IEEE 100
The Authoritative Dictionary of IEEE Standards Terms

Seventh Edition

Published by
Standards Information Network
IEEE Press

Trademarks and disclaimers

IEEE believes the information in this publication is accurate as of its publication date; such information is subject to change without notice. IEEE is not responsible for any inadvertent errors.

Other tradenames and trademarks in this document are those of their respective owners.

The Institute of Electrical and Electronics Engineering, Inc.
3 Park Avenue, New York, NY, 10016-5997, USA

To order IEEE Press publications, call 1-800-678-IEEE.

Print: ISBN 0-7381-2601-2 *SP1122*

See other standards and standards-related product listings at: http://standards.ieee.org/

The publisher believes that the information and guidance given in this work serve as an enhancement to users, all parties must rely upon their own skill and judgement when making use of it. The publisher does not assume any liability to anyone for any loss or damage caused by any error or omission in the work, whether such error or omission is the result of negligence or any other cause. Any and all such liability is disclaimed.

This work is published with the understanding that the IEEE is supplying information through this publication, not attempting to render engineering or other professional services. If such services are required, the assistance of an appropriate professional should be sought. The IEEE is not responsible for the statements and opinions advanced in this publication.

Library of Congress Cataloging-in-Publication Data

IEEE 100 : the authoritative dictionary of IEEE standards terms.—7th ed.
p. cm.
ISBN 0-7381-2601-2 (paperback : alk. paper)
1. Electric engineering—Dictionaries. 2. Electronics—Dictionaries. 3. Computer engineering—Dictionaries. 4. Electric engineering—Acronyms. 5. Electronics—Acronyms. 6. Computer engineering—Acronyms. I. Institute of Electrical and Electronics Engineers.

TK9 .I28 2000
621.3′03—dc21 00-050601

Contents

Introduction

IEEE standards establish an authoritative common language that defines quality and sets technical criteria. By guaranteeing consistency and conformity through open consensus, IEEE standards add value to products, facilitate trade, drive markets, and ensure safety. That's why leading companies, organizations, and industries around the globe rely on them.

Critical components of this common language are the terms and definitions that are at the foundation of the vast body of IEEE standards. In the past decade alone, hundreds of terms—describing the latest tools, techniques, and best practices—have been added to the lexicon of IEEE standards.

In this newly updated *Authoritative Dictionary of IEEE Standards Terms,* professional experts and students alike will gain an in-depth understanding and appreciation for the breadth of coverage of IEEE standards terms and definitions not found in any other single source.

The seventh edition of IEEE 100 has been revised to include nearly 35 000 technical terms and definitions from over 800 standards—covering areas such as power and energy, communications, information technology, and transportation systems. In addition to an extensive list of widely used acronyms and abbreviations, this new edition also contains detailed abstracts of each term's associated standard(s). What's more, all definitions are augmented by a combination of indispensable information, including:

- Preferred and popular usage of each term
- Variations in meanings among different technical specialties
- Cross-indexing to related works
- Key explanatory notes for further term clarification

In preparing this latest edition of the Dictionary, we realized that the standards community desired more than just a compilation of IEEE standardized terms and definitions. They needed an authoritative resource created by the organization that develops and produces the standards from which the terms and definitions are derived—the IEEE. In addition, we determined the Dictionary needed to be not only user friendly, but also rich in information. In other words, it needed to be the *Authoritative Dictionary of IEEE Standards Terms*.

Susan K. Tatiner
Director, IEEE Standards Publishing Programs

IEEE Standards Project Editors for the seventh edition:
Kim Breitfelder
Don Messina

Additional assistance was provided by the IEEE Standards editorial staff.

How to Use This Dictionary

The terms defined in the Dictionary are listed in *letter-by-letter* alphabetical order. Spaces are ignored in this style of alphabetization, so *cable value* will come before *cab signal*. Descriptive categories associated with the term in earlier editions of the Dictionary will follow the term in parentheses. New categories appear after the definitions (see Categories, below), followed by the designation of the standard or standards that include the definition. If a standard designation is followed by the letter *s,* it means that edition of the standard was superseded by a newer revision and the term was not included in the revision. If a designation is followed by the letter *w,* it means that edition of the standard was withdrawn and not replaced by a revision. A bracketed number refers to the non-IEEE standard sources given in the back of the book.

Abstracts of the current set of approved IEEE standards are provided in the back of the book. It should be noted that updated information about IEEE standards can be obtained at any time from the IEEE Standards World Wide Web site at http://standards.ieee.org/.

Categories

The category abbreviations that are used in this edition of the Dictionary are defined below. This information is provided to help elucidate the context of the definition. Older terms for which no category could be found have had the category *Std100* assigned to them. Note that terms from sources other than IEEE standards, such as the National Electrical Code® (NEC®) or the National Fire Protection Association, may not be from the most recent editions; the reader is cautioned to check the latest editions of all sources for the most up-to-date terminology.

Categories sorted by abbreviation

AES	aerospace and electronic systems
AHDL	computer—Analog Hardware Descriptive Language
AMR	automatic meter reading and energy management
AP	antennas and propagation
ATL	computer—Abbreviated Test Language for All Systems
BA	computer—bus architecture
BT	broadcast technology
C	computer
CAS	circuits and systems
CE	consumer electronics
CHM	components, hybrids, and manufacturing technology
COM	communications
CS	control systems
DA	computer—design automation
DEI	dielectrics and electrical insulation
DESG	dispersed energy storage and generation
DIS	computer—distributed interactive simulation
ED	electron devices
EDU	education
EEC	electrical equipment and components
ELM	electricity metering
EM	engineering management
EMB	engineering in medicine and biology
EMC	electromagnetic compatibility
GRS	geoscience and remote sensing
GSD	graphic symbols and designations
IA	industry applications
IE	industrial electronics
II	information infrastructure
IM	instrumentation and measurement
IT	information theory

IVHS	intelligent vehicle highway systems
LEO	lasers and electro-optics
LM	computer—local and metropolitan area networks
MAG	magnetics
MIL	military
MM	computer—microprocessors and microcomputers
MTT	microwave theory and techniques
NEC	National Electrical Code
NESC	National Electrical Safety Code
NFPA	National Fire Protection Association
NI	nuclear instruments
NIR	non-ionizing radiation
NN	neural networks
NPS	nuclear and plasma sciences
ODM	computer—optical disk and multimedia platforms
OE	oceanic engineering
PA	computer—portable applications
PE	power engineering
PEL	power electronics
PQ	power quality
PSPD	power surge protective devices
PV	photovoltaics
QUL	quantities, units, and letter symbols
R	reliability
RA	robotics and automation
REM	rotating electrical machinery
RL	roadway lighting
S&P	computer—security and privacy
SB	stationary batteries
SE	computer—software engineering
SMC	systems, man, and cybernetics
SP	signal processing
Std100	Standard 100 legacy data
SUB	substations
SWG	power switchgear
T&D	transmission and distribution
TF	time and frequency
TRR	transformers, regulators, and reactors
TT	test technology
UFFC	ultrasonics, ferroelectrics, and frequency control
VT	vehicular technology

Categories sorted by name

aerospace and electronic systems	AES
antennas and propagation	AP
automatic meter reading and energy management	AMR
broadcast technology	BT
circuits and systems	CAS
communication	COM
components, hybrids, and manufacturing technology	CHM
computer	C
computer—Abbreviated Test Language for All Systems	ATL
computer—Analog Hardware Descriptive Language	AHDL
computer—bus architecture	BA
computer—design automation	DA
computer—distributed interactive simulation	DIS
computer—local and metropolitan area networks	LM
computer—microprocessors and microcomputers	MM
computer—optical disk and multimedia platforms	ODM
computer—portable applications	PA
computer—security and privacy	S&P
computer—software engineering	SE
consumer electronics	CE

control systems	CS
dielectrics and electrical insulation	DEI
dispersed energy storage and generation	DESG
education	EDU
electrical equipment and components	EEC
electricity metering	ELM
electromagnetic compatibility	EMC
electron devices	ED
engineering in medicine and biology	EMB
engineering management	EM
geoscience and remote sensing	GRS
graphic symbols and designations	GSD
industrial electronics	IE
industry applications	IA
information infrastructure	II
information theory	IT
instrumentation and measurement	IM
intelligent vehicle highway systems	IVHS
lasers and electro-optics	LEO
magnetics	MAG
microwave theory and techniques	MTT
military	MIL
National Electrical Code	NEC
National Electrical Safety Code	NESC
National Fire Protection Association	NFPA
neural networks	NN
non-ionizing radiation	NIR
nuclear and plasma sciences	NPS
nuclear instruments	NI
oceanic engineering	OE
photovoltaics	PV
power electronics	PEL
power engineering	PE
power quality	PQ
power surge protective devices	PSPD
power switchgear	SWG
quantities, units, and letter symbols	QUL
reliability	R
roadway lighting	RL
robotics and automation	RA
rotating electrical machinery	REM
signal processing	SP
Standard 100 legacy data	Std100
stationary batteries	SB
substations	SUB
systems, man, and cybernetics	SMC
test technology	TT
time and frequency	TF
transformers, regulators, and reactors	TRR
transmission and distribution	T&D
ultrasonics, ferroelectrics, and frequency control	UFFC
vehicular technology	VT

Trademarks

The following is a list of trademarks that may be used in *IEEE 100: The Authoritative Dictionary of IEEE Standards Terms.*

802 is a registered trademark of the Institute of Electrical and Electronics Engineers, Inc.

ABBET is a registered trademark of the Institute of Electrical and Electronics Engineers, Inc.

Adobe is a trademark of Adobe Systems Incorporated.

Analog devices is a trademark of Analog Devices, Inc.

Appletalk is a registered trademark of Apple Computer, Inc.

BOCA is a registered trademark of Building Officials and Code Administrators International, Inc.

BOOM is a registered trademark of Fakespace, Inc.

Centronics is a registered trademark of Genicom Corporation.

CompactPCI is a registered trademark of the PCI Industrial Computer Manufacturers Group.

Cray is a registered trademark of Cray Research, Inc.

DSSI is a registered trademark of Discreet Surveillance Systems.

Futurebus+ is a registered trademark of the Institute of Electrical and Electronics Engineers, Inc.

IBM is a registered trademark of International Business Machines, Inc.

Intel386 is a trademark of Intel Corporation.

Life Safety Code is a registered trademark of the National Fire Protection Association.

Mylar is a registered trademark of E.I. du Pont de Nemours and Company.

National Electrical Code is a registered trademark of the National Fire Protection Association.

NEC is a registered trademark of the National Fire Protection Association.

National Electrical Safety Code is a registered trademark and service mark of the Institute of Electrical and Electronics Engineers, Inc.

NESC is a registered trademark and service mark of the Institute of Electrical and Electronics Engineers, Inc.

Netbios is a registered trademark of International Business Machines, Inc.

Nomex is a registered trademark of E. I. Dupont de Nemours and Company.

NuBus is a registered trademark of Texas Instruments, Inc.

OpenBoot is a trademark of Sun Microsystems, Inc.

Open Software Foundation, OSF, and the OSF logo are registered trademarks of the Open Software Foundation, Inc.

PostScript is a trademark of Adobe Systems Incorporated.

POSIX is a registered certification mark of the Institute of Electrical and Electronics Engineers, Inc.

PS/2 is a registered trademark of International Business Machines, Inc.

SCSI is a registered trademark of SCSI Solutions.

SDI is a registered trademark of Maurice Siebenberg.

SPARC is a registered trademark of SPARC International, Inc.

SPARCstation is a trademark of SPARC International, Inc.

SPAsystem is a registered trademark of the Institute of Electrical and Electronics Engineers, Inc.

Stylized 8 (Futurebus+) is a registered trademark of the Institute of Electrical and Electronics Engineers, Inc.

Sun Microsystems is a registered trademark of Sun Microsystems, Inc.

TURBOchannel is a registered trademark of Digital Equipment Corporation.

Uniform Building Code is a trademark of the International Conference of Building Officials (ICBO).

UNIX is a registered trademark in the United States and other countries, licensed exclusively through X/Open Company Limited.

VAX is a registered trademark of Digital Equipment Corporation.

VAXBI is a registered trademark of Digital Equipment Corporation.

Verilog is a registered trademark of Cadence Design Systems, Inc.

Velcro is a registered trademark of Velcro Industries B. V.

Windows is a trademark of Microsoft Corporation.

X/Open is a registered trademark and the “X” device is a trademark of X/Open Company, Ltd.

A

***aa* auxiliary switch** *See: aa* contact; auxiliary switch.

AAAC Concentric-lay-stranded all aluminum alloy conductor. (T&D/PE) 524-1992r

AAC Concentric-lay-stranded all aluminum conductor. (T&D/PE) 524-1992r

***aa* contact** A contact that is open when the operating mechanism of the main device is in the standard reference position and that is closed when the operating mechanism is in the opposite position. *See also:* standard reference position. (SWG/PE) C37.100-1992

AACSR *See:* aluminum alloy conductor, steel reinforced.

***A* and *R* display (radar)** An *A*-display, any portion of which may be expanded. *See also:* navigation. (AES/RS) 686-1982s

AAU *See:* alternate access unit.

***a* auxiliary switch** *See: a* contact; auxiliary switch.

abampere The unit of current in the centimeter-gram-second (cgs) electromagnetic system. The abampere is 10 A. (Std100) 270-1966w

abandoned call (telephone switching systems) A call during which the calling station goes on-hook prior to its being answered. (COM) 312-1977w

ABASIC A dialect of the BASIC programming language. (C) 610.13-1993w

***A* battery** A battery designed or employed to furnish current to heat the filaments of the tubes in a vacuum-tube circuit. *See also:* battery. (EEC/PE) [119]

ABBET application An end-use program constructed using one or more ABBET components. (ATLAS) 1226-1993s

ABBET component An implementation of the services defined in an IEEE ABBET component standard. (ATLAS) 1226-1993s

ABBET implementation The installation and utilization of one or more ABBET applications. (ATLAS) 1226-1993s

ABBET layer A natural grouping to the ABBET services that is recognized by the ABBET layer model. (ATLAS) 1226-1993s

abbreviated dialing (telephone switching systems) A feature permitting the establishment of a call with an input of fewer digits than required under the numbering plan. (COM) 312-1977w

abbreviated ringing A short, variable burst of power ringing that is required to establish a temporary communications path in certain types of network pair-gain equipment. The switch is instructed, via trunk signals, to output this abbreviated ringing on the end user's line. (SCC31/AMR) 1390.3-1999, 1390.2-1999, 1390-1995

Abbreviated Test Language for All Systems (ATLAS) (1) A standard abbreviated English language used in the preparation and documentation of test procedures or test programs that can be implemented either manually or with automatic or semiautomatic test equipment. (ATLAS/SCC20) 1232-1995, 1226-1998, 771-1984s, 993-1997
(2) A test language used by test engineers in controlling automatic test equipment. (C) 610.13-1993w

A Broad-Based Environment for Test (ABBET) A set of international standards that define language-independent interfaces to industry standards regarding automatic testing and integrated diagnositcs. (SCC20) 993-1997

abbreviation A shortened form of a word or expression. *See also:* reference designation; symbol for a unit; letter combination; graphic symbol; mathematical symbol; symbol for a quantity; functional designation. (GSD) 267-1966

abend *See:* abnormal end.

ability A mode that a device can advertise using Auto-Negotiation. For modes that represent a type of data service, a device shall be able to operate that data service before it may advertise this ability. A device may support multiple abilities. (C/LM) 802.3-1998

abnormal decay The dynamic decay of multiply written, superimposed (integrated) signals whose total output amplitude changes at a rate distinctly different from that of an equivalent singly written signal. *Note:* Abnormal decay is usually very much slower than normal decay and is observed in bombardment-induced conductivity type of tubes. *See also:* charge-storage tube. (ED) 158-1962w

abnormal end Termination of a process prior to completion. *Synonym:* abend. *See also:* exception; abort. (C) 610.12-1990

abnormal glow discharge (gas tube) The glow discharge characterized by the fact that the working voltage increases as the current increases. *See also:* discharge. (Std100) [31]

abnormality Any deviation from the pre-established test conditions, including the tolerance limits, that may affect the outcome of the test. (PE/IC) 1407-1998

abnormal preamble A preamble that does not match the synchronization pattern resulting in a packet error. (C) 610.7-1995

abort (1) (software) To terminate a process prior to completion. *See also:* abend; exception. (C) 610.12-1990
(2) To terminate the transmission of a frame before it has been completely transmitted. (EMB/MIB) 1073.4.1-2000

abort completion point A point at which the execution of an aborted construct must complete. (C) 1003.5-1999

abort deferred operation An operation that always continues to completion without being affected by an abort. Certain operations are required by the Ada language to be abort deferred. (C) 1003.5-1999

abortive release An abrupt termination of a network connection that may result in the loss of data. (C) 1003.5-1999

abort sequence A sequence transmitted by an originating ring station that terminates the transmission of a frame prematurely. It also causes the ring station receiving this frame to terminate the frame's reception. (C/LM) 8802-5-1998

above threshold firing time (microwave switching tubes) (nonlinear, active, and nonreciprocal waveguide components) The time to establish an above-threshold discharge in the gas tube after the application of radio frequency power. This time delay is responsible for the spike in the leading edge of the output leakage waveform. *See also:* duplexer; gas tube. (MTT) 457-1982w

abrupt junction (nonlinear, active, and nonreciprocal waveguide components) (semiconductor) A semiconductor crystal having an *n*-region containing a near-constant net concentration of donor impurities adjoining a *p*-region with a near-constant net concentration of acceptors; used primarily in microwave frequency multipliers, dividers, and parametric circuits. (MTT) 457-1982w

ABS (cable systems in power generating stations) Conduit fabricated from acrylonitrile-butadiene-styrene. (PE/SUB/EDPG) 422-1977, 525-1992r

ABSBH load *See:* average busy season busy-hour load.

absolute accuracy Accuracy as measured from a reference that must be specified. (IA/EEC) [61], [74]

absolute address (software) An address that is permanently assigned to a device or storage location and that identifies the device or location without the need for translation or calculation. *Synonyms:* specific address; explicit address. *Contrast:* symbolic address; relative address; relocatable address. *See also:* absolute assembler; absolute code; absolute loader; absolute instruction. (C) 610.12-1990
(2) (A) (computers) An address that is assigned by the machine designer to a physical storage location. **(B) (computers)** A pattern of characters that identifies a unique storage loca-

tion without further modification. *See also:* machine address. (C) [20], [85]

(3) **(A)** An address that is permanently assigned to a device or storage location and that identifies the device or location without the need for translation or calculation. **(B)** The actual complete address of a device or storage location. *Synonyms:* specific address; address reference; machine address; explicit address. *See also:* relocatable address; virtual address; symbolic address; base address; relative address. (C) 610.10-1994

absolute altimeter (1) (electronic navigation) A device that measures altitude above local terrain. (AES/RS) 686-1982s, [42]

(2) (navigation aid terms) A device that measures altitude above local terrain. In its usual form, it does this by measuring the time interval between transmission of a signal and the return of its echo, or by measuring the phase difference between the transmitting signal and the echo. (AES/GCS) 172-1983w

absolute assembler An assembler that produces absolute code. *Contrast:* relocating assembler. (C) 610.12-1990

absolute block (automatic train control) A block governed by the principle that no train shall be permitted to enter the block while it is occupied by another train. (EEC/PE) [119]

absolute code (1) (microprocessor object modules) Data or executable machine code in memory or an image thereof. *Contrast:* relocatable code. (MM/C) 695-1985s

(2) (software) Code in which all addresses are absolute addresses. *Synonym:* specific code. *Contrast:* relocatable code. (C) 610.12-1990

absolute delay (1) (loran) The interval of time between the transmission of a signal from the master station and transmission of the next signal from the slave station. *See also:* navigation. (AES/RS/GCS) 686-1982s, 172-1983w, [42]

(2) (telecommunications) The interval of time between the transmission of a signal and the reception of the same signal at a different point in the circuit. (COM/TA) 1007-1991r

(3) The interval of time between the transmission of a signal and the reception of the same signal (or its associated signal, if in a different domain) at a different point in a circuit. *Synonym:* transmission delay or propagation delay. (COM/TA) 743-1995

(4) The time elapsed between transmission of a signal and reception of the same signal. (COM/TA) 1007-1991r

absolute deviation integral (automatic control) The time integral of the absolute value of the system deviation following a stimulus specified as to location, magnitude, and time pattern. *Note:* The stimulus commonly employed is a step input. (PE/EDPG) [3]

absolute dimension A dimension expressed with respect to the initial zero point of a coordinate axis. *See also:* coordinate dimension word. (IA/EEC) [61], [74]

absolute error (A) The amount of error expressed in the same units as the quantity containing the error. *Contrast:* relative error. **(B)** Loosely, the absolute value of the error; i.e., the magnitude of the error without regard to its algebraic sign. (C) 1084-1986

absolute gain *See:* gain.

absolute instruction A computer instruction in which all addresses are absolute addresses. *See also:* indirect instruction; effective instruction; direct instruction; immediate instruction. (C) 610.12-1990, 610.10-1994w

absolute loader (1) (microprocessor object modules) A process that can load one or more sections of absolute code only at the locations specified by the sections. *See also:* relocating loader. (C/MM) 695-1985s

(2) (software) A loader that reads absolute machine code into main memory, beginning at the initial address assigned to the code by the assembler or compiler, and performs no address adjustments on the code. *Contrast:* relocating loader. (C) 610.12-1990

absolute luminance threshold (illuminating engineering) Luminance threshold for a bright object like a disk on a totally dark background. (ED) [127]

absolute machine code (software) Machine language code that must be loaded into fixed storage locations at each use and may not be relocated. *See also:* relocatable machine code. (C/SE) 729-1983s

absolute path If the underlying system is based upon a conforming implementation of POSIX.1; then a pathname that begins with /; otherwise, *absolute path* is implementation defined. (C/PA) 1387.2-1995

absolute permissive block (automatic train control) A term used for an automatic block signal system on a track signaled in both directions. For opposing movements, the block is from siding to siding and the signals governing entrance to this block indicate stop. For following movements, the section between sidings is divided into two or more blocks, and train movements into these blocks, except the first one, are governed by intermediate signals usually displaying stop; then the trains proceed at restricted speed, as their most restrictive indication. (EEC/PE) [119]

absolute photocathode spectral response (diode-type camera tube) The ratio of the photocathode current, measured in amperes, to the radiant power incident on the photocathode face, measured in watts, as a function of the photon energy, frequency, or wavelength. Units: amperes/watt^{-1} (A/W^{-1}). (ED) 503-1978w

absolute refractory state (medical electronics) The portion of the electrical recovery cycle during which a biological system will not respond to an electric stimulus. (EMB) [47]

absolute Seebeck coefficient The integral, from absolute zero to the given temperature, of the quotient of the Thomson coefficient of the material by the absolute temperature. *See also:* thermoelectric device. (ED) [46]

absolute stability Global asymptotic stability maintained for all nonlinearities within a given class. *Note:* A typical problem to which the concept of absolute stability has been applied consists of a system with dynamics described by the vector differential equation

$$\dot{\mathrm{x}} = A\mathrm{x} + \mathrm{b}f(\sigma)$$

$$\sigma = \mathrm{c}^{\mathrm{T}}\mathrm{x}$$

with a nonlinearity class defined by the conditions

$$f(0) = 0$$

$$k_1 \le f(\sigma)/\sigma \le k_2$$

The solution $\mathrm{x}(t) = 0$ is said to be absolutely stable if it is globally asymptotically stable for all nonlinear functions $f(\sigma)$ in the above class. *See also:* control system. (CS/IM) [120]

absolute steady-state deviation (control) The numerical difference between the ideal value and the final value of the directly controlled variable (or another variable, if specified). *See also:* deviation; percent steady-state deviation. (IA/IAC) [60]

absolute system deviation (control) At any given point on the time response, the numerical difference between the ideal value and the instantaneous value of the directly controlled variable (or another variable, if specified). *See also:* deviation. (IA/IAC) [60]

absolute threshold The luminance threshold or minimum perceptible luminance (photometric brightness) when the eye is completely dark-adapted. *See also:* visual field. (ED) [127]

absolute transient deviation (control) The numerical difference between the instantaneous value and the final value of the directly controlled variable (or another variable, if specified). *See also:* percent transient deviation; deviation. (IA/IAC) [60]

absolute value The magnitude of a quantity without regard to its algebraic sign. (C) 1084-1986w

absolute-value circuit A transducer or circuit employed in analog computers that produces an output signal equal in magnitude to the input signal but always of one polarity. (C) 610.10-1994w, 165-1977w

absolute-value device A transducer that produces an output signal equal in magnitude to the input signal but always of one polarity. *See also:* electronic analog computer. (C) 165-1977w

absorbed dose The energy imparted to the material by the incident radiation (usually abbreviated to "dose"). It is dependent on the magnitude of the radiation field and on the degree of interaction between the radiation and the material. The SI unit of absorbed dose is the gray (Gy), which equals one joule per kilogram. A special unit of absorbed dose, the rad (rd), is also widely used. One gray equals 100 rd (10 Gy = 1 Mrd). (DEI/RE) 775-1993w

absorbed dose rate The increment of absorbed dose in a given time interval (usually abbreviated to "dose rate"). The SI unit is grays per second. Special units of rads per second or per hour are also widely used at present. (DEI/RE) 775-1993w

absorbed electrolyte Electrolyte in a VRLA cell that has been immobilized in absorbent separators. (SB) 1189-1996

absorbed electrolyte cell A valve-regulated lead-acid (VRLA) cell whose electrolyte has been immobilized in absorbent separator (normally, glass or polymeric fiber). *Synonyms:* absorbed glass mat cell; starved electrolyte cell. (IA/PSE) 446-1995

absorbed glass mat cell *See:* absorbed electrolyte cell.

absorber-lined chamber (ALC) A room or enclosure (either shielded or unshielded) with all of its surfaces lined with radio-frequency (RF) absorber material. Commonly referred to as an anechoic chamber. (EMC) 1128-1998

absorber-lined open-area test site (ATS) An open-area test site (OATS) in which the ground plane is covered with radio-frequency (RF) absorber to suppress ground reflections. *See also:* open-area test site. (EMC) 1128-1998

absorbing clamp A portable testing device that is effective at detecting electromagnetic radiation. The absorbing clamp has a great capacity for electromagnetic compatibility cable measurements in the frequency range of 30–1000 MHz, and is non-destructive to the specimen. The test fixture clamps over the sample cable and inductively detects signal leakage. (PE/IC) 1143-1994r

absorbing Markov chain model A Markov chain model that has at least one absorbing state and in which from every state it is possible to get to at least one absorbing state. (C) 610.3-1989w

absorbing state In a Markov chain model, a state that cannot be left once it is entered. *Contrast:* nonabsorbing state. (C) 610.3-1989w

absorptance (illuminating engineering) The ratio of the absorbed flux to the incident flux. *Note:* The sum of the hemispherical reflectance, the hemispherical transmittance, and the absorptance is one. (ED) [127]

absorption (1) (fiber optics) In an optical waveguide, that portion of attenuation resulting from conversion of optical power into heat. *Note:* Intrinsic components consist of tails of the ultraviolet and infrared absorption bands. Extrinsic components include impurities, for example, the OH^- ion and transition metal ions and, defects; for example, results of thermal history and exposure to nuclear radiation. *See also:* attenuation. (Std100) 812-1984w

(2) (illuminating engineering) A general term for the process by which incident flux is converted to another form of energy, usually and ultimately to heat. *Note:* All of the incident flux is accounted for by the processes of reflection, transmission, and absorption. (ED) [127]

(3) (laser maser) The transfer of energy from a radiation field to matter. (LEO) 586-1980w

(4) The process of converting electromagnetic energy to heat. (AP/PROP) 211-1997

absorption band A band of frequencies for which a medium is considered to be absorbing. (AP/PROP) 211-1997

absorption coefficient (κ_a) (1) (power station noise control) The ratio of the energy absorbed by the surface to the energy incident upon it. (PE/EDPG) 640-1985w

(2) (of a medium) The rate of decrease of power density of a wave per unit distance, due to absorption. For a homogeneous medium with relative complex permittivity ε_r and the permeability of free space μ_0:

$$\kappa_a = -4\pi \mathrm{Im}\{\sqrt{\varepsilon_r}\}/\lambda_0$$

where

λ_0 = the free-space wavelength
$\exp(+j\omega t)$= the time factor

For inhomogeneous media. *See also:* extinction coefficient. (AP/PROP) 211-1997

absorption cross-section (σ_a) (of a lossy body) The ratio of power absorbed by the body, P_a, to the power density of an incident plane wave, S_i:

$$\sigma_a = P_a/S_i$$

See also: extinction cross-section. (AP/PROP) 211-1997

absorption current (1) (rotating machinery) (or component) A reversible component of the measured current, which changes with time of voltage application, resulting from the phenomenon of "dielectric absorption" within the insulation when stressed by direct voltage. (PE/EM) 95-1977r

(2) (electric submersible pump cable) Current resulting from charge absorbed in the dielectric as a result of polarization. (IA/PE/PC/IC/TR) 1017-1985s, 400-1991, C57.19.03-1996

(3) A current resulting from molecular polarizing and electron drift, which decays with time of voltage application at a decreasing rate from a comparatively high initial value to nearly zero, and depends on the type and condition of the bonding material used in the insulation system. (PE/EM) 43-2000

absorption, deviative *See:* deviative absorption.

absorption frequency meter (reaction frequency meter) (waveguide) A one-port cavity frequency meter that, when tuned, absorbs electromagnetic energy from a waveguide. *See also:* waveguide. (AP/ANT) [35]

absorption loss (data transmission) The loss of signal energy in a communication circuit that results from coupling to a neighboring circuit or conductor. (PE) 599-1985w

absorption modulation A method for producing amplitude modulation of the output of a radio transmitter by means of a variable-impedance (principally resistive) device inserted in or coupled to the output circuit. (BT) 182A-1964w

absorptive attenuator *See:* resistive attenuator.

absorptive loss *See:* arc loss.

abstract class (1) An OM class of OM objects of which instances are forbidden. (C/PA) 1328.2-1993w, 1326.2-1993w, 1224.2-1993w, 1327.2-1993w

(2) A class, instances of which are forbidden unless they belong to one of its concrete subclasses. (C/PA) 1328-1993w, 1224.1-1993w, 1327-1993w, 1238.1-1994w, 1224-1993w

(3) A class that cannot be instantiated independently, i.e., instantiation must be accomplished via a subclass. A class for which every instance must also be an instance of a subclass in the cluster (i.e., a total cluster) is called an abstract class with respect to that cluster. (C/SE) 1320.2-1998

abstract data type (1) A data type for which only the properties of the data and the operations to be performed on the data are specified, without concern for how the data will be represented or how the operations will be implemented. (C) 610.12-1990

(2) A data type for which the user can create instances and operate on those instances, but the range of valid operations available to the user does not depend in any way on the internal representation of the instances or the way in which the operations are realized. The data is "abstract" in the sense that values in the extent, i.e., the concrete values that represent the instances, are any set of values that support the operations and are irrelevant to the user. An abstract data type defines the operations on the data as part of the definition of the data and separates what can be done (interface) from how it is done (realization). (C/SE) 1320.2-1998

abstraction (A) A view of an object that focuses on the information relevant to a particular purpose and ignores the remainder of the information. *See also:* data abstraction. **(B)** The process of formulating a view as in (A). (C) 610.12-1990

abstract machine (A) (software) A representation of the characteristics of a process or machine. **(B) (software)** A module that processes inputs as though it were a machine. *See also:* module; process. (C/SE) 729-1983

abstract quantity *See:* mathematico-physical quantity.

abstract symbol A symbol whose meaning and use have not been determined by a general agreement but have to be defined for each application of the symbol. (C) 1084-1986w

Abstract Syntax Notation One (ASN.1) A notation that both enables complicated types to be defined and also enables values of these types to be specified. (C/PA) 1328.2-1993w, 1224.2-1993w, 1327.2-1993w, 1326.2-1993w

ac (alternating current) *See:* alternating current.

AC *See:* acoustic coupler.

ACA *See:* adjacent-channel attenuation.

academic simulation *See:* instructional simulation.

ac analog computer An analog computer in which electrical signals are of the form of carrier signals where the absolute value of a mathematical variable is represented by the amplitude of the carrier and the sign of the mathematical variable is represented by the phase (0 or 180 degrees) of the carrier relative to the computer. (C) 610.10-1994w, 165-1977w

ACAR *See:* aluminum conductor, aluminum alloy reinforced.

ac breakdown voltage (gas tube surge-protective device) The minimum root-mean-square value of sinusodial voltage at frequencies between 15 Hz and 62 Hz that results in arrester sparkover. (SPD/PE) C62.31-1981s

ac cable (armored cable) A fabricated assembly of insulated conductors in a flexible metallic enclosure. (NESC/NEC) [86]

accelerated aging The application of intensified aging stress or stresses in order to increase the degradation rate above that expected in service. (DEI/RE) 775-1993w

accelerated life test (test, measurement, and diagnostic equipment) A test in which certain factors, such as voltage, temperature, and so forth, are increased or decreased beyond normal operating values to obtain observable deterioration in a reasonable period of time, and thereby afford some measure of the probable life under normal operating conditions or some measure of the durability of the equipment when exposed to the factors being aggravated. (MIL) [2]

accelerated test (evaluation of thermal capability) (thermal classification of electric equipment and electrical insulation) A functional test in which one or more factors of influence are increased in magnitude or frequency of application so as to decrease the time needed for the test. (EI) 1-1986r

accelerating (rotating machinery) The process of running a motor up to speed after breakaway. *See also:* asynchronous machine. (PE) [9]

accelerating device (power system device function numbers) A device that is used to close or to cause the closing of circuits that are used to increase the speed of a machine. (PE/SUB) C37.2-1979s

accelerating electrode An electrode to which a potential is applied to increase the velocity of the electrons or ions in the beam. (NPS) 61-1971w, 398-1972r

accelerating grid *See:* accelerating electrode.

accelerating relay A programming relay whose function is to control the acceleration of rotating electrical equipment. (SWG/PE) C37.100-1992

accelerating time (control) The time in seconds for a change of speed from one specified speed to a higher specified speed while accelerating under specified conditions. *See also:* electric drive. (IA/ICTL/APP/IAC) [69], [60]

accelerating torque (rotating machinery) Difference between the input torque to the rotor (electromagnetic for a motor or mechanical for a generator) and the sum of the load and loss torques; the net torque available for accelerating the rotating parts. *See also:* rotor. (PE) [9]

accelerating voltage (oscilloscopes) The cathode-to-viewing-area voltage applied to a cathode-ray tube for the purpose of accelerating the electron beam. *See also:* oscillograph. (IM/HFIM) [40]

acceleration (electric drive) Operation of raising the motor speed from zero or a low level to a higher level. *See also:* electric drive. (IA/IAC) [60]

acceleration factor The ratio between the times necessary to obtain the same stated proportion of failures in two equal samples under two different sets of stress conditions involving the same failure modes and mechanisms. (R) [29]

acceleration-forced response (automatic control) The total (transient plus steady-state) time response resulting from a sudden increase in the rate of the rate of change of input from zero to some finite value. (PE/EDPG) [3]

acceleration-insensitive drift rate (gyros) The component of systematic drift rate that has no correlation with acceleration. *See also:* systematic drift rate. (AES/GYAC) 528-1994

acceleration, programmed *See:* programmed acceleration.

acceleration-sensitive drift rate (gyros) Those components of systematic drift rates that are correlated with the first power of linear acceleration applied to the gyro case. The relationship of these components of drift rate to acceleration can be stated by means of coefficients having dimensions of angular displacement per unit time per unit acceleration for accelerations along each of the principal axes of the gyro (for example, drift rate caused by mass unbalance). *See also:* systematic drift rate. (AES/GYAC) 528-1994

acceleration space (velocity-modulated tube) The part of the tube following the electron run in which the emitted electrons are accelerated to reach a determined velocity. *See also:* velocity-modulated tube. (ED) [45], [84]

acceleration-squared-sensitive drift rate (gyros) Those components of systematic drift rates that are correlated with the second power or product of linear accelerations applied to the gyro case. The relationship of these components of drift rate to acceleration can be stated by means of coefficients having dimensions of angular displacement per unit time per unit acceleration squared for accelerations along each of the principal axes of the gyro and angular displacement per unit time per the product of accelerations along combinations of two principal axes of the gyro (for example, drift rate caused by anisoelasticity). (AES/GYAC) 528-1994

acceleration time The part of access time that is required to bring a storage device, typically a tape or disk drive, to the speed at which data can be read or written. *Synonym:* start time. *Contrast:* deceleration time. (C) 610.10-1994w

acceleration, timed *See:* timed acceleration.

accelerator (1) An Xt Intrinsics facility that allows the binding of a widget event to a keyboard action or a series of actions. (C) 1295-1993w
(2) A circuit or device that accelerates some unit in a computer, as in an accelerator board. *See also:* hardware accelerator. (C) 610.10-1994w

accelerator board A printed circuit board that replaces or augments the computer's main processor with a faster processor. (C) 610.10-1994w

accelerometer A device that senses the inertial reaction of a proof mass for the purpose of measuring linear or angular acceleration. *Note:* In its simplest form, an accelerometer consists of a case-mounted spring and mass arrangement in which displacement of the mass from its rest position, relative to the case, is proportional to the total nongravitational acceleration experienced along the instrument's sensitive axes. (AES/GYAC/GCS) 528-1994, 172-1983w

accent lighting (illuminating engineering) Directional lighting to emphasize a particular object or draw attention to a part of the field of view. (ED) [127]

accept The condition assumed by an LLC upon accepting a correctly received PDU for processing. (C/LM/CC) 8802-2-1998

acceptability criteria A set of standards, established by the modeling and simulation (M&S) application sponsor or accreditation authority, that a particular model or simulation must meet to be accredited for a given use. The criteria will be unique to each problem and will give key insights to potential solutions. (C/DIS) 1278.4-1997

acceptable (1) (diesel-generator unit) Demonstrated to be adequate by the safety analysis of the plant. (PE/NP) 387-1995

(2) Demonstrated to be adequate by the safety analysis of the station. (PE/NP) 603-1998

acceptable deviation In the context of evaluating specific test-case post conditions, a deviation is permitted based on an informed decision to specify that deviation as noncritical. (C/PA) 2000.2-1999

acceptable energized background noise level (1) (A) Energized background noise level present during test that is considered acceptable. **(B)** (partial discharge measurement in liquid-filled power transformers and shunt reactors) Energized background noise level present during test that is considered acceptable. It should not exceed 50% of the acceptable terminal discharge level and in any case should be below 100 pC. (PE/TR) C57.113-1988

(2) (dry-type transformers) The acceptable energized background noise level present during test should not exceed 50% of the acceptable terminal discharge level, and in any case should be below 100 pC (5 pC if an acceptable terminal discharge level of 10 pC is required.). (PE/TR) C57.124-1991r

(3) Energized background noise level present during test that does not exceed 50% of the acceptable partial discharge level of the test specimen. Spurious noise, however, can exceed this level if identified as not emanating from the specimen. This may require extending the period of voltage application. (SWG/PE) 1291-1993r

acceptable terminal partial discharge level (1) (dry-type transformers) The acceptable terminal partial discharge level is that specified maximum terminal partial discharge value for which measured terminal partial discharge values exceeding said value are considered unacceptable. The method of measurement and the test voltage for a given test object must be specified with the acceptable terminal partial discharge level. (PE/TR) C57.124-1991r

(2) The specified maximum terminal partial discharge level for which measured terminal partial discharge values exceeding this value are considered unacceptable. This level may be defined by the appropriate apparatus test standard or may be a level agreed to by the user and manufacturer. The method of measurement and the test voltage for a given test object must be specified with respect to the acceptable terminal partial discharge level. (SWG/PE) 1291-1993r

(3) That specified maximum terminal partial discharge value for which measured terminal partial discharge values exceeding said value are considered unacceptable. The method of measurement and the test voltage for a given test object should be specified with the acceptable terminal partial discharge level. (PE/TR) C57.113-1988s

acceptance An action by an authorized representative of the acquirer by which the acquirer assumes ownership of software products as partial or complete performance of a contract. (C/SE) J-STD-016-1995

acceptance angle (fiber optics) Half the vertex angle of that cone within which optical power may be coupled into bound modes of an optical waveguide. *Notes:* 1. Acceptance angle is a function of position on the entrance face of the core when the refractive index is a function of radius in the core. In that case, the local acceptance angle is

$$\arcsin \sqrt{n^2(r) - n^2{}_2}$$

where $n(r)$ is the local refractive index and n_2 is the minimum refractive index of the cladding. The sine of the local acceptance angle is sometimes referred to as the local numerical aperture. 2. Power may be coupled into leaky modes at angles exceeding the acceptance angle. *See also:* launch numerical aperture; power-law index profile. (Std100) 812-1984w

acceptance criteria (1) (nuclear power quality assurance) Specified limits placed on characteristics of an item, process, or service defined in codes, standards, or other requirement documents. (PE/NP) [124]

(2) (software) The criteria that a system or component must satisfy in order to be accepted by a user, customer, or other authorized entity. *See also:* requirement; test criteria. (C) 610.12-1990

acceptance proof test (rotating machinery) A test applied to new insulated winding before commercial use. It may be performed at the factory or after installation, or both. (PE/EM) 95-1977r

acceptance quality level (aql) The maximum percent defective (maximum number of defects per 100 units) that, for the purpose of a sampling inspection, can be considered satisfactory as a process average. (PE/T&D) C135.61-1997

acceptance test (1) (A) (general) A test to demonstrate the degree of compliance of a device with purchaser's requirements. **(B) (general)** A test demonstrating the quality of the units of a consignment, without implication of contractual relations between buyer and seller. *Note:* American National Standards should use the term "conformance test" as directed by the Standards Council of ANSI, rather than the term acceptance test. Use of the term "conformance test" avoids the implication of contractual relations between buyer and seller. *See also:* test; acceptance testing; routine test; conformance tests. (SWG/SPD/PE) 32-1972, C37.100-1981

(2) (power cable systems) A test made after installation but before the cable system is placed in normal service. This test is intended to detect shipping or installation damage and to show any gross defects or errors in workmanship or splicing and terminating. (PE/IC) 400-1991

(3) A constant current or power capacity test made on a new battery to determine that it meets specifications or manufacturer's ratings. (PE/EDPG/NP) 1106-1995, 450-1995, 380-1975w

(4) (electric submersible pump cable) Test intended to detect damage prior to the initial installation of new cable. (IA/PC) 1017-1985s

(5) (battery) Capacity test made on a new battery to determine that it meets specifications or manufacturer's ratings. (SB) 1188-1996

acceptance testing (1) (A) (software) Formal testing conducted to determine whether or not a system satisfies its acceptance criteria and to enable the customer to determine whether or not to accept the system. **(B) (software)** Formal testing conducted to enable a user, customer, or other authorized entity to determine whether to accept a system or component. *Contrast:* development testing. *See also:* operational testing; qualification testing. (C) 610.12-1990

(2) (nuclear power plants) Evaluation or measurement of performance characteristics to verify that certain stated specifications and contractual requirements are met. (NI) N42.17B-1989r, N42.20-1995

(3) Testing conducted in an operational environment to determine whether a system satisfies its acceptance criteria (i.e., initial requirements and current needs of its user) and to enable the customer to determine whether to accept the system. (C/SE) 1012-1998

Acceptance Test or Launch Language A test language used to test applications on the Apollo launch vehicle. (C) 610.13-1993w

acceptance tests *See:* conformance tests.

accepted test A test on a system or model system that simulates the electrical, thermal, and mechanical stresses occurring in service. (IA/PC) 1068-1996

acceptor *See:* semiconductor.

access **(1)** **(A)** The process of obtaining data from or placing data into a storage device. *Synonym:* storage access. *See also:* access method; access mode. **(B)** To obtain data from or place data into a storage device as in definition (A). *See also:* random access; partitioned access; direct access; indexed sequential access; serial access; sequential access; indexed access. (C) 610.5-1990, 610.10-1994
(2) **(A)** Any means of establishing logical or physical communication with a computer or communications system. **(B)** Any means of obtaining the use of such a system. **(C)** Any actions that result in a flow of information involving such a system. **(D)** That part of a public network connecting the customer premises to the public network switching system (central office). (C) 610.7-1995
(3) To obtain data from or plate data into a storage device as in definition (A). *Synonym:* storage access. (C) 610.10-1994w

access arm In a magnetic disk device, an arm that supports and positions one or more magnetic heads. *See also:* voice-coil actuator. (C) 610.10-1994w

access code (telephone switching systems) One or more digits required in certain situations in lieu of or preceding an area or office code. (COM) 312-1977w

access control **(1)** The prevention of unauthorized use of a resource, including the prevention of use of a resource in an unauthorized manner. (LM/C) 802.10-1992
(2) The means to allow authorized entry and prevent unauthorized entry of persons, vehicles, and materials into an area. (PE/NP) 692-1997
(3) The prevention of unauthorized usage of resources. (C/LM) 8802-11-1999

access control field The *protocol control information* in a *slot,* which is used to support the *access control function.* (LM/C) 8802-6-1994

access control function The generic name for the *Queued Arbitrated (QA) Access* and *Pre-Arbitrated (PA) Access functions* in the *DQDB Layer* that control access to the medium in this part of ISO/IEC 8802. (LM/C) 8802-6-1994

access-control mechanism (software) Hardware or software features, operating procedures, or management procedures designed to permit authorized access to a computer system. *See also:* hardware; computer system; procedure; software. (C/SE) 729-1983s

access coupler (fiber optics) A device placed between two waveguide ends to allow signals to be withdrawn from or entered into one of the waveguides. *See also:* optical waveguide coupler. (Std100) 812-1984w

access credentials Data that are transferred to establish the claimed identity of a roadside equipment (RSE) application. (SCC32) 1455-1999

access fitting A fitting permitting access to the conductors in a raceway at locations other than at a box. *See also:* raceway. (EEC/PE) [119]

accessibility **(1)** **(software)** The extent to which software facilitates selective use or maintenance of its components. *See also:* software; maintenance; components. (C/SE) 729-1983s
(2) **(telephone switching systems)** The ability of a given inlet to reach the available outlets. (COM) 312-1977w
(3) **(A)** (As applied to equipment). Admitting close approach; not guarded by locked doors, elevation, or other effective means. *See also:* readily accessible. **(B)** (As applied to wiring methods.) Capable of being removed or exposed without damaging the building structure or finish or not permanently closed in by the structure or finish of the building. *See also:* exposed; concealed. (NESC/NEC) [86]
(4) **(power and distribution transformers)** Admitting close approach because not guarded by locked doors, elevation, or other effective means. (PE/TR) C57.12.80-1978r
(5) **(wiring methods)** Not permanently closed in by the structure or finish of the building; capable of being removed without disturbing the building structure or finish. (PE/EEC) [119]
(6) Admitting close approach to contact by persons due to lack of locked doors, elevation, or other effective safeguards. (PE/TR) C57.12.80-1978r

accessible object An object for which the client possesses a valid designator or handle. (C/PA) 1328-1993w, 1224-1993w, 1327-1993w

accessible, readily *See:* readily accessible.

accessible voltage drop Voltage difference between any two points accessible to workers at the work site. (T&D/PE) 1048-1990

access list A list of user IDs and group IDs of those users and groups authorized to place jobs in a queue. An access list is associated with a queue. A batch server uses the access list of a queue as one of the criteria in deciding to put a job in a queue. (C/PA) 1003.2d-1994

access mechanism A mechanism that is responsible for moving an access arm. *Synonym:* actuator. (C) 610.10-1994w

access method **(1)** **(LANs)** A communication technique where data is allowed or disallowed access to a communication system. (LM/C) 802.7-1989r
(2) **(data management)** A method for logically structuring data so that the storage location of any specific data item is well-defined. *Synonym:* access technique. *See also:* direct access method; basic access method. (C) 610.5-1990w

access mode **(1)** A technique that is used to access logical records within a file. *See also:* indexed sequential access mode; sequential access mode; file access mode; direct access mode. (C) 610.5-1990w
(2) A form of access permitted to a file. (PA/C) 9945-1-1996
(3) A form of access to a file or an attribute of a file indicating the kind of operations that may be performed on the file. (C) 1003.5-1999

accessories **(1)** **(power and distribution transformers) (general)** Devices that perform a secondary or minor duty as an adjunct or refinement to the primary or major duty of a unit of equipment. (SWG/PE/TR) C37.100-1992, C57.12.80-1978r
(2) **(raceway) (raceway systems for Class 1E circuits for nuclear power generating stations)** Devices that are used to supplement the functions of raceway systems. These include such items as dropouts, covers, conduit adapters, fastening devices (items such as conduit clamps, support connections, and cable tray cover clamps), adjustable connectors, and dividers. (PE/NP) 628-1987r

accessory **(1)** **(test, measurement, and diagnostic equipment)** An assembly of a group of parts or a unit that is not always required for the operation of a test set or unit as originally designed, but serves to extend the functions or capabilities of the test set; similarly, as headphones for a radio set supplied with a loudspeaker; a vibrator power unit for use with a set having a built-in power supply, or a remote control unit for use with a set having integral controls. (MIL) [2]
(2) **(electric and electronics parts and equipment)** A basic part, subassembly, or assembly designed for use in conjunction with or to supplement another assembly, unit, or set, contributing to the effectiveness thereof without extending or varying the basic function of the assembly or set. An accessory may be used for testing, adjusting, or calibrating purposes. Typical examples: test instrument, recording camera for radar set, headphones, emergency power supply. (GSD) 200-1975w

(3) **(power line maintenance)** A removable device attached to a major or primary operating tool allowing diversified operations. Example: universal tool. (T&D/PE) 516-1987s

accessory equipment (Class 1E motor) (nuclear power generating station) Devices other than the principal motor components that are furnished with or built as a part of the motor structure and are necessary for the operation of the motor. (PE/NP) 334-1974s

access path The manner in which related data items are linked to one another to permit access. (C) 610.5-1990w

access point (AP) **(1)** The point at which an abstract service is obtained. (C/PA) 1327.2-1993w, 1224.2-1993w, 1326.2-1993w, 1328.2-1993w
(2) Any entity that has station functionality and provides access to the distribution services, via the wireless medium (WM) for associated stations. (C/LM) 8802-11-1999

access technique *See:* access method.

access time **(1)** A time interval that is characteristic of a storage device, and is essentially a measure of the time required to communicate with that device. *Note:* Many definitions of the beginning and ending of this interval are in common use. (C) 162-1963w
(2) The elapsed time required to read from or write to a storage device after the proper controls and address have been applied. *See also:* latency; seek time; acceleration time; mean access time; transfer time. (C) 610.10-1994w
(3) **(A)** The time interval between the instant at which data are called for from a storage device and the instant delivery is completed, that is, the read time. **(B)** The time interval between the instant at which data are requested to be stored and the instant at which storage is completed; that is, the write time. (C) [85]
(4) **(A)** **(acousto-optic deflector)** The minimum time to randomly deflect the light beam from one spot position to another. It is given by the time it takes the acoustic beam to cross the optical beam; viz, $\tau = S/V$, with τ the access time, S the optical beam dimension, and V the acoustic velocity. **(B)** **(acoustically tunable optical filter)** The minimum time to randomly tune the filter from one wavelength to another. It is given by the time it takes the acoustic beam to cross the optical beam; namely: $\tau = S \cdot V$, with S the length of the optical beam along the acoustic beam direction and V the acoustic velocity. (UFFC) [17]

access tools (relaying) (switchgear assembly) (tamper-resistant switchgear assembly) Keys or other special accessories with unique characteristics that make them suitable for gaining access to the tamper-resistant switchgear assembly. (SWG/PE) C37.100-1981s, C37.20-1968w

access type *See:* file access mode.

access unit (AU) **(1)** The abstraction of the device that provides the IEEE 802.9 functionality, i.e., the integrated set of services, to stations connected across the IEEE 802.9 interface. (LM/C/COM) 8802-9-1996
(2) The functional unit in a *node* that performs the *DQDB Layer* functions to control access to both *buses*. Access units attach to each bus via a *write* connection and a *read* tap placed *upstream* of the write connection. (LM/C) 8802-6-1994

accident An unplanned event or series of events that results in death, injury, illness, environmental damage, or damage to or loss of equipment or property. (C/SE) 1228-1994

accommodation **(1)** **(general) (illuminating engineering)** The process by which the eye changes focus from one distance to another. *See also:* visual field. (ED) [127]
(2) **(laser maser)** The ability of the eye to change its power and thus focus for different object distances. (LEO) 586-1980w

accommodation, electrical *See:* electrical accomodation.

accommodation spaces Spaces provided for passengers and crew members that are used for berthing, dining rooms, mess spaces, offices, private baths, toilets and showers, lounges, and similar spaces. (IA/MT) 45-1998

accompanying documents Documents accompanying equipment or an accessory and containing all important documentation for the user, operator, installer, or assembler of equipment, particularly regarding safety. (EMB/MIB) 1073.4.1-2000

ac converter (self-commutated converters) A converter for changing alternating current (ac) power of a given voltage, frequency, and phase number to ac power in which one or more of these parameters are different. (IA/SPC) 936-1987w

accounting machine A device that reads data from external storage media, such as cards or tapes, and automatically produces accounting records or tabulation, usually on continuous forms. (C) 610.10-1994w

accounting management In networking, a management function defined for collecting and processing of data to evaluate resource consumption. (C) 610.7-1995

accreditation **(1)** The systematic and objective determination of a laboratory's competence to perform its services according to specific test methods/standards, by a qualified accreditation body, and issuance of a certificate attesting to that competence by the body. (NI) N42.23-1995
(2) *See also:* distributed simulation accreditation; model accreditation. (DIS/C) 1278.3-1996
(3) **(A)** Model/simulation accreditation is the official certification that a model or simulation is acceptable for use for a specific purpose. **(B)** Distributed simulation accreditation is the official certification that a distributed simulation is acceptable for use for a specific purpose. (C/DIS) 1278.4-1997

accredited standards committee A standards developing committee whose procedures have been determined to meet ANSI's requirements for fairness, openness, and other attributes necessary for developing a consensus position on a proposed ANSI standard relating to a specific technology area. (C) 610.7-1995

accredited standards development organization An organization recognized as a standards development organization by ISO, IEC, ITU-T, or recognized as a standards development organization by one of the member bodies of one of these three organizations. (C/PA) 14252-1996

accredited testing laboratory A testing laboratory that has been accredited by an authoritative body with respect to its qualifications to perform verification tests on the type of instruments covered by this standard. (NI) N42.20-1995

accumulated jitter The jitter at a PHY entity in the ring measured against the transmit clock of the active monitor. It is the total jitter accumulated by all the stations from the active monitor to the measurement point. It is typically used to determine the required size of the elastic buffer. (C/LM) 8802-5-1998

accumulated service years The length of time the transformer is operating from its in-service date until it is retired from service. It is suggested that de-energized time of three months or more not be considered in-service time. (PE/TR) C57.117-1986r

accumulating stimulus (electrotherapy) A current that increases so gradually in intensity as to be less effective than it would have been if the final intensity had been abruptly attained. *See also:* electrotherapy. (EMB) [47]

accumulator **(1)** **(A)** A device that retains a number (the augend), adds to it another number (the addend), and replaces the augend with the sum. **(B)** Sometimes only the part of definition (A) that retains the sum. *Note:* The term is also applied to devices that function as described but that also have other properties. (C) 162-1963
(2) Container that stores hydraulic oil under pressure as a source of fluid energy. (PE/EDPG) 1020-1988r
(3) A register or storage location in which the result of an operation is formed. (C) 610.10-1994w

accumulator function *See:* supervisory control functions.

accumulator point interfaces Master station or RTU (or both) element(s) that accept(s) a pulsing digital input signal to accumulate a total of pulse counts. (SUB/PE) C37.1-1994

accumulator SCADA function The capability of a supervisory system to accept and totalize digital pulses and make them available for display or recording, or both. (SUB/PE) C37.1-1994

accuracy (1) The quality of freedom from mistake or error, that is, of conformity to truth or to a rule. *Notes:* 1. Accuracy is distinguished from precision as in the following example: A six-place table is more precise than a four-place table. However, if there are errors in the six-place table, it may be more or less accurate than the four-place table. 2. The accuracy of an indicated or recorded value is expressed by the ratio of the error of the indicated value to the true value. It is usually expressed in percent. Since the true value cannot be determined exactly, the measured or calculated value of highest available accuracy is taken to be the true value or reference value. Hence, when a meter is calibrated in a given echelon, the measurement made on a meter of a higher-accuracy echelon usually will be used as the reference value. Comparison of results obtained by different measurement procedures is often useful in establishing the true value. *See also:* static accuracy; dynamic accuracy; measurement system; electronic analog computer. (IM) [38]

(2) (A) (analog computer) Conformity of a measured value to an accepted standard value. **(B) (analog computer)** A measure of the degree by which the actual output of a device approximates the output of an ideal device nominally performing the same function. *See also:* electronic analog computer. (C) 165-1977

(3) (power supply) Used as a specification for the output voltage of power supplies, accuracy refers to the absolute voltage tolerance with respect to the stated nominal output. (AES) [41]

(4) (numerically controlled machines) Conformity of an indicated value to the true value, that is, an actual or an accepted standard value. *Note:* Quantitatively, it should be expressed as an error or an uncertainty. The property is the joint effect of method, observer, apparatus, and environment. Accuracy is impaired by mistakes, by systematic bias such as abnormal ambient temperature, or by random errors (imprecision). The accuracy of a control system is expressed as the system deviation (the difference between the ultimately controlled variable and its ideal value), usually in the steady state or at sampled instants. *See also:* reproducibility; precision. (IA) [61]

(5) (electronic navigation) Generally, the quality of freedom from mistake or error; that is, of conformity to truth or a rule. Specifically, the difference between the mean value of a number of observations and the true value. *Note:* Often refers to a composite character including both accuracy and precision. *See also:* precision; navigation. (AES/RS) 686-1982s, [42]

(6) (signal-transmission system) Conformity of an indicated value to an accepted standard value or true value. *Note:* Quantitatively, it should be expressed as an error or uncertainty. The accuracy of a determination is affected by the method, observer, environment, and apparatus, including the working standard used for the determination. *See also:* signal. (IE) [43]

(7) (indicated or recorded value) The accuracy of an indicated or recorded value is expressed by the ratio of the error of the indicated value to the true value. It is usually expressed in percent. (EEC/PE) [119]

(8) (instrument transformers) The extent to which the current or voltage in the secondary circuit reproduces the current or voltage of the primary circuit in the proportion stated by the marked ratio, and represents the phase relationship of the primary current or voltage. (PE/TR) C57.12.80-1978r

(9) The extent to which the current in the secondary circuit reproduces the current in the primary circuit in the proportion stated by the marked ratio, and represents the phase relationship of the primary current. (PE/PSR) C37.110-1996

(10) (test, measurement, and diagnostic equipment) The degree of correctness with which a measured value agrees with the true value. *See also:* precision. (MIL) [2]

(11) (electrothermic power meters) The degree of correctness with which a measurement device yields the true value of a measured quantity; quantitatively expressed by uncertainty. *See also:* uncertainty. (IM) 544-1975w

(12) (nuclear power generating station) The quality of freedom from mistake or error. (PE/NP) 498-1985s

(13) (pulse measurement) The degree of agreement between the result of the application of a pulse measurement process and the true magnitude of the pulse characteristic, property, or attribute being measured. (IM/WM&A) 181-1977w

(14) (metric practice) The degree of conformity of a measured or calculated value to some recognized standard or specified value. This concept involves the systematic error of an operation, which is seldom negligible. *See also:* precision. (QUL) 268-1982s

(15) (measuring and test equipment) (nuclear power generating station) A measure of the degree by which the actual output of a device approximates the output of an ideal device nominally performing the same function. (PE/NP) 498-1985s

(16) (excitation systems for synchronous machines) The degree of correspondence between the controlled variable and the desired value under specified conditions such as load changes, ambient temperature, humidity, frequency, and supply voltage variations. Quantitatively, it is expressed as the ratio of difference between the controlled variable and the desired value to the desired value. (PE/EDPG) 421.1-1986r

(17) (A) (software) A qualitative assessment of correctness, or freedom from error. **(B) (software)** A quantitative measure of the magnitude of error. *See also:* precision. (C) 610.12-1990, 1084-1986

(18) ("dose calibrator" ionization chambers) Usually described in terms of overall uncertainty, accuracy is the estimate of the overall possible deviation from the stated value. The overall uncertainty is a total of the estimated error plus the random uncertainty of the measurement. (NI) N42.13-1986

(19) The extent to which a given measurement agrees with the defined value. (ELM) C12.1-1988

(20) (mathematics of computing) A qualitative assessment of correctness, or freedom from error. Contrast with precision. (C) 1084-1986w

(21) The degree of agreement of the measured value with the true value of the quantity being measured. (NI) N42.12-1994, N317-1980r, N323-1978r, N42.18-1980r

(22) (nuclear power generating station) A measure of the degree by which the actual output of a device approximates the output of an ideal device nominally performing the same function. (PE/NP) 498-1985s

(23) The degree of agreement of the observed value with the conventionally true value of the quantity being measured. (NI) N42.17B-1989r

(24) The degree of agreement between a measured value and the true value. (PE/PSIM) 4-1995

(25) The degree of agreement of the observed value with the conventionally true value of the quantity being measured. This degree of agreement can be quantified by computing the difference between the indicated value of a quantity and the correct (conventionally true) value of the quantity at the point of measurement. In the case of dose equivalent, it is expressed as $H_i - H_t$, where H_i is the indicated value and H_t is the conventionally true value. (NI) N42.20-1995

(26) The degree of exactness of an approximation or measurement. *Note:* Accuracy denotes the absolute quality of computed results; precision refers to the amount of detail used in representing those results. *See also:* precision. (C) 610.10-1994w

(27) A concept employed to describe the dispersion of measurements with respect to a known value. A measurement with small systematic uncertainties is said to have high accuracy. (NI) N42.23-1995

(28) The degree of conformity of a measured or calculated value to some reference value, which may be specified or unknown. This concept includes the systematic error of an operation, which is seldom negligible or known exactly. *See also:* precision. (SCC14) SI 10-1997

accuracy class The limits of transformer correction factor, in terms of percent error, that have been established to cover specific performance ranges for line power factors between 1.0 and 0.6 lag. (ELM) C12.11-1987

accuracy classes for metering (instrument transformers) Limits of a transformer correction factor, in terms of percent error, that have been established to cover specific performance ranges for line power-factor conditions between 1.0 and 0.6 lag. (PE/TR) [57], [117]

accuracy classes for relaying (instrument transformers) Limits, in terms of percent ratio error, that have been established. (PE/TR/PSR) [117], C37.110-1996

accuracy control character A control character used to indicate whether the data with which it is associated are in error, are to be disregarded, or cannot be represented on a particular device. *Synonym:* error control character. (C) 610.5-1990w

accuracy rating (1) (general) (electric instruments) The accuracy classification of the instrument. It is given as the limit, usually expressed as a percentage of full-scale value, that errors will not exceed when the instrument is used under reference conditions. *Notes:* 1. The accuracy rating is intended to represent the tolerance applicable to an instrument in an "as-received condition." Additional tolerances for the various influences are permitted when applicable. It is required that the accuracy, as received, be directly in terms of the indications on the scale and without the application of corrections from a curve, chart, or tabulation. Over that portion of the scale where the accuracy tolerance applies, all marked division points shall conform to the stated accuracy class. 2. Generally, the accuracy of electrical indicating instruments is stated in terms of the electrical quantities to which the instrument responds. In instruments with the zero at a point other than one end of the scale, the arithmetic sum of the end-scale readings to the right and to the left of the zero point shall be used as the full-scale value. *Exceptions:*

— The accuracy of frequency meters shall be expressed on the basis of the percentage of actual scale range. Thus, an instrument having a scale range of 55 Hz to 65 Hz would have its error expressed as a percentage of 10 Hz.

— The accuracy of a power-factor meter shall be expressed as a percentage of scale length.

— The accuracy of instruments that indicate derived quantities, such as series type ohmmeters, shall be expressed as a percentage of scale length.

3. In the case of instruments having nonlinear scales, the stated accuracy only applies to those portions of the scale where the divisions are equal to or greater than two-thirds the width they would be if the scale were even divided. The limit of the range at which this accuracy applies may be marked with a small isosceles triangle whose base marks the limit and whose point is directed toward the portion of the scale having the specified accuracy. 4. Instruments having an accuracy rating of 0.1% are frequently referred to as laboratory standards. Portable instruments having an accuracy rating of 0.25% are frequently referred to as portable standards. (EEC/AII) [102]

(2) (automatic control system) The limit that the system deviation will not exceed under specified operating conditions. (PE/EDPG) [3]

accuracy ratings for relaying The relay accuracy class is described by a letter denoting whether the accuracy can be obtained by calculation or must be obtained by test, followed by the minimum secondary terminal voltage that the transformer will produce at 20 times rated secondary current with one of the standard burdens without exceeding the relay accuracy class limit. (This is usually taken as 10%.) (PE/PSR) C37.110-1996

accuracy ratings of instrument transformers Means of classifying transformers in terms of percent error limits under specified conditions of operation. (PE/TR) C57.13-1978s

accuracy, synchronous-machine regulating system The degree of correspondence (or ratio) between the actual and the ideal values of a controlled variable of the synchronous-machine regulating system under specified conditions, such as load changes, drift, ambient temperature, humidity, frequency, and supply voltage. (PE) [9]

accuracy test (instrument transformers) A test to determine the degree to which the value of the quantity obtained from the secondary reflects the value of the quantity applied to the primary. (PE/TR) [57]

ACD *See:* automatic call distribution.

ac-dc general-use snap switch A form of general-use snap switch suitable for use on either ac or dc circuits for controlling the following:

a) Resistive loads not exceeding the ampere rating of the switch at the voltage applied.
b) Inductive loads not exceeding 50% of the ampere rating of the switch at the applied voltage. Switches rated in horsepower are suitable for controlling motor loads within their rating at voltage applied.
c) Tungsten-filament lamp loads not exceeding the ampere rating of the switch at the applied voltage if "T" rated.

(NESC/NEC) [86]

ACE (area control error) *See:* area control error.

ac electric field strength The electric field strength produced by ac power systems as defined by its space components along three orthogonal axes. For steady-state sinusoidal fields, each component can be represented by a complex number or phasor. The magnitudes of the components are expressed by their rms values in volts per meter, and their phases need not be the same. *Notes:* 1. A phasor is a complex number expressing the magnitude and phase of a time-varying quantity. Unless otherwise specified, it is used only within the context of linear systems driven by steady-state sinusoidal sources. In polar coordinates, it can be written as $Ae^{j\phi}$ where A is the amplitude or magnitude (usually rms but sometimes indicated as peak value) and ϕ is the phase angle. The phase angle should not be confused with the space angle of a vector. 2. The space components (phasors) are not vectors. The space components have a time-dependent angle while vectors have space angles. For example, the sinusoidal electric field strength, $\vec{E}$, can be expressed in rectangular coordinates as

$$\vec{E} = \vec{a}_x E_x + \vec{a}_y E_y + \vec{a}_z E_z$$

where, for example, the x component is

$$E_x = \mathrm{Re}\,(E_{x0}\, e^{j\phi x}\, e^{j\omega t}) = E_{x0} \cos\,(\phi_x + \omega t)$$

The magnitude, phase, and time-dependent angle are given by E_{x0}, ϕ_x, and $(\phi_x + \omega t)$, respectively. In this representation, the space angle of the x component is specified by the unit vector $\vec{a}_x$. An alternative general representation of a steady-state sinusoidal electric field can be derived algebraically from equation 1 above and is perhaps more useful in characterizing power-line fields because the fields along the direction of the line are small and can usually be neglected. It is a vector rotating in a plane where it describes an ellipse whose major semi-axis represents the magnitude and direction of the maximum value of the electric field, and whose minor semi-axis represents the magnitude and direction of the field a quarter-cycle later. As mentioned above, the electric field in the direction perpendicular to the plane of the ellipse is assumed to be zero. *See also:* single-phase ac fields; polyphase ac fields. (T&D/PE) 539-1990

ac electric field strength meter (1) A meter designed to measure the power-frequency electric field. Two types of electric field strength meters are in common use. (T&D/PE) 539-1990
(2) A meter designed to measure ac electric fields. Three types of electric field strength meters are available—free-body meter, ground-reference meter, and electro-optic meter. (T&D/PE) 1308-1994

acetate disks Mechanical recording disks, either solid or laminated, that are made of various acetate compounds. (SP) [32]

ACF (access control field) *See:* access control field.

ac filter Resistor-capacitor circuits connected in three-phase wye or delta on the ac terminals of a converter. (IA/ID) 995-1987w

ac general-use snap switch A form of general-use snap switch suitable only for use on alternating-current circuits for controlling the following:

a) Resistive and inductive loads, including electric-discharge lamps, not exceeding the ampere rating of the switch at the voltage involved.
b) Tungsten-filament lamp loads not exceeding the ampere rating of the switch at 120 V.
c) Motor loads not exceeding 80% of the ampere rating of the switch at its rated voltage.

(NESC/NEC) [86]

achromatic locus (television) (achromatic region) A region including those points in a chromaticity diagram that represent, by common acceptance, arbitrarily chosen white points (white references). *Note:* The boundaries of the achromatic locus are indefinite, depending on the tolerances in any specific application. Acceptable reference standards of illumination (commonly referred to as white light) are usually represented by points close to the locus of Planckian radiators having temperatures higher than about 2000°K. While any point in the achromatic locus may be chosen as the reference point for the determination of dominant wavelength, complementary wavelength, and purity for specification of object colors, it is usually advisable to adopt the point representing the chromaticity of the illuminator. Mixed qualities of illumination and luminators with chromaticities represented very far from the Planckian locus require special consideration. After a suitable reference point is selected, dominant wavelength may be determined by noting the wavelength corresponding to the intersection of the spectrum locus with the straight line drawn from the reference point through the point representing the sample. When the reference point lies between the sample point and the intersection, the intersection indicates the complementary wavelength. Any point within the achromatic locus chosen as a reference point may be called an achromatic point. Such points have also been called white points. (BT/AV) 201-1979w

acid-resistant So constructed that it will not be injured readily by exposure to acid fumes. (IA/ICTL/IAC/APP) [60], [75]

ACK (acknowledge character) *See:* acknowledge character.

ack cycle A cycle in which a slave responds to a master and terminates a transaction. (C/MM) 1196-1987w

acknowledge (1) An acknowledge packet. (C/MM) 1394-1995
(2) Operator action to indicate awareness of an event or alarm. (PE/NP) 692-1997

acknowledge bit A bit used by IEEE 802.3 Auto-Negotiation to indicate that a station has successfully received multiple identical copies of the Link Code Word. This bit is only set after an identical Link Code Word has been received three times in succession. (C/LM) 802.3-1998

acknowledge character (A) A transmission control character transmitted by a station as an affirmative response to the station with which the connection has been set up. **(B)** A transmission control character transmitted by a receiver as an affirmative response to a sender. An acknowledge character may also be used as an accuracy control character. *See also:* negative acknowledge character. (C) 610.5-1990

acknowledge gap The period of idle bus between the end of a packet and the start of an acknowledge. (C/MM) 1394-1995

acknowledge packet (1) A link-layer packet returned by a destination node back to a source node in response to most primary packets. An acknowledge packet is always exactly 8 bits long. (C/MM) 1394-1995
(2) The first packet returned by an individually addressed S-module that conveys to the M-module that the appropriate S-module is responding and indicates the current status of the responding S-module. (TT/C) 1149.5-1995
(3) An 8-bit packet that may be transmitted in response to the receipt of a primary packet. The most and least significant nibbles are the one's complement of each other. *Synonym:* acknowledge. (C/MM) 1394a-2000

acknowledger (forestaller) A manually operated electric switch or pneumatic valve by means of which, on a locomotive equipped with an automatic train stop or train control device, an automatic brake application can be forestalled, or by means of which, on a locomotive equipped with an automatic cab signal device, the sounding of the cab indicator can be silenced. (PE/EEC) [119]

acknowledging (forestalling) The operating by the engineman of the acknowledger associated with the vehicle-carried equipment of an automatic speed control or cab signal system to recognize the change of the aspect of the vehicle-carried signal to a more restrictive indication. The operation stops the sounding of the warning whistle, and in a locomotive equipped with speed control, it also forestalls a brake application. (PE/EEC) [119]

acknowledging device *See:* acknowledger.

acknowledging switch *See:* acknowledger.

acknowledging whistle An air-operated whistle that is sounded when the acknowledging switch is operated. Its purpose is to inform the fireman that the engineman has recognized a more restrictive signal indication. (EEC/PE) [119]

acknowledgment (1) (of a message) A reply transmitted by a receiving station to inform the sending station that a message has arrived and the message is error-free. *Contrast:* negative acknowledgment. (C) 610.7-1995
(2) A signal that is used to reply to a message or signal originator that its message or signal was received. (IM/ST) 1451.2-1997

ACL *See:* audit command language.

***a* contact** A contact that is open when the main device is in the standard reference position and that is closed when the device is in the opposite position. *Notes:* 1. *a* contact has general application. However, this meaning for front contact is restricted to relay parlance. 2. For indication of the specific point of travel at which the contact changes position, an additional letter or percentage figure may be added to *a*. *See also:* standard reference position. (SWG/PE) C37.100-1992

acoustic absorber Material with high acoustic loss placed on any part of the substrate for acoustic absorption purposes. (UFFC) 1037-1992w

acoustical depth finder *See:* echo sounder.

acoustically tunable optical filter An optical filter that is driven by an acoustic wave and that is tunable by varying the acoustic frequency. (UFFC) [17]

acoustic coupler (1) A type of data communication equipment that has sound transducers that permit the use of a telephone handset as a connection to a voice communication system for the purpose of data transmission. (LM/COM) 168-1956w
(2) A modem that interconnects a communicating device with a telephone handset. (C) 610.7-1995

acoustic delay line (1) A delay line whose operation is based on the time of propagation of sound waves. (C) [20], [85]

(2) A delay line whose operation is based on the time of propagation of sound waves within a given medium. *Synonym:* sonic delay line. *See also:* mercury storage. (C) 610.10-1994w

acoustic echo canceller (AEC) A circuit or algorithm designed to eliminate acoustic echoes and prevent howling due to acoustic feedback from loudspeaker to microphone. (COM/TA) 1329-1999

acoustic echo path (1) In a telephone set, the coupling from the receiver to the microphone (or transmitter). (COM/TA) 269-1992
(2) In a handset or headset system, the coupling from the receiver to the microphone (or transmitter). (COM/TA) 1206-1994

acoustic-gravity wave In the atmosphere, a low-frequency wave whose restoring forces are compressional, gravitational, and buoyant. (AP/PROP) 211-1997

acoustic input (1) The free-field sound pressure level developed by an artificial mouth at the mouth reference point. *See also:* sound pressure level. (COM/TA) 269-1992, 1206-1994
(2) The free-field sound pressure level developed by a mouth simulator at the mouth reference point. *See also:* sound pressure level. (COM/TA) 1329-1999

acoustic interferometer An instrument for the measurement of wavelength and attenuation of sound. Its operation depends upon the interference between reflected and direct sound at the transducer in a standing-wave column. *See also:* instrument. (EEC/PE) [119]

acoustic memory *See:* acoustic storage.

acoustic monitoring The detection of sound patterns emitted by equipment to determine its operating condition for predictive monitoring. (PE/NP) 933-1999

acoustic noise *See:* audible noise.

acoustic output (1) The sound pressure level developed in an artificial ear. *See also:* sound pressure level. (COM/TA) 269-1992, 1206-1994
(2) The sound pressure level developed at the measuring microphone. *See also:* sound pressure level. (COM/TA) 1329-1999

acoustic propagation loss Amplitude decay of the acoustic wave due to material damping; scattering caused by defects, surface finish, or electrodes; and acoustic bulk-wave radiation into the ambient environment. Specifically, this is the ratio of the power transmitted in a surface acoustic wave (SAW) beam to the power received, expressed in dB. (UFFC) 1037-1992w

acoustic radiator A means for radiating acoustic waves. (EEC/PE) [119]

acoustic regeneration The generation of a secondary acoustic wave by the potential variations of an electrode caused by a primary surface acoustic wave passing under it. (UFFC) 1037-1992w

acoustic storage A type of storage consisting of acoustic delay lines. (C) 610.10-1994w

acoustic tablet A data tablet on which the position of the sensor or stylus is determined by acoustic sensing techniques. (C) 610.10-1994w

acoustic wave filter A filter designed to separate acoustic waves of different frequencies. *Note:* Through electroacoustic transducers, such a filter may be associated with electric circuits. *See also:* filter. (EEC/PE) [119]

acoustic waveguide A perturbation along the direction of propagation of a surface acoustic wave to produce a decreased phase velocity, and hence, transverse concentration and guiding of the surface acoustic wave. (UFFC) 1037-1992w

acousto-optic device A device that is used to modulate light in amplitude, frequency, phase, polarization, or spatial position by virtue of optical diffraction from an acoustically generated diffraction grating. (UFFC) [23]

acousto-optic effect (fiber optics) A periodic variation of refractive index caused by an acoustic wave. *Note:* The acousto-optic effect is used in devices that modulate and deflect light. *See also:* modulation. (Std100) 812-1984w

ac power-line fields Power frequency electric and magnetic fields produced by ac power lines. (T&D/PE) 539-1990

acquirer (1) An organization that procures software products for itself or another organization. (C/SE) J-STD-016-1995
(2) The individual or organization that specifies requirements for and accepts delivery of a new or modified software product and its documentation. The acquirer may be internal or external to the supplier organization. Acquisition of a software product may involve, but does not necessarily require, a legal contract or a financial transaction between acquirer and supplier. (C/SE) 1058-1998
(3) A person or organization that acquires or procures a system or software product (which may be part of a system) from a supplier. (C/SE) 1062-1998

acquisition (1) The process of establishing a stable track on a target that is designated in one or more coordinates. A search of a limited given volume of coordinate space is usually required because of errors or incompleteness of the designation. (AES) 686-1997
(2) The process of obtaining a system or software product. (C/SE) 1062-1998

acquisition phase The final phase of the arbitration operation entered after determining that an agent has the highest priority and the bus is available. *See also:* arbitration operation; agent. (C/MM) 1296-1987s

acquisition probability The probability of establishing a stable track on a designated target. (AES) 686-1997

acquisition start time The start time of the acquisition of the histogram data, as

DD/MM/YR_HH:NN:SS_

where the '_' (underscore character) is an ASCII space; DD is the day; MM is the month; YR is the year; HH is the hours; NN is the minutes; and SS is the seconds. (NPS/NID) 1214-1992r

ac reactor (thyristor converter) An inductive reactor that is inserted between the transformer and the thyristor converter for the purpose of controlling the rate of rise of current in the thyristor and possibly the magnitude of fault current. (IA/IPC) 444-1973w

acronym A contrived reduction of nomenclature yielding mnemonics (ACRONYM). (C/MM) 1394a-2000

across-the-line starter A device that connects the motor to the supply without the use of a resistance or autotransformer to reduce the voltage. It may consist of a manually operated switch or a master switch, which energizes an electromagnetically operated contactor. (IA/MT) 45-1998

across-the-line starting (rotating machinery) The process of starting a motor by connecting it directly to the supply at rated voltage. (PE) [9]

ACSE/Presentation Service (APS) Environment The collection of information, associated with a particular APS instance, necessary to initiate and maintain an association with another application entity. (C/PA) 1351-1994w

ACSL *See:* Advanced Continuous Simulation Language.

ACSR (aluminum conductor, steel reinforced, aluminum cable steel reinforced) *See:* aluminum conductor, steel reinforced; aluminum cable steel reinforced.

act Abbreviation for ACTUAL, indicating the programmed functional capabilities of an end device. (AMR/SCC31) 1377-1997

acting stress (1) (seismic design of substations) Maximum applied or expected stress in the material during normal operation of the apparatus of which it is a part, including the stresses caused by wind, seismic or short-circuit loading, acting either independently or simultaneously, as determined by the user. (PE/SUB) 693-1984s

(2) (gas-insulated substations) The maximum applied or expected stress in a material during operation of the apparatus of which it is a part and including the stresses caused by gas pressure, wind, ice or loading. (SUB/PE) C37.122-1983s
(3) (working stress) The maximum applied or expected mechanical stress in a material during operation of the apparatus of which it is a part and including the stresses caused by seismic and other loading, acting independently or simultaneously as determined by the user.
(SUB/PE) C37.122.1-1993
(4) The maximum applied or expected stress in a material during operation of the apparatus of which it is a part and including the stresses caused by gas pressure, wind, ice, or seismic loading. *Synonym:* working stress.
(SWG/PE) C37.100-1992

Action An instance of the class `IEEE1451_Action` or of a subclass thereof. (IM/ST) 1451.1-1999

action A step a user takes to complete a task; a step that cannot be subdivided further. A single user action may invoke one or more functions but need not invoke any.
(C/SE) 1063-1987r

action potential (1) (medical electronics) The instantaneous value of the potential observed between excited and resting portions of a cell or excitable living structure. *Note:* It may be measured direct or through a volume conductor.
(EMB) [47]
(2) (overhead power lines) The electrical response of an excitable membrane that leads to the propagation of a nerve impulse; a nerve impulse. (T&D/PE) 539-1990

action spike (medical electronics) The greatest in magnitude and briefest in duration of the characteristic negative waves seen during the observation of action potentials.
(EMB) [47]

activate (A) The action of applying signals to a group of bus lines. **(B)** The state of a group of bus lines when they carry signals. (C/BA) 896.10-1997

activation (1) (thermionics) (cathode) The treatment applied to a cathode in order to create or increase its emission. *See also:* electron emission. (ED) [45], [84]
(2) One occurrence of a function's transformation of some subset of its inputs into some subset of its outputs.
(C/SE) 1320.1-1998

activation constraint A function's requirement for the presence of a nonempty object set in a particular arrow role as a precondition for some activation of the function.
(C/SE) 1320.1-1998

activation distance The distance traveled by a fall arrester or the amount of line payed out by a self-retracting lanyard from the point of onset of a fall to the activation point where the fall arrester begins to apply a braking or stopping force. This activation point is where the fall arrester engages the lifeline or, in the case of a self retracting lanyard, where an internal brake engages. Activation distance is part of the free fall distance experienced in a fall. (PE/T&D) 1307-1996

activation polarization The difference between the total polarization and the concentration polarization. *See also:* electrochemistry. (EEC/PE) [119]

activation time *See:* turn-on time.

active (1) (power system measurement) (electric generating unit reliability, availability, and productivity) The state in which a unit is in the population of units being reported on.
(PE/PSE) 762-1987w
(2) (696 interface devices) (signals and paths) A signal in its logically true state. (MM/C) 696-1983w
(3) (broadband local area networks) A cable plant component that consumes electrical power to perform its intended function. Examples of active devices include status monitors and amplifiers. (LM/C) 802.7-1989r
(4) Pertaining to a record or file that has been accessed by one or more transactions during a given processing cycle. *See also:* inactive; logically deleted; purged. (C) 610.2-1987
(5) When associated with a logic level (e.g., in the word active-low), this term identifies the logic level to which a signal shall be set to cause a defined action to occur. When referring to an output driver (e.g., in the phrase an active driver), this term describes the mode in which the driver is capable of determining the voltage of the network to which it is connected. (TT/C) 1149.1-1990, 1149.5-1995

active air terminal An air terminal which has been modified to lower its corona inception gradient. (PE/T&D) 1243-1997

active area (solar cells) The illuminated area normal to light incidence, usually the face area less the contact area. *Note:* For the purpose of determining efficiency, the area covered by collector grids is considered a part of the active area. *See also:* semiconductor. (AES/SS) 307-1969w

active array antenna system An array in which all or part of the elements are equipped with their own transmitter or receiver, or both. *Notes:* 1. Ideally, for the transmitting case, amplitudes and phases of the output signals of the various transmitters are controllable and can be coordinated in order to provide the desired aperture distribution. 2. Often it is only a stage of amplification or frequency conversion that is actually located at the array elements, with the other stages of the receiver or transmitter remotely located.
(AP/ANT) 145-1993

active current (rotating machinery) The component of the alternating current that is in phase with the voltage. *See also:* asynchronous machine. (PE) [9]

active-current compensator (rotating machinery) A compensator that acts to modify the functioning of a voltage regulator in accordance with active current. (PE) [9]

active data dictionary A data dictionary that ensures its own consistency with a system by limiting the data items that may be used by a process to those that are defined in the data dictionary. *Synonym:* embedded data dictionary. *Contrast:* passive data dictionary. (C) 610.5-1990w

active dimension (charged-particle detectors) (of a position-sensitive detector) A dimension (length, width) of that region of a position-sensitive detector that is depleted.
(NPS) 300-1988r

active electric network An electric network containing one or more sources of power. *See also:* network analysis.
(EEC/PE) [119]

active electrode (A) (electrobiology) A pickup electrode that, because of its relation to the flow pattern of bioelectric currents, shows a potential difference with respect to ground or to a defined zero, or to another (reference) electrode on related tissue. **(B) (electrobiology)** Any electrode, in a system of stimulating electrodes, at which excitation is produced. **(C) (electrobiology)** A stimulating electrode (different electrode) applied to tissue for stimulation and distinguished from another (inactive, dispersive, diffuse, or indifferent) electrode by having a smaller area of contact, thus affording a higher current density. *See also:* electrobiology. (EMB) [47]

active file (A) A file that is in current use. **(B)** A file with an expiration date that has not yet been reached.
(C) 610.5-1990

active filter (A) A filter network containing one or more voltage-dependent or current-dependent sources in addition to passive elements. **(B)** A filter containing energy generating elements. (CAS) [13]

active fire protection The minimizing of fire hazards in electrical systems by the use of fuses, circuit breakers, and other devices. (DEI) 1221-1993w

active-high signal A signal for which the logical-true (activated) state is represented by the high electrical state, and the logical-false (deactivated) state is represented by the low electrical state. (C/MM) 959-1988r

active homing guidance (navigation aid terms) A system of homing guidance wherein both the source of illuminating the target and the receiver for detecting the energy reflected from the target, as a result of illuminating the target, are carried within the vehicle. (AES/GCS) 172-1983w

active impedance (of an array element) The ratio of the voltage across the terminals of an array element to the current flowing at those terminals when all array elements are in place and excited. (AP/ANT) 145-1993

active laser medium (fiber optics) The material within a laser, such as crystal, gas, glass, liquid, or semiconductor, that emits coherent radiation (or exhibits gain) as the result of stimulated electronic or molecular transitions to lower energy states. *Synonym:* laser medium. *See also:* laser; optical cavity. (Std100) 812-1984w

active-low signal A signal for which the logical-false (deactivated) state is represented by the high electrical state, and the logical-true (activated) state is represented by the low electrical state. (C/MM) 959-1988r

active maintenance time The time during which maintenance actions are performed on an item, either manually or automatically. *Notes:* 1. Delays inherent in the maintenance operation (for example, those due to design or to prescribed maintenance procedures) shall be included. 2. Active maintenance may be carried out while the item is performing its intended function. (R) [29]

active materials (storage battery) The materials of the plates that react chemically to produce electric energy when the cell discharges and that are restored to their original composition, in the charged condition, by oxidation and reduction processes produced by the charging current. *See also:* battery. (PE/EEC) [119]

active monitor A station on the ring that is performing certain functions to ensure proper operation of the ring. These functions include 1) establishing clock reference for the ring; 2) assuring that a usable token is available; 3) initiating the neighbor notification cycle; 4) preventing circulating frames and priority tokens. In normal operation only one station on a ring may be the active monitor at any instance in time. (C/LM) 8802-5-1998

active package The package, if any, whose methods are accessible by name to the command interpreter, and to which newly created methods andproperties are added. (C/BA) 1275-1994

active port A connected, enabled port that observes bias and is capable of detecting all Serial Bus signal states and participating in the reset, tree identify, self-identify, and normal arbitration phases. (C/MM) 1394a-2000

active power (1) (rotating machinery) A term used for power when it is necessary to distinguish among apparent power, complex power, and its components, active and reactive power. *See also:* asynchronous machine. (PE) [9]
(2) (metering) The time average of the instantaneous power over one period of the wave. *Notes:* 1. For sinusoidal quantities in a two-wire circuit, it is the product of the voltage, the current, and the cosine of the phase angle between them. For nonsinusoidal quantities, it is the sum of all the harmonic components, each determined as above. In a polyphase circuit, it is the sum of the active powers of the individual phases. *See also:* active power. (ELM) C12.1-1982s
(3) (A) At the terminals of entry of a polyphase circuit into a delimited region, the algebraic sum of the active powers for the individual terminals of entry when the voltages are all determined with respect to the same arbitrarily selected common reference point in the boundary surface (which may be the neutral terminal of entry). *Notes:* 1. The active power for each terminal of entry is determined by considering each conductor and the common reference point as a single-phase two-wire circuit and finding the active power for each in accordance with the definition of "power, active (single-phase two-wire circuit)." If the voltages and currents are sinusoidal and of the same period, the active power P for a three-phase circuit is given by

$$P = E_aI_a \cos(\alpha_a - \beta_a) + E_bI_b \cos(\alpha_b - \beta_b) + E_cI_c \cos(\alpha_c - \beta_c)$$

where the symbols have the same meaning as in "power, instantaneous (polyphase circuit)." 2. If there is no neutral conductor and the common point for voltage measurement is selected as one of the phase terminals of entry, the expression will be changed in the same way as that for "power, instantaneous (polyphase circuit)." 3. If both the voltages and the currents in the preceding equations constitute symmetrical sets of the same phase sequences

$$P = 3E_aI_a \cos(\alpha_a - \beta_a)$$

4. In general the active power P at the ($m = 1$) terminals of entry of a polyphase circuit of m phases to a delimited region, when one of the terminals is the neutral terminal of entry, is expressed by the equation

$$P = \sum_{s=1}^{s=m} \sum_{r=1}^{r=\infty} E_{sr}I_{sr} \cos(\alpha_{sr} - \beta_{sr})$$

where E_{sr} is the root-mean-square amplitude of the rth harmonic of the voltage e_s, from phase conductor to neutral. I_{sr} is the root-mean-square amplitude of the rth harmonic of the current i_s through terminal s. α_{sr} is the phase angle of the rth harmonic of e_s with respect to a common reference. β_{sr} is the phase angle of the rth harmonic of i_s with respect to the same reference as the voltages. The indexes s and r have the same meaning as in "power, instantaneous (polyphase circuit)." 5. The active power can also be stated in terms of the root-mean-square amplitudes of the symmetrical components of the voltages and currents as

$$P = m \sum_{k=0}^{k=m-1} \sum_{r=1}^{r=\infty} E_{kr}I_{kr} \cos(\alpha_{kr} - \beta_{kr})$$

where m is the number of phase conductors, k denotes the number of the symmetrical component, and r denotes the number of the harmonic component. 6. When the voltages and currents are quasi-periodic and the amplitudes of the voltages and currents are slowly varying, the active power for the circuit of each conductor may be determined for this condition as in "ower, active (single-phase two-wire circuit)." The active power for the polyphase circuit is the sum of the active power values for the individual conductors. The active power is also the time average of the instantaneous power for the polyphase circuit. 7. Mathematically, the active power at any time t_0 is

$$P = \frac{1}{T}\int_{t_0-T/2}^{t_0+T/2} p\,dt$$

where p is the instantaneous power and T is the period. This formulation may be used when the voltage and current are periodic or quasi-periodic so that the period is defined. The active power is expressed in watts when the voltages are in volts and the currents in amperes. **(B)** At the terminals of entry of a single-phase, two-wire circuit into a delimited region, when the voltage and current are periodic or quasi-periodic, the time average of the values of the instantaneous power, the average being taken over one period. *Notes:* 1. Mathematically, the active power P at a time t_0 is given by the equation

$$P = \frac{1}{T}\int_{t_0-T/2}^{t_0+T/2} p\,dt$$

where T is the period, and p is the instantaneous power. 2. If both the voltage and current are sinusoidal and of the same period the active power P is given by

$$P = EI \cos(\alpha - \beta)$$

in which the symbols have the same meaning as in "power, instantaneous (two-wire circuit)." 3. If both the voltage and current are sinusoidal, the active power P is also equal to the real part of the product of the phasor voltage and the conjugate of the phasor current, or to the real part of the product of the conjugate of the phasor voltage and the phasor current. Thus,

$$P = \mathrm{Re}\mathbf{EI}^* = \mathrm{Re}\mathbf{E}^*\mathbf{I} = \frac{1}{2}[\mathbf{EI}^* + \mathbf{E}^*\mathbf{I}]$$

in which **E** and **I** are the root-mean-square phasor voltage and root-mean-square phasor current, respectively (see "phasor quantity"), and the * denotes the conjugate of the phasor to which it is applied. 4. If the voltage is an alternating voltage and the current is an alternating current (see "alternating voltage and alternating current"), the active power is given by the equations

$$P = E_1 I_1 \cos(\alpha_1 - \beta_1) + E_2 I_2 \cos(\alpha_2 - \beta_2) + \ldots$$

$$= \sum_{r=1}^{r=\infty} \mathrm{E}_r \mathrm{I}_r \cos(\alpha_r - \beta_r)$$

$$= Re \sum_{r=1}^{r=\infty} E_r I_r$$

$$= \frac{1}{2} \sum_{r=1}^{r=\infty} [E_r I_r + E_r I_r]$$

in which r is the order of the harmonic component of the voltage (see "harmonic components (harmonics)") and r is also the order of the harmonic component of the current. E_r and I_r are the phasors corresponding to the rth harmonic of the voltage and current, respectively. 5. If the voltage and current are quasi-periodic functions of the form given in "power, instantaneous (two-wire circuit)," the integral over the period T will not result in the simple expressions that are obtained when E_r and I_r are constant. However, if the relative rates of change of the quantities are so small that each may be considered to be constant during any one period, but to have slightly different values in successive periods, the active power at any time t is very closely approximated by

$$P = \sum_{r=1}^{r=\infty} E_r(t) I_r(t) \cos(\alpha_r - \beta_r)$$

which is analogous to the preceding expression. When the amplitudes of voltage and current are slowly changing, the active power may be represented by this expression. 6. Active power is expressed in watts when the voltage is in volts and the current in amperes. 7. With reference to "power," when it is clear that "average power" and not "instantaneous power" is meant, "power" is often used for "active power." (Std100) 270-1966

(4) The average power consumed by a unit. For a two terminal device with current voltage waveforms $i(t)$ and $v(t)$, which are periodic T, the real or active power is

$$P = \frac{1}{T} \int_0^T v(t) i(t) dt$$

(PEL) 1515-2000

active-power relay (general) A power relay that responds to active power. *See also:* power relay; relay; reactive power relay. (SWG/PE/PSR) C37.100-1992, C37.90-1978s

active preventive maintenance time That part of the active maintenance time in which preventive maintenance is carried out. *Notes:* 1. Delays inherent in the preventive maintenance operation (for example, those due to design or prescribed maintenance procedures) shall be included. 2. Active preventive maintenance time does not include any time taken to maintain an item that has been replaced. (R) [29]

active redundancy (1) (computers) In fault tolerance, the use of redundant elements operating simultaneously to prevent, or permit recovery from, failures. *Contrast:* standby redundancy. *See also:* active redundancy. (C) 610.12-1990

(2) That redundancy wherein all means for performing a given function are operating simultaneously. (R) [29]

active reflection coefficient (of an array element) The reflection coefficient at the terminals of an array element when all array elements are in place and excited. (AP/ANT) 145-1993

active region A region of a detector in which charge created by ionizing radiation contributes significantly to the output signal. (NPS) 325-1996

active repair time The time during which corrective maintenance actions are performed on an item either manually or automatically. *Notes:* 1. Delays inherent in the repair operation (for example, those due to design or to prescribed maintenance procedures) shall be included. 2. Active repair time does not include any time taken to repair an item that has been replaced as part of the corrective maintenance action under consideration. (R) [29]

active requester *See:* requester.

active retimed concentrator A type of token ring concentrator that performs an embedded repeater function in the lobe port's data path, thereby providing ring segment boundaries at the concentrator lobe port connector (CMIC). (C/LM) 8802-5-1998

active segment interconnect A segment interconnect is said to be active if it is asserting AS=1 on the far-side segment. (NID) 960-1993

active sensor (test, measurement, and diagnostic equipment) A sensor requiring a source of power other than the signal being measured. (MIL) [2]

active sounding The remote sensing of atmospheric or ionospheric parameters by transmission and reception of radio signals. (AP/PROP) 211-1997

active speech level A period of time during which speech spurt intervals are followed by speech pause intervals. (COM/TA) 743-1995

active storage Storage that holds data that is being processed. (C) 610.10-1994w

active test An on-going test that is invoked by a write to the TEST_START register. The node is in the testing state (STATE_CLEAR.state is equal to testing) while an active test is in progress. (C/MM) 1212-1991s

active testing (test, measurement, and diagnostic equipment) The process of determining equipment static and dynamic characteristics by performing a series of measurements during a series of known operating conditions. Active testing may require an interruption of normal equipment operations, and it involves measurements made over the range of equipment operation. *See also:* interference testing. (MIL) [2]

active topology At any time, the set of communication paths in a Bridged Local Area Network that can be used in transferring data between end stations on the LANs. (C/LM) 802.1G-1996

active transducer A transducer whose output waves are dependent upon sources of power, apart from that supplied by any of the actuating waves, which power is controlled by one or more of the waves. *Note:* The definition of active transducer is a restriction of the more general active network: that is, one in which there is an impressed driving force. *See also:* transducer. (Std100) 270-1966w

activities Events in the software life cycle for which effort data is collected and reported. (C/SE) 1045-1992

Activity A defined body of work to be performed, including its required Input and Output Information. *See also:* Activity Group. (C/SE) 1074-1997

activity (1) The expected number of spontaneous nuclear decays (transformations) in unit time from a specified energy state (excluding prompt decays from a lower nuclear level) for a given amount of a radionuclide. Its standard unit (SI) is the becquerel (Bq), where one Bq equals one decay per second. Activity has often been expressed in curies (Ci), where 3.7×10^{10} Bq equals 1 Ci, exactly. (NI) N42.14-1991

(2) (computers) In modeling and simulation, a task that consumes time and resources and whose performance is necessary for a system to move from one event to the next. (C) 610.3-1989w

(3) A set of tasks that relate to the performance of a specific function in a plant phase. Information is compiled throughout the Plant Information Network (PIN) at the activity level. (PE/EDPG) 1150-1991w

(4) A constituent task of a Process. *See also:* task. (C/SE) 1074-1995s
(5) *See also:* function. (C/SE) 1320.1-1998

activity-based simulation A discrete simulation that represents the components of a system as they proceed from activity to activity; for example, a simulation in which a manufactured product moves from station to station in an assembly line. (C) 610.3-1989w

activity coordinator A person who is an expert in the methodology and development of the activity documentation packages and who is responsible for coordination and development of the activity documentation packages with the activity technical contacts. (PE/EDPG) 1150-1991w

activity data list A list that itemizes the major data items used by the activity, gives a brief description of each data item, and lists other activities that provide or receive each data item. (PE/EDPG) 1150-1991w

activity dip For a vibrating beam accelerometer (VBA), the phenomenon where at certain frequencies the resonator vibration amplitude decreases due to parasitic resonances within itself or with the surrounding structure. (AES/GYAC) 1293-1998

activity documentation package A summary of the results of the activity investigation which includes the activity description, activity process diagram, activity data list, activity entity-relationship diagram, and the activity support modules. (PE/EDPG) 1150-1991w

activity description An overview of the activity that briefly describes the activity and its scope, and delineates the boundaries and major tasks of the activity. (PE/EDPG) 1150-1991w

activity entity-relationship (E-R) diagram A diagram that defines the data contents of the activity by identifying its data entities, associated data attributes, and the relationships among data entities. (PE/EDPG) 1150-1991w

activity fractioning monitor An instrument that separates airborne radioactivity into two or more specific fractions and monitors each fraction. (NI) N42.17B-1989r

Activity Group A set of related Activities. *See also:* Activity. (C/SE) 1074-1997

activity list A list containing names of activities that define the work processes of a generating plant from conception to decommissioning. There are approximately 400 such activities that comprise the power-plant life cycle and are mainline activities that are common throughout the industry. The list also separately contains brief descriptions of the activities. (PE/EDPG) 1150-1991w

activity process diagram A diagram that shows the relationships and flow of tasks within an activity and represents the process required to complete an activity. (PE/EDPG) 1150-1991w

activity ratio The ratio of active records to the total number of records in a file. (C) 610.2-1987

activity response (sodium iodide detector) The net number of counts registered by the detector system per unit of time divided by the activity of the radionuclide that is being measured during the same unit of time. (NI) N42.12-1980s

activity support modules Procedures or computer programs that operate on the data associated with the activity. (PE/EDPG) 1150-1991w

activity technical contact A person who is knowledgeable about the functions, tasks, data, and details related to an activity. (PE/EDPG) 1150-1991w

ACTOR An object-oriented language designed to facilitate development of SAA-compliant systems. (C) 610.13-1993w

Actor The local entity in a Link Aggregation Control Protocol exchange. (C/LM) 802.3ad-2000

actual address The real or designed address that is built into the computer by the manufacturer as a storage location or register. (C) 610.10-1994w

actual ESD events Non-simulated electrostatic discharges that occur in the intended environment of the electronic equipment. (EMC) C63.16-1993

actual generation (electric generating unit reliability, availability, and productivity) The energy that was generated by a unit in a given period. Actual generation can be expressed as gross actual generation (GAAG) or net actual generation (NAAG). (PE/PSE) 762-1987w

actual instruction* *See:* effective instruction.

* Deprecated.

actual key A key that directly expresses the physical location of a logical record on a storage medium. (C) 610.5-1990w

actual parameter *See:* argument.

actual time *See:* real time.

actual time to crest The time interval from the start of the transient to the time when the maximum amplitude is reached. (PE/TR) C57.12.90-1999

actual transient recovery voltage (TRV) **(1)** The TRV (transient recovery voltage) that actually occurs across the terminals of a pole of a switching device following current interruption. *Note:* The actual TRV may differ from the inherent TRV due to the modifying effects of device impedance and arc-circuit interaction.
(SWG) C37.04E-1985w, C37.4D-1985w, C37.100B-1986w
(2) That which actually occurs across the terminals of a pole of a switching device following current interruption. *Note:* The actual TRV may differ from the inherent TRV due to the modifying effects of device impedance and arc-circuit interaction. (SWG/PE) C37.100-1992

actual weight The measured weight of a finished, ready-to-run vehicle; the tare weight. *Synonym:* empty weight. (VT) 1475-1999

actuated equipment **(1) (nuclear power plants)** A component or assembly of components that performs, or directly contributes to the performance of, a protective function such as reactor trip, containment isolation, or emergency coolant injection. The following are examples of actuated equipment: an entire control rod with its release or drive mechanism, a containment isolation valve with its operator, and a safety injection pump with its prime mover. (PE/NP) 380-1975w, 381-1977w
(2) The assembly of prime movers and driven equipment used to accomplish a protective action. *Note:* Examples of prime movers are: turbines, motors, and solenoids. Examples of driven equipment are: control rods, pumps, and valves. (PE/NP) 603-1998

actuating current (of an automatic line sectionalizer) The rms current that actuates a counting operating or an automatic operation. (SWG/PE) C37.100-1992

actuating device (protective signaling) A manually or automatically operated mechanical or electric device that operates electric contacts to effect signal transmission. *See also:* protective signaling. (EEC/PE) [119]

actuating signal **(1)** The reference input signal minus the feedback signal. *See also:* feedback control system. (IA/ICTL/APP/IAC) [69], [60]
(2) A particular input pulse in the control circuitry of a computer. (C) 610.10-1994w

actuation device **(1) (nuclear power plants)** A component or assembly of components (or module) that directly controls the motive power (electricity, compressed air, etc.) for actuated equipment. The following are examples of an actuation device: a circuit breaker, a relay, a valve (with its operator) used to control compressed air to the operator of a containment isolation valve, (and a module containing such equipment). (PE/NP) 381-1977w
(2) A component or assembly of components that directly controls the motive power (electricity, compressed air, hydraulic fluid, etc.) for actuated equipment. *Note:* Examples of actuation devices are: circuit breakers, relays, and pilot valves. (PE/NP) 603-1998

actuation time, relay *See:* relay actuation time.

actuator (1) (automatic train control) A mechanical or electric device used for automatic operation of a brake valve. (EEC/PE) [119]
(2) **(A)** A mechanism that moves an object in order to access a storage device. For example, the device that selects a laser disk in a jukebox, or an access arm in a magnetic disk drive. **(B)** In robotics, a motor or transducer that uses electrical, hydraulic, or pneumatic energy to effect motion in a robot. (C) 610.10-1994
(3) A transducer that accepts an electrical signal and converts it into a physical action. (IM/ST) 1451.2-1997
(4) A component that provides a physical output in response to a stimulating variable or signal. (IM/ST) 1451.1-1999

actuator, centrifugal *See:* centrifugal actuator.

actuator, relay *See:* relay actuator.

actuator valve An electropneumatic valve used to control the operation of a brake valve actuator. (EEC/PE) [119]

acutance A measure of the sharpness of the edges in an image. (C) 610.4-1990w

acute care Short-term care, i.e., less than 30 days. (EMB/MIB) 1073-1996

acute exposure Exposure to a large dose during a relatively short time. (T&D/PE) 539-1990

ac winding *See:* alternating-current winding.

acyclic machine (rotating machinery) A direct-current machine in which the voltage generated in the active conductors maintains the same direction with respect to those conductors. *Synonym:* homopolar machine. (PE) [9]

A/D Acronym for analog-to-digital, as in A/D converter. (C) 610.10-1994w

Ada A programming language designed, developed, and primarily used by the United States Department of Defense. The original design of Ada was based on Pascal, with more complex features such as private data types, synchronized rendezvous for multi-tasking environments, and exception handlers. *Note:* Named after Ada Lovelace, an early pioneer in computing. *See also:* extensible language; HAL; block-structured language. (C) 610.13-1993w

Ada83 The original Ada language standard, approved by ANSI in 1983 and by ISO/IEC in 1987. (C/PA) 1003.5b-1995

Ada I/O The input/output operations defined in Ada RM and further defined in IEEE Std 1003.5b-1995. (C/PA) 1003.5b-1995

Ada95 The 1995 Ada language standard, item 1 in 1.2, used in contrast to Ada 83. (C/PA) 1003.5b-1995

adaptability *See:* flexibility.

adaptation (illuminating engineering) The process by which the retina becomes accustomed to more or less light than it was exposed to during an immediately preceding period. It results in a change in the sensitivity to light. *Note:* Adaptation is also used to refer to the final state of the process, as reaching a condition of adaptation to this or that level of luminance. *See also:* photopic vision; chromatic adaptation; scotopic vision. (ED) [127]

adaptation data Data used to adapt a program to a given installation site or to given conditions in its operational environment. (C) 610.12-1990

adaptation parameter A variable that is given a specific value to adapt a program to a given installation site or to given conditions in its operational environment; for example, the variable Installation_Site_Latitude. (C) 610.12-1990

adapter (1) (general) A device for connecting parts that will not mate. An accessory to convert a device to a new or modified use. (IM/HFIM) [40]
(2) A device, or series of devices, designed to provide a compatible connection between the unit under test (UUT) and the test equipment. It may include proper stimuli or loads not contained in the test equipment. (MIL/SCC20) [2], 993-1997
(3) A device or series of devices designed to provide a compatible connection between the test subject and the test equipment. *Synonyms:* interface device; interface test adapter; test adapter. (SCC20) 1226-1998

adapter kit (test, measurement, and diagnostic equipment) A kit containing an assortment of cables and adapters for use with test or support equipment. (MIL) [2]

adapter, standard *See:* standard adapter.

adapter, waveguide *See:* waveguide adapter.

adapting *See:* self-adapting.

adaptive antenna system An antenna system having circuit elements associated with its radiating elements such that one or more of the antenna properties are controlled by the received signal. (AP/ANT) 145-1993

adaptive coding The application of two or more image compression techniques to a single image, based on properties of different parts of the image. (C) 610.4-1990w

adaptive color shift (illuminating engineering) The change in the perceived object's color caused solely by the change of the state of chromatic adaptation. *See also:* state of chromatic adaptation. (ED) [127]

adaptive control system A control system within which automatic means are used to change the system parameters in a way intended to improve the performance of the control system. *See also:* feedback control system. (IA/IM/PE/ICTL/APP/EDPG/IAC) [69], [120], [3], [60]

adaptive equalization (data transmission) A system that has a means of monitoring its own frequency response characteristics and a means of varying its own parameters by closed-loop action to obtain the desired overall frequency response. (PE) 599-1985w

adaptive equalizer An electronic device for maximizing the signal quality on a transmission channel by monitoring the signal and adjusting the equalization. *Synonym:* automatic equalizer. (C) 610.7-1995

adaptive maintenance (1) (software) Software maintenance performed to make a computer program usable in a changed environment. *Contrast:* corrective maintenance; perfective maintenance. (C) 610.12-1990
(2) Modification of a software product performed after delivery to keep a computer program usable in a changed or changing environment. (C/SE) 1219-1998

adaptive relay A relay that can change its setting and/or relaying logic upon the occurrence of some external signal or event. (PE/PSR) C37.113-1999

adaptive relaying A protection philosophy that permits, and seeks to make adjustments automatically, in various protection functions to make them more attuned to prevailing power conditions. (PE/PSR) C37.113-1999

adaptive routing A routing strategy that dynamically adjusts path selection based on current network parameters. (C) 610.7-1995

adaptive system A system that has a means of monitoring its own performance and a means of varying its own parameters by closed-loop action to improve its performance. *See also:* system science. (SMC) [63]

ADC *See:* analog-to-digital converter.

ADC conversion gain The number of channels over which the full amplitude span can be spread; usually 2048−8192 channels are used for Ge gamma-ray spectrometry. (NI) N42.14-1991

ADC number A four-character number identifying the ADC (analog to digital converter) used for the data. Leading spaces are interpreted as leading zeros. Normally, the ADC numbers would start at 1 and go up in sequence for a given system. Different systems in a specific laboratory could use non-sequential numbers, e.g., 1 to 4, and 11 to 14, for different types of equipment. (NPS/NID) 1214-1992r

Adcock antenna A pair of vertical antennas separated by a distance of one-half wavelength or less, and connected in phase opposition to produce a radiation pattern having the shape of the figure eight in all planes containing the centers of the two antennas. (AP/ANT) 145-1993

add To insert a record into an existing file. (C) 610.5-1990w

add-and-subtract relay *See:* bidirectional relay.

added source statements The count of source statements that were created specifically for the software product. (C/SE) 1045-1992

addend A number to be added to another number (the augend) to produce a result (the sum). (C) 1084-1986w

adder (1) A device whose output is a representation of the sum of the two or more quantities represented by the inputs. *See also:* half-adder; electronic analog computer. (C/MIL) 162-1963w, [2]
(2) A device whose output data is the arithmetic sum of the two or more quantities presented as input data. *Contrast:* subtracter. *See also:* summer; half adder; quarter adder; serial adder; adder-subtracter; full adder; parallel adder. (C) 610.10-1994w

adder-subtracter A device that acts either as an adder or subtracter depending upon the control signal received. *Note:* The adder-subtracter may be constructed so as to yield the sum and the difference at the same time. (C) 610.10-1994w

add file A file containing records that are being added or are to be added to a master file. (C) 610.5-1990w

addition agent (electroplating) A substance that, when added to an electrolyte, produces a desired change in the structure or properties of an electrodeposit, without producing any appreciable change in the conductivity of the electrolytes, or in the activity of the metal ions or hydrogen ions. *See also:* electroplating. (PE/EEC) [119]

addition without carry* *See:* exclusive OR.
* Deprecated.

additive A chemical compound or compounds added to an insulating fluid for the purpose of imparting new properties or altering those properties that the fluid already has. (PE/TR) 637-1985r

add-on board *See:* expansion board.

add record A record that is to be added or that has been added to a master file. *Contrast:* deletion record. (C) 610.5-1990w

address (1) (semiconductor memory) Those inputs whose states select a particular cell or group of cells. (TT/C/AMR/SCC31) 662-1980s, 1377-1997
(2) (A) (electronic computation) An identification, as represented by a name, label, or number, for a register, location in storage, or any other data source or destination such as the location of a station in a communication network. **(B) (electronic computation)** Loosely, any part of an instruction that specifies the location of an operand for the instruction. **(C) (electronic computation) (electronic machine-control system)** A means of identifying information or a location in a control system. Example: The *x* in the command *x* 12345 is an address identifying the numbers 12345 as referring to a position on the *x* axis. (C) [85]
(3) (test pattern language) The identification of a specific memory word, usually expressed in *x*-, *y*-, and *z*-coordinates, and in binary code. (TT/C) 660-1986w
(4) A character or group of characters that identifies a register, a particular part of storage, or some other data source or destination. (ED/ED) 641-1987w, 1005-1998
(5) (STEbus) The reference to a unit of data or the value represented by the address lines while ADRSTB* is active. (MM/C) 1000-1987r
(6) An identifier that tells where a *service access point (SAP)* may be found (ISO 7498). (LM/C) 8802-6-1994
(7) (A) A number, character or group of character that identifies a given device or storage location. **(B)** To refer to a device, data item or storage location by an identifying number, character, or group of characters, known as its address, as in definition (A). *Synonym:* address reference. *See also:* relative address; relocatable address; indirect address; virtual address; absolute address; implied addressing; effective address. (C) 610.10-1994
(8) An identifying name, label, or number for a data terminal, source, or storage location calculation. (SUB/PE) 999-1992w
(9) An unambiguous name, label, or number that identifies the location of a particular entity or service. (C/PA) 1328.2-1993w, 1326.2-1993w, 1327.2-1993w, 1224.2-1993w
(10) A character or group of characters that identifies a register, a particular part of storage, or some other data source or destination. (IM/ST) 1451.2-1997
(11) *See also:* primary address. (NID) 960-1993

addressable memory A region of memory that can be located by an address. *Synonym:* addressed memory. (C) 610.10-1994w

addressable point In computer graphics, a position on a device that can be specified by coordinates. *See also:* pixel. (C) 610.6-1991w

addressable register A register with a fixed location and address. (C) 610.10-1994w

address broadcast The phase of a bus cycle that selects one slave as the responding slave and zero or more slaves as participating slaves. During the address broadcast the active master broadcasts the addressing information and then asserts an address strobe. After the slaves acknowledge the address broadcast, the master terminates the address broadcast. (C/MM) 1096-1988w

address bus A bus used to carry an address from the processor to memory or to a peripheral device. (C) 610.10-1994w

address calculation sort An insertion sort in which each of the items to be sorted is inserted into one of several lists, according to an address calculated from its value, and the resulting lists are then merged. *Synonym:* multiple list insertion sort. (C) 610.5-1990w

address cycle *See:* primary address cycle.

address/data bus signal group A set of thirty-six (36) signals, consisting of 32 address/data signals and four parity signals that are used for address and data transfers. (C/MM) 1296-1987s

addressed board A board that recognizes its address while ADRSTB* is active. (C/MM) 1000-1987r

addressed memory *See:* addressable memory.

addressed refresh A RAM-refresh protocol, in which the controller-provided read0-requests schedule the timing and specify addresses for RAM refresh cycles. (C/MM) 1596.4-1996

address, effective *See:* effective address.

address error An error that occurs when a node recognizes its own address in a packet's improper source or destination information. (C) 610.7-1995

address_error An error-status code returned to the requester when a transaction is directed to a non-existing address; on some buses, this has been called a NACK (negative acknowledge). The address_error status is generally returned if a valid address acknowledgement is not observed within a fixed timeout period. (C/MM) 1212-1991s

address field (1) The part of a *protocol data unit (PDU)* that contains an *address* that identifies one or more addressable entities. (The address may be a single-source address, single-destination address, or multiple-destination address [:*multicast*]:). (LM/C/EMB/MIB) 8802-6-1994, 1073.3.1-1994
(2) A sequence of bits that identifies the intended destination or receiver of a transmission. *Note:* May be single-source, single-destination, or multiple-destination address. (C) 610.7-1995
(3) Any of the fields of a computer instruction that contain addresses, information necessary to derive either other addresses, or values of operands. *Synonym:* address part. *See also:* operation field; operand field. (C) 610.10-1994w, 610.12-1990

address fields The ordered pair of service access point (SAP) addresses at the beginning of an LLC PDU that identifies the LLC(s) designated to receive the protocol data unit (PDU) and LLC sending the PDU. Each address field is one octet in length. (C/LM/CC) 8802-2-1998

address format (computers) The arrangement of the address parts of an instruction. *Note:* The expression plus-one is frequently used to indicate that one of the addresses specifies the location of the next instruction to be executed, such as one-plus-one, two-plus-one, three-plus-one, four-plus-one. (C) [20], [85]
(2) (A) (computers) The number and arrangement of address fields in a computer instruction. *See also:* n–address instruction; *n*–plus-one address instruction. **(B) (computers)** The number and arrangement of elements within an address, such as the elements needed to identify a particular channel, device, disk sector, and record in magnetic disk storage. (C) 610.12-1990
(3) (A) The number and arrangement of elements within an address, such as the elements needed to identify a particular channel, device, disk sector or record on a storage device. *Note:* The expression "plus-one" is frequently used to indicate that one of the addresses specifies the location of the next instruction to be executed; for example in an three-plus-one address format, an instruction contains three addresses of operands for the present operation, plus one address representing the next instruction to be executed. Such an instruction is known as a "three-plus-one address instruction." *See also:* two-address instruction; three-address instruction; four-address instruction; n-plus-one address instruction format; instruction format; one-address instruction. **(B)** The number and arrangement of elements within an address, such as the elements within an address, such as the elements needed to identify a particular channel, device, disk sector or record on a storage device. (C) 610.10-1994

addressibility The ability to locate an item in storage using an address. (C) 610.10-1994w

address ID (broadband local area networks) A unique digital identification sequence that is used to identify a device on a network. (LM/C) 802.7-1989r

addressing *See:* extended addressing (32-bit); fixed addressing (64-bit); extended addressing (64-bit).

addressing exception (software) An exception that occurs when a program calculates an address outside the bounds of the storage available to it. *See also:* protection exception; data exception; underflow exception; operation exception; overflow exception. (C) 610.12-1990

addressing mode (1) (microprocessor assembly language) The manner in which an operand is to be accessed during execution of an instruction. (C/MM) 695-1985s
(2) A means of combining information in an instruction, in registers, or in memory to define the location of a datum; for example, direct addressing, immediate addressing; implied addressing; indirect addressing; indexed addressing; relative addressing; symbolic addressing; virtual addressing. (C) 610.10-1994w

address invariance When a multi-byte data item is transferred over the bus, bytes with the same relative memory address are always mapped to the same bus lanes, regardless whether the processor is big or little endian and regardless of the significance of the bytes. (C/BA) 896.3-1993w

address-invariant A convention for defining the byte-ordering of DMA messages and data. The first byte in a contiguous set corresponds to the lowest order address, the second byte corresponds to the next address, etc. This is independent of how these addresses are sequenced onto a serial bus or assigned to physical positions and sequenced on a parallel bus. When different bus types are connected within a system, the bridge is expected to map the respective byte lanes and sequential positions to maintain address-invariance. (C/MM) 1212.1-1993

addressless instruction *See:* zero-address instruction.

address locked operation (FASTBUS acquisition and control) An operation directed to a single primary address containing a mixture of read and write cycles, possibly including block transfers as well. (NID) 960-1993

address mark A mark on a disk that is used to identify the specific areas on the disk such as an index, or free storage. *See also:* index mark. (C) 610.10-1994w

address modification (software) Any arithmetic, logical, or syntactic operation performed on an address. *See also:* effective address; relative address; relocatable address; indexed address. (C) 610.12-1990, 610.10-1994w

address-only cycle (1) (VMEbus) A data transfer bus (DTB) cycle that consists of an address broadcast, but does not have a data transfer. Slaves do not acknowledge address-only cycles and masters terminate the cycle without waiting for an acknowledgment. (C/BA) 1014-1987
(2) (VSB) The DTB cycle that consists of an address broadcast, but no data transfer. The active master terminates the cycle after the slaves acknowledge the address broadcast. (C/MM) 1096-1988w

address-only transaction(s) A bus transaction that does not include a data phase. The only information transferred is contained within the connection phase and, in some cases, the disconnection phase. (C/BA) 10857-1994, 1014.1-1994w, 896.3-1993w, 896.4-1993w

address part (1) A part of an instruction that usually is an address, but that may be used in some instructions for another purpose. (C) 162-1963w
(2) (software) *See also:* address field. (C) 610.12-1990

address reference *See:* address.

address register (1) (computers) A register in which an address is stored. (C) [20], [85]
(2) A register in which an address is stored. *Note:* An address register is generally used in an operand field of a processor instruction and contains a pointer to the address holding the data value to be used by the instruction. *See also:* base address register; instruction address register. (C) 610.10-1994w

address space (1) (A) The range of addresses that a computer program can access. *Note:* In some systems, this may be the set of physical storage locations that a program can access, disjoint from other programs, together with the set of virtual addresses referring to those storage locations which may be accessible by other programs. **(B)** The number of memory locations that a central processing unit can address. *See also:* virtual address space. (C) 610.10-1994, 610.12-1990
(2) The memory locations that can be referenced by a process. (C/PA) 1003.5-1999, 9945-2-1993
(3) The memory locations that can be referenced by the threads of a process. (C/PA) 9945-1-1996

address space identifier (ASI) An 8-bit field appended to the address by the **integer unit.** It identifies the address space being accessed and typically encodes whether the processor is in user or supervisor mode. (C/MM) 1754-1994

address stop An address that, when it is encountered by a program, causes the program to halt execution. *See also:* breakpoint instruction; instruction address stop. (C) 610.10-1994w

address table sorting (data management) A sorting technique in which a table of addresses that point to the items to be sorted is manipulated instead of moving the items themselves. *See also:* list sorting; key sorting. (C) 610.5-1990w

address, tag *See:* symbolic address.

address trace (A) To monitor references made to a particular address. **(B)** A list of addresses of previously executed instructions, in the order in which they were executed. *Note:* Generally used for debugging. (C) 610.10-1994

address track A track that contains addresses that may be used to locate data on other tracks of the same data medium. *Note:* Usually refers to disk drives. (C) 610.10-1994w

add transaction A transaction that causes a new record to be added to a master file. *See also:* update transaction; delete transaction; null transaction; change transaction. (C) 610.2-1987

address transfer (MULTIBUS II) The passing of address information over the multiplexed address/data bus from the bus owner in order to select the replying agent(s). *See also:* bus owner; replying agent. (C/MM) 1296-1987s
MULTIBUS is a registered trademark of Intel corporation.

address translator (A) A device that transforms the address of an instruction to the address in main storage at which it is to be loaded or relocated. **(B)** In virtual storage, a device that transforms the address of an item of data or instruction from its virtual address into its real address. (C) 610.10-1994

add time The elapsed time required to perform one addition operation, not including the time required to obtain the operands or to return the result to storage. *Contrast:* subtract time; multiply time. (C) 610.10-1994w

Adel'son-Velskii and Landis tree (data management) A height-balanced binary tree in which the difference in height of the two subtrees of any node is at most 1. *Note:* Also referred to as a HB tree; a height-balanced 1-tree. (C) 610.5-1990w

ADF *See:* automatic direction finder.

adhesion (coefficient of) During rolling contact, the ratio between the longitudinal tangential force at the wheel-rail/running surface interface and the normal force. (VT) 1475-1999

ad hoc (data management) Pertaining to an item such as a computer program or database used for a particular and specific purpose; for example, an ad hoc query. *Note:* Usually the item is used for a relatively short time, then discarded. (C) 610.5-1990w

ad hoc network A network composed solely of stations within mutual communication range of each other via the wireless medium (WM). An ad hoc network is typically created in a spontaneous manner. The principal distinguishing characteristic of an ad hoc network is its limited temporal and spatial extent. These limitations allow the act of creating and dissolving the ad hoc network to be sufficiently straightforward and convenient so as to be achievable by nontechnical users of the network facilities; i.e., no specialized "technical skills" are required and little or no investment of time or additional resources is required beyond the stations that are to participate in the ad hoc network. The term *ad hoc* is often used as slang to refer to an independent basic service set (IBSS). (C/LM) 8802-11-1999

ad hoc query (data management) A query that is used for a particular and specific purpose. *Note:* Such a query is usually used once or twice, then discarded. (C) 610.5-1990w

adiabatic atmosphere of refraction *See:* refractive index gradient.

A-display A display in which targets appear as vertical deflections from a horizontal line representing a time base. Time delay, or target range is indicated by the horizontal position of the deflection from one end of the time base. The vertical deflection is a function of signal amplitude.

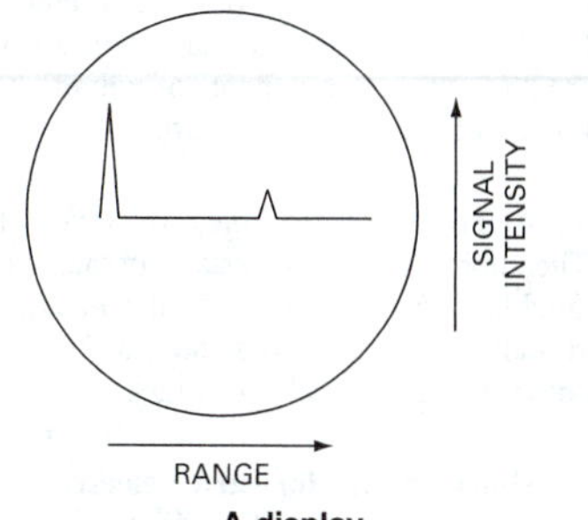

A-display

(AES) 686-1997

adjacency In character recognition, a condition in which the character spacing reference lines of two consecutive characters printed on the same line are separated by less than a specified distance. (C) 610.2-1987

adjacent bridges Two Local or Remote Bridges are termed adjacent if both are attached to the same LAN or Remote Bridge Group. (C/LM) 802.1G-1996

adjacent channel A channel whose frequency band is adjacent to that of another channel, known as the reference channel. (C/PE) 610.10-1994w, 599-1985w

adjacent-channel attenuation *See:* selectance.

adjacent-channel interference (data transmission) Interference, in a reference channel, caused by the operation of an adjacent channel. (PE) 599-1985w

adjacent-channel selectivity and desensitization (receiver performance) (receiver) A measure of the ability to discriminate against a signal at the frequency of the adjacent channel. Desensitization occurs when the level of any off-frequency signal is great enough to alter the usable sensitivity. (VT) [37]

adjoint system (1) (analog computer) A method of computation based on the reciprocal relation between a system of ordinary linear differential equation and its adjoint. *Note:* By solution of the adjoint system, it is possible to obtain the weighting function (response to a unit impulse) $W(T, t)$ of the original system for fixed T (the time of observation) as a function of t (the time of application of the impulse). Thus, this method has particular application to the study of systems with time-varying coefficients. The weighting function then may be used in convolution to give the response of the original system to an arbitrary input. *See also:* electronic analog computer. (C) 165-1977w
(2) For a system whose state equations are $dx(t)/dt = f(x(t),u(t),t)$, the adjoint system is defined as that system whose state equations are $dy(t)/dt = -y(t)$, where A^* is the conjugate transpose of the matrix whose i,j element is $\partial f_i/\partial x_j$. *See also:* control system. (CS/IM) [52]

adjust (1) (instrument) Change the value of some element of the mechanism, or the circuit of the instrument or of an auxiliary device, to bring the indication to a desired value, within a specified tolerance for a particular value of the quantity measured. *See also:* instrument. (EEC/PE) [119]
(2) (airborne radioactivity monitoring) To alter the response by means of a variable, built-in control such as a potentiometer. (NI) N42.17B-1989r
(3) To alter the reading of an instrument by means of a variable (hardware or software) control. (NI) N42.20-1995

adjustable (As applied to circuit breakers.) A qualifying term indicating that the circuit breaker can be set to trip at various values of current and/or time within a predetermined range. (NESC/NEC) [86]

adjustable constant-speed motor A motor, the speed of which can be adjusted to any value in the specified range, but when adjusted, the variation of speed with load is a small percentage of that speed. For example, a direct-current shunt motor with field-resistance control designed for a specified range of speed adjustment. *See also:* asynchronous machine. (PE) [9]

adjustable impedance-type ballast (illuminating engineering) A reference ballast consisting of an adjustable inductive reactor and a suitable adjustable resistor in series. These two components are usually designed so that the resulting combination has sufficient current-carrying capacity and range of impedance to be used with a number of different sizes of lamps. The impedance and power factor of the reactor-resistor combination are adjusted and checked each time the unit is used. (EEC/LB) [97]

adjustable-speed drive An electric drive designed to provide easily operable means for speed adjustment of the motor, within a specified speed range. *See also:* electric drive. (IA/ICTL/IAC) [60]

adjustable-speed motor A motor whose speed can be varied gradually over a range of speeds, but when once adjusted remains practically unaffected by the load, such as a dc shunt-wound motor with field resistance control designed for a range of speed adjustments. (IA/MT) 45-1998

adjustable varying-speed motor A motor whose speed can be adjusted gradually, but when once adjusted for a given load will vary with change in load; such as a dc compound-wound motor adjusted by field control or a wound-rotor induction motor with speed control. (IA/MT) 45-1998

adjustable varying-voltage control A form of armature-voltage control obtained by impressing a voltage that may be changed by small increments on the armature of the motor, but that, when adjusted for a given load, will vary considerably with change in load with a consequent change in speed, such as may be obtained from a differentially compound-wound generator with adjustable field current or by means of an adjustable resistance in the armature circuit. *See also:* control. (IA/ICTL/IAC) [60]

adjustable voltage control A form of armature-voltage control obtained by impressing on the armature of the motor a voltage that may be changed in small increments; but when adjusted, it, and consequently the speed of the motor, are practically unaffected by a change in load. *Note:* Such a voltage may be obtained from an individual shunt-wound generator with adjustable field current for each motor. *See also:* control. (IA/IAC) [60]

adjusted NEXT loss The NEXT loss in decibels of a channel plus $15\log F/F_{ref}$, where F is the measured frequency and F_{ref} is a reference frequency (4 MHz at 4 Mbit/s and 16 Mbit/s). It is used to determine the NEXT to interference (NIR) ratio of a channel. (C/LM) 8802-5-1998

adjusted speed The speed obtained intentionally through the operation of a control element in the apparatus or system governing the performance of the motor. *Note:* The adjusted speed is customarily expressed in percent (or per unit) of base speed (for direct-current shunt motors). *See also:* electric drive. (IA/ICTL/IAC) [60]

adjuster A means to shorten or lengthen a strap, webbing or rope. (T&D/PE) 1307-1996

adjust line mode In text formatting, an operating mode in which line endings are automatically adjusted to comply with the current margin setting. *See also:* text end adjustment; word wrap. (C) 610.2-1987

adjustment (test, measurement, and diagnostic equipment) The act of manipulating the equipment's controls to achieve a specified condition. (MIL) [2]

adjustment accuracy (direct-current instrument shunts) The limit of error, expressed as a percentage of the rated output voltage, in the initial adjustment of the shunt made when employing a low-current measurement method. (PE/PSIM) 316-1971w

adjustment accuracy of instrument shunts (electric power system) The limit of error, expressed as a percentage of the rated voltage drop, of the initial adjustment of the shunt by resistance or low-current methods. (PE/PSIM) [55]

Adler tube* *See:* beam parametric amplifier.

* Deprecated.

administrative application A program that is concerned with managing operational aspects of the Media Management System (MMS), and typically does not itself own media. An administrative application may include an interface for administrative users. Examples of administrative applications include those that allow the addition and removal of applications, drives, libraries, media, and computer systems from the MMS, as well as those concerned with allocation and policy management of an installation. (C/SS) 1244.1-2000

administrative authority (1) The governmental authority exercising jurisdiction over application of this guide. (T&D/PE) 1307-1996
(2) The governmental authority exercising jurisdiction over application of this code. (NESC) C2-1997

administrative controls Rules, orders, instructions, procedures, policies, practices, and designations of authority and responsibility. (PE/NP) 603-1998

administrative data processing The use of computers for administrative applications such as personnel, payroll, and accounting functions. (C) 610.2-1987

administrative downtime Downtime caused by administrative or maintenance activities. If an activity that restores service for a few customers interferes with service to a larger number, then the outage of the larger number shall be included. Outages due to maintenance or administrative activities that can be scheduled to minimize interference with the customer may be weighted differently from the contribution of randomly occurring outages. Program software up-dates ordinarily fall into this category. (COM/TA) 973-1990w

administrative security Management constraints, operational procedures, and other administrative controls to enforce a security policy. (C/BA) 896.3-1993w

admissible control input set A set of control inputs that satisfy the control constraints. *See also:* control system. (CS/IM) [52]

admittance (data transmission) The reciprocal of impedance. (PE) 599-1985w
(2) (A) (linear constant-parameter system) The corresponding admittance function with p replaced by j_ω in which ω is real. **(B) (linear constant-parameter system)** The ratio of the phasor equivalent of a steady-state sine-wave current or current-like quantity (response) to the phasor equivalent of the corresponding voltage or voltage-like quantity (driving force). *Note:* Definitions (A) and (B) are equivalent. (Std100) 270-1966

admittance, effective input (electron tube or valve) The quotient of the sinusoidal component of the control-grid current by the corresponding component of the control voltage, taking into account the action of the anode voltage on the grid current; it is a function of the admittance of the output circuit and the interelectrode capacitance. *Note:* It is the reciprocal of the effective input impedance. *See also:* electron-tube admittances. (ED) [44], [84]

admittance, effective output (electron tube or valve) The quotient of the sinusoidal component of the anode current by the corresponding component of the anode voltage, taking into account the output admittance and the interelectrode capacitance. *Note:* It is the reciprocal of the effective output impedance. *See also:* electron-tube admittances. (ED) [44], [84]

admittance, electrode *See:* electrode admittance.

admittance matrix, short-circuit (multiport network) A matrix whose elements have the dimension of admittance and, when multiplied into the vector of port voltages, gives the vector of port currents. (CAS) [13]

admittance, short-circuit (A) (general) An admittance of a network that has a specified pair or group of terminals short-circuited. **(B)** (four-terminal network or line) The input, output, or transfer admittance parameters y_{11}, y_{22}, and y_{12} of a four-terminal network when the far end is short-circuited. (CAS) [13]

admittance, short-circuit driving-point (*j*th terminal of an *n*-terminal network). The driving-point admittance between that terminal and the reference terminal when all other terminal shave zero alternating components of voltage with respect to the reference point. *See also:* electron-tube admittances. (ED) 161-1971w

admittance, short-circuit feedback (electron-device transducer) The short-circuit transfer admittance from the physically available output terminals to the physically available input terminals of a specified socket, associated filters, and electron device. *See also:* electron-tube admittances. (ED) 161-1971w

admittance, short-circuit forward (electron-device transducer) The short-circuit transfer admittance from the physically available output terminals of a specified socket, associated filters, and electron device. *See also:* electron-tube admittances. (ED) 161-1971w

admittance, short-circuit input (electron-device transducer) The driving-point admittance at the physically available input terminals of a specified socket, associated filters, and tube. All other physically available terminals are short-circuited. *See also:* electron-tube admittances. (ED) 161-1971w

admittance, short-circuit output (electron-device transducer) The driving-point admittance at the physically available output terminals of a specified socket, associated filters, and tube. All other physically available terminals are short-circuited. *See also:* electron-tube admittances. (ED) 161-1971w

admittance, short-circuit transfer (from the *j*th terminal to the *l*th terminal of an *n*-terminal network) The transfer admittance from terminal *j* to terminal *l* when all terminals except *j* have zero complex alternating components of voltage with respect to the reference point. *See also:* electron-tube admittances. (ED) 161-1971w

ADP (automatic data processing, administrative data processing) *See:* automatic data processing; automated data processing; administrative data processing.

ADSIM *See:* Applied Dynamics International Simulation Language.

advance ball (mechanical recording) A rounded support (often sapphire) attached to a cutter that rides on the surface of the recording medium so as to maintain a uniform depth of cut and to correct for small irregularities of the disk surface. (SP) [32]

Advanced Continuous Simulation Language A simulation language used for continuous simulation applications. (C) 610.13-1993w

advanced z transform (data processing) The advanced z transform of $f(t)$ is the z transform of $f(t + \Delta T)$; that is,

$$\sum_{n=0}^{\infty} f(nT + \Delta T)z^{-n}$$

$$0 < \Delta < 1$$

(IM) [52]

adverse water conditions (power operations) Water conditions that limit hydroelectric energy production. (PE/PSE) 858-1987s

adverse weather (1) (electric power system) Weather conditions that cause an abnormally high rate of forced outages for exposed components during the periods such conditions persist. *Note:* Adverse weather conditions can be defined for a particular system by selecting the proper values and combinations of weather: thunderstorms, tornadoes, wind velocities, precipitation, temperature, etc. *See also:* outage. (PE/PSE) [54], 859-1987w
(2) (generating station) Designates weather conditions that cause an abnormally high rate of forced outages for exposed components during the periods such conditions persist, but do not qualify as major storm disasters. Adverse weather conditions can be defined for a particular system by selecting the proper values and combinations of conditions reported by the Weather Bureau: thunderstorms, tornadoes, wind velocities, precipitation, temperature, etc. *Note:* This definition derives from transmission and distribution applications and does not necessarily apply to generation outages. *See also:* major storm disaster. (PE/PSE) 346-1973w

adverse-weather lamps *See:* fog lamps.

advertised ability An operational mode that is advertised using Auto-Negotiation. (C/LM) 802.3-1998

AEEC (Airlines Electronic Engineering Committee) *See:* Airlines Electronic Engineering Committee.

aeolian flexure Flexure of cables caused by the wind. (PE/IC) 1143-1994r

aeolight (optical sound recording) A glow lamp employing a cold cathode and a mixture of permanent gases in which the intensity of illumination varies with the applied signal voltage. (SP) [32]

AEP *See:* application environment profile.

aeration cell *See:* differential aeration cell.

Aerex *See:* explosives.

aerial belt A single D-ring belt designed for attachment when in an aerial bucket or platform. (T&D/PE) 1307-1996

aerial cable (1) A cable for installation on a pole line or similar overhead structure that may be self-supporting or installed on a supporting messenger (cable) and is designed to resist solar radiation and precipitation. A self-supported aerial cable is one that includes a messenger cable that has an outer jacket that covers the messenger and the shield. The messenger is available for support, gripping, pulling, and tensioning. (PE/PSC) 789-1988w
(2) An assembly of insulated conductors installed on a pole or similar overhead structures; it may be self-supporting or installed on a supporting messenger cable. *See also:* cable. (T&D/PE) [10]

aerial device A vehicular mounted articulating device or telescoping boom-type personal lift device, or both, equipped with one or more buckets or a platform used to position a worker. (T&D/PE) 516-1995

aerial lug *See:* external connector.

aerial platform A device designed to be attached to the boom tip of a crane or aerial lift and support a worker in an elevated working position. Platforms may be constructed with surrounding railings that are fabricated from aluminum, steel, or fiber reinforced plastic. Occasionally, a platform is suspended from the load line of a large crane. *Synonyms:* platform; cage. (T&D/PE) 524-1992r

aerial work (power line maintenance) Work performed on equipment used for the transmission and distribution of electricity, which is performed in an elevated position on various structures, conductors, or associated equipment. (T&D/PE) 516-1995

aerodrome beacon (illuminating engineering) An aeronautical beacon used to indicate the location of an aerodrome. *Note:* An aerodrome is any defined area on land or water—including any buildings, installations, and equipment—intended to be used either wholly or in part for the arrival, departure, and movement of aircraft. (ED) [127]

aerometeorgraph (navigation aid terms) A self-recording instrument for the simultaneous recording of atmospheric pressure, temperature, and humidity. (AES/GCS) 172-1983w

aeronautical beacon (illuminating engineering) An aeronautical ground light visible at all azimuths, either continuously or intermittently, to designate a particular location on the surface of the earth. (ED) [127]

aeronautical ground light (illuminating engineering) Any light specially provided as an aid to air navigation, other than a light displayed on an aircraft. (ED) [127]

aeronautical light (illuminating engineering) Any luminous sign or signal that is specially provided as an aid to air navigation. (ED) [127]

Aeronautical Radio Incorporated (ARINC) An organization concerned with providing services to airlines, including sponsorship of voluntary standardization among airlines and airframe and avionics manufacturers. (ATLAS) 771-1989s

aerophare *See:* navigation; radio beacon.

aerophase *See:* radio beacon.

aerosol (1) (laser maser) A suspension of small solid or liquid particles in a gaseous medium. Typically, the particle sizes may range from 100 μm to 0.01 μm or less. (LEO) 586-1980w
(2) (nuclear power plants) Suspension of solid or liquid particles in a gas. (NI) N42.17B-1989r

aerosol development (electrostatography) Development in which the image-forming material is carried to the field of the electrostatic image by means of a suspending gas. *See also:* electrostatography. (ED) [46]

aerospace support equipment (test, measurement, and diagnostic equipment) All equipment (implements, tools, test equipment, devices [mobile or fixed], and so forth), both airborne and ground, required to make an aerospace system (aircraft, missile, and so forth) operational in its intended environment. Aerospace support equipment includes ground support equipment. (MIL) [2]

AEW *See:* airborne early warning.

AF *See:* analog-to-frequency converter.

AFC (automatic frequency control) *See:* automatic frequency control.

AFCS (automatic flight control system) *See:* automatic flight control system.

afferent Pertaining to a flow of data or control from a subordinate module to a superordinate module in a software system. *Contrast:* efferent. (C) 610.12-1990

affiliate A remote convergence protocol entity (CPE) whose CPE address is known to the local CPE. (LM/C) 15802-2-1995

affiliation A state that exists if both remote and local CPEs know each other's CPE addresses. (LM/C) 15802-2-1995

affirmative response An input string that matches one of the responses acceptable to the LC_MESSAGES category keyword yesexpr, matching an ERE in the current locale. (C/PA) 9945-2-1993

AFIPS *See:* American Federation of Information Processing Societies.

afterimage (illuminating engineering) A visual response that occurs after the stimulus causing it has ceased. (ED) [127]

afterpulse (photo multipliers) A spurious pulse induced in a photomultiplier by a previous pulse. *See also:* phototube. (NPS) 398-1972r

AGC (automatic generation control, automatic gain control) *See:* automatic generation control; automatic gain control.

agent (1) A physical unit that has an interface to the parallel system bus, for example, a single-board computer. (C/MM) 1296-1987s

(2) A switch or switch-like component or bridge between the requester and the responder. During normal operation the agent's intervention is transparent to the requester and responder. (C/MM) 1596-1992

(3) An active switch, switch-like component, or bridge, between the requester and responder. During normal system operation, the agent is transparent to the requester and responder. (C/MM) 1212-1991s

(4) An active component or bridge that acts on behalf of the real target for an action. For example, a DMA queue could be placed in a bus bridge in order to perform special operations when crossing address or protection domain boundaries. (C/MM) 1212.1-1993

(5) A switch or switch-like component between a RamLink controller and a RamLink slave. During normal operation, the agent has two behaviors: from the higher-level controller's perspective, the agent behaves like a RamLink slave, and from the lower-level slave's perspective, the agent behaves like a controller. (C/MM) 1596.4-1996

(6) Refers to the managed nodes in a network. Managed nodes are those nodes that contain a network management entity (NME), which can be used to configure the node and/or collect data describing operation of that node. The agent is controlled by a network control host or manager that contains both an NME and network management application (NMA) software to control the operations of agents. Agents include systems that support user applications as well as nodes that provide communications services such as front-end processors, bridges, and routers. (C/LM) 802.3-1998

agent code A term used to refer to network management entity software residing in a node that can be used to remotely configure the host system based on commands received from the network control host, collect information documenting the operation of the host, and communicate with the network control host. (C/LM) 802.3-1998

agent error An agent status that indicates an error condition in a replying agent. (C/MM) 1296-1987s

agent status The condition of the replying agent, transmitted during the reply phase of a transfer operation. *See also:* transfer operation; reply phase. (C/MM) 1296-1987s

aggregate A group of entities or a group of other aggregates. The substitution of the word "unit" is used to avoid phrases like "aggregate aggregate." (C/DIS) 1278.1a-1998

aggregate responsibility A broadly stated responsibility that is eventually refined as specific properties and constraints. (C/SE) 1320.2-1998

aggregation The process of changing the resolution of an aggregate to represent it in less detail. (C/DIS) 1278.1a-1998

Aggregation Key A parameter associated with each port and with each aggregator of an Aggregation System identifying those ports that can be aggregated together. Ports in an Aggregation System that share the same Aggregation Key value are potentially able to aggregate together. (C/LM) 802.3ad-2000

Aggregation Link An instance of a MAC-Physical Layer-Medium Physical Layer-MAC entity between a pair of Aggregation Systems. (C/LM) 802.3ad-2000

Aggregation Port An instance of a MAC-Physical Layer entity within an Aggregation System. (C/LM) 802.3ad-2000

Aggregation System A uniquely identifiable entity comprising (among other things) an arbitrary grouping of one or more ports for the purpose of aggregation. An instance of an aggregated link always occurs between exactly two Aggregation Systems. A physical device may comprise a single Aggregation System or more than one Aggregation System. (C/LM) 802.3ad-2000

aggressive carbon dioxide Free carbon dioxide in excess of the amount necessary to prevent precipitation of calcium as calcium carbonate. (IA) [59]

agile device A device that supports automatic switching between multiple Physical Layer technologies. (C/LM) 802.3-1998

aging (1) (Class 1E battery chargers and inverters) The change with passage of time in physical, chemical, or electrical properties of components or equipment under design range operating conditions that may result in degradation of significant performance characteristics. (PE/NP) 650-1979s

(2) (nuclear power generating station) The effect of operational, environmental, and system conditions on equipment during a period of time up to, but not including, design basis events, or the process of simulating these events. (SWG/PE/NP) 382-1985, 627-1980r, C37.100-1992, 323-1974s

(3) (thermal classification of electric equipment and electrical insulation) The irreversible change (usually degradation) that takes place with time. (EI) 1-1986r

aging acceleration factor For a given hottest-spot temperature, the rate at which transformer insulation aging is accelerated compared with the aging rate at a reference hottest-spot temperature. The reference hottest-spot temperature is 110°C for 65°C average winding rise and 95°C for 55°C average winding rise transformers (without thermally upgraded insulation). For hottest-spot temperatures in excess of the reference hottest-spot temperature the aging acceleration factor is greater than 1. For hottest-spot temperatures lower than the reference hottest-spot temperature, the aging acceleration factor is less than 1. (PE/TR) C57.91-1995

aging assessment Evaluation of appropriate information for determining the effects of aging on the current and future ability of systems, structures, and components to function within acceptance criteria. (PE/NP) 1205-1993

aging degradation Gradual deterioration in the physical characteristics of a system, structure, or component, that is due to aging mechanisms, that occurs with time or use under preservice or service conditions, and could impair its ability to perform any of its design functions. (PE/NP) 1205-1993

aging factor **(1)** (thermal classification of electric equipment and electrical insulation) A factor of influence that causes aging. (EI) 1-1986r
(2) A quantitative factor expressing the degradation in the ability of the battery, due to usage, to deliver electrical energy under specified operating conditions such as, but not limited to, operating ambient temperature, cycling, depth of discharge, and maintenance practices. (VT) 1476-2000

aging mechanism **(1)** **(nuclear power generating station)** Any process attributable to service conditions that results in degradation of an equipment's ability to perform its Class 1E functions. (PE/NP) 649-1980s
(2) A specific process that gradually changes the characteristics of a system, structure, or component with time or use. (PE/NP) 1205-1993
(3) The microscopic or molecular level process or processes (such as chain scission, cross-linking, oxidation, evaporation, or diffusion) that produce changes in the material. (DEI/RE) 775-1993w

agitator (hydrometallurgy) (electrowinning) A receptacle in which ore is kept in suspension in a leaching solution. *See also:* electrowinning. (PE/EEC) [119]

aided tracking A tracking technique in which the manual correction of the tracking error automatically corrects the rate of motion of the tracking mechanism. (AES) 686-1997

aid to navigation *See:* navigational aid.

AI-ESTATE (Artificial Intelligence - Expert System Tie to ATE) *See:* Artificial Intelligence and Expert System Tie to Automatic Test Equipment.

aiming symbol A circle or other pattern of light projected by a light pen onto a display surface to aid in positioning the pen or to describe the light pen's field of view. (C) 610.6-1991w

air- (Used as a prefix). Applied to a device that interrupts an electric circuit; this prefix indicates that the interruption occurs in air. (IA/ICTL/IAC/APP) [60], [75]

AI radar *See:* airborne-intercept radar.

air, ambient *See:* ambient air.

air-blast circuit breaker *See:* circuit breaker.

airborne early warning (radar) (navigation aid terms) An early-warning radar carried by an airborne or spaceborne vehicle. *See also:* early-warning radar. (AES/GCS) 172-1983w, 686-1997

airborne-intercept radar A fire-control radar for use in interceptor aircraft. (AES) 686-1997

airborne moving-target indication radar (AMTI radar) An MTI radar flown in an aircraft or other moving platform with corrections applied for the effects of platform motion, which include the changing clutter Doppler frequency and the spread of the clutter Doppler spectrum. *See also:* displaced phase center antenna; space-time adaptive processing; time-averaged-clutter coherent airborne radar. (AES) 686-1997

airborne radioactivity Radioactivity in any chemical or physical form that is dissolved, mixed, suspended, or otherwise entrained in air. (NI) N42.17B-1989r

air cell A gas cell in which depolarization is accomplished by atmospheric oxygen. *See also:* electrochemistry. (EEC/PE) [119]

air circuit breaker *See:* circuit breaker.

air conditioning The process of treating air so as to simultaneously control temperature, humidity, and distribution to the conditioned space. (IA/PSE) 241-1990r

air-conditioning equipment All of that equipment intended or installed for the purpose of processing the treatment of air so as to control simultaneously its temperature, humidity, cleanliness, and distribution to meet the requirements of the conditioned space. (NESC/NEC) [86]

air conduction (hearing) The process by which sound is conducted to the inner ear through the air in the outer ear canal as part of the pathway. (SP) [32]

air-core inductance (winding inductance) The effective self-inductance of a winding when no ferromagnetic materials are present. *Note:* The winding inductance is not changed when ferromagnetic materials are present. (CHM) [51]

air-core reactor A reactor that does not include a magnetic core or magnetic shield. (PE/TR) C57.16-1996

aircraft aeronautical light (illuminating engineering) Any aeronautical light specially provided on an aircraft. (ED) [127]

aircraft bonding The process of electrically interconnecting all parts of the metal structure of the aircraft as a safety precaution against the buildup of isolated static charges and as a means of reducing radio interference. (EEC/PE) [119]

aircraft electric machine An electric machine designed for operation aboard aircraft. *Note:* Minimum weight and size and extreme reliability for a specified (usually short) life are required while operating under specified conditions of coolant temperature and flow, and for air-cooled machines, pressure and humidity. (PE) [9]

aircraft hangar A location used for storage or servicing of aircraft in which gasoline, jet fuels, or other volatile flammable liquids or flammable gases are used. (NESC/NEC) [86]

air data system (navigation aid terms) A set of aerodynamic and thermodynamic sensors, and a computer which provide flight parameters such as airspeed, static pressure, air temperature, and Mach number. (AES/GCS) 172-1983w

air-derived navigation data (navigation aid terms) Data obtained from measurements made on an airborne vehicle. *See also:* navigation. (AES/RS/GCS) 686-1982s, [42], 172-1983w

air discharge method **(1)** A method of ESD testing in which the charged electrode of the ESD simulator approaches the EUT or coupling plane regardless of the conductivity of the ESD receptor. The discharge is actuated by a spark in air to the EUT or coupling plane. (EMC) C63.16-1993
(2) A method of ESD testing in which the charged electrode of the ESD simulator approaches the Unit Under Test (UUT) or coupling plane. The discharge is actuated by a spark in the air to the UUT or to the coupling plane. (SPD/PE) C62.38-1994r

air equivalent radiation dose (valve actuators) The energy that is absorbed per unit mass of air at the geometric center of the volume occupied by the specimen if it were replaced with air and a uniform flux were incident at the boundary of the volume, directed toward the center. (PE/NP) 382-1985

air failure A failure in the cable above the waterline but below the termination. (PE/IC) 1407-1998

air-floating head *See:* floating head.

air gap The space between the magnetic shunt and the core, used to establish the required reluctance of the shunt flux path. (PEL) 449-1998

air-gap field voltage (excitation systems for synchronous machines) The synchronous machine field voltage required to produce rated voltage on the air-gap line of the synchronous machine with its field winding at 75°C for field windings designed to operate at rating with a temperature rise of 60°C or less; or 100°C for field windings designed to operate at rating with a temperature rise greater than 60°C. *Note:* This defines one per unit excitation system voltage for use in computer representation of excitation systems. (PE/EDPG) 421.1-1986r

air-gap line (excitation systems for synchronous machines) The extended straight line part of the no-load saturation curve of the synchronous machine. (PE/EDPG) 421.1-1986r

air gap, relay *See:* relay air gap.

air-gap surge arrester (low-voltage air-gap surge-protective devices) A gap or gaps in air at ambient atmospheric pressure, designed to protect apparatus and personnel, or both, from high transient voltages. (SPD/PE) C62.32-1981s

air-gap surge protector (low-voltage air-gap surge-protective devices) A protective device, consisting of one or more air-gap surge arresters; optional fuses, short-circuiting devices, etc.; and a mounting assembly, for limiting surge voltages on low voltage (600 V rms or less) electrical and electronic equipment or circuits. (SPD/PE) C62.32-1981s

air horn A horn having a diaphragm that is vibrated by the passage of compressed air. *See also:* protective signaling. (EEC/PE) [119]

Airlines Electronic Engineering Committee The Aeronautical Radio Incorporated (ARINC) committee that originated the Abbreviated Test Language for All Systems (ATLAS). (ATLAS) 771-1989s

air mass The mass of air between a surface and the sun that affects the spectral distribution and intensity of sunlight. *See also:* air mass one; air mass zero. (AES/SS) 307-1969w

air mass one A term that specifies the spectral distribution and intensity of sunlight on earth at sea level with the sun directly overhead and passing through a standard atmosphere. *See also:* air mass; air mass zero. (AES/SS) 307-1969w

air mass zero A term that specifies the spectral distribution and intensity of sunlight in near-earth space without atmospheric attenuation. *Note:* The air mass must be specified when reporting the efficiency of solar cells; for example, 10% efficient at air mass zero, 60°C. *See also:* air mass one; air mass. (AES/SS) 307-1969w

air navigation (navigation aid terms) The navigation of aircraft. (AES/GCS) 172-1983w

airport beacon *See:* aerodrome beacon.

airport surface detection equipment (ASDE) (1) A ground-based radar for observation of the positions of aircraft and other vehicles on the surface of an airport. (AES/GCS) 172-1983w
(2) A high-resolution radar usually located on the airport control tower or other high point and used for observation of the positions of aircraft and other vehicles on the surface of an airport. (AES) 686-1997

airport-surveillance radar (ASR) (1) (navigation aid terms) A medium-range (for example, 60 nautical miles [nmi]) surveillance radar used to control aircraft in the vicinity of an airport. (AES/GCS) 172-1983w
(2) A medium-range (e.g., 100 km) surveillance radar used to control aircraft in the vicinity of an airport. (AES) 686-1997

air position indicator (API) (navigation aid terms) An airborne computing system that presents a continuous indication of the aircraft's position on the basis of aircraft heading, airspeed, and elapsed time. (AES/RS/GCS) 686-1982s, 172-1983w

air, recirculated *See:* recirculated air.

air, return *See:* return air.

air-route surveillance radar (ARSR) (1) (navigation aid terms) A long-range (for example, 200 nautical miles [nmi]) surveillance radar used to control aircraft on airways beyond the coverage of airport surveillance radar (ASR). (AES/GCS) 172-1983w
(2) A long-range (e.g., 350 km) surveillance radar used to control aircraft on airways beyond the coverage of airport-surveillance radar (ASR). (AES) 686-1997

airspeed (navigation aid terms) The rate of motion of a vehicle relative to the air mass. (AES/RS/GCS) 686-1982s, 172-1983w

airspeed indicator (navigation aid terms) An instrument for measuring airspeed. (AES/RS/GCS) 686-1982s, 172-1983w

air-surveillance radar A surveillance radar whose function is to detect and track aircraft over a volume of space. (AES) 686-1997

air switch (1) (high-voltage switchgear) A switch with contacts that separate in air. (SWG/PE) C37.40-1993
(2) A switching device designed to close and open one or more electric circuits by means of guided separable contacts that separate in air. The switching device may be equipped with arcing horns. (SWG/PE) C37.36b-1990r
(3) A switching device designed to close and open one or more electric circuits by means of guided separable contacts that separate in air. (SWG/PE) C37.100-1992

air terminal (lightning protection) The combination of an elevation rod and brace, or footing placed on upper portions of structures, together with tip or point, if used. (PE/T&D) 1243-1997

air ventilation The amount of supply air required to maintain the desired quality of air within a designated space. (IA/PSE) 241-1990r

airway beacon (illuminating engineering) An aeronautical beacon used to indicate a point on the airway. (ED) [127]

AIS *See:* alarm indication signal.

alarm (1) (power generating stations) A signal for attracting attention to some abnormal condition. Alarms associated with electric heat tracing systems can signal high temperature, low temperature, loss of heater circuit voltage, etc. *Synonym:* alarm signal. (PE/EDPG) 622A-1984r, 622B-1988r
(2) An audible and/or visible signal activated when the instrument reading exceeds a preset value or falls outside of a preset range. (NI) N42.17B-1989r, N42.20-1995
(3) A signal generated by a slave and received by the controller, which is typically used to interrupt the processor, or to activate processing of the slave's request/response packet queues. (C/MM) 1596.4-1996
(4) A signal for attracting attention to some abnormal condition. A warning of danger, safeguard threat, equipment failure, or other condition requiring attention. (PE/NP) 692-1997

alarm, blue* *See:* alarm indication signal.
* Deprecated.

alarm checking The identification of an alarm from a remote location by communicating with its point of origin. (COM) 312-1977w

alarm condition (1) (supervisory control, data acquisition, and automatic control) A predefined change in the condition of equipment or the failure of equipment to respond correctly. Indication may be audible or visual, or both. (SUB/PE) C37.1-1994
(2) A predefined change in the condition of equipment or the failure of equipment to respond correctly. Indication may be audible, visual, or both. (SWG/PE) C37.100-1992

alarm function *See:* supervisory control functions.

alarm indication signal A signal that replaces the normal traffic signal when a maintenance alarm indication has been activated. (COM/TA) 1007-1991r

alarm point (power-system communication) A supervisory control status point considered to be an alarm. (PE) 599-1985w

alarm point interfaces Master station or RTU (or both) element(s) that input(s) a signal to the alarm function. (SUB/PE) C37.1-1994

alarm, red *See:* red alarm.

alarm relay (1) (signal) A monitoring relay whose function is to operate an audible or visual signal to announce the occurrence of an operation or a condition needing personal attention, and usually provided with a signaling cancellation device. *See also:* relay. (SWG/PE/PSR) C37.90-1978s
(2) (power system device function numbers) A relay other than an annuciator, as covered under device function 30, [annunciator relay], that is used to operate, or to operate in connection with, a visual or audible alarm. (SUB/PE) C37.2-1979s
(3) A monitoring relay whose function is to operate an audible or visual signal to announce the occurrence of an operation or a condition needing personnel attention, and which is usually provided with a signaling cancellation device. (SWG/PE) C37.100-1992

alarm SCADA function The capability of a supervisory system to accomplish a predefined action in response to an alarm condition. (SUB/PE) C37.1-1994

alarm sending (telephone switching systems) The extension of alarms from an office to another location. (COM) 312-1977w

alarm signal A signal for attracting attention to some abnormal condition. *See also:* alarm. (COM/PE/EDPG) [48], 622B-1988r

alarm summary printout (sequential events recording systems) The recording of all inputs currently in the alarm state. (PE/EDPG) [5], [1]

alarm switch An auxiliary switch that actuates a signaling device upon the automatic opening of the circuit breaker with which it is associated. (IA/PSP) 1015-1997

alarm system (protective signaling) An assembly of equipment and devices arranged to signal the presence of a hazard requiring urgent attention. *See also:* protective signaling. (EEC/PE) [119]

alarm, yellow *See:* yellow alarm.

albedo (photovoltaic power system) The reflecting power expressed as the ratio of light reflected from an object to the total amount falling on it. (AES) [41]
(2) **(A)** In astronomy (where the sizes of the objects/surfaces are extremely large in comparison to a wavelength), the ratio of the total radiation reflected (scattered) from an object to the total incident power. **(B)** In transport theory or particle scattering (where the size of the object is not extremely large), the ratio of the total scattering cross-section to the sum of the scattering and absorption cross-sections. (AP/PROP) 211-1997

ALC *See:* automatic load (level) control.

alert (1) To cause the terminal of the user to give some audible or visual indication that an error or some other event has occurred. When the standard output is directed to a terminal device, the method for alerting the terminal user is unspecified. When the standard output is not directed to a terminal device, the alert shall be accomplished by writing the ⟨`alert`⟩ character to standard output (unless the utility description indicates that the use of standard output produces undefined results in this case). (C/PA) 9945-2-1993
(2) A notification to be watchful that shall not be considered the same priority as an alarm. (PE/NP) 692-1997

alert level A probability value placed on equipment failure rates to identify when systems, trains, or components are not achieving their target availability or reliability values. (PE/NP) 933-1999

alertness function A device or system that monitors the operator for signs of incapacitation, usually by requiring movement or response to take place within a prescribed period of time. (VT) 1475-1999

⟨`alert`⟩ A character that in the output stream shall indicate that a terminal should alert its user via a visual or audible notification. The ⟨`alert`⟩ shall be the character designated by `'\a'` in the C-language binding. It is unspecified whether this character is the exact sequence transmitted to an output device by the system to accomplish the alert function. (C/PA) 9945-2-1993

alert tone A non-power ringing tone, or combination of tones, used to request the telemetry interface unit (TIU) or customer premise equipment (CPE) to become active. (AMR/SCC31) 1390-1995, 1390.2-1999, 1390.3-1999

alert tone code (1) A data byte, from the utility controller, that identifies which alert tone is to be used by the central office service unit (COSU). (AMR/SCC31) 1390-1995, 1390.2-1999
(2) A data byte that identifies which alert tone is to be used by the central office service unit (COSU). (SCC31) 1390.3-1999

Alford loop antenna A multi-element antenna having approximately equal amplitude currents that are in phase and uniformly distributed along each of its peripheral elements and producing a substantially circular radiation pattern in its principal E-plane. *Note:* This antenna was originally developed as a four-element, horizontally polarized, UHF loop antenna. (AP/AES/ANT/GCS) 145-1993, 172-1983w

Alfvén velocity (radio-wave propagation) The characteristic velocity of an Alfven wave, given by:

$$V_a = H_0 \left[\frac{\mu}{\rho}\right]^{1/2}$$

where μ is the permeability, H_o is the static magnetic field strength, and ρ is the mass density of the conducting fluid. (AP/PROP) 211-1990s

Alfvén wave (radio-wave propagation) In a homogeneous magneto-ionic medium, the magneto-hydrodynamic wave that propagates in the direction of the static magnetic field, with associated electric and magnetic fields and fluid particle velocities oriented perpendicular to the direction of propagation. (AP/PROP) 211-1990s

algebraic coding function In hashing, a hash function that returns the result of evaluating some polynomial in which selected digits of the original key are used as coefficients. For example, in the function below, the first three digits of the original key are evaluated as *a*, *b*, and *c*, respectively, in the polynomial $a + b\,x + c\,x^2$ with $x = 14$.

Original key	Calculation	Hash value
964721	9 + 6(14) + 4(14)2 = 877	877
864765	8 + 6(14) + 4(14)2 = 876	876

(C) 610.5-1990w

algebraic language A programming language that permits the construction of statements resembling algebraic expressions, such as $Y = X + 5$. For example, NOMAD or FORTRAN. *See also:* algorithmic language; logic programming language; list processing language; functional language. (C) 610.13-1993w, 610.12-1990

algebraic manipulation The processing of mathematical expressions without concern for the numeric values of the symbols that represent numbers. (C) 1084-1986w

algebraic sum The answer arrived at when adding two operands numerically. For example: 01102 + 01012 = 10112. *Contrast:* logical sum. (C) 610.10-1994w

ALGOL (ALGOrithmic Language or ALGebraic Oriented Language). A high-order programming language suitable for expressing solutions to problems requiring numeric computations, algorithms, or mathematical formulas; its many elegant features and formal syntactic definition have inspired much research in programming language theory. *Note:* Jointly developed by the United States and European communities, ALGOL 60 was the first language standard to be adopted as an ISO standard. As of this writing, ALGOL 68 is the dialect accepted as the latest standard language. *See also:* extensible language; EULER; GLYPNIR; block-structured language. (C) 610.13-1993w

ALGOL 58 A dialect of ALGOL developed as an IEEE standard language in 1958. (C) 610.13-1993w

ALGOL 60 A dialect of ALGOL that was the first version to be adopted as an ISO language standard for ALGOL. *See also:* EL1; MP; SIMULA. (C) 610.13-1993w

ALGOL 68 A dialect of ALGOL characterized by being the first instance of a complete formal definition language. (C) 610.13-1993w

algorithm (general) A prescribed set of well-defined rules or processes for the solution of a problem in a finite number of steps; for example, a full statement of an arithmetic procedure for evaluating sin*x* to a stated precision. *See also:* heuristic. (MIL/C) [2], [20], [85]
(2) **(A)** **(software) (mathematics of computing)** A finite set of well-defined rules for the solution of a problem in a finite number of steps; for example, a complete specification of a sequence of arithmetic operations for evaluating sine *x* to a given precision. **(B)** **(software)** Any sequence of operations for performing a specific task. (C) 610.12-1990, 1084-1986

algorithm analysis (software) The examination of an algorithm to determine its correctness with respect to its intended use, to determine its operational characteristics, or to understand it more fully in order to modify, simplify, or improve it. *See also:* algorithm. (C/SE) 729-1983s

algorithmic language (1) (software) A programming language designed for expressing algorithms; for example, ALGOL. *See also:* logic programming language; functional language; list processing language; algebraic language. (C) 610.12-1990, 610.13-1993w

(2) (test, measurement, and diagnostic equipment) A language designed for expressing algorithms. (MIL) [2]

alias (1) (A) (software) An additional name for an item. **(B) (software)** An alternate label. For example, a label and one or more aliases may be used to refer to the same data element or point in a computer program. *See also:* data; label; computer program; alternate name. (C/SE) 729-1983, 610.5-1990

(2) An alternate name for a directory object, provided by the use of one or more alias entries in the DIT. *Synonym:* alias name. (C/PA) 1328.2-1993w, 1224.2-1993w, 1327.2-1993w, 1326.2-1993w

(3) An alternate name for an IDEF1X model construct (class, responsibility, entity, or domain). (C/SE) 1320.2-1998

alias entry A Directory entry, of Object Class "alias," containing information used to provide an alternative name for an object. (C/PA) 1327.2-1993w, 1326.2-1993w, 1328.2-1993w, 1224.2-1993w

aliasing The visual misrepresentation that occurs when an image or model contains more detail than the display device's resolution can present. *Note:* A result of aliasing is jagged stairstepping of slanted lines. (C) 610.6-1991w

alias name In the shell command language, a word consisting solely of underscores, digits, and alphabetics from the portable character set and any of the following characters:

! % , @

Implementations may allow other characters within alias names as an extension. *See also:* alias. (C/PA) 9945-2-1993

align (test, measurement, and diagnostic equipment) To adjust a circuit, equipment, or system so that its functions are properly synchronized or its relative positions properly oriented. For example, trimmers, padders, or variable inductances in tuned circuits are adjusted to give a desired response for fixed tuned equipment or to provide tracking for tunable equipment. (MIL) [2]

aligned A term that refers to the constraints placed on the address of the data; the address is constrained to be a multiple of the data format size. (C/MM) 1596.5-1993

aligned address This is an integer multiple of the data block size. The maximum data block size that can be transferred by an implementation under test (IUT) Master is the product of data width and data length. (C/BA) 896.4-1993w

aligned bundle (fiber optics) A bundle of optical fibers in which the relative spatial coordinates of each fiber are the same at the two ends of the bundle. *Note:* The term "coherent bundle" is often employed as a synonym, and should not be confused with phase coherence or spatial coherence. *Synonym:* coherent bundle. *See also:* fiber bundle. (Std100) 812-1984w

aligned-grid tube (or valve) A vacuum multigrid tube or valve in which at least two of the grids are aligned, one behind the other, so as to obtain a particular effect (canalizing an electron beam, suppressing noise, etc.). *See also:* electron tube. (ED) [45], [84]

alignment (1) (data transmission) In communication practice, alignment is the process of adjusting a plurality of components of a system for proper interrelationship. The term is applied especially to the adjustment of the tuned circuits of an amplifier for desired frequency response, and the synchronization of the components of a system. (PE) 599-1985w

(2) (inertial navigation equipment) (navigation aid terms) The orientation of the measuring axes of the inertial components with respect to the coordinate system in which the equipment is used. *Note:* Inertial alignment refers to the result of either the process of bringing the measuring axis into a desired orientation or the computation of the angles between the measuring axis and the desired orientation with respect to the coordinate system in which the equipment is used. The initial alignment can be accomplished by the use of noninertial sensors. *See also:* gyrocompass alignment; transfer alignment. (AES/GCS) 172-1983w

(3) (communication practice) The process of adjusting a plurality of components of a system for proper interrelationship. *Note:* The term is applied especially to the adjustment of the tuned circuits of an amplifier for desired frequency response, and the synchronization of components of a system. *See also:* radio transmission. (PE) 599-1985w

(4) (computers) Pertaining to data that are stored beginning at certain machine-dependent boundaries. Such data is said to be "aligned," otherwise it is said to be "unaligned;" for example, a four-bit data item is aligned if it begins on a full-word boundary of eight-bit words. *Synonym:* boundary alignment. (C) 610.5-1990w

(5) The suitability of particular addresses for accessing particular types of data. For example, some processors require even addresses for accessing 16-bit data items. (C/BA) 1275-1994

(6) *See also:* input-axis misalignment. (AES/GYAC) 528-1994

alignment error (1) An error that occurs when a packet is not a multiple of eight bits. *Note:* It is only applicable to specific protocols. (C) 610.7-1995

(2) The deviation of the recovered clock from the ideal recovered clock embedded by the transmitter. The deviation from the ideal sampling point may be caused by static timing errors in the timing recovery circuit, internal jitter generated in the timing recovery circuit, and the inability to track exactly the jitter on the received data signal. (C/LM) 8802-5-1998

alignment jitter The jitter measured against the clock of the upstream adapter. This is not a type of jitter per se; rather, it is a way to measure jitter. When "zero transferred jitter" is specified, the jitter measured is alignment jitter. (LM/C) 802.5-1989s

alignment kit (test, measurement, and diagnostic equipment) A kit containing all the instruments or tools necessary for the alignment of electrical or mechanical components. (MIL) [2]

alignment tool (test, measurement, and diagnostic equipment) A small screwdriver, socket wrench, or special tool used for adjusting electronic, mechanical, or optical units, usually constructed of nonmagnetic materials. (MIL) [2]

alive (1) (electric systems) Electrically connected to a source of potential difference, or electrically charged so as to have a potential different from that of the ground. *Note:* The term "alive" is sometimes used in place of the term current-carrying, where the intent is clear to avoid repetitions of the longer term. *Synonym:* live. *See also:* energized; insulated. (T&D) C2.2-1960

(2) *See also:* energized. (T&D/PE) 524-1992r

alkaline cleaning (electroplating) Cleaning by means of alkaline solutions. *See also:* electroplating. (EEC/PE) [119]

alkaline storage battery A storage battery in which the electrolyte consists of an alkaline solution, usually potassium hydroxide. *See also:* battery. (EEC/PE) [119]

Allan deviation *See:* two-sample deviation.

Allan variance The average of the variance of adjacent pairs of elements in a contiguous time series of data versus the averaging time used to generate the elements. The term "Allan variance" is also used to refer to its square root versus averaging time, although "square root of Allan variance" would be more proper usage. (AES/GYAC) 1293-1998

Allan variation *See:* two-sample variance.

alligator *See:* running board.

allocated baseline In configuration management, the initial approved specifications governing the development of configuration items that are part of a higher level configuration item. *Contrast:* developmental configuration; product baseline; functional baseline. *See also:* allocated configuration identification. (C) 610.12-1990

allocated configuration identification In configuration management, the current approved specifications governing the development of configuration items that are part of a higher level configuration item. Each specification defines the functional characteristics that are allocated from those of the higher level configuration item, establishes the tests required to demonstrate achievement of its allocated functional characteristics, delineates necessary interface requirements with other associated configuration items, and establishes design constraints, if any. *Contrast:* functional configuration identification; product configuration identification. *See also:* allocated baseline. (C) 610.12-1990

allocated storage Portions of storage that are assigned or reserved for active instructions or for data. (C) 610.10-1994w

allocation (1) (A) (software) The process of distributing requirements, resources, or other entities among the components of a system or program. **(B) (software)** The result of the distribution in definition (A). (C) 610.12-1990
(2) (broadband local area networks) The assignment of specific broadcast frequencies by a national organization (such as the FCC) for various communications uses (e.g., commercial television and radio, land-mobile radio, defense communications, microwave links). This divides the available spectrum between competing services and minimizes interference between them. The manager of a broadband network must allocate the available bandwidth of the cable among different services for the same reason. (LM/C) 802.7-1989r
(3) (computers) *See also:* storage allocation. (C) [20], [85]
(4) The decision to assign a function or decision to hardware, software, or humans. Allocation may be made entirely to one of these three system element types or to some combination to be resolved upon further functional decomposition. (C/SE) 1220-1998

allocation protocols The protocols used to allocate resources that are shared by multiple nodes. These include bandwidth allocation protocols and queue allocation protocols. (C/MM) 1596-1992

allotting (telephone switching systems) The preselecting by a common control of an idle circuit. (COM) 312-1977w

allowable continuous current (of a fuse link, fuse unit or refill unit) The maximum rms current in amperes at rated frequency and at a specific ambient temperature, which a device will carry continuously without exceeding the allowable total temperature. (SWG/PE) C37.40-1993, C37.41-1981s

allowable continuous-current class designation (of an air switch) A code that identifies the composite curve relating the loadability factor *LF* of the switch to the ambient temperature θ_A as determined by the limiting switch part class designations. (SWG/PE) C37.30-1992s, C37.37-1996

allowable stress (seismic design of substations) The maximum stress permitted by applicable standards or codes, or both. (PE/SUB) 693-1984s, C37.122.1-1993

alloy junction (semiconductor) A junction formed by recrystallization on a base crystal from a liquid phase of one or more components and the semiconductor. *See also:* semiconductor. (IA) 59-1962w, [12]

alloy plate An electrodeposit that contains two or more metals codeposited in combined form or in intimate mixtures. *See also:* electroplating. (EEC/PE) [119]

all-pass filter A filter designed to introduce phase shift or delay over a band of frequencies without introducing appreciable attenuation distortion over those frequencies. (CAS) [13]

all-pass function (linear passive networks) A transmittance that provides only phase shift, its magnitude characteristic being constant. *Notes:* 1. For lumped-parameter networks, this is equivalent to specifying that the zeros of the function are the negatives of the poles. 2. A realizable all-pass function exhibits non-decreasing phase lag with increasing frequency. 3. A trivial all-pass function has zero phase at all frequencies. (CAS) 156-1960w

all-pass network A network designed to introduce phase shift or delay without introducing appreciable attenuation at any frequency. *Synonym:* all-pass transducer. *See also:* network analysis. (EEC/PE) [119]

all-pass transducer *See:* all-pass network.

all-purpose computer *See:* general-purpose computer.

all-relay system An automatic telephone switching system in which all switching functions are accomplished by relays. (EEC/PE) [119]

all routes explorer (ARE) A frame that traverses every path and combination of paths through a bridged network. (C/LM/CC) 8802-2-1998

all-segments broadcast The transmission of a frame to all interconnected segments of a local area network. *See also:* all-stations broadcast. (C) 610.7-1995

all-stations broadcast The transmission of a frame to all stations on a given local area network segment. *See also:* all-segments broadcast. (C) 610.7-1995

all terrain vehicle (ATV) *See:* off-road vehicle.

all-weather distribution A distribution of corona-effect data collected under all weather conditions. Such data are usually obtained from long-term recording stations. Weather conditions are defined in the next section. (T&D/PE) 539-1990

almanac (navigation aid terms) A periodic publication of astronomical data useful to a navigator. (AES/GCS) 172-1983w

ALOHA network A telecommunication network that uses a multi-access contention protocol, first developed for use in Hawaii. (C) 610.7-1995

ALP *See:* automated language processing.

ALPHA An extension to PL/1 providing BNF (backus naur form) parsing capabilities. *Note:* Semantic routines are defined in PL/1 and invoked during the parse. (C) 610.13-1993w

alpha *See:* alphabetic; alphanumeric.

alphabet (1) (computers) An ordered set of all the letters or symbols used in a language, including letters with diacritical signs where appropriate, but not including punctuation marks. (C) 610.5-1990w
(2) A character set arranged in certain order. *Note:* Character sets are finite quantities of letters of the normal alphabet, digits, punctuation marks, control signals, such as carriage return and other ideographs. Characters are usually represented by letters (graphics) or technically realized in the form of combinations of punched holes, sequences of electric pulses, etc. (COM) [49]

alphabetic Pertaining to data that consist solely of letters from the same alphabet. For example, (AaBbCcDdEe...) plus the space character. *Note:* IEEE Std 610.5 deprecates the use of "alpha" as an abbreviation for "alphabetic". *Synonym:* alpha. *See also:* alphanumeric; character. (C) 610.5-1990w

alphabetic character set A character set that contains alphabetic characters and that may contain control characters, special characters, and the space character, but not digits. (C) 610.5-1990w

alphabetic code A code that uses alphabetic characters to represent data. (C) 610.5-1990w

alphabetic shift A control for selecting the alphabetic character set on an keyboard or printer. *Contrast:* numeric character. *See also:* shift character. (C) 610.5-1990w

alphabetic string A character string consisting solely of alphabetic characters. (C) 610.5-1990w

alphabetic word (A) A word consisting solely of letters from the same alphabet; for example, the word "CIRCUS." **(B)** A word that consists of letters and associated special characters, but not digits; for example, the word "HEAVY-DUTY." (C) 610.10-1994

alpha end (1) The end of the module nearest the lowest-numbered connector contact. (C/BA) 1101.3-1993
(2) The end of the module nearest the lowest-numbered contact. (C/BA) 1101.4-1993, 1101.7-1995

alpha key The connector keying pin located at the alpha end of the module connector. (C/BA) 1101.3-1993

alphameric *See:* alphanumeric.

alphanumeric (1) (computers) Pertaining to data that contain the letters of an alphabet (AaBbCcDdEeFfGgHh...), the decimal digits (0123456789), and may contain control characters, special characters and the space character. *Synonym:* alphameric. (C) 610.5-1990w
(2) Pertaining to a character set that contains both letters and digits, but usually some other characters such as punctuation symbols. *Synonym:* alphameric. (C) [20]

alphanumeric character set A character set that contains alphanumeric characters. (C) 610.5-1990w

alphanumeric code A code that uses alphanumeric characters to represent data. (C) 610.5-1990w

alphanumeric display device *See:* character display device.

alpha profile *See:* power-law index profile.

ALT *See:* alternate key.

alter (A) To insert, delete, or modify a data record. **(B)** To change a logical relationship or physical structure of a database. *See also:* modify. (C) 610.5-1990

alteration (elevators) Any change or addition to the equipment other than ordinary repairs or replacements. *See also:* elevator. (EEC/PE) [119]

alternate ac (AAC) source An ac power source that is available to and located at or nearby a nuclear power plant and that meets the following requirements: It can be connected to (but is not normally connected to) the offsite or onsite emergency ac power systems; It has minimum potential for common mode failure with offsite power or onsite emergency ac power sources; It is available in a timely manner after the onset of a station blackout; It has sufficient capacity and reliability to operate all systems required for both coping with a station blackout and for the time needed to bring the plant to and maintain the plant in a safe shutdown (nondesign basis accident). (PE/NP) 765-1995

alternate access unit (AAU) Type of unit architecture that defines access between multiple IEEE Standard-compliant buses (e.g., Futurebus+ and Serial Bus) when the buses share a common module. (C/BA) 896.3-1993w

alternate-channel interference (second-channel interference) Interference caused in one communication channel by a transmitter operating in a channel next beyond an adjacent channel. *See also:* radio transmission. (EEC/PE) [119]

alternate display (oscillography) A means of displaying output signals of two or more channels by switching the channels in sequence. *See also:* oscillograph. (IM/HFIM) [40]

alternate function key A function key that, when used in conjunction with the alternate key, performs a different function or command than when it is used alone. (C) 610.10-1994w

alternate hierarchical routing A routing strategy in which the traffic is routed through the lowest available level of the network hierarchy. *Note:* It uses a tree like structure of five classes: class 1—regional center, class 2—sectional center, class 3—primary center, class 4—toll center, and class 5—end office. *Synonym:* alternative hierarchical routing. (C) 610.7-1995

alternate index An index that uses alternate keys to reference indexed data. *See also:* secondary index. (C) 610.5-1990w

alternate key (1) (A) In a relation, a candidate key that is not chosen to be the primary key for that relation. **(B)** A secondary key for an indexed sequential file. *See also:* alternate index; prime key. (C) 610.5-1990
(2) A control key that controls the interpretation of other keys. That is, when used in conjunction with another key it causes a different interpretation of that key than when the key is used alone. *See also:* shift key. (C) 610.10-1994w
(3) Any candidate key of an entity other than the primary key. (C/SE) 1320.2-1998

alternate mark inversion code *See:* bipolar signal.

alternate name Any name besides the data element name by which a data item is known. *Note:* Often stored in data dictionaries. *Synonym:* alias. (C) 610.5-1990w

alternate power source One or more generator sets intended to provide power during the interruption of the normal electrical service or the public utility electrical service intended to provide power during interruption of service normally provided by the generating facilities on the premises. (NESC/NEC) [86]

alternate root directory A pathname other than / for managing installed software. (C/PA) 1387.2-1995

alternate route (data transmission) A secondary communications path used to reach a destination if the primary path is unavailable. (PE) 599-1985w

alternate-route trunk group (telephone switching systems) A trunk group that accepts alternate-routed traffic. (COM) 312-1977w

alternate routing (1) (telephone switching systems) A means of selectively distributing traffic over a number of routes ultimately leading to the same destination. (COM) 312-1977w
(2) A routing strategy that assigns a secondary communications path to a destination when the primary path is busy or unavailable. *Synonym:* alternative routing. (C) 610.7-1995

alternate track On a disk, a spare track that is used in place of a normal track in the event that the latter is damaged or inoperable. *Synonym:* alternative track. (C) 610.10-1994w

alternating charge characteristic (nonlinear capacitor) The function relating the instantaneous values of the alternating component of transferred charge, in a steady state, to the corresponding instantaneous values of a specified applied periodic capacitor-voltage. *Note:* The nature of this characteristic may depend upon the nature of the applied voltage. *See also:* nonlinear capacitor. (ED) [46]

alternating current (ac) (1) An electric current that reverses direction at regularly recurring intervals of time. *Contrast:* direct current. (C) 610.10-1994w
(2) A periodic current with an average value over a period of time of zero. (Unless distinctly specified otherwise, the term refers to a current that reverses at regularly recurring intervals of time and that has alternately positive and negative values.). (IA/MT) 45-1998

alternating-current and direct-current ringing Ringing in which alternating current activates the ringer and direct current controls the removal of ringing upon answer. (COM) 312-1977w

alternating-current arc welder transformer A transformer with isolated primary and secondary windings and suitable stabilizing, regulating, and indicating devices required for transforming alternating current from normal supply voltages to an alternating-current output suitable for arc welding. (EEC/AWM) [91]

alternating-current breakdown voltage (gas-tube surge protective devices) The minimum root-mean-square (rms) value of a sinusoidal voltage at frequencies between 15 Hz and 62 Hz that results in arrester sparkover. (SPD/PE) C62.31-1984s

alternating-current circuit A circuit that includes two or more interrelated conductors intended to be energized by alternating current. (Std100) 270-1966w

alternating-current circuit breaker (power system device function numbers) A device that is used to close and interrupt an ac power circuit under normal conditions or to interrupt this circuit under fault or emergency conditions. (PE/SUB) C37.2-1979s

alternating-current commutator motor An alternating-current motor having an armature connected to a commutator and included in an alternating-current circuit. *See also:* asynchronous machine. (PE) [9]

alternating-current component *See:* symmetrical component.

alternating-current–direct-current general-use snap-switch A form of general-use snap-switch suitable for use on either direct- or alternating-current circuits for controlling the following:

a) Resistive loads not exceeding the ampere rating at the voltage involved.
b) Inductive loads not exceeding one-half the ampere rating at the voltage involved, except that switches having a marked horsepower rating are suitable for controlling motors not exceeding the horse-power rating of the switch at the voltage involved.
c) Tungsten filament lamp loads not exceeding the ampere rating at 125 V, when marked with the letter "T." Alternating-current-direct-current general use snap-switches are not generally marked alternating-current-direct-current, but are always marked with their electrical rating.

See also: switch. (NESC) [86]

alternating-current–direct-current ringing Ringing in which a combination of alternating and direct currents is utilized, the direct current being provided to facilitate the functioning of the relay that stops the ringing. (EEC/PE) [119]

alternating-current directional overcurrent relay (power system device function numbers) A relay that functions on a desired value of ac overcurrent flowing in a predetermined direction. (SUB/PE) C37.2-1979s

alternating-current distribution The supply to points of utilization of electric energy by alternating current from its source to one or more main receiving stations. *Note:* Generally, a voltage is employed that is not higher than that which could be delivered or utilized by rotating electric machinery. Step-down transformers of a capacity much smaller than that of the line are usually employed as links between the moderate voltage of distribution and the lower voltage of the consumer's apparatus. (T&D/PE) [10]

alternating-current electric locomotive An electric locomotive that collects propulsion power from an alternating-current distribution system. *See also:* electric locomotive. (EEC/PE) [119]

alternating-current erasing head A head that uses alternating current to produce the magnetic field necessary for erasing. *Note:* Alternating-current erasing is achieved by subjecting the medium to a number of cycles of a magnetic field of a decreasing magnitude. The medium is, therefore, essentially magnetically neutralized. (SP/MR) [32]

alternating-current floating storage-battery system A combination of alternating-current power supply, storage battery, and rectifying devices connected so as to charge the storage battery continuously and at the same time to furnish power for the operation of signal devices. (EEC/PE) [119]

alternating-current general-use snap-switch A form of general-use snap-switch suitable only for use on alternating-current circuits for controlling the following:

a) Resistive and inductive loads (including electric discharge lamps) not exceeding the ampere rating at the voltage involved.
b) Tungsten filament lamp loads not exceeding the ampere rating at 120 V.
c) Motor loads not exceeding 80% of the ampere rating of the switches at the rated voltage.

Note: All alternating-current general-use snap-switches are marked ac in addition to their electrical rating. *See also:* switch. (NESC) [86]

alternating-current generator A generator for the production of alternating-current power. (PE) [9]

alternating current-linked ac converter (self-commutated converters) A converter comprising two cascaded frequency changers in which the intermediate link is usually a high-frequency tank circuit. (IA/SPC) 936-1987w

alternating current-linked dc converter (self-commutated converters) A converter comprising an inverter and a rectifier, with an intermediate ac link. (IA/SPC) 936-1987w

alternating-current magnetic biasing Magnetic biasing accomplished by the use of an alternating current, usually well above the signal-frequency range. *Note:* The high-frequency linearizing (biasing) field usually has a magnitude approximately equal to the coercive force of the medium. (SP/MR) [32]

alternating-current motor An electric motor for operation by alternating current. (PE) [9]

alternating-current pulse An alternating-current wave of brief duration. *See also:* pulse. (EEC/PE) [119]

alternating-current reclosing relay (power system device function numbers) A relay that controls the automatic reclosing and locking out of an ac circuit interrupter. (SUB/PE) C37.2-1979s

alternating-current relay A relay designed for operation from an alternating-current source. *See also:* relay. (EEC/REE) [87]

alternating current root-mean-square voltage rating (semiconductor rectifiers) The maximum root-mean-square value of applied sinusoidal voltage permitted by the manufacturer under stated conditions. *See also:* semiconductor rectifier stack. (IA) [62]

alternating-current saturable reactor (power and distribution transformers) A reactor whose impedance varies cyclically with the alternating current (or voltage). (PE/TR) C57.12.80-1978r

alternating-current signal A time-varying signal whose polarity varies with a period of time T, and whose average value is zero. (PEL) 1515-2000

alternating-current standby power (low voltage varistor surge arresters) Varistor ac power dissipation measured at rated root-mean-square (rms) voltage. (PE) [8]

alternating-current time overcurrent relay (power system device function numbers) A relay that operates when its ac input current exceeds a predetermined value, and in which the input current and operating time are inversely related through a substantial portion of the performance range. (PE/SUB) C37.2-1979s

alternating-current transmission (1) The transfer of electric energy by alternating current from its source to one or more main receiving stations for subsequent distribution. *Note:* Generally, a voltage is employed that is higher than that which would be delivered or utilized by electric machinery. Transformers of a capacity comparable to that of the line are usually employed as links between the high voltage of transmission and the lower voltage used for distribution or utilization. *See also:* alternating-current distribution. (T&D/PE) [10]

(2) (television) That form of transmission in which a fixed setting of the controls makes any instantaneous value of signal correspond to the same value of brightness only for a short time. *Note:* Usually, this time is not longer than one field period and may be as short as one line period. *See also:* television. (EEC/PE) [119]

alternating-current winding (1) (power and distribution transformers) (of a rectifier transformer) The primary winding that is connected to the alternating-current circuit and usually has no conductive connection with the main electrodes of the rectifier. (PE/TR) C57.12.80-1978r

(2) **(thyristor converter)** The winding of a thyristor converter transformer that is connected to the ac circuit and usually has no conductive connection with the thyristor circuit elements. *Synonym:* primary winding. (IA/IPC) 444-1973w

alternating function A periodic function whose average value over a period is zero. For instance, $f(t)$ = **B** sin wt is an alternating function (w,**B** assumed constants). (Std100) 270-1966w

alternating voltage *See:* alternating current.

alternative (electric power system) (generating stations electric power system) A qualifying word identifying a power circuit equipment, device, or component available to be connected (or switched) into the circuit to perform a function when the preferred component has failed or is inoperative. *See also:* reserve. (PE/EDPG) 505-1977r

alternative hierarchical routing *See:* alternate hierarchical routing.

alternative routing *See:* alternate routing.

alternative track *See:* alternate track.

alternator (rotating electric machinery) An alternating-current generator. (PE/EM) 11-1980r

alternator-rectifier exciter (1) (excitation systems for synchronous machines) An exciter whose energy is derived from an alternator and converted to direct current by rectifiers. The exciter includes an alternator and power rectifiers, which may be either noncontrolled or controlled, including gate circuitry. It is exclusive of input control elements. The alternator may be driven by a motor, prime mover, or by the shaft of the synchronous machine. The rectifiers may be stationary or rotating with the alternator shaft. (PE/EDPG) 421.1-1986r
(2) (synchronous machines) An exciter whose energy is derived from an alternator and converted to direct current by rectifiers. *Notes:* 1. The exciter includes an alternator and power rectifiers which may be either noncontrolled or controlled, including gate circuitry. 2. It is exclusive of input control elements. 3. The alternator may be driven by a motor, prime mover, or by the shaft of the synchronous machine. 4. The rectifiers may be stationary or rotating with the alternator shaft. (PE/EDPG) 421-1972s

alternator transmitter A radio transmitter that utilizes power generated by a radio-frequency alternator. *See also:* radio transmitter. (AP/BT/ANT) 145-1983s, 182-1961w

altimeter (navigation aid terms) An instrument which determines the height of an object with respect to a fixed level, such as sea level. There are two common types: an aneroid, or barometric altimeter, and the radio, or radar altimeter. (AES/GCS) 172-1983w

altitude (1) (illuminating engineering) The angular distance of a heavenly body measured on that great circle that passes, perpendicular to the plane of the horizon, through the body and through the zenith. It is measured positively from the horizon to the zenith, from 0 to 90 degrees. (ED) [127]
(2) (A) (navigation aid terms) Angular distance above the horizon—the arc of a vertical circle between the horizon and a point on the celestial sphere. **(B) (navigation aid terms)** Vertical distance above a given datum. (AES/GCS) 172-1983
(3) (series capacitor) The elevation of the series capacitor above mean sea level. (T&D/PE) [26]

altitude-treated current-carrying brush A brush specially fabricated or treated to improve its wearing characteristics at high altitudes (over 6000 m). (EEC/PE) [119]

ALU *See:* arithmetic and logic unit.

aluminum alloy conductor, steel reinforced (AACSR) A composite conductor made up of a combination of aluminum alloy and coated steel wires. In the usual construction, the aluminum wires surround the steel. (T&D/PE) 524-1992r

aluminum cable steel reinforced A composite conductor made up of a combination of aluminum wires surround the steel. *See also:* conductor. (PE/T&D) [10]

aluminum conductor A conductor made wholly of aluminum. *See also:* conductor. (T&D/PE) [10]

aluminum conductor, aluminum alloy reinforced (ACAR) A composite conductor made up of a combination of aluminum and aluminum alloy wires. In the usual construction, the aluminum wires surround the aluminum alloy. (T&D/PE) 524-1992r

aluminum conductor, steel reinforced A composite conductor made up of a combination of aluminum and coated steel wires. In the usual construction, the aluminum wires surround the steel. (T&D/PE) 524-1992r

aluminum-covered steel wire (power distribution, underground cables) A wire having a steel core, to which is bonded a continuous outer layer of aluminum. (PE) [4]

aluminum sheath (aluminum sheathing for power cables) An impervious aluminum or aluminum alloy tube, either smooth or corrugated, which is applied over a cable core to provide mechanical protection. (PE/IC) 635-1989r

always_swap A bus transaction that atomatically writes a new value to an address and returns the previous value. (C/MM) 1212.1-1993

AM *See:* amplitude modulation.

AMA *See:* automatic message accounting.

amalgam (electrolytic cells) The product formed by mercury and another metal in an electrolytic cell. (EEC/PE) [119]

amateur band (overhead-power-line corona and radio noise) Frequency bands assigned for the transmission of signals by amateur radio operators. *Note:* The amateur bands may differ from country to country. The bands presently in use in the United States under 300 MHz are 1.8–2.0 MHz, 3.5–4.0 MHz, 7.0–7.3 MHz, 10.1–10.15 MHz, 14.00–14.35 MHz, 21.00–21.45 MHz, 24.89–24.99 M.Hz, 28.0–29.7 MHz, 50–54 MHz; 144–148 MHz, and 220–225 MHz. (T&D/PE) 539-1990

ambient air (1) The air surrounding or occupying a space or object. (IA/PSE) 241-1990r
(2) The general air in the area of interest (e.g., the general room atmosphere) distinct from a specific stream or volume of air that may have different properties. (NI) N42.17B-1989r

ambient air temperature (relaying) (metal enclosed bus) The temperature of the surrounding air that comes in contact with equipment. *Note:* Ambient air temperature, as applied to enclosed bus or switchgear assemblies, is the average temperature of the surrounding air that comes in contact with the enclosure. (SWG/PE/SWG-OLD) C37.20-1968w, C37.20.1-1993r, C37.20.2-1993, C37.20.3-1996, C37.21-1985r, C37.23-1969s, C37.100-1992

ambient background Those counts that can be observed, and thereby allowed for, by measuring a sample that is identical to the unknown sample in all respects except for the absence of radioactivity. These counts are attributable to environmental radioactivity in the detector itself, the detector shielding material, and the sample container; cosmic rays; electronic noise pulses; etc. (NI) N42.12-1994

ambient conditions Characteristics of the environment, for example, temperature, humidity, pressure. *See also:* measurement system. (MIL/IM/HFIM) [2], [40]

ambient level The values of radiated and conducted signal and noise existing at a specific test location and time when the test sample is not activated. *See also:* electromagnetic compatibility. (EMC/CHM) C63.5-1988, [51], C63.4-1991

ambient noise (1) (mobile communication) The average radio noise power in a given location that is the integrated sum of atmospheric, galactic, and man-made noise. *See also:* telephone station. (VT) [37]
(2) The all-encompassing noise associated with a given environment, usually a composite of contributions from many sources near and far. (T&D/PE) 539-1990

ambient operating-temperature range The range of environmental temperatures in which a power supply can be safely operated. For units with forced-air cooling, the temperature is measured at the air intake. (AES/PE) [41], [78]

ambient radio noise *See:* ambient level.

ambient sound pressure level The sound pressure level measured at the test facility or at the substation without the transformer energized. (PE/TR) C57.12.90-1999

ambient temperature (1) (electrical heating systems) The environmental temperature surrounding the object under consideration. For objects enclosed in thermal insulation, the ambient temperature is the temperature external to the thermal insulation. (IA/PC) 844-1991
(2) The temperature surrounding the object under consideration. Where electrical heating cable is enclosed in thermal insulation, the ambient temperature is the temperature exterior to the thermal insulation.
(BT/IA/AV/PC) 152-1953s, 515.1-1995, 515-1997
(3) (electric equipment) The temperature of the ambient medium.
(4) (shunt power capacitors) (power and distribution transformers) (neutral grounding devices) The temperature of the medium such as air, water, or earth into which the heat of the equipment is dissipated. *Notes:* 1. For self-ventilated equipment, the ambient temperature is the average temperature of the air in the immediate vicinity of the equipment. 2. For air- or gas-cooled equipment with forced ventilation or secondary water cooling, the ambient temperature is taken as that of the ingoing air or cooling gas. 3. For self-ventilated enclosed (including oil-immersed) equipment considered as a complete unit, the ambient temperature is the average temperature of the air outside of the enclosure in the immediate neighborhood of the equipment.
(SPD/PE/T&D/TR) 32-1972r, 18-1992, C57.12.80-1978r
(5) (light-emitting diodes) (free air temperature) The air temperature measured below a device, in an environment of substantially uniform temperature, cooled only by natural air convention and not materially affected by reflective and radiant surfaces. (IE/EEC) [126]
(6) (nuclear power generating station) The average of air temperature readings at several locations in the immediate neighborhood of the equipment. (PE/NP) 649-1980s
(7) (packaging machinery) The temperature of the surrounding cooling medium, such as gas or liquid, that comes into contact with the heated parts of the apparatus.
(IA/PKG) 333-1980w
(8) The temperature of the surrounding medium that comes in contact with the device or equipment.
(SWG/PE) C37.40-1993, C37.100-1992
(9) The temperature of the surrounding air that comes in contact with the bushing and device or equipment in which the bushing is mounted. (PE/TR) C57.19.03-1996, 21-1976
(10) The temperature of the medium such as air, gas, or water, into which the heat of the equipment is dissipated.
(T&D/PE) 824-1994
(11) The temperature of the medium, usually air, surrounding the battery charger. (IA/PSE) 602-1996
(12) The temperature of the medium, such as air or water, into which the heat generated in the equipment is dissipated.
(PE/TR) C57.15-1999
(13) Temperature of the ambient air immediately surrounding the unit under test. (PEL) 1515-2000
(14) The temperature of the cooling air surrounding a smoothing reactor. (PE/TR) 1277-2000

ambient temperature time constant At a constant operating resistance, the time required for the change in (bolometer unit) bias power to reach 63% of the total change in bias power after an abrupt change in ambient temperature.
(IM) 470-1972w

ambiguity (1) (navigation aid terms) (navigation) The condition obtained when navigation coordinates define more than one point, direction, line of position, or surface of position.
(AES/RS/GCS) 686-1982s, 172-1983w
(2) In fault isolation, an ambiguity that exists when the failure(s) in a system have not been localized to a single diagnostic unit for a repair level. (ATLAS) 1232-1995

ambiguity function The squared magnitude $|\chi(\tau, f_d)|^2$ of the function that describes the response of a radar receiver to targets displaced in range delay τ and Doppler frequency f_d from a reference position, where $|\chi(0,0)|$ is normalized to unity. Mathematically:

$$\chi(\tau, f_d) = \int u(t)u^*(t + \tau)\exp(2\pi j f_d t)dt$$

where

$u(t)$ = the transmitted waveform, suitably normalized
positive τ = a target beyond the reference delay
positive f_d = an approaching target

Ambiguity function is used to examine the suitability of radar waveforms for achieving accuracy, resolution, freedom from ambiguities, and reduction of unwanted clutter.
(AES) 686-1997

ambiguity group The collection of all diagnostic units that are in ambiguity. (ATLAS) 1232-1995

AM broadcast array One or more towers fed the same broadcast signal but at different current levels and with different delays. By carefully choosing the height, location, current level and delay for each tower, a far-field pattern can be constructed to broadcast strongly in some directions and weakly in others. (T&D/PE) 1260-1996

American Federation of Information Processing Societies A national (American) association of computing and information-related organizations that represents the United States in the International Federation of Information Processing (IFIP) organization. (C) 610.10-1994w

American Standard Code for Information Interchange (ASCII) (1) A seven-bit code that standardizes a set of characters representing letters and numbers for international use.
(PE/SUB) 1379-1997
(2) A binary code in which 128 letters, numbers, and special characters are represented by seven-bit numerals. *Note:* Some systems make use of an eight-bit binary code, called ASCII-8, in which 256 symbols are represented. (C) 1084-1986w

American National Standards Institute An organization that establishes and maintains standards for the information processing industry within the United States.
(C) 610.7-1995, 610.10-1994w

American Wire Gauge (AWG) A system of measurement of the thickness and current carrying capacity of wire.
(EMB/MIB) 1073.4.1-2000

αmin *See:* limiting angular subtense.

ammeter (1) (general) An instrument for measuring the magnitude of an electric current. *Note:* It is provided with a scale, usually graduated in either amperes, milliamperes, microamperes, or kiloamperes. If the scale is graduated in milliamperes, microamperes, or kiloamperes, the instrument is usually designated as a milliammeter, a microammeter, or a kiloammeter. *See also:* instrument. (EEC/PE) [119]
(2) An instrument for measuring electric current in amperes.
(CAS) [13]

amnesia address The module address ('FA' HEX) to which a module will respond as though uniquely addressed if that module implements the ability to detect when it cannot determine its address unambiguously and detects that it cannot determine its address unambiguously. (TT/C) 1149.5-1995

amortisseur bar (rotating machinery) A single conductor that is a part of an amortisseur winding or starting winding. *Synonym:* damper bar. *See also:* rotor; stator. (PE) [9]

amortisseur winding A permanently short-circuited winding used for starting induction motors consisting of conductors embedded in the pole shoes of a synchronous machine and connected together at the ends of the poles, but not necessarily connected between poles. (IA/MT) 45-1998

amp *See:* amplifier; ampere.

ampacity (1) (power and distribution transformers) Current-carrying capacity, expressed in amperes, of a wire or cable under stated thermal conditions. (PE/TR) C57.12.80-1978r
(2) Current-carrying capacity of electric conductors expressed in amperes. (NESC/NEC) [86]
(3) (packaging machinery) Current-carrying capacity expressed in amperes. (IA/PKG) 333-1980w
(4) The current-carrying capacity, expressed in amperes, of an electric conductor under stated thermal conditions. (NESC) C2-1997

ampacity correction factor A numeric value equal to one minus the ampacity derating factor. (PE/IC) 848-1996

ampacity derating factor A numeric value representing the fractional reduction from a base ampacity cable rating. Ampacity derating factors are associated with specific installation conditions not presently addressed in the base ampacity. (PE/IC) 848-1996

ampere (metric practice) The constant current that, if maintained in two straight parallel conductors of infinite length, of negligible circular cross section, and placed one meter apart in vacuum, would produce between these conductors a force equal to 2×10^{-7} newton per meter of length. (QUL) 268-1982s

ampere-hour capacity (storage battery) The number of ampere-hours that can be delivered under specified conditions as to temperature, rate of discharge, and final voltage. *See also:* battery. (PE/EEC) [119]

ampere-hour efficiency (storage battery) (storage cell) The electrochemical efficiency expressed as the ratio of the ampere-hour output to the ampere-hour input required for the recharge. *See also:* charge. (PE/EEC) [119]

ampere-hour meter An electricity meter that measures and registers the integral, with respect to time, of the current of the circuit in which it is connected. *Note:* The unit in which this integral is measured is usually the ampere-hour. *See also:* electricity meter. (EEC/PE) [119]

ampere rating (protection and coordination of industrial and commercial power systems) The current that the fuse will carry continuously without deterioration and without exceeding temperature rise limits specified for that fuse. (IA/PSP) 242-1986r

Ampere's law *See:* magnetic field strength produced by an electric current.

ampere-turn per meter The unit of magnetic field strength in SI units (International System of Units). The ampere-turn per meter is the magnetic field strength in the interior of an elongated uniformly wound solenoid that is excited with a linear current density in its winding of one ampere per meter of axial distance. (Std100) 270-1966w

ampere-turns (electrical heating systems) The product of the number of turns and the alternating-current (ac) amperes flowing in an induction heating coil. (IA/PC) 844-1991

& Logical AND. (C/BA) 14536-1995

amphiboly A logical expression that has more than one meaning. For example, the expression A OR B AND C might mean (A OR B) AND C or A OR (B AND C) depending upon the rules of interpretation used. (C) 1084-1986w

amplification (mechanical) The relationship between response acceleration and ground acceleration. (SUB/PE) C37.122.1-1993

amplification, current *See:* current amplification.

amplification, voltage *See:* voltage amplification.

amplified spontaneous emission (laser maser) The radiation resulting from amplification of spontaneous emission. (LEO) 586-1980w

amplifier (1) (analog computer) A device that enables an input signal to control a source of power, and thus is capable of delivering at its output a reproduction or analytic modification of the essential characteristics of the signal. (C) 165-1977w
(2) (data transmission) A unidirectional device that is capable of delivering an enlargement of the waveform of the electric current, voltage, or power supplied to it. (PE) 599-1985w
(3) A device that enables an input signal to control power from a source independent of the signal and thus be capable of delivering an output that bears some relationship to, and is generally greater than, the input signal. (AP/ANT) 145-1983s
(4) (photomultipliers for scintillation counting) A device whose output is an enlarged reproduction of the essential features of an input signal and that draws power from a source other than the input signal. (NPS) 398-1972r
(5) (A) An apparatus or device used to increase the amplitude or the power of an input signal by means of energy drawn from an external source. **(B)** In an analog computer, a device that enables an input signal to control a source of power and thus is capable of delivering at its output an enlarged reproduction or analytical modification of the essential characteristics of the signal. *See also:* high-gain dc amplifier; buffer amplifier; unloading amplifier; relay amplifier; operational amplifier; servo amplifier. (C) 610.10-1994

amplifier, balanced *See:* balanced amplifier.

amplifier, bridging *See:* bridging amplifier.

amplifier, buffer *See:* isolating amplifier.

amplifier, carrier *See:* carrier amplifier.

amplifier, chopper *See:* chopper amplifier.

amplifier class ratings (1) (A) (electron tube) Class-A amplifier. An amplifier in which the grid bias and alternating grid voltages are such that anode current in a specific tube flows at all times. *Note:* The suffix 1 is added to the letter or letters of the class identification to denote that grid current does not flow during any part of the input cycle. The suffix 2 is used to denote that current flows during some part of the cycle. *See also:* amplifier. **(B) (electron tube) Class-AB amplifier.** An amplifier in which the grid bias and alternating grid voltages are such that anode current in a specific tube flows for appreciably more than half but less than the entire electrical cycle. *Note:* The suffix 1 is added to the letter or letters of the class identification to denote that grid current does not flow during any part of the input cycle. The suffix 2 is used to denote that current flows during some part of the cycle. *See also:* amplifier. **(C) (electron tube) Class-B amplifier.** An amplifier in which the grid bias is approximately equal to the cutoff value so that the anode current is approximately zero when no exciting grid voltage is applied, and so that anode current in a specific tube flows for approximately one half of each cycle when an alternating grid voltage is applied. *Note:* The suffix 1 is added to the letter or letters of the class identification to denote that grid current does not flow during any part of the input cycle. The suffix 2 is used to denote that current flows during some part of the cycle. *See also:* amplifier. **(D) (electron tube) Class-C amplifier.** An amplifier in which the grid bias is appreciably greater than the cutoff value so that the anode current in each tube is zero when no alternating grid voltage is applied, and so that anode current in a specific tube flows for appreciably less than one half of each cycle when an alternating grid voltage is applied. *Note:* The suffix 1 is added to the letter or letters of the class identification to denote that grid current does not flow during any part of the input cycle. The suffix 2 is used to denote that current flows during some part of the cycle. *See also:* amplifier. (ED/AP/ANT) 161-1971, 145-1983
(2) A device that increases the amplitude of an electrical signal. Amplifiers are placed in a cable system to strengthen signals weakened by cable and component attenuation. (LM/C) 802.7-1989r

amplifier, difference *See:* differential amplifier.

amplifier, differential *See:* differential amplifier.

amplifier ground *See:* receiver ground.

amplifier, horizontal *See:* horizontal amplifier.

amplifier, integrating *See:* integrating amplifier.

amplifier, intensity *See:* intensity amplifier.

amplifier, inverting *See:* inverting amplifier.

amplifier, isolating *See:* isolating amplifier.

amplifier, isolation *See:* isolation amplifier.

amplifier, line *See:* line amplifier.

amplifier, monitoring *See:* monitoring amplifier.

amplifier noise *See:* noise referred to the input.

amplifier, peak limiting *See:* peak limiter.

amplifier, program *See:* line amplifier.

amplifier, relay *See:* relay amplifier.

amplifier shaping time A nonspecific indication of the shaped-pulse width issuing from a linear pulse amplifier. *See also:* shaping index. (NPS) 325-1996

amplifier, servo *See:* servo amplifier.

amplifier, summing *See:* summing amplifier.

amplifier time constant A misnomer for the width of the shaped pulse issuing from a linear pulse amplifier. *See also:* shaping index. (NPS) 325-1996

amplifier, vertical *See:* vertical amplifier.

amplifier, *x*-axis *See:* horizontal amplifier.

amplifier, *y*-axis *See:* vertical amplifier.

amplifier, *z*-axis *See:* intensity amplifier; Z-axis amplifier.

amplistat reactor A reactor conductively connected between the direct-current winding of a rectifier transformer and rectifier circuit elements that when operating in conjunction with other similar reactors, provides a relatively small controlled direct-current voltage range at the rectifier output terminals. *See also:* reactor. (PE/TR) [57]

amplitude (1) The strength or volume of a periodic signal, usually measured in decibels. (C) 610.7-1995

(2) The maximum or peak value of a periodically varying quantity. *Note:* Sometimes the term complex amplitude is used to denote a phasor. *See also:* magnitude. (AP/PROP) 211-1997

amplitude-comparison monopulse A form of monopulse in which the angular deviation of the target from the antenna axis is measured as the amplitude ratio of the target as received by two antenna patterns. The patterns may be a pair of beams displaced on opposite sides of the antenna axis, or a difference-channel beam having odd symmetry about the axis and a sum beam having even symmetry. In the latter case the ratio may have positive and negative values (0° or 180° phase shift, or in some cases +90° and −90°). Distinguished from phase-comparison monopulse, in which the relative phase of the two patterns carries the information on target displacement. *See also:* phase-comparison monopulse. (AES) 686-1997

amplitude deviation ($\varepsilon(t)$) Instantaneous amplitude departure from a nominal amplitude. (SCC27) 1139-1999

amplitude discriminator (radar) A circuit whose output is a function of the relative magnitudes of two signals. *See also:* navigation. (AES/RS) 686-1982s, [42]

amplitude distortion (data transmission) Distortion caused by a deviation from a desired linear relationship between specified measures of the output and input of a system. *Note:* The related measures need not be output and input values of the same quantity; for example, in a linear detector, the desired relation is between the output signal voltage and the input modulation envelope, or the modulation of the input carrier and the resultant detected signal. (PE) 599-1985w

amplitude factor (of transient recovery voltage) The ratio of the highest peak of the transient recovery voltage to the peak value of the normal-frequency recovery voltage. *Note:* In tests made under one condition to simulate duty under another, as in single-phase tests made to simulate duty on three-phase ungrounded faults, the amplitude factor is expressed in terms of the duty being simulated. (SWG/PE) C37.100-1992

amplitude flatness The variation in output amplitude as a function of frequency in response to a constant amplitude sine wave input. (IM/WM&A) 1057-1994w

amplitude fluctuation *See:* target fluctuation.

amplitude-frequency response (1) (data transmission) The variation of gain, loss, amplification, or attenuation as a function of frequency. *Note:* This response is usually measured in the region of operation in which the transfer characteristic of the system or transducer is essentially linear. (PE) 599-1985w

(2) (high voltage testing) The amplitude frequency response $G(f)$ of a measuring system is the ratio as a function of the frequency f of the output amplitude to the input amplitude of the system when the input is a sinusoid. A convenient form is "the normalized frequency response $g(f)$" in which the constant value of the output amplitude is denoted as unity when that amplitude, multiplied by the scale factor of the system, equals the input amplitude. (PE/PSIM) 4-1978s

amplitude gate *See:* slicer.

amplitude instability ($S_a(f)$) One-sided spectral density of the fractional amplitude deviation. (SCC27) 1139-1999

amplitude jitter A short term instability in the amplitude of a transmission signal. *See also:* phase jitter. (C) 610.7-1995

amplitude locus (control system feedback) (for a nonlinear system or element whose gain is amplitude dependent) A plot of the describing function, in any convenient coordinate system. *See also:* feedback control system. (PE/EDPG) [3]

amplitude-modulated transmitter A transmitter that transmits an amplitude-modulated wave. *Note:* In most amplitude-modulated transmitters, the frequency is stabilized. *See also:* radio transmitter. (AP/BT/ANT) 145-1983s, 182-1961w

amplitude modulation (1) (data transmission) The process by which a continuous high-frequency wave (carrier) is caused to vary in amplitude by the action of another wave containing information. The usual procedure is to key the carrier wave on and off in accordance with the data to be transmitted. For example, a 1170 Hz tone (the carrier) could be off for "space" and on for "mark." This method has several disadvantages. It does not use bandwidth efficiently, since two sidebands of the carrier are produced, and unlike single sideband voice communication methods, the carrier and one sideband cannot be completely eliminated and still do a satisfactory job. The information carrying characteristic of an AM signal is its amplitude. (PE) 599-1985w

(2) (signal-transmission system) The process, or the result of the process, whereby the amplitude of one electrical quantity is varied in accordance with some selected characteristic of a second quantity, which need not be electrical in nature. *See also:* signal. (IE) [43]

(3) Modulation in which the amplitude of a wave is the characteristic varied. (AP/ANT) 145-1983s

(4) (overhead-power-line corona and radio noise) Modulation in which the amplitude of a carrier is caused to depart from its reference value by an amount proportional to the instantaneous value of the modulating signal. 539-1990

(5) A modulation technique in which a data signal is sent onto a carrier at a fixed frequency by raising and lowering the amplitude of the carrier. *See also:* pulse amplitude modulation. (C) 610.7-1995

(6) Modulation in which the amplitude of a carrier is caused to depart from its reference value by an amount proportional to the instantaneous value of the modulating wave. (T&D/PE) 1260-1996

amplitude-modulation noise The noise produced by undesired amplitude variations of a radio-frequency signal. *See also:* radio transmission. (BT) 182-1961w

amplitude-modulation noise level The noise level produced by undesired amplitude variations of a radio frequency signal in the absence of any intended modulation. (AP/ANT) 145-1983s

amplitude noise Used variously to describe target fluctuation and scintillation error. Use of one of these specific terms is recommended to avoid ambiguity. (AES) 686-1997

amplitude pattern *See:* radiation pattern.

amplitude permeability (magnetic core testing) The value of permeability at a stated value of field strength (or induction), the field strength varying periodically with time and with no static magnetic field being present.

$$\mu_a = \frac{1}{\mu_0}\frac{B}{H}$$

μ_a = relative amplitude permeability. Maximum permeability is the maximum value of the amplitude permeability as a function of the field strength (or of the induction). (MAG) 393-1977s

amplitude probability distribution (APD) (1) (electromagnetic site survey) A distribution showing the probability (commonly percentage of time) that an amplitude is exceeded as a function of the amplitude. (EMC) 473-1985r
(2) The fraction of the total time interval for which the envelope of a function is above a given level *x*. (EMC) C63.12-1987

amplitude range (electroacoustics) The ratio, usually expressed in decibels, of the upper and lower limits of program amplitudes that contain all significant energy contributions. (SP) 151-1965w

amplitude ratio *See:* subsidence ratio; gain.

amplitude reference level (pulse techniques) The arbitrary reference level from which all amplitude measurements are made. *Note:* The arbitrary reference level normally is considered to be at an absolute amplitude of zero but may, in fact, have any magnitude of either polarity. If this arbitrary reference level is other than zero, its value and polarity must be stated. *See also:* pulse. (IM/HFIM) [40]

amplitude resonance Resonance in which amplitude is stationary with respect to frequency. (Std100) 270-1966w

amplitude response (camera tubes) The ratio of the peak-to-peak output from the tube resulting from a spatially periodic test pattern, to the difference in output corresponding to large-area blacks and large-area whites, having the same illuminations as the test pattern minima and maxima, respectively. *Note:* The amplitude response is referred to as modulation transfer (sine-wave response) when a sinusoidal test pattern is used and as square-wave response when the pattern consists of alternate black and white bars of equal width. *See also:* camera tube. (ED) [45]

amplitude response characteristic (camera tubes) The relation between amplitude response and television line number (camera tubes) or (image tubes) test-pattern spatial frequency, usually in line pairs per millimeter. *See also:* camera tube. (ED) [45]

amplitude selection In an analog computer, a summation of one or more variables with a constant resulting in a sudden change in rate or level at the output of a computing element as the sum changes sign. *See also:* electronic analog computer. (C) 610.10-1994w, 165-1977w

amplitude shift keying A modulation technique that encodes data by transmitting a signal at two different amplitudes representing binary digit one and binary digit zero. *See also:* binary phase shift keying; frequency shift keying. (C) 610.7-1995

amplitude suppression ratio (frequency modulation) The ratio of the undesired output to the desired output of a frequency-modulation receiver when the applied signal has simultaneous amplitude modulation and frequency modulation. *Note:* This ratio is generally measured with an applied signal that is amplitude modulated 30% at a 400-Hz rate and is frequency modulated 30% of maximum system deviation at a 1000-Hz rate. *See also:* frequency modulation. (EEC/PE) [119]

AM radio broadcast band (overhead-power-line corona and radio noise) A band of frequencies assigned for amplitude-modulated broadcasting to the general public. *Note:* In the United States and Canada, the frequency band is 535−1605 kHz. This is one of the International Telecommunications Union (ITU) frequency allocations, on a worldwide basis, for broadcasting. (T&D/PE) 539-1990

AM to FS converter *See:* transmitting converter; facsimile.

anaerobic Free of uncombined oxygen. (IA) [59]

analog (1) (analog computer) Pertaining to representation by means of continuously variable physical quantities; for example, to describe a physical quantity, such as voltage or shaft position, that normally varies in a continuous manner, or devices such as potentiometers and synchros that operate with such quantities. (C) 165-1977w
(2) (data transmission) Used to describe a physical quantity, such as voltage or shaft position, that normally varies in a continuous manner. (PE) 599-1985w
(3) Pertains to information content that is expressed by signals dependent upon magnitude. *See also:* feedback control system. (IA/ICTL/IAC) [60]
(4) (computers) Pertaining to data in the form of continuously variable physical quantities. *Contrast:* digital. *See also:* analog computer. (C) 610.10-1994w, 1084-1986w

analog and digital data Analog data implies continuity, as contrasted to digital data, that is concerned with discrete states. *Note:* Many signals can be used in either the analog or digital sense, the means of carrying the information being the distinguishing feature. The information content of an analog signal is conveyed by the value or magnitude of some characteristics of the signal such as the amplitude, phase, or frequency of a voltage, the amplitude or duration of a pulse, the angular position of a shaft, or the pressure of the fluid. To extract the information, it is necessary to compare the value or magnitude of the signal to a standard. The information content of the digital signal is concerned with discrete states of the signal, such as the presence or absence of a voltage, a contact in the open or closed position, or a hole or no hole in certain positions on a card. The signal is given meaning by assigning numerical values or other information to the various possible combinations of the discrete states of the signal. *See also:* analog; digital. (EEC) [74]

analog boundary module (ABM) A circuit module connected between the core circuit and an analog function pin to provide facilities for test in a mixed-signal integrated circuit. *Note:* An ABM may be attached to a digital function pin in order to provide analog measurement capability to the pin. *See also:* core circuit; function pin; mixed-signal circuit. (C/TT) 1149.4-1999

analog channel (1) (data transmission) A channel on which the information transmitted can take any value between the limits defined by the channel. Voice channels are analog channels. (PE) 599-1985w
(2) A channel in which transmitted information can take any value between the defined limits of the channel. *Note:* The limits for an analog channel are usually the upper and lower frequencies which will pass through the channel. (C) 610.10-1994w

analog computer (1) (A) (general) An automatic computing device that operates in terms of continuous variation of some physical quantities, such as electrical voltages and currents, mechanical shaft rotations, or displacements, and that is used primarily to solve differential equations. The equations governing the variation of the physical quantities have the same or very nearly the same form as the mathematical equations under investigation and therefore yield a solution analogous to the desired solution of the problem. Results are measured on meters, dials, oscillograph recorders, or oscilloscopes. *See also:* simulator. **(B) (direct current)** An analog computer in which computer variables are represented by the instantaneous values of voltages. **(C) (alternating current)** An analog computer in which electrical signals are in the form of amplitude modulated suppressed carrier signals where the absolute value of a computer variable is represented by the amplitude of the carrier and the sign of a computer variable is represented by the phase (0° or 180°) of the carrier relative to the reference alternating-current signal. (C) 165-1977

(2) A computer that processes analog data. *Synonym:* electronic analog computer. *Contrast:* digital computer; hybrid computer. *See also:* direct-current analog computer; ac analog computer. (C) 610.10-1994w

analog cut* *See:* clipping.

* Deprecated.

analog data (1) Date represented by scalar values. (SUB/PE) 999-1992w

(2) Data in the form of continuous numerical properties represented by physical variables. *Contrast:* digital data. (C) 610.7-1995

(3) Data that represents a variable that is mathematically continuous in the domain of the application. For example, the measurements of time, velocity, pressure, are all continuous variables, excluding quantum effects. (IM/ST) 1451.1-1999

analog device A device that operates with variables represented by continuously measured quantities such as voltages, resistances, rotations, and pressures. (SWG/PE/SUB) C37.100-1992, C37.1-1994

analog divider A divider whose output analog variable is proportional to the quotient of the input analog variables. (C) 610.10-1994w

analog function *See:* supervisory control functions.

analog function check Monitor a reference quantity. A check of master and remote station equipment by exercising a predefined component or capability. (SUB/PE) C37.1-1994

analog multiplier A multiplier whose output analog variable is proportional to the product of two input analog variables. *Note:* This term may also be applied to a device that can perform more than one multiplication, such as a servo multiplier. *Contrast:* analog divider. *See also:* quarter-squares multiplier. (C) 610.10-1994w

analog output One type of continuously variable quantity used to represent another; for example, in temperature measurement, an electric voltage or current output represents temperature input. *See also:* signal. (IE) [43]

analog pin A pin on an integrated circuit or other component that is intended to pass information represented as a current or voltage that can have any value between the limits defined by the driver or receiver to which it is connected. *Notes:* 1. Analog pins can have several forms. In addition to input, output, and bidirectional pins, which are analogous to corresponding digital forms, it is possible to have pins that do not readily fit into any of these categories (e.g., those supporting compensation elements for operational amplifiers). Any such pin that has no identifiable drive capability should be regarded for the purposes of this standard as an input pin. 2. An analog pin may be put into a state in which no signals can pass in either direction between the pin and the core circuit. 3. An analog pin can pass digital data, using discrete levels that lie within its analog range. *Contrast:* digital pin. *See also:* high-Z; core circuit; core disconnect. (C/TT) 1149.4-1999

analog plotter A plotter that presents analog data in the form of a two-dimensional graphic representation. *Contrast:* digital plotter; raster plotter. (C) 610.10-1994w

analog point interfaces Master station or RTU (or both) element(s) that input(s) or output(s) an analog quantity. (SUB/PE) C37.1-1994

analog quantity (A) (station control and data acquisition) A variable represented by a scalar value. **(B) (supervisory control, data acquisition, and automatic control)** A continuous variable that is typically digitized and represented as a scalar value. (SWG/SUB/PE) C37.1-1994, C37.100-1992

analog representation (mathematics of computing) The representation of numerical quantities by means of continuous physical variables such as translation, rotation, voltage, or resistance. *Contrast:* digital representation. (C) 1084-1986w

analog/RF modules Modules whose electronics are primarily analog or radio frequency (RF) rather than digital. These modules often utilize some digital logic for control and test and may interface to digital buses and interconnects. Analog modules often require shielding from EMI produced by digital modules. (C/BA) 14536-1995

analog SCADA function The capability of a supervisory system to accept, record, or display, or do all of these, an analog quantity as presented by a transducer or external device. The transducer may or may not be a part of the supervisory control system. (SUB/PE) C37.1-1994

analog signal (1) (control) A signal that is solely dependent upon magnitude to express information content. *See also:* feedback control system. (IA/ICTL/APP/IAC) [69], [60]

(2) A continuously changing signal. *Contrast:* digital signal. *See also:* digitize. (C) 610.7-1995

analog signaling A means of communicating between devices that uses continuously variable signals. (SUB/PE) 999-1992w

analog simulation (A) A simulation that is designed to be executed on an analog system. **(B)** A simulation that is designed to be executed on a digital system but that represents an analog system. **(C)** A simulation of an analog circuit. *Contrast:* digital simulation. *See also:* hybrid simulation. (C) 610.3-1989

analog switch (telephone loop performance) A switch capable of switching analog and digital signals without converting them into a set digital format. Most analog end office switches are two-wire systems that have simple interfaces with the loop. (COM/TA) 820-1984r

analog switching (telephone switching systems) Switching of continuously varying-level information signals. (COM) 312-1977w

analog telemetering (station control and data acquisition) Telemetering in which some characteristic of the transmitter signal is proportional to the quantity being measured. (SWG/PE/SUB) C37.100-1992, C37.1-1994

analog telephone set A telephone set where the two-way voice communication interface to the network is in an analog format. (COM/TA) 269-1992

analog test access port (ATAP) A set of two mandatory and two optional pins on a mixed-signal integrated circuit. The pins are connected to a bus allowing automatic test equipment to gain access to the components' on-chip analog test facilities. The mandatory pins are labelled AT1 and AT2; the optional pins (labelled AT1N and AT2N) are normally used for differential testing. (C/TT) 1149.4-1999

analog-to-digital (A/D) conversion Production of a digital output corresponding to the value of an analog input quantity. (SWG/PE/SUB) C37.100-1992, C37.1-1994

analog-to-digital converter (1) (data processing) A device that converts a signal that is a function of a continuous variable into a representative number sequence. (MIL) [2]

(2) A circuit whose input is information in analog form and whose output is the same information in digital form. *See also:* digital; analog. (PE) 599-1985w

(3) (hybrid computer linkage components) Provides the means of obtaining a digital number representation of a specific analog voltage value. (C) 166-1977w

(4) (x-ray energy spectrometers) A device whose input is information in analog form and whose output is the same information in digital form. (NPS/NID) 759-1984r

(5) An electronic device used to convert the amplitude of a voltage pulse from analog to digital format. (NI) N42.14-1991

(6) (ADC or A/D converter) A device that provides the means to obtain a digital number representation from a specific analog value. *Contrast:* digital-to-analog converter. (C) 610.10-1994w

(7) A circuit whose input is information in analog form and whose output is the same information in digital form. (IM/ST) 1451.2-1997

(8) A device or a group of devices that converts an analog quantity or analog position input signal into some type of numerical output signal or code. The input signal is either the measure and or a signal derived from it. (SWG/PE) C37.100-1992

analog-to-frequency converter A circuit whose input is information in an analog form other than frequency and whose output is the same information as a frequency proportional to the magnitude of the information. *See also:* analog. (PE) 599-1985w

analog voice frequency circuits abbreviations (A) dBm. Decibels relative to one milliwatt. This is the customary unit worldwide for measurement of communications signal power. **(B)** dBm0. Decibels relative to one milliwatt, referred to a zero transmission level point (0 TLP). **(C)** dBrn. Decibels to one picowatt reference noise level. This is the customary North American unit for measurement of noise power in communications signal circuits. **(D)** dBrnC. Decibels relative to one picowatt reference noise level, measured with C-message or C-notch frequency weighting. **(E)** TLP. Transmission level point. The symbol TLP is preceded by a number that indicates, for a particular point in a transmission system, the design signal level in decibels (dB) relative to the level at a reference point (0 TLP). (COM/TA) 743-1984

analysis (1) (electric penetration assemblies) A process of mathematical or other logical reasoning that leads from stated premises to the conclusion concerning the qualification of an assembly or components. (PE/NP) 317-1983r
(2) (safety systems equipment in nuclear power generating stations) (valve actuators) A course of reasoning showing that a certain result is a consequence of assumed premises. (PE/NP) 382-1985, 627-1980r
(3) (Class 1E battery chargers and inverters) (nuclear power generating systems) A process of mathematical or other logical reasoning that leads from stated premises to the conclusion concerning specific capabilities of equipment and its adequacy for a particular application. *See also:* numerical analysis. (PE/NP) 380-1975w, 323-1974s, 650-1979s, 933-1999
(4) Examination for the purpose of understanding. (C/SE) 1074-1995s
(5) The process of studying a system by partitioning the system into parts (functions, components, or objects) and determining how the parts relate to each other. (C/SE) 1362-1998

analysis phase (1) The steps a software administration utility performs, before modifying the target, while attempting to ensure that the execution of operations on the target will succeed. (C/PA) 1387.2-1995
(2) (software) *See also:* requirements phase. (C/SE) 729-1983s

analyst A member of the technical community (such as a systems engineer or business analyst, developing the system requirements) who is skilled and trained to define problems and to analyze, develop, and express algorithms. (C/SE) 1233-1998

analyte The particular radionuclide(s) to be determined in a sample of interest. As a matter of clarity when interpreting various sections of IEEE Std N42.23-1995, a gamma-ray spectral analysis is considered one analyte. (NI) N42.23-1995

analytical engine A device from which modern digital computers are descended, invented in the mid 1800's by Charles Babbage, a British mathematician, to solve mathematical problems. (C) 610.10-1994w

analytical limit Limit of a measured or calculated variable established by the safety analysis to ensure that a safety limit is not exceeded. (PE/NP) 603-1998

analytical model (1) (software) A representation of a process or phenomenon by a set of solvable equations. *See also:* process; simulation. (C/SE) 729-1983s
(2) (modeling and simulation) A model consisting of a set of solvable equations; for example, a system of solvable equations that represents the laws of supply and demand in the world market. (C) 610.3-1989w

analyzer *See:* network analyzer; digital differential analyzer; differential analyzer.

ANC *See:* ancillary logic.

ancestor (1) Relative to a given node *x* within a tree, any node *y* for which *x* is a descendent node of *y*. (C) 610.5-1990w
(2) (of a class) A generic ancestor of the class or a parent of the class or an ancestor of a parent of the class. *Contrast:* generic ancestor; reflexive ancestor. (C/SE) 1320.2-1998

ancestral box A box related to a specific diagram by a hierarchically consecutive sequence of one or more parent/child relationships. (C/SE) 1320.1-1998

ancestral diagram A diagram that contains an ancestral box. (C/SE) 1320.1-1998

anchor (conductor stringing equipment) A device that serves as a reliable support to hold an object firmly in place. The term anchor is normally associated with cone, plate, screw or concrete anchors, but the terms snub, deadman, and anchor log are usually associated with pole stubs or logs set or buried in the ground to serve as temporary anchors. The latter are often used at pull and tension sites *Synonyms:* snub structure; snub; deadman; anchor log. (T&D/PE) 1048-1990, 524-1992r, 524a-1993r

anchorage (1) (raceway) (raceway systems for Class 1E circuits for nuclear power generating stations) The connection between the building structure and the raceway support. (PE/NP) 628-1987r
(2) A secure point of attachment to which the fall protection system is connected. (NESC/PE/T&D) C2-1997, 1307-1996

anchor guy guard A protective cover over the guy, often a length of sheet metal or plastic shaped to a semicircular or tubular section and equipped with means of attachment to the guy. It may also be of wood. *See also:* tower. (T&D/PE) [10]

anchor light (illuminating engineering) An aircraft light designed for use on a seaplane or amphibian to indicate its position when at anchored or moored. (ED) [127]

anchor log (1) A piece of rigid material such as timber, metal, or concrete, usually several feet in length, buried in earth in a horizontal position and at right angles to anchor rod attachment. *Synonym:* dead man. *See also:* tower; anchor. (T&D/PE) [10]
(2) *See also:* anchor. (T&D/PE) 524-1992r

anchor point *See:* dead-end.

anchor rod A steel or other metal rod designed for convenient attachment to a buried anchor and also to provide for one or more guy attachments above ground. *See also:* tower. (T&D/PE) [10]

anchor site The location along the line where anchors are installed to temporarily hold the conductors in facilitating splicing, pulling, or tensioning. (T&D/PE) 524a-1993r

ancillary data Optional, protocol-specific or local-system-specific information. The information can be both local or end-to-end significant. It can be header information or part of the data portion. It can be protocol-specific and implementation- or system-specific. (C) 1003.5-1999

ancillary equipment (1) (test, measurement, and diagnostic equipment) Equipment that is auxiliary or supplementary to an automatic test equipment installation. Ancillary equipment usually consists of standard off-the-shelf items such as an oscilloscope and distortion analyzer. (MIL) [2]
(2) Auxiliary or accessory equipment (e.g., thermometer, liquid level gauge, pressure gauge). (SWG/PE) C37.10-1995

ancillary logic (ANC) Logic required for each segment but not part of any device. Operations associated with arbitration, geographical addressing, system handshake and run/halt control are carried out in the ancillary logic. The circuit board containing the ANC may also contain segment terminators. (NID) 960-1993

AND (mathematics of computing) A Boolean operator having the property that if P is a statement, Q is a statement, R is a statement,. . ., then the AND of P,Q,R,. . . is true if and only if all statements are true. *Note:* P AND Q is often represented by P·Q, P&Q, P∧ Q, or PQ.

P	*Q*	*P∧Q*
0	0	0
0	1	0
1	0	0
1	1	1

AND Truth Table

Synonyms: conjunction; meet; logic multiply; intersection; logical multiply; collation; Boolean multiplication. (C) 1084-1986w

AND-circuit *See:* AND gate.

AND element *See:* AND gate.

Anderson bridge A six-branch network in which an outer loop of four arms is formed by three nonreactive resistors and the unknown inductor, and an inner loop of three arms is formed by a capacitor and a fourth resistor in series with each other and in parallel with the arm that is opposite the unknown inductor, the detector being connected between the junction of the capacitor and the fourth resistor and that end of the unknown inductor that is separated from a terminal of the capacitor by only one resistor, while the source is connected to the other end of the unknown inductor and to the junction of the capacitor with two resistors of the outer loop. *Note:* Normally used for the comparison of self-inductance with capacitance. The balance is independent of frequency. *See also:* bridge.

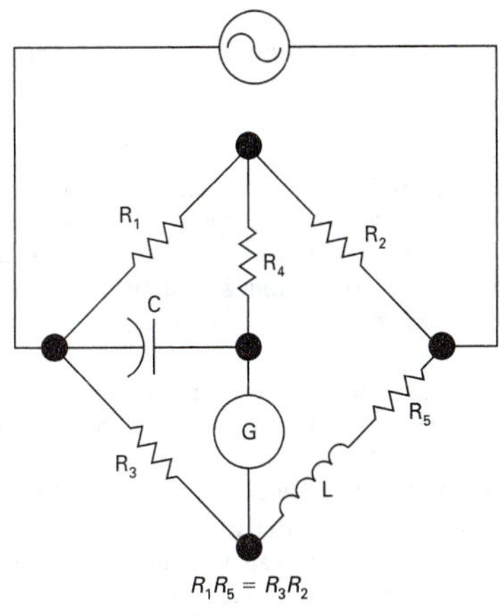

$$R_1R_5 = R_3R_2$$

$$L = CR_3\left[R_4\left(1 + \frac{R_2}{R_1}\right) + R_2\right]$$

Anderson bridge

(EEC/PE) [119]

AND gate (1) (general) A combinational logic element such that the output channel is in its ONE state if and only if each input channel is in its ONE state. (C) 162-1963w
(2) A gate that implements the logic AND operator. (C) [85]
(3) A gate that performs the Boolean operation of conjunction. *Synonym:* AND element. (C) 610.10-1994w

AND-NOT* *See:* exclusion.

* Deprecated.

AND-parallelism Pertaining to the performance of multiple predicate operations concurrently; the successful completion of which results in a true response. *Contrast:* OR-parallelism. (C) 610.10-1994w

anechoic chamber An enclosure especially designed with boundaries that absorb sufficiently well the sound incident thereon to create an essentially free-field condition in the frequency range of interest. (SP) [32]

anechoic enclosure (radio frequency) An enclosure whose internal walls have low reflection characteristics. *See also:* electromagnetic compatibility. [53]

anelectrotonus (electrobiology) Electrotonus produced in the region of the anode. *See also:* excitability. (EMB) [47]

anemometer (navigation aid terms) An instrument for measuring the speed of wind. (AES/GCS) 172-1983w

aneroid altimeter *See:* barometric altimeter.

anesthetizing location (health care facilities) Any area in which it is intended to administer any flammable or nonflammable inhalation anesthetic agents in the course of examination or treatment and includes operating rooms, delivery rooms, emergency rooms, anesthetizing rooms, corridors, utility rooms and other areas that are intended for induction of anesthesia with flammable or nonflammable anesthetizing agents. (NESC/NEC) [86]

anesthetizing-location receptacle (health care facilities) A receptacle designed to accept the attachment plugs listed for use in such locations. (NESC/NEC) [86]

angel *See:* angel echo.

angel echo (1) Radar returns caused by atmospheric inhomogeneities, refractive index discontinuities, insects, birds, or unknown sources. *Note:* Originally, when some physical target could not be identified through direct visual observation, echoes from such unknown causes were designated as "angels." (AP/PROP) 211-1997
(2) A radar echo caused by birds, insects, and atmospheric clear-air turbulence not usually visible to the eye. Often a term applied to any unknown radar echo that does not appear to be related to conventional targets. *Synonym:* angel. (AES) 686-1997

angle (of a waveform). The phase of a periodic or approximately periodic waveform. *See also:* phase; phase angle. (IT) [7]

angle brackets The characters "⟨" (*left-angle-bracket*) and "⟩" (*right-angle-bracket*). When used in the phrase "enclosed in angle brackets" the symbol "⟨" shall immediately precede the object to be enclosed, and "⟩" shall immediately follow it. When describing these characters, the names ⟨`less-than-sign`⟩ and ⟨`greater-than-sign`⟩ are used. (C/PA) 9945-2-1993

angle, bunching *See:* bunching angle.

angle, effective bunching *See:* effective bunching angle.

angle, flow *See:* flow angle.

angle, maximum-deflection *See:* maximum-deflection angle.

angle modulation (1) Modulation in which the angle of a sine-wave carrier is the characteristic varied from its reference value. *Notes:* 1. Frequency modulation and phase modulation are particular forms of angle modulation; however, the term "frequency modulation" is often used to designate various forms of angle modulation. 2. The reference value is usually taken to be the angle of the unmodulated wave. *See also:* modulation index. (AP/ANT) 145-1983s
(2) (data transmission) The process of causing the angle of the carrier wave to vary in accordance with the signal wave. Phase and frequency modulation are two particular types of angle modulation. (PE) 599-1985w

angle noise The noise-like variation in the apparent angle of arrival of a signal received from a target, caused by changes in phase and amplitude of multiple, unresolved target-scattering sources. *Note:* Includes both glint and scintillation errors. *See also:* glint; scintillation error. (AES) 686-1997

angle of advance (1) (power inverter) The time interval in electrical degrees by which the beginning of anode conduction leads the moment at which the anode voltage would attain a negative value equal to that of the succeeding anode in the commutating group. *See also:* rectification. (EEC/PCON) [110]
(2) (semiconductor power converter) (semiconductor rectifiers) The angle by which forward conduction is advanced by the control means only, in the incoming circuit element, ahead of the instant in the cycle at which the incoming com-

mutating voltage passes through zero in the direction to produce forward conduction in the outgoing circuit element. *See also:* rectification; semiconductor rectifier stack. (IA) [62]

angle-of-approach lights (illuminating engineering) Aeronautical ground lights arranged so as to indicate a desired angle of descent during an approach to an aerodrome runway. *Synonym:* optical glide path lights. (ED) [127]

angle of arrival (of a wave) The angle between the negative of the propagation vector and a reference direction. (AP/PROP) 211-1997

angle of attack (navigation aid terms) The angle between the mean chord or the wing and the line of flow of the air past the aircraft. (AES/GCS) 172-1983w

angle of climb (navigation aid terms) The angle between a climbing aircraft's flight path and the horizontal. (AES/GCS) 172-1983w

angle of collimation (illuminating engineering) The angle subtended by a luminaire on an irradiated surface. (ED) [127]

angle of cut (navigation aid terms) (navigation) The angle at which two lines of position intersect. *Synonym:* crossing angle. (AES/GCS) 172-1983w

angle of descent (navigation aid terms) The angle between a descending aircraft's flight path and the horizontal. (AES/GCS) 172-1983w

angle of deviation (fiber optics) In optics, the net angular deflection experienced by a light ray after one or more refractions or reflections. *Note:* The term is generally used in reference to prisms, assuming air interfaces. The angle of deviation is then the angle between the incident ray and the emergent ray. *See also:* refraction; reflection. (Std100) 812-1984w

angle of extinction The phase angle of the stopping (extinction) instant of anode-current flow in a glass tube with respect to the starting instant of the corresponding positive half cycle of the anode voltage of the tube. *See also:* electronic controller. (IA/ICTL/IAC) [60]

angle of ignition The phase angle of the starting instant of anode-current flow in a gas tube with respect to the starting instant of the corresponding positive half cycle of the anode voltage of the tube. *See also:* electronic controller. (IA/ICTL/IAC) [60]

angle of incidence (1) (acousto-optic device) The angle in air between the acoustic wavefront and the normal to the optical wavefront. For operation in the Bragg region, maximum diffraction into the first order occurs when the angle of incidence is equal to the Bragg angle, θ_B, which is given by the equation $\sin \theta_B = \lambda_0/2\Lambda$. (UFFC) [23]
(2) (fiber optics) The angle between an incident ray and the normal to a reflecting or refracting surface. (Std100) 812-1984w
(3) At a point on a surface, the angle between the negative of the incident propagation vector and the outward normal to this surface. (AP/PROP) 211-1997

angle of protection (lightning) The angle between the vertical plane and a plane through the ground wire, within which the line conductors must lie in order to ensure a predetermined degree of protection against direct lightning strokes. *See also:* surge arrester. (PE) [8], [84]

angle of retard (1) (thyristor) The interval in electrical angular measure by which the trigger pulse is delayed in relation to operation that would occur with continuous gated control elements and a resistive load. (IA/IPC) 428-1981w
(2) (semiconductor rectifiers) The angle by which forward conduction is delayed by the control means only, beyond the instant in the cycle at which the incoming commutating voltage passes through zero in the direction to produce forward conduction in the incoming circuit element. *See also:* semiconductor rectifier stack. (IA) [62]

angle of retard unbalance (thyristor) (tracking unbalance) The load voltage/current unbalance due to unequal angles of retard either between positive and negative half cycles of a single ac wave or between two or more phases in a three-phase system. (IA/IPC) 428-1981w

angle optimum bunching *See:* optimum bunching.

angle or phase (sine wave) The measure of the progression of the wave in time or space from a chosen instant or position or both. *Notes:* 1. In the expression for a sine wave, the angle or phase is the value of the entire argument of the sine function. 2. In the representation of a sine wave by a phasor or rotating vector, the angle or phase is the angle through which the vector has progressed. *See also:* wavefront. (AP/ANT) 145-1983s

angle, overlap *See:* overlap angle.

angle random walk *See:* random walk.

angle, roll over *See:* roll over angle.

angle tower A tower located where the line changes horizontal direction sufficiently to require special design of the tower to withstand the resultant pull of the wires and to provide adequate clearance. *See also:* tower. (T&D/PE) [10]

angle tracking *See:* tracking.

angle, transit *See:* transit angle.

angstrom† **(fiber optics)** A unit of optical wavelength.

$$1\text{Å} = 10^{-10}\ \text{m}$$

Note: The angstrom has been used historically in the field of optics, but it is not an SI (International System) unit. (Std100) 812-1984w

† Obsolete.

angular acceleration sensitivity (A) (accelerometer) The output (divided by the scale factor) of a linear accelerometer that is produced per unit of angular acceleration input about a specified axis. **(B) (gyros)** The ratio of drift rate due to angular acceleration about a gyro axis divided by the angular acceleration causing it. *Note:* In single-degree-of-freedom gyros, it is nominally equal to the effective moment of inertia of the gimbal assembly divided by the angular momentum. (AES/GYAC) 528-1994

angular accelerometer A device that senses angular acceleration about an input axis. An output signal is produced by the reaction of the moment of inertia of a proof mass to an angular acceleration input. The output is usually an electrical signal proportional to applied angular acceleration. (AES/GYAC) 528-1994

angular accuracy (radar) The degree to which the measurement of the angular location of a target with respect to a given reference represents the true angular location of the target with respect to this reference. (AES/RS) 686-1982s

angular-case-motion sensitivity (dynamically tuned gyro) The drift rate resulting from an oscillatory angular input about an axis normal to the spin axis at twice the rotor spin frequency. This effect is due to the single-degree-of-freedom of the gimbal relative to the support shaft and is proportional to the input amplitude and phase relative to the flexure axes. *See also:* two-N (2N) angular sensitivity. (AES/GYAC) 528-1994

angular dependence The dependence of the response of an instrument upon the direction of the incident radiation. (NI) N42.20-1995

angular deviation loss (acoustic transducer) The ratio of the response in a specific direction to the response on the principal axis, usually expressed in decibels. (SP) [32]

angular deviation sensitivity (navigation aid terms) The ratio of change of course indication to the change of angular displacement from the course line. (AES/RS/GCS) 686-1982s, 172-1983w

angular displacement (polyphase transformer) The phase angle expressed in degrees between the line-to-neutral voltage of the reference identified high-voltage terminal and the line-to-neutral voltage of the corresponding identified low-voltage of terminal. *Note:* The preferred connection and arrangement of terminal markings for polyphase transformers are those that have the smallest possible phase-angle displacements and are measured in a clockwise direction from the line-to-neutral voltage of the reference identified high-voltage terminal. Thus, standard three-phase transformers have angular displacements of either zero or 30 degrees. *See also:* routine test. (PE/TR) C57.12.80-1978r

angular displacement of polyphase regulator (A) The time angle, expressed in degrees, between the line-to-neutral voltage of the reference identified source voltage terminal S_1 and the line-to-neutral voltage of the corresponding identified load voltage terminal L_1. **(B)** The connection and arrangement of terminal markings for three-phase regulators in a wye connection has an angular displacement of zero degrees. **(C)** The connection and arrangement of terminal markings for three-phase regulators in a delta connection has an angular displacement of zero degrees when the regulator is on the neutral tap position. When the regulator is on a tap position other than neutral, the angular displacement will be other than zero degrees. The angular displacement with the regulator connected in delta will be less than $\pm 5°$ for a $\pm 10\%$ range of regulation. (PE/TR) C57.15-1999

angular frequency (ω) (of a sinusoidal wave) 2π times the frequency. *Synonym:* radian frequency. (AP/PROP) 211-1997

angular misalignment loss (fiber optics) The optical power loss caused by angular deviation from the optimum alignment of source to optical waveguide, waveguide to waveguide, or waveguide to detector. *See also:* extrinsic joint loss; intrinsic joint loss; lateral offset loss; gap loss. (Std100) 812-1984w

angular power spectrum Constituted by the mean squared magnitudes of the plane wave spectrum of an electromagnetic field as a function of the direction cosines k_x/k and k_y/k. *Note:* The angular power spectrum and the mutual coherence function are Fourier transform pairs. *See also:* mutual coherence function. (AP/PROP) 211-1997

angular resolution The ability to distinguish between two targets solely by the measurement of their angle, usually expressed in terms of the minimum angle separation by which two targets at a given range can be distinguished. *Note:* The required separation should be specified for targets of given relative power level at the receiver. Equal powers are often assumed, but where resolution of targets of different powers is important it may be necessary to specify the separation at two or more power ratios. (AES) 686-1997

angular spectrum An electromagnetic field that is source-free in the homogeneous half space $z > 0$, can be represented in this half space by a superposition of plane waves. The complex amplitude of these plane waves, as a function of their direction cosines, constitute the angular spectrum. *Notes:* 1. The direction cosines are defined by k_x/k and k_y/k where k_x and k_y are the x and y components of the wave vector $\vec{k}$. 2. The angular spectrum is the Fourier transform of the field in the plane $z = 0$ or any other plane z = constant > 0. (AP/PROP) 211-1990s

angular swing (acousto-optic deflector) The center-to-center angular separation between the deflected light beams obtained upon application of the maximum and minimum acoustic frequency of the frequency range. (UFFC) [17]

angular-velocity-sensitivity (accelerometer) (inertial sensors) The output (divided by the scale factor) of a linear accelerometer that is produced per unit of angular velocity input about a specified axis. (AES/GYAC) 528-1984s

angular vibration sensitivity (inertial sensors) (gyros) The ratio of the change in output due to angular vibration about a sensor axis divided by the amplitude of the angular vibration causing it. (AES/GYAC) 528-1994

angular width *See:* course width.

ANI *See:* automatic number identification.

animation A technique that presents a logical sequence of images in such a manner as to create an illusion of motion. (C) 610.6-1991w

anisoelasticity (gyros) The inequality of compliance of a structure in different directions. *See also:* principal axis of compliance; acceleration-squared-sensitive drift rate. (AES/GYAC) 528-1994

anisoinertia (A) (accelerometer) A relationship among the principal axis moments of inertia of an accelerometer pendulum in which the moment of inertia about the output axis differs from the difference of the moments of inertia about the other two principal axes. This inequality causes the effective centers of mass for angular velocity and for angular acceleration to be physically separated. In a system in which the accelerometer is modeled as though it were located at the effective center of mass for angular acceleration, there will be an offset in accelerometer output proportional to the product of the angular rates about the input and pendulous axes. Anisoinertia may be expressed as the magnitude of the actual separation in units of length, or as a compensation term in units of μg/(rad/s) squared. Anisoinertia, in this usage, differs from standard physical definitions, but it describes a real effect that is closely analgous to the effect of the same name in gyros. **(B) (gyros)** The inequality of the moments of inertia about the gimbal principal axes. When the gyro is subjected to angular rates about the input and spin axes, and the moments of inertia about these axes are unequal, a torque is developed about the output axis that is proportional to the difference of the inertias about the input and spin axes multiplied by the product of the rates about these two axes. (AES/GYAC) 528-1994

anisotropic (1) (fiber optics) Pertaining to a material whose electrical or optical properties are different for different directions of propagation or different polarizations of a traveling wave. *See also:* isotropic. (Std100) 812-1984w **(2) (wood transmission structures)** Of unequal physical properties along different axes. (T&D/PE) 751-1990

anisotropic medium A medium that is not isotropic, i.e., whose constitutive parameters depend on the polarization and direction of wave propagation of the electric and magnetic fields. (AP/PROP) 211-1997

anisotropic substrate A substrate whose electric or magnetic properties, or both, are directionally dependent. (MTT) 1004-1987w

annealing (metal-nitride-oxide field-effect transistor) In the context of metal oxide semiconductor (MOS) device properties under irradiation, annealing refers to the sometimes observed reduction of the radiation-induced threshold voltage change over a period of seconds to hours after exposure to radiation has ceased. (ED) 581-1978w

annotation Further documentation accompanying a requirement such as background information and/or descriptive material. (C/SE) 1233-1998

annoyance shock An electric shock from a steady-state or a discharge current for which a person would consider the sensation to be a mild irritant if it were to occur repeatedly. (PE/T&D) 539-1990

annual cycle One complete execution of a data processing function that must be performed once a year. *Synonym:* yearly cycle. *See also:* monthly cycle; weekly cycle; daily cycle. (C) 610.2-1987

annular slot antenna A slot antenna with the radiating slot having the shape of an annulus. (AP/ANT) 145-1993

annul bit A bit in a delayed control-transfer instruction that can cause the delay instruction to have no effect. (C/MM) 1754-1994

annunciator (thyristor) A visual signal device consisting of a number of pilot lights or drops, each one indicating the condition that exists or has existed in an associated circuit, accordingly labeled. (IA/IPC) 428-1981w

annunciator relay (power system device function numbers) A nonautomatically reset device that gives a number of separate visual indications upon the functioning of protective devices, and which may also be arranged to perform a lockout function. (SUB/PE) C37.2-1979s

anode (1) An electrode through which current enters any conductor of the nonmetallic class. Specifically, an electrolytic anode is an electrode at which negative ions are discharged, or positive ions are formed, or at which other oxidizing reactions occur. (EEC/PE) [119]
(2) An electrode or portion of an electrode at which a net oxidation-reaction occurs. *See also:* electrochemical cell. (AES/IA/APP) [41], [59], [73]
(3) (thyristor) The electrode by which current enters the thyristor, when the thyristor is in the ON state with the gate open-circuited. *Note:* This term does not apply to bidirectional thyristors. (IA/ED) 223-1966w, [46], [12], [62]
(4) (electron tube or valve) An electrode through which a principal stream of electrons leaves the interelectrode space. *See also:* electrode. (ED/NPS) 161-1971w, 398-1972r
(5) (semiconductor rectifier diode) The electrode from which the forward current flows within the cell. *See also:* semiconductor. (IA) 59-1962w, [12]
(6) (x-ray tubes) *See also:* target.
(7) (light-emitting diodes) The electrode from which the forward current is directed within the device. (IE/EEC) [126]

anode butt A partially consumed anode. *See also:* fused electrolyte. (EEC/PE) [119]

anode characteristic *See:* anode-to-cathode voltage-current characteristic.

anode circuit A circuit that includes the anode-cathode path of an electron tube in series connection with other elements. *See also:* electronic controller. (IA/ICTL/IAC) [60]

anode circuit breaker (1) (power system device function numbers) A device used in the anode circuits of a power rectifier circuit if an arc-back should occur. (PE/SUB) C37.2-1979s
(2) A low-voltage power circuit breaker that is designed for connection in an anode of a mercury-arc power rectifier unit, that trips automatically only on reverse current and starts reduction of a current in a specified time when the arc-back occurs at the end of the forward current conduction, and that substantially interrupts the arc-back current within one cycle of the fundamental frequency after the beginning of the arc-back. *Note:* The specified time in present practice is 0.008 s or less (at an ac frequency of 60 Hz). (SWG/PE) C37.100-1992

anode cleaning (electroplating) (reverse-current cleaning) Electrolytic cleaning in which the metal to be cleaned is made the anode. *See also:* battery. (PE/EEC) [119]

anode corrosion efficiency The ratio of the actual corrosion of an anode to the theoretical corrosion calculated from the quantity of electricity that has passed. (IA) [59]

anode current *See:* electrode current; electronic controller.

anode dark space (gas tube) (gas) A narrow dark zone next to the surface of the anode. *See also:* discharge. (Std100) [84]

anode differential resistance *See:* anode resistance.

anode effect A phenomenon occurring at the anode, characterized by failure of the electrolyte to wet the anode and resulting in the formation of a more or less continuous gas film separating the electrolyte and anode and increasing the potential difference between them. *See also:* fused electrolyte. (EEC/PE) [119]

anode efficiency The current efficiency of a specified anodic process. *See also:* electrochemistry. (EEC/PE) [119]

anode, excitation *See:* excitation anode.

anode fall (gas) The fall of potential due to the space charge near the anode. *See also:* discharge. (ED) [45], [84]

anode firing The method of initiating conduction of an ignitron by connecting the ignitor through a rectifying element to the anode of the ignitron to obtain power for the firing current pulse. *See also:* electronic controller. (IA/ICTL/IAC) [60]

anode glow (gas tube) A very bright narrow zone situated at the near end of the positive column with respect to the anode. *See also:* discharge. (ED) [45], [84]

anode layer A molten metal or alloy, serving as the anode in an electrolytic cell, that floats on the fused electrolyte or upon which the fused electrolyte floats. *See also:* fused electrolyte. (EEC/PE) [119]

anode, main *See:* main anode.

anode mud *See:* anode slime.

anode paralleling reactor (power and distribution transformers) A reactor with a set of mutually coupled windings connected to anodes operating in parallel from the same transformer terminal. (PE/TR) C57.12.80-1978r

anode power supply (electron tube) (plate power supply) The means for supplying power to the plate at a voltage that is usually positive with respect to the cathode. *See also:* power pack. (PE/EEC) [119]

anode-reflected-pulse rise time The rise time of a pulse reflected from the anode. *Note:* This time can be measured with a time-domain reflectometer. (NPS) 398-1972r

anode region (gas tube) The group of regions comprising the positive column, anode glow, and anode dark space. *See also:* discharge. (ED) [45], [84]

anode relieving (gas tube) (pool-cathode tube) An anode that provides an alternative conducting path to reduce the current to another electrode. *See also:* electrode. (ED) [45], [84]

anode resistance The quotient of a small change in anode voltage by a corresponding small change of the anode current, all the other electrode voltages being maintained constant. It is equal to the reciprocal of the anode conductance. *See also:* ON period. (ED) [45], [84]

anode scrap That portion of the anode remaining after the scheduled period for the electrolytic refining of the bulk of its metal content has been completed. *See also:* electrorefining. (EEC/PE) [119]

anode slime Finely divided insoluble metal or compound forming on the surface of an anode or in the solution during electrolysis. (EEC/PE) [119]

anode sputtering *See:* cathode sputtering.

anode strap (magnetrons) A metallic connector between selected anode segments of a multicavity magnetron, principally for the purpose of mode separation. *See also:* magnetron. (ED) 161-1971w, [45]

anode supply voltage The voltage at the terminals of a source of electric power connected in series in the anode circuit. *See also:* electronic controller. (IA/ICTL/IAC) [60]

anode terminal (1) (semiconductor devices) The terminal by which current enters the device. *See also:* semiconductor device; semiconductor. (ED) 216-1960w
(2) (thyristor) The terminal that is connected to the anode. *Note:* This term does not apply to bidirectional thyristors. *See also:* anode. (ED) [46]

anode-to-cathode voltage (thyristor) The voltage between the anode terminal and the cathode terminal. *Note:* It is called positive when the anode potential is higher than the cathode potential and called negative when the anode potential is lower than the cathode potential. *Synonym:* anode voltage. *See also:* electronic controller; principal voltage-current characteristic. (ED) [46]

anode-to-cathode voltage-current characteristic (thyristor) A function, usually represented graphically, relating the anode-to-cathode voltage to the principal current with gate current, where applicable, as a parameter. *Note:* This term does not apply to bidirectional thyristors. *Synonym:* anode characteristic. (ED) [46]

anode voltage *See:* electrode voltage; electronic controller; anode-to-cathode voltage.

anode voltage drop (glow-discharge cold-cathode tube) The main gap voltage drop after conduction is established in the main gap. (ED) [45]

anode voltage, forward, peak *See:* peak forward anode voltage.

anode voltage, inverse, peak *See:* peak inverse anode voltage.

anodic polarization Polarization of an anode. *See also:* electrochemistry. (EEC/PE) [119]

anolyte The portion of an electrolyte in an electrolytic cell adjacent to an anode. If a diaphragm is present, it is the portion of electrolyte on the anode side of the diaphragm. *See also:* electrolytic cell. (IA) [59]

anomaly (1) (software verification and validation plans) Anything observed in the documentation or operation of software that deviates from expectations based on previously verified software products or reference documents. (C/SE) 1012-1986s, 610.12-1990

(2) (data management) An irregularity that arises when processing an improperly structured database. For example, in order to retrieve all the SUPPLIERS from the database in the figure below, one would have to search sequentially through all the PARTS INVENTORY segments.

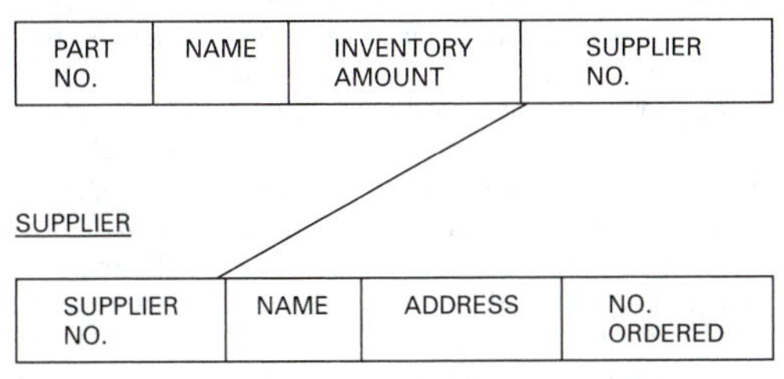

anomaly

(C) 610.5-1990w

(3) Any deviation from requirements, expected or desired behavior, or performance of the software. (C/SE) 1074-1995s

(4) Irregularity; deviation from usual behavior. (ATLAS) 1232-1995

(5) Any condition that deviates from expectations based on requirements specifications, design documents, user documents, standards, etc., or from someone's perceptions or experiences. Anomalies may be found during, but not limited to, the review, test, analysis, compilation, or use of software products or applicable documentation. (C/SE) 1044.1-1995, 1044-1993, 1028-1997

(6) Deviation from the normal behavior of a test subject. Faults (e.g., output stuck high, gain low) and manufacturing defects (e.g., missing or incorrect components, incorrectly installed components) are kinds of anomalies. (SCC20) 1226-1998

A-N radio range (navigation aid terms) A radio range providing four radial lines of position identified aurally as a continuous tone resulting from the interleaving of equal amplitude A and N of international Morse code. The sense of deviation from these lines is indicated by deterioration of the steady tone into audible A or N code signals. (AES/GCS) 172-1983w

ANSI *See:* American National Standards Institute.

ANSI standard A standard approved by The American National Standards Institute. Examples of ANSI standards include programming languages (C, FORTRAN, or COBOL), media formats (Hollerith cards), and interface standards (SCSI interfaces, device drivers). (C) 610.7-1995, 610.10-1994w

ANSI C A standardized version of C established by ANSI. (C) 610.13-1993w

answer To respond to a calling station, either automatically, under program control, or manually, to establish a connection between stations. (C) 610.7-1995

answered call (1) (telephone switching systems) A call on which an answer signal occurred. (COM) 312-1977w

(2) (public telephone service) The called party off-hook supervision duration exceeds the minimum chargeable duration (MCD) after an allowance equal to the worst possible inaccuracy known about the timing sensor has been applied. *See also:* charge delay. (COM/TA) 973-1990w

answering plug and cord A plug and cord used to answer a calling line. (EEC/PE) [119]

answer signal (telephone switching systems) A signal that indicates that the call has been answered. (COM) 312-1977w

answer supervision delay The time interval between the transition of the called line equipment from on-hook to off-hook and transfer of the answer signal to the originating system during a multi-office call. *Note:* To transmit an answer signal to an originating switching system in per-trunk-signaling, the terminating system sends an on-hook-to-off-hook transition and maintains the off-hook state until disconnect. (COM/TA) 973-1990w

antenna (1) (general) That part of a transmitting or receiving system that is designed to radiate or to receive electromagnetic waves. (AP/ANT) 145-1993

(2) (data transmission) A means for radiating or receiving radio waves. *See also:* horn antenna; effective height; dipole antenna; effective area antenna; loop antenna; helical antenna; slot antenna. (PE) 599-1985w

(3) (overhead-power-line corona and radio noise) A device used to radiate or receive electromagnetic waves. (T&D/PE) 539-1990

(4) A device used for transmitting or receiving electromagnetic signals or power. As such, it is designed to maximize its coupling to the electromagnetic field; as a receiver it is made to intercept as much of the field as possible. Those devices that are made to measure the power level of the electromagnetic field rather than its field components are included in this category. (EMC) 1309-1996

(5) A device used to send or receive radio waves. (SCC32) 1455-1999

antenna [aperture] illumination efficiency The ratio, usually expressed in percent, of the maximum directivity of an antenna [aperture] to its standard directivity. *Note:* For planar apertures, the standard directivity is calculated by using the projected area of the actual antenna in a plane transverse to the direction of its maximum radiation intensity. *Synonym:* normalized directivity. *See also:* standard directivity. (AP/ANT) 145-1993

antenna array *See:* array antenna.

antenna correction factor (land-mobile communications transmitters) A factor usually supplied with the antenna, which, when properly applied to the meter reading of the measuring instrument, yields the electric field in volts/meters (V/m) or the magnetic field strength in amperes/meters (A/m). *Notes:* 1. This factor includes the effects of antenna effective length and impedance mismatch plus transmission line losses. 2. The factor for electric field strength is not necessarily the same as the factor for the magnetic field strength. (EMC) 377-1980r

antenna effect (1) (radio direction finding) (navigation aid terms) The presence of output signals having no directional information and caused by the directional array acting as simple nondirectional antenna; the effect is manifested by angular displacement of the nulls, or a broadening of the nulls. (AES/GCS) 172-1983w

(2) (loop antenna) Any spurious effect resulting from the capacitance of the loop to ground. *See also:* antenna. (AP/ANT) 145-1983s

antenna, effective area *See:* effective area antenna.

antenna, effective height *See:* effective height antenna.

antenna, effective height base station *See:* effective height base station antenna.

antenna, effective length *See:* effective length antenna.

antenna efficiency (of an aperture-type anenna) For an antenna with a specified planar aperture, the ratio of the maximum effective area of the antenna to the aperture area. (AP/ANT) 145-1993

antenna factor (1) Quantity relating the strength of the field in which the antenna is immersed to the output voltage across the load connected to the antenna. (EMC) C63.5-1988, 1128-1998
(2) A factor that, when properly applied to the meter reading of the measuring instrument, yields the electric field strength in volts/meter or the magnetic field strength in amperes/meter. *Notes:* 1. This factor includes the effects of antenna effective length and mismatch and may include transmission line losses. 2. The factor for electric field strength is not necessarily the same as the factor for the magnetic field strength. (EMC) [53], C63.4-1991
(3) A factor that, when properly applied to the voltage meter reading of the measuring instrument, yields the electric field strength in volts/meter or the magnetic field strength in amperes/meter. *Notes:* 1. This factor includes the effects of antenna effective length and mismatch and may include transmission line loss. 2. The factor for the electric field strength is not necessarily the same as the factor for the magnetic field strength. (EMC) 1128-1998

antenna figure of merit (communication satellite) An antenna performance parameter equaling the antenna gain G divided by the antenna noise temperature T, measured at the antenna terminals. It can be expressed as a ratio, $M = G/T$, or logarithmically,

$$M(\text{dB}) = 10 \log_{10} G - 10 \log_{10} T$$

(COM) [25]

antenna pattern *See:* radiation pattern.

antenna-pattern loss *See:* beamshape loss.

antenna resistance (1) (general) The real part of the input impedance of an antenna. (AP/ANT) 145-1993
(2) (test procedures for antennas) The ratio of the power accepted by the entire antenna circuit to the mean-square antenna current referred to a specified point. *Note:* Antenna resistance is made up of such components as radiation resistance, ground resistance, radio-frequency resistance of conductors in the antenna circuit, and equivalent resistance due to corona, eddy currents, insulator leakage, and dielectric power loss. *See also:* antenna. (AP) 149-1979r

antenna sensitivity-test input (amplitude-modulation broadcast receivers) The sensitivity input is the least signal-input voltage of a specified carrier frequency, modulated 30% at 400 cycles and applied to the receiver through a standard dummy antenna, which results in normal test output when all controls are adjusted for greatest sensitivity. It is expressed in decibels below one volt, or in microvolts. (CE) 186-1948w

antenna temperature The temperature of a blackbody that, when placed around a matched, loss-free antenna similar to the actual antenna, produces the same available noise power, in a specified frequency range, as the actual antenna in its normal electromagnetic environment. *See also:* blackbody. (AP/PROP) 211-1997

antenna terminal conducted interference Any undesired voltage or current generated within a receiver, transmitter, or their associated equipment appearing at the antenna terminals. *See also:* electromagnetic compatibility. (EMC) [53]

anti-aliasing (1) By the Nyquist Theorem, the maximum reproducible frequency is one-half the sampling rate. Aliasing is caused when frequencies higher than one half of the sampling rate are present. This results in the higher frequencies being "aliased" down to look like lower frequency components. Anti-aliasing is providing low pass filtering to block out frequencies higher than those that can be accurately reproduced by the given sampling rate. (PE/PSR) 1344-1995
(2) A technique that reduces the visual effects of aliasing. (C) 610.6-1991w

anticathode *See:* anode.

anticipatory buffering A buffering technique in which data are stored in a buffer in anticipation of a need for the data. *See also:* simple buffering; dynamic buffering. (C) 610.12-1990

anticipatory paging A storage allocation technique in which pages are transferred from auxiliary storage to main storage in anticipation of a need for those pages. *Contrast:* demand paging. (C) 610.12-1990

anticlutter circuits Circuits that attenuate undesired reflections from the natural environment (clutter) to permit detection of targets otherwise obscured by such reflections. *See also:* clutter. (AES) 686-1997

anticlutter gain control (nonlinear, active, and nonreciprocal waveguide components) (radar) A device that automatically and smoothly increases the gain of a radar receiver from a low level to the maximum, within a specified period after each transmitter pulse, so that short-range echoes producing clutter are amplified less than long-range echoes. (MTT) 457-1982w

anticoincidence (radiation counter) The occurrence of a count in a specified detector unaccompanied simultaneously or within an assignable time interval by a count in one or more other specified detectors. (ED) [45]

anticoincidence circuit (pulse techniques) A circuit that produces a specified output pulse when one (frequently predesignated) of two inputs receives a pulse and the other receives no pulse within an assigned time interval. (NPS) 398-1972r

anticollision light (illuminating engineering) A flashing aircraft aeronautical light or system of lights designed to provide a red signal throughout 360° of azimuth for the purpose of giving long-range indication of an aircraft's location to pilots of other aircraft. (ED) [127]

antiferroelectric material A material that exhibits structural phase changes and anomalies in dielectric permittivity, as do ferroelectrics, but has zero net spontaneous polarization, and hence, exhibits no hysteresis phenomena. *Note:* In some cases, it is possible to apply electric fields sufficiently high to produce a structural transition to a ferroelectric phase as evidenced by appearance of a double hysteresis loop. *See also:* paraelectric region; ferroelectric material. (UFFC) [21]

antifouling Pertaining to the prevention of marine organism attachment and growth on a submerged metal surface (through the effects of chemical action). (IA) [59]

antifreeze pin, relay *See:* relay antifreeze pin.

anti-lock means (laser gyro) A means of preventing lock-in, such as magneto-optical (Faraday cell, magnetic mirrors), mechanical (dither, rate biasing), or mechanical-optical (mirror dither). *See also:* lock-in. (AES/GYAC) 528-1994

anti-lock residual (laser gyro) Output noise remaining after compensation for anti-lock means. (AES/GYAC) 528-1994

antinoise microphone A microphone with characteristics that discriminate against acoustic noise. *See also:* microphone. (EEC/PE) [119]

anti-overshoot The effect of a control function or a device that causes a reduction in the transient overshoot. *Note:* Anti-overshoot may apply to armature current, armature voltage, field current, etc. *See also:* feedback control system. (IA/ICTL/IAC) [60]

antioxidant *See:* oxidation inhibitor.

antiplugging protection The effect of a control function or a device that operates to prevent application of counter torque by the motor until the motor speed has been reduced to an acceptable value. *See also:* feedback control system. (IA/ICTL/IAC) [60]

antipodal focusing Ionospheric focusing sometimes observed in the vicinity of the antipodal point or region. (AP/PROP) 211-1997

antipump device A device that prevents reclosing after an opening operation as long as the device initiating closing is maintained in the position for closing. *Synonym:* pump-free device. (SWG/PE) C37.100-1992

antireflection coating (fiber optics) A thin, dielectric or metallic film (or several such films) applied to an optical surface to reduce the reflectance and thereby increase the transmittance. *Note:* The ideal value of the refractive index of a single layered film is the square root of the product of the refractive indices on either side of the film, the ideal optical thickness being one quarter of a wavelength. *See also:* dichroic filter; Fresnel reflection; transmittance; reflectance. (Std100) 812-1984w

antiresonant frequency Usually in reference to a crystal unit or the parallel combination of a capacitor and inductor. The frequency at which, neglecting dissipation, the impedance of the object under consideration is infinite. (CAS) [13]

antisidetone induction coil An induction coil designed for use in an antisidetone telephone set. *See also:* telephone station. (EEC/PE) [119]

antisidetone telephone set A telephone set that includes a balancing network for the purpose of reducing sidetone. *See also:* telephone station; sidetone. (EEC/PE) [119]

anti-single-phase tripping device A device that operates to open all phases of a circuit by means of a polyphase switching device, in response to the interruption of the current in one phase. *Notes:* 1. This device prevents single phasing of connected equipment resulting from the interruption of any one phase of the circuit. 2. This device may sense operation of a specific single-phase interrupting device or may sense loss of single-phase potential. (SWG/PE) C37.100-1992

antistatic (1) (health care facilities) Adjective describing that class of materials that includes conductive materials and also those materials that, throughout their stated life, meet the requirements of 4-6. 6. 3 and 4-6. 6. 4 of NFPA-56A, 1978. (EMB) [47]

(2) A property of materials that resist triboelectric charging. (SPD/PE) C62.47-1992r

antisubmarine warfare radar A radar used in antisubmarine warfare (ASW). It includes radars for the detection of submarines and submarine effects as well as radars on ships or aircraft employed in ASW operations to obtain situational awareness of the surface ships and aircraft in the vicinity of ASW operations. (AES) 686-1997

anti-transmit-receive switch A radio-frequency switch that automatically decouples the transmitter from the antenna during the receiving period. *Note:* The ATR switch is employed when a common transmitting and receiving antenna is used. (AES) 686-1997

anti-transmit-receive tube (electron tube) A gas-filled radio-frequency switching tube used to isolate the transmitter during the interval for pulse reception. *Synonym:* ATR tube. *See also:* gas tube. (ED/MTT) 161-1971w, 457-1982w

***A* operator** An operator assigned to an *A* switchboard. (EEC/PE) [119]

APC *See:* automatic phase control.

APD *See:* amplitude probability distribution; avalanche photodiode.

aperiodically sampled equivalent time format (pulse measurement) A format that is identical to the aperiodically sampled real time format except that the time coordinate is equivalent to and convertible to real time. Typically, each datum point is derived from a different measurement on a different wave in a sequence of waves. *See also:* sampled format. (IM/WM&A) 181-1977w

aperiodically sampled real time format (pulse measurement) A format that is identical to the periodically sampled real time format except that the sampling in real time is not periodic and wherein the data exists as coordinate point pairs t_1, m_1; t_2, m_2; . . .; t_n, m_n. *See also:* sampled format. (IM/WM&A) 181-1977w

aperiodic antenna An antenna that, over an extended frequency range, does not exhibit a cyclic behavior with frequency of either its input impedance or its pattern. *Note:* This term is often applied to an electrically small monopole or loop, containing an active element as an integral component, with impedance and pattern characteristics varying but slowly over the extended frequency range. (AP/ANT) 145-1993

aperiodic circuit A circuit in which it is not possible to produce free oscillations. *See also:* oscillatory circuit. (PE) 599-1985w, [84]

aperiodic component (rotating machinery) (of short-circuit current) The component of current in the primary winding immediately after it has been suddenly short-circuited when all components of fundamental and higher frequencies have been subtracted. *See also:* asynchronous machine. (PE) [9]

aperiodic damping *See:* overdamping.

aperiodic tasks Tasks that arrive at irregular intervals and have only soft (not rigid) deadlines, but a good response time is typically desirable. (C/BA) 896.3-1993w

aperiodic time constant (rotating machinery) The time constant of the aperiodic component when it is essentially exponential, or of the exponential that can most nearly be fitted. *See also:* asynchronous machine; direct-current commutating machine. (PE) [9]

aperture (1) A surface, near or on an antenna, on which it is convenient to make assumptions regarding the field values for the purpose of computing fields at external points. *Notes:* 1. In some cases the aperture may be considered as a line. 2. In the case of a unidirectional antenna, the aperture is often taken as that portion of a plane surface near the antenna, perpendicular to the direction of maximum radiation, through which the major part of the radiation passes. *See also:* antenna. (AP) 149-1979r

(2) (data transmission) For a unidirectional antenna, that portion of a plane surface near the antenna perpendicular to the direction of maximum radiation through which the major part of the radiation passes. (PE) 599-1985w

(3) (of an antenna) A surface, near or on an antenna, on which it is convenient to make assumptions regarding the field values for the purpose of computing fields at external points. *Note:* The aperture is often taken as that portion of a plane surface near the antenna, perpendicular to the direction of maximum radiation, through which the major part of the radiation passes. (AP/ANT) 145-1993

(4) An opening in a data medium or device such as the opening in the an aperture card, or an opening in a multiaperture core. (C) 610.10-1994w

(5) Maximum interdigital transducer finger overlap length, which is expressed in length units or normalized in terms of wavelength. (UFFC) 1037-1992w

aperture averaging The reduction in output signal variation when the size of the antenna is large compared to the decorrelation distance of the incident field across the aperture. *Note:* The beamwidth of the antenna is much smaller than the angular spectrum of the incoming wave. *See also:* angular spectrum. (AP/PROP) 211-1997

aperture blockage A condition resulting from objects lying in the path of rays arriving at or departing from the aperture of an antenna. *Note:* For example, the feed, subreflector, or support structure produce aperture blockage for a symmetric reflector antenna. (AP/ANT) 145-1993

aperture card A punch card of standard dimensions into which microfilm frames may be inserted. (C) 610.10-1994w

aperture compensation *See:* aperture equalization.

aperture correction *See:* aperture equalization.

aperture dimension (1) The space or opening between features. (C/BA) 1301.2-1993

(2) The usable height, width, or depth between features. (C/MM) 1301.3-1992r

(3) The usable space between features. (C/MM) 1301.1-1991

(4) The usable space between standardized features. (C/BA) 1301.4-1996

aperture distribution (excitation systems) The field over the aperture as described by amplitude, phase, and polarization distributions. *Synonym:* aperture illumination. (AP/ANT) 145-1993

aperture efficiency (for an antenna aperture) The ratio of its directivity to the directivity obtained when the aperture illumination is uniform. *See also:* antenna. (AP/ANT) [35], 145-1983s

aperture equalization (television) Electrical compensation for the distortion introduced by the size of a scanning aperture. *See also:* television. (BT/AV) [34]

aperture illumination (excitation systems) The field over the aperture as described by amplitude, phase, and polarization distributions. *Synonym:* aperture illumination. (AP/ANT) 145-1993

aperture uncertainty The standard deviation of the sample instant in time. *Synonyms:* timing phase noise; short-term timing instability; timing jitter. (IM/WM&A) 1057-1994w

API *See:* application program interface; air position indicator.

APL *See:* A Programming Language; average picture level.

apoapsis (communication satellite) The most distant point from the center of a primary body (or planet) to an orbit around it. (COM) [19]

apodization Response weighting produced by the change of finger overlap along the length of the interdigital transducer. (UFFC) 1037-1992w

apogee (1) (navigation aid terms) That orbital point farthest from the earth, when the earth is the center of attraction. (AES/GCS) 172-1983w
(2) (communication satellite) The most distant point from the center of the earth to an orbit around it. (COM) [19]

apparatus (1) (power and distribution transformers) A general designation for large electrical equipment such as generators, motors, transformers, circuit breakers, etc. (PE/TR) C57.12.80-1978r
(2) A device or system of devices that performs a distinct function within a basic operating unit, including a device or system of devices whose principal function is data communications. (VT) 1473-1999

apparatus insulator (cap and pin, post) An assembly of one or more apparatus-insulator units, having means for rigidly supporting electric equipment. *See also:* insulator. (EEC/IEPL) [89]

apparatus insulator unit The assembly of one or more elements with attached metal parts, the function of which is to support rigidly a conductor, bus, or other conducting elements on a structure or base member. *See also:* tower. (T&D/PE) [10]

apparatus interoperability The ability of any specific apparatus to communicate with other apparatuses in such a way that it can successfully replace another apparatus of the same type without any requirement for manual configuration other than the address or unique identifier of the replacement apparatus. (VT) 1473-1999

apparatus termination A termination designed for use in sealed enclosures where the external dielectric strength is dependent upon liquid or special gaseous dielectric and where the ambient temperature of the medium immediately surrounding the termination may reach 55°C. (PE/IC) 48-1996

apparatus thermal device (power system device function numbers) A device that functions when the temperature of the shunt field, or the amortisseur winding of a machine, or that of a load limiting or load shifting resistor, or of a liquid or other medium exceeds a predetermined value; or if the temperature of the protected apparatus, such as a power rectifier, or of any medium decreases below a predetermined value. (SUB/PE) C37.2-1979s

apparatus type A predefined configuration that, when adhered to by a given apparatus, makes it possible for that apparatus to achieve apparatus interoperability, without restriction on the internal constructional details of the apparatus concerned. (VT) 1473-1999

apparent altitude (navigation aid terms) That sextant altitude corrected for reading and reference level inaccuracies. (AES/GCS) 172-1983w

apparent bearing (navigation aid terms) (direction finding systems) A bearing from a direction-finder site to a target transmitter determined by averaging the readings made on a calibrated direction-finder test standard; the apparent bearing is then used in the calibration and adjustment of other direction finders at the same site. (AES/GCS) 172-1983w

apparent candlepower (extended source) At a specified distance, the candlepower of a point source that would produce the same illumination at that distance. (ED) [127]

apparent charge (1) That charge that, if it could be injected instantaneously between the terminals of the test object, would momentarily change the voltage between its terminals by the same amount as the partial discharge itself. The apparent charge should not be confused with the charge transferred across the discharging cavity in the dielectric medium. Apparent charge is expressed in coulombs (C). One pC is equal to 10^{-12} C. (PE/PSIM/TR) 62-1995, C57.113-1991
(2) (terminal charge) A charge that, if injected instantaneously between the terminals of the test object, would momentarily change the voltage between its terminals by the same amount as the partial discharge itself. The apparent charge should not be confused with the charge transferred across the overstressed insulation in the dielectric medium. Apparent charge within the terms of this document is expressed in picocoulombs, which is abbreviated as pC (10^{-12} Coulombs). (SWG/PE) 1291-1993r
(3) (dielectric tests) That charge of a partial discharge which, if injected instantaneously between the terminals of the test object, would momentarily change the voltage between its terminals by the same amount as the partial discharge itself. The apparent charge should not be confused with the charge transferred across the discharging cavity in the the dielectric. Apparent charge is expressed in coulombs. *Synonym:* terminal charge. (PE/PSIM) 454-1973w

apparent dead time *See:* dead time.

apparent discharge magnitude (corona measurement) The charge transfer measured at the terminals of a sample caused by a corona pulse in a sample. (MAG/ET) 436-1977s

apparent horizon (navigation aid terms) Visible horizon. (AES/GCS) 172-1983w

apparent impedance The impedance to a fault as seen by a distance relay is determined by the applied current and voltage. It may be different from the actual impedance because of current outfeed or current infeed at some point between the relay and the fault. (PE/PSR) C37.113-1999

apparent inductance The reactance between two terminals of a device or circuit divided by the angular frequency at which the reactance was determined. This quantity is defined only for frequencies at which the reactance is positive. *Note:* Apparent inductance includes the effects of the real and parasitic elements that comprise the device or circuit and is therefore a function of frequency and other operating conditions. (CHM) [51]

apparent output power (self-commutated converters) (converters having ac output) The product of fundamental current and fundamental phase voltage summed for all phases of the circuit. (IA/SPC) 936-1987w

apparent (phasor) power

$$S = VI$$

where S is the apparent power, V is the rms value of the voltage, and I is the rms value of the current. (PE/PSIM) 120-1989r

apparent power (1) (rotating machinery) The product of the root-mean-square current and the root-mean-square voltage. *Notes:* 1. It is a scalar quantity equal to the magnitude of the phasor power. *See also:* asynchronous machine. (PE) [9]

(2) (metering) For sinusoidal quantities in either single-phase or polyphase circuits, apparent power is the square root of the sum of the squares of the active and reactive powers. *Note:* This is, in general, not true for nonsinusoidal quantities. (ELM) C12.1-1982s

(3) (A) (polyphase circuit) At the terminals of entry of a polyphase circuit, a scalar quantity equal to the magnitude of the vector power. *Note:* In determining the apparent power, the reference terminal for voltage measurement shall be taken as the neutral terminal of entry, if one exists, otherwise as the true neutral point. 2. If the ratios of the components of the vector power, for each of the terminals of entry, to the corresponding apparent power are the same for every terminal of entry, the total apparent power is equal to the arithmetic apparent power for the polyphase circuit: otherwise the apparent power is less than the arithmetic apparent power. 3. If the voltages have the same wave form as the corresponding currents, the apparent power is equal to the amplitude of the phasor power. 4. Apparent power is expressed in volt-amperes when the voltages are in volts and the currents in amperes. **(B) (single-phase two-wire circuit)** At the two terminals of entry of a single-phase two-wire circuit into a delimited region, a scalar equal to the product of the root-mean-square voltage between one terminal of entry and the second terminal of entry, considered as the reference terminal, and the root-mean-square value of the current through the first terminal. *Note:* Mathematically, the apparent power U is given by the equation

$$U = EI$$
$$= (\pm)\,(E_1^2 + E_2^2 + \ \ldots\ + E_r^2 + \ \ldots\ )^{1/2}$$
$$\times\,(I_1^2 + I_r^2 + \ \ldots\ + I_q^2 + \ \ldots\ )^{1/2}$$

in which E and I are the root-mean-square amplitudes of the voltage and current, respectively. E_r and I_q are the root-mean-square amplitudes of the rth harmonic of voltage and the qth harmonic of current, respectively. 2. If both the voltage and current are sinusoidal and of the same period, so that the distortion power is zero, the apparent power becomes

$$U = EI = E_1I_1$$

in which E_1 and I_1 are the root-mean-square amplitudes of voltage and current of the primitive period. The apparent power is equal to the amplitude of the phasor power. 3. If the voltage and current are quasiperiodic and the amplitude of the voltage and current components are slowly varying, the apparent power at any instant may be taken as the value derived from the amplitudes of the components at that instant. 4. Apparent power is expressed in volt-amperes when the voltage is in volts and the current in amperes. Because apparent power has the property of magnitude only and its sign is ambiguous, it does not have a definite direction of flow. For convenience, it is usually treated as positive. (Std100) 270-1966

apparent-power loss (volt-ampere loss) (electric instruments) Of the circuit for voltage-measuring instruments, the product of end-scale voltage and the resulting current: and for current-measuring instruments, the product of the end-scale current and the resulting voltage. *Notes:* 1. For other than current-measuring or voltage-measuring instruments, for example, wattmeters, the apparent power loss of any circuit is expressed at a stated value of current or of voltage. 2. Computation of loss: For the purpose of computing the loss of alternating-current instruments having current circuits at some selected value other than that for which it is rated, the actual loss at the rated current is multiplied by the square of the ratio of the selected current to the rated current. Example: A current transformer with a ratio of 500:5 amperes is used with an instrument having a scale of 0−300 amperes and, therefore, a 3-ampere field coil, and the allowable loss at end scale is as stated on the Detailed Requirement Sheet. The allowable loss of the instrument referred to a 5-amperes basis is as follows: Allowable loss in volt-amperes equals (allowable loss end-scale volt-amperes). *See also:* accuracy rating. (EEC/AII) [102]

apparent resistance (insulation testing) Ratio of the voltage across the electrodes in contact with the specimen to the current between them as measured under the specified test conditions and specified electrification time. (PE) 402-1974w

apparent sag (1) (A) (wire in a span) The maximum departure in the vertical plane of the wire in a given span from the straight line between the two points of support of the span, at 60°F, with no wind loading. *Note:* Where the two supports are at the same level this will be the sag. *See also:* tower. **(B) (wire in a span)** The departure in the vertical plane of the wire at the particular point in the span from the straight line between the two points of support. (T&D/PE) [10], C2.2-1960

(2) (of a span) The maximum distance between the wire in a given span and the straight line between the two points of support of the wire, measured perpendicularly from the straight line. (NESC) C2-1997

(3) (of a span) The maximum departure of the wire in a given span from the straight line between the two points of support of the span. (T&D) C2.2-1960

apparent sag at any point in the span (1) The departure of the wire at the particular point in the span from the straight line between the two points of support of the span. (T&D) C2.2-1960

(2) The distance, at the particular point in the span, between the wire and the straight line between the two points of support of the wire, measured perpendicularly from the straight line.

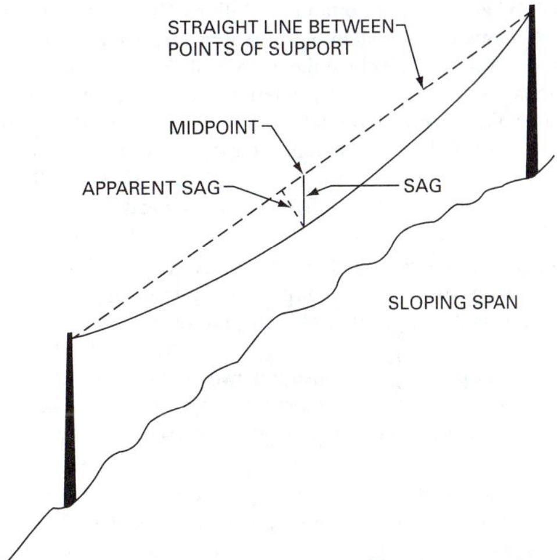

sag and apparent sag

(NESC) C2-1997

apparent time constant The time required for 63% of the change in output electromotive force to occur after an abrupt change in the input quantity to a new constant value. *Synonym:* 63% response time. *See also:* characteristic time; thermal converter; response. (EEC/AII) [102]

apparent vertical (navigation aid terms) The direction of the vector sum of the gravitational and all other accelerations. (AES/GCS) 172-1983w

apparent visual angle (laser maser) The angular subtense of the source as calculated from the source size and distance from the eye. It is not the beam divergence of two sources. (LEO) 586-1980w

appliance (1) (electric) A utilization item of electric equipment, usually complete in itself, generally other than industrial, normally built in standardized sizes or types, that transforms electric energy into another form, usually heat or mechanical motion, at the point of utilization. For example, a toaster, flatiron, washing machine, dryer, hand drill, food mixer, air conditioner. (IA/APP) [80]

(2) Utilization equipment, generally other than industrial, normally built in standardized sizes or types, which is in-

stalled or connected as a unit to perform one or more functions such as clothes washing, air conditioning, food mixing, deep frying, etc. (NESC/NEC) [86]
(3) Current-conducting, energy-consuming equipment, fixed or portable; for example, heating, cooling, and small motor-operated equipment. (NESC/T&D) C2-1977s, C2.2-1960

appliance branch circuit (1) A branch circuit supplying energy to one or more outlets to which appliances are to be connected; such circuits to have no permanently connected lighting fixtures not a part of an appliance. (NESC/NEC) [86]
(2) A circuit that supplies energy to one or more outlets to which appliances are connected. These circuits have no permanently connected lighting fixtures that are not a part of an appliance. (IA/MT) 45-1998

appliance, fixed *See:* fixed appliance.

appliance outlet (household electric ranges) An outlet mounted on the range and to which a portable appliance may be connected by means of an attachment plug cap. (IA/APP) [90]

appliance, portable *See:* portable appliance.

appliance, stationary *See:* stationary appliance.

application (1) The use to which a computer system is put; for example, a payroll application, an airline application, or a network application. (C) 610.2-1987, 610.5-1990w
(2) The use of capabilities provided by an information system specific to the satisfaction of a set of user requirements. *Note:* These capabilities include hardware, software, and data. (C/PA) 14252-1996
(3) When the User Portability Utilities Option is supported, requirements associated with the term *application* also shall be interpreted to include the actions of the user who is interacting with the system by entering shell command language statements from a terminal. (C/PA) 2003.2-1996
(4) A software program consisting of one or more processes and supporting functions. (PE/SUB) 1379-1997
(5) A computer program that performs some desired function. (C) 1003.5-1999

application-association (1) A cooperative relationship between two applications for the purpose of communication of information and coordination of their joint operations. (C/PA) 1351-1994w
(2) A cooperative relationship between two application-entities, formed by their exchange of application-protocol-control-information through their use of presentation services. (C/PA) 1238.1-1994w

application engineering The process of constructing or refining application systems by reusing assets. (C/SE) 1517-1999

application entity The aspects of an application process pertinent to OSI. (C/PA) 1238.1-1994w

application entity title In OSI, a title that unambiguously identifies an application entity. An application entity title is composed of an application process title and an application entity qualifier. (C) 1003.5-1999

application entity qualifier In OSI, a component of an application entity title that is unambiguous within the scope of the application process. (C) 1003.5-1999

application environment The physical environment of a backplane serial bus. This includes the bus itself, the modules, and the system that contains them. This environment may be a standardized host backplane (e.g., a Futurebus+ profile) that describes signal requirements, transceivers, mechanical arrangement of the modules, and temperature range over which operation is guaranteed. (C/MM) 1394-1995

application environment profile (aep, AEP) (1) A document that describes functional requirements and points to existing standards, selecting and binding options within those standards. An implementer who then designs a specific module and/or system should be reasonably assured that another designer's (manufacturer's or supplier's) modules will properly function within the same system. This includes all aspects of definition: mechanical, electrical, protocol, environmental, and system considerations. (C/BA) 896.2-1991w, 896.3-1993w, 896.4-1993w, 896.10-1997
(2) A profile specifying a complete and coherent specification of the Open System Environment (OSE), in which the standards, options, and parameters chosen are necessary to support a class of applications. (C/PA) 14252-1996

application generator A code generator that produces programs to solve one or more problems in a particular application area; for example, a payroll generator. (C) 610.12-1990

application identifier (AID) An identifier that defines the category of dedicated short-range communications (DSRC) applications to which a specific application belongs. (SCC32) 1455-1999

application interface The programming access mechanism to the communication resources of a network. (DIS/C) 1278.2-1995

application layer (1) (Layer 7) The layer of the OSI reference model (ISO 7498: 1984) that provides the means for simulation applications to access and use the network's communications resources. (DIS/C) 1278.1-1995, 1278.2-1995
(2) The seventh and highest layer of the seven-layer OSI model providing the only interface between the user and the application program. *Note:* It hides from the user the physical distribution of processors, communications media, and data resources while maximizing the utility of those resources. *See also:* entity layer; logical link control sublayer; session layer; client layer; data link layer; presentation layer; physical layer; transport layer; sublayer; network layer; medium access control sublayer. (C) 610.7-1995

application logic That portion of a module that excludes the MTM-Bus interface logic. *See also:* module. (TT/C) 1149.5-1995

application-oriented language A programming language with facilities or notations applicable primarily to a single application area; for example, a language for computer-assisted instruction or hardware design. *See also:* simulation language; specification language; authoring language. (C) 610.13-1993w, 610.12-1990

application platform (1) A set of resources, including hardware and software, that support the services on which application software will run. The application platform provides services at its interfaces that, as much as possible, make the specific characteristics of the platform transparent to the application software. (C/PA) 14252-1996
(2) A set of resources on which an application will run. (C/PA) 1003.13-1998

application process title In OSI, a title that unambiguously identifies an application process. An application process title is a single name, which, for convenience, may be structured internally. (C) 1003.5-1999

application program (1) A computer program that is used for a specific application. (C) 610.5-1990w
(2) A program executed with the processor in **user mode.** *Note:* Statements made in this document regarding application programs may be inapplicable to programs (for example, debuggers) that have access to privileged processor state (e.g., as stored in a memory-image dump). (C/MM) 1754-1994

application program interface (API) The interface between the application software and the application platform, across which all services are provided. (C/PA) 14252-1996

application-service-element The part of an application-entity that provides an OSI environment, using underlying services when appropriate. (C/PA) 1238.1-1994w

application software (1) Software designed to fulfill specific needs of a user; for example, software for navigation, payroll, or process control. *Contrast:* support software; system software. (C) 610.12-1990
(2) Software that is specific to an application and is composed of programs, data, and documentation. (C/PA) 14252-1996

application-specific data dictionary A data dictionary specific to a particular implementation of an Intelligent Transportation Systems (ITS) application. (SCC32) 1489-1999

application valve (brake application valve) An air valve through the medium of which brakes are automatically applied. (EEC/PE) [119]

application view *See:* logical database.

applicative order A property of a programming language or procedure: the arguments to a procedure call are evaluated before the procedure is invoked, and the result of each evaluation is passed to the procedure in place of its argument expression. (C/MM) 1178-1990r

applicator (dielectric heating) (electrodes) Appropriately shaped conducting surfaces between which is established an alternating electric field for the purpose of producing dielectric heating. (IA) 54-1955w

Applied Dynamics International Simulation Language A simulation language designed for use in dynamic simulation applications. (C) 610.13-1993w

applied-fault protection A protective method in which, as a result of relay action, a fault is intentionally applied at one point in an electrical system in order to cause fuse blowing or further relay action at another point in the system. (SWG/PE/PSR) C37.100-1992, C37.90-1978s

applied-potential tests (electric power) Dielectric tests in which the test voltages are low-frequency alternating voltages from an external source applied between conducting parts, and between conducting parts and ground. (SPD/PE) 32-1972r

applied voltage (corona measurement) Voltage that is applied across insulation. Applied voltage may be between windings or from winding(s) to ground. (MAG/ET) 436-1977s

applied voltage tests (power and distribution transformers) Dielectric tests in which the test voltages are low-frequency alternating voltages from an external source applied between conducting parts and ground without exciting the core of the transformer being tested. (PE/TR) C57.12.80-1978r

approach circuit A circuit used to announce the approach of trains at block or interlocking stations. (EEC/PE) [119]

approach indicator A device used to indicate the approach of a train. (EEC/PE) [119]

approach-light beacon (illuminating engineering) An aeronautical ground light placed on the extended centerline of the runway at a fixed distance from the runway threshold to provide an early indication of position during an approach to a runway. *Note:* The runway threshold is the beginning of the runway usable for landing. (ED) [127]

approach lighting An arrangement of circuits so that the signal lights are automatically energized by the approach of a train. (EEC/PE) [119]

approach-lighting relay A relay used to close the lighting circuit for signals upon the approach of a train. (EEC/PE) [119]

approach lights (illuminating engineering) A configuration of aeronautical ground lights located in extension of a runway or channel before the threshold to provide visual approach and landing guidance to pilots. (ED) [127]

approach locking (electric approach locking) Electric locking effective while a train is approaching, within a specified distance, a signal displaying an aspect to proceed, and that prevents, until after the expiration of a predetermined time interval after such signal has been caused to display its most restrictive aspect, the movement of any interlocked or electrically locked switch, movable-point frog, or derail in the route governed by the signal, and that prevents an aspect to proceed from being displayed for any conflicting route. *See also:* interlocking. (EEC/PE) [119]

approach navigation (navigation aid terms) Navigation during the time that the approach to a dock, runway, or other terminal facility is of immediate importance. (AES/GCS) 172-1983w

approach path (navigation aid terms) That portion of the flight path between the point at which the descent for landing is normally started and the point at which the aircraft touches down on the runway. (AES/GCS) 172-1983w

approach signal A fixed signal used to govern the approach to one or more other signals. (EEC/PE) [119]

approach speed The rate at which the intruder approaches the receptor. (SPD/PE) C62.47-1992r

appropriate privileges (1) An implementation-defined means of associating privileges with a implementation defined process with regard to the function calls and function call options defined in ISO/IEC 9945 that need special privileges. There may be zero or more such means. (C/PA) 9945-1-1996, 1003.5-1992r, 9945-2-1993

(2) An implementation defined means of associating privileges with a process with regard to the subprogram calls and options defined in this standard that need special privileges. There may be zero or more such means. (C/PA) 1003.5b-1995

approval Written notification by an authorized representative of the acquirer that a developer's plans, design, or other aspects of the project appear to be sound and can be used as the basis for further work. Such approval does not shift responsibility from the developer to meet contractual requirements. (C/SE) J-STD-016-1995

approval plate A label that the United States Bureau of Mines requires manufacturers to attach to every completely assembled machine or device sold as permissible mine equipment. *Note:* By this means, the manufacturer certifies to the permissible nature of the machine or device. (PE/EEC/MIN) [119]

approval test (metering) (acceptance tests) A test of one or more meters or other items under various controlled conditions to ascertain the performance characteristics of the type of which they are a sample. (ELM) C12.1-1982s

approved (1) (general) Approved by the enforcing authority. (EEC/PE) [119]

(2) Acceptable to the authority having jurisdiction. (NEC/NESC/DEI) 1221-1993w, [86]

approved supplier (replacement parts for Class 1E equipment in nuclear power generating stations) A supplier whose quality assurance (QA) system has been evaluated and found to meet the owner's QA requirements for the item or service to be purchased. (PE/NP) 934-1987w

approximate value (metric practice) A value that is nearly but not exactly correct or accurate. (QUL) 268-1982s

A Programming Language An interactive programming language with a concise syntax that is well-suited for solving mathematical problems requiring intricate vector or matrix manipulations. *Notes:* 1. Requires a special keyboard configuration due to its extended character set. 2. Standardized by ISO/IEC. (C) 610.13-1993w

APSE *See:* Automatic Programming and Scaling of Equations.

APS Instance An instantiation of an APS API service provider, including the APS environment and internal state information. (C/PA) 1351-1994w

APT (automatically programmed tools) *See:* automatic programmed tools; Automatically Programmed Tools.

aramid A manufactured material in which the base polymer is a long-chain synthetic polyamide with at least 85% of the amide linkages attached directly to two aromatic rings. Paper and transformerboard are made from this material and have been shown to be suitable for use in high-temperature and hybrid high-temperature insulation systems. (PE/TR) 1276-1997

arbiter (1) A functional module that accepts bus requests from requester modules and grants control of the data transfer bus (DTB) to one requester at a time. (C/BA) 1014-1987

(2) When implementing the serial arbitration method, the arbiter module accepts requests for the DTB from requesters and grants control of the DTB to one requester at a time.

There is one and only one active arbiter in the serial arbitration scheme, and it is always located in slot 1. An arbiter is not required in the parallel arbitration method. (C/MM) 1096-1988w

(3) The module that is performing the arbitration. (C/BA) 896.4-1993w

arbitrary sequence computer A computer in which each instruction explicitly specifies the location of the next instruction to be executed. *Contrast:* consecutive sequence computer. *See also:* nonsequential computer. (C) 610.10-1994w

arbitrated message (1) A number broadcast on the arbitrated message bus lines to all modules on the bus. (C/BA) 10857-1994, 896.3-1993w

(2) An event number broadcast on the arbitrated message bus lines to all modules on the bus. (C/BA) 896.4-1993w

arbitration (1) The process of determining which requesting device will gain access to a resource. (C/MM) 959-1988r

(2) The means whereby masters compete for control of the bus and the process by which a master is granted control of the bus. (C/MM) 1000-1987r

(3) A collection of mechanisms that allow masters to access the bus without conflicting with each other. (C/MM) 1196-1987w

(4) The process of selecting the next bus master. (C/BA) 10857-1994, 896.3-1993w, 896.4-1993w, 1014.1-1994w

(5) The process by which nodes compete for ownership of the bus. The cable environment uses a hierarchical point-to-point algorithm, while the backplane environment uses the bit-serial process of transmitting an arbitration sequence. At the completion of an arbitration contest, only one node will be able to transmit a data packet. (C/MM) 1394-1995

(6) In 1000BASE-X, Auto-Negotiation process that ensures proper sequencing of configuration information between link partners using the Physical Coding Sublayer (PCS) Transmit and Receive functions. (C/LM) 802.3-1998

(7) The process by which nodes compete for control of the bus. Upon completion of arbitration, the winning node is able to transmit a packet or initiate a short bus reset. (C/MM) 1394a-2000

arbritration bus One of the four buses provided by the backplane. This bus allows an arbiter module and several requester modules to coordinate use of the DTB. (C/BA) 1014-1987

arbitration clock rate The rate used to define a number of timing requirements within the backplane physical layer. It is 49.152 MHz × 100 ppm, regardless of the backplane interface technology. (C/MM) 1394-1995

arbitration contest This is the core mechanism to resolve bus ownership between one or more competing masters. It takes two bus periods. (C/MM) 1196-1987w

arbitration cycle (1) (FASTBUS acquisition and control) The process by which the next master to be granted bus mastership is determined. It is initiated by the arbitration timing controller and is complete when the winning master assumes bus mastership. (NID) 960-1993

(2) (VSB) A cycle that is initiated by the active requester in response to a bus request, after its associated active master no longer needs the bus. This cycle is used to select the master that will be granted use of the DTB. If the active requester detects a request for the bus, and if its associated master no longer needs the bus, it initiates an arbitration cycle. During the arbitration cycle, all contending requesters drive an arbitration ID on the bus. This ID is a combination of the geographical address of the board that is supplied by the backplane slot, and a priority code that is supplied by user-defined on-board logic. At the end of the arbitration cycle, one of the contending requesters becomes the active requester. (MM/C) 1096-1988w

arbitration locked sequence A sequence of operations by one master, directed to a number of different primary addresses, which is not interruptable by any other master because the originating master does not allow bus arbitration to take place. (NID) 960-1993

arbitration operation The bus operation in which agents attempt to gain exclusive access to the parallel system bus. (C/MM) 1296-1987s

arbitration reset gap (1) The minimum period of idle bus that has to occur after a source using the fairness protocol has won an arbitration contest before it can once again compete for bus mastership. This is longer than a normal subaction gap. (C/MM) 1394-1995

(2) The minimum period of idle bus (longer than a normal subaction gap) that separates fairness intervals. (C/MM) 1394a-2000

arbitration sequence For the backplane environment, a set of bits transmitted by nodes that wish to transmit packets that is used to determine which node will be able to transmit next. (C/MM) 1394-1995

arbitration signal Bidirectional signal exchanged between nodes during arbitration. One of the PDUs for the physical layer (the other is the data bit). (C/MM) 1394-1995

arbitration signaling A protocol for the exchange of bidirectional, unclocked signals between nodes during arbitration. (C/MM) 1394a-2000

arbitration timing control (ATC) Logic associated with each segment for the purpose of supervising and generating the arbitration control signals, run/halt control, and broadcast system handshake. (This is part of the ancillary logic.). (NID) 960-1993

arc (1) (A) (computer graphics) A continuous portion of a circle. **(B) (computer graphics)** A finite set of pixels representing a portion of a curve. **(C) (overhead power lines)** A continuous luminous discharge of electricity across an insulating medium, usually accompanied by the partial volatilization of the electrodes.

illustrations of arc

(C/PE/T&D) 610.4-1990, 539-1990

(2) A discharge of electricity through a gas, normally characterized by a voltage drop in the immediate vicinity of the cathode approximately equal to the ionization potential of the gas. *See also:* gas tube. (ED/T&D/PE) 161-1971w, 539-1990

(3) *See also:* timing arc. (C/DA) 1481-1999

arc-back (gas tube) A failure of the rectifying action, which results in the flow of a principal electron stream in the reverse direction, due to the formation of a cathode spot on an anode. *See also:* gas tube; rectification. (ED) 161-1971w

arc cathode (gas tube) A cathode whose electron emission is self-sustaining, with a small voltage drop approximately equal to the ionization potential of the gas. (Std100) [84]

arc chute (of a switching device) A structure affording a confined space or passageway, usually lined with arc-resisting material, into or through which an arc is directed to extinction. (SWG/PE) C37.100-1992

arc, clockwise *See:* clockwise arc.

arc converter A form of negative-resistance oscillator utilizing an electric arc as the negative resistance. *See also:* radio transmission. (BT) 182-1961w

arc, counterclockwise *See:* counterclockwise arc.

arc current (gas-tube surge protective devices) (gas tube surge arresters) The current that flows after breakdown when the circuit impedance allows a current that exceeds the glow-to-arc transition current. *Synonym:* arc mode current. (PE/SPD) C62.31-1987r, [8]

arc discharge (1) (illuminating engineering) An electric discharge characterized by high cathode current densities and a low voltage drop at the cathode. (ED) [127]
(2) (nonlinear, active, and nonreciprocal waveguide components) Commonly refers to weakly ionized plasma created by a radio-frequency (rf) discharge in gas tubes, receiver protectors, or duplexers. (MTT) 457-1982w

arc-discharge tube (valve) A gas-filled tube or valve in which the required current is that of an arc discharge. (PE/PSR) C37.90-1978s

arc-drop loss (gas tube) The product of the instantaneous values of the arc-drop voltage and current averaged over a complete cycle of operation. *See also:* gas tube. (ED) 161-1971w

arc-drop voltage (gas tube) The voltage drop between the anode and cathode of a rectifying device during conduction. *See also:* tube voltage drop; electrode voltage. (ED) [45]

arc-extinguishing medium (1) Material included in the fuse to facilitate current interruption. *Synonym:* fuse filler. (SWG/PE) C37.40-1993
(2) (of a fuse) Material included in the fuse to facilitate current interruption. (SWG/PE) C37.100-1992

arc furnace An electrothermic apparatus, the heat energy for which is generated by the flow of electric current through one or more arcs internal to the furnace. *See also:* electrothermics. (EEC/PE) [119]

arc gap *See:* resonant gap.

arching time (of a mechanical switching device) The interval of time between the instant of the first initiation of the arc and the instant of final arc extinction in all poles. *Note:* For switching devices that embody switching resistors, a distinction should be made between the arcing time up to the instant of the extinction of the main arc, and the arcing time up to the instant of the breaking of the resistance current. (SWG/PE) C37.100-1992

architecture (1) (computers) The organizational structure of a system or component. *See also:* system architecture; component; program architecture; subprogram; routine; module. (C) 610.12-1990
(2) (of AI-ESTATE) The elements of AI-ESTATE and their interrelationships. (ATLAS) 1232-1995
(3) In computer hardware, the organizational structure and interrelationship between the parts of a computing system, including the arrangement, design and interconnection of components. *Note:* This term is sometimes taken to mean the "instruction set" of a computer, since the physical architecture of a computer is often very tightly coupled with the instruction set of a computer. *See also:* computer architecture; network architecture. (C) 610.10-1994w
(4) The organizational structure of a system or a software item, identifying its components, their interfaces, and a concept of execution among them. (C/SE) J-STD-016-1995

architecture design (A) The process of defining a collection of hardware and software components and their interfaces to establish the framework for the development of a computer system. **(B)** The result of the process in definition (A). *See also:* functional design. (C) 610.10-1994, 610.12-1990

archival database A copy of a database saved for later reference or use. (C) 610.5-1990w

archival pages On-line data that is no longer maintained, is not expected to change, and may not be readily renderable by future tools. (C) 2001-1999

arcing chamber (expulsion-type arrester) The part of an expulsion-type arrester that permits the flow of discharge current to the ground and interrupts the follow current. *See also:* surge arrester. (IA/ICTL) 74-1958w

arcing contacts (1) The contacts of a switching device on which the arc is drawn after the main (and intermediate, where used) contacts have parted. (SWG/PE/TR) C57.12.44-1994
(2) The contacts of a switching device on which the arc is drawn after the main (and intermediate, where used) contacts have parted. (SWG/PE) C37.100-1992

arcing distance The shortest external tight-string distance measured over the insulating envelope between the metal parts at line potential and ground. Formerly referred to as striking distance or flashover distance. (PE/TR) C57.19.03-1996

arcing horn (1) One of a pair of diverging electrodes on which an arc is extended to the point of extinction after the main contacts of the switching device have parted. *Synonym:* arcing runners. (SWG/PE) C37.36b-1990r
(2) One of a pair of diverging electrodes on which an arc is extended to the point of extinction after the main contacts of the switching device have parted. *Note:* Arcing horns are sometimes referred to as arcing runners. (SWG/PE) C37.100-1992

arcing runners *See:* arcing horn.

arcing switch A switching device used in conjunction with a tap selector to carry, make, and break current in circuits that have already been selected. (PE/TR) C57.131-1995

arcing tap switch A switching device capable of carrying current and also breaking and making current while selecting a tap position. It, thereby, combines the duties of an arcing switch and a tap selector. (PE/TR) C57.131-1995

arcing time (1) (protection and coordination of industrial and commercial power systems) The arcing time of a fuse is the time elapsing from the melting of the current-responsive element (such as the link) to the final interruption of the circuit. This time will be dependent upon such factors as voltage and reactance of the circuit. (IA/PSP) 242-1986r
(2) (mechanical switching device). The interval of time between the instant of the first initiation of the arc and the instant of final arc extinction in all poles. *Note:* For switching devices that embody switching resistors, a distinction should be made between the arcing time up to the instant of the extinction of the main arc, and the arcing time up to the instant of the breaking of the resistance current. (SWG)
(3) (of a fuse) The time elapsing from the severance of the current-responsive element to the final interruption of the circuit. (PE/SWG-OLD) C37.100-1992, C37.40-1993

arc loss (1) (nonlinear, active, and nonreciprocal waveguide components) Power absorbed in an active nonlinear device during above-threshold switching or limiting in gas tubes, duplexers, ferrite limiters, or diode limiters. *Synonym:* absorptive loss. (MTT) 457-1982w
(2) (switching tubes) The decrease in radio-frequency power measured in a matched termination when a fired tube, mounted in a series or shunt junction with a waveguide, is inserted between a matched generator and the termination. *Note:* In the case of a pretransmit-receive tube, a matched output termination is also required for the tube. *See also:* gas tube. (ED) 161-1971w

arc mode current *See:* arc current.

arc mode voltage *See:* arc voltage.

arc reach The distance from a point midway between the arc extremities to the most remote point of the arc at the time of its maximum length. (SWG/PE) C37.36b-1990r, C37.100-1992

arc resistance The impedance of an arc that is resistive by nature; it is a function of the current magnitude and arc length. (PE/PSR) C37.113-1999

arc-shunting-resistor-current arcing time The interval between the parting of the secondary arcing contacts and the extinction of the arc-shunting-resistor current. (SWG/PE) C37.100-1992

arc suppression (rectifier) The prevention of the recurrence of conduction, by means of grid or ignitor action, or both, during the idle period, following a current pulse. *See also:* rectification. (EEC/PCON) [110]

arc-through (gas tube) A loss of control resulting in the flow of a principal electron stream through the rectifying element in the normal direction during a scheduled nonconducting period. *See also:* rectification. (ED) 161-1971w

arc-tube relaxation oscillator *See:* gas-tube relaxation oscillator.

arc voltage (gas-tube surge protective devices) (gas tube surge arresters) The voltage drop across the arrester during arc current flow. *Synonym:* arc mode voltage. (SPD/PE) C62.31-1987r, [8]

arc welder generator (generator, alternating-current arc welder) An alternating-current generator with associated reactors, regulators, control, and indicating devices required to produce alternating current suitable for arc welding.
(2) (A) (generator-rectifier, direct-current arc welder) A combination of static rectifiers and the associated alternating-current generator, reactors, regulators, controls, and indicating devices required to produce direct current suitable for arc welding. **(B) (generator, direct-current arc welder)** A direct-current generator with associated reactors, regulators, control, and indicating devices required to produce direct current suitable for arc welding. (EEC/AWM) [91]

arc-welding engine generator A device consisting of an engine mechanically connected to and mounted with one or more arc-welding generators. (EEC/AWM) [91]

arc-welding motor-generator A device consisting of a motor mechanically connected to and mounted with one or more arc-welding generators. (EEC/AWM) [91]

ARE *See:* all routes explorer.

area (1) (A) (data management) In CODASYL, a part of a database that can be opened or closed as a unit. *Note:* This term was used in early CODASYL documents, but is now considered deprecated. **(B) (data management)** A named collection of records within a database. *Note:* May contain occurrences of one or more record types, and a record type may have occurrences in one or more area. *Synonym:* realm. (C) 610.5-1990
(2) (image processing) The number of pixels in a region. (C) 610.4-1990w
(3) *See also:* equivalent flat plate area of a scattering object; partial effective area; effective area antenna. (AP/ANT) 145-1993

area assist action (electric power system) A control feature that bypasses economic control and that controls all available units while the area control error violates a preset limit. (PE/PSE) 94-1991w

area code (telephone switching systems) A one-, two-, or three-digit number that, for the purpose of distance dialing, designates a geographical subdivision of the territory covered by a separate national or integrated numbering plan. (COM) 312-1977w

area control error (1) (electric power system) A quantity reflecting the deficiency or excess of power within a control area. (PE/PSE) 858-1993w, 94-1991w
(2) (isolated-power system consisting of a single control area) The frequency deviation (of a control area on an interconnected system) is the net interchange minus the biased scheduled net interchange. *Note:* The above polarity is that which has been generally accepted by electric power systems and is in wide use. It is recognized that this is the reverse of the sign of control error generally used in servomechanism and control literature, which defines control error as the reference quantity minus the controlled quantity. (PE/PSE) [54]

area fill *See:* fill.

area frequency-response characteristic (control area) The sum of the change in total area generation caused by governor action and the change in total area load, both of which result from a sudden change in system frequency, in the absence of automatic control action. (PE/PSE) 94-1970w

areal beamwidth For pencil-beam antennas the product of the two principal half-power beamwidths. *See also:* principal half-power beamwidths. (AP/ANT) 145-1993

area load-frequency characteristic (control area) The change in total area load that results from a change in system frequency. (PE/PSE) 94-1970w

areal object A synthetic environment object that is geometrically anchored to the terrain with a set of at least three points that come to a closure. (C/DIS) 1278.1a-1998

area moving target indication A method of MTI based upon amplitude changes in corresponding resolution cells for radar returns obtained at different times. (AES) 686-1997

area supplementary control (electric power system) The control action applied, manually or automatically, to area generator speed governors in response to changes in system frequency, tie-line loading, or the relation of these to each other, so as to maintain the scheduled system frequency and/or the established net interchange with other control areas within predetermined limits. (PE/PSE) 94-1970w

area tie line (electric power system) A transmission line connecting two control areas. *Note:* Similar to interconnection tie. *See also:* transmission line. (PE/PSE) 94-1970w

argand plane A graphical representation of a vector used in complex number notation. (SCC20) 771-1998

Argument A value of type `Argument`. (IM/ST) 1451.1-1999

argument (1) (A) An independent variable; for example, the variable m in the equation $E = mc^2$. **(B)** A specific value of an independent variable; for example, the value $m = 24$ kg. **(C)** A constant, variable, or expression used in a call to a software module to specify data or program elements to be passed to that module. *Synonym:* actual parameter. *Contrast:* formal parameter. (C) 610.12-1990
(2) A parameter passed to a utility as the equivalent of a single string in the *argv* array created by one of the POSIX.1 *exec* functions. An argument is one of the options, option-arguments, or operands following the command name. (C/PA) 9945-2-1993
(3) Information that is passed to an interface operation or a directory operation. (C/PA) 1328.2-1993w, 1327.2-1993w, 1224.2-1993w, 1326.2-1993w
(4) An expression occurring as the actual value in a function call or procedure call. (C/DA) 1076.3-1997
(5) The value or the address of a data item passed to a function or procedure by the caller. (C/DA) 1481-1999
(6) The usual mathematical meaning. (IM/ST) 1451.1-1999

Argument Array A value of type `ArgumentArray`. (IM/ST) 1451.1-1999

ARINC *See:* Aeronautical Radio Incorporated.

arithmetic Pertaining to data that has the characteristics of base, scale, mode, and precision. *Note:* Used to represent numbers. *Contrast:* string. *See also:* decimal picture data; coded arithmetic data; binary picture data; numeric data. (C) 610.5-1990w

arithmetic and logic unit A functional component of a computer system that performs arithmetic and logical operations. *Synonym:* arithmetic-logic unit. *See also:* logic unit; arithmetic unit; exponent arithmetic and logic unit; register-arithmetic and logic unit. (C) 610.10-1994w

arithmetic check *See:* mathematical check.

arithmetic element *See:* arithmetic unit.

arithmetic expression An expression containing any combination of variables and constants joined by one or more arithmetic operators such that the expression can be reduced to a single numerical result. (C) 1084-1986w

arithmetic instruction An instruction in which the operation field specifies an arithmetical operation; for example, an add instruction or a multiply instruction. *Contrast:* logic instruction. (C) 610.10-1994w

arithmetic-logic unit *See:* arithmetic and logic unit.

arithmetic mean The numerical result obtained by dividing the sum of two or more quantities by the number of quantities. *Notes:* 1. Strictly speaking, arithmetic means of corona-effect data expressed in decibels cannot be taken unless the numbers are converted back to real units such as microvolts per meter (μV/m) or micropascals (μPa). 2. An arithmetic mean that is commonly used in audible noise investigations is the energy average of the quantities. The units in decibels above 20 μPa are converted to energy units such as microwatts (μW), which are then averaged. (T&D/PE) 539-1990

arithmetic operation **(1)** An operation for which the VHDL operator is +, -, *, /, **mod, rem, abs,** or **. (C/DA) 1076.3-1997
(2) An operation that is performed in accordance with the rules of ordinary arithmetic. (C) 1084-1986w
(3) **(test, measurement, and diagnostic equipment)** Operations in which numerical quantities form the elements of the calculation. (MIL) [2]

arithmetic overflow *See:* overflow.

arithmetic point *See:* radix point.

arithmetic power factor The ratio of the active power to the arithmetic apparent power. The arithmetic power factor is expressed by the equation

$$F_{pa} = \frac{P}{U_a}$$

where
F_{pa} = arithmetic power factor
P = active power
U = arithmetic apparent power.

Normally power factor, rather than arithmetic power factor, will be specified, but in particular cases, especially when the determination of the apparent power for a polyphase circuit is impracticable with the available instruments, arithmetic power factor may be used. When arithmetic power factor and power factor differ, arithmetic power factor is the smaller. (Std100) 270-1966w

arithmetic reactive factor The ratio of the reactive power to the arithmetic apparent power. (Std100) 270-1966w

arithmetic register A register that holds the operands or the results of operations such as arithmetic operations, logic operations, and shift operations. (C) 610.10-1994w

arithmetic shift **(1)** **(mathematics of computing)** A shift that affects all digit positions in a register, word, or numeral but does not affect the sign position. For example, +231.702 shifted two places to the left becomes +170.200. *Note:* The result is equivalent to multiplication or division by an integral power of the radix, except for the truncation effects. *Synonym:* numerical shift. *Contrast:* logical shift. (C) 1084-1986w
(2) **(general)** **(A)** A shift that does not affect the sign position. **(B)** A shift that is equivalent to the multiplication of a number by a positive or negative integral power of the radix. (MIL/C) [2], [85]

arithmetic underflow *See:* underflow.

arithmetic unit **(1)** The unit of a computing system that contains the circuits that perform arithmetic operations. (MIL/C) [2], [20], [85]
(2) A functional component of a computer system that performs arithmetic operations. *Note:* The term is also sometimes used for an arithmetic and logic unit. *Synonym:* arithmetic element. *See also:* vector unit; scalar unit. (C) 610.10-1994w

arm *See:* network analysis; branch.

arm a timer To start a timer measuring the passage of time, enabling the notification of a process when the specified time or time interval has passed. (C) 1003.5-1999

armature (of a relay) The moving element of an electromechanical relay that contributes to the designed response of the relay and that usually has associated with it a part of the relay contact assembly. (SWG/PE) C37.100-1992

armature band (rotating machinery) A thin circumferential structural member applied to the winding of a rotating armature to restrain and hold the coils so as to counteract the effect of centrifugal force during rotation. *Note:* Armature bands may serve the further purpose of archbinding the coils. They may be on the end windings only or may be over the coils within the core. (PE) [9]

armature band insulation (rotating machinery) An insulation member placed between a rotating armature winding and an armature band. *See also:* armature. (PE) [9]

armature bar (rotating machinery) (half coil) Either of two similar parts of an armature coil, comprising an embedded coil side and two end sections, that when connected together form a complete coil. *See also:* armature. (PE) [9]

armature coil (rotating machinery) A unit of the armature winding composed of one or more insulated conductors. *See also:* asynchronous machine; armature. (PE/EEC) [119]

armature core (rotating machinery) A core on or around which armature windings are placed. *See also:* armature. (PE) [9]

armature I^2R loss (synchronous machines) The sum of the I^2R losses in all of the armature current paths. *Note:* The I^2R loss in each current path shall be the product of its resistance in ohms, as measured with direct current and corrected to a specified temperature, and the square of its current in amperes. (PE/REM) [9], [115]

armature quill *See:* armature spider.

armature reaction (rotating machinery) The magnetomotive force due to armature-winding current. (PE) [9]

armature-reaction excited machine A machine having a rotatable armature, provided with windings and a commutator, whose load-circuit voltage is generated by flux that is produced primarily by the mag-netomotive force of currents in the armature winding. *Notes:* 1. By providing the stationary member of the machine with various types of windings, different characteristics may be obtained, such as a constant-current characteristic or a constant-voltage characteristic. 2. The machine is normally provided with two sets of brushes, displaced around the commutator from one another, so as to provide primary and secondary circuits through the armature. 3. The primary circuit carrying the excitation armature current may be completed externally by a short-circuit connection, or through some other external circuit, such as a field winding or a source of power supply; and the secondary circuit is adapted for connection to an external load. (EEC/PE) [119]

armature sleeve *See:* armature spider.

armature spider A support upon which the armature laminations are mounted and which in turn is mounted on the shaft. *Synonyms:* armature sleeve; armature quill. (EEC/PE) [119]

armature terminal (rotating machinery) A terminal connected to the armature winding. *See also:* armature. (PE) [9]

armature to field transfer function (G[s]) (synchronous machine parameters by standstill frequency testing) (standstill frequency response testing). The ratio of the Laplace transform of the direct-axis armature flux linkages to the Laplace transform of the field voltage, with the armature open-circuited. (PE/EM) 115A-1987

armature-voltage control A method of controlling the speed of a motor by means of a change in the magnitude of the voltage impressed on its armature winding. *See also:* control. (IA/IAC) [60]

armature winding (rotating machinery) The winding in which alternating voltage is generated by virtue of relative motion with respect to a magnetic flux field. *See also:* asynchronous machine. (PE) [9]

armature winding cross connection *See:* armature winding equalizer.

armature winding equalizer (rotating machinery) An electric connection to normally equal-potential points in an armature circuit having more than two parallel circuits. *Synonym:* armature winding cross connection. *See also:* armature. (PE) [9]

armed sweep *See:* single sweep.

armor clamp (wiring methods) A fitting for gripping the armor of a cable at the point where the armor terminates or where the cable enters a junction box or other piece of apparatus. (T&D/PE) [10]

armored cable (interior wiring) A fabricated assembly of insulated conductors and a flexible metallic covering. *Note:* Armored cable for interior wiring has its flexible outer sheath or armor formed of metal strip, helically wound and with interlocking edges. Armored cable is usually circular in cross section but may be oval or flat and may have a thin lead sheath between the armor and the conductors to exclude moisture, oil, etc., where such protection is needed. *See also:* nonmetallic-sheathed cable. (EEC/PE) [119]

arm, thermoelectric *See:* thermoelectric arm.

arm, thermoelectric, graded *See:* thermoelectric, graded arm.

arm, thermoelectric, segmented A thermoelectric arm composed of two or more materials having different compositions. *See also:* thermoelectric device. (ED) [46], 221-1962w

ARQ *See:* automatic repeat request.

Array Denotes the IEEE 1451.1 array datatype. (IM/ST) 1451.1-1999

array (1) (photovoltaic converter) A combination of panels coordinated in structure and function. (AES) [41]
(2) (solar cells) A combination of solar-cell panels or paddles coordinated in structure and function. (AES/SS) 307-1969w
(3) (data management software) An *n*-dimensional ordered set of data items identified by a single name and one or more indices, so that each element of the set is individually addressable. (C) 610.5-1990w, 610.12-1990
(4) The language-independent syntax for a family of datatypes constructed from a base datatype and an index datatype. The base datatype may be any datatype, the index datatype shall be a finite ordered datatype. (C/PA) 1351-1994w
(5) A datatype constructed from a base datatype and an index datatype. The base datatype may be any datatype; the index datatype shall be a finite, ordered datatype. (C/PA) 1224.1-1993w
(6) A group of memory cells that are arranged in a pattern. (ED) 1005-1998

array antenna An antenna comprised of a number of identical radiating elements in a regular arrangement and excited to obtain a prescribed radiation pattern. *Notes:* 1. The regular arrangements possible include ones in which the elements can be made congruent by simple translation or rotation. 2. This term is sometimes applied to cases where the elements are not identical or arranged in a regular fashion. For those cases qualifiers shall be added to distinguish from the usage implied in this definition. For example, if the elements are randomly located, one may use the term random array antenna. *Synonym:* antenna array. (AP/ANT) 145-1993

array control (terrestrial photovoltaic power systems) All electrical and mechanical controls that ensure proper electric and thermal performance of the array field. (PV) 928-1986r

array element In an array antenna, a single radiating element or a convenient grouping of radiating elements that have fixed relative excitations. (AP/ANT) 145-1993

array factor The radiation pattern of an array antenna when each array element is considered to radiate isotropically. *Note:* When the radiation patterns of individual array elements are identical, and the array elements are congruent under translation, then the product of the array factor and the element radiation pattern gives the radiation pattern of the entire array. (AP/ANT) 145-1993

array field (terrestrial photovoltaic power systems) The aggregate of all array subfields. *See also:* array control. (PV) 928-1986r

array processor A processor capable of executing instructions in which the operands may be arrays rather than scalar data elements. *Synonym:* vector processor. (C) 610.10-1994w

array source A common diffusion that provides the source of electrons for the cell in the read mode or, in the case of a cell that programs via channel hot elecrons (CHE), provides the source of electrons for programming. (ED) 1005-1998

array subsystem (terrestrial photovoltaic power systems) The array field and the controls that together produce dc electric and thermal energy. Associated thermal energy may be utilized or dissipated. *See also:* array control. (PV) 928-1986r

arrester *See:* surge arrester.

arrester alternating sparkover voltage The root-mean-square value of the minimum 60-Hz sine-wave voltage that will cause sparkover when applied between its line and ground terminals. *See also:* surge arrester; current rating, 60-hertz. (PE) [8]

arrester, dead-front type *See:* dead-front type arrester.

arrester discharge capacity The crest value of the maximum current of specified wave shape that the arrester can withstand without damage to any of its parts. *See also:* surge arrester. (PE) [8]

arrester discharge current The current that flows through an arrester resulting from an impinging surge. (SPD/PE) C62.22-1997

arrester discharge voltage The voltage that appears across the terminals of an arrester during the passage of discharge current. (SPD/PE) C62.22-1997

arrester discharge voltage-current characteristic The variation of the crest values of discharge voltage with respect to discharge current. *Note:* This characteristic is normally shown as a graph based on three or more current surge measurements of the same wave shape but of different crest values. *See also:* surge arrester; lightning; current rating, 60-hertz. (PE) 28-1974, [8]

arrester discharge voltage-time curve A graph of the discharge voltage as a function of time while discharging a current surge of given wave shape and magnitude. *See also:* surge arrester. (PE) [8]

arrester disconnector A means for disconnecting an arrester in anticipation of, or after, a failure in order to prevent a permanent fault on the circuit and to give indication of a failed arrester. *Note:* Clearing of the fault current through the arrester during disconnection is generally done by the nearest source side over-current-protective device. (SPD/PE) C62.11-1999

arrester, distribution, normal duty class *See:* distribution arrester.

arrester duty cycle rating The designated maximum permissible root-mean-square (rms) value of power-frequency voltage between its line and ground terminals at which it is designed to perform its duty cycle. (SPD/PE) C62.22-1997

arrester, expulsion-type *See:* expulsion-type surge arrester.

arrester ground An intentional electric connection of the arrester ground terminal to the ground. *See also:* surge arrester. (PE) [8]

arresters, classification of Arrester classification is determined by prescribed test requirements. These classifications are: station valve arrester; intermediate valve arrester; secondary valve arrester; protector tube. (PE) [8]

arrester unit Any section of a multiunit arrester. (SPD/PE) C62.11-1999

arrester, valve-type An arrester having a characteristic element consisting of a resistor with a nonlinear volt-ampere characteristic that limits the follow current to a value that the series gap can interrupt. *Note:* If the arrester has no series gap, the characteristic element limits the follow current to a magnitude that does not interfere with the operation of the system. *See also:* surge arrester; nonlinear resistor-type arrester. (PE) [8]

arrow A directed line, composed of one or more connected arrow segments in a single diagram from a single source (box or diagram boundary) to a single use (box or diagram boundary). *See also:* arrow segment; boundary arrow; internal arrow. (C/SE) 1320.1-1998

arrow button A visual user interface control that is a button labeled with an arrow pointing in a specified direction that represents an action associated with that direction. The action is invoked if the user clicks the mouse within the button. (C) 1295-1993w

arrow label A noun or noun phrase associated with an arrow segment to signify the arrow meaning of the arrow segment. Specifically, an arrow label identifies the object type set that is represented by an arrow segment. (C/SE) 1320.1-1998

arrow meaning The object types (e.g., a physical thing, a data element) of an object type set, regardless of how these object types may be collected, aggregated, grouped, bundled, or otherwise joined within the object type set. (C/SE) 1320.1-1998

arrow reference *See:* ICOM code.

arrow role The relationship between an object type set represented by an arrow segment and the activity represented by the box to which the arrow segment is attached. There are four arrow roles: input, control, output, and mechanism. (C/SE) 1320.1-1998

arrow segment A directed line that originates at a box side, arrow junction (branch or join), or diagram boundary and terminates at the next box side, arrow junction (branch or join), or diagram boundary that occurs in the path of the line. (C/SE) 1320.1-1998

ARSR *See:* air-route surveillance radar.

articulated part A visible part of a simulated entity that may not move relative to the entity, but is able to move relative to the entity itself. (DIS/C) 1278.1-1995

articulated unit substation (1) (A) power and distribution transformer. A unit substation in which the incoming, transforming, and outgoing sections are manufactured as one or more subassemblies intended for connection in the field. **(B)** radial type. One that has a single stepdown transformer and that has an outgoing section for the connection of one or more outgoing radial (stub end) feeders. **(C)** distributed-network type. One that has a single step-down transformer having its outgoing side connected to a bus through a circuit breaker equipped with relays that are arranged to trip the circuit breaker on reverse power flow to the transformer and to reclose the circuit breaker upon the restoration of the correct voltage, phase angle, and phase sequence at the transformer secondary. The bus has one or more outgoing radial (stub end) feeders and one or more tie connections to a similar unit substation. **(D)** spot-network type. One that has two step-down transformers, each connected to an incoming high-voltage circuit. The outgoing side of each transformer is connected to a common bus through circuit breakers equipped with relays which are arranged to trip the circuit breaker on reverse power flow to the transformer and to reclose the circuit breaker upon the restoration of the correct voltage, phase angle and phase sequence at the transformer secondary. The bus has one or more outgoing radial (stub end) feeders. **(E)** secondary-selective type (low-voltage-selective type). One which has two stepdown transformers, each connected to an incoming high-voltage circuit. The outgoing side of each transformer is connected to a separate bus through a suitable switching and protective device. The two sections of bus are connected by a normally open switching and protective device. Each bus has one or more outgoing radial (stub end) feeders. **(F)** duplex type (breaker-and-a-half arrangement). One that has two stepdown transformers, each connected to an incoming high-voltage circuit. The outgoing side of each transformer is connected to a radial (stub end) feeder. These feeders are joined on the feeder side of the power circuit breakers by a normally open-tie circuit breaker. (SWG/PE/TR) C57.12.80-1978
(2) A unit substation in which the incoming, transforming, and outgoing sections are manufactured as one or more subassemblies intended for connection in the field. (SWG/PE) C37.100-1992

articulation (percent articulation) and intelligibility (percent intelligibility) The percentage of the speech units spoken by a talker or talkers that is correctly repeated, written down, or checked by a listener or listeners. *Notes:* 1. The word "articulation" is used when the units of speech material are meaningless syllables or fragments; the word "intelligibility" is used when the units of speech material are complete, meaningful words, phrases, or sentences. 2. It is important to specify the type of speech material and the units into which it is analyzed for the purpose of computing the percentage. The units may be fundamental speech sounds, syllables, words, sentences, etc. 3. The percent articulation or percent intelligibility is a property of the entire communication system: talker, transmission equipment or medium, and listener. Even when attention is focused upon one component of the system (for example, a talker, a radio receiver), the other components of the system should be specified. 4. The kind of speech material used is identified by an appropriate adjective in phrases such as syllable articulation, individual sound articulation, vowel (or consonant) articulation, monosyllabic word intelligibility, discrete word intelligibility, discrete sentence intelligibility. *See also:* volume equivalent. (SP) [32]

articulation equivalent (complete telephone connection) A measure of the articulation of speech reproduced over it. The articulation equivalent of a complete telephone connection is expressed numerically in terms of the trunk loss of a working reference system when the latter is adjusted to give equal articulation. *Note:* For engineering purposes, the articulation equivalent is divided into articulation losses assignable to the station set, subscriber line, and battery supply circuit that are on the transmitting end, the station set, subscriber line, and battery supply circuit that are on the receiving end, the trunk, and interaction effects arising at the trunk terminals. *See also:* volume equivalent. (EEC/PE) [119]

artificial antenna A device that has the necessary impedance characteristics of an antenna and the necessary power-handling capabilities, but that does not radiate or receive radio waves. *Synonym:* dummy antenna. *See also:* antenna. (AP/ANT) 145-1983s

artificial dielectric A medium containing a distribution of scatterers, usually metallic, that react as a dielectric to radio waves. *Notes:* 1. The scatterers are usually small compared to a wavelength and embedded in a dielectric material whose effective permittivity and density are intrinsically low. 2. The scatterers may be in either a regular arrangement or a random distribution. (AP/ANT) 145-1993

artificial ear (1) (transmission performance of telephone sets) A device for the measurement of the acoustic output of telephone-set receivers. It presents to the receiver an acoustic impedance approximating the impedance presented by the average human ear. (COM/TA) 269-1983s
(2) (general) A device for the measurement of the acoustic output of earphones in which the artificial ear presents to the earphone an acoustic impedance approximating the impedance presented by the average human ear and is equipped with a microphone for measurement of the sound pressures developed by the earphone. (SP/COM) [32], [50]

artificial hand A device simulating the impedance between an electric appliance and the local earth when the appliance is grasped by the hand. *See also:* electromagnetic compatibility. (EMC/INT) [53], [70]

artificial horizon (A) (navigation aid terms) A device for indicating the horizontal, as a bubble, gyroscope, pendulum, or the flat surface of a liquid. **(B) (navigation aid terms)** A gyroscopic flight instrument that shows the pitching and banking attitudes of a vehicle with respect to a reference line horizon. (AES/GCS) 172-1983

artificial intelligence (A) The study of designing computer systems exhibiting the characteristics associated with intelligence in human behavior including understanding language, learning, reasoning from incomplete or uncertain information, and solving problems. **(B)** The discipline for developing computer systems capable of passing the *Turing Test* in which behavior of the computer system is indistinguishable from human behavior. **(C)** The study of problem solving using computational models. (ATLAS) 1232-1995

Artificial Intelligence and Expert System Tie to Automatic Test Equipment (AI-ESTATE) (1) A set of specifications that defines the software interface of artificial intelligence and expert systems to system test and diagnosis. *Note:* The AI-ESTATE set of standards currently includes IEEE Std 1232-1995, IEEE P1232.1, IEEE P1232.2, and IEEE P1347. (ATLAS) 1232-1995
(2) Provides standard representation of dependency information used to drive intelligent diagnostic reasoning systems. (ATLAS) 1226-1993s

artificial language *See:* formal language.

artificial line (1) (data transmission) An electric network that simulates the electrical characteristics of a line over a desired frequency range. *Note:* Although the term basically is applied to the case of simulation of an actual line, by extension it is used to refer to all periodic lines that may be used for laboratory purposes in place of actual lines, but that may represent no physically realizable line. For example, an artificial line may be composed of pure resistances. *See also:* network analysis. (AP/PE/ANT) 145-1983s, 599-1985w
(2) (waveguide) A network that simulates the electrical characteristics of a transmission line over a given frequency range. (MTT) 146-1980w

artificial load A dissipative but essentially nonradiating device having the impedance characteristics of an antenna, transmission line, or other practical utilization circuit. (AP/PE/ANT) 145-1983s, 599-1985w

artificial mains network A network inserted in the supply mains lead of the apparatus to be tested, which provides a specified measuring impedance for interference voltage measurements and isolates the apparatus from the supply mains at radio frequencies. *See also:* line-impedance stabilization network; electromagnetic compatibility. (EMC/INT) [53], [70]

artificial mouth (transmission performance of telephone sets) An electroacoustic transducer that produces a sound field simulating that of a typical human talker. The reference point for the handset and the headset is the center of the circular plane of contact of the handset ear-cap and the ear. If the handset ear-cap is not circular or has no external plane of contact, an effective center and an effective plane of contact are determined. (COM/TA) 269-1983s

artificial pupil (illuminating engineering) A device or arrangement for confining the light passing through the pupil of the eye to an area smaller than the natural pupil. (ED) [127]

artificial test head A fixture containing an artificial mouth and an artificial ear located in a specified relationship with each other. *See also:* loudness rating guard-ring position. (COM/TA) 269-1992, 1206-1994

artificial voice A sound source for microphone measurements consisting of a small loudspeaker mounted in a shaped baffle proportioned to simulate the acoustic constants of the human head. *See also:* close-talking pressure-type microphones. (SP) 258-1965w

as-built curve (rotating electric machinery) A curve that is found on an individual machine during testing. (PE/EM) 11-1980r

as-built drawings A complete set of drawings which, in addition to the original drawings, includes all drawings that accurately record changes made to the equipment or subsystem to indicate the final installation and commissioning. (PE/SUB) 1303-1994

ASC *See:* accredited standards committee.

***A*-scan** *See: A* and *R* display.

ascender The portion of a graphic character that extends above the main part of the character; for example, the upper portion of the letters b and h. *Contrast:* descender. (C) 610.2-1987

ascending node (communication satellite) The point on the line of nodes that the satellite passes through as the satellite travels from below to above the equatorial plane. (COM) [19]

ASCII *See:* American Standard Code for Information Interchange.

ASCIIz An ASCII string concatenated with a NULL character. (C/MM) 1284.1-1997

A-scope A cathode-ray oscilloscope arranged to present an A-display. (AES/RS) 686-1990

ASC T1 A standards committee accredited by the American National Standards Institute organization that recommends standards for telecommunication. (C) 610.7-1995

ASC X3 A standards committee accredited by the American National Standards Institute organization that recommends standards for computers and information processing systems. (C) 610.7-1995

ASDE *See:* airport surface detection equipment.

ash layer porosity (fly ash resistivity) (cm^3/cm^3) The ratio of the ash layer void volume to the test cell volume in a test cell used for the laboratory measurement of fly ash resistivity. (PE/EDPG) 548-1984w

ASI *See:* address space identifier.

A64 module A module whose address space is limited to a 64-bit width. (C/BA) 14536-1995

askarel (1) (handling and disposal of transformer grade insulating liquids containing PCBs) A generic term for a group of synthetic, fire-resistant, chlorinated aromatic hydrocarbons used as electrical insulating liquid. They have a property under arcing conditions such that any gases produced will consist predominantly of noncombustible hydrogen chloride with lesser amounts of combustible gases. Askarel does not necessarily contain polychlorinated biphenyl (PCBs). (LM/C) 802.2-1985s
(2) A generic term for a group of nonflammable synthetic chlorinated hydrocarbons used as electrical insulating media. Askarels of various compositional types are used. The gases produced under arcing conditions, while consisting predominantly of noncombustible hydrogen chloride, can include varying amounts of combustible gases depending upon the askarel type. (NESC/NEC) [86]
(3) (power and distribution transformers) A generic term for a group of synthetic, fire-resistant, chlorinated, aromatic hydrocarbons used as electrical insulating liquids. They have a property under arcing conditions such that any gases produced will consist predominantly of noncombustible hydrogen chloride with lesser amounts of combustible gases. (PE/TR) 637-1985r, C57.12.80-1978r

ASN.1 *See:* Abstract Syntax Notation One.

aspect ratio (1) (television) The ratio of the frame width to the frame height. The ratio of the frame width to the frame height as defined by the active picture. (BT/AV) 201-1979w, 202-1954w
(2) The ratio of the height to the width of a rectangle such as is found in a display surface, window, viewport, or character space. (C) 610.6-1991w

asphalt (rotating machinery) A dark brown to black cementitious material, solid or semisolid in consistency, in which the predominating constituents are bitumens that occur in nature as such or are obtained as residue in refining of petroleum. (PE) [9]

ASR *See:* airport-surveillance radar.

assay (sodium iodide detector) The determination of the activity of a radionuclide in a sample. (NI) N42.12-1994

assemble (1) (software) To translate a computer program expressed in an assembly language into its machine language equivalent. *Contrast:* disassemble; compile; interpret. (C) 610.12-1990
(2) The process of constructing from parts one or more identified pieces of software. (C/SE) 1517-1999

assemble-and-go An operating technique in which there are no stops between the assembling, linking, loading, and execution of a computer program. (C) 610.12-1990

assembled origin The address of the initial storage location assigned to a computer program by an assembler, a compiler, or a linkage editor. *Contrast:* loaded origin. *See also:* starting address; offset. (C) 610.12-1990

assembler (1) (microprocessor assembly language) A utility program that translates symbolic assembly language instructions into machine instructions or data on a one-to-one basis. (C/MM) 695-1985s

(2) (software) A computer program that translates programs expressed in assembly language into their machine language equivalents. *Contrast:* interpreter; compiler. *See also:* relocating assembler; absolute assembler; cross-assembler. (C) 610.12-1990

(3) (test, measurement, and diagnostic equipment) A computer program that is one step more automatic than a translator; it translates not only operations but also data and input-output quantities from symbolic to machine language form in a one-to-one ratio. An assembler program may have the capability to assign locations within a storage device. (MIL) [2]

assembler code *See:* assembly code.

assembler language *See:* assembly language.

assembly (GIS) (1) (electric and electronics parts and equipment) A number of basic parts or subassemblies, or any combination thereof, joined together to perform a specific function. The application, size, and construction of an item may be factors in determining whether an item is regarded as a unit, an assembly, a subassembly, or a basic part. A small electric motor might be considered as a part if it is not normally subject to disassembly. The distinction between an assembly and a subassembly is not always exact: An assembly in one instance may be a subassembly in another where it forms a portion of an assembly. Typical examples are: electric generator, audio-frequency amplifier, power supply. (GSD) 200-1975w

(2) (nuclear power generating station) (seismic qualification of Class 1E equipment) Two or more devices sharing a common mounting or supporting structure. *Note:* Examples are control panels and diesel generators. (PE/NP) 380-1975w, 344-1975s

(3) A collection of GIS components that are interconnected and ready for insertion as a subassembly in a GIS, such as a breaker bay shipping assembly. The term is also used to describe a complete GIS. (SUB/PE) C37.122.1-1993, C37.122-1993

(4) Gas-insulated substation fully erected. (SWG/PE) C37.100-1992

(5) (A) An element of the physical or system architecture, specification tree, and system breakdown structure that is a subordinate element to a subsystem and is comprised of two or more components. It represents a consumer product (automatic brake system) of a subsystem (braking system of an automobile); or a life-cycle process product (control system) of a subsystem (flight controls of a simulator) related to a life-cycle process (training) that supports an assembly or group of assemblies. **(B)** The act of fitting together fabricated or manufactured elements into a larger element. (C/SE) 1220-1994

assembly code Computer instructions and data definitions expressed in a form that can be recognized and processed by an assembler. *Synonym:* assembler code. *Contrast:* compiler code; machine code; interpretive code. (C) 610.12-1990

assembly language A symbolic programming language that corresponds closely to the instruction set of a given computer, allows symbolic naming of operations and addresses, and usually results in a one-to-one translation of program instructions into machine instructions. *Synonyms:* assembler language; second generation language. *Contrast:* high-order language; machine language; fifth generation language; fourth generation language. *See also:* META 5. (C) 610.13-1993w, 610.12-1990

assembly, microelectronic device (electric and electronics parts and equipment) An assembly of inseparable parts, circuits, or a combination thereof. Typical examples are: microcircuit, integrated-circuit package, micromodule. (GSD) 200-1975w

assert (1) (signals and paths) To cause a signal line to make a transition from its logically false (inactive) state to its logically true (active) state. The true or active state is either a high or low state, as specified for each signal. (C/MM) 696-1983w

(2) (STD bus) To place a signal in its logic 1 state. (C/MM) 961-1987r

(3) (A) For a set of parallel signals of the same function, to place the desired logic state pattern on the bus, which may include both one and zero values. **(B)** For a single signal, to drive a signal to the one ("1"), or asserted, logic state. (C/BA) 1496-1993

(4) (A) The action of applying a logic one signal to a bus line. **(B)** The state of a bus line when the signal it carries represents a logic one. (C/BA) 10857-1994, 896.10-1997, 896.2-1991, 896.4-1993, 896.3-1993

(5) To change the value of a bus signal from logic 0 (released) to logic 1 (asserted) or ensure that such a signal remains at a logic 1. (TT/C) 1149.5-1995

asserted (1) The state of a signal line. Since all lines are active low signals, this state is the low state for all bus lines. (C/MM) 1196-1987w

(2) Having a current value equal to logic 1 (said of any signal). (TT/C) 1149.5-1995

assertion (1) (software) A logical expression specifying a program state that must exist or a set of conditions that program variables must satisfy at a particular point during program execution. Types include input assertion, loop assertion, output assertion. *See also:* proof of correctness; invariant. (C) 610.12-1990

(2) A statement of functionality or behavior for a POSIX element that is derived from the POSIX standard being tested and that is true for a conforming POSIX implementation. (C/PA) 13210-1994, 2003.1-1992

(3) A statement that is derived from the standard to which conformance is being measured, that is true for a conforming implementation, and that pertains either to functionality or behavior of a functional interface or namespace allocation or to the documentation associated with the implementation being tested. (C/PA) 1328.2-1993w, 1326.2-1993w, 1224-1993w, 1327-1993w, 1224.1-1993w, 1326.1-1993w, 1328-1993w

(4) The specification for testing a conformance requirement in an IUT. It defines what to test and is **TRUE** for a conforming implementation. Assertions are the basic entities for test method specifications and test method standards. (C/PA) 2003-1997

assertion identifier The identifier assigned to an assertion. The name of the element and the assertion identifier together shall uniquely identify an assertion within a test method specification. (C/PA) 2003-1997

assertion number The numeric identifier assigned to an assertion. The name of the element and the assertion number together uniquely identify an assertion. (C/PA) 13210-1994, 2003.1-1992

assertion test (1) The software or procedural methods that ascertain the conformance of a POSIX implementation to an assertion. (C/PA) 13210-1994, 2003.1-1992

(2) The software or procedural methods that generate the test result codes used for assessment of conformance to an assertion. (C/PA) 2003-1997

assessed failure rate The failure rate of an item determined by a limiting value or values of the confidence interval associated with a stated confidence level, based on the same data as the observed failure rate of nominally identical items. *Notes:* 1. The source of the data shall be stated. 2. Results can be accumulated (combined) only when all conditions are similar. 3. The assumed underlying distribution of failures against time shall be stated. 4. It should be stated whether a one-sided or a two-sided interval is being used. 5. Where one limiting value is given this is usually the upper limit. (R) [29]

assessed mean active maintenance time The active maintenance time determined as the limit or the limits of the confidence interval associated with a stated confidence level, and based on the same data as the observed mean active maintenance time or nominally identical items. *Notes:* 1. The source of the data shall be stated. 2. Results can be accumulated (combined) only when all conditions are similar. 3. It should be stated whether a one-sided or two-sided interval is being used. 4. The assumed underlying distribution of mean active maintenance times shall be stated with the reason for the assumption. 5. When one value is given, this is usually the upper limit. (R) [29]

assessed mean life (non-repaired items) The mean life of an item determined by a limiting value or values of the confidence interval associated with a stated confidence level, based on the same data as the observed mean life of nominally identical items. *Notes:* 1. The source of the data shall be stated. 2. Results can be accumulated (combined) only when all conditions are similar. 3. The assumed underlying distribution of failures against time shall be stated. 4. It should be stated whether a one-sided or a two-sided interval is being used. 5. Where one limiting value is given, this is usually the lower limit. (R) [29]

assessed reliability The reliability of an item determined by a limiting value or values of the confidence interval associated with a stated confidence level, based on the same data as the observed reliability of nominally identical items. *Notes:* 1. The source of the data shall be stated. 2. Results can be accumulated (combined) only when all conditions are similar. 3. The assumed underlying distribution of failures against time shall be stated. 4. It should be stated whether a one-sided or a two-sided interval is being used. 5. Where one limiting value is given, this is usually the lower limit. (R) [29]

assessment A planned and documented activity performed to determine whether various elements within a quality management system are effective in achieving stated quality objectives. (NI) N42.23-1995

asset **(1)** The items of interest that are stored in a reuse library, such as design documentation, specifications, source code, documentation, test suites, etc., or any other unit of information of potential value to a reuser. (C/SE) 1420.1-1995, 1420.1a-1996
(2) A class in the BIDM. *Asset* will always be capitalized when referring to the class Asset. (C/SE) 1420.1-1995
(3) An item, such as design, specifications, source code, documentation, test suites, manual procedures, etc., that has been designed for use in multiple contexts. (C/SE) 1517-1999

Asset Certification Framework A technique and associated data model used for organizing, selecting, communicating, and guiding the process of certifying assets. (C/SE) 1420.1a-1996

assigned indexing Automatic indexing in which appropriate keywords are assigned from a list of preselected keywords rather than from the text of the document or information being indexed. *Synonym:* assignment indexing. *Contrast:* derivative indexing. (C) 610.2-1987

assigned value The best estimate of the value of a quantity. The assigned value may be from an instrument reading, a calibration result, a calculation, etc. (IM) 470-1972w, 544-1975w

assignment *See:* variable assignment.

assignment indexing *See:* assigned indexing.

assignment reference The occurrence of a literal or other expression as the waveform element of a signal assignment statement or as the right-hand side expression of a variable assignment statement. (C/DA) 1076.3-1997, 1076.6-1999

assignment statement (software) A computer program statement that assigns a value to a variable; for example, $Y = X - 5$. *Contrast:* control statement; declaration. *See also:* initialize; reset; clear. (C) 610.12-1990

assistance call (telephone switching systems) A call to an operator for help in making a call. (COM) 312-1977w

associated circuits **(1) (nuclear power generating station)** (design and installation of cable systems for Class 1E circuits in nuclear power generating stations) Non-Class 1E circuits that share power supplies, signal sources, enclosures, or raceways with Class 1E circuits or are not physically separated or electrically isolated from Class 1E circuits by acceptable separation distance, barriers, or isolation devices. *Note:* Circuits include the interconnecting cabling and the connected loads. *See also:* circuit. (EDPG) 690-1984r
(2) Non-Class 1E circuits that are not physically separated or are not electrically isolated from Class 1E circuits by acceptable separation distance, safety class structures, barriers, or isolation devices. (PE/NP) 603-1998

associated equipment (packaging machinery) Any attachment or component part that is not necessarily located on the packaging machine but is directly associated with the performance of the machine. Limit switches and photoelectric devices are examples. (IA/PKG) 333-1980w

associate developer An organization that is neither prime contractor nor subcontractor to the developer, but who has a development role on the same or related system or project. (C/SE) J-STD-016-1995

associated structural parts **(1) (insulation system)** Includes items such as slot wedges, space blocks, and ties used to position the coil ends and connections, any nonmetallic supports for the winding, and field-coil flanges. *See also:* insulation system. (IA/PC) 1068-1996
(2) (insulation systems of synchronous machines) The associated structural parts of the installation system include the field collars, the slot wedges, the filler strips under the support ring insulation, the nonmetallic support for the winding, the space blocks used to separate the coil ends and connections, the lead cleats, and the terminal boards. (REM) [115]

association **(1)** In data management, a relationship established in a data model to represent a connection between entities that is not reflected solely by the attributes inherent in the entities. (C) 610.5-1990w
(2) The service used to establish access point/station (AP/STA) mapping and enable STA invocation of the distribution system services (DSSs). (C/LM) 8802-11-1999
(3) *See also:* application-association. (C/PA) 1351-1994w

associative class A class introduced to resolve a many-to-many relationship. (C/SE) 1320.2-1998

associative literal A literal that denotes an instance in terms of its value. The form of expression used to state an associative literal is `className with propertyName: propertyValue`. (C/SE) 1320.2-1998

associative lookup Table lookup performed on a table that is stored in associative memory. (C) 610.5-1990w

associative memory A type of memory whose locations are identified by their contents or by a part of their contents, rather than by their names or positions. *Synonyms:* search memory; content addressable storage. (C) 610.10-1994w

associative storage A storage device in which storage locations may be identified by specifying part or all of their contents. *Synonym:* content addressed storage. (C) 162-1963w

assumed binary point The position in a binary numeral at which the binary point is assumed to be located; usually at the right unless otherwise specified. *Synonym:* implied binary point. (C) 1084-1986w

assumed decimal point The position in a decimal numeral at which the decimal point is assumed to be located; usually at the right unless otherwise specified. *Synonym:* implied decimal point. (C) 1084-1986w

assumed position (navigation aid terms) A point at which a craft is assumed to be located. (AES/GCS) 172-1983w

assumed radix point The position in a numeral at which the radix point is assumed to be located; usually at the right unless otherwise specified. *Synonym:* implied radix point. (C) 1084-1986w

assumptions Conditions and/or resource requirements that are mandatory for process completion. (C/SE) 1209-1992w

assured access protocol (FASTBUS acquisition and control) A potential master is operating in the assured access protocol if, on detecting an arbitration request inhibit (AI) assertion, it will not assert arbitration request (AR) and thus will not participate in subsequent arbitration cycles until all devices currently asserting AR have obtained bus mastership and completed their operations. (NID) 960-1993

assured disruptive discharge voltage The prospective value of the test voltage that causes disruptive discharge under specified conditions. (PE/PSIM) 4-1995

asterisk (*) (1) When appended to a signal's name, the suffix "*" (asterisk) indicates that the logic one state of the signal is such that it will override the logic zero state applied by any other module on the line. (C/BA) 896.4-1993w, 14536-1995, 896.2-1991w, 10857-1994
(2) When appended to a signal's name, the suffix "*" (asterisk) indicates that the logic 1 state of the signal is represented by a less positive voltage than the logic 0 state. (C/BA) 896.10-1997

astern (navigation aid terms) Bearing approximately 180° relative. (AES/GCS) 172-1983w

aster rectifier circuit A circuit that employs 12 or more rectifying elements with a conducting period of 30 electrical degrees plus the commutating angle. *See also:* rectification. (EEC/PE) [119]

astigmatism (electron optical) In an electron-beam tube, a focus defect in which electrons in different axial planes come to focus at different points. *See also:* oscillograph. (ED) 161-1971w

ASTM The American Society for Testing and Materials. Founded in 1898, the ASTM is a scientific and technical organization formed for the development of standards on characteristics and performance of materials, products, systems, and service; and the promotion of related knowledge. (T&D/PE) 524a-1993r

Aston dark space (gas) The dark space in the immediate neighborhood of the cathode, in which the emitted electrons have a velocity insufficient to excite the gas. *See also:* discharge. (ED) [45], [84]

astrocompass (navigation aid terms) An instrument that, when oriented to the horizontal and the celestial sphere, indicates horizontal reference direction relative to the earth. It is used to obtain true heading by reference to celestial bodies. (AES/GCS) 172-1983w

astrodynamics (communication satellite) Engineering application of celestial mechanics. (COM) [19]

astro-inertial navigation equipment *See:* celestial-inertial navigation equipment.

astronomical position (A) (navigation aid terms) A point on the earth where coordinates have been determined as a result of the observation of celestial bodies. **(B) (navigation aid terms)** A point on the earth defined in terms of astronomical latitude and longitude. (AES/GCS) 172-1983

astronomical unit (communication satellite) Abbreviated AU: the mean distance between the centers of the sun and the earth, 149.6 x 10^6 kilometers, 92.98 × 10^6, miles or 80.78 × 10^6 nautical miles. (COM) [19]

astronomical unit of distance The length of the radius of the unperturbed circular orbit of a body of negligible mass moving around the sun with a sidereal angular velocity of 0.017 202 098 950 radian per day of 86 400 ephemeris seconds. In the system of astronomical constants of the International Astronomical Union, the value adopted for it is 1 AU = 149 600 × 10^6 m. (QUL) 268-1982s

astrotracker (navigation aid terms) An automatic sextant that has the ability to sight on and track selected stars throughout the day and night, providing heading and position data. The tracker may be optical or radiometric. *Synonym:* star tracker. (AES/GCS) 172-1983w

ASW *See:* antisubmarine warfare radar.

***A* switchboard (telephone switching systems)** A telecommunications switchboard in a local central office, used primarily to extend calls received from local stations. (COM) 312-1977w

asymmetrical cell A cell in which the impedance to the flow of current in one direction is greater than in the other direction. (EEC/PE) [119]

asymmetric multiprocessor A multiprocessor in which the processors are not assigned equal tasks. *Note:* Typically one processor is in charge of assigning tasks to processors and controlling I/O for them all. *Contrast:* symmetric multiprocessor. (C) 610.10-1994w

asymmetric terminal voltage Terminal voltage measured with a delta network between the midpoint of the resistors across the mains lead and ground. *See also:* electromagnetic compatibility. (EMC/INT) [53], [70]

asymptomatic stability (of a solution $\phi(x(t_0);t)$) The solution is (1) Lyapunov stable, (2) such that

$$\lim_{t\to\infty} \|\Delta\phi\| = 0$$

where $\Delta\phi$ is a change in the solution due to an initial state perturbation. See "stability" for explanation of symbols. *Note:* The solution x = 0 of the system x = ax is asymptotically stable for a < 0, but not for a = 0. In this case

$$\varphi(x(t_0);t) = x(t0) \exp(-a(t - t_0))$$

In some cases the rate of convergence to zero depends on both the initial state $x(t0)$ and the initial time t_0. See **stability, equiasymptotic** for stability concepts where this rate of convergence is independent of either $\mathbf{x}(t_0)$ or t_0. *See also:* stability; control system. (CS/IM) [120]

async-cancel-safe function A function that may be safely invoked by an application while the asynchronous form of cancellation is enabled. No function in this standard is async-cancel safe unless explicitly described as such. *Note:* Section 18 for further clarifications of the meaning of this term. (C/PA) 9945-1-1996

asynchronous (1) A transmission process such that between any two significant instants in the same group, there is always an integral number of unit intervals. Between two significant instants located in different groups, there is not always an integral number of unit intervals. *Synonym:* nonsynchronous. (COM/TA) 1007-1991r
(2) Protocol operation in which more than one exchange between a given pair of entities can be handled simultaneously. (LM/C) 15802-2-1995
(3) Describes an activity initiated by a function that is not necessarily complete when the function returns. (C/MM) 855-1990

asynchronous circuit A logic circuit in which the timing of the result is not related to a clock. *Contrast:* synchronous circuit. *See also:* double-rail logic. (C) 610.10-1994w

Asynchronous Client Port An instance of the class `IEEE1451_AsynchronousClientPort` or of a subclass thereof. (IM/ST) 1451.1-1999

asynchronous communication In the IEEE 1451.1 client-server model, refers to a communication in which the client does not block. State is maintained to allow the client to determine whether the return has been received from the server and to permit the client to retrieve the return. (IM/ST) 1451.1-1999

asynchronous completion (1) Completion of an asynchronous directory operation. An asynchronous directory operation is complete when a corresponding synchronous directory op-

eration would complete and any associated status fields have been updated.
(C/PA) 1328.2-1993w, 1326.2-1993w, 1327.2-1993w, 1224.2-1993w

(2) The state of an asynchronous read or write operation when a corresponding synchronous read or write would have completed and any associated status attributes have been updated.
(C) 1003.5-1999

asynchronous computer (1) A computer in which each event or the performance of each operation starts as a result of a signal generated by the completion of the previous event or operation, or by the availability of the parts of the computer required for the next event or operation. (C) [20], [85]
(2) A computer in which each event or operation is performed upon receipt of a signal generated by the completion of a previous event or operation, or upon availability of the system resources required by the event or operation. *Contrast:* synchronous computer. (C) 610.10-1994w

asynchronous errored second A one-second interval during which one or more errors are received, which is measured by detecting errors within seconds defined by a clock that is independent of the error occurrence. (COM/TA) 1007-1991r

asynchronous events Events that occur independently of the execution of the application. (C) 1003.5-1999

asynchronous impedance (rotating machinery) The quotient of the voltage, assumed to be sinusoidal and balanced, supplied to a rotating machine out of synchronism, and the same frequency component of the current. *Note:* The value of this impedance depends on the slip. *See also:* asynchronous machine. (PE) [9]

asynchronous I/O completion (1) For an asynchronous read or write operation, when a corresponding synchronous read or write would have completed and when any associated status fields have been updated. (C/PA) 9945-1-1996
(2) An asynchronous read or write operation is complete when a corresponding synchronous read or write would have completed and any associated status attributes have been updated. (C/PA) 1003.5b-1995

asynchronous I/O operation (1) An I/O operation that does not of itself cause the thread requesting the I/O to be blocked from further use of the processor. This implies that the thread and the I/O operation may be running concurrently.
(C/PA) 9945-1-1996
(2) An I/O operation that does not of itself cause the task requesting the I/O to be blocked. This implies that the requesting task and the I/O operation may be running concurrently. (C/PA) 1003.5b-1995

asynchronously generated signal (1) A signal that is not attributable to a specific thread. *Note:* Examples are: signals sent via *kill*( ), signals sent from the keyboard, and signals delivered to process groups. Being asynchronous is a property of how the signal was generated and not a property of the signal number. All signals may be generated asynchronously.
(C/PA) 9945-1-1996
(2) An occurrence of a signal that is generated by some mechanism external to the task receiving the signal; for example, via a call to `POSIX_Signals.Send_Signal` by another process or via the keyboard. Being asynchronous is a property of how an occurrence of the signal was generated and not a property of the signal. All signals may be generated asynchronously. *Note:* Only asynchronously generated signal occurrences are visible to an Ada application as signals. All signal occurrences that are not generated asynchronously are translated into exceptions. (C) 1003.5-1999

asynchronous machine (1) (rotating machinery) An ac machine in which the rotor does not turn at synchronous speed.
(PE) [9]
(2) A machine in which the speed of operation is not proportional to the frequency of the system to which it is connected.
(IA/MT) 45-1998

asynchronous operation (1) (rotating machinery) Operation of a machine where the speed of the rotor is other than synchronous speed. *See also:* asynchronous machine. (PE) [9]
(2) An operation that occurs without a regular or predictable time relationship to a specified event; for example, an interrupt. (C) 610.10-1994w
(3) An operation that does not of itself cause the process requesting the operation to be blocked from further use of the CPU. This implies that the process and the operation are running concurrently.
(C/PA) 1328.2-1993w, 1326.2-1993w, 1224.2-1993w, 1327.2-1993w
(4) An I/O operation that does not of itself cause the task requesting the I/O to be blocked. An asynchronous I/O operation and the requesting task may be running concurrently.
(C) 1003.5-1999

asynchronous packet (1) A primary packet that contains the bus_ID of the destination in the first quadlet. It is sent as the request subaction and/or response subaction of a transaction.
(C/MM) 1394-1995
(2) A primary packet transmitted in accordance with asynchronous arbitration rules (outside of the isochronous period).
(C/MM) 1394a-2000

asynchronous reactance (rotating machinery) The quotient of the reactive component of the average voltage at rated frequency, assumed to be sinusoidal and balanced, applied to the primary winding of a machine rotating out of synchronism, and the average current component at the same frequency.
(PE) [9]

asynchronous receiver/transmitter *See:* universal asynchronous receiver/transmitter.

asynchronous resistance (rotating machinery) The quotient of the active component of the average voltage at rated frequency assumed to be sinusoidal and balanced, applied to the primary winding of a machine rotating out of synchronism, and the average current component at the same frequency.
(PE) [9]

asynchronous transfer mode A LAN WAN communications architecture that switches or relays small fixed length (53 octets) packets called cells. *Note:* Each cell has a 5 octet header and 48 octets of data. (C) 610.7-1995

asynchronous transmission (1) A transmission in which each information character, word, or block is individually synchronized, and the transmission is controlled by start and stop bits at the beginning and end of each character. *Synonym:* start-stop transmission. *Contrast:* synchronous transmission.
(C) 610.7-1995
(2) *See also:* nonsynchronous transmission.

async-signal-safe function A function that may be invoked, without restriction, from signal-catching functions. No function in this standard is async-signal safe unless explicitly described as such. (C/PA) 9945-1-1996

ATC *See:* arbitration timing control.

ATE *See:* automatic test equipment.

A32 module A module whose address space is limited to a 32-bit width. (C/BA) 14536-1995

ATLAS *See:* Abbreviated Test Language for All Systems.

ATLAS compiler Software that converts ATLAS statements into executable machine code which may involve more than one operation. (SCC20) 771-1998

ATLAS translator Software that converts ATLAS statements into another language. This language may be a computer language or a restricted and structured version of a natural language. (SCC20) 771-1998

ATLAS vocabulary The range of words and symbols used in standard ATLAS. (SCC20) 771-1998

ATM *See:* asynchronous transfer mode; automated teller machine.

atmospheric absorption (1) (general) The loss of energy in transmission of radio waves, due to dissipation in the atmosphere. *See also:* radiation. (EEC/PE) [119]
(2) (communication satellite) Absorption, by the atmosphere, of electromagnetic energy traversing it.
(COM) [25]

atmospheric condition monitor (power system device function numbers) A device that functions upon the occurrence of an abnormal atmospheric condition, such as damaging fumes, explosive mixtures, smoke, or fire. (SUB/PE) C37.2-1979s

atmospheric correction factor A factor applied to account for the difference between the atmospheric conditions in service and the standard atmospheric conditions. (In terms of this standard, it applies to external insulation only). (PE/C) 1313.1-1996

atmospheric duct *See:* atmospheric radio duct.

atmospheric noise (communication satellite) Noise radiated by the atmosphere into a space communications receiver antenna. (COM) [25]

atmospheric paths (atmospheric correction factors to dielectric tests) Paths entirely through atmospheric air, such as along the porcelain surface of an outdoor bushing. 579-1975w

atmospheric radio duct A layer in the atmosphere within which radio waves propagate with low attenuation. (AP/PROP) 211-1997

atmospheric radio noise (control of system electromagnetic compatibility) Noise having its source in natural atmospheric phenomena. *See also:* electromagnetic compatibility. (EMC) [53], C63.12-1987

atmospheric radio wave A radio wave that is propagated by reflection in the atmosphere. *Note:* It may include either the ionospheric wave or the tropospheric wave, or both. *See also:* radiation. (EEC/PE) [119]

atmospherics Transient bursts of electromagnetic radiation arising from natural electrical disturbances in the lower atmosphere. *Notes:* 1. In the past, the term *static* was used to include atmospherics and other radio noise. The term *sferics* is in current use. 2. Below 1 Hz, noise is primarily of geomagnetic origin; above 1 Hz it is due to lightning. (AP/PROP) 211-1997

atmospheric transmissivity The ratio of the directly transmitted flux incident on a surface after passing through unit thickness of the atmosphere to the flux which would be incident on the same surface if the flux had passed through a vacuum. (ED) [127]

ATOLL *See:* Acceptance Test or Launch Language.

atomic A term that describes the constraints placed on accesses of a data format; the read, write, and lock accesses are performed indivisibly. (C/MM) 1596.5-1993

atomic condition The basic qualification condition in a query, consisting of the name of a data item, a logical operation, and a value; for example, LASTNAME = 'Jones'. *See also:* item condition; record condition. (C) 610.5-1990w

atomic data element A data element that cannot be broken into constituent data elements. *Contrast:* composite data element. (C) 610.5-1990w

atomic transaction(s) **(1)** Transactions that are indivisible with respect to other transactions. Other concurrent transactions will either see the state of memory as it was before the atomic transaction or after, but not a mixture. For example, a write4 transaction is atomic if no observer of the same address can see a mixture of old and new byte values, and two write4's for the same address are atomic with respect to each other if one overwrites the other without mixing bytes. (C/MM) 1212.1-1993

(2) A transaction is atomic (or indivisible) if the only system states on which other transactions can act (which are visible to other transactions) are states prior to the commencement of the transaction and states after the completion of the transaction. (C/BA) 896.3-1993w

atomic type A data type, each of whose members consists of a single, nondecomposable data item. *Synonym:* primitive type. *Contrast:* composite type. (C) 610.12-1990

ATR switch *See:* anti-transmit-receive switch.

ATR tube *See:* anti-transmit-receive tube.

attach The term *attach* is used consistently to express the relationship between an end station, a Bridge, or a Bridge Port and the physical or logical communications elements by which it is interconnected with the other Bridges in a Bridged Local Area Network. *Note:* In particular: an end station, a Local Bridge, or a LAN Port is said to attach to a LAN; a Remote Bridge is said to attach to a LAN, a Group, or a Cluster, a Virtual Port is said to attach to a Group, a Subgroup, or a Cluster. Where other uses are encountered, they are to be interpreted in a similar sense. (C/LM) 802.1G-1996

attached A worker is connected to an anchorage when utilizing a fall protection system to prevent or arrest a fall. (T&D/PE) 1307-1996

attached part A visible part of a simulated entity that may not move relative to the entity, but that may or may not be present. For example, a bomb on an aircraft wing station. (DIS/C) 1278.1-1995

attached peer PHY A peer cable PHY at the other end of a particular physical connection from the local PHY. (C/MM) 1394-1995

attached slave A slave that in the previous primary address cycle recognized its address and address type and as a result will participate in the ensuing data cycles. (NID) 960-1993

attachment (electric and electronics parts and equipment) A basic part, subassembly, or assembly designed for use in conjunction with another assembly, unit, or set, contributing to the effectiveness thereof by extending or varying the basic function of the assembly, unit, or set. A typical example is an ultra-high-frequency (UHF) converter for a very-high-frequency (VHF) receiver. (GSD) 200-1975w

attachment plug A device that, by insertion in a receptacle, establishes connection between the conductors of an attached cord and the conductors connected permanently to the receptacle. (IA/MT) 45-1998

attachments Accessories to be attached to switchgear apparatus, as distinguished from auxiliaries. (SWG/PE) C37.100-1992

attachment unit interface (AUI) **(1)** **(broadband local area networks)** The cable, connectors, and transmission reception circuitry used to interconnect the physical layer signaling and MAU. (LM/C) 802.7-1989r

(2) In a local area network, the interface between the medium attachment unit (MAU) and the data terminal equipment (DTE) within a data station. *Note:* The AUI carries encoded signals and provides for duplex data transmission. (C) 610.7-1995

(3) In 10 Mb/s CSMA/CD, the interface between the medium attachment unit (MAU) and the data terminal equipment (DTE) within a data station. Note that the AUI carries encoded signals and provides for duplex data transmission. (LM/C) 802.3-1998

attachment unit interface cable A cable that connects a workstation to a transceiver. *Note:* This term is contextually specific to IEEE Std 802.3, clause 7. *See also:* drop cable; coaxial cable; transceiver cable. (C) 610.7-1995

attack time **(1)** **(electroacoustics)** The interval required, after a sudden increase in input signal amplitude to a system or transducer, to attain a stated percentage (usually 63%) of the ultimate change in amplification or attenuation due to this increase. (SP) 151-1965w

(2) *See also:* build-up time. (COM/TA) 1329-1999

attempted domain The MD to which a tracing domain attempts but fails to transfer a communique or report. (C/PA) 1224.1-1993w

attempted MTA The MTA to which a tracing MTA attempts but fails to transfer a communique or report. (C/PA) 1224.1-1993w

attempts Occur when the system recognizes an off-hook that is long enough to be a valid originating or incoming call request. The system usually returns a wink on trunks with per-trunk-signaling and returns dial tones to lines. (COM/TA) 973-1990w

attendant (telephone switching systems) A private branch exchange or centrex operator. (COM) 312-1977w

attended operation A central office that normally has maintenance staff on duty. (C) 610.7-1995

attention cycle A single cycle in which a master indicates the start and acknowledge in the same cycle. (C/MM) 1196-1987w

attention key (ATTN) A control key that causes an attention or input-output interrupt signal to be generated, thereby causing the processing unit to cease processing. *Synonyms:* break key; program attention key. *See also:* escape key. (C) 610.10-1994w

attenuating pad *See:* pad.

attenuation (1) (data transmission) A general term used to denote a decrease in signal magnitude in transmission from one point to another. Attenuation may be expressed as a scalar ratio of the input magnitude to the output magnitude or in decibels. (PE) 599-1985w
(2) (fiber optics) In an optical waveguide, the diminution of average optical power. *Note:* In optical waveguides, attenuation results from absorption, scattering, and other radiation. Attenuation is generally expressed in decibels (dB). However, attenuation is often used as a synonym for attenuation coefficient, expressed as dB/km. This assumes the attenuation coefficient is invariant with length. *See also:* transmission loss; extrinsic joint loss; attenuation coefficient; Rayleigh scattering; equilibrium mode distribution; leaky mode; coupling loss; material scattering; waveguide scattering; spectral window; macrobend loss; differential mode attenuation; microbend loss. (Std100) 812-1984w
(3) (laser maser) The decrease in the radiant flux as it passes through an absorbing or scattering medium. (LEO) 586-1980w
(4) (quantity associated with a traveling waveguide or transmission-line wave) (waveguide) The decrease with distance in the direction of propagation. *Note:* Attenuation of power is usually measured in terms of decibels or decibels per unit length. *See also:* loss. (MTT) 147-1979w, 146-1980w
(5) (A) A decrease in signal magnitude between two points or between two frequencies. **(B)** The reciprocal of gain. *Note:* It may be expressed as a scalar ratio or in decibels as 20 times the log of that ratio. A decrease with time is usually called "damping" or "subsidence." *See also:* subsidence ratio. (CS/PE/EDPG) [3]
(6) (broadband local area networks) The quantity of reduction of a defined parameter, expressed in dB. (LM/C) 802.7-1989r
(7) (germanium spectrometers) The net loss at the detector of primary photons of a given energy resulting from their interaction with matter either due to the occurrence of scattering or absorption in the sample or in material between the sample and the detector crystal. (NI) N42.14-1991
(8) A loss or decrease of signal power in a transmission, usually measured in decibels. (C) 610.7-1995
(9) (A) A decrease in the magnitude of current, voltage, or power of a transmitted signal due to loss through a communication medium. *See also:* equalization. **(B)** A decrease in the magnitude of current, voltage, or power of a signal during transmission from one point to another. (C) 610.10-1994
(10) A general term used to denote a decrease in signal magnitude in transmission from one point to another. (PE/EDPG/PSC) 1050-1996, 789-1988w
(11) (of an electromagnetic wave) The decrease in magnitude of a field with distance or with changes in the path in excess of the decrease due to a geometrical spreading factor. *See also:* spreading factor. (AP/PROP) 211-1997

attenuation band (uniconductor waveguide) Rejection band. *See also:* waveguide. (MTT) 147-1979w, 146-1980w

attenuation coefficient (1) (fiber optics) The rate of diminution of average optical power with respect to distance along the waveguide. Defined by the equation

$$P(z) = P(0)10^{\,(\alpha z/10)}$$

where $P(z)$ is the power at distance z along the guide and $P(0)$ is the power at $z = 0$; α is the attenuation coefficient in dB/km if z is in km. From this equation,

$$\alpha z = -10 \log_{10}[P(z)/P(0)].$$

This assumes that α is independent of z; if otherwise, the definition must be given in terms of incremental attenuation as:

$$P(z) = P(0)10^{-\left[\int_0^z \frac{\alpha(z)dz}{10}\right]}$$

or, equivalently,

$$\alpha(z) = -10d/dz \log_{10}[P(z)/P(0)]$$

See also: axial propagation constant; attenuation; attenuation constant. (Std100) 812-1984w
(2) *See also:* attenuation constant. (AP/PROP) 211-1997

attenuation constant (α) (1) (waveguide) The rate of decrease in amplitude of a field component (or of voltage or current) of a traveling wave in a uniform transmission line at a given frequency in the direction of propagation as a function of distance; the real part of the propagation constant. (MTT) 146-1980w
(2) The magnitude of the attenuation vector. *Synonym:* attenuation coefficient. *See also:* propagation vector. (AP/PROP) 211-1997

attenuation, current *See:* current attenuation.

attenuation distortion (1) (A) (frequency distortion) (data transmission) Either a departure in a circuit or system from uniform amplification or attenuation over the frequency range required for transmission, or the effect of such departure on a transmitted signal. **(B) (frequency response)** The relative attenuation at any frequency with respect to that at a reference frequency. (PE/COM/TA) 599-1985, 1007-1991
(2) The change in attenuation at any frequency with respect to that at a reference frequency. (COM/TA) 743-1995

attenuation equalizer (data transmission) A corrective network that is designed to make the absolute value of the transfer impedance, with respect to two chosen pairs of terminals, substantially constant for all frequencies within a desired range. (PE) 599-1985w

attenuation-limited operation (fiber optics) The condition prevailing when the received signal amplitude (rather than distortion) limits performance. *See also:* distortion-limited operation; bandwidth-limited operation. (Std100) 812-1984w

attenuation range (a_H) The difference in level in dB between maximum inserted switched loss and the full removal of that switched loss. If the send to receive is not the same as send to transmit, then the larger of the two is the attenuation range. (COM/TA) 1329-1999

attenuation ratio (radio-wave propagation) The magnitude of the propagation ratio. (AP) 211-1977s

attenuation vector ($\vec{\alpha}$) The imaginary part of the propagation vector, $\vec{k}$. The attenuation vector points in the direction of maximum decrease in magnitude. *See also:* propagation vector. (AP/PROP) 211-1997

attenuation vector in physical media The real part of the propagation vector. (AP/ANT) 145-1983s

attenuation, voltage *See:* voltage attenuation.

attenuator A two-port device that provides a fixed amount of signal loss over a wide range of frequencies (attenuation). These devices have identical attenuation in either direction. (LM/C) 802.7-1989r

attenuator tube (electron tube) A gas-filled radio-frequency switching tube in which a gas discharge, initiated and regulated independently of radio-frequency power, is used to control this power by reflection or absorption. *See also:* gas tube. (ED) 161-1971w, [45]

attenuator, waveguide *See:* waveguide attenuator.

attitude **(1) (navigation aid terms)** The position of a body as determined by the inclination of the axes to some frame of reference. (AES/GCS) 172-1983w
(2) (communication satellite) Orientation of a satellite vehicle with respect to a reference coordinate system. Deviations of the satellite axes from the reference system are called roll, pitch, and yaw. The reference system is generally an orbital reference system with the x-axis (roll axis) in the orbital plane in direction of the satellite motion, the y-axis (pitch axis) normal to the orbital plane, and the z-axis (yaw axis) in the orbital plane in direction of the center of the earth. (COM) [19]

attitude control (navigation aid terms) Devices or system that automatically regulates and corrects attitude. (AES/GCS) 172-1983w

attitude-effect error (navigation aid terms) A manifestation of polarization error; an error in indicated bearing that is dependent upon the attitude of the vehicle with respect to the direction of signal propagation. *See also:* heading-effect error. (AES/GCS) 172-1983w

attitude gyro-electric indicator An electrically driven device that provides a visual indication of an aircraft's roll and pitch attitude with respect to the earth. *Note:* It is used in highly maneuverable aircraft and differs from the gyro-horizon electric indicator in that the gyro is not limited by stops and has complete freedom about the roll and pitch axes. (EEC/PE) [119]

attitude stabilized satellite (communication satellite) A satellite with at least one axis maintained in a specified direction, namely, toward the center of the earth, the sun, or a specified point in space. (COM) [19]

attitude storage (gyros) The transient deviation of the output of a rate-integrating gyro from that of an ideal integrator when the gyro is subjected to an input rate. It is a function of the gyro characteristic time. *See also:* float storage; torque-command storage.

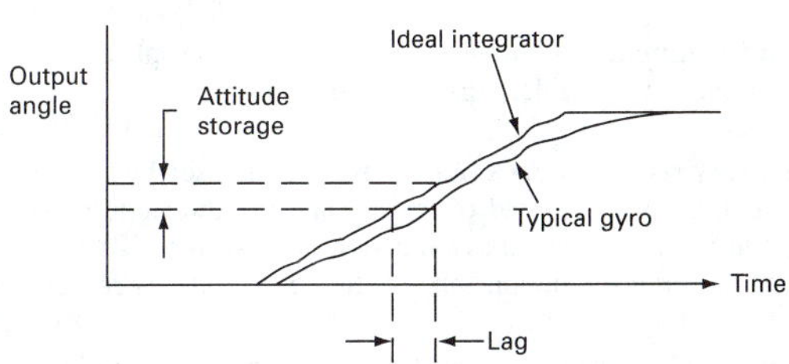

attitude storage
(AES/GYAC) 528-1994

ATTN *See:* attention key.

attribute **(1) (data management)** In a relation, a named characteristic, property, or description of an entity. *Note:* Also known as data element, data field, data item or column in a table. *Synonym:* data field. *See also:* column. (C) 610.5-1990w
(2) (computer graphics) A characteristic of an item; for example, the item's color, size, or type. *Note:* In computer graphics, an attribute may be represented by the text or numeric data associated with the item. (C) 610.6-1991w, 610.5-1990w, 610.12-1990
(3) A property of a managed object or a property of an association among OSI entities. An attribute has an associated value, which may have a simple or complex structure. (LM/C) 802.10-1992
(4) A predefined characteristic that provides a property of a class. Properties are inherited from a class to its subclasses. (C/SE) 1420.1-1995
(5) A single piece of information stored in the APS environment. (C/PA) 1351-1994w
(6) A component of an object, possessing a name and one or more values. (PA/C) 1387.2-1995
(7) A property or fact about the entities in an entity set. (PE/EDPG) 1150-1991w
(8) A component of an object, comprised of an integer denoting the type of the attribute and an ordered sequence of one or more attribute values, each accompanied by an integer denoting the syntax of the value. (C/PA) 1328-1993w, 1238.1-1994w, 1224.1-1993w, 1327-1993w, 1224-1993w
(9) Used alone, means directory attribute. (C/PA) 1224.2-1993w, 1327.2-1993w, 1326.2-1993w, 1328.2-1993w
(10) A measurable physical or abstract property of an entity. (C/SE) 1061-1998
(11) (A) A kind of property associated with a set of real or abstract things (people, objects, places, events, ideas, combinations of things, etc.) that is some characteristic of interest. An attribute expresses some characteristic that is generally common to the instances of a class. **(B)** An attribute is a function from the instances of a class to the instances of the value class of the attribute. **(C)** The name of the attribute is the name of the role that the value class plays in describing the class, which may simply be the name of the value class (as long as using the value class name does not cause ambiguity). (C/SE) 1320.2-1998
(12) A documenting characteristic of an entity. (SCC32) 1489-1999
(13) an inherent characteristic of an item ascribing an inherent quality. (SCC32) 1488-2000

attribute data element A data element within a record that represents a property, feature, or characteristic of the subject of that record; for example, the data element "date of birth" in a record containing "name," "address," and "date of birth" of a person. *Contrast:* primary data element. (C) 610.5-1990w

attribute name A role name for the value class of the attribute. (C/SE) 1320.2-1998

attributes Measurable characteristics of a primitive. (C/SE) 1045-1992

attribute syntax A definition of the set of values that a directory attribute may assume. It includes the datatype, in ASN.1, and, usually, one or more matching rules by which values may be compared. *Synonym:* directory syntax. (C/PA) 1328.2-1993w, 1327.2-1993w, 1224.2-1993w, 1326.2-1993w

attribute type Used alone, means directory attribute type. *See also:* OM attribute; OM attribute type. (C/PA) 1328.2-1993w, 1224.2-1993w, 1326.2-1993w, 1327.2-1993w

attribute value **(1)** Used alone, means directory attribute value. *See also:* OM attribute value. (C/PA) 1328.2-1993w, 1327.2-1993w, 1326.2-1993w, 1224.2-1993w
(2) (C/PA) 1224.2-1993w

attribute value assertion **(1)** A proposition, which may be true, false, or undefined, according to the specified matching rules for the type, concerning the presence in an entry of an attribute value (or a distinguished value) of a particular type. *Note:* An attribute value assertion consists of an attribute type and a single value. Loosely, it is true if one of the values of the given attribute in the entry matches the given value. (C/PA) 1327.2-1993w, 1328.2-1993w, 1224.2-1993w
(2) A proposition, which may be true, false, or undefined, about the values (or perhaps only the distinguished values) of an entry. *Note:* An attribute value assertion consists of an attribute type and a single value. Loosely, it is true if one of the values of the given attribute in the entry matches the given value. (C/PA) 1326.2-1993w

attribute value syntax *See:* OM syntax.

ATV *See:* all terrain vehicle.

AU *See:* access unit.

audible busy signal An audible signal connected to the calling line to indicate that the called line is in use. *Synonym:* busy tone. (EEC/PE) [119]

audible cab indicator A device (usually an air whistle, bell, or buzzer) located in the cab of a vehicle equipped with cab signals or continuous train control designed to sound when the cab signal changes to a more restrictive indication. (EEC/PE) [119]

audible noise Any undesired sound. *Synonym:* acoustic noise. (T&D/PE) 539-1990

audible signal device (protective signaling) A general term for bells, buzzers, horns, whistles, sirens, or other devices that produce audible signals. *See also:* protective signaling. (EEC/PE) [119]

audience Persons who are expected to need a given software user document. (C/SE) 1063-1987r

audio (data transmission) Pertaining to frequencies corresponding to a normally audible sound wave. *Note:* These frequencies range roughly from 15 Hz to 20 000 Hz. (PE) 599-1985w

audio frequency (1) (general) Any frequency corresponding to a normally audible sound wave. *Notes:* 1. Audio frequencies range roughly from 15 Hz to 20 000 Hz. 2. This term is frequently shortened to audio and used as a modifier to indicate a device or system intended to operate at audio frequencies; for example, an audio amplifier. (AP/CHM/SP/ANT) 145-1983s, [51], [32]
(2) (interference terminology) Components of noise having frequencies in the audio range. *See also:* signal. (IE) [43]
(3) (overhead power lines) Any frequency corresponding to a normally audible sound wave. This usually covers the range from 20 Hz to 20 kHz. (T&D/PE) 539-1990

audio-frequency distortion The form of wave distortion in which the relative magnitudes of the different frequency components of the wave are changed on either a phase or amplitude basis. (VT) [37]

audio-frequency harmonic distortion The generation in a system of integral multiples of a single audio-frequency input signal. *See also:* modulation. (AP/ANT) 145-1983s

audio-frequency noise Any unwanted disturbance in the audio-frequency range. (SP) 151-1965w

audio-frequency oscillator (audio oscillator) A nonrotating device for producing an audio-frequency sinusoidal electric wave, whose frequency is determined by the characteristics of the device. *See also:* oscillatory circuit. (SP) 151-1965w

audio-frequency peak limiter A circuit used in an audio-frequency system to cut off peaks that exceed a predetermined value. (AP/ANT) 145-1983s

audio-frequency response (receiver performance) The measure of the relative departure of all audio-frequency signal levels within a specified bandwidth, from a specified reference frequency signal power level. (VT) 184-1969w

audio-frequency spectrum (audio spectrum) The continuous range of frequencies extending from the lowest to the highest audio frequency. (SP) 151-1965w

audio-frequency transformer A transformer for use with audio-frequency currents. (CHM) [51]

audiogram (threshold audiogram) A graph showing hearing level as a function of frequency. (SP) [32]

audio input power (transmitter performance) The input power level to the modulator, expressed in decibels referred to a one milliwatt power level. *See also:* audio-frequency distortion. (VT) [37]

audio input signal (transmitter performance) That composite input to the transmitter modulator that consists of frequency components normally audible to the human ear. *See also:* audio-frequency distortion. (VT) [37]

audiometer An instrument for measuring hearing level. *See also:* instrument. (SP) [32]

audio oscillator *See:* audio-frequency oscillator.

audio output power (receiver) The audio-frequency power dissipated in a load across the output terminals. (VT) [37]

audio power output (receiver performance) The measure of the audio-frequency energy dissipated in a specified output load. (VT) 184-1969w

audio response device An output device capable of generating spoken language. (C) 610.10-1994w

audio-tone channel *See:* voice-frequency carrier telegraph.

audit (1) (software) An independent examination of a work product or set of work products to assess compliance with specifications, standards, contractual agreements, or other criteria. *See also:* physical configuration audit; functional configuration audit. (C) 610.12-1990
(2) A planned and documented activity performed to determine by investigation, examination, or evaluation of objective evidence, the adequacy of and compliance with established procedures, instructions, drawings, and other applicable documents, and the effectiveness of implementation. An audit should not be confused with surveillance or inspection activities performed for the sole purpose of process control or product acceptance. (NI/PE/NP) N42.23-1995, [124]
(3) An independent examination of a software product, software process, or set of software processes to assess compliance with specifications, standards, contractual agreements, or other criteria. (C/SE) 1028-1997

auditable data (1) (valve actuators) Information which is documented and organized in a readily understandable and traceable manner that permits independent verification of inferences or conclusions based on the information. *Note:* Examples of information include product catalog information, dimensional drawings, bills of material, engineering specifications, installation and calibration instructions and manuals, maintenance manuals, test reports, and analyses. (PE/NP) 382-1985, 627-1980r
(2) (nuclear power generating station) Technical information that is documented and organized in a readily understandable and traceable manner that permits independent auditing of the inferences or conclusions based on the information. (PE/NP) 323-1974s

audit command language (ACL) A high-order programming language used widely in audit applications. (C) 610.13-1993w

auditory sensation area (A) The region enclosed by the curves defining the threshold of feeling and the threshold of audibility, each expressed as a function of frequency. **(B)** The part of the brain (temporal lobe of the cortex) that is responsive to auditory stimuli. (SP) [32]

audit trail A manual or computerized record that can be used to trace the transactions affecting the contents of a record or a file. (C) 610.2-1987

AUDYSIM *See:* Autodynamics simulation language.

augend A number to which another number (the addend) is added to produce a result (the sum). (C) 1084-1986w

augment (information processing) An independent variable, for example, in looking up a quantity in a table, the number or any of the numbers, that identifies the location of the desired value. (C) [20]

AUI (Attachment Unit Interface) *See:* attachment unit interface.

AUI cable *See:* attachment unit interface cable.

***A* unit** A motive power unit so designed that it may be used as the leading unit of a locomotive, with adequate visibility in a forward direction, and that includes a cab and equipment for full control and observation of the propulsion power and brake applications for the locomotive and train. *See also:* electric locomotive. (EEC/PE) [119]

aural harmonic A harmonic generated in the auditory mechanism. (SP) [32]

aural radio range *See:* A-N radio range.

aural transmitter The radio equipment used for the transmission of the aural (sound) signals from a television broadcast station. *See also:* television. (AP/BT/ANT) 145-1983s, 182A-1964w

aurora Collective name of optical, electrical, and magnetic phenomena, generally at high latitudes, resulting from direct excitation of the upper atmosphere by energetic particles. (AP/PROP) 211-1997

auroral (power fault effects) Electrical voltages and currents on or around the earth, due to emission of particle energy from the sun. *See also:* susceptibility. (PE/PSC) 367-1979

auroral absorption The increased attenuation of radio waves propagating through the D and E regions of the ionosphere when additional ionization is produced by precipitating charged particles usually associated with the visual aurora. (AP/PROP) 211-1997

auroral attenuation (radio-wave propagation) The attenuation of radio waves propagating through the D and E regions of the ionosphere when additional ionization is produced by an aurora. (AP/PROP) 211-1990s

auroral effects Electrical voltages and currents on the earth due to emission of particle energy from the sun. (PE/PSC) 367-1996

auroral hiss Audio-frequency electromagnetic noise associated with auroras. (AP/PROP) 211-1997

auroral oval *See:* auroral zone.

auroral zone An annular region situated between approximately 60° and 70° geomagnetic latitude, north or south, in which auroras are frequently present. *Synonym:* auroral oval. (AP/PROP) 211-1997

austenitic The face-centered cubic crystal structure of ferrous metals. (IA) [59]

authentication (1) The process of validating a user or process to verify that the user or process is not a counterfeit. (C/PA) 1003.2d-1994
(2) The service used to establish the identity of one station as a member of the set of stations authorized to associate with another station. (C/LM) 8802-11-1999
(3) *See also:* data origin authentication. (LM/C) 802.10-1998

author and keyword in context index A variation of a keyword in context (KWIC) index in which author and keyword entries are combined and presented in a KWIC format. *Contrast:* word and author index. (C) 610.2-1987

authoring language (1) A high level programming language used to develop courseware for computer-assisted instruction. *See also:* authoring system. (C) 610.2-1987, 610.12-1990
(2) An application-oriented programming language used to develop courseware for computer-assisted instruction. (C) 610.13-1993w

authoring system A programming system that incorporates an authoring language. (C) 610.2-1987, 610.12-1990

authorities (monitoring radioactivity in effluents) Any governmental agencies or recognized scientific bodies that by their charter define regulations or standards dealing with radiation protection. (NI) N42.18-1980r

authority A geographical or political division created specifically for the purpose of providing transportation service. (VT) 1476-2000, 1475-1999

authority having jurisdiction (1) The organization, office, or individual that has the responsibility and authority for approving equipment, installations, or procedures. (IA) 515-1997
(2) That entity that defines the contractual (including specification) requirements for the procurement. (VT/RT) 1473-1999, 1475-1999, 1474.1-1999, 1476-2000

authorization The process of verifying that a user or process has permission to use a resource in the manner requested. To assure security, the user or process would also need to be authenticated before granting access. (C/PA) 1003.2d-1994

authorized bandwidth (mobile communication) The frequency band containing those frequencies upon which a total of 99% of the radiated power appears. *See also:* mobile communication system. (VT) [37]

auto alarm A radio receiver that automatically produces an audible alarm when a prescribed radio signal is received. (EEC/PE) [119]

autoanswer A capability of a terminal, modem, computer, or similar device to respond to an incoming call over the switched network, and to establish a data connection with a remote device without operator intervention. *See also:* autodial. (C) 610.7-1995

autochanger *See:* jukebox.

Autocoder An early symbolic programming language developed for programming computers. (C) 610.13-1993w

autocondensation (electrotherapy) A method of applying alternating currents of frequencies exceeding 100 kilohertz to limited areas near the surface of the human body through the use of one very large and one small electrode, the patient becoming part of the capacitor. *See also:* electrotherapy. (EMB) [47]

autoconduction (electrotherapy) A method of applying alternating currents, of frequencies exceeding 100 kHz for therapeutic purposes, by electromagnetic induction, the patient being placed inside a large solenoid. *See also:* electrotherapy. (EMB) [47]

autodial A capability of a terminal, modem, computer, or similar device to place a call over the switched network, and to establish a data connection without operator intervention. *See also:* autoanswer. (C) 610.7-1995

Autodynamics simulation language A simulation language used in dynamic simulation applications. (C) 610.13-1993w

autodyne reception A system of heterodyne reception through the use of a device that is both an oscillator and a detector. (EEC/PE) [119]

autoerection (gyros) The process by which gimbal axis friction causes the spin axis of a free gyro to tend to align with the axis about which the case is rotated. The resulting drift rate is a function of the angular displacement between the spin axis and the rotation axis. (AES/GYAC) 528-1994

auto ignition Ignition without a pilot source. (DEI) 1221-1993w

automated data medium *See:* machine-readable medium.

automated data processing *See:* automatic data processing.

automated design tool (software) A software tool that aids in the synthesis, analysis, modeling, or documentation of a software design. Examples include simulators, analytic aids design representation processors, and documentation generators. *See also:* simulator; design; documentation. (C/SE) 729-1983s

automated dictionary In machine-aided translation, an automated lexicon in which entries are single words. *Contrast:* automated glossary. (C) 610.2-1987

automated glossary In machine-aided translation, an automated lexicon in which entries may consist of multiple words. *Synonym:* terminology bank. *Contrast:* automated dictionary. (C) 610.2-1987

automated guideway transit Any guided transit mode with fully automated operation (i.e. no crew on the train). The term usually refers, however, only to guided modes with small and medium-sized vehicles that operate on exclusive right-of-way. (VT) 1475-1999

automated language processing The application of data processing, word processing, and machine-aided translation to the processing or translation of natural languages. (C) 610.2-1987

automated lexicon A computer-resident table of source language and target language equivalents that serves as the central component in a machine-aided translation system. *See also:* automated dictionary; automated glossary. (C) 610.2-1987

automated library A robotic (electromechanical) library. (C/SS) 1244.1-2000

automated office *See:* electronic office.

automated teller machine (ATM) An unattended terminal-type device that offers simple banking services such as cash withdrawals, transfer of funds between accounts, and account balance inquiry. *Synonym:* customer-bank communication terminal. (C) 610.2-1987

automated test case generator *See:* automated test generator.

automated test data generator *See:* automated test generator.

automated test generator (software) A software tool that accepts as input a computer program and test criteria, generates test input data that meet these criteria, and, sometimes, determines the expected results. *See also:* data; computer program. (C/SE) 729-1983s

automated thesaurus In machine-aided translation, a computer-resident thesaurus used in conjunction with an automated lexicon to handle words with multiple meanings. (C) 610.2-1987

automated verification system (A) (software) A software tool that accepts as input a computer program and a representation of its specification, and produces, possibly with human help, a proof or disproof of the correctness of the program. **(B) (software)** Any software tool that automates part or all of the verification process. (C) 610.12-1990

automated verification tools (software) A class of software tools used to evaluate products of the software development process. These tools aid in the verification of such characteristics as correctness, completeness, consistency, traceability, testability, and adherence to standards. Examples are design analyzers, automated verification systems, static analyzers, dynamic analyzers, and standards enforcers. *See also:* tool; verification; automated verification system; testability; dynamic analyzer; static analyzer; software development process; correctness; design analyzer. (C/SE) 729-1983s

automatic (1) Pertaining to a function, operation, process, or device that, under specified conditions, functions without intervention by a human operator. (C/SUB/PE) 610.2-1987, 610.10-1994w, C37.1-1987s
(2) Self-acting, operating by its own mechanism when actuated by some impersonal influence—as, for example, a change in current strength; not manual; without personal intervention. Remote control that requires personal intervention is not automatic, but manual. (NESC/T&D) C2-1997, C2.2-1960
(3) Self-acting, operating by its own mechanism when actuated by some impersonal influence, as, for example, a change in current strength, pressure, temperature, or mechanical configuration. *See also:* nonautomatic. (NESC/NEC/IA/ICTL/IAC) [60], [86]
(4) Pertaining to a process or device that, under specified conditions, functions without intervention by a human operator. (SWG/PE) C37.100-1992

automatic abstracting In library automation, the automatic selection of words and phrases from a document to produce an abstract. (C) 610.2-1987

automatic acceleration (1) (automatic train control) Acceleration under the control of devices that function automatically to maintain, within relatively close predetermined values or schedules, current passing to the traction motors, the tractive force developed by them, the rate of vehicle acceleration, or similar factors affecting acceleration. *See also:* multiple-unit control; electric drive. (EEC/PE) [119]
(2) Acceleration under the control of devices that function automatically to raise the motor speed. *See also:* multiple-unit control; electric drive. (IA/IAC) [60]

Automatically Programmed Tools (APT) (1) A problem-oriented programming language used for programming numerically controlled machine tools. (C) 610.13-1993w
(2) A programming system using English-like symbolic descriptions of part and tool geometry and tool motion for numerical control. (C) 610.2-1987

automatically regulated (rotating machinery) Applied to a machine that can regulate its own characteristics when associated with other apparatus in a suitable closed-loop circuit. (PE) [9]

automatically reset relay *See:* self-reset relay.

automatic approach control A system that integrates signals, received by localizer and glide path receivers, into the automatic pilot system, and guides the airplane down the localizer and glide path beam intersection. (EEC/PE) [119]

automatic bias nulling A circuit or system technique for setting the mean value of sensor output, averaged over a defined time period, to zero, or to some defined value. (AES/GYAC) 528-1994

automatic block signal system A series of consecutive blocks governed by block signals, cab signals, or both, operated by electric, pneumatic, or other agency actuated by a train or by certain conditions affecting the use of a block. *See also:* block-signal system. (EEC/PE) [119]

automatic cab signal system A system that provides for the automatic operation of cab signals. *See also:* automatic train control. (EEC/PE) [119]

automatic calendar A component of some office automation systems that allows users to store their appointments in a database and to set up meetings by requesting a search for an available meeting time in each of the participants' calendars. (C) 610.2-1987

automatic call distribution (ACD) A service that evenly distributes calls among incoming end user lines. (AMR/SCC31) 1390-1995, 1390.2-1999, 1390.3-1999

automatic call distributor (telephone switching systems) The facility for allotting incoming traffic to idle operators or attendants. (COM) 312-1977w

automatic capacitor control equipment A piece of equipment that provides automatic control for functions related to capacitors, such as their connection to and disconnection from a circuit in response to predetermined conditions such as voltage, load, or time. (SWG/PE) C37.100-1992

automatic carriage (1) A control mechanism for a typewriter or other output device that can automatically control the feeding, spacing, skipping and ejecting of paper and preprinted forms. (C) [20], 610.10-1994w
(2) Pertaining to a function, operation, process, or device that, under specified conditions, functions without intervention by a human operator. (C) 610.10-1994w

automatic chart-line follower (navigation aid terms) A device that automatically derives error signals proportional to the deviation of the position of a vehicle from a predetermined course line drawn on a chart. (AES/GCS) 172-1983w

automatic check A check that is built into a device in order to verify the accuracy of information transmitted, manipulated, or stored by that device. *Synonyms:* built-in check; hardware check. (C) 610.5-1990w, 610.10-1994w

automatic circuit closer (supervisory control, data acquisition, and automatic control) A self-controlled device for automatically interrupting and reclosing an alternating-current circuit, with a predetermined sequence of opening and reclosing followed by resetting, hold-closed, or lockout operation. (SUB/PE) C37.1-1987s

automatic circuit recloser A self-controlled device for automatically interrupting and reclosing an alternating-current circuit, with a predetermined sequence of opening and reclosing followed by resetting, hold-closed, or lockout operation. *Note:* When applicable, it includes an assembly of control elements required to detect overcurrents and control the recloser operation. (SWG/SUB/PE) C37.1-1987s, C37.100-1992

automatic combustion control A method of combustion control that is effected automatically by mechanical or electric devices. (T&D/PE) [10]

automatic component interconnection matrix A hardware system for connecting inputs and outputs of parallel computing components according to a predetermined program. *Note:* This system, which may consist of a matrix of mechanical and/or electronic switches, replaces the manual program patch boards and patch cords on analog computers. *Synonym:*

autopatch. *See also:* problem board. (C) 610.10-1994w, 165-1977w

automatic computer* A computer that can perform a sequence of operations without intervention by a human operator. (C) [20], 610.10-1994w

* Deprecated.

automatic control (1) (excitation systems for synchronous machines) In excitation control system usage, automatic control refers to maintaining synchronous machine terminal voltage without operator action, over the operating range of the synchronous machine within its capabilities. *Note:* Voltage regulation under automatic control may be modified by the action of reactive or active load compensators or by var control elements; or may be constrained by the action of various limiters included in the excitation system. *See also:* control. (EDPG) 421.1-1986r

(2) (analog computer) In an analog computer, a method of computer operation using auxiliary automatic equipment to perform computer-control state selections, switching operations, or component adjustments in accordance with previously selected criteria. Such auxiliary automatic equipment usually consists of programmable digital logic that is part of the analog, a separate digital computer, or both. The case of the digital computer controlling the analog computer is an example of a hybrid computer. (C) 165-1977w

(3) (electrical controls) An arrangement that provides for switching or otherwise controlling, or both, in an automatic sequence and under predetermined conditions, the necessary devices comprising an equipment. *Note:* These devices thereupon maintain the required character of service and provide adequate protection against all usual operating emergencies. (SWG/PE) [56]

(4) (computers) Describes a control system capable of operating without external or human intervention. *See also:* process control; numerical control. (C) 610.2-1987

(5) An arrangement of electrical controls that provides for switching or otherwise controlling or both in an automatic sequence and under predetermined conditions the necessary devices comprising a piece of equipment. These devices thereupon maintain the required character of service and provide adequate protection against all unusual operating emergencies. (SWG/PE/SUB) C37.100-1992, C37.1-1987s

automatic control equipment (1) (station control and data acquisition) Equipment that provides automatic control of power apparatus in response to predetermined conditions. (SUB/PE) C37.1-1979s

(2) Equipment that provides automatic control for a specified type of power circuit or apparatus. (SWG/PE) C37.100-1992

automatic controller (electrical heating applications to melting furnaces and forehearths in the glass industry) (process control) A device that operates automatically to regulate a controlled variable in response to a command and a feedback signal. *Note:* The term originated in process control usage. Feedback elements and final control elements may also be part of the device. *See also:* automatic controller. (IA/PE/PSE/EDPG) 446-1987s, 668-1987w, [3]

automatic control system A control system that operates without human intervention. *See also:* feedback control system. (IM/PE/EDPG) [120], [3]

automatic current limiting (power supplies) An overload protection mechanism that limits the maximum output current to a preset value, and automatically restores the output when the overload is removed. *See also:* short-circuit protection. (AP/ANT) [35]

automatic data processing Data processing performed by a computer system. *Synonyms:* automated data processing; electronic data processing. (C) 610.2-1987

automatic detection and tracking (ADT) In a surveillance radar, the computer-based ADT of targets based on target locations obtained from scan-to-scan. *See also:* track-while-scan. (AES) 686-1997

automatic direct-control telecommunications system (telephone switching systems) A system in which the connections are set directly in response to pulsing from the originating calling device. (COM) 312-1977w

automatic direction finder (navigation aid terms) A direction finder that automatically and continuously provides a measure of the direction of arrival of the received signal. Data are usually displayed visually. (AES/GCS) 172-1983w

automatic dispatching system (electric power system) A controlling means for maintaining the area control error or station control error at zero by automatically loading generating sources; it also may include facilities to load the sources in accordance with a predetermined loading criterion. (PE/PSE) 94-1970w

automatic equalizer *See:* adaptive equalizer.

automatic equipment (for a specified type of power circuit or apparatus) Equipment that provides automatic control. (SWG/PE) C37.100-1981s

automatic extraction or induction turbine, or both—condensing or noncondensing (control systems for steam turbine-generator units) Steam is extracted from or inducted into one or more stages with means for controlling the pressure(s) of the extraction or induction steam, or both. (PE/EDPG) 122-1985s

automatic extraction turbine (control systems for steam turbine-generator units) (condensing or noncondensing) Steam is extracted from one or more stages with means for controlling the pressure(s) of the extracted steam. (PE/EDPG) 122-1985s

automatic feedback control system A feedback control system that operates without human intervention. *See also:* feedback control system. (IM/PE/EDPG) [120], [3]

automatic-feed punch A card punch or keypunch into which cards are fed automatically. *Contrast:* hand-feed punch. (C) 610.10-1994w

automatic fire-alarm system A fire-alarm system for automatically detecting the presence of fire and initiating signal transmission without human intervention. *See also:* protective signaling. (EEC/PE) [119]

automatic fire detector (fire protection devices) A device designed to detect the presence of fire and initiate action. (NFPA) [16]

automatic flight control system An autopilot or automatic pilot. A system that controls the attitude, direction, and speed of a vehicle and directs it to travel along a selected course in response to manual or electronic commands. Stabilizes the dynamic response of the vehicle. (AES/GCS) 172-1983w

automatic frequency control (data transmission) An arrangement whereby the frequency of an oscillator or the tuning of a circuit is automatically maintained within specified limits with respect to a reference frequency. (PE) 599-1985w

automatic-frequency-control synchronization A process for locking the frequency (phase) of a local oscillator to that of an incoming synchronizing signal by the use of a comparison device whose output continuously corrects the local-oscillator frequency (phase). (EEC/PE) [119]

automatic gain control (AGC) (1) (general) A process or means by which gain is automatically adjusted in a specified manner as a function of input or other specified parameters. (SP/BT/AV) 151-1965w, [34]

(2) (data transmission) A method of automatically obtaining a substantially constant output of some amplitude characteristic of the signal over a range of variation of that characteristic at the input. The term is also applied to a device for accomplishing this result. (PE) 599-1985w

(3) A circuit or algorithm that varies gain as a function of the input signal amplitude. (COM/TA) 1329-1999

automatic generation control (1) Any supplementary control that automatically adjusts the power output levels of electric generators within a control area. Automatic generation control schemes can include one or more control subsystem(s),

such as load frequency control, economic dispatch, environmental dispatch control, security dispatch control, and the like. (PE/PSE) 858-1993w, 94-1991w
(2) The regulation of the power output of electric generators within a prescribed area in response to change in system frequency, tie-line loading, or the relation of these to each other, so as to maintain the scheduled system frequency or the established interchange with other areas within predetermined limits or both. (PE/PSE) [54]

automatic grid bias Grid-bias voltage provided by the difference of potential across resistance(s) in the grid or cathode circuit due to grid or cathode current or both. *See also:* radio receiver. (AP/ANT) 145-1983s

automatic hold (analog computer) Attainment of the hold condition automatically through amplitude comparison of a problem variable or through an overload condition. *See also:* electronic analog computer. (C) 165-1977w

automatic holdup alarm system An alarm system in which the signal transmission is initiated by the action of the robber. *See also:* protective signaling. (EEC/PE) [119]

automatic hyphenation In text formatting, hyphenation in which all line-ending and word break decisions are made automatically. Word break decisions may be made using syllabication algorithms or a dictionary containing commonly used words and their syllables. *See also:* manual hyphenation; semi-manual hyphenation. (C) 610.2-1987

automatic-identified outward dialing (telephone switching systems) A method of automatically obtaining the identity of a calling station from a private branch exchange over a separate data link for use in automatic message accounting. (COM) 312-1977w

automatic index An index produced by automatic indexing. *See also:* selective listing in combination index; word index; permutation index; keyword out of context index; keyword in context index. (C) 610.2-1987

automatic indexing Automated production of an index by selecting keywords and organizing them according to the type of index being produced. *Note:* Methods include assigned indexing and derivative indexing. *See also:* word index; keyword in context index; permutation index; selective listing in combination index; keyword out of context index. (C) 610.2-1987

automatic indirect-control telecommunications system (telephone switching systems) A system in which the pulsing from the originating calling device is stored in a register temporarily associated with the call, for the subsequent establishing of connections. (COM) 312-1977w

automatic interlocking An arrangement of signals, with or without other signal appliances, that functions through the exercise of inherent powers as distinguished from those whose functions are controlled manually, and that are so interconnected by means of electric circuits that their movements must succeed one another in proper sequence. *See also:* interlocking. (EEC/PE) [119]

automatic keying device A device that, after manual initiation, controls automatically the sending of a radio signal that actuates the auto alarm. *Note:* The prescribed signal is a series of twelve dashes, each of four seconds duration, with one-second intervals between dashes, transmitted on the radiotelegraph distress frequency in the medium-frequency band. This signal is used only to proceed distress calls or urgent warnings. (EEC/PE) [119]

automatic line sectionalizer (1) (supervisory control, data acquisition, and automatic control) A self-contained circuit-opening device that automatically opens the main electrical circuit through it after sensing and responding to a predetermined number of successive main current impulses equal to or greater than a predetermined magnitude. It opens while the main electrical circuit is de-energized. It may also have a provision to be manually operated to interrupt loads. (SWG/PE/SUB) C37.1-1987s
(2) A self-contained circuit-opening device that automatically opens the main electrical circuit through it after sensing and responding to a predetermined number of successive main current impulses equal to or greater than a predetermined magnitude. It opens while the main electrical circuit is de-energized. It may also have provision to be manually operated to interrupt loads. *Note:* When applicable, it includes an assembly of control elements required to detect overcurrents and control the sectionalizer operation. (SWG/PE) C37.100-1992

automatic load (armature current division) The effect of a control function or a device to automatically divide armature currents in a prescribed manner between two or more motors or two or more generators connected to the same load. *See also:* feedback control system. (IA/ICTL/IAC/APP) [60], [75]

automatic load (level) control (ALC) (power-system communication) A method of automatically maintaining the peak power of a single-sideband suppressed-carrier transmitter at a constant level. *See also:* radio transmitter. (PE) 599-1985w

automatic load throwover equipment (1) (supervisory control, data acquisition, and automatic control) (transfer or switchover) Equipment that automatically transfers a load to another source of power when the original source to which it has been connected fails, and that automatically restores the load to the original source under desired conditions. *Note:* The restoration of the load to the preferred source from the emergency source upon reenergization of the preferred source after an outage may be of the continuous circuit restoration type or interrupted circuit restoration type.

a) *Equipment of the nonpreferential type.* Equipment that automatically restores the load to the original source only when the other source, to which it has been connected, fails.
b) *Fixed preferential type.* Equipment in which the original source always serves as the preferred source and other source as the emergency source. The automatic transfer equipment will restore the load to the preferred source upon its reenergization.
c) *Selective preferential source.* Equipment in which either source may serve as the preferred or the emergency source of preselection as desired, and which will restore the load to the preferred source upon its reenergization.
d) *Semiautomatic load throwover equipment.* Equipment that automatically transfers a load to another (emergency) source of power when the original (preferred) source to which it has been connected fails, but requires manual restoration of the load to the original source.

(PE/SUB) C37.1-1987s
(2) Equipment that automatically transfers a load to another source of power when the original source to which it has been connected fails, and that automatically restores the load to the original source under desired conditions. *Note:* The restoration of the load to the preferred source from the emergency source upon reenergization of the preferred source after an outage may be of the continuous circuit restoration type or interrupted circuit restoration type. *See also:* equipment of the selective preferential type; equipment of the nonpreferential type; semiautomatic load throw-over equipment; equipment of the fixed preferential type. (SWG/PE) C37.100-1992

automatic machine control equipment Equipment that provides automatic control for functions related to rotating machines or power rectifiers. (SWG/PE) C37.100-1992

automatic message accounting (AMA) (telephone switching systems) An arrangement for automatically collecting, recording, and processing information relating to calls for billing purposes. (COM) 312-1977w

automatic number identification (ANI) (1) (telephone switching systems) The automatic obtaining of a calling station directory or equipment number for use in automatic message accounting. (COM) 312-1977w

(2) The local access and transport area (LATA) or interLATA billing number of the calling party. (AMR/SCC31) 1390-1995, 1390.2-1999
(3) A network service that delivers the phone number/billing number of the calling party. (SCC31) 1390.3-1999

automatic opening (1) (supervisory control, data acquisition, and automatic control) (station control and data acquisition) The opening of a switching device under predetermined conditions without operator intervention. (SUB/PE) C37.1-1987s, C37.1-1979s
(2) (tripping) The opening of a switching device under predetermined conditions without the intervention of an attendant. (SWG/PE) C37.100-1992

automatic operation (1) (elevators) Operation wherein the starting of the elevator car is effected in response to the momentary actuation of operating devices at the landing, and/or of operating devices in the car identified with the landings, and/or in response to an automatic starting mechanism, and wherein the car is stopped automatically at the landings. *See also:* control. (EEC/PE) [119]
(2) (of a switching device) The ability to complete an assigned sequence of operations by automatic control without the assistance of an attendant. (SWG/PE) C37.100-1992

automatic outage An outage occurrence that results from automatic operation of switching devices. (PE/PSE) 859-1987w

automatic pagination In text formatting, the automatic arrangement or rearrangement of text according to preset page layout parameters such as margin width and lines per page. *Note:* May also include the assignment and placement of page numbers on the pages. *Synonyms:* repagination; pagination. (C) 610.2-1987

automatic phase control (television) A process or means by which the phase of an oscillator signal is automatically maintained within specified limits by comparing its phase to the phase of an external reference signal and thereby supplying correcting information to the controlled source or a device for accomplishing this result. *Note:* Automatic phase control is sometimes used for accurate frequency control and under these conditions is often called automatic frequency control. *See also:* television. (EEC/PE) [119]

automatic pilot (electronic navigation) Equipment that automatically controls the attitude of a vehicle about one or more of its rotational axes (pitch, roll, and yaw), and may be used to respond to manual or electronic commands. *See also:* navigation. (AES) [42]

automatic-pilot servo motor A device that converts electric signals to mechanical rotation so as to move the control surfaces of an aircraft. (EEC/PE) [119]

automatic programmed tools (APT) (numerically controlled machines) A computer-based numerical control programming system that uses English-like symbolic descriptions of part and tool geometry and tool motion. (PE/EEC/TR) [57], [74]

automatic programming (analog computer) A method of computer operation using auxiliary automatic equipment to perform computer control state selections, switching operations, or component adjustments in accordance with previously selected criteria. *See also:* electronic analog computer. (C) 165-1977w

Automatic Programming and Scaling of Equations A programming language similar to FORTRAN, characterized by its ability to describe equation-oriented specifications used in continuous simulation models. (C) 610.13-1993w

automatic punch *See:* card punch.

automatic reclosing equipment (1) (supervisory control, data acquisition, and automatic control) (station control and data acquisition) Equipment that initiates automatic closing of a switching device under predetermined conditions without operator intervention. (PE/SUB) C37.1-1987s, C37.1-1979s
(2) Automatic equipment that provides for reclosing a switching device as desired after it has opened automatically under abnormal conditions. *Note:* Automatic reclosing equipment may be actuated by conditions sensed on either or both sides of the switching device as designed. (SWG/PE) C37.100-1992

automatic repeat request A protocol that uses positive or negative acknowledgment with retransmission techniques to ensure reliability. *Note:* The sender automatically repeats the request if it does not receive an answer. (C) 610.7-1995

automatic reset A function which operates to automatically reestablish specific conditions. (SWG/PE) C37.100-1981s

automatic-reset manual release of brakes (control) A manual release that, when operated, will maintain the braking surfaces in disengagement but will automatically restore the braking surfaces to their normal relation as soon as electric power is again applied. *See also:* feedback control system. (IA/ICTL/IAC) [60]

automatic reset relay (A) A stepping relay that returns to its home position either when it reaches a predetermined contact position, or when a pulsing circuit fails to energize the driving coil within a given time. May either pulse forward or be spring reset to the home position. **(B)** An overload relay that restores the circuit as soon as an overcurrent situation is corrected. (EEC/REE) [87]

automatic-reset thermal protector (rotating machinery) A thermal protector designed to perform its function by opening the circuit to or within the protected machine and then automatically closing the circuit after the machine cools to a satisfactory operating temperature. *See also:* starting-switch assembly. (PE) [9]

automatic reversing Reversing of an electric drive, initiated by automatic means. *See also:* electric drive. (IA/ICTL/IAC) [60]

automatic selective control or transfer relay (power system device function numbers) A relay that operates to select automatically between certain sources or conditions in an equipment, or performs a transfer operation automatically. (SUB/PE) C37.2-1979s

automatic send/receive (ASR) (A) A teletypewriter with a keyboard, printer, and paper tape punch/reader, allowing tape to be produced and edited off-line for automatic transmission. **(B)** A keyboard/printer device that uses asynchronous set-serial connection to a computer. (C) 610.10-1994

automatic signal A signal controlled automatically by the occupancy or certain other conditions of the track area that it protects. (EEC/PE) [119]

automatic smoke alarm system An alarm system designed to detect the presence of smoke and to transmit an alarm automatically. *See also:* protective signaling. (EEC/PE) [119]

automatic speed adjustment Speed adjustment accomplished automatically. *See also:* electric drive; automatic. (IA/ICTL/IAC) [60]

automatic starter A starter in which the influence directing its performance is automatic. (IA/MT) 45-1998

automatic station (1) (supervisory control, data acquisition, and automatic control) (station control and data acquisition) A station that operates in automatic control mode. *Note:* An automatic station may go in and out of operation in response to predetermined voltage, load, time, or other conditions, or in response to a remote or locally manually operated control device. (SWG/SUB/PE) C37.1-1987s, C37.100-1992
(2) A station (usually unattended) that under predetermined conditions goes into operation by an automatic sequence; that thereupon by automatic means maintains the required character of service within its capability; that goes out of operation by automatic sequence under other predetermined conditions; and includes protection against the usual operating emergencies. *Note:* An automatic station may go in and out of operation in response to predetermined voltage, load, time, or other conditions, or in response to supervisory control or to

a remote or local manually operated control device. (SWG/PE) C37.100-1992

automatic switchboard A switchboard in which the connections are made by apparatus controlled from remote calling devices. (COM) [48]

automatic switching system (telephone switching systems) The switching entity for an automatic telecommunication system. (COM) 312-1977w

automatic system A system in which the operations are performed by electrically controlled devices without the intervention of operators. (COM) [49]

automatic telecommunications exchange (telephone switching systems) A telecommunications exchange in which connections between stations are automatically established as a result of signals produced by calling devices. (COM) 312-1977w

automatic telecommunications system (telephone switching systems) A system in which connections between stations are automatically established as a results of signals produced by calling devices. (COM) 312-1977w

automatic telegraphy That form of telegraphy in which transmission or reception of signals, or both, are accomplished automatically. *See also:* telegraphy. (COM) [49]

automatic test equipment (ATE) (1) (test, measurement, and diagnostic equipment) Equipment that is designed to conduct analysis of functional or static parameters to evaluate the degree of performance degradation and may be designed to perform fault isolation of unit malfunctions. The decision-making, control, or evaluative functions are conducted with minimum reliance on human intervention. (MIL/SCC20) [2], 993-1997
(2) A system providing a test capability for the automatic testing of one or more units under test (UUTs). The ATE system consists of a controller, test resource devices, and peripherals. The controller directs the testing process and interprets the results. The test resource devices provide stimuli, measurements, and physical interconnections. The peripherals, such as displays, keyboards, printers, mass storage, etc., supply the necessary capability for information management. (ATLAS) 1232-1995
(3) Equipment on which an implementation of the tests, test methods, and test sequences to be performed on a unit under test (UUT) to verify conformance with its test specification with or without fault diagnosis may be executed with minimum reliance on human intervention. (SCC20) 771-1998
(4) Equipment that is designed to conduct analysis of functional or static parameters to evaluate the degree of performance degradation and that may be designed to perform fault isolation of unit malfunctions. (SCC20) 1226-1998

automatic test equipment control software (test, measurement, and diagnostic equipment) Software used during execution of a test program, which controls the nontesting operations of the ATE. This software is used to execute a test procedure but does not contain any of the stimuli or measurement parameters used in testing the unit under test (UUT). Where test software and control software are combined in one inseparable program, that program will be treated as test software, not control software. (MIL) [2]

automatic test equipment oriented language (test, measurement, and diagnostic equipment) A computer language used to program an automatic test equipment to test units under test (UUTs), whose characteristics imply the use of a specific ATE system or family of ATE systems. (MIL) [2]

automatic test equipment support software (test, measurement, and diagnostic equipment) Computer programs that aid in preparing, analyzing, and maintaining test software. Examples are: ATE compilers, translation analysis programs, and punch/print programs. (MIL) [2]

automatic test pattern generator (ATPG) Any tool that generates test information for a device based on structural analysis of the device. (C/TT) 1450-1999

automatic test system (ATS) Includes the automatic test equipment (ATE) as well as all support equipment, software, test programs, and adapters. (SCC20) 1226-1998

automatic threshold variation A constant-false-alarm-rate (CFAR) technique in which the detection decision threshold is varied continuously in proportion to the incoming IF and video noise level. (AES/RS) 686-1990

automatic throw-over equipment *See:* automatic transfer equipment.

automatic throw-over equipment of the fixed preferential type *See:* automatic transfer equipment.

automatic throw-over equipment of the nonpreferential type *See:* automatic transfer equipment.

automatic throw-over equipment of the selective-preferential type *See:* automatic transfer equipment.

automatic ticketing (telephone switching systems) An arrangement for automatically recording information relating to calls, for billing purposes. (COM) 312-1977w

automatic track follower *See:* automatic chart-line follower.

automatic tracking (1) Tracking in which a system employs some feedback mechanism, for example a servo or computer, to follow automatically some characteristic of a signal or target, such as range angle, Doppler frequency, or phase. *See also:* tracking radar; tracking. (AES/GCS) 172-1983w
(2) Tracking with the use of electronic circuitry rather than a human operator in which a system employs some feedback mechanism, (e.g., a servo or computer) to follow automatically some characteristic of a signal or target, such as range, angle, Doppler frequency, or phase. *See also:* tracking; tracking radar. (AES) 686-1997

automatic train control (ATC) (1) (train control) (automatic speed control) A system or an installation so arranged that its operation on failure to forestall or acknowledge will automatically result in either one or the other or both of the following conditions:

a) Automatic train stop: The application of the brakes until the train has been brought to a stop; and
b) Automatic speed control: The application of the brakes when the speed of the train exceeds a prescribed rate and continued until the speed has been reduced to a predetermined and prescribed rate.

(EEC/PE) [119]
(2) The system for automatically controlling train movement, enforcing train safety, and directing train operations. ATC must include automatic train protection (ATP) and may include automatic train operation (ATO) and/or automatic train supervision (ATS). (VT/RT) 1475-1999, 1474.1-1999

automatic train control application An application of the brake by the automatic train control device. (EEC/PE) [119]

automatic train operation (ATO) The subsystem within the automatic train control system that performs any or all of the functions of speed regulation, programmed stopping, door control, performance level regulation, or other functions otherwise assigned to the train operator. (VT/RT) 1474.1-1999, 1475-1999

automatic train protection (ATP) The subsystem within the automatic train control (ATC) system that maintains fail-safe protection against collisions, excessive speed, and other hazardous conditions through a combination of train detection, train separation, and interlocking. (VT/RT) 1475-1999, 1474.1-1999

automatic train stop A wayside system that works in conjunction with equipment installed on the vehicle to apply the brakes at designated restrictions or on a dispatcher's signal, should the operator not respond properly. (VT) 1475-1999

automatic train supervision (ATS) The subsystem within the automatic train control (ATC) system that monitors trains, adjusts the performance of individual trains to maintain schedules, and provides data to adjust service to minimize the inconveniences otherwise caused by irregularities. *Note:* The

ATS subsystem also typically includes manual and automatic routing functions. (VT/RT) 1475-1999, 1474.1-1999

automatic transfer equipment (A) Equipment that automatically transfers a load to another source of power when the original source to which it has been connected fails, and that will automatically retransfer the load to the original source under desired conditions. *Notes:* 1. It may be of the nonpreferential, fixed-preferential, or selective-preferential type. 2. Compare with transfer switch where transfer is accomplished without current interruption. **(B)** (of the fixed preferential type) Automatic transfer equipment in which the original source always serves as the preferred source and the other source as the emergency source. The automatic transfer equipment will retransfer the load to the preferred source upon its reenergization. *Note:* The restoration of the load to the preferred source from the emergency source upon the reenergization of the preferred source after an outage may be of the continuous-circuit restoration type or the interrupted-circuit restoration type. **(C)** (of the non-preferential type) Automatic transfer equipment that automatically retransfers the load to the original source only when the other source, to which it has been connected, fails. **(D)** (of the selective-preferential type) Automatic transfer equipment in which either source may serve as the preferred or the emergency source of preselection as desired, and which will retransfer the load to the preferred source upon its reenergization. *Note:* The restoration of the load to the preferred source from the emergency source upon the reenergization of the preferred source after an outage may be of the continuous-circuit restoration type or the interrupted-circuit restoration type. (SWG/PE) C37.100-1992

automatic transfer switch (emergency and standby power) Self-acting equipment for transferring one or more load conductor connections from one power source to another. (IA/PSE) 446-1995

automatic transformer control equipment Equipment that provides automatic control for functions relating to transformers, such as their connection, disconnection or regulation in response to predetermined conditions such as system load, voltage or phase angle. (SWG/PE) C37.100-1992

automatic triggering (oscilloscopes) A mode of triggering in which one or more of the triggering-circuit controls are preset to conditions suitable for automatically displaying repetitive waveforms. *Note:* The automatic mode may also provide a recurrent trigger of recurrent sweep in the absence of triggering signals. *See also:* oscillograph. (IM/HFIM) [40]

automatic tripping *See:* automatic opening.

automatic video noise leveling A constant-false-alarm-rate (CFAR) technique in which the receiver gain is readjusted to maintain a constant video noise level. The noise level is sampled at the receiver output at the end of each range sweep, prior to the next transmission. The resulting receiver gain is fixed throughout the next sweep. Under some jamming conditions, a fixed video noise can be maintained at the display. (AES) 686-1997

automatic voltage-current crossover (power supplies) The characteristic of a power supply that automatically changes the method of regulation from constant voltage to constant current (or vice versa) as dictated by varying load conditions. (AES) [41]

automatic volume control (data transmission) A method of automatically obtaining a substantially constant audio output volume over a range of input volume. The term is also applied to a device for accomplishing this result. (PE) 599-1985w

automation (1) (A) The implementation of a process by automatic means. **(B)** The theory, art, or technique of making a process more automatic. **(C)** The investigation, design, development, and application of methods of rendering processes automatic, self-moving, or self-controlling. (C) 610.2-1987

(2) Computerization of data or of a process that uses that data. (PE/EDPG) 1150-1991w

autonavigator (navigation aid terms) Navigation equipment that includes means for coupling the output navigational data derived from the navigation sensors to the control system of the vehicle. (AES/GCS) 172-1983w

auto-negotiation (1) A link pulse signalling scheme at the PMD sublayer, whereby devices at each end of a link segment can determine the modes of operation of devices at the other end. (C/LM) 802.9a-1995w

(2) The algorithm that allows two devices at either end of a link segment to negotiate common data service functions. (C/LM) 802.3-1998

autonomous system A collection of gateways and networks administered by one administrative entity. (C) 610.7-1995

autopatch *See:* automatic component interconnection matrix.

autopilot *See:* automatic flight control system; automatic pilot.

autopilot coupler (electronic navigation) (navigation) (navigation aid terms) The means used to link the navigation system-receiver output to the automatic pilot. (AES/RS/GCS) 686-1982s, 172-1983w

autoradar plot (electronic navigation) A particular chart comparison unit using a radar presentation of position. *See also:* navigation. (AES) [42]

autorecovery The process of restoring installed software to the state it was in prior to the invocation, and subsequent failure during execution, of the `swinstall` utility. (C/PA) 1387.2-1995

autorefresh A RAM-refresh protocol in which controller-provided *refreshNow* signals schedule the timing for RAM refresh cycles and RAM-local hardware specifies refresh-cycle addresses. *See also:* refresh. (C/MM) 1596.4-1996

autoregulation induction heater An induction heater in which a desired control is effected by the change in characteristics of a magnetic charge as it is heated at or near its Curie point. *See also:* coupling; induction heating; dielectric heater. (IA) 54-1955w

autoselect The automatic selection, within a utility, of software beyond that directly specified by the user in order to meet the dependencies of the user-specified software. (C/PA) 1387.2-1995

autotrack (communication satellite) The capability of a space communications receiver antenna to automatically track an orbiting satellite vehicle, for example, by using a monopulse system. (COM) [19]

autotransformer A transformer in which part of one winding is common to both the primary and the secondary circuits associated with that winding. (PE/TR) C57.15-1999

autotransformer, individual-lamp *See:* specialty transformer.

autotransformer starter A starter that includes an autotransformer to furnish a reduced voltage for starting a motor. It includes the necessary switching mechanism and is frequently called a compensator or autostarter. (IA/MT) 45-1998

autotransformer starting (rotating machinery) The process of starting a motor at reduced voltage by connecting the primary winding to the supply initially through an autotransformer and reconnecting the winding directly to the supply at rated voltage for the running conditions. *See also:* asynchronous machine. (PE) [9]

auxiliaries (1) (collective) (generating stations electric power system) For more than one auxiliary, that is, auxiliaries bus, auxiliaries power transformer, etc. (PE/EDPG) 505-1977r

(2) Accessories to be used with switchgear apparatus but not attached to it, as distinguished from attachments. (SWG/PE) C37.100-1992

auxiliary (1) (controller) (thyristor) Apparatus peripheral to the main power flow but necessary for the operation of the controller. (IA/IPC) 428-1981w

(2) (generating stations electric power system) Any item not directly a part of a specified component or system but required for its functional operation. (PE/EDPG) 505-1977r

auxiliary anode An anode located adjacent to the pool cathode in an ignitron to facilitate the maintenance of a cathode spot under conditions adverse to its maintenance by the main anode circuit. *See also:* electronic controller. (IA/ICTL/IAC) [60]

auxiliary branch (self-commutated converters) (converter circuit elements) A branch other than a principal branch. *Note:* Examples of auxiliary branches are regenerative branches and turn-off branches. (IA/SPC) 936-1987w

auxiliary building (radiological monitoring instrumentation) Building(s) near or adjacent to the reactor containment building in which primary system support equipment is housed. (NI) N320-1979r

auxiliary burden (capacitance potential device) A variable burden furnished, when required, for adjustment purposes. *See also:* outdoor coupling capacitor. (PE/EM) 43-1974s

auxiliary bus A relatively low capacity narrow bus used for miscellaneous functions for which the primary bus is not suited, or for which an alternate module access is needed. (C/BA) 14536-1995

auxiliary capacitance (capacitance potential devices) The capacitance between the network connection and ground, if present. *Synonym:* shunt capacitance. *See also:* outdoor coupling capacitor. (PE/EM) 43-1974s

auxiliary circuit breaker (ac high-voltage circuit breakers) The circuit breaker used to disconnect the current circuit from direct connection with the test circuit breaker. (SWG/PE) C37.081-1981r, C37.083-1999, C37.100-1992

auxiliary circuits (1) All control, indicating, and measuring circuits. (SUB/PE) C37.122-1993, C37.122.1-1993
(2) All control, indicating and measuring circuits. (SWG/PE) C37.100-1992

auxiliary compartment That portion of the switchgear assembly that is assigned to the housing of auxiliary equipment, such as potential transformers, control power transformers, or other miscellaneous devices. (SWG/PE) C37.20.1-1993r, C37.20.2-1993

auxiliary console In a computer system with more than one console, an alternate console used primarily to supervise operations within the computer. *Contrast:* master console. (C) 610.10-1994w

auxiliary device (1) (auxiliary devices for motors) Components installed either integrally within the motor, located adjacent to or mounted on the motor, or attached to its terminals for the purpose of monitoring the operating conditions or protecting the motor. (IA/PC) 303-1984s
(2) Any electrical device other than motors and motor starters necessary to fully operate the machine or equipment. (IA/PKG) 333-1980w

auxiliary device to an instrument A separate piece of equipment used with an instrument to extend its range, increase its accuracy, or otherwise assist in making a measurement or to perform a function additional to the primary function of measurement. (EEC/PE) [119]

auxiliary equipment (Class 1E motor) (nuclear power generating station) Equipment that is not part of the motor but is necessary for the operation of the motor and will be installed within the containment. *See also:* ancillary equipment. (PE/NP) 380-1975w

auxiliary function (numerically controlled machine) A function of a machine other than the control of the coordinates of a workpiece or tool. Includes functions such as miscellaneous, feed, speed, tool selection, etc. *Note:* Not a preparatory function. (IA) [61]

auxiliary generator A generator, commonly used on electric motive power units, for serving the auxiliary electric power requirements of the unit. *See also:* traction motor. (EEC/PE) [119]

auxiliary generator set A device usually consisting of a commonly mounted electric generator and a gasoline engine or gas turbine prime mover designed to convert liquid fuel into electric power. *Note:* It provides the aircraft with an electric power supply independent of the aircraft propulsion engines. (EEC/PE) [119]

auxiliary ground electrode A ground electrode with certain design or operating constraints. Its primary function may be other than conducting the ground fault current into the earth. (PE/SUB) 80-2000

auxiliary lead (rotating machinery) A conductor joining an auxiliary terminal to the auxiliary device. (PE) [9]

auxiliary means A system element or group of elements that changes the magnitude but not the nature of the quantity being measured to make it more suitable for the primary detector. In a sequence of measurement operations, it is usually placed ahead of the primary detector. *See also:* measurement system. (EEC/PE) [119]

auxiliary motor or motor generator (power system device function numbers) One used for operating auxiliary equipment, such as pumps, blowers, exciters, rotating magnetic amplifiers, etc. (SUB/PE) C37.2-1979s

auxiliary operation Any operation that is performed by equipment that is not under continuous control of the central processing unit. (C) [20], 610.10-1994w, [85]

auxiliary power (thyristor) The power used by the controller to perform its various auxiliary functions, as opposed to the principal power. (IA/IPC) 428-1981w
(2) (A) (thyristor power converter) (General) The power required for fans or blowers, relays, breaker control, phase loss detection, etc. **(B) (thyristor power converter)** Input power used by the thyristor converter to perform its various auxiliary functions as opposed to the power that may be flowing between the ac supply and the load. (IA/IPC) 444-1973

auxiliary power supply A power source supplying power other than load power as required for the proper functioning of a device. *See also:* electronic controller. (IA/ICTL/IAC) [60]

auxiliary power transformer A transformer having a fixed phase position used for supplying excitation for the rectifier station and essential power for the operation of rectifier equipment auxiliaries. *See also:* transformer. (Std100) C57.18-1964w

auxiliary relay A relay whose function is to assist another relay or control device in performing a general function by supplying supplementary actions. *Notes:* 1. Some of the specific functions of an auxiliary relay are as follows: (a) Reinforcing contact current-carrying capacity of another relay or device. (b) Providing circuit seal-in functions. (c) Increasing available number of independent contacts. (d) Providing circuit-opening instead of circuit-closing contacts or vice-versa. (e) Providing time delay in the completion of a function. (f) Providing simple functions for interlocking or programming. 2. The operating coil of the contacts of an auxiliary relay may be used in the control circuit of another relay or other control device. *Example:* An auxiliary relay may be applied to the auxiliary contact circuits of a circuit breaker in order to coordinate closing and tripping control sequences. 3. A relay that is auxiliary in its functions even though it may derive its driving energy from the power system current or voltage is a form of auxiliary relay. *Example:* A timing relay operating from current or potential transformers. 4. Relays that, by direct response to power system input quantities, assist other relays to respond to such quantities with greater discrimination are NOT auxiliary relays. *Example:* Fault detector relay. 5. Relays that are limited in function by a control circuit, but are actuated primarily by system input quantities, are NOT auxiliary relays. *Example:* Torque-controlled relays. (SWG/PE) C37.100-1992

auxiliary relay contacts Contacts of lower current capacity than the main contacts: used to keep the coil energized when the original operating circuit is open, to operate an audible or visual signal indicating the position of the main contacts, or to establish interlocking circuits, etc. (EEC/REE) [87]

auxiliary relay driver A circuit that supplies an input to an auxiliary relay. (SWG/PE) C37.100-1992

auxiliary rope-fastening device (elevators) A device attached to the car or counterweight or to the overhead dead-end rope-hitch support that will function automatically to support the car or counterweight in case the regular wire-rope fastening fails at the point of connection to the car or counterweight or at the overhead dead-end hitch. *See also:* elevator. (EEC/PE) [119]

auxiliary secondary terminals The auxiliary secondary terminals provide the connections to the auxiliary secondary winding, when furnished. *See also:* auxiliary secondary winding. (PE/EM) 43-1974s

auxiliary secondary winding (capacitance potential device) The auxiliary secondary winding is an additional winding that may be provided in the capacitance potential device when practical considerations permit. *Note:* It is a separate winding that provides a potential that is substantially in phase with the potential of the main winding. The primary purpose of this winding is to provide zero-sequence voltage by means of a broken delta connection of three single-phase devices. *See also:* outdoor coupling capacitor; auxiliary secondary terminals. (PE/EM) 43-1974s

auxiliary storage (1) (computers) A storage that supplements another storage. (C) [20], [85]
(2) A type of secondary storage that is available to a processor only through input-output channels; for example, storage on magnetic tape or a disk drive. *Synonym:* peripheral storage. *Contrast:* main storage. *See also:* paging device. (C) 610.10-1994w

auxiliary supporting features (1) (nuclear power generating station) Installed systems or components that provide services, such as cooling, illumination, and energy supply and that are required by the post accident monitoring instrumentation to perform its functions. (PE/NP) 497-1981w
(2) Systems or components that provide services (such as cooling, lubrication, and energy supply) required for the safety systems to accomplish their safety functions. (PE/NP) 603-1998

auxiliary switch (1) A switch mechanically operated by the main device for signaling, interlocking, or other purposes. *Note:* Auxiliary switch contacts are classed as follows: *a, b, aa, bb,* LC, etc., for the purpose of specifying definite contact positions with respect to the main device. (SWG/PE) C37.100-1992
(2) A switch that is mechanically operated by the main switching device for signaling, interlocking, or other purposes. *Note:* Auxiliary switch contacts are classified as a, b, aa, bb, LC, etc., for the purpose of specifying definite contact positions with respect to the main device. (IA/PSP) 1015-1997

auxiliary terminal (rotating machinery) A termination for parts other than the armature or field windings. (PE) [9]

auxiliary wayside system A back-up or secondary train control system capable of providing full or partial automatic train protection for trains not equipped with trainborne communications-based train control (CBTC) equipment, and/or trains with partially or totally inoperative trainborne CBTC equipment. The auxiliary wayside system may include trainborne equipment and may also provide broken rail detection. (VT/RT) 1474.1-1999

auxiliary winding (single-phase induction motor) A winding that produces poles of a magnetic flux field that are displaced from those of the main winding, that serves as a means for developing torque during starting operation, and that, in some types of design, also serves as a means for improvement of performance during running operation. An auxiliary winding may have a resistor or capacitor in series with it and may be connected to the supply line or across a portion of the main winding. *See also:* asynchronous machine. (PE) [9]

availability (1) (emergency and standby power) The fraction of time within which a system is actually capable of performing its mission. (IA/SMC/C/BA/PSE) 446-1995, [63], 896.9-1994w
(2) (supervisory control, data acquisition, and automatic control) The ratio of uptime to total time (uptime plus downtime). (PE/SUB) C37.1-1994
(3) (software) The degree to which a system or component is operational and accessible when required for use. Often expressed as a probability. (C) 610.12-1990
(4) (nuclear power generating station) The characteristic of an item expressed by the probability that it will be operational at a randomly selected future instant in time. (PE/NP) 380-1975w, 352-1975s, 577-1976r, 933-1999
(5) Relates to the accessibility of information to the operator on a "continuous," "sequence," or "as called for" basis. (PE/NP) 566-1977w
(6) (telephone switching systems) The number of outlets of a group that can be reached from a given inlet in a switching stage or network. (COM) 312-1977w
(7) The ability of an item—under combined aspects of its reliability, maintainability, and maintenance support—to perform its required function at a stated instant of time or over a stated period of time. *Note:* The term "availability" is also used as an availability characteristic denoting either the probability of performing at a stated instant of time or the probability related to an interval of time. (R) [29]
(8) Availability = service time/reporting period time. (PE/PSE) 859-1987w
(9) The probability that a system will be able to execute a function accurately at any given time. (C/BA) 896.3-1993w
(10) The ratio of uptime and uptime plus downtime. (SWG/PE) C37.100-1992
(11) As applied either to the performance of individual components or to that of a system, it is the long-term average fraction of time that a component or system is in service and satisfactorily performing its intended function. An alternative and equivalent definition for availability is the steady-state probability that a component or system is in service. (IA/PSE) 493-1997, 399-1997

availability factor The ratio of the time a generating unit or piece of equipment is ready for, or in service to, the total time interval under consideration. *See also:* generating station. (T&D/PE) [10]

availability model (software) A model used for predicting, estimating, or assessing availability. *See also:* availability. (C/SE) 729-1983s

available (power system measurement) (electric generating unit reliability, availability, and productivity) The state in which a unit is capable of providing service, whether or not it is actually in service and regardless of the capacity level that can be provided. (PE/PSE) 762-1987w

available accuracy (noise temperature of noise generators) An accuracy that is readily available to the public at large, such as may be announced in calibration service bulletins or instrument catalogs. This term shall not include accuracies that may be obtainable at any echelon by employing special efforts and expenditures over and above those invested in producing the advertised or announced accuracies, nor shall it include accuracies of calibration or measurement services that are not readily available to any and all customers and clients. (IM) 294-1969w

available capacity (1) (electric generating unit reliability, availability, and productivity) The dependable capacity, modified for equipment at any time. (PE/PSE) 762-1987w
(2) The capacity for a given discharge time and end-of-discharge voltage that can be withdrawn from a cell under the specific conditions of operation. (PE/EDPG) 1115-1992

available conversion power gain (conversion transducer) The ratio of the available output-frequency power from the output terminals of the transducer to the available input-

frequency power from the driving generator with terminating conditions specified for all frequencies that may affect the result. *Notes:* 1. This applies to outputs of such magnitude that the conversion transducer is operating in a substantially linear condition. 2. The maximum available conversion power gain of a conversion transducer is obtained when the input termination admittance, at input frequency, is the conjugate of the input-frequency driving-point admittance of the conversion transducer. *See also:* transducer.
(ED) 161-1971w, 196-1952w

available current (A) The current that would flow if each pole of the breaking device under consideration were replaced by a link of negligible impedance without any change of the circuit or the supply. *Synonym:* prospective current. *See also:* contactor. **(B)** (of a circuit with respect to a switching device situated therein) The current that would flow in that circuit if each pole of the switching device were to be replaced by a link of negligible impedance without any other change in the circuit or the supply. *Synonym:* prospective current.
(CAS/SWG/IA/PE/ICTL/IAC) [60], [84], C37.100-1981

available generation (electric generating unit reliability, availability, and productivity) The energy that could have been generated by a unit in a given period if operated continuously at its available capacity. (PE/PSE) 762-1987w

available hours (electric generating unit reliability, availability, and productivity) The number of hours a unit was in the available state. *Note:* Available hours is the sum of service hours and reserve shutdown hours, or may be computed from period hours minus unavailable hours. *See also:* unavailable hours. (PE/PSE) 762-1987w

available line The portion of the scanning line that can be used specifically for picture signals. (COM) 168-1956w

available power (1) (hydraulic turbines) (at a port) The maximum power that can be transferred from the port to a load. *Note:* At a specified frequency, maximum power transfer will take place when the impedance of the load is the conjugate of that of the source. The source impedance must have a positive real part. *See also:* network analysis.
(ED) 161-1971w, [45]

(2) (audio and electroacoustics) The maximum power obtainable from a given source by suitable adjustment of the load. *Note:* For a source this is equivalent to a constant sinusoidal electromotive force in series with an impedance independent of amplitude, the available power is the mean-square value of the electromotive force divided by four times the resistive part of the impedance of the source.
(SP) 151-1965w, 196-1952w, 270-1966w

(3) (of a sound field with a given object placed in it) The power that would be extracted from the acoustic medium by an ideal transducer having the same dimensions and the same orientation as the given object. The dimensions and the orientation with respect to the sound field must be specified. *Note:* The acoustic power available to an electroacoustic transducer, in a plane-wave sound field of given frequency, is the product of the free-field intensity and the effective area of the transducer. For this purpose the effective area of an electroacoustic transducer, for which the surface velocity distribution is independent of the manner of excitation of the transducer, is set 1.4 times the product of the receiving directivity factor and the square of the wavelength of a free progressive wave in the medium. If the physical dimensions of the transducer are small in comparison with the wavelength, the directivity factor is near unity, and the effective area varies inversely as the square of the frequency. If the physical dimensions are large in comparison with the wavelength, the directivity factor is nearly proportional to the square of the frequency, and the effective area approaches the actual area of the active face of the transducer. (SP) [32]

(4) (signal generators) The power at the output port supplied by the generator into a specified load impedance. *See also:* signal generator. (IM/HFIM) [40]

available power efficiency (electroacoustics) Of an electroacoustic transducer used for sound reception, the ratio of the electric power available at the electric terminals of the transducer to the acoustic power available to the transducer. *Notes:* 1. For an electroacoustic transducer that obeys the reciprocity principle, the available power efficiency in sound reception is equal to the transmitting efficiency. 2. In a given narrow frequency band, the available power efficiency is numerically equal to the fraction of the open-circuit mean-square thermal noise voltage present at the electric terminals that contributed by thermal noise in the acoustic medium. *See also:* microphone. (SP) [32]

available power gain (1) (two-port linear transducer) At a specified frequency, the ratio of the available signal power from the output port of the transducer, to the available signal power from the input source. *Note:* The available signal power at the output port is a function of the match between the source impedance and the impedance of the input port. *See also:* network analysis. (ED) 161-1971w

(2) The maximum power gain that can be obtained from a signal source. For a source of internal impedance Z_s, $R_s jX_s$, the maximum power gain is obtained when the source is connected to a conjugate matched load; i.e., if Z_2, $R_s - jX_s$. It is sometimes called completely matched power gain or available gain. *See also:* network analysis; transducer. (CAS) [13]

available power response (electroacoustics) (electroacoustic transducer used for sound emission) The ratio of the mean-square sound pressure apparent at a distance of one meter in a specified direction from the effective acoustic center of the transducer to the available electric power from the source. *Notes:* 1. The sound pressure apparent at a distance of one meter can be found by multiplying the sound pressure observed at a remote point where the sound field is spherically divergent by the number of meters from the effective acoustic center to that point. 2. The available power response is a function not only of the transducer but also of some source impedances, either actual or nominal, the value of which must be specified. (SP) [32]

available (prospective) current (of a circuit with respect to a switching device situated therein) The current that would flow in that circuit if each pole of the switching device were to be replaced by a link of negligible impedance without any other change in the circuit or the supply.
(SWG/PE) C37.100-1992

available short-circuit current (at a given point in a circuit) The maximum current that the power system can deliver through a given circuit point to any negligible-impedance short circuit applied at the given point, or at any other point that will cause the highest current to flow through the given point. *Notes:* 1. This value can be in terms of either symmetrical or asymmetrical: peak or root-mean-square current, as specified. 2. In some resonant circuits, the maximum available short-circuit current may occur when the short circuit is placed at some other point than the given one where the available current is measured.
(SWG/PE/IA/PSP) C37.40-1981s, C37.100-1992, 1015-1997

available (prospective) short-circuit test current (at the point of test) The maximum short-circuit current for any given setting of a test circuit that the test power source can deliver at the point of test, with the test circuit short-circuited by a link of negligible impedance at the line terminals of the device to be tested. *Note:* This value can be in terms of either symmetrical or asymmetrical; peak or rms current, as specified.
(SWG/PE) C37.100-1992, C37.40-1981s

available signal-to-noise ratio (at a point in a circuit) The ratio of the available signal power at that point to the available random noise power. *See also:* signal-to-noise ratio.
(EEC/PE) [119]

available state A state that occurs when all of the following are true:

a) Bit-error ratio better than one in ten to the *n* power, for a specific number of consecutive observation periods of fixed duration.

b) Block-error ratio better than one in ten to the *n* power, for a specific number of consecutive observation periods of fixed duration.

c) A specific number of consecutive observation periods of fixed duration without a severely errored unit of time.

Note: the consecutive observation periods in a), b), and c) above are different for each case. (COM/TA) 1007-1991r

available time (electric drive) The period during which a system has the power turned on, is not under maintenance, and is known or believed to be operating correctly or capable of operating correctly. *See also:* electric drive. (IA/ICTL/IAC) [60]

(2) **(A)** The time during which a functional unit is on and is operating correctly or is ready to use. **(B)** In time-sharing computer systems, the time during which a system or system component is performing tasks for the user. *Contrast:* unavailable time. *See also:* makeup time. (C) 610.10-1994

avalanche The cumulative process in which charged particles accelerated by an electric field produce additional charged particles through collision with neutral gas molecules or atoms. It is therefore a cascade multiplication of ions. (NI/NPS) 309-1999

avalanche breakdown (1) (germanium gamma-ray detectors) (charged-particle detectors) (of a semiconductor device) A breakdown that is caused by the cumulative multiplication of charge carriers through field-induced impact ionization. (ED/NPS) 216-1960w, 300-1988r

(2) A breakdown caused by the cumulative multiplication of charge carriers through electric-field-induced impact ionization. (NPS) 325-1996

avalanche impedance *See:* breakdown impedance; semiconductor.

avalanche photodiode (fiber optics) A photodiode designed to take advantage of avalanche multiplication of photocurrent. *Note:* As the reverse-bias voltage approaches the breakdown voltage, hole-electron pairs created by absorbed photons acquire sufficient energy to create additional hole-electron pairs when they collide with ions; thus, a multiplication (signal gain) is achieved. *See also:* photodiode; PIN diode. (Std100) 812-1984w

average absolute burst magnitude (audio and electroacoustics) The average of the instantaneous burst magnitude taken over the burst duration. *See also:* burst; burst duration. (SP) [32]

average absolute pulse amplitude The average of the absolute value of the instantaneous amplitude taken over the pulse duration. (IM/WM&A) 194-1977w

average absolute value

$$y_{\mathrm{AAV}} = \frac{1}{T}\int_{a}^{a+T} |y|\, dt$$

where y_{AAV} is the average absolute value (often called simply the average) of y, a is any instant of time, and T is the period. (PE/PSIM) 120-1989r

average active power

$$P = \frac{1}{T}\int_{t_0-T/2}^{t_0+T/2} p\, dt$$

where P is the average active power at any time t_0, $p = vi$ is the instantaneous power, v and i are the instantaneous values of voltage and current, and T is the period. If both the voltage and current are sinusoidal and of the same period, P is given by

$$P = VI \cos \theta$$

where V and I are the rms values of voltage and current respectively and θ is the phase angle separating V and I. The general expression for polyphase active power in a system with m phases and n harmonics involves a summation over all harmonies in accordance with the above equation and a summation over all phases. (PE/PSIM) 120-1989r

average bundle gradient (overhead-power-line corona and radio noise) For a bundle of two or more subconductors, the arithmetic mean of the average gradients of the individual subconductors. (PE/T&D) 539-1990

average busy season *See:* average busy season busy-hour load; busy season; time-consistent traffic measures.

average busy season busy-hour load The busy-hour traffic level averaged across the busy season. Busy season data excludes Mother's Day, Christmas, or extremely high-traffic days that can be attributed to unusually severe weather or catastrophic events and are not reasonably expected to recur from year to year. Generally, but not always, the busy season data excludes weekends. *See also:* busy season; time-consistent traffic measures; busy hour. (COM/TA) 973-1990w

average crossing rate (1) (electromagnetic site survey) The average number of crossings in the positive direction of a given level *v* 1 per unit time. (EMC) 473-1985r

(2) (control of system electromagnetic compatibility) The average rate at which a specified level (zero if not specified) is crossed in the positive-going direction. (EMC) C63.12-1987

average current (periodic current) The value of the current averaged over a full cycle unless otherwise specified. *See also:* rectification. (IA) 59-1962w, [12]

average demand The consumption (e.g., energy, volume) recorded during the integration period divided by the integration time period. (AMR/SCC31) 1377-1997

average detector (overhead-power-line corona and radio noise) A detector, the output voltage of which is the average value of the magnitude of the envelope of an applied signal or noise. *Notes:* 1. This detector function is often identified on radio noise meters as *field intensity* (FI). [The term "field intensity" is deprecated; *field strength* should be used.] 2. FI (field strength) setting on some radio noise meters produces a reading proportional to the average value of the logarithmic detector output on the meter scale. 3. Radio noise meters of modern design do not have the detector function identified as "FI" or "field intensity." Also, modern radio noise meters have true average detector functions, but a few still have average logarithm (sometimes called "carrier") detector functions. (T&D/PE) 539-1990

average discharge current (dielectric tests) The sum of the rectified charge quantities passing through the terminals of the test object due to partial discharges during a time interval, divided by this interval. The average discharge current is expressed in coulombs per second (amperes). (PE/PSIM) 454-1973w

average electrode current (electron tube) The value obtained by integrating the instantaneous electrode current over an averaging time and dividing by the averaging time. *See also:* electrode current. (ED) 161-1971w

average forward current rating (rectifier circuit element) The maximum average value of forward current averaged over a full cycle, permitted by the manufacturer under stated conditions. (IA) [62]

average information content, per symbol (information rate from a source, per symbol) The average of the information content per symbol emitted from a source. *Note:* The term "entropy rate" is also used to designate average information content. *See also:* information theory. (IT) [7]

average inside air temperature (of enclosed switchgear) The average temperature of the surrounding cooling air that comes in contact with the heated parts of the apparatus within the enclosure. (SWG/PE) C37.100-1992

average luminance (illuminating engineering) Luminance is the property of a geometric ray. Luminance as measured by conventional meters is averaged with respect to two independent variables, area and solid angle; both must be defined for a complete description of a luminance measurement. (ED) [127]

average magner *See:* magner.

average maximum bundle gradient (overhead-power-line corona and radio noise) For a bundle of two or more subconductors, the arithmetic mean of the maximum gradients of the individual subconductors. For example, for a three-conductor bundle with individual maximum subconductor gradients of 16.5, 16.9, and 17.0 kV/cm, the average maximum bundle gradient would be (1/3) (16.5 + 16.9 + 17.0) = 16.8 kV/cm. (T&D/PE) 539-1990

average mutual information (output symbols and input symbols) Mutual information averaged over the ensemble of pairs of transmitted and received symbols. *See also:* information theory. (IT) [7]

average noise factor *See:* average noise figure.

average noise figure (1) (average noise factor) (transducer) The ratio of total output noise power to the portion thereof attributable to thermal noise in the input termination, the total noise being summed over frequencies from zero to infinity, and the noise temperature of the input termination being standard (290 kelvins). *See also:* signal-to-noise ratio; noise figure. (EEC/PE) [119]
(2) (communication satellite) Of a two-port transducer the ratio of the total noise power to the input noise power, when the input termination is at the standard temperature of 290°K. *See also:* noise factor. (COM) [25]

average noise temperature The noise temperature of an antenna averaged over a specified frequency band. (AP/ANT) 145-1983s

average outgoing quality limit (AOQL) A statistical measure of outgoing quality. (ED) 1005-1998

average phasor power (single-phase two-wire, or polyphase circuit) A phasor of which the real component is the average active power and the imaginary component is the average reactive power. The amplitude of the phasor power is

$$S_{av} = [(P_{av})^2 + (Q_{av})^2]^{1/2}$$

where P_{av} and Q_{av} are the active and the reactive power, respectively. (Std100) 270-1966w

average picture level (television) The average signal level, with respect to the blanking level, during the active picture scanning time (averaged over a frame period, excluding blanking intervals) expressed as a percentage of the difference between the blanking and reference white levels. *See also:* television. (BT/AV) [34]

average power (1) (in a waveguide) For a periodic wave, the time-average of the power passing through a given transverse section of the waveguide in a time interval equal to the fundamental period. *See also:* average active power. (MTT) 146-1980w
(2) The time-averaged rate of energy transfer.

$$P_{\text{avg}} = \frac{1}{t_2 - t_1}\int_{t_1}^{t_2} P(t)dt$$

where
$P(t)$ = instantaneous power
t_1 = initial time
t_2 = final time of the interval over which $P(t)$ is averaged
(NIR) C95.1-1999

average power density The instantaneous power density integrated over a source repetition period. (NIR) C95.1-1999

average power output (amplitude-modulated transmitter) The radio-frequency power delivered to the transmitter output terminals averaged over a modulation cycle. *See also:* radio transmitter. (AP/ANT) 145-1983s

average relative bias The average relative bias for a test category is calculated from the individual relative biases B_{ri} and defined as:

$$B_r = \sum_{i=1}^{N} B_{ri}/N$$

where
N = the number of test samples measured by an individual service laboratory in a given test category. The sample size N shall be at least five. (NI) N42.23-1995

average sensing rms detector A detector circuit that rectifies the signal from the probe and is calibrated to give the correct rms value of a sinusoidal field at some given frequency. *Note:* If there are harmonics in the field, a field meter with an average sensing rms detector will not indicate the true rms value of the field if the signal from the probe is proportional to the time derivative of the field. If the detector contains a stage of integration, the error is reduced. The error will also be a function of the phase relation between the harmonic and fundamental field components. (T&D/PE) 1308-1994

average single-conductor (or subconductor) gradient The value E_{av}, obtained from

$$E_{\text{av}} = \frac{1}{2\pi}\int_0^{2\pi} E(\theta)\text{d}(\theta)$$

Approximately, the average single-conductor gradient is given by

$$E_{\text{av}} = \frac{\lambda}{2\pi\epsilon_0 r}$$

where
λ = Total charge on conductor per unit length
ϵ_0 = Permittivity of free space
r = Radius of conductor

For practical cases, the average conductor gradient is approximately equal to the arithmetic mean of the maximum and minimum conductor gradients. (T&D/PE) 539-1990

average test value ($\bar{X}_n$) $\bar{X}_n = (X_1 + X_2 + X_3 + \ldots + X_n)/n$

where
$X_1, X_2, \ldots, X_n$ are individual test values and n is the total number of units tested. (PE/T&D) C135.61-1997

average voltage (rotating electric machinery) The value declared by the user to be the average of the system described, where externally powered. (PE/EM) 11-1980r

average water conditions (power operations) Precipitation and runoff conditions that provide water for hydroelectric energy production approximating the average amount and distribution available over a long time period, usually the period of record. (PE/PSE) 858-1987s

average winding temperature The average temperature of the winding as determined from the ohmic resistance measured across the terminals of the winding, in accordance with the cooling curve procedure specified in IEEE Std C57.12.90-1993. (PE/TR) 1276-1997

average winding temperature rise (1) The arithmetic difference between the average winding temperature and the average temperature of the air surrounding the transformer. (PE/TR) 1276-1997
(2) The arithmetic difference between the average winding temperature and the ambient temperature as determined from the change in the ohmic resistance measured across the terminals of the winding. (PE/TR) C57.134-2000

averaging time (T_{avg}) The appropriate time period over which exposure is averaged for purposes of determining compliance with a maximum permissible exposure (MPE). For exposure durations less than the averaging time, the maximum exposure, *MPE'*, in any time interval equal to the averaging time is found from

$$MPE' = MPE\left(\frac{T_{\text{avg}}}{T_{\text{exp}}}\right)$$

where
T_{exp} = exposure duration in that interval expressed in the same units as T_{avg}. (Restrictions on peak power density limit T_{exp}.) (NIR) C95.1-1999

averaging time, electrode current *See:* electrode-current averaging time.

aversive shock An electric shock from a steady-state or a discharge current that after one exposure would motivate people to avoid situations that they felt would lead to similar experiences. (T&D/PE) 539-1990

AVL tree *See:* Adel'son-Velskii and Landis tree.

A-weighted sound level (1) (speech quality measurements) (airborne sound measurements on rotating electric machinery) A weighted sound pressure level obtained by the use of a metering characteristic and the weighting "*A*," specified in USAS S1. 4-1961 (*General Purpose Sound Level Meters*). (PE/EM) 297-1969w, 85-1973w
(2) The representation of the sound pressure level that has as much as 40 dB of the sound below 100 Hz and a similar amount above 10 000 Hz filtered out. This level best approximates the response of the average young ear when listening to most ordinary, everyday sounds. Generally designated as dBA. (PE/SUB) 1127-1998
(3) Loudness that is measured with a sound level meter using the A-weighted response filter that is built into the meter circuitry. The A-weighting filter is commonly used to measure community noise, and it simulates the frequency response of the human ear. (PE/TR) C57.12.90-1999

A-weighted sound power level (airborne sound measurements on rotating electric machinery) The A-weighted sound power level, in decibels, is equal to the sound power level determined by weighting each of the frequency bands. (PE/EM) 85-1973w

AWG American Wire Gage. Also known as the Brown and Sharp gage, AWG was devised in 1857 by J. R. Brown. This gage has the property such that its sizes represent approximately the successive steps in the process of wire drawing. Also, its numbers are retrogressive; a larger number denotes a smaller wire corresponding to the operations of drawing. These gage numbers are not arbitrarily chosen, but follow the mathematical law upon which the gage is founded. (T&D/PE) 524a-1993r

AWK A computer language designed for file processing applications. *Note:* AWK originated in the UNIX environment and was named after its originators, Aho, Wienberger, and Kernighan. (C) 610.13-1993w

axial flow (hydroelectric power plants) Used to describe any turbine, such as a propeller type with an inlet that directs the water axially toward the runner, as contrasted with radial entry to the runner. (PE/EDPG) 1020-1988r

axial magnetic centering force (rotating machinery) The axial force acting between rotor and stator resulting from the axial displacement of the rotor from magnetic center. *Note:* Unless other conditions are specified, the value of axial magnetic centering force will be for no-load and rated voltage, and for rated no-load field current and rated frequency as applicable. (PE) [9]

axial mode (laser maser) The mode in a beamguide or beam resonator that has one or more maxima for the transverse field intensities over the cross-section of the beam. (LEO) 586-1980w

axial propagation constant (fiber optics) The propagation constant evaluated along the axis of a waveguide (in the direction of transmission). *Note:* The real part of the axial propagation constant is the attenuation constant while the imaginary part is the phase constant. *Synonym:* axial propagation wave number. *See also:* attenuation coefficient; attenuation; propagation constant; attenuation constant. (Std100) 812-1984w

axial propagation wave number *See:* axial propagation constant.

axial ratio (1) (waveguide) The ratio of the axes of the polarization ellipse. *Note:* The shape of the ellipse is defined by the axial ratio, which is the major axis/minor axis. Sometimes, the ratio is defined as the reciprocal of the above, that is, minor axis. (MTT) 146-1980w
(2) (of a polarization ellipse) The ratio of the major to minor axes of a polarization ellipse. *Note:* The axial ratio sometimes carries a sign that is taken as plus if the sense of polarization is right-handed and minus if it is left-handed. *See also:* sense of polarization. (AP/ANT) 145-1993

axial ratio pattern A graphical representation of the axial ratio of a wave radiated by an antenna over a radiation pattern cut. (AP/ANT) 145-1993

axial ray (fiber optics) A light ray that travels along the optical axis. *See also:* geometric optics; meridional ray; fiber axis; skew ray; paraxial ray. (Std100) 812-1984w

axial slab interferometry *See:* slab interferometry.

axially extended interaction tube (microwave tubes) A klystron tube utilizing an output circuit having more than one gap. (ED) [45]

axis *See:* direct axis; quadrature axis; magnetic axis.

axis-of-freedom (gyros) The axis about which a gimbal provides a degree-of-freedom. (AES/GYAC) 528-1994

axle bearing A bearing that supports a portion of the weight of a motor or a generator on the axle of a vehicle. *See also:* traction motor; bearing. (EEC/PE) [119]

axle-bearing cap The member bolted to the motor frame supporting the bottom half of the axle bearing. *See also:* bearing. (EEC/PE) [119]

axle-bearing-cap cover A hinged or otherwise applied cover for the waste and oil chamber of the axle bearing. *See also:* bearing. (EEC/PE) [119]

axle circuit The circuit through which current flows along one of the track rails to the train, through the wheels and axles of the train, and returns to the source along the other track rail. (EEC/PE) [119]

axle current The electric current in an axle circuit. (EEC/PE) [119]

axle generator An electric generator designed to be driven mechanically from an axle of a vehicle. *See also:* axle-generator system. (EEC/PE) [119]

axle-generator pole changer A mechanically or electrically actuated changeover switch for maintaining constant polarity at the terminals of an axle generator when the direction of the rotation of the armature is reversed due to a change in direction of movement of a vehicle on which the generator is mounted. *See also:* axle-generator system. (EEC/PE) [119]

axle-generator regulator A control device for automatically controlling the voltage and current of a variable-speed axle generator. *See also:* axle-generator system. (EEC/PE) [119]

axle-generator system A system in which electric power for the requirements of a vehicle is supplied from an axle generator carried on the vehicle, supplemented by a storage battery. (EEC/PE) [119]

axle-hung generator *See:* axle-hung motor.

axle-hung motor A traction motor (or generator), a portion of the weight of which is carried directly on the axle of a vehicle by means of axle bearings. *Synonym:* axle-hung generator. *See also:* traction motor. (EEC/PE) [119]

A-0 A programming language developed in 1953 for UNIVAC computers; uses three-address code instructions for solving mathematical problems. *Note:* Developed by Grace Hopper, A-0 was the first computer language for which a compiler was developed. (C) 610.13-1993w

A-0 context diagram The only context diagram that is a required for a valid IDEF0 model, the **A-0** diagram contains one box, which represents the top-level function being modeled, the inputs, controls, outputs, and mechanisms attached to this box, the full model name, the model name abbreviation, the model's purpose statement, and the model's viewpoint statement. (C/SE) 1320.1-1998

A-0 diagram *See:* A-0 context diagram.

azimuth (1) (A) (navigation aid terms) The direction of a celestial point from a terrestrial point, expressed as the angle in

the horizontal plane between a reference line and the horizontal projection of the line joining the two points. *Note:* True north is usually, but not always, implied where no reference direction is stated. **(B) (navigation aid terms)** The angle between horizontal reference direction and the horizontal of the direction of boresight of the antenna. **(C) (navigation aid terms)** Bearing. (AES/GCS) 172-1983
(2) (illuminating engineering) The angular distance between the vertical plane containing a given line or celestial body and the plane of the meridian. (ED) [127]
(3) The angle between a horizontal reference direction (usually north) and the horizontal projection of the direction of interest, measured clockwise. (AES/PE/T&D) 686-1997, 1260-1996

azimuth discrimination *See:* angular resolution.

azimuth markers *See:* azimuth marks.

azimuth marks (1) (radar) (navigation aid terms) (markers) Calibration marks for azimuth. (AES/GCS) 172-1983w
(2) Calibration marks used on a display for azimuth. *Synonym:* azimuth markers. (AES) 686-1997

azimuth-stabilized plan-position indicator A plan-position indicator (PPI) on which the reference direction remains fixed with respect to the indicator, regardless of the vehicle orientation. (AES) 686-1997

B

b *See:* bit.

B A procedural language used in non-numerical computations; primarily designed for systems programming. *Note:* B is based on BCPL and is a precursor to C. (C) 610.13-1993w

babble The aggregate crosstalk from a large number of interfering channels. *See also:* signal-to-noise ratio. (SP) 151-1965w

back (motor or generator) (turbine or drive end) The end that carries the largest coupling or driving pulley. *See also:* armature. (PE) [9]

back-annotation The annotation of information from further downstream steps (toward fabrication) in the design process. *See also:* back-annotation file. (C/DA) 1481-1999

back-annotation file A file containing information to be read by a tool for the purpose of back-annotation, for example Physical Design Exchange Format (PDEF) and Standard Parasitic Exchange Format (SPEF) files. *See also:* back-annotation; timing annotation. (C/DA) 1481-1999

backbone network A network designed to interconnect lower speed distribution channels, devices, or clusters of dispersed users. (C) 610.7-1995

back-connected device A device in which the current-carrying conductors are fastened to the studs in the rear of the mounting base. (SWG/PE) C37.100-1992

back-connected fuse (high-voltage switchgear) A fuse in which the current-carrying conductors are fastened to the studs in the rear of the mounting base. (SWG/PE) C37.40-1993

back-connected switch A switch in which the current-carrying conductors are connected to studs in back of the mounting base. (SWG/PE) C37.100-1981s

back contact (1) (electric power apparatus relaying) A contact that is closed when the relay is reset. *Synonym: b* contact. (SWG/PE/PSR) C37.90-1978s, C37.100-1981s
(2) (utility-consumer interconnections relaying) A contact that is closed when the relay is de-energized. (PE/PSR) C37.95-1973s

back course (navigation aid terms) [instrument landing system (ILS)] The course that is located on the opposite side of the localizer from the runway. (AES/GCS) 172-1983w

back edge By convention, the edge of the module closest to the backplane. (C/MM) 1101.2-1992

backed stamper (phonograph techniques) (mechanical recording) A thin metal stamper that is attached to a backing material, generally a metal disk of desired thickness. *See also:* phonograph pickup. (NESC/SP) [32], [86]

backend Pertaining to one part of a process which has two parts, the frontend and the backend; the frontend usually denotes what the user sees and the backend denotes some special process. *Contrast:* backend; frontend. *See also:* backend computer. (C) 610.10-1994w

backend computer A specialized computer that is attached to another computer, known as a frontend, or host, computer that handles the interface to the users while the backend computer performs functions such as database access, simulation, or vector processing. *Synonyms:* backend processor; backend machine. *Contrast:* front end computer. *See also:* bifunctional machine. (C) 610.10-1994w

backend machine *See:* backend computer.

backend processor *See:* backend computer.

backfeed To energize a section of a power network that is supplied from a source other than its normal source. (SWG/PE) C37.100-1992

backfill Materials such as sand, crushed stone, or soil, that are placed to fill an excavation. (NESC) C2-1997

back filter A filter inserted in the power line feeding an equipment to be surge tested; this filter has a dual purpose

— of preventing the applied surge from being fed back to the power source where it may [*might*, according to the word usage in this guide] cause damage.
— of eliminating loading effects of the power source on the surge generator.

See also: decoupling network. (SPD/PE) C62.45-1992r

backfire antenna An antenna consisting of a radiating feed, a reflector element, and a reflecting surface such that the antenna functions as an open resonator, with radiation from the open end of the resonator. (AP/ANT) 145-1993

back flashover (lightning) A flashover of insulation resulting from a lightning stroke to part of a network or electric installation that is normally at ground potential. *See also:* direct-stroke protection. (T&D/PE/SPD) [10], C62.23-1995, 1243-1997, 1410-1997

back-flashover rate The annual outage rate on a circuit or tower-line length basis caused by back flashover on a transmission line. (PE/T&D) 1243-1997

back focal length (laser maser) The distance from the last optical surface of a lens to the focal point. (LEO) 586-1980w

background (1) (x-ray energy spectrometers) (associated with a spectral peak from a semiconductor detector) Non-ideal spectral response that results from radiation that is not part of the monoenergetic line of interest. (NPS/NID) 759-1984r
(2) (test, measurement, and diagnostic equipment) Those effects present in physical apparatus or surrounding environment that limit the measurement or observation of low-level signals or phenomena; commonly referred to as noise (background acoustical noise, background electromagnetic radiation, background ionizing radiation). (MIL) [2]
(3) (nuclear power plants) Spectral data including peaks not caused by the source but rather resulting from radioactive decay occurring in the surrounding environment or resulting from cosmic-ray interactions in or adjacent to the detector. *See also:* baseline. (NI) N42.14-1991
(4) (software) (job scheduling) The computing environment in which low-priority processes or those not requiring user interaction are executed. *Contrast:* foreground. *See also:* background processing. (C) 610.12-1990
(5) (micrographics) The portion of a document that does not contain lettering or other information. (C) 610.2-1987
(6) (image processing and pattern recognition) A connected component of a region's complement such that the connected component completely surrounds the region.

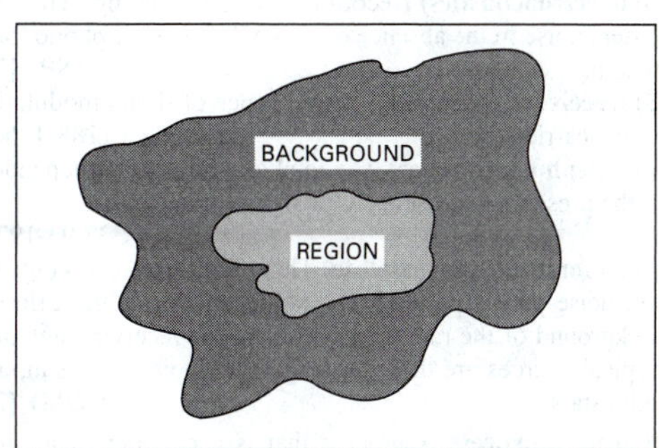

illustration of background

(C) 610.4-1990w

(7) (ambient) The spectrum of X or gamma rays originating from materials other than the radionuclide being measured. (NPS) 325-1996

(8) (under a peak) The background from all sources under a peak being measured, including Compton and degraded-energy counts from higher energy and ambient background events. (NPS) 325-1996

(9) Ambient signal response, recorded by measuring instruments, that is independent of radioactivity contributed by the radionuclides being measured in the sample. (NI) N42.23-1995

background check source A sealed vial of liquid-scintillation solution containing no added radioactive material. (NI) N42.15-1990, N42.16-1986

background count rate (liquid-scintillation counters) (in radioactivity counters) Count rate recorded by the instrument when measuring a background check source. (NI) N42.15-1990, N42.16-1986

background counts Counts caused by ionizing radiation coming from sources other than that to be measured and by any electronic disturbance in the circuitry that is used to record the counts. (NI/NPS) 309-1999

background fields Any electric or magnetic field that does not originate from the VDT under test. (EMC) 1140-1994r

background image The part of a display image that can not be modified. *Contrast:* foreground image. *See also:* form overlay. (C) 610.6-1991w

background ionization voltage (surge arresters) A high-frequency voltage appearing at the terminals of the apparatus to be tested that is generated by ionization extraneous to the apparatus. *Note:* While this voltage does not add arithmetically to the radio influence or internal ionization voltage, it affects the sensitivity of the test. *See also:* surge arrester. (PE/IA/APP) [8], [79]

background job *See:* background process group.

background level (sound measurement) Any sound at the points of measurement other than that of the machine being tested. It also includes the sound of any test support equipment. (PE/EM) 85-1973w

background noise (1) (A) (radio noise from overhead power lines and substations) The total system noise independent of the presence or absence of radio noise from the power line or substation. *Note:* Background noise is presumed to be reduced to a level of insignificance. **(B)** The total of all sources of interference in a system used for the production, detection, measurement, or recording of a signal, independent of the presence of the signal. *Note:* Ambient noise detected, measured, or recorded with the signal becomes part of the background noise. *See also:* ambient noise. (T&D/PE) 430-1986, 539-1990

(2) (data transmission) Noise due to audible disturbances or periodic random occurrence, or both. (PE) 599-1985w

(3) Noise due to audible disturbances of periodic and/or random occurrence. *See also:* modulation. (AP/ANT) 145-1983s

(4) (electroacoustics) (recording and reproducing) The total system noise in the absence of a signal. *See also:* phonograph pickup. (SP) [32]

(5) (receivers) The noise in the absence of signal modulation on the carrier. 188-1952w

(6) (telephone practice) The total system noise independent of the presence or absence of a signal. (PE/PSR) C37.93-1976s

(7) (communication satellite) That part of the receiving system noise power produced by noise sources in the celestial background of the radiation pattern of the receiving antenna. Typical sources are the galaxy (galactic noise), the sun, and radio stars. (COM) [25]

background process A process that is a member of a background process group. (C/PA) 9945-1-1996, 9945-2-1993, 1003.5-1999

background process group Any process group, other than a foreground process group, that is a member of a session that has established a connection with a controlling terminal. *Synonym:* background job. (C/PA) 9945-1-1996, 9945-2-1993, 1003.5-1999

background processing (software) The execution of a low-priority process while higher priority processes are not using computer resources, or the execution of processes that do not require user interaction. *Contrast:* foreground processing. (C) 610.12-1990

background response (radiation detectors) Response caused by ionizing radiation coming from sources other than that to be measured. *See also:* ionizing radiation. (NPS) 398-1972r

background return *See:* clutter.

backing (rotating machinery) (planar structure) A fabric, mat, film, or other material used in intimate conjunction with a prime material and forming a part of the composite for mechanical support or to sustain or improve its properties. (PE) [9]

backing lamp *See:* backup lamp.

backlash (1) (general) A relative movement between interacting mechanical parts, resulting from looseness. *See also:* industrial control; feedback control system. (IA/PE/EDPG/APP/IAC) [61], [93], [69], [60]

(2) (signal generators) The difference in actual value of a parameter when the parameter is set to an indicated value by a clockwise rotation of the indicator, and when it is set by a counterclockwise rotation. *See also:* signal generator. (IM/HFIM) [40]

(3) (tunable microwave tube) The amount of motion of the tuner control mechanism (in a mechanically tuned oscillator) that produces no frequency change upon reversal of the motion. (ED) 158-1962w, [45]

back light (illuminating engineering) Illumination from behind the subject in a direction substantially parallel to a vertical plane through the optical axis of the camera. (EEC/IE) [126]

back lobe A radiation lobe whose axis makes an angle of approximately 180 degrees with respect to the beam axis of an antenna. *Note:* By extension, a radiation lobe in the half-space opposed to the direction of peak directivity. (AP/ANT) 145-1993

back office application (BOA) Intelligent transportation system (ITS) or other application that resides and executes on the back office equipment (BOE). BOAs exchange messages with the onboard equipment (OBE) via the resource manager. (SCC32) 1455-1999

back office equipment (BOE) Computer equipment that hosts the applications and data required for some intelligent transportation system (ITS) function. Data is exchanged with the OBE via the vehicle-to-roadside communications (VRC) controller. (SCC32) 1455-1999

back out *See:* roll back.

back pitch (rotating machinery) The coil pitch at the nonconnection end of a winding (usually in reference to a wave winding). (PE) [9]

backplane (1) The printed circuit board that contains the connectors and interconnect traces. (C/MM) 961-1987r

(2) A printed circuit board (pcb) on which connectors are mounted, into which boards or plug-in units are inserted. (C/MM) 1000-1987r

(3) A printed-circuit board (pcb) with 96-pin connectors and signal paths that bus (connect corresponding) connector pins. Some systems have a single pcb, J1 backplane. It provides the signal paths needed for basic operation. Other systems also have an optional second pcb, J2 backplane. It provides the additional 96-pin connectors and signal paths needed for wider data and address transfers. Still others have a single pcb, J1/J2 backplane, which provides the signal conductors and connectors of the J1 and J2 backplanes. (C/BA) 1014-1987

(4) A circuit board with one or more bus connectors that provides signals for communication between bus modules, and provides certain resources to the connected modules. (C/MM) 1196-1987w

(5) The physical mechanism by which signals are routed between agents. (C/MM) 1296-1987s
(6) A board that holds the connectors into which SCI modules can be plugged. In ring-based SCI systems, the backplane may contain wiring that connects the output link of one module to the input link of the next. In switch-based SCI systems, the backplane may merely provide mechanical mounting for connectors that are connected by cables to the switch circuitry; or, part of the switch circuitry may be implemented on the backplane. Usually the backplane provides power connections, power status information and physical position information to the module. (C/MM) 1596-1992
(7) An electronic circuit board and connectors used to interconnect modules together electrically. The backplane connects selected pins of the connectors, thus providing the medium for the transfer of signals needed for the operation of the bus.
(C/BA) 1014.1-1994w, 896.4-1993w, 896.2-1991w, 896.3-1993w
(8) An assembly, typically a PCB, with 96-pin connectors and signal paths that bus the connector pins. VXIbus systems will have up to three sets of bussed connectors, called the J1, J2, and J3 backplanes. (MM/C) 1155-1992
(9) A subassembly that holds the connectors into which one or more boards can be plugged. In addition to providing bus signal connections, the backplane usually provides power connections, power status information, and physical position information to the board. (C/MM) 1212-1991s
(10) Motherboard comprising connectors for the modules of a system and wiring interconnecting those modules. The intermodule wiring of the MTM-Bus is expected to be on this motherboard. (C/TT) 1149.5-1995
(11) Circuit board (typically printed) at the rear of a crate which, by means of its attached connectors, mates with the modules and constitutes the crate segment.
(NID) 960-1993
(12) (A) The circuitry and mechanical elements used to connect the circuit boards within a computer system. *Note:* This circuitry is usually limited to terminating the bus signals, and sometimes generating central clocks or providing an arbiter. (B) The main circuit board of a computer into which other circuit boards are plugged. *Contrast:* motherboard.
(C) 610.10-1994
(13) An electronic and/or fiber optic interconnected set of connectors used to connect modules together electrically and/or optically. (The backplane connects selected pins of the connectors, thus providing the medium for the transfer of signals needed for the operation of the bus. Note that a backplane may contain zero or more backplane buses as well as other communications and power interconnects.).
(C/BA) 14536-1995
(14) A board that holds the connectors into which SCI modules can be plugged. In ring-based SCI systems, the backplane may contain wiring that connects the output link of one module to the input link of the next. Usually the backplane provides power connections, power status information, and physical position information to the module.
(C/MM) 1596.3-1996
(15) An electronic circuit board and connectors used to interconnect modules together electrically. The backplane connects selected pins of the connectors, thus providing the medium for the transfer of signals needed for the operation of the bus. (C/BA) 896.10-1997

backplane bus (1) A means to connect circuit modules using common signal traces on a backplane and a standard set of rules.
(C/BA) 896.3-1993w, 896.2-1991w, 896.4-1993w, 896.10-1997
(2) A means in a backplane to connect corresponding signals of circuit modules using a standard set of electrical, timing, and logical rules. (C/BA) 14536-1995

backplane capacity The minimum and maximum number of backplane slots permitted. (C/BA) 896.2-1991w

backplane interface logic (1) Special logic that takes into account the characteristics of the backplane; its signal line impedance, propagation time, termination values, etc. The specification prescribes certain rules for the design of this logic based on the maximum length of the backplane and its maximum number of printed-circuit board (pcb) slots.
(C/BA) 1014-1987
(2) (VSB bus structure) Special interface logic that takes into account the characteristics of the backplane. The VSB specification prescribes certain requirements for the design of this logic, which takes into account the signal line impedance, propagation times, termination values, the maximum length of the backplane, and the number of slots allowed.
(MM/C) 1096-1988w

backplane interface standard A set of specifications that define physical and electrical attributes, and some functional and protocol properties, of electronic modules for interconnection to a common backplane interface.
(C/BA) 14536-1995

backplane PHY *See:* backplane physical layer.

backplane physical layer The version of the physical layer applicable to the Serial Bus backplane environment.
(C/MM) 1394-1995

backplane slot The backplane connections devoted to a single module. (C/BA) 14536-1995

back plate (camera tubes) (signal plate) The electrode in an iconoscope or orthicon camera tube to which the stored charge image is capacitively coupled. *See also:* television.
(BT/AV) [34]

back porch (1) (monochrome composite picture signal). The portion that lies between the trailing edge of a horizontal synchronizing pulse and the trailing edge of the corresponding blanking pulse.
(2) (National Television System Committee composite color-picture signal). The portion that lies between the color burst and the trailing edge of the corresponding blanking pulse. *See also:* television. (BT/AV) [34]

backquote The character "`", also known as a *grave accent.*
(C/PA) 9945-2-1993

back relay contacts Sometimes used for relay contacts, normally closed. (EEC/REE) [87]

backscatter (1) The scattering of waves back toward the source.
(AP/PROP) 211-1997
(2) Energy reflected or scattered in a direction opposite to that of the incident wave. (AES) 686-1997

backscatter coefficient A normalized measure of radar return from a distributed scatterer. For area targets, such as ground or sea clutter, it is defined as the average *monostatic* radar cross section per unit surface area. *Note:* The backscatter coefficient for area targets, often expressed in decibels and denoted by σ^0, is dimensionless but is sometimes written in units of m^2/m^2 for clarity. For volume scatter, such as that from rain, chaff, or deep snow cover, it is defined as the average monostatic radar cross section per unit volume and is expressed in units of m^2/m^3 or m^{-1}. The volume backscatter coefficient is often expressed in decibels and denoted by the symbol η_v. (AES) 686-1997

backscatter cross section *See:* radar cross section.

backscattering (fiber optics) The scattering of light into a direction generally reverse to the original one. *See also:* reflection; Rayleigh scattering; reflectance. (Std100) 812-1984w

backscattering coefficient (echoing area) (data transmission) Of an object for an incident plane wave is 4π the ratio of the reflected power per unit solid angle ϕ_r in the direction of the source to the power per unit area (W_1) in the incident wave:

$$B = 4\pi \frac{\phi_r}{W_1} = 4\pi r^2 \frac{W_r}{W_1}$$

where (W_r) is the power per unit area at distance r. *Note:* For large objects, the backscattering coefficient of an object is approximately the product of its interception area and its scattering gain in the direction of the source, where the interception area is the projected geometrical area and the scattering gain is the reradiated power gain relative to an isotropic radiator. (PE) 599-1985w

backscattering cross section (1) The scattering cross-section in the direction towards the source. *See also:* backscattering coefficient; radar cross section. (RS/ANT)
(2) (radar) *See also:* monostatic cross section; radar cross section. (AES/AP/RS/ANT/PROP) 686-1990, 145-1993, 211-1997

back-shunt keying A method of keying a transmitter in which the radio-frequency energy is fed to the antenna when the telegraph key is closed and to an artificial load when the key is open. *See also:* radio transmission. (AP/BT/ANT) 145-1983s, 182A-1964w

backslash The character "\", also known as a *reverse solidus.* (PA/C) 9945-2-1993

⟨**backspace**⟩ A character that normally causes printing (or displaying) to occur one column position previous to the position about to be printed. The ⟨`backspace`⟩ shall be the character designated by `'\b'` in the C-language binding. It is unspecified whether this character is the exact sequence transmitted to an output device by the system to accomplish the backspace function. The ⟨`backspace`⟩ character defined here is not necessarily the ERASE special character defined in POSIX.1. (C/PA) 9945-2-1993

backspace character (BS) A format effector character that causes the print or display position to move one position backward along the line without producing the printing or display of any graphic. (C) 610.5-1990w

backstop, relay *See:* relay backstop.

backswing (pulse transformers) (last transition overshoot) The maximum amount by which the instantaneous pulse value is below the zero axis in the region following the fall time. It is expressed in amplitude units or as a percentage of A_M. *See also:* input pulse shape. (PEL/MAG/ET) 390-1987r, 391-1976w

back-to-back capacitor bank switching Switching a capacitor bank with and in close electrical proximity to one or more other capacitor banks.power systems relaying. (PE) C37.99-2000

back-to-back switching The switching of a capacitor bank that is connected in parallel with one or more other capacitor banks. (T&D/PE) 1036-1992

back-to-back test A test of a bipolar station in which the transmission terminals of two converters are temporarily jumpered in the station. One converter is run as rectifier while the other converter is run as inverter. (PE/SUB) 1378-1997

back-to-back testing Testing in which two or more variants of a program are executed with the same inputs, the outputs are compared, and errors are analyzed in case of discrepancies. *See also:* mutation testing. (C) 610.12-1990

backup (supervisory control, data acquisition, and automatic control) Provision for an alternate means of operation if the primary system is not available. (SWG/PE/SUB) C37.100-1992, C37.1-1994
(2) (A) (software) A system, component, file, procedure, or person available to replace or help restore a primary item in the event of a failure or externally-caused disaster. **(B)** To create or designate a system, component, file, procedure, or person as in definition (A) above. (C) 610.5-1990, 610.12-1990

backup air-gap device An air-gap device connected in parallel with a sealed gas-tube device, having a higher breakdown voltage than the gas tube, which provides a secondary means of protection in the event of a venting to atmosphere by the primary gas-tube device. (SPD/PE) C62.31-1987r, C62.32-1981s

backup current-limiting fuse (1) A fuse capable of interrupting all currents from the maximum rated interrupting current down to the rated minimum interrupting current. *See also:* function Class-A (back-up) current-limiting fuse. (SWG/PE) C37.40-1993
(2) A fuse capable of interrupting all currents from the rated maximum interrupting current down to the rated minimum interrupting current. (SWG/PE) C37.100-1992

backup, degraded *See:* degraded backup.

backup gap (series capacitor) A supplementary gap that may be set to sparkover at a voltage level higher than the protective level of the primary protective device, and that is normally placed in parallel with the primary protective device. (PE/T&D) 824-1985s

backup lamp (illuminating engineering) A lighting device mounted on the rear of a vehicle for illuminating the region near the back of the vehicle while moving in reverse. It normally can be used only while backing up. (EEC/IE) [126]

backup overcurrent protective device or apparatus (nuclear power generating station) A device or apparatus that performs the circuit interrupting function in the event the primary protective device or apparatus fails or is out of service. (PE/NP) 317-1976s

backup path Secondary transmission path in trunk cabling and concentrator, normally used for token ring signal transmission only when there is a failure on the main ring path. (C/LM) 8802-5-1998

backup programmer (software) The assistant leader of a chief programmer team; responsibilities include contributing significant portions of the software being developed by the team, aiding the chief programmer in reviewing the work of other team members, substituting for the chief programmer when necessary, and having an overall technical understanding of the software being developed. *See also:* chief programmer. (C) 610.12-1990

backup protection (as applied to a relay system) A form of protection that operates independently of specified components in the primary protective system. It may duplicate the primary protection or may be intended to operate only if the primary protection fails or is temporarily out of service. (SWG/PE/PSR) C37.100-1992, C37.90-1978s

backup zone The protected zone of a relay that is not the primary protection. It is usually time delayed (e.g., zones 2 and 3 of a distance relay). In addition, the backup zone will usually remove more of the system elements than required by the operation of the primary zone of protection. (PE/PSR) C37.113-1999

Backus-Naur form A recursive metalanguage used to specify or describe the syntax of a language in which each symbol, by itself, represents a set of strings of symbols. *Note:* Developed by John Backus and Peter Naur, BNF was one of the first formal systems developed to specify languages. *See also:* ALPHA. (C/ATLAS) 610.13-1993w, 771-1989s

Backus-Naur format (BNF) A particular metalanguage developed by Backus and Naur. (SCC20) 771-1998

Backus-Naur notation (BNN) A general term relating to metalanguages that use the concepts developed by Backus and Naur. (SCC20) 771-1998

backus normal *See:* Backus-Naur form.

backward-acting regulator A transmission regulator in which the adjustment made by the regulator affects the quantity that caused the adjustment. *See also:* transmission regulator. (EEC/PE) [119]

backward channel A channel, associated with the forward channel, used for supervisory or error control signals, but with a direction of transmission opposite to that of the forward channel in which user information is being transferred. *Note:* In the case of simultaneous transfer of user information in both directions, this definition applies with respect to the data source under consideration. *Synonym:* reverse channel. (C) 610.10-1994w

backward diode (nonlinear, active, and nonreciprocal waveguide components) A semiconductor device used primarily as a detector or mixer. Quantum-mechanical tunneling in this diode results in a current-voltage characteristic in which the reverse current is greater than the forward current for equal applied voltages of opposite polarity. (MTT) 457-1982w

backward execution *See:* reversible execution.

backward read To read data from a sequential storage medium in a reverse direction; for example, to read a magnetic tape from the end to the beginning. (C) 610.10-1994w

backward recovery (1) The reconstruction of a file to a given state by reversing all changes made to the file since it was in that state. *Contrast:* forward recovery; inline recovery. (C) 610.5-1990w, 610.12-1990
(2) A type of recovery in which a system program database or other system resource is restored to a previous state in which it can perform required functions. (C) 610.12-1990

backward supervision The use of supervisory sequences from a secondary station or node to a primary station or node. *Contrast:* forward supervision. (C) 610.7-1995

backward wave (traveling-wave tubes) A wave whose group velocity is opposite to the direction of electron-stream motion. *See also:* amplifier. (ED) 161-1971w

backward-wave oscillator *See:* carcinotron.

backward-wave structure (BW) (microwave tubes) A slow-wave structure whose propagation is characterized on an ω/β diagram (sometimes called a Brillouin diagram) by a negative slope in the region $0 < \beta < \pi$ (in which the phase velocity is therefore of opposite sign to the group velocity). (ED) [45]

back wave A signal emitted from a radio telegraph transmitter during spacing portions of the code characters and between the code characters. *See also:* radio transmission. (BT) 182A-1964w

bactericidal effectiveness (illuminating engineering) The capacity of various portions of the ultraviolet spectrum to destroy bacteria, fungi, and viruses. *Synonym:* germicidal effectiveness. (EEC/IE) [126]

bactericidal efficiency of radiant flux (illuminating engineering) The ratio of the bactericidal effectiveness of that wavelength to that of wavelength 265.0 nm (nanometers), which is rated as unity. (EEC/IE) [126]

bactericidal exposure (illuminating engineering) The product of bactericidal flux density on a surface and time. It usually is measured in bactericidal microwatt-minutes per square centimeter or bactericidal watt-minutes per square foot. *Synonym:* germicidal exposure. (EEC/IE) [126]

bactericidal flux (illuminating engineering) Radiant flux evaluated according to its capacity to produce bactericidal effects. It is usually measured in microwatts of ultraviolet radiation weighted in accordance with its bactericidal efficiency. Such quantities of bactericidal flux would be in bactericidal microwatts. *Note:* Ultraviolet radiation of wavelength 253.7 nm (nanometers) is usually referred to as "ultraviolet microwatts" or "UV watts." These terms should not be confused with "bactericidal microwatts" because the radiation has not been weighted in accordance with the values given in the table under erythemal flux density. *Synonym:* germicidal flux. (EEC/IE) [126]

bactericidal flux density (illuminating engineering) The bactericidal flux per unit area of the surface being irradiated. It is equal to the quotient of the incident bactericidal flux divided by the area of the surface when the flux is uniformly distributed. It is usually measured in microwatts per square centimeter or watts per square foot of bactericidally weighted ultraviolet radiation (bactericidal microwatts per square centimeter or bactericidal watts per square foot). *Synonym:* germicidal flux density. (EEC/IE) [126]

badge number A numeric character code assigned to a badge. (PE/NP) 692-1997

badge reader A reader capable of reading information on specially coded badges or cards. (C) 610.10-1994w

baffle (1) (audio and electroacoustics) A shielding structure or partition used to increase the effective length of the transmission path between two points in an acoustic system; as, for example, between the front and back of an electroacoustic transducer. *Note:* In the case of a loudspeaker, a baffle is often used to increase the acoustic loading of the diaphragm. (SP) [32]
(2) (illuminating engineering) A single opaque or translucent element to shield a source from direct view at certain angles, or to absorb unwanted light. (EEC/IE) [126]
(3) (gas tube) An auxiliary member, placed in the arc path and having no separate external connection. *Note:* A baffle may be used for:
a) Controlling the flow of mercury vapor or mercury particles,
b) Controlling the flow of radiant energy,
c) Forcing a distribution of current in the arc path,
d) Deionizing the mercury vapor following conduction. It may be of either conducting or insulating material.
See also: electrode. (ED) [45]

bag A kind of collection class whose members are unordered but in which duplicates are meaningful. *Contrast:* list; set. (C/SE) 1320.2-1998

bag-type construction (dry cell) (primary cell) A type of construction in which a layer of paste forms the principal medium between the depolarizing mix, contained within a cloth wrapper, and the negative electrode. *See also:* electrolytic cell. (EEC/PE) [119]

baker board *See:* lineperson's platform.

balance beam (of a relay) A lever form of relay armature, one end of which is acted upon by one input and the other end restrained by a second input. (SWG/PE/PSR) C37.100-1992, C37.90-1978s

balance check In an analog computer, the computer-control state in which all amplifier summing junctions are connected to the computer zero reference level (usually signal ground) to permit zero balance of the operational amplifiers. (C) 610.10-1994w, 165-1977w

balanced (1) (general) Used to signify proper relationship between two or more things, such as stereophonic channels.
(2) (data transmission) In communication practice, signifies electrically alike and symmetrical with respect to ground, or arranged to provide conjugate conductors between certain sets of terminals. (PE) 599-1985w
(3) (to ground) The state of impedance on a two-wire circuit when the impedance-to-ground of one wire is equal to the impedance-to-ground of the other wire. *Contrast:* unbalanced. *See also:* balun. (C) 610.7-1995
(4) Pertaining to a relationship between two or more objects that are alike or symmetrical in some respect. *Contrast:* unbalanced. (C) 610.10-1994w

balanced amplifier (push-pull amplifier) An amplifier in which there are two identical signal branches connected so as to operate in phase opposition and with input and output connections each balanced to ground. (AP/BT/PE/ANT) 145-1983s, 182-1961w, 599-1985w

balanced cable (1) A cable consisting of one or more metallic symmetrical cable elements (twisted pairs or quads).local area networks. (LM/C) 802.3u-1995s, 8802-12-1998
(2) A cable consisting of one or more metallic symmetrical cable elements (twisted pairs or quads).

balanced capacitance (between two conductors) (mutual capacitance between two conductors) The capacitance between two conductors when the changes in the charges on the two are equal in magnitude but opposite in sign and the potentials of the other $n - 2$ conductors are held constant. *See also:* direct capacitances. (IM/HFIM) [40]

balanced circuit (1) (measuring longitudinal balance of telephone equipment operating in the voice band) A circuit in which two branches are electrically alike and symmetrical with respect to a common reference point, usually ground. (COM/TA) 455-1985w
(2) (signal-transmission system) A circuit, in which two branches are electrically alike and symmetrical with respect to a common reference point, usually ground. *Note:* For an

applied signal difference at the input, the signal relative to the reference at equivalent points in the two branches must be opposite in polarity and equal in amplitude. (IM/HFIM) [40]
(3) (electric power system) A circuit in which there are substantially equal currents, either alternating or direct, in all main wires and substantially equal voltages between main wires and between each main wire and neutral (if one exists). *See also:* center of distribution. (T&D/PE) [10]

balanced conditions (1) (rotating machinery) (time domain) A set of polyphase quantities (phase currents, phase voltages, etc.) that are sinusoidal in time, that have identical amplitudes, and that are shifted in time with respect to each other by identical phase angles.
(2) (space domain) In space, a set of coils (for example, of a rotating machine) each having the same number of effective turns, with their magnetic axes shifted by identical angular displacements with respect to each other. *Notes:* 1. The impedance (matrix) of a balanced machine is balanced. A balanced set of currents will produce a balanced set of voltage drops across a balanced set of impedances. 2. If all sets of windings of a machine are balanced and if the magnetic structure is balanced, the machine is balanced. *See also:* asynchronous machine. (PE) [9]

balanced currents (waveguide) (on a balanced line) Currents flowing in the two conductors of a balanced line, which, at every point along the line, are equal in magnitude and opposite in direction. (MTT) 146-1980w

balanced duplexer (radar) (nonlinear, active, and nonreciprocal waveguide components) A dualized network using two quadrature hybrids on each side of a pair of self-switching elements used to interconnect the transmitter, receiver, and antenna in a radar. *See also:* duplexer. (MTT) 457-1982w

balanced error (A) A set of error values in which the maximum and minimum are opposite in sign and equal in magnitude. *Contrast:* unbalanced error. **(B)** A set of error values whose average is zero. *Contrast:* unbalanced error. (C) 1084-1986

balanced line (waveguide) (two conductor) A transmission line consisting of two conductors in the presence of ground capable of being operated in such a way that the voltages on the two conductors at all transverse planes are equal in magnitude and opposite in direction. The ground may be a conducting sheath, forming a shielded transmission line. (MTT) 146-1980w

balanced line system (waveguide) A system consisting of a generator and a balanced line, and load-adjusted so that the voltages of the two conductors at all transverse planes are equal in magnitude and opposite in polarity with respect to ground. (MTT) 146-1980w

balanced merge A merge in which the subsets to be merged are equally distributed among half of the available storage, then the subsets are merged onto the other half of storage. *Contrast:* unbalanced merge. (C) 610.5-1990w

balanced merge sort A merge sort in which the sorted subsets created by internal sorts are equally distributed among half of the available storage, the subsets are merged onto the other half of the available storage, and this process is repeated until all the items are in one sorted set. *Contrast:* unbalanced merge sort. (C) 610.5-1990w

balanced mixer (1) (single, double) A type of mixer that forms from two signals A & B a third signal C having the form C = (a+A)(b+B). "Single balanced" implies $a = 0$, $b \neq 0$; "double balanced" implies $a = b = 0$. *Note:* Such mixers can suppress a RF carrier and/or a local oscillator in their output spectrum. *Synonym:* balanced modulator. (CAS) [13]
(2) A hybrid junction with crystal receivers in one pair of uncoupled arms the arms of the remaining pair being fed from a signal source and a local oscillator. *Note:* The resulting intermediate-frequency signals from the crystals are added in such a manner that the effect of local-oscillator noise is minimized. *See also:* radio receiver; hybrid junction; converter; waveguide. (AP/ANT) [35], [84]

balanced modulator (signal-transmission system) A modulator, specifically a push-pull circuit, in which the carrier and modulating signal are so introduced that after modulation takes place the output contains the two sidebands without the carrier. *See also:* modulation. (AP/ANT) 145-1983s

balanced oscillator An oscillator in which, at the oscillator frequency, the impedance centers of the tank circuit are at ground potential and the voltages between either end and their centers are equal in magnitude and opposite in phase. *See also:* oscillatory circuit. (AP/BT/ANT) 145-1983s, 182A-1964w

balanced polyphase load A load to which symmetrical currents are supplied when it is connected to a system having symmetrical voltages. *Note:* The term "balanced polyphase load" is applied also to a load to which two currents having the same wave form and root-mean-square value and differing in phase by 90 electrical degrees are supplied when it is connected to a quarter-phase (or two-phase) system having voltages of the same wave form and root-mean-square value. *See also:* generating station. (T&D/PE) [10]

balanced polyphase system A polyphase system in which both the currents and voltages are symmetrical. *See also:* alternating-current distribution. (T&D/PE) [10]

balanced relay armature An armature that is approximately in equilibrium with respect to both static and dynamic forces. (EEC/REE) [87]

balanced telephone-influence factor (three-phase synchronous machine) The ratio of the square root of the sum of the squares of the weighted root-mean-square values of the fundamental and the nontriple series of harmonics to the root-mean-square value of the normal no-load voltage wave. (PE) [9]

balanced termination (system or network having two output terminals) A load presenting the same impedance to ground for each of the output terminals. *See also:* network analysis. (MTT) 146-1980w

balanced three-wire system A three-wire system in which no current flows in the conductor connected to the neutral point of the supply. *See also:* three-wire system; alternating-current distribution. (T&D/PE) [10]

balanced tree *See:* height-balanced tree.

balanced voltages (1) (waveguide) (on a balanced line) Voltages relative to ground on the two conductors of a balanced line which, at every point along the line, are equal in magnitude and opposite in polarity. (MTT) 146-1980w
(2) (signal-transmission system) The voltages between corresponding points of a balanced circuit (voltages at a transverse plane) and the reference plane relative to which the circuit is balanced. *See also:* signal. (IE) [43]

balanced wire circuit (data transmission) One whose two sides are electrically alike and symmetrical with respect to ground and other conductors. The term is commonly used to indicate a circuit whose two sides differ only by chance. (PE) 599-1985w

balancer That portion of a direction-finder that is used for the purpose of improving the sharpness of the direction indication. *See also:* radio receiver. (EEC/PE) [119]

balance relay A relay that operates by comparing the magnitudes of two similar input quantities. *Note:* The balance may be effected by counteracting electromagnetic forces on a common armature, or by counteracting magnetomotive forces in a common magnetic circuit, or by similar means, such as springs, levers, etc. (SWG/PE) C37.100-1992

balance test (rotating machinery) A test taken to enable a rotor to be balanced within specified limits. *See also:* rotor. (PE) [9]

balancing (1) Adjusting the gains and losses in each path of a system to achieve proper cable plant characteristics. (LM/C) 802.7-1989r

(2) (analog computer) (of an operational amplifier) The act of adjusting the output level of an operational amplifier to coincide with its input reference level, usually ground or zero voltage, in the "balance check" computer-control state. This operation may not be required in some amplifiers, and there may be no provision for performing it. (C) 165-1977w

balancing network An electric network designed for use in a circuit in such a way that two branches of the circuit are made substantially conjugate; that is, such that an electromotive force inserted in one branch produces no current in the other branch. *See also:* network analysis. (PE) 599-1985w

ballast (1) (fluorescent lamps or mercury lamps) Devices that by means of inductance, capacitance, or resistance, singly or in combination, limit the lamp current of fluorescent or mercury lamps to the required value for proper operation, and also, where necessary, provide the required starting voltage and current and, in the case of ballasts for rapid-start lamps, provide for low-voltage cathode heating. *Note:* Capacitors for power-factor correction and capacitor-discharge resistors may form part of such a ballast. (EEC/LB) [95], [94]
(2) (fixed-impedance type) (reference ballast) Designed for use with one specific type of lamp that, after adjustment during the original calibration, is expected to hold its established impedance through normal use. (EEC/LB) [96], [97]
(3) (variable-impedance type) An adjustable inductive reactor and a suitable adjustable resistor in series. *Note:* These two components are usually designed so that the resulting combination has sufficient current-carrying capacity and range of impedance to be used with a number of different sizes of lamps. The impedance and power factor of the reactor-resistor combination are adjusted, or rechecked, each time the unit is used. (EEC/LB) [96], [97]
(4) (illuminating engineering) A device used with an electric-discharge lamp to obtain the necessary circuit conditions (voltage, current, and wave form) for starting and operating. (EEC/IE) [126]
(5) (electric power systems in commercial buildings) An electrical device that is used with one or more discharge lamps to supply the appropriate voltage to a lamp for starting, to control lamp current while it is in operation, and, usually, to provide for power factor correction. (IA/PSE) 241-1990r

ballast factor (illuminating engineering) The fractional loss of task illuminance due to use of a ballast other than the standard one. (EEC/IE) [126]

ballast leakage The leakage of current from one rail of a track circuit to another through the ballast, ties, earth, etc. (EEC/PE) [119]

ballast resistance The resistance offered by the ballast, ties, earth, etc., to the flow of leakage current from one rail of a track circuit to another. (EEC/PE) [119]

ballast section (railroads) The section of material, generally trap rock, that provides support under railroad tracks. (NESC) C2-1997

ballast tube (ballast lamp) A current-controlling resistance device designed to maintain substantially constant current over a specified range of variation in the applied voltage or the resistance of a series circuit. (EEC/PE) [119]

ball bearing (rotating machinery) A bearing incorporating a peripheral assembly of balls. *See also:* bearing. (PE) [9]

ball burnishing Burnishing by means of metal balls. *See also:* electroplating. (EEC/PE) [119]

ballistic deficit (germanium gamma-ray detectors) The loss in signal amplitude that occurs when the charge collection time in a detector is a significant fraction of the amplifier's differentiating time constant. (NPS) 325-1996

ballistic focusing (microwave tubes) A focusing system in which static electric fields cause an initial convergence of the beam, and the electron trajectories are thereafter determined by momentum and space charge forces only. (ED) [44]

ballistic munition Any munition that follows a ballistic trajectory. (DIS/C) 1278.1-1995

ball lightning A type of lightning discharge reported from visual observations to consist of luminous, ball-shaped regions of ionized gases. *Note:* In reality ball lightning may or may not exist. *See also:* direct-stroke protection. (T&D/PE) [10]

balun (1) In networking, a passive device with distributed electrical constants used to couple a balanced system or device to an unbalanced system or device. For example, a transformer used to connect balanced twisted-pair cables to unbalanced coaxial cables. *Note:* Derived from "balance to unbalance" transformer. (C/CHM) 610.7-1995, [51]
(2) A network for the transformation from an unbalanced transmission line or system to a balanced line or system, or vice versa. (IM/HFIM) [40]

BAM *See:* basic access method.

banana plug A single-conductor plug with a spring metal tip that somewhat resembles a banana in shape. (IM) [120]

band (1) A group of circular recording tracks, on a moving storage device such as a drum or disc. *See also:* channel. (C/MIL) [2]
(2) (data transmission) Range of frequency between two defined limits. (PE) 599-1985w
(3) A group of tracks on a magnetic drum or a magnetic disk which are read or written as a group. (C) 610.10-1994w

band I The frequency band 5 Hz to 2 kHz. (EMC) 1140-1994r

band II The frequency band 2 kHz to 400 kHz. (EMC) 1140-1994r

band-edge The highest or lowest frequency passed for a defined range of frequencies. The band-edge frequencies are normally identified to be the half power points of a frequency band. (LM/C) 802.7-1989r

band, effective *See:* effective band.

band-elimination filter (1) A network designed to eliminate a band or frequencies. Its frequency response has a single pass band bounded by two attenuation bands. (CAS) [13]
(2) (signal-transmission system) A filter that has a single attenuation band, neither of the cutoff frequencies being zero or infinite. *See also:* filter; rejection filter. (SP/PE) 151-1965w, 599-1985w

band gap (charged-particle detectors) (in a semiconductor) The energy difference between the bottom of the conduction band and the top of the valence band. (NPS) 325-1996, 300-1988r

banding insulation (rotating machinery) Insulation between the winding overhang and the binding bands. (PE) [9]

band of regulated voltage (synchronous machines) The band or zone, expressed in percent of the rated value of the regulated voltage, within which the excitation system will hold the regulated voltage of an electric machine during steady or gradually changing conditions over a specified range of load. (PE/EDPG) 421-1972s, 421.1-1986r

band-pass (broadband local area networks) A range of frequencies that express the difference between the lowest and highest frequencies of interest. The band-pass frequencies are normally associated with frequencies that define the half power points. (LM/C) 802.7-1989r

band-pass filter (1) (data transmission) A wave filter that has a single transmission band, neither of the cutoff frequencies being zero or infinite. *See also:* optical filter. (PE) 599-1985w
(2) (broadband local area networks) A filter that allows passage of a desired range of frequencies and attenuates frequencies outside the desired range. (LM/C) 802.7-1989r

band-pass tube *See:* broadband tube.

band pressure level For a specified frequency band, the sound pressure level for the sound contained within the restricted band. The reference pressure must be specified. *Note:* The band may be specified by its lower and upper cutoff frequencies or by its geometric center frequency and bandwidth. The width of the band may be indicated by a modifying prefix, e.g., octave band (sound pressure) level, half-octave band level, third-octave band level, 50-Hz band level. (PE/T&D) 539-1990

band printer An element printer in which type slugs are carried on a flexible band. (C) 610.10-1994w

band spreading (A) The spreading of tuning indicators over a wide scale range to facilitate tuning in a crowded band of frequencies. *See also:* radio receiver. **(B)** The method of double-sideband transmission in which the frequency band of the modulating wave is shifted upward in frequency so that the sidebands produced by modulation are separated in frequency from the carrier by an amount at least equal to the bandwidth of the original modulating wave, and second-order distortion products may be filtered from the demodulator output. *See also:* radio receiver. (EEC/PE) [119]

band-stop filter (broadband local area networks) A band-stop or band reject filter attenuates a desired range of frequencies and passes frequencies that are higher and lower than the rejection band. (LM/C) 802.7-1989r

band switch A switch used to select any one of the frequency bands in which an electric transmission apparatus may operate. (EEC/PE) [119]

bandwidth (1) (amplitude-modulation broadcast receivers) As applied to the selectivity of a radio receiver, the bandwidth is the width of a selectivity graph at a specified level on the scale of ordinates. (CE) 186-1948w
(2) (device) The range of frequencies within which performance, with respect to some characteristic, falls within specific limits. *See also:* radio receiver. (T&D/PE/VT) 539-1990, [37]
(3) (signal-transmission system) The range of frequencies within which performance, with respect to some characteristic, falls within specific limits. *Notes:* 1. For systems capable of transmitting at zero frequency the frequency at which the system response is less than that at zero frequency by a specified ratio. For carrier-frequency systems: the difference in the frequencies at which the system response is less than that at the frequency of reference response by a specified ratio. For both types of systems, bandwidth is com m only defined at the points where the response is three decibels less than the reference value (0.707 root-mean-square voltage ratio). *See also:* equivalent noise bandwidth. (IE) [43]
(4) (wave) The least frequency interval outside of which the power spectrum of a time-varying quantity is everywhere less than some specified fraction of its value at a reference frequency. Warning: This definition permits the spectrum to be less than the specified fraction within the interval. *Note:* Unless otherwise stated, the reference frequency is that at which the spectrum has its maximum value. 188-1952w
(5) (burst) (burst measurements). The smallest frequency interval outside of which the integral of the energy spectrum is less than some designated fraction of the total energy of the burst. *See also:* burst. (SP) 265-1966w
(6) (of an antenna) The range of frequencies within which the performance of the antenna, with respect to some characteristics, conforms to a specified standard. (AP/ANT) [35], 145-1993
(7) (facsimile) The difference in hertz between the highest and the lowest frequency components required for adequate transmission of the facsimile signals. *See also:* facsimile. (COM) 168-1956w
(8) (excitation systems) The interval separating two frequencies between which both the gain and the phase difference (of sinusoidal output referred to sinusoidal input) remain within specified limits. *Note:* For control systems and many of their components, the lower frequency often approaches zero. *See also:* feedback control system. (IA/IM/PE/ICTL/APP/EDPG/IAC) [69], [120], [93], [60], 421A-1978s
(9) (pulse terminology) The two portions of a pulse waveform that represents the first nominal state from which a pulse departs and to which it ultimately returns. Typical closed-loop frequency response of an excitation control system with the synchronous machine open circuited.
(10) (oscilloscopes) The difference between the upper and lower frequency at which the response is 0.707 (−3 dB) of the response at the reference frequency. Usually both upper and lower limit frequencies are specified rather than the difference between them. When only one number appears, it is taken as the upper limit. *Notes:* 1. The reference frequency shall be at least 20 times greater for the lower bandwidth limit and at least 20 times less for the upper bandwidth limit than the limit frequency. The upper and lower reference frequencies are not required to be the same. In cases where exceptions must be made, they shall be noted. 2. This definition assumes the amplitude response to be essentially free of departures from a smooth roll-off characteristic. 3. If the lower bandwidth limit extends to zero frequency, the response at zero frequency shall be equal to the response at the reference frequency, not −3 dB from it. (IM/HFIM) [40]
(11) (dispersive and nondispersive delay lines) A specified frequency range over which the amplitude response does not vary more than a defined amount. *Note:* Typically, amplitude range is 1 dB bandwidth, 3 dB bandwidth. (UFFC) [22]
(12) (A) (analog computer) Of a signal, the difference between the limiting frequencies encountered in the signal. **(B) (analog computer)** Of a device, the range of frequencies within which performance in respect to some characteristic falls within specific limits. (C) 165-1977
(13) (data transmission) The range of frequencies within which performance, with respect to some characteristic, falls within specific limits. Bandwidth is commonly defined at the points where the response is three decibels less than the reference value. (PE) 599-1985w
(14) (broadband local area networks) The frequency range that a component, circuit, or system passes or uses. For example, voice transmission by telephone requires a bandwidth of about 3000 Hz (3 kHz). A television channel occupies a bandwidth of 6 000 000 Hz (6 MHz). Cable systems occupy 5−300 MHz or higher of the electromagnetic spectrum. (LM/C) 802.7-1989r
(15) The range of frequencies, expressed in hertz, that can pass over a given channel. *See also:* pass band. (C) 610.7-1995
(16) A specified frequency range over which the amplitude response does not vary more than a defined amount. *Note:* Typically, amplitude variations to specify bandwidth are 1 dB or 3 dB (dispersive and nondispersive delay lines). (UFFC) 1037-1992w
(17) (fiber optics) *See also:* fiber bandwidth. 812-1984w

bandwidth allocation protocols The protocols used to allocate bandwidth on a ringlet. This involves inhibiting send-packet transmissions from one or more nodes when another node is being starved (never gets an opportunity to transmit its send packet). (C/MM) 1596-1992

bandwidth balancing mechanism A procedure to facilitate effective sharing of the bandwidth, whereby a node occasionally skips the use of empty Queued Arbitrated (QA) slots. (LM/C) 8802-6-1994

bandwidth, coherent *See:* dispersive bandwidth.

bandwidth, dispersive *See:* dispersive bandwidth.

bandwidth, effective *See:* effective bandwidth.

bandwidth, frequency selective *See:* frequency selective bandwidth.

bandwidth-limited operation (fiber optics) The condition prevailing when the system bandwidth, rather than the amplitude (or power) of the signal, limits performance. The condition is reached when the system distorts the shape of the waveform beyond specified limits. For linear systems, bandwidth-limited operation is equivalent to distortion-limited operation. *See also:* attenuation-limited operation; distortion-limited operation. (Std100) 812-1984w

bandwidth reuse A ring segmentation feature that multiplies the overall data throughput capacity of the spaceborne fiber-optic data bus (SFODB) network by allowing independent ring segments to use the same dedicated data bandwidth. (C/BA) 1393-1999

bang snuffer (nonlinear, active, and nonreciprocal waveguide components). A switch used in radar receivers to suppress carrier leakage during the transmit period. *See also:* gate. (MTT) 457-1982w

bank (A) (navigation) Lateral inclination of an aircraft in flight. *See also:* list. **(B)** An aggregation of similar devices (for example, transformers, lamps, etc.) connected together and used in cooperation. *Note:* In automatic switching, a bank is an assemblage of fixed contacts over which one or more wipers or brushes move in order to establish electric connections. *See also:* relay level. (AES/EEC/PE/GCS) 172-1983, [119]

(2) **(A)** One or more disk drives lined up in a row. **(B)** Any group of similar devices that are connected together for use as a single device. For example, a row of light-emitting diodes connected to form a display. **(C)** A contiguous section of addressable memory. For example, eight memory devices, each of which is 64 kB by 1; forming a 64 kB × 8 memory bank. (C) 610.10-1994

bank-and-wiper switch (telephone switching systems) A switch in which an electromagnetic ratchet or other mechanisms are used, first, to move the wipers to a desired group of terminals, and second, to move the wipers over the terminals of this group to the desired bank contacts. (EEC/PE) [119]

banked winding *See:* bank winding.

bank winding (banked winding) A compact multilayer form of coil winding, for the purpose of reducing distributed capacitance, in which single turns are wound successively in each of two or more layers, the entire winding proceeding from one end of the coil to the other, without return. (IM) [120]

bar (1) (illuminating engineering) (of lights) A group of three or more aeronautical ground lights placed in a line transverse to the axis, or extended axis, of the runway. (EEC/IE) [126]

(2) The darker element of a bar code. (PE/TR) C57.12.35-1996

bar code (1) An identification code consisting of a pattern of vertical bars whose width and spacing identifies the item marked. *Note:* The code is meant to be read by an optical input device, such as a bar code scanner. Applications include retail product pricing labels, identification of library documents, and railroad box car identification. *Synonym:* optical bar code. *See also:* universal product code. (C) 610.2-1987, 610.10-1994w

(2) An array of rectangular marks and spaces in a predetermined pattern. (PE/TR) C57.12.35-1996

bar code reader *See:* bar code scanner.

bar code symbol An array of rectangular bars and spaces which are arranged in a predetermined pattern following specific rules to represent elements of data that are referred to as characters. A bar code symbol typically contains a leading quiet zone, start character, data character(s) including a check character (if any), stop character, and a trailing quiet zone. (PE/TR) C57.12.35-1996

bar code scanner An optical scanner used to read a bar-code using reflected light. *Synonym:* bar code reader. *See also:* light pen. (C) 610.10-1994w

bare conductor A conductor having no covering or electrical insulation whatsoever. *See also:* covered conductor. (NESC/NEC) [86]

barehand work A technique of performing live maintenance on energized wires and equipment whereby one or more line workers work directly on an energized part after having been raised and bonded to the same potential as the energized wire or equipment. These line workers are normally supported by an insulating ladder, nonconductive rope, insulating aerial device, helicopter, or the energized wires or equipment being worked on. Most barehand work includes the use of insulating live tools. (T&D/PE) 516-1995

bare lamp (illuminating engineering) A light source with no shielding. *Synonym:* exposed lamp. (EEC/IE) [126]

barette (illuminating engineering) A short bar in which the lights are closely spaced so that from a distance they appear to be a linear light. *Note:* Barettes are usually less than 4.6 m (15 ft) in length. (EEC/IE) [126]

bar generator (television) A generator of pulses that are uniformly spaced in time and are synchronized to produce a stationary bar pattern on a television screen. *See also:* television. 188-1952w

Barker code A binary phase code used for pulse compression, in which a long pulse is divided into n subpulses with the phase of each subpulse being 0 or π radians. Barker coded pulses have the property that after matched filter processing there are $(n - 1)/2$ sidelobes, or $n/2$ for n even, on each side of the main response, each at a voltage level $1/n$ relative to the main response. Barker codes exist with $n = 2, 3, 4, 5, 7, 9$, and 13. *See also:* coded pulse. (AES) 686-1997

Barkhausen-Kurz oscillator An oscillator of the retarding-field type in which the frequency of oscillation depends solely upon the electron transit-time within the tube. *See also:* oscillatory circuit. (AP/ANT) 145-1983s

Barkhausen tube *See:* positive-grid oscillator tube.

barometric altimeter (navigation aid terms) Essentially an aneroid barometer, an instrument which determines atmospheric pressure and is graduated in feet above sea level. (AES/GCS) 172-1983w

barothermograph (navigation aid terms) An instrument which automatically records pressure and temperature. (AES/GCS) 172-1983w

bar pattern (television) A pattern of repeating lines or bars on a television screen. When such a pattern is produced by pulses that are equally separated in time, the spacing between the bars on the television screen can be used to measure the linearity of the horizontal or vertical scanning systems. *See also:* television. (EEC/PE) [119]

bar printer An element printer in which the members of the character set are carried on a type bar. (C) 610.10-1994w

barrel connector A double-sided male coupling that interconnects two coaxial cables. *Contrast:* end connector. (C) 610.7-1995

barrel distortion (1) A defect in a display surface that causes parallel lines to bow away from each other, causing a distorted image. *See also:* pin-cushion distortion. (C) 610.6-1991w

(2) A distortion that results in a progressive decrease in radial magnification in the reproduced image away from the axis of symmetry of the electron optical system. *Note:* For a camera tube, the reproducer is assumed to have no geometric distortion. (ED) 161-1971w

barrel plating Mechanical plating in which the cathodes are kept loosely in a container that rotates. *See also:* electroplating. (EEC/PE) [119]

barrel shifter A circuit which will shift a word a certain number of bits in either direction within a single clock cycle. (C) 610.10-1994w

barretter (waveguide components) A form of bolometer element having a positive temperature coefficient of resistivity which typically employs a power-absorbing wire or thin metal film. (MTT) 147-1979w

barrier (1) A partition for the insulation or isolation of electric circuits or electric arcs. (SWG/PE) C37.40-1993, C37.100-1992

(2) (Class 1E equipment and circuits) A device or structure interposed between redundant Class 1E equipment or circuits, or between Class 1E equipment or circuits and a potential source of damage to limit damage to Class 1E systems to an acceptable level. (PE/NP) 384-1992r

(3) Any product whose sole purpose is to act as an obstruction to the path of the animal. A barrier may have electrical insulating properties, but by design and application, its use is limited to blocking an animal's passage or an animal's contact with energized conductors or equipment. (SUB/PE) 1264-1993

(4) An obstruction composed of suitable construction and materials or a time delay mechanism that imposes a delay for an intended purpose. (PE/NP) 692-1997

barrier grid (charge-storage tubes) A grid, close to or in contact with a storage surface, which establishes an equilibrium voltage for secondary-emission charging and serves to minimize redistribution. *See also:* charge-storage tube. (ED) 158-1962w, [45]

barrier layer (fiber optics) In the fabrication of an optical fiber, a layer that can be used to create a boundary against OH^- ion diffusion into the core. *See also:* core. (Std100) 812-1984w

barrier transaction (1) Transaction that is guaranteed to become visible to other observers after all transactions created before it have become visible. (C/BA) 896.3-1993w

(2) A transaction that ensures that all previously generated write transactions have the global appearance of having been written to memory. This is used before signaling another noncoherent unit, or one in a different coherence domain, that the data is available. In some systems, this is an explicit bus transaction that will be treated specially by the bus bridges (e.g., that may not return a response until all write buffers for the unit are flushed). For buses that delay the write-response until write bus transactions have been adequately completed, a separate barrier transaction is not needed since the effect of a barrier can be achieved by waiting for all outstanding write-responses. *Synonym:* write barrier. (C/MM) 1212.1-1993

barrier wiring techniques (coupling in control systems) Those wiring techniques which obstruct electric or magnetic fields, excluding or partially excluding the fields from a given circuit. Barrier techniques are often effective against electromagnetic radiation also. In general, these techniques change the coupling coefficients between wires connected to a noise source and the signal circuit. *Example*: placement of signal lines within steel conduit to isolate them from an existing magnetic field. *See also:* suppressive wiring techniques; compensatory wiring techniques. (IA/ICTL) 518-1982r

barring hole (rotating machinery) A hole in the rotor to permit insertion of a pry bar for the purpose of turning the rotor slowly or through a limited angle. *See also:* rotor. (PE) [9]

bar, rotor *See:* rotor bar.

bar-type current transformer One that has a fixed and straight single primary winding turn passing through the magnetic circuit. The primary winding and secondary winding(s) are insulated from each other and from the core(s) and are assembled as an integral structure. (PE/TR) C57.13-1993, [57]

base (1) (number system) An integer whose successive powers are multiplied by coefficients in a positional notation system. *See also:* radix; positional notation. (C) 162-1963w

(2) (rotating machinery) A structure, normally mounted on the foundation, that supports a machine or a set of machines. In single-phase machines rated up through several horsepower, the base is normally a part of the machine and supports it through a resilient or rigid mounting to the end shields. (PE) [9]

(3) (electron tube or valve) The part attached to the envelope, carrying the pins or contacts used to connect the electrodes to the external circuit and that plugs into the holder. *See also:* electron tube. (ED) [45], [84]

(4) (electroplating) (basis or base metals) The object upon which the metal is electroplated. *See also:* electroplating. (PE/EEC) [119]

(5) (transistor) A region that lies between an emitter and a collector of a transistor and into which minority carriers are injected. *See also:* transistor. (ED/IA) 216-1960w, [12]

(6) (high-voltage fuse) The supporting member to which the insulator unit or units are attached. (SWG/PE) C37.40-1993, C37.100-1992

(7) (pulse terminology) The two portions of a pulse waveform which represents the first nominal state from which a pulse departs and to which it ultimately returns. (IM/WM&A) 194-1977w

base active power (synchronous generators and motors) The total (generator) output or (motor) input power at base voltage and base current with a power factor of unity.

base address (1) (computers) An address used as a reference point to which a relative address is added to determine the address of the storage location to be accessed. *See also:* indexed address; relative address; self-relative address. (C) 610.12-1990, 610.10-1994w

(2) A given address from which an absolute address is derived by combination with a relative address. *Synonyms:* reference address; presumptive address; constant address. (C) [20], 610.10-1994w, [85]

base address register A register used in an operand field of a processor instruction with a specified offset, the sum of which points to a data value within a data structure to be used by the instruction. *See also:* base register. (C) 610.10-1994w

base ambient temperature (power distribution, underground cables) (cable or duct) The no-load temperature in a group with no load on any cable or duct in the group. (PE) [4]

base apparent power (1) (ac rotating machinery) (basic per-unit quantities for ac rotating machines) A reference value expressing an electrical power rating of the machine. *Notes:* 1. Base apparent power may be either input or output power, and the numerical value may be either real power—watts (W)—or total apparent electrical power—voltamperes (VA)—depending upon machine type. Base apparent power is usually expressed in voltamperes, but any consistent set of units may be used. For synchronous generators, induction generators, and synchronous motors, base apparent power is the total apparent electrical at rated voltage and rated current. In induction motors (preferred method), base apparent power is numerically equal to the rated power output. For induction motors (alternate method), base apparent power is the total apparent electrical power at rated voltage and rated current. 2. When the alternate method is used it should be identified as "input voltampere based." (EM/PE) 86-1987w

(2) (synchronous generators and motors) The total rated apparent power at rated voltage and rated current. *Note:* Base apparent power is usually expressed in volt-amperes, but any consistent set of units may be used. 86-1961

base assertion An assertion that is required to be tested for required features and for implemented conditional feaures. (C/PA) 1326.2-1993w, 1328-1993w, 13210-1994, 2003.1-1992, 1328.2-1993w

baseband (carrier or subcarrier wire or radio transmission system) The band of frequencies occupied by the signal before it modulates the carrier (or subcarrier) frequency to form the transmitted line or radio signal. *Note:* The signal in the baseband is usually distinguished from the line or radio signal by ranging over distinctly lower frequencies, which at the lower end relatively approach or may include direct current (zero frequency). In the case of a facsimile signal before modulation on a subcarrier, the baseband includes direct current. *See also:* facsimile transmission. (BT/COM/PE/AV) [34], 168-1956w, 599-1985w

baseband coaxial system (1) A baseband system employing coaxial cables as a data transmission medium. At any point on the medium only one information signal at a time can be present without disruption. *Contrast:* baseband twisted-pair system. (C) 610.7-1995

(2) A system whereby information is directly encoded and impressed upon the transmission medium. At any point on the medium only one information signal at a time can be present without disruption. (C/LM) 802.3-1998

baseband-multiplexed (data transmission) The frequency band occupied by the aggregate of the transmitted signals applied to the facility interconnecting the multiplexing and line equipment. The multiplex baseband is also defined as the frequency band occupied by the aggregate of the received signals obtained from the facility interconnecting the line and the multiplex equipment. (PE) 599-1985w

baseband response function *See:* transfer function.

baseband signaling The transmission of a signal at its original frequency, that is, not changed by modulation. *Note:* It can be an analog or a digital signal. *Contrast:* broadband signaling. (C) 610.7-1995

baseband system A system used for networking in which information is encoded, modulated, and impressed directly on the transmission medium. *Note:* Generally used for limited distance. *Contrast:* broadband system. *See also:* baseband twisted-pair system; baseband coaxial system. (C) 610.7-1995

baseband twisted-pair system A baseband system employing twisted-pair wiring cables as the transmission medium. *Contrast:* baseband coaxial system. (C) 610.7-1995

Base Client Port An instance of a subclass of `IEEE1451_BaseClientPort`. (IM/ST) 1451.1-1999

base complement *See:* radix complement.

base current (ac rotating machinery) (basic per-unit quantities for ac rotating machines) The value of phase current corresponding to the value of base apparent power, base voltage, and the number of phases. *Note:* Base current is usually expressed in amperes, but any consistent set of units may be used. Base current equals the base apparent power divided by the product of base voltage and the number of phases. (EM/PE) 86-1987w

base electrode (transistor) An ohmic or majority-carrier contact to the base region. (IA) [12]

base font The font that is used by a printer or other peripheral device when no font is specified. *Synonym:* default font. (C) 610.10-1994w

base group address The group address (GP) value that is used for geographical addressing on a segment. Normally the lowest GP assigned to the segment. (NID) 960-1993

base impedance (ac rotating machinery) (basic per-unit quantities for ac rotating machines) The value of impedance corresponding to the value of the base voltage divided by the value of the base current. *Note:* Base impedance is usually expressed in ohms (Ω), but any consistent set of units may be used. (EM/PE) 86-1987w

basic insulation Insulation applied to live parts to provide basic protection against electric shock. (EMB/MIB) 1073.4.1-2000

base light (illuminating engineering) A uniform, diffuse illumination approaching a shadowless condition, which is sufficient for a television picture of technical acceptability, and which may be supplemented by other lighting. (EEC/IE) [126]

baseline (1) (germanium gamma-ray detectors) (x-ray energy spectrometers) (charged-particle detectors) (at pulse peak) The instantaneous value that the voltage would have had at the time of the pulse peak in the absence of that pulse. (NPS/NID) 759-1984r

(2) (navigation) The line joining the two points between which electrical phase or time is compared in determining navigation coordinates. For two ground stations, this is normally the great circle joining the two stations, and, in the case of a rotation collector system, it is the line joining the two sides of the collector. (AES/GCS) 172-1983w

(3) (pulse techniques) That amplitude level from which the pulse waveform appears to originate. (IM/HFIM) [40]

(4) (A) (software) A specification or product that has been formally reviewed and agreed upon, that thereafter serves as the basis for further development, and that can be changed only through formal change control procedures. **(B) (software)** A document or a set of such documents formally designated and fixed at a specific time during the life cycle of a configuration item. *Note:* Baselines, plus approved changes from those baselines, constitute the current configuration identification. *See also:* product baseline; functional baseline; developmental configuration; allocated baseline. **(C) (software)** Any agreement or result designated and fixed at a given time, from which changes require justification and approval. (C) 610.12-1990

(5) (charged-particle detectors) The average of the levels from which a pulse departs and to which it returns in the absence of a following overlapping pulse. (NPS) 300-1988r, 325-1996

(6) (radiation instrumentation) The part of the pulse-height distribution lying underneath a peak, including contributions associated with the source, detector, and measuring conditions that affect the spectral shape. (NI) N42.14-1991

(7) The agreed specification, or software item, which has been uniquely identified and becomes the focus for further development, and which can only be altered under strict control procedures. (C/SE) 1298-1992w

(8) A work product that has been formally reviewed and accepted by the involved parties. A baseline should be changed only through formal configuration management procedures. Some baselines may be project deliverables while others provide the basis for further work. (C/SE) 1058-1998

(9) A specification or system that has been formally reviewed and agreed upon, that thereafter serves as the basis for further development and can be changed only through formal change control procedures. (C/SE) 1233-1998

baseline clipper-intensifier (spectrum analyzer) A means of changing the relative brightness between the signal and baseline portion of the display. (IM) 748-1979w

baseline data (1) (electric pipe heating systems) Information retained for the purpose of evaluation against repeated information in order to establish trends in parameters. (PE/EDPG) 622-1979s

(2) (nuclear power generating station) Reference data that may be used to show acceptable functioning of the equipment during qualification testing. (PE/NP) 649-1980s

(3) Initial data needed to show acceptable functioning of the equipment during qualification testing. (SWG/PE) C37.100-1992

baseline delay (navigation aid terms) The time interval needed for a signal from a loran master station to travel to the slave station. (AES/GCS) 172-1983w

baseline management In configuration management, the application of technical and administrative direction to designate the documents and changes to those documents that formally identify and establish baselines at specific times during the life cycle of a configuration item. (C) 610.12-1990

baseline offset (pulse techniques) The algebraic difference between the amplitude of the baseline and the amplitude reference level. *See also:* pulse. (IM/HFIM) [40]

baseline overshoot *See:* pulse distortion.

baseline restoration (x-ray energy spectrometers) Appropriate linear or nonlinear technique(s), or their combination, used to accelerate the return of a voltage to its baseline. (NPS/NID) 759-1984r

baseline restorer A circuit that rapidly restores the baseline following an amplifier's output pulse (or train of pulses) to the level that existed before the pulse. (NPS) 325-1996

Base Link Code Word The first 16-bit message exchanged during IEEE 802.3 Auto-Negotiation. (C/LM) 802.3-1998

base load (power operations) (electric power utilization) The minimum load over a given period of time. *See also:* generating station. (PE/PSE) 858-1987s, 346-1973w

base load control (electric generating unit or station) For an electric generating unit or station, a mode of operation in which the unit or station generation is held constant. (PE/PSE) 94-1991w

base magnitude (pulse terminology) The magnitude of the base as obtained by a specified procedure or algorithm. (IM/WM&A) 194-1977w

basement The rock region underlying the overburden largely comprising aged rock types, often crystalline and of low conductivity. (COM) 365-1974w

base-minus-one complement *See:* diminished-radix complement.

base-minus-ones complement A number representation that can be derived from another by subtracting each digit from one less than the base. Nines complements and ones complements are base-minus-ones complements. (C) 162-1963w

base-mounted electric hoist A hoist similar to an overhead electric hoist except that it has a base or feet and may be mounted overhead, on a vertical plane, or in any position for which it is designed. *See also:* hoist. (EEC/PE) [119]

basename The final, or only, filename in a pathname. (C/PA) 9945-2-1993

base notation *See:* radix notation.

base number *See:* radix.

Base Page *See:* Base Link Code Word.

base page address (microprocessor assembly language) An address of reduced size which references a pre-specified portion of memory (which might be an on-board RAM). (C/MM) 695-1985s

base point *See:* radix point.

Base Port An instance of a subclass of `IEEE1451_BasePort`. (IM/ST) 1451.1-1999

Base Publisher Port An instance of a subclass of `IEEE1451_BasePublisherPort`. (IM/ST) 1451.1-1999

base rate (1) (telephone switching systems) A fixed amount charged each month for any one of the classes-of-service that is provided to a customer. (COM) 312-1977w
(2) The lowest data rate used by the Serial Bus in a particular cable environment. In multiple speed environments, all nodes have to be able to receive and transmit at the base rate. The base rate for the cable environment is 98.304 MHz × 100 ppm. (C/MM) 1394-1995
(3) The lowest data rate used by Serial Bus in a backplane or cable environment. In multiple speed environments, all nodes are able to receive and transmit at the base rate. The base rate for the cable environment is 98.304 MHz ± 100 ppm. (C/MM) 1394a-2000

base-rate area (telephone switching systems) The territory in which the tariff applies. (COM) 312-1977w

base region (transistor) The interelectrode region of a transistor into which minority carriers are injected. *See also:* transistor. (EEC/PE) [119]

base register *See:* base address register.

base relation A relation that is not derivable from other base relations in a given data-base. *Contrast:* derived relation. (C) 610.5-1990w

base repetition rate *See:* basic repetition frequency.

base resistivity The electrical resistivity of the material composing the base of a semiconductor device. (AES/SS) 307-1969w

base speed (1) (ac rotating machinery) (basic per-unit quantities for ac rotating machines) The rated synchronous speed. *Note:* Synchronous speed equals 120 times the value of line frequency, divided by the number of poles. Base speed is usually expressed in revolutions per minute (r/min), but any consistent set of units may be used. (EM/PE) 86-1987w
(2) The lowest speed obtained at rated load and rated voltage at the specified temperature rise. (IA/MT) 45-1998

base standard (1) An approved international standard, technical report, ITU-T Recommendation, or national standard. (C/PA) 14252-1996
(2) The standard for which a test method specification is written and/or a test method implementation is developed. (C/PA) 2003-1997

base station (mobile communication) A land station in the land-mobile service carrying on a radio communication service with mobile and fixed radio stations. *See also:* mobile communication system. (VT) [37]

base torque (ac rotating machinery) (basic per-unit quantities for ac rotating machines) The value of torque corresponding to the value of base apparent power and base synchronous speed. The value of base torque in pound-force feet (lbf · ft) is 7.043 times the value of the base apparent power—in voltamperes (VA), divided by the value of base speed in revolutions per minute (r/min). The value of base torque in newton meters per radian (N · m/rad) is 9.549 times the value of the base apparent power (in voltamperes), divided by the value of the base speed in revolutions per minute. *Note:* Base torque has conventionally been expressed in pound-force feet or in newton meters (N · m). To avoid confusion with the unit of energy, which is also the newton meter, the designation newton meter per radian is recommended. (EM/PE) 86-1987w

Base Transducer Block An instance of a subclass of the class `IEEE1451_BaseTransducerBlock`. (IM/ST) 1451.1-1999

base value (rotating machinery) A normal or nominal or reference value in terms of which a quantity is expressed in per unit or percent. *See also:* asynchronous machine; direct-current commutating machine. (PE) [9]

base voltage (ac rotating machinery) (basic per-unit quantities for ac rotating ma-chines) The rated phase voltage. *Note:* The value of the base voltage is the value of the rated line voltage for a delta-connected machine, and is the value of the rated line voltage divided by $\sqrt{3}$ for a wye-connected machine. Base voltage is usually expressed in volts (V), but any consistent set of units may be used. (EM/PE) 86-1987w

BASIC *See:* Beginner's All-purpose Symbolic Instruction Code.

basic access method (BAM) An access method in which each input or output statement invokes a corresponding machine operation. For example, when reading a file with 10 records, exactly 10 READ operations will be invoked. *Contrast:* direct access method. *See also:* basic partitioned access method; basic sequential access method; basic indexed sequential access method; basic direct access method. (C) 610.5-1990w

basic alternating voltage (power rectifier) The sustained sinusoidal voltage that must be impressed on the terminal of the alternating-current winding of the rectifier transformer, when set on the rated voltage tap, to give rated output voltage at rated load with no phase control. *See also:* rectification. (EEC/PE) [119]

basic control element (thyristor) The basic thyristor or thyristor/diode circuit configuration, or both, employed as the principal means of power control. (IA/IPC) 428-1981w

basic current range (watthour meter) The current range of a multirange standard watthour meter designated by the manufacturer for the adjustment of the meter (normally the five-ampere range). (ELM) C12.1-1982s

BASIC definitions (real-time BASIC for CAMAC) The syntax definitions make use of a metalanguage that is the usual extension to BNF notation. The meta-language contains symbols such as { or ::= or] or < which do not occur in BASIC. In addition, the meta-language contains words in angle brackets; for example, <numeric-expression>, where the meta-language symbols < and > indicate that the word between (in this case numeric-expression) is in the meta-language, not in BASIC. The symbols of the meta-language are listed below, together with their meanings. Any other symbols not enclosed by angle brackets stand for themselves and are part of BASIC.

::=	means is defined by. It separates the left part from the right part of a definition
<	opens a character string that constitutes a meta-language symbol
>	terminates a character string that constitutes a meta-language symbol
/	separates alternatives in the right part of a definition

[opens an option—that is, the syntactic units enclosed by square brackets are optionally present
] terminates an option
{ opens a group of elements that are to be considered a single syntactic unit for the purposes of the definition
} terminates a group of elements to be considered as a single syntactic unit
. . . means that the preceding syntactic unit may be repeated zero or more times
.is. used in place of ::= in the formal semantic definition of a terminal symbol

Notes: 1. Concatenation takes precedence over alternation; for example: F<integer>/<null> is equivalent to {F<integer>}/<null>. 2. The statement number is omitted in the formal definitions. It is mandatory for statements that form part of a program, but it may be omitted to indicate immediate mode execution of single statements in the usual way. 3. Tabulation, blanks, and new lines are used in the syntax definitions to make them easier to read, but they have no other significance. A program must follow the rules for the implementation concerning blanks. (NPS) 726-1982r

basic device *See:* common device.

basic direct access method (BDAM) A variation on the basic access method that allows direct access to the data. (C) 610.5-1990w

basic element (measurement system) A measurement component or group of components that performs one necessary and distinct function in a sequence of measurement operations. *Note:* Basic elements are single-purpose units and provide the smallest steps into which the measurement sequence can be classified conveniently. Typical examples of basic elements are: a permanent magnet, a control spring, a coil, and a pointer and scale. *See also:* measurement system. (EEC/PE) [119]

basic encoding rules A specific set of rules used to encode ASN.1 values as strings of octets. (C/PA) 1327.2-1993w, 1326.2-1993w, 1224.2-1993w, 1328.2-1993w

basic frequency Of an oscillatory quantity having sinusoidal components with different frequencies, the frequency of the component considered to be the most important. *Note:* In a driven system, the basic frequency would, in general, be the driving frequency, and in a periodic oscillatory system, it would be the fundamental frequency. (SP) [32]

basic functions (controller) The functions of those of its elements that govern the application of electric power to the connected apparatus. *See also:* electric controller. (IA/ICTL/IAC) [60]

basic impulse insulation level (BIL, bil) (1) (electric power) Reference levels expressed in impulse crest voltage with a standard wave not longer than 1.5 × 40 μs wave. *See also:* insulation. (SWG/SPD/PE) 28-1974, 32-1972r, [8], [98], [99], [100]
(2) (surge arresters) (rated impulse withstand voltage) A reference impulse insulation strength expressed in terms of the crest value of withstand voltage of a standard full impulse voltage wave. (SWG/PE/SPD/T&D/SWG-OLD) C37.40-1993, C62.11-1993s, C62.1-1981s, C37.100-1992, 1410-1997
(3) (outdoor apparatus bushings) A reference insulation level expressed as the impulse crest voltage of the 1.2 × 50 microsecond wave which the bushing will withstand when tested in accordance with specified conditions. (PE/TR) 21-1976
(4) (power cable systems) Impulse voltage that electrical equipment is required to withstand without failure or disruptive discharge when tested under specified conditions of temperature and humidity. Basic impulse levels (BILs) are designated in terms of the crest voltage of a 1.2 · 50 μs full-wave impulse voltage test. (PE/IC) 400-1991

basic indexed sequential access method (BISAM) A variation on the basic access method that allows indexed sequential access to the data. (C) 610.5-1990w

Basic Interoperability Data Model (BIDM) Defines the minimal set of information that reuse libraries should be able to exchange about assets in order to interoperate. (C/SE) 1420.1-1995

Basic Language for Implementation of System Software A programming language designed for writing systems software such as compilers and operating systems. (C) 610.13-1993w

basic lightning impulse insulation level (BIL) (1) (power and distribution transformers) A specific insulation level expressed in kilovolts of the crest value of a standard lightning impulse. (PE/TR) C57.13-1993, C57.12.80-1978r
(2) The electrical strength of insulation expressed in terms of the crest value of a standard lightning impulse under standard atmospheric conditions. BIL may be expressed as either statistical or conventional. (PE/C) 1313.1-1996
(3) (A) The electrical strength of insulation expressed in terms of the crest value of a standard lightning impulse under standard atmospheric conditions. BIL may be expressed as either statistical or conventional. **(B)** A specific insulation level expressed as the crest value of a standard lightning impulse.

— **BIL (conventional):** Applicable specifically to non-self-restoring insulations. The crest value of a standard lightning impulse for which the insulation does not exhibit disruptive discharge when subjected to a specific number of applications of this impulse under specified conditions.
— **BIL (statistical)**: Applicable specifically to self-restoring insulations. The crest value of a standard lightning impulse for which the insulation exhibits a 90% probability of withstand (or a 10% probability of failure) under specified conditions.

(SPD/PE) C62.11-1999

basic metallic rectifier One in which each rectifying element consists of a single metallic rectifying cell. *See also:* rectification. (EEC/PE) [119]

basic numbering plan USA (telephony) The plan whereby every telephone station is identified for nationwide dialing by a code for routing and a number of digits. (COM) [48]

basic operating unit (A) A single vehicle designed for independent operation. **(B)** A permanent or semipermanent combination, designed for independent operation, consisting of two or more vehicles of one or more types. (VT/RT) 1477-1998, 1473-1999, 1475-1999, 1474.1-1999

basic part (electric and electronics parts and equipment) One piece, or two or more pieces joined together, which are not normally subject to disassembly without destruction of designed use. The application, size, and construction of an item may be factors in determining whether an item is regarded as a unit, an assembly, a subassembly, or a basic part. A small electric motor might be considered as a part if it is not normally subject to disassembly. Typical examples: electron tube, resistor, relay, power transformer, microelectronic device. (GSD) 200-1975w

basic partitioned access method (BPAM) A variation on the basic access method that allows partitioned access to the data. (C) 610.5-1990w

basic planned derating (electric generating unit reliability, availability, and productivity) The planned derating that is originally scheduled and of predetermined duration. *See also:* planned derating. (PE/PSE) 762-1987w

basic planned outage (electric generating unit reliability, availability, and productivity) The planned outage state that is originally scheduled and of a predetermined duration. (PE/PSE) 762-1987w

basic reference designation (electric and electronics parts and equipment) The simplest form of a reference designation, consisting only of a class letter portion and a number (namely, without mention of the item within which the reference-designated item is located). The reference designation for a unit consists of only a number. (GSD) 200-1975w

basic reference standards (metering) Those standards with which the values of the electrical units are maintained in the laboratory, and which serve as the starting point of the chain of sequential measurements carried out in the laboratory. (ELM) C12.1-1982s

basic regular expression (BRE) A pattern (sequence of characters or symbols) constructed according to the rules defined in POSIX.2. (C/PA) 9945-2-1993

basic repetition frequency (navigation) (loran) The lowest pulse repetition frequency of each of the several sets of closely spaced repetition frequencies employed. (AES/GCS) 172-1983w

basic repetition rate *See:* basic repetition frequency.

basic sequential access method (BSAM) A variation of the basic access method that allows sequential access to the data. *See also:* indexed sequential access method; virtual sequential access method. (C) 610.5-1990w

basic series ferroresonant voltage regulator This regulator consists of a series connection of a saturating inductor and a capacitor connected across the source. The load is inductively or conductively coupled to the saturating inductor. See the figure below. *Note:* Applications of this circuit are limited by the requisite large ratio of reactive to real powers.

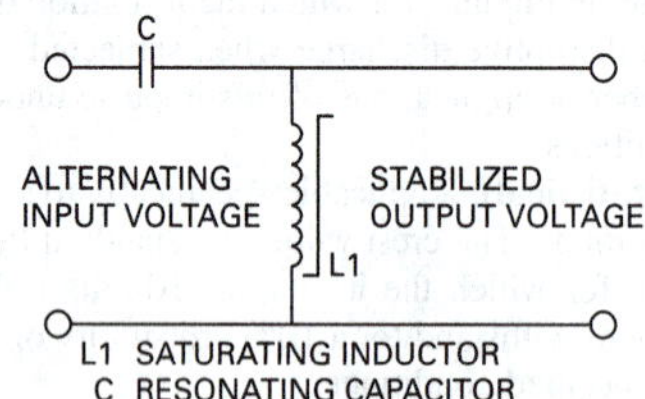

Basic series ferroresonant voltage regulator

Basic series ferroresonant voltage regulator

(PEL) 449-1998

basic series parallel ferroresonant voltage regulator This regulator consists of an essentially linear inductor connected in series with a parallel combination of a nonlinear inductor and a capacitor. This combination is connected across the source as shown in the figure below. Load voltage is derived by inductive or conductive coupling to the nonlinear inductor.

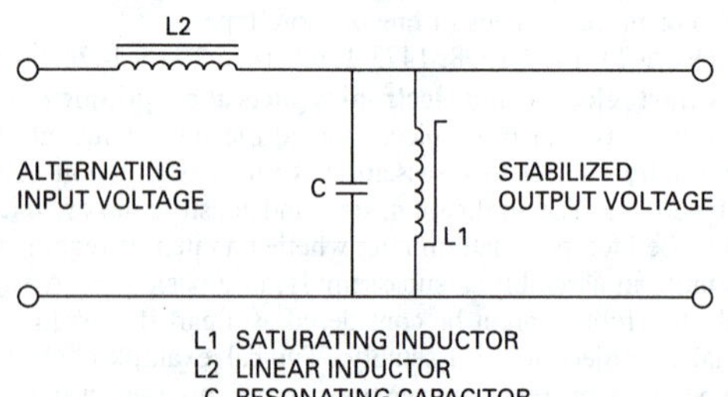

Basic series parallel ferroresonant voltage regulator

Basic series parallel ferroresonant voltage regulator

(PEL) 449-1998

basic service area (BSA) The conceptual area within which members of a basic service set (BSS) may communicate. (C/LM) 8802-11-1999

basic service set (BSS) A set of stations controlled by a single coordination function. (C/LM) 8802-11-1999

basic service set basic rate set The set of data transfer rates that all the stations in a BSS will be capable of using to receive frames from the wireless medium (WM). The BSS basic rate set data rates are preset for all stations in the BSS. (C/LM) 8802-11-1999

basic status The capability of an LLC to send or receive a PDU containing an information field. (C/LM/CC) 8802-2-1998

basic switching impulse insulation level (BSL) (1) (power and distribution transformers) A specific insulation level expressed in kilovolts of the crest value of a standard switching impulse. (PE/TR) C57.12.80-1978r
(2) The electrical strength of insulation expressed in terms of the crest value of a standard switching impulse. BSL may be expressed as either statistical or conventional. (PE/C) 1313.1-1996
(3) (A) The electrical strength of insulation expressed in terms of the crest value of a standard switching impulse. BSL may be expressed as either statistical or conventional. **(B)** A specific insulation level expressed as the crest value of a standard switching impulse.

— **BSL (conventional)**: Applicable specifically to non-self-restoring insulations. The crest value of a standard switching impulse for which the insulation does not exhibit disruptive discharge when subjected to a specific number of impulses under specified conditions.
— **BSL (statistical)**: Applicable specifically to self-restoring insulations. The crest value of a standard switching impulse for which the insulation exhibits a 90% probability of withstand (or a 10% probability of failure) under specified conditions.

(SPD/PE) C62.11-1999

basic voltage range (watthour meter) The voltage range of a multirange standard watthour meter designated by the manufacturer for the adjustment of the meter (normally the 120-volt range). (ELM) C12.1-1982s

basket *See:* bucket; woven wire grip.

bass boost An adjustment of the amplitude-frequency response of a system or transducer to accentuate the lower audio frequencies. (EEC/PE) [119]

batch Pertaining to a system or mode of operation in which inputs are collected and processed all at one time, rather than being processed as they arrive, and a job, once started, proceeds to completion without additional input or user interaction. *Contrast:* real time; interactive; conversational. (C) 610.12-1990

batch administrator A person who is authorized to use all restricted batch services. (C/PA) 1003.2d-1994

batch client A computational entity that utilizes batch services by making requests of batch servers. Batch clients often provide the means by which users access batch services, although a batch server may act as a batch client by virtue of making requests of another batch server. *Synonym:* client. (C/PA) 1003.2d-1994

Batcher's parallel sort (data management) A merge sort in which corresponding items in two ordered subsets are simultaneously compared and, if necessary, exchanged; the resulting subsets are divided in half and interleaved with one another, and these steps are repeated until the merge is complete. *Note:* This algorithm is particularly appropriate for parallel processing. *Synonyms:* odd-even sort; merge exchange sort. *See also:* bitonic sort. (C) 610.5-1990w

batch job A set of computational tasks for a computing system. Batch jobs are managed by batch servers. Once created, a batch job may be executing or pending execution. A batch job that is executing has an associated session leader (a process) that initiates and monitors the computational tasks of the job. *Synonym:* job. (C/PA) 1003.2d-1994

batch job attribute A named data type whose value affects the processing of a batch job. The values of the attributes of a batch job affect the processing of that job by the batch server that manages the job. The attributes defined for a batch job are called the batch job attributes. (C/PA) 1003.2d-1994

batch operator A person who is authorized to use some, but not all, restricted batch services. *Synonym:* operator. (C/PA) 1003.2d-1994

batch node A host containing part or all of a batch system. A *batch node* is a host meeting at least one of the following conditions:

— Is capable of executing a batch client
— Contains a routing queue
— Contains an execution queue

Synonym: node. (C/PA) 1003.2d-1994

batch queue A manageable object that represents a set of batch jobs and is managed by a single batch server. *Note:* Each batch job managed by a batch server is a member of a single batch queue managed by that server. Such a set of batch jobs is called a queue largely for historical reasons. Jobs are selected from the queue for execution based on attributes such as priority, resource requirements, and hold conditions. *Synonym:* queue. (C/PA) 1003.2d-1994

batch queue attribute A named data type whose value affects the processing of all jobs that are members of the queue. A batch queue has attributes that affect the processing of jobs that are members of the queue. The attributes defined for a batch queue are called the batch queue attributes. (PA/C) 1003.2d-1994

batch server A computational entity that provides batch services. *Synonym:* server. (C/PA) 1003.2d-1994

batch service Computational and organizational services performed by a batch system on behalf of batch jobs. Batch services are of two types: *requested* and *deferred.* (C/PA) 1003.2d-1994

batch server name A string that identifies a specific server in a network. A string of characters in the portable character set used to specify a particular server in a network. (C/PA) 1003.2d-1994

batch system A collection of one or more batch servers. *Synonym:* system. (C/PA) 1003.2d-1994

batch user A person who is authorized to make use of batch services. (C/PA) 1003.2d-1994

bathtub curve (software) A graph of the number of failures in a system or component as a function of time. The name is derived from the usual shape of the graph: a period of decreasing failures (the early-failure period), followed by a relatively steady period (the constant-failure period), followed by a period of increasing failures (the wearout-failure period). (C) 610.12-1990

bath voltage The total voltage between the anode and cathode of an electrolytic cell during electrolysis. It is equal to the sum of

a) equilibrium reaction potential,
b) IR drop,
c) anode polarization, and
d) cathode polarization.

See also: tank voltage; electrolytic cell. (EEC/PE) [119]

bathythermograph (navigation aid terms) A recording thermometer for determining the temperature of the sea at various depths. (AES/GCS) 172-1983w

battery (primary or secondary) Two or more cells electrically connected for producing electric energy. [Common usage permits this designation to be applied also to a single cell used independently. In this document, IEEE Std 100, unless otherwise specified, the term "battery" will be used in this dual sense.] (IA/PE/EEC/PSE) 446-1995, [119]

battery-and-ground pulsing (telephone switching systems) Dial pulsing using battery-and-ground signaling. (COM) 312-1977w

battery-and-ground signaling (telephone switching systems) A method of loop signaling, used to increase the range, in which battery and ground at both ends of the loop are poled oppositely. (COM) 312-1977w

battery cabinet A structure used to support and enclose a group of cells. (SB) 1188-1996

battery carry-over (magnetic tape pulse recorders for electricity meters) A device that maintains actual time of the interval recording from a standby power source for a specified period when the principal power source is inoperative. (ELM) C12.14-1982r

battery charger As defined in IEEE Std 602-1996, static equipment that is capable of restoring and maintaining the charge in a storage battery. (IA/PSE) 602-1996

battery chute A small cylindrical receptacle for housing track batteries and so set in the ground that the batteries will be below the frost line. (EEC/PE) [119]

battery-current regulation (generator) That type of automatic regulation in which the generator regulator controls only the current used for battery charging purposes. *See also:* axle-generator system. (EEC/PE) [119]

battery duty cycle The loads a battery is expected to supply for specified time periods. (SCC29) 485-1997

battery, electric *See:* electric battery.

battery eliminator A device that provides direct-current energy from an alternating-current source in place of a battery. *See also:* battery. (PE) 599-1985w

battery feed (telephone loop performance) The direct current (dc) supply and coupling circuit powering the loop. (COM/TA) 820-1984r

battery, power station *See:* power station battery.

battery rack (A) (lead storage batteries) A structure used to support a group of cells. **(B) (lead storage batteries) (nuclear power generating station)** A rigid structure used to accommodate a group of cells. (PE/IA/SB/EDPG/PSE) 450-1987, 446-1995, 1188-1996

battery voltage Voltage that is provided within specified limits by the low voltage power supply (or, in its absence, the control voltage on-board battery). (VT) 1475-1999

baud (1) (supervisory control, data acquisition, and automatic control) The signaling speed, that is, keying rate of the modem. The signaling speed in baud is equal to the reciprocal of the shortest element duration in seconds to be transmitted. The terms *bit rate* and *baud* are not synonymous and shall not be interchanged in usage. Preferred usage is bit rate, with baud used only when the details of a communication modem or channel are specified. (PE/SUB/SWG-OLD) C37.100-1992, C37.1-1994

(2) (general) A unit of signalling speed equal to the number of discrete conditions or signal events per second. For example, one baud equals one half dot cycle per second in Morse code, one bit per second in a train of binary signals, and one 3-bit value per second in a train of signals each of which can assume one of 8 different states. *See also:* telegraphy. (C) [85]

(3) (telegraphy) The unit of telegraph signaling speed, derived from the duration of the shortest signaling pulse. A telegraphic speed of one baud is one pulse per second. *Note:* The term "unit pulse" is often used for the same meaning as "baud." A related term, "dot cycle," refers to an ON-OFF or MARK-SPACE cycle in which both mark and space intervals have the same length as the unit pulse. (AP/ANT) 145-1983s

(4) (data transmission) A unit of signaling speed equal to the number of discrete conditions or signal events per second, or the reciprocal of the time of the shortest signal element in a character.local area networks. (LM/PE/C) 599-1985w, 8802-12-1998

(5) A unit of signaling speed, expressed as the number of times per second the signal can change the electrical state of the transmission line or other medium. *Note:* Depending on the encoding strategies, a signal event may represent a single bit, more, or less, than one bit. *Contrast:* bits per second; bit rate. (C/Std100/LM/EMB/MIB) 610.7-1995, 610.10-1994w, 802.3-1998, 1073.3.2-2000

Baudot code A code for the transmission of data in which five data bits represent one character. (C) 610.7-1995

baud rate (1) The rate of signal transitions per unit time, usually expressed in baud. *Note:* Often confused with bit rate. *Contrast:* bit rate; bits per second. *See also:* data signaling rate. (C) 610.7-1995, 610.10-1994w

(2) For a given encoding scheme and data rate, the maximum number of signal transitions transmitted on a serial interface in a 1 s period. (EMB/MIB) 1073.4.1-2000

***b* auxiliary switch** *See:* auxiliary switch; *b* contact.

bay *See:* patch bay; electronic analog computer.

Bayliss distribution (A) Circular. A continuous distribution over a circular planar aperture that yields a difference pattern with a sidelobe structure similar to that of a sum pattern produced by a Taylor circular distribution. **(B)** Linear. A continuous distribution of a line source that yields a difference pattern with a side-lobe structure similar to that of a sum pattern produced by a Taylor linear distribution. (AP/ANT) 145-1993

***B* battery** A battery designed or employed to furnish the plate current in a vacuum-tube circuit. *See also:* battery. (EEC/PE) [119]

***bb* auxiliary switch** *See: bb* contact; auxiliary switch.

***bb* contact** A contact that is closed when the operating mechanism of the main device is in the standard reference position and that is open when the operating mechanism is in the opposite position. *See also:* standard reference position. (SWG/PE) C37.100-1992

B-box *See:* index register.

BCC *See:* bedside communications controller; block check character.

BCD *See:* binary coded decimal; borderline between comfort and discomfort.

BCD real data *See:* binary coded decimal real data.

B channel A channel that provides 64 kbit/s, full-duplex, isochronous access. B channels support all ISDN bearer services. The information on a B channel may be nonswitched or either circuit or packet switched depending on user request and network capabilities. (C/LM/COM) 802.9a-1995w, 8802-9-1996

BCNF *See:* Boyce/Codd Normal form.

***b* contact** A contact that is closed when the main device is in the standard reference position and that is open when the device is in the opposite position. *Notes:* 1. *b* contact has general application. However, this meaning for back contact is restricted to relay parlance. 2. For indication of the specific point of travel at which the contact changes position, an additional letter or percentage figure may be added to *b*. *See also:* standard reference position. (SWG/PE) C37.100-1992

BCPL *See:* Bootstrap Combined Programming Language.

BDAM *See:* basic direct access method.

B-display A rectangular display in which each target appears as an intensity-modulated blip, with azimuth indicated by the horizontal coordinate and range by the vertical coordinate.

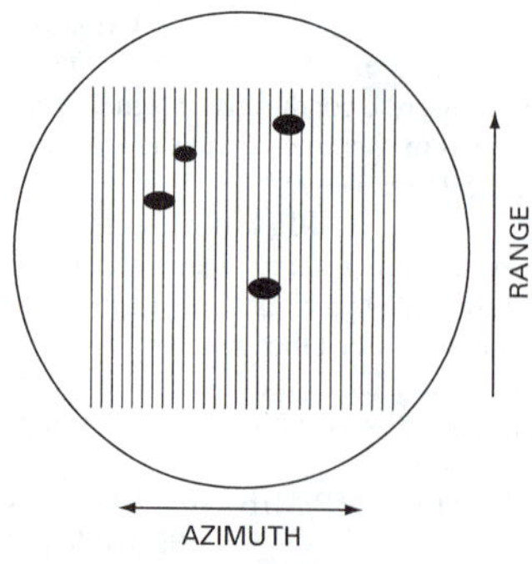

B-display

(AES) 686-1997

BDP *See:* business data processing.

beacon (1) (A) (navigation aid terms) A fixed aid to navigation. *See also:* racon; fan-marker beacon; marker beacon; z-marker beacon; lighted beacon; radio beacon; identification beacon; landing beacon; radar beacon; homing beacon. **(B) (navigation aid terms)** An unlighted aid to navigation. *See also:* marker beacon; radio beacon; fan-marker beacon; homing beacon; radar beacon; lighted beacon; landing beacon; identification beacon; z-marker beacon; racon. **(C) (navigation aid terms)** Anything serving as a signal or conspicuous indication, either for guidance or warning. *See also:* radio beacon; lighted beacon; landing beacon; identification beacon; z-marker beacon; marker beacon; racon; fan-marker beacon; radar beacon; homing beacon. **(D) (navigation aid terms)** In radar, a transponder used for replying to interrogations from a radar. *See also:* radar beacon; homing beacon; fan-marker beacon; radio beacon; z-marker beacon; landing beacon; lighted beacon; identification beacon; marker beacon; racon. (AES/GCS) 172-1983
(2) A roadside system at which dedicated short-range communications (DSRC) can be accomplished. A beacon typically consists of a reader and an antenna. (SCC32) 1455-1999
(3) *See also:* radar beacon. (AES) 686-1997

beacon equation An equation that gives the maximum detection range of a transponder or secondary radar as a function of system parameters for a given set of conditions. It is the one-way counterpart of the two-way radar equation. (AES) 686-1997

beaconing A ring state that occurs when a station on the ring has detected a ring failure. The frame transmitted by the station to alert the other stations on the ring of the failure is called a beacon frame. (C/LM) 8802-5-1998

beacon receiver A radio receiver for converting waves, emanating from a radio beacon, into perceptible signals. *See also:* radio beacon; radio receiver. (EEC/PE) [119]

beacon reconfigure A beacon (Type 1) used in the reconfiguration protocols. (LM/C) 802.5c-1991r

beacon service table (BST) A data structure created and transmitted by a beacon. It contains data such as the application identifier (AID) relevant for initiating communication with an onboard equipment (OBE) transponder. The reception of the BST by an OBE transponder results in a vehicle service table (VST) being sent back to the beacon. (SCC32) 1455-1999

beam (1) (laser maser) A collection of rays that may be parallel, divergent, or convergent. (LEO) 586-1980w
(2) (of an antenna) The major lobe of the radiation pattern of an antenna. (AP/ANT) 145-1993

beam alignment (camera tubes) An adjustment of the electron beam, performed on tubes employing low-velocity scanning, to cause the beam to be perpendicular to the target at the target surface. (ED) 161-1971w

beam angle *See:* scan angle.

beam axis (1) (of a pencil-beam antenna) The direction, within the major lobe of a pencil-beam antenna, for which the radiation intensity is a maximum. (AP/ANT) 145-1993
(2) (illuminating engineering) (of a projector) A line midway between two lines that intersect the candlepower distribution curve at points equal to a stated percent of its maximum (usually 50%). (EEC/IE) [126]

beam bending (camera tubes) Deflection of the scanning beam by the electrostatic field of the charges stored on the target. (ED) 161-1971w

beam compressors Structures on the surface of a substrate that increase the power density in a surface acoustic wave device by decreasing its lateral extent, such as the following: **horn:** A tapered structure of reduced velocity to produce gradual reduction of transverse width of beam; **multistrip beam compressor:** A multistrip coupler with spacing of the strips chosen so that one track (path) is appreciably wider than the other; **lenses:** Regions of decreased phase velocity so shaped as to produce focusing of an incident surface acoustic wave beam. (UFFC) 1037-1992w

beam coverage solid angle (of an antenna over a specified surface) The solid angle, measured in steradians, subtended at the antenna by the footprint of the antenna beam on a specified surface. *Contrast:* beam solid angle. *See also:* footprint. (AP/ANT) 145-1993

beam current (1) (storage tubes) The current emerging from the final aperture of the electron gun. *See also:* storage tube. (ED) 158-1962w
(2) (computer graphics) The flow of electrons from an electron gun onto the phosphor-coated screen of a cathode ray tube. (C) 610.6-1991w

beam-deflection tube An electron-beam tube in which current to an output electrode is controlled by the transverse movement of an electron beam. (ED) 161-1971w

beam diameter (1) (fiber optics) The distance between two diametrically opposed points at which the irradiance is a specified fraction of the beam's peak irradiance; most commonly applied to beams that are circular or nearly circular in cross section. *Synonym:* beamwidth. *See also:* beam divergence. (Std100) 812-1984w
(2) (laser maser) The distance between diametrically opposed points in that cross section of a beam where the power per unit area is $1/e$ times that of the peak power per unit area. (LEO) 586-1980w

beam divergence (laser maser) The full angle of the beam spread between diametrically opposed $1/e$ irradiance points; usually measured in mrad (one mrad ≃ Δ 3.4 minutes of arc). (LEO) 586-1980w
(2) (A) (fiber optics) For beams that are circular or nearly circular in cross section, the angle subtended by the far-field beam diameter. *See also:* collimation; far-field region; beam diameter. **(B) (fiber optics)** For beams that are not circular or nearly circular in cross section, the far-field angle subtended by two diametrically opposed points in a plane perpendicular to the optical axis, at which points the irradiance is a specified fraction of the beam's peak irradiance. Generally, only the maximum and minimum divergences (corresponding to the major and minor diameters of the far-field irradiance) need be specified. *See also:* beam diameter; far-field region; collimation. (Std100) 812-1984

beam error (navigation aids) (navigational systems using directionally propagated signals) The lateral or angular distance between the mean direction of the actual course and the desired course direction. (AES/GCS) 172-1983w

beam expander (laser maser) A combination of optical elements that will increase the diameter of a laser beam. (LEO) 586-1980w

beam finder (oscilloscopes) A provision for locating the spot when it is not visible. (IM) 311-1970w

beamguide (laser maser) A set of beam-forming elements spaced in such a way as to conduct a well-defined beam of radiation. Analogs are waveguides and fiber optic filaments. (LEO) 586-1980w

beam-indexing color tube A color-picture tube in which a signal, generated by an electron beam after deflection, is fed back to a control device or element in such a way as to provide an image in color. (ED) 161-1971w

beam landing error (camera tubes) A signal non-uniformity resulting from beam electrons arriving at the target with a spatially varying component of velocity parallel to the target. *See also:* camera tube. (ED) [45]

beam locator *See:* beam finder.

beam modulation, percentage (image orthicons) One hundred times the ratio of the signal output current for highlight illumination on the tube to the dark current. (ED) 161-1971w

beam noise (navigation aids) (navigational systems using directionally propagated signals) Extraneous disturbances tending to interfere with ideal system performance. *Note:* Beam noise is the aggregate effect of bends, scalloping, roughness, etc. (AES/GCS) 172-1983w

beam parametric amplifier A parametric amplifier that uses a modulated electron beam to provide a variable reactance. *See also:* parametric device. (ED) [46]

beam pattern *See:* directional response pattern.

beam pointing (communication satellite) The ability to orient the beam of a high gain antenna into a specific direction in a coordinate system. (COM) [19]

beam position *See:* current position.

beam power tube An electron-beam tube in which use is made of directed electron beams to contribute substantially to its power-handling capability, and in which the control grid and the screen grid are essentially aligned. (ED) 161-1971w

beam resonator (laser maser) A resonator that serves to confine a beam of radiation to a given region of space without continuous guidance along the beam. (LEO) 586-1980w

beam rider guidance That form of missile guidance wherein a missile, through a self-contained mechanism, automatically guides itself along a beam. *See also:* guided missile. (EEC/PE) [119]

beamshape loss A loss factor included in the radar equation to account for the use of the peak antenna gain in the radar equation instead of the effective gain that results when the received train of pulses is modulated by the two-way pattern of a scanning antenna. *Synonym:* antenna-pattern loss. (AES) 686-1997

beam shaping (communication satellite) Controlling the shape of an antenna beam, by design of the surfaces of the antenna or by controlling the phasing of the signals radiated from the antenna. (COM) [25]

beam solid angle The solid angle through which all the radiated power would stream if the power per unit solid angle were constant throughout this solid angle and at the maximum value of the radiation intensity. (AP/ANT) 145-1993

beamsplitter (fiber optics) A device for dividing an optical beam into two or more separate beams; often a partially reflecting mirror. (Std100) 812-1984w

beam splitter (laser maser) An optical device which uses controlled reflection to produce two beams from a single incident beam. (LEO) 586-1980w

beam spot size *See:* spot size.

beam spread (1) (illuminating engineering) (in any plane) The angle between the two directions in the plane in which the intensity is equal to a stated percentage of the maximum beam intensity. The percentage typically is 10% for floodlights and 50% for photographic lights. (EEC/IE) [126]
(2) (light-emitting diodes) (source of light, θ *y*, where *y* is the stated percent.) See definition (1) above. (ED) [127]

beam steering (1) Changing the direction of the major lobe of a radiation pattern. *See also:* radiation. (AP/ANT) [35], 145-1993
(2) Surface acoustic wave propagation phenomena in anisotropic materials described by a nonzero angle of power flow. (UFFC) 1037-1992w

beam waveguide A quasioptical structure consisting of a sequence of lenses or mirrors used to guide an electromagnetic wave. (MTT) 146-1980w

beamwidth *See:* beam diameter; half-power beamwidth.

bearer channel protocol intervention level The highest protocol level at which a private switching network (PSN) provides protocol termination on a given bearer channel. (LM/C/COM) 8802-9-1996

bearer service A telecommunication service that provides the capability for the transmission of signals between user-network interfaces. (C/LM/COM) 802.9a-1995w, 8802-9-1996

bearing (A) (navigation aid terms) The horizontal direction of one terrestrial point from another, expressed as the angle in the horizontal plane between a reference line and the horizontal projection of the line joining two points. **(B) (navigation aid terms)** Azimuth. A bearing is often designated as true, magnetic, compass, grid, or relative, and is dependent upon the reference direction. (AES/GCS) 172-1983
(2) (A) (rotating machinery) A stationary member or assembly of stationary members in which a shaft is supported and may rotate. **(B) (rotating machinery)** In a ball or roller

bearing, a combination (frequently preassembled) of stationary and rotating members containing a peripheral assembly of balls or rollers, in which a shaft is supported and may rotate. (PE) [9]

bearing accuracy, instrumental (A) (direction finding systems) The difference between the indicated and the apparent bearings in a measurement of the same signal source. *See also:* navigation. **(B) (direction finding systems)** As a statement of overall system performance, a difference between indicated and correct bearings whose probability of being exceeded in any measurement made on the system is less than some stated value. *See also:* navigation. (AES) [42]

bearing bracket (rotating machinery) A bracket which supports a bearing, but including no part thereof. A bearing bracket is not specifically constructed to provide protection for the windings or rotating parts. (PE) [9]

bearing cap (rotating machinery) (or bearing bracket cap) A cover for the bearing enclosure of a bearing bracket type machine or the removable upper half of the enclosure for a bearing. *See also:* bearing. (PE) [9]

bearing cartridge (rotating machinery) A complete enclosure for a ball or roller bearing, separate from the bearing bracket or end shield. *See also:* bearing. (PE) [9]

bearing clearance (A) (rotating machinery) The difference between the bearing inner diameter and the journal diameter. *See also:* bearing. **(B) (rotating machinery)** The total distance for axial movement permitted by a double-acting thrust bearing. *See also:* bearing. (PE) [9]

bearing distance heading indicator (navigation aid terms) A display device which presents continuous references as to course and distance to destination. (AES/GCS) 172-1983w

bearing dust-cap (rotating machinery) A removable cover to prevent the entry of foreign material into the bearing. *See also:* bearing. (PE) [9]

bearing error curve (A) (navigation aid terms) [DF (direction finder) equipment]. A plot of the instrumental bearing errors versus either indicated or correct bearing. **(B) (navigation aid terms)** (in DF installations) A plot of the combined instrumental bearing error (of the equipment) and site error versus indicated bearings. (AES/GCS) 172-1983

bearing housing (rotating machinery) A structure supporting the actual bearing liner or ball or roller bearing in a bearing assembly. *See also:* bearing. (PE) [9]

bearing insulation (rotating machinery) Insulation that prevents the circulation of stray currents by electrically insulating the bearing from its support. *See also:* bearing. (PE) [9]

bearing liner (rotating machinery) The assembly of a bearing shell together with its lining. *See also:* bearing. (PE) [9]

bearing lining (rotating machinery) The element of the journal bearing assembly in which the journal rotates. *See also:* bearing. (PE) [9]

bearing locknut (rotating machinery) A nut that holds a ball or roller bearing in place on the shaft. *See also:* bearing. (PE) [9]

bearing lock washer (rotating machinery) A washer between the bearing locknut and the bearing that prevents the locknut from turning. *See also:* bearing. (PE) [9]

bearing offset, indicated (electronic navigation) (direction finding systems) The mean difference between the indicated and apparent bearings of a number of signal sources, the sources being substantially uniformly distributed in azimuth. *See also:* navigation. (AES/GCS) 173-1959w, [42]

bearing oil seal *See:* oil seal.

bearing oil system (rotating machinery) (oil-circulating system) All parts that are provided for the flow, treatment, and storage of the bearing oil. *See also:* oil cup. (PE) [9]

bearing pedestal (rotating machinery) A structure mounted from the bedplate or foundation of the machine to support a bearing, but not including the bearing. *See also:* bearing. (PE) [9]

bearing-pedestal cap (rotating machinery) The top part of a bearing pedestal. *See also:* bearing. (PE) [9]

bearing plates Plates of large surface area attached to the structure below ground surface to prevent uplift or to increase the bearing capability in unstable soils. (T&D/PE) 751-1990

bearing protective device (power system device function numbers) A device that functions on excessive bearing temperature, or on other abnormal mechanical conditions associated with the bearing, such as undue wear, which may eventually result in excessive bearing temperature or failure. (SUB/PE) C37.2-1979s

bearing reciprocal *See:* reciprocal bearing.

bearing reservoir (rotating machinery) (oil tank) (oil well) A container for the oil supply for the bearing. It may be a sump within the bearing housing. *See also:* oil cup. (PE) [9]

bearing seal *See:* oil seal.

bearing seat (rotating machinery) The surface of the supporting structure for the bearing shell. *See also:* bearing. (PE) [9]

bearing sensitivity (electronic navigation) The minimum field strength input to a direction-finder system to obtain repeatable bearings within the bearing accuracy of the system. *See also:* navigation. (AES) 270-1966w, [42]

bearing shell (rotating machinery) The element of the journal bearing assembly that supports the bearing lining. *See also:* bearing. (PE) [9]

bearing shoe *See:* segment shoe.

bearing-temperature detector (rotating machinery) A temperature detector whose sensing element is mounted at or near the bearing surface. *See also:* bearing. (PE) [9]

bearing-temperature relay (rotating machinery) A relay whose temperature sensing element is mounted at or near the bearing surface. *Synonym:* bearing thermostat. *See also:* bearing. (PE) [9]

bearing thermometer (rotating machinery) A thermometer whose temperature sensing element is mounted at or near the bearing surface. *See also:* bearing. (PE) [9]

bearing thermostat *See:* bearing-temperature relay.

beat An event that begins with the transition on a synchronization line by the master, followed by the release of an acknowledge line by one or more slaves. Command and data information may be transferred from the master to one or more slaves in the first half of the beat. During the second half of the beat the slaves may transfer capability, status, and data information back to the master. (C/BA) 10857-1994, 896.4-1993w, 896.3-1993w

beating (data transmission) A phenomenon in which two or more periodic quantities of different frequencies produce a resultant having pulsations of amplitude. (PE) 599-1985w

beat note The wave of difference frequency created when two sinusoidal waves of different frequencies are supplied to a nonlinear device. *See also:* radio receiver. 188-1952w

beat reception *See:* heterodyne reception.

beats (1) (general) Periodic variations that result from the superposition of waves having different frequencies. *Note:* The term is applied both to the linear addition of two waves, resulting in a periodic variation of amplitude, and to the nonlinear addition of two waves, resulting in new frequencies, of which the most important usually are the sum and difference of the original frequencies. *See also:* signal wave. (COM) 312-1977w

(2) (data transmission) Periodic variations that result from the superposition of waves having difference frequencies. *Note:* The term is applied both to the linear addition of two waves, resulting in a periodic variation of amplitude, and to the nonlinear addition of two waves, resulting in new frequencies, of which the most important usually are the sum and difference of the original frequencies. (PE/EDPG) 599-1985w, [3]

becquerel (metric practice) The activity of a radionuclide decaying at the rate of one spontaneous nuclear transition per second. (QUL) 268-1982s

bedside Those medical devices that directly interact with, monitor, provide treatment to, or are in some way associated with a single patient. (EMB/MIB) 1073.4.1-1994s, 1073-1996, 1073.3.1-1994

bedside communications controller (BCC) A communications controller, typically located at a patient bedside, that serves to interface between one or more medical devices. The BCC may be embedded into local display, monitoring, or control equipment. Alternatively, it may be part of a communications router to a remote hospital host computer system. (EMB/MIB) 1073.4.1-2000, 1073.3.2-2000

bedside environment Encompassing a particular patient, bed or treatment area which is specific to one patient, and usually including those systems and personnel which are involved in the acute monitoring and treatment of the patient. (EMB/MIB) 1073-1996

bedside medical device A medical device that directly interacts with, monitors, provides treatment to, or is in some way associated with a single patient. (EMB/MIB) 1073.4.1-2000

beep A brief audible warning emitted by the terminal. (C) 1295-1993w

Beer-Lambert Law Also called Beer's Law or Bouger's Law, this law, valid for discrete random media, relates the intensity of an electromagnetic wave at one point to the intensity at another point in the direction of propagation. The intensity decreases exponentially with distance and the attenuation coefficient is equal to the product of the concentration of particles and the extinction cross-section per particle. Consequently, the application of Beer's Law is restricted to weakly scattering media. (AP/PROP) 211-1997

begin-end block (software) A sequence of design or programming statements bracketed by "begin" and "end" delimiters and characterized by a single entrance and a single exit. *See also:* design. (C/SE) 729-1983s

Beginner's All-purpose Symbolic Instruction Code (BASIC) A general-purpose programming language designed for writing programs in scientific and business applications. *Notes:* 1. Originally developed on a mainframe computer in 1964 at Dartmouth College, BASIC was later implemented as the first high-order language available for a microcomputer. 2. Numerous implementations of BASIC have been developed for various computers. Examples include ABASIC, MBASIC, S-BASIC, and ZBASIC. *See also:* common language. (C) 610.13-1993w

beginning-of-file label (BOF) An internally-recorded label that identifies a file, marks its location, and contains information for use in file control. *Synonym:* header label. *Contrast:* end-of-file label. (C) 610.10-1994w

beginning of frame (BOF) An octet specified by infrared link access protocol (IrLAP). (EMB/MIB) 1073.3.2-2000

beginning-of-tape marker (BOT) A marker on a magnetic tape used to indicate the beginning of the permissible recording area. *Note:* It might be a photo reflective strip, a unique data pattern, or a transparent section of tape *Contrast:* end-of-tape marker. *See also:* load point. (C) 610.10-1994w

beginning-of-volume label (BOV) An internally-recorded label that identifies a volume and which indicates the beginning of the recording area on that volume. *Synonyms:* volume header; volume label. *Contrast:* end-of-volume label. (C) 610.10-1994w

behavior (1) A formal representation of the characteristics that describe the operation, function, relationships, control, or static properties of a test entity. (SCC20) 1226-1998
(2) The aspect of an instance's specification that is determined by the state-changing operations it can perform. (C/SE) 1320.2-1998
(3) A statement of the externally visible response and internal change of state of an object to invoked operations or internal events, given its current internal state. (IM/ST) 1451.1-1999

behavioral analysis The analysis of the logical (stimulus/response) and design (resource consumption, event timing, throughput, etc.) execution of a system to assess the functional and design architectures. (C/SE) 1220-1998

behavioral design The design of how an overall system or software item will behave, from a user's point of view, in meeting its requirements, ignoring the internal implementation of the system or software item. This design contrasts with architectural design, which identifies the internal components of the system or software item, and with the detailed design of those components. (C/SE) J-STD-016-1995

behavioral model *See:* black box model.

BEL *See:* bell character.

bel (1) The fundamental division of a logarithmic scale for expressing the ratio of two amounts of power, the number of bels denoting such a ratio being the logarithm to the base 10 of this ratio. *Note:* With P_1 and P_2 designating two amounts of power and N the number of bels denoting their ratio, $N =$ log10(P_1/P_2) bels. (AP/ANT) 145-1983s
(2) The fundamental unit in a logarithmic scale for expressing the ratio of two amounts of power. *Notes:* 1. The number of bels is equal to the $\log_{10}(P_1/P_2)$, where P_1 is the power level being considered and P_2 is an arbitrary reference level. 2. The decibel, a more commonly used unit, is equal to 0.1 bel. (C) 610.10-1994w

bell box (ringer box) An assemblage of apparatus, associated with a desk stand or hand telephone set, comprising a housing (usually arranged for wall mounting) within which are those components of the telephone set not contained in the desk stand or hand telephone set. These components are usually one or more of the following: induction coil, capacitor assembly, signaling equipment, and necessary terminal blocks. In a magneto set a magneto and local battery may also be included. *See also:* telephone station. (EEC/PE) [119]

bell character (BEL) A control character that is used when there is a need to call for human attention and that may activate an alarm or other attention devices. (C) 610.5-1990w

bell crank A lever with two arms placed at an angle diverging from a given point, thus changing the direction of motion of a mechanism. (SWG/PE) C37.100-1992

bell crank hanger A support for a bell crank. (SWG/PE) C37.100-1992

Bell Laboratories' Low-level Linked List Language (L) A list processing language that allows programmers to specify list sizes and types. (C) 610.13-1993w

belt (rotating machinery) A continuous flexible band of material used to transmit power between pulleys by motion. (PE) [9]

belt, aerial *See:* aerial belt.

belt, bucket *See:* aerial belt.

belt-drive machine (elevators) An indirect-drive machine having a single belt or multiple belts as the connecting means. *See also:* driving machine. (PE/EEC) [119]

belted-type cable A multiple-conductor cable having a layer of insulation over the assembled insulated conductors. (PE/T&D) [10]

belt insulation (rotating machinery) A form of overhang packing inserted circumferentially between adjacent layers in the winding overhang. *See also:* stator; rotor. (PE) [9]

belt leakage flux (rotating machinery) The low-order harmonic airgap flux attributable to the phase belts of a winding. The magnitude of this leakage flux varies with winding pitch. *See also:* rotor; stator. (PE) [9]

belt printer An element printer in which the type slugs are carried on a flexible belt. (C) 610.10-1994w

belt-type conveyor A conveyor consisting of an endless belt used to transport material from one place to another. *See also:* conveyor. (EEC/PE) [119]

benchboard A combination of a control desk and a vertical switchboard in a common assembly. (SWG/PE) C37.100-1992, C37.21-1985r

benchmark **(A)** A standard against which measurements or comparisons can be made. *See also:* benchmark program; benchmark problem. **(B)** A procedure, problem, or test that can be used to compare systems or components to each other or to a standard as in definition (A). **(C)** A recovery file. (C) 610.12-1990, 610.10-1994

benchmark problem (computers) A problem used to evaluate the performance of computers relative to each other. (C) [85]
(2) **(A)** A problem used to evaluate the performance of hardware, software, or both. **(B)** A problem used to evaluate the performance of several computer systems relative to one another, or relative to system specification. (C) 610.10-1994

benchmark program A standard program that can be used to evaluate the performance of a computer system. *See also:* synthetic benchmark program; kernel benchmark program; local benchmark program. (C) 610.10-1994w

bend (navigation) A departure of the course line from the desired direction at such a rate that it can be followed by the vehicle. (AES/GCS) 172-1983w

bend amplitude (navigation) The measured maximum amount of course deviation due to bend; measurement is made from the nominal or bend-free position of the course. (AES/GCS) 172-1983w

bend frequency (navigation) The frequency at which the course indicator oscillates when the vehicle track is straight and the course contains bends; bend frequency is a function of the vehicle velocity. (AES/GCS) 172-1983w

bend radius The radial distance of any arc formed by a bent cable, measured to the geometric center of the cable. *See also:* minimum bend radius. (C) 610.7-1995

bend ratio (cable plowing) The radius of a bend (segment of a circle) divided by the outside diameter of a cable, pipe, etc. (T&D/PE) 590-1977w

bend-reduction factor (navigation) The ratio of bend amplitude existing before the introduction of bend-reducing features to that existing afterward. (AES/GCS) 172-1983w

bend, waveguide *See:* waveguide bend.

benign failure Failure whose penalties are of the same order of magnitude as the benefit provided by correct service delivery. (C/BA) 896.9-1994w

BER *See:* bit error rate; bit error ratio.

BERT *See:* bit error rate testing.

best effort service A communication service in which transmitted data is not acknowledged. Such data typically arrives in order, complete and without errors. However, if an error occurs, or a packet is not delivered, nothing is done to correct it (e.g., there is no retransmission). (DIS/C) 1278.1-1995, 1278.2-1995

beta The ratio of the collector current to the base current of a bipolar transistor, commonly referred to as either the common-emitter current gain or the current amplification factor. (CAS) [13]

beta circuit (feedback amplifier) That circuit that transmits a portion of the amplifier output back to the input. *See also:* feedback. (EEC/PE) [119]

beta end **(1)** The end of the module farthest from the lowest-numbered connector contact. (C/BA) 1101.3-1993
(2) The end of the module nearest the highest-numbered contact. (C/BA) 1101.4-1993, 1101.7-1995

beta figure of merit (β) (nonlinear, active, and nonreciprocal waveguide components) A figure of merit for parametric amplifier varactors that relates to capacitive nonlinearity. Historically, for silicon varactors,

$$\beta = \frac{C_J(+1\mu A)}{C_{J-3}} = \frac{C_{JV_S}}{C_{J-3}}$$

and for GaAs varactors,

$$\beta = \frac{C_{J+0.5}}{C_{J-3}}$$

where
C_{JV} = junction capacitance at voltage V
V_s = voltage at which the forward current is 1μV
(MTT) 457-1982w

beta key The connector keying pin located at the beta end of the module connector. (BA/C) 1101.3-1993

betatron An electric device in which electrons revolve in a vacuum enclosure in a circular or a spiral orbit normal to a magnetic field and have their energies continuously increased by the electric force resulting from the variation with time of the magnetic flux enclosed by their orbits. (ED) [45]

bevatron A synchrotron designed to produce ions of a billion (10^9) electron-volts energy or more. (ED) [45]

beveled brush corners Where material has been removed from a corner, leaving a triangular surface. *See also:* brush. (EEC/EM/LB) [101]

beveled brush edges The removal of an edge to provide a slanting surface from which a shunt connection can be made or for clearance of pressure fingers or for any other purpose. *See also:* brush. (EEC/EM/LB) [101]

beveled brush ends and toes The angle included between the beveled surface and a plane at right angles to the length. The toe is the uncut or flat portion on the beveled end. When a brush has one or both ends beveled, the front of the brush is the short side of the side exposing the face level. *See also:* brush. (EEC/EM/LB) [101]

Beverage antenna A directional antenna composed of a system of parallel horizontal conductors from one-half to several wavelengths long, terminated to ground at the far end in its characteristic impedance. *Synonym:* wave antenna. (AP/ANT) 145-1993

bezel (cathode-ray oscilloscopes) The flange or cover used for holding an external graticule or cathode-ray tube cover in front of the cathode-ray tube. It may also be used for mounting a trace recording camera or other accessory item. (IM) 311-1970w

BF *See:* ballistic focusing.

BI *See:* buffered interconnect.

bias **(1)** **(germanium gamma-ray detectors)** The voltage applied to a detector to produce the electric field to sweep out the signal charge. (NPS/NID) 325-1986s, 759-1984r
(2) **(telegraph transmission)** A uniform displacement of like signal transitions resulting in a uniform lengthening or shortening of all marking signal intervals. *See also:* telegraphy. (COM) [49]
(3) **(A)** (of a semiconductor radiation detector) (in a biased amplifier). The applied threshold voltage (or current) below which the gain is zero. **(B)** (in a detector) The polarizing electric field that causes charge to be collected. (NPS) 300-1988
(4) **(A)** **(computers)** A systematic deviation of a value from a reference value. *Synonym:* bias error. **(B)** **(computers)** The amount by which the average of a set of values departs from a reference value. *Synonym:* bias error. (C) 610.10-1994, 1084-1986
(5) **(A)** **(accelerometer)** The average component of accelerometer output, which has no correlation with input acceleration. Bias is typically expressed in units of gravity. **(B)** **(gyros)** The average component of gyro output, which has no correlation with input rotation or acceleration. Bias is typically expressed in degrees per hour (°/h). (AES/GYAC) 528-1994
(6) **(A)** The deviation of the expected value of a random variable from a corresponding stated (correct or known) value. **(B)** A fixed deviation from the true value that remains constant over replicated measurements within the statistical precision of the measurement. *Synonyms:* fixed error; systematic error; deterministic error. (NI) N42.23-1995

(7) The time difference between the data arrival time and a specified signal edge (e.g., of a clock). Also the BIAS clause used in a CHECK statement. (C/DA) 1481-1999

bias current or power The direct and/or alternating current or power required to operate a bolometer at a specified resistance under specified ambient conditions. (IM) 470-1972w

bias, detector *See:* detector bias.

bias distortion (data transmission) A measure of the difference in the pulse width of the positive and negative pulses of a dotting signal. Usually expressed in percent of a full signal. (PE) 599-1985w

biased amplifier (1) (charged-particle detectors) (semiconductor radiation detectors) An amplifier giving essentially zero output for all inputs below a threshold and having constant incremental gain for all inputs above the threshold up to a specified maximum amplitude. (NPS/NID) 325-1986s, 301-1976s, 759-1984r
(2) (semiconductor charged-particle detectors) An amplifier giving zero output for input signals below an adjustable threshold and having a constant incremental gain above that threshold up to a specified maximum output. (NPS) 300-1988r

biased exponent (1) (binary floating-point arithmetic) The sum of the exponent and a constant (bias) chosen to make the biased exponent's range nonnegative. (C/MM) 754-1985r
(2) (mathematics of computing) In floating-point arithmetic, the sum of the exponent and a constant (bias) chosen to make the biased exponent's range nonnegative. (C) 1084-1986w

biased scheduled net interchange The scheduled net interchange power plus the algebraic sum of frequency bias, time-error bias, and other control-area biases. (PE/PSE) 94-1991w

bias error (1) A systematic error, whether due to equipment or propagation conditions. A nonzero mean component of a random error. (AES) 686-1997
(2) (mathematics of computing) *See also:* bias. (C) 1084-1986w

bias, grid, direct *See:* direct grid bias.

biasing (1) (laser gyro) The action of intentionally imposing a real or artificial rate into a laser gyro to avoid the region in which lock-in occurs. (AES/GYAC) 528-1994
(2) A technique used in memory mapping whereby the translation from a logical address to a physical address is performed by simply adding a bias to the logical address to determine the physical address. *Synonym:* relocation. *See also:* segmenting. (C) 610.10-1994w

bias instability (gyros) The random variation in bias as computed over specified finite sample time and averaging time intervals. This non-stationary (evolutionary) process is characterized by a 1/f power spectral density. It is typically expressed in degrees per hour (°/h). (AES/GYAC) 528-1994

bias magnet (magnetic tape pulse recorders for electricity meters) A device that provides a magnetic field used to orient the direction of magnetization on the magnetic tape to a predetermined polarity. (ELM) C12.14-1982r

bias resistor (1) (charged-particle detectors) (germanium gamma-ray detectors) (x-ray energy spectrometers) (of a semiconductor radiation detector) The resistor through which bias voltage is applied to a detector. (NPS/NID) 759-1984r, 301-1976s
(2) (semiconductor charged-particle detectors) The resistor through which the polarizing voltage is applied to a detector. (NPS) 300-1988r
(3) The resistor through which the bias voltage is applied to a detector. (NPS) 325-1996

bias spectrum At reference ambient, a specification of the fractions of total bias power in the dc and ac components and the frequency of the ac component. *Note:* The polarity of the dc component should also be given. (IM) 470-1972w

bias statistic An estimation of bias calculated from a finite sample of data using a specified formula. (NI) N42.23-1995

bias telegraph distortion (1) Distortion in which all mark pulses are lengthened (positive bias) or shortened (negative bias). It may be measured with a steady stream of unbiased reversals, square waves having equal-length mark and space pulses. The average lengthening or shortening gives true bias distortion only if other types of distortion are negligible. *See also:* modulation. (AP/ANT) 145-1983s
(2) (relay) *See also:* relay bias winding.

biaxial test (1) (seismic testing of relays) The relay under test is subjected to acceleration in one principal horizontal axis and the vertical axis simultaneously. (PE/PSR) C37.98-1977s
(2) The specimen under test is subject to acceleration in one principal horizontal axis and the vertical axis simultaneously. (SWG/PE) C37.100-1992, C37.81-1989r
(3) Simultaneously testing in one horizontal and the vertical direction. (PE/SUB) 693-1997

BIB *See:* Bridge Interconnect Bus.

biconical antenna (1) (overhead power lines) An antenna consisting of two conical conductors that have a common axis and vertex and are excited or connected to the receiver at the vertex. When the vertex angle of one of the cones is 180°, the antenna is called a discone. (T&D/PE) 539-1990
(2) An antenna consisting of two conical conductors having a common axis and vertex. (AP/ANT) 145-1993

biconical reflectance (illuminating engineering) Ratio of reflected flux collected through a conical solid angle to the incident flux limited to a conical solid angle. *Note:* The directions and extent of each cone must be specified.

biconical reflectance

(EEC/IE) [126]

biconical transmittance (illuminating engineering) Ratio of transmitted flux, collected over an element of solid angle surrounding the direction, to the incident flux limited to a conical solid angle. *Note:* The directions and extent of each cone must be specified.

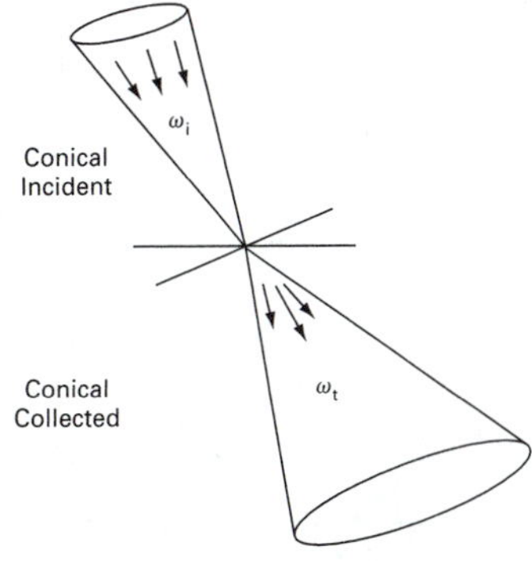

biconical transmittance

(EEC/IE) [126]

BICS *See:* bus implementation conformance statement.

BICS pro forma *See:* bus implementation conformance statement pro forma.

bicycle *See:* cable car.

BIDI *See:* bidirectional bus.

bidirectional (1) Providing for information transfer in both directions between master and remote terminals (of a communication channel). (SUB/PE) 999-1992w
(2) A pin or port that can place logic signals onto an interconnect and receive logic signals from it (i.e., act both as a driver and a receiver). (C/DA) 1481-1999

bidirectional antenna An antenna having two directions of maximum response. *See also:* antenna. (AP/ANT) [35]

bidirectional bar code symbol A bar code symbol format that permits decoding of the contents whether scanned in one direction or the reverse direction. (PE/TR) C57.12.35-1996

bidirectional bus (BIDI) **(1)** A bus which provides a communication path in either direction between two or more devices; for example, between a central processor and peripheral devices. (C) 610.10-1994w
(2) (programmable instrumentation) A bus used by any individual device for two-way transmission of messages; that is, both input and output. (IM/C/MM/AIN) 488.1-1987r, 696-1983w

bidirectional diode-thyristor (thyristor ac power controllers) A two-terminal thyristor having substantially the same switching behavior in the first and third quadrants of the principal voltage-current characteristic. (IA/IPC) 428-1981w

bidirectional operation When the peripheral and host communicate using both forward and reverse data channels. As defined in this standard, Nibble and Byte Modes provide reverse channel communication and are used in conjunction with Compatibility Mode to provide bidirectional operation. Extended Capabilities Port (ECP) and Enhanced Parallel Port (EPP) Modes support bidirectional communication. (C/MM) 1284-1994

bidirectional pin A component pin that can either drive or receive signals from external connections. (TT/C) 1149.1-1990

bidirectional printer A printer that can print in two directions, that is, left-to-right and right-to-left. *Synonym:* reverse printer. (C) 610.10-1994w

bidirectional pulse *See also:* bidirectional pulses.

bidirectional pulses Pulses, some of which rise in one direction and the remainder in the other direction. *See also:* pulse. (AP/ANT) 145-1983s

bidirectional reflectance (illuminating engineering) Ratio of reflected flux collected over an element of solid angle surrounding the given direction to essentially collimated incident flux. *Note:* (1) The directions of incidence and collections and the size of the solid angle "element" of collection must be specified. (2) In each case of conical incidence or collection, the solid angle is not restricted to a right circular cone, but may be of any cross section, including rectangular, a ring, or a combination of two or more solid angles.

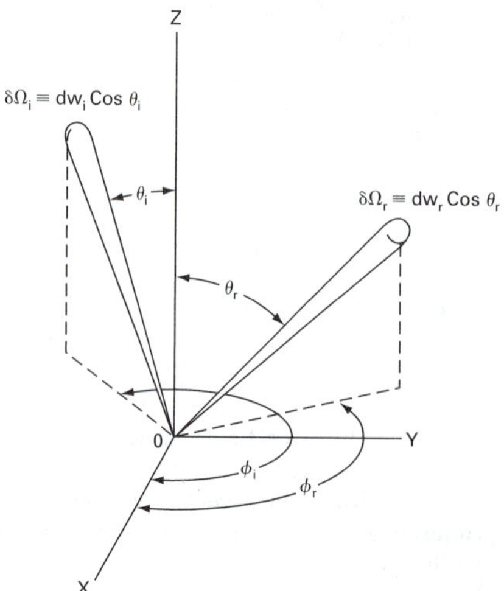

bidirectional reflectance

(EEC/IE) [126]

bidirectional reflectance-distribution function (f_r) The ratio of the differential luminance of a ray $dL_r(\theta_r,\phi_r)$ reflected in a given direction (θ_r,ϕ_r) to the differential luminous flux density $dE_i(\theta_i,\phi_i)$ incident from a given direction of incidence (θ_r,ϕ_r) that produces it.

$$f_r(\theta_i,\phi_i;\theta_r,\phi_r) \equiv dL_r(\theta_r,\phi_r)/dE_i(\theta_i,\phi_i)(\mathrm{sr})^{-1}$$
$$= dL_r(\theta_r,\phi_r)/L_i(\theta_i,\phi_i)d\Omega_i$$

where
$d\Omega \equiv d\omega \cdot \cos\theta.$

Notes: 1. This distribution function is the basic parameter for describing (geometrically) the reflecting properties of an opaque surface element (negligible internal scattering). 2. It may have any positive value and will approach infinity in the particular direction for ideally specular reflectors. 3. The spectral and polarization aspects must be defined for complete specification, since the BRDF as given above only defines the geometric aspects.

bidirectional reflectance—distribution function

(EEC/IE) [126]

bidirectional relay A stepping relay in which the rotating wiper contacts may move in either direction. *Synonym:* add-and-subtract relay. (IM) [120]

bidirectional signal line A signal line that may be defined in either direction across an interface. The direction is determined by control signals for each operation. (C/MM) 959-1988r

bidirectional transducer (bilateral transducer) A transducer that is not a unidirectional transducer. *See also:* transducer. (Std100) 270-1966w

bidirectional transmission (fiber optics) Signal transmission in both directions along an optical waveguide or other component. (Std100) 812-1984w

bidirectional transmittance (illuminating engineering) Ratio of incident flux collected over an element of solid angle surrounding the given direction to essentially collimated incident flux. *Note:* The directions of incidence and collection, and the size of the solid angle "elements" must be specified. (EEC/IE) [126]

bidirectional triode-thyristor A three-terminal thyristor having substantially the same switching behavior in the first and third quadrants of the principal voltage-current characteristic. (IA/IPC) 428-1981w

BIDM *See:* Basic Interoperability Data Model.

bifilar suspension A suspension employing two parallel ligaments, usually of conducting material, at each end of the moving element. (EEC/AII) [102]

bifunctional machine A computer that can perform either the host computer or backend computer functions. (C) 610.10-1994w

bifurcated feeder A stub feeder that connects two loads in parallel to their only power source. (SWG/PE) C37.100-1992

bigAdd A bus transaction that adds an integer *addend* argument to a specified data address and returns the previous data value from that address. All values in this transaction are assumed to be big-endian integers. In the CSR Architecture this is called a fetch_add transaction. (C/MM) 1596.5-1993

big addressan Bus that multiplexes the most significant byte of the address with the data byte that has the lowest address. (C/BA) 896.3-1993w

big addressian A term used to describe the physical location of data-byte addresses on a multiplexed address/data bus. On a big-addressian bus, the data byte with the largest address is multiplexed (in time or space) with the least-significant byte of the address. (C/MM) 1212-1991s

big-bang testing A type of integration testing in which software elements, hardware elements, or both are combined all at once into an overall system, rather than in stages. (C) 610.12-1990

big endian (1) The most significant byte of a data item, it has the lowest relative memory address. Correspondingly, less significant bytes have higher relative memory addresses. (C/BA) 896.3-1993w

(2) A representation of multibyte numerical values in which bytes with greater numerical significance appear at lower memory addresses. (C/BA) 1275-1994

(3) A term used to describe the arithmetic significance of data-byte addresses within a multibyte register. Within a big endian register or register set, the data byte with the largest address is the least significant. (MM/C) 1212-1991s

(4) A multibyte data value that is stored in memory with the most significant data byte through least significant data byte in the lowest through highest memory addresses, respectively. (C/MM) 1212.1-1993

(5) A specified ordering of bytes within a data structure where the low-order byte (byte 0) is placed in the most significant byte lane of that data structure. (C/BA) 1014.1-1994w

(6) A term used to imply that bytes in a word and words in a double word are transmitted most significant byte or word first in a serial stream of bytes. (C/MM) 1284.1-1997

(7) A method of storing multibyte data in a byte-addressable memory such that the most significant byte of the data is stored at the lowest address. (C/BA) 896.10-1997

(8) A data format where the most significant byte of a multibyte object is at the lowest address and the least-significant byte is at the highest address. (C/MM) 1284.4-2000

big endian processor A processor architecture that is optimized for the processing of big endian data values, as opposed to little endian data values. (C/MM) 1212.1-1993

bignum A multiple-precision computer representation for very large integers. *See also:* fixnum. (C/MM) 1178-1990r

bihemispherical reflectance (illuminating engineering) Ratio of reflected flux collected over the entire hemisphere to the flux incident from the entire hemisphere. *See also:* hemispherical reflectance.

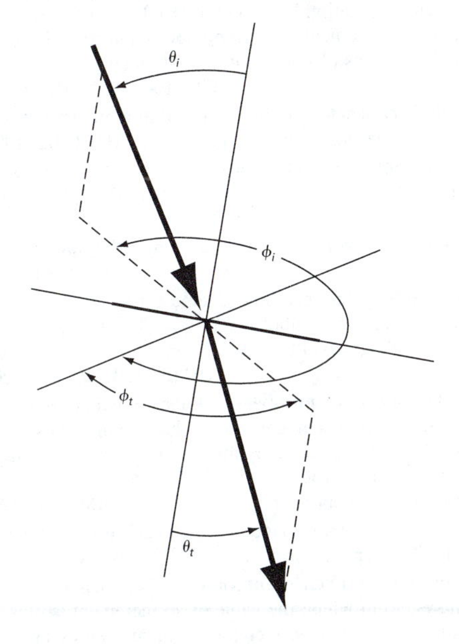

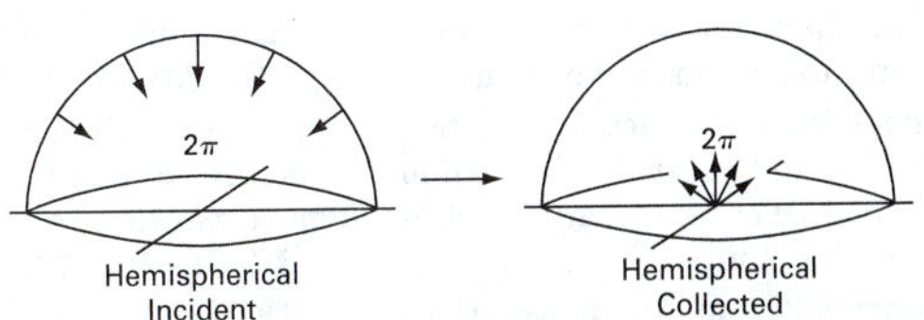

bihemispherical reflectance

(EEC/IE) [126]

bihemispherical transmittance (illuminating engineering) Ratio of transmitted flux collected over the entire hemisphere to the incident flux from the entire hemisphere.

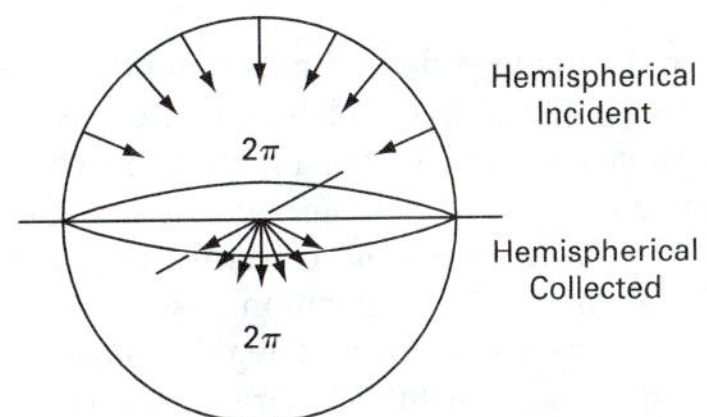

bihemispherical transmittance

(EEC/IE) [126]

bil *See:* basic impulse insulation level.

BIL *See:* preferred basic impulse insulation level; basic impulse insulation level.

bilateral-area track (electroacoustics) A photographic sound track having the two edges of the central area modulated according to the signal. *See also:* phonograph pickup. (SP) [32]

bilateral network (network analyzers) Network capable of transmission in both directions, not necessarily equal or symmetrical. (IM/HFIM) 378-1986w

bilateral transducer A transducer capable of transmission simultaneously in both directions between at least two terminations. 196-1952w

BILBO *See:* built-in logic block observer.

bilevel operation Operation of a storage tube in such a way that the output is restricted to one or the other of two permissible levels. *See also:* storage tube. (ED) 158-1962w

billing demand (power operations) The demand that is used to determine the demand charges in accordance with the provisions of a rate schedule or contract. (PE/PSE) 858-1987s, 346-1973w

billing error (switching systems in telecommunications environments) Occurs when a call is billed incorrectly. Billing errors may be measured as the number of incorrectly billed calls per 10 000 billable calls. Billing errors are the aggregate of the following specific billing errors. The billing accuracy definitions for automatic message accounting (AMA) data collection apply from the point at which information is submitted to the sensor up to the point at which the information is sent on to the revenue accounting office. Passage of answer supervision and number identification information outside of the AMA data collection system is considered separately. The AMA data collection system should be designed such that a single equipment failure does not cause an unrecoverable loss of more than a given number of call records. (COM/TA) 973-1990w

bill of materials A report showing the material costs of a single unit of product; listing of all unit components with part numbers, quantities, and supplier prices. (SCC22) 1346-1998

bill to wrong party (switching systems in telecommunications environments) Occurs when a call is charged to the wrong customer. (COM/TA) 973-1990w

bimetallic element An actuating element consisting of two strips of metal with different coefficients of thermal expansion bound together in such a way that the internal strains caused by temperature changes bend the compound strip. *See also:* relay. (EEC/REE) [87]

bimetallic thermometer A temperature-measuring instrument comprising an indicating pointer and appropriate scale in a protective case and a bulb having a temperature-sensitive bimetallic element. The bimetallic element is composed of two or more metals mechanically associated in such a way that relative expansion of the metals due to temperature change produces motion. (PE/PSIM) 119-1974w

bin *See:* pocket.

binary (A) (data transmission) Pertaining to a characteristic or property, involving a selection, choice or condition in which there are two possibilities. **(B) (data transmission)** Pertaining to the numeration system with a radix of two. (PE) 599-1985

(2) (A) (data management) (mathematics of computing) Pertaining to a selection in which there are two possible outcomes. *Synonyms:* straight binary; pure binary; natural binary; ordinary binary; normal binary; regular binary; standard binary. **(B) (mathematics of computing) (data management)** Pertaining to the numeration system with a radix of two. *Synonyms:* standard binary; regular binary; normal binary; straight binary; natural binary; ordinary binary; pure binary. (C) 610.5-1990, 1084-1986

binary arithmetic operation (mathematics of computing) An arithmetic operation in which the operands and the results are represented in the binary numeration system. (C) 1084-1986w

binary Boolean operation* *See:* dyadic Boolean operation.
* Deprecated.

binary card A punch card that is to contain information in column binary or row binary form. (C) 610.10-1994w

binary cell (1) An elementary unit of storage that can be placed in either of two stable states. *Note:* It is therefore a storage cell of one binary digit capacity, for example, a single-bit register. (C) [20], [85]
(2) A storage cell that can hold one binary digit. For example, a single-bit register. (C) 610.10-1994w

binary chop* *See:* binary search.
* Deprecated.

binary code (1) A code in which each code element may be either of two distinct kinds or values, for example, the presence or absence of a pulse. (PE) 599-1985w
(2) A code that makes use of members of an alphabet containing exactly two characters, usually 0 and 1. The binary number system is one of many binary codes. *See also:* reflected binary code; pulse; information theory. (C/IA) [20], [61]
(3) (mathematics of computing) A code that uses exactly two distinct characters, usually 0 and 1, to represent data or instructions. (C) 610.5-1990w, 1084-1986w

binary coded decimal (BCD) (mathematics of computing) Pertaining to a number representation system in which each decimal digit is represented by a unique arrangement of binary digits (usually four); for example, the number 23 is represented as 0010 0011, whereas in binary notation, 23 is represented as 10111. *Synonym:* coded decimal. (C) 1084-1986w

binary coded decimal character set A character set containing all 64 characters that can be represented as permutations of six bits. (C) 610.5-1990w

binary-coded-decimal number (or BCD number). The representation of the cardinal numbers 0 through 9 by 10 binary codes of any length. Note that the minimum length is four and that there are over 29×10^9 possible four-bit binary-coded-decimal codes. *Note:* An example of 8-4-2-1 binary-coded decimal code follows for number 1 through 9.

Number	*r*4	*r*3	*r*2	*r*1
0	0	0	0	0
1	0	0	0	1
2	0	0	1	0
3	0	0	1	1
4	0	1	0	0
5	0	1	0	1
6	0	1	1	0
7	0	1	1	1
8	1	0	0	0
9	1	0	0	1

Where *r*1 is termed the **least significant binary digit (bit).** *See also:* digital. (PE) 599-1985w

binary coded decimal real data A technique for assigning numeric characters such that each decimal digit is represented by a unique arrangement of binary digits with an implied radix point at a specified position.

decimal	163.3_{10}
BCD real	$10001\ 0110\ 0011\ .\ 0011_2$

(C) 610.5-1990w

binary-coded digit A digit of any number representation system that is represented as a fixed number of binary digits. For example, the decimal digit 9 is represented as 1001. (C) 1084-1986w

binary-coded octal Pertaining to a three-bit binary code in which the octal digits 0−7 are represented by the binary numerals 000−111. (C) 1084-1986w

binary data Numeric data used to represent binary digits. *See also:* packed binary data; binary picture data; fixed-point binary data. (C) 610.5-1990w

binary digit (1) (A) (computers) A unit of information that can be represented by either a zero or a one. *See also:* word; byte; binary element. **(B) (computers)** An element of computer storage that can hold a unit of information as in definition (A). *See also:* binary element; byte; word. **(C) (computers)** A numeral used to represent one of the two digits in the binary numeration system; zero (0) or one (1) *See also:* byte; word; binary element. (C) 610.5-1990, 610.12-1990, 1084-1986, 610.6-1991
(2) (data transmission) A character used to represent one of the two digits in the numeration system with a radix of two. Abbreviated "bit." (PE) 599-1985w
(3) A character used to represent one of the two digits in the binary number system and the basic unit of information in a two-state device. The two states of a binary digit are usually represented by "0" and "1". *Synonym:* bit. (SUB/PE) 999-1992w
(4) (A) A unit of information that can be represented by either a zero or a one. *See also:* word; clocking bit; start bit; block; byte; synchronization bit; stop bit. **(B)** The fundamental unit of digital communication; information is transmitted over networks as streams of units as in definition (A). (C) 610.7-1995, 610.10-1994

binary digit character A character within a picture specification that represents a binary digit. (C) 610.5-1990w

binary element A data element that can assume either of two possible values or states. *See also:* binary digit; binary variable. (C) 610.5-1990w, 1084-1986w

binary element string A string consisting solely of binary elements. (C) 1084-1986w

binary encoding An encoding scheme for serial communications in which the symbol for a logic "0" or "1" assumes a specified voltage level for the full duration of a signalling bit period. (EMB/MIB) 1073.4.1-2000

binary floating point number (or binary floating-point arithmetic) A bit-string characterized by three components: a sign, a signed exponent, and a significand. Its numerical value, if any, is the signed product of its significand and two raised to the power of its exponent. (C/MM) 754-1985r

binary image A digital image in which each pixel is assigned a value of either zero or one. (C) 610.4-1990w

binary incremental representation An incremental representation system in which the value of an increment is plus one or minus one. *Synonym:* incremental binary representation. (C) 1084-1986w

binary information (microprocessor object modules) Bit patterns to be loaded into memory. (C/MM) 695-1985s

binary insertion sort An insertion sort in which each item in the set to be sorted is inserted into its proper position in the sorted set using a binary search algorithm. *Contrast:* two-way insertion sort. (C) 610.5-1990w

binary notation Any notation that uses the binary digits and the radix 2. *Synonyms:* binary scale; two-scale. (C) 1084-1986w

binary number (A) A quantity that is expressed by using the binary numeration system. **(B)** Loosely, a binary numeral. (C) 1084-1986, [20], [85]

binary number system *See:* positional notation; binary numeration system.

binary numeral (1) The binary representation of a number; for example, 101 is the binary numeral and V is the Roman numeral of the number of fingers on one hand. (C) [20], [85]
(2) (mathematics of computing) A numeral in the binary numeration system. For example, the binary numeral 101 is equivalent to the decimal numeral 5. (C) 1084-1986w

binary numeration system The numeration system that uses the binary digits and the radix 2. *Note:* The use of "binary number system" as a synonym for this term is deprecated. *Synonyms:* pure binary numeration system; binary system. *See also:* positional notation. (C) 1084-1986w

binary one The "true" binary state, usually represented as 1 or T. *Contrast:* binary zero. (C) 1084-1986w

binary operation* *See:* dyadic operation; Boolean operation.
* Deprecated.

binary operator *See:* dyadic operator.

binary phase shift keying (Binary PSK, BPSK) (1) A specific form of PSK that defines two states of carrier phase that are digitally encoded in a binary data stream. The states have a change in phase of 180° that corresponds to the 0 or 1 binary state. (LM/C) 802.7-1989r
(2) A form of modulation in which binary data are transmitted by changing the carrier phase by 180°. *See also:* frequency shift keying; amplitude shift keying. (LM/C) 610.7-1995, 802.3-1998

binary picture data Arithmetic data that is associated with a picture specification that allows binary digit characters, a radix point, exponent characters, and sign characters. *Synonym:* numeric bit data. *Contrast:* decimal picture data. (C) 610.5-1990w

binary point (1) (mathematics of computing) The radix point in the binary numeration system. *See also:* radix point; point. 10.1-1990
(2) *See also:* point.

Binary PSK *See:* binary phase shift keying.

binary pulse width modulation torquing (digital accelerometer) A torquing technique in which the time between (positive, negative) torquing transitions is constant. (AES/GYAC) 530-1978r

binary radix trie search A radix trie search using a binary trie in which only one bit is considered on each branch. *See also:* multiway radix trie search. (C) 610.5-1990w

binary relation A relation with two attributes. (C) 610.5-1990w

binary scale *See:* binary notation.

binary search (1) A dichotomizing search in which, at each step of the search, the remaining set of items is partitioned into two equal parts. *Synonyms:* logarithmic search; binary chop; bisection. *Contrast:* Fibonacci search; interpolation search. *See also:* binary search tree; binary tree search. (C) 610.5-1990w
(2) A search in which a set of items is divided into two parts, one part is rejected, and the process is repeated on the accepted part until those items with the desired property are found. *See also:* dichotomizing search. (C) [20], [85]

binary search tree A search tree of order 2. (C) 610.5-1990w

binary signaling A means of communicating between devices that uses two-state signals. Where multiple binary data bits are to be transferred, either multiple signaling paths ("parallel binary") or a time series of individual data bits ("serial binary") transmission methods are to be used. (SUB/PE) 999-1992w

binary-state variable *See:* binary variable.

binary symmetric channel A channel designed to convey messages consisting of binary characters and which has the property that the conditional probabilities of changing any one character to the other character are equal. (C) 610.10-1994w

binary system *See:* binary numeration system.

binary-to-decimal conversion The process of converting a binary numeral to an equivalent decimal numeral. For example, binary 10001011.01 is converted to decimal 139.25. (C) 1084-1986w

binary-to-hexadecimal conversion The process of converting a binary numeral to an equivalent hexadecimal numeral. For example, binary 10001011.01 is converted to hexadecimal 8B.4. (C) 1084-1986w

binary-to-octal conversion The process of converting a binary numeral to an equivalent octal numeral. For example, binary 10001011.01 is converted to octal 213.2. (C) 1084-1986w

binary torquing (1) (digital accelerometer) System with two stable torquing states (for example, positive and negative). (AES/GYAC) 530-1978r
(2) (accelerometer) (gyros) A torquing mechanization that uses only two torquer current levels, usually positive and negative of the same magnitude; no sustained zero current or off condition exists. The positive and negative current periods can be either discrete pulses or duration-modulated pulses. In the case of zero input (acceleration or angular rate), a discrete pulse system will produce an equal number of positive and negative pulses. A pulse-duration-modulated system will produce positive and negative current periods of equal duration for zero input. Binary torquing delivers constant power to a sensor torquer (as compared to variable power ternary torquing) and results in stable thermal gradients for all inputs. (AES/GYAC) 528-1994

binary tree A tree in which each nonterminal node has at most two subtrees. *Note:* B tree is sometimes used incorrectly in reference to a binary tree. This usage is considered deprecated. *See also:* complete binary tree; weight-balanced tree; n-ary tree; binary search tree; full binary tree. (C) 610.5-1990w

binary tree search A search in which the items in the set to be searched are placed in a binary tree, and the tree is traversed making key comparisons until the argument is found, or the end of the tree is encountered. *See also:* digital tree search; binary search. (C) 610.5-1990w

binary variable A variable that can assume either of two values or logic states: binary zero (false) or binary one (true). *Synonyms:* two-valued variable; two-state variable; Boolean variable; binary-state variable. (C) 1084-1986w

binary zero The "false" binary state, usually represented as 0 or F. *Contrast:* binary one. (C) 1084-1986w

bind (1) To assign a value to an identifier. For example, to assign a value to a parameter or to assign an absolute address to a symbolic address in a computer program. *See also:* static binding; dynamic binding. (C) 610.12-1990
(2) To assign a network address to an endpoint. (C) 1003.5-1999

binder *See:* binder load.

binder (bond) (1) (rotating machinery) A solid, liquid, or semiliquid composition that exhibits marked ability to act as an adhesive, and that, when applied to wires, insulation components, or other parts, will solidify, hold them in position, and strengthen the structure. (PE) [9]
(2) (electroacoustics) A resinous material that causes the various materials of a record compound to adhere to one another. *See also:* phonograph pickup. (SP) [32]

binder-hole card A punch card that contains one or more holes used to bind the cards together. (C) 610.10-1994w

binder load A toggle device designed to secure loads in a desired position. It is normally used to secure loads on mobile equipment. *Synonyms:* chain binder; binder. (T&D/PE) 524-1992r

binding (1) (software) The assigning of a value or referent to an identifier; for example, the assigning of a value to a parameter, the assigning of an absolute address, virtual address, or device identifier to a symbolic address or label in a computer program. *See also:* label; dynamic binding; identifier; computer program; parameter; static binding. (C/SE) 729-1983s

(2) A defined mapping from the syntax and semantics of a software specification language into the syntax and semantics of a general purpose programming language. The purpose of such mapping to allow either a human or a machine to translate a program specified in the former language into a compilable and executable program in the latter. (SCC20) 1226-1998

binding band (rotating machinery) A band of material, encircling stator or rotor windings to restrain them against radial movement. *See also:* rotor. (PE) [9]

binding post *See:* binding screw.

binding screw A screw for holding a conductor to the terminal of a device or equipment. *Synonyms:* clamping screw; binding post; terminal screw. (EEC/PE) [119]

binnacle (navigation aids) (marine navigation) The stand in which a compass is mounted. (AES/GCS) 172-1983w

binocular (navigation aids) An optical instrument for use with both eyes simultaneously. (AES/GCS) 172-1983w

binocular portion of the visual field (illuminating engineering) That portion of space where the fields of the two eyes overlap. (EEC/IE) [126]

binomial array A linear array in which the currents in successive elements are made proportional to the binomial coefficients of $(x+ y)^{n-1}$ for the purpose of reducing minor lobes. *See also:* antenna. (AP/ANT) [35]

bioelectric null (zero lead) (medical electronics) A region of tissue or other area in the system, which has such electric symmetry that its potential referred to infinity does not significantly change. *Note:* This may or may not be ground potential. (EMB) [47]

biological electrode impedance The ratio between two vectors, the numerator being the vector that represents the potential difference between the electrode and biological material, and the denominator being the vector that represents the current between the electrode and the biological material. *See also:* polarization reactance; loss angle; polarization capacitance; polarization resistance. (EMB) [47]

biological electrode potential The potential between an electrode and biological material. (EMB) [47]

biological variability A range in the degree of response to internal and external stimuli that organisms normally exhibit because of genetic makeup and environmental conditioning. This biological or normal variability must be considered when determining the effect of any one specific factor; e.g., an electric field. (T&D/PE) 539-1990

Biomedical Statistics Package A computer language used widely in biomedical statistical applications. (C) 610.13-1993w

bionics A branch of technology relating the functions, characteristics, and phenomena of living systems to the development of mechanical systems. (C) 610.2-1987

biophysical study One approach used to assess the potential for biological effects of artificial electric or magnetic fields. The magnitude of induced body currents and fields is compared with levels known to cause biological effects by certain physical mechanisms; e.g., heating of tissues. (T&D/PE) 539-1990

Biot-Savart law *See:* magnetic field strength produced by an electric current.

biparting door (elevators) A vertically sliding or a horizontally-sliding door, consisting of two or more sections so arranged that the sections or groups of sections open away from each other and so interconnected that all sections operate simultaneously. *See also:* hoistway. (PE/EEC) [119]

bipolar (power supplies) Having two poles, polarities, or directions. *Note:* Applied to amplifiers or power supplies, it means that the output may vary in either polarity from zero; as a symmetrical program it need not contain a direct-current component. *See also:* unipolar. (PE) [78]

(2) (A) Having two opposite states, such as positive and negative; For example, in computer logic, a value of true is represented by an electrical voltage polarity opposite to that representing a value of false. *Contrast:* unipolar. **(B)** Pertaining to a semiconductor technology in which transistors are built from alternating layers of positively and negatively doped semiconductor material. *See also:* diode-transistor logic; transistor-transistor logic; emitter-coupled logic. (C) 610.10-1994

bipolar code violation A violation that occurs whenever two consecutive nonzero elements of the same polarity occur in an AMI signal. A bipolar violation is a code violation if (A) it occurs in an AMI signal, or (B) it occurs in a B*n*ZS signal separate from a zero substitution code. Such a code violation indicates an error in transmission. (COM/TA) 1007-1991r

bipolar device An electronic device whose operation depends on the transport of both holes and electrons. (CAS) [13]

bipolar electrode An electrode, without metallic connection with the current supply, one face of which acts as an anode surface and the opposite face as a cathode surface when an electric current is passed through the cell. *See also:* electrolytic cell. (EEC/PE) [119]

bipolar electrode system (electrobiology) Either a pickup or stimulating system consisting of two electrodes whose relation to the tissue currents is roughly symmetrical. *See also:* electrobiology. (EMB) [47]

bipolar pulse (1) (pulse terminology) Two pulse waveforms of opposite polarity that are adjacent in time and that are considered or treated as a single feature. (IM/WM&A) 194-1977w

(2) A signal pulse having two lobes, one above and the other below the baseline. When produced by a linear filter network, the two lobes have the same area but not necessarily the same peak amplitude. (NPS) 325-1996

bipolar signal A line code that employs a ternary signal to convey binary digits in which successive binary ones are represented by signal elements that are normally of alternating positive and negative polarity but equal in amplitude, and in which binary zeros are represented by signal elements that have zero amplitude. (COM/TA) 1007-1991r

bipolar video A radar video signal whose amplitude can have both positive and negative values; derived from a synchronous phase detection process. *Note:* Coherent detection produces one type of bipolar video. *See also:* coherent signal processing. (AES) 686-1997

bipolar violation A nonzero signal element in an alternate mark inversion signal that has the same polarity as the previous nonzero element. (COM/TA) 1007-1991r

biquinary (1) (mathematics of computing) Pertaining to a two-part representation of decimal digits consisting of a binary portion with values 0 or 5, and a quinary portion with values 0 through 4. For example, the decimal digit 7 is coded as 12, which implies 5 + 2. (C) 1084-1986w

(2) (information processing) Pertaining to the number representation system in which each decimal digit N is represented by the digit pair AB, where $N = 5A + B$, and where $A =$ 0 or 1 and B = 0, 1, 2, 3, or 4; for example, decimal 7 is represented by biquinary 12. This system is sometimes called a mixed-radix system having the radices 2 and 5. (C) [85]

biquinary code A two-part representation of decimal digits consisting of a binary portion with values 0 or 5 and a quinary portion with values 0 through 4. For example, decimal digit 7 is coded as 12. (C) 1084-1986w

biquinary coded decimal Pertaining to a number representation system in which each decimal digit is represented by a biquinary code. (C) 1084-1986w

biquinary notation Any notation that uses the biquinary code to represent numbers. (C) 1084-1986w

biquinary numeration system A numeration system that alternately uses 2 and 5 as bases. *Note:* The abacus uses a biquinary system. (C) 1084-1986w

bird *See:* running board.

birdie *See:* running board.

birefringence *See:* birefringent medium.

birefringent medium (fiber optics) A material that exhibits different indices of refraction for orthogonal linear polarizations of the light. The phase velocity of a wave in a birefringent medium thus depends on the polarization of the wave. Fibers may exhibit birefringence. *See also:* refractive index. (Std100) 812-1984w

BISAM *See:* basic indexed sequential access method.

bisection *See:* binary search.

BIST Built-in self-test. (C/BA) 896.2-1991w

bistable (1) (general) The ability of a device to assume either of two stable states. (IA/IAC) [60]
(2) Pertaining to a device capable of assuming either one or two stable states. *See also:* feedback control system. (C/ICTL) [85]
(3) Pertaining to a circuit or device that is capable of assuming one of two stable states. *See also:* monostable. (C) 610.10-1994w

bistable amplifier An amplifier with an output that can exist in either of two stable states without a sustained input signal and can be switched abruptly from one state to the other by specified inputs. *See also:* feedback control system. (IA/ICTL/IAC) [60]

bistable logic function A sequential logic function that has two and only two internal output states. *Synonym:* flip-flop. (GSD) 91-1984r

bistable operation Operation of a charge-storage tube in such a way that each storage element is inherently held at either of two discrete equilibrium potentials. *Note:* Ordinarily this is accomplished by electron bombardment. *See also:* charge-storage tube. (ED) 158-1962w

bistatic cross section The scattering cross section in any specified direction other than back toward the source. *See also:* monostatic cross section; radar cross section. (AP/ANT) 145-1993

bistatic radar A radar using antennas for transmission and reception at sufficiently different locations that the angles or ranges from those locations to the target are significantly different. (AES) 686-1997

bistatic reflectivity The reflectivity when the reflected wave is in any specified direction other than back toward the transmit antenna. The transmit and receive antennas are at different locations. (EMC) 1128-1998

bistatic-scatter cross section *See:* radar cross section.

bistatic scattering coefficient The scattering coefficient when the transmitter and receiver are not collocated. *See also:* scattering coefficient. (AP/PROP) 211-1997

BIT *See:* built-in test.

bit (b) (1) (microprocessor operating systems) A contraction of the term "binary digit;" a unit of information represented by either a zero or a one. (C/MM) 855-1990
(2) A single binary integer. A set bit represents to a binary "1." A cleared bit represents a binary "0." (C/MM) 1284.1-1997
(3) (A) An abbreviation of binary digit. **(B)** A single occurrence of a character in a language employing exactly two distinct kinds of characters. **(C)** A unit of storage capacity. The capacity, in bits, of a storage device is the logarithm to the base two of the number of possible states of the device. *See also:* storage capacity.
(4) A unit of information content equal to the information content of a message the *a priori* probability of which is one-half. *Note:* If, in the definition of information content, the logarithm is taken to the base two, the result will be expressed in bits. One bit equals $\log_{10} 2$ hartley. *See also:* information theory; parity bit; check bit. (IT) [7]
(5) A binary digit. (C/MM) 1296-1987s
(6) A unit of information in the binary numeration system. Also a unit of storage capacity of a memory. (ED) 641-1987w
(7) The smallest unit of information in the binary system of notation. (DIS/C) 1278.1-1995
(8) *See also:* least significant bit; most significant bit. (SWG/SUB/PE) C37.1-1987s, C37.100-1992
(9) *See also:* binary digit. (C) 610.5-1990w, 610.12-1990, 1084-1986w, 610.6-1991w
(10) *See also:* binary digit; block; byte; clocking bit; start bit; stop bit. (C) 610.7-1995
(11) (A) An abbreviation of binary digit. **(B)** A single occurrence of a character in a language that employs exactly two distinct kinds of characters. **(C)** A unit of storage capacity. The capacity, in bits, of a storage device is the logarithm to the base two of the number of possible states of the device. (ED) 1005-1998

bit blitter circuit A circuit that performs bit block transfer operations. *Synonym:* blitter. *See also:* blt chip. (C) 610.10-1994w

bit block transfer (A) To move information, optionally with a masking step, from one storage location to another. *Note:* Used extensively in bit mapped displays. Often abbreviated as bit blt, bitblt, or just blt (pronounced "blit"). *See also:* bit blitter circuit. **(B)** The transfer or combination of the pixel values in rectangular regions of bit maps. (C) 610.10-1994

bitblt *See:* bit block transfer.

bit blt (bitblt) *See:* bit block transfer.

bit cell The time interval used for the transmission of a single data (CD0 or CD1) or control (CVH or CVL) symbol. (C/LM) 802.3-1998

bit clock A clock recovered from the input Rx Serial Symbol stream which is bit synchronous with less than 5 degrees of single-sided phase noise. (C/BA) 1393-1999

bit density *See:* recording density.

BITE *See:* built-in test equipment.

bit error A bit is said to be in error when it is transferred from the source to the destination within the assigned timeslot, but the delivered bit is of a different value than that sent from the source. *See also:* error rate. (COM/TA) 1007-1991r

bit-encoded byte A byte with a definition for each bit. (C/MM) 1284.1-1997

bit-encoded word A word with a definition for each bit. (C/MM) 1284.1-1997

bit error rate (BER) (1) The ratio of errors to the total number of bits being sent in a data transmission from one location to another. (LM/C/BA) 802.7-1989r, 1355-1995
(2) A measurement of error rate stated as a ratio of the number of bits with an error to the total number of bits passing a given point on the ring. A BER of 10^{-6} indicates that an average of one bit per million bits is in error. (C/LM) 8802-5-1998

bit error rate testing The process of testing a data transmission channel using some predictable bit pattern so that the bits can be compared before and after the transmission to detect errors. *See also:* block error rate testing. (C) 610.7-1995

bit error ratio (BER) (1) The ratio of the number of bits received in error to the total number of bits received. (C/LM) 802.3-1998
(2) The ratio of the number of bit errors to the total number of bits transmitted in a given time interval. BER may be measured directly by detecting errors in a known signal, or approximated from code violations or framing bit errors. Numerical values of error ratio should be expressed in the form $n \cdot 10^{-p}$, where p is an integer greater than zero. When n is omitted, the implied value is 1. (COM/TA) 1007-1991r

bit-line The line that connects the memory cell drain to the sense amplifier during the read cycle and to a data line or latch during a write cycle. (ED) 1005-1998

bit map In computer graphics, a block of memory that stores a raster image of pixels in a device-specific format, in which the characteristics of each pixel are determined by a set of bits. *Synonyms:* refresh buffer; frame buffer.

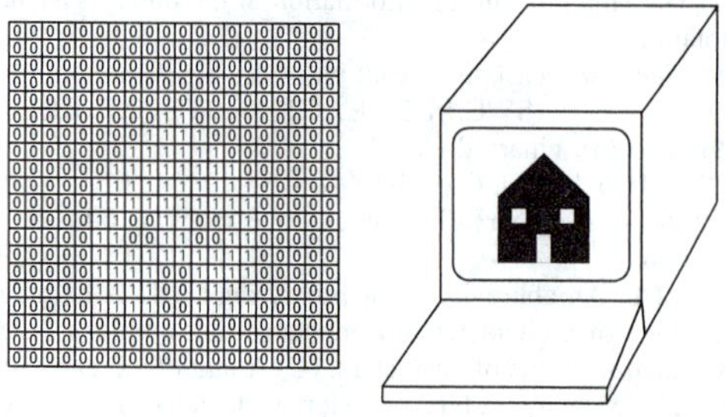

bit map

(C) 610.6-1991w

(2) A data structure that stores information about entities in the form of a series of one-bit entries, each of which describes the state of the corresponding entity; for example, in graphics, a block of memory that stores a raster image in a device-specific format in which the characteristics of each pixel are determined by a set of bits. *See also:* bit plane; bit-mapped. (C) 610.10-1994w

bit map font A font defined in the form of a bit map that specifies the bit pattern which makes up each character. *Synonym:* intrinsic font. *Contrast:* derived font; vector font; outline font. (C) 610.10-1994w

bit-mapped Pertaining to a display screen on which a character or image is generated from a bit map in memory. (C) 610.10-1994w

bitonic merge *See:* bitonic sort.

bitonic sort A variation on Batcher's parallel sort in which one of the two ordered subsets begins in reverse order and the items to be compared and exchanged are selected from the same subset. *Synonym:* bitonic merge. (C) 610.5-1990w

bit pad *See:* data tablet.

bit parallel Pertaining to a method for simultaneously processing all bits as a contiguous set of bits over separate wires, one wire for each bit. (C) 610.10-1994w

bit-parallel (programmable instrumentation) (696 interface devices) (signals and paths) A set of concurrent data bits present on a like number of signal lines used to carry information. Bit-parallel data bits may be acted upon concurrently as a group or independently as individual data bits. (IM/MM/C/AIN) 488.1-1987r, 696-1983w

bit pattern The image created on the screen of a display device by the mapping of the bit map onto the screen. (C) 610.10-1994w

bit plane A portion of a bit map that stores one bit of every pixel of a raster image. *Note:* Several bit planes are combined to make the full image.

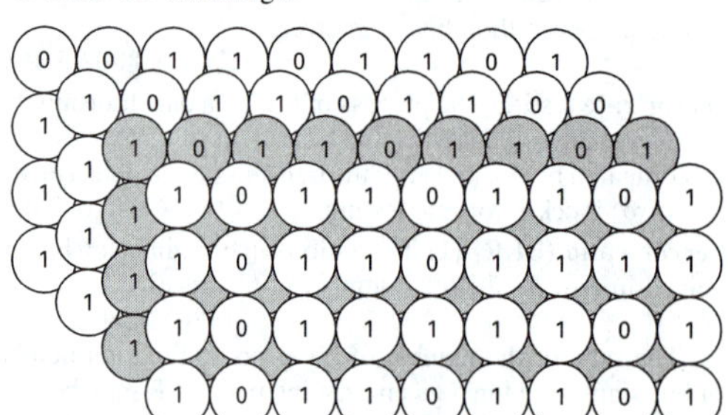

bit plane

(C) 610.10-1994w

bit rate (BR) (1) (station control and data acquisition) (supervisory control) (data acquisition and automatic control) The number of bits transferred in a given time interval. Bits per second is a measure of the rate at which bits are transmitted. (SWG/PE/SUB) C37.100-1992, C37.1-1994

(2) (data transmission) The speed at which bits are transmitted; usually expressed in bits per second. (PE) 599-1985w

(3) The number of bits transmitted per unit of time, usually expressed in bits per second (bps). (COM/TA) 1007-1991r

(4) The rate of data throughput on the medium in bits per second or hertz, whichever is more appropriate to the context. *Synonym:* bit transfer rate. *Contrast:* baud rate. (C) 610.7-1995

(5) The rate at which data are transmitted, expressed in bits per unit time. *Synonym:* bit transfer rate. *See also:* baud rate. (C) 610.10-1994w

(6) The total number of bits per second transferred to or from the Media Access Control (MAC). For example, 100BASE-T has a bit rate of one hundred million bits per second (10^8 b/s). (C/LM) 802.3-1998

bit rate/2 One-half of the BR in hertz. (C/LM) 802.3-1998

bit-retention The time interval from the writing of either state into a specified memory location to the earliest appearance of an incorrect state from such memory location, as measured on the specified output terminal(s). (ED) 1005-1998

bit-retention time The retention time for one address location. (ED) 641-1987w

bit serial Pertaining to a method of sequentially processing a contiguous set of bits one at a time over a single wire, according to a fixed sequence. (C) 610.10-1994w

bit slice Pertaining to a device consisting of an n-bit functional component, such as an arithmetic and logic unit (ALU), or a sequencer, which may be cascaded with one or more identical devices to expand the width of its function by multiples of n. *See also:* bit slice device; bit slice processor. (C) 610.10-1994w

bit slice architecture An architecture in which a section of the register and the arithmetic and logic unit in a computer is placed into one package. *See also:* bit slice processor. (C) 610.10-1994w

bit slice device A device that uses bit slice technology. (C) 610.10-1994w

bit slice microprocessor *See:* bit slice device; bit slice processor.

bit slice processor A processor that is built from multiple bit slices to any given word-size. (C) 610.10-1994w

bits per second (bps) A unit of data transmission speed, expressed as the number of bits transmitted per second. *Note:* IEEE Std 260.1-1993 specifies b/s as the SI unit symbol for bits per second. *Contrast:* baud rate. (C) 610.7-1995

bits per unit time (test, measurement, and diagnostic equipment) Operating number of bits, handled by a device in a given unit of time, under specified conditions. (MIL) [2]

bit steering A microprogramming technique in which the meaning of a field in a microinstruction is dependent on the value of another field in the microinstruction. *Synonym:* immediate control. *Contrast:* residual control. *See also:* two-level encoding. (C) 610.12-1990

bit stream A continuous stream of bits transmitted over a channel with no separators between the character groups. (C) 610.7-1995, 610.10-1994w

bit string (1) A sequence of binary digits; for example, the bit string 0101001. *See also:* character string. (C) 610.5-1990w

(2) An ordered sequence of zero or more bits. (C/PA) 1328-1993w, 1327-1993w, 1224-1993w

bit stuffing A method to insert extra bits in a bit stream to achieve transparency throughout the bit stream. (C) 610.7-1995

bit time (BT) The duration of one bit as transferred to and from the MAC. The bit time is the reciprocal of the bit rate. For example, for 100BASE-T the bit rate is 10^{-8} s or 10 ns. (LM/C) 8802-3-1993s, 802.3-1998

bit transfer rate *See:* bit rate.

BIU *See:* bus interface unit.

bivariant function generator A function generator having two input variables. *See also:* electronic analog computer. (C) 165-1977w

BIXIT *See:* bus implementation extra information for testing.

BIXIT pro forma *See:* bus implementation extra information for testing pro forma.

black Pertains to the parts of a computer or communications system in which data being transmitted or manipulated is encrypted. *Contrast:* red. (C) 610.7-1995

black and white *See:* monochrome.

blackbody (1) (A) (fiber optics) A totally absorbing body (which reflects no radiation). *Note:* In thermal equilibrium, a blackbody absorbs and radiates at the same rate; the radiation will just equal absorption when thermal equilibrium is maintained. *See also:* emissivity. **(B) ([planckian] locus [illuminating engineering])** The locus of points on a chromaticity diagram representing the chromaticities of blackbodies having various (color) temperatures. **(C) (illuminating engineering)** A temperature radiator of uniform temperature whose radiant exitance in all parts of the spectrum is the maximum obtainable from any temperature radiator at the same temperature. Such a radiator is called a blackbody because it will absorb all the radiant energy that falls upon it. All other temperature radiators may be classed as nonblackbodies. They radiate less in some or all wavelength intervals than a blackbody of the same size and the same temperature. *Note:* The blackbody is practically realized over limited solid angles in the form of a cavity with opaque walls at a uniform temperature and with a small opening for observation purposes. (Std100/EEC/IE) 812-1984, [126]
(2) An ideal material that absorbs all incident radiation. *Note:* Under thermal equilibrium, a blackbody is a perfect emitter with its emissivity and absorptivity equal to unity. The radiation spectrum of a blackbody is given by Planck's radiation law. (AP/PROP) 211-1997

black box (A) A system or component whose inputs, outputs, and general function are known but whose contents or implementation are unknown or irrelevant. *See also:* encapsulation. **(B)** Pertaining to an approach that treats a system or component as in definition (A). *Contrast:* glass box. *See also:* encapsulation. (C) 610.12-1990

black box model A model whose inputs, outputs, and functional performance are known, but whose internal implementation is unknown or irrelevant; for example, a model of a computerized change-return mechanism in a vending machine, in the form of a table that indicates the amount of change to be returned for each amount deposited. *Synonyms:* input/output model; behavioral model. *Contrast:* glass box model. (C) 610.3-1989w

black-box testing *See:* functional testing.

black compression (television) The reduction in gain applied to a picture signal at those levels corresponding to dark areas in a picture with respect to the gain at that level corresponding to the mid-range light value in the picture. *Note:* (1) The gain referred to in the definition is for a signal amplitude small in comparison with the total peak-to-peak picture signal involved. A quantitative evaluation of this effect can be obtained by a measurement of differential gain. (2) The overall effect of black compression is to reduce contrast in the low lights of the picture as seen on a monitor. *Synonym:* black saturation. *See also:* television. (BT/AV) [34]

black level (television) The level of the picture signal corresponding to the maximum limit of black peaks. *See also:* television. (BT/AV) [34]

black light (illuminating engineering) The popular term for ultraviolet energy near the visible spectrum. *Note:* For engineering purposes the wavelength range 320–400 nm (nanometers) has been found useful for rating lamps and their effectiveness upon fluorescent materials (excluding phosphors used in fluorescent lamps). By confining "black light" applications to this region, germicidal, and erythemal effects are, for practical purposes, eliminated. (EEC/IE) [126]

black light flux (illuminating engineering) Radiant flux within the wavelength range 320−400 nm (nanometers). It is usually measured in milliwatts. *Note:* The fluoren is used as a unit of "black light" flux and is equal to one milliwatt of radiant flux in the wavelength range 320−400 nm. Because of the variability of the spectral sensitivity of materials irradiated by "black light" in practice, no attempt is made to evaluate "black light" flux according to its capacity to produce effects. (EEC/IE) [126]

black light flux density (illuminating engineering) "Black light" flux per unit area of the surface being irradiated. It is equal to the incident "black light" flux divided by the area of the surface when the flux is uniformly distributed. It usually is measured in milliwatts per unit area of "black light" flux. (EEC/IE) [126]

black peak (television) A peak excursion of the picture signal in the black direction. (BT/AV) [34]

black recording (A) (amplitude-modulation facsimile system) The form of recording in which the maximum received power corresponds to the maximum density of the record medium. **(B) (frequency-shift facsimile system)** The form of recording in which the lowest received frequency corresponds to the maximum density of the record medium. *See also:* recording. (COM) 168-1956

black saturation *See:* black compression.

black signal (at any point in a facsimile system) The signal produced by the scanning of a maximum-density area of the subject copy. *See also:* facsimile signal. (COM) 168-1956w

black transmission (A) (amplitude-modulation facsimile system) The form of transmission in which the maximum transmitted power corresponds to the maximum density of the subject copy. **(B) (frequency-modulation facsimile system)** The form of transmission in which the lowest transmitted frequency corresponds to the maximum density of the subject copy. *See also:* facsimile transmission. (COM) 168-1956

blade (1) (of a switching device) The moving contact member that enters or embraces the contact clips. *Notes:* 1. In cutouts the blade may be a fuse carrier or fuseholder on which a nonfusible member has been mounted in place of a fuse link. When so used the nonfusible member alone is also called a blade in fuse parlance. 2. In distribution cutouts the blade may be a nonfusible member for mounting on a fuse carrier in place of a fuse link, or in a fuse support, in place of a fuse holder. *Synonym:* disconnecting blade. (SWG/PE) C37.30-1971s
(2) (disconnecting blade of a switch or disconnecting cutout) The moving contact member that enters or embraces the contact clips. *Note:* In distribution cutouts, the blade may be a non-fusible member for mounting on a fuse carrier in place of a fuse link, or in a fuse support, in place of a fuseholder. (SWG/PE) C37.40-1993
(3) (of a switching device) The moving contact member that enters or embraces the contact clips. *Note:* In cutouts, the blade may be a fuse carrier or fuseholder on which a nonfusible member has been mounted in place of a fuse link. When so used, the nonfusible member alone is also called a blade in fuse parlance. *Synonym:* disconnecting blade. (SWG/PE) C37.100-1992

blade antenna A form of monopole antenna that is blade-shaped for strength and low aerodynamic drag. (AP/ANT) 145-1993

blade control deadband (hydraulic turbines) The magnitude of the change in the blade control cam follower position required to reverse the travel of the blade control servomotor. The deadband is expressed in percent of the change in cam follower position required to move the blades from extreme "flat" to extreme "steep." (PE/EDPG) 125-1977s

blade guide An attachment to ensure proper alignment of the blade and contact clip when closing the switch. (SWG/PE) C37.100-1992

blade latch A latch used on a stick operated switch to hold the switch blade in the closed position. (SWG/PE) C37.100-1992, C37.40-1981s

⟨`blank`⟩ One of the characters that belong to the blank character class as defined via the LC_CTYPE category in the current locale. In the POSIX Locale, a ⟨`blank`⟩ is either a ⟨`tab`⟩ or a ⟨`space`⟩. (PA/C) 9945-2-1993

blank (A) (test, measurement, and diagnostic equipment) A place of storage where data may be stored. *Synonym:* space. **(B) (test, measurement, and diagnostic equipment)** A character, used to indicate an output space on a printer in which nothing is printed. **(C) (test, measurement, and diagnostic equipment)** A condition of no information at all in a given column of a punched card or in a given location on perforated tape. (MIL) [2]

blank character (1) A character used to produce a character space on an output medium. (C) [20], [85]
(2) (computer graphics) A graphic representation of the space character. (C) 610.5-1990w

blanked picture signal (television) The signal resulting from blanking a picture signal. *Note:* (1) Adding synchronizing signal to the blanked picture signal forms the composite picture signal. (2) This signal may or may not contain setup. A blanked picture signal with setup is commonly called a noncomposite signal. *See also:* television. (BT/AV) [34], [27]

blanket *See:* conductor cover.

blanketing The action of a powerful radio signal or interference in rendering a receiving set unable to receive desired signals. *See also:* radiation. (EEC/PE) [119]

blank groove *See:* unmodulated groove.

blanking (1) (general) The process of making a channel or device noneffective for a desired interval.
(2) (television) The substitution for the picture signal, during prescribed intervals, of a signal whose instantaneous amplitude is such as to make the return trace invisible. *See also:* television. (EEC/PE) [119]
(3) (oscilloscopes) Extinguishing of the spot. Retrace blanking is the extinction of the spot during the retrace portion of the sweep waveform. The term does not necessarily imply blanking during the holdoff interval or while waiting for a trigger in a triggered sweep system. (IM/HFIM) [40]

blanking, chopped *See:* chopping transient blanking.

blanking, chopping transient *See:* chopping transient blanking.

blanking, deflection *See:* deflection blanking.

blanking level (television) That level of a composite picture signal which separates the range containing picture information from the range containing synchronizing information. *Note:* This term should be used for controls performing this function. (BT) [27]

blanking signal (television) A wave constituted of recurrent pulses, related in time to the scanning process, used to effect blanking. *Note:* In television, this signal is composed of pulses at line and field frequencies, which usually originate in a central synchronizing generator and are combined with the picture signal at the pickup equipment in order to form the blanked picture signal. The addition of synchronizing signal completes the composite picture signal. The blanking portion of the composite picture signal is intended primarily to make the return trace on a picture tube invisible. The same blanking pulses or others of somewhat shorter duration are usually used to blank the pickup device also. *See also:* television. 337

blanking, transient *See:* chopping transient blanking.

blank line A line consisting solely of zero or more ⟨`blank`⟩s terminated by a ⟨`newline`⟩. *See also:* empty line. (C/PA) 9945-2-1993

blank medium A data medium on which neither marks of reference, nor user data are recorded; For example, an unformatted floppy disk. *See also:* empty medium; virgin medium. (C) 610.10-1994w

blaster *See:* blasting unit.

blasting circuit A shot-firing cord together with connecting wires and electric blasting caps used in preparation for the firing of a blast in mines, quarries, and tunnels. *See also:* blasting unit. (EEC/PE) [119]

blasting switch A switch used to connect a power source to a blasting circuit. *Note:* A blasting switch is sometimes used to short-circuit the leading wires as a safeguard against premature blasts. *See also:* blasting unit. (EEC/PE) [119]

blasting unit A portable device including a battery or a hand-operated generator designed to supply electric energy for firing explosive charges in mines, quarries, and tunnels. *Synonyms:* exploder; blaster. (EEC/PE) [119]

bleaching (laser maser) The decrease of optical absorption produced in a medium by radiation or by external forces. (LEO) 586-1980w

bleeder A resistor connected across a power source to improve voltage regulation, to drain off the charge remaining in capacitors when the power is turned off, or to protect equipment from excessive voltages if the load is removed or substantially reduced. (EEC/PE) [119]

blemish (television) A small area brightness gradient in the reproduced picture, not present in the original scene. (BT) [27]

blemish charge (storage tubes) A localized imperfection of the storage assembly that produces a spurious output. *See also:* storage tube. (ED) 158-1962w, [45], 161-1971w

blending The combination of two or more modes of braking (e.g., rheostatic electric brake, regenerative electric brake, and friction brake) to produce the desired total retarding effort. (VT) 1475-1999

BLERT *See:* block error rate testing.

blinder A relay having a characteristic on an *R-X* diagram of one or more essentially straight lines, usually positioned at 75° to 90° from the *R*-axis and displaced from the origin. (SWG/PE) C37.100-1992

blinder characteristic A nondirectional distance relay characteristic in which the threshold of operation substantially plots as a straight line on an *R-X* diagram with the reach essentially resistive and largely independent of the reactance value. Generally this threshold of operation is positioned at an angle of 75° to 90° from the *R* axis. See figure below.

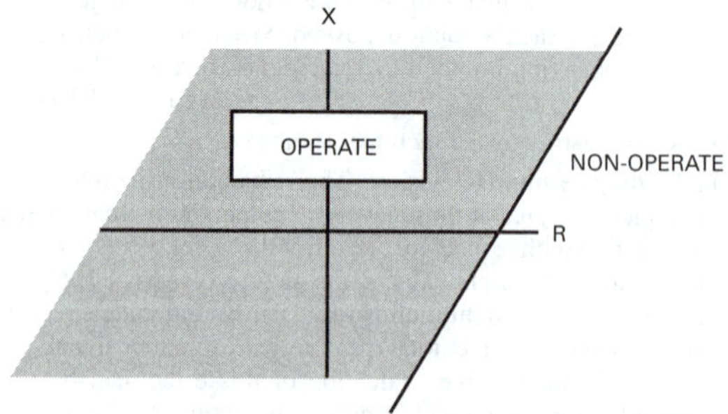

blinder characteristic

(SWG/PE) C37.100-1992

blinding glare (illuminating engineering) Glare that is so intense that for an appreciable length of time after it has been removed, no object can be seen. (EEC/IE) [126]

blind interrogation Access to a facility (e.g., the device identification register) without prior knowledge of the test logic operation of the specific component being accessed. (TT/C) 1149.1-1990

blind matrix spike A matrix spike sample sent through normal processing wherein the processor knows that the sample is of QA origin but does not know the nuclide or nuclide concentration. The terminology "Single Blind Matrix Spike" is also commonly used. (NI) N42.23-1995

blind phase In moving-target indication (MTI) radars, when the echo of interest is in quadrature to the reference signal. It occurs in systems that detect only the in-phase signal component. *See also:* moving-target indication. (AES) 686-1997

blind quality control sample Also referred to as a single blind quality control sample. Similar to a spike sample except that a blind sample is presented to the laboratory or analyst by whom the sample may be recognized as a quality control sample but the quantity and identity of the analytes are unknown. (NI) N42.23-1995

blind range A range corresponding to the time delay of an integral multiple of the interpulse period plus a time less than or equal to the transmitted pulse length. *Note:* A radar usually cannot detect targets at a blind range because of interference by a subsequent transmitted pulse. *See also:* eclipsing. (AES) 686-1997

blind replicate A replicate sample unknown to the analyst. (NI) N42.23-1995

blind replicate reference standard A sample of known concentration prepared from a purchased standard reference material and submitted as a blind sample into the laboratory. (NI) N42.23-1995

blind speed [radar using moving-target indication (MTI)] Radial velocity of a target with respect to the radar for which the MTI response is approximately zero. *Note:* In a coherent MTI system using a uniform repetition rate, a blind speed is a radial velocity at which the target changes its distance by one-half wavelength, or a multiple thereof, during each pulse-repetition interval. (AES) 686-1997

blind spot A limited range within the total domain of application of a device, generally at values inferior to the maximum rating. Operation of the equipment or of the protective device might fail in that limited range despite the device's demonstration of satisfactory performance at maximum ratings. (SPD/PE) C62.45-1992r

blind study *See:* single-blind study; double-blind study.

B-line *See:* index register.

blink A technique in which a display element is alternately blanked and displayed. *See also:* highlight. (C) 610.6-1991w

blinking (pulse systems) (navigation aids) A method of providing information by modifying the signal at its source so that the signal presentation on the display at the receiver alternately appears and disappears, for example, in loran, blinking is used to indicate that the signals of a pair of stations are out of synchronization. (AES/GCS) 172-1983w

blip (1) (navigation aids) (radar) A deflection or a spot of contrasting luminescence on a radar display caused by the presence of a target. (AES/GCS) 172-1983w, 686-1997 **(2)** *See also:* document mark. (C) 610.2-1987

blip-scan ratio The fraction of scans for which a blip is observed at a given range. *Synonym:* single-scan probability of detection. (AES) 686-1997

BLISS *See:* Basic Language for Implementation of System Software.

blitter *See:* bit blitter circuit; bit block transfer.

Block An instance of a subclass of `IEEE1451_Block`. (IM/ST) 1451.1-1999

block (1) (A) (data transmission) A set of things, such as words, characters, or digits handled as a unit. **(B) (data transmission)** A collection of contiguous records recorded as a unit. **(C) (data transmission)** In data communications, a group of contiguous characters formed for transmission purposes. **(D)** A circuit assemblage that functions as a unit. For example, a logic block within a sequential circuit. (PE/C) 599-1985, 610.7-1995, 610.10-1994 **(2) (railway practice)** A length of track of defined limits on which the movement of trains is governed by block signals, cab signals, or both. *See also:* absolute block. (EEC/PE) [119] **(3) (A) (software)** A group of contiguous storage locations, computer program statements, records, words, characters, or bits that are treated as a unit. **(B) (software)** To form a group as in definition (A). *Contrast:* deblock. (C) 610.7-1995, 610.5-1990, 610.10-1994, 610.12-1990 **(4)** In text editing and text formatting, one or more contiguous characters or lines of text. (C) 610.2-1987 **(5)** (city, town, or village) A square or portion of a city, town, or village enclosed by streets and including the alleys so enclosed but not any street. (NESC/NEC) [86] **(6)** A device designed with one or more single sheaves, a wood or metal shell, and an attachment hook or shackle. When rope is reeved through two of these devices, the assembly is commonly referred to as A *block and tackle.* A *set of fours* refers to a block and tackle arrangement utilizing two 4 in double sheave blocks to obtain four load-bearing lines. Similarly, a *set of fives* or a *set of sixes* refers to the same number of load-bearing lines obtained using two 5 in or two 6 in double sheave blocks, respectively. (PE/T&D) 524-1992r **(7)** (as applied to static relay design) An output signal of constant amplitude and specified polarity derived from an alternating input and with the duration controlled by the polarity of the input quantity. (SWG/PE) C37.100-1992 **(8)** A group of data that is contiguous in nature. *Synonym:* sector. (C/MM/ED) 855-1990, 1005-1998

block allocation *See:* paging.

Block and List Manipulator A programming language, based on LISP, but containing an ALGOL-like syntax, data types such as vectors and strings, and the ability to write macro-instructions. (C) 610.13-1993w

block and tackle *See:* rope block.

block average demand An average value occurring over a demand period specified by the end device (e.g., watthours/hours). The value may be saved by the end device for maximum or minimum registration. (AMR/SCC31) 1377-1997

block–block element A signal element in which two blocks are compared as to coincidence or sequence. (SWG/PE) C37.100-1992

block cable (communication practice) A distribution cable installed on poles or outside building walls, in the interior of a block, including cable run within buildings from the point of entrance to a cross-connecting box, terminal frame, or point of connection to house cable. *See also:* cable. (EEC/PE) [119]

block character *See:* end of transmission block character.

block check character In longitudinal redundancy checking and cyclic redundancy checking, a character that is transmitted by the sender after each message block and is compared with a character computed by the receiver to determine if the transmission was successful. (C) 610.7-1995

block clear Operation that sets a block of data within a memory to a common "1" state without affecting any other memory block. *Note:* In the field of nonvolatile memories, clear conventionally means "set to a '1' state." (ED) 1005-1998

Block Cookie The value of the cookie of a specific Block. (IM/ST) 1451.1-1999

block copy (1) In text editing, an operation that copies a block of text from one point to another within a file or between files, leaving the original block of text intact. (C) 610.2-1987 **(2)** A series of read or write transactions to sequential memory locations. (C/BA) 896.3-1993w **(3)** A block copy operation is characterized by a long series of read or write transactions to sequential memory locations. (C/BA) 10857-1994 **(4)** This operation is characterized by a long series of read or write transactions to sequential memory locations. (C/BA) 896.4-1993w

block count readout Display of the number of blocks that have been read from the tape derived by counting each block as it is read. *See also:* sequence-number readout. (IA) [61]

block delete In text editing, an operation that removes a block of text from a file. (C) 610.2-1987

block_descriptor A cell in a block_vector containing a pointer to the first byte of a contiguous storage block, the size of the block, and a flag. The flag indicates whether the storage block contains data or an extension to the block_vector. (C/MM) 1212.1-1993

block diagram (software) A diagram of a system, computer, or device in which the principal parts are represented by suitably annotated geometrical figures to show both the functions of the parts and their functional relationships. *Synonym:* configuration diagram; system resources chart. *See also:* structure chart; input-process-output chart; box diagram; graph; bubble chart.

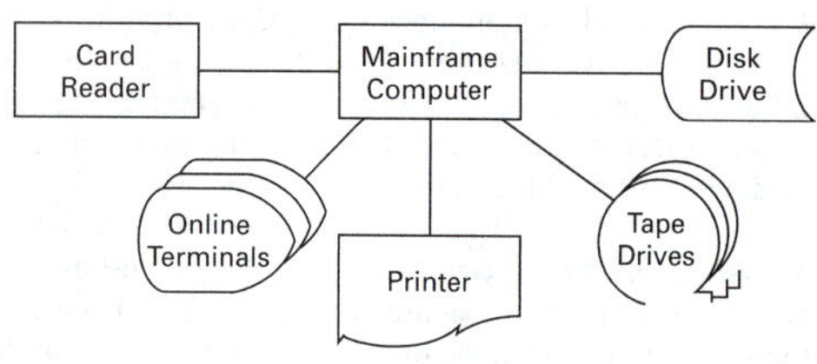

block diagram

(C) 610.12-1990

blocked channel A channel with at least one blocked endpoint. (C/MM) 1284.4-2000

blocked conductor A stranded conductor whose interstices are filled with a compound that prevents the migration of moisture along the interstices. (PE/IC) 1142-1995

blocked dial tone Occurs when the customer line cannot access any digit receiver because of overload, particularly in the front-end concentrator. In some systems, overload is the major reason for extremely long dial-tone delays. (COM/TA) 973-1990w

blocked endpoint An endpoint that is unable to send data on a connection due to lack of credit. A blocked endpoint becomes unblocked when it is granted credit to send data on that connection. (C/MM) 1284.4-2000

blocked impedance (transducer) The input impedance of the transducer when its output is connected to a load of infinite impedance. *Note:* For example, in the case of an electromechanical transducer, the blocked electric impedance is the impedance measured at the electric terminals when the mechanical system is blocked or clamped: the blocked mechanical impedance is measured at the mechanical side when the electric circuit is open-circuited. *See also:* self-impedance. (SP) [32]

blocked process A process that is waiting for some condition (other than the availability of a processor) to be satisfied before it can continue execution. (C/PA) 1003.1b-1993s

blocked record A record that is contained in a block that contains at least one other record. *See also:* unblocked record; spanned record. (C) 610.5-1990w

blocked task An Ada task that is not running or ready to run. A task is either blocked or ready to run. While ready, a task competes for the available execution resources that it requires to run. An operation that causes a task to become blocked is said to *block* the task, and an operation that causes a task to no longer be blocked is said to *unblock* the task. (C) 1003.5-1999

blocked thread A thread that is waiting for some condition (other than the availability of a processor) to be satisfied before it can continue execution. (C/PA) 9945-1-1996

block erase (A) Signal (command) that causes the erasing of a block of data within a memory without affecting any other memory block. **(B)** The operation of erasing a block of data. (ED) 1005-1998

block error (data transmission) A discrepancy of information in a block as detected by a checking code or technique. (PE) 599-1985w

block error rate testing The process of testing a data transmission channel using groups of information arranged into transmission blocks in a given message for error checking. *See also:* bit error rate testing. (C) 610.7-1995

block-error ratio The ratio of the blocks in error received in a specified period to the total number of blocks received in the same period. (COM/TA) 1007-1991r

block errors A block is said to be in error when one or more bit errors occur in that block when it is transferred from the source to the destination within the timeslot assigned. (COM/TA) 1007-1991r

block gap* *See:* interblock gap.

* Deprecated.

block ground *See:* traveler ground.

block, hold-down *See:* hold-down block.

block indicator A device used to indicate the presence of a train in a block. (EEC/PE) [119]

blocking (1) (tube rectifier) The prevention of conduction by means of grid or ignitor action, or both, when forward voltage is applied across a tube. (IA/CEM) [58]
(2) (semiconductor rectifiers) The action of a semiconductor rectifier cell that essentially prevents the flow of current. *See also:* rectification. (IA) 59-1962w, [12]
(3) (rotating machinery) A structure or combination of parts, usually of insulating material, formed by hold coils in relative position for mechanical support. *Note:* Usually inserted in the end turns to resist forces during running and abnormal conditions. *See also:* stator. (PE) [9]
(4) (telephone switching systems) The inability of a telecommunication system to establish a connection due to the unavailability of paths. (COM) 312-1977w
(5) (computers) The process of creating a block from one or more records. (C) 610.5-1990w, 610.12-1990
(6) A relay function that prevents action that would otherwise be initiated by the relay system. (SWG/PE) C37.100-1992
(7) Executing with `POSIX_IO.Non_Blocking` not set. *See also:* nonblocking. (C) 1003.5-1999

blocking behavior The effect on other tasks in the same partition when a task is blocked by a POSIX operation. Certain POSIX operations are required to block the calling task under defined conditions. For implementation-defined reasons a blocked task may prevent other tasks from executing. (C) 1003.5-1999

blocking capacitor (1) A capacitor that introduces a comparatively high series impedance for limiting the current flow of low-frequency alternating current or direct current without materially affecting the flow of high-frequency alternating current. *Synonym:* blocking condenser. (IM) [120]
(2) (check valve) An asymmetrical cell used to prevent flow of current in a specified direction. (PE/EEC) [119]

blocking condenser *See:* blocking capacitor.

blocking contact (of a semiconductor radiation detector) That contact from which depletion proceeds into the semiconductor material under conditions of reverse bias. (NPS) 300-1988r

blocking factor The number of records, words, characters, or bits in a block. (C) 610.12-1990, 610.5-1990w

blocking interval (circuit properties) (self-commutated converters) An interval during which voltage is impressed across a switching element in its off-state. (IA/SPC) 936-1987w

blocking oscillator (1) A relaxation oscillator consisting of an amplifier (usually single-stage) with its output coupled back to its input by means that include capacitance, resistance, and mutual inductance. *See also:* oscillatory circuit. (EEC/PE) [119]
(2) (squegging oscillator) An electron-tube oscillator operating intermittently with grid bias increasing during oscillation to a point where oscillations stop, then decreasing until

oscillation is resumed. *Note:* Squegge rhymes with wedge. *See also:* oscillatory circuit. (AP/ANT) 145-1983s

blocking period (1) (rectifier circuit element) The part of an alternating-voltage cycle during which reverse voltage appears across the rectifier-circuit element. *Note:* The blocking period is not necessarily the same as the reverse period because of the effect of circuit parameters and semiconductor rectifier cell characteristics. *See also:* rectifier circuit element. (IA) 59-1962w

(2) (gas tube) The part of the idle period corresponding to the commutation delay due to the action of the control grid. (ED) [45], [84]

blocking relay (1) (power system device function numbers) A relay that initiates a pilot signal for blocking of tripping on external faults in a transmission line or in other apparatus under predetermined conditions, or cooperates with other devices to block tripping or to block reclosing on an out-of-step condition or on power swings. (SUB/PE) C37.2-1979s

(2) A relay whose function is to render another relay or device ineffective under specified conditions. (SWG/PE) C37.100-1992

blocking signal A logic signal that is transmitted in a pilot scheme to prevent tripping. (PE/PSR) C37.113-1999

blocking switching network (telephone switching systems) A switching network in which a given outlet cannot be reached from any given inlet under certain traffic conditions. (COM) 312-1977w

blocking voltage The maximum voltage that can be applied to current-limiting pairs of terminals of a surge protector containing one or more current-protective devices without degradation of the surge protector. (SPD/PE) C62.36-1994

block, input *See:* input block.

block-interval demand meter *See:* integrated-demand meter; demand meter.

block-interval demand register (mechanical demand registers). A demand register that indicates or registers the maximum demand obtained by arithmetically averaging the meter registration over a regularly repeated time interval. (ELM) C12.4-1984

block length The number of units in a block. *Synonym:* block size. (C) 610.5-1990w

block-mode terminal A terminal device operating in a mode incapable of the character-at-a-time input and output operations described by some of the standard utilities. (C/PA) 9945-2-1993

block move In text editing, an operation that moves a block of text from one point to another within a file or between files, deleting the block of text from its original location. *Synonym:* block movement. (C) 610.2-1987

block movement *See:* block move.

block operation In text editing, an operation that affects a block of text. For example, block copy, block delete, block move. (C) 610.2-1987

block-organized random-access memory (BORAM) A memory arrangement that permits random access to blocks of memory cells that are read using serial transmission methods. *Note:* A block is a singly addressed large number of memory cells (greater than needed for a single computer word operation). (ED) 641-1987w

block overhead Any information, besides the actual data, that is stored with a block; for example, the size and location of the records within the block is considered overhead. *See also:* loading factor. (C) 610.5-1990w

block parity A parity check system capable of detecting and correcting a single error in a binary message. *See also:* parity check. (C) 1084-1986w

block read cycle A data transfer bus (DTB) cycle that is used to transfer a block of bytes ranging in number from 1 to 256 bytes from a to a master. This transfer is executed by using a string of l-, 2-, or 4-byte data transfers. Once the block transfer is initiated, the master does not release the DTB until all of the bytes have been transferred. This operation differs from a string of read cycles insofar as the master broadcasts only one address and one address modifier (at the beginning of the cycle). The slave then increments this address on each transfer so that the data for the next transfer is retrieved from the next higher location. (C/BA) 1014-1987

block read transaction An address beat followed by a block of one or more data read transfers from a set of contiguous addresses beginning with the address in the address beat. This is terminated by the appropriate style of end beat. (MM/C) 896.1-1987s

block select An input terminal to which a signal must be applied in order to permit the device to read or write a block of data. (ED) 1005-1998

block signal A fixed signal installed at the entrance of a block to govern trains entering and using that block. (EEC/PE) [119]

block-signal system A method of governing the movement of trains into or within one or more blocks by block signals or cab signals. (EEC/PE) [119]

block size *See:* block length.

block special file A file that refers to a device. A block special file is normally distinguished from a character special file by providing access to the device in a manner such that the hardware characteristics of the device are not visible. (C/PA) 9945-1-1996, 9945-2-1993, 1003.5-1999

block-spike element A signal element in which a block and a spike are compared as to coincidence. (SWG/PE) C37.100-1992

block, splice release *See:* hold-down block.

Block State Machine The state machine specified for the referenced Block. (IM/ST) 1451.1-1999

block station A place at which manual block signals are displayed. (EEC/PE) [119]

block-structured language A design language or programming language in which sequences of statements, called blocks, are defined, usually with begin and end delimiters, and variables or labels defined in one block are not recognized outside that block. Examples include Ada, ALGOL, C, PL/1, Pascal, MENTOR, and Modula II. *See also:* structured programming language. (C) 610.13-1993w, 610.12-1990

block transfer (1) (FASTBUS acquisition and control) The portion of a FASTBUS operation in which a master either sends data to or receives data from an attached slave on every transition of data sync. The slave acknowledges receipt of or sends data with every transition of data acknowledge. (NID) 960-1993

(2) (STEbus) A sequence of data transfers, in the same direction, that occur during a single bus transaction. (MM/C) 1000-1987r

(3) (NuBus) A transaction in which a single address is conveyed by the master and multiple data items from sequential addresses are then communicated between the master and the slave. (C/MM) 1196-1987w

®NuBus is a registered trademark of Texas Instruments

block-transfer read cycle A DTB cycle that is used to transfer a block of bytes from the responding slave to the active master, and possibly to participating slaves. This transfer is done using a number of 1, 2, or 4-byte data transfers. It differs from a series of single-transfer read cycles in that the master broadcasts the address only once, at the beginning of the cycle. It is the responsibility of the selected slaves to control the address for each subsequent data transfer. (C/MM) 1096-1988w

block-transfer write cycle A DTB cycle that is used to transfer a block of bytes from the active master to the selected slaves. This transfer is done using a series of 1, 2, or 4-byte data transfers. It differs from a series of single-transfer write cycles in that the master broadcasts the address only once, at the beginning of the cycle. It is the responsibility of the selected slaves to control the address for each subsequent data transfer. (C/MM) 1096-1988w

block_vector An effective_length parameter and an array of block_descriptors referencing storage locations for application data or unit-dependent information. The referenced blocks are physically segmented but logically contiguous. The descriptor array may also be segmented and one flagged descriptor per array segment may be used to point to an extension of the vector. The effective_length parameter limits the transfer of data bytes. (C/MM) 1212.1-1993

block write cycle A data transfer bus (DTB) cycle used to transfer a block of bytes ranging in number from 1 to 256 bytes from a master to a slave. It uses a string of 1-, 2-, or 4-byte data transfers. Once the block transfer is initiated, the master does not release the DTB until all of the bytes have been transferred. It differs from a string of write cycles insofar as the master broadcasts only one address and one address modifier (at the beginning of the cycle). The slave then increments this address on each transfer so that the data from the next transfer is stored in the next higher location. (C/BA) 1014-1987

block write transaction An address beat followed by a block of one or more data write transfers to a set of contiguous addresses beginning with the address in the address beat. This is terminated by the appropriate style of end beat. (C/MM) 896.1-1987s

Blondel diagram (rotating machinery) A phasor diagram intended to illustrate the currents and flux linkages of the primary and secondary windings of a transformer, and the components of flux due to primary and secondary winding currents acting alone. *Note:* This diagram is also useful as an aid in visualizing the fluxes in an induction motor. *See also:* asynchronous machine. (PE) [9]

blooming (1) (A) (diode-type camera tube) The increase in the size of the displayed image of a bright source when its irradiance is sufficient to cause overload of the mosaic target. It is measured in the display of the video output as the ratio of the enlarged spot size to the dimension of the active raster diagonal. **(B) (diode-type camera tube)** The ratio of the image device generated spot size at overload to the size of the active raster diagonal. The actual spot size imaged upon the device photosensitive surface is chosen as one percent of the active raster diagonal. (ED) 503-1978
(2) An increase in the blip size on the display as a result of an increase in signal intensity or duration. (AES) 686-1997

blowback In micrographics, an enlargement. (C) 610.2-1987

blower blade (rotating machinery) An active element of a fan or blower. *See also:* fan. (PE) [9]

blower housing *See:* fan housing.

blowoff valve (gas turbines) A device by means of which a part of the air flow bypasses the turbine(s) and/or the regenerator to reduce the rate of energy input to the turbine(s). *Note:* It may be used in the speed governing system to control the speed of the turbine(s) at rated speed when fuel flow permitted by the minimum fuel limiter would otherwise cause the turbine to operate at a higher speed. *See also:* asynchronous machine. (PE/EDPG) 282-1968w, [5]

blowout coil An electromagnetic device that establishes a magnetic field in the space where an electric circuit is broken and helps to extinguish the arc by displacing it, for example, into an arc chute. (IA/EEC/IAC/REE) [60], [84], [87]

blowout magnet A permanent-magnet device that establishes a magnetic field in the space where an electric circuit is broken and helps to extinguish the arc by displacing it. *See also:* relay. (EEC/REE) [87]

blt *See:* bit block transfer.

blt chip An integrated circuit whose purpose is to perform bit block transfer operations. (C) 610.10-1994w

blue alarm *See:* alarm indication signal.

blue dip (electroplating) A solution containing a mercury compound, and used to deposit mercury upon an immersed metal, usually prior to silver plating. *See also:* electroplating. (EEC/PE) [119]

blur (navigation aid terms) [null type direction finder (DF) systems] The output (including noise) at the bearing of minimum response expressed as a percentage of the output at the bearing of maximum response. (AES/GCS/RS) 173-1959w, 172-1983w, 686-1982s

blurred Pertaining to elements in an image that are indistinct or not readily discernable. *Contrast:* sharp. (C) 610.4-1990w

BMDP *See:* Biomedical Statistics Package.

B-message (analog voice frequency circuits) A frequency-weighting characteristic, used for measurement of noise in voice-frequency communications circuits and designed to weight noise frequencies in proportion to their perceived annoyance effect in telephone service. (COM/TA) 743-1984s

BNF *See:* Backus-Naur form.

BNR *See:* beacon reconfigure.

B*n*ZS code A bipolar line code with *n*-zero substitution. (COM/TA) 1007-1991r

board (1) (STEbus). A printed circuit board (pcb) that complies with IEEE Std 1000-1987. *See also:* problem board. (C/MM) 1000-1987r
(2) (VMEbus) A printed-circuit board (pcb), its collection of electronic components, and either one or two 96-pin connectors that can be plugged into backplane connectors. (BA/C) 1014-1987
(3) (VSB) A printed circuit (pc) board, its collection of electronic components, and at least one 96-pin connector. (C/MM) 1096-1988w
(4) (NuBus) A device connected to a bus. Usually constructed from a printed circuit board. Also referred to as a module. (C/MM) 1196-1987w
(5) A physical component that is inserted into one of the backplane slots. Note that a board may contain two nodes. (C/BA) 896.3-1993w
(6) The physical component of an SCI module that is inserted into one of the backplane slots. Note that a board may contain multiple nodes, and that nodes can be implemented without using boards or modules. (C/MM) 1596-1992
(7) A blank PCB. (C/MM) 1155-1992
(8) An electronic circuit assembly that connects to a single slot on the backplane. It is removable from and replaceable to a backplane assembly via connectors. This is standard terminology for VME64, while Futurebus+ uses module synonymously. (C/BA) 1014.1-1994w
(9) The physical component that is inserted into one of the backplane slots. (C/MM) 1212-1991s
(10) A device connected to a bus. Usually constructed from a printed circuit board. Also referred to as a module. (C/BA) 896.9-1994w
(11) A generic term used as an abbreviation for circuit board. (C) 610.10-1994w
(12) The physical component that is inserted into one of the backplane slots. Note that a board may contain multiple nodes. (C/MM) 1596.3-1996

®NuBus is a registered trademark of Texas Instruments Inc.

board assembly A board and its associated electrical components and connectors. (C/MM) 1155-1992

boatswain's chair (conductor stringing equipment) A seat designed to be suspended on a line reeved through a block and attached to a pulling device to hoist a workman to an elevated position. *Synonym:* bosun's chair. (T&D/PE) 524-1992r

bobbin (1) (primary cell) A body in a dry cell consisting of a depolarizing mix molded around a central rod of carbon and constituting the positive electrode in the assembled cell. *See also:* electrolytic cell. (EEC/PE) [119]
(2) (rotating machinery) Spool-shaped ground insulation fitting tightly on a pole piece, into which field coil is wound or placed. *See also:* rotor; stator. (PE) [9]

bobbin core A tape-wound core in which ferromagnetic tape has been wrapped on a form or bobbin that supplies mechanical support to the tape. *Note:* The dimensions of a bobbin are illustrated in the accompanying figure. Bobbin I.D. is the center-hole diameter (D) of the bobbin. Bobbin O.D. is the over-all diameter (E) of the bobbin. The bobbin height is the over-all axial dimension (F) of the bobbin. Groove diameter is the diameter (G) of the center portion of the bobbin on which the first tape wrap is placed. The groove width is the axial dimension (H) of the bobin measured inside the groove at the groove diameter.

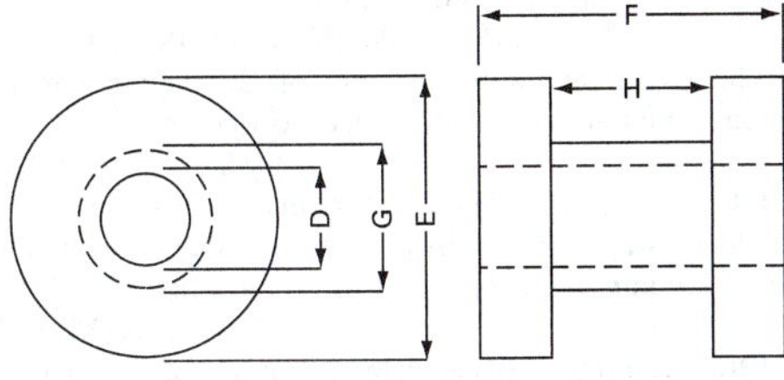

dimensions of a bobbin

bobbin core

(Std100) 163-1959w

bobbin height *See:* bobbin core; tape-wound core.

bobbin I.D. *See:* bobbin core; tape-wound core.

bobbin O.D. *See:* bobbin core; tape-wound core.

Bode diagram (automatic control) A plot of log-gain and phase-angle values on a log-frequency base, for an element transfer function $G(j\omega)$, a loop transfer function $GH(j\omega)$, or an output transfer function $G(j\omega)/[1 + GH(j\omega)]$. The generalized Bode diagram comprises similar plots of functions of the complex variable $s = \sigma + j\omega$. *Note:* Except for functions containing lightly damped quadratic factors, the gain characteristic may be approximated by asymptotic straight-line segments that terminate at corner frequencies. The ordinate may be expressed as a gain, a log-gain, or in decibels as 20 times log-gain; the abscissa as cycles per unit time, radians per unit time, or as the ratio of frequency to an arbitrary reference frequency. *See also:* feedback control system. (IM/PE/EDPG) [120], [3]

body *See:* housing.

body capacitance Capacitance introduced into an electric circuit by the proximity of the human body. (EEC/PE) [119]

body-capacitance alarm system A burglar alarm system for detecting the presence of an intruder through his or her body capacitance. *See also:* protective signaling. (EEC/PE) [119]

body effect (metal-nitride-oxide field-effect transistor) This effect occurs when the potential in the substrate of a (p-channel) insulated-gate field-effect transistor (IGFET) is more positive than the source potential. It can be expressed as an increment that increases the threshold voltage of an IGFET. The effect occurs routinely in integrated circuits. (ED) 581-1978w

body/finger ESD An electrostatic discharge from an intruding human finger or hand. Also called body/finger discharge. (SPD/PE) C62.47-1992r

body generator suspension A design of support for an axle generator in which the generator is supported by the vehicle body. *See also:* axle-generator system. (EEC/PE) [119]

body/metal discharge *See:* hand/metal ESD.

body/metal ESD *See:* hand/metal ESD.

body resistance (1) Determined from the ratio of voltage applied to current flowing in a human body, neglecting capacitive and inductive effects. (T&D/PE) 524a-1993r
(2) Determined from the ratio of voltage applied to current flowing in a body, neglecting capacitive and inductive effects. That value impeding the current flow through the common body resulting from contact with an energized line. (T&D/PE) 1048-1990

BOF *See:* beginning-of-file label.

bog anchor A heavy anchor of large surface area for use in unstable soils. (PE/T&D) 751-1990

bog shoe A piece of material, such as a section of pole or railroad tie, attached horizontally below the ground surface to increase the bearing area in unstable soils. Usually four shoes are installed, one on each side of the pole, at ninety degrees to each other. (T&D/PE) 751-1990

boilerplate text In word processing, standardized previously-stored textual material that may be used to create a new document. *Synonym:* stored paragraph. (C) 610.2-1987

bole The main stem of a tree of substantial diameter. Roughly capable of yielding sawn timber, veneer logs, or poles. (PE/T&D) 751-1990

bolometer (1) (fiber optics) A device for measuring radiant energy by measuring the changes in resistance of a temperature-sensitive device exposed to radiation. *See also:* radiant energy; radiometry. (Std100) 812-1984w
(2) (waveguide components) A term commonly used to denote the combination of a bolometer element and a bolometer mount; sometimes used imprecisely to refer to a bolometer element. (MTT) 147-1979w
(3) (laser maser) A radiation detector of the thermal type in which absorbed radiation produces a measurable change in the physical property of the sensing element. The change in state is usually that of electrical resistance. (LEO) 586-1980w

bolometer bridge A bridge circuit with provisions for connecting a bolometer in one arm and for converting bolometer-resistance changes to indications of power. *See also:* bolometric power meter. (IM/HFIM) [40]

bolometer bridge, balanced A bridge in which the bolometer is maintained at a prescribed value of resistance before and after radio-frequency power is applied, or after a change in radio-frequency power, by keeping the bridge in a state of balance. *Note:* The state of balance can be achieved automatically or manually by decreasing the bias power when the radio-frequency power is applied or increased and by increasing the bias power when the radio-frequency power is turned off, or decreased. The change in the bias power is a measure of the applied radio-frequency power. *See also:* bolometric power meter. (IM) 470-1972w

bolometer bridge, unbalanced A bridge in which the resistance of the bolometer changes after the radio-frequency power is applied and unbalances the bridge. The degree of bridge unbalance is a measure of the radio-frequency power dissipated in the bolometer. *See also:* bolometric power meter. (IM/HFIM) [40]

bolometer-coupler unit A directional coupler with a bolometer unit attached to either the side arm or the main arm, normally used as a feed-through power-measuring system. *Note:* Typically, a bolometer unit is attached to the side arm of the coupler so that the radio-frequency power at the output port can be determined from a measurement of the substitution power in the side arm. This system can be used as a terminating power meter by terminating the output port of the directional coupler. *See also:* bolometric power meter. (IM) 470-1972w

bolometer element (waveguide components) (bolometric detector) A power-absorbing element that uses the resistance change related to the temperature coefficient of resistivity (either positive or negative) as a means of measuring or detecting the power absorbed by the element. (MTT) 147-1979w

bolometer mount (1) (general) A waveguide or transmission-line termination that houses a bolometer element(s). *Note:* It normally contains internal matching devices or other reactive elements to obtain specified impedance conditions when a bolometer element is inserted and appropriate bias power is applied. Bolometer mounts may be subdivided into tunable, fixed-tuned, and broad-band untuned types. *See also:* bolometric power meter. (IM) 470-1972w

(2) (waveguide components) A waveguide or transmission line termination that can house a bolometer element. (MTT) 147-1979w

bolometer unit An assembly consisting of a bolometer element or elements and bolometer mount in which they are supported. *See also:* bolometric power meter. (IM) 470-1972w

bolometer unit, dual element An assembly consisting of two bolometer elements and a bolometer mount in which they are supported. *Note:* The bolometer elements are effectively in series to the bias power and in parallel to the radio frequency power. (IM/HFIM) [40]

bolometric detector (bolometers) The primary detector in a bolometric instrument for measuring power or current and consisting of a small resistor, the resistance of which is strongly dependent on its temperature. *Notes:* 1. Two forms of bolometric detector are commonly used for power or current measurement: The barretter that consists of a fine wire or metal film, and the thermistor that consists of a very small bead of semiconducting material having a negative temperature-coefficient of resistance; either is usually mounted in a waveguide or coaxial structure and connected so that its temperature can be adjusted and its resistance measured. 2. Bolometers for measuring radiant energy usually consist of blackened metal-strip temperature-sensitive elements arranged in a bridge circuit including a compensating arm for ambient temperature compensation. *See also:* instrument; bolometric instrument. (IM/HFIM) [40]

bolometric instrument (bolometers) An electrothermic instrument in which the primary detector is a resistor, the resistance of which is temperature sensitive, and that depends for its operation on the temperature difference maintained between the primary detector and its surroundings. Bolometric instruments may be used to measure nonelectrical quantities, such as gas pressure or concentration, as well as current and radiant power. *See also:* instrument. (EEC/PE) [119]

bolometric power meter A device consisting of a bolometer unit and associated bolometer-bridge circuit(s). (IM) 470-1972w

bolometric technique (power measurement) A technique wherein the heating effect of an unknown amount of radio-frequency power is compared with that of a measured amount of direct-current or audio-frequency power dissipated within a temperature sensitive resistance element (bolometer). *Note:* The bolometer is generally incorporated into a bridge network, so that a small change in its resistance can be sensed. This technique is applicable to the measurement of low levels of radio-frequency power, that is, below 100 mW. (IM/HFIM) [40]

bolted fault (1) (generating station grounding) A short circuit or electrical contact between two conductors at different potentials, in which the impedance or resistance between the conductors is essentially zero. (PE/EDPG) 665-1987s
(2) A short-circuit condition that assumes zero impedance exists at the point of the fault. (SWG/PE) C37.100-1992

Boltzmann's constant (fiber optics) The number k that relates the average energy of a molecule to the absolute temperature of the environment. k is approximately 1.38×10^{-23} joules/kelvin. (Std100) 812-1984w

bombardment-induced conductivity (storage tubes) An increase in the number of charge carriers in semiconductors or insulators caused by bombardment with ionizing particles. *See also:* storage tube. (ED) 158-1962w

bomb-control switch A switch that closes an electric circuit, thereby tripping the bomb-release mechanism of an aircraft, usually by means of a solenoid. (EEC/PE) [119]

bond A reliable connection to assure the required electrical conductivity between conductive parts required to be electrically connected. (IA/PC) 463-1993w

bonded (conductor stringing equipment) (power line maintenance, grounding) The mechanical interconnection of conductive parts to maintain a common electrical potential. *Synonym:* connected. *See also:* bonding. (T&D/PE) 524a-1993r, 1048-1990, 516-1995, 524-1992r

bonded motor (rotating machinery) A complete motor in which the stator and end shields are held together by a cement, or by welding or brazing. (PE) [9]

bonded sheath Cable shielding that is bonded to a plastic jacket by means of a plastic coating on the shielding. (PE/IC) 1143-1994r

bonding (1) (generating station grounding) The permanent joining of metallic parts to form an electrically conductive path that will ensure electrical continuity and the capacity to conduct safely any current likely to be imposed. (PE/IA/EDPG/PSE) 665-1995, 1100-1999
(2) The electrical interconnecting of conductive parts, designed to maintain a common electrical potential. (IA/PSE) 1100-1999
(3) (electric cables) The electric interconnecting of cable sheaths or armor to sheaths or armor of adjacent conductors. *See also:* continuity cable bond; cross cable bond; cable bond. (T&D/PE) [10]
(4) (data management) A technique used in database design, in which two or more data items are defined and physically stored together; for example, one might bond data items FIRST-NAME and LAST-NAME. (C) 610.5-1990w

bonding jumper A reliable conductor to ensure the required electrical conductivity between metal parts that need to be electrically connected. (PE/EDPG) 665-1995

bone conduction (hearing) The process by which sound is conducted to the inner ear through the cranial bones. (SP) [32]

Bookmaster A text-formatting language developed by IBM; a superset of DCF and GML that allows for elaborate markup of simple text into complex books, with a large degree of output device independence. (C) 610.13-1993w

Boolean (1) (mathematics of computing) Pertaining to the rules of logic formulated by the Irish mathematician George Boole in 1847. (C) 1084-1986w
(2) The language-independent syntax for a datatype with two values, names "TRUE" and "FALSE", and a set of logical operations: NOT, OR, AND, and so on. (C/PA) 1351-1994w
(3) (general) (A) Pertaining to the processes used in the algebra formulated by George Boole. (B) Pertaining to the operations of formal logic. (C) [85]
(4) A datatype with two values, named "TRUE" and "FALSE", and a set of logical operations: NOT, OR, AND, etc. (C/PA) 1224.1-1993w
(5) A variable that can assume only two states, true or false. (SCC20) 771-1998

Boolean add *See:* OR.

Boolean algebra The binary system of algebra formulated by George Boole, dealing with binary variables and employing the basic logical operators AND, OR, NOT, etc. *Synonyms:* Boolean logic; Boolean math. (C) 1084-1986w

Boolean calculus An extension of Boolean algebra that includes time-dependent operators such as BEFORE, DURING, AFTER. (C) 1084-1986w

Boolean complementation *See:* NOT.

Boolean connective *See:* Boolean operator.

Boolean function A switching function in which the number of possible values of the function and each of its independent variables is two. (C) 1084-1986w

Boolean logic *See:* Boolean algebra.

Boolean math *See:* Boolean algebra.

Boolean multiplication *See:* AND.

Boolean operation Any operation in which each of the operands and the result take one of two values. (C) 1084-1986w

Boolean operation table *See:* truth table.

Boolean operator An operator whose operands and results are binary variables. *Synonym:* Boolean connective. (C) 1084-1986w

Boolean value The value of a binary variable; either binary zero or binary one. (C) 1084-1986w

Boolean variable *See:* binary variable.

boolean vector machine A special type of attractor neural network that uses binary values for its connectivity-states matrix. (C) 610.10-1994w

boost (1) The act of increasing the power output capability of an operational amplifier by circuit modification in the output stage. *See also:* electronic analog computer. (C) 165-1977w
(2) In an analog computer, to increase the power output capability of an operational amplifier by circuit modification in the output stage. (C) 610.10-1994w

boost charge (storage battery) A partial charge, usually at a high rate for a short period. *Synonym:* quick charge. *See also:* charge. (EEC/PE) [119]

booster An electric generator inserted in series in a circuit so that it either adds to or subtracts from the voltage furnished by another source. (EEC/PE) [119]

booster coil An induction coil utilizing the aircraft direct-current supply to provide energy to the spark plugs of an aircraft engine during its starting period. (EEC/PE) [119]

booster dynamotor A dynamotor having a generator mounted on the same shaft and connected in series for the purpose of adjusting the output voltage. *See also:* converter. (EEC/PE) [119]

boot (1) To initialize a computer system by clearing memory and reloading the operating system. Derived from **bootstrap.** (C) 610.12-1990
(2) To load and execute a client program. (C/BA) 1275-1994

bootleg (railway techniques) A protection for track wires when the wires leave the conduit or ground near the rail. (EEC/PE) [119]

bootstrap (1) (A) (software) A short computer program that is permanently resident or easily loaded into a computer and whose execution brings a larger program, such as an operating system or its loader, into memory. *Synonym:* initial program load. **(B) (software)** To use a program as in definition (A). *Synonym:* initial program load. (C) 610.12-1990
(2) (metal nitrite oxide semiconductor arrays) A circuit design technique in which a junction point (node) is capacitively driven to a voltage of greater magnitude than that available from the device power supply. (ED) 641-1987w

bootstrap circuit (1) (general) A single-stage electron-tube amplifier circuit in which the output load is connected between cathode and ground or other common return, the signal voltage being applied between the grid and the cathode. *Note:* The name "bootstrap" arises from the fact that a change in grid voltage changes the potential of the input source with respect to ground by an amount equal to the output signal. (BT) 182A-1964w
(2) A circuit in which an increment of the applied input signal is partially fed back across the input impedance resulting in a higher effective input impedance. (CAS) [13]

Bootstrap Combined Programming Language A recursive computer language used primarily for compiler writing and systems programming. *See also:* CINEMA; B. (C) 610.13-1993w

bootstrap loader (software) A short computer program used to load a bootstrap. (C) 610.12-1990

bootstrap SAIDs Four SAID values that are reserved for the purpose of establishing initial communication with key management or system management when an SAID has not already been negotiated. These SAID values have a preestablished security association. (C/LM) 802.10-1998

boot-up The process an NCAP and its operating system perform, usually on application of power, in preparation for executing operations related to the application and application visible components of the system. System level network visible actions may be accomplished as well, for example, the publication of `PSK_NCAPBLOCK_ANNOUNCEMENT`. (IM/ST) 1451.1-1999

***B* operator** An operator assigned to a *B* switchboard. *See also:* telephone system. (EEC/PE) [119]

BORAM *See:* block-organized random-access memory.

border The set of pixels in a region of a digital image that are adjacent to pixels in the region's complement. *Synonym:* boundary. *Contrast:* interior. *See also:* perimeter; edge. (C) 610.4-1990w

border delineation *See:* border detection.

border detection Any image segmentation technique that identifies borders within a digital image. *Synonym:* border delineation. (C) 610.4-1990w

borderline between comfort and discomfort (BCD) (illuminating engineering) The average luminance of a source in a field of view which produces a sensation between comfort and discomfort. (EEC/IE) [126]

bore (1) (rotating machinery) The surface of a cylindrical hole (for example, stator bore). *See also:* stator. (PE) [9]
(2) The inside diameter of a spool of magnetic tape. (C) 610.10-1994w

borehole cable A cable designed for vertical suspension in a borehole or shaft and used for power circuits in mines. *See also:* mine feeder circuit. (PE/EEC/MIN) [119]

bore-hole lead insulation (rotating machinery) Special insulation surrounding connections that pass through a hollow shaft. *See also:* rotor. (PE) [9]

boresight *See:* electrical boresight; reference boresight.

boresight error The angular deviation of the electrical boresight of an antenna from its reference boresight. *See also:* antenna. (AP/ANT) 149-1979r, 145-1993

boresighting (1) (navigation aids) The process of aligning or determining the angle of the electrical or mechanical axes of a navigation system to a set of vehicle reference axes. Usually accomplished by an optical procedure. (AES/GCS) 172-1983w
(2) The process of aligning the electrical and mechanical axes of a directional antenna system, usually by an optical procedure. (AES) 686-1997

Born approximation A single-scattering approximation in which the exciting field is assumed to be equal to the incident field. (AP/PROP) 211-1997

borrow (general math) In direct subtraction, a carry that arises when the result of the subtraction in a given digit place is less than zero. (C) 162-1963w
(2) (A) (mathematics of computing) A mathematical process used in subtraction, in which, when the difference in a digit place would be arithmetically negative, the subtraction in that digit place is preceded by increasing the digit in the minuend by the value of the radix, and decreasing the digit in the next higher digit place by one. **(B) (mathematics of computing)** The value added to the digit place in definition (A). **(C) (mathematics of computing)** To perform the process defined in definition (A). (C) 1084-1986

Bose-Chaudhuri-Hocquenghem (BCH) Code A class of security code that is relatively simple to implement in hardware and that provides a high degree of immunity to transmission errors for a small reduction in communication efficiency. (SUB/PE) 999-1992w

bosun's chair *See:* boatswain's chair.

BOT *See:* beginning-of-tape marker.

bottom (A) In a queue, the position of the item that has been in the queue for the shortest time. **(B)** In a stack, the position of the item that has been in the stack for the longest time. (C) 610.5-1990

bottom-car clearance (elevators) The clear vertical distance from the pit floor to the lowest structural or mechanical part, equipment, or device installed beneath the car platform, except guide shoes or rollers, safety jaw assemblies, and platform aprons or guards, when the car rests on its fully compressed buffers. *See also:* hoistway. (EEC/PE) [119]

bottom-coil slot (rotating machinery) (radially outer-coil side) The coil side of a stator slot farthest from the bore of the stator or from the slot wedge. *See also:* stator. (PE) [9]

bottom-connected electromechanical watthour meter An electromechanical watthour meter having a bottom-connection terminal assembly. (ELM) C12.10-1987

bottom edge By convention, that edge of the module that is seen counterclockwise from the faceplate when viewing the component side. (C/MM) 1101.2-1992

bottom-half bearing (rotating machinery) The bottom half of a split-sleeve bearing. *See also:* bearing. (PE) [9]

bottom-terminal landing (elevators) The lowest landing served by the elevator that is equipped with a hoistway door and hoistway-door locking device that permits egress from the hoistway side. *See also:* elevator landing. (EEC/PE) [119]

bottom-up (software) Pertaining to an activity that starts with the lowest-level components of a hierarchy and proceeds through progressively higher levels; for example, bottom-up design; bottom-up testing. *Contrast:* top-down. *See also:* critical piece first. (C) 610.12-1990

bottom-up design (software) The design of a system starting with the most basic or primitive components and proceeding to higher level components that use the lower level ones. *See also:* components; top-down design; system; design. (C/SE) 729-1983s

bounce (television) A transient disturbance affecting one or more parameters of the display and having duration much greater than the period of one frame. *Note:* The term is usually applied to changes in vertical position or in brightness. *See also:* television. (BT/AV) [34]

boundary *See:* border.

boundary alignment *See:* alignment.

boundary arrow An arrow with one end (source or use) not connected to any box in a diagram. *Contrast:* internal arrow. (C/SE) 1320.1-1998

boundary condition The values assumed by the variables in a system, model, or simulation when one or more of them is at a limiting value or a value at the edge of the domain of interest. *Contrast:* final condition; initial condition. (C) 610.3-1989w

boundary ICOM code An ICOM code that maps an untunneled boundary arrow in a child diagram to an arrow attached to the parent box that is detailed by that diagram. (C/SE) 1320.1-1998

boundary lights (illuminating engineering) Aeronautical ground lights delimiting the boundary of a land aerodrome without runways. (EEC/IE) [126]

boundary marker (navigation aid terms) [instrument landing system (ILS)] A radio-transmitting station near the approach end of the landing runway that provides a fix on the localizer course. (AES/GCS) 172-1983w

boundary node A node with two or more ports, at least one of which is active and another suspended. (C/MM) 1394a-2000

boundary, *p-n* (semiconductor) A surface in the transition region between *p*-type and *n*-type material at which the donor and acceptor concentrations are equal. *See also:* semiconductor; transistor. (ED) 216-1960w

boundary potential The potential difference, of whatever origin, across any chemical or physical discontinuity or gradient. *See also:* electrobiology. (EMB) [47]

boundary value A data value that corresponds to a minimum or maximum input, internal, or output value specified for a system or component. *See also:* stress testing. (C) 610.12-1990

bounded-input-bounded-output stability (A) Driven stability when the solution of interest is the output solution. *See also:* control system. **(B) (excitation systems)** A system exhibits bounded input-bounded output (BIBO) stability if the output is bounded for every bounded input. *Note:* BIBO stability is also known as stability in the sense of Liapunov and it refers to force systems. In nonlinear systems, a bounded limit cycle appearing in the output signal is an example of BIBO stability. (CS/PE/EDPG) 421A-1978

bounded scheduling A scheduling algorithm used by a simple controller. Rather than specifying the exact time for a slave to return its response, a time window is provided where the time window is longer than the expected request-processing delay. (C/MM) 1596.4-1996

bounding volume The six-sided, rectangular enclosing space whose width, length, and height are aligned with those of the entity. (DIS/C) 1278.1-1995

bound mode (fiber optics) In an optical waveguide, a mode whose field decays monotonically in the transverse direction everywhere external to the core and that does not lose power to radiation. Specifically a mode for which $N(a)k = \beta = n(0)k$ where β is the imaginary part (phase constant) of the axial propagation constant, $n(a)$ is the refractive index at $r = a$, the core radius, $n(0)$ is the refractive index at $r = 0$, k is the free-space wave number, $2\pi/\lambda$, and λ is the wavelength. Bound modes correspond to guided rays in the terminology of geometric optics. *Note:* Except in a monomode fiber, the power in bound modes is predominantly contained in the core of the fiber. *Synonyms:* guided mode; trapped mode. *See also:* unbound mode; cladding mode; leaky mode; mode; guided ray; normalized frequency. (Std100) 812-1984w

bound ray *See:* guided ray.

bounds register A register which holds an address specifying a storage boundary. *Note:* An access outside the boundary results in an error. (C) 610.10-1994w

Bourdon A closed and flattened tube formed in a spiral, helix, or arc, which changes in shape when internal pressure changes are applied. *Note:* Bourdon tube, or simply Bourdon, has at times been used more restrictively to mean only the C-shaped member invented by Bourdon. (PE/PSIM) 119-1974w

BOV *See:* beginning-of-volume label.

bowl (illuminating engineering) An open top diffusing glass or plastic enclosure used to shield a light source from direct view and to redirect or scatter the light. (EEC/IE) [126]

box (1) A mechanical unit which contains links; the links may either remain inside the box, connecting internal devices, or may leave the box in order to connect internal devices to external ones. A box is assumed to be an EMC compliant enclosure and to operate under a single electrical environment. (C/BA) 1355-1995
(2) (electronic) A protective enclosure to house modules, backplane(s), I/O connector assemblies, internal cables, and other electronic, mechanical, and thermal devices. *Synonyms:* box; rack; cabinet. (BA/C) 14536-1995
(3) A rectangle containing a box name, a box number, and possibly a box detail reference and representing a function in a diagram. (C/SE) 1320.1-1998

box-car detector† In radar, a detector whose output is held at the amplitude of the last sample until the next sample arrives. *Note:* Functionally the same as a sample and hold. (AES) 686-1997

† Obsolete.

box detail reference A square enclosure encompassing a box number, which indicates that the box is decomposed or detailed by a child diagram. (C/SE) 1320.1-1998

box diagram A control flow diagram consisting of a rectangle that is subdivided to show sequential steps, if-then-else conditions, repetition, and case conditions. *Synonyms:* Chapin chart; Nassi-Shneiderman chart; program structure diagram. *See also:* block diagram; graph; flowchart; program structure diagram; structure chart; bubble chart.

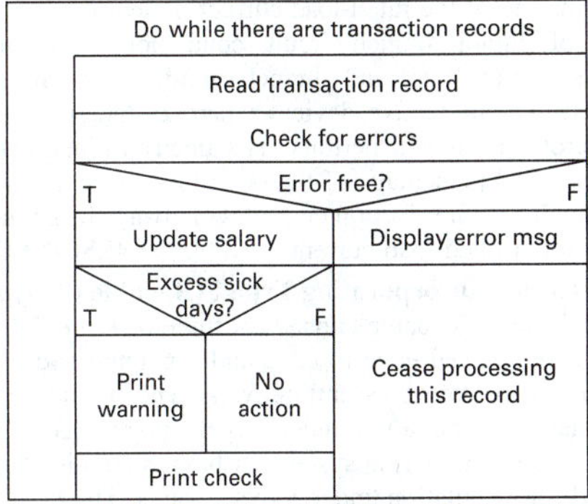

box diagram

(C) 610.12-1990

box frame (rotating machinery) A stator frame in the form of a box with ends and sides and that encloses the stator core. *See also:* rotor. (PE) [9]

box ICOM code An ICOM code that maps a tunneled boundary arrow to an arrow attached to some ancestral box. (C/SE) 1320.1-1998

boxing glove *See:* conductor lifting hook.

box name The verb or verb phrase placed inside a box that names the modeled function. A box takes as its box name the function name of the function represented by the box. *See also:* function name. (C/SE) 1320.1-1998

box number A single digit (*0, 1, 2, . . ., 9*) placed in the lower right corner of a box to uniquely identify that box in a diagram. The only box that may be numbered 0 is the box that represents the **A0** function in **A-0** and **A-1** context diagrams. (C/SE) 1320.1-1998

Boyce/Codd Normal form (BCNF) Developed by R. F. Boyce and E. F. Codd, one of the forms used to characterize relations; a relation is said to be in Boyce/Codd Normal form if every determinant in the relation is or contains a candidate key; that is, no attribute is transitively dependent on any key. *Note*: This is an extension of third normal form. (C) 610.5-1990w

BPAM *See:* basic partitioned access method.

bps *See:* bits per second.

BPSK *See:* binary phase shift keying.

BR *See:* bit rate.

BR/2 *See:* bit rate/2.

braces The characters "{" (*left brace*) and "}" (*right brace*), also known as *curly braces*. When used in the phrase "enclosed in (curly) braces" the symbol "{" shall immediately precede the object to be enclosed, and "}" shall immediately follow it. When describing these characters, the names <left-brace> and <right-brace> are used. (C/PA) 9945-2-1993

bracket (1) (illuminating engineering) An attachment to a lamp post or pole from which a luminaire is suspended. *Synonym:* mast arm. (EEC/IE) [126]
(2) (rotating machinery) A solid or skeletal structure usually consisting of a central hub and a plurality of arms extending (often radially) outward from the hub to a supporting structure. The supporting structure usually is the stator frame when the axis of the shaft is horizontal. When the axis of the shaft is vertical, the stator usually supports the upper bracket and the foundation supports the lower bracket. *See also:* bearing bracket. (PE) [9]

bracket arm (rotating machinery) One of several structural members (beams) extending from the hub portion of a bracket to the supporting structure. The arms may be individual or parallel pairs extending radially or near-radially from the hub. (PE) [9]

brackets The characters "[" (*left bracket*) and "]" (*right bracket*), also known as *square brackets*. When used in the phrase "enclosed in (square) brackets" the symbol "[" shall immediately precede the object to be enclosed, and "]" shall immediately follow it. When describing these characters, the names <left-square-bracket> and <right-square-bracket> are used. (C/PA) 9945-2-1993

bracket-type handset telephone *See:* hang-up hand telephone set.

Bragg angles When an incident plane wave is diffracted by a periodic structure into discrete directions, the angles these directions of travel make with respect to the normal of the mean boundary. (AP/PROP) 211-1997

Bragg region (acousto-optic device) The region that occurs when the length of the acoustic column in the direction of light propagation, L, satisfies the inequality $L > n\,\Lambda^2\,\lambda_0$, with n the index of refraction at wavelength λ_0 and Λ the acoustic wavelength. (UFFC) [23]

Bragg resonant scattering Originally described the scattering in discrete directions by spatially periodic boundaries or constitutive parameter(s), where the scattering directions are determined by the resonance condition in which two source-to-scatter-to-receiver path lengths differ by an integer multiple of 2π radians. This same physical mechanism has been found to apply to some randomly rough planar interfaces and random fluctuations of spatially continuous constitutive parameter(s). In these cases, there is a continuum of scattering angles provided there is either a continuous surface roughness or a continuous constitutive parameter fluctuation spectrum that satisfies the proper Bragg resonance condition. (AP/PROP) 211-1997

braided shield Cable shield that consists of groups of metallic strands, one set woven in a clockwise direction and interwoven with another set in a counter-clockwise direction. Braided shields provide superior structural integrity, while maintaining good flexibility and flex life. (PE/IC) 1143-1994r

brake *See:* bullwheel tensioner.

brake assembly (rotating machinery) All parts that are provided to apply braking to the rotor. *See also:* rotor. (PE) [9]

brake control The provision for controlling the operation of an electrically actuated brake. *Note:* Electrical energizing of the brake may either release or set the brake, depending upon its design. *See also:* electric controller. (IA/ICTL/IAC) [60]

brake drum *See:* brake ring.

brake ring (rotating machinery) A rotating ring mounted on the rotor that provides a bearing surface for the brake shoes. *Synonym:* brake drum. *See also:* rotor. (PE) [9]

brakes applied An indication that all friction brakes are applied to some agreed-upon preset level. (VT) 1475-1999

brake service (maximum) A nonemergency brake application that obtains the (maximum) brake rate that is consistent with the design of the brake system, retrievable under the control of master control. (VT/RT) 1474.1-1999

braking The control function of retardation by dissipating the kinetic energy of the drive motor and the driven machinery. *See also:* electric drive. (IA/ICTL/IAC) [60]

braking effort Longitudinal retarding force generated by the friction brake system or the propulsion system (in electric brake). (VT) 1475-1999

braking magnet *See:* retarding magnet.

braking resistor A resistor commonly used in some types of dynamic braking systems, the prime purpose of which is to convert the electric energy developed during dynamic braking into heat and to dissipate this energy to the atmosphere. *See also:* dynamic braking. (EEC/PE) [119]

braking test (A) (rotating machinery) A test in which the mechanical power output of a machine acting as a motor is determined by the measurement of the shaft torque, by means of a brake, dynamometer, or similar device, together with the rotational speed. *See also:* asynchronous machine; direct-current commutating machine. **(B) (rotating machinery)** A test performed on a machine acting as a generator, by means of a dynamometer or similar device, to determine the mechanical power input. *See also:* asynchronous machine; direct-current commutating machine. (PE) [9]

braking torque (synchronous motor) Any torque exerted by the motor in the same direction as the load torque so as to reduce its speed. (PE) [9]

branch (1) (A) (software) A computer program construct in which one of two or more alternative sets of program statements is selected for execution. *See also:* go to; jump; if-then-else; case. **(B) (software)** A point in a computer program at which one of two or more alternative sets of program statements is selected for execution. *Synonym:* branchpoint. **(C) (software)** Any of the alternative sets of program statements in definition (A). **(D) (software)** To perform the selection in definition (A). (C) 610.12-1990
(2) (local area networks) A cable distribution line in a broadband coaxial network that is connected to a trunk line. (LM/C) 802.7-1989r
(3) (network analysis) A line segment joining two nodes, or joining one node to itself. *See also:* network analysis; directed branch. (CAS) 155-1960w
(4) (A) A set of instructions that are executed between two successive decision instructions. **(B)** To select a branch as in definition (A). **(C)** Loosely, a conditional jump. *See also:* conditional jump. (C) [85]
(5) A portion of a network consisting of one or more two-terminal elements, comprising a section between two adjacent branch-points. *See also:* principal branch; auxiliary branch. (CAS) [13]
(6) *See also:* turn-off branch; principal branch; regenerative branch; auxiliary branch; subtree.
(7) A junction at which a root arrow segment (going from source to use) divides into two or more arrow segments. May denote unbundling of arrow meaning, i.e., the separation of object types from an object type set. Also refers to an arrow segment into which a root arrow segment has been divided. (C/SE) 1320.1-1998

branch cable In 10BROAD36, the Attachment Unit Interface (AUI) cable interconnecting the data terminal equipment and Medium Attachment Unit (MAU) system components. (C/LM) 802.3-1998

branch circuit (1) (electrical heating applications to melting furnaces and forehearths in the glass industry). One, two, or more circuits whose main power is connected through the same main switch. (IA) 668-1987w
(2) The circuit conductors between the final overcurrent device protecting the circuit and the outlet(s). *See also:* thermal cutout; thermal relay. (NESC/NEC) [86]
(3) (packaging machinery) That portion of a wiring system extending beyond the final overcurrent device protecting the circuit. (A device not approved for branch circuit protection, such as a thermal cutout or motor overload protective device, is not considered as the overcurrent device protecting the circuit.). (IA/PKG) 333-1980w
(4) (or final subcircuit) That portion of a wiring system that extends beyond the final overcurrent device protecting the circuit. (IA/MT) 45-1998

branch-circuit distribution center A distribution center at which branch circuits are supplied. *See also:* distribution center. (EEC/PE) [119]

branch circuit, general purpose A branch circuit that supplies a number of outlets for lighting and appliances. (NESC) [86]

branch circuit, individual A branch circuit that supplies only one utilization equipment. (NESC) [86]

branch-circuit load The load on that portion of a wiring system extending beyond the final overcurrent device protecting the circuit. (IA/PSE) 241-1990r

branch circuit, multiwire A circuit consisting of two or more ungrounded conductors having a potential difference between them, and an identified grounded conductor having equal potential difference between it and each ungrounded conductor of the circuit and that is connected to the neutral conductor of the system. (NESC) [86]

branch-circuit selection current The value in amperes to be used instead of the rated-load current in determining the ratings of motor branch-circuit conductors, disconnecting means, controllers and branch-circuit short-circuit and ground-fault protective devices wherever the running overload protective device permits a sustained current greater than the specified percentage of the rated-load current. The value of branch-circuit selection current will always be greater than the marked rated-load current. (NESC/NEC) [86]

branch circuits incorporating Type FCC cable (A) *type FCC cable.* Type FCC cable consists of three or more flat copper conductors placed edge to edge and separated and enclosed within an insulating assembly. *Note:* The wiring system is designed for installation under carpet squares. **(B)** *FCC system.* A complete wiring system for branch circuits that is designed for installation under carpet squares. The FCC system includes Type FCC cable and associated shielding, connectors, terminators, adapters, boxes, and receptacles. **(C)** *cable connector.* A connector designed to join Type FCC cables without using a junction box. **(D)** *insulating end.* An insulator designed to electrically insulate the end of a Type FCC cable. **(E)** *top shield.* A grounded metal shield covering under carpet components of the FCC system for the purposes of providing electrical safety and protection against physical damage. **(F)** *bottom shield.* A shield mounted on the floor under the FCC system to provide protection against physical damage. **(G)** *transition assembly.* An assembly to facilitate connection of the FCC system to other approved wiring systems, incorporating: (1) a means of electrical interconnection; and (2) a suitable box or covering for providing electrical safety and protection against physical damage. **(H)** *metal shield connections.* Means of connection designed to electrically and mechanically connect a metal shield to another metal shield, to a receptacle housing or self-contained device or to a transition assembly. (NESC) [86]

branch conductor (lightning protection) A conductor that branches off at an angle from a continuous run of conductor. (EEC/PE) [119]

branch input signal (network analysis) The signal *xj* at the input end of branch *jk*. (CAS) 155-1960w

branch instruction (1) An instruction in the program that provides a choice between alternative subprograms in accordance with the test logic. (MIL) [2]
(2) A computer instruction that changes the sequence in which computer instructs are performed. *Note:* A branch instruction generally specifies the next instruction in terms of a relative address based on the program counter. *Synonym:* decision instruction. *See also:* conditional branch instruction; jump instruction. (C) 610.10-1994w

branch joint (1) (general) A joint used for connecting a branch conductor or cable to a main conductor or cable, where the latter continues beyond the branch. *Note:* A branch joint may be further designated by naming the cables between which it is made; for example, single-conductor cables, three-conductor main cable to single-conductor branch cable, etc. With the term "multiple joint" it is customary to designate the various kinds as 1-way, 2-way, 3-way, 4-way, etc., multiple joint. *See also:* reducing joint; straight joint; cable joint. (T&D/PE) [10]
(2) (power cable joints) A cable joint used for connecting one or more cables to a main cable. *Note:* A branch joint may be further designated by naming the cables between which it is made, for example, single conductor cable, three conductor cable, three conductor main cable to single conductor branch, etc. It is customary to designate the various kinds as *Y* joint, *T* joint, *H* joint, cross joint, etc. (PE/IC) 404-1986s

branch metric The result of dividing the total number of modules in which every branch has been executed at least once by the total number of modules. *Note:* This definition assumes that the modules are essentially the same size. (C/SE) 730-1998

branch node *See:* nonterminal node.

branch number (*b*) (subroutines for CAMAC) The symbol *b* represents an integer which is the branch number component of a CAMAC address. It may represent a physical highway number in multiple highway systems, or it may represent sets of crates grouped together for functional or other reasons. In some systems it may be ignored, although it must be included in the parameter list for the sake of compatibility. (NPS) 758-1979r

branch output signal (network analysis) (branch *jk*) The component of signal *xk* contributed to node *k* via branch *jk*. (CAS) 155-1960w

branchpoint *See:* node; branch.

branch point (1) (electric networks) A junction where more than two conductors meet. *See also:* network analysis; node. (PE/EEC) [119]

(2) (computers) A place in a routine where a branch is selected. *See also:* network analysis. (C) [85]

branch testing Testing designed to execute each outcome of each decision point in a computer program. *Contrast:* path testing; statement testing. (C) 610.12-1990

branch, thermoelectric Alternative term for thermoelectric arm. *See also:* thermoelectric device. (ED/ED) [46], 221-1962w

branch transmittance (network analysis) The ratio of branch output signal to branch input signal. (CAS) 155-1960w

BRE *See:* basic regular expression.

breadboard An experimental model of a circuit, usually roughly conceived, that can be used as a prototype for planning, design, and feasibility evaluation. (C) 610.10-1994w

breadboard construction (communication practice) An arrangement in which components are fastened temporarily to a board for experimental work. (EEC/PE) [119]

break (1) (circuit-opening device) The minimum distance between the stationary and movable contacts when these contacts are in the open position.

a) The length of a single break is as defined above.
b) The length of a multiple break (breaks in series) is the sum of two or more breaks.

See also: contactor. (IA/IAC) [60]

(2) (communication circuits) For the receiving operator or listening subscriber to interrupt the sending operator or talking subscriber and take control of the circuit. *Note:* The term is used especially in connection with half-duplex telegraph circuits and two-way telephone circuits equipped with voice-operated devices. *See also:* telegraphy. (EEC/PE) [119]

breakaway The condition of a motor at the instant of change from rest to rotation. (PE) [9]

breakaway starting current (rotating machinery) (alternating-current motor) The highest root mean square current absorbed by the motor when at rest, and when it is supplied at the rated voltage and frequency. *Note:* This is a design value and transient phenomena are ignored. (PE) [9]

breakaway torque (rotating machinery) The torque that a motor is required to develop to break away its load from rest to rotation. *See also:* asynchronous machine. (PE) [9]

break distance (of a switching device) The minimum open-gap distance between the main-circuit contacts, or live parts connected thereto, when the contacts are in the open position. *Note:* In a multiple-break device, it is the sum of the breaks in series. (SWG/PE) C37.100-1992, C37.40-1993

breakdown (1) (gas-tube surge protective devices) (low-voltage air-gap surge-protective devices) The abrupt transition of the gap resistance from a practically infinite value to a relatively low value. In the case of a gap, this is sometimes referred to as sparkover or ignition. *See also:* sparkover. (PE/SPD) C62.31-1981s, C62.32-1987r

(2) (germanium gamma-ray detectors) (x-ray energy spectrometers) (charged-particle detectors) (of a semiconductor diode) A phenomenon occurring in a reverse-biased semiconductor diode, the initiation of which is observed as a transition from a region of high dynamic resistance to a region of substantially lower dynamic resistance for increasing magnitude of reverse current. (NPS/NID) 759-1984r, 300-1988r

(3) (rotating machinery) The condition of operation when a motor is developing breakdown torque. *See also:* asynchronous machine. (PE) [9]

(4) (thyristor converter) A failure that permanently deprives a rectifier diode or a thyristor of its property to block voltage in the reverse direction (reverse breakdown) or a thyristor in the forward direction (forward breakdown). (IA/IPC) 444-1973w

(5) A disruptive discharge occurring through a dielectric. (PE/IC) 48-1996

(6) A phenomenon occurring in a reverse-biased semiconductor diode that appears as an increase in noise, reverse current, or both when the bias is increased beyond a certain value. (NPS) 325-1996

breakdown current (semiconductor) The current at which the breakdown voltage is measured. (IA) [12]

breakdown impedance (semiconductor diode) The small-signal impedance at a specified direct current in the breakdown region. *See also:* semiconductor. (ED) 216-1960w

breakdown maintenance Those repair actions that are conducted after a failure in order to restore equipment or systems to an operational condition. (IA/PSE) 902-1998

breakdown region (germanium gamma-ray detectors) (x-ray energy spectrometers) (of a semiconductor diode characteristic) (charged-particle detectors) That entire region of the voltage-current characteristic beyond the initiation of breakdown for increasing magnitude of reverse current. (NPS/NID) 325-1986s, 300-1988r, 759-1984e5r

breakdown strength *See:* dielectric strength.

breakdown torque (1) (rotating machinery) The maximum shaft-output torque that an induction motor (or a synchronous motor operating as an induction motor) develops when the primary winding is connected for running operation, at normal operating temperature, with rated voltage applied at rated frequency. *Note:* A motor with a continually increasing torque as the speed decreases to standstill is not considered to have a breakdown torque. (PE) [9]

(2) The maximum torque a motor will develop, with rated voltage applied at rated frequency, without an abrupt drop in speed. (IA/MT) 45-1998

breakdown-torque speed (rotating machinery) The speed of rotation at which a motor develops breakdown torque. *See also:* asynchronous machine. (PE) [9]

breakdown transfer characteristic (gas tube) A relation between the breakdown voltage of an electrode and the current to another electrode. *See also:* gas tube. (ED) 161-1971w

breakdown voltage (1) (diode) (nonlinear, active, and nonreciprocal waveguide components) The reverse voltage at which there is a conduction of current due to the Zener effect or the avalanche multiplication process. This voltage is usually specified at 10 μA of reverse current. (MTT) 457-1982w

(2) (germanium gamma-ray detectors) (charged-particle detectors) (x-ray energy spectrometers) (of a semiconductor diode) The voltage measured at a specified current in the breakdown region. (NPS/NID) 759-1984r, 300-1988r

(3) (rotating machinery) The voltage at which a disruptive discharge takes place through or over the surface of the insulation. (PE/EM) 95-1977r

(4) (gas) The voltage necessary to produce a breakdown. *See also:* gas tube. (ED) [45]

(5) (electrode of a gas tube) The voltage at which breakdown occurs to that electrode. *Notes:* 1. The breakdown voltage is a function of the other electrode voltages or currents and of

the environment. 2. In special cases where the breakdown voltage of an electrode is referred to an electrode other than the cathode, this reference electrode shall be indicated. 3. This term should be used in preference to pickup voltage, firing voltage, starting voltage, etc., which are frequently used for specific types of gas tubes under specific conditions. *See also:* critical grid voltage. (ED) 161-1971w
(6) **(A)** (ac). The minimum rms value of a sinusoidal voltage at frequencies between 15 Hz and 62 Hz that results in arrester sparkover. **(B)** (dc). The minimum slowly rising dc voltage that causes breakdown or sparkover when applied across the terminals of an arrester. (SPD/PE) C62.31-1987r
(7) The voltage measured at a specified current in the breakdown region. (NPS) 325-1996

breakdown voltage, ac *See:* alternating-current breakdown voltage.

breakdown voltage alternating current (gas tube surge arresters) The minimum root-mean-square (rms) value of sinusoidal voltage at frequencies between 15 Hz and 62 Hz that results in arrester sparkover. (PE) [8]

breakdown voltage, dc *See:* direct-current breakdown voltage.

breaker failure The failure of a circuit breaker to operate or to interrupt a fault. (PE/PSR) C37.113-1999

break indication **(A)** The state where the physical layer (PHY) is unable to recover data from the incoming signal, or the incoming signal power level is less than a defined threshold. **(B)** The state where the Physical Layer (PHY) is unable to recover data from the incoming signal, or the incoming signal power level is less than a defined threshold. (LM/C) 802.5c-1991

breaking capacity (interrupting capacity) The current that the device is capable of breaking at a stated recovery voltage under prescribed conditions of use and behavior. *See also:* control. (IA/ICTL/IAC) [60], [84]

breaking current (pole of a breaking device) The current in that pole at the instant of contact separation, expressed as a root-mean-square value. *See also:* interrupting current; contactor. (SWG/PE/IA/ICTL/IAC) C37.100-1981s, [60], [84]

breaking point (transmission system or element thereof). A level at which there occurs an abrupt change in distortion or noise that renders operation unsatisfactory. *See also:* level. (EEC/PE) [119]

break key *See:* attention key.

break link A weak section of rope connected between the cable pulling attachment and the pull rope that is intended to break when the pulling tension exceeds a certain limit. (PE/IC) 1185-1994

break-make relay contacts A contact form in which one contact opens its connection to another contact and then closes its connection to a third contact. (EEC/REE) [87]

breakover current (thyristor) The principal current at the breakover point. *See also:* principal current. (ED/IA) [46], [62]

breakover point (thyristor) Any point on the principal voltage-current characteristic for which the differential resistance is zero and where the principal voltage reaches a maximum value. *See also:* principal voltage-current characteristic. (ED) [46]

breakover voltage (thyristor) The principal voltage at the breakover point. *See also:* principal voltage-current characteristic. (ED/IA) [46], [62]

break, % break (dial-pulse address signaling systems) (telephony) In dial-pulse signaling, that portion of the signal in which the dialing contacts are open (broken). % break is the ratio of break time to the total pulse period; (make + break) time. (COM/TA) 753-1983w

breakpoint (1) (A) (computer routine) Pertaining to a type of instruction, instruction digit, or other condition used to interrupt or stop a computer at a particular place in a routine when manually requested. **(B) (computer routine)** A place in a routine where such an interruption occurs or can be made to occur. (C) 162-1963
(2) (software) A point in a computer program at which execution can be suspended to permit manual or automated monitoring of program performance or results. Types include code breakpoint, data breakpoint, dynamic breakpoint, epilog breakpoint, programmable breakpoint, prolog breakpoint, static breakpoint. *Note:* A breakpoint is said to be set when both a point in the program and an event that will cause suspension of execution at that point are defined; it is said to be initiated when program execution is suspended. (C) 610.12-1990
(3) A position within a pattern set where the pattern may be segmented into multiple independent bursts while still achieving predictable behavior of the device. (C/TT) 1450-1999

breakpoint halt *See:* breakpoint instruction.

breakpoint instruction **(A)** A computer instruction that causes program flow to be halted. *See also:* address stop. **(B)** A computer instruction that causes program flow to be redirected to a monitor or debugging program. *Synonym:* breakpoint halt; dynamic stop. (C) 610.10-1994

breakthrough (thyristor converter) The failure of the forward-blocking action of an arm of a thyristor connection during a normal off-state period with the result that it allows on-state current to pass during a part of this period. *Note:* Breakthrough can occur in rectifier operation as well as inverter operation and for various reasons, for example, excessive virtual junction temperature, voltage surges in excess of rated peak off-state voltage, excessive rate of rise of off-state voltage, advance gating, or forward breakdown. (IA/IPC) 444-1973w

breather A device fitted in the wall of an explosion-proof compartment, or connected by piping thereto, that permits relatively free passage of air through it, but that will not permit the passage of incendiary sparks or flames in the event of gas ignition inside the compartment. (EEC/PE) [119]

breathing (carbon microphones) The phenomenon manifested by a slow cyclic fluctuation of the electric output due to changes in resistance resulting from thermal expansion and contraction of the carbon chamber. *See also:* close-talking pressure-type microphones. (SP) 258-1965w

breezeway (television synchronizing waveform for color transmission) The time interval between the trailing edge of the horizontal synchronizing pulse and the start of the color burst. (BT/AV) [34]

b-register *See:* index register.

BRE (ERE) matching a single character A basic or extended regular expression that matches either a single character or a single collating element. Only a BRE or ERE of this type that includes a bracket expression can match a collating element. (C/PA) 9945-2-1993

BRE (ERE) matching multiple characters A basic or extended regular expression that matches a concatenation of single characters or collating elements. Such a BRE or ERE is made up from a *BRE ERE matching a single character* and *BRE ERE special characters.* (C/PA) 9945-2-1993

Brewster angle The angle of incidence of a wave on the planar bounding surface of a lossless medium for which the reflection coefficient for parallel polarization is zero. *Note:* For a lossy medium, the pseudo-Brewster angle is that angle at which the modulus of the reflection coefficient is a minimum. (AP/PROP) 211-1997

Brewster's angle (fiber optics) For light incident on a plane boundary between two regions having different refractive indices, that angle of incidence at which the reflectance is zero for light, and that has its electric field vector in the plane defined by the direction of propagation and the normal to the surface. For propagation from medium 1 to medium 2, Brewster's angle is $\arctan(n_2 n_1)$. *See also:* reflectance; refractive index; angle of incidence. (Std100) 812-1984w

Bridge An interconnect between two or more buses that provides signal and logical protocol translation from one bus to another. The buses may adhere to different bus standards for mechanical, electrical, and logical operation (such as a bus

Bridge from Futurebus+ to VME64). (C/BA) 1014.1-1994w

bridge **(1)** **(A)** **(data transmission)** A network with minimum of two ports or terminal pairs capable of being operated in such a manner that when power is fed into one port, by suitable adjustment of the elements in the network or the element connected to one or more other ports, zero output can be obtained at another port. Under these conditions the bridge is balanced. **(B)** **(data transmission)** An instrument or intermediate means in a measurement system that embodies all or part of a bridge circuit, and by means of which one or more of the electrical constants of a bridge may be measured. (PE) 599-1985

(2) **(protection and coordination of industrial and commercial power systems)** That narrowed portion of a fuse link that is expected to melt first. One link may have two or more bridges in parallel and in series as well. The shape and size of the bridge is a factor in determining the fuse characteristics under overload and fault current conditions. (IA/PSP) 242-1986r

(3) A pair of communicating nodes, each of which selectively (based on target address) accepts certain packets for retransmission by the other. For example, a symmetric bridge may be used to connect two SCI ringlets. Such a bridge (the simplest kind of switch) acts as an agent, taking the place of the target on one ringlet and of the source on the other. It acts like a node that has many addresses. Bridges may also connect dissimilar systems, such as SCI and VME. Such bridges are generally much more complex, because they must translate protocols. (C/MM) 1596-1992

(4) A hardware adapter that forwards transactions between buses. (C/MM) 1212-1991s

(5) A functional unit that interconnects two subnetworks that use a single Logical Link Control (LLC) procedure but may use different Medium Access Control (MAC) procedures. Local area networks (LANs and metropolitan area networks (MANs) are examples of the subnetworks that a bridge may interconnect. (LM/C) 8802-6-1994

(6) In networking, a device that connects two systems using the similar or identical data link layer protocols. *Note:* Bridges are independent of the protocol of the network layer and above. *Contrast:* gateway. *See also:* learning bridge; router; mail bridge; brouter. (C) 610.7-1995

(7) An interface between heterogeneous memory-mapped buses. Due to the distinct capabilities of the attached buses, a bridge is expected to perform address translations, fragmentation (one transaction is decomposed into several smaller transactions), concatenation (multiple writes are combined into one), or prefetching (one read initiates speculative reads of likely-to-be-used addresses). (C/MM) 1596.4-1996

(8) A layer 2 interconnection device that does not form part of a CSMA/CD collision domain but conforms to the ISO/IEC 15802-3: 1998 [ANSI/IEEE 802.1D, 1998 Edition] International Standard. A bridge does not form part of a CSMA/CD collision domain but, rather appears as a Media Access Control (MAC) to the collision domain. (C/LM) 802.3-1998

(9) A system element that converts from one data format to another. Bridging can be incorporated in most Year 2000 remediations to interpret date-data formats. This may be helpful in transferring dates between date formats for remediated system elements and those used in the original system. There may also be situations in which multiple remediation techniques requiring different date formats are used, creating a need for bridges between them. (C/PA) 2000.2-1999

(10) A layer 2 interconnection device that conforms to ISO/IEC DIS 15802-3.local area networks. (C) 802.12c-1998

(11) An intermediary mechanism that converts data passed between system elements. (C/PA) 2000.1-1999

bridge circuit A circuit of elements that is arranged such that when an elecromotive force is present in one branch, the response of a detecting device in another branch can be zeroed by adjusing the electrical constants of the other branches. *See also:* bridge limiter. (C) 610.10-1994w

bridge control Apparatus and arrangement providing for direct control from the bridge or wheelhouse of the speed and direction of a vessel. (EEC/PE) [119]

bridge current (power supply) The circulating control current in the comparison bridge. *Note:* Bridge current equals the reference voltage divided by the reference resistor. Typical values are 1 milliampere and 10 milliamperes, corresponding to control ratios of 1000 ohms per volt and 100 ohms per volt, respectively. (AES) [41]

Bridged Local Area Network **(1)** A concatenation of individual Local Area Networks interconnected by MAC Bridges. (C/LM) 10038-1993, 802.1G-1996

(2) A concatenation of individual IEEE 802 LANs interconnected by MAC Bridges. (C/LM) 802.1D-1998

bridged tap (telephone loop performance) Any portion of a metallic circuit that is not in the path between the end office and the customer. The bridged tap may be connected at an intermediate location or be an extension of the circuit beyond the customer location. The pair associated with the bridged tap introduces a frequency-dependent bridging loss in the loop. (COM/TA) 820-1984r

bridged-T network A T network with a fourth branch connected across the two series arms of the T, between an input terminal and an output terminal. *See also:* network analysis.

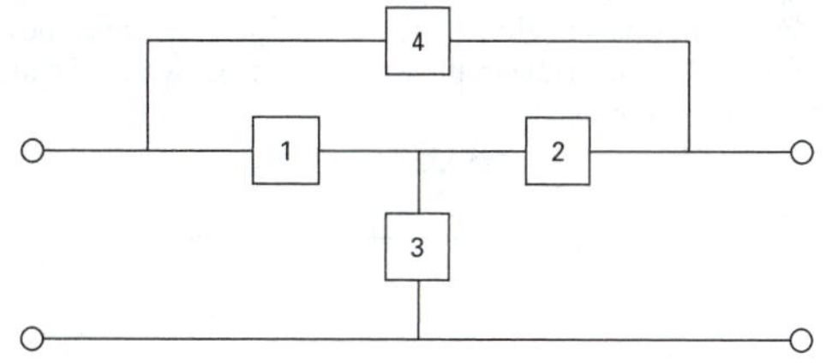

bridged-T network

(Std100) 106-1972

bridge duplex system A duplex system based on the Wheatstone bridge principle in which a substantial neutrality of the receiving apparatus to the sent currents is obtained by an impedance balance. *Note:* Received currents pass through the receiving relay that is bridged between the points that are equipotential for the sent currents. *See also:* telegraphy. (EEC/PE) [119]

Bridge Interconnect Bus The medium used to connect two or more Bridge modules together. (C/BA) 1014.1-1994w

bridge limiter **(1)** A bridge circuit that is used as a limiter circuit. (C) 610.10-1994w

(2) *See also:* limiter circuit.

Bridge Port A LAN Port or Virtual Port. (C/LM) 802.1G-1996

bridger (or bridging amplifier) The point of amplification of signals between a trunk and a feeder cable, usually consisting of an additional amplifier module fitted into a trunk amplifier station. (LM/C) 802.7-1989r

bridge rectifier (power semiconductor) A rectifier unit which makes use of a bridge-rectifier circuit. (IA) [12]

bridge rectifier circuit A full-wave rectifier with four rectifying elements connected as the arms of a bridge circuit. *See also:* single-way rectifier circuit; rectifier; double-way rectifier circuit. (AP/ANT) 145-1983s

bridge semiconverter A bridge in which one commutating group uses thyristors and the other uses diodes. (IA/IPC) 444-1973w

bridge thyristor converter (double-way) A bridge thyristor converter in which the current between each terminal of the alternating-voltage circuit and the thyristor converter circuit elements conductively connected to it flows in both directions. *Note:* The terms single-way and double-way (bridge) provide a means for describing the effect of the thyristor converter circuit on current flow in the transformer windings connected to the converter. Most thyristor converters may be classified into these two general types. The term bridge relates back to the single-phase "bridge" which resembles the Wheatstone bridge. (IA/IPC) 444-1973w

bridge transition A method of changing the connection of motors from series to parallel in which all of the motors carry like currents throughout the transfer due to the Wheatstone bridge connection of motors and resistors. *See also:* multiple-unit control. (EEC/PE) [119]

bridging (1) (signal circuits) The shunting of one signal circuit by one or more circuits usually for the purpose of deriving one or more circuit branches. *Note:* A bridging circuit often has an input impedance of such a high value that it does not substantially affect the circuit bridged. (SP) 151-1965w
(2) (soldered connections) Solder that forms an unwanted conductive path. (EEC/AWM) [105]
(3) (relays) *See also:* relay bridging.

bridging amplifier An amplifier with an input impedance sufficiently high so that its input may be bridged across a circuit without substantially affecting the signal level of the circuit across which it is bridged. *See also:* amplifier. (SP) 151-1965w

bridging connection (data transmission) A parallel connection by means of which some of the signal energy in a circuit may be withdrawn, frequently with imperceptible effect on the normal operation of the circuit. (PE) 599-1985w

bridging gain (data transmission) The ratio of the signal power a transducer delivers to its load (Z_B) to the signal power dissipated in the main circuit load (Z_M) across which the input transducer is bridged.

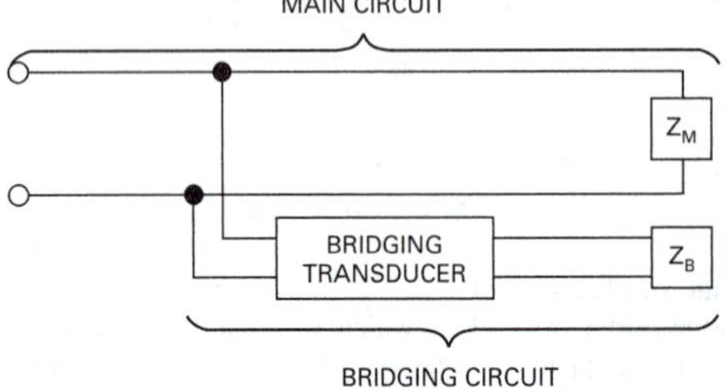

bridging gain

(PE) 599-1985w

bridging loss (A) (data transmission) The ratio of the signal power delivered to that part of the system following the bridging point, before the insertion of the bridging element to this signal power delivered to the same part after the bridging. *Note:* Bridging loss is the inverse of bridging gain, and is usually expressed in decibels. **(B) (data transmission)** The ratio of the power dissipated in a load B across which the input of a transducer delivers to its load A. *Note:* Bridging loss is the inverse of bridging gain, and is usually expressed in decibels. (PE) 599-1985

bridging relay contacts A contact form in which the moving contact touches two stationary contacts simultaneously during transfer. (EEC/REE) [87]

bridle wire Insulated wire for connecting conductors of an open wire line to associated pole-mounted apparatus. (EEC/PE) [119]

bright dip (electroplating) A dip used to produce a bright surface on a metal. *See also:* electroplating. (PE/EEC) [119]

brightener (electroplating) An addition agent used for the purpose of producing bright deposits. *See also:* electroplating. (PE/EEC) [119]

brightness (1) (fiber optics) An attribute of visual perception, in accordance with which a source appears to emit more or less light; obsolete. *Notes:* 1. Usage should be restricted to nonquantitative reference to physiological sensations and perceptions of light. 2. "Brightness" was formerly used as a synonym for the photometric term "luminance" and (incorrectly) for the radiometric term "radiance." *See also:* radiance; radiometry. (BT/AV) 201-1979w, 812-1984w
(2) (illuminating engineering) (of a perceived aperture color) The attribute by which an area of color of finite size is perceived to emit, transmit, or reflect a greater or lesser amount of light. No judgement is made as to whether the light comes from a reflecting, transmitting, or self-luminous object. (EEC/IE) [126]
(3) (computer graphics) A measure of the visible light intensity of the image displayed on the surface of a display device. (C) 610.6-1991w
(4) (image processing) A value associated with a point of an image, representing the amount of light projected from a scene in a given direction. (C) 610.4-1990w
(5) (electric power systems in commercial buildings) The subjective attribute of any light sensation, including the entire scale of the qualities "bright," "light," "brilliant," "dim," and "dark." (IA/PSE) 241-1990r
(6) *See also:* spectral brightness. (AP/PROP) 211-1997

brightness channel *See:* television.

brightness contrast threshold (illuminating engineering) When two patches of color are separated by a brightness contrast border as in the case of a bipartite photometric field or in the case of a disk shaped object surrounded by its background, the border between the two patches is a brightness contrast border. The contrast which is just detectable is known as the brightness contrast threshold. (EEC/IE) [126]

brightness control (television) A control, associated with a picture display device, for adjusting the average luminance of the reproduced picture. *Note:* In a cathode-ray tube the adjustment is accomplished by shifting bias. This affects both the average luminance and the contrast ratio of the picture. In a color-television system, saturation and hue are also affected. (EEC/PE) [119]

brightness of surface (radio-wave propagation) The power radiated per unit area, per unit bandwidth, per unit solid angle. (AP/PROP) 211-1990s

brightness signal *See:* luminance signal.

brightness temperature For a region or an extended source at a given wavelength, the temperature of an equivalent black-body radiator that has the same brightness. (AP/PROP) 211-1997

bright signal* *See:* luminance signal.

* Deprecated.

brine A salt solution. (EEC/PE) [119]

brittle metal component A metallic component that fails at an elongation of less than 10% in 5 cm (2 in). (PE/SUB) 693-1997

British thermal unit The quantity of heat required to raise one pound of water 1°F. (IA/PSE) 241-1990r

broadband (1) In general, wide bandwidth equipment or systems that can carry signals occupying a large portion of the electromagnetic spectrum. A broadband communication system can simultaneously accommodate television, voice, data, and many other services. (LM/C) 802.7-1989r
(2) (electrical noise) Electrical noise that contains energy covering a wide frequency range. *Contrast:* narrow-band electrical noise. (PE/INT/IC) 1143-1994r

broadband coaxial system A broadband system employing coaxial cables as a data transmission medium. (C) 610.7-1995

broadband interference (measurement) A disturbance that has a spectral energy distribution sufficiently broad, so that the response of the measuring receiver in use does not vary significantly when tuned over a specified number of receiver bandwidths. *See also:* electromagnetic compatibility. (EMC) [53]

broadband local area network A local area network (LAN) in which information is transmitted on modulated carriers, allowing coexistence of multiple simultaneous services on a single physical medium by frequency division multiplexing. (LM/C) 802.3-1998, 610.7-1995

broadband radio noise Radio noise having a spectrum broad in width as compared to the nominal bandwidth of the measuring instrument, and whose spectral components are sufficiently close together and uniform so that the measuring instrument cannot resolve them. (EMC) C63.4-1988s

broadband response spectrum (seismic qualification of Class 1E equipment for nuclear power generating stations) A response spectrum that describes motion in which amplified response occurs over a wide (broad) range of frequencies. (PE/NP) 344-1987r

broadband signaling The transmission of a signal in an analog form that may use frequency division multiplexing to allow multiple channels. *Contrast:* baseband signaling. (C) 610.7-1995

broadband spurious emission (land-mobile communications transmitters) The term as used in IEEE Std 377-1980 is applicable to modulation products near the carrier frequency generated as a result of the normal modulation process of the transmitter and appearing in the spectrum outside the authorized bandwidth (FCC). The products may result from over-deviation or internal distortion and noise and may have a Gaussian distribution. (EMC) 377-1980r

broadband system A system used for networking in which information is encoded, modulated onto a carrier, and pass band filtered or otherwise constrained to occupy only a limited frequency spectrum on the transmission medium. *Note:* Generally used for large amounts of voice, data, and video signals. *Contrast:* baseband system. (C) 610.7-1995

broadband tube (microwave gas tubes) A gas-filled fixed-tuned tube incorporating a bandpass filter of geometry suitable for radio-frequency switching. *See also:* transmit-receive tube; pretransmit-receive tube; gas tube. (ED) 161-1971w, [45]

broadcast (1) (FASTBUS acquisition and control) (broadcast operation) An operation directed to one or more slaves on one or more segments. (NID) 960-1993
(2) A mode of information transfer in which a single message is transmitted simultaneously to multiple receivers. (SUB/PE) 999-1992w
(3) A transmission mode in which a single message is sent to all network destinations, (i.e., one-to-all). Broadcast is a special case of multicast. (DIS/C) 1278.2-1995
(4) A mode of operation of the MTM-Bus in which an MTM-Bus Master transmits data to all connected S-modules simultaneously throughout a message. Also, a message transmitted in this mode. (TT/C) 1149.5-1995
(5) A technique that allows copies of a single packet from one node on a LAN to be passed to all possible nodes on a LAN. *Contrast:* multicast. (C) 610.7-1995
(6) The act of sending a frame addressed to all stations. (C/LM) 8802-5-1998
(7) The transfer of data from one endpoint to several endpoints. (C) 1003.5-1999

broadcast address (1) (FASTBUS acquisition and control) A primary address asserted by a master during a broadcast. (NID) 960-1993
(2) A predefined destination address that denotes the set of all service access points (SAPs) within a given layer. (LM/C) 8802-6-1994
(3) A predefined address that denotes the set of all stations on a given local area network. *Note:* This allows a message to be "broadcast" to all users simultaneously. (C) 610.7-1995
(4) A special address consisting of all 1's indicating all end nodes on the network.local area networks. (C) 8802-12-1998
(5) A unique multicast address that specifies all stations. (C/LM) 8802-11-1999

broadcast, global *See:* global broadcast.

broadcast, linear *See:* linear broadcast.

broadcast, local *See:* local broadcast.

broadcast message A sequence of one or more data transfers from the bus owner to all replying agents, with uninterrupted bus ownership. (C/MM) 1296-1987s

broadcast mode Beacon-initiated transmissions that are intended for all onboard equipment (OBE) in the communications zone. (SCC32) 1455-1999

Broadcast/Multicast Received (BMR) bit A bit in the Bus Error register of all S-modules. An S-module sets this bit to indicate that the last broadcast or multicast message was received without error. (TT/C) 1149.5-1995

broadcast_physical_ID A physical_ID with a value of 111111_2. (C/MM) 1394-1995

broadcast transaction (1) A transaction that may be processed by more than one responder. Although a broadcast transaction is distributed to all nodes on the ringlet, it is only accepted by nodes that support the broadcast option. Broadcast transactions are flow-controlled, and bridges or switches may forward these transactions to other ringlets in the system. Only *move* transactions can be broadcast, so higher-level protocols are needed to confirm when all broadcast transactions have completed in a multiple-ringlet system. (C/MM) 1596-1992
(2) A transaction that is distributed to all nodes on a bus. (C/MM) 1212-1991s

broadcast transmission (token ring access method) A transmission addressed to all stations. (LM/C) 802.5-1989s

broadside array antenna A linear or planar array antenna whose direction of maximum radiation is perpendicular to the line or plane, respectively, of the array. (AP/ANT) 145-1993

bronze conductor A conductor made wholly of an alloy of copper with other than pure zinc. *Note:* The copper may be alloyed with tin, silicon, cadmium, manganese, or phosphorus, for instance, or several of these in combination. *See also:* conductor. (T&D/PE) [10]

bronze leaf brush (rotating machinery) A brush made up of thin bronze laminations. *See also:* brush. (PE) [9]

brother *See:* sibling node.

brouter A device that performs router and bridging functions. Also known as a routing bridge. *See also:* router; gateway; bridge. (C) 610.7-1995

browsing Attempts by a user or intruder to access information to which read access is not authorized or intended. Browsing includes the threat of inadvertent access to sensitive information by users and nonusers (e.g., over displays visible to others, hardcopy output at printers, misrouted electronic mail). Browsing could violate the principle of least privilege, need-to-know requirements, or clearance authorizations and could result in the unauthorized disclosure of sensitive or classified information. (C/BA) 896.3-1993w

brush (1) A conductor, usually composed in part of some form of the element carbon, serving to maintain an electric connection between stationary and moving parts of a machine or apparatus. *Note:* Brushes are classified according to the types of material used, as follows: carbon, carbon-graphite, electrographitic, graphite, and metal-graphite. (PE/EM) [9]
(2) (relay) *See also:* relay wiper.

brush box (rotating machinery) The part of a brush holder that contains a brush. *See also:* brush. (EEC/LB) [101]

brush-by An electrostatic discharge from the human torso, such as from the hip or shoulder. (SPD/PE) C62.47-1992r

brush chamfer The slight removal of a sharp edge. *See also:* brush. (EEC/EM/LB) [101]

brush contact loss (rotating machinery) The I2R loss in brushes and contacts of the field collector ring or the direct-current armature commutator. *See also:* brush. (PE) [9]

brush convex and concave ends Partially cylindrical surfaces of a given radius. *Note:* When concave bottoms are applied to bevels, both bevel angle and radius shall be given. *See also:* brush. (EEC/EM/LB) [101]

brush corners The point of intersection of any three surfaces. *Note:* They are designated as top or face corners. *See also:* brush. (EEC/EM/LB) [101]

brush diameter The dimension of the round portion that is at right angles to the length. *See also:* brush. (EEC/EM/LB) [101]

brush edges The intersection of any two brush surfaces. *See also:* brush. (EEC/EM/LB) [101]

brush ends The surface defined by the width and thickness of the brush. *Note:* They are designated as top or holder end and bottom or commutator end. The end that is in contact with the commutator or ring is also knows as the brush face. *See also:* brush. (EEC/EM/LB) [101]

brush friction loss (rotating machinery) The mechanical loss due to friction of the brushes normally included as part of the friction and windage loss. *See also:* brush. (PE) [9]

brush hammer, lifting, or guide clips (electric machines) Metal parts attached to the brush that serve to accommodate the spring finger or hammer or to act as guides. *Note:* Where these serve to prevent the wear of the carbon due to the pressure finger, they are called hammer or finger clips. Rotary converter brushes may have clips that serve the dual purpose of lifting the brushes and of preventing wear from the spring finger. These are generally called lifting clips. *See also:* brush. (EEC/LB) [101]

brush holder (rotating machinery) A structure that supports a brush and that enables it to be maintained in contact with the sliding surface. *See also:* brush. (PE) [9]

brush-holder bolt insulation (rotating machinery) A combination of members of insulating materials that insulate the brush yoke mounting bolts, brush yoke, and brush holders. *See also:* brush. (PE) [9]

brush-holder insulating barriers (rotating machinery) Pieces of sheet insulation installed in the brush yoke assembly to provide longer leakage paths between live parts and ground, or between live parts of different polarities. *See also:* brush. (PE) [9]

brush-holder spindle insulation (rotating machinery) Insulation members that (when required by design) insulate the spindle on which the brush spring is mounted from the brush yoke and the brush holder. *See also:* brush. (PE) [9]

brush-holder spring That part of the brush holder that provides pressure to hold the brush against the collector ring or commutator. *See also:* brush. (PE) [9]

brush holder stud (rotating machinery) An intermediate member between the brush holder and the supporting structure. *See also:* brush. (PE) [9]

brush-holder-stud insulation (rotating machinery) An assembly of insulating material that insulates the brush holder or stud from the supporting structure. *See also:* brush. (PE) [9]

brush-holder support (rotating machinery) The intermediate member between the brush holder or holders and the supporting structure. *Note:* This may be in the form of plates, spindles, studs, or arms. *See also:* brush. (PE) [9]

brush-holder yoke A rocker arm, ring, quadrant, or other support for maintaining the brush holders or brush-holder studs in their relative positions. *See also:* brush. (PE) [9]

brush length The maximum overall dimension of the carbon only, measured in the direction in which the brush feeds to the commutator or collector ring. *See also:* brush. (EEC/EM/LB) [101]

brushless (rotating machinery) Applied to machines with primary and secondary or field windings that are constructed such that all windings are stationary, or in which the conventional brush gear is eliminated by the use of transformers having both moving and stationary windings, or by the use of rotating rectifiers. *See also:* brush. (PE) [9]

brushless exciter (1) (control of small hydroelectric power plants) Direct-connected ac generator with shaft-mounted rotating rectifiers and without a commutator and brushes. (PE/EDPG) 1020-1988r

(2) (excitation systems for synchronous machines) An alternator-rectifier exciter employing rotating rectifiers with a direct connection to the synchronous machine field, thus eliminating the need for field brushes. (PE/EDPG) 421.1-1986r

(3) An ac (rotating armature type) exciter whose output is rectified by semiconductor devices to provide excitation to an electric machine. The semiconductor devices are mounted on, and rotate with, the ac exciter armature. (IA/MT) 45-1998

brushless synchronous machine A synchronous machine that has a brushless exciter with its rotating armature and semiconductor devices on a common shaft with the field of the main machine. This type of machine has no collector, commutator, or brushes. (IA/MT) 45-1998

brush-operating device (power system device function numbers) (or slip-ring short-circuiting device) A device for raising, lowering, or shifting the brushes of a machine, or for short-circuiting its slip rings, or for engaging or disengaging the contacts of a mechanical rectifier. (SUB/PE) C37.2-1979s

brush or sponge plating (electroplating) A method of plating in which the anode is surrounded by a brush or sponge or other absorbent to hold electrolyte while it is moved over the surface of the cathode during the plating operation. *See also:* electroplating. (PE/EEC) [119]

brush rigging (rotating machinery) The complete assembly of parts whose main function is to position and support all of the brushes for a commutator or collector. *See also:* brush. (PE) [9]

brush rocker (rotating machinery) The structure from which the brush holders are supported and fixed relative to each other and so arranged that the whole assembly may be moved circumferentially. *Synonym:* brush yoke. *See also:* brush. (PE) [9]

brush-rocker gear (rotating machinery) The worm wheel or other gear by means of which the position of the brush rocker may be adjusted. *See also:* brush. (PE) [9]

brush shoulders When the top of the brush has a portion cut away by two planes at right angles to each other, this is designated as a shoulder. *See also:* brush. (EEC/EM/LB) [101]

brush shunt (rotating machinery) The stranded cable or other flexible conductor attached to a brush to connect it electrically to the machine or apparatus. *Note:* Its purpose is to conduct the current that would otherwise flow from the brush to the brush holder or brush-holder finger. *See also:* brush. (PE) [9]

brush shunt length The distance from the extreme top of the brush to the center of the hole or slot in the terminal, or the center of the inserted portion of a plug terminal or, if there is no terminal, to the end of the shunt. *See also:* brush. (PE/EM) [9]

brush sides (A) Front and back (bounded by width and length). *Note:* If the brush has one or both ends beveled, the short side of the brush is the front. *See also:* brush. **(B)** If there are no top or bottom bevels and width is greater than thickness and there is a top clip, the side to which the clip is attached is the back, except in the case of angular clips where the front or short side is determined by the slope of the clip and not by the side to which it is attached. *See also:* brush. **(C)** Left side and right side (bounded by thickness and length). *See also:* brush. (EEC/EM/LB) [101]

brush slots, grooves, and notches Hollows in the brush. *See also:* brush. (EEC/EM/LB) [101]

brush spring (rotating machinery) The portion of a brush holder that exerts pressure on the brush to hold it in contact with the sliding surface. (PE) [9]

brush thickness The dimension at right angles to the length in the direction of rotation. *See also:* brush. (EEC/EM/LB) [101]

brush width The dimension at right angles to the length and to the direction of rotation. *See also:* brush. (EEC/EM/LB) [101]

brush yoke *See:* brush rocker.

BS *See:* backspace character.

BSAM *See:* basic sequential access method.

B-scope A cathode-ray oscilloscope arranged to present a B-display. (AES/RS) 686-1990

BSE *See:* Bus Error.

BSL *See:* basic switching impulse insulation level.

***B* stage** An intermediate stage in the reaction of certain thermosetting resin in which the material swells when in contact with certain liquids and softens when heated, but may not entirely dissolve or fuse. *Note:* The resin in an uncured thermosetting moulding compound is usually in this stage. (PE) [9]

***B* switchboard (telephone switching systems)** A telecommunications switchboard in a local central office, used primarily to complete calls received from other central offices. (COM) 312-1977w

BT The time required for one data bit to cross the Medium Independent Interface (MII)—Bit Time = 1/TxClk.local area networks. (C) 8802-12-1998

***B*-trace (navigation aid terms) (loran)** The second (lower) trace on the scope display. (AES/GCS) 172-1983w

B-tree (A) A height-balanced search tree of order n in which each node contains keys $\{k_1, k_2, ...k_m\}$ in ascending order, where $m \leq n - 1$. The ith subtree of that node contains all the key values falling between $k_i - _1$ and k_i, with the first subtree containing all key values less than k_1 and the last subtree containing all key values greater than k_m. For example, in the B-tree in the figure below, the lowest notes contain "values less than 10," "11−19," "20−44," "45−59," and "values greater than 60," respctively. *Note:* The height balance of a B-tree is zero. *Synonym:* B-tree index. *See also:* B*-tree; B′-tree. **(B)** A B-tree as in definition (A) in which every nonterminal node except the root has at least $n/2$ subtrees. *Note:* When a node overflows, it is split into two separate nodes, with the parent node updated accordingly. *See also:* binary tree.

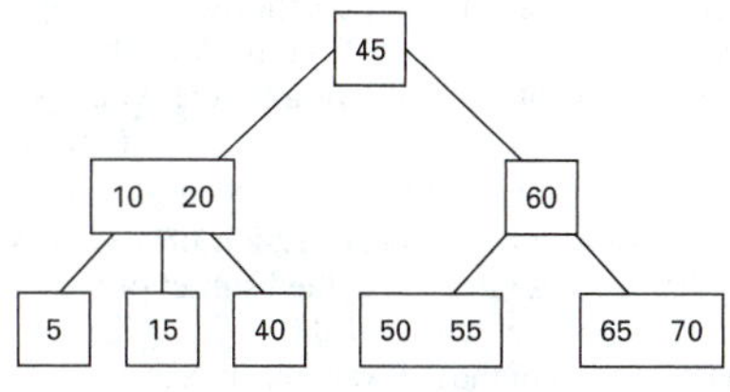

B-tree of order 3

(C) 610.5-1990

B′-tree A modified B-tree in which identifiers for all nodes are stored in terminal nodes.

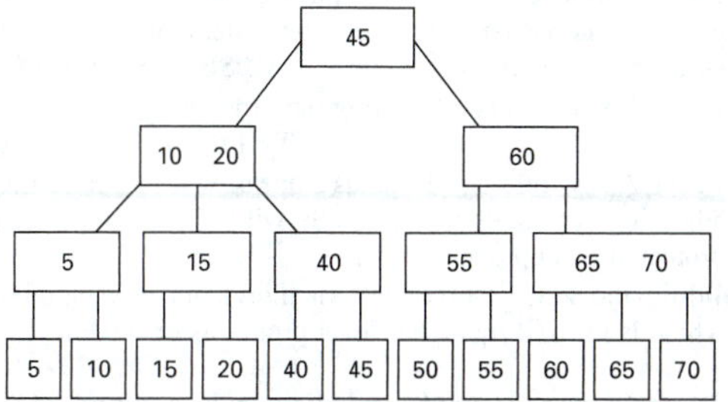

B′-tree of order 3

(C) 610.5-1990w

B*-tree A B-tree in which the root node has between 2 and 2+1 descendants, and each remaining node has between (2m−1)/3 and m descendants. That is, two-thirds of the available space in each node is used. *Note:* When a node overflows, keys from that node are moved into one of its sibling nodes if possible; otherwise the node, together with one of its sibling nodes, is split into three nodes.

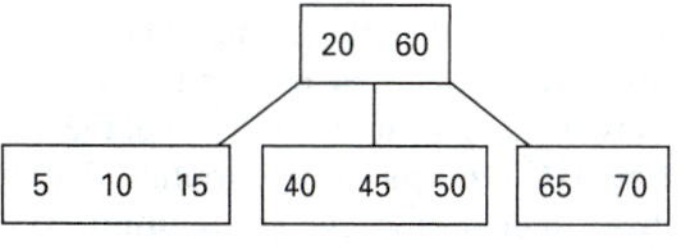

B*-tree of order 3

(C) 610.5-1990w

B-tree index *See:* B-tree.

bubble chart A data flow, data structure, or other diagram in which entities are depicted with circles (bubbles) and relationships are represented by links drawn between the circles. *See also:* box diagram; input-process-output chart; block diagram; flowchart; graph; structure chart.

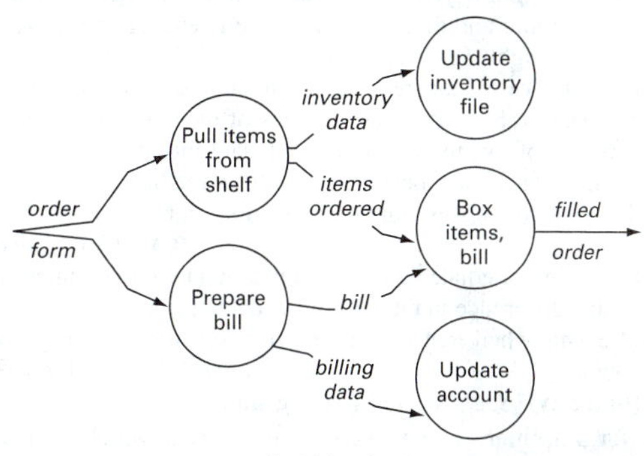

bubble chart

(C) 610.12-1990

bubble memory A type of nonvolatile storage that uses magnetic fields to create regions of magnetization; a pulsed field breaks the regions into isolated bubbles, free to move along the surface and the presence or absence of a bubble represents digital information. *Synonym:* magnetic bubble memory. (C) 610.10-1994w

bubble sort An exchange sort in which adjacent pairs of items are compared and exchanged, if necessary, and all passes through the set proceed in the same direction. *Synonyms:* propagation sort; sifting sort; exchange selection sort. *Contrast:* cocktail shaker sort. (C) 610.5-1990w

Buchmann-Meyer pattern *See:* light pattern.

buck arm A crossarm placed approximately at right angles to the line crossarm and used for supporting branch or lateral conductors or turning large angles in line conductors. *See also:* tower. (T&D/PE) [10]

bucket (1) (A) (data management) An area of storage that may contain more than one record and that is referenced as a whole by some addressing technique. **(B) (data management)** In hashing, a section of a hash table that can hold all records with identical hash values. (C) 610.5-1990
(2) A device designed to be attached to the boom tip of a line truck, crane, or aerial lift and used to support workers in an elevated working position. It is normally constructed of fiberglass to reduce its physical weight, maintain strength, and obtain good dielectric characteristics *Synonym:* basket. (T&D/PE) 516-1995, 524-1992r
(3) A colloquial reference for an area of storage that may contain more than one record and that is referenced as a whole by some addressing technique. (C) 610.10-1994w

bucket belt *See:* aerial belt.

buffalo *See:* conductor grip.

buffer (1) (A) (supervisory control, data acquisition, and automatic control) (buffer storage) A device in which data are stored temporarily, in the course of transmission from one point to another; used to compensate for a difference in the flow of data, or time of occurrence of events, when transmitting data from one device to another. **(B) (supervisory con-**

trol, data acquisition, and automatic control) (buffer storage) An isolating circuit used to prevent a driven circuit from influencing a driving circuit. **(C) (computers)** A device or storage area used to store data temporarily to compensate for differences in rates of data flow, time of occurrence of events, or amounts of data that can be handled by the devices or processes involved in the transfer or use of the data. *Synonym:* input buffer. **(D) (computers)** A routine that accomplishes the objectives in definition (A). **(E) (computers)** To allocate, schedule, or use devices or storage areas as in definition (A). *See also:* simple buffering; anticipatory buffering; dynamic buffering. (SWG/SUB/PE) C37.1-1987
(2) (data processing) A storage device used to compensate for a difference in rate of flow of information or time of occurrence of events when transmitting information from one device to another. (C) 162-1963w
(3) (elevators) A device designed to stop a descending car or counterweight beyond its normal limit of travel by storing or by absorbing and dissipating the kinetic energy of the car or counterweight. *See also:* elevator. (PE/EEC) [119]
(4) A device or storage area used to store data temporarily to compensate for differences in rates of data flow, time or occurrence of events, or amounts of data that can be handled by the devices or processes involved in the transfer or use of the data. *Synonyms:* input-output area; output buffer; input buffer. (C) 610.10-1994w
(5) An intermediate data storage location used to compensate for the difference in rate of flow of data or time of occurrence of events when transmitting information from one device to another. (IM/ST) 1451.2-1997
(6) (relay) *See also:* relay spring stud.

buffer amplifier (1) (general) An amplifier in which the reaction of output-load-impedance variation on the input circuit is reduced to a minimum for isolation purposes. *See also:* unloading amplifier; amplifier. (AP/C/ANT) 145-1983s, 165-1977w
(2) An amplifier employed in analog computers that produces an output signal equal in magnitude to the input signal but always of one polarity. *Note:* This isolates a preceding circuit from the effects of the following circuit. *See also:* unloading amplifier. (C) 610.10-1994w

buffered channel A channel in which the data is placed into a buffer prior to a trigger event and then transmitted or acted upon following that trigger event. This contrasts with an unbuffered channel in which the data is not taken by, or available to, the channel until following the trigger event. (IM/ST) 1451.2-1997

buffered computer A computer that can perform input-output and process operations simultaneously by using input and output buffers. (C) 610.10-1994w

buffered input Input that is received using buffers. (C) 610.5-1990w

buffered interconnect (BI) A device that implements an intersegment connection such that the FASTBUS protocol (FBP) on one segment is not synchronized with that on the other. (NID) 960-1993

buffered write A write transaction that appears to complete when the request is queued in the agent or responder. A buffered-write transaction returns an optimistic (done_correct) status before the responder's completion status (which could report an error) is available. (C/MM) 1212-1991s

buffering The process of using a buffer. *See also:* dynamic buffering. (C) 610.10-1994w

buffer memory (sequential events recording systems) The memory used to compensate for the difference in rate of flow of information or time of occurrence of events when transmitting information from one device to another. *See also:* storage; buffer; event. (PE/EDPG) [1]

buffer pool A collection of buffers that can be allocated and used as needed. (C) 610.5-1990w

buffer prefix An area contained within a buffer that is used to store control information for the buffer. (C) 610.10-1994w

buffer register *See:* data buffer register; input buffer register.

buffers (buffer salts) Salts or other compounds that reduce the changes in the pH of a solution upon the addition of an acid or alkali. *See also:* ion. (EEC/PE) [119]

buffer salts *See:* buffers.

buffer storage (1) An intermediate storage medium between data input and active storage. (IA) [61]
(2) (data management) A storage device that is used as a buffer. *Synonym:* buffer store. (C) 610.5-1990w
(3) (telecommunications) Memory provided in a digital switching system or digital facility interface (DFI) to compensate for timing drift and frame registration differences between a DFI and the switching system. Reduces the probability of slips caused by environmentally produced phase modulation, such as those resulting from diurnal temperature variations. The mechanism for absorbing slips in the DFI of a local digital switch could consist of several single frame stores that are alternately written and read. This scheme allows the two clocks to drift within the limits of the buffer storage. In addition, a type of hysteresis should be provided at the DFI whereby a buffer that was involved in a slip is protected against an immediate slip in the reverse direction. Enough buffering should be used to minimize such occurrences. (COM/TA) 973-1990w
(4) (A) A type of storage that is used as temporary storage; to compensate for differences in data rate and data flow. *See also:* dynamic buffering. **(B)** A portion of main storage that is assigned to temporary storage as in definition (A). (C) 610.10-1994

buffer store *See:* buffer storage.

buffing (electroplating) The smoothing of a metal surface by means of flexible wheels, to the surface of which fine abrasive particles are applied, usually in the form of a plastic composition or paste. *See also:* electroplating. (EEC/PE) [119]

bug (1) (telegraphy) A semiautomatic telegraph key in which movement of a lever to one side produces a series of correctly spaced dots and movement to the other side produces a single dash. *See also:* fault; error. (EEC/PE) [119]
(2) In computer hardware, a recurring physical problem that prevents a system or system component from working together properly. (C) 610.10-1994w

bugduster An attachment used on shortwall mining machines to remove cuttings (bugdust) from back of the cutter and to pile them at a point that will not interfere with operation. (EEC/PE) [119]

bug seeding *See:* fault seeding; error seeding.

build (software) An operational version of a system or component that incorporates a specified subset of the capabilities that the final product will provide. (C) 610.12-1990
(2) (A) A version of the software that meets a specified subset of the requirements that the completed software will meet. **(B)** The period of time during which such a version is developed. *Note:* The relationship of the terms "build" and "version" is up to the developer; for example, it may take several versions to reach a build, a build may be released in several parallel versions (such as to different sites), or the terms may be used as synonyms. (C/SE) J-STD-016-1995

builder The entity manufacturing the product. (VT) 1475-1999, 1476-2000

building A structure which stands alone or which is cut off from adjoining structures by fire walls with all openings therein protected by approved fire walls. (NESC/NEC) [86]

building block (1) (software) An individual unit or module which is utilized by higher-level programs or modules. (C/SE) 729-1983s
(2) (test, measurement, and diagnostic equipment) Any programmable measurement or stimulus device, such as multimeter, power supply switching unit, frequency meter, installed as an integral part of the automatic test equipment. (MIL) [2]

building bolt (rotating machinery) A bolt used to insure alignment and clamping of parts. (PE) [9]

building component Any subsystem, subassembly, or other system designed for use in or integral with or as part of a structure, which can include structural, electrical, mechanical, plumbing and fire protection systems and other systems affecting health and safety. (NESC/NEC) [86]

building out (communication practice) The addition to an electric structure of an element or elements electrically similar to an element or elements of the structure, in order to bring a certain property of characteristics to a desired value. *Note:* Examples are building-out capacitors, building-out sections of line, etc. (PE/EEC) [119]

building-out capacitor A capacitor employed to increase the capacitance of an electric structure to a desired value. *Note:* The use of "building-out condenser" as a synonym for this term is deprecated. *Synonym:* building-out condenser. (IM) [120]

building-out condenser* *See:* building-out capacitor.
* Deprecated.

building-out network An electric network designed to be connected to a basic network so that the combinations will simulate the sending-end impedance, neglecting dissipation, of a line having a termination other than that for which the basic network was designed. *See also:* network analysis. (EEC/PE) [119]

building pin (rotating machinery) A dowel used to insure alignment of parts. (PE) [9]

building system Plans, specifications, and documentation for a system of manufactured building or for a type or a system of building components, which can include structural, electrical, mechanical, plumbing, and fire protection systems, and other systems affecting health and safety, and including such variations thereof as are specifically permitted by regulation, and which variations are submitted as part of the building system or amendment thereto. (NESC/NEC) [86]

building up (electroplating) Electroplating for the purpose of increasing the dimensions of an article. *See also:* electroplating. (PE/EEC) [119]

buildup or decay (diode-type camera tube). The response to the camera tube to a positive or negative step in irradiance. (ED) 503-1978w

build-up time (T_R) (1) (automatic control) In a continuous step-forced response, the fictitious time interval, which would be required for the output to rise from its initial to its ultimate value, assuming that the entire rise were to take place at the maximum rate. *Note:* It can be evaluated as π/ω_O, where ω_O is the cut-off frequency of an ideal low-pass filter. (PE/EDPG) [3]
(2) Time from the input signal going above the threshold level until the time at which the output level reaches 3 dB below the complete removal of the insertion loss. *Synonyms:* attack time; rise time. (COM/TA) 1329-1999

built-in *See:* built-in utility.

built-in ballast (mercury lamp) A ballast specifically designed to be built into a lighting fixture. (EEC/LB) [95]

built-in check *See:* automatic check.

built-in class A class that is a primitive in the IDEF1X metamodel. (C/SE) 1320.2-1998

built-in device A device that is either permanently attached to the computer system, not easily removable, or present in all system configurations (i.e., not optional). (C/BA) 1275-1994

built-in font *See:* internal font.

built-in logic block observer (BILBO) A shift-register based structure used in some forms of self-testing circuit design. (TT/C) 1149.1-1990

built-in self-test (BIST) A test paradigm that incorporates circuitry in the device for executing and resolving test information about the device. (C/TT) 1450-1999

built-in simulation (computers) A special-purpose simulation provided as a component of a simulation language; for example, a simulation of a bank that can be made specific by stating the number of tellers, number of customers, and other parameters. (C) 610.3-1989w

built-in simulator (computers) A simulator that is built-in to the system being modeled; for example, an operator training simulator built into the control panel of a power plant such that the system can operate in simulator mode or in normal operating mode. (C) 610.3-1989w

built-in test (BIT) (1) An integral capability of the test subject used to provide self-test capability. (SCC20) 1226-1998
(2) A test approach using built-in-test equipment (BITE) or self-test hardware or software to test all or part of the unit under test (UUT). *See also:* built-in test equipment. (ATLAS/MIL) 1232-1995, [2]

built-in test equipment (BITE) (1) (test, measurement, and diagnostic equipment) Any device that is part of an equipment or system and is used for the express purpose of testing the equipment or system. BITE is an identifiable unit of the equipment or system. *See also:* self-test. (MIL/ATLAS) [2], 1232-1995
(2) Hardware included solely for the built-in test function. (SCC20) 1226-1998

built-in transformer A transformer specifically designed to be built into a luminaire. (EEC/LB) [98]

built-in utility A utility implemented within a shell. The utilities referred to as *special built-ins* have special qualities. *Synonym:* built-in. (C/PA) 9945-2-1993

built-up connection A toll call that has been relayed through one or more switching points between the originating operator and the receiving exchange. *See also:* telephone system. (EEC/PE) [119]

bulb (A) (electron tubes and electric lamp) The glass envelope used in the assembly of an electron tube or an electric lamp. **(B) (electron tubes and electric lamp)** The glass component part used in a bulb assembly. (EEC/GB) [106]

bulb unit Propeller turbine and generator, with the generator in a bulbous enclosure in the water passageway. *Note:* The term "bulb turbine" has no meaning. (PE/EDPG) 1020-1988r

bulk erase Operation of removing electrons from all of the bits of an array. (ED) 1005-1998

bulkhead mounting (of a filter) Installation in which the metallic case of the filter is bolted directly to a metallic bulkhead that is at reference or ground potential. (EMC) C63.13-1991

bulk parameters Complex permittivity, complex permeability, and conductivity properties of the bulk material used in the radio-frequency (RF) absorber. The conductivity may be included in the imaginary part of the complex permittivity. (EMC) 1128-1998

bulk power system (power operations) An interconnected system for the movement or transfer of electric energy in bulk on transmission levels. (PE/PSE) 858-1987s

bulk storage (test, measurement, and diagnostic equipment) A supplementary large volume memory or storage device. (MIL) [2]

bulk-storage plant A location where gasoline or other volatile flammable liquids are stored in tanks having an aggregate capacity of one carload or more, and from which such products are distributed (usually by tank truck). (NESC/NEC) [86]

bullet *See:* connector link.

bulletin board *See:* electronic bulletin board.

bull line A high-strength line, normally synthetic fiber rope, used for pulling and hoisting large loads. *Synonyms:* bull rope; pulling line. (T&D/PE) 524-1992r

bull ring A metal ring used in overhead construction at the junction point of three or more guy wires. *See also:* tower. (T&D/PE) [10]

bull rope *See:* bull line.

bullwheel (conductor stringing equipment) A wheel incorporated as an integral part of a bullwheel puller or tensioner to generate pulling or braking tension on conductors or pulling lines, or both, through friction. A puller or tensioner normally has one or more pairs of wheels arranged in tandem incorporated in its design. The physical size of the wheels will vary for different designs, but 17 in (43 cm) face widths and diameters of 5 ft (150 cm) are common. The wheels are power driven or retarded and lined with single or multiple groove neoprene or urethane linings. Friction is accomplished by reeving the pulling line or conductor around the groove of each pair. (T&D/PE) 524-1992r

bullwheel puller (conductor stringing equipment) A device designed to pull pulling lines and conductors during stringing operations. It normally incorporates one or more pairs of urethane- or neoprene-lined, power-driven, single- or multiple-groove bullwheels where each pair is arranged in tandem. Pulling is accomplished by friction generated against the pulling line, which is reeved around the grooves of a pair of the bullwheels. The puller is usually equipped with its own engine which drives the bullwheels mechanically, hydraulically, or through a combination of both. Some of these devices function as either a puller or tensioner. *Synonym:* puller. (T&D/PE) 524a-1993r, 524-1992r

bullwheel tensioner (conductor stringing equipment) A device designed to hold tension against a pulling line or conductor during the stringing phase. Normally, it consists of one or more pairs of urethane- or neoprene-lined, power-braked, single- or multiple-groove bullwheels where each pair is arranged in tandem. Tension is accomplished by friction generated against the conductor which is reeved around the grooves of a pair of the bullwheels. Some tensioners are equipped with their own engines which retard the bullwheels mechanically, hydraulically, or through a combination of both. Some of these devices function as either a puller or tensioner. Other tensioners are equipped with only friction type retardation. *Synonyms:* retarder; brake; tensioner. (T&D/PE/T&D/PE) 524a-1993r, 524-1992r

bump *See:* pulse distortion.

bumper (elevators) A device other than an oil or spring buffer designed to stop a descending car or counterweight beyond its normal limit of travel by absorbing the impact. *See also:* elevator. (PE/EEC) [119]

buncher space (velocity-modulated tube) The part of the tube following the acceleration space where there is a high-frequency field, due to the input signal, in which the velocity modulation of the electron beam occurs. *Note:* It is the space between the input resonator grids. *See also:* velocity-modulated tube. (ED) [45], [84]

bunching The action in a velocity-modulated electron stream that produces an alternating convection-current component as a direct result of differences of electron transit time produced by the velocity modulation. *See also:* overbunching; reflex bunching; electron device; space-charge debunching; optimum bunching; underbunching. (ED) 161-1971w

bunching angle (electron stream) (given drift space) The average transit angle between the processes of velocity modulation and energy extraction at the same or different gaps. *See also:* electron device; effective bunching angle. (ED) 161-1971w

bunching, optimum *See:* optimum bunching.

bunching, parameter *See:* parameter bunching.

bunching time, relay *See:* relay bunching time.

bundle (1) (conductor stringing equipment) A circuit phase consisting of more than one conductor. Each conductor of the phase is referred to as a subconductor. A two-conductor bundle has two subconductors per phase. These may be arranged in a vertical or horizontal configuration. Similarly, a three-conductor bundle has three subconductors per phase. These are usually arranged in a triangular configuration with the vertex of the triangle up or down. A four-conductor bundle has four subconductors per phase. These are normally arranged in a square configuration. Although other configurations are possible, those listed are the most common. *Synonyms:* quad-bundle; twin-bundle; tri-bundle. *See also:* fiber bundle. (T&D/PE) 524a-1993r, 524-1980s, 1048-1990
(2) A software object, which is a grouping of other software objects, such as all or parts of other bundles and products. (C/PA) 1387.2-1995
(3) A group of signals that have a common set of characteristics and differ only in their information content. (C/LM) 802.3-1998
(4) **(A)** (As a verb) To combine separate arrow meanings into a composite arrow meaning, expressed by joining arrow segments, i.e., the inclusion of multiple object types into an object type set. **(B)** (As a noun) An arrow segment that collects multiple meanings into a single construct or abstraction, i.e., an arrow segment that represents an object type set that includes more than one object type. (C/SE) 1320.1-1998

bundled cable A cable consisting of multiple twisted pairs.local area networks. (C) 8802-12-1998

bundled conductor An assembly of two or more conductors used as a single conductor and employing spacers to maintain a predetermined configuration. The individual conductors of this assembly are called subconductors. (NESC/T&D) C2-1997, C2.2-1960

bundle table A workstation-dependent table specifying the attributes of a display element. (C) 610.6-1991w

bundle, two-conductor, three-conductor, four-conductor, multiconductor A circuit phase consisting of more than one conductor. Each conductor of the phase is referred to as a *subconductor*. A two-conductor bundle has two subconductors per phase. These may be arranged in a vertical or horizontal configuration. Similarly, a three-conductor bundle has three subconductors per phase. These usually are arranged in a triangular configuration with the vertex of the triangle up or down. A four-conductor bundle has four subconductors per phase. These normally are arranged in a square configuration. Although other configurations are possible, those listed are the most common. *Synonyms:* twin-bundle; tri-bundle; quad-bundle. (T&D/PE) 524-1992r

***B* unit** A motive power unit designed primarily for use in multiple with an *A* unit for the purpose of increasing locomotive power, but not equipped for use as the leading unit of a locomotive or for full observation of the propulsion power and brake applications for a train. *Note: B* units are normally equipped with a single control station for the purpose of independent movement of the unit only, but are not usually provided with adequate instruments for full observation of power and brake applications. *See also:* electric locomotive. (EEC/PE) [119]

bunker material Expendable material used for protecting commodities during shipping or hauling. (T&D/PE) 751-1990

buoy (navigation aids) A floating object, other than a lightship, moored or anchored to the bottom of the sea, which is an aid to navigation. *See also:* combination buoy; danger buoy; lighted buoy; radio-beacon buoy; buoys classified to location; sound buoy; sonobuoy. (AES/GCS) 172-1983w

buoys classified to location (navigation aid terms) Channel, mid-channel, turning, fairway, bifurcation, junction, sea. *See also:* buoy. (AES/GCS) 172-1983w

burden (1) (metering) (instrument transformers) The impedance of the circuit connected to the secondary winding. *Note:* For voltage transformers it is convenient to express the burden in terms of the equivalent volt-amperes and power factor at a specified voltage and frequency. (ELM) C12.1-1982s
(2) Load imposed by a relay on an input circuit, expressed in ohms or volt-amperes. *See also:* relay. (PE/PSR) [6]
(3) That property of the circuit connected to the secondary winding that determines the real and reactive power at the secondary terminals. It is expressed either as total impedance with effective resistance and reactance components or as the total voltamperes and power factor at the specified value of current and frequency. (PEL/ET) 389-1990
(4) (of a relay) Load impedance imposed by a relay on an input circuit expressed in ohms and phase angle at specified conditions. *Note:* If burden is expressed in other terms such

as volt-amperes, additional parameters such as voltage, current, and phase angle must be specified. (SWG/PE/PSR) C37.100-1992, C37.110-1996

(5) (of an instrument transformer) That property of the circuit connected to the secondary winding that determines the active and reactive power at the secondary terminals. *Note:* The burden is expressed either as total ohms impedance with the effective resistance and reactance components, or as the total voltamperes and power factor at the specified value of current or voltage, and frequency. (PE/TR/PSR) C57.13-1993, C37.110-1996

burden regulation (capacitance potential devices) Refers to the variation in voltage ratio and phase angle of the secondary voltage of the capacitance potential device as a function of burden variation over a specified range, when energized with constant, applied primary line-to-ground voltage. *See also:* outdoor coupling capacitor. (PE/EM) 43-1974s

burglar-alarm system An alarm system signaling an entry or attempted entry into the area protected by the system. *See also:* protective signaling. (EEC/PE) [119]

burial depth (cable plowing) The depth of soil cover over buried cable, pipe, etc., measured on level ground. (T&D/PE) 590-1977w

buried cable (1) (direct buried) A cable for installation under the surface of the earth in such a manner that it cannot be removed without disturbing the soil. It is designed for direct burial in the earth to withstand submersion in ground water, the pressure of backfill material, and in special applications, to withstand the gnawing of burrowing rodents. (PE/PSC) 789-1988w

(2) A cable installed under the surface of the ground in such a manner that it cannot be removed without disturbing the soil. (EEC/PE) [119]

burn in (1) (station control and data acquisition) (supervisory control, data acquisition, and automatic control) A period, prior to on-line operation, during which equipment is continuously energized for the purpose of forcing infant mortality failures. (PE/SUB) C37.1-1994

(2) (Class 1E battery chargers and inverters) The operation of components or equipment, prior to type test or ultimate application, intended to stabilize their characteristics and to identify early failures. (PE/NP) 650-1979s

(3) The operation of items prior to their ultimate application, intended to stabilize their characteristics and to identify early failures. *See also:* reliability. (R) [29]

(4) The process of running a device for a period in order to identify early failures caused by "infant mortality." *Note:* This is not the same as programming, or "burning" electrically programmable read-only memory. (C) 610.10-1994w

(5) A test performed for the purpose of screening out devices with inherent defects or defects resulting from manufacturing aberrations that cause time and stress dependent failures. Burn-in is intended to eliminate infant mortality and early lifetime failures by stressing devices at or above maximum operating conditions. (C/BA) 1156.4-1997

burn-in period *See:* early-failure period.

burnishing The smoothing of metal surfaces by means of a hard tool or other article. *See also:* electroplating. (EEC/PE) [119]

burnishing surface (mechanical recording) The portion of the cutting stylus directly behind the cutting edge, that smooths the groove. *See also:* phonograph pickup. (SP) [32]

burnout (nonlinear, active, and nonreciprocal waveguide components) The point at which a sensitive receiving device suffers a specified permanent degradation of noise figure or equivalent increase in noise temperature. (MTT) 457-1982w

burnt deposit A rough or noncoherent electrodeposit produced by the application of an excessive current density. *See also:* electroplating. (EEC/PE) [119]

burnup, nuclear *See:* nuclear burnup.

burst (1) (pulse techniques) A wave or waveform composed of a pulse train or repetitive waveform that starts as a prescribed time and/or amplitude, continues for a relatively short duration and/or number of cycles, and upon completion returns to the starting amplitude. *See also:* pulse. (IM/HFIM) [40]

(2) (audio and electroacoustics) An excursion of a quantity (voltage, current, or power) in an electric system that exceeds a selected multiple of the long-time average magnitude of this quantity taken over a period of time sufficiently long that increasing the length of this period will not change the result appreciably. This multiple is called the upper burst reference. *Notes:* 1. If measurements are made at different points in a system, or at different times, the same quantity must be measured consistently. 2. The excursion may be an electrical representation of a change of some other physical variable such as pressure, velocity, displacement, etc. (SP) [32]

(3) (radio-wave propagation) A transient increase in intensity of radiation over a short period, such as is observed from the sun. (AP) 211-1977s

(4) (A) To read or write data in such a manner that does not require or permit an interruption to occur. **(B)** To separate the pages of a continuous form, often by means of a device called a burster. (C) 610.10-1994

(5) A set of stimulus patterns and related unit under test (UUT) responses that are set up, applied, and read as a group. A test program may employ more than one burst to provide the stimuli and responses necessary to test the UUT. (SCC20) 1445-1998

(6) Tester execution of a pattern or set of patterns. Generally controlled by "start" and "stop" definitions. (C/TT) 1450-1999

(7) *See also:* pulse burst. (AES) 686-1997

burst4 Exactly four consecutive signal elements of the same polarity. (C/LM) 8802-5-1998

burst5 Five or more consecutive signal elements of the same polarity. (C/LM) 8802-5-1998

burst6 Six or more consecutive signal elements of the same polarity. (C/LM) 8802-5-1998

burst build-up interval The time interval between the burst leading-edge time and the instant at which the upper burst reference is first equaled. (SP) 257-1964w, [32]

burst corona (overhead-power-line corona and radio noise) Corona mode that may be considered as the initial stage of positive glow. It occurs at a positive electrode with electric field strengths at and slightly above the corona inception voltage gradient. Burst corona appears as a bluish film of velvet-like glow adhering closely to the electrode surface. The current pulses of burst corona are of low amplitude and may last for periods of milliseconds. (T&D/PE) 539-1990

burst decay interval (audio and electroacoustics) The time interval between the instant at which the peak burst magnitude occurs and the burst trailing-edge time. *See also:* burst; burst duration. (SP) 257-1964w, [32]

burst duration (audio and electroacoustics) The time interval during which the instantaneous magnitude of the quantity exceeds the lower burst reference, disregarding brief excursions below the reference, provided the duration of any individual excursion is less than a burst safeguard interval of selected length. *Notes:* 1. If the duration of an excursion is equal to or greater than the burst safeguard interval, the burst has ended. 2. These terms, as well as those defined below, are illustrated in the accompanying figure.

a) A burst is found with the aid of a "window" that is slid horizontally to the right with its base resting on the lower burst reference. The width of the window equals the burst safeguard interval and the height of the window equals the difference between the upper and lower burst references. The window is slid to the right until the trace crosses the top of the window. The upper burst reference has then been reached and a burst has occurred.

b) The burst leading-edge time is found by sliding the window to the left until the trace disappears from the window.

The right-hand side of the window marks the burst leading-edge time.

c) The burst trailing edge time is found by a similar procedure. The window is slid to the right past its position in (A) until the trace disappears from the window. The left-hand side of the window marks the burst trailing-edge time.

d) Terms used in defining a burst: burst leading-edge time, t_1; burst build-up interval, $t_2 - t_1$; burst rise interval, $t_3 - t_1$; burst trailing-edge time, t_5; burst decay interval, $t_5 - t_3$; burst fall-off interval, $t_5 - t_4$; burst duration, $t_5 - t_1$; upper burst reference, U; lower burst reference, L; long-time average power, P.

See also: burst.

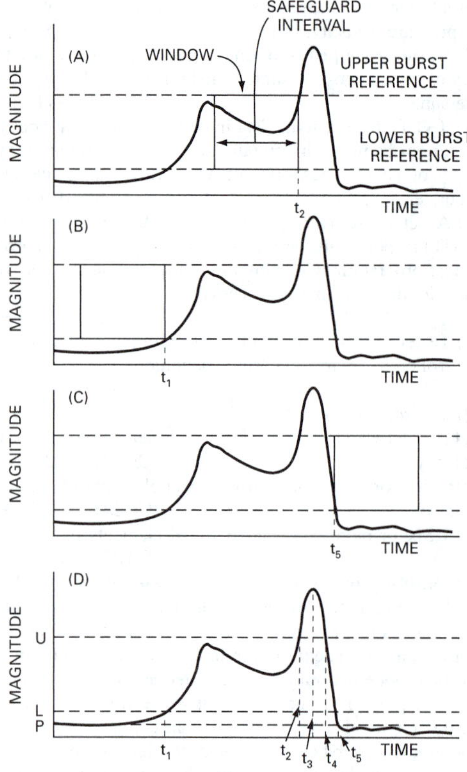

Plot of instantaneous magnitude versus time to illustrate terms used in defining a burst.

burst duration

(SP) 257-1964w, [32]

burst duty factor (audio and electroacoustics) The ratio of the average burst duration to the average spacing. *Note:* This is equivalent to the product of the average burst duration and the burst repetition rate. *See also:* burst. (SP) 257-1964w, [32]

burst error In data communications, a series of consecutive errors in data transmission that tend to be grouped together, with a longer time interval separating multiple bursts. (C) 610.7-1995

burst fall-off interval (audio and electroacoustics) The time interval between the instant at which the upper burst reference is last equaled and the burst trailing edge time. *See also:* burst. (SP) 257-1964w, [32]

burst flag (television) A keying or gating signal used in forming the color burst from a chrominance subcarrier source. *See also:* television. (BT/AV) [34]

burst gate (television) A keying or gating device or signal used to extract the color burst from a color picture signal. *See also:* television. (BT/AV) [34]

burst keying signal *See:* television; burst flag.

burst leading-edge time (audio and electroacoustics) The instant at which the instantaneous burst magnitude first equals the lower burst reference. *See also:* burst. (SP) 257-1964w, [32]

burst measurements *See:* energy-density spectrum.

burst mode (1) A mode of transmission by which a system can send a burst of data at higher speed for some period of time. (C) 610.7-1995

(2) An operational mode in which an end node may send one or more packets each time it is granted permission to transmit.local area networks. (C) 802.12c-1998

burst-quiet interval (audio and electroacoustics) The time interval between successive bursts during which the instantaneous magnitude does not equal the upper burst reference. *See also:* burst. (SP) 257-1964w, [32]

burst repetition rate (audio and electroacoustics) The average number of bursts per unit of time. *See also:* burst. (SP) 257-1964w, [32]

burst rise interval (audio and electroacoustics) The time interval between the burst leading-edge time and the instant at which the peak burst magnitude occurs. *See also:* burst. (SP) 257-1964w, [32]

burst safeguard interval (audio and electroacoustics) A time interval of selected length during which excursions below the lower burst reference are neglected; it is used in determining those instants at which the lower burst references are first and last equaled during a burst. *See also:* burst. (SP) 257-1964w, [32]

burst spacing (audio and electroacoustics) The time interval between the burst leading-edge times of two consecutive bursts. *See also:* burst. (SP) 257-1964w, [32]

burst trailing-edge time (audio and electroacoustics) The instant at which the instantaneous burst magnitude last equals the lower burst reference. *See also:* burst. (SP) 257-1964w, [32]

burst train (audio and electroacoustics) A succession of similar bursts having comparable adjacent burst-quiet intervals. *See also:* burst. (SP) 257-1964w, [32]

bus (1) A three-phase junction common to two or more ways. (SWG/PE) C37.71-1984r

(2) (signals and paths) (microcomputer system bus) A signal line or a set of lines used by an interface system to connect a number of devices and to transfer data. (MM/C/IM/AIN) 796-1983r, 488.1-1987r, 1000-1987r, 696-1983w, 959-1988r

(3) One or more conductors used for transmitting signals or power from one or more sources to one or more destinations. (C) 162-1963w

(4) (simple 32-bit backplane bus) A set of signal lines to which a number of devices are connected and over which information is transferred between them. (MM/C) 1196-1987w

(5) (hydroelectric power plants) A conductor or group of electrical conductors serving as common connections between circuits, generally in the form of insulated cable, rigid rectangular or round bars, or stranded overhead cables held under tension. (PE/EDPG) 1020-1988r

(6) The concatenation of the *transmission links* between *nodes* and the data path within nodes that provides unidirectional transport of the digital bit stream from the *Head of Bus function* past the *access unit (AU)* of each node to the end of bus. (LM/C) 8802-6-1994

(7) One or more conductors that are used for the transmission of signals, data, or power. *See also:* address bus; data chain bus; data bus; memory bus; control bus; bidirectional bus; time-multiplexed bus. (C) 610.10-1994w

(8) A conductor, or group of conductors, that serves as a common connection for two or more circuits. (SWG/PE) C37.100-1992

(9) In PDEF, a physical collection of nets and/or pnets, or of pins and/or nodes. If the items collected in the PDEF bus are logical, the PDEF bus may or may not correspond to a logical bus described in the netlist. (C/DA) 1481-1999

bus address A label used to define a communications path to a device in a bus environment where multiple devices share a common data path. (SCC20) 993-1997

bus bar A common metallized region that connects the individual interdigital transducer fingers and provides a contact area for external circuit connection via bonding or other means. (UFFC) 1037-1992w

bus-based architecture A computer architecture in which the components such as processors, peripheral devices and memory are interconnected by one or more busses. *Contrast:* non-bus-based architecture. (C) 610.10-1994w

bus bridge A bus bridge is an interconnect between two or more buses that provides signal and protocol translation from one bus to another. The buses may adhere to different bus standards for mechanical, electrical, and logical operation (such as a bus bridge from Futurebus+ to VMEbus or to Multibus II). (C/BA) 10857-1994, 896.2-1991w, 896.3-1993w, 896.4-1993w, 896.10-1997

bus clock cycle An amount of time equal to one bus clock period, nominally 100 ns. (C/MM) 1296-1987s

bus cycle (1) (general system) (microcomputer system bus) The process whereby digital signals effect the transfer of data bytes or words across the interface by means of an interlocked sequence of control signals. Interlocked denotes a fixed sequence of events in which one event must occur before the next event can occur. (MM/C) 796-1983r
(2) (696 interface devices) (signals and paths) The basic sequence of electrical events required to complete a transfer of data on the bus. A bus cycle contains at least three bus states. (MM/C) 696-1983w

bus-dependent (1) A term used to describe parameters that may vary among different bus standards, but are defined by them. Although the CSR Architecture may constrain the definition of these fields, their detailed definition is provided by the appropriate bus standard. (C/MM) 1212-1991s
(2) This term is used to describe technology-dependent parameters. Although the CSR Architecture may specify the size and address of these parameters, their format and definition is provided by the appropriate bus standards. (C/BA) 896.4-1993w

bus driver (A) A device capable of providing sufficient current to drive all loads connected to a bus. *See also:* bus slave. **(B)** A device that controls access to a bus. (C) 610.10-1994

Bus Error BSE bit A bit in the Slave Status register of every S-module that is set by the S-module when a Bus Error is recorded in the Bus Error register. (TT/C) 1149.5-1995

Bus Error register A status register that is required to be implemented in the MTM-Bus interface circuitry of every S-module. Bits in this register provide the S-module with the ability to record error conditions associated with message transmission. The register may be interrogated by the M-module. Some bits in the register are reserved for application-specific uses. (TT/C) 1149.5-1995

bushing (1) (rotating machinery) (electrical) Insulator to permit passage of a lead through a frame or housing. (PE) [9]
(2) An insulating structure including a through conductor, or providing a passageway for such a conductor, with provision for mounting on a barrier, conducting or otherwise, for the purpose of insulating the conductor from the barrier and conducting current from one side of the barrier to the other. (SWG/PE/NESC/TR/PSIM) C37.100-1992, C2-1997, C57.12.80-1978r, 62-1995
(3) (relay) *See also:* relay spring stud.

bushing condenser The component within a capacitive graded bushing in which the grading element is embedded in the major insulation. (PE/TR) C57.19.03-1996

bushing insert (separable insulated connectors) A connector component intended for insertion into a bushing well. (T&D/PE) 386-1995

bushing potential tap (outdoor apparatus bushings) An insulated connection to one of the conducting layers of a bushing providing a capacitance voltage divider to indicate the voltage applied to the bushing. (PE/TR) 21-1976

bushing, rotor *See:* rotor bushing.

bushings (A) (for combined voltage application) A bushing applied to the valve winding side of a converter transformer or a bushing applied to the converter transformer side of a dc converter valve. This bushing is exposed to a large ac stress superimposed on a dc bias. **(B)** (for pure DC application) A bushing applied to thedc side of a dc converter valve or a bushing applied on a dc smoothing reactor. This bushing is exposed to dc stress with a small AC ripple. (PE/TR) C57.19.03-1996

bushing tap (partial discharge measurement in liquid-filled power transformers and shunt reactors) Connection to a capacitor foil in a capacitively graded bushing designed for voltage or power factor measurement that also provides a convenient connecting point for partial discharge measurement. The tap-to-phase capacitance is generally designated as C_1 and the tap-to-ground capacitance is designated as C_2. *See also:* bushing test tap; capacitance; bushing potential tap. (SWG/PE/TR) 1291-1993r, C57.113-1988s

bushing test tap (1) (outdoor apparatus bushings) An insulated connection to one of the conduction layers of a bushing for the purpose of making insulation power factor tests. (PE/TR) 21-1976
(2) A connection to one of the conducting layers of a capacitance graded bushing for measurement of partial discharge, power factor, and capacitance values. (PE/TR) C57.19.03-1996

bushing type current transformer A current transformer (CT) that has an annular core with a secondary winding insulated from and permanently assembled on the core but has no primary winding or insulation for a primary winding. This type of CT is for use with a fully insulated conductor as a primary winding. A bushing type ct is usually used in equipment where the primary conductor is a component part of other apparatus. *Note:* This type of ct has very low leakage flux and is also known as a Low Inductance Type CT. (PE/PSR/TR) C37.110-1996, C57.13-1993, C57.12.80-1978r

bushing voltage tap A connection to one of the conducting layers of a capacitance graded bushing providing a capacitance voltage divider. *Note:* Additional equipment can be designed, connected to this tap and calibrated to indicate the voltage applied to the bushing. This tap can also be used for measurement of partial discharge, power factor and capacitance values. (PE/TR) C57.19.03-1996

bushing well (separable insulated conductors) An apparatus bushing having a cavity for insertion of a connector component, such as a bushing insert. (PE/T&D) 386-1995

bus_ID A 10-bit number uniquely specifying a particular bus within a system of multiple interconnected buses. (C/MM) 1394-1995

busied A status indication returned in an echo packet that indicates to the sender that the send packet was not accepted (and was discarded), probably because there was no room in the destination queue. The sender should retransmit the packet later. (C/MM) 1596-1992

bus implementation conformance statement (BICS) This is a completed BICS pro forma questionnaire. (C/BA) 896.4-1993w

bus implementation conformance statement pro forma (BICS pro forma) A questionnaire that lists implementation capabilities. (C/BA) 896.4-1993w

bus implementation extra information for testing (BIXIT) A completed BIXIT pro forma. (BA/C) 896.4-1993w

bus implementation extra information for testing pro forma (BIXIT pro forma) This questionnaire provides extra information about the module that might be necessary to configure and perform the tests. (C/BA) 896.4-1993w

business area (BA) The logical subdivision of an enterprise into areas of similar business directions, e.g., finance, sales, and marketing. (C/PA) 1003.23-1998

business data processing The use of computers for processing information to support the operational, logistical, and functional activities performed by an organization. (C) 610.2-1987

business function A set of processes that support the attainment of a particular business goal. (C/PA) 1003.23-1998

business graphics The use of computer graphics to display business data; for example, bar charts, histograms, pie charts. (C) 610.6-1991w

business information system *See:* management information system.

business system requirement (BSR) The enterprise-driven requirement for a business system, i.e., a set of processes, procedures, and documentation supported by technology to deliver either a major CSF or a KPI in the measurement of the attainment of the enterprise business goals and vision. *See also:* critical success factor; key performance indicator. (C/PA) 1003.23-1998

bus interface unit (BIU) The logic on a module that converts bus signals to and from signals that are compatible with the functional logic of the module. (C/BA) 896.3-1993w

bus line (1) (railways) A continuous electric circuit other than the electric train line, extending through two or more vehicles of a train, for the distribution of electric energy. *See also:* multiple-unit control. (EEC/PE) [119]
(2) Signal transmission line, that may be driven by several modules simultaneously using drivers with wire-OR capability. Therefore, a signal carried by a bus line is the combination of signals applied to that line from each module. (C/BA) 896.3-1993w
(3) The medium for the transmission of signals. Since Futurebus+ uses open collector drivers, a bus line may be driven by several boards simultaneously. Therefore, the signal carried by the bus line is the combination of signals applied to that line from each board. (C/BA) 896.2-1991w
(4) The medium for the transmission of signals. Since Futurebus+ requires drivers with wire-OR capability, a bus line may be driven by several modules simultaneously. Therefore, the signal carried by the bus line is the combination of signals applied to that line from each module. (C/BA) 10857-1994, 896.4-1993w
(5) The medium for the transmission of signals. Futurebus+ Spaceborne Profile uses both open collector and push-pull drivers to match whether a signal is expected to be driven by more than one board simultaneously. A bus line is driven by only one board for those signals that are push-pull. A bus line may be driven by several boards simultaneously for those signals that are open collector. Therefore, the signal carried by the bus line is the combination of signals applied to that line from each board in the open collector case or from the one board in the push-pull case. (C/BA) 896.10-1997

bus lock Method of a master ensuring continued tenure of the bus. Not identical to resource lock. (C/MM) 1196-1987w

bus loss The amount of time required for a valid signal transition to occur at every point on the backplane. This value is equivalent to two bus propagation delays plus the clock skew. (C/MM) 1296-1987s

bus manager (1) The node that provides advanced power management, optimizes Serial Bus performance, describes the topology of the bus, and cross-references the maximum speed for data transmission between any two nodes on the bus. The bus manager node may also be the isochronous resource manager node. (C/MM) 1394-1995
(2) The node that provides power management, sets the gap count in the cable environment, and publishes the topology of the bus and the maximum speed for data transmission between any two nodes on the bus. The bus manager node may also be the isochronous resource manager node. (C/MM) 1394a-2000

bus master A device connected to a bus which controls all other devices connected to the same bus. *Note:* The bus master controls which slave devices may, and when they may, place data on the bus. *Contrast:* bus slave. (C) 610.10-1994w

bus mouse A mouse that connects to the computer system using a bus, generally contained within a special expansion board. *Contrast:* serial mouse. (C) 610.10-1994w

bus node In the device tree, a descendant node that represents the interface, or "bridge," between an SBus and its parent (which could be another bus). (C/BA) 1275.2-1994w, 1275.4-1995

bus operation The basic unit of processing whereby digital signals effect the transfer of data across an interface by means of a sequence of control signals and an integral number of bus clock cycles. (C/MM) 1296-1987s

bus owner The agent that enters the acquisition phase of the arbitration operation and initiates one or more transfer operations. *See also:* transfer operation; arbitration operation; acquisition phase. (C/MM) 1296-1987s

bus reactor (power and distribution transformers) A current-limiting reactor for connection between two different buses or two sections of the same bus for the purpose of limiting and localizing the disturbance due to a fault in either bus. *See also:* reactor. (PE/TR) C57.12.80-1978r, [57]

bus request sequence A set of one or more arbitration operations in which all agents that simultaneously request the bus become the bus owner, one at a time. *See also:* bus owner; arbitration operation. (C/MM) 1296-1987s

bus-ring topology A topology where the stations are physically wired as a bus but logically act like a ring. Every station on the bus knows its logical predecessor and successor. Transmissions can be broadcast to all stations on the bus or addressed to another individual station. *See also:* bus topology; star-ring topology; ring topology; star topology; star-bus topology; loop topology; tree topology. (C) 610.7-1995

Bus Sizing The dynamic modification of the data transfer width to meet the SBus Slave's bus width requirements. (C/BA) 1496-1993w

bus slave (A) A device which responds to signals on a bus. *Contrast:* bus master. **(B)** A device connected to a bus which cannot put data onto the bus until given permission by the bus driver or bus master. (C) 610.10-1994

bus standard An abbreviated notation used throughout this document, rather than the more exact "bus standard document that claims conformance to this specification." (C/MM) 1212-1991s

bus state (696 interface devices) (signals and paths) A bus state is one clock cycle long and begins and ends just before the rising edge of ϕ. There are at least three bus states in every bus cycle. (C/MM) 696-1983w

bus structure An assembly of bus conductors, with associated connection joints and insulating supports. (PE/SUB) 605-1998

bus support (1) An insulating support for a bus. It includes one or more insulator units with fittings for fastening to the mounting structure and for receiving the bus. (SWG/PE) C37.100-1992
(2) An insulating support for a bus. *Note:* A bus support includes one or more insulator units with fittings for fastening to the mounting structure and for receiving the bus. (PE/SUB) 605-1998

bus tenure The duration of a master's control of the bus; i.e., the time during which a module has the right to initiate and execute bus transactions. (C/BA) 1014.1-1994w, 896.4-1993w, 896.3-1993w, 10857-1994

bus tie reactor A current limiting reactor for connection between two different buses or two sections of the same bus for the purpose of limiting and localizing the disturbance due to a fault in either bus. (PE/TR) C57.16-1996

bus timer A functional module that measures the time each data transfer takes on the DTB and terminates the DTB cycle when the transfer time is not within reason. Without this module, when the master attempts to transfer data to or from a nonexistent slave location it could wait forever. The bus timer prevents this delay by terminating the cycle. (C/BA) 1014-1987

bus topology A topology in which stations are attached to a common transmission medium, known as a bus; data propagate the length of the medium and are received by all stations. *See also:* star-bus topology; tree topology; loop topology; ring topology; bus-ring topology; star topology; star-ring topology. (C) 610.7-1995

bus transaction An event initiated with a connection phase and terminated with a disconnection phase. Data may or may not be transferred during a bus transaction. *See also:* transaction (C/BA) 1014.1-1994w, 896.4-1993w, 10857-1994, 896.3-1993w

bus-type shunt (direct-current instrument shunts) An instrument shunt for switchboard use so that it can be installed in the bus or connection bar structure of the circuit whose current is to be measured. (PE/PSIM) 316-1971w

busway A grounded metal enclosure containing factory-mounted, bare or insulated conductors that are usually copper or aluminum bars, rods, or tubes. *See also:* cable bus. (NESC/NEC) [86]

busy (1) Pertaining to a system or component that is operational, in service, and in use. *See also:* up; down; idle. (C) 610.12-1990
(2) If a slave is unable to accept a bus transaction from a master, it may issue a busy status to the master of the transaction. The master must relinquish the bus and may reacquire the bus and retry the transaction after a suitable time interval. (C/BA) 10857-1994, 896.4-1993w, 896.3-1993w, 1014.1-1994w

busy hour (1) (telephone switching systems) That uninterrupted period of 60 min during the day when the traffic offered is a maximum. (COM) 312-1977w
(2) (data transmission) The peak 60 min period during a 24 h period when the largest volume of communication traffic is handled. (PE) 599-1985w
(3) (telecommunications) The hour having the highest average traffic for the three highest traffic months. A "busy hour" determination study uses only about two weeks worth of hour-by-hour data collected just in advance of the expected high-traffic months. *Synonym:* time-consistent busy hour. *See also:* time-consistent traffic measures. (COM/TA) 973-1990w
(4) An hour-long window during which the communication traffic load is at its maximum for a given 24 h period. (C) 610.7-1995

busy season The three months, not necessarily consecutive, that have the highest average traffic in the busy hour. *See also:* average busy season busy-hour load. (COM/TA) 973-1990w

busy slot A slot that contains information and is not available for Queued Arbitrated (QA) access. (LM/C) 8802-6-1994

busy test (telephone switching systems) A test made to determine if certain facilities, such as a line, link, junctor, trunk, or other servers, are available for use. (COM) 312-1977w, [48]

busy time In computer performance engineering, the period of time during which a system or component is operational, in service, and in use. *See also:* idle time; down time; setup time. (C) 610.12-1990

busy tone *See:* audible busy signal.

busy verification (telephone switching systems) A procedure for checking whether or not a called station is in use or out-of-order. (COM) 312-1977w

butt contacts An arrangement in which relative movement of the cooperating members is substantially in a direction perpendicular to the surface of contact. *See also:* contactor. (IA/ICTL/IAC) [60], [84]

Butterworth filter A filter whose pass-band frequency response has a maximally flat shape brought about by the use of Butterworth polynomials as the approximating function. (CAS) [13]

butt ground *See:* structure base ground.

butt joint (waveguides) A connection between two waveguides or transmission lines that provides physical contact between the ends of the waveguides in order to maintain electric continuity. *See also:* waveguide. (MTT) 147-1979w

button A generic term for any control that initiates an action when pressed. (C) 1295-1993w

button device *See:* choice device.

buyer (A) An individual or organization responsible for acquiring a product or service (for example, a software system) for use by themselves or other users. *See also:* customer. **(B)** The person or organization that accepts the system and pays for the project. (C/SE) 1362-1998

buzz A disturbance of relatively short duration, but longer than a specified value as measured under specified conditions. *Note:* For the specified values and conditions, guidance should be found in documents of the International Special Committee on Radio Interference. *See also:* electromagnetic compatibility. (EMC/IM) [53], [76]

buzzer A signaling device for producing a buzzing sound by the vibration of an armature. (EEC/PE) [119]

buzz stick A device for testing suspension insulator units for fault when the units are in position on an energized line. *Note:* It consists of an insulating stick, on one end of which are metal prongs of the proper dimensions for spanning and short-circuiting the porcelain of one insulator unit at a time, and thereby checking conformity to normal voltage gradient. *See also:* tower. (T&D/PE) [10]

BW *See:* backward-wave structure.

by-link (telephone switching systems) A temporary connection between trunks and registers set up before the normal connection between them can be established. (COM) 312-1977w

bypass (1) (hydroelectric power plants) A means to pass the flow of water around a turbine to the same discharge outlet. *See also:* jumper. (PE/EDPG) 1020-1988r
(2) The state of the station attachment when the TCU does not route the station signals onto the trunk ring. Instead, the station signals will be returned to the station for lobe testing, and trunk signals will continue along the trunk. (C/LM) 11802-4-1994
(3) *See also:* jumper. (T&D/PE) 516-1995

BYPASS A defined instruction for the test logic defined by IEEE Std 1149.1-1990. (TT/C) 1149.1-1990

bypass capacitor A capacitor for providing an alternating-current path of comparatively low impedance around some circuit element. *Synonym:* bypass condenser. (EEC/PE) [119]

bypass condenser* *See:* bypass capacitor.

* Deprecated.

bypass contacts For reactance-type load tap changers (LTCs), a set of through current-carrying contacts that commutates the current to the transfer contacts without any arc. (PE/TR) C57.131-1995

bypass current The current flowing through the bypass switch, protective device, or other devices, in parallel with the series capacitor. (PE/T&D) 824-1994

bypass device (series capacitor) A device such as a switch or circuit breaker used in parallel with a series capacitor and its protective device to shunt line current for some specified time, or continuously. This device may also have the capability of inserting the capacitor into a circuit and carrying a specified level of current. (T&D/PE) 824-1985s

bypass gap A gap, or systems of gaps, to protect either the capacitor against overvoltage or the varistor against thermal overload, by carrying load or fault current around the protected equipment for some specified time. The bypass gap normally consists of a power gap and a trigger circuit. (T&D/PE) 824-1994

bypass interlocking device A device that requires all three phases of the switch to be in the same open or closed position. (T&D/PE) 824-1994

bypass/isolation switch (emergency and standby power) A manually operated device used in conjunction with an automatic transfer switch to provide a means of directly connecting load conductors to a power source and of disconnecting the automatic transfer switch. (IA/PSE) 446-1995

bypass key A signalling pattern sent by a fibre optic station to leave the ring and enter the BYPASS state. This pattern consists of a low light-level detected at the FOTCU for greater than 4 ms. (C/LM) 11802-4-1994

bypass switch A device such as a switch or circuit breaker used in parallel with a series capacitor and its protective device to shunt line current for some specified time or continuously. This device may also have the capability of inserting and bypassing the capacitor into a circuit carrying a specified level of current. (T&D/PE) 824-1994

byproduct energy (power operations) Electric energy produced as a byproduct incidental to some other operation. (PE/PSE) 858-1987s

byte (1) (programmable instrumentation) A group of adjacent binary digits operated on as a unit and usually shorter than a computer word (frequently connotes a group of eight bits). (IM/C/Std100/AIN) 488.1-1987r, 610.5-1990w, 1084-1986w, 610.12-1990, 610.7-1995

(2) (signals and paths) (microcomputer system bus) (MULTIBUS) A group of eight adjacent bits operated on as a unit. (MM/C) 796-1983r, 1296-1987s

(3) (software) An element of computer storage that can hold a group of bits. *See also:* bit; word. (C/Std100) 610.12-1990, 610.7-1995

(4) (696 interface devices) (signals and paths) A set of bit-parallel signals corresponding to binary digits operated on as a unit. Connotes a group of eight bits where the most significant bit carries the subscript 7 and the least significant bit carries the subscript 0. (MM/C) 696-1983w

(5) (STEbus) A set of eight signals, individually referred to as bits, which are operated on as a unit. (C/MM) 1000-1987r

(6) (NuBus) A set of 8 signals or bits taken as a unit. (C/MM) 1196-1987w

(7) A set of eight adjacent binary digits. (C/BA) 896.2a-1994w, 10857-1994, 896.3-1993w, 896.4-1993w

(8) A group of adjacent binary digits operated on as a unit. Usually 8 b. (SUB/PE/MM/C) 999-1992w, 959-1988r

(9) A unit of computer data consisting of 8 bits. (BA/C) 1275-1994

(10) A set of eight bit-parallel signals corresponding to binary digits operated on as a unit. The most significant bit carries the index value 7 and the least significant bit carries the index value 0. (C/BA) 1496-1993w

(11) Eight consecutive bits of data. *Note:* A byte is not necessarily equivalent to a character. (C/MM) 1754-1994

(12) An ordered set of eight binary digits (bits). (C/BA) 1014.1-1994w

(13) Eight bits of data. (C/MM/IM/ST) 1212-1991s, 1596.3-1996, 1394-1995, 1596.5-1993, 1596-1992, 1451.2-1997

(14) An individually addressable unit of data storage that is equal to or larger than an octet, used to store a character or a portion of a character.A byte is composed of a contiguous sequence of bits, the number of which is implementation defined. The least significant bit is called the *low-order* bit; the most significant is called the *high-order* bit. *Note:* This definition of *byte* is actually from the C Standard because POSIX.1 merely references it without copying the text. It has been reworded slightly to clarify its intent without introducing the C Standard terminology "basic execution character set," which is inapplicable to this standard. It deviates intentionally from the usage of *byte* in some other standards, where it is used as a synonym for *octet* (always 8 b). On a POSIX.1 system, a byte may be larger than 8 b so that it can be an integral portion of larger data objects that are not evenly divisible by 8 b (such as a 36 b word that contains four 9 b bytes). (C/PA) 1003.2-1992s

(15) A sequence of bits transmitted over a serial line. (C/PA) 1003.5-1992r

(16) A unit of machine storage containing an ordered sequence of 8 b. (C/PA) 1328-1993w, 1224-1993w, 1327-1993w

(17) A set of eight signals or bits taken as a unit. (C/MM) 1596.4-1996

(18) By common usage the term "byte" usually refers to eight bits of data, but within this context the size of a byte is implementation defined subject to the constraints given in IEEE Std 1003.5b-1996. The size of a byte is given by the constant POSIX.Byte_Size. *Note:* In the context of serial I/O, transmitting a byte of data may require the transmission of more bits than the size of a byte in memory, since for example stop bits and parity bits might be included. (C/PA) 1003.5b-1995

(19) A binary bit string that is operated on as a unit and is usually eight bits long and capable of holding one character in the local character set. (C/MM/ED) 855-1990, 1005-1998

(20) An entity composed of 8 bits, used to define a unit element of memory or transmitted data. It is capable of describing integers in the decimal range −128 to 127. (C/MM) 1284.1-1997

(21) A group of eight adjacent bits that function as a single unit. *See also:* octet. (PE/SUB) 1379-1997

(22) An 8 bit value. (C/MM) 1284.4-2000

byte clear Operation that sets all bits in an addressed byte of memory to a common "1" state. (ED) 1005-1998

byte clock A clock derived from the bit clock with a period equal to 10 clock cycles. (C/BA) 1393-1999

byte lane (1) A data path formed by eight data lines and one parity line, used to carry a single byte among the system modules. (C/BA) 10857-1994, 896.4-1993w, 896.3-1993w

(2) A data path formed by eight data lines and one parity line (if parity is implemented) that is used to carry a single byte of information among the system modules. (C/BA) 1014.1-1994w

byte select transistor The transistor, controlled by the wordline or row select line, that isolates each byte from the other bytes along the control line (each row of bytes). (ED) 1005-1998

Byte Mode An asynchronous, byte-wide reverse (peripheral-to-host) channel using the eight data lines of the interface for data and the control/status lines for handshaking. Byte Mode is used with Compatibility Mode to implement a bidirectional channel, with transfer direction controlled by the host when the host and peripheral both support bidirectional use of the data lines. The two modes cannot be active simultaneously. (C/MM) 1284-1994

byte serial (programmable instrumentation) (696 interface devices) (signals and paths) A sequence of bit-parallel data bytes used to carry information over a common bus. (IM/MM/C/AIN) 488.1-1987r, 696-1983w

byte Slave An SBus Slave having a data path only through bits D[31:24] of the data bus. (C/BA) 1496-1993w

C

c *See:* centi.

C A programming language, standardized by ANSI and ISO, designed for systems programming but also well-suited for general problem solving. Features include concise expressions, well-designed control flow and data structures, and a broad range of operators. *Note:* B is an ancestor of C. *See also:* C++; ANSI C; block-structured language. (C) 610.13-1993w

C++ A general-purpose programming language based on C, characterized by having facilities for performing object-oriented programming. (C) 610.13-1993w

cabinet (1) An enclosure designed either for surface or flush mounting and provided with a frame, mat, or trim in which a swinging door or doors are or may be hung. (NESC/NEC) [86]
(2) (power system communication equipment) An enclosure provided with an internal equipment mounting rack and hinged doors. (PE/PSC) 281-1984w
(3) (electronic) A protective enclosure to house modules, backplane(s), I/O connector assemblies, internal cables, and other electronic, mechanical, and thermal devices. *Synonyms:* rack; box; rack; box. (BA/C) 14536-1995

cabinet interface The cabinet interface provides within a cabinet the standardized mechanisms for mounting modules, backplane(s), and I/O connector assemblies, and for mounting other electronic, mechanical and thermal devices. The cabinet interface provides heat exchanging facilities, with standardized capabilities for transferring heat from modules and other heat sources within the cabinet, to external heat sink(s). (C/BA) 14536-1995

cabinet for safe (burglar-alarm system) Usually a wood enclosure, having protective linings on all inside surfaces and traps on the doors, built to surround a safe and designed to produce an alarm condition in a protection circuit if an attempt is made to attack the safe. *See also:* protective signaling. (EEC/PE) [119]

cable (1) (signal-transmission system) A transmission line or group of transmission lines mechanically assembled into a complex flexible form. *Note:* The conductors are insulated and are closely spaced and usually have a common outer cover which may be an electric portion of the cable. This definition also includes a twisted pair. (MTT) 146-1980w
(2) (communication and control cables) An insulated conductor or combination of electric conductors that are insulated from each other. A shield is usually provided. (PE/PSC) 789-1988w
(3) An assembly of one or more conductors within an enveloping protective sheath, constructed to allow use of the conductors separately or in groups. *See also:* transceiver cable; coaxial cable; drop cable; trunk cable; attachment unit interface cable; twinaxial cable; optical cable. (C) 610.7-1995
(4) A conductor with insulation, or a stranded conductor with or without insulation and other coverings (single-conductor cable) or a combination of conductors insulated from one another (multiple-conductor cable). *See also:* spacer cable. (NESC/T&D) C2-1997, C2.2-1960
(5) (fiber optics) *See also:* optical cable. 812-1984w
(6) *See also:* conductor. (PE/T&D) 524-1992r

cable accessories (power cable systems) Those components of a cable system which cannot be readily disconnected from the cable and which will be subjected to the full test voltage applied to the cable system. (PE) 400-1980s

cable armor A metallic element or envelope inserted in or around a cable sheath to provide mechanical protection against rodents, severe installation conditions, etc. (PE/PSC) 789-1988w

cable assembly *See:* optical cable assembly; multifiber cable.

cable attenuation *See:* cable tilt.

cable bedding (power distribution, underground cables) A relatively thick layer of material, such as a jute serving, between two elements of a cable to provide a cushion effect, or gripping action, as between the lead sheath and wire armor of a submarine cable. (PE) [4]

cable bond An electric connection across a joint in the armor or lead sheath of a cable, or between the armor or lead sheath and the earth, or between the armor or sheath of adjacent cables. *See also:* continuity cable bond; cross cable bond. (EEC/PE) [119]

cable buggy *See:* conductor car.

cable bus An approved assembly of insulated conductors with fittings and conductor terminations in a completely enclosed, ventilated protective metal housing. The assembly is designed to carry fault current and to withstand the magnetic forces of such current. Cablebus shall be permitted at any voltage or current for which the spaced conductors are rated. Cablebus is ordinarily assembled at the point of installation from components furnished or specified by the manufacturer in accordance with instructions for the specific job. (NESC/NEC) [86]

cable car A seat or basket-shaped device, designed to be suspended by a framework, and two or more sheaves arranged in tandem to enable a workman to ride a single conductor, wire, or cable. *Synonyms:* conductor car; bicycle; cable trolley. (T&D/PE) 524-1992r

cable charging current Current supplied to an unloaded cable. *Note:* Current is expressed in rms amperes. (SWG/PE) C37.100-1992

cable clamp A device designed to clamp cables together. It consists of a "U" bolt threaded on both ends, two nuts, and a base, and is commonly used to make temporary *bend back* eyes on wire rope. *Synonym:* Crosby clip. (T&D/PE) 524-1992r

cable complement (communication practice) A group of pairs in a cable having some common distinguishing characteristic. *See also:* cable. (EEC/PE) [119]

cable connection assembly The combination of the cable termination with the cable connection enclosure, GIS conductor end and removable conductor link. (PE/IC) 1300-1996

cable connection enclosure The part of the GIS which surrounds the cable termination. (PE/IC) 1300-1996

cable core (1) (A) (cable) The portion lying under other elements of a cable. **(B)** The core of a cable is the cylindrical center consisting of insulated conductors usually twisted together in pairs and in pairs arranged together in groups that lie under the sheath. (T&D/PE/PSC) [10], 789-1988
(2) The portion of a cable that includes the conductor, the conductor shield, the insulation, and the extruded insulation shield. (PE/IC) 1142-1995

cable core binder A wrapping of tapes or cords around the several conductors of a multiple-conductor cable used to hold them together. *Note:* Cable core binder is usually supplemented by an outer covering of braid, jacket, or sheath. (T&D/PE) [10]

cable coupler (rotating machinery) A form of termination in which the ends of the machine winding are connected to the supply leads by means of a plug-and-socket device. (PE) [9]

cable entrance fitting (pothead) A fitting used to seal or attach the cable sheath or armor to the pothead. *Note:* A cable entrance fitting is also used to attach and support the cable sheath or armor where a cable passes into a transformer removable cable terminating box without the use of potheads. *See also:* transformer. (PE/TR) [107], [108]

cable fill The ratio of the number of pairs in use to the total number of pairs in a cable. *Note:* The maximum cable fill is the percentage of pairs in a cable that may be used safely and economically without serious interference with the availability and continuity of service. *See also:* cable. (EEC/PE) [119]

cable filler (1) The material used in multiple-conductor cables to occupy the interstices formed by the assembly of the insulated conductors, thus forming a cable core of the desired shape (usually circular). (T&D/PE) [10]
(2) (communication and control cables) The material used in multiple-pair cables to occupy the interstices formed by the assembly of the insulated conductors, and to form a cable core of the desired shape (usually circular). A material may be used that resists the entrance of water (nonhydroscopic) and ionizes at a higher voltage than air. (PE/PSC) 789-1988w

cable-fire break Material, devices, or an assembly of parts installed in a cable system, other than at a cable penetration of a fire-resistive barrier, to prevent the spread of fire along the cable system. (PE/SUB/EDPG) 690-1984r, 525-1992r

cableheads *See:* submersible entrance terminals.

cable in free air That portion of a cable not routed in either a raceway or an enclosure. (PE/NP) 384-1992r

cable in the zone of influence (wire-line communication facilities) A high dielectric cable which provides high-voltage insulation between conductors and between conductors and shield. (PE/PSC) 487-1980s

cable jacket (1) A protective covering over the insulation, core, or sheath of a cable. (NESC) C2-1997
(2) (communication and control cables) A thermoplastic or thermosetting covering that is extruded over a cable to provide physical protection and electrical insulation. *(A) inner jacket.* A jacket that is extruded over the cable core covering to provide additional dielectric strength when it is needed between the conductors and the shield. An inner jacket may be used in cables that are used for direct burial and also where high ground potential rise is to be withstood. *(B) outer jacket.* A jacket that is extruded over the cable shield. It also may be extruded over both the shield and a supporting messenger cable. (PE/PSC) 789-1988w

cable joint (1) (cable splice) A connection between two or more separate lengths of cable with the conductors in one length connected individually to conductors in other lengths and with the protecting sheaths so connected as to extend protection over the joint. *Note:* Cable joints are designated by naming the conductors between which the joint is made, for example, 1 single-conductor to 2 single-conductor cables; 1 single-conductor to 3 single-conductor cables; 1 concentric to 2 concentric cables; 1 concentric to 1 single-conductor cable; 1 concentric to 2 single-conductor cables; 1 concentric to 4 single-conductor cables; 1 three-conductor to 3 single-conductor cables. *See also:* branch joint; reducing joint. (T&D/PE) [10]
(2) (power cable joints) A complete insulated splice or group of insulated splices contained within a single protective covering or housing. In some designs, the insulating material may also serve as the protective covering. Insulated end caps are considered joints in this context. *See also:* straight joint; branch joint; transition joint; insulating (isolating) joint. (PE/IC) 404-1986s

cable Morse code A three-element code, used mainly in submarine cable telegraphy, in which dots and dashes are represented by positive and negative current impulses of equal length, and a space by absence of current. *See also:* telegraphy. (EEC/PE) [119]

cable penetration (cable-penetration fire stops, fire breaks, and system enclosures) (nuclear power generating station) An assembly or group of assemblies for electrical conductors to enter and continue through a fire-rated structural wall, floor, or floor-ceiling assembly. (PE/SUB/EDPG) 690-1984r, 525-1992r, 634-1978w

cable-penetration fire stop Material, devices, or an assembly of parts providing cable penetrations through fire-rated walls, floors, for floor-ceiling assemblies, while maintaining required fire rating. (PE/SUB/IC/EDPG) 848-1996, 690-1984r, 525-1992r

cable PHY *See:* cable physical layer.

cable physical layer The version of the physical layer applicable to the Serial Bus cable environment. (C/MM) 1394-1995

cable powered Supplying power to active CATV equipment (for example, amplifiers) from the coaxial cable. This ac or dc power does not interfere with the RF information signal. (LM/C) 802.7-1989r

cable pullback The pulling of one or more cables out of a conduit system for the express purpose of repulling the cables into the same conduit. *Note:* Cable pullback is normally performed to allow relocation of a portion of a conduit system or to avoid pullbys during the installation of additional cables. (PE/IC) 1185-1994

cable pullby The pulling of cable(s) into a conduit that already contains one or more cables. (PE/IC) 1185-1994

cable rack A device usually secured to the wall of a manhole, cable raceway, or building to provide support for cables. (T&D/PE) [10]

cable reel A drum on which conductor cable is wound, including one or more collector rings and associated brushes, by means of which the electric circuit is made between the stationary winding on the locomotive or other mining device and the trailing cable that is wound on the drum. *Note:* The drum may be driven by an electric motor, a hydraulic motor, or mechanically from an axle on the machine. *See also:* mine feeder circuit. (PE/EEC/MIN) [119]

cable segment (FASTBUS acquisition and control) A FASTBUS segment consisting of a cable together with appropriate connectors for mating with devices. (NID) 960-1993

cable separator (power distribution, underground cables) A serving of threads, tapes, or films to separate two elements of the cable, usually to prevent contamination or adhesion. (PE) [4]

cable sheath (1) A tubular impervious metallic protective covering applied directly over the cable core. (PE/T&D/IC) [4], [10]
(2) (communication and control cables) The outer covering over the insulated conductors to provide mechanical and electrical protection for the conductors. In telephone-type cables, the sheath usually includes a shield, and may include armor. (PE/PSC) 789-1988w
(3) A conductive protective covering applied to cables. *Note:* A cable sheath may consist of multiple layers, of which one or more is conductive. (NESC/T&D) C2-1997, C2.2-1960

cable sheath insulator (pothead) An insulator used to insulate an electrically conductive cable sheath or armor from the metallic parts of the pothead or transformer removable cable terminating box in contact with the supporting structure for the purpose of controlling cable sheath currents. *See also:* transformer; transformer removable cable-terminating box. (PE/TR) [107], [108]

cable shield (communication and control cables) A conducting envelope, composed of metal strands, ribbon or sheet metal that encloses a wire, group of wires, or cable, so constructed that substantially every point on the surface of the underlying insulation or core wrap is at ground potential or at some predetermined potential with respect to ground. *See also:* duct edge fair-lead. (PE/PSC) 789-1988w

cable shielding (nuclear power generating station) (cable systems in power generating stations) (shielding and shield grounding) A nonmagnetic metallic material applied over the insulation of the conductor or conductors to confine the electric field of the cable to the insulation of the conductor or conductors. (PE/EDPG) 422-1977, 690-1984r

cable splicer A short piece of tubing or a specially formed band of metal generally used without solder in joining ends of portable cables for mining equipment. *See also:* mine feeder circuit. (PE/EEC/MIN) [119]

cable spreading room (cable systems) The cable spreading room is normally the area adjacent to the control room where cables leaving the panels are dispersed into various cable trays for routing to all parts of the plant. (PE/EDPG) 422-1977

cable-system enclosure (nuclear power generating station) (cable-penetration fire stops, fire breaks, and system enclosures) An assembly installed around a cable system to maintain circuit integrity, for a specified time, of all circuits within the enclosure when it is exposed to the most severe fire that may be expected to occur in the area. (PE/SUB/EDPG) 690-1984r, 525-1992r

cable terminal (1) A device that provides insulated egress for the conductors. *Synonyms:* termination. (NESC) C2-1997
(2) (power work) A device that seals the end of a cable and provides insulated egress for the conductors. *Synonyms:* pothead; end bell. (PE/T&D) [10]

cable termination Parts assembled onto the end of the cable to provide the electrical and mechanical interface into the gas-insulated environment. Typically this includes a solid insulation barrier between the cable/cable fluid and the gas insulation of the GIS. (PE/IC) 1300-1996

cable tilt (loss) The amount of RF signal attenuation by a given coaxial cable. Cable attenuation is mainly a function of signal frequency, cable length, and diameter. Cables attenuate higher frequency signals more than lower frequency signals (tilt). Cable losses are usually referenced to the highest frequency carried (greatest loss) on the cable. (LM/C) 802.7-1989r

cable tray (1) (raceway systems for Class 1E circuits for nuclear power generating stations) A prefabricated metal raceway with or without covers consisting of siderails and bottom support sections. Bottom support sections may be ladder, trough, or solid. (PE/NP) 628-1987r
(2) (electric power systems in commercial buildings) A unit or assembly of units or sections, and associated fittings, made of metal or other noncombustible material forming a continuous rigid structure used to support cables. (IA/PSE) 241-1990r
(3) A raceway resembling a ladder and usually constructed of metal. Other styles of trays include solid-bottom and channel type. (PE/IC) 848-1996
(4) A continuous rigid structure used to support cables. Cable trays include ladders, troughs, channels and other similar structures. Conduits are not included in this category. (PE/IC) 817-1993w

cable tray system (raceway systems for Class 1E circuits for nuclear power generating stations) An assembly of metallic cable tray sections, fittings, supports, anchorages, and accessories that form a structural system to support wire and cables. (PE/NP) 628-1987r

cable trolley *See:* cable car.

cable TV A communication system that simultaneously distributes several different channels of broadcast programs and other information to customers via a coaxial cable. Previously called community antenna television (CATV). (LM/C) 802.7-1989r

cable type (nuclear power generating station) A cable type for purposes of qualification testing shall be representative of those cables having the same materials, similar construction, and service rating, as manufactured by a given manufacturer. (PE/NP) 380-1975w

cable value *See:* manhole.

cab signal (1) A signal located in the engineman's compartment or cab indicating a condition affecting the movement of a train or engine and used in conjunction with interlocking signals and in conjunction with or in lieu of block signals. *See also:* automatic train control. (EEC/PE) [119]
(2) (system) A signal located in the cab, indicating a condition affecting the movement of a train and used in conjunction with interlocking signals and in conjunction with or in lieu of block signals. (VT) 1475-1999

cache (1) A buffer inserted between one or more processors and the bus, used to hold currently active copies of blocks from main memory. (C/BA) 896.3-1993w
(2) A small portion of high-speed memory used for temporary storage of frequently-used data, instuctions, or operands. *See also:* instruction cache; disk cache; high-speed buffer; caching; cache architecture; data cache; cache memory. (C) 610.10-1994w
(3) *See also:* copy. (C/PA) 1328.2-1993w, 1224.2-1993w

cache coherence A system of caches is said to be coherent with respect to a cache line if each cache and main memory in the coherence domain observes all modifications of that same cache line. A modification is said to be observed by a cache when any subsequent read would return the newly written value. (C/BA) 1014.1-1994w, 10857-1994, 896.3-1993w, 896.4-1993w

cache agent A module that uses split transactions to assume all the rights and responsibilities of some number of remote cache modules. (C/BA) 896.4-1993w

cache line (1) Often called simply a "line." The unit of data on which coherence checks are performed, and for which coherence tag information is maintained. In SCI, a line consists of 64 data bytes. (MM/C) 1596-1992
(2) Often called simply a "line." The block of memory (sometimes called a "sector") that is managed as a unit for coherence purposes; i.e., cache tags are maintained on a per-line basis. SCI directly supports only one line size, 64 bytes. (C/MM) 1596.5-1993
(3) Often simply called a "line," the block of memory (sometimes called a sector) that is managed as a unit for coherence purposes; i.e., cache tags are maintained on a per-line basis. Although the SCI line size influenced the RamLink packet sizes, coherence protocols are beyond the scope of this standard. (C/MM) 1596.4-1996

cache architecture (A) A computer architecture that employs an extremely high-speed memory block, called a cache, in which data is stored. **(B)** The organization of cache memory; for example, direct mapped cache, two-way set associative cache. (C) 610.10-1994

cache hit *See:* hit.

caching The process of accessing a cache. (C) 610.10-1994w

cache memory (1) A buffer memory inserted between one or more processors and the bus, which is used to hold currently active copies of blocks of information from main memory. (C/BA) 1014.1-1994w
(2) A buffer memory inserted between one or more processors and the bus, used to hold currently active copies of blocks from main memory. Cache memories exploit spatial locality by what is brought into a cache. Temporal locality is exploited by the strategy employed for determining what is removed from the cache. (C/BA) 10857-1994, 896.4-1993w

CAD *See:* computer-aided design.

CADD *See:* computer-aided design and drafting.

CADEM *See:* computer-aided engineering; computer-aided manufacturing; computer-aided design.

CADF *See:* commutated antenna direction finder.

CADM *See:* computer-aided manufacturing; computer-aided design.

CAE *See:* computer-aided engineering; computer-aided education.

cage (1) A system of conductors forming an essentially continuous conducting mesh or network over the object protected and including any conductors necessary for interconnection to the object protected and an adequate ground. *See also:* Faraday cage. (EEC/PE) [119]
(2) emptydef;. *See also:* aerial platform. (T&D/PE) 524-1992r

cage antenna A multi-wire element whose wires are so disposed as to resemble a cylinder, in general of circular cross section; for example, an elongated cage. (AP/ANT) 145-1993

cage synchronous motor (rotating machinery) A salient pole synchronous motor having an amortisseur (damper) winding embedded in the pole shoes, the primary purpose of this winding being to start the motor. (PE) [9]

cage winding *See:* squirrel-cage winding.

caging (gyros) The process of orienting and mechanically locking one or more gyro axes or gimbal axes to a reference position. (AES/GYAC) 528-1994

CAI *See:* computer-assisted instruction; computer-aided inspection; computer-aided instruction.

CAL *See:* Conversational Algebraic Language; computer-assisted learning; computer-augmented learning.

calc algorithm *See:* hash function.

calc chain *See:* collision chain.

calculating punch A calculator, with card reader and card punch, that reads data from a punch card, performs some arithmetic operations or logic operations on the data, and punches the results on the same or another punch card. *Synonym:* multiplying punch. (C) 610.10-1994w

calculations (International System of Units (SI)) Errors in calculations can be minimized if the base and the coherent derived International System (SI) units are used and the resulting numerical values are expressed in power-of-ten notation instead of using prefixes. *See also:* prefixes and symbols; units and letter symbols. (QUL) 268-1982s

calculator (1) (A) A device capable of performing arithmetic. **(B)** A calculator as in definition (A) that requires frequent manual intervention. **(C)** Generally and historically, a device for carrying out logic and arithmetic digital operations of any kind. (C) [85]
(2) A device that is suitable for performing logic and arithmetic digital operations, but that requires manual intervention to initiate each operation. *See also:* calculating punch. (C) 610.10-1994w

calibrate (1) (monitoring radioactivity in effluents) Adjustment of the system and the determination of system accuracy using one or more sources traceable to the National Bureau of Standards (NBS). (NI) N42.18-1980r
(2) (radiological monitoring instrumentation) (plutonium monitoring) To determine the response or reading of an instrument relative to a series of known radiation values over the range of the instrument. (NI) N317-1980r, N320-1979r
(3) (radiation protection) To determine the response or reading of an instrument relative to a series of known radiation values over the range of the instrument or the strength of a radiation source relative to a standard. (NI) N323-1978r
(4) (airborne radioactivity monitoring) To adjust or determine or both: The response or reading of an instrument relative to a series of conventionally true values; or The strength of a radiation source relative to a standard or conventionally true value. (NI) N42.17B-1989r

calibrated Checked for proper operation at selected points on the operating characteristic. (IA/ICTL/IAC) [60]

calibrated-driving-machine test (rotating machinery) A test in which the mechanical input or output of an electric machine is calculated from the electric input or output of a calibrated machine mechanically coupled to the machine on test. *See also:* direct-current commutating machine; asynchronous machine. (PE) [9]

calibrated Marinelli beaker standard source (germanium semiconductor detector) A calibrated MBSS is an MBSS that has been calibrated by comparing its photon emission rate to that of a certified MBSS. *Note:* The photon emission rate as used in this standard is the number of photons per second resulting from the decay of radionuclides in the source and is thus higher than the detected rate at the surface. (NPS) 680-1978w

calibrated-solution Marinelli beaker standard source (germanium semiconductor detector) A calibrated-solution MBSS is a standard beaker that contains as its radioactive filling material a solution that has been calibrated by comparing its photon emission rate at specified energies to that of a certified solution. *Note:* The photon emission rate as used in this standard is the number of photons per second resulting from the decay of radionuclides in the source and is thus higher than the detected rate at the surface. (NPS) 680-1978w

calibrated source *See:* radioactivity standard source.

calibration (1) (nuclear power generating station) Comparison of items of measuring and test equipment with reference standards or with items of measuring and test equipment of equal or closer tolerance to detect and quantify inaccuracies and to report or eliminate those inaccuracies. (PE/NP) 498-1985s
(2) (supervisory control, data acquisition, and automatic control) Adjustment of a device so that the output is within a specific range for particular values of the input. (PE/SUB) C37.1-1994
(3) The adjustment of a device to have the designed operating characteristics, and the subsequent marking of the positions of the adjusting means, or the making of adjustments necessary to bring operating characteristics into substantial agreement with standardized scales or marking. (SWG/PE/PSR) C37.100-1992, C37.90-1978s, [56], [6]
(4) (metering) Comparison of the indication of the instrument under test, or registration of the meter under test, with an appropriate standard. (ELM) C12.1-1988
(5) The process of determining the numerical relationship, within an overall stated uncertainty, between the observed output of a measurement system and the value, based on standard sources, of the physical quality being measured. (EMC/NI) 1140-1994r, N42.13-1986
(6) (germanium spectrometers) The determination of a value that converts a measured number into a desired physical quantity (e.g., pulse height into photon energy, or counts per second into emission rate). (NI) N42.14-1991

calibration error (1) (electric pipe heating systems) In operation, the departure under specified conditions of actual performance from performance indicated by scales, dials, or other markings on the device. *See also:* alarm signal. (PE/EDPG) 622A-1984r, 622B-1988r
(2) In the operation of a device, the departure, under specified conditions, of actual performance from performance indicated by scales, dials, or other markings on the device. *Note:* The indicated performance may be by calibration markings in terms of input or performance quantities (amperes, ohms, seconds, etc.) or by reference to a specific performance data recorded elsewhere. *See also:* setting error. (SWG/PE) C37.100-1992

calibration factor (A) (bolometer-coupler unit) The ratio of the substitution power in the bolometer attached to the side arm of the directional coupler to the microwave power incident on a nonreflecting load connected to the output port of the main arm of the directional coupler. *Notes:* 1. If the bolometer unit is attached to the main arm of the directional coupler, the calibration factor is the ratio of the substitution power in the bolometer unit attached to the main arm of the directional coupler to the microwave power incident upon a nonreflecting load connected to the output port of the side arm of the directional coupler. **(B) (bolometer units)** The ratio of the substitution power to the radio-frequency power incident upon the bolometer unit. The ratio of the bolometer-unit calibration factor to the effective efficiency is determined by the reflection coefficient of the bolometer unit. The two terms are related as follows:

$$K_b/\eta_e = 1 - |\Gamma|^2$$

where K_b, η_e, and Γ are the calibration factor's effective efficiency, and reflection coefficient of the bolometer unit, respectively. **(C) (calibration)** (loosely called antenna factor) The factor or set of factors that, at given frequency, expresses the relationship between the field strength of an electromagnetic wave impinging upon the antenna of a field-strength

meter and the indication of the field-strength meter. *Note:* The composite of antenna characteristics, balun and transmission line effects, receiver sensitivity and linearity, etc. *See also:* measurement system. (IM/HFIM) [40], 284-1968
(2) (A) (electrothermic unit) The ratio of the substituted reference power (dc, audio, or rf) in the electrothermic unit to the power incident upon the electrothermic unit for the same dc output voltage from the electrothermic unit at a prescribed temperature. *Notes:* 1. Calibration factor and effective efficiency are related as in the equation above, where K_b, η_e, and Γ are the calibration factor, effective efficiency, and reflection coefficient of the electrothermic unit, respectively. 2. The reference frequency is to be supplied with the calibration factor. **(B) (electrothermic-coupler unit)** The ratio of the substituted reference power (dc, audio, or rf) in the electrothermic unit attached to the side arm of the directional coupler to the power incident upon a nonreflecting load connected to the output port of the main arm of the directional coupler for the same dc output voltage from the electrothermic unit is attached to the main arm of the directional coupler, the calibration factor is the ratio of the substituted reference (dc, audio, or rf) power in the electrothermic unit attached to the main arm of the directional coupler to the power incident upon a nonreflecting load connected to the output port of the side arm of the directional coupler for the same dc output voltage from the electrothermic unit at a prescribed temperature. *Note:* The reference frequency is to be supplied with the calibration factor. (IM) 544-1975

calibration interval (test, measurement, and diagnostic equipment) The maximum length of time between calibration services during which each standard and test and measuring equipment is expected to remain within specific performance levels under normal conditions of handling and use. (BT) 511-1979w

calibration level (signal generators) The level at which the signal generator output is calibrated against a standard. *See also:* signal generator. (IM/HFIM) [40]

calibration markers *See:* calibration marks.

calibration marks (radar) (navigation aids) Indications superimposed on a display to provide a numerical scale of the parameters displayed. (AES/GCS) 686-1997, 172-1983w

calibration procedure (test, measurement, and diagnostic equipment) A document which outlines the steps and operations to be followed by standards and calibration laboratory and field calibration activity personnel in the performance of an instrument calibration. (MIL) [2]

calibration programming (power supplies) Calibration with reference to power-supply programming describes the adjustment of the control-bridges current to calibrate the programming ratio in ohms per volt. *Note:* Many programmable supplies incorporate a calibrate control as part of the reference resistor that performs this adjustment. (AES/PE) [41], [78]

calibration scale A set of graduations marked to indicate values of quantities, such as current, voltage, or time at which an automatic device can be set to operate. (SWG/PE) C37.100-1992

calibration voltage The voltage applied during the adjustment of a meter. *See also:* test. (ELM) C12.1-1982s

calibrator (oscilloscopes) The signal generator whose output is used for purposes of calibration, normally either amplitude or time or both. (IM) 311-1970w

caliche (cable plowing) Common sedimentary rock normally formed from ancient marine life. (T&D/PE) 590-1977w

call (1) (computers) The action performed by the calling party, or the operations necessary in making a call, or the effective use made of a connection between two stations. (COM/C) [85]
(2) (telephone switching systems) A demand to set up a connection. (COM) 312-1977w
(3) (A) (software) A transfer of control from one software module to another, usually with the implication that control will be returned to the calling module. *Contrast:* go to. **(B) (software)** A computer instruction that transfers control from one software module to another as in definition (A) and, often, specifies the parameters to be passed to and from the module. **(C) (software)** To transfer control from one software module to another as in definition (A) and, often, to pass parameters to the other module. *Synonym:* cue. *See also:* call list; call by name; call by reference; call by value; calling sequence. (C) 610.12-1990
(4) (telecommunications) A measure of traffic intensity or event data. *Synonym:* call attempts per hour. *See also:* CCS; time-consistent traffic measures. (COM/TA) 973-1990w

call announcer (automatic telephone office) A device for receiving pulses and audibly reproducing the corresponding number in words so that it may be heard by a manual operator. (EEC/PE) [119]

call arrow An arrow that enables the sharing of detail between IDEF0 models (linking them together) or within an IDEF0 model. The tail of a call arrow is attached to the bottom side of a box. One or more page references are attached to a call arrow. (C/SE) 1320.1-1998

call attempts per hour A measure of traffic intensity or event data. *Synonym:* call. *See also:* time-consistent traffic measures; CCS. (COM/TA) 973-1990w

callback A function written as part of the application, associated with a specific widget resource, that is invoked as a result of a specific change of state associated with that widget. For example, the `XmNactivateCallback` resource of the `PushButton` widget points to the callback function that is called when the button is pushed. (C) 1295-1993w

call back A security procedure that verifies the identity of a terminal accessing a computer system by terminating the original connection and then reestablishing it by placing a new call to the terminal. (C) 610.7-1995

call by address *See:* call by reference.

call by location *See:* call by reference.

call by name A method for passing parameters, in which the calling module provides to the called module a symbolic expression representing the parameter to be passed, and a service routine evaluates the expression and provides the resulting value to the called module. *Note:* Because the expression is evaluated each time its corresponding formal parameter is used in the called module, the value of the parameter may change during the execution of the called module. *Contrast:* call by reference. (C) 610.12-1990

call by reference A method for passing parameters, in which the calling module provides to the called module the address of the parameter to be passed. *Note:* With this method, the called module has the ability to change the value of the parameter stored by the calling module. *Synonyms:* call by address; call by location. *Contrast:* call by value; call by name. (C) 610.12-1990

call by value A method of passing parameters, in which the calling module provides to the called module the actual value of the parameter to be passed. *Note:* With this method, the called module cannot change the value of the parameter as stored by the calling module. *Contrast:* call by name; call by reference. (C) 610.12-1990

call capacity The number of call attempts per busy hour that can be processed without exceeding various service standards. The capacity may alternatively be expressed in terms of originating-plus-incoming (O + I) calls that can be processed during the busy hour. (COM/TA) 973-1990w

call circuit (manual switching) A communication circuit between switching points used by the traffic forces for the transmission of switching instructions. (EEC/PE) [119]

call count Occurs when one digit of the called number is received by the system after dial tone or wink is sent. (COM/TA) 973-1990w

called diagram A decomposition diagram invoked by a calling box and identified by a page reference attached to a call arrow. (C/SE) 1320.1-1998

called-line release (telephone switching systems) Release under the control of the line to which the call was directed. (COM) 312-1977w

call forwarding (telephone switching systems) A feature that permits a customer to instruct the switching equipment to transfer calls intended for his or her station to another station. (COM) 312-1977w

call graph A diagram that identifies the modules in a system or computer program and shows which modules call one another. *Note:* The result is not necessarily the same as that shown in a structure chart. *Synonym:* tier chart; call tree. *Contrast:* structure chart. *See also:* data structure diagram; control flow diagram; state diagram; data flow diagram.

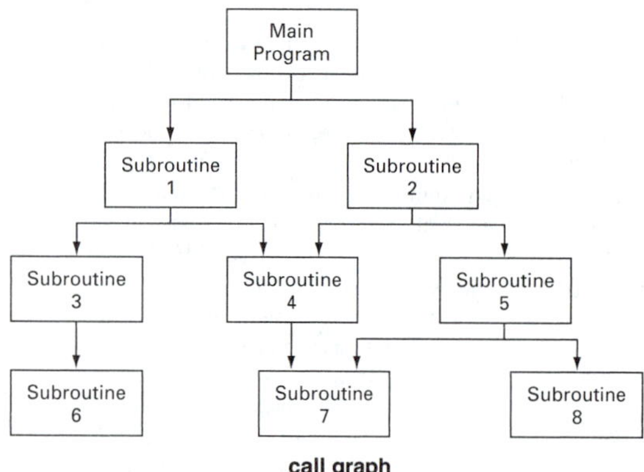

call graph

(C) 610.12-1990

call indicator A device for receiving pulses from an automatic switching system and displaying the corresponding called number before an operator at a manual switchboard. (EEC/PE) [119]

calling box A box that is detailed by a decomposition diagram that is not the box's child diagram. A call arrow is attached to the bottom of a calling box. (C/SE) 1320.1-1998

calling device (telephone switching systems) An apparatus that generates the signals required for establishing connections in an automatic switching system. (COM) 312-1977w

calling line identification (telephone switching systems) Means for automatically identifying the source of calls. (COM) 312-1977w

calling-line release (telephone switching systems) Release under the control of the line from which the call originated. (COM) 312-1977w

calling-line timed release (telephone switching systems) Timed release initiated by the calling line. (COM) 312-1977w

calling plug and cord A plug and cord that are used to connect to a called line. (EEC/PE) [119]

calling sequence (1) (computers) A specified arrangement of instructions and data necessary to set up and call a given subroutine. (C) [20], [85]
(2) (software) A sequence of computer instructions and, possibly, data necessary to perform a call to another module. (C) 610.12-1990

call list The ordered list of arguments used in a call to a software module. (C) 610.12-1990

call packing (telephone switching systems) A method of selecting paths in a switching network according to a fixed hunting sequence. (COM) 312-1977w

call rate (telephone switching systems) The number of calls per unit of time. (COM) 312-1977w

call reference A page reference attached to a call arrow. (C/SE) 1320.1-1998

call splitting (telephone switching systems) Opening the transmission path between the parties of a call. (COM) 312-1977w

call tone (telephone switching systems) A tone that indicates to an operator or attendant that a call has reached the position or console. (COM) 312-1977w

call trace *See:* subroutine trace.

call tracing (telephone switching systems) A means for manually identifying the source of calls. (COM) 312-1977w

call tree *See:* call graph.

call-type information (CTI) digits Digits sent to the switch from the central office service unit (COSU) via signaling on the utility telemetry trunk, per call, which specify the customer premise equipment (CPE) transmission interface (i.e., on-hook or off-hook operation). These information digits are assigned, on a trunk group basis, through the switch administration procedures. (SCC31/AMR) 1390.3-1999, 1390.2-1999, 1390-1995

call waiting (telephone switching systems) A feature providing a signal to a busy called line to indicate that another call is waiting. (COM) 312-1977w

call-waiting tone (telephone switching systems) A tone used in the call-waiting feature. (COM) 312-1977w

calomel electrode *See:* calomel half-cell.

calomel half-cell (calomel electrode) A half-cell containing a mercury electrode in contact with a solution of potassium chloride of specified concentration that is saturated with mercurous chloride of which an excess is present. *See also:* electrochemistry. (EEC/PE) [119]

calorie The quantity of heat required to raise one gram of water 1°F. (IA/PSE) 241-1990r

calorimeter (laser maser) A device for measuring the total amount of energy absorbed from a source of electromagnetic radiation. (LEO) 586-1980w

calorimetric test (rotating machinery) A test in which the losses in a machine are deduced from the heat produced by them. The losses are calculated from the temperature rises produced by this heat in the coolant or in the surrounding media. *See also:* asynchronous machine. (PE) [9]

CAM *See:* computer-aided management; computer-aided manufacturing.

CAMA *See:* centralized accounting, automatic message.

CAMAC *See:* computer automated measurement and control.

CAMAC branch driver *See:* CAMAC parallel highway driver.

CAMAC branch highway *See:* CAMAC parallel highway.

CAMAC compatible crate A mounting unit for CAMAC plug-in units that does not conform to the full requirements for a CAMAC crate but in which CAMAC modules can be mounted and operated in accordance with the dataway requirements of IEEE Std 583-1975. (NPS) 583-1975s

CAMAC crate A mounting unit for CAMAC plug-in units that includes a CAMAC dataway and conforms to the mandatory requirements for a CAMAC crate as specified in IEEE Std 583-1975. (NPS) 583-1975s

CAMAC crate assembly An assembly of a CAMAC crate controller and one or more CAMAC modules mounted in a CAMAC crate (or CAMAC compatible crate), and operable in conformity with the dataway requirements of IEEE Std 583-1975. (NPS) 583-1975s

CAMAC crate controller A functional unit that when mounted in the control station and one or more normal stations of a CAMAC crate (or CAMAC compatible crate) communicates with the dataway in accordance with IEEE Std 583-1975. (NPS) 583-1975s

CAMAC data array (subroutines for CAMAC) The symbol *intc* represents an array of CAMAC data words. Each element of *intc* has the same form as the CAMAC data word variable *int*. The length of *intc* is given by the value of the first element of *cb* at the time the subroutine is executed. *See also:* control block; CAMAC data word. (NPS) 758-1979r

CAMAC dataway An interconnection between CAMAC plug-in units which conforms to the mandatory requirements for a CAMAC dataway as specified in IEEE Std 583-1975. (NPS) 583-1975s

CAMAC data word (subroutines for CAMAC) The symbol *int* represents a CAMAC data word stored in computer memory. The form is not specified, but the word must be stored in an addressable storage entity capable of containing twenty-four bits. In a computer or programming system which does not have an addressable unit of storage which can contain twenty-four bits, multiple units must be used. (NPS) 758-1979r

CAMAC external addresses (subroutines for CAMAC) The symbol *exta* represents an array of integers each of which is a CAMAC register address. The form and information content of each element of *exta* must be identical to the form and information content of the quantity *ext*. The length of *exta* is given by the value of the first element of *cb* at the time the subroutine is executed. *See also:* control block; external address. (NPS) 758-1979r

CAMAC module A CAMAC plug-in unit that when mounted in one or more normal stations of a CAMAC crate is compatible with IEEE Std 583-1975. (NPS) 583-1975s

CAMAC parallel highway A standard highway (for a CAMAC system) in which the data is transferred in parallel and that conforms to the requirements of IEEE Std 596-1982. *Synonym:* CAMAC branch highway. (NPS) 583-1975s

CAMAC parallel highway driver A unit that communicates via the CAMAC parallel highway with up to seven CAMAC crates and conforms to the requirements as specified in IEEE Std 596-1982. *Synonym:* CAMAC branch driver. (NPS) 583-1975s

CAMAC plug-in unit A functional unit that conforms to the mandatory requirements for a plug-in unit as specified in IEEE Std 583-1975. (NPS) 583-1975s

CAMAC serial highway A standard highway (for a CAMAC system) in which the data is transferred in bit or byte serial and which conforms to the requirements of IEEE Std 595-1982. (NPS) 583-1975s

CAMAC system A system including at least one CAMAC crate assembly. (NPS) 583-1975s

CAMAL *See:* CAMbridge ALgebra system.

cambium A layer of delicate meristematic tissue between the inner bark and the wood that produces all secondary growth in plants and is responsible for the annual rings of wood. (T&D/PE) 751-1990

CAMbridge ALgebra system A programming language used to perform large scale formal algebraic manipulation, particularly in celestial mechanics and general relativity. (C) 610.13-1993w

cam contactor (cam switch) A contactor or switch actuated by a cam. *See also:* control switch. (VT/LT) 16-1955w

camera storage tube A storage tube into which the information is introduced by means of electromagnetic radiation, usually light, and read at a later time as an electric signal. *See also:* storage tube. (ED) 158-1962w, [45]

camera tube (television) A tube for conversion of an optical image into an electrical signal. (BT/AV) 201-1979w

cam-operated switch A switch consisting of fixed contact elements and movable contact elements operated in sequence by a camshaft. *See also:* switch. (IA/ICTL/IAC) [60], [84]

camping trailer A vehicular portable unit mounted on wheels and constructed with collapsible partial side walls that fold for towing by another vehicle and unfold at the campsite to provide temporary living quarters for recreational, camping, or travel use. *See also:* recreational vehicle. (NESC/NEC) [86]

camp-on busy (telephone switching systems) A feature whereby a call encountering a busy condition can be held and subsequently connected automatically when the busy condition is required. (COM) 312-1977w

cam-programmed (test, measurement, and diagnostic equipment) **(A)** A programming technique that uses a rotating shaft, having specifically oriented, eccentric projections that control a series of switches that set up the proper circuits for a test. **(B)** A cam-follower system used to set positions or values of a shafted instrument for programming instructions to the test system. (MIL) [2]

camshaft position (electric power system) The angular position of the main shaft directly operating the governor-controlled valves. (PE/PSE) 94-1991w

CAN *See:* cancel character.

can (1) (dry cell) A metal container, usually zinc, in which the cell is asserted and that serves as its negative electrode. *See also:* electrolytic cell. (PE/EEC) [119]
(2) An indication of a permissible optional feature or behavior available to the application; the implementation shall support such features or behaviors as mandatory requirements. (C/PA) 2003.2-1996

cancel (numerically controlled machines) A command that will discontinue any fixed cycles or sequence commands. (IA/EEC) [61], [74]

cancel character (CAN) **(A)** A control character used by some convention to indicate that the data with which it is associated are in error or are to be disregarded. **(B)** An accuracy control character used to indicate that the data with which it is associated are in error or are to be disregarded. (C) 610.5-1990

canceled video In moving-target indication (MTI), the video output remaining after the cancellation process. *See also:* clutter residue; moving-target indication. (AES) 686-1997

canceler That portion of the system in which unwanted signals, such as clutter, fixed targets, and other interference, are suppressed by a process of linear subtraction. (AES) 686-1997

cancellation ratio **(A)** In moving-target indication (MTI), the ratio of canceler voltage amplification for fixed-target echoes received with a fixed antenna, to the gain for a single pulse passing through the unprocessed channel of the canceler. *Note:* This measure of MTI performance, in which high performance is represented by a low numerical value, has been largely replaced by the MTI improvement factor. *See also:* moving-target indication. **(B)** In interference-reduction techniques other than MTI, the cancellation ratio is the ratio of interference output power in the absence of the technique to that when the technique is applied, when the system gains for the two cases are adjusted to provide equal noise outputs. *Note:* High performance is indicated by a high numerical value, generally expressed in decibels to avoid ambiguity between power and voltage ratios. (AES) 686-1997

cancelled video (radar moving-target indicator) The video output remaining after the cancellation process. *See also:* navigation. (AES/RS) 686-1982s, [42]

cancer promoter An agent that advances carcinogenesis after its initiation. (T&D/PE) 539-1990

candela (1) (illuminating engineering) The SI unit of luminous intensity. One candela is one lumen per steradian (lm/sr). Formerly, candle. *Notes:* 1. The fundamental luminous intensity definition in the SI is the candela in terms of a complete (blackbody) radiator. From this relation K_m and K_m, and consequently the lumen, are determined. One candela is defined as the luminous intensity of 1/600 000 of one square meter of projected area of a blackbody radiator operating at the temperature of solidification of platinum, at a pressure of 101 325 newtons per square meter (N/m^2 = PA). From 1909 until the introduction of the present photometric system on January 1, 1948, the unit of luminous intensity in the United States, as well as in France and Great Britain, was the "international candle," which was maintained by a group of carbon-filament vacuum lamps. For the present unit as defined above, the internationally accepted term is candela. The difference between the candela and the old international candle is so small that only measurements of high precision are affected. The following resolution was adopted at the Seizièeme Conférence Générale des Poids et Mesures (the Sixteenth General Conference on Weights and Measures) on October 11, 1979. The Conference had decided: (1) The candela is the luminous intensity, in a given direction, of a source emitting monochromatic radiation of frequency 540 × 10^{12} Hz and whose radiant intensity in this direction is 1/683 watt per steradian. (2) The candela so defined is the base unit applicable to photopic quantities, scotopic quantities, and quantities to be defined in the mesopic domain. *See also:* luminous flux. (EEC/IE) [126]

(2) (metric practice) The luminous intensity, in the perpendicular direction, of a surface of 1/600 000 square meters of blackbody at the temperature of freezing plantinum under a pressure of 101 325 newtons per square meter (adopted by the 13th General Conference on Weights and Measures 1967). (QUL) 268-1982s

(3) (television) The luminous intensity, in the perpendicular direction, of a 1/600 000 square meter surface of a blackbody at the freezing temperature of plantinum under a pressure of 101 325 pascals. *Notes:* 1. Values for standards having other spectral distributions are derived by the use of accepted spectral luminous efficiency data for photopic vision. 2. From 1909 until the introduction of the present photometric system on January 1, 1948, the unit of luminous intensity in the United States, as well as in France and Great Britain, was the international candle, which was maintained by a group of carbon-filament vacuum lamps. For the present unit as defined above, the internationally accepted term is candela. The difference between the candela and the old international candle is so small that only measurements of high precision are affected. (BT/AV) 201-1979w

candidate key (1) In a relational data model, any minimal set of attributes within a relation that forms a key that is a determinant of all attributes in the relation. *Note:* In normalization, one of the candidate keys of each relation is chosen as the primary key and the others are known as alternate keys. *See also:* compound key. (C) 610.5-1990w

(2) An attribute, or combination of attributes, of an entity for which no two instances agree on the values. (C/SE) 1320.2-1998

candle *See:* candela.

candlepower (illuminating engineering) (television) Luminous intensity expressed in candelas. (BT/EEC/IE/AV) 201-1979w, [126]

can loss (rotating machinery) Electric losses in a can used to protect electric components from the environment. *See also:* asynchronous machine. (PE) [9]

canned (rotating machinery) Completely enclosed and sealed by a metal sheath. (PE) [9]

canned cycle *See:* fixed cycle.

canonical input processing The processing of terminal input in the form of text lines. (C) 1003.5-1999

canonical model A data model that represents the inherent structure of the data, independent of any specific implementations. (C) 610.5-1990w

canonical synthesis A technique for generating a canonical model from the relations in a database. (C) 610.5-1990w

cant hook A tool similar to a peavey (except that it has a blunt end) used to turn a pole or stabilize it if necessary during installation. (T&D/PE) 751-1990

capability (1) (power operations) The maximum load-carrying capability expressed in kilovolt-amperes (kVA) or kilowatts (kW) of generating equipment, other electrical apparatus, or system under specified conditions for a given time interval. *See also:* installed incremental transfer capability; dependable capability; second contingency incremental transfer capability; maximum capability; extended capability; system assured capability; first contingency incremental transfer capability; total for load capability; normal transfer capability; steam capability; system margin capability; pumped-storage hydro capability; hydro capability.

(2) A set of functions that are logically related by the common set of resources on which they operate. (C/MM) 855-1990

capability margin (electric power supply) The difference between the "total capability for load" and the "system load responsibility." It is the margin of capability available to provide for scheduled maintenance, emergency outages, adverse system operating requirements, and unforeseen loads. *See also:* generating station. (PE/PSE) [54]

capability module A set of functions within a capability that provide a class of support. Each module is intended to be provided in its entirety. (C/MM) 855-1990

capacitance (1) (semiconductor diode) (semiconductor radiation conductor) The small-signal capacitance measured between the terminals of the diode or detector under specified conditions of bias and frequency. *See also:* rectification; semiconductor device; semiconductor. (IM/ED/NPS/HFIM/NID) 314-1971w, 216-1960w, 325-1986s, 301-1976s

(2) A property expressible by the ratio of the time integral of the flow rate of a quantity, such as heat, or electric charge to or from a storage, divided by the related potential change. *Note:* Typical units are microfarads, Btu/°F, lb/psi, gal/ft. (CS/PE/EDPG) [3]

(3) (A) (outdoor apparatus bushings) The main capacitance, C_1, of a condenser bushing is the value in picofarads between the high-voltage conductor and the potential tap or the test tap. **(B) (outdoor apparatus bushings)** The tap capacitance, C_2, of a condenser bushing is the value in picofarads between the potential tap and mounting flange (ground). **(C) (outdoor apparatus bushings)** The capacitance, C, of a bushing without a potential or test tap is the value in picofarads between the high-voltage conductor and the mounting flange (ground). (PE/TR) 21-1976

(4) (VLF insulation testing) Capacitance, as used here, and distinguished from power-frequency capacitance, is that value which would result from a measurement at VLF, that is, 0.1 Hz ± 25%. In magnitude, it would tend to be greater than the power-frequency capacitance, to the extent of increased contributions made by dipole and interfacial polarizations. (PE/EM) 433-1974r

(5) (of a semiconductor radiation detector) The small-signal capacitance measured between terminals of the detector under specified conditions of bias and frequency. (NPS) 300-1988r

(6) The ratio of a conductor's electrostatic charge to the potential difference between conductors (required to maintain that charge). (PE/IC) 1143-1994r

(7) (A) (of bushing) The main capacitance, C_1, of a bushing is the capacitance between the high-voltage conductor and the voltage tap or the test tap. **(B)** (of bushing) The tap capacitance, C_2, of a capacitance graded bushing is the capacitance between the voltage tap and mounting flange (ground). **(C)** (of bushing) The capacitance, C, of a bushing without a voltage or test tap is the capacitance between the high-voltage conductor and the mounting flange (ground). (PE/TR) C57.19.03-1996

(8) That property of a system of conductors and dielectrics that permits the storage of electricity when potential differences exist between the conductors. Its value is expressed as the ratio of a quantity of electricity to a potential difference. A capacitance value is always positive. (IA/MT) 45-1998

capacitance between two conductors, balanced *See:* balanced capacitance.

capacitance coupling *See:* coupling capacitance.

capacitance current (1) (electric submersible pump cable) Current required to charge the capacitor formed by the dielectric of the cable under test. (IA/PC) 1017-1985s

(2) (rotating machinery) (or component) A reversible component of the measured current on charge or discharge of the winding that is due to the geometrical capacitance; that is, the capacitance as measured with alternating current of power or higher frequencies. With high direct voltage this current has a very short time constant and so does not affect the usual measurements. (PE/EM) 95-1977r

(3) (power cable systems) Current that charges the capacitor that is formed by the capacitance of the cable under test. (PE/IC) 400-1991

capacitance, detector *See:* detector capacitance.

capacitance, discontinuity *See:* discontinuity capacitance.

capacitance, effective *See:* effective capacitance.

capacitance graded bushing A bushing in which metallic or non-metallic conducting layers are arranged within the insulating material for the purpose of controlling the distribution of the electric field of the bushing, both axially and radially. (PE/TR) C57.19.03-1996

capacitance, input *See:* input capacitance.

capacitance meter An instrument for measuring capacitance. *Note:* If the scale is graduated in microfarads the instrument is usually designated as a microfaradmeter. *See also:* instrument. (EEC/PE) [119]

capacitance, nonlinear element *See:* nonlinear element capacitance.

capacitance, output *See:* output capacitance.

capacitance potential device A voltage-transforming equipment or network connected to one conductor of a circuit through a capacitance, such as a coupling capacitor or suitable high-voltage bushing, to provide a low voltage such as required for the operation of instruments and relays. *Notes:* 1. The term "potential device" applies only to the network and is exclusive of the coupling capacitor or high-voltage bushing. 2. The term "capacitance potential device" indicates use with any type of capacitance coupling. 3. Capacitance potential devices and their associated coupling capacitors or bushings are designed for line-to-ground connection, and not line-to-line connection. The potential device is a single-phase device, and, in combination with its coupling capacitor or bushing, is connected line-to-ground. The low voltage thus provided is a function of the line-to-ground voltage and the constants of the capacitance potential device. Two or more capacitance potential devices, in combination with their coupling capacitors or bushings, may be connected line-to-ground on different high-voltage phases to provide low voltages of other desired phase relationships. 4. Zero-sequence voltage may be obtained from the broken-delta connection of the auxiliary windings or by the use of one device with three coupling capacitors or bushings. In the latter case, the three operating-tap connection-points are joined together and one device connected between this common point and ground. Although used in combination with three coupling capacitors or bushings, the device output and accuracy rating standards are based on the single-phase conditions. *See also:* outdoor coupling capacitor. 341

capacitance ratio (nonlinear capacitor) The ratio of maximum to minimum capacitance over a specified voltage range, as determined from a capacitance characteristic, such as a differential capacitance characteristic, or a reversible capacitance characteristic. *See also:* nonlinear capacitor. (ED) [46]

capacitance, short-circuit input *See:* short-circuit input capacitance.

capacitance, short-circuit output *See:* short-circuit output capacitance.

capacitance, short-circuit transfer *See:* short-circuit transfer capacitance.

capacitance, signal electrode *See:* electrode capacitance.

capacitance, stray *See:* stray capacitance.

capacitance-switching transient overvoltage ratio (high voltage air switches, insulators, and bus supports) The ratio of the peak value of voltage above ground, during the transient conditions resulting from the operation of the switch, to the peak value of the steady-state line-to-neutral voltage. *Note:* It is measured at either terminal of the switch, whichever is higher, and is expressed in multiples of the peak values of the operating line-to-ground voltages at the switch with the capacitance connected. (SWG/PE) C37.30-1971s, C37.100-1992

capacitance, target *See:* target capacitance.

capacitance unbalance detection function The detection of objectionable unbalance in capacitance between capacitor groups within a phase, such as that caused by blown capacitor fuses or faulted capacitors, and to initiate an alarm, the closing of the capacitor bypass switch, or both. (T&D/PE) 824-1994

capacitance unbalance, pair to ground The unbalance that exists between the capacitance of each conductor of pair *ab* to the grounded shield with all the other conductors connected to the shield. This is: $C_{UB,PG} = (Cag + Cap) - (Cbg + Cbp)$, where *Cag* and *Cbg* are capacitances between each conductor (*a* and *b*) to ground and *Cap* and *Cbp* are capacitances between each conductor (*a* and *b*) and all other pairs connected together and grounded. (PE/PSC) 789-1988w

capacitance unbalance, pair to pair The unbalance in capacitance that exists between each conductor in a pair (*ab*) to each conductor in another pair (*cd*). This is: $C_{UB,PP} = (Cad + Cbc) - (Cac + Cbd)$, where *Cac*, *Cad*, *Cbc*, and *Cbd* are direct capacitances between conductors. (PE/PSC) 789-1988w

capacitator braking (rotating machinery) A form of dynamic braking for induction motors in which a capacitator is used to magnetize the motor. *See also:* asynchronous machine. (PE) [9]

capacitive coupling *See:* electrical coupling.

capacitive current (1) (rotating machinery) (or component) A reversible component of the measured current on charge or discharge of the winding that is due to the geometrical capacitance; that is, the capacitance as measured with alternating current of power or higher frequencies. With high direct voltage this current has a very short time constant and so does not affect the usual measurements. (PE/EM) 95-1977r
(2) (maintenance of energized power lines) The component of the measured current that leads the applied voltage by 90° due to the geometrical capacitance of the tool or equipment. (T&D/PE) 516-1995

capacitive gap (nonlinear, active, and nonreciprocal waveguide components) (microwave receiver protectors) The distance between cone apexes (apices) in a waveguide resonant structure. *See also:* resonant gap. (MTT) 457-1982w

capacitive load A lumped capacitance that is switched as a unit. (SWG/PE) C37.100-1992

capacitive weighting Response weighting by change of capacitance between fingers and bus bar, where the capacitance value is varied from electrode to electrode along the interdigital transducer. (UFFC) 1037-1992w

capacitor (1) (series capacitor) An assembly of one or more capacitor elements in a single container, with one or more insulated terminals brought out. (T&D/PE) 824-1994
(2) An element within a circuit consisting of two conductors, each with an extended surface exposed to that of the other, but separated by a layer of insulating material called the dielectric. *Note:* The dielectric is designed so the electric charge on one conductor is equal in value but opposite in polarity to that of the other conductor. *See also:* storage capacitor. (C) 610.10-1994w
(3) A device with the primary purpose of introducing capacitance into an electric circuit. Capacitors are usually classified, according to their dielectrics, as air capacitors, mica capacitors, paper capacitors, etc. (IA/MT) 45-1998

capacitor antenna (condenser antenna) An antenna consisting of two conductors or systems of conductors, the essential characteristic of which is its capacitance. *See also:* antenna. (CHM) [51]

capacitor bank An assembly at one location of capacitors and all necessary accessories, such as switching equipment, protective equipment, controls, etc., required for a complete operating installation. It may be a collection of components assembled at the operating site or may include one or more piece(s) of factory-assembled equipment.power systems relaying. (T&D/PE) 18-1992, C37.99-2000

capacitor bank overcurrent protection Common name for all or part of the overcurrent protective equipment at a capacitor installation. (SWG/PE) C37.40b-1996

capacitor bus (series capacitor) The main conductors that serve to connect the capacitor assemblies in series with the line. (T&D/PE) 824-1994

capacitor bushing (outdoor apparatus bushings) A bushing in which cylindrical conducting layers are arranged coaxially with the conductor within the insulating material for the purpose of controlling the electric field of the bushing. *Synonym:* condenser bushing. (PE/TR) 21-1976

capacitor bypass switch (series capacitor) A switch device with moving and stationary contacts that functions as a means of bypassing the capacitor. This switch may also have the capability of inserting the capacitor against a specified level of current. (T&D/PE) [26]

capacitor-bypass-switch interlocking devices (series capacitor) Devices that perform the function of having all three integral bypass switches of a capacitor step take the same open or close position. (T&D/PE) [26]

capacitor control The device required to automatically switch shunt power capacitor banks.power systems relaying. (T&D/PE) 1036-1992, C37.99-2000

capacitor element (1) (series capacitor) An individual part of a capacitor unit consisting of coiled conductors separated by dielectric material. (T&D/PE) [26]
(2) The smallest unit of a capacitor consisting of metallic foil plates separated by a dielectric film made typically of a polymer, paper, or combination of the two materials. (T&D/PE) 824-1994
(3) (power systems relaying) A device consisting essentially of two electrodes separated by a dielectric. (PE) C37.99-2000

capacitor enclosure The case in which the capacitor is mounted. (EEC/PE) [119]

capacitor equipment (shunt power capacitors) An assembly of capacitors with associated accessories, such as fuses, switches, etc., all mounted on a common frame for handling, transportation, and operation as a single unit. (PE/T&D) 18-1992

capacitor fuse (series capacitor) A capacitor fuse that provides an externally visible indication of fuse operation. (T&D/PE) 824-1985s, 824-1994

capacitor group (series capacitor) An assembly of more than one capacitor connected in parallel between two buses or terminals. (SWG/PE/T&D) C37.82-1971s, 824-1994

capacitor group fuse *See:* capacitor line fuse.

capacitor, ideal *See:* ideal capacitor.

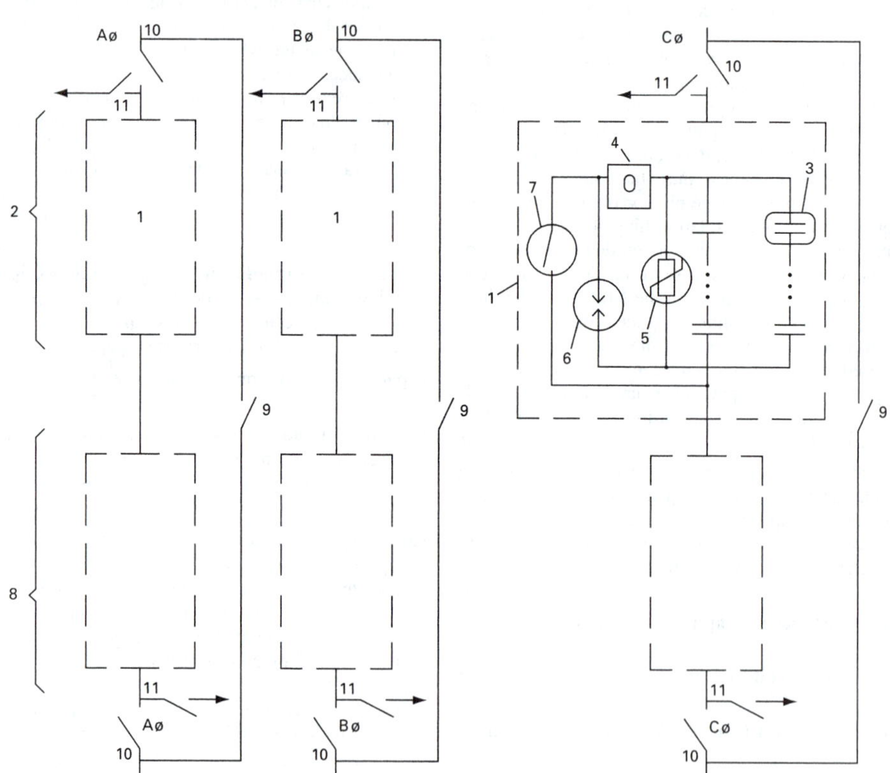

Typical series capacitor bank nomenclature

capacitor group

capacitor indicating fuse (series capacitor) A capacitor fuse that provides an externally visible indication of fuse operation. (T&D/PE) 824-1985s, 824-1994

capacitor inrush current The transient charging current that flows in a capacitor when a capacitor bank is initially connected to a voltage source.power systems relaying. (T&D/PE) 1036-1992, C37.99-2000

capacitor line fuse (capacitor group fuse) (power systems relaying) A fuse applied to disconnect a faulted phase of a capacitor bank from a power system. (SWG/PE/T&D) C37.40b-1996, 1036-1992, C37.99-2000

capacitor loudspeaker *See:* electrostatic loudspeaker.

capacitor microphone *See:* electrostatic microphone.

capacitor motor A single-phase induction motor with a main winding arranged for direct connection to a source of power and an auxiliary winding connected in series with a capacitor. The capacitor may be directly in the auxiliary circuit or connected into it through a transformer. *See also:* two-value capacitor motor; permanent-split capacitor motor; capacitor-start motor; asynchronous machine. (PE) [9]

capacitor mounting strap A device by means of which the capacitor is affixed to the motor. (EEC/PE) [119]

capacitor outrush current The high-frequency, high-magnitude current discharge of one or more capacitors into a short circuit, such as into a failed capacitor unit connected in parallel with the discharging units, or into a breaker closing into a fault.power systems relaying. (PE/T&D) C37.99-2000, 1036-1992

capacitor pickup A phonograph pickup that depends for its operation upon the variation of its electric capacitance. *See also:* phonograph pickup. (EEC/ACO) [109]

capacitor platform A structure that supports the capacitor rack assemblies and all associated equipment and protective devices, and is supported on insulators compatible with line-to-ground insulation requirements. (T&D/PE) 824-1994

capacitor rack (series capacitor) A frame that supports one or more capacitors. (T&D/PE) 824-1994

capacitor segment A single-phase assembly of groups of capacitors that has its own voltage-limiting device and relays to protect the capacitors from overvoltages and overloads. *See also:* capacitor group. (T&D/PE) 824-1994

capacitor start-and-run motor A capacitor motor in which the auxiliary primary winding and series-connected capacitors remain in circuit for both starting and running. *See also:* asynchronous machine; permanent-split capacitor motor. (PE) [9]

capacitor-start motor A capacitor motor in which the auxiliary winding is energized only during the starting operation. *Note:* The auxiliary-winding circuit is open-circuited during running operation. *See also:* asynchronous machine. (PE) [9]

capacitor storage A type of storage that uses the capacitive properties of certain materials. (C) 610.10-1994w

capacitor(s) stored energy The value of energy, measured in Joules, that is stored in a capacitor or group of capacitors at a given instantaneous value of voltage.

$$E = \frac{CV^2}{2}$$

where
E = Energy in Joules
C = Capacitance in microfarads
V = instantaneous voltage in kilovolts

(SWG/PE) C37.40b-1996

capacitor switch A switch capable of making and breaking capacitive currents of capacitor banks. (SWG/PE) C37.100-1992

capacitor switching step A three-phase function that consists of one or more capacitor segments per phase with capacitor bypass devices connected in parallel control devices, and provision for interlocked operation of the single-phase or three-phase switches when bypassing or inserting the capacitor segments. This is sometimes referred to as a capacitor module. (T&D/PE) 824-1994

capacitor unbalance protection A protective system sensitive to unbalanced voltages and/or currents in a normally balanced capacitor bank. The imbalance may be the result of blown fuses or due to an insulation failure within the capacitor bank. (SWG/PE) C37.40b-1996

capacitor unit (1) (general) A single assembly of dielectric and electrodes in a container with terminals brought out. *See also:* indoor; outdoor; alternating-current distribution. (T&D/PE) 18-1980s
(2) (series capacitor) An assembly of one or more capacitor elements in a single container, with one or more insulated terminals brought out. (T&D/PE) [26]
(3) An assembly of dielectric and electrodes in a container (case), with terminals brought out, that is intended to introduce capacitance into an electric power circuit.power systems relaying. (PE) C37.99-2000

capacitor unit fuse A fuse applied to disconnect an individual faulted capacitor from its bank. *Synonyms:* individual capacitor fuse; capacitor fuse. (SWG/PE/T&D) C37.40b-1996, 1036-1992

capacitor voltage The voltage across two terminals of a capacitor. (ED) [46]

capacity (C) (1) (A) (data transmission) The number of digits or characters in a machine word regularly handled in a computer. **(B) (data transmission)** The upper and lower limits of the numbers which may be regularly handled in a computer. **(C) (data transmission)** The maximum number of binary digits that can be transmitted by a communications channel in one second. (PE) 599-1985
(2) (nuclear power generating station) Maximum output of a turbine generator unit. (PE/EDPG) 1020-1988r
(3) A measure of the ability to generate electric power, usually expressed in megawatts or kilowatts. Capacity can refer to the output of a single generator, a plant, an item of electrical equipment, an entire electric system, or a power pool. (PE/PSE) 858-1993w
(4) The combined weight for which the component is designed to be used. Combined weight includes the user's body weight and clothing, tools, and other objects borne or carried by the user. (T&D/PE) 1307-1996
(5) Generally, the total number of ampere-hours that can be withdrawn from a fully charged battery at a specific discharge rate and electrolyte temperature, and to a specific cutoff voltage. (SCC21) 937-2000

capacity charge The charge for generation or transmission capacity used or reserved on the seller's system. *Synonyms:* demand charge; reservation charge. (PE/PSE) 858-1993w

capacity emergency The operating situation that exists when a system is unable to supply its firm demand and regulating requirements. (PE/PSE) 858-1993w

capacity factor The ratio of the average load on a machine or equipment for the period of time considered to the capacity of the machine or equipment. *See also:* generating station. (T&D/PE/PSE) [10], 346-1973w

capacity, firm, purchases or sales (electric power supply) Firm capacity that is purchased, or sold, in transactions with other systems and which is not from designated units, but is from the overall system of the seller. *Note:* It is understood that the seller treats this type of transaction as a load obligation. (PE/PSE) 346-1973w

capacity, heat *See:* heat capacity.

capacity limits (major) There are three major capacity categories that may limit the ultimate size of a switching system. The categories are termination capacity, call capacity, and traffic usage capacity. The fundamental hardware and software architecture of the switching system determines these major capacity limits. *See also:* traffic usage capacity; call capacity; termination capacity. (COM/TA) 973-1990w

capacity, specific unit, purchases or sales (electric power supply) Capacity that is purchased, or sold, in transactions with other systems and which is from a designated unit on the system of the seller. (PE/PSE) 346-1973w

capacity test (battery) A discharge of a battery at a constant current or a constant power to a specified voltage. (SB/PE/EDPG) 1188-1996, 450-1995, 1106-1995

cap-and-pin insulator An assembly of one or more shells with metallic cap and pin, having means for direct and rigid mounting. *See also:* insulator. (EEC/IEPL) [89]

capstan A rotating shaft within a tape drive that pulls the tape across the read or write heads. (C) 610.10-1994w

capture effect (1) (modulation systems) The effect occurring in a transducer (usually a demodulator) whereby the input wave having the largest magnitude controls the output. (Std100) 270-1964w
(2) The tendency of a receiver to suppress the weaker of two time-coincident signals within its passband. (AES) 686-1997

capturing (accelerometer) (gyros) The use of a torquer (forcer) in a servo loop to restrain a gyro gimbal, rotor, or accelerometer proof mass to a specified reference position. (AES/GYAC) 528-1994

car *See:* vehicle.

carabiner A connector component generally comprised of a trapezoidal or oval shaped body with a normally closed gate or similar arrangement which may be opened to permit the body to receive an object, and when released, automatically closes to retain the object. (T&D/PE) 1307-1996

carabiner, locking *See:* locking carabiner.

carabiner, manual locking *See:* manual locking carabiner.

carabiner, non-locking *See:* nonlocking carabiner.

car annunciator An electric device in the car that indicates visually the landings at which an elevator-landing signal-registering device has been actuated. *See also:* elevator. (EEC/PE) [119]

carbon-arc lamp (illuminating engineering) An electric-discharge lamp employing an arc discharge between carbon electrodes. One or more of these electrodes may have cores of special chemicals that contribute importantly to the radiation. (EEC/IE) [126]

carbon block protector An assembly of two or three carbon blocks and air gaps designed to a specific breakdown voltage. These devices are normally connected to telecommunication circuits to provide overvoltage protection and a current path to ground during such overvoltage. (PE/PSC) 487-1992

carbon brush (A) (motors and generators) A specific type of brush composed principally of amorphous carbon. *Note:* This type of brush is usually hard and is adapted to low speeds and moderate currents. **(B) (motors and generators)** A broader classification of brush, containing carbon in appreciable amount. *See also:* brush. (PE) [9]

carbon-consuming cell (carbon-combustion cell) A cell for the production of electric energy by galvanic oxidation of carbon. *See also:* electrochemistry. (PE/EEC) [119]

carbon-contact pickup A phonograph pickup that depends for its operation upon the variation in resistance of carbon contacts. *See also:* phonograph pickup. (EEC/PE) [119]

carbon-dioxide system (rotating machinery) A fire-protection system using carbon-dioxide gas as the extinguisher. (PE) [9]

carbon-graphite brush A carbon brush to which graphite is added. This type of brush can vary from medium hardness to very hard. It can carry only moderate currents and is adapted to moderate speeds. *See also:* brush. (EEC/EM/LB) [101]

carbon noise (carbon microphones) The inherent noise voltage of the carbon element. *See also:* close-talking pressure-type microphones. (SP) 258-1965w

carbon-pressure recording facsimile That type of electromechanical recording in which a pressure device acts upon carbon paper to register upon the record sheet. *See also:* recording. (COM) 168-1956w

carbon telephone transmitter A telephone transmitter that depends for its operation upon the variation in resistance of carbon contacts. *See also:* telephone station. (EEC/PE) [119]

car builder (1) The entity assembling or manufacturing the vehicle. (VT) 1475-1999
(2) The entity manufacturing the vehicle. (VT) 1476-2000

car, cable *See:* cable car.

carcinogen An agent that tends to produce cancer. (T&D/PE) 539-1990

carcinotron (microwave tubes) (m-type backward-wave oscillator) A crossed-field oscillator tube in which an electron stream interacts with a backward wave on the nonreentrant circuit. The oscillation frequency is a function of anode-to-solve voltage. (ED) [45]

car, conductor *See:* conductor car.

card (1) (STD bus) A printed circuit board and components that make up the modules that plug in to the bus backplane. (C/MM) 961-1987r
(2) (A) A generic term used as an abbreviation for a circuit board. **(B)** An input medium made of paperboard, formed in a uniform size and shape such that it may be punched or marked and sensed electronically. *See also:* magnetic card; mark-sensing card; punch card. (C) 610.10-1994
(3) (computers) *See also:* magnetic card; tape to card; punched card.

card cage A chassis in which a printed circuit board may be mounted. (C) 610.10-1994w

card code The set or combination of punched holes in a punch card that represent a character. (C) 610.10-1994w

card column A single vertical line of punch positions on a punch card. *Contrast:* card row. (C) 610.10-1994w

card deck A group of punch cards. (C) 610.10-1994w

card duplicator *See:* card reproducing punch.

card extender A device that provides access to components on a circuit card for testing purposes while maintaining all the electrical connections to the card. (SWG/PE) C37.100-1992

card feed (1) (test, measurement, and diagnostic equipment) The mechanism that moves cards serially into a machine. (MIL) [2]
(2) A mechanism that moves cards one at a time from the card hopper to the card path. (C) 610.10-1994w

card field (test, measurement, and diagnostic equipment) An area (one or more columns) of a card that is regularly assigned for the same information item. (MIL) [2]

card hopper (1) (computers) A device that holds cards and makes them available to a card-feed mechanism. *See also:* card stacker. (C) [85]
(2) The part of a card-processing device that holds the cards to be processed and makes them available to the card feed mechanism. *Synonym:* punched card holder. *Contrast:* card stacker. (C) 610.10-1994w

card image (1) (computers) A one-to-one representation of the contents of a punched card, for example, a matrix in which a 1 represents a punch and a 0 represents the absence of a punch. (C) [85]
(2) A representation of the hole patterns found in a punched card, for example, a matrix in which a one represents a punch and a zero represents the absence of a punch. (C) 610.10-1994w

cardinality (1) (A) The number of elements in a set. **(B)** In a relational data model, the number of tuples in a relation. (C) 610.5-1990
(2) The numeric relationship between entity sets, labeled as one to one (1-1), many to one (M-1), or many to many (M-N), which indicates the number of items in one entity set that could possibly be associated with the items in another entity set. (PE/EDPG) 1150-1991w
(3) A specification of how many instances of a first class may or must exist for each instance of a second (not necessarily distinct) class, and how many instances of a second class may

or must exist for each instance of a first class. For each direction of a relationship, the cardinality can be constrained. *See also:* cardinality constraint. (C/SE) 1320.2-1998

cardinality constraint (A) A kind of constraint that limits the number of instances that can be associated with each other in a relationship. *See also:* cardinality. **(B)** A kind of constraint that limits the number of members in a collection. *See also:* collection cardinality. (C/SE) 1320.2-1998

cardinal plane For an infinite planar array whose elements are arranged in a regular lattice, any plane of symmetry normal to the planar array and parallel to an edge of a lattice cell. *Notes:* 1. This term can be applied to a finite array, usually one containing a large number of elements, by the assumption that it is a subset of an infinite array with the same lattice arrangement. 2. This term is used to relate the regular geometrical arrangement of the array elements to the radiation pattern of the antenna. (AP/ANT) 145-1993

cardiogram *See:* electrocardiogram.

cardiovascular effect Effect pertaining to the system comprised of the heart and the blood vessels. (T&D/PE) 539-1990

car door (elevators) The sliding portion of the car or the hinged or sliding portion in the hoistway enclosure that closes the opening giving access to the car or to the landing. *See also:* hoistway. (PE/EEC) [119]

car-door contact An electric device, the function of which is to prevent operation of the driving machine by the normal operating device unless the car door or gate is in the closed position. *See also:* elevator. (EEC/PE) [119]

car-door closer A device or assembly of devices that closes a manually opened car door or gate by power other than by hand, gravity, springs, or the movement of the car. *See also:* elevator. (EEC/PE) [119]

card path In a card-processing device, a path along which cards are moved and guided. *See also:* card feed. (C) 610.10-1994w

card-processing device Any device that can read or write data to punch cards. *See also:* card reader; card reproducing punch; card punch. (C) 610.10-1994w

card-programmed (test, measurement, and diagnostic equipment) The capability of performing a sequence of tests according to instructions contained in one or a deck of punched cards. (MIL) [2]

card punch An output device that produces a record of data as hole patterns in punch cards. *Synonym:* automatic punch. *See also:* card reproducing punch; keypunch. (C) 610.10-1994w

card reader (1) (test, measurement, and diagnostic equipment) A mechanism that senses and obtains information from punched cards. (MIL) [2]
(2) An input device that reads or senses hole patterns in a punch card, transforming the data from hole patterns to electrical signals. *Synonym:* punched card reader. *See also:* card stacker; card hopper; card track; paper tape reader. (C) 610.10-1994w
(3) A device used to read a coded credential at an entry point. (PE/NP) 692-1997

card, relay *See:* relay armature card.

card reproducer *See:* card reproducing punch.

card reproducing punch A card-processing device that prepares one punch card, copying all or part of the data from another punch card. *Synonyms:* card reproducer; card duplicator. (C) 610.10-1994w

card row A single horizontal line of punch positions on a punch card. *Contrast:* card column. (C) 610.10-1994w

card set function generator (analog computer) A diode function generator whose values are stored and set by means of a punched card and a mechanical card-reading device. (C) 165-1977w

card sorter A sorting device that deposits punch cards in pockets selected according to the hole patterns in the cards. (C) 610.10-1994w

card stacker (1) (computers) An output device that accumulates punched cards in a deck. *See also:* card hopper. (MIL/C) [2], [20], [85]
(2) The part of a card-processing device that receives the cards after they have been processed. *Contrast:* card hopper. (C) 610.10-1994w

card tester (test, measurement, and diagnostic equipment) An instrument for testing and diagnosing printed circuit cards. (MIL) [2]

card-to-disk converter An input device that converts data from punch cards to disk storage. *See also:* card-to-tape converter; key-to-disk converter. (C) 610.10-1994w

card-to-tape converter An input device that converts data from punch cards to magnetic or paper tape. *See also:* card-to-disk converter; key-to-tape converter. (C) 610.10-1994w

card track That part of a card-processing device that moves and guides the card along the card path. (C) 610.10-1994w

car enclosure (elevators) Consists of the top and the walls resting on and attached to the car platform. *See also:* elevator. (EEC/PE) [119]

car-frame sling The supporting frame to which the car platform, upper and lower sets of guide shoes, car safety and hoisting ropes or hoisting-rope sheaves, or the plunger of a direct-plunger elevator are attached. *See also:* elevator. (EEC/PE) [119]

cargo vessel A vessel that carries bulk, containerized, or roll-on/roll-off dry cargo, and no more than 12 passengers. Research vessels, search and rescue vessels, and tugs are also considered to be cargo vessels. (IA/MT) 45-1998

car platform (elevators) The structure that forms the floor of the car and that directly supports the load. *See also:* hoistway. (EEC/PE) [119]

car retarder A braking device, usually power operated, built into a railway track and used to reduce the speed of cars by means of brake shoes that when set in braking position press against the sides of the lower portions of the wheels. *See also:* master controller; control machine; switch machine; trimmer signal. (EEC/PE) [119]

carriage The mechanism in a typewriter or other printing device that holds the paper and moves it past the printing position. *See also:* automatic carriage. (C) 610.10-1994w

carriage control tape (A) A tape that is used to control vertical tabulation of printing or display positions. **(B)** A tape that contains line feed and form feed control data for a printing device. (C) 610.10-1994

carriage restore key *See:* carriage return key.

⟨`carriage-return`⟩ A character that in the output stream shall indicate that printing should start at the beginning of the same physical line in which the ⟨`carriage-return`⟩ occurred. The ⟨`carriage-return`⟩ shall be the character designated by `'\r'` in the C-language binding. It is unspecified whether this character is the exact sequence transmitted to an output device by the system to accomplish the movement to the beginning of the line. (C/PA) 9945-2-1993

carriage return (CR) (1) (typewriter) The operation that causes the next character to be printed at the left margin. (C) [85]
(2) A command or signal sent to a printer to instruct it to move to the beginning of the writing line. *Note:* Often used in conjunction with a line feed to move to the beginning of the next writing line. (C) 610.10-1994w

carriage return character (CR) A format effector that causes the print or display position to move to the first position on the same line. *Synonym:* new-line character. (C) 610.5-1990w

carriage return key A control key on a keyboard that initiates a carriage return. *Note:* Often used to terminate a command or to request its execution. *Synonyms:* carriage restore key; enter key. (C) 610.10-1994w

carried traffic (telephone switching systems) A measure of the calls served during a given period of time. (COM) 312-1977w

carrier (1) (A) (data transmission) A wave having at least one characteristic that may be varied from a known reference value by modulation. **(B) (data transmission)** That part of the modulated wave that corresponds in a specified manner to the unmodulated wave, having, for example, the carrier-frequency spectral components. *Note:* Examples of carriers are a sine wave and a recurring series of pulses. (PE/AP/ANT) 599-1985, 145-1993
(2) (overhead-power-line corona and radio noise) A continuous electromagnetic wave having a repeating variation in time and at least one characteristic that may be varied from a known reference value by modulation. *Note:* Examples of carriers are a sine wave and a recurring series of pulses. (T&D/PE) 539-1990
(3) (semiconductor) A mobile conduction electron or hole. *See also:* semiconductor device. (MIL) [2]
(4) (electrostatography) The substance in a developer that conveys a toner, but does not itself become a part of the viewable record. *See also:* electrostatography. (ED) 224-1965w, [46]
(5) (A) A continuous frequency capable of being modulated or impressed with a signal. *Synonym:* carrier wave. **(B)** An alternating current that oscillates at a fixed frequency, used to transmit a signal. (C) 610.7-1995

carrier amplifier (signal-transmission system) An alternating current amplifier capable of amplifying a prescribed carrier frequency and information side-bands relatively close to the carrier frequency. *See also:* signal. (IE) [43]

carrier-amplitude regulation The change in amplitude of the carrier wave in an amplitude-modulated transmitter when modulation is applied under conditions of symmetrical modulation. *Note:* The term "carrier shift," often applied to this effect, is deprecated. (AP/ANT) 145-1983s

carrier beat (facsimile) The undesirable heterodyne of signals each synchronous with a different stable reference oscillator causing a pattern in received copy. *Note:* Where one or more of the oscillators is fork controlled, this is called fork beat. *See also:* facsimile transmission. (COM) 168-1956w

carrier chrominance signal *See:* chrominance signal.

carrier-controlled approach system (CCA) (1) An aircraft carrier radar system providing information by which aircraft approaches may be directed via radio communication. (AES/GCS) 172-1983w
(2) An aircraft carrier radar system providing information that facilitates direction of aircraft via radio communication. (AES) 686-1997

carrier current The current associated with a carrier wave. *See also:* carrier. (PE) 599-1985w

carrier-current choke coil (capacitance potential devices) A reactor or choke coil connected in series between the potential tap of the coupling capacitor and the potential device transformer unit, to present a low impedance to the flow of power current and a high impedance to the flow of carrier-frequency current. Its purpose is to limit the loss of carrier-frequency current through the potential-device circuit. *See also:* outdoor coupling capacitor. (PE/EM) 43-1974s

carrier-current grounding-switch and gap Consists of a protective gap for limiting the voltage impressed on the carrier-current equipment and the line turning unit (if used); and a switch that, when closed, solidly grounds the carrier equipment for maintenance or adjustment without interrupting either high-voltage line or potential-device operation. *See also:* outdoor coupling capacitor. (PE/EM) 43-1974s

carrier detect A dc electrical signal presented by a modem to its associated terminal equipment when the modem is receiving a modulatory signal. (SUB/PE) 999-1992w

carrier extension The addition of nondata symbols to the end of frames that are less than slotTime bits in length so that the resulting transmission is at least one slotTime in duration. (C/LM) 802.3-1998

carrier frequency (1) (A) (data transmission) (periodic carrier) The reciprocal of its period. *Note:* The frequency of a periodic pulse carrier is often called the pulse repetition frequency in a signal transmission system. **(B) (data transmission)** (modulated amplifier) The frequency that is used to modulate the input signal for amplification. (PE) 599-1985
(2) A unique frequency of a carrier that is used to carry data. *Note:* It is measured in cycles per second or hertz. (C) 610.7-1995

carrier-frequency pulse A carrier that is amplitude modulated by a pulse. *Notes:* 1. The amplitude of the modulated carrier is zero before and after the pulse. 2. Coherence of the carrier (with itself) is not implied. (IM/WM&A) 194-1977w

carrier frequency range (transmitter) The continuous range of frequencies within which the transmitter may be adjusted for normal operation. A transmitter may have more than one carrier-frequency range. *See also:* radio transmitter. (AP/BT/ANT) 145-1983s, 182-1961w

carrier frequency stability (radio transmitters) (transmitter performance) The measure of the ability to remain on assigned channel as determined on both a short-term (1 second) and a long-term (24 hour) basis. (VT) [37]

carrier, fuse *See:* fuse carrier.

carrier group (data transmission) The frequency band, 60 kHz to 108 kHz, containing twelve voice channels which serves as the basic building block of a larger system. (PE) 599-1985w

carrier group alarm A combination of a carrier failure alarm and trunk conditioning, which is the process of disconnecting any trunk-group connections and making trunks busy. (COM/TA) 1007-1991r

carrier isolating choke coil An inductor inserted, in series with a line on which carrier energy is applied, to impede the flow of carrier energy beyond that point. (IM) [120]

carrier modulation (data transmission) A process whereby a high-frequency carrier wave is altered by a signal containing the information to be transmitted. (PE) 599-1985w

carrier noise level (residual modulation) The noise produced by undesired variations of radio-frequency signal in the absence of any intended modulation. (AP/ANT) 145-1983s

carrier or pilot-wire receiver relay (power system device function numbers) A relay that is operated or restrained by a signal used in connection with carrier-current or direct-current (dc) pilot-wire fault relaying. (SUB/PE) C37.2-1979s

carrier-pilot-protection A form of pilot protection in which the communication means between relays is a carrier current channel. (SWG/PE) C37.100-1992

carrier power output (transmitter performance) The radio-frequency power available at the antenna terminal when no modulating signal is present. *See also:* audio-frequency distortion. (VT) [37]

carrier relaying protection A form of pilot protection in which high-frequency current is used over a metallic circuit (usually the line protected) for the communicating means between the relays at the circuit terminals. (SWG/PE) C37.100-1981s

carrier sense In a local area network, an ongoing activity of a data station to detect whether another station is transmitting. *Note:* The carrier sense signal indicates that one or more DTEs are currently transmitting. (LM/C) 610.7-1995, 8802-3-1990s, 802.3-1998

carrier sense multiple access with collision detection A local area network access technique. When a station wants to gain access to the network, it listens for conflicting traffic and checks to see if the network is free. If the network is not free, it waits for a small amount of time and retries. (C) 610.7-1995

carrier shift (1) (data transmission) The difference between the steady state, mark, and space frequencies in a system utilizing frequency shift modulation. (PE) 599-1985w
(2) *See also:* carrier-amplitude regulation. (AP/ANT) 145-1983s

carrier suppression (radio communication) The method of operation in which the carrier wave is not transmitted. *See also:* modulation. (AP/ANT) 145-1983s

carrier system (1) (data transmission) A communication system using frequency multiplexing to a number of channels over a single path by modulating each channel upon a different carrier frequency and demodulating at the receiving point to restore the signals to their original form. (PE) 599-1985w
(2) A means for obtaining a number of channels over a single path, known as a carrier. (C) 610.7-1995

carrier tap choke coil A carrier-isolating choke coil inserted in series with a line tap. (IM) [120]

carrier telegraphy The form of telegraphy in which, in order to form the transmitted signals, alternating-current is supplied to the line after being modulated under the control of the transmitting apparatus. *See also:* telegraphy. (COM) [49]

carrier telephone channel A telephone channel employing carrier transmission. *See also:* channel. (EEC/PE) [119]

carrier-to-noise ratio (1) (A) The ratio of the powers of the carrier and the noise after specified band limiting and before any nonlinear process such as amplitude limiting and detection. *Note:* This ratio is expressed in many different ways, for example, in terms of peak values in the case of impulse noise and in terms of mean-squared values for other types of noise. **(B)** A combination of transmission media and equipment capable of accepting signals at one point and delivering related signals at another point. *See also:* amplitude modulation; information theory. (IT) [7]
(2) (data transmission) The ratio of the magnitude of the carrier to that of the noise after selection and before any nonlinear process, such as amplitude limiting and detection. The bandwidth used for measurement of the noise should be specified when using this ratio. (PE) 599-1985w
(3) (broadband local area networks) The carrier-to-noise ratio expresses the relationship between signaling and noise power on a communications medium. The ratio is normally referenced to a specific noise bandwidth that corresponds to the signaling bandwidth. (LM/C) 802.7-1989r

carrier transmission That form of electric transmission in which the transmitted electric wave is a wave resulting from the modulation of a single-frequency wave by a modulating wave. *See also:* carrier. (EEC/PE) [119]

carrier velocity (semiconductor) (nonlinear, active, and nonreciprocal waveguide components) The average velocity of the random thermal motion of electrons in *n*-type semiconductors and of holes in *p*-type semiconductors. (MTT) 457-1982w

carrier wave *See:* carrier.

carry (A) A character or characters, produced in connection with an arithmetic operation on one digit place of two or more number representations in positional notation, and forwarded to another digit place for processing there. **(B)** The number represented by the character or characters in definition (A). **(C)** Usually, a signal or expression as defined in definition (A) which arises in adding, when the sum of two digits in the same digit place equals or exceeds the base of the number system in use. *Note:* If a carry into a digit place will result in a carry out of the same digit place, and if the normal adding circuit is bypassed when generating this new carry, it is called a high-speed carry, or standing-on-nines carry. If the normal adding circuit is used in such a case, the carry is called a cascaded carry. If a carry resulting from the addition of carries is not allowed to propagate (for example, when forming the partial product in one step of a multiplication process), the process is called a partial carry. If it is allowed to propagate, the process is called a complete carry. If a carry generated in the most-significant-digit place is sent directly to the least-significant place (for example, when adding two negative numbers using nines complements) that carry is called and end-around carry. **(D)** A carry, in direct subtraction, is a signal or expression as defined in definition (A) that arises when the difference between the digits is less than zero. Such a carry is frequently called a borrow. **(E)** To carry is the action for forwarding a carry. **(F)** A carry is the command directing a carry to be forwarded. *See also:* high-speed carry; end-around carry; standing-on-nines carry; complete carry; partial carry; cascaded carry. (C) 162-1963
(2) (A) (mathematics of computing) A mathematical process used in addition and subtraction, in which a value is generated when a sum or product in a digit place exceeds the largest number that can be represented in that digit place, and the value is transferred to the next higher digit place for processing there. *See also:* complete carry; partial carry; cascaded carry; standing-on-nines carry; half carry; high-speed carry; end-around carry. **(B) (mathematics of computing)** The value generated in definition (A). **(C) (mathematics of computing)** To perform the process defined in definition (A). (C) 1084-1986

car safety (counterweight safety) A mechanical device attached to the car frame or to an auxiliary frame, or to the counterweight frame, to stop and hold the car or counterweight in case of predetermined overspeed of free fall, or if the hoisting ropes slacken. *See also:* elevator. (EEC/PE) [119]

car-switch automatic floor-stop operation (elevators) Operation in which the stop is initiated by the operator from within the car with a definite reference to the landing at which it is desired to stop, after which the slowing down and stopping of the elevator is effected automatically. *See also:* control. (PE/EEC) [119]

car-switch operation (elevators) Operation wherein the movement and direction of travel of the car are directly and solely under the control of the operator by means of a manually operated car switch of continuous-pressure buttons in the car. *See also:* control. (PE/EEC) [119]

cartridge (1) A container holding some form of data medium such that the medium can be accessed without separating it from the container; for example, a magnetic tape cartridge or a font cartridge. *See also:* cassette. (C) 610.10-1994w
(2) A unit of physical media that contains one or more sides. (C/SS) 1244.1-2000

cartridge font A font that is stored on a font cartridge. (C) 610.10-1994w

cartridge fuse A low-voltage fuse consisting of a current-responsive element inside a fuse tube with terminals on both ends. (SWG/PE) C37.100-1992

cartridge size (of a cartridge fuse) The range of voltage and ampere ratings assigned to a fuse cartridge with specific dimensions and shape. (SWG/PE) C37.100-1992

cartridge-type bearing (rotating machinery) A complete ball or roller bearing assembly consisting of a ball or roller bearing and bearing housing that is intended to be inserted into a machine endshield. *See also:* bearing. (PE) [9]

cart, splicing *See:* splicing cart.

car-wiring apparatus *See:* wire; multiple-unit control.

cascade (1) (electrolyte cells) A series of two or more electrolytic cells or tanks so placed that electrolyte from one flows into the next lower in the series, the flow being favored by differences in elevation of the cells, producing a cascade at each point where electrolyte drops from one cell to the next. *See also:* tandem; electrowinning. (EEC/PE) [119]
(2) (broadband local area networks) The number of amplifiers connected in series. (LM/C) 802.7-1989r
(3) A multilevel repeater topology in which higher-level repeaters are connected through their local ports to the cascade port of lower-level repeaters.local area networks. (C) 8802-12-1998

cascade button A choice on a menu that, when activated, presents another menu with additional related choices. (C) 1295-1993w

cascade connection (cascade) A tandem arrangement of two or more similar component devices in which the output of one is connected to the input of the next. *See also:* tandem. (PE/EEC) [119]

cascade control (1) (street lighting system) A method of turning street lights on and off in sections, each section being controlled by the energizing and de-energizing of the preceding section. *See also:* direct-current distribution; alternating-current distribution. (T&D/PE) [10]
(2) (automatic control) *See also:* cascade control system. (PE/EDPG) [3]

cascade control system A control system in which the output of one subsystem is the input for another subsystem. *See also:* feedback control system. (IM/PE/EDPG) [120], [3]

cascaded carry (mathematics of computing) A carry process in which the addition of two numerals results in a partial-sum numeral and a carry numeral that are in turn added together, this process being repeated until no new carries are generated.

Cascaded Carry

Augend	289594
Addend	320607
First Partial Sum	509191
First Partial Carry	101010
Second Partial Sum	600101
Second Partial Carry	010100
True Sum	610201

Contrast: high-speed carry. *See also:* partial carry; carry; partial sum. (C) [20], 1084-1986w

cascade development (electrostatography) Development in which the image-forming material is carried to the field of the electrostatic images by means of gravitational forces, usually in combination with a granular carrier. *See also:* electrostatography. (ED) 224-1965w, [46]

cascaded thermoelectric device A thermoelectric device having two or more stages arranged thermally in series. *See also:* thermoelectric device. (ED) [46], 221-1962w

cascade matrix *See:* chain matrix.

cascade merge sort An unbalanced merge sort in which the distribution of the sorted subsets is based on the cascade numbers. *See also:* polyphase merge sort. (C) 610.5-1990w

cascade node (network analysis) A node (branch) not contained in a loop. (CAS) 155-1960w

cascade port The repeater port that enables a cascade connection to a higher-level repeater.local area networks. (C) 8802-12-1998

cascade rectifier (cascade rectifier circuit) A rectifier in which two or more similar rectifiers are connected in such a way that their direct voltages add, but their commutations do not coincide. *Note:* When two or more rectifiers operate so that their commutations coincide, they are said to be in parallel if the direct currents add, and in series if the direct voltages add. *See also:* power rectifier; rectifier circuit element; rectification. (IA/EEC/PCON) 59-1962w, [110]

cascade summing of X and gamma rays The simultaneous detection of two or more photons originating from a single nuclear disintegration that results in only one observed (summed) pulse. *Synonym:* coincidence summing. (NI) N42.14-1991

cascade thyristor converter A thyristor converter in which two or more simple converters are connected in such a way that their direct voltages add, but their commutations do not coincide. (IA/IPC) 444-1973w

cascade transitions Gamma rays in the radioactive decay of a single atom that are emitted sequentially and within the resolving time of the spectrometer. (NI) N42.14-1991

cascade-type voltage transformer (1) An insulated-neutral terminal type voltage transformer with the primary distributed on two or more cores electromagnetically coupled by coupling windings. The secondary is on the core at the neutral end of the primary. Each core is insulated from the other cores and is maintained at a fixed percentage of the voltage between the primary terminal and the neutral terminal. (PE/TR) [57]
(2) (instrument transformers) (power and distribution transformers) An insulated-neutral terminal type voltage transformer with the primary winding distributed on several cores with the cores electromagnetically coupled by coupling windings. The secondary winding is on the core at the neutral end of the high-voltage winding. Each core of this type of transformer is insulated from the other cores and is maintained at a fixed voltage with respect to ground and the line-to-ground voltage. (PE/TR) C57.12.80-1978r
(3) A voltage transformer that has an insulated-neutral or grounded-neutral terminal and that has the primary winding subdivided into two or more (usually equal) series connected sections, mounted on one or more magnetic cores, and that has the secondary winding located about the core at the neutral end of the primary winding. The sections of the primary winding are coupled by "coupling windings." The cores, if more than one, are insulated from each other and connected to definite voltage levels along the primary winding. (PE/TR) C57.13-1993

cascading (of switching devices) The application of switching devices in which the devices nearest the source of power have interrupting ratings equal to, or in excess of, the available short-circuit current, while devices in succeeding steps further from the source, have successively lower interrupting ratings. (SWG/PE) C37.100-1992

CASE *See:* computer-aided software engineering.

case (1) (storage battery) (storage cell) A multiple compartment container for the elements and electrolyte of two or more storage cells. Specifically, wood cases are containers for cells in individual jars. *See also:* battery. (EEC/PE) [119]
(2) (electrotyping) A metal plate to which is attached a layer of wax to serve as a matrix. (EEC/PE) [119]
(3) (semiconductor devices) The housing of a semiconductor device. (IA) [12]
(4) (software) A single-entry, single-exit multiple-way branch that defines a control expression, specifies the processing to be performed for each value of the control expression, and returns control in all instances to the statement immediately following the overall construct. *Synonym:* multiple exclusive selective construct. *Contrast:* if-then-else; go to; jump. *See also:* multiple inclusive selective construct.

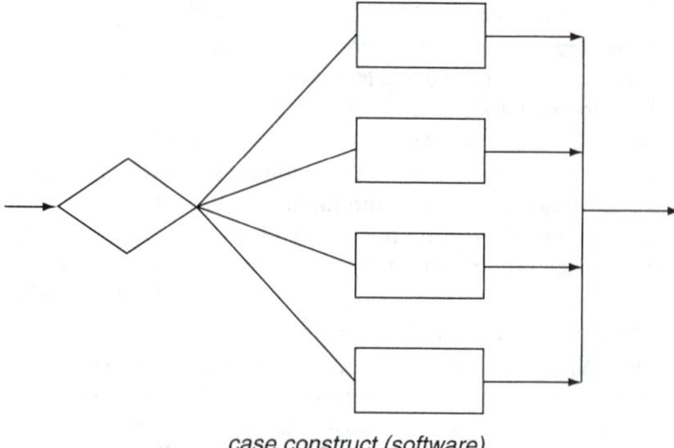

case construct (software)

case

(C) 610.12-1990
(5) (accelerometer) (gyros) The structure that provides the mounting surfaces and establishes the reference axes. (AES/GYAC) 528-1994

case capacitance (nonlinear, active, and nonreciprocal waveguide components) (semiconductor) The fixed capacitance of an empty enclosure (neither semiconductor chip nor connecting wires or straps are present). (MTT) 457-1982w

case ground protection Overcurrent relay protection used to detect current flow in the ground or earth connection of the equipment or machine. (SWG/PE) C37.100-1992

case shift (telegraphy) The change-over of the translating mechanism of a telegraph receiving machine from letters-case to figures-case or vice versa. *See also:* telegraphy. (COM) [49]

CASE tool A software tool that aids in software engineering activities, including, but not limited to, requirements analysis and tracing, software design, code production, testing, document generation, quality assurance, configuration management, and project management. *Note:* A CASE tool may provide support in only selected functional areas or in a wide variety of functional areas. (C/SE) 1209-1992w, 1348-1995

CASE tool evaluation A process wherein various aspects of a CASE tool are measured against defined criteria and the results are recorded for future use. (C/SE) 1209-1992w

CASE tool selection A process wherein the data from one or more CASE tool evaluations are weighted and compared against defined criteria to determine whether one or more of the CASE tools can be recommended for selection. (C/SE) 1209-1992w

Cassegrainian feed (communication satellite) A feed system used for parabolic reflector antennas, where a small hyperbolic subreflector is placed near the focus of the paraboloid. The Cassegrainian feed system prevents spillover to the back of the reflector; thus, a better noise performance is achieved. (COM) [24]

Cassegrain reflector antenna A paraboloidal reflector antenna with a convex subreflector, usually hyperboloidal in shape, located between the vertex and the prime focus of the main reflector. *Notes:* 1. To improve the aperture efficiency of the antenna, the shapes of the main reflector and the subreflector are sometimes modified. 2. There are other alternate forms that are referred to as Cassegrainean. Examples include the following: one in which the subreflector is surrounded by a reflecting skirt and one that utilizes a concave hyperboloidal reflector. When referring to these alternate forms the term shall be modified in order to differentiate them from the antenna described in the definition. (AP/ANT) 145-1993

cassette A container holding some form of data medium on reels which are driven at their axis at a variable speed which allows the tape to be accessed without separating it from the container. *See also:* cartridge; magnetic tape cassette. (C) 610.10-1994w

cast To treat an object of one type as an object of another type. *Contrast:* coerce. (C/SE) 1320.2-1998

casting (electrotyping) The pouring of molten electrotype metal upon tinned shells. (PE/EEC) [119]

casting out nines A method of checking addition, subtraction, or multiplication results by dividing decimal values by nine and comparing the remainders. *Synonym:* nines check. (C) 1084-1986w

cat *See:* tractor, crawler; crawler tractor.

CAT *See:* computerized axial tomography; computer-assisted tomography; computer-aided testing.

catalog (1) (A) A directory of the location of files within a system. *See also:* file directory. **(B)** The set of all indices used to reference a file, database, or system. **(C)** The index to all other indices. **(D)** To enter information about a file, database, or system as in definitions (A) and (B). (C) 610.5-1990
(2) The metadata describing all the software objects that are a part of a single software_collection (distribution or installed_software object). Catalogs exist both in distributions and for installed software, although storage of catalogs for installed software is undefined within this standard. A catalog in a distribution shall always use the exported catalog structure, since it is required to be stored in a portable or exported catalog structure. A catalog for installed software shall use the exported catalog structure when information is listed with `swlist -v`. (C/PA) 1387.2-1995

catastrophic failure (1) (software) A failure of critical software. (C) 610.12-1990
(2) Failure that is both sudden and complete. (R) [29]
(3) A failure of any portion of the GIS due to internal or external faults that results in sufficient damage to that portion of the GIS that it cannot be returned to service without major repairs. (PE/SUB) C37.123-1996

catcher *See:* output resonator.

catcher space (velocity-modulated tube) The part of the tube following the drift space, and where the density modulated-electron beam excites oscillations in the output resonator. It is the space between the output-resonator grids. *See also:* velocity-modulated tube. (ED) [45], [84]

categorization *See:* generalization.

category (1) In pattern recognition, a synonym for **pattern class.** (C) 610.4-1990w
(2) An attribute of an anomaly to which a group of classifications belongs. (C/SE) 1044-1993
(3) An attribute of an anomaly to which a group of classifications belongs. A specifically defined division in a system of classification; class. (C/SE) 1044.1-1995

category cluster *See:* subclass cluster.

category discriminator *See:* discriminator.

category entity An entity whose instances represent a subtype or subclassification of another entity (generic entity). *Synonyms:* subclass; subtype. (C/SE) 1320.2-1998

Category 3 balanced cabling Balanced 100 Ω and 120 Ω cables and associated connecting hardware whose transmission characteristics are specified up to 16 MHz. Commonly used by IEEE 802.3 10BASE-T installations. (C/LM) 802.3-1998

Category 4 balanced cabling Balanced 100 Ω and 120 Ω cables and associated connecting hardware whose transmission characteristics are specified up to 20 MHz. (C/LM) 802.3-1998

Category 5 balanced cabling Balanced 100 Ω and 120 Ω unshielded twisted-pair (UTP) cables and associated connecting hardware whose transmission characteristics are specified up to 100 MHz. (C/LM) 802.3-1998

catelectrotonus (electrobiology) Electrotonus produced in the region of the cathode. *See also:* excitability. (EMB) [47]

catenate *See:* concatenate.

cathode (1) (electron tube or valve) An electrode through which a primary stream of electrons enters the interelectrode space. *See also:* electrode. (ED) 161-1971w, [45]
(2) (semiconductor rectifier diode) The electrode to which the forward current flows within the cell. *See also:* semiconductor. (IA) 59-1962w, [12]
(3) (electrolytic) An electrode through which current leaves any conductor of the nonmetallic class. Specifically, an electrolytic cathode is an electrode at which positive ions are discharged, or negative ions are formed, or at which other reducing reactions occur. *See also:* electrolytic cell. (EEC/PE) [119]
(4) (thyristor) The electrode by which currents leaves the thyristor when the thyristor is in the ON state with the gate open-circuited. *Note:* This term does not apply to bidirectional thyristors. (ED/IA) [46], [62]

cathode border (gas) (gas tube) The distinct surface of separation between the cathode dark space and the negative glow. *See also:* discharge. (Std100) [84]

cathode cleaning (electroplating) Electrolytic cleaning in which the metal to be cleaned is the cathode. *See also:* battery. (EEC/PE) [119]

cathode coating impedance (electron tube) The impedance excluding the cathode interface (layer) impedance, between the the base metal and emitting surface of a coated cathode. (ED) 161-1971w, [45]

cathode, cold *See:* cold cathode.

cathode current *See:* electrode current; electronic controller.

cathode current, peak (A) (fault) The highest instantaneous value of a nonrecurrent pulse of cathode current occurring under fault conditions. *See also:* electrode current. **(B) (steady-state)** The maximum instantaneous value of a periodically recurring cathode current. *See also:* electrode current. **(C) (surge)** The highest instantaneous value of a randomly recurring pulse of cathode current. *See also:* electrode current. (ED) [45]

cathode dark space (gas tube) The relatively nonluminous region in a glow-discharge cold-cathode tube between the cathode glow and the negative glow. *Synonym:* Crookes dark space. *See also:* gas tube. (ED) [45]

cathode efficiency The current efficiency of a specified cathodic process. *See also:* electrochemistry.

cathode fall (gas) The difference of potential due to the space charge near the cathode. *See also:* discharge. (ED) [45], [84]

cathode follower A circuit in which the output load is connected in the cathode circuit of an electron tube and the input is applied between the control grid and the remote end of the cathode load, which may be at ground potential. *Note:* The circuit is characterized by low output impedance, high input impedance, gain less than unity, and negative feedback. (AP/BT/ANT) 145-1983s, 182A-1964w

cathode glow (gas tube) The luminous glow that covers all, or part, of the surface of the cathode in a glow-discharge cold-cathode tube, between the cathode and the cathode dark space. *See also:* gas tube. (ED) 161-1971w, [45]

cathode heating time (vacuum tubes) The time required for the cathode to attain a specified condition; for example: a specified value of emission, or a specified rate of change of emission. *Note:* All electrode voltages are to remain constant during measurement. The tube elements must all be at room temperature at the start of the test. *See also:* operation time. (ED) 161-1971w, [45]

cathode interface (layer) capacitance (electron tube) A capacitance that, in parallel with a suitable resistance, forms an impedance approximating the cathode interface impedance. *Note:* Because the cathode interface impedance cannot be represented accurately by the two-element resistance-capacitance circuit, this value of capacitance is not unique. (ED) 161-1971w, [45]

cathode interface (layer) impedance (electron tube) An impedance between the cathode base and coating. *Note:* This impedance may be the result of a layer of high resistivity or a poor mechanical bond between the cathode base and coating. (ED) 161-1971w, [45]

cathode interface (layer) resistance (electron tube) The low-frequency limit of cathode interface impedance. (ED) 161-1971w, [45]

cathode, ionic-heated *See:* ionic-heated cathode.

cathode layer A molten metal or alloy forming the cathode of an electrolytic cell and which floats on the fused electrolyte, or upon which fused electrolyte floats. *See also:* fused electrolyte. (EEC/PE) [119]

cathode luminous sensitivity (photocathodes) The quotient of photoelectric emission current from the photocathode by the incident luminous flux under specified conditions of illumination. *Notes:* 1. Since cathode luminous sensitivity is not an absolute characteristic but depends on the spectral distribution of the incident flux, the term is commonly used to designate the sensitivity to radiation from a tungsten filament lamp operating at a color temperature of 2870 kelvins. 2. Cathode luminous sensitivity is usually measured with a collimated beam at normal incidence. *See also:* phototube. (ED/NPS) 161-1971w, 398-1972r

cathode modulation Modulation produced by application of the modulating voltage to the cathode of any electron tube in which the carrier is present. *Note:* Modulation in which the cathode voltage contains externally generated pulses is called cathode pulse modulation. (EEC/PE) [119]

cathode, pool *See:* pool cathode.

cathode, preheating time *See:* preheating time cathode.

cathode pulse modulation Modulation produced in an amplifier or oscillator by application of externally generated pulses to the cathode circuit. (AP/ANT) 145-1983s

cathode radiant sensitivity (photocathodes) The quotient of the photoelectric emission current from the photocathode by the incident radiant flux at a given wavelength under specified conditions of irradiation. *Note:* Cathode radiant sensitivity is usually measured with a collimated beam at normal incidence. (ED/NPS) 161-1971w, 398-1972r

cathode-ray charge-storage tube A charge-storage tube in which the information is written by means of cathode-ray beam. *Note:* Dark-trace tubes and cathode-ray tubes with a long persistence are examples of cathode-ray storage tubes that are not charge-storage tubes. Most television camera tubes are examples of charge-storage tubes that are not cathode-ray storage tubes. *See also:* charge-storage tube. (ED) 158-1962w

cathode-ray oscillograph An oscillograph in which a photographic or other record is produced by means of the electron beam of a cathode-ray tube. *Note:* The term "cathode-ray oscillograph" has frequently been applied to a cathode-ray oscillograph, but this usage is deprecated. *See also:* oscillograph. (EEC/PE) [119]

cathode-ray oscilloscope An oscilloscope that employs a cathode-ray tube as the indicating device. *See also:* oscillograph. (EEC/PE) [119]

cathode ray storage A type of matrix storage in which a cathode ray beam is used to access data. (C) 610.10-1994w

cathode-ray storage tube A storage tube in which the information is written by means of a cathode-ray beam. *See also:* storage tube. (ED) 158-1962w

cathode-ray tube (1) (supervisory control, data acquisition, and automatic control) (station control and data acquisition) A display device in which controlled electron beams are used to present alphanumeric or graphical data on an electroluminescent screen. (SUB/PE/SWG-OLD) C37.1-1987s, C37.100-1992
(2) (computer graphics) An evacuated glass tube in which a beam of electrons is emitted and focused onto the phosphor-coated display surface of the tube. A beam deflection system moves the beam as required to generate an image.

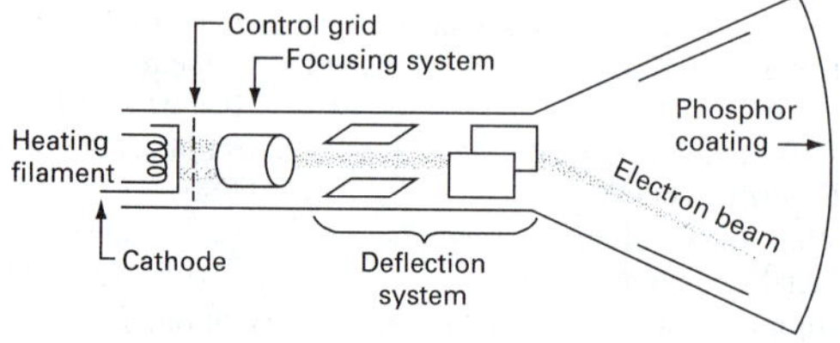

cathode-ray tube

(C) 610.6-1991w
(3) An evacuated glass tube in a well-defined and controllable beam of electrons is focused onto a phosphor-coated display surface of the tube causing the phosphors to emit light. *Note:* A beam deflection system moves the beam as required to generate an image. *See also:* Williams-tube storage; storage tube. (C) 610.10-1994w

cathode-ray-tube display area *See:* graticule area.

cathode ray tube display device A display device that presents data in visual form by means of controlled electron beams within a cathode ray tube. (C) 610.6-1991w

cathode region (gas) (gas tube) The group of regions that extends from the cathode to the Faraday dark space inclusively. *See also:* discharge. (EEC/ACO) [109], [84]

cathode spot An area on the cathode of an arc from which electron emission takes place at a current density of thousands of amperes per square centimeter and where the temperature of the electrode is too low to account for such currents by thermionic emission. *See also:* gas tube. (EEC/ACO) [109]

cathode sputtering (gas) The emission of fine particles from the cathode (or anode) produced by positive ion (or electron) bombardment. *Synonym:* anode sputtering. *See also:* discharge. (ED) [45], [84]

cathode terminal (1) (semiconductor devices) The terminal from which forward current flows to the external circuit. *Note:* In the semiconductor rectifier components field, the cathode terminal is normally marked positive. *See also:* semiconductor rectifier cell; semiconductor. (IA) [62], [12], 59-1962w

(2) **(thyristor)** The terminal that is connected to the cathode. *Note:* The term does not apply to bidirectional thyristors. *See also:* anode. (IA/ED) 223-1966w, [12], [46]

cathodic corrosion An increase in corrosion of a metal by making it cathodic. *See also:* stray-current corrosion. (IA) [71], [59]

cathodic polarization Polarization of a cathode. *See also:* electrochemistry. (IA) [59], [71]

cathodic protection Reduction or prevention of corrosion by making a metal the cathode in a conducting medium by means of a direct electric current (which is either impressed or galvanic). (IA) [59], [71]

catholyte The portion of an electrolyte in an electrolytic cell adjacent to a cathode. If a diaphragm is present, it is the portion of electrolyte on the cathode side of the diaphragm. *See also:* electrolytic cell. (EEC/PE) [119]

cation A positively charged ion or radical that migrates toward the cathode under the influence of a potential gradient. *See also:* ion. (IA) [59], [71]

CATV *See:* community antenna television.

CATV-type broadband medium *See:* community antenna television (CATV)-type broadband medium.

cat whisker A small, sharp-pointed wire used to make contact with a sensitive point on the surface of a semiconductor. (EEC/PE) [119]

caustic A point in space where geometric or ray optics theory predicts infinite field strength. (AP/PROP) 211-1997

caustic embrittlement Stress-corrosion cracking in alkaline solutions. (IA) [59], [71]

caustic soda cell A cell in which the electrolyte consists primarily of a solution of sodium hydroxide. *See also:* electrochemistry. (EEC/PE) [119]

caution Advisory in a software user document that performing some action may lead to consequences that are unwanted or undefined. *Contrast:* warning. (C/SE) 1063-1987r

cavitation (1) (liquid) Formation, growth, and collapse of gaseous and vapor bubbles due to the reduction of pressure of the cavitation point below the vapor pressure of the fluid at the working temperature. (SP) [32]

(2) (hydroelectric power plants) With respect to an operating hydraulic turbine, the formation of vapor-filled bubbles in high velocity, low pressure regions of the water passage—for example, around the turbine runner. The rapid collapse of the bubbles as they are propelled out of the low pressure region produces a pressure wave, which can erode nearby material. (PE/EDPG) 1020-1988r

cavitation damage Deterioration caused by formation and collapse of cavities in a liquid. (IA) [59]

cavity *See:* optical cavity; unloaded applicator impedance.

cavity dumpers (acousto-optic device) Generally, a fast rise-time pulse modulator used intracavity. (UFFC) [17]

cavity ratio (CR) A number indicating cavity proportions.

$$CR = \frac{5 \times (\text{height of C}) \times (\text{C} + \text{C width})}{(\text{C length}) \times (\text{C width})}$$

For cavities of irregular shape,

$$CR = \frac{2.5 \times (\text{C height}) \times (\text{C perimeter})}{(\text{area of C base})}$$

Note: The relationship between "cavity ratio" and "room coefficient" should be noted. If the entire room is considered as a cavity, the room height becomes the cavity height and CR $= 10K_r$. (EEC/IE) [126]

cavity resonator (1) A space normally bounded by an electrically conducting surface in which oscillating electromagnetic energy is stored, and whose resonant frequency is determined by the geometry of the enclosure. *See also:* waveguide. (AP/ANT) 145-1983s

(2) (waveguide components) A resonator formed by a volume of propagating medium bounded by reflecting surfaces. *See also:* waveguide resonator. (MTT) 147-1979w

cavity resonator frequency meter (waveguide components) A cavity resonator used to determine frequency. *See also:* cavity resonator. (MTT) 147-1979w

CAX *See:* unattended automatic exchange.

C-band A radar-frequency band between 4 GHz and 8 GHz, usually in the International Telecommunications Union (ITU) allocated band 5.2–5.9 GHz. (AES) 686-1997

C battery A battery designed or employed to furnish voltage used as a grid bias in a vacuum-tube circuit. *See also:* battery. (EEC/PE) [119]

CBCT *See:* customer-bank communication terminal.

CBE *See:* computer-based education.

CBEMA *See:* Computer and Business Equipment Manufacturers Association.

CBL *See:* computer-based learning.

CCA *See:* carrier-controlled approach system.

CCB *See:* change control board; configuration control board.

CCD *See:* charge-coupled device.

C channel (1) A channel that provides an integer multiple of 64 kbit/s, full-duplex, isochronous clear channels. The "C" shall be used to indicate that this is a circuit-switched channel. C_m stands for a channel (m) times 64 kbit/s in size. In general, a C channel has the same characteristics as an ISDN B or H channel, except that it can be any multiple of 64 kbit/s in size, rather than only the ITU-T approved rates. There may be multiple C_m channels of varying sizes ($n \times C_m$). For ISLAN4-T/ISLAN16-T, the following also applies: The maximum composite bandwidth allocated for the C channel(s) shall be limited by the application requirements for the P channel. That is, the available bandwidth that may be assigned for C channels shall be limited to the difference between the TDM payload field minus the bandwidth required for the P channel. In practice, the C channel size(s) will be effected by the interconnection facility(ies) between the AU and the ISDN wide area network (WAN). (C/LM) 802.9a-1995w

(2) A channel that provides an integer multiple of 64 kb/s, full duplex, isochronous clear channels. The "C" is used to indicate that this is a circuit-switched channel. $C sub m$ stands for a channel (m) times 64 kb/s in size. In general, a C channel has the same characteristics as an ISDN B or H channel, except that it can be any multiple of 64 kb/s in size, rather than only the CCITT approved rates. There may be multiple C channels of varying sizes ($n \times C$). Note the following equivalencies:

$C_1 = B = 64$ kb/s

$C_6 = H_0 = 384$ kb/s

$C_{24} = H_{11} = 1.536$ Mb/s

$C_{30} = H12 = 1.920$ Mb/s

The maximum composite bandwidth allocated for the C channel(s) shall be limited by the application requirements for the P channel. That is, the available bandwidth that may be assigned for C channels shall be limited to the difference between the TDM payload field minus the bandwidth required for the P channel. In practice, the C channel size(s) will be effected by the interconnection facility(ies) between the AU and the ISDN wide area network (WAN). (LM/C/COM) 8802-9-1996

CCITT (Comité Consultatif International de Télégraphique et Téléphonique) *See:* Consultative Committee on International Telegraphy and Telephony; Comité Consultatif Internationale Télégraphique et Téléphone.

CCITT Standard A standard recommended by CCITT (Consultative Committee on International Telegraphy and Telephony). *Notes:* 1. CCITT recommendations are published by lettered series. For example, the X-series, for equipment and protocols used with computer networks. 2. CCITT Standards are now designated as ITU-T Standards. *See also:* Consultative Committee on International Telegraphy and Telephony. (C) 610.7-1995

C conditioning A North American term for a type of conditioning that controls attenuation, distortion, and delay distortion, thus making transmission impairments of a circuit lie within specified limits. *See also:* D conditioning. (C) 610.7-1995

CCR *See:* condition code register.

CCS Hundreds of call seconds per hour. A measure of traffic intensity or event data. *Synonym:* call attempts per hour. *See also:* time-consistent traffic measures. (COM/TA) 973-1990w

CD *See:* compact disc; collision detection.

C-display A rectangular display in which each target appears as an intensity-modulated blip with azimuth indicated by the horizontal coordinate and angle of elevation by the vertical coordinate.

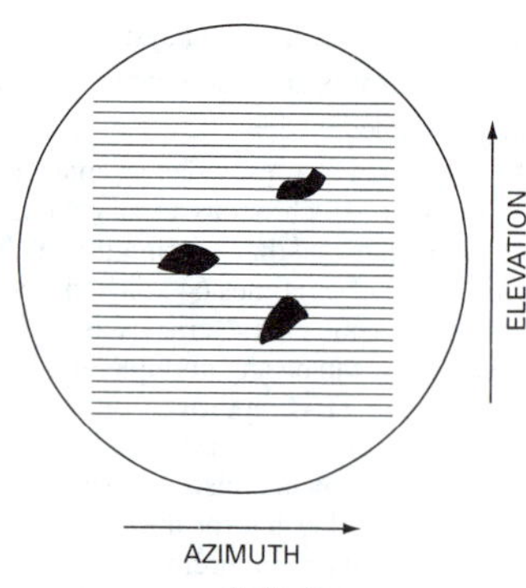

C-display

(AES) 686-1997

CDL *See:* Computer Design Language.

CDO *See:* unattended automatic exchange.

CD1 *See:* clocked data one.

CDP *See:* commercial data processing.

CDR *See:* critical design review.

CD-ROM *See:* compact disc read-only memory.

CD-ROM storage A read-only form of optical storage employing compact discs to store information. (C) 610.10-1994w

CD state The state of an analog pin when it is isolated from the core circuit and all test circuits. *Note:* When a pin is in the CD state, there may be residual elements to which it remains connected. *See also:* core disconnect; residual element. (C/TT) 1149.4-1999

CD0 *See:* clocked data zero.

C-effective A capacitance value, often computed as an approximation to a resistor/inductor/capacitor (RLC) network or a π-model, that characterizes the admittance of an interconnect structure at a particular driver. The reduction of real parasitics and pin capacitances to a C-effective allows the calculation of delay and slew values from cell characterization data which assumes a pure capacitive output load. *Synonym:* effective capacitance. (C/DA) 1481-1999

CEI (television) The initials of the official French name, Commission Éléctrotechnique Internationale, of the International Electrotechnical Commission (IEC). (BT/AV) 201-1979w

ceiling The result obtained by rounding a number up to the nearest integer. For example, the ceiling of 5.3 is 6. *Contrast:* floor. (C) 1084-1986w

ceiling area lighting (illuminating engineering) A general lighting system in which the entire ceiling is, in effect, one large luminaire. *Note:* Ceiling area lighting includes luminous ceilings and louvered ceilings. *See also:* luminous ceiling; louvered ceiling. (EEC/IE) [126]

ceiling cavity ratio (illuminating engineering) For a cavity formed by the ceiling, the plane of the luminaire, and the wall surfaces between these two planes, the CCR is computed by using the distance from the plane of the luminaire to the ceiling (h_c as the cavity height in the equations given in the definition of "cavity ratio." (EEC/IE) [126]

ceiling current (excitation systems for synchronous machines) The maximum direct current that the excitation system is able to supply from its terminals for a specified time. (PE/EDPG) 421.2-1990, 421.1-1986r

ceiling direct voltage (direct potential rectifier unit) The average direct voltage at rated direct current with rated sinusoidal voltage applied to the alternating-current line terminals, with the rectifier transformer set on rated voltage tap and with voltage regulating means set for maximum output. *See also:* rectification; power rectifier. (IA/EEC/PCON) [62], [110]

ceiling projector (illuminating engineering) A device designed to produce a well-defined illuminated spot on the lower portion of a cloud for the purpose of providing a reference mark for the determination of the height of that part of the cloud. (EEC/IE) [126]

ceiling ratio (illuminating engineering) The ratio of the luminous flux which reaches the ceiling directly to the upward component of the luminaire. (EEC/IE) [126]

ceiling voltage (1) (excitation systems for synchronous machines) The maximum direct voltage that the excitation system is able to supply from its terminals under defining conditions. *Notes:* 1. The no-load ceiling voltage is determined with the excitation system supplying no current. 2. The ceiling voltage under load is determined with the excitation system supplying ceiling current. 3. For excitation systems whose supply depends on the synchronous machine voltage and (if applicable) current, the nature of power system disturbance and specific design parameters of the excitation system and the synchronous machine influence the excitation system output. For such systems, the ceiling voltage is determined considering an appropriate voltage drop and (if applicable) current increase. 4. For excitation systems employing a rotating exciter, the ceiling voltage is determined at rated speed. (PE/EDPG) 421.1-1986r
(2) (excitation systems) The maximum direct voltage that the excitation system is able to supply from its terminals under defined conditions. (PE/EDPG) 421.2-1990

ceiling voltage, exciter nominal The ceiling voltage of an exciter loaded with a resistor having an ohmic value equal to the resistance of the field winding to be excited and with this field winding at a temperature of (A) 75°C for field windings designed to operate at rating with a temperature rise of 60°C or less; or (B) 100°C for field windings designed to operate at rating with a temperature rise greater than 60°C. (PE/EDPG) 421-1972s

ceilometer (navigation aid terms) An instrument for measuring the height of clouds. (AES/GCS) 172-1983w

celestial fix (navigation aid terms) A position fix established by observation of celestial bodies. (AES/GCS) 172-1983w

celestial-inertial navigation equipment (navigation aid terms) An equipment employing both celestial and inertial sensors. *Synonyms:* astro-inertial navigation equipment; stellar-inertial navigation equipment. (AES/GCS) 172-1983w

celestial mechanics (communication satellite) The mechanics of motion of celestial bodies, including satellites. (COM) [19]

celestial navigation (navigation aid terms) Navigation with the aid of celestial bodies. Applied principally to the measurement of the altitudes of a celestial body. (AES/GCS) 172-1983w

cell (1) (lead-acid batteries for photovoltaic systems) The basic electrochemical unit, characterized by an anode and a cathode used to receive, store, and deliver electrical energy. For a lead-acid system, the cell is characterized by a nominal two-volt potential. (PV) 937-1987s
(2) (batteries for photovoltaic systems) The basic electrochemical unit, characterized by an anode, a cathode, and electrolyte, used to receive, store, and deliver electrical energy. *Notes:* 1. For a nickel-cadmium cell, the nominal voltage is 1.2 V. 2. For a lead-acid cell, the nominal voltage is 2.0 V. (PV) 1013-1990, 1144-1996, 1145-1990s

(3) (test pattern language) The element of a memory in which one bit is stored. (TT/C) 660-1986w
(4) (information storage) An elementary unit of storage; for example, binary cell, decimal cell. (C) 162-1963w, 270-1966w
(5) A single, enclosed tubular space in a cellular metal floor member, the axis of the cell being parallel to the axis of the metal floor member. (NESC/NEC) [86]
(6) The primary unit of information in the architecture of a Forth System. (C/BA) 1275-1994
(7) A packet with fixed length. *Notes:* 1. Each cell has a 5 octet header and 48 octets of data. 2. This definition is specific to asynchronous transfer mode (ATM). (C) 610.7-1995
(8) (A) A module used in assembling application-specific integrated circuits. **(B)** The storage position of one unit of information, such as a character, a bit, or a word. *See also:* storage cell. (C) 610.10-1994
(9) A primitive in an integrated circuit library. Primitive means the timing properties of the cell are directly described in the DPCM without reference back to the application for the internal structure of the cell. This primitiveness typically is a result of the characterization of that cell by the semiconductor vendor, but it may instead be a result of the construction of a timing model for a subcircuit by the application, and its loading into the DPCM at run-time. For PDEF, cell refers to a logical instance, and a library primitive is called a gate. The term cell can arise in the context of the abstraction of a type of cell available in the library or in the concrete selection and placement of a cell in the final design. If the context is not clear, the terms cell type and cell instance (or just instance) shall be used. *See also:* cell type; instance. (C/DA) 1481-1999
(10) (data management) *See also:* data item; data element. (C) 610.5-1990w
(11) *See also:* memory cell. (ED) 1005-1998

cell cavity (electrolysis) The container formed by the cell lining for holding the fused electrolyte. *See also:* fused electrolyte. (PE/EEC) [119]

cell connector (storage cell) An electric conductor used for carrying current between adjacent storage cells. *See also:* battery. (EEC/PE) [119]

cell constant (electrolytic cells) The resistance in ohms of that cell when filled with a liquid of unit resistivity. (PE/EEC) [119]

cell cover The transparent medium (glass, quartz, etc.) that protects the solar cells from space particulate radiation. (AES/SS) 307-1969w

cell instance A particular, concrete appearance of a cell in the fully expanded (flattened, unfolded, elaborated) design description of an integrated circuit; also referred to elsewhere as an "occurrence." An instance is a "leaf" of the unfolded design hierarchy. In Physical Design Exchange Format (PDEF), this is a physical cluster or a logical cell. *See also:* cell; cell type; cluster; hierarchical instance. (C/DA) 1481-1999

cell library A collection of cells used to design and lay out application-specific integrated circuits in accordance with the functional requirements of particular end users. (C) 610.10-1994w

cell line An assembly of electrically interconnected electrolytic cells supplied from a source of dc power. (NESC/IA/PC) 463-1993w, [86]

cell line attachments and auxiliary equipment Include, but are not limited to: auxiliary tanks; process piping; duct work, structural supports, exposed cell line conductors; conduits and other raceways; pumps, positioning equipment, and cell cutout or by-pass electrical devices. Auxiliary equipment includes tools, welding machines, crucibles, and other portable equipment used for operation and maintenance within the electrolytic cell line working zone. In the cell line working zone, auxiliary equipment includes the exposed conductive surfaces of the ungrounded cranes and crane-mounted cell-servicing equipment. (NESC/NEC) [86]

cell line potential (electrolytic cell line) The dc voltage applied to the positive and negative buses supplying power to a cell line. (IA/PC) 463-1993w, 463-1977s

cell line voltage (electrolytic cell line) The dc voltage applied to the positive and negative buses supplying power to a cell line. (IA/PC) 463-1993w, 463-1977s

cell line working zone The space envelope where operation or maintenance is normally performed on or in the vicinity of exposed energized surfaces of electrolytic cell lines or their attachments. (IA/PC) 463-1993w

cell-organized raster display device A raster display device on which an image is constructed by a collection of rectangular characters, each character represented by an n-by-m set of illuminated control indicators. (C) 610.10-1994w

cell potential (electrolytic cells) The dc voltage between the positive and negative termials of one electrolytic cell. (IA/PC) 463-1977s

cell relay A fast packet switching technology that provides a virtual circuit service for the transfer of cells. For example, asynchronous transfer mode is the most common type of cell relay. *See also:* frame relay. (C) 610.7-1995

cell size The rated capacity of a lead-acid cell or the number of positive plates in a cell. (SCC29) 485-1997

cell switching A technique used in data communications, in which messages are broken into fixed-size packets and forwarded to another party over the network. *Contrast:* packet switching. (C) 610.7-1995

cell type (1) (Class 1E lead storage batteries) Cells of identical design (for example, plate size, alloy, construction details), but that may have differences in the number of plates and spacers, quantity of electrolyte, or length of container. (PE/EDPG) 535-1979s
(2) Name used to identify a particular cell in the library. (C/DA) 1481-1999

cell-type tube (microwave gas) A gas-filled radio-frequency switching tube that operates in an external resonant circuit. *Note:* A tuning mechanism may be incorporated in either the external resonant circuit or the tube. *See also:* gas tube. (ED) 161-1971w

cellular metal floor raceway The hollow spaces of cellular metal floors, together with suitable fittings, that may be approved as enclosures for electric conductors. (NESC/NEC) [86]

cellulose Unbleached kraft insulation material from which paper and transformerboard are made, that is suitable for use in 65°C average winding temperature rise insulation systems. (PE/TR) 1276-1997

cell voltage (electrolytic cell line working zone) The dc voltage between the positive and negative terminals of one electrolytic cell. (IA/PC) 463-1993w

cent The interval between two sounds whose basic frequency ratio is the twelve-hundredth root of 2. *Note:* The interval, in cents, between any two frequencies is 1200 times the logarithm to the base of 2 of the frequency ratio. Thus, 1200 cents equal 12 equally tempered semitones equal 1 octave. (SP/ACO) 157-1951w, [32]

center In text formatting, to format one or more lines of text so that the left and right margins are equal in size. (C) 610.2-1987

center-break switching device A mechanical switching device in which both contacts are movable and engage at a point substantially midway between their supports. (SWG/PE) C37.100-1992

center-center point A point on a VDT screen [cathode ray tube (CRT) face panel] that is both the horizontal and vertical midpoint. The center-center point is represented by the bisector of the horizontal and vertical center lines on the VDT screen. (EMC) 1140-1994r

center frequency (1) (frequency modulation) The average frequency of the emitted wave when modulated by a symmetrical signal. *See also:* frequency modulation. (AP/ANT) 145-1983s

(2) **(burst measurements)** The arithmetic mean of the two frequencies that define the bandwidth of a filter. *See also:* burst. (SP) 265-1966w

(3) **(spectrum analyzer) (non-real time spectrum analyzer)** That frequency which corresponds to the center of a frequency span. (IM) 748-1979w

center-frequency delay (1) The nominal group delay of the device at the center frequency f_o, generally expressed in microseconds, where f_o is defined as the lower band-edge frequency plus one-half the bandwidth (dispersive delay line). (UFFC) 1037-1992w

(2) The frequency delay of the device at the center frequency, F_0, generally expressed in microseconds. (UFFC) [22]

centerline (navigation aid terms) The lows of the points equidistant from two reference points or lines, as the perpendicular bisector of the baseline of a hyperbolic system of navigation, such as loran. (AES/GCS) 172-1983w

center of distribution (primary distribution) The point from which the electric energy must be supplied if the minimum weight of conducting material is to be used. *Note:* The center of distribution is commonly considered to be that fixed point that, in practice, most nearly meets the ideal conditions stated above. (PE/T&D) [10]

center pivot irrigation machines A center pivot irrigation machine is a multi-motored irrigation machine that revolves around a central pivot and employs alignment switches or similar devices to control individual motors. (NESC/NEC) [86]

center wavelength (1) The wavelength that is the arithmetic mean of the half-maximum spectral intensity points of the transmitter. If the spectral intensity distribution is symmetric and singly peaked, the center wavelength is at maximum intensity. (C/BA) 1393-1999

(2) The average of two optical wavelengths at which the spectral radiant intensity is 50% of the maximum value. (C/LM) 802.3-1998

centi (mathematics of computing) A prefix indicating one hundredth (10^{-2}). (C) 1084-1986w

centimeter-gram-second electromagnetic system of units A system in which the basic units are the centimeter, gram, second, and abampere. *Notes:* 1. The abampere is a derived unit defined by assigning the magnitude one to the unrationalized magnetic constant (sometimes called the permeability of space). 2. Most electrical units of this system are designated by prefixing the syllable "ab-" to the name of the corresponding unit in the mksa system. Exceptions are the maxwell, gauss, oersted, and gilbert. (Std100) 270-1966w

centimeter-gram-second electrostatic system of units The system in which the basic units are the centimeter, gram, second, and statcoulomb. *Notes:* 1. The statcoulomb is a derived unit defined by assigning the magnitude 1 to the unrationalized electric constant (sometimes called the permittivity of space). 2. Each electrical unit of this system is commonly designated by prefixing the syllable "stat-" to the name of the corresponding unit in the International System of Units. (Std100) 270-1966w

centimeter-gram-second system of units A system in which the basic units are the centimeter, gram, and second. (Std100) 270-1966w

central alarm station (CAS) A continuously manned location that provides primary security system monitoring and communications functions. (PE/NP) 692-1997

central arbiter requester A module that requests access to the bus on a system that uses central arbitration. (C/BA) 896.4-1993w

central computer *See:* host computer.

central control room (nuclear power generating station) A continuously manned, protected enclosure from which actions are normally taken to operate the nuclear generating station under normal and abnormal conditions. (PE/NP) 567-1980w

central distribution frame grounding A type of grounding system where all signal grounds are referenced to a central point rather than at their respective signal sources. (PE/EDPG) 1050-1996

centralized accounting, automatic message (CAMA) (telephone switching systems) An arrangement at an intermediate office for collecting automatic message accounting information. (COM) 312-1977w

centralized computer network (1) A computer network configuration in which a central node provides computing power, control, or other services. *See also:* decentralized computer network. (LM/COM) 168-1956w

(2) A computer network in which a central node provides all network control functions and services to other nodes. *Contrast:* decentralized computer network. *See also:* distributed computer network. (C) 610.7-1995

centralized control (1) (electric pipe heating systems) A common (central) point where multiple control, alarm, or both signals or functions are brought together. With respect to electric pipe heating systems, centralized control/alarm stations usually consist of cabinets or panels where remote control, alarm, or both signals are brought together for a common output signal to the generating unit control room. (PE/EDPG) 622A-1984r

(2) **(electric heat tracing systems)** A common (central) point where multiple control, alarm, or both signals or functions are brought together. With respect to electric heat tracing systems, centralized control/alarm stations usually consist of cabinets or panels where remote control, alarm, or both signals are brought together for a common output signal to the generating unit control room. (PE/EDPG) 622B-1988r

(3) The local, remote, or programmed operations of more than one equipment or system from one central control station in a common area or room by a broadly skilled operator or supervisory system. (IA/MT) 45-1998

centralized polling A polling technique in which a single, central authority controls access to transmission medium. Each station is invited to transmit periodically according to a scheme or list. *Contrast:* distributed polling. (C) 610.7-1995

centralized test system (test, measurement, and diagnostic equipment) A test system that processes, records, or displays at a central location information gathered by test point data sensors at more than one remotely located equipment or system under test. (MIL) [2]

centralized traffic-control machine (railway practice) A control machine for operation of a specific type of traffic control system of signals and switches. *See also:* centralized traffic-control system. (EEC/PE) [119]

centralized traffic-control system (railway practice) A specific type of traffic control system in which the signals and switches for a designated section of track are controlled from a remotely located centralized traffic control machine. *See also:* electropneumatic interlocking machine; control machine; centralized traffic-control machine; block-signal system. (EEC/PE) [119]

central office (CO) (1) (telephone loop performance) The building, one or more switching systems, and related equipment contained therein that provide telephone service. (COM/TA) 820-1984r

(2) **(data transmission)** The place where communications common carriers terminate customer lines and locate the equipment which interconnects those lines. Usually the junction point between metallic pair and carrier system. (PE) 599-1985w

(3) **(telephone switching systems)** A switching entity that has one or more office codes and a system control serving a telecommunication exchange. (COM) 312-1977w

(4) A physical location where communications common carriers terminate customer lines and locate the switching devices that interconnect these lines. *Synonyms:* telephone exchange; local central office; exchange. *See also:* end office. (C) 610.7-1995

central office diagram (telephone switching systems) A simplified switching network plan for a given installation, specifying types and quantities of equipment and trunk groups and other parameters. (COM) 312-1977w

central office exchange (data transmission) The place where a communication common carrier locates the equipment which interconnects incoming subscribers and circuits. (PE) 599-1985w

central office service unit (COSU) A telephone company controller resident in a central office that connects to the utility controller and, via the utility telemetry trunk, to the switch. The COSU provides the function of originating and terminating calls to and from telemetry interface units (TIUs). For the COSU access method, the COSU performs a security check with the utility controller and places a call to the end user in response to the information sent to it by the utility controller. The COSU performs a security check and initiates a connection to the utility controller when called by the TIU. The COSU also provides a multiplexing interface between the utility controller and COSU and the COSU and TIU(s). The COSU may also provide traffic measurements. (AMR/SCC31) 1390-1995, 1390.2-1999, 1390.3-1999

central office service unit (COSU) access method An access method that utilizes the switched telephone network, comprised of a COSU, switch, and other network elements. This method provides for automatically invoking/ignoring certain switch-based telemetry communications capabilities and establishes a communications path between a utility/enhanced service provider (ESP) and a telemetry interface unit (TIU). (SCC31/AMR) 1390.3-1999, 1390.2-1999, 1390-1995

central processing unit (CPU) **(1)** The unit of a computing system that includes the circuits controlling the interpretation of instructions and their execution. (C/C) [20], [85]
(2) Describes that part of a computer that does the primary computational functions. Loosely describes the computer system other than connected input and output devices. (C/BA) 1496-1993w
(3) That unit of a computer system which fetches, decodes and executes programmed instructions and maintains the status of results as the program is executed. *Synonym:* central processor. *See also:* uniprocessor; processor. (C) 610.10-1994w

central processor *See:* central processing unit.

central services module A specific module that is required in all systems using the parallel system bus. Its services, such as starting certain bus operations and guaranteeing uniform initialization of all agents, are required by all agents on the parallel system bus. It is always located in a specific slot in the system backplane. *See also:* parallel system bus. (C/MM) 1296-1987s

central station (protective signaling) An office to which remote alarm and supervisory signaling devices are connected, where operators supervise the circuits, and where guards are maintained continuously to investigate signals. *Note:* Facilities may be provided for transmission of alarms to police and fire departments or other outside agencies. *See also:* protective signaling. (EEC/PE) [119]

central station equipment (protective signaling) The signal receiving, recording, or retransmission equipment installed in the central station. *See also:* protective signaling. (EEC/PE) [119]

central station switchboard (protective signaling) That portion of the central station equipment on or in which are mounted the essential control elements of the system. *See also:* protective signaling. (EEC/PE) [119]

central station system (central office system) (protective signaling) A system in which the operations of electric protection circuits and devices are signaled automatically to, recorded in, maintained, and supervised from a central station having trained operators and guards in attendance at all times. *See also:* protective signaling. (EEC/PE) [119]

central vision (illuminating engineering) The seeing of objects in the central or foveal part of the visual field, approximately two degrees in diameter. It permits seeing much finer detail than does peripheral vision. *Synonym:* foveal vision. (EEC/IE) [126]

central visual field (illuminating engineering) That region of the visual field that corresponds to the foveal portion of the retina. (EEC/IE) [126]

centrex CO (telephone switching systems) (company) The provision of centrex service by switching, station equipment, and attendant facilities located on the premises of the customer. (COM) 312-1977w

centrex CU (customer) (telephone switching systems) The provision of centrex service by switching, station equipment, and attendant facilities located on the premises of the customer. (COM) 312-1977w

centrex service (telephone switching systems) A service that provides direct inward dialing and identified outward dialing in accordance with the national numbering plan for stations served as they would be by a private branch exchange. (COM) 312-1977w

centrifugal actuator (rotating machinery) Rotor-mounted element of a centrifugal starting switch. *See also:* centrifugal starting switch. (PE) [9]

centrifugal-mechanism pin (governor pin) A component of the linkage between the centrifugal mechanism weights and the short-circuiting device. *See also:* centrifugal starting switch. (EEC/PE) [119]

centrifugal-mechanism spring (governor spring) A spring that opposes the centrifugal action of the centrifugal-mechanism weights in determining the motor speed at which the switch or short-circuiting device is actuated. *See also:* centrifugal starting switch. (EEC/PE) [119]

centrifugal-mechanism weights (governor weights) Moving parts of the centrifugal-mechanism assembly that are acted upon by centrifugal force. *See also:* centrifugal starting switch. (EEC/PE) [119]

centrifugal relay An alternating-current frequency-selective relay in which the contacts are operated by a fly-ball governor or centrifuge driven by an induction motor. (EEC/PE) [119]

centrifugal starting switch (rotating machinery) A centrifugally operated automatic mechanism used to perform a circuit-changing function in the primary winding of a single-phase induction motor after the rotor has attained a predetermined speed, and to perform the reverse circuit-changing operation prior to the time the rotor comes to rest. *Notes:* 1. One of the circuit changes that is usually performed is to open or disconnect the auxiliary winding circuit. 2. In the usual form of this device, the part that is mounted to the stator frame or end shield is the starting switch, and the part that is mounted on the rotor is the centrifugal actuator. (PE) [9]

Centronics The popular name for the parallel printer port used as the parallel interface for most printers and supported by most "MS-DOS compatible" PCs. The name is derived from the printer manufacturer that introduced this interface, Centronics Data Computer Corporation. This interface has never been formalized. Despite a basic similarity, many variations of this interface have been implemented in different peripherals and hosts. This specification describes the more prevalent variations of the "Centronics" interface and defines a family of signaling methods that are backward compatible with the typical "Centronics" interface. (C/MM) 1284-1994

Centronics connector The popular name for the 36-pin ribbon contact type connector commonly used for the parallel port on printers. (C/MM) 1284-1994

CEP *See:* circular probable error.

ceramic insulator Insulators made from porcelain or glass or a general class of rigid material. (T&D/PE) 957-1995

ceraunic level *See:* keraunic level.

certificate A statement that an asset has been assessed according to specified certification criteria. (C/SE) 1420.1a-1996

certificate of conformance (replacement parts for Class 1E equipment in nuclear power generating stations) (nuclear power quality assurance) A document signed by an authorized individual, certifying the degree to which items or services meet specified requirements. (PE/NP) 934-1987w, [124]r

certification (1) (nuclear power quality assurance) The act of determining, verifying, or attesting in writing to the qualifications of personnel, processes, procedures, or items in accordance with specified requirements. (PE/NP) [124]
(2) (A) (software) A written guarantee that a system or component complies with its specified requirements and is acceptable for operational use. For example, a written authorization that a computer system is secure and is permitted to operate in a defined environment. **(B) (software)** A formal demonstration that a system or component complies with its specified requirements and is acceptable for operational use. **(C) (software)** The process of confirming that a system or component complies with its specified requirements and is acceptable for operational use. (C) 610.12-1990
(3) (test, measurement, and diagnostic equipment) Attestation that a support test system is capable, at the time of certification demonstration, of correctly assessing the quality of the items to be tested. This attestation is based on an evaluation of all support test system elements and establishment of acceptable correlation among similar test systems. (MIL) [2]

certification artifact The tangible results from a certification process (e.g., inspection checklists, metrics, problem reports). (C/SE) 1420.1a-1996

certification criteria A set of standards, rules, or properties to which an asset must conform in order to be certified to a certain level. Certification criteria are defined by a certification policy. Certification criteria may be specified as a set of certification properties that must be met. (C/SE) 1420.1a-1996

certification levels The step-wise organization of a certification policy by a reuse library into increasingly more stringent certification processes and criteria. (C/SE) 1420.1a-1996

certification method A documented technique applied as part of a certification process (e.g., inspections, static analysis, testing, formal verification). (C/SE) 1420.1a-1996

certification policy The statement of a reuse library's standard process for asset certification, the levels of certification, the properties and criteria for each level and the methods employed. (C/SE) 1420.1a-1996

certification process The process of assessing whether an asset conforms to predetermined certification criteria appropriate for that class of asset. (C/SE) 1420.1a-1996

certification property A statement about some feature or characteristic of an asset that may be assessed as being true or false during a certification process. Properties may relate to what an asset is, what it does, or how it relates to its operating environment. An assessment of a certification quality factor is accomplished by assessing the underlying certification properties. (C/SE) 1420.1a-1996

certification quality factor A high level aspect of an asset (e.g., completeness, correctness, reusability) that is assessed during a certification process. A certification quality factor is a manifestation of one or more certification properties. (C/SE) 1420.1a-1996

certification tests (1) (surge arresters) Tests made, when required, to verify selected performance characteristics of a product or representative samples thereof. (SPD/PE) C62.1-1981s
(2) Tests run on a regular, periodic basis to verify that selected key performance characteristics of a product, or representative samples thereof, have remained within performance specifications. (SPD/PE) C62.11-1999, C62.62-2000

certified design test (station control and data acquisition) A test performed on a production model specimen of a generic type of equipment to establish a specific performance parameter of that genre of equipment. The condition and results of the test are described in a document that is signed and attested to by the testing engineer and other appropriate, responsible individuals. (SWG/PE/SUB) C37.100-1992, C37.1-1994

certified Marinelli beaker standard source (germanium semiconductor detector) An MBSS that has been calibrated as to photon emission rate at specified energies by a laboratory recognized as a country's national standardizing laboratory for radioactivity measurements and has been so certified by the calibrating laboratory. *Notes:* 1. The photon emission rate as used in this standard is the number of photons per second resulting from the decay of radionuclides in the source and is thus higher than the detected rate at the surface. 2. For the United States, the US National Bureau of Standards is the National Standardizing Laboratory. *Synonym:* certified MBSS. *See also:* certified solution; Marinelli beaker standard source; certified-solution Marinelli beaker standard source; Marinelli beaker. (NPS) 680-1978w

certified MBSS *See:* certified Marinelli beaker standard source.

certified radioactivity standard source (germanium detectors) A calibrated radioactive source, with stated accuracy, whose calibration is certified by the source supplier as traceable to the National Radioactivity Meas-urements System. (PE/EDPG) 485-1983s

certified reference material A reference material one or more of whose property values are certified by a technically valid procedure, accompanied by or traceable to a certificate or other documentation which is used by a certifying body. (NI) N42.23-1995

certified solution (germanium semiconductor detector) A liquid radioactive filling material that has been calibrated by a laboratory (for the United States, the US National Bureau of Standards) recognized as a country's National Standardizing Laboratory for radioactivity measurements and has been so certified by the calibrating laboratory. (NPS) 680-1978w

certified-solution Marinelli beaker standard source (germanium semiconductor detector) A standard beaker that contains a certified solution as its radioactive filling material. *See also:* certified Marinelli beaker standard source; Marinelli beaker; certified solution; Marinelli beaker standard source. (NPS) 680-1978w

certified-solution MBSS *See:* certified-solution Marinelli beaker standard source.

certified sources Sources that have been certified for radionuclide activity (Bq) concentration (Bq/g) or alpha, beta, x-, or gamma-ray emission rate ($s-1$). (NI) N42.22-1995

certified unit (test, measurement, and diagnostic equipment) A unit whose demonstrated ability to perform in accordance with preestablished criteria has been attested. (MIL) [2]

certifying agency Organization that validates that equipment meets tests and standards. (IA) 515-1997

CFAR *See:* constant-false-alarm rate.

C_{50} The difference in dB between the first 50 ms of reverberant decay energy and the remaining decay energy from 50 ms and later. It is a member of the class of acoustic measurements commonly known as early-to-late ratios. (COM/TA) 1329-1999

CGF *See:* computer generated force.

CGI *See:* Computer Graphics Interface.

CGM *See:* Computer Graphics Metafile.

cgs electromagnetic system of units *See:* centimeter-gram-second electromagnetic system of units.

cgs electrostatic system of units *See:* centimeter-gram-second electrostatic system of units.

cgs system of units *See:* centimeter-gram-second system of units.

chad (1) The piece of material removed when forming a hole or notch in a storage medium such as punched tape or punched cards. (MIL/C) [2], [85], [20]

(2) The bit of material resulting from punching a hole in a paper card or tape. *Synonym:* chip. (C) 610.10-1994w

chadded Pertaining to the punching of tape in which chad results. (C) [20], [85]

chadless Pertaining to the punching of tape in which chad does not result. (MIL/C) [2], [85], [20]

chadless tape (1) A punched tape wherein only partial perforation is completed and the chad remains attached to the tape. (IA) [61]
(2) Perforated tape that has been punched in such a way that chad is not formed. (C) 610.10-1994w

chaff Strips of lightweight metal or metallized material that are dispensed in large numbers (bundles) so as to simulate a true target, or, more usually, to create a large clutter signal that masks the detection of wanted targets. *Notes:* 1. Each bundle may contain thousands of individual reflectors whose lengths are related to the wavelength of the radar. 2. Chaff for use at HF and VHF frequencies is sometimes called "rope." 3. During WWII, chaff was called "window" in Great Britain and "Dueppel" in Germany. (AES) 686-1997

chafing strip *See:* drive strip.

chain (1) (navigation aids) A network of similar stations operating as a group for determination of position or for furnishing navigational information. (AES/GCS) 172-1983w
(2) (mathematics of computing) A sequence of bits used to construct a binary code. *See also:* chain code. (C) 1084-1986w
(3) (data management) *See also:* linked list. (C) 610.5-1990w

chain binder *See:* binder load.

chain code An arrangement in a cyclic sequence of some or all of the possible different *n*-bit words in which adjacent words are related such that each word is derivable from its neighbor by displacing the bits one place to the left or right, dropping the leading bit, and inserting a bit at the end. The value of the inserted bit needs only to meet the requirement that a word must not recur before the cycle is complete. For example, 000 001 010 101 011 111 110 100 000. . . (C) [20], 1084-1986w, [85]

chain-drive machine (elevators) An indirect-drive machine having a chain as the connecting means. *See also:* driving machine. (EEC/PE) [119]

chained list (1) (software) A list in which the items may be dispersed but in which each item contains an identifier for locating the next item. *Synonym:* linked list. *See also:* list; identifier. (C/SE) 729-1983s
(2) (data management) *See also:* linked list. (C) 610.5-1990w

chain field *See:* link field.

chain hoist *See:* hoist.

chaining (1) A method for storing records in which each record has a link field that is used to access subsequent records. (C) 610.5-1990w
(2) A mode of interaction optionally used by a DSA that cannot perform an directory operation itself; the DSA chains by invoking a directory operation of another DSA and then relaying the outcome to the original requester. (C/PA) 1328.2-1993w, 1326.2-1993w, 1327.2-1993w, 1224.2-1993w

chaining search A search in which each item contains a means for locating the next item to be considered in the search. (C) 610.5-1990w

chain matrix The 2 × 2 matrix relating voltage and current at one port of a two-port network to voltage and current at the other port.

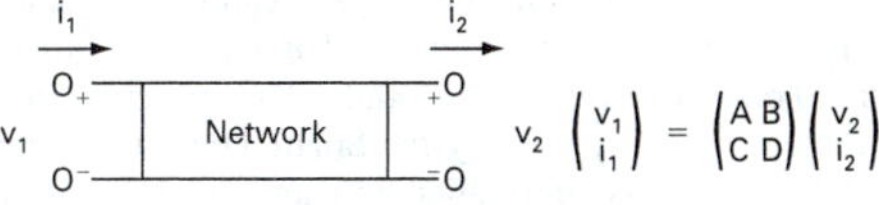

chain matrix

(CAS) [13]

chain printer An element printer in which the type slugs are carried by the links of a revolving chain, called a print chain. (C) 610.10-1994w

chain tugger *See:* hoist.

chain-type conveyor A conveyor using a driven endless chain or chains, equipped with flights that operate in a trough and move material along the trough. *See also:* conveyor. (EEC/PE) [119]

chair, boatswain *See:* boatswain's chair.

chalking (1) (composite insulators) The powdered surface on weathersheds consisting of particles of filler resulting from ultraviolet exposure. (T&D/PE) 987-1985w
(2) The development of loose removable powder at or just beneath a coating surface. (IA) [59]
(3) The powdered surface on the polymeric insulator consisting of particles of filler resulting from ultraviolet exposure or leakage current activity. (PE/IC) 48-1996

challenge (navigation aids) To cause an interrogator to transmit a signal which puts a transponder into operation. *See also:* interrogation. (AES/GCS) 172-1983w

change control *See:* configuration control.

change control board *See:* configuration control board.

change detection An image processing technique in which the pixels of two registered images are compared to detect differences. (C) 610.4-1990w

change dump A selective dump of those storage locations whose contents have changed since some specified time or event. *Synonym:* differential dump. *See also:* dynamic dump; snapshot dump; static dump; selective dump; memory dump; postmortem dump. (C) 610.12-1990

change-of-frame alignment A state that occurs when an off-line framer realigns the receiver to the proper frame alignment signal. *See also:* out-of-frame condition. (COM/TA) 1007-1991r

change of state (COS) A significant change (as defined by a particular system) in the condition of a point being monitored, for example, a change in flow rate, temperature, voltage, etc. Usually associated with dual-status (that is, alarm/normal conditions). (PE/SUB) 1379-1997

change-over selector A device designed to carry, but not to make or break current, used in conjunction with a tap selector or arcing tap switch to enable its contacts, and the connected taps, to be used more than once when moving from one extreme position to the other. (PE/TR) C57.131-1995

change recording *See:* nonreturn-to-zero (change) recording.

changeover switch A switching device for changing electric circuits from one combination to another. *Note:* It is usual to qualify the term "changeover switch" by stating the purpose for which it is used, such as a series-parallel changeover switch, trolley-shoe changeover switch, etc. *See also:* multiple-unit control. (EEC/PE) [119]

change transaction A transaction that causes information in a master file to be changed. *See also:* delete transaction; add transaction; update transaction; null transaction. (C) 610.2-1987

CHANHI Abbreviation for upper channel corresponding to the half-amplitude point of a distribution. (NPS) 398-1972r

CHANLO Abbreviation for lower channel corresponding to the half-amplitude point of a distribution. (GSD) 200-1975w

Channel A control or data path established between two buses that allows information to flow from one bus to the other. The Channels specific to this standard are the CSR channels, the F2V and V2F data channels, the event channel, and the dual port memory Channel. (C/BA) 1014.1-1994w

channel (1) (A) (electric communication) A single path for transmitting electric signals, usually in distinction from other parallel paths. **(B) (electric communication)** A band of frequencies. *Note:* The word "path" is to be interpreted in a broad sense to include separation by frequency division or time division. The term "channel" may signify either a one-way path, providing transmission in one direction only, or a two-way path, providing transmission in two directions. (EEC/PE) [119], 599-1985

(2) **(A)** A path along which signals can be sent, for example, data channel, output channel. **(B)** The portion of a storage medium that is accessible to a given reading station. *See also:* track. (C) 162-1963

(3) A combination of transmission media and equipment capable of receiving signals at one point and delivering related signals at another point. *See also:* information theory. (IT) 171-1958w

(4) (illuminating engineering) An enclosure containing the ballast, starter, lamp holders, and wiring for a fluorescent lamp, or a similar enclosure on which filament lamps (usually tubular) are mounted. (EEC/IE) [126]

(5) (nuclear power generating station) An arrangement of components and modules as required to generate a single protective action signal when required by a generating station condition. A channel loses its identity where single protective action signals are combined. (PE/NP) 379-1994, 338-1987r, 603-1998

(6) (metal-nitride-oxide field-effect transistor) A surface layer of carriers connecting source and drain in an insulated-gate field-effect transistor (IGFET). This channel was formed by inversion with the help of a gate voltage, or by the presence of charges in the gate insulator, or by deliberate doping of the region. (ED) 581-1978w

(7) A band of frequencies dedicated to a certain service transmitted on the broadband medium. (LM/C) 610.7-1995, 802.3u-1995s

(8) (broadband local area networks) The bandwidth required for the transportation of a signal. The bandwidth will vary according to the information being transported. A band of frequencies dedicated to a certain service transmitted on a broadband medium. (LM/C) 802.7-1989r

(9) (speech telephony) A means of one-way or two-way transmission provided by a vendor between two defined interface points. (The customer can realize a connection by connecting together channels from one or more vendors.) *Notes:* 1. "Provided by a vendor" means responsibility for the service and does not necessarily mean ownership of facilities. 2. Channels are provided either dedicated or switched. A dedicated channel may be a non-switched channel for the exclusive use of a customer for a contracted time period. A switched channel may be a channel established (set up and released) under customer control. 3. A channel may consist of two or more equipment items, such as transmission facilities, switching systems, etc. *See also:* connection. (COM/TA) 823-1989w

(10) A logically independent data path between two Functions. Multiple channels can be used to reach different Functions or to represent independent instances of inter-unit communication (e.g., X.25 connections, I/O operations on different discs, datagrams to different network SAPs, etc.). Channels can either be provided by physically separate queues or by multiplexing a shared queue. (C/MM) 1212.1-1993

(11) **(A)** A one-way path for transmission of signals between two or more points; for example, a data channel. *See also:* line; circuit; link. **(B)** In data transmission, either one-way path, providing transmission in one direction only, or two-way path, providing transmission in two directions. *Synonyms:* path. (C) 610.7-1995

(12) **(A)** A one-way path for transmission of signals between two or more points; for example, an output channel or a data channel. *Synonyms:* link; path; line. *See also:* circuit. **(B)** The portion of a storage medium that is accessible to a given reading or writing station, such as a track, or a band. **(C)** A two-way communications path between the central processor and its peripheral devices. (C) 610.10-1994

(13) The tester electronics associated with a digital input/output (I/O) pin that either drives or senses a particular node on the unit under test (UUT). (SCC20) 1445-1998

(14) The data path from any transmitting MIC to the next downstream receiving MIC. (LM/C) 8802-5-1998

(15) A single flow path for digital data or an analog signal, usually in distinction from other parallel paths. An IEEE 1451.2 channel provides a path for a single commodity or logical state, either real or virtual, using a single data model and a single set of physical units. (IM/ST) 1451.2-1997

(16) An instance of medium use for the purpose of passing protocol data units (PDUs) that may be used simultaneously, in the same volume of space, with other instances of medium use (on other channels) by other instances of the same physical layer (PHY), with an acceptably low frame error ratio due to mutual interference. Some PHYs provide only one channel, whereas others provide multiple channels. Examples of channel types are as shown in the following table:

Single channel	n-channel
Narrowband radio-frequency (RF) channel	Frequency division multiplexed channels
Baseband infrared	Direct sequence spread spectrum (DSSS) with code division multiple access

(C/LM) 8802-11-1999

(17) A physical or logical communication link to a single transducer or to a group of transducers considered as a single transducer. (IM/ST) 1451.1-1999

(18) (overhead power lines) *See also:* frequency band. (T&D/PE) 539-1990

(19) *See also:* communication channel. (SUB/PE) 999-1992w

channel address The portion of a full data transport address that specifies the channel to which the read or write is directed. (IM/ST) 1451.2-1997

channel-attached terminal A terminal that is connected directly to the computer by wires or cables. *Synonym:* locally-attached terminal. *Contrast:* link-attached terminal. (C) 610.10-1994w

channel bank A device that multiplexes high-speed communication circuits into lower-speed communication channels; used primarily to digitize analog voice transmission. (C) 610.7-1995

channel-busy tone (telephone switching systems) A tone that indicates that a server other than a destination outlet is either busy or not accessible. (COM) 312-1977w

channel calibration The adjustment of channel output such that it responds, with acceptable range and accuracy, to known values of the parameter that the channel measures, and the performance of a functional test. (PE/NP) 338-1987r

channel capacity (1) (data transmission) The maximum possible information rate through a channel subject to the constraints of that channel. *Note:* Channel capacity may be either per second or per symbol. (PE) 599-1985w

(2) (software) The maximum amount of information that can be transferred on a given channel per unit of time; usually measured in bits per second or in baud. *See also:* memory capacity; storage capacity. (C) 610.12-1990

channel check A qualitative assessment of performance carried out at designated intervals to determine if all elements of the channel are operating within their designated limits. (PE/NP) 338-1987r

channel failure alarm (power-system communication) A circuit to give an alarm if a communication channel should fail. *See also:* power-line carrier. (PE) 599-1985w

channel group (data transmission) A number of channels regarded as a unit. *Note:* The term is especially used to designate part of a larger number of channels. (PE) 599-1985w

channel groupings Manufacturer specifications that define the inherent relationships between the channels of a multichannel Smart Transducer Interface Module. This grouping information is not normally used by the Smart Transducer Interface Module itself. This information will normally be used by Network Capable Application Processor applications to properly compose human readable displays or in formulating other computations. For example, channel groupings can be used to indicate which channels represent the three vector axes of a three-axis vector measurement. (IM/ST) 1451.2-1997

channel hot electrons (CHE) Electrons that are generated by sufficiently high electric fields in the channel with energies that exceed the thermal equilibrium energy. (ED) 1005-1998

channel insertion loss The static loss of a link between a transmitter and receiver. It includes the loss of the fiber, connectors, and splices. (C/LM) 802.3-1998

channeling, lattice *See:* lattice channeling.

channel lights (illuminating engineering) Aeronautical ground lights arranged along the sides of a channel of a water aerodrome. (EEC/IE) [126]

channel load factor (1) The fraction of channel operating time used to transfer the required volume of information between its terminals. (SUB/PE) C37.1-1994

(2) The percent of channel capacity in bits per second required to support the effective data rate for information exchange. (SWG/PE) C37.100-1992

channel, melting *See:* melting channel.

channel multiplier A tubular electron-multiplier with a continuous interior surface of secondary-electron emissive material. *See also:* camera tube; amplifier. (ED) [45]

channel_number A system-dependent, system-global value that is used by communicating Functions to designate a channel. (C/MM) 1212.1-1993

channel path The routing, switching and line links between an input-output channel and some peripheral device. *Note:* There may be multiple channel paths between a channel and a device. (C) 610.7-1995, 610.10-1994w

channel, radio *See:* radio channel.

channel router A machine used to determine a path between two points. *Note:* Often used in the design and layout of integrated circuits and printed circuit boards. (C) 610.10-1994w

channel spacing (radio communication) The frequency increment between the assigned frequency of two adjacent radio-frequency channels. *See also:* radio transmission; two-frequency simplex operation; single-frequency simplex operation; dispatch operation. (VT) [37]

channel service unit (CSU) A device that performs transmit and receive filtering, signal shaping, longitudinal balance, voltage isolation, equalization, and remote loopback testing in a digital communications environment. *Synonym:* digital modem. *See also:* data service unit. (C) 610.7-1995

channel supergroup (data transmission) A number of channel groups regarded as a unit. *Note:* The term is especially used to designate part of a larger number of channels. (PE) 599-1985w

channel, surface *See:* surface channel.

channel timeslots A timeslot that occupies a specified position(s) in a frame and is allocated to a particular time-derived channel. (COM/TA) 1007-1991r

channel utilization index The ratio of the information rate (per second) through a channel to the channel capacity (per second). *See also:* information theory. (IT) 171-1958w

Chapin chart *See:* box diagram.

char The name of a data-type in the C programming language that stands for character, or a group of eight bits that function as a single unit. (PE/SUB) 1379-1997

character (1) (A) An elementary mark or event that may be combined with others, usually in the form of a linear string, to form data or represent information. If necessary to distinguish from definition (B) below, such a mark may be called a "character event." **(B)** A class of equivalent elementary marks or events as in definition (A) having properties in common, such as shape or amplitude. If necessary to distinguish from definition (A) above, such a class may be called a "character design." There are usually only a finite set of character designs in a given language. *Notes:* 1. In "bookkeeper" there are six character designs and ten character events, while in "1010010" there are two character designs and seven character events. 2. A group of characters, in one context, may be considered as a single character in another, as in the binary-coded-decimal system. *See also:* control character; numerical control; special character; check character; escape character. (C/MIL) 162-1963, [2]

(2) (data transmission) One of a set of elementary symbols which normally include both alpha and numeric codes plus punctuation marks and any other symbol which may be read, stored, or written and is used for organization, control, or representation of data. (PE) 599-1985w

(3) (computers) A letter, digit, or other symbol that is used to represent information. *See also:* alphanumeric; alphabetic. (C) 610.7-1995, 1084-1986w, 610.5-1990w, 610.12-1990

(4) A group of consecutive bits used to represent control or data information. Characters are of two types: normal characters (N_chars) or link characters (L_chars). *See also:* link character; normal character; control character; data character. (C/BA) 1355-1995

(5) A sequence of one or more bytes representing a single graphic symbol. *Note:* This term corresponds in the C Standard to the term *multibyte character,* noting that a single-byte character is a special case of multibyte character. Unlike the usage in the C Standard, *character* here has no necessary relationship with storage space, and byte is used when storage space is discussed. (PA/C) 9945-1-1996, 9945-2-1993

(6) A letter, digit, or other special form that is used as part of the organization, control, or representation of data. A character is often in the form of a spatial arrangement of adjacent or connected strokes. (PE/TR) C57.12.35-1996

(7) A sequence of one or more values of type `POSIX.POSIX_Character`. *Note:* This definition of the term *character* applies when it is used by itself. It does not apply to qualified phrases containing the word character, such as POSIX character, graphic character, Ada character, and character special file. (C) 1003.5-1999

character-at-a-time printer A printer that prints a single character at a time. *Synonym:* serial printer. *Contrast:* page printer; line printer. (C) 610.10-1994w

character attribute A characteristic of a single text character. For example, expansion factor, spacing, height, up vector. (C) 610.6-1991w

character-based user interface A user interface in which commands must be expressed in characters entered on a keyboard. *Synonym:* text-based user interface. *Contrast:* graphical user interface. (C) 610.10-1994w

character boundary In character recognition, the largest rectangle, with a side parallel to the document reference edge, whose sides are tangential to a given character outline. (C) 610.2-1987

character class A named set of characters sharing an attribute associated with the name of the class. The classes and the characters that they contain are dependent on the value of the LC_CTYPE category in the current locale. (PA/C) 9945-2-1993

character code A code that uses unique numeric values to represent character data; for example, in ASCII the hexadecimal value 40 is used to represent the character "@." (C) 610.5-1990w

character-deletion character A character within a line of terminal input specifying that it and the immediately preceding character are to be removed from the line; for example, if "\" is the character-deletion character in the string "ABD\C," the following would appear on the terminal: "ABC." *See also:* line-deletion character. (C) 610.5-1990w

character density (test, measurement, and diagnostic equipment) The number of characters that can be stored per unit area or length. (MIL) [2]

character device A printer or other peripheral device that receives data character by character. (C) 610.10-1994w

character display device A display device that provides a representation of data only in the form of characters. *Synonyms:* readout device; alphanumeric display device. *Contrast:* graphic display device. (C) 610.10-1994w

character distortion (data transmission) The normal and predictable distortion of data bit produced by characteristics of a given circuit at a particular transmission speed. (PE) 599-1985w

character fill To insert into a storage medium, as often as necessary, the representation of a specified filler character that does not itself convey data but that may delete unwanted data or initialize storage. *See also:* zero fill. (C) 610.5-1990w

character font (1) A set of graphic characters that are of the same size and style. *Synonym:* type font. *See also:* handprinted character font; OCR-A; OCR-B; optical font; font disk. (C) 610.2-1987
(2) A family or related set of graphic characters that are of the same style of type. For example, Courier Bold Oblique. (C) 610.10-1994w

character form (microprocessor object modules) The printable character representation of binary information as opposed to bit pattern information. (C/MM) 695-1985s

character generator (1) A device that uses predefined character patterns to generate characters on a display surface. *See also:* stroke character generator; matrix character generator. (C) 610.6-1991w
(2) A device that forms character images on a display device or printer. (C) 610.10-1994w

character-indicator tube A blow-discharge tube in which the cathode glow displays the shape of a character, for example, letter, number, or symbol. (ED) [45]

character interval (data transmission) In start-stop operation the duration of a character expressed as the total number of unit intervals (including information, error checking and control bits, and the start and stop elements) required to transmit any given character in any given communication system. (PE) 599-1985w

characteristic (1) (A) (mathematics of computing) The integer part of a logarithm. *Contrast:* mantissa. **(B) (mathematics of computing)** For floating point arithmetic. *See also:* exponent. (C) 1084-1986
(2) (software) *See also:* data characteristic; software characteristic. (C) 610.12-1990
(3) (nuclear power quality assurance) Any property or attribute of an item, process, or service that is distinct, describable, or measurable. (PE/NP) [124]
(4) (semiconductor devices) An inherent and measurable property of a device. Such a property may be electrical, mechanical, thermal, hydraulic, electro-magnetic or nuclear and can be expressed as a value for stated or recognized conditions. A characteristic may also be a set of related values, usually shown in graphical form. (IA) [12]

characteristic curve (1) (illuminating engineering) A curve that expresses the relationship between two variable properties of a light source, such as candlepower and voltage, flux and voltage, etc. (EEC/IE) [126]
(2) (Hall generator) A plot of Hall output voltage versus control current, magnetic flux density, or the product of magnetic flux density and control current. (MAG) 296-1969w

characteristic curves (rotating machinery) The graphical representation of the relationships between certain quantities used in the study of electric machines. *See also:* asynchronous machine. (PE) [9]

characteristic distortion (telegraphy) A displacement of signal transitions resulting from the persistence of transients caused by preceding transitions. *See also:* telegraphy. (COM) [49]

characteristic element (surge arresters) The element that in a valve-type arrester determines the discharge voltage and the follow current. *See also:* surge arrester. (PE) [8]

characteristic equation (feedback control system) The relation formed by equating to zero the denominator of a transfer function of a closed loop. *See also:* feedback control system. (IM) [120]

characteristic harmonic (1) (converter characteristics) (self-commutated converters) Those harmonics produced by semiconductor converter equipment in the course of normal operation. In a six-pulse converter, the characteristic harmonics are the nontriplen odd harmonics, for example, the 5th, 7th, 11th, 13th, etc. (IA/SPC) 936-1987w
(2) Those harmonics produced by semiconductor converter equipment in the course of normal operation. In a six-pulse converter, the characteristic harmonics are the nontriple odd harmonics, for example, the 5th, 7th, 11th, 13th, etc.

$h = kq \pm 1$

k = any integer

q = pulse number of converter

(IA/SPC) 519-1992

characteristic impedance (1) (A) (data transmission) *(two-conductor transmission line for a traveling transverse electromagnetic wave).* The ratio of the complex voltage between the conductors to the complex current on the conductors in the same transverse plane with the sign so chosen that the real part is positive. **(B) (data transmission)** *(coaxial transmission line).* The driving impedance of the forward-traveling transverse electromagnetic wave. (PE) 599-1985
(2) (circular waveguide) For a traveling wave in the dominant ($TE_{1,1}$) mode of a lossless circular waveguide at a specified frequency above the cutoff frequency,

a) the ratio of the square of the root-mean-square voltage along the diameter where the electric vector is a maximum to the total power flowing when the guide is match terminated.
b) the ratio of the total power flowing to the square of the total root-mean-square longitudinal current flowing in one direction when the guide is match terminated.
c) the ratio of the root-mean-square voltage along the diameter where the electric vector is a maximum to the total root-mean-square longitudinal current flowing along the half surface bisected by the diameter when the guide is match terminated.

Note: Under definition (a) the power $W = V^2/Z_{(W,V)}$ where V is the voltage and $Z_{(W,V)}$ is the characteristic impedance defined in (a). Under definition (b) the power $W = I^2Z_{(W,I)}$ where I is the current and $Z_{(W,I)}$ is the characteristic impedance defined in (b). The characteristic impedance $Z_{(V,I)}$ as defined in (c) is the geometric mean of the values given by (a) and (b). Definition (c) can be used also below the cutoff frequency. *See also:* waveguide; self-impedance.
(3) (rectangular waveguide) For a traveling wave in the dominant ($TE_{1,0}$) mode of a lossless rectangular waveguide at a specified frequency above the cutoff frequency,

a) the ratio of the square of the root-mean-square voltage between midpoints of the two conductor's faces normal to the electric vector, to the total power flowing when the guide is match terminated.
b) the ratio of the total power flowing to the square of the root-mean-square longitudinal current, flowing on one face normal to the electric vector when the guide is match terminated.
c) the ratio of the root-mean-square voltage, between midpoints of the two conductor faces normal to the electric vector, to the total root-mean-square longitudinal current, flowing on one face when the guide is match terminated.

Note: Under definition (a) the power $W = V^2/Z_{(W,I)}$ where V is the voltage, and $Z_{(W,V)}$ the characteristic impedance defined in definition (a). Under definition (b) the power $W = 2/Z_{(W,I)}$ where I is the current and $Z_{(W,V)}$ the characteristic impedance defined in definition (b). The characteristic impedance $Z_{(V,I)}$

as defined in definition (c) is the geometric mean of the values given by definition (a) and definition (b). Definition (c) can be used also below the cutoff frequency. *See also:* self-impedance; waveguide.

(4) (two-conductor transmission line) (for a traveling transverse electromagnetic wave). The ratio of the complex voltage between the conductors to the complex current on the conductors in the same transverse plane with the sign so chosen that the real part is positive. *See also:* waveguide; self-impedance; transmission line.

(5) (coaxial transmission line) The driving impedance of the forward-traveling transverse electromagnetic wave. *See also:* self-impedance; transmission line. (MTT) 146-1980w

(6) (surge arresters) (surge impedance) The driving-point impedance that the line would have if it were of infinite length. *Note:* It is recommended that this term be applied only to lines having approximate electric uniformity. For other lines or structures the corresponding term is "iterative impedance." *See also:* self-impedance. (PE) [8]

(7) (overhead-power-line corona and radio noise) The ratio of the complex voltage of a propagation mode to the complex current of the same propagation mode in the same transverse plane with the sign so chosen that the real part is positive. *Note:* The characteristic impedance of a line with losses neglected is known as the surge impedance. *See also:* propagation mode. (T&D/PE) 539-1990

(8) The ratio of the complex value of voltage between the conductors to the complex value of current on the conductors in the same transverse plane with the sign so chosen that the real part is positive. (PE/PSC) 789-1988w

(9) (planar transmission lines) A parameter having the dimensions of impedance (volt per ampere = ohm) that characterizes a mode of propagation. For a transverse electromagnetic (TEM) mode propagating in a single direction on a two-conductor transmission line,

Z_0 = the ratio of voltage to current at any cross section or

$Z_0 = R(Z/Y)$

where
Z = series inductance per unit length
Y = shunt admittance per unit length of the transmission line

$$Z_0 = V^2/2P = 2P/I^2$$

where
P = time-average power transmitted through any cross section
V = amplitude of the voltage
I = amplitude of the current

For modes other than TEM, different definitions will not in general provide the same numerical value; in those cases, the definition of the characteristic impedance is dictated by custom and by its usefulness in the specific application in question. For most planar transmission lines, the fundamental propagation mode is not purely TEM and a characteristic impedance cannot be defined unambiguously. Techniques for defining and calculating impedances in these cases can be found in the literature. (MTT) 1004-1987w

(10) (of a symmetrical pair of coupled lines) The geometric mean of the even and odd mode characteristic impedances. (MTT) 1004-1987w

(11) (A) For a transmission line, the ratio of the complex voltage between the conductors to the complex current on the conductors, taken at a common reference plane for a single transverse electromagnetic (TEM) propagating wave. **(B)** For a wave guide, the ratio of the complex transverse electric field component at any point in the wave guide to the complex magnetic field component measured perpendicular to the electric field at the same point in the wave guide for a single propagating wave guide mode. **(C)** For an unbounded homogeneous medium, the ratio of the complex transverse electric field component at any point to the complex magnetic field component measured perpendicular to the electric field at the same point. *Note:* For example, for a linearly polarized transverse electromagnetic (TEM) wave propagating in the z-direction of an isotropic medium E_x/H_y is the characteristic impedance. (AP/PROP) 211-1997

characteristic insertion loss (1) (waveguide and transmission line) The insertion loss in a transmission system that is reflectionless looking toward both the source and the load from the inserted transducer. *Notes:* 1. This loss is a unique property of the inserted transducer. 2. The frequency, internal impedance, and available power of the source and the impedance of the load have the same value before and after the transducer is inserted. *See also:* waveguide. (IM/HFIM) [40]

(2) (fixed and variable attenuators)

P_{INPUT} = Incident power from Z_0 source

P_{OUTPUT} = Net power into Z_0 load

Characteristic insertion loss =

$$10\log_{10}\frac{P_{INPUT}}{P_{OUTPUT}}\text{ (dB)}$$

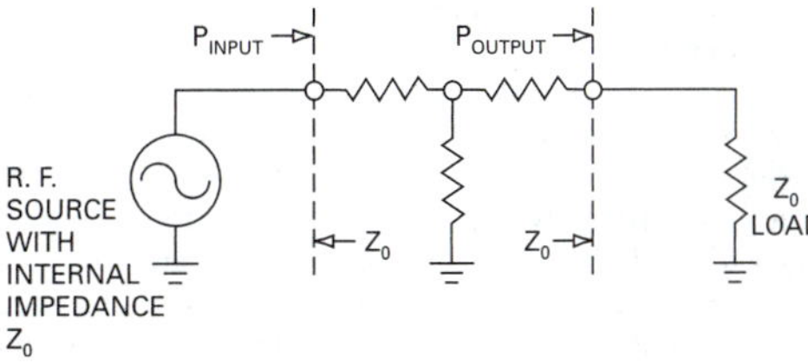

P_{INPUT} P_{OUTPUT} R. F. SOURCE WITH INTERNAL IMPEDANCE Z_0 Z_0 Z_0 Z_0 LOAD

characteristic insertion loss

(NPS/NID) 309-1970s

characteristic insertion loss, incremental The change in the characteristic insertion loss of an adjustable device between two settings. *See also:* waveguide. (IM/HFIM) [40]

characteristic insertion loss, residual The characteristic insertion loss of an adjustable device at an indicated minimum position. *See also:* waveguide. (IM/HFIM) [40]

characteristic insertion phase shift (network analyzers) (waveguide and transmission line) The phase shift occurring upon insertion of a device in a transmission system that is reflectionless looking toward both the source and the load from the insertion plane. *Notes:* 1. The frequency, incident power from the source port, and impedance of the load port are the same before and after the device is inserted. 2. The connectors of source and load ports mate directly. The device can be inserted and its connectors can mate directly with the connectors of the source and load ports. (IM/HFIM) 378-1986w

characteristic overflow *See:* exponent overflow.

characteristic phase shift For a two-port device inserted into a stable, nonreflecting system between the generator and its load, the magnitude of the phase change of the voltage wave incident upon the load before and after insertion of the device, or change of the device from initial to final condition. *Note:* The following conditions apply:

— The frequency, the load impedance, and the generator characteristics, internal impedance and available power, initially have the same values as after the device is inserted;
— the joining devices, connectors, or adapters belonging to the system conform to some set of standard specifications—the same specifications to be used by different laboratories, if measurements are to agree precisely;
— the nonreflecting conditions are to be obtained in uniform, standard sections of waveguide on the system sides of the connectors at the place of insertion.

See also: measurement system. (IM) [38]

characteristics Those inherent factors of software development that may have a significant impact on productivity. (C/SE) 1045-1992

characteristics related to the voltage collapse during chopping (high voltage testing) (chopped impulses) The characteristics of the voltage collapse during chopping are defined in terms of two points C and D at 70% and 10% of the voltage at the instant of chopping. The virtual duration of the voltage collapse is 1.67 times the time interval between points C and D. The virtual steepness of the voltage collapse is the ratio of the voltage at the instant of chopping to the virtual duration of voltage collapse. *Note:* The use of points C and D is for definition purposes only; it is not implied that the duration and steepness of chopping can be measured with any degree of accuracy using conventional measuring circuits. (PE/PSIM) 4-1978s

characteristic telegraph distortion Distortion that does not affect all signal pulses alike, the effect on each transistion depending upon the signal previously sent, due to remnants of previous transitions or transients that persist for one or more pulse lengths. *Note:* Lengthening of the mark pulse is positive, and shortening, negative. Characteristic distortion is measured by transmitting biased reversals, square waves having unequal mark and space pulses. The average lengthening or shortening of mark pulses, expressed in percent of unit pulse length, gives a true measure of characteristic distortion only if other types of distortion are negligible. *See also:* modulation. (AP/ANT) 145-1983s

characteristic time (gyros) The time required for the output to reach 63% of its final value for a step input. *Note:* For a single-degree-of-freedom, rate-integrating gyro, characteristic time is numerically equal to the ratio of the float moment of inertia to the damping coefficient about the output axis. For certain fluid-filled sensors, the float moment of inertia may include other effects, such as that of transported fluid. (AES/GYAC) 528-1994

characteristic underflow *See:* exponent underflow.

characteristic wave A wave that propagates in a homogeneous anisotropic medium with unchanging polarization. *See also:* extraordinary wave; ordinary wave. (AP/PROP) 211-1997

characteristic wave impedance The wave impedance of a traveling wave, with the sign so chosen that the real part is positive. *Note:* In a given mode, in a homogeneously filled waveguide, this is constant for all points and all cross-sections. (MTT) 146-1980w

character layer The layer of the protocol stack which specifies the representation of characters in terms of groups of consecutive bits. The character layer provides the service to the higher layers of the transmission of a continuous sequence of characters on a link. (C/BA) 1355-1995

character outline In character recognition, the graphic pattern established by the stroke edges of a character. (C) 610.2-1987

character printer A printer that can print only character text. *Contrast:* graphic printer. (C) 610.10-1994w

character reader A reader that can recognize hand-written or printed characters using character recognition. *See also:* optical character reader; page reader; magnetic ink character reader. (C) 610.10-1994w

character recognition The use of pattern recognition techniques to identify characters by automatic means. *See also:* magnetic ink character recognition; single-font character recognition; omni-font character recognition; optical character recognition. (C) 610.2-1987

character representation system *See:* character set.

character set **(1)** **(A)** The set of all characters that is defined for a given system. **(B)** A finite set of unique characters upon which agreement has been reached and that is considered complete for some purpose; for example, all the letters, numbers, and symbols used in a language. **(C)** A finite set of unique representations called characters, made to denote and distinguish data; for example, the 26 letters of the English alphabet; 0 and 1 of the boolean alphabet; the set of signals in the Morse code alphabet; and the 128 ASCII characters. *See also:* coded character set; numeric character set; alphanumeric character set; alphabetic character set. (C) 610.5-1990

(2) Those characters that are available for encoding within the bar code symbol. (PE/TR) C57.12.35-1996

character spacing reference line In character recognition, a vertical line that is used to evaluate the horizontal spacing of characters. *Note:* It may be a line that equally divides the distance between the sides of a character boundary or that coincides with the centerline of a vertical stroke. (C) 610.2-1987

character special file A file that refers to a device. One specific type of character special file is a terminal device file. (C/PA) 9945-1-1996, 1003.5-1999, 9945-2-1993

character special file for use with XTI calls A file of a particular type that is used for process-to-process communication. A character special file for use with XTI calls corresponds to a communications endpoint that uses a specified family of communications protocols. (C) 1003.5-1999

character string **(1)** A sequence of characters; for example, the character string 72ZABC. *See also:* bit string. (C) 610.5-1990w

(2) An ordered sequence of zero or more characters. (C/PA) 1328-1993w, 1224-1993w, 1327-1993w

character string picture data Picture data that is associated with a picture specification that specifies at least one alphanumeric character. (C) 610.5-1990w

character stroke In optical character recognition, a line, point, arc, or other mark used as a portion of a graphic character. For example, the dot over the letter i or the cross of the letter t. (C) 610.2-1987

character type A data type whose members can assume the values of specified characters and can be operated on by character operators, such as concatenation. *Contrast:* enumeration type; logical type; real type; integer type. (C) 610.12-1990

character variable A variable that may assume values of any character within some character set. (C) 610.5-1990w

charactron A CRT display device that incorporates a metallic foil into which characters are embossed. The electron beam is directed to the location of a desired character on the foil and its image focused onto the display surface. (C) 610.10-1994w

charge **(1)** **(power operations)** The amount paid for a service rendered or facilities used or made available for use. *See also:* terminations charge; facilities charge; wheeling charge; customer charge; connection charge; energy charge; demand charge. (PE/PSE) 858-1987s

(2) **(storage battery) (storage cell)** The conversion of electric energy into chemical energy within the cell or battery. *Note:* This restoration of the active materials is accomplished by maintaining a unidirectional current in the cell or battery in the opposite direction to that during discharge; a cell or battery that is said to be charged is understood to be fully charged. (EEC/PE) [119]

(3) **(electric power supply)** The amount paid for a service rendered or facilities used or made available for use. (PE/PSE) 346-1973w

(4) The conversion of electrical energy into chemical energy within the battery. (IA/PSE) 602-1996

(5) **(induction and dielectric heating usage)** *See also:* load.

charge, apparent *See:* apparent charge.

charge carrier **(1)** **(x-ray energy spectrometers) (charged-particle detectors) (germanium gamma-ray detectors)** (of a semiconductor) A mobile conduction electron or mobile hole. (IM/ED/NID) 314-1971w, 216-1960w, 300-1988r, 301-1976s

(2) A mobile electron or hole. (NPS) 325-1996

charge collection time (1) (x-ray energy spectrometers) (charged-particle detectors) (germanium gamma-ray detectors) (semiconductor) The time interval, after the passage of an ionizing particle, for the integrated current flowing between the terminals of the detector to increase from 10% to 90% of its final value. *Synonym:* charge sweep-out time. (IM/NPS/HFIM/NID) 314-1971w, 759-1984r, 300-1988r, 325-1996

(2) (semiconductor radiation detectors) The time interval, after the passage of an ionizing particle, for the integrated current flowing between the terminals of the detector to increase from 10% to 93% of its final value. (NID) 301-1976s

charge, connection *See:* connection charge.

charge-coupled device A storage device in which individual semiconductor components are connected to each other so that the electrical charge at the output of one device provides the input to the next. (C) 610.10-1994w

charge, customer *See:* customer charge.

charged aerosol Ion comprised of charged particles, liquid or solid, suspended in air. Typical radius is in the range of 2×10^{-8} m to 2×10^{-7} m. Mobility is in the range of 10^{-9} m to 10^{-7} m^2/Vs. *Note:* Historically, these have been referred to as large or Langevin ions. The use of the term "charged aerosols" is encouraged. (T&D/PE) 1227-1990r

charge delay The time delay after answer supervision is recognized before the beginning of charge recording. In public telephone service (PTS), a call is defined as answered when the called party off-hook supervision duration exceeds the minimum chargeable duration (MCD) after an allowance equal to the worst possible inaccuracy known about the timing sensor has been applied. Should called supervision return on-hook before the MCD has elapsed, MCD timing may start again with the next called party off-hook transition. (COM/TA) 973-1990w

charge-delay interval (telephone switching systems) The recognition time for a valid answer signal in message charging. (COM) 312-1977w

charge, demand *See:* demand charge.

charge, energy *See:* energy charge.

charge, facilities *See:* facilities charge.

charge pump Circuitry that is used to create an on-chip voltage that is greater in magnitude than the voltage available from the device power supply. This voltage is typically used for the write operation. (ED) 1005-1998

charge-resistance furnace A resistance furnace in which the heat is developed within the charge acting as the resistor. *See also:* electrothermics.

charge-sensitive preamplifier An amplifier preceding the main amplifier in which the output amplitude is proportional to the charge injected at the input. *See also:* voltage-sensitive preamplifier. (NPS) 325-1996

charge, space *See:* space charge.

charge storage (semiconductor) (nonlinear, active, and nonreciprocal waveguide components) An electrical property of step recovery, dual mode, and p-i-n diodes. As the diode is driven into forward conduction by the first half-cycle of the incident signal, it stores a charge and appears as a low impedance. As the polarity of the incident signal reverses, the charge is extracted, and the diode remains in its low-impedance state until virtually all of the charge is removed, whereupon the diode rapidly switches to a high-impedance state. (MTT) 457-1982w

charge-storage tube (electrostatic memory tube) A storage tube in which the information is retained on the storage surface in the form of a pattern of electric charges. (ED) 158-1962w

charge sweep-out time *See:* charge collection time.

charge, termination *See:* termination charge.

charge-to-third-number call (telephone switching systems) A call for which the charges are billed to a number other than that of the calling or called number. (COM) 312-1977w

charge transfer The process of charge movement, especially that occurring during a transient discharge. (PE/T&D) 539-1990

charge transit time *See:* transit time.

charge voltage The voltage difference between the intruder and the receptor just prior to an ESD. (SPD/PE) C62.47-1992r

charge, wheeling *See:* wheeling charge.

charging (1) (overhead power lines) The process, or the result of any process, by which an atom, molecule, molecular cluster, or aerosol acquires either a positive or a negative charge. (PE/T&D) 539-1990

(2) (electrostatography) *See also:* sensitizing; electrostatography.

charging (capacitance) current Current resulting from charge absorbed by the capacitor formed by the capacitance of the bushing. (PE/TR) C57.19.03-1996

charging circuit (surge generator) (surge arresters) The portion of the surge generator connections through which electric energy is stored up prior to the production of a surge. *See also:* surge arrester. (PE) [8], 64

charging current (1) (transmission lines) The current that flows in the capacitance of a transmission line when voltage is applied at its terminals. *See also:* transmission line. (T&D/PE) [10]

(2) The maximum continuous current at any charge voltage that may flow at the ESD simulator probe tip as measured to the return path of the simulator through a 1500 Ω resistor that is connected to the probe tip. (EMC) C63.16-1993

charging inductor An inductive component used in the charging circuit of a pulse-forming network. (MAG) 306-1969w

charging rack A device used for holding batteries for mining lamps and for connecting them to a power supply while the batteries are being recharged. *See also:* mine feeder circuit. (EEC/PE/MIN) [119]

charging rate (1) (storage battery) (storage cell) The current expressed in amperes at which a battery is charged. *See also:* charge. (EEC/PE) [119]

(2) The output current expressed in amperes at which the battery is charged. (IA/PSE) 602-1996

charles or kino gun *See:* end injection.

chart (1) (navigation aids) A map intended primarily for navigation use. (AES/GCS) 172-1983w

(2) (recording instrument) The paper or other material upon which the graphic record is made. *See also:* moving element. (EEC/PE) [119]

chart-comparison unit (navigation aids) A device for the simultaneous viewing of a navigational chart in such a manner that one appears superimposed upon the other. (AES/GCS) 172-1983w

chart mechanism (recording instrument) The parts necessary to carry the chart. *See also:* moving element. (EEC/ERI) [111]

chart scale (recording instrument) The scale of the quantity being recorded, as marked on the chart. *Note:* Independent of and generally in quadrature with the chart scale is the time scale that is graduated and marked to correspond to the principal rate at which the chart is advanced in making the recording. This quadrature scale may also be used for quantities other than time. *See also:* moving element. (EEC/PE) [119]

chart scale length (recording instrument) The shortest distance between the two ends of the chart scale. *See also:* instrument. (EEC/PE) [119]

chassis (1) (printed-wiring boards) (frame connection: equivalent chassis connection) A conducting connection to a chassis or frame, or equivalent chassis connection of a printed-

wiring board. The chassis or frame (or equivalent chassis connection of a printed-wiring board) may be at substantial potential with respect to the earth or structure in which this chassis or frame (or printed-wiring board) is mounted. (GSD) 315-1975r
(2) A subrack that is in accordance with IEC 50. (C/BA) 1101.3-1993
(3) A subrack as specified in IEC 50. (C/BA) 1101.4-1993

chassis shield A shield that resides between two modules and attaches to the mainframe. (C/MM) 1155-1992

chatter A condition that results when transceiver electronics fail to shut down and the transceiver floods the network with random signals. *Synonym:* transceiver chatter. (C) 610.7-1995

chatter, relay *See:* relay chatter time; relay contact chatter.

CHDL *See:* computer hardware description language.

Chebyshev filter A filter whose pass-band frequency response has an equal-ripple shape brought about by the use of Chebyshev cosine polynomials as the approximating function. (CAS) [13]

check (1) (monitoring radioactivity in effluents) The use of a source to determine if the detector and all electronic components of the system are operating correctly. (NI) N42.18-1980r
(2) (radiological monitoring instrumentation) To determine if the detector and all electronic components of a system are operating satisfactorily by determining consistent response to the same source. (NI) N320-1979r
(3) (instrument or meter) Ascertain the error of its indication, recorded value, or registration. *Note:* The use of the word "standardize" in place of "adjust" to designate the operation of adjusting the current in the potentiometer circuit to balance the standard cell is deprecated. *See also:* test. (EEC/PE) [119]
(4) (computer-controlled machines) A process of partial or complete testing of either the "correctness of machine operations" or "the existence of certain prescribed conditions within the computer." A check of any of these conditions may be made automatically by the equipment or may be programmed. (C) 162-1963w, 270-1966w
(5) (nuclear power generating station) The use of a source to determine if the detector and all electronic components of the system are operating correctly. (PE/NP) 380-1975w
(6) (A) (transmission line supporting structures) A separation along the grain of the wood, the separation occurring across the annual rings. **(B) (transmission line supporting structures)** A lengthwise separation of the wood that usually extends across the rings of annual growth and commonly results from stresses set up in wood during seasoning. (T&D/PE) 751-1990
(7) (software) To verify the accuracy of data transmitted, manipulated, or stored by any unit or device in a computer. *See also:* automatic check; sequence check; check key; echo check; consistency check; check character. (C) 610.5-1990w

checkback (1) The retransmission from the receiving end to the initiating end of a coded signal or message to verify, at the initiating end, the initial message before proceeding with the transmitting of data or a command. (SWG/PE) C37.100-1992
(2) *See also:* check before operate. (PE) 599-1985w

check back The retransmission from the receiving end to the initiating end of a coded signal or message to verify, at the initiating end, the initial message before proceeding with the transmitting of data or a command. (SWG/PE) C37.100-1981s

checkback message The response from the receiving end to the initiating end of a coded signal or message. *See also:* complete checkback message; partial checkback message. (SWG/PE/SUB) C37.100-1992, C37.1-1987s

check before operate (data transmission) A message and control technique providing for confirmation of control request before operation. *Synonym:* checkback. (PE) 599-1985w

check bit A binary check digit. For example, a parity bit. (C) 1084-1986w

check bits (data transmission) Associated with a code character or block for the purpose of checking the absence of error within the code character or block. *See also:* data processing. (COM) [49]

check box A visual user interface control used to set and reset parameters that have only two (binary) values (e.g., True/False, On/Off, Active/Inactive). When the control is set, a visual indication is provided to indicate its state (e.g., the check box is filled). The user can reset the parameter by selecting the check box again. A check box can be within a group of check boxes. Normally a group of check boxes are not mutually exclusive. (C) 1295-1993w

check card A punch card so formatted as to be suitable for use as a negotiable bank check; for example, a U.S. series E bond. (C) 610.10-1994w

check character (1) A character used for the purpose of performing a check, but often otherwise redundant. (C) [20]
(2) (A) (data management) A character used for the purpose of performing a check. **(B) (data management)** A single character from a check key. (C) 610.5-1990
(3) A character added to a group of characters to provide data redundancy to permit error detection and error correction. (C) 610.7-1995
(4) A calculated character often included within a bar code symbol whose value is used for performing a mathematical check of the validity of the decoded data. (PE/TR) C57.12.35-1996

check code *See:* security code.

check digit (1) A digit used for the purpose of performing a check, but often otherwise redundant. *See also:* forbidden combination; check. (C) [20]
(2) (mathematics of computing) One of a set of redundant digits in a word, byte, character, or message that depends upon the remaining digits in such a fashion that if a digit changes, the error can be detected. (C) 1084-1986w

checking or interlocking relay (power system device function numbers) A relay that operates in response to the position of a number of other devices (or to a number of predetermined conditions) in an equipment, to allow an operating sequence to proceed, or to stop, or to provide a check of the position of these devices or of these conditions for any purpose. (PE/SUB) C37.2-1979s

check key A key that is used for the purpose of performing a check; for example, in the following example the check key is equal to the sum of the first and last digit in field x; this check key could be used to ensure that field x is accurate and complete.

record number	field x	check key
1	0125	5
2	1136	7
3	2228	10

(C) 610.5-1990w

checkout (1) (test, measurement, and diagnostic equipment) A sequence of tests for determining whether or not a device or system is capable of, or is actually performing, a required operation or function. (MIL) [2]
(2) (software) Testing conducted in the operational or support environment to ensure that a software product performs as required after installation. (C) 610.12-1990

checkout equipment (test, measurement, and diagnostic equipment) Electric, electronic, optical, mechanical, hydraulic, or pneumatic equipment, either automatic, semiautomatic, or any combination thereof, which is required to perform the checkout function. (MIL) [2]

checkout time (test, measurement, and diagnostic equipment) Time required to determine whether designated characteristics of a system are within specified values. (MIL) [2]

checkpoint (1) (electronic computation) A place in a routine where a check, or a recording of data for restart purposes, is performed.
(2) (electronic navigation) *See also:* way point. (AES/C/RS) 686-1982s, [85], [20]
(3) (software) A point in a computer program at which program state, status, or results are checked or recorded. (C) 610.12-1990

check problem (A) (electronic computation) A routine or problem that is designed primarily to indicate whether a fault exists in the computer, without giving detailed information on the location of the fault. *Synonym:* check routine. *See also:* programmed check; test; diagnostic. **(B)** A test or problem that is chosen to determine whether an operations or computer program is operating properly. (C) 162-1963, 610.5-1990

check, programmed *See:* programmed check.

check, redundant *See:* redundant check.

check routine *See:* check problem.

check, selection *See:* selection check.

check solution (analog computer) A solution to a problem obtained by independent means to verify a computer solution. (C) 165-1977w

check source (1) (radiological monitoring instrumentation) (liquid-scintillation counting) (radiation protection) (sodium iodide detector) (germanium detectors) A radioactivity source, not necessarily calibrated, that is used to confirm the continuing satisfactory operation of an instrument. Four types of check sources that are of the vial type may be used:

1) Flame-sealed glass (activity known)
2) Flame-sealed glass (activity unknown)
3) Screw-capped glass or plastic (activity known)
4) Screw-capped glass or plastic (activity unknown)

Check sources of the type (1) can be used for all measurements described in the standards listed below. Such sources are available from instrument manufacturers of radiochemicals. They are often designated as "unquenched standards." (PE/EDPG) N320-1979r, N323-1978r, 485-1983s
(2) (liquid-scintillation counting) A radioactive source, not necessarily calibrated, that is used to confirm the continuing satisfactory operation of an instrument. (NI/NPS) N42.15-1990, N42.12-1994, 309-1999

checksum (1) (microprocessor object modules) A deterministic function of a file's contents. If a file is copied and the checksum of the copy is different from the original, there has been an error in copying. (MM/C) 695-1985s
(2) A sum obtained by adding the digits in a numeral, or a group of numerals, usually without regard to meaning, position, or significance. This sum may be compared with a previously computed value to verify that no errors have occurred. *See also:* summation check; sideways sum. (C) 610.7-1995, 1084-1986w

check summation (data transmission) A redundant check in which groups of digits are summed usually without regard for overflow, and that sum checked against a previously computed sum to verify accuracy. *See also:* checksum. (PE) 599-1985w

check, transfer *See:* transfer check.

check valve *See:* blocking capacitor.

cheek, field-coil flange *See:* collar.

cheese antenna A reflector antenna having a cylindrical reflector enclosed by two parallel conducting plates perpendicular to the cylinder, spaced more than one wavelength apart. *Contrast:* pillbox antenna. (AP/ANT) 145-1993

chemical conversion coating A protective or decorative coating produced in situ by chemical reaction of a metal with a chosen environment. (IA) [59]

chemical vapor deposition technique (CVD) (fiber optics) A process in which deposits are produced by heterogeneous gas-solid and gas-liquid chemical reactions at the surface of a substrate. *Note:* The CVD method is often used in fabricating optical waveguide preforms by causing gaseous materials to react and deposit glass oxides. Typical starting chemicals include volatile compounds of silicon, germanium, phosphorus, and boron, which form corresponding oxides after heating with oxygen or other gases. Depending upon its type, the preform may be processed further in preparation for pulling into an optical fiber. *See also:* preform. (Std100) 812-1984w

Chicago grip *See:* conductor grip.

chief programmer (software) The leader of a chief programmer team; a senior-level programmer whose responsibilities include producing key portions of the software assigned to the team, coordinating the activities of the team, reviewing the work of the other team members, and having an overall technical understanding of the software being developed. *See also:* backup programmer; chief programmer team. (C) 610.12-1990

chief programmer team (software) A software development group that consists of a chief programmer, a backup programmer, a secretary/librarian, and additional programmers and specialists as needed, and that employs procedures designed to enhance group communication and to make optimum use of each member's skills. *See also:* chief programmer; egoless programming; backup programmer. (C) 610.12-1990

child A widget that is the immediate inferior of the current widget in the widget instance hierarchy. *Synonym:* descendant. (C) 1295-1993w

child box A box in a child diagram. (C/SE) 1320.1-1998

child diagram A decomposition diagram related to a specific box by exactly one child/parent relationship. (C/SE) 1320.1-1998

child entity The entity in a specific relationship whose instances can be related to zero or one instance of the other entity (parent entity). (C/SE) 1320.2-1998

Child-Langmuir equation (thermionics) An equation representing the cathode current of a thermionic diode in a space-charge-limited-current state.

$I = GV^{3/2}$

where I is the cathode current, V is the anode voltage of a diode or the equivalent diode of a triode or of a multi-electrode value or tube, and G is a constant (perveance) depending on the geometry of the diode or equivalent diode. *See also:* electron emission. (ED) [45], [84]

child node In a tree, a descendant node having a given node as its parent node. *Synonym:* daughter; son. *Contrast:* parent node. *See also:* sibling node.

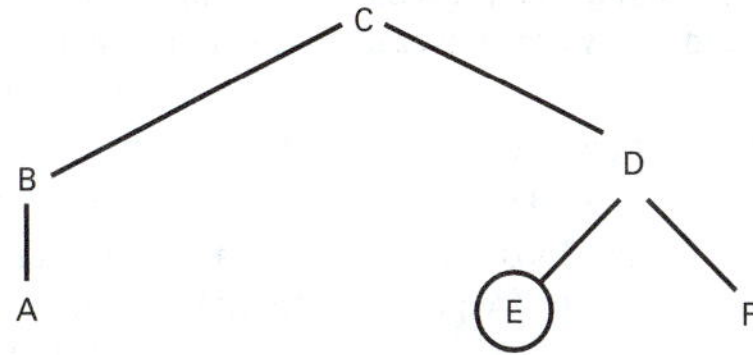

E is a child node of node D

child node

(C) 610.5-1990w
(2) A node that "descends" from another node, i.e., all nodes except the root node. *See also:* parent node. (BA/C) 1275-1994
(3) In the device tree, a descendant node that represents a device plugged into an Sbus. (C/BA) 1275.4-1995, 1275.2-1994w

child process *See:* POSIX process.

child segment In a hierarchical database, a segment that has a parent segment and that is dependent on that segment for its existence. *Note:* If the parent segment is deleted, the child segment must be deleted. *Contrast:* parent node. *See also:* logical child segment; physical child segment; dependent segment; twin segment. (C) 610.5-1990w

CHILL A high-order language, standardized by CCITT (Consultative Committee on International Telephone & Telegraph, Geneva), used for communication applications. (C) 610.13-1993w

Chinese binary *See:* column binary.

Chinese finger *See:* woven wire grip.

chip (1) (mechanical recording) The material removed from the recording medium by the recording stylus while cutting the groove. (SP) [32]
(2) (nonlinear, active, and nonreciprocal waveguide components) (semiconductor) A small unpackaged functional element made by subdividing a wafer of semiconductor material. Sometimes referred to as a "die." (MTT) 457-1982w
(3) A small piece of silicon or other semiconductive material on which circuits can be placed. (C) 610.10-1994w
(4) A small unpackaged functional element made by subdividing a wafer of semiconductor material. Sometimes referred to as a die. Also used as a modifier to indicate an operation that applies to the entire chip as in chip enable or chip clear. *Synonym:* integrated circuit. (ED) 1005-1998

chip clear (A) Operation that causes the writing of all memory elements to a common "1" state. **(B)** Terminal to which the clear signal is applied (preferred usage). (ED) 1005-1998

chip density The number of transistors implemented on a single integrated circuit. (C) 610.10-1994w

chip enable (semiconductor memory) The inputs that when true permit input, internal transfer, manipulation, refreshing, and output of data, and when false cause the memory to be in a reduced power standby mode. *Note:* Chip enable is a clock or strobe that significantly affects the power dissipation of the memory. Chip select is a logical function that gates the inputs and outputs. For example, chip enable may be the cycle control of a dynamic memory or a power reduction input on a static memory. (TT/C) 662-1980s

chip erase *See:* bulk erase.

chip-on-board testing A test of a component after it has been assembled onto a printed circuit board or other substrate; for example, using the facilities defined by IEEE Std 1149.1-1990. (TT/C) 1149.1-1990

chip select (semiconductor memory) The inputs that when false prohibit writing into the memory and disable the output of the memory. *Note:* Chip enable is a clock or strobe that significantly affects the power dissipation of the memory. Chip select is a logical function that gates the inputs and outputs. For example, chip enable may be the cycle control of a dynamic memory or a power reduction input on a static memory. (TT/C) 662-1980s

chirp A form of pulse compression that uses frequency modulation (usually linear) during the pulse. (AES) 686-1997

chirp filter A filter whose group delay is a nonconstant function of the instantaneous frequency of the input signal. (UFFC) 1037-1992w

chirping (fiber optics) A rapid change (as opposed to long-term drift) of the emission wavelength of an optical source. Chirping is often observed in pulsed operation of a source. (Std100) 812-1984w

choice (1) (telephone switching systems) The position of an outlet in a group with respect to the order of selection. (COM) 312-1977w
(2) The language-independent syntax for a family of datatypes constructed from a sequence of base datatypes, each associated with a name. A value of choice datatype contains, for exactly one name, a value of the corresponding datatype. (C/PA) 1351-1994w
(3) A datatype constructed from a sequence of base datatypes, each associated with a name. A choice value contains, for exactly one name, a value of the corresponding datatype. (C/PA) 1224.1-1993w

choice device A logical input device used to make a selection from a set of predefined menu options in a graphics system. A typical physical device is a function keyboard or a set of function keys. *Synonym:* button device. (C) 610.6-1991w, 610.10-1994w

choke (waveguide) A device for preventing energy within a waveguide in a given frequency range from taking an undesired path. *See also:* waveguide. (AP/ANT) [35]

choke coil An inductor used in a special application to impede the current in a circuit over a specified frequency range while allowing relatively free passage of the current at lower frequencies. (CHM) [51]

choke flange (1) (microwave technique) A flange in whose surface is cut a groove so dimensioned that the flange may form part of a choke joint. *See also:* waveguide. (AP/ANT) [35]
(2) (waveguide components) A flange designed with auxiliary transmission-line elements to form a choke joint when used with a cover flange. (MTT) 147-1979w

choke joint (waveguide components) A connection designed for essentially complete transfer of power between two waveguides without metallic contact between the inner walls of the waveguides. It typically consists of one cover flange and one choke flange. (MTT) 147-1979w

choke piston (waveguide) A piston in which there is no metallic contact with the walls of the waveguide at the edges of the reflecting surface; the short-circuit to high-frequency currents is achieved by a choke system. *Note:* This definition covers a number of configurations: dumbbell; Z-slot; inverted bucket; etc. *Synonyms:* choke plunger; noncontact plunger. *See also:* waveguide. (AP/ANT) [35]

choke plunger *See:* choke piston.

choker *See:* traveler sling.

chop A sudden cessation of the flow of arc current during circuit interruption. (PE/IC) 1143-1994r

chopped display (oscilloscopes) A time-sharing method of displaying output signals of two or more channels with a single cathode-ray-tube gun, at a rate that is higher than and not referenced to the sweep rate. *See also:* oscillograph. (IM/HFIM) [40]

chopped frequency *See:* chopping rate.

chopped impulses (high voltage testing) Generally, chopping of an impulse is characterized by an initial discontinuity, decreasing the voltage, which then falls to zero with or without oscillations. See figures 1 and 3 to definition of "virtual origin O_1." *Note:* With some test objects or test arrangements, there may be a flattening of the crest or a rounding off of the voltage before the final voltage collapse. Similar effects may also be observed, due to imperfections of the measuring system. Exact determination of the parameters related to the chopping then requires special consideration. (PE/PSIM) 4-1978s

chopped impulse voltage A transient voltage derived from a full impulse voltage that is interrupted by the disruptive discharge of an external gap or the external portion of the test specimen causing a sudden collapse in the voltage, practically to zero value. *Note:* The collapse can occur on the front, at the peak, or on the tail. *See also:* test voltage.

chopped impulse wave (surge arresters) An impulse wave that has been caused to collapse suddenly by a flashover. (PE) [8], [84]

chopped lightning impulse A prospective full lightning impulse during which any type of discharge causes a rapid collapse of the voltage. (PE/PSIM) 4-1995

chopped wave A voltage impulse that is terminated intentionally by sparkover of a gap. (PE/SPD/T&D) 32-1972r, [10]

chopped-wave lightning impulse test (power and distribution transformers) A voltage impulse that is terminated intentionally by sparkover of a gap, which occurs subsequent to the maximum crest of the impulse wave voltage, with a specified minimum crest voltage, and a specified minimum time to flashover. (PE/TR) C57.12.80-1978r

chopper (1) (analog computer) A mechanical, electrical, or electromechanical device that converts dc into a square wave. As applied to a direct-coupled operational amplifier, it is a modulator used to convert the dc at the summing junction to ac for amplifier and reinsertion as a correcting voltage to reduce offset. (C) 165-1977w, 610.10-1994w
(2) A device for interrupting a current or a light beam at regular intervals. Choppers are frequently used to facilitate amplification. (COM/PE/EEC) [119]
(3) (capacitance devices) A special form of pulsing relay having contacts arranged to rapidly interrupt, or alternately reverse, the direct-current polarity input to an associated circuit. 31-1944w

chopper amplifier (signal-transmission system) A modulated amplifier in which the modulation is achieved by an electronic or electromechanical chopper, the resultant wave being substantially square. *See also:* signal. (IE) [43]

chopping frequency *See:* chopping rate.

chopping rate (oscilloscopes) (cathode-ray oscilloscopes) The rate at which channel switching occurs in chopped-mode operation. *See also:* oscillograph. (IM/HFIM) [40], 311-1970w

chopping transient blanking The process of blanking the indicating spot during the switching periods in chopped display operation. (IM) 311-1970w

chroma *See:* Munsell chroma.

chromatic adaptation (illuminating engineering) The process by which the chromatic properties of the visual system are modified by the observation of stimuli of various chromaticities and luminances. (EEC/IE) [126]

chromatic color (illuminating engineering) Perceived color possessing a hue. In everyday speech, the word "color" is often used in this sense in contradistinction to white, gray, or black. (EEC/IE) [126]

chromatic dispersion *See:* dispersion.

chromaticity (1) (general) The color quality of light definable by its chromaticity coordinates, or by its dominant (or complementary) wavelength and its purity, taken together. *See also:* color. (BT/AV) [34], [84]
(2) (illuminating engineering) The dominant or complementary wavelength and purity aspects of the color taken together, or of the aspects specified by the chromaticity coordinates of the color taken together. (EEC/IE) [126]
(3) (television) That color attribute of light definable by its chromaticity coordinates. *Note:* When a specific white, the value of dominant (or complementary) wavelength, and saturation are given, there will be a corresponding set of unique chromaticity coordinates. (BT/AV) 201-1979w
(4) (electric power systems in commercial buildings) The measure of the warmth or coolness of a light source, which is expressed in the Kelvin (K) temperature scale. (IA/PSE) 241-1990r

chromaticity coordinates of a color, *x*, *y*, *z* (illuminating engineering) The ratio of each of the tristimulus values of the color to the sum of the three tristimulus values. *See also:* tristimulus values of a light, ***X***, ***Y***, ***Z***. (IE/EEC/BT/AV) [126], 201-1979w

chromaticity diagram (1) (illuminating engineering) A plane diagram formed by plotting one of the three chromaticity coordinates against another. (EEC/IE) [126]
(2) (television) A plane diagram formed by plotting one chromaticity coordinate against another. *Notes:* 1. A commonly used chromaticity diagram is the 1931 CIE (*x*,*y*) diagram. 2. Another chromaticity diagram coming into use is defined in the 1960 CIE (*u*,*v*) uniform chromaticity system (UCS). In contrast with the CIE (x,y) diagram, chromaticities that have just noticeable differences (*j*, *n*, *d*) are spaced by essentially equal distances over the entire diagram. Coordinate values in the two systems are related by the transformations:

$$u = \frac{4x}{-2x + 12y + 3}$$

$$v = \frac{6y}{-2x + 12y + 3}$$

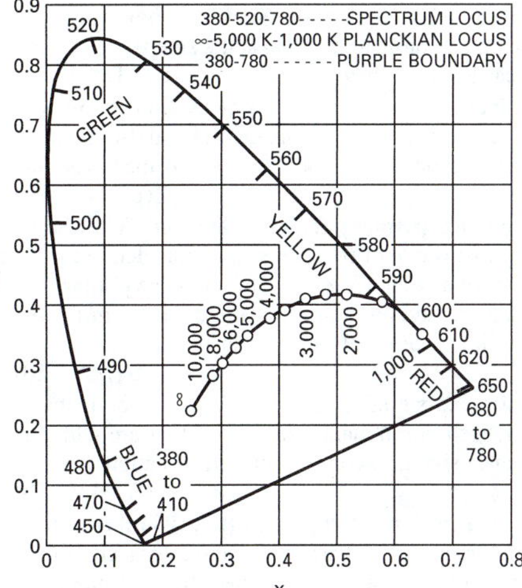

1931 CIE (x, y) chromaticity diagram

chromaticity diagram

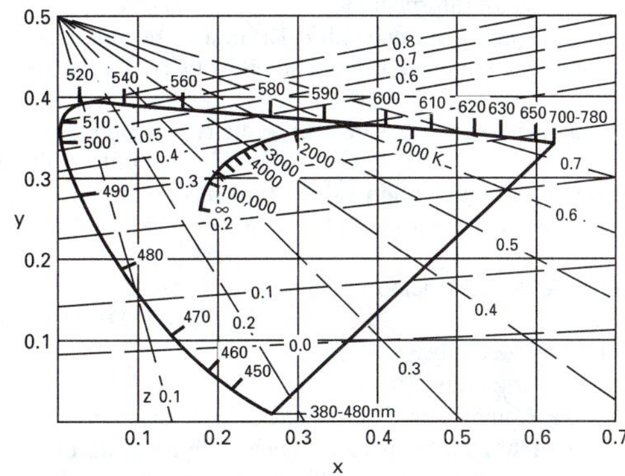

1960 CIE-UCS (u, v) chromaticity diagram

chromaticity diagram

(BT/AV) 201-1979w

chromaticity difference thresholds (illuminating engineering) The smallest difference in chromaticity, between two colors of the same luminance, that makes them perceptibly different. The difference may be a difference in hue or saturation, or a combination of the two. (EEC/IE) [126]

chromaticity flicker (television) The flicker that results from fluctuation of chromaticity only. (BT/AV) 201-1979w

chrominance (1) (television) The colorimetric difference between any color and a reference color of an equal luminance, the reference color having a specified chromaticity. *Notes:* 1. In three-dimensional color space, chrominance is a vector that lies in a plane of constant luminance. In that plane, it may be resolved into components called chrominance components. 2. In color television transmission, for example, the chromaticity of the reference color may be that of a specified white. (BT/AV) 201-1979w
(2) (broadband local area networks) The portion of a video signal that contains color information. (LM/C) 802.7-1989r

chrominance channel (color television) Any path that is intended to carry the chrominance signal. (BT/AV) 201-1979w

chrominance channel bandwidth (color television) The bandwidth of the path intended to carry the chrominance signal. (BT/AV) 201-1979w

chrominance components *See:* chrominance.

chrominance demodulator (color television) A demodulator used for deriving video-frequency chrominance components from the chrominance signal and a sine wave of chrominance subcarrier frequency. (BT/AV) 201-1979w

chrominance modulator (color television) A modulator used for generating the chrominance signal from the video-frequency chrominance components and the chrominance subcarrier. (BT/AV) 201-1979w

chrominance primary (color television) A transmission primary that is one of two whose amounts determine the chrominance of a color. *Notes:* 1. Chrominance primaries have zero luminance and are nonphysical. 2. This term is obsolete because it is useful only in a linear system. (BT/AV) 201-1979w

chrominance signal (color television) The sidebands of the modulated chrominance subcarrier that are added to the luminance signal to convey color information. (BT/AV) 201-1979w

chrominance signal component (television) A signal resulting from suppressed-carrier modulation of a chrominance subcarrier voltage at a specified phase, by a chrominance primary signal such as the I Video Signal or the Q Video Signal. (BT) 204-1961w

chrominance subcarrier (color television) The carrier whose modulation sidebands are added to the luminance signal to convey color information. (BT/AV) 201-1979w

chronaxie (medical electronics) The minimum duration of time required to stimulate with a current of twice the rheobase. (EMB) [47]

chronic exposure Exposure over a relatively long time. (T&D/PE) 539-1990

chronometer (navigation aids) A time piece with a nearly constant rate. Set approximately to Greenwich Mean Time. (AES/GCS) 172-1983w

chunk A block of memory, typically 64 bits. (ED) 1005-1998

chute *See:* feed tube.

CI *See:* configuration item.

CIE *See:* Commission Internationale de l'Eclairage.

CIE (L*a*b*) uniform color space (illuminating engineering) A transformation of CIE tristimulus values X, Y, Z into three coordinates that define a space in which equal distances are more nearly representative of equal magnitudes of perceived color difference. This space is specially useful in cases of colorant mixtures, for example, dye-stuffs, paints. (EEC/IE) [126]

CIE (L*u*v*) uniform color space (illuminating engineering) A transformation of CIE tristimulus values X, Y, Z into three coordinates that define a space in which equal distances are more nearly representative of equal magnitudes of perceived color difference. This space is specially useful in cases where colored lights are mixed additively for example, color television. (EEC/IE) [126]

CIE standard chromaticity diagram (illuminating engineering) One in which the *x* and *y* chromaticity coordinates are plotted in rectangular coordinates. *Note:* The diagram may be based on the CIE 1931 Standard Observer or on the CIE 1964 Supplementary Standard Observer. *See also:* color matching functions. (EEC/IE) [126]

CIE standard colorimetric observer, 1931 Receptor of radiation whose colorimetric characteristics correspond to the distribution coefficients $\bar{x}_\lambda$, $\bar{y}_\lambda$, $\bar{z}_\lambda$ adopted by the International Commission on Illumination in 1931. *See also:* color. (BT/AV) [34], [84]

CIGRE *See:* Conférence Internationale Des Grands Réseaux Electriques.

CIM *See:* computer input microfilm; computer-integrated manufacturing.

CINEMA A hardware description language with a compiler written in BCPL; contains normal control statements and also statements providing parallel execution of program statements. (C) 610.13-1993w

cine-oriented image In micrographics, an image appearing on a roll of microfilm in such a manner that the top edge of the image is perpendicular to the long edge of the film. *Synonyms:* portrait image; motion picture display. *Contrast:* comic-strip oriented image. (C) 610.2-1987

C interface The C language binding, defined in terms of ISO/IEC 9899: 1990. (C/PA) 1328-1993w, 1327-1993w

ciphertext Data produced through the use of encipherment, the semantic content of which is not available. *Note:* Ciphertext may itself be input to encipherment, producing superenciphered data. (C/LM) 802.10-1998

circadian rhythm Oscillation of biological processes with an approximate 24 h period regulated by external stimuli. (PE/T&D) 539-1990

circle diagram (A) (rotating machinery) Circular locus describing performance characteristics (current, impedance, etc.) of a machine or system. In case of rotating machinery, the term "circle diagram" has, in addition, some specific usages: The locus of the armature current phasor of an induction machine, or of some other type of asynchronous machine, displayed on the complex plane, with the shaft speed as the variable (parameter), when the machine operates at a constant voltage and at a constant frequency. **(B) (rotating machinery)** The locus of the current vector(s) of a nonsalient-pole synchronous machine, displayed in a synchronously rotating reference frame (Park transform, *d-q* coordinates), with the active component of the load, hence with the rotor displacement angle, as the variable (parameter), when the machine operates at a constant field current. **(C) (rotating machinery)** The locus of the current phasor(s) of (2) *See also:* asynchronous machine. (PE) [9]

circling guidance lights (illuminating engineering) Aeronautical ground lights provided to supply additional guidance during a circling approach when the circling guidance furnished by the approach and runway lights is inadequate. (EEC/IE) [126]

circuit (1) (A) The physical medium on which signals are carried across the AUI. The data and control circuits consist of an A circuit and a B circuit forming a balanced transmission system so that the signal carried on the B circuit is the inverse of the signal carried on the A circuit. **(B) (data transmission)** A network providing one or more closed paths. **(C)** An arrangement of interconnected components that has at least one input and one output terminal, and whose purpose is to produce at the output terminals a signal that is a function of the signal at the input terminals. *Synonyms:* physical circuit; network. *See also:* expansion board; channel; telecommunication circuit. **(D)** An arrangement of interconnected electronic components that can perform specific functions upon application of proper voltages and signals. *See also:* logic circuit; integrated circuit. (LM/C/COM/PE/TA) 8802-3-1990, 455-1985, 599-1985, 610.10-1994

(2) A conductor or system of conductors through which an electric current is intended to flow. (NESC/PE) C2-1997, 599-1985w

(3) (machine winding) The element of a winding that comprises a group of series-connected coils. A single-phase winding or one phase of a polyphase winding may comprise one circuit or several circuits connected in parallel. (PE) [9]

(4) An interconnection of electrical elements. *See also:* network. (CAS) [13]

(5) In networking, a means of communication of electrical or electronic signals between two points. *Synonym:* network. *See also:* telecommunication circuit; dial-up circuit; simplex circuit; four-wire circuit; two-wire circuit; foreign exchange circuit; leased circuit; channel. (C) 610.7-1995

(6) The physical medium on which signals are carried across the Attachment Unit Interface (AUI) for 10BASE-T or Media Independent Interface (MII) for 100BASE-T. For 10BASE-T, the data and control circuits consist of an A circuit and a B circuit forming a balanced transmission system so that the signal carrier on the B circuit is the inverse of the signal carried on the A circuit. (C/LM) 802.3-1998

circuit analyzer (multimeter) The combination in a single enclosure of a plurality of instruments or instrument circuits for use in measuring two or more electrical quantities in a circuit. *See also:* instrument. (EEC/PE) [119]

circuit, balanced *See:* balanced circuit.

circuit board A flat piece of insulating material, often multilayered, constituted of epoxy-glass or phenolic resin, on which electrical components are mounted and interconnected by etched copper foil so patterned as to form a circuit. *Note:* Sometimes referred to as a "board" or a "card." *See also:* printed circuit board. (C) 610.10-1994w

circuit bonding jumper The connection between portions of a conductor in a circuit to maintain required ampacity of the circuit. (NESC/NEC) [86]

circuit breaker (1) (general) (thyristor) A device designed to open and close a circuit by nonautomatic means, and to open the circuit automatically on a predetermined overload of current, without injury to itself when properly applied within its rating. *Notes:* 1. A circuit breaker is usually intended to operate infrequently, although some types are suitable for frequent operation. (NESC/IA/IPC/PKG) 428-1981w, [86], 333-1980w
(2) A switching device capable of making, carrying, and breaking currents under normal circuit conditions and also making, carrying for a specified time, and breaking currents under specified abnormal conditions such as those of short circuit. (NESC) C2-1997
(3) (packaging machinery) An automatic device designed to open under abnormal conditions a current-carrying circuit without damage to itself. (IA/PKG) 333-1980w
(4) (hydroelectric power plants) A fast-acting switching device used to close and open an electric circuit and capable of interruption of fault currents. (PE/EDPG) 1020-1988r
(5) A mechanical switching device, capable of making, carrying, and breaking currents under normal circuit conditions and also, making and carrying for a specified time and breaking currents under specified abnormal circuit conditions such as those of short circuit. *Notes:* 1. A circuit breaker is usually intended to operate infrequently, although some types are for frequent operation. 2. The medium in which circuit interruption is performed may be designated by suitable prefix, that is, airblast circuit breaker, air circuit breaker, compressed-air circuit breaker, gas circuit breaker, oil circuit breaker, vacuum circuit breaker, oilless circuit breaker, etc. 3. Circuit breakers are classified according to their application or characteristics and these classifications are designated by the following modifying words or clauses delineating the several fields of application, or pertinent characteristics: High-voltage power—Rated above 1000 V ac. Molded-case—See separate definition. Low-voltage power—Rated 1000 V ac or below, or 300 V dc and below, but not including molded-case circuit breakers. Direct-current low-voltage power circuit breakers are subdivided according to their specified ability to limit fault-current magnitude by being called general purpose, high-speed, semi-high-speed, rectifier or anode. For specifications of these restrictions see latest revision of the applicable standard. (SWG/PE) C37.100-1992

circuit-breaker compartment (1) That portion of a switchgear assembly that contains one circuit breaker and the associated primary conductors and secondary control connection devices including current transformers. (SWG/PE) C37.20.1-1993r
(2) That portion of the switchgear assembly that contains one circuit breaker or other removable primary interrupting device and the associated primary conductors. (SWG/PE) C37.20.2-1993

circuit breaker downtime Time from the discovery of the failure until the breaker is returned to service. (SWG/PE) C37.10-1995

circuit breaker, field discharge A circuit breaker having main contacts for energizing and deenergizing the field of a generator, motor, synchronous condenser, or rotating exciter, and having discharge contacts for short-circuiting the field through the discharge resistor at the instant preceding the opening of the circuit-breaker main contacts. The discharge contacts also disconnect the field from the discharge resistor at the instant following the closing of the main contacts. For direct-current generator operation, the discharge contacts may open before the main contacts close. *Note:* When used in the main field circuit of an alternating- or direct-current generator, motor, or synchronous condenser, the circuit breaker is designated as a main field discharge circuit breaker. When used in the field circuit of the rotating exciter of the main machine, the circuit breaker is designated as an exciter field discharge circuit breaker. *See also:* field discharge circuit breaker. (SWG/PE) C37.100-1992, C37.18-1979r

circuit breaker, general purpose low-voltage dc power A circuit breaker that, during interruption, does not limit the current peak of the available (prospective) fault current and may not prevent the fault current from rising to its sustained value. (SWG/PE) C37.100-1992, C37.14-1999

circuit-breaker grouping The three poles of a circuit breaker grouped in adjacent configuration along the line of the same row. (SWG/SUB/PE) C37.122-1983s, C37.100-1992

circuit breaker, high-speed low-voltage dc power A circuit breaker that, during interruption, limits the current peak to a value less than the available (prospective) fault current. (SWG/PE) C37.100-1992, C37.14-1999

circuit breaker interrupting rating For an unfused circuit breaker, the designated limit of available (prospective) current at which the circuit breaker is required to perform its short-circuit current duty cycle at rated maximum voltage under the prescribed test conditions. This current is expressed as the rms symmetrical value envelope at a time 1/2 cycle after short-circuit is initiated. (For dc breakers, the rated interrupting current is the maximum value of direct current.). (IA/MT) 45-1998

circuit breaker, rectifier low-voltage dc power A circuit breaker that carries the normal current output of one rectifier and that, during fault conditions, functions to withstand and/or interrupt abnormal current as required. (SWG/PE) C37.100-1992, C37.14-1999

circuit breaker, semi-high-speed low-voltage dc power A circuit breaker that, during interruption, does not limit the current peak of the available (prospective) fault current on circuits with minimal inductance but that does limit current to a value less than the sustained current available on higher-inductance circuits. (SWG/PE) C37.100-1992, C37.14-1999r

circuit bypass means (bypass) An assembly of parts which, when properly operated, closes the circuit between the line and load jaws. (ELM) C12.7-1993

circuit-commutated turn-off time (thyristor) The time interval between the instant when the principal current has decreased to zero after external switching of the principal voltage circuit and the instant when the thyristor is capable of supporting a specified principal voltage without turning on. *See also:* principal voltage-current characteristic. (IA/ED) 223-1966w, [46], [62], [12]

circuit components (thyristor) Those electrical controller devices that may conduct current during some part of the cycle. Instrumentation is excluded. *Note:* This definition may include devices within the controller that are used for the suppression of voltage and current transients. (IA/IPC) 428-1981w

circuit controller A device for closing and opening electric circuits. (EEC/PE) [119]

circuit efficiency (output circuit of electron tubes) The ratio of the power at the desired frequency delivered to a load at the output terminals of the output circuit of an oscillator or amplifier to the power at the desired frequency delivered by the electron stream to the output circuit. *See also:* network analysis. (ED) 161-1971w, [45]

circuit element A basic constituent part of a circuit, exclusive of interconnections. (EEC/PE) [119]

circuit interrupter (packaging machinery) A manually operated device designed to open under abnormal conditions a current-carrying circuit without damage to itself. (IA/PKG) 333-1980w

circuit limiter *See:* limiter circuit.

circuit malfunction analysis (test, measurement, and diagnostic equipment) The logical, systematic examination of circuits and their diagrams to identify and analyze the probability and consequence of potential malfunctions and for determining related maintenance and support requirements to investigate effects of failures. (MIL) [2]

circuit, multipoint *See:* multipoint circuit.

circuit noise meter An instrument for measuring circuit noise level. Through the use of a suitable frequency-weighting network and other characteristics, the instrument gives equal readings for noises that are approximately equally interfering. The readings are expressed as circuit noise levels in decibels above reference noise. *Synonym:* noise measuring set. *See also:* instrument. (EEC/PE) [119]

circuit pack A printed circuit board (PCB) populated with components, i.e., a PCB assembly. Also called a feature card. (SPD/PE) C62.38-1994r

circuit properties (thyristor) Those conditions which exist, or actions which take place, inside the controller during its operating cycle. (IA/IPC) 428-1981w

circuit simulator A software program that predicts a circuit's response to a given stimulus. (SCC20) 1445-1998

circuit switch (data transmission) A communications switching system which completes a circuit from sender to receiver at the time of transmission (as opposed to a message switch). (PE) 599-1985w

circuit switcher (1) A mechanical switching device with an integral interrupter, suitable for making, carrying, and interrupting currents under normal circuit conditions. It is also suitable for interrupting specified short-circuit current that may be less than its close and latch, momentary, and short-time current ratings. *Note:* This device may be suitable for transformer protection where the majority of faults are limited by the transformer and system impedance. (SWG/PE) C37.100-1992

(2) A circuit interrupting device with a limited interrupting rating as compared with a circuit breaker. It is often integrated with a disconnecting switch. Its design usually precludes the integration of current transformers (CTs). (PE/PSR) C37.113-1999

circuit-switched network A switched network having the capability to switch lines in different configurations to establish a continuous pathway between the sender and the recipient. (C) 610.7-1995

circuit switching (1) A method of communications where an electrical connection between calling and called stations is established on demand for exclusive use of the circuit until the connection is released. *See also:* store-and-forward switching; packet switching; message switching. (LM/COM) 168-1956w

(2) In data communications, a method of communication in which a dedicated communications path is set up between two devices through one or more intermediate switching nodes. *Synonym:* line switching. *See also:* message switching; space-division switching. (C) 610.7-1995

circuit switching element (inverters) A group of one or more simultaneously conducting thyristors, connected in series or parallel or any combination of both, bounded by no more than two main terminals and conducting principal current between these main terminals. *See also:* self-commutated inverters. (IA) [62]

circuit switching system (telephone switching systems) A switching system providing through connections for the exchange of messages. (COM) 312-1977w

circuit transient recovery voltage *See:* inherent transient recovery voltage.

circuit voltage class (electric power system) A phase-to-phase reference voltage that is used in the selection of insultation class designations for neutral grounding devices. (SPD/PE) 32-1972r

circular array An array of elements whose corresponding points lie on a circle. *Note:* Practical circular arrays may include arrangements of elements that are congruent under translation or rotation. *Synonym:* ring array. (AP/ANT) 145-1993

circular Bayliss distribution *See:* Bayliss distribution.

circular electric wave (waveguide) A transverse electric wave for which the lines of electric force form concentric circles. (MTT) 146-1980w

circular grid array An array of elements whose corresponding points lie on coplanar concentric circles. (AP/ANT) 145-1993

circular interpolation (numerically controlled machines) A mode of contouring control that uses the information contained in a single block to produce an arc of a circle. *Note:* The velocities of the axes used to generate this arc are varied by the control. (IA/EEC) [61], [74]

circular list *See:* circularly-linked list.

circularly-linked list A linked list in which the last item contains a pointer to the first item. *Synonyms:* circular list; ring; chain. (C) 610.5-1990w

circularly polarized field vector At a point in space, a field vector whose extremity describes a circle as a function of time. *Note:* Circular polarization may be viewed as a special case of elliptical polarization where the axial ratio has become equal to one. (AP/ANT) 145-1993

circularly polarized plane wave A plane wave whose electric field vector is circularly polarized. (AP/ANT) 145-1993

circularly polarized wave (1) (general) An elliptically polarized wave in which the ellipse is a circle in a plane perpendicular to the direction of propagation. *See also:* radiation. (PE/EEC) [119]

(2) An electromagnetic wave for which the locus of the tip of the instantaneous electric field vector is a circle in a plane orthogonal to the wave normal. This circle is traced at a rate equal to the angular frequency of the wave with a left-hand or right-hand sense of rotation. *See also:* left-hand polarized wave; right-hand polarized wave. (AP/PROP) 211-1997

circular magnetic wave (waveguide) A transverse magnetic wave for which the lines of magnetic force form concentric circles. (MTT) 146-1980w

circular mil A unit of area equal to $\pi/4$ of a square mil (0.7854 square mil). The cross-sectional area of a circle in circular mils is therefore equal to the square of its diameter in mils. A circular inch is equal to one million circular mils. *Note:* A mil is one-thousandth part of an inch. There are 1974 circular mils in a square millimeter. (SWG) 341-1980

circular orbit (communication satellite) An orbit of a satellite in which the distance between the centers of mass of the satellite and of the primary body is constant. (COM) [19]

circular probable error (navigation aids) In two-dimensional error distribution, the radius of a circle encompassing half of all errors. (AES/GCS) 172-1983w

Circular Queue A software array of message storage locations for which the last cell is logically adjacent to the first. It is a first-in-first-out (FIFO) queue where the producer leads the consumer through the array, wrapping back to the first cell when the end is reached. (C/MM) 1212.1-1993

circular scanning (1) (radio) Scanning in which the direction of maximum response generates a plane or a right circular cone whose vertex angle is close to 180 degrees. (AP/ANT) [35]

(2) Scanning where the beam axis of the antenna generates a conical surface. *Note:* This can include the special case where the cone degenerates to a plane. (AP/ANT) 145-1993

circular shift (mathematics of computing) A variation of a logical shift in which the digits moved out of one end of a register, word, or numeral are returned at the other end. For example, + 231.702 shifted two places to the left becomes 3170.2+2. *Note:* A circular shift may be applied to the multiple precision representation of a number. *Synonyms:* end-around shift; cyclic shift; rotate; end-around carry shift; ring shift. (C) 1084-1986w

circular Taylor distribution *See:* circular Taylor distribution.

circulating current The current that flows through the transition impedance as a result of two taps being bridged during a tap change operation for resistance-type LTCs or being in the bridging position for reactance-type LTCs. (PE/TR) C57.131-1995

circulating current fault (thyristor converter) A circulating current in excess of the design value. *Note:* In a double converter precaution must be taken to control circulating direct current between the forward and reverse sections. (IA/IPC) 444-1973w

circulating memory *See:* circulating register.

circulating-power test Operation of a bipolar HVDC transmission system so that power flows in the opposite direction in each pole. High power levels up to system rating can be used for test purposes with losses and net reactive only being supplied from the ac network. *Synonym:* round-power test. (PE/SUB) 1378-1997

circulating register (1) (data processing) A register that retains data by inserting it into a delaying means, and regenerating and reinserting the data into the register. (C) 162-1963w
(2) Shift register in which data that are moved out of one end of the register are reentered into the other end, as in a closed loop. *See also:* cyclic shift. (C) [20], [85], 610.10-1994w

circulating storage Dynamic storage in the form of a closed loop. *Synonym:* cyclic storage. *See also:* regenerative track. (C) 610.10-1994w

circulation of electrolyte A constant flow of electrolyte through a cell to facilitate the maintenance of uniform conditions of electrolysis. *See also:* electrorefining. (EEC/PE) [119]

circulator (waveguide system) A passive waveguide junction of three or more arms in which the arms can be listed in such an order that when power is fed into any arm it is transferred to the next arm on the list, the first arm being counted as following the last in order. *See also:* transducer; waveguide. (AP/ANT) [35], [84]

circulator coupled (isolated) port (nonlinear, active, and nonreciprocal waveguide components) With reference to a particular port of the circulator, a port to which waves pass from the reference port with low (high) insertion loss. (MTT) 457-1982w

circum-aural receiver A receiver that surrounds and encloses the ear pinna without making intimate contact with it. (COM/TA) 1206-1994

circumflex The character "^". (C/PA) 9945-2-1993

CISC *See:* complex-instruction-set computer.

CISPR International Special Committee on Radio Interference.

citizens bands Frequency bands allocated for short-distance personal or business radio communication, radio signaling, and control of remote devices by radio. *Note:* The frequency bands may differ from country to country. The bands presently in use in the United States are 26.965–27.405 MHz; 72–76 MHz and 462.550–467.425 MHz. (T&D/PE) 539-1990

CIU *See:* computer interface unit.

civil speed limit The maximum speed authorized for each section of track, as determined primarily by the alignment, profile, and structure. (VT/RT) 1474.1-1999

civil twilight (illuminating engineering) (morning and evening) Civil twilight ends in the evening when the center of the sun's disk is six degrees below the horizon and begins in the morning when the center of the sun's disk is six degrees below the horizon. *See also:* night. (EEC/IE) [126]

cladding (1) (fiber optics) The dielectric material surrounding the core of an optical waveguide. *See also:* tolerance field; core; optical waveguide; normalized frequency. (Std100) 812-1984w

(2) A layer that surrounds the glass core of an optical fiber. (C) 610.7-1995

cladding center (fiber optics) The center of the circle that circumscribes the outer surface of the homogeneous cladding, as defined under tolerance field. *See also:* cladding; tolerance field. (Std100) 812-1984w

cladding diameter (fiber optics) The length of the longest chord that passes through the fiber axis and connects two points on the periphery of the homogeneous cladding. *See also:* core diameter; tolerance field; cladding. (Std100) 812-1984w

cladding mode (fiber optics) A mode that is confined by virtue of a lower index medium surrounding the cladding. Cladding modes correspond to cladding rays in the terminology of geometric optics. *See also:* mode; bound mode; leaky mode; unbound mode; cladding ray. (Std100) 812-1984w

cladding mode stripper (fiber optics) A device that encourages the conversion of cladding modes to radiation modes; as a result, the cladding modes are stripped from the fiber. Often a material having a refractive index equal to or greater than that of the waveguide cladding. *See also:* cladding mode; cladding. (Std100) 812-1984w

cladding ray (fiber optics) In an optical waveguide, a ray that is confined to the core and cladding by virtue of reflection from the outer surface of the cladding. Cladding rays correspond to cladding modes in the terminology of mode descriptors. *See also:* cladding mode; guided ray; leaky ray. (Std100) 812-1984w

claiming A ring state that occurs when a station detects that the active monitor functions are not being performed and at least one station is contending to become active monitor. (C/LM) 8802-5-1998

clamp (converter circuit elements) (self-commutated converters) An auxiliary circuit element or combination of elements employed to limit the peak voltage or current of a semiconductor device. *See also:* clamping circuit. (IA/SPC) 936-1987w

clamp, cable *See:* cable clamp.

clamper (data transmission) When used in broadband transmissions, it reinserts low frequency signal components that were not faithfully transmitted. (PE) 599-1985w

clamp, grounding *See:* grounding clamp.

clamp, hose *See:* strand restraining clamp.

clamping (1) (control) A function by which the extreme amplitude of a waveform is maintained at a given level. *See also:* feedback control system. (IA/ICTL/IAC) [60]
(2) (pulse operations) A process in which a specified instantaneous magnitude of a pulse is fixed at a specified magnitude. Typically, after clamping, all instantaneous magnitudes of the pulse are offset, the pulse shape remaining unaltered.

clamping circuit (1) (electronic circuits) A circuit that adds a fixed bias to a wave at each occurrence of some predetermined feature of the wave so that the voltage or current of the feature is held at, or "clamped," to some specified level. The level may be fixed or adjustable. *See also:* clamp; clamper. (EEC/PE) [119]
(2) A circuit used in analog computers to provide automatic hold and reset action electronically for the purpose of switching or supplying repetitive operation. *See also:* limiter circuit. (C) 610.10-1994w, 165-1977w

clamping-in *See:* clipping-in.

clamping screw *See:* binding screw.

clamping voltage (1) (low voltage varistor surge arresters) Peak voltage across the varistor measured under conditions of a specified peak pulse current and specified waveform. *Note:* Peak voltage and peak current are not necessarily coincident in time. (PE) [8]
(2) The maximum magnitude of voltage across a surge-protective device during the passage of a specified surge current (e.g., 100 A, 8/20 µs waveshape). (T&D/PE) 1250-1995
(3) The peak voltage across the surge-protective device measured under conditions of a specified surge current and specified current waveform. (SPD/PE) C62.62-2000

clamp, strand restraining *See:* strand restraining clamp.

clapper An armature that is hinged or pivoted. (EEC/REE) [87]

class (1) A category into which objects are placed on the basis of both their purpose and their internal structure. (C/PA) 1238.1-1994w, 1224.1-1993w, 1224-1993w
(2) A template for the creation of an object instance. The class defines the properties of an object. (SCC20) 1226-1998
(3) An abstraction of the knowledge and behavior of a set of similar things. Classes are used to represent the notion of "things whose knowledge or actions are relevant." (C/SE) 1320.2-1998
(4) A collection of objects that share common characteristics and features. (IM/ST) 1451.1-1999
(5) *See also:* object class; pattern class; class of messages; OM class.

Class *See:* accuracy rating.

Class 0 unplanned outage (electric generating unit reliability, availability, and productivity) An outage that results from the unsuccessful attempt to place the unit in service. *See also:* starting failure. (PE/PSE) 762-1987w

Class 1, Division 2 locations (auxiliary devices for motors) Basically those in which there may be flammable gas present due to a failure of a process system. Under normal conditions a flammable mixture of gas is not present. (IA/PC) 303-1984s

Class 1E The safety classification of the electric equipment and systems that are essential to emergency reactor shutdown, containment isolation, reactor core cooling, and containment and reactor heat removal, or are otherwise essential in preventing significant release of radioactive material to the environment. (PE/NP) 603-1998

Class 1E circuits (design and installation of cable systems for Class 1E circuits in nuclear power generating stations) The safety classification of circuits that are essential to emergency reactor shutdown, containment isolation, reactor core cooling, and containment and reactor heat removal, or are otherwise essential in preventing a significant release of radioactive material to the environment. (PE/EDPG) 690-1984r

Class 1E control board, panel, or rack A control board, panel, or rack fitted with Class 1E equipment. (PE/NP) 420-1982

Class 1E electric systems *See:* nuclear power generating stations, class ratings.

Class 1 electric equipment *See:* nuclear power generating stations, class ratings.

Class I equipment Equipment in which protection against electric shock is achieved by using basic insulation, and providing a means of connecting the conductive parts, which are otherwise capable of assuming hazardous voltages if the basic insulation fails, to the protective ground conductor in the building wire. (EMC) 1140-1994r

Class I repeater A type of 100BASE-T repeater set with internal delay such that only one repeater set may exist between any two DTEs within a single collision domain when two maximum length copper cable segments are used. (C/LM) 802.3-1998

Class 1 structures and equipment *See:* nuclear power generating stations, class ratings.

Class 1 termination Provides electric stress control for the cable insulation shield terminus; provides complete external leakage insulation between the cable conductor(s) and ground; and provides a seal to the end of the cable against the entrance of the external environment and main tains the pressure, if any, of the cable system. (PE/IC) 48-1990s

Class 1 unplanned outage (electric generating unit reliability, availability, and productivity) (immediate) An outage that requires immediate removal from the existing state. *Note:* A Class 1 unplanned outage can be initiated from either the in-service or shutdown states. A Class 1 unplanned outage can also be initiated from the planned outage state. *See also:* extended planned outage. (PE/PSE) 762-1987w

Class II equipment Equipment in which protection against electric shock does not rely on basic insulation only, but also on the provision of additional safety precautions, such as double insulation or reinforced insulation. There is no provision for protective grounding or reliance upon installation conditions. (EMC) 1140-1994r

Class II repeater A type of IEEE 802.3 100BASE-T repeater set with internal delay such that only two or fewer such repeater sets may exist between any two DTEs within a single collision domain when two maximum length copper cable segments are used. (C/LM) 802.3-1998

Class 2 structures and equipment *See:* nuclear power generating stations, class ratings.

Class 2 termination Provides electric stress control for the cable insulation shield terminus; and provides complete external leakage insulation between the cable conductor (s) and ground. (PE/IC) 48-1990s

Class 2 transformer (power and distribution transformers) A step-down transformer of the low-secondary-voltage type, suitable for use in class 2 remote-control low-energy circuits. It shall be of the energy-limiting type, or of a non-energy-limiting type equipped with an overcurrent device. *Note:* "Low-secondary-voltage," as used here, has a value of approximately 24 V. (PE/TR) C57.12.80-1978r

Class 2 unplanned outage (electric generating unit reliability, availability, and productivity) (delayed) An outage that does not require immediate removal from the in-service state but requires removal within 6 hours. (PE/PSE) 762-1987w

Class 3 structures and equipment *See:* nuclear power generating stations, class ratings.

Class 3 termination Provides electric stress control for the cable insulation shield terminus. *Note:* Some cables do not have an insulation shield. Termination for such cables would not be required to provide electric stress control. In such cases, this provision would not be part of the definition. (PE/IC) 48-1990s

Class 3 unplanned outage (electric generating unit reliability, availability, and productivity) (postponed) An outage that can be postponed beyond six hours but requires that a unit be removed from the in-service state before the end of the next weekend. *Note:* Classes 2 and 3 can only be initiated from the in-service state. (PE/PSE) 762-1987w

Class 4 unplanned outage (electric generating unit reliability, availability, and productivity) (deferred) An outage that will allow a unit outage to be deferred beyond the end of the next weekend but requires that a unit be removed from the available state before the next planned outage. (PE/PSE) 762-1987w

Class 90 insulation *See:* class ratings insulation.

Class 105 insulation *See:* class ratings insulation.

Class 105 insulation system Materials or combinations of materials such as cotton, silk, and paper when suitably impregnated or coated or when immersed in a dielectric liquid. *Note:* Other materials or combinations may be included in this class if by experience or accepted tests the insulation system can be shown to have comparable themal life at 105°C. (PE/TR) C57.12.80-1978r

Class 120 insulation system Materials or combinations of materials such as cotton, silk, and paper when suitably impregnated or coated or when immersed in a dielectric liquid; and which possess a degree of thermal stability which allows them to be operated at a temperature 15°C higher than temperature index 105 materials. *Note:* Other materials or combinations may be included in this class if by experience or accepted tests the insulation system can be shown to have comparable thermal life at 120°C. (PE/TR) C57.12.80-1978r

Class 130 insulation *See:* class ratings insulation.

Class 150 insulation system Materials or combinations of sealed dry-type transformer, self-cooled (class GA) (power and distribution transformer) materials such as mica, glass fiber, asbestos, etc., with suitable bonding substances. *Note:* Other materials or combinations of materials may be included in this class if by experience or accepted tests the insulation system can be shown to have comparable thermal life at 150°C. (PE/TR) C57.12.80-1978r

Class 155 insulation *See:* class ratings insulation.

Class 180 insulation *See:* class ratings insulation.

Class 185 insulation system Materials or combinations of materials such as silicone elastometer, mica, glass fiber, asbestos, etc, with suitable bonding substances such as appropriate silicone resins. *Note:* Other materials or combinations of materials may be included in this class if by experience or accepted tests the insulation system can be shown to have comparable thermal life at 220°C. (PE/TR) C57.12.80-1978r

Class 220 insulation *See:* class ratings insulation.

Class 220 insulation system Materials or combinations of materials such as silicone elastomer, mica, glass fiber, asbestos, etc, with suitable bonding substances such as appropriate silicone resins. *Note:* Other materials or combinations of materials may be included in this class if by experience or accepted tests, the insulation system can be shown to have comparable thermal life at 220°C. (PE/TR) C57.12.80-1978r

Class A amplifier *See:* amplifier class ratings.

Class A component (seismic design of substations) Any component or system whose failure, malfunction, or need for repair prevents the proper operation of the substation during or after the design earthquake. (SUB/PE) 693-1984s

Class A modulator A Class-A amplifier that is used specifically for the purpose of supplying the necessary signal power to modulate a carrier. *See also:* modulation. (AP/BT/ANT) 145-1983s, 182-1961w

Class A operation *See:* amplifier class ratings.

Class A push-pull sound track A Class-A push-pull photographic sound track consists of two single tracks side by side, the transmission of one being 180° out of phase with the transmission of the other. Both positive and negative halves of the sound wave are linearly recorded on each of the two tracks. *See also:* phonograph pickup. (SP) [32]

Class A seismic component A component or system whose failure, malfunction, or need for repair prevents the proper operation of the gas-insulated substation during or after the design earthquake. (SWG/SUB/PE) C37.122-1983s, C37.122.1-1993, C37.100-1992

Class AB amplifier *See:* amplifier class ratings.

Class AB operation *See:* amplifier class ratings.

Class B amplifier *See:* amplifier class ratings.

Class B component (seismic design of substations) Any component or system whose failure, malfunction, or need for repair does not prevent the operation of the substation during or after the design earthquake. (SUB/PE) 693-1984s

Class B modulator A class-B amplifier that is used specifically for the purpose of supplying the necessary signal power to modulate a carrier. *Note:* In such a modulator the class-B amplifier is normally connected in push-pull. *See also:* modulation. (AP/BT/ANT) 145-1983s, 182-1961w

Class B operation *See:* amplifier class ratings.

Class B push-pull sound track A class-B push-pull photographic sound track consists of two tracks side by side, one of which carries the positive half of the signal only, and the other the negative half. *Note:* During the inoperative half-cycle, each track transmits little or no light. *See also:* phonograph pickup. (SP) [32]

Class B seismic component A component or system whose failure, malfunction, or need for repair does not prevent the proper operation of the gas-insulated substation during or after the design earthquake. Class B components are designed to meet either normal building codes and national standards in force at the site or another lower-level design earthquake. Application of further design requirements is left to the discretion of the user. (SWG/SUB/PE) C37.122-1983s, C37.100-1992, C37.122.1-1993

Class C amplifier *See:* amplifier class ratings.

Class C operation *See:* amplifier class ratings.

class designation (watthour meter) The maximum of the load range in amperes. *See also:* load range; watthour meter—class designation. (ELM) C12.1-1981

classes of grounding (neutral grounding in electrical utility systems) A specific range of degree of grounding; for example, effectively and noneffectively. (PE/SPD) C62.92-1987r

classes of insulation systems (insulation systems of synchronous machines) The insulation systems usually employed in synchronous machines covered by this standard are defined below. These definitions, in general, correspond with the principles set forth in IEEE Std 1-1969, *IEEE Standard General Principles for Temperature Limits in the Rating of Electric Equipment and for the Evaluation of Electrical Insulation,* which is also the accepted basis for interpretation. (REM) [115]

class hierarchy (1) The tree-structured organization of classes. A widget class in this tree always supports all of the operations supported by its ancestor classes closer to the root of the tree, but might support them with different implementations and might also add new operations. (C) 1295-1993w **(2)** An ordering of classes, in which a subclass is a specialization of its superclass. A class inherits attributes and relationships from its superclass and can define additional attributes and relationships of its own. (C/SE) 1420.1-1995

classical maximum usable frequency *See:* maximum usable frequency.

Class ID All classes have a descriptive header entry "Class ID," the value of which is the Class ID (note capitalization). For any object of class `IEEE1451_Root`, the operation `GetClassID` returns a value, `class_id`, that has the same value as Class ID. (IM/ST) 1451.1-1999

classification (1) A choice within a category. (C/SE) 1044-1993 **(2)** A choice within a category to describe the category (attribute) of an anomaly. (C/SE) 1044.1-1995 **(3)** A grouping of objects on the basis of common characteristics. This is the normal method for grouping objects within an object-oriented programming environment. (SCC20) 1226-1998 **(4)** The manner in which the assets are organized for ease of search and extraction within a reuse library. (C/SE) 1517-1999

classification current (metal-oxide surge arresters for ac power circuits) The designated current used to perform the classification tests. (SPD/PE) C62.11-1993s, C62.11-1987s

classification lamp (classification light) A signal lamp placed at the side of the front end of a train or vehicle, displaying light of a particular color to identify the class of service in which the train or vehicle is operating. (PE/NP) 344-1975s

classification light *See:* classification lamp.

classification of arresters (1) Arrester classification is determined by prescribed test requirements. These classifications are: station valve arrester, intermediate valve arrester, distri-

bution valve arrester, secondary valve arrester, protector tube. (PE/SPD) [8], C62.1-1981s
(2) Arrester classification is determined by prescribed test requirements. These classifications are: station, intermediate, distribution heavy duty, distribution normal duty, distribution light duty, secondary. (SPD/PE) C62.11-1993s

classification of insulation *See:* self-restoring insulation; internal insulation; external insulation; non-self-restoring insulation.

classification process The classification process is a series of activities, starting with the recognition of an anomaly through to its closure. The process is divided into four sequential steps interspersed with three administrative activities. The sequential steps are as follows: Step 1: Recognition; Step 2: Investigation; Step 3: Action; Step 4: Disposition. The three administrative activities applied to each sequential step are as follows: Recording; Classifying; Identifying impact. (C/SE) 1044-1993

classification scheme A scheme for the arrangement or division of entities into groups based on properties that the entities have in common. (SCC32) 1489-1999

classifier *See:* decision rule.

classifying current The designated current used to perform the classification tests. (SPD/PE) C62.11-1999

class-level attribute A mapping from the class itself to the instances of a value class. (C/SE) 1320.2-1998

class-level operation A mapping from the (cross product of the) class itself and the instances of the input argument types to the (cross product of the) instances of the other (output) argument types. (C/SE) 1320.2-1998

class-level responsibility A kind of responsibility that represents some aspect of the knowledge, behavior, or rules of the class as a whole. For example, the total `registered-VoterCount` would be a class-level property of the class `registeredVoter`; there would be only one value of `registeredVoterCount` for the class as a whole. *Contrast:* instance-level responsibility. (C/SE) 1320.2-1998

Class Name All classes have a descriptive header entry "Class," the value of which is the Class Name (note capitalization). For any Object, the operation `GetClassName` returns a value, `class_name`, that has the same value as Class Name. (IM/ST) 1451.1-1999

class of messages A group of messages having in the Command fields of their respective HEADER packets command codes of commands belonging to a single class of commands. The name, *C*, of a class of messages is the same as the name of the class of commands that defines the class *C*. (TT/C) 1149.5-1995

class-of-service indication (telephone switching systems) An indication of the features assigned to a switching network termination. (COM) 312-1977w

class-of-service tone (telephone switching systems) A tone that indicates to an operator that a certain class-of-service is appropriate to a call. (COM) 312-1977w

Class O insulation *See:* class ratings insulation.

Class over-220 insulation *See:* class ratings insulation.

Class over-220 insulation system Materials consisting entirely of mica, porcelain, glass quartz, and similar inorganic materials. *Note:* Other materials or combinations of materials may be included in this class if by experience or accepted tests the insulation system can be shown to have the required thermal life at temperatures over 220°C. (PE/TR) C57.12.80-1978r

class ratings insulation These temperatures are, and have been in most cases over a long period of time, benchmarks descriptive of the various classes of insulating materials, and various accepted test procedures have been or are being developed for use in their identification. They should not be confused with the actual temperatures at which these same classes of insulating materials may be used in the various specific types of equipment, nor with the temperatures on which specified temperature rise in equipment standards are based. **(A)** In the following definitions the words accepted tests are intended to refer to recognized test procedures established for the thermal evaluation of materials by themselves or in simple combinations. Experience or test data, used in classifying insulating materials, are distinct from the experience or test data derived for the use of materials in complete insulation systems. The thermal endurance of complete systems may be determined by test procedures specified by the responsible technical committees. A material that is classified as suitable for a given temperature may be found suitable for a different temperature, either higher or lower, by an insulation system test procedure. For example, it has been found that some materials suitable for operation at one temperature in air may be suitable for a higher temperature when used in a system operated in an inert gas atmosphere. Likewise some insulating materials when operated in dielectric liquids will have lower or higher thermal endurance than in air. **(B)** It is important to recognize that other characteristics, in addition to thermal endurance, such as mechanical strength, moisture resistance, and corona endurance, are required in varying degrees in different applications for the successful use of insulating materials. A) *class 90 insulation.* Materials or combinations of materials such as cotton, silk, and paper without impregnation. *Note:* Other materials or combinations of materials may be included in this class if by experience or accepted tests they can be shown to have comparable thermal life at 90 degrees Celsius. B) *class 105 insulation.* Materials or combinations of materials such as cotton, silk, and paper when suitably impregnated or coated or when immersed in a dielectric liquid. *Note:* Other materials or combinations may be included in this class if by experience or accepted tests they can be shown to have comparable thermal life at 105 degrees Celsius. **(C)** *class 130 insulation.* Materials or combinations of materials such as mica, glass fiber, asbestos, etc., with suitable bonding substances. *Note:* Other materials or combinations of materials may be included in this class if by experience or accepted tests they can be shown to have comparable thermal life at 130 degrees Celsius. **(D)** *class 155 insulation.* Materials or combinations of materials such as mica, glass fiber, asbestos, etc., with suitable bonding substances. *Note:* Other materials or combinations of materials may be included in this class if by experience or accepted tests they can be shown to have comparable thermal life at 155 degrees Celsius. **(E)** *class 180 insulation.* Materials or combinations of materials such as silicone elastomer, mica, glass fiber, asbestos, etc., with suitable bonding substances such as appropriate silicone resins. *Note:* Other materials or combinations of materials may be included in this class if by experience or accepted tests they can be shown to have comparable thermal life at 180 degrees Celsius. **(F)** *class 220 insulation.* Materials or combinations of materials which by experience or accepted tests can be shown to have the required thermal life at 220 degrees Celsius. **(G)** *class over-220 insulation.* Materials consisting entirely of mica, porcelain, glass, quartz, and similar inorganic materials. *Note:* Other materials or combinations of materials may be included in this class if by experience or accepted tests they can be shown to have comparable thermal life at 90 degrees Celsius. B) class 105 insulation. Materials or combinations of materials such as cotton, silk, and paper when suitably impregnated or coated or when immersed in a dielectric liquid. *Note:* Other materials or combinations may be included in this class if by experience or accepted tests they can be shown to have the required thermal life at temperatures over 220 degrees Celsius.
(2) **(A)** *class O insulation. See also:* Class 90 insulation. **(B)** *class A insulation.*

- Cotton, silk, paper, and similar organic materials when either impregnated or immersed in a liquid dielectric.
- Molded and laminated materials with cellulose filler, phenolic resins, and other resins of similar properties.
- Films and sheets of cellulose acetate and other cellulose derivatives of similar properties.
- Varnishes (enamel) as applied to conductors.

Note: An insulation is considered to be impregnated when a suitable substance replaces the air between its fibers, even if this substance does not completely fill the spaces between the insulated conductors. The impregnating substances, in order to be considered suitable, must have good insulating properties; must entirely cover the fibers and render them adherent to each other and to the conductor; must not produce interstices within itself as a consequence of evaporation of the solvent or through any other cause; must not flow during the operation of the machine at full working load or at the temperature limit specified; and must not unduly deteriorate under prolonged action of heat. **(C)** *class B insulation.* Mica, asbestos, glass fiber, and similar inorganic materials in built-up form with organic binding substances. *Note:* A small proportion of class A materials may be used for structural purposes only. Glass fiber or asbestos magnet-wire insulations are included in this temperature class. These may include supplementary organic materials, such as polyvinyl acetal or polyamide films. The electrical and mechanical properties of the insulated winding must not be impaired by application of the temperature permitted for class B material. (The word "impaired" is here used in the sense of causing any change that could disqualify the insulating material for continuous service.) The temperature endurance of different class B insulation assemblies varies over a considerable range in accordance with the percentage of class A materials employed, and the degree of dependence placed on the organic binder for maintaining the structural integrity of the insulation. **(D)** *class H insulation.* Insulation consisting of:

- mica, asbestos, glass fiber, and similar inorganic materials in built-up form with binding substances composed of silicone compounds or materials with equivalent properties.
- silicone compounds in rubbery or resinous forms or materials with equivalent properties.

Note: A minute proportion of class A materials may be used only where essential for structural purposes during manufacture. The electrical and mechanical properties of the insulated winding must not be impaired by the application of the hottest-spot temperature permitted for the specific insulation class. The word "impaired' is here used in the sense of causing any change that could disqualify the insulating materials for continuously performing its intended function, whether creepage spacing, mechanical support, or dielectric barrier action. **(E)** *class C insulation.* Insulation consisting entirely of mica, porcelain, glass, quartz, and similar inorganic materials. **(F)** *class F insulation.* Materials or combinations of materials such as mica, glass fiber, asbestos, etc., with suitable bonding substances. Other materials or combinations of materials, not necessarily inorganic, may be included in this class if by experienced or accepted tests they can be shown to be capable of operation at 155°C. (IA/MT) 45-1983

cleaner (electroplating) A compound or mixture, used in degreasing, that is usually alkaline. (PE/EEC) [119]

cleaning (electroplating) The removal of grease or other foreign material from a metal surface, chiefly by physical means. *See also:* electroplating. (EEC/PE) [119]

cleanse instruction A cleanse (cache-control) instruction converts a line to the clean state (the data in cache and memory are the same). (C/MM) 1596-1992

cleanup data cycle A data cycle that is not accompanied by data transfer but is for the purpose of turning off the slave's AD, PA, and PE drivers. (NID) 960-1993

clear (1) (A) To preset a storage or memory device to a prescribed state, usually that denoting zero. **(B)** To place a binary cell in the zero state. *See also:* nonlinear capacitor; reset. (C/ED) [46]

(2) (software) To set a variable, register, or other storage location to zero, blank, or other null value. *See also:* initialize; reset. (C) 610.12-1990

(3) To force the contents of one or more storage elements to the logic 0 state. (TT/C) 1149.5-1995

(4) To replace a variable, register, or other storage location with a zero, blank, or null value. (C) 610.10-1994w

(5) To preset a storage or memory device to a prescribed state, usually that denoting zero. *Note:* In the field of nonvolatile memories, "clear" conventionally means to set the outputs of the memory to the high logic level. (ED) 1005-1998

(6) Operator action to remove specific displays. (PE/NP) 692-1997

(7) The action that removes the outgoing signal from a link (i.e., clears the link) and prepares the Physical Medium Dependent (PMD) to receive a packet.local area networks. (C) 8802-12-1998

clear algorithm The timed sequence of signals necessary to clear the memory for a flash electrically erasable programmable read-only memory (EEPROM). (ED) 1005-1998

clearance (1) (A) (navigation aids) [instrument landing system (ILS)] The difference in depth of modulation (DDM) in excess of that required to produce full-scale deflection of the course-deviation indicator in flight areas outside the on-course sector; when the DDM is too low the indicator falls below full-scale deflection and the condition of low clearance exists. **(B) (navigation aids)** (air traffic control) Permission by a control facility to the pilot to proceed in a mutually understood manner. (AES/GCS) 172-1983

(2) The minimum separation between two conductors, between conductors and supports or other objects, or between conductors and ground. *See also:* tower. (PE/T&D) [10]

(3) (fence safety clearances in electric-supply stations) The separation between two conductors, between conductors and supports or other objects, or between conductors and ground. (PE/SUB) 1119-1988w

(4) (A) The condition in which a circuit has been deenergized to enable work to be performed more safely. A clearance is normally obtained on a circuit presenting a source of hazard prior to starting work. *Synonyms:* restriction; outage; permit. **(B)** The minimum separation between two conductors, between conductors and supports or other objects, or between conductors and ground or the clear space between any objects. (T&D/PE) 524-1992

(5) The clear distance between two objects measured surface to surface. (NESC) C2-1997

(6) (maintenance of energized power lines) *See also:* work permit. (T&D/PE) 516-1995

clearance antenna array (directional localizer) (navigation aid terms) The antenna array that radiates a localizer signal on a separate frequency within the pass band of the receiver and provides the required signals in the clearance sectors as well as a back course. (AES/GCS) 172-1983w

clearance lamps (illuminating engineering) Lighting devices for the purpose of indicating the width and height of a vehicle. (EEC/IE) [126]

clearance point The location on a turnout at which the carrier's specified clearance is provided between tracks. (EEC/PE) [119]

clearances The separation between two conductors, between conductors and supports or other objects, or between conductors and ground. (PE/SUB) 1268-1997

clearance sector (instrument landing systems) (navigation aids) The sector extending around either side of the localizer from the course sector to the back course sector, and within which the deviation indicator provides the required offcourse indication. (AES/GCS) 172-1983w

clear area In character recognition, a specified area that is to be kept free of printing or other markings that are not related to machine readings. (C) 610.2-1987

clear channel assessment function That logical function in the physical layer (PHY) that determines the current state of use of the wireless medium (WM). (C/LM) 8802-11-1999

clear disturb The corruption of data in one location caused by the clearing of data at another location. (ED) 1005-1998

clearing (low-voltage air-gap surge-protective devices) (low voltage surge protective devices) The characteristic of some types of air gap surge arresters to exhibit a low resistance and then to revert to a high resistance state as a result of an external influence. (SPD/PE) C62.32-1981s, [8]

clearing circuit A circuit used for the operation of a signal in advance of an approaching train. (EEC/PE) [119]

clearing-out drop (cord circuit or trunk circuit) A drop signal that is operated by ringing current to attract the attention of the operator. (EEC/PE) [119]

clearing source (low-voltage air-gap surge-protective devices) (low voltage surge protective devices) A defined electrical source which is intentionally applied as a clearing stimulus to an air gap surge protective device under laboratory test conditions. This stimulus is intended to simulate conditions encountered during normal usage. (SPD/PE) C62.32-1981s, [8]

clearing time (fuse) The time elasping from the beginning of an overcurrent to the final circuit interruption. *Note:* The clearing time is equal to the sum of melting time and arcing time. *Synonym:* total clearing time. (SWG/PE) [56]
(2) (A) (mechanical switching device). The interval between the time the actuating quantity in the main circuit reaches the value causing actuation of the release and the instant of final arc extinction on all poles of the primary arcing contacts. *Note:* Clearing time is numerically equal to the sum of contact parting time and arcing time. **(B)** (total clearing time of a fuse). The time elapsing from the beginning of a specified overcurrent to the final circuit interruption, at rated maximum voltage. *Note:* The clearing time is equal to the sum of melting time and the arcing time. (SWG/PE/NP) C37.100-1992, 308-1980, C37.40-1993

clearly discernable Capable of being noticed easily and without close inspection. (SUB/PE) C37.123-1996

clear packet A packet used during initialization to empty linc buffers and initialize the linc. CSR state is unaffected; e.g., the node's address is unchanged by a "clear." Clear may be sent by any node that has lost synchronization in order to trigger reinitialization. (C/MM) 1596-1992

clear sky (illuminating engineering) A sky that has less than 30% cloud cover. (EEC/IE) [126]

cleartext Intelligible data, the semantic content of which is available. (LM/C) 802.10-1992

cleat An assembly of two pieces of insulating material provided with grooves for holding one or more conductors at a definite spacing from the surface wired over and from each other, and with screw holes for fastening in position. *See also:* raceway. (EEC/PE) [119]

clerestory (illuminating engineering) That part of a building which rises clear of the roofs or other parts and whose walls contain windows for lighting the interior. (EEC/IE) [126]

CLI *See:* cumulative leakage index.

click (1) A disturbance of a duration less than a specified value as measured under specified conditions. *See also:* electromagnetic compatibility. (EMC/IM) [53], C63.4-1991, [76]
(2) The act of pressing and releasing a mouse button without moving the mouse pointer. (C) 1295-1993w

client (1) (MULTIBUS) An agent that requests services of a server. *See also:* server. (C/MM) 1296-1987s
(2) Software that uses the interface. (C/PA) 1351-1994w, 1224-1993w, 1327-1993w, 1328-1993w
(3) In networking, a station or program requesting a service. *Contrast:* server. (C) 610.7-1995
(4) Software that uses an interface. (C/PA) 1224.1-1993w
(5) Refers to the software component on one device that uses the services provided by a server on another device. (C/MM) 1284.4-2000
(6) *See also:* batch client.
®MULTIBUS is a registered trademark of Intel Corporation.

client application An application program that makes use of Media Management System (MMS) services to manage its media. Examples of client applications include a backup program, a hierarchical storage manager, and an application that allows individual users to mount their own tapes. (C/SS) 1244.1-2000

Client Cached Block Cookie A particular Block Cookie associated with a Base Client. (IM/ST) 1451.1-1999

client execution environment The machine state that exists when a client program begins execution. (C/BA) 1275-1994

client interface A set of data and procedures giving a client program access to client interface services. (C/BA) 1275-1994

client instance A manifestation of the client that shares the input and output queues of the client with other instances. (C/PA) 1224.1-1993w

client interface handler A mechanism by which control and data are transferred from a client program to the firmware, and subsequently returned, for the purpose of providing client interface services. (C/BA) 1275-1994

client layer In the OSI model, refers to the data link and physical layers. *See also:* transport layer; network layer; presentation layer; physical layer; sublayer; data link layer; entity layer; application layer; session layer; logical link control sublayer; medium access control sublayer. (C) 610.7-1995

client interface services Those services that Open Firmware provides to client programs, including device tree access, memory allocation, mapping, console I/O, mass storage, and network I/O. (C/BA) 1275-1994

Client object Any object that invokes operations on other objects. (IM/ST) 1451.1-1999

Client Port An instance of the class `IEEE1451_ClientPort` or of a subclass thereof. (IM/ST) 1451.1-1999

client program A software program that is loaded and executed by Open Firmware (or a secondary boot program). (The client program may use services provided by the Open Firmware client interface.). (C/BA) 1275-1994

client role The location where the software is actually executed or used (as opposed to the target where it is actually installed). The configuration of software is performed by this role. (C/PA) 1387.2-1995

client-server In a communications network, the client is the requesting device and the server is the supplying device. For example, the user interface could reside in the client workstation while the storage and retrieval functions could reside in the server database. (C) 610.7-1995

client-server communication A communication pattern, where a specific object, the client, communicates in a one-to-one fashion with a specific server object, the server. (IM/ST) 1451.1-1999

climber in training A worker who is in training to become a qualified climber. (T&D/PE) 1307-1996

climbing The vertical movement (ascending and descending) and horizontal movement to access or depart the worksite. (NESC/T&D/PE) C2-1997, 1307-1996

climbing space The vertical space reserved along the side of a pole or structure to permit ready access for linemen to equipment and conductors located on the pole structure. (T&D/PE) 196-1951w, [10], C2.2-1960

clinometer (navigation aids) An instrument for indicating the degree of slope of the angle of roll or pitch of a vehicle, according to the plane in which it is mounted. (AES/GCS) 172-1983w

clip (1) (charged-particle detectors) (x-ray energy spectrometers) (radiation detectors) A limiting operation, such as the use of a high-pass filter or a nonlinear operation such as diode limiting of pulse amplitude. *Synonym:* clipping. *See also:* fuse clips; differentiated; contact clip. (NPS/NID) 325-1971w, 759-1984r, 301-1976s
(2) (charged-particle detectors) A limiting operation, such as the use of a high-pass filter (differentiator) or a nonlinear operation to limit the amplitude of a pulse. The first usage is archaic. *Synonym:* clipping. (NPS) 300-1988r
(3) *See also:* cable clamp. (PE/T&D) 524-1992r

clipboard A software storage device that is used to store an object that is cut or copied from the screen and to retrieve an object that is pasted. (C) 1295-1993w

clipper (data transmission) A device that automatically limits the instantaneous value of the output to a predetermined maximum value. *Note:* The term is usually applied to devices which transmit only portions of an input wave lying on one side of an amplitude boundary. (PE) 599-1985w

clipper limiter A transducer that gives output only when the input lies above a critical value and a constant output for all inputs above a second higher critical value. *Synonym:* amplitude gate. *See also:* transducer. (AP/ANT) 145-1983s

clipping (1) (voice-operated telephone circuit) The loss of initial or final parts of words or syllables due to nonideal operation of the voice-operated devices. (EEC/PE) [119]
(2) (computer graphics) A computer graphics technique in which display elements lying totally outside a view area are made invisible and display elements lying partially inside a view area are scissored to remove the parts outside the view area before they are mapped to the display image. *Note:* In two-dimensional graphics, this view area is called the window; in three-dimensional graphics, it is called the view volume. *See also:* view volume; window; wrap-around; scissoring.

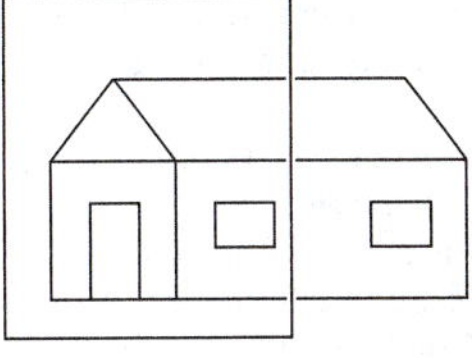

clipping

(C) 610.6-1991w
(3) *See also:* clip. (NPS) 300-1988r
(4) *See also:* clipping-in; semiconductor; chip. (PE/T&D) 524-1992r

clipping-in (conductor stringing equipment) The transferring of sagged conductors from the travelers to their permanent suspension positions and the installing of the permanent suspension clamps. *Synonyms:* clipping; clamping-in. (T&D/PE) 524a-1993r, 524-1992r

clipping offset (conductor stringing equipment) A calculated distance, measured along the conductor from the plumb mark to a point on the conductor at which the center of the suspension clamp is to be placed. When stringing in rough terrain, clipping offsets may be required to balance the horizontal forces on each suspension structure. (T&D/PE) 524-1992r

clips *See:* fuse clips; contact clips.

CLIST A command language used in the IBM MVS environment. (C) 610.13-1993w

CLK A fixed-frequency clock signal. The main SBus timing signal. (C/BA) 1496-1993w

clock (1) (A) A device that generates periodic signals used for synchronization. **(B)** A device that measures and indicates time. *See also:* timer; real-time clock; master clock; wall clock; time-of-day clock. **(C)** A register whose content changes at regular intervals in such a way as to measure time. (C) [20], 610.10-1994
(2) A signal, the transitions of which (between the low and high logic level [or vice versa]) are used to indicate when a stored-state device, such as a flip-flop or latch, may perform an operation. (TT/C) 1149.5-1995, 1149.1-1990
(3) An object that measures the passage of time. The current value of the time measured by a clock can be queried and, possibly, set to a value within the legal range of the clock. (C/PA) 9945-1-1996, 1003.5-1999
(4) (A) A device that generates periodic, accurately spaced signals used for such purposes as timing, regulation of the operations of a processor, or generation of interrupts. **(B)** To trigger a circuit to perform an operation, such as to accept data into a register. (C) 610.10-1994
(5) A device that generates periodic signals used for synchronization. A device that measures and indicates time. A register whose content changes at regular intervals in such a way as to measure time. (AMR/SCC31) 1377-1997
(6) *See also:* dynamometer. (T&D/PE) 516-1987s, 524-1992r

clock accuracy The deviation from absolute accuracy per unit of time. In a hierarchical, master-slave synchronization plan, with one primary and at least one backup reference being designated for each local digital switch, the clock rate of the local switch is controlled by the master. Under that method of operation, the local digital switch should operate at zero nominal slips. If the link connecting the master switch to the slave switch is broken, the number of slips will depend on clock accuracy. (COM/TA) 973-1990w

clock cycle One period of the CLK signal, beginning with the rising edge of the signal and ending on the following rising edge of the signal. (C/BA) 1496-1993w

clocked data one (CD1) A Manchester-encoded data 1. A CD1 is encoded as a LO for the first half of the bit-cell and a HI for the second half of the bit-cell. (C/LM) 802.3-1998

clocked data zero (CD0) A Manchester-encoded data 0. A CD0 is encoded as a HI for the first half of the bit-cell and a LO for the second half of the bit-cell. (C/LM) 802.3-1998

clocked logic (power-system communication) The technique whereby all the memory cells (flip-flops) of a logic network are caused to change in accordance with logic input levels but at a discrete time. *See also:* digital. (PE) 599-1985w

clocked violation HI (CVH) A symbol that deliberately violates Manchester-encoding rules, used as a part of the Collision Presence signal. A CVH is encoded as a transition from LO to HI at the beginning of the bit cell, HI for the entire bit cell, and a transition from HI to LO at the end of the bit cell. (C/LM) 802.3-1998

clocked violation LO (CVL) A symbol that deliberately violates Manchester-encoding rules, used as a part of the Collision Presence signal. A CVL is encoded as a transition from HI to LO at the beginning of the bit cell, LO for the entire bit cell, and a transition from LO to HI at the end of the bit cell. (C/LM) 802.3-1998

clocking (data transmission) The generation of periodic signals used for synchronization. *See also:* data processing. (COM) [49]

clocking bit (1) In asynchronous transmission, a bit that signals a synchronization event. (C) 610.7-1995
(2) A bit containing an encoded signal, preceding the data within a data stream, or on a separate channel; used for establishing timing intervals. *See also:* synchronization bit; clock track. (C) 610.10-1994w

clock pulse *See:* clock signal.

clock reference (digital accelerometer) Basic system timing reference. (MTT) 457-1982w

clock, reference *See:* reference clock.

clock register *See:* timer.

clock signal A periodic signal used for synchronizing events. *Synonyms:* clock pulse; timing pulse. (C) 610.10-1994w

clockStrobe signal A packet that causes a node to record its time-of-day registers (if any) when it is received, and to record the duration of the propagation of the packet within the node. Used for precisely synchronizing multiple time-of-day clocks within a system. (C/MM) 1596-1992

clock tick An interval of time. A number of these occur each second. Clock ticks are one of the units that may be used to express a value found in type *clock_t*. (C/PA) 9945-1-1996

clock track A track on which a pattern of signals, known as synchronization bits, is recorded to provide a timing reference. *Synonym:* timing track. *See also:* clocking bit. (C) 610.10-1994w

clockwise arc (numerically controlled machines) An arc generated by the coordinated motion of two axes in which curvature of the path of the tool with respect to the workpiece is clockwise when viewing the plane of motion in the negative direction of the perpendicular axis. (IA) [61]

CLOS *See:* Common LISP Object System.

close To destroy a package instance. (C/BA) 1275-1994

close and latch The capability of a switching device to close (allow current flow) and immediately thereafter latch (remain closed) and conduct a specified current through the device under specified conditions. (SWG/PE) C37.100-1992

close coupling Any degree of coupling greater than the critical coupling. *Synonym:* tight coupling. *See also:* critical coupling; coupling. (EEC/PE) [119]

closed air circuit (rotating machinery) A term referring to duct-ventilated apparatus used in conjunction with external components so constructed that while it is not necessarily airtight, the enclosed air has no deliberate connection with the external air. *Note:* The term must be qualified to describe the means used to circulate the cooling air and to remove the heat produced in the apparatus. (PE) [9]

closed amortisseur An amortisseur that has the end connections connected together between poles by bolted or otherwise separable connections. (EEC/PE) [119]

closed architecture An architecture for which design parameters and specifications are not available to anyone except the manufacturer of the system. *Contrast:* open architecture. (C) 610.10-1994w

closed-circuit cooling (rotating machinery) A method of cooling in which a primary coolant is circulated in a closed circuit through the machine and if necessary a heat exchanger. Heat is transferred from the primary coolant to the secondary coolant through the structural parts or in the heat exchanger. (PE) [9]

closed-circuit principle The principle of circuit design in which a normally energized electric circuit, on being interrupted or de-energized, will cause the controlled function to assume its most restrictive condition. (EEC/PE) [119]

closed-circuit signaling (data transmission) That type of signaling in which current flows in the idle condition, and a signal is initiated by increasing or decreasing the current. (PE) 599-1985w

closed-circuit transition As applied to reduced-voltage controllers, including star-delta controllers, a method of starting in which the power to the motor is not interrupted during the starting sequence. *See also:* electric controller. (IA/ICTL/IAC) [60]

closed-circuit transition auto-transformer starting (rotating machinery) The process of auto-transformer starting whereby the motor remains connected to the supply during the transition from reduced to rated voltage. (PE) [9]

closed-circuit voltage (batteries) The voltage at its terminals when a specified current is flowing. *See also:* battery. (EEC/PE) [119]

closed construction Any building, building component, assembly or system manufactured in such a manner that all concealed parts of processes of manufacture cannot be inspected before installation at the building site without disassembly, damage, or destruction. (NESC/NEC) [86]

closed curve (image processing and pattern recognition) A curve whose beginning and ending points are the same point.

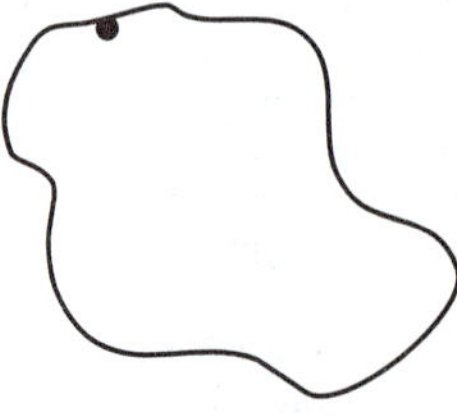

closed curve

(C) 610.4-1990w

closed loop (1) (automatic control) A signal path that includes a forward path, a feedback path, and a summing point and that forms a closed circuit. *See also:* feedback loop. (IA/ICTL/APP/IAC) [69], [60]
(2) (software) A loop that has no exit and whose execution can be interrupted only by intervention from outside the computer program or procedure in which the loop is located. *Contrast:* WHILE; UNTIL. (C) 610.12-1990

closed-loop control (1) (station control and data acquisition) A type of automatic control in which control actions are based on signals fed back from the controlled equipment or system. For example, RTUs can manage local voltage conditions by control of load tap changers and volt amperes reactive (VAR) control compensation equipment. (PE/SUB/SWG-OLD) C37.100-1992, C37.1-1994
(2) Pertaining to a control system in which the output is measured and compared with a standard representing the acceptable range, and any deviation from the standard is fed back into the system in a way that will reduce the deviation. *Synonym:* feedback control. *Contrast:* open-loop control. (C) 610.2-1987

closed-loop control system (1) (control system feedback) A control system in which the controlled quantity is measured and compared with a standard representing the desired performance. *Note:* Any deviation from the standard is fed back into the control system in such a sense that it will reduce the deviation of the controlled quantity from the standard. *See also:* control; network analysis. (PE/PEL/PSE/ET) 94-1970w, 111-1984w
(2) (high-power wide-band transformers) A system in which the controlled quantity is measured and compared with a standard representing the desired performance. Any deviation from the standard is fed back into the control system in such sense that it will reduce the deviation of the controlled quantity from the standard. (MAG) 264-1977w

closed-loop gain (operational gain) (power supplies) The gain, measured with feedback, is the ratio of voltage appearing across the output terminal pair to the causative voltage required at the input resistor. If the open-loop gain is sufficiently large, the closed-loop gain can be satisfactorily approximated by the ratio of the feedback resistor to the input resistor. *See also:* open-loop gain. (SP/EEC/PE) [32], [119]

closed-loop series street lighting system Street lighting system that employs two-wire series circuits in which the return wire is always adjacent. *See also:* alternating-current distribution; direct-current distribution. (EEC/PE) [119]

closed-loop testing (test, measurement, and diagnostic equipment) Testing in which the input stimulus is controlled by the equipment output monitor. (MIL) [2]

closed network A network that prevents outside access by eliminating external connections and external entry or use. *Contrast:* open network. (C) 610.7-1995

closed-numbering plan (telephone switching systems) A numbering plan in which a fixed number of digits is always dialed. (COM) 312-1977w

closed subroutine (1) A subroutine that can be stored at one place and can be connected to a routine by linkages at one or more location. *See also:* open subroutine. (C) [20], [85]
(2) (software) A subroutine that is stored at one given location rather than being copied into a computer program at each place that it is called. *Contrast:* open subroutine. (C) 610.12-1990

closed user group A specified group of network users who are permitted communications among themselves but not with other network users. (C) 610.7-1995

close-open operation (of a switching device) A close operation followed immediately by an open operation without purposely delayed action. *Note:* The letters CO signify this operation: Close-Open. (SWG/PE) C37.100-1992

close operation (of a switching device) The movement of the contacts from the normally open to the normally closed position. *Note:* The letter C signifies this operation: Close. (SWG/PE) C37.100-1992

close-talking microphone A microphone designed particularly for use close to the mouth of the speaker. *See also:* microphone. (EEC/PE) [119]

close-talking pressure-type microphones An acoustic transducer that is intended for use in close proximity to the lips of the talker and is either hand-held or boom-mounted. *Notes:* 1. Various types of microphones are currently used for close-talking applications. These include carbon, dynamic, magnetic, piezoelectric, electrostrictive, and capacitor types. Each of these microphones has only one side of its diaphragm exposed to sound waves, and its electric output substantially corresponds to the instantaneous sound pressure of the impressed sound wave. 2. Since a close-talking microphone is used in the near sound field produced by a person's mouth, it is necessary when measuring the performance of such microphones to utilize a sound source that approximates the characteristics of the human sound generator. (SP) 258-1965w

close-time delay-open operation (of a switching device) A close operation followed by an open operation after a purposely delayed action. *Note:* The letters CTO signify this operation: Close-Time Delay-Open. (SWG/PE) C37.100-1992

closing coil (of a switching device) A coil used in the electromagnet that supplies power for closing the device. *Note:* In an air-operated, or other stored-energy-operated device, the closing coil may be the coil used to release the air or other stored energy that in turn closes the device. (SWG/PE) C37.100-1992

closing operating time The interval during which the contacts move from the fully open position to the fully closed position. (SWG/PE) C37.100-1992

closing relay A form of auxiliary relay used with an electrically operated device to control the closing and opening of the closing circuit of the device so that the main closing current does not pass through the control switch or other initiating device. (SWG/PE/PSR) C37.100-1992, [56], [6]

closing time (of a mechanical switching device) The interval of time between the initiation of the closing operation and the instant when metallic continuity is established in all poles. *Notes:* 1. It includes the operating time of any auxiliary equipment that is necessary to close the switching device, and that forms an integral part of the switching device. 2. For switching devices that embody switching resistors, a distinction should be made between the closing time up to the instant of establishing a circuit at the secondary arcing contacts, and the closing time up to the establishment of a circuit at the main or primary arcing contacts, or both. (SWG/PE) C37.100-1992

cloud chamber smoke detector (fire protection devices) A device which is a form of sampling detector. The air pump draws a sample of air into a high humidity chamber within the detector. After the air is in the humidity chamber, the pressure is lowered slightly. If smoke particles are present, the moisture in the air condenses on them forming a cloud in the chamber. The density of this cloud is then measured by the photoelectric principle. When the density is greater than a predetermined level, the detector responds to the smoke. (NFPA) [16]

cloud pulse (charge-storage tubes) The output resulting from space-charge effects produced by the turning on or off of the electron beam. *See also:* charge-storage tube. (ED) 158-1962w, 161-1971w

clouds *See:* fog.

cloudy sky (illuminating engineering) A sky that has more than 70% cloud cover. (EEC/IE) [126]

CLR *See:* trunk circuit, combined line and recording; combined-line-recording trunk; recording-completing trunk.

cluster (1) (A) In image processing, a set of pixels in a digital image that are close to one another and similar in some way. **(B)** In pattern recognition, a set of points in a feature space that are similar in some way. (C) 610.4-1990
(2) One or more contiguous sectors on a magnetic disk. *See also:* Remote Bridge. (C/LM) 610.10-1994w, 802.1G-1996
(3) *See also:* subclass cluster. (C/SE) 1320.2-1998
(4) A grouping of cell instances and/or clusters that are constrained to each other due to physical location or some other shared characteristic(s). It is not valid to have a cell instance explicitly made a member of more than one cluster. (C/DA) 1481-1999

cluster analysis (A) The detection and description of clusters in a digital image. **(B)** The detection and description of clusters in a feature space. (C) 610.4-1990

clustered word processing Word processing performed on a system composed of multiple work stations, each with its own memory but operating under the control of a master work station. *Contrast:* stand-alone word processing; shared-logic word processing; shared-resource word processing; dedicated word processing. (C) 610.2-1987

clutter (1) (navigation aids) Atmospheric noise, extraneous signals, etc., that tend to obscure the reception of a desired signal. (AES/GCS) 172-1983w
(2) Unwanted echoes, typically from the ground, sea, rain or other precipitation, chaff, birds, insects, meteors, and aurora. *Synonym:* background return. (AES) 686-1997

clutter attenuation (CA) In moving-target indication (MTI) or Doppler radar, the ratio of the clutter-to-noise ratio at the input to the processor, to the clutter-to-noise ratio at the output. *Note:* In MTI, a single value of CA will be obtained, while in Doppler radar the value will generally vary over the different target Doppler filters. In MTI, CA will be equal to MTI improvement factor if the targets are assumed uniformly distributed in velocity. *See also:* MTI improvement factor. (AES) 686-1997

clutter cancellation *See:* clutter attenuation.

clutter detectability factor The predetection signal-to-clutter ratio that provides stated probability of detection for a given false alarm probability in an automatic detection circuit. *Note:* In MTI systems, it is the ratio after cancellation or Doppler filtering. (AES) 686-1997

clutter fence A barrier surrounding a ground-based radar to serve as an artificial horizon and suppress ground clutter. (AES) 686-1997

clutter filter A filter or group of filters (filter bank) included in a radar for the purpose of rejecting clutter returns and passing target returns at Doppler frequencies different from the Doppler frequencies of clutter. *Note:* Moving-target indication (MTI) and pulsed-Doppler processors are examples. (AES) 686-1997

clutter improvement factor *See:* moving-target indication improvement factor.

clutter map Computer-stored values of radar-measured clutter for each range-azimuth resolution cell or local region, used to set thresholds for each cell in a constant-false-alarm rate (CFAR) detection system or to adjust other processing parameters. (AES) 686-1997

clutter-referenced MTI A type of noncoherent MTI that uses clutter as a reference. *See also:* noncoherent MTI. (AES/RS) 686-1990

clutter reflectivity The backscatter coefficient of clutter. *See also:* backscatter coefficient. (AES) 686-1997

clutter residue The uncanceled clutter power remaining at the output of an moving-target indication (MTI) or Doppler signal processor. *See also:* canceled video. (AES) 686-1997

clutter visibility factor The predetection signal-to-clutter ratio that provides stated probability of detection for a given false alarm probability on a display. *Note:* In moving-target indication (MTI) systems, it is the ratio after cancellation or Doppler filtering. (AES) 686-1997

CM *See:* configuration management.

CMC *See:* code for magnetic characters.

C-Message noise The noise on an idle channel or circuit, i.e., a channel or circuit with a termination and no signal (holding tone) at the transmitting end, measured through a C-Message weighting. The noise is expressed in dBrnC. (COM/TA) 743-1995

C-Message weighting (1) (data transmission) A noise weighting used in a noise measuring set to measure noise on a line that is terminated by a subset with a number 500 receiver or a similar subset. The meter scale readings are in dBrn (C-Message). (PE) 599-1985w
(2) (voice-frequency electrical-noise test) A weighting derived from listening tests to indicate the relative annoyance or speech impairment by an interfering signal of frequency *f* as heard through a "500-type" telephone set. The result, called "C-Message waiting," is shown in the corresponding figure. (COM/TA) 469-1988w
(3) A frequency-weighting characteristic used for measurement of noise in voice-frequency communications circuits and designed to weight noise frequencies in proportion to their perceived annoyance effect in telephone service. The C-Message weighting is used to evaluate noise to the human ear, using a 500-type telephone set. (COM/TA) 743-1995

CMI *See:* computer-managed instruction.

CMOS *See:* complementary metal-oxide semiconductor.

CMRR *See:* common-mode rejection ratio.

CNC *See:* computer numerical control.

C network A network composed of three impedance branches in series, the free ends being connected to one pair of terminals, and the junction points being connected to another pair of terminals. *See also:* network analysis. (EEC/PE) [119]

C-Notch (analog voice frequency circuits) The measure of noise on a channel when a signal is present. A very narrow band-elimination filter (notch filter) is used with a C-Message filter to eliminate the holding tone at the measuring end of the circuit. *See also:* holding tone. (COM/TA) 743-1984s

C-Notched filter (telephone loop performance) A filter used in front of the noise detector in conjunction with the measurement of noise in certain systems. A tone is transmitted in these systems to activate signal-dependent noise sources, but the tone power should not be included in the measurement. The C-Notched filter has a C-Message weighting transfer function with a sharp notch which removes this tone from the received signal before its power is measured. (COM/TA) 820-1984r

C-Notched noise The noise power on a channel with a holding tone (signal) at the transmit end, measured through a C-Message weighting and a 1010 Hz notch filter in tandem. (COM/TA) 743-1995

CO *See:* central office; unit operation; close-open operation.

coagulating current *See:* Tesla current.

coal cleaning equipment Equipment generally electrically driven, to remove impurities from the coal as mined, such as slate, sulphur, pyrite, shale, fire clay, gravel, and bone. (EEC/PE) [119]

coalesce To combine two or more sets into one set. *See also:* merge; collate. (C) 610.5-1990w

coarse change-over selector A change-over selector that connects the tap winding to a coarse winding, a main winding, or to portions of the main winding. (PE/TR) C57.131-1995

coarse chrominance primary* *See:* Q chrominance signal.

* Deprecated.

coarse/fine operation A winding arrangement in which a coarse change-over selector connects the tap winding to the coarse or main winding, and allows the use of the taps twice when travelling through the tapping range. (PE/TR) C57.131-1995

coarse-fine control system A control system that uses some elements to reduce the difference between the directly controlled variable and its ideal value to a small value and that uses other elements to reduce the remaining difference to a smaller value. (IA/ICTL/IAC) [60]

coarse-grain parallel architecture Parallel architecture that uses between 2 and 16 processors. *Contrast:* fine-grain parallel architecture; medium-grain parallel architecture. (C) 610.10-1994w

coarse winding A winding that extends the regulating range beyond the range of the finely tapped winding. (PE/TR) C57.131-1995

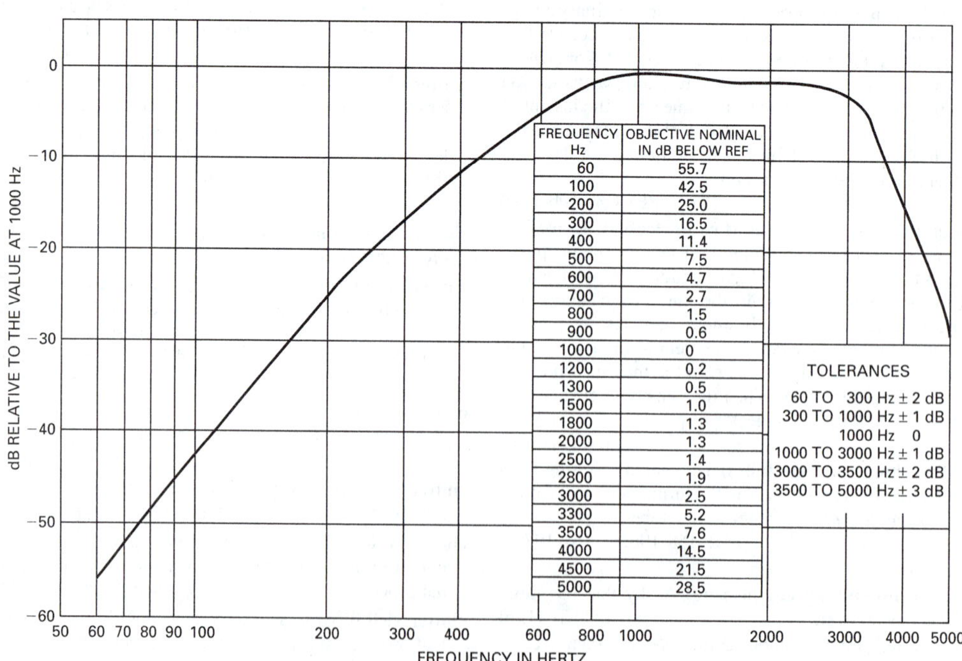

FREQUENCY Hz	OBJECTIVE NOMINAL IN dB BELOW REF
60	55.7
100	42.5
200	25.0
300	16.5
400	11.4
500	7.5
600	4.7
700	2.7
800	1.5
900	0.6
1000	0
1200	0.2
1300	0.5
1500	1.0
1800	1.3
2000	1.3
2500	1.4
2800	1.9
3000	2.5
3300	5.2
3500	7.6
4000	14.5
4500	21.5
5000	28.5

Response in decibels indicating relative interfering effect, C-Message weighting.

C-Message weighting

coast (1) The mode of operation of a vehicle or train in which both tractive effort from the propulsion system and braking effort from the propulsion and friction brake systems are zero. *Note:* The inherent design characteristics of some propulsion systems require that a negligible level of electric brake be present in the coast mode. (VT) 1475-1999
(2) A radar memory feature that causes the range or angle tracking systems to continue to move in the same direction and at the same rate that an original target was moving. *Note:* Coast is invoked manually or automatically when the tracked target approaches a stronger echo (target or clutter) to prevent capture of the track by that echo or to maintain track over brief periods of signal loss. (AES) 686-1997

coast time *See:* run-down time.

coated card *See:* edge-coated card.

coated fabric A fabric or mat in which the elements and interstices may or may not in themselves be coated or filled but that has a relatively uniform compound or varnish finish on either one or both surfaces. *Synonym:* coated mat. *See also:* stator; rotor. (PE) [9]

coated magnetic tape (magnetic powder-coated tape) A tape consisting of a coating of uniformly dispersed, powdered ferromagnetic material (usually ferromagnetic oxides) on a nonmagnetic base. *See also:* magnetic tape; phonograph pickup. (SP) [32]

coated mat *See:* coated fabric.

coating (1) (electroplating) The layer deposits by electroplating. *See also:* electroplating. (PE/EEC) [119]
(2) As defined by the Steel Structures Painting Council, a generic term for paints, lacquer, enamels, etc. A liquid, liquefiable, or mastic composition that has been converted to a solid protective, decorative, or functional adherent film after application as a thin layer. (SUB/PE) 1264-1993

coax A colloquial reference to coaxial cable. (C) 610.7-1995

coaxial antenna An antenna comprised of an extension to the inner conductor of a coaxial line and a radiating sleeve that in effect is formed by folding back the outer conductor of the coaxial line. *Contrast:* sleeve-dipole antenna. (AP/ANT) 145-1993

coaxial cable (1) A two-conductor (center conductor, shield system), concentric, constant impedance transmission line used as the trunk medium in the baseband system. (LM/C/LM/C) 802.3-1998, 8802-3-1990s
(2) (broadband local area networks) A cable with two conductors where one completely surrounds the other. Coax cables are unbalanced transmission lines that have an outer conductor that shields the center conductor from electrostatic interference. The two conductors are spaced by an insulating dielectric that, depending on the mechanical and material configuration, affects the speed, attenuation, and impedance of transmission. (LM/C) 802.7-1989r
(3) A cable consisting of a central conductor and an outer, concentric conductor. *Contrast:* twinaxial cable. *See also:* transceiver cable; trunk cable; shield; core; drop cable; attachment unit interface cable. (C) 610.7-1995

coaxial cable interface The electrical and mechanical interface to the shared coaxial cable medium either contained within or connected to the Medium Attachment Unit (MAU). Also known as the Medium Dependent Interface (MDI). (C/LM) 802.3-1998

coaxial cable section A single length of coaxial cable, terminated at each end with a male BNC connector. Cable sections are joined to other cable sections via BNC plug/receptacle barrel or Type T adapters. (LM/C) 8802-3-1990s, 802.3-1998

coaxial cable segment (medium attachment units and repeater units) A length of coaxial cable made up from one or more coaxial cable sections and coaxial connectors, and terminated at each end in its characteristic impedance. (LM/C) 8802-3-1990s, 802.3-1998

coaxial conductor An electric conductor comprising outgoing and return current paths having a common axis, one of the paths completely surrounding the other throughout its length. (IA) 54-1955w

coaxial detector A detector in which all or part of the two electrical contacts on the detector element are substantially coaxial. Typically one end of each contact configuration is closed (closed-ended coaxial detector), but both ends may be open (open-ended coaxial detector). (NPS) 325-1996

coaxial detector, conventional-electrode geometry (germanium gamma-ray detectors) Conventional-electrode geometry. A coaxial detector in which the outer contact is an n-type layer. (NPS) 325-1996

coaxial detector, reverse-electrode geometry (germanium gamma-ray detectors) A coaxial detector in which the outer contact is a p-type layer. (NPS) 325-1996

coaxial relay A relay that opens and closes an electric contact switching high-frequency current as required to maintain minimum losses. *See also:* relay. (SWG) 341-1986

coaxial stop filter A tuned movable filter set round a conductor in order to limit the radiating length of the conductor for a given frequency. *See also:* electromagnetic compatibility. (EMC/INT) [53], [70]

coaxial stub A short length of coaxial that is joined as a branch to another coaxial. *Note:* Frequently a coaxial stub is short-circuited at the outer end and its length is so chosen that a high or low impedance is presented to the main coaxial in a certain frequency range. *See also:* waveguide. (EEC/PE) [119]

coaxial switch A switch used with and designed to simulate the critical electric properties of coaxial conductors. (PE/PSE) 346-1973w

coaxial transmission line (waveguide) A transmission line consisting of two essentially concentric cylindrical conductors. (MTT) 146-1980w

COBOL 85 A dialect of COBOL; developed as a standard language in 1985, and standardized by IEEE, ISO, and ANSI. (C) 610.13-1993w

co-channel interference Interference caused in one communication channel by a transmitter operating in the same channel. *See also:* radio transmission. (BT/AV) [34]

cock A pneumatic device having two positions, closed/shut and open/through. (VT) 1475-1999

cocktail shaker sort An exchange sort in which adjacent pairs of items are compared and exchanged, if necessary, and alternate passes through the set proceed in opposite directions. *Contrast:* bubble sort. (C) 610.5-1990w

CODASYL *See:* Conference on Data Systems Languages.

CODASYL database A database that adheres to the standards established by the Database Task Group of CODASYL. *Note:* A network database is generally accepted to be synonymous with a CODASYL database. (C) 610.5-1990w

CODASYL model A network database model defined by the CODASYL organization. The CODASYL model is based on sets that are used to specify associations between different record types that exist in a database. *Synonym:* flex model. (C) 610.5-1990w

CODASYL set *See:* set.

code (microprocessor object modules) Data or executable machine code. *See also:* relocatable code; absolute code. (MM/C) 695-1985s
(2) (A) (computer terminology) A character or bit pattern that is assigned a particular meaning; for example, a status code. **(B)** The characters or expressions of an originating or source language, each correlated with its equivalent expression in an intermediate or target language, for example, alphanumeric characters correlated with their equivalent six-bit expressions in a binary machine language. *Note:* For punched or magnetic tape; a predetermined arrangement of possible locations of holes or magnetized areas and rules for interpreting the various possible patterns. **(C)** Frequently, the set of expressions in the target language that represent the set of characters of the source language. **(D)** To encode is to express

given information by means of a code. **(E)** To translate the program for the solution of a problem on a given computer into a sequence of machine-language or pseudo instructions acceptable to that computer. (C) 610.5-1990, 610.12-1990, 162-1963

(3) (A) (computer terminology) In software engineering, computer instructions and data definitions expressed in a programming language or in a form output by an assembler, compiler, or other translator. *See also:* source code; machine code; object code; microcode. **(B) (computer terminology)** To express a computer program in a programming language. (C) 610.12-1990

(4) (A) (computer terminology) A set of rules used to convert data from one form of representation to another. *Synonym:* coding scheme; data element tag; data code. **(B) (computer terminology)** To represent data in symbolic form. **(C) (computer terminology)** Data that have been expressed in symbolic form. (C) 610.5-1990, 1084-1986

(5) (A) (computer terminology) Data that have been converted from one form of representation to another, using a set of rules as in definition (5A). *Synonym:* encoded data. *See also:* coded representation; symbol; code set. **(B) (computer terminology)** To convert data from one form of representation to another, using a set of rules as in definition (5A). *See also:* encode; decode. (C) 610.5-1990

code audit (software) An independent review of source code by a person, team, or tool to verify compliance with software design documentation and programming standards. Correctness and efficiency may also be evaluated. *See also:* tool; inspection; code; audit; efficiency; walk-through; correctness; static analysis. (C/SE) 729-1983s

code bin *k* A digital output that corresponds to a particular set of input values.

Code Transition Level	Code Bin	Code Bin Width
	2N– 1	
T[2N– 1]		
	2N– 2	W[2N– 2]
T[2N– 2]		
•	•	•
•	•	•
•	•	•
T[k + 2]		
	k + 1	W[k + 1]
T[k + 1]		
	k	W[k]
T[k]		
	k – 1	W[k – 1]
T[k – 1]		
•	•	•
•	•	•
•	•	•
T[2]		
	1	W[1]
T[1]		
	0	

Definitions pertaining to input quantization

code bin K

(IM/WM&A) 1057-1994w

code bin width *W[k]* The difference of the code transition levels that delimit the bin.

$W[k] = T[k + 1] - T[k]$

(IM/WM&A) 1057-1994w

code-bit In 100BASE-T, the unit of data passed across the Physical Medium Attachment (PMA) service interface, and the smallest signaling element used for transmission on the medium. A group of five code-bits constitutes a code-group in the 100BASE-X Physical Coding Sublayer (PCS). (C/LM) 802.3-1998

code breakpoint A breakpoint that is initiated upon execution of a given computer instruction. *Synonym:* control breakpoint. *Contrast:* data breakpoint. *See also:* prolog breakpoint; dynamic breakpoint; static breakpoint; epilog breakpoint; programmable breakpoint. (C) 610.12-1990

codec A combination of a coder and decoder operating in different directions of transmission in the same equipment. (COM/TA) 1007-1991r

code character A particular arrangement of code elements representing a specific symbol or value. (COM/PE) [49], 599-1985w

code classes (safety systems equipment in nuclear power generating stations) Levels of structural integrity and quality commensurate with the relative importance of the individual mechanical components of the nuclear power generating station. *Note:* For the recognized code classes, refer to the following documents: ANSI N18.2-1973, *Nuclear Safety Criteria for the Design of Stationary Pressurized Water Reactor Plants;* ANSI/ANS 51.8, *Nuclear Safety Criteria for the Design of Stationary Pressurized Water Reactor Plants;* ANSI/ASME BPV-III, *Boiler and Pressure Vessel Cod* and its latest addenda, Section III; ANSI/ANS 52.1-1980, *Nuclear Safety Criteria for Design of Stationary BWR Plants.* (PE/NP) 627-1980r

code conversion (telephone switching systems) The substitution of a routing code for a destination code. (COM) 312-1977w

code converter A converter that changes the representation of data from one code to another. (C) 610.10-1994w, 610.5-1990w

coded arithmetic data Data stored in a form that is acceptable for arithmetic calculations without conversion to an intermediate form; for example, data stored in integer form. (C) 610.5-1990w

coded character set A set of characters for which coded representations exist. *Synonyms:* coded representation; code set. (C) 610.5-1990w

coded decimal *See:* binary coded decimal.

coded-decimal code The decimal number system with each decimal digit expressed by a code. (IA/EEC) [61], [74]

code-decode table A table that identifies a correspondence between encoded and decoded data items. *Synonym:* encode-decode table. (C) 610.5-1990w

code density The number of characters that can appear per unit of length. (PE/TR) C57.12.35-1996

coded fire-alarm system A local fire-alarm system in which the alarm signal is sounded in a predetermined coded sequence. *See also:* protective signaling. (EEC/PE) [119]

code distance *See:* hamming distance.

coded pulse A pulse compression waveform in which a long pulse is divided into many subpulses, with the phase of each subpulse assuming a discrete value (often 0 or π radians) chosen in a deterministic manner (as in Barker codes, which result in all time sidelobes being equal) or chosen in a pseudorandom manner (such as with linear recursive or maximal-length sequences). *See also:* Barker code. (AES) 686-1997

coded representation The result of applying a code to a particular item of data. For example, the designation ORY for Paris International Airport, obtained by applying the international three-letter code for airports. *Synonym:* code value. *See also:* coded character set. (C) 610.5-1990w

coded track circuit A track circuit in which the energy is varied or interrupted periodically. (EEC/PE) [119]

code element One of the discrete conditions or events in a code, for example, the presence or absence of a pulse. *See also:* data processing; information theory. (COM) [49]

code extension character Any control character used to indicate that one or more of the succeeding coded representations are to be interpreted according to a different code or according to a different coded character set. (C) 610.5-1990w

code for magnetic characters (CMC) A set of rules used in magnetic ink character recognition. *See also:* magnetic ink character recognition. (C) 610.2-1987

code generator (A) (software) A routine, often part of a compiler, that transforms a computer program from some intermediate level of representation (often the output of a root compiler or parser) into a form that is closer to the language of the machine on which the program will execute. **(B) (software)** A software tool that accepts as input the requirements or design for a computer program and produces source code that implements the requirements or design. *Synonym:* source code generator. *See also:* application generator. (C) 610.12-1990

code-group A set of encoded symbols representing encoded data or control information. For 100BASE-T4, a set of six ternary symbols that, when representing data, conveys an octet. For 100BASE-TX and 100BASE-FX, a set of five code-bits that, when representing data, conveys a nibble. For 100BASE-T2, a pair of PAM5×5 symbols that, when representing data, conveys a nibble. For 1000BASE-X, a set of ten bits that, when representing data, conveys an octet. (C/LM) 802.3-1998

code-group alignment In 1000BASE-X, the receiver action that resets the existing code-group boundary to that of the comma or K28.5 character currently being received. (C/LM) 802.3-1998

code-group slipping In 1000BASE-X, the receiver action to align the correct receive clock and code-group containing a comma. (C/LM) 802.3-1998

code, idle channel *See:* idle channel code.

code inspection *See:* inspection.

code letter (locked-rotor kilovolt-amperes) A letter designation under the caption "code" on the nameplate of alternating-current motors (except wound-rotor motors) rated 1/20 horsepower and larger to designate the locked-rotor kilovolt-amperes per horsepower as measured at rated voltage and frequency. (PE) [9]

code of ethics standard A standard that describes the characteristics of a set of moral principles dealing with accepted standards of conduct by, within, and among professionals. (C) 610.12-1990

code, peak *See:* peak code.

coder (1) (general) A device that sets up a series of signals in code form. (EEC/PE) [119]
(2) (code transmitter) A device used to interrupt or modulate the track or line current periodically in various ways in order to establish corresponding controls in the other apparatus. (EEC/PE) [119]

code review A meeting at which software code is presented to project personnel, managers, users, customers, or other interested parties for comment or approval. *Contrast:* formal qualification review; test readiness review; design review. (C) 610.12-1990

code ringing (telephone switching systems) Ringing wherein the number of rings or the duration, or both, indicate which system on a party line is being called. (COM) 312-1977w

Code Rule Violation (CRV) An analog waveform that is not the result of the valid Manchester-encoded output of a single optical transmitter. The collision of two or more 10BASE-FB optical transmissions will cause multiple CRVs. The preamble encoding of a single 10BASE-FP optical transmission contains a single CRV. (C/LM) 802.3-1998

coder offset The difference between the code that is supposed to result from a zero-voltage input to the encoder and the code that actually occurs. (COM/TA) 1007-1991r

code set The complete set of coded representations used by a particular code. For example, the set of three-letter codes used to represent airports. (C) 610.5-1990w, 1084-1986w

code symbol A 4B/5B encoded sequence of five bits representing a unique pattern. (C/LM) 802.9a-1995w

code system A system of control of wayside signals, cab signals, train stop or continuous train control in which electric currents of suitable character are supplied to control apparatus, each function being controlled by its own distinctive code. *See also:* block-signal system. (EEC/PE) [119]

code trace *See:* execution trace.

code transition level The boundary between two adjacent code bins. (IM/WM&A) 1057-1994w

code transition level *T[k]* The value of the recorder input parameter at the transition point between two given, adjacent code bins. The transition point is defined as the input value that causes 50% of the output codes to be greater than or equal to the upper code of the transition, and 50% to be less than the upper code of the transition. The transition level $T[k]$ lies between code bin $k-1$ and code bin k. *See also:* transition point. (IM/WM&A) 1057-1994w

code translator *See:* digital converter.

code value *See:* coded representation.

code violation Violation of a coding rule; e.g., the AMI coding rule where the code is corrupted by bipolar violation(s). (COM/TA) 1007-1991r

code walk-through *See:* walk-through.

coding (1) The process of transforming messages or signals in accordance with a definite set of rules. (COM) 270-1964w
(2) (computers) Loosely, a routine. *See also:* relative coding; straight-line coding; symbolic coding. (C) [20]
(3) (test, measurement, and diagnostic equipment) A part of the programming process in which a completely defined, detailed sequence of operation is translated into computer-entry language. (MIL) [2]
(4) (A) (software) In software engineering, the process of expressing a computer program in a programming language. **(B) (software)** The transforming of logic and data from design specifications (design descriptions) into a programming language. *See also:* software development process. (C) 610.12-1990
(5) Coding is the translation from the original set of bits (character) to a new set of bits (coded character) suitable for serial transmission. *See also:* decoding. (C/BA) 1355-1995

coding delay* (navigation aids) (loran) An arbitrary time delay in the transmission of pulse signals from the slave station to permit the resolution of ambiguities; the term "suppressed time delay" more accurately represents what is being accomplished and should be used instead of "coding delay." (AES/GCS) 172-1983w

* Deprecated.

coding fan *See:* electrode radiator.

coding scheme *See:* code.

coding siren A siren having an auxiliary mechanism to interrupt the flow of air through the device, thereby enabling it to produce a series of sharp blasts as required in code signaling. *See also:* protective signaling. (EEC/PE) [119]

coefficient of attenuation (illuminating engineering) The decrement in flux per unit distance in a given direction within a medium. It is defined by the relation: $\Phi_0 e$ − ux where Φ_x is the flux at any distance x from a reference point having flux Φ_0. (EEC/IE) [126]

coefficient of beam utilization (illuminating engineering) The ratio of the luminous flux (lumens) reaching a specified area directly from a floodlight or projector to the total beam luminous flux (lumens). (EEC/IE) [126]

coefficient of coupling *See:* coupling coefficient.

coefficient of grounding (COG) (1) (surge arresters) (power and distribution transformers) The ratio (E_{LG}/E_{LL}) expressed as a percentage, of the highest root-mean-square line-to-ground power-frequency voltage (E_{LG}) on a sound phase,

at a selected location, during a fault to earth affecting one or more phases to the line-to-line power-frequency voltage (E_{LL}) that would be obtained, at the selected location, with the fault removed. *Notes:* 1. Coefficients of grounding for three-phase systems are calculated from the phase-sequence impedance components as viewed from the selected location. For machines use the subtransient reactance. 2. The coefficient of grounding is useful in the determination of an arrester rating for a selected location. 3. A value not exceeding 80% is obtained approximately when for all system conditions the ratio of zero-sequence reactance to positive-sequence reactance is positive and less than three and the ratio of zero-sequence resistance to positive-sequence reactance is positive and less than one. (PE/TR) C57.12.80-1978r, [8]
(2) The ratio E_{LG}/E_{LL}, expressed as a percentage, of the highest root-mean-square line-to-ground power-frequency voltage E_{LG} on a sound phase, at a selected location, during a fault to ground affecting one or more phases to the line-to-line power-frequency voltage E_{LL} that would be obtained, at the selected location, with the fault removed. *Notes:* 1. Coefficients of grounding for three-phase systems are calculated from the phase-sequence impedance components as viewed from the selected location. For machines, use the subtransient reactance. 2. The coefficient of grounding is useful in the determination of an arrester rating for a selected location. 3. A value not exceeding 80 percent is obtained approximately when for all system conditions the ratio of zero-sequence reactance to positive-sequence reactance is positive and less than three, and the ratio of zero-sequence resistance to positive-sequence reactance is positive and less than one. 28-1974
(3) The ratio, ELG/ELL (expressed as a percentage), of the highest root-mean-square (rms) line-to-ground power-frequency voltage ELG on a sound phase, at a selected location, during a fault to ground affecting one or more phases to the line-to-line power-frequency voltage ELL that would be obtained at the selected location with the fault removed. (SPD/PE) C62.22-1997

coefficient of performance (1) (A) (thermoelectric cooling couple) The quotient of the net rate of heat removal from the cold junction by the thermoelectric couple by the electric power input to the thermoelectric couple. *Note:* This is an idealized coefficient of performance assuming perfect thermal insulation of the thermoelectric arms. *See also:* thermoelectric device. **(B) (thermoelectric cooling device)** The quotient of the rate of heat removal from the cooled body by the electric power input to the device. *See also:* thermoelectric device. **(C) (thermoelectric heating device)** The quotient of the rate of heat addition to the heated body by the electric power input to the device. *See also:* thermoelectric device. **(D) (thermoelectric heating couple)** The quotient of the rate of heat addition to the hot junction by the thermoelectric couple by the electric power input to the thermoelectric couple. *Note:* This is an idealized coefficient of performance assuming perfect thermal insulation of the thermoelectric arms. *See also:* thermoelectric device. (ED) [46], 221-1962
(2) (heat pump) Ratio of heating effect produced to the energy supplied. (IA/PSE) 241-1990r

coefficient of performance, reduced (thermoelectric device) The ratio of "a specified coefficient of performance" to "the corresponding coefficient of performance of a Carnot cycle." *See also:* thermoelectric device. (ED) [46], 221-1962w

coefficient of trip point repeatability *See:* trip-point repeatability coefficient.

coefficient of utilization (1) (electric power systems in commercial buildings) For a specific room, the ratio of the average lumens delivered by a luminaire to a horizontal work plane to the lumens generated by the luminaire's lamps alone. (IA/PSE) 241-1990r
(2) (illuminating engineering) The ratio of luminous flux (lumens) calculated as received on the work-plane to the rated luminous flux (lumens) emitted by the lamps alone. (It is equal to the product of "room utilization factor" and "luminaire efficiency."). (EEC/IE) [126]

coefficient of variation The standard deviation, expressed as a percentage of the mean [i.e., (standard deviation/$\bar{x}$)(100)]. (NI) N42.17B-1989r

coefficient of zero error *See:* environmental coefficient.

coefficient potentiometer A parameter potentiometer that is used to represent a coefficient. (C) 610.10-1994w

coefficient sensitivity *See:* sensitivity coefficient.

coenetic variable In modeling, a variable that affects both the system under consideration and that system's environment. (C) 610.3-1989w

coerce To treat an object of one type as an object of another type by using a different object. *Contrast:* cast. (C/SE) 1320.2-1998

coercive field (E_c) (primary ferroelectric terms) The electric field required to switch the polarization from $P = \pm P_R$ to $P = 0$. The coercive field of a ferroelectric crystal depends on its thermal and electrical history, temperature, pressure, type of electrodes, magnitude, and waveshape of the applied switching voltage (that is, E_c increases as a function of the rate of polarization reversal). (UFFC) 180-1986w

coercive force (H_c) (magnetic core testing) The magnetic field strength at which the magnetic induction is zero, when the core material is in a symmetrically cyclically magnetized condition, with a specified maximum value of field strength (that is, loci of points on the hysteresis curve when $B = 0$). (MAG) 393-1977s

cofactor *See:* path factor.

coffer (illuminating engineering) A recessed panel or dome in the ceiling. (EEC/IE) [126]

Coffing *See:* hoist.

Coffing hoist *See:* hoist.

coffin hoist *See:* hoist.

COG *See:* coefficient of grounding.

cogeneration The generation of electric energy and commercial or industrial quality heat or steam from a single facility. (SUB/PE) 1109-1990w

cogging (rotating machinery) Variations in motor torque at very low speeds caused by variations in magnetic flux due to the alignment of the rotor and stator teeth at various positions of the rotor. *See also:* rotor; stator. (PE) [9]

cognitive process An internal human activity that receives, manipulates, and stores knowledge or information, or that controls actions according to this knowledge. (PE/NP) 1082-1997

COGO *See:* COordinate GeOmetry.

cohered video (in radar moving-target indicator) Video-frequency signal output employed in a coherent system. *See also:* navigation. (AES/RS) 686-1982s

coherence (1) (metric practice) A characteristic of a coherent system. In such a system the product or quotient of any two unit quantities is the unit of the resulting quantity. The SI base units, supplementary units, and derived units form a coherent set. (QUL) 268-1982s
(2) (computer graphics) The property that neighboring pixels tend to possess similar attributes. (C) 610.6-1991w
(3) The correlation between electromagnetic fields at points separated in space, time, or both. (AP/LEO/PROP) 211-1997, 586-1980w

coherence area (1) (fiber optics) The area in a plane perpendicular to the direction of propagation over which light may be considered highly coherent. Commonly the coherence area is the area over which the degree of coherence exceeds 0.88. *See also:* coherent; degree of coherence. (Std100) 812-1984w
(2) (laser maser) A quantitative measure of spatial coherence. The largest cross-sectional area of a light beam, such that light from this area (passing through any two pin holes placed in this area) will produce interference fringes. (LEO/PE) 586-1980w, [9]

coherence domain **(1)** A region in a multiple-cache system, inside of which, cache consistency measures are enforced. In a system that contains bus bridges, a coherence domain may or may not be extensible beyond the local bus through a bus bridge to remote buses. (C/BA) 10857-1994, 896.3-1993w, 896.4-1993w
(2) A coherence domain is a region in a multiple-cache system in which cache consistency measures are enforced. (C/BA) 1014.1-1994w

coherence function **($R(\vec{\Delta}, \tau)$)** **(1)** **(radio-wave propagation)** The expected value of the product of a component of the complex field (F_x) at a given location ($\vec{r}$) and time (t) and the complex conjugate of that field component (F^*_x) at a different location ($\vec{r} + \vec{\Delta}$) and time ($t + \tau$): *Notes:* 1. This definition assumes that the statistics of the fields are homogeneous and stationary. 2. The normalized coherence function, also called the mutual coherence function, is the coherence function divided by the expected value of the square of the magnitude of the field. (AP/PROP) 211-1997
(2) **(seismic qualification of Class 1E equipment for nuclear power generating stations)** Defines a comparative relationship between two time histories. It provides a statistical estimate of how much two motions are related, as a function of frequency. The numerical range is from zero for unrelated, to 1.0 for related motions. (PE/NP) 344-1987r

coherence length **(1)** **(fiber optics)** The propagation distance over which a light beam may be considered coherent. If the spectral linewidth of the source is $\Delta\lambda$ and the central wavelength is λ_0, the coherence length in a medium of refractive index n is approximately $\lambda^2_0/n\Delta\lambda$. *See also:* degree of coherence; spectral width. (Std100) 812-1984w
(2) The distance between two wavefronts of an electromagnetic wave, measured in the direction of propagation, over which the phase of these wavefronts remains sufficiently correlated to result in observable interference between them. (AP/PROP) 211-1997

coherence line A data block for which cache consistency attributes are maintained. (C/BA) 10857-1994, 896.4-1993w

coherence time **(τ_0)** **(1)** **(fiber optics)** The time over which a propagating light beam may be considered coherent. It is equal to coherence length divided by the phase velocity of light in a medium; approximately given by $\lambda^2_0/c\Delta\lambda$, where λ_0 is the central wavelength, $\Delta\lambda$ is the spectral linewidth and c is the velocity of light in vacuum. *See also:* phase velocity; coherence length. (Std100) 812-1984w
(2) **(laser maser)** A quantitative measure of temporal coherence. The maximum delay time which can be introduced between the two beams in a Michelson interferometer before the interference fringes disappear. (LEO) 586-1980w
(3) The time over which the mutual coherence function has decreased to $1/e$ at a given location. (AP/PROP) 211-1997

coherent **(1)** **(fiber optics)** Characterized by a fixed phase relationship between points on an electromagnetic wave. *Note:* A truly monochromatic wave would be perfectly coherent at all points in space. In practice, however, the region of high coherence may extend only a finite distance. The area on the surface of a wavefront over which the wave may be considered coherent is called the coherence area or coherence patch; if the wave has an appreciable coherence area, it is said to be spatially coherent over that area. The distance parallel to the wave vector along which the wave may be considered coherent is called the coherence length; if the wave has an appreciable coherence length, it is said to be phase or length coherent. The coherence length divided by the velocity of light in the medium is known as the coherence time; hence a phase coherent beam may also be called time (or temporally) coherent. *See also:* coherence time; coherence length; degree of coherence; coherence area; monochromatic. (Std100) 812-1984w
(2) **(laser maser)** A light beam is said to be coherent when the electric vector at any point in it is related to that at any other point by a definite, continuous sinusoidal function. (LEO) 586-1980w

coherent bandwidth *See:* frequency selective bandwidth.

coherent bundle *See:* aligned bundle.

coherent data-access operation A data-access operation, when used to access coherently cached data. (C/MM) 1596.5-1993

coherent field In situations where the magnitude, phase, and/or vector direction of an electromagnetic field are random variables, the result of averaging the field over all random characteristics. Also called the mean or average field. (AP/PROP) 211-1997

coherent integration Integration of radio frequency (RF), intermediate frequency (IF), or bipolar envelope signals over an interval in which phase or polarity is preserved. *Note:* Sometimes called "predetection integration." *See also:* integration. (AES) 686-1997

coherent interrupted waves Interrupted continuous waves occurring in wave trains in which the phase of the waves is maintained through successive wave trains. *See also:* wavefront. (EEC/PE) [119]

coherent moving-target indication The usual form of MTI in which a moving target is separated from large clutter echoes as a result of a pulse-to-pulse change in echo phase relative to the phase of a coherent reference oscillator. *See also:* moving-target indication. (AES) 686-1997

coherent MTI *See:* coherent moving-target indication.

coherent processing interval The time during which the radar signal is received and processed coherently. *Note:* Such processing is usually for Doppler filtering. (AES) 686-1997

coherent pulse operation The method of pulse operation in which a fixed phase relationship is maintained from one pulse to the next. *See also:* pulse. (EEC/PE) [119]

coherent radiation *See:* coherent.

coherent sampling Sampling of a periodic waveform in which there is an integer number of cycles in the data record. In other words, coherent sampling occurs when the following relationship exists:

$$f_s \cdot M_c = f_0 \cdot M$$

where
f_s is the sampling frequency.
M_c is the integer number of cycles in the data record.
f_0 is the frequency of the input.
M is the number of samples in the record.
(IM/WM&A) 1057-1994w

coherent signal processing Integration, filtering, or detection of an echo signal using the amplitude of the received signal and its phase referred to that of a reference oscillator or to the transmitted signal. (AES) 686-1997

coherent transaction A transaction (typically read or write) that provides protocols for checking and maintaining consistency with other caches. Coherent transactions are expected to address a cache-line. For example, tightly coupled multiprocessors are expected to use coherent transactions when accessing shared-memory resident data. (C/MM) 1212-1991s

coherent video Bipolar video obtained from a synchronous (coherent) detector. (AES) 686-1997

cohesion **(software)** The manner and degree to which the tasks performed by a single software module are related to one another. Types include coincidental, communicational, functional, logical, procedural, sequential, and temporal. *Synonym:* module strength. *Contrast:* coupling. (C) 610.12-1990

coho A term derived from coherent oscillator, coho designates an oscillator used in a coherent radar to provide a reference phase by which changes in the phase of successively received pulses may be recognized. *Note:* In practice, a coho usually operates at the receiver intermediate frequency. (AES) 686-1997

coil (1) (general) An assemblage of successive convolutions of a conductor.
(2) (rotating machinery) A unit of a winding consisting of one or more insulated conductors connected in series and surrounded by common insulation, and arranged to link or produce magnetic flux. *See also:* rotor; stator. (PE) [9]

coil brace (1) A structure for the support or restraint for one or more coils. *Synonym:* coil support.
(2) (v wedge, salient-pole construction) A trapezoidal insulated insert clamped between field poles, to provide radial restraint for the field coil turns against centrifugal force and to brace the coils tangentially. *See also:* stator. (PE) [9]

coil end-bracing *See:* stator; end-winding support; rotor.

coil insulation (rotating machinery) The main insulation to ground or between phases surrounding a coil, additional to any conductor or turn insulation. *See also:* rotor; stator. (PE) [9]

coil insulation with its accessories (1) (insulation systems of synchronous machines) The coil insulation comprises all of the insulating materials that envelope the current-carrying conductors and their component turns and strands and form the insulation between them and the machine structure, and includes the armor tape, the tying cord, slot fillers, slot tube insulation, pole body insulation, and rotor-retaining ring insulation. (REM) [115]
(2) (repair and rewinding of motors) Comprises all of the insulating materials that envelop and separate the current-carrying conductors and their component turns and strands and form the insulation between them and the machine structure; includes wire coatings, varnish, encapsulants, slot insulation, slot fillers, tapes, phase insulation, pole-body insulation, and retaining ring insulation when present. *See also:* insulation system. (IA/PC) 1068-1996

coil lashing (rotating machinery) The binding used to attach a coil end to the supporting structure. *See also:* rotor; stator. (PE) [9]

coil loading Loading in which inductors, commonly called loading coils, are inserted in a line at intervals. *Note:* The loading coils may be inserted either in series or in shunt. As commonly understood, coil loading is a series loading in which the loading coils are inserted at uniformly spaced recurring intervals. *See also:* loading. (EEC/PE) [119]

coil pitch (rotating machinery) The distance between the two active conductors (coil sides) of a coil, usually expressed as a percentage of the pole pitch. *See also:* armature. (PE) [9]

coil probe A magnetic flux density sensor comprised of a coil of wire that produces an induced voltage proportional to the time derivative of the magnetic flux density. *Notes:* 1. To eliminate effects due to electric field induction, it is essential that the coil of wire be shielded. 2. Since the induced voltage is proportional to the time derivative of the magnetic flux density, the detector circuit of the sensor often contains an integrating stage to recover the waveform of the magnetic field. The integrating stage is also desirable, particularly for measurements of magnetic field strength with harmonic content, since this stage (i.e., its integrating property) eliminates the excessive weighting of the harmonic components in the voltage signal produced by the probe. 3. This probe can also be used to measure static (dc) magnetic flux density if the probe is rotated at a known rate. (T&D/PE) 539-1990, 1308-1994

coil *Q* (dielectric heating) Ratio of reactance to resistance measured at the operating frequency. *Note:* The loaded-coil Q is that of a heater coil with the charge in position to be heated. Correspondingly, the unloaded-coil Q is that of a heater coil with the charge removed from the coil. (IA) 54-1955w

coil Q power factor *See:* coil Q.

coil section (rotating machinery) The basic electrical element of a winding comprising an assembly of one or more turns insulated from one another. (PE) [9]

coil shape factor (dielectric heating) A correction factor for the calculation of the inductance of a coil based on its diameter and length. (IA) 54-1955w

coil side (rotating machinery) Either of the two normally straight parts of a coil that lie in the direction of the axial length of the machine. *See also:* rotor; stator. (PE) [9]

coil-side separator (rotating machinery) Additional insulation used to separate embedded coil sides. *See also:* rotor; stator. (PE) [9]

coil space factor The ratio of the cross-sectional area of the conductor metal in a coil to the total cross-sectional area of the coil. *Note:* If the overall insulation, such as spool bodies or stop linings, is omitted from consideration when the space factor is calculated, the omission should be specifically stated. *See also:* asynchronous machine. (EEC/PE) [119]

coil span *See:* stator; coil pitch; rotor.

coil support *See:* coil brace.

coil support bracket (rotating machinery) A bracket used to mount a coil support ring or binding band. *See also:* stator; rotor. (PE) [9]

coin box A telephone set equipped with a device for collecting coins in payment for telephone messages. *See also:* telephone station. (EEC/PE) [119]

coin call (telephone switching systems) A call in which a coil collection device is used. (COM) 312-1977w

coincidence (radiation counters) The practically simultaneous production of signals from two or more counter tubes. *Note:* A genuine or true coincidence is due to signals from related events (passage of one particle or of two or more related particles through the counter tubes); an accidental, spurious, or chance coincidence is due to unrelated signals that coincide accidentally. *See also:* anticoincidence. (ED) [45]

coincidence circuit A circuit that produces a specified output pulse when and only when a specified number (two or more) or a specified combination of input terminals receives pulses within an assigned time interval. *See also:* anticoincidence; pulse. (NPS) 398-1972r

coincidence factor (electric power utilization) The ratio of the maximum coincident total demand of a group of consumers to the sum of the maximum power demands of individual consumers comprising the group both taken at the same point of supply for the same time. *See also:* generating station. (T&D/PE) [10]

coincidence summing *See:* cascade summing of X and gamma rays.

coincidental cohesion A type of cohesion in which the tasks performed by a software module have no functional relationship to one another. *Contrast:* temporal cohesion; sequential cohesion; procedural cohesion; logical cohesion; communicational cohesion; functional cohesion. (C) 610.12-1990

coincident-current selection The selection of a magnetic cell for reading or writing, by the simultaneous application of two or more currents. (C) [20]

coincident demand (1) (A) (radio-wave propagation) Any demand that occurs simultaneously with any other demand. **(B) (radio-wave propagation)** The sum of any set of coincident demands. (IA/PE/PSE) 241-1990, 858-1993
(2) Any demand that occurs simultaneously with any other demand, also the sum of any set of coincident demands. Information on these factors for the various loads and groups of loads is useful in designing the system. For example, the sum of the connected loads on a feeder, multiplied by the demand factor of these loads, will give the maximum demand that the feeder must carry. The sum of the individual maximum demands on the circuits associated with a load center or panelboard, divided by the diversity factor of those circuits, will give the maximum demand at the load center and on the circuit supplying it. The sum of the individual maximum demands on the circuits from a transformer, divided by the diversity factor of those circuits, will give the maximum demand on the distribution transformer. The sum of the maximum demand on all distribution transformers, divided

by the diversity factor of the transformer loads, will give the maximum demand on their primary feeder. By the use of the proper factors, as outlined, the maximum demands on the various parts of the system from the load circuits to the power source can be estimated. Allowances should also be made for future load expansion in these calculations. (IA/PSE) 141-1993r

(3) (electric power utilization) Any demand that occurs simultaneously with any other demand; also the sum of any set of coincident demands. *See also:* alternating-current distribution. (PE/PSE) 346-1973w

coin-control signal (telephone switching systems) On a coin call, one of the signals used for collecting or returning coins. (COM) 312-1977w

coin-denomination tone (telephone switching systems) The tone that indicates the value of coins when they are deposited in a coin telephone. (COM) 312-1977w

coin tone (telephone switching systems) A class-of-service tone that indicates to an operator that the call has originated from a coin telephone. (COM) 312-1977w

cold cathode A cathode that functions without the application of heat. *See also:* electrode. (EEC/PE) [119]

cold-cathode glow-discharge tube (glow tube) A gas tube that depends for its operation on the properties of a glow discharge. (ED) [45]

cold-cathode lamp (illuminating engineering) An electric-discharge lamp whose mode of operation is that of a glow discharge, and which has electrodes so spaced that most of the light comes from the positive column between them. (EEC/IE) [126]

cold-cathode stepping tube A glow discharge tube having several main gaps with or without associated auxiliary gaps, and in which the main discharge has two or more stable positions and can be made to step in sequence, when a suitable shaped signal is applied to an input electrode, or a group of input electrodes. (ED) [45]

cold-cathode tube An electron tube containing a cold cathode. (ED) 161-1971w

cold-end termination (electrical heat tracing for industrial applications) The termination applied to the end of a heating cable where the power is supplied. (BT/AV) 152-1953s

cold lead (electrical heat tracing for industrial applications) An electrically insulated conductor used to connect a heating conductor to the branch-circuit conductors and designed so as not to produce any appreciable heat. (BT/AV) 152-1953s

cold-lead connection An electrically insulated conductor used to connect a heating conductor to the branch-circuit conductors and designed so as not to produce appreciable heat. (IA) 515-1997

cold reserve Thermal generating capacity available for service but not maintained at operating temperature. (T&D/PE) [10]

cold shrink A joint that consists of a tube or a series of tubes that are applied over the conductor and reduced in diameter over the cable without the use of heat. (PE/IC) 404-1993

cold side *See:* unexposed side.

"cold" standby redundant UPS configuration Consists of two independent, nonredundant modules with either individual module batteries or a common battery. (IA/PSE) 241-1990r

cold-start A sequence of events performed on the application of power that ensures a uniform initialization period for all agents, giving them the ability to begin operation from a known state. (C/MM) 1296-1987s

cold test *See:* passive test.

collaboration The cooperative exchange of requests among classes and instances in order to achieve some goal. (C/SE) 1320.2-1998

collapsing loss The increase in required input signal-to-noise ratio to maintain given probability of detection for a given false alarm probability when resolution cells or samples containing only noise are integrated along with those containing signal and noise. *Note:* This type of loss occurs, for example, when radar returns containing range, azimuth, and elevation information are constrained to a two-dimensional display. (AES) 686-1997

collapsing ratio The ratio

$(m + n)/n$

where

m = number of noise-only samples

n = number of signal-plus-noise samples

(AES) 686-1997

collar (rotating machinery) (washer) Insulation between the field coil and the pole shoe (top collar) and between the field coil and the member carrying the pole body (bottom collar). *See also:* rotor; field-coil flange. (PE) [9]

collate (1) (mathematics of computing) To compare and merge two or more similarly ordered or sequenced sets onto one ordered set. For example, to arrange the set 1, 4, 9, 12, 18 and the set 2, 5, 10, 19 as the single set 1, 2, 4, 5, 9, 10, 12, 18, 19. (C) [20], [85], 1084-1986w

(2) (data management) To arrange items from two or more ordered subsets into one or more other subsets. The resulting subsets will commonly contain at least one item from each of the original subsets and may be ordered in some specified order that is not necessarily the order of any of the original subsets. *Contrast:* decollate. *See also:* merge; collating sequence; coalesce. (C) 610.5-1990w

collating element The smallest entity used to determine the logical ordering of strings. A collating element shall consist of either a single character, or two or more characters collating as a single entity. The value of the LC_COLLATE category in the current locale determines the current set of collating elements. *See also:* collation sequence. (C/PA) 9945-2-1993

collating sequence An ordering assigned to a set of items, such that any two sets in that assigned order can be collated. (C) [20], [85]

(2) (A) (data management) A sequence assigned to a set of items such that any two sets that are in that assigned order can be collated. **(B) (data management)** A specified arrangement of the items in a set used in sequencing. *Synonym:* sequence. *See also:* order. (C) 610.5-1990

collating significance Any attribute of a set that may be used to define a specified arrangement to be used in collating. (C) 610.5-1990w

collation *See:* AND.

collation sequence The relative order of collating elements as determined by the setting of the LC_COLLATE category in the current locale. The character order, as defined for the LC_COLLATE category in the current locale defines the relative order of all collating elements such that each element occupies a unique position in the order. This is the order used in ranges of characters and collating elements in REs and pattern matching. In addition, the definition of the collating weights of characters and collating elements uses collating elements to represent their respective positions within the collation sequence. Multilevel sorting is accomplished by assigning one or more collation weights to elements, up to the limit

$$\left\{\text{COLL\_WEIGHTS\_MAX}\right\}.$$

On each level, elements may be given the same weight (at the primary level, this is called an *equivalence class*) or may be omitted from the sequence. Strings that collate equally using the first assigned weight (primary ordering) are then compared using the next assigned weight (secondary ordering), and so on. (PA/C) 9945-2-1993

collator (1) A device to collate sets of punched cards or other documents into a sequence. (C) [20], [85]

(2) A device that compares and merges sets of punch cards or other documents into a sequence. (C) 610.10-1994w

collect call (telephone switching systems) A call for which the called customer agrees to pay. (COM) 312-1977w

collecting pit A pit built under oil-filled equipment to collect any accidental discharge of oil from that piece of equipment. (SUB/PE) 980-1994

collection cardinality A specification, for a collection-valued property, of how many members the value of the property, i.e., the collection, may or must have for each instance. *See also:* cardinality constraint. (C/SE) 1320.2-1998

collection class A kind of class in which each instance is a group of instances of other classes. (C/SE) 1320.2-1998

collection configuration In a collection of objects, the configuration is the enumeration of the members of the collection, and the specification of the allowed communications between the members. To configure a collection means to make the necessary changes to the collection and its members to make real the defined enumeration of members and the specification of allowed communication between members. (IM/ST) 1451.1-1999

collection efficiency (quantum yield) The number of carriers crossing the p-n junction per incident photon. (AES/SS) 307-1969w

collection property *See:* collection-valued property.

collection-valued A value that is complex, i.e., having constituent parts. *Contrast:* scalar. (C/SE) 1320.2-1998

collection-valued class A class in which each instance is a collection of values. *Contrast:* scalar-valued class. (C/SE) 1320.2-1998

collection-valued property A property that maps to a collection class. *Contrast:* scalar-valued property. (C/SE) 1320.2-1998

collector (1) (rotating machinery) An assembly of collector rings, individually insulated, on a supporting structure. *See also:* asynchronous machine. (PE) [9]
(2) (electron tube) An electrode that collects electrons or ions that have completed their functions within the tube. *See also:* electrode. (ED) 161-1971w
(3) (transistor) A region through which primary flow of charge carriers leaves the base. (ED) 216-1960w

collector grid (solar cells) A pattern of conducting material making ohmic contact to the active surface of a solar cell to reduce the series resistance of the device by reducing the mean path of the current carriers within the semiconductor. (AES/SS) 307-1969w

collector junction (semiconductor devices) A junction normally biased in the high-resistance direction, the current through which can be controlled by the introduction of minority carriers. *Note:* The polarity of the voltage across the junction reverses when a switching occurs. *See also:* semiconductor; transistor; semiconductor device. (ED/IA) 216-1960w, 270-1966w, [12]

collector plates Metal inserts embedded in the cell lining to minimize the electric resistance between the cell lining and the current leads. *See also:* fused electrolyte. (EEC/PE) [119]

collector ring (1) A metal ring suitably mounted on an electric machine that (through stationary brushes bearing thereon) conducts current into or out of the rotating member. *Synonym:* slip ring. *See also:* asynchronous machine. (EEC/PE) [119]
(2) An assembly of slip rings for transferring electrical energy from a stationary to a rotating member. (NESC/NEC) [86]

collector-ring lead insulation (rotating machinery) Additional insulation, applied to the leads that connect the collector rings to the windings of the rotating member, to prevent grounding to the metallic parts of the rotating members, and to provide electrical separation between leads. *See also:* rotor. (PE) [9]

collector-ring shaft insulation (rotating machinery) The combination of insulating members that insulate the collector rings from the parts of the structure that are mounted on the shaft. *See also:* rotor. (PE) [9]

collimate (storage tubes) To modify the paths of electrons in a flooding beam or of various rays of a scanning beam in order to cause them to become more nearly parallel as they approach the storage assembly. *See also:* storage tube. (ED) 158-1962w

collimated beam (laser maser) Effectively, a parallel beam of light with very low divergence or convergence. (LEO) 586-1980w

collimating lens (storage tubes) An electron lens that collimates an electron beam. *See also:* storage tube. (ED) 158-1962w

collimation (fiber optics) The process by which a divergent or convergent beam of radiation is converted into a beam with the minimum divergence possible for that system (ideally, a parallel bundle of rays). *See also:* beam divergence. (Std100) 812-1984w

collimator (laser maser) An optical device for converting a diverging or converging beam of light into a collimated or parallel one. (LEO) 586-1980w

collinear array antenna A linear array of radiating elements, usually dipoles, with their axes lying in a straight line. (AP/ANT) 145-1993

collision (1) An unwanted condition that results from concurrent transmission on the physical medium. (LM/C) 8802-3-1990s
(2) (data management) In hashing, the occurrence of the same hash value for two or more different keys. *Synonym:* hash clash. *See also:* synonym. (C) 610.5-1990w
(3) The condition occurring when two MTM-Bus modules are simultaneously MTM-Bus drivers and are attempting to drive a MTM-Bus signal to complementary values (one driving logic 1, one driving logic 0). (TT/C) 1149.5-1995
(4) The condition when multiple packets/signals are observed simultaneously at a single point on the medium where the "listening" station is unable to function properly due to multiple signals being present. *See also:* forced collision; collision detect signal; contention; collision detection. (C) 610.7-1995
(5) A condition that results from concurrent transmissions from multiple data terminal equipment (DTE) sources within a single collision domain. (LM/C) 802.3-1998, 610.7-1995

collision-avoidance system (navigation aids) A system providing the means of detection and prevention of impending collision between vehicles. The system performs one or more of the following functions: detection of intruders in surrounding vicinity, evaluation of miss distance of a collision hazard, determination of precise maneuver needed to avoid the hazard, and specification of when an avoidance maneuver should be initiated. (AES/GCS) 172-1983w

collision chain A list used in hashing to hold all the keys for which the hash address is identical. *Synonym:* calc chain. (C) 610.5-1990w

collision detection The ability of a node to detect collision. *Note:* This term is contextually specific to IEEE Std 802.3. (C) 610.7-1995

collision detect signal A signal provided by the physical layer to the data link layer, to indicate collision detection. *Note:* This term is contextually specific to IEEE Std 802.3. (C) 610.7-1995

collision enforcement The emission of an encoded sequence by the transmitting node after a collision is detected, to ensure that all other transmitting nodes detect the collision. *Note:* This term is contextually specific to IEEE Std 802.3. (C) 610.7-1995

collision domain A single, half duplex mode CSMA/CD network. If two or more Media Access Control (MAC) sublayers are within the same collision domain and both transmit at the same time, a collision will occur. MAC sublayers separated by a repeater are in the same collision domain. MAC sublayers separated by a bridge are within different collision domains. (C/LM) 802.3-1998

collision error In networking, an indication that two or more nodes have attempted to transmit within the same time slot. *Note:* This term is contextually specific to IEEE Std 802.3. (C) 610.7-1995

collision frequency In a plasma, the average number of collisions per second of a particle of a given species with particles of another or of the same species. (AP/PROP) 211-1997

collision presence (1) The signal provided by the physical signaling sublayer to the physical medium attachment sublayer (within the data link layer) to indicate that multiple stations are contending for access to the transmission medium. (C) 610.7-1995
(2) A signal generated within the Physical Layer by an end station or hub to indicate that multiple stations are contending for access to the transmission medium. (C/LM) 802.3-1998

collision resolution In hashing, the process of applying further calculations or other means to resolve a collision. Methods include open-address hashing, separate chaining, and the use of buckets. *Synonym:* rehashing. (C) 610.5-1990w

colloidal ions Ions suspended in a medium, that are larger than atomic or molecular dimensions but sufficiently small to exhibit Brownian movement. (EEC/PE) [119]

colon definition A command defined as a sequence of previously existing commands. (C/BA) 1275-1994

color (1) (television) That characteristic of visual sensation in the photopic range that depends on the spectral composition of light entering the eye. (BT/AV) 201-1979w
(2) (illuminating engineering) (of a physical stimulus) One of the ways in which the word "color" may be used is to designate the property of light falling on the retina which causes it to generate an impression perceived as having or lacking a quality such as whiteness, redness, greenness, and the like. This property of light is determined by its spectral power distribution and may be specified in terms of its chromaticity and luminance. This same property of light may be imputed to a to a beam of light being propagated through space or originating at a distal stimulus. The distal stimulus itself may be described as colored because it gives off colored light. (EEC/IE) [126]

color breakup (color television) Any transient or dynamic distortion of the color in a television picture. *Note:* This effect may originate in videotape equipment, in a television camera, or in a receiver. In videotape recording or playback, it occurs as intermittent misphasing or loss of the chrominance signal. In a field-sequential system, it may be caused at the camera by rapid motion of the image on the camera sensor through motion of either the camera or the subject. It may be caused at the receiver by rapid changes in viewing conditions such as blinking or motion of the eyes. (BT/AV) 201-1979w

color burst (color television) The portion of the composite or noncomposite color-picture signal, comprising a few cycles of a sine wave of chrominance subcarrier frequency, that is used to establish a reference for demodulating the chrominance signal. (BT/AV) 201-1979w

color-burst flag keying signal (television) A keying signal used to form the color burst from a color-subcarrier signal source. *See also:* burst flag. (EEC/PE) [119]

color-burst gate (television) A keying or gating signal used to extract the color burst from a color-television signal. *See also:* television; burst gate. (PE/EEC) [119]

color-burst keying signal *See:* color-burst flag keying signal.

color carrier *See:* chrominance subcarrier.

color cell (repeating pattern of phosphors on the screen of a color-picture tube) The smallest area containing a complete set of all the primary colors contained in the pattern. *Note:* If the cells are described by only one dimension as in the line type of screen, the other dimension is determined by the resolution capabilities of the tube. (ED) 161-1971w

color center (color picture tubes) A point or region (defined by a particular color-selecting electrode and screen configuration) through which an electron beam must pass in order to strike the phosphor array of one primary color. *Note:* This term is not to be used to define the color-triad center of a color-picture tube screen. (ED) 161-1971w

color code (1) (electrical) A system of standard colors adopted for identification of conductors for polarity, etc., and for identification of external terminals of motors and starters to facilitate making power connections between them. *See also:* mine feeder circuit. (EEC/PE) [119]
(2) (communication and control cables) A system of standard colors used for identification of conductors. Colors identify the tip and ring conductors in pairs of a communications cable. Combinations of colors identify the pair numbers. Pair groups are bound with threads or tapes that are identified with color bands or with unit numbers and the names of the colors. (PE/PSC) 789-1988w

color coder *See:* color encoder.

color comparison (illuminating engineering) The judgement of equality, or of the amount and character of difference, of the color of two objects viewed under identical illumination. *Synonym:* color grading. (EEC/IE) [126]

color contamination (color television) An error of color rendition caused by incomplete separation of paths carrying different color components of the picture. *Note:* Such errors can arise in the optical, electronic, or mechanical portions of a color television system as well as in the electrical portions. (BT/AV) 201-1979w

color coordinate transformation (color television) Computation of the tristimulus values of colors in terms of one set of primaries from the tristimulus values of the same colors in another set of primaries. *Note:* This computation may be performing electrically in a color television system. (BT/AV) 201-1979w

color correction (illuminating engineering) (of a photograph or printed picture) The adjustment of a color reproduction process to improve the perceived-color conformity of the reproduction to the original. (EEC/IE) [126]

color decoder (color television) An apparatus for deriving the signals for the color display device from the color picture signal and the color burst. (BT/AV) 201-1979w

color-difference signal (color television) An electrical signal that, when added to the luminance signal, produces a signal representative to one of the tristimulus values (with respect to a stated set of primaries) of the transmitted color. (BT/AV) 201-1979w

color difference thresholds (illuminating engineering) The difference in chromaticity or luminance or both, between two colors, that makes them just perceptibly different. The difference may be a difference in hue, saturation, or brightness (lightness for surface colors), or a combination of the three. (EEC/IE) [126]

color discrimination (illuminating engineering) The perception of differences between two or more colors. (EEC/IE) [126]

color display device A display device that can display more than one color, in addition to the background color. *Contrast:* monochrome display device. *See also:* red, green, blue display device. (C) 610.6-1991w, 610.10-1994w

color encoder (National Television System Committee color television) An apparatus for generating either the noncomposite or the composite color picture signal and the color burst from camera signals (or equivalents) and the chrominance subcarrier. (BT/AV) 201-1979w

color-field corrector (electron tube) A device located external to the tube producing an electric or magnetic field that affects the beam after deflection as an aid in the production of uniform color fields. (ED) 161-1971w

color flicker (television) The flicker that results from fluctuation of both chromaticity and luminance. (BT/AV) 201-1979w

color fringing (color television) Spurious chromaticity at boundaries of objects in the picture. *Note:* Color fringing can be caused by a change in relative position of the televised object from field to field (in a field-sequential system), or by misregistration in either camera or receiver; in the case of small objects, it may cause them to appear separated into different color. (BT/AV) 201-1979w

colorfulness (illuminating engineering) (of a perceived color) The attribute according to which it appears to exhibit more or less chromatic color. For a stimulus of a given chromaticity, colorfulness normally increases as the absolute luminance is increased. (EEC/IE) [126]

color grading *See:* color comparison.

colorimetric purity (illuminating engineering) (of a light) The ratio L_1/L_2, where L_1 is the luminance of the single frequency component which must be mixed with a reference standard to match the color of the light and L_2 is the luminance of the light. (EEC/IE) [126]

colorimetric shift (illuminating engineering) The change of chromaticity and luminance factor of an object color due to change of the light source. *See also:* resultant color shift; adaptive color shift. (EEC/IE) [126]

colorimetry (1) (illuminating engineering) The measurement of color. (EEC/IE) [126]
(2) (television) The techniques for the measurement of color and for the interpretation of the results of such measurements. *Note:* The measurement of color is made possible by the properties of the eye, and is based on a set of conventions. (BT/AV) 201-1979w

coloring (1) (electroplating) (chemical) The production of desired colors on metal surfaces by appropriate chemical action. *See also:* electroplating. (PE/EEC) [119]
(2) (buffing) Light buffing of metal surfaces, for the purpose of producing a high luster. *See also:* electroplating. (EEC/PE) [119]

color light signal A fixed signal in which the indications are given by the color of a light only. (EEC/PE) [119]

color look-up table A workstation-dependent table in which the entries specify the red, green, and blue intensity values that define the color of a pixel on the display surface. *Synonyms:* video look-up table; color map. *See also:* color mapping. (C) 610.6-1991w

color map *See:* color look-up table.

colormap A set of entries *(colorcells)* controlling the display of colors for a pixel by assigning numeric intensity values to each of red, blue, and green attributes of the pixel. (C) 1295-1993w

color mapping The use of the color look-up table to produce color output on the screen of a cathode ray tube. (C) 610.6-1991w

color match (colorimetry) (television) The condition in which the two halves of a structureless photometric field are judged by the observer to have exactly the same appearance. *Note:* A color match for the standard observer may be calculated. (BT/AV) 201-1979w

color matching (illuminating engineering) Action of making a color appear the same as a given color. (EEC/IE) [126]

color matching functions (illuminating engineering) (spectral tristimulus values) The tristimulus value per unit wavelength interval and unit spectral radiant flux. *Notes:* 1. Color-matching functions have been adopted by the International Commission on Illumination. They are tabulated as functions of wavelength throughout the spectrum and are the basis for the evaluation of radiant energy as light and color. The $\bar{y}$ values are identical with the values of spectral luminous efficiency for photopic vision. 2. The $\bar{x}$, $\bar{y}$, and $\bar{z}$ values for the 1931 Standard Observer are based on a two degree bipartite field, and are recommended for predicting matches for stimuli subtending between one degree and four degrees. Supplementary data based on a ten-degree bipartite field were adopted in 1964 for use for angular subtenses greater than four degrees. *See also:* values of spectral luminous efficiency for photopic vision. (EEC/IE) [126]

color mixture (television) Color produced by the combination of lights of different colors. *Notes:* 1. The combination may be accomplished by successive presentation of the components, provided the rate of alternation is sufficiently high; or the combination may be accomplished by simultaneous presentation, either in the same area or on adjacent areas, provided they are small enough and close enough together to eliminate pattern effects. 2. A color mixture as here defined is sometimes denoted as an additive color mixture to distinguish it from combinations of dyes, pigments, and other absorbing substances. Such mixtures of substances are sometimes called subtractive color mixtures, but might more appropriately be called colorant mixtures. (BT/AV) 201-1979w

color mixture data *See:* tristimulus values.

color picture signal* *See:* noncomposite color picture signal; composite color picture signal.
* Deprecated.

color-picture tube An electron tube used to provide an image in color by the scanning of a raster and by varying the intensity of excitation of phosphors to produce light of the chosen primary colors. (ED) 161-1971w

color plane (multibeam color-picture tubes) A surface approximating a plane containing the color centers. (ED) 161-1971w

color-position light signal A fixed signal in which the indications are given by the color and the position of two or more lights. (EEC/PE) [119]

color preference index (illuminating engineering) (of a light source) Measure appraising a light source for enhancing the appearance of an object or objects by making their colors tend toward people's preferences. Judd's flattery index is an example. (EEC/IE) [126]

color printer A printer that utilizes multi-colored ribbons, pens, or ink supplies, allowing it to print in more than one color. (C) 610.10-1994w

color-purity magnet A magnet in the neck region of a color-picture tube to alter the electron beam path for the purpose of improving color purity. (ED) 161-1971w

color rendering index (illuminating engineering) (of a light source) Measure of the degree of color shift objects undergo when illuminated by the light source as compared with the color of those same objects when illuminated by a reference source of comparable color temperature. (EEC/IE) [126]

color-selecting-electrode system A structure containing a plurality of openings mounted in the vicinity of the screen of a color-picture tube (electron tubes), the function of this structure being to cause electron impingement on the proper screen area by using either masking, focusing, deflection, reflection, or a combination of these effects. *See also:* shadow mask. (ED) 161-1971w

color-selecting-electrode system transmission (electron tube) The fraction of incident primary electron current that passes through the color-selecting-electrode system. (ED) 161-1971w

color signal (color television) Any signal at any point for wholly or partially controlling the chromaticity values of a color television picture. *Note:* This is a general term that encompasses many specific connotations such as those conveyed by the words "color picture signal" (either composite or noncomposite), "chrominance signal color carrier signal," "luminance signal" (in color television). (BT/AV) 201-1979w

color sync signal (color television) A signal used to establish and maintain the same color relationships that are transmitted. *Note:* In Rules Governing Radio Broadcast Services, Part 3, of the Federal Communications Commission, the color sync signal consists of a sequence of color bursts that recur every line except for a specified time interval during the vertical interval, each burst occurring on the back porch. (BT/AV) 201-1979w

color temperature (1) (television) The absolute temperature of the full (blackbody) radiator for which the ordinates of the spectral distribution curve of emission are proportional (or approximately so) in the visible regions, to those of the distribution curve of the radiation considered, so that both radiations have the same chromaticity. *Note:* In certain countries, by extension, the term "color temperature" is used in the case of a selective radiator when, for the colorimetric standard observer, this radiator has the same color (or at least approximately the same color) as a full radiator at a certain temperature; this temperature is then called the color temperature of the selective radiator. (BT/AV) 201-1979w
(2) (illuminating engineering) (of a light source) The absolute temperature of a blackbody radiator having a chromaticity equal to that of the light source. (EEC/IE) [126]

color tracking (television) (A) The degree to which color balance is maintained over the complete range of the achromatic (neutral gray) scale. **(B)** A qualitative term indicating the degree to which constant chromaticity within the achromatic region in the chromaticity diagram is achieved on a color-display device over the range of luminances produced from a monochrome signal. *See also:* television. (BT/AV) [34]

color transmission (color television) The transmission of a signal wave for controlling both the luminance values and the chromaticity values in a picture. (BT/AV) 201-1979w

color triad (phosphor-dot screen) A color cell of a three-color phosphor-dot screen. (ED) 161-1971w

color triangle (television) A triangle drawn on a chromaticity diagram, representing the entire range of chromaticities obtainable as additive mixtures of three prescribed primaries represented by the corners of the triangle. (BT/AV) 201-1979w

Colpitts oscillator An electron tube or solid state circuit in which the parallel-tune tank circuit is connected between grid and plate, the capacitive portion of the tank circuit being comprised of two series elements, the connection between the two being at cathode potential with the feedback voltage obtained across the grid-cathode portion of the capacitor. *See also:* radio-frequency generator. (BT) 182A-1964w

column (1) (A) (positional notation) A vertical arrangement of characters or other expressions. **(B)** Loosely, a digital place. *See also:* place. **(C) (test pattern language)** A group of words or bits in a memory, identified by a common *Y*-address. **(D) (metal nitrite oxide semiconductor arrays)** A group of memory cells having a common sense amplifier that detects the state of the cell being addressed. **(E) (data management)** A vertically corresponding set of entries in a table. *Contrast:* row. *See also:* attribute.
(ED/C/TT) 641-1987, 610.5-1990, 660-1986, 162-1963
(2) In a Physical Design Exchange Format (PDEF) datapath cluster, a cluster of cell, spare_cell, and/or cluster instances placed or constrained to be placed in the vertical (*Y*-axis) direction. *See also:* datapath; row. (C/DA) 1481-1999

column binary (1) Pertaining to the binary representation of data on punched cards in which adjacent positions in a column correspond to adjacent bits of data, for example, each column in a 12-row card may be used to represent 12 consecutive bits of a 36-bit word. (C) [20], [85]
(2) (mathematics of computing) Pertaining to the binary representation of data in which adjacent positions in a column correspond to adjacent binary digits. For example, each column in a 12-row card may be used to represent 12 consecutive bits of a binary word. *Synonym:* Chinese binary. *Contrast:* row binary. (C) 1084-1986w
(3) Pertaining to the binary representation of data on punch cards in which the weights of punch positions are assigned along card columns, for example, each column in a 12-row card may be used to represent 12 consecutive bits. *Synonym:* Chinese binary. *Contrast:* row binary. *See also:* binary card. (C) 610.10-1994w

column enable (semiconductor memory) The input used to strobe in the column address in multiplexed address random access memories (RAM). (TT/C) 662-1980s

column-major order A method for storing the elements of a matrix in computer memory, in which elements are ordered in a column-by-column manner; that is, all elements of column 1, followed by all elements of column 2, etc. *Contrast:* row-major order. (C) 610.5-1990w

column position A unit of horizontal measure related to characters in a line. It is assumed that each character in a character set has an intrinsic column width independent of any output device. Each printable character in the portable character set has a column width of one. The standard utilities, when used as described in this standard, assume that all characters have integral column widths. The column width of a character is not necessarily related to the internal representation of the character (numbers of bits or octets). The column position of a character in a line is defined as one plus the sum of the column widths of the preceding characters in the line. Column positions are numbered starting from 1. (C/PA) 9945-2-1993

column, positive *See:* positive column.

column select line The line that is determined by the column addresses (output of the Y decoder) that are used to select the appropriate access transistors during a read or write. (ED) 1005-1998

column select transistor The transistor, controlled by the column select line, that accesses the appropriate bit-line during a read or write cycle. (ED) 1005-1998

column sort *See:* distribution sort.

column split The capability of a punch card device to read or punch two parts of a card column independently. (C) 1084-1986w

column vector A matrix with only one column. That is, a matrix of size *m*-by-1. *Contrast:* row vector. (C) 610.5-1990w

COM *See:* computer output microfilm.

comb filter A filter whose insertion loss forms a sequence of narrow pass bands or narrow stop bands centered at multiples of some specified frequency. (CAS) [13]

combination An unordered sequence of items chosen from a set. *Contrast:* permutation. *See also:* forbidden combination. (C) 610.5-1990w

combinational Pertaining to a logic whose output values at any given instant depend only upon the input values at that time. *Contrast:* sequential. (C) 610.10-1994w

combinational circuit A logic circuit whose output values at any given instant depend only upon the input values at that time. *Synonym:* combinatorial circuit. *Contrast:* sequential circuit. *See also:* gate. (C) 610.10-1994w

combinational logic element (A) A device having zero or more input channels and one output channel, each of which is always in one of exactly two possible physical states, except during switching transients. *Note:* On each of the input channels and the output channel, a single state is designated arbitrarily as the "one" state, for that input channel or output channel, as the case may be. For each input channel and output channel, the other state may be referred to as the "zero" state. The device has the property that the output channel state is determined completely by the comtemporaneous input-channel-state combination, to within switching transients. **(B)** By extension, a device similar to that in definition (A), except that one or more of the input channels or the output channel, or both, have a finite number, but more than two, possible physical states each of which is designated as a distinct logic state. The output channel state is determined completely by the contemporaneous input-channel-state combination, to within switching transients. **(C)** A device similar to that of definition (A) or (B), except that it has more than one output channel. *See also:* OR gate; AND gate. (C) 162-1963

combinational logic function A logic function in which there exists one and only one resulting combination of states of the outputs for each possible combination of input states. *Note:* The terms "combinative" and "combinatorial" have also been used to mean "combinational." (GSD) 91-1984r

combination buoy (navigation aids) A buoy that has more than one means of conveying intelligence. *See also:* buoy. (AES/GCS) 172-1983w

combination controller A full magnetic or semimagnetic controller with additional externally operable disconnecting means contained in a common enclosure. The disconnecting means may be a circuit breaker or a disconnect switch. *See also:* electric controller. (IA/ICTL/IAC) [60]

combination current and voltage regulation That type of automatic regulation in which the generator regulator controls both the voltage and current output of the generator. *Note:* This type of control is designed primarily for the purpose of ensuring proper charging of storage batteries on cars or locomotives. *See also:* axle-generator system. (EEC/PE) [119]

combination detector (fire protection devices) A device that either responds to more than one of fire phenomena (heat, smoke, or flame) or employs more than one operating principle to sense one of these phenomena. (NFPA) [16]

combination effect An electric disturbance not caused by one of the following mechanisms, but to some extent by a combination of them: normal-mode noise (transverse or differential), common-mode noise (longitudinal), and common-mode to normal-mode conversions. *See also:* normal-mode. (PE/IC) 1143-1994r

combination electric locomotive An electric locomotive, the propulsion power for which may be drawn from two or more sources, either located on the locomotive of elsewhere. *Note:* The prefix "combination" may be applied to cars, buses, etc., of this type. *See also:* electric locomotive. (EEC/PE) [119]

combination lighting and appliance branch circuit A circuit supplying energy to one or more lighting outlets and to one or more appliance outlets. *See also:* branch circuit. (EEC/PE) [119]

combination microphone A microphone consisting of a combination of two or more similar or dissimilar microphones. Examples: Two oppositely phased pressure microphones acting as a gradient microphone; a pressure microphone and velocity microphone acting as a unidirectional microphone. *See also:* microphone. (EEC/PE) [119]

combination monopulse A form of monopulse employing amplitude comparison in one angular coordinate plane and phase comparison in the orthogonal coordinate plane. (AES) 686-1997

combination rubber tape The assembly of both rubber and friction tape into one tape that provides both insulation and mechanical protection for joints. (EEC/PE) [119]

combinations of pulses and waveforms *See:* bipolar pulse; double pulse; staircase.

combination starter (packaging machinery) A starter having manually operated disconnecting means built into the same enclosure with the magnetic contactor. (IA/PKG) 333-1980w

combination support (raceway systems for Class 1E circuits for nuclear power generating stations) A support that serves either raceways or different types of raceway(s) and other mechanical or electric systems such as heating, ventilating, and air-conditioning (HVAC) ducts, piping, and lighting fixtures. (PE/NP) 628-1987r

combination surge *See:* combination wave.

combination thermoplastic tape An adhesive tape composed of a thermoplastic compound that provides both insulation and mechanical protection for joints. (EEC/PE) [119]

combination-type surge protective device A surge protective device that incorporates both voltage-switching-type components and voltage-limiting-type components may exhibit voltage switching, voltage limiting, or both voltage-switching and voltage-limiting behavior, depending upon the characteristics of the applied voltage. (SPD/PE) C62.48-1995

combination watch-report and fire-alarm system A coded manual fire-alarm system, the stations of which are equipped to transmit a single watch-report signal or repeated fire-alarm signals. *See also:* protective signaling. (EEC/PE) [119]

combination wave (1) The combination wave is delivered by a generator that applies a 1.2/50 voltage impulse across an open circuit and an 8/20 impulse current into a short circuit. The voltage and current and wave forms that are delivered to the surge protective device (SPD) are determined by the generator and the impedance of the SPD to which the surge is applied. The ratio of open-circuit voltage to peak short-circuit current is 2 ω. (PE) C62.34-1996
(2) A surge delivered by an instrument that has the inherent capability of applying a 1.2/50-voltage wave across an open-circuit, and delivering an 8/20-current wave into a short circuit. The instantaneous impedance to which the combination wave is applied determines the exact wave that is delivered. The peak magnitudes of the voltage or current wave shall be specified. *Synonym:* combination surge. (SPD/PE) C62.62-2000

combinatorial circuit *See:* combinational circuit.

combined head *See:* read/write head.

combined-line-recording trunk (CLR) (telephone switching) A one-way trunk for operator recording and extending of toll calls. (COM) 312-1977w

combined mechanical and electrical strength (insulators) The loading in pounds at which the insulator fails to perform its function either electrically or mechanically, voltage and mechanical stress being applied simultaneously. *Note:* The value will depend upon the conditions under which the test is made. *See also:* insulator; tower. (T&D/PE) [10]

combined-stress aging A form of accelerated aging in which several stresses are applied simultaneously. Ideally, the relative levels of the stresses are adjusted to produce the anticipated effects of the operational and environmental stresses in service. (DEI/RE) 775-1993w

combined telephone set A telephone set including in a single housing all the components required for a complete telephone set except the handset which it is arranged to support. *Note:* Wall hand telephone sets are of this type, but the term is usually reserved for a self-contained desk telephone set to distinguish it from desk telephone sets requiring an associated bell box. A desk local-battery telephone set may be referred to as a combined set if it includes in its mounting all components except its associated local batteries. *See also:* telephone station. (EEC/PE) [119]

combined uncertainty The uncertainty resulting from combining category A and category B uncertainties, as defined by the Bureau International des Poids et Mésures (BIPM), using standard statistical methods. Category A uncertainties are evaluated by applying statistical methods to a series of repeated measurements and are characterized by the estimated standard deviation, s_A; category B uncertainties are assigned to quantities whose variation is not explicitly observed. Category B uncertainties are determined by estimating from other information an approximation to a corresponding "standard deviation," s_B, whose existence is assumed. They are combined as if they are all standard deviations. (NI) N42.14-1991

combined voltage and current influence (wattmeter) The percentage change (of full-scale value) in the indication of an instrument that is caused solely by a voltage and current departure from specified references while constant power at the selected scale point is maintained. *See also:* accuracy rating. (EEC/AII) [102]

combustible Capable of undergoing combustion in air, at pressures and temperatures that might occur during a fire in a building, or in a more severe environment when specified. (DEI) 1221-1993w

combustible materials (power and distribution transformers) Materials which are external to the apparatus and made of or surfaced with wood, compressed paper, plant fibers, or other materials that will ignite and support flame. (PE/TR) C57.12.80-1978r

combustion A chemical process of oxidation that occurs at a rate fast enough to produce heat and usually light, either as a glow or flame. (DEI) 1221-1993w

combustion control The regulation of the rate of combination of fuel with air in a furnace. (T&D/PE) [10]

COM device *See:* computer output microfilmer.

come-along *See:* conductor grip.

comic-strip oriented image In micrographics, an image appearing on a roll of microfilm in such a manner that the top edge of the image is parallel to the long edge of the film. *Synonym:* landscape image. *Contrast:* cine-oriented image. (Std100) 10.2-1987

COMIT One of the first languages designed to manipulate text strings; provides pattern matching and substitution capabilities. (C) 610.13-1993w

Comité Consultatif Internationale Télégraphique et Téléphone (CCITT) **(1)** **(data transmission)** An advisory committee established under the United Nations in accordance with the International Tele-Communications Convention (Geneva 1959) Article 13, to study and recommend solutions for questions on technical operation and tariffs. The organization is attempting to establish standards for intercountry operation on a worldwide basis. (PE) 599-1985w
(2) An international organization that studies and issues recommendations on issues related to communication technology. *Note:* Also know in English as International Telegraph and Telephone Consultative Committee. (C) 610.10-1994w

comma In 1000BASE-X, the seven-bit sequence that is part of an 8B/10B code-group that is used for the purpose of code-group alignment. (C/LM) 802.3-1998

comma- In 1000BASE-X, the seven-bit sequence (1100000) of an encoded data stream. (C/LM) 802.3-1998

comma+ In 1000BASE-X, the seven-bit sequence (0011111) of an encoded data stream. (C/LM) 802.3-1998

command **(1)** **(logical link control)** In data communications, an instruction represented in the control field of a protocol data unit (PDU) and transmitted by a logical link control (LLC). It causes the addressed LLC(s) to execute a specific data link control function. (LM/PE/C/TR/CC) 799-1987w, 8802-2-1998
(2) **(A)** **(electronic computation)** One of a set of several signals (or groups of signals) that occurs as a result of interpreting an instruction; the commands initiate the individual steps that form the process of executing the instruction's operation. **(B)** **(electronic computation)** Loosely: an instruction in machine language. **(C)** **(electronic computation)** Loosely: a mathematical or logic operator. **(D)** **(electronic computation)** Loosely: an operation. (MIL/C/Std100) [2], [20], [85], 162-1963
(3) An input variable established by means external to, and independent of, the feedback (automatic) control system. It sets, is equivalent to, and is expressed in the same units as the ideal value of the ultimately controlled variable. *See also:* feedback control system; set point. (IA/ICTL/IAC) [60]
(4) **(software)** An expression that can be input to a computer system to initiate an action or affect the execution of a computer program; for example, the "log on" command to initiate a computer session. (C) 610.12-1990
(5) A pulse, signal, or set of signals initiating one step in the performance of a controlled operation. (SUB/PE) 999-1992w
(6) A procedure in the Forth programming language. The execution of a command performs some operation, usually affecting the state of one or more system resources in a predefined way. (New commands may be defined as sequences of previously defined commands. Most commands have human-readable names expressed as a sequence of textual characters.) *See also:* word name; Forth word. (C/BA) 1275-1994
(7) Any communication from a commander to a message-based servant, consisting of a write to the servant's data low register, possibly preceded by a write to the data high or data high and data extended registers. (C/MM) 1155-1992
(8) A directive to the shell to perform a particular task. (C/PA) 9945-2-1993
(9) **(A)** In hardware, a control signal. **(B)** An expression that can be input to a computer system to initiate aan action or affect the execution of a computer program; for example, the (log on(command to initiate a computer session. **(C)** Loosely, a mathematical or logic operator. **(D)** Loosely, a computer instruction. (C) 610.10-1994
(10) A message from the host directed to the printer that may or may not include print data. (C/MM) 1284.1-1997
(11) A package of information transmitted from the roadside to the vehicle that requests that the transponder on the vehicle perform a specific action. (SCC32) 1455-1999
(12) The instruction sent from an initiator to a target directing the target to execute a specified process. (C/MM) 1284.4-2000

command character *See:* control character.

command control (electric power system) An automatic generation control methodology that reduces unit control error irrespective of area control error. (PE/PSE) 94-1991w

command-driven Pertaining to a system or mode of operation in which the user directs the system through commands. *Contrast:* menu-driven. (C) 610.12-1990

commander A message-based device that is also a bus master and can control one or more servants. (C/MM) 1155-1992

command group A set of commands with defined behaviors, the group as a whole providing some particular capability (for example, one command group is concerned with client program debugging). (C/BA) 1275-1994

command guidance (navigation aid terms) Guidance in which information transmitted to a craft from an outside source causes it to follow a prescribed path. (AES/GCS) 172-1983w

command interpreter The portion of a Forth system that processes user input and Forth language source code by accepting a sequency of textual characters representing Forth word names and executing the corresponding Forth words. (C/BA) 1275-1994

command key Any control key on a keyboard used to represent a particular machine command. (C) 610.10-1994w

command language **(1)** **(software)** A language used to express commands to a computer system. *See also:* command-driven. (C) 610.12-1990
(2) A computer language used to express commands to a computer system and to control their execution. For example, job control language, or REXX. *Synonym:* command-level language. *See also:* interactive language; declarative language; rule-based language. (C) 610.13-1993w
(3) A type of dialog in which a user composes entries to evoke a system response. (PE/NP) 1289-1998

command language interpreter *See:* shell.

command-level language *See:* command language.

command line interface A means of invoking utilities by issuing commands from within a POSIX.2 shell, implying that neither graphics nor windows are required. (C/PA) 1387.2-1995

command link (communication satellite) A data transmission link (generally earth to spacecraft or satellite) used to command a satellite or spacecraft in space. (COM) [24]

command protocol data unit (PDU) **(1)** **(logical link control)** All PDU's transmitted by a logical link control (LLC) in which the C/R (command/response) bit is equal to "O." (PE/TR) 799-1987w

(2) All PDUs sent by an LLC in which the C/R bit in the SSAP address field is equal to "0." (C/LM/CC) 8802-2-1998

command rate (gyros) The input rate equivalent of a torquer command signal. (AES/GYAC) 528-1994

command readout (numerically controlled machines) Display of absolute position as derived from position command. *Note:* In many systems the readout information may be taken directly from the dimension command storage. In others it may result from the summation of command departures. (IA) [61]

command reference (power supplies) (servo or control system) The voltage or current to which the feedback signal is compared. As an independent variable, the command reference exercises complete control over the system output. *See also:* operational programming. (AES) [41]

command_reset An initialization event that is initiated by a write to the RESET_START CSR. (C/BA) 896.2-1991w, 896.10-1997

Command Resource Unavailable (CRU) bit A bit in the Bus Error register of all S-modules. An S-module sets this bit to indicate that resources required to complete execution of a command were not available and that the command was not executed. (TT/C) 1149.5-1995

commands, class of One of the groups of MTM-Bus commands. Every MTM-Bus command is assigned to a command class. (TT/C) 1149.5-1995

Command Sequence Error (CSE) bit A bit in the Bus Error register of all S-modules. An S-module sets this bit to indicate that the module has received a command that requires a previous enabling command without receipt of such an enabling command. (TT/C) 1149.5-1995

command set A field in the Device ID message identifying the type of data expected by the peripheral. For example, a printer might use this field to report which page description language(s) it supports. (C/MM) 1284-1994

command transfer The passing of command information over the system control signal group, from the bus owner to the replying agent(s), during the request phase of a transfer operation. Command information includes parameters for the impending transfer operation, as well as additional address space information not transmitted with the address transfer. *See also:* system control signal group; request phase. (C/MM) 1296-1987s

command & US core;_reset An initialization event that is initiated by a write to the RESET_START register. (C/MM) 1212-1991s

command *X*, receipt of Error-free receipt of the HEADER packet containing in its Command field the command code of *X*. (TT/C) 1149.5-1995

comment (software) Information embedded within a computer program, job control statements, or a set of data, that provides clarification to human readers but does not affect machine interpretation. (C) 610.12-1990

comment source statements Source statements that provide information to people reading the software source code and are ignored by the compiler. (C/SE) 1045-1992

commercial character (A) One of the set of characters used commonly in commercial operations; for example, CR (credit) and DB (debit). **(B)** A character within a picture specification that represents one of the characters as in (A). (C) 610.5-1990

commercial data processing Data processing performed to support a commercial organization or function. (C) 610.2-1987

commercial grade dedication A process of evaluating (which includes testing) and accepting commercial grade items to obtain adequate confidence of their suitability for safety application. (PE/NP) 7-4.3.2-1993

commercial grade item An item satisfying a), b), and c) below:

a) Not subject to design or specification requirements that are unique to nuclear facilities
b) Used in applications other than nuclear facilities
c) Ordered from the manufacturer/supplier on the basis of specifications set forth in the manufacturer's published product description (for example, catalog)

. (PE/NP) 7-4.3.2-1993

commercial grade part (replacement parts for Class 1E equipment in nuclear power generating stations) A part that is:

a) Not subject to design or specification requirements that are unique to nuclear power plants;
b) Used in applications other than nuclear power plants;
c) Ordered from the manufacturer/supplier on the basis of specifications set forth in the manufacturer's published product description (for example, a catalog).

(PE/NP) 934-1987w

commercial-off-the-shelf (COTS) Software defined by a market-driven need, commercially available, and whose fitness for use has been demonstrated by a broad spectrum of commercial users. (C/SE) 1062-1998

commercial operation The acceptance, by the user, of the static var compensator (SVC) from the supplier. (PE/SUB) 1031-2000

commercial power (1) (emergency and standby power) Power furnished by an electric power utility company; when available, it is usually the prime power source. However, when economically feasible, it sometimes serves as an alternative or standby source. *Synonym:* utility power. (IA/PSE) 446-1995
(2) Power furnished by an electric power utility company. (IA/PSE) 1100-1999

commercial, residential, and institutional buildings All buildings other than industrial buildings and residential dwellings. (IA/PSE) 241-1990r

commercial tank (electrorefining) An electrolytic cell in which the cathode deposit is the ultimate electrolytically refined product. *See also:* electrorefining. (EEC/PE) [119]

commercial zone A zone that includes offices, shops, hotels, motels, service establishments, or other retail/commercial facilities as defined by local ordinances. (PE/SUB) 1127-1998

commissioning The process of providing to the appropriate components, the information necessary for the designed communication between components. (IM/ST) 1451.1-1999

Commission Internationale de l'Eclairage The initials CIE are the initials of the official French name of the International Commission on Illumination. This translated name is approved for usage in English-speaking countries, but at its 1951 meeting the Commission recommended that only the initials of the French name be used. The initials ICI, which have been used commonly in this country, are deprecated because they conflict with an important trademark registered in England and because the initials of the name translated into other languages are different. (BT/AV) 201-1979w

commissioning tests (rotating machinery) Tests applied to a machine at site under normal service conditions to show that the machine has been erected and connected in a correct manner and is able to work satisfactorily. *See also:* asynchronous machine. (PE) [9]

common *See:* common storage.

common ancestor constraint A kind of constraint that involves two or more relationship paths to the same ancestor class and states either that a descendent instance must be related to the *same* ancestor instance through each path or that it must be related to a *different* ancestor instance through each path. (C/SE) 1320.2-1998

common area *See:* common storage.

common-battery central office *See:* common-battery office.

common-battery office (telephone switching systems) A central office that supplies transmitter and signaling currents for its associated stations and current for the central office equipment from a power source located in the central office. (COM) 312-1977w

common-battery signaling (data transmission) A method of actuating a line or supervisory signal at the distant end of a telephone line by the closure of a direct-current (dc) circuit with the exchange providing the feeding current. (PE) 599-1985w

common-battery switchboard A telephone switchboard for serving common-battery telephone sets. (COM) [48]

common block *See:* common storage.

common bonding network (CBN) **(A)** The principal means for affecting bonding and earthing inside a building. **(B)** The set of metallic components that are intentionally or incidentally interconnected to form the (earthed) bonding network (a mesh) in a building. These components include structural steel or reinforcing rods, metallic plumbing, ac power conduit, equipment grounding conductors, cable racks, and bonding conductors. The CBN always has a mesh topology and is connected to the grounding electrode system. *Note:* The CBN may also be known in the public telephone network as an integrated ground plane. (IA/PSE) 1100-1999

COmmon Business-Oriented Language (COBOL) A high-order programming language standardized by ANSI and ISO, designed for business applications. *See also:* common language; general-purpose programming language; IDS/1. (C) 610.13-1993w

common carrier **(1)** In telecommunications, a public utility company that is recognized by an appropriate regulatory agency as having a vested interest and responsibility in furnishing communication services to the general public. *See also:* value-added service; specialized common carrier. (LM/COM) 168-1956w

(2) *See also:* communications common carrier. (C) 610.7-1995

common-cause failure **(1) (reliability data for pumps and drivers, valve actuators, and valves)** Two or more redundant component failures due to a single cause. The common-cause events that cause multiple failures are usually secondary events or events that exceed the design envelope of the component. (PE/NP) 500-1984w

(2) (nuclear power generating station safety systems) Multiple failures attributable to a common cause. (PE/NP) 379-1994, 603-1998, 933-1999

common-channel interoffice signaling (telephone switching systems) The use of separate paths between switching entities to carry the signaling associated with a group of communication paths. (COM) 312-1977w

common class Defines those aspects of different software objects that are the same. The common classes for this standard are software_collections, software, and software_files. The names of these classes are also used to generically describe any object that shares that common class. (C/PA) 1387.2-1995

common control (telephone switching systems) An automatic switching arrangement in which the control equipment necessary for the establishment of connections is shared, being associated with a given call only during the period required to accomplish the control function. (COM) 312-1977w

common coupling *See:* common-environment coupling.

common data *See:* global data.

common device (of a supervisory system) A device in either the master or remote station that is required for the basic operation of the supervisory system and is not part of the equipment for the individual points. *Synonym:* basic device. (SWG/PE) C37.100-1992

common-environment coupling A type of coupling in which two software modules access a common data area. *Synonym:* common coupling. *Contrast:* pathological coupling; content coupling; data coupling; control coupling; hybrid coupling. (C) 610.12-1990

common equipment That complement of either the master or remote station supervisory equipment that interfaces with the interconnecting channel and is otherwise basic to the operation of the supervisory system, but is exclusive of those elements that are peculiar to and required for the particular applications and uses of the equipment. (SWG/PE/SUB) C37.100-1992, C37.1-1987s

common language Any programming language that is used widely on a variety of computers; For example, BASIC, C, COBOL, and FORTRAN. *See also:* general-purpose programming language. (C) 610.13-1993w

Common LISP A dialect of LISP that is widely accepted as the standard language for LISP. *See also:* CLOS. (C) 610.13-1993w

Common LISP Object System An object-oriented language based on Common LISP. (C) 610.13-1993w

common-mode **(1) (general)** The instantaneous algebraic average of two signals applied to a balanced circuit, both signals referred to a common reference. *See also:* oscillograph. (IM/HFIM) [40]

(2) (medical electronics) (in-phase signal) A signal applied equally and in phase to the inputs of a balanced amplifier or other differential device. (EMB) [47]

common-mode conversion (interference terminology) The process by which differential-mode interference is produced in a signal circuit by a common-mode interference applied to the circuit.Common-mode currents are converted to differential-mode voltages by impedances R_1, R_2, R_3, R_4, R_S, R_R, and c. The differential-mode voltage at the receiver resulting from the conversion is the algebraic summation of the voltage drops produced by the various currents in these impedances. Various of the impedances may be neglected at particular frequencies. At direct current,

$$V_{CM} = I_r R_r \approx I_{CM1}(R_S + R_1 + R_2) - I_{CM2}(R_3 + R_4)$$

At

$$f > \frac{I}{c(R_1 + R_3 + R_S)}$$

$$V_{CM} \approx I_c X_c \frac{R_R}{R_2 + R_4 + R_R}$$

See also: interference.

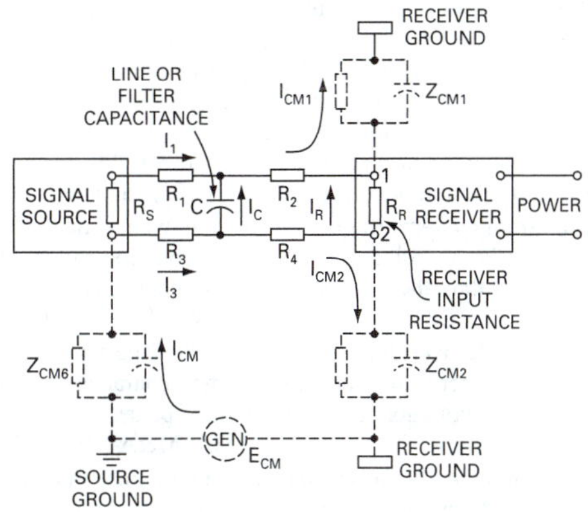

common-mode conversion

(IE) [43]

common-mode failure (nuclear power generating station) (safety systems equipment in nuclear power generating stations) Multiple failures attributable to a common cause.

(SWG/PE/NP) 627-1980r, 650-1979s, 649-1980s, 308-1980s, C37.100-1992

common-mode interference **(1) (automatic null-balanced electrical instruments)** Interference that appears between both signal leads and a common reference plane (ground) and causes the potential of both sides of the transmission path to be changed simultaneously and by the same amount relative to the common reference plane (ground). *See also:* interference. (IE/EMC/PE/SUB) [43], C63.13-1991, C37.1-1994

(2) A form of interference that appears between any measuring circuit terminal and ground. *See also:* accuracy rating.

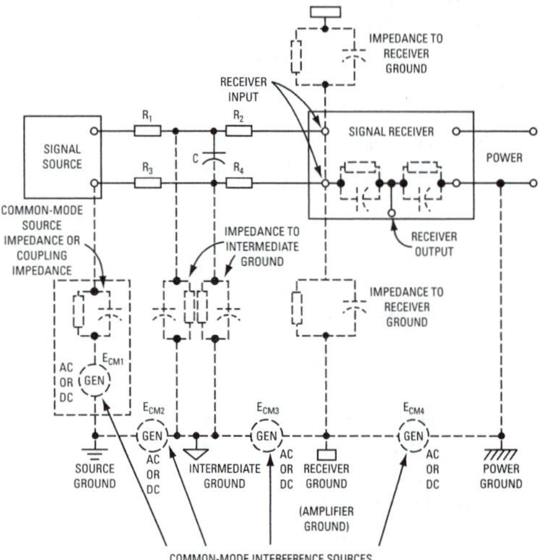

Common-mode interference sources and current paths. The common-mode voltage V_{CM} in any path is equal to the sum of the common-mode generator voltages in that path; for example, in the source-receiver path.

$$V_{CM} = E_{CM1} + E_{CM2} + E_{CM3}.$$

common-mode interference
(EEC/SUB/PE/EMI) [112], C37.1-1994

common-mode noise (longitudinal) The noise voltage that appears equally, and in phase, from each current carrying conductor to ground. (IA/PSE) 1100-1999

common-mode outage *See:* common-mode outage event.

common-mode outage event A related multiple outage event consisting of two or more primary outage occurrences initiated by a single incident or underlying cause where the outage occurrences are not consequences of each other. *Note:* Primary outage occurrences in a common-mode outage event are referred to as common-mode outage occurrences or simply common-mode outages. Examples of common-mode outage events are a single lightning stroke causing tripouts of both circuits on a common tower, and an external object causing the outage of two circuits on the same right-of-way. *See also:* related multiple outage event. (PE/PSE) 859-1987w

common-mode overvoltage A signal level whose magnitude is less than the specified maximum safe common-mode signal but greater than the maximum operating common-mode signal. (IM/WM&A) 1057-1994w

common-mode overvoltage recovery time The time required for the recorder to return to its specified characteristics after the end of a common-mode overvoltage pulse.
(IM/WM&A) 1057-1994w

common-mode radio noise Conducted radio noise that appears between a common reference plane (ground) and all wires of a transmission line causing their potentials to be changed simultaneously and by the same amount relative to the common reference plane (ground). (EMC) C63.4-1991

common-mode rejection (in-phase rejection) The ability of certain amplifiers to cancel a common-mode signal while responding to an out-of-phase signal. (EMB) [47]

common-mode rejection quotient (in-phase rejection quotient) The quotient obtained by dividing the response to a signal applied differentially by the response to the same signal applied in common mode, or the relative magnitude of a common-mode signal that produces the same differential response as a standard differential input signal. (EMB) [47]

common-mode rejection ratio (CMRR) (1) (signal-transmission signal) The ratio of the common-mode interference voltage at the input terminals of the system to the effect produced by the common-mode interference, referred to the input terminals for an amplifier. For example,

$$\text{CMRR} = \frac{V_{\text{CM}}\ \text{(root-mean-square) at input}}{\text{effect at output/amplifier gain}}$$

See also: interference. (IE) [43]
(2) (oscilloscopes) The ratio of the deflection factor for a common-mode signal to the deflection factor for a differential signal applied to a balanced-circuit input. *See also:* oscillograph.
(3) The ratio of the input common-mode signal to the effect produced at the output of the recorder in units of the input.
(IM/WM&A) 1057-1994w

common mode signal The average value of the signals at the positive and negative inputs of a differential input waveform recorder. If the signal at the positive input is designated V_+, and the signal at the negative input is designated V_-, then the common mode signal (V_{cm}) is

$$V_{cm} = \frac{V+ + V-}{2}$$

(IM/WM&A) 1057-1994w

common-mode to normal-mode conversion In addition to the common-mode voltages which are developed in the single conductors by the general environmental sources of electrostatic and electromagnetic fields, differences in voltage exist between different ground points in a facility due to the flow of ground currents. These voltage differences are considered common mode when connection is made to them either intentionally or accidentally, and the currents they produce are common mode. These common-mode currents can develop normal-mode noise voltage across unequal circuit impedances. (PE/SUB/EDPG) 422-1977, 525-1992r

common-mode voltage (1) The voltage that, at a given location, appears equally and in phase from each signal conductor to ground. (PE/PSR) C37.90.1-1989r
(2) The instantaneous algebraic average of two signals applied to a balanced circuit, with both signals referenced to a common reference. *Synonym:* longitudinal voltage.
(LM/C) 802.3-1998
(3) (surge withstand capability tests) The voltage common to all conductors of a group as measured between that group at a given location and an arbitrary reference (usually earth).
(SWG/PE/PSR) C37.100-1992, C37.90-1978s
(4) The instantaneous algebraic mean of two signals applied to a balanced circuit, where both signals are referred to a common reference.local area networks. (C) 8802-12-1998

common return A return conductor common to several circuits. *See also:* center of distribution. (T&D/PE) [10]

common services Data type and functional declarations defining an application procedural interface used to access and manipulate numerical and digital data, physical data units, mathematical functions and constants, unit conversion factors, and functions used by two or more interfaces.
(SCC20) 1226-1998

common spectrum multiple access (communication satellite) A method of providing multiple access to a communication satellite in which all of the participating earth stations use a common time-frequency domain. Signal processing is employed to detect a wanted signal in the presence of others. Three typical approaches utilizing these techniques are spread spectrum, frequency-time matrix, and frequency-hopping.
(COM) [19]

common storage A portion of main storage that can be accessed by two or more modules in a software system. *Synonyms:* common area; common block. *See also:* global data.
(C) 610.12-1990, 610.10-1994w

common trunk (telephone switching systems) A trunk, link, or junctor accessible from all input groups of a grading.
(COM) 312-1977w

common use Simultaneous use by two or more utilities of the same kind. (NESC) C2-1997

common winding (power and distribution transformers) (autotransformer) That part of the autotransformer winding that is common to both the primary and secondary circuits. *Synonym:* shunt winding. (PE/TR) C57.15-1999, C57.12.80-1978r

communication (1) (data transmission) (electric systems) (telecommunications) The transmission of information from one point to another by means of electromagnetic waves. (PE) 599-1985w
(2) The flow of information from one point, known as the source, to another, the receive. (C) 610.10-1994w

communication access Passive and active attacks on information transmitted over communication channels and on the system's communication services themselves. This threat area assumes that an intruder has access to the communication media or the components (both hardware and software) that provide the communication services. (C/BA) 896.3-1993w

communication channel A facility that permits signaling between terminals. (SUB/PE) 999-1992w

communicational cohesion A type of cohesion in which the tasks performed by a software module use the same input data or contribute to producing the same output data. *Contrast:* coincidental cohesion; functional cohesion; temporal cohesion; sequential cohesion; procedural cohesion; logical cohesion. (C) 610.12-1990

communication band *See:* frequency band of emission.

communication circuits Electrical circuits supplying equipment and systems for voice, sound, or data transmission, such as telephone, engine order telegraph, data communication, interior communication, paging systems, wired music systems, fire and general alarm systems, smoke and fire detection systems, closed circuit television, navigational equipment, and microprocessor based automated alarm and control systems. (IA/MT) 45-1998

communication conductor (measuring longitudinal balance of telephone equipment operating in the voice band) A conductor used in a communication network. (COM/TA) 455-1985w

communication control character (1) A functional character intended to control or facilitate transmission over data networks. Control characters form the basis for character-oriented communications control procedures. (LM/COM) 168-1956w
(2) (data management) *See also:* transmission control character. (C) 610.5-1990w

communication facility (data transmission) Anything used or available for use in the furnishing of communication service. (PE) 599-1985w

communication interface That part of the API devoted to communications with other application software, external data transport facilities, and devices. (C/PA) 14252-1996

communication line *See:* telecommunication line.

communication lines (1) The conductors and their supporting or containing structures that are used for public or private signal or communication service, and which operate at potentials not exceeding 400 V to ground or 750 V between any two points of the circuit, and the transmitted power of which does not exceed 150 W. When operating at less than 150 V, no limit is place on the transmitted power of the system. Under specified conditions, communication cables may include communication circuits exceeding the preceding limitation where such circuits are also used to supply power solely to communication equipment. *Note:* Telephone, telegraph, railroad-signal, data, clock, fire, police-alarms, cable television and other systems conforming with the above are included. Lines used for signaling purposes, but not included under the above definition, are considered as supply lines of the same voltage and are to be so installed. *See also:* electric supply lines. (NESC/T&D) C2.2-1960
(2) The conductors and their supporting or containing structures that are used for public or private signal or communications service, and which operate at potentials not exceeding 400 V to ground or 750 V between any two points of the circuit, and the transmitted power of which does not exceed 150 W. When operating at less than a nominal voltage of 90 V, no limit is placed on the transmitted power of the system. Under specified conditions, communication cables may include communication circuits exceeding the preceding limitation where such circuits are also used to supply power solely to communications equipment. *Note:* Telephone, telegraph, railroad-signal, data, clock, fire, police-alarm, cable-television, and other systems conforming with the above are included. Lines used for signaling purposes, but not included under the above definition, are considered as supply lines of the same voltage and are to be so installed. (NESC) C2-1997

communication provider A component of the system that provides the communications service through an endpoint. (C) 1003.5-1999

communication reliability (mobile communication) A specific criterion of system performance related to the percentage of times a specified signal can be received in a defined area during a given interval of time. *See also:* mobile communication system. (VT) [37]

communications architecture The hardware and software structure that facilitates the communications operations. (C) 610.7-1995

communication satellite A satellite used for communication between two or more ground points by transmitting the messages to the satellite and retransmitting them to the participating ground station. (COM) [24]

communications-based train control A continuous automatic train control system utilizing high-resolution train location determination, independent of track circuits; continuous, high capacity, bidirectional train-to-wayside data communications; and trainborne and wayside processors capable of implementing vital functions. (VT/RT) 1474.1-1999

communications cable A cable that carries a low level of electric energy used for the transmission of communication frequencies. A telephone-type cable consists of two or more solid, insulated, twisted, paired and/or quadded, shielded or unshielded conductors ranging from No 19 to No 26 AWG, with either a shielded or unshielded sheath. (PE/PSC) 789-1988w

communications common carrier (1) (data transmission) A company recognized by an appropriate regulatory agency as having a vested interest in furnishing communications services to the public at large. (PE) 599-1985w
(2) In telecommunication, a public utility company that is recognized by an appropriate regulatory agency as having a vested interest in and responsibility for furnishing communications services to the general public. *Synonym:* common carrier. (C) 610.7-1995

communications computer A computer that is specially designed to be an interface between another computer or terminal and a network, or to control data flow in a network. *See also:* switching computer; front-end computer; concentrator. (C/COM) 610.7-1995, 168-1956w, 610.10-1994w

communications controller A dedicated computer that checks and manages data traffic through a network. (C) 610.7-1995

communication security Protective measures for information transmitted between system components, over telecommunication links, and through networks to provide data confidentiality, integrity, andauthenticity. (C/BA) 896.3-1993w

communications endpoint *See:* endpoint.

communication services interface (CSI) The boundary across which access to services for interaction between internal application software entities and application platform external entities is provided. (C/PA) 14252-1996

communications interface equipment (relays and relay systems associated with electric power apparatus) A portion of a relay system that transmits information from the relay logic to a communications link, or conversely to logic, for example, audio tone equipment, a carrier transmitter-receiver when an integral part of the relay system. (SWG/PE/PSR) C37.100-1992, C37.90-1978s

communications link (relays and relay systems associated with electric power apparatus) Any of the communications media, for example, microwave, power line carrier, wire line. (PE/PSR) C37.90-1978s

communications network A network of communication circuits managed as a single unit. *See also:* computer network; value-added network. (C) 610.7-1995

communications processor A computer that performs protocol (terminates one or more protocols layers) or network management functions. (C) 610.7-1995

communications user An application that uses process-to-process communication services. (C) 1003.5-1999

communications zone The area of space within which a beacon can communicate with transponders in or on passing vehicles. (SCC32) 1455-1999

communications security The use of administrative, technical, or physical measures to deny unauthorized persons information from a computer or a communications network and to ensure the authenticity and integrity of such communications. (C) 610.7-1995

communication theory (data transmission) The mathematical theory underlying the communication of messages from one point to another. (PE) 599-1985w

community antenna television (CATV) *See:* cable TV.

community antenna television (CATV)-type broadband medium A broadband system comprising coaxial cables, taps, splitters, amplifiers, and connectors the same as those used in CATV or cable television installations. (C/LM) 802.3-1998

community dial office (telephone switching systems) A small automatic central office that serves a separate exchange area that ordinarily has no permanently assigned central office operating or maintenance forces. (COM) 312-1977w

community-of-interest (telephone switching systems) A characteristic of traffic resulting from the calling habits of the customers. (COM) 312-1977w

commutated antenna direction finder (CADF) (navigation aid terms) A system using a multiplicity of antennas in a circular array and a receiver which is connected to the antennas in sequence through a commutating device for finding the direction of arrival of radio waves; the directional sensing is related to phase shift that occurs as a result of the communication. (AES/GCS) 172-1983w

commutating angle (1) (rectifier circuits) The time, expressed in degrees, during which the current is commutated between two rectifying elements. *See also:* rectifier circuit element; rectification. (IA) [62]

(2) (thyristor converter circuit) (μ) The time, expressed in degrees (one cycle of the ac wave form—360°), during which the current is commutated between two thyristor converter circuit elements. (IA/IPC) 444-1973w

commutating capacitor (converter circuit elements) (self-commutated converters) A capacitor that provides commutating voltage for circuit-commutated thyristors in a self-commutated converter. (IA/SPC) 936-1987w

commutating-field winding An assembly of field coils, located on the commutating poles, that produces a field strength approximately proportional to the load current. The commutating field is connected in direction and adjusted in strength to assist the reversal of current in the armature coils for successful commutation. This field winding is used along, or supplemented by, a compensating winding. *See also:* asynchronous machine. (EEC/PE) [119]

commutating group (1) (rectifier circuits) A group of rectifier-circuit elements and the alternating-voltage supply elements conductively connected to them in which the direct current of the group is commutated between individual elements that conduct in succession. *See also:* rectifier; rectification; circuit element. (IA) [62]

(2) A group of thyristor converter circuit elements and the alternating-voltage supply elements conductively connected to them in which the direct current of the group is commutated between individual elements that conduct in succession. (IA/IPC) 444-1973w

commutating impedance (1) (rectifier transformer) The impedance that opposes the transfer of current between two direct-current winding terminals of a commutating group, or a set of commutating groups. *See also:* rectifier transformer. (Std100) C57.18-1964w

(2) (rectifier transformer) The impedance that opposes the transfer of current between two secondary winding terminals of a commutating group, or a set of commutating groups. (PE/TR) C57.18.10-1998

commutating period (inverters) The time during which the current is commutated. *See also:* self-commutated inverters. (IA) [62]

commutating pole (interpole) An auxiliary pole placed between the main poles of a commutating machine. Its exciting winding carries a current proportional to the load current and produces a flux in such a direction and phase as to assist the reversal of the current in the short-circuited coil. (EEC/PE) [119]

commutating reactance (thyristor converter) The reactance that effectively opposes the transfer of current between thyristor converter circuit elements of a commutating group or set of commutating groups. *Note:* For convenience, the reactance from phase to neutral, or one half the total reactance in the commutating circuit, is the value usually employed in computations, and is designated as the commutating reactance. (IA/IPC) 444-1973w

commutating reactance factor (rectifier circuits) The line-to-neutral commutating reactance in ohms multiplied by the direct current commutated and divided by the effective (root-mean-square) value of the line-to-neutral voltage of the transformer direct-current winding. *See also:* circuit element; rectification. (IA) [62]

commutating reactance transformation constant A constant used in transforming line-to-neutral commutating reactance in ohms on the direct-current winding to equivalent line-to-neutral reactance in ohms referred to the alternating-current winding. *See also:* rectification. (IA) [62]

commutating reactor (1) (converter circuit elements) (commutating inductor) (self-commutated converters) An inductor having one or more windings that modifies or couples the transient current produced by the commutating voltage. (IA/SPC) 936-1987w

(2) (power and distribution transformers) A reactor used primarily to modify the rate of current transfer between rectifying elements. (PE/TR) C57.12.80-1978r

commutating resistance (rectifier transformer) The resistance component of the commutating impedance. *See also:* rectifier transformer. (Std100) C57.18-1964w

commutating voltage (1) (self-commutated converters) (circuit properties) The voltage that causes the current to commutate from one switching branch to another. *Notes:* 1. In an internally commutated converter, the commutating voltage is supplied by an ac (alternating current) source outside the converter. 2. In a self-commutated converter using switching devices that have turn-off capability, such as power transistors or gate turn-off thyristors, the commutating voltage results from the interruption of current in the outgoing device branch. 3. In a self-commutated converter using circuit-commutated thyristors, the commutating voltage is usually supplied by capacitors. (IA/SPC) 936-1987w

(2) (ac adjustable-speed drives) The voltage that causes the current to commutate from one switching branch to another. *Note:* In an externally commutated converter, the commutating voltage is supplied by an ac source outside the converter. (IA/ID) 995-1987w

commutation (1) (circuit properties) (ac adjustable-speed drives) (self-commutated converters) The transfer of

current from one converter switching branch to another. (IA/SPC/ID) 936-1987w, 995-1987w
(2) (harmonic control and reactive compensation of static power converters) The transfer of unidirectional current between thyristor (or diode) converter circuit elements that conduct in succession. (IA/IPC/SPC) 428-1981w, 519-1992, 444-1973w

commutation elements (semiconductor rectifiers) The circuit elements used to provide circuit-commutated turnoff time. *See also:* semiconductor rectifier stack. (IA) [62]

commutation factor (1) (rectifier circuits) The product of the rate of current decay at the end of conduction, in amperes per microsecond, and the initial reverse voltage, in kilovolts. *See also:* element; rectification. (IA) [62]
(2) (gas tube) The product of the rate of current decay and the rate of the inverse voltage rise immediately following such current decay. *Note:* The rates are commonly stated in amperes per microsecond and volts per microsecond. *See also:* rectification; heterodyne conversion transducer; gas tube. (ED) 161-1971w

commutation interval (self-commutated converters) (circuit properties) The time interval between the application of commutating voltage to a pair of commutating branches and the cessation of the resulting transient currents. *Note:* The commutation interval is the same as the overlap interval in an externally commutated converter in which the commutating voltage is supplied by the ac (alternating current) line. (IA/SPC) 936-1987w

commutation shrink ring A member that holds the commutator-segment assembly together and in place by being shrunk on an outer diameter of and insulated from the commutator-segment assembly. *See also:* commutator. (EEC/PE) [119]

commutator (rotating machinery) An assembly of conducting members insulated from one another, in the radial-axial plane, against which brushes bear, used to enable current to flow from one part of the circuit to another by sliding contact. (PE) [9]

commutator bars *See:* commutator segments.

commutator bore Diameter of the finished hole in the core that accommodates the armature shaft. *See also:* commutator. (EEC/PE) [119]

commutator brush track diameter That diameter of the commutator segment assembly that after finishing on the armature is in contact with the brushes. *See also:* commutator. (EEC/PE) [119]

commutator core The complete assembly of all of the retaining members of a commutator. *See also:* commutator. (EEC/PE) [119]

commutator-core extension That portion of the core that extends beyond the commutator segment assembly. *See also:* commutator. (EEC/PE) [119]

commutator inspection cover A hinged or otherwise attached part that can be moved to provide access to commutator and brush rigging for inspection and adjustment. *See also:* commutator. (EEC/PE) [119]

commutator insulating segments (rotating machinery) The insulation between commutator segments. (PE) [9]

commutator insulating tube (rotating machinery) The insulation between the underside of the commutator segment assembly and the core. *See also:* commutator. (PE) [9]

commutator motor meter (seismic qualification of Class 1E equipment for nuclear Power generating stations) A motor type of meter in which the rotor moves as a result of the magnetic reaction between two windings, one of which is stationary and the other assembled on the rotor and energized through a commutator and brushed. *See also:* electricity meter. (PE/NP) 344-1975s

commutator nut The retaining member that is used in combination with a vee ring and threaded shell to clamp the segment assembly. *See also:* commutator. (EEC/PE) [119]

commutator riser (rotating machinery) A conducting element for connecting a commutator segment to a coil. *See also:* commutator. (PE) [9]

commutator-segment assembly A cylindrical ring or disc assembly of commutator segments and insulating segments that are bound and ready for installation. *Note:* The binding used may consist of wire, temporary assembly rings, shrink rings, or other means. *See also:* commutator. (EEC/PE) [119]

commutator segments (commutator bars) Metal current-carrying members that are insulated from one another by insulting segments and that make contact with the brushes. *See also:* commutator. (EEC/PE) [119]

commutator shell The support on which the component parts of the commutator are mounted. *Note:* The commutator may be mounted on the shaft, on a commutator spider, or it may be integral with a commutator spider. *See also:* commutator. (EEC/PE) [119]

commutator-shell insulation (rotating machinery) The insulation between the under (or in the case of a disc commutator, the back) side of the commutator assembled segments and the commutator shell. *See also:* commutator. (PE) [9]

commutator vee ring The retaining member that, in combination with a commutator shell, clamps or binds the commutator segments together. *See also:* commutator. (EEC/PE) [119]

commutator vee ring insulation (rotating machinery) The insulation between the V-ring and the commutator segments. (PE) [9]

commutator vee-ring insulation extension (rotating machinery) The portion of the vee-ring insulation that extends beyond the commutator segment assembly. *See also:* commutator. (PE) [9]

commuter rail A passenger railroad service that operates within metropolitan areas on trackage that usually is part of the general railroad system. The operations, primarily for commuters, are generally run as part of a regional system that is publicly owned, or by a railroad company as part of its overall service. (VT/RT) 1474.1-1999

compact disc An optical disk that is compact in size, generally 4 to 5 inches in diameter. *See also:* CD-ROM storage; laser disk. (C) 610.10-1994w

compact disc read-only memory *See:* CD-ROM storage.

compact disc storage *See:* CD-ROM storage.

compaction (software) In microprogramming, the process of converting a microprogram into a functionally equivalent microprogram that is faster or shorter than the original. (C) 610.12-1990

companding (data transmission) A process in which compression is followed by expansion. *Note:* Companding is often used for noise reduction, in which case the compression is applied before the noise exposure and the expansion after the exposure. (PE) 599-1985w, 270-1964w

compandor (data transmission) A combination of a compressor at one point in a communication path for reducing the amplitude range of signals followed by an expander at another point for a complementary increase in the amplitude range. *Note:* The purpose of a compandor is to improve the ratio of the signal to the interference entering in the path between the compressor and expander. (PE) 599-1985w

company_id A 24-bit binary value used to identify a company within the context of the CSR Architecture. The company_id values are expected to be uniquely assigned to each company. (C/MM) 1212-1991s

comparative tests (test, measurement, and diagnostic equipment) Comparative tests compare end item signal or characteristic values with a specified tolerance band and present the operator with a go/no-go readout; a go for signals within tolerances, and a no-go for signals out of tolerance. (MIL) [2]

comparator (1) A circuit for performing amplitude selection between either two variables or between a variable and a constant. (C) [20]

(2) (test, measurement, and diagnostic equipment) A device capable of comparing a measured value with predetermined limits to determine if the value is within these limits. (MIL) [2]

(3) (analog computer) A circuit, having only two logic output states, for comparing the relative amplitudes of two analog variables, or of a variable and a constant, such that the logic signal output of the comparator uniquely determines which variable is the larger at all times. (C) 165-1977w

(4) (software) A software tool that compares two computer programs, files, or sets of data to identify commonalities or differences. Typical objects of comparison are similar versions of source code, object code, data base files, or test results. (C) 610.12-1990

compare (1) (mathematics of computing) To examine a quantity for the purpose of determining its relationship to zero. (C) 1084-1986w

(2) (data management) To examine two items to determine their relative magnitudes, their relative positions in a given sequence, or whether they are identical. (C) 610.5-1990w

comparer A signal element that performs an AND logic function. (SWG/PE) C37.100-1992

compare&swap A data-access operation that conditionally stores a *next* value to a specified data type and returns the previous data value. The store occurs when the addressed memory value and a second *test* value are equal. When accessing uncached data, this data-access operation generates a compareSwap bus transaction. (C/MM) 1596.5-1993

compare_swap A bus transaction that takes a test value and a new value as inputs, compares the test value to the current contents of an address and, if equal, atomically writes the new value to the address and returns the old. (C/MM) 1212.1-1993

compareSwap A bus that conditionally stores a next argument to a specified data address and returns the previous data value from that address. The store occurs when the addressed memory value and a second *test* value are equal. In the CSR Architecture, this is called a compare_swap transaction. (C/MM) 1596.5-1993

comparison (A) (data management) The process of examining two or more items for identity, similarity, equality, relative magnitude, or for order in a sequence. **(B) (data management)** The result of such an examination as in (A). (C) 610.5-1990

comparison amplifier (power supplies) A high-gain non-inverting direct-current amplifier that, in a bridge-regulated power supply, has as its input the voltage between the null junction and the common terminal. The output of the comparison amplifier drives the series pass elements. (AES) [41]

comparison bridge (power supplies) A type of voltage-comparison-circuit whose configuration and principle of operation resemble a four-arm electric bridge. *See also:* error signal. (AES) [41]

comparison lamp (luminous standards) (illuminating engineering) A light source having a constant, but not necessarily known, luminous intensity with which standard and test lamps are successively compared. (IE/EEC) [126]

compartment (GIS) (1) (packaging machinery) A space within the base, frame, or column of the industrial equipment. (IA/PKG) 333-1980w

(2) Any gas section of the gas-insulated substation assembly that provides gas isolation. (SUB/PE) C37.122-1993, C37.122.1-1993

(3) Any gas section of the gas-insulated substation assembly that can be isolated from the system by internal or external means. (SWG/PE) C37.100-1992

compass (navigation aid terms) An instrument for indicating a horizontal reference direction relative to the earth. (AES/GCS) 172-1983w

compass bearing (navigation aid terms) Bearing relative to compass north. (AES/GCS) 172-1983w

compass-controlled directional gyro A device that uses the earth's magnetic field as a reference to correct a directional gyro. *Note:* The direction of the earth's field is sensed by a remotely located compass that is connected electrically to the gyro. (EEC/PE) [119]

compass course (navigation aid terms) Course relative to compass north. (AES/GCS) 172-1983w

compass declinometer *See:* declinometer.

compass deviation *See:* magnetic deviation.

compass heading (navigation aid terms) Heading relative to compass north. (AES/GCS) 172-1983w

compass locator *See:* nondirectional beacon.

compass north (navigation aid terms) The direction north as indicated by a magnetic compass. (AES/GCS) 172-1983w

compass repeater (navigation aid terms) That part of a remote-indicating compass system which repeats, at a distance, the indications of the master compass. (AES/GCS) 172-1983w

compass rose (navigation aid terms) A compass used to assist in aircraft magnetic compass compensation. (AES/GCS) 172-1983w

compatibility (1) (696 interface devices) (general system) The degree to which devices may be interconnected and used without modification, when designed to conform to IEEE Std 696-1983. (MM/C) 696-1983w

(2) (microcomputer system bus) The degree to which devices may be interconnected and used without modification, when designed as defined in IEEE Std 796-1983. (C/MM) 796-1983r

(3) (color television) The property of a color television system that permits substantially normal monochrome reception of the transmitted signal by typical unaltered monochrome receivers. (BT/AV) 201-1979w

(4) (programmable instrumentation) The degree to which devices may be interconnected and used, without modification, when designed as defined throughout IEEE Std 488.1-1987 (for example, mechanical, electrical, or functional compatibility). (IM/AIN) 488.1-1987r

(5) (A) (software) The ability of two or more systems or components to perform their required functions while sharing the same hardware or software environment. **(B) (software)** The ability of two or more systems or components to exchange information. *See also:* interoperability. (C) 610.12-1990

(6) (SBX bus) The degree to which devices may be interconnected and used without modification. (C/MM) 959-1988r

(7) (STEbus) The degree to which boards may be interconnected and used without modification when designed according to the specifications contained within IEEE Std 1000-1987. (C/MM) 1000-1987r

compatibility interfaces (1) (medium attachment units and repeater units) The medium dependent interface (MDI) coaxial cable interface and the attachment unit interface (AUI) branch cable interface, the two points at which hardware compatibility is defined to allow connection of independently designed and manufactured components to the baseband transmission system. (LM/C) 8802-3-1990s

(2) The Medium Dependent Interface (MDI) cable, the Attachment Unit Interface (AUI) branch cable, and the Media Independent Interface (MII); the three points at which hardware compatibility is defined to allow connection of independently designed and manufactured components to a baseband transmission medium. (C/LM) 802.3-1998

Compatibility Mode An asynchronous, byte-wide forward (host-to-peripheral) channel with data and status lines used according to their original definitions. Compatibility Mode is backward compatible with many existing devices, including the PC parallel port, and is the base mode common to all compliant interfaces. (C/MM) 1284-1994

Compatibility Mode forward data transfer phase Begins when the host asserts nStrobe and ends following data hold time and nStrobe de-assertion. (Note that the host is not free to send the next data byte until the peripheral acknowledges the transfer using nAck.) The host may not initiate negotiation to a new operating mode until the interface returns to Compatibility Mode forward idle phase. (C/MM) 1284-1994

Compatibility Mode forward idle phase When the interface is in Compatibility Mode with no data transfer in progress. The host may initiate a data transfer in Compatibility Mode or may initiate negotiation to a new operating mode. (C/MM) 1284-1994

compatible Pertaining to a computer system or system component that is capable of handling data and programs intended for use with some other system or component. *See also:* downward compatible; upward compatible. (C) 610.10-1994w

compatible device A device that supports any of a specified range of popular variants of the "Centronics" interface. Compatible devices will interoperate with compliant devices in Compatibility Mode only. (C/MM) 1284-1994

compelled data transfer protocol A technology-independent transfer mechanism in which the slave is compelled to provide a response before the master proceeds to the next transfer. (C/BA) 10857-1994, 1014.1-1994w, 896.4-1993w

compensated control system (control systems for steam turbine-generator units) An interconnected system that controls two or more variables (speed, load, pressure, etc.) with compensation designed to minimize the interaction between the controlled variables. (PE/EDPG) 122-1985s

compensated-loop direction-finder A direction-finder employing a loop antenna and a second antenna system to compensate polarization error. *See also:* radio receiver. (EEC/PE) [119]

compensated repulsion motor A repulsion motor in which the primary winding on the stator is connected in series with the rotor winding via a second set of brushes on the commutator in order to improve the power factor and commutation. (PE) [9]

compensated semiconductor (charged-particle detectors) (germanium gamma-ray detectors) (x-ray energy spectrometers) A semiconductor in which one type of impurity or imperfection (for example, donor) partially cancels the electric effects of the other type of impurity or imperfection (for example, acceptor). (IM/NPS/ED/HFIM/NID) 314-1971w, 325-1996, 300-1988r, 301-1976s, 216-1960w

compensated series-wound motor A series-wound motor with a compensating-field winding. The compensating-field winding and the series-field winding may be combined into one field winding. *See also:* asynchronous machine. (PE) 224-1965w, [9]

compensating-field winding (rotating machinery) Conductors embedded in the pole shoes and their end connections. It is connected in series with the commutating-field winding and the armature circuit. *Note:* A compensating-field winding supplements the commutating-field winding, and together they function to assist the reversal of current in the armature coils for successful commutation. *See also:* asynchronous machine. (PE/EEC) [119]

compensating-rope sheave switch A device that automatically causes the electric power to be removed from the elevator driving-machine motor and brake when the compensating sheave approaches its upper or lower limit of travel. *See also:* hoistway. (EEC/PE) [119]

compensation (control system feedback) A modifying of supplementary action (also, the effect of such action) intended to improve performance with respect to some specified characteristic. *Note:* In control usage, this characteristic is usually the system deviation. Compensation is frequently qualified as series, parallel, feedback, etc., to indicate the relative position of the compensating element. *See also:* equalization; feedback control system. (PE/EDPG) 421-1972s, [3]

compensation theorem States that if an impedance is inserted in a branch of a network, the resulting current increment produced in any branch in the network is equal to the current that would be produced at that point by a compensating voltage, acting in series with the modified branch, whose values is, where I is the original current that flowed where the impedance was inserted before the insertion was made. (EEC/PE) [119]

compensator (1) (rotating machinery) An element or group of elements that acts to modify the functioning of a device in accordance with one or more variables. *See also:* asynchronous machine. (PE) [9]
(2) (radio direction-finders) That portion of a direction-finder that automatically applies to the direction indication all or a part of the correction for the deviation. *See also:* radio receiver. (EEC/PE) [119]
(3) (excitation systems) A feedback element of the regulator that acts to compensate for the effect of a variable by modifying the function of the primary detecting element. *Notes:* 1. Examples are reactive current compensator and active current compensator. A reactive current compensator is a compensator that acts to modify the function of a voltage regulator in accordance with reactive current. An active current compensator is a compensator that acts to modify the function of a voltage regulator in accordance with active current. 2. Historically, terms such as equalizing reactor and cross-current compensator have been used to describe the function of a reactive compensator. These terms are deprecated. 3. Reactive compensators are generally applied with generator voltage regulators to obtain reactive current sharing among generators operating in parallel. They function in the following two ways.

1) Reactive droop compensation is the more common method. It creates a droop in generator voltage proportional to reactive current and equivalent to that which would be produced by the insertion of a reactor between the generator terminals and the paralleling point.
2) Reactive differential compensation is used where droop in generator voltage is not wanted. It is obtained by a series differential connection of the various generator current transformer secondaries and reactive compensators. The difference current for any generator from the common series current creates a compensating voltage in the input to the particular generator voltage regulator which acts to modify the generator excitation to reduce to minimum (zero) its differential reactive current.
3) Line drop compensators modify generator voltage by regulator action to compensate for the impedance drop from the machine terminals to a fixed point. Action is accomplished by insertion within the regulator input circuit of a voltage equivalent to the impedance drop. The voltage drops of the resistance and reactance portions of the impedance are obtained, respectively, in per unit quantities by an active compensator and a reactive compensator.

(PE/EDPG) 421-1972s

(4) (as applied to relaying) A transducer with an air-gapped core that produces an output voltage proportional to input current. The voltage modifies (or *compensates*) the voltage applied to the relay. (SWG/PE) C37.100-1992

compensatory leads Connections between an instrument and the point of observation so contrived that variations in the properties of leads, such as variations of resistance with temperature, are so compensated that they do not affect the accuracy of the instrument readings. *See also:* auxiliary device to an instrument. (EEC/PE) [119]

compensatory wiring techniques (coupling in control systems) Those writing techniques which result in a substantial cancellation or counteracting of the effects of rates of change of electric or magnetic fields, without actually obstructing or altering the intensity of the fields. If the signal wires are considered to be part of the control circuit, these techniques change the susceptibility of the circuit. Example: twisting of signal and return wires associated with a susceptable instrument so as to cancel the voltage difference between wires caused by an existing varying magnetic field. *See also:* suppressive wiring techniques; barrier wiring techniques. (IA/ICTL) 518-1982r

competent person One who, because of training, experience, and authority is capable of identifying and correcting hazardous or dangerous conditions in the fall arrest system or any component thereof under consideration, as well as its application and use with related equipment. (T&D/PE) 1307-1996

competitor (1) (NuBus) A master that participates in a particular arbitration contest. (C/MM) 1196-1987w
(2) A module actively participating in the current control acquisition cycle of the arbitration process. (C/BA) 10857-1994, 896.3-1993w, 896.4-1993w

compile (software) To translate a computer program expressed in a high-order language into its machine language equivalent. *Contrast:* interpret; assemble; decompile. (C) 610.12-1990

compile-and-go (software) An operating technique in which there are no stops between the compiling, linking, loading, and execution of a computer program. (C) 610.12-1990

compiler (software) A computer program that translates programs expressed in a high-order language into their machine language equivalents. *Contrast:* interpreter; assembler. *See also:* incremental compiler; cross-compiler; root compiler. (C) 610.12-1990

compiler code (software) Computer instructions and data definitions expressed in a form that can be recognized and processed by a compiler. *Contrast:* interpretive code; machine code; assembly code. (C) 610.12-1990

compiler directive source statements Source statements that define macros, or labels, or direct the compiler to insert external source statements (for example, an *include* statement), or direct conditional compilation, or are not described by one of the other type attributes. (C/SE) 1045-1992

compiler generator (software) A translator or interpreter used to construct part or all of a compiler. *Synonym:* metacompiler. (C) 610.12-1990

compiler specification language A specification language used to develop compilers. *See also:* LEX. (C) 610.13-1993w

complement (1) (mathematics of computing) A numeral derived from a given numeral by a specified subtraction rule. Often used to represent the negative of the number represented by the given numeral. *See also:* radix complement; diminished-radix complement. (C) 1084-1986w
(2) (image processing and pattern recognition) All points in an image that do not belong to a given subset of the image.

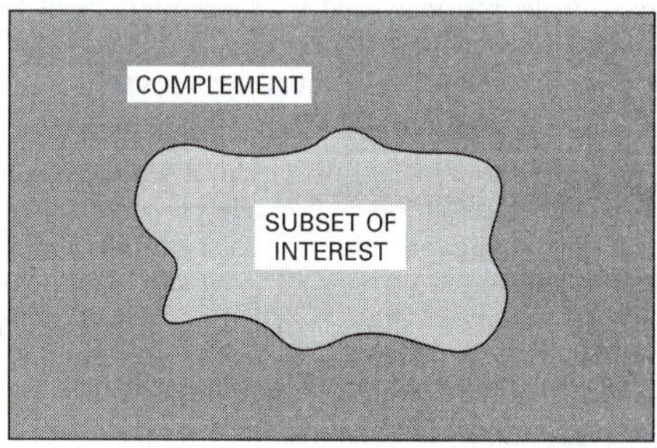

complement

(C) 610.4-1990w

(3) (test pattern language) Another number in which each zero bit has been replaced by a one and each one bit has been replaced by a zero. Ones complement is formed by interchanging all ones and zeros. This is equivalent to logical inversion. (TT/C) 660-1986w

complementary commutation (circuit properties) (self-commutated converters) Commutation occurs from one to the other of a complementary pair of principal switching branches arranged as a two-pulse group that conduct in alternate but not necessarily equal time intervals. The commutation may be direct or indirect. *Note:* An example of a converter employing complementary commutation is given in the figures below.

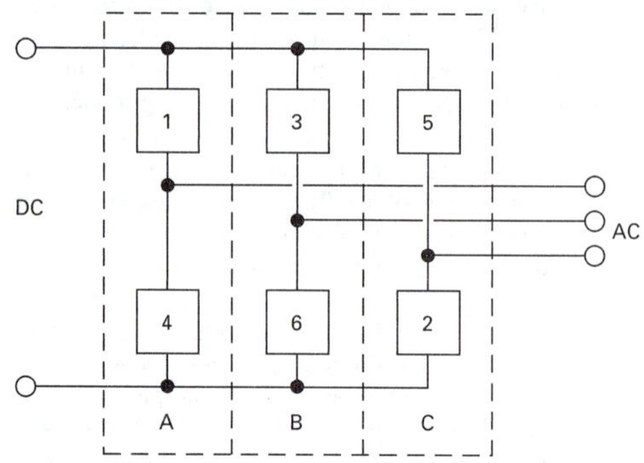

Note: The principle switching branches 1–6 are numbered in the order in which they begin conduction.

a) Three 2-pulse commutating groups: A, B, C

complementary commutation

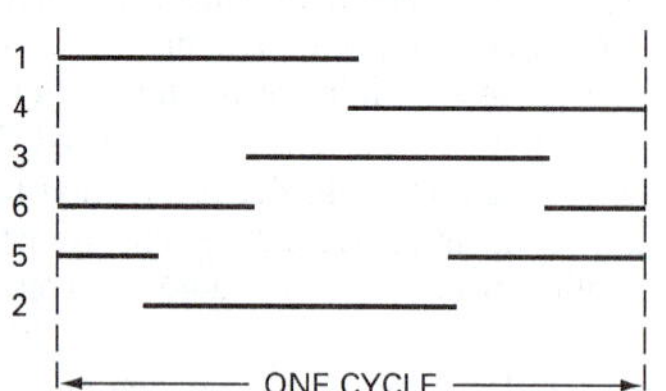

b) Conducting intervals of principle switching branches 1-6

(IA/SPC) 936-1987w

complementary function (automatic control) The solution of a homogeneous differential equation, representing a system or element, which describes a free motion. (PE/EDPG) [3]

complementary functions Two driving-point functions whose sum is a positive constant. (CAS) 156-1960w

complementary metal-oxide semiconductor A semiconductor technology in which circuits are composed of paired NMOS and PMOS devices; characterized by extremely low power dissipation when not changing states. (C) 610.10-1994w

complementary operation Two Boolean operations are complementary if the result of one operation is the negation of the result of the other, for all combinations of operands. For example, the AND and NAND operations are complementary. *Contrast:* dual operation. (C) 1084-1986w

complementary operator *See:* NOT.

complementary tracking (power supplies) A system of interconnection of two regulated supplies in which one (the master) is operated to control the other (the slave). The slave supply voltage is made equal (or proportional) to the master supply voltage and of opposite polarity with respect to a common point. (AES) [41]

complementary wavelength (1) (television) (color) The wavelength of a spectrum light that, when combined in suitable proportions with the light considered, yields a match with the specified achromatic light. *See also:* dominant wavelength. (BT/AV) 201-1979w
(2) (illuminating engineering) (of a light) The wavelength of radiant energy of a single frequency that, when combined in suitable proportion with the light, matches the color of a reference standard. (EEC/IE) [126]

complementation The process of obtaining a complement. (C) 1084-1986w

complement base The numeral from which a given numeral is subtracted to obtain its complement. (C) 1084-1986w

complemented representation A positional notation system in which negative numbers are represented by their complements and positive numbers are represented in their usual form. *See also:* twos-complement notation. (C) 1084-1986w

complementer A device whose output data are a representation of the complements of the numbers represented by its input data. (C) 610.10-1994w

complement on *n* *See:* radix complement.

complement on nine *See:* nines complement.

complement on *n* − 1 *See:* diminished-radix complement.

complement on one *See:* ones complement.

complement on ten *See:* tens complement.

complement on two *See:* twos complement.

complete binary tree A complete tree of order 2. *Note:* The nodes in the tree can be read sequentially from left to right; top to bottom. *Synonym:* full binary tree.

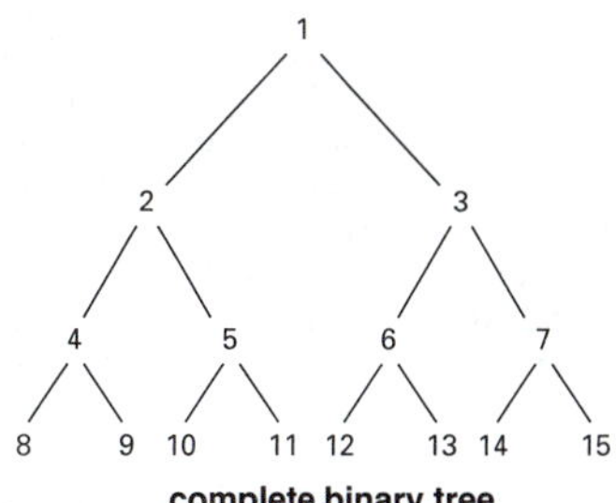

complete binary tree

(C) 610.5-1990w

complete carry **(1)** A carry process in which a carry resulting from addition of carries is allowed to propagate. Contrasted with partial carry. *See also:* carry. (C) 162-1963w **(2) (mathematics of computing)** A carry process in which the carry digits are transferred and processed as they occur. *Contrast:* partial carry. (C) 1084-1986w

complete checkback message Message from the initiating end is interpreted by the receiving end. A new message is sent to the initiating end to verify error-free transmission and proper interpretation of the message. (SWG/PE/SUB) C37.100-1992, C37.1-1987s

complete cluster *Contrast:* incomplete cluster. *See also:* total cluster. (C/SE) 1320.2-1998

complete connection An association of channels, switching systems, other functional units, and telephone sets set up to provide means to allow telephone users to converse. (COM/TA) 823-1989w

completed call (telephone switching systems) An answered call that has been released. (COM) 312-1977w

complete diffusion (illuminating engineering) That in which the diffusing medium completely redirects the incident flux by scattering, that is, no incident flux can remain in an image-forming state. (EEC/IE) [126]

complete failure **(1)** Failure of equipment that is both sudden and total. *Synonym:* catastrophic failure. (PE/NP) 933-1999 **(2)** *See also:* failure.

complete ICOM code A diagram feature reference in which dot notation joins an ICOM code to a diagram reference. (C/SE) 1320.1-1998

completely immersed bushing A bushing in which both ends are intended to be immersed in an insulating medium such as oil or gas. (PE/TR) C57.19.03-1996

completely polarized wave A wave with no randomly polarized content. (AP/PROP) 211-1997

complete operating test equipment (test, measurement, and diagnostic equipment) Equipment together with the necessary detail parts, accessories, and components, or any combination thereof, required for the testing of a specified operational function. (MIL) [2]

complete outage state The component or unit is completely de-energized or is connected so that it is not serving any of its functions within the power system. (PE/PSE) 859-1987w

complete reference designation (electric and electronics parts and equipment) A reference designation that consists of a basic reference designation and, as prefixes, all the reference designations that apply to the subassemblies or assemblies within which the item is located, including those of the highest level needed to designate the item uniquely. The reference designation for a unit consists of only a number. (GSD) 200-1975w

complete tree A tree of order *n* in which each node has exactly *n* subtrees. *Synonym:* full tree. *See also:* heap. (C) 610.5-1990w

completion code A code communicated to a job stream processor by a batch program to influence the execution of succeeding steps in the input stream. (C) 610.12-1990

completion of a call The execution of a construct or entity is complete when the end of that execution has been reached, or when a transfer of control causes it to be abandoned. Completion due to reaching the end of execution, or due to the transfer of control of an `exit`, `return`, `goto`, `requeue`, or of the selection of a `terminate` alternative is normal completion. Completion is abnormal when control is transferred out of a construct due to abort or the raising of an exception. (C) 1003.5-1999

completion queue A DMA queue that is used primarily to pass I/O transaction-completion messages. (C/MM) 1212.1-1993

complex capacitivity *See:* relative complex dielectric constant.

complex conductivity For isotropic media, at a particular point, and for a particular frequency, the ratio of the complex amplitude of the total electric current density to the complex amplitude of the electric field strength. *Note:* The electric field strength and total current density are both expressed as phasors, with the latter composed of the conduction current density plus the displacement current density. (AP/ANT) 145-1993

complex data Numeric data used to represent complex numbers. (C) 610.5-1990w

complex data structure *See:* nonprimitive data structure.

complex dielectric constant The complex permittivity of a physical medium in ratio to the permittivity of free space. *See also:* relative complex dielectric constant. (AP/ANT) 145-1993

complex electrical ground and test device A device with two terminal sets and a manually operated terminal selector switch for connecting either terminal set to the device ground connection system through a power-operated ground-making switch, complete with necessary isolation barriers and suitable interlocking. Voltage test ports may be provided. (SWG/PE) C37.20.6-1997

complex-instruction-set computer A computer with a very expansive and robust instruction set, incorporating several types of addressing modes and varying length instruction words. *Note:* Such instructions are usually stored in microcode. (C) 610.10-1994w

complexity **(1) (A) (software)** The degree to which a system or component has a design or implementation that is difficult to understand and verify. *Contrast:* simplicity. **(B) (software)** Pertaining to any of a set of structure-based metrics that measure the attribute in definition (A). (C) 610.12-1990 **(2) (magnetic core testing)** Under stated conditions, the complex quotient of vectors representing induction and field strength inside the core material. One of the vectors is made to vary sinusoidally and the other referenced to it. (MAG) 393-1977s

complex number A number consisting of a real part (a) and an imaginary part (b), expressed in the form $a + bi$, where $i^2 = -1$. (C) 610.5-1990w, 1084-1986w

complex permeability (μ) A macroscopic material property of a medium that relates the magnetic flux density, $\vec{B}$, to the magnetic field, $\vec{H}$, in the medium. For a monochromatic wave in a linear medium, that relationship is described by the (phasor) equation:

$$\vec{B} = \overline{\mu} \cdot \vec{H}$$

where μ', the complex permeability, is a tensor that is generally frequency dependent. For an isotropic medium, the tensor reduces to a complex scalar:

$$\mu = \mu' - j\mu''$$

where $\overline{\mu}$ is the real part of the permeability and μ'' accounts for losses.

(AP/PROP) 211-1990s

complex permittivity (1) For isotropic media, the ratio of the complex amplitude of the electric displacement density to the complex amplitude of the electric field strength. *See also:* relative complex dielectric constant. (AP/ANT) 145-1993
(2) A macroscopic material property of the medium that relates the electric field, $\vec{E}$ to the electric flux density, $\vec{D}$, in the medium. For a monochromatic wave in a linear medium, that relationship is described by the (phasor) equation:

$$\vec{D} = \overline{\varepsilon} \cdot \vec{E}$$

where $\overline{\varepsilon}$, the complex permittivity, is a complex-valued tensor, generally frequency dependent. For an isotropic medium, the tensor reduces to a complex scalar:

$$\varepsilon = \varepsilon' - j\varepsilon''$$

where ε' is the real part of the permittivity and ε'' accounts for losses. (AP/PROP) 211-1990s

complex plane (automatic control) A plane defined by two perpendicular reference axes, used for plotting a complex variable or functions of this variable, such as a transfer function. (PE/EDPG) [3]

complex polarization ratio For a given field vector at a point in space, the ratio of the complex amplitudes of two specified orthogonally polarized field vectors into which the given field vector has been resolved. *Note:* For these amplitudes to define definite phase angles, particular unitary vectors (basis vectors) must be chosen for each of the orthogonal polarizations. *See also:* plane wave; polarization vector.
(AP/ANT) 145-1993

complex power *See:* phasor power.

complex refractive index A dimensionless complex quantity, characteristic of a medium and so defined that its real part is the ratio of the phase velocity in free space to the phase velocity in the medium. The product of the imaginary part of the refractive index and the free space propagation constant is the attenuation constant in the medium.
(AP/PROP) 211-1990s

complex target A target composed of more than one scatterer within a single radar resolution cell. A target may be both complex and distributed. *See also:* distributed target.
(AES) 686-1997

complex tone (A) A sound containing simple sinusoidal components of different frequencies. **(B)** A sound sensation characterized by more than one pitch. (SP) [32]

complex variable (automatic control) A convenient mathematical concept having a complex value, that is having a real part and an imaginary part. *Note:* In control systems, the pertinent independent variable is a generalized frequency $s = \sigma + j\omega$ used in the Laplace transform. (PE/EDPG) [3]

complex waveforms *See:* waveforms produced by continuous time superposition of simpler waveforms; waveforms produced by operations on waveforms; waveforms produced by noncontinuous time superposition of simpler waveforms; combinations of pulses and waveforms; waveforms produced by magnitude superposition.

compliance A property reciprocal to stiffness. *See also:* control system, feedback; feedback control system.
(IA/ICTL/IAC) [60]

compliance extension (power supply) A form of master/slave interconnection of two or more current-regulated power supplies to increase their compliance voltage range through series connection. *See also:* compliance voltage. (AES) [41]

compliance voltage (power supplies) The output voltage of a direct-current power supply operating in constant-current mode. *Note:* The compliance range is the range of voltages needed to sustain a given value of constant current throughout a range of load resistances. (AES) [41]

compliant device A device that supports either the Level 1 or Level 2 electrical interface, plus Compatible and Nibble Mode operation, as well as the negotiation phases necessary to transition between these two modes.
(C/MM) 1284-1994

Component An instance of a subclass of `IEEE1451_Component`. (IM/ST) 1451.1-1999

component (1) (reliability data for pumps and drivers, valve actuators, and valves) The largest entity of hardware for which data are most generally collected and expected to be reliable (for example, pump with motor, valve with operator, amplifier, pressure transmitter). It is generally an off-the-shelf item procured by the system designer as a basic building block for his system. It should be distinguished from seals, materials, nuts, bolts, and other piece parts from which the component is made. (PE/NP) 500-1984w
(2) (seismic design of substations) The devices and equipment which are assembled at the erection site, or readily removed or accessed for maintenance, and which perform a function (for example, power circuit breakers, disconnect switches, relays, sensors). (SUB/PE) 693-1984s
(3) (unique identification in power plants) A part or assembly of parts that is viewed as an entity for purposes of design, operation, and reporting. (PE/EDPG) 803-1983r
(4) (unique identification in power plants and related facilities) A part or assembly of parts considered an entity for purpose of design, operation, and reporting.
(PE/EDPG) 804-1983r
(5) (software) One of the parts that make up a system. A component may be hardware or software and may be subdivided into other components. *Notes:* 1. The terms "module," "component," and "unit" are often used interchangeably or defined to be subelements of one another in different ways depending upon the context. The relationship of these terms is not yet standardized. (C) 610.12-1990, 610.10-1994w
(6) Any part, assembly, or subdivision of a computer, such as a resistor, amplifier, power supply or rack.
(C) 610.10-1994w
(7) (electrical transmission facilities) A device that performs a major operating function and that is regarded as an entity for purposes of recording and analyzing data on outage occurrences. *Notes:* 1. Some examples of components are line sections, transformers, ac/dc converters, series capacitors or reactors, shunt capacitors or reactors, circuit breakers, line protection systems, and bus sections. 2. Sometimes it is necessary to subdivide a line section into segments to allow proper calculation of failure rates and exposure data. For example, if a line section is composed of an overhead line segment and an underground line segment, failure and exposure data for each line segment may be recorded separately.
(PE/PSE) 859-1987w
(8) (electric utility power systems) A part within or associated with a transformer that is viewed as an entity. This is usually a replaceable part; for example, main winding, tap changer motor, etc. (PE/TR) C57.117-1986r
(9) A piece of electrical or mechanical equipment, a line or circuit, or a section of a line or circuit, or a group of items that is viewed as an entity for the purposes of reliability evaluation. (IA/PSE) 493-1997
(10) A model, simulation, or database used or considered for use in a Distributed Interactive Simulation (DIS) exercise.
(C/DIS) 1278.4-1997
(11) A piece of equipment, a line or circuit, or a section of a line or circuit, or a group of items that is viewed as an entity for purposes of reliability evaluation. (IA) 399-1997

component assembly (1) (unique identification in power plants) An assembly of components, physically contiguous, which is viewed as a single entity for purposes of procurement, for example, boric-acid control panel. (PE/EDPG) 803-1983r
(2) (unique identification in power plants and related facilities) An assembly of contiguous components, considered as a single entity for purpose of procurement, that is, boric acid control panel. (PE/EDPG) 804-1983r

component data element A component of a data structure. *Synonym:* element. (C) 610.5-1990w

component failure Malfunctions in the system hardware (e.g., failures in equipment, electronics, input/output device, distribution system), software (e.g., due to deadlock conditions, exceptions, error conditions), or media that are not precipitated through design flaws. Component failures could be manifested because of faulty equipment, unanticipated system events, or environmental effects (e.g., power surge, humidity, heat), and could result in denial of service conditions. However, failure of components that implement security mechanisms could result in the violation of the system security policy through unauthorized disclosure or modification of information, unauthorized receipt of services, or denial of service to legitimate users or critical functions. (C/BA) 896.3-1993w

component function (1) (unique identification in power plants) The action performed by a component within a system. (PE/EDPG) 803-1983r
(2) (unique identification in power plants and related facilities) The primary function performed by a component (element) within a system. (PE/EDPG) 804-1983r

component function identifier (1) (unique identification in power plants and related facilities) A one to four character alpha-numeric code that identifies the function the component performs within the system. (PE/EDPG) 804-1983r
(2) A one to four (1 to 4) character alphanumeric code that identifies the function that will be performed by a component. (Reference codes have been established for the basic functions performed by the principal components currently in use in nuclear and fossil-fueled power plants and related facilities.). (PE/EDPG) 803.1-1992

Component Group An instance of the class `IEEE1451_ComponentGroup` or of a subclass thereof. (IM/ST) 1451.1-1999

component hazard (reliability data) The instantaneous failure rate of a component or its conditional probability of failure versus time. (PE/NP) 500-1977s

component profile A profile that is made up of a formally defined subset of a single standard. (C/PA) 14252-1996

components (1) (safety systems equipment in nuclear power generating stations) Items from which equipment is assembled (for example, attachments, bearings, bolts, capacitors, connectors, governors, inspection access ports, instrument sensors, locking devices, position indicators, resistors, seals, sight glasses, springs, switches, transistors, tubes, wires, etc. *Note:* Certain items, for example, instrument sensors, may satisfy the definition of the term component or the term equipment as used in IEEE Std 627-1980. Where such items are included within defined boundaries of equipment items, they are correctly referred to as components. Where such items are installed outside of defined boundaries for equipment items and perform independent functions, they are correctly referred to as equipment. (PE/NP) 627-1980r
(2) (switchgear assemblies for Class 1E applications in nuclear power generating stations) Items from which the switchgear assemblies are made (for example, power circuit breakers, instrument transformers, protective relays, control switches, primary insulation, etc.). (SWG/PE) C37.100-1992, C37.82-1971s
(3) (accident monitoring instrumentation) Discrete items from which a system is assembled. (PE/NP) 497-1981w
(4) (electric heat tracing systems) Items from which a system is assembled; for example, resistors, capacitors, wires, connectors, transistors, switches, etc. (PE/EDPG/NP) 622B-1988r, 650-1979s, 622A-1984r, 323-1974s
(5) (nuclear power generating station) Items from which equipment is assembled. (For example, a component is a resistor, capacitor, wire, connector, spring, terminal block, bus support, etc.). (PE/NP) 649-1980s
(6) Items from which the equipment is assembled. (For motors, typical components include slator coils, rotor bars, bearings, bolts, capacitors, internal thermal overload relays, connectors, instrument sensors, locking devices, seals, sight glasses, springs, switches, etc.). (PE/NP) 334-1994r
(7) Discrete items from which a system is assembled. *Note:* Examples of components are: wires, transistors, switches, motors, relays, solenoids, pipes, fittings, pumps, tanks, valves, computer programs, computer hardware, or computer firmware. (PE/NP) 603-1998
(8) Items from which the system is assembled. (PE/NP) 933-1999

component side By convention, the side of the module seen furthest from row A of the connector. On single-sided modules, this is the side populated with circuit components. This is the right side when looking at an IEEE 1101.1 system through the front door. (C/MM) 1101.2-1992

component standard (1) (software) A standard that describes the characteristics of data or program components. (C) 610.12-1990
(2) One standard, within a set of standards, that is developed in accordance with the architecture, terminology, guidelines, and requirements set forth in the base document of the same set of standards, and that provides a detailed definition of a component part of the architecture. (ATLAS) 1232-1995

component testing (1) (software) Testing of individual hardware or software components or groups of related components. *Synonym:* module testing. *See also:* system testing; integration testing; interface testing; unit testing. (C) 610.12-1990
(2) Testing conducted to verify the correct implementation of the design and compliance with program requirements for one software element (e.g., unit, module) or a collection of software elements. (C/SE) 1012-1998

composite bushing (outdoor apparatus bushings) A bushing in which the major insulation consists of several coaxial layers of different insulation materials. (PE/TR) 21-1976

composite cable (communication practice) A cable in which conductors of different gauges or types are combined under one sheath. *Note:* Differences in length of twist are not considered here as constituting different types. *See also:* cable. (PE/EEC) [119]

composite color picture signal (National Television System Committee color television) The electric signal that represents complete color picture information and all sync signals. (BT/AV) 201-1979w

composite color signal (color television) The color-picture signal plus blanking and all synchronizing signals. (BT/AV) 201-1979w

composite color sync [National Television System Committee (NTSC) color television] The signal comprising all the sync signals necessary for proper operations of a color receiver. *Note:* This includes the deflection sync signals to which the color sync signal is added in the proper time relationship. (BT/AV) 201-1979w

composite conductor A composite conductor consists of two or more strands consisting of two or more materials. *See also:* conductor. (T&D/PE) [10]

composite controlling voltage (electron tube) The voltage of the anode of an equivalent diode combining the effects of all individual electrode voltages in establishing the space-charge-limited currents. *See also:* excitation. (ED) 161-1971w, [45]

composite data element A data element that contains two or more data elements that can be referred to either collectively or individually; for example, a data element named "date of birth" containing data elements "year," "month," and "day." *Synonyms:* molecular data element; data chain. *Contrast:* atomic data element. *See also:* data aggregate. (C) 610.5-1990w

composited circuit A circuit that can be used simultaneously for telephony and direct-current telegraphy or signaling, separation between the two being accomplished by frequency discrimination. *See also:* transmission line. (EEC/PE) [119]

composite error The maximum deviation of the output data from a specified output function. Composite error is due to the composite effects of hysteresis, resolution, nonlinearity, nonrepeatability, and other uncertainties in the output data. It is generally expressed as a percentage of half the output span. (AES/GYAC) 528-1994

composite key A key comprised of two or more attributes. (C/SE) 1320.2-1998

composite lens characteristic A modification of an impedance or mho characteristic in which the operating area on an *R-X* diagram is inherently restricted in the plus and minus *R* directions. The common area between two overlapping circles produces such a characteristic. (SWG/PE) C37.100-1992

composite level (measuring the performance of tone address signaling systems) In two-tone signaling systems, the total power of the two tones comprising a specific signal present condition. (COM/TA) 752-1986w

composite overhead groundwire with optical fibers (OPGW) Concentric-lay-stranded composite conductor for use as overhead groundwire with telecommunication capability. The conductor is constructed with a central optical fiber core surrounded by helically laid aluminum-clad wires, aluminum alloy wires, galvanized steel wires, or combinations thereof. (T&D/PE) 524-1992r

composite picture signal (television) The signal that results from combining a blanked picture signal with the asynchronizing signal. *See also:* television. (BT/AV) [34]

composite plate (electroplating) An electrodeposit consisting of two or more layers of metals deposited separately. *See also:* electroplating. (PE/EEC) [119]

composite pulse (navigation aid terms) (pulse navigational systems) A pulse composed of a series of overlapping pulses received from the same signal source but by way of different paths. (AES/GCS) 172-1983w

composite set An assembly of apparatus designed to provide one end of a composited circuit. (EEC/PE) [119]

composite signal A signal that is composed of both ac and dc components. (PEL) 1515-2000

composite signaling (telephone switching systems) A form of polar-duplex signaling capable of simultaneously serving a number of circuits using low-pass filters to separate the signaling currents from the voice currents. (COM) 312-1977w

composite supervision The use of a composite signaling channel for transmitting supervisory signals between two points in a connection. (EEC/PE) [119]

composite tomato characteristic A modification of an impedance or mho characteristic in which the operating area on an *R-X* diagram is inherently expanded in the plus and minus direction. The total area of two overlapping circles produces such a characteristic. (SWG/PE) C37.100-1992

composite triple beat distortion The combination of all possible third-order beat frequencies ($F_1 \pm F_2 \pm F_3$) that occurs within a channel of the cable plant. *See also:* intermodulation distortion. (LM/C) 802.7-1989r

composite type A data type each of whose members is composed of multiple data items. For example, a data type called PAIRS whose members are ordered pairs (*x,y*). *Contrast:* atomic type. (C) 610.12-1990

composite video signal The complete video signal. For monochrome systems, it comprises the picture, blanking, and synchronizing signals. For color systems it includes additional color synchronizing signals and color picture information. (LM/C) 802.7-1989r

composite waveform (pulse terminology) A waveform that is, or that for analytical or descriptive purposes is treated as, the algebraic summation of two or more waveforms.

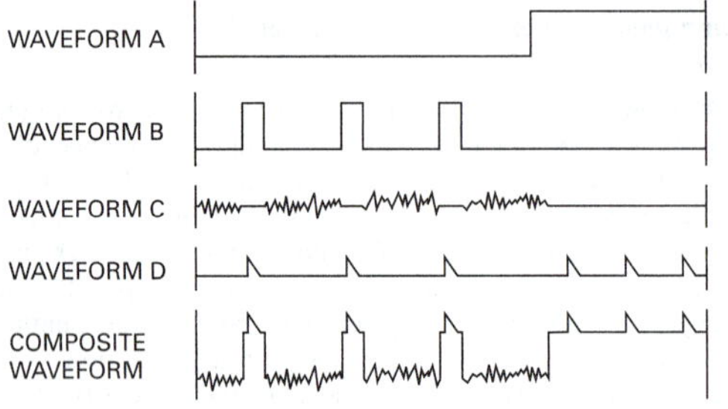

composite waveform

(IM/WM&A) 194-1977w

composite widget A parent widget that physically contains other widgets. (C) 1295-1993w

compound (A) (rotating machinery) A definite substance resulting from the combination of specific elements or radicals in fixed proportions: distinguished from mixture. **(B) (rotating machinery)** The intimate admixture of resin with ingredients such as fillers, softeners, plasticizers, catalysts, pigments, or dyes. *See also:* rotor. (PE) [9]

compound cartridge An ordered set of cartridges that may be treated atomically. (C/SS) 1244.1-2000

compound circular horn antenna A horn antenna of circular cross section with two or more abrupt changes of flare angle or diameter. (AP/ANT) 145-1993

compound-filled (grounding device) (reactor, transformer) Having the coils/windings encased in an insulating fluid that becomes solid or remains slightly plastic at normal operating temperatures. *See also:* instrument transformer; reactor. (SPD/PE) 32-1972r

compound-filled bushing (outdoor electric apparatus) A bushing in which the space between the inside surface of the porcelain/weather casing and the major insulation (or conductor where no major insulating is used) is filled with compound. (PE/TR) 21-1976

compound filled joints (power cable joints) Joints in which the joint housing is filled with an insulating compound that is non-fluid at normal operating temperatures. (PE/IC) 404-1986s

compound-filled transformer (power and distribution transformers) A transformer in which the windings are enclosed with an insulating fluid which becomes solid, or remains slightly plastic, at normal operating temperatures. *Note:* The shape of the compound-filled transformer is determined in large measure by the shape of the contain or mold used to contain the fluid before solidification. (PE/TR) C57.12.80-1978r

compound horn antenna *See:* compound circular horn antenna; compound rectangular horn antenna.

compounding curve (direct-current generator) A regulation curve of a compound-wound direct-current generator. *Note:* The shunt field may be either self or separately excited. *See also:* direct-current commutating machine. (EEC/PE) [119]

compound interferometer system An antenna system consisting of two or more interferometer antennas whose outputs are combined using nonlinear circuit elements such that grating lobe effects are reduced. (AP/ANT) 145-1993

compound key A candidate key consisting of more than one attribute. (C) 610.5-1990w

compound list *See:* list structure.

compound microstrip A microstrip line in which the substrate consists of two or more layers of different electromagnetic properties. (MTT) 1004-1987w

compound rectangular horn antenna A horn antenna of rectangular cross section in which at least one pair of opposing sides has two or more abrupt changes of flare angle or spacing. (AP/ANT) 145-1993

compound source-rectifier exciter (1) (excitation systems for synchronous machines) An exciter whose energy is derived from the currents and potentials of the ac terminals of the synchronous machine and converted to direct current by rectifiers. The exciter included the power transformers (current and potential), reactors, and rectifiers which may be either noncontrolled or controlled, including gate circuitry. It is exclusive of input control elements. (PE/EDPG) 421.1-1986r **(2) (synchronous machines)** An exciter whose energy is derived from the currents and potentials of the alternating current terminals of the synchronous machine and converted to direct current by rectifiers. *Notes:* 1. The exciter includes the power transformers (current and potential), power reactors, and power rectifiers which may be either noncontrolled or controlled, including gate circuitry. 2. It is exclusive of input control elements. (PE/EDPG) 421-1972s

compound target* This term has been used to mean either complex target or distributed target. Because of its ambiguity, it is deprecated. (AES/RS) 686-1990

* Deprecated.

compound-wound A qualifying term applied to a direct-current machine to denote that the excitation is supplied by two types of windings, shunt and series. *Note:* When the electromagnetic effects of the two windings are in the same direction, it is termed cumulative compound wound; when opposed, differential compound wound. *See also:* direct-current commutating machine. (EEC/PE) [119]

compound-wound generator A dc generator that has two separate field windings. One supplies the predominating excitation, and is connected in parallel with the armature circuit. The other supplies only partial excitation and is connected in series with the armature circuit. It is proportioned to require an equalizer connection for satisfactory parallel operation. (IA/MT) 45-1998

compound-wound motor A dc motor that has two separate field windings: one, usually the predominating field, connected in parallel with the armature circuit, and the other connected in series with the armature circuit. Speed and torque characteristics are between those of shunt and series motors. (IA/MT) 45-1998

compressed-air circuit breaker *See:* circuit breaker.

compressed file A file that has been transformed in a manner intended to reduce its size without loss of information. (C/PA) 1387.2-1995

compression (1) (data transmission) A process in which the effective gain applied to a signal is varied as a function of the signal magnitude, the effective gain being greater for small rather than for large signals. (PE) 599-1985w **(2) (television)** The reduction in gain at one level of a picture signal with respect to the gain at another level of the same signal. *Note:* The gain referred to in the definition is for a signal amplitude small in comparison with the total peak-to-peak picture signal involved. A quantitative evaluation of this effect can be obtained by a measurement of differential gain. *See also:* white compression; black compression; television. (BT/AV) [34] **(3) (oscillography)** An increase in the deflection factor usually as the limits of the quality area are exceeded. *See also:* oscillograph. (IM/HFIM) [40] **(4) (image processing and pattern recognition)** *See also:* image compression. (C) 610.4-1990w

compressional wave A wave in an elastic medium that is propagated by fluctuations in elemental volume, accompanied by velocity components along the direction of propagation only. *Note:* A compressional plane wave is a longitudinal wave. (SP) [32]

compression gain 10log of the ratio of the magnitude of the peak power of a compressed pulse to the RMS noise power measured. For an unweighted chirp pulse compression system, the value is 10log (TB), where TB is the time bandwidth product (in decibels). (UFFC) 1037-1992w

compression joint (conductor stringing equipment) A tubular compression fitting designed and fabricated from aluminum, copper, or steel to join conductors or overhead ground wires. It is usually applied through the use of hydraulic or mechanical presses. However, in some cases, automatic, wedge, and explosive type joints are utilized. *Synonyms:* splice; sleeve; conductor splice. (T&D/PE) 524a-1993r, 524-1992r

compression point (nonlinear, active, and nonreciprocal waveguide components) The level of the output signal at which the gain of a device is reduced by a specified amount, usually expressed in decibels, as in the 1 dB compression point. (MTT) 457-1982w

compression ratio (gain or amplification) The ratio of (1) the magnitude of the gain (or amplification) at a reference signal level to (2) its magnitude at a higher stated signal level. *See also:* amplifier. (ED) 161-1971w

compressor (data transmission) A transducer, which for a given amplitude range of input voltages, produces a smaller range of output voltages. One important type of compressor employs the envelope of speech signals to reduce their volume range by amplifying weak signals and attenuating strong signals. (PE) 599-1985w

compressor-stator-blade-control system (gas turbines) A means by which the turbine compressor stator blades are adjusted by vary the operating characteristics of the compressor. *See also:* speed-governing system. (PE/EDPG) [5]

COM printer *See:* computer output microfilm printer.

compromise A violation of the security of a system such that an unauthorized disclosure of sensitive information may have occurred. (LM/C) 802.10-1992

computation *See:* implicit computation.

computational bandwidth The maximum number of operations per second a machine can perform. (C) 610.10-1994w

computational data *See:* fixed-point data.

computational model A model consisting of well-defined procedures that can be executed on a computer; for example, a model of the stock market, in the form of a set of equations and logic rules. (C) 610.3-1989w

compute-bound Pertaining to programs that have an abundance of computations. *Synonym:* process bound. *Contrast:* input-output bound. (C) 610.10-1994w

computed tomography (CT) A medical diagnostic technique in which a computer is used to produce an image of cross-sections of the human body by using measured attenuation of X rays through a cross-section of the body. *Synonym:* computer-assisted tomography. *See also:* computer-aided testing; computerized axial tomography. (C) 610.2-1987

computer (1) (A) (emergency and standby power) A machine for carrying out calculations. **(B) (emergency and standby power)** By extension, a machine for carrying out specified transformations on information. (IA/C/PSE) 446-1987, 165-1977 **(2) (A) (software)** A functional unit that can perform substantial computation, including numerous arithmetic operations, or logic operations without intervention by a human operator during a run. **(B) (software)** A functional programmable unit that consists of one or more associated processing units and peripheral equipment, that is controlled by internally stored programs, and that can perform substantial computation, including numerous arithmetic operations or logic operations, without human intervention. *See also:* program. (C/SE) 729-1983 **(3)** A device that consists of one or more associated processing units and peripheral units, that is controlled by internally stored programs, and that can perform substantial computations, including numerous arithmetic operations, or logic operations, without human intervention during a run. *Note:* May be stand alone, or may consist of several interconnected units. (C) 610.10-1994w

computer-aided design (CAD) (computer graphics) The use of computers to aid in design layout and analysis. May include modeling, analysis, simulation, or optimization of designs for production. Often used in combinations such as CAD/CAM. *See also:* computer-aided engineering; computer-aided manufacturing; computer-aided design and drafting; design automation. (C) 610.2-1987, 610.6-1991w

computer-aided design and drafting (CADD) The use of computers to aid in design layout, drafting, and analysis. Often used as a synonym for computer-aided design. (C) 610.6-1991w

computer-aided education (CAE) *See:* computer-assisted instruction.

computer-aided engineering (CAE) (1) (computer graphics) The use of computers to aid in engineering analysis and design. May include solution of mathematical problems, process control, numerical control, and execution of programs performing complex or repetitive calculations. *See also:* computer-aided manufacturing; computer-aided design. (C) 610.2-1987, 610.6-1991w
(2) The application of computers to the engineering process. The term now commonly applies to any computer system or program that manipulates data for the purpose of assisting engineering, design, procurement, maintenance, etc. (PE/EDPG) 1150-1991w
(3) A computer-based set of tools to assist in the design and development of integrated circuits. (C/TT) 1450-1999

computer-aided inspection (CAI) The use of computers to inspect manufactured parts. *Synonym:* mechanical inspection. (C) 610.2-1987

computer-aided instruction (CAI) The use of computers to present instructional material and to accept and evaluate student responses. *See also:* computer-assisted instruction; computer-based instruction. (C) 610.2-1987

computer-aided management (CAM) The application of computers to business management activities. For example, database management, control reporting, and information retrieval. *See also:* decision support system; management information system. (C) 610.2-1987

computer-aided manufacturing (CAM) (computer graphics) The use of computers and numerical control equipment to aid in manufacturing processes. May include robotics, automation of testing, management functions, control, and product assembly. Often used in combinations such as CAD/CAM. *See also:* computer-aided design; computer-aided engineering. (C) 610.2-1987, 610.6-1991w

computer-aided page makeup The use of computers to automate the formation of text and graphics into discrete camera-ready pages. *See also:* computer-aided typesetting; photocomposition. (C) 610.2-1987

computer-aided software engineering (CASE) The use of computers to aid in the software engineering process. May include the application of software tools to software design, requirements, tracing, code production, testing, document generation, and other software engineering activities. (C/SE) 1348-1995, 610.12-1990

computer-aided testing (CAT) The use of computers to test manufactured parts. (C) 610.2-1987

computer-aided typesetting The use of computers at any stage of the document composition process. This may involve text formatting, input from a word processing system, or computer-aided page makeup. *Synonym:* computer typesetting. (C) 610.2-1987

Computer and Business Equipment Manufacturers Association The Secretariat for ASC X3-series standards on information technology. (C) 610.7-1995, 610.10-1994w

computer architecture The organizational structure of a computer system, including the hardware and the software. *Contrast:* computer network architecture. (C) 610.10-1994w

computer-assisted instruction (CAI) The use of computers to present instructional material and to accept and evaluate student responses. *Synonyms:* computer-assisted learning; computer-aided instruction; computer-aided education; computer-augmented learning. *See also:* computer-based instruction. (C) 610.2-1987, 610.6-1991w

computer-assisted learning (CAL) *See:* computer-assisted instruction.

computer-assisted system A system that utilizes separate and standalone computers or processors for arithmetic computational and logic functions. All data manipulation and evaluation (e.g., alarm condition annunciation) functions are performed by the system. (IA/MT) 45-1998

computer-assisted tester (test, measurement, and diagnostic equipment) A test not directly programmed by a computer but that operates in association with a computer by using some arithmetic functions of the computer. (MIL) [2]

computer-assisted tomography (CAT) *See:* computed tomography.

computer-augmented learning (CAL) *See:* computer-assisted instruction.

computer automated measurement and control (CAMAC) **(1)** A standard modular instrumentation and digital interface system. (NPS) 583-1982r
(2) (FASTBUS acquisition and control) An internationally standardized modular instrumentation and digital interface system as defined in IEEE Std 583-1982, *IEEE Standard Modular Instrumentation and Digital Interface System (CAMAC),* and the corresponding documents EUR 4100-1972, *CAMAC: A Modular Instrumentation System for Data Handling,*and IEC Pub 516-1975, *A Modular Instrumentation System for Data Handling; CAMAC System.*; Compiler Automated Measurement and Control. (NID) 960-1986s

computer-based education (CBE) *See:* computer-based instruction.

computer-based instruction The use of computers to support any process involving human learning. *Synonyms:* computer-based education; computer-based learning. (C) 610.2-1987

computer-based learning (CBL) *See:* computer-based instruction.

computer-based simulation A simulation that is executed on a computer. *Synonym:* machine-centered simulation. *Contrast:* human-centered simulation. (C) 610.3-1989w

computer-based system A system that utilizes one or more embedded computers or processors to perform its functions. (IA/MT) 45-1998

computer channel *See:* input-output channel.

computer code A machine code for a specific computer. (C) [20], [85]

computer component (analog computer) Any part, assembly, or subdivision of a computer, such as resistor, amplifier, power supply, or rack. (C) 165-1977w

computer conferencing A form of teleconferencing that allows one or more users to exchange messages on a computer network. *See also:* video conferencing. (C) 610.2-1987

computer control (electric power system) (physical process) A mode of control wherein a computer, using as input the process variables, produces outputs that control the process. *See also:* power system. (PE/PSE) [54]

computer-control state (1) (analog computer) One of several distinct and selectable conditions of the computer-control circuits. *See also:* potentiometer set; hold; reset; operate; balance check; static test. (C) 165-1977w
(2) In an analog computer, one of several distinct and selectable conditions of the control circuit. *See also:* operate; balance check; static test; hold; reset; potentiometer set. (C) 610.10-1994w

computer control unit *See:* instruction control unit.

computer data (software) Data available for communication between or within computer equipment. Such data can be external (in computer-readable form) or resident within the computer equipment and can be in the form of analog or digital signals. *See also:* computer. (C/SE) 729-1983s

computer database *See:* database.

computer description language *See:* hardware description language.

Computer Design Language A design language for describing or designing computer architectures at the register level. (C) 610.13-1993w

computer diagram (analog computer) A functional drawing showing interconnections between computing elements, such interconnections being specified for the solution of a particular set of equations. *See also:* computer program; problem board. (C) 165-1977w

computer equation (machine equation) (analog computer) An equation derived from a mathematical model for use on a computer which is equivalent or proportional to the original equation. *See also:* scale factor. (C) 165-1977w

computer generated force (CGF) Simulation of entities on the virtual battlefield. CGF entities may be fully autonomous (needing no human direction) or semi-autonomous (requiring some direction by a human controller who is not a participant in the virtual events). CGF entities represent friendly, opposing forces (OPFOR), and neutral battlefield participants not portrayed by manned simulators. (DIS/C) 1278.3-1996

computer graphics (A) The branch of computer science concerned with methods of creating, modifying, or analyzing pictorial data. **(B)** The use of a computer in any discipline to create, modify, or analyze images. (C) 610.6-1991

Computer Graphics Interface (CGI) (A) A computer graphics standard that provides a method for exchanging device-independent data between graphics systems or device-dependent parts of a graphics system. It is under development by the American National Standards Institute (ANSI) and the International Standards Organization (ISO). **(B)** A method for exchanging device-independent data between graphics systems or device-dependent parts of a graphics system. (C) 610.6-1991

Computer Graphics Metafile (CGM) (A) A computer graphics standard that provides a method for recording graphical information in a metafile. It was developed by the American National Standards Institute (ANSI) and the International Standards Organization (ISO). **(B)** A method for recording graphical information in a metafile. (C) 610.6-1991

computer hardware Devices capable of accepting and storing computer data, executing a systematic sequence of operations on computer data, or producing control outputs. Such devices can perform substantial interpretation, computation, communication, control, or other logical functions. (C/SE) J-STD-016-1995

computer hardware description language *See:* hardware description language.

computer input microfilm (CIM) The input to a process that converts data contained on microform into machine-readable data. (C) 610.2-1987

computer instruction A machine instruction for a specific computer. (C) [20], [85]

(2) (A) (software) A statement in a programming language, specifying an operation to be performed by a computer and the addresses or values of the associated operands; for example, Move A to B. *See also:* instruction set; instruction format. **(B) (software)** Loosely, any executable statement in a computer program. (C) 610.12-1990, 610.10-1994

(3) (A) A statement in a computer language; specifying an operation to be performed by a computer and the address or values of the associated operands; for example, MOVE A to B. *See also:* machine instruction; operation field; operand field; address field. **(B)** An instruction expressed in machine language. (C) 610.10-1994

computer instruction code A code used to represent the instruction within an instruction set. *See also:* machine code. (C) 610.10-1994w

computer instruction set The collection of computer instructions possible on a given computer. *Synonym:* machine instruction set. (C) 610.10-1994w

computer-integrated manufacturing (CIM) Use of an integrated system of computer-controlled manufacturing centers. The centers may use robotics, design automation, or CAD/CAM (computer-aided design/computer-aided manufacturing) technologies. *See also:* flexible manufacturing system. (C) 610.2-1987

computer interface equipment (1) (surge withstand capability) A device that interconnects a protective relay system to an independent computer, for example, an analog to digital converter, a scanner, a buffer amplifier. (PE/PSR) C37.90-1978s

(2) A device that interconnects a protective relay system to an independent computer, for example, a scanner or a buffer amplifier. (SWG/PE) C37.100-1992

computer interface unit A device used to connect peripheral devices with a computer. (C) 610.10-1994w

computerized axial tomography (CAT) *See:* computed tomography.

computerized healthcare information systems *See:* patient care information system.

computer language A language designed to enable humans to communicate with computers. *See also:* system profile; workload model; programming language; design language. (C) 610.12-1990

(2) (A) A language designed to enable humans to communicate with computers and computer systems. **(B)** A language that is used to control, design, or define a computer or computer program. (C) 610.13-1993, 610.10-1994

computer literacy An understanding of the capabilities, operation, and applications of computers. (C) 610.2-1987

computer-managed instruction (CMI) The use of computers for management of student progress. Activities may include record keeping, progress evaluation, and lesson assignment. *See also:* computer-based instruction. (C) 610.2-1987

computer network (1) (software) A complex consisting of two or more interconnected computers. *See also:* computer. (C/SE) [20], 729-1983s, [85]

(2) An interconnection of assemblies of computer systems, terminals and communications facilities. (LM/COM) 168-1956w

(3) A structured connection of computer systems and peripheral devices that exchange data as necessary to perform the specific function of the network. *See also:* hierarchical computer network; homogeneous computer network; heterogeneous computer network; centralized computer network; decentralized computer network; distributed computer network. (C) 610.7-1995, 610.10-1994w

computer network architecture The logical structure and the operating principles, including those concerning services, functions, nd protocols, of a computer network. *Contrast:* computer architecture. (C) 610.7-1995, 610.10-1994w

computer numerical control (CNC) Numerical control in which one or more machines that produce manufactured parts are linked together via a single computer. (C) 610.2-1987

computer operation (A) An operation which can be performed by a computer with a single instruction. **(B)** An operation performed by a functional unit within a computer. For example: an instruction fetch, or an addition. *Synonym:* machine operation. (C) 610.10-1994

computer output microfilm (COM) The end result of a process that converts and records data from a computer directly to a microform. (C) 610.2-1987

computer output microfilmer A device for producing computer output microfilm. *Synonym:* COM device. (C) 610.2-1987

computer output microfilm printer A page printer that produces a microimage of each page on a photographic film. (C) 610.10-1994w

computer performance evaluation (software) An engineering discipline that measures the performance of computer systems and investigates methods by which that performance can be improved. *See also:* throughput; utilization. (C) 610.12-1990

computer program (1) (general) A plan or routine for solving a problem on a computer, as contrasted with such terms as fiscal program, military program, and development program. (MIL/C) [2], [20], [85]
(2) (analog computer) That combination of computer diagram, potentiometer list, amplifier list, trunk list, switch list, scaled equations, and any other documentation that defines the analog configuration for the particular problem to be solved. This term sometimes is used to include the problem patch board as well, and, in some loose usage, the computer program may be (incorrectly) used to refer solely to the program patch panel. (C) 165-1977w
(3) (programmable digital computer systems in safety systems of nuclear power generating stations) A schedule or plan that specifies actions that may or may not be taken, expressed in a form suitable for execution by a programmable digital computer. 7432-1982w
(4) (computer terminology) A combination of computer instructions and data definitions that enable computer hardware to perform computational or control functions. *See also:* software.
(C/SE) J-STD-016-1995, 610.12-1990, 610.5-1990w, 610.10-1994w

computer program abstract (software) A brief description of a computer program that provides sufficient information for potential users to determine the appropriateness of the program to their needs and resources. (C) 610.12-1990

computer program annotation *See:* comment.

computer program certification *See:* certification.

computer program component* (CPC) *See:* computer software component.
* Deprecated.

computer program configuration identification *See:* configuration identification.

computer program configuration item (CPCI) *See:* computer software configuration item.

computer program development plan *See:* software development plan.

computer program validation *See:* validation.

computer program verification *See:* verification.

computer resource allocation The assignment of computer resources to current and waiting jobs; for example, the assignment of main memory, input/output devices, and auxiliary storage to jobs executing concurrently in a computer system. *See also:* dynamic resource allocation; storage allocation. (C) 610.12-1990

computer resources The computer equipment, programs, documentation, services, facilities, supplies, and personnel available for a given purpose. *See also:* computer resource allocation. (C) 610.12-1990, 610.10-1994w

computer security Protection of information, system resources, and system services through controls provided by hardware and software mechanisms, including access controls, user authentication mechanisms, and audit facilities. (C/BA) 896.3-1993w

computer security object An information object used to maintain a condition of security in computerized environments. Examples include: representations of computer or communications systems resources, security label semantics, modes of operation for cryptographic algorithms, and one-way hashing functions. (C/LM) 802.10g-1995

computer simulation A simulation of the operation of a computer. *See also:* computer-based simulation. (C) 610.3-1989w

computer software *See:* software.

computer software component (CSC) A functionally or logically distinct part of a computer software configuration item, typically an aggregate of two or more software units. (C) 610.12-1990

computer software configuration item (CSCI) An aggregation of software that is designated for configuration management and treated as a single entity in the configuration management process. *Contrast:* hardware configuration item. *See also:* configuration item. (C) 610.12-1990

computer system (1) (software) A system containing one or more computers and associated software. (C) 610.12-1990
(2) A system containing one or more computers, peripheral devices and associated software. *Synonym:* computing system. (C) 610.7-1995, 610.10-1994w

Computer System Simulation II (CSS/II) A simulation language that is based on the concepts used in GPSS, but specialized for use in modeling computer systems. (C) 610.13-1993w

computer time *See:* time.

computer typesetting (CTS) *See:* computer-aided typesetting.

computer variable (1) A dependent variable as represented on the computer. *See also:* time. (C) 165-1977w
(2) (machine variable) *See also:* scale factor. (C) 165-1977w

computer word (1) A sequence of bits or characters treated as a unit and capable of being stored in one computer location. (C) [20], [85]
(2) A unit of storage, typically a set of bits, that is suitable for processing by a given computer; for example, two bytes. *Synonyms:* machine word; fullword. *See also:* double word; word. (C) 610.10-1994w
(3) (computer terminology) *See also:* word. (C) 610.5-1990w, 610.12-1990, 1084-1986w

computing center A facility designed to provide computer services to a variety of users through the operation of computers and auxiliary hardware and through services provided by the facility's staff. (C) 610.12-1990

computing elements (analog computer) A computer component that performs a mathematical operation required for problem solution. It is shown explicitly in computer diagrams, or computer programs. (C) 165-1977w

computing system *See:* computer system.

concatenate To append one item to the end of another so as to form a single unit in a contiguous pattern. For example, if we concatenate 'AP' with 'PLE,' the result is 'APPLE.' *Synonym:* catenate. (C) 610.5-1990w

concatenated key (A) (data management) A key derived from the concatenation of two or more keys. *Synonyms:* multifield key; fully concatenated key. **(B) (data management)** A concatenation of the keys for the first *N*segments found in a hierarchical path. For example, in the structure below, the concatenated key for segment *x*is "AFINANCE006."

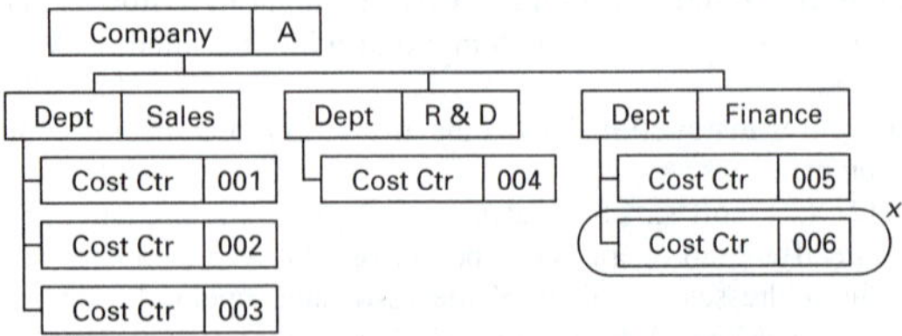

concatenated key

(C) 610.5-1990

concatenated transaction (1) A transaction where the request and response subactions are directly concatenated without a gap between the acknowledge of the request and the response packet. (C/MM) 1394-1995
(2) A split transaction comprised of concatenated subactions. (C/MM) 1394a-2000

concatenation (fiber optics) (of optical waveguides) The linking of optical waveguides, end to end. (Std100) 812-1984w

concave (image processing and pattern recognition) Pertaining to a region for which at least one straight line segment between two points of the region is not entirely contained within the region. *Contrast:* convex.

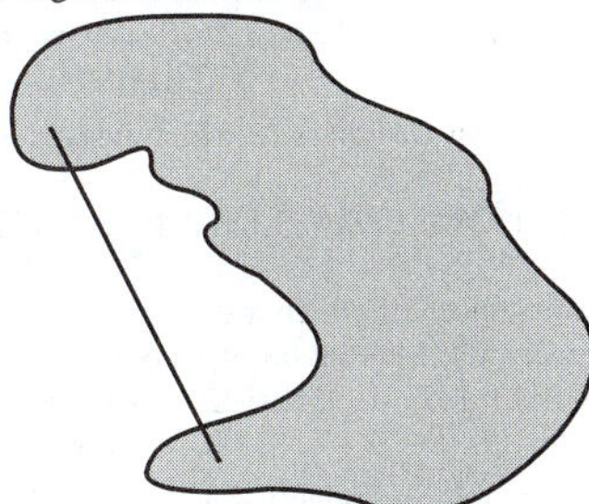

illustration of concave

(C) 610.4-1990w

concealed Rendered inaccessible by the structure or finish of the building. Wires in concealed raceways are considered concealed, even though they may become accessible by withdrawing them. (NESC/NEC) [86]

concealed knob-and-tube wiring A wiring method using knobs, tubes, and flexible nonmetallic tubing for the protection and support of single insulated conductors concealed in hollow spaces of walls and ceilings of buildings. (NESC/NEC) [86]

concentrate (metallurgy) The product obtained by concentrating disseminated or lean ores by mechanical or other processes thereby eliminating undesired minerals or constituents. *See also:* electrowinning. (EEC/PE) [119]

concentrated winding (rotating machinery) A winding, the coils of which occupy one slot pole: or a field winding mounted on salient poles. *See also:* direct-current commutating machine; asynchronous machine. (PE) [9]

concentration The quantity of radioactive material stated in terms of activity (or mass) per unit of volume or mass of a medium. (NI) N42.23-1995

concentration cell (1) An electrolyte cell, the electromotive force of which is due to differences in composition of the electrolyte at anode and cathode areas. (IA) [59]
(2) A cell of the two-fluid type in which the same dissolved substance is present in differing concentrations at the two electrodes. *See also:* electrochemistry. (EEC/PE) [119]

concentration polarization (1) That part of the total polarization that is caused by changes in the activity of the potential-determining components of the electrolyte. *See also:* electrochemistry. (EEC/PE) [119]
(2) That portion of the polarization of an electrode produced by concentration changes at the metal-environment interface. (IA) [59]

concentrator (1) (telephone switching systems) A switching entity for connecting a number of inlets to a smaller number of outlets. (COM) 312-1977w
(2) A device that provides communications capability between many low-speed, usually asynchronous channels and fewer high-speed, usually synchronous channels the sum of whose data rates is (usually) less than the sum of the data rates of the low-speed channels. (LM/COM) 168-1956w
(3) (A) A device that combines incoming messages into a single message or that extracts individual messages from the data set in a single transmission sequence. *Note:* The former process is called "concentration" and the latter, "deconcentration." **(B)** A communications computer that provides communications capability between many low speed asynchronous channels and one or more high-speed synchronous channels. *See also:* multiplexer; data concentrator. **(C)** A device in token ring networks that contains multiple interconnected trunk coupling units. (C) 610.7-1995, 610.10-1994
(4) A device that contains multiple interconnected trunk coupling units (TCUs). The concentrator contains two ports, referred to as *ring in* and *ring out*, to interface trunk cable. (C/LM) 8802-5-1998

concentrator concentric electrode system (coaxial electrode system) (electrobiology) An electrode system that is geometrically coaxial but electrically unsymmetrical. Example: One electrode may have the form of a cylindrical shell about the other so as to afford electrical shielding. *See also:* electrobiology. (EMB) [47]

concentricity error (fiber optics) When used in conjunction with a tolerance field to specify core/cladding geometry, the distance between the center of the two concentric circles specifying the cladding diameter and the center of the two concentric circles specifying the core diameter. *See also:* core diameter; cladding diameter; tolerance field; core; cladding. (Std100) 812-1984w

concentricity of coaxial connectors (fixed and variable attenuators) Total indicator runout between the diameter of outer conductor and that diameter of that portion of inner conductor which engages with the corresponding diameters of mating connector. *Note:* This does not apply to precision connectors with only butt contacts. (IM/HFIM) 474-1973w

concentric-lay cable A multiple-conductor cable composed of a central core surrounded by one or more layers of helically laid insulated conductors. *See also:* concentric-lay conductor. (T&D/PE) [10]

concentric-lay conductor A conductor composed of a central core surrounded by one or more layers of helically laid wires. *Note:* In the most common type of concentric-lay conductor, all wires are of the same size and the central core is a single wire. *See also:* conductor. (T&D/PE) [10]

concentric resonator (laser maser) A beam resonator comprising a pair of spherical mirrors having the same axis of rotational symmetry and positioned so that their centers of curvature coincide on this axis. (LEO) 586-1980w

concentric winding (rotating machinery) A winding in which the two coil sides of each coil of a phase belt, or of a pole of a field winding, are symmetrically located so as to be equidistant from a common axis. *See also:* asynchronous machine. (PE) [9]

concentric windings (power and distribution transformers) (of a transformer) An arrangement of transformer windings where the primary and secondary windings, and the tertiary winding, if any, are located in radial progression about a common core. (PE/TR) C57.12.80-1978r

concentric-wound relay coil A coil with two or more insulated windings, wound one over the other. (EEC/REE) [87]

concept A unit of thought constituted through abstraction on the basis of characteristics common to a group of entities. (SCC32) 1489-1999

concept analysis The derivation of a system concept through the application of analysis. *See also:* analysis. (C/SE) 1362-1998

concept level The level of verification activities at which vital functions and vital implementation requirements, imposed on the system's design and implementation by the safety assurance concept selected, are determined and identified. (VT/RT) 1483-2000

concept of operations document (ConOps document) A user-oriented document that describes a system's operational characteristics from the end user's viewpoint. *Synonym:* operational concept description. (C/SE) 1362-1998

concept phase (software) The period of time in the software development cycle during which the user needs are described and evaluated through documentation (for example, statement of needs, advance planning report, project initiation memo, feasibility studies, system definition, documentation, regulations, procedures, or policies relevant to the project). (C/SE) 1012-1986s, 610.12-1990, 982.1-1988, 982.2-1988

conceptual data definition language A data definition language used to describe the format layout and contents of all data stored in a database, the result of which is a conceptual schema. *Note:* May also include authorization levels and validation procedures. (C) 610.5-1990w

conceptual design The process of developing a conceptual schema for a database. (C) 610.5-1990w

conceptual model (1) A simulation implementation-independent representation of the exercise architect's understanding of the exercise objectives, requirements, and environment. The model includes logic and algorithms and explicitly recognizes assumptions and limitations. (C/DIS) 1278.4-1997 **(2)** A model of the concepts relevant to some endeavor. (C/SE) 1320.2-1998

conceptual population (results from a measurement process) The set of measurements that would result from infinite repetition of a measurement process in a state of statistical control. (IM) 470-1972w

conceptual record A record within a conceptual view. (C) 610.5-1990w

conceptual schema (A) A description of the format and layout of the entire data contents of a database. *Note:* The schema is written using a conceptual data definition language. It may include authorization levels and validation procedures. **(B)** The comprehensive, logical description of the information environment in which an enterprise exists, free of both the physical structure and application systems considerations. *Synonym:* enterprise view; conceptual model. *Contrast:* internal schema; external view. (C) 610.5-1990

conceptual switch A circuit feature, acting under the control of a digital control signal, that allows two circuit nodes to be electrically connected or isolated, as though there were a switch between them. *Note:* Depending on physical constraints, such as size, power consumption, and electrical characteristics, it may be possible in any particular application to implement all, some, or none of the conceptual switches by conventional complimentary metal oxide silicon (CMOS) transmission gates. *See also:* switch. (C/TT) 1149.4-1999

conceptual view The format and layout of the entire data content of a database, as described in a conceptual schema. *Note:* There may be many external views of a database, but only one conceptual view. (C) 610.5-1990w

concrete A class, instances of which are permitted either by direct instantiation or the instantiation of its concrete subclasses. (C/PA) 1328-1993w, 1224.1-1993w, 1327-1993w, 1238.1-1994w, 1224-1993w

concrete class (1) An OM class of which instances are permitted. (C/PA) 1328.2-1993w, 1326.2-1993w, 1327.2-1993w, 1224.2-1993w **(2)** A class, instances of which are permitted either by direct instantiation or the instantiation of its concrete subclasses. (C/PA) 1327-1993w, 1224.1-1993w, 1328-1993w, 1224-1993w, 1238.1-1994w

concrete-encased ground electrode A grounding electrode completely encased within concrete, located within, and near the bottom of, a concrete foundation or footing or pad, that is in direct contact with the earth. *Synonym:* ufer ground. (IA/PSE) 1100-1999

concrete model A model in which at least one component represented is a tangible object; for example, a physical replica of a building. (C) 610.3-1989w

concrete pole structures Structures consisting of one or more concrete poles. Other members of the structure may be reinforced concrete or other materials (i.e., wood, steel, aluminum). These structures are prefabricated, as opposed to being cast-in-place. Concrete pole structures may be manufactured in a variety of ways. A few examples are: Hollow or solid members; Different cross-sectional shape (i.e., round or square); Spun or statically cast (see figure below); Pretensioned, posttensioned, or nontensioned reinforcing steele); and Single-piece or multipiece poles. Combination of the above may be used to achieve the desired results (see figure below). (T&D/PE) 1025-1993r

concrete quantity *See:* physical quantity.

STATIC CAST CONCRETE POLE

OUTSIDE DIM.
PARTIAL VOID SOMETIMES PRESENT
A
A
CONCRETE COVER
SPIRAL REINFORCEMENT
CROSS SECTION
PRESTRESSING STRAND
SECTION A–A

SPUN CAST CONCRETE POLE

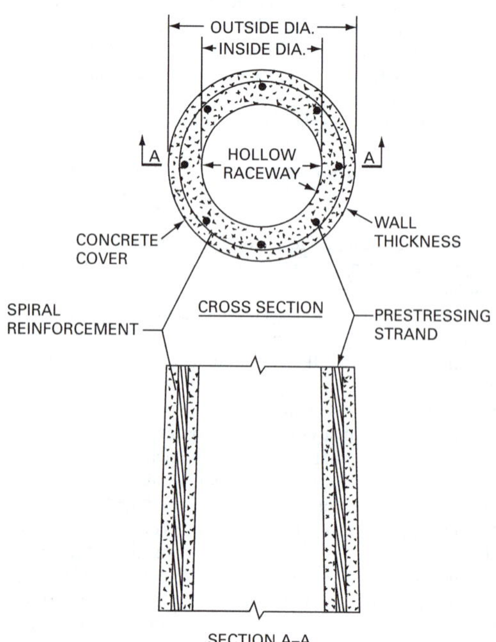

various concrete pole sections

concrete-tight fitting (for conduit) A fitting so constructed that embedment in freshly mixed concrete will not result in the entrance of cement into the fitting. (EEC/REWS) [113]

concurrency The process of multiple users accessing and manipulating a data item simultaneously, with the data-base management system linking the transaction of each user so that the access appears to be sequential. (PE/EDPG) 1150-1991w

concurrent (software) Pertaining to the occurrence of two or more activities within the same interval of time, achieved either by interleaving the activities or by simultaneous execution. *Synonym:* parallel. *Contrast:* simultaneous. (C) 610.12-1990

concurrent bus operation In dual bus systems, "concurrent bus operation capable" describes buses capable of conducting simultaneous unrelated bus transactions. "Non-concurrent bus operation capable" describes buses capable of conducting a transaction on only one bus at a time. (C/BA) 14536-1995

concurrent engineering The simultaneous engineering of products and life cycle processes to ensure usability, producibility, and supportability, and to control life cycle and total ownership costs. (C/SE) 1220-1998

concurrent execution Functions that suspend the execution of the calling thread shall not cause the execution of other threads to be indefinitely suspended. (C/PA) 9945-1-1996

concurrent processes (software) Processes that may execute in parallel on multiple processors or asynchronously on a single processor. Concurrent processes may interact with each other, and one process may suspend execution pending receipt of information from another process or the occurence of an external event. *See also:* sequential processes; execution. (C/SE) 729-1983s

concurrent reorganization (data management) Database reorganization in which users have access to the reorganized portion of the database while one or more reorganization processes are modifying other portions of the database. (C) 610.5-1990w

condensed-mercury temperature (mercury-vapor tube) The temperature measured on the outside of the tube envelope in the region where the mercury is condensing in a glass tube or at a designated point on a metal tube. *See also:* gas tube. (ED) [45]

condenser *See:* capacitor; fuse condenser.

condenser antenna *See:* capacitor antenna.

condenser box *See:* subdivided capacitor.

condenser bushing *See:* capacitor bushing.

condenser loudspeaker *See:* electrostatic loudspeaker.

condenser microphone *See:* electrostatic microphone.

condition (1) (modeling and simulation) The values assumed at a given instant by the variables in a system, model, or simulation. *See also:* initial condition; boundary condition; state; final condition. (C) 610.3-1989w
(2) (data management) *See also:* item condition; record condition; atomic condition. (C) 610.5-1990w
(3) *See also:* initial condition. (C)

condition adverse to quality An all-inclusive term used in reference to any of the following: failures, malfunctions, deficiencies, defective items and non-conformances. A significant condition adverse to quality is one which, if uncorrected, could have a serious effect on safety, quality, or operability. (NI/PE/NP) N42.23-1995, [124]

conditional branch* *See:* conditional jump.

* Deprecated.

conditional branch instruction A branch instruction that specifies conditions and, if those conditions are met, changes the program flow to a new location. *See also:* branch instruction; conditional jump instruction. (C) 610.10-1994w

condition code register A flag register used to hold the status bits used to decide conditional branches. *Note:* These bits generally include: zero, negative, and overflow. (C) 610.10-1994w

conditional control structure A programming control structure that allows alternative flow of control in a program depending upon the fulfillment of specified conditions, for example, case, if. . .then. . .else. . . . *See also:* program; flow of control; case; control structure. (C/SE) 729-1983s

conditional control transfer instruction *See:* conditional jump instruction.

conditional event A sequentially dependent event that will occur only if some other event has already taken place. *See also:* time-dependent event. (C) 610.3-1989w

conditional feature A feature or behavior referred to in a POSIX standard that need not be present on all conforming implementations. (C/PA) 13210-1994, 2003.1-1992

conditional implication *See:* implication.

conditional jump (1) To cause, or an instruction that causes, the proper one of two (or more) addresses to be used in obtaining the next instruction, depending upon some property of one or more numerical expressions or other conditions. *Synonym:* branch. *See also:* jump. (C) 162-1963w
(2) (software) A jump that takes place only when specified conditions are met. *Contrast:* unconditional jump. (C) 610.12-1990

conditional jump instruction A jump instruction that specifies conditions and, if those conditions are met, changes program flow to a new location. *Contrast:* unconditional jump instruction. *See also:* conditional branch instruction. (C) 610.10-1994w

conditionally invalid date-component value A date-component value that is improperly produced or improperly accepted by a system element dependent on other date-component values. *Note:* In the Gregorian calendar, the following are the conditionally invalid date-component values for dates following the adoption of the calendar:

— Values of the day-of-month equal to 29, 30, or 31 in February in years that are not leap years.
— Values of the day-of-month equal to 30 or 31 in February in years that are leap years.
— Values of the day-of-month equal to 31 in April, June, September, and November.
— Values of the day-of-year equal to 366 in years that are not leap years.

Normalization of invalid date-component values to valid date-component values does not constitute improper acceptance of a date-component. (C/PA) 2000.1-1999

conditional stability (linear feedback control system) A property such that the system is stable for prescribed operating values of the frequency-invariant factor of the loop gain and becomes unstable not only for higher values, but also for some lower values. *See also:* feedback control system. (PE/EDPG) 421A-1978s

conditional transfer instruction *See:* conditional jump instruction.

condition code *See:* status code.

conditioning (1) The addition of equipment to or selection of communication facilities to provide the performance characteristics required for certain types of data transmission. (LM/COM) 168-1956w
(2) (replacement parts for Class 1E equipment in nuclear power generating stations) Any additional work or process imposed upon a part that makes it different from nominally similar parts. *Note:* Conditioning may include calibration, adjustment, tuning, selection testing, "burn-in," heat treatment, machining, and similar processes. For example, if several parts are selected to test one that displays a special characteristic, the selected part is conditioned because it then displays a characteristic that makes it unique from parts with the same nominal description. (PE/NP) 934-1987w
(3) In telecommunication, a means to improve the performance of a line by reducing distortion and amplifying weak signals. For example, in telecommunication, line conditioning will bring attenuation, impedance, and delay characteristics

to within set limits. *Synonym:* line conditioning. *See also:* D conditioning; C conditioning. (C) 610.7-1995

conditioning stimulus (medical electronics) A stimulus of given configuration applied to a tissue before a test stimulus. (EMB) [47]

condition monitoring Observation, measurement, or trending of condition or functional indicators with respect to some independent parameter (usually time or cycles) to indicate the current and future ability to function within acceptance criteria. (PE/NP) 933-1999

condition variable (1) A synchronization object that allows a thread to suspend execution, repeatedly, until some associated predicate becomes true. A thread whose execution is suspended on a condition variable is said to be *blocked on* the condition variable. (C/PA) 9945-1-1996
(2) A synchronization object that allows a task to become blocked until it is unblocked by some event. The unblocking may occur spontaneously or as a result of a timeout or another task performing a condition-signaling operation on the condition variable. In use, condition variables are always associated with mutexes. (C) 1003.5-1999

Condition Variable Service An instance of the class `IEEE1451_ConditionVariableService` or of a subclass thereof. (IM/ST) 1451.1-1999

conductance (A) That physical property of an element, device, branch, network or system, that is the factor by which the mean square voltage must be multiplied to give the corresponding power lost by dissipation as heat or as other permanent radiation or loss of electromagnetic energy from the circuit. **(B)** The real part of admittance. *Note:* (A) and (B) are not equivalent but are supplementary. In any case where confusion may arise, specify the definition being used. (IM/HFIM) [40]

conductance coupling (interference terminology) The type of coupling in which the mechanism is conductance between the interference source and the signal system. *See also:* raceway; interference.

conductance, electrode *See:* electrode conductance.

conductance for rectification (electron tube) The quotient of (A) the electrode alternating current of low frequency by (B) the in-phase component of the electrode alternating voltage of low frequency, a high frequency sinusoidal voltage being applied to the same or another electrode and all other electrode voltages being maintained constant. *See also:* rectification factor. (ED) [45]

conductance relay A mho relay for which the center of the operating characteristic on the *R-X* diagram is on the *R*-axis. *Note:* The equation that describes such a characteristic is

$Z - K \cos\theta$

where K is a constant and θ is the phase angle by which the input voltage leads the input current. (SWG/PE) C37.100-1992

conducted emissions test site A site meeting specified requirements suitable for measuring radio interference voltages and currents emitted by an equipment under test (EUT). (EMC) C63.4-1991

conducted heat The thermal energy transported by thermal conduction. *See also:* thermoelectric device. (ED) [46]

conducted interference Interference resulting from conducted radio noise or unwanted radio signals entering a transducer (receiver) by direct coupling. *See also:* electromagnetic compatibility. (EMC/T&D/PE) [53], 539-1990

conducted radio noise (1) Radio noise produced by equipment operation, which exists on the power line of the equipment and is measurable under specified conditions as a voltage or current. (EMC) C63.4-1988s
(2) Radio noise propagated along circuit conductors. *Note:* It may enter a transducer (receiver) by direct coupling or by an antenna as by subsequent radiation from some circuit element. *See also:* electromagnetic compatibility. (EMC) [53]
(3) (overhead power lines) Radio noise that is propagated by conduction from a source through electrical connections. (T&D/PE) 539-1990

conducted spurious emission power (land-mobile communications transmitters) Any part of the spurious emission power output conducted over a tangible transmission path. Radiation is not considered a tangible path. (EMC) 377-1980r

conducted spurious transmitter output (land-mobile communications transmitters) Any spurious output of a radio transmitter conducted over a tangible transmission path. *Note:* Power lines, control leads, radio frequency transmission lines and waveguides are all considered as tangible paths in the foregoing definition. Radiation is not considered a tangible path in this definition. (EMC) 377-1980r

conducting (conduction) period (1) (rectifier circuit element) (semiconductor) That part of an alternating voltage cycle during which the current flows in the forward direction. *Note:* The forward period is not necessarily the same as the conducting period because of circuit parameters and semiconductor rectifier diode characteristics. (IA) [12]
(2) (gas tube) That part of an alternating-voltage cycle during which a certain arc path is carrying current. (ED) [45]

conducting element (of a fuse) The conducting means, including the current-responsive element, for completing the electric circuit between the terminals of a fuse-holder or fuse unit. *Synonym:* fuse link. (SWG/PE) C37.100-1992, C37.40-1993

conducting ground plane A conducting flat surface or plate that is used as a common reference point for circuit returns and electric or signal potentials, and that reflects electromagnetic waves. (EMC) C63.4-1991

conducting interval (self-commutated converters) (circuit properties) An interval during which the principal current flows through a blocking element. (IA/SPC) 936-1987w

conducting material A material, such as a metal, that has a very large number of free electrons that can easily be put into motion to create an electric current. *Contrast:* insulating material; semiconducting material. (C) 610.10-1994w

conducting mechanical joint The juncture of two or more conducting surfaces held together by mechanical means. *Note:* Parts jointed by fusion processes, such as welding, brazing, or soldering, are excluded from this definition. (SWG/PE) C37.100-1992

conducting paint (rotating machinery) A paint in which the pigment or a portion of pigment is a conductor of electricity and the composition is such that when it is converted to a solid film, the electric conductivity of the film approaches that of metallic substances. (PE) [9]

conducting parts The parts that are designed to carry current or that are conductively connected therewith. (IA/ICTL/IAC) [60]

conducting salts Salts that, when added to a plating solution, materially increase its conductivity. *See also:* electroplating. (EEC/PE) [119]

conduction band (semiconductor) A range of states in the energy spectrum of a solid in which electrons can move freely. *See also:* semiconductor. (ED) 216-1960w

conduction current (I_G) (1) (A) The component of the measured current in phase with the applied voltage that is delivered to the volume of the tool or equipment, due to the physical resistance of the material comprising the tool or equipment. **(B) (electric submersible pump cable) (leakage current)** Current resulting from conduction through the cable insulating medium or over surfaces. Corona discharge from external energized elements will be indicated as conduction current. (IA/PE/T&D/PC/IC) 1017-1985, 516-1995, 400-1991
(2) Current in the specimen under steady-state conditions. *Notes:* 1. This is sometimes called "leakage" current. 2. Absorption and capacitive effects are assumed to have been made negligible under steady-state conditions. 3. Surface

leakage current is assumed excluded from the measured current. (PE) 402-1974w

(3) A current that is constant in time, that passes through the bulk insulation from the grounded surface to the high-voltage conductor, and that depends on the type of bonding material used in the insulation system. (PE/EM) 43-2000

conduction electrons (semiconductor) The electrons in the conduction band of a solid that are free to move under the influence of an electric field. *See also:* semiconductor. (AES) [41]

conduction-through (thyristor converter) The failure to achieve forward blocking, during inverter operation, of an arm of a thyristor connection at the end of the normally conducting period, thus enabling the direct current to continue to pass during the period when the thyristor is normally in the off state. *Note:* A conduction-through occurs, for example, when the margin angle is too small or because of a misgating in the succeeding arm. (IA/IPC) 444-1973w

conductive (health care facilities) Adjective describing not only those materials, such as metals, which are commonly considered as electrically conductive, but also that class of materials which, when tested in accordance with NFPA standard 56A, 1978, have a resistance not exceeding 1 000 000 ohms. Such materials are required where electrostatic interconnection is necessary. (EMB) [47]

conductive clothing Clothing made of natural or synthetic material that is either conductive or interwoven with conductive thread to provide mitigation of effects of the electric fields of high-voltage energized electrical conductors and equipment. (T&D/PE) 516-1995

conductive coating (rotating machinery) Conducting paint applied to the slot portion of a coil-side, to carry capacitive and leakage currents harmlessly between insulation and grounded iron. (PE) [9]

conductive coupling (1) (overhead power lines) The process of generating voltages and/or currents in conductive objects and electric circuits, otherwise unenergized, due to deposition of charge. (T&D/PE) 539-1990

(2) (interference terminology) *See also:* coupling; conductance coupling.

conductive heat release The energy released from a burning material to whatever is in direct contact with it. (DEI) 1221-1993w

conductivity (σ) (1) (material) A factor such that the conduction-current density is equal to the electric-field intensity in the material multiplied by the conductivity. *Note:* In the general case it is a complex tensor quantity. *See also:* transmission line. (IM/HFIM) [40]

(2) A macroscopic material property that relates the conduction current density (*J*) to the electric field ($\vec{E}$) in the medium. *Note:* For a monochromatic wave in a linear medium, that relationship is described by the (phasor) equation:

$\vec{J} = \sigma\!= \cdot \vec{E}$

where

$\sigma\!=$ = a tensor, generally frequency dependent

$\vec{J}$ = in phase with $\vec{E}$

For an isotropic medium, the tensor conductivity reduces to a complex scalar conductivity σ, in which case $\vec{J} = \sigma\vec{E}$. (AP/PROP) 211-1997

conductivity chamber An instrument that determines the conductivity of the air. (T&D/PE) 539-1990

conductivity in physical media The real part of the complex conductivity. *See also:* complex permittivity. (AP/ANT) 145-1983s

conductivity modulation (semiconductor) The variation of the conductivity of a semiconductor by variation of the charge-carrier density. *See also:* semiconductor device; semiconductor. (ED) 216-1960w

conductivity-modulation transistor A transistor in which the active properties are derived from minority-carrier modulation of the bulk resistivity of a semiconductor. *See also:* semiconductor; transistor. (ED) 216-1960w

conductivity, *n*-type (semiconductor) The conductivity associated with conduction electrons in a semiconductor. *See also:* semiconductor. (ED) 216-1960w

conductivity, *p*-type (semiconductor) The conductivity associated with holes in a semiconductor. *See also:* semiconductor. (AES/IA/ED) [41], [12], 270-1966w, 216-1960w

conductivity, thermal *See:* thermal conductivity.

conductor (1) (A) (general) A substance or body that allows a current of electricity to pass continuously along it. **(B) (general)** The portion of a lightning-protection system designed to carry the lightning discharge between air terminal and ground. (PE/NFPA) [9], [114]

(2) A material, usually in the form of a wire, cable, or bus bar, suitable for carrying an electric current. *See also:* open conductor; bundled conductor; covered conductor; grounded conductor; lateral conductor; insulated conductor; grounding conductor. (NESC/T&D) C2-1997, C2.2-1960

(3) (power line maintenance) A wire or combination of wires not insulated from one another, suitable for carrying an electrical current. However, it may be bare or insulated. *Synonyms:* wire; cable.

(T&D/PE) 524a-1993r, 30-1937w, 516-1995, 524-1992r, 1048-1990

(4) (substation grounding) A metallic substance that allows a current of electricity to pass continuously along it. As used in IEEE Std 837-1989, a conductor includes cable (wire), rods (electrodes), and metallic structures. (SUB/PE) 837-1989r

(5) A device made from conducting material; for example, a metal wire. (C) 610.10-1994w

conductor, bare *See:* bare conductor.

conductor car (conductor stringing equipment) A device designed to carry workmen and ride on sagged bundle conductors, thus enabling them to inspect the conductors for damage and install spacers and dampers where required. These devices may be manual or powered. *Synonyms:* cable buggy; cable car; spacer cart; spacing bicycle; spacer buggy. (PE/T&D) 524-1992r

conductor, coaxial *See:* coaxial conductor.

conductor combination (substation grounding) The various conductors that may be joined by a connector. (SUB/PE) 837-1989r

conductor-cooled (rotating machinery) A term referring to windings in which coolant flows in close contact with the conductors so that the heat generated within the principal portion of the windings reaches the cooling medium without flowing through the major ground insulation. (Std100) [84]

conductor cover Electrical protection equipment designed specifically to cover conductors *Synonyms:* hose; hard cover; eel; snake. *See also:* cover-up equipment. (T&D/PE) 516-1995

conductor, covered *See:* covered conductor.

conductor current connection interface The connection for transfer of conductor current from the GIS conductor to the cable termination. (PE/IC) 1300-1996

conductor grip A device designed to permit the pulling of a conductor without splicing on fittings, eyes, etc. It permits the pulling of a continuous conductor where threading is not possible. The designs of these grips vary considerably. Grips such as the Klein (Chicago) and Crescent utilize an open-sided, rigid body with opposing jaws and swing latch. In addition to pulling conductors, this type of grip is commonly used to tension guys and, in some cases, to pull wire rope. The design of the come-along (pocket-book, suitcase, four bolt, etc.) incorporates a bail attached to the body of a clamp that folds to completely surround and envelop the conductor. Bolts are then used to close the clamp and obtain a grip *Synonyms:* buffalo; six bolt; crescent; Kellem grip; four bolt; Kellem; Klein; seven bolt; suitcase; slip-grip; grip; pocketbook; Chicago grip; come-along. *See also:* conductor grip. (T&D/PE) 524a-1993r, 524-1992r, 516-1995

conductor hook *See:* conductor lifting hook.

conductor insulation (rotating machinery) The insulation on a conductor or between adjacent conductors. (PE) [9]

conductor lifting hook (power line maintenance) A device resembling an open boxing glove designed to permit the lifting of conductors from a position above the conductors. Normally used during clipping-in operations. Suspension clamps are sometimes used for this purpose. (T&D/PE) 516-1987s

conductor loading (mechanical) The combined load per unit length of a conductor due to the weight of the wire plus the wind and ice loads. *See also:* tower. (T&D/PE) [10]

conductor-loop resistance (telephone switching systems) The series resistance of the conductors of a line or trunk loop, excluding terminal equipment or apparatus. (COM) 312-1977w

conductor loss That contribution to the attenuation constant of a propagating mode on a planar transmission line that represents losses attributed to the finite conductivity of the conductors involved. (MTT) 1004-1987w

conductor payout station *See:* tension site.

conductor safety A sling arranged in a vertical basket configuration, with both ends attached to the supporting structure and passed under the clipped-in conductor(s). These devices, when used, are normally utilized with bundled conductors to act as a safety device in case of insulator failure while workers in conductor cars are installing spacers between the subconductors, or as an added safety measure when crossing above energized circuits. These devices may be fabricated from synthetic fiber rope or wire rope. (T&D/PE) 524-1992r

conductor shielding (1) (power distribution, underground cables) A conducting or semiconducting element in direct contact with the conductor and in intimate contact with the inner surface of the insulation so that the potential of this element is the same as the conductor. Its function is to eliminate ionizable voids at the conductor and provide uniform voltage stress at the inner surface of the insulating wall. (PE) [4]
(2) (cable systems) A conducting material applied in manufacture directly over the surface of the conductor and firmly bonded to the inner surface of the insulation. (PE/EDPG) 422-1977
(3) An envelope that encloses the conductor of a cable and provides an equipotential surface in contact with the cable insulation. (NESC) C2-1997

conductor splice *See:* compression joint.

conductor support box A box that is inserted in a vertical run of raceway to give access to the conductors for the purpose of providing supports for them. *See also:* cabinet. (EEC/PE) [119]

conductor temperature (1) (electrical heat tracing for industrial applications) The temperature of the heat-producing element. (BT/AV) 152-1953s
(2) The temperature of a conductor. *Note:* The conductor is assumed to be isothermal (i.e., no axial or radial temperature variation) for all steady-state calculations and for all transient calculations where the time period of interest exceeds 1 min or the conductor consists of a single material. With transient calculations for times less than 1 min with nonhomogeneous ACSR conductors (i.e., aluminum conductor steel reinforced), the aluminum strands are isothermal; but the heat capacity of the steel core is assumed to be zero. (T&D/PE) 738-1993

conduit (1) (aircraft) An enclosure used for the radio shielding or the mechanical protection of electric wiring in an aircraft. *Note:* It may consist of either rigid or flexible, metallic or nonmetallic tubing. Conduit differs from pipe and metallic tubing in that it is not normally used to conduct liquids or gases. *See also:* flexible metal conduit; rigid metal conduit. (EEC/PE) [119]
(2) (packaging machinery) A tubular raceway for holding wires or cables, which is designed expressly for, and used solely for, this purpose. (IA/PKG) 333-1980w
(3) A structure containing one or more ducts. *Note:* Conduit may be designated as iron-pipe conduit, tile conduit, etc. If it contains only one duct it is called *single-duct conduit;* if it contains more than one duct it is called *multiple-duct conduit,* usually with the number of ducts as a prefix, for example, *two-duct multiple conduit.* (NESC/T&D/PE) C2-1997, C2.2-1960, [10]

conduit body A separate portion of a conduit or tubing system that provides access through a removable cover(s) to the interior of the system at a junction of two or more sections of the system or at a terminal point of the system. (NESC/NEC) [86]

conduit fitting An accessory that serves to complete a conduit system, such as bushings and access fittings. *See also:* raceway. (EEC/REWS) [113]

conduit knockout *See:* knockout.

conduit run *See:* duct bank.

conduit system (raceway systems for Class 1E circuits for nuclear power generating stations) Any assembly of conduit sections, fittings, supports, anchorages, and accessories that form a structural system to support wire and cable. . (PE/NP) 628-1987r
(2) Any combination of duct, conduit, conduits, manholes, handholes, and vaults joined to form an integrated whole. (NESC) C2-1997

cone (1) (cathode-ray tubes) The divergent part of the envelope of the tube. *See also:* cathode-ray tube. (ED) [45]
(2) (vision) Retinal elements that are primarily concerned with the perception of detail and color by the light-adapted eye. *See also:* retina. (EEC/IE) [126]

cone of ambiguity (navigation aid terms) A generally conical volume of airspace above a navigation aid within which navigational information from that facility is unreliable. (AES/GCS) 172-1983w

cone of nulls A conical surface formed by directions of negligible radiation. *See also:* antenna. (AP/ANT) [35]

cone of protection (lightning) The space enclosed by a cone formed with its apex at the highest point of a lightning rod or protecting tower, the diameter of the base of the cone having a definite relation to the height of the rod or tower. *Note:* This relation depends on the height of the rod and the height of the cloud above the earth. The higher the cloud, the larger the radius of the base of the protecting cone. The ratio of radius of base to height varies approximately from one to two. When overhead ground wires are used, the space protected is called a zone of protection or protected zone. (T&D/PE) [10]

cone of silence (navigation aid terms) A conically shaped region above an antenna where the field strength is relatively weak because of the configuration of the antenna system. (AES/GCS) 172-1983w

cone, leader *See:* leader cone.

cones (illuminating engineering) Retinal receptors which dominate the retinal response when the luminance level is high and provide the basis for the perception of color. (EEC/IE) [126]

conference call (telephone switching systems) A call in which communication is provided among more than two main stations. (COM) 312-1977w

conference connection A special connection for a telephone conversation among more than two stations. (EEC/PE) [119]

Conférence Internationale Des Grands Réseaux Electriques An international organization concerned with large high voltage electric power systems. (PE) 599-1985w

Conference on Data Systems Languages (CODASYL) An organization that establishes industry standards for database structures. (C) 610.5-1990w

confidence limit The uncertainty associated with the estimate of a time- or frequency-domain instability measure from a finite number of measurements. (SCC27) 1139-1999

confidence test (test, measurement, and diagnostic equipment) A test primarily performed to provide a high degree of certainty that the unit under test is operating acceptably. (MIL) [2]

confidence tester (test, measurement, and diagnostic equipment) Any test equipment, either automatic, semiautomatic, or manual, which is used expressly for performing a test or series of tests to increase the degree of certainty that the unit under test is operating acceptably. (MIL) [2]

confidentiality The property of information that is not made available or disclosed to unauthorized individuals, entities, or processes. (C/LM) 8802-11-1999

Configurable Startup Set A Component Group Object owned by an NCAP Block Object for use in bringing the system to a known state. (IM/ST) 1451.1-1999

configuration (1) (A) (software) The arrangement of a computer system or component as defined by the number, nature, and interconnections of its constituent parts. **(B) (software)** In configuration management, the functional and physical characteristics of hardware or software as set forth in technical documentation or achieved in a product. *See also:* version. **(C)** The physical and logical elements of an information processing system, the manner in which they are organized and connected, or both. *Note:* May refer to hardware configuration or software configuration. (C) 610.7-1995, 610.12-1990, 610.10-1994
(2) A collection of capability modules. (C/MM) 855-1990
(3) A process often synonymous with commissioning. Often includes the selection of attributes of an object that change its appearance or performance characteristics as opposed to its communication properties. (IM/ST) 1451.1-1999

configuration audit *See:* physical configuration audit; functional configuration audit.

configuration baseline The configuration at a point in time recorded in documentation that fully describes the functional, performance, usability, interface requirement, and design characteristics, as appropriate to the stage of the life cycle. (C/SE) 1220-1998

configuration control (software) An element of configuration management, consisting of the evaluation, coordination, approval or disapproval, and implementation of changes to configuration items after formal establishment of their configuration identification. *Synonym:* change control. *Contrast:* configuration status accounting; configuration identification. *See also:* engineering change; interface control; notice of revision; waiver; configuration item; deviation; specification change notice. (C) 610.12-1990

configuration control board (software) A group of people responsible for evaluating and approving or disapproving proposed changes to configuration items, and for ensuring implementation of approved changes. *Synonym:* change control board. *See also:* configuration control. (C) 610.12-1990

configuration control function The function that ensures that the resources of all *nodes* of a DQDB *subnetwork* are configured into a correct Dual Bus topology. The resources that are managed are the Head of Bus function, the *External Timing Source function,* and the *Default Slot Generator function.* (LM/C) 8802-6-1994

configuration diagram *See:* block diagram.

configuration factor (illuminating engineering) The ratio of illuminance on a surface at point 2 (due to flux directly received from lambertian surface 1) to the exitance of surface 1. It is used in flux transfer theory $C_{1-2} = (E_2)/(M_1)$. (EEC/IE) [126]

configuration identification (A) (software) An element of configuration management, consisting of selecting the configuration items for a system and recording their functional and physical characteristics in technical documentation. *Contrast:* configuration status accounting; configuration control. **(B) (software)** The current approved technical documentation for a configuration item as set forth in specifications, drawings, associated lists, and documents referenced therein. *See also:* functional configuration identification; allocated configuration identification; product configuration identification; baseline. (C) 610.12-1990

configuration index A document used in configuration management, providing an accounting of the configuration items that make up a product. *See also:* configuration item development record; configuration status accounting. (C) 610.12-1990

configuration information The data or information that defines the operational limits and characteristics of a particular device. Depending on the device, this information is either manually downloaded into NVRAM or EEPROM, or is pre-programmed into EPROM. (PE/SUB) 1379-1997

configuration item (software) An aggregation of hardware, software, or both, that is designated for configuration management and treated as a single entity in the configuration management process. *See also:* configuration identification; hardware configuration item; critical item; computer software configuration item. (C) 610.12-1990

configuration item development record A document used in configuration management, describing the development status of a configuration item based on the results of configuration audits and design reviews. *See also:* configuration index; configuration status accounting. (C) 610.12-1990

configuration management (1) (software) A discipline applying technical and administrative direction and surveillance to: identify and document the functional and physical characteristics of a configuration item, control changes to those characteristics, record and report change processing and implementation status, and verify compliance with specified requirements. *See also:* configuration control; configuration status accounting; configuration audit; configuration identification; baseline. (C) 610.12-1990
(2) In networking, a management function that identifies, controls, collects data from, and provides data to, open network systems. (C) 610.7-1995

configuration registers A device's A16 registers that are required for the system configuration process. (C/MM) 1155-1992

configuration report server (CRS) A function that monitors and controls the stations of the ring. It receives configuration information from the stations on the ring and either forwards it to the network manager or uses it to maintain a configuration of the ring. It can also, when requested by a network manager, check the status of stations on the ring, change operational parameters of stations on the ring, and request that a station remove itself from the ring. (C/LM) 8802-5-1998

configuration status accounting (software) An element of configuration management, consisting of the recording and reporting of information needed to manage a configuration effectively. This information includes a listing of the approved configuration identification, the status of proposed changes to the configuration, and the implementation status of approved changes. *Contrast:* configuration identification; configuration control. *See also:* configuration item development record; configuration index. (C) 610.12-1990

configuration variable A named parameter, whose value is stored in nonvolatile memory, that controls some aspect of the firmware's behavior. (C/BA) 1275-1994

configure To initialize a device so that it operates in a particular way. For instance, a customer may configure a device so the device never requests data link confirmations, using a variety of mechanisms (e.g., parameters in NVRAM, parameters in ROM, dip switches, or hardware jumpers). (PE/SUB) 1379-1997

confinement (A) (software) Prevention of unauthorized alteration, use, destruction, or release of data during authorized access. **(B) (software)** Restriction on programs and processes so that they do not access or have influence on data, programs, or processes other than that allowed by specific authorization. *See also:* data; integrity. (C/SE) 729-1983

confirm A primitive provided from one layer entity to another layer entity to verify that the entity generating the confirm primitive responded to a request primitive originating from the entity that received the confirm primitive.
(EMB/MIB) 1073.4.1-2000

conflict_error An error-status code that is returned when a transaction has been transmitted successfully, but a queue or usage conflict inhibits the transaction completion. A conflict _error status is returned to the original requester, which is expected to retry the transaction. This is different than a bus-dependent delay (wait or busy status), which delays the forwarding of a transaction or subaction across the bus.
(C/MM) 1212-1991s

confocal resonator (laser maser) A beam resonator comprising a pair of spherical mirrors having the same axis of rotational symmetry and positioned so that their focal points coincide on this axis. (LEO) 586-1980w

conformal antenna An antenna (an array) that conforms to a surface whose shape is determined by considerations other than electromagnetic; for example, aerodynamic or hydrodynamic. *Synonym:* conformal array. (AP/ANT) 145-1993

conformal array *See:* conformal antenna.

conformance document (CD) **(1)** A document provided by an implementor that contains implementation details.
(C/PA) 1003.5b-1995
(2) The conformance document required by a standard that meets the requirements specified in that standard for such a document. (C/PA) 2003-1997

Conformance Documentation Audit The process of reviewing a conformance document to ascertain that it meets the requirements of a base standard as specified by documentation assertions. (C/PA) 2003-1997

conformance requirement A requirement stated in a base standard that identifies a specific requirement in a finite, measurable, and unambiguous manner. A conformance requirement by itself or in conjunction with other conformance requirements corresponds to an assertion. *Note:* Behavior and/or capabilities imposed upon an implementation by the base standard for the implementation to conform to that base standard.
(C/PA) 2003-1997

conformance test interface connector (CTIC) A defined connector for the purpose of conformance testing.
(C/LM) 11802-4-1994

Conformance Test Procedure (CTP) Manual procedures used in conjunction with other test methods to measure conformance. (C/PA) 2003-1997

conformance tests (1) (power and distribution transformers) (general) Tests that are specifically made to demonstrate conformity with applicable standards. (SWG/PE) [56]
(2) (mechanical switching device) (X-radiation limits for ac high-voltage power vacuum interrupters used in power switchgear). Those tests that are specifically made to demonstrate the conformity of switchgear or its component parts with applicable standards.
(SWG/PE) C37.40-1981s, C37.85-1972w, C37.60-1981r
(3) Tests that are made by agreement between the manufacturer and the purchaser at the time the order is placed. In some cases, by mutual agreement, certain Design Tests may be made as Conformance Tests. (PE/TR) C57.12.80-1978r
(4) Tests that demonstrate compliance with the applicable standards. The test specimen is normally subjected to all planned production tests prior to initiation of the conformance test program. *Note:* The conformance tests may, or may not, be similar to certain design tests. Demonstration of margin (capabilities) beyond the standards is not required.
(SWG/PE) C37.20.1-1993r, C37.20.4-1996, C37.20.2-1993, C37.20.3-1996, C37.100-1992
(5) Tests made, when required, to demonstrate selected performance characteristics of a product or representative samples thereof. (SPD/PE) C62.11-1999
(6) (instrument transformers) A performance test to demonstrate compliance with the applicable standard(s). The test specimen is normally subjected to all planned routine tests prior to initiation of the conformance test program. *Note:* The conformance tests may, or may not, be similar to certain design tests. Demonstration of margin (capabilities beyond the standard) is not required.
(SWG/PE/TR) C57.13.2-1986s, C37.21-1985r
(7) A test that is specifically made to demonstrate the continuing conformance of equipment with the applicable standard.
(SUB/PE) C37.122-1983s

Conformance Test Software (CTS) Test software used to ascertain conformance to standards. (C/PA) 2003-1997

conforming test result codes The complete list of test result codes associated with each assertion that a CTS can report for a conforming implementation. (C/PA) 2003-1997

conformity (1) (potentiometer) The accuracy of its output: used especially in reference to a function potentiometer.
(C) 165-1977w, 166-1977w
(2) (automatic control) (curve) The closeness with which it approximates the specified functional curve (for example logarithmic, parabolic, cubic, etc.). *Note:* It is usually expressed in terms of a nonconformity, for example the maximum deviation. For "independent conformity," any shift or rotation is permissible to reduce this deviation. For "terminal conformity," the specified functional curve must be drawn to give zero output at zero input and maximum output at maximum input, but the actual deviation at these points is not necessarily zero. (PE/EDPG) [3]

confounding variable *See:* uncontrolled variable.

congenital effect Existing at or from birth.
(T&D/PE) 539-1990

congestion In networking, a condition that occurs when the traffic exceeds the capacity of the network. *See also:* source quench; fair queuing. (C) 610.7-1995

congruencing An image processing technique in which two images of the same scene are transformed so that the size, shape, position, and orientation of all objects in one image are the same as those in the other image. (C) 610.4-1990w

conical array A two-dimensional array of elements whose corresponding points lie on a conical surface.
(AP/ANT) 145-1993

conical-directional reflectance Ratio of reflected flux collected over an element of solid angle surrounding the given direction to the incident flux limited to a conical solid angle. See the corresponding figure. *Note:* The direction and extent of the cone must be specified and the direction of collection and size of the solid angle "element" must be specified.

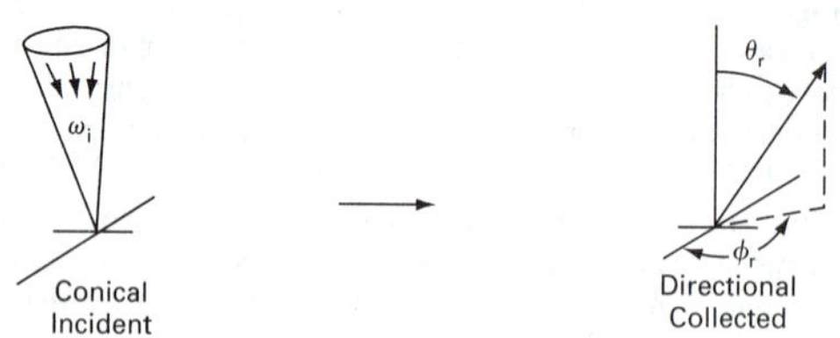

conical-directional reflectance

(EEC/IE) [126]

conical-directional transmittance Ratio of transmitted flux, collected over an element of solid angle surrounding the direction, to the incident flux to a conical solid angle. *Note:* The direction and extent of the cone must be specified and the direction of collection and size of the solid angle "element" must be specified. (EEC/IE) [126]

conical-hemispherical reflectance, $\rho(\omega_i; 2\pi)$ (illuminating engineering) Ratio of reflected flux collected over the entire hemisphere to the incident flux limited to a conical solid angle. *Note:* The direction and extent of the cone must be specified. (EEC/IE) [126]

conical-hemispherical transmittance, $\tau(\omega_t; 2\pi)$ (illuminating engineering) Ratio of transmitted flux collected over the entire hemisphere to the incident flux limited to a conical solid angle. (See the corresponding figure.). *Note:* The direction and extent of the cone must be specified.

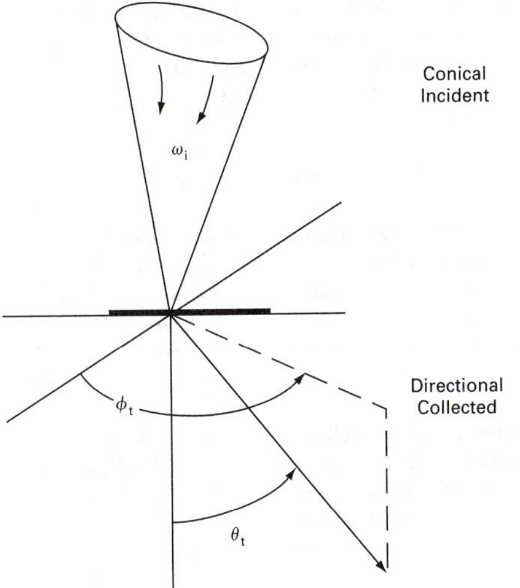

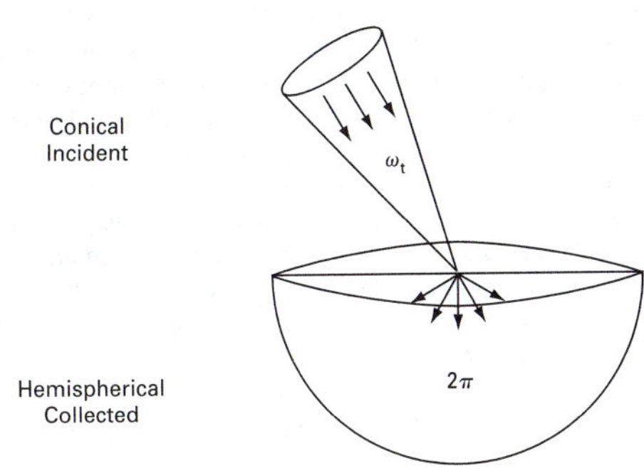

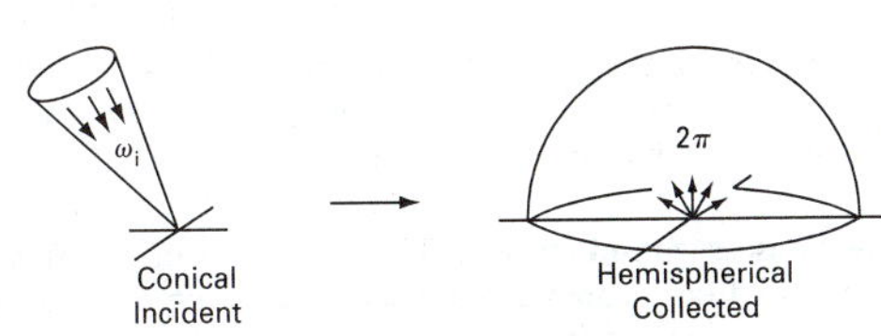

conical-hemispherical transmittance

(EEC/IE) [126]

conical horn A horn whose cross-sectional area increases as the square of the axial length. (SP) [32]

conical scanning (1) A form of sequential lobing in which the direction of maximum radiation generates a cone whose vertex angle is of the order of the antenna half-power beamwidth. *Note:* Such scanning may be either rotating or nutating according to whether the direction of polarization rotates or remains unchanged. *See also:* antenna. (AP/ANT) 149-1979r, 145-1993

(2) A form of angular tracking in which the antenna beam is offset from the tracking axis of the antenna. Rotation of the beam about the axis generates a cone whose vertex angle is of the order of the beamwidth. *Note:* Such scanning may be either rotating or nutating, according to whether the direction of polarization rotates or remains unchanged. The variation of signal amplitude as the beam scans provides information on the amount and direction of displacement of the target from the axis of rotation. (AES) 686-1997

conical-scan-on-receive-only (COSRO) A method of angle tracking in which only the receiving beam is conically scanned. (AES) 686-1997

conical wave (radio-wave propagation) A wave whose equiphase surfaces asymptotically form a family of coaxial circular cones. (AP) 211-1977s

coning effect (gyros) The apparent drift rate caused by motion of an input axis in a manner that generally describes a cone. This usually results from a combination of oscillatory motions about the gyro principal axes. The apparent drift rate is a function of the amplitudes and frequencies of oscillations present and the phase angles between them. (AES/GYAC) 528-1994

conjugate bridge The detector circuit and the supply circuit are interchanged as compared with a normal bridge of the given type. *See also:* bridge. (EEC/PE) [119]

conjugate impedance An impedance the value of which is the complex conjugate of a given impedance. *Note:* For an impedance associated with an electric network, the complex conjugate is an impedance with the same resistance component and a reactance component the negative of the original. (SP/IM/HFIM) 151-1965w, 270-1966w, [40]

conjugate termination A termination whose input impedance is the complex conjugate of the output impedance of the source or network to which it is connected. *See also:* transmission line. (IM/HFIM) [40]

conjunction (1) The logical 'AND' operator. (C) 610.5-1990w, 1084-1986w

(2) The result of joining two conditions by the logical 'AND' operator. *Contrast:* disjunction. *See also:* conjunctive query. (C) 610.5-1990w

(3) The Boolean operation whose result has the value 1 if and only if each operand has the value 1. *Contrast:* nonconjunction. *See also:* AND gate. (C) 610.10-1994w

conjunctive query A database query formed by using one of the logical operators 'AND' and 'OR.'. (C) 610.5-1990w

connect (1) The capability for communication, across a Group or LAN, between adjacent Bridges, or between the Bridge Ports of adjacent Bridges. *Contrast:* attach. (C/LM) 802.1G-1996

(2) Indicates that a device communications controller (DCC) has been physically attached to a bedside communications controller (BCC) port, by means of an interconnecting cable. Connect does not necessarily imply that communications have been logically established between the BCC and the DCC. (EMB/MIB) 1073.4.1-2000

connected (1) (networks) A network is connected if there exists at least one path, composed of branches of the network, between every pair of nodes of the network. *See also:* network analysis. (BT) 153-1950w

(2) (graph) A graph is connected if there exists at least one path between every pair of its vertices. (CAS) [13]

(3) (image processing and pattern recognition) Pertaining to a subset of an image, any two points of which can be joined by an arc that is entirely contained within the subset.

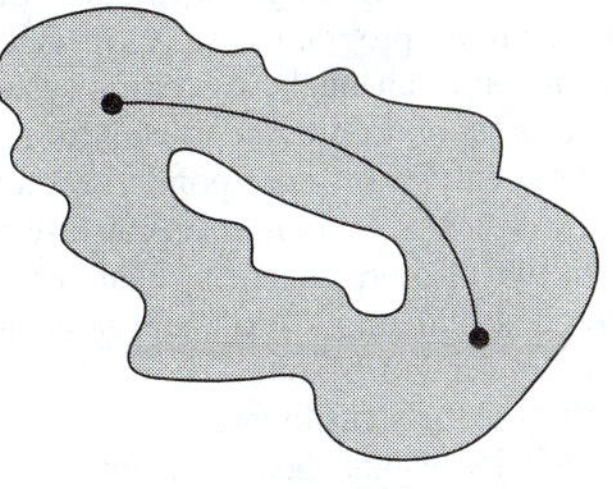

connected

(C) 610.4-1990w

(4) (maintenance of energized power lines) *See also:* bonded. (T&D/PE) 516-1987s, 524-1992r

connected load (1) The sum of the continuous ratings of the load-consuming apparatus connected to the system or any part thereof. *See also:* generating station. (T&D/PE) [10]

(2) (electric power systems in commercial buildings) The sum of the continuous ratings of the power consuming ap-

paratus connected to the system or any part thereof in watts, kilowatts, or horsepower. (IA/PSE) 241-1990r
(3) The connected transformer kVA, peak load, or metered demand (to be clearly specified when reporting) on the circuit or portion of circuit that is interrupted. When reporting, the report should state whether it is based on an annual peak or on a reporting period peak. (PE/T&D) 1366-1998

connected PHY A peer cable PHY at the other end of a particular physical connection from the local PHY. (C/MM) 1394-1995

connected position (of a switchgear-assembly removable element) That position of the removable element in which both primary and secondary disconnecting devices are in full contact. (SWG/PE) C37.100-1992

connected slave A slave that is permitted to participate actively in the data transfer handshake. A selected slave that is not disabled (including a diverted slave), a reflecting slave, and an intervening slave are connected slaves. A disabled slave or, an unselected slave that is not intervening or reflecting, is not a connected slave. (MM/C) 896.1-1987s

connected system All segments of a connected system are capable of communicating directly with one another through SIs. Note that because of the route map table implementation of message paths, segments of a system that are connected electrically by SIs are not necessarily also logically connected in the sense used here. (NID) 960-1993

connected transaction A transaction in which both the request and response are performed within the same bus transaction. (C/BA) 1014.1-1994w, 896.4-1993w, 10857-1994, 896.3-1993w

connecting rod (high voltage air switches, insulators, and bus supports) A component of a switch operating mechanism designed to transmit motion from an offset bearing or bell crank to a switch pole unit. *Synonym:* connecting shaft. (SWG/PE) C37.30-1971s, C37.100-1992

connecting shaft *See:* connecting rod.

connecting wire A wire generally of smaller gauge than the shot-firing cord and used for connecting the electric blasting-cap wires from one drill hole to those of an adjoining one in mines, quarries, and tunnels. *See also:* mine feeder circuit. (EEC/PE/MIN) [119]

connection (1) (rotating machinery) Any low-impedance tie between electrically conducting components. (PE) [9]
(2) (nuclear power generating station) (cable, field splice, and connection qualification) (design and installation of cable systems for Class 1E circuits in nuclear power generating stations) A cable terminal, splice, or hostile environment boundary seal at the interface of cable and equipment. (PE/NP/EDPG) 383-1974r, 690-1984r
(3) (A) (software) A reference in one part of a program to the identifier of another part (that is something found elsewhere). **(B) (software)** An association established between functional units for conveying information. *See also:* functional unit; identifier; program. (C/SE) 729-1983
(4) An association of channels, switching systems, and other functional units set up to provide means for a transfer of information between two or more points in a telecommunications network. *Notes:* 1. A connection can also be one channel or a series of two or more channels regardless of the method used to interconnect them. 2. A transmission path that is considered as a connection by one vendor may be resold as a channel to the vendor's customers.
(5) An association established by a *layer* between two or more users of the layer service for the transfer of information. (C) 8802-6-1994
(6) A parallel interface state that is outside the scope of this standard and is not defined herein. This state is indicative of the state of the physical or logical connection between a host and the printer. Only in this state can data be transferred between a host and the printer. (C/MM) 1284.1-1997
(7) An association established between two or more endpoints or the transfer of data. (C) 1003.5-1999
(8) A persistent communications path between two endpoints. (C/MM) 1284.4-2000

connection and winding support insulation (1) (insulation systems of synchronous machines) The connection and winding support insulation includes all of the insulation materials that envelope the connections, which carry current from coil to coil or from bar to bar, and from field and armature coil terminals to the points of external circuit and attachment; and also the insulation of metallic supports for the winding. (REM) [115]
(2) (repair and rewinding of motors for the petroleum and chemical industry) Includes all of the insulation materials that envelop the connections that carry current from coil to coil, and from stationary or rotating coil terminals to the points of external circuit attachment; and the insulation of any metallic supports for the winding. *See also:* insulation system. (IA/PC) 1068-1996

connection assembly (Class 1E connection assemblies) Any connector or termination combined with related cables or wires as an assembly. This assembly may include environmental seals, but excludes fire stops, in-line splices, and containment electric penetration assemblies. (PE/NP) 572-1985r

connection charge (power operations) The amount paid by a customer for connecting the customer's facilities to the supplier's facilities. (PE/PSE) 858-1987s, 346-1973w

connection diagram (1) A diagram that shows the connection of an installation or its component devices, controllers, and equipment. *Notes:* 1. It may cover internal or external connections, or both, and shall contain such detail as is needed to make or trace connections that are involved. It usually shows the general physical arrangement of devices and device elements and also accessory items such as terminal blocks, resistors, etc. 2. A connection diagram excludes mechanical drawings, commonly referred to as wiring templates, wiring assemblies, cable assemblies, etc. (PE/IA/TR/IAC) [116], [60]
(2) (packaging machinery) A diagram showing the electrical connections between the parts comprising the control and indicating the external connections. (IA/PKG) 333-1980w
(3) A diagram showing the relation and connections of devices and apparatus of a circuit or a group of circuits. (SWG/PE) C37.100-1992

connection insulation (rotating machinery) (joint insulation) The insulation at an electric connection such as between turns or coils or at a bushing connection. *See also:* stator. (PE) [9]

connectionless confidentiality The protection of (N)-service data units from unauthorized disclosure during transmission from one (N+1)-entity to one or more (N+1)-entities, where each entity has an association with the physical layer, and no association is established for the transmission of data or for the application of the confidentiality service between the layer peer-entities themselves. (C/LM) 802.10-1998

connectionless integrity A service providing for the integrity of a single SDU. It may take the form of determining whether or not the received SDU has been modified. (C/LM) 802.10-1998

connectionless mode The transfer of data other than in the context of a connection. *See also:* connection mode; datagram. (C) 1003.5-1999

connectionless service (1) A kind of delivery service that treats each packet as a separate entity. Each packet contains all protocol layers and destination address at each intermediate node in the network. *Note:* Order of arrival of packets is not necessarily the same as order of transmission. *Synonym:* datagram. (C) 610.7-1995
(2) A kind of delivery service offered by most hardware that treats each packet or datagram as a separate entity containing the source and destination address. (C) 610.10-1994w

connection mode The transfer of data in the context of a connection. *See also:* connectionless mode. (C) 1003.5-1999

connection-oriented confidentiality The protection of all (N)-service data units from unauthorized disclosure during communications from one (N+1)-entity to one or more (N+1)-entities for which a security association is established for the transfer of data and for the application of confidentiality service between the entities themselves and between each entity and the physical layer. (C/LM) 802.10-1998

connection-oriented integrity A service providing for the integrity of all (N)-service data on a security association and detecting any modification, insertion, deletion, or replay of any data within an entire SDU sequence. (C/LM) 802.10-1998

connection-oriented service A kind of delivery service where different virtual circuit configurations are used to transmit messages. *Synonym:* virtual circuit service. (C) 610.7-1995

connection phase A beat that begins with the assertion of the address synchronization line followed by the release of an address acknowledge line. It is used to broadcast the address and command information. Modules determine whether they wish to take part in the transaction based on this information. (C/BA) 10857-1994, 896.4-1993w, 896.3-1993w

connections In-line and tee splices that are designed to join heating cables or heating cables and power leads. (IA/PC) 515.1-1995

connections of polyphase circuits *See:* star-connected circuit; zig-zag connection of polyphase circuits; mesh-connected circuit.

connector **(1)** A coupling device employed to connect conductors of one circuit or transmission element with those of another circuit or transmission element. *See also:* auxiliary device to an instrument. (IM/ELM/HFIM) [40], C12.8-1981r
(2) (wires) A device attached to two or more wires or cables for the purpose of connecting electric circuits without the use of permanent splices. (VT/LT) 16-1955w
(3) (splicing sleeve) A metal sleeve, that is slipped over and secured to the butted ends of the conductors in making up a joint. (T&D/PE) [10]
(4) (waveguides) A mechanical device, excluding an adapter, for electrically joining separable parts of a waveguide or transmission-line system. (AP/ANT) [35]
(5) (power cable joints) A metallic device of suitable electric conductance and mechanical strength, used to splice the ends of two or more cable conductors, or as a terminal connector on a single conductor. Connectors usually fall into one of the following types: solder, welded, mechanical, and compression or indent. Conductors are sometimes spliced without connectors, by soldering, brazing or welding. (PE/IC) 404-1986s
(6) (substation grounding) A metallic device of suitable electric conductance and mechanical strength used to connect conductors. (SUB/PE) 837-1989r
(7) (fiber optics) *See also:* optical waveguide connector. 812-1984w
(8) *See also:* connector link. (PE/T&D) 524-1992r

connector base (motor plug) (motor attachment plug cap) A device, intended for flush or surface mounting on an appliance, that serves to connect the appliance to a cord connector. (EEC/PE) [119]

connector insertion loss *See:* insertion loss.

connector link (conductor stringing equipment) A rigid link designed to connect pulling lines and conductors together in series. It will not spin and relieve torsional forces. *Synonyms:* bullet; link; link; slug; connector; connector. (T&D/PE) 524-1992r

connector, precision *See:* precision connector.

connector, rope *See:* rope connector.

connector switch (connector) A remotely controlled switch for connecting a trunk to the called line. (EEC/PE) [119]

connector thermal capacity (substation grounding) The ability of a connector to withstand the amount of current required to produce a specified temperature on the control conductor without increasing the resistance of the connector beyond that specified in IEEE Std 837-1989. (SUB/PE) 837-1989r

connector, waveguide *See:* waveguide connector.

connect time In time-sharing computer systems, the time that a terminal or user is connected and able to communicate with a computer. *See also:* CPU time. (C) 610.10-1994w

conopulse A tracking radar that uses two simultaneous squinted beams that are rotated around the antenna boresight to produce, on a time-shared basis, monopulse angle-error signals in two orthogonal coordinates (such as azimuth and elevation). *Notes:* 1. Only two receivers are required rather than three as in the usual monopulse tracker. 2. Also called "konopulse" and "scan with compensation." (AES) 686-1997

consecutive Pertaining to the occurrence of two sequential events or items without the intervention of any other event or item; that is, one immediately after the other. (C) 610.12-1990

consecutive sequence computer A type of computer in which instructions are executed in an implicitly defined sequence unless a jump instruction specifies the storage location of the next instruction to be executed. *Contrast:* arbitrary sequence computer. (C) 610.10-1994w

consecutive spill method *See:* linear probing.

consequences The result(s) of (i.e., events that follow and depend upon) a specified event. (PE/NP) 1082-1997

conservation of radiance (fiber optics) A basic principle stating that no passive optical system can increase the quantity Ln^{-2} where L is the radiance of a beam and n is the local refractive index. Formerly called "conservation of brightness" or the "brightness theorem." *See also:* brightness; radiance. (Std100) 812-1984w

conservator (expansion tank system) (power and distribution transformers) A system in which the oil in the main tank is sealed from the atmosphere, over the temperature range specified, by means of an auxiliary tank partly filled with oil and connected to the completely filled main tank. (PE/TR) C57.12.80-1978r

conservator/diaphragm system (power and distribution transformers) A system in which the oil in the main tank is completely sealed from the outside atmosphere, and is connected to an elastic diaphragm tank contained inside a tank mounted at the top of the transformer. As oil expands and contracts within a specified temperature range the system remains completely sealed with an approximately constant pressure. (PE/TR) C57.12.80-1978r

conservator system An oil preservation system in which the oil in the main tank is sealed from the atmosphere, over the temperature range specified, by means of an ancillary tank partly filled with oil and connected to the completely filled main tank. *Synonym:* expansion tank system. (PE/TR) C57.15-1999

consist The makeup or composition (number and specific identity) of individual units on a train. (VT/RT) 1477-1998, 1475-1999, 1474.1-1999

consistency (software) The degree of uniformity, standardization, and freedom from contradiction among the documents or parts of a system or component. *See also:* traceability. (C) 610.12-1990

consistency check A check that verifies that an item of data is compatible with certain rules specified for that data. For example, one might wish to check the consistency between two data elements ORDER-DATE and DELIVER-DATE such that DELIVER-DATE may not be earlier than ORDER-DATE. *See also:* limit check; crossfooting check; range check; validity check. (C) 610.5-1990w

consol (navigation aid terms) A keyed C-W (continuous wave) short-baseline-radio navigation system operating in the L/MF (low- and medium frequency) band, generally useful to about 1500 nmi (nautical miles) (2800 kilometers [km]), and using three radiators to provide a multiplicity of overlapping lobes of dot-and-dash patterns which form equisignal hyperbolic lines of position. These lines of position are moved slowly in azimuth by changing rf (radio frequency) phase, thus allowing a simple listening and counting of timing operation to be used to determine a line of position within the sector bounded by any pair of equisignal lines. (AES/GCS) 172-1983w

consolan (navigation aid terms) A form of consol using two radiators instead of three. (AES/GCS) 172-1983w

console (1) (telephony) A control cabinet located apart from the associated switching equipment arranged to control those functions for which an attendant or an operator is required. (SWG/COM) [48]
(2) (telephony) The part of a computer used for communication between the operator or maintenance engineer and the computer. (C) [85]
(3) (telephone switching systems) (telephony) A desk or desk-top cordless switchboard which may include display elements in addition to those required for supervisory purposes is required. (COM) 312-1977w
(4) (supervisory control, data acquisition, data control) (station control and data acquisition) (telephony) That component of the system that provides facilities for observation and control of the system (e.g., operator's console, maintenance console). *See also:* control panel; control; panel. (SWG/PE/SUB) C37.100-1992, C37.1-1994
(5) A device used as the primary means of communication with a human being, consisting of an input device, used for receiving information supplied by the human, and an output device, used for sending information to the human. (Typically, a console is either an ASCII terminal connected to a serial port or the combination of a text/graphics display device and a keyboard.). (C/BA) 1275-1994
(6) A functional unit used for communication between the computer operator and the computer. *Note:* May provide special-purpose keys, input devices, and display devices employed to operate and control the computer. *Synonyms:* console display; display console. *See also:* control panel. (C) 610.10-1994w

console display *See:* console.

console language The human language in which information is to be displayed on local or remote consoles. (C/MM) 1284.1-1997

consonant articulation (percent consonant articulation) The percent articulation obtained when the speech units considered are consonants (usually combined with vowels into meaningless syllables). *See also:* volume equivalent. (EEC/PE) [119]

conspicuity (illuminating engineering) The capacity of a signal to stand out in relation to its background so as to be readily discovered by the eye. (EEC/IE) [126]

constancy *See:* residual probe pickup.

constant (1) (computers) A quantity or data item whose value cannot change; for example, the data item FIVE, with an unchanging value of 5. *Contrast:* variable. *See also:* literal; figurative constant. (C) 610.12-1990
(2) (A) (As a noun) An instance whose identity is known at the time of writing. The identity of a constant state class instance is represented by #K, where K is an integer or a name. **(B)** (As an adjective) The specification that an attribute or participant property value, once assigned, may not be changed, or that an operation shall always provide the same output argument values given the same input argument values. (C/SE) 1320.2-1998
(3) *See also:* time constant of integrator.

constant address *See:* base address.

constant-amplitude recording (mechanical recording) A characteristic wherein, for a fixed amplitude of a sinusoidal signal, the resulting recorded amplitude is independent of frequency. *See also:* phonograph pickup. (SP) [32]

constant available power source (transmission performance of telephone sets) A signal source with a purely resistive internal impedance and a constant open-circuit terminal voltage, independent of frequency. *Note:* Receiving is tested under conditions of constant available power. A generator having an open-circuit voltage, E, and an internal voltage, R_θ, both constant with frequency, provides a constant available power of $E^2/4R_\theta$. When the load impedance is a resistance equal to R_θ, an impedance match exists and the maximum available power of $E^2/4R_\theta$, is dissipated in the load. When a generator with a constant open-circuit voltage and a constant internal resistance of the required value is not available, suitable test conditions for this type of measurement can be provided by a generator whose voltage is maintained at a constant value, E_θ, across its terminals in series with a resistance, R_θ. The generator and the resistance, R_θ, in series are then equivalent to a source of constant available power. (COM/TA) 269-1983s

constant-current (Heising) modulation A system of amplitude modulation wherein the output circuits of the signal amplifier and the carrier-wave generator or amplifier are directly and conductively coupled by means of a common inductor that has ideally infinite impedance to the signal frequencies and that therefore maintains the common plate-supply current of the two devices constant. *Note:* The signal-frequency voltage thus appearing across the common inductor appears also as modulation of the plate supply to the carrier generator or amplifier with corresponding modulation of the carrier output. (AP/ANT) 145-1983s

constant-current arc-welding power supply A power supply that has characteristically drooping volt-ampere curves producing relatively constant current with a limited change in load voltage. *Note:* This type of supply is conventionally used in connection with manual-stick-electrode or tungsten-inert-gas arc welding. (EEC/AWM) [91]

constant-current characteristic (electron tube) The relation, usually represented by a graph, between the voltages of two electrodes, with the current to one of them as well as all other voltages maintained constant. (ED) 161-1971w

constant-current charge (storage battery) (storage cell) A charge in which the current is maintained at a constant value. *Note:* For some types of lead-acid batteries this may involve two rates called the starting and finishing rates. *See also:* charge. (EEC/PE) [119]

constant-current (series) incandescent filament lamp transformer *See:* incandescent-filament-lamp transformer.

constant current loads A load that demands constant current even when the input voltage varies. *Note:* Typical of such loads is lighting when driven from an inverter ballast configuration. (VT) 1476-2000

constant-current (series) mercury-lamp transformer A transformer that receives power from a current-regulated series circuit and transforms the power to another circuit at the same or different current from that in the primary circuit. *Note:* It also provides the required starting and operating voltage and current for the specified lamp. Further, it provides protection to the secondary circuit, casing, lamp, and associated luminaire from the high voltage of the primary circuit. (EEC/LB) [98]

constant-current power supply A power supply that is capable of maintaining a preset current through a variable load resistance. *Note:* This is achieved by automatically varying the load voltage in order to maintain the ratio V_{load}/R_{load} constant. (AES) [41]

constant-current regulation (generator) That type of automatic regulation in which the regulator maintains a constant-current output from the generator. *See also:* axle-generator system. (EEC/PE) [119]

constant current retention (metal-nitride-oxide field-effect transistor) Retention inherent in the metal-nitride-oxide-semiconductor (MNOS) transistor when gate and drain are biased to result in a constant drain current during information storage. The time period is defined by the intersection of the high conduction (HC) threshold voltage curve obtained under constant current condition, with the low conduction (LC) threshold voltage curve obtained under zero bias condition, when both are plotted against the logarithm of trd, the time elapsed between writing and the threshold voltage measurement. (ED) 581-1978w

constant-current street-lighting system (series street-lighting system) A street-lighting system employing a series circuit in which the current is maintained substantially constant. *Note:* Special generators or rectifiers are used for direct current while suitable regulators or transformers are used for alternating current. *See also:* alternating-current distribution; direct-current distribution. (T&D/PE) [10]

constant-current transformer (power and distribution transformers) A transformer that automatically maintains an approximately constant current in its secondary circuit under varying conditions of load impedance when supplied from an approximately constant-voltage source. *See also:* rated primary voltage of a constant current transformer; rated kilowatts; rated secondary current of a constant-current transformer; current regulation. (PE/TR) C57.12.80-1978r

constant cutting speed (numerically controlled machines) The condition achieved by varying the speed of rotation of the workpiece relative to the tool inversely proportional to the distance of the tool from the center of rotation. (EEC) [74]

constant-delay discriminator *See:* pulse decoder.

constant-failure period (software) The period of time in the life cycle of a system or component during which hardware failures occur at an approximately uniform rate. *Contrast:* wearout-failure period; early-failure period. *See also:* bathtub curve. (C) 610.12-1990

constant failure rate period That possible period during which the failures occur at an approximately uniform rate. *Note:* The curve in the figure below shows the failure rate pattern when the terms of minor failure, early failure period, and constant failure rate period all apply to the item.

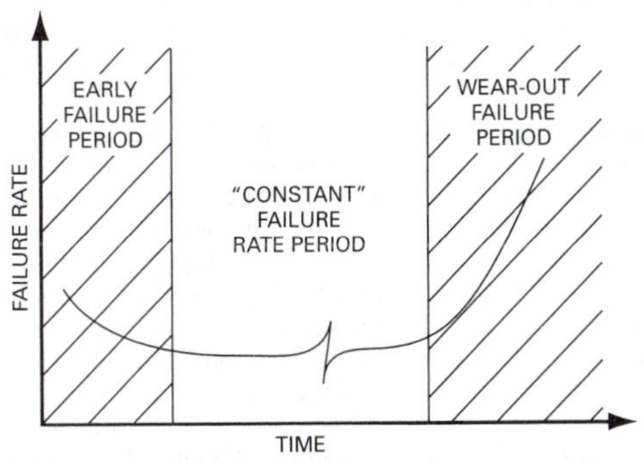

constant failure rate period

(R/VT) [37]

constant-false-alarm rate A property of threshold or gain control devices that maintain an approximately constant rate of false target detections when the noise, and/or clutter levels, and/or ECM (electronic countermeasures) into the detector are variable. (AES/RS) 686-1990

constant-false-alarm rate receiver A radar receiver that maintains the output false-alarm rate constant in spite of the varying nature of the receiver noise level, echoes from the clutter environment, or from electronic countermeasures. *Note:* CFAR is usually achieved by establishing a threshold level that varies according to the local noise and/or clutter environment measured in the near vicinity of the target echo. (AES) 686-1997

constant-fraction discriminator (1) A discriminator in which the threshold is set at a fixed fraction of the input signal (instead of to a fixed amplitude). This is one of a class of timing discriminators. (NPS) 325-1996
(2) An amplitude discriminator in which the triggering threshold is set at a constant percentage of the input pulse amplitude independent of its size. (NPS) 300-1988r

constant-frequency control (power system) For a power system, a mode of operation under load-frequency control in which the area control error is directly proportional to the frequency error. (PE/PSE) 94-1991w

constant-horsepower motor *See:* constant-power motor.

constant-horsepower range (electric drive) The portion of its speed range within which the drive is capable of maintaining essentially constant horsepower. *See also:* electric drive. (IA/IAC) [60]

constant-linear-bit recording A method for recording information on a storage device whereby the rotational speed is kept constant, but the data rate (density) is varied with the track to ensure that data is stored with the same number of bits per inch in all tracks. *Contrast:* constant-linear-velocity recording. (C) 610.10-1994w

constant-linear-velocity recording A method for recording information on a circular disk whereby the rotational speed is varied so that the speed of the storage medium past the recording head is constant for all tracks on the disk. *Note:* Since the outer-most tracks are longer than the inner-most tracks, this allows the device to store more information there. *Synonym:* group code recording. *Contrast:* constant-linear-bit recording. (C) 610.10-1994w

constant-luminance transmission (color television) A type of transmission in which the sole control of luminance is provided by the luminance signal, and no control of luminance is provided by the chrominance signal. *Notes:* 1. In such a system, noise signals falling within the bandwidth of the chrominance channel produce only chromaticity variations at the outputs of the chromiance demodulators. Coarse-structured chromaticity variations thus produced are subjectively less objectionable than correspondingly coarse-structured luminance variations. 2. Because of the use of gamma correction in the camera, these ideal conditions are not completely realized, especially for colors of high saturation. (BT/AV) 201-1979w, [34]

constant multiplier (computers) A computing element that multiplies a variable by a constant factor. *See also:* electronic analog computer; multiplier. (C) 165-1977w, 166-1977w

constant-net-interchange control (power system) A mode of operation under load-frequency control in which the area control error is proportional to the net interchange error. (PE/PSE) 94-1991w

constant potential charge A charge in which the voltage at the output terminals of the charger is held to a constant value. (IA/PSE) 602-1996

constant power load A load that demands constant power from the source even when the voltage value drops such as when switching from the low-voltage power supply to the battery. *Note:* Typical of such loads are those that have their own built-in regulator such as propulsion control power supplies. (VT) 1476-2000

constant-power motor (constant-horsepower motor) A multispeed motor that develops the same related power output at all operating speeds. The torque then is inversely proportional to the speed. *See also:* asynchronous machine. (PE) [9]

constant-resistance (conductance) network A network having at least one driving-point impedance (admittance) that is a positive constant. (EEC/PE) [119]

constant-speed motor (electric installations on shipboard) A motor, the speed of normal operation of which is constant or practically constant. For example, a synchronous motor, an induction motor with small slip, or an ordinary dc shunt-wound motor. (IA/MT) 45-1983s

constant-torque motor Multispeed motor that is capable of developing the same torque for all design speeds. The rated power output varies directly with the speed. *See also:* asynchronous machine. (PE) [9]

constant-torque range (electric drive) The portion of its speed range within which the drive is capable of maintaining essentially constant torque. *See also:* electric drive. (IA/IAC) [60]

constant-torque resistor A resistor for use in the armature or rotor circuit of a motor in which the current remains practically constant throughout the entire speed range. (IA/MT) 45-1998

constant-torque speed range The portion of the speed range of a drive within which the drive is capable of maintaining essentially constant torque. (IA/ICTL/IAC) [60]

constant-velocity recording (mechanical recording) A characteristic wherein for a fixed amplitude of a sinusoidal signal, the resulting recorded amplitude is inversely proportional to the frequency. *See also:* phonograph pickup. (SP) [32]

constant-voltage arc-welding power supply Power supply (arc welder) that has characteristically flat volt-amperer curves producing relatively constant voltage with a change in load current. This type of power supply is conventionally used in connection with welding processes involving consumable electrodes fed at a constant rate. (EEC/AWM) [91]

constant-voltage charge (storage battery) (storage cell) A charge in which the voltage at the terminals of the battery is held at a constant value. *See also:* charge. (PE/EEC) [119]

constant-voltage regulation (generator) That type of automatic regulation in which the regulator maintains constant voltage of the generator. (PE/EEC) [119]

constant voltage retention (metal-nitride-oxide field-effect transistor) Retention inherent in the metal-nitride-oxide-semiconductor (MNOS) transistor when source, drain, and substrate are grounded, and a fixed read bias VGR is maintained at the gate. The time period is defined by the intersection of the two high conduction (HC) and low conduction (LC) threshold voltage curves obtained under this condition when plotted versus the logarithm of t_{rd}, the time elapsed between writing and threshold voltage measurement. (ED) 581-1978w

constant-voltage transformer (power and distribution transformers) A transformer that maintains an approximately constant voltage ratio over the range from zero to rated output. (PE/TR) C57.12.80-1978r

constitutive parameters The permittivity and permeability of a medium. *See also:* permeability; permittivity. (AP/PROP) 211-1997

constitutive relations (radio-wave propagation) Constraints imposed by the medium on the relationships between electric and magnetic field vectors and their respective flux density vectors. (AP) 211-1977s

constrained painting *See:* grid constraint.

constraint (1) A restriction on software life cycle process (SLCP) development. (C/SE) 1074-1997
(2) A limitation or implied requirement that constrains the design solution or implementation of the systems engineering process, is not changeable by the enterprise, and is generally nonallocable. (C/SE) 1220-1998
(3) An externally imposed limitation on system requirements, design, or implementation or on the process used to develop or modify a system. (C/SE) 1362-1998
(4) A statement that expresses measurable bounds for an element or function of the system. That is, a constraint is a factor that is imposed on the solution by force or compulsion and may limit or modify the design changes. (C/SE) 1233-1998
(5) **(A)** A kind of **responsibility** that is a statement of facts that are required to be true in order for the constraint to be met. Classes have constraints, expressed in the form of logical sentences about property values. An instance conforms to the constraint if the logical sentence is true. Some constraints are inherent in the modeling constructs; other constraints are specific to a particular model and are stated in the specification language. **(B)** A rule that specifies a valid condition of data. (C/SE) 1320.2-1998
(6) A timing property of a design that is supplied as a goal or objective to an electronic design automation (EDA) tool, such as logic synthesis, floorplanning, or layout. The tool shall not start out with a fixed design implementation; it shall build or modify the design to meet the constraint. *See also:* timing check. (C/DA) 1481-1999

constraints (1) Limits on the ranges of variables or system parameters because of physical or system requirements. *See also:* system. (SMC) [63]
(2) A restriction placed on the control signal, control law, or state variables. *See also:* control system. (CS/IM) [120]
(3) Conditions and/or resource requirement limitations affecting the process. (C/SE) 1209-1992w
(4) Restrictions, resources, rules, etc., that limit Software Life Cycle Model selection, project planning, and management. (C/SE) 1074.1-1995

construction The process of writing, assembling, or generating assets. (C/SE) 1517-1999

construction agency *See:* constructor.

construction department *See:* constructor.

construction diagram A diagram that shows the physical arrangement of parts, such as wiring, buses, resistor units, etc. Example: A diagram showing the arrangement of grids and terminals in a grid-type resistor. (IA/ICTL/IAC) [60]

construction stage The time related to the installation or modification of fixtures or structures, including services, foundations, steel, conductors, buildings, and grounding. (PE/SUB) 1402-2000

construction test (Class 1E power systems and equipment) A test to verify proper installationand operation of individual components in a system prior to operation of the system as an entity. It is assumed that the construction test does not verify the interconnected-system equipment external to that component. For example, the protective relays are bench tested by simulating fault conditions to verify conformance with approved characteristics. During bench tests the alarm, trip, and permissive-interlock functions that the protective relay circuits are to perform are not verified. (PE/NP) 415-1986w

construction testing Performing required inspections and tests to ensure that completed installations are in accordance with contract requirements and the latest engineering and design information. (PE/EDPG) 1248-1998

constructor (1) A party who undertakes the assembly and erection of a transmission structure for an owner, or an owner who undertakes all or part of a project alone. *Synonyms:* installer; contractor; construction department; construction agency. (T&D/PE) 1025-1993r
(2) A party who undertakes the assembly and erection of a transmission structure. The constructor can be an owner or an agent acting for an owner. *Synonym:* construction department; construction agency; installer; contractor. (T&D/PE) 951-1996

Consultative Committee on International Telegraphy and Telephony (CCITT) An international organization that studies and issues recommendations on issues related to communication technology. *Note:* In March 1993, the CCITT was reorganized and renamed to be the International Telecommunication Union (ITU) Telecommunications Standardization Sector (TSS). (C) 610.7-1995

consumer (1) The node on a ringlet that strips a send packet from the ringlet and creates the echo packet that is returned to the producer. (C/MM) 1596-1992
(2) A unit that removes messages from a DMA queue. (C/MM) 1212.1-1993

consumption of idles Idle symbols arriving at a node may be discarded (after saving certain information) while other symbols that arrived earlier and were stored in the bypass FIFO are being transmitted. Consuming idles thus reduces the number of symbols stored in the bypass FIFO. (C/MM) 1596-1992

consumer product An end item delivered to a customer. (C/SE) 1220-1994s

contact (A) (general). A conducting part that co-acts with another conducting part to make or break a circuit. **(B)** (of a relay). A conducting part that acts with another conducting part to make or break a circuit. (SWG/SWG/PE/PSR) C37.30-1971, C37.100-1992, C37.90-1978

contact area (1) (photoelectric converter) The area of ohmic contact provided on either the p or n faces of a photoelectric converter for electric circuit connections. *See also:* semiconductor. (AES) [41]
(2) (solar cells) That area of ohmic contact provided on either the p or n surface of a solar cell for electric circuit connections. (AES/SS) 307-1969w

contact bounce (dial-pulse address signaling systems) (telephony) The intermittent and undesired opening of contacts during the closure of open contacts or opening of closed contacts. An irregular wavefront during transition from one state to the other is implied. (COM/TA) 753-1983w

contact chatter, relay *See:* relay contact chatter.

contact clip (of a mechanical switching device) The clip that the blade enters or embraces. (SWG/PE) C37.100-1992

contact clips *See:* fuse clips.

contact conductor (electric traction) The part of the distribution system other than the track rails, that is in immediate electric contact with current collectors of the cars or locomotives. *See also:* contact wire; trolley; underground collector or plow; multiple-unit control. (VT/LT) 16-1955w

contact converter (as applied to relaying) A buffer element used to produce a prescribed output as the result of the opening or closing of a contact. (SWG/PE) C37.100-1992

contact corrosion *See:* crevice corrosion.

contact current-carrying rating (of a relay) The current that can be carried continuously or for stated periodic intervals without impairment of the contact structure or interrupting capability. (SWG/PE/PSR) C37.100-1992, C37.90-1989r

contact current-closing rating (of a relay) The current that the device can close successfully with prescribed operating duty and circuit conditions without significant impairment of the contact structure. (SWG/PE) C37.100-1992

contact discharge method (1) A method of ESD testing in which the electrode of the ESD simulator is in firm contact with a conductive surface of the EUT or coupling plane prior to discharge. The discharge is actuated by a switching device (i.e., a relay) within the simulator. (EMC) C63.16-1993
(2) A method of ESD testing in which the electrode of the ESD simulator is in firm conductive contact with the UUT or coupling plane prior to and during the discharge. The discharge is actuated by a switching device, such as a relay, within the simulator. (SPD/PE) C62.38-1994r

contact flange (waveguide components) A flat flange used in conjunction with another flat flange to provide a contact joint. (MTT) 147-1979w

contact follow-up (relays, switchgear, and industrial control) The distance between the position one contact face would assume, were it not blocked by the second (mating) contact, and the position the second contact removed, when the actuating member is fixed in its final contact-closed position. *See also:* electric controller. (IA/ICTL/IAC) 74-1958w, [60]

contact gap (break) The final length of the isolating distance of a contact in the open position. *See also:* contactor. (IA/ICTL/IAC) [84], [60]

contact high recombination rate (semiconductor) A semiconductor-semiconductor or metal-semiconductor contact at which thermal equilibrium charge-carrier concentrations are maintained substantially independent of current density. *See also:* semiconductor device; semiconductor. (ED) 216-1960w

contact interrupting rating (of a relay) The current that the device can interrupt successfully with prescribed operating duty and circuit conditions without significant impairment of the contact structure. (SWG/PE/PSR) C37.100-1992, C37.90-1978s

contact, ion implanted *See:* ion-implanted contact.

contact joint (waveguide components) (contact coupling) A connection designed for essentially complete transfer of power between two waveguides by means of metallic contact between the inner walls of the waveguides. It typically consists of two contact flanges. (MTT) 147-1979w

contactless vibrating bell A vibrating bell whose continuous operation depends upon application of alternating-current power without circuit-interrupting contacts. *See also:* protective signaling. (EEC/PE) [119]

contact-making clock demand meter (metering) A device designed to close momentarily an electric circuit to a demand meter at periodic intervals. (ELM) C12.1-1982s

contact mechanism (demand meter) A device for attachment to an electricity meter or to a demand-totalizing relay for the the purpose of providing electric impulses for transmission to a demand meter relay. *See also:* demand meter. (EEC/PE) [119]

contact, ohmic *See:* ohmic contact.

contact opening time (of a relay) The time a contact remains closed while in process of opening following a specified change of input. (SWG/PE) C37.100-1992

contactor (1) (thyristor) A device for repeatedly establishing and interrupting an electric power circuit. (IA/PE/IPC/TR) 428-1981w, C57.12.80-1978r
(2) A device which upon receipt of an electrical signal establishes or opens repeatedly an electrical circuit with a nominal current rating of 5 amperes minimum for its main contacts. (VT/LT) 16-1955w

contactor, load *See:* load switch or contactor.

contactor or unit switch A device operated other than by hand for repeatedly establishing and interrupting an electric power circuit under normal conditions. *See also:* control switch. (VT/LT) 16-1955w

contact parting time (of a mechanical switching device) The interval between the time when the actuating quantity in the release circuit reaches the value causing actuation of the release and the instant when the primary arcing contacts have parted in all poles. *Note:* Contact parting time is the numerical sum of release delay and opening time. (SWG/PE) C37.100-1992

contact piston (waveguide) (contact plunger) A piston with sliding metallic contact with the walls of a waveguide. *See also:* waveguide. (AP/ANT) [35]

contact plating The deposition, without the application of an external electromotive force, of a metal coating upon a base metal, by immersing the latter in contact with another metal in a solution containing a compound of the metal to be deposited. *See also:* electroplating. (EEC/PE) [119]

contact plunger *See:* contact piston.

contact position indicator A device that is located at or near the operating mechanism to indicate whether the main contacts are in the closed or open position. Typically, colors are used to indicate a closed or open position; red shall signify closed and green shall signify open. (SWG/PE) C37.100-1992

contact potential The difference in potential existing at the contact of two media or phases. *See also:* positive after-potential; negative after-potential; depolarization; electrolytic cell; depolarization front. (EEC/PE) [119]

contact-potential difference The difference between the work functions of two materials divided by the electronic charge. (ED) 161-1971w

contact pressure, final *See:* final contact pressure.

contact pressure, initial *See:* initial contact pressure.

contact race A circuit design condition wherein two or more independently operated contacts compete for the control of a circuit which they will open and close. (PE/PSR) [6]

contact rectifier A rectifier consisting of two different solids in contact, in which rectification is due to greater conductivity across the contact in one direction than in the other. *See also:* rectifier. (AP/ANT) 145-1983s

contacts Conducting parts which co-act to complete or to interrupt a circuit. (EEC/PE) [119]
(2) (A) (nonoverlapping) Combinations of two sets of contacts, actuated by a common means, each set closing in one of two positions, and so arranged that the contacts of one set open before the contacts of the other set close. *See also:* electric controller. **(B) (switching device) (auxiliary)** Contacts in addition to the main circuit contacts that function with the movement of the latter. *See also:* contactor. **(C) (overlapping, industrial control)** Combinations of two sets of contacts, actuated by a common means, each set closing in one of two positions, and so arranged that the contacts of one set open after the contacts of the other set have been closed. *See also:* electric controller. (IA/ICTL/IAC) [60]

contact surface That surface of a contact through which current is transferred to the coacting contact. (SWG/PE) C37.100-1992

contact voltage (human safety) A voltage accidentally appearing between two points with which a person can simultaneously make contact. (PE) [8], [84]

contact-wear allowance The total thickness of material that may be worn away before the co-acting contacts cease to perform adequately. *See also:* contactor. (SWG/PE/ICTL) C37.30-1971s, C37.100-1992

contact wire (trolley wire) A flexible contact conductor, customarily supported above or to one side of the vehicle. *See also:* contact conductor. (VT/LT) 16-1955w

container (1) A parent widget that defines a region containing zero or more subobjects of a given type. (C) 1295-1993w
(2) An ordered set of 1, 2, 4, or 8 contiguous bytes fully packed with one or more signed or unsigned field formats. (C/MM) 1596.5-1993

container class A class that defines an object that holds other objects. (SCC20) 1226-1998

containment (1) (valve actuators) (safety systems equipment in nuclear power generating stations) That portion of the engineered safety features designed to act as the principal barrier, after the reactor system pressure boundary, to prevent the release, even under conditions of a reactor accident, of unacceptable quantities of radioactive material beyond a controlled zone. (PE/NP) 382-1985, 627-1980r, 323-1974s, 383-1974r, 334-1974s, 650-1979s
(2) (radiological monitoring instrumentation) A structure or vessel which encloses the components of the reactor coolant pressure boundary or which serves as a leakage limiting barrier to radioactive material that could be released from the reactor coolant pressure boundary, or both. (NI) N320-1979r
(3) (data management) The result of placing all occurrences of a repeating group within the same logical record. (C) 610.5-1990w
(4) A relationship between two objects such that one is said to belong to, or form part of, the other. All objects except software_collection objects shall be contained within exactly one object. (C/PA) 1387.2-1995

contamination (rotating machinery) This deteriorates electrical insulation by actually conducting current over insulated surfaces, or by attacking the material reducing its electrical insulating quality or its physical strength, or by thermally insulating the material forcing it to operate at higher than normal temperatures. *Note:* Included here are: wetness or extreme humidity, oil or grease, conducting dusts and particles, non-conducting dusts and particles, and chemicals of industry. (PE/EM) 432-1976s

contend To actively and simultaneously vie for the attention of the MTM-Bus Master module (said of a group or one or more S-modules). (TT/C) 1149.5-1995

contending requester *See:* requester.

contending slave *See:* interrupt-acknowledge cycle.

content addressable storage *See:* associative memory.

content addressed storage *See:* associative storage.

content, average information *See:* information theory; average information content, per symbol.

content, conditional information *See:* information theory.

content coupling A type of coupling in which some or all of the contents of one software module are included in the contents of another module. *Contrast:* control coupling; hybrid coupling; common-environment coupling; data coupling; pathological coupling. (C) 610.12-1990

contention (1) (data transmission) A condition on a multipoint communication channel when two or more locations try to transmit at the same time. (PE) 599-1985w
(2) (station control and data acquisition) An operational condition in which two or more devices simultaneously try to use the same resource (e.g., communication channel, disk, memory). (SUB/PE) C37.1-1994
(3) A condition on a communications channel when two or more stations may try to seize the channel at the same time. (LM/COM) 168-1956w
(4) A condition that occurs when two or more devices simultaneously request the services of another device, network medium, or resource that can handle only one request at a time. *See also:* collision. (C) 610.7-1995
(5) An operational condition on a data communication channel in which no station is designated a master station. In contention, each station on the channel shall monitor the signals on the channel and wait for a quiescent condition before initiating a bid for circuit control. (SWG/PE) C37.100-1992

contention interval *See:* slot time.

contention resolution The management of contention for a communications resource so as to minimize collisions, resolve access order, and maximize utilization. (C) 610.7-1995

contents list In word processing, a list of stored information available for user selection. (C) 610.2-1987

context (1) Reflects the intended scope of a set of tests. Examples of context include manufacturing process test, maintenance test, design verification test, screening test, etc. (SCC20) 1226-1998
(2) The immediate environment in which a function (or set of functions in a diagram) operates. (C/SE) 1320.1-1998

context diagram A diagram that presents the context of the top-level function of an IDEF0 model, whose diagram number is ***A-n***, where ***0≤n≤9***. The one-box A-0 context diagram is a required context diagram; those with diagram numbers ***A-1***, ***A-2***, . . ., ***A-9*** are optional context diagrams. (C/SE) 1320.1-1998

context editing A method of line editing in which the line to be viewed or altered is identified by specifying part or all of its contents. (C) 610.2-1987

context free The mode of API operation in which the underlying FTAM initiator establishes an FTAM Regime for the sole purpose of executing the requested operation, closing the regime once the operation is complete. (C/PA) 1238.1-1994w

context sensitive The mode of API operation in which the underlying FTAM initiator performs the requested operation, using a pre-existing FTAM Regime that is established and maintained independently of individual operation invocations. (C/PA) 1238.1-1994w

contiguous allocation A storage allocation technique in which programs or data to be stored are allocated a block of storage of equal or greater size, so that logically contiguous programs and data are assigned physically contiguous storage locations. *Contrast:* paging. (C) 610.12-1990

contiguous memory An area of storage that occupies consecutive or adjacent address locations. (C) 610.10-1994w

contingency The unexpected failure or outage of a system component(s) (generator, transmission line, breaker, switch, etc.). (PE/PSE) 858-1993w

continuation reference A reference that describes how the performance of all or part of a directory operation can be continued at a different DSA or DSAs. *See also:* referral. (C/PA) 1328.2-1993w, 1327.2-1993w, 1326.2-1993w, 1224.2-1993w

continuous rating (of diesel-generator unit) (of diesel-generator unit) The electric power output capability that the diesel-generator unit can maintain in the service environment for 8760 h of operation per year with only scheduled outages for maintenance. (PE/NP) 387-1995

continuing current (lightning) The low-magnitude current that may continue to flow between components of a multiple stroke. *See also:* direct-stroke protection. (T&D/PE) [10]

continuity cable bond A cable bond used for bonding of cable sheaths and armor across joints between continuous lengths of cable. *See also:* cable bond; cross cable bond. (T&D/PE) [10]

continuity test (1) (test, measurement, and diagnostic equipment) A test for the purpose of detecting broken or open connections and ground circuits in a network or device. (MIL) [2]
(2) (battery) A test on a cell/unit or battery to determine the integrity of its conduction path. (SB) 1188-1996

continuity tester (test, measurement, and diagnostic equipment) An electrical tester used to determine the presence and location of broken or open connections and grounded circuits. (MIL) [2]

continuity transfer relay contacts Sometimes used for relay contacts, make-break. (EEC/REE) [87]

continuous air monitor (cam) An instrument used to continuously sample and measure airborne radioactivity concentrations. (NI) N42.17B-1989r

continuous change model *See:* continuous model.

continuous corona (1) (corona measurement) Corona discharges that recur at regular intervals; for example, on approximately every cycle of an applied alternating voltage or at least once per minute for an applied direct voltage. (MAG/ET) 436-1977s
(2) (overhead-power-line corona and radio noise) Corona discharge that is either steady or recurring at regular intervals (approximately every cycle of an applied alternating voltage or at least several times per minute for an applied direct voltage). (T&D/PE) 539-1990

continuous cumulative demand The sum of (A) all previous billing period maximum demands, and (B) the highest demand to date for the present billing period. (ELM) C12.15-1990

continuous current (1) The maximum constant rms power frequency current that can be carried continuously without causing further measurable increase in temperature rise under prescribed conditions of test, and within the limitations of established standards. (PE/TR) C57.16-1996
(2) A current that is expected to continue for three hours or more. (IA/PSP) 1015-1997

continuous current rating The designated rms alternating or direct current that the connector can carry continuously under specified conditions. (T&D/PE) 386-1995

continuous-current tests Tests made at rated current, until temperature rise ceases, to determine that the device or equipment can carry its rated continuous current without exceeding its allowable temperature rise. (SWG/PE) C37.100-1992

continuous data Data of which the information content can be ascertained continuously in time. (IA) [61]

continuous duty (1) Operation at a substantially constant load for an indefinitely long time. (NESC/NEC) [86]
(2) Operation at a substantially constant load for an indefinitely long time. (For motors, the constant load is to be within the nameplate rating of the motor.) (PE/NP) 334-1994r
(3) A requirement of service that demands operation at a substantially constant current for an extended period of time. (PE/TR) C57.16-1996
(4) A requirement of service that demands operation at a constant load for an indefinite period of time. (IA/MT) 45-1998

continuous-duty current rating The rating in amperes that a meter socket will carry continuously under stated conditions, without exceeding the allowable temperature rise. A multiposition trough socket has an additional current rating that denotes the maximum ampere capacity of the line buses. (ELM) C12.7-1993

continuous-duty rating The rating applying to operation for an indefinitely long time. (AP/ANT) 145-1983s

continuous electrode A furnace electrode that receives successive additions in lengths at the end remote from the active zone of the furnace to compensate for the length consumed therein. *See also:* electrothermics. (EEC/PE) [119]

continuous enclosure (1) (generating station grounding) A type of isolated-phase bus in which the enclosure is electrically continuous over the full length of the bus. All enclosures are electrically tied together at each end of the bus. (PE/EDPG) 665-1987s
(2) A bus enclosure in which the consecutive sections of the enclosure are electrically bonded together to provide a continuous current path through the entire enclosure length. (SUB/PE) C37.122-1993, C37.122.1-1993
(3) A bus enclosure in which the consecutive sections of the enclosure for the same phase conductor are electrically bonded together to provide a continuous current path throughout the entire enclosure length. *Note:* Cross-connections to the other phase enclosures are made only at the extremities of the installation and at selected intermediate points. (SWG/PE) C37.100-1992
(4) A bus enclosure in which the consecutive sections of the housing along the same phase conductor are bonded together to provide an electrically continuous current path throughout the entire enclosure length. Cross-bondings, connecting the other phase enclosures, are made only at the extremities of the installation and at a few selected intermediate points. (PE/SUB) 80-2000

continuous exposure Exposure for durations exceeding the corresponding averaging time. Exposure for less than the averaging time is called *short-term exposure*. (NIR) C95.1-1999

continuous feed A mechanism enabling a printer to employ continuous form paper using friction feed or tractor feed. *Contrast:* single-sheet feed. (C) 610.10-1994w

continuous form (A) A series of connected paper forms, each divided by a tear-off perforation, that feeds continuously through a printing device. *Synonym:* Z-fold paper. *Contrast:* cut form. *See also:* burst. **(B)** Pertaining to a series of cards or paper as in (A). For example, continuous form cards or continuous form paper.

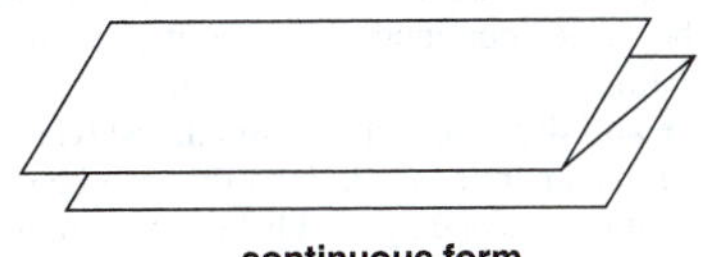

continuous form

(C) 610.10-1994

continuous inductive train control *See:* continuous train control.

continuous iteration A loop that has no exit. (C) 610.12-1990

continuous lighting (railway practice) An arrangement of circuits so that the signal lights are continuously energized. (EEC/PE) [119]

continuous load A load where the current continues for 3 h or more. (ELM) C12.7-1993

continuous load rating (power inverter unit) Defines the maximum load that can be carried continuously without exceeding established limitations under prescribed conditions of test, and within the limitations of established standards. *See also:* self-commutated inverters. (IA) [62]

continuously acting regulator (synchronous machines) A regulator that initiates a corrective action for a sustained infinitesimal change in the controlled variable.
(PE/EDPG) 421.1-1986r, 421-1972s

continuously adjustable inductor An adjustable inductor in which the inductance can have every possible value within its range. *Synonym:* variable inductor.
(Std100) 270-1966w

continuous model (A) A mathematical or computational model whose output variables change in a continuous manner; that is, in changing from one value to another, a variable can take on all intermediate values; for example, a model depicting the rate of air flow over an airplane wing. *Synonym:* continuous-variable model. **(B)** A model of a system that behaves in a continuous manner. *Contrast:* discrete model.
(C) 610.3-1989

continuous monitoring (1) The process of sampling the state of some phenomenon at a time interval shorter than the time constant of the phenomenon.
(SUB/PE) C37.122-1993, C37.122.1-1993
(2) The process of sampling the state of some phenomenon either continuously or at a sample interval of one second or less. (SWG/PE) C37.100-1992

continuous noise Noise, the effect of which is not resolvable into a succession of discrete impulses. *See also:* electromagnetic compatibility. (EMC) [53]

continuous operating current The current flowing through the surge protective device when energized at the maximum continuous operating voltage. (PE) C62.34-1996

continuous periodic rating The load that can be carried for the alternate periods of load and rest specified in the rating and repeated continuously without exceeding the specified limitation. (IA/ICTL/IAC) [60]

continuous-pressure operation (elevators) Operation by means of buttons or switches in the car and at the landings, any one of which may be used to control the movement of the car as long as the button or switch is manually maintained in the actuating position. *See also:* control.
(PE/EEC) [119]

continuous-progression code *See:* unit-distance code.

continuous pulse (thyristor) A gate signal applied during the desired conducting interval, or parts thereof, as a dc signal.
(IA/IPC) 428-1981w

continuous rating (1) (nuclear power generating station) (of diesel-generator unit) The electric power output capability that the diesel-generator unit can maintain in the service environment for 8760 hours of operation per (common) year with only scheduled outages for maintenance.
(PE/NP) 387-1984s
(2) (packaging machinery) The rating that defines the load that can be carried continuously without exceeding the temperature rating. (IA/PKG) 333-1980w
(3) (power and distribution transformers) (electric equipment) The maximum constant load that can be carried continuously without exceeding established temperature-rise limitations under prescribed conditions. *See also:* rectification; duty. (PE/TR) C57.12.80-1978r, C57.15-1968s
(4) (rotating electric machinery) The output that the machine can sustain for an unlimited period under the conditions of IEEE Std 11-1980 without exceeding the limits of temperature rise. (PE/EM) 11-1980r

continuous service current (A) (thyristor converter) The value of direct current that a converter unit or section can supply to its load for unlimited time periods under specified conditions. **(B)** (long-time) The rms value and duration (minutes) of direct current which may be applied to the converter unit or section within the service current profile. *Note:* This value establishes point B on the service current profile and it may be identical to the long-time test current. **(C)** (profile) The time-current profile that defines the allowable rms currents the converter section can sustain. *Note:* The profile is defined for times from zero to infinity, and the rms current derived from any current-time diagram must not exceed this profile. **(D)** (short time) The peak rms value and duration (seconds) of direct current which may be applied to the converter unit or section within the service current profile. *Note:* This value establishes point C on the service current profile.
(IA/IPC) 444-1973

continuous simulation A simulation that uses a continuous model. (C) 610.3-1989w

continuous simulation language A simulation language designed for use in describing continuous simulations.
(C) 610.13-1993w

continuous-speed adjustment Refers to an adjustable-speed drive capable of being adjusted with small increments, or continuously, between minimum and maximum speed. *See also:* electric drive. (IA/ICTL/IAC) [60]

continuous-stream printer A printer that can print processed data off-line in a continuous form. (C) 610.10-1994w

Continuous System Modeling Program III (CSMP III) A simulation language used to simulate the dynamics of continuous systems that use ordinary differential equations.
(C) 610.13-1993w

Continuous Systems Simulation Language (CSSL) A statement-oriented simulation language used to simulate the dynamics of continuous systems that are describable by ordinary differential equations. (C) 610.13-1993w

continuous test (batteries) A service test in which the battery is subjected to an uninterrupted discharge until the cutoff voltage is reached. *See also:* battery; cutoff voltage.
(EEC/PE) [119]

continuous test current (thyristor converter) The value of direct current that a converter unit or section can supply to its load for unlimited time periods under specified conditions.
(IA/IPC) 444-1973w

continuous thermal burden (metering) (voltage transformer) The volt-ampere burden that the voltage transformer will carry continuously at rated voltage and frequency without causing the specified temperature limitations to be exceeded.
(ELM) C12.1-1982s

continuous thermal current rating factor (1) (instrument transformers) (RF) The factor by which the rated primary current of a current transformer is multiplied to obtain the maximum primary current that can be carried continuously without exceeding the limiting temperature rise from 30°C average ambient air temperature. The RF of tapped-secondary or multi-ratio current transformers applies to the highest ratio, unless otherwise stated. (When current transformers are incorporated internally as parts of larger transformers or power circuit breakers, they shall meet allowable average winding and hot spot temperature limits under the specific conditions and requirements of the larger apparatus).
(PE/TR) [57], C57.13-1993
(2) (power and distribution transformers) (RF) The specified factor by which the rated primary current of a current transformer can be multiplied to obtain the maximum primary current that can be carried continuously without exceeding the limiting temperature rise from 30°C ambient air temperature. (When current transformers are incorporated internally as parts of larger transformers or power circuit breakers, they shall meet allowable average winding and hot-spot temperatures under the specific conditions and requirements of the larger apparatus.).
(PE/PSR/TR) C37.110-1996, C57.12.80-1978r

continuous train control (continuous inductive train control) A type of train control in which the locomotive apparatus is constantly in operative relation with the track circuit and is immediately responsive to a change in the character of the current flowing in the track circuit of the track on which the locomotive is traveling. *See also:* automatic train control.
(EEC/PE) [119]

continuous-type control (electric power system) A control mode that provides a continuous relation between the deviation of the controlled variable and the position of the final controlling element. *See also:* speed-governing system. (PE/PSE) 94-1970w

continuous update *See:* supervisory control.

continuous update supervisory system (station control and data acquisition) A system in which the remote station continuously updates indication and telemetering to the master station regardless of action taken by the master station. The remote station may interrupt the continuous data updating to perform a control operation. (SWG/PE/SUB) C37.100-1992, C37.1-1994

continuous-variable model *See:* continuous model.

continuous-voltage-rise test (rotating machinery) A controlled overvoltage test in which voltage is increased in continuous function of time, linear or otherwise. *See also:* asynchronous machine. (PE) [9]

continuous wave (CW) (1) (data transmission) Waves, the successive oscillations of which are identical under steady-state conditions. (PE) 599-1985w

(2) (laser maser) The output of a laser which is operated in a continuous rather than pulsed mode. In this standard, a laser operating with a continuous output for a period greater than 0. 25 s is regarded as a CW laser. (LEO) 586-1980w

(3) A sinusoidal wave that has reached a steady state value. Continuous wave noise would be noise at a single frequency (e.g., 60 Hz "hum"). (PE/IC) 1143-1994r

(4) A carrier that is not modulated or switched. (C/LM) 802.3-1998

continuous wave Doppler radar A radar that transmits a continuous-wave signal and discriminates desired targets from other targets or clutter on the basis of the Doppler shift due to radial motion. *See also:* Doppler radar; continuous wave radar. (AES) 686-1997

continuous wave radar A radar that transmits a continuous-wave signal. *See also:* continuous wave Doppler radar. (AES) 686-1997

continuum *See:* baseline.

contour analysis In optical character recognition, a technique for locating the outline of a character by searching around its exterior edges with a spot of light. (C) 610.2-1987

contoured beam antenna A shaped-beam antenna designed in such a way that when its beam intersects a given surface, the lines of equal power flux density incident upon the surface form specified contours. *See also:* footprint. (AP/ANT) 145-1993

contour encoding An image compression technique in which a region that has a constant gray level is encoded by specifying only its border. (C) 610.4-1990w

contouring control system (numerically controlled machines) A system in which the controlled path can result from the coordinated, simultaneous motion of two or more axes. (EEC) [74]

contract (1) (diode-type camera tube) The ratio of the difference between the peak and minimum values of irradiance to the peak irradiance of an image or specified portion of an image.

$$C = \frac{E_p - E_m}{E_p} \times 100 \text{ (percent)}$$

(ED) 503-1978w

(2) A legally binding document agreed upon by the customer and supplier. This includes the technical and organizational requirements, cost, and schedule for a product. A contract may also contain informal but useful information such as the commitments or expectations of the parties involved. (C/SE) 830-1998

(3) In project management, a legally binding document agreed upon by the customer and the hardware or software developer or supplier; includes the technical, organizational, cost, and/or scheduling requirements of a project. (C/SE) 1362-1998

(4) A binding agreement between two parties, especially enforceable by law or similar internal agreement wholly within an organization, for supply of service or for the supply, development, production, operation, or maintenance of a software product. (C/SE) 1062-1998

contract curve (rotating electric machinery) A specified machine characteristic curve that becomes part of the contract. (PE/EM) 11-1980r

contract demand (power operations) The demand that the supplier of electric service agrees to have available for delivery. *See also:* alternating-current distribution. (PE/PSE) 858-1993w, 346-1973w

contractor (1) (hydroelectric power plants) A device used for repetitive opening and closing operation of an electric circuit, and that has load-current interrupting capability. It has no fault-current interrupting capability. (PE/EDPG) 1020-1988r

(2) (metal transmission structures) *See also:* constructor. (T&D/PE) 951-1988s

(3) (power and distribution transformers) A device for repeatedly establishing and interrupting an electric power circuit. (PE/TR) C57.12.80-1978r

contract start The date a contract to supply a static var compensator (SVC) becomes effective, and the user has given notice to proceed. (PE/SUB) 1031-2000

contractual requirements Customer-imposed performance, logistics, and other requirements and commitments governing the scope of software development, delivery, or support. (C/SE) 1074-1995s

contrast (1) (image processing and pattern recognition) The difference between the average brightness of two subsets of an image. (C) 610.4-1990w

(2) (computer graphics) The relationship between the highest and lowest intensity levels of a display image, usually expressed as the ratio of light to dark. (C) 610.6-1991w

(3) (display presentation) The subjective assessment of the difference in appearance of two parts of a field of view seen simultaneously or successively. (Hence: luminosity contrast, lightness contrast, color contrast, simultaneous contrast, successive contrast). *See also:* television; photometry. (BT/AV) [34], [84]

(4) (electric power systems in commercial buildings) Indicates the degree of difference in light reflectance of the details of a task compared with its background. (IA/PSE) 241-1990r

contrast control A control, associated with a picture-display device, for adjusting the contrast ratio of the reproduced picture. *Note:* The contrast control is normally an amplitude control for the picture signal. In a monochrome-television system, both average luminance and the contrast ratio are affected. In a color-television system, saturation and hue also may be affected. *See also:* television. (EEC/PE) [119]

contrast ratio (1) (television) The ratio of the maximum to the minimum luminance values in a television picture or a portion thereof. *Note:* Generally, the entire area of the picture is implied, but smaller areas may be specified as in detail contrast. (BT/AV) 201-1979w

(2) (amplitude, frequency, and pulse modulation) For any diffraction order, the ratio of the maximum light intensity to the minimum light intensity in the order, so that $C = I_{max}/I_{min}$, where C is the contrast ratio. *Note:* In the limiting case when the depth of modulation is equal to 1, the minimum light intensity is due to background light, so that $C = I_{max}/I_b$. In the other extreme, when $m = 0$, the contrast ratio is equal to 1. (UFFC) [17]

(3) (acoustically tunable optical filter) The ratio of the dynamic transmission at a given acoustic frequency and power level to the dynamic transmission with no applied acoustic power. *Note:* The contrast ratio is a measure of light leakage through the device. It should be specified for either a monochromatic or white light source input, and the angular spread of the input light. (UFFC) [17]

contrast rendition factor (illuminating engineering) The ratio of visual task contrast with a given lighting environment to the contrast with sphere illumination. (EEC/IE) [126]

contrast sensitivity (illuminating engineering) The ability to detect the presence of luminance differences. Quantitatively, it is equal to the reciprocal of the brightness contrast threshold. *See also:* brightness contrast threshold. (EEC/IE) [126]

contrast stretching An image enhancement technique in which the contrast between image subsets and their complements is increased. (C) 610.4-1990w

contrast transfer function square-wave response (diode-type camera tube) The contrast transfer function or CTF represents the response of the imaging system in the spatial frequency domain to a square-wave input. A bar pattern represents a one-dimensional input to a two-dimensional imaging sensor. CTF is synonymous with the square-wave amplitude response, R_{sq} (N). (ED) 503-1978w

contravariance A rule governing the overriding of a property and requiring that the set of values acceptable for an input argument in the overriding property shall be a superset (includes the same set) of the set of values acceptable for that input argument in the overridden property, and the set of values acceptable for an output argument in the overriding property shall be a subset (includes the same set) of the set of values acceptable for that output argument in the overridden property. (C/SE) 1320.2-1998

contributing cause A cause that, of itself, may not result in failure. (SWG/PE) C37.10-1995

control (1) (A) (electronic computation) Usually, those parts of a digital computer that effect the carrying out of instructions in proper sequence, the interpretation of each instruction, and the application of the proper signals to the arithmetic unit and other parts in accordance with this interpretation. **(B) (electronic computation)** In some business applications of mathematics, a mathematical check. (C) 162-1963
(2) (cryotron) An input element of a cryotron. (ED) [46]
(3) (packaging machinery) A device or group of devices that serves to govern in some predetermined manner the electric power delivered to the apparatus to which it is connected. (IA/PKG) 333-1980w
(4) (electric power systems in commercial buildings) Any device used for regulation of a system or component. (IA/PSE) 241-1990r
(5) (overhead power lines) In experiments, establishment of an untreated group of animals, plants, cells, etc., that serve as the basis for comparing responses of a similar, but treated, group that has been subjected (exposed) to some agent (i.e., an electric field). (T&D/PE) 539-1990
(6) A visual user interface element that is defined by IEEE Std 1295-1993. (C) 1295-1993w
(7) The execution of a system change by manual means, remote means, automatic means, or partially automatic means. (SWG/PE/SUB) C37.100-1992, C37.1-1987s
(8) In an IDEF0 model, a condition or set of conditions required for a function to produce correct output. (C/SE) 1320.1-1998

control accuracy The degree of correspondence between the final value and the ideal value of the directly controlled variable. *See also:* feedback control system. (IA/ICTL/IAC) [60]

control action (automatic control) Of a control element or a controlling system, the nature of change of the output effected by the input. *Note:* The output may be a signal or the value of a manipulated variable. The input may be the control loop feedback signal when the command is constant, an actuating signal, or the output of another control element. One use of control action is to effect compensation. *See also:* compensation. (PE/EDPG) [3]

control acquisition (1) The total of all bus activity associated with acquiring exclusive control of the bus by a module. (C/BA) 896.3-1993w
(2) The total of all bus activity associated with acquiring exclusive control of the bus. (C/BA) 10857-1994, 896.4-1993w

control action, derivative *See:* derivative control action.

control action, proportional (1) Control action in which there is a continuous linear relation between the output and the input. *Note:* This condition applies when both the output and input are within their normal operating ranges. (IA/ICTL/IAC) [60]
(2) (automatic control) Action in which there is a linear relation between the output and the input of the controller. *Note:* The ratio of the change in output produced by the proportional control action to the change in input is defined as the proportional gain. (PE/PSE) 94-1970w

control action, proportional plus derivative Control action in which the output is proportional to a linear combination of the input and the time rate-of-change of input. *Note:* In the practical embodiment of proportional plus derivative control action the relationship between output and input, neglecting high frequency terms, is

$$\frac{Y}{X} = \pm P \frac{\frac{I}{s} + 1 + Ds}{\frac{bI}{s} + 1 + \frac{Ds}{a}} \qquad \frac{a > 1}{0 \le b \ll 1}$$

where
a = derivative action gain
D = derivative action time constant
P = proportional gain
s = complex variable
X = input transform
Y = output transform

Synonym: P.D. (CS/PE/EDPG) [3]

control action, proportional plus integral *See:* proportional plus integral control action.

control action, proportional plus integral plus derivative *See:* proportional plus integral plus derivative control action.

control and instrumentation cables (cable systems in substations) Insulated electrical conductors utilized to convey information or to intermittently operate devices controlling power switching or conversion equipment. The cross-sectional areas of the conductors are generally No. 6 American Wire Gage (AWG) or smaller, and the duty cycle is such that conductor heating is insignificant. 382

control and status register (CSR) (1) A memory-mapped register that is accessed through read and write transactions and is used to observe the state of a node or to control its operation. (C/BA) 896.9-1994w
(2) A register used to control the operation of a device and/or record the status of an operation. It is accessible through a separate address space in a FASTBUS device. CSR#0, mandatory for all devices, contains the manufacturer's ID for the device and a number of device status bits as well as some user-defined bits. (NID) 960-1993
(3) A register, storage location, or address that is used to control buses, interconnects, and multiple processor systems. (C/BA) 14536-1995

control and status registers (CSR) A set of registers, storage locations, and addresses that are used to control buses, interconnects, and multiple processor systems. (C/BA) 896.5-1993s

control and status register space (FASTBUS acquisition and control) A FASTBUS primary address cycle may specify with a code on the mode select (MS) control lines one of two separate address spaces in a device; CSR space and data space. CSR space contains registers for control of and status reporting registers for the device. Its allocation and usage is part of the FASTBUS specification. *Synonym:* CSR space. *See also:* data space. (NID) 960-1986s

control apparatus A set of control devices used to accomplish the intended control functions. *See also:* control. (IA/IAC) [60]

control area (1) (electric power) A power system, a part of a power system, or a combination of several power systems under common control for which a single area control error is defined. (PE/PSE) 94-1991w
(2) A storage area used to hold information necessary for the control of a task, function, or operation. (C) 610.10-1994w
(3) A power system, a part of a power system, or a combination of several power systems under common control, that uses tie-line bias control if it is part of an interconnected system. (PE/PSE) 858-1993w

control arrow An arrow or arrow segment that expresses IDEF0 control, i.e., an object type set whose instances establish a condition or set of conditions required for a function to produce correct output. The arrowhead of a control arrow is attached to the top side of a box. (C/SE) 1320.1-1998

control ball An input device consisting of a ball, rotatable about its center and recessed into a surface, used as a locator. *Synonym:* track ball. (C) 610.6-1991w

control battery A battery used as a source of energy for the control of an electrically operated device. (IA/ICTL/IAC) [60]

control block (1) (subroutines for CAMAC) The symbol *cb* represents an integer array having four elements. The contents of these elements are:

element 1	Repeat Count
element 2	Tally
element 3	LAM Identification
element 4	Channel Identification

The repeat count specifies the number of individual CAMAC actions or the maximum number of data words to be transferred. Some multiple action and block transfer subroutines permit termination of the sequence upon a signal from the addressed module. In such cases the repeat count represents an upper limit. The tally is the number of actions usually performed or the number of CAMAC data words actually transferred. If the block transfer or multiple action is terminated by the controller due to exhaustion of the repeat count, the tally will be equal to the repeat count; otherwise it may be less. The LAM identification is an integer value having the same form and information content as the variable *lam*. The channel identification is an integer value which identifies system-dependent facilities which may be necessary to perform the block transfer or multiple action. This number, if it is required, has the same form and content as the parameter *chan* and can be created by the subroutine CDCHN. (NPS) 758-1979r
(2) The circuitry within a computer that performs control functions such as decoding microinstructions and generating the internal control signals that perform requested operations. (C) 610.10-1994w

control board (control boards, panels, and racks) An assembly of panels on which are installed components and modules for monitoring, measuring, and controlling remotely operated systems and equipment. It provides a visual and physical interface between the operator and the systems. *Synonyms:* control switchboard; console; benchboard; control panel. (PE/NP) 420-1982

control breakpoint *See:* code breakpoint.

control bus (1) A bus that carries signals that regulate system operations. (C) 610.10-1994w
(2) A bus used to distribute power for operating electrically controlled devices. (SWG/PE) C37.100-1992

control cable (1) (cable systems in power generating stations) Cable applied at relatively low current levels or used for intermittent operation to change the operating status of a utilization device of the plant auxiliary system. (PE/EDPG) 422-1977
(2) (communication and control cables) A cable that usually carries relatively low current levels used for indication purposes to change the operating status of a utilization device of a plant auxiliary system. A control cable usually consists of two or more insulated, unpaired, shielded or unshielded conductors. Sizes may be No 22, 20, 19, 18, or 16 AWG solid and No 14, 12, 10, or 9 AWG stranded or solid conductor. Control cable conductor insulation usually has voltage ratings of 300, 600, or 1000 V rms, 50–60 Hz. (PE/PSC) 789-1988w
(3) Cable used in a control function application, e.g., interconnection of control switches, indicating lights, relays, solenoids, etc. Generally the cable construction is 600 V or 1000 V, single or multiple conductors, typically in wire sizes 14 AWG (2.08 mm^2), 12 AWG (3.31 mm^2), 10 AWG (5.26 mm^2), 9 AWG (6.63 mm^2), or 8 AWG (8.37 mm^2). (PE/IC) 1185-1994

control capacitor The element by which voltage is coupled to the floating gate for reading or writing. (ED) 1005-1998

control card A punch card containing input parameters for controlling the execution of a program or job. (C) 610.10-1994w

control center (1) (generating stations electric power system) An assembly of devices for the purpose of switching and protecting a number of load circuits. The control center may contain transformers, contactors, circuit breakers, protective and other devices intended primarily for energizing and deenergizing load circuits. (PE/EDPG) 505-1977r
(2) The facility from which instructions and signals are issued for controlling the bulk electric system, and, in some instances, the distribution system as well. (PE/PSE) 858-1993w

control character (1) A character whose occurrence in a particular context initiates, modifies, or stops a control operation, for example, a character to control carriage return. (C) [20], [85]
(2) (A) (data management) A character whose occurrence in a particular context initiates, modifies, or stops a control operation. A control character may be recorded for use in a subsequent action, and it may have a graphic representation in some circumstances. *Synonyms:* operational character; functional character; command character. *See also:* transmission control character; accuracy control character; print control character; device control character. **(B) (data management)** A character that initiates some kind of physical control action but is not printed on the output page. For example, line feed, tab, form feed. (C) 610.5-1990
(3) A character used for signaling purposes by the exchange, packet or transaction layers of the stack. Both N_chars and L_chars are used as control characters. *See also:* link character; normal character. (C/BA) 1355-1995

control characteristic (gas tube) A relation, usually shown by a graph, between critical grid voltage and anode voltage. *See also:* gas tube. (ED) 161-1971w

control circuit (1) (packaging machinery) The circuit that carries the electric signals directing the performance of the controller but does not carry the main power circuit. (IA/PKG) 333-1980w
(2) The circuit that carries the electric signals of a control apparatus or system directing the performance of the controller but that does not carry the main power circuit. (IA/MT) 45-1998

control circuit failure Failure attributed to the inability of the electrical control circuit to perform its function. (SWG/PE) C37.10-1995

control-circuit limit switch A limit switch the contacts of which are connected only into the control circuit. *See also:* control; switch. (IA/IAC) [60]

control circuit transformer (packaging machinery) A voltage transformer utilized to supply a voltage suitable for the operation of control devices. (IA/PKG) 333-1980w

control circuit voltage (packaging machinery) The voltage provided for the operation of shunt coil magnetic devices. (IA/PKG) 333-1980w

control compartment (packaging machinery) A space within the base, frame, or column of the machine, used for mounting the control panel. (IA/PKG) 333-1980w

control conductor (substation grounding) The conductor that is utilized to measure equivalent changes in temperature, size, etc., that are occurring in at least one of the conductors joined by the connector under test. (PE/SUB) 837-1989r

control coupling A type of coupling in which one software module communicates information to another module for the explicit purpose of influencing the latter module's execution. *Contrast:* hybrid coupling; data coupling; pathological coupling; content coupling; common-environment coupling. (C) 610.12-1990

control current (Hall effect devices) The current through the Hall plate that by its interaction with a magnetic flux density generates the Hall voltage. (MAG) 296-1969w

control current sensitivity (Hall effect devices) The ratio of the voltage across the Hall terminals to the control current for a given magnitude of magnetic flux density. (MAG) 296-1969w

control current terminals (Hall effect devices) The terminals through which the control current flows. (MAG) 296-1969w

control cut-out switch (land transportation vehicles) An isolating switch that isolates the control circuits of a motor controller from the source of energy. (VT/LT) 16-1955w

control data (software) Data that select an operating mode, direct the sequential flow of a program, or otherwise directly influence the operation of software; for example, a loop control variable. (C) 610.12-1990

control designation symbol A symbol that identifies the particular manner, permissible or required, in which an input variable (possibly in combination with other variables) causes the logic element to perform according to its defined function. (GSD) 91-1973s

control desk (console) A control switchboard consisting of one or more relatively short horizontal or inclined panels mounted on an assembly of such a height that the panel-mounted devices are within convenient reach of an attendant. (SWG/PE) C37.100-1992, C37.21-1985r

control device An individual device used to execute a control function. *See also:* control. (IA/IAC) [60]

control dial An input device, consisting of one or more rotating knobs or levers, that provides coordinate input data. (C) 610.6-1991w

control direction In T101, control direction is transmission from the controlling station (master/RTU) to the controlled station (RTU/IED). (PE/SUB) 1379-1997

control directory The directory below which the control_files for filesets and products are stored within exported catalogs for distributions and installed software. (C/PA) 1387.2-1995

control electrode (electron tube) An electrode used to initiate or vary the current between two or more electrodes. *See also:* electrode. (ED) 161-1971w

control-electrode discharge recovery time (attenuator tubes) The time required for the control-electrode discharge to deionize to a level such that a specified fraction of the critical high-power level is required to ionize the tube. *See also:* gas tube. (ED) 161-1971w

control enclosure (packaging machinery) The metal housing for the control panel, whether mounted on the industrial equipment or separately mounted. (IA/PKG) 333-1980w

control exciter (rotating machinery) An exciter that acts as a rotary amplifier in a closed-loop circuit. *See also:* asynchronous machine. (PE) [9]

control field (C) **(1)** The sequence of eight (or sixteen, if extended) bits immediately following the address field of a frame. This octet identifies the HDLC frame type. (EMB/MIB) 1073.3.1-1994
(2) A sequence of bits that identifies the type of frame being transmitted, and optionally, contains sequence or acknowledgment numbers. (C) 610.7-1995
(3) The field immediately following the DSAP and SSAP address fields of a PDU. The content of the control field is interpreted by the receiving destination LLC(s) designated by the DSAP address field:

1) As a command, from the source LLC designated by the SSAP address field, instructing the performance of some specific function; or
2) As a response, from the source LLC designated by the SSAP address field.

(C/LM/CC) 8802-2-1998

control_files The control scripts executed by the utilities, the INFO file describing the files in a fileset, and other files associated with a software object. (C/PA) 1387.2-1995

control flow The sequence in which operations are performed during the execution of a computer program. *Synonym:* flow of control. *Contrast:* data flow. (C) 610.12-1990

control flow architecture A computer architecture in which execution is controlled by the need for a particular result; that is, an instruction is executed only when its result is needed by another process. *Synonym:* Von Neumann architecture. *Contrast:* data flow architecture; Harvard class architecture. (C) 610.10-1994w

control flow diagram A diagram that depicts the set of all possible sequences in which operations may be performed during the execution of a system or program. Types include box diagram, flowchart, input-process-output chart, state diagram. *Contrast:* data flow diagram. *See also:* call graph; structure chart. (C) 610.12-1990

control flow trace *See:* execution trace.

control function *See:* supervisory control functions.

control function check Control and indication from a control-check relay. A check of master and remote station equipment by exercising a predefined component or capability. (SUB/PE) C37.1-1994

control gate A form of the control capacitor in which the top polysilicon gate, above the floating gate, of the memory transistor is connected to the word-line or control line. (ED) 1005-1998

control generator A generator, commonly used on electric motive power units for the generation of electric energy in proportion to vehicle speed, prime mover speed, or some similar function, thereby serving as a guide for initiating appropriate control functions. *See also:* traction motor. (EEC/PE) [119]

control grid (electron tube) A grid, ordinarily placed between the cathode and an anode, for use as a control electrode. *See also:* electrode; grid. (ED) 161-1971w

control hole *See:* designation hole.

control host The spaceborne fiber-optic data bus (SFODB) network management node. (C/BA) 1393-1999

control initiation The function introduced into a measurement sequence for the purpose of regulating any subsequent control operations in relation to the quantity measured. *Note:* The system element comprising the control initiator is usually included in the end device but may be associated with the primary detector or the intermediate means. *See also:* measurement system. (EEC/PE) [119]

control interaction factors In a proportional plus integral plus derivative control action unit, the ratio of the effective values to the values that would be measured when the product (integral action rate) (derivative action time constant) is zero. Example: Assume a control unit composed of elements whose ratios of output to input are $1 + D's$ and $P'(I'/s + 1)$ connected so that the output of one is the input of the other. The ratio of output to input of the combination is

$$\frac{Y}{X} = P'(1 + I'D')\left[\frac{I'/s}{1 + I'D'} + 1 + \frac{D's}{1 + I'D'}\right]$$

By comparison with the equation

$$\frac{Y}{X} = P\left[\frac{1}{s} + 1 + Ds\right]$$

it is seen that the effective values are

$P = P'(1 + I'D')$ = proportional gain
$I = I'/(1 + I'D')$ = integral action rate
$D = D'/(1 + I'D')$ = derivative action time constant.

When either I' or D' is set equal to zero the factor $1 + I'D'$ equals unity and the measured values are P', I' and D'. Consequently, $1 + I'D'$ is the "proportional interaction factor" and $1/(1 + I'D')$ is both the "integral action rate interaction factor" and "derivative action time interaction factor. ". (CS/PE/EDPG) [3]

control interface The interface to a device through which the operation and response of the device is controlled either by manual operation or by a system controller. (SCC20) 993-1997

control key (CTRL) Any key on a keyboard that is used to control a process. *Note:* The control key, usually labelled "CTRL" is said to represent a control character, and when used in conjunction with another key, such as "C", the combination is said to represent the control character "CONTROL C" or "C". *Contrast:* typing key. *See also:* attention key; alternate key; cursor control key; escape key; enter key; shift key; function key; command key. (C) 610.10-1994w

controllability In comparison of processes, a qualitative term indicating the relative ease with which they can be controlled. *Note:* The type of disturbance for which the comparison is made should be specified. (CS/PE/EDPG) [3]

controllable A property of a component of a state whereby, given an initial value of the component at a given time, there exists a control input that can change this value to any other value at a later time. *See also:* control system. (IM) [120]

controllable, completely The property of a plant whereby all components of the state are controllable within a given time interval. *See also:* control system. (IM) [120]

control language *See:* job control language.

control law A function of the state of a plant and possibly of time, generated by a controller to be applied as the control input to a plant. *See also:* control system. (IM) [120]

control law, closed-loop A control law specified in terms of some function of the observed state. *See also:* control system. (IM) [120]

control law, open-loop A control law specified in terms of the initial state only and possibly of time. *See also:* control system. (IM) [120]

controlled access (communication satellite) A mode of operation of a communication satellite in which an earth station desiring access to the system must request and obtain access to the system via a network management facility. (COM) [19]

controlled area (laser maser) An area where the occupancy and activity of those within is subject to control and supervision for the purpose of protection from radiation hazards. (LEO) 586-1980w

controlled-avalanche rectifier diode (semiconductor) A rectifier diode that has specified maximum and minimum breakdown-voltage parameters and is specified to operate under steady-state conditions in the breakdown region of its reverse characteristic. *See also:* breakdown. (IA) [12]

controlled carrier (floating carrier) (variable carrier) A system of compound modulation wherein the carrier is amplitude modulated by the signal frequencies in any conventional manner, and, in addition, the carrier is simultaneously amplitude modulated in accordance with the envelope of the signal so that the percentage of modulation, or modulation factor, remains approximately constant regardless of the amplitude of the signal. (AP/BT/ANT) 145-1983s, 182A-1964w

controlled ESD environment One in which an attempt is made to maintain charge levels on humans and objects below a certain level. Typical control measures include humidity controls, equipment earth grounding, use of antistatic materials, ionized air, and high-resistance discharge paths for humans. (EMC) C63.16-1993

controlled ferroresonant regulators A regulator consisting basically of an inductor connected in series with a parallel combination of a capacitor and controllable simulated inductor. This combination is connected across the source as shown in the figure below. Stabilized output voltage is derived by inductive or conductive coupling to the parallel combination of C and the controllable simulated inductor. In a controlled ferroresonant regulator the controllable simulated inductor can be a combination of switching devices (such as thyristors or transistors) and linear or saturating inductors. This circuit, in combination with a control input to the simulated inductor, controls the flux swing (or simulated flux swing) in the saturated (or simulated saturating) inductor, thereby controlling the stabilized output voltage.

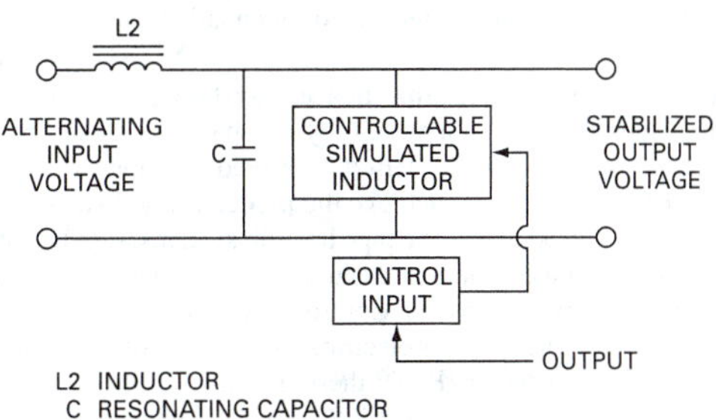

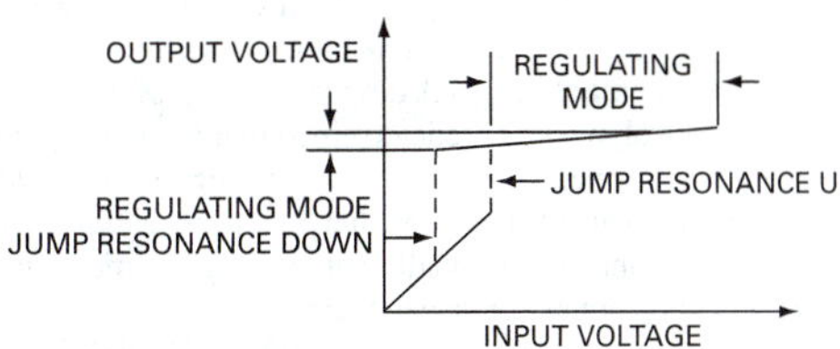

Controlled ferroresonant regulator schematic

Controlled ferroresonant regulators

(PEL) 449-1998

controlled list (A) A list whose access is controlled in some way; for example, access to an array is controlled by its index variable. **(B)** A list that can contain a finite number of entries. (C) 610.5-1990

controlled list data element A data element that is contained in a controlled list. (C) 610.5-1990w

controlled manual block signal system A series of consecutive blocks governed by block signals, controlled by continuous track circuits, operated manually upon information by telegraph, telephone, or other means of communication, and so constructed as to require the cooperation of the signalmen at both ends of the block to display a clear or permissive block signal. *See also:* block-signal system. (EEC/PE) [119]

controlled overvoltage test (rotating machinery) (dc leakage, measured current, or step voltage test) A test in which the increase of applied direct voltage is controlled and measured currents are continuously observed for abnormalities with the intention of stopping the test before breakdown occurs. (PE/EM) 95-1977r

controlled plasma switch (nonlinear, active, and nonreciprocal waveguide components) A triggered gas switch that uses an electron-beam-excited gaseous plasma in a waveguide to limit or switch radio frequency (rf) power. (MTT) 457-1982w

controlled rectifier A rectifier in which means for controlling the current flow through the rectifying devices is provided. *See also:* rectification; electronic controller. (EEC/PE) [119]

controlled slip The occurrence of a replication or deletion of all the information bits in a frame at the receiving terminal. (COM/TA) 1007-1991r

controlled-speed axle generator An axle generator in which the speed of the generator is maintained approximately constant at all vehicle speeds above a predetermined minimum. *See also:* axle-generator system. (EEC/PE) [119]

controlled system (automatic control) The apparatus, equipment, or machine used to effect changes in the value of the ultimately controlled variable. *See also:* control system. (PE/EDPG) [3]

controlled vented power fuse (installations and equipment operating at over 600 volts, nominal) A fuse with provision for controlling discharge circuit interruption such that no solid material may be exhausted into the surrounding atmosphere. The discharge gases shall not unite or damage insulation in the path of the discharge nor shall these gases propagate a flashover to or between grounded members or conduction members in the path of the discharge when the distance between the vent and such insulation or conduction members conforms to manufacturer's recommendations. (NESC/NEC) [86]

controller (1) (electric pipe heating systems) A device that regulates the state of a system by comparing a signal from a sensor located in the system with a predetermined value and adjusting its output to achieve the predetermined value. Controllers, as used in electric pipe heating systems, regulate temperatures on the system and can be referred to as temperature controllers or thermostats. Controller sensors can be mechanical (bulb, bimetallic) or electrical (thermocouple, resistance-temperature detector [RTD] thermistor). (PE/EDPG) 622A-1984r, 622B-1988r
(2) A device or group of devices that serves to govern, in some predetermined manner, the electric power delivered to the apparatus to which it is connected. (NESC/NEC) [86]
(3) (packaging machinery) A device or group of devices that serves to control in some predetermined manner the apparatus to which it is connected. (IA/PKG) 333-1980w
(4) The component of a system that functions as the system controller. A controller typically sends program messages to and receives response messages from devices. (IM/AIN) 488.2-1992r
(5) (A) A functional unit in a computer system that controls one or more units of the peripheral equipment. *Synonym:* peripheral control unit. *See also:* input-output controller; dual-channel controller. **(B)** In robotics, a processor that takes as input desired and measured position, velocity or other pertinent variables and whose output is a drive signal to a controlling motor or activator. **(C)** A device through which one can introduce commands to a control system. (C) 610.10-1994
(6) The entity that initiates RamLink transactions. There is exactly one controller on each RamLink ringlet. (C/MM) 1596.4-1996
(7) A device or group of devices used to control in a predetermined manner the electric power delivered to the apparatus to which it is connected. (IA/MT) 45-1998
(8) (CAMAC system) *See also:* CAMAC crate.
(9) *See also:* SBus Controller. (C/BA) 1496-1993w

Controller *See:* SBus Controller.

controller, automatic *See:* automatic controller.

controller characteristics (thyristor) The electrical characteristics of an ac power controller measured or observed at its input or output terminal. (IA/IPC) 428-1981w

controller current (thyristor) The current flowing through the terminals of the controller. (IA/IPC) 428-1981w

controller diagram (electric-power devices) A diagram that shows the electric connections between the parts comprising the controller and that shows the external connections. (IA/IAC) 270-1966w, [60]

controller equipment (thyristor) An operative unit for ac power control comprising one or more thyristor assemblies together with any input or output transformers, filters, other switching devices and auxiliaries required by the thyristor ac power controller to function. (IA/IPC) 428-1981w

controller faults (thyristor) A fault condition exists if the conduction cycles of some semiconductors are abnormal. (IA/IPC) 428-1981w

controller ON-state interval (thyristor) The time interval in which the controller conducts. *Note:* It is assumed that the starting instant of the controller ON-state interval is coincident with the starting instant of the trigger pulse. (IA/IPC) 428-1981w

controller power transformer (thyristor) A transformer within the controller employed to provide isolation or the transformation of voltage or current, or both. (IA/IPC) 428-1981w

controller section (thyristor) That part of a controller circuit containing the basic control elements necessary for controlling the load voltage. (IA/IPC) 428-1981w

controller, self-operated *See:* self-operated controller.

controllers for steel-mill accessory machines Controllers for machines that are not used directly in the processing of steel, such as pumps, machine tools, etc. *See also:* electric controller. (IA/IAC) [60]

controllers for steel-mill auxiliaries Controllers for machines that are used directly in the processing of steel, such as screwdowns and manipulators but not cranes and main rolling drives. *See also:* electric controller. (IA/IAC) [60]

controller, time schedule *See:* time schedule controller.

control line The line, connected to the memory transistor control element, that provides the reference voltage to the memory cell during a read and may provide a high voltage during a write cycle. (ED) 1005-1998

controlling element, final *See:* final controlling element.

controlling elements The functional components of a controlling system. *See also:* feedback control system. (IM/PE/EDPG) [120], [3]

controlling elements, forward *See:* forward controlling elements.

controlling means (of an automatic control system) Consists of those elements that are involved in producing a corrective action. (PE/PSE) 94-1970w

controlling section A length of track consisting of one or more track circuit sections, by means of which the roadway elements or the device that governs approach to or movement within a block are controlled. (EEC/PE) [119]

controlling system (1) (automatic control system without feedback) That portion of the control system that manipulates the controlled system. (IM/PE/EDPG) [120], [3]
(2) (control system feedback) The portion that compares functions of a directly controlled variable and a command and adjusts a manipulated variable as a function of the difference. *Note:* It includes the reference input elements; summing point; forward and final controlling elements; and feedback elements. *See also:* feedback control system. (IM/PE/EDPG) [120], [3]

controlling voltage, composite *See:* composite controlling voltage.

control loopback Loopback of output from one function to be control for another function in the same diagram. *Synonym:* feedback. (C/SE) 1320.1-1998

control machine (A) (railroad practice) An assemblage of manually operated levers or other devices for the control of signals, switches, or other units, without mechanical interlocking, usually including a track diagram with indication lights. *See also:* car retarder. **(B) (railroad practice)** A group of levers or equivalent devices used to operate the various mechanisms and signals that constitute the car retarder installation. *See also:* centralized traffic-control system; car retarder. (EEC/PE) [119]

control, manual *See:* manual control.

control mechanism (control systems for steam turbine-generator units) Includes all systems, devices, and mechanisms between a controller and the controlled valves. (PE/EDPG) 122-1985s

control metering point (1) (tie line) The location of the metering equipment that is used to measure power on the tie line for the purpose of control. *See also:* power system; center of distribution. (PE/PSE) [54]
(2) (electric power system) The actual or equivalent location of power flow measurement on an area tie line. (PE/PSE) 94-1991w

control mode (thyristor) The starting instant of the controller ON-state interval is periodic. The control mode is defined only for steady state operation. *Note:* It is possible to combine several control modes, for example, ON-OFF control and phase control. *See also:* operation modes. (IA/IPC) 428-1981w

control operator In the shell command language, a token that performs a control function. A control operator is one of the following symbols:

&
&&
(
)
;
;;
⟨newline⟩
|
||

The end-of-input indicator used internally by the shell is also considered a control operator.On some systems, the symbol ((is a control operator; its use produces unspecified results. (C/PA) 9945-2-1993

control panel (1) (supervisory control, data acquisition, and automatic control) (station control and data acquisition) An assembly of man/machine interface devices. (PE/SUB) C37.1-1987s
(2) The part of a console that contains switches, pushbuttons and indicators. (C) 610.10-1994w

control point (project control point) A project agreed on point in time or times when specified agreements or controls are applied to the software configuration items being developed, e.g., an approved baseline or release of a specified document/code. (C/SE) 828-1998

control point interfaces Master station or RTU (or both) element(s) that operate(s) to perform a control function. (SUB/PE) C37.1-1994

control point selector (test, measurement, and diagnostic equipment) A device capable of selecting and controlling the proper stimuli, power of loads, and applying it to the unit under test, in accordance with instructions from the programming device. (MIL) [2]

control position electric indicator A device that provides an indication of the movement and position of the various control surfaces or structural parts of an aircraft. It may be used for wing flaps, cowl flaps, trim tabs, oil-cooler shutters, landing gears, etc. (EEC/PE) [119]

control positioning accuracy, precision, or reproducibility (numerically controlled machines) Accuracy, precision, or reproducibility of position sensor or transducer and interpreting system and including the machine positioning servo. *Note:* May be the same as machine positioning accuracy, precision, or reproducibility in some systems. (IA) [61]

control power disconnecting device (power system device function numbers) A disconnecting device, such as a knife switch, circuit breaker, or pull-out fuse block, used for the purpose of respectively connecting and disconnecting the source of control power to and from the control bus or equipment. *Note:* Control power is considered to include auxiliary power which supplies such apparatus as small motors and heaters. (SUB/PE) C37.2-1979s

control-power winding (power and distribution transformers) The winding (or transformer) that supplies power to motors, relays, and other devices used for control purposes. *Synonym:* control-power transformer. *See also:* high-voltage and low-voltage windings. (PE/TR) C57.12.80-1978r, [57]

control-power transformer *See:* control-power winding.

control precision Precision evidenced by either the directly or the indirectly controlled variable, as specified. (CS/PE/EDPG) [3]

control procedure The means used to control the orderly communication of information between stations on a data link. (LM/COM) 168-1956w

controlling process The session leader that established the connection to the controlling terminal. Should the terminal subsequently cease to be a controlling terminal for this session, the session leader shall cease to be the controlling process. (PA/C) 9945-1-1996, 1003.5-1999

controlling terminal A terminal that is associated with a session. Each session may have at most one controlling terminal associated with it, and a controlling terminal is associated with exactly one session. Certain input sequences from the controlling terminal cause signals to be sent to all processes in the process group associated with the controlling terminal. (C/PA) 9945-1-1996, 1003.5-1999

control program *See:* supervisory program.

control punch *See:* designation hole.

control range The total inductive plus capacitive range of reactive current or megavar variation of the static var compensator (SVC), at the point of connection. (PE/SUB) 1031-2000

control ratio (1) (gas tube) The ratio of the change in anode voltage to the corresponding change in grid voltage, with all other operating conditions maintained constant. *See also:* gas tube. (ED) 161-1971w
(2) (power supplies) The required charge in control resistance to produce a one-volt change in the output voltage. The control ratio is expressed in ohms per volt and is reciprocal of the bridge current. (AES) [41]

control read-only memory (CROM) A type of read-only storage in the control block of some processors such that the ROM has been programmed to decode the control logic. (C) 610.10-1994w

control register A register in a computer or peripheral device, the contents of which control the operations of the computer or peripheral. *See also:* program counter; device register. (C) 610.10-1994w

control relay An auxiliary relay whose function is to initiate or permit the next desired operation in a control sequence. (SWG/PE) C37.100-1992

control ring *See:* grading ring.

control room complex (nuclear power generating station) The complex that houses and protects plant operating personnel and control and instrumentation equipment. It includes the central control room, adjacent rooms that house supporting control equipment and instrumentation (sometimes known as the auxiliary equipment room), ventilation and life support equipment, and the cable spreading areas serving the equipment therein. (PE/NP) 567-1980w

control SCADA function The capability of a supervisory system to selectively perform manual or automatic, or both, operation (singularly or in selected groups) of external devices. Control may be either analog (magnitude or duration) or digital. (SUB/PE) C37.1-1994

control script A control_file associated with a software object that is executed by the software administration utilities. (C/PA) 1387.2-1995

control sequence table (electric-power devices) A tabulation of the connections that are made for each successive position of the controller. (IA/IAC) [60]

control signal Any signal that purposely affects the recording, processing, transmission or interpretation of data by a system element. (C) 610.10-1994w

control signal one (CS1) An encoded control signal used on the Control In and Control Out circuits. A CS1 is encoded as a signal at half the bit rate (BR)/2. (C/LM) 802.3-1998

control signal zero (CS0) An encoded control signal used on the Control In and Control Out circuits. A CS0 is encoded as a signal at the bit rate (BR). (C/LM) 802.3-1998

control space A dedicated area or compartment provided with equipment such as control consoles, gauge boards, control bench boards, switchboards, instrumentation, displays, control switches, communications, and other equipment for the local, remote, or programmed control and monitoring of equipment. The control system equipment may be operated by one or more individuals acting together or independently. (IA/MT) 45-1998

control span *See:* sag span.

control statement (software) A program statement that selects among alternative sets of program statements or affects the order in which operations are performed. For example, if-then-else, case. *Contrast:* assignment statement; declaration. (C) 610.12-1990

control station (1) (mobile communication) A base station, the transmission of which is used to control automatically the emission or operation of another radio station. *See also:* mobile communication system. (VT) [37]
(2) A facility that provides the individual responsible for controlling the simulation and that provided the capability to implement simulation control as PDUs on the DIS network. (DIS/C) 1278.3-1996

control store (software) In a microprogrammed computer, the computer memory in which microprograms reside. *See also:* microword; nanostore. (C) 610.12-1990, 610.10-1994w

control structure (software) A construct that determines the flow of control through a computer program. *See also:* conditional control structure; flow of control; computer program. (C/SE) 729-1983s

control switch A manually operated switching device for controlling power-operated devices. *Note:* It may include signaling, interlocking, etc., as dependent functions. (SWG/PE) C37.100-1992

control switchboard A type of switchboard including control, instrumentation, metering, protective (relays) or regulating equipment for remotely controlling other equipment. Control switchboards do not include the primary power circuit-switching devices or their connections. (SWG/PE) C37.100-1992, C37.21-1985r

control-switching point (telephone switching systems) A switching entity arranged for routing and control in the distance dialing network, at which intertoll trunks are interconnected. (COM) 312-1977w

control system (1) (broadly) An assemblage of control apparatus coordinated to execute a planned set of controls. *See also:* control. (IA/IAC) [60]
(2) A system in which a desired effect is achieved by operating on the various inputs to the system until the output, which is a measure of the desired effect, falls within an acceptable range of values. *See also:* network analysis; control; open-loop control system; transfer function; closed-loop control system. (MAG/PEL/ET) 264-1977w, 111-1984w
(3) (automatic control) A system in which deliberate guidance or manipulation is used to achieve a prescribed value of a variable. *Note:* It is subdivided into a controlling system and a controlled system. (PE/EDPG) [3]
(4) A system in which a desired effect is achieved by operating on inputs until the output, which is a measure of the desired effect, falls within an acceptable range of values. *See also:* open-loop control; automatic control; closed-loop control. (C/MAG) 610.2-1987, 264-1977w

control system, adaptive *See:* adaptive control system.

control system, automatic *See:* automatic control system.

control system, automatic feedback *See:* automatic feedback control system.

control system, cascade *See:* cascade control system.

control system, closed-loop *See:* closed-loop control system.

control system, coarse-fine *See:* coarse-fine control system.

control system, dual-mode *See:* dual-mode control system.

control system, duty factor *See:* duty factor control system.

control system, feedback *See:* feedback control system.

control system, floating *See:* floating control system.

control system, multiple-speed floating *See:* multiple-speed floating control system.

control system, multi-step *See:* step control system.

control system, on-off *See:* on-off control system.

control system, positioning *See:* positioning control system.

control system, ratio *See:* ratio control system.

control system, sampling *See:* sampling control system.

control system, single-speed floating *See:* single-speed floating control system.

control system, step *See:* step control system.

control system, two-step *See:* two-step control system.

control system, two-step neutral zone *See:* two-step control system.

control system, two-step single-point *See:* two-step control system.

control tape *See:* carriage control tape.

control terminal (mobile communication) (base station) Equipment for manually or automatically supervising a multiplicity of mobile and/or radio stations including means for calling or receiving calls from said stations. *See also:* mobile communication system. (VT) [37]

control total *See:* hash total.

control track (electroacoustics) A supplementary track usually placed on the same medium with the record carrying the program material. *Note:* Its purpose is to control, in some respect, the reproduction of the program, or some related phenomenon. Ordinarily, the control track contains one or more tones, each of which may be modulated either as to amplitude, frequency, or both. *See also:* phonograph pickup. (SP) [32]

control transfer instruction *See:* jump instruction.

control transformers (power and distribution transformers) Step-down transformers generally used in circuits which are characterized by low power levels and which contribute to a control function, such as in heating and air conditioning, printing, and general industrial controls. (PE/TR) C57.12.80-1978r

control unit (1) (digital computers) The parts that effect the retrieval of instructions in proper sequence, the interpretation of each instruction, and the application of the proper signals to the arithmetic unit and other parts in accordance with this interpretation. (C) [20], [85]
(2) (mobile communication) (mobile station) Equipment including a microphone and/or handset and loudspeaker together with such other devices as may be necessary for controlling a mobile station. *See also:* mobile communication system. (VT) [37]
(3) A functional unit of a computer that interprets and executes the instructions of a program in a prescribed sequence. *See also:* instruction control unit; main control unit. (C) 610.10-1994w

control valve (control systems for steam turbine-generator units) Those valves that control the energy input to the turbine and that are actuated by a controller through the control mechanism. (PE/EDPG) 122-1985s

control variable *See:* loop-control variable.

control voltage (1) The voltage applied to the operating mechanism of a device to actuate it, usually measured at the control power terminals of the mechanism. (SWG/PE) C37.100-1992
(2) Voltage that is provided for operating the controlled elements of the vehicle. *Note:* Control voltage may or may not be the same nominal potential as the battery voltage. *See also:* battery voltage. (VT) 1475-1999

control winding (1) (rotating machinery) An excitation winding that carries a current controlling the performance of a machine. *See also:* asynchronous machine. (PE) [9]

(2) (saturable reactor) A winding by means of which a controlling magnetomotive force is applied to the core. *See also:* magnetic amplifier. (EEC/PE) [119]

convection current In an electron stream, the time rate at which charge is transported through a given surface. *See also:* electron emission. (ED) 161-1971w, [45]

convection-current modulation The time variation in the magnitude of the convection current passing through a surface, or the process of directly producing such a variation. *See also:* electron emission. (ED) 161-1971w

convection heater A heater than dissipates its heat mainly by convection and conduction. (EEC/PE) [119]

convective discharge (medical electronics) (effluve) (electrical wind) (static breeze) The movement of a visible or invisible stream of particles carrying away charges from a body that has been charged to a sufficiently high voltage. (EMB) [47]

convective heat release The heat contained in the hot gases produced in a fire. (DEI) 1221-1993w

convenience outlet *See:* receptacle.

convention Any practice that is not formally standardized, but which is adopted by a group in a given situation. For example, programmers usually adopt the convention of indenting subordinate instructions in a routine so that the structure of the program is more easily visualized. *See also:* standard. (C) 610.7-1995, 610.10-1994w

conventional BIL (basic lightning impulse insulation level) The crest value of a standard lightning impulse for which the insulation shall not exhibit disruptive discharge when subjected to a specific number of applications of this impulse under specified conditions, applicable specifically to nonself-restoring insulations. (PE/SPD/C) C62.22-1997, 1313.1-1996

conventional BSL (basic switching impulse insulation level) The crest value of a standard switching impulse for which the insulation does not exhibit disruptive discharge when subjected to a specific number of impulses under specified conditions, applicable to nonself-restoring insulations. (PE/SPD/C) C62.22-1997, 1313.1-1996

conventional deviation of the disruptive discharge voltage (*z*) The difference between the 50% and 16% disruptive discharge voltages. (PE/PSIM) 4-1995

conventional-electrode coaxial detector (germanium gamma-ray detectors) Conventional-electrode geometry. A coaxial detector in which the outer contact is an n-type layer. (NPS) 325-1996

conventionally (true value of a quantity) The commonly accepted best estimate of the value of that quantity. This and its associated uncertainty will normally be determined by a national or transfer standard, or by a reference instrument that has been calibrated against a national or transfer standard, or by measurement quality assurance (MQA) with a national laboratory or qualified secondary laboratory. (NI) N42.17B-1989r

conventionally cooled (rotating machinery) A term referring to windings in which the heat generated within the principal portion of the windings must flow through the major ground insulation before reaching the cooling medium. (PE/REM) [9], [115]

conventionally true value The best estimate of the value determined by a primary or secondary standard, or by a reference instrument that has been calibrated against a primary or secondary standard. (NI) N42.20-1995

conventional withstand voltage (1) The voltage that an insulation system is capable of withstanding without failure or disruptive discharge under specified test conditions. (PE/C) 1313.1-1996

(2) The voltage that an insulation is capable of withstanding with a 0% probability of failure. (SPD/PE) C62.22-1997

conventions (1) (software) Requirements employed to prescribe a disciplined uniform approach to providing consistency in a software product, that is, uniform patterns or forms for arranging data. *See also:* practices. (C/SE) 610.12-1990, 983-1986w

(2) Accepted guidelines employed to prescribe a disciplined, uniform approach to providing consistency in a software item, for example, uniform patterns or forms for arranging data. (C/SE) 730.1-1995

convergence (multibeam cathode-ray tubes) A condition in which the electron beams intersect at a specified point. (ED) 161-1971w

convergence, dynamic *See:* dynamic convergence.

convergence electrode (multibeam cathode-ray tubes) An electrode whose electric field converges two or more electron beams. (ED) 161-1971w

convergence function A function or procedure that provides sufficient additional services to enable a *layer or sublayer* to provide the services expected by a particular higher layer user. (For example, the MAC Convergence Function enables the capabilitites of the *Queued Arbitrated access function* to be enhanced to provide the *Medium Access Control (MAC) Sublayer* service to the *Logical Link Control Sublayer.*). (LM/C) 8802-6-1994

convergence magnet (multibeam cathode-ray tubes) A magnet assembly whose magnetic field converges two or more electron beams. (ED) 161-1971w

convergence plane (multibeam cathode-ray tubes) A plane containing the points at which the electron beams appear to experience a deflection applied for the purpose of obtaining convergence. (ED) 161-1971w

convergence protocol (1) A protocol that provides the convergence service for the provision of enhancements to an underlying service in order to provide for the specific requirements of the convergence service user. (LM/C/COM) 8802-9-1996

(2) A protocol that provides the convergence service. (LM/C) 15802-2-1995

convergence service A service that provides enhancements to an underlying service in order to provide for the specific requirements of the convergence service user. (LM/C) 15802-2-1995

convergence surface (multibeam cathode-ray tubes) The surface generated by the point of intersection of two or more electron beams during the scanning process. (ED) 161-1971w

convergence time (T_c) The time required to reach within 3 dB of maximum echo return loss, or 25 dB loss, whichever occurs first. (COM/TA) 1329-1999

Conversation A set of MAC frames transmitted from one end station to another, where all of the MAC frames form an ordered sequence, and where the communicating end stations require the ordering to be maintained among the set of MAC frames exchanged. (C/LM) 802.3ad-2000

conversational (software) Pertaining to an interactive system or mode of operation in which the interaction between the user and the system resembles a human dialog. *Contrast:* batch. *See also:* online; interactive; real time. (C) 610.12-1990

Conversational Algebraic Language (CAL) A general-purpose programming language used in time-sharing environments for solving numerical problems. (C) 610.13-1993w

conversational compiler *See:* incremental compiler.

converse inorder traversal The process of traversing a binary tree in a recursive fashion as follows: the right subtree is traversed, then the root is visited, then the left subtree is traversed. *Contrast:* converse postorder traversal; converse preorder traversal. *See also:* inorder traversal. (C) 610.5-1990w

converse postorder traversal The process of traversing a binary tree in a recursive fashion as follows: the right subtree is traversed, then the left subtree is traversed, then the root is visited. *Contrast:* converse inorder traversal; converse preorder traversal. *See also:* postorder traversal. (C) 610.5-1990w

converse preorder traversal The process of traversing a binary tree in a recursive fashion as follows: the root is visited, then the right subtree is traversed, then the left subtree is traversed. *Contrast:* converse inorder traversal; converse postorder traversal. *See also:* preorder traversal. (C) 610.5-1990w

conversion (1) (software) Modification of existing software to enable it to operate with similar functional capability in a different environment; for example, converting a program from Fortran to Ada, converting a program that runs on one computer to run on another. (C) 610.12-1990
(2) A general term covering the process of altering existing power switchgear equipment. (SWG/PE) C37.100-1992
(3) The process of altering existing power switchgear equipment from the original manufacturers design. (SWG/PE) C37.59-1996

conversion efficiency (1) (electrical conversion) In alternating-current to direct-current conversion equipment, the ratio of the product of output direct-current and voltage to input watts expressed in percent. *Note:* It reflects alternating-current power capacity required for a given voltage and current output and does not necessarily reflect watts lost.

$$= \frac{(E_{dc})(I_{dc})}{P}(100 \text{ percent})$$

(AES) [41]
(2) (overall) (photoelectric converter) The ratio of available power output to total incident radiant power in the active area for photovoltaic operation. *Note:* This depends on the spectral distribution of the source and junction temperature. *See also:* semiconductor. (AES) [41]
(3) (klystron oscillator) The ratio of the high-frequency output power to the direct-current power supplied to the beam. *See also:* velocity-modulated tube. (Std100) [84]
(4) (solar cells) The ratio of the solar cell's available power output (at a specified voltage) to the total incident radiant power. The cell active area shall be used in this calculation; that is, ohmic contact (but no grid lines) areas on the irradiated side shall be deducted from the total irradiated cell area to determine active area. The spectral distribution of the source and the junction temperature must be specified. (AES/SS) 307-1969w

conversion factor *See:* calibration factor.

conversion loss (nonlinear, active, and nonreciprocal waveguide components) In a frequency converter (mixer), the ratio of the output power at the converted frequency to the available input power at the signal frequency; often expressed in decibels. (MTT) 457-1982w

conversion rate (A) (hybrid computer linkage components) (analog-to-digital converter) The maximum rate at which the start conversion commands can be applied to the converter, to which the converter will respond by providing the desired signal at the output to within a given accuracy. **(B) (analog-to-digital converter with multiplexor with sample and hold)** The maximum rate at which the start sample commands can be applied to the system to which the system will respond by providing the desired signal at the output to within a given accuracy. (Pre-selected channel). (C) 166-1977

conversion time (A) (hybrid computer linkage components) (analog-to-digital converter) That time required from the instant at which a conversion command is received and a final digital representation is available for external output to within a given accuracy. **(B) (analog-to-digital converter with multiplexor with sample and hold)** That time required from the time at which a sample command is received and a final digital representation is available for external output to within a given accuracy. (Pre-selected channel). (C) 166-1977

conversion transconductance (heterodyne conversion transducer) The quotient of (1) the magnitude of the desired output-frequency component of currents by (2) the magnitude of the input-frequency (signal) component of voltage when the impedance of the output external termination is negligible for all of the frequencies that may affect the result. *Note:* Unless otherwise stated, the term refers to the cases in which the input-frequency voltage is of infinitesimal magnitude. All direct electrode voltages, and the magnitude of the local-oscillator voltage, must remain constant. *See also:* transducer; modulation. (ED) 161-1971w

conversion transducer (1) (general) A transducer in which the signal undergoes frequency conversion. *Note:* The gain or loss of a conversion transducer is specified in terms of the useful signal. *See also:* transducer. (PE/EEC) [119]
(2) An electric transducer in which the input and the output frequencies are different. *Note:* If the frequency-changing property of a conversion transducer depends upon a generator of frequency different from that of the input or output frequencies, the frequency and voltage or power of this generator are parameters of the conversion transducer. *See also:* heterodyne conversion transducer. (ED) 161-1971w

conversion voltage gain (conversion transducer) The ratio of the magnitude of the output-frequency voltage across the output termination, with the transducer inserted between the input-frequency generator and the output termination" to the magnitude of the input-frequency voltage across the input termination of the transducer. (ED) [45]

convert (data processing) To change the representation of data from one form to another, for example, to change numerical data from binary to decimal or from cards to tape. (C) [20], [85]

converter (1) (general) A machine or device for changing alternating-current power to direct-current power or vice versa. (PE) [9]
(2) (A) (heterodyne reception) (frequency converter) The portion of the receiver that converts the incoming signal to the intermediate frequency. **(B) (data transmission)** A device for changing one form of information language to another, so as to render the language acceptable to a different machine (that is, card to tape conversion). (PE) 599-1985
(3) (facsimile) A device that changes the type of modulation. *See also:* facsimile. (COM) 168-1956w
(4) A network or device for changing the form of information or energy. (IA/ICTL/APP/IAC) [69], [60]
(5) (test measurement and diagnostic equipment) A device that changes the manner of representing information from one form to another. (MIL) [2]
(6) A device that changes electrical energy from one form to another, as from alternating current to direct current. (NESC/NEC) [86]
(7) A device that changes electrical energy from one form to another. A semiconductor converter is a converter that uses semiconductors as the active elements in the conversion process. (IA/SPC) 519-1992
(8) A machine or device for changing dc power to ac power, for changing ac power to dc power, or for changing from one frequency to another. This definition covers several different power conversion functions, each of which is known by a separate term. *See also:* dc-dc converter; frequency converter; inverter; rectifier. (PEL/ET) 388-1992r
(9) A device capable of converting impulses from one mode to another, such as analog to digital, parallel to serial, or from one code to another. *See also:* code converter; digital-to-analog converter; power supply. (C) 610.10-1994w
(10) (self-commutated converters) (ac adjustable-speed drives) An operative unit for electronic power conversion, comprising one or more electronic switching devices and any associated components, such as transformers, filters, commutation aids, controls, and auxiliaries. *Synonym:* converter equipment. (IA/SPC/ID) 936-1987w, 995-1987w
(11) A type of repeater that converts the data signal from one media to another. (C/LM) 8802-5-1998

converter, analog-to-digital *See:* analog-to-digital converter.

converter, digital-to-analog *See:* digital-to-analog converter.

converter equipment *See:* converter.

converter, reversible power *See:* reversible power converter.

converter, static solid state *See:* static, solid-state converter.

converter switching element (ac adjustable-speed drives) A part of the converter circuit, bounded by two principal terminals, containing one or more semiconductor devices having the property of controllable or noncontrollable conduction in at least one direction. (IA/ID/SPC) 995-1987w, 936-1987w

converter tube An electron tube that combines the mixer and local-oscillator functions of a heterodyne conversion transducer. *See also:* heterodyne conversion transducer. (ED) 161-1971w, [45]

converting station (power operations) A station where machinery is used for changing alternating-current (ac) power to direct-current power or vice versa, or from one frequency to another. (PE/PSE) 858-1987s

convex Pertaining to a region for which a straight line segment between any two points of the region is entirely contained within the region. *Contrast:* concave.

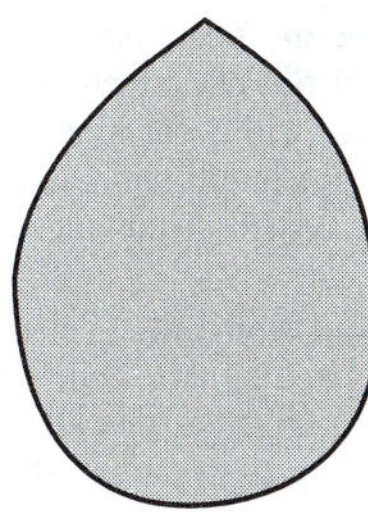

convex

(C) 610.4-1990w

convex programming In operations research, a particular type of nonlinear programming in which the function to be maximized or minimized and the constraints to be applied are appropriately convex or concave functions, respectively. (C) 610.2-1987

conveyor A mechanical contrivance, generally electrically driven, that extends from a receiving point to a discharge point and conveys, transports, or transfers material between those points. *See also:* chain-type conveyor; shaker-type conveyor; vibrating-type conveyor; belt-type conveyor. (EEC/PE) [119]

conveyor, belt-type *See:* belt-type conveyor.

conveyor, chain-type *See:* chain-type conveyor.

conveyor, shaker-type *See:* shaker-type conveyor.

conveyor, vibrating-type *See:* vibrating-type conveyor.

convolution efficiency The ratio of the power density of the desired output at frequency (ω_1 and ω_2) to the product of the power densities of the two inputs at frequencies ω_1 and ω_2, respectively, in the convolution region, expressed in decibels. (UFFC) 1037-1992w

convolution function (burst measurements) The integral of the function x (τ) multiplied by another function y ($-\tau$) shifted in time by t

$$\int_{-\infty}^{\infty} x(\tau)y(t-\tau)d\tau$$

See also: burst. (SP) 265-1966w

convolution integral (automatic control) A mathematical integral operation that is used to describe the time response of a linear element to an input function in terms of the weighting function of the element. The integral generally takes the form $\int_0^t f(x)g(t-x)dx$ where $f(x)$ is an arbitrary input, and g $(t-x)$ is a weighting function that extends backward from instant t through x as far as zero. (PE/EDPG) [3]

convolver A three-port device whose output signal is the convolution of two time waveforms applied simultaneously to the input ports; convolution is achieved by physically passing the input signals over one another at the output transducer. (UFFC) 1037-1992w

cookie A quantity used to indicate or signal to a recipient of data, significant changes in the state of the entity supplying the data. (IM/ST) 1451.1-1999

cooking unit, counter-mounted *See:* counter-mounted cooking unit.

coolant A fluid, usually air, hydrogen, or water, used to remove heat from a machine or from certain of its components. *Synonym:* cooling medium. (PE/EM) 67-1990r, [9]

cooled-input FET preamplifier (germanium gamma-ray detectors) A preamplifier in which the input field-effect transistor (FET) is cooled to achieve a reduction in noise. (NPS) 325-1986s

cooler (heat exchanger) (rotating machinery) A device used to transfer heat between two fluids without direct contact between them. (PE) [9]

Coolidge tube An X-ray tube in which the needed electrons are produced by a hot cathode. (ED) [45]

cooling (power supplies) The cooling of regulator elements refers to the method used for removing heat generated in the regulating process. *Note:* Methods include radiation, convection, and conduction or combination thereof. (AES) [41]

cooling coil (rotating machinery) A tube through whose wall, heat is transferred between two fluids without direct contact between them. (PE) [9]

cooling, convection *See:* convection cooling.

convection cooling (power supplies) A method of heat transfer that uses the natural upward motion of air warmed by the heat dissipators. (AES/PE) [41], [78]

cooling duct *See:* ventilating duct.

cooling fin A metallic part of fin extending the cooling area to facilitate the dissipation of the heat generated in the device. *See also:* electron device. (ED) [45]

cooling, lateral force-air (power supplies) An efficient method of heat transfer by means of side-to-side circulation that employs blower movement of air through or across the heat dissipators. (AES/PE) [41], [78]

cooling medium *See:* coolant.

cooling system (1) (rectifier) Equipment, that is, parts and their interconnections, used for cooling a rectifier. *Note:* It includes all or some of the following: rectifier water jacket, cooling oils or fins, heat exchanger, blower, water pump, expansion tank, insulating pipes, etc. *See also:* rectification. (IA) [62]

(2) (thyristor) Any equipment, that is, parts and their interconnections, used for cooling a thyristor controller. It includes all or some of the following; thyristor heat sink, cooling coils or fins, heat exchanger, fan or blower, water pump, expansion tank, insulating pipes, equipment enclosure, etc. (IA/IPC) 428-1981w

(3) (thyristor converter) Equipment, that is, parts and their interconnections, used for cooling a thyristor converter. *Note:* It includes all or some of the following: thyristor heat sink, cooling coils or fins, heat exchanger, fan or blower, water pump, expansion tank, insulating pipes, etc. (IA/IPC) 444-1973w

cooling system, direct raw-water (thyristor converter) A cooling system in which water, received from a constantly available supply, such as a well or water system, is passed directly over the cooling surfaces of the thyristor converter and discharged. (IA/IPC) 444-1973w

cooling system, direct raw-water, with recirculation (thyristor converter) A direct raw-water cooling system in which part of the water passing over the cooling surfaces of the thyristor converter is recirculated and raw water is added as needed to maintain the required temperature, the excess being discharged. (IA/IPC) 444-1973w

cooling system, forced-air (thyristor converter) An air cooling system in which heat is removed from the cooling surfaces of the thyristor converter by means of a flow of air produced by a fan or blower. (IA/IPC) 444-1973w

cooling system, heat-exchanger (thyristor converter) A cooling system in which the coolant, after passing over the cooling surfaces of the thyristor converter, is cooled in heat exchanger and recirculated. *Note:* Heat may be removed from the thyristor converter cooling surfaces by liquid or air using the following types of heat exchangers: water-to-water, water-to-air, air-to-water, air-to-air, and refrigeration cycle. The liquid in the closed system may be other than water, and the gas in the closed system may be other than air. (IA/IPC) 444-1973w

cooling system, natural-air (thyristor converter) An air cooling system in which heat is removed from the cooling surfaces of the thyristor converter only by the natural action of the ambient air. (IA/IPC) 444-1973w

cooling system regulating equipment (thyristor) Any equipment used for heating and cooling the thyristor controller, together with the devices for controlling and indicating its temperature. (IA/IPC) 428-1981w

cooling-water system (rotating machinery) All parts that are provided for the flow, treatment, or storage of cooling water. (PE) [9]

coordinate dimension word (numerically controlled machines) A word defining an absolute dimension. (IA/EEC) [61], [74]

coordinated operation **(1)** Operation of generation and transmission facilities of two or more interconnected systems to achieve greater reliability and economy. (PE/PSE) 858-1993w

(2) **(hydro plants)** Operation of a group of hydro plants and storage reservoirs so as to obtain optimum power benefits with due consideration to all other uses. (PE/PSE) 346-1973w

(3) **(electric power supply)** Operation of generation and transmission facilities of two or more interconnected electrical systems to achieve greater reliability and economy. (PE/PSE) 346-1973w

coordinated operation of hydroplants (power operations) Operation of a group of hydroplants and storage reservoirs so as to obtain optimum power benefits with due consideration to all other uses. (PE/PSE) 858-1987s

coordinated transpositions (electric supply or communication circuits) Transpositions that are installed for the purpose of reducing inductive coupling, and that are located effectively with respect to the discontinuities in both the electric supply and communication circuits. *See also:* inductive coordination. (PE/EEC) [119]

COordinate GeOmetry (COGO) A problem-oriented programming language used to solve coordinate geometry problems in civil engineering applications. (C) 610.13-1993w

coordinates A set of data values that specify a location. (C) 610.6-1991w

coordinate system (pulse terminology) Throughout the following, a rectangular Cartesian coordinate system is assumed in which, unless otherwise specified:

1) Time (t) is the independent variable taking alone the horizontal axis, increasing in the positive sense from left to right.
2) Magnitude (m) is the dependent variable taken along the vertical axis, increasing the positive sense or polarity from bottom to top.
3) The following additional symbols are used:
   a) e—The base of natural logarithms.
   b) a, b, c, etc.—Real constants that, unless otherwise specified, may have any value and either sign.
   c) n—A positive integer.

(Std100)

coordinating entity **(1)** That part of an end system or interworking unit (IWU) responsible for the coordination and synchronization of functions belonging to the data and signalling subentities of the layer entity implementing a PSN access protocol. (C/LM) 802.9a-1995w

(2) That part of the Network Layer within an end system or interworking unit responsible for the coordination and synchronization of functions belonging to the Data and Signalling subentities of the layer entity implementing a PSN access protocol. (LM/C/COM) 8802-9-1996

coordination dimension A reference dimension used to coordinate mechanical interfaces. This is not a manufacturing dimension with a tolerance. (C/BA/MM) 1301.2-1993, 1301.4-1996, 1301.1-1991, 1301.3-1992r

coordination function The logical function that determines when a station operating within a basic service set (BSS) is permitted to transmit and may be able to receive protocol data units (PDUs) via the wireless medium (WM). The coordination function within a BSS may have one point coordination function (PCF) and will have one distributed coordination function (DCF). (C/LM) 8802-11-1999

coordination function pollable A station able to respond to a coordination function poll with a data frame, if such a frame is queued and able to be generated, and interpret acknowledgments in frames sent to or from the point coordinator. (C/LM) 8802-11-1999

coordination of protection The process of choosing settings or time delay characteristics of protective devices, such that operation of the devices will occur in a specified order to minimize customer service interruption and power system isolation due to a power system disturbance. (PE/PSR) C37.113-1999

coplanar strip transmission line A planar transmission line consisting of two parallel thin conducting strips of finite width, separated by a finite gap and affixed to the same plane surface of an insulating substrate of arbitrary thickness. (MTT) 1004-1987w

coplanar waveguide A planar transmission line consisting of a single thin conducting strip of finite width situated between two semi-infinite ground planes and separated from them by finite gaps, which are all affixed to the same plane surface of an insulating substrate of arbitrary thickness. (MTT) 1004-1987w

co-polarization That polarization that the antenna is intended to radiate [receive]. *See also:* polarization pattern. (AP/ANT) 145-1993

co-polar (radiation) pattern A radiation pattern corresponding to the co-polarization. *See also:* co-polarization. (AP/ANT) 145-1993

copper brush (rotating machinery) A brush composed principally of copper. *See also:* brush. (PE) [9]

copper-clad aluminum conductors Conductors drawn from a copper-clad aluminum rod with the copper metallurgically bonded to an aluminum core. The copper forms a minimum of 10 percent of the cross-sectional area of a solid conductor or each strand of a stranded conductor. (NESC/NEC) [86]

copper-clad steel Steel with a coating of copper welded to it, as distinguished from copper-plated or copper-sheathed material. (EEC/PE) [119]

copper-covered steel wire A wire having a steel core to which is bounded a continuous outer layer of copper. *See also:* conductor. (PE) [4], 64

copper losses *See:* load losses.

coprocessor (CP) **(1)** An optional processing unit (impl. dep. #4). (C/MM) 1754-1994

(2) A processor used in conjunction with a central processing unit, designed to perform specific functions that may not be executed efficiently by the central processing unit, for example: a floating-point coprocessor. (C) 610.10-1994w

coprocessor operate (CPop) instructions Instructions that perform coprocessor calculations, as defined by the CPop1 and CPop2 opcodes. CPop instructions do not include CBccc instructions, nor loads and stores between memory and the coprocessor. (C/MM) 1754-1994

copy **(1)** **(A)** **(software)** To read data from a source, leaving the source data unchanged, and to write the same data elsewhere in a physical form that may differ from that of the source. For example, to copy data from a magnetic disk onto a magnetic

tape. *Contrast:* move. **(B) (software)** The result of a copy process as in definition (A). For example, a copy of a data file. *See also:* soft copy; display; hard copy. (C) 610.2-1987, 610.12-1990
(2) (A) (electronic data processing) To reproduce data leaving the original data unchanged. *See also:* transfer. **(B) (electronic data processing)** To produce a sequence of character events equivalent, character by character, to another sequence of character events. *See also:* transfer. **(C) (electronic data processing)** The sequence of character events produced in (B). *See also:* transfer. (C) 162-1963
(3) To duplicate text or graphic objects from the screen to the clipboard. (C) 1295-1993w
(4) (A) A copy of an entry stored in other DSA(s) through bilateral agreement. **(B)** A locally and dynamically stored copy of an entry resulting from a request (a cache copy). (C/PA) 1328.2-1993, 1224.2-1993, 1327.2-1993, 1326.2-1993

copyback cache A cache memory scheme with the attribute that data written from the processor is normally written to the cache rather than the main memory. Modified data in the cache is written to the main memory to avoid loss of the data when a cache line flush or replacement occurs. (C/BA) 10857-1994, 896.4-1993w

copyright The exclusive right granted to the owner of an original work of authorship, which is fixed in any tangible medium of expression, to reproduce, publish, perform, and/or sell the work. (C/SE) 1420.1b-1999

cord One or a group of flexible insulated conductors, enclosed in a flexible insulating covering and equipped with terminals. (EEC/PE) [119]

cord adjuster A device for altering the pendant length of the flexible cord of pendant. *Note:* This device may be a rachet reel, a pulley and counterweight, a tent-rope stick, etc. (EEC/PE) [119]

cord circuit (telephone switching systems) A connecting circuit, usually terminating in a plug at one or both ends, used at switchboard positions in establishing telephone connections. (COM) 312-1977w

cord-circuit repeater A repeater associated with a cord circuit so that it may be inserted in a circuit by an operator. *See also:* repeater. (EEC/PE) [119]

cord connector A plug receptacle provided with means for attachment to flexible cord. (EEC/PE) [119]

cord grip (strain relief) A device by means of which the flexible cord entering a device or equipment is gripped in order to relieve the terminals from tension in the cord. (EEC/PE) [119]

cordless switchboard (telephone switching systems) A telecommunications switchboard in which manually operated keys are used to make connections. (COM) 312-1977w

core (1) (power and distribution transformers) An element made of magnetic material, serving as part of a path for magnetic flux. (PE/TR) C57.12.80-1978r
(2) (electronic information storage) *See also:* digital computer.
(3) (mechanical recording) The central layer or basic support of certain types of laminated media. (SP) [32]
(4) (electromagnet) The part of the magnetic structure around which the magnetizing winding is place. (Std100) 270-1966w
(5) (fiber optics) The central region of an optical waveguide through which light is transmitted. *See also:* normalized frequency; cladding; optical waveguide. (Std100) 812-1984w
(6) (composite insulators) The axially aligned glass fiber reinforced resin rod that forms the mechanically load-bearing component of the insulator. (T&D/PE) 987-1985w
(7) (A) The central conductor element of a coaxial cable. *Note:* It is usually constructed of copper. **(B)** Single conductor in a cable (British usage). (C) 610.7-1995

core area (fiber optics) The cross sectional area enclosed by the curve that connects all points nearest the axis on the periphery of the core where the refractive index of the core exceeds that of the homogeneous cladding by k times the difference between the maximum refractive index in the core and the refractive index of the homogeneous cladding, where k is a specified positive or negative constant $k1$. *See also:* cladding; homogeneous cladding; core; tolerance field. (Std100) 812-1984w

core center (fiber optics) A point on the fiber axis. *See also:* fiber axis; optical axis. (Std100) 812-1984w

core circuit The part of the circuitry in an integrated circuit that provides the intended data manipulation function (as distinct from dedicated test circuitry). *Note:* Residual elements, which are permanently connected to the pin (i.e., in test mode as well as in normal function mode), are not regarded as being part of the core circuit. *See also:* residual element. (C/TT) 1149.4-1999

core diameter (fiber optics) The diameter of the circle that circumscribes the core area. *See also:* core; tolerance field; core area; cladding. (Std100) 812-1984w

core disconnect (CD) A facility provided within an analog boundary module (ABM) (usually mediated by a conceptual switch) that allows a pin to be disconnected from the core circuit so that the signal at the pin can be driven to any value within the pin's normal functional range without affecting the core circuit, and no value generated in the core circuit will affect the pin. *Notes:* 1. The core disconnect facility could be provided as part of the functional driver or receiver attached to the pin, e.g., by implementing a driver with high-Z capability. 2. It is necessary to document all residual elements that remain connected to the pin when it enters the CD state. *See also:* conceptual switch; core circuit; high-Z; residual element; CD state. (C/TT) 1149.4-1999

core duct (rotating machinery) The space between or through core laminations provided to permit the radial or axial flow of coolant gas. *See also:* rotor. (PE) [9]

core dump* *See:* memory dump.

* Deprecated.

core end plate (rotating machinery) A plate or structure at the end of a laminated core to maintain axial pressure on the laminations. (PE) [9]

core-form transformer (power and distribution transformers) A transformer in which those parts of the magnetic circuit surrounded by the windings have the form of legs with two common yokes. (PE/TR) C57.12.80-1978r

core length (rotating machinery) The dimension of the stator, or rotor, core measured in the axial direction. *See also:* stator; rotor. (PE) [9]

core loss (1) The power dissipated in a magnetic core subjected to a time-varying magnetizing force. *Note:* Core loss includes hysteresis and eddy-current losses of the core. (PE/TR) C57.12.80-1978r
(2) (synchronous machines) The difference in power required to drive the machine at normal speed, when excited to produce a voltage at the terminals on open circuit corresponding to the calculated internal voltage, and the power required to drive the unexcited machine at the same speed. *Note:* The internal voltage shall be determined by correcting the rated terminal voltage for the resistance drop only. (PE) [9], [84]
(3) (electronic power transformer) The measured power loss, expressed in watts, attributable to the material in the core and associated clamping structure, of a transformer that is excited, with no connected load, at a core flux density and frequency equal to that in the core when rated voltage and frequency is applied and rated load current is supplied. (PEL/ET) 295-1969r
(4) (power and distribution transformers) The power dissipated in a magnetic core subjected to a time-varying magnetizing force. Core loss includes hysteresis and eddy-current losses of the core. (PE/TR) C57.12.80-1978r

core-loss current The in-phase component (with respect to the induced voltage) of the exciting current supplied to a coil. *Note:* It may be regarded as a hypothetical current, assumed to flow through the equivalent core-loss resistance. (CHM) [51]

core loss, open-circuit (rotating machinery) The difference in power required to drive a machine at normal speed, when excited to produce a specified voltage at the open-circuited armature terminals, and the power required to drive the unexcited machine at the same speed. (Std100) [84]

core-loss test (rotating machinery) A test taken on a built-up (usually unwound) core of a machine to determine its loss characteristic. *See also:* stator. (PE) [9]

core memory *See:* magnetic core.

core package (rotating machinery) The portion of core lying between two adjacent vent ducts or between an end plate and the nearest vent duct. (PE) [9]

corequisite The specification in a software object such that another software object shall be installed, in conjunction with the installation of the first, and configured in conjunction with the configuration of the first. (C/PA) 1387.2-1995

core, relay *See:* relay core.

core specification Synonym for IEEE Std 1275-1994, i.e., the standard that specifies the system-independent and bus-independent requirements for Open Firmware. (C/BA) 1275.1-1994w, 1275.2-1994w

core storage A type of storage in which the data medium consists of magnetic cores. *Contrast:* semiconductor storage. (C) 610.10-1994w

CORE System A prototype computer graphics standard that contains common concepts and practices of graphics programming. It was developed by an Association for Computing Machinery (ACM) Special Interest Group on Computer Graphics (SIGGRAPH) committee. (C) 610.6-1991w

core test (rotating machinery) FRA test taken on a built-up (usually unwound) core of a machine to determine its loss characteristics or its magnetomotive force characteristics, or to locate short-circuited laminations. *See also:* stator; rotor. (PE) [9]

Core Test Information Model (CTIM) The fundamental information entities required to describe tests, test specifications, test requirements, and other test entities within the ABBET™ domain. The CTIM contains the significant types from which the component standards may be derived. (SCC20) 1226-1998

core-type transformer A transformer in which those parts of the magnetic-circuit surrounded by the windings have the form of legs with two common yokes. (PE/TR) C57.12.80-1978r

Coriolis acceleration (inertial sensors) That increment of acceleration relative to inertial space that arises from the velocity of a particle relative to a rotating coordinate system. Sometimes, the term Coriolis acceleration is also used to describe the apparent acceleration relative to a rotating coordinate system of a force-free moving particle. (AES/GYAC) 528-1994

coriolis correction (navigation) (navigation aid terms) An acceleration correction that must be applied to measurements of acceleration with respect to a coordinate system relative to inertial space. (AES/GCS) 172-1983w

cornea (laser maser) The transparent outer coat of the human eye that covers the iris and the crystalline lens. It is the main refracting element of the eye. (LEO) 586-1980w

corner (waveguide technique) An abrupt change in the direction of the axis of a waveguide. *Synonym:* elbow. *See also:* waveguide. (AP/ANT) [35], [84]

corner frequency (1) The frequency at which the skin depth is equal to the thickness of the shield. (PE/IC) 1143-1994r
(2) (asymptotic form of Bode diagram) (control system feedback) The frequency indicated by a breakpoint, that is, the junction of two confluent straight lines asymptotic to the log gain curve. *Note:* One breakpoint is associated with each distinct real root of the characteristic equation, one with each set of repeated roots, and one with each pair of complex roots. For a single real root, corner frequency (in radians per second) is the reciprocal of the corresponding time constant (in seconds), and the corresponding phase angle is halfway between the phase angles belonging to the asymptotes extended to infinity. *See also:* feedback control system. (IM/PE/EDPG) [120], [3]

corner reflector (1) A reflecting object consisting of two or three mutually intersecting conducting flat surfaces. *Note:* Dihedral forms of corner reflectors are frequently used in antennas; trihedral forms with mutually perpendicular surfaces are more often used as radar targets. (AP/ANT) 145-1993
(2) (A) (antenna) A reflecting object consisting of two or three mutually intersecting conducting flat surfaces. *Note:* Dihedral forms of corner reflectors are frequently used in antennas; trihedral forms are more often used as radar targets. **(B) (radar)** Two (dihedral) or three (trihedral) mutually intersecting conducting surfaces designed to return electromagnetic radiation towards its source. Also used as calibration devices. (AP/PROP) 211-1997
(3) Two (dihedral) or three (trihedral) orthogonal conducting surfaces, designed to return an incident electromagnetic wave toward its source. *Note:* A corner reflector is often used to provide a conspicuous radar target as a safety measure for a small sailboat, to enhance the detectability of a radar target on which it is mounted, or to calibrate a radar. (AES) 686-1997

corner reflector antenna An antenna consisting of a feed and a corner reflector. (AP/ANT) 145-1993

corner, waveguide *See:* waveguide bend.

cornice lighting (illuminating engineering) Lighting comprising sources shielded by a panel parallel to the wall and attached to the ceiling, and distributing light over the wall. (EEC/IE) [126]

corona (1) (air) A luminous discharge due to ionization of the air surrounding a conductor caused by a voltage gradient exceeding a certain critical value. *See also:* tower. (T&D/PE/TR) [10], C57.19.03-1996
(2) (gas) A discharge with slight luminosity produced in the neighborhood of a conductor, without greatly heating it, and limited to the region surrounding the conductor in which the electric field exceeds a certain value. *See also:* partial discharge; discharge. (ED) [45], [84]
(3) (overhead-power-line corona and radio noise) A luminous discharge due to ionization of the air surrounding an electrode caused by a voltage gradient exceeding a certain critical value. *Note:* For the purpose of IEEE Std 539-1990, electrodes may be conductors, hardware, accessories, or insulators. (PE/T&D) 539-1990
(4) (partial discharge) (corona measurement) A type of localized discharge resulting from transient gaseous ionization in an insulation system when the voltage stress exceeds a critical value. The ionization is usually localized over a portion of the distance between the electrodes of the system. (MAG/ET) 436-1977s
(5) (dc electric-field strength and ion-related quantities) A luminous discharge due to ionization of the air surrounding an electrode caused by a voltage gradient exceeding a certain critical value. (T&D/PE) 1227-1990r
(6) (non-preferred term) (power and distribution transformers) *See also:* partial discharge. (PE/TR) C57.12.80-1978r
(7) (dry-type transformers) *See also:* partial discharge. (PE/TR) C57.124-1991r

corona charging (electrostatography) Sensitizing by means of gaseous ions of a corona. *See also:* electrostatography. (ED) [46]

corona-discharge tube A low-current gas-filled tube utilizing the corona-discharge properties. (ED) [45]

corona extinction gradient (overhead-power-line corona and radio noise) The gradient on that part of an electrode surface at which continuous corona last persists as the applied voltage is gradually decreased. (PE/T&D) 539-1990

corona extinction voltage (1) (corona measurement) The highest voltage at which continuous corona of specified pulse amplitude no longer occurs as the applied voltage is gradually decreased from above the corona inception value. Where the applied voltage is sinusoidal, the CEV is expressed as

$$1/\sqrt{2}$$

of the peak voltage. (MAG/ET) 436-1977s
(2) (overhead-power-line corona and radio noise) The voltage applied to the electrode to produce the corona extinction gradient. (T&D/PE) 539-1990

corona inception gradient (overhead-power-line corona and radio noise) The gradient on that part of an electrode surface at which continuous corona first occurs as the applied voltage is gradually increased. *See also:* continuous corona. (T&D/PE) 539-1990

corona inception test *See:* discharge inception test.

corona inception voltage (A) (corona measurement) The lowest voltage at which continuous corona of specified pulse amplitude occurs as the applied voltage is gradually increased. Where the applied voltage is sinusoidal, the CIV is expressed as

$$1/\sqrt{2}$$

of the peak voltage. **(B)** The voltage applied to the electrode to produce the corona inception gradient. (MAG/T&D/PE/ET) 436-1977, 539-1990

corona level *See:* ionization extinction voltage.

corona loss Power lost due to corona process. On overhead power lines, this loss is expressed in watts per meter (W/m) or kilowatts per kilometer (kW/km). (T&D/PE) 539-1990

corona modes (overhead-power-line corona and radio noise) Two principal modes of corona are the glow mode and the streamer mode. Their characteristics and occurrence depend on the polarity of the electrode, the basic ionization characteristics of the ambient air, and the magnitude, as well as the distribution of the electric field. Thus, the geometry of the electrodes, the ambient weather conditions, and the magnitude, as well as the polarity of the applied voltage, are the main factors determining corona modes. Corona modes that are possible during alternating half-cycles of the alternating-voltage waveform are essentially similar to those of corresponding direct-voltage corona modes when effects of space charges left behind from each preceding half-cycle are taken into account. *See also:* streamer mode; glow mode. (PE/T&D) 539-1990

corona, overhead power lines Coronas occurring at the surfaces of power-line conductors and their fittings under the positive or negative polarity of the power-line voltage. *Notes:* 1. Surface irregularities such as stranding, nicks, scratches, and semiconducting or insulating protrusions are usual corona sites. 2. Dry or wet airborne particles in the proximity of power-line conductors and their fittings may cause corona discharges. 3. Weather has a pronounced influence on the occurrence and characteristics of overhead power-line coronas. (T&D/PE) 539-1990

corona pulse (1) (corona measurement) A voltage or current pulse which occurs at some location in a transformer as a result of a corona discharge. (MAG/ET) 436-1977s
(2) (overhead-power-line corona and radio noise) A voltage or current pulse that occurs at some designated location in a circuit as a result of corona discharge. (T&D/PE) 539-1990

corona shielding (rotating machinery) (corona grading) A means adapted to reduce potential gradients along the surface of coils. *See also:* asynchronous machine; direct-current commutating machine. (PE) [9]

corona voltmeter A voltmeter in which the crest value of voltage is indicated by the inception of corona. *See also:* instrument. (EEC/PE) [119]

coroutine (software) A routine that begins execution at the point at which operation was last suspended, and that is not required to return control to the program or subprogram that called it. *Contrast:* subroutine. (C) 610.12-1990

coroutines (software) Two or more modules that can call each other, but are not in a superior to subordinate relationship. *See also:* module. (C/SE) 729-1983s

corrected-compass course *See:* magnetic course.

corrected-compass heading *See:* magnetic heading.

correcting signal *See:* synchronizing signal.

correction (1) (mathematics of computing) (digital computers) A quantity (equal in absolute value to the error) added to a calculated or observed value to obtain the true value. *See also:* accuracy rating; error. (C) 162-1963w, 1084-1986w
(2) The evaluation of a multinomial function using information from the Calibration Transducer Electronic Data Sheet together with data from one or more channels. (IM/ST) 1451.2-1997
(3) (analog computer) *See also:* error. (C) 165-1977w

correction angle* (navigation aid terms) The angular difference between heading and course of a vehicle. Preferably called drift-correction angle. (AES/GCS) 172-1983w

* Deprecated.

correction factor (metering) (instrument transformers) The factor by which the reading of a wattmeter or the registration of a watthour meter must be multiplied to correct for the effects of the error in ratio and the phase angle of the instrument transformer. This factor is the product of the ratio and phase-angle correction factors for the existing conditions of operation. (ELM) C12.1-1982s

correction rate The velocity at which the control system functions to correct error in register. (IA/ICTL/IAC) [60]

corrective action (1) (nuclear power quality assurance) Measures taken to rectify conditions adverse to quality and, where necessary, to preclude repetition. (PE/NP) [124]
(2) Intended to eliminate anomalies. Corrective actions include repair, replacement, calibration, alignment, and other services. *See also:* maintenance. (SCC20) 1226-1998

corrective maintenance (1) (availability, reliability, and maintainability) The maintenance carried out after a failure has occurred and intended to restore an item to a state in which it can perform its required function. (R) [29]
(2) (test, measurement, and diagnostic equipment) Actions performed to restore a failed or degraded equipment. It includes fault isolation, repair or replacement of defective units, alignment and checkout. (MIL) [2]
(3) (software) Maintenance performed to correct faults in hardware or software. *Contrast:* perfective maintenance; adaptive maintenance. (C) 610.12-1990
(4) Maintenance that is performed specifically to overcome existing faults. *Contrast:* preventive maintenance. (C) 610.10-1994w
(5) Reactive modification of a software product performed after delivery to correct discovered faults. (C/SE) 1219-1998

corrective network An electric network designed to be inserted in a circuit to improve its transmission properties, its impedance properties, or both. *See also:* network analysis. (EEC/PE) [119]

correctness (software) (A) The degree to which a system or component is free from faults in its specification, design, and implementation. (B) The degree to which software, documentation, or other items meet specified requirements. (C) The degree to which software, documentation, or other items meet user needs and expectations, whether specified or not. (C) 610.12-1990

correctness proof *See:* proof of correctness.

correct relaying-system performance The satisfactory operation of all equipment associated with the protective-relaying function in a protective-relaying system. It includes the satisfactory presentation of system input quantities to the relaying equipment, the correct operation of the relays in response to these input quantities, and the successful operation of the assigned switching device or devices.
(SWG/PE/PSR) C37.100-1992, C37.90-1978s

correct relay operation An output response by the relay that agrees with the operating characteristic for the input quantities applied to the relay. *See also:* correct relaying-system performance.
(SWG/PE/PSR) C37.100-1992, C37.90-1978s

correlation length The direction-dependent distance over which the mutual coherence function for fields or the covariance function for statistical properties of a medium or surface decreases to $1/e$ of its maximum value.
(AP/PROP) 211-1997

correlated color temperature (illuminating engineering) (of a light source) The absolute temperature of a blackbody whose chromaticity most nearly resembles that of the light source. (EEC/IE) [126]

correlated gamma ray summing The simultaneous detection of two or more gamma rays originating from a single atom disintegration. (NI) N42.12-1994

correlated jitter The portion of the total jitter that is related to the data pattern. Since every PHY receives the same pattern, this jitter is correlated among all similarly configured PHYs receiving the same data pattern and therefore may grow in a systematic way along the ring. Also referred to as *pattern jitter* or *systematic jitter*. (C/LM) 8802-5-1998

correlated photon summing (germanium detectors) The simultaneous detection of two or more photons originating from a single nuclear disintegration.
(PE/NI/EDPG) 485-1983s, N42.12-1980s

correlation (test, measurement, and diagnostic equipment) That portion of certification which establishes the mutual relationships between similar or identical support test systems by comparing test data collected on specimen hardware or simulators. (MIL) [2]

correlation coefficient function (seismic qualification of Class 1E equipment for nuclear power generating stations) Defines a comparative relationship between two time histories. It provides a statistical estimate of how much two motions are related, as a function of time delay. The numerical range is from zero for unrelated, to +1.0 for related motions.
(PE/NP) 344-1987r

correlation detection (modulation systems) Detection based on the averaged product of the received signal and a locally generated function possessing some known characteristic of the transmitted wave. *Notes:* 1. The averaged product can be formed, for example, by multiplying and integrating, or by the use of a matched filter whose impulse response, when reversed in time, is the locally generated function. 2. Strictly, the foregoing definition applies to detection based on cross correlation. The term correlation detection may also apply to detection involving autocorrelation, in which case the locally generated function is merely a delayed form of the received signal. (Std100) [123]

correlator A filter whose impulse response is the time-reversed complex conjugate of the coded waveform intended to be received. (UFFC) 1037-1992w

correspondence *See:* fidelity.

corrosion (1) The deterioration of a substance (usually a metal) because of a reaction with its environment. (IA) [59]
(2) A process of gradual weakening or destruction, usually by a chemical action. (DEI) 1221-1993w

corrosion fatigue Reduction in fatigue life in a corrosive environment. (IA) [59]

corrosion fatigue limit The maximum repeated stress endured by a metal without failure in a stated number of stress applications under defined conditions of corrosion and stressing.
(IA) [59]

corrosion rate The rate at which corrosion proceeds.
(IA) [59]

corrosion-resistant (power and distribution transformers) So constructed, protected, or treated that corrosion will not exceed specified limits under specified test conditions.
(PE/TR) C57.12.80-1978r

corrosion-resistant parts (A) (electric installations on shipboard) General. Where essential to minimize deterioration due to marine atmospheric corrosion, corrosion-resisting materials, or other materials treated in a satisfactory manner to render them adequately resistant to corrosion should be used. **(B) (electric installations on shipboard)** Corrosion-resisting materials. Silver, corrosion-resisting steel, copper, brass, bronze, copper-nickel, certain nickel-copper alloys, and certain aluminum alloys are considered satisfactory corrosion-resisting materials within the intent of the foregoing. **(C) (electric installations on shipboard)** Corrosion-resistant treatments. The following treatments, when properly done and of a sufficiently heavy coating, are considered satisfactory corrosion-resistant treatments within the intent of the foregoing. Electroplating of: cadmium, chromium, copper, nickel, silver, and zinc, sheradizing, galvanizing dipping and painting. (Phosphate or suitable cleaning, followed by the application of zinc chromate primer or equivalent.) **(D) (electric installations on shipboard)** Application. These provisions should apply to the following components: (1) Parts. Interior small parts which are normally expected to be removed in service, such as bolts, nuts, pins, screws, cap screws, terminals, brushholder studs, springs, etc. (2) Assemblies, subassemblies, and other units. Where necessary due to the unit function, or for interior protection, such as shafts within a motor or generator enclosure, and surface of stator and rotor. (3) Enclosures and their fastenings and fittings. Enclosing cases for control apparatus, outer cases for signal and communication systems (both outside and inside), and similar items together with all their fastenings and fittings which would be seriously damaged or rendered ineffective by corrosion. (IA/MT) 45-1983

corrugated horn (antenna) A hybrid-mode horn antenna produced by cutting narrow transverse grooves of specified depth in the interior walls of the horn. *See also:* hybrid-mode horn.
(AP/ANT) 145-1993

corrupt data error (A) An error condition that results when hardware components fail or an external impulse enters into the system upsetting at least one data bit. **(B)** A condition that results from erratic hardware performance, characterized by introduction of a high degree of random errors in the data.
(C) 610.10-1994

cosecant-squared antenna A shaped-beam antenna in which the radiation intensity over a part of its pattern in some specified plane (usually the vertical) is proportional to the square of the cosecant of the angle measured from a specified direction in that plane (usually the horizontal). *Note:* Its purpose is to lay down a uniform field along a line that is parallel to the specified direction but that does not pass through the antenna. *See also:* antenna. (AP/ANT) [35]

cosecant-squared beam antenna A shaped-beam antenna whose pattern in one principal plane consists of a main beam with well-defined side lobes on one side, but with the absence of nulls over an extended angular region adjacent to the peak of the main beam on the other side, with the radiation intensity in this region designed to vary as the cosecant-squared of the angle variable. *Note:* The most common applications of this antenna are for use in ground-mapping radars and target acquisition radars, since the cosecant-squared coverage provides constant signal return for targets with the same radar cross section at different ranges but a common height. (AP/ANT) 145-1993

cosecant-squared pattern A vertical-plane antenna pattern in which the transmitting and receiving power gains vary as the square of the cosecant of the elevation angle. *Note:* The unique property of this pattern is that it results in the received echo signal being independent of range if

— The target is of constant radar cross section
— The target moves at constant altitude
— The earth's surface can be considered flat

See also: modified cosecant-squared antenna pattern. (AES) 686-1997

cosine-cubed law (illuminating engineering) An extension of the cosine law in which the distance d between the source and surface is replaced by $h/\cos\theta$, where h is the perpendicular distance of the source from the plane in which the point is located. It is expressed by $E=(I\ \cos^3\theta)/h^2$. (See figure below.). *See also:* cosine law.

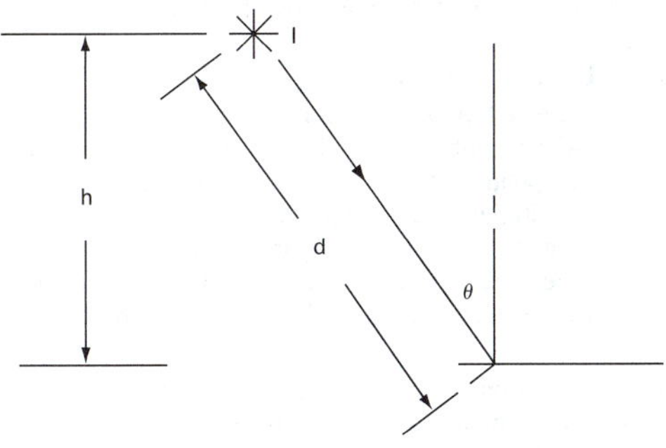

cosine-cubed law

(EEC/IE) [126]

cosine emission law *See:* Lambert's cosine law.

cosine law (illuminating engineering) A law stating that the illuminance on any surface varies as the cosine of the angle of incidence. The angle of incidence θ is the angle between the normal to the surface and the direction of the incident light. The inverse-square law and the cosine law can be combined as $E=(I\cos\theta)/d_2$. *See also:* inverse-square law. (EEC/IE) [126]

cosmic noise Noise-like radio waves originating from extragalactic sources. (AP/PROP) 211-1997

cosmic radio waves *See:* cosmic noise.

COSRO *See:* conical-scan-on-receive-only.

Costas code A frequency-hopping pulse compression waveform in which a long pulse is divided into n subpulses with the frequency of each subpulse chosen from n contiguous frequencies in a manner first suggested by John P. Costas. (AES) 686-1997

costate The state of the adjoint system. *See also:* control system. (CS/IM) [120]

cost of incremental fuel (electric power system) The ultimate replacement cost of the fuel that would be consumed to supply an additional increment of generation (usually expressed in cents per million British thermal units). (PE/PSE) 94-1970w

costs (power operations) Monies associated with investment or use of electrical plant. *See also:* fixed investment costs. (PE/PSE) 858-1987s

COSU *See:* central office service unit.

COTS *See:* commercial-off-the-shelf.

coulomb The unit of electric charge in SI units (International System of Units). The coulomb is the quantity of electric charge that passes any cross section of a conductor in one second when the current is maintained constant at one ampere. (Std100) 270-1966w

Coulomb's law (electrostatic attraction) The force of repulsion between two like charges of electricity concentrated at two points in an isotropic medium is proportional to the product of their magnitudes and inversely proportional to the square of the distance between them and to the dielectric constant of the medium. *Note:* The force between unlike charges is an attraction. (Std100) 270-1966w

coulometer (voltameter) An electrolytic cell arranged for the measurement of a quantity of electricity by the chemical action produced. *See also:* electricity meter. (PE/EEC) [119]

count A single response of the counting system. *See also:* tube count. (NI/NPS) 309-1999

count-down (transponder) The ratio of the number of interrogation pulses not answered to the total number of interrogation pulses received. (AES/RS) 686-1982s, [42]

counter (1) (test, measurement, and diagnostic equipment) **(A)** A device such as a register or storage location used to represent the number of occurrences of an event. **(B)** An instrument for storing integers, permitting these integers to be increased or decreased sequentially by unity or by an arbitrary integer, and capable of being reset to zero or to an arbitrary integer. (MIL) [2]
(2) (software) A variable used to record the number of occurrences of a given event during the execution of a computer program; for example, a variable that records the number of times a loop is executed. (C) 610.12-1990
(3) (A) A device with a finite number of states each of which represents a number which, upon receipt of an appropriate signal, can be incremented or decremented by a given constant. *Note:* The device may be capable of being set to a particular state such as zero. *See also:* reversible counter; modulo-n counter; keystroke counter; line counter. **(B)** A register or storage location used to accumulate the number of occurrences of some event. *See also:* program counter. (C) 610.10-1994

counter beam system Tunnel lighting system or luminaires having a light distribution that is greater in the opposite direction of travel. (RL) C136.27-1996

counter cells *See:* counter-electromotive-force cells.

counterclockwise arc (numerically controlled machines) An arc generated by the coordinated motion of two axes in which curvature of the path of the tool with respect to the workpiece is counterclockwise, when viewing the plane of motion in the negative direction of the perpendicular axis. (IA) [61]

counter electromotive force (any system) The effective electromotive force within the system that opposes the passage of current in a specified direction. (EEC/PE) [119]

counter-electromotive-force cells (counter cells) Cells of practically no ampere-hour capability used to oppose the battery voltage. *See also:* battery. (EEC/PE) [119]

counter-mounted cooking unit **(A)** A cooking appliance designed for mounting in or on a counter and consisting of one or more heating elements, internal wiring, and build-in or separately mountable controls. *See also:* wall-mounted oven. **(B)** An assembly of one or more domestic surface heating elements for cooking purposes, designed for flush mounting in, or supported by, a counter, and which assembly is complete with inherent or separately mountable controls and internal wiring. (NESC/NEC/C2) [86]

counterpoise (1) A system of conductors, elevated above and insulated from the ground, forming a lower system of conductors of an antenna. *Note:* The purpose of a counterpoise is to provide a relatively high capacitance and thus a relatively low impedance path to earth. The counterpoise is sometimes used in medium- and low-frequency applications where it would be more difficult to provide an effective ground connection. (AP/ANT) 145-1993
(2) A conductor or system of conductors arranged beneath the line; located on, above, or most frequently below the surface of the earth; and connected to the grounding systems of the towers or poles supporting the transmission lines.
(PE/T&D/SPD/PSIM) 1243-1997, 81-1983, C62.23-1995
(3) *See also:* ground grid. (T&D/PE) 524-1992r

counter, radiation *See:* radiation counter.

counter tube (A) (externally quenched) A radiation-counter tube that requires the use of an external quenching circuit to inhibit reignition. **(B)** (gas-filled) A gas tube used for detection of radiation by means of gas ionization. **(C)** (gas-flow) A radiation-counter tube in which an appropriate gas-fill concentration is maintained by a flow of gas through the tube. **(D)** (Geiger-Mueller) A radiation-counter tube operated in the Geiger-Mueller region. **(E)** (self-quenched) A radiation-counter tube in which reignition of the discharge is inhibited by internal processes. (NI/NPS) 309-1999

counting channel (liquid-scintillation counting) A region of the pulse-height spectrum that is defined by upper and lower boundaries set by discriminators. (NI) N42.15-1990

counting efficiency (1) (radiation counter tubes) The average fraction of the number of ionizing particles or quanta incident on the sensitive area that produce tube counts. *Note:* The operating conditions of the counter and the condition of irradiation must be specified. (ED) 161-1971w
(2) (scintillation counters) The ratio of the average number of photons or particles of ionizing radiation that produce counts to the average number incident on the sensitive area. *Note:* The operating conditions of the counter and the conditions of irradiation must be specified. *See also:* scintillation counter. (NPS) 398-1972r
(3) (liquid-scintillation counting) The ratio of the count rate to the disintegration rate, usually expressed as a percentage:

$$E = (R/A) \times 100.$$

E = counting system efficiency
R = net count rate in an individual measurement, counts per minute
A = activity of the radionuclide contained in the check source.

(NI) N42.15-1990
(4) The ratio of the number of observed counts to the total number of ionizing particles impinging upon the counter surface when the counting rate is so low that dead-time correction is unnecessary. (NI/NPS) 309-1999

counting mechanism (of an automatic line sectionalizer or automatic circuit recloser) A device that counts the number of electrical impulses and, following a predetermined number of successive electrical impulses, actuates a releasing mechanism. It resets if the total predetermined number of successive impulses do not occur in a predetermined time. (SWG/PE) C37.100-1992

counting operation (of an automatic line sectionalizer or automatic circuit recloser) Each advance of the counting mechanism towards an opening operation. (SWG/PE) C37.100-1992

counting operation time (of an automatic line sectionalizer) The time between the cessation of a current above the minimum actuating current value and the completion of a counting operation. (SWG/PE) C37.100-1992

counting rate (1) Number of counts per unit time. *See also:* anticoincidence. (ED) [45]
(2) (germanium spectrometers) The rate at which detector pulses are being registered in a selected voltage interval. The unit is reciprocal seconds (i.e., s^{-1}). (NI) N42.14-1991

counting-rate meter (pulse techniques) A device that indicates the time rate of occurrence of input pulses averaged over a time interval. *See also:* scintillation counter. (NPS) 398-1972r

counting rate versus voltage characteristic (gas-filled radiation counter tube) The counting rate as a function of applied voltage for a given constant average intensity of radiation.

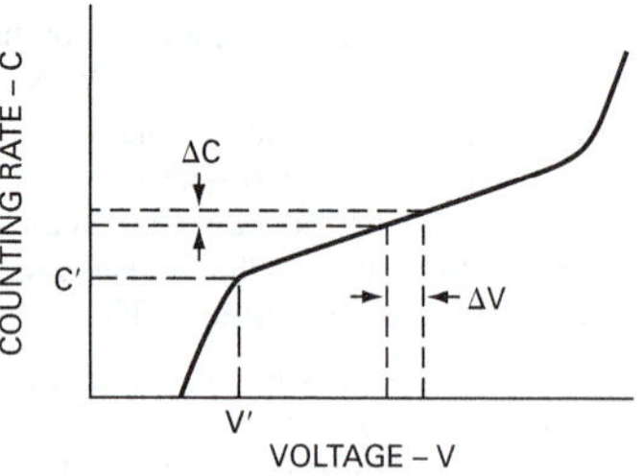

Counting rate-voltage characteristic in which

$$\text{relative plateau slope} = 100\frac{\Delta C/C}{\Delta V}$$

$$\text{normalized plateau slope} = \frac{\Delta C/\Delta V}{C'/V'} = \frac{\Delta C/C'}{\Delta V/V'}$$

counting rate versus voltage characteristic

(ED/NI/NPS) 161-1971w, 309-1999

counting region A region that identifies the first and last memory location of a contiguous series to be summed in a multichannel analyzer. (NI) N42.15-1990

country beam *See:* upper (driving) beams.

country code (telephone switching systems) The one-, two-, or three-digit number that, in the world numbering plan, identifies each country or integrated numbering plan area in the world. The initial digit is always the world-zone number. Any subsequent digits in the code further define the designated geographical area normally identifying a specific country. On an international call, this code is dialed ahead of the national number. (COM) 312-1977w

counts, tube, multiple *See:* multiple tube counts.

counts, tube, spurious *See:* spurious tube counts.

couple (1) (storage cell) An element of a storage cell consisting of two plates, one positive and one negative. *Note:* The term couple is also applied to a positive and a negative plate connected together as one unit for installation in adjacent cells. *See also:* battery. (PE/EEC) [119]
(2) (thermoelectric) A thermoelectric device having two arms of dissimilar composition. *Note:* The term thermoelement is ambiguously used to refer to either a thermoelectric arm or to a thermoelectric couple, and its use is therefore not recommended. *See also:* thermoelectric device. (ED) [46]

coupled fine A transmission line with multiple guiding members whose propagating waves interact with each other. (MTT) 1004-1987w

coupled modes (fiber optics) Modes whose energies are shared. *See also:* mode. (Std100) 812-1984w

coupler (1) (navigation aid terms) That portion of a navigational system which receives signals of one type from a sensor and transmits signals of a different type to an actuator. *See also:* autopilot coupler. (AES/GCS) 172-1983w
(2) (surge testing for equipment connected to low-voltage ac power circuits) A device, or combination of devices, used to feed a surge from a generator to powered equipment while limiting the flow of current from the power source into the generator. *See also:* coupling network. (SPD/PE) C62.45-1992r
(3) (fiber optics) *See also:* optical waveguide coupler. 812-1984w

coupler interface That facility of a basic operating unit that is designed to provide convenient connection to, and disconnection from, any other basic operating unit without requiring disassembly of any constituent part of either basic operating unit. This includes standardized mechanical, electrical, electronic, pneumatic, and other interfaces as required. (VT) 1473-1999

coupler, optical *See:* optical directional coupler.

coupling (1) (ground systems) The association of two or more circuits or systems in such a way that power or signal information may be transferred from one to another. *Note:* Coupling is described as close or loose. A close-coupled process has elements with small phase shift between specified variables; close-coupled systems have large mutual effect shown mathematically by cross-products in the system matrix. (PE/PSIM) 81-1983

(2) (rotating machinery) A part or combination of parts that connects two shafts for the purpose of transmitting torque or maintaining alignment of the two shafts. (PE) [9]

(3) (software) The manner and degree of interdependence between software modules. Types include common-environment coupling, content coupling, control coupling, data coupling, hybrid coupling, and pathological coupling. *Contrast:* cohesion. (C) 610.12-1990

(4) (waveguide) The power transfer from one transmission path to a particular mode or form in another. *Note:* Small, undesired coupling is sometimes called isolation, decoupling, or cross coupling. (MTT) 146-1980w

(5) (instrumentation and control equipment grounding in generating stations) The mechanism by which an interference source produces interference in a signal circuit. (PE/EDPG) 1050-1996

(6) The mode of propagation of disturbing energy from a power system to a telecommunications system. There are three forms of coupling between the two systems: magnetic (inductive) coupling, electric (capacitive) coupling, and conductive (resistive) coupling. In addition, coupling by electromagnetic radiation exists and is associated with propagation of radiation fields, e.g., radio frequency interference (RFI), electromagnetic pulse (EMP), and corona. (PE/PSC) 487-1992

(7) The association of two or more circuits or systems in such a way that power or signal information may be transferred from one system or circuit to another. (IA/PE/PSE) 1100-1999, 599-1985w

coupling aperture (coupling hole, coupling slot) (waveguide components) An aperture in the bounding surface of a cavity resonator, waveguide, transmission line, or waveguide component which permits the flow of energy to or from an external circuit. (MTT) 147-1979w

coupling capacitance (1) (ground systems) The association of two or more circuits with one another by means of capacitance mutual to the circuits. (PE/PSIM) 81-1983

(2) (interference terminology) The type of coupling in which the mechanism is capacitance between the interference source and the signal system; that is, the interference is induced in the signal system by an electric field produced by the interference source. *See also:* interference. (IE) [43]

coupling-capacitor voltage transformer (metering) A voltage transformer comprised of a capacitor divider and an electromagnetic unit so designed and interconnected that the secondary voltage of the electromagnetic units is substantially proportional to, and in phase with, the primary voltage applied to the capacitor divider for all values of secondary burdens within the rating of the coupling-capacitor voltage transformer. (ELM) C12.1-1988

coupling coefficient (1) (coefficient of coupling) The ratio of impedance of the coupling to the square root of the product of the total impedances of similar elements in the two meshes. *Notes:* 1. Used only in the case of resistance, capacitance, self-inductance, and inductance coupling. 2. Unless otherwise specified, coefficient of coupling refers to inductance coupling, in which case it is equal to $M/(L_1L_2)^{1/2}$, where M is the mutual inductance, L_1 the total inductance of one mesh, and L_2 the total inductance of the other. *See also:* network analysis. (IM/HFIM) [40]

(2) (planar transmission lines) A number used as a measure of the degree of interaction between the members of a coupled line. One commonly used definition of the coupling coefficient of a symmetrical coupled pair of transmission lines is K, a voltage or field ratio:

$$\frac{\frac{Z_{0e}}{Z_{0o}}-1}{\frac{Z_{0e}}{Z_{0o}}+1}$$

where Z_{0e} and Z_{0o} = even- and odd-mode characteristic impedances. (MTT) 1004-1987w

coupling coefficient, small-signal (electron stream) The ratio of (A) the maximum change in energy of an electron traversing the interaction space to (B) the product of the peak alternating gap voltage by the electronic charge. *See also:* electron emission; coupling coefficient; coupling. (ED) 161-1971w

coupling, conductance *See:* conductance coupling.

coupling efficiency (fiber optics) The efficiency of optical power transfer between two optical components. *See also:* coupling loss. (Std100) 812-1984w

coupling, electric *See:* electric coupling.

coupling factor (1) (lightning) The ratio of the induced voltage to the inducing voltage on parallel conductors. *See also:* direct-stroke protection. (T&D/PE) [10]

(2) (directional coupler) The ratio of the incident power fed into the main port, and propagating in the preferred direction, to the power output at an auxiliary port, all ports being terminated by reflectionless terminations. *See also:* waveguide. (IM/HFIM) [40]

(3) The ratio of the induced voltage to the inducing voltage on parallel conductors. For example, at the tower, the shield or coupling wires and tower crossarms are at practically the same potential (because of lightning stroke travel time). The stress across the insulator string is one minus the coupling factor multiplied by the tower top potential.

$$\text{Stress} = (1.0 - K_{\text{fc}}) \times V_{\text{TT}}$$

where
K_{fc} is the coupling factor
V_{TT} is the tower top voltage

(PE/SPD) C62.23-1995

coupling flange (rotating machinery) The disc-shaped element of a half coupling that permits attachment to a mating half coupling. *Synonym:* flange. *See also:* rotor. (PE) [9]

coupling function A mathematical, graphical, or tabular statement of the influence that one element or subsystem has on another element or subsystem, expressed as the effect/cause ratio of related variables or their transforms. *Note:* For a multi-terminal system described by m differential equations and having m input transforms $R_1 \ldots R_m$ and m output transforms $C_1 \ldots C_m$, the coupling functions consist of all effect/cause ratios which can be formed from transforms bearing unlike-numbered subscripts. (CS/PE/EDPG) [3]

coupling hole *See:* coupling aperture.

coupling, hysteresis *See:* hysteresis coupling.

coupling, inductance *See:* inductance coupling.

coupling, induction *See:* induction coupling.

coupling loop (waveguide components) A conducting loop that permits the flow of energy between a cavity resonator, waveguide, transmission line, or waveguide component and an an external circuit. (MTT) 147-1979w

coupling loss (fiber optics) The power loss suffered when coupling light from one optical device to another. *See also:* lateral offset loss; insertion loss; gap loss; extrinsic joint loss; angular misalignment loss; intrinsic joint loss. (Std100) 812-1984w

coupling, magnetic friction *See:* magnetic friction coupling.

coupling, magnetic-particle *See:* magnetic-particle coupling.

coupling network Electrical circuit for the purpose of transferring energy from one circuit to another. *See also:* coupler. (SPD/PE) C62.45-1992r

coupling plane A metal plate to which discharges are applied to simulate electrostatic discharge to objects adjacent (vertically or horizontally) to the EUT. (EMC) C63.16-1993

coupling probe (waveguide components) A probe that permits the flow of energy between a cavity resonator, waveguide, transmission line, or waveguide component and an external circuit. (MTT) 147-1979w

coupling, radiation *See:* radiation coupling.

couplings (pothead) Entrance fittings which may be provided with a rubber gland to provide a hermetic seal at the point where the cable enters the box and may have, in addition, a threaded portion to accommodate the conduit used with the cable or have an armor clamp to clamp and ground the armored sheath on armor-covered cable. (PE/TR) [108]

coupling slat *See:* coupling aperture.

coupling, synchronous *See:* synchronous coupling.

coupling wire A conductor attached to the transmission line structure and below the phase wires, with proper clearance, and connected to the grounding system of the towers or the pole supporting the line. (SPD/PE) C62.23-1995

course (A) (navigation aids) The intended direction of travel, expressed as an angle in the horizontal plane between a reference line and the course line, usually measured clockwise from the reference line. **(B) (navigation aids)** The intended direction of travel as defined by a navigational facility. **(C) (navigation aids)** Common usage for "course line." (AES/GCS) 172-1983

course-deviation indicator *See:* course-line deviation indicator.

course line (navigation aids) The projection in the horizontal plane of a path (proposed path of travel). (AES/GCS) 172-1983w

course linearity (navigation aids) (instrument landing systems) A term used to describe the change in DDM (difference in depth of modulation) of the two modulation signals with respect to displacement of the measuring position from the course line but within the course sector. *Synonyms:* flight path; desired track. (AES/GCS) 172-1983w

course-line computer (navigation aids) A device, usually carried aboard a vehicle, to convert navigational signals such as VOR/DME (very high-frequency omnidirectional range/distance measuring equipment) into course extending between any desired points regardless of their orientation with respect to the source of the signals. (AES/GCS) 172-1983w

course-line deviation (navigation aids) The amount by which the track of a vehicle differs from its course line, expressed in terms of either an angular or linear measurement. (AES/GCS) 172-1983w

course-line deviation indicator (course deviation indicator) (navigation aids) A device providing a visual display of the direction and amount of deviation from the intended course. *Synonym:* flight-path-deviation indicator. (AES/GCS) 172-1983w

course made good (navigation aids) The direction from the point of departure to the position of the vehicle on the horizontal plane. (AES/GCS) 172-1983w

course push (pull) (navigation aids) An erroneous deflection of the indicator of a navigational aid, produced by altering the attitude of the receiving antenna. *Note:* This effect is a manifestation of polarization error and results in an apparent displacement of the course line. (AES/GCS) 172-1983w

course roughness (navigation aids) A term used to describe the imperfections in a visually indicated course when such imperfections cause the course indicator to make rapid erratic movements. *See also:* scalloping. (AES/GCS) 172-1983w

course scalloping *See:* scalloping.

course section width (instrument landing systems) The transverse dimension at a specified distance, or the angle in degrees between the sides of the course sector. *See also:* navigation. (AES/RS) 686-1982s, [42]

course sector (instrument landing systems) (navigation aid terms) A wedge-shaped section of airspace containing the course line and spreading with distance from the ground station; it is bounded on both sides by the loci of points at which the DDM (difference in depth of modulation) is a specified amount, usually the DDM giving full-scale deflection of the course-deviation indicator. (AES/GCS) 172-1983w

course-sector width (navigation aids) (instrument landing systems) The transverse dimension at a specified distance, or the angle in degrees, between the sides of the course sector. (AES/GCS) 172-1983w

course sensitivity (navigation systems) (navigation aids) The relative response of a course-line deviation indicator to the actual or simulated departure of the vehicle from the course line. In VOR (very high-frequency) omnidirectional range), Tacan (tactical air navigation), or similar omnirange systems, course sensitivity is often taken as the number of degrees through which the omnibearing selector must be moved to change the deflection of the course-line deviation indicator from full scale on one side to full scale on the other, while the receiver omnibearing-input signal is held constant. (AES/GCS) 172-1983w

course softening (navigation aids) The intentional decrease in course sensitivity upon approaching a navigational aid such than the ratio of indicator deflection to linear displacement from the course line tends to remain constant. (AES/GCS) 172-1983w

courseware Instructional materials, such as software and student documentation, designed for use in computer-based instruction. (C) 610.2-1987

course width (navigation aids) Twice the displacement (of the vehicle), in degrees, to either side of a course line, which produces a specifed indication on the course deviation indicator (usually the specified indication is full scale). (AES/GCS) 172-1983w

Coursewriter A programming language used to write instructional programs for computer-assisted instruction. (C) 610.13-1993w

cove lighting (illuminating engineering) Lighting comprising light sources shielded by a ledge or horizontal recess, and distributing light over the ceiling and upper wall. (EEC/IE) [126]

cover (power system communication equipment) A protective covering used to enclose or partially enclose equipment that may be mounted in a rack. (PE/PSC) 281-1984w

coverage Measure of the representative nature of situations to which a system is submitted during its validation compared to the actual situations it will be confronted with during its operational life. (C/BA) 896.9-1994w

coverage area (1) (mobile communication) The area surrounding the base station that is within the signal-strength contour that provides a reliable communication service 90 percent of the time. *See also:* mobile communication system. (VT) [37]

(2) The area surrounding the broadcast array that is within the signal strength contour that provides adequate reception. (T&D/PE) 1260-1996

covered conductor A conductor covered with a dielectric having no rated insulating strength or having a rated insulating strength less than the voltage of the circuit in which the conductor is used. (NESC/T&D/PE) C2-1997, [10], C2.2-1960

covered line An idealized planar transmission line with two conducting ground planes, parallel to the strip conductor. One of the ground planes has only a minor effect on the propagation properties of the line. (MTT) 1004-1987w

covered plate (storage cell) A plate bearing a layer of oxide between perforated sheets. *See also:* battery. (PE/EEC) [119]

cover flange (waveguide components) A flat flange used in conjunction with a choke flange to provide a choke joint. (MTT) 147-1979w

cover-up equipment (power line maintenance) Equipment designed to protect persons from energized parts in a specific work area. Many different types are available to cover conductors, insulators, dead-end assemblies, structures, and apparatus. Cover-up material may be either flexible or rigid. (T&D/PE) 516-1995

CP *See:* collision presence; coprocessor.

CPC *See:* computer program component.

CPCI *See:* computer program configuration item.

CPE *See:* circular probable error.

CPE active state A state in which the CPE performs a communications functions. (AMR) 1390-1995

CPE inactive state A state in which the CPE does not perform a communications function. (AMR) 1390-1995

CPE address The LSAP address at which the CPE may be reached. (LM/C) 15802-2-1995

CPE instance identifier The tuple of CPE address and CPE instance number that uniquely identifies a CPE instance within the LAN/MAN environment, within the limits of uniqueness of the CPE address and instance values used. (LM/C) 15802-2-1995

CPE instance number A number, allocated to the CPE at instantiation time, that distinguishes a CPE instance from all other CPE instances, past and present, associated with a particular CPE address. (LM/C) 15802-2-1995

CPM *See:* critical path method.

CPU *See:* central processing unit.

CPU busy time *See:* CPU time.

CPU time In time-sharing computer systems, the time devoted by the central processing unit to the execution of instructions of a particular process, task, or user. *Synonym:* CPU busy time. *See also:* connect time. (C) 610.10-1994w

CPU timer A feature of some computer systems that measures elapsed CPU time and that causes an interrupt when a previously specified amount of time has elapsed. (C) 610.10-1994w

CR *See:* carriage return character; carriage return; cavity ratio.

crab angle* *See:* drift angle; drift correction angle.

* Deprecated.

cracking (1) Rupture of the polymeric insulator material to depths equal to or greater than 0.1 mm. (PE/IC) 48-1996
(2) Rupture of the weathershed material to depths greater than 0.1 mm. (SPD/PE) C62.11-1999

cradle base (rotating machinery) A device that supports the machine at the bearing housings. (PE) [9]

crane A machine for lifting or lowering a load and moving it horizontally, in which the hoisting mechanism is an integral part of the machine. *Note:* It may be driven manually or by power and may be fixed or a mobile machine. (EEC/PE) [119]

crash (1) The sudden and complete failure of a computer system or component. *See also:* hard failure; disk crash; head crash. (C) 610.12-1990, 610.10-1994w
(2) To fail as in definition (A). (C) 610.10-1994w

crate (1) (CAMAC system) *See also:* CAMAC crate.
(2) (FASTBUS crate) The mechanical housing for FASTBUS modules in a crate segment. (NID) 960-1993

crate number (c) (subroutines in CAMAC) The symbol c represents an integer which is the crate number component of a CAMAC address. Crate number in this context can be either the physical crate number or it can be an integer symbol which is interpreted by the computer system software to produce appropriate hardware access information. (NPS) 758-1979r

crate segment (FASTBUS acquisition and control) A FASTBUS segment that consists of a backplane mounted on a FASTBUS crate and having connectors to mate with a multiplicity of FASTBUS modules. (NID) 960-1993

Crawford cell *See:* transverse-electromagnetic cell.

crawler *See:* crawler tractor.

crawler tractor (conductor stringing equipment) A tracked unit employed to pull pulling lines, sag conductor, level or clear pull and tension sites, and miscellaneous other work. It is also frequently used as a temporary anchor. Sagging winches on this unit are usually arranged in a vertical configuration. *Synonyms:* tractor; cat; crawler. (T&D/PE) 524-1992r, 524a-1993r

crawling (rotating machinery) The stable but abnormal running of a synchronous or asynchronous machine at a speed near to a submultiple of the synchronous speed. *See also:* asynchronous machine. (PE) [9]

crazing (1) (composite insulators) Surface microfractures of the weathershed material to depths less than 0. 1 millimeter resulting from ultraviolet exposure. (T&D/PE) 987-1985w
(2) The small internal cracking around a point of mechanical stress that sometimes occurs in plastics. (PE/EDPG) 1184-1994
(3) Surface microfractures of the insulator material to depths less than 0.1 mm resulting from ultraviolet exposure. (PE/IC) 48-1996

CRC (1) violation If the transmitted and received CRC codes are not identical, a CRC violation has occurred, meaning one or more errors has occurred in transmission. (COM/TA) 1007-1991r
(2) The cyclic redundancy code used for error detection on each packet. (C/MM) 1596-1992
(3) *See also:* cyclic redundancy check. (C) 610.7-1995

CR differentiator A high-pass electrical filter section consisting of a capacitor in series with the signal path followed by a resistor across the path. (NPS) 325-1996

credentials Information supplied to authenticate a communication. (SCC32) 1455-1999

credit-card call (telephone switching systems) A call in which a credit-card identity is used for billing purposes. (COM) 312-1977w

creep (1) Continued deformation of material under stress. (IA/PSE) 241-1990r
(2) *See also:* watthour meter—creep. (ELM) C12.1-1988

creepage The travel of electrolyte up the surface of electrode or other parts of the cell above the level of the main body of electrolyte. *See also:* electrolytic cell. (EEC/PE) [119]

creepage distance (power and distribution transformers) The shortest distance between two conducting parts measured along the surface or joints of the insulating material between them. (SWG/PE/TR) C37.100-1992, C57.12.80-1978r

creepage surface (rotating machinery) An insulating-material surface extending across the separating space between components at different electric potential, where the physical separation provides the electrical insulation. *See also:* asynchronous machine. (PE) [9]

creep distance (1) (outdoor apparatus bushings) The distance measured along the external contour of the weather casing separating the metal parts which have the operating line-to-ground voltage between them. (PE/TR) 21-1976
(2) The shortest distance measured along the external contour of the insulating envelope that separates the metal part operating at line voltage and the metal flange at ground potential. (PE/TR) C57.19.03-1996

creeping stimulus *See:* accumulating stimulus.

creeping wave A wave propagating along a smooth convex surface that has diffracted into the shadow region. (AP/PROP) 211-1997

crescent *See:* conductor grip.

coordination of insulation (1) (lightning insulation strength) The steps taken to prevent damage to electric equipment due to overvoltages and to localize flashovers to points where they will not cause damage. *Note:* In practice, coordination consists of the process of correlating the insulating strengths of electric equipment with expected overvoltages and with the characteristics of protective devices. (PE/EEC) [8], [74]
(2) The selection of insulation strength consistent with expected overvoltages to obtain an acceptable risk of failure. (SPD/PE) C62.22-1997

crest factor (1) (germanium gamma-ray detectors) (x-ray energy spectrometers) (semiconductor radiation detectors) (charged-particle detectors) (of an average reading or root-mean-square voltmeter) The ratio of the peak voltage value that an average reading or root-mean-square voltmeter will accept without overloading to the full scale value of the range being used for measurement. (NPS/NID) 759-1984r, 301-1976s

(2) (ac voltmeter) The highest ratio of peak to rms voltage that can be applied to an ac voltmeter before overload sets in. The crest factor may depend upon the full-scale setting of the meter. (NPS) 300-1988r

(3) (of a periodic function) The ratio of its crest (peak, maximum) value to its root-mean-square (rms) value. (PE/TR) C57.12.80-1978r

(4) (electrical measurements in power circuits) (of a periodic function) The ratio of the peak value to the rms value $cf = y_p/y_{rms}$. (PE/PSIM) 120-1989r

(5) (of a periodic function) The ratio of the peak value of a periodic function (y_{peak}) to the rms value (y_{rms}); $cf = y_{peak}/y_{rms}$. (IA/PSE) 1100-1999

(6) (pulse carrier) The ratio of the peak pulse amplitude to the root-mean-square amplitude. *See also:* carrier. (IM/AP/WM&A/ANT) 194-1977w, 145-1983s

(7) (of an rms voltmeter) The highest ratio of peak to rms voltage that can be applied to an ac voltmeter before overload sets in. The crest factor may depend on the full-scale setting of the voltmeter. (NPS) 325-1996

(8) The ratio of the peak value to the rms value of an ac waveform measured under steady-state conditions. It is unitless, and the ratio for a pure sine wave is equal to $\sqrt{2}$.

$$cf = \frac{V_{in,pk}}{V_{in,rms}}$$

where

V_{in} = the voltage at the user input terminals. (PEL) 1515-2000

crest value (1) (peak value) (power and distribution transformers) The maximum absolute value of a function when such a maximum exists. (PE/C/TR) 1313.1-1996, C57.12.80-1978r

(2) (of a wave, surge, or impulse) The maximum value that a wave, surge, or impulse attains. (SPD/PE) C62.11-1999, C62.62-2000

(3) (surge arresters) The maximum value that an impulse attains. *Synonym:* peak value. (SPD/PE) C62.22-1997, C62.1-1981s, C62.11-1987s, 2-1978w

crest voltmeter A voltmeter depending for its indications upon the crest or maximum value of the voltage applied to its terminals. *Note:* Crest voltmeters should have clearly marked on the instrument whether readings are in equivalent root-mean-square values or in true crest volts. It is preferred that the marking should be root-mean-square values of the sinusoidal wave having the same crest value as that of the wave measured. *See also:* instrument. (EEC/PE) [119]

crest working line voltage (vlwm) (thyristor) The highest instantaneous value of the line voltage excluding all repetitive and nonrepetitive transient voltages, but including voltage variations. (IA/IPC) 428-1981w

crest working voltage (semiconductor rectifiers) (between two points) The maximum instantaneous difference of voltage, excluding oscillatory and transient overvoltages, that exists during normal operation. *See also:* semiconductor rectifier stack; rectification. (IA/EEC/PCON) [62], [110]

crevice corrosion Localized corrosion as a result of the formation of a crevice between a metal and a nonmetal, or between two metal surface. (IA) [59], [71]

criteria Parameters against which the CASE tool is evaluated, and upon which selection decisions are made. (C/SE) 1209-1992w

critical angle (fiber optics) When light propagates in a homogeneous medium of relatively high refractive index (n_{high}) onto a planar interface with a homogeneous material of lower index (n_{low}), the critical angle is defined byarcsin(n_{low}/n_{high}). *Note:* When the angle of incidence exceeds the critical angle, the light is totally reflected by the interface. This is termed "total internal reflection." *See also:* step index profile; acceptance angle; angle of incidence; reflection; refractive index; total internal reflection. (Std100) 812-1984w

critical anode voltage *See:* gas tube; breakdown voltage.

critical branch (health care facilities) A subsystem of the Emergency System consisting of feeders and branch circuits supplying energy to task illumination, special power circuits, and selected receptacles serving areas and functions related to patient care, and which can be connected to alternate power sources by one or more transfer switches during interruption of normal power source. (NEC/NESC/EMB) [47], [86]

critical build-up resistance (rotating machinery) The highest resistance of the shunt winding circuit supplied from the primary winding for which the machine voltage builds up under specified conditions. (PE) [9]

critical build-up speed (rotating machinery) The limiting speed below which the machine voltage will not build up under specified condition of field-circuit resistance. *See also:* direct-current commutating machine. (PE) [9]

critical characteristics (1) (replacement parts for Class 1E equipment in nuclear power generating stations) (equipment) Those properties or attributes that are essential for performance of an equipment's safety function. (PE/NP) 934-1987w

(2) (replacement parts for Class 1E equipment in nuclear power generating stations) (parts) Those properties or attributes of the part that are essential to the safety function of the equipment in which the part is installed. *Note:* Typical critical characteristics are attributes such as dimensions, materials, electrical and temperature parameters, output tolerances, and fluid viscosity. (PE/NP) 934-1987w

critical components Equipment whose failure will result in complete system or functional failure. (PE/NP) 933-1999

critical component temperature The temperature of semiconductor components that are most susceptible to malfunction from high temperature. (C/BA) 14536-1995

critical control command An MTM-Bus command that has significant effect on the operation of a module to a degree that, for added security, a message conveying such a command should be difficult to send unintentionally. This Standard provides that a message containing a critical control command has to be proceeded by an Enable Module Control (EMC) message. If this procedure is not followed, a Command Sequence Error will occur. (TT/C) 1149.5-1995

critical controlling current (cryotron) The current in the control that just causes direct-current resistance to appear in the gate, in the absence of gate current and at a specified temperature. *See also:* superconductivity. (SPD/PE) 32-1972r

critical coupling That degree of coupling between two circuits, independently resonant to the same frequency, that results in maximum transfer of energy at the resonance frequency. *See also:* coupling. (EEC/PE) [119]

critical current (1) (superconductor) The current in a superconductive material above which the material is normal and below which the material is superconducting, at a specified temperature and in the absence of external magnetic fields. *See also:* superconductivity. (ED) [46]

(2) The first-stroke lightning current to a phase conductor which produces a critical impulse flashover voltage wave. (PE/T&D) 1243-1997

critical damping The least amount of viscous damping that causes a single-degree-of-freedom system to return to its original position without oscillation after initial disturbance. (PE/SUB) 693-1997

critical design review (CDR) (A) A review conducted to verify that the detailed design of one or more configuration items satisfy specified requirements; to establish the compatibility among the configuration items and other items of equipment, facilities, software, and personnel; to assess risk areas for each configuration item; and, as applicable, to assess the results of producibility analyses, review preliminary hardware product specifications, evaluate preliminary test planning, and evaluate the adequacy of preliminary operation and support documents. *See also:* preliminary design review; system design review. **(B)** A review as in (A) of any hardware or software component. (C) 610.12-1990

critical dimension (waveguide) The dimension of the cross-section that determines the cutoff frequency. *See also:* waveguide. (EEC/PE) [119]

critical event simulation A simulation that is terminated by the occurrence of a certain event; for example, a model depicting the year-by-year forces leading up to a volcanic eruption, that is terminated when the volcano in the model erupts. *See also:* time-slice simulation. (C) 610.3-1989w

critical failure *See:* failure.

critical field (1) (magnetrons) The smallest theoretical value of steady magnetic flux density, at a steady anode voltage, that would prevent an electron emitted from the cathode at zero velocity from reaching the anode. *See also:* magnetron. **(2) (nonlinear, active, and nonreciprocal waveguide components)** In a gyromagnetic material that radio-frequency (rf) magnetic field levelabove which transfer of energy occurs from the uniform precession mode to spin waves; that is the field corresponding to nonlineal loss threshold. (MTT) 457-1982w

critical flashover voltage (CFO) The amplitude of voltage of a given waveshape that, under specified conditions, causes flashover through the surrounding medium on 50% of the voltage applications. (SPD/PE) C62.22-1997

critical freeze protection (electric pipe heating systems) The use of electric pipe heating systems to prevent the temperature of fluids from dropping below the freezing point of the fluid in important or critical outdoor (usually) piping systems at nuclear generating stations. An example of a critical freeze protection system is the heating for the nuclear service water system. (PE/EDPG) 622A-1984r, 622B-1988r

critical frequency (1) (data transmission) In radio propagation (by way of the ionosphere) the limiting frequency below which a wave component is reflected by, and above which it penetrates through, an ionospheric layer of vertical incidence. *Note:* The existence of the critical frequency is the result of electron limitation, that is, the inadequacy of the existing number of free electrons to support reflection at higher frequencies. (PE) 599-1985w
(2) (network or system) A pole or zero of a transfer or driving-point function. (CAS) [13]
(3) (of an ionospheric layer) The limiting frequency below which a normally-incident magneto-ionic wave component is returned by, and above which it penetrates through, an ionospheric layer. (AP/PROP) 211-1997

critical grid voltage (multielectrode gas tubes) The grid voltage at which anode breakdown occurs. *Note:* The critical grid voltage is a function of the other electrode voltages or currents and of the environment. *See also:* breakdown voltage. (ED) 161-1971w

critical head (power operations) The head at which the full-gate output of the hydroturbine equals the nameplate generator capacity. (PE/PSE) 858-1987s

critical heat flux The heat flux below which ignition is not possible. (DEI) 1221-1993w

critical high-power level (attenuator tubes) The radio-frequency power level at which ionization is produced in the absence of a control-electrode discharge. (ED) 161-1971w

critical humidity The relative humidity above which the atmospheric corrosion rate of a given metal increases sharply. (IA) [59]

critical hydro period (power operations) (electric power supply) Period when the limitations of hydroelectric energy supply due to water conditions are most critical with respect to system load requirements. (PE/PSE) 858-1987s, 346-1973w

critical impulse (of a relay) The maximum impulse in terms of duration and input magnitude that can be applied suddenly to a relay without causing pickup. (SWG/PE) C37.100-1992

critical impulse flashover voltage (CFO) (insulators) The crest value of the impulse wave that, under specified conditions, causes flashover through the surrounding medium on 50% of the applications. *See also:* impulse flashover voltage. (PE/T&D/SPD) 1410-1997, 32-1972r, 1243-1997

critical impulse time (of a relay) The duration of a critical impulse under specified conditions. (SWG/PE/PSR) C37.100-1992, C37.90-1978s

critical item (software) In configuration management, an item within a configuration item that, because of special engineering or logistic considerations, requires an approved specification to establish technical or inventory control at the component level. (C) 610.12-1990

criticality (1) (power operations) The state of an assembly of fissionable material in which a stable, self-sustaining chain reaction exists. At this condition a nuclear reactor will produce energy at a constant rate and the effective multiplication factor keff is exactly equal to 1. (PE/PSE) 858-1987s
(2) (software) The degree of impact that a requirement, module, error, fault, failure, or other item has on the development or operation of a system. *Synonym:* severity. (C) 610.12-1990
(3) A subjective description of the intended use and application of the system. Software criticality properties may include safety, security, complexity, reliability, performance, or other characteristics. (C/SE) 1012-1998

criticality analysis A structured evaluation of the software characteristics (e.g., safety, security, complexity, performance) for severity of impact of system failure, system degradation, or failure to meet software requirements or system objectives. (C/SE) 1012-1998

critical jamming ratio The ratio of conduit diameter (D) to cable diameter (d) that could result in the cable wedging or jamming in the conduit during the cable pull. (PE/IC) 1185-1994

critical load (1) That part of the load that requires continuous quality electric power for its successful operation. (IA/PSE) 241-1990r
(2) Devices and equipment whose failure to operate satisfactorily jeopardizes the health or safety of personnel, and/or results in loss of function, financial loss, or damage to property deemed critical by the user. (IA/PSE) 1100-1999

critical magnetic field (superconductor) The field below which a superconductor material is superconducting and above which the material is normal, at a specified temperature and in the absence of current. *See also:* superconductivity. (ED) [46]

critical mating dimension (standard connector) Those longitudinal and transverse dimensions assuring nondestructive mating with a corresponding standard connector. (IM/HFIM) 474-1973w

critical overtravel time (of a relay) The time following a critical impulse until movement of the responsive element ceases just short of pickup. (SWG/PE/PSR) C37.100-1992, C37.90-1978s

critical path In the critical path method, a path whose sum of activity times is greater than or equal to the sum of activity times for any other path through the network. *Note:* This sum of activity times is the shortest possible completion time of the overall project. (C) 610.2-1987

critical path method (CPM) A project management technique in which the activities that constitute a project are identified, dependencies among the activities are determined, a network of parallel and sequential activities is produced, an estimated time is assigned to each activity, and a sequence of activities taking the longest time (a critical path) is identified, determining the shortest possible completion time for the overall project. *See also:* program evaluation and review technique. (C) 610.2-1987

critical period That portion of the duty cycle that is the most severe, or the specified time period of the battery duty cycle. (PE/EDPG) 450-1995

critical piece first (software) A system development approach in which the most critical aspects of a system are implemented first. The critical piece may be defined in terms of services provided, degree of risk, difficulty, or other criteria. *See also:* top-down; bottom-up. (C) 610.12-1990

critical point (1) (feedback control system) (Nichols chart) The bound of stability for the *GH* (*j*ω) plot; the intersection of |*GH*| = 1 with ang *GH* = −180°.
(2) (Nyquist diagram) The bound of stability for the locus of the loop trasfer function *GH*(*j*ω); the (-1.*j*0) point. (PE/IM/EDPG) [3], [120]

critical process control (electric pipe heating systems) The use of electric heat tracing systems to increase or maintain, or both, the temperature of fluids (or processes) in important or critical mechanical piping systems including pipes, pumps, valves, tanks, instrumentation, etc., in nuclear power generating stations. An example of an important or critical mechanical piping system is the safety injection system. (PE/EDPG) 622A-1984r, 622B-1988r

critical range Metric values used to classify software into the categories of acceptable, marginal, or unacceptable. (C/SE) 1061-1998

critical rate-of-rise of OFF-state voltage (thyristor) The minimum value of the rate of rise of principal voltage which will cause switching from the OFF-state to the ON-state. (IA/IPC) 428-1981w

critical rate-of-rise of ON-state current (thyristor) The maximum value of the rate-of-rise of ON-state current that a thyristor can withstand without deleterious effect. *See also:* principal current. (ED) [46]

critical section (software) A segment of code to be executed mutually exclusively with some other segment of code which is also called a critical section. Segments of code are required to be executed mutually exclusively if they make competing uses of a computer resource or data item. *See also:* segment; computer; code; data; execute. (C/SE) 729-1983s

critical service loads Station auxiliary loads that are sensitive to power supply disturbances and that have an immediate effect upon power transmission or whose outages could cause damage to the equipment. (SUB/PE) 1158-1991r

critical short-circuit ratio (CSCR) The SCR corresponding to the operation at maximum available power (MAP); for typical inverter design, CSCR = 2. *Note:* The following operational characteristics are associated with CSCR:

— CSCR represents the borderline between “stable” and “unstable” operating regions. For SCR values lower than CSCR, the operation is in the “unstable” region of the ac voltage/dc power characteristic.

— If the operation is at unity power factor for systems at CSCR (i.e., the operation is at MAP), then the fundamental component of the temporary overvoltage (TOV_{fc}) at full load rejection would be near to $\sqrt{2}$.

— A resonance near the second harmonic will occur for systems operating at CSCR.

(PE/T&D) 1204-1997

critical software (software verification and validation plans) (software) Software whose failure could have an impact on safety, or could cause large financial or social loss. (C/SE) 1012-1986s, 610.12-1990, 730-1998

critical speed (rotating machinery) A speed at which the amplitude of the vibration of a rotor due to shaft transverse vibration reaches a maximum value. *See also:* rotor. (PE) [9]

critical stroke amplitude The amplitude of the current of the lightning stroke that, upon terminating on the phase conductor, would raise the voltage of the conductor to a level at which flashover is likely. (SUB/PE) 998-1996

critical success factor (CSF) A business system performance measurement that combines with other CSFs to form a key performance indicator (KPI). (C/PA) 1003.23-1998

critical system (health care facilities) A system of feeders and branch circuits in nursing homes and residential custodial care facilities arranged for connection to the alternate power source to restore service to certain critical receptacles, task illumination and equipment. (EMB) [47]

critical temperature (superconductor) The temperature below which a superconductive material is superconducting and above which the material is normal, in the absence of current and external magnetic fields. *See also:* superconductivity. (ED) [46]

critical torsional speed (rotating machinery) A speed at which the amplitude of the vibration of a rotor due to shaft torsional vibration reaches a maximum value. *See also:* rotor. (PE) [9]

critical travel (of a relay) The amount of movement of the responsive element of a relay during a critical impulse, but not subsequent to the impulse. (SWG/PE/PSR) C37.100-1992, C37.90-1978s

critical value Metric value of a validated metric that is used to identify software that has unacceptable quality. (C/SE) 1061-1998

critical voltage (1) (magnetrons) The highest theoretical value of steady anode voltage, at a given steady magnetic flux density, at which electrons emitted from the cathode at zero velocity would fail to reach the anode. (ED) 161-1971w
(2) (relay) *See also:* relay critical voltage.

critical-voltage parabola (magnetrons) (cutoff parabola) The curve representing in Cartesian coordinates the variation of the critical voltage as a function of the magnetic induction. *See also:* magnetron. [84]

critical withstand current (surge) (impulse) The highest crest value of a surge of given waveshape and polarity that can be applied without causing disruptive discharge on the test specimen. (PE) [8]

CROM *See:* control read-only memory.

Crookes dark space *See:* cathode dark space.

Crosby *See:* cable clamp.

Crosby clip *See:* cable clamp.

cross acceleration (accelerometer) The acceleration applied in a plane normal to an accelerometer input reference axis. (AES/GYAC) 528-1994

crossarm A horizontal member (usually wood or steel) attached to a pole, post, tower or other structure and equipped with means for supporting the conductors. *Note:* The crossarm is placed at right angles to conductors on straight line poles, but splits the angle on light corners. *See also:* tower. (T&D/PE) [10]

crossarm guy A tensional support for a crossarm used to offset unbalanced conductor stress. (T&D/PE) [10]

cross-assembler (software) An assembler that executes on one computer but generates machine code for a different computer. (C) 610.12-1990

cross-axis sensitivity (accelerometer) The proportionality constant that relates a variation of accelerometer output to cross acceleration. This sensitivity varies with the direction of cross acceleration, and is primarily due to misalignment. (AES/GYAC) 528-1994

crossband transponder (navigation) A transponder that replies in a different frequency band from that of the received interrogation. *See also:* navigation. (AES/GCS/RS) 172-1983w, 686-1982s, [42]

crossbar switch (1) A switch having a plurality of vertical paths, a plurality of horizontal paths, and electromagnetically-operated mechanical means for interconnecting any one of the vertical paths with any one of the horizontal paths. (PE/EDPG) [3]
(2) A switch having vertical and horizontal paths and an electromagnetically operated mechanical means for interconnection of any one vertical path with any one horizontal path. *See also:* step-by-step switch. (C) 610.7-1995

cross bar switch A relay-operated device that makes a connection between a line in a set of lines and a line in another set, where the two sets are arranged along adjacent sides of a matrix of contacts or switch points. (C) 610.10-1994w

crossbar system An automatic switching system in which the selecting mechanisms are crossbar switches, common circuits select and test the switching paths and control the operation of the selecting mechanism, and the method of operation is one in which the switching information is received and stored by controlling mechanisms that determine the operations necessary in establishing a telephone connection. *See also:* step-by-step system; electronic switching system. (C/PE/EEC) 610.7-1995, [119]

cross cable bond A cable bond used for bonding between the armor or lead sheath of adjacent cables. *See also:* continuity cable bond; cable bond. (T&D/PE) [10]

cross-category services A set of tools or features or both that has a direct effect on the operation of one or more components of the OSE, but is not in and of itself a stand-alone component. (C/PA) 14252-1996

cross check To test for accuracy by comparing the results of two different methods of computation. (C) 1084-1986w

cross-compiler (software) A compiler that executes on one computer but generates machine code for a different computer. (C) 610.12-1990

cross connect A group of connection points, often wall- or rack-mounted in a wiring closet, used to mechanically terminate and interconnect twisted-pair building wiring.local area networks. (LM/C) 802.3-1998, 8802-12-1998

cross connection (telephone switching systems) Easily changed or removed wire that is run loosely between equipment terminals to establish an electrical association. (COM) 312-1977w

cross-correlation (excitation systems) The cross-correlation of two random signals $x_1(t)$ and $x_2(t)$ is $R_{12}(t)$ defined by

$$R_{12}(t) = \int_0^t x_1(t - \tau)x_2(\tau)d\tau$$

If $x_1(t)$ is a random input to a linear stationary system and $x_2(t)$ is the response, then $R_{12}(t)$ is the inverse Laplace transform of the transfer function of the system. (PE/EDPG) 421A-1978s

cross coupling (transmission medium) A measure of the undesired power transferred from one channel to another. *See also:* transmission line; coupling. (MTT) 146-1980w

cross-coupling coefficient (accelerometer) The proportionality constant that relates a variation of accelerometer output to the product of acceleration applied normal and parallel to an input reference axis. This coefficient can vary, depending on the direction of cross acceleration. (AES/GYAC) 528-1994

cross-coupling errors (inertial sensors) (gyros) The errors in the gyro output resulting from gyro sensitivity to inputs about axes normal to an input reference axis. (AES/GYAC) 528-1994

crossed-field amplifier (microwave tubes) A crossed-field tube or valve, with a nonreentrant slow-wave structure, used as an amplifier. (ED) [45]

crossed-field tube (microwave) A high-vacuum electron tube in which a direct, alternating, or pulsed voltage is applied to produce an electric field perpendicular both to a static magnetic field and to the direction of propagation of a radio-frequency delay line. *Note:* The electron beam interacts synchronously with a slow wave on the delay line. (ED) [45]

cross fire (data transmission) An interfering current in one telegraph or signaling channel resulting from telegraph or signaling current in another channel. (PE) 599-1985w

crossfooting check A consistency check in which two totals obtained by adding the same set of numbers in different sequences are compared. (C) 610.5-1990w

cross-grained wood Wood in which the fibers deviate from a line parallel to the sides of the piece. Cross-grain may be either diagonal or spiral grain or a combination of the two. (T&D/PE) 751-1990

crosshairs A set of two intersecting perpendicular lines whose intersection indicates a position on a graphical display device. *See also:* thumbwheel. (C) 610.6-1991w

crosshatch (A) A series of evenly spaced parallel lines within a closed boundary on a display surface. *See also:* hatch. **(B)** To insert a series of evenly spaced parallel lines as in (A).

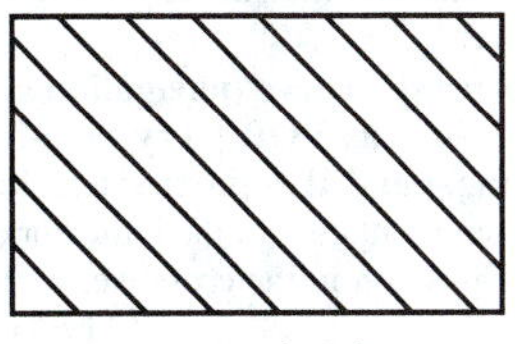

crosshatch

(C) 610.6-1991

cross-index A link between two files containing related data. For example, in a library, the subject card catalog is a cross-index for the title and author card catalogs. *See also:* inverted file; cross-indexed file. (C) 610.5-1990w

cross-indexed file A file whose contents are linked with another file through a cross-index. *See also:* inverted file. (C) 610.5-1990w

cross-indexing (A) The process of linking entities in two files to facilitate searches performed on data contained in those files. **(B)** A method of linking entities as in (A). (C) 610.5-1990

crossing angle *See:* angle of cut.

crossing structure A structure built of poles and, sometimes, rope nets. It is used whenever conductors are being strung over roads, power lines, communications circuits, highways, or railroads, and is normally constructed in such a way as to prevent the conductor from falling onto or into any of these facilities in the event of equipment failure, broken pulling lines, loss of tension, etc. *Synonyms:* guard structure; temporary structure; H-frame; rider structure. (T&D/PE) 524-1992r

cross light (illuminating engineering) Equal illumination in front of the subject from two directions at substantially equal and opposite angles with the optical axis of the camera and a horizontal plane. (EEC/IE) [126]

cross modulation A type of intermodulation due to the modulation of the carrier of the desired signal by an undesired signal wave. (AP/PE/ANT) 145-1983s, 599-1985w

cross-modulation distortion (1) (nonlinear, active, and nonreciprocal waveguide components) A third-order distortion product that can occur when nonlineal devices are exposed simultaneously to two carriers of different frequency where modulation is present. It is measured by using two separate carriers and providing 100% modulation on one. The cross modulation is defined as the power in one sideband of the unmodulated carrier below the power of the carrier. (MTT) 457-1982w

(2) (broadband local area networks) Cross-modulation is the process where the modulation of one carrier is imposed onto another carrier. The exchange of modulation information involves the change of an amplifier's transfer characteristic brought about by a change in amplifier loading. The non-ideal amplifier has an associated compression and expansion characteristic that is dependent on amplifier loading. The change in load caused by the total power of signals and their modulation will create a variation in transfer gain. The change in load due to modulation change is normally the primary cause of cross-modulation. Broadband networks have data devices that have switched carrier transmissions that express cross-modulation as a change in transmission characteristics dependent on channel loading and traffic. (LM/C) 802.7-1989r

cross neutralization A method of neutralization used in push-pull amplifiers whereby a portion of the plate-cathode alternating voltage of each tube is applied to the grid-cathode circuit of the other tube. *See also:* feedback; amplifier. (AP/BT/ANT) 145-1983s, 182A-1964w

cross-office delay For a tandem call switched cross-office, the interval between starting time point and ending point. *See also:* ending point; starting point. (COM/TA) 973-1990w

crossover (cathode-ray tubes) The first focusing of the beam that takes place in the electron gun. *See also:* cathode-ray tube. (ED) [45], [84]

crossover, automatic voltage-current *See:* automatic voltage-current crossover.

crossover characteristic curve (navigation aid terms) (navigation systems such as VOR [very high-frequency omnidirectional range] and ILS [instrument landing system]) The graphical representation of the indicator current variation with change of position in the crossover region. (AES/GCS) 172-1983w

crossover frequency (1) (frequency-dividing networks) The frequency at which equal power is delivered to each of two adjacent channels when all channels are properly terminated. *See also:* transition frequency. (SP) [32]
(2) (automatic control) *See also:* phase-crossover frequency; gain-crossover frequency.

crossover loss For a tracker that uses an offset beam, such as a conical scan tracker, the reduction in signal-to-noise ratio for a target on the tracking axis relative to that for a target on the peak two-way antenna gain of the beam. *Note:* The crossover loss factor is the ratio of the signal-to-noise ratio for a target on the peak two-way antenna gain to that for a target on the tracking axis. *See also:* conical scanning. (AES) 686-1997

crossover network *See:* dividing network.

crossover region (navigation systems) A loosely defined region in space containing the course line and within which a transverse flight yields information useful in determining course sensitivity and "flyability." (AES/GCS) 172-1983w

crossover spiral *See:* lead-over groove.

crossover time (1) (germanium gamma-ray detectors) (charged-particle detectors) The instant at which the waveform of a bipolar pulse passes through a designated level. (NID) 301-1976s
(2) (semiconductor radiation detectors) The instant at which the transition between the two lobes of a bipolar pulse passes through a designated level (usually the baseline). (NPS) 300-1988r
(3) (of a bipolar pulse) The instant at which the transition between the two lobes passes through a designated level (usually the baseline). At this level, the timing with respect to the initiating events is nearly invariant with pulse amplitude. (NPS) 325-1996

crossover transition (liquid scintillation counting systems) A gamma ray occurring between two nonadjacent nuclear levels. (NI) N42.14-1991

crossover voltage, secondary-emission *See:* secondary-emission crossover voltage.

crossover walk (1) (germanium gamma-ray detectors) (charged-particle detectors) (of a pulse) The deviation of the crossover time for some variable, such as (pulse) amplitude. (NPS/NID) 325-1986s, 301-1976s
(2) (semiconductor radiation detectors) (of a pulse) A deviation of crossover time of a bipolar pulse due to a change in peak amplitude. *See also:* walk. (NPS) 300-1988r

crosspoint (telephone switching systems) A controlled device used in extending a transmission or control path. (COM) 312-1977w

crosspoint switch (local area networks) A switching function that provides normal connections, wrap connections, breaks in the transmit, and local loopback for lobe testing in the wrap mode. (LM/C) 802.5c-1991r

cross polarization (1) In a specified plane containing the reference polarization ellipse, the polarization orthogonal to a specified reference polarization. *Note:* The reference polarization is usually the copolarization. (AP/ANT) 145-1993
(2) (waveguide) The polarization orthogonal to a reference polarization. *Note:* Two fields have orthogonal polarization if their polarization ellipses have the same axial ratio, major axes at right angles, and opposite senses of rotation. *See also:* orthogonal polarization. (MTT/AP/PROP) 146-1980w, 211-1997
(3) (protective relaying) The polarization of a relay for directionality using some proportion of the voltage from a healthy (unfaulted) phase(s). One example of this is quadrature polarization. In this case, the polarizing voltage is in quadrature to the faulted phase voltage. (PE/PSR) C37.113-1999

cross-polarization discrimination The ratio of the power level at the output of a receiving antenna, nominally co-polarized with the transmitting antenna, to the output of a receiving antenna of the same gain but nominally orthogonally polarized to the transmitting antenna. (AP/PROP) 211-1997

cross polarization electronic countermeasures An ECM technique that transmits with a polarization orthogonal to the principal polarization of the victim radar. *Note:* Since cross-polarized antenna patterns are often very different from the normal, copolarized patterns, tracking might be disrupted. *See also:* electronic countermeasures. (AES) 686-1997

cross-polarization isolation The ratio of the wanted power to the unwanted power in the same receiver channel when the transmitting antenna is radiating nominally orthogonally polarized signals at the same frequency and power level. (AP/PROP) 211-1997

cross-polar radiation pattern A radiation pattern corresponding to the polarization orthogonal to the co-polarization. *See also:* co-polarization. (AP/ANT) 145-1993

cross-polar side lobe level, relative The maximum relative partial directivity (corresponding to the **cross polarization**) of a side lobe with respect to the maximum partial directivity (corresponding to the **co-polarization**) of the antenna. *Note:* Unless otherwise specified, the cross-polar side lobe level shall be taken to be that of the highest side lobe of the cross-polar radiation pattern. (AP/ANT) 145-1993

cross product *See:* vector product.

cross protection An arrangement to prevent the improper operation of devices from the effect of a cross in electric circuits. (EEC/PE) [119]

cross rectifier circuit A circuit that employs four or more rectifying elements with a conducting period of 90 electrical degrees plus the commutating angle. *See also:* rectification. (EEC/PE) [119]

cross-reference generator (software) A software tool that accepts as input the source code of a computer program and produces as output a listing that identifies each of the program's variables, labels, and other identifiers and indicates which statements in the program define, set, or use each one. *Synonym:* cross-referencer. (C) 610.12-1990

cross-reference list (software) A list that identifies each of the variables, labels, and other identifiers in a computer program and indicates which statements in the program define, set, or use each one. (C) 610.12-1990

cross-referencer *See:* cross-reference generator.

cross section (1) (image processing and pattern recognition) The intersection of an n-dimensional image or region with an $(n-1)$-dimensional object. For example, the intersection of a two-dimensional image or region with a straight line.

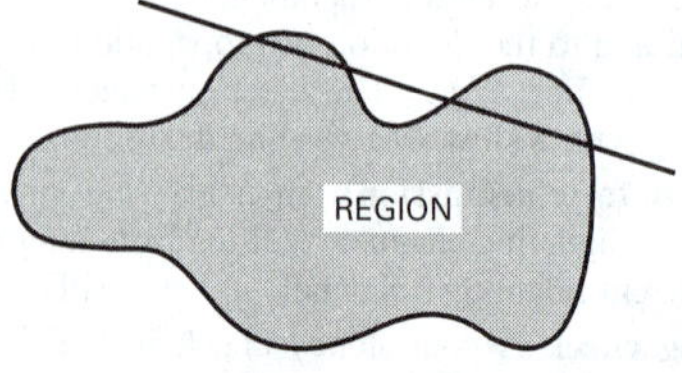

cross section

(C) 610.4-1990w
(2) Often used as a shortened form of radar cross section. *Note:* This term should be avoided when there is a possibility of confusion with geometric cross section. (AES) 686-1997

(3) *See also:* radar cross section; scattering cross section; bistatic cross section; monostatic cross section; backscattering cross section. (ANT)

cross-sectional area (conductor) (cross section of a conductor) The sum of the cross-sectional areas of its component wires, that of each wire being measured perpendicular to its individual axis. (PE) 599-1985w

crosstalk (1) (cable systems in power generating stations) The noise or extraneous signal caused by ac or pulse-type signals in adjacent circuits. (PE/SUB/EDPG) 422-1977, 525-1992r
(2) (A) (data transmission) Undesired energy appearing in one signal path as a result of coupling from other signal paths. *Note:* Path implies wires, waveguides, or other localized or constrained transmission systems. **(B) (data transmission)** (electroacoustics) The unwanted sound reproduced by an electroacoustic receiver associated with a given transmission channel resulting from cross coupling to another transmission channel carrying sound-controlled electric waves or, by extension, the electric waves in the disturbed channel that result in such sound. *Note:* In practice, crosstalk may be measured either by the volume of the overheard sounds or by the magnitude of the coupling between the disturbed and the disturbing channels. In the latter case, to specify the volume of the overheard sounds, the volume in the disturbing channel must also be given. (PE) 599-1985
(3) Unwanted electric signals injected into a circuit by stray coupling. (ELM) C12.1-1981
(4) (instrumentation and control equipment grounding in generating stations) The noise or extraneous signal caused by ac or dc pulse-type signals in adjacent circuits. (PE/EDPG) 1050-1996
(5) (telecommunications) Undesired energy appearing in one path as a result of coupling from another path. Crosstalk is further classified as near-end, far-end, intelligible, or unintelligible. (COM/TA) 1007-1991r
(6) (telecommunications) Unwanted coupling between any two paths through the switching system. There are two situations of interest. In the first the crosstalk coupling loss of the system is linear; that is it is independent of applied disturbing signal level and is unaffected by other things, such as circuit noise. The other situation of interest arises with a digital switching system where the crosstalk coupling loss is nonlinear and is a function of applied disturbing signal level circuit noise, and encoder bias. Such systems will, under some conditions of circuit noise and coder bias, display crosstalk coupling loss that tends to decrease with decreasing disturbing signal level. *See also:* equal-level crosstalk coupling loss; crosstalk coupling loss. (COM/TA) 973-1990w
(7) (multichannel) The ratio of the signal induced in one channel to a common signal applied to all other channels. (IM/WM&A) 1057-1994w
(8) The noise or extraneous signal caused by ac or pulse-type signals in adjacent circuits (measurement of power frequency magnetic fields). (T&D/PE) 644-1994, 1308-1994
(9) An electromagnetic field in the space surrounding a cable circuit created by an electrical signal. This field induces currents and electromotive forces in other circuits located close enough to the disturbing cable circuit to be affected. (PE/IC) 1143-1994r
(10) (A) A type of noise characterized by unwanted coupling of a signal or the interaction of signals on two adjacent channels. *See also:* near end crosstalk. **(B)** Undesired energy appearing in one signal path as a result of coupling from other signal paths. (C) 610.7-1995
(11) Crosstalk is undesired energy appearing in one signal path as a result of coupling from other signal paths. (C/LM) 8802-5-1998

crosstalk coupling (crosstalk loss) Cross coupling between speech communication channels or their component parts. *Note:* Crosstalk coupling is measured between specified points of the disturbing and disturbed circuits and is preferably expressed in decibels. *See also:* coupling. (PE/EEC) [119]

crosstalk coupling loss (telecommunications) The loss of the crosstalk path. *See also:* crosstalk. (COM/TA) 973-1990w

crosstalk, electron beam *See:* electron beam crosstalk.

crosstalk loss *See:* crosstalk coupling.

crosstalk unit Crosstalk coupling is sometimes expressed in crosstalk units through the relation

$$\text{Crosstalk units} = 10^{[6-(L/20)]}$$

where L = crosstalk coupling in decibels. *Note:* For two circuits of equal impedance, the number of crosstalk units expresses the current in the disturbed circuit as millionths of the current in the disturbing circuit. *See also:* coupling. (EEC/PE) [119]

crowbar A protective circuit in a power distribution circuit that rapidly shorts the output voltage to ground when an overvoltage or other error condition occurs. (C) 610.10-1994w

CR-RC shaping (1) (germanium gamma-ray detectors) (x-ray energy spectrometers) (charged-particle detectors) The pulse shaping present in an amplifier that has a simple high-pass filter consisting of a capacitor and a resistor together with a simple low-pass filter, separated by impedance isolation. (Pulse shaping in such an amplifier cuts off at 6 dB (decibels) per octave at both ends of the band.). (NPS/NID) 301-1976s, 759-1984r
(2) (semiconductor charged-particle detectors) In an amplifier, the pulse shaping produced by a single-section high-pass network followed by a single-section low-pass network, with the two networks separated by an isolating stage and with both networks having the same time constant. (NPS) 300-1988r
(3) In an amplifier, the pulse shape produced by m CR high-pass filter sections (differentiators) in conjunction with n RC low-pass filter sections (integrators), all with the same time constant. If the input signal is a step function and no other high-pass sections are in the signal path, the pulse shape is unipolar if m = 1, bipolar if m = 2. For unipolar pulses, the waveform is described by

$$Kt\,{}^{n}_{e} - t/\tau$$

where K is a constant, τ is time, and t is the time constant of the differentiator. (NPS) 325-1996

CRS *See:* configuration report server.

CRT *See:* cathode-ray tube.

CRT display device A display device that displays data onto a phosphor coated display screen using controlled electron beams within a CRT. *Note:* Raster display devices and random-scan display devices are two major categories of CRT display devices. *See also:* penetration CRT display device; dark trace tube display device; raster display device; storage tube display device. (C) 610.10-1994w

crude metal Metal that contains impurities in sufficient quantities to make it unsuitable for specified purposes or that contains more valuable metals in sufficient quantities to justify their recovery. *See also:* electrorefining. (EEC/PE) [119]

crush loaded weight The weight of a vehicle when loaded with crew, all seats occupied, and standees to a specified maximum number. (VT) 1475-1999

crust A layer of solidified electrolyte. *See also:* fused electrolyte. (EEC/PE) [119]

CRV *See:* Code Rule Violation.

cryogenics (1) (general) The study and use of devices utilizing properties of materials near absolute-zero temperature. (C) [20], [85]
(2) (laser maser) The branch of physics dealing with very low temperatures. (LEO) 586-1980w
(3) A branch of technology concerned with devices that make use of the properties assumed by materials at temperatures near absolute zero. (C) 610.2-1987

cryogenic storage A type of storage that uses the superconductive and magnetic properties of certain materials at temperatures near absolute zero. (C) 610.10-1994w

cryotron (1) A superconductive device in which current in one or more input circuits magnetically controls the superconducting-to-normal transition in one or more output circuits, provided the current in each output circuit is less than its critical value. *See also:* superconductivity. (ED) [46]
(2) A device that makes uses of the effects of extremely low temperatures on conductive materials such that small magnetic field changes can control large current changes. (C) 610.10-1994w

cryptography The discipline embodying principles, means, and methods for the transformation of data in order to hide its information content, prevent its undetected modification, and/or prevent its unauthorized use. (LM/C) 802.10-1992

crystal (A) (communication practice) A piezoelectric crystal. **(B) (communication practice)** A piezoelectric crystal plate. **(C) (communication practice)** A crystal rectifier. (PE/EEC) [119]

crystal-controlled oscillator *See:* crystal oscillator.

crystal diode A rectifying element comprising a semiconducting crystal having two terminals designed for use in circuits in a manner analogous to that of electron-tube diodes. *See also:* rectifier. (EEC/PE) [119]

crystal loudspeaker (piezoelectric loudspeaker) A loudspeaker in which the mechanical displacements are produced by piezoelectric action. (EEC/PE) [119]

crystal microphone (piezoelectric microphone) A microphone that depends for its operation of the generation of an electric charge by the deformation of a body (usually crystalline) having piezoelectric properties. *See also:* microphone. (EEC/PE) [119]

crystal mixer (mixer) A crystal receiver that can be fed simultaneously from a local oscillator and signal source, for the purpose of frequency changing. *See also:* waveguide. (Std100) [84]

crystal oscillator (crystal-controlled oscillator) An oscillator in which the principal frequency-determining factor is the mechanical resonance of a piezoelectric crystal. *See also:* oscillatory circuit. (BT) 182A-1964w

crystal pickup (piezoelectric pickup) A phonograph pickup that depends for its operation on the generation of an electric charge by the deformation of a body (usually crystalline) having piezoelectric properties. *See also:* phonograph pickup. (EEC/PE) [119]

crystal pulling A method of crystal growing in which the developing crystal is gradually withdrawn from a melt. (IA) [12]

crystal receiver A waveguide incorporating a crystal detector for the purpose of rectifying received electromagnetic signals. *See also:* waveguide. (AP/ANT) [35], [84]

crystals hierarchy (primary ferroelectric terms) Depending on their geometry, crystals are commonly classified into seven systems: triclinic (the least symmetrical), monoclinic, orthorhombic, tetragonal, trigonal, hexagonal, and cubic. The seven systems in turn are divided into point groups (crystal classes) according to their symmetry with respect to a point. There are 32 such crystal classes; 20 are piezoelectric. Piezoelectric crystals have the following property: if stress is applied along certain directions in the crystals, they develop an electric polarization whose magnitude is (within limits) proportional to the applied stress. Conversely, when an electric field is applied along certain directions in a piezoelectric crystal, the crystal is strained by an amount proportional to the applied field. Each crystal system contains at least one piezoelectric class. Ten of the 20 piezoelectric classes possess spontaneous electrical polarization; that is, they have a nonvanishing dipole moment per unit volume and are called polar. The ten polar crystal classes are designated 1, 2, m, 2mm, 4, 4mm, 3, 3m, 6, 6mm in the notation of Hermann and Maugin, C_1, C_2, C_{1h}, C_{2v}, C_4, C_{4v}, C_3, C_{3v}, C_6, C_{6v}, respectively in the notation of Schoenflies. (UFFC) 180-1986w

crystal spots Spots produced by the growth of metal sulfide crystals upon metal surfaces with a sulfide finish and lacquer coating. The appearance of crystal spots is called spotting in. *See also:* electroplating. (EEC/PE) [119]

crystal-stabilized transmitter A transmitter employing automatic frequency control, in which the reference frequency is that of a crystal oscillator. *See also:* radio transmitter. (AP/ANT) 145-1983s

crystal systems The term "crystal" is applied to a solid in which the atoms are arranged in a single pattern repeated throughout the body. In a crystal the atoms may be thought of as occurring in small groups, all groups being exactly alike, similarly oriented, and regularly aligned in all three dimensions. Each group can be regarded as bounded by a parallelepiped and each parallelepiped regarded as one of the ultimate building blocks of the crystal. The crystal is formed by stacking together in all three dimensions replicas of the basic parallelepiped without any spaces between them. Such a building block is called a unit cell. Since the choice of a particular set of atoms to form a unit cell is arbitrary, it is evident that there is a wide range of choices in the shapes and dimensions of the unit cell. In practice, that unit cell is selected which is most simply related to the actual crystal faces and X-ray reflections, and which has the symmetry of the crystal itself. Except in a few special cases, the unit cell has the smallest possible size. In crystallography the properties of a crystal are described in terms of the natural coordinate system provided by the crystal itself. The axes of this natural system, indicated by the letters a, b, and, c, are the edges of the unit cell. In a cubic crystal, these axes are of equal length and are mutually perpendicular; in a triclinic crystal they are of unequal lengths and no two are mutually perpendicular. The faces of any crystal are all parallel to planes whose intercepts on the a, b, c axes are small multiples of unit distances or else infinity, in order that their reciprocals, when multiplied by a small common factor, are all small integers or zero. These are the indices of the planes. In this nomenclature we have, for example, faces (100), (010), (001), also called the a, b, c faces, respectively. In the orthorhombic, tetragonal, and cubic systems, these faces are normal to the a, b, c axes, 100, etc. Even in the monoclinic and triclinic systems, these faces contain respectively, the b and c, a and c, and, a and b axes. As referred to the set of rectangular axes X, Y, Z, these indices are in general irrational except for cubic crystals. Depending on their degrees of symmetry, crystals are commonly classified into seven systems: triclinic (the least symmetrical), monoclinic, orthorhombic, tetragonal, trigonal, hexagonal, and cubic. The seven systems, in turn, are divided into point groups (classes) according to their symmetry with respect to a point. There are 32 such classes, eleven of which contain enantiomorphous forms. Twelve classes are of too high a degree of symmetry to show piezoelectric properties. Thus twenty classes can be piezoelectric. Every system contains at least one piezoelectric class. (UFFC) 176-1978s

crystal-video receiver A receiver consisting of a crystal detector and a video amplifier. (EEC/PE) [119]

CSC *See:* computer software component.

CSCI *See:* computer software configuration item.

C-scope A cathode-ray oscilloscope arranged to present a C-display. (AES/RS) 686-1990

CSI *See:* communication services interface.

CSMA/CD LAN (1) Any local area network using the CSMA/CD access protocol. (LM/C) 802.1H-1995
(2) An IEEE 802.3 LAN is a CSMA/CD LAN, as is an Ethernet LAN. Most CSMA/CD networks are hybrids, carrying both Ethernet and IEEE 802 style frames. (LM/C) 802.1H-1995

CSMP III *See:* Continuous System Modeling Program III.

CSR *See:* control and status register.

CSR Architecture (1) IEEE Std 1212-1991, IEEE Standard Control and Status Register (CSR) Architecture for Microcomputer Buses. (C/MM) 1596-1992

(2) Refers to IEEE Std 1212-1991, the parent document of this series. It specifies the overall architecture of a node on conformant standard buses and specifically the structure and use of its CSRs and standard ROM locations. (C/MM) 1212.1-1993

(3) ISO/IEC 13213: 1994 [ANSI/IEEE Std 1212, 1994 Edition], Information technology-Microprocessor systems-Control and Status Registers (CSR) Architecture for microcomputer buses. (C/MM) 1394-1995

(4) Refers to IEEE Std 1212-1991. (C/MM) 1596.5-1993, 1212-1991s

(5) ISO/IEC 13213: 1994 [ANSI/IEEE Std 1212, 1994 Edition] Information technology—Microprocessor systems—Control and Status Register (CSR) Architecture for microcomputer buses. (C/MM) 1596.4-1996

CSR space *See:* control and status register space.

CSR unit architecture The logical component of a node that is accessed by I/O driver software. After the node is initialized and configured, the units normally operate independently. Note that one node could have multiple units (for example, processor, memory, and SCSI controller). (C/BA) 896.3-1993w

CSS/II *See:* Computer System Simulation II.

CSSL *See:* Continuous Systems Simulation Language.

CSU *See:* channel service unit.

CT *See:* current transformer; computed tomography.

CTIC *See:* conformance test interface connector.

CTRL *See:* control key.

CTS *See:* computer typesetting.

CTS build system The hardware and software used to compile and configure a CTS. (C/PA) 2003-1997

CTS execution system The hardware and system software on which the CTS is executed. (C/PA) 2003-1997

cube tap *See:* multiple plug.

cubic meters per second (m^3/s) (cubic meters per second) Volume of water or liquid discharged per second under standard conditions. (T&D/PE) 957-1995, 957-1987s

cubic natural spline (pulse terminology) A catenated piecewise sequence of cubic polynominal functions p(1, 2), p(2, 3),. . ., p($n-1$, n) between knots t_1m_1 and t_2m_2, t_2 m_2 and t_3m_3, . . ., t_{n-1} m_{n-1} and t_nm_n, respectively, wherein: (1) At all knots the first and second derivatives of the adjacent polynominal functions are equal, and (2) For all values of t less than t_1 and greater than t_n the function is linear. *See also:* waveforms produced by operations on waveforms. (IM/WM&A) 194-1977w

cue *See:* call.

cumulative amplitude probability distribution (control of system electromagnetic compatibility) A cumulative distribution showing the probability that all amplitudes equal to, or above, a stated value are exceeded as a function of that value. C63.12-1984

cumulative compound (rotating machinery) Applied to a compound machine to denote that the magnetomotive forces of the series and the shunt field windings are in the same direction. *See also:* magnetomotive force. (PE) [9]

cumulative demand An indicating demand meter in which the accumulated total of maximum demands during the preceding periods is indicated during the period after the meter has been reset and before it is reset again. *Note:* The maximum demand for any one period is equal or proportional to the difference between the accumulated readings before and after reset. (AMR/SCC31) 1377-1997

cumulative demand meter (or register) An indicating demand meter in which the accumulated total of maximum demands during the preceding periods is indicated during the period after the meter has been reset and before it is reset again. *Note:* The maximum demand for any one period is equal or proportional to the difference between the accumulated readings before and after reset. *See also:* electricity meter. (ELM) C12.1-1982s

cumulative demand register (metering) A register that indicates the sum of the previous maximum demand readings prior to reset. When reset, the present reading is added to the previous accumulated readings. The maximum demand for the present reading period is the difference between the present and previous readings. (ELM) C12.1-1982s

cumulative detection probability The probability that a target is detected on at least one of n successive scans or detection opportunities of a surveillance radar. (AES) 686-1997

cumulative latency The time it takes for a signal element to travel from the active monitor's transmitter output to its receiver input. (C/LM) 8802-5-1998

cumulative leakage index (broadband local area networks) A measurement of cumulative RF signal leakage of a cable system. The measurement is usually specified for a given area. (LM/C) 802.7-1989r

cumulative probability distribution *See:* probability distribution function.

cuprous chloride cell A primary cell in which depolarization is accomplished by cuprous chloride. *See also:* electrochemistry. (EEC/PE) [119]

Curie temperature (electrical heating systems) The temperature at which the magnetic properties of a substance change from ferromagnetic to paramagnetic. (IA/PC) 844-1991

Curie-Weiss temperature (primary ferroelectric terms) The intercept $_q$ of the linear portion of the plot of 1/k versus T, in the region above the ferroelectric Curie point, where k is the small-signal relative dielectric permittivity measured at zero bias field along the polar axis, and T is the absolute temperature. *Note:* In many ferroelectrics, k follows the Curie-Weiss relation. (UFFC) 180-1986w

curl A vector that has a magnitude equal to the limit of the quotient of the circulation around a surface element on which the point is located by the area of the surface, as the area approaches zero, provided the surface is oriented to give a maximum value of the circulation: the positive direction of this vector is that traveled by a right hand screw turning about an axis normal to the surface element when an integration around the element in the direction of the turning of the screw gives a positive value to the circulation. If the vector A of a vector field is expressed in terms of its three rectangular components A_x, A_y, and A_z, so that the values of A_x, A_y, and A_z are each given as a function of x, y, and z, the curl of the vector field (abbreviated curl **A** or $\nabla \times \mathbf{A}$) is the vector sum of the partial derivatives of each perpendicular to it, or curl

$$\text{curl}\,\mathbf{A} = \nabla \times \mathbf{A} = \begin{vmatrix} \mathbf{i} & \mathbf{j} & \mathbf{k} \\ \frac{\delta}{\delta x} & \frac{\delta}{\delta y} & \frac{\delta}{\delta z} \\ \mathbf{A}_x & \mathbf{A}_y & \mathbf{A}_z \end{vmatrix}$$

$$= \mathbf{i}\left\{\frac{\delta A_z}{\delta y} - \frac{\delta A_y}{\delta z}\right\} + \mathbf{j}\left\{(\frac{\delta A_x}{\delta z} - \frac{\delta A_z}{\delta x}\right\} + \mathbf{k}\left\{\frac{\delta A_y}{\delta x} - \frac{\delta A_x}{\delta y}\right\}$$

$$= \mathbf{i}(D_yA_z - D_zA_y) + \mathbf{j}(D_zA_x - D_xA_z) + \mathbf{k}(D_xA_y - D_yA_z)$$

where **i**, **j**, and **k** are unit vectors along the x, y, and z axes, respectively. Example: The curl of the linear velocity of points in a rotating body is equal to twice the angular velocity. The curl of the magnetic field strength at a point within an electric conductor is equal to k times the current density at the point where k is a constant depending on the system of units. (Std100) 270-1966w

currency symbol character A character within a picture specification that represents the currency sign. *Note:* $ is commonly used. (C) 610.5-1990w

current (1) The flow of electrons within a wire or a circuit; measured in amperes. (C) 610.7-1995

(2) (general) A generic term used when there is no danger of ambiguity to refer to any one or more of the currents specifically described. *Note:* (1) For example, in the expression "the current in a simple series circuit, " the word current refers to the conduction current in the wire of the inductor and the displacement current between the plates of the capacitor. (2)

A direct current is a unidirectional current in which the changes in value are either zero or so small that they may be neglected. A given current would be considered a direct current in some applications, but would not necessarily be so considered in other applications. (Std100) 270-1966w
(3) The use of certain adjectives before "current" is often convenient, as in convection current, anode current, electrode current, emission current, etc. The definition of conducting current usually applies in such cases and the meaning of adjectives should be defined in connection with the specific applications. (Std100) 270-1966w
(4) Sum of the polarization and conductance currents. (PE) 402-1974w

current amplification (1) The ratio of the output current to the cathode current due to photoelectric emission at constant electrode voltages. *Notes:* 1. The term output current and photocathode current as here used does not include the dark current. 2. This characteristic is to be measured at levels of operation that will not cause saturation. *See also:* phototube. (ED/NPS) 161-1971w, 398-1972r
(2) (magnetic amplifier) The ratio of differential output current to differential control current. (MAG) 107-1964w
(3) The ratio of the signal output current to the current applied to the input. *See also:* amplifier. (ED) [45]

current, anode *See:* electrode current.

current attenuation Either a decrease in signal current magnitude, in transmission from one point to another, or the process thereof, or of a transducer, the scalar ratio of the signal input current to the signal output current. *Note:* By incorrect extension of the term "decibel," this ratio is sometimes expressed in decibels by multiplying its common logarithm by 20. It may be correctly expressed in decilogs. *See also:* decibel; attenuation. (SP) 151-1965w

current, average discharge *See:* average discharge current.

current balance ratio The ratio of the metallic-circuit current or noise-metallic (arising as a result of the action of the longitudinal-circuit induction from an exposure on unbalances outside the exposure) to the longitudinal circuit current or noise-longitudinal in sigma at the exposure terminals. It is expressed in microamperes per milliampere or the equivalent. *See also:* inductive coordination. (EEC/PE) [119]

current-balance relay A balance relay that operates by comparing the magnitudes of two current inputs. (SWG/PE/PSR) C37.100-1992, C37.90-1978s

current-balancing device (thyristor) Device used to achieve satisfactory division of current among parallel connected semiconductor devices, for example, reactor, resistor, impedance. (IA/IPC) 428-1981w

current-balancing reactor A reactor used in semiconductor rectifiers to achieve satisfactory division of current among parallel-connected semicondutor diodes. *See also:* reactor. (PE/TR) [57]

current balancing transformer *See:* sharing transformer and current balancing transformer.

current carrier In a semiconductor, a mobile conduction electron or hole. (AES/SS) 307-1969w

current-carrying *See:* energized.

current-carrying capacity The maximum current that a contact is able to carry continuously or for a specified period of time. *See also:* contactor. (IA/IAC) [60], [84]

current-carrying part A conducting part intended to be connected in an electric circuit to a source of voltage. *Note:* Non-current-carrying parts are those not intended to be so connected.
(SWG/NESC/T&D/PE) C2-1997, 516-1995, C37.100-1992, C37.40-1993, C37.30-1971s

current circuit (1) (ac high-voltage circuit breakers) That part of the synthetic test circuit from which the major part of the power frequency current is obtained. (SWG/PE) C37.081-1981r, C37.083-1999
(2) (relays) An input circuit to that is applied a voltage or a current which is a measure of primary current. (PE/PSR) C37.90.1-1989r

current clamp (self-commutated converters) (converter circuit elements) A clamp that limits the current through a semiconductor device. (IA/SPC) 936-1987w

current comparator (metering) A device by which the ratio of two currents and the phase angle between them can be measured precisely. *Note:* A common form of current comparator relies on a balance on ampere-turns produced by currents in two or more windings on one or more magnetic cores. (ELM) C12.1-1988

current compensator (excitation systems for synchronous machines) An element of the excitation system that acts to compensate for synchronous machine load current effects. *Notes:* 1. Examples are reactive current compensator and active current compensator. A reactive current compensator is a compensator that acts to modify the regulated voltage in accordance with reactive current. An active current compensator is a compensator that acts to modify the regulated voltage in accordance with active current. 2. Historically, terms such as equalizing reactor and cross current compensator have been used to describe the function of a reactive compensator. These terms are deprecated. 3. Reactive compensators are generally applied with synchronous machine voltage regulators to obtain reactive current sharing among synchronous machines operating in parallel. They function in the following two ways:

a) Reactive droop compensation is the more common mathod. It creates a droop in synchronous machine terminal voltage proportional to reactive current and equivalent to that which would be produced by the insertion of a reactor between the synchronous machine terminals and the paralleling point.
b) Reactive differential compensation is used where droop in synchronous machine voltage is not wanted. It is obtained by a series differential connection of the various synchronous machine, current transformer secondaries, and reactive compensators. The difference current for any synchronous machine from the common series current creates a compensating voltage in the input to the particular synchronous machine voltage regulator which acts to modify the synchronous machine excitation to reduce to minimum (zero) its differential reactive current.

4. Line drop compensators modify synchronous machine terminal voltage by regulator action to compensate for the impedance drop from the machine terminals to a fixed point in the external circuit. Action is accomplished by insertion of a voltage equivalent to the impedance drop within the regulator input circuit. The voltage drops of the resistance and reactance portions of the impedance are obtained, respectively, by an active compensator and a reactive compensator. (PE/EDPG) 421.1-1986r

current, conduction *See:* conduction current.

current crest factor The ratio of the peak value of lamp current to the root-mean-square value of lamp current. (EEC/LB) [97]

current cutoff (power supplies) An overload protective mechanism designed into certain regulated power supplies to reduce the load current automatically as the load resistance is reduced. This negative resistance characteristic reduces overload dissipation to negligible proportions and protects sensitive loads. (AP/ANT) [35]

current cycle loop (substation grounding) The combination of conductors and connectors that carries the current of the circuit under test. (SUB/PE) 837-1989r

current delay angle (thyristor) The interval in electrical angular measure by which the starting instant of conduction is delayed in relation to operation that would occur with contin-

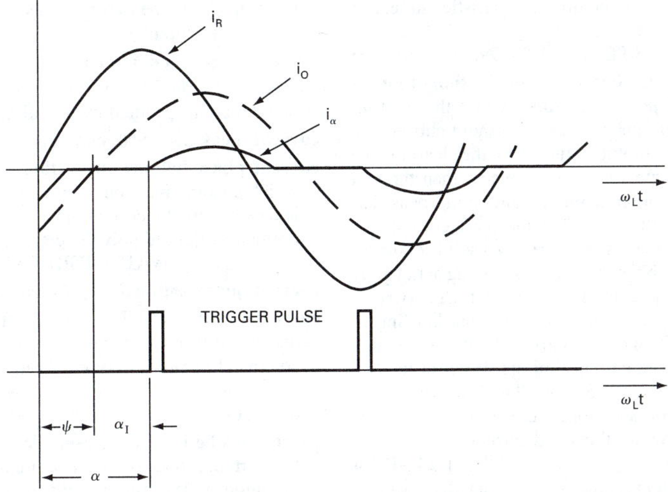

i_R = current in a control element with all control elements continuously gated and a resistive load

NOTE: In the case of a single phase controller i_R is in phase with the line voltage. The latter may be used as a convenient reference voltage to measure α

i_O = current in a control element with all control elements continuously gated and at the specified load

i_α = current in a control element with a trigger delay angle of α and at the specified load

α = angle of retard

α_I = current delay angle

ψ = angle between i_R and i_O

NOTE: In the case of a single phase controller, ψ is identical with the load power factor angle.

current delay angle

uously gated control elements. See α_I of the corresponding figure. (IA/IPC) 428-1981w

current density (1) A generic term used where there is no danger of ambiguity to refer either to conduction-current density or to displacement-current density, or to both. (Std100) 270-1966w
(2) A vector-point function describing the magnitude and direction of charge flow per unit area. The preferred unit is amperes per square meter (A/m^2). (T&D/PE) 1227-1990r

current derived voltage A voltage produced by a combination of currents. *Note:* (1) The element used to create this voltage in a pilot system is popularly referred to as a filter. A typical example is a filter that is supplied three-phase currents and produces an output voltage proportional to the symmetrical component content of these currents. (For example, $V_F = K_1\, IA_1 + K_2\, IA_2 + K_0\, IA_0$ where IA_1, IA_2, and IA_0 are the symmetrical components of the *A* phase current and the *K* are weighting factors.). (PE/PSR) C37.95-1973s

current differential relay A relay designed to detect faults by measuring the current magnitude and phase angle difference between relay terminals of a transmission line. (PE/PSR) C37.113-1999

current efficiency (specified electrochemical process) The proportion of the current that is effective in carrying out that process in accordance with Faraday's law. *See also:* electrochemistry. (EEC/PE) [119]

current extent *See:* extensional set.

current generator (signal-transmission system) A two-terminal circuit element with a terminal current substantially independent of the voltage between its terminals. *Note:* An ideal current generator has zero internal admittance. *See also:* network analysis; signal. (ED) 161-1971w

current injection method A synthetic test method in which the voltage circuit is applied to the test circuit breaker before power frequency current zero.
(SWG/PE) C37.100-1992, C37.083-1999, C37.081-1981r

current instance The package instance whose private data is currently accessible. (C/BA) 1275-1994

current limit (1) The maximum output of the battery charger delivered to a discharged battery and load, usually stated as a percentage of output rating and with nominal input voltage supplied to the charger. (IA/PSE) 602-1996
(2) A control function that prevents a current from exceeding its prescribed limits. *Note:* Current-limit values are usually expressed as percent of rated-load value. If the current-limit circuit permits the limit value to increase somewhat instead of being a single value, it is desirable to provide either a curve of the limit value of current as a function of some variable such as speed or to give limit values at two or more conditions of operation. (IA/APP/IAC) [69], [60]

current-limit acceleration (electric drive) A system of control in which acceleration is so governed that the motor armature current does not exceed an adjustable maximum value. *See also:* electric drive. (IA/ICTL/IAC) [60]

current-limit control (electric drive) A system of control in which acceleration, or retardation, or both, are so governed that the armature current during speed changes does not exceed a predetermined value. *See also:* electric drive. (EEC/PE) [119]

current limiter (protection and coordination of industrial and commercial power systems) A device intended to function only on fault currents of high magnitude and that may not successfully open on lesser overcurrents regardless of time. Such a device should always be used in series with a fuse, contactor, or circuit breaker to protect against overloads and low-level short circuits. Current limiters are typically added to molded-case circuit breakers, power circuit breakers, or instantaneous circuit protectors. (IA/PSP) 242-1986r

current limiting, automatic *See:* automatic current limiting.

current-limiting characteristic curve (of a current-limiting fuse) A curve showing the relationship between the maximum peak current passed by a fuse and the correlated root-mean-square available current magnitudes under specified voltage and circuit impedance conditions. *Synonyms:* peak let-through characteristic curve; cutoff characteristic curve. *See also:* current-limiting fuse. (SWG/PE) C37.100-1992, C37.40-1993

current-limiting fuse (protection and coordination of industrial and commercial power systems) A fuse that will interrupt all available currents above its threshold current and below its maximum interrupting rating, limit the clearing time at rated voltage to an interval equal to or less than the first major or symmetrical loop duration, and limit peak let-through current to a value less than the peak current that would be possible with the fuse replaced by a solid conductor of the same impedance. Note that current-limiting action only becomes effective at a specific value of current. Underwriters Laboratories (UL) only recognizes and permits labeling of Classes G, J, L, R, CC, and T as current limiting, although Class K fuses are, in fact, current limiting. Refer to the National Electrical Code (NEC), Section 240-60b, which prohibits fuse clips for current-limiting fuses accepting noncurrent-limiting fuses. *See also:* threshold current. (IA/PSP) 242-1986r

current-limiting fuse unit (1) A fuse unit that, when its current-responsive element is melted by a current within the fuse's specified current-limiting range, abruptly introduces a high resistance to reduce current magnitude and duration, resulting in subsequent current interruption. *Notes:* 1. The values specified in standards for the threshold ratio, peak let-through current, and I^2t characteristics are used as the measures of current-limiting ability. 2. There are two classes of current-limiting fuse units—power and distribution. They are differentiated from one another by current ratings and minimum melting time-current characteristic. (SWG/PE) C37.40-1993
(2) A fuse unit that, when it is melted by a current within its specified current-limiting range, abruptly introduces a high resistance to reduce the current magnitude and duration. *Note:* There are two classes of current-limiting fuse units—power and distribution. They are differentiated one from the other by current ratings and minimum melting time current characteristics. 2. The values specified in standards for the threshold ratio, peak let-through current, and 1^2t characteristics are used as the measures of current-limiting ability. (SWG/PE) C37.100-1992

current-limiting overcurrent protective device A device that, when interrupting currents in its current-limiting range, will reduce the current flowing in the faulted circuit to a magnitude substantially less than that obtainable in the same circuit if the device were replaced with a solid conductor having comparable impedance. (NESC/NEC) [86]

current-limiting range (of a current-limiting fuse) That specified range of currents between the threshold current and the rated interrupting current within which current limitation occurs. (SWG/PE) C37.100-1992

current-limiting reactor (1) (power and distribution transformers) A reactor intended for limiting the current that can flow in a circuit under short-circuit conditions, or under other operating conditions such as starting, synchronizing, etc. *See also:* reactor. (PE/TR) C57.12.80-1978r, [57]
(2) A reactor connected in series with the phase conductors for limiting the current that can flow in a circuit under short-circuit conditions, or under other operating conditions, such as capacitor switching, motor starting, synchronizing, arc stabilization, etc. (PE/TR) C57.16-1996

current-limiting resistor A resistor inserted in an electric circuit to limit the flow of current to some predetermined value. *See also:* feedback control system. (EEC/PE/ICTL) [119]

current loss (electric instruments) In a voltage-measuring instrument, the value of the current when the applied voltage corresponds to nominal end-scale indication. *Note:* In other instruments it is the current in the voltage circuit at rated voltage. *See also:* accuracy rating. (EEC/AII) [102]

current margin The difference between the steady-state currents flowing through a receiving instrument, corresponding, respectively, to the two positions of the telegraph transmitter. *See also:* telegraphy. (EEC/PE) [119]

current master *See:* master.

current of traffic The movement of trains on a main track in one direction specified by the rules. (EEC/PE) [119]

current, peak *See:* peak current.

current phase-balance protection A method of protection in which an abnormal condition within the protected equipment is detected by the current unbalance between the phases of a normally balanced polyphase system. (SWG/PE/PSR) C37.100-1992, C37.90-1978s

current, polarization *See:* polarization current.

current position (A) The position of the electron beam on a display surface following the most recently executed display command. *Synonym:* beam position. **(B)** A CORE System value that defines the current drawing location in world coordinates. (C) 610.6-1991

current probe Used to measure dc, ac, or composite currents. DC current probes should measure dc and composite currents to within ± 1% with a probe calibrator and ± 3% without the calibrator. AC current probes should measure ac currents to within ± 5%. This accuracy should be maintained up to the worst case expected peak current. Proper bandwidth should also be ensured. (PEL) 1515-2000

current pulsation (rotating machinery) The difference between maximum and minimum amplitudes of the motor current during a single cycle corresponding to one revolution of the driven load expressed as a percentage of the average value of the current during this cycle. *See also:* asynchronous machine. (PE) [9]

current, rated *See:* rated current.

current rating (of a relay) The current at specified frequency that may be sustained by the relay for an unlimited period without causing any of the prescribed limitations to be exceeded. (PE/SWG-OLD/PSR) C37.100-1992, C37.90-1978s

current rating, 60-hertz (arresters) A designation of the range of the symmetrical root-mean-square fault currents of the system for which the arrester is designed to operate. *Notes:* 1. An expulsion arrester is given a maximum current rating and may also have a minimum current rating. 2. The designation of the maximum and minimum current ratings of an expulsion arrester not only specifies the useful operating range of the arrester between those extreme values for symmetrical root-mean-square short-circuit current, but indicates that at the point of application of the arrester the root-mean-square short-circuit current for the system should neither be greater than the maximum nor less than the minimum current rating. (PE/EEC) [119]

current ratio (series transformer) (mercury lamp) The ratio of the (root-mean-square) primary current to the root-mean-square secondary current under specified conditions of load. (EEC/LB) [98]

current-recovery ratio (arc-welding apparatus) With a welding power supply delivering current through a short-circuited resistor whose resistance is equivalent to the load setting on the power supply, and with the short-circuit suddenly removed, the ratio of (A) the minimum transient value of current upon the removal of the short-circuit to (B) the final steady-state value is the current-recovery ratio. (EEC/AWM) [91]

current regulation (1) (constant-current transformer) The maximum departure of the secondary current from its rated value, with rated primary voltage at rated frequency applied, and at rated secondary power factor, and with the current variation taken between the limits of a short-circuit and rated load. *Note:* This regulation may be expressed in per unit, or percent, on the basis of the rated secondary current. (PE/TR) C57.12.80-1978r
(2) (thyristor) The method whereby the current is controlled to a specified value. (IA/IPC) 428-1981w

current relay (1) (general) A relay that functions at a predetermined value of current. *Note:* It may be an overcurrent, undercurrent, or reverse-current relay. *See also:* relay. (EI) 1-1978s

(2) A relay that responds to current. (SWG/PE/PSR) C37.100-1992, C37.90-1978s

current-responsive element (of a fuse) That part with predetermined characteristics, the melting and severance or severances of which initiate the interrupting function of the fuse. *Note:* The current-response element may consist of one or more fusible elements combined with a strain element or other component(s), or both, that affect(s) the current-responsive characteristic. (SWG/PE) C37.100-1992, C37.40-1993

current ripple (rotating electric machinery) Current ripple, for the purposes of this standard, is defined as

$$[(I_{max} - I_{min})/(I_{max} + I_{min})] \times 100$$

expressed in percent where I_{max} and I_{min} are the maximum and minimum values of the current waveform, provided that the current is continuous. (PE/EM) 11-1980r

current-sensing resistor (power supplies) A resistor placed in series with the load to develop a voltage proportional to load current. A current-regulated direct-current power supply regulates the current in the load by regulating the voltage across the sensing resistor. (AES/PE) [41], [78]

current sensitivity (nonlinear, active, and nonreciprocal waveguide components) (diode) The output current developed by a diode detector for a specified load resistance per unit available radio-frequency (rf) input power. It is expressed in mA/mW. (MTT) 457-1982w

current, short-circuit *See:* short-circuit current.

current tap *See:* plug adapter lampholder.

current terminals (direct-current instrument shunts) Those terminals which are connected into the line whose current is to be measured and that will carry the current of the shunt. (PE/PSIM) 316-1971w

current terminals of instrument shunts (electric power system) Those terminals that are connected into the line whose current is to be measured and that will carry the current of the shunt. (PE/PSIM) [55]

current-transformation ratio The ratio of the rms value of the primary current to the rms value of the secondary current under specified conditions. (PEL/ET) 389-1990

current transformer (CT) (1) An instrument transformer that is intended to have its primary winding connected in series with the conductor carrying the current to be measured or controlled. (In window-type cts, the primary winding is provided by the line conductor and is not an integral part of the transformer.). (PE/TR/PSR) C57.13-1993, C57.12.80-1978r, C37.110-1996

(2) (metering) An instrument transformer designed for use in the measurement or control of current. *Note:* Its primary winding, which may be a single turn or bus bar, is connected in series with the load. *See also:* continuous thermal current rating factor; phase angle; instrument transformer. (ELM) C12.1-1982s

current turn-off time (gas tube surge-protective device) The time required for the arrester to restore itself to a nonconducting state following a period of conduction. This definition applies only to a condition where the arrester is exposed to a continuous specified dc potential under a specified circuit condition. (SPD/PE) C62.31-1987r, C62.32-1981s, [8]

current-type telemeter A telemeter that employs the magnitude of a single current as the translating means. (SWG/PE) C37.100-1992

current window The block of 24 *r* registers to which the current window pointer (CWP) points. (C/MM) 1754-1994

current withstand rating The maximum allowable current, either instantaneous or for a specified period of time, that a device can withstand without damage, or without exceeding the criteria of an applicable safety or performance standard. (IA/PSE) 446-1995

current working directory *See:* working directory.

curriculum standard A standard that describes the characteristics of a course of study on a body of knowledge that is offered by an educational institution. (C) 610.12-1990

cursor (1) A moveable icon or spot of light on the screen of a display device that indicates the currently selected object or character. (C) 610.2-1987

(2) (computer graphics) A cross, flashing underscore, or other symbol that represents a position on a display surface. (C) 610.6-1991w

(3) (A) In computer graphics, a cross, flashing underscore, or other symbol that represents a position on a graphics display surface. **(B)** A moveable icon or spot of light on the screen of a display device that indicates a particular object or character. **(C)** A moveable mark as in that (A) indicates the position on which the next operation will occur. (C) 610.10-1994

cursor control The ability to modify the position of a cursor by explicit commands. (C) 610.2-1987

cursor control device An inut device used to control the position of the cursor on a display device. *Synonym:* pointing device. *See also:* paddle; cursor control keypad; mouse; joystick. (C) 610.10-1994w, 610.6-1991w

cursor control key Any key on the keyboard that may be used to control a cursor function such as moving the cursor up or down a line. *See also:* home key; carriage return key. (C) 610.10-1994w

cursor control keypad A keypad comprising a set of cursor control keys such as in the following diagram of a standard inverted "T" cursor control keypad.

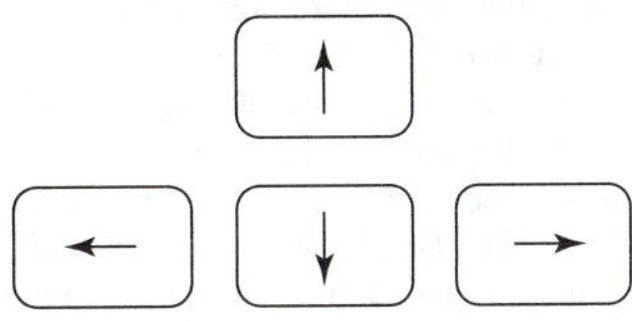

cursor control keypad

(C) 610.10-1994w

cursor position The line and column position on the screen denoted by the cursor of the terminal. (C/PA) 9945-2-1993

curvature (coaxial transmission line) The radial departure from a straight line between any two points on the external surface of a conductor. *See also:* waveguide. (EEC/REWS) [92]

curvature loss *See:* macrobend loss.

curve (A) (computer graphics) The path traced by a point moving continuously in space. *See also:* arc; closed curve. **(B) (computer graphics)** A finite set of pixels representing a path as in (A). *See also:* closed curve; arc.

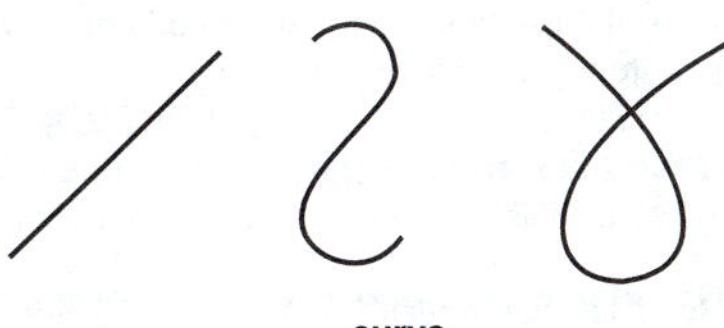

curve

(C) 610.4-1990

curve follower *See:* curve-follower function generator.

curve-follower function generator (analog computer) A function generator that operates by automatically following a curve f(x) drawn or constructed on a surface, as the input x varies over its range. (C) 165-1977w

curve, integrated energy *See:* integrated energy curve.

curve, load duration *See:* load duration curve.

curve, monthly peak duration *See:* monthly peak duration curve.

curve, reservoir operating rule *See:* reservoir operating rule curve.

cushion clamp (rotating machinery) A device for securing the cushion to the supporting member. (PE/EEC) [119]

cushioning time (speed governing of hydraulic turbines) The elapsed time during which the (closing) rate of servomotor travel is retarded by the slow closure device. *See also:* slow-closure device. (PE/EDPG/EDPG) 125-1977s, [5]

customer (1) The individual or organization that specifies and accepts the project deliverables. The customer may be internal or external to the parent organization of the project, and may or may not be the end user of the software product. A financial transaction between customer and developer is not necessarily implied. (C/SE) 1058.1-1987s
(2) The entity or entities for whom the requirements are to be satisfied in the system being defined and developed. This can be an end user of the completed system, an organization within the same company as the developing organization (e.g., System Management), a company or entity external to the developing company, or some combination of all of these. This is the entity to whom the system developer must provide proof that the system developed satisfies the system requirements specified. (C/SE) 1233-1998
(3) A person or organization ordering, purchasing, receiving, or affected by a product or process provided by the enterprise. Customers include developers, manufacturers, testers, distributors, operators, supporters, trainers, disposers, and the general public. (C/SE) 1220-1998
(4) **(A)** An individual or organization who specifies the requirements for and formally accepts delivery of a new or modified hardware or software product and its documentation; the customer may or may not be the ultimate user of the system. There are potentially many levels of customers, each with a different level of requirements to satisfy. The customer may be internal or external to the development organization for the project. *See also:* user. **(B)** An individual or organization who acts for the ultimate user of a new or modified hardware or software product to acquire the product and its documentation. *See also:* buyer. (SE/C) 1362-1998
(5) The person, or persons, for whom the product is intended, and usually (but not necessarily) who decides the requirements. (C/SE) 1219-1998
(6) The purchaser and/or user of a product or service supplied by a service provider or utility. (AMR/SCC31) 1377-1997
(7) *See also:* user. (PE/SUB) 1109-1990w

customer alert (1) (watthour meters) A switching output used to indicate a time-of-use period. (ELM) C12.13-1985s
(2) (watthour meters) A switching output used to indicate events or conditions. (ELM) C12.15-1990

customer-bank communication terminal (CBCT) *See:* automated teller machine.

customer charge (power operations) The amount paid periodically by a customer without regard to demand or energy consumption. (PE/PSE) 858-1987s, 346-1973w

customer count The number of customers or number of meters. The number of customers is the preferred item to count if the counting system is not already in place. (PE/T&D) 1366-1998

customer generation reserve (power operations) The operating reserve available through startup of customer generation. (PE/PSE) 858-1987s

customer load (power operations) Total of loads including distribution system load and losses but excluding station service, transmission losses, and pumping load. (PE/PSE) 858-1987s

customer premise equipment (CPE) (1) Equipment located on the customer's premises that is connected to the telephone line [e.g., telemetry interface units (TIUs), telephones, answering machines, and modems]. (AMR/SCC31) 1390-1995, 1390.2-1999, 1390.3-1999
(2) (telephone loop performance) Any equipment connected by customer premises wiring to the customer side of the network interface. (COM/TA) 820-1984r
(3) Any equipment connected by customer premises wiring to the customer side of the demarcation point (network interface). (IA/PSE) 1100-1999

customer premise equipment (CPE) active state Occurs when the CPE has been alerted and is performing a communication function. (SCC31) 1390.3-1999, 1390.2-1999

customer premise equipment (CPE) inactive state Occurs when the CPE has completed its communication function. (SCC31) 1390.3-1999, 1390.2-1999

customizing Modifying, or adding to, the structure of data in the PIN to tailor the general data requirements to an needs of an organization. Customizing can result in an organization-specific data model that should be cross-referenced to the PIN. (PE/EDPG) 1150-1991w

cut To duplicate text or graphic objects from the screen to the clipboard and then delete them from the screen. (C) 1295-1993w

cutback technique (fiber optics) A technique for measuring fiber attenuation or distortion by performing two transmission measurements. One is at the output end of the full length of the fiber. The other is within 1 to 3 meters of the input end, access being had by "cutting back" the test fiber. *See also:* attenuation. (Std100) 812-1984w

cut form A series of individual paper forms that feed into a printing device. *Contrast:* continuous form.

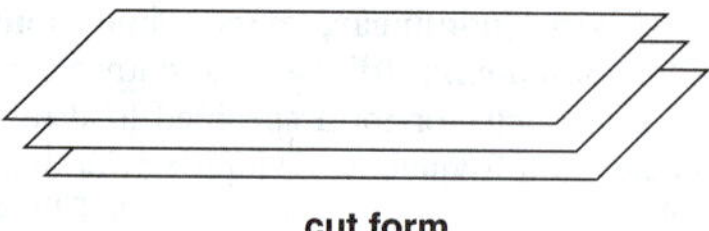

cut form

(C) 610.10-1994w

cut-in loop A circuit on the roadway energized to automatically cut in the train control or cab signal apparatus on a passing vehicle. (EEC/PE) [119]

cutoff *See:* cutoff frequency.

cutoff angle (illuminating engineering) (of a luminaire) The angle, measured up from nadir, between the vertical axis and the first line of sight at which the bare source is not visible. (EEC/IE) [126]

cutoff attenuator An adjustable length of waveguide used below its cutoff frequency to introduce variable nondissipative attenuation. *See also:* waveguide.

cutoff calls Occurs when an established connection is terminated for some reason other than an on-hook by one of the parties. In a switching system, cutoffs can be caused by hardware or software failures, procedural errors, or (in the case of time-division switching systems) digital-signal impairments such as slips, misframes, and errors. The performance measure for cutoff calls is the proportion of cutoff calls to total calls based on a holding time of 3 min. (COM) 973-1990w

cutoff characteristic *See:* current-limiting characteristic curve.

cutoff characteristic curve *See:* current-limiting characteristic curve.

cutoff frequency (1) (general) The frequency that is identified with the transition between a pass band and an adjacent attenuation band of a system or transducer. *Note:* It may be either a theoretical cutoff frequency or an effective cutoff frequency. (SP/PE/MTT/EDPG) 151-1965w, 1050-1996, 146-1980w
(2) (seismic qualification of Class 1E equipment for nuclear power generating stations) The frequency in the response spectrum where the zero period acceleration asymptote begins. This is the frequency beyond which the single-degree-of-freedom oscillators exhibit no amplification of motion, and indicate the upper limit of the frequency content of the waveform being analyzed. (PE/NP) 344-1987r
(3) (planar transmission lines) For a given mode in a planar transmission line, the frequency below which the mode is not guided by the line. For a shielded transmission line (with a finite cross section), the mode phase constant is very small

compared to the attenuation constant below cutoff, and is zero if the line is nondissipative. For an open transmission line (infinite cross section), the mode is guided above cutoff and radiating below cutoff. (MTT) 1004-1987w

(4) (nonlinear, active, and nonreciprocal waveguide components) A figure of merit for a varactor diode. It is the frequency at which Q equals 1. Its relationship to diode series resistance and junction capacitance is given by

$$f_c = \frac{1}{2\pi R_s C_j}$$

where

f_c = cutoff frequency at bias voltage V_R

R_S = series resistancce at bias voltage V_R

Cj = junction capacitance at bias voltage V_R

(MTT) 457-1982w

(5) The frequency below which a waveguide fails to transmit a signal in the differential mode. (PE/IC) 1143-1994r

(6) The frequency in the response spectrum where the zero period acceleration asymptote begins. This is the frequency beyond which the single-degree-of-freedom oscillators exhibit no amplification of input motion and which indicates the upper limit of the frequency content of the waveform being analyzed. (PE/SUB) 693-1997

(7) (of a waveguide) For a given transmission mode in a nondissipative waveguide, the frequency at which the propagation constant is 0. (PE/MTT/EDPG) 1050-1996, 146-1980w

cutoff frequency, effective *See:* effective cutoff frequency.

cutoff mode (waveguide) A nonpropagating waveguide mode such that the variation of phase along the direction of the guide is negligible. (MTT) 146-1980w

cutoff relay (telephony) A relay associated with a subscriber line, that disconnects the line relay from the line when the line is called or answered. (PE/EEC) [119]

cutoff voltage (1) (batteries) The prescribed voltage at which the discharge is considered complete. *Note:* The cutoff or final voltage is usually chosen so that the useful capacity of the battery is realized. The cutoff voltage varies with the type of battery, the rate of discharge, the temperature, and the kind of service. The term cutoff voltage is applied more particularly to primary batteries, and final voltage to storage batteries. (PE/EEC) [119]

(2) (electron tube) The electrode voltage that reduces the value of the dependent variable of an electron-tube characteristic to a specified low value. *Note:* A specific cutoff characteristic should be identified as follows: current versus grid cutoff voltage, spot brightness versus grid cutoff voltage, etc. *See also:* electrode voltage. (ED/ED) 161-1971w, [45]

(3) (magnetrons) *See also:* critical voltage.

cutoff waveguide A waveguide used as a frequency below its cutoff frequency. *See also:* waveguide. (AP/ANT) [35], [84]

cutoff wavelength (1) That wavelength, in free space or in the unbounded guide medium, as specified, above which a traveling wave in that mode cannot be maintained in the guide. *Note:* For $TE_{m,n}$ or $TM_{m,n}$ waves in hollow rectangular cylinders

$$\lambda_c = 2/[(m/a)^2 + (n/b)^2]^{1/2}$$

where a is the width of the waveguide along the x coordinate and b is the height of the waveguide along the y coordinate. *See also:* guided wave; waveguide. (AP/ANT) [35], [84]

(2) (uniconductor waveguide) The ratio of the velocity of electromagnetic waves in free space to the cutoff frequency. *See also:* waveguide. (MTT) 146-1980w

(3) (fiber optics) That wavelength greater than which a particular waveguide mode ceases to be a bound mode. *Note:* In a single mode waveguide, concern is with the cutoff wavelength of the second order mode. *See also:* mode. (Std100) 812-1984w

(4) (of a waveguide) The free-space wavelength corresponding to the cutoff frequency of the waveguide. (MTT) 146-1980w

cut out The state of being disabled by the conscious use of a cutout device or function. (VT) 1475-1999

cutout (1) (general) An electric device used manually or automatically to interrupt the flow of current through any particular apparatus or instrument. (PE/EEC) [119]

(2) An assembly of a fuse support with either a fuseholder, fuse carrier, or disconnecting blade. The fuseholder or fuse carrier may include a conducting element (fuse link), or may act as a disconnecting blade by the inclusion of a nonfusible member. *Note:* The term *cutout,* as defined here, is restricted in practice to equipment used on distribution systems. *See also:* distribution; power; distribution cutout; power fuse. (SWG/PE) C37.100-1992, [56]

(3) A device or function whose purpose is deliberately to disable a specified device or function, e.g., "dynamic brake cutout." (VT) 1475-1999

cutout base *See:* fuseholder.

cutout box (interior wiring) An enclosure designed for surface mounting and having swinging doors or covers secured directly to and telescoping with the walls of the box proper. *See also:* cabinet. (NESC/NEC/EEC/PE) [86], [119]

cutout loop (railway practice) A circuit in the roadway that cooperates with vehicle-carried apparatus to cut out the vehicle train control or cab signal apparatus. (EEC/PE) [119]

cutout type sectionalizers A single-phase automatic line sectionalizer that is very similar in outward appearance to a distribution open dropout type fuse cutout and is used in a distribution cutout mounting. (SWG/PE) C37.63-1997

cut paraboloidal reflector A reflector that is not symmetrical with respect to its axis. *See also:* antenna. (AP/ANT) 145-1983s

cut-section A location within a block other than a signal location where two adjacent track circuits end. (EEC/PE) [119]

cut-set (networks) A set of branches of a network such that the cutting of all the branches of the set increases the number of separate parts of the network, but the cutting of all the branches except one does not. *See also:* network analysis. (Std100) 270-1966w

cut-sheet feed A mechanism enabling a printer to print on multiple sheets of paper. *See also:* single-sheet feed. (C) 610.10-1994w

cutter (audio and electroacoustics) An electromechanical transducer that transforms an electric input into a mechanical output, that is typified by mechanical motions that may be inscribed into a recording medium by a cutting stylus. *See also:* phonograph pickup. (SP) [32]

cutter compensation (numerically controlled machines) Displacement, normal to the cutter path, to adjust for the difference between actual and programmed cutter radii or diameters. (EEC/IA) [74], [61]

cut-through A transmission path through the switched telephone network to an end user. (AMR/SCC31) 1390-1995, 1390.3-1999, 1390.2-1999

cutting down (electroplating) Polishing for the purpose of removing roughness or irregularities. *See also:* electroplating. (PE/EEC) [119]

cutting stylus (electroacoustics) A recording stylus with a sharpened tip that, by removing material, cuts a groove into the recording medium. *See also:* phonograph pickup. (SP) [32]

CVH *See:* clocked violation HI.

CVD *See:* chemical vapor deposition technique.

CVL *See:* clocked violation LO.

CW *See:* continuous wave.

C-weighted sound level Loudness that is measured with a sound level meter using the C-weighted filter that is built into the sound level meter. The C-weighting has only little dependence on frequency over the greater part of the audible frequency range. (PE/TR) C57.12.90-1999

cybernetics **(1)** A branch of technology concerned with the comparative study of communication and control in living organisms and in machines. (C) 610.2-1987
(2) *See also:* system science.

cycle **(1)** **(A)** An interval of space or time in which one set of events or phenomena is completed. **(B)** Any set of operations that is repeated regularly in the same sequence. The operations may be subject to variations on each repetition. (C) [20], [85]
(2) **(pulse terminology)** The complete range of states or magnitudes through which a periodic waveform or a periodic feature passes before repeating itself identically. (IM/WM&A) 194-1977w
(3) **(A)** **(data transmission)** An interval of space or time in which one set of events or phenomena is completed; any set of operations that is related regularly in the same sequence. The operations may be subject to variations on each repetition. **(B)** **(data transmission)** The complete set of values of a periodic quantity that occurs during a period. *Note:* It is one complete set of positive and negative values of an alternating current. (PE) 599-1985
(4) **(test pattern language)** A complete operation, such as writing or reading, performed by a memory. *Synonym:* period. (TT/C) 660-1986w
(5) **(A)** **(software)** A period of time during which a set of events is completed. *See also:* software life cycle; software development cycle. **(B)** **(software)** A set of operations that is repeated regularly in the same sequence, possibly with variations in each repetition; for example, a computer's read cycle. *See also:* pass. (C) 610.12-1990
(6) **(NuBus)** One period of the bus clock, from rising edge to the next rising edge. (C/MM) 1196-1987w
(7) A battery discharge followed by a complete recharge. A deep (or full) cycle is described as the removal and replacement of 80% or more of the cell's design capacity. (PE/EDPG) 1184-1994
(8) **(A)** In an ac voltage or current, exactly one complete set of positive and negative values. **(B)** Any set of operations that is repeated regularly in the same sequence. *See also:* machine cycle; instruction cycle; cycle time; read cycle; write cycle. **(C)** To perform, or cause to perform, one set of operations as in definition (B). **(D)** An interval of space or time in which one set of operations as in definition (B) is completed. (C) 610.10-1994
(9) The complete series of values of a periodic quantity that occurs during a period. (It is one complete set of positive and negative values of an alternating current.). (IA/MT) 45-1998

®NuBus is a registered trademark of Texas Instruments, Inc.

cycle counter *See:* index register.

cycle life The number of cycles (discharges and recharges), under specified conditions, that a battery can undergo before failing to meet its specified end-of-life capacity. (PV) 1013-1990, 1144-1996

cycle master **(1)** The node that generates the periodic cycle start. (C/MM) 1394-1995
(2) The node that generates the periodic cycle start packet 8000 times a second. (C/MM) 1394a-2000

cycle of operation **(1)** The discharge and subsequent recharge of the cell or battery to restore the initial conditions. *See also:* charge. (EEC/PE) [119]
(2) The movement of the LTC from one end of its range to the other and back to its original position. (PE/TR) C57.131-1995

cycle start A primary packet sent by the cycle master that indicates the start of an isochronous cycle. (C/MM) 1394-1995

cycle start packet A primary packet sent by the cycle master that indicates the start of an isochronous period. (C/MM) 1394a-2000

cycle stealing The process of suspending the operation of a central processing unit for one or more cycles to permit the occurrence of other operations, such as transferring data from main storage in response to an output request from an input-output controller. (C) 610.10-1994w, 610.12-1990

cycle termination The phase of a cycle during which the master terminates the cycle, and slaves acknowledge this termination by establishing the intercycle state of bus signals. (C/MM) 1096-1988w

cycle time The minimum amount of time between the start of successive read or write cycles of a storage device. *See also:* write cycle time; read cycle time. (C) 610.10-1994w

cyclically magnetized condition A condition of a magnetic material when, under the influence of a magnetizing force that is a cyclic (but not necessarily periodic) function of time having one maximum and one minimum per cycle, it follows identical hysteresis loops on successive cycles. (Std100) 270-1966w

cyclic binary code *See:* Gray code.

cyclic code *See:* Gray code.

cyclic code error detection (power-system communication) The process of cyclically computing bits to be added at the end of a word such that an identical computation will reveal a large portion of errors that may have been introduced in transmission. *See also:* digital. (PE) 599-1985w

cyclic decimal code A binary code in which sequential decimal digits are represented by four-bit BCD expressions, each of which differs from the preceding expression in one place only. *Note:* This is an example of unit-distance code. (C) 1084-1986w

cyclic duration factor (rotating machinery) The ratio between the period of loading including starting and electric braking, and the duration of the duty cycle, expressed as a percentage. *See also:* asynchronous machine; direct-current commutating machine. (PE) [9]

cyclic function A function that repetitively assumes a given sequence of values at an arbitrarily varying rate. *Note:* That is, if y is a periodic function of x and x in turn is a monotonic nondecreasing function of t, then y is said to be a cyclic function of t. (Std100) 270-1966w

cyclic irregularity (rotating machinery) The periodic fluctuation of speed caused by irregularity of the prime-mover torque. *See also:* direct-current commutating machine; asynchronous machine. (PE) [9]

cyclic permuted code *See:* unit-distance code.

cyclic redundancy (check) code Defined for some digital transmission formats (usually stated with the number of bits in the code; e.g., CRC6, CRC9, etc.). The CRC is the result of a calculation carried out on the set of transmitted bits by the transmitter. The CRC is encoded into the transmitted signal with the data. At the receiver, the calculation creating the CRC may be repeated, and the result compared to that encoded in the signal. The calculations are chosen to optimize the error detection capability. (COM/TA) 1007-1991r

cyclic redundancy check (CRC) **(1)** A form of error check used to ensure the accuracy of transmitting a message. *Note:* The CRC is the result of a calculation carried out on the set of transmitted bits by the transmitter. The CRC is encoded into the transmitted signal with the data. At the receiver, the calculation creating the CRC may be repeated, and the result compared to that encoded in the signal. The calculations are chosen to optimize the error detection capability. *Contrast:* parity check; parity. *See also:* frame check sequence; frame check sequence error. (C) 610.7-1995
(2) An error-detection scheme that checks the integrity of a transmitted message for errors introduced during transmission. (PE/SUB) 1379-1997
(3) The result of a calculation carried out on the octets within an IrLAP frame; also called a frame check sequence. The CRC is appended to the transmitted frame. At the receiver, the calculation creating the CRC may be repeated, and the

result compared to that encoded in the signal. *Synonym:* frame check sequence. (EMB/MIB) 1073.3.2-2000

cyclic search A storage allocation technique in which each search for a suitable block of storage begins with the block following the one last allocated. (C) 610.12-1990

cyclic shift (1) An operation that produces a word whose characters are obtained by a cyclic permutation of the characters of a given word. (C) 162-1963w, 270-1966w
(2) A shift in which the data moved out of one end of the storing register are reentered into the other end, as in a closed loop. *See also:* circulating register. (C) [20], [85]
(3) *See also:* circular shift. (C) 1084-1986w

cyclic storage *See:* circulating storage.

cycling The repeated charge/discharge cycle of a storage battery. Some batteries are rated by their ability to withstand repeated, deep discharge cycles. (PE/EDPG) 1184-1994

cyclize To drive a tester, data must be provided in uniform, consistent, repeatable collections. These collections are termed "cycles" or "tester cycles." The process of constructing these collections, generally from simulation environments, is called "cyclizing." (C/TT) 1450-1999

cycloconverter/synchroconverter A converter using controlled rectifier or transistor devices that has the capability of adjusting the frequency and proportional voltage of the output waveform to provide speed control of motors. (IA/MT) 45-1998

cyclometer register A set of four or five wheels numbered from zero to nine inclusive on their edges, and so enclosed and connected by gearing that the register reading appears as a series of adjacent digits. *See also:* watthour meter. (EEC/PE) [119]

cyclotron A device for accelerating positively charged particles (for example, protons, deuterons, etc.) to high energies. The particles in an evacuated tank are guided in spiral paths by a static magnetic field while they are accelerated many times by an electric field of mixed frequency. (ED) [45]

cyclotron frequency *See:* gyro-frequency.

cyclotron-frequency magnetron oscillations Those oscillations whose frequency is substantially the cyclotron frequency. (ED) 161-1971w

cyclotron, frequency-modulated *See:* frequency-modulated cyclotron.

cylinder In an assembly of magnetic disks, the set of all tracks that can be accessed by all the magnetic heads at a given fixed position. (C) 610.10-1994w

cylindrical antenna* *See:* cylindrical dipole; cylindrical array.

* Deprecated.

cylindrical array A two-dimensional array of elements whose corresponding points lie on a cylindrical surface. (AP/ANT) 145-1993

cylindrical dipole (antenna) A dipole, all of whose transverse cross sections are the same, the shape of a cross section of a cylinder being circular. (AP/ANT) 145-1993

cylindrical reflector A reflector that is a portion of a cylindrical surface. *Note:* The cylindrical surface is usually parabolic, although other shapes may be employed. (AP/ANT) 145-1993

cylindrical-rotor generator An alternating-current generator driven by a high-speed turbine (usually steam) and having an exciting winding embedded in a cylindrical steel rotor. (PE) [9]

cylindrical wave A wave whose equiphase surfaces form a family of coaxial cylinders. (AP/PROP) 211-1997

CYPHERTEXT A text-formatting language commonly used for typesetting. (C) 610.13-1993w

cytac (navigation aid terms) The designation of loran C in an earlier stage of development. (AES/GCS) 172-1983w

D

D/A *See:* digital-to-analog converter.

DA *See:* data administrator; design automation; destination address.

DAA *See:* direct access arrangement.

DAC *See:* digital-to-analog converter.

D/A converter *See:* digital-to-analog converter.

daily cycle One complete execution of a data processing function that must be performed once a day. *See also:* monthly cycle; annual cycle; weekly cycle. (C) 610.2-1987

daisy-chain **(1)** A special type of signal line that is used to propagate a signal level from printed-circuit board (pcb) to pcb, starting with the first slot and ending with the last slot. There are four bus grant daisy-chains and one interrupt acknowledge daisy-chain on the backplane. (C/BA) 1014-1987

(2) A special type of signal line that is used to propagate bus grants from board to board, starting with the board installed in the first slot and ending with the one installed in the last slot. (C/MM) 1096-1988w

(3) (FASTBUS acquisition and control) A backplane connection between adjacent module stations that allows information to flow between modules independent of the FASTBUS protocol. (NID) 960-1993

daily load curves (electric power supply) Curves of net 60-minute integrated demand for each clock hour of a 24-hour day. *See also:* generating station. (PE/PSE) [54]

daisy wheel printer A wheel printer in which the type slugs are mounted on a "daisy wheel," a central hub with numerous spring fingers each of which is embossed with one character. (C) 610.10-1994w

DAM *See:* digital-to-analog multiplier; direct access method.

damage Attacks against system components (e.g., hardware, distribution system, media, software) todestroy the component or render it useless. Damage could be perpetrated inadvertently or intentionally using physical, electrical, or environmental (e.g., fire, water exposure) means, and could result in denial of service conditions. (C/BA) 896.3-1993w

damped filter A filter generally consisting of combinations of capacitors, inductors, and resistors that have been selected in such a way as to present a low impedance over a broad range of frequencies. The filter usually has a relatively low *Q* (*X*/*R*). (IA/SPC) 519-1992

damped frequency (automatic control) The apparent frequency of a damped oscillatory time response of a system resulting from a nonoscillatory stimulus. *Note:* The value of the frequency in a particular system depends somewhat on the subsidence ratio. *See also:* feedback control system. (IM/PE/EDPG) [120], [3]

damped harmonic system (1) (linear system with one degree of freedom) A physical system in which the internal forces, when the system is in motion, can be represented by the terms of a linear differential equation with constant coefficients, the order of the equation being higher than the first. Example: The differential equation of a damped system is often of the form

$$M\frac{d^2x}{dt^2} + F\frac{dx}{dt} + Sx = f(t)$$

where M, F, and S are positive constants of the system: x is the dependent variable of the system (displacement in mechanics, quantity of electricity, etcetera): and $f(t)$ is the applied force. Examples: A tuning fork is a damped harmonic system in which M represents a mass, F a coefficient of damping, S a coefficient of restitution, and x a displacement. Also, an electric circuit containing constant inductance, resistance, and capacitance is a damped harmonic system, in which case M represents the self-inductance of the circuit, F the resistance of the circuit, S the reciprocal of the capacitance, and x the charge that has passed through a cross section of the circuit.

(2) (critical damping) The name given to that special case of damping that is the boundary between underdamping and overdamping. A damped harmonic system is critically damped if F^2, $4Ms$. *See also:* network analysis.

(3) (overdamping) (aperiodic damping) The special case of a damping in which the free oscillation does not change sign. A damped harmonic system is overdamped if $F^2 > 4$ MS. *See also:* network analysis.

(4) (periodic damping) (underdamping) The special case of damping in which the free oscillation changes sign at least once. A damped harmonic system is underdamped if $F^2 < 4$ MS. 270-1966w

(5) (underdamped) Damped insufficiently to prevent oscillation of the output following an abrupt input stimulus. *Note:* In an underdamped linear second-order system, the roots of the characteristic equation have complex values. *See also:* feedback control system; critical damping; control; damping. (IA/APP/IAC) [69], [60]

(6) (anode) (x-ray tubes) (anticathode) An electrode, or part of an electrode, on which a beam of electrons is focused and from which x rays are emitted. *See also:* radar; electrode. (ED) [45]

damper bar *See:* amortisseur bar.

damper segment (rotating machinery) One portion of a short-circuiting end ring (of an amortisseur winding) that can be separated into parts for mounting or removal without access to a shaft end. *See also:* rotor. (PE) [9]

damper winding (rotating machinery) (amortisseur winding) A winding consisting of a number of conducting bars short-circuited at the ends by conducting rings or plates and distributed on the field poles of a synchronous machine to suppress pulsating changes in magnitude or position of the magnetic field linking the poles. (PE) [9]

damping (1) The temporal decay of the amplitude of a free oscillation of a system, associated with energy loss from the system.

(2) Pertaining to or productive of damping. *Note:* The damping of many physical systems is conveniently approximated by a viscous damping coefficient in a second-order linear differential equation (or a quadratic factor in a transfer function). In this case the system is said to be critically damped when the time response to an abrupt stimulus is as fast as possible without overshoot; underdamped (oscillatory) when overshoot occurs; overdamped (aperiodic) when response is slower than critical. The roots of the quadratic are, respectively, real and equal: complex, and real and unequal.

(3) (relative underdamped system) A number expressing the quotient of the actual damping of a second-order linear system or element by its critical damping. *Note:* For any system whose transfer function includes a quadratic factor $s^2 + 2z\omega_n s + \omega_n^2$, relative damping is the value of z, since $z + 1$ for critical damping. Such a factor has a root — $\sigma + j\omega$ in the complex s plane, from which $z = \sigma/\omega_n = \sigma/(\sigma^2 + \omega^2)^{1/2}$.

(4) (coulomb) That due to Coulomb friction.

(5) (instrument) Term applied to the performance of an instrument to denote the manner in which the pointer settles to its steady indication after a change in the value of the measured quantity. Two general classes of damped motion are distinguished as follows: periodic (underdamped) in which the pointer oscillates about the final position before coming to rest, and aperiodic (overdamped) in which the pointer comes to rest without overshooting the rest position. The point of change between periodic and aperiodic damping is called critical damping. *Note:* An instrument is considered for practical purposes to be critically damped when overshoot is

present but does not exceed an amount equal to one-half the rated accuracy of the instrument. (IA/IAC) [60]
(6) The generic term ascribed to the numerous complex energy dissipating mechanisms in a system. As an identifying parameter of a specific seismic response spectrum, the percent of critical damping is assumed to be constant. (SWG/PE) C37.81-1989r
(7) A dynamic property of a vibrating structure that indicates its ability to dissipate mechanical energy. The phenomenon of damping is represented by a quantity called the damping factor, which is expressed as a percentage of critical damping. After being forced to deflect and allowed to freely vibrate, structures with zero damping will vibrate with a harmonic motion indefinitely. Structures with critical damping will creep back to their static or neutral position with no velocity reversal. (SWG/PE) C37.100-1992
(8) An energy dissipation mechanism that reduces the response amplification and broadens the vibratory response over frequency in the region of resonance. Damping is usually expressed as a percentage of critical damping. *See also:* critical damping. (PE/SUB/NP) 693-1997, 344-1987r

damping amortisseur An amortisseur the primary function of which is to oppose rotation or pulsation of the magnetic field with respect to the pole shoes. (EEC/PE) [119]

damping fluid (accelerometer) (gyros) (inertial sensors) A fluid that provides viscous damping forces or torques to the inertial sensing element. *See also:* flotation fluid. (AES/GYAC) 528-1994

damping magnet A permanent magnet so arranged in conjunction with a movable conductor such as a sector or disk as to produce a torque (or force) tending to oppose any relative motion between them. *See also:* moving element. (EEC/PE) [119]

damping torque (synchronous machines) The torque produced, such as by action of the amortisseur winding, that opposes the relative rotation, or changes in magnitude, of the magnetic field with respect to the rotor poles. (PE) [9]

damping torque coefficient (synchronous machines) A proportionality constant that, when multiplied by the angular velocity of the rotor poles with respect to the magnetic field, for specified operating conditions, results in the damping torque. (PE) [9]

damp location Partially protected locations under canopies, marquees, roofed open porches, and like locations, and interior locations subject to moderate degrees of moisture, such as some basements, some barns, and some cold-storage warehouses. (NESC/NEC) [86]

danger buoy (navigation aid terms) Classified as: obstruction, wreck, telegraph, cable, fish net, dredging. *See also:* buoy. (AES/GCS) 172-1983w

dap To cut and form a recess in timbers for making a joint. (T&D/PE) 751-1990

DARE *See:* Differential Analyzer Replacement.

dark adaptation (illuminating engineering) The process by which the retina becomes adapted to a luminance less than about 0.034 cd/m^2, (2.2 $\times$ 10^{-5} cd/in^2), (0.01 fL). (EEC/IE) [126]

dark current (1) (diode-type camera tube) The current that flows in the output lead of the target in the absence of any external irradiation. Units: amperes. (ED) 503-1978w
(2) (fiber optics) The external current that, under specified biasing conditions, flows in a photosensitive detector when there is no incident radiation. (Std100) 812-1984w
(3) (photoelectric device) The current flowing in the absence of irradiation. *See also:* electrode; photoelectric effect; dark current; dark-current pulses. (ED/NPS) 161-1971w, 398-1972r, [84]

dark-current pulses (phototubes) Dark-current excursions that can be resolved by the system employing the phototube. *See also:* phototube. (NPS) 175-1960w

darkening (electroplating) The production by chemical action, usually oxidation, of a dark colored film (usually a sulfide) on a metal surface. *See also:* electroplating. (PE/EEC) [119]

dark pulses Pulses observed at the output electrode when the photomultiplier is operated in total darkness. These pulses are due primarily to electrons originating at the photocathode. (NPS) 398-1972r

dark space, cathode *See:* cathode dark space.

dark space, Crookes *See:* cathode dark space.

dark-trace screen (cathode-ray tubes) A screen giving a spot darker than the remainder of the surface. *See also:* cathode-ray tube. (Std100) [84]

dark-trace tube (1) (electronic navigation) A cathode-ray tube having a special screen that changes color but does not necessarily luminesce under electron impact, showing, for example, a dark trace on a bright background. *See also:* cathode-ray tube; navigation. (Std100) [84]
(2) A type of cathode-ray tube having a bright face, on which signals are displayed as dark traces or dark blips; sometimes used as a storage tube or long-persistence display because the dark traces remain on the screen until erased by heat or electron bombardment. This device is now obsolete. *Synonym:* skiatron. (AES/RS) 686-1990

dark trace tube display device A CRT display device whose electron beam causes the display surface of the tube to darken rather than to brighten. For example, the image may be viewed by illumination from the rear as a reverse image against the otherwise transparent or translucent face of the tube. (C) 610.10-1994w

D'Arsonval current (medical electronics) (solenoid current) The current of intermittent and isolated trains of heavily damped oscillations of high frequency, high voltage, and relatively low amperage. *See also:* D'Arsonvalization. (EMB) [47]

D'Arsonvalization (medical electronics) The therapeutic use of intermittent and isolated trains of heavily damped oscillations of high frequency, high voltage, and relatively low amperage. *Note:* This term is deprecated because it was initially ill-defined and because the technique is not of contemporary interest. (EMB) [47]

dart leader The downward leader of a subsequent stroke of a multiple-stroke lightning flash. (SUB/PE) 998-1996

DASD *See:* direct-access storage device.

D* *See:* D-star.

data (1) (programmable digital computer systems in safety systems of nuclear power generating stations) A representation of facts, concepts, or instructions in a formalized manner suitable for communication, interpretation or processing by a programmable digital computer. (C) 610.10-1994w
(2) (supervisory control, data acquisition, and automatic control) (station control and data acquisition) Any representation of a digital or analog quantity to which meaning has been assigned. (SWG/SUB/PE) 999-1992w, C37.1-1994, C37.100-1992
(3) (test pattern language) The binary information that is stored in or read out of a memory array. (TT/C) 660-1986w
(4) (A) (data management) (software) A representation of facts, concepts, or instructions in a manner suitable for communication, interpretation, or processing by humans or by automatic means. *Note:* "Data" is plural for datum, but is often used as a collective noun, as in "The data is in this file." *See also:* pointer data; logical data; numeric data; null data; data type. **(B) (data management) (software)** Anything observed in the documentation or operation of software that deviates from expectations based on previously verified software products or reference documents. *Synonym:* documentation. (C) 610.5-1990
(5) A representation of facts, concepts, or instructions in a formalized manner suitable for communication, interpretation, or processing by humans or by automatic means. (C/DIS) 1278.4-1997

(6) Representations of static or dynamic entities in a formalized manner suitable for communication, interpretation, or processing by humans or by machines. (SCC32) 1489-1999

data abstraction (A) (software) The process of extracting the essential characteristics of data by defining data types and their associated functional characteristics and disregarding representation details. *See also:* encapsulation; information hiding. **(B) (software)** The result of the process in definition (A). (C) 610.12-1990

data-access operation A processor-initiated load, store, or lock that involves a data-format copy and (for lock operations) a data-update action (such as swap or add). (C/MM) 1596.5-1993

data access register A register that is used for arithmetic associated with random-access of data. (C) 610.10-1994w

data acquisition (supervisory control, data acquisition, and automatic control) (station control and data acquisition) The collection of data. (SWG/PE/SUB) 999-1992w, C37.1-1994, C37.100-1992

data acquisition system (1) (supervisory control, data acquisition, and automatic control) (station control and data acquisition) A system that receives data from one or more locations. *See also:* telemetering. (PE/SUB) C37.1-1994
(2) A centralized system that receives data from one or more remote points—a telemetering system. Data may be transported by either analog or digital telemetering. (SWG/PE) C37.100-1992

data administrator An individual who is responsible for the definition, organization, supervision, and protection of data within some organization. *See also:* database administrator. (C) 610.5-1990w

data aggregate A collection of two or more data items that are treated as a unit. *Synonyms:* aggregate; group item. *See also:* composite data element. (C) 610.5-1990w

data attribute A characteristic of a unit of data. (C) 610.5-1990w

data bank (A) A collection of data libraries. *Note:* A record contains one or more items, a file contains one or more records, a library contains one or more files, and a data bank contains one or more libraries. **(B)** A collection of data relating to a particular subject area. *Note:* The data may or may not be machine-readable. (C) 610.5-1990

data bar polling An end-of-write indicator. (ED) 1005-1998

database (DB) (1) (A) (data management) (software) A collection of logically related data stored together in one or more computerized files. *Note:* Each data item is identified by one or more keys. *See also:* database management system. **(B) (data management) (software)** In CODASYL, the collection of all the record occurrences, set occurrences, and areas controlled by a specific schema. (C) 610.5-1990
(2) A collection of data fundamental to a system. (C/SE) 1074-1995s
(3) A collection of related data stored in one or more computerized files in a manner that can be accessed by users or computer programs via a database management system. (C/SE) J-STD-016-1995
(4) A collection of interrelated data, often with controlled redundancy, organized according to a schema to serve one or more applications; the data are stored so that they can be used by different programs without concern for the data structure or organization. A common approach is used to add new data and to modify and retrieve existing data. (C/DIS) 1278.4-1997

database access method A technique for organizing and storing a physical database in computer storage. (C) 610.5-1990w

database administration (DBA) The responsibility for the definition, operation, protection, performance, and recovery of a database. (C) 610.5-1990w

database administrator (DBA) An individual who is responsible for the definition, operation, protection, performance, and recovery of a database. *See also:* data administrator. (C) 610.5-1990w

database command language (DBCL) A procedural data manipulation language used to access a database through a database management system. *See also:* database manipulation language. (C) 610.5-1990w

database creation The process of naming, allocating space, formatting, and defining a database. *See also:* database definition; database design. (C) 610.5-1990w

database definition (A) The process of translating a conceptual schema for a database into a data storage schema. *See also:* redefinition; database design; database creation. **(B)** The result of such a translation. (C) 610.5-1990

database description language (DBDL) *See:* data definition language.

database design (A) The process of developing a conceptual schema for a database that will meet a user's requirements. *Synonym:* implementation design. *See also:* database creation; database definition. **(B)** The result of the process in definition (A). (C) 610.5-1990

database engine A software engine that is specially designed for database applications; performs low-level database operations such as record creation, editing, and deletion. *See also:* relational engine. (C) 610.10-1994w

database extract A file, each record of which contains data items selected from a database based on a particular criterion. (C) 610.5-1990w

database integrity The degree to which the data in a database are current, consistent and accurate. *See also:* data integrity; database security; integrity. (C) 610.5-1990w

database key A field in a database that identifies a record in that database. (C) 610.5-1990w

database management system (DBMS) (1) A computer system involving hardware, software, or both that provides a systematic approach to creating, storing, retrieving and processing information stored in a database. A DBMS acts as an interface between computers' programs and data files as well as between users and the database. It may include backup/recovery, checkpoint processing, and ad-hoc query capability. (C) 610.5-1990w
(2) An integrated set of computer programs that provide the capabilities needed to establish, modify, make available, and maintain the integrity of a database. (C/SE) J-STD-016-1995

database manipulation language (DBML) *See:* data manipulation language.

database organization The manner in which a database is structured; for example, a hierarchical organization, a relational organization. *See also:* reorganization. (C) 610.5-1990w

database record (A) A collection of data elements that are stored in a database. *See also:* record. **(B)** A collection of hierarchically dependent segments (one root and all its descendants) within a hierarchical database. *See also:* record. (C) 610.5-1990

database reorganization *See:* reorganization.

database security The degree to which a database is protected from exposure to accidental or malicious alteration or destruction. *See also:* database integrity; data security. (C) 610.5-1990w

database segment *See:* segment.

database server On a network, a server that provides access to a database at the record level; that is, the server sends and locks only the records affected by a particular requestor. *See also:* file server; disk server; mail server; terminal server; network server; print server. (C) 610.7-1995

database sublanguage *See:* data sublanguage.

database system A software system that supports multiple applications using a common database. (C) 610.5-1990w

Database Task Group (DBTG) A task group of the CODASYL Programming Language Committee that established a set of standards for specification and design of network database structures. *See also:* CODASYL database. (C) 610.5-1990w

data bit (1) The smallest signaling element used by the physical layer for transmission of packet data on the medium. One of the PDUs for the physical layer (the other is the arbitration signal). (C/MM) 1394-1995
(2) A single entity of information that is transmitted across a serial signalling medium. A bit assumes one of two values: logic "0" or logic "1." A data bit may convey control, address, information, or frame check sequence (FCS) data. (EMB/MIB) 1073.4.1-2000

data block *See:* block.

data-break *See:* direct memory access.

data breakpoint A breakpoint that is initiated when a specified data item is accessed. *Synonym:* storage breakpoint. *Contrast:* code breakpoint. *See also:* dynamic breakpoint; static breakpoint; prolog breakpoint; programmable breakpoint; epilog breakpoint. (C) 610.12-1990

data broadcall An operation wherein participating slaves capture the data that are placed on the data lines by the responding slave during a read cycle. (C/MM) 1096-1988w

data broadcast An operation wherein participating slaves capture the data that are placed on the data lines by the active master during a write cycle. (C/MM) 1096-1988w

data buffer register A register in a central processing unit or peripheral device capable of receiving or transmitting data at different data transfer rates. *See also:* input buffer register. (C) 610.10-1994w

data bus A bus used to communicate data to and from a processing unit or a storage device. *See also:* bidirectional bus. (C) 610.10-1994w

data cache An area of high-speed buffer storage, used to store data and operands. *Contrast:* instruction cache. (C) 610.10-1994w

data card A punch card that contains data to be used by a computer program. *See also:* source data card. (C) 610.10-1994w

data carrier Material that serves as a data medium or to which a data medium is applied and that facilitates the transport of data; for example, a punch card, a disk, or a plastic card with a magnetic surface that serves as the data medium. *See also:* data medium. (C) 610.5-1990w, 610.10-1994w

data cell *See:* storage cell.

data certification The determination that data have been verified and validated. *See also:* data user certification; data producer certification. (DIS/C) 1278.3-1996

data chain *See:* composite data element.

data chain bus A connection by which electrical signals are transmitted and/or received at multiple circuit elements. (C) 610.10-1994w

data change An event in which at least one bit of data is caused to change. *Note:* This event may be used as a unit of endurance for erasable programmable read-only memories. (ED) 1005-1998

data channel *See:* input-output channel.

data channels (test pattern language) All memory devices have one or more (up to 16) independent data inputs or outputs. Each of these is called a data channel. (TT/C) 660-1986w

data character A character used for packet payload or packet header. A data character represents one of the values of a byte, i.e., 0–255 (decimal). Only N_chars are used as data characters. *See also:* link character; normal character. (C/BA) 1355-1995

data characteristic (software) (software unit testing) An inherent, possibly accidental, trait, quality, or property of data (for example, arrival rates, formats, value ranges, or relationships between field values). (C/SE) 610.12-1990, 1008-1987r

data circuit A circuit used to transmit data. *Synonym:* duplex circuit. (C) 610.7-1995

data circuit-terminating equipment (DCE) (1) A device that provides the signal conversion and coding between the data terminal equipment (DTE) and the network carrier facility. Note that in the context of an ITU-T X.25 network, for example, the DCE performs functions at the network end of an access line to the network. (C/LM/COM) 802.9a-1995w, 8802-9-1996
(2) A device that interfaces between the data terminal equipment (DTE) and the line. (C) 610.7-1995

data code *See:* code.

data-collection application (DCA) A software application that acquires data from and sends data to various intelligent electronic devices. (PE/SUB) 1379-1997

data collection station *See:* data input station.

data communication equipment (1) The equipment that provides the functions required to establish, maintain, and terminate a connection, as well as the signal conversion, and coding required for communication between data terminal equipment and data circuit. (COM) 168-1956w
(2) An equipment that transmits data from one point to another. (C) 610.7-1995

data communications (1) (data transmission) The movement of encoded information by means of communications techniques. (PE) 599-1985w
(2) A data transfer between data source and data destination via one or more data links. (C) 610.7-1995

data compaction Any technique used to encode data in order to reduce the amount of storage it requires. *Contrast:* data compression. (C) 610.5-1990w

data compression Any technique used to reduce the amount of storage required to store data. *Contrast:* data compaction. (C) 610.5-1990w

data concentrator A concentrator that permits a common transmission medium to serve more data sources than there are channels available within the transmission medium. (C) 610.7-1995

data concept Any of a group of data dictionary structures defined in this standard (e.g., data element, data element concept, entity type, property, value domain, and generic property domain) referring to abstractions or things in the natural world that can be identified with explicit boundaries and meaning, and whose properties and behavior all follow the same rules. (SCC32) 1489-1999

data connection The interconnection of two or more data circuits by means of switching equipment to enable data transmission to take place between DTEs. *See also:* virtual data connection. (C) 610.7-1995

data conversion (1) To change data from one form of representation to another; for example, to convert data from an ASCII representation to an EBCDIC representation. (C) 610.5-1990w
(2) The translation of data from one numeric form into another (e.g., converting a digital-to-analog converter input bit stream into a voltage). (IM/ST) 1451.2-1997

data converter A device whose purpose it is to convert data from one representation to an equivalent representation. (C) 610.10-1994w

data coupling A type of coupling in which output from one software module serves as input to another module. *Synonym:* input-output coupling. *Contrast:* control coupling; hybrid coupling; content coupling; common-environment coupling; close coupling. (C) 610.12-1990

data cycle (1) (A) (FASTBUS acquisition and control) The portion of a FASTBUS operation in which a master either sends data to or receives data from an attached slave. It begins with the master causing a data sync transition and terminates with the master receiving a data acknowledge transition from the slave. **(B)** A period in which data are valid and are acknowledged. This occurs when acknowledge is asserted at the end of a transaction and on intermediate acknowledges during a block transfer. (C/MM/NID) 960-1993, 1196-1987

(2) A cycle in which each bit changes to its opposite state and back to its original state. *Notes:* 1. These changes may occur for all bits in parallel or series, e.g., by page, block, word, byte, or bit. 2. This cycle may be used as a unit of endurance for erasable programmable read-only memories. (ED) 1005-1998

data deciphering key A key used for the decipherment of an (N)-layer SDU. (It is not used to decipher other keys.). (C/LM) 802.10-1998

data declaration source statements Source statements that reserve or initialize memory at compilation time. (C/SE) 1045-1992

data definition A description of the format, structure, and properties of a data item, data element, or data structure. (C) 610.5-1990w

data definition language (DDL) **(A)** A language for describing the organization of data within a database. *Note:* In some software, the logical organization is described; in some, both the logical and physical organizations are described. **(B)** A language used to describe the logical structure of a database. *Synonyms:* database description language; schema language; data description language; schema definition language. *Contrast:* data manipulation language. *See also:* database manipulation language. (C) 610.5-1990

data density The amount of data that can be stored in one unit of data medium. For example, the number of bits stored in an inch of magnetic tape medium. (C) 610.5-1990w

data description language *See:* data definition language.

data dictionary **(1)** **(A)** **(software)** A collection of the names of all data items used in a software system, together with relevant properties of those items, for example, length of data item, representation, etcetera. **(B)** **(software)** A set of definitions of data flows, data elements, files, data bases, and processes referred to in a leveled data flow diagram set. *See also:* data; file; data flow diagram; database. (C/SE) 729-1983
(2) **(data management)** A collection of entries specifying the name, source, usage, and format of each data element used in a system or set of systems. *Synonym:* data element dictionary. *See also:* data directory. (C) 610.5-1990w
(3) An information technology for documenting, storing, and retrieving the syntactical form (i.e., representational form) and some semantics of data elements and other data concepts. (SCC32) 1489-1999

data dictionary/directory (DD/D) *See:* data dictionary.

data dictionary system A software system that maintains and manages a data dictionary. (C) 610.5-1990w

data directory A collection of entries specifying the data name, source, location, ownership, usage and format of each data element used in some system or set of systems. *See also:* data dictionary. (C) 610.5-1990w

data element **(1)** **(A)** A uniquely named and defined component of a data definition; a data "cell" into which data items (actual values) can be placed. For example, the data element AGE, into which data items 1, 2, ... can be placed. *Note*: The terms **data element** and **data item** are often used interchangeably or with the reverse definitions from those given here. No standard of use exists at this time. *Synonym:* cell. *See also:* data item; attribute. **(B)** A data definition as in definition (A) that cannot be divided into other individually named data definitions. *See also:* data item; attribute. (C) 610.5-1990
(2) A syntactically formal representation of some single unit of information of interest (such as a fact, proposition, or observation), with a singular instance value at any point in time, about some entity of interest (e.g., a person, place, process, property, object, concept, association, state, event). A data element is considered indivisible in a certain context. *Note:* An example of a data element is "ROUTE.BUS_Stop_location". (SCC32) 1489-1999

data element concept An expression of the inherent concept embodied in a data element without regard to the value domain(s) by which it can be physically represented. *Note:* An example of a data element concept is "ROUTE.BUS_Stop". (SCC32) 1489-1999

data element dictionary *See:* data dictionary.

data element tag *See:* code.

data enciphering key A key used for the encipherment of an (N)-layer SDU. (It is not used to encipher other keys.). (C/LM) 802.10-1998

data encryption The changing of the form of a data stream such that only the intended recipient can read or alter the information and detect unauthorized messages. (AMR/SCC31) 1377-1997

data encryption standard (DES) A private key cryptography standard promoted by the United States government for use with private or unclassified information. (C/BA) 896.3-1993w

data entry To input data into a computer system. (C) 610.5-1990w

data entry device An input device used to prepare data so that a computer can accept it. For example, a keyboard, or barcode scanner. (C) 610.10-1994w

data exception An exception that occurs when a program attempts to use or access data incorrectly. *See also:* protection exception; addressing exception; overflow exception; operation exception; underflow exception. (C) 610.12-1990

data exchange **(A)** The use of data by more than one computer program or system. **(B)** The movement of data between two or more programs or systems. *See also:* exchange data; data interchange. (C) 610.5-1990

data field *See:* attribute.

data file *See:* file.

data flow The sequence in which data transfer, use, and transformation are performed during the execution of a computer program. *Contrast with:* control flow. (C) 610.12-1990

data flow architecture A computer architecture in which execution is controlled only by the data needed for that operation and not the order in which instructions are stored in memory. *Contrast:* control flow architecture. (C) 610.10-1994w

data flowchart *See:* data flow diagram.

data flow diagram (DFD) **(software)** A diagram that depicts data sources, data sinks, data storage, and processes performed on data as nodes, and logical flow of data as links between the nodes. *Synonyms:* data flow graph; data flowchart. *Contrast:* data structure diagram; control flow diagram.

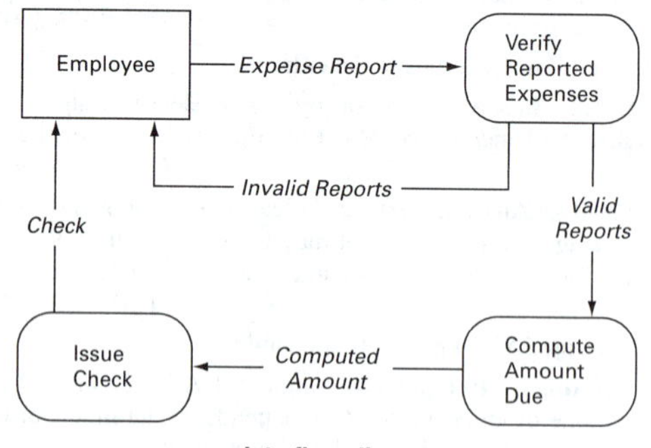

data flow diagram

(C) 610.12-1990

data flow graph *See:* data flow diagram.

data flowpath A segment of the overall plant data flow that represents one attribute sent from a providing activity to a receiving activity. (PE/EDPG) 1150-1991w

data flow trace *See:* variable trace.

data frame **(1)** Consists of the Destination Address, Source Address, Length Field, logical link control (LLC) Data, PAD, and Frame Check Sequence. (C/LM) 802.3-1998
(2) A grouping of data elements primarily for the purpose of referring to the group with a single name, and thereby efficiently reusing groups of data elements that commonly appear together (as an ASN.1 Sequence, Sequence Of, Set, Set Of or

Choice) in a message body specification. This data concept type may, however, be used to specify groups of data elements for other purposes as well. (SCC32) 1488-2000

data glossary A collection of entries specifying a data definition and a specification of its uses. (C) 610.5-1990w

datagram (1) A unit of data that is transferred as a single, nonsequenced, unacknowledged unit. (DIS/C) 1278.2-1995
(2) A unit of data transferred from one endpoint to another in connectionless mode service. (C) 1003.5-1999
(3) *See also:* connectionless service. (C) 610.7-1995

data hierarchy A set of directed relationships between two or more units of data, such that each unit has one and only one owner. *See also:* hierarchy. (C) 610.5-1990w

data-hold (data processing) A device that converts a sampled function into a function of a continuous variable. The output between sampling instants is determined by an extrapolation rule or formula from a set of past inputs. (IM) [52]

data host The spaceborne fiber-optic data bus (SFODB) source and destination nodes. (C/BA) 1393-1999

data independence The degree to which the logical view of a database is immune to changes in the physical structure of the database. (C) 610.5-1990w

data input (semiconductor memory) The inputs whose states determine the data to be written into the memory. (TT/C) 662-1980s

data input/output (semiconductor memory) The ports that function as data input during write operations and as data output during read operations. (TT/C) 662-1980s

data input sheet User documentation that describes, in a worksheet format, the required and optional input data for a system or component. *See also:* user manual. (C) 610.12-1990

data input station A workstation that is used primarily as an input device. *Synonyms:* data collection station; input workstation. (C) 610.10-1994w

data integrity (1) The degree to which a collection of data is complete, consistent, and accurate. *Synonym:* data quality. *See also:* database integrity; integrity; data security. (C) 610.5-1990w
(2) The condition or state in which data has not been altered or destroyed in an unauthorized manner. (LM/C) 802.10-1992

data interchange The use of data by two or more different systems. *See also:* data exchange. (C) 610.5-1990w

data interchange format A standarized data file format allowing data interchange between software packages on personal computers. For example, data interchange between an electronic spread sheet and a word processorcould be accomplished by converting the spread sheet data to data interchange format, then to the format required for the word processor. (C) 610.2-1987

data invariance A technique for mapping together big endian and little endian domains by mapping together bits with the same significance. This shall be done with some reference data width, typically either 32 bits or the implementation-dependent local data bus width. Futurebus+ is address invariant, but not necessarily data invariant. (C/BA) 896.3-1993w

data item A value contained in a data element; for example the data element AGE might contain data items 1, 2, ... *Note:* The terms **data element** and **data item** are often used interchangeably or with the reverse definitions from those given here. No standard of use exists at this time. *See also:* data element. (C) 610.5-1990w

Data Language I (DL/I) A database manipulation language used with IMS hierarchical databases. (C) 610.13-1993w

data latch Latch to temporarily store data to be subsequently written into a floating gate memory. (ED) 1005-1998

data library A set of related files, tables, or sets. (C) 610.5-1990w

data line *See:* bit-line.

data link (1) An assembly of data terminals and the interconnecting circuits operating according to a particular method that permits information to be exchanged between the terminals. (LM/COM) 168-1956w
(2) (test, measurement, and diagnostic equipment) Any information channel used for connecting data processing equipment to any input, output, display device, or other data processing equipment, usually at a remote location. (MIL) [2]
(3) (data management) The physical means of connecting two computers together for the purpose of transmitting and receiving data. (C) 610.5-1990w
(4) The services required to allow a local entity to establish a connection-oriented communications channel with a remote entity and exchange packets over the channel. (EMB/MIB) 1073.3.1-1994
(5) The assembly of parts of two data terminals that are controlled by a link protocol and an interconnecting data circuit to enable data to be transferred from a data source to a data sink. (C) 610.7-1995
(6) An assembly of two or more terminal installations and the interconnecting communications channel operating according to a particular method that permits information to be exchanged; in this context the term *terminal installation* does not include the data source and the data sink. (C/LM/CC) 8802-2-1998

data link escape character A transmission control character that changes the meaning of a limited number of contiguously following characters or coded representations to provide supplementary transmission control characters. (C) 610.7-1995

data link layer (1) The second layer of the OSI seven-layer model; provides error-free communication across the physical link. *Note:* This layer takes a bit stream from the physical layer, frames it into a data packet, appends leading and trailing headers for detection and correction of damaged packets and moves it to the network layer. It also performs the inverse operation on packets received from the network layer. *See also:* medium access control sublayer; application layer; client layer; network layer; logical link control sublayer; entity layer; session layer; physical layer; sublayer; presentation layer; transport layer. (C) 610.7-1995
(2) The conceptual layer of control or processing logic existing in the hierarchical structure of a station that is responsible for maintaining control of the data link. The data link layer functions provide an interface between the station higher layer logic and the data link. These functions include address/control field interpretation, channel access, and command PDU/response PDU generation, sending, and interpretation. (C/LM/CC) 8802-2-1998

data link service access point (DLSAP) The point at which logical connection between Data Link layers occurs. In IEEE 1073, both BCCs and DCCs have DLSAPs. (EMB/MIB) 1073.3.1-1994

data logger (1) (power-system communication) A system to measure a number of variables and make a written tabulation and.or record in a form suitable for computer input. *See also:* digital. (PE) 599-1985w
(2) A device that accepts PDUs from the network and stores them for later replay according to either the time sequence in which they were originally received or the time sequence as indicated by their time stamps. (DIS/C) 1278.3-1996

data logging (1) (supervisory control, data acquisition, and automatic control) The recording of selected data on suitable media. (SUB/PE) C37.1-1994
(2) An arrangement for the alphanumerical representation of selected quantities on log sheets; papers, magnetic tape, or the like, by means of an electric typewriter or other suitable devices. (SWG/PE) C37.100-1992

data-logging equipment Equipment for numerical recording of selected quantities on log sheets or paper or magnetic tape or the like, by means of an electric typewriter or other suitable device. (SWG/PE) C37.100-1981s, [56]

data management The function of controlling the acquisition, analysis, storage, retrieval, and distribution of data. (C) 610.5-1990w

data manipulation language (DML) A language used to retrieve, insert, delete, or modify the data in a database. *Synonym:* database manipulation language. *Contrast:* data definition language. *See also:* MODEL 204; dBASE; INQUIRE; FOCUS; SQL; RAMIS; Datatrieve; DL/I; Easytrieve; NATURAL. (C) 610.5-1990w, 610.13-1993w

data medium A material in or on which data are or may be represented. *See also:* data carrier; prerecorded data medium; media. (C) 610.10-1994w, 610.5-1990w

data model **(1)** A description of data that consists of all entities represented in a data structure or database and the relationships that exist among them. *See also:* physical data model; view; schema; logical data model. (C) 610.5-1990w
(2) A conceptual representation of the information requirements, data flows, and data relationships for an organization, facility, activity, or process. (PE/EDPG) 1150-1991w
(3) The numeric format in which the Smart Transducer Interface Module will output or accept data. (IM/ST) 1451.2-1997
(4) A graphical and textual representation of analysis that identifies the data needed by an organization to achieve its mission, functions, goals, objectives, and strategies and to manage and rate the organization. A data model identifies the entities, domains (attributes), and relationships (associations) with other data and provides the conceptual view of the data and the relationships among data. (C/SE) 1320.2-1998

data mover A system program through which a client application accesses the data on media. A data mover is not part of the Media Management System (MMS), nor is a data mover required for operation of the MMS. If a data mover is present, the MMS provides appropriate interfaces and facilities for it to operate. (C/SS) 1244.1-2000

data multiplexer A device that permits two or more data sources to share a common transmission medium. (C) 610.7-1995

data name One or more characters used to identify a data element. (C) 610.5-1990w

data normalization *See:* normalization.

data origin authentication The corroboration that the source of data received is as claimed. This service, when provided by the (N)-layer, provides the corroboration to an (N+1)-entity that the source of the data is the claimed peer (N+1)-entity. (LM/C) 802.10-1992

data output (semiconductor memory) The outputs whose states represent the data read from the memory. (TT/C) 662-1980s

Data Overrun Error (DOR) bit A bit in the Bus Error register of all S-modules. An S-module sets this bit to indicate that the module has received input data from the M-module when the S-module was not ready to receive it. (TT/C) 1149.5-1995

DATA packet Any packet other than a HEADER, PACKET COUNT, or ACKNOWLEDGE packet. (TT/C) 1149.5-1995

datapath **(1)** A type of Physical Design Exchange Format (PDEF) cluster that contains rows and/or columns of cluster, cell, skip, and/or spare_cell instances. A PDEF datapath typically corresponds to structured logic. *See also:* column; row. (C/DA) 1481-1999
(2) Signal lines on a bus associated with data. (C/MM) 959-1988r

data phase A period within a transaction used to transfer data. (C/BA) 10857-1994, 896.3-1993w, 896.4-1993w

data processing **(1)** The systematic performance of operations upon data, such as data manipulation, merging, sorting, and computing. *Synonym:* information processing. *See also:* mechanical data processing; business data processing; office automation; distributed data processing; integrated data processing; commercial data processing; automatic data processing; remote-access data processing; administrative data processing. (C) 610.2-1987
(2) (emergency and standby power) Pertaining to any operation or combination of operations on data. (IA/PSE) 446-1987s

data processing cycle *See:* processing cycle.

data processing system A system, including computer systems and associated personnel, that performs input, processing, storage output, and control functions to accomplish a sequence of operations on data. *See also:* information system. (C) 610.2-1987, 610.10-1994w

data processor* **(1)** **(A)** A processor capable of performing operations on data. For example: a desk calculator or tabulating machine, or a computer. **(B)** A person who operates a computer. (C) 610.10-1994
(2) Any device capable of being used to perform operations on data, for example, a desk calculator, tape recorder, analog computer, or digital computer. (IA/PSE) 446-1987s
* Deprecated.

data producer certification The determination by the data producer that data have been verified and validated against documented standards of criteria. (DIS/C) 1278.3-1996

data quality *See:* data integrity.

data quality objective The qualitative and quantitative statements that specify the quality of data required to support decisions for any process requiring radiochemical analysis (radioassay). (NI) N42.23-1995

data rate **(1)** The rate at which a data path (e.g., a channel) carries data, measured in bits per second. (EMB/MIB) 1073.4.1-2000
(2) *See also:* transfer rate. (C) 610.7-1995

data reconstruction (date processing) The conversion of a signal defined on a discrete-time argument to one defined on a continuous-time argument. (IM) [52]

data record *See:* record.

data recorder The device used to record any type of data. (VT) 1475-1999

data recording The act of recording any type of data. (VT) 1475-1999

data reduction **(1)** The transformation of raw data into a more useful form, for example, smoothing to reduce noise. (MIL) [2]
(2) (data management) Any technique used to transform data from raw data into a more useful form of data. For example, grouping, summing, or averaging related data. (C) 610.5-1990w

data registry A data dictionary that contains not only data about data elements in terms of their names, representational forms, and usage in applications, but also substantial data about the semantics or meaning associated with the data elements as concepts that describe or provide information about real or abstract entities. A data registry may contain abstract data concepts that do not get directly represented as data elements in any application system, but that help in information interchange and reuse both from the perspective of human users and for machine interpretation of data elements. (SCC32) 1489-1999

data resource A purposely organized body of data that is of use to some person or group of people. (C) 610.5-1990w

data retention time *See:* retention time.

data rewrite An operation in which data is written into any array and that includes one data cycle or at least one data change. (ED) 1005-1998

data security The degree to which a collection of data is protected from exposure to accidental or malicious alteration or destruction. *See also:* database security; data integrity. (C) 610.5-1990w

data-sensitive fault A fault that causes a failure in response to some particular pattern of data. *Synonym:* pattern-sensitive fault. *Contrast:* program-sensitive fault. (C) 610.12-1990

data sequence sensor A sensor that samples data independent of any triggers from a network capable application processor. (IM/ST) 1451.2-1997

data service unit A device that provides bipolar conversion functions to ensure proper signal shaping and adequate signal strength in a digital communications environment. *See also:* channel service unit. (C) 610.7-1995

data set (1) (data management) A named collection of related records. *Synonym:* file. *See also:* partitioned data set. (C) 610.5-1990w
(2) (data transmission) A modem serving as a conversion element and interface between a data machine and communication facilities. *See also:* modem. (PE) 599-1985w

data sheet A set of information on a device that defines the parameters of operation and conditions of usage (usually produced by the device's manufacturer). (IM/ST) 1451.2-1997

data signaling rate The rate of data transmission, generally expressed as bits per second. *See also:* baud rate. (C) 610.7-1995, 610.10-1994w

data sink (1) (data transmission) The equipment which accepts data signals after transmission. (PE) 599-1985w
(2) The functional unit that accepts transmitted data. *Contrast:* data source. (C) 610.7-1995

data source (1) (data transmission) The equipment which supplies data signals that enter into a data link. (PE) 599-1985w
(2) The functional unit that originates data for transmission. *Contrast:* data sink. (C) 610.7-1995

data space The address space which devices may have that is recommended for use in data operations. There are few constraints applied to data space uses. *See also:* CSR space. (NID) 960-1993

data stabilization (vehicle-borne navigation systems) (navigation aid terms) The stabilization of the output signals with respect to a selected reference invariant with vehicle orientation. (AES/GCS) 172-1983w

data stack A stack that may be used for passing parameters between Forth definitions. (C/BA) 1275-1994

data station *See:* station.

data storage description language A language used to define the organization of stored data in terms that are independent of any particular storage device or operating system. (C) 610.5-1990w

data storage schema A data structure that describes the manner in which data items are physically stored in storage. *See also:* database definition. (C) 610.5-1990w

data stream (A) All data that is transmitted through an input-output channel in a single read or write operation. **(B)** A continuous stream of data elements being transmitted, or intended for transmission. (C) 610.10-1994

data striping RAID storage A form of RAID storage, known as level 0, in which data is striped across the multiple drives by system block size. *Note:* No parity check is performed. (C) 610.10-1994w

data structure (1) (data management) (software) A physical or logical relationship among data elements, designed to support specific data manipulation functions. *Synonym:* logical structure. (C) 610.5-1990w, 610.12-1990
(2) A group of digital data fields organized in some logical order for some specific purpose. A two-dimensional paper version of a data structure is an empty fill-in-the-blanks form or an empty tabular chart with organized column and row headings. A data structure is the template by which data is stored in computer memory. (IM/ST) 1451.2-1997

data structure-centered design A software design technique in which the architecture of a system is derived from analysis of the structure of the data sets with which the system must deal. *See also:* structure clash; modular decomposition; input-process-output; structured design; transform analysis; object-oriented design; rapid prototyping; transaction analysis; stepwise refinement. (C) 610.12-1990

data structure diagram A diagram that depicts a set of data elements, their attributes, and the logical relationships among them. *Contrast:* data flow diagram. *See also:* entity-relationship diagram.

<table>
<tr><th colspan="11">Employee Record</th></tr>
<tr><td rowspan="2">Emp. No. (4I)</td><td colspan="3">Emp. Name</td><td colspan="4">Emp. Address</td><td rowspan="2">Dept. No. (3I)</td><td rowspan="2">Emp. Sal. (4I)</td></tr>
<tr><td>First (10C)</td><td>Mid. (1C)</td><td>Last (16C)</td><td>Street (20C)</td><td>City (20C)</td><td>State (2C)</td><td>Zip (9I)</td></tr>
</table>

I = Integer C = Character

data structure diagram

(C) 610.12-1990

data sublanguage (DSL) A subset of another language, called the host language, that is used to perform database operations. *Synonym:* database sublanguage. (C) 610.5-1990w

data submodel *See:* external schema.

data switch A switch device that is designed to handle data communications rather than voice communications. (C) 610.7-1995

data switching exchange In networking, the equipment installed at a single location to provide circuit switching, packet switching, or both functions. (C) 610.7-1995

data tablet A graphical input device, used as a locator, consisting of a flat surface with a sensing apparatus, such as a grid of wires, and a pointing device such as a mouse, puck, or stylus to indicate tablet locations. *Synonyms:* writing tablet; bit pad. *See also:* graphic tablet; locator; stylus; digitizer; acoustic tablet. (C) 610.10-1994w, 610.6-1991w

data terminal (data transmission) A device which modulates or demodulates data between one input-output device and a data transmission link, or both. (PE) 599-1985w

data terminal equipment (DTE) (1) The equipment comprising the data source, the data sink, or both. (LM/COM) 168-1956w
(2) A device that serves as a data source and/or a data sink. (C/LM/COM) 802.9a-1995w, 610.7-1995, 8802-9-1996
(3) Any source or destination of data connected to the local area network (LAN). (C/LM) 802.3-1998

data test (A) (station control and data acquisition) The recorded results of test. **(B) (station control and data acquisition)** A set of data developed specifically to test the adequacy of a computer run or system. They may be actual data taken from previous operations or artificial data created for this purpose. (SWG/PE/SUB) C37.100-1992, C37.1-1979, C37.1-1994

data trace *See:* variable trace.

data transfer (1) The passing of data over the multiplexed address/data bus, between the bus owner and the replying agent(s), during the reply phase of a transfer operation. (C/MM) 1296-1987s
(2) The phase of a cycle during which data are transferred between the master and the selected slaves. It starts when the active master asserts the data strobe and ends after the responding slave acknowledges the transfer and all participating slaves indicate that they are ready to participate in a new cycle. (C/MM) 1096-1988w

data transfer bus (DTB) (1) One of the four buses provided by the backplane. The data transfer bus allows masters to direct the transfer of binary data among themselves and slaves. (C/BA) 1014-1987
(2) One of the two subbuses defined in the VSB specification. It allows masters to direct the transfer of binary data to and from slaves. The DTB contains 32 multiplexed address/data lines and the associated control signals that are required to execute cycles on the VSB. (C/MM) 1096-1988w

data-transfer-bus cycle A sequence of level transitions on the signal lines of the DTB that result in the transfer of an address or an address and data between a master and a slave. The data-transfer-bus cycle is divided into two portions: the address broadcast and zero or more data transfers. (C/BA) 1014-1987

data-transfer interface An interface that enables a connection between a computer and a peripheral unit such as a magnetic disk. *See also:* enhanced small device interface; small computer systems interface; ST-506 interface. (C) 610.10-1994w

data translation The modification of the physical representation of data used in one hardware/software environment so that it is compatible with a different hardware/software environment. (C) 610.5-1990w

data transmission The sending of data from one place to another. (C/PE/PSCC) 610.7-1995, 599-1985w

Datatrieve A database manipulation language used primarily for database applications under Digital's VAX/VMS environment. (C) 610.13-1993w

data type (1) (software) A class of data, characterized by the members of the class and the operations that can be applied to them. For example, character type, enumeration type, integer type, logical type, real type. *See also:* strong typing. (C) 610.12-1990

(2) **(A)** A categorization of an abstract set of possible values, characteristics, and set of operations for an attribute. Integers, real numbers, and character strings are examples of data types. **(B)** A set of values and operations on those values. The set of values is called the *extent* of the type. Each member of the set is called an instance of the type. (C/SE) 1320.2-1998

(3) A classification of the collection of letters, digits, and/or symbols used to encode values of a data element based upon the operations that can be performed on the data element. (SCC32) 1489-1999

datatype A collection of distinguished values, together with a collection of characterizing operations on those values. (C/PA) 1351-1994w, 1328-1993w, 1224-1993w, 1224.1-1993w, 1327-1993w

data unit The smallest unit of the contents of a file that the filestore actions can manipulate. (C/PA) 1238.1-1994w

data user certification The determination by the application sponsor or designated agent that data have been verified and validated as appropriate for the specific Modeling and Simulation (M&S) usage. (DIS/C) 1278.3-1996

data validation (1) The documented assessment of data by subject area experts and its comparison to known or best-estimate values.

— *Data producer validation.* That documented assessment within stated criteria and assumptions.
— *Data user validation.* That documented assessment of data as appropriate for use in an intended M&S.

(DIS/C) 1278.3-1996

(2) The process of checking data-entry items for correct content or format. (PE/NP) 1289-1998

data value The actual value that is stored in a data item. For example, the numeric value of the data item SALARY may be 20 000. *Synonym:* value. (C) 610.5-1990w

data verification The use of techniques and procedures to ensure that data meets specified constraints defined by data standards and business rules.

— *Data producer verification.* The use of techniques and procedures to ensure that data meets constraints defined by data standards and business rules derived from process and data modeling.
— *Date user verification.* The use of techniques and procedures to ensure that data meets user specified constraints defined by data standards and business rules derived from process and data modeling and to ensure that data are transformed and formatted properly.

(DIS/C) 1278.3-1996

data verification, validation, and certification (VV&C) The process of verifying the internal consistency and correctness of data, validating that it represents real-world entities appropriate for its intended purpose or expected range of purposes, and certifying it as having a specified level of quality or as being appropriate for a specified use, type of use, or range of uses. The process is conducted from two perspectives: (1) the data producer ensures the data produced satisfy the appropriate standards and (2) each data user ensures the data selected are appropriate for the specific application. (C/DIS) 1278.4-1997

data volatility The rate of change, over a specified period of time, in the values of stored data items. (C) 610.5-1990w

dataway *See:* CAMAC dataway.

data window A set of coefficients by which corresponding samples in the data record are multiplied to more accurately estimate certain properties of the signal, particularly frequency domain properties. Generally, the coefficient values increase smoothly toward the center of the record. (IM/WM&A) 1057-1994w

date-component Any sub-element of a calendar system or date format. (C/PA) 2000.1-1999

date-component overflow A condition in which an operation involving a date component produces a result too large to be contained in the location or register allotted to it. (C/PA) 2000.1-1999

date data overflow *See:* date-component overflow.

date data processing The processing of date data within a system element, which may include receiving, manipulating, and providing date data. (C/PA) 2000.1-1999

date-insensitivity A type of year-insensitivity in which year and date are not maintained or represented, but day indicator or time-of-day may be. (C/PA) 2000.1-1999

date interval The inclusive interval of time between two specified dates. Unless otherwise specified, a date interval spans the fully inclusive interval specified by its two endpoints. Intervals using exact dates as endpoints include the first instant of the inception date through the last instant of the terminal date. Intervals using years as endpoints include the first instant of the first day of the inception year through the last instant of the last day of the terminal year. (C/PA) 2000.1-1999

date-invariance range Specifies the date interval able to be represented and consistently manipulated within a system element. (C/PA) 2000.1-1999

date overflow Synonym for date-component overflow. (C/PA) 2000.1-1999

date-specifier An instantiation of a date, using a specific date format. (C/PA) 2000.1-1999

date/time-of-day clock A clock that shall be maintained by the BCC, providing an indication of time, from year to milliseconds, with a resolution of 1 ms. (EMB/MIB) 1073.3.1-1994

datum (1) Singular for data. (C) 610.5-1990w, 610.12-1990
(2) Singular form for data. (C) 610.10-1994w

daughter *See:* child node.

daughter board A printed circuit board that attaches to another, often the main system board, or motherboard, to provide additional functionality or performance. *Synonym:* piggyback board. (C) 610.10-1994w

davit (power line maintenance) An assembly attached to a support or assembled on a structure to provide a rigging point for rope blocks, chains, or hoists so as to manipulate various pieces of apparatus. The davit is a rigid assembly and does not swivel. (T&D/PE) 516-1995, 458-1990

davit arm A rigid upswept cantilever arm used to support an insulator string. (T&D/PE) 751-1990

day-insensitivity A type of year-insensitivity in which year, date, and day indicator are not maintained or represented, but time-of-day may be. (C/PA) 2000.1-1999

daylight factor (illuminating engineering) A measure of daylight illuminance at a point on a given plane expressed as a ratio of the illuminance on the given plane at that point to the simultaneous exterior illuminance on a horizontal plane from the whole of an unobstructed sky of assumed or known luminance distribution. Direct sunlight is excluded from both interior and exterior values of illuminance. (EEC/IE) [126]

day-night sound level (L_{dn}) The L_{dn} rating is the average A-weighted sound level, in decibels, integrated over a 24 h period. A 10 db(A) penalty is applied to all sound occuring between 10 P.M. and 7 A.M. *Notes:* 1. L_{dn} is intended to improve upon the L_{eq} rating by adding a correction for night-time noise intrusions because people are more sensitive to such intrusions. 2. The L_{dn} can be derived from daytime and nighttime L_{eq} values as follows:

$$L_{dn} = 10\log\left(\frac{1}{24}\right)\left[15\text{ antilog }\frac{L_d}{10} + 9\text{ antilog }\frac{L_n + 10}{10}\right]$$

where

L_d = The L_{eq} for the 15 daytime hours

L_n = The L_{eq} for the 9 nighttime hours

(T&D/PE) 656-1992, 539-1990

days of battery reserve (1) The number of days a fully-charged battery can satisfy the load with no contribution from the PV array or auxiliary power source. (PV) 1013-1990 **(2)** The number of days a fully charged battery can satisfy the load with no contribution from the photovoltaic array or auxiliary power source. (PV) 1144-1996

dB *See:* decibel.

DB-9 The designation of a standard plug and jack set used in EIA/TIA-530-A [formerly RS-449] wiring. It has a 9-pin connector. *Note:* The connector is specific to ISO standard. (C) 610.7-1995

DB-25 The designation of a standard plug and jack set used in EIA/TIA-232-E [formerly RS-232-C] wiring. It has a 25-pin connector, with 13 pins in the top row and 12 in the bottom row. *Note:* The connector is specific to ISO standard. (C) 610.7-1995

DB-37 The designation of a standard plug and jack set used in EIA/TIA-530-A [formerly RS-449] wiring. It has a 37-pin connector. *Note:* The connector is specific to ISO standard. (C) 610.7-1995

dBASE A database manipulation language used commonly for database applications in the PC DOS Environment. *Note:* Several version have been published with Roman numeral suffixes, as in dBASE III, dBASE IV. (C) 610.13-1993w

DBA *See:* database administration; database administrator.

DBCL *See:* database command language.

DBDL *See:* database description language.

dBDSX A representation of signal level expressed by a measure of base-to-peak signal voltage at the digital signal crossconnect (DSX) point. (COM/TA) 1007-1991r

DBE *See:* design basis events.

DBL A programming language used in conjunction with DIBOL for developing applications within the UNIX and DOS environments. (C) 610.13-1993w

dBm (1) (A) Decibels relative to 1 milliwatt. **(B) (data transmission)** A unit for expression of power level in decibels with reference to a power of one milliwatt. (LM/C/PE) 802.7-1989, 599-1985 **(2)** Decibel with reference to one milliwatt. (COM/TA) 1007-1992

DBML *See:* database manipulation language.

DBMS *See:* database management system.

dBmV (1) Decibels relative to 1 mV across 75 Ω. Zero dBmV is defined as 1 mV across 75 Ω.

dBmV = 20 log 10 (V_1/V_2)

V_1 is the measurement of voltage at a point having identical impedance to V_2 (0.001 V across 75 Ω). (LM/C) 802.7-1989r

(2) Decibels referenced to 1.0 mV measured at the same impedence. Used to define signal levels in Community Antenna Television (CATV)-type broadband systems. (C/LM) 802.3-1998

D-board *See:* lineperson's platform.

dBV *See:* voltage level.

DbNC *See:* distributed numerical control.

DBS *See:* Doppler beam sharpening.

DBTG *See:* Database Task Group.

dc *See:* direct current.

DC *See:* dynamic configuration.

DCA *See:* digital coefficient attenuator.

D cable A two-conductor cable, each conductor having the shape of the capital letter D with insulation between the conductors themselves and between conductors and sheath. (T&D/PE) [10]

dc-ac transfer standards (metering) Instruments used to establish the equality of a root-mean-square (rms) current or voltage (or the average values of alternating power) with the corresponding steady-state direct-current (dc) quantity. (ELM) C12.1-1982s

DCC *See:* device communications controller.

dc-dc converter A machine, device, or system, typically combining the functions of inversion and rectification, for changing dc at one voltage to dc at a different voltage. (PEL/ET) 388-1992r

DC device A device whose logical address can be programmed by a dynamic configuration (DC) resource manager. (C/MM) 1155-1992

DCE Formerly the Open Software Foundation (OSF) Distributed Computing Environment (DCE), and now the Open Group's DCE. (C/SS) 1244.1-2000

DCDL *See:* Digital Control Design Language.

DCF *See:* Document Composition Facility.

D channel (for the ISLAN16-T) A channel that provides a 64 kbit/s full-duplex packet access. The D channel provides ISDN call control services via the ITU-T Q.93x family of protocols. The use of the Dchannel may be restricted to conveying signalling information in some applications. The Dchannel is capable of supporting packet mode information. All information on Dchannels is packetized. (LM/C/COM) 802.9a-1995w, 8802-9-1996

DCL *See:* Digital Command Language.

D conditioning A North American term for a type of conditioning that controls harmonic distortion and signal-to-noise ratio, thus making transmission impairments lie within specified limits. *See also:* C conditioning. (C) 610.7-1995

DC resource manager A resource manager that supports dynamic configuration. (C/MM) 1155-1992

DC system A VXIbus system that includes a DC resource manager and one or more DC devices. (C/MM) 1155-1992

DCTL *See:* direct-coupled transistor logic.

DDA *See:* digital differential analyzer.

DDD *See:* direct distance dialing.

DD/D *See:* data dictionary/directory.

D dimension (motors) The standard designation of the distance from the centerline of the shaft to the plane through the mounting surface bottom of the feet, in National Electrical Manufacturers Association approved designations. (PE) [9]

D-display Similar to a C-display, but composed of a series of horizontal stripes representing successive elevation angles. Each stripe is a miniature B-display with compressed vertical scale. Horizontal position of a blip represents azimuth, the gross vertical scale (the stripe in which the blip appears) represents elevation, and vertical position within the stripe represents range.

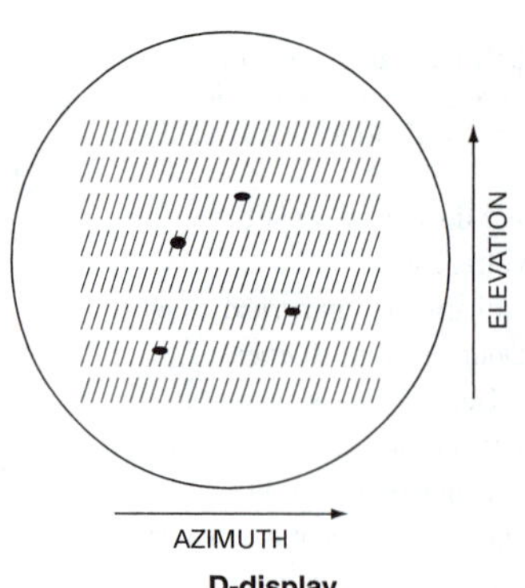

D-display

(AES) 686-1997

DDL *See:* data definition language.

DDM *See:* difference in depth of modulation.

DDP *See:* dispersed data processing; distributed data processing.

deactivate (696 interface devices) (signals and paths) To cause a signal to transition from its logically true (active) state to its logically false (inactive) state. Opposite of assert. (C/MM) 696-1983w

deactivated shutdown (power system measurement) (electric generating unit reliability, availability, and productivity) The state in which a unit is unavailable for service for an extended period of time because of its removal for economy or reasons not related to the equipment. Under this condition, a unit generally requires weeks of preparation to make it available. (PE/PSE) 762-1987w

deactivated shutdown hours (electric generating unit reliability, availability, and productivity) The number of hours a unit was in a deactivated shutdown state. (PE/PSE) 762-1987w

deactivation The process of removing active constituents from a corroding liquid (as removal of oxygen from water). (IA) [59]

deactivation date (electric generating unit reliability, availability, and productivity) The date a unit was placed into the deactivated shutdown state. (PE/PSE) 762-1987w

dead (1) *See also:* de-energized. (PE/T&D) 524-1992r
(2) A circuit that has been de-energized so that the circuit has been disconnected from all intended electrical sources. However, it could be electrically charged through induction from energized circuits in proximity to it, particularly when the circuits are parallel. *See also:* de-energized. (T&D/PE) 516-1995

dead band (1) (supervisory control, data acquisition, and automatic control) The range through which an analog quantity can vary without initiating response. (SUB/PE/NP) C37.1-1994, 381-1977w
(2) (accelerometer) (gyros) A region between the input limits within which variations in the input produce output changes of less than 10% (or other small value) of those expected based on the nominal scale factor. (AES/GYAC) 528-1994
(3) The range through which an input can be varied without initiating response. (SWG/PE) C37.100-1992

dead band differential (electric pipe heating systems) The difference in degrees between the OFF and the ON stage of temperature controllers. (PE/EDPG) 622A-1984r, 622B-1988r

dead-band rating The limit that the dead band will not exceed when the instrument is used under rated operating conditions. *See also:* accuracy rating. (EEC/EMI) [112]

dead-break connector (1) (power and distribution transformers) A separable insulated connector designed to be separated and engaged on de-energized circuits only. (PE/TR) C57.12.80-1978r
(2) (separable insulated connectors) A connector designed to be separated and engaged on de-energized circuits only. (T&D/PE) 386-1995

dead-end The point at which mechanical force (primarily) and longitudinal strain is applied to a reliable support. *Synonyms:* termination; strain attachment; anchor point. (T&D/PE) 516-1995

dead-end board *See:* lineperson's platform.

dead-end cover *See:* insulator cover.

dead-end guy An installation of line or anchor guys to hold the pole at the end of a line. *See also:* tower; pole guy. (T&D/PE) [10]

dead-end loop *See:* jumper.

dead-end platform *See:* lineperson's platform.

dead-end tower A tower designed to withstand unbalanced pull from all of the conductors in one direction together with wind and vertical loads. *See also:* tower. (T&D/PE) [10]

dead front (1) (A) Without live parts exposed to a person on the operating side of the equipment. **(B)** (As applied to switches, circuit breakers, switchboards, and distribution panelboards.) So designed, constructed and installed that no current-carrying parts are normally exposed on the front. (NESC/NEC) [86]
(2) (power and distribution transformers) So constructed that there are no exposed live parts on the front of the assembly. (PE/TR) C57.12.80-1978r

dead-front mounting (of a switching device) A method of mounting in which a protective barrier is interposed between all live parts and the operator, and all exposed operating parts are insulated or grounded. *Note:* The barrier is usually grounded metal. (SWG/PE) C37.100-1992

dead-front pad-mounted switchgear (PMSG) A switchgear assembly in which all energized parts are insulated and completely enclosed within a grounded shield system when separable connectors are in place. The overall enclosure is of suitable environmental and tamper-resistant construction for outdoor above-ground installation. The term *front* refers specifically to any side of the enclosure that provides access to enclosed accessories external to the ground shield system. (SWG/PE) C37.100-1992

dead-front switchboard A switchboard that has no exposed live parts on the front. *Note:* The switchboard panel is normally grounded metal and provides a barrier between the operator and all live parts. (SWG/PE) C37.100-1992

dead-front type arrester An arrester assembled in a shielded housing providing system insulation and conductive ground shield, intended to be installed in an enclosure for the protection of underground and padmounted distribution equipment and circuits. (SPD/PE) C62.22-1997, C62.11-1999

dead-tank switching device A switching device in which a vessel(s) at ground potential surrounds and contains the interrupter(s) and the insulating medium. (SWG/PE) C37.100-1992

dead layer (1) (of a semiconductor detector) An inactive region (layer) in which the energy absorbed from the passage of monoenergetic charged particles does not significantly contribute to the resulting full energy peak. (NPS) 300-1988r
(2) In a semiconductor detector, a layer (frequently associated with a contact region) in which no significant part of the energy lost by photons or particles can contribute to the resulting signal. (NPS) 325-1996

dead layer thickness (x-ray energy spectrometers) (of a semiconductor radiation detector) The thickness of an inactive region (in the form of a layer) through which the incident radiation must pass to reach the sensitive volume. *Synonym:* window. (NPS/NID) 759-1984r, 301-1976s

dead leg (1) A segment of process piping that is not in the normal flow pattern. (IA/BT/AV) 515-1997, 152-1953s
(2) A portion of a piping system, without flow, used to simulate the overall system conditions for control sensing. (IA/PC) 844-1991

deadlock (1) A situation in which computer processing is suspended because two or more devices or processes are each awaiting resources assigned to the other(s). For example, a situation in which computer program A, with an exclusive lock on record X, asks for a lock on record Y, which is allocated to computer program B. Likewise, program B is waiting for exclusive control over record X before giving up control of record Y. *Synonym:* deadly embrace. *See also:* lockout. (C) 610.5-1990w, 610.12-1990

(2) A state that occurs when modules are awaiting actions that can only be performed by those waiting, and those waiting cannot perform the actions. (C/BA) 10857-1994, 896.3-1993w, 1014.1-1994w, 896.4-1993w

deadly embrace *See:* deadlock.

dead man *See:* anchor log.

deadman A pressure- or activity-actuated alertness device to detect inattention or disability of a train operator. *Note:* A deadman can be contained within the master controller main handle grip, obtained by a separate foot switch, or obtained through an alertness type function. The device, when not properly maintained in an operational condition, will result in an emergency or full service brake application. (VT) 1475-1999

deadman's handle A handle of controller or master switch that is designed to cause the controller to assume a preassigned operating condition if the force of the operator's hand on the handle is released. *See also:* electric controller. (IA/ICTL/IAC) [60]

deadman's release The effect of that feature of a semiautomatic or nonautomatic control system that acts to cause the controlled apparatus to assume a preassigned operating condition if the operator becomes incapacitated. *See also:* control. (IA/ICTL/IAC) [60]

dead-metal part (power and distribution transformers) A part, accessible or inaccessible, which is conductively connected to the grounded circuit under conditions of normal use of the equipment. (PE/TR) C57.12.80-1978r

dead reckoning (DR) (1) (navigation aid terms) The determining of the position of a vehicle at one time with respect to its position at a different time by the application of vectors representing courses and distances. (AES/GCS) 172-1983w

(2) A method for the estimation of the position/orientation of an entity based on a previously known position/orientation and estimates of time and motion. (DIS/C) 1278.1-1995

dead room (audio and electroacoustics) Those locations completely within the coverage area where the signal strength is below the level needed for reliable communication. *See also:* mobile communication system. (VT/ACO) [37]

dead state A node state that is reflected by the value of 3 in the STATE_CLEAR.state field. A node enters the dead state when a fatal error has been detected and the node is connected but no longer operational. Note that the severest errors could leave the node in a broken state, with its registers undefined, rather than indicating a dead state. (C/MM) 1212-1991s

dead tank switching device A switching device in which a vessel(s) at ground potential surrounds and contains the interrupter(s) and the insulating medium. (SWG/PE) C37.100-1981s

dead time (1) (A) (of a circuit breaker on a reclosing operation) The interval between interruption in all poles on the opening stroke and reestablishment of the circuit on the reclosing stroke. *Notes:* 1. In breakers using arc-shunting resistors, the following intervals are recognized and the one referred to should be stated: Dead time from interruption on the primary arcing contacts to reestablishment through the primary arcing contacts; Dead time from interruption on the primary arcing contacts to reestablishment through the secondary arcing contacts; Dead time from interruption on the secondary arcing contacts to reestablishment on the primary arcing contacts; Dead time from interruption on the secondary arcing contacts to reestablishment on the secondary arcing contacts. 2. The dead time of an arcing fault on a reclosing operation is not necessarily the same as the dead time of the circuit breakers involved, since the dead time of the fault is the interval during which the faulted conductor is de-energized from all terminals. **(B)** The time during which the ADC is unable to process input pulses because it is processing a previous pulse. (SWG/PE/NI) C37.100-1992, N42.14-1991

(2) (navigation) The time interval in an equipment's cycle of operating during which the equipment is prevented from providing normal response. For example, in a radar display, the portion of the interpulse interval which is not displayed; or, in secondary radar, the interval immediately following the transmission of a pulse relay during which the transponder is insensitive to interrogations. (AES/RS) 686-1982s

(3) (sodium iodide detector) The time after a triggering pulse during which the system is unable to retrigger. (NI) N42.12-1994

(4) The time interval after the start of an essentially full-amplitude pulse during which a radiation detector is insensitive to further ionizing events. *See also:* recovery time. (NI/NPS) 309-1999

dead-time correction (radiation counters) A correction to the observed counting rate to allow for the probability of the occurrence of events within the dead time of the system. *See also:* anticoincidence. (ED) [45]

dead time tt (sodium iodide detector) The time after a triggering pulse during which the system is unable to retrigger. (NI) N42.12-1980s, N42.12-1994

dead zone The period(s) in the operating cycle of a machine during which corrective functions cannot be initiated. (IA/ICTL/IAC) [60]

deassembler* *See:* disassembler.

* Deprecated.

deauthentication The service that voids an existing authentication relationship. (C/LM) 8802-11-1999

deblock To separate the parts of a block. *Synonym:* unblock. *Contrast:* block. (C) 610.12-1990, 610.5-1990w

deblurring *See:* sharpening.

debug (1) To examine or test a procedure, routine, or equipment for the purpose of detecting and correcting errors. (COM) [49]

(2) (computers) To detect, locate, and remove mistakes from a routine or malfunctions from a computer. *See also:* troubleshoot. (C) [20], [85]

(3) (software) To detect, locate, and correct faults in a computer program. Techniques include use of breakpoints, desk checking, dumps, inspection, reversible execution, single-step operation, and traces. (C) 610.12-1990

debugging (1) The operation of an equipment or complex item prior to use to detect and replace parts that are defective or expected to fail, and to correct errors in fabrication or assembly. *See also:* reliability. (MIL/R) [2], [29]

(2) (software) The process of locating, analyzing, and correcting suspected faults. *See also:* testing. (C/SE) 729-1983s

debugging model *See:* error model.

Debye length (L_D) That distance in a plasma over which a free electron may move under its own kinetic energy before it is pulled back by the electrostatic restoring forces of the polarization (ion) cloud surrounding it. Over this distance, a net charge density can exist in an ionized gas. The Debye length is given by:

$$L_D = 6.9\sqrt{\frac{T_e}{N_e}}$$

where

T_e = the electron temperature

N_e = the electron number density

(AP/PROP) 211-1997

deca A prefix indicating ten. *Synonym:* deka. (C) 1084-1986w

decade (1) A functional grouping of tables by application into groups of ten. The tables are numbered "X0" through "X9," with "X" representing the decade number. (AMR/SCC31) 1377-1997
(2) A range of values for which the upper limit is a power of ten above the lower limit. (NI) N42.17B-1989r, N323-1978r

decalescent point (induction heating usage) The temperature at which there is a sudden absorption of heat when metals are raised in temperature. *See also:* recalescent point; induction heating; coupling. (ED) 161-1971w

decay (storage tubes) A decrease in stored information by any cause other than erasing or writing. *Note:* Decay may be caused by an increase, a decrease, or a spreading of stored charge. *See also:* storage tube. (ED) 161-1971w, 158-1962w

decay characteristic *See:* persistence characteristic.

decay, dynamic *See:* dynamic decay.

decay, static *See:* static decay.

decay [test] method A test that determines the power dissipation characteristics of a damper by the measurement of the decay rate of the amplitude of motion of a span following a period of forced vibration at a natural frequency and a fixed test amplitude. (T&D/PE) 664-1993

decay time (1) (storage tubes) The time interval during which the stored information decays to a stated fraction of its initial value. *Note:* Information may not decay exponentially. *See also:* storage tube. (ED/ED) 158-1962w, 161-1971w
(2) *See also:* hang-over time. (COM/TA) 1329-1999

decay time constant (1) (A) (germanium gamma-ray detectors) (semiconductor radiation detectors) (x-ray energy spectrometers) The time for a true single-exponential waveform to decay to a value of 1/e of the original step height. **(B)** With a signal that decays with a single exponential waveform, the time required for it to fall to 1/*e* from any selected level on the asymptote of the last transition. (NPS/NID) 759-1984, 301-1976, 300-1988
(2) In the absence a low-pass network preceding the point of observation, the time for a single-exponential waveform to decay to the fraction 1/e of its original amplitude. (NPS) 325-1996

Decca (navigation aid terms) A radio navigation system transmitting on several related frequencies near 100 kHz (kilohertz), useful to about 200 nmi (nautical miles) (370 km [kilometers]) in which sets of hyperbolic lines of position are determined by comparison of the phase of (A) one reference continuous wave signal from a centrally located master with (B) each of several continuous wave signals from slave transmitters located in a star pattern, each about 70 nmi (130 km) from the master. (AES/GCS) 172-1983w

decelerating electrode (electron-beam tubes) An electrode the potential of which provides an electric field to decrease the velocity of the beam electrons. (ED/NPS) 161-1971w, 398-1972r

decelerating relay A relay that functions automatically to maintain the armature current or voltage within limits, when decelerating from speeds above base speed, by controlling the excitation of the motor field. *See also:* relay. (IA/ICTL/IAC/APP) [60], [75]

decelerating time The time in seconds for a change of speed from one specified speed to a lower specified speed while decelerating under specified conditions. *See also:* feedback control system. (IA/ICTL/IAC) [60]

deceleration *See:* retardation.

deceleration distance The additional vertical distance a falling worker travels, excluding lifeline elongation and free fall distance, before stopping, from the point at which the energy absorbing device begins to operate. It is measured as the distance between the location of a line-worker's body belt, aerial belt, or full body harness attachment point at the moment of activation (at the onset of fall arrest forces) of the energy absorbing device during a fall, and the location of that attachment point after the worker comes to a stop. (T&D/PE) 1307-1996

deceleration, programmed *See:* programmed deceleration.

decelerating device A device that is used to close or to cause the closing of circuits that are used to decrease the speed of a machine. (PE/SUB) C37.2-1979s

deceleration time The time that is required to slow a storage device, typically a tape or disk drive, to a stop after data has been read or written. *Synonym:* stop time. *Contrast:* acceleration time. (C) 610.10-1994w

deceleration, timed *See:* timed deceleration.

decentralized computer network (1) A computer network in which control functions are distributed over several network nodes. *Contrast:* centralized computer network. *See also:* distributed computer network. (C/COM) 610.7-1995, 168-1956w
(2) A computer network, where some of the computing power and network control functions are distributed over several network nodes. *See also:* centralized computer network. (LM/COM) 168-1956w

deci A prefix indicating one tenth. (C) 1084-1986w

decibel (1) (A) (power station noise control) Ten times the logarithm to base 10 of a ratio of two powers. **(B)** A standard unit for expressing the ratio between two parameters using logarithms to the base 10. Decibels provide a convenient format to express voltages or powers that range several orders of magnitude for a given system. **(C)** One-tenth of a bel, the number of decibels denoting the ratio of two amounts of power being ten times the common logarithm of this ratio. *Note:* The abbreviation dB is commonly used for the term decibel. With P_1 and P_2 designating two amounts of power, and n the number of decibels denoting their ratio:

$$n = 10 \log (P_1/P_2) \text{ dB}$$

When the conditions are such that ratios of currents or ratios of voltages (or analogous quantities in other disciplines) are the square roots of the corresponding power ratios, the number of decibels by which the corresponding powers differ is expressed by the following equations:

$$n = 20 \log (I_1/I_2) \text{ dB}$$

$$n = 20 \log (V_1/V_2) \text{ dB}$$

where I_1/I_2 = The given current ratio V_1/V_2 = The given voltage ratio. By extension, these relations between numbers of decibels and ratios of currents or voltages are sometimes applied where these ratios are not the square roots of the corresponding power ratios; to avoid confusion, such usage should be accompanied by a specific statement of the application in question. **(D)** A unit of measurement for the relative strength of a signal parameter such as power or voltage. **(E)** The standard unit for expressing transmission gain or loss and relative power levels. Decibels indicate the ratio of power input to power output: dB = $10 \log_{10}(P_{out}/P_{in})$. *Note:* One decibel is 0.1 bel.
(LM/PE/C/T&D/AP/EDPG/ANT) 640-1985, 802.7-1989, 539-1990, 599-1985, 145-1993, 610.7-1995, 610.10-1994
(2) (automatic control) A logarithmetic scale unit relating a variable x (e.g., angular displacement) to a specified reference level x_0; dB = $20 \log x/x_0$. *Note:* The relation is strictly applicable only where the ratio x/x_0 is the square root of the power ratio P/P_0, as is true for voltage or current ratios. The value dB = $10 \log_{10} P/P_0$ originated in telephone engineering, and is approximately equivalent to the old "transmission unit." (PE/EDPG) [3], 421A-1978s

decibel meter An instrument for measuring electric power level in decibels above or below an arbitrary reference level. *See also:* instrument. (EEC/PE) [119]

decile (A) (electromagnetic site survey) (D_u) The ratio of the upper decile value (the value of x exceeded 10% of the time) of the random variable x to its median value, expressed in decibels. **(B) (D_l) (electromagnetic site survey)** The ratio of the lower decile value (the value of x exceeded by 90% of the time) of the random variable x to its median value, expressed in decibels. (EMC) 473-1985

decilog (data transmission) A division of the logarithmic scale used for measuring the logarithm of the ratio of two values of any quantity. Its value is such that the number of decilogs is equal to 10 times the logarithm to the base 10 of the ratio. One decilog, therefore, corresponds to a ratio of 100.1 (that is, 1.25892 +). (PE) 599-1985w

decimal (A) (mathematics of computing) Pertaining to a selection in which there are ten possible outcomes. **(B) (mathematics of computing)** Pertaining to the numeration system with a radix of ten. *Synonym:* denary. (C) 1084-1986

decimal alignment An operation in which two or more decimal numbers are arranged such that their decimal points, whether explicit or implicit, are aligned vertically; for example,

unaligned	aligned
1.4	1.4
5	5
.067	.067

See also: radix alignment. (C) 610.2-1987, 610.5-1990w

decimal character string A sequence of characters from the set of decimal digits the first of which shall not be the digit zero. Decimal character strings shall consist only of the following characters:

0 1 2 3 4 5 6 7 8 9

Within software definition files of exported catalogs, all such strings shall be encoded using IRV. (C/PA) 1387.2-1995

decimal code A code in which each allowable position has one of 10 possible states. The conventional decimal number system is a decimal code. (IA) [61], [84]

decimal-coded digit A digit or character defined by a set of decimal digits, such as a pair of decimal digits specifying a letter or special character in a system of notation. (C) 1084-1986w

decimal data Data used to represent decimal numbers; that is, numeric values expressed in base 10. *See also:* binary coded decimal real data; packed decimal data; unsigned packed decimal data; zoned decimal data; decimal picture data. (C) 610.5-1990w

decimal digit A numeral used to represent one of the ten digits in the decimal numeration system; 0, 1, 2, 3, 4, 5, 6, 7, 8, or 9. (C) 1084-1986w

decimal notation Any notation that uses the decimal digits and the radix 10. (C) 1084-1986w

decimal number (A) A quantity that is expressed using the decimal numeration system. **(B)** Loosely, a decimal numeral. (C) 1084-1986

decimal number system *See:* positional notation; decimal numeration system.

decimal numeral A numeral in the decimal numeration system. For example, the decimal numeral 12 is equivalent to the Roman numeral XII. (C) 1084-1986w

decimal numeration system The numeration system that uses the decimal digits and the radix 10. *Synonyms:* decimal system; decimal number system. (C) 1084-1986w

decimal picture data Arithmetic picture data that is associated with a picture specification that allows decimal digit characters, a radix point, zero-suppression characters, sign characters, currency symbol characters, insertion characters, commercial characters, and exponent characters. *Synonym:* numeric character data. *Contrast:* binary picture data. (C) 610.5-1990w

decimal point (mathematics of computing) The radix point in the decimal numeration system. *See also:* point. (C) 1084-1986w

decimal system *See:* decimal numeration system.

decimal-to-BCD conversion The process of converting a decimal numeral to an equivalent BCD numeral. For example, decimal 139 is converted to BCD 1 0011 1001. (C) 1084-1986w

decimal-to-binary conversion The process of converting a decimal numeral to an equivalent binary numeral. For example, decimal 139.25 is converted to binary 10001011.01. (C) 1084-1986w

decimal-to-hexadecimal conversion The process of converting a decimal numeral to an equivalent hexadecimal numeral. For example, decimal 139.25 is converted to hexadecimal 8B.4. (C) 1084-1986w

decimal-to-octal conversion The process of converting a decimal numeral to an equivalent octal numeral. For example, decimal 139.25 is converted to octal 213.2. (C) 1084-1986w

decineper One-tenth of a neper. (AP/ANT) 145-1983s

decipherment The reversal of a corresponding reversible encipherment. (LM/C) 802.10-1992

decision gate (navigation aid terms) A specified point near the lower end of an ILS (instrument landing system) approach at which a pilot must make a decision either to complete the landing or to execute a missed-approach procedure. (AES/GCS) 172-1983w

decision instruction (computers) An instruction that effects the selection of a branch of a program, for example, a conditional jump instruction. (C) [20], [85]

decision level concentration (DLC) Quantity of analyte at or above which an *a priori* decision is made that a positive quantity of the analyte is present. For IEEE Std N42.23-1995, the probability of a Type I error (probability of erroneously reporting a detectable nuclide in an appropriate blank or sample) is set at 0.05. (NI) N42.23-1995

decision point metric The result of dividing the total number of modules in which every decision point has had 1) all valid conditions, and 2) at least one invalid condition, correctly processed, by the total number of modules. *Note:* This definition assumes that the modules are essentially the same size. (C/SE) 730-1998

decision rule A rule or algorithm used in pattern classification to assign an observed unit of image data to a pattern class based on features extracted from the image. *Synonym:* classifier. (C) 610.4-1990w

decision support services (DSS) (A) The services provided by a decision support system. For example, software components for model building, forecasting, statistical analysis, ad hoc model interrogation, report generation, and graphics. **(B)** A computer system that supports decision making by performing such functions as modeling, forecasting, and statistical analysis. *See also:* management information system; computer-aided management. **(C)** The services provided by the staff of an information center. (C) 610.2-1987

decision support software Interactive software used in a decision support system. For example, software components for model building, forecasting, statistical analysis, ad hoc model interrogation, report generation, and graphics. (C) 610.2-1987

decision support system (DSS) A computer system that supports decision making by performing such functions as modeling, forecasting, and statistical analysis. *See also:* management information system; computer-aided management. (C) 610.2-1987

decision support system generator A package of decision support software that enables users to develop customized decision support systems for specific applications. (C) 610.2-1987

decision table (1) A matrix-providing program branching which may be a complex function of a number of variables. (ATLAS) 771-1980s
(2) (software) A table used to show sets of conditions and the actions resulting from them. (C) 610.12-1990

Decision Table Translator (D-TRAN) A computer language developed as a preprocessor that converts decision table constructs into conventional programming language code. (C) 610.13-1993w

deck (computers) A collection of punched cards. (C) [20], [85]

declaration A non-executable program statement that affects the assembler or compiler's interpretation of other statements in the program. For example, a statement that identifies a name, specifies what the name represents, and, possibly, assigns it an initial value. *Contrast:* control statement; assignment statement. *See also:* pseudo-instruction. (C) 610.12-1990

declarative language (1) A nonprocedural language that permits the user to declare a set of facts and to express queries or problems that use these facts. *See also:* interactive language; command language; rule-based language. (C) 610.12-1990, 610.13-1993w
(2) A programming language that can be understood without reference to the behavior of any particular computer system. (C) 610.13-1993w

declared curve (rotating electric machinery) A characteristic curve of the machine type, as obtained by averaging the results of testing four to ten machines, of which at least two shall have had a type test. (PE/EM) 11-1980r

declination rate of ON-state current (thyristor) Average rate of declination or fall of ON-state current measured from 50 percent IF to 0. (IA/IPC) 428-1981w

declinometer (navigation aid terms) An instrument for measuring magnetic declination. (AES/GCS) 172-1983w

decode (1) To produce a single output signal from each combination of a group of input signals. *See also:* translate; matrix. (C) 162-1963w
(2) (data management) To convert data by reversing the effect of previous encoding. *Contrast:* encode. (C) 610.5-1990w

decoder (1) (electronic computation) A matrix of logic elements that selects one or more output channels according to the combination of input signals present. (C) [20], [85]
(2) (telecommunications) A device that performs decoding. (COM/TA) 1007-1991r
(3) (data management) A device or system that decodes data. *Contrast:* encoder. (C) 610.5-1990w
(4) (A) A device that has a number of input lines such that any number may carry signals and a number of output lines such that no more than one at a time may carry a signal. *Note:* the combination of input signals serves as a code to indicate which output line carries the signal. *Synonyms:* decoder matrix; many-to-one decoder. **(B)** A device that can decode data. (C) 610.10-1994

decoder matrix *See:* decoder.

decoding Decoding is the translation from the coded set of bits (coded character) to the original set of bits (character). *See also:* coding. (C/BA) 1355-1995

decollate (1) To divide the items in a set into unique subsets. *Contrast:* collate. (C) 610.5-1990w
(2) To separate the parts of a multipart form, often by means of a device called a decollator. *Synonym:* deleave. *See also:* burst. (C) 610.10-1994w

decompile To translate a compiled computer program from its machine language version into a form that resembles, but may not be identical to, the original high-order language program. *Contrast:* compile. (C) 610.12-1990

decompiler (1) A software tool that decompiles computer programs. (C) 610.12-1990
(2) A software component that takes one or more compiled Forth commands and generates the equivalent text representation for those commands. (C/BA) 1275-1994

decomposition The partitioning of a modeled function into its component functions. (C/SE) 1320.1-1998

decomposition diagram A diagram that details its parent box. (C/SE) 1320.1-1998

decomposition potential (decomposition voltage) The minimum potential (excluding IR drop) at which an electrochemical process can take place continuously at an appreciable rate. *See also:* electrochemistry. (EEC/PE) [119]

decorrelation distance The direction-dependent distance over which the mutual coherence function falls to $1/e$ of its maximum value. (AP/PROP) 211-1990s

decorrelation time The time required for the mutual coherence function to decay to 1/e of its maximum value. (AP/PROP) 211-1990s

decoupled architecture A computer architecture in which a program is divided into two or more instruction streams, and a number of processors cooperate in the execution of the task. (C) 610.10-1994w

decoupling (1) The reduction of coupling. *See also:* coupling. (EEC/PE) [119]
(2) (software) The process of making software modules more independent of one another to decrease the impact of changes to, and errors in, the individual modules. *See also:* coupling. (C) 610.12-1990

decoupling network Electrical circuit for the purpose of preventing an electrical fast transient (EFT) signal applied to the equipment under test (EUT) from affecting other devices, equipment or systems that are not under test. *See also:* back filter. (SPD/PE) C62.45-1992r

decrement (test-pattern language) The action of reducing the arithmetic value of a counter by one. (TT/C) 660-1986w
(2) (A) (mathematics of computing) The quantity by which a variable is decreased. **(B) (mathematics of computing)** To decrease the value of a variable. *Contrast:* increment. **(C) (mathematics of computing)** To decrease the value of a variable by one. *Contrast:* increment. (C) 610.10-1994, 1084-1986
(3) (A) To decrease the value of a variable. **(B)** To decrease the value of a variable by one. *Contrast:* increment. (C) 610.10-1994

decremental energy cost The cost avoided by reducing the production of electric energy below some base level. (PE/PSE) 858-1993w

decrement factor An adjustment factor used in conjunction with the symmetrical ground fault current parameter in safety-oriented grounding calculations. It determines the rms equivalent of the asymmetrical current wave for a given fault duration, t_f, accounting for the effect of initial dc offset and its attenuation during the fault. (PE/SUB) 80-2000

dectra (navigation aid terms) An adaptation of the Decca low frequency (lf) radio navigation system in which two pairs of continuous wave (cw) transmitters are oriented so that the center lines of both pairs are along and at opposite ends of the same great circle path, to provide course guidance along and adjacent to the great circle path. Distance along track may be indicated by synchronized signals from one transmitter from each pair. (AES/GCS) 172-1983w

dedicated cable A cable containing only pairs servicing an electric power station. It is installed at any station with ground potential rise (GPR) above a given level (for example, 300 V rms), and will have a core and jacket dielectric capability suitable to withstand worst fault-produced voltage stress. (PE/PSC) 789-1988w

dedicated circuit *See:* leased circuit.

dedicated line *See:* leased line.

dedicated computer A special-purpose computer that can be used exclusively for one purpose, such as a dedicated word processing system or a numerical control system for machine tooling. (C) 610.10-1994w

dedicated service A CSMA/CD network in which the collision domain consists of two and only two DTEs so that the total network bandwidth is dedicated to supporting the flow of information between them. (C/LM) 802.3-1998

dedicated word processing Word processing performed on a system used exclusively for that purpose. *Contrast:* stand-alone word processing; clustered word processing; shared-resource word processing; shared-logic word processing. (C) 610.2-1987

de-emphasis (1) (data transmission) The use of an amplitude-frequency characteristic complementary to that used for pre-emphasis earlier in the system. (PE) 599-1985w
(2) (post emphasis) (post equalization) The use of an amplitude-frequency characteristic complementary to that used for pre-emphasis earlier in the system. *See also:* pre-emphasis. (PE/EEC) [119]

de-emphasis network A network inserted in a system in order to restore the pre-emphasized frequency spectrum to its original form. (AP/ANT) 145-1983s

de-energize (relay) To disconnect the relay from its power source. (EEC/REE) [87]

de-energized Free from any electrical connection to a source of potential difference and from electric charge; not having a potential different from that of the earth. *Note:* The term is used only with reference to current-carrying parts that are sometimes energized (alive). *Synonym:* dead.
(NESC/T&D/PE) C2-1997, 524a-1993r, 1048-1990, 516-1995, C2.2-1960, 524-1992r

deep-bar rotor A squirrel-cage induction-motor rotor having a winding that is narrow and deep giving the effect of varying secondary resistance, large at standstill and decreasing as the speed rises. *See also:* rotor. (PE) [9]

deep space (communication satellite) Space at distances from the earth approximately equal to or greater than the distance between the earth and the moon. (COM) [19]

deep space instrumentation facility (communication satellite) A ground network of worldwide communication stations (earth terminals) maintained for providing communications to and from lunar and inter-planetary spacecraft and deep space probes. Each earth terminal utilizes large antennas, low-noise receiving systems and high-power transmitters.
(COM) [24]

deepwell pump An electrically-driven pump located at the low point in the mine to discharge the water accumulation to the surface. (EEC/PE) [119]

de-excitation (excitation systems for synchronous machines) The removal of an excitation of a synchronous machine, main exciter, or pilot exciter. *Note:* De-excitation may be accomplished by various means, such as a dc field breaker, alternating-current (ac) supply breaker, static switches, phase-back control of controlled rectifiers, or a combination of these. (PE/EDPG) 421.1-1986r

de facto standard A standard that is developed informally when a single entity develops a product or technology and, through success and imitation, that product or technology becomes so widely used that deviation causes compatibility problems or limits marketability. One such example is the Hayes modem handshake protocol. (C) 610.7-1995, 610.10-1994w

default (1) (A) Pertaining to a value, attribute, or option that is assumed in place of a value, attribute or option when one is required, but not specified explicitly; for example, the default value for a field called MARITAL STATUS might be M (for married). **(B)** To assign the value, attribute, or option as in definition (A). (C) 610.5-1990
(2) The choice used when no specification is given.
(IM/AIN) 488.2-1992r

default font *See:* base font.

default option The value for an extended option as defined in a defaults file. (C/PA) 1387.2-1995

defaults file A system-specific or user-specific file that contains the default values for extended options used by the software administration utilities. (C/PA) 1387.2-1995

default slot generator function The function that defines the identity (i.e., Bus A or Bus B) for each bus of a Dual Bus subnetwork. Additionally, the function that provides Head of Bus functions for both Bus A and Bus B in a looped Dual Bus subnetwork. (If the looped Dual Bus subnetwork is reconfigured to an open Dual Bus subnetwork due to transmission link faults, then the Head of Bus functions for either one or both buses are assigned to other nodes for the duration of the fault.). (LM/C) 8802-6-1994

defeater *See:* interlocking deactivating means.

defect (1) (A) (software) A product anomaly. Examples include such things as omissions and imperfections found during early life cycle phases; and faults contained in software sufficiently mature for test or operation. *See also:* fault. **(B) (solar cells)** A localized deviation of any type from the regular structure of the atomic lattice of a single crystal. *See also:* fault.
(C/SE/AES/SS) 982.2-1988, 982.1-1988, 307-1969
(2) Imperfection or partial lack of performance that can be corrected without taking a transformer out of service.
(PE/TR) C57.117-1986r
(3) Imperfection in the state of an item (or inherent weakness) which can result in one or more failures of the item itself or of another item under the specific service or environmental or maintenance conditions for a stated period of time.
(SWG/PE) C37.10-1995
(4) Any nonconformance with specified requirements of the tested unit of product. (PE/T&D) C135.61-1997

deference A process by which a data station delays its transmission when the channel is busy to avoid collision with ongoing transmissions. (C) 610.7-1995

deferred address *See:* indirect address.

deferred batch service A service that is performed as a result of events that are asynchronous with respect to requests. *Note:* Once a batch job has been created, it is subject to deferred services. (C/PA) 1003.2d-1994

`defer` word A Forth word whose name has been entered into the dictionary (by the defining word `defer`), but whose action was left unresolved and may be resolved at a later time.
(C/BA) 1275-1994

deficiency A deviation from the accepted procedures, practices, or standards, or a defect in an item that could lead to degradation of quality. (NI) N42.23-1995

defined pulse width (semiconductor radiation detectors) The time elapsed between the first and final crossings of the defined zero level for the maximum rated output pulse amplitude. (NID) 301-1976s

defined reference pulse waveform (pulse measurement) A reference pulse waveform which is defined without reference to any practical or derived pulse waveform. Typically, a defined reference pulse waveform is an ideal pulse waveform.
(IM/WM&A) 181-1977w

defined zero (semiconductor radiation detectors) An arbitrarily chosen voltage level at the amplifier output resolvable from zero by the measuring apparatus. (NID) 301-1976s

definite-minimum-time relay An inverse-time relay in which the operating time becomes substantially constant at high values of input.
(SWG/PE/PSR) C37.100-1992, C37.90-1978s

definite-purpose circuit breaker (1) A circuit breaker that has been designed, tested, and rated in accordance with general-purpose circuit breaker requirements of applicable standards and that has been designed, tested, and rated in accordance with the requirements of one or more specific performance requirements for a definite-purpose circuit breaker.
(SWG/PE) C37.100-1992
(2) A definite purpose circuit breaker is one that is designed specifically for capacitance current switching.
(SWG) 341-1972w

definite-purpose controller Any controller having ratings, operating characteristics, or mechanical construction for use under service conditions other than usual or for use on a definite type of application. *See also:* electric controller.
(IA/ICTL/IAC) [60]

definite-purpose motor Any motor designated, listed, and offered in standard ratings with standard operating characteristics or mechanical construction for use under service con-

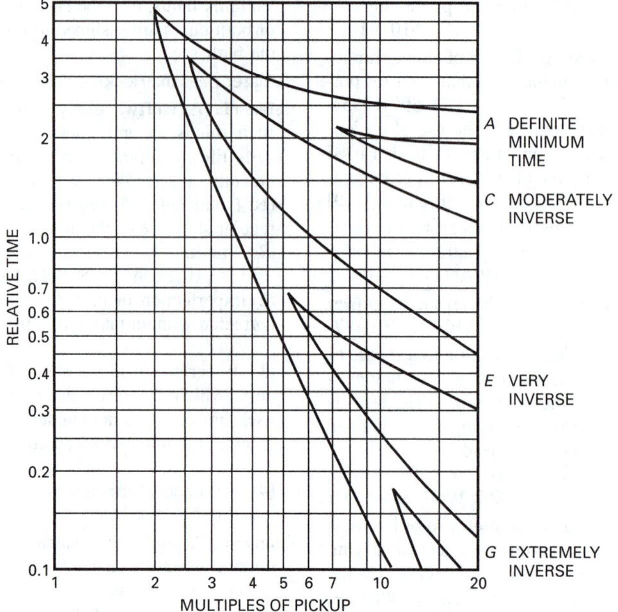

typical operating characteristic curve shapes of various inverse-time relays

ditions other than usual or for use on a particular type of application. *Note:* Examples: crane, elevator, and oil-burner motors. *See also:* asynchronous machine. (EEC/PE) [119]

definite time (relays) A qualifying term indicating that there is purposely introduced a delay action, which delay remains substantially constant regardless of the magnitude of the quantity that causes the action. *See also:* relay. (SWG/PE) C37.100-1981s

definite-time acceleration (electric drive) A system of control in which acceleration proceeds on a definite-time schedule. *See also:* electric drive. (IA/ICTL/IAC) [60]

definite-time delay A qualifying term indicating that there is purposely introduced a delay in action, which delay remains substantially constant regardless of the magnitude of the quantity that causes the action. (SWG/PE) C37.100-1992

definite-time relay A relay in which the operating time is substantially constant regardless of the magnitude of the input quantity. *See also:* relay. (SWG/PE/PSR) C37.100-1992, C37.90-1978s

definition (1) (facsimile) Distinctness or clarity of detail or outline in a record sheet, or other reproduction. (COM) 168-1956w
(2) (data management) *See also:* database definition. (C) 610.5-1990w

definition phase *See:* requirements phase.

definitions of classes (insulation systems of synchronous machines) Insulation systems are those which by service experience or accepted comparative tests with service proven systems can be shown to be capable of continuous operation with the limiting observable temperature rise or hottest spot total temperature as specified in the appropriate American National Standard, C50.13-1977 or C50.14-1977. Insulation systems of synchronous machines shall be classified as Class A, Class B, Class F, or Class H. (REM) [115]

deflecting electrode (electron-beam tubes) An electrode the potential of which provides an electric field to produce deflection of an electron beam. *See also:* electrode. (ED) 161-1971w

deflecting force (direct-current recording instrument) At any part of the scale (particularly full scale), the force for that position, measured at the marking device, and produced by the electrical quantity to be measured, acting through the mechanism. *See also:* accuracy rating. (PE/EEC) [119]

deflecting voltage (cathode-ray tubes) Voltage applied between the deflector plates to create the deflecting electric field. *See also:* cathode-ray tube. (Std100) [84]

deflecting yoke An assembly of one or more coils that provide a magnetic field to produce deflection of an electron beam. (ED) [45]

deflection axis, horizontal *See:* horizontal deflection axis.

deflection axis, vertical *See:* vertical deflection axis.

deflection blanking (oscilloscopes) Blanking by means of a deflection structure in the cathode-ray tube electron gun that traps the electron beam inside the gun to extinguish the spot, permitting blanking during retrace and between sweeps regardless of intensity setting. *See also:* oscillograph. (IM/HFIM) [40], 311-1970w

deflection center (electron-beam tubes) The intersection of the forward projection of the electron path prior to deflection and backward projection of the electron path in the field-free space after deflection. (ED) 161-1971w

deflection coefficient *See:* deflection factor.

deflection defocusing (cathode-ray tubes) A fault of a cathode-ray tube characterized by the enlargement, usually nonuniform, of the deflected spot which becomes progressively greater as the deflection is increased. *See also:* cathode-ray tube. (Std100) [84]

deflection factor (1) (inverse sensitivity) (general) The reciprocal of sensitivity. *Note:* It is, for example, often used to describe the performance of a galvanometer by expressing this in microvolts per millimeter (or per division) and for a mirror galvanomter at a specified scale distance, usually 1 meter. *See also:* accuracy rating. (ED) 161-1971w
(2) (oscilloscopes) The ratio of the input signal amplitude to the resultant displacement of the indicating spot, for example, volts per division. *See also:* oscillograph. (ED) 161-1971w
(3) (spectrum analyzer) The ratio of the input signal amplitude to the resultant output indication. The ratio may be in terms of volts root-mean-square (rms) per division, describes decibels (dB) per division, watts per division, or any other specified factor. (IM) 748-1979w

deflection plane (cathode-ray tubes) A plane perpendicular to the tube axis containing the deflection center. (ED) [45]

deflection polarity (oscilloscopes) The relation between the polarity of the applied signal and the direction of the resultant displacement of the indicating spot. *Note:* Conventionally a positive-going voltage causes upward deflection or deflection from left to right. *See also:* oscillograph. (IM/HFIM) [40]

deflection sensibility (1) (oscilloscopes) The number of trace widths per volt that can be simultaneously resolved anywhere within the quality area. *See also:* oscillograph. (IM/HFIM) [40]
(2) (oscilloscopes) The number of trace widths per volt of input signal that can be simultaneously resolved anywhere within the quality area. (IM) 311-1970w

deflection sensitivity (1) (magnetic-deflection cathode-ray tube and yoke assembly) The quotient of the spot displacement by the change in deflecting-coil current. (ED) 161-1971w, [45]
(2) (oscilloscopes) The reciprocal of the deflection factor (for example, divisions.volt). (IM) 311-1970w

deflection system A system that causes an electron beam to be aimed at a specific position on the screen of a cathode ray tube by creating an electrostatic or magnetic field around the beam. (C) 610.6-1991w

deflection yoke (television) An assembly of one or more coils whose magnetic field deflects an electron beam. (BT/ED/AV) 201-1979w, 161-1971w

deflection-yoke pull-back (cathode-ray tubes) (A) (color). The distance between the maximum possible forward position of the yoke and the position of the yoke to obtain optimum color purity. (2) (monochrome). The maximum distance the yoke can be moved along the tube axis without producing neck shadow. (ED) 161-1971w

deflector (1) (acousto-optics) A device which directs a light beam to an angular position in space upon application of an acoustic frequency. (UFFC) [17]
(2) A means for directing the flow of gas discharge from the vent of the arrester. (SPD/PE) C62.11-1999

deflector plates *See:* deflecting electrode.

defocusing The failure of rays to converge. (AP/PROP) 211-1997

defruiter Equipment that deletes random asynchronous unintentional returns in a beacon system. *Note:* Commonly used in secondary surveillance radars. (AES) 686-1997

degassing (electron tube) (rectifier) The process of driving out and exhausting occluded and remanent gases within the vacuum tank or tube, anodes, cathode, etcetera, that are not removed by evacuation alone. *See also:* rectification. (EEC/PE) [119]

degauss To apply a variable, alternating current field for the purpose of demagnetizing magnetic media or devices. *See also:* delete. (C) 610.10-1994w

degausser A device that removes unwanted magnetization from objects; commonly used to demagnetize read/write heads, and to erase infromation from magnetic storage media. (C) 610.10-1994w

degaussing (navigation aid terms) Neutralization of the strength of the magnetic field of a vessel. (AES/GCS) 172-1983w

degaussing coil A single conductor or a multiple-conductor cable so disposed that passage of current through it will neutralize or bias the magnetic polarity of a ship or portion of a ship. *Note:* Continuous application of current, with adjustment to suit changes of position or heading of the ship, is required to maintain a degaussed condition. (EEC/PE) [119]

degaussing generator An electric generator provided for the purpose of supplying current to a degaussing coil or coils. (EEC/PE) [119]

degeneracy (resonant device) The condition where two or more modes have the same resonance frequency. *See also:* waveguide. (AP/ANT) [35]

degenerate gas A gas formed by a system of particles whose concentration is very great, with the result of the Maxwell-Boltzmann law does not apply. Example: An electronic gas made up of free electrons in the interior of the crystal lattice of a conductor. *See also:* discharge. (Std100) [84]

degenerate mode (waveguide) In a uniform waveguide, one of a set of modes having the same exponential variation along the direction of the guide, but having different configurations in the transverse plane. In a cavity, one of a set of modes having the same natural frequency. (MTT) 146-1980w

degenerate parametric amplifier (nonlinear, active, and nonreciprocal waveguide components) An inverting parametric device for which the two signal frequencies are identical and equal to one-half the frequency of the pump. *Note:* This exact but restrictive definition is often relaxed to include cases where the signals occupy frequency bands that overlap. *See also:* parametric device. (MTT) 457-1982w

degenerate tree A tree in which each nonterminal node has exactly one subtree. (C) 610.5-1990w

degeneration *See:* feedback; negative feedback.

degradation failure Failure that is both gradual and partial. *Note:* In time, such a failure may develop into a complete failure. (IA/PSE) 1100-1999

degraded A failure that is gradual, partial, or both; for example, the equipment degrades to a level that, in effect, is a termination of the ability to perform its required function. (PE/NP) 933-1999

degraded backup (supervisory control, data acquisition, and automatic control) (station control and data acquisition) A backup capability that does not perform all of the functions of the primary system. (PE/SUB) C37.1-1994, C37.100-1992

degraded minute Sixty seconds of available time that has a BER worse that 10^{-6} and better than 10^{-3}. Degraded minutes are expressed as an integer. (COM/TA) 1007-1991r

degraded mode When part of the central alarm station equipment is inoperable, causing less then optimal operational conditions. (PE/NP) 692-1997

degraded voltage condition A voltage deviation, above or below normal, to a level that, if sustained, could result in unacceptable performance of, or damage to, the connected loads and/or their control circuitry. (SWG/PE/NP) 741-1997, C37.40-1981s

degree (A) With regard to a relation, the number of attributes in that relation. **(B)** With regard to a given node in a tree, the number of subtrees within that node. (C) 610.5-1990

degree Celsius (metric practice) It is equal to the kelvin and is used in place of the kelvin for expressing Celsius temperature (symbol *t*) defined by the equation $t = T - T_0$ where T is the thermodynamic temperature and $T_0 = 273.15$ K by definition. (QUL) 268-1982s

degree day A unit based upon temperature difference and time, which is used for estimating fuel consumption and for specifying nominal heating loads of buildings during the heating season. Degree days = number of degrees (°F) that the mean temperature is below 65°F × days. (IA/PSE) 241-1990r

degree of asymmetry (of a current at any time) The ratio of the direct-current component to the peak value of the symmetrical component determined from the envelope of the current wave at that time. *Note:* This value is 100% when the direct-current component equals the peak value of the symmetrical component. (SWG/PE) C37.100-1992

degree of coherence (fiber optics) A measure of the coherence of a light source; the magnitude of the degree of coherence is equal to the visibility, V, of the fringes of a two-beam interference experiment, where

$$V = \frac{I_{max} - I_{min}}{I_{max} + I_{min}}$$

I_{max} is the intensity at a maximum of the interference pattern, and I_{min} is the intensity at a minimum. *Note:* Light is considered highly coherent when the degree of coherence exceeds 0.88, partially coherent for values less than 0.88, and incoherent for "very small" values. *See also:* coherence length;

coherent; interference; coherence area. (Std100) 812-1984w

degree of distortion (data transmission) At the digital interface for binary signals, a measure of the time displacement of the transitions between signal states from their ideal instants. The degree of distortion is generally expressed as a percentage of the unit interval. (PE) 599-1985w

degree-of-freedom (gyros) An allowable mode of angular motion of the spin axis with respect to the case. The number of degrees-of-freedom is the number of orthogonal axes about which the spin axis is free to rotate. (AES/GYAC) 528-1994

degree of gross start-stop distortion (data transmission) The degree of start-stop distortion determined using the unit interval which corresponds to the actual mean modulation rate of the signal involved. (PE) 599-1985w

degree of individual distortion (of a particular signal transition) (data transmission) The ratio to the unit interval of the displacement, expressed algebraically, of this transition from its ideal instant. This displacement is considered positive when the transition occurs after its ideal instant (late). (PE) 599-1985w

degree of isochronous distortion (A) (data transmission) The ratio to the unit interval of the maximum measured difference, irrespective of sign, between the actual and the theoretical intervals separating any two transitions of modulation (or of restitution), these transitions not being necessarily consecutive. **(B) (data transmission)** The algebraical difference between the highest and lowest value of individual distortion affecting the transitions of an isochronous modulation. (This difference is independent of the choice of the reference ideal instant.) The degree of distortion (of an isochronous modulation or restitution) is usually expressed as a percentage. (PE) 599-1985

degree of longitudinal balance (measuring longitudinal balance of telephone equipment operating in the voice band) The ratio of the disturbing longitudinal voltage V_s and the resulting metallic voltage V_m of the network under test expressed in decibels, as follows:

$$\text{longitudinal balance} = 20\log|V_s/V_m| \text{ (dB)}$$

where V_s and V_m are of the same frequency. *Note:* Here, log is assumed to mean log to the base 10. (COM/TA) 455-1985w

degree of polarization The fraction of the total power in a wave that is completely polarized. *Note:* Sometimes the definition is further restricted to a given polarization state, as in degree of linear polarization. (AP/PROP) 211-1997

degree of start-stop distortion (data transmission) The ratio to the unit interval of the maximum measured difference (irrespective of sign) between the actual interval and the theoretical interval (the appropriate integral multiple of unit intervals) separating any transition from the start transition preceding it. (PE) 599-1985w

dehumidification Condensation of water vapor from the air by cooling below the dew point, or removal of water vapor from air by physical or chemical means. (IA/PSE) 241-1990r

deionization time (gas tube) The time required for the grid to regain control after anode-current interruption. *Note:* To be exact the dionization time of a gas tube should be presented as a family of curves relating such factors as condensed-mercury temperature, anode and grid currents, anode and grid voltages, and regulation of the rid current. *See also:* gas tube. (ED) 161-1971w

deionizing grid (gas tube) A grid accelerating deionization in its vicinity in a gas-filled valve or tube, and forming a screen between two regions within the envelope. (Std100) [84]

deka *See:* deca.

DEL *See:* delete character.

delay (1) (protection and coordination of industrial and commercial power systems) The opening time of a fuse when in excess of one cycle, where the time may vary considerably between types and makes and still be within established standards. This word, in itself, has no specific meaning other than in manufacturers' claims unless published standards specify delay characteristics. *See also:* time delay.
(2) (A) (phase) (data transmission) The time interval by which a pulse is time retarded with respect to a reference time. **(B) (time delay) (data transmission)** (General). The time interval between the manifestation of a signal at one point and the manifestation or detection of the same signal at another point. *Notes:* 1. Generally, the term time delay is used to describe a process whereby an output signal has the same form as an input signal causing it, but is delayed in time; that is, the amplification of all frequency components of the output are related by a single constant to those of corresponding input frequency components but each output component lags behind the corresponding input component by a phase angle proportional to the frequency of the component. 2. Transport delay is synonymous with time delay but usually is reserved for applications that involve the flow of material. (PE) 599-1985
(3) The amount of time by which a signal or event lags due to an external condition or event. *See also:* time delay; rotational delay; duration of unscheduled interrupt; delay line; propagation delay. (C) 610.10-1994w
(4) The time taken for a digital signal to propagate between two points. (C/DA) 1481-1999

delay, absolute *See:* absolute delay.

delay and power calculation module (DPCM) A software component compliant with the delay and power calculation system (DPCS) supplied by a semiconductor vendor that is responsible for computing instance-specific timing data under control of an EDA application. The DPCM is loaded into memory at run-time and linked to the application via the procedural interface (PI). A DPCM typically is created from DCL subrules compiled by the DCL compiler and linked together with run-time support modules. (C/DA) 1481-1999

delay and power calculation system (DPCS) The DCL language, the procedural interface (PI) for delay and power calculations, and text formats for physical design and parasitic information. (C/DA) 1481-1999

delay angle (thyristor converter) The time, expressed in degrees (1 cycle of the ac waveform = 360°), by the starting point of commutation is delayed by phase control in relation to rectifier operationing without phase control, including possible inherent delay angle. (IA/IPC) 444-1973w

delay angle, inherent *See:* inherent delay angle.

delay arc *See:* timing arc.

delay calculation language (DCL) The programming language used to calculate instance-specific timing data. DCL contains high-level constructs that can refer to the aspects of the design topology that influence timing and also express the sequence of calculations necessary to compute the desired delay and timing check limit values. (C/DA) 1481-1999

delay calculation language compiler A software program, used in conjunction with a C compiler, that reduces DCL from ASCII text to computer executable format. *See also:* delay and power calculation module. (C/DA) 1481-1999

delay circuit (pulse techniques) A circuit that produces an output signal that is delayed intentionally with respect to the input signal. *See also:* pulse. (NPS) 398-1972r

delay coincidence circuit (pulse techniques) A coincidence circuit that is actuated by two pulses, one of which is delayed by a specified time interval with respect to the other. *See also:* pulse. (NPS) 398-1972r

delay dispersion (1) The change in phase delay over a specified operating frequency range (dispersive delay line). (UFFC) 1037-1992w, [22]
(2) A delay line that has a transfer characteristic with a constant modulus and an argument (phase) that is a nonlinear function of frequency. The phase characteristic of devices of common interest is a quadratic function of frequency (linear FM), but in general may be represented by higher-order pol-

ynomials and/or other nonlinear functions. (UFFC) 1037-1992w

delay distortion (1) (data transmission) Phase delay distortion (also called phase distortion) which is either departure from flatness in the phase delay of a circuit, or system over the frequency range required for transmission or the effect of such departure on a transmitted signal, or envelope delay distortion, which is either departure from flatness in the envelope delay of a circuit or system over the frequency range required for transmission, or the effect of such departure on a transmitted signal. *See also:* envelope delay distortion; distortion. (PE) 599-1985w

(2) A distortion on communication lines due to different speeds of signals at different frequencies in a given transmission medium. *Synonym:* phase distortion. (C) 610.7-1995

delayed application (railway practice) The application of the brakes by the automatic train control equipment after the lapse of a predetermined interval of time following its initiation by the roadway apparatus. *See also:* automatic train control. (EEC/PE) [119]

delayed overcurrent trip *See:* overcurrent release; delayed release.

delayed-plan-position indicator A plan-position indicator in which the initiation of the time base is delayed. (AES/RS) 686-1990

delayed release (trip) A release with intentional delay introduced between the instant when the activating quantity reaches the release setting and the instant when the release operates. (SWG/PE) C37.100-1992

delayed sweep (A) (oscilloscopes) A sweep that has been delayed either by a predetermined period or by a period determined by an additional independent variable. **(B) (oscilloscopes)** A mode of operation of a sweep, as defined in definition (A). *See also:* radar. (IM/HFIM) [40]

delayed test A service test of a battery made after a specified period of time, which is usually made for comparison with an initial test to determine shelf depreciation. *See also:* battery. (EEC/PE) [119]

delayed z transform (data processing) The delayed z transform of $f(t)$, denoted $F(z, \Delta)$, is the z transform of $f(t - \Delta T)u(T - \Delta T)$, where $u(t)$ is the unit step function: that is,

$$F(z, \Delta) = \sum_{n=0}^{\infty} f(nT - \Delta T)u(nT - \Delta T)z^{-n}$$

$$0 < \Delta < 1$$

(IM) [52]

delay electric blasting cap An electric blasting cap with a delay element between the priming and detonating composition to permit firing of explosive charges in sequence with but one application of the electric current. *See also:* blasting unit. (EEC/PE) [119]

delay, envelope *See:* envelope delay.

delay equalizer (data transmission) A corrective network that is designed to make the phase delay or envelope delay of a circuit, or system, substantially constant over a desired frequency range. (PE) 599-1985w

delay equation Any mathematical expression describing cell delay or interconnect delay. (C/DA) 1481-1999

delaying (operations on a pulse) (pulse terminology) A process in which a pulse is delayed in time by active circuitry or by propagation. (IM/WM&A) 194-1977w

delaying sweep (oscilloscopes) A sweep used to delay another sweep. *See also:* delayed sweep; oscillograph. (IM/HFIM) [40]

delay instruction The instruction following a delayed control-transfer instruction. The delay instruction is always fetched, even if the delayed control transfer is an unconditional branch. However, the **annul bit** in the delayed control-transfer instruction can cause the delay instruction to be annulled (that is, to have no effect) if the branch is not taken (or in the branch-always case, if the branch is taken). (C/MM) 1754-1994

delay line (1) (A) (data transmission) Originally a device utilizing wave propagation for producing a time delay of a signal. **(B) (scintillation counting)** Commonly, a real or artificial transmission line or equivalent device designed to introduce delay. (PE/NPS) 599-1985, 398-1972

(2) (digital computers) A sequential logic element or device with one input channel in which the output-channel state at a given instant t is the same as the input-channel state at the instant t-n, that is, the input sequence undergoes a delay of n units. There may be additional taps yielding output channels with smaller values of n. *See also:* acoustic delay line; magnetic delay line; electromagnetic delay line; sonic delay line; pulse. (C/PE/PSCC) 162-1963w, 599-1985w

(3) A line or circuit designed to introduce a desired delay in the transmission of a signal. *Synonym:* delay unit. *See also:* magnetic delay line; electromagnetic delay line; electric delay line; acoustic delay line. (C) 610.10-1994w

(4) A device that operates over some defined range of electrical and environmental conditions as a linear passive circuit element. The transfer characteristic has a modulus and argument (phase) that may be constant or a function of frequency. (UFFC/UFFC) 1037-1992w, [22]

delay line, digital *See:* digital delay line.

delay line, dispersive *See:* dispersive delay line.

delay-line memory *See:* delay-line storage.

delay line, nondispersive *See:* nondispersive delay line.

delay-line storage (1) (electronic computation) (delay-line memory) A storage or memory device consisting of a delay line and means for regenerating and reinserting information into the delay line. (MIL) [2]

(2) A storage technique in which data are stored by sending them through a circuit loop having a data capacity (measured in bits) equal to its propagation delay for one complete pass around the loop, measured in bit times past the read position. *Synonym:* time delay register. (C) 610.10-1994w

delay, phase *See:* phase delay.

delay pickoff (oscilloscopes) A means of providing an output signal when a ramp has reached an amplitude corresponding to a certain length of time (delay interval) since the start of the ramp. The output signal may be in the form of a pulse, a gate, or simply amplification of that part of the ramp following the pickoff time. (IM/HFIM) [40]

delay, pulse *See:* pulse delay.

delay pulsing (telephone switching systems) A method of pulsing control and trunk integrity check wherein the sender delays the sending of the address pulses until it receives from the far end an off-hook signal (terminating register not yet attached), followed by a steady on-hook signal (terminating register attached). (COM) 312-1977w

delay relay A relay having an assured time interval between energization and pickup or between de-energization and dropout. *See also:* relay. (PE/EM) 43-1974s

delay, signal *See:* signal delay.

delay slope (1) (dispersive delay line) The ratio of the delay dispersion to the dispersive bandwidth. (UFFC) [22]

(2) The second derivative of radian phase with respect to radian frequency $\partial^2\phi/\partial\omega^2$, which is the instantaneous rate of change of the group delay with respect to frequency. This generally refers to the ideal or best-fit phase, ignoring undesired ripple. For a linear FM dispersive delay line, this is equal to the ratio of the delay dispersion to the dispersive bandwidth (dispersive delay line). (UFFC) 1037-1992w

delay spread *See:* time delay spread.

delay switching system (telephone switching systems) A switching system in which a call is permitted to wait until a path becomes available. (COM) 312-1977w

delay time (1) (railway practice) The period or interval after the initiation of an automatic train-control application by the roadway apparatus and before the application of the brakes becomes effective. *See also:* automatic train control. (EEC/PE) [119]

(2) (nondispersive delay line) The transit time of the envelope of an RF tone burst. (UFFC) [22]

(3) In data communications, the wait time period between a signal sent and a signal received. *See also:* time delay; propagation delay. (C) 610.7-1995

(4) The transit time of the envelope of an RF tone burst (nondispersive delay line). (UFFC) 1037-1992w

delay unit *See:* delay line.

deleave *See:* decollate.

delete (A) (data management) To remove data from a storage device or data medium. *Synonym:* erase. *See also:* update; read; write. **(B) (data management)** To render data unretrievable, although it may continue to be physically present on a storage device. (C) 610.5-1990, 610.10-1994

delete access A type of access to data in which the data may be deleted. *See also:* read/write access; read-only access; write access; update access. (C) 610.5-1990w

delete character (DEL) (data management) A control character used to obliterate an erroneous or unwanted character. *Note:* On a perforated tape, this character consists of a card hole in each punch position. *Synonyms:* rub-out character; erase character. (C) 610.5-1990w

delete transaction A transaction that causes a record to be deleted from a master file. *See also:* change transaction; null transaction; update transaction; add transaction. (C) 610.2-1987

deleted source statements Source statements that are removed or modified from an existing software product as a new product is constructed. (C/SE) 1045-1992

deletion record (A) (data management) A record that indicates that data is to be deleted. *Contrast:* add record. **(B) (data management)** A record that has been deleted from a master file. *Contrast:* add record. (C) 610.5-1990

delimiter (1) (data management) (software) A bit, character, or set of characters used to denote the beginning or end of a group of related bits, characters, words, or statements. For example, the ampersand "&" in the character string "&APPLE&." (C) 610.5-1990w, 610.12-1990

(2) A character that provides punctuation in an ATLAS statement. (SCC20) 771-1998

(3) A nondata symbol transmitted on a serial link to delineate the beginning or end of a frame of data. (EMB/MIB) 1073.4.1-2000

delink (data management) To retrieve and delete an item from a linked list. (C) 610.5-1990w

deliverable software product A software product that is required by the contract to be delivered to the acquirer or other designated recipient. (C/SE) J-STD-016-1995

delivered source statements Source statements that are incorporated into the product delivered to the customer. (C/SE) 1045-1992

delivery (software) Release of a system or component to its customer or intended user. *See also:* system life cycle; software life cycle. (C) 610.12-1990

delivery queue One of two alternative databases that the service uses to convey objects to the client of the MA interface. (C/PA) 1224.1-1993w

Dellinger effect *See:* radio fadeout.

DeLoach measurement (nonlinear, active, and nonreciprocal waveguide components) A method used in the characterization of varactor diodes that involves the measurement of device resonance parameters in a reduced-height waveguide transmission line. Parameters such as cutoff frequency, Q, capacitance, and inductance can be realized from this measurement. (MTT) 457-1982w

delta The difference between a partial-select output of a magnetic cell in a ONE state and a partial-select output of the same cell in ZERO state. *See also:* coincident-current selection. (Std100) 163-1959w

delta B *See:* delta induction.

delta connection (power and distribution transformers) So connected that the windings of a three-phase transformer (or the windings for the same rated voltage of single-phase transformers associated in a three-phase bank) are connected in series to for a closed circuit. (PE/TR) C57.12.80-1978r

delta-function light source (scintillation counting) A light source whose rise time, fall time, and FWHM (full width at half maximum) are not more than one third of the corresponding parameters of the output pulse of the photomultiplier. (NPS) 398-1972r

delta induction (toroidal magnetic amplifier cores) The change in induction (flux density) when a core is in a cyclically magnetized condition. (Std100) 106-1972

delta, minimum *See:* minimum delta.

delta network (1) A network or that part of a network that consists of three branches connected among three terminals. (CAS) [13]

(2) An artificial mains network of specified symmetric and asymmetric impedance used for two-wire mains operation and comprising resistors connected in delta formation between the two conductors, and each conductor and earth. *See also:* electromagnetic compatibility. (EMC/CHM) [51]

delta N_J (parametric mode) (nonlinear, active, and nonreciprocal waveguide components) A figure of merit for parametric amplifier varactors that relates to capacitance nonlinearity.

delta tan delta The increment in the dielectric dissipation factor (tan) of the insulation measured at two designated voltages. *Note:* When the values of power factors or dissipation factors are in the 0-0·10 range (see dielectric dissipation factor), the value of delta tan delta may be used as the equivalent of the power-factor tip-up value. (EM/PE) 286-1975w

delta_V_{gnd} Voltage differential of 0 V dc rail between any two worst-case points of use on the backplane, including terminating resistors and backplane capacitors. (C/BA) 896.2-1991w

demagnetization (magnetic tape pulse recorders for electricity meters) The removal of the residual magnetization by application of a demagnetizing alternating current (ac) field of sufficient initial magnitude. (ELM) C12.14-1982r

demand (1) (electric power systems in commercial buildings) The electrical load at the receiving terminals averaged over a specified interval of time. Demand is expressed in kilowatts, kilovoltamperes, kilovars, amperes, or other suitable units. The interval of time is generally 15 minutes, 30 minutes, or 60 minutes. *Note:* If there are two 50 hp motors (which drive 45 hp loads) connected to the electric power system, but only one load is operating at any time, the demand load is only 45 hp but the connected load is 100 hp. *Synonym:* demand load. (IA/PSE) 241-1990r, 141-1993r

(2) The average value of power or a related quantity over a specified interval of time. *Note:* Demand is expressed in kilowatts, kilovolt-amperes, kilovars, or other suitable units. (ELM) C12.1-1988

(3) The rate at which electric energy is being used. (PE/PSE) 858-1993w

(4) The rate of consumption, e.g., power, volume/hour. (AMR/SCC31) 1377-1997

demand, billing *See:* billing demand.

demand charge (power operations) That portion of the charge for electric service based upon a customer's demand. (PE/PSE) 858-1987s, 346-1973w

demand clause, ratchet *See:* ratchet demand clause.

demand, coincident *See:* coincident demand.

demand constant (pulse receiver) The value of the measured quantity for each received pulse, divided by the demand interval, expressed in kilowatts per pulse, kilovars per pulse, or other suitable units. (ELM) C12.1-1988

demand, contract *See:* contract demand.

demand deviation The difference between the indicated or recorded demand and the true demand, expressed as a percentage of the full-scale value of the demand meter or demand register. *See also:* pulse-count deviation. (ELM) C12.1-1982s

demand factor (1) (power operations) The ratio of the maximum coincident demand of a system, or part of a system, to the total connected load of the system, or part of the system, under consideration. (PE/PSE) 858-1987s
(2) (electric power systems in commercial buildings) The ratio of the maximum demand of a system to the total connected load of the system. *Notes:* 1. Since demand load cannot be greater than the connected load, the demand factor cannot be greater than unity. 2. Those demand factors permitted by the NEC (for example, services and feeders) must be considered in sizing the electric system (with few exceptions, this is 100%); otherwise, the circuit may be sized to support the anticipated load. (IA/PSE) 241-1990r
(3) The ratio of the maximum coincident demand of a system, or part of a system, to the total connected load of the system, or part of the system, under consideration. The resultant is always 1 or less and can range from 0.8 to 1 to as low as 0.15 to 0.25 for some plants with very low diversity. (IA/PSE) 141-1993r
(4) The ratio of the operating load demand of a system or part of a system to the total connected load of the system or part of the system under consideration. (IA/MT) 45-1998

demand failure rate (reliability data for pumps and drivers, valve actuators, and valves) The probability (per demand) of failure that a component will fail to operate upon demand when required to start, change state, or function. (PE/NP) 500-1984w

demand, instantaneous *See:* instantaneous demand.

demand, integrated *See:* integrated demand.

demand interval (1) (demand meter or register) The length of the interval of time upon which the demand measurement is based. *Note:* The demand interval of a block-interval demand meter is a specific period of time such as 15, 30, or 60 minutes during which the electric energy flow is average. The demand interval of a lagged-demand meter is the time required to indicate 90 percent of the full value of a constant load suddenly applied. Some meters record the highest instantaneous load. *See also:* demand meter; electricity meter. (ELM/PE/PSE) [54]
(2) (block-interval demand meter) The specified interval of time on which a demand measurement is based. *Note:* Intervals such as 15, 30, or 60 minutes are commonly specified. *See also:* demand meter—time characteristic. (ELM) C12.1-1988
(3) The period during which the demand is integrated. (PE/PSE) 858-1993w

demand-interval deviation (metering) The difference between the measured demand interval and the specified demand interval, expressed as a percentage of the specified demand interval. (ELM) C12.1-1988

demand load *See:* demand.

demand—maximum The highest demand measured over a selected period of time, such as 1 month. (ELM) C12.1-1981

demand meter (metering) A metering device that indicates or records the demand, maximum demand, or both. *Note:* Since demand involves both an electrical factor and a time factor, mechanisms responsive to each of these factors are required, as well as an indicating or recording mechanism. These mechanisms may be either separate from or structurally combined with one another. (ELM) C12.1-1988

demand meter—contact-making clock A device designed to close momentarily an electric circuit to a demand meter at periodic intervals. (ELM) C12.1-1988

demand meter—indicating A demand meter equipped with a readout that indicates demand, maximum demand, or both. (ELM) C12.1-1988

demand meter—integrating (block-interval) A meter that integrates power or a related quantity over a fixed-time interval, and indicates or records the average. (ELM) C12.1-1988

demand meter—lagged A meter that indicates demand by means of thermal or mechanical devices having an approximately exponential response. (ELM) C12.1-1988

demand meter—time characteristic (lagged-demand meter) The nominal time required for 90% of the final indication, with constant load suddenly applied. *Note:* The time characteristic of lagged-demand meters describes the exponential response of the meter to the applied load. The response of the lagged-demand meter to the load is continuous and independent of the selected discrete time intervals. (ELM) C12.1-1988

demand meter—timing deviation The difference between the elapsed time indicated by the timing element and the true elapsed time, expressed as a percentage of the true elapsed time. (ELM) C12.1-1988

demand, native system *See:* native system demand.

demand, noncoincident *See:* noncoincident demand.

demand paging (software) A storage allocation technique in which pages are transferred from auxiliary storage to main storage only when those pages are needed. *Contrast:* anticipatory paging. (C) 610.12-1990

demand priority A round-robin arbitration method to provide LAN access based on message priority level.local area networks. (C) 8802-12-1998

demand register A mechanism, for use with an integrating electricity meter, that indicates maximum demand and also registers electric energy (or other integrated quantity). (ELM) C12.1-1988

demand register—cumulative A register that indicates the sum of the previous maximum demand readings prior to reset. When reset, the present reading is added to the previous accumulated readings. The maximum demand for the present reading period is the difference between the present and previous readings. (ELM) C12.1-1988

demand register—multiple-pointer form An indicating demand register from which the demand is obtained by reading the position of the multiple pointers relative to their scale markings. The multiple pointers are resettable to zero. (ELM) C12.1-1988

demand register—single-pointer form An indicating demand register from which the demand is obtained by reading the position of a pointer relative to the markings on a scale. The single pointer is resettable to zero. (ELM) C12.1-1988

demand-totalizing relay A device designed to receive and totalize electric pulses from two or more sources for transmission to a demand meter or to another relay. *See also:* demand meter. (ELM) C12.1-1982s

demarcation strip (data transmission) The terminals at which the telephone company's service ends and the customer's equipment is connected. (PE) 599-1985w

demineralization The process of removing dissolved minerals (usually by chemical means). (Std100) [71]

demodulate To receive signals transmitted over a communications computer; and to convert them into electrical pulses that can serve as inputs to a computer system. *Contrast:* modulate. (C) 610.7-1995

demodulation (1) A modulation process wherein a wave resulting from previous modulation is employed to derive a wave having substantially the characteristics of the original modulating wave. *Note:* The term is sometimes used to describe the action of a frequency converter or mixer, but this practice is deprecated except in the case of shifting a single-sideband signal to baseband. (IT) [123]
(2) (data transmission) A modulation process wherein a wave resulting from previous modulation is employed to derive a wave substantial to the characteristics of the original modulating wave. *Note:* The term is sometimes used to describe the action of a frequency converter or mixer, but this practice is deprecated. (PE) 599-1985w

(3) **(overhead power lines)** The process by which the signal is recovered from a modulated carrier. (PE/T&D) 539-1990

(4) The reconversion of a modulated signal back into its original form by extracting the data from the modulated carrier. *Contrast:* modulation. (C) 610.7-1995

demodularization In software design, the process of combining related software modules, usually to optimize system performance. *See also:* lateral compression; downward compression; upward compression. (C) 610.12-1990

demodulator A device to effect the process of demodulation. *See also:* demodulation. (Std100) 270-1964w

demodulator-modulator *See:* modem.

demonstration (1) (safety systems equipment in nuclear power generating stations) The provision of evidence to support the conclusion derived from assumed premises. (PE/NP) 627-1980r

(2) (Class 1E battery chargers and inverters) A course of reasoning showing that a certain result is a consequence of assumed premises; an explanation or illustration, as in teaching by use of examples. (PE/NP) 650-1979s

(3) (software) A dynamic analysis technique that relies on observation of system or component behavior during execution, without need for post-execution analysis, to detect errors, violations of development standards, and other problems. *See also:* testing. (C) 610.12-1990

demultiplexer An electronic switch with one input and several outputs. Encoded selection signals control which output is connected to the input. *Contrast:* multiplexer. (C) 610.7-1995

demultiplexing The separation from a common input into several outputs. For example, hardware may demultiplex signals from a transmission line based on time or carrier frequency to allow multiple, simultaneous transmissions across a single physical cable. *Contrast:* multiplexing. (C) 610.7-1995

denary *See:* decimal.

denial of message service Preventing or delaying the performance of legitimate or critical communication services. Denial of message service attacks may be perpetrated at any point in the communication architecture (e.g., data link, network, transport, application), and could result in denial of service conditions. (C/BA) 896.3-1993w

denormalized number (1) (mathematics of computing) A nonzero floating-point number whose exponent has a reserved value, usually the format's minimum, and whose explicit or implicit significand digit is zero. (C) 1084-1986w

(2) (binary floating-point arithmetic) A nonzero floating-point number whose exponent has a reserved value, usually the format's minimum, and whose explicit or implicit leading significand bit is zero. (MM/C) 754-1985r

dense binary code A binary code in which all possible bit combinations are used. (C) 1084-1986w

dense list *See:* packed array.

densitometer (illuminating engineering) A photometer for measuring the optical density (common logarithm of the reciprocal of the transmittance or reflectance) of materials. (EEC/IE) [126]

density (1) (facsimile) A measure of the light-transmitting or reflecting properties of an area. *Notes:* 1. It is expressed by the common logarithm of the ratio of incident to transmitted or reflected light flux. 2. There are many types of density that will usually have different numerical values for a given material; for example, diffuse density, double diffuse density, specular density. The relevant type of density depends upon the geometry of the optical system in which the material is used. *See also:* scanning. (COM) 168-1956w

(2) (electron or ion beam) The density of the electron or ion current of the beam at any given point. (Std100) [84]

(3) (computers) *See also:* packing density.

(4) (A) On an integrated circuit, the number of logic gates per unit area of usable surface. *See also:* surface density; chip density. **(B)** A measure of the number of characters per inch on an output medium such as paper. *See also:* recording density. (C) 610.10-1994

density coefficient *See:* environmental coefficient.

density-modulated tube (space-charge-control tube) (microwave tubes) Microwave tubes or valves characterized by the density modulation of the electron stream by a gating electrode. *Note:* The electron stream is collected on those electrodes that form a part of the microwave circuit, principally the anode. These electrodes are often small compared to operating wavelength so that for this reason space-charge-control tubes or valves are often not considered to be microwave tubes even though they are used at microwave frequencies. (ED) [45]

density modulation (electron beams) The process whereby a desired time variation in density is impressed on the electrons of a beam. *See also:* velocity-modulated tube. (Std100) [84]

density-tapered array antenna *See:* space-tapered array antenna.

denuder That portion of a mercury cell in which the metal is separated from the mercury. (EEC/PE) [119]

dependability (of a relay or relay system) The facet of reliability that relates to the degree of certainty that a relay or relay system will operate correctly. (SWG/PE/PSR) C37.100-1992, C37.90-1978s

dependable capability (power operations) The maximum generation, expressed in kilowatt-hours per hour (kWh/h) which a generating unit, station, power source, or system can be depended upon to supply on the basis of average operating conditions. (PE/PSE) 858-1987s

dependable capacity The maximum capacity modified for ambient limitations that a generating unit, power plant, item of electrical equipment, or system can sustain over a specified period of time. (PE/PSE) 858-1993w, 762-1987w

dependency (1) A logical relationship between two tests or between a test and an element of the unit under test (UUT) (either an actual part or a failure mode of a part). A diagnostic test is said to be dependent on a particular diagnostic element or test if the failed outcome of the test implies the failed condition of the diagnostic element or test on which the dependent test relies. (ATLAS) 1232-1995

(2) *See also:* multivalued dependency; functional dependency; join dependency; nontransitive dependency; transitive dependency.

dependency model A diagram, list, or topological graph indicating which events depend on related elements and other events. (ATLAS) 1232-1995

dependency notation A means of obtaining simplified symbols for complex elements by denoting the relationships between inputs, outputs, or inputs and outputs, without actually showing all the elements and interconnections involved. (GSD) 91-1984r

dependency_spec A `software_spec` that describes a dependency. (C/PA) 1387.2-1995

dependent A software object that specifies a prerequisite, corequisite or exrequisite on another software object. (C/PA) 1387.2-1995

dependent biaxial test A test in which the horizontal and the vertical acceleration components are derived from a single-input signal. (SWG/PE/PSR) C37.100-1992, C37.98-1977s

dependent contact A contacting member designed to complete any one of two or three circuits, depending on whether a two- or a three-position device is considered. (EEC/PE) [119]

dependent entity An entity for which the unique identification of an instance depends upon its relationship to another entity. Expressed in terms of the foreign key, an entity is said to be dependent if any foreign key is wholly contained in its primary key. *Synonym:* identifier-dependent entity. *Contrast:* independent entity. (C/SE) 1320.2-1998

dependent manual operation An operation solely by means of directly applied manual energy, such that the speed and force of the operation are dependent upon the action of the attendant. (SWG/PE) C37.100-1992

dependent node (network analysis) A node having one or more incoming branches. (CAS) 155-1960w

dependent power operation An operation by means of energy other than manual, where the completion of the operation is dependent upon the continuity of the power supply (to solenoids, electric or pneumatic motors, etc.). (SWG/PE) C37.100-1992

dependent segment In a hierarchical data-base, a segment that is not the root segment. (C) 610.5-1990w

dependent state class A class whose instances are, by their very nature, intrinsically related to certain other state class instance(s). It would not be appropriate to have a dependent state class instance by itself and unrelated to an instance of another class(es) and, furthermore, it makes no sense to change the instance(s) to which it relates. *Contrast:* independent state class. (C/SE) 1320.2-1998

dependent variable A variable whose value is dependent on the values of one or more independent variables. *Contrast:* independent variable. (C) 610.3-1989w

deperm To remove, as far as practicable, the permanent magnetic characteristic of a ship's hull by powerful external demangetizing coils. (EEC/PE) [119]

depletion (metal-nitride-oxide field-effect transistor) The state of the silicon surface in the insulated-gate field-effect transistor (IGFET) structure when a gate voltage of such polarity has been applied that all majority carriers have been repelled. The space charge region so formed is depleted of all mobile majority carriers. (ED) 581-1978w

depletion layer (x-ray energy spectrometers) (region) (in a semiconductor) A region in which the charge-carrier charge density is insufficient to neutralize the net fixed charge density of donors and acceptors. In a diode-type semiconductor radiation detector the depletion region is the sensitive region of the device. (NPS/NID) 759-1984r

depletion mode transistor (metal-nitride-oxide field-effect transistor) An insulated-gate field-effect transistor (IGFET) where the channel connecting source and drain is a preexisting thin layer of the same conductivity type as the source and drain. (ED) 581-1978w

depletion region (1) (in a semiconductor) A region in which the charge-carrier charge density is insufficient to neutralize the net fixed charge density of ionized donors and acceptors. In a diode-type semiconductor radiation detector the depletion region is the sensitive region of the device. (NPS/NID) 300-1988r, 301-1976s

(2) A region in which the mobile charge-carrier density is insufficient to neutralize the net fixed charge density of donors and acceptors. In a diode-type semiconductor radiation detector the depletion region is the active (sensitive) region. (NPS) 325-1996

depletion voltage (1) The voltage at which a junction detector becomes fully depleted. (NPS) 325-1996

(2) (charged-particle detectors) (germanium gamma-ray detectors) (of a semiconductor radiation detector) The voltage at which a junction detector becomes fully depleted. (NPS) 300-1988r

depolarization (1) (electrochemistry) A decrease in the polarization of an electrode at a specified current density. *See also:* electrochemistry. (EEC/PE) [119]

(2) (medical electronic biology) A reduction of the voltage between two sides of a membrane or interface below an initial value. *See also:* electrochemistry; contact potential. (EMB) [47]

(3) The conversion of power from a reference polarization into the cross polarization. (AP/ANT) 145-1993

(4) A process by which the polarization state of a wave is altered. (AP/PROP) 211-1997

depolarization field (primary ferroelectric terms) A self-generated electric field that opposes the spontaneous polarization. In a crystal of finite dimensions there will be a discontinuity of P_s at the crystal surfaces, which gives rise to a bound polarization charge ($Q = P_s a$) of surface density P_s. This charge gives rise to an electric field called the depolarizing field, which opposes the spontaneous polarization. The magnitude of this depolarizing field depends on the shape of the crystal. In real materials the depolarizing field is neutralized by the flow of free charge through the crystal or from the environment. For electrically insulating crystals in an insulating environment, this process of neutralization may be very slow. (UFFC) 180-1986w

depolarization front (medical electronics) The border of a wave of electric depolarization, traversing an excitable tissue that has appreciable width and thickness as well as length. *See also:* contact potential. (EMB) [47]

depolarizer (1) A substance or a means that produces depolarization. (EEC/PE) [119]

(2) (primary cell) A cathodic depolarizer that is adjacent to or a part of the positive electrode. *See also:* electrolytic cell; electrochemistry. (EEC/PE) [119]

depolarizer, optical *See:* optical depolarizer.

depolarizing mix (primary cell) A mixture containing a depolarizer and a material to improve conductivity. *See also:* electrolytic cell. (EEC/PE) [119]

deposit attack (deposition corrosion) Pitting corrosion resulting from deposits on a metallic surface. (IA) [59]

deposited-carbon resistor A resistor containing a thin coating of carbon deposited on a supporting material. (PE/EM) 43-1974s

deposition corrosion *See:* deposit attack.

depot maintenance (test, measurement, and diagnostic equipment) Maintenance performed on material requiring major overhaul or a complete rebuild of parts, subassemblies, and end items, including the manufacture of parts, modification, testing, and reclamation as required. Depot maintenance serves to support lower categories of maintenance by providing technical assistance and performing that maintenance beyond their responsibility. Depot maintenance provides stocks of serviceable equipment by using more extensive facilities for repair than are available in lower level maintenance activities. (MIL) [2]

depth *See:* height.

depth control (cable plowing) The means used to maintain a predetermined plowing depth. (T&D/PE) 590-1977w

depth cueing (computer graphics) The simulation of three dimensions on a two-dimensional graphical display device by modulation of line intensity, use of perspective, use of stereopsis and kinetic depth effects, and other techniques. *Synonym:* intensity cueing. (C) 610.6-1991w

depth-finder (navigation aid terms) An instrument for determining the depth of water, particularly an echo sounder. (AES/GCS) 172-1983w

depth-first search (DFS) (data management) A search of a tree using preorder traversal. (C) 610.5-1990w

depth of current penetration (induction heating usage) The thickness of a layer extending inward from the surface of a conductor which has the same resistance to direct current as the conductor as a whole has to alternating current of a given frequency. *Note:* About 87 percent of the heating energy of an alternating current is dissipated in the so-called depth of penetration. *See also:* induction heating. (IA) 54-1955w, 169-1955w

depth of discharge (DOD) (1) (batteries) The ampere-hours removed from a fully-charged battery, expressed as a percentage of its rated capacity at the applicable discharge rate. (PV) 1013-1990

(2) The ampere-hours removed from a fully charged battery, expressed as a percentage of its rated capacity at the applicable discharge rate. (PV) 1144-1996

depth of focus In synthetic-aperture radar (SAR), the range interval over which the cross-range resolution is maintained without introducing or changing the focusing correction in signal processing. (AES) 686-1997

depth of heating (dielectric heating) The depth below the surface of a material in which effective dielectric heating can be confined when the applicator electrodes are applied adjacent to one surface only. (IA) 54-1955w, 169-1955w

depth of modulation (amplitude, frequency, and pulse modulation) The ratio of the maximum minus minimum light intensity to the sum of the maximum and minimum light intensity, namely: $m = (I_{max} - I_{min}) / (I_{max} + I_{min})$. This applies to either the diffracted or the zero order. (UFFC) [17]

depth of penetration *See:* depth of current penetration.

depth of velocity modulation (electron beams) The ratio of the amplitude of a stated frequency component of the varying velocity of an electron beam, to the average beam velocity. (ED) 161-1971w, [45]

deque *See:* double-ended queue.

dequeue To retrieve and delete an item from a queue. *Contrast:* enqueue. (C) 610.5-1990w

derail detector A device so arranged as to detect a derailment condition. (VT) 1475-1999

derated generation (electric generating unit reliability, availability, and productivity) The generation that was not available due to unit deratings. DG = equivalent unit derated hours · maximum capacity = EUNDH · MC. (PE/PSE) 762-1987w

derated operation Use of equipment or a system at a more restricted performance level than that for which the equipment or system was originally designed. Derated operation is usually implemented either to forestall failures or as a result of system component failure. (SUB/PE) 1303-1994

derating The intentional reduction of stress/strength ratio in the application of an item, usually for the purpose of reducing the occurrence of stress-related failures. *See also:* reliability. (R) [29]

derivative control action (1) (automatic generation control) A control mode designed to provide a controller output proportional to the rate of change of the input. *See also:* derivative control action. (PE/PSE) 94-1991w
(2) The component of control action for which the output is proportional to the rate of change to the input. *See also:* control system, feedback; feedback control system. (IA/ICTL/IAC) [60]

derivative indexing (software) Automatic indexing in which the keywords are extracted directly from the text of the document or information being indexed. For example, keyword in context index. *Synonyms:* derived-term indexing; extraction indexing. *Contrast:* assigned indexing. *See also:* uniterm indexing. (C) 610.2-1987

derivative time (speed governing of hydraulic turbines) The derivative time, T_n, of a derivative element is also the derivative gain, G_n. The derivative gain is the ratio of the element's percent output to the time derivative of the element's percent input (input slope with respect to time).

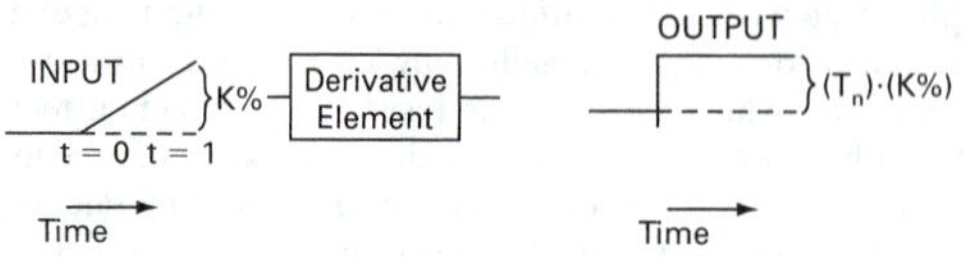

INPUT } K% — Derivative Element — OUTPUT } $(T_n)\cdot(K\%)$
t = 0 t = 1
Time → Time →

derivative time

(CS/PE/EDPG) 125-1977s

derived attribute *See:* derived property.

derived data (data management) Data that is computed or otherwise obtained from other data by application of a specified procedure. (C) 610.5-1990w

derived data element (data management) A data element whose entries are obtained from those in another data element by application of a specified procedure; for example, entries in a data element "age" could be derived from entries in the data element "date of birth." (C) 610.5-1990w

derived envelope (navigation aid terms) (loran C) The waveform equivalent to the summation of the video envelope of the rf (radio frequency) received pulse and the negative of its derivative, in proper proportion; the resulting envelope has a zero crossing at a standard point (for example, 25 μs) from the pulse beginning, serving as an accurate reference point for envelope time-difference measurements, and as a gating point in rejecting the latter part of the received pulse which may be contaminated by skywave transmissions. (AES/GCS) 172-1983w

derived font A font that has been scaled, or modifed from a scalable font. *Contrast:* bit map font. (C) 610.10-1994w

derived gauge The computed difference between values of a counter type attribute sampled in successive scan intervals. (LM/C) 802.1F-1993r

derived MAC protocol data unit (DMPDU) The Protocol Data Units (PDUs) of a length of 48 octets formed by the addition of protocol control information (including message identifier and error protection information) to each of the 44-octet segmentation units created from the segmentation of an Initial MAC Protocol Data Unit (IMPDU). Each DMPDU is carried as the payload of a Queued Arbitrated (QA) segment. (LM/C) 8802-6-1994

derived participant property *See:* derived property; participant property.

derived property The designation given to a property whose value is determined by computation. The typical case of a derived property is as a derived attribute although there is nothing to prohibit other kinds of derived property. (C/SE) 1320.2-1998

derived pulse (navigation aid terms) (loran C) A pulse derived by summing the received rf (radio frequency) pulse and an oppositely phased rf pulse so that it has an envelope which is the derivative of the received rf pulse envelope; the resultant envelope has a zero point and an rf phase reversal at a standard interval (for example, 25 μs) from the pulse beginning and it serves as an accurate reference for cycle time-difference measurements and as a gating point in rejecting the latter part of the received pulse which may be contaminated by sky-wave transmissions. (AES/GCS) 172-1983w

derived reference pulse waveform (pulse measurement) A reference pulse waveform which is derived by a specified procedure or algorithm from the pulse waveform which is being analyzed in a pulse measurement process. (IM/WM&A) 181-1977w

derived relation (data management) A relation in a database that can be entirely obtained from previously defined base relations by applying some sequence of relational operators. *Contrast:* base relation. (C) 610.5-1990w

derived requirement A requirement deduced or inferred from the collection and organization of requirements into a particular system configuration and solution. (C/SE) 1233-1998

derived-term indexing *See:* derivative indexing.

derived traceable sources Sources prepared or derived from certified sources that have been calibrated in accordance with this standard. Examples are dilutions of a liquid standard and special geometries such as charcoal cartridges, marinelli (reentrant) beakers, and filters, manufacturer. Any organization that produces and distributes sources that are certified with respect to radionuclide activity or a radiation emission rate. (NI) N42.22-1995

derived type (software) A data type whose members and operations are taken from those of another data type according to some specified rule. *See also:* subtype. (C) 610.12-1990

derrick An apparatus consisting of a mast or equivalent members held at the top by guys or braces, with or without a boom, for use with a hoisting mechanism and operating ropes. *See also:* elevator. (EEC/PE) [119]

DES *See:* data encryption standard.

descendant *See:* child.

descendant node (data management) In a tree, a node that is in a subtree of a given node. *See also:* parent node; ancestor.

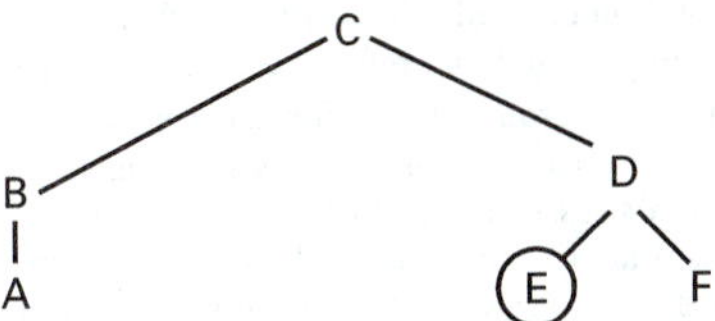

Node E Is a descendant node of nodes C and D

descendant node

(C) 610.5-1990w

descendent box A box in a descendent diagram. (C/SE) 1320.1-1998

descendent diagram A decomposition diagram related to a specific box by a hierarchically consecutive sequence of one or more child/parent relationships. (C/SE) 1320.1-1998

descender The portion of a graphic character that extends below the main part of the character; for example the lower portion of the letters "g" and "y." *Contrast:* ascender. (C) 610.2-1987

descending node (communication satellite) The point of the line of nodes that the satellite passes through as the satellite travels from above to below the equatorial plane. (COM) [19]

describing function (control system feedback) (nonlinear element under periodic input) A transfer function based solely on the fundamental, ignoring other frequencies. *Note:* This equivalent linearization implies amplitude dependence with or without frequency dependence. *See also:* feedback control system. (IM/PE/EDPG) [120], [3]

description standard (software) A standard that describes the characteristics of product information or procedures provided to help understand, test, install, operate, or maintain the product. (C) 610.12-1990

descriptive adjectives (A) (pulse terminology) *major (minor).* Having or pertaining to greater (lesser) importance, magnitude, time, extent, or the like, than another similar feature(s). **(B) (pulse terminology)** *ideal.* Of or pertaining to perfection in, or existing as a perfect exemplar or, a waveform or a feature. **(C) (pulse terminology)** *reference.* Of or pertaining to a time, magnitude, waveform, feature, or the like which is used for comparison with, or evaluation of, other times, magnitudes, waveforms, features, or the like. A reference entity may, or may not, be an ideal entity. (IM/WM&A) 194-1977

descriptive model A model used to depict the behavior or properties of an existing system or type of system; for example, a scale model or written specification used to convey to potential buyers the physical and performance characteristics of a computer. *Synonym:* representational model. *Contrast:* prescriptive model. (C) 610.3-1989w

descriptor (1) The means by which the client and service exchange an OM attribute value, and the integers that denote its representation, type, and syntax. (C/PA) 1224.2-1993w
(2) The portion of the routing information field that indicates the individual segment and bridge of the network path. A series of descriptors therefore describe a path through the network. (C/LM/CC) 8802-2-1998
(3) *See also:* keyword.

descriptor list An ordered sequence of descriptors that is used to represent several OM attribute types and values. (C/PA) 1328.2-1993w, 1326.2-1993w, 1327.2-1993w, 1224.2-1993w

desensitizing relay A relay that prevents tripping of a network protector on transient power reversals, which neither exceed a predetermined value nor persist for a predetermined time. (PE/TR) C57.12.44-1994

deserialization The assembly of a coded character from the sequence of serial bits. *See also:* serialization. (C/BA) 1355-1995

design (n, v) (1) (A) (data management) (software) The process of defining the architecture, components, interfaces, and other characteristics of a system or component. *See also:* database design; detailed design; preliminary design. **(B) (software) (data management)** The result of the process in definition (A). (C) 610.5-1990, 610.12-1990
(2) The results of the synthesis process that provide sufficient details, drawings, or other pertinent information for a physical or software element that permits further development, fabrication, assembly, and integration, or production of a product element. (C/SE) 1220-1994s
(3) The act of preparing drawings or other pertinent information for a physical or software element during synthesis within the systems engineering process. (C/SE) 1220-1994s
(4) Those characteristics of a system or a software item that are selected by the developer in response to the requirements. Some will match the requirements; others will be elaborations of requirements, such as definitions of all error messages in response to a requirement to display error messages; others will be implementation related, such as decisions about what software units and logic to use to satisfy the requirements. (C/SE) J-STD-016-1995
(5) (pulse terminology) *See also:* logic design. (IM/WM&A) 194-1977w

design A motor An integral-horsepower polyphase squirrel-cage induction motor designed for full-voltage starting with normal values of locked-rotor torque and breakdown torque, and with locked-rotor current higher than that specified for design B, C, and D motors. (PE) [9]

design analysis (1) (software) The evaluation of a design to determine correctness with respect to stated requirements, conformance to design standards, system efficiency, and other criteria.
(2) (software) The evaluation of alternative design approaches. *See also:* design; preliminary design; requirement; correctness. (C/SE) 729-1983s

design analyzer (software) An automated design tool that accepts information about a program's design and produces such outputs as module hierarchy diagrams, graphical representations of control and data structure, and lists of access data blocks. *See also:* automated design tool; design; data structure; data block. (C/SE) 729-1983s

design architecture An arrangement of design elements that provides the design solution for a product or life cycle process intended to satisfy the functional architecture and the requirements baseline. (C/SE) 1220-1998

designated employee *See:* designated person.

designated person A qualified person designated to perform specific duties under the conditions existing. *Synonym:* designated employee. (NESC) C2-1997

designated representative (nuclear power quality assurance) An individual or organization authorized by the purchaser to perform functions in the procurement process. (PE/NP) [124]

designation Selection of a particular target and transmission of its approximate coordinates from some external source to a radar. *Note:* Designation is usually intended to initiate tracking. (AES) 686-1997

designation hole A hole that has been punched in a punch card to indicate the nature of the data on the card, or the functions that a machine is to perform. *Synonyms:* control punch; control hole. (C) 610.10-1994w

designation number In logic design, the bottom line of a truth table written such that the values of the variables equal the binary number of the state. For example, the designation number for a two-variable exclusive-OR function is as follows:

State	0	1	2	3	
Variable A	0	1	0	1	
Variable B	0	0	1	1	
Exclusive OR	0	1	1	0	Designation number

(C) 1084-1986w

design automation The use of computers to automate the design process. *See also:* computer-aided design. (C) 610.2-1987

design basis earthquake (nuclear power generating station) That earthquake producing the maximum vibratory ground motion that the nuclear power generating station is designed to withstand without functional impairment to those features necessary to shut down the reactor, maintain the station in a safe condition, and prevent undue risk to the health and safety of the public. (AES/RS) 686-1982s

design basis event conditions (nuclear power generating station) Conditions calculated to occur as a result of the design basis events. (PE/NP) 380-1975w, 334-1974s

design basis events (1) (standby power supplies) (diesel-generator unit) Postulated events used in the design to establish the performance requirements of the structures and systems. 387-1995

(2) (Class 1E motor control centers) (valve actuators) (safety systems equipment in nuclear power generating stations) Postulated events specified by the safety analysis of the station used in the design to establish the acceptable performance requirements of the structures and systems. (SWG/PE/NP/EDPG) 382-1985, 649-1980s, 627-1980r, 690-1984r, C37.100-1992, 323-1974s, 650-1979s

(3) Postulated abnormal events used in the design to establish the acceptable performance requirements of the structures and systems, and components. (PE/NP) 384-1992r, 323-1974s

(4) Postulated events used in the design to establish the acceptable performance requirements for the structures, systems, and components. (PE/NP) 603-1998

design B motor An integral-horsepower polyphase squirrel-cage induction motor, designed for full-voltage starting, with normal locked-rotor and breakdown torque and with locked-rotor current not exceeding specified values. (PE) [9]

design characteristic The design attributes or distinguishing features that pertain to a measurable description of a product or process. (C/SE) 1220-1998

design C motor An integral-horsepower polyphase squirrel-cage induction motor, designed for full-voltage starting with high locked-rotor torque and with locked-rotor current not exceeding specified values. (PE) [9]

design current (glow lamp) The value of current flow through the lamp upon which rated-life values are based. (EEC/EL) [104]

design description (software) A document that describes the design of a system or component. Typical contents include system or component architecture, control logic, data structures, input/output formats, interface descriptions, and algorithms. *Synonyms:* design specification; design document. *Contrast:* requirements specification. *See also:* product specification. (C) 610.12-1990

design D motor An integral-horsepower poly−phase squirrel-cage induction motor with rated-load slip of at least 5 percent and designed for full-voltage starting with locked-rotor torque at least 275 percent of rated-load torque and with locked-rotor currents not exceeding specified values. (PE) [9]

design document *See:* design description.

design earthquake The greatest earthquake postulated during the life of the gas-insulated substation that the user wishes the gas-insulated substation to survive in operating condition.

(SWG/SUB/PE) 693-1984s, C37.100-1992, C37.122.1-1993

designed availability (nuclear power generating station) The probability that an item will be operable when needed as determined through the design analyses. (PE/NP) 380-1975w, 338-1977s

design element (software) (Ada as a program design language) A basic component or building block in a design. (C/SE) 610.12-1990, 990-1987w

design entity An element (component) of a design that is structurally and functionally distinct from other elements and that is separately named and referenced. (C/SE) 1016-1998

design family A group of transformer designs that share common characteristics such as design arrangement, materials, and design stresses to meet performance characteristics such as temperature rise, impedance, losses, and seismic capability. Due to different ratings, the transformers may have dimensional differences. (PE/TR) 638-1992r

design input (nuclear power quality assurance) Those criteria, parameters, bases, or other design requirements upon which detailed final design is based. (PE/NP) [124]

design inspection *See:* inspection.

design language (software) A specification language with special constructs and, sometimes, verification protocols, used to develop, analyze, and document a hardware or software design. Types include hardware design language, program design language. *See also:* requirements specification language; CDL. (C/C) 610.12-1990, 610.13-1993w

design letters Terminology established by the National Electrical Manufacturers Association to describe a standard range of characteristics. The characteristics covered are slip at rated load, locked-rotor and breakdown torque, and locked-rotor current. (PE) [9]

design level The design decomposition of the software item (e.g., system, subsystem, program, or module). (C/SE) 829-1998

design life (1) (nuclear power generating station) The time during which satisfactory performance can be expected for a specific set of service conditions. *Note:* The life may be specified in calendar time. However, operating time, number of operating cycles or other performance interval, as appropriate may be used to determine the time. (PE/NP) 323-1974s

(2) (safety systems equipment in nuclear power generating stations) (valve actuators) The time during which satisfactory performance can be expected for a specific set of service conditions. *Note:* The time may be specified in real time, operating time, number of operating cycles or other performance interval, as appropriate. (PE/NP/EDPG) 382-1985, 422-1986w, 627-1980r, 649-1991r

(3) (Class 1E battery chargers and inverters) The time during which satisfactory performance can be expected for a specific set of service conditions, based upon component selection and applications. (PE/NP) 650-1979s

(4) The time during which satisfactory performance can be expected for a specific set of service conditions. (SWG/PE) C37.100-1992

(5) (of a substation) The time during which satisfactory substation performance can be expected for a specific set of operating conditions. (SUB/PE) 525-1992r

design limits Design aspects of the instrument in terms of certain limiting conditions to which the instrument may be subjected without permanent physical damage or impairment of operating characteristics. *See also:* instrument. (EEC/EMI) [112]

design L motor An integral-horsepower single-phase motor, designed for full-voltage starting with locked-rotor current not exceeding specified values, which are higher than those for design M motors. (PE) [9]

design load That combination of electric loads (kW and kvar), having the most severe power demand characteristic, which is provided with electric energy from a diesel-generator unit for the operation of engineered safety features and other systems required during and following shutdown of the reactor. (PE/NP) 387-1995

design margin Additional capacity above requirements to allow for unforeseen additions to the dc system and less than optimum operating conditions due to improper maintenance, recent discharge, or ambient conditions lower than anticipated. (PE/EDPG) 1106-1987s

design method A definition of a set of essential entities. It includes a systematic procedure for the set of models with rules or heuristics used to determine the entities and entity attributes of the models, and a notation to represent the entities and the attributes expressed in each model. (C/SE) 1016.1-1993w

design methodology (1) (software) A systematic approach to creating a design, consisting of the ordered application of a specific collection of tools, techniques, and guidelines. *See also:* tool; design. (C/SE) 729-1983s
(2) A guideline identifying how to design software. As a process, a methodology is a practical set of procedures that facilitate the design of software. (C/SE) 1016.1-1993w

design M motor An integral-horsepower single-phase motor, designed for full-voltage starting with locked-rotor current not exceeding specified values, which are lower than those for design L motors. (PE) [9]

design N motor A fractional-horsepower single-phase motor, designed for full-voltage starting with locked-rotor current not exceeding specified values, which are lower than those for design O motors. (PE) [9]

design O motor A fractional-horsepower single-phase motor, designed for full-voltage starting with locked-rotor current not exceeding specified values, which are higher than those for design N motors. (PE) [9]

design output (nuclear power quality assurance) Documents, such as drawings, specifications, and other documents, defining technical requirements of structures, systems, and components. (PE/NP) [124]

design phase (software verification and validation plans) (software) The period of time in the software life cycle during which the designs for architecture, software components, interfaces, and data are created, documented, and verified to satisfy requirements. *See also:* data; preliminary design; detailed design; component; architecture; requirement; software life cycle; design. (SE/C) 1012-1986s, 610.12-1990

design pressure (1) (working pressure) The maximum steady-state gas pressure to which a gas-insulated substation enclosure is subjected under normal operating conditions. (SUB/PE) C37.122-1983s, C37.122.1-1993
(2) The maximum gas pressure to which a gas-insulated substation enclosure will be subjected under normal operating conditions. *Synonym:* working pressure. (SWG/PE) C37.100-1992

design process (nuclear power quality assurance) Technical and management processes that commence with identification of design input and that lead to and include the issuance of design output documents. (PE/NP) [124]

design qualification (safety systems equipment in nuclear power generating stations) The generation and maintenance of evidence to demonstrate that equipment can perform within its specification requirements. *Note:* In the context of IEEE Std 627-1980, design qualification is synonymous with equipment qualification or qualification. Normal production testing and preoperational testing performed after installation and acceptance of the equipment is outside the scope of this definition and standard. (PE/NP) 627-1980r

design requirement (software) A requirement that specifies or constrains the design of a system or system component. *Contrast:* functional requirement; implementation requirement; physical requirement; performance requirement; interface requirement. (C) 610.12-1990

design review (software) A process or meeting during which a system, hardware, or software design is presented to project personnel, managers, users, customers, or other interested parties for comment or approval. Types include critical design review, preliminary design review, system design review. *See also:* code review; requirements review; test readiness review; formal qualification review. (C) 610.12-1990

design service conditions (electric penetration assemblies) The service conditions used as the basis for ratings and for the design qualification of electric penetration assemblies. (PE/NP) 317-1983r

design specification *See:* design description.

design standard (software) A standard that describes the characteristics of a design or a design description of data or program components. (C) 610.12-1990

design test A test made on an LTC or the components of an LTC, or a range of LTCs or components all based on the same design, to prove compliance with this standard. *Note:* A range of LTCs is a number of LTCs based on the same design and having the same characteristics, with the exception of the insulation levels to ground and possibly between phases, the number of steps and the value of the transition impedance. (PE/TR) C57.131-1995

design tests (1) (electric penetration assemblies) Tests performed to verify that an electric penetration assembly meets design requirements. (PE/NP) 317-1983r
(2) (metal-enclosed bus and calculating losses in isolated-phase bus) Those tests made to determine the adequacy of a particular type, style, or model of metal-enclosed bus or its component parts to meet its assigned ratings and to operate satisfactorily under normal service conditions or under special conditions if specified. *Note:* Design tests are made only on representative apparatus to substantiate the ratings assigned to all other apparatus of basically the same design. These tests are not intended to be used as a part of normal production. The applicable portion of these design tests may also be used to evaluate modifications of a previous design and to assure that performance has not been adversely affected. Test data from previous similar designs may be used for current designs, where appropriate. (SWG/PE) C37.23-1987r
(3) (general) Those tests made to determine the adequacy of the design of a particular type, style, or model of equipment or its component parts to meet its assigned ratings and to operate satisfactorily under normal service conditions or under special conditions if specified. *Note:* Design tests are made only on representative apparatus to substantiate the ratings assigned to all other apparatus of basically the same design. These tests are not intended to be used as a part of normal production. The applicable portion of these design tests may also be used to evaluate modifications of a previous design and to assure that performance has not been adversely affected. Test data from previous similar designs may be used for current designs, where appropriate. (SWG/TR) C37.40-1981s, C57.12.80-1978r
(4) Tests made by the manufacturer to obtain data for design or application, or to obtain information on the performance of each type of high-voltage cable termination. (PE/IC) 48-1996
(5) Tests made by the manufacturer to determine the adequacy of the design of a particular type, style, or model of equipment or its component parts to meet its assigned ratings and to operate satisfactorily under normal service conditions or under special conditions if specified, and may be used to demonstrate compliance with applicable standards of the industry. *Notes:* 1. Design tests are made on representative apparatuses or prototypes to verify the validity of design analyses and calculation methods and to substantiate the ratings assigned to all other apparatuses of basically the same design. These tests are not intended to be made on every design variation or to be used as part of normal production. The applicable portion of these design tests may also be used to evaluate modifications of a previous design and to ensure that performance has not been adversely affected. Test data from previous similar designs may also be used for current designs, where appropriate. Once made, the tests need not be repeated unless the design is changed so as to modify performance. 2. Design tests are sometimes called type tests. (SWG/PE/SWG-OLD) C37.20.1-1993r, C37.21-1985r, C37.20.2-1993, C37.20.3-1996, C37.20.4-1996, C37.20.6-1997

(**6**) **(power cable joints)** Tests made on typical joint designs to obtain data to substantiate the design. These tests are of such nature that after they have once been made, they need not be repeated unless significant changes are made in the material or design which may change the performance of the joint. (PE/IC) 404-1986s
(**7**) Those tests made to determine the adequacy of the equipment comprising a gas-insulated substation, which enables it to operate satisfactorily under specified service conditions. *Note:* Design tests are made on representative samples of the same type and rating and are not intended to be used as a pan of normal production. The applicable portions of these design tests may also be used to evaluate modifications of a previous design and to ensure that performance has not been adversely affected. Test data from previous similar designs may be used for current designs, where appropriate. (SUB/PE) C37.122-1983s
(**8**) Tests that are required on new designs of insulators. (T&D/PE) 1024-1988w
(**9**) Those tests made to determine the adequacy of a particular type, style, or model of equipment with its component parts to meet its assigned ratings and to operate satisfactorily under normal service conditions or under special conditions if specified. *Note:* Design tests are made only on representative apparatus to substantiate the ratings assigned to all other apparatus of basically the same design. These tests are not intended to be used as a part of normal production. The applicable portion of these design tests may also be used to evaluate modifications of a previous design and to assure that performance has not been adversely affected. Test data from previous similar designs may be used for current designs, where appropriate. (SWG/PE) C37.100-1992
(**10**) Those tests made to determine the adequacy of the design of a particular type, style, or model generator circuit breaker to meet its assigned ratings and to operate satisfactorily under usual service conditions or under unusual conditions, if specified. Design tests are made only on representative circuit breakers of basically the same design, i.e., the same interrupters operating at the same contact speeds, and having at least the same dielectric strength. These tests are not intended to be used as a part of normal production. The applicable portions of these design tests may also be used to evaluate modifications of a previous design and to assure that performance has not been adversely affected. Test data from previous similar designs may be used for current designs, where appropriate. (SWG/PE) C37.013-1997
(**11**) Tests made on each design to establish the performance characteristics and to demonstrate compliance with the appropriate standards of the industry. Once made they need not be repeated, unless the design is changed so as to modify performance. (SPD/PE) C62.11-1999
(**12**) Tests made by the manufacturer on each design to establish the performance characteristics and to demonstrate compliance with the appropriate standards of the industry. Once made, they need not be repeated unless the design is changed (so as to modify performance). (SPD/PE) C62.62-2000

design to cost Cost goals for the production cost (design to unit production cost) and life cycle cost (design to life cycle cost) of the products used to make the design converge on cost targets, and is equivalent to any other performance parameter. (C/SE) 1220-1998

design unit (Ada as a program design language) A logically related collection of design elements. In an Ada program design language (PDL), a design unit is represented by an Ada compilation unit. (C/SE) 990-1987w, 610.12-1990

design verification (**1**) A general term covering the overall qualification of any conversion to standards by means of design testing supported by justified technical evaluation. (SWG/PE) C37.100-1992
(**2**) The process of overall qualification, in accordance with all appropriate standards, of any conversion by means of design testing and/or evaluation, supported by justified technical evaluation and documentation. (SWG/PE) C37.59-1996
(**3**) The process of evaluating whether or not the requirements of the design architecture are traceable to the verified functional architecture and satisfy the validated requirement baseline. (C/SE) 1220-1998

design view A subset of design entity attribute information that is specifically suited to the needs of a software project activity. (C/SE) 1016-1998

design voltage The voltage at which the device is designed to draw rated watts input. *See also:* appliance outlet. (IA/APP) [90]

design walk-through *See:* walk-through.

desired polarization (navigation aid terms) The polarization of the radio wave for which an antenna system is designed. (AES/GCS) 172-1983w

desired track *See:* course line.

desired value *See:* ideal value.

desk checking (software) A static analysis technique in which code listings, test results, or other documentation are visually examined, usually by the person who generated them, to identify errors, violations of development standards, or other problems. *See also:* walk-through; inspection. (C) 610.12-1990

deskstand A movable pedestal or stand (adapted to rest on a desk or table) that serves as a mounting for the transmitter of a telephone set and that ordinarily includes a hook for supporting the associated receiver when not in use. *See also:* telephone station. (EEC/PE) [119]

deskstand telephone set A telephone set having a deskstand. *See also:* telephone station. (EEC/PE) [119]

desktop computer (**1**) A computer designed for use on a desk or table. (C) 610.2-1987
(**2**) A computer designed to be placed on a desk or table. (C) 610.10-1994w

desktop management interface (DMI) A facility, normally host resident, for handling and translating defined interfaces for component information, event information, stored management information format (MIF) structures and management information. In this context, a component is an integral device or product, such as a printer; and event is an asynchronous alert or trap. The management information is used by an application such as a user management program, or may be converted for use by a network management facility. The desktop management interface is controlled by the Desktop Management Task Force. (C/MM) 1284.1-1997

Desktop Management Task Force (DMTF) An association of software developers, host computer, and peripheral device manufacturers that promulgate a platform-independent interface standard for the management of desktop computers and the peripheral devices attached to such computers. In addition to defining the interfaces between components, the host-based service layer and management applications, the association supports the development of consistent management information format (MIF) structures that define the significant manageable and status attributes of various components incorporated into desktop computers. (C/MM) 1284.1-1997

despun antenna On a rotating vehicle, an antenna whose beam is scanned such that, with respect to fixed reference axes, the beam is stationary. (AP/ANT) 145-1993

dest_id *See:* destination_identifier.

destination (**1**) **(binary floating-point arithmetic)** The location for the result of a binary or unary operation. A destination may be either explicitly designated by the user or implicitly supplied by the system (for example, intermediate results in subexpressions or arguments for procedures). Some languages place the results of intermediate calculations in destinations beyond the user's control. Nonetheless, IEEE Std 754-1985 defines the result of an operation in terms of that destination's format and the operands' values. (C/MM) 754-1985r

(2) (radix-independent floating-point arithmetic) The location for the result of a binary or unary operation. A destination may be either designated by the user or implicitly supplied by the system (that is, intermediate results in subexpressions or arguments for procedures). Some languages place the results of intermediate calculations in destinations beyond the user's control. Nevertheless, ANSI/IEEE Std 854-1987 defines the result of an operation in terms of that destination's precision as well as the operand's values. (MM/C) 854-1987r
(3) (mathematics of computing) The location for the result of a binary or unary operation. (C) 1084-1986w
(4) (software) The address of the device or storage location to which data is to be transferred. *Contrast:* source address. (C) 610.12-1990
(5) The batch server in a batch system to which a batch job should be sent for processing. Acceptance of a job at a destination is the responsibility of a receiving batch server. A destination may consist of a batch server specific portion, a network wide portion, or both. The batch server specific portion is referred to as the queue. The network wide portion is referred to as a batch server name. (C/PA) 1003.2d-1994
(6) One or more destination_identifiers, identifying the destination node(s) to which the packet is to be transmitted. *See also:* destination_identifier. (C/BA) 1355-1995
(7) A node that is addressed by a packet. If the destination is individually addressed by a source, then it has to return an acknowledge packet. (C/MM) 1394-1995

destination address (DA) (1) The address of a device or storage location to which data are to be transferred. (C) 610.10-1994w
(2) A field in the packet format identifying the end node(s) to which the packet is being sent.local area networks. (C) 8802-12-1998

destination address. The address of a device or storage location to which data are to be transferred. *Contrast:* source address. (C) 610.10-1994w

destination code (telephone switching systems) A combination of digits providing a unique termination address in a communication network. (COM) 312-1977w

destination-code routing (telephone switching systems) The means of using the area and office codes to direct a call to a particular destination regardless of its point of origin. (COM) 312-1977w

destination identifier A string that identifies a specific destination. A string of characters in the portable character set used to specify a particular destination. (C/PA) 1003.2d-1994

destination_identifier (dest_id) An implementation dependent identity of the/a destination node for a packet. (C/BA) 1355-1995

destination node The terminal node(s) which is/are to receive a particular packet. *See also:* node. (C/BA) 1355-1995

destructive addition (mathematics of computing) Computer addition in which the sum is placed in the storage location, register, or accumulator previously occupied by an operand, usually the augend, which is then lost. *Contrast:* nondestructive addition. (C) 1084-1986w

destructive read (1) (A) (computers) A read process that also erases the data in the source. **(B)** A read operation that alters the data in the accessed location. (C) [20], [85], 610.12-1990
(2) A read operation that deletes the data being read. *Contrast:* nondestructive read. (C) 610.10-1994w

destructive reading (charge-storage tubes) Reading that partially or completely erases the information as it is being read. *See also:* charge-storage tube. (ED) 158-1962w

destructive testing (A) (test, measurement, and diagnostic equipment) Prolonged endurance testing of equipment or a specimen until it fails in order to determine service life or design weakness. **(B) (test, measurement, and diagnostic equipment)** Testing in which the preparation of the test specimen until it fails in order to determine service life or design weakness. **(C) (test, measurement, and diagnostic equipment)** Testing in which the preparation of the test specimen or the test itself may adversely affect the life expectancy of the unit under test or render the sample unfit for its intended use. (MIL) [2]

destructive backspace In word processing, an operation that moves the cursor back one character and deletes the character that was in the cursor's new location. (C) 610.2-1987

detachable (socket-mounted) electromechanical watthour meters A detachable electromechanical watthour meter is one having bayonet-type (blade) terminals arranged on the back side of the meter for insertion into matching jaws of a meter socket (or detachable meter-mounting device). (ELM) C12.10-1987

detachable ladder Detachable ladders are those that are not permanently installed on a structure but are the normal means for accessing the structure and attached facilities. (T&D/PE) 1307-1996

detachable step Detachable steps are those that are not permanently installed on a structure but are the normal means for accessing the structure and attached facilities. (T&D/PE) 1307-1996

detail contrast *See:* resolution response.

detailed-billed call (telephone switching systems) A call for which there is a record, including the calling and called line identities, that will appear in a customer's billing statement. (COM) 312-1977w

detailed design (A) (software) The process of refining and expanding the preliminary design of a system or component to the extent that the design is sufficiently complete to be implemented. *See also:* software development process. **(B)** The result of the process in definition (A). (C) 610.12-1990

detail file *See:* transaction file.

detailed-record call (telephone switching systems) A call for which there is a record including the calling and called line identities that may be used in the billing process as well as for other purposes. (COM) 312-1977w

detectability A segment attribute that determines if a display element or segment can be identified by a pick device. (C) 610.6-1991w

detectability factor (1) (continuous-wave radar) The ratio of single-look signal energy to noise power per unit bandwidth, using a filter matched to the time on target. (AES) [42]
(2) In pulsed radar, the ratio of single-pulse signal energy to noise power per unit bandwidth that provides stated probability of detection for a given false alarm probability, measured in the intermediate-frequency amplifier bandwidth and using an intermediate-frequency filter matched to the single pulse and followed by optimum video integration. (AES) 686-1997

detectable element A display element that can be identified by a pick device. (C) 610.6-1991w

detectable failures Failures that can be identified through periodic testing or can be revealed by alarm or anomalous indication. Component failures that are detected at the channel, division, or system level are detectable failures. *Note:* Identifiable, but nondetectable, failures are failures identified by analysis that cannot be detected through periodic testing or cannot be revealed by alarm or anomalous indication. (PE/NP) 603-1998

detected error Error recognized as such by a detection algorithm or mechanism. (C/BA) 896.9-1994w

detecting element *See:* primary detecting element.

detecting means The first system element or group of elements that responds quantitatively to the measured variable and performs the initial measurement operation. The detecting means performs the initial conversion or control of measurement energy. *See also:* instrument. (EEC/EMI) [112]

detection (A) Determination of the presence of a signal. **(B)** Demodulation. The process by which a wave corresponding to the modulating wave is obtained in response to a modulated wave. *See also:* square-law detection; power detection; linear detection. (AP/ANT) 145-1983

detection efficiency The ratio between the number of selected pulses recorded per unit time to the number of photons emitted by the source per unit time. *See also:* total efficiency. (NI) N42.14-1991

detection, error *See:* error detection.

detection limit (radioactivity monitoring instrumentation) The extreme of quantification for the radiation of interest by the instrument as a whole or by an individual readout scale or decade. The lower detection limit is the minimum quantifiable instrument response or reading. The upper detection limit is the maximum quantifiable instrument response or reading. Quantifiable, in this case, means within the specified accuracy. (NI) N42.17B-1989r, N323-1978r

detection probability The probability that a signal, when actually present at the input to the receiver, will be correctly declared a target signal based on observation of the receiver output. *See also:* false-alarm probability. (AES) 686-1997

detection zone Any area equipped to sense the presence of an intruder. (PE/NP) 692-1997

detectivity (fiber optics) The reciprocal of noise equivalent power (NEP). *See also:* noise equivalent power. (Std100) 812-1984w

detector (1) (monitoring radioactivity in effluents) Any device for converting radiation flux to a signal suitable for observation and measurement. (NI) N42.18-1980r
(2) (electromagnetic energy) A device for the indication of the presence of electromagnetic fields. *Note:* In combination with an instrument, a detector may be employed for the determination of the complex field amplitudes. *See also:* auxiliary device to an instrument. (IM/HFIM) [40]
(3) (overhead-power-line corona and radio noise) A device that performs detection (extraction of signal or noise from a modulated input) and weighting (extraction of a particular characteristic of the signal or noise). *Note:* In a radio noise receiver, the voltage applied to the detector depends upon the nature of the noise and the bandwidth of the filters used in the intermediate frequency stages. To furnish calibrations that are independent of the bandwidth and can be made with readily available equipment, an unmodulated carrier is used. With such input, all detectors (peak, quasi-peak, average, or rms) will indicate the same value of radio noise. (T&D/PE) 539-1990
(4) (radiation protection) A device or component which produces an electronically measurable quantity in response to ionizing radiation. (NI) N323-1978r
(5) (airborne radioactivity monitoring) That portion of an instrument system sensitive to and used for the quantification of ionizing radiation. (NI) N42.17B-1989r

detector, average *See:* average detector.

detector bias The voltage applied to a detector to produce the electric field that sweeps out the signal charge. (NPS) 325-1996

detector capacitance The small-signal electrical capacitance measured between terminals of the detector under specified conditions of bias and frequency. (NPS) 325-1996

detector, coaxial *See:* coaxial detector.

detector, diffused-junction *See:* diffused-junction detector.

detector element The semiconductor crystal including its contacts. (NPS) 325-1996

detector element geometry The physical configuration of a detector element. *See also:* detector element. (NPS) 325-1996

detector figure of merit (nonlinear, active, and nonreciprocal waveguide components) A measure of the performance of a diode detector. It can be expressed quantitatively as the ratio of the open-circuit voltage sensitivity to the square root of the video resistance. (MTT) 457-1982w

detector geometry (x-ray energy spectrometers) (detector jargon) (x-ray energy spectrometers) (semiconductor radiation detectors) The physical configuration of a solid-state detector. (NPS/NID) 325-1971w, 300-1988r, 301-1976s, 759-1984r

detector, germanium gamma-ray *See:* germanium gamma-ray detector.

detector, p-i-n *See:* p-i-n detector.

detector, Schottky-barrier *See:* Schottky-barrier detector.

detector, semiconductor radiation *See:* semiconductor radiation detector.

detector, surface barrier *See:* surface barrier detector.

detector, transmission *See:* transmission detector; differential d*E*/d*x* detector.

detector, well-type coaxial *See:* well-type coaxial detector.

determinant (1) A square array of numbers or elements bordered on either side by a straight line. The value of the determinant is a function of its elements. (CAS) [13]
(2) (data management) Within a relation, an attribute on which some other attribute is functionally dependent. (C) 610.5-1990w

deterministic Pertaining to a process, model, or variable whose outcome, result, or value does not depend on chance. *Contrast:* stochastic. (C) 610.3-1989w

deterministic error *See:* bias.

deterministic model A model in which the results are determined through known relationships among the states and events, and in which a given input will always produce the same output; for example, a model depicting a known chemical reaction. *Contrast:* stochastic model. (C) 610.3-1989w

deterministic routing A network routing strategy where the choice of destination drives the decision at each node, regardless of changing conditions in the network. (C) 610.7-1995

DETOL *See:* Directly Executable Test Oriented Language.

detuners Devices attached to a structure which alter the impedance at the connection point such that a minimum of current at the design frequency (frequencies) flows in the structure. (T&D/PE) 1260-1996

developed source statements Source statements that are newly created for, added to, or modified for a software product. (C/SE) 1045-1992

developer (1) (electrostatography) A material or materials that may be used in development. (ED) 224-1965w, [45]
(2) An organization that develops software products; "develops" may include new development, modification, reuse, reengineering, maintenance, or any other activity that results in software products, and includes the testing, quality assurance, configuration management, and other activities applied to these products. *Synonym:* supplier. (C/SE) J-STD-016-1995, 1362-1998
(3) A person or organization that performs development activities (including requirements analysis, design, testing through acceptance) during the software life cycle process. (C/SE) 1062-1998

developer role Where software is developed, tested, and maintained. (C/PA) 1387.2-1995

development (1) (electrostatography) The act of rendering an electrostatic image viewable. *See also:* electrostatography. (ED) [46]
(2) All activities that are carried out to create a software product. (C/SE) 1298-1992w

developmental baseline* *See:* developmental configuration.
* Deprecated.

developmental configuration In configuration management, the software and associated technical documentation that define the evolving configuration of a computer software configuration item during development. *Note:* The developmental configuration is under the developer's control, and therefore is not called a baseline. *Contrast:* allocated baseline; product baseline; functional baseline. (C) 610.12-1990

development cycle *See:* software development cycle.

development life cycle *See:* software development cycle.

development methodology (software) A systematic approach to the creation of software that defines development phases and specifies the activities, products, verification procedures, and completion criteria for each phase. *See also:* software. (C/SE) 729-1983s

development platform A system used to prepare an application for execution. Such a system is possibly distinct from the system on which the application will execute. (C/PA) 1003.13-1998

development specification *See:* requirements specification.

development system (1) The computer system used to compile and configure a PCTS.1. (C/PA) 2003.1-1992
(2) The computer system used to compile and configure a PCTS. (C/PA) 13210-1994

development testing Formal or informal testing conducted during the development of a system or component, usually in the development environment by the developer. *Contrast:* acceptance testing; operational testing. *See also:* qualification testing. (C) 610.12-1990

deviation (1) (A) (software) A departure from a specified requirement. *Contrast:* waiver; engineering change. *See also:* configuration control. **(B) (software)** A written authorization, granted prior to the manufacture of an item, to depart from a particular performance or design requirement for a specific number of units or a specific period of time. *Note:* Unlike an engineering change, a deviation does not require revision of the documentation defining the affected item. *Contrast:* waiver; engineering change. *See also:* configuration control. **(C) (navigation aid terms)** The angle between the magnetic meridian and the axis of a compass card. Indicates the offset of the compass card from magnetic north. (C/AES/GCS) 610.12-1990, 172-1983
(2) (automatic control) Any departure from a desired or expected value or pattern. (IA/PE/APP/EDPG/IAC) [69], [3], [60]
(3) (nuclear power quality assurance) A departure from specified requirements. (PE/NP) [124]
(4) Departure from a specified dimension or design requirement, usually defining upper and lower limits. *See also:* tolerance. (SCC14/QUL) SI 10-1997, 268-1982s

deviation distortion (data transmission) Distortion in an FM receiver due to inadequate bandwidth and inadequate amplitude modulation rejection, or inadquate discriminator linearity. (PE) 599-1985w

deviation factor (1) (rotating machinery) (wave) The ratio of the maximum difference between corresponding ordinates of the wave and of the equivalent sine wave when the waves are superposed in such a way as to make this maximum difference as small as possible. *Note:* The equivalent sine wave is defined as having the same frequency and the same root-mean-square value as the wave being tested. *See also:* direct-axis synchronous impedance. (PE) [9]
(2) (electrical measurements in power circuits) The deviation factor is the ratio of the maximum difference between corresponding ordinates of the wave and of the equivalent sine wave to the maximum ordinate of the equivalent sine wave when the waves are superposed in such a way as to make this maximum difference as small as possible. The equivalent sine wave is defined as having the same frequency and the same rms value as the wave being tested. (PE/PSIM) 120-1989r

deviation, frequency *See:* frequency deviation.

deviation from a sine wave (harmonic control and reactive compensation of static power converters) (converter characteristics) (self-commutated converters) A single number measure of the distortion of a sinusoid due to harmonic components. It is equal to the ratio of the absolute value of the maximum difference between the distorted wave and the fundamental to the crest value of the fundamental. *See also:* maximum theoretical deviation from a sine wave. (IA/SPC) 936-1987w, 519-1992

deviation integral, absolute *See:* absolute deviation integral.

deviation ratio (frequency-modulation systems) (data transmission) The ratio of the maximum frequency deviation to the maximum modulating frequency of the system. (PE) 599-1985w

deviation sensitivity (1) (navigation aid terms) The rate of change of course indication with respect to the change of displacement from the course line. (AES/GCS) 172-1983w
(2) (frequency-modulation receivers) The least frequency deviation that produces a specified output power. 188-1952w

deviation, steady-state *See:* steady-state deviation.

deviation system (control) The instantaneous value of the ultimately controlled variable minus the command. *Note:* The use of system error to mean a system deviation with its sign changed is deprecated. *Synonym:* system overshoot. *See also:* deviation. (PE/IA/EDPG/IAC) 421-1972s, [60]

deviation, transient *See:* transient deviation.

device (1) (FASTBUS acquisition and control) (FASTBUS device) Any equipment capable of connecting to a segment and responding to the mandatory features of the FASTBUS protocol. (NID) 960-1993
(2) (696 interface devices) (general system) A circuit or logical group of circuits resident on one or more boards capable of interacting with other such devices through the bus. (C/MM) 696-1983w
(3) (nuclear power generating station) An item of electric equipment that is used in connection with, or as an auxiliary to, other items of electric equipment. (For example, as used in IEEE Std 649-1980, a device is a starter, contactor, circuit breaker, relay, etc.). (PE/COM/TA/NP) 649-1980s, 455-1985w, 344-1975s
(4) (programmable instrumentation) A component of a system that does not function as the system-controller but typically receives program messages from and sends response messages to the controller. A device may optionally have the capability to receive control from the controller and become the controller-in-charge of the system. A device meets all the requirements stated in IEEE Std 488.2-1987. (IM/AIN) 488.2-1992r
(5) (packaging machinery) A unit of an electrical system which is intended to carry but not consume electrical energy. (IA/PKG) 333-1980w
(6) A medical instrument or other device used to generate data on a particular patient. (EMB/MIB) 1073.3.1-1994
(7) A hardware unit that is capable of performing some specific function. (C/BA) 1275-1994
(8) A component of an VXIbus system. Normally, a device will consist of one VXIbus board. However, multiple-slot devices and multiple-device modules are permitted. Some examples of devices are computers, multimeters, multiplexers, oscillators, operator interfaces, and counters. (C/MM) 1155-1992
(9) In networking, a unit that provides a means for inputting and outputting data over the transmission medium. (C) 610.7-1995
(10) (software) A mechanism or piece of equipment designed to serve a purpose or perform a function. (C) 610.10-1994w, 610.12-1990
(11) A computer peripheral or an object that appears to the application as such. (C/PA) 9945-1-1996, 1003.5-1999
(12) (electrical equipment) An operating element such as a relay, contactor, circuit breaker, switch, valve, or governor used to perform a given function in the operation of electrical equipment. (SWG/PE/SUB) C37.100-1992, C37.1-1994
(13) Any independent test resource. A test resource may be either manually or automatically controlled. Devices can generate stimuli, measure response, or provide switching control. Examples include voltmeters, counters, and power supplies. (SCC20) 993-1997

(14) A reference to an integrated circuit or other design structure. (C/TT) 1450-1999

device address The (32-m)-bit identifying number assigned to a FASTBUS device that is compared with the signals on the AD lines during a logical primary address cycle of a FASTBUS operation. The device address is formed by the group and module address fields. The (remaining) low-order m bits are assigned to the internal address field. (NID) 960-1993

device alias A shorthand representation for a device path. (C/BA) 1275-1994

device arguments The component of a node name that is provided to a package's `open` method to provide addtional device-specific information. (C/BA) 1275-1994

device class-broadcast Selective broadcast-class specified by CSR#7. Controls device response to subsequent cycles within the broadcast. (NID) 960-1993

device communications controller (DCC) A communications interface associated with a medical device. A DCC may support one or more physically distinct devices acting as a single network communications unit. Its purpose is to provide a point-to-point serial communication link to a bedside communications controller (BCC). (EMB/MIB) 1073.4.1-2000, 1073.3.2-2000

device control character (data management) A control character used for the control of auxiliary devices associated with a data processing system or data communication system; for example, a control character for switching such devices on or off. (C) 610.5-1990w

device control language A language used to monitor and/or control the state of a device. (C/MM) 1284.4-2000

device coordinate system (computer graphics) A device-dependent coordinate system in which the coordinates of addressable points are expressed in integer addressable units. *Note:* A device driver maps normalized device coordinates or world coordinates to actual device coordinates. (C) 610.6-1991w

device-dependent (computer graphics) Pertaining to that which can be used only on a particular device. *Contrast:* device-independent. (C) 610.6-1991w

device driver (1) (computer graphics) The software that translates device-independent commands into device-specific commands. (C) 610.6-1991w
(2) The software responsible for managing low-level I/O operations for a particular hardware device or set of devices. Contains all the device-specific code necessary to communicate with a device and provides a standard interface to the rest of the system. *See also:* firmware device driver; operating system device driver. (C/BA) 1275-1994
(3) A program that runs on the host and manages the sending and receiving of information from the peripheral. The driver utilizes the link level interface defined in this standard to communicate data between the application program and the peripheral personality. (C/MM) 1284-1994
(4) A software component that permits a system to control and communicate with a peripheral device. *See also:* printer driver; disk driver. (C) 610.10-1994w

Device ID A structured, variable length ASCII message identifying the manufacturer, command set, and model of the peripheral. The message is provided by the peripheral in response to a request issued by the host during the negotiation phase. Provided that the peripheral supports the bidirectional mode requested by the host, this message is provided in the requested mode. The Device ID is intended to assist the host in selecting the device and/or peripheral driver appropriate to the peripheral. (C/MM) 1284-1994

device-independent (computer graphics) Pertaining to that which can be used on a variety of devices. *Contrast:* device-dependent. (C) 610.6-1991w

device interface One of the interfaces specified in this standard that allows devices to be identified, characterized, and used to assist other Open Firmware functions such as booting. (C/BA) 1275-1994

device media control language (data management) A language that may be used to describe the physical layout and organization of data within some physical storage media. (C) 610.5-1990w

device node A particular entry in the device tree, usually describing a single device or bus, consisting of properties, methods, and private data. (A device node may have multiple child nodes and has exactly one parent node. The root node has no parent node.). (C/BA) 1275-1994

device path A textual name identifying a device node by showing its position in the device tree. (C/BA) 1275-1994

device register (A) An addressable register used to store information describing the device. *See also:* control register. **(B)** An addressable register used to store status and control information, and data for transmission to or from a device. *Synonym:* device status word. (C) 610.10-1994

device rise time (photomultipliers for scintillation counting) The mean time difference between the 10- and 90-percent amplitude points on the output waveform for full cathode illumination and delta-function excitation. DRT is measured with a repetitive delta-function light source and a sampling oscilloscope. The trigger signal for the oscilloscope may be derived from the device output pulse, so that light sources such as the the scintillator light source may be employed. (NPS) 398-1972r

device space (computer graphics) The area defined by the addressable points of a display device. (C) 610.6-1991w

device specifier Either a device path, a device alias, or a hybrid path that begins with a device alias and ends with a device path. (C/BA) 1275-1994

device status word *See:* device register.

device tree A hierarchical data structure representing the physical configuration of the system. (The device tree describes the properties of the system's devices and the devices' relationships to one another. Most Open Firmware elements [devices, buses, libraries of software procedures, etc.] are named and located by the device tree.). (C/BA) 1275-1994

dew point The temperature at which the water vapor in the gas begins to condense, expressed in degrees Fahrenheit (°F) or Celsius (°C). (PE/IC) 1125-1993

device port The physical connection points through which signals flow into or out of a device or where timing, synchronization, and triggering control are accomplished. (SCC20) 993-1997

device type Identifies the set of properties and package classes that a node is expected to implement. Specified by the "`device_type`" property. (C/BA) 1275-1994

device under test (DUT) The device to be placed in a test fixture and tested. (C/TT) 1450-1999

dew point temperature The temperature at which condensation of water vapor begins in a space. (IA/PSE) 241-1990r

dew withstand voltage test A test to determine the ability of the insulating system to withstand specified overvoltages for a specified time without flashover or puncture while completely covered with dew. (SWG/PE) C37.100-1992, C37.23-1987r

dezincification Parting of zinc from an alloy (parting is the preferred term). *Note:* Other terms in this category, such as denickelification, dealuminification, demolybdenization, etcetera, should be replaced by the term parting. *See also:* parting. (IA) [59]

DF *See:* direction finder.

DF antenna *See:* direction finder antenna system.

DFD *See:* data flow diagram.

D Filter A 300 Hz to 3400 Hz bandpass filter used for measuring noise, impulse noise, or data modem signal power. Noise measured through the D-Notched filter is used to evaluate its effect on the performance of a data modem. (COM/TA) 743-1995

D flip-flop A flip-flop that has one data input, one trigger, and an output which assumes the state of the data input when the trigger is received. (C) 610.10-1994w

DFS *See:* depth-first search.

DF sensitivity *See:* direction finder sensitivity.

dg *See:* decilog.

diad (mathematics of computing) A group of two closely related items or digits. (C) 1084-1986w

diagnosis (1) The conclusion(s) resulting from tasks, tests, observations, or other information. (ATLAS/SCC20) 1232-1995, 1226-1998
(2) A cognitive assessment of the state of the system. (PE/NP) 1082-1997

diagnosis, fault *See:* fault diagnosis.

diagnostic (1) (software) Pertaining to the detection and isolation of faults or failures; for example, a diagnostic message, a diagnostic manual. (C) 610.12-1990
(2) A process by which hardware malfunctions may be detected. (SUB/PE) 999-1992w

diagnostic controller The agent (this could be from an expert/reasoner system or from an operator) that invokes test procedures in the sequence required to achieve test goals. (SCC20) 1226-1998

diagnostic factor (thermal classification of electric equipment and electrical insulation) (evaluation of thermal capability) A variable or fixed stress, which can be applied periodically or continuously during an accelerated test, to measure the degree of aging without in itself influencing the aging process. (EI) 1-1986r

diagnostic field tests and measurements (power apparatus) Procedures that are performed on site on the complete apparatus or parts thereof in order to determine its suitability for service. *Note:* The parameters measured differ from apparatus to apparatus and may include electrical, mechanical, chemical, thermal, etc., quantities. Interpretation of the results is usually based on a change in the measured characteristics and/or by comparison with pre-established criteria. The tests are normally carried out at regular intervals based on users' experience and/or manufacturers' recommendations. These tests may also be performed on defective apparatus in order to determine the location and/or cause of failure. (PE/PSIM) 62-1995

diagnostic knowledge Provides the information required to support the diagnostic process. This knowledge defines the relationships between possible test outcomes and anomalies that may cause these outcomes. (SCC20) 1226-1998

Diagnostic Machine Aid—Digital (DMAD) A test language used for functional testing of digital devices; allows device description in terms of registers, signal names, and functional operators, such as logical operators and Boolean operations. (C) 610.13-1993w

diagnostic manual (software) A document that presents the information necessary to execute diagnostic procedures for a system or component, identify malfunctions, and remedy those malfunctions. Typically described are the diagnostic features of the system or component and the diagnostic tools available for its support. *See also:* installation manual; support manual; user manual; programmer manual; operator manual. (C) 610.12-1990

diagnostic procedure A structured combination of tasks, tests, observations, and other information used to localize a fault or faults. (ATLAS) 1232-1995

diagnostic process A structured combination of tasks, tests, observations, and other information used to localize a fault or faults. (SCC20) 1226-1998

diagnostic resolution The ability to trace a trouble to a minimum number of replaceable elements. (COM/TA) 973-1990w

diagnostic routine (1) A routine designed to locate either a malfunction in the computer or a mistake in coding. *See also:* programmed check. (C) 270-1966w
(2) (test, measurement, and diagnostic equipment) A logical sequence of tests designed to locate a malfunction in the unit under test. (MIL) [2]

diagnostics, self *See:* self diagnostics.

diagnostic test (1) (test, measurement, and diagnostic equipment) A test performed for the purpose of isolating a malfunction in the unit under test or confirming that there actually is a malfunction. (MIL) [2]
(2) A test, or collection of tests, that is invoked by writing to the TEST_START register. There are four forms of diagnostic tests: initialization tests, extended tests, manual tests, and system tests. (C/MM) 1212-1991s
(3) A test applied to a unit under test (UUT) for the purpose of isolating a fault to a lower level of assembly. (SCC20) 771-1998

diagnostic tests Comparative tests or measurements of one or more of the characteristic parameters of a circuit breaker to verify that it performs its functions. *Note:* The result from diagnostic tests can lead to the decision of carrying out overhaul. (SWG/PE) C37.10-1995

diagnostic unit (recursive) A collection of one or more diagnostic conclusions and diagnostic units. It represents a conclusion that might be drawn through the process of diagnosis and is related to repair by physical mapping and repair actions. (ATLAS) 1232-1995

diagonally integrated microprocessor A microprocessor in which diagonal microinstructions can be performed. *Contrast:* vertically integrated microprocessor; horizontally integrated microprocessor. (C) 610.10-1994w

diagonal microinstruction (1) A microinstruction capable of specifying a limited number of simultaneous operations needed to carry out a machine language instruction. *Note:* Diagonal microinstructions fall, in size and functionality, between horizontal microinstructions and vertical microinstructions. The designation "diagonal" refers to this compromise rather than to any physical characteristic of the microinstruction. *Contrast:* horizontal microinstruction; vertical microinstruction. (C) 610.12-1990
(2) A microinstruction capable of specifying a limited number of simultaneous operations needed to carry out a machine language instruction. *Contrast:* vertical microinstruction; horizontal microinstruction. (C) 610.10-1994w

diagram An instantiation of the formal diagram structure that consists only of semantically and syntactically valid IDEF0 graphical statements. Each diagram is a single unit of an IDEF0 model that presents the top-level function that is the subject of the model (the A-0 context diagram), presents the context of the subject function (other context diagrams), or presents the details of a box (decomposition diagrams). (C/SE) 1320.1-1998

diagram boundary An edge of a diagram in a diagram page. (C/SE) 1320.1-1998

diagram feature An element of a diagram. Diagram features include boxes, arrow segments, arrow labels, ICOM codes, ICOM labels, model notes, and reader notes. (C/SE) 1320.1-1998

diagram feature reference An expression that unambiguously identifies a diagram feature within an IDEF0 model. (C/SE) 1320.1-1998

diagram number That part of a diagram reference that corresponds to a diagram's parent function's node number. The diagram number refers to the diagram that details or decomposes the function designated by the same node number. (C/SE) 1320.1-1998

diagram page A model page that contains a context diagram or a decomposition diagram. (C/SE) 1320.1-1998

diagram reference An expression that unambiguously identifies a diagram and specifies the diagram's position in a specific model hierarchy; a diagram reference is composed of a model name abbreviation and a diagram number. (C/SE) 1320.1-1998

diagram title A verb or verb phrase that describes the overall function presented by a diagram; the diagram title of a child diagram is the box name of its parent box. (C/SE) 1320.1-1998

dial (1) A plate or disc, suitably marked, that served to indicate angular position, as for example the position of a handwheel. (IA/ICTL/IAC) [60]
(2) (automatic control) A type of calling device used in automatic switching that, when wound up and released, generates pulses required for establishing connections. (EMB) [47]

dialect (A) In computer languages, a variation of a particular language. *Synonyms:* variation; variant; version. **(B)** A form of a particular language, peculiar to a specific population or group, differing from some standard language in some significant manner. *See also:* extension; subset. (C) 610.13-1993

dialing (telephone switching systems) The act of using a calling device. (COM) 312-1977w

dialing pattern (telephone switching systems) The implementation of a numbering plan with reference to an individual automatic exchange. (COM) 312-1977w

dial-mobile telephone system (mobile communication) A mobile communication system that can be interconnected with a telephone network by dialing, or a mobile communication system connected on a dial basis with a telephone network. *See also:* mobile communication system. (VT) [37]

dialog (dialogue) Computer-human interaction in which the responses provided by the computer are highly responsive to the questions, answers, and directives given by the user. *Synonym:* online dialog. (C) 610.2-1987

dialogue window A window, such as a `DialogBox`, that pops up to perform a specific function and is then dismissed. (C) 1295-1993w

dial pulse (1) (dial-pulse address signaling systems) (telephony) A momentary interruption or change in the direct-current path of a signalling system to provide address information. (COM/TA) 753-1983w
(2) A means of pulsing that consists of regular, momentary interruptions of a direct or alternating current path in which the number of interruptions corresponds to the value of the digit or carrier. (C) 610.7-1995

dial-pulse signaling An address signaling method using the opening and closing of contacts to represent the dialed phone number. The digits are represented as a string of pulses closely spaced (a few milliseconds), and consecutive digits are separated by a longer period without pulsing. (COM/TA) 973-1990w

dial pulsing (dial-pulse address signaling systems) (telephony) A means of transmitting the address telephone number over a direct-current path. The current is interrupted, at the transmitting end, in a regular, momentary pattern. The number of interruptions corresponds to the digit being transmitted. (COM/TA) 753-1983w

dial pushing (telephone switching systems) A means of pulsing consisting of regular, momentary interruptions of a direct or alternating current path at the sending end in which the number of interruptions corresponds to the value of the digit or character. (COM) 312-1977w

dial tone (telephone switching systems) The tone that indicates that the switching equipment is ready to receive signals from a calling device. (COM) 312-1977w

dial-tone delay The time it takes for a telephone switching system to return a dial tone to an originating line after the customer goes off-hook. (COM/TA) 973-1990w

dial train (register) All the gear wheels and pinions used to interconnect the dial pointers. *See also:* watthour meter. (EEC/PE) [119]

dial-up circuit A telecommunication circuit that is established and broken, under human or machine control, using the public switched network as the routing and transmission medium. *See also:* simplex circuit; two-wire circuit; foreign exchange circuit; leased circuit; four-wire circuit. (C) 610.7-1995

dial-up line A line established on a circuit-switched network for public use. *Contrast:* leased line. (C) 610.7-1995

diameter (computer graphics) In image processing, the maximum distance between any two points in a subset of an image.

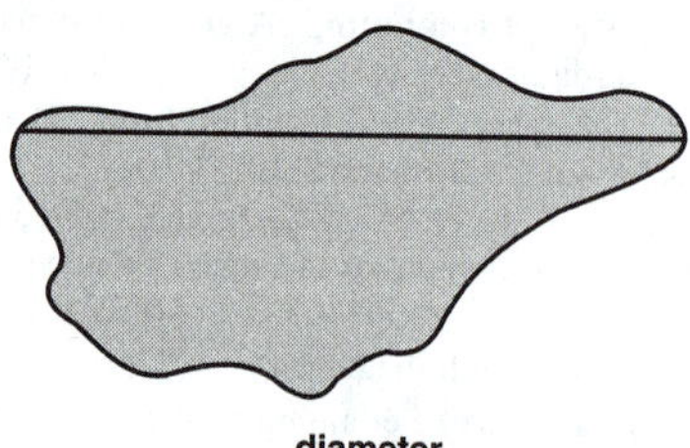

diameter

(C) 610.4-1990w

diametric rectifier circuit A circuit that employs two or more rectifying elements with a conducting period of 180 electrical degrees plus the commutating angle. *See also:* rectification. (EEC/PE) [119]

diamond winding (rotating machinery) A distributed winding in which the individual coils have the same shape and coil pitch. (PE) [9]

diaphragm (electrolytic cells) A porous or permeable membrane separating anode and cathode compartments of an electrolytic cell from each other or from an intermediate compartments for the purpose of preventing admixture of anolyte and catholyte. *See also:* electrolytic cell. (PE/EEC) [119]

diathermy (medical electronics) The therapeutic use of alternating currents to generate heat within some part of the body, the frequency being greater than the maximum frequency for neuromuscular response. (EMB) [47]

dibit (1) (data transmission) Two bits; two binary digits. (PE) 599-1985w
(2) (data management) Two bits. (C) 610.5-1990w

DIBOL *See:* Digital Business Oriented Language.

dicap storage A type of storage that uses an array of diodes to control current directed to storage capacitors. (C) 610.10-1994w

dice Multiple pieces of silicon, each of which contains one or more circuits and is or will be packaged as a unit. *Note:* This is the plural form of die. (C) 610.10-1994w

dichotomizing search (data management) A search in which an ordered set of items is partitioned into two parts, one of which is rejected, and the process is repeated on the accepted part until the search is completed. *See also:* interpolation search; Fibonacci search; binary search. (C) 610.5-1990w

dichotomy (mathematics of computing) A division into two classes that are mutually exclusive and dual in nature. For example, all zero and all nonzero, or all true and all false. (C) 1084-1986w

dichroic filter (fiber optics) An optical filter designed to transmit light selectively according to wavelength (most often, a high-pass or low-pass filter). *See also:* optical filter. (Std100) 812-1984w

dichroic mirror (fiber optics) A mirror designed to reflect light selectively according to wavelength. *See also:* dichroic filter. (Std100) 812-1984w

dichromate cell A cell having an electrolyte consisting of a solution of sulphuric acid and a dichromate. *See also:* electrochemistry. (EEC/PE) [119]

Dicke fix An electronic counter-countermeasures (ECCM) technique designed to counter impulsive jamming and some types of swept-frequency jamming. *Note:* The usual configuration is a broadband intermediate frequency (IF) amplifier followed by a limiter and then an IF amplifier of optimum bandwidth for the radar signal. (AES) 686-1997

dictionary A list of data items and information about those items, used both to describe and to reference the items. *Contrast:* directory. *See also:* index; data dictionary; table. (C) 610.5-1990w

dictionary/directory *See:* data directory; data dictionary.

die **(1)** A single piece of silicon that contains one or more circuits and is or will be packaged as a unit. *Note:* This is the plural form of die. *See also:* dice. (C) 610.10-1994w
(2) *See also:* semiconductor; clipping-in; chip. (T&D/PE) 524-1992r

dielectric (surge arresters) A medium in which it is possible to maintain an electric field with little or no supply of energy from outside sources. (PE) [8]

dielectric constant **(1)** **(dielectric)** That property which determines the electrostatic energy stored per unit volume for unit potential gradient. *Note:* This numerical value usually is given relative to a vacuum. (IA) 54-1955w
(2) The real part of the complex dielectric constant. (AP/ANT) 145-1993
(3) Relative permittivity (possilby complex). (AP/PROP) 211-1997

dielectric dissipation factor **(A)** The cotangent of the dielectric phase angle of a dielectric material or the tangent of the dielectric loss angle. **(B)** The ratio of the loss index ento the relative dielectric constant e. *See also:* relative complex dielectric constant. (EM/PE) 286-1975

dielectric filling factor *See:* filling factor.

dielectric filter *See:* interference filter.

dielectric guide A waveguide in which the waves travel through solid dielectric material. *See also:* waveguide. (EEC/PE) [119]

dielectric heater A device for heating normally insulating material by applying an alternating-current field to cause internal losses in the material. *Note:* The normal frequency range is above 10 megahertz. *See also:* interference. (IE) [43]

dielectric lens A lens made of dielectric material and used for refraction of radio-frequency energy. *See also:* waveguide; antenna. (Std100) [84]

dielectric loss (planar transmission lines) That contribution to the attenuation constant of a propagating mode on a planar transmission line that represents losses associated with the dielectric properties of the substrates (and overlays) materials involved, which may also include conduction mechanisms. (MTT) 1004-1987w

dielectric loss angle d (rotating machinery) The angle whose tangent is the dissipation factor. (EM/PE) 286-1975w

dielectric loss factor The factor by which the product of a sinusoidal alternating voltage applied to a dielectric and the component of the resulting current having the same period as the voltage have to be multiplied in order to obtain the power dissipated in the dielectric. *See also:* loss factor. (PE/PSIM) 4-1995

dielectric loss filling factor *See:* filling factor.

dielectric phase angle **(A)** The angular difference in phase between the sinusoidal alternating voltage applied to a dielectric and the component of the resulting alternating current having the same period as the voltage. **(B)** The angle whose contangent is the dissipation factor, or arc cot $\varepsilon''/\varepsilon'$. *See also:* relative complex dielectric constant; dielectric dissipation factor. (EM/PE) 286-1975

dielectric power factor The cosine of the dielectric phase angle (or the sine of the dielectric loss angle). (EM/PE) 286-1975w

dielectric rod antenna An antenna that employs a shaped dielectric rod as the electrically significant part of a radiating element. *Note:* The polyrod rod antenna is a notable example of the dielectric rod antenna when constructed of polystyrene. (AP/ANT) 145-1993

dielectric strength (general) (material) (electric strength-breakdown strength) The potential gradient at which electric failure or breakdown occurs. To obtain the true dielectric strength the actual maximum gradient must be considered, or the test piece and electrodes must be designed so that uniform gradient is obtained. The value obtained for the dielectric strength in practical tests will usually depend on the thickness of the material and on the method and conditions of test. (PE) [8]

dielectric tests **(1)** **(general)** Tests which consist of the application of a voltage higher than the rated voltage, for a specified time to assure the withstand strength of insulation materials and spacing. These various types of dielectric tests have been developed to allow selectivity testing the various insulation components of a transformer, without overstressing other components; or to simulate transient voltages which transformers may encounter in service. *See also:* applied voltage tests; induced voltage tests; impulse test. (PE/TR) C57.12.80-1978r
(2) **(high voltage air switches)** Tests that consist of the application of a standard test voltage for a specific time and are designed to determine the adequacy of insulating materials and spacing. (SWG/PE) C37.34-1971s
(3) **(neutral grounding devices)** Tests that consists of the application of a voltage, higher than the rated voltage, for a specified time to prove compliance with the required voltage class of the device. (SPD/PE) 32-1972r
(4) Tests that consist of the application of a voltage higher than the rated voltage for a specified time for the purpose of determining the adequacy of insulating materials against breakdown, and for spacing under normal conditions. (PE/TR) C57.15-1999

dielectric waveguide A waveguide consisting of a dielectric structure. (MTT) 146-1980w

dielectric withstand voltages The maximum voltage that may be applied between line or phase terminals and the neutral or ground terminals (or the device enclosure) without causing electrical failure or breakdown of the device insulation. (SPD/PE) C62.62-2000

dielectric withstand voltage tests (x-radiation limits for ac high-voltage power vacuum interrupters used in power switchgear) Tests made to determine the ability of insulating materials and spacings to withstand specified overvoltages for a specified time without flashover or puncture. (SWG/PE/TR) C37.40-1981s, C57.12.80-1978r, 553-1981, C37.100-1992

diesel-electric drive A self-contained system of power generation and application in which the power generated by a diesel engine is transmitted electrically by means of a generator and a motor (or multiples of these) for propulsion purposes. *Note:* The prefix diesel-electric is applied to ships, locomotives, cars, buses, etcetera, that are equipped with this drive. *Synonym:* oil-electric drive. *See also:* electric locomotive. (EEC/PE) [119]

diesel-generator unit An independent source of standby electrical power that consists of a diesel-fueled internal combustion engine (or engines) coupled directly to an electrical generator (or generators); the associated mechanical and electrical auxiliary systems; and the control, protection, and surveillance systems. (PE/NP) 387-1995

DIF *See:* data interchange format.

difference **(1)** **(mathematics of computing)** The result of a subtraction operation. (C) 1084-1986w
(2) **(data management)** A relational operator that combines two relations having identical attributes and results in a relation containing the tuples that are in the first but not the second relation. *Synonyms:* set difference; minus. *See also:* join; projection; union; selection; product; intersection.

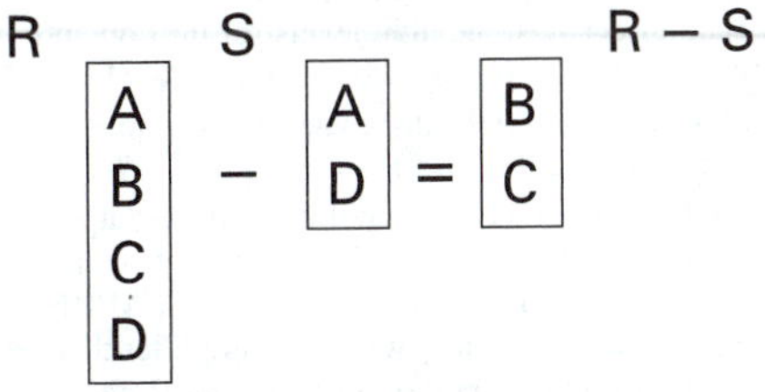

difference

(C) 610.5-1990w

difference amplifier *See:* differential amplifier.

difference channel (monopulse radar) A receiving channel in which the response, as a function of a given radar coordinate, approximates the first derivative of the response of the main (sum) channel, to indicate the displacement of the target from the center of the main channel. *Note:* This term was originally applied to monopulse radar, in which the difference between two offset beams or antenna phase centers was used to generate an error signal for tracking. In more modern radars the difference channel in angle can be generated by a feed network producing an aperture illumination function with odd symmetry. Similar channels in range and Doppler coordinates can be generated by suitable gates and filters. (AES) 686-1997

difference detector A detector circuit in which the output is a function of the difference of the peak amplitudes or root-mean-square amplitudes of the input waveforms. *See also:* navigation. (AES) [42]

difference frequency (parametric device) The absolute magnitude of the difference between a harmonic nfp of the pump frequency fp and the signal frequency fs, where n is a positive integer. *Note:* Usually n is equal to one. *See also:* parametric device. (ED) 254-1963w, [46]

difference-frequency parametric amplifier* *See:* inverting parametric device.

* Deprecated.

difference in depth of modulation [directive systems employing overlapping lobes with modulated signals such as instrument landing system (ILS)] A fraction obtained by subtracting from the percentage of modulation of the smaller signal and dividing by 100. (AES/GCS) 172-1983w

difference limen (differential threshold) (just noticeable difference) The increment in a stimulus that is just noticeable in a specified fraction of trials. *Note:* The relative difference limen is the ratio of the difference limen to the absolute magnitude of the stimulus to which it is related. *See also:* phonograph pickup. (SP) [32]

difference pattern (1) A radiation pattern characterized by a pair of main lobes of opposite phase, separated by a single null, plus a family of side lobes, the latter usually desired to be at a low level. *Note:* Antennas used in many radar applications are capable of producing a sum pattern and two orthogonal difference patterns. The difference patterns can be employed to determine the position of a target in a right/left and up/down sense by antenna pattern pointing, which places the target in the null between the twin lobes of each difference pattern. *Contrast:* sum pattern. (AP/ANT) 145-1993
(2) The curve of antenna gain versus angle for the difference channel of a monopulse antenna. *See also:* difference channel. (AES) 686-1997

difference signal *See:* differential signal.

difference slope In a monopulse radar, the slope of the difference-pattern voltage (normalized with respect to the sum-pattern voltage) as a function of target angle from the tracking axis. *Note:* The slope is usually specified at the point on the curve where the difference-pattern voltage is zero, which corresponds to the tracking axis. In range and Doppler coordinates, it is the corresponding slope of the difference channel voltage normalized to that of the sum channel. (AES) 686-1997

differential (photoelectric lighting control) The difference in foot-candles between the light levels for turn-on and turn-off operation. *See also:* photoelectric control. (IA/IAC) [60]

differential aeration cell An oxygen concentration cell. *See also:* electrolytic cell. (IA) [59], [71]

differential amplifier (1) An amplifier whose output signal is proportional to the algebraic difference between two input signals. *See also:* amplifier. (IM/HFIM) [40]
(2) (signal-transmission system) An amplifier that produces an output only in response to a potential difference between its input terminals (differential-mode signal) and in which outputs from common-mode interference voltages on its input terminals are suppressed. *Note:* An ideal differential amplifier produces neither a differential-mode nor a common-mode output in response to a common-mode interference input. *See also:* amplifier; signal. (IE) [43]
(3) An amplifier with two input circuits that amplifies the difference between the two input signals. (C) 610.10-1994w

differential analyzer (1) (analog computer) A computer designed primarily for the convenient solution of differential equations. (C) 165-1977w
(2) An analog computer that uses interconnected integrators to solve differential equations. (C) 610.10-1994w

Differential Analyzer Replacement (DARE) A series of continuous simulation languages for use in batch and on-line applications. (C) 610.13-1993w

differential capacitance (nonlinear capacitor) The derivative with respect to voltage of a charge characteristics, such as an alternating charge characteristic or a mean charge characteristic, at a given point on the characteristic. (ED) [46]

differential-capacitance characteristic (nonlinear capacitor) The function relating differential capacitance to voltages. *See also:* nonlinear capacitor. (ED) [46]

differential capacitance voltage The difference in magnitudes of the rms system normal frequency line-to-neutral voltage multiplied by the square root of two, with and without the capacitance connected. *Note:* This can be calculated from the equations:

$$\Delta V = \sqrt{2}E_S \frac{X_L}{X_C - X_L}$$

or

$$\Delta V = \sqrt{2}E_S \frac{\text{kvar}}{\text{kVA}_{SC} - \text{kvar}}$$

where
ΔV = Differential capacitance voltage in volts
E_S = System phase-to-neutral voltage in volts rms
X_L = Source inductive reactance to point of application, in ohms per phase
X_C = Capacitive reactance of bank being switched in ohms per phase
kvar = Size of bank being switched (three phase)
kVA_{SC} = System short-circuit kVA at point of capacitor application (symmetrical three phase)

(SWG/PE/SWG-OLD) C37.100-1992, C37.30-1971s

differential compounded (rotating machinery) Applied to a compound machine to denote that the magnetomotive forces of the series field winding is opposed to that of the shunt field winding. (PE) [9]

differential control A system of load control for self-propelled electrically driven vehicles wherein the action of a differential field wound on the field poles of a main generator (or of an exciter) and connected in circuit between the main generator and the traction motors, serves to limit the power demand from the prime mover. *See also:* multiple-unit control. (EEC/PE) [119]

differential control current (magnetic amplifier) The total absolute change in current in a specified control winding necessary to obtain differential output voltage when the control current is varied very slowly (a quasistatic characteristic). (MAG) 107-1964w

differential control voltage (magnetic amplifier) The total absolute change in voltage across the specified control terminals necessary to obtain differential output voltage when the control voltage is varied very slowly (a quasistatic characteristic). (MAG) 107-1964w

differential d*E*/d*x* detector (1) A transmission detector in which the thickness is small compared to the range of the incident particle. (NPS) 300-1988r
(2) A detector in which the thickness is small compared to the range of the incident particle and in which the entrance and exit dead layers are small compared to the thickness of the detector. (NPS) 325-1996

differential Doppler frequency (radio-wave propagation) The time rate of change of difference in phase path at two frequencies in a dispersive medium. *Note:* Sometimes called the dispersive Doppler frequency. (AP) 211-1977s

differential dump *See:* change dump.

differential duplex system A duplex system in which the sent currents divide through two mutually inductive sections of the receiving apparatus, connected respectively to the line and to a balancing artificial line, in opposite directions so that there is substantially no net effect on the receiving apparatus; whereas the received currents pass mainly through one section; or though the two sections in the same direction, and operate the apparatus. *See also:* telegraphy. (EEC/PE) [119]

differential gain (video transmission system) The difference between (a) the ratio of the output amplitudes of a small high-frequency sine-wave signal at two stated levels of a low frequency signal on which it is superimposed, and (b) unity. *Notes:* 1. Differential gain may be expressed in percent by multiplying the above difference by 100. 2. Differential gain may be expressed in decibels by multiplying the common logarithm of the ratio described in (a) above by 20. 3. In this definition, level means a specified position on an amplitude scale applied to a signal wave-form. 4. The low- and high-frequency signals must be specified. *See also:* television. (BT/AV) [34], 206-1960w

differential-gain control (gain sensitivity control) A device for altering the gain of a radio receiver in accordance with an expected change of signal level, to reduce the amplitude differential between the signals at the output of the receiver. *See also:* radio receiver. (EEC/PE) [119]

differential gain-control circuit (electronic navigation) The circuit of a receiving system that adjusts the gain of a single radio receiver to obtain desired relative output levels from two alternately applied or sequentially unequal input signals. *Note:* This may be accomplished automatically or or manually; if automatic, it is referred to as automatic differential-gain control. Example: Loran circuits that adjust gain between successive pulses from different ground stations. (AES/RS) 686-1982s, [42]

differential-gain-control range The maximum ratio of signal amplitudes (usually expressed in decibels), at the input of a single receiver, over which the differential-gain-control circuit can exercise proper control and maintain the desired output levels. *See also:* navigation. (AES/RS) 686-1982s, [42]

differential gap *See:* neutral zone.

differential gear A mechanism used for addition and subtraction in an analog computer in which the angles of rotation of three shafts are related to each other such that the algebraic sum of the rotation of two shafts is equal to twice the rotation of the third. (C) 610.10-1994w

differential input impedance to ground The impedance between either the positive input and ground or the negative input and ground. (IM/WM&A) 1057-1994w

differential interconnect A pair of connections carrying signals from a transmitter on one component to a receiver on another component, where the transmitted information (which may be either analog or digital) is represented by the difference between two signals rather than by either signal individually. *See also:* simple interconnect; extended interconnect. (C/TT) 1149.4-1999

differential Manchester encoding A signaling method used to encode clock and data bit information into bit symbols. Each bit symbol is split into two halves, or signal elements, where the second half is the inverse of the first half. A 0 bit is represented by a polarity change at the start of the bit time. A 1 bit is represented by no polarity change at the start of the bit time. Differential Manchester encoding is polarity-independent. (C/LM) 8802-5-1998

differential mode attenuation (fiber optics) The variation in attenuation among the propagating modes of an optical fiber. (Std100) 812-1984w

differential mode delay (fiber optics) The variation in propagation delay that occurs because of the different group velocities of the modes of an optical fiber. *Synonym:* multimode group delay. *See also:* mode; group velocity; multimode distortion. (Std100) 812-1984w

differential-mode interference (1) (emi power line filters for commercial use) (signal-transmission system) Interference that causes the potential of one side of the signal transmission path to be changed relative to the other side. *Note:* That type of interference in which the interference current path is wholly in the signal transmission path. (EMC/IE/SUB/PE) C63.13-1991, [43], C37.1-1994
(2) (interference terminology) *See also:* accuracy rating; interference; normal-mode interference.

differential-mode noise (1) (normal or transverse) The noise voltage that appears differentially between two signal wires and acts on the signal sensing circuit in the same manner as the desired signal. Normal mode noise may be caused by one or more of the following: Electrostatic induction and differences in distributed capacitance between the signal wires and the surroundings; Electromagnetic induction and magnetic fields linking unequally with the signal wires; Junction or thermal potentials due to the use of dissimilar metals in the connection system; Common mode to normal mode noise conversion. (PE/EDPG) 1050-1996
(2) *See also:* transverse-mode noise. (IA/PSE) 1100-1999

differential-mode radio noise Conducted radio noise that causes the potential of one side of the signal transmission path to be changed relative to another side. (EMC) C63.4-1988s

differential-mode voltage (1) The instantaneous algebraic difference between the potential of two signals applied to the two sides of a balanced circuit. Also called *metallic voltage* in the telephone industry. (C/LM) 802.3-1998
(2) The instantaneous algebraic difference of two signals applied to a balanced circuit, where both signals are referred to a common reference.local area networks. (C) 8802-12-1998
(3) *See also:* transverse-mode voltage. (PE/PSR) C37.90.1-1989r

differential nonlinearity (1) (semiconductor radiation detectors) (percent) The percentage departure of the slope of the plot of output versus input from the slope of a reference line. (NID) 301-1976s
(2) The difference between a specified code bin width and the average code bin width, divided by the average code bin width. (IM/WM&A) 1057-1994w

differential nonreversible output voltage *See:* differential output voltage.

differential output current (magnetic amplifier) The ratio of differential output voltage to rated load impedance. (MAG) 107-1964w

differential output voltage (1) (magnetic amplifier) (nonreversible output) The voltage equivalent to the algebraic difference between maximum test output voltage and minimum test output voltage. (MAG) 107-1964w
(2) (reversible output) The voltage equivalent to the algebraic difference between positive maximum test output voltage and negative maximum test output voltage. (MAG) 107-1964w

differential permeability (magnetic core testing) The rate of change of the induction with respect to the magnetic field strength.

$$\mu_{dif} = \frac{1}{\mu_0}\frac{dB}{dH}$$

where

μ_{dif} = relative differential permeability
dH = infinitely small change in field strength
dB = corresponding change induction

(MAG) 393-1977s

differential permittivity (primary ferroelectric terms) The slope of the hysteresis loop (electric displacement versus electric field) at any point. Differential permittivity is usually measured at low frequency (60 Hz) due to the self-heating produced on cycling through a hysteresis loop. The value of differential permittivity is often different from the small-signal permittivity measured under equivalent bias conditions. (UFFC) 180-1986w

differential phase (video transmission system) The difference in output phase of a small high-frequency sine-wave signal at the two stated levels of a low-frequency signal on which it is superimposed. *Note:* Notes C and D appended to differential gain apply also to differential phase. *See also:* television. (BT/AV) 206-1960w

differential phase shift (A) (nonlinear, active, and nonreciprocal waveguide components) The difference between insertion phase changes, resulting from a change of configuration or material state, in the two opposite directions of propagation between two ports of a junction. **(B) (nonlinear, active, and nonreciprocal waveguide components)** Differential insertion phase. (MTT) 457-1982

differential-phase-shift keying (DPSK) (1) A method for encoding a signal in which the value of a bit stream is encoded on the differences between the phase of adjacent signals; that is, if the signals are in phase, the bit is a one; if not, the bit is a zero. (C) 610.10-1994w
(2) A form of phase-shift keying in which the reference phase for a given keying interval is the phase of the signal during the preceding keying interval. (Std100) 270-1964w

differential position (loran, omega, [global positioning system]) (navigation aid terms) The difference between position axis determined by separated receivers or antennas. Close proximity error sources are minimized, thereby greatly enhancing the accuracy of this parameter. (AES/GCS) 172-1983w

differential protection A method of apparatus protection in which an internal fault is identified by comparing electrical conditions at all terminals of the apparatus. (SWG/PE/PSR) C37.100-1992, C37.90-1978s

differential protective relay (power system device function numbers) A protective relay that functions on a percentage or phase angle or other quantitative difference of two currents or of some other electrical quantities. (SUB/PE) C37.2-1979s

differential quantum efficiency (fiber optics) In an optical source or detector, the slope of the curve relating output quanta to input quanta. (Std100) 812-1984w

differential relay (1) A relay that by its design or application is intended to respond to the difference between incoming and outgoing electrical quantities associated with the protected apparatus. (SWG/PE/PSR) C37.100-1992, C37.90-1978s
(2) A relay with multiple windings that functions when the power developed by the individual windings is such that pickup or dropout results from the algebraic summation of the fluxes produced by the effective windings. (PE/EM) 43-1974s

differential resistance (semiconductor rectifiers) The differential change of forward voltage divided by a stated increment of forward current producing this change. *See also:* semiconductor rectifier stack. (IA) [62]

differential reversible output voltage *See:* differential output voltage.

differential signal (1) The instantaneous, algebraic difference between two signals. *See also:* oscillograph. (IM/HFIM) [40]
(2) A signal that is conveyed between two separate conductors, instead of one active conductor and signal ground. The magnitude of the differential signal is the difference between the two signals, rather than the voltages between the two individual signals and ground. (EMB/MIB) 1073.4.1-2000

differential skew The difference in time between the midpoint voltage crossings of the true and complement components of a differential signal. (C/LM) 802.3-1998

differential threshold *See:* difference limen.

differential trip signal (magnetic amplifier) The absolute magnitude of the difference between trip OFF and trip ON control signal. (MAG) 107-1964w

differential voltage signal The voltage difference between the true and complementary signals from a driver with two single-ended outputs whose signals always complement each other. Differential signals are also referred to as "balanced signals." (C/MM) 1596.3-1996

differentiated (1) (germanium gamma-ray detectors) (pulse amplifier) (charged-particle detectors) (pulse) (semiconductor radiation detectors) (x-ray energy spectrometers) A pulse is differentiated when it is passed through a high-pass network, such a a CR filter. (NPS/NID) 325-1986s, 759-1984r, 301-1976s
(2) (pulse) (pulse amplifier) A pulse that is passed through a high-pass network, such as a CR filter. (NPS) 300-1988r

differentiating network *See:* differentiator.

differentiator (1) (electronic circuits) A device whose output function is reasonably proportional to the derivative of the input function with respect to one or more variables, for example, a resistance-capacitance network used to select the leading and trailing edges of a pulse signal. (C) [85]
(2) (differentiating circuit) (modulation circuits and industrial control) (differentiating network) A transducer whose output waveform is substantially the time derivative of its input waveform. *Note:* Such a transducer preceding a frequency modulator makes the combination a phase modulator; or following a phase detector makes the combination a frequency detector. Its ratio of output amplitude to input amplitude is proportional to frequency and its output phase leads its input phase by 90 degrees. (PE/AP/IA/SWG-OLD/ANT/IAC) C37.100-1992, 145-1983s, [60]
(3) A circuit or device whose output signal is proportional to the derivative of its input signal with respect to one or more variables, usually time. For example, a resistor-capacitor circuit used to detect the edges of a pulse. (C) 610.10-1994w
(4) A high-pass filter network in which the waveform of the output signal approximates the mathematical derivative of the input waveform. (NPS) 325-1996
(5) A high-pass network, usually comprising a series capacitor and a shunt resistor, for the purpose of reducing the duration of a signal. (NI/NPS) 309-1999

diffracted wave (1) When a wave in a medium of certain propagation characteristics is incident upon a discontinuity or a second medium, the diffracted wave is the wave component that results in the first medium in addition to the incident wave and the waves corresponding to the reflected rays of geometrical optics. *See also:* radiation. (EEC/PE) [119]
(2) (audio and electroacoustics) A wave whose front has been changed in direction by an obstacle or other nonhomogeneity in a medium, rather than by reflection or refraction. *See also:* radiation. (SP) [32]
(3) An electromagnetic wave that has been modified by an obstacle or spatial inhomogeneity in the medium by means other than reflection or refraction. (AP/PROP) 211-1997

diffraction (1) (general) A process that produces a diffracted wave. *See also:* radiation. (SP) [32]
(2) (laser maser) Deviation of part of a beam, determined by the wave nature of radiation, and occurring when the radiation passes the edge of an opaque obstacle. (LEO) 586-1980w
(3) (fiber optics) The deviation of a wavefront from the path predicted by geometric optics when a wavefront is restricted by an opening or an edge of an object. *Note:* Diffraction is usually most noticeable for openings of the order of a wavelength. However, diffraction may still be important for apertures many orders of magnitude larger than the wavelength. *See also:* near-field diffraction pattern; far-field diffraction pattern. (Std100) 812-1984w

(4) The deviation of the direction of energy flow of a wave, not attributable to reflection or refraction, when it passes an obstacle, a restricted aperture, or other inhomogeneities in a medium. (AP/PROP) 211-1997

diffraction angle (acousto-optic device) The angle between the *N*th order diffraction beam and the zeroth order beam. It is given by the ratio of the optical wavelength λ_0 to the acoustic wavelength, times the order of the diffracted beam $N = \pm 1, \pm 2, \pm 3, \ldots$ so that ${}_{\theta}N = N\lambda_0/\wedge$. (UFFC) [23]

diffraction efficiency (acousto-optic device) For the *N*th order, the percent ratio of the light intensity diffracted into the *N*th order divided by the light intensity in the zeroth order with the acoustic drive power off, thus

$$\eta_N = (I_N/I_O) \times 100$$

For a device of fixed design, the diffraction efficiency will depend on the optical wavelength, beam diameter, angle of incidence, and acoustic drive power. (UFFC) [23]

diffraction grating (fiber optics) An array of fine, parallel, equally spaced reflecting or transmitting lines that mutually enhance the effects of diffraction to concentrate the diffracted light in a few directions determined by the spacing of the lines and the wavelength of the light. *See also:* diffraction. (Std100) 812-1984w

diffraction limited (fiber optics) A beam of light is diffraction limited if the far-field beam divergence is equal to that predicted by diffraction theory, or in focusing optics, the impulse response or resolution limit is equal to that predicted by diffraction theory. *See also:* diffraction; beam divergence. (Std100) 812-1984w

diffraction loss (laser maser) That portion of the loss of power in a propagating wave (beam) which is due to diffraction. (LEO) 586-1980w

diffused junction (semiconductor) A junction that has been formed by the diffusion of an impurity within a semiconductor crystal. *See also:* semiconductor. (ED) 216-1960w

diffused-junction detector A semiconductor detector in which the rectifying junction is produced by diffusion of acceptor (p) or donor (n) defects. (NPS) 325-1996, 300-1982s

diffused lighting (illuminating engineering) Lighting provided on the work-plane or on an object light that is not incident predominantly from any particular direction. (EEC/IE) [126]

diffuse ferroelectrics (diffuse ferroelectric single crystals or polycrystalline solid solutions) (primary ferroelectric terms) Materials whose small-signal permittivity, when measured as a function of temperature, indicates by its width a broad or diffuse phase transition between the ferroelectric and nonferroelectric phases. These materials exhibit weak ferroelectric properties in the temperature range of the diffuse phase transition. Diffuse ferroelectrics are noted for their almost anhysteretic P versus E behavior and have been called weak or dilute ferroelectrics, slim-loop materials, penferroelectric, and quasiferroelectric or alpha-phase materials. (UFFC) 180-1986w

diffuse field The non-coherent component of the scattered field. *Note:* The diffuse electromagnetic field has a zero average value, (i.e., it is a zero-mean process). (AP/PROP) 211-1997

diffuse intensity Power density associated with the diffuse field. (AP/PROP) 211-1997

diffuse multipath Propagation between radar and target for which one path is direct and the other(s) involve scattering from a rough surface or an atmospheric volume. (AES) 686-1997

diffuser (illuminating engineering) A device to redirect or scatter the light from a source, primarily by the process of diffuse transmission. (EEC/IE) [126]

diffuse reflectance (illuminating engineering) The ratio of the flux leaving a surface or medium by diffuse reflection to the incident flux. (EEC/IE) [126]

diffuse reflection (1) (illuminating engineering) That process by which incident flux is redirected over a range of angles. *See also:* diffusing surfaces and media. (EEC/IE) [126]
(2) (laser maser) Change of the spatial distribution of a beam of radiation when it is reflected in many directions by a surface or by a medium. (LEO) 586-1980w
(3) (fiber optics) *See also:* reflection. (LEO) 586-1980w
(4) *See also:* diffuse scattering. (AP/PROP) 211-1997

diffuse scattering (1) Scattering of incident electromagnetic energy over a range of angles other than the specular direction. (EMC) 1128-1998
(2) The generation of non-coherent (diffuse) fields caused by scattering of an incident electromagnetic wave by a random rough surface or a medium randomly varying with time and/or space. (AP/PROP) 211-1997

diffuse sound field A sound field in which the time average of the mean-square sound pressure is everywhere the same and the flow of energy in all directions is equally probable. (SP) [32]

diffuse transmission (illuminating engineering) That process by which the incident flux passing through a surface or medium is scattered. (EEC/IE) [126]

diffuse transmission density The value of the photographic transmission density obtained when the light flux impinges normally on the sample and all the transmitted flux is collected and measured. (SP) [32]

diffuse transmittance (illuminating engineering) The ratio of the diffusely transmitted flux leaving a surface or medium to the incident flux. *Note:* Provision for exclusion of regularly transmitted flux must be clearly described. (EEC/IE) [126]

diffusing panel (illuminating engineering) A translucent material covering the lamps in a luminaire in order to reduce the brightness by distributing the flux over an extended area. (EEC/IE) [126]

diffusing surfaces and media (illuminating engineering) Those surfaces and media that redistribute at least some of the incident flux by scattering. (EEC/IE) [126]

diffusion (laser maser) Change of the spatial distribution of a beam of radiation when it is deviated in many directions by a surface or by a medium. (LEO) 586-1980w

diffusion approximation For wave propagation in lossy media, it corresponds to neglecting the displacement current. (AP/PROP) 211-1997

diffusion capacitance (semiconductor) (nonlinear, active, and nonreciprocal waveguide components) A capacitance enhancement effect associated with p-n junctions. Because the diffusion of electrons and holes takes time, there is a storage effect that is equivalent to adding additional capacitance in shunt with the junction. In the forward bias state, the diffusion capacitance becomes predominant over the space-charge capacitance to the point that the latter can be neglected at lower microwave frequencies. Diffusion capacitance varies inversely with frequency, but increases as minority carrier lifetime increases and must be dealt with in many frequency multiplier and parametric amplifier designs. (MTT) 457-1982w

diffusion charging Charging of aerosols by small ions in collisions resulting from thermal motion of the small ions. (T&D/PE) 539-1990

diffusion constant (charge carrier) (homogeneous semiconductor) The quotient of diffusion current density by the charge-carrier concentration gradient. It is equal to the product of the drift mobility and the average thermal energy per unit charge of carriers. *See also:* semiconductor. (ED) 216-1960w

diffusion depth *See:* junction depth.

diffusion length, charge-carrier (homogeneous semiconductor) The average distance to which minority carriers diffusion length is equal to the square root of the product of the charge-carrier diffusion constant and the volume lifetime. *See also:* semiconductor. (AES/IA/ED) [41], [12], 270-1966w, 216-1960w

digit (1) (metric practice) One of the ten Arabic numerals (0 to 9). (QUL) 268-1982s
(2) (A) (positional notation) (notation) A character that stands for an integer; Loosely, the integer that the digit stands for; Loosely, any character. **(B) (positional notation) (notation)** A character used to represent one of the nonnegative integers smaller than the radix, for example, in decimal notation one of the characters 0 to 9. (C) [85]
(3) (mathematics of computing) (data management) A symbol or character that represents one of the non-negative integers smaller than the radix; for example, in decimal notation, a digit is one of the characters 0 1 2 3 4 5 6 7 8 9. *Synonym:* numeric character. (C) 610.5-1990w, 1084-1986w
(4) One of the ten numerals (0 to 9) in the decimal number system. A position in a number. (SCC14) SI 10-1997

digit absorption (telephone switching systems) The interpretation and rejection of those digits received, but not required, in the setting of automatic direct control system crosspoints. (COM) 312-1977w

digital (1) Pertaining to data in the form of digits. *See also:* analog. (C) 162-1963w
(2) (mathematics of computing) Pertaining to quantities in the form of discrete, integral values. *Contrast:* analog. (C) 1084-1986w, 610.10-1994w

digital-analog converter *See:* digital-to-analog converter.

digital automatic test program generator (DATPG) A program, often based on simulation, that aids in the development of test patterns and diagnostic information from the model of a unit under test (UUT). (SCC20) 1445-1998

digital bit rate The number of bits per unit of time. (COM/TA) 1007-1991r

digital boundary module (DBM) A circuit module connected between the digital core circuit and a digital function pin to provide facilities for test in a digital or mixed-signal component. *Notes:* 1. A DBM may also be interposed at the boundary between the digital and analog portions of the core circuit. 2. A control-and-observe DBM includes a switching function in the serial data path between pin and core; an observe-only DBM captures data from the data path without interrupting the path. (C/TT) 1149.4-1999

Digital Business Oriented Language (DIBOL) A problem-oriented programming language developed by Digital Equipment Company; used to develop business applications. *See also:* DBL. (C) 610.13-1993w

digital coefficient attenuator (1) (hybrid computer linkage components) Essentially the same as a digital-to-analog multiplier (DAM). This term is generally reserved for those components that are used as the high speed hybrid replacement for manual and servo potentiometers. *Synonym:* digital potentiometer. (C) 166-1977w
(2) A component that is used as a high-speed hybrid replacement for manual and servo potentiometers in analog computers. *Synonym:* digital potentiometer. *See also:* digital-to-analog multiplier. (C) 610.10-1994w

Digital Command Language (DCL) A command language used under Digital's VAX/VMS environments. (C) 610.13-1993w

digital computer (1) (information processing) A computer that operates on discrete data by performing arithmetic and logic processes on these data. *Contrast:* analog computer. (C) [20], [85]
(2) (test, measurement, and diagnostic equipment) A computer in which discrete quantities are represented in digital form and which generally is made to solve mathematical problems by iterative use of the fundamental processes of addition, subtraction, multiplication, and division. (MIL) [2]
(3) (A) A computer that consists of one or more associated processing units and that is controlled by internally-stored programs. **(B)** A computer that utilizes digital circuitry to perform calculations and logical instructions, and to control sequencing of operations. *Contrast:* analog computer; hybrid computer. (C) 610.10-1994

Digital Control Design Language (DCDL) A simulation language for use in designing digital computer systems. (C) 610.13-1993w

digital controller (data processing) A controller that accepts an input sequence of numbers and processes them to produce an output sequence of numbers. (IM) [52]

digital converter A device, or group of devices, that converts an input numerical signal or code of one type into an output numerical signal or code of another type. *Synonym:* code translator. (SWG/PE) C37.100-1992

digital data (1) (data transmission) Pertaining to data in the form of digits or interval quantities. *Contrast:* analog data. (PE) 599-1985w
(2) Data in the form of discrete integral values. *Contrast:* analog data. (C) 610.7-1995
(3) Data that does not admit to representations obeying the laws of arithmetic. Examples of digital data are digital state vectors, image bit maps, and proximity switch outputs. (IM/ST) 1451.1-1999

digital data circuit Any circuit that transfers data in a digitally encoded form which is essential for the proper operation of the relay system. (PE/PSR) C37.90.1-1989r

digital delay line A delay line designed specifically to accept digital (video) electrical signals. The signals are specified usually as bipolar, RZ, or NRZ. The definitions are based on the output signal being a doublet generated by an input step function. (UFFC) [22]

digital device (1) (control equipment) A device that operates on the basis of discrete numerical techniques in which the variables are represented by coded pulses or states. (PE/PSE) 94-1970w
(2) (radio-noise emissions) An information technology equipment (ITE) that falls into the class of unintentional radiators that uses digital techniques and generates and uses timing signals or pulses at a rate in excess of 9000 pulses per second. (EMC) C63.4-1991

digital differential analyzer A special-purpose digital computer consisting of many parallel computing elements, that performs integration by means of a suitable integration code on incremental quantities and that can be programmed for the solution of differential equations in a manner similar to an analog computer. (C) 610.10-1994w, 165-1977w

digital filter A device that produces a predetermined digital output in response to a digital input. For example, a digital filter may use arithmetic or delays in order to obtain the desired transfer function. (C) 610.10-1994w

digital image An image that has been converted into an array of pixels, each of which has an associated value called its gray level. *Note:* A digital image may be referred to as an image when the intended meaning is clear from the context. *Synonym:* digitized image. *See also:* digitization. (C) 610.4-1990w

digital interface A set of wires and a protocol for transferring information by binary means only. (IM/ST) 1451.2-1997

digital line link (digital line path) A digital link that comprises a digital line section or a number of tandem-connected digital line sections. (COM/TA) 1007-1991r

digital line section A digital section implemented on a single type of manufactured transmission medium, such as symmetric cable pair, coaxial, or fiber. (COM/TA) 1007-1991r

digital link The method of digital transmission of a digital signal of specified rate between two digital distribution frames (or equivalent). *Notes:* 1. A digital link comprises one or more digital sections and may include multiplexing or demultiplexing, with the rule that the digital signal exiting the link must not differ in information content from the signal entering the link. 2. The term may be qualified to indicate the transmission medium used, for example, "digital satellite link." 3. The term always applies to the combination of "forward" and "return" directions of transmission, unless stated otherwise. 4. The term *digital path* is sometimes used to describe one or more digital links connected in tandem, especially between equipment at which the signals of the specified rate originate and terminate. (COM/TA) 1007-1991r

digital logic (A) Any logic used for digital integrated circuits and systems. **(B)** Circuitry that produces two or more distinct states, which can be used for logical operations. (C) 610.10-1994

digital logic elements (analog computer) In an analog computer, a number of digital functional modules, consisting of logic gates, registers, flip-flops, timers, etcetera, all operating in parallel, either synchronously or asynchronously, and whose inputs and outputs are interconnected, according to a "logic program," via patch cards, on a patch board. (C) 165-1977w

digital loop carrier (DLC) Equipment that increases the number of end users served by existing loopside pairs through the use of digital multiplexing. These concentration systems are often called pair-gain devices. Both universal and integrated DLCs may be used. (AMR/SCC31) 1390-1995, 1390.2-1999, 1390.3-1999

digitally-controlled function generator (1) (analog computer) A hybrid component using DAC's and DAM's to insert the linear segment approximation values to the desired arbitrary function. The values are stored in a self-contained digital core memory, which is accessed by the DAC's and DAM's at digital-computer speeds (microseconds). (C) 165-1977w

(2) A hybrid component employed in analog computers to insert linear segment approximation values into a desired arbitrary function. (C) 610.10-1994w

digital milliwatt (1) The repetitive transmission of a sequence of codes in a given channel, which will be decoded in a receiving terminal as a 0 dBm, 1000.0 Hz signal. The digital milliwatt, which has no quantizing noise, is generally not transmitted over a telecommunication facility because it can cause false framing on D4-formatted DS1 facilities. (COM/TA) 743-1995

(2) The repetitive transmission of the following sequence of codes in a given channel will be decoded in a receiving terminal as 0 dBm0, 1000.0 Hz signal:

Digit number	1	2	3	4	5	6	7	8
	0	0	0	1	1	1	1	0
	0	0	0	0	1	0	1	1
	0	0	0	0	1	0	1	1
	0	0	0	1	1	1	1	0
	1	0	0	1	1	1	1	0
	1	0	0	0	1	0	1	1
	1	0	0	0	1	0	1	1
	1	0	0	1	1	1	1	0

(COM/TA) 1007-1991r

digital modem *See:* channel service unit.

digital multimeter (DMM) A DMM is used to measure electrical quantities such as dc or ac voltage, ac or dc current, resistance, etc. The input resistance/impedance should be at least 1000 times the resistance/impedance of the circuit being measured. (PEL) 1515-2000

digital offset The offset is the position of the intersection point of the zero voltage input value on the channel number axis. The digital offset value is subtracted from the ADC output value before storage. This corresponds to the first channel in the stored spectrum. It is expressed as six characters with leading spaces interpreted as leading zeros. The digital offset is used where the low channel part of the data does not contain useful information and is digitally discarded before storage in the memory. This number is added to the stored data channel number to obtain the ADC output value. This allows various spectra to be compared even if they are incomplete. (NPS/NID) 1214-1992r

digital optical disk *See:* optical disk.

digital path The whole of the means of transmitting and receiving a digital signal of specified rate between those two digital signal crossconnect frames (DSX) (or equivalent) at which terminal equipment or switches will be connected. A digital path comprises one or more sections. *See also:* digital link. (COM/TA) 1007-1991r

digital phase lock loop (communication satellite) A circuit for synchronizing the received waveform, by means of discrete corrections. (COM) [24]

digital pin A pin on an integrated circuit or other component that is intended to pass data represented as a voltage or current that can have one of two discrete values. *Notes:* 1. In addition to the set of discrete data values, a digital pin may be put into a state in which its driver is disabled so that it cannot actively sink or source current and, therefore, cannot influence the state of the attached net. 2. If analog data are applied externally to a digital input pin, the internal circuitry will normally interpret the data as digital. Use of the pin in this way may lead to possible adverse effects on the internal circuitry (e.g., power consumption). 3. A digital function pin may have additional circuitry to allow analog signals to be applied or monitored for test purposes. This additional circuitry would not affect its status as a digital pin. *Contrast:* analog pin. *See also:* high-Z; net. (C/TT) 1149.4-1999

digital plotter A plotter that presents digital data in the form of a two-dimensional graphic representation. *Contrast:* analog plotter; raster plotter. (C) 610.10-1994w

digital potentiometer *See:* digital coefficient attenuator.

digital quantity (1) (station control and data acquisition) A variable represented by coded pulses (for example, bits) or states. (SUB/PE) 999-1992w

(2) (supervisory control, data acquisition, and automatic control) A variable represented by a number of bits. (SUB/PE) C37.1-1994

(3) A variable represented by a number of discrete units. (SWG/PE) C37.100-1992

digital readout clock A clock that gives (usually with visual indication) a voltage or contact closure pattern of electrical circuitry for a readout of time. A digital readout calendar clock also includes a readout of day, month, and year, usually, also with indication. (SWG/PE) C37.100-1992

Digital Reference Sequence (DRS) A 797-byte representation, in A-law or μ-law coding of one of three levels (0, −10, or −13 dBm) of a 1013.8-Hz sine wave. With digital bit integrity, the 797-byte sequence may also be used for bit error ratio testing through the network. (COM/TA) 743-1995

digital reference signal A sequence of bits that represents a 1002 Hz to a 1020 Hz signal that, when decoded, will produce a sinusoidal signal of the same power as the digital milliwatt. (COM/TA) 1007-1991r

digital representation The representation of numerical quantities by means of digits, or discrete values. *Contrast:* analog representation. (C) 1084-1986w

digital search tree A search tree in which the order of the keys is representational of the data contained in the tree. For example, a thumb-index of a dictionary that organizes the alphabet by groups of three or four letters; ABC, DEF, GHI, ... WXYZ. (C) 610.5-1990w

digital section The whole of the means of digital transmission of a digital signal of specified rate between two consecutive digital signal crossconnect (DSX) frames or equivalent. *Notes:* 1. A digital section forms either a part or the whole of a digital link, and includes terminating equipment at both ends, but excludes multiplexers. 2. Where appropriate, the digital rate or multiplex order should qualify the title. 3. The definition applies to the combination of "forward" and "return" directions of transmission, unless stated otherwise. (COM/TA) 1007-1991r

digital signal A discrete and/or discontinuous signal. *Contrast:* analog signal. *See also:* digitize. (C) 610.7-1995

digital signal crossconnect frame A device for circuit rearrangement, patching, and testing purposes. The digital signal crossconnect is designated DSX*n*, where *n* indicates the hierarchical level of the digital signal network (DSN) interconnected at that crossconnect. (COM/TA) 1007-1991r

digital signal level A hierarchy of transmission formats of digital signals. (C) 610.7-1995

digital signal level four A digital transmission format in which six digital signal level three are time-division multiplexed together. *See also:* T4. (C) 610.7-1995

digital signal level one A digital transmission format in which twenty-four digital signal level zero are time-division multiplexed together. *See also:* T1. (C) 610.7-1995

digital signal level 1C A digital transmission format in which forty eight digital signal level zero are time-division multiplexed together. *See also:* T1C. (C) 610.7-1995

digital signal level three A digital transmission format in which seven digital signal level two are time-division multiplexed together. *See also:* T3. (C) 610.7-1995

digital signal level two A digital transmission format in which four digital signal level one are time-division multiplexed together. *See also:* T2. (C) 610.7-1995

digital signal level zero Fundamental transmission rate of the "digital signal level." *Note:* The data rate is 64 kb/s. (C) 610.7-1995

digital signature Information supplied with a transmission that allows the receiver to verify that the transmission has not been modified after initial generation. (SCC32) 1455-1999

digital simulation (A) A simulation that is designed to be executed on a digital system. **(B)** A simulation that is designed to be executed on an analog system but that represents a digital system. **(C)** A simulation of a digital circuit. *Contrast:* analog simulation. *See also:* hybrid simulation. (C) 610.3-1989

Digital Simulation Language (DSL) A simulation language that represents blocks, switching functions, and function generators, similar to those available with an analog computer. (C) 610.13-1993w

digital sort A radix sort in which base 10 notation is used. (C) 610.5-1990w

digital subscriber line (DSL) (failure rate) The expected frequency of outages a DSL can experience due to switching system and subsystem malfunctions. Problems that may occur in the station or wiring, outside plant, or loop electronics are excluded. The DSL failure rate may be given for hardware faults alone. (COM/TA) 973-1990w

digital sum variation The difference between the number of logical 1s and the logical 0s transmitted by a link output since commencing operation. *See also:* running disparity. (C/BA) 1355-1995

digital switch (telephone loop performance) A switch that, internally, performs switching only of digital signals of a set format. It is inherently a four-wire entity, requiring a two-wire to four-wire hybrid at the channel interface when accepting analog signals from two-wire channels. (COM/TA) 820-1984r

digital switching (1) (telephone switching systems) Switching of discrete-level information signals. (COM) 312-1977w
(2) A switching process in which connections are established by operations directly on the digital signals. (C) 610.7-1995

digital synchronization slip Occurs when the frame rate of an incoming signal received at the SPCS is different from the frame rate of the SPCS itself. A higher or lower incoming frame rate causes the receiver buffer to overflow or underflow, resulting in a lost or repeated frame. Slip occurrences are minimized by requiring that the incoming signal be synchronized by the synchronization plan of the switch. However, overflow and underflow of the receive synchronizer in the digital facility interface may still result from imperfect synchronization, independent operation of switching office clocks, or temperature variations in the cable. (COM/TA) 973-1990w

digital telemeter indicating receiver A device that receives the numerical signal transmitted from a digital telemeter transmitter and gives a visual numerical display of the quantity measured. (SWG/PE) C37.100-1992

digital telemeter receiver A device that receives the numerical signal transmitted by a digital telemeter transmitter and stores it or converts it to a usable form, or both, for such purposes as recording, indication, or control. (SWG/PE) C37.100-1992

digital telemeter transmitter A device that converts its input signal to a numerical form for transmission to a digital telemeter receiver over an interconnecting channel. (SWG/PE) C37.100-1992

digital telemetering Telemetering in which a numerical representation, as for example some form of pulse code, is generated and transmitted; the number being representative of the quantity being measured. (SWG/PE) C37.100-1992

digital telephone set A telephone set where the two-way voice communication interface to the network is in a digital format. (COM/TA) 269-1992

digital termination system A form of local loop that connects private homes and/or business locations to the common carrier switching facility. (C) 610.7-1995

digital-to-analog (d/a) conversion (1) Production of an analog signal whose magnitude is proportional to the value of a digital input. (SWG/PE) C37.100-1992
(2) Production of an analog signal whose magnitude is proportional to the value of a digital quantity. (SUB/PE) C37.1-1994

digital-to-analog converter (1) (data processing) A device that converts an input number sequence into a function of a continuous variable. (IM) [52]
(2) (hybrid computer linkage components) A circuit or device whose input is information in digital form and whose output is the same information in analog form. In a hybrid computer, the input is a number sequence (or word) coming from the digital computer, while the output is an analog voltage proportional to the digital number. (C) 166-1977w
(3) A converter that converts an input number sequence (digital) into a function of a continuous variable (analog). *Synonyms:* D/A converter; d-to-a converter; digital-analog converter. *Contrast:* analog-to-digital converter. *See also:* digital-to-analog multiplier. (C) 610.10-1994w
(4) A device, or group of devices, that converts a numerical input signal or code into an output signal some characteristic of which is proportional to the input. (SWG/PE) C37.100-1992
(5) A circuit that converts an input number sequence (digital) into a function of a continuous variable (analog). (IM/ST) 1451.2-1997

digital-to-analog multiplier (DAM) (hybrid computer linkage components) A device which provides the means of obtaining the continuous multiplication of a specific digital value with a changing analog variable. The product is represented by a varying analog voltage. *Synonym:* multiplying-digital-to-analog converter. (C) 166-1977w

digital tree search A radix search in which the items in the set to be searched are placed in a tree according to the digital representation of the search keys. *Note:* The tree is traversed in a top-down fashion making comparisons on the bit representations until a match is found for the search argument or the lowest-level node of the tree is encountered. *See also:* binary tree search. (C) 610.5-1990w

digital variable* *See:* binary variable; integer variable.
* Deprecated.

digit deletion In the processing of a call, the elimination of a portion of the destination code. (COM) 312-1977w

digitization (1) (mathematics of computing) The process of converting analog data to digital data. (C) 1084-1986w
(2) (image processing and pattern recognition) The process of converting an image into a digital image. *See also:* quantization; sampling. (C) 610.4-1990w

digitize (1) (mathematics of computing) To express analog data in digital form. (C) 1084-1986w
(2) (computer graphics) To convert graphical data, such as a drawing, into digital data that can be processed by a computer graphics system. (C) 610.6-1991w
(3) To convert an analog signal to a digital signal. (C) 610.7-1995

digitized image *See:* digital image.

digitizer A graphic input device that converts analog data, such as those derived from a drawing, into digital form. Examples include optical scanners and graphic tablets. *Synonym:* quantizer. *See also:* data tablet. (C) 610.10-1994w, 610.6-1991w

digit place (A) In a positional notation system, a position corresponding to a given power of the radix. **(B)** A location in which a digit may occur in a numeral. *Synonyms:* position; digit position; symbol rank; place. (C) 1084-1986

digit position *See:* digit place.

digit punch A punch in one of the punch rows designated as 1 through 9 of a twelve-row punch card. *Contrast:* zone punch. (C) 610.10-1994w

digit transformation function A hash function that returns a permutation of the original key with one or more digits removed. For example, in the function below, every other digit is dropped from the original key.

Original key	Hash value
964721	942
78394	734

(C) 610.5-1990w

digraph *See:* directed graph.

dim Abbreviation for DIMENSION, indicating the maximum functional capability designed into an end device. *See also:* function-limiting control table. (AMR/SCC31) 1377-1997

dimension (1) (metric practice) A geometric element in a design, such as length or angle, or the magnitude of such a quantity. (QUL/SCC14) 268-1982s, SI 10-1997
(2) The number of degrees of freedom of a physical quantity. For example, an electric field, which is a spatial vector quantity, has dimension 3. (IM/ST) 1451.1-1999

dimension, critical mating *See:* critical mating dimension.

diminished-radix complement (mathematics of computing) The complement obtained by subtracting each digit of a given numeral from the largest digit in the numeration system. For example, ones complement in binary notation, nines complement in decimal notation. *Synonyms:* complement on $n - 1$; radix-minus-one complement; base-minus-one complement. *Contrast:* radix complement. (C) 1084-1986w

diminishing increment sort An insertion sort in which the items in the set to be sorted are divided into subsets, each containing *N* items; the corresponding items in the subsets are ordered using an insertion sort; and this process is repeated using subsets of diminishing size until the subsets are of size 1. *Synonyms:* Shell's method; Shell sort. (C) 610.5-1990w

dimmed A visual effect on an object, such as a reduction in brightness or color, indicating that it cannot be selected or accept input. (C) 1295-1993w

dimming reactor (thyristor) A reactor that may be inserted in a lamp circuit at will for reducing the luminous intensity of the lamp. *Note:* Dimming reactors are normally used to dim headlamps, but may be applied to other circuits, such as gauge lamp circuits. (EEC/PE) [119]

diode (1) (electron tube) A two-electrode electron tube containing an anode and a cathode. *See also:* equivalent diode. (ED) 161-1971w
(2) (semiconductor) A semiconductor device having two terminals and exhibiting a nonlinear voltage-current characteristic; in more-restricted usage, a semicondutor device that has the asymmetrical voltage-current characteristic exemplified by a single p-n junction. *See also:* semiconductor. (ED) 216-1960w
(3) A semiconducting device used to permit current flow in one direction and to inhibit current flow in the other direction. *Synonym:* rectifier. (C) 610.10-1994w

diode array An integrated circuit that contains two or more diodes. (C) 610.10-1994w

diode characteristic (multielectrode tube) The composite electrode characteristic taken with all electrodes except the cathode connected together. (ED) 161-1971w

diode-current-balancing reactor A reactor with a set of mutually coupled windings that, operating in conjunction with other similar reactors, forces substantially equal division of current among the parallel paths of a rectifier circuit element. *See also:* reactor. (PE/TR) [57]

diode equivalent The imaginary diode consisting of the cathode of a triode or multigrid tube and a virtual anode to which is applied a composite controlling voltage such that the cathode current is the same as in the triode or multigrid tube. (ED) 161-1971w

diode function generator (analog computer) A function generator that uses the transfer characteristics of resistive networks containing biased diodes. The desired function is approximated by linear segments whose values are manually inserted by means of potentiometers and switches. (C) 165-1977w

diode fuses (semiconductor rectifiers) Fuses of special characteristics connected in series with one or more semiconductor rectifier diodes to disconnect the semiconductor rectifier diode in case of failure and protect the other components of the rectifier. *Note:* Diode fuses may also be employed to provide coordinated protection in case of overload or short-circuit. *See also:* semiconductor rectifier stack. (IA) [62]

diode laser *See:* injection laser diode.

diode-transistor logic (DTL) A family of bipolar integrated circuit logic formed by diodes, transistors, and resistors; characterized by medium speed, low power dissipation, high drive capability and low cost. (C) 610.10-1994w

dip (1) (electroplating) A solution used for the purpose of producing a chemical reaction upon the surface of a metal. *See also:* electroplating. (EEC/PE) [119]
(2) *See also:* sag. (SCC22) 1346-1998

DIP *See:* dual in-line package.

diplex filter A filter having a low pass and high pass filter that divide the frequency spectrum into two separate frequency bands that do not overlap. The conventional designation assigns the low band of frequencies to the inbound path, and the high band of frequencies to the outbound path. The diplex filter allows the placement of duplex signals onto a cable by the use of frequency division multiplexing. (LM/C) 802.7-1989r

diplex operation (data transmission) The simultaneous transmission or reception of two signals using a specified common feature, such as a single antenna or a single carrier. (PE) 599-1985w

diplex radio transmission The simultaneous transmission of two signals using a common carrier wave. *See also:* radio transmission. (AP/ANT) 145-1983s

dip needle A device for indicating the angle between the magnetic field and the horizontal. *See also:* magnetometer. (EEC/PE) [119]

dipole *See:* folded dipole; Hertzian electric dipole; dipole antenna; electrically short dipole; half-wave dipole; Hertzian magnetic dipole; magnetic dipole; microstrip dipole.

dipole antenna (overhead-power-line corona and radio noise) Any one of a class of antennas having a radiation pattern approximating that of an elementary electric dipole. *Note:* Common usage considers the dipole antenna to be a metal radiating or receiving structure that supports a line current distribution similar to that of a thin straight wire, a half-wavelength long, so that the current has a node at each end of the antenna. *Synonym:* doublet. (T&D/PE/AP/ANT) 539-1990, 145-1993, 599-1985w

dipole modulation *See:* non-polarized return-to-zero recording.

dipole molecule A molecule that possesses a dipole moment as a result of the permanent separation of the centroid of positive charge from the centroid of negative charge for the molecule as a whole. (Std100) 270-1966w

dip plating *See:* immersion plating.

dip soldering (soldered connections) The process whereby assemblies are brought in contact with the surface of molten solder for the purpose of making soldered connections. (EEC/AWM) [105]

DIP switch One or more two-position switches housed in a dual in-line package (DIP); used on a circuit board to control certain functions or to specify particular operating characteristics. (C) 610.10-1994w

direct access Pertaining to the process of storing and retrieving data using direct access mode. *Contrast:* sequential access. *See also:* indexed sequential access; indexed access. (C) 610.5-1990w

direct access arrangement A circuit, typically used in modems, which allows a device to be connected to telephone lines. (C) 610.7-1995

direct-access merge sort An external merge sort in which the auxiliary storage used is direct-access storage. *See also:* tape merge sort. (C) 610.5-1990w

direct access method (DAM) An access method in which the access time required to access data is effectively independent of the location of the data. *Contrast:* basic access method. *See also:* direct data set; basic direct access method. (C) 610.5-1990w

direct access mode An access mode in which data records are stored and retrieved from storage in such a way that the location of a data record can be derived from the value of some element or elements in the record, regardless of the contents of other data records. *Contrast:* sequential access mode; indexed sequential access mode. (C) 610.5-1990w

direct-access storage device An auxiliary storage device that can provide direct access to the data stored on the device. *See also:* immediate access storage. (C) 610.10-1994w

direct ac converter (self-commutated converters) (cycloconverter) The alternating current (ac) conversion is accomplished directly, without an intermediate link having different power characteristics, such as direct current (dc) or high-frequency ac. (IA/SPC) 936-1987w

direct-acting machine-voltage regulator A machine-voltage regulator having a voltage-sensitive element that acts directly without interposing power-operated means to control the excitation of an electric machine. (SWG/PE) C37.100-1992

direct-acting overcurrent trip device (1) A release or tripping system that is completely self-contained on a circuit breaker and requires no external power or control circuits to cause it to function. (SWG/PE) C37.100-1992
(2) *See also:* direct release; overcurrent release; indirect release.

direct-acting overcurrent trip device current rating (trip devices for ac and general-purpose dc low-voltage power circuit breakers) The value of current designated by the manufacturer on which trip element calibration marks are based. (PE) C37.100-1992, C37.17-1972w

direct-acting recording instrument A recording instrument in which the marking device is mechanically connected to, or directly operated by, the primary detector. *See also:* instrument. (EEC/PE) [119]

direct address (1) (A) (computers) An address that specifies the location of an operand. *See also:* one-level address. **(B)** An address that identifies the storage location of an operand. *Contrast:* *n*–level address; indirect address; immediate data. *See also:* one-level address; direct instruction. (C) [20], [85], 610.12-1990
(2) An address that explicitly specifies the location of an operand. *Synonym:* one-level address; first-level address. (C) 610.10-1994w

direct addressing (1) (microprocessor assembly language) An addressing mode in which the address is treated as a constant. It is to be distinguished from relative addressing. *See also:* relative addressing. (C/MM) 695-1985s
(2) An addressing mode in which the address field of an instruction contains a direct address. (C) 610.10-1994w

direct air-impingement cooling A way to implement cooling, using forced air across the components. (C/BA) 1101.4-1993

direct-arc furnace An arc furnace in which the arc is formed between the electrodes and the charge. (EEC/PE) [119]

direct axis (synchronous machines) The axis that represents the direction of the plane of symmetry of the no-load magnetic-flux density, produced by the main field winding current, normally coinciding with the radial plane of symmetry of a field pole. *See also:* direct-axis synchronous reactance. (PE) [9]

direct-axis armature to field transfer function *See:* armature to field transfer function.

direct-axis component of armature current That component of the armature current that produces a magnetomotive force distribution that is symmetrical about the direct axis. (EEC/PE) [119]

direct-axis component of armature voltage That component of the armature voltage of any phase that is in time phase with the direct-axis component of current in the same phase. *Note:* A direct-axis component of voltage may be produced by:

1) Rotation of the quadrature-axis component of magnetic flux,
2) Variation (if any) of the direct-axis component of magnetic flux,
3) Resistance drop caused by flow of the direct-axis component of armature current. As shown in the phasor diagram, the direct-axis component of terminal voltage, assuming no field magnetization in the quadrature-axis, is given by

$$E_{ad} = -RI_{ad} - jX_q I_{aq}$$

(EEC/PE) [119]

direct-axis component of magnetomotive force (rotating machinery) The component of magnetomotive force that is directed along the direct axis. *See also:* asynchronous machine; direct-current commutating machine; direct-axis synchronous impedance. (PE) [9]

direct-axis current (rotating machinery) The current that produces direct-axis magnetomotive force. *See also:* direct-axis synchronous reactance. (PE) [9]

direct-axis magnetic-flux component (rotating machinery) The magnetic-flux component directed along the direct axis. *See also:* direct-axis synchronous reactance. (PE) [9]

direct-axis operational inductance [Ld(s)] (synchronous machine parameters by standstill frequency testing) The ratio of the Laplace transform of the direct-axis armature flux linkages to the Laplace transform of the direct-axis current, with the field winding short-circuited. (PE/EM) 115A-1987

direct-axis subtransient impedance (rotating machinery) The magnitude obtained by the vector addition of the value for armature resistance and the value for direct-axis subtransient reactance. *Note:* The resistance value to be applied in this case will be a function of frequency depending on rotor iron losses. (PE) [9]

direct-axis subtransient open-circuit time constant The time in seconds required for the rapidly decreasing component (negative) present during the first few cycles in the quadrature-axis component of symmetrical armature voltage under suddenly removed symmetrical shot-circuit condition, with the machine running at rated speed to decrease to 1/e D 0.368 of its initial value. *Note:* If the rotor is made of slid steel no single subtransient time constant exists but a spectrum of time constants will appear in the subtransient region. *See also:* direct-axis synchronous impedance. (PE) [9]

direct-axis subtransient reactance (rotating machinery) The quotient of the initial value of a sudden change in that fundamenatal alternating-current component of armature voltage, which is produced by the total direct-axis primary flux, and the value of this simultaneous change in fundamental alternating-current component of direct-axis armature current, the machine running at rated speed. *Note:* The rated current value is obtained from the tests for the rated current value of direct-axis transient reactance. The rated voltage value is that obtained from a sudden short-circuit test at the terminals of the machine at rated armature voltage, no load. *See also:* direct-axis synchronous reactance. (PE) [9]

direct-axis subtransient short-circuit time constant The time required for the rapidly changing component, present during the first few cycles in the direct-axis alternating component of a shot-circuit armature current, following a sudden change in operating conditions, to decrease to 1/e approx 0.368 of its initial value, the machine running at rated sped. *Note:* The rated current value is obtained from the test for the rated current value of the direct-axis transient reactance. The rated voltage value is obtained from the test for the rated voltage value of the direct-axis transient reactance. *See also:* direct-axis synchronous reactance. (PE) [9]

direct-axis subtransient voltage (rotating machinery) The direct-axis component of the terminal voltage which appears immediately after the sudden opening of the external circuit when the machine is running at a specified load, before ay flux variation in the excitation and damping circuits has taken place. (PE) [9]

direct-axis synchronous impedance (rotating machinery) (synchronous machines) The magnitude obtained by the vector addition of the value for armature resistance and the value for direct-axis synchronous reactance. (PE) [9]

direct-axis synchronous reactance (rotating machinery) The quotient of a sustained value of that fundamental alternating-current component of armature voltage that is produced by the total direct-axis flux due to direct-axis armature current and the value of the fundamental alternating-current component of this current, the machine running at rated speed. Unless otherwise specified, the value of synchronous reactance will be that corresponding to rated armature current. For most machines, the armature resistance is negligibly small compared to the synchronous reactance. Hence the synchronous reactance may be taken also as the synchronous impedance. (PE) [9]

direct-axis transient impedance (rotating machinery) The magnitude obtained by the vector addition of the value for armature resistance and value for direct-axis transient reactance. (PE) [9]

direct-axis transient open-circuit time constant (rotating machinery) The time in seconds required for the root-mean-square alternating-current value of the slowly decreasing component present in the quadrature axis (T'do) component of symmetrical armature voltage on open-circuit to decrease to 1/e = 0.368 of its initial value when the field winding is suddenly short-circuited with the machine running at rated speed. *See also:* direct-axis synchronous impedance. (PE) [9]

direct-axis transient reactance (rotating machinery) The quotient of the initial value of a sudden change in that fundamental alternating-current component of armature voltage, which is produced by the total direct-axis flux, and the value of the simultaneous change in fundamental alternating-current component of direct-axis armature current, the machine running at rated speed and the high-decrement components during the first cycles being excluded. *Note:* The rated current value is that obtained from a three-phase sudden short-circuit test at the terminals of the machine at no load, operating at a voltage such as to give an initial value of the alternating component of current, neglecting the rapidly decaying component of the first few cycles, equal to the rated current. This requirement means that the pr-unit test voltage is equal to the rated current value of transient reactance (per unit). In actual practice, the test voltage will seldom result in initial transient current of exactly rated value, and it will usually be necessary to determine the reactance from a curve of reactance plotted against voltage. The rated voltage value is that obtained from a three-phase sudden short-circuit test at the terminals of the machine at rated voltage, no load. *See also:* direct-axis synchronous reactance; direct-axis synchronous impedance. (PE) [9]

direct-axis transient short-circuit time constant The time in seconds required for the root-mean-square value of the slowly decreasing component present in the direct-axis component of the alternating-current component of the armature current under suddenly applied symmetrical short-circuit conditions with the machine running at rated speed, to decrease to 1/e = 0.368 of its initial value.

direct-axis transient voltage (rotating machinery) The direct-axis component of the armature voltage that appears immediately after the sudden opening of the external circuit when running at a specified load, the components that decay very fast during the first few cycles, if any, being neglected. *See also:* direct-axis synchronous reactance. (PE) [9]

direct-axis voltage (rotating machinery) The component of voltage that would produce direct-axis current when resistance-limited. *See also:* direct-axis synchronous reactance. (PE) [9]

direct-buried transformer (power and distribution transformers) A transformer designed to be buried in the earth with connecting cables. (PE/TR) C57.12.80-1978r

direct capacitances (system of conductors) The direct capacitances of a system of n conductors such as that considered in coefficients of capacitance (system of conductors) are the coefficients in the array of linear equations that express the charges on the conductors in terms of their differences in potential, instead of potentials relative to ground.

$$Q_1 = O + C_{12}(V_1 - V_2) + C_{13}(V_1 - V_3) + \ldots + C_{1(n-1)}(V_1 - V_{n-1}) + C_{10}V_1$$

$$Q_2 = C_{21}(V_2 - V_1) + O + C_{23}(V_2 - V_3) + \ldots + C_{2(n-1)}(V_2 - V_{n-1}) + C_{20}V_2$$

$$Q_{n-1} = C_{(n-1)1}(V_{n-1} - V_1) + C_{(n-1)2} (V_{n-1} - V_2) + \ldots + O + C_{(n-1)0}V_{n-1}$$

with $C_{rp} = C_{rp}$ and C_{re} not involved but defined as zero. *Note:* The coefficients of capacitance c are related to the direct capacitances C as follows $c_{rp} = -C_{rp}$, for $r \neq p$ and

$$c_{rr} = \sum_{p=1}^{p=n} C_{rp}$$

direct chaining *See:* separate chaining.

direct commutation (self-commutated converters) (circuit properties) A commutation between two principal switching branches without the involvement of other switching branches. In converters using devices such as power transistors or gate turn-off thyristors, this is accomplished by turning off the switch in the outgoing branch and turning on the switch in the incoming branch. In converters using circuit-commutated thyristors, commutating capacitors coupled to the switching branches turn off the outgoing switch when the incoming switch is turned on. (IA/SPC) 936-1987w

direct component (illuminating engineering) That portion of the light from a luminaire which arrives at the work-plane without being reflected by room surfaces. (EEC/IE) [126]

direct-connected exciter (rotating machinery) An exciter mounted on or coupled to the main machine shaft so that both machines operate at the same speed. *See also:* asynchronous machine. (PE) [9]

direct-connected system *See:* headquarters system.

direct-connect end node An end-node configuration that allows the node to be directly connected to another end node in a two-node, repeaterless network. A direct-connect node may be connected to another direct-connect node (allowing either full- or half-duplex packet transmission), to an end node without direct-connect capability (allowing half-duplex packet transmission), or to a repeater (allowing half-duplex packet transmission as an ordinary end node).local area networks. (C) 802.12c-1998

direct coupled (electrical heating applications to melting furnaces and forehearths in the glass industry) The power modulation device is conductively connected directly to the electrodes carrying current into the molten glass. (IA) 668-1987w

direct-coupled amplifier (signal-transmission system) A direct-current amplifier in which all signal connections between active channels are conductive. *See also:* signal. (IE) [43]

direct-coupled attenuation (transmit-receive, pretransmit-receive, and attenuator tubes) The insertion loss measured with the resonant gaps, or their functional equivalent, short-circuited. 160-1957w

direct-coupled transistor logic A family of circuit logic in which only transistors and resistors are used, with transistors directly connected to each other. (C) 610.10-1994w

direct coupling The association of two or more circuits by means of self-inductance, capacitance, resistance, or a combination of these that is common to the circuits. (PE/PSIM) 81-1983

direct current (dc) **(1)** An electric current that flows in one direction. (C) 610.10-1994w
(2) A unidirectional current in which the changes in value (polarity) are either zero or so small that they may be neglected. (As ordinarily used, the term designates a practically nonpulsating current.). (IA/MT) 45-1998

direct-current amplifier **(1)** An amplifier capable of amplifying waves of infinitesimal frequency. *See also:* amplifier. (AP/ANT) 145-1983s
(2) (signal-transmission system) An amplifier capable of producing a sustained single-valued, unidirectional output in response to a similar but smaller input. *Note:* It generally employs between stages either resistance coupling alone or resistance coupling alone or resistance coupling combined with other forms of coupling. *See also:* amplifier. (IE/PE) [43], 599-1985w

direct-current analog computer (analog computer) An analog computer in which computer variables are represented by the instantaneous values of voltages. (C) 165-1977w, 610.10-1994w

direct-current balance (amplifiers) An adjustment to avoid a change in direct-current level when changing gain. *See also:* amplifier. (IM/HFIM) [40]

direct-current balancer A machine that comprises two or more similar dc machines (usually with shunt or compound excitation) directly coupled to each other and connected in series across the outer conductors of a multiple wire system of distribution, for the purpose of maintaining the potentials of the intermediate conductors of the system, which are connected to the junction points between the machines. (IA/MT) 45-1998

direct-current blocking voltage rating (rectifier circuit element) The maximum continuous direct-current reverse voltage permitted by the manufacturer under stated conditions. *See also:* rectifier circuit element. (IA) 59-1962w, [62], [12]

direct-current breakdown voltage (gas-tube surge protective devices) (low-voltage air-gap surge-protective devices) The minimum slowly rising direct-current (dc) voltage that will cause breakdown or sparkover when applied across the terminals of an arrester. (SPD/PE) C62.31-1984s, C62.32-1981s

direct-current bushing A bushing carrying a dc stress, i.e., bushings applied to the valve winding side of a converter transformer, bushing applied to a dc smoothing reactor or a bushing applied to a dc converter valve. (PE/TR) C57.19.03-1996

direct-current circuit breaker (power system device function numbers) A circuit breaker that is used to close and interrupt a dc power circuit under normal conditions or to interrupt this circuit under fault or emergency conditions. (SUB/PE) C37.2-1979s

direct-current commutating machine A machine that comprises a magnetic field excited from a dc source or formed of permanent magnets, an armature and a commutator connected therewith. Specific types of dc commutating machines are dc generators, motors, synchronous converters, boosters, balancers, and dynamotors. (IA/MT) 45-1998

direct-current compensator *See:* direct-current balancer.

direct-current component **(1)** (of a total current) That portion of the total current which constitutes the asymmetry. (SWG/PE) C37.100-1992
(2) (television) (of a composite picture signal, blanked picture signal, or picture signal) The difference in level between the average value, taken over a specified time interval, and the peak value in the black direction, which is taken as zero. *Note:* The averaging period is usually one line interval or greater. *See also:* television. (BT/AV) [34]

direct current converter (self-commutated converters) A converter for changing dc power at a given voltage to dc power at a higher or lower voltage. (IA/SPC) 936-1987w

direct-current converter load rejection Tripping, blocking, or ramping HVDC converters to cause a significant reduction in transmitted power. (PE/SUB) 1378-1997

direct-current distribution The supply, to points of utilization, of electric energy by direct current from its point of generation or conversion. (T&D/PE) [10]

direct-current dynamic short-circuit ratio The ratio of the maximum transient value of a current, after a suddenly applied short circuit, to the final steady-state value. (EEC/AWM) [91]

direct-current electric-field strength (dc electric field) The time-invariant electric field, produced by dc power systems and space charge, defined by its space components along three orthogonal axes. The magnitudes of the components are expressed in volts per meter (V/m). *Notes:* 1. The convention in the discussion of electric fields near HVDC transmission lines has been to designate the electric field into the ground as positive; i.e., the electric field under a positive conductor is denoted as positive. 2. The "time-invariant" electric field in this definition is an idealization of actual conditions. Typically, climatic effects will perturb the space charge, which will then perturb the electric-field strength. (T&D/PE) 1227-1990r, 539-1990

direct-current electric-field-strength meter A meter designed to measure dc electric field. Two types of dc field strength meters are in common use. (T&D/PE) 539-1990

direct-current erase disturb Sometimes called deprogramming or gate disturb. Floating gate to control gate charge loss when the word-line (and control line) is high for programming another location on the same word-line (and control line). (ED) 1005-1998

direct-current filter A resistor-capacitor circuit connected between the positive and negative terminals of the dc link. (IA/ID) 995-1987w

direct-current electric locomotive An electric locomotive that collects propulsion power from a direct-current distribution system. *See also:* electric locomotive. (EEC/PE) [119]

direct-current electron-stream resistance (electron tube) The quotient of electron-stream potential and the direct-current component of stream current. *See also:* electrode current. (ED) 161-1971w

direct-current equipment grounding conductor (DCEG) (telecommunications) The conductor used to connect the metal parts of equipment, raceways, and other enclosures to the system grounded conductor (battery return), the conductor providing the system ground reference, or both, at the source of a direct current system (dc power plant). (IA/PSE) 1100-1999

direct-current erasing head One that uses direct current to produce the magnetic field necessary for erasing. *Note:* Direct-current erasing is achieved by subjecting the medium to a unidirectional field. Such a medium is, therefore, in a different magnetic state than one erased by alternating current. (SP/MR) [32]

direct current form factor (self-commutated converters) (converter characteristics) Of a periodic function, the ratio of the rms (root-mean-square) value to the mean value, averaged over a full period of the function. (IA/SPC) 936-1987w

direct-current generator A generator for production of direct-current power. (PE) [9]

direct-current generator-commutator exciter (excitation systems for synchronous machines) An exciter whose energy is derived from a dc generator. The exciter includes a dc generator with its commutator and brushes. It is exclusive of input control elements. The exciter may be driven by a motor, prime mover, or by the shaft of the synchronous machine. (PE/EDPG) 421.1-1986r

direct-current holdover (gas-tube surge protective devices) (low-voltage air-gap surge-protective devices) In applications where direct-current (dc) voltage exists on a line, a holdover condition is one in which a surge-protective device continues to conduct after it is subjected to an impulse large enough to cause breakdown. Factors that affect the time required to recover from the conducting state include the dc voltage and the current. (SPD/PE) C62.31-1987r, C62.32-1981s

direct-current holdover voltage (gas-tube surge protective devices) (low-voltage air-gap surge-protective devices) The maximum dc voltage across the terminals of an arrester under which it may be expected to clear and return to the high-impedance state after the passage of a surge, under specified circuit conditions. (PE/SPD) C62.31-1987r, C62.32-1981s

direct-current input current (self-commutated converters) (converters having dc input) The mean value of the direct current into the input terminals, taken over one period of the ripple current into those terminals. (IA/SPC) 936-1987w

direct-current input power (converters having dc input) (self-commutated converters) The product of the dc supply voltage and the dc input current (mean values of both as defined in dc supply voltage and dc input current). (IA/SPC) 936-1987w

direct-current insulation resistance (1) The resistance offered by its insulation to the flow of current resulting from an impressed direct voltage. *See also:* conductor. (PE/IC) 599-1985w
(2) (between two electrodes in contact with or embedded in a specimen) The ratio of the direct voltage applied to the electrodes to the total current between them. *Note:* It is dependent upon both the volume and surface resistances of the specimen. (Std100) 270-1966w
(3) (rotating machinery) The quotient of a specified direct voltage maintained on an insulation system divided by the resulting current at a specified time after the application of the voltage under designated conditions of temperature, humidity, and previous charge. *Note:* If steady state has not been reached the apparent resistance will be affected by the rate of absorption by the insulation of electric charge. *See also:* asynchronous machine. (PE) [9]

direct-current interface (terrestrial photovoltaic power systems) The connections between the array subsystem, the dc auxiliary power source, and the input of the power conditioning subsystem at the input terminals of the power conditioning system (PCS). *See also:* array control. (PV) 928-1986r

direct-current leakage *See:* controlled overvoltage test.

direct-current level The input or output voltage level on a dc-coupled instrument when there are no pulses present. For Ge spectrometer systems, the dc level of an output signal from a linear amplifier should be matched to the input requirements for dc levels of an ADC if the gain conversion scale is to include the origin. (NI) N42.14-1991

direct-current level unmodulated direct-current power Average dc level of an optical signal. For example, if 100% effective power is 3 mW (4.8 dBm) and 0% effective power is 1 mW (0 dBm) the dc level is 2 mW (3 dBm). (C/BA) 1393-1999

direct-current-linked alternating-current converter (self-commutated converters) A converter comprising a rectifier and an inverter, with an inmtermediate dc link. *Note:* This definition is intended to include only those circuits in which the dc link is readily identified or explicit, and not those circuits having an implicit dc link but no single pair of conductors that can be identified as the dc link. (IA/SPC/ID) 936-1987w, 995-1987w

direct-current magnetic biasing Magnetic biasing accomplished by the used of direct current. *See also:* phonograph pickup. (SP/MR) [32]

direct-current neutral grid A network of neutral conductors, usualy grounded, formed by connecting together within a given area of all the neutral conductors of a low-voltage direct-current supply system. *See also:* center of distribution. (T&D/PE) [10]

direct-current offset (1) (amplifiers) A direct-current level that may be added to the input signal, referred to the input terminals. *See also:* amplifier. (IM/HFIM) [40]
(2) The difference between the symmetrical current wave and the actual current wave during a power system transient condition. Mathematically, the actual fault current can be broken into two parts: a symmetrical alternating component and a unidirectional dc component, either or both with decreasing magnitudes (usually both). The unidirectional component can be of either polarity, but will not change polarity during its decay period, and will reach zero at some predetermined time. (PE/PSC) 367-1996
(3) Difference between the symmetrical current wave and the actual current wave during a power system transient condition. Mathematically, the actual fault current can be broken into two parts, a symmetrical alternating component and a unidirectional (dc) component. The unidirectional component can be of either polarity, but will not change polarity, and will decrease at some predetermined rate. (PE/SUB) 80-2000

direct-current offset factor The ratio of the peak asymmetrical fault current to the peak symmetrical value. (PE/PSC) 367-1996

direct-current overcurrent relay (power system device function numbers) A relay that functions when the current in a dc circuit exceeds a given value. (PE/SUB) C37.2-1979s

direct-current power supply (alternating-current to direct-current) Generally, a device consisting of a transformer, rectifier, and filter for converting alternating current to a prescribed direct voltage or current. (AES) [41]

direct-current program disturb Sometimes called word-line disturb. Floating gate charge gain when the word-line (and control line) is high for programming another location on the same word-line (and control line). (ED) 1005-1998

direct-current quadruplex system A direct-current telegraph system that affords simultaneous transmission of two messages in each direction over the same line, operation being obtained by superposing neutral telegraph upon polar telegraph. *See also:* telegraphy. (EEC/PE) [119]

direct current, rated *See:* rated direct current.

direct-current reactor (thyristor converter) An inductive reactor between the dc output of the thyristor converter and the load in order to limit the magnitude of fault current and also, in some cases, to limit the magnitude of ripple current in the load. In this latter case, it is called a ripple reactor. (IA/IPC) 444-1973w

direct-current reclosing relay (power system device function numbers) A relay that controls the automatic closing and reclosing of a dc circuit interrupter, generally in response to load circuit conditions. (SUB/PE) C37.2-1979s

direct-current regulated power supply A direct-current power supply whose output voltage is automatically controlled to remain within specified limits for specified variations in supply voltage and load current. *See also:* direct-current power supply. (Std100) [123]

direct-current relay A relay designed for operation from a direct-current source. *See also:* relay. (EEC/REE) [87]

direct-current restoration (television) The reestablishment of the dc and low-frequency components of a video signal that have been lost by ac transmission. (BT/AV) 201-1979w

direct-current restorer (television) A device for accomplishing dc restoration. (BT/AV) 201-1979w, 204-1961w

direct-current self-synchronous system A system for transmitting angular position or motion, comprising a transmitter and one or more receivers. The transmitter is an arrangement of resistors that furnishes the receiver with two or more voltages that are functions of transmitter shaft position. The receiver has two or more stationary coils that set up a magnetic field causing a rotor to take up an angular position corresponding to the angular position of the transmitter shaft. *See also:* synchro system. (EEC/PE) [119]

direct-current shift (oscilloscopes) A deviation of the displayed response to an input step, occurring over a period of several seconds after the input has reached its final value. (IM) 311-1970w

direct-current side short-circuit test A test in which the dc terminals of a converter are temporarily jumpered. *Synonym:* zero direct current voltage test. (PE/SUB) 1378-1997

direct-current signal A signal whose polarity and amplitude do not vary with time. (PEL) 1515-2000

direct-current standby current Voltage across the varistor measured at a specified pulsed direct-current (dc) current of specific duration. (PE) [8]

direct-current static tests Voltage and current tests to determine the input levels, output levels, and power dissipation characteristics of a device. (ED) 1005-1998

direct-current supply voltage (converters having dc input) (self-commutated converters) The mean value of the direct voltage between the input terminals, taken over one period of the ripple voltage appearing between the input terminals. (IA/SPC) 936-1987w

direct-current telegraphy That form of telegraphy in which, in order to form the transmitted signals, direct current is supplied to the line under the control of the transmitting apparatus. *See also:* telegraphy. (EEC/PE) [119]

direct-current transmission (1) (electric energy) The transfer of electric energy by direct current from its source to one or more main receiving stations. *Note:* For transmitting large blocks of power, high voltage may be used such as obtained with generators in series, rectifiers, etc. *See also:* direct-current distribution. (T&D/PE) [10]
(2) (television) A form of transmission in which the dc component of the video signal is transmitted. *Note:* In an amplitude-modulated signal with dc transmission, the black level is represented always by the same value of envelope. In a frequency-modulated signal with dc transmission, the black level is represented always by the same value of the instantaneous frequency. (BT/AV) 201-1979w, 204-1961w

direct-current winding (1) (thyristor converter) The winding of a thyristor converter transformer that is conductively connected to the thyristor converter circuit elements and that conducts the direct current of the thyristor converter. *See also:* secondary winding. (IA/IPC) 444-1973w
(2) (power and distribution transformers) (of rectifier transformer) The secondary winding that is conductively connected to the main electrodes of the rectifier, and that conducts the direct current of the rectifier. (PE/TR) C57.12.80-1978r

direct data set A file that is accessed using the direct access method. *Contrast:* sequential file. (C) 610.5-1990w

direct direct-current converter (self-commutated converters) (dc chopper) The dc conversion is accomplished directly, without an intermediate ac (alternating current) link. (IA/SPC) 936-1987w

direct dial access method Utilizes the switched telephone network, comprised of a switch and other network elements that establish a communication path between a utility/enhanced service provider (ESP) and a telemetry interface unit (TIU). (AMR/SCC31) 1390-1995, 1390.2-1999, 1390.3-1999

direct digital control (electric power system) A mode of control wherein digital computer outputs are used to directly control a process. *See also:* power system. (PE/PSE) 94-1970w, [54]

direct distance dialing (1) (telephone switching systems) The automatic establishing of toll calls in response to signals from the calling device of a customer. (COM) 312-1977w
(2) A telephone exchange service that enables the telephone user to call long distance subscribers without operator assistance. *Synonym:* subscriber trunk dialing. (C) 610.7-1995

direct-drive machine An electric driving machine the motor of which is directly connected mechanically to the driving sheave, drum, or shaft without the use of belts or chains, either with or without intermediate gears. (EEC/PE) [119]

directed address *See:* unicast frame.

directed branch (network analysis) A branch having an assigned direction. *Note:* In identifying the branch direction, the branch *jk* may be thought of as outgoing from node j and incoming at node *k*. Alternatively, branch *jk* may be thought of as originating or having its input at node *j* and terminating or having its output at node *k*. The assigned direction is conveniently indicated by an arrow pointing from node *j* toward node *k*. (CAS) 155-1960w

directed flow (oil-immersed forced-oil-cooled transformers) Indicates that the principal part of the pumped oil from heat exchangers or radiators is forced to flow through the windings. (PE/TR) C57.91-1995

directed graph (software) A graph in which direction is implied in the internode connections. *Synonym:* digraph. *Contrast:* undirected graph.

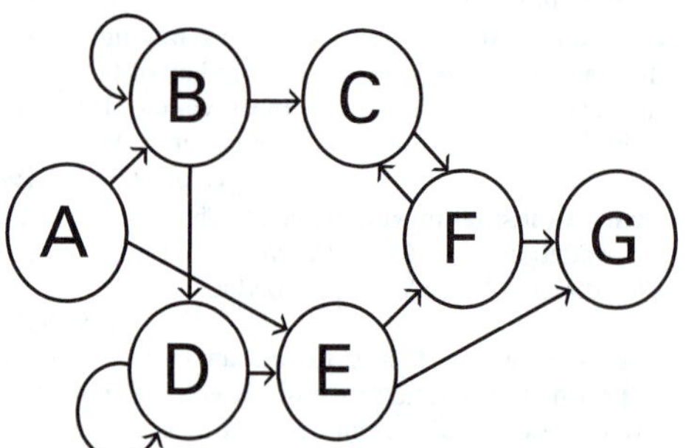

directed graph

(C) 610.5-1990w, 610.12-1990

directed reference flight (navigation aid terms) That type of stabilized flight which obtains control information from external signals which may be varied as necessary to direct the flight; for example, flight of a guided missile or a target aircraft. (AES/GCS) 172-1983w

directed rounding The process of approximating an exact value by a digital numeral such that the resulting error is known to be either non-positive or non-negative. The resulting number is therefore guaranteed to be an upper or lower bound. (C) 1084-1986w

directed transaction (1) A transaction that is processed by one and only one responder. The read, write, and lock transactions are always directed transactions. (C/MM) 1596-1992
(2) A transaction that is processed by one and only one responder. The read and write transactions are always directed. The event transactions may be directed or broadcast. (C/MM) 1596.4-1996

direct ESD event An ESD event that takes place between an intruder and a receptor in which the intruder or the receptor, or both, is an equipment victim. (SPD/PE) C62.47-1992r

direct ESD test A test in which ESD is applied directly to the surface or structure of the EUT. (EMC) C63.16-1993

direct feeder A feeder that connects a generating station, substation, or other supply point to one point of utilization. *See also:* center of distribution; auxiliary device to an instrument; radial feeder. (EEC/PE) [119]

direct glare (illuminating engineering) Glare resulting from high luminances or insufficiently shielded light sources in the field of view. It usually is associated with bright areas, such as luminaires, ceilings, and windows which are outside the visual task or region being viewed. (EEC/IE) [126]

direct grid bias The direct component of grid voltage. *See also:* electrode voltage. (ED) 161-1971w

direct-indirect lighting (illuminating engineering) A variant of general diffuse lighting in which the luminaires emit little or no light at angles near the horizontal. (EEC/IE) [126]

direct insert subroutine *See:* open subroutine.

direct instruction (1) A computer instruction that contains the direct addresses of its operands. *Contrast:* immediate instruction; indirect instruction. *See also:* absolute instruction; effective instruction. (C) 610.12-1990
(2) A computer instruction that contains the direct addresses of its operands. *Contrast:* indirect instruction. (C) 610.10-1994w

direct interelectrode capacitance (electron tube) The direct capacitance between any two electrodes excluding all capacitance between either electrode and any other electrode or adjacent body. (Std100) [84]

direct inward dialing (telephone switching systems) A private automatic branch exchange or centrex service feature that permits outside calls to be dialed directly to the stations. (COM) 312-1977w

direct I/O An operation that attempts to circumvent a system performance optimization for the optimization of the individual I/O operation. (C/PA) 9945-1-1996

direction (navigation aid terms) The position of one point in space relative to another without reference to the distance between them; direction may be either three dimensional or two dimensional, and it is not an angle, but is often indicated in terms of its angular difference from a reference direction. *Note:* Five terms used in navigation—azimuth, bearing, course, heading, and track—involve measurement of angles from reference directions. To specify the reference directions, certain modifiers are used. These are: true, magnetic, compass, relative, grid and gyro. (AES/GCS) 172-1983w

directional antenna An antenna having the property of radiating or receiving electromagnetic waves more effectively in some directions than others. *Note:* This term is usually applied to an antenna whose maximum directivity is significantly greater than that of a half-wave dipole. (AP/ANT) 145-1993

directional-comparison protection A form of pilot protection in which the relative operating conditions of the directional units at the line terminals are compared to determine whether a fault is in the protected line section. (SWG/PE/PSR) C37.100-1992, C37.90-1978s

directional conical reflectance Ratio of reflected flux collected through a conical solid angle to essentially collimated incident flux. *Note:* The direction of incidence must be specified, and the direction and extent of the cone must be specified. (EEC/IE) [126]

directional conical transmittance Ratio of transmitted flux collected through a conical solid angle to essentially collimated incident flux. *Note:* The direction of incidence must be specified, and the direction and extent of the cone must be specified. (EEC/IE) [126]

directional constraint A technique used in computer graphics to force an input line to be parallel with a particular two-dimensional or three-dimensional axis. (C) 610.6-1991w

directional control (as applied to a protective relay or relay scheme) A qualifying term that indicates a means of controlling the operating force in a nondirectional relay so that it will not operate until the two or more phasor quantities used to actuate the controlling means (directional relay) are in a predetermined band of phase relations with a reference input. (SWG/PE) C37.100-1992

directional coupler (1) (A) (transmission lines) A transmission coupling device for separately (ideally) sampling (through a known coupling loss for measuring purposes) either the forward (incident) or the backward (reflected) wave in a transmission line. *Notes:* 1. Similarly, it may be used to excite in the transmission line either a forward or backward wave. 2. A unidirectional coupler has available terminals or connections for sampling only one direction of transmission; a bidirectional coupler has available terminals for sampling both directions. *See also:* auxiliary device to an instrument. **(B) (tap)** A passive device used in cable systems to divide and combine RF signals. It has at least three ports: line in, line out, and the tap. The signal passes between line in and line out ports with loss referred to as the insertion loss. A small portion of the signal power applied to the line in port passes to the tap port. A signal applied to the tap port is passed to the line in port less the tap attenuation value. The tap signals are isolated from the line out port to prevent reflections. A signal applied to the line out port passes to the line in port and is isolated from the tap port. Some devices provide more than one tap output line (multi-taps). (LM/EEC/PE/C) [119], 802.7-1989
(2) (waveguide components) A four port junction consisting of two waveguides coupled together in such a manner that a single traveling wave in either guide will induce a single traveling wave in the other, the direction of the latter wave being determined by the direction of the former. (MTT) 147-1979w
(3) (fiber optics) *See also:* tee coupler.

directional-current tripping *See:* directional-overcurrent protection; directional-overcurrent relay.

directional gain directivity index (transducer) (audio and electroacoustics) In decibels, 10 times the logarithm to the base 10 of the directivity factor. (SP) [32]

directional-ground relay A directional relay used primarily to detect single-phase-to-ground faults, but also sensitive to double-phase-to-ground faults. *Note:* This type of relay is usually operated from the zero-sequence components of voltage and current, but is sometimes operated from negative-sequence quantities. (SWG/PE/PSR) C37.100-1992, C37.90-1978s

directional gyro A two-degree-of-freedom gyro with a provision for maintaining the spin axis approximately horizontal. In this gyro, an output signal is produced by gimbal angular displacement that corresponds to the angular displacement of the case about an axis that is nominally vertical. (AES/GYAC) 528-1994

directional gyro electric indicator An electrically driven device for use in aircraft for measuring deviation from a fixed heading. (EEC/PE) [119]

directional-hemispherical reflectance $[\rho(\theta_i, \phi_i 2\pi)]$ Ratio of reflected flux collected over the entire hemisphere to essentially collimated incident flux. (See figure below). *Note:* The direction of incidence must be specified.

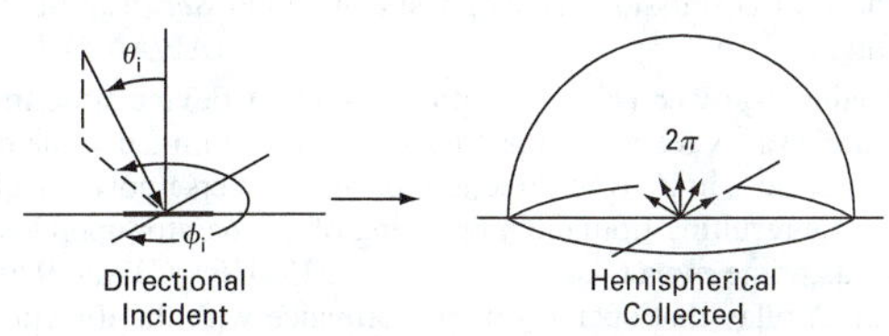

directional-hemispherical reflectance

(EEC/IE) [126]

directional-hemispherical transmittance [$\tau(\theta_i, \phi_i; 2\pi)$] Ratio of transmitted flux collected over the entire hemisphere to essentially collimated incident flux. (See figure below). *Note:* The direction of incidence must be specified.

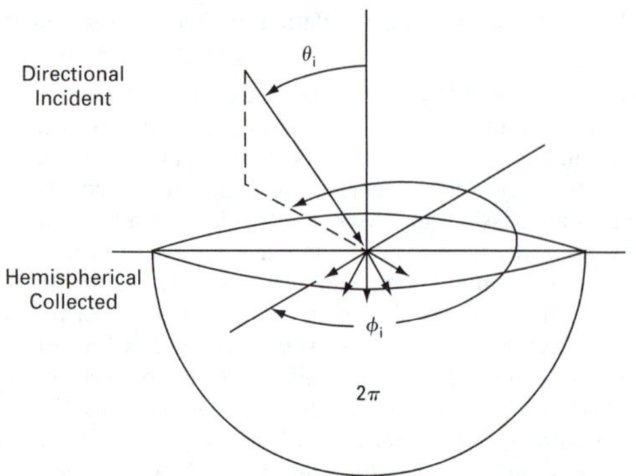

directional-hemispherical transmittance

(EEC/IE) [126]

directional homing (navigation aid terms) The process of homing wherein the navigational quantity maintained constant is the bearing. (AES/GCS) 172-1983w

directional lighting (illuminating engineering) Lighting provided on the work plane or on an object. Light that is predominantly from a preferred direction. (EEC/IE) [126]

directional localizer (navigation aid terms) (instrument landing systems) A localizer in which maximum energy is directed close to the runway centerline, thus minimizing extraneous reflections. (AES/GCS) 172-1983w

directional microphone A microphone the response of which varies significantly with the direction of sound incidence. *See also:* microphone. (EEC/PE) [119]

directional-null A sharp minimum in a radiation pattern that has been produced for the purpose of direction-finding or the suppression of unwanted radiation in a specified direction. (AP/ANT) 145-1993

directional-null antenna An antenna whose radiation pattern contains one or more directional nulls. *See also:* null-steering antenna system. (AP/ANT) 145-1993

directional-overcurrent protection A method of protection in which an abnormal condition within the protected equipment is detected by the current being in excess of a predetermined amount and in a predetermined band of phase relations with a reference input.

(SWG/PE/PSR) C37.100-1992, C37.90-1978s

directional-overcurrent relay A relay consisting of an overcurrent unit and a directional unit combined to operate jointly. (SWG/PE/PSR) C37.100-1992, C37.90-1978s

directional pattern (radiation pattern) The directional pattern of an antenna is a graphical representation of the radiation or reception of the antenna as a function of direction. *Note:* Cross sections in which directional patterns are frequently given are vertical planes and the horizontal planes or the principal electric and magnetic polarization planes. *See also:* antenna. (EEC/PE) [119]

directional phase shifter (directional phase changer) (nonreciprocal phase shifter) A passive changer in which the phase change for transmission in one direction differs from that for transmission in the opposite direction. *See also:* transmission line. (AP/ANT) [35]

directional-power relay (1) (power system device function numbers) A relay that operates on a predetermined value of power flow in a given direction, or upon reverse power such as that resulting from the monitoring of a generator upon loss of its prime mover. (PE/SUB) C37.2-1979s

(2) A relay that operates in conformance with the direction of power. (SWG/PE) C37.100-1992

directional-power tripping *See:* directional-power relay.

directional relay A relay that responds to the relative phase position of a current with respect to another current or voltage reference. *Note:* The above definition, which applies basically to a single-phase directional relay, may be extended to cover a polyphase directional relay.

(SWG/PE/PSR) C37.100-1992, C37.90-1978s

directional response pattern (beam pattern) (electroacoustics) (transducer used for sound emission or reception) A description, often presented graphically, of the response of the transducer as a function of the direction of the transmitted or incident sound waves in a specified plane and at a specified frequency. *Notes:* 1. A complete description of the directional response pattern of a transducer would require a three-dimensional presentation. 2. The directional response pattern is often shown as the response relative to the maximum response. (SP) [32]

direction finder (DF) *See:* radio direction-finder.

direction-finder antenna Any antenna used for radio direction finding. *See also:* navigation.

(AES/AES/RS) 686-1982s, [42]

direction finder antenna system (navigation aid terms) One or more DF antennas, their combining circuits and feeder systems, together with the shielding and all electrical and mechanical items up to the termination at the receiver-input terminals. (AES/GCS) 172-1983w

direction finder bearing sensitivity The minimum field-strength input to a direction-finder system to obtain repeatable bearings within the bearing accuracy of the system.

(AES/GCS) 172-1983w

direction finder deviation (navigation aid terms) The amount by which an observed radio bearing differs from the corrected bearing. (AES/GCS) 172-1983w

direction-finder noise level (in the absence of the desired signals) The average power or root-mean-square voltage at any specified point in a direction-finder system circuit. *Note:* In radio-frequency and audio channels, the direction-finder noise level is usually measured in terms of the power dissipated in suitable termination. In a video channel, it is customarily measured in terms of voltage across a given impedance or of the cathode-ray deflection. *See also:* navigation.

(AES/GCS/RS) 173-1959w, [42], 686-1982s

direction finder noise level (navigation aid terms) In the absence of the desired signals, the average power or rms (root-mean-square) voltage at any specified point in a direction finder system circuit. *Note:* In rf (radio frequency) and audio channels, the direction finding (DF) noise level is usually measured in terms of the power dissipated in suitable termination. In a video channel, it is customarily measured in terms of voltage across a given impedance, or of the cathode-ray deflection. (AES/RS/GCS) 686-1982s, 172-1983w

direction finder sensitivity (navigation aid terms) That field strength at the DF antenna, in microvolts per meter, which produces a ratio of signal-plus-noise to noise, equal to 20 dB (decibels) in the receiver output, the direction of arrival of the signal being such as to produce maximum pickup in the DF antenna system.

(AES/RS/GCS) 686-1982s, 172-1983w

direction-finder sensitivity The field strength at the direction-finder antenna, in microvolts per meter, that produces a ratio of signal-plus-noise to noise equal to 20 decibels in the receiver output, the direction of arribal of the signal being such as to produce maximum pickup in the direction-finder antenna system. *See also:* navigation.

(AES/GCS/RS) 173-1959w, [42], 686-1982s

direction finding *See:* radio direction-finder.

direction finding antenna Any antenna used for radio direction finding. *Synonym:* DF antenna. (AES/RS) 686-1982s

direction finding antenna system One or more DF antennas, their combining circuits and feeder systems, together with the shielding and all electrical and mechanical items up to the termination at the receiver input terminals.

(AES/RS) 686-1982s

direction of lay (cable) The lateral direction, designated as left-hand or right-hand, in which the elements of a cable run over the top of the cable as they recede from an observer looking along the axis of the cable. (PE/T&D) [4], [10]

direction of polarization (of an elliptically polarized wave) The direction of the major axis of the electric vector ellipse. *See also:* elliptically polarized wave. (AP/PROP) 211-1997

direction of propagation (1) (A) (point in a homogeneous isotropic medium) The normal to an equiphase surface taken in the direction of increasing phase lag. *See also:* radiation. **(B) (point in a homogeneous isotropic medium)** The direction of time-average energy flow. *Notes:* 1. In a uniform waveguide the direction of propagation is often taken along the axis.X. 2. The directional response pattern is often shown as the response relative to the maximum response. *See also:* waveguide; radiation. (PE/EEC/MTT) [119], 146-1980 **(2) (waveguide)** The direction of time average energy flow in a given mode. *Note:* In the case of a uniform lossless waveguide, the direction of propagation at every point is parallel to the axis. (MTT) 146-1980w **(3)** At any point in a medium, the direction of the time-averaged energy flow. *See also:* Poynting vector. (AP/PROP) 211-1997

direction of rotation of phasors (power and distribution transformers) Phasor diagrams should be drawn so that an advance in phase of one phasor with respect to another is in the counterclockwise direction. In the following figure, phasor 1 is 120 degrees in advance of phasor 2, and the phase sequence is 1, 2, 3.

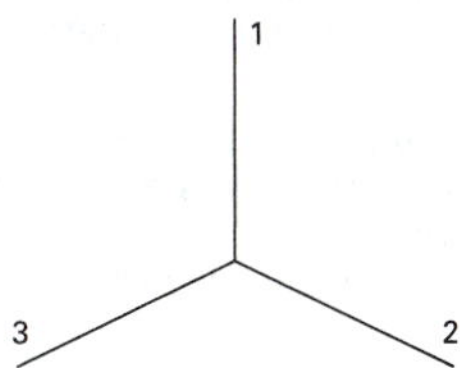

direction of rotation of phasors

(PE/TR) C57.12.80-1978r

direction operation Operation by means of a mechanism connected directly to the main operating shaft or an extension of the same. (SWG/PE) [56], C37.100-1992, C37.30-1971s, C37.100-1981s

directive A line from a file that is interpreted by the batch server. The line is usually in the form of a comment and is an additional means of passing options to the `qsub` utility. (C/PA) 1003.2d-1994

directive gain* *See:* directivity.

* Deprecated.

directive gain in physical media In a given direction and at a given point in the far field, the ratio of the power flux per unit area from an antenna to the power flux per unit area from an isotropic radiator at a specified location delivering the same power from the antenna to the medium. *Note:* The isotropic radiator must be within the smallest sphere containing the antenna. Suggested locations are antenna terminals and points of symmetry if such exist. (AP/ANT) 145-1983s

directivity (1) (gain) The value of the directive gain in the direction of its maximum value. *See also:* antenna. (AP) 149-1979r **(2) (directional coupler)** The ratio of the power output at an auxiliary port, when power is fed into the main waveguide or transmission line in the preferred direction, to the power output at the same auxiliary port when power is fed into the main guide or line in the opposite direction, the incident power fed into the main guide or line being the same in each case, and reflectionless terminations being connected to all ports. *Note:* The ratio is usually expressed in decibels. (IM/HFIM) [40] **(3)** (of an antenna) (in a given direction) The ratio of the radiation intensity in a given direction from the antenna to the radiation intensity averaged over all directions. *Notes:* 1. The average radiation intensity is equal to the total power radiated by the antenna divided by 4π. 2. If the direction is not specified, the direction of maximum radiation intensity is implied. (AP/ANT) 145-1993

directivity factor (A) (transducer used for sound emission) (audio and electroacoustics) The ratio of the sound pressure squared, at some fixed distance and specified direction, to the mean-square sound pressure at the same distance averaged over all directions from the transducer. *Notes:* 1. The distance must be great enough so that the sound pressure appears to diverge spherically from the effective acoustic center of the transducer. Unless otherwise specified, the reference direction is understood to be that of a maximum response. The frequency must be stated. **(B) (transducer used for sound reception)** The ratio of the square of the open-circuit voltage produced in response to sound waves arriving in a specified direction to the mean-square voltage that would be produced in a perfectly diffused sound field of the same frequency and mean-square sound pressure. *Notes:* 1. This definition may be extended to cover the case of finite frequency bands whose spectrum may be specified. 2. The average free-field response may be obtained in various ways, such as by the use of a spherical integrator, by numerical integration of a sufficient number of directivity patterns corresponding to different planes, or by integration of one or two directional patterns whenever the pattern of the transducer is known to possess adequate symmetry. *See also:* microphone. (SP) [32]

directivity, partial *See:* partial directivity.

direct lighting (illuminating engineering) Lighting involving luminaires which distribute 90 to 100 percent of the emitted light in the general direction of the surface to be illuminated. The term usually refers to light emitted in a downward direction. (EEC/IE) [126]

direct liquid cooling system (semiconductor rectifiers) A cooling system in which a liquid, received from a constantly available supply, is passed directly over the cooling surfaces of the semiconductor power converter and discharged. *See also:* semiconductor rectifier stack. (IA) [62]

direct liquid cooling system with recirculation (semiconductor rectifiers) A direct liquid cooling system in which part of the liquid passing over the cooling surfaces of the semiconductor power converter is recirculated and additional liquid is added as needed to maintain the required temperature, the excess being discharged. *See also:* semiconductor rectifier stack. (IA) [62]

direct lookup A table lookup in which the position of an entry is computed as a function of its key value. (C) 610.5-1990w

directly controlled system That portion of the controlled system that is directly guided or restrained by the final controlling element to achieve a prescribed value of the directly controlled variable. *See also:* feedback control system. (IM/IA/IAC) [120], [60]

directly controlled variable (automatic control) The variable in a feedback control system whose value is sensed to originate the primary feedback signal. *See also:* feedback control system. (PE/ICTL/EDPG) 421-1972s

Directly Executable Test Oriented Language (DETOL) A test language used to control a specific type of automatic test equipment. (C) 610.13-1993w

directly grounded *See:* grounded solidly.

direct manipulation User manipulation of symbols in the display by direct interaction with the symbol. It is generally performed through the use of a display structure, such as a pointer, and a cursor control device, such as a mouse. (PE/NP) 1289-1998

direct memory access (DMA) (1) Access to data by which data is transferred directly between main memory and storage devices. (C) 610.5-1990w **(2)** Ability of I/O controller modules to independently access memory. An I/O controller with DMA capabilities can access

commands, fetch data, and report status by accessing memory directly. (C/BA) 896.3-1993w
(3) A method for transferring data between an external device and memory without interrupting program flow or requiring CPU intervention. *Note:* The interface device takes control of the memory and transfers the data. *See also:* programmed input-output; direct memory transfer. (C) 610.10-1994w
(4) This refers to the ability of I/O controller modules to independently access memory. An I/O controller with DMA capabilities can access commands, fetch data, and report status by accessing memory directly. (C/BA) 896.4-1993w

direct memory access controller (DMAC) The block transfer processor used to implement direct memory access. (C) 610.5-1990w

direct memory control (DMC) *See:* direct memory transfer.

direct memory transfer (DMT) A method for transferring data between an external device and memory without interrupting program flow. *Note:* The CPU microcode flow is changed to a routine which transfers the data. *Synonym:* direct memory control. *See also:* direct memory access; programmed input-output. (C) 610.10-1994w

direct metric A metric that does not depend upon a measure of any other attribute. (C/SE) 1061-1998

direct metric value A numerical target for a quality factor to be met in the final product. For example, mean time to failure (MTTF) is a direct metric of final system reliability. (C/SE) 1061-1998

direct numerical control Numerical control in which a dedicated computer controls the operation of the parts programs in a single numerical control machine. (C) 610.2-1987

direct on-line starting (rotating machinery) The process of starting a motor by connecting it directly to the supply at a rated voltage. *See also:* asynchronous machine. (PE) [9]

direct operation Operation by means of a mechanism connected directly to the main operating shaft or an extension of the same. (SWG/PE) C37.30-1971s, C37.100-1992, [56], C37.100-1981s

direct orbit (communication satellite) An inclined orbit with an inclination between zero and ninety degrees. (COM) [19]

director element (1) (data transmission) A parasitic element located forward of the driven element of an antenna, intended to increase the directive gain of the antenna in the forward direction. (PE) 599-1985w
(2) A parasitic element located forward of the driven element of an antenna, intended to increase the directivity of the antenna in the forward direction. (AP/ANT) 145-1993

directory (1) (A) A list of data items and information about those items, used to reference the items; for example, the directory for each user's personal disk space contains an entry for each file within that space, and a reference to its physical location. *Contrast:* dictionary. *See also:* data directory; file directory. **(B)** *See also:* index. (C) 610.5-1990, 610.12-1990
(2) A contiguous collection of one or more entries, which is contained within the node's ROM. (C/MM) 1212-1991s
(3) A file that contains directory entries. No two directory entries in the same directory shall have the same name. (C/PA) 9945-1-1996, 9945-2-1993
(4) A repository of information about objects that provides directory services to its users, which allows its users access to the information. (PA/C) 1328.2-1993w, 1224.2-1993w, 1327.2-1993w, 1326.2-1993w
(5) A file that contains directory entries. No two entries in a directory shall have the same filename. (C) 1003.5-1999

directory-assistance call (telephone switching systems) A call placed to request the directory number of a customer. (COM) 312-1977w

directory attribute The information of a particular type concerning an object and appearing in an entry describing the object in the DIB. *Synonym:* attribute. (C/PA) 1328.2-1993w, 1327.2-1993w, 1224.2-1993w, 1326.2-1993w

directory attribute type That component of a directory attribute that indicates the class of information given by that attribute. It is an object identifier, and so completely unique. *Synonym:* attribute type. (C/PA) 1328.2-1993w, 1224.2-1993w, 1327.2-1993w, 1326.2-1993w

directory attribute value A particular instance of the class of information indicated by a directory attribute type. *Synonym:* attribute value. (C/PA) 1328.2-1993w, 1224.2-1993w, 1327.2-1993w, 1326.2-1993w

directory class *See:* object class.

directory entry (1) A ROM entry that specifies the address of another ROM directory. (C/MM) 1212-1991s
(2) An object that associates a filename with a file. Several directory entries can associate names with the same file. *Synonym:* link. (C/PA) 9945-1-1996, 9945-2-1993
(3) An object that associates a filename with a file. Several directory entries can associate different filenames with the same file. (C) 1003.5-1999

directory information base The complete set of information to which the directory provides access and that includes all of the pieces of information that can be read or manipulated using the operations of the directory. (C/PA) 1328.2-1993w, 1224.2-1993Ow, 1326.2-1993w, 1327.2-1993w

directory information tree The directory information base, considered as a tree, whose vertices (other than the root) are the directory entries. (C/PA) 1328.2-1993w, 1326.2-1993w, 1327.2-1993w, 1224.2-1993w

directory medium A medium that contains a distribution in a POSIX.1 hierarchical file system format. (C/PA) 1387.2-1995

directory number (telephone switching systems) The full complement of digits required to designate a customer in a directory. (COM) 312-1977w

directory-numbering plan (telephone switching systems) The arrangement whereby each customer is identified by an office and main-station code. (COM) 312-1977w

directory object Anything in some "world," generally the world of telecommunications and information processing or some part thereof, that is identifiable (can be named) and that it is of interest to hold information on in the directory information base. (C/PA) 1326.2-1993w, 1327.2-1993w, 1224.2-1993w, 1328.2-1993w

directory operation Processing performed within the directory to provide a service, such as a read directory operation; it is given some arguments as input, performs some processing, and returns some results. An application process causes a directory operation to be performed by invoking an interface operation. (C/PA) 1328.2-1993w, 1327.2-1993w, 1326.2-1993w, 1224.2-1993w

directory syntax *See:* attribute syntax.

directory system agent An OSI application process that is part of the directory. (C/PA) 1328.2-1993w, 1326.2-1993w, 1327.2-1993w, 1224.2-1993w

directory user agent An OSI application process that represents a user accessing the directory. *Note:* This may be composed of an arbitrary number of system processes and application processes, including the one or more that are the user(s). (C/PA) 1328.2-1993w, 1327.2-1993w, 1224.2-1993w, 1326.2-1993w

direct outward dialing (telephone switching systems) A private automatic branch exchange or centrex service feature that permits stations to dial outside numbers without intervention of an attendant. (COM) 312-1977w

direct-plunger driving machine (elevators) A machine in which the energy is applied by a plunger or piston directly attached to the car frame or platform and that operates in a cylinder under hydraulic pressure. *Note:* It includes the cylinder and plunger or piston. *See also:* driving machine. (PE/EEC) [119]

direct-plunger elevator A hydraulic elevator having a plunger or piston directly attached to the car frame or platform. *See also:* elevator. (EEC/PE) [119]

direct polarity indication The designation of the internal state produced by the external level of an input, or producing the external level of an output, by the presence or absence of the polarity symbol. *See also:* mixed logic. (GSD) 91-1984r

direct radiation monitors Those monitors that provide monitoring of ambient airborne radioactivity by measurement of the external radiation field associated with the airborne radioactivity. (NI) N42.17B-1989r

direct ratio (illuminating engineering) The ratio of the luminous flux which reaches the work-plane directly to the downward component from the luminaire. (EEC/IE) [126]

direct raw-water cooling system (1) (rectifier) A cooling system in which water, received from a constantly available supply, such as a well or water system, is passed directly over the cooling surfaces of the rectifier and discharged. *See also:* direct liquid cooling system; rectification. (IA) [62]
(2) (thyristor controller) A cooling system in which water, received from a constantly available supply, such as a well or water system, is passed directly over the cooling surfaces of the thyristor controller components and discharged. (IA/IPC) 428-1981w

direct raw-water cooling system with recirculation A direct raw-water cooling system in which part of the water passing over the cooling surfaces of the rectifier is recirculated and raw water is added as needed to maintain the required temperature, the excess being discharged. *See also:* rectification; direct liquid cooling system with recirculation. (IA) [62]

direct-reading ammeters Ammeters that employ a shunt and are connected in series and carry some of the line current through them for measurement purposes. They are part of the circuit being measured. (IA/PSE) 1100-1999

direct-recording (facsimile) That type of recording in which a visible record is produced, without subsequent processing, in response to the received signals. *See also:* recording. (COM) 168-1956w

direct reference address A virtual address that is not modified by indirect addressing. *Note:* It can be modified by indexing. (C) 610.10-1994w

direct release (series trip) A release directly energized by the current in the main circuit of a switching device. (SWG/PE) C37.100-1992

direct staff-hour The amount of effort directly expended in creating a specific output product. (C/SE) 1045-1992

direct stroke A lightning stroke direct to any part of a network or electric installation. (PE/T&D) [84], [8], 1410-1997

direct-stroke protection (lightning) Lightning protection designed to protect a network or electric installation against direct strokes. (T&D/PE) [10], 1243-1997

direct support maintenance *See:* intermediate maintenance.

direct test (ac high-voltage circuit breakers) A test in which the applied voltage, current, and recovery voltage is obtained from a single power source, which may be comprised of generators, transformers, networks, or combinations of these. (SWG/PE) C37.081-1981r, C37.083-1999, C37.100-1992

direct vacuum-tube current (medical electronics) A current obtained by applying to the part to be treated an evacuated glass electrode connected to one terminal of a generator of high-frequency current (100 to 1000 kilohertz), the other terminal being grounded. *Note:* Deprecated as confusing and as representing an ill-defined and obsolescent produce. (EMB) [47]

direct-view storage tube (DVST) A type of random-scan display device whose screen surface will retain an image for a long period of time, so that the image, once generated, does not need to be continuously refreshed to remain visible and avoid flicker. *Contrast:* refresh display device. (C) 610.6-1991w

direct-voltage high-potential test (rotating machinery) A test that consists of the application of a specified unidirectional voltage higher than the rated root-mean-square value for a specified time for the purpose of determining the adequacy against breakdown of the insulation system under normal conditions, or the resistance characteristic of the insulation system. (PE) [9]

direct wave A wave propagated over an unobstructed ray path from a source to a point. (AP/PROP) 211-1997

direct-wire circuit (one-wire circuit) A supervised circuit, usually consisting of one metallic conductor and a ground return, and having signal receiving equipment responsive to either an increase or a decrease in current. *See also:* protective signaling. (EEC/PE) [119]

Dirichlet boundary condition A boundary condition applied to the solution of a partial differential equation in which the function is specified as a constant at the boundary. *Note:* When applied to the wave equation for electromagnetic fields, it requires continuity of the tangential field components across the boundary. (AP/PROP) 211-1997

dirty read To access data from a storage device or data medium while that same data is being modified by another process. *Synonym:* transient read. (C) 610.5-1990w

DIR violation A word-serial protocol error that occurs when a servant receives a DIR synchronized command (such as *byte available*) and is unable to process that command because the DIR bit is zero (0). (C/MM) 1155-1992

DIS *See:* Distributed Interactive Simulation.

disability glare (illuminating engineering) Glare which reduces visual performance and visibility and often is accompanied by discomfort. *See also:* veiling brightness. (EEC/IE) [126]

disability glare factor (illuminating engineering) A measure of the visibility of a task in a given lighting installation in comparison with its visibility under reference lighting conditions, expressed in terms of the ratio of luminance contrasts having an equivalent effect upon task visibility. The value of DGF takes account of the equivalent veiling luminance produced in the eye by the pattern of luminances in the task surround. (EEC/IE) [126]

disable A command or condition that prohibits some specific event from proceeding. (SWG/PE/SUB) C37.100-1992, C37.1-1994

disabled port A port configured to neither transmit, receive, or repeat Serial Bus signals. A disabled port shall be reported as disconnected in a physical layer's (PHY's) self-ID packet(s). (C/MM) 1394a-2000

disabled slave (backplane bus specification for multiprocessor architectures) A selected slave that detects that an intervening slave is participating in the data transfer in its place. A disabled slave treats the transaction as an address-only transaction and does not participate in the data transfer phase. A slave may be disabled only during single-slave mode transactions. (C/MM) 896.1-1987s

disaggregation The process of changing the resolution of an aggregate to represent it in more detail. (C/DIS) 1278.1a-1998

disarm a timer To stop a timer from measuring the passage of time, thereby disabling any future process notifications (until the timer is armed again). (C/PA) 1003.5-1999, 9945-1-1996

disassemble To translate an assembled computer program from its machine language version into a form that resembles, but may not be identical to, the original assembly language program. *Contrast:* assemble. (C) 610.12-1990

disassembler **(1)** A software tool that disassembles computer programs. *Synonym:* deassembler. (C) 610.12-1990
(2) A program that translates machine code into an equivalent human-readable assembly-language representation. (C/BA) 1275-1994

disassociation The service that removes an existing association. (C/LM) 8802-11-1999

disaster, major storm *See:* major storm disaster.

disc *See:* magnetic disk; disk recorder; disk.

discharge **(1)** **(A)** **(storage cell)** The conversion of the chemical energy of the battery into electric energy. *See also:* charge. **(B)** Water flow, exiting from a turbine. It is generally measured in cubic feet per second or cubic meters per second. (PE/EEC/EDPG) [119], 1020-1988
(2) **(gas)** The passage of electricity through a gas. (ED) [45], [84]
(3) The passage of electricity through gaseous, liquid, or solid insulation. (PE/PSIM) 4-1995

discharge capacity *See:* arrester discharge capacity.

discharge circuit **(surge generator)** That portion of the surge-generator connections in which exist the current and voltage variations constituting the surge generated. (T&D/PE) [10], [8]

discharge counter A means for recording the number of arrester discharge operations. (SPD/PE) C62.11-1999

discharge current **(1)** **(surge arresters)** The surge current that flows through an arrester when sparkover occurs. (PE/SPD) C62.1-1981s
(2) The current that flows through an arrester when sparkover occurs. (SPD/PE) C62.31-1987r, C62.32-1981s
(3) The surge current that flows through an arrester. (SPD/PE) C62.11-1999
(4) The surge current that flows through the surge-protective device when conduction occurs. (SPD/PE) C62.62-2000

discharge current limiting device A reactor or equivalent device to limit the current magnitude and frequency of the discharge of the capacitors during closing operations of the bypass switch or gap. (T&D/PE) 824-1994

discharge current withstand rating The specified magnitude and wave shape of a discharge current that can be applied to a surge-protective device a specified number of times without causing damage or degradation beyond specified limits. (SPD/PE) C62.62-2000

discharge detector **(rotating machinery) (ionization or corona detector)** An instrument that can be connected in or across an energized insulation circuit to detect current or voltage pulses produced by electric discharges within the circuit. *See also:* instrument. (PE) [9]

discharge device **(1)** An internal or external device intentionally connected in shunt with the terminals of a capacitor for the purpose of reducing the residual voltage after the capacitor is disconnected from an energized line. (PE/CAP) 18-1992, C37.99-2000
(2) An internal or external device permanently connected in parallel with the terminals of a capacitor for the purpose of reducing the residual voltage after the capacitor is disconnected from an energized line. (T&D/PE) 824-1994

discharge energy **(overhead power lines)** The energy transferred during a transient discharge. (T&D/PE) 539-1990

discharge-energy test **(rotating machinery)** A test for determining the magnitude of the energy dissipated by a discharge or discharges within the insulation. *See also:* asynchronous machine; direct-current commutating machine. (PE) [9]

discharge extinction voltage **(rotating machinery) (ionization or corona extinction voltage)** The voltage at which discharge pulses that have been observed in an insulation system, using a discharge detector of specified sensitivity, cease to be detectable as the voltage applied to the system is decreased. *See also:* asynchronous machine; direct-current commutating machine. (PE) [9]

discharge inception test **(rotating machinery) (corona inception test)** A test for measuring the lowest voltage at which discharges of the specified magnitude recur in successive cycles when an increasing alternating voltage is applied to insulation. *See also:* asynchronous machine; direct-current commutating machine. (PE) [9]

discharge inception voltage **(1)** **(rotating machinery)** (ionization or corona inception voltage) The voltage at which discharge pulses in an insulation system become observable with a discharge detector of specified sensitivity, as the voltage applied to the system is raised. *See also:* direct-current commutating machine; asynchronous machine. (PE) [9]
(2) **(surge arresters)** The root-mean-square value of the power-frequency voltage at which discharges start, the measurement of their intensity being made under specified conditions. (PE) [8], [84]

discharge indicator A means for indicating that the arrester has discharged. (SPD/PE) C62.11-1999

discharge opening **(rotating machinery)** A port for the exit of ventilation air. (PE) [9]

discharge oscillations **(laser gyro)** Periodic variations in voltage and current at the terminals of a dc discharge tube that are supported by the negative resistance of the discharge tube itself. (AES/GYAC) 528-1994

discharge probe **(rotating machinery) (ionization or corona probe)** A portable antenna, safely insulated, and designed to be used with a discharge detector for locating sites of discharges in an energized insulation system. *See also:* instrument. (PE) [9]

discharge rate The rate, in amperes, at which current is delivered by a battery. *See also:* hour rate. (PV) 1013-1990, 1144-1996

discharge resistor **(1)** **(excitation systems for synchronous machines)** A resistor that, upon interruption of excitation source current, is connected across the field windings of a synchronous machine or an exciter to limit the transient voltage in the field circuit and to hasten the decay of field current of the machine. (PE/EDPG) 421.1-1986r
(2) A resistor that, upon interruption of excitation source current, is connected across the field windings of a generator, motor, synchronous condenser, or an exciter to limit the transient voltage in the field circuit and to hasten the decay of field current of these machines. (SWG/PE) C37.100-1992

discharge tube An evacuated enclosure containing a gas at low pressure that permits the passage of electricity through the gas upon application of sufficient voltage. *Note:* The tube is usually provided with metal electrodes, but one form permits an electrodeless discharge with induced voltage. (EEC/PE) [119]

discharge voltage **(1)** The voltage that appears across the terminals of an arrester during passage of discharge current. (SPD/PE) C62.11-1999
(2) The voltage that appears across other terminals of a surge-protective device during passage of discharge current. (SPD/PE) C62.62-2000

discharge voltage-current characteristic The variation of the crest values of discharge voltage with respect to discharge current. *Note:* This characteristic is normally shown as a graph based on three or more current-surge measurements of the same wave shape, but of different crest values. (SPD/PE) C62.11-1999, C62.62-2000

discharge voltage-time curve *See:* arrester discharge voltage-time curve.

discharge withstand current The specified magnitude and wave shape of a discharge current that can be applied to an arrester a specified number of times without causing damage to it. (SPD/PE) C62.11-1999

disclaimer A notice that renounces or repudiates a legal claim or right. (C/SE) 1420.1b-1999

discomfort glare (illuminating engineering) Glare which produces discomfort. It does not necessarily interfere with visual performance or visibility. (EEC/IE) [126]

discomfort glare factor† (illuminating engineering) The numerical assessment of the capacity of a single source of brightness, such as a luminaire, in a given visual environment for producing discomfort. *Note:* This term is obsolete and is retained for reference and literature searches. (EEC/IE) [126]

† Obsolete.

discomfort glare rating (illuminating engineering) A numerical assessment of the capacity of a number of sources of luminance, such as luminaires, in a given visual environment for producing discomfort. It is the net effect of the individual values of index of sensation M, for all luminous areas in the field of view

$$\mathrm{DGR} = \Sigma(M)^{a}$$

where

$a = n^{-0.914}$

n = number of sources (ceiling elements) in the field of view.

(EEC/IE) [126]

DIS compatible Two or more simulations/simulators that are Distributed Interactive Simulation (DIS) compliant and whose models and data that send and interpret protocol data units (PDUs) support the realization of a common operational environment among the systems (i.e., they are coherent in time and space). (C/DIS) 1278.4-1997

DIS compliant A simulation/simulator that can send or receive protocol data units (PDUs) in accordance with IEEE Std 1278.1-1995 and IEEE Std 1278.2-1995. A specific statement must be made regarding the qualifications of each PDU. (C/DIS) 1278.4-1997

discone antenna A biconical antenna with one cone having a vertex angle of 180°. *See also:* biconical antenna. (AP/ANT) 145-1993

disconnect (1) (A) (release) (telephony) To disengage the apparatus used in a telephone connection and to restore it to its condition when not in use. **(B)** A signal that indicates that the call is over and the line should be returned to the on-hook state. The signal's duration is what distinguishes between true disconnects and shorter on-hook signals. The hit timing interval is provided so that a short, random on-hook signal does not initiate any call processing functions. An on-hook signal that lasts longer than a hit may be either a flash or a disconnect. If the customer does not have calling features that require flashes, the signal should be considered a disconnect. An on-hook state that lasts beyond a hit or a flash is considered to be a disconnect signal. *See also:* flash; hit, flash, and disconnect timing. (COM/TA) [48], 973-1990
(2) (watthour meter) A conductor, bar, or nut used to open an electrical circuit for isolation purposes. (ELM) C12.8-1981r

disconnectable device A grounding device that can be disconnected from ground by the operation of a disconnecting switch, circuit breaker, or other switching device. (SPD/PE) 32-1972r

disconnected position (of a switchgear-assembly removable element) That position in which the primary and secondary disconnecting devices of the removable element are separated by a safe distance from the stationary element contacts. *Note:* Safe distance, as used here, is a distance at which the equipment will meet its withstand ratings, both power frequency and impulse, between line and load stationary terminals and phase-to-phase and phase-to-ground on both line and load stationary terminals with the switching device in the closed position. (SWG/PE) C37.100-1992

disconnected state A state in which the node no longer responds to bus transactions. Since the node no longer responds to bus transactions, a power_reset is required to change to another node state. (C/MM) 1212-1991s

disconnecting blade *See:* blade.

disconnecting cutout A distribution cutout, having a disconnecting blade, that is used for closing, opening, or changing the connections in a circuit or system, or for isolating purposes. *Note:* Some load-break ability is inherent in the device, but it has no load-break rating. This ability can best be evaluated by the user, based on experience under operating conditions. (SWG/PE) C37.100-1992

disconnecting device (packaging machinery) A device whereby the conductors of a circuit can be disconnected from their source of supply. (IA/PKG) 333-1980w

disconnecting fuse *See:* fuse-disconnecting switch.

disconnecting means (1) (A) A device, or group of devices, or other means by which the conductors of a circuit can be disconnected from their source of supply. **(B) (recreational vehicles)** The necessary equipment usually consisting of a circuit breaker or switch and fuses, and their accessories, located near the point of entrance of supply conductors in a recreational vehicle and intended to constitute the means of cutoff for the supply to that recreational vehicle. (NESC) [86]
(2) A device or group of devices used to disconnect the conductors of a circuit from the source of supply. (IA/MT) 45-1998

disconnecting switch (1) A switch used for changing the connections in a circuit, or for isolating a circuit or equipment from the source of power. *Note:* The switch is required to carry normal load current continuously and also abnormal or short-circuit currents for short intervals as specified. It is also required to open or close circuits either when negligible current is broken or made, or when no significant change in the voltage across the terminals of each of the switch poles occurs. Some disconnecting switches have some inherent load-break ability which can best be evaluated by the user, based on experience under operating conditions. (SWG/PE) C37.40-1993
(2) A mechanical switching device used for changing the connections in a circuit, or for isolating a circuit or equipment from the source of power. *Note:* It is required to carry normal load current continuously, and also abnormal or short-circuit currents for short intervals as specified. It is also required to open or close circuits when negligible current is broken or made, or when no significant change in the voltage across the terminals of each of the switch poles occurs. *Synonym:* isolating switch. (SWG/PE/NESC/IA/IPC) C37.100-1992, C2-1997, 428-1981w

disconnection (control) Connotes the opening of a sufficient number of conductors to prevent current flow. (IA/ICTL/IAC) [60]

disconnection phase (1) Period at the end of a transaction used to return the bus signals to their quiescent state. In addition, this phase might be used to transfer additional information required to perform or abort the requested operation. (C/BA) 896.3-1993w
(2) A period within a transaction used to return the bus signals to their quiescent state. In addition, this phase might be used to transfer additional information required to perform or abort the requested operation. (C/BA) 10857-1994, 896.4-1993w

disconnector A switch that is intended to open a circuit only after the load has been thrown off by some other means. *Note:* Manual switches designed for opening loaded circuits are usually installed in circuit with disconnectors, to provide a safe means for opening the circuit under load. (BT/AV) [34]

disconnected port A port whose connection detect circuitry detects no peer physical layer (PHY) at the other end of a cable. It is not important whether the peer PHY is powered or the peer port is enabled. (C/MM) 1394a-2000

disconnect signal (telephony) A signal transmitted from one end of a subscriber line or trunk to indicate that the relevant party has released. (COM) [48]

disconnect timing *See:* hit; disconnect; flash; hit, flash, and disconnect timing.

disconnect-type pothead A pothead in which the electric continuity of the circuit may be broken by physical separation of the pothead parts, part of the pothead being on each conductor end after the separation. *See also:* pothead. (PE) 48-1975s

discontinuity **(1)** **(A)** An abrupt nonuniformity in a uniform waveguide or transmission line that causes reflected waves. *See also:* discontinuity capacitance; waveguide. **(B)** An abrupt change in the cross section of the planar transmission line. Abrupt refers usually to a change in dimensions or material over a length short compared to a wavelength. (IM/MTT/HFIM) [40], 1004-1987
(2) **(inductive coordination)** An abrupt change at a point in the physical relations of electric supply and communication circuits or in electrical parameters of either circuit, that would materially affect the coupling. *Note:* Although technically included in the definition, transpositions are not rated as discontinuities because of their application to coordination. *See also:* inductive coordination. (EEC/PE) [119]

discontinuity capacitance (waveguide or transmission line) The shunt capacitance that, when inserted in a uniform waveguide or transmission line, would cause reflected waves of the dominant mode equal to those resulting from the given discontinuity. *See also:* waveguide. (COM) [49]

discrete bar code symbology A bar code symbology where the spaces between characters (intercharacter gap) are not part of the encoding scheme. (PE/TR) C57.12.35-1996

discrete change model *See:* discrete model.

discrete component An electrical element that is mounted on a printed circuit board or other substrate and connected to the interconnect wiring system, but is not integrated to any significant extent with other elements and does not contain boundary module test circuitry adjacent to its pins. *See also:* interconnect. (C/TT) 1149.4-1999

discrete data Data that represents a variable that can be mathematically represented in integer form. For example, a count of whole items processed is a discrete variable. (IM/ST) 1451.1-1999

discrete event model *See:* discrete model.

discrete model **(A)** A mathematical or computational model whose output variables take on only discrete values; that is, in changing from one value to another, they do not take on the intermediate values; for example, a model that predicts an organization's inventory levels based on varying shipments and receipts. *Synonyms:* discrete change model; discrete event model. *Contrast:* continuous model. *See also:* state machine. **(B)** A model of a system that behaves in a discrete manner. *Contrast:* continuous model. *See also:* state machine. (C) 610.3-1989

discrete programming *See:* integer programming.

discrete sentence intelligibility The percent intelligibility obtained when the speech units considered are sentences (usually of simple form and content). *See also:* volume equivalent. (EEC/PE) [119]

discrete simulation A simulation that uses a discrete model. (C) 610.3-1989w

discrete-state system (system, finite-state) A system whose state is defined only for discrete values of time and amplitude. *See also:* control system. (CS/IM) [120]

discrete type A data type whose members can assume any of a set of distinct values. A discrete type may be an enumeration type or an integer type. (C) 610.12-1990

discrete system A system whose signals are inherently discrete. *See also:* control system. (CS/IM) [120]

discrete variable model *See:* discrete model.

discrete word intelligibility The percent intelligibility obtained when the speech units considered are words (usually presented so as to minimize the contextual relation between them). *See also:* volume equivalent. (EEC/PE) [119]

discretionary hyphen In word processing, a hyphen inserted into a word by the user to indicate the desired position for a break, if required by justification. *Note:* If the hyphen is not needed, it does not appear in the formatted text. *Synonyms:* soft hyphen; syllable hyphen; ghost hyphen. *Contrast:* required hyphen. (C) 610.2-1987

discrimination **(1)** **(any system of transducer)** The difference between the losses at specified frequencies, with the system or transducer terminated in specified impedances. *See also:* transmission loss. (EEC/PE) [119]
(2) Separation or identification of the differences between similar (but not identical) signals. (AES) 686-1997

discrimination instruction One of a class of instructions that comprises branch instructions and conditional instructions. (C) 610.10-1994w

discriminator **(1)** A circuit having a threshold below which signals applied to the input will not cause an output signal and above which the input signal will trigger an output having an amplitude independent of the input signal height. (NPS) 325-1996
(2) A circuit in which the output is dependent upon how an input signal differs in some aspect from a standard or from another signal. (AES) 686-1997
(3) **(A)** A property of a superclass, associated with a cluster of that superclass, whose value identifies to which subclass a specific instance belongs. Since the value of the discriminator (when a discriminator has been declared) is equivalent to the identity of the subclass to which the instance belongs, there is no requirement for a discriminator in identity-style modeling. **(B)** An attribute in the generic entity (or a generic ancestor entity) of a category cluster whose values indicate which category entity in the category cluster contains a specific instance of the generic entity. All instances of the generic entity with the same discriminator value are instances of the same category entity. (C/SE) 1320.2-1998

discriminator, amplitude *See:* pulse-height discriminator.

discriminator, constant-fraction pulse-height A pulse-height discriminator in which the threshold changes with input amplitude in such a way that the triggering point corresponds to a constant fraction of the input pulse height. (NPS) 398-1972r

discriminator, pulse-height *See:* pulse-height discriminator.

DIS exercise **(A)** The total process of designing, assembling, testing, conducting, evaluating, and reporting on an activity. **(B)** One or more sessions involving two or more interacting simulation applications with a common objective and accreditation. Participating simulations share a common identifying number called the exercise identifier and use correlated representations of the synthetic environment in which they operate. (C/DIS) 1278.4-1997

dish **(1)** **(radio practice)** **(data transmission)** A reflector, the surface of which is concave, as, for example, a part of a sphere or of a paraboloid of revolution. (PE) 599-1985w
(2) A colloquial term for the reflecting surface of a paraboloidal-reflector antenna. (AES) 686-1997

DIS interoperable Two or more simulations/simulators that, for a given exercise, are Distributed Interactive Simulation (DIS) compliant and DIS compatible and whose performance characteristics support the fidelity required for the exercise. (C/DIS) 1278.4-1997

disjunction **(1)** The logical 'OR' operator. (C) 610.5-1990w, 1084-1986w
(2) The result of joining two conditions by the logical 'OR' operator. *Contrast:* conjunction. *See also:* conjunctive query. (C) 610.5-1990w
(3) The Boolean operation whose result has the value 0 if and only if each operand has the value 0. *Synonym:* logical add. (C) 610.10-1994w

disk A generic term for any storage medium in the form of circular, flat recording surface. *Note:* Also known as a platter. *Synonym:* platter. *See also:* magnetic disk; diskette; magneto-optical disk; optical disk. (C) 610.10-1994w

disk array Multiple disks arranged in such a manner as to increase storage capacity or to provide redundant data for disaster recovery. *See also:* RAID storage. (C) 610.10-1994w

disk cache A cache consisting of random-access memory, used by a disk driver as intermediate storage between a rotating disk and main storage. *Note:* The disk cache minimizes access to the rotating disk by storing recently-used data or adjacent data in the random-access memory. (C) 610.10-1994w

disk cartridge An assembly of one or more magnetic disks that is removable from the disk drive, but which cannot be separated from its associated container. *Contrast:* disk pack. *See also:* removable storage. (C) 610.10-1994w

disk crash The sudden and complete failure of a disk drive. *See also:* head crash. (C) 610.10-1994w

disk drive An electromechanical device that reads from and writes to disks. *Contrast:* tape drive. *See also:* disk storage device; magnetic disk drive; WORM drive; head-per-track disk drive; disk pack; full-height disk drive; half-height disk drive. (C) 610.10-1994w

disk driver A device driver that supports a specific class of disk drives. (C) 610.10-1994w

diskette A magnetic disk enclosed in a protective container. *See also:* floppy diskette; double-sided disk; disk. (C) 610.10-1994w

diskette compatibility The ability of a diskette to be accessed by one or more systems such that data exchange can take place. (C) 610.2-1987

disk file A file, typically containing data, residing on a magnetic or optical disk. (C) 610.10-1994w

disk label Contains descriptive information, usually in a well-known location such as physical block zero, about the device and the media and may include logical partitioning information. (C/BA) 1275-1994

diskless workstation A workstation with no storage capacity, intended to be used in conjunction with another workstation networked in such a way that the two workstations can share the storage. (C) 610.10-1994w

disk mirroring *See:* double storage.

disk pack An assembly of one or more magnetic disks that is removable from the disk drive together with its container, however the disks must be separated from the container when they are in use. *Contrast:* disk cartridge. *See also:* volume. (C) 610.10-1994w

disk recorder (phonograph techniques) A mechanical recorder in which the recording medium has the geometry of a disk. *See also:* phonograph pickup. (SP) [32]

disk server On a network, a server that allows access to a disk storage device at the disk sector level; that is, the server sends absolute disk sectors to the requestor. *See also:* database server; print server; mail server; file server; terminal server; network server. (C) 610.7-1995

disk storage device *See:* disk drive.

dismiss To remove a menu or popup window from the screen. (C) 1295-1993w

disparity The difference between the number of logical 1s and logical 0s in a character. A positive or negative disparity indicates an excess of 1s or 0s, respectively. (C/BA) 1355-1995

dispatch Issue a fetched instruction to one or more functional units for execution. (C/MM) 1754-1994

dispatcher The software that implements the service interface functions using workspace interface functions. (C/PA) 1328-1993w, 1327-1993w

dispatching system (mining practice) A system employing radio, telephone, and/or signals (audible or light) for orderly and efficient control of the movements of trains of cars in mines. *See also:* mine radio telephone system; mine-fan signal system. (PE/EEC) [119]

Dispatch List A DMA model where the DMA queues are linked lists of dispatch_items. Each dispatch_item contains a pointer to the next dispatch_item and a message being passed to the consumer. (C/MM) 1212.1-1993

dispatch operation (radio-communication circuit) A method for permitting a maximum number of terminal devices to have access to the same two-way radio communication circuit. *See also:* channel spacing. (VT) [37]

dispenser cathode (electron tube) A cathode that is not coated but is continuously supplied with suitable emission material from a separate element associated with it. *See also:* electron tube. (Std100) [84]

dispersed data processing (DDP) *See:* distributed data processing.

dispersed magnetic power tape *See:* magnetic-powder-impregnated tape.

dispersed power An electric power generation source (or sources) not directly under established electric utility ownership and control. (SUB/PE) 1109-1990w

dispersion **(1)** The property of a planar transmission line whereby the phase velocity of the mode of propagation is frequency dependent, or equivalently, the phase constant is not proportional to frequency. (MTT) 1004-1987w
(2) (of a wave) The variation of the phase velocity with frequency. (AP/PROP) 211-1997

dispersion relation The functional relationship between the angular frequency, ω, and the wave vector, $\bar{k}$, for waves in a source-free medium. For a dispersionless medium, the components of $\bar{k}$ are linearly proportional to ω. (AP/PROP) 211-1997

dispersion slope The rate of change of the chromatic dispersion of a fiber with wavelength. (C/LM) 802.3-1998

dispersive bandwidth **(1)** The operating frequency range over which the delay dispersion is defined (dispersive delay line). *Synonyms:* dispersive bandwidth; frequency selective bandwidth. (UFFC) 1037-1992w, [22]
(2) *See also:* frequency selective bandwidth. (AP/PROP) 211-1997

dispersive delay line A delay line which has a transfer characteristic with a constant modulus and an argument (phase) which is a nonlinear function of frequency. The phase characteristic of devices of common interest is a quadratic function of frequency, but in general may be represented by higher order polynominals and/or other nonlinear functions. (UFFC) [22]

dispersive Doppler frequency *See:* differential Doppler frequency.

dispersive medium A medium in which one or more of the constitutive parameters vary with frequency. *Note:* As a result, the phase velocity of propagating waves in a dispersive medium depends on frequency. (AP/PROP) 211-1997

displaced phase center antenna (DPCA) An antenna and signal processing method used in airborne moving-target indication (AMTI) radar to compensate for the spread of the clutter Doppler spectrum caused by platform motion. *Note:* An example of a popular DPCA method is to employ two squinted antenna beams, take their sum and their difference on each of two successive pulses, and combine them in such a manner that the radar antenna appears to be stationary from pulse to pulse (a stationary antenna does not cause widening of the clutter spectrum). *See also:* space-time adaptive processing; airborne moving-target indication radar. (AES) 686-1997

displacement current The time rate of change of the electric flux density. (AP/PROP) 211-1997

display **(1)** **(A)** To present data visually. **(B)** The result of a display process. *See also:* copy. (C/Std100) 610.2-1987
(2) **(navigation aid terms)** The visual representation of output data. (AES/RS/GCS) 686-1990, 172-1983w
(3) **(watthour meters)** A means for visually identifying and presenting measured or calculated quantities and other information. (ELM) C12.15-1990, C12.13-1985s

(4) (test, measurement, and diagnostic equipment) A mechanical, optical, electro-mechanical, or electronic device for presenting information to the operator or maintenance technician about the state or condition of the unit under test or the checkout equipment itself. (MIL) [2]
(5) (oscilloscopes) The visual presentation on the indicating device of an oscilloscope. (IM) 311-1970w
(6) (A) (computer graphics) A visual presentation of graphics or other data such as text. *See also:* display device. **(B) (computer graphics)** To visually present graphics or other data. *See also:* display device. (C) 610.6-1991
(7) To output to the terminal of the user. If the output is not directed to a terminal, the results are undefined. (C/PA) 9945-2-1993
(8) To make a popup menu visible. (C) 1295-1993w
(9) Visual depiction of a single, integrated, organized set of information. A display may be an integration of several display formats (such as a system mimic that includes bar charts, trend graphs, and data fields). (PE/NP) 1289-1998

display address space The range of addresses available for display of a graphical image. (C) 610.6-1991w

display buffer A storage device or area of memory that contains the graphics display commands and coordinate data used to create images on a display device. *See also:* image memory. (C) 610.10-1994w, 610.6-1991w

display command An instruction that is processed by a graphical display device or display processor. *Synonym:* display order. (C) 610.6-1991w

display console *See:* console.

display controller *See:* display generator.

display cycle The time needed to completely refresh a graphical display image. *Synonym:* refresh cycle. (C) 610.6-1991w

display data A collection of data that defines a graphical image, or a portion thereof, for display. *Synonym:* graphics data. (C) 610.6-1991w

display device (1) In computer graphics, an output device on which display images can be represented. For example, cathode ray tube display device, plotter, hard copy unit. *See also:* graphical display device. (C) 610.6-1991w
(2) An output device that gives a visual representation of data. *Note:* The representation is usually temporary, however arrangements may be made for producing a hard copy of the representation. *Synonyms:* display screen; display unit; display monitor; display station. *See also:* monitor; display surface. (C) 610.10-1994w
(3) The hardware used to provide the display to users. An example is a video display unit. *See also:* display. (PE/NP) 1289-1998

display element (1) A basic computer graphics element used in the construction and rendering of a display image. For example, line, arc, circle, text string. *Synonyms:* primitive; output primitives. *See also:* selectable element. (C) 610.6-1991w
(2) The individual component(s) of a display, such as labels, abbreviations, acronyms, icons, symbols, numbers, color, graph lines, coding, highlighting, and background. *See also:* display. (PE/NP) 1289-1998

display flatness (spectrum analyzer) The total variation in displayed amplitude over a specified span, decibel (dB). *Note:* Display flatness is closely related to frequency response. The main difference is that the tuning control of the spectrum analyzer is not readjusted to center the display. (IM) 748-1979w

display file *See:* display list.

display frequency (spectrum analyzer) The input frequency as indicated by the spectrum analyzer (Hz). (IM) 748-1979w

display format Method of data presentation, such as a trend plot, bar chart, graph, table, or cross-plot. *See also:* display. (PE/NP) 1289-1998

display formatting A word processing capability that presents the formatted version of a document on a display device. (C) 610.2-1987

display frame A single display image analogous to a frame in a motion picture. *Synonym:* frame. (C) 610.6-1991w

display generator A processing unit that will produce the corresponding image on the display device when presented with a display instruction. *Synonym:* display controller. (C) 610.10-1994w

display graphic A hardware device (crt, plasma panel, arrays of lamps, or light-emitting diodes) used to present pictorial information. (SWG/PE) C37.100-1992

display group *See:* segment.

display head A head within a display device that employs the signals obtained from the various function generators to control the display of information on the display device. (C) 610.10-1994w

display image The portion of an image that is displayed on a graphical display device. *Synonym:* screen image. (C) 610.6-1991w, 610.10-1994w

display law (spectrum analyzer) The mathematical law that defines the input-output function of the instrument.

1) *Linear.* A display in which the scale divisions are a linear function of the input voltage.
2) *Square law (power).* A display in which the scale divisions are a linear function of the input power.
3) *Logarithmic.* A display in which the scale divisions are a logarithmic function of the input signal.

(IM) 748-1979w

display line The writing line on a display device. *See also:* display position. (C) 610.10-1994w

display list The display commands and data required to generate a graphical image. *Synonym:* display file. *See also:* metafile; display buffer. (C) 610.6-1991w

display menu *See:* menu.

display monitor *See:* display device.

display order *See:* display command.

display panel (A) A set of indicators on the front of a device or component, used to indicate its status. **(B)** A special display area on a processing unit that is used to show the contents of the display register and is activated by a display switch. (C) 610.10-1994

display position One character position on a display line. (C) 610.10-1994w

display primaries (color television) The colors of constant chromaticity and variable luminance produced by the receiver or any other display device that, when mixed in proper proportions, are used to produce other colors. *Note:* Usually the three primaries used are red, green, and blue. (BT/AV) 201-1979w

display processor A hardware device that executes a sequence of display commands to create a display image. *Synonym:* graphics processor. (C) 610.6-1991w

display reference level (spectrum analyzer) A designated vertical position representing specified input levels. The level may be expressed in dBm, volts, or any other units. (IM) 748-1979w

display register A register with corresponding indicators on the display panel, used to display the contents of the register selected by the display switch. (C) 610.10-1994w

displays (nuclear power generating station) Devices that convey information to the operator. (PE/NP) 566-1977w

display screen The surface of a display device on which the visual representation of data is displayed. For example, the phosphor-coated portion of a CRT display device. *See also:* display device. (C) 610.10-1994w

display segment *See:* segment.

display space The portion of the device space on a graphical display device that is available for displaying images. *Synonym:* image space. (C) 610.6-1991w

display station A generic term for a terminal or a console. (C) 610.10-1994w

display storage tube A storage tube into which the information is introduced as an electric signal and read at a later time as a visible output. *See also:* storage tube. (ED) 158-1962w

display surface **(1)** The surface of an output unit, such as a display device or plotting unit, on which the output is displayed. (C) 610.6-1991w
(2) The surface of an output device such as a display device or plotting unit. *See also:* display screen. (C) 610.10-1994w

display switch A switch used to select the register that is to be shown on the display panel. (C) 610.10-1994w

display tube A tube, usually a cathode-ray tube, used to display data. (C) [20], [85]

display unit *See:* display device.

disposal (liquid-filled power transformers) Intentionally or accidentally to discard, throw away, or otherwise complete or terminate the useful life of polychlorinated biphenyls (PCBs) and PCB items. Disposal includes spills, leaks, and other uncontrolled discharges of PCBs as well as actions related to containing, transporting, destroying, degrading, decontaminating, or confining PCBs and PCB items. Examples: Spill, chemical dechlorination, landfill, incineration.

disruptive discharge **(1)** A discharge that completely bridges the insulation under test, reducing the voltage between the electrodes practically to zero. (PE/PSIM) 4-1995
(2) A term that relates to phenomena associated with the breakdown of insulation under electrical stress, in which the discharge completely bridges the insulation under test, reducing the voltage between the electrodes to zero or nearly to zero. It applies to electrical breakdown in solid, liquid, and gaseous dielectrics and combinations of these. (SWG/PE) C37.100-1992
(3) The sudden and large increase in current through an insulating medium due to the complete failure of the medium under electrical stress. (SPD/PE) C62.22-1997, C62.11-1999
(4) **(A)** The sudden and large increase in current through an insulating medium, due to the complete failure of the medium under the electrostatic stress. **(B)** An increase that causes explosive mechanical or electrical failure in one of the failure modes. (SPD/PE) C62.62-2000

disruptive discharge probability ***(p)*** The probability that one application of a prospective voltage of a given shape and type will cause a disruptive discharge. (PE/PSIM) 4-1995

disruptive discharge voltage The voltage causing the disruptive discharge for tests with direct voltage, alternating voltage, and impulse voltage chopped at or after the peak; the voltage at the instant when the disruptive discharge occurs for impulses chopped on the front. (PE/PSIM) 4-1995
(2) **(A)** **(surge arresters)** (fifty percent) The voltage that has a 50-percent probability of producing a disruptive discharge. *Note:* The term applies mostly to impulse tests and has significance only in cases when the loss of electric strength resulting from a disruptive discharge is temporary. **(B)** **(surge arresters)** (one hundred percent) The specified voltage that is to be applied to a test object in a 100-percent disruptive discharge test under specified conditions. *Note:* The term applies mostly to impulse tests and has significance only in cases when the loss of electric strength resulting from a disruptive discharge is temporary. During the test, in general, all voltage applications should cause disruptive discharge. (Std100/PE/ELECTEC) [8], [84]

disruptive test A test that is invoked through a write to the TEST_START register and disrupts the node's operation by temporarily moving the node to the testing state. (C/MM) 1212-1991s

dissector *See:* image dissector.

dissector tube A camera tube having a continuous photocathode on which is formed a photoelectric-emission pattern that is scanned by moving its electron optical image over an aperture. *See also:* image dissector tube. (EEC/PE) [119]

dissipation **(1)** **(electrical energy)** Loss of electric energy as heat. (CHM) [51]
(2) **(waveguide)** The power reduction in a transmission path caused by resistive or conductive loss, or both. (MTT) 146-1980w

dissipation, electrode *See:* electrode dissipation.

dissipation factor **(1)** **(A)** The ratio of energy dissipated to the energy stored in an element for one cycle. **(B)** The loss tangent of an element. **(C)** The inverse of Q. *See also:* dielectric dissipation factor. (CAS) [13]
(2) **(dielectric)** The cotangent of the phase angle between a sinusoidal voltage applied across a dielectric (or combinations of dielectrics) and the resulting current through the dielectric system. (PE/PSIM) 62-1995
(3) The tangent of the dielectric loss angle. *Note:* For small values of dielectric loss angle dissipation factor is virtually equal to the insulation power factor. (PE/TR) C57.19.03-1996

dissipation-factor test *See:* loss-tangent test.

dissolved gas Implies that the gas is dissolved in the dielectric fluid. In solution, these gases lose all semblance of gaseous matter. There are no gas bubbles in the fluid under normal circumstances, and there is little tendency for the gases to respond to gravitational forces. Dissolved gases will not accumulate at high points, and they generally do not pose any threat to the dielectric function of the fluid. (PE/IC) 1406-1998

dissolved gas content The amount of gas dissolved in the fluid expressed as a percent of, or in parts-per-million relative to, the volume of the fluid, where the volume of the gas is defined as that occupied by the gas at a pressure of one atmosphere and temperature of 0°C. Dissolved gas content can refer to a specific gas, the total amount of all gases, or the total amount of combustible gases (TCG) only. (PE/IC) 1406-1998

dissymmetrical transducer Dissymmetrical with respect to a specified pair of terminations when the interchange of that pair of terminations will affect the transmission. *See also:* transducer. (EEC/PE) [119]

distal stimuli (illuminating engineering) In the physical space in front of the eye one can identify points, lines, surfaces, and three dimensional arrays of scattering particles which constitute the distal physical stimuli which form optical images on the retina. Each element of a surface or volume to which an eye is exposed subtends a solid angle at the entrance pupil. Such elements of solid angle make up the field of view and each has a specifiable luminance and chromaticity. Points and lines are specific cases which have to be dealt with in terms of total candlepower and candlepower per unit length. Distal stimuli are sometimes referred to simply as lights or colors. (EEC/IE) [126]

distance *See:* signal distance; hamming distance.

distance clearance D The minimum separation between two conductors, between conductors and supports or other objects, or between conductors and ground, or the clear space between any objects. (T&D/PE) 516-1995

distance dialing (telephone switching systems) The automatic establishing of toll calls by means of signals from the calling device of either a customer or an operator. (COM) 312-1977w

distance mark (range mark) (on a radar display) A calibration marker used on a cathode-ray screen in determining target distance. *See also:* radar. (AES) [42]

distance measuring equipment (DME) (navigation aid terms) A radio aid to navigation which provides distance information by measuring total round-trip time of transmission from an interrogator to a transponder and return. (AES/GCS) 172-1983w

distance protection A method of line protection in which an abnormal condition within a predetermined electrical distance of a line terminal on the protected circuit is detected by measurement of system conditions at that terminal. (SWG/PE/PSR) C37.100-1992, C37.90-1978s

distance relay (1) A generic term covering those forms of protective relays in which the response to the input quantities is primarily a function of the electrical circuit distance between the relay location and the point of fault. (SWG/PE/PSR) C37.90-1978s

(2) (power system device function numbers) A relay that functions when the circuit admittance, impedance, or reactance increases or decreases beyond a predetermined value. (SUB/PE) C37.2-1979s

(3) A generic term covering those forms of measuring relays in which the response to the input quantities is a function of the electric circuit distance (impedance) between the point of measurement and the point of fault. *Note:* A distance relay response characteristic, when presented on an *R-X* impedance diagram, will have an operating area dependent on the manner in which the input quantities are processed and compared. (SWG/PE) C37.100-1992

(4) A protective relay in which the response to the input quantities is primarily a function of the electrical circuit distance between the relay location and the point of fault. (PE/PSR) C37.113-1999

distance relay characteristic The defined threshold between the operate and nonoperate response of a distance relay, generally referred to as reach and presented on an *R-X* impedance diagram. (SWG/PE) C37.100-1992

distance resolution (radar) The ability to distinguish between two targets solely by the measurement of distances; generally expressed in terms of the minimum distance by which two targets of equal strength at the same azimuth and elevation angles must be spaced to be separately distinguishable and measureable. (AES/RS) 686-1982s

distance teaching Instruction in which the teacher and student are not in face-to-face contact. Communication is made through correspondence, radio, television, or computer-assisted instruction. (C) 610.2-1987

distant source An emitter far enough away from a shielding enclosure for its energy to illuminate uniformly an entire shielding face. (EMC) 299-1991s

distinguished encoding (1) Restrictions to the basic encoding rules designed to ensure a unique encoding of each ASN.1 value. (C/PA) 1327.2-1993w, 1224.2-1993w

(2) Restrictions to the basic encoding rules designed to ensure a unique encoding of each ASN.1 value, defined in ISO/IEC 9594-8. (C/PA) 1328.2-1993w, 1327.2-1993w

(3) Restrictions to the basic encoding rules designed to ensure a unique encoding of each ASN.1 value, defined in ISO/IEC 9594-8: 1990. (C/PA) 1326.2-1993w

distinguished name One of the names of an object, formed from the sequence of RDNs of its object entry and each of its superior entries. (C/PA) 1328.2-1993w, 1326.2-1993ow, 1327.2-1993w, 1224.2-1993w

distinguished value An attribute value in an entry that has been designated to appear in the relative distinguished name of the entry. (C/PA) 1328.2-1993w, 1224.2-1993w, 1327.2-1993w, 1326.2-1993w

distorted Born approximation *See:* extended Born approximation.

distorted current The current through the test circuit breaker that is influenced by the arc voltage of both the test and auxiliary circuit breakers during the high-current interval. (SWG/PE) C37.083-1999, C37.081-1981r, C37.100-1992

distortion (1) (data transmission) An undesired change in waveform. The principal sources of distortion are: A nonlinear relation between input and output at a given frequency; Nonuniform transmission at different frequencies; Phase shift not proportional to frequency. (PE) 599-1985w

(2) (fiber optics) A change of signal waveform shape. *Note:* In a multimode fiber, the signal can suffer degradation from multimode distortion. In addition, several dispersive mechanisms can cause signal distortion in an optical waveguide: waveguide dispersion, material dispersion, and profile dispersion. *See also:* profile dispersion; dispersion. (Std100) 812-1984w

(3) (broadband local area networks) The creation of additional or undesired effects due to nonlinearities in the system. *See also:* cross-modulation distortion; time distortion; intermodulation distortion; composite triple beat distortion. (LM/C) 802.7-1989r

(4) The change in waveform caused by outside interferences. *See also:* phase distortion; delay distortion; end distortion; intermodulation distortion. (C) 610.7-1995

(5) The rms value of the ac signal exclusive of the fundamental component. It may include various harmonic and interharmonic components. In a dc system, distortion is the rms value of the ac (ripple) component on the (fundamental) dc level. (Harmonics are sinusoidal distortion components that occur at integer multiples of the fundamental frequency. Inter-harmonics are distortion components that occur at noninteger multiples of the fundamental frequency.). (PEL) 1515-2000

distortion, amplitude-frequency *See:* distortion.

distortion, barrel *See:* barrel distortion.

distortion, envelope delay *See:* envelope delay distortion.

distortion factor (1) (power system communication equipment) The ratio of the root-mean-square value of the residue of a voltage wave after the elimination of the fundamental to the root-mean-square value of the original wave. (PE/PSC) 281-1984w

(2) (rotating machinery) (wave) The ratio of the root-mean-square value of the residue of a voltage wave after the elimination of the fundamental to the root-mean-square value of the original wave.

$$df = \left[\frac{\text{sum of squares of amplitudes of all harmonics}}{\text{square of amplitude of fundamental}}\right]^{1/2} \cdot 100\%$$

(PE) [9]

(3)

$$df = \left[\sum_{k=2}^{n} \frac{A_k^2}{A_1^2}\right]^{1/2}$$

where A_1 is the rms value of the fundamental component and A_2 to A_n are the rms values of the harmonic components. (PE/PSIM) 120-1989r

(4) (harmonic factor) The ratio of the root-mean-square of the harmonic content to the root-mean-square value of the fundamental quantity, expressed as a percent of the fundamental.

$$DF = \sqrt{\frac{\text{sum of squares of amplitudes of all harmonics}}{\text{square of amplitude of fundamental}}} \cdot 100\%$$

(IA/SPC) 519-1992

(5) The ratio of the root square value of the harmonic content to the root square value of the fundamental quantity, expressed as a percent of the fundamental. *Note:* Also referred to as *total harmonic distortion.* (IA/PSE) 1100-1999

(6) The ratio of the rms value of an ac signal, exclusive of the fundamental, to the rms value of the fundamental component of the ac signal, expressed as a percent. For example, voltage distortion factor may be expressed as:

$$D_V = \sqrt{\left(\frac{V_2}{V_1}\right)^2 + \left(\frac{V_3}{V_1}\right)^2 + \ldots \left(\frac{V_n}{V_1}\right)^2}$$

with

$< f_n \leq f_m$

where

D_V = voltage distortion factor in percent,
V_1 = rms value of the fundamental frequency component,
V_n = rms value of an individual non-fundamental frequency component,
f_1 = the fundamental frequency,
f_n = frequency of other individual waveform components, including harmonics and inter-harmonics (non-integral multiples of the fundamental),
f_{max} = the maximum frequency of measurement.

(PEL) 1515-2000

distortion, field-time waveform *See:* field-time waveform distortion.

distortion, intermodulation *See:* intermodulation distortion.

distortion, keystone *See:* keystone distortion.

distortion-limited operation (fiber optics) The condition prevailing when the distortion of the received signal, rather than its amplitude (or power), limits performance. The condition is reached when the system distorts the shape of the waveform beyond specified limits. For linear systems, distortion-limited operation is equivalent to bandwidth-limited operation. *See also:* multimode distortion; bandwidth-limited operation; distortion; attenuation-limited operation. (Std100) 812-1984w

distortion, linear *See:* linear distortion.

distortion, linear TV waveform *See:* linear TV waveform distortion.

distortion, line-time waveform *See:* line-time waveform distortion.

distortion, pattern *See:* pattern distortion.

distortion, percent harmonic *See:* percent harmonic distortion.

distortion, phase delay *See:* phase delay distortion.

distortion, pin-cushion *See:* pin-cushion distortion.

distortion power (A) (single-phase two-wire circuit) At the two terminals of entry of a single-phase two-wire circuit into a delimited region, a scalar quantity having an amplitude equal to the square root of the difference of the squares of the apparent power and the amplitude of the phasor power. *Note:* Mathematically, the amplitude of the distortion power D is given by the equation

$$D = (U^2 - S^2)^{1/2} = (U^2 - P^2 - Q^2)^{1/2}$$

$$= \left[\sum_{r=1}^{r=\infty} \sum_{q=1}^{q=\infty} \left\{E_r^2 I_q^2 E_r E_q I_r I_q \cdot \cos\left[(\alpha_r - \beta_r) - (\alpha_q - \beta_q)\right]\right\}\right]^{1/2}$$

where the symbols are as in power, apparent (single-phase two-wire circuit). If the voltage and current are quasi-periodic and the amplitudes are slowly varying, the distortion power at any instant may be taken as the value derived from the amplitude of the apparent power and phasor power at that instant. By this definition the sign of distortion power is not definitely determined, and it may be given either sign. In the absence of other definite information, it is to be taken the same as for the active power. Distortion power is expressed in volt-amperes when the voltage is in volts and the current in amperes. The distortion power is zero if the voltage and the current have the same waveform. This condition is fulfilled when the voltage and current are sinusoidal and have the same period, or when the circuit consists entirely of noninductive resistors. *See also:* apparent power. **(B) (polyphase circuit)** At the terminals of entry of a polyphase circuit, equal to the sum of the distortion powers for the individual terminals of entry. *Notes:* 1. The distortion power for each terminal of entry is determined by considering each phase conductor, in turn, with the common reference point as a single-phase, two- wire circuit and finding the distortion power for each in accordance with the definition of distortion power (single-phase two-wire circuit). The common reference terminal shall be taken as the neutral terminal of entry, if one exists, otherwise as the true neutral point. The sign given to the distortion power for each single-phase current, and therefore to the total for the polyphase circuit, shall be the same as that of the total active power. 2. Distortion power is expressed in volt-amperes when the voltages are in volts and the current in amperes. 3. The distortion power is zero if each voltage has the same wave form as the corresponding current. This condition is fulfilled, of course, when all the currents and voltages are sinusoidal. (Std100/Std100) 270-1966

distortion, pulse *See:* pulse distortion.

distortion, short-time waveform *See:* short-time waveform distortion.

distortion, single frequency *See:* single frequency distortion.

distortion, spiral *See:* spiral distortion.

distortion tolerance (telegraph receiver) The maximum signal distortion that can be tolerated without error in reception. *See also:* telegraphy. (COM) [49]

distortion, total *See:* total distortion.

distortion, total harmonic *See:* total harmonic distortion.

distortion, waveform *See:* waveform distortion.

distributed Spread out over an electrically significant length or area. (EEC/PE) [119]

distributed architecture A computer architecture characterized by its suitability for distributed processing. (C) 610.10-1994w

distributed constant (waveguide) A circuit parameter that exists along the length of a waveguide or transmission line. *Note:* For a transverse electromagnetic wave on a two-conductor transmission line, the distributed constants are series resistance, series inductance, shunt conductance, and shunt capacitance per unit length of line. *See also:* waveguide. (MTT) 146-1980w

distributed computer network A network in which all node pairs are connected either directly or through redundant paths through intermediate nodes. *See also:* decentralized computer network; centralized computer network. (C) 610.7-1995

distributed control system A system comprised of software, hardware, cabling, sensors, and activators, which is used to control and monitor equipment. (PE/EDPG) 1050-1996

distributed coordination function (DCF) A class of coordination function where the same coordination function logic is active in every station in the basic service set (BSS) whenever the network is in operation. (C/LM) 8802-11-1999

distributed database (A) A database that is not stored in its entirety at a single physical location, but is dispersed over a network of interconnected computers. **(B)** A database under the overall control of a central database management system, whose storage devices are not all attached to the same processor. (C) 610.5-1990

distributed data processing (DDP) The use of computers for processing information within a distributed system. (C) 610.2-1987

distributed element (waveguide) (for a transmission line) A circuit element that exists continuously along a transmission line. *Note:* For a transverse electromagnetic wave on a two-conductor transmission line, the distributed elements are series resistance, series inductance, shunt conductance, and shunt capacitance per unit length of line. (MTT) 146-1980w

distributed element circuit (microwave tubes) A circuit whose inductance and capacitance are distributed over a physical distance that is comparable to a wavelength. (ED) [45]

distributed function (dot logic, wired logic) A logic function (either AND or OR) implemented by connecting together outputs of the appropriate type; these outputs are the inputs of the logic function thus formed; the joined connection is the output. (GSD) 91-1984r

Distributed Interactive Simulation (DIS) A time-and-space-coherent synthetic representation of world environments designed for linking the interactive, free play activities of people in operational exercises. The synthetic environment is created through real-time exchange of data units between distributed, computationally autonomous simulation applications in the form of simulations, simulators, and instrumented equipment interconnected through standard computer communicative services. The computational simulation entities may be present in one location or may be distributed geographically. (C/DIS) 1278.4-1997, 1278.2-1995, 1278.1-1995

distributed-network type A unit substation which has a single stepdown transformer having its outgoing side connected to a bus through a circuit breaker equipped with relays which are arranged to trip the circuit breaker on reverse power flow to the transformer and to reclose the circuit breaker upon the restoration of the correct voltage, phase angle, and phase sequence at the transformer secondary. The bus has one or more outgoing radial (stub-end) feeders and one or more tie connections to a similar unit substation. (PE/TR) C57.12.80-1978r

distributed numerical control (DbNC) Numerical control in which a computer controls one or more remote numerical control machines. (C) 610.2-1987

distributed polling A polling technique in which all stations participate equally in the control of access to the transmission medium. *Contrast:* centralized polling. (C) 610.7-1995

distributed processing (1) (supervisory control, data acquisition, and automatic control) A design in which data is processed by more than one processor. (PE/SUB) C37.1-1994
(2) A design in which all data is not processed in one processor. Multiple processors in the master station or in the remote stations, or both, share the functions. (SWG/PE) C37.100-1992

distributed queue The medium access control (MAC) procedure in ISO/IEC 8802 for queued arbitrated (QA) access. (LM/C) 8802-6-1994

distributed resource islanding An islanding condition in which the distributed resource(s) supplying the loads within the island are not within the direct control of the power system operator. (SCC21) 929-2000

distributed simulation accreditation The official certification that a distributed simulation is acceptable for use for a specific purpose. (DIS/C) 1278.3-1996

distributed simulation validation The process of determining the degree to which a distributed simulation is an accurate representation of the real world from the perspective of its intended use(s) as defined by the requirements. (DIS/C) 1278.3-1996

distributed simulation verification the process of determining that an implementation of a distributed simulation accurately represents the developer's conceptual description and specifications. (DIS/C) 1278.3-1996

distributed system A computer system in which several interconnected computers share the computing tasks assigned to the system. (C) 610.2-1987, 610.10-1994w

distributed target A target composed of a number of scatterers, where the target extent in any dimension is greater than the radar resolution in that dimension. *See also:* complex target. (AES) 686-1997

distributed winding (rotating machinery) A winding, the coils of which occupy several slots per pole. *See also:* stator; armature; rotor. (PE) [9]

distributing cable *See:* distribution cable.

distributing frame (1) A structure for terminating permanent wires of a central office, private branch exchange, or private exchange and for permitting the easy change of connections between them by means of cross-connecting wires. (COM) [48]
(2) A structure for terminating permanent wires of a central office for permitting the easy change of connections between them by means of cross-connect wires. (C) 610.7-1995

distributing valve The element of the governor-control actuator that controls the flow of hydraulic fluid to the turbine-control servomotor(s). (PE/EDPG) 125-1988r

distribution (1) (used as an adjective) A general term used, by reason of specific physical or electrical characteristics, to denote application or restriction of the modified term, or both, to that part of an electrical system used for conveying energy to the point of utilization from a source or from one or more main receiving stations. *Notes:* 1. From the standpoint of a utility system, the area described is between the generating source or intervening substations and the customer's entrance equipment. 2. From the standpoint of a customer's internal system, the area described is between a source or receiving station within the customer's plant and the points of utilization. (SWG/PE/PE) C37.40-1993, C37.100-1992
(2) A software_collection containing software in the software packaging layout. (C/PA) 1387.2-1995
(3) The service that, by using association information, delivers medium access control (MAC) service data units (MSDUs) within the distribution system (DS). (C/LM) 8802-11-1999
(4) *See also:* probability distribution function. (T&D/PE) 539-1990

distribution amplifier A high-gain amplifier used to overcome high losses encountered in signal distribution. The generic term referring to any amplifier in a broadband coaxial system. In this document, the term signifies an amplifier that is used to operate at the higher levels that are normally associated with the distribution portion of the cable plant. (LM/C) 802.7-1989r

distribution arrester (A) Light duty class: An arrester generally installed on and used to protect underground distribution systems where the major portion of the lightning stroke current is discharged by an arrester located at the overhead line/cable junction. **(B) Normal duty class:** An arrester generally used to protect overhead distribution systems exposed to normal lightning currents. **(C) Heavy duty class:** An arrester most often used to protect overhead distribution systems exposed to severe lightning currents. (SPD/PE) C62.22-1997, C62.11-1999

distribution box (mine type) A portable piece of apparatus with enclosure by means of which an electric circuit is carried to one or more machine trailing cables from a single incoming feed line, each trailing cable circuit being connected through individual overcurrent protective devices. *See also:* mine feeder circuit; distributor box. (EEC/PE) [119]

distribution catalog The catalog of metadata for a distribution software_collection. Unlike a catalog for an installed_software object, a distribution catalog is stored in a particular exported catalog structure that is part of the software packaging layout. (C/PA) 1387.2-1995

distribution cable (communication practice) (distributing cable) A cable extending from a feeder cable into a specific area for the purpose of providing service to that area. *See also:* cable. (EEC/PE) [119]

distribution center Enclosed apparatus that consists of automatic protective devices connected to bus bars, to subdivide the feeder supply and provide control and protection of subfeeders or branch circuits. (IA/MT) 45-1998

distribution coefficients (television) (color) The tristimulus values of monochromatic radiations of equal power. *Note:* Generally represented by overscored, lowercase letters such as $\bar{x}$, $\bar{y}$, $\bar{z}$ in the CIE system. (BT/AV) 201-1979w

distribution counting sort An insertion sort in which the sort keys of the items to be sorted fall within some finite range of values, and by counting the number of sort keys having each value, the position of the items in the sorted set can be determined. (C) 610.5-1990w

distribution current-limiting fuse A fuse consisting of a fuse support and a current-limiting fuse unit. *Note:* In addition, the distribution current-limiting fuse is identified by the following characteristics:

a) Dielectric withstand basic impulse insulation level (BIL) strengths at distribution levels
b) Application primarily on distribution feeders and circuits

c) Operating voltage limits correspond to distribution system voltage

(SWG/PE) C37.40-1993, C37.100-1992

distribution cutout A fuse or disconnecting device consisting of any one of the following assemblies:

a) A fuse support and fuseholder that may or may not include the conducting element (or fuse link)
b) A fuse support and disconnecting blade
c) A fuse support and fuse carrier that may or may not include the conducting element (fuse link) or disconnecting blade

Note: In addition, the distribution cutout is identified by the following characteristics: Dielectric withstand (basic impulse insulation level) strengths at distribution levels; Application primarily on distribution feeders and circuits; Mechanical construction basically adapted to pole or crossarm mounting except for the distribution oil cutout; Operating voltage limits correspond to distribution system voltage.

(SWG/PE) C37.40-1993, C37.100-1992

distribution disconnecting cutout (1) A distribution cutout having a disconnecting blade that is used for closing, opening, or changing the connections in a circuit or system, or for isolating purposes. *Note:* Some load-break ability is inherent in the device but it has no load-break rating. This ability can best be evaluated by the user based on experience under operating conditions. (SWG/PE) C37.40-1993
(2) *See also:* distribution cutout; disconnecting cutout.

(SWG/PE) C37.100-1981s

distribution enclosed single-pole air switch (distribution enclosed air switches) A single-pole disconnecting switch in which the contacts and blade are mounted completely within an insulated enclosure (cannot be converted into a distribution cutout or disconnecting fuse). *Notes:* 1. The distribution enclosed air switch is identified by the following characteristics: Dielectric withstand basic impulse insulation level (BIL) strengths at distribution level; Application primarily on distribution feeders and circuits; Mechanical construction basically adapted to crossarm mounting; Operating voltage limits correspond to distribution voltages; Unless incorporating load-break means, it has no interrupting load-break current rating. 2. Some load-break ability is inherent in the device. This ability can best be evaluated by the user based on experience under operating conditions.

(SWG/PE) C37.40-1993, C37.100-1992

distribution factor (rotating machinery) A factor related to a distributed winding, taking into account the spatial distribution of the slots in which the winding considered is laid, that is the decrease in the generated voltage, as a result of a geometrical addition of the corresponding representative vectors.

(PE) [9]

distribution feeder *See:* primary distribution feeder; secondary distribution feeder; distribution center.

distribution function The probability that a parameter is less than a given value *x*. (EMC) C63.12-1987

distribution fuse cutout (1) A distribution cutout having a fuseholder or fuse carrier and fuse link or a fuse unit. *Note:* A fuse cutout is a fuse disconnecting switch. It has some inherent load-break ability but does not have a load-break rating. The load-break ability can best be evaluated by the user based on experience under operating conditions.

(SWG/PE) C37.40-1993

(2) *See also:* fuse cutout; distribution cutout.

(SWG/PE) C37.100-1981s

distribution line Electric power lines which distribute power from a main source substation to consumers, usually at a voltage of 34.5 kV or less. (T&D/PE) 751-1990, 1410-1997

distribution main *See:* primary distribution mains; secondary distribution mains; center of distribution.

distribution network *See:* secondary distribution network; center of distribution.

distribution oil cutout *See:* oil cutout; distribution; distribution cutout.

distribution open cutout *See:* open cutout; distribution; distribution cutout.

distribution open-link cutout *See:* distribution; distribution cutout; open-link cutout.

distribution panel A panel that receives energy from a switchboard or a distribution center and distributes power to energy-consuming devices. (IA/MT) 45-1998

distribution panelboard A single panel or group of panel units designed for assembly in the form of a single panel; including buses, and with or without switches and/or automatic overcurrent protective devices for the control of light, heat or power circuits of small individual as well as aggregate capacity; designed to be placed in a cabinet or cutout box placed in or against a wall or partition and accessible only from the front. (NESC/NEC) [86]

distribution path The pathname below which the catalog describing the distribution is located. If the distribution is on a single medium, all software for it is located below this path.

(C/PA) 1387.2-1995

distribution sort A sort in which the set to be sorted is divided into two or more subsets such that all items within each subset are within some exclusive range. Each subset is sorted, then the subsets are joined in the correct order. *Synonyms:* distributive sort; column sort; separation sort; pocket sort. *Contrast:* merge sort. *See also:* radix sort; digital sort.

(C) 610.5-1990w

distribution station (power operations) A transforming station where the transmission is linked to the distribution system.

(PE/PSE) 858-1987s

distribution switchboard A power switchboard used for the distribution of electric energy at the voltages common for such distribution within a building. *Note:* Knife switches, air circuit breakers, and fuses are generally used for circuit interruption on distribution switchboards, and voltages seldom exceed 600. However, such switchboards often include switchboard equipment for a high-tension incoming supply circuit and a step-down transformer.

(SWG/PE) C37.100-1992

distribution system (DS) (1) That portion of an electric system that transfers electric energy from the bulk electric system to the customers. (PE/PSE) 858-1993w
(2) That portion of an electric system that delivers electric energy from transformation points on the transmission system to the customer. *Note:* The distribution system is generally considered to be anything from the distribution substation fence to the customer meter. Often the initial overcurrent protection and voltage regulator are within the substation fence.

(PE/T&D) 1366-1998

(3) A system used to interconnect a set of basic service sets (BSSs) and integrated local area networks (LANs) to create an extended service set (ESS). (C/LM) 8802-11-1999

distribution system medium (DSM) The medium or set of media used by a distribution system (DS) for communications between access points (APs) and portals of an extended service set (ESS). (C/LM) 8802-11-1999

distribution system service (DSS) The set of services provided by the distribution system (DS) that enable the medium access control (MAC) to transport MAC service data units (MSDUs) between stations that are not in direct communication with each other over a single instance of the wireless medium (WM). These services include transport of MSDUs between the access points (APs) of basic service sets (BSSs) within an extended service set (ESS), transport of MSDUs between portals and BSSs within an ESS, and transport of MSDUs between stations in the same BSS in cases where the MSDU has a multicast or broadcast destination address or where the destination is an individual address, but the station sending the MSDU chooses to involve DSS. DSSs are provided between pairs of IEEE 802.11 MACs.

(C/LM) 8802-11-1999

distribution temperature (illuminating engineering) (of a light source) The absolute temperature of a blackbody whose relative spectral distribution is most nearly the same in the visible region of the spectrum as that of the light source. (EEC/IE) [126]

distribution transformer (power and distribution transformers) A transformer for transferring electrical energy from a primary distribution circuit to a secondary distribution circuit or consumer's service circuit. *Note:* Distribution transformers are usually rated in the order of 5–500 kVA. (PE/TR) C57.12.80-1978r

distribution trunk line *See:* primary distribution trunk line; center of distribution.

distributive sort *See:* distribution sort.

distributor box A box or pit through which cables are inserted or removed in a draw-in system of mains. It contains no links, fuses, or switches and its usual function is to facilitate tapping into a consumer's premises. *See also:* tower; distribution box. (T&D/PE) [10]

distributor duct A duct installed for occupancy of distribution mains. *See also:* service pipe. (T&D/PE) [10]

distributor suppressor (internal-combustion engine terminology) A suppressor designed for direct connection to the high-voltage terminals of a distributor cap. (INT) [53], [70]

disturbance (1) (A) (communication practice) Any irregular phenomenon associated with transmission that tends to limit or interfere with the interchange of intelligence. **(B) (electromagnetic)** Any electromagnetic phenomenon that may degrade the performance of a device, equipment, or system, or adversely affect living or inert matter. *Note:* An electromagnetic disturbance may be a noise, an unwanted signal, or a change in the propagation medium itself. **(C)** Any perturbation to the electric system. **(D)** An unexpected change in area control error (ACE) that exceeds three times L_d, and which is caused by a sudden loss of generation or interruption of load. (EEC/PE/EMC/PE/PSE) [119], C63.4-1991, 858-1993
(2) An undesired input variable that may occur at any point within a feedback control system. *See also:* feedback control system. (IA/ICTL/IAC) [60]
(3) (storage tubes) That type of spurious signal generated within a tube that appears as abrupt variations in the amplitude of the output signal. *Notes:* 1. These variations are spatially fixed with reference to the target area. 2. The distinction between this and shading. 3. A blemish, a mesh pattern, and moire present in the output are forms of disturbance. Random noise is not a form of disturbance. *See also:* storage tube. (ED) 158-1962w
(4) (interference terminology) *See also:* interference.

disturbance testing Tests performed to verify the dc system response to ac faults, dc line faults, loss of one pole (if applicable), and other dynamic response criteria. (PE/SUB) 1378-1997

disturbed-ONE output (magnetic cell) A ONE output to which partial-read pulses have been applied since that cell was last selected for writing. *See also:* coincident-current selection. (Std100) 163-1959w

disturbed-ZERO output (magnetic cell) A ZERO output to which partial-write pulses have been applied since that cell was last selected for reading. (Std100) 163-1959w

dit *See:* hartley.

dither (control circuits) A useful oscillation of small amplitude, introduced to overcome the effects of friction, hysteresis, or clogging. *See also:* feedback control system. (IA/IAC) [60]

dithering A technique for displaying an image with many colors or gray levels on a device having fewer colors or gray levels than the image by simulating the colors or gray levels with a group of closely spaced, equal-sized dots. *See also:* halftoning. (C) 610.6-1991w

dither spillover *See also:* anti-lock residual. (AES/GYAC) 528-1994

diurnal Pertaining to daytime, as in the daily biological rhythm. (T&D/PE) 539-1990

divergence (1) (vector field at a point) A scalar equal to the limit of the quotient of the outward flux through a closed surface that surrounds the point by the volume within the surface, as the volume approaches zero. *Note:* If the vector **A** of a vector field is expressed in terms of its three rectangular components A_x, A_y, A_z, so that the values of A_x, A_y, A_z are each given as a function of x, y, and z, the divergence of the vector field **A** (abbreviated $\nabla \cdot$ A) is the sum of the three scalars obtained by taking the derivatives of each component in the direction of its axis, or

$$\text{div A} \equiv \nabla \cdot \text{A} = \frac{\partial A_x}{\partial_x} + \frac{\partial A_y}{\partial_y} + \frac{\partial A_z}{\partial_z}$$

Examples: If a vector field **A** represents velocity of flow such as the material flow of water or a gas or the imagined flow of heat or electricity, the divergence of **A** at any point is the net outward rate of flow per unit volume and pr unit time. It is the time rate of decrease in density of the fluid at that point. Because the density of an incompressible fluid is always zero. The divergence of the flow of heat at a point in a body is equal to the rate of generation of heat per unit volume at the point The divergence of the electric field strength at a point is proportional to the volume density of charge at the point. (Std100) 270-1966w
(2) (fiber optics) *See also:* beam divergence. 812-1984w

divergence loss (acoustic wave) The part of the transmission loss that is due to the divergence or spreading of the sound rays in accordance with the geometry of the system (for example, spherical waves emitted by a point source). (SP) [32]

diverse redundancy *See:* diversity.

diversity (1) (software) In fault tolerance, realization of the same function by different means. For example, use of different processors, storage media, programming languages, algorithms, or development teams. *See also:* software diversity. (C) 610.12-1990
(2) *See also:* load diversity.

diversity factor (1) The ratio of the sum of the individual noncoincident maximum demands of various subdivisions of the system to the maximum demand of the complete system. The diversity factor is always 1 or greater. The (unofficial) term *diversity*, as distinguished from *diversity factor* refers to the percent of time available that a machine, piece of equipment, or facility has its maximum or nominal load or demand (i.e., a 70% diversity means that the device in question operates at its nominal or maximum load level 70% of the time that it is connected and turned on). (IA/PSE) 141-1993r
(2) The ratio of the sum of the individual maximum demands of the subdivisions of the system to the maximum demand of the complete system. (IA/T&D/PE/PSE) 241-1990r, [10]

diversity gain The reduction in predetection signal-to-interference energy ratio required to achieve a given level of performance, relative to that of a nondiversity radar, resulting from the use of diversity in frequency, polarization, space, or other characteristics. (AES) 686-1997

diversity, seasonal *See:* seasonal diversity.

diversity, time zone *See:* time zone diversity.

diverted slave A selected slave that detects that a reflecting slave is participating in the data transfer in its place. A diverted slave participates in the transfer by reading the data being transferred between the master and the reflecting slave. A slave may be diverted only during single-slave mode transactions. (C/MM) 896.1-1987s

divide-and-conquer sort *See:* radix exchange sort.

divide check An indicator that denotes that an invalid division has been attempted or has occurred. (C) 1084-1986w

divided code ringing (divided ringing) A method of code ringing that provides partial ringing selectivity by connecting one-half of the ringers from one side of the line to ground and the other half from the other side of the line to the ground. This term is not ordinarily applied to selective and semiselective ringing systems. (PE/EEC) [119]

divided ringing *See:* divided code ringing.

dividend A number to be divided by another number (the divisor) to produce a result (the quotient), and perhaps a remainder. (C) 1084-1986w

divider (1) (A) (analog computer) A device capable of dividing one variable by another. **(B) (analog computer)** A device capable of attenuating a variable by a constant or adjustable amount, as an attenuator. (C) 165-1977
(2) A device capable of dividing one variable into another. *Contrast:* multiplier. *See also:* analog divider. (C) 610.10-1994w

dividing network (loadspeaker dividing network) (crossover network) A frequency selective network that divides the spectrum into two or more frequency bands for distribution to different loads. (SP) [32]

diving board *See:* lineperson's platform.

division (nuclear power generating station) (accident monitoring instrumentation) The designation applied to a given system or set of components that enables the establishment and maintenance of physical, electrical, and functional independence from other redundant sets of components. *Notes:* 1. The terms division, train, channel, separation group, safety group, or load group, when used in this context, are interchangeable. 2. A division can have one or more channels. (PE/NP) 497-1981w, 308-1991, 384-1992r, 603-1998

Division 1 Terminology used for classification of an industrial area in which flammable gases or combustible dusts can be present under normal conditions, or from frequent breakdowns, or where failure of equipment could release materials and create simultaneous failure of electrical equipment. (IA) 515-1997

division transformation function A hash function that returns the remainder from the division of some value into the original key. For example, in the function below, the original key is divided by 997.

Original Key	Calculation	Hash Value
35721	35721/997 = 35 R 826	826
87452	87452/997 = 87 R 713	713

(C) 610.5-1990w

Division 2 Terminology used for classification of an industrial area in which flammable gases or combustible dusts will only be present under abnormal conditions. (IA) 515-1997

divisor A number by which another number (the dividend) is divided to produce a result (the quotient), and perhaps a remainder. (C) 1084-1986w

DL *See:* data link.

DL/I *See:* Data Language I.

D layer (radio-wave propagation) An ionized layer in the D region. (AP/PROP) 211-1990s

DLE character *See:* data link escape character.

DLC *See:* decision level concentration; digital loop carrier.

DLSAP *See:* data link service access point.

DMA *See:* direct memory access.

DMAC *See:* direct memory access controller.

DMAD *See:* Diagnostic Machine Aid—Digital.

DMA Framework Refers to IEEE Std 1212.1-1993: a description of the fundamental structures and protocols for the overall exchange of messages and data via shared memory. In this framework, a given DMA model is not a complete interface specification, but rather a skeleton. To it must be added conventions for message contents, semantics, and allowed sequences in order to define application-dependent I/O Unit and Function interface standards. (C/MM) 1212.1-1993

DMC *See:* direct memory control.

DME *See:* distance measuring equipment.

DML *See:* data manipulation language.

DMPDU (derived MAC protocol data unit (ARCHIVE), derived MAC protocol data unit (ARCHIVED)) *See:* derived MAC protocol data unit.

DMT *See:* direct memory transfer.

DNC *See:* direct numerical control.

D noise The noise on an idle channel or circuit, i.e., a channel or circuit with a termination and no signal (holding tone) at the transmitting end, measured through a D Filter. The noise is expressed in dBrnD. (COM/TA) 743-1995

D-Notched noise The noise power on a channel with a holding tone (signal) at the transmit end, measured through a D Filter and a 1010 Hz notch filter in tandem. May also be measured as a signal-to-D notched noise ratio (S/DNN in dB). (COM/TA) 743-1995

document (1) (A) (information processing) A medium and the data recorded on it for human use, for example, a report sheet, a book. **(B) (information processing)** By extension, any record that has permanence and that can be read by man or machine. (C) [20], [85]
(2) (nuclear power quality assurance) Any written or pictorial information describing, defining, specifying, reporting, or certifying activities, requirements, procedures, or results. A document is not considered to be a quality assurance record until it satisfies the definition of a quality assurance record. *See also:* quality assurance record. (PE/NP) [124]
(3) (A) (software) A medium, and the information recorded on it, that generally has permanence and can be read by a person or a machine. Examples in software engineering include project plans, specifications, test plans, user manuals. **(B) (software)** To create a document as in definition (A). **(C) (software)** To add comments to a computer program. *See also:* document cycle. (C) 610.12-1990
(4) A collection of data, regardless of the medium on which it is recorded, that generally has permanence and can be read by humans or machines. *Synonym:* documentation. (C/SE) J-STD-016-1995
(5) An encoded, electronically transmittable image, set of images, or image-related information, which is handled by the printer interface control unit. (C/MM) 1284.1-1997

document assembly In word processing, the assembly of new documents from previously recorded documents or boilerplate text in accordance with specified variables such as names and addresses in iterative documents, or sales figures embedded in the text of a document. *Synonym:* document merge. (C) 610.2-1987

documentation (nuclear power generating station) Any written or pictorial information describing, defining, specifying, reporting or certifying activities, requirements, procedures, or results. (PE/EDPG) 690-1984r
(2) (A) (software) A collection of documents on a given subject. **(B) (software)** Any written or pictorial information describing, defining, specifying, reporting, or certifying activities, requirements, procedures, or results. **(C) (software)** The process of generating or revising a document. **(D) (software)** The management of documents, including identification, acquisition, processing, storage, and dissemination. (C) 610.12-1990

documentation assertion (1) An assertion that pertains to the documentation associated with the implementation being tested. (C/PA) 1328.2-1993w, 1326.2-1993w, 1328-1993w
(2) An assertion generated by a requirement in the base standard being tested that a specific feature or behavior be documented. (C/PA) 2003-1997

documentation level *See:* level of documentation.

documentation tree A diagram that depicts all of the documents for a given system and shows their relationships to one another. *See also:* specification tree

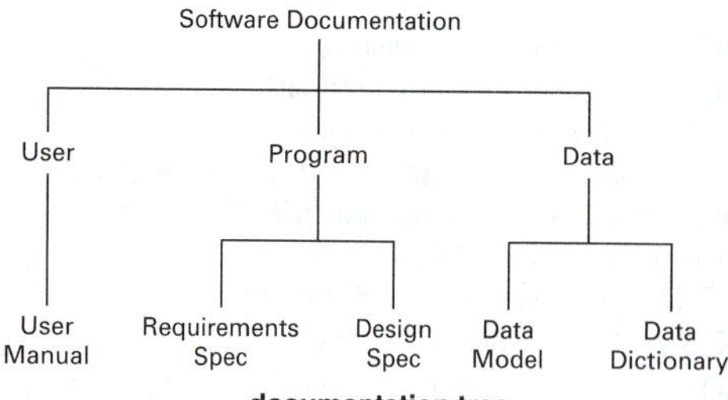

documentation tree

(C) 610.12-1990

Document Composition Facility (DCF) A text-formatting language, developed by IBM, for use as an engine for GML to generate formatted documents. *See also:* SCRIPT. (C) 610.13-1993w

document cycle The steps involved in the creation and handling of a document. *Note:* The cycle typically includes origination, production, reproduction, distribution, filing, and storage. *See also:* information traffic. (C) 610.2-1987

document editor A text editor used to enter, alter, and view documents. *Synonym:* manuscript editor. *Contrast:* program editor. (C) 610.2-1987

document mark In micrographics, a mark on microfilm, used for counting images or film frames automatically. *Synonym:* blip. (C) 610.2-1987

document merge *See:* document assembly.

document page count The total number of nonblank pages contained in a hard-copy document. (C/SE) 1045-1992

document reader A device that can be used to sense and interpret information contained on documents such as punch cards or paper. For example, a card reader, or scanner. (C) 610.10-1994w

document reference edge In character recognition, a specified document edge with respect to which the alignment of characters is defined. (C) 610.2-1987

documents (1) (nuclear power generating station) Drawings and other records significant to the design, construction, testing, maintenance, and operation of Class 1E equipment and systems for nuclear power generating stations. *Note:* Documents include: Drawings such as instrument diagrams, functional control diagrams, one line diagrams, schematic diagrams, equipment arrangements, cable and tray lists, wiring diagrams; Instrument data sheets; Design specifications; Instruction manuals; Test specifications, procedures, and reports; Device lists. Not to be included as documents are: project schedules, financial reports, meeting minutes, correspondence such as letters and memoranda, and equipment procurement documentation covered by quality assurance programs. (PE/NP) 380-1975w
(2) Hard copy, screen images, text, and graphics used to convey information to people. (C/SE) 1045-1992

document screen count The total number of page images for electronically displayed documents. (C/SE) 1045-1992

document set Document or group of documents that offers an audience the information it needs about a software product. (C/SE) 1063-1987r

document traffic *See:* information traffic.

document type The specification of a class of documents, which states their necessary semantics, abstract syntaxes, and dynamics. (C/PA) 1238.1-1994w

DOD *See:* depth of discharge.

Doherty amplifier A particular arrangement of a radio-frequency linear power amplifier wherein the amplifier is divided into two sections whose inputs and outputs are connected by quarter-wave (90-degree) networks and whose operating parameters are so adjusted that, for all values of input signal voltage up to one-half maximum amplitude, Section No. 2 is inoperative and Section No. 1 delivers all the power to the load, which presents an impedance at the output of Section No. 1 that is twice the optimum for maximum output. At one-half maximum input level, Section No. 1 is operating at peak efficiency, but is beginning to saturate. Above this level, Section No. 2 comes into operation, thereby decreasing the impedance presented to Section No. 1, which causes it to deliver additional power into the load until, at maximum signal input, both sections are operating at peak efficiency and each section is delivering one-half the total output power to the load. *See also:* amplifier. (AP/ANT) 145-1983s

dollar sign The character "$". (C/PA) 9945-2-1993

dolly *See:* traveler.

dolly car *See:* traveler rack.

Dolph-Chebyshev array antenna* *See:* Dolph-Chebyshev distribution.

* Deprecated.

Dolph-Chebyshev distribution A set of excitation coefficients for an equispaced linear array antenna such that the array factor can be expressed as a Chebyshev polynomial. (AP/ANT) 145-1993

Domain A publication-specific scoping mechanism that defines the set of entities that can potentially communicate via the publish-subscribe model. Specifically, the term Domain shall denote a value having type `PubSubDomain`. (IM/ST) 1451.1-1999

domain (1) (A) The set of all possible values that can be taken on by an independent variable. **(B)** In a relational data model, the set of all possible values that can be taken on by some attribute. For example, the domain of a three-digit positive integer is [001, 002, 003, ..., 999]. (C) 610.5-1990
(2) A problem space. (C/SE) 1517-1999
(3) *Synonym:* value class. (C/SE) 1320.2-1998

domain analysis (A) The analysis of systems within a domain to discover commonalities and differences among them. **(B)** The process by which information used in developing software systems is identified, captured, and organized so that it can be reused to create new systems, within a domain. **(C)** The result of the process in (A) and (B). (C/SE) 1517-1999

domain architecture A generic, organizational structure or design for software systems in a domain. The domain architecture contains the designs that are intended to satisfy requirements specified in the domain model. The domain architecture documents design, whereas the domain model documents requirements. A domain architecture: 1) can be adapted to create designs for software systems within a domain, and 2) provides a framework for configuring assets within individual software systems. (C/SE) 1517-1999

domain definition The process of determining the scope and boundaries of a domain. The result of the process. (C/SE) 1517-1999

domain engineer A party that performs domain engineering activities (including domain analysis, domain design, asset construction, and asset maintenance). (C/SE) 1517-1999

domain engineering A reuse-based approach to defining the scope (i.e., domain definition), specifying the structure (i.e., domain architecture), and building the assets (e.g., requirements, designs, software code, documentation) for a class of systems, subsystems, or applications. Domain engineering may include the following activities: domain definition, domain analysis, developing the domain architecture, and domain implementation. (C/SE) 1517-1999

domain expert An individual who is intimately familiar with the domain and can provide detailed information to the domain engineers. (C/SE) 1517-1999

domain metric The result of dividing the total number of modules in which one valid sample and one invalid sample of every class of input data items (external messages, operator inputs, and local data) have been correctly processed, by the total number of modules. *Note:* This definition assumes that the modules are essentially the same size. (C/SE) 730-1998

domain model A product of domain analysis that provides a representation of the requirements of the domain. The domain model identifies and describes the structure of data, flow of information, functions, constraints, and controls within the domain that are included in software systems in the domain. The domain model describes the commonalities and variabilities among requirements for software systems in the domain. (C/SE) 1517-1999

domain of applicability For a system element, the range, set, or set of ranges of intrinsic dates that can be represented, stored, manipulated, and processed invariantly within the system element. (C/PA) 2000.1-1999

dominant mode (waveguide) In a uniform waveguide, the propagating mode with the lowest cutoff frequency. (MTT) 146-1980w

dominant mode of propagation The mode of propagation with the lowest cutoff frequency. For a planar transmission line containing at least two disjoint conductors, the cutoff frequency can be zero. In this limit, the dominant mode reduces to a static electromagnetic field pattern and is also called the fundamental mode. For a planar transmission line with more than two conductors, there is more than one fundamental mode. (MTT) 1004-1987w

dominant wave (uniconductor waveguide) The guided wave having the lowest cutoff frequency. *Note:* It is the only wave that will carry energy when the excitation frequency is between the lowest cutoff frequency and the next higher cutoff frequency. *See also:* guided wave; waveguide.

dominant wavelength (1) (television) (colored light, not purple) The wavelength of the spectrum light that, when combined in suitable proportions with the specified achromatic light, yields a match with the light considered. *Note:* When the dominant wavelength cannot be given (this applies to purples), its place is taken by the complementary wavelength. (BT/AV) 201-1979w

(2) (of a light) The wavelength of radiant energy of a single frequency that, when combined in suitable proportion with the radiant energy of a reference standard, matches the color of light. (EEC/IE) [126]

done_correct A status code that is returned when a transaction is completed without errors. On many buses, the done_correct status is implicitly assumed when no error-status codes are observed. (C/MM) 1212-1991s

Donnan potential (electrobiology) The potential difference across an inert semipermeable membrane separating mixtures of ions, attributed to differential diffusion. *See also:* electrobiology. (EMB) [47]

donor *See:* semiconductor.

don't care value (1) The enumeration literal '-' of the type STD_ULOGIC defined by IEEE Std 1164-1993. (C/DA) 1076.3-1997

(2) The enumeration literal '-' of the type STD_ULOGIC (or subtype STD_LOGIC) defined by IEEE Std 1164-1993. (C/DA) 1076.6-1999

do-nothing instruction *See:* dummy instruction.

do-nothing operation *See:* no-operation.

door (gate closer) A device that closes a manually opened hoistway door, a car door, or gate by means of a spring or by gravity. *See also:* hoistway. (EEC/PE) [119]

door contact (burglar-alarm system) An electric contacting device attached to a door frame and operated by opening or closing the door. *See also:* protective signaling. (PE/EEC) [119]

door power operator A device, or assembly of devices, that opens a hoistway door and/or a car door or gate by power other than by hand, gravity, springs, or the movement of the car; and that closes them by power other than by hand, gravity, or the movement of the car. *See also:* hoistway. (EEC/PE) [119]

doors closed A state, as given by trainline signal indication, in which doors are fully closed and locked. (VT) 1475-1999

doors locked The condition reached in the door-closing cycle when the drive has achieved a latching condition that will hold doors closed mechanically until a door-opening cycle is initiated. (VT) 1475-1999

doors open A state, as given by trainline signal indication, in which doors are not fully closed and locked. (VT) 1475-1999

dopant (1) (semiconductor) (acceptor) An impurity that may induce hole conduction. *Synonym:* impurity. *See also:* semiconductor device.

(2) (donor) (semiconductor) An impurity that may induce electron conduction. *See also:* semiconductor device; impurity. (IA) [12]

doped junction (semiconductor) A junction produced by the addition of an impurity to the melt during crystal growth. *See also:* semiconductor. (ED) 216-1960w

Dopen-loop phase angle *See:* loop phase angle.

Dopen-shut control system *See:* on-off control system.

doping (semiconductor) Addition of impurities to a semiconductor or production of a deviation from stoichiometric composition, to achieve a desired characteristic. *See also:* semiconductor. (ED) 216-1960w

doping compensation (semiconductor) Addition of donor impurities to a p-type semiconductor or of acceptor impurities to an n-type semiconductor. *See also:* semiconductor. (IA) [12]

Doppler beam sharpening (DBS) A form of squint-mode synthetic-aperture radar (SAR) employed in a sector-scanning air-to-ground radar. *Note:* It usually has less resolution than a conventional SAR, since it employs a shorter processing (integration) time and varies this time as a function of beam squint angle so as to keep the resolution constant. Often displayed in near-real time on a plan-position indicator (PPI). (AES) 686-1997

Doppler effect (1) (data transmission) The phenomenon changing the observed frequency of a wave in a transmission system caused by a time rate of change in the effective length of the path of travel between the source and the point of observation. (PE) 599-1985w

(2) For an observer, the apparent change in frequency of a wave when there is relative motion between the source and the observer. (AP/PROP) 211-1997

Doppler filter A filter used in continuous wave (CW) radar, moving-target indication (MTI), or pulsed-Doppler radars for the purpose of separating moving target echo signals from stationary clutter echo signals. Also may be used to separate one moving target from another moving target due to a different Doppler frequency shift. (AES) 686-1997

Doppler-inertial navigation equipment (navigation aid terms) Hybrid navigation equipment which employs both Doppler navigation radar and inertial sensors. (AES/GCS) 172-1983w

Doppler frequency (radio-wave propagation) Of a wave traveling between a source and a point, the shift in frequency of the wave caused by the change of phase path with time. *Note:* The change may be due to variations in the separation of source and the point or in the refractive index of the intervening medium. (AP) 211-1977s

Doppler navigator (navigation aid terms) A self-contained dead reckoning navigation aid transmitting two or more beams of electromagnetic or acoustic energy outward and downward from the vehicle and utilizing the Doppler effect of the reflected energy, a reference direction, and the relationship of the beams to the vehicle to determine speed and direction of motion over the reflecting surface. (AES/GCS) 172-1983w

Doppler radar A radar that utilizes the Doppler effect to determine the radial component of relative radar-target velocity or to select targets having particular radial velocities. (AES/GCS) 172-1983w, 686-1997

Doppler shift The magnitude of the change in the observed frequency of a wave due to the Doppler effect. The unit is the hertz. (SP) [32]

Doppler spread ($\sigma\lambda$) The spreading in the frequency domain of the power spectrum of a wave. *Note:* The Doppler spread is inversely related to the decorrelation time (τ_0):

$$\sigma_\lambda = (2\pi\tau_0)^{-1}$$

(AP/PROP) 211-1997

Doppler tracking (communication satellite) A method of determining the position of an observer on earth using the known (exact) satellite transmission frequency and the known satellite ephemeris and measuring the Doppler frequency shift of the signal received from the satellie. (COM) [19]

Doppler VOR (navigation aid terms) (very high-frequency omnidirectional range) A very high frequency radio range, operationally compatible with conventional VOR, less susceptible to siting difficulties because of its increased aperture. In it the variable signal (the signal producing azimuthal information) is developed by sequentially feeding a radio frequency signal to a multiplicity of antennas disposed in a ring-shaped array; the array usually surrounds the central source of reference signal. (AES/GCS) 172-1983w

dormancy (accelerometer) (gyros) The state wherein a device is connected to a system in the normal operational configuration and experiences below-normal, often periodic structural, mechanical, electrical, or environmental stresses for prolonged periods before being used in a mission. Dormancy consists of a long, predominantly inactive, period where material and component degradation effects due to age and/or storage environment dominate. (AES/GYAC) 528-1994

DOR violation A word-serial protocol error that occurs when a servant receives a DOR synchronized command (such as *byte request*) and is unable to process that command because the DOR bit is zero (0). (C/MM) 1155-1992

dose (A) (photovoltaic power system) The radiation delivered to a specified area of the whole body. *Note:* Units of dose are rads or roentgens for X or gamma rays and rads for beta rays and protons. *See also:* photovoltaic power system. **(B)** The amount of a chemical or other agent delivered to an organism; usually normalized to the mass of an organism. (AES/T&D/PE) [41], 539-1990

dose equivalent The product, H, of the absorbed dose, D, and the quality factor, Q, at the point of interest in tissue.

$$H = DQ$$

The shallow and deep dose equivalents, H_s and H_d, are the dose equivalents at depths in tissue of 0.07 mm and 10 mm, respectively. The SI unit of dose equivalent has been given the special name of sievert (Sv).

$$1\text{Sv} = 1\text{J/kg}$$

Note: For photon and beta radiation, Q may be taken as equal to unity for external radiations. (NI) N42.20-1995

dose equivalent rate The quotient ($\dot{H}$) of dH by dt, where dH is the increment of dose equivalent in the time interval dt.

$$\dot{H} = dH/dt$$

(NI) N42.20-1995

dose rate (1) (A) (metal-nitride-oxide field-effect transistor) The time rate of deposition of radiation expressed in rads in SI per second. The integration of dose rate over a specified period of time represents the total dose. **(B)** The time rate of deposition of radiation expressed in grays in silicon per second. *Note:* The integration of dose rate over a specified period of time represents the total dose. (ED) 581-1978, 641-1987
(2) (photovoltaic power system) Radiation dose delivered per unit time. *See also:* photovoltaic power system. (AES) [41]
(3) The time rate of deposition of ionizing radiation. (ED) 1005-1998

dose-response relationship The relationship between dose and magnitude (or frequency) of response. (T&D/PE) 539-1990

dosimetry (A) The determination of dose or dose rate arising from an experimental or environmental exposure. **(B)** The comparison of dose or dose rate in one experiment (or situation) with that in another. (T&D/PE) 539-1990

dot (1) The filename consisting of a single dot character (.). In the context of shell special built-in utilities. *See also:* pathname resolution. (C/PA) 9945-2-1993
(2) The filename consisting of a single dot character (.). *See also:* pathname resolution. (C/PA) 9945-1-1996, 1003.5-1999

dot cycle (data transmission) One cycle of a periodic alternation between two signaling conditions, with each condition having unit duration. *Note:* Thus, in two-condition signaling, it consists of a dot, or marking element followed by a spacing element. (PE) 599-1985w

dot-dot (1) The filename consisting solely of two dot characters (..). *See also:* pathname resolution. (C/PA) 9945-1-1996, 9945-2-1993
(2) The filename consisting of (..). (C) 1003.5-1999

Dot4 Transducer Block An instance of the class `IEEE1451_Dot4TransducerBlock` or of a subclass thereof. (IM/ST) 1451.1-1999

dot matrix printer An impact printer in which each character is represented by a pattern of dots selected from a matrix of available dot positions. *Synonyms:* wire printer; dot printer; matrix printer. *Contrast:* formed character printer. (C) 610.10-1994w

dot notation A technique for naming that joins the name of a parent class to the name of a dependent class with the period character. For example, the diagram feature reference ***ABC/A31.3*** uses dot notation to join the page reference of the parent diagram ***ABC/A31*** to the feature reference for box 3 in that diagram. (C/SE) 1320.1-1998

dot printer *See:* dot matrix printer.

dot product *See:* scalar product.

dot-product line integral *See:* line integral.

dot sequential (color television) Sampling of primary colors in sequence with successive picture elements. (BT/AV) 201-1979w

dot signal (data transmission) A series of binary digits having equal and opposite states, such as a series of alternate 1 and 0 states. (PE) 599-1985w

Dot3 Transducer Block An instance of the class `IEEE1451_Dot3TransducerBlock` or of a subclass thereof. (IM/ST) 1451.1-1999

Dot2 Transducer Block Denotes an instance of the class `IEEE1451_Dot2TransducerBlock` or of a subclass thereof. (IM/ST) 1451.1-1999

DotX Transducer Block A Transducer Block implementing one of the IEEE 1451.X family of standards. (IM/ST) 1451.1-1999

double address *See:* indirect address.

double-address instruction *See:* two-address instruction.

double aperture seal (electric penetration assemblies) Two single aperture seals in series. (PE/NP) 317-1983r

double-blind study To reduce possible effects of bias, a study in which neither the experimenter nor the subject knows what treatment an individual subject receives until after the experiment has been completed. (T&D/PE) 539-1990

double blind matrix spike A matrix spike sample sent through normal processing wherein the processor has no knowledge that the sample is of quality assurance origin. (NI) N42.23-1995

double blind quality assurance sample Similar to a blind sample except that the identity as well as the composition (quantity and identity of analytes) are unknown to the analyst. (NI) N42.23-1995

double break relay contacts A contact form in which one contact is normally closed in simultaneous connection with two other contacts. (EEC/REE) [87]

double-break switch One that opens a conductor of a circuit at two points. (SWG/PE) C37.100-1992

double bridge *See:* Kelvin bridge.

double-buffered (hybrid computer linkage components) A digital-to-analog converter (DAC) or a digital-to-analog multiplier (DAM) with two registers in cascade, one a holding register, and the other the dynamic register. (C) 166-1977w

double-circuit system (protective signaling) A system of protective wiring in which both the positive and the negative sides of the battery circuit are employed, and that utilizes either an open or a short circuit in the wiring to initiate an alarm. *See also:* protective signaling. (EEC/PE) [119]

double click To press and release a mouse button twice in rapid succession without moving the pointer. (C) 1295-1993w

double connection (telephone switching systems) A fault condition whereby two separate calls are connected together. (COM) 312-1977w

double crucible method (fiber optics) A method of fabricating an optical waveguide by melting core and clad glasses in two suitably joined concentric crucibles and then drawing a fiber from the combined melted glass. *See also:* chemical vapor deposition technique. (Std100) 812-1984w

double-current generator A machine that supplies both direct and alternating currents from the same armature winding. (PE) [9]

double-density disk A floppy disk that is capable of storing information at twice the density of a single-sided disk. *See also:* high-density disk. (C) 610.10-1994w

double diode An electron tube or valve containing two diode systems. *See also:* multiple tube. (ED) [45]

double electric conductor seal (electric penetration assemblies) Two single electric conductor seals in series. (PE/NP) 317-1983r

double-end control A control system in which provision is made for operating a vehicle from either end. *See also:* multiple-unit control. (EEC/PE) [119]

double-ended queue (deque) (deque is pronounced "deck") A list whose contents may be changed by adding or removing items at either end. *Note:* This data structure is inaccurately named because it contradicts the definition of queue. (C) 610.5-1990w

double-faced tape Fabric tape finished on both sides with a rubber or synthetic compound. (PE/T&D) [10]

double-fed asynchronous machine (rotating machinery) An asynchronous machine of which the stator winding and the rotor winding are fed by supply frequencies each of which may be either constant or variable. *See also:* asynchronous machine. (PE) [9]

double-gun cathode-ray tube A cathode-ray tube containing two separate electron-gun systems. (ED) [45]

double hashing Open-address hashing in which collision resolution is handled by using a second hash function to determine a fixed increment and adding multiples of this increment to the original hash value until an empty position is found in the hash table. (C) 610.5-1990w

double insulation Insulation comprising both basic insulation and supplementary insulation. (EMB/MIB) 1073.4.1-2000

double-integrating gyro A single-degree-of-freedom gyro having no intentional elastic or viscous restraint of the gimbal about the output axis so that the dynamic behavior is primarily established by the inertial properties of the gimbal. In this gyro, an output signal is produced by gimbal angular displacement, relative to the case, which is proportional to the double integral of the angular rate of the case about the input axis. (AES/GYAC) 528-1984s

double length Pertaining to twice the normal length of a unit of data or a storage device in a given computing system. *Note:* For example, a double-length register would have the capacity to store twice as much data as a single-length or normal register; a double-length word would have twice the number of characters or digits as a normal or single-length word. *See also:* double precision. (C) 162-1963w, 610.5-1990w

double-length register Two registers that function as a single register that may be used for storing the product of multiplication, storing the partial quotient in division, or for accessing the left-hand or the right-hand portions in character string manipulation. *Synonym:* double register. *See also:* triple-length register; n-tuple length register; quadruple-length register. (C) 610.10-1994w

double make relay contacts A contact form in which one contact, which is normally open, makes simultaneous connection when closed with two other independent contacts. (EEC/REE) [87]

double modulation The process of modulation in which a carrier wave of one frequency is first modulated by a signal wave and a resultant wave is then made to modulate a second carrier wave of another frequency. (EEC/PE) [119]

double-operand instruction *See:* two-address instruction.

double pole-piece magnetic head (electroacoustics) A magnetic head having two separate pole pieces in which pole faces of opposite polarity are on opposite sides of the medium. *Note:* One or both of these pole pieces may be provided with an energizing winding. *See also:* phonograph pickup. (SP) [32]

double-pole relay A term applied to a contact arrangement to denote that it includes two separate contact forms: that is, two single-pole contact assemblies. (EEC/REE) [87]

double precision (mathematics of computing) Pertaining to the use of two computer words to represent a number in order to preserve or gain precision. *Synonym:* double length. *Contrast:* multiple precision; triple precision; single precision. (C) 610.5-1990w, 1084-1986w

double-precision addition Computer addition performed with operands that are expressed in double-precision representation. (C) 1084-1986w

double-precision arithmetic Computer arithmetic performed with operands that are expressed in double-precision representation. (C) 1084-1986w

double pulse (pulse terminology) Two pulse waveforms of the same polarity which are adjacent in time and which are considered or treated as a single feature. (IM/WM&A) 194-1977w

double-pulse recording A variation of phase-modulation recording resulting in unmagnetized regions on each side of the magnetized regions. (C) 610.10-1994w

double quote The character "″", also known as *quotation mark*. *Note:* The "double" adjective in this term refers to the two strokes in the character glyph. This standard never uses the term *double quote* to refer to two apostrophes or quotation marks. (C/PA) 9945-2-1993

double-rail logic A family of circuit logic in which each logic variable is represented by two electrical lines which together can take on three meaningful states: zero, one, and undecided. *Synonyms:* dual-rail logic; two-rail logic. (C) 610.10-1994w

double register *See:* double-length register.

double-secondary current transformer One that has two secondary windings each on a separate magnetic circuit with both magnetic circuits excited by the same primary winding. (PE/TR) C57.13-1993, [57]

double-secondary voltage transformer (1) (instrument transformers) (power and distribution transformers) One that has two secondary windings on the same magnetic circuit insulated from each other and the primary. C57.12.80-1978r **(2)** One that has two secondary windings on the same magnetic circuit with the secondary winding insulated from each other. (PE/TR) C57.13-1993

double sideband transmitter (data transmission) A transmitter that transmits the carrier frequency and both sidebands resulting from the modulation of the carrier by the modulating signal. (PE) 599-1985w

double-sided board A board with components on both sides. (C/BA) 14536-1995

double-sided disk A floppy disk that utilizes both of its sides for information storage. *Synonym:* flippy. *Contrast:* single-sided disk. (C) 610.10-1994w

double squirrel cage (rotating machinery) A combination of two squirrel-cage windings mounted on the same induction-motor rotor, one at a smaller diameter than the other. *Note:* It is common but not essential for the two windings to have thesame number of slots. In any case, each bar of the lower (smaller-diameter) winding is located at the bottom of a slot containing a bar of the upper winding. A narrow portion of the slot (called the leakage slot) is provided in the radial separation between the two bars. *See also:* asynchronous machine. (PE) [9]

double (split) electrode Quarter-wavelength-spaced (center-to-center) fingers, typically one-eighth wavelength wide, used to reduce reflections from transducers. (UFFC) 1037-1992w

double storage A method for storing and recovering information from storage in which duplicate copies of the data are stored in physically independent memory units; if one unit fails, the data can be retrieved from the second. *Synonym:* disk mirroring. (C) 610.10-1994w

double-superheterodyne reception (triple detection) The method of reception in which two frequency converters are employed before final detection. *See also:* radio receiver. (EEC/PE) [119]

doublet (1) A group of two adjacent digits operated upon as a unit. (C) 610.5-1990w, 1084-1986w
(2) A set of adjacent bytes. (C/BA) 896.3-1993w
(3) A unit of computer data consisting of 16 bits. (C/BA) 1275-1994
(4) Two bytes (16 bits) of data. (C/MM) 1754-1994, 1596-1992, 1596.4-1996, 1394-1995, 13213-1994, 1394a-2000
(5) A set of two adjacent bytes. (C/BA) 10857-1994, 896.4-1993w
(6) An ordered set of two adjacent bytes. (C/BA) 1014.1-1994w
(7) A byte composed of two bits. *Synonym:* two-bit byte. (C) 610.10-1994w
(8) A data format or data type that is 2 bytes in size. (C/MM) 1596.5-1993

double talk (DT) Two talkers speaking simultaneously in opposite transmission directions. (COM/TA) 1329-1999

doublet antenna *See:* dipole antenna.

double-throw (as applied to a mechanical switching device) A qualifying term indicating that the device can change the circuit connections by utilizing one or the other of its two operating positions. *Note:* A double-throw air switch changes circuit connections by moving the switchblade from one of two sets of contact clips into the other. (SWG/PE) C37.100-1992

double-throw relay A term applied to a contact arrangement to denote that each contact form included is a break-make. (EEC/REE) [87]

double thyristor converter unit Two converters connected to a common dc circuit such that this circuit can accept or give up energy with direct current in both directions. *Note:* The converters may be supplied from separate cell windings on a common transformer, from common cell windings, or from separate transformers, The converter connections may be single way or symmetrical double way. Where two converters are involved, the designated forward converter arbitrarily operates in quadrant I. Quadrant I implies motoring torque in the agreed-upon forward direction. (IA/IPC) 444-1973w

doublet impulse (automatic control) An impulse having equal positive and negative peaks. (PE/EDPG) [3]

double-tuned amplifier An amplifier of one or more stages in which each stage utilizes coupled circuits having two frequencies of resonance, for the purpose of obtaining wider bands than those obtainable with single tuning. *See also:* amplifier. (EEC/PE) [119]

double-tuned circuit A circuit whose response is the same as that of two single-tuned circuits coupled together. (EEC/PE) [119]

double-way rectifier (power semiconductor) A rectifier unit which makes use of a double-way rectifier circuit. (IA) [12]

double-way rectifier circuit A rectifier circuit in which the current between each terminal of the alternating-voltage circuit and the rectifier circuit elements conductively connected to it flows in both directions. *Note:* The terms single-way and double-way provide a means for describing the effect of the rectifier circuit on current flow in transformer windings connect to rectifier circuit elements. Most rectifier circuits may be classified into these two general types. Double-way rectifier circuits are also referred to as bridge rectifier circuits. *See also:* single-way rectifier circuit; power rectifier; bridge rectifier circuit; rectifier circuit element; rectification. (IA) [62]

double-winding synchronous generator A generator that has two similar windings, in phase with one another, mounted on the same magnetic structure but not connected electrically, designed to supply power to two independent external circuits. (PE) [9]

double word (1) A sequence of contiguous bits or characters that comprise two computer words and that may be addressed as a unit. (C) 610.10-1994w, 610.5-1990w
(2) An aligned octlet. *Note:* The definition of this term is architecture-dependent, and so may be different from that used in other processor architectures. (C/MM) 1754-1994
(3) Eight bytes or 64 bits operated on as a unit. The most significant byte carries the index value 0 and the least significant byte carries the index value 7. (C/BA) 1496-1993w
(4) A field composed of two words. In a message, the most significant word is transmitted/received first. It is capable of describing integers in the decimal range −2 147 483 648 to 2 147 483 647. (C/MM) 1284.1-1997

doubly-chained tree *See:* doubly-threaded tree.

doubly-linked list A linked list in which each item contains two pointers, one pointing forward to the next item in the list, and one pointing backward to the previous item in the list. *Synonym:* two-way chain. (C) 610.5-1990w

doubly-threaded tree A binary tree in which each node contains two link fields; one each for its successor and predecessor nodes with respect to some traversal. *Synonym:* doubly-chained tree. *Contrast:* triply-threaded tree. (C) 610.5-1990w

doughnut *See:* toroid.

dovetail projection A tenon, commonly flared; used for example, to fasten a pole to the spider. *See also:* stator. (PE) [9]

dovetail slot (A) A recess along the side of a coil slot into which a coil-slot wedge is inserted. *See also:* stator. **(B)** A flaring slot into which a dovetail projection is engaged; used for example, to fasten a pole to the spider. *See also:* stator. (PE) [9]

dowel A pin fitting with close tolerance into a hole in abutting pieces to establish and maintain accurate alignment of parts. Frequently designed to resist a shear load at the interface of the abutting pieces. (PE) [9]

down (1) Pertaining to a system or component that is not operational or has been taken out of service. *Contrast:* up. *See also:* idle; busy; crash. (C) 610.12-1990
(2) A colloquial expression used in reference to a system or a system component that is not functioning. *Contrast:* up. (C) 610.10-1994w

downconverter (nonlinear, active, and nonreciprocal waveguide components) A heterodyne frequency conversion device that converts an input signal to a lower frequency output signal. (MTT) 457-1982w

downdate Installation of software with a revision older than that of the software currently installed in the same location. This is also referred to as downgrading or reverting. (C/PA) 1387.2-1995

down lead (lightning protection) The conductor connecting an overhead ground wire or lightning conductor with the grounding system. *See also:* direct-stroke protection. (T&D/PE) [10]

downlight (illuminating engineering) A small direct lighting unit which directs the light downward and can be recessed, surface mounted, or suspended. (EEC/IE) [126]

downlink The transmission medium between a repeater and a connected end node or lower level repeater, as viewed from the local repeater. *Contrast:* uplinklocal area networks. (C) 8802-12-1998

down link (communication satellite) A transmission link carrying information from a satellite or spacecraft to earth. Typically down links carry telemetry, data and voice. (COM) [24]

download (A) To transfer some collection of data from a computer memory to another storage location. **(B)** To transfer some collection of data from the memory of one computer to the memory of a second computer that is relatively smaller than the first; for example, to transfer data from a mainframe computer to a microcomputer. (C) 610.5-1990

downloadable font A font that must be downloaded from a computer to a printer each time the font is to be used. *Synonym:* soft font. *Contrast:* online font. (C) 610.10-1994w

downstream The direction of data flow along a bus, i.e., away from the Head of Bus function. (LM/C) 8802-6-1994

down time (1) (A) (supervisory control, data acquisition, and automatic control) The time during which a device or system is not capable of meeting its functional requirements. **(B) (software)** The period of time during which a system or component is not operational or has been taken out of service. *See also:* busy time; mean time to repair; idle time; setup time. **(C) (broadband local area networks)** The time interval during which a network is not available for access by the user. (LM/PE/SUB/C) C37.1-1994, 610.12-1990, 802.7-1989
(2) The period during which a device cannot be operated due to a fault within itself or within the environment. *Synonym:* fault time. *See also:* inoperable time; environmental loss time; unavailable time. (C) 610.10-1994w
(3) The time during which a device or system is not capable of meeting performance requirements. (SWG/PE) C37.100-1992

downward compatible (1) Pertaining to hardware or software that is compatible with an earlier or less complex version of itself; for example, a program that handles files created by an earlier version of itself. *Contrast:* upward compatible. (C) 610.12-1990
(2) Pertaining to hardware or software that is compatible with an earlier or less complex version of itself. For example, if an early version of a program can handle files from a later version, the later version is said to be "downward compatible." *Contrast:* upward compatible. (C) 610.10-1994w, 610.12-1990

downward component (illuminating engineering) That portion of the luminous flux from a luminaire that is emitted at angles below the horizontal. (EEC/IE) [126]

downward compression In software design, a form of demodularization in which a superordinate module is copied into the body of a subordinate module. *Contrast:* lateral compression; upward compression. (C) 610.12-1990

downward modulation Modulation in which the instaneous amplitude of the modulated wave is never greater than the amplitude of the unmodulated carrier. 188-1952w

DP *See:* dial pulse; data processing.

DPCA *See:* displaced phase center antenna.

DPSK *See:* differential-phase-shift keying.

DQDB layer The sublayer in this part of ISO/IEC 8802 that uses the services of the Physical Layer to provide the following: Medium Access Control (MAC) Sublayer service to the Logical Link Control (LLC) Sublayer, and isochronous service, and connection-oriented data service. (LM/C) 8802-6-1994

DR *See:* dead reckoning.

draft gauge (navigation aid terms) A hydrostatic instrument installed in vessels to indicate the depth to which a vessel is submerged. (AES/GCS) 172-1983w

draft quality Pertaining to printed output that is readable but not of extremely high quality. *Note:* May be used for internal communication and rough drafts. *Contrast:* near-letter quality; letter-quality. (C) 610.10-1994w

drag-in (electroplating) The quantity of solution that adheres to cathodes when they are introduced into a bath. *See also:* electroplating. (PE/EEC) [119]

dragging The process of moving a selected display element on a display surface from one position to another with a locator. (C) 610.6-1991w

drag magnet *See:* retarding magnet.

drag-out (electroplating) The quantity of solution that adheres to cathodes when they are removed from a bath. *See also:* electroplating. (EEC/PE) [119]

drain (1) (general) The current supplied by a cell or battery when in service. *See also:* battery. (EEC/PE) [119]
(2) (metal-nitride-oxide field-effect transistor) Region in the device structure of an insulated-gate field-effect transistor (IGFET) which contains the terminal into which charge carriers flow from the source through the channel. It has the potential which is more attractive than the source for the carriers in the channel. (ED) 581-1978w

drainage Conduction of current (positive electricity) from an underground metallic structure by means of a metallic conductor. (IA) [59]

drainage reactors *See:* drainage units.

drainage units Center-tapped inductive devices designed to relieve conductor-to-conductor and conductor-to-ground voltage stress by draining extraneous currents to ground. They are also designed to serve the purpose of a mutual drainage reactor forcing near-simultaneous protector-gap operation. *Synonym:* drainage reactors. (PE/PSC) 487-1992

drain erase disturb Floating gate to drain charge loss, on an erased cell, when the bit-line is high for programming or erasing another location on the same bit-line. (ED) 1005-1998

drain line (rotating machinery) (bearing oil system) A return pipe line using gravity flow. *See also:* oil cup. (PE) [9]

drawbar pull (cable plowing) The effective pulling force delivered. (T&D/PE) 590-1977w

drawbridge coupler *See:* movable bridge coupler.

drawdown (power operations) The distance that the water surface of a reservoir is lowered from a given elevation as the result of the withdrawal of water. (PE/PSE) 858-1987s

draw-lead bushing A bushing that will allow the use of a draw-lead conductor. (PE/TR) C57.19.03-1996

draw-lead conductor A cable or solid conductor that has one end connected to the transformer or reactor winding and the other end drawn through the central tube of the bushing and connected to the top of the bushing. (PE/TR) C57.19.03-1996

drawout circuit breaker Circuit breaker equipped with slides or rollers and quick disconnect means to facilitate removal and replacement. (PE/EDPG) 1020-1988r

drawout-mounted device One having disconnecting devices and in which the removable portion may be removed from the stationary portion without the necessity of unbolting connections or mounting supports. *Contrast:* stationary-mounted device. (SWG/PE) C37.100-1992

D region The region of the terrestrial ionosphere between about 50 km and 90 km altitude. *Note:* The D region is responsible for most of the daytime attenuation of LF, MF and HF radio waves, and it forms the upper boundary of the Earth-ionosphere waveguide for VLF waves. (AP/PROP) 211-1997

dribble bits Extra bits added to the end of a packet that allow extra synchronization in implementations. (C/MM) 1394-1995

drift (1) (rotating machinery) A long-time change in synchronous-machine resulting system error resulting from causes such as aging of components, self-induced temperature changes, and random phenomena. *Note:* Maximum acceptable drift is normally a specified change for a specified period of time, for specified conditions. (PE) [9]
(2) An undesired but relatively slow change in output over a period of time, with a fixed reference input. *Note:* Drift is usually expressed in percent of the maximum rated value of the variable being measured. *See also:* control system; feedback. (IA/ICTL/APP/IAC) [69], [60]
(3) (sound recording and reproducing) Frequency modulation of the signal in the range below approximately 0.5 Hz resulting in distortion that may be perceived as a slow changing of the average pitch. *Note:* Measurement of drift is not covered by this definition. (SP) 193-1971w
(4) (analog computer) In an analog computer, a slowly varying error in an integrator, caused by the integration of offset errors at the inputs, capacitor leakage, or both. Also, any slowly varying error in a computer component. (C) 165-1977w
(5) Change in readout, usually gradual, without concomitant change in the influence quantity. (NI) N42.17B-1989r
(6) The latent tendency of control system output to digress from the desired effect. (C) 610.2-1987
(7) The unwanted change of the value of an output signal of a device over a specified period of time when the values of all input signals of the device are kept constant. *See also:* zero drift. (C) 610.10-1994w
(8) (A) Drift angle. **(B) (navigation aid terms)** Component of a vehicle's ground speed perpendicular to heading. **(C)** Distance a craft is moved by current and wind. (AES/GCS) 172-1983
(9) (oscilloscopes) *See also:* stability.
(10) (electronic navigation) *See also:* G drift.

drift angle (navigation aid terms) The angular difference between the heading and the track. (AES/GCS) 172-1983w

drift band of amplification (magnetic amplifier) The maximum change in amplification due to uncontrollable causes for a specified period of time during which all controllable quantities have been held constant. *Note:* The units of this drift band are the amplification units per the time period over which the drift band was determined. (MAG) 107-1964w

drift compensation The effect of a control function, device, or means to decrease overall systems drift by minimizing the drift in one or more of the control elements. *Note:* Drift compensation may apply to feedback elements, reference input, or other portions of a system. *See also:* feedback control system. (IA/ICTL/IAC) [60]

drift correction angle (navigation aid terms) The angular difference between the course and the heading. *Synonyms:* crab angle; correction angle. (AES/GCS) 172-1983w

drift, direct-current *See:* drift.

drift, kinematic *See:* misalignment drift.

drift mobility (homogeneous semiconductor) The ensemble average of the drift velocities of the charge-carriers per unit electric field. *Note:* In general, the mobilities of electrons and holes are different. *See also:* semiconductor device. (ED) 216-1960w

drift offset (magnetic amplifier) The change in quiescent operating point due to uncontrollable causes over a specified period of time when all controllable quantities are held constant. (MAG) 107-1964w

drift rate (1) (voltage regulators or reference tubes) The slope at a stated time of the smoothed curve of tube voltage drop with time at constant operating conditions. (ED) 161-1971w
(2) (gyros) The time rate of output deviation from the desired output. It consists of random and systematic components and is expressed as an equivalent input angular displacement per unit time with respect to inertial space. It is typically expressed in degrees per hour (°/h). (AES/GYAC) 528-1994
(3) (of a clock) The rate at which the time measured by a clock deviates from the actual passage of real time. A positive drift rate causes a clock to gain time with respect to real time; a negative drift rate causes a clock to lose time with respect to real time. (C/PA) 9945-1-1996

drift space (electron tube) A region substantially free of externally applied alternating fields, in which a relative repositioning of the electrons takes place. (ED) 161-1971w

drift, stability *See:* stability drift.

drift stabilization (1) (analog computer) Any automatic method used to minimize the drift of a dc amplifier. (C) 165-1977w
(2) In an analog computer, any automatic method used to minimize the drift of an amplifier. (C) 610.10-1994w

drift tunnel (velocity-modulated tube) A piece of metal tubing, held at a fixed potential, that forms the drift space. *Note:* The drift tunnel may be divided into several parts, which constitute the drift electrodes. *See also:* velocity-modulated tube. (EEC/ACO) [109], [84]

drift velocity (semiconductor) (nonlinear, active, and nonreciprocal waveguide components) A velocity component in a doped semiconductor that occurs when an electric field exists within the semiconductor material. This component is superimposed upon the thermal carrier's random motion. For electrons, the drift velocity will have a direction opposite to the electric field. (MTT) 457-1982w

drift, zero *See:* zero drift.

drill and practice interaction An instruction method employed by some computer-assisted instruction systems, in which the student is asked repeatedly to perform the same or similar tasks. (C) 610.2-1987

D-ring A connector used integrally in a line-worker's body belt, aerial belt, or full body harness as an attachment element and in lanyards, energy absorbers, lifelines and, non-permanent boom attachments. To be in accordance with ASTM F 887-91a. (T&D/PE) 1307-1996

dripproof So constructed or protected that successful operation is not interfered with when faling drops of liquid or solid particles strike or enter the enclosure at any angle from 0 to 15 degrees from the downward vertical unless otherwise specified. *See also:* asynchronous machine. (SWG/IA/PE/MT) 45-1983s, [9], C37.100-1981s

dripproof enclosure (1) (metal-enclosed bus and calculating losses in isolated-phase bus) An enclosure usually for indoor application, so constructed or protected that falling drops of liquid or solid particles that strike the enclosure at any angle not greater than 15 degrees from the vertical will not be able to interfere with the successful operation of the metal-enclosed bus. (SWG/PE) C37.23-1987r
(2) (power and distribution transformers) An enclosure, usually for indoor application, so constructed or protected that falling drops of liquid or solid particles that strike the enclosure at any angle within a specified variation from the vertical shall not interfere with the successful operation of the enclosed equipment. (PE/TR) C57.12.80-1978r
(3) An enclosure in which the openings are so constructed that drops of liquid or solid particles falling on the enclosure at any angle not greater than 15° from the vertical either cannot enter the enclosure, or if they do enter the enclosure, they will not prevent the successful operation of, or cause damage to, the enclosed equipment. (IA/MT) 45-1998

dripproof machine An open machine in which the ventilating openings are so constructed that drops of liquid or solid particles falling on the machine at any angle not greater than 15 degrees from the vertical cannot enter the machine either directly or by striking and running along a horizontal or inwardly inclined surface. *See also:* asynchronous machine. (EEC/PE) [119]

driptight So constructed or protected as to exclude falling dirt or drops of liquid, under specified test conditions. (PE/TR) C57.12.80-1978r

driptight enclosure An enclosure so constructed that falling drops of liquid or solid particles striking the enclosure at any angle within a specified variation from the vertical cannot enter the enclosure either directly or by striking and running along a horizontal or inwardly inclined surface.
(SWG/PE/TR) C37.30-1971s, C57.12.80-1978r, C37.100-1981s

drive (1) The equipment used for converting available power into mechanical power suitable for the operation of a machine. *See also:* electric drive.
(PE/IA/ICTL/NP/APP) 317-1976s, [75]
(2) (NuBus) A module activity causing a bus signal line to be in a particular sate. (C/MM) 1196-1987w
(3) A piece of hardware that allows access to data that is stored on media. A set of devices such as RAID may be considered as a single drive. (C/SS) 1244.1-2000
(4) (electronic computation) *See also:* tape drive.
NuBus is a registered trademark of Texas Instruments, Inc.

drive mechanism The means by which the LTC is actuated.
(PE/TR) C57.131-1995

driven element (1) A radiating element coupled directly to the feed line. *See also:* antenna. (AP/ANT) [35]
(2) A radiating element coupled directly to the feed line of an antenna. (AP/ANT) 145-1993

driven stability (solution $\phi(u;t)$). For each bounded system input perturbation $\Delta u(t)$ the output perturbation $\Delta\phi$ is also bounded for $t \geq t_0$. *Note:* A necessary and sufficient condition for a solution of a linear system to be driven-stable is that the solution be uniformly asymptotically stable. See **stability** for explanation of symbols. *See also:* control system.
(CS/IM) [120]

drive pattern (facsimile) Density variation caused by periodic errors in the position of the recording spot. When caused by gears this is called gear pattern. *See also:* recording.
(COM) 167-1966w

drive pin (disk recording) A pin similar to the center pin, but located to one side thereof, that is used to prevent a disc record from slipping on the turntable. *See also:* phonograph pickup. (SP) [32]

drive-pin hole (disk recording) A hole in a disc record that accommodates the turntable drive pin. (SP) [32]

drive pulse (static magnetic storage) A pulsed magnetomotive force applied to a magnetic cell from one or more sources.
(Std100) 163-1959w

driver (1) (communication practice) An electronic circuit that supplies input to another electronic circuit.
(PE/EEC) [119]
(2) (A) (software) A software module that invokes and, perhaps, controls and monitors the execution of one or more other software modules. **(B) (software)** A computer program that controls a peripheral device and, sometimes, reformats data for transfer to and from the device. *See also:* test driver.
(C) 610.12-1990
(3) A program, circuit or device used to power or control other programs, circuits or devices. *See also:* bus driver; device driver. (C) 610.10-1994w
(4) An electrical circuit whose purpose is to signal a binary state for transmitting information. Also referred to as a "generator" in international standards. (MM/C) 1596.3-1996
(5) A pin of a cell instance that, in the current context, is placing or can place a signal onto an interconnect structure.
(C/DA) 1481-1999

driver name The component of a node name that corresponds to the value of the device's "`name`" property.
(C/BA) 1275-1994

drive strip (rotating machinery) (chafing strip) An insulating strip located in the coil slots between the wedge and the top of the slot armor or the top coil side, to provide protection during assembly of the wedges. *See also:* stator. (PE) [9]

driving agent The agent that is permitted to assert or negate a signal on the bus. (C/MM) 1296-1987s

driving edge A time corresponding to a rising edge of the bus clock. (C/MM) 1196-1987w

driving machine The power unit that applies the energy necessary to raise and lower an elevator or dumbwaiter car or to drive an escalator or a private-residence inclined lift.
(EEC/PE) [119]

driving-point admittance (between the *j*th terminal and the reference terminal of an *n*-terminal network) The quotient of the complex alternating component I_j of the current flowing to the *j*th terminal from its external termination by the complex alternating component V_j of the voltage applied to the *j*th terminal with respect to the reference point when all other terminals have arbitrary external terminations. *Note:* In specifying the driving-point admittance of a given pair of terminals of a network or transducer having two or more pairs of terminals, no two pairs of which contain a common terminal, all other pairs of terminals are connected to arbitrary admittances. *See also:* network analysis; electron-tube admittances.
(ED) 161-1971w

driving-point function (linear passive networks) A response function for which the variables are measured at the same port (terminal pair). (CAS) 156-1960w

driving-point impedance (networks) At any pair of terminals the ratio of an applied potential difference to the resultant current at these terminals, all terminals being terminated in any specified manner. *See also:* self-impedance.
(EEC/PE) [119]

driving power, grid *See:* grid driving power.

driving signals (television) Signals that time the scanning at the pickup point. *Note:* Two kinds of driving signals are usually available from a central synchronizing generator. One is composed of pulses at line frequency and the other is composed of pulses at field frequency. *See also:* television.
(BT/AV) [34]

driving test circuit (measuring longitudinal balance of telephone equipment operating in the voice band) A test circuit used to convert an exciting test voltage into balanced longitudinal voltages on tip and ring leads.
(COM/TA) 455-1985w

droop* *See:* pulse distortion.
* Deprecated.

droop, frequency *See:* frequency droop.

drop (data transmission) A connection made between a through transmission circuit and a local terminal unit.
(PE) 599-1985w

drop-away The electrical value at which the movable member of an electromagnetic device will move ot its de-energized position. (EEC/PE) [119]

drop cable (1) The cable assembly that connects a distribution tap to a user outlet. The cable is usually a flexible RG type 75 Ω coax. (LM/C) 802.7-1989r
(2) The small diameter flexible coaxial cable of the broadband medium that connects to a medium access unit. *See also:* attachment unit interface cable; transceiver cable; coaxial cable; trunk cable. (C) 610.7-1995
(3) In 10BROAD36, the small diameter flexible coaxial cable of the broadband medium that connects to a Medium Attachment Unit (MAU). *See also:* trunk cable.
(C/LM) 802.3-1998

drop-in In the storage and retrieval of data from a magnetic storage device, an error revealed by the reading of a binary character not previously recorded. *Note:* Usually caused by defects or the present of particles in the magnetic surface layer. (C) 610.10-1994w

drop line The type or specific cable that is used in a drop cable.
(LM/C) 802.7-1989r

dropout (1) (protective relaying of utility-consumer interconnections) Contact operation (opening or closing) as a relay just departs from pickup. The value at which dropout occurs is usually stated as a percentage of pickup. For example, dropout ratio of a typical instantaneous overvoltage relay is 90 percent. (PE/PSR) C37.95-1973s

(2) (of a relay) A term for contact operation (opening or closing) as a relay just departs from pickup. Also identifies the maximum value of an input quantity that will allow the relay to depart from pickup. (SWG/PE) C37.100-1992
(3) A loss of equipment operation (discrete data signals) due to noise, voltage sags, or interruption. (IA/PSE) 1100-1999

drop-out In the storage and retrieval of data from a magnetic storage device, an error due to the failure to read a binary character. *Note:* Usually caused by defects or the presence of particles in the magnetic surface layer. (C) 610.10-1994w

dropout fuse A fuse in which the fuseholder or fuse unit automatically drops into an open position after the fuse has interrupted the circuit. (SWG/PE) C37.40-1993, C37.100-1992

dropout ratio (of a relay) The ratio of dropout to pickup of an input quantity. *Note:* This term has been used mostly with relays for which reset is not differentiated from dropout. Hence a similar term, reset ratio, the ratio of reset to pickup, is not generally used, though technically correct. (SWG/PE) C37.100-1992

dropouts (1) (data transmission) A loss of discrete data signals due to noise or attenuation hits. (PE) 599-1985w
(2) A subset of gain hit, dropouts are a loss of data caused by a decrease in signal, usually greater than 12 dB and lasting longer than 4 ms. Dropouts interrupt the information flow between two modems. Furthermore, even when the signal returns, some modems will take additional time to recover. Experience suggests that dropouts occur less often than the other types of transients. However, each dropout causes modem errors. *See also:* gain hit or change; phase hit or change. (PE/IC) 1143-1994r

dropout time (of a relay) The time interval to dropout following a specified change of input conditions. *Note:* When the change of input conditions is not specified it is intended to be a sudden change from pickup value of input to zero input. (SWG/PE) C37.100-1992

dropout voltage (1) (or current) The voltage (or current) at which a magnetically operated device will release to its de-energized position. It is a level of voltage (or current) that is insufficient to maintain the device in an energized state. (IA/PSE) 446-1995
(2) The voltage at which a device will revert to its de-energized position, i.e., the voltage at which a device fails to operate. (IA/PSE) 1100-1999

drop, voltage, anode *See:* anode voltage drop.

drop, voltage, starter *See:* starter voltage drop.

drop, voltage, tube *See:* tube voltage drop.

drop wire (drop) (data transmission) A wire suitable for extending an open wire or cable pair from a pole or cable terminal to a building. (PE) 599-1985w

DRP *See:* eardrum reference point.

DRS *See:* Digital Reference Sequence.

DRT box *See:* anti-transmit-receive switch.

drum *See:* magnetic drum.

drum controller (electric installations on shipboard) An electric controller which utilizes a drum switch as the main switching element. A drum controller usually consists of a drum switch and a resistor. (IA/ICTL/MT/IAC) 45-1983s, [60]

drum factor (facsimile) The ratio of usable drum length to drum diameter. *Note:* Before a picture is transmitted, it is necessary to verify that the ratio of used transmitter drum length to transmitter drum diameter is not greater than the receiver drum factor if the receiver is of the drum type. *See also:* facsimile. (COM) [49]

drum plotter A plotter that draws an image on a display surface mounted on a rotating drum. (C) 610.10-1994w

drum puller (conductor stringing equipment) A device designed to pull conductors(s) during stringing operations. It is normally equipped with its own engine which drives the drum(s) mechanically, hydraulically, or through a combination of both. It may be equipped with synthetic fiber rope or wire rope to be used as the pulling line. The pulling line is payed out from the unit, pulled through the travelers in the sag section, and attached to the conductor. The conductor is then pulled in by winding the pulling line back onto the drum. This unit is sometimes used with synthetic fiber rope acting as a pilot line to pull heavier pulling lines across canyons, rivers, etc. *Synonyms:* hoist; tugger; hoist, single drum; winch, single-drum. (PE/T&D) 524a-1993r, 524-1992r

drum printer An element printer in which a full set of type slugs, placed on a rotating print drum, is made available for each printing position. (C) 610.10-1994w

drum reference point (DRP) A point located at the end of the ear canal, corresponding to the eardrum position. (COM/TA) 1329-1999

drum speed (facsimile) The angular speed of the transmitter of recorder drum. *Note:* This speed is measured in revolutions per minute. *See also:* recording; scanning. (COM) 168-1956w

drum switch A switch in which the electric contacts are made of segments or surfaces on the periphery of a rotating cylinder or sector, or by the operation of a rotating cam. *See also:* switch; control switch. (IA/ICTL/IAC) [60]

dry-arcing distance (insulators) The shortest distance through the surrounding medium between terminal electrodes, or the sum of the distances between intermediate electrodes, whichever is the shorter, with the insulator mounted for dry flash-over test. *See also:* insulator. (EEC/IEPL) [89]

dry bulb temperature The temperature of a gas, or a mixture of gases, that is indicated by an accurate thermometer after correction for radiation. (IA/PSE) 241-1990r

dry cell A cell in which the electrolyte is immobilized. *See also:* electrochemistry. (EEC/PE) [119]

dry-charged battery (lead-acid batteries for photovoltaic systems) A battery in which the electrolyte has been removed for ease in shipping or storage or both. (PV) 937-1987s

dry-charged cell A cell that does not contain electrolyte for ease in shipping or storage, or both. (SCC21) 937-2000

dry circuit relay Erroneously used for a relay with either dry or low-level contacts. *See also:* low-level relay contacts. (EEC/REE) [87]

dry contact One through which no direct current flows. (COM) 312-1977w, [48]

dry location A location not normally subject to dampness or wetness. A location classified as dry may be temporarily subject to dampness or wetness, as in the case of a building under construction. (NEC/NESC/IA/PC) [86], 515.1-1995

dry-niche lighting fixtures A lighting fixture intended for installation in the wall of the pool in a niche that is sealed against the entry of pool water by a fixed lens. (NESC/NEC) [86]

dry reed relay A reed relay with dry (nonmercury-wetted) contacts. (PE/EM) 43-1974s

dry relay contacts (A) Contacts which neither break nor make current. **(B)** Erroneously used for relay contacts, low level. (EEC/REE) [87]

dry-run test A test of switching or tripping sequences with equipment not energized or "dry." (PE/SUB) 1378-1997

dry-type (1) (grounding device) (current-limiting reactor) Having the coils immersed in an insulating gas. *See also:* reactor. (PE/SPD) 32-1972r, C57.16-1958w
(2) (regulator) Having the core and coils not immersed in an insulting liquid. *See also:* voltage regulator. (PE/TR) C57.15-1968s
(3) Having the core and coils neither impregnated with an insulating fluid nor immersed in an insulating oil. *See also:* dry-type transformer. (PE/TR) [57]

dry-type encapsulated water-cooled transformer (electrical heating applications to melting furnaces and forehearths in the glass industry) Dry-type (nonoil-cooled) transformer in which the windings are made from hollow conducting tubes that are cooled by transferring heat to water flowing in the tubes. Since the cooling is internal, the windings are usually totally encapsulated with epoxy resin. (IA) 668-1987w

dry-type forced-air-cooled transformer (power and distribution transformers) (class AFA) A dry-type transformer which derives its cooling by the forced circulation of air. (PE/TR) C57.12.80-1978r

dry-type instrument transformer *See:* dry-type.

dry-type nonventilated self-cooled transformer (power and distribution transformers) (class ANV) A dry-type self-cooled transformer which is so constructed as to provide no intentional circulation of external air through the transformer, and operating air through the transformer, and operating at zero gauge pressure. (PE/TR) C57.12.80-1978r, C57.15-1968s

dry-type self-cooled/forced-air-cooled transformer (power and distribution transformers) (Class AA/FA) A dry-type transformer which has a self-cooled rating with cooling obtained by the natural circulation of air and a forced-air-cooled rating with cooling obtained by the forced circulation of air. (PE/TR) C57.12.80-1978r

dry-type self-cooled shunt reactor (shunt reactors over 500 kVA) (class AA) A dry-type shunt reactor which is cooled by the natural circulation of the cooling air. (PE/TR) C57.21-1981s

dry-type self-cooled transformer (power and distribution transformers) (class AA) A dry-type transformer that is cooled by the natural circulation of air. (PE/TR) C57.12.80-1978r

dry-type shunt reactor (shunt reactors over 500 kVA) One in which the coils and magnetic circuit are neither impregnated with an insulating fluid nor immersed in an insulating oil. (PE/TR) C57.21-1981s

dry-type transformer (A) (electrical heating applications to melting furnaces and forehearths in the glass industry) A transformer that relies on convection or forced air rather than liquid, such as oil or water, for cooling. **(B) (power and distribution transformers)** A transformer in which the core and coils are in a gaseous or dry compound insulating medium. *See also:* ventilated dry-type transformer; sealed transformer; nonventilated dry-type transformer; compound-filled transformer; gas-filled transformer. (IA/PE/TR) 668-1987, C57.12.80-1978

dry vault A ventilated, enclosed area not subject to flooding. (SWG/PE) C37.100-1992

DS *See:* digital signal.

DSAP *See:* address fields.

D-scan *See:* D-display.

D-scope A cathode-ray oscilloscope arranged to present a D-display. (AES/RS) 686-1990

DS/DD *See:* double-density disk; double-sided disk.

DSE *See:* data switching exchange.

DS-4 *See:* digital signal level four.

DS/HD *See:* high-density disk; double-sided disk.

DS1 frame The DS1 frame consists of 193 bit positions, the first of which is the frame overhead bit position. The remaining 192 bits are available as payload, and can be divided into 24 blocks (channels) of 8 bits each. (COM/TA) 1007-1991r

D64 module A module whose data space is limited to a 64-bit width. (C/BA) 14536-1995

DSK *See:* Dvorak keyboard.

DSL *See:* data sublanguage; digital subscriber line.

DS-1 *See:* digital signal level one.

DSS *See:* decision support system; decision support services.

D-star (D*) (fiber optics) A figure of merit often used to characterize detector performance, defined as the reciprocal of noise equivalent power (NEP), normalized to unit area and unit bandwidth.

$$D^* = \frac{\sqrt{(A\Delta f)}}{\text{NEP}}$$

where A is the area of the photosensitive region of the detector and (Δf) is the effective noise bandwidth. *Synonym:* specific detectivity. *See also:* noise equivalent power; detectivity. (Std100) 812-1984w

DS-3 *See:* digital signal level three.

DS-2 *See:* digital signal level two.

DSU *See:* data service unit.

DS-0 *See:* digital signal level zero.

DTB *See:* data transfer bus.

DTE (Data Terminal Equipment) *See:* data terminal equipment.

DTG *See:* dynamically tuned gyro.

D32 module A module whose data space is limited to a 32-bit width. (C/BA) 14536-1995

DTL *See:* diode-transistor logic.

DTMF *See:* dual tone multifrequency.

DTMF signaling *See:* dual tone multifrequency signaling.

d-to-a converter *See:* digital-to-analog converter.

D-TRAN *See:* Decision Table Translator.

DTS *See:* digital termination system.

dual alternate routing (data transmission) (When applied to the routing of two circuits) The transmission facility assigned to one circuit is geographically separated from the transmission facility assigned to the other circuit throughout their entire length. To meet the aforementioned criteria, dual alternate routes are constructed in such a manner that an interruption on one route will not result in an interruption on the other. (PE) 599-1985w

dual-beam oscilloscope An oscilloscope in which the cathode-ray tube produces two separate electron beams that may be individually or jointly controlled. *See also:* multibeam oscilloscope; oscillograph. (IM/WM&A) 311-1970w, 181-1977w

dual benchboard A combination assembly of a benchboard and a vertical hinged panel switchboard placed back to back (no aisle) and enclosed with a top and ends. (SWG/PE) C37.100-1992, C37.21-1985r

dual bus A pair of buses carrying digital bit streams flowing in opposite directions. One bus is referred to as Bus A and the other bus as Bus B. (LM/C) 8802-6-1994

dual buses Two functional buses (primary and secondary) on the same backplane, each possessing the signals required to transfer data between boards. (C/BA) 896.3-1993w

dual cable A type of broadband coaxial cable system that uses separate cables to carry the inbound and outbound signals. (LM/C) 802.7-1989r

dual-channel controller A controller that enables reading from and writing to a device to occur simultaneously. (C) 610.10-1994w

dual coding (software) A development technique in which two functionally identical versions of a program are developed by different programmers or different programming teams from the same specification. The resulting source code may be in the same or different languages. The purpose of dual coding is to provide for error detection, increase reliability, provide additional documentation, or reduce the probability of systematic programming errors or compiler errors influencing the end result. *See also:* documentation; program; compiler; reliability; specification; error; code. (C/SE) 729-1983s

dual control A term applied to signal appliances provided with two authorized methods of operation. (EEC/PE) [119]

dual diversity (data transmission) The term applied to the simultaneous combining of four signals and their detection through the use of space, frequency or polarization characteristics. (PE) 599-1985w

dual duplex A signaling system that supports simultaneous duplex communication over two cabling pairs. (C/LM) 802.3-1998

dual-element bolometer unit An assembly consisting of two bolometer elements and a bolometer mount in which they are supported. *Note:* The bolometer elements are effectively in series to the bias power and in parallel to the RF power. (IM) 470-1972w

dual-element electrothermic unit An assembly consisting of two thermopile elements and an electrothermic mount in which they are supported. The thermopile elements are effectively in series to the output voltage and in parallel to the RF power. The thermopiles also serve as the power absorber. (IM) 544-1975w

dual-element fuse (1) (protection and coordination of industrial and commercial power systems) A cartridge fuse having two or more current-responsive of different fusing characteristics in series in a single cartridge. This is a construction/design technique frequently used to obtain a time-delay response characteristic. Labeling a fuse as dual element means this fuse meets UL time delay requirements (can carry five times rated current for a minimum of 10 seconds (s) for Class H, K, J, and R fuses) and in this case defines a time-current response characteristic and not necessarily a dual-element construction technique. (IA/PSP) 242-1986r
(2) A fuse having current-responsive elements of two different fusing characteristics in series in a single fuse. (SWG/PE) C37.100-1992

dual-element substitution effect (error) A component of substitution error, peculiar to dual-element bolometer units, that can cause the effective efficiency to vary with RF input power level. *Note:* This component, usually very small, is included in the effective efficiency correction for substitution error only with reference conditions for input RF power level and frequency. It results from a different division of RF and bias powers between the two elements. (IM) 470-1972w

dual headlighting system (illuminating engineering) Two double headlighting units, one mounted on each side of the front end of a vehicle. Each unit consists of two sealed-beam lamps mounted in a single housing. The upper or outer lamps have two filaments which supply the lower beam and part of the upper beam, respectively. The lower or inner lamps have one filament which provides the primary source of light for the upper beam. (EEC/IE) [126]

dual in-line package A common type of integrated circuit package in which the circuit leads or pins extend symmetrically outward and downward from opposite sides of the rectangular package body. (C) 610.10-1994w

dual mode (semiconductor) (nonlinear, active, and nonreciprocal waveguide components) A class of semiconductor devices used in frequency multipliers. Multiplication results not only from the diode's voltage variable reactive properties, but also from its ability to recover rapidly after storing charge. (MTT) 457-1982w

dual-mode agent An agent that supports both memory-mode and message-mode communication on the parallel system bus. *See also:* message-mode agent; memory-mode agent. (C/MM) 1296-1987s

dual-mode control system A control system in which control alternates between two predetermined modes. *Note:* The condition for change from one mode to the other is often a function of the actuating signal. One use of dual-mode action is to provide rapid recovery from large deviations without incurring large overshoot. *See also:* feedback control system. (PE/EDPG) [3]

dual-mode system A system that supports both memory-mode and message-mode communication. A mixture of both communication types is used on the parallel system bus. (C/MM) 1296-1987s

dual modulation (facsimile) The process of modulting a common carrier wave or subcarrier by two different types of modulation. For example, amplitude and frequency modulation, each conveying separate information. *See also:* facsimile transmission. (COM) 168-1956w

dual networks *See:* structurally dual networks.

dual operation Two Boolean operations are dual if the result of one operation is the negation of the result of applying the other operation to the negated operands, for all combinations of operands. For example, the AND and NOR operations are dual operations. *Contrast:* complementary operation. (C) 1084-1986w

dual overcurrent trip *See:* overcurrent release; dual release.

dual-pitch printer A printer that can print two or more type sizes by using different character spacing. (C) 610.10-1994w

dual polarization The polarization of a relay using current and voltage sources. (PE/PSR) C37.113-1999

Dual Port Memory A set of registers (or memory) that are asynchronously accessible from each of two ports (or buses, in the context of a Bridge) without the requirement for arbitration on one bus while accessing the memory from the other bus. (C/BA) 1014.1-1994w

dual race (display device) A multitrace operation in which a single beam in a cathode-ray tube is shared by two signal channels. *See also:* oscillograph; chopped display; alternate display; multitrace. (IM/WM&A) 181-1977w, 311-1970w

dual-rail logic *See:* double-rail logic.

dual-range single-pointer-form demand register (mechanical demand registers) An indicating demand register having an arrangement for changing the full-scale capacity from one value to another, usually by reversing the scale plate. For example, Scale Class 1/2; Scale Class 2/6. An interlock assures proper scale and scale-class relation. (ELM) C12.4-1984

dual release (trip) A release that combines the function of a delayed and an instantaneous release. (SWG/PE) C37.100-1992

dual ring A topology in which stations are linked by link pairs and, from any one station to another, there are exactly two distinct paths, where a path is defined as a sequence of consecutive links in which no link pair is traversed more than once. (LM/C) 802.5c-1991r

dual ring management The management functions of a dual ring station responsible for dual ring reconfiguration. (LM/C) 802.5c-1991r

dual service (plural service) Two separate services, usually of different characteristics, supplying one consumer. *Note:* A dual service might consist of an alternating-current and direct-current service, or of 208Y/120 volt 3-phase, 4-wire service for light and some power and a 13.2-kilovolt service for power, etcetera. *See also:* loop service; emergency service; duplicate service; service. (PE/T&D) [10]

dual simplex A link segment configuration containing two simultaneous signal paths, one in each direction. (STP and fibre optic links are configured as dual simplex.)local area networks. (C) 8802-12-1998

dual ring station A station that attaches link pairs that have opposite directions of data flow with respect to the adjacent dual ring stations. It consists of MACs, a crosspoint function, and dual ring management. It may have other attachments. (LM/C) 802.5c-1991r

dual switchboard A control switchboard with front and rear panels separated by a comparatively short distance and enclosed at both ends and top. The panels on at least one side are hinged for access to the panel wiring. (SWG/PE) C37.100-1992, C37.21-1985r

dual tone multifrequency (DTMF) *See:* dual tone multifrequency signaling.

dual tone multifrequency signaling (1) (telephone loop performance) Voiceband signaling by simultaneous transmission of two tones, one from a low-frequency and one from a high-frequency group. Each of these groups consists of four voiceband frequency tones no two of which are harmonically related. Only 12 of the 16 combinations are currently in use for customer address signaling. (COM/TA) 820-1984r

(2) An address signaling method for PTS using 16 pairs of frequencies to represent digits and other characters. Although it is most commonly used by a station set to signal into a switching system, it may be used for signaling from a local switching system to another system for certain services. The DTMF codes are pairs of frequencies, each consisting of one out of four frequencies from a low group and one out of three or four frequencies from a mutually exclusive higher group. Performance is measured as tolerances for each frequency signaling level twist and timing. (COM/TA) 973-1990w

dual-tone multifrequency pulsing (telephone switching systems) A means of pulsing utilizing a simultaneous combination of one of a lower group of frequencies and one of a higher group of frequencies to represent each digit or character. (COM) 312-1977w

dubbing (electroacoustics) A term used to describe the combining of two or more sources of sound into a complete recording at least one of the sources being a recording. *See also:* phonograph pickup; rerecording. (SP) [32]

duck tape Tape of heavy cotton fabric, such as duck or drill, that may be impregnated with an asphalt, rubber, or synthetic compound. (T&D/PE) [10]

duct (1) A single enclosed raceway for conductors or cable. (T&D/NESC) C2-1997

(2) (underground electric systems) A single enclosed runway for conductors or cables. (T&D/PE) [10]

duct bank (conduit run) An arrangement of conduit providing one or more continuous ducts between two points. *Note:* An underground runway for conductors or cables, large enough for workmen to pass through, is termed a gallery or tunnel. (T&D/PE) [10]

duct edge fair-lead (cable shield) A collar or thimble, usually flared, inserted at the duct entrance in a manhole for the purpose of protecting the cable sheath or insulation from being worn away by the duct edge. (PE/T&D) [4], [10]

duct entrance The opening of a duct at a manhole, distributor box, or other accessible space. (T&D/PE) [10]

ductility factor (seismic design of substations) The ratio of the maximum displacement (ultimate) to the displacement that corresponds to initiation of the yielding. (PE/SUB) 693-1984s

ducting (1) Guided propagation of radio waves inside an atmospheric or tropospheric radio duct. *See also:* atmospheric radio duct. (AP/PROP) 211-1997

(2) Confinement of near-horizontally directed electromagnetic waves to a restricted horizontal layer in the atmosphere, resulting from a sufficiently steep negative vertical gradient of the atmospheric refractive index in a limited altitude region. *Note:* The region of steep gradient is not necessarily identical to the dimensions of the duct. *Synonyms:* trapping; super-refraction. (AES) 686-1997

duct rodding (rodding a duct) The threading of a duct by means of a jointed rod of suitable design for the purpose of pulling in the cable-pulling rope, mandrel, or the cable itself. (T&D/PE) [10]

duct sealing The closing of the duct entrance for the purpose of excluding water, gas, or other undesirable substances. (T&D/PE) [10]

duct spacer (rotating machinery) (vent finger) A spacer between adjacent packets of laminations to provide a radial ventilating duct. (PE) [9]

duct system A continuous passageway for the transmission of air which, in addition to ducts, may include duct fittings, dampers, plemums, fans, and accessory air handling equipment. (NESC/NEC) [86]

duct ventilated (rotating machinery) (pipe ventilated) A term applied to apparatus that is so constructed that a cooling gas can be conveyed to or from it through ducts. (PE) [9]

DUI *See:* duration of unscheduled interrupt.

dumb terminal A terminal that can only send and receive information; that is, one that is lacking in local processing capability and built-in logic. *Contrast:* intelligent terminal. (C) 610.10-1994w

dumbwaiter A hoisting and lowering mechanism equipped with a car that moves in guides in a substantially vertical direction, the floor area of which does not exceed 9 square feet, whose total inside height whether or not provided with fixed or removable shelves does not exceed 4 feet, the capacity of which does not exceed 500 pounds, and which is used exclusively for carrying materials. (EEC/PE) [119]

dummy Pertaining to a nonfunctioning item used to satisfy some format or logic requirement or to fulfill prescribed conditions. For example, a dummy report containing only titles and column headings with place-holding data instead of real data. (C) 610.5-1990w

(2) (A) Pertaining to a nonfuctional item used to satisfy some format or logic requirement or to fulfill prescribed conditions. *See also:* dummy instruction; dummy address. **(B)** Pertaining to an item such as a character, data item or statement that has the appearance of a specified item, but not the capacity to function as such. *Synonym:* placeholder. (C) 610.10-1994

dummy address A nonfunctional address used for illustration or instruction purposes. (C) 610.10-1994w

dummy antenna A device that has the necessary impedance characteristics of an antenna and the necessary power-handling capabilities, but that does not radiate or receive radio waves. *Note:* In receiver practice, that portion of the impedance not included in the signal generator is often called dummy antenna. *See also:* radio receiver. 188-1952w

dummy-antenna system An electric network that simulates the impedance characteristics of an antenna system. *See also:* navigation. (AES/GCS) 173-1959w

dummy coil (rotating machinery) A coil that is not required electrically in a winding, but that is installed for mechanical reasons and left unconnected. *See also:* rotor; stator. (PE) [9]

dummy data Data that is used to satisfy some format or logic requirement or to fulfill prescribed conditions. For example, an artificial character used as a placeholder variable within a program. (C) 610.5-1990w

dummy finger A passive electrode that may be included in an interdigital transducer in order to suppress wavefront distortion. (UFFC) 1037-1992w

dummy instruction (A) An item of data, in the form of an instruction, that requires modification before being executed. *Synonyms:* do-nothing instruction; no-op instruction. **(B)** An item of data, in the form of an instruction, that is inserted into a sequence of instructions, but that is not intended to be executed. (C) 610.10-1994

dummy load (radio transmission) A dissipative but essentially nonradiating substitute device having impedance characteristics simulating those of the substituted device. *See also:* artificial load; radio transmission. (IM/HFIM) [40]

dummy parameter *See:* formal parameter.

dump (A) (computers) To copy the contents of all or part of a storage, usually from an internal storage into an external storage. **(B) (computers)** A process as in definition (A). **(C) (computers)** The data resulting from the process as in definition (A). *See also:* static dump; selective dump; dynamic dump; snapshot dump. (MIL) [2]

(2) (A) (software) A display of some aspect of a computer program's execution state, usually the contents of internal storage or registers. Types include change dump, dynamic dump, memory dump, postmortem dump, selective dump, snapshot dump; static dump. **(B) (software)** A display of the contents of a file or device. **(C) (software)** To copy the contents of internal storage to an external medium. **(D) (software)** To produce a display or copy as in definitions (A), (B), or (C). (C) 610.12-1990

dump energy (1) Energy generated from any source that cannot be stored and that is beyond the immediate needs of the electric system producing the energy. (PE/PSE) 858-1993w

(2) **(electric power supply)** Energy generated from water, as, wind, or other source which cannot be stored and which is beyond the immediate needs of the electric system producing the energy. (PE/PSE) 346-1973w

dump power Power generated from water, gas, wind, or other source that cannot be stored or conserved and that is beyond the immediate needs of the electric system producing the power. *See also:* generating station. (T&D/PE) [10]

dunnage material *See:* bunker material.

duodecimal **(A)** Pertaining to a characteristic or property involving a selection, choice, or condition in which there are twelve possibilities. **(B)** Pertaining to the numeration system with a radix of twelve. (C) [20], [85]
(2) **(A)** **(mathematics of computing)** Pertaining to a selection in which there are 12 possible outcomes. **(B)** **(mathematics of computing)** Pertaining to the numeration system with a radix of 12. (C) 1084-1986

duolater coil *See:* honeycomb coil.

duosexadecimal **(A)** Pertaining to a selection in which there are 32 possible outcomes. **(B)** Pertaining to the numeration system with a radix of 32. *Synonym:* duotricenary. (C) 1084-1986

duotricenary *See:* duosexadecimal.

duplex **(1)** A simultaneous, two-way, independent transmission in both directions. *Synonym:* full duplex. (SUB/PE) 999-1992w
(2) **(data transmission)** Pertaining to a simultaneous two-way independent transmission in both directions. (PE) 599-1985w
(3) A type of printing that involves the process of creating images or impressions on both sides of the printing media. (C/MM) 1284.1-1997

duplex artificial line (balancing network) A balancing network, simulating the impedance of the real line and distant terminal apparatus, that is employed in a duplex circuit for the purpose of making the receiving device unresponsive to outgoing signal currents. *See also:* telegraphy. (PE/EEC) [119]

duplex benchboard A combination assembly of a benchboard and a vertical control switchboard placed back to back and enclosed with a top and ends (not grille). Access space with entry doors is provided between the benchboard and vertical control switchboard. (SWG/PE) C37.100-1992, C37.21-1985r

duplex cable A cable composed of two insulated single-conductor cables twisted together. *Note:* The assembled conductors may or may not have a common covering of binding or protecting material. (T&D/PE) [10], 30-1937w

duplex channel A communications channel capable of simultaneous duplex communication. (C/LM) 802.3-1998

duplex circuit *See:* data circuit.

duplex current-limiting reactor A center tapped reactor used in two circuit branches fed by a common circuit and wound in such a way as to employ negative coupling under normal operating conditions to reduce circuit impedance and positive coupling under fault conditions to increase circuit impedance. (PE/TR) C57.16-1996

duplex data circuit A pair of associated transmit and receive channels that provide a means of two-way data communications. *See also:* virtual circuit. (C) 610.7-1995

duplexer (nonlinear, active, and nonreciprocal waveguide components) (radar) A device that utilizes the finite delay between the transmission of a pulse and the echo thereof so as to permit the connection of the transmitter and receiver to a common antenna. A duplexer commonly employs either a circulator and receiver protector or a balanced network of transmit-receive switches and a receiver protector. (MTT) 457-1982w

duplexing *See:* double storage.

duplexing assembly, radar *See:* transmit-receive switch.

duplex lap winding (rotating machinery) A lap winding in which the number of parallel circuits is equal to twice the number of poles. (PE) [9]

duplex operation **(1)** **(data transmission)** **(A)** **(general)** The operation of transmitting and receiving apparatus at one location in conjunction with associated transmitting and receiving equipment at another location; the processes of transmission and reception being concurrent. **(B)** **(radio communication) (two-way radio communication circuit)** The operation utilizing two radio-frequency channels, one for each direction of transmission, in such manner that intelligence may be transmitted concurrently in both directions. (AP/PE/VT/ANT) 145-1983s, 599-1985w, [37]
(2) A mode of operation of a data link or a data circuit in which data is transmitted in both directions simultaneously. (C) 610.7-1995

duplex signaling (telephone switching systems) A form of polar-duplex signaling for a single physical circuit. (COM) 312-1977w

duplex switchboard A control switchboard consisting of panels placed back to back and enclosed with a top and ends (not grille). Access space with entry doors is provided between the rows of panels. (SWG/PE) C37.100-1992, C37.21-1985r

duplex system A telegraph system that affords simultaneous independent operation in opposite directions over the same. *See also:* telegraphy. (EEC/PE) [119]

duplex transmission Transmission in which data may be sent simultaneously in both directions on a transmission medium. *Contrast:* half-duplex transmission; simplex transmission. (C) 610.7-1995

duplex type (breaker-and-a-half arrangement) A unit substation which has two stepdown transformers, each connected to an incoming high-voltage circuit. The outgoing side of each transformer is connected to a radial (stub-end) feeder. These feeders are joined on the feeder side of the power circuit breakers by a normally open-tie circuit breaker. (PE/TR) C57.12.80-1978r

duplex wave winding (rotating machinery) A wave winding in which the number of parallel circuits is four, whatever the number of poles. (PE) [9]

duplicate (data management) To copy data from a source to a destination that has similar physical form as the source. *Synonym:* reproduce. *See also:* copy. (C) 610.5-1990w

duplicate lines (power transmission) Lines of substantially the same capacity and characteristics, normally operated in parallel, connecting the same supply point with the same distribution point. *See also:* center of distribution. (T&D/PE) [10]

duplicate service (power transmission) Two services, usually supplied from separate sources, of substantially the same capacity and characteristics. *Note:* The two services may be operated in parallel on the consumer's premises, but either one alone is of sufficient capacity to carry the entire load. *See also:* loop service; dual service; service; emergency service. (T&D/PE) [10]

duplication check **(1)** A check based on the consistency of two independent performances of the same task. (C/EEC/IE) [20], [126]
(2) **(data management)** A check that requires that the results of two independent performances of the same operations be identical. (C) 610.5-1990w

duration (pulse terminology) The absolute value of the interval during which a specified waveform or feature exists or continues. (IM/WM&A) 194-1977w

duration of unscheduled interrupt The length of the delay caused by an unscheduled or unexpected interrupt. (C) 610.10-1994w

dust-ignition-proof (class II locations) Enclosed in a manner that will exclude ignitable amounts of dusts or amounts that might affect performance or rating and that, where installed and protected in accordance with this Code, will not permit arcs, sparks, or heat otherwise generated or liberated inside of the enclosure to cause ignition of exterior accumulations or atmospheric suspensions of a specified dust on or in the vicinity of the enclosure. (NEC/NESC) [86]

dust-ignition proof machine A totally enclosed machine whose enclosure is designed and constructed in a manner that will exclude ignitable amounts of dusts or amounts that might affect performance or rating, and that, when installation and protection are in conformance with the National Electrical Code (ANSI CI-1975; section 502-1), will not permit arcs, sparks, or heat otherwise generated or liberated inside of the enclosure to cause ignition of exterior accumulations or atmospheric suspensions of a specific dust on or in the vicinity of the enclosure. *See also:* asynchronous machine. (PE/IA/APP) [9], [82]

dustproof (1) (general) So constructed or protected that the accumulation of dust will not interfere with successful operation. *See also:* luminaire. (SWG/PE/IA/NP/IAC) C37.30-1971s, 308-1980s, [60]
(2) (enclosure) An enclosure so constructed or protected that any accumulation of dust that may occur within the enclosure will not prevent the successful operation of, or cause damage to, the enclosed equipment. (IA/MT) 45-1983s
(3) (luminaire) Luminaire so constructed or protected that dust will not interfere with its successful operation. *See also:* luminaire. (EEC/IE) [126]
(4) So constructed or protected that dust will not interfere with its successful operation. (NESC/NEC) [86]

dustproof enclosure An enclosure so constructed or protected that any accumulation of dust that may occur within the enclosure will not prevent the successful operation of, or cause damage to, the enclosed equipment. (IA/MT) 45-1998

dust seal (rotating machinery) A sealing arrangement intended to prevent the entry of a specified dust into a bearing. *See also:* direct-current commutating machine; asynchronous machine. (PE) [9]

dust-tight enclosure (1) (electric installations on shipboard) An enclosure constructed so that dust cannot enter the enclosing case. (SWG/PE) C37.30-1971s
(2) An enclosure constructed so that dust cannot enter. (IA/MT) 45-1998
(3) (power and distribution transformers) An enclosure so constructed that dust will not enter the enclosing case under specified conditions. (NESC/SWG/EEC/IE/PE/TR) [126], C57.12.80-1978r, [86], C37.100-1981s

duty (1) (general) A statement of loads including no-load and rest and de-energized periods, to which the machine or apparatus is subjected including their duration and sequence in time. (PE) [9]
(2) (rating of electric equipment) A statement of the operating conditions to which the machine or apparatus is subjected, their respective durations, and their sequence in time. (Std100/EI) 96-1969w, [83]
(3) (power and distribution transformers) A requirement of service that defines the degree of regularity of the load. (PE/ICTL/TR) C57.12.80-1978r
(4) (excitation systems) Those voltage and current loadings imposed by the synchronous machine upon the excitation system including short circuits and al conditions of loading. *Note:* The duty will include the action of limiting devices to maintain synchronous machine loading at or below that defined by American National Standard Requirements for Cylindrical Rotor Synchronous Generators. (REM/IM/WM&A) C50.13-1965s, 181-1977w
(5) A requirement of service that defines the degree of regularity of the current through the reactor. (PE/TR) C57.16-1996

duty continuous (thyristor converter) A duty where the converter equipment carries a direct current of fixed value for an interval sufficiently long for the components of the converter to reach equilibrum temperatures corresponding to the said value of current. (IA/IPC) 444-1973w

duty cycle (1) (general) The time interval occupied by a device on intermittent duty in starting, running, stopping, and idling. (AP/ANT) 145-1983s
(2) (rotating machinery) A variation of load with time which may or may not be repeated, and in which the cycle time is too short for thermal equilibrium to be attained. *See also:* asynchronous machine. (PE) [9]
(3) (pulse systems) The ratio of the sum of all pulse durations to the total period, during a specified period of continuous operation. *See also:* navigation. (AES) [42]
(4) (welding) The percentage of the time during which the welder is loaded. For instance, a spot welder supplied by a 60-Hertz system (216 000 cycles per hour) making four hundred 15-cycle welds per hour would have a duty cycle of 2.8 percent (400 multiplied by 15, divided by 216 000, multiplied by 100). A seam welder operating 2 cycles "on" and 2 cycles "off" would have a duty cycle of 50 percent. (NESC/NEC) [86]
(5) (large lead storage batteries) The load currents a battery is expected to supply for specified time periods. (PE/EDPG) 450-1995, 1115-1992, 485-1983s
(6) The ratio of the sum of all pulse durations to the total period during a specified period of continuous operation. (PEL/ET) 388-1992r
(7) A prescribed sequence of operations for a specific time with specified time intervals between sequences. (SWG/PE) C37.100-1992
(8) (of a valve actuator motor) The duty cycle of a valve actuator motor (VAM) consists of the number of consecutive valve strokes required for the valve's intended service. Plant operation or testing may require successive stroking of the VAM. (PE/NP) 1290-1996
(9) The ratio of the active or ON time within a specified period to the duration of the specified period. *Note:* For a pulsed radar, the ratio of transmitted pulse width to pulse repetition interval. (AES) 686-1997
(10) (radar) *See also:* duty factor.

duty-cycle rating (rotating machinery) The statement of the loads and conditions assigned to the machine by the manufacturer, at which the machine may be operated on duty cycles. *See also:* asynchronous machine. (PE) [9]

duty cycle test A test to determine if a device can repeatedly function, extinguish follow current, and avoid thermal runaway. (PE) C62.34-1996

duty cycle time [of a valve actuator motor (VAM)] A VAM's duty cycle time is the time for one stroke multiplied by the duty cycle. (PE/NP) 1290-1996

duty-cycle voltage rating The designated maximum permissible voltage between its terminals at which an arrester is designed to perform its duty cycle. (SPD/PE) C62.11-1999

duty factor (1) (pulse techniques) The ratio of the pulse duration to the pulse period of a periodic pulse train. *See also:* pulse carrier; pulse. (COM) [25]
(2) (electron tube) The ratio of the ON period to the total period during which an electronic valve or tube is operating. *See also:* pulse; ON period. (Std100) [84]
(3) (automatic control) The ratio of working time to the time taken for the complete sequence of a duty cycle. (PE/EDPG) [3]
(4) (waveguide) The ratio of the average power to the peak pulse power passing through the transverse section of the waveguide. *Synonym:* duty cycle. (MTT) 146-1980w
(5) The ratio of pulse duration to the pulse period of a periodic pulse train. A duty factor of 1.0 corresponds to continuous-wave (CW) operation. (NIR) C95.1-1999
(6) *See also:* duty cycle. (AES) 686-1997

duty factor control system (automatic control) A control system in which the signal to the final controlling element consists of periodic pulses whose duration is varied to relate, in some prescribed manner, the time average of the signal to the actuating signal. *Note:* This mode of control differs from two-step control in that the period of the pulses in duty-factor control is predetermined. (PE/EDPG) [3]

duty, peak load *See:* peak load duty.

duty ratio (pulse system) The ratio of average to peak pulse power. *See also:* navigation. (AES) [42]
(2) (A) In a pulsed-radar system, the ratio of average-to-peak pulse power. **(B)** Same as duty factor. (AES/RS) 686-1990

Dvorak keyboard A keyboard layout in which the letters of the alphabet are arranged according to their frequency of use. *Note:* First patented in 1932 by Dr. August Dvorak. *Synonym:* Dvorak simplified keyboard. *See also:* QWERTY keyboard. (C) 610.10-1994w

Dvorak simplified keyboard *See:* Dvorak keyboard.

DVST *See:* direct-view storage tube.

dwell (numerically controlled machines) A timed delay of programmed or established duration, not cyclic or sequential, that is, not an interlock or hold. (IA) [61]

dwelling unit One or more rooms for the use of one or more persons as a housekeeping unit with space for eating, living, and sleeping, and permanent provisions for cooking and sanitation. (NESC/NEC) [86]

dwell time The time a transit unit (vehicle or train) spends at a station or stop, measured as the interval between its stopping and starting. (VT/RT) 1474.1-1999

dyadic (mathematics of computing) Pertaining to an operation involving two operands. *Contrast:* monadic. (C) 1084-1986w

dyadic Boolean operation A logical operation involving two operands. For example, the equivalence operation. *Contrast:* monadic Boolean operation. (C) 1084-1986w

dyadic operation An operation involving two operands. *Contrast:* monadic operation. (C) 1084-1986w

dyadic operator An operator that specifies an operation on two operands. *Synonym:* binary operator. *Contrast:* monadic operator. (C) 1084-1986w

dyadic selective construct An if-then-else construct in which processing is specified for both outcomes of the branch. *Contrast:* monadic selective construct. (C) 610.12-1990

dynamic (1) (software) Pertaining to an event or process that occurs during computer program execution; for example, dynamic analysis, dynamic binding. *Contrast:* static. (C) 610.12-1990
(2) (excitation systems) A state in which one or more quanitities exhibit appreciable change within an arbitrarily short time interval. *Note:* For excitation control systems, this time interval encompasses up to 15-20 sec.; that is, sufficient time to ascertain whether oscillations are decaying or building up with time. *See also:* feedback control system. (PE/ICTL/EDPG) 421A-1978s

dynamic accuracy Accuracy determined with a time-varying output. *Contrast:* static accuracy. *See also:* electronic analog computer. (BT/C) 185-1975w, 165-1977w

dynamic allocation *See:* dynamic resource allocation.

dynamically programmable logic gate A gate in a field programmable gate array, the function of which can be changed while it is in the circuit. *Note:* This is a function that is available in some RAM-based field programmable gate arrays. (C) 610.10-1994w

dynamically tuned gyro (DTG) (inertial sensors) A two-degree-of-freedom gyro in which a dynamically tuned flexure and gimbal mechanism both supports the rotor and provides angular freedom about axes perpendicular to the spin axis. *See also:* dynamic tuning. (AES/GYAC) 528-1994

dynamic analysis (software) The process of evaluating a system or component based on its behavior during execution. *Contrast:* static analysis. *See also:* testing; demonstration. (C) 610.12-1990

dynamic analyzer (software) A software tool that aids in the evaluation of a computer program by monitoring execution of the program. Examples include instrumentation tools, software monitors, and tracers. *See also:* computer program; execution; tracer; software monitor; program; static analyzer. (C/SE) 729-1983s

dynamic binding (software) Binding performed during the execution of a computer program. *Contrast:* static binding. (C) 610.12-1990

dynamic braking (rotating machinery) A system of electric braking in which the excited machine is disconnected from the supply system and connected as a generator, the energy being dissipated in the winding and, if necessary, in a separate resistor. (PE) [9]

dynamic braking envelope A curve that defines the dynamic braking limits in terms of speed and tractive force as restricted by such factors asmaximum current flow, maximum permissible voltage, minimum field strength, etcetera. *See also:* dynamic braking. (EEC/PE) [119]

dynamic breakpoint A breakpoint whose predefined initiation event is a runtime characteristic of the program, such as the execution of any twenty source statements. *Contrast:* static breakpoint. *See also:* data breakpoint; epilog breakpoint; code breakpoint; programmable breakpoint; prolog breakpoint. (C) 610.12-1990

dynamic buffering (1) A buffering technique in which the buffer allocated to a computer program varies during program execution, based on current need. *Contrast:* simple buffering. (C) 610.12-1990
(2) Buffering in which buffer storage is allocated in the sizes and at the times as required by an application. (C) 610.10-1994w

dynamic bus sizing The ability of some microprocessors to adjust the number and the size of data transfers to the amount of data that the responding board can access in one transfer. During the address broadcast portion of the cycle, the slave informs the master how many data lines it actually drives or receives. This information is made available to on-board logic that can then adjust the amount of data that it access during the data transfer to the capabilities of the slave. (C/MM) 1096-1988w

dynamic characteristic *See:* load characteristic.

dynamic check *See:* problem check.

dynamic computer check *See:* problem check.

dynamic configuration (DC) An optional, alternative method of automatically assigning logical addresses to VXIbus devices at system power-on or other configuration times, using the MODID lines. It allows for each slot to contain one or more devices as well as different devices within a slot to share address decoding hardware. (C/MM) 1155-1992

dynamic convergence (multibeam cathode-ray tubes) The process whereby the locus of the point of convergence of electron beams is made to fall on a specified surface during scanning. (ED) 161-1971w, [45]

dynamic cutoff frequency (semiconductor) (nonlinear, active, and nonreciprocal waveguide components) A figure of merit used for varactor diodes. Unlike fixed cutoff frequency measurements at specific bias voltages, dynamic cutoff frequency is a measure of the varactor's total change in Q from a slight forward bias current to reverse breakdown voltage. This dynamic or total figure of merit is useful in evaluating the frequency multiplier performance of fully driven multipliers. (MTT) 457-1982w

dynamic decay (storage tubes) Decay caused by an action such as that of the reading beam, ion currents, field emission, or holding beam. *See also:* storage tube. (ED/C60) 158-1962w

dynamic dump A dump that is produced during the execution of a computer program. *Contrast:* static dump. *See also:* selective dump; memory dump; snapshot dump; change dump; postmortem dump. (C) 610.12-1990

dynamic dumping (test, measurement, and diagnostic equipment) The printing of diagnostic information without stopping the program being tested. (MIL) [2]

dynamic electrode potential An electrode potential when current is passing between the electrode and the electrolyte. *See also:* electrochemistry. (EEC/PE) [119]

dynamic energy sensitivity *See:* dynamic sensitivity.

dynamic error (1) (analog computer) An error in a time-varying signal resulting from imperfect dynamic response of a transducer. (C) 165-1977w
(2) (software) An error that is dependent on the time-varying nature of an input. *Contrast:* static error. (C) 610.12-1990

dynamic focusing (picture tubes) The process of focusing in accordance with a specified signal in synchronism with scanning. (ED) 161-1971w

dynamic holding brake A braking system designed for the purpose of exerting maximum braking force at a fixed speed only and used primarily to assist in maintaining this fixed speed when a train is descending a grade, but not to effect a deceleration. *See also:* dynamic braking. (EEC/PE) [119]

dynamic impedance (low voltage varistor surge arresters) A measure of small signal impedance at a given operating point, described as the rate of change of varistor voltage with respect to varistor current at the operating point. (PE) [8]

dynamicizer *See:* serializer.

dynamic load line The locus of all simultaneous values of total instantaneous output electrode current and voltage for a fixed value of load impedance. (ED) [45]

dynamic loudspeaker *See:* moving-coil loudspeaker.

dynamic microphone *See:* moving-coil microphone.

dynamic memory *See:* dynamic storage.

dynamic model (1) A model of a system in which there is change, such as the occurrence of events over time or the movement of objects through space; for example, a model of a bridge that is subjected to a moving load to determine characteristics of the bridge under changing stress. *Contrast:* static model. (C) 610.3-1989w
(2) A kind of model that describes individual requests or patterns of requests among objects. *Contrast:* static model. (C/SE) 1320.2-1998

dynamic patterns A set of controlled, time-variant patterns within a time interval. (SCC20) 1445-1998

dynamic performance test Testing of the transient performance of a dc system either on the actual system or on a system simulator. Such testing includes fault recovery performance, step responses, and ac system interaction. (PE/SUB) 1378-1997

dynamic problem check *See:* problem check.

dynamic programming In operations research, a procedure for optimizing a multi-stage problem solution, in which a number of decisions are available at each stage of the process. (C) 610.2-1987

dynamic radiation test (1) A test of the instantaneous effects of radiation electrical and memory properties of interest continuously, during and immediately after exposure to a transient radiation pulse. (ED) 641-1987w
(2) Test of the instantaneous effects of radiation that are obtained by monitoring electrical and memory properties of interest continuously during and immediately after exposure to a transient radiation pulse. (ED) 1005-1998

dynamic random-access memory A dynamic form of random-access memory that uses, as its memory elements, capacitors that are built into the integrated circuit. *Note:* Since the capacitors lose their charge, this type of storage requires periodic refreshing. *Contrast:* static random-access memory. (C) 610.10-1994w

dynamic range (DR) (1) (nonlinear, active, and nonreciprocal waveguide components) (parametric amplifier) The ratio, usually expressed in decibels, of the maximum to the minimum signal input power levels over which the amplifier can operate within some specified range of performance. The minimum level is usually determined by the noise level of the amplifier, while the maximum level is usually set by the maximum tolerable nonlinear effects. (MTT) 457-1982w
(2) (general) The difference, in decibels, between the overload level and the minimum acceptable signal level in a system or transducer. *Note:* The minimum acceptable signal level of a system or transducer is ordinarily fixed by one or more of the following: noise level, low-level distortion, interference, or resolution level. (C) 165-1977w
(3) (control system or element) The ratio of two instantaneous signal magnitudes, one being the maximum value consistent with specified criteria of performance, the other the maximum value of noise. (PE/EDPG) [3]
(4) (spectrum analyzer) The maximum ratio of two signals simultaneously present at the input which can be measured to a specified accuracy. **(A) (harmonic dynamic range)** The maximum ratio of two harmonically related sinusoidal signals simultaneously present at the input which can be measured with a specified accuracy. **(B) (nonharmonic dynamic range)** The maximum ratio of two nonharmonically related sinusoidal signals simultaneously present at the input which can be measured with a specified accuracy. **(C) (display dynamic range)** The maximum ratio of two nonharmonically related sinusoids each of which can be simultaneously measured on the screen to a specified accuracy. (IM) 748-1979w
(5) (analog computer) The ratio of the specified maximum signal level capability of a system or component to its noise or resolution level usually expressed in decibels. Also the ratio of the maximum to minimum amplitudes of a variable during a computer solution. (C) 165-1977w
(6) (accelerometer) (gyros) The ratio of the input range to the resolution. (AES/GYAC) 528-1994
(7) The range of amplitudes over which the receive system operates linearly. The DR is numerically equal to the difference between the maximum and minimum signal amplitudes when both terms are expressed in decibels. For a shielding effectiveness (SE) measurement, the important portion of the DR is from the reference level to the noise floor. (EMC/STCOORD) 299-1997

dynamic range, reading (storage tubes) The range of output levels that can be read, from saturation level to the level of the minimum discernible output signal. *See also:* storage tube. (ED) 158-1962w

dynamic range, writing (storage tubes) The range of input levels that can be written under any stated condition of scanning, from the input that will write the minimum usable signal. *See also:* storage tube. (ED) 158-1962w

dynamic register (hybrid computer linkage components) The register that produces the analog equivalent voltage or coefficient. (C) 166-1977w

dynamic regulation Expresses the maximum or minimum output variations occurring during transient conditons, as a percentage of the final value. *Note:* Typical transient conditions are instantaneous or permanent input or load changes.

$$\text{dynamic regulation} = \frac{E_{\text{max}} - E_{\text{final}}}{E_{\text{final}}}\left(100\%\right)$$

(SP) [32]

dynamic regulator A transmission regulator in which the adjusting mechanism is in self-equilibrium at only one or a few settings and requires control power to maintain it at any other setting. (EEC/PE) [119]

dynamic relocation Relocation of a computer program during its execution. (C) 610.12-1990

dynamic resource allocation A computer resource allocation technique in which the resources assigned to a program vary during program execution, based on current need. (C) 610.12-1990

dynamic response *See:* time response; feedback control system.

dynamic restructuring (software) (data management) The process of restructuring a database, data structure, computer program, or set of system components during program execution. For example, concurrent reorganization of a database. (C) 610.5-1990w, 610.12-1990

dynamics characteristics test *See:* forced response [test] method.

dynamic sensitivity (phototubes) The quotient of the modulated component of the output current by the modulated component of the incident radiation at a stated frequency of modulation. *Note:* Unless otherwise stated the modulation wave shape is sinusoidal. *See also:* phototube. (ED) 158-1962w, [45]

dynamic short-circuit output current (converters having ac output) (self-commutated converters) The transient current that flows from the converter into a short-circuit across the output terminals. (IA/SPC) 936-1987w

dynamic signal to noise ratio (digital delay line) The ratio of the minimum peak output signal to the maximum peak noise output when operated with a random bit sequence at a specified clock frequency. (UFFC) [22]

dynamic slowdown Dynamic braking applied for slowing down, rather than stopping, a drive. *See also:* electric drive. (IA/ICTL/IAC) [60]

dynamic stop *See:* breakpoint instruction.

dynamic storage (A) A type of storage in which data is stored and retrieved from a moving data medium. *See also:* hold time. **(B)** A type of storage that requires periodic refreshment for retention of data. *Synonym:* dynamic memory. *Contrast:* static storage. *See also:* dynamic random-access memory. (C) 610.10-1994

dynamic storage allocation A storage allocation technique in which the storage assigned to a computer program varies during program execution, based on the current needs of the program and of other executing programs. (C) 610.12-1990

dynamic test (test, measurement, and diagnostic equipment) A test of one or more of the signal properties or characteristics of the equipment or of any of its constituent items performed while the equipment is energized. (MIL) [2]

dynamic time constant (A) (accelerometer) The delay time between an input ramp and the output after steady-state is reached. For a second-order system it has a value of twice the damping ratio divided by the natural frequency in radians/second. **(B) (dynamically tuned gyro)** The time required for the rotor to move through an angle equal to 63% of its final value following a step change in case angular position about an axis normal to the spin axis with the gyro operating open loop. The value depends on the gimbal and rotor damping and drag forces, and is inversely proportional to quadrature spring rate. (AES/GYAC) 528-1994

dynamic torque (electric coupling) That torque of an electric coupling developed or transmitted at a specified value or range of speed differential between input and output members and at specified excitation and other applicable conditions. (EM/PE) 290-1980w

dynamic transmission (acoustically tunable optical filter) The ratio of the intensity of the light transmitted by the device at the wavelength to be filtered to the light intensity at this this wavelength incident on the device, namely: $T(\lambda) = [I(\lambda)/I_o(\lambda)]$. It includes all static losses as well as the diffraction efficiency of the interaction. For a given design, the dynamic transmission is a function of acoustic drive power. (UFFC) [17]

dynamic tuning (dynamically tuned gyro) The adjustment of the gimbal inertia, flexure spring rate, or the spin rate of a rotor suspension system to achieve a condition in which the dynamically induced (negative) spring rate cancels the spring rate of the flexure suspension. (AES/GYAC) 528-1994

dynamic variable brake A dynamic braking system designed to allow the operator to select (within the limits of the electric equipment) the braking force best suited to the operation of a train descending a grade and to increase or decrease this braking force for the purpose of reducing or increasing train speed. *See also:* dynamic braking. (EEC/PE) [119]

dynamic vertical *See:* apparent vertical.

dynamite *See:* explosives.

DYNAMO A programming language for continuous simulations; used to construct large multisector models of economic, industrial, and social systems, as well as other continuous closed-loop information feedback systems. *Note:* DYNAMO was written entirely in FORTRAN. (C) 610.13-1993w

dynamometer (conductor stringing equipment) (power line maintenance) A device designed to measure loads or tension on conductors. Various models of these devices are used to tension guys or sag conductors. *Synonyms:* load cell; clock. (T&D/PE) 516-1987s, 524-1992r, 524a-1993r

dynamometer, electric (rotating machinery) An electric generator, motor or eddy-current load absorber equipped with means for indicating torque. *Note:* When used for determining power input or output of a coupled machine, means for indicating speed are also provided. (PE) [9]

dynamometer test (rotating machinery) A braking or motoring test in which a dynamometer is used. *See also:* asynchronous machine; braking test. (PE) [9]

dynamotor A form of converter that combines both motor and generator action, with one magnetic field and with two armatures or with one armature having separate windings. (IA/MT) 45-1998

dynatron effect (electron tube) (dynatron characteristic) An effect equivalent to a negative resistance, which results when the electrode characteristic (or transfer characteristic) has a negative slope. Example: Anode characteristic of a tetrode, or tetrode-connected valve or tube. (ED) [45], [84]

dynatron oscillation Oscillation produced by negative resistance due to secondary emission. *See also:* oscillatory circuit. (AP/ANT) 145-1983s

dynatron oscillator A negative-resistance oscillator in which negative resistance is derived between plate and cathode of a screen-grid tube operating so that secondary electrons produced at the plate are attracted to the higher potential screen grid. *See also:* oscillatory circuit. (AP/ANT) 145-1983s

dyne The unit of force in the cgs (centimeter-gram-second) systems. The dyne is 10^{-5} newton. (Std100) 270-1966w

dynode (electron tube) An electrode that performs a useful function, such as current amplification, by means of secondary emission. *See also:* electrode; electron tube. (ED/NPS) 161-1971w, 398-1972r

dynode spots (image orthicons) A spurious signal caused by variations in the secondary-emission ratio across the surface of a dynode that is scanned by the electron beam. (ED) 161-1971w

E

EACK *See:* extended acknowledgment.

E&M signaling (telephone switching systems) A technique for transferring information between a trunk circuit and a separate signaling circuit over leads designated "E" and "M." The "M" lead transmits to the signaling circuit and the "E" lead transmits to the trunk circuit. (COM) 312-1977w

EAM *See:* electrical accounting machine.

eardrum reference point (DRP) A point located at the end of the ear canal, corresponding to the eardrum position. (COM/TA) 1206-1994

early decay time (EDT) Reverberation time based on the first 10 dB of sound decay in a room. (COM/TA) 1329-1999

early-failure period That possible early period, beginning at a stated time and during which the failure rate decreases rapidly in comparison with that of subsequent period. *See also:* constant failure rate period.

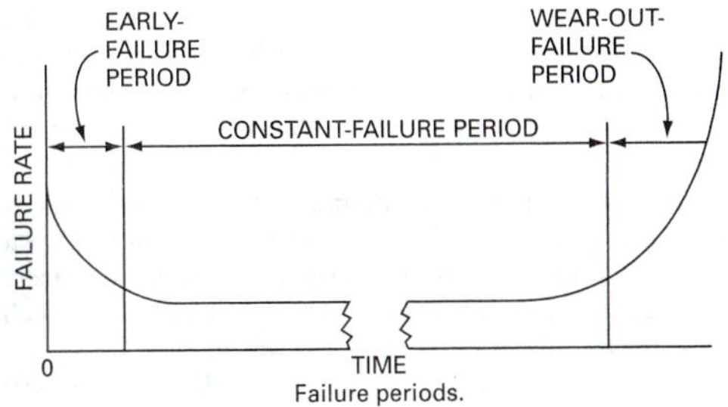

early failure period

(R) [29]

(2) (software) The period of time in the life cycle of a system or component during which hardware failures occur at a decreasing rate as problems are detected and repaired. *Synonym:* burn-in period. *Contrast:* wearout-failure period; constant-failure period. *See also:* bathtub curve. (C) 610.12-1990

early mode The very first edge that propagates through a given cone of logic. (C/DA) 1481-1999

early relay contacts Sometimes used for relay contacts, preliminary. (EEC/REE) [87]

early-warning radar Radar employed to search for distant enemy aircraft or missiles. (AES) 686-1997

ear mold A receiver-to-ear coupling device consisting of a short length of tubing (sound-pipe) and a fitting, usually made of hard plastic. The fitting is custom molded to an individual's concha and ear canal entrance. It occludes that portion of the ear canal into which the fitting extends. (COM/TA) 1206-1994

EAROM *See:* electrically alterable read-only memory.

earphone (receiver) An electroacoustic transducer intended to be closely coupled acoustically to the ear. *Note:* The term receiver should be avoided when there is risk of ambiguity. (SP) [32]

earphone coupler A cavity of predetermined size and shape that is used for the testing of earphones. The coupler is provided with a microphone for the measurement of pressures developed in the cavity. *Note:* Couplers generally have a volume of six cubic centimeters for testing regular earphones and a volume of two cubic centimeters for testing insert earphones. (SP) [32]

ear reference point (ERP) (1) A point on the artificial test head where the intersection of the receiver ear-cap axis with the external ear-cap plane is placed for testing. (COM/TA) 269-1992

(2) A virtual point for geometric reference located at the entrance to the listener's ear, traditionally used for calculating telephonometric loudness ratings. (COM/TA) 1206-1994, 1329-1999

(3) The intersection of the receiver ear-cap axis with the external ear-cap plane (planar area containing points of the receiver-end of the handset which, in normal handset use which, in normal handset use, rest against the ear) as shown below.

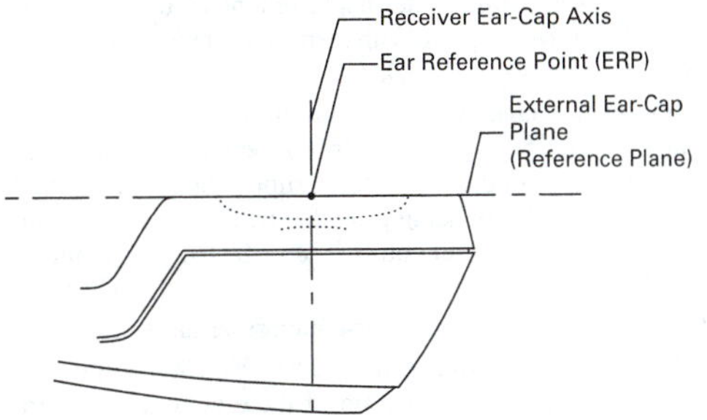

ear reference point

(COM/TA) 1027-1996

earth, effective radius (radio-wave propagation) A value for the radius of the earth that is used in place of the geometrical radius to correct approximately for atmospheric refraction when the index of refraction in the atmosphere changes linearly with height. *Note:* Under conditions of standard refraction the effective radius of the earth is 8.5×10^6 meters, or 4/3 the geometrical radius. *See also:* radio-wave propagation; radiation. (AP/PROP) [36]

earth-fault protection *See:* ground protection.

earth inductor *See:* generating magnetometer.

earth rate The angular velocity of the earth with respect to inertial space. Its magnitude is $7.292 \cdot 10^{-5}$ rad/s (15.041°/h). This vector quantity is usually expressed as two components in local level coordinates, north (or horizontal) and up (or vertical). (AES/GYAC) 528-1994

earth resistivity The measure of the electrical impedance of a unit volume of soil. The commonly used unit is the ohm-meter (Ω-m) that refers to the impedance measured between opposite faces of a cubic meter of soil. (PE/PSC) 367-1996

earth's rate correction (1) (navigation aid terms) A rate applied to a gyroscope to compensate for the apparent precession of the spin axis caused by the rotation of the earth. (GCS) 172-1983w

(2) (gyros) A command rate applied to a gyro to compensate for the rotation of the earth with respect to the gyro input axis. (AES/GYAC) 528-1994

earth, remote *See:* remote earth.

earth station (communication satellite) A ground station designed to transmit to and receive transmission from communication satellites. (COM) [24]

earth terminal *See:* ground terminal.

earth wire *See:* overhead ground wire.

ear tip A receiver-to-ear coupling device consisting of a short length of tubing (sound-pipe) and a soft bulb-shaped fitting. It occludes that portion of the ear canal into which the ear tip extends. (COM/TA) 1206-1994

Easytrieve A database manipulation language used for extracting data from data files and databases. (C) 610.13-1993w

EBCDIC *See:* extended binary coded decimal interchange code.

E bend (E-plane bend) (waveguide technique) A smooth change in the direction of the axis of a waveguide, throughout which the axis remains in a plane parallel to the direction of polarization. *See also:* waveguide. (PE/EEC) [119]

EBR *See:* electron beam recording.

ECAP II *See:* Electronic Circuit Analysis Program II.

ECC *See:* error-correcting code; error-correction coding.

eccentric groove (disc recording) (eccentric circle) A locked groove whose center is other than that of the disc record (generally used in connection with mechanical control of phonographs). *See also:* phonograph pickup. (SP) [32]

eccentricity (1) (general) (power distribution, underground cables) The ratio of the difference between the minimum and average thickness to the average thickness of an annular element, expressed in percent. (PE) [4]
(2) (disc recording) The displacement of the center of the recording groove spiral, with respect to the record center hole. *See also:* phonograph pickup. (SP) [32]

Eccles-Jordan circuit A flip-flop circuit consisting of a two-stage resistance-coupled electron-tube amplifier with its output similarly coupled back to its input, the two conditions of permanent stability being provided by the alternate biasing of the two stages beyond cutoff. *See also:* trigger circuit. (EEC/PE) [119]

ECCM *See:* electronic counter-countermeasures.

electronic counter-countermeasures (ECCM) improvement factor (EIF) The power ratio of the electronic countermeasures (ECM) signal level required to produce a given output signal-to-interference ratio from a receiver using an ECCM technique to the ECM signal level producing the same output signal-to-interference ratio from the same receiver without the ECCM technique. *Notes:* 1. The principal application is the representation of the performance of certain ECCM techniques in the analysis of radar performance in hostile electromagnetic environments. 2. The EIF is not a measure of the overall efficacy of an ECCM technique. (AES) 686-1997

EDC *See:* error-detection coding.

echelon (calibration) A specific level of accuracy of calibration in a series of levels, the highest of which is represented by an accepted national standard. *Note:* There may be one or more auxiliary levels between two successive echelons. *See also:* measurement system. (IM) 285-1968w, 294-1969w, [38]

echo (1) (A) (supervisory control, data acquisition, and automatic control) A communication technique assuring that a word received at the termination point in a system is the same as the word originally transmitted. The received word is retransmitted to the sending device and matched to ensure that the original message was received properly. **(B) (software)** To return a transmitted signal to its source, often with a delay to indicate that the signal is a reflection rather than the original. **(C) (software)** A returned signal, as in definition (A). **(D) (data transmission) (general)** A wave which has been reflected or otherwise returned with sufficient magnitude and delay to be perceived in some manner as a wave distinct from that directly transmitted. *Note:* Echoes are frequently measured in decibels relative to the directly transmitted wave. **(E) (computer graphics)** The immediate notification of the current values provided by an input device to an operator at a display console; for example, by displaying the input value. (SWG/PE/SUB/C) C37.1-1987, 610.7-1995, C37.100-1992, 610.12-1990, 599-1985, 610.6-1991
(2) (facsimile) A wave that has been reflected at one or more points with sufficient magnitude and time difference to be perceived in some manner as a wave distinct from that of the main transmission. (COM) 168-1956w
(3) The second subaction packet. This 8-byte packet reports the status of the queueing of the corresponding send packet. (C/MM) 1596-1992
(4) In a communication channel, noise characterized by undesired return of the transmitted signal back to the sender after a delay interval corresponding to the round-trip transmission time, caused by improper echo suppression or impedance mismatch. (C) 610.7-1995
(5) The portion of energy of the radar signal that is reflected to a receiver. (AES) 686-1997

echo area, effective *See:* effective echo area.

echo attenuation (data transmission) In a four-wire or two-wire circuit in which the two directions of transmission can be separated from each other, the attenuation of the echo currents (which return to the input of the circuit under consideration) is determined by the ratio of the transmitted power to the echo power received expressed in decibels. (PE) 599-1985w

echo box A high-Q resonant cavity that stores part of the transmitted pulse power and feeds the resulting exponentially decaying power into the receiver after completion of the pulse transmission. (AES) 686-1997

echo canceler disabling tone A 1350 ms, 2100-Hz tone sequence, which includes two phase reversals. It is used to disable echo cancelers and echo suppressors. (COM/TA) 743-1995

echo check (1) A method of checking the accuracy of transmission of data in which the received data are returned to the sending end for comparison with the original data. (MIL) [2]
(2) (data management) A check in which information that has been transmitted is returned to the information source and compared with the original information to ensure accuracy of the transmission. *Synonym:* read-back check. (C) 610.5-1990w
(3) An error control technique in which the receiving terminal or computer returns the original message to the sender to verify that the message was received correctly. *See also:* echoplex. (C) 610.7-1995

echo checkback message A communication technique assuring that a message received at the termination point in a system is the same message as originally transmitted. The received message is retransmitted to the sending device and matched to ensure that the original message was received properly. (SUB/PE) C37.1-1994

echo path delay (EPD) (1) The time difference between an incident signal and the returned reflection of that incident signal. There can be multiple echo paths with different delays. (COM/TA) 743-1995
(2) The total delay of the echo path from the receive electrical test point to the send electrical test point, excluding any delay in the test equipment. (COM/TA) 1329-1999

echo path loss (EPL) The difference in dB between an incident signal and the returned reflection of that signal. There can be multiple echo paths with different losses. (COM/TA) 743-1995

echo path response The output at the send electrical test point due to an input at the receive electrical test point. It is a measure of acoustic, vibration, and electrical coupling from the receive circuit to the send circuit. (COM/TA) 1329-1999

echoplex An echo check applied to network terminals operating in duplex transmission to assure that data is received correctly at the other end. (C) 610.7-1995

echo radar (navigation aid terms) The portion of energy of the transmitted pulse which is reflected to a receiver. (AES/GCS) 172-1983w

echo ranging (navigation aid terms) The process of determination of distance by measuring the time interval between transmission of a radiant energy source, usually sound, and the return of its echo. *See also:* radio-acoustic ranging. (AES/GCS) 172-1983w

echo return loss (ERL) (1) The frequency weighted average of the return losses over the middle of the voice band, with the far end terminated with a specified impedance. The 3 dB down frequencies of the weighting are 560 Hz to 1965 Hz. *See also:* singing return loss. (COM/TA) 743-1995
(2) A frequency-weighted average, over the middle of the voice band, of the return losses RL(f) at any point in a channel, with the output of the channel terminated with a specified standard impedance. The weighting is given in IEEE Std 743-1984. The 3 dB bandwidth of the weighting is 560 Hz to 1965 Hz. (COM/TA) 1007-1991r
(3) *See also:* terminal coupling loss. (COM/TA) 1329-1999

echo, second-time-around *See:* second-time-around echo.

echo sounder (navigation aid terms) An instrument used for echo sounding. (AES/GCS) 172-1983w

echo sounding (navigation aid terms) Determination of the depth of water by measuring the time interval between emissions of a sonic or ultrasonic signal and the return of its echo from the bottom. (AES/GCS) 172-1983w

echo sounding system (depth finer) A system for determination of the depth of water under a ship's keel, based on the measurement of clapsed time between the propagation and projection through the water of a sonic or supersonic signal, and reception of the echo reflected from the bottom. (PE/EEC) [119]

echo suppressor (1) (navigation aid terms) (navigation) A circuit component that desensitizes the receiving equipment for a period after the reception of one pulse, for the purpose of rejecting pulses arriving later over indirect reflection paths. (AES/GCS) 172-1983w
(2) (data transmission) A voice-operated device for connection to a two-way telephone circuit to attenuate echo current in one direction caused by telephone current in the other direction. (PE) 599-1985w

ECL *See:* emitter-coupled logic.

eclipsing The loss of information on radar echoes at ranges when the receiver is blanked because of the occurrence of a transmitter pulse. Numerous such blankings can occur in radars having high pulse-repetition frequencies. *See also:* blind range. (AES) 686-1997

ECM *See:* electronic countermeasures.

ecology The interrelation between organisms and their environment or the division of biology concerned with the study of such relationships. (T&D/PE) 539-1990

Econometric Software Package (ESP) A programming language used for statistical analysis of time series and other data by regression and more sophisticated econometric techniques. Includes data editing, transformation and display, matrix manipulation, and a variety of complex forecasting procedures. (C) 610.13-1993w

economic dispatch (electric power system) The optimization of the incremental cost of delivered power by allocating generating requirements among the on-control units with consideration of such factors as incremental generating costs and incremental transmission losses. (PE/PSE) 94-1991w

economic dispatch control (ED) (electric power system) An automatic generation control subsystem designed to allocate unit generation to minimize the incremental cost of delivered power. (PE/PSE) 94-1991w

economy energy (power operations) Energy produced in one system and substituted for less economical energy in another system. *See also:* generating station. (PE/PSE) 858-1987s, 346-1973w

economy power Power produced from a more economical source in one system and substituted for less economical power in another system. *See also:* generating station. (T&D/PE) [10]

ECP *See:* engineering change proposal.

ECR *See:* electronic cash register.

ECSA *See:* Exchange Carriers' Standards Association.

ECSS II *See:* Extendible Computer System Simulator II.

ED *See:* economic dispatch control.

EDD *See:* envelope delay distortion.

eddy current(s) (1) (electrical heating systems) Current that circulates in a metallic material as a result of electromotive forces induced by a variation of magnetic flux. (IA/PC) 844-1991
(2) The currents that are induced in the body of a conducting mass by the time variation of magnetic flux. (PE/TR) C57.12.80-1978r

eddy-current braking (rotating machinery) A form of electric braking in which the energy to be dissipated is converted into heat by eddy currents produced in a metallic mass. *See also:* asynchronous machine. (PE) [9]

eddy-current loss (1) (parts, hybrids, and packaging) Power dissipated due to eddy currents. *Note:* The eddy-current loss of a magnetic device includes the eddy-current losses in the core, windings, case, and associated hardware. (CHM) [51]
(2) (power and distribution transformers) The energy loss resulting from the flow of eddy currents in a metallic material. (PE/TR) C57.12.80-1978r

edge (1) (image processing and pattern recognition) In image processing, a set of pixels belonging to an arc and having the property that pixels on opposite sides of the arc have differing gray levels. (C) 610.4-1990w
(2) A logic state transition that is considered instantaneous for a given pattern in the simulation process. (SCC20) 1445-1998

edge detection (image processing and pattern recognition) An image segmentation technique in which edge pixels are identified by examining their neighborhoods. *See also:* edge linking. (C) 610.4-1990w

edge diffraction Diffraction by a transverse obstacle with a relatively sharp profile, located between the transmission and reception points. Diffraction over a very sharp profile is frequently called knife-edge diffraction. (AP/PROP) 211-1997

edge enhancement (image processing and pattern recognition) An image enhancement technique in which edges are sharpened by increasing the contrast between the gray levels of the pixels on opposite sides of the edge. (C) 610.4-1990w

edge image An image in which each pixel is labeled as either an edge pixel or a non-edge pixel. (C) 610.4-1990w

edge linking An image processing technique in which neighboring pixels labeled as edge pixels are connected to form an edge. (C) 610.4-1990w

edge operator A neighborhood operator that determines which pixels in an image are edge pixels. (C) 610.4-1990w

edge pixel A pixel that lies on an edge. (C) 610.4-1990w

EDIF *See:* Electronic Design Interchange Format.

edge-coated card A punch card that has been strengthened by treating one or more of its edges with a special coating. (C) 610.10-1994w

edge-notched card A punch card into which notches representing data are punched around the edges. *See also:* edge-punched card. (C) 610.10-1994w

edge-punched card A punch card that is punched with hole patterns in tracks along the edges. *Synonym:* verge-punched card. (C) 610.10-1994w

edge sensitive Pertaining to a circuit that responds to a transition, usually in one direction, of an input signal; for example, responding to the rising edge of a signal. *Contrast:* level sensitive. (C) 610.10-1994w

edge-sensitive signal Signals whose leading and/or trailing edges are used to strobe information contained on level sensitive signals. (C/BA) 896.9-1994w

edge-sensitive storage element A storage element mapped to by a synthesis tool that

a) Propagates the value at the data input whenever an appropriate value is detected on a clock control input, and
b) Preserves the last value propagated at all other times, except when any asynchronous control inputs become active.

(For example, a flip-flop.) (C/DA) 1076.6-1999

Edison distribution system A three-wire direct-current system, usually about 120−240 volts, for combined light and power service from a single set of mains. *See also:* direct-current distribution. (T&D/PE) [10]

Edison effect *See:* thermionic emission.

Edison storage battery An alkaline storage battery in which the positive active material is nickel oxide and thenegative an iron alloy. *See also:* battery. (EEC/PE) [119]

E-display A rectangular display in which targets appear as intensity-modulated blips with range indicated by the horizontal coordinate and elevation angle by the vertical coordinate. *Note:* The term "E-display" has also been applied to a display in which height or altitude is the vertical coordinate. This usage is deprecated because of ambiguity. The preferred term for such a display is "range-height indicator (RHI)."

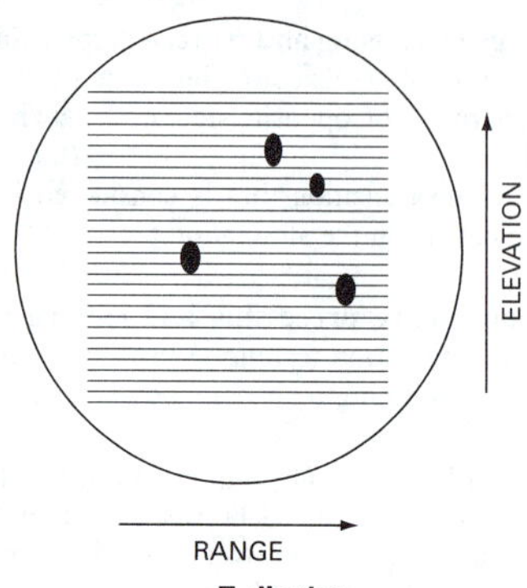

E-display

(AES) 686-1997

edit (1) (computers) To modify the form or format of data, for example, to insert or delete characters such as page numbers or decimal points. (MIL) [2]
(2) (software) To modify the form or format of computer code, data, or documentation; for example, to insert, rearrange, or delete characters. (C) 610.12-1990

editing symbol In micrographics, a symbol on microfilm that is human readable without magnification and that provides cutting, loading, or other preparation instructions. (C) 610.2-1987

editor *See:* linkage editor; text editor.

EDP *See:* electronic data processing.

EDR *See:* electrodermal reaction.

EEI *See:* external environment interface.

eel *See:* conductor cover.

EEPROM *See:* electronically erasable programmable read-only memory.

EEPROM redundancy *See:* redundancy.

effect A change in an organism or in a specific biological parameter as a result of application of some treatment (e.g., chemical). Also, a difference in some parameter between a control and treatment group that is biologically and/or statistically significant. (PE/T&D) 539-1990

effective address (1) (microprocessor assembly language) The result of evaluating an address in accordance with its addressing mode. (C/MM) 695-1985s
(2) (software) The address that results from performing any required indexing, indirect addressing, or other address modification on a specified address. *Note:* If the specified address requires no modification, it is also the effective address. *See also:* relative address; indirect address; generated address. (C) 610.12-1990, 610.10-1994w
(3) (computers) The address that is derived by applying any specified rules (such as rules relating to an index register or indirect address) to the specified address and that is actually used to identify the current operand. (C) [20]

effective aperture (1) (EM-radiation collection device) (radar) Synonymous with effective area for an antenna (IEEE Std 145-1973, Definitions of Terms for Antennas); also, the effective area of other EM-radiation collecting devices, such as lenses. (AES/RS) 686-1982s
(2) Normalized beamwidth of the SAW generated at center frequency and normalized to the corresponding wavelength. (UFFC) 1037-1992w

effective area (of an antenna) (in a given direction) In a given direction, the ratio of the available power at the terminals of a receiving antenna to the power flux density of a plane wave incident on the antenna from that direction, the wave being polarization matched to the antenna. *Notes:* 1. If the direction is not specified, the direction of maximum radiation intensity is implied. 2. The effective area of an antenna in a given direction is equal to the square of the operating wavelength times its gain in that direction divided by 4π. *See also:* polarization match. (AP/ANT) 145-1993

effective area antenna (data transmission) The ratio of the power available at the terminals of an antenna to the incident power density of a plane wave from that direction polarized, coincident with the polarization that the antenna would radiate. *See also:* antenna. (PE/AP/ANT) 599-1985w, [35]

effective area, partial *See:* partial effective area.

effective asymmetrical fault current (1) (safety in ac substation grounding) The root-mean-square (rms) value of asymmetrical current wave, integrated over the entire interval of fault duration. See corresponding figure. *Note:* It can be expressed as

$$I_F = D_f(t_f)I_f$$

where

I_F = effective asymmetrical current in A
I_f = (initial) symmetrical ground fault current in amperes
$D_f(t_f)$ = decrement factor accounting for the effect of a dc offset during the subtransient period of fault current wave on an equivalent time basis of the entire fault duration, t_f, for t_f given in s.2T

(2) The rms value of asymmetrical current wave, integrated over the interval of fault duration.

$$I_F = D_f \times I_f$$

where

I_F = the effective asymmetrical fault current in A
I_f = the rms symmetrical ground fault current in A
D_f = the decrement factor

(PE/SUB) 80-2000

effective band (facsimile) The frequency band of a facsimile signal wave equal in width to that between zero frequency and maximum keying frequency. *Note:* The frequency band occupied in the transmission medium will in general be greater than the effective band. (COM) 168-1956w

effective bandwidth (bandpass filter in a signal transmission system) The width of an assumed rectangular bandpass filter having the same transfer ratio at a reference frequency and passing the same mean square of a hypothetical current and voltage having even distribution of energy over all frequencies. *Note:* For a nonlinear system, the bandwidth at a specified input level. *See also:* signal; network analysis. (IE/AP/ANT) [43], 145-1983s

effective bunching angle (reflex klystrons) In a given drift space, the transit angle that would be required in a hypothetical drift space in which the potentials vary linearly over the same range as in the given space and in which the bunching action is the same as in the given space. (ED) 161-1971w

effective capacitance The imaginary part of a capacitive admittance divided by the angular frequency. (COM) [49]

effective ceiling cavity reflectance (illuminating engineering) A number giving the combined reflectance effect of the wall and ceiling reflectance of the ceiling cavity. *See also:* ceiling cavity ratio. (EEC/IE) [126]

effective center (radiation protection) The point within a detector that produces, for a given set of irradiation conditions, an instrument response equivalent to that which would be produced if the entire detector were located at the point. (NI) N42.17B-1989r, N323-1978r

effective center of mass (accelerometer) That point defined by the intersection of the pendulous axis and an axis parallel to the output axis about which angular acceleration results in minimum accelerometer output. *See also:* spin-offset coefficient. (AES/GYAC) 528-1984s

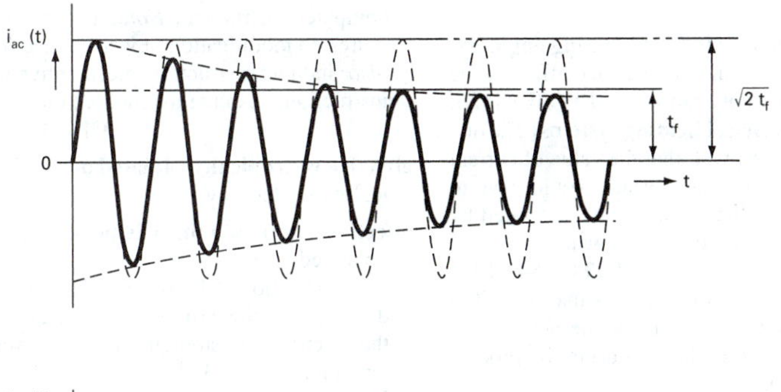

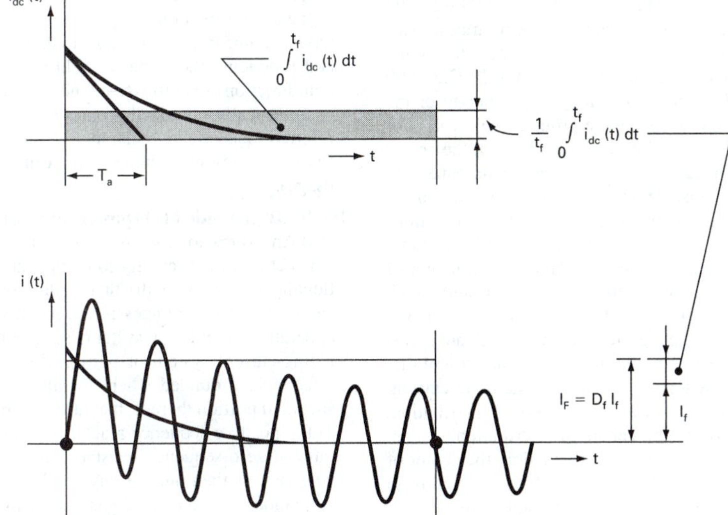

Relationship between actual values of fault current and values of I_F, I_f, and D_f for fault duration t_f

effective asymmetrical fault current

effective center-of-mass for angular acceleration (accelerometer) That point defined by the intersection of the pendulous axis and an axis parallel to the output axis, about which angular acceleration results in a minimum accelerometer output. (AES/GYAC) 528-1994

effective center-of-mass for angular velocity (accelerometer) (inertial sensors) That point defined by the intersection of the pendulous axis and an axis of constant speed rotation approximately parallel to the input axis, for which the offset due to spin becomes independent of orientation. *See also:* spin-offset coefficient. (AES/GYAC) 528-1994

effective cutoff frequency A frequency at which its insertion loss between specified terminating impedances exceeds by some specified amount the loss at some reference point in the transmission band. *Note:* The specified insertion loss is usually three decibels. *See also:* cutoff frequency; network analysis. (EEC/PE) [119]

effective dc inertia constant (H_{dc}) The rotational ac system inertia constant H converted to the base of dc power. (PE/T&D) 1204-1997

effective dielectric constant A parameter frequently used to characterize the phase velocity of modes propagating on planar transmission lines with inhomogeneous or anisotropic media. It is the square of the ratio of actual to free space propagation constant of a mode. It is the dielectric constant of an equivalent line filled by a nonmagnetic, homogeneous, isotropic medium in which a TEM wave propagates with the same phase velocity. (MTT) 1004-1987w

effective echo area (radar) The area of a fictitious perfect electromagnetic reflector that would reflect the same amount of energy back to the radar as the target. *See also:* navigation. (AES) [42]

effective echoing area *See:* radar cross section.

effective efficiency (1) (bolometer units) The ratio of the substitution power to the total RF power dissipated within the bolometer unit. *Notes:* 1. Effective efficiency includes the combined effect of the direct-current-radio-frequency substitution error and bolometer unit efficiency. *See also:* bolometric power meter. (IM) 470-1972w, [38]
(2) (electrothermic unit) The ratio of the substituted reference power (direct current, audio or radio frequency) in the electrothermic unit to the power dissipated within the electrothermic unit for the same direct current output voltage from the electrothermic unit at a prescribed frequency, poer level, and temperature. *Notes:* 1. Calibration factor and effective efficiency are related as follows:

$$\frac{K_b}{\eta_c} = 1 - |\Gamma|^2$$

where K_b, η_c and Γ are the calibration factor, effective efficiency, and reflection coefficient of the electrothermic unit, respectively. 2. The reference frequency is to be supplied with the calibration factor. (IM) 544-1975w

effective energy (radiation survey instruments) The energy of monochromatic photons which undergoes the same percentage attenuation in a specified filter as the heterogeneous beam under consideration. Aluminum is the filter specified for photon energies less than, or equal to, 100 kiloelectronvolts (keV), copper for photon energies between 100 keV and 1.5 megaelectronvolts (MeV), and lead for photons with energies greater than 1.5 MeV. (NI) N13.4-1971w

effective field *See:* rms field.

effective floor cavity reflectance, rfc (illuminating engineering) A number giving the combined reflectance effect of the floor cavity. *See also:* floor cavity ratio. (EEC/IE) [126]

effective flux penetration (electrical heating systems) The distance into a pipeline or a vessel wall that the value of current induced by the magnetic field at the surface would have to penetrate in order to generate the same heat as generated by the actual induced current distribution in the wall. (IA/PC) 844-1991

effective group ID (1) An attribute of a process that is used in determining various permissions, including file access permissions. This value is subject to change during the process lifetime. *See also:* group ID. (C/PA) 9945-1-1996, 9945-2-1993
(2) An attribute of a process that is used in determining various permissions. This value is subject to change during the process lifetime. *See also:* group ID. (C) 1003.5-1999

effective height (A) *High-frequency usage.* The height of the antenna center of radiation above the ground level. *Note:* For an antenna with symmetrical current distribution, the center of radiation is the center of distribution. For an antenna with asymmetrical current distribution, the center of radiation is the center of current moments when viewed from directions near the direction of maximum radiation. **(B) (data transmission)** The effective height of an antenna is the height of its center of radiation above the effective ground level. **(C) (data transmission)** In low-frequency applications the term effective height is applied to loaded or nonloaded vertical antennas and is equal to the moment of the current distribution in the vertical section, divided by the input current. *Note:* For an antenna with symmetrical current distribution, the center of radiation is the center of distribution. For an antenna with asymmetrical current distribution, the center of radiation is the center of current moments when viewed from directions near the direction of maximum radiation. (AP/PE/ANT) 145-1993, 599-1985

effective height antenna (1) (A) (data transmission) The height of its center of radiation above the effective ground level. **(B) (data transmission)** In low-frequency applications, the term "effective height" is applied to loaded or nonloaded vertical antennas and is equal to the moment of the current distribution in the vertical section, divided by the input current. *Note:* For an antenna with symmetrical current distribution, the center of radiation is the center of distribution. For an antenna with asymmetrical current distribution, the center of radiation is the center of current moments when viewed from directions near the direction of maximum radiation. *See also:* antenna. (PE) 599-1985
(2) (mobile communication) The height of the center of a vertical antenna (of at least 1.4 wavelength) above the effective ground plane of the vehicle on which the antenna is mounted. *See also:* mobile communication system; antenna. (VT) [37]

effective inductance (1) (general) The imaginary part of an inductive impedance divided by the angular frequency. (IM/HFIM) [40]
(2) (winding) The self-inductance at a specified frequency and voltage level, determined in such a manner as to exclude the effects of distributed capacitance and other parasitic elements of the winding but not the parasitic elements of the core. (CHM) [51]

effective induction area of the control current loop (Hall effect devices) The effective area of the loop enclosed by the control current leads and the relevant conductive path through the Hall element. (MAG) 296-1969w

effective induction area of the output loop (Hall effect devices) The effective induction area of the loop enclosed by the leads to the Hall terminals and the relevant conductive path through the Hall plate. (MAG) 296-1969w

effective instruction (software) The computer instruction that results from performing any required indexing, indirect addressing, or other modification on the addresses in a specified computer instruction. *Note:* If the specified instruction requires no modification, it is also the effective instruction. *See also:* indirect instruction; presumptive instruction; immediate instruction; absolute instruction; direct instruction. (C) 610.12-1990, 610.10-1994w

effective isotropically radiated power *See:* equivalent isotropically radiated power.

effective length of a linearly polarized antenna For a linearly polarized antenna receiving a plane wave from a given direction, the ratio of the magnitude of the open circuit voltage developed at the terminals of the antenna to the magnitude of the electric field strength in the direction of the antenna polarization. *Notes:* 1. Alternatively, the effective length is the length of a thin straight conductor oriented perpendicularly to the given direction and parallel to the antenna polarization, having a uniform current equal to that at the antenna terminals and producing the same far-field strength as the antenna in that direction. 2. In low-frequency usage, the effective length of a vertically polarized ground-based antenna is frequently referred to as effective height. Such usage should not be confused with **effective height of an antenna** (high-frequency usage). (AP/ANT) 145-1993

effectively grounded (1) (power and distribution transformers) An expression that means grounded through a grounding connection of sufficiently low impedance (inherent or intentionally added, or both) that fault grounds that may occur cannot build up voltages in excess of limits established for apparatus, circuits, or systems so grounded. 1. 1. An alternating-current system or portion thereof may be said to be effectively grounded when, for all points on the system or specified portion thereof, the ratio of zero-sequence reactance to the positive-sequence reactance is less than three and the ratio of zero-sequence resistance to positive-sequence reactance is less than one for any condition of operation and for any amount of connected generator capacity. (PE/TR) C57.12.80-1978r
(2) Intentionally connected to earth through a ground connection or connections of sufficiently low impedance and having sufficient current-carrying capacity to limit the buildup of voltages to levels below that which may result in undue hazard to persons or to connected equipment. (NESC) C2-1997
(3) (A) (grounding of industrial and commercial power systems) An expression that means grounded through a grounding connection of sufficiently low impedance (inherent or intentionally added or both) that ground fault that may occur cannot build up voltages in excess of limits established for apparatus, circuits, or systems so grounded. *Notes:* 1. An alternating-current system or portion thereof may be said to be effectively grounded when, for all points on the system or specified portion thereof, the ratio of zero-sequence reactance to positive-sequence reactance is not greater than three and the ratio of zero-sequence resistance to positive-sequence reactance is not greater than one for any condition of operation and for any amount of connected generator capacity. 2. This definition is basically used in the application of line-to-neutral surge arresters. surge arresters with less than line-to-line voltage ratings are applicable on effectively grounded systems. *See also:* ground; grounded. **(B) (system grounding)** Grounded through a sufficiently low impedance such that for all system conditions the ratio of zero-sequence reactance (X_0 $/X_1$) is positive and less than 3, and the ratio of zero-sequence resistance to positive-sequence reactance (R_0 $/X_1$) is positive and less than 1. *Note:* The effectively grounded system permits the application of surge arresters with less than line-to-line voltage ratings. Ground fault currents will be approximately of the same magnitude as three-phase fault currents. (IA/PSE) 142-1982

effectively grounded system A system in which the neutral points are connected directly to the ground through a connection in which no impedance has been inserted intentionally. (PE/C) 1313.1-1996

effective ground plane (mobile communication) The height of the average terrain above mean sea level as measured for a distance of 100 meters out from the base of the antenna in the desired direction of communication. It may be considered the same as ground level only in open flat country. *See also:* mobile communication system. (VT) [37]

effective input impedance (1) (electron tube or valve) (output) The quotient of the sinusoidal component of the control-electrode voltage (output-electrode voltage) by the corresponding component of the current for the given electrical conditions of all the other electrodes. *See also:* ON period. (ED) [45], [84]
(2) (output) The quotient of voltage by current at the input port of a device when it is operating normally (usually steady-state). (CAS) [13]

effective height base station antenna (mobile communication) The height of the physical center of the antenna above the effective ground plane. *See also:* mobile communication system. (VT) [37]

effective length antenna (1) (A) (general) For an antenna radiating linearly polarized waves, the length of a thin, straight conductor oriented perpendicular to the direction of maximum radiation, having a uniform current equal to that at the antenna terminals and producing the same far field strength as the antenna. **(B) (general)** Alternatively, for the same antenna receiving linearly polarized waves from the same direction, the ratio of the open-circuit voltage developed at the terminals of the antenna to the component of the electric field strength in the direction of antenna polarization. *Notes:* 1. The two definitions yield equal effective lengths. 2. In low-frequency usage, the effective length of a ground-based antenna is taken in the vertical direction and is frequently referred to as effective height. Such usage should not be confused with effective height (of an antenna, high-frequency usage). *See also:* antenna. (AP/ANT) [35]
(2) The ratio of the antenna open-circuit voltage to the strength of the field component being measured. *See also:* electromagnetic compatibility. (EMC) [53]

effective medium The replacement of an inhomogeneous medium by an equivalent homogeneous medium having complex constitutive parameters derived from the propagation of the coherent (i.e., mean) field in the actual medium. The equivalent medium describes only the coherent field. (AP/PROP) 211-1997

effective mode volume (fiber optics) The square of the product of the diameter of the near-field pattern and the sine of the radiation angle of the far-field pattern. The diameter of the near-field radiation pattern is defined here as the full width at half maximum and the radiation angle at half maximum intensity. *Note:* Effective mode volume is proportional to the breadth of the relative distribution of power amongst modes in a multimode fiber. It is not truly a spatial volume but rather an "optical volume" equal to the product of area and solid angle. *See also:* radiation pattern; mode volume. (Std100) 812-1984w

effective multiplication factor The ratio of the average number of neutrons produced by nuclear fission in each generation to the total number of corresponding neutrons absorbed or leaking out of the system. If $k_{eff} = 1$, that is, the number of neutrons produced is equal to the number being absorbed or leaking out of the system, a stable, self-sustaining chain reaction exists and the assembly is said to be critical. If $k_{eff} > 1$, the chain reaction is not self-sustaining and will terminate, such a system is said to be subcritical. If $k_{eff} < 1$, the chain reaction is divergent and the system is supercritical. (PE/PSE) 858-1987s

effectiveness analysis An analysis of how well a design solution will perform or operate given anticipated operational scenarios. (C/SE) 1220-1998

effectiveness assessment The evaluation of the design solution with respect to manufacturing, test, distribution, operations, support, training, environmental impact, cost effectiveness, and life cycle cost. (C/SE) 1220-1998

effectiveness criteria The measure of value used to determine the success or failure of a design solution. (C/SE) 1220-1998

effective power The difference, expressed in dBm, between the absolute optical power, measured in milliwatts, at the midpoint in time of a high optical signal vs. the midpoint in time of a low optical signal. (C/BA) 1393-1999

effective radiated power (ERP) (1) In a given direction, the relative gain of a transmitting antenna with respect to the maximum directivity of a half-wave dipole multiplied by the net power accepted by the antenna from the connected transmitter. *Synonym:* equivalent radiated power. *Contrast:* equivalent isotropically radiated power. (AP/ANT) 145-1993
(2) (mobile communication) The product in a given direction of the effective gain of the antenna in that direction over a half-wave dipole antenna, and the antenna power input. *See also:* mobile communication system. (VT) [37]

effective radius of the Earth (1) (radio-wave propagation) An effective value for the radius of the Earth that is used in place of the actual radius to correct approximately for atmospheric refraction. *Note:* Under conditions of standard refraction, the effective radius of the Earth is 8.5×10^6 m, or 4/3 the geometrical radius. (AP/PROP) 211-1997
(2) (data transmission) In radio transmission, a value which is used in place of the geometrical radius to correct for atmospheric refraction when the index of refraction in the atmosphere changes linearly with height. *Note:* Under conditions of standard refraction, the effective radius is 1.33 the geometrical radius. (PE) 599-1985w

effective range of measurement The range of values of the quantity to be measured by which the performance of a dosimeter meets the requirements of this standard. (NI) N42.20-1995

effective relative permeability A parameter frequently used to characterize the phase velocity of modes propagating on planar transmission lines containing a magnetic material as compared to a transmission line of the same configuration and equal dielectric constant, but with free space magnetic properties. It is the square of the ratio of the propagation constants of these two transmission lines. (MTT) 1004-1987w

effective relay actuation time The sum of the initial actuation time and the contact chatter intervals following such actuation. (EEC/REE) [87]

effective resistance The effective resistance, or ac resistance, of a series reactor is derived by dividing the total losses, as defined in (1), (2) and (3) above, by the current squared. (PE/TR) C57.16-1996

effective resistivity A factor such that the conduction current density is equal to the electric field in the material divided by the resistivity. (PE/PSIM) 81-1983

effective shielding That which permits lightning strokes no greater than those of critical amplitude (less design margin) to reach phase conductors. (SUB/PE) 998-1996

effective sound pressure (root-mean-square sound pressure) At a point over a time interval, the root-mean-square value of the instantaneous sound pressure at the point under consideration. In the case of periodic sound pressures, the interval must be an integral number of periods or an interval long compared to a period. In the case of nonperiodic sound pressures, the interval should be long enough to make the value obtained essentially independent of small changes in the length of the interval. *Note:* The term effective sound pressure is frequently shortened to sound pressure. (SP) [32]

effective speed of transmission Speed, less than rated, of information transfer that can be averaged over a significant period of time and that reflects effects of control codes, timing codes, error detection, retransmission, tabbing, hand keying, etc. (COM) [49]

effective strip width The strip width of an idealized planar transmission line introduced to model an actual physical structure having equivalent electrical characteristics of interest. Two examples are: effective width of microstrip (conductor thickness correction). The width of a zero thickness microstrip introduced to represent the additional fringing capacitance associated with the finite thickness of the strip conductor; and effective width of microstrip (parallel plate waveguide model). The width of a parallel plate waveguide having magnetic wall boundaries and a height equal to the substrate thickness and having the same phase constant and characteristic impedance as the microstrip mode considered. (MTT) 1004-1987w

effective surface acoustic wave coupling coefficient K_s The electromechanical coupling coefficient $K_s^2 = 2|\Delta v/v|$ where $\Delta v/v$ is the relative velocity change produced by short-circulating the surface potential from the open circuit condition. (UFFC) 1037-1992w

effective synchronous reactance An assumed value of synchronous reactance used to represent a machine in a system study calculation for a particular operating condition. (PE) [9]

effective temperature (1) (laser maser) The temperature that must be used in the Boltzmann formula to describe the relative populations of two energy levels that may or may not be in thermal equilibrium. (LEO) 586-1980w
(2) An arbitrary index that combines, into a single value, the effects of temperature, humidity, and air movement on the sensation of hot or cold felt by the human body. (IA/PSE) 241-1990r

effective thermal resistance (semiconductor devices) (semiconductor rectifiers) The effective temperature rise per unit power dissipation of a designated junction, above the temperature of a stated external reference point under conditions of thermal equilibrium. *Note:* Thermal impedance is the temperature rise of the junction above a designated point on the case, in degrees Celsius per watt of heat dissipation. *See also:* rectification; semiconductor rectifier stack; semiconductor. (ED) 216-1960w

effective turns per phase (rotating machinery) The product of the number of series turns of each coil by the number of coils connected in series per phase and the winding factor. (PE) [9]

effective user ID An attribute of a process that is used in determining various permissions, including file access permissions. This value is subject to change during the process lifetime. *See also:* user ID. (C/PA) 9945-2-1993, 9945-1-1996, 1003.5-1999

effective voltage overshoot (arc-welding apparatus) The area under the transient voltage curve during the time that the transient voltage exceeds the steady-state value. *See also:* voltage recovery time. (EEC/AWM) [91]

effective X/R ratio The value of X/R as seen from the fault location looking back into the power system far enough to include the reduction of the X/R ratio due to the effects of the terminal apparatus. (PE/PSC) 367-1996

efferent Pertaining to a flow of data or control from a superordinate module to a subordinate module in a software system. *Contrast:* afferent. (C) 610.12-1990

efficacy *See:* lumens per watt.

efficiency (1) (x-ray energy spectrometers) (of a semiconductor radiation conductor for a monoenergetic radiation source) The ratio of the number of events in the spectral distribution to the total number of photons incident on the active detector volume during the same time interval. (NPS/NID) 759-1984r
(2) (converter characteristics) (of power conversion) The ratio of active (real) output power and active (real) input power. *Note:* Both powers are to be taken as the total average power as given by the formula:

$$P = \frac{1}{T}\int_0^T ei \quad dt$$

where T = the period for ac (alternating current) and the ripple period for dc (direct current). (IA/SPC) 936-1987w
(3) (rotating machinery) (from total loss) The method of indirect calculation of efficiency from the measurement of total loss. *See also:* direct-current commutating machine; asynchronous machine. (PE) [9]
(4) (rectification) Ratio of the direct-current component of the rectified voltage at the input terminals of the apparatus to the maximum amplitude of the applied sinusoidal voltage in the specified conditions. (ED) [45]
(5) (software) The degree to which a system or component performs its designated functions with minimum consumption of resources. *See also:* storage efficiency; execution efficiency. (C) 610.12-1990
(6) (rotating machinery) (by direct calculation) The method by which the efficiency is calculated from the input and output, these having been measured directly. *See also:* asynchronous machine; direct-current commutating machine. (PE) [9]
(7) (electric power systems in commercial buildings) The power (kW) output divided by the power (kW) input at rated output. (IA/PSE) 241-1990r
(8) The net number of counts registered by the detector system per unit of time, divided by the number of photons of interest originating in the radioactive source during the same unit of time. (NI) N42.12-1994
(9) (power and distribution transformers) (of a transformer) The ratio of the useful power output of a transformer to the total power input. (PE/TR) C57.12.80-1978r, C57.12.90-1999
(10) The output real power divided by the input real power. (IA/PSE) 1100-1999
(11) *See also:* counting efficiency. (NI/NPS) 309-1999

efficiency, effective *See:* effective efficiency.

efficiency, generator *See:* generator efficiency.

efficiency, generator, overall *See:* over-all generator efficiency.

efficiency, generator, reduced *See:* reduced generator efficiency.

efficiency, load circuit *See:* load-circuit efficiency.

efficiency, overall electrical *See:* over-all electrical efficiency.

efficiency, quantum *See:* quantum efficiency.

effluent (1) (monitoring radioactivity in effluents) The liquid or gaseous waste streams released to the environment. (NI) N42.18-1980r
(2) (radiological monitoring instrumentation) Liquid or airborne radioactive materials released to the environs. (NI) N320-1979r

effluve *See:* convective discharge.

E Filter A 1 kHz to 50 kHz bandpass filter used for measuring the power of a digital data signal, noise, or impulse noise on an ISDN basic access digital subscriber line. (COM/TA) 743-1995

EFTS *See:* electronic funds transfer system.

EGM *See:* electrogeometric model.

egoless programming (software) A software development technique based on the concept of team, rather than individual, responsibility for program development. Its purpose is to prevent individual programmers from identifying so closely with their work that objective evaluation is impaired. (C) 610.12-1990

egress The process whereby signals exit the cable system; i.e., signal leakage. (LM/C) 802.7-1989r

EHF *See:* extremely high frequency.

E-H tee (waveguide components) A junction composed of E- and H-plane tee junctions wherein the axes of the arms intersect at a common point in the main guide. *Note:* Compare to "hybrid tee." (MTT) 147-1979w

E-H tuner (waveguide components) An E-H tee having E and H arms terminated in movable open- or short-circuit terminations. (MTT) 147-1979w

EHV *See:* extra-high voltage.

EI *See:* end injection.

EIA *See:* Electronic Industries Association.

EIA-232-D *See:* RS-232-C; EIA/TIA-232-E.

EIA-422-A An EIA standard that specifies electrical characteristics for balanced transmission in which each of the main circuits has its own ground lead. *Note:* There is a 10 Mb/s limit on speed. *Synonym:* RS-422-A. (C) 610.7-1995

EIA-423-A An EIA standard that specifies electrical characteristics for unbalanced circuits using common or shared grounding techniques. *Note:* There is a 300 kb/s limit on speed. *Synonym:* RS-423-A. (C) 610.7-1995

EIA-530 An EIA standard which uses the 25-pin connector commonly associated with EIA-232-D. Note. Represents high-speed electrical characteristics of EIA-422-A and 423-A. (C) 610.10-1994w

EIA/TIA-232-E An EIA/TIA standard for asynchronous serial data communications between terminal devices, such as printers; computers; and communications devices, such as modems. *Note:* IEEE Std 610.7-1995 defines a 25-pin (DB-25) connector and certain electrical and mechanical characteristics for interfacing computer equipment. There is a 20 kb/s limit on speed and 15 m (50 ft) cable limit. *Synonyms:* RS-232; RS-232-C; EIA-232-D. (C) 610.7-1995

EIA/TIA-530-A An EIA/TIA physical and mechanical standard that specifies cabling and connectors for EIA-422-A and EIA-423-A interfaces. IEEE Std 610.7-1995 defines a 37-pin (DB-37) and 9-pin (DB-9) connector. *Synonym:* RS-449. (C) 610.7-1995

EIF *See:* electronic counter-countermeasures (ECCM) improvement factor.

EIFFEL An object-oriented programming language. (C) 610.13-1993w

8B/10B encoding A byte-oriented encoding scheme developed by IBM which encodes 8-bit data into "dc balanced" 10-bit symbols. 8B/10B encoding guarantees a minimum of four (4) non-return-to-zero (NRZ) transitions per 10-bit symbol and, with "running disparity," produces a bit stream with zero dc offset. (C/BA) 1393-1999

eight-bit byte *See:* octet.

eight-hour rating (magnetic contactor) The rating based on its current-carrying capacity for eight hours, starting with new clean contact surfaces, under conditions of free ventilation, with full-rated voltage on the operating coil, and without causing any of the established limitations to be exceeded. (IA/IAC) [60]

eight-pin modular An eight-wire connector. (LM/C) 802.3u-1995s

einschleichender stimulus *See:* accumulating stimulus.

Einstein's law (photoelectric device) The law according to which the absorption of a photon frees a photo-electron with a kinetic energy equal to that of the photon less the work function

$$\frac{1}{2}mv^2 = hv - p \text{ (if } hv > p)$$

See also: photoelectric effect. (Std100) [84]

EIRP *See:* equivalent isotropically radiated power.

EIS *See:* executive information system.

EITHER-OR* *See:* OR.

* Deprecated.

either-way operation *See:* two-way alternate operation.

eject (A) To remove, either manually or under software control, a storage medium, from the storage device; for example, to eject a diskette from a disk drive. **(B)** To advance a printer to the top of the next page to be printed. *Note:* This is commonly called a "form feed." (C) 610.10-1994

EL1 An extensible language that includes most of the concepts of ALGOL 60 and LISP, but with a syntax similar to ALGOL. (C) 610.13-1993w

elapsed time Counting time uncompensated for periods in which an instrument might be unable to respond. Elapsed time of a count equals live time plus dead time. (NI) N42.14-1991

elapsed time printout (sequential events recording systems) The recording of time interval between first and successive detected events. (PE/EDPG) [1]

elastances (system of conductors) (coefficients of potential−maxwell) A set of n conductors of any shape that are insulated from each other and that are mounted on insulating supports within a conducting shell, or on one side of a conducting sheet of infinite extent or above the surface of the earth constitutes a system of n capacitors having mutual elastances and capacitances. *Note:* If the shell (or the earth) is regarded as the electrode common to all n capacitors and the transfers of charge as taking place between shell and the individual electrodes, the sum of the charges on the conductors will be equal and opposite in sign to the charge on the common electrode. The shell (or the earth) is taken to be at zero potential. Let Q_r represent the value of the charge that has been transferred from the shell to the other electrode of the rth capacitor,and let V_r represent the algebraic value of the potential of this electrode resulting from the charges in all n capacitors. If the charges are known the values of the potentials can be computed from the equations:

$$V_1 = S_{11}Q_1 + S_{12}Q_2 + S_{13}Q_3 + \dots$$

$$V_2 = S_{21}Q_1 + S_{22}Q_2 + S_{23}Q_3 + \dots$$

$$V_3 = S_{31}Q_1 + S_{32}Q_2 + S_{33}Q_3 + \dots$$

$$V_r = \sum_{c=l}^{c=n} S_{r,c}Q_c$$

The multiplying operators $S_{r,r}$ are the self-elastances and the multipliers $S_{r,c}$ are the mutual elastances of the system. Maxwell termed them the coefficients of potential of the system. Their values can be measured by noting that the defining equation for the mutual elastance $S_{r,c}$ is

$$S_{r,c} \text{ (reciprocal farad*)} = \frac{V_r\text{(volt)}}{Q_c\text{(coulomb)}}$$

(every Q except Q_c being zero)

It can be shown that $S_{r,c} = S_{c,r}$ and that under the conventions stated all the elastances have positive values. Formerly, sometimes called the daraf. (Std100) 270-1966w

elastic buffer A variable delay element inserted in the ring by the active monitor to ensure that ring latency remains constant when the cumulative latency changes. (C/LM) 8802-5-1998

elasticity buffer A first-in-first-out (FIFO) buffer in the network repeater that can provide temporary storage for a message packet during retransmission delays. The buffer acts as a shift register or delay line, and does not need to hold an entire, full-length packet. *See also:* store-and-forward bufferlocal area networks. (C) 8802-12-1998

elastic-restraint coefficient (inertial sensors) (gyros) The ratio of gimbal restraining torque about an output axis to the output angle. (AES/GYAC) 528-1994

elastic-restraint drift rate (gyros) The component of systematic drift rate that is proportional to the angular displacement of a gyro gimbal about an output axis. The relationship of this component of drift rate to gimbal angle can be stated by means of a coefficient having dimensions of angular displacement per unit time per unit angle. This coefficient is equal to the elastic-restraint coefficient divided by angular momentum. (AES/GYAC) 528-1994

elastomer (rotating machinery) Macromolecular material that returns rapidly to approximately the initial dimensions and shape after substantial deformation by a weak stress and release of the stress. *See also:* asynchronous machine. (PE) [9]

E layer An ionized layer in the E region. The ionization within the E region is highly correlated with the incident solar flux. Therefore, the normal E layer is present only during daytime. (AP/PROP) 211-1997

elbow (separable insulated connectors) A connector component for connecting a power cable to a bushing, so designed that when assembled with the bushing, the axes of the cable and bushing are perpendicular. *See also:* corner. (T&D/PE) 386-1995

electric (1) Containing, producing, arising from, actuated by, or carrying electricity, or designed to carry electricity and capable of so doing. Examples: Electric eel, energy, motor, vehicle, wave. *Note:* Some dictionaries indicate electric and electrical as synonymous but usage in the electrical engineering field has in general been restricted to the meaning given in the definitions above. It is recognized that there are borderline cases wherein the usage determines the selection. *See also:* electrical. (SWG/PE) [56]
(2) Containing, producing, arising from, actuated by, or carrying electricity, or designed to carry electricity and capable of so doing. *Examples:* Electric eel, energy, motor, vehicle, wave. (SWG/PE) C37.100-1992

electric air-compressor governor A device responsive to variations in air pressure that automatically starts or stops the operation of a compressor for the purpose of maintaining air pressure in a reservoir between predetermined limits. (EEC/PE) [119]

electrical (1) (general) Related to, pertaining to, or associated with electricity, but not having its properties or characteristics. Examples: Electrical engineer, handbook, insulator, rating, school, unit. *Note:* Some dictionaries indicate electric and electrical as synonymous but usage in the electrical engineering field has in general been restricted to the meaning given in the definitions above. It is recognized that there are borderline cases wherein the usage determines the selection. *See also:* electric. (SWG/PE) [56]
(2) Related to, pertaining to, or associated with electricity but not having its properties or characteristics. *Examples:* Electrical engineer, handbook, insulator, rating, school, unit. (SWG/PE) C37.100-1992

electrical accomodation (electrobiology) (biology) A rise in the stimulation threshold of excitable tissue due to its electrical environment, often observed following a previous stimulation cycle. *See also:* excitability. (EMB) [47]

electrical accounting machine A machine that is predominantly electromechanical in nature. Examples include keypunches, mechanical sorters, collators, and tabulators. (C) 610.10-1994w

electrical anesthesia (medical electronics) More or less complete suspension of general or local sensibility produced by electric means. (EMB) [47]

electrical arc (gas) A discharge characterized by a cathode drop that is small compared with that in a glow discharge. *Note:* The electron emission of a cathode is due to various causes (thermionic emission, high-field emission, etc.) acting simultaneously or separately, but secondary emission plays only a small part. *See also:* discharge. (Std100) [84]

electrical back-to-back test *See:* pump-back test.

electrical boresight *See:* electric boresight.

electrical center *See:* electric center.

electrical codes (1) (general) A compilation of rules and regulations covering electric installations.
(2) (official electrical code) One issued by a municipality, state, or other political division, and which may be enforced by legal means.
(3) (unofficial electrical code) One issued by other than political entities such as engineering societies, and the enforcement of which depends on other than legal means.
(4) The code of rules and regulations as recommended by the National Fire Protection Association (NFPA) and approved by the American National Standards Institute (ANSI). *Note:* This code is the accepted minimum standard for electric installations and has been accepted by many political entities as their official code, or has been incorporated in whole or in part in their official codes.
(5) A set of rules, prepared by the National Electrical Safety Code committee (secretariat held by the Institute of Electrical and Electronics Engineers) and approved by the American National Standards Institute governing: Methods of grounding; Installation and maintenance of electric-supply stations and equipment; Installation and maintenance of overhead supply and communication lines; Installation and maintenance of underground and electric-supply and communications lines; Operation of electric-supply and communication lines and equipment (Work Rules).

electrical conductor seal, double (nuclear power generating station) An assembly of two single electrical conductor seals in series and arranged in such a way that there is a double pressure barrier seal between the inside and the outside of the containment structure along the axis of the conductors. (PE/NP) 380-1975w

electrical conductor seal, single (nuclear power generating station) A mechanical assembly providing a single pressure barrier between the electrical conductors and the electrical penetration assembly. (PE/NP) 380-1975w

electrical coupling Electrical charges in conductors of a disturbed circuit formed by electrical induction. Since the ratio of a conductor's electrostatic charge to the potential difference between conductors (required to maintain that charge) is the general definition of capacitance, electrical coupling is also called capacitive coupling. Its magnitude depends on the cable geometry and the cable insulation properties: dielectric constant and dissipative losses. Magnetic coupling introduces electromotive force in the disturbed circuit due to magnetic induction. This electromotive force opposes the change in the current that generated it. *Synonym:* capacitive coupling. (PE/IC) 1143-1994r

electrical degree (rotating machinery) The 360th part of the angle subtended, at the axis of a machine, by two consecutive field poles of like polarity. One mechanical degree is thus equal to as many electrical degrees as there are pairs of poles in the machine. *See also:* direct-current commutating machine. (PE) [9]

electrical distance (navigation aid terms) The distance between two points expressed in terms of the duration of travel of an electromagnetic wave in free space between the two points. *Note:* An often used unit of electrical distance is the light-microsecond, approximately 300 m (983 ft). (AES/GCS) 172-1983w

electrical equipment (1) A general term that is applied to materials, fittings, devices, fixtures, and apparatus that are a part of, or are used in connection with, an electrical installation. This includes the electrical power generating system; substations; distribution systems including cable and wiring; utilization equipment; and associated control, protective, and monitoring devices. (IA/PSE) 902-1998
(2) A general term including materials, fittings, devices, appliances, fixtures, apparatus, machines, etc., used as a part of, or in connection with, an electric installation. (IA/PSE) 493-1997

electrical failure (of a circuit breaker) Failure attributable to the application of electrical stresses to the main circuit of the circuit-breaker. (SWG/PE) C37.10-1995

electrical installation, insulation (cable) (electric pipe heating systems) A part that is relied upon to insulate the conductor from other conductors or conducting parts or from ground. Electrical insulation as related to electric pipe heating systems includes that part of a heater that electrically insulates the current carrying conductor(s) from the sheath material. (PE/EDPG) 622-1979s

electrical insulating material (thermal classification of electric equipment and electrical insulation) A substance in which the electrical conductivity is very small (approcahing zero) and provides electric isolation. (EI) 1-1986r

electrical insulation A dielectric material that insulates each conductor from other conductors or from conductive parts at or near earth potential. (IA) 515-1997

electrical insulation system (thermal classification of electric equipment and electrical insulation) An insulating material or a suitable combination of insulating materials specifically designed to perform the functions needed in electric and electronic equipment. (EI) 1-1986r

electrical interchangeability (of fuse links or fuse units) The characteristic that permits the designs of various manufacturers to be used interchangeably so as to provide a uniform degree of overcurrent protection and fuse coordination. (SWG/PE) C37.100-1992, C37.40-1993

electrical length (1) (A) (two-port network at a specified frequency) The length of an equivalent lossless reference waveguide or reference air line (which in the ideal case would be evacuated) introducing the same total phase shift as the two-port when each is terminated in a reflectionless termination. *Note:* It is usually expressed in fractions or multiples of waveguide wavelength. When expressed in radians or degrees it is equal to the phase angle of the transmission coefficient + $2n\pi$. *See also:* waveguide. **(B) (waveguide)** For a traveling wave of a given frequency, a distance in a transmission or guiding medium expressed in wavelengths of the wave in the medium. *Note:* Electrical length is sometimes expressed in radians or degrees. (MTT) 146-1980
(2) For a wave of a given frequency, a distance between field-points, expressed in wavelengths of the wave in the medium. *Note:* The electrical length is sometimes expressed in radians or degrees. *See also:* phase path length. (AP/PROP) 211-1997

electrical load (power operations) Electric power used by devices connected to an electrical generating system. (PE/PSE) 858-1987s

electrically connected Connected by means of a conducting path or through a capacitor, as distinguished from connection merely through electromagnetic induction. *See also:* inductive coordination. (EEC/PE) [119]

electrically alterable read-only memory A type of read-only memory that can be erased electrically. *See also:* electrically erasable programmable read-only memory. (C) 610.10-1994w

electrically erasable programmable read-only memory (EEPROM) (1) A type of read-only memory that can be erased and reprogrammed by electronic methods. (C) 610.10-1994w
(2) A reprogrammable read-only memory in which the cells at each address can be erased electrically and reprogrammed electrically. *Note:* These devices are characterized by slower write times than read times. The memory is used for applications that specify a maximum number of write operations. (ED) 1005-1998, 641-1987w

electrically heated airspeed tube A Pitot-static or Pitot-Venturi tube utilizing a heating element for deicing purposes. (EEC/PE) [119]

electrically heated flying suit A garment that utilizes sewn-in heating elements energized by electric means designed to cover the torso and all or part of the limbs. *Note:* It may be a one-piece garment or consist of a coat, trousers, and the like. The lower portion of the one-piece suit is in trouser form. (EEC/PE) [119]

electrically interlocked manual release of brakes (control) A manual release provided with a limit switch that is operated when the braking surfaces are disengaged manually. *Note:* The limit switch may operate a signal, open the control circuit, or perform other safety functions. *See also:* switch. (IA/ICTL/IAC) [60]

electrically operated valve (power system device function numbers) An electrically operated, controlled or monitored valve used in a fluid line. *Note:* The functions of the valve may be indicated by the use of the suffixes in 3.3 of IEEE Std C37.2-1979. (SUB/PE) C37.2-1979s

electrically programmable Pertaining to any memory in which binary digits may be entered electrically using a special programming device. This process is often referred to as "burning." (C) 610.10-1994w

electrically release-free (as applied to an electrically operated switching device) A term indicating that the release can open the device even though the closing control circuit is energized. *Note:* Electrically release-free switching devices are usually arranged so that they are also anti-pump. With such an arrangement, the closing mechanism will not reclose the switching device after opening until the closing control circuit is opened and again closed. *Synonym:* electrically trip-free. (SWG/PE) C37.100-1992

electrically reset relay A relay that is so constructed that it remains in the picked-up condition even after the input quantity is removed; an independent electrical input is required to reset the relay. (SWG/PE) C37.100-1992

electrically short dipole A dipole whose total length is small compared to the wavelength. *Note:* For the common case that the two arms are collinear, the radiation pattern approximates that of a Hertzian dipole. (AP/ANT) 145-1993

electrically small antenna An antenna whose dimensions are such that it can be contained within a sphere whose diameter is small compared to a wavelength at the frequency of operation. (AP/ANT) 145-1993

electrically suspended gyro (ESG) A free gyro in which the main rotating element—the inertial member—is suspended by an electrostatic or an electromagnetic field within an evacuated enclosure. (AES/GYAC) 528-1994

electrically trip-free *See:* electrically release-free.

electrical metallic tubing A thin-walled metal raceway of circular cross section constructed for the purpose of the pulling in or the withdrawing of wires or cables after it is installed in place. *See also:* raceway. (EEC/PE) [119]

electrical noise (1) Unwanted voltage or current, or both, that appears in an electrical system. For given system characteristics, electrical noise may or may not impair proper functioning. The unwanted noise can have effects that range from totally undetectable to system malfunction or even damage or destruction. (PE/IC) 1143-1994r
(2) Unwanted electrical signals that produce undesirable effects in the circuits of the control systems in which they occur. (IA/PSE/ICTL) 1100-1999, 518-1982r

electrical null (accelerometer) (gyros) The minimum electrical output. It may be specified in terms of rms, peak-to-peak, quadrature component, or other electrical parameters. (AES/GYAC) 528-1994

electrical null position (gyros) (accelerometer) The angular or linear position of a pickoff corresponding to electrical null. (AES/GYAC) 528-1994

electrical objective loudness rating (loudness ratings of telephone connections) For a network

$$\text{EOLR} = -20 \log_{10} \frac{V_T}{1/2V_W}$$

where

V_W = open-circuit voltage of the electric source (in millivolts)

V_T = output voltage of the network (in millivolts)

(COM/TA) 661-1979r

electrical operation Power operation by electric energy. (SWG/PE) C37.100-1992

electrical penetration assembly (nuclear power generating station) An electrical penetration assembly provides the means to allow passage of one or more electrical circuits through a single aperture (nozzle or other opening) in the containment pressure barrier, while maintaining the integrity of the pressure barrier. (PE/NP) 380-1975w

electrical penetration assembly current capacity (nuclear power generating station) The maximum current that each conductor in the assembly is specified to carry for its duty cycle in the design service environment without causing stabilized temperatures of the conductors or the penetration nozzle-concrete interface (if applicable) to exceed their design limits. (PE/NP) 380-1975w

electrical penetration assembly short-time overload rating (nuclear power generating station) The limiting overload current that any one third of the conductors (but in no case less than three of the conductors) in the assembly can carry, for a specified time, in the design service environment, while all remaining conductors carry rated continuous current, without causing the conductor temperatures to exceed those values recommended by the insulated conductor manufacturer as the short-time overload conductor temperature and without causing the stabilized temperature of the penetration nozzle-concrete interface (if applicable) to exceed its design limit. (PE/NP) 380-1975w

electrical pitch The distance between two adjacent connections to the electrical backplane. (C/BA) 14536-1995

electrical preventive maintenance A system of planned inspection, testing, cleaning, drying, monitoring, adjusting, corrective modification, and minor repair of electrical equipment to minimize or forestall future equipment operating problems or failures, which, depending upon equipment type, may require exercising or proof testing. (IA/PSE) 493-1997

electrical range The range expressed in equivalent electrical units. *See also:* instrument; electrical distance. (EEC/EMI) [112]

electric rate tariff *See:* electric rate schedule.

electrical reference plane (standard connector) A transverse plane of the waveguide or transmission line on the drawing standardizing the critical mating dimensions shown in relation to the mechanical reference plane. *Notes:* 1. The electrical reference planes of two mating standard connectors forming a mated standard connector pair nearly coincide. 2. The electrical and mechanical reference planes of standard connectors do not necessarily coincide except for precision coaxial connectors complying with IEEE Std 287-1968[w], Precision Coaxial Connectors, and many connectors for uniconductor waveguides. (IM/HFIM) 474-1973w

electrical reserve (power operations) (electric power supply) The capacity in excess of that required to carry the system load. (PE/PSE) 858-1987s, 346-1973w

electrical resistance heat tracing (1) The utilization of electric heating cables, other electric heating devices, and support components that are externally applied and used to reduce or eliminate ice build-up, to prevent the freezing of pipes or surfaces, or to maintain a pipe or surface at a prescribed temperature. (IA/PC) 515.1-1995
(2) The utilization of electric heating cables, other electric heating devices, and support components that are externally applied and used to maintain or raise the temperature of fluids/materials in piping and associated equipment. (IA) 515-1997

electrical system The existing utility network consisting of interconnected and synchronized generation, transmission, and distribution facilities. (SUB/PE) 1109-1990w

electrical utility (terrestrial photovoltaic power systems) An organization that provides and distributes electric energy to consumers. In the utility interconnected configuration, solar photovoltaic (PV) systems may be interactive with the utility distribution network to permit the interchange of electric power and energy. *See also:* array control. (PV) 928-1986r

electrical zero* *See:* electrical null position.
* Deprecated.

electric back-to-back test *See:* pump-back test.

electric battery A device that transforms chemical energy into electric energy. *See also:* battery. (PE) 599-1985w

electric bell An audible signal device consisting of one or more gongs and an electromagnetically actuated striking mechanism. *Note:* The gong is the resonant metallic member that produces an audible sound when struck. However, the term going is frequently applied to the complete electric bell. (EEC/PE) [119]

electric bias, relay *See:* relay electric bias.

electric blasting cap A device for detonating charges of explosives electrically. *See also:* blasting unit.

electric boresight The tracking axis as determined by an electric indication, such as the null direction of a conical-scanning or monopulse antenna system, or the beam-maximum direction of a highly directive antenna. *See also:* reference boresight. (AP/ANT) 145-1983s

electric brake A mode of operation of the propulsion system in which retardation is provided. *Note:* Although generally considered synonymous with dynamic brake, electric brake is a more global term, in that it includes the possibility of providing retardation by drawing power from the line or by other means not dependent on conversion of kinetic energy into retarding power, which is the key element of dynamic braking. (VT) 1475-1999

electric braking A system of braking wherein electric energy, either converted from the kinetic energy of vehicle movement or obtained from a separate source, is one of the principal agents for the braking of the vehicle or train. *See also:* regenerative braking; magnetic track braking; electropneumatic brake. (EEC/PE) [119]

electric bus A passenger vehicle operating without track rails, the propulsion of which is effected by electric motors mounted on the vehicle. *Note:* A prefix diesel-electric, gas-electric, etc., may replace the word electric. *See also:* trolley coach. (EEC/PE) [119]

electric-cable-reel mine locomotive An electric mine locomotive equipped with a reel for carrying an electric conductor cable that is used to conduct power to the locomotive when operating beyond the trolley wire. *See also:* electric mine locomotive. (EEC/PE) [119]

electric capacitance altimeter An altimeter, the indications of which depend on the variation of an electric capacitance with distance from the earth's surface. (EEC/PE) [119]

electric center (of a power system out of synchronism) A point at which the voltage is zero when a machine is 180° out of phase with the rest of the system. *Note:* There may be one or more electrical centers depending on the number of machines and the interconnections among them. (SWG/PE) C37.100-1992

electric charge time constant (detector) The time required, after the instantaneous application of a sinusoidal input voltage of constant amplitude, for the output voltage across the load capacitor of a detector circuit to reach 63% of its steady-state value. *See also:* electromagnetic compatibility. (INT) [53], [70]

electric coal drill An electric motor-driven drill designed for drilling holes in coal for placing blasting charges. (EEC/PE) [119]

electric components (generating stations electric power system) The electric equipment, assemblies, and conductors that to-gether form the electric power systems. (PE/EDPG) 505-1977r

electric conduction and convection current density At any point at which there is a motion of electric charge, a vector quantity whose direction is that of the flow of positive charge at this point, and whose magnitude is the limit of the time rate of flow of net (positive) charge across a small plane area perpendicular to the motion, divided by this area as the area taken approaches zero in a macroscopic sense, so as to always include this point. *Note:* The flow of charge may result from the movement of free electrons or ions but is not, in general, except in microscopic studies, taken to include motions of charges resulting from the polarization of the dielectric. (Std100) 270-1966w

electric conductivity The property of a material or medium permitting flow of electricity through its volume, expressed as the ratio of electric current density to electric field strength in a material or medium. For isotropic homogeneous media, the conductivity is a scalar quantity, with the preferred unit siemens per meter (S/m); 1 S/m = 1 mho/m. (T&D/PE) 539-1990, 1227-1990r

electric console lift An electrically driven mechanism for raising and lowering an organ console and the organist. *See also:* elevator. (EEC/PE) [119]

electric constant (permittivity or capacitivity of free space pertinent to any system of units) The scalar ε_0 that in that system relates the electric flux density D, in empty space, to the electric field strength $E (D = \varepsilon_0 E)$. *Notes:* 1. It also relates the mechanical force between two charges in empty space to their magnitudes and separation. Thus, in the equation

$$F = Q_1 Q_2 / (n \varepsilon_0 r^2)$$

for the force F between charges Q_1 and Q_2 separated by a distance r, ε_0 is the electric constant, and n is a dimensionless factor that is unity in unrationalized systems and 4π in a rationalized system. 2. In the International System of Units (SI), the magnitude of ε_0 is that of $10^7/(4\pi c^2)$ and the dimension is $(L^{-3} M^{-1} T^4 I^2)$. Here, c is the speed of light expressed in the appropriate system of units. (Std100) 270-1966w

electric contact The junction of conducting parts permitting current to flow. (EEC/PE) [119]

electric controller A device (or group of devices) that serves to govern, in some predetermined manner, the electric power delivered to the apparatus to which it is connected. (IA/MT) 45-1998

electric-controller rail car A trail car used in a multiple-unit train, provided at one or both ends with a master controller and other apparatus necessary for controlling the train. *See also:* electric trail car; electric motor car. (EEC/PE) [119]

electric coupler (1) A group of devices (plugs, receptacles, cable, etc.) that provides for readily connecting or disconnecting electric circuits. (EEC/PE) [119]
(2) A device used to allow trainline signals to be transmitted from vehicle to vehicle or unit to unit in a train, with the connection of trainlines performed automatically when vehicles are coupled. (VT) 1475-1999

electric coupler plug The removable portion of an electric coupler. (EEC/PE) [119]

electric coupler receptacle (electric coupler socket) The fixed portion of an electric coupler. (EEC/PE) [119]

electric coupler socket *See:* electric coupler receptacle.

electric coupling (1) A device for transmitting torque by means of electromagnetic force in which there is no mechanical torque contact between the driving and driven members. The slip type electric coupling has poles excited by direct current on one rotating member, and an armature winding, usually of the double squirrel cage type, on the other rotating member. (IA/MT) 45-1998
(2) *See also:* coupling. (PE/PSC) 487-1992

electric course recorder A device that operates, under control of signals from a master compass, to make a continuous record of a ship's heading with respect to time. (EEC/PE) [119]

electric crab-reel mine locomotive An electric mine locomotive equipped with an electrically driven winch, or crab reel, for the purpose of hauling cars by means of a wire rope from places beyond the trolley wire. *See also:* electric mine locomotive. (EEC/PE) [119]

electric current The flow of electric charge. The preferred unit is the ampere (A). (T&D/PE) 539-1990

electric current density A vector-point function describing the magnitude and direction of charge flow per unit area. The preferred unit is A/m^2. (T&D/PE) 539-1990

electric delay line *See:* electromagnetic delay line.

electric depth recorder A device for continuously recording, with respect to time, the depth of water determined by an echo sounding system. (EEC/PE) [119]

electric design automation application Any software program that interacts with the delay and power calculation module (DPCM) through the procedural interface (PI) to compute instance specific timing values. Examples include batch delay calculators, synthesis tools, floorplanners, static timing analyzers, etc. *See also:* delay and power calculation module; procedural interface. (C/DA) 1481-1999

electric dipole (1) (general) An elementary radiator consisting of a pair of equal and opposite oscillating electric charges an infinitesimal distance apart. *Note:* It is equivalent to a linear current element. (AP/ANT) 149-1979r, [35], 145-1983s
(2) The limit of an electric doublet as the separation approaches zero while the moment remains constant. (Std100/ANT) 270-1966w
(3) *See also:* Hertzian electric dipole. (AP/ANT) 145-1983s

electric dipole moment (two point charges, q and $-q$, a distance a apart) A vector at the midpoint between them, whose magnitude is the product qa and whose direction is along the line between the charges from the negative toward the positive charge. (Std100) 270-1966w

electric-discharge lamp (gas discharge) (illuminating engineering) A lamp in which light (or radiant energy near the visible spectrum) is produced by the passage of an electric currrent through a vapor or a gas. *Note:* Electric-discharge lamps may be named after the filling gas or vapor which is responsible for the major portion of the radiation; for example, mercury lamps, sodium lamps, neon lamps, argon lamps, etc. A second method of designating electric-discharge lamps is by physical dimensions or operating parameters; for example, short-arc lamps, high-pressure lamps, low-pressure lamps, etc. A third method of designating electric-discharge lamps is by their application; in addition to lamps for illumination there are photochemical lamps, bactericidal lamps, blacklight lamps, sun lamps, etc. (EEC/IE) [126]

electric-discharge time constant (detector) The time required, after the instantaneous removal of a sinusoidal input voltage of constant amplitude, for the output voltage across the load capacitor of the detector circuit to fall to 37% of its initial value. *See also:* electromagnetic compatibility. (INT) [53], [70]

electric displacement *See:* electric flux density.

electric displacement density *See:* electric flux density.

electric drive A system consisting of one or several electric motors and of the entire electric control equipment designed to govern the performance of these motors. The control equipment may or may not include various rotating electric machines. (IA/ICTL/IAC) [60]

electric driving machine A machine where the energy is applied by an electric motor. *Note:* It includes the motor and brake and the driving sheave or drum together with its connecting gearing, belt, or chain, if any. *See also:* driving machine. (EEC/PE) [119]

electric elevator A power elevator where the energy is applied by means of an electric motor. *See also:* elevator. (EEC/PE) [119]

electric energy (1) The electric energy delivered by an electric circuit during a time interval is the integral with respect to time of the instaneous power at the terminals of entry of the circuit to a delimited region. *Note:* If the reference direction for energy flow is selected as into the region when the sign of the energy is positive and out of the region when the sign is negative. If the reference direction is selected as out of the region, the reverse will apply. Mathematically,

$$W = \int_{t_0}^{t+t_0} p \, dt$$

where
W = electric energy
p = instantaneous power
t = time during which energy is determined.

When the voltages and currents are periodic, the electric energy is the product of the active power and the time interval, provided the time interval is one or more complete periods or is quite long in comparison with the time of one period. The energy is expressed by

$$W = pt$$

where
P = active power
t = time interval

If the instantaneous power is constant, as is true when the voltages and currents form polyphase symmetrical sets, there is no restriction regarding the relation of the time interval to the period. If the voltages and currents are quasi-periodic and amplitudes of the voltages and currents are slowly varying, the electric energy is the integral with respect to time of the active power, provided the integration is for a time that is one or more complete periods or that is quite long in comparison with the time of one period. Mathematically,

$$W = \int_{t_0}^{t+t_0} P\, dt$$

where P = active power determined for the condition of voltages and currents having slowly varying amplitudes. Electric energy is expressed in joules (watt-seconds) or watthours when the voltages are in volts and the currents in amperes, and the time interval is in seconds or hours, respectively. (Std100/EDPG) 270-1966w

(2) Usually, electric demand integrated over the period of one hour. (PE/PSE) 858-1993w

electric explosion-tested mine locomotive An electric mine locomotive equipped with explosion-tested equipment. *See also:* electric mine locomotive. (EEC/PE) [119]

electric field ($\vec{E}$) (1) (general) A vector field of electric field strength or of electric flux density. *Note:* The term is also used to denote a region in which such vector fields have a significant magnitude. *See also:* vector field. (Std100) 270-1966w

(2) (signal-transmission system) A state of a medium characterized by spatial potential gradients (electric field vectors) caused by conductors at different potentials, that is, the field between conductors at different potentials that have capacitance between them. *See also:* signal. (IE) [43]

(3) The field surrounding a charged object. *See also:* magnetic field. (PE/IC) 1143-1994r

(4) The electric force that acts on a unit electric charge independent of the velocity of that charge. (AP/PROP) 211-1997

electric field induction (1) (capacitive coupling) The process of generating voltages or currents or both in a conductive object or electric circuit by means of time-varying electric fields. *Notes:* 1. "Electric field induction" is preferred over "electric induction" because the latter may be taken to mean electric flux density. 2. "Electric field induction" was formerly called "electrostatic induction." This usage is deprecated because electrostatic fields are time invariant. (T&D/PE) 1048-1990

(2) (grounding of power lines) The induction process that results from time-varying quasi-static electric fields. *Notes:* 1. The term "electric field induction" is preferred over "electric induction" because the latter may be taken to mean electric flux density. 2. Electric field induction was formerly called "electrostatic induction." This usage is deprecated because electrostatic fields are time invariant. (T&D/PE) 539-1990

(3) (capacitive coupling) The process of generating voltages and/or currents in conductive objects or electrical circuits by the induction process that results from time-varying quasi-static electric fields. *Notes:* 1. The term "electric field induction" is preferred over "electric induction" because the latter may be taken to mean electric flux density. 2. Electric field induction was formerly called electrostatic induction. This usage is deprecated because electrostatic fields are time invariant. (T&D/PE) 524a-1993r

electric field integral equation An integrodifferential equation having the form of a Fredholm integral equation of the first kind for the electric current density and its spatial derivative along the surface S of a perfect electric conductor. *Note:* The tangential component of the incident electric field acts as the source for the current, hence the name. The equation is as follows:

$$\hat{n} \times \vec{E}^i = \frac{j}{\omega\varepsilon_0}\hat{n} \times \int_{S_0} \{k_0^2 \vec{J}_S g - (\nabla_{S_0} \cdot \vec{J}_S)\nabla_0 g\} dS_0$$

where
$\hat{n}$ = unit normal to S
$\vec{E}^i$ = incident electric field
$j = \sqrt{-1}$
$k_0^2 = \omega^2\mu_0\varepsilon_0$
$\exp(j\omega t)$ = time convention
$g = \exp[-jk_0|\vec{r} - \vec{r}_0|]/4\pi|\vec{r} - \vec{r}_0|$
∇_0 = gradient evaluated on the surface
∇_{S_0} = gradient in the direction tangential to the surface

(AP/PROP) 211-1997

electric field strength (E) (1) (radio-wave propagation) (fly ash resistivity) (electric field) (kv/cm) (measurement of power frequency electric and magnetic fields from ac power lines) At a given point in space, the ratio of force on a positive test charge placed at the point to the magnitude of the test charge, in the limit that the magnitude of the test charge goes to zero. The electric field strength (E-field) at a point in space is a vector defined by its space components along three orthogonal axes. For steady-state sinusoidal fields, each space component is a complex number or phasor. The magnitudes of the components, expressed by their root-mean-square (rms) values in volts per meter (V/m), and the phases need not be the same. *Note:* The space components (phasors) are not vectors. The space components have a time dependent angle, while vectors have space angles. For example, the sinusoidal electric field E can be expressed in rectangular coordinates as

$$\vec{E} = \hat{a}_x E_x + \hat{a}_y E_y + \hat{a}_z E_z$$

The space component in the x-direction is

$$E_x = \mathrm{Re}\,(E_{x0}\, e^{j\varphi_x}\, e^{j\omega t}) = E_{x0} \cos(\varphi_x + \omega t)$$

The magnitude, phase angle, and time dependent angle are given by E_{x0}, φ_x, and $(\varphi_x + \omega t)$, respectively. In this representation the space angle of the x-component is specified by the unit vector $\hat{a}_x$. An alternative general representation of a steady-state sinusoidal E-field, derivable algebraically from the above equation, and perhaps more useful in characterizing power line fields, is a vector rotating in a plane where it describes an ellipse whose semimajor axis represents the magnitude and direction of the maximum value of the electric field, and whose semiminor axis represents the magnitude and direction of the field a quarter cycle later. The electric field in the direction perpendicular to the plane of the ellipse is zero. *See also:* single-phase ac fields; phasor; polyphase ac fields. (T&D/PE) 644-1994

(2) The ratio of the applied voltage to the ash layer thickness in a test cell used for the laboratory measurement of electrical resistivity of fly ash. bulk density (g/cm^3) (fly ash resistivity). The ratio of ash layer is the ratio of the mass of the particulate in the test cell to the cell volume in a test cell used for the laboratory measurement of electrical resistivity of fly ash. (PE/EDPG) 548-1984w

(3) (overhead power lines) A vector field, often denoted as $\vec{E}$ at a specific point. In a zero magnetic field, it is numerically equal to the force on a motionless unit positive test charge placed at that point. *Note:* In a zero magnetic field, the force $\vec{F}$ is given by $\vec{F} = q\vec{E}$. The magnitudes of the field components are expressed in volts per meter (V/m) (which dimensionally is the same as Newton/Coulomb). *Synonym:* electric field. *See also:* voltage gradient. (T&D/PE) 539-1990

(4) (waveguide) The magnitude of the electric (or magnetic) field vector. *Synonym:* magnetic field strength. (MTT) 146-1980w

(5) The magnitude of the electric field vector $\vec{E}$. The units of electric field strength are in volts per meter. (AP/PROP) 211-1997

(6) A field vector quantity that represents the force (*F*) on a positive test charge (*q*) at a point divided by the charge.

$$E = \frac{F}{q}$$

Electric field strength is expressed in units of volts per meter (V/m). (NIR) C95.1-1999

electric field strength meter An instrument used to measure electric field strength. (T&D/PE) 539-1990

electric field vector (1) (at a point in an electric field) The force on a stationary positive charge per unit charge. *Notes:* 1. This may be measured either in newtons per coulomb or in volts per meter. This term is sometimes called the electric field intensity, but such use of the word intensity is deprecated since intensity connotes power in optics and radiation. *See also:* waveguide; radio-wave propagation. (MTT) 146-1980w

(2) (radio-wave propagation) At a point in an electric field, the force per unit charge acting on a stationary positive charge. *Notes:* 1. This may be expressed either in newtons coulomb or in volts/meter. This term has sometimes been called the electric field intensity, but such use of the word "intensity" is deprecated in favor of field strength since intensity connotes power in optics and radiation. 2. This term has sometimes been called the electric-field intensity, but such use of the word "intensity" is deprecated in favor of field strength, since intensity connotes power in optics and radiation. *Synonyms:* electric vector; electric field strength. (AP) 211-1977s

electric flux density ($\vec{D}$) A vector quantity related to the charge displaced within the medium by an electric field. The electric flux density is that function whose divergence is the charge density. *Note:* Using phasor notation, the electric flux density is given by:

$$\vec{D} = \varepsilon = \cdot \vec{E}$$

where

$\vec{D}$ = the electric flux density
$\varepsilon=$ = the permittivity in the medium
$\vec{E}$ = the electric field

In an isotropic medium, ε is a scalar and $\vec{D}$ is parallel to $\vec{E}$. In an anisotropic medium, $\varepsilon=$ is a tensor and $\vec{D}$ and $\vec{E}$ are not necessarily parallel. The units of electric flux density are in coulombs per meter squared. *Synonym:* electric displacement. (AP/PROP) 211-1997

electric focusing (microwave tubes) The combination of electric fields that acts upon the electron beam in addition to the forces derived from momentum and space charge. (ED) [45]

electric freight locomotive An electric locomotive, commonly used for hauling freight trains and generally designed to operate at higher tractive force values and lower speeds than a passenger locomotive of equal horsepower capacity. *Note:* A prefix diesel-electric, gas-electric, turbine-electric, etc., may replace the word electric. *See also:* electric locomotive. (EEC/PE) [119]

electric gathering mine locomotive An electric mine locomotive, the chief function of which is to move empty cars into, and remove loaded cars from, the working places. *See also:* electric mine locomotive. (EEC/PE) [119]

electric generator A machine that transforms mechanical power into electric power. (IA/MT) 45-1998

electric gun heater An electrically heated element attached to the gun breech to prevent the oil from congealing or the gun mechanism from freezing. (EEC/PE) [119]

electric haulage mine locomotive An electric mine locomotive used for hauling trains of cars, that have been gathered from the working faces of the mine, to the point of delivery of the cars. *See also:* electric mine locomotive. (EEC/PE) [119]

electric horn A horn having a diaphragm that is vibrated electrically. *See also:* protective signaling. (EEC/PE) [119]

electric-hydraulic governor (hydraulic turbines) A governor in which the control signal is proportional to speed error and the stabilizing signals are developed electrically, summed by appropriate electrical networks, and are then hydraulically amplified. Electrical signals may be derived by analog or digital means. (PE/EDPG) 125-1988r

electric hygrometer An instrument for indicating by electric means the humidity of the ambient atmosphere. *Note:* Electric hygrometers usually depend for their operation on the relation between the electric conductance of a film of hygroscopic material and its moisture content. *See also:* instrument. (EEC/PE) [119]

electric incline railway A railway consisting of an electric hoist operating a single car with or without counterweights, or two cars in balance, which car or cars travel on inclined tracks. *See also:* elevator. (EEC/PE) [119]

electric indication lock An electric lock connected to a lever of an interlocking machine to prevent the release of the level or latch until the signals, switches, or other units operated, or directly affected by such lever, are in the proper position. *See also:* interlocking. (EEC/PE) [119]

electric indication locking Electric locking adapted to prevent manipulation of levers that would bring about an unsafe condition for a train movement in case a signal, switch, or other operated unit fails to make a movement corresponding with that of its controlling lever; or adapted directly to prevent the operation of one unit in case another unit to be operated first, fails to make the required movement. *See also:* interlocking. (EEC/PE) [119]

electric induction *See:* electric field induction.

electric interlocking machine An interlocking machine designed for the control of electrically operated functions. *See also:* interlocking. (EEC/PE) [119]

electricity meter A device that measures and registers the integral of an electrical quantity with respect to time. (ELM) C12.1-1988

electric larry car A burden-bearing car for operation on track rails used for short movements of materials, the propulsion of which is effected by electric motors mounted on the vehicle. *Note:* A prefix (diesel-electric, gas-electric, etc.) may replace the word "electric." *See also:* electric motor car. (EEC/PE) [119]

electric loading (rotating machinery) The average ampere-conductors of the primary winding per unit length of the air-gap periphery. *See also:* stator; rotor. (PE) [9]

electric lock A device to prevent or restrict the movement of a lever, a switch, or a movable bridge unless the locking member is withdrawn by an electric device such as an electromagnet, solenoid, or motor. *See also:* interlocking. (CAS) 156-1960w

electric locking The combination of one or more electric locks and controlling circuits by means of which levers of an interlocking machine, or switches, or other units operated in connection with signaling and interlocking, are secured against operation under certain conditions, as follows:

1) Approach locking
2) Indication locking
3) Switch-lever locking
4) Time locking
5) Traffic locking

See also: interlocking. (EEC/PE) [119]

electric locomotive A vehicle on wheels, designed to operate on a railway for haulage purposes only, the propulsion of which is effected by electric motors mounted on the vehicle. *Note:* While this is a generic term covering any type of locomotive driven by electric motors, it is usually applied to locomotives receiving electric power from a source external to the locomotive. The prefix electric may also be applied to cars, buses, etc., driven by electric motors. A prefix (i.e.,

diesel-electric, etc.) may replace the word "electric."
(EEC/PE) [119]

electric machine An electric apparatus depending on electromagnetic induction for its operation and having one or more component members capable of rotary and/or linear movement. *See also:* asynchronous machine. (PE) [9]

electric-machine regulating system (rotating machinery) A feedback control system that includes one or more electric machines and the associated control. (PE) [9]

electric-machine regulator (rotating machinery) A specified element or a group of elements that is used within an electric-machine regulating system to perform a regulating function by acting to maintain a designated variable (or variables) at a predetermined value, or to vary it according to a predetermined plan. (PE) [9]

electric mechanism (demand meter) That portion, the action of which, in response to the electric quantity to be measured, gives a measurement of that quantity. *Note:* For example, the electric mechanism of certain demand meters is similar to the ordinary ammeter of wattmeter of the deflection type; in others it is a watt-hour meter or other integrating meter; and in still others it comprises an electric circuit that heats temperature-responsive elements, such as bimetallic spirals, that deflect to move the indicating means. The electrical quantity may be measured in kilowatts, kilowatt-hours, kilovolt-amperes, kilovolt-ampere-hours, amperes, ampere-hours, kilovars, kilovar-hours, or other suitable units. *See also:* demand meter. (EEC/PE) [119]

electric mine locomotive (1) (general) An electric locomotive designed for use underground; for example, in such places as coal, metal, gypsum, and salt mines, tunnels, and in subway construction.

(2) (storage-battery type) An electric locomotive that receives its power supply from a storage battery mounted on the chassis of the locomotive.

(3) (trolley type) An electric locomotive that receives its power supply from a trolley-wire distribution system.

(4) (combination type) An electric locomotive that receives power either from a trolley-wire distribution system or from a storage battery carried on the locomotive.

(5) (separate tandem) An electric mine locomotive consisting of two locomotive units that can be coupled together or operated from one controller as a single unit, or else separated and operated as two independent units.

(6) (permanent tandem) A locomotive consisting of two locomotive units permanently connected together and provided with one set of controls so that both units can be operated by a single operator. (EEC/PE) [119]

electric motive power unit A self-contained electric traction unit, comprising wheels and a superstructure capable of independent propulsion from a power supply system, but not necessarily equipped with an independent control system. *Note:* While this is a generic term covering any type of motive power driven by electric motors, it is usually applied to locomotives receiving electric power from an external source. A prefix (diesel-electric, gas-electric, turbine-electric, etc.) may replace the word "electric." *See also:* electric locomotive. (EEC/PE) [119]

electric motor (1) (packaging machinery) A device that converts electrical energy into rotating mechanical energy.
(IA/PKG) 333-1980w

(2) A machine that transforms electric power into mechanical power. (IA/MT) 45-1998

electric motor car A vehicle for operating on track rails, used for the transport of passengers or materials, the propulsion of which is effected by electric motors, mounted on the vehicle. *Note:* A prefix (diesel-electric, gas-electric, etc.) may replace the word "electric." (EEC/PE) [119]

electric motor controller A device or group of devices that serve to govern, in some predetermined manner, the electric power delivered to the motor. *Note:* An electric motor controller is distinct functionally from a simple disconnecting means whose principal purpose in a motor circuit is to disconnect the circuit, together with the motor and its controller, from the source of power. *See also:* electric controller.
(IA/IAC) [60]

electric movable-bridge (drawbridge) lock A device used to prevent the operation of a movable bridge until the device is released. *See also:* interlocking. (EEC/PE) [119]

electric network *See:* network.

electric noise (1) (general) Unwanted electrical energy other than crosstalk present in a transmission system.

(2) (interface terminology) A form of interference introduced into a signal system by natural sources that constitutes for that system an irreducible limit on its signal-resolving capability. *Note:* Noise is characterized by randomness of amplitude and frequency distribution and therefore cannot be eliminated by band-rejection filters tuned to preselected frequencies. *See also:* distortion; interference.
(PE/AP/ANT) [9], 145-1983s

electric operation Power operation by electric energy.
(SWG/PE) C37.100-1981s

electric orchestra lift An electrically driven mechanism for raising and lowering the musicians' platform and the musicians. *See also:* elevator. (EEC/PE) [119]

electric parachute-flare-launching tube A tube mounted on an aircraft through which a metal container carrying a parachute flare is launched, the tube being so designed that as the parachute-flare container passes through the tube, an electric circuit is completed that ignites a slow-burning fuse in the container, the fuse being so designed as to permit the container to clear the aircraft before it ignites the parachute flare.
(EEC/PE) [119]

electric passenger locomotive An electric locomotive, commonly used for hauling passenger trains and generally designed to operate at higher speeds and lower tractive-force values than a freight locomotive of equal horsepower capacity. *Note:* A prefix diesel-electric, gas-electric, turbine-electric, etc., may replace the word electric. *See also:* electric locomotive. (EEC/PE) [119]

electric penetration assembly (electric penetration assemblies) An assembly of insulated electric conductors, conductor seals, module seals (if any), and aperture seals that provides the passage of the electric conductors through a single aperture in the nuclear containment structure, while providing a pressure barrier between the inside and the outside of the containment structure. The electric penetration assembly includes terminal (junction) boxes, terminal blocks, connectors and cable supports, and splices which are designed and furnished as an integral part of the assembly.
(PE/NP) 317-1983r

electric permissible mine locomotive An electric locomotive carrying the official approval plate of the United States Bureau of Mines. *See also:* electric mine locomotive.
(EEC/PE) [119]

electric pin-and-socket coupler (connector) A readily disconnective assembly used to connect electric circuits between components of an aircraft electric system by means of mating pins and sockets. (PE/EEC) [119]

electric pipe heating system (electric pipe heating systems) A system of components and devices consisting of electric heaters, controllers, sensors, dedicated power system components such as transformers, panelboards, cables and systems alarm devices (as required), which, when taken together as a system, is used to increase or maintain the temperature of fluids in mechanical pipes, valves, pumps, tanks, instrumentation, etc. (PE/EDPG) 622-1979s

electric polarizability Of an isotropic medium for which the direction of electric polarization and electric field strength are the same at any point in the medium, the magnitude P of the electric polarization at that point divided by the electric field strength there, **E**. *Note:* In a rationalized system, the electric polarizability $P_e = P/\mathbf{E} = \varepsilon_0(\varepsilon - 1)$. (Std100) 270-1966w

electric polarization (electric field) At any point, the vector difference between the electric flux density at that point and the electric flux density that would exist at that point for the same electric field strength there, if the medium were a vacuum there. *Note:* Electric polarization is the vector limit of the quotient of the vector sum of electric dipole moments in a small volume surrounding a given point, and this volume, as the volume approaches zero in a microscopic sense. (Std100) 270-1966w

electric port (optoelectronic device) A port where the energy is electric. *Note:* A designated pair of terminals may serve as one or more electric ports. *See also:* optoelectronic device. (ED) [46]

electric potential The potential difference between the point and some equipotential surface, usually the surface of the earth, which is arbitrarily chosen as having zero potential (remote earth). *Note:* A point which has a higher potential than a zero surface is said to have a positive potential; one having a lower potential has a negative potential. (PE/PSIM) 81-1983

electric potential difference The line integral of the scalar product of the electric field strength vector and the unit vector along any path from one point to the other, in an electric field resulting from a static distribution of electric charge. (T&D/PE) 539-1990

electric power cable shielding The practice of confining the electric field of the cable to the insulation surrounding the conductor by means of conducting or semiconducting layers, or both, that are in intimate contact or bonded to the inner and outer surfaces of the insulation. (IA/PSE) 241-1990r

electric power distribution panel A metallic or nonmetallic, open or enclosed, unit of an electric system. The operable and the indicating components of an electric system, such as switches, circuit breakers, fuses, indicators, etc., usually are mounted on the face of the panel. Other components, such as terminal strips, relays, capacitors, etc., usually are mounted behind the panel. (EEC/PE) [119]

electric propulsion apparatus Electric apparatus (generators, motors, control apparatus, etc.) provided primarily for ship's propulsion. *Note:* For certain applications, and under certain conditions, auxiliary power may be supplied by propulsion apparatus. *See also:* electric propulsion system. (EEC/PE) [119]

electric propulsion system A system providing transmission of power by electric means from a prime mover to a propeller shaft with provision for control, partly or wholly by electric means, of speed and direction. *Note:* An electric coupling (which see) does not provide electric propulsion. (EEC/PE) [119]

electric rate schedule (electric power supply) A statement of an electric rate (charges) and the terms and conditions governing its application. *Synonym:* electric rate tariff. (PE/PSE) 858-1993w, 346-1973w

electric reset relay (1) A relay that is so constructed that it remains in the picked-up condition even after the input quantity is removed: an independent electric input is required to reset the relay. (SWG/PE/PSR) C37.100-1981s, [6], [56]
(2) A relay that may be reset electrically after an operation. (EEC/REE) [87]

electric resistance-type temperature indicator A device that indicates temperature by means of a resistance bridge circuit. (EEC/PE) [119]

electric road locomotive An electric locomotive designed primarily for hauling dispatched trains over the main or secondary lines of a railroad. *Note:* A prefix diesel-electric, gas-electric, turbine-electric, etc., may replace the word electric. *See also:* electric locomotive. (EEC/PE) [119]

electric road-transfer locomotive An electric locomotive designed primarily so that it may be used either for hauling dispatched trains over the main or secondary lines of a railroad or for transferring relatively heavy cuts of cars for short distances within a switching area. *Note:* A prefix diesel-electric, gas-electric, turbine-electric, etc., may replace the word electric. *See also:* electric locomotive. (EEC/PE) [119]

electric shock Stimulation of the nerves and possible convulsive contraction of the muscle caused by the passage of an electric current through the human or the animal body. (T&D/PE) 539-1990

electric sign A fixed, stationary, or portable self-contained, electrically illuminated utilization equipment with words or symbols designed to convey information or attract attention. (NESC/NEC) [86]

electric-signal storage tube A storage tube into which the information is introduced as an electric signal and read at a later time as an electric signal. *See also:* storage tube. (ED) 158-1962w

electric sounding machine A motor-driven reel with wire line and weight for determination of depth of water by mechanical sounding. (EEC/PE) [119]

electric squib A device similar to an electric blasting cap but containing a gunpowder composition that simply ignites but does not detonate an explosive charge. *See also:* blasting unit. (EEC/PE) [119]

electric stage lift An electrically driven mechanism for raising and lowering various sections of a stage. (EEC/PE) [119]

electric storage subsystem (terrestrial photovoltaic power systems) The subsystem that stores electric energy. *See also:* array control. (PV) 928-1986r

electric strength (rotating machinery) (dielectric strength) The maximum potential gradient that the material can withstand without rupture. (PE/EM) 95-1977r

electric stroboscope An instrument for observing rotating or vibrating objects or for measuring rotational speed or vibration frequency, or similar periodic quantities, by electrically produced periodic changes in illumination. *See also:* instrument. (EEC/PE) [119]

electric submersible pump (electric submersible pump cable) Deep-well electric submersible pumps as commonly used to lift fluids from subsurface formations. (IA/PC) 1017-1985s

electric submersible pump cable Three-conductor power cable installed in the well for the purpose of transmitting power from the surface to the motor lead extension cable. (IA/PC) 1017-1985s

electric-supply equipment Equipment that produces, modifies, regulates, controls, or safeguards a supply of electric energy. *Synonym:* supply equipment. (PE/SUB/NESC) 1268-1997, C2-1997, 1119-1988w

electric supply lines Those conductors used to transmit electric energy and their necessary supporting or containing structures. Signal lines of more than 400 V are always supply lines within the meaning of the rules, and those of less than 400 V may be considered as supply lines, if so run and operated throughout. *Synonym:* supply lines. (NESC/T&D) C2-1997, C2.2-1960

electric supply station (1) Any building, room, or separate space within which electric supply equipment is located and the interior of which is accessible, as a rule, only to qualified persons. This includes generating stations and substations, including their associated generator, storage battery, transformer, and switchgear rooms or enclosures, but does not include facilities such as pad-mounted equipment and installations in manholes and vaults. (NESC) C2-1997
(2) Any building, room, or separate space within which electric-supply equipment is located and the interior of which is accessible, as a rule, only to properly qualified persons. This includes generating stations and substations, including their associated generator, storage battery, transformer, and switchgear rooms. (PE/SUB) 1268-1997
(3) Any building, room, or separate space within which electric-supply equipment is located and the interior of which is accessible, as a rule, only to properly qualified persons. This includes generating stations, substations and generator, storage battery, and transformer rooms. (SUB/PE) 1119-1988w

electric surges (nuclear power generating station) Any spurious voltage or current pulses conducted into the module from external sources. (PE/NP) 381-1977w

electric susceptibility Of an isotropic medium, for which the direction of electric polarization and electric field strength are the same, at any point in the medium, the magnitude of the electric polarization at that point of the medium, divided by the electric flux density that would exist at that point for the same electric field strength, if the medium there were a vacuum. *Note:* In a rationalized system the electric susceptibility $\chi_{\varepsilon} = P/D(\varepsilon - 1)$. (Std100) 270-1966w

electric switching locomotive An electric locomotive designed for yard movements of freight or passenger cars, its speed and continuous electrical capacity usually being relatively low. *Note:* A prefix diesel-electric, gas-electric, turbine-electric, etc., may replace the word electric. *See also:* electric locomotive. (EEC/PE) [119]

electric switch-lever lock An electric lock used to prevent the movement of a switch lever or latch in an interlocking machine until the lock is released. *See also:* interlocking. (EEC/PE) [119]

electric switch-lever locking A general term for route or section locking. *See also:* interlocking. (EEC/PE) [119]

electric switch lock An electric lock used to prevent the operation of a switch or a switch movement until the lock is released. *See also:* interlocking. (EEC/PE) [119]

electric system loss (1) Total electric energy losses in the electric system. It consists of transmission, transformer, and distribution losses between the supply and receiving points. (PE/PSE) 858-1993w

(2) Total electric energy loss in the electric system. It consists of transmission, transformation, and distribution losses between sources of supply and points of delivery. (PE/PSE) 346-1973w

electric tachometer (marine usage) An instrument for measuring rotational speed by electric means. *See also:* instrument. (EEC/PE) [119]

electric telegraph A telegraph having the relationship of the moving parts of the transmitter and receiver maintained by the use of self-synchronous motors or equivalent devices. (EEC/PE) [119]

electric telemeter The measuring, transmitting, and receiving apparatus, including the primary detector, intermediate means (excluding the channel) and end devices for electric telemetering. *Note:* A telemeter that measures current is called a teleammeter; voltage, a televoltmeter; power, a telewattmeter; one that measures angular or linear position, a position telemeter. The names of the various component parts making up the telemeter are, in general, self-defining; for example, the transmitter, receiver, indicator, etc. (SWG/PE) C37.100-1992

electric telemetering (electric telemetry) Telemetering performed by an electrical translating means separate from the measured. (SWG/PE) C37.100-1992

electric thermometer (rotating electric machinery) An instrument that utilizes electric means to measure temperature. Electric thermometers include thermocouples and resistance temperature detectors. (PE/EM) 11-1980r

electric tower car A rail vehicle, the propulsion of which is effected by electric means and that is provided with an elevated platform, generally arranged to be raised and lowered, for the installation, inspection, and repair of a contact wire system. *Note:* A prefix diesel-electric, gas-electric, etc., may replace the word electric. *See also:* electric motor car. (EEC/PE) [119]

electric trail car (electric trailer) A car not provided with motive power that is used in a train with one or more electric motor cars. *Note:* A prefix diesel-electric, gas-electric, etc., may replace the word electric to identify the motor cars. *See also:* electric motor car. (EEC/PE) [119]

electric transducer A transducer in which all of the waves concerned are electric. *See also:* transducer. (Std100) 196-1952w, 270-1966w

electric transfer locomotive An electric locomotive designed primarily for transferring relatively heavy cuts of cars for short distances within a switching area. *Note:* A prefix diesel-electric, gas-electric, turbine-electric, etc., may replace the word electric. *See also:* electric locomotive. (EEC/PE) [119]

electric-tuned oscillator An oscillator whose frequency is determined by the value of a voltage, current, or power. Electric tuning includes electronic tuning, electrically activated thermal tuning, electromechanical tuning, and tuning methods in which the properties of the medium in a resonant cavity are changed by an external electric means. An example is the tuning of a ferrite-filled cavity by changing an external magnetic field. (ED) [45], 158-1962w

electric turn-and-bank indicator A device that utilizes an electrically driven gyro for turn determination and a gravity-actuated inclinometer for bank determination. (EEC/PE) [119]

electric valve operator (nuclear power generating station) An electric-powered mechanism for opening and closing a valve, including all electric and mechanical components that are integral to the mechanism and are required to operate and control valve action. (PE/NP) 380-1975w, 382-1980s

electric vector *See:* electric field vector.

electric wind *See:* convective discharge.

electrification by friction *See:* triboelectrification.

electrification time (cable-insulation materials) Time during which a steady direct voltage is applied to electrical insulating materials before the current is measured. (PE) 402-1974w

electrified track A railroad track suitably equipped in association with a contact conductor or conductors for the operation of electrically propelled vehicles that receive electric power from a source external to the vehicle. *See also:* electric locomotive. (EEC/PE) [119]

electroacoustical reciprocity theorem For an electroacoustic transducer satisfying the reciprocity principle, the quotient of the magnitude of the ratio of the open-circuit voltage at output terminals (or the short-circuit current) of the transducer, when used as a sound receiver, to the free-field sound pressure referred to an arbitrarily selected reference point on or near the transducer, divided by the magnitude of the ratio of the sound pressure apparent at a distance d from the reference point to the current flowing at the transducer input terminals (or the voltage applied at the input terminals), when used as a sound emitter, is a constant, called the reciprocity constant, independent of the type or constructional details of the transducer. *Note:* The reciprocity constant is given by

$$\left|\frac{M_O}{S_S}\right| = \left|\frac{M_S}{S_S}\right| = \left|\frac{2\delta}{\rho f}\right|$$

where

M_O = open free-field voltage response, as a sound receiver, in open-circuit volts per newton per square meter, referred to the arbitrary reference point on or near the transducer.

M_S = free-field current response in short-circuit amperes per newton per square meter, referred to the arbitrary reference point on or near the transducer

S_O = sound pressure in newtons per square meter per ampere of input current produced at a distance *d* meters from the arbitrary reference point

S_S = sound pressure in newtons per square meter per volt applied at the input terminals produced at a distance *d* meters from the arbitrary reference point

f = frequency in hertz

ρ = density of the medium in kilograms per cubic meter

δ = distance in meters from the arbitrary reference point on or near the transducer to the point in which the sound pressure established by the transducer when emitting is evaluated.

(SP) [32]

electroacoustic transducer (electric systems) A transducer for receiving waves and delivering waves to an acoustic system, or vice versa. *See also:* transducer. (SP) [32]

electrobiology The study of electrical phenomena in relation to biological systems. (EMB) [47]

electrocardiogram The graphic record of the variation with time of the voltage associated with cardiac activity. *See also:* spindle wave; electrodermogram; vector electrocardiogram; electrocorticogram. (EMB) [47]

electrocardiographic waves, P, Q, R, S, and, T (medical electronics) (in electrocardiograms obtained from differential electrodes placed on the right arm and left leg) The characteristic tracing consists of five consecutive waves: P, a prolonged, low, positive wave; Q, brief, low, negative; R, brief, high, positive; S, brief, low, negative; and T, prolonged, low, positive. (EMB) [47]

electrocautery (electrotherapy) An instrument for cauterizing the tissues by means of a conductor brought to a high temperature by an electric current. *See also:* electrotherapy. (EMB) [47]

electrochemical cell A system consisting of an anode, cathode, and an electrolyte plus such connections (electric and mechanical) as may be needed to allow the cell to deliver or receive electric energy. (AES/IA/APP) [41], [73]

electrochemical equivalent: element, compound, radical, or ion (1) (general) The weight of that substance involved in a specified electrochemical reaction during the passage of a specified quantity of electricity, such as a faraday, ampere-hour, or coulomb. (EEC/PE) [119]
(2) (oxidation) The weight of an element or group of elements oxidized or reduced at 100% efficiency by a unit quantity of electricity. *See also:* electrochemistry. (IA) [59]

electrochemical recording (facsimile) Recording by means of a chemical reaction brought about by the passage of signal-controlled current through the sensitized portion of the record sheet. *See also:* recording. (COM) 168-1956w

electrochemical series *See:* electromotive force series.

electrochemical valve An electric valve consisting of a metal in contact with a solution or compound across the boundary of which current flows more readily in one direction than in the other direction and in which the valve action is accompanied by chemical changes. (EEC/PE) [119]

electrochemical valve metal A metal or alloy having properties suitable for use in an electrochemical valve. *See also:* electrochemical valve. (EEC/PE) [119]

electrochemistry That branch of science and technology that deals with interrelated transformations of chemical and electric energy. (EEC/PE) [119]

electrochromeric display device A display device that uses materials that change from transparent to opaque under the control of an electric field. For example, a liquid crystal display device. (C) 610.10-1994w

electrocoagulation (medical electronics) The clotting of tissue by heat generated within the tissue by impressed electric currents. (EMB) [47]

electrocorticogram (medical electronics) A graphic record of the variation with time of voltage taken from exposed cortex cerebra. (EMB) [47]

electroculture (medical electronics) The stimulation of growth, flowering, or seeding by electric means. (EMB) [47]

electrocution The destruction of life by means of electric current. (EMB) [47]

electrode (1) (electrochemistry) An electric conductor for the transfer of charge between the external circuit and the electroactive species in the electrolyte. *Note:* Specifically, in an electrolytic cell, an electrode is a conductor at the surface of which a change occurs from conduction by electrons to conduction by ions or colloidal ions. *See also:* electrolytic cell; electrochemical cell. (AES) [41]
(2) (electron tube) A conducting element that performs one or more of the functions of emitting, collecting, or controlling by an electric field the movements of electrons or ions. (ED) 161-1971w
(3) (A) (biological electronics) (reference, inactive, diffuse, dispersive, indifferent electrode) A pickup electrode that, because of averaging, shunting, or other aspects of the tissue-current pattern to which it connects, shows potentials not characteristic of the region near the active electrode. **(B) (biological electronics) (reference, inactive, diffuse, dispersive, indifferent electrode)** Any electrode, in a system of stimulating electrodes, at which due to its dispersive action, excitation is not produced. **(C) (biological electronics) (reference, inactive, diffuse, dispersive, indifferent electrode)** An electrode of relatively large area applied to some inexcitable or distant tissue in order to complete the circuit with the active electrode that is used for stimulation. (EMB) [47]

electrode, accelerating *See:* accelerating electrode.

electrode admittance (*j*th electrode of an *n*-electrode electron tube) The short-circuit driving-point admittance between the jth electrode and the reference point measured directly at the jth electrode. *Note:* To be able to determine the intrinsic electronic merit of an electron tube, the driving-point and transfer admittances must be defined as if measured directly at the electrodes inside the tube. The definitions of electrode admittance and electrode impedance are included for this reason. *See also:* electron-tube admittances. (ED) 161-1971w

electrode alternating-current resistance The real component of the electrode impedance. *See also:* self-impedance. (ED) [45]

electrode bias (electron tube) The voltage at which an electrode is stabilized under operating conditions with no incoming signal, but taking into account the voltage drops in the connected circuits. *See also:* electrode voltage. (Std100) [84]

electrode capacitance (*n*-terminal electron tube) The capacitance determined from the short-circuit driving-point admittance at that electrode. *See also:* electron-tube admittances. (ED) 161-1971w

electrode characteristic A relation, usually shown by a graph, between the electrode voltage and the current of an electrode, all other electrode voltages being maintained constant. (ED) 161-1971w

electrode conductance The real part of the electrode admittance. (ED) 161-1971w

electrode, control *See:* control electrode.

electrode current (electron tube) The current passing to or from an electrode through the interelectrode space. *Note:* The terms cathode current, grid current, anode current, plate current, etc., are used to designate electrode currents for these specific electrodes. Unless otherwise stated, an electrode current is measured at the available terminal. (ED) [45]

electrode current, average *See:* average electrode current.

electrode-current averaging time (electron tube) The time interval over which the current is averaged in defining the operating capabilities of the electrode. *See also:* electrode current. (ED) [45]

electrode dark current (1) (phototubes) The component of electrode current remaining when ionizing radiation and optical photons are absent. *Notes:* 1. Optical photons are photons with energies corresponding to wavelengths between 2000 and 1500 angstroms. 2. Since the dark current may change considerably with temperature, the temperature should be specified. *See also:* phototube. (ED) [45]
(2) (camera tubes) The current from an electrode in a photoelectric tube under stated conditions of radiation shielding. *See also:* camera tube. (BT/ED/AV) [34], [45]

electrode dissipation The power dissipated in the form of heat by an electrode as a result of electron or ion bombardment, or both, and radiation from other electrodes. *See also:* grid driving power. (ED) [45]

electrode drop (arc-welding apparatus) The voltage drop in the electrode due to its resistance (or impedance). (EEC/AWM) [91]

electrode impedance The reciprocal of the electrode admittance. *See also:* electron-tube admittances. (ED) [45]

electrode impedance, biological *See:* biological electrode impedance.

electrode, pad *See:* pad electrode.

electrode potential, biological *See:* biological electrode potential.

electrode radiator (electron tube) (cooling fin) A metallic piece, often of large area, extending the electrode to facilitate the dissipation of the heat generated in the electrode. *See also:* electron tube. (ED) [45], [84]

electrode reactance The imaginary component of the electrode impedance. *See also:* self-impedance. (ED) [45]

electrode resistance (1) (general) The reciprocal of the electrode conductance. *Note:* This is the effective parallel resistance and is not the real component of the electrode impedance. (ED) [45]
(2) (at a stated operating point) The quotient of the direct electrode voltage by the direct electrode current. *See also:* self-impedance. (ED) [45]

electrodermal reaction (EDR) (medical electronics) The change in electric resistance of the skin during emotional stress. (EMB) [47]

electrodermogram (electromyogram) (electrobiology) (electroretinogram) A graphic record of the variation with time of voltage taken from the given and anatomical structure (skin, muscle, and retina, respectively). *See also:* electrocardiogram. (EMB) [47]

electrodesiccation (fulguration) The superficial destruction of tissue by electric sparks from a movable electrode. *See also:* electrotherapy. (EMB) [47]

electrode, signal *See:* signal electrode.

electrode susceptance The imaginary component of the electrode admittance. *See also:* self-impedance. (ED) [45]

electrode voltage The voltage between an electrode and the cathode or a specified point of a filamentary cathode. *Note:* The terms grid voltage, anode voltage, plate voltage, etc., are used to designate the voltage between these specific electrodes and the cathode. Unless otherwise stated, electrode voltages are understood to be measured at the available terminals. (ED) [45], 161-1971w

electrodiagnosis The study of functional states of parts of the body either by studying their responses to electric stimulation or by studying the electric potentials (or currents) that they spontaneously produce. (EMB) [47]

electroencephalogram (medical electronics) A graphic record of the changes with time of the voltage obtained by means of electrodes applied to the scalp over the cerebrum. (EMB) [47]

electroendosmosis effect A phenomenon occasionally observed, more often on older windings, when, in the presence of moisture, different insulation resistance values may be obtained when the polarity of the tester leads are reversed. Typically for older wet windings, the insulation resistance for reverse polarity, where the ground lead is connected to the winding and the negative voltage lead to ground, is much higher than for normal polarity. (PE/EM) 43-2000

electrographic recording (electrostatography) The branch of electrostatic electrography that employs a charge transfer between two or more electrodes to form directly electrostatic-charge patterns on an insulating medium for producing a viewable record. *See also:* electrostatography. (ED) [46]

electrographitic brush (rotating machinery) A brush composed of selected amorphous carbon that, in the process of manufacturer, is carried to a temperature high enough to convert the carbon to the graphitized form. *Note:* This type of brush is exceedingly versatile in that it can be made soft or very hard, also nonabrasive or slightly abrasive. Grades of brushes of this type have a high current-carrying capacity, but differ greatly in operating speed from low to high. *See also:* brush. (PE) [9]

electrohydraulic elevator A direct-plunger elevator where liquid is pumped under pressure directly into the cylinder by a pump driven by an electric motor. *See also:* elevator. (EEC/PE) [119]

electrokinetic potential (zeta potential) (medical electronics) A set of four electric or velocity potentials that accompany relative motion between solids and liquids. (EMB) [47]

electroluminescence (1) (illuminating engineering) The emission of light from a phosphor excited by an electromagnetic field. (EEC/IE) [126]
(2) (light-emitting diodes) The emission of light from a material (phosphor or semiconductor) where the exciting mechanism is the application of an electromagnetic field. (ED) [127]
(3) (fiber optics) Nonthermal conversion of electrical energy into light. One example is the photon emission resulting from electron-hole recombination in a pn junction such as in a light emitting diode. *See also:* injection laser diode. (Std100) 812-1984w

electroluminescent display device An optoelectronic device with a multiplicity of electric ports, each capable of independently producing an optic output from an associated electroluminator element. *See also:* optoelectronic device. (ED) [46]

electroluminescent display panel A thin, usually flat, electroluminescent display device. *See also:* optoelectronic device. (ED) [46]

electrolysis (underground structures) The destructive chemical action caused by stray or local electric currents to pipes, cables, and other metalwork. *See also:* corrosion. (T&D/PE) [10]

electrolyte A conducting medium in which the flow of electric current takes place by migration of ions. *Note:* Many physical chemists define electrolyte as a substance that when dissolved in a specified solvent, usually water, produces an ionically conducting solution. *See also:* electrolytic cell. (Std100) 270-1966w

electrolyte cells *See:* cascade.

electrolytic *See:* cathode.

electrolytic cell (1) A receptacle or vessel in which electrochemical reactions are caused by applying electrical energy for the purpose of refining or producing usable materials. (NESC/NEC) [86]
(2) A receptacle or vessel in which electrochemical reactions are caused by applying electrical energy for the purpose of refining or producing materials. (IA/PC) 463-1993w

electrolytic cell line working zone The cell line working zone is the space envelope wherein operation or maintenance is normally performed on or in the vicinity of exposed energized surfaces of electrolytic cell lines or their attachments. (NESC/NEC) [86]

electrolytic cleaning The process of degreasing or descaling a metal by making it an electrode in a suitable bath. (Std100) [71]

electrolytic recording (facsimile) That type of electrochemical recording in which the chemical change is made possible by the presence of an electrolyte. *See also:* recording. (COM) 168-1956w

electrolytic tank A vessel containing a poorly conducting liquid, in which are inserted conductors that are scale models of an electrode system. *Note:* It is used to obtain potential diagrams. *See also:* electron optics. (ED) [45], [84]

electrolyzer An electrolytic cell for the production of chemical products.

electromagnet A device consisting of a ferromagnetic core and a coil, that produces appreciable magnetic effects only when an electric current exists in the coil. 270-1966w

electromagnetic compatibility (EMC) (1) (supervisory control, data acquisition, and automatic control) (station con-

trol and data acquisition) A measure of equipment tolerance to external electromagnetic fields. (SWG/PE/SUB) C37.100-1992, C37.1-1994
(2) (control of system electromagnetic compatibility) The ability of a device, equipment, or system to function satisfactorily in its electromagnetic environment without introducing intolerable electromagnetic disturbances to anything in that environment. (EMC/EMB/MIB) C63.12-1987, 1073.3.2-2000
(3) (equipment) The capability of electronic equipment or systems to be operated in the intended operational electromagnetic environment at designed levels of efficiency. (PE/EDPG) 1050-1996
(4) (of an electrical system) An electrical system's ability to perform its specified functions in the presence of electrical noise generated either internally or externally by other systems. The goal of EMC is to minimize the influence of electrical noise. (PE/IC) 1143-1994r
(5) The requirements for electromagnetic emission and susceptibility dictated by the physical environment and regulatory governing bodies within whose jurisdiction a piece of equipment is operated. (EMB/MIB) 1073.4.1-2000

electromagnetic delay line (computers) (information processing) A delay line whose operation is based on the time of propagation of electromagnetic waves through distributed or lumped capacitance and inductance. *Synonym:* electric delay line. (C) [20], [85], 610.10-1994w

electromagnetic disturbance (1) An electromagnetic phenomenon that may be superimposed on a wanted signal. *See also:* electromagnetic compatibility. (EMC) [53]
(2) (overhead power lines) Any electromagnetic phenomenon that may degrade the performance of a device, a piece of equipment, or a system. *Notes:* 1. An electromagnetic disturbance may be electromagnetic noise, an unwanted signal, or a change in the propagation medium itself. 2. The term "system" is used here in its generic sense, which may include inert and living matter. (T&D/PE) 539-1990

electromagnetic environment The electromagnetic field(s) and or signals existing in a transmission medium. *See also:* electromagnetic compatibility . [53]

electromagnetic field (1) The energy field radiating from a source and containing both electric and magnetic field components. *See also:* magnetic field; electric field. (PE/IC) 1143-1994r
(2) A time-varying field, associated with the electric or magnetic forces and described by Maxwell's equations. (AP/PROP) 211-1997

electromagnetic field induction (1) (electro-magnetic coupling) The induction process that includes both electric and magnetic fields. (T&D/PE) 1048-1990
(2) (overhead power lines) (electromagnetic coupling) The induction process that results from time-varying electromagnetic fields. (T&D/PE) 539-1990, 524a-1993r

electromagnetic induction (1) The production of an electromotive force in a circuit by a change in the magnetic flux linking with that circuit. (CHM) [51]
(2) *See also:* electromagnetic interference. (PE/IC) 1143-1994r

electromagnetic interference (EMI) (1) (station control and data acquisition) A measure of electromagnetic radiation from equipment. (SWG/PE/SUB) C37.100-1992, C37.1-1994
(2) (overhead power lines) Degradation of the performance of a device, a piece of equipment, or a system caused by an electromagnetic disturbance. *Note:* The English words "interference" and "disturbance" are often used indiscriminately. (T&D/PE) 539-1990
(3) Impairment of a wanted electromagnetic signal by an electromagnetic disturbance. (PE/IA/EDPG/PSE) 1050-1996, 241-1990r
(4) Electromagnetic energy from sources external or internal to electrical or electronic equipment that adversely affects equipment by creating undesirable responses (degraded performance or malfunctions). EMI can be divided into two classes: continuous wave (CW) and transient. *Synonym:* electromagnetic induction. *See also:* continuous wave; electrical noise; transient electrical noise. (PE/IC) 1143-1994r
(5) Signals emanating from external sources (e.g., power supplies, transmitters) or internal sources (e.g., adjacent electronic components, energy sources) that disrupt or prevent operation of electronic systems. (EMB/C/BA/MIB) 1073.3.2-2000, 896.3-1993w
(6) Any disturbance that interrupts, obstructs, or otherwise impairs the performance of electronic equipment. (PEL) 1515-2000

electromagnetic lens A three-dimensional structure, through which electromagnetic waves can pass, possessing an index of refraction that may be a function of position and a shape that is chosen so as to control the exiting aperture illumination. (AP/ANT) 145-1993

electrogeometric model (EGM) A geometrical representation of a facility, that, together with suitable analytical expressions correlating its dimensions to the current of the lightning stroke, is capable of predicting if a lightning stroke will terminate on the shielding system, the earth, or the element of the facility being protected. (SUB/PE) 998-1996

electrogeometric model theory The theory describing the electrogeometric model together with the related quantitative analyses including the correlation between the striking distances to the different elements of the model and the amplitude of the first return stroke. (SUB/PE) 998-1996

electromagnetic noise (1) An unwanted electromagnetic disturbance that is not of a sinusoidal character. *See also:* electromagnetic compatibility. (EMC/INT) [53], [70]
(2) (overhead power lines) A time-varying electromagnetic phenomenon that apparently does not convey information and that may be superimposed on or combined with a wanted signal. (T&D/PE) 539-1990

electromagnetic radiation (1) (radio frequency radiation hazard warning symbol) The term is restricted to that part of the spectrum commonly defined as the radio frequency region, which for the purpose of this sxtandard includes microwave frequencies. (NIR) C95.2-1982r
(2) The emission of electromagnetic energy from a finite region in the form of unguided waves. (AP/ANT) 145-1993
(3) (laser maser) The flow of energy consisting of orthogonally vibrating electric and magnetic fields lying transverse to the direction of propagation. X rays, ultraviolet, visible, infrared, and radio waves occupy various portions of the electromagnetic spectrum and differ only in frequency and wavelength. (LEO) 586-1980w

electromagnetic relay (1) An electromechanical relay that operates principally by action of an electromagnetic element that is energized by the input quantity. (SWG/PE) C37.100-1992
(2) A relay, controlled by electromagnetic means, that opens and closes electric contacts. *See also:* relay. (EEC/REE) [87]

electromagnetic signal The intelligence, message, or effect to be conveyed over a communication system or broadcasting system via electromagnetic waves. (T&D/PE) 539-1990

electromagnetic spectrum The spectrum of electromagnetic radiation, as shown in the following table:

Spectral region	Wavelength
Gamma rays	< 0.006 nm
X-rays	0.006–5 nm
Ultraviolet rays	5 nm–0.4 μm
Visible light	0.4–0.7 μm
Infrared	0.7 μm–0.1 mm
Radio	> 0.1 mm

See also: radio spectrum. (AP/PROP) 211-1997

electromagnetic pulse (EMP) (1) A high-energy electromagnetic pulse initiated by a nuclear reaction (e.g., upper atmospheric detonation of a nuclear weapon). (PE/IC) 1143-1994r

(2) An intense transient electromagnetic field. *Note:* EMP is commonly associated with nuclear explosions in or near the Earth's atmosphere; however, electromagnetic pulses can arise from other sources, such as lightning. (AP/PROP) 211-1997

electromagnetic waves (1) Waves characterized by variations of electric and magnetic fields. *Note:* Electromagnetic waves are known as radio waves, heat rays, light rays, etc., depending on the frequency. *See also:* radio wave; propagation; waveguide. (MTT) 146-1980w
(2) Waves characterized by temporal and spatial variations of electric and magnetic fields. Electromagnetic waves are known as radio waves, infrared waves, light waves, etc., depending on the frequency. (AP/PROP) 211-1997

electromechanical device (control equipment) A device that is electrically operated and has mechanical motion such as relays, servos, etc. (PE/PSE) 94-1970w

electromechanical recording (facsimile) Recording by means of a signal-actuated mechanical device. *See also:* recording. (COM) 168-1956w

electromechanical relay A relay that operates by physical movement of parts resulting from electromagnetic, electrostatic, or electrothermic forces created by the input quantities. (SWG/PE) C37.100-1992

electromechanical switching system (telephone switching systems) An automatic switching system in which the control functions are performed principally by electromechanical devices. (COM) 312-1977w

electromechanical transducer A transducer for receiving waves from an electric system and delivering waves to a mechanical system, or vice versa. *See also:* transducer. (T&D/PE/SP) 590-1977w, [32]

electromechanical watthour meter designation The maximum continuous load in amperes at which a watthour meter meets the accuracy of ANSI C12.1-1988, or the latest revision thereof. (ELM) C12.10-1987

electrometer tube A vacuum tube having a very low control-electrode conductance to facilitate the measurement of extremely small direct current or voltage. (ED) 161-1971w, [45]

electromotive force *See:* voltage.

electromotive force series A list of elements arranged according to their standard electrode potentials. (IA) [59], [71]

electromyograph (medical electronics) An instrument for recording action potentials or physical movements of muscles. (EMB) [47]

electron (1) An elementary particle containing the smallest negative electric charge. *Note:* The mass of the electron is approximately equal to 1/1837 of the mass of the hydrogen atom. (Std100) [84]
(2) Operated by, containing, or producing electrons. Examples: Electron tube, electron emission, and electron gun. *See also:* electronics; electronic. (EEC/PE) [119]

electron accelerator, linear *See:* linear electron accelerator.

electronarcosis The production of transient insensibility by means of electric current applied to the cranium at intensities insufficient to cause generalized convulsions. *See also:* electrotherapy. (EMB) [47]

electron beam A beam of electrons (ions) emitted from a single source and moving in neighboring paths that are confined to a desired region. (Std100) [84]

electron beam crosstalk (charge-storage tubes) Any spurious output signal that arises from scanning or from the input of information. *See also:* charge-storage tube. (ED) 158-1962w

electron beam recording (EBR) In micrographics, a specific method of producing computer output microfilm in which a beam of electrons is directed onto an energy-sensitive microfilm. (C) 610.2-1987

electron-beam tube An electron tube, the performance of which depends upon the formation and control of one or more electron beams. (ED) 161-1971w

electron collector (microwave tubes) The electrode that receives the electron beam at the end of its path. *Note:* The power of the beam is used to produce some desired effect before it reaches the collector. *See also:* velocity-modulated tube. (Std100) [84]

electron-coupled oscillator An oscillator employing a multigrid tube with the cathode and two grids operating as an oscillator in any conventional manner, and in which the plate circuit load is coupled to the oscillator through the electron system. *See also:* oscillatory circuit. (AP/BT/ANT) 145-1983s, 182-1961w

electron (proton) damage coefficient The change in a stated quantity (such as minority carrier inverse squared diffusion length) of a given material per unit particle fluence of a stated energy spectrum. (AES/SS) 307-1969w

electron device A device in which conduction is principally by electrons moving through a vacuum, gas, or semiconductor. (ED) 161-1971w

electron-device transducer *See:* short-circuit forward admittance.

electron emission The liberation of electrons from an electrode into the surrounding space. *Note:* Quantitatively, it is the rate at which electrons are emitted from an electrode. (ED) 161-1971w

electron gun (1) (electron tube) An electrode structure that produces and may control, focus, deflect, and converge one or more electron beams. *See also:* electrode. (ED) 161-1971w
(2) (computer graphics) A device in a cathode ray tube that emits a stream of electrons that is directed by the deflection system toward the phosphor-coated screen. *See also:* flood gun. (C) 610.6-1991w

electron-gun density multiplication (electron tube) The ratio of the average current density at any specified aperature through which the stream passes to the average current density at the cathode surface. (ED) 161-1971w

electronic Of, or pertaining to, devices, circuits, or systems utilizing electron devices. Examples: Electronic control, electronic equipment, electronic instrument, and electronic circuit. *See also:* electron device; electronics. (ED) 161-1971w

electronically de-spun antenna (communication satellite) A directional antenna, mounted to a rotating object (namely spin stabilized communication satellite), with beam switching and phasing such that the antenna beam points into the same direction in space regardless of its mechanical rotation. (COM) [24]

electronically erasable programmable read-only memory (EEPROM) A type of memory chip designed to be programmed more than once. The chips are functionally the same as EPROMs, but are erased using a particular electrical voltage. (PE/SUB) 1379-1997

electronic analog computer An automatic computing device that operates in terms of continuous variation of some physical quantities, such as electric voltages and currents, mechanical shaft rotations, or displacements, and that is used primarily to solve differential equations. *Note:* The equations governing the variation of the physical quantities have the same or very nearly the same form as the mathematical equations under investigation and therefore yield a solution analogous to the desired solution of the problem. Results are measured on meters, dials, oscillograph recorders, or oscilloscopes. (C) 165-1977w

electronic bulletin board In an electronic mail system, a storage area shared by several users, each having access to all messages left in that area. (C) 610.2-1987

electronic cash register (ECR) A device that functions as both a cash register and a point-of-sale terminal to a central computer performing inventory control, price updating, and other retail sales functions. (C) 610.2-1987

Electronic Circuit Analysis Program II (ECAP II) A simulation language used for modeling and analyzing electrical networks, allowing synthesis of device models using a function generator. (C) 610.13-1993w

electronic contactor A contactor whose function is performed by electron tubes. *See also:* contactor. (IA/ICTL/IAC) [60]

electronic controller An electric controller in which the major portion or all of the basic functions are performed by electron tubes. (IA/ICTL/IAC) [60]

electronic counter-countermeasures (ECCM) Any electronic technique designed to make a radar less vulnerable to electronic countermeasures (ECM). (AES) 686-1997

electronic counter-countermeasures improvement factor (radar) The power ratio of the electronic countermeasures (ECM) signal level required to produce a given output signal from a receiver using an ECCM technique to the ECM signal level producing the same output from the same receiver without the ECCM technique. (AES/RS) 686-1982s

electronic countermeasures (ECM) Any electronic technique designed to deny detection or accurate information to a radar. *Note:* Screening with noise, confusion with false targets, and deception by affecting tracking circuits are typical ECM. (AES) 686-1997

electronic data processing (EDP) *See:* automatic data processing.

electronic data sheet A data sheet stored in some form of electronically readable memory (as opposed to a piece of paper). (IM/ST) 1451.2-1997

Electronic Design Interchange Format (EDIF) An industry standard for transfer of schematic and structured connectivity information for electronic design automation. (ATLAS) 1232-1995

electronic direct-current motor controller A phase-controlled rectifying system using tubes of the vapor- or gas-filled variety for power conversion to supply the armature circuit or the armature and shunt-field circuits of a direct-current motor, to provide adjustable-speed, adjustable- and regulated-speed characteristics. *See also:* electronic controller. (IA/ICTL/IAC) [60]

electronic direct-current motor drive The combination of an electronic direct-current motor controller with its associated motor or motors. *See also:* electronic controller. (IA/ICTL/IAC) [60]

electronic efficiency (electron tube) The ratio of the power at the desired frequency delivered by the electron stream to the circuit in an oscillator or amplifier to the average power supplied to the stream. (ED) 161-1971w

electronic funds transfer system A data collection and telecommunication system that electronically transports information about the movement of funds between accounts managed by financial institutions. (C) 610.2-1987

electronic gun A device in a cathode ray tube that emits of electrons directed by the deflection system toward the phosphor-coated screen, thereby causing the phosphor to emit light. *See also:* flood gun. (C) 610.10-1994w

electronic keying A method of keying whereby the control is accomplished solely by electronic means. *See also:* telegraphy. (AP/ANT) 145-1983s

electronic line scanning (facsimile) That method of scanning that provides motion of the scanning spot along the scanning line by electronic means. *See also:* scanning. (COM) 168-1956w

electronic mail (A) The generation, transmission, and display of correspondence and documents by electronic means. *Synonym:* mailbox service. *See also:* electronic bulletin board; electronic mailbox. **(B)** The concepts and technologies employed for the electronic communication of textual material. (C) 610.2-1987

(2) **(A)** A networking service that electronically provides all the basic services of traditional mail. *See also:* mail exploder. **(B)** A computerized store-and-forward system for electronic delivery of text memos and messages. (C) 610.7-1995

electronic mailbox A storage area used to hold all messages addressed to a particular user of an electronic mail system. (C) 610.2-1987

electronic microphone A microphone that depends for its operation on a change in the terminal electrical characteristic of an active device when a force is applied to some part of the device. *See also:* microphone. (SP) [32]

electronic multiplier An all-electronic device capable of forming the product of two variables. *Note:* Examples are a time-division multiplier, a square-law multiplier, an amplitude-modulation-frequency-modulation (AM-FM) multiplier, and a triangular-wave multiplier. *See also:* electronic analog computer. (C) 165-1977w

electronic navigation *See:* navigation.

electronic office An office that makes use of office automation. *Synonyms:* automated office; office of the future. *See also:* paperless office. (C) 610.2-1987

electronic pen A pick device that detects a display element or segment by sensing electronic pulses. (C) 610.10-1994w

electronic position indicator (navigation aid terms) A radio navigation system used in hydrographic surveying that provides circular lines of position. (AES/GCS) 172-1983w

electronic power converter Electronic devices for transforming electric power. *See also:* rectification. (EEC/PE) [119]

electronic raster scanning (facsimile) That method of scanning in which motion of the scanning spot in both dimensions is accomplished by electronic means. *See also:* scanning. (COM) 168-1956w

electronic rectifier A rectifier in which electron tubes are used as rectifying elements. *See also:* rectification of an alternating current; electronic controller. (IA/ICTL/IAC) [60]

electronics (1) Of, or pertaining to, the field of electronics. Examples: Electronics engineer, electronics course, electronics laboratory, and electronics committee. *See also:* electronic; electron.

(2) That field of science and engineering that deals with electron devices and their utilization. *See also:* electron device. (ED) [45]

(3) That branch of science and technology that relates to devices in which conduction is principally by electrons moving through a vacuum, gas, or semiconductor. (IA/MT) 45-1998

electronic scanning Scanning an antenna beam by electronic or electric means without moving parts. *Synonym:* inertialess scanning. *See also:* radiation. (AP/ANT) 145-1993

Electronic Industries Association (EIA) An organization that establishes and maintains standards for the electronics industries in the United States. (C/C) 610.7-1995, 610.10-1994w

electronic signatures The use of encryption techniques to authenticate a message as originating from a specific source, often utilizing a public key system. (C) 610.7-1995

electronic spread sheet (A) A computer program that enables the user to set up a display of rows and columns in which some entries are manually entered and others are calculated automatically using formulas supplied by the user. *Synonym:* spread sheet. **(B)** The display of rows and columns produced by a computer program as in (A). (C) 610.2-1987

electronic storage register (1) (watthour meters) An electronic circuit, which is an integral part of the time-of-use register, where data are stored for display or retrieval, or both. (ELM) C12.13-1985s

(2) (electromechanical watthour meters) An electronic circuit, which is an integral part of the solid-state register, where data are stored for display and/or retrieval. (ELM) C12.15-1990

electronic switching system (1) (telephone switching systems) An automatic switching system in which the control functions are performed principally by electronic devices. (COM) 312-1977w

(2) A type of telephone switching system that uses a special-purpose computer to direct and control the switching operation. *See also:* crossbar system; step-by-step system. (C) 610.7-1995

electronic thermal conductivity The part of the thermal conductivity resulting from the transport of thermal energy by electrons and holes. *See also:* thermoelectric device. (ED) [46]

electronic transformer (power and distribution transformers) Any transformer intended for use in a circuit or system utilizing electron or solid-state devices. *Note:* Mercury-arc rectifier transformers and luminous-tube transformers are normally excluded from this classification. (PE/TR) C57.12.80-1978r

electronic trigger circuit A network containing electron tubes in which the output changes abruptly with an infinitesimal change in input at one or more points in the operating range. (IA/ICTL/IAC) [60]

electronic trip unit A self-contained portion of a circuit breaker that senses the condition of the circuit breaker electronically and that actuates the mechanism that opens the circuit breaker contacts automatically. (IA/PSP) 1015-1997

electronic tuning The process of changing the operating frequency of a system by changing the characteristics of a coupled electron stream. Characteristics involved are, for example: velocity, density, or geometry. *See also:* oscillatory circuit. (ED) 161-1971w

electronic tuning range The frequency range of continuous tuning between two operating points of specified minimum power output for an electronically tuned oscillator. *Note:* The reference points are frequently the half-power points, but should always be specified. *See also:* oscillatory circuit. (ED) 161-1971w

electronic tuning sensitivity At a given operating point, the rate of change of oscillator frequency with the change of the controlling electron stream. For example, this change may be expressed in terms of an electrode voltage or current. *See also:* oscillatory circuit; pushing figure. (ED) 161-1971w

electronic-warfare support measures (ESM) Actions taken to search for, intercept, locate in angle, record, and analyze radiated electromagnetic energy for the purpose of exploiting such radiations in support of military operations. (AES) 686-1997

electron injector The electron gun of a betatron. (ED) [45]

electron lens A device for the purpose of focusing an electron (ion) beam. *See also:* electron optics. (Std100) [84]

electron microscope An electron-optical device that produces a magnified image of an object. *Note:* Detail may be revealed by virtue of selective transmission, reflection, or emission of electrons by the object. (ED) [45]

electron mirror An electronic device causing the total reflection of an electron beam. *See also:* electron optics. (ED) [45]

electron multiplier A structure, within an electron tube, that employs secondary electron emission from solids to produce current amplification. *See also:* amplifier; electron emission. (ED/NPS) 161-1971w, 398-1972r

electron multiplier transit time That portion of photomultiplier transit time corresponding to the time delay between an electron packet leaving the first dynode and the multiplier packet striking the anode. (NPS) 398-1972r

electron optics The branch of electronics that deals with the operation of certain electronic devices, based on the analogy between the path of electron (ion) beams in magnetic or electric fields and that of light rays in refractive media. (ED) [45], [84]

electron-ray indicator tube An elementary form of cathode-ray tube used to indicate a change of voltage. *Note:* Such a tube used to indicate the tuning of a circuit is sometimes called a magic eye. *See also:* cathode-ray tube. (ED) [45], [84]

electron resolution The ability of the electron multiplier section of the photomultiplier to resolve inputs consisting of n and $n + 1$ electrons. This may be expressed as a fractional full width at half maximum of the nth peak, as the peak to valley ratio of the nth peak to the valley between the nth and $n \times 1$th peaks. (NPS) 398-1972r

electrons, conduction *See:* conduction electrons.

electron sheath (gas) A film of electrons (or of ions) that has formed on or near a surface that is held at a potential different from that of the discharge. *Synonym:* ion sheath. *See also:* discharge. (ED) [45], [84]

electron-stream potential (electron tube) (any point in an electron stream) The time average of the potential differential difference between that point and the electron-emitting surface. *See also:* electron emission. (ED) 161-1971w

electron-stream transmission efficiency (electron tube) (electrode through which the electron stream passes) The ratio of the average stream current through the electrode to the average stream current approaching the electrode. *Note:* In connection with multitransit tubes, the term electron stream should be taken to include only electrons approaching the electrode for the first time. *See also:* electron emission. (ED) 161-1971w

electron telescope An optical instrument for astronomy including an electronic image transformer associated with an optical telescope. *See also:* electron optics. (ED) [45], [84]

electron tube An electron device in which conduction by electrons takes place through a vacuum or gaseous medium within a gastight envelope. *Note:* The envelope may be either pumped during operation or sealed off. (ED) 161-1971w

electron-tube admittances The cross-referenced terms generalize the familiar electron-tube coefficients so that they apply to all types of electron devices operated at any frequency as linear transducers. *Note:* The generalizations include the familiar low-frequency tube concepts. In the case of a riode, for example, at relatively low frequencies the short-circuit input admittance reduces to substantially the grid admittance, the short-circuit output admittance reduces to substantially the plate admittance, the short-circuit forward admittance reduces to substantially the grid-plate transconductance, and the short-circuit feedback admittance reduces to substantially the admittance of the grid-plate capacitance. When reference is made to alternating-voltage or -current components, the components are understood to be small enough so that linear relations hold between the various alternating voltages and currents. Consider a generalized network or transducer having n available terminals to each of which is flowinga complex alternating component I_j of the current and between each of which and a reference point (which may or may not be one of the n network terminals) is applied a complex alternating voltage V_j. This network represents an n-terminal electron device in which each one of the terminals is connected to an electrode. (EEC/PE) [119]

electron-tube amplifier An amplifier that obtains its amplifying properties by means of electron tubes. (IA/IAC) [60]

electron-wave tube An electron tube in which mutually interacting streams of electrons having different velocities cause a signal modulation to change progressively along their length. (ED) 161-1971w

electronvolt The kinetic energy acquired by an electron in passing through a potential difference of 1V in vacuum; $1\ \text{eV} = 1.602\,19 \times 10^{-19}$ J approximately. (QUL) 268-1982s

electro-optic effect (fiber optics) A change in the refractive index of a material under the influence of an electric field. *Notes:* 1. Pockels and Kerr effects are electro-optic effects that are respectively linear and quadratic in the electric field strength. 2. "Electro-optic" is often erroneously used as a synonym for "optoelectronic." *See also:* optoelectronic. (Std100) 812-1984w

electro-optic field meter A meter that measures changes in the transmission of light through a fiber or crystal due to the influence of the electric field. *Note:* While there are several electro-optic methods that can be used for measuring electric fields, e.g., the Pockels effect, the Kerr effect, and interferometric techniques, this recommended practice only considers electro-optic field meters that utilize the Pockels effect. (T&D/PE) 1308-1994

electroosmosis The movement of fluids through diaphragms that is as a result of the application of an electric current. (EEC/PE) [119]

electroosmotic potential (electrobiology) The electrokinetic potential gradient producing unit velocity of liquid flow through a porous structure. *See also:* electrobiology. (EMB) [47]

electrophonic effect The sensation of hearing produced when an alternating current of suitable frequency and magnitude from an external source is passed through an animal. (SP) [32]

electrophoresis A movement of colloidal ions as a result of the application of an electric potential. *See also:* ion. (EEC/PE) [119]

electrophoretic potential (electrobiology) The electrokinetic potential gradient required to produce unit velocity of a colloidal or suspended material through a liquid electrolyte. *See also:* electrobiology. (EMB) [47]

electroplating The electrodeposition of an adherent coating upon an object for such purposes as surface protection or decoration. (EEC/PE) [119]

electropneumatic brake An air brake that is provided with electrically controlled valves for control of the application and release of the brakes. *Note:* The electric control is usually in addition to a complete air brake equipment to provide a more prompt and synchronized operation of the brakes on two or more vehicles. *See also:* electric braking. (EEC/PE) [119]

electropneumatic contactor (1) A contactor actuated by air pressure. *See also:* contactor. (IA/ICTL/IAC) [60]
(2) (electropneumatic unit switch) A contactor or switch controlled electrically and actuated by air pressure. *See also:* control switch; contactor. (VT/LT) 16-1955w

electropneumatic controller An electrically supervised controller having some or all of its basic functions performed by air pressure. *See also:* multiple-unit control; electric controller. (IA/IAC) [60]

electropneumatic interlocking machine An interlocking machine designed for electric control of electropneumatically operated functions. *See also:* centralized traffic-control system. (EEC/PE) [119]

electropneumatic valve An electrically operated valve that controls the passage of air. (EEC/PE) [119]

electropolishing (electroplating) The smoothing or brightening of a metal surface by making it anodic in an appropriate solution. *See also:* electroplating. (EEC/PE) [119]

electrorefining The process of electrodissolving a metal from an impure anode and depositing it in a more pure state. (EEC/PE) [119]

electroretinogram *See:* electrodermogram.

electroscope An electrostatic device for indicating a potential difference or an electric charge. *See also:* instrument. (EEC/PE) [119]

electrosensitive printer A nonimpact printer in which images are generated on specially coated paper by an electric stylus. (C) 610.10-1994w

electroshock therapy The production of a reaction in the central nervous system by means of electric current applied to the cranium. *See also:* electrotherapy. (EMB) [47]

electrostatic actuator An apparatus constituting an auxiliary external electrode that permits the application of known electrostatic forces to the diaphragm of a microphone for the purpose of obtaining a primary calibration. *See also:* microphone. (SP) [32]

electrostatic coupling *See:* signal.

electrostatic deflection (cathode-ray tubes) Deflecting an electron beam by the action of an electric field. *See also:* cathode-ray tube. (ED) [45]

electrostatic discharge (ESD) (1) Electrical discharges of static electricity that build up on personnel or equipment, generated by interaction of dissimilar materials. (PE/IC) 1143-1994r
(2) The sudden transfer of charge between bodies of differing electrostatic potentials. (EMC) C63.16-1993
(3) The sudden transfer of charge between bodies of differing electrostatic potentials that may produce voltages or currents that could destroy or damage electrical components. (EMB/MIB) 1073.3.2-2000

electrostatic electrography The branch of electrostatography that employs an insulating medium to form, without the aid of electromagnetic radiation, latent electrostatic-charge patterns for producing a viewable record. *See also:* electrostatography. (ED) [46]

electrostatic electron microscope An electron microscope with electrostatic lenses. *See also:* electron optics. (ED) [45]

electrostatic electrophotography The branch of electrostatography that employs a photoresponsive medium to form, with the aid of electromagnetic radiation, latent electrostatic-charge patterns for producing a viewable record. *See also:* electrostatography. (ED) [46]

electrostatic focusing (electron beams) A method of focusing an electron beam by the action of an electric field. (ED) 161-1971w

electrostatic induction* **(electric coupling)** A common misnomer. The term "static" implies "at rest" or not varying with time. Therefore, this term may be construed to mean induced potential or current resulting from an object being placed in a dc electric field, but often the term is loosely used to include ac field effects. (T&D/PE) 1048-1990

* Deprecated.

electrostatic instrument An instrument that depends for its operation on the forces of attraction and repulsion between bodies charged with electricity. *See also:* instrument. (EEC/PE) [119]

electrostatic lens An electron lens in which the result is obtained by an electrostatic field. *See also:* electron optics.

electrostatic loudspeaker A loudspeaker in which the mechanical forces are produced by the action of electrostatic fields. *Synonyms:* condenser loudspeaker; capacitor loudspeaker. (SP) [32]

electrostatic microphone A microphone that depends for its operation upon variations of its electrostatic capacitance. *Synonyms:* condenser microphone; capacitor microphone. *See also:* microphone. (SP) [32]

electrostatic plotter A raster plotter in which images are drawn by attracting toner particles to a static charge on the surface of a photoconductor, then transferring the image to a sheet of paper. (C) 610.10-1994w

electrostatic printer A nonimpact printer in which images are generated by attracting toner particles to a static charge on the surface of a photoconductor, then transferring the image to a sheet of paper. *Synonym:* optical printer. (C) 610.10-1994w

electrostatic recording (facsimile) Recording by means of a signal-controlled electrostatic field. *See also:* recording. (COM) 168-1956w

electrostatic relay (1) A relay in which operation depends upon the application or removal of electrostatic charge. 341-1956
(2) A relay in which the actuator element consists of nonconducting media separating two or more conductors that change their relative positions because of the mutual attraction or repulsion by electric charges applied to the conductors. *See also:* relay. (EEC/REE) [87]

electrostatics The branch of science that treats of the electric phenomena associated with electric charges at rest in the frame of reference. (Std100) 270-1966w

electrostatic storage A type of storage that uses electrically charged areas on a dielectric surface layer. *See also:* Williams-tube storage. (C) [20], 610.10-1994w, [85]

electrostatic voltmeter A voltmeter depending for its action upon electric forces. An electrostatic voltmeter is provided with a scale, usually graduated in volts or kilovolts. *See also:* instrument. (EEC/PE) [119]

electrostatic wave *See:* longitudinal wave.

electrostatography The formation and utilization of latent electrostatic-charge patterns for the purpose of recording and reproducing patterns in viewable form. (ED) 224-1965w, [46]

electrostenolysis The discharge of ions or colloidal ions in capillaries through the application of an electric potential. *See also:* ion. (EEC/PE) [119]

electrostrictive relay A relay in which an electrostrictive dielectric serves as the actuator. *See also:* relay. (EEC/REE) [87]

electrotaxis (galvanotaxis) (electrobiology) The act of a living organism in arranging itself in a medium in such a way that its axis bears a certain relation to the direction of the electric current in the medium. *See also:* electrobiology. (EMB) [47]

electrotherapy The use of electric energy in the treatment of disease. (EMB) [47]

electrothermal efficiency The ratio of energy usefully employed in a furnace to the total energy supplied. *See also:* electrothermics. (EEC/PE) [119]

electrothermal recording (facsimile) That type of recording that is produced principally by signal-controlled thermal action. *See also:* recording. (COM) 168-1956w

electrothermic-coupler unit (electrothermic power meters) A three-port directional coupler with an electrothermic unit attached to either the side arm or the main arm which is normally used as a feed-through power measuring system. Typically, an electrothermic unit is attached to the side arm of the coupler so that the power at the output port of the main arm can be determined from a measurement of the power in the side arm. This system also can be used as a terminating powermeter by terminating the output port of the directional coupler. (IM) 544-1975w

electrothermic element (electrothermic power meters) A power absorber and a thermocouple (or thermopile) which are either two separate units or where the thermocouple (or thermopile) is also the power absorber. (IM) 544-1975w

electrothermic instrument An instrument that depends for its operation on the heating effect of a current or currents. *Note:* Among the several possible types are the expansion type, including the hot-wire and hot-strip instruments; the thermocouple type; and the bolometric type. *See also:* instrument. (EEC/PE) [119]

electrothermic mount (electrothermic power meters) A waveguide or transmission line structure which is designed to accept the electrothermic element. It normally contains internal matching devices or other reactive elements to obtain specified impedance conditions at its input terminal when an electrothermic element is installed. It usually contains a means of protecting the electrothermic element and the immediate environment from thermal gradients which would cause an undesirable thermoelectric output. (IM) 544-1975w

electrothermic power indicator (electrothermic power meters) An instrument that may or may not amplify the low level dc output voltage from the electrothermic unit and provides a display, usually in the form of the D'Arsonval type indication or a digital readout. (IM) 544-1975w

electrothermic power indicator error (electrothermic power meters) Ability of the metering circuitry to indicate exactly the substituted power within an electrothermic unit. Included are such factors as meter calibration, open loop gain, meter linearity, tracking errors, range switching errors, line voltage errors, and temperature compensation errors. (IM) 544-1975w

electrothermic power meter This consists of an electrothermic unit and an electrothermic power indicator. (IM) 544-1975w

electrothermics The branch of science and technology that deals with the direct transformation of electric energy and heat. (EEC/PE) [119]

electrothermic substitution power (electrothermic power meters) The power at a reference frequency which, when dissipated in the electrothermic element, produces the same dc electrothermic output voltage that the element produces when subjected to radio frequency power. (IM) 544-1975w

electrothermic technique of power measurement (electrothermic power meters) A technique wherein the heating effect of power dissipated in an electrothermic element (which consists of an energy absorber and a thermocouple or thermopile) is used to generate a dc voltage. The power is dissipated either in a separate absorber or in the resistance of the electrothermic element. The resultant heat causes a temperature rise in a portion of the element. This temperature rise is sensed by the thermocouple which generates a dc output voltage proportional to the power. (IM) 544-1975w

electrothermic unit (electrothermic power meters) An assembly consisting of the electrothermic element installed in the electrothermic mount. (IM) 544-1975w

electrotonic wave (electrobiology) A brief nonpropagated change of potential on an excitable membrane in the vicinity of an applied stimulus; it is often accompanied by a propagated response and always by electrotonus. *See also:* excitability. (EMB) [47]

electrotonus (A) (physical) The change in distribution of membrane potentials in nerve and muscle during or after the passage of an electric current. *See also:* excitability. **(B) (physiological)** The change in the excitability of a nerve or muscle during the passage of an electric current. *See also:* excitability. (EMB) [47]

electrotyping The production or reproduction of printing plates by electroforming. (EEC/PE) [119]

electrowinning The electrodeposition of metals or compounds from solutions derived from ores or other materials using insoluable anodes. (EEC/PE) [119]

element (1) A representation of all or part of a logic function within a single outline, which may, in turn, be subdivided into smaller elements representing subfunctions of the overall function. Alternatively, the function so represented. (GSD) 91-1984r

(2) (measuring longitudinal balance of telephone equipment operating in the voice band) Any electric device (such as inductor, resistor, capacitor, generator, or line) with terminals at which it may be directly connected to other devices, elements, or apparatus. (COM/TA) 455-1985w

(3) (electron tube) A constituent part of the tube that contributes directly to its electrical operation. (ED) 161-1971w

(4) (semiconductor devices) Any integral part that contributes to its operation. (ED) 216-1960w

(5) (integrated circuit) A constituent part of the integrated circuit that contributes directly to its operation. (ED) 274-1966w, [46]

(6) (storage cell) Consists of the positive and negative groups with separators, or separators and retainers, assembled for one cell. (EEC/PE) [119]

(7) (primary detecting) That portion of the feedback elements which first either utilizes or transforms energy from the controlled medium to produce a signal that is a function of the value of the directly controlled variable. (PE/EDPG) 421-1972s

(8) A product, subsystem, assembly, component, subcomponent or subassembly, or part of a physical or system architecture, specification tree, or system breakdown structure, including the system itself. (C/SE) 1220-1994s

(9) A functional interface or a namespace allocation. Examples of elements are C functions or utility programs. Examples of namespace allocation include headers or error return value constants. (C/PA) 2003.1-1992, 13210-1994

(10) A functional interface or a namespace allocation. Examples of elements are functions and utility programs. Examples of namespace allocation include headers and error return value constants. (C/PA) 2003-1997

(11) A component of a circuit, such as a resistor or capacitor. (C) 610.10-1994w

(12) Any of the bits of a bit string, the octets of an octet string, or the bytes by means of which the characters of a character string are represented. (C/PA) 1224-1993w, 1327-1993 w, 1328-1993w

(13) Within AI-ESTATE, element refers to the smallest entity of a model. For example, in a particular model, the smallest test, the smallest diagnosis, and the no-fault conclusion are all elements. (SCC20) 1232.1-1997

(14) A component of a system; may include equipment, a computer program, or a human. (C/SE) 1233-1998

(15) *See also:* relay element. (SWG/PE) C37.100-1981s

(16) (computing system) *See also:* combinational logic element; logic element; threshold element.

(17) *See also:* array element; director element; driven element; linear electric current element; linear magnetic current element; multiwire element; parasitic element; radiating element; reflector element. (AP/ANT) 145-1993

(18) (data management) *See also:* binary element; data element. (C) 610.5-1990w

elemental area (facsimile) Any segment of the scanning line of the subject copy the dimension of which along the line is exactly equal to the nominal line width. *Note:* Elemental area is not necessarily the same as the scanning spot. *See also:* scanning. (COM) 168-1956w

elementary diagram (packaging machinery) A diagram using symbols and a plan of connections to illustrate, in simple form, the scheme of control. (IA/PKG) 333-1980w

element cell (of an array antenna) In an array having a regular arrangement of elements that can be made congruent by translation, an element and a region surrounding it that, when repeated by translation, covers the entire array without gaps or overlay between cells. *Note:* There are many possible choices for such a cell. Some may be more convenient than others for analytic purposes. (AP/ANT) 145-1993

element conduction interval (thyristor) That part of an operating cycle in which ON-state current flows in the basic control element. (IA/IPC) 428-1981w

element identifier (EID) A numeric identifier generated by the application on the onboard equipment (OBE) transponder. An EID is unambiguous within the context of a complete exchange of messages with an application on the roadside equipment (RSE). Multiple EIDs can be maintained between the RSE and OBE over a single data link session (LID). (SCC32) 1455-1999

element linear *See:* linear system or element.

element, measuring *See:* measuring element.

element nonconduction interval (thyristor) That part of an operating cycle during which no ON-state current flows in the basic control element. (IA/IPC) 428-1981w

element of a fix (navigation aid terms) The specific values of the navigation coordinates necessary to define a position. (AES/GCS) 172-1983w

element, primary detecting *See:* primary detecting element.

element printer An impact printer that generates characters using interchangeable print elements such as daisy wheels or thimbles, each of which contains a full character set. *See also:* chain printer; band printer; wheel printer; stick printer; bar printer; drum printer. (C) 610.10-1994w

elements The discrete pieces that make up an asset (e.g., documents, requirements specifications, test cases, source code, installation information, and read me files). (C/SE) 1420.1a-1996

elements, feedback *See:* feedback elements.

elements, forward *See:* forward elements.

elements, loop *See:* loop elements.

elements, reference-input *See:* reference-input elements.

element type A category or class of elements. (C/SE) 1420.1a-1996

ELEPL *See:* equal level echo path loss.

elevated duct A tropospheric radio duct in which the lower boundary is above the surface of the Earth. (AP/PROP) 211-1997

elevated-zero range A range where the zero value of the measured variable, measured signal, etc., is greater than the lower range value. *Note:* The zero may be between the lower and upper range values, at the upper range value, or above the upper range value. For example: −20 to 100; −40 to 0; and −50 to −10. *See also:* instrument. (EEC/EMI) [112]

elevation (illuminating engineering) The angle between the axis of a searchlight drum and the horizontal. For angles above the horizontal, elevation is positive, and below the horizontal negative. (EEC/IE) [126]

elevation angle (1) Complement of the angle of incidence. May also refer to the angle of radiation measured above the horizon from a source. *See also:* grazing angle. (AP/PROP) 211-1997

(2) In radar, the angle between the line-of-sight in the direction of interest and a horizontal reference plane, measured upwards. (AES) 686-1997

elevation rod (lightning protection) The vertical portion of conductor in an air terminal by means of which it is elevated above the object to be protected. (EEC/PE) [119]

elevator A hoisting and lowering mechanism equipped with a car or platform that moves in guides in a substantially vertical direction, and that serves two or more floors of a building or structure. (EEC/PE) [119]

elevator automatic dispatching device A device, the principal function of which is to either operate a signal in the car to indicate when the car should leave a designated landing; or actuate its starting mechanism when the car is at a designated landing. *See also:* control. (EEC/PE) [119]

elevator automatic signal transfer device A device by means of which a signal registered in a car is automatically transferred to the next car following, in case the first car passes a floor, for which a signal has been registered, without making a stop. *See also:* control. (EEC/PE) [119]

elevator car The load-carrying unit including its platform, car frame, enclosure, and car door or gate. *See also:* elevator. (EEC/PE) [119]

elevator car bottom runby (elevators) The distance between the car buffer striker plate and the striking surface of the car buffer when the car floor is level with the bottom terminal landing. *See also:* elevator. (EEC/PE) [119]

elevator-car flash signal device One providing a signal light, in the car, that is illuminated when the car approaches the landings at which a landing-signal-registering device has been actuated. *See also:* control. (EEC/PE) [119]

elevator car-leveling device Any mechanism that will, either automatically or under the control of the operator, move the car within the leveling zone toward the landing only, and automatically stop it at the landing. *Notes:* 1. Where controlled by the operator by means of up-and-down continuous-pressure switches in the car, this device is known as an inching device. 2. Where used with a hydraulic elevator to correct automatically a change in car level caused by leakage in the hydraulic system, this device is known as an anticreep device. *See also:* leveling zone; two-way automatic maintaining leveling device; one-way automatic leveling device; two-way automatic nonmaintaining leveling device; elevator. (EEC/PE) [119]

elevator counterweight bottom runby The distance between the counterweight buffer striker plate and the striking surface of the counterweight buffer when the car floor is level with the top terminal landing. *See also:* elevator. (EEC/PE) [119]

elevator landing That portion of a floor, balcony, or platform used to receive and discharge passengers or freight. *See also:* landing zone; bottom-terminal landing; top terminal landing. (EEC/PE) [119]

elevator-landing signal registering device A button or other device, located at the elevator landing that, when actuated by a waiting passenger, causes a stop signal to be registered in the car. *See also:* control. (EEC/PE) [119]

elevator-landing stopping device A button or other device, located at an elevator landing that, when actuated, causes the elevator to stop at that floor. *See also:* control. (EEC/PE) [119]

elevator parking device An electric or mechanical device, the function of which is to permit the opening, from the landing side, of the hoistway door at any landing when the car is within the landing zone of that landing. The device may also be used to close the door. *See also:* control. (EEC/PE) [119]

elevator pit That portion of a hoistway extending from the threshold level of the lowest landing door to the floor at the bottom of the hoistway. *See also:* elevator. (EEC/PE) [119]

elevator separate-signal system A system consisting of buttons or other devices located at the landings that, when actuated by a waiting passenger, illuminate a flash signal or operate an annunciator in the car, indicating floors at which stops are to be made. *See also:* control. (EEC/PE) [119]

elevator signal-transfer switch A manually operated switch, located in the car, by means of which the operator can transfer a signal to the next car approaching in the same direction, when he desires to pass a floor at which a signal has been registered in the car. *See also:* control. (EEC/PE) [119]

elevator starter's control panel An assembly of devices by means of which the starter may control the manner in which an elevator, or group of elevators, functions. *See also:* control. (EEC/PE) [119]

elevator truck zone The limited distance above an elevator landing within which the truck-zoning device permits movement of the elevator car. *See also:* control. (EEC/PE) [119]

elevator truck-zoning device A device that will permit the operator in the car to move a freight elevator, within the truck zone, with the car door or gate and a hoistway door open. *See also:* control. (EEC/PE) [119]

eleven punch A zone punch in punch row eleven (second from the top) in a twelve-row punch card. *Synonym:* X punch. *See also:* zero punch; twelve punch. (C) 610.10-1994w

ELF *See:* extremely low frequency.

elliptically polarized field vector At a point in space, a field vector whose extremity describes an ellipse as a function of time. *Note:* Any single-frequency field vector is elliptically polarized if "elliptical" is understood in the wide sense as including circular and linear. Often, however, the expression is used in the strict sense meaning noncircular and nonlinear. (AP/MTT/ANT) 145-1993, 146-1980w

elliptically polarized plane wave A plane wave whose electric field vector is elliptically polarized. (AP/ANT) 145-1993

elliptically polarized wave (1) (given frequency) An electromagnetic wave for which the component of the electric vector in a plane normal to the direction of propagation describes an ellipse. *See also:* electromagnetic waves; radiation; waveguide; radio transmitter. (MTT) 146-1980w
(2) An electromagnetic wave for which the locus of the tip of the electric field vector is an ellipse in a plane orthogonal to the wave normal. This ellipse is traced at the rate in radians equal to the angular frequency of the wave. *See also:* left-hand polarized wave; right-hand polarized wave. (AP/PROP) 211-1997

elliptical orbit (communication satellite) An orbit of a satellite in which the distance between the centers of mass of the satellite and of the primary body is not constant. The general type of orbit is a special case. (COM) [19]

elliptic filter A filter having an equiripple pass band and an equiminima stop band. (CAS) [13]

ellipticity *See:* waveguide; axial ratio.

EM *See:* end-of-medium character.

E-Mail *See:* electronic mail.

emanations attacks Passive collection, analysis, and interpretation of electromagnetic emissions from and the direction of electromagnetic interference at electrical and electronic equipment. Emanations attacks may be perpetrated at several points in a system (e.g., processors, input/output devices, communication channels, power sources), and could result in unauthorized disclosure or modification of information, denial of service to legitimate users or critical functions, or destruction of sensitive electronic components (e.g., from an electromagnetic pulse created by a nuclear explosion). (C/BA) 896.3-1993w

emanations security Protection of electromagnetic emanations of systems from unauthorized interception and analysis, and protection of the system from electromagnetic interference. *See also:* tempest; electromagnetic interference. (C/BA) 896.3-1993w

embedded coil side (rotating machinery) That part of a coil side which lies in a slot between the ends of the core. (PE) [9]

embedded computer system (software) A computer system that is part of a larger system and performs some of the requirements of that system; for example, a computer system used in an aircraft or rapid transit system. (C) 610.12-1990, 610.10-1994w

embedded data dictionary *See:* active data dictionary.

embedded hyphen *See:* required hyphen.

embedded software Software that is part of a larger system and performs some of the requirements of that system; for example, software used in an aircraft or rapid transit system. (C) 610.12-1990

embedded temperature detector (1) (rotating machinery) An element, usually a resistance thermometer or thermocouple, built into apparatus for the purpose of measuring temperature. *Notes:* 1. This is ordinarily installed in a stator slot between coil sides at a location at which the highest temperature is anticipated. 2. Examination or replacement of an embedded detector after the apparatus is placed in service is usually not feasible. (SWG/PE) C37.30-1971s, [9]
(2) A resistance thermometer or thermocouple built into a machine for the purpose of measuring the temperature. (IA/MT) 45-1998

embossing stylus A recording stylus with a rounded tip that displaces the material in the recording medium to form a groove. *See also:* phonograph pickup. (SP) [32]

embrittlement Severe loss of ductility of a metal or alloy. (IA) [59]

EMC *See:* electromagnetic compatibility.

emergency announcing system A system of microphones, amplifier, and loud speakers (similar to a public address system) to permit instructions and orders from a ship's officers to passengers and crew in an emergency and particularly during abandon-ship operations. (EEC/PE) [119]

emergency brake Fail-safe, open-loop braking to a complete stop with an assured maximum stopping distance, considering all relevant factors. Once the brake application is initiated, it is irretrievable, i.e., it cannot be released until the train has stopped or a predetermined time has passed. (VT/RT) 1475-1999, 1474.1-1999

emergency cells (storage cell) End cells that are held available for use exclusively during emergency discharges. *See also:* battery. (PE/EEC) [119]

emergency egress A path or route that provides an immediate exit path or way out of an area in the event of a sudden, unexpected, or dangerous occurrence. (PE/NP) 692-1997

emergency electric system (marine) All electric apparatus and circuits the operation of which, independent of ship's service supply, may be required under casualty conditions for preservation of a ship or personnel. (PE/EEC) [119]

emergency generator (marine) An internal-combustion-engine-driven generator so located in the upper part of a vessel as to permit operation as long as the ship can remain afloat, and capable of operation, independent of any other apparatus on the ship, for supply of power to the emergency electric system upon failure of a ship's service power. *See also:* emergency electric system. (PE/EEC) [119]

emergency lighting (illuminating engineering) Lighting designed to supply illumination essential to the safety of life and property in the event of failure of the normal supply. (EEC/IE) [126]

emergency lighting storage battery A storage battery for instant supply of emergency power, upon failure of a ship's service supply, to certain circuits of special urgency principally temporary emergency lighting. *See also:* emergency electric system. (EEC/PE/MT) [119]

emergency maintenance Unscheduled corrective maintenance performed to keep a system operational. (C/SE) 1219-1998

emergency message An arbitration cycle with a special high arbitration number, which is selected from a set of numbers assigned to emergency messages. (C/MM) 896.1-1987s

emergency operations area (nuclear power generating station) Functional area(s) allocated for the displays used to assess the status of safety systems and the controls for manual operations required during emergency situations. (PE/NP) 566-1977w

emergency power (power operations) (electric power system) Power required by a system to make up a deficiency between the current firm power demand and the immediately available generating capability. (PE/PSE) 858-1987s, 346-1973w

emergency power feedback An arrangement permitting feedback of emergency-generator power to a ship's service system for supply of any apparatus on the ship within the limit of the emergency-generator rating. *See also:* emergency electric system. (EEC/PE) [119]

emergency power system An independent reserve source of electric energy that, upon failure or outage of the normal source, automatically provides reliable electric power within a specified time to critical devices and equipment whose failure to operate satisfactorily would jeopardize the health and safety of personnel or result in damage to property. (IA/PSE) 446-1995

emergency rating (1) The level of power flow in excess of the normal rating that a facility can carry for the time sufficient for adjustment of transfer schedules or generation dispatch in an orderly manner, with acceptable loss of life to the facility involved. (PE/PSE) 858-1993w

(2) (generating station) Capability of installed equipment for a short time interval. (PE/PSE) 346-1973w

emergency service An additional service intended only for use under emergency conditions. *See also:* service; duplicate service; loop service; dual service. (T&D/PE) [10]

emergency stop switch (elevators) A device located in the car that, when manually operated, causes the electric power to be removed from the driving-machine motor and brake of an electric elevator or from the electrically operated valves and/or pump motor of a hydraulic elevator. *See also:* control. (EEC/PE) [119]

emergency switchboard (1) A switchboard for control of sources of emergency power and for distribution to all emergency circuits. *See also:* emergency electric system. (EEC/PE) [119]

(2) A switchgear and control assembly that receives energy from the emergency generating plant and distributes directly or indirectly to all emergency loads. (IA/MT) 45-1998

emergency system (1) (health care facilities) A system of feeders and branch circuits meeting the requirements of Article 700 of NFPA 70-1978, National Electrical Code, and intended to supply alternate power to a limited number of prescribed functions vital to the protection of life and safety, with automatic restoration of electrical power within 10 seconds of power interruption. (EMB) [47]

(2) A system of feeders and branch circuits meeting the requirements of Article 700, connected to alternate power sources by a transfer switch and supplying energy to an extremely limited number of prescribed functions vital to the protection of life and patient safety, with automatic restoration of electrical power within 10 seconds of power interruption. (NESC/NEC) [86]

emergency-terminal stopping device (elevators) A device that automatically causes the power to be removed from an electric elevator driving-machine motor and brake, or from a hydraulic elevator machine, at a predetermined distance from the terminal landing, and independently of the functioning of the operating device and the normal-terminal stopping device, if the normal-terminal stopping device does not slow down the car as intended. *See also:* control. (PE/EEC) [119]

emergency transfer capability (electric power supply) The maximum amount of power that can be transmitted following a loss of transmission or generation capacity without causing additional transmission outages. (PE/PSE) 346-1973w

emergency voltage limit (power operations) The voltage range that is acceptable without serious system consequences, for the time sufficient for system adjustments to be made. *See also:* normal voltage limit. (PE/PSE) 858-1987s

emerging standard A specification that is under consideration by an accredited standards development organization, but that has not completed the process of approval by the sponsoring body. Emerging standards are often subject to significant change prior to approval. (C/PA) 14252-1996, 1003.23-1998l

E meter *See:* electricity meter.

EMF *See:* electromotive force.

EMI *See:* electromagnetic interference.

emission (1) (laser maser) The transfer energy from matter to a radiation field. (LEO) 586-1980w

(2) (radio-noise emissions) (electromagnetic) The phenomenon by which electromagnetic energy emanates from a source. (EMC) C63.4-1991

emission characteristic A relation, usually shown by a graph, between the emission and a factor controlling the emission (such as temperature, voltage, or current of the filament or heater). *See also:* electron emission. (ED) 161-1971w

emission current The current resulting from electron emission. (ED) [45], [84]

emission current, field-free *See:* field-free emission current.

emission efficiency (thermionics) The quotient of the saturation current by the heating power absorbed by the cathode. *See also:* electron emission. (ED) [45], [84]

emission probability per decay The probability that a radioactive decay will be followed by the emission of the specified radiation. Gamma-ray emission probabilities are often expressed per 100 decays. (NI) N42.14-1991

emissivity (1) (fiber optics) The ratio of power radiated by a substance to the power radiated by a blackbody at the same temperature. Emissivity is a function of wavelength and temperature. *See also:* blackbody. (Std100) 812-1984w

(2) (photovoltaic power system) The emittance of a specimen of material with an optically smooth, clean surface and sufficient thickness to be opaque. *See also:* photovoltaic power system. (AES) [41]

(3) The ratio of power (per unit surface area, per unit solid angle, over a specified bandwidth) radiated by a material body to the power radiated by a blackbody at the same temperature. (AP/PROP) 211-1997

emittance (1) (illuminating engineering) ε The ratio of radiance in a given direction (for directional emittance) or radiant

exitance (for hemispherical emittance) of a sample of a thermal radiator to that of a blackbody radiator at the same temperature. Formerly, exitance. The use of exitance with this meaning is deprecated. (EEC/IE) [126]
(2) (photovoltaic power system) The ratio of the radiant flux-intensity from a given body to that of a black body at the same temperature. *See also:* photovoltaic power system. (AES) [41]

emitter (1) (transistor) A region from which charge carriers that are minority carriers in the base are injected into the base. (ED/IA) 216-1960w, [12], 270-1966w
(2) A device that is able to discharge detectable electromagnetic, seismic, or acoustic energy. (DIS/C) 1278.1-1995

emitter-coupled logic A family of non-saturated, very high speed, bipolar logic devices that are commonly used in high performance processors, which dissipate relatively large amounts of power. (C) 610.10-1994w

emitter junction (semiconductor devices) A junction normally biased in the low-resistance direction to inject minority carriers into an interelectrode region. *See also:* semiconductor; transistor. (Std100) 270-1966w

emitter, majority *See:* majority emitter.

emitter, minority *See:* minority emitter.

emitting sole (microwave tubes) An electron source in crossed-field amplifiers that is extensive and parallel to the slow-wave circuit and that may be a hot or cold electron-emitter. (ED) [45]

EMP *See:* electromagnetic pulse.

emperor The processor that has the responsibility for initialization of an entire multiprocessor system. (C/MM) 1596-1992

emperor processor (1) The monarch processor that is selected to initialize and configure the system. On a single-bus system, the monarch and emperor processor are always the same. On a multiple-bus system, the single emperor processor is selected from the available monarch processors. (C/MM) 1212-1991s
(2) The monarch processor selected to direct the configuration and initialization of an entire system with multiple interconnected logical buses. (C/BA) 896.4-1993w, 10857-1994

emphasis Highlighting, color change, or another visual indication of the condition of an object or choice, and the effect of that condition on the ability of the user to interact with that object or choice. Emphasis can also give the user additional information about the state of an object or choice. (C) 1295-1993w

empirical Pertaining to information that is derived from observation, experiment, or experience. (C) 610.3-1989w

empirical propagation model A propagation model that is based solely on measured path-loss data. *See also:* electromagnetic compatibility. (EMC) [53]

empty directory (1) A directory that contains, at most, directory entries for dot and dot-dot. (C/PA) 9945-1-1996, 9945-2-1993
(2) A directory that contains, at most, entries for dot and dot-dot. (C) 1003.5-1999

empty line A line consisting of only a ⟨`newline`⟩ character. *See also:* blank line. (C/PA) 9945-2-1993

empty medium (1) A data medïum that does not contain data. (C) 610.5-1990w
(2) A data medium that contains only marks of reference and no user data; For example, a formatted floppy disk. *See also:* blank medium; virgin medium. (C) 610.10-1994w

empty queued arbitrated slot A Queued Arbitrated (QA) slot that was designated by the Head of Bus function as being available for transfer of a QA segment, and that does not contain a QA segment. (LM/C) 8802-6-1994

empty string (1) A character array whose first element is a null character. *Synonym:* null string. (C/PA) 9945-1-1996, 9945-2-1993
(2) A zero-length array whose components are of some character type. *Synonym:* null string. (C) 1003.5-1999

empty weight *See:* actual weight.

EMT *See:* electrical metallic tubing.

emulate To represent a system by a model that accepts the same inputs and produces the same outputs as the system represented. For example, to emulate an 8-bit computer with a 32-bit computer. *See also:* simulate. (C) 610.3-1989w

emulation (A) (software) A model that accepts the same inputs and produces the same outputs as a given system. *See also:* simulation. **(B) (software)** The process of developing or using a model as in (A). (C) 610.3-1989, 610.12-1990

emulator (modeling and simulation) (software) A device, computer program, or system that performs emulation. (C) 610.3-1989w, 610.12-1990

enable (1) (supervisory control, data acquisition, and automatic control) (station control and data acquisition) A command or condition that permits some specific event to occur. (PE/SUB) C37.1-1994
(2) A command or condition that permits some specific event to proceed. (SWG/PE) C37.100-1992

enable high only (local area networks) A link control signal from an upper repeater to a lower repeater pre-empting a lower repeater's normal- priority round-robin control cycle. (C) 8802-12-1998

enabling pulse (1) (navigation) A pulse that prepares a circuit for some subsequent action. (AES/RS) 686-1982s, [42]
(2) A pulse that opens an electric gate normally closed, or otherwise permits an operation for which it is a necessary but not a sufficient condition. *See also:* pulse. (EEC/PE) [119]

enamel (1) (general) A paint that is characterized by an ability to form an especially smooth film. (PE/IA/PC) [9], [65]
(2) (rotating machinery) (wire) A smooth film applied to wire usually by a coating process. *See also:* rotor; stator. (PE) [9]

encapsulated (rotating machinery) A machine in which one or more of the windings is completely encased by molded insulation. *See also:* asynchronous machine. (PE) [9]

encapsulation (1) (germanium gamma-ray detectors) (of a semiconductor radiation detector) The packaging of a detector for protective or mounting purposes, or both. (NPS) 325-1986s
(2) (software) A software development technique that consists of isolating a system function or a set of data and operations on those data within a module and providing precise specifications for the module. *See also:* information hiding; data abstraction. (C) 610.12-1990
(3) In the context of AI-ESTATE, the act of specifying a test or collection of tests together with associated preconditions and post conditions. (ATLAS) 1232-1995
(4) In the context of AI-ESTATE, the act of specifying a test or collection of tests together with the associated preconditions and postconditions. Alternately, the process of hiding all of the details of an object that do not contribute to the essential characteristics. (SCC20) 1232.1-1997
(5) A technique used by layered protocols to carry foreign protocols in a network. (C) 610.7-1995
(6) The grouping of data and operations upon that data into a single object. (SCC20) 1226-1998
(7) The concept that access to the names, meanings, and values of the responsibilities of a class is entirely separated from access to their realization. (C/SE) 1320.2-1998
(8) In 1000BASE-X, the process by which a MAC packet is enclosed within a PCS code-group stream. (C/LM) 802.3-1998

encipherment The cryptographic transformation of data to produce ciphertext. *See also:* cryptography. (LM/C) 802.10-1992

enclosed Surrounded by case, cage, or fence designed to protect the contained equipment and limit the likelihood, under normal conditions, of dangerous approach or accidental contact by persons or objects. (NESC) C2-1997

enclosed brake A brake that is provided with an enclosure that covers the entire brake, including the brake actuator, the brake shoes, and the brake wheel. *See also:* electric drive. (IA/ICTL/IAC/APP) [60], [75]

enclosed capacitor (shunt power capacitors) A capacitor having enclosed terminals. The enclosure is provided with means for connection to a rigid or flexible conduit. (T&D/PE) 18-1992

enclosed cutout A cutout in which the fuse clips and fuseholder or disconnecting blade are mounted completely within an insulating enclosure. (SWG/PE) C37.40-1993, C37.100-1992

enclosed relay A relay that has both coil and contacts protected from the surrounding medium. *See also:* relay. (EEC/REE) [87]

enclosed self-ventilated machine A machine that has openings for the ventilating air circulated by means integral with the machine, the machine being otherwise totally enclosed. These openings are so arranged that inlet and outlet ducts or pipes may be connected. (IA/MT) 45-1998

enclosed separately ventilated machine A machine that has openings for ventilating air circulated by means external to and not a part of the machine, the machine being otherwise totally enclosed. These openings are so arranged that inlet and outlet duct pipes may be connected to them. (IA/MT) 45-1998

enclosed switch (safety switch) A switch either with or without fuse holders, meter-testing equipment, or accommodation for meters, having all current-carrying parts completely enclosed in metal, and operable without opening the enclosure. *See also:* switch. (IA/ICTL/IAC) [60]

enclosed switchboard A dead-front switchboard that has an overall sheet-metal enclosure (not grille) covering back and ends of the entire assembly. *Note:* Access to the interior of the enclosure is usually provided by doors or removable covers. The top may or may not be covered. (SWG/PE) C37.100-1992, C37.21-1985r

enclosed switches (indoor or outdoor) Switches designed for service within a housing restricting heat transfer to the external medium. (SWG/PE) C37.100-1992

enclosed switchgear assembly An assembly that is enclosed on all sides and top. (SWG/PE) C37.100-1992

enclosed ventilated (rotating machinery) A term applied to an apparatus with a substantially complete enclosure in which openings are provided for ventilation only. *See also:* asynchronous machine. (PE) [9]

enclosed ventilated apparatus Apparatus totally enclosed except that openings are provided for the admission and discharge of the cooling air. *Note:* These openings may be so arranged that inlet and outlet ducts or pipes may be connected to them. An enclosed ventilated apparatus or machine may be separately ventilated or self-ventilated. (EEC/PE) [119]

enclosure (1) (power system communication equipment) A surrounding case or housing to protect the contained equipment against external conditions and to prevent personnel from accidentally contacting live parts. (PE/PSC/TR) 281-1984w, C57.12.80-1978r

(2) The case or housing of apparatus, or the fence or walls surrounding an installation to prevent personnel from accidentally contacting energized parts, or to protect the equipment from physical damage. (NESC/NEC) [86]

(3) A surrounding case or housing used to protect the contained equipment and to prevent personnel from accidentally contacting live parts. (SWG/PE) C37.23-1987r, C37.100-1992, [56]

(4) An identifiable housing, such as a cubicle, compartment, terminal box, panel, or enclosed raceway, used for electrical equipment or cables. (PE/NP) 384-1992r

enclosure currents (1) Currents that result from the voltages induced in the metallic enclosure by effects of currents flowing in the enclosed conductors. (SWG/PE/SUB) C37.122-1983s, C37.100-1992, C37.122.1-1993

(2) Currents that result from the voltages induced in the metallic enclosure by the current(s) flowing in the enclosed conductor(s). (PE/SUB) 80-2000

encode (1) (general) To express a single character or a message in terms of a code. (C) 162-1963w

(2) (electronic control) To produce a unique combination of a group of output signals in response to each of a group of input signals. (C) 162-1963w

(3) (computers) To apply the rules of a code. *See also:* matrix; translate; code; decode. (C) [20], [85]

(4) (modeling and simulation) To represent data in symbolic form using a code or a coded character set such that reconversion to the original form is possible. *Note:* Sometimes used when complete reconversion is not possible. *Contrast:* decode. *See also:* code. (Std100) 270-1966w

encoded data *See:* code.

encode/decode Encoding is the mapping of typed information from its internal datatype format into the types allowed by the signatures of the `Perform-`, `Execute-`, and `Publish-`like operations. Decoding is the mapping from the types allowed by the signatures of the `Perform-`, `Execute-`, and subscription callback-like operations into the datatypes used internally. (IM/ST) 1451.1-1999

encode-decode table *See:* code-decode table.

encoded symbol A 10-bit symbol created from 8-bit data using the 8B/10B encoding scheme. (C/BA) 1393-1999

encoder (1) A network or system in which only one input is excited at a time and each input produces a combination of outputs. *See also:* matrix. (Std100) 270-1966w

(2) A device that performs encoding. (COM/TA) 1007-1991r

(3) A device or system that encodes data. *Contrast:* decoder. (C) 610.5-1990w

encoding (1) A means of producing a unique combination of bits (a code) in response to an analog input signal. (COM/TA) 1007-1991r

(2) The representation of data bits and nondata information for signal transmission across a serial communications medium. Nondata information includes indications of start and end of octets and frame transmission. (EMB/MIB) 1073.4.1-2000

encoding law An algorithm for encoding; i.e., "μ-law" or "A law." (COM/TA) 1007-1991r

end-application The portion of a computer program that is separate from the communications stack. Specifically, an end-application does not include the Application Layer of the OSI Reference Model, nor any of the layers below that. (SCC32) 1488-2000

end-around carry (1) (computers) A carry generated in the most significant place and forwarded directly to the least significant place, for example, when adding two negative numbers, using nines complement. *See also:* carry. (C) 162-1963w

(2) (mathematics of computing) A carry process in which a carry digit generated in the most significant digit place is added directly to the least significant digit place. For example, when adding two negative numbers using nines complement. (C) 1084-1986w

end-around carry shift *See:* circular shift.

end-around shift *See:* circular shift.

end bell *See:* cable terminal.

end bracket (rotating machinery) A beam or bracket attached to the frame of a machine and intended for supporting a bearing. (PE) [9]

end capacitor A conducting element or group of conducting elements, connected at the end of a radiating element of an antenna, to modify the current distribution on the antenna, thus changing its input impedance. (AP/ANT) 145-1993

end cells (storage battery) (storage cell) Cells that may be cut in or cut out of the circuit for the purpose of adjusting the battery voltage. *See also:* battery. (PE/EEC) [119]

end closure The degree of accuracy with which two separate lines, defined to end at the same point, actually meet. (C) 610.6-1991w

end connector A female coupling which attaches to the ends of a coaxial cable section to interconnect sections. *Contrast:* barrel connector. (C) 610.7-1995

end device (1) (of a telemeter) The final system element that responds quantitatively to the measurand through the translating means and performs the final measurement operation. *Note:* An end device performs the final conversion of measurement energy to an indication, record, or the initiation of control. (SWG/PE) C37.100-1992
(2) The closest device to the sensor or control point within a metering application communication system that is compliant with the utility industry end device data tables. (AMR/SCC31) 1377-1997

end distortion (1) (data transmission) The shifting of the end of all marking pulses from their proper positions in relation to the beginning of the start pulse, of telegraph signals. (PE) 599-1985w
(2) A distortion in the end of all marking pulses of start-stop teletypewriter signals from their proper positions in relation to the beginning of the start pulse. (C) 610.7-1995

end finger (rotating machinery) (outside space-block) A radially extending finger piece at the end of a laminated core to transfer pressure from an end clamping plate or flange to a tooth. *See also:* rotor; stator. (PE) [9]

end-fire array antenna A linear array antenna whose direction of maximum radiation lies along the line of the array. (AP/ANT) 145-1993

end fittings (composite insulators) The insulator attachment hardware that is connected to the core. (T&D/PE) 987-1985w

ending point (A) (for CCS outgoing trunk) Transmittal of IAM. **(B)** (for per-trunk-signaling outgoing trunk). Transmittal of connect signal to next office. *See also:* cross-office delay; starting point. (COM/TA) 973-1990

end injection (microwave tubes) A gun used in the presence of crossed electric and magnetic fields to inject an electron beam into the end of a slow-wave structure. *Synonym:* charles or kino gun. (ED) [45]

end item An entity identified with an element of the system breakdown structure. An end item is represented by one or more of the following: equipment (hardware and software), data, facilities, material, services, and/or techniques. (C/SE) 1220-1998

end mark A mark that indicates the end of a word or another unit of data. (C) 610.10-1994w

end node (local area networks) A physical device that may be attached to a LAN link segment for the purpose of transmitting and receiving information on that link medium. For example, an end node may be a user station, a bridge, or a LAN analyzer. It is identified by a unique 48-bit address. (C) 8802-12-1998

end-of-block signal (numerically controlled machines) A symbol or indicator that defines the end of one block of data. (EEC/NFPA) [74], [114]

end-of-copy signal (facsimile) A signal indicating termination of the transmission of a complete subject copy. *See also:* facsimile signal. (COM) 168-1956w

end-of-demand-interval indicator An indicator for the end of the demand interval for nonrolling-interval demand, or the end of the sub-interval for rolling-interval demand. (ELM) C12.15-1990

end-of-dialing determination The use of code interpretation and digit counting on critical interdigital timing to determine if additional dialed digits are to be expected. The critical interdigital timing interval is a specified time interval. In these cases, dialing should be considered complete if a potentially complete code has been received and if no additional character is received within the critical interdigital timing interval. It is desirable to avoid the use of timing whenever possible since this delays call completion and is a potential source of misdirected calls. (COM/TA) 973-1990w

end office (1) (telephone switching systems) A local office that is part of the toll hierarchy of World Zone 1. An end office is classified as a Class 5 office. *See also:* office class. (COM) 312-1977w
(2) Class 5 office in the North American hierarchical routing plan; a switching center where subscriber's loops are terminated and where toll calls are switched through to called lines. *Synonyms:* local exchange; wire center. *See also:* sectional center; central office; regional center; toll center; primary center. (C) 610.7-1995
(3) A switching system to which customer premises equipment is directly connected by loops. The switch connects loops to loops and loops to trunks. (COM/TA) 820-1984r

end of file (EOF) An internal label, immediately following the last record of a file, signalling the end of that file. *Synonym:* end-of-file. (C) 610.5-1990w

end-of-file *See:* end of file.

end-of-file label (EOF) An internally-recorded label that indicates the end of a file and that may contain information for use in file control. *Synonym:* trailer label. *Contrast:* beginning-of-file label. (C) 610.10-1994w

end-of-medium character A control character that is used to identify the physical end of the data medium, the end of the used portion of the medium, or the end of the wanted portion of the data recorded on the medium. (C) 610.10-1994w

end-of-message packet The last packet of a message in the data stream. (C/MM) 1284.4-2000

End_of_Packet Delimiter (EPD) In 1000BASE-X, a defined sequence of three single code-group 8B/10B ordered_sets used to delineate the ending boundary of a data transmission sequence for a single packet. (C/LM) 802.3-1998

end_of_packet marker A control character which indicates the end of a packet. *See also:* packet. (C/BA) 1355-1995

end of program (numerically controlled machines) A miscellaneous function indicating completion of workpiece. *Note:* Stops spindle, coolant, and feed after completion of all commands in the block. Used to reset control and/or machine. Resetting control may include rewinding of tape or progressing a loop tape through the splicing leader. The choice for a particular case must be defined in the format classification sheet. (IA/EEC) [61], [74]

end-of-stream delimiter (ESD) (1) (local area networks) Patterns that identify the end of an MII data stream. (C) 8802-12-1998
(2) A code-group pattern used to terminate a normal data transmission. For 100BASE-T4, the ESD is indicated by the transmission of five predefined ternary code-groups named eop1-5. For 100BASE-X, the ESD is indicated by the transmission of the code-group/T/R. For 100BASE-T2, the ESD is indicated by two consecutive pairs of predefined PAM5×5 symbols which are generated using unique Start-of-Stream Delimiter (SSD)/ESD coding rules. (C/LM) 802.3-1998

end of tape (numerically controlled machines) A miscellaneous function that stops spindle, coolant, and feed after completion of all commands in the block. *Note:* Used to reset control and/or machine. Resetting control will include rewinding of tape, progressing a loop tape through the splicing leader, or transferring to a second tape reader. The choice for a particular case must be defined in the format classification sheet. (MIL) [2]

end-of-tape marker (EOT) A marker on a magnetic tape used to indicate the end of the permissible recording area. *Note:* It might be a photoreflective strip, a unique data pattern, or a transparent section of tape. *Contrast:* beginning-of-tape marker. (C) 610.10-1994w

end of transfer status A handshake status that indicates the last data transfer of the transfer operation. *See also:* handshake status. (C/MM) 1296-1987s

end of transmission block character A transmission control character that indicates the end of a transmission block of data. *Synonyms:* block character; transmission block character. (C) 610.7-1995

end-of-volume label (EOV) **(1)** An internal label that precedes and initiates the beginning of the data contained in that volume. *Synonym:* volume label. (C) 610.5-1990w
(2) An internally-recorded label that indicates the end of the recording area contained in a volume. *Contrast:* beginning-of-volume label. (C) 610.10-1994w

endogenous variable A variable whose value is determined by conditions and events within a given model. *Synonym:* internal variable. *Contrast:* exogenous variable. (C) 610.3-1989w

end-on armature relay *See:* relay.

end-on relay armature An armature whose motion is in the direction of the core axis, with the pole face at the end of the core and perpendicular to this axis. (EEC/REE) [87]

endorder traversal *See:* postorder traversal.

endothermic Characterized by or formed with the absorption of heat. (DEI) 1221-1993w

end plate, rotor *See:* rotor end plate.

end-play washers (rotating machinery) Washers of various thicknesses and materials used to control axial position of the shaft. (PE) [9]

endpoint **(1)** A measurable response of interest in a biological experiment. (T&D/PE) 539-1990
(2) An object that is created and maintained by a communications provider and used by applications for sending and receiving data; endpoints are used by the communications providers to identify the sources and destinations of data. (C) 1003.5-1999
(3) A point at each end of a channel, line, or a circuit. (C) 610.7-1995
(4) A consumer or producer of data on a communication link. (C/MM) 1284.4-2000

end-point criterion (thermal classification of electric equipment and electrical insulation) (evaluation of thermal capability) A value of property or property degradation (either absolute or percentage change) that defines failure in a functional test. (EI) 1-1986r

end rail (rotating machinery) A rail on which a bearing pedestal can be mounted. *See also:* bearing. (PE) [9]

end ring, rotor *See:* rotor end ring.

end-scale value (electric instruments) The value of the actuating electrical quantity that corresponds to end-scale indication. *Notes:* 1. When zero is not at the end or at the electrical center of the scale, the higher value is taken. 2. Certain instruments such as power-factor meters, ohmmeters, etc., are necessarily excepted from this definition. 3. In the specification of the range of multiple-range instruments, it is preferable to list the ranges in descending order, as 750/300/150. *See also:* instrument; accuracy rating. (EEC/AII) [102]

end shield **(1) (rotating machinery)** A solid or skeletal structure, mounted at one end of a machine, for the purpose of providing a specified degree of protection for the winding and rotating parts or to direct the flow of ventilating air. *Note:* Ordinarily, a machine has an end shield at each end. For certain types of machine, one of the end shields may be constructed as an integral part of the stator frame. The end shields may be used to align and support the bearings, oil deflectors, and, for a hydrogen-cooled machine, the hydrogen seals. (PE) [9]
(2) (magnetrons) A shield for the purpose of confining the space charge to the interaction space. *See also:* magnetron. (ED) 161-1971w

end-shift frame (rotating machinery) A stator frame so constructed that it can be moved along the axis of the machine shaft for purposes of inspection. *See also:* stator. (PE) [9]

end station A system attached to a LAN that is an initial source or a final destination of MAC frames transmitted across that LAN. A Network layer router is, from the perspective of the LAN, an end station; a MAC Bridge, in its role of forwarding MAC frames from one LAN to another, is not an end station. (C/LM) 802.3ad-2000

end termination The termination applied to the end of the heating cable, opposite the power supply end. (IA/PC) 515.1-1995

end termination connection The termination applied to the end of a heating cable that may be heat producing, opposite where the power is supplied. (IA) 515-1997

end-to-end test **(1)** A test series of all performance requirements with the dc system under normal operating conditions and, as conditions permit, under contingency operating conditions. (PE/SUB) 1378-1997
(2) A test sequence to establish pass (functioning properly) or fail (not functioning properly) conditions. *Synonym:* go/no-go test. (SCC20) 1445-1998

endurance **(1) (metal-nitride-oxide field-effect transistor)** The number of write-high write-low cycles accumulated before any defined unacceptable changes in device properties occur. (ED) 581-1978w
(2) (metal-nitrite-oxide semiconductor arrays) The minimum number of data alteration cycles possible without catastrophic failure or degradation beyond the specified performance characteristics of any memory cell within an array. (ED) 641-1987w
(3) The ability of a reprogrammable read-only memory to withstand data rewrites and still comply with its specifications. (ED) 1005-1998

endurance limit The maximum stress a metal can withstand without failure during a specified large number (usually 10 million) cycles of stress. (IA) [59], [71]

endurance test An experiment carried out to investigate how the properties of an item are affected by the application of stresses and the elapse of time. (R) [29]

end user The person or persons who will ultimately be using the system for its intended purpose. (C/SE) 1233-1998

end user computing The performance of system development and data processing tasks by the user of a computer system. *Synonym:* user-driven computing. (C) 610.2-1987

end winding (rotating machinery) That portion of a winding extending beyond the slots. *Note:* It is outside the major flux path and its purpose is to provide connections between parts of the winding within the slots of the magnetic circuit. *See also:* stator; asynchronous machine; rotor; direct-current commutating machine. (PE) [9]

end-winding cover (rotating machinery) (winding shield) A cover to protect an end winding against mechanical damage and/or to prevent inadvertent contact with the end winding. (PE) [9]

end-winding support (rotating machinery) The structure by which coil ends are braced against gravity and electromagnetic forces during start-up (for motors), running, and abnormal conditions such as sudden short-circuit, for example, by blocking and lashings between coils and to brackets or rings. *See also:* stator. (PE) [9]

end-window counter tube (radiation) A counter tube designed for the radiation to enter at one end. *See also:* anticoincidence. (ED) [45]

end-wire insulation (rotating machinery) Insulation members placed between the end wires of individual coils such as between main and auxiliary windings. *See also:* rotor; stator. (PE) [9]

end wire, winding *See:* winding end wire.

energization test Any test requiring that system voltage be applied to the equipment. (SUB/PE) 1303-1994

energized **(1) (conductor stringing equipment) (power line maintenance)** Electrically connected to a source of potential difference, or electrically charged so as to have a potential different from that of the ground. *Synonyms:* alive; hot; live. *See also:* alive. (T&D/PE) 524a-1993r, 1048-1990, 516-1995, C2.2-1960, 524-1992r

(2) Electrically connected to a source of potential difference, or electrically charged so as to have a potential significantly different from that of earth in the vicinity. *Synonym:* alive; live. (NESC) C2-1997

energized background noise level (1) (liquid-filled power transformers) (dry-type transformers) Stated in pC (one pC = 10^{-12} Coulombs), the residual response of the partial discharge measurement system to background noise of any nature after the test circuit has been calibrated and the test object is energized at 50% of its nominal operating voltage. (PE/TR) C57.113-1988s, C57.124-1991r

(2) Stated in pC, the residual response of the partial discharge measurement system to background noise of any nature after the test circuit has been calibrated and energized at 100% of the test voltage without the test object connected. (SWG/PE) 1291-1993r

energy (1) (power operations) That which does work or is capable of doing work. As used by electric utilities, it is generally a reference to electrical energy and is measured in kilowatt hours (kWh). *See also:* incremental energy cost; economy energy; dump energy; off-peak energy; net system energy; fuel replacement energy; energy loss; byproduct energy; energy control center; interchange energy; potential hydro energy; on-peak energy. (PE/PSE) 858-1987s

(2) (metering) The integral of active power with respect to time. (ELM) C12.1-1988

(3) (laser maser) (Q) The capacity for doing work. Energy content is commonly used to characterize the output from pulsed lasers and is generally expressed in joules. (LEO) 586-1980w

(4) (system) The available energy is the amount of work that the system is capable of doing. *See also:* electric energy. (Std100) 270-1966w

energy and channel pairs The energy (keV) of the corresponding channel is stored as energy-channel pairs. Each member of the pair is stored as a 16-character floating point number, with unused pairs being ASCII spaces or zeros. They are stored as ordered pairs, i.e., the first entry is the energy, the second is the channel at that energy, the third is the energy, the fourth is the channel at that energy, and then to the next record. This is intended to provide sufficient numbers of channel pairs to allow for an adequate reconstruction of the energy-channel function by the analysis program. (NPS/NID) 1214-1992r

energy and efficiency pairs The detection efficiency at the corresponding energy is stored as energy-efficiency pairs. Each member of the pair is stored as a 16-character floating point number, with unused pairs being ASCII spaces or zeros. They are stored as ordered pairs, i.e., the first entry is the energy, the second is the efficiency at that energy, the third is the energy, the fourth is the efficiency at that energy, and then to the next record. This is intended to provide sufficient numbers of efficiency pairs to allow for an adequate reconstruction of the efficiency function by the analysis program. (NPS/NID) 1214-1992r

energy and resolution pairs The detector resolution at the corresponding energy is stored as energy-resolution pairs. Each member of the pair is stored as a 16-character floating point number, with unused pairs being ASCII spaces or zeros. They are stored as ordered pairs, i.e., the first entry is the energy, the second is the resolution at that energy, the third is the energy, the fourth is the resolution at that energy, and then to the next record. This is intended to provide sufficient numbers of resolution pairs to allow for an adequate reconstruction of the resolution function by the analysis program. (NPS/NID) 1214-1992r

energy and torque (International System of Units (SI)) The vector product of force and moment arm is widely designated by the unit newton meter. This unit for bending moment of torque results in confusion with the unit for energy, which is also newton meter. If torque is expressed as newton meter per radian, the relationship to energy is clarified, since the product of torque and angular rotation is energy:

$$(\mathrm{N \cdot m/rad}) \cdot \mathrm{rad} = \mathrm{N \cdot m}$$

See also: units and letter symbols. (QUL) 268-1982s

energy calibration (sodium iodide detector) The relationship between the height of the amplifier output pulse and the energy of the photons originating in the radioactive source. (NI) N42.12-1994

energy calibration coefficients The energy (E) (in units of keV) versus channel number (Ch) coefficients as

$$E = A + B \cdot Ch + C \cdot Ch^2 + D \cdot Ch^3$$

with the coefficients, A, B, C, and D stored as four successive 14-character numbers including the decimal point. Leading spaces are interpreted as zeros. Any values not used or calculated should be set to all spaces. The A term is usually called the offset or zero intercept. The B term is usually called the slope of the energy-channel curve. The C term is called the quadratic component of the energy-channel curve. The D term is called the cubic component of the energy-channel curve. (NPS/NID) 1214-1992r

energy capacity The energy, usually expressed in watthours (Wh), that a fully charged battery can deliver under specified conditions. (PV) 1013-1990, 1144-1996

energy charge (1) The charge for electric service based upon the electric energy delivered or billed. (PE/PSE) 858-1993w

(2) (electric power utilization) That portion of the charge for electric service based upon the electric energy consumed or billed. (PE/PSE) 346-1973w

energy consumption Telecommunications switching systems may be characterized by their peak power consumption and by their long-term average power consumption. Peak power consumption is measured in watts and will determine the size of the power plant required. Long-term average power is a measure of energy. It is measured in watthours for a period of time such as a day, a month, or a year. This will determine the air-conditioning load and the cost of energy. Energy is sometimes measured as watts per line averaged over a year. However, both peak power and energy are better described as a function of line size, traffic, and other variations in central office equipment. (COM/TA) 973-1990w

energy control center *See:* power control center.

energy cost, incremental *See:* incremental energy cost.

energy count-rate product *See:* energy rate limit.

energy density (1) (audio and electroacoustics) (point in a field) The energy contained in a given infinitesimal part of the medium divided by the volume of that part of the medium. *Notes:* 1. The term energy density may be used with prefatory modifiers such as instantaneous, maximum, and peak. 2. In speaking of average energy density in general, it is necessary to distinguish between the space average (at a given instant) and the time average (at a given point). (SP) [32]

(2) The electromagnetic energy contained in an infinitesimal volume divided by that volume. (SP/NIR) C95.1-1999

energy-density spectrum (burst measurements) (finite energy signal) The square of the magnitude of the Fourier transform of a burst. *See also:* network analysis; burst. (IT) [7]

energy dependence (radiation protection) A change in instrument response with respect to radiation energy for a constant exposure or exposure rate. (NI) N323-1978r

energy distribution (solar cells) The distribution of the flux or fluence of particles with respect to particle energy. (AES/SS) 307-1969w

energy, dump *See:* dump energy.

energy, economy *See:* economy energy.

energy efficiency (specified electrochemical process) The product of the current efficiency and the voltage efficiency. *See also:* electrochemistry. (EEC/PE) [119]

energy-equivalent sound level (L_{eq}) (1) (audible noise measurements) The equivalent sound level L_{eq} is the energy average of the level (usually A-weighted) of a varying sound over a specified period of time. The term "equivalent" signifies that the average of the fluctuating sound would have the same sound-energy level as a steady sound having the same level. The term "energy" is used because the sound amplitude is averaged on a root-mean-square (rms) pressure-squared basis, and pressure-squared is proportional to energy. Mathematically, the equivalent sound level is defined as:

$$L_{eq} = 10\log\left[\frac{1}{(t_2 - t_1)}\int_{t_1}^{t_2}\frac{p^2(t)}{p_{inf}^2}\,dt\right]\ \text{dB}$$

where

$p(t)$ = The time-varying A-weighted sound level, in μPa

p_{ref} = The reference pressure, 20 μPa

$(t_2 - t_1)$ = The time period of interest

If the cumulative probability distribution of a noise is known, then L_{eq} can be estimated by:

$$L_{eq} = 10\log\left[\frac{1}{100}\sum_{0}^{n}(P_x - P_{x-1})\ \text{antilog}\,\frac{L_x}{10}\right]\ \text{dB}$$

where

L_x = The highest noise level in each step

P_x, P_{x-1} = Selected adjacent steps along the probability scale, expressed in percent probability.

(2) The average of the sound energy level (usually A-weighted) of a varying sound over a specified period of time. *Notes:* 1. The simplest and most popular method for rating intermittent or fluctuating noise intrusions is to rely upon some measure of the average sound-level magnitude over time. The most common such average is the equivalent sound level, L_{eq}, expressed in decibels. 2. The term "equivalent" signifies that a steady sound having the same level as the L_{eq} would have the same sound energy as the fluctuating sound. The term "energy" is used because the sound amplitude is averaged on an rms-pressure-squared basis, and the square of the pressure is proportional to energy. For example, two sounds, one of which contains 24 times as much energy as the other but lasts for 1 h instead of 24 h, would have the same energy-equivalent sound level. 3. Mathematically, the equivalent sound level is defined as

$$L_{eq} = 10\log\left[\frac{1}{(t_2 - t_1)}\int_{t_1}^{t_2}\frac{p^2(t)}{p_{ref}^2}\,dt\right]$$

where

$p(t)$ = The time-varying A-weighted sound level, in μPa

p_{ref} = The reference pressure, 20 μPa

$(t_2 - t_1)$ = The time period of interest

If the cumulative probability distribution of a noise is known, then L_{eq} can be estimated by

$$L_{eq} = 10\log\left[\frac{1}{100}\sum_{0}^{n}(P_x - P_{x-1})\ \text{antilog}\,\frac{L_x}{10}\right]$$

where

P_x, P_{x-1} = Selected adjacent steps along the probability scale, expressed in percent (%)

L_x = The highest noise level in each step

x = The step number

n = The total number of steps

(T&D/PE) 539-1990, 656-1992

energy flux (audio and electroacoustics) The average rate of flow of energy per unit time through any specified area. *Note:* For a sound wave in a medium of density r and for a plane or spherical free wave having a velocity of propagation c, the sound-energy flux through the area S corresponding to an effective sound pressure p is

$$J = \frac{p^2 S}{\rho c}\cos\theta$$

where q is the angle between the direction of propagation of the sound and the normal to the area S. (SP) [32]

energy, fuel replacement *See:* fuel replacement energy.

energy gap (semiconductor) The energy range between the bottom of the conduction band and the top of the valence band. *See also:* semiconductor device. (ED) 216-1960w

energy, interchange *See:* interchange energy.

energy-limiting transformer (power and distribution transformers) A transformer that is intended for use on an approximately constant-voltage supply circuit and that has sufficient inherent impedance to limit the output current to a thermally safe maximum value. *See also:* specialty transformer. (PE/TR) C57.12.80-1978r, [57]

energy loss (power operations) The difference between energy input and output as a result of transfer of energy between two points. (T&D/PE/PSE) 858-1993w, 346-1973w

energy metering point (electric power system) For a tie line, the actual or equivalent location of a power flow measurement that is integrated to produce an energy transfer value. (PE/PSE) 94-1991w

energy, net system *See:* net system energy.

energy, nuclear *See:* nuclear energy.

energy, off-peak *See:* off-peak energy.

energy, on-peak *See:* on-peak energy.

energy, partial discharge *See:* partial discharge energy.

energy, potential hydro *See:* potential hydro energy.

energy, Q *See:* Q energy.

energy rate The average energy per event times the number of events per second. (NPS) 325-1996

energy rate limit In a preamplifier dc-coupled to a detector, the highest energy rate in units of MeV per second that causes no more than a specified fraction of the pulses (usually 1%) to overload the preamplifier. (NPS) 325-1996

energy ratio The ratio of signal energy to noise power spectral density in the receiver, at a point where the noise factor has been established and prior to filtering that would exclude components of the input signal. *Note:* Energy ratio also equals the maximum output signal-to-noise power ratio for a matched-filter system. (AES) 686-1997

energy resolution (1) (A) In keV: the FWHM of a spectral line in units of keV. **(B)** In percent: 100 · (FWHM/*E*) where FWHM and energy *E* are expressed in the same units. *See also:* full width at half maximum. (NPS) 300-1988

(2) (full width at half maximum) (fwhm) (x-ray energy spectrometers) (of a semiconductor radiation detector) The detector's contribution (including detector leakage current noise), expressed in units of energy, to the FWHM of a pulse-height distribution corresponding to an energy spectrum. (NPS/NID) 759-1984r

(3) One hundred times the energy resolution divided by the energy for which the resolution is specified. (NPS/NID) 759-1984r, 301-1976s

(4) The full width at half maximum of a peak in a spectrum*f*after subtracting the background under the peak*f*expressed in units of energy, usually keV, or as a percentage of the energy corresponding to that peak. (NPS) 325-1996

energy resolution, full width at half-maximum (fwhm) The width of a peak at half of the maximum peak height with the baseline removed. (NI) N42.14-1991

energy resolution, full width at tenth maximum (full width at half maximum) The width of a peak at one-tenth of the maximum peak height with the baseline removed. [For a normal (Gaussian) distribution, FWTM is 1.823 times its FWHM]. (NI) N42.14-1991

energy (shock) absorber A component whose primary function is to dissipate energy and limit deceleration forces which the system imposes on the body during fall arrest. Such devices may employ various principles such as deformation, friction, tearing of materials or breaking of stitches to accomplish energy absorption. An energy absorber causes an increase in the deceleration distance. An energy absorber may be borne by the user (personal) or be a part of a horizontal lifeline subsystem or a vertical lifeline subsystem. (T&D/PE) 1307-1996

energy spectrum A differential distribution of the intensity of radiation as a function of energy. (NPS) 325-1996

energy straggling (1) (semiconductor radiation detectors) The random fluctuations in energy loss whereby those particles having the same initial energy lose different amounts of energy when traversing a given thickness of matter. (This process may lead to the broadening of spectral lines.) (NPS/NID) 325-1996, 300-1988r, 301-1976s
(2) *See also:* energy straggling. (NPS) 300-1988r

engine A dedicated processor, architecture, or system component that is used for a single and special purpose; for example, an inferencing co-processor (inferencing engine), floating-point processor, a print engine in a laser printer, or a database engine (software engine). (C) 610.10-1994w

engine-driven generator for aircraft A generator mechanically, hydraulically, or pneumatically coupled to an aircraft propulsion engine to provide power for the electric and electronic systems of an aircraft. It may be classified as follows: 1) Engine-mounted; 2) Remote-driven; A) Flexible-shaft-driven; B) Variable- ratio-driven; C) Air-turbine-driven. (EEC/PE) [119]

engine equilibrium temperature (1) (periodic testing of diesel-generator units applied as standby power supplies in nuclear power generating stations) The condition at which the jacket water and lube oil temperatures are both within ± 10°F (5.5°C) of their normal operating temperatures established by the engine manufacturer. (PE/NP) 749-1983w
(2) The condition at which the jacket water and lube oil temperatures are both within ± 5.5°C (10°F) of their normal operating temperatures established by the engine manufacturer. (PE/NP) 387-1995

engineered safety features (nuclear power generating station) Features of a unit, other than reactor trip or those used only for normal operation, that are provided to prevent, limit, or mitigate the release of radioactive material. 627-1980r, 308-1991

engineered system A fall protection system which is designed and will operate to withstand the maximum expected impact load while maintaining a specified overload capacity factor (OCF). (T&D/PE) 1307-1996

engineering The application of a systematic, disciplined, quantifiable approach to structures, machines, products, systems, or processes. (C) 610.12-1990

engineering change In configuration management, an alteration in the configuration of a configuration item or other designated item after formal establishment of its configuration identification. *Contrast:* deviation. *See also:* waiver; configuration control; engineering change proposal. (C) 610.12-1990

engineering change proposal In configuration management, a proposed engineering change and the documentation by which the change is described and suggested. *See also:* configuration control. (C) 610.12-1990

engineering units (O/M) (1) (supervisory control, data acquisition, and automatic control) A unit of physical measurement (e.g., volts, amperes). (PE/SUB) C37.1-1994
(2) A unit of measure for use by operating/maintenance personnel usually provided by scaling the input quantity for display (meter, stripchart, or crt). (SWG/PE) C37.100-1992

engine-generator system (electric power supply) A system in which electric power for the requirements of a railway vehicle (other than propulsion) is supplied by an engine-driven generator carried on the vehicle, either as an independent source of electric power or supplemented by a storage battery. *See also:* axle-generator system. (PE/EEC) [119]

engine-room control Apparatus and arrangement providing for control in the engine room, on order from the bridge, of the speed and direction of a vessel. (EEC/PE) [119]

engine synchronism indicator A device that provides a remote indication of the relative speeds between two or more engines. (EEC/PE) [119]

engine-temperature thermocouple-type indicator A device that indicates temperature of an aircraft engine cylinder by measuring the electromotive force of a thermocouple. (EEC/PE) [119]

engine-torque indicator A device that indicates engine torque in pound-feet. *Note:* It is usually converted to horsepower with reference to engine revolutions per minute. (EEC/PE) [119]

English language programming (test, measurement, and diagnostic equipment) A technique of programming which allows the programmer to write programs and routines in English language statements. (MIL) [2]

English unit of luminance (illuminating engineering) (USA unit of luminance) Candela per square foot(cd/ft^2) also lumen per steradian, square foot (lm/(sr × ft^2). Another unit is candela per square inch (cd/in^2); also, lumen per steradian, square inch (lm/(sr × in^2). *See also:* lambert; footlambert; lambertian units of luminance. (EEC/IE) [126]

enhanced backscatter Stronger than expected backscattered signal due to resonant surface or internal waves in the target region. (AP/PROP) 211-1997

enhanced parallel port mode (EPP mode) An asynchronous, byte-wide, bidirectional channel controlled by the host device. This mode provides separate address and data cycles over the eight data lines of the interface. (C/MM) 1284-1994

enhanced service provider (ESP) A service provider offering services through the telephone network using the telemetry transport capabilities to deliver their services. (AMR/SCC31) 1390-1995, 1390.2-1999, 1390.3-1999

enhanced small device interface (ESDI) A data-transfer interface characterized by improved seek times and greater throughput than its predecessor, the ST-506 interface. (C) 610.10-1994w

enhanced solar radiation (radio-wave propagation) The electromagnetic radiation of the sun under other than quiet conditions. *See also:* quiet sun. (AP/PROP) 211-1990s

enhancement *See:* image enhancement.

enhancement mode transistor (metal-nitride-oxide field-effect transistor) An insulated-gate field effect transistor (IGFET) where the channel connecting source and drain was formed by the effects of an applied gate voltage. (ED) 581-1978w

enqueue To append an item to a queue. *Contrast:* dequeue. (C) 610.5-1990w

enter key A control key that signals the end of input to a computer. *See also:* carriage return key. (C) 610.10-1994w

enterprise The organization that performs specified tasks. (C/SE) 1220-1998

enterprise service (telephone switching systems) A service in which calls from certain designated exchanges are completed and billed to a number in another exchange. (COM) 312-1977w

enterprise view *See:* conceptual view.

entire RE The concatenated set of one or more BREs or EREs that make up the pattern specified for string selection. (C/PA) 9945-2-1993

Entity An instance of a subclass of `IEEE1451_Entity`. (IM/ST) 1451.1-1999

entity (1) (software) In computer programming, any item that can be named or denoted in a program. For example, a data item, program statement, or subprogram. (C) 610.12-1990
(2) (data management) A distinguishable object, either real or abstract, about which data are recorded; for example, a person such as a CUSTOMER, or a concept, such as SALES-REVENUE, about which data is stored in a data structure. *Synonym:* entity instance. (C) 610.5-1990w
(3) In an open system, an element in a hierarchical division. *Note:* It has attributes that describe it, a name that identifies it, and an interface that provides management operations. (C) 610.7-1995

(4) A group of like items or subjects that can be individually identified and about which information is recorded. Examples include cable, drawing, modification, and system. *Synonym:* entity set. (PE/EDPG) 1150-1991w
(5) An active element in an open system. (C/LM) 802.10g-1995
(6) Any component in a system that requires explicit representation in a model. Entities possess attributes denoting specific properties. *See also:* simulation entity. (C/DIS) 1278.3-1996
(7) **(A)** The representation of a concept, or meaning, in the minds of the people of the enterprise. **(B)** The representation of a set of real or abstract things (people, objects, places, events, ideas, combination of things, etc.) that are recognized as the same type because they share the same characteristics and can participate in the same relationships. (C/SE) 1320.2-1998
(8) Signifies the hardware/software embodiment of an object. (IM/ST) 1451.1-1999
(9) Anything of interest (such as a person, place, process, property, object, concept, association, state, or event) within a given domain of discourse (in this case, within the ITS domain of discourse). (SCC32) 1489-1999
(10) *See also:* simulation entity. (DIS/C) 1278.1-1995

entity attribute (1) A named characteristic or property of a design entity that provides a systematic procedure for the statement of fact about the entity. (C/SE) 1016.1-1993w
(2) A named characteristic or property of a design entity. It provides a statement of fact about the entity. (C/SE) 1016-1998

entity/attribute matrix A representation of a relation in the form of a matrix such that each row represents an entity and each column represents an attribute of the entity.

attributes → R

No.	Name	Grade	Homeroom
15	Mary	4	26A
20	Joe	6	43
21	Harry	4	27
27	Michael	5	25
30	Susan	5	25
42	Mickey	6	41

entity ↓

students

entity/attribute matrix

(C) 610.5-1990w

entity class *See:* entity set.

entity coordinate system A system whereby location with respect to a simulation entity is described by an entity coordinate system. (DIS/C) 1278.1-1995

entity instance One of a set of real or abstract things represented by an entity. Each instance of an entity can be specifically identified by the value of the attribute(s) participating in its primary key. (C/SE) 1320.2-1998

entity layer In the OSI model, one of a collection of network-processing functions representing a level of a hierarchy of functions. *See also:* session layer; logical link control sublayer; client layer; physical layer; application layer; client layer; entity layer; presentation layer; sublayer; data link layer; network layer; transport layer; medium access control sublayer. (C) 610.7-1995

entity-relationship *See:* entity-relationship map; entity-relationship data model; entity-relationship diagram.

entity-relationship data model A logical view of data within a system, representing the entities in the system as well as relationships among the entities, attributes of the entities, and attributes of the relationships. (C) 610.5-1990w

entity-relationship diagram (E-R diagram) A diagram that depicts a set of real-world entities and the logical relationships among them. *Synonym:* entity-relationship map. *See also:* data structure diagram. (C) 610.12-1990

entity-relationship map *See:* entity-relationship diagram.

entity set A collection of entities that have similar properties, such as a set of CUSTOMERS. *Synonyms:* entity type; entity class. (C) 610.5-1990w

entity type The construct used to represent an entity in Intelligent Transportation Systems (ITS) data dictionaries. (SCC32) 1489-1999

entrance *See:* routine entry point.

entrance terminal (for distribution oil cutouts) (distribution oil cutouts) A terminal with an electrical connection to the fuse contact and suitable insulation where the connection passes through the housing. (SWG/PE) C37.40-1993, C37.100-1992

entrant A live inserted module in the process of aligning itself with the arbitration protocol. (C/BA) 10857-1994, 896.4-1993w

entry (1) (data management) An element of information in a data structure, that describes an identifiable entity; for example, a member of a table, list, or queue. *See also:* data entry. (C) 610.5-1990w
(2) A component of a directory, which is located within the node's ROM. An entry may contain information, or a pointer to another directory or leaf. *See also:* ROM. (C/MM) 1212-1991s
(3) A part of the DIB that contains information about an object. Each entry is made up of attributes. *Synonym:* object entry. (C/PA) 1328.2-1993w, 1326.2-1993w, 1224.2-1993vw, 1327.2-1993Aw
(4) (software) *See also:* routine entry point. (C) 610.12-1990

entry point identifier (label) (CAMAC system) The symbol label represents an entry point into a programmed procedure. Such a procedure will typically be executed in response to the recognition of a LAM, and it may interrupt the process being executed at the time of recogniton of the LAM. Under these circumstances the procedure must be capable of saving and restoring the state of the computer so that the interrupted process can be resumed. At least one value of labels should identify a system error procedure which deals with LAMs not linked to user processes. (NPS) 758-1979r

entry point, routine *See:* routine entry point.

enumeration The listing of the meaning associated with each binary numeric value possible in a data field's storage. Binary numbers are usually expressed in decimal terms for human convenience. Not all possible numeric values need have a specific meaning. Values without meaning are declared to be unused or reserved for future use. Enumeration is the process of declaring the encoding of human interpretable information in a manner convenient for digital electronic machine storage and interchange. The subclause that defines each transducer electronic data sheet data field that is of data type *enumeration* shall contain a table that defines the meaning of the data field for each binary number possible. The meanings encoded in each data field shall be specific and unique to that data field and only that data field. The value becomes meaningless if not associated with the data field and its defining table. (IM/ST) 1451.2-1997

enumeration type A discrete data type whose members can assume values that are explicitly defined by the programmer. For example, a data type called COLORS with possible values RED, BLUE, and YELLOW. *Contrast:* logical type; real type; integer type; character type. (C) 610.12-1990

envelope (1) (wave) (general) The boundary of the family of curves obtained by varying a parameter of the wave. For the special cas

$$y = E(t) \sin (\omega t + \theta)$$

variation of the parameter q yields $E(t)$ as the envelope. (Std100) 270-1966w
(2) (wave) (automatic control) Another wave composed of the instantaneous peak values of the original wave of an

alternating quantity, and which indicates the variation in amplitude undergone by that quantity. (PE/EDPG) [3]

envelope amplitude distribution A cumulative distribution of the impulse response positive crossing rates of a bandpass filter at different spectrum amplitudes. (EMC) C63.12-1987

envelope delay **(1)** The time of propagation, between two points, of the envelope of a wave. *Note:* It is equal to the rate of change with angular frequency of the difference in phase between these two points. It has significance over the band of frequencies occupied by the wave only if this rate is approximately constant over that band. If the system distorts the envelope, the envelope delay at a specified frequency is defined with reference to a modulated wave that occupies a frequency bandwidth approaching zero. *See also:* radio-wave propagation; facsimile transmission; television. (COM/PE) 168-1956w, 599-1985w
(2) The time that the envelope of a modulated signal takes to pass from one point in a network (or transmission system) to a second point in the network. *Note:* Envelope delay is often defined the same as group delay, that is, as the rate of change, with angular frequency, of the phase shift between two points in a network. *See also:* time delay; group delay time. (CAS) [13]
(3) **(non-real time spectrum analyzer)** The display produced on a spectrum analyzer when the resolution bandwidth is greater than the spacing of the individual frequency components. (IM) [14]
(4) **(PCM telecommunications circuits and systems)** The propagation time between two points for the envelope of a wave. (COM/TA) 1007-1991r
(5) The time that the envelope of a modulated signal takes to pass from one point in a network (or transmission system) to a second point in the network. Envelope delay is often defined as the rate of change, with angular frequency, of the phase shift between two points in a network. Examples of envelope delay are as follows: The time it takes the envelope of a carrier frequency amplitude modulated by 83 1/3 Hz to pass between two points in a network. The rate of change, with angular frequency, of the phase shift between adjacent pairs of tones (frequency difference of 156.25 Hz) in the 23-tone test signal. The rate of change, with angular frequency, of the phase shift measured between frequencies spaced 31.25 Hz apart in the Network Impulse Response test signal. (COM/TA) 743-1995
(6) The time of propagation of the envelope of a wave between two points provided that the envelope is not significantly distorted. *Synonym:* group delay. *See also:* group velocity. (AP/PROP) 211-1997

envelope delay distortion (EDD) **(1)** The difference between the envelope delay at one frequency and the envelope delay at a reference frequency, which is usually taken as the frequency of minimum envelope delay. (COM/TA) 1007-1991r
(2) **(facsimile)** That form of distortion which occurs when the rate of change of phase shift with frequency of a circuit or system is not constant over the frequency range required for transmission. *Note:* In facsimile, envelope delay distortion is usually expressed as one-half the difference in microseconds between the maximum and the minimum envelope delays existing between the two extremes of frequency defining the channel used. *See also:* facsimile transmission. (COM) 168-1956w
(3) The difference between the envelope delay at a test frequency and the envelope delay at a reference frequency. (COM/TA) 743-1995
(4) **(general)** Of a system or transducer, the difference between the envelope delay at one frequency and the envelope delay at at a reference frequency. (SP) 151-1965w
(5) *See also:* envelope delay distortion.

envelope delay, round trip *See:* round-trip envelope delay.

envelope display (spectrum analyzer) The display produced on a spectrum analyzer when the resolution bandwidth is greater than the spacing of the individual frequency components. (IM) 748-1979w

envelope, vacuum *See:* vacuum envelope.

envelope voltage (electromagnetic site survey) The magnitude of the complex representation of the observed instantaneous voltage. *Note:* Envelope voltage is always a positive quantity permitting the logarithmic operation to be performed upon the value. (EMC) 473-1985r

environment **(1)** The universe within which the system must operate. All the elements over which the designer has no control and that affect the system or its inputs and outputs. (SMC) [63]
(2) **(Class 1E battery chargers and inverters)** The external conditions and influences such as temperature, humidity, altitude, shock and vibration which may affect the life and function of the components or equipment. (PE/NP) 650-1979s
(3) **(overhead power lines)** The combined external factors that affect the health, growth, reproduction, and survival of an organism. (T&D/PE) 539-1990
(4) **(modeling and simulation)** The external objects, conditions, and processes that influence the behavior of a system. (C) 610.3-1989w
(5) The natural (weather, climate, ocean conditions, terrain, vegetation, dust, etc.) and induced (electromagnetic, interference, heat, vibration, etc.) conditions that constrain the design solutions for consumer products and their life-cycle processes. (C/SE) 1220-1994s
(6) The circumstances, objects, and conditions that surround a system to be built; includes technical, political, commercial, cultural, organizational, and physical influences as well as standards and policies that govern what a system must do or how it will do it. (C/SE) 1362-1998
(7) The circumstances, objects, and conditions that will influence the completed system; they include political, market, cultural, organizational, and physical influences as well as standards and policies that govern what the system must do or how it must do it. (C/SE) 1233-1998
(8) A concept space, i.e., an area in which a concept has an agreed-to meaning and one or more agreed-to names that are used for the concept. (C/SE) 1320.2-1998
(9) **(A)** A general term relating to everything that supports a system or the performance of a function. **(B)** The conditions that affect the performance of a system or function. (C) 610.12-1990

environmental application factor (reliability data for pumps and drivers, valve actuators, and valves) (reliability data) A multiplicative constant used to modify a failure rate to incorporate the effects of other normal or abnormal environments. *Note:* When available these factors are included in Appendix D of IEEE Std 500-1984 P&V in the appropriate chapter prefaces. (PE/NP) 500-1984w

environmental change of amplification (magnetic amplifier) The change in amplification due to a specified change in one environmental quantity while all other environmental quantities are held constant. *Note:* Use of a coefficient implies a reasonable degree of linearity of the considered quantity with respect to the specified environmental quantity. If significant deviations from linearity exist within the environmental range over which the amplifier is expected to operate, particularly if the amplification, for example, is not a monotonic function of the environmental quantity, the existence of such deviations should be noted. (MAG) 107-1964w

environmental coefficient **(A)** (output from a control system or element having a specified input). The ratio of a change of output to the change in the specified environment (temperature, pressure, humidity, vibration, etc.), measured from a specified reference level, which causes it; in a linear system, it includes the "coefficient of sensitivity," and the "coefficient of zero error." **(B)** (sensitivity). The ratio of a change in sensitivity to the change in the specified environment (measured from a specified reference level) which causes it. **(C)** (zero error). The ratio of a change in zero error to the change in the specified environment (measured from a specified reference level) which causes it. (CS/PE/EDPG) [3]

environmental coefficient of amplification (magnetic amplifier) The ratio of the change in amplification to the change in the specified environmental quantity when all other environmental quantities are held constant. *Note:* The units of this coefficient are the amplification units per unit of environmental quantity. (MAG) 107-1964w

environmental coefficient of offset (magnetic amplifier) The ratio of the change in quiescent operating point to the change in the specified environmental quantity when all other environmental quantities are held constant. *Note:* The units of this coefficient are the output units per unit of environmental quantity. (MAG) 107-1964w

environmental coefficient of trip-point stability (magnetic amplifier) The ratio of the change in trip point to the change in the specified environmental quantity when all other environmental quantities are held constant. *Notes:* 1. The units of this coefficient are the control signal units per unit of environmental quantity. 2. Use of a coefficient implies a reasonable degree of linearity of the considered quantity with respect to the specified environmental quantity. If significant deviations from linearity exist within the environmental range over which the amplifier is expected to operate, particularly if the amplification, for example, is not a monotonic function of the environmental quantity, the existence of such deviations should be noted. (MAG) 107-1964w

environmental conditions (electric penetration assemblies) Physical service conditions external to the electric penetration assembly such as ambient temperature, pressure, radiation, humidity, vibration, chemical or demineralized water spray and submergence expected as a result of normal operating requirements, and postulated conditions appropriate for the design basis events applicable to the electric penetration assembly. (PE/NP) 317-1983r

environmental dispatch control An automatic generation control subsystem that allocates unit generation levels within a control area based upon environmental considerations. (PE/PSE) 94-1991w

environmental impact A change in existing conditions due to a natural or artificial cause, whether beneficial or adverse, that affects an organism and its surroundings. (T&D/PE) 539-1990

environmental loss time The part of down-time that is due to a fault in the computer environment. *Synonym:* external loss time. (C) 610.10-1994w

environmental offset (magnetic amplifier) The change in quiescent operating point due to a specified change in one environmental quantity (such as line voltage) while all other environmental quantities are held constant. (MAG) 107-1964w

environmental radio noise (control of system electromagnetic compatibility) The total electromagnetic disturbance complex in which an equipment, subsystem, or system may be immersed, exclusive of its own electromagnetic contribution. (EMC) C63.12-1987

environmental seal (Class 1E connection assemblies) A device or system that restricts the passage of a gas or liquid through a boundary in conjunction with related cables or wires as an assembly. This does not include fire stops, in-line splices, or containment electric penetrations. (PE/NP) 572-1985r

environmental simulation (modeling and simulation) A simulation that depicts all or part of the natural or man-made environment of a system; for example, a simulation of the radar equipment and other tracking devices that provide input to an aircraft tracking system. (C) 610.3-1989w

environmental temperature (separable insulated connectors) The temperature of the surrounding medium, such as air, water, and earth, into which the heat of the connector is dissipated directly, including the effect of heat dissipation from associated cables and apparatus. (T&D/PE) 386-1995

environmental trip-point stability (magnetic amplifier) The change in the magnitude of the trip point (either trip OFF or trip ON, as specified) control signal due to a specified change in one environmental quantity (such as line voltage) while all other environmental quantities are held constant. (MAG) 107-1964w

environment glossary *See:* glossary.

environment task The anonymous task whose execution elaborates the library items of the declarative part of an active partition, and then calls the main subprogram, if there is one. (C) 1003.5-1999

environs (radiological monitoring instrumentation) The uncontrolled area at or near the site boundary. (NI) N320-1979r

EOF *See:* end-of-file label; end of file.

EOL-3 *See:* Expression-Oriented Language 3.

EOT *See:* end-of-tape marker.

EOV *See:* end-of-volume label.

EPC-40 Electrical plastic conduit for type II applications, fabricated from PE; or for type II and III applications, fabricated from PVC. (SUB/PE) 525-1992r

EPC-80 Electrical plastic conduit for type IV applications, fabricated from PVC. (SUB/PE) 525-1992r

EPD *See:* echo path delay.

ephapse The electric junction of two parallel or crossing nerve fibers at which there may occur phenomena similar to those occurring at a synapse. (EMB) [47]

ephemeris (communication satellite) The position vector of a satellite or spacecraft in space with respect to time. (COM) [19]

epidemiology The study of the frequency and distribution of a disease, or a physiological condition in human populations, and of the factors that influence its frequency and distribution. (T&D/PE) 539-1990

epilog breakpoint A breakpoint that is initiated upon exit from a given program or routine. *Synonym:* postamble breakpoint. *Contrast:* prolog breakpoint. *See also:* data breakpoint; code breakpoint; static breakpoint; programmable breakpoint; dynamic breakpoint. (C) 610.12-1990

EPL *See:* echo path loss.

E-plane bend (waveguide components) A waveguide bend (corner) in which the longitudinal axis of the guide remains in a plane parallel to the electric field vector throughout the bend (corner). (MTT) 147-1979w

E-plane line A rectangular waveguide containing one or more planar conducting structures, with or without dielectric backings, which are oriented in the plane defined by the electric field and the direction of propagation of the dominant waveguide mode. The guiding structures consist of one or more thin conducting strips, each having one edge extending to the broad wall of the enclosure. (MTT) 1004-1987w

E-plane, principal *See:* principal E-plane.

E-plane tee junction (waveguide components) (series tee) A waveguide tee junction in which the electric field vector of the dominant mode in each arm is parallel to the plane of the longitudinal axes of the guides. (MTT) 147-1979w

Epoch **(1)** The time 0 hours, 0 minutes, 0 seconds, January 1, 1970, Coordinated Universal Time. *See also:* seconds since the Epoch. (C/PA) 9945-2-1993, 9945-1-1996
(2) A base reference time defined as 0 hours, 0 minutes, 0.0 seconds, 1 January 1970, Universal Coordinated Time. (C) 1003.5-1999

epoch **(1)** A base reference time defined as 0 hours, 0 minutes, 0.0 seconds, 1 January 1970, Universal Coordinated Time. (C/PA) 1003.5b-1995
(2) The reference time defining the origin of the time scale used in a particular measurement. (IM/ST) 1451.1-1999

EPROM *See:* erasable programmable read-only memory.

EPT Electrical plastic tubing for type I applications, fabricated from PVC. (SUB/PE) 525-1992r

equal-energy source (light) (television) A light source from which the emitted power per unit of wavelength is constant throughout the visible spectrum. (BT/AV) 201-1979w

equal interval (isophase) light (illuminating engineering) A rhythmic light in which the light and dark periods are equal. (EEC/IE) [126]

equal interval quantizing A quantization technique in which the range of gray levels in an image is divided into intervals of equal length and the quantization level assigned to each pixel is the same for all pixels whose original gray levels fall within the same interval. *Synonym:* linear quantizing. (C) 610.4-1990w

equality *See:* equivalence.

equality relation A VHDL relational expression in which the relational operator is =. (C/DA) 1076.3-1997

equalization (1) (transmission performance of telephone sets) The function a telephone set performs when it automatically adjusts transmitting or receiving, or both, so as to compensate for loop loss. (COM/TA) 269-1983s

(2) (data transmission) The process of reducing frequency or phase distortion, or both, of a circuit by the introduction of networks to compensate for the difference in attenuation or time delay, or both, at the various frequencies in the transmission band. (PE) 599-1985w

(3) (feedback control system) Any form of compensation used to secure a closed-loop gain characteristic which is approximately constant over a desired range of frequencies. *See also:* compensation. (PE/EDPG) [3]

(4) (broadband local area networks) A technique used to modify the frequency response of an amplifier or network to compensate for variations in the frequency response across the network bandwidth. The ideal result is a flat overall response. This slope compensation is often done by a module within an amplifier enclosure. (LM/C) 802.7-1989r

(5) The process of reducing the frequency and/or phase distortion of a circuit to compensate for the difference in attenuation and/or delay distortion. (C) 610.7-1995

(6) (electroacoustics) *See also:* frequency-response equalization.

equalizer (1) (substation grounding) A device to provide equipotential planes for resistance measurements. (SUB/PE) 837-1989r

(2) (rotating machinery) A connection made between points on a winding to minimize any undesirable potential voltage between these points. *See also:* direct-current commutating machine; asynchronous machine. (PE) [9]

(3) A device, such as a capacitor or resistor, inserted in a transmission line to improve its frequency response and thus compensate for distortion introduced by transmission facilities. (C) 610.7-1995

equalizer circuit breaker (power system device function numbers) A breaker that serves to control or to make and break the equalizer or the current-balancing connections for a machine field, or for regulating equipment, in a multiple-unit installation. (SUB/PE) C37.2-1979s

equalize voltage A voltage approximately 10% higher than the float voltage. This higher voltage is used for periodic equalizing of lead-acid and nickel-cadmium batteries. Equalize voltage is expressed in volts/cell. (IA/PSE) 602-1996

equalizing charge (1) (storage battery) (storage cell) An extended charge to a measured end point that is given to a storage battery to insure the complete restoration of the active materials in all the plantes of all the cells. *See also:* charge. (PE/EEC) [119]

(2) A prolonged charge, at a rate higher than the normal float voltage, to correct any inequalities of voltage and specific gravity that may have developed between the cells during service. (SCC29) 485-1997

equalizing pulses (pulse terminology) Pulse trains in which the pulse-repetition frequency is twice the line frequency and that occur just before and just after a vertical synchronizing pulse. *Note:* The equalizing pulses minimize the effect of line-frequency pulse on the interlace. (IM/WM&A) 194-1977w

equalizing resistor The resistor connected across the circuit element to equalize the off state voltage across elements that are connected in series. (IA/ID) 995-1987w

equalizing voltage The voltage, higher than float, applied to a battery to correct inequalities among battery cells (voltage or specific gravity) that may develop in service. (PE/SCC21/EDPG) 450-1995, 937-2000

equal-level crosstalk coupling loss The path loss measured between points at the same transmission level on the disturbing and disturbed circuits. *See also:* crosstalk. (COM/TA) 973-1990w

equal level echo path loss (ELEPL) The measure of echo path loss at a four-wire interface that is corrected by the difference in dB, between the transmit and receive TLPs. (COM/TA) 743-1995

equally tempered scale A series of notes selected from a division of the octave (usually) into 12 equal intervals, with a frequency ratio between any two adjacent notes equal to the twelfth root of two.

Equally Tempered Intervals

Name of Interval	Frequency Ratio	Cents
Unison	1:1	0
Minor Second or Semitone	1.059463:1	100
Major Second or Whole Tone	1.122462:1	200
Minor Third	1.189207:1	300
Major Third	1.259921:1	400
Perfect Fourth	1.334840:1	500
Augmented Fourth / Diminished Fifth	1.414214:1	600
Perfect Fifth	1.498307:1	700
Minor Sixth	1.587401:1	800
Major Sixth	1.681793:1	900
Minor Seventh	1.781797:1	1000
Major Seventh	1.887749:1	1100
Octave	2:1	1200

(SP) [32]

*The frequency ratio is $[(2)^{1/12}]^n$ where *n* equals the number of the interval. (The number of the interval is its value in cents divided by 100.)

equal probability quantizing A quantization technique in which the range of gray levels in an image is divided into contiguous intervals such that the frequency of occurrence of each quantization level is the same. (C) 610.4-1990w

equal vectors Two vectors are equal when they have the same magnitude and the same direction. (Std100) 270-1966w

equation *See:* computer equation.

equational format (pulse measurement) One or more algebraic equations which specify a waveform wherein, typically, a first equation specifies the waveform from t_0 to t_1, a second equation specifies the waveform from t_1 to t_2, etc. The equational format is typically used to specify hypothetical, ideal, or reference waveforms. (IM/WM&A) 181-1977w

equatorial orbit (communication satellite) An inclined orbit with an inclination of zero degrees. The plane of an equatorial orbit contains the equator of the primary body. (COM) [19]

equiasymptotic stability Asymptotic stability where the rate of convergence to zero of the perturbed-state solution is independent of all initial states in some region $\|\Delta x(t_0)\| \leq v$. *See also:* control system. (CS/IM) [120]

equilibrium *See:* steady state.

equilibrium condition *See:* final condition.

equilibrium coupling length *See:* equilibrium length.

equilibrium electrode potential A state electrode potential when the electrode and electrolyte are in equilibrium with respect to a specified electrochemical reaction. *See also:* electrochemistry. (EEC/PE) [119]

equilibrium length (fiber optics) For a specific excitation condition, the length of multimode optical waveguide necessary to attain equilibrium mode distribution. *Note:* The term is sometimes used to refer to the longest such length, as would result from a worst-case, but undefined excitation. *Synonyms:* equilibrium coupling length; equilibrium mode distribution length. *See also:* mode coupling; equilibrium mode distribution. (Std100) 812-1984w

equilibrium mode distribution (fiber optics) The condition in a multimode optical waveguide in which the relative power distribution among the propagating modes is independent of length. *Synonym:* steady-state condition. *See also:* mode coupling; equilibrium length; mode. (Std100) 812-1984w

equilibrium mode distribution length *See:* equilibrium length.

equilibrium mode simulator (fiber optics) A device or optical system used to create an approximation of the equilibrium mode distribution. *See also:* mode filter; equilibrium mode distribution. (Std100) 812-1984w

equilibrium point A point in state space of a system where the time derivative of the state vector is identically zero. *See also:* control system. (CS/PE/EDPG) [3]

equilibrium potential The electrode potential at equilibrium. (IA) [59]

equilibrium reaction potential The minimum voltage at which an electrochemical reaction can take place. *Note:* It is equal to the algebraic difference of the equilibrium potentials of the anode and cathode with respect to the specified reaction. It can be computed from the free energy of the reaction. Thus,

$$\Delta F = -nFE$$

where ΔF is the free energy of the reaction, n is the number of chemical equivalents involved in the reaction, F is the value of the Faraday expressed in calories per volt gram-equivalent (23 060.5) and E is the equilibrium reaction potential (in volts). *See also:* electrochemistry. (EEC/PE) [119]

equilibrium temperature (thyristor power converter) The steady-state temperature reached by a component of a thyristor converter under specified conditions of load and cooling. *Note:* The steady-state temperatures are, in general, different for different components. The times necessary to establish steady-state temperatures are also different and proportional to the thermal time constants. (IA/IPC) 444-1973w

equilibrium voltage *See:* storage-element equilibrium voltage; storage tube.

equiphase surface Any surface over which the field vectors of a time harmonic wave have the same phase. (AP/PROP) 211-1997

equiphase zone (navigation aid terms) The region in space within which difference in phase of two radio signals is indistinguishable. (AES/GCS) 172-1983w

equipment (1) (nuclear power generating station) An assembly of components designed and manufactured to perform specific functions. *Note:* Examples of equipment are motors, transformers, valve operators, and instrumentation and control devices. (PE/NP) 323-1974s
(2) (safety systems equipment in nuclear power generating stations) An assembly of components designed and manufactured to perform specific functions. *Note:* Certain items which satisfy the definition of the term equipment as used in IEEE Std 627-1980 are those referred to as components in the ASME Boiler and Pressure Vessel Code and its latest addenda, Section III (IEEE BPV-III), for example, pumps and valves. Other examples of equipment are motors, transformers, and instrumentation and control devices. Structures and structural support items are not included in the definition of equipment. (PE/NP) 627-1980r
(3) (power and distribution transformers) A general term including material, fittings, devices, appliances, fixtures, apparatus, and the like, as a part of, or in connection with, an electrical installation. (NEC/NESC/PE/TR) C57.12.80-1978r, [86]
(4) A general term relating to devices and functional units that are part of an electrical installation. (C) 610.10-1994w
(5) A general term including fittings, devices, appliances, fixtures, apparatus, and similar terms used as part of or in connection with an electric supply or communications system. (NESC) C2-1997

equipment bonding jumper The connection between two or more portions of the equipment grounding conductor. (NESC/NEC) [86]

equipment certification An act or process resulting in documentation that attests to product performance. (T&D/PE) 1307-1996

equipment ground (1) (general) A ground connection to non-current-carrying metal parts of a wiring installation or of electric equipment, or both. *See also:* ground. (T&D/PE) [10]
(2) For the purposes of IEEE Std 1050-1996, it is the safety ground connection to the conductive, non current-carrying parts of electrical equipment. (PE/EDPG) 1050-1996

equipment grounding conductor (1) The conductor used to connect the non-current-carrying metal parts of equipment, raceways, and other enclosures to the service equipment, the service power source(s) ground, or both. (PE/SPD/EDPG) 665-1995, C62.45-1992r
(2) The conductor used to connect the non-current-carrying parts of conduits, raceways, and equipment enclosures to the grounding electrode at the service equipment (main panel) or secondary of a separately derived system (e.g., isolation transformer). (IA/PSE) 1100-1999

equipment noise (sound recording and reproducing system) The noise output that is contributed by the elements of the equipment during recording and reproducing, excluding the recording medium, when the equipment is in normal operation. *Note:* Equipment noise usually comprises hum, rumble, tube noise, and component noise. *See also:* noise. 191-1953w

equipment number (telephone switching systems) A unique, physical or other identification of an input or output termination of a switching network. (COM) [48]

equipment of the fixed preferential type Equipment in which the original source always serves as the preferred source and the other source as the emergency source. The automatic transfer equipment will restore the load to the preferred source upon its reenergization. (SWG/PE) C37.100-1992

equipment of the nonpreferential type Equipment that automatically restores the load to the original source only when the other source, to which it has been connected, fails. (SWG/PE) C37.100-1992

equipment of the selective preferential type Equipment in which either source may serve as the preferred or the emergency source of preselection as desired, and that will restore the load to the preferred source upon its reenergization. (SWG/PE) C37.100-1992

equipment outage (relay systems) The electrical isolation of equipment from the electric system such that it can no longer perform usefully for the duration of such isolation. *Note:* Since the term "outage" can also refer to service as well as equipment, it should always carry the appropriate modifier. (PE/PSR) C37.90-1978s

equipment qualification (1) (Class 1E battery chargers and inverters) The generation and maintenance of evidence to assure that the equipment will meet the system performance requirements. (PE/NP) 649-1980s, 650-1979s, 323-1974s
(2) The generation and maintenance of evidence to assure that the equipment will operate on demand, to meet the system performance requirements. (SWG/PE) C37.100-1992

equipment signature (test, measurement, and diagnostic equipment) The special characteristics of an equipment's response to, or reflection of, impinging impulsive energy, or of its electromagnetic, infrared or acoustical emissions. (MIL) [2]

equipment system (health care facilities) A system of feeders and branch circuits arranged for delayed, automatic or manual connection to the alternate power source and which serves primarily three-phase power equipment. (NESC/NEC/EMB) [86], [47]

equipment under test (EUT) (1) (radio-noise emissions) A device or system used for evaluation that is representative of a product to be marketed. (EMC) C63.4-1991
(2) A representative component, unit, or system to be used for evaluation purposes. (SPD/PE) C62.45-1992r
(3) The equipment being measured or tested, as opposed to support or ancillary equipment. (EMC) 1128-1998

equipment victim The electronic equipment or subassembly that is subjected to the effects associated with an ESD event. It may be the intruder or receptor, or it may be in proximity to the discharge between the intruder and receptor and therefore subjected to the stress of ESD- related electromagnetic fields. (SPD/PE) C62.47-1992r

equipotential (conductor stringing equipment) (power line maintenance) An identical state of electrical potential for two or more items. (T&D/PE) 524a-1993r, 524-1992r, 1048-1990, 516-1995

equipotential line or contour The locus of points having the same potential at a given time. (PE/PSIM) 81-1983

equipotential work zone (area, site) A work zone (area, site) where all equipment is interconnected by jumpers, grounds, ground rods, and/or grids that will provide acceptable potential differences between all parts of the zone under worst-case conditions of energization. (T&D/PE) 524a-1993r

equisignal localizer (navigation aid terms) A localizer in which the localizer on-course line is established as an equality of the amplitudes of two signals. (AES/GCS) 172-1983w

equisignal zone The region in space within which the difference in amplitude of two radio signals (usually emitted by a single station) is indistinguishable. *See also:* radio navigation. (PE/EEC/RN) [119]

equivalence (1) (mathematics of computing) A dyadic Boolean operator having the property that if P is a statement and Q is a statement, then the equivalence of P and Q is true if and only if both statements are true or both statements are false. *Note:* The equivalence of P and Q is often represented by P≡Q.

Equivalence Truth Table

P	Q	P≡Q
0	0	1
0	1	0
1	0	0
1	1	1

Synonyms: IF-AND-ONLY-IF; exclusive NOR. (C) 1084-1986w
(2) The dyadic Boolean operation whose result has the Boolean value 1 if and only if the operands have the same Boolean value. *Synonym:* equality. *Contrast:* nonequivalence. *See also:* IF-AND-ONLY-IF gate. (C) 610.10-1994w

equivalence class A set of collating elements with the same primary collation weight. Elements in an equivalence class are typically elements that naturally group together, such as all accented letters based on the same base letter. The collation order of elements within an equivalence class is determined by the weights assigned on any subsequent levels after the primary weight. (C/PA) 9945-2-1993

equivalent binary digit(s) (1) (mathematics of computing) The number of binary digits required to represent a number expressed in another numeration system with no loss of precision. *Note:* This number is approximately 3-1/3 times the number of decimal digits. *Synonym:* equivalent binary digit factor. (C) 1084-1986w
(2) (computers) The number of binary places required to count the elements of a given set. (C) [20], [85]

equivalent binary digit factor *See:* equivalent binary digit(s).

equivalent circuit (1) (general) An arrangement of circuit elements that has characteristics, over a range of interest, electrically equivalent to those of a different circuit or device. *Note:* In many useful applications, the equivalent circuit replaces (for convenience of analysis) a more-complicated circuit or device. *See also:* network analysis. (Std100) 270-1966w
(2) (piezoelectric crystal unit) An electric circuit that has the same impedance as the unit in the frequency region of resonance. *Note:* It is usually represented by an inductance, capacitance, and resistance in series, shunted by the direct capacitance between the terminals of the crystal unit. *See also:* crystal. (PE/EEC) [119]

equivalent concentration (ion type) The concentration equal to the ion concentration divided by the valency of the ion considered. *See also:* ion. (PE/EEC) [119]

equivalent conductance (1) (acid, base, or salt) The conductance of the amount of solution that contains one gram equivalent of the solute when measured between parallel electrodes that are one centimeter apart and large enough in area to include the necessary volume of solution. *Note:* Equivalent conductance is numerically equal to the conductivity multiplied by the volume in cubic centimeters containing one gram equivalent of the acid, base, or salt. *See also:* electrochemistry. (EEC/PE) [119]
(2) (microwave gas tubes) The normalized conductance of the tube in its mount measured as its resonance frequency. *Note:* Normalization is with respect to the characteristic impedance of the transmission line at its junction with the tube mount. *See also:* electron-tube admittances; element. (ED) 161-1971w

equivalent continuous rating (rotating machinery) The statement of the load and conditions assigned to the machine for test purposes, by the manufacturer, at which the machine may be operated until thermal equilibrium is reached, and which is considered to be equivalent to the duty or duty type. (PE) [9]

equivalent contrast ($\tilde{C}$) (illuminating engineering) A numerical description of the relative visibility of a task. It is the contrast of the standard visibility reference task giving the same visibility as that of a task whose contrast has been reduced to threshold when the background luminances are the same. *See also:* visual task evaluator. (EEC/IE) [126]

equivalent contrast, $\tilde{C}_e$ (illuminating engineering) The actual equivalent contrast in a real luminous environment with nondiffuse illumination. The actual equivalent contrast C_e is less than the equivalent contrast due to veiling reflection. $C_e = C \times \text{CRF}$. *See also:* contrast rendition factor. (EEC/IE) [126]

equivalent core-loss resistance A hypothetical resistance, assumed to be in parallel with the magnetizing inductance, that would dissipate the same power as that dissipated in the core of the transformer winding for a specified value of excitation. (CHM) [51]

equivalent dark-current input (phototubes) The incident luminous (or radiant) flux required to give a signal output current equal to the output electrode dark current. *Note:* Since the dark current may change considerably with temperature, the temperature should be specified. *See also:* phototube. (ED) [45]

equivalent diode *See:* diode equivalent.

equivalent diode voltage *See:* composite controlling voltage.

equivalent faults Two or more faults that result in the same failure mode. (C) 610.12-1990

equivalent flat plate area of a scattering object For a given scattering object, an area equal to the wavelength times the square root of the ratio of the monostatic cross section to 4π. *Note:* A perfectly reflecting plate parallel to the incident wavefront and having this area, if it is large compared to the wavelength, will have approximately the same monostatic cross section as the object. (AP/ANT) 145-1993

equivalent 4-wire (data transmission) Use of different frequency bands to form a "high group" and "low group" for the two directions of transmission, thereby permitting operation over a single pair of conductors. (PE) 599-1985w

equivalent hours (electric generating unit reliability, availability, and productivity) The number of hours a unit was in a time category involving unit derating, expressed as equivalent hours of full outage at maximum capacity. Both unit derating and maximum capacity shall be expressed on a consistent basis, gross or net. Equivalent hours can be calculated for each of the time categories—unit derated hours, in-service unit derated hours, reserve shutdown unit derated hours, planned derated hours, in-service planned derated hours, reserve shutdown planned derated hours, unplanned derated hours, in-service unplanned derated hours, reserve shutdown unplanned derated hours, forced derated hours, in-service forced derated hours, reserve shutdown forced derated hours, maintenance derated hours, in-service maintenance derated hours, reserve shutdown maintenance derated hours, and seasonal derated hours. The symbol designation for the equivalent hours is formed by adding an E in front of the symbol for the corresponding time designation (for example, equivalent unit derated hours is designated EUNDH). Equivalent hours can be calculated from the following equation:

$$E(\ ) = \frac{\Sigma D(\ )_i T_i}{\text{MC}}$$

where $E(\ )$ = equivalent hours in the time category represented by the parentheses, which can be any one of the time categories in sections 5.11 through 5.16 in IEEE Std 762-1987. D = the derating for the time category shown in parentheses, after the ith change in either available capacity (unit deratings) or dependable capacity (seasonal deratings). *Note:* In order to apportion equivalent hours among the various time categories, appropriate ground rules are established in the reporting system so that after each change in either available capacity or dependable capacity, the sum of all subcategories of unit derating is equal to the unit derating. T_i = the number of hours accumulated in the time category of interest between the ith and the $(i + 1)$th change in either available capacity (unit deratings) or dependable capacity (seasonal deratings) MC = maximum capacity. (PE/PSE) 762-1987w

equivalent input noise sensitivity (spectrum analyzer) The average level of a spectrum analyzer's internally generated noise referenced to the input. *See also:* input signal level sensitivity; sensitivity. (IM) 748-1979w

equivalent isotropically radiated power (EIRP) In a given direction, the gain of a transmitting antenna multiplied by the net power accepted by the antenna from the connected transmitter. *Synonym:* effective isotropically radiated power. (AP/ANT) 145-1993

equivalent load reflection coefficient *See:* reflection coefficient.

equivalent luminous intensity of an extended source at a specified distance (illuminating engineering) The intensity of a point source which would produce the same illuminance at that distance. Formerly, apparent luminous intensity of an extended source. (EEC/IE) [126]

equivalent network A network that, under certain conditions of use, may replace another network without substantial effect on electrical performance. *Note:* If one network can replace another network in any system whatsoever without altering in any way the electrical operation of that portion of the system external to the networks, the networks are said to be networks of general equivalence. If one network can replace another network only in some particular system without altering in any way the electrical operation of that portion of the system external to the networks, the networks are said to be networks of limited equivalence. Examples of the latter are networks that are equivalent only at a single frequency, over a single band, in one direction only, or only with certain terminal conditions (such as H and T networks). *See also:* network analysis. (Std100) 270-1966w

equivalent noise bandwidth (interference terminology) (signal system) The frequency interval, determined by the response-frequency characteristics of the system, that defines the noise power transmitted from a noise source of specified characteristics. *Note:* For Gaussian noise

$$\Delta f = \int_0^\infty y(f)^2 df$$

where $y(f) = Y(0)/Y(f)$ is the relative frequency dependent response characteristic. *See also:* interference. (IE) [43]

equivalent noise conductance (interference terminology) A quantitative representation in conductance units of the spectral density of a noise-current generator at a specified frequency. *Notes:* 1. The relation between the equivalent noise noise conductance G_n and the spectral density W_i of the noise-current generator is

$$G_n = \pi W_i/(kT_0)$$

where k is Boltzmann's constant and T_0 is the standard noise temperature (290 kelvins) and $kT_0 = 4.00 \times 10^{-21}$ watt-seconds. 2. The equivalent noise conductance in terms of the mean-square noise-generator current i^2 within a frequency increment δf is

$$G_n = i^2/(4kT_0\Delta f)$$

See also: electron-tube admittances. (IE/ED) [43], [45]

equivalent noise current (electron tube) (interference terminology) A quantitative representation in current units of the spectral density of a noise current generator at a specified frequency. *Notes:* 1. The relation between the equivalent noise current I_n and the spectral density W_i of the noise-current generator is

$$I_n = (2\pi W_i)/e$$

where e is the magnitude of the electron charge. 2. The equivalent noise current in terms of the mean-square noise-generator current $\bar{I}^2$ within a frequency increment δf is

$$I_n = i^2/(2e\Delta f)$$

See also: interference; signal-to-noise ratio. (IE/ED) [43], [45]

equivalent noise input (phototubes) The value of incident luminous (or radiant) flux that, when modulated in a stated manner, produces a root-mean-square signal output current equal to the root-mean-square dark-current noise both in the same specified bandwidth (usually one hertz). *See also:* phototube. (ED) 158-1962w

equivalent noise referred to input (germanium gamma-ray detectors) (x-ray energy spectrometers) (of a linear amplifier) The value of noise at the input that would produce the same value of noise at the output as does the actual noise source. (NPS/NID) 325-1986s, 759-1984r, 301-1976s

equivalent noise resistance (1) (A) (charged-particle detectors) (parallel noise). In a hypothetically noise-free amplifier, that resistance which when placed across the input terminals of the amplifier will produce an output signal attributable to the observed parallel-noise component. This definition applies only to noise with a constant spectral density (white noise). **(B) (charged-particle detectors)** (series noise). In a hypothetically noise-free amplifier, that resistance which when connected between the signal source and the amplifier will produce an output signal attributable to the observed series-noise component. This definition applies only to noise with a constant spectral density (white noise). (NPS) 300-1988

(2) (electron tube) A quantitative representation in resistance units of the spectral density of a noise voltage generator at a specified frequency. *Notes:* 1. The relation between the equivalent noise resistance R_n and the spectral density W_e of the noise-voltage generator is

$$R_n = (\pi W_e)/(kT_0)$$

where k is Boltzmann's constant and T_0 is the standard noise temperature (290 kelvins) and $kT_0 = 4.00 \times 10^{-21}$ watt-seconds. 2. The equivalent noise resistance in terms of the mean-square noise-generator voltage $\bar{e}^2$ within a frequency increment δf is

$R_n = \bar{e}^2/(4kT_0\Delta f)$

See also: signal-to-noise ratio; interference.

(ED) 161-1971w

equivalent noise resistance referred to input (1) (germanium gamma-ray detectors) (charged-particle detectors) (of a linear amplifier) That value of resistor which, when applied to the input of a hypothetical noiseless amplifier with the same gain and bandwidth, would produce the same output noise.

(NPS) 325-1986s

(2) (linear amplifier) (semiconductor radiation detectors) That value of resistor which when applied to the input of a hypothetical noiseless amplifier with the same gain and bandwidth would produce the same output noise.

(NID) 301-1976s

equivalent noise resistance referred to the input of an amplifier This is an ambiguous term*f*the qualifiers "series" or "parallel" must be specified. See equivalent series {parallel} noise resistance referred to the input of an amplifier.

(NPS) 325-1996

equivalent 1-megaelectronvolt electron flux The flux of electrons of 1-megaelectronvolt energy that changes a stated physical quantity (such as minority carrier diffusion length) of a given material or device to the same value as would the flux of penetrating particles of another stated energy spectrum. (AES/SS) 307-1969w

equivalent parallel circuit elements (magnetic core testing) Under stated conditions of excitation and coil configuration, the values of inductance and resistance connected parallel so that they give representation to the real permeability of the core (μ'_s) and the total losses in the core (μ''_s)

$L_p, \mu'_p L_0$

$R_p = \omega\mu''_p L_0$

$$\frac{1}{Z} = \frac{1}{j\omega L_p} + \frac{1}{R_p} = \frac{1}{j\omega\bar{\mu}L_0}$$

where

$\bar{\mu}$ = complex relative permeability

μ'_p = real complement of $\bar{\mu}$ parallel representation

μ''_p = imaginary component of $\bar{\mu}$, parallel representation

L_0 = self-inductance of coil with a core of unit relative permeability, but with the same flux distribution as with a ferromagnetic core

L_p = parallel equivalent self-inductance of the coil with a core of $\bar{\mu}$ permeability

R_p = parallel equivalent loss resistance of the core

ω = angular frequency in radians/sec.

(MAG) 393-1977s

equivalent periodic line (uniform line) A periodic line having the same electrical behavior, at a given frequency, as the uniform line when measured at its terminals or at corresponding section junctions. *See also:* transmission line.

(Std100) 270-1966w

equivalent radiated power *See:* effective radiated power.

equivalent reflection coefficient (ERC) The measure of the reflection coefficient of an actual radio-frequency (RF) absorber-lined reflecting surface. The ERC includes not only the RF absorber reflection, but other effects such as mounting fixtures, adhesive, and any air space between the RF absorber and the reflecting surface. (EMC) 1128-1998

equivalent salt-deposit density (equivalent salt-deposit density) A measure of contamination level.

(PE/T&D) 957-1995, 957-1987s

equivalent series circuit elements (magnetic core testing) Under stated conditions of excitation and coil configuration, values of a reactance and a resistance connected in series so that they give representation to the real permeability of the core (μ'_s) and to the total losses in the core (μ''_s)

$L_s = \mu'_s L_0$

$R_s = \omega\mu''_s L_0$

$Z = R_s + j\omega L_s = j\omega\bar{\mu}L_0$

where

L_s = self-inductance of oil with a core of $\bar{\mu}$ permeability; series equivalent inductance

R_s = equivalent weries resistance of coil in ohms with a core of $\bar{\mu}$ permeability

ω = angular frequency in radians/sec.

(MAG) 393-1977s

equivalent series {parallel} noise resistance referred to the input of an amplifier In a hypothetically noise-free amplifier, the value of resistor that, when connected in series with {shunted across} its input, will produce the same output noise spectrum as is observed in the real amplifier. The gain and bandwidth of the real and hypothetical amplifiers must be the same for this definition to be valid. (NPS) 325-1996

equivalent source reflection coefficient (network analyzers) The reflection coefficient equal to that caused by the source impedance Z_s

$$\Gamma_s = \frac{Z_s - Z_0}{Z_s + Z_0}$$

where the source impedance Z_s is the Thevenin impedence and is only considered in the linear range of the source. The Thevenin impedance is the impedance in Thevenin's Theorem. The impedance, Z_0, is the characteristic impedance of the transmission system. *Notes:* 1. In order to approximate a Z_0 source impedance, that is, $\Gamma_s = 0$, a directional coupler or suitable power splitter can be used as part of a feedback control circuit to maintain a constant incident power at its main-arm output port independent of the source impedance of the radio-frequency source connected to the main-arm input port of the coupler. 2. At lower frequencies, in order to approximate a Z_0 source impedance, a Z_0 impedance can be put in series with a constant voltage source that is maintained at zero impedance by means of a feedback control circuit independent of the source impedance of the radio-frequency source.

(IM/HFIM) 378-1986w

equivalent sources *See:* Huygens' sources.

equivalent sphere illumination (1) (electric power systems in commercial buildings) The measure of the effectiveness with which a practical lighting system renders a task visible compared with the visibility of the same task that is lit inside a sphere of uniform luminance. (IA/PSE) 241-1990r

(2) (illuminating engineering) The level of sphere illumination that would produce task visibility equivalent to that produced by a specific lighting environment.

(EEC/IE) [126]

equivalent test alternating voltage (charging inductors) A sinusoidal root-mean-square test voltage equal to 0.707 times the power-supply voltage of the network-charging circuit and having a frequency equal to the resonance frequency of charging. *Note:* This is the alternating component of the voltage that appears across the charging inductor in a resonance-charging circuit of the pulse forming network.

(MAG) 306-1969w

equivalent two-winding kVA rating (power and distribution transformers) The equivalent two-winding rating of multi-one-half the sum of the kVA ratings of all windings. *Note:* It is customary to base this equivalent two-winding kVA rating on the self-cooled rating of the transformer.

(PE/TR) C57.12.80-1978r

equivocation The conditional information content of an input symbol given an output symbol, averaged over all input-output pairs. *See also:* information theory. (IT) [123]

erasable programmable read-only memory (EPROM) **(1)** Same as EAROM, except erasure is implemented by exposure to ultraviolet light. (ED) 641-1987w

(2) A type of programmable read-only memory that can be erased and reprogrammed using ultraviolet light.

(C) 610.10-1994w

(3) A reprogrammable read-only memory in which all cells can be simultaneously erased using ultraviolet light and in which the cells at each address can be reprogrammed electrically. (ED) 1005-1998

(4) A type of memory chip designed to be programmed more than once, using special erasing procedures involving ultraviolet light. The processor can only read but not alter the data, considered as permanent memory. (PE/SUB) 1379-1997

erasable read-only-memory (EROM) *See:* erasable programmable read-only memory.

erasable storage A type of storage whose contents can be erased or modified. *Note:* This is generally applied only to nonvolatile storage. *Contrast:* permanent storage. (C) 610.10-1994w

erase (1) (charge-storage tubes) To reduce by a controlled operation the amount of stored information. (ED) 158-1962w, 161-1971w
(2) (computer graphics) To remove one or more display elements from the screen of a cathode ray tube. (C) 610.6-1991w
(3) In the field of electrically erasable programmable read-only memories, the removal of electrons from the floating gate of the memory cell. (ED) 1005-1998

erase algorithm The timed sequence of signals necessary to erase the memory for a flash electrically erasable programmable read-only memory (EEPROM). (ED) 1005-1998

erase character* *See:* delete character.
* Deprecated.

erase disturb The corruption of data in one location caused by the erasing of data at another location. (ED) 1005-1998

erase head Any magnetic head used to erase information from magnetic storage media. (C) 610.10-1994w

erase margin The minimum measured difference between the erased states and the sensing level for the array. (ED) 1005-1998

erase-program cycle The event of writing a memory cell from the erased state to the programmed state and back to the erased state. *Note:* This event may be used as a unit of measurement for endurance. Within a sequence, erase-program cycles are indistinguishable from program-erase cycles. *Contrast:* program-erase cycle. (ED) 1005-1998

erasing head A device for obliterating any previous magnetic recordings. *See also:* direct-current erasing head; alternating-current erasing head; permanent-magnet erasing head; phonograph pickup. (SP) [32]

erasing rate (charge-storage tubes) The time rate of erasing a storage element line or area, from one specified level to another. Note the distinction between this and erasing speed. *See also:* storage tube. (ED) 158-1962w

erasing, selective *See:* selective erasing.

erasing speed (charge-storage tubes) The linear scanning rate of the beam across the storage surface in erasing. Note the distinction between this and erasing rate. *See also:* storage tube. (ED) 158-1962w, 161-1971w

erasing time, minimum usable (storage tubes) The time required to erase stored information from one specified level to another under stated conditions of operation and without rewriting. *Note:* The qualifying adjectives minimum usable are frequently omitted in general usage when it is clear that the minimum usable erasing time is implied. *See also:* storage tube. (ED) 158-1962w

E-R diagram *See:* entity-relationship diagram.

erection (gyros) The process of aligning, by precession, a reference axis with respect to the vertical. (AES/GYAC) 528-1994

erection cut-out (gyros) The feature wherein the signal supplying the erection torque is disconnected in order to minimize vehicle maneuver effects. (AES/GYAC) 528-1994

erection or slaving rate (gyros) The angular rate at which the spin axis is precessed to a reference position. It is expressed as angular displacement per unit time. (AES/GYAC) 528-1994

E region The region of the terrestrial ionosphere between about 90 km and 150 km altitude. (AP/PROP) 211-1997

erg The unit of work and of energy in the centimeter-gram-second systems. The erg is 10^{-7} joule. (Std100) 270-1966w

Ergodic hypothesis For stationary random processes, the equivalence of spatial or temporal average with ensemble average. (AP/PROP) 211-1997

ERL *See:* echo return loss.

erlang (1) (telephone switching systems) Unit of traffic intensity, measured in number of arrivals per mean service time. For carried traffic measurements, the number of erlangs is the average number of simultaneous connections observed during a measurement period. (COM) 312-1977w
(2) (data transmission) A term used in message loading of telephone leased facilities. One erlange is equal to the number of call-seconds divided by 3600 and is equal to a fully loaded circuit over a one-hour period. (PE) 599-1985w

EROM *See:* erasable read-only-memory; erasable programmable read-only memory.

erosion (1) (composite insulators) The loss of material by leakage current or corona discharge. (T&D/PE) 987-1985w
(2) Deterioration by the abrasive action of fluids, usually accelerated by the presence of solid particles of matter in suspension. When deterioration is further increased by corrosion, the term erosion-corrosion is often used. (IA) [59]

ERP *See:* effective radiated power; ear reference point.

erroneous execution The term *erroneous execution* is used in this standard as defined in [Ada RM {1} 1.1.5]. (C/PA) 1003.5b-1995

error (1) (mathematics) Any discrepancy between a computed, observed, or measured quantity and the true, specified, or theoretically correct value or condition. *Notes:* 1. A positive error denotes that the indication of the instrument is greater than the true value. Error = Indication − True. *See also:* absolute error; correction; inherited error. (PE/EDPG) 421-1972s
(2) Any incorrect step, process, or result. *Note:* In the computer field the term commonly is used to refer to a machine malfunction as a machine error (or computer error) and to a human mistake as a human error (or operator error). Frequently it is helpful to distinguish between these errors as follows; an error results from incorrect programming, coding, data transcription, manual operation, etc., a malfunction results from a failure in the operation of a machine component such as a gate, a flip-flop, or an amplifier. *See also:* dynamic error; resolution error; electronic analog computer; loading error; linearity error; static error. (MIL/C) [2], 270-1966w, [20]
(3) (A) (analog computer) In science, the difference between the true value and a calculated or observed value. A quantity (equal in absolute magnitude to the error) added to a calculated, indicated, or observed value to obtain the true value is called a correction. **(B) (analog computer)** In a computer or data processing system, any incorrect step, process, or result. In the computer field, the following terms are commonly used: a machine malfunction is a "machine error" (or "computer error"); an incorrect program is a "program error"; and a human mistake is a "human error" (or "operator error"). Frequently it is helpful to distinguish among these errors as follows: an error results from approximations used in numerical methods or imperfections in analog components; a mistake results from incorrect programming, coding, data transcription, manual operation, etc; a malfunction results from a failure in the operation of a machine component such as a gate, flip-flop, or an amplifier. (C) 165-1977
(4) (automatic control) An indicated value minus an accepted standard value, or true value. *Note:* ASA C85 deprecates use of the term as the negative of deviation. *See also:* accuracy; precision. (PE/EDPG) [3]
(5) (unbalanced transmission-line impedance) "In any measurement of a particular quantity, the difference between the measurement concerned and the true value of the magnitude of this quantity, taken positive or negative accordingly as the measurement is greater or less than the true value"

(Churchill Eisenhart, "Realistic Evaluation of the Precision and Accuracy of Instrument Calibration Systems," Journal of Research of the National Bureau of Standards, Vol. 67C, No. 2, April-June 1963). (IM/HFIM) 314-1971w

(6) (measurement) The algebraic difference between a value that results from measurement and a corresponding true value. (PE/PSE) 94-1970w

(7) (pascal computer programming language) A violation by a program of the requirements of IEEE 770X3.97-1983 that a processor is permitted to leave undetected. *Notes:* 1. If it is possible to construct a program in which the violation or non-violation of this standard requires knowledge of the data read by the program or the implementation definition of implementation-defined features, then violation of that requirement is classified as an error. Processors may report on such violations of the requirement without such knowledge, but there always remain some cases that require execution or simulated execution, or proof procedures with the required knowledge. Requirements that can be verified without such knowledge are not classified as errors. 2. Processors should attempt the detection of as many errors as possible. Permission to omit detection is provided for implementations in which the detection would be an excessive burden. (Std100) 812-1984w

(8) (A) (software) The difference between a computed, observed, or measured value or condition and the true, specified, or theoretically correct value or condition. For example, a difference of 30 meters between a computed result and the correct result. *See also:* syntactic error; dynamic error; semantic error; transient error; indigenous error; static error; fatal error. **(B)** An incorrect step, process, or data definition. For example, an incorrect instruction in a computer program. **(C)** An incorrect result. For example, a computed result of 12 when the correct result is 10. **(D)** A human action that produces an incorrect result. For example, an incorrect action on the part of a programmer or operator. *Note:* While all four definitions are commonly used, one distinction assigns definition A to the word "error," definition B to the word "fault," definition C to the word "failure," and definition D to the word "mistake." (C/Std100) 610.12-1990

(9) (software reliability) Human action that results in software containing a fault. Examples include omission or misinterpretation of user requirements in a software specification, incorrect translation, or omission of a requirement in the design specification. (SE/C) 982.2-1988, 982.1-1988

(10) Manifestation of a failure in a system. (C/BA) 896.9-1994w, 896.3-1993w

(11) The difference between the measured value of a quantity and the true value of that quantity under specified conditions. (PE/PSIM) 4-1995

(12) The difference between a computed, observed, or measured value or condition and the true, specified, or theoretically correct value or condition. For example: A difference of 30 m between a computed result and the correct result. *See also:* parity error; frame check sequence error; burst error. (C) 610.7-1995

error analysis (A) (software) The process of investigating an observed software fault with the purpose of tracing the fault to its source. **(B) (software)** The process of investigating an observed software fault to identify such information as the cause of the fault, the phase of the development process during which the fault was introduced, methods by which the fault could have been prevented or detected earlier, and the method by which the fault was detected. (C) The process of investigating software errors, failures, and faults to determine quantitative rates and trends. *See also:* fault; failure. (C/SE) 729-1983

error and correction The difference between the indicated value and the true value of the quantity being measured. *Note:* It is the quantity that algebraically subtracted from the indicated value gives the true value. A positive error denotes theat the indicated value of the instrument is greater than the true value. The correction has the same numerical value as the error of the indicated value, but the opposite sign. It is the quantity that algebraically added to the indicated value gives the true value. If T, I, E, and C represent, respectively, the true value, the indicated value, the error, and the correction, the following equations hold: $E = I - T$; $C = T - I$ Example: a voltmeter reads 112 volts when the voltage applied to its terminals is actually 110 volts. *See also:* accuracy rating. (EEC/PE) [119]

error band (accelerometer) (gyros) A specified band about the specified output function that contains the output data. The error band contains the composite effects of nonlinearity, resolution, nonrepeatability, hysteresis, and other uncertainties in the output data. (AES/GYAC) 528-1994

error, bit *See:* bit error.

error bit A bit in a status register of an S-module that is associated with detection of some error detected by that S-module. Such bits may be found in the Bus Error register, the optional Module Status register, or in an Additional Status register. Error bits of the Bus Error register affect the value of the BSE bit of the Slave Status register. Error bits of the optional Module Status register or of an Additional Status register are permitted to affect the value of either the BSE bit or EVO bit of the Slave Status register. (TT/C) 1149.5-1995

error burst (1) (data transmission) A group of bits in which two successive erroneous bits are always separated by less than a given number x of correct bits. The last erroneous bit in the burst and the first erroneous bit in the following burst are accordingly separated by x correct bits or more. Number x should be specified when describing an error burst. (PE) 599-1985w

(2) (mathematics of computing) A group of bits in which two erroneous bits are separated by fewer than a specified number of correct bits. (C) 1084-1986w

error category (software) One of a set of classes into which an error, fault, or failure might fall. Categories may be defined for the cause, criticality, effect, life cycle phase when introduced or detected, or other characteristics of the error, fault, or failure. *See also:* failure; fault; software; criticality. (C/SE) 729-1983s

error character A control character used to indicate that an error exists in the data or has occurred during transmission. (C) 610.5-1990w

error code of a task An attribute of a task that ordinarily specifies information about the most recent error that caused `POSIX_Error` to be raised. (C) 1003.5-1999

error coefficient (control system feedback) The real number C_n by which the n th derivative of the reference input signal is multiplied to give the resulting n th component of the actuating signal. *Note:* The error coefficients may be obtained by expanding in a Maclaurin series the error transfer function as follows:

$$\frac{1}{1 + GH(s)} = C_0 + C_1 s + C_2 s^2 + \ldots + C_n s^n$$

See also: feedback control system. (IM) [120]

error compensation Form of error processing when the erroneous state contains enough redundancy to enable correct service delivery. (C/BA) 896.9-1994w

error constant (control system feedback) The real number K_n by which the nth derivative of the reference input signal is divided to give the resulting nth component of the actuating signal. *Note:* $K_n = 1/C_n$; $K_0 = 1 + K_p$, where K_p is position constant; $K_1 = K_v$ velocity constant; $K_2 = K_a$ acceleration constant; $K_3 = K_j$ jerk constant. In some systems these constants may equal infinity. *See also:* feedback control system. (PE/EDPG) [3]

error control (1) Any of a variety of techniques employed to detect and/or correct transmission errors that occur on a communication channel. (SUB/PE) 999-1992w

(2) A technique used to detect the presence of errors and add refinements to correct the detected errors. *See also:* echo check. (C) 610.7-1995

error control character *See:* accuracy control character.

error-correcting code (1) A code in which each telegraph or data signal conforms to specific rules of construction so that departures from this construction in the received signals can be automatically detected, and permits the automatic correction, at the received terminal, or some or all of the errors. *Note:* Such codes require more signal elements than are necessary to convey the basic information. *See also:* error-detecting system; error-detecting code; error-detecting and feedback system. (COM) [49]
(2) (mathematics of computing) A code containing redundant information that can be used to detect certain classes of errors and to restore a word, byte, character, quantity, or message to its correct representation. *Synonym:* error-detecting and correcting code. (C) 610.7-1995, 1084-1986w

error-correction coding An encoding of data and redundant check bits that enables decoding hardware to reconstruct the original data in the presence of a data-bit or check-bit error. (C/MM) 1596.4-1996

error count The number of detected errors in the operation of some device. For communication channels, separate error counts may be maintained for several different error types, e.g., no response, invalid response, and multiple retries, to simplify determination of the error source(s). (SUB/PE) 999-1992w

error data (software) A term commonly (but not precisely) used to denote information describing software problems, faults, failures, and changes, their characteristics, and the conditions under which they are encountered or corrected. *See also:* fault; failure; software. (C/SE) 729-1983s

error-detecting and correcting code *See:* error-correcting code.

error-detecting and feedback system A system employing an error-detecting code and so arranged that a character or block detected as being in error automatically initiates a request for retransmission of the signal detected as being in error. (COM) [49]

error-detecting code (1) A code in which each expression conforms to specific rules of construction, so that if certain errors occur in an expression the resulting expression will not conform to the rules of construction and thus the presence of the errors is detected. *Note:* Such codes require more signal elements than are necessary to convey the fundamental information. *See also:* check; forbidden combination; error-correcting code. (C/COM) [85], [49]
(2) (mathematics of computing) A code containing redundant information that can be used to detect certain classes of errors in a word, byte, character, quantity, or message. *Synonym:* self-checking code. (C) 610.7-1995, 1084-1986w

error-detecting system (data transmission) A system employing an error-detecting code and so arranged that any signal detected as being in error is either deleted from the data delivered to the receiver, in some cases with an indication that such deletion has taken place, or delivered to the receiver together with an indication that it has been detected as being in error. (COM) [49]

error detection The action of identifying that a system state is erroneous. (C/BA) 896.9-1994w

error-detection coding An encoding of data and redundant check bits, such that in the presence of a data-bit or check-bit error decoding hardware can detect the error, but cannot reconstruct the original data. (C/MM) 1596.4-1996

error, dynamic *See:* dynamic error.

errored second A one-second interval during which one or more errors are received. (COM/TA) 1007-1991r

errored second, asynchronous *See:* asynchronous errored second.

errored second, severely *See:* severely errored second.

errored second, synchronous *See:* synchronous errored second.

error, fractional *See:* fractional error.

error-free second A one-second interval during which no error occurs. (COM/TA) 1007-1991r

error, linearity *See:* linearity error.

error log Memory space specifically allocated for recording errors. (C/BA) 896.3-1993w

error logging The recording of an error condition detected during the execution of a service. (SCC20) 1226-1998

error, logical *See:* logical error.

error, matching *See:* matching error.

error message metric The result of dividing the total number of error messages that have been formally demonstrated, by the total number of error messages. (C/SE) 730-1998

error model (A) (modeling and simulation) (software) A model used to estimate or predict the extent of deviation of the behavior of an actual system from the desired behavior of the system; for example, a model of a communications channel, used to estimate the number of transmission errors that can be expected in the channel. **(B) (modeling and simulation) (software)** In software evaluation, a model used to estimate or predict the number of remaining faults, required test time, and similar characteristics of a system. *Synonym:* error prediction model. (C) 610.3-1989, 610.12-1990

error prediction (software) A quantitative statement about the expected number or nature of faults in a system or component. *See also:* error model; error seeding. (C) 610.12-1990

error prediction model *See:* error model.

error, random *See:* random error.

error range The difference between the highest and lowest error values. (C) [85], [20]

error rate (1) (data transmission) Ratio of the number of characters of a message incorrectly received to the number of characters of the message received. (PE) 599-1985w
(2) The probability of an error occurring in the course of data manipulation. For serial binary channels, the error rate is usually expressed as the "bit error rate," i.e., the probability that an individual bit will be received in error. (SUB/PE) 999-1992w
(3) The ratio of the number of characters of a message incorrectly received to the total number of characters of the message received. (C) 610.7-1995

error recovery Form of error processing where an error-free state is substituted for an erroneous state. (C/BA) 896.9-1994w

error report A summary, either in full or in part, of the error log. (C/BA) 896.9-1994w, 896.3-1993w

error, resolution *See:* resolution error.

error seeding The process of intentionally adding known faults to those already in a computer program for the purpose of monitoring the rate of detection and removal, and estimating the number of faults remaining in the program. *Synonyms:* fault seeding; bug seeding. *See also:* indigenous error. (C) 610.12-1990

error signal (1) (excitation systems for synchronous machines) In a control system the error signal is the difference between a sensing signal and a constant reference signal. *Note:* In excitation control systems sensing signals may be proportional to synchronous machine terminal voltage, the ratio of terminal voltage to frequency, active or reactive armature current, active or reactive power, power factor, terminal frequency, shaft speed, generator field voltage or field current, and exciter field voltage or field current. (PE/EDPG) 421.1-1986r
(2) (automatic control device) A signal whose magnitude and sign are used to correct the alignment between the controlling and the controlled elements. (EEC/PE) [119]
(3) (power supplies) The difference between the output voltage and a fixed reference voltage compared in ratio by the two resistors at the null junction of the comparison bridge. The error signal is amplified to drive the pass elements and correct the output. (AES/PE) [41], [78]

(4) (control system feedback) (closed loop) The signal resulting from subtracting a particular return signal from its corresponding input signal. (See the corresponding figure.) *See also:* feedback control system.

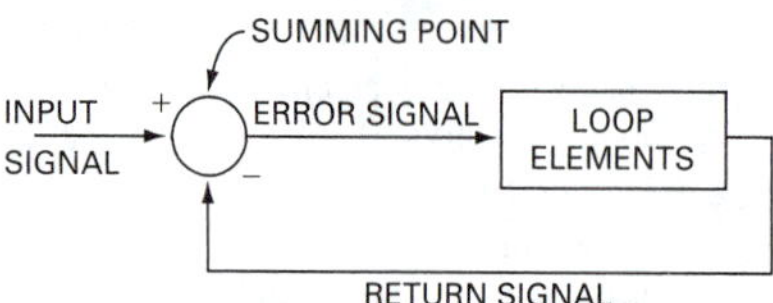

block diagram of a closed loop

signal, error

(PE/EDPG) 421-1972s, [3]

error, static *See:* static error.

error strategy Methodology targeted at dealing with temporary errors. (C/BA) 896.3-1993w

error, systematic *See:* systematic error.

error tolerance The ability of a system or component to continue normal operation despite the presence of erroneous inputs. *See also:* fault tolerance; robustness. (C) 610.12-1990

error transfer function (closed loop) (control system feedback) The transfer function obtained by taking the ratio of the Laplace transform of the error signal to the Laplace transform of its corresponding input signal. *See also:* feedback control system. (IM/PE/EDPG) [120], [3]

erythema (illuminating engineering) The temporary reddening of the skin produced by exposure to ultraviolet energy. *Note:* The degree of erythema is used as a guide to dosages applied in ultraviolet therapy. (EEC/IE) [126]

erythemal effectiveness (illuminating engineering) The capacity of various portions of the ultraviolet spectrum to produce erythema. (EEC/IE) [126]

erythemal efficiency of radiant flux (illuminating engineering) (for a particular wavelength) The ratio of the erythemal effectiveness of that wavelength to that of wavelength 296.7 nm (nanometers), which is rated as unity. *Note:* This term formerly was called "relative erythemal factor." (EEC/IE) [126]

erythemal exposure (illuminating engineering) The product of erythemal flux density on a surface and time. It usually is measured in erythemal microwatt-minutes per square centimeter. *Note:* For average untanned skin a minimum perceptible erythema requires about 300 microwatt-minutes per square centimeter of radiation at 296.7 nm (nanometers). (EEC/IE) [126]

erythemal flux (illuminating engineering) Radiant flux evaluated according to its capacity to produce erythema of the untanned human skin. It usually is measured in microwatts of ultraviolet radiation weighted in accordance with its erythemal efficiency. Such quantities of erythemal flux would be in erythemal microwatts. *Note:* A commonly used practical unit of erythemal flux is the erythemal unit (EU) or E-viton (erytheme) which is equal to the amount of radiant flux which will produce the same erythemal effect as 10 microwatts of radiant flux at wavelength 296.7 nm (nanometers). (EEC/IE) [126]

erythemal flux density (illuminating engineering) The erythemal flux per unit area of the surface being irradiated. It is equal to the quotient of the incident erythemal flux divided by the area of the surface when the flux is uniformly distributed. It usually is measured in microwatts per square centimeter of erythemally weighed ultraviolet radiation (erythemal microwatts per square centimeter). *Note:* A suggested practical unit of erythemal flux density is the Finsen which is equal to one E-viton per square centimeter. (EEC/IE) [126]

ESC *See:* escape character; escape key.

escalator A power-driven, inclined, continuous stairway used for raising or lowering passengers. *See also:* elevator. (EEC/PE) [119]

escape character (ESC) (1) (computers) A character used to indicate that the succeeding one or more characters are expressed in a code different from the code currently in use. (C) [85], [20]

(2) (modeling and simulation) A code extension character used, in some cases with one or more succeeding characters, to indicate by some convention or agreement that the coded representations following the character or the group of characters are to be interpreted according to a different code or according to a different coded character set. (C) 610.5-1990w

escape key (ESC) (A) A special key on a keyboard that is used to represent the escape character. **(B)** A command key that is used to terminate a process or transfer from one mode of operation to another. *See also:* attention key. (C) 610.10-1994

escapement The relative movement by one increment between the printing medium and the printing position. (C) 610.10-1994w

escape ratio (charge-storage tubes) The average number of secondary and reflected primary electrons leaving the vicinity of a storage element per primary electron entering that vicinity. *Note:* The escape ratio is less than the secondary-emission ratio when, for example, some secondary electrons are returned to the secondary-emitting surface by a retarding field. *See also:* charge-storage tube. (ED) 158-1962w

E-scope A cathode-ray oscilloscope arranged to present an E-display. (AES/RS) 686-1990

ESD *See:* end-of-stream delimiter; electrostatic discharge.

ESD current wave The waveform of the discharge current between an intruder and a receptor. (SPD/PE) C62.47-1992r

ESDD (equivalent salt-deposit density) A measure of contamination level. (T&D/PE) 957-1987s, 957-1995

ESD event (1) The occurrence of a single ESD. (EMC) C63.16-1993

(2) An interval that includes the ESD current, electromagnetic fields, and corona effects before and during an ESD. (SPD/PE) C62.47-1992r

ESD receptor The surface (or target) of the object at rest being subjected to the ESD event. (EMC) C63.16-1993

ESD response The EUT reaction to ESD. (EMC) C63.16-1993

ESD simulator A testing device used to simulate a human or furniture ESD event. (EMC) C63.16-1993

ESD simulator ground The pulse-current return connection of the ESD simulator. (SPD/PE) C62.38-1994r

ESD test voltage The amplitude (usually expressed in kV) of the initial electrostatic voltage that exists prior to discharge. (EMC) C63.16-1993

ESG *See:* electrically suspended gyro.

ESDI *See:* enhanced small device interface.

Es layer *See:* sporadic E layer.

ESM *See:* electronic-warfare support measures.

ESONE A multi-national committee representing European nuclear laboratories. It produced the initial CAMAC specification and collaborates with NIM in the maintenance and extension of CAMAC and in the development of FASTBUS. (NID) 960-1993

ESP *See:* enhanced service provider; Econometric Software Package.

ESP cable *See:* electric submersible pump cable.

ESS *See:* electronic switching system.

essential electrical systems (health care facilities) Systems comprised of alternate sources of power, transfer switches, overcurrent protective devices, distribution cabinets, feeders, branch circuits, motor controls, and all connected electrical equipment, designed to provide designated areas with continuity of electrical service during disruption of normal power sources and also designed to minimize the interruptive effects of disruption within the internal wiring system. (NESC/NEC) [86]

essential freeze protection The use of electric heat tracing systems to prevent the temperature of fluids from dropping below the freezing point of the fluid in desirably available or essential outdoor (usually) piping systems at fossil fueled generating stations. An example of an essential freeze protection system is the heat tracing for the feedwater system. (PE/EDPG) 622B-1988r

essential loads Those station auxiliary loads necessary to maintain full output of the station. (SUB/PE) 1158-1991r

essentially zero source impedance (electronic power transformer) Implies that the source impedance is low enough so that the test currents under consideration would cause less than five (5) percent distortion (instantaneous) in the voltage amplitude or waveshape at the load terminals. (PEL/ET) 295-1969r

essential performance requirements (nuclear power generating station) Requirements that must be met if a component, module, or channel is to carry out its part in the implementation of a protective function. (PE/NP) 379-1977s

essential process control (1) (electric pipe heating systems) The use of electric pipe heating systems to increase or maintain or both, the temperature of fluids (or processes) in desirably available or essential mechanical piping systems including pipes, pumps, valves, tanks, instrumentation, etc., in fossil-fueled generating stations. An example of an essential process control system is the heating for the fuel oil system. (PE/EDPG) 622A-1984r
(2) (electric heat tracing systems) The use of electric heat tracing systems to increase, maintain, or both, the temperature of fluids (or processes) in desirably available or essential mechanical piping systems including pipes, pumps, valves, tanks, instrumentation, etc, in fossil-fueled generation stations. An example of an essential process control system is the heating for the fuel oil system. (PE/EDPG) 622B-1988r

Estelle A specification language for telecommunications and distributed systems based on extended state transitions. (C) 610.13-1993w

estimated entry search *See:* interpolation search.

estimated life (thermal classification of electric equipment and electrical insulation) (performance) The expected useful service life based upon service experience or the results of tests performed in accordance with appropriate evaluation procedures established by the responsible technical committee, or both. (EI) 1-1986r

estimated maximum load The calculated maximum heat transfer that a heating or cooling system will be called upon to provide. (IA/PSE) 241-1990r

estimated position (navigation aid terms) The most probable position of a craft determined from incomplete data or data of questionable accuracy. (AES/GCS) 172-1983w

ETB character *See:* end of transmission block character.

ETC *See:* Extendible Compiler.

etched circuit *See:* printed circuit.

Ethernet LAN A CSMA/CD LAN that does *not* use LLC headers on its frames but instead encodes a protocol type field directly after the source address. (LM/C) 802.1H-1995

Ethernet Type-encoding The use of the Type interpretation of an IEEE 802.3 Length/Type field value in a frame as a protocol identifier associated with the MAC Service user data carried in the frame. *Note:* Ethernet Type-encoding can be used with MAC Service user data carried on non-IEEE 802.3 MACs by means of the SNAP-based encapsulation techniques specified in ISO/IEC 11802-5, IETF RFC 1042, and IETF RFC 1390. (C/LM) 802.1Q-1998

EU *See:* erythemal flux.

EULER An experimental programming language that is a generalization of the formal definition of ALGOL. (C) 610.13-1993w

Euler angles A set of three angles used to describe the orientation of an entity as a set of three successive rotations about three different orthogonal axes (x, y, and z). The order of rotation is first about z by angle psi (ψ), then about the new y by angle theta (θ), then about the newest x by angle phi (ϕ). Angles ψ and ϕ range between $\pm\pi$, while angle θ ranges only between $\pm\pi/2$ radians. These angles specify the successive rotations needed to transform from the world coordinate system to the entity coordinate system. The positive direction of rotation about an axis is defined by the right-hand rule. (DIS/C) 1278.1-1995

E-unit *See:* execution unit.

EUT *See:* equipment under test.

evacuating equipment The assembly of vacuum pumps, instruments, and other parts for maintaining and indicating the vacuum. *See also:* rectification. (EEC/PE) [119]

evaluation (1) Interpretation of measurements and observations, including determination of compliance with applicable specification. (NI) N42.17B-1989r
(2) Determination of fitness for use. (C/SE) 1074-1995s
(3) The process of determining whether an item or activity meets specified criteria. (C/SE) J-STD-016-1995

evaluation stack In a stack-based processor, a memory structure in which operands are stored before and after computations. (C) 610.10-1994w

evaluators Those who execute the evaluation portion of the process described in this recommended practice. They may also act in other roles (for example, selector). (C/SE) 1209-1992w

evanescent field (1) (fiber optics) A time varying electromagnetic field whose amplitude decreases monotonically, but without an accompanying phase shift, in a particular direction is said to be evanescent in that direction. (Std100) 812-1984w
(2) An electromagnetic field for which, as one moves away from a boundary, the phase is spatially invariant and the magnitude decays exponentially. *Notes:* 1. An evanescent field is a special case of an inhomogeneous plane wave. 2. Fields in a waveguide beyond cutoff are evanescent. (AP/PROP) 211-1997

evanescent mode (cutoff mode) (1) (waveguide) A field configuration in a waveguide such that the amplitude of the field diminishes along the waveguide, but the phase is unchanged. The frequency of this mode is less than the critical frequency. *See also:* waveguide. (AP/ANT) [35]
(2) *See also:* cutoff mode. (MTT) 146-1980w

evanescent waveguide *See:* cutoff waveguide.

EVE *See:* extreme value engineering.

even and odd mode characteristic impedances The characteristic impedances associated with the even and odd modes of a propagation of a symmetrical pair of coupled transmission lines with respect to ground. These impedances are a function of the degree of coupling between the lines. (MTT) 1004-1987w

even and odd modes The modes of propagation on a symmetrical planar transmission-line structure whose electric field distribution in the transverse cross section in even or odd with respect to reflections in the plane of symmetry of the structure. *Notes:* 1. A symmetrical coupled pair of transmission lines can support two fundamental modes—an even mode and an odd mode. 2. A single planar transmission line can support only one fundamental mode, which may be even or odd, depending on the structure of the transmission line. For example, the fundamental mode on a single microstrip line is an even mode. The first higher order mode is odd. On a single slot line, the fundamental mode is an odd mode. (MTT) 1004-1987w

even-odd check *See:* parity check.

even parity (1) An error detection method in which the number of ones in a binary word, byte, character, or message is maintained as an even number. (C) 1084-1986w
(2) The property possessed by a binary word, byte, character, or message that has an even number of ones. (C) 1084-1986w

event (1) (sequential events recording systems) A change in a process or a change in operation of equipment that is detected by bistable sensors. (PE/EDPG) [1]
(2) (A) (modeling and simulation) An occurrence that causes a change of state in a simulation. **(B) (modeling and simulation)** The instant in time at which a change in some variable occurs. (C/Std100) 610.3-1989
(3) A semantic construct associated with a point in time that may result in an instance of processing or state transitions on the part of the receiver. Events are usually carried between entities by DMA messages. For example, an inbound DMA event message may indicate an asynchronous error requiring Processor attention. (C/MM) 1212.1-1993
(4) An occurrence that may require reporting by the utilities defined in this standard. The reporting of an event may cause data to be written to stdout, stderr, or to a log file. (C/PA) 1387.2-1995
(5) A discrete change of state (status) of a system or device. (SWG/PE/SUB) C37.100-1992, C37.1-1987s
(6) **(A)** Any change in conditions or performance of interest. **(B)** An occurrence at a specific point in time. (PE/NP) 1082-1997
(7) Change of status or condition. (PE/NP) 692-1997
(8) An abstraction of the mechanism by which asynchronously generated signals or conditions are generated and represented. (IM/ST) 1451.1-1999

event-based planning An approach to establishing engineering plans, tasks, and milestones based upon satisfying significant accomplishments associated with key events rather than calendar-oriented milestones. (C/SE) 1220-1998

event data *See:* time-consistent traffic measures; traffic intensity.

event-driven simulation *See:* event-oriented simulation.

event, event command A command contained within an *event packet.* (C/MM) 1596.4-1996

event flag (1) A Boolean associated with a session and maintained by the service that is used to signal the arrival of objects in the delivery, retrieval, or input queue. (C/PA) 1224.1-1993w
(2) A single bit variable used to represent the occurrence of a particular event. (C/MM) 855-1990

Event Generator Publisher Port An instance of the class `IEEE1451_EventGeneratorPublisherPort` or of a subclass thereof. (IM/ST) 1451.1-1999

event horizon The earliest future date on which a system element will fail to perform date data processing consistently and predictably. (C/PA) 2000.1-1999

event management The mechanism that enables applications to register for and be made aware of external events such as data becoming available for reading. (C) 1003.5-1999

Event Occurrence EVO bit A bit in the Slave Status register of every S-module that is set by the S-module when a module-application-related condition requiring an interrupt has occurred. (TT/C) 1149.5-1995

event-oriented simulation A simulation in which attention is focused on the occurrence of events and the times at which those events occur; for example, a simulation of a digital circuit that focuses on the time of state transition. *Synonyms:* event-sequenced simulation; event-driven simulation. (C) 610.3-1989w

event packet A short, four-byte packet containing an event command that is directed to one slave or broadcast to all. Device state is affected by the event command, but no response is returned to the controller. (C/MM) 1596.4-1996

event recognition (sequential events recording systems) The capability to detect and process changes of state of one or more inputs. (PE/EDPG) [1]

event recorder (1) On-board device/system with crashworthy, nonvolatile memory, which records data to support accident/incident analysis. (VT) 1482.1-1999
(2) An on-board device/system with crashworthy memory that records data to support accident/incident analysis. (VT) 1475-1999

events Signals or interrupts generated by a device to notify another device of an asynchronous event. The contents of events are device dependent. (C/MM) 1155-1992

event-sequenced simulation *See:* event-oriented simulation.

event sequence sensor A sensor that detects a change of state in the physical world. The instant in time of the change of state, not the state value, is the "measurement." (IM/ST) 1451.2-1997

event tree A graphical representation of the logical progression of the possible scenarios through a multiple series of events that may or may not occur. (PE/NP) 1082-1997

everyday load (composite insulators) The bare conductor weight and wind load that predominates for the greatest period of time over the life of a line. (T&D/PE) 987-1985w

evh *See:* extra-high voltage.

E-viton *See:* erythemal flux.

evoked potential The electrical response of a neuron or neurons elicited by electrical or natural (i.e., auditory, visual, etc.) stimulus. To be contrasted with spontaneous activity, such as that recorded by the EEG. (T&D/PE) 539-1990

evolving fault A change in the current during interruption whereby the magnitude of current increases to a fault current or to a higher value of fault current in one or more phases. (SWG/PE) C37.100-1992

EV traffic measures *See:* extreme value traffic measures.

EW Acronym for early warning; electronic warfare. (AES) 686-1997

exact scheduling A scheduling algorithm used by the controller to predict the time delay needed by a slave to generate response packets. The controller reserves a time slot for the response, where the time slot is the exact size needed and is at the precise time for the expected response packet. (C/MM) 1596.4-1996

exalted carrier reception *See:* reconditioned carrier reception.

examination An inspection with the addition of partial dismantling, as required, supplemented by diagnostic tests in order to reliably evaluate the condition of the circuit breaker. (SWG/PE) C37.10-1995

exceedance level A statistical descriptor that is often used in expressing levels of quantities. For example, in acoustics, the L_{10} is the A-weighted sound level exceeded for 10% of the time over a specified time period (and for corona noise over a specified weather condition). For the other 90% of the time, the sound level is less than the L_{10}. Similarly, the L_{50} is the sound level exceeded 50% of the time; the L_{90} is the sound level exceeded 90% of the time, etc. The concept of exceedance levels can also be used as a statistical term for other corona effects such as radio noise, corona loss, and dc fields and ions. Any exceedance level canbe easily obtained from distributions that have been plotted on probability paper. (T&D/PE) 539-1990

exception (1) (software) An event that causes suspension of normal program execution. Types include addressing exception, data exception, operation exception, overflow exception, protection exception, underflow exception. (C) 610.12-1990
(2) (MULTIBUS) An abnormal condition on the bus caused by either a bus parity error, a bus time-out, a protocol violation, or a bus owner reply phase termination. (C/MM) 1296-1987s

®Multibus is a registered trademark of Intel Corporation.

exception condition (1) A condition assumed by a secondary or remote station when it receives a command that it cannot execute, or when it receives data it cannot process. (C) 610.7-1995
(2) The condition assumed by an LLC upon receipt of a command PDU that it cannot execute due to either a transmission error or an internal processing malfunction. (C/LM/CC) 8802-2-1998

exception operation (MULTIBUS) A bus operation in which an agent places an error indication on the parallel system bus. The error indication causes all bus agents to terminate arbitration and transfer operations. (C/MM) 1296-1987s

®Multibus is a registered trademark of Intel Corporation.

exception reporting An information processing technique that screens large amounts of computerized data and produces a report containing only the data that require action. *See also:* information overload. (C) 610.2-1987

Exception Window A time interval during which the impedance of a mated connector and associated transmission line is allowed to exceed the impedance tolerance specification for signals passed through that connector. (C/LM) 802.3-1998

except operation* *See:* exclusion.

* Deprecated.

excess gain The value of the positive gain (in decibels) at any specified frequency for the open oscillator loop measured under small signal conditions (no limiting action). The source and load impedance must be specified. (UFFC) 1037-1992w

excess-fifty code A binary code in which a decimal number *n* is represented by the binary equivalent of $n + 50$. *Synonym:* excess-fifty representation. (C) 1084-1986w

excess-fifty representation *See:* excess-fifty code.

excess insertion loss (fiber optics) In an optical waveguide coupler, the optical loss associated with that portion of the light which does not emerge from the nominally operational ports of the device. *See also:* optical waveguide coupler. (Std100) 812-1984w

excess-sixty-four code A binary code in which a decimal number *n* is represented by the binary equivalent of $n + 64$. *Synonym:* excess-sixty-four representation. (C) 1084-1986w

excess-sixty-four representation *See:* excess-sixty-four code.

excess-three BCD *See:* excess-three code.

excess-three code (A) A BCD code in which a decimal digit *n* is represented by the four-bit binary equivalent of $n + 3$.

Excess-Three Code

DECIMAL DIGIT:	0	1	2	3
EXCESS 3 CODE:	0011	0100	0101	0110

Synonyms: excess-three representation; excess-three BCD. **(B) (electronic computation)** Number code in which the decimal digit *n* is represented by the four-bit binary equivalent of $n + 3$. Specifically:

decimal digit	excess-three code
0	0011
1	0100
2	0101
3	0110
4	0111
5	1000
6	1001
7	1010
8	1011
9	1100

(C) 1084-1986, 162-1963

excess-three representation *See:* excess-three code.

excess meter An electricity meter that measures and registers the integral, with respect to time, of those portions of the active power in excess of the predetermined value. *See also:* electricity meter. (EEC/PE) [119]

excess reactivity (power operations) More reactivity than that needed to achieve criticality. In order to avoid frequent reactor shutdowns to replace fuel that has been consumed and to compensate for the accumulation of fission products that have high neutron absorption cross sections and negative temperature coefficients, excess reactivity is provided in a reactor by including additional fuel in the core at startup. *See also:* reactivity. (PE/PSE) 858-1987s

exchange *See:* private automatic exchange; private branch exchange; exchange service; private automatic branch exchange; central office; central office exchange.

exchangeable power (per unit bandwidth, at a port) The extreme value of the power flow per unit bandwidth from or to a port under arbitrary variations of its terminating impedance. *Notes:* 1. The exchangeable power pe at a port with a mean-square open-circuit voltage spectral density e^2 and an internal impedance with a real part *R* is given by the relation

$$p_e = \frac{\overline{e^2}}{4R}$$

2. The exchangeable power is equal to the available power when the internal impedance of the port has a positive real part. *See also:* waveguide; signal-to-noise ratio. (ED) [45]

exchangeable power gain (two-port linear transducer) At a pair of selected input and output frequencies, the ratio of the exchangeable signal power of the output port of the transducer to the exchangeable signal power of the source connected to the input port. *Note:* The exchangeable power gain is equal to the available power gain when the internal impedances of the source and the output port of the transducer have positive real parts. *See also:* signal-to-noise ratio; waveguide. (ED) [45]

exchange area (1) (telephone switching systems) The territory included within the boundaries of a telecommunications exchange. (COM) 312-1977w
(2) In North America, an area within which there is a single uniform set of charges for telephone service. An exchange area may be served by a number of end offices. *Note:* In Europe, the area of service of a single end office is an exchange area. A call between any two points within an exchange area is a local call. (C) 610.7-1995

Exchange Carriers' Standards Association The Secretariat for ASC T1, which develops the ASC T1's series of standards on telecommunication. (C) 610.7-1995, 610.10-1994w

exchange, central office *See:* central office exchange.

exchange data Data that is received or transmitted via data exchange in an appropriate format. *See also:* data exchange. (C) 610.5-1990w

eXchange IDentification (XID) A frame exchange used during the logical connection (initialization) sequence for the purpose of transferring Data Link layer and upper layer parameters between the primary station (BCC) and the secondary station (DCC). (EMB/MIB) 1073.3.1-1994

exchange layer The exchange layer describes the procedure of the node-to-node exchange of characters to ensure the proper functioning of the link. The exchange layer provides the service to the higher layers of the transmission of an indefinite sequence of N_chars. (C/BA) 1355-1995

exchange selection sort *See:* bubble sort.

exchange service (1) (data transmission) A service permitting interconnection of any two customers' telephones through the use of a switching equipment. (PE) 599-1985w
(2) In data communications, a service that permits interconnection of any two customers' stations through the use of the exchange system. (C) 610.7-1995

exchange sort A sort in which pairs of items in a set are examined in some sequence, pairs found out of order are exchanged, and the process is repeated until all items are in the correct order. Multiple passes are usually required. *See also:* bubble sort; radix exchange sort; cocktail shaker sort; Batcher's parallel sort. (C) 610.5-1990w

exchange system In data communications, a system that controls the connection of incoming and outgoing lines. (C) 610.7-1995

excitability (1) (irritability) (electrobiology) The inherent ability of a tissue to start its specific reaction in response to an electric current. (EMB) [47]
(2) (overhead power lines) The sensitivity of an excitable membrane to a stimulus. (T&D/PE) 539-1990

excitability curve (medical electronics) A graph of the excitability of a given tissue as a function of time, where excitability is expressed either as the reciprocal of the intensity of an electric current just sufficient at a given instant to start the specific reaction of the tissue, or as the quotient of the initial (or conditioning) threshold intensity for the tissue by subsequent threshold intensities. (EMB) [47]

excitable membrane The membrane of nerve or muscle cells having an electrochemical property that results in sudden, major changes in ionic permeability when excited by an appropriate stimulus. (T&D/PE) 539-1990

excitation (1) (control of small hydroelectric plants) A source of direct current for the synchronous generator field. (PE/EDPG) 1020-1988r
(2) (array antenna) For an array of radiating elements, the specification, in amplitude and phase, of either the voltage applied to each element or the input current to each element. (AP/ANT) 145-1993

excitation anode (pool-cathode rectifier tube) An electrode that is used to maintain an auxiliary arc in the vacuum tank. *See also:* electrode; rectification. (EEC/PE) [119]

excitation coefficients The relative values, in amplitude and phase, of the excitation currents or voltages of the radiating elements of an array antenna. *Synonym:* feeding coefficients. (AP/ANT) 145-1993

excitation control system A feedback control system that includes the synchronous machine and its excitation system. (PE/EDPG) 421.4-1990, 421-1972s

excitation control system stabilizer (synchronous machines) An element or group of elements that modify the forward signal by either series or feedback compensation to improve the dynamic performance of the excitation control system. (PE/EDPG) 421-1972s

excitation current (1) (no-load current) (power and distribution transformers) The current that flows in any winding used to excite the transformer when all other windings are open-circuited. It is usually expressed in percent of the rated current of the winding in which it is measured. (PE/TR) C57.12.80-1978r
(2) (voltage regulators) The current that maintains the excitation of the regulator. *Note:* It is usually expressed in per unit or in percent of the rated series-winding current of the regulator. *See also:* voltage regulator; efficiency. (PE/TR) C57.15-1968s
(3) The current supplied to unloaded transformers or similar equipment. (SWG/PE) C37.100-1992
(4) The current that maintains the excitation of the regulator. It may be expressed in amperes, per unit, or percent of the rated current of the regulator. (PE/TR) C57.15-1999

excitation current interrupting rating (of an interrupter switch) The highest rms current in amperes between zero and 0.1 power factor lagging that a device is required to interrupt, without requiring maintenance, at its rated maximum voltage and at rated frequency, for a number of operations equal to the life expectancy of the switch. (SWG/PE) C37.30-1992s

excitation current switching capability (of a generator circuit breaker) The highest magnetizing current that a generator circuit breaker shall be required to switch at any voltage up to rated maximum voltage at power frequency without causing an overvoltage exceeding the levels agreed upon between the user and the manufacturer. (SWG/PE) C37.013-1997

excitation equipment (rectifier) The equipment for starting, maintaining, and controlling the arc. *See also:* rectification. (PE/EEC) [119]

excitation losses *See:* no-load losses **(series transformer)** The losses in the transformer with the secondary winding open-circuited when the primary winding is excited at rated frequency and at a voltage that corresponds to the primary voltage obtained when the transformer is operating at nominal rated load. *Note:* The measurement should be made with a constant voltage source of supply with not more than 3-percent harmonic deviation from sine wave. (EEC/LB) [98]
(2) (instrument transformers) The watts required to supply the energy necessary to excite the transformer which include the dielectric watts, the core watts, and the watts in the excited winding due to the excitation current. (PE/TR) C57.12.80-1978r, C57.13-1978s

excitation losses for an instrument transformer The power (usually expressed in watts) required to excite the transformer at its primary terminals. *Note:* Excitation losses include core, dielectric, and winding losses due to the excitation current. (PE/TR) C57.13-1993

excitation power current transformer (excitation systems for synchronous machines) The elements in a compound source-rectifier excitation system which transfer electrical energy from the synchronous machine armature current to the excitation system at a magnitude and phase relationship required by the excitation system. (PE/EDPG) 421.1-1986r

excitation power potential transformer (excitation systems for synchronous machines) The element or elements in a compound source-rectifier excitation system which transfer electrical energy from the synchronous machine armature terminals to the excitation system at a magnitude and phase relationship required in the excitation system. Also, the element or elements in a potential source-rectifier excitation system which transfer electrical energy either from the machine terminals or from an auxiliary bus to the excitation system at a magnitude level required by the excitation system. (PE/EDPG) 421.1-1986r

excitation purity (1) (light) (television) The ratio of the distance from the reference point to the point representing the sample to the distance along the same straight line from the reference point to the spectrum locus or to the purple boundary, both distances being measured (in the same direction from the reference point) on the CIE chromaticity diagram. *Note:* When giving excitation purity and dominant (or complimentary) wavelength as a pair of values to determine the chromaticity coordinates, the reference point must be the same in all cases, and it must represent the reference standard light (specified achromatic light) mentioned in the definitions of dominant wavelength. (BT/AV) 201-1979w
(2) (illuminating engineering) (of a light) The ratio of the distance on the CIE chromaticity diagram between the reference point and the light point to the distance in the same direction between the reference point and the spectrum locus or the purple boundary. (EEC/IE) [126]

excitation-regulating winding (power and distribution transformers) (two-core regulating transformer) In some designs, the main unit will have one winding operating as an autotransformer which performs both functions listed under excitation and regulating windings. Such a winding is called the "excitation-regulating winding." (PE/TR) C57.12.80-1978r

excitation response *See:* exciter voltage response.

excitation system (1) (excitation systems for synchronous machines) The equipment providing field current for a synchronous machine, including all power, regulating, control, and protective elements. (PE/EDPG) 421.4-1990, 421.1-1986r
(2) (rotating machinery) The source of field current for the excitation of a principal electric machine, including means for its control. *See also:* direct-current commutating machine. (PE) [9]

excitation system ceiling voltage (excitation systems) The maximum dc component of system output voltage that may be attained by an excitation system under specified conditions. *Note:* In some excitation systems, ceiling voltage may have both positive and negative values. Also, in some special applications, the excitation system is capable of supplying both positive and negative field current to the synchronous machine. (PE/EDPG) 421A-1978s

excitation system duty cycle (excitation systems for synchronous machines) An initial operating condition and a subsequent sequence of events of specified duration to which the excitation system will be exposed. *Note:* The duty cycle usually involves a three-phase fault of specified duration which is located electrically close to the synchronous machine. Its primary purpose is to specify the duty that the excitation system components can withstand without incurring maloperation or damage. (PE/EDPG) 421.1-1986r

excitation system, high initial response An excitation system having an excitation system voltage response time of 0.1 second or less. (PE/EDPG) 421-1972s

excitation system nominal ceiling voltage (excitation systems) The ceiling voltage attained by an excitation system under the following conditions: (A) The exciter loaded with a resistor having an ohmic value equal to the resistance of the filed winding to be excited and with this field winding at a temperature of 75°C for field windings designed to operate at rating with a temperature rise of 60°C or less; or 100°C for field windings designed to operate at rating with a temperature rise greater than 60°C. For rectifier exciters nominal ceiling voltage should be determined with the exciter loaded with a load having resistance as specified above and sufficient inductance so that regulation effects and voltage and current waveforms can be properly duplicated. For test purposes, providing such a load may often be impractical. In such cases, analytical means may be used to predict performance under actual loading and conditions. (B) For excitation systems employing a rotating exciter, the ceiling should be determined at rated speed. (C) For potential-source rectifier excitation systems, the ceiling should be determined with rated (100 percent) potential applied unless otherwise specified. (D) In compound-rectifier excitation systems both generator voltage and current inputs are utilized as the source of power for the excitation system. The nominal ceiling voltage will be determined under specified reduced generator terminal voltage and increased generator terminal current conditions as would be encountered during power system faults and other disturbances. For some applications where relay coordination is a consideration, the ceiling voltage will be determined by a requirement that the generator produce a specific value of steady-state three-phase short circuit current. (PE/EDPG) 421A-1978s

excitation system nominal response (excitation systems for synchronous machines) The rate of increase of the excitation system output voltage determined from the excitation system voltage response curve, divided by the rated field voltage. This rate, if maintained constant, would develop the same voltage-time area as obtained from the actual curve over the first half-second interval (unless a different time interval is specified). (PE/EDPG) 421.1-1986r

excitation system output terminals (excitation systems for synchronous machines) The place of output from the equipmant comprising the excitation system. These terminals may be identical with the field winding terminals. (PE/EDPG) 421.1-1986r

excitation system rated current (excitation systems for synchronous machines) The direct current at the excitation system output terminals which the excitation system can supply under defined conditions of its operation. This current is at least that value required by the synchronous machine under the most demanding continuous operating conditions (generally resulting from synchronous machine voltage frequency variations and power factor variations). (PE/EDPG) 421.1-1986r

excitation system rated voltage (excitation systems for synchronous machines) The direct voltage at the excitation system output terminals which the excitation system can provide when delivering excitation system rated current under rated continuous load conditions of the synchronous machine with its field winding at 75°C for field windings designed to operate at rating with a temperature rise of 60°C or less; or 100°C for field windings designed to operate at rating with a temperature rise greater than 60°C. (PE/EDPG) 421.1-1986r

excitation-system stability (rotating machinery) (synchronous machines) The ability of the excitation system to control the field voltage of the principal electric machine so that transient changes in the regulated voltage are effectively suppressed and sustained oscillations in the regulated voltage are not produced by the excitation system during steady-load conditions or following a change to a new steady-load condition. *Note:* It should be recognized that under some system conditions it may be necessary to use power system stabilizing signals as additional inputs to excitation control systems to achieve stability of the power system including the excitation system. *See also:* direct-current commutating machine. (PE/EDPG) [9], 421-1972s

excitation system stabilizer (1) (excitation systems for synchronous machines) An element or group of elements that modify the forward signal by either series or feedback compensation to improve the dynamic performance of the excitation control system. (PE/EDPG) 421.1-1986r
(2) (excitation systems for synchronous machines) A control element that is used to stabilize the excitation control system. (PE/EDPG) 421.4-1990

excitation system voltage response (1) (excitation systems) The rate of increase or decrease of the excitation system output voltage determined from the excitation system voltage-time response curve, which rate if maintained constant, would develop the same voltage-time area as obtained from the curve for a specified period. The starting point for determining the rate of voltage change is the initial value of the excitation system voltage time response curve. Referring to the figure below, the excitation system voltage response is illustrated by line ac. This line is determined by establishing the area acd equal to area abd. *Notes:* 1. The starting point for determining the rate of voltage change is the initiation of the disturbance, that is, the excitation system voltage time response should include any delay time that may be present. 2. A system having an excitation system voltage response time of 0.1 s or less is defined as a high response excitation system. (PE/EDPG) 421A-1978s
(2) (synchronous machines) The rate of increase or decrease of the excitation system output voltage determined from the excitation system voltage-time response curve, that if maintained constant would develop the same voltage-time area as obtained from the curve for a specified period. The starting point for determining the rate of voltage change is the initial value of the excitation system voltage-time response curve. *Notes:* 1. Similar definitions can be applied to the excitation system major components such as the exciter and regulator. 2. A system having an excitation system voltage response time of 0.1 second or less is defined as a "high initial response excitation system."

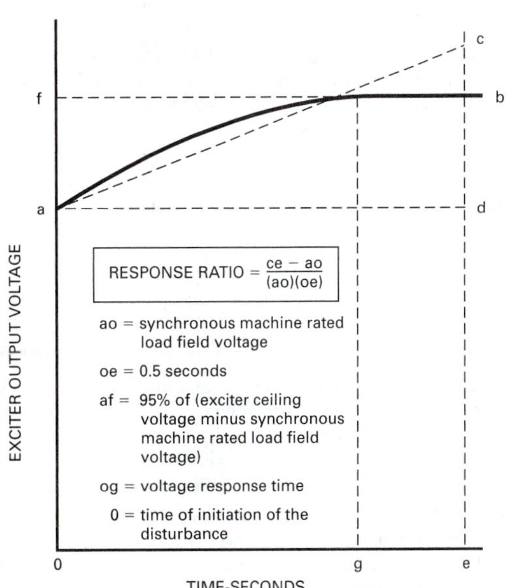

excitation system voltage response
(PE/EDPG) 421-1972s

excitation system voltage response ratio (1) (excitation systems) The numerical value that is obtained when the excitation system voltage response, in volts per second measured over the first 1/2 S interval, unless otherwise specified, is divided by the rated load field voltage of the synchronous machine. *Notes:* 1. Referring to the figure to the definition of **excitation system voltage response,** the excitation system response ratio, unless otherwise specified, applies apply only to the increase in excitation system voltage. 2. Response ratio is determined with the exciter voltage initially equal to the rated load field voltage of the synchronous machine to which the exciter is applied, and then suddenly establishing circuit conditions required to obtain nominal exciter ceiling voltage. Excitation system response ratio is determined by suddenly reducing the voltage sensed by the synchronous machine voltage regulator from 100 percent to 80 percent unless otherwise specified. 3. Unless otherwise specified, excitation system response ratio should be determined with the exciter loaded as specified in 3.2.2. If, for practical considerations, the test is performed at no load, analytical means may be utilized to predict the performance under load. 4. For excitation systems employing a rotating exciter, the response ratio should be determined at rated speed. 5. For potential-source rectifier excitation systems, the nature of a power system disturbance greatly affects the available power supply voltages. The ceiling voltage available and the voltage respose time are more meaningful parameters. To specify a response ratio implies equivalence with other systems whose output is not adversely affected by such depressed voltage conditions. Therefore, response ratio is not recommended as a specification parameter for these excitation systems. 6. For compound-rectifier excitation systems, the nature of the power system disturbance and the specific design prameters of the exciter and the synchronous generator influence the performance of the exciter output voltage. For equivalence with rotating exciters, the response ratio should be based on performance under specified reduced generator terminal voltage and increased generator stator current conditions as would be encountered during power system faults and disturbances. (PE/EDPG) 421A-1978s
(2) (synchronous machines) The numerical value that is obtained when the excitation system voltage response, in volts per second, measured over the first half-second interval, unless otherwise specified, is divided by the rated-load field voltage of the synchronous machine. *Note:* Unless otherwise specified, the excitation system voltage response ratio shall apply only to the increase in excitation system voltage. (PE/EDPG) 421-1972s

excitation system voltage response time (excitation systems) The time in seconds for the excitation voltage to attain 95% of the difference between ceiling voltage and rated field voltage under specified conditions. (PE/EDPG) 421.2-1990, 421.1-1986r

excitation system voltage time response (excitation systems for synchronous machines) The excitation system output voltage expressed as a function of time, under specified conditions. *Note:* A similar definition can be applied to the excitation system major components, the exciter and regulator, separately. (PE/EDPG) 421.2-1990, 421.1-1986r

excitation voltage The nominal voltage of the excitation circuit. (R) [29]

excitation winding (power and distribution transformers) (two-core regulating transformer) The winding of the main unit, which draws power from the system to operate the two-core transformer. *See also:* field winding. (PE/TR) C57.12.80-1978r

excite (rotating machinery) To initiate or develop a magnetic field in (such as in an electric machine). *See also:* asynchronous machine. (PE) [9]

excited-field loudspeaker A loudspeaker in which the steady magnetic field is produced by an electromagnet. (EEC/PE) [119]

excited-state maser (laser maser) A maser in which the terminal level of the amplifying transition is not appreciably populated at thermal equilibrium for the ambient temperature. (LEO) 586-1980w

excited winding (power and distribution transformers) (two-core regulating transformer) The winding of the series unit which is excited from the regulating winding of the main unit. (PE/TR) C57.12.80-1978r

exciter (1) (excitation systems for synchronous machines) The equipment providing the field current for the excitation of a synchronous machine. (PE/EDPG) 421.1-1986r
(2) (excitation systems) The equipment that provides the field current for the excitation of a synchronous machine. (PE/EDPG) 421.4-1990
(3) (rotating machinery) The source of all or part of the field current for the excitation of an electric machine. *Note:* Familar sources include direct-current commutator machines; alternating-current generators whose output is rectified; and batteries. *See also:* direct-current commutating machine; synchronous machine. (PE/IA/MT) [9], 45-1998
(4) (data transmission) In antenna practice, the portion of a transmitting array, (of the type which includes a reflector or director), which is directly connected with the source of power. (COM/PE) 599-1985w

exciter, alternator-rectifier *See:* alternator-rectifier exciter.

exciter-ceiling voltage (rotating machinery) (field discharge circuit breakers) The maximum voltage that may be attained by an exciter under specified conditions. (SWG/PE) C37.100-1992, C37.18-1979r

exciter ceiling voltage, nominal (rotating machinery) The ceiling voltage of an exciter loaded with a resistor having an ohmic value equal to the resistance of the field winding to be excited and with this field winding at a temperature of: (1) 75 degrees Celsius for field windings designed to operate at rating with temperature rise of 60 degrees Celsius or less. (2) 100 degrees Celsius for field windings designed to operate at rating with a temperature rise greater than 60 degrees Celsius. (EEC/PE) [119]

exciter, compound-rectifier *See:* compound source-rectifier exciter.

exciter, direct current generator-commutator (synchronous machines) An exciter whose energy is derived from a direct current generator. *Notes:* 1. The exciter includes a direct current generator with its commutator and brushes. It is exclusive of input control elements. 2. The exciter may be driven by a motor, prime mover, or by the shaft of the synchronous machine. (PE/EDPG) 421-1972s

exciter dome (rotating machinery) Exciter housing for a vertical machine. (PE) [9]

exciter losses (synchronous machines) The total of the electric and mechanical losses in the equipment supply excitation. (PE) [9], [84]

exciter, main *See:* main exciter.

exciter or direct-current generator relay (power system device function numbers) A relay that forces the dc machine field excitation to build up during starting or which functions when the machine voltage has built up to a given value. (SUB/PE) C37.2-1979s

exciter, pilot *See:* pilot exciter.

exciter platform (rotating machinery) A deck on which to stand while inspecting the exciter. (PE) [9]

exciter, potential source-rectifier *See:* potential source-rectifier exciter.

exciter response *See:* exciter voltage response.

exciter response ratio, main (synchronous machines) The numerical value obtained when the response, in volts per second, is divided by the rated-load field voltage, which response, if maintained constant, would develop, in one half-second, the same excitation voltage-time area as attained by the actual exciter. *Note:* The response is determined with no load on the exciter voltage initially equal to the the rated-load field voltage, and then suddenly establishing circuit conditions which would be used to obtain nominal exciter ceiling voltage. For a rotating exciter, response should be determined at rated speed. This definition does not apply to main exciters having one or more series field, except a light differential series field, or to electronic exciters. (PE/EDPG) 421-1972s

exciter voltage response The rate of increase or decrease of the exciter voltage when a change in this voltage is demanded. It is the rate determined from the exciter voltage response curve that if maintained constant would develop the same exciter voltage-time area as is obtained from the curve for a specified period. The starting point for determining the rate of voltage change shall be the initial value of the exciter voltage-time response curve. *See also:* asynchronous machine. (PE) [9]

exciting current The total current applied to a coil that links a ferromagnetic core. The component of the primary current of a transformer that is sufficient by itself to cause the counter electromotive force to be induced in the primary winding. (CHM) 270-1966w, [51]

exciting field The total field responsible for the waves scattered by a particle or elemental surface area. In the case of single scattering, the exciting field consists solely of the incident field. In the multiple scattering case, the exciting field consists of the incident field plus the fields scattered by and among all other particles or elemental surface areas. *See also:* incident field. (AP/PROP) 211-1997

excitron A single-anode pool tube provided with means for maintaining a continuous cathode spot. (ED) [45]

exclusion (1) (mathematics of computing) A dyadic Boolean operator having the property that if P is a statement and Q is a statement, then the expression P exclusion Q is true if and only if P is true and Q is false. *Note:* P exclusion Q is often represented by a combination of AND and NOT symbols such as P∧Q.

Exclusion Truth Table

P	Q	P∧~Q
0	0	0
0	1	0
1	0	1
1	1	0

Synonym: NOT-IF-THEN. (C) 1084-1986w
(2) The dyadic Boolean operation whose result has the Boolean value 1 if and only if the first operand has the Boolean value 1 and the second has the Boolean value 0. *See also:* NOT-IF-THEN gate. (C) 610.10-1994w

exclusive lock A lock that grants the holder sole access to the locked data. No other process can access the data for either read or write purposes. *Contrast:* shared lock. (C) 610.5-1990w

exclusive modified An attribute assigned to a cache line if there is an up-to-date copy of the line in the module's cache and the module has the only valid copy in the system. (C/BA) 896.4-1993w

exclusive NOR (XNOR) *See:* equivalence.

exclusive-NOR element *See:* exclusive-NOR gate.

exclusive-NOR gate A gate that performs the Boolean operation of equivalence. *Synonyms:* exclusive-NOR element; IF-AND-ONLY-IF element. (C) 610.10-1994w

exclusive OR (XOR) (mathematics of computing) A dyadic Boolean operator having the property that if P is a statement and Q is a statement, then P exclusive-OR Q is true if and only if either, but not both, is true. *Note:* P exclusive OR Q is often represented by P⊕Q or P∪Q.

Exclusive OR Truth Table

P	Q	P⊕Q
0	0	0
0	1	1
1	0	1
1	1	0

Synonyms: modulo-two sum; nonequivalence; inequivalence. *Contrast:* OR. (C) 1084-1986w

exclusive-OR element *See:* exclusive-OR gate.

exclusive-OR gate A gate that performs the Boolean operation of nonequivalence. *Synonym:* exclusive-OR element. (C) 610.10-1994w

exclusive unmodified An attribute assigned to a cache line if there is an up-to-date copy of the line in the module's cache and the module is to assume that no other copies of the line are valid in any other cache in the system. (C/BA) 896.4-1993w

excursion *See:* reference excursion.

EXEC A command language used in IBM's VM/CMS environment to carry out command level processing. *Note:* EXEC was superseded by EXEC2 and is superseded by REXX. (C) 610.13-1993w

EXEC2 *See:* EXEC.

Exec family of operations The collection of operations that cause a new program to be executed, *i.e.*, the `Exec` and `Exec_Search` procedures in `POSIX_Unsafe_Process_Primitives` and the `Start_Process` and `Start_Process_Search` procedures in `POSIX_Process_Primitives`. (C) 1003.5-1999

executable instruction (A) An instruction that is in the instruction set for a given computer and can be executed in its current form. **(B)** A word or words containing the complete machine code for a computer operation. (C) 610.10-1994

executable file A regular file acceptable as a new process image file by the equivalent of the POSIX.1 *exec* family of functions, and thus usable as one form of a utility. The standard utilities described as compilers can produce executable files, but other unspecified methods of producing executable files may also be provided. The internal format of an executable file is unspecified, but a conforming application shall not assume an executable file is a text file. (C/PA) 9945-2-1993

executable source statements Source statements that direct the actions of the computer at run time. (C/SE) 1045-1992

execute (1) To carry out an instruction, process, or computer program. (C) 610.12-1990
(2) To perform the actions described in 3.9.1.1. *See also:* invoke. (C/PA) 9945-2-1993

execute features The electrical and mechanical equipment and interconnections that perform a function, associated directly or indirectly with a safety function, upon receipt of a signal from the sense and command features. The scope of the execute features extends from the sense and command features output to and including the actuated equipment-to-process coupling. *Note:* In some instances, protective actions may be performed by execute features that respond directly to the process conditions (for example, check valves, self-actuating relief valves). (PE/NP) 603-1998

execution (software) The process of carrying out an instruction or the instructions of a computer program by a computer. *See also:* instruction; computer program. (C/SE) 729-1983s

execution efficiency The degree to which a system or component performs its designated functions with minimum consumption of time. *See also:* storage efficiency; execution time. (C) 610.12-1990

execution monitor *See:* monitor.

execution phase The operations a software administration utility performs that modify the target. (C/PA) 1387.2-1995

execution time (software) The amount of elapsed time or processor time used in executing a computer program. *Note:* Processor time is usually less than elapsed time because the processor may be idle (for example, awaiting needed computer resources) or employed on other tasks during the execution of a program. *Synonyms:* run time; running time. *See also:* overhead time. (C) 610.12-1990

execution time theory (software) A theory that uses cumulative execution time as the basis for estimating software reliability. *See also:* software reliability; execution time. (C/SE) 729-1983s

execution trace A record of the sequence of instructions executed during the execution of a computer program. Often takes the form of a list of code labels encountered as the program executes. *Synonym:* code trace. *See also:* subroutine trace; variable trace; retrospective trace; symbolic trace. (C) 610.12-1990

execution unit In a pipelined machine, the portion of the computer that actually performs the operation specified by an instruction. (C) 610.10-1994w

executive *See:* supervisory program.

executive information system *See:* management information system.

executive program *See:* supervisory program.

executive routine (computers) A routine that controls the execution of other routines. *See also:* supervisory routine. (MIL/C) [2], [20], [85]

executive state *See:* supervisor state.

exercise (1) (test, measurement, and diagnostic equipment) To operate an equipment in such a manner that it performs all its intended functions to allow observation, testing, measurement and diagnosis of its operational condition. (MIL) [2]
(2) (A) One or more sessions with a common objective and accreditation. **(B)** The total process of designing, assembling, testing, conducting, evaluation, and reporting on an activity. *See also:* simulation exercise. (DIS/C) 1278.3-1996
(3) *See also:* simulation exercise. (DIS/C) 1278.1-1995

exerciser (1) Physical activity generator designed to manifest defects as errors. (C/BA) 896.3-1993w
(2) A device used to operate the EUT. (EMC) C63.16-1993

exfoliation A thick layer-like growth of corrosion product. (IA) [59]

existence constraint A kind of constraint stating that an instance of one entity cannot exist unless an instance of another related entity also exists. (C/SE) 1320.2-1998

existence dependency A kind of constraint between two related entities indicating that no instance of one can exist without being related to an instance of the other. The following association types represent existence dependencies: identifying relationships, categorization structures and mandatory nonidentifying relationships. (C/SE) 1320.2-1998

existing installation (elevators) An installation, prior to the effective date of a code: A) all work of installation was completed, or B) the plans and specifications were filed with the enforcing authority and work begun not later than three months after the approval of such plans and specifications. *See also:* elevator. (EEC/PE) [119]

exit (software) A point in a software module at which execution of the module can terminate. *Contrast:* routine entry point. *See also:* return. (C) 610.12-1990

exit routine A routine that receives control when a specified event, such as an error, occurs. (C) 610.12-1990

exogenous variable A variable whose value is determined by conditions and events external to a given model. *Synonym:* external variable. *Contrast:* endogenous variable. (C) 610.3-1989w

exothermic Characterized by or formed with the release of heat. (DEI) 1221-1993w

expand In the shell command language, when not qualified, the act of applying all the expansions described in 3.6. (C/PA) 9945-2-1993

expandability (1) (supervisory control, data acquisition, and automatic control) The capability of a system to be increased in capacity or provided with additional functions. (SWG/PE/SUB) C37.100-1992, C37.1-1994
(2) *See also:* extendability. (C) 610.12-1990

expanded sweep A sweep of the electron beam of a cathode-ray tube in which the movement of the beam is speeded up during a part of the sweep. *See also:* magnified sweep; radar. (EEC/PE) [119]

expander (data transmission) A transducer which, for a given amplitude range of input voltages, produces a larger range of output voltages. One important type of expander employs the envelope of speech signals to expand their volume range. (PE) 599-1985w

expandor (telephone switching systems) A switching entity for connecting a number of inlets to a greater number of outlets. (COM) 312-1977w

expansion (1) (modulation systems) A process in which the effective gain applied to a signal is varied as a function of the signal magnitude, the effective gain being greater for large than for small signals. (PE) [4]
(2) (oscillograph) A decrease in the deflection factor, usually as the limits of the quality area are exceeded. (Std100) [123]
(3) (data transmission) A process in which the effective gain applied to a signal is varied as a function of the signal magnitude, the effective gain being greater for large than for small signals; (in a switching stage), a switching stage in which the number of inputs is smaller than the number of outputs. (PE) 599-1985w

expansion board A circuit board that can be installed in an expansion slot in a computer; often used to increase the memory capabilities of the computing system. *Synonym:* add-on board. (C) 610.10-1994w

expansion chamber (for an oil cutout) A sealed chamber separately attachable to the vent opening to provide additional air space into which the gases developed during circuit interruption can expand and cool. (SWG/PE) C37.100-1992, C37.40-1993

expansion element *See:* module retainer.

expansion orbit (electronic device) The last part of the electron path that terminates at the target. It is outside the equilibrium orbit. *See also:* electron device. (ED) [45]

expansion slot An area within a computer that is reserved for an expansion board. (C) 610.10-1994w

expansion tank system *See:* conservator system.

expected data (test pattern language) The binary data that is expected to be read out of a memory array. It is identified by the symbol "Q." (TT/C) 660-1986w

expected failure duration The expected or long-term average duration of a single failure event. (IA/PSE) 493-1997

expected interruption duration The expected, or average, duration of a single-load interruption event. (IA/PSE) 493-1997, 399-1997

expedited traffic Traffic that requires preferential treatment as a consequence of jitter, latency, or throughput constraints, or as a consequence of management policy. (C/LM) 802.1D-1998

expendable cap (of an expendable-cap cutout) A replacement part or assembly for clamping the button head of a fuse link and closing one end of the fuseholder. It includes a pressure-responsive section that opens to relieve the pressure within the fuseholder, when a predetermined value is exceeded during circuit interruption. (SWG/PE) C37.100-1992, C37.40-1981s

expendable-cap cutout An open cutout having a fuse support designed for, and equipped with, a fuseholder having an expendable cap. (SWG/PE) C37.40-1993, C37.100-1992

experience Successful operation for a long time under actual operating conditions of machines designed with temperature rise at or near the temperature rating limit. (IA/PC) 1068-1996

experience or accepted test (insulation systems of synchronous machines) In accordance with IEEE Std 1-1969: "Experience," as used in ANSI C50.10-1977, means successful operation for a long time under actual operating conditions of machines designed with temperatures at or near the temperature limits. "Accepted test" as used in this standard means a test on a system or model system that simulates the electrical, thermal, and mechanical stresses occurring in service. (REM) [115]

expert system A computer system designed to solve a specific problem or class of problems by processing information specific to the problem domain. (Typically, information processed by an expert system corresponds to rules or procedures applied by human experts to solve similar problems.) (ATLAS) 1232-1995

explicit address *See:* absolute address.

exploder *See:* blasting unit.

explosion-proof apparatus (1) (explosionproof) (mine apparatus) Apparatus capable of withstanding explosion tests as established by the United State Bureau of Mines, namely, internal explosions of methane-air mixtures, with or without coal dust present, without ignition of surrounding explosive methane-air mixtures and without damage to the enclosure or discharge of flame. *See also:* distribution center; luminaire. (NESC/SWG/BT/PE/IA/AV/PC/APP) [34], [56], [11], [86], [82]
(2) Apparatus enclosed in a case that is capable of withstanding an explosion of a specified gas or vapor that may occur within it, and of preventing the ignition of a specified gas or vapor surrounding the enclosure by sparks, flashes, or explosion of the gas or vapor within. The apparatus operates at such an external temperature that a surrounding flammable atmosphere will not be ignited. (SWG/PE/NESC) C37.100-1992, C2-1984s

explosion-proof enclosure An enclosure designed and constructed to withstand an explosion of a specified flammable gas or vapor that may occur within it, and to prevent the ignition of flammable gas or vapor in the atmosphere surrounding the enclosure by sparks, flashes, or explosions of the specified gas or vapor that may occur within the enclosure. *Note:* Explosionproof apparatus should bear a nationally recognized independent testing laboratory approval rating of the proper class and group consonant with the spaces in which flammable volatile liquids, flammable gases, mixtures, or highly flammable substances may be present. (IA/MT) 45-1998

explosion-proof fuse A fuse, so constructed or protected, that for all current interruptions within its rating shall not be damaged nor transmit flame to the outside of the fuse. (SWG/PE) C37.40-1993, C37.100-1992

explosion-proof luminaire (illuminating engineering) A luminaire which is completely enclosed and capable of withstanding an explosion of a specific gas or vapor which may occur within it, and preventing the ignition of a specific gas or vapor surrounding the enclosure by sparks, flashes or explosion of the gas or vapor within. It must operate at such an external temperature that a surrounding flammable atmosphere will not be ignited thereby. (EEC/IE) [126]

explosion-tested equipment Equipment in which the housings for the electric parts are designed to withstand internal explosions of methane-air mixtures without causing ignition of such mixtures surrounding the housings. (EEC/PE) [119]

explosion-tested shuttle car A shuttle car equipped with explosion-tested equipment. (EEC/PE) [119]

explosives Mixtures of solids, liquids, or a combination of the two that, upon detonation, transform almost instantaneously into other products that are mostly gaseous and that occupy much greater volume than the original mixtures. This transformation generates heat, which rapidly expands the gases, causing them to exert enormous pressure. Dynamite and Primacord are explosives as manufactured. Aerex, Triex, and Quadrex are manufactured in two components and are not true explosives until mixed. Explosives are commonly used to build construction roads, blast holes for anchors, structure footings, etc. *Synonyms:* Quadrex; fertilizer; Primacord; powder; Aerex; dynamite; Triex. (T&D/PE) 524-1992r

exponent (1) (binary floating-point arithmetic) The component of a binary floating-point number that normally signifies the integer power to which two is raised in determining the value of the represented number. Occasionally the exponent is called the signed or unbiased exponent. (C/MM) 754-1985r
(2) (A) (mathematics of computing) A superscript indicating the number of times a number is to be used as a factor. **(B) (mathematics of computing)** The component of a floating-point number that normally signifies the integer power to which the radix is raised in determining the value of the represented number. *Synonyms:* floating-point coefficient; characteristic; exrad. *Contrast:* significand. (C) 1084-1986
(3) (radix-independent floating-point arithmetic) The component of a floating-point number that normally signifies the integer power to which the radix is raised in determining the value of the represented number. Occasionally, the exponent is called the signed or unbiased exponent. (C/MM) 854-1987r

exponent arithmetic and logic unit A special-purpose arithmetic and logic unit for handling exponent calculations or floating-point operands. (C) 610.10-1994w

exponent character (A) A character within a picture specification that represents the beginning of the exponent within a floating point number. *Note:* K and E are commonly used. **(B)** A character within a picture specification that represents the scaling factor for a decimal number. Specified with an integer constant, it indicates the number of decimal positions the decimal point is to be moved from its assumed position to the right (if the constant is positive) or to the left (if the constant is negative). *Note:* F is commonly used. (C) 610.5-1990

exponential minus cosine (−cosine) envelope (1) (exponential minus cosine) (of a transient recovery voltage) A voltage-versus-time curve that represents the maximum at any time of the 1 − cosine (1 minus cosine) envelope and the exponential envelope. (SWG/PE) C37.100-1992
(2) (transient recovery voltage) (exponential minus cosine) The greater at any instant of: (A) The curve traced by the multiple exponential, transient voltage across Z when a switch is closed on the circuit shown below. It reaches its crest E_1 at $t = \infty$. (B) The 1-cosine curve with its initial crest at $P.E_1$ represents the alternating current driving or ceiling voltage which is considered at its peak at the time of a current zero and remains practically constant during that portion of the transient defined by the first curve. Hence, it can be considered as a direct current source during this time. The voltage application is simulated by the closing of the switch. e represents the transient voltage across the circuit breaker pole unit. L represents the equivalent effective inductance on the source side of the circuit breaker. Z represents the equivalent surge impedance of associated transmission lines. C represents the equivalent lumped capacitance on the source side of the breaker and modifies the ex-cos envelope by what may be considered as a slight initial time delay, $T_1 \cdot R$ is the transient recovery voltage rate. Besides forming a basis of rating the above definition is also useful in discussing the changes of transient voltage caused by varying the parameters. *Note:* The ex-cos curve is the standard envelope for rating circuit breaker transient recovery voltage performance for circuit breakers rated 121 kV and above.

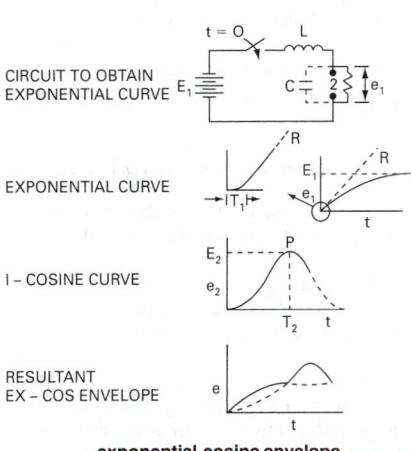

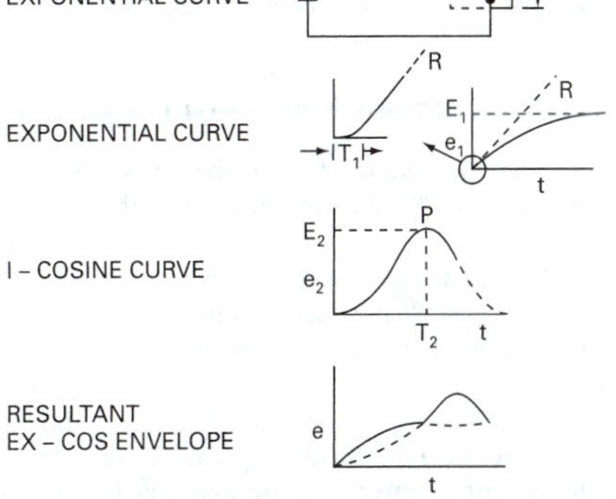

exponential-cosine envelope

exponential (ex) envelope (of a transient recovery voltage) A voltage-versus-time curve of the general exponential form e_1 E_1 [1-ex (t/T)] in which e_1 represents the transient voltage across a switching device pole unit, reaching its crest E_1 at infinite time. *Note:* In practice, this envelope curve is derived from a circuit in which a voltage E_1 charges, by means of a switch, a circuit with inductance L in series with impedance Z and capacitance C in parallel. The voltage of e_1 is measured across Z. E_1 represents the ac driving or ceiling voltage that is considered at its peak at the time of a current zero and remains practically constant during that portion of the transient defined by the first curve. Hence, it can be considered as dc source during this time. The voltage application is simulated by the closing of the switch. e_1 represents the transient voltage across the circuit-breaker pole unit. L represents the equivalent effective inductance on the source side of the circuit breaker. Z represents the equivalent surge impedance of associated transmission lines. C represents the equivalent lumped capacitance on the source side of the breaker and modifies the exponential envelope by what may be considered as a slight initial time delay, T_1. R is the transient recovery voltage rate, corresponding to the initial slope of the exponential envelope. (PE) C37.100-1992

exponential function One of the form $y = ae_{bx}$, where a and b are constants and may be real or complex. An exponential function has the property that its rate of change with respect to the independent variable is proportional to the function, or $dy/dx = by$. (Std100) 270-1966w

exponential horn A horn the cross-sectional area of which increases exponentially with axial distance. *Note:* If S is the area of a plane section normal to the axis of the horn at a distance x from the throat of the horn, S_0 is the area of a plane section normal to the axis of the horn at the throat, and m is a constant that determines the rate of taper or flare of the horn, then

$$S = S_0 e^{mx}$$

(SP) [32]

exponential lag *See:* lag.

exponentially damped sine function A generalized sine function of the form $Ae^{-bx} \sin (x + a)$ where $b > 0$. (Std100) 270-1966w

exponentially weighted moving average (EWMA) algorithm A specific metric algorithm whose behaviour emphasizes recent observation values and that can also, depending upon initialization, transparently pass through the behaviour of the observed gauge or derived gauge type attribute. (LM/C) 802.1F-1993r

exponential reference atmosphere A mathematical model of atmospheric refraction in which the refractivity is approximated by an exponential function of height:

$$N = N_S \exp(-c_e h)$$

where

N = refractivity = $(n - 1) \times 10^6$
n = atmospheric refractive index
N_s = value of N at the surface
h = height in km above the surface

In Bean and Thayer, the exponential coefficient c_e is given by

$$c_e = -\ln \left[1 - \frac{7.32}{N_S} \exp(0.005577 N_S) \right]$$

Note: The average value of N_s in the United States is 313, and the value for a "4/3 earth radius" is 301. (AES) 686-1997

exponential transmission line A tapered transmission line whose characteristic impedance varies exponentially with electrical length along the line. *See also:* waveguide; transmission line. (MTT) 146-1980w

exponent overflow A condition that occurs in floating-point arithmetic if an attempt is made to create an exponent greater than the largest positive number that can be processed or stored. *Synonym:* characteristic overflow. (C) 1084-1986w

exponent spill A condition that occurs in floating-point arithmetic when the exponent of a computed result lies outside the range that can be processed or stored. (C) 1084-1986w

exponent underflow A condition that occurs in floating-point arithmetic if an attempt is made to create a negative exponent greater in absolute value than the smallest nonzero number that can be processed or stored. *Synonym:* characteristic underflow. (C) 1084-1986w

exported catalog Refers to information organized in the exported catalog structure of the standard packaging layout. It is used for distribution catalogs as well as exporting installed software catalogs using `swlist -c` *catalog*. Within software definition files of an exported catalog, all data that can be encoded using IRV, shall be. Any such data that cannot be so encoded shall be transformed using UTF-8. (C/PA) 1387.2-1995

exposed (1) Not isolated or guarded. (T&D/NESC/SUB/PE) C2-1997, 1119-1988w, 1268-1997
(2) (wiring methods) Not concealed. (EEC/PE) [119]
(3) (communication circuits) The circuit is in such a position that in case of failure of supports or insulation, contact with another circuit may result. (NESC) [86]

exposed conductive surfaces (health care facilities) Those surfaces which are capable of carrying electric current and which are unprotected, unenclosed or unguarded, permitting personal contact. Paint, anodizing and similar coatings are not considered suitable insulation, unless they are approved for the purpose. (NESC/NEC) [86]

exposed installation (lightning) An installation in which the apparatus is subject to overvoltages of atmospheric origin. *Note:* Such installations are usually connected to overhead transmission lines either directly or through a short length of cable. *See also:* surge arrester. (PE) [8], [84]

exposed lamp *See:* bare lamp.

exposure (1) (overhead power lines) An expression of the quantity of some material, agent, etc., that is incident on an organism. (T&D/PE) 539-1990
(2) (laser maser) The product of an irradiance and its duration. (LEO) 586-1980w
(3) The subjection of a person to electric, magnetic, or electromagnetic fields or to contact currents other than those originating from physiological processes in the body and other natural phenomena. (NIR) C95.1-1999
(4) (A) (operations) The number of operations during which a component or components within a unit are exposed to failures of response functions. *Note:* For example, the number of commands to open the breaker is the exposure parameter in the case of the circuit breaker failure mode failure to open on command. **(B) (time)** The aggregate time during which a component or components that make up a unit are exposed to failures of continuously required functions. *Note:* Exposure time may include only service time, or it may also include outage time, depending on the type of component or unit and mode of failure. Time is the major measure of exposure for most failure modes (open circuit, short circuit, etc.) of lines and transformers and the switching equipment failure modes "opening without command" and "closing without command." (PE/PSE) 859-1987

exposure fire (Class 1E equipment and circuits) Fire initiated by other than electrical means or supported by fuel other than cable insulation. (PE/NP) 384-1981s

exposure meter A device for measuring the amount of a quantity, to which the device has been exposed, over a period of time. (T&D/PE) 539-1990

exposure time (1) The time during which a component is performing its intended function and is subject to failure. (IA/PSE) 493-1997
(2) The time during which a component is performing its intended function and is subject to failure. Usually expressed in years. (IA) 399-1997

exposure-to-dose-equivalent conversion factor The numerical quantity that relates the exposure (or air kerma) in air to the dose equivalent at a specified depth in tissue. (NI) N42.20-1995

EXPRESS A standard information modeling language being developed by ISO 10303-11:1994. (ATLAS) 1232-1995

expression (1) An ordered set of one or more characters. (C) 162-1963w, [20]
(2) (mathematics of computing) A sequence of constants, variables, and functions connected by operators to indicate a desired computation. (C) 1084-1986w

Expression-Oriented Language 3 (EOL-3) A programming language used to manipulate strings of characters. (C) 610.13-1993w

expulsion arrester (surge arresters) An arrester that includes an expulsion element. (PE/SPD) C62.1-1981s

expulsion fuse (high-voltage switchgear) A vented fuse or fuse unit in which the expulsion effect of gases produced by the arc and lining of the fuseholder, either alone or aided by a spring, extinguishes the arc. *Synonym:* expulsion fuse unit. (SWG/NESC/PE) C37.100-1992, [86], C37.40-1993

expulsion fuse unit *See:* expulsion fuse.

expulsion-type surge arrester An arrester having an arcing chamber in which the follow-current arc is confined and brought into contact with gas-evolving or other arc-extinguishing material in a manner that results in the limitation of the voltage at the line terminal and the interruption of the follow current. *Note:* The term "expulsion arrester" includes any external series-gap or current-limiting resistor if either or both are used as a part of the complete device as installed for service. (Std100) [84]

exrad *See:* exponent.

exrequisite The specification in a software object such that it shall not be installed if one or more specific software objects are installed. (C/PA) 1387.2-1995

extendability The ease with which a system or component can be modified to increase its storage or functional capacity. *Synonyms:* extensibility; expandability. *See also:* flexibility; maintainability. (C) 610.12-1990

extended acknowledgment *See:* selective acknowledgment.

extended addressing (32-bit) The address model implemented by bus standards supporting 32-bit addresses. The 32-bit extended addressing model is a subset of the 64-bit extended addressing model. (C/MM) 1212-1991s

extended addressing (64-bit) An address model implemented by bus standards supporting 64-bit and 32-bit addresses. The 32-bit extended addressing model is a subset of the 64-bit extended addressing model. (C/MM) 1212-1991s

extended assertion An assertion that is not required to be tested. (C/PA) 1328.2-1993w, 13210-1994, 2003.1-1992, 1328-1993w, 1326.2-1993w

extended back-to-back test A test of a bipolar station in which the transmission terminals of two converters are temporarily jumpered at the remote end of the transmission line. One converter is run as rectifier while the other converter is run as inverter. (PE/SUB) 1378-1997

extended binary coded decimal interchange code (EBCDIC) A binary code in which 256 letters, numbers, and special characters are represented by eight-bit numerals. (C) 1084-1986w

extended binary tree A full binary tree in which all terminal nodes contain data. (C) 610.5-1990w

extended Born approximation The extended Born approximation takes the exciting field to be the incident field after propagating through the average medium. *Synonym:* distorted Born approximation. (AP/PROP) 211-1997

Extended Capabilities Port (ECP) Mode An asynchronous, byte-wide, bidirectional channel. An interlocked handshake replaces the minimum timing requirements of Compatibility Mode. A control line is provided to distinguish between command and data transfers. A command may optionally be used to indicate single-byte data compression or channel address. (C/MM) 1284-1994

extended capability (power operations) The generating capability increment in excess of dependable capability which can be obtained under emergency operating procedures. (PE/PSE) 858-1987s

extended delta connection (power and distribution transformers) A connection similar to a delta, but with a winding extension at each corner of the delta, each of which is 120 degrees apart in phase relationship. *Note:* The connection may be used as an autotransformer to obtain a voltage change or a phase shift, or a combination of both. (PE/TR) C57.12.80-1978r

extended devices A device that has VXIbus configuration registers and a subclass register. This category is intended to allow for definition of additional device types. (C/MM) 1155-1992

extended Huygen's-Fresnel principle An integral relationship between a scalar wave over one plane in an extended random medium and the wave over a parallel plane located a distance away. In this formulation, the random effects are explicitly represented by the change in log amplitude and phase of a spherical wave propagating between the two planes. (AP/PROP) 211-1997

extended interconnect A connection pathway consisting of two or more nets that includes one or more discrete components, where the test requirement is to verify both the integrity of the connections and the values of the components. *Contrast:* simple interconnect. *See also:* discrete component; net. (C/TT) 1149.4-1999

extended longword serial A form of word-serial communication that allows 48-bit data transfers between commanders and servants. (C/MM) 1155-1992

extended option The options that can be specified with the -x option. These options may be defined in defaults files or options files. (C/PA) 1387.2-1995

extended memory space An extended address space on a node that provides a RAM-access window for a memory-controller unit architecture. The base address and upper bound of the extended memory space are specified through writes to the node's MEMORY_BASE and MEMORY_BOUND registers. The extended memory space is not relevant to bus standards implementing 64-bit fixed addressing. (C/MM) 1212-1991s

extended packet A version of a command *packet* that is generated by a controller to initiate a directed transaction with a specified slave. (C/MM) 1596.4-1996

extended planned derating (electric generating unit reliability, availability, and productivity) The planned derating that is the extension of the basic planned derating beyond its predetermined duration. (PE/PSE) 762-1987w

extended planned outage (electric generating unit reliability, availability, and productivity) The planned outage state that is the extension of the basic planned outage beyond its predetermined duration. *Note:* Extended planned outage applies only when planned work exceed predetermined duration. The extension, due to a condition discovered during the planned outage, is to be classified as Class 1 unplanned outage (see Class 1 unplanned outage [immediate]). Start-up failure would result in Class 0 unplanned outage (see Class 0 unplanned outage [starting failure]). (PE/PSE) 762-1987w

extended precision *See:* multiple precision.

extended rate set (ERS) The set of data transfer rates supported by a station (if any) beyond the extended service set (ESS) basic rate set. This set may include data transfer rates that will be defined in future physical layer (PHY) standards. (C/LM) 8802-11-1999

extended regular expression A pattern (sequence of characters or symbols) constructed according to the rules defined in IEEE Std 1003.2-1992. (C/PA) 9945-2-1993

extended return to bias (magnetic tape pulse recorders for electricity meters) A method whereby a recording head current, which results in a magnetic field polarity opposite that of the bias magnet, is applied to the magnetic tape for a portion of the interval in order to record a pulse. (ELM) C12.14-1982r

extended round trip envelope delay The round-trip envelope delay of a network as measured by a small shift in envelope modulation frequency to extend the range of measurement beyond the normal range limit of 1/fmodul. (COM/TA) 743-1995

extended security controls **(1)** A concept of the underlying system, as follows. The access control mechanisms have been defined to allow implementation-defined extended security controls. These permit an implementation to provide security mechanisms to implement different security policies than those described in POSIX.1. These mechanisms shall not alter or override the defined semantics of any of the functions in POSIX.1. *See also:* appropriate privileges; file access permissions. (C/PA) 9945-2-1993
(2) The access control and privilege mechanisms have been defined to allow implementation-defined extended security controls. These permit an implementation to provide security mechanisms to implement different security policies than described in this part of ISO/IEC 9945. These mechanisms shall not alter or override the defined semantics of any of the functions in this part of ISO/IEC 9945. *See also:* file access permissions; appropriate privileges. (C/PA) 9945-1-1996

extended segment A multiplicity of crate segments accessed by the same group address. Unlike operations on segments linked by segment interconnects, independent operations on each of the segments that are part of an extended segment never proceed concurrently. Depending on the method of implementation, some restrictions may exist as to the placement of masters. Depending on the disposition of modules on the extended segment, some broadcast operations may not be useable or may require special interpretation. (NID) 960-1993

extended service area (ESA) **(1) (telephone switching systems)** That part of the local service area that is outside of the boundaries of the exchange area of the calling customer. (COM) 312-1977w
(2) The conceptual area within which members of an extended service set (ESS) may communicate. An ESA is larger than or equal to a basic service area (BSA) and may involve several basic service sets (BSSs) in overlapping, disjointed, or both configurations. (C/LM) 8802-11-1999

extended service set (ESS) A set of one or more interconnected basic service sets (BSSs) and integrated local area networks (LANs) that appears as a single BSS to the logical link control layer at any station associated with one of those BSSs. (C/LM) 8802-11-1999

extended source (laser maser) An extended source of radiation can be resolved by the eye into a geometrical image, in contrast to a point source of radiation, which cannot be resolved into a geometrical image. (LEO) 586-1980w

Extended Spanning Tree Protocol The protocol specified in sections 9 and 10 of IEEE Std 802.1G-1995, for optional use among the Remote Bridges of a Group in determining how the Virtual Ports attaching them to the Group are to participate in the active topology. (C/LM) 802.1G-1996

extended superframe format A structure consisting of 24 DS1 frames. The frame overhead bit positions are shared between an extended superframe frame alignment signal, a cyclic redundancy check (CRC), and a data link. (COM/TA) 1007-1991r

extended test A test or collection of tests that shall perform bus transactions and use an external memory buffer. An extended test is invoked by writing to the TEST_START register. (C/MM) 1212-1991s

extended-time rating (grounding device) A rated time in which the period of time is greater than the time required for the temperature rise to become constant but is limited to a specified average number of days operation per year. (PE/SPD) 32-1972r

extended units space An extended address space on a node that provides an access window for unit architectures. The base address and upper bound of the extended units space are specified through writes to the node's UNITS_BASE and UNITS_BOUND registers. The extended units space is not relevant to bus standards implementing 64-bit fixed addressing. (C/MM) 1212-1991s

Extendible Compiler (ETC) An extensible language whose extended language is similar to PL/1. Contains provisions for a programmer to code machine-dependent statements to maximize efficiency. (C) 610.13-1993w

Extendible Computer System Simulator II (ECSS II) An extension of SIMSCRIPT providing statements and data structures for simulating computer hardware configurations, software, and work load. (C) 610.13-1993w

extensibility *See:* extendability.

extensible language A computer language that can be altered or can alter itself to provide a programmer with additional user-specified functions or capabilities. Examples include Ada, ALGOL, FORTH, and LOGO, because each can be used in a building block fashion to construct increasingly complex functions. *See also:* MP; EL1; MADCAP; ETC. (C) 610.13-1993w

extension **(1) (Pascal computer programming language)** A modification to Section 6 of IEEE 770X3.97-1983 that does not invalidate any program complying with this standard, as defined by Section 5.2, except by prohibiting the use of one or more particular spellings of identifiers. (Std100) 812-1984w
(2) A dialect of a particular language that varies from its referenced standard language such that the extension language has all the capabilities of the referenced language plus some additional capabilities. For example, ALPHA is an extension of PL/1. *Contrast:* subset. (C) 610.13-1993w

extensional set The set containing the currently existing instances of a class. The instances in the extensional set correspond to the database and data modeling notion of *instance*. *Synonym:* current extent. (C/SE) 1320.2-1998

extension bit A bit decoded from the received carrier stream that does not map into the data space but nonetheless denotes the presence of carrier for the purposes of CSMA/CD. (C/LM) 802.3-1998

extension bits One to four bits that are appended to the last octet of a MAC frame to complete the last data quintet.local area networks. (C) 8802-12-1998

extension cord An assembly of a flexible cord with an attachment plug on one end and a cord connector on the other. (EEC/PE) [119]

extension station A telephone station associated with a main station through connection to the same subscriber line and having the same call number designation as the associated main station. *See also:* telephone station. (EEC/PE) [119]

extent **(1) (scheme programming language)** A period of time, usually referring to the lifetime of an object. Once created, an object with unlimited extent exists forever. (C/MM) 1178-1990r
(2) (data management) A continuous area of storage on a direct access data medium, occupied by or reserved for a particular file. *See also:* secondary space allocation; primary space allocation. (C) 610.5-1990w

external Not associated with the equipment design. (PEL) 1515-2000

external address (subroutines for CAMAC) The symbol *ext* represents an integer that is used an identifier on an external CAMAC address. The address may represent a register that can be read or written, a complete CAMAC address that can be accessed by control or test functions, or a crate address. The value of *ext* is explicitly defined to be an integer. Normally, it can be expected to be an encoded version of the address components, in which the coding has been selected for the most efficient execution of CAMAC actions on the interface to which the implementation applies. Other possibilities are allowed, however. For example, *ext* may be an index or a point into a data structure in which the actual CAMAC address components are stored. (NPS) 758-1979r

external addresses (subroutines for CAMAC) The symbol *extb* represents an array of integers containing external CAMAC addresses. The array has two elements: (1) The starting address for an Address Scan multiple action; (2) The final address that can be permitted to participate in the Address Scan sequence. Each element has the same form and information content as the parameter *ext*. *See also:* external address. (NPS) 758-1979r

external audit (nuclear power quality assurance) An audit of those portions of another organization's quality assurance program not under the direct control or within the organizational structure of the auditing organization. (PE/NP) [124]

external capacitor fuse A fuse external to, and in series with, a capacitor unit or group of units. (SWG/PE) C37.40b-1996

external chaining *See:* separate chaining.

external connector (aerial lug) A connector that joins the external conductor to the current-carrying parts of a cable termination. (PE/IC) 48-1996

external data file (A) A data file that is sorted on an external storage medium such as a magnetic tape. **(B)** A data file that is stored apart from the system using the data. (C) 610.5-1990

external data model A data model depicting entities within a specific application or type of application in an organization. *Contrast:* internal model. (C) 610.5-1990w

external data submodel *See:* external schema.

external dc short circuit (thyristor power converter) A short circuit on the dc side outside the converter. *Note:* External short circuits may require different protecting means, depending on the character of the short circuit. Complete dc short circuit occurs when the short-circuit impedance is negligible compared to internal impedance of the converter. Limited dc short circuit occurs when the short-circuit impedance is large enough to limit the fault current. Feeder dc short circuit is a short circuit in a feeder with a separate protective device with much lower rating than the feeding converter (multimotor drives). (IA/IPC) 444-1973w

external device (A) A unit of processing equipment in a computer system external to the central processing unit. **(B)** In a personal computer, a device that is not physically contained within the main cabinet. *Note:* Examples include external disk drives and external modems. (C) 610.10-1994

external entry search *See:* interpolation search.

external environment A set of entities external to the application platform with which services are provided. External entities include people, exchangeable media that is not mounted in the platform, communication wiring, and other platforms. (C/PA) 14252-1996

external environment interface (EEI) The interface between the application platform and the external environment across which services are provided. The EEI is defined primarily in support of systems and application interoperability. The primary services present at the EEI are

— Human/computer interface
— Information
— Communications

(C/PA) 1003.23-1998, 14252-1996

external field influence (electric instruments) The percentage change (of full-scale value) in indication caused solely by a specified external field. Such a field is produced by a standard method with a current of the same kind and frequency as that which actuates the mechanism. This influence is determined with the most unfavorable phase and position of the field in relation to the instrument. *Note:* The coil used in the standard method shall be approximately 40 inches in diameter not over 5 inches long, and carrying sufficient current to produce the required field. The current to produce a field to an accuracy of ± 1 percent in air shall be calculated without the instrument in terms of the specific dimensions and turns of the coil. In this coil, 400 ampere-turns will produce a field of approximately 5 oersteds. The instrument under test shall be placed in the center of the coil. *See also:* accuracy rating. (EEC/ERI/AII) [111], [102]

external insulation (1) (surge arresters) (power and distribution transformers) (apparatus) The external insulating surfaces and the surrounding air. *Note:* The dielectric strength of external insulation is dependent on atmospheric conditions. (SWG/PE/TR) C57.12.80-1978r, [8]
(2) The air insulation and the exposed surfaces of solid insulation of equipment, which are both subject to dielectric stresses and to the effects of atmospheric and other external conditions such as contamination, humidity, vermin, etc. (C/PE/PSIM) 1313.1-1996, 4-1995
(3) Insulation that is designed for use outside of buildings and for exposure to the weather. (SWG/PE) C37.100-1992

external label (1) A label, usually not machine-readable, attached to a data medium container; for example, a paper sticker attached to the outside of a reel of magnetic tape. *Contrast:* internal label. (C) 610.5-1990w, 610.10-1994w
(2) A marking on the exterior of a cartridge that identifies it. The external label may be human-readable, machine-readable, or both. A bar-code label is an example of an external label. (C/SS) 1244.1-2000

external line fault A fault that occurs on lines or equipment other than the transmission line that includes the series capacitor installation. (T&D/PE) 824-1994

external logic state A logic state assumed to exist outside a symbol outline, either on an input line prior to any external qualifying symbol at that input, or on an output line beyond any external qualifying symbol at that output. (GSD) 91-1984r

external loss time *See:* environmental loss time.

externally commutated converter (self-commutated converters) A converter in which the commutating voltages are supplied by the ac supply lines, the ac load, or some other ac source outside the converter. (IA/SPC/ID) 936-1987w, 995-1987w

externally commutated inverters An inverter in which the means of commutation is not included within the power inverter. *See also:* self-commutated inverters. (IA) [62]

externally operable Capable of being operated without exposing the operator to contact with live parts. (NESC/NEC) [86]

externally programmed automatic test equipment (test, measurement, and diagnostic equipment) An automatic tester using any programming technique in which the programming instructions are not read directly from within the ATE (automatic test equipment), but from a medium which is added to the equipment such as punched tape, punched cards, and magnetic tape. (MIL) [2]

externally quenched counter tube A radiation counter tube that requires the use of an external quenching circuit to inhibit reignition. (ED) [45]

externally ventilated machine (rotating machinery) A machine that is ventilated by means of a separate motor-driven blower. The blower is usually mounted on the machine enclosure but may be separately mounted on the foundation for large machines. *See also:* separately ventilated machine; open pipe-ventilated machine. (PE) [9]

externally visible Actions or values that can be seen by other Units, as opposed to those that are unit-private. Externally visible conditions are specified by this document, unit-private conditions are not. (C/MM) 1212.1-1993

external merge sort A merge sort that makes use of auxiliary storage. *Contrast:* internal merge sort. *See also:* balanced merge sort; tape merge sort; unbalanced merge sort; direct-access merge sort; multiway merge sort. (C) 610.5-1990w

external node *See:* terminal node.

external record A record within an external view. (C) 610.5-1990w

external remanent residual voltage (Hall effect devices) That portion of the zero field residual voltage which is due to remanent magnetic flux density in the external electromagnetic core. (MAG) 296-1969w

external schema (A) A description of the format, layout, and contents of the data, within a database, to be employed by a user or application program. *Note:* The schema is written using the data definition portion of the data sublanguage. *Synonyms:* data submodel; logical view. *Contrast:* conceptual schema. *See also:* internal schema. **(B)** A logical description of an organization or enterprise. *Note:* The external schema may differ from the conceptual schema in that some entities, attributes, or relationships may be omitted, renamed, or otherwise transformed. **(C)** A description of the user's view of data. *Synonym:* external data submodel; view; subschema. (C) 610.5-1990

external series gap (expulsion-type arrester) An intentional gap between spaced electrodes, in series with the gap or gaps in the arcing chamber. (PE) [8], [84]

external sort A sort that requires the use of auxiliary storage. *Contrast:* internal sort. (C) 610.5-1990w

external storage (test, measurement, and diagnostic equipment) Information storage off-line in media such as magnetic tape, punched tape, and punched cards. (MIL) [2]

external stress An aging stress that is derived from the environment to which insulation is exposed, such as temperature, gas composition, or radiation. (DEI/RE) 775-1993w

external system interfaces The system or product interfaces to other systems, platforms, or products that influence the design solutions for consumer products and their life-cycle processes. (C/SE) 1220-1994s

external temperature influence (direct-current instrument shunts) The percentage change in the output voltage of a shunt (expressed in terms of rated output and measured with low current) when the ambient temperature is changed from 25°C to 100°C. (PE/PSIM) 316-1971w

external termination (*j*th terminal of an *n*-terminal network) The passive or active two-terminal network that is attached externally between the *j* th terminal and the reference point. *See also:* electron-tube admittances. (ED/ED) 161-1971w, [45]

external test (EXTEST) A defined instruction for the test logic defined by 1149.1-1990. (TT/C) 1149.1-1990

external timing source function The function of providing the primary point of synchronization of the DQDB subnetwork to some external timing reference, for example, that provided by a public network operator. (LM/C) 8802-6-1994

external trigger (oscilloscopes) A triggering signal introduced into the trigger circuit from an external source. (IM) 311-1970w

external variable *See:* exogenous variable.

external view The format, layout, and contents of the data in a database that a user or application program uses, as described in an external schema. *Contrast:* conceptual schema. *See also:* external record. (C) 610.5-1990w

EXTEST *See:* external test.

extinction The decrease of power flux density of an electromagnetic wave due to absorption and scattering. (AP/PROP) 211-1997

extinction coefficient (κ_e) (of a medium) The rate of decrease of power density of a wave, per unit distance, due to absorption and scattering. (AP/PROP) 211-1997

extinction cross-section (σ_e) (of a body) The ratio of power absorbed (P_a) and scattered (P_s) by the body to the power density of an incident plane wave, S_i:

$$\sigma_e = \frac{(P_a + P_s)}{S_i} = \sigma_a + \sigma_{ts}$$

where

σ_a = the absorption cross-section of the body

σ_{ts} = the total scattering cross-section of the body

(AP/PROP) 211-1997

extinction matrix ($\kappa{=}_e$) (of a medium) Vector analog of extinction coefficient for polarized waves propagating in an anisotropic medium. (AP/PROP) 211-1997

Extinction Ratio The ratio of the low optical power level to the high optical power level on an optical segment. (C/LM) 802.3-1998

extinction voltage (gas tube) The anode voltage at which the discharge ceases when the supply voltage is decreasing. (ED) [45], [84]

extinguishing voltage (drop-out voltage) (glow lamp) Dependent upon the impedance in series with the lamp, the voltage across the lamp at which an abrupt decrease in current between operating electrodes occurs and is accompanied by the disappearance of the negative glow. *Note:* In recording or specifying extinguishing voltage, the impedance must be specified. (EEC/EL) [104]

extraband spurious transmitter output (land-mobile communications transmitters) Spurious output of a transmitter outside of its specified band of transmission. (EMC) 377-1980r

extracameral (radiation protection) (radiological monitoring instrumentation) Pertaining to that portion of the instrument exclusive of the detector. (NI) N320-1979r, N323-1978r

extracameral effect (monitoring radioactivity in effluents) Apparent response of an instrument caused by radiation on any other portion of the system than the detector. (NI) N42.18-1980r

extracameral response (plutonium monitoring) An instrument response arising from the action of the radiation field on parts of the instrument other than the intended radiosensitive element. (NI) N317-1980r

extract (A) (electronic computation) To form a new word by juxtaposing selected segments of given words. **(B)** To pick, from a set of items, all items that meet a particular criterion. *See also:* select; database extract. (Std100/C) 270-1966, 610.5-1990

extraction indexing *See:* derivative indexing.

extract instruction (1) (electronic digital computation) An instruction that requests the formation of a new expression from selected parts of given expressions. (C) 162-1963w, [20], [85]
(2) An instruction that creates a new data item from parts of one or more other data items. (C) 610.10-1994w

extraction liquor The solvent used in hydrometallurgical processes for extraction of the desired constituents from ores or other products. *See also:* electrowinning. (EEC/PE) [119]

extragalactic radio waves Radio waves from beyond our galaxy. *See also:* cosmic noise. (AP/PROP) 211-1997

extra-high voltage (EHV) (1) (power operations) A term applied to voltage levels that are higher than 230 000 V. (PE/PSE) 858-1987s
(2) A term applied to voltage levels that are greater than 240 000 V. (PE/T&D) 516-1995

extra-high voltage aluminum-sheathed power cable (aluminum sheaths for power cables) Cable used in an electric system having a maximum phase-to-phase rms ac voltage above 242 000 V, the cable having an aluminum sheath as a major component in its construction. (PE/IC) 635-1989r

extra-high-voltage system (electric power) An electric system having a maximum root-mean-square alternating-current voltage above 240 000 volts to 800 000 volts. *See also:* low-voltage system; medium-voltage system; high-voltage system. (IA/PSE) 570-1975w

extranet A set of intranets connected for specific objectives. (C) 2001-1999

extraordinary load (composite insulators) The ice or wind load, or both, that may last for as long as one week, recurring as often as once per year. (T&D/PE) 987-1985w

extraordinary wave The magneto-ionic wave component in which the electric vector rotates in the opposite sense to that for the ordinary wave component. *Synonym:* X wave. *See also:* ordinary wave. (AP/PROP) 211-1997

extrapolated failure rate Extension by a defined extrapolation or interpolation of the observed or assessed failure rate for durations and/or conditions different from those applying to the observed or assessed failure rate. *Note:* The validity of the extrapolation shall be justified. (R) [29]

extrapolated mean life (non-repaired items) Extension by a defined extrapolation or interpolation of the observed or assessed mean life for stress conditions different from those applying in the observed or assessed mean life. *Note:* The validity of the extrapolation shall be justified. (R) [29]

extrapolated range for electrons (solar cells) The distance of travel in a material by electrons of a given energy, at which the flux of primary electrons extrapolates to zero. (AES/SS) 307-1969w

extrapolated reliability Extension by a defined extrapolation or interpolation of the observed or assessed reliability for durations and/or conditions different from those applying to the observed or assessed reliability. *Note:* The validity of the extrapolation shall be justified. (R) [29]

extra work (extras) Work performed by the contractor that has to be added to the contract for unforeseen conditions or changes in the scope of work. (IA/PSE) 241-1990r

extreme environment One in which ambient temperature or humidity or both fall outside a specified range of values. (NI) N42.17B-1989r

extreme load (composite insulators) The greatest load to occur on the line in a 50-year period. It may last as long as one day. (T&D/PE) 987-1985w

extremely low frequency (ELF) 3 Hz to 3 kHz. *See also:* radio spectrum. (AP/PROP) 211-1997

extreme low frequency range Frequency range from 3 Hz to 3 kHz. (T&D/PE) 539-1990

extremely high frequency (EHF) 30–300 GHz. *See also:* radio spectrum. (AP/PROP) 211-1997

extreme operating conditions (automatic null-balancing electric instrument) The range of operating conditions within which a device is designed to operate and under which operating influences are usually stated. *See also:* measurement system. (EEC/EMI) [112]

extreme value engineering (EVE) *See:* traffic engineering limits.

extreme value traffic measures Modern data collection systems that can collect and process large amounts of data have made it possible to observe an entire year's traffic data and then to engineer on only a subset of this (e.g., THDBH and HDBH). Though engineering on the peak data alone provides greater protection from overloads, peak data can be very volatile since it is based on only a small number of observations. Statistically based estimates of peak traffic hours can replace single busy-hour observations, in order to reduce volatility of peak estimates. These estimates reduce the risk of overengineering on the basis of outlying data. These methods applied to PTS are called EV methods because they provide estimates of "extreme" traffic loads, such as HDBH and THDBH, that are more accurate than observations of actual traffic. These estimates are more accurate because they are based on a large amount of the busy season's daily peak data, rather than only the highest day or 10 highest days of the year. Further advances in data collection systems have made it possible to observe the entire day's data for every day of the busy season, not just the busy hour. The newer systems can automatically pick out the single hour of data when the office is busiest. (COM/TA) 973-1990w

extrinsic joint loss (fiber optics) That portion of joint loss that is not intrinsic to the fibers (that is, loss caused by imperfect jointing). *See also:* gap loss; angular misalignment loss; intrinsic joint loss; lateral offset loss. (Std100) 812-1984w

extrinsic properties (semiconductor) The properties of a semiconductor as modified by impurities or imperfections within the crystal. *See also:* semiconductor; semiconductor device. (ED) 216-1960w

extrinsic semiconductor (1) A semiconductor whose charge-carrier concentration is dependent upon impurities. *See also:* semiconductor. (IA/ED) 59-1962w, 216-1960w, [12]
(2) (power semiconductor) A semiconductor in which the concentrations of holes and electrons are unbalanced by the introduction of impurities. (PE/EDPG) [93]

extruded A joint in which both cables are insulated with extruded dielectrics. The dielectrics may or may not be of the same material type. (PE/IC) 404-1993

eye bolt (rotating machinery) A bolt with a looped head used to engage a lifting hook. (PE) [9]

eyelet A hollow tube inserted in a printed circuit or terminal board to provide electric connection or mechanical support for component leads. *See also:* soldered joints. (EEC/AWM) [105]

eye light (illuminating engineering) Illumination on a person to provide a specular reflection from the eyes (and teeth) without adding a significant increase in light on the subject. (EEC/IE) [126]

eye-opening penalty The difference, in dB, between (a) the optical power measured at the center of the data eye, and (b) the optical power measured at a point defined by the total worst-case peak-to-peak jitter at the receiver. (C/LM) 802.3-1998

eye pattern (data transmission) An oscilloscope display of the detector voltage waveform in a data modem. This pattern gives a convenient representation of cross-over distortion that is indicated by a closing of the center of the eye. (PE) 599-1985w

F

FA *See:* functional address; oil-immersed transformer.

fabric (1) (rotating machinery) A planar structure comprising two or more sets of fiber yarns interlaced in such a way that the elements pass each other essentially at right angles and one set of elements is parallel to the fabric axis. (IA/PE/TFF) [66], [9]
(2) A device or a collection of devices which provides a general routing capability, constructed from one or more switches using links. *See also:* link; switch. (C/BA) 1355-1995

fac *See:* facsimile.

faceplate (1) The large transparent end of the envelope through which the image is viewed or projected. (ED) [45], [84]
(2) By convention, the edge of the module that is furthest from the backplane, also known as the *front panel* or *front plane*. (C/MM) 1101.2-1992

faceplate controller An electric controller consisting of a resistor and a faceplate switch in which the electric contacts are made between flat segments, arranged on a plane surface, and a contact arm. *See also:* electric controller. (IA/ICTL/IAC) [60]

faceplate rheostat A rheostat consisting of a tapped resistor and a panel with fixed contact members connected to the taps, and a lever carrying a contact rider over the fixed members for adjustment of the resistance. (IA/ICTL/IAC) [60]

face validation The process of determining whether a model or simulation based on performance seems reasonable to people knowledgeable about the system under study. The process does not review software code or logic, but rather reviews the inputs and outputs to assure that they appear realistic or representative. (DIS/C) 1278.3-1996

facilitation The brief rise of excitability above normal either after a response of after a series of subthreshold stimuli. (EMB) [47]

facilities charge The amount paid by the customer as a lump sum, or periodically, as reimbursement for facilities furnished. The charge may include operation and maintenance as well as fixed costs. (PE/PSE) 858-1993w, 346-1973w

facility, communication *See:* communication facility.

facing (rotating machinery) (planar structure) A fabric, mat, film, or other material used in intimate conjunction with a prime material and forming a relatively minor part of the composite for the purpose of protection, handling, or processing. *See also:* direct-current commutating machine; asynchronous machine. (PE) [9]

facsimile (1) (electrical communication) (data transmission) The process, or the result of the process, by which fixed graphic material including pictures or images is scanned and the information converted into signal waves which are used either locally or remotely to produce in record form a likeness (facsimile) of the subject copy. (PE) 599-1985w
(2) (A) An exact copy or likeness. **(B)** The process by which fixed graphic images are scanned, transmitted electronically, and reproduced either locally or remotely. **(C)** The result of the process in definition (B). *See also:* microfacsimile. (C) 610.2-1987
(3) A process by which textual or pictorial images are communicated, typically but not exclusively, over telephone lines. The images may be coded in raster or compressed raster format (such as CCITT group 3) or in a page description language such as Adobe™ PostScript™. Facsimile typically operates down to the physical link level and includes protocols providing control and addressing mechanisms specific to the media being used. This is distinguished from the process of communicating similarly encoded images over local or wide area networks. However, both may be considered implementations of remote printing. (C/MM) 1284.1-1997

facsimilie receiving converter (frequency-shift to amplitude-modulation converter) A device which changes the type of modulation from frequency shift to amplitude. *See also:* facsimile transmission. (COM) 168-1956w

facsimile signal (picture signal) A signal resulting from the scanning process. (COM) 168-1956w

facsimile-signal level The maximum facsimile signal power or voltage (root-mean-square or direct-current) measured at any point in a facsimile system. *Note:* It may be expressed in decibels with respect to some standard value, such as one milliwatt. *See also:* facsimile signal. (COM) 168-1956w

facsimile system An integrated assembly of the elements used for facsimile. *See also:* facsimile. (COM) 168-1956w

facsimile telegraphy A facsimile transmission system designed specifically for the transmission of photographic images. The reproduction may be in two significant states only (for example, black and white), may contain intermediate shades, or may be colored. *Synonym:* telephotography. (C) 610.2-1987

facsimile terminal A terminal used in facsimile transmission. (C) 610.2-1987, 610.10-1994w

facsimile transient A damped oscillatory transient occurring in the output of the system as a result of a sudden change in input. *See also:* facsimile transmission. (COM) 168-1956w

facsimile transmission (1) The transmission of signal waves produced by the scanning of fixed graphic material, including pictures, for reproduction in record form. (AP/COM/ANT) 145-1983s, 168-1956w
(2) The use of a telecommunication system to transmit fixed graphic images. *Synonym:* telefax. *See also:* facsimile telegraphy; facsimile. (C) 610.2-1987

facsimile transmitter The apparatus employed to translate the subject copy into signals suitable for delivery to the communication system. *See also:* facsimile. (COM) 168-1956w

factor (1) (A) Any of the operands in a multiplication operation. **(B)** A number used as a multiplier to cause a set of quantities to fall within a given range of values. *Synonym:* factor scale. (C) 1084-1986
(2) *See also:* quality factor. (C/SE) 1061-1992s

factoring (A) The process of decomposing a system into a hierarchy of modules. *See also:* modular decomposition. **(B)** The process of removing a function from a module and placing it into a module of its own. (C) 610.12-1990

factor of assurance (wire or cable insulation) The ratio of the voltage at which completed lengths are tested to that at which they are used. (T&D/PE) [10]

factor of influence (thermal classification of electric equipment and electrical insulation) A specific physical stress imposed by operation, environment, or test that influences the performance of an insulating material, insulation system, or electric equipment. (EI) 1-1986r

factor sample A set of factor values that is drawn from the metrics database and used in metrics validation. (C/SE) 1061-1992s

factor scale *See:* factor.

factor value A value of the direct metric that represents a factor. *See also:* metric value. (C/SE) 1061-1992s

factory fabricated (1) A heating cable assembled by the manufacturer, including terminations and connections. (BT/IA/AV/PC) 152-1953s, 515.1-1995
(2) A heating cable or surface heating device assembled by the manufacturer, including terminations and connections. (IA) 515-1997

factory-renewable fuse unit A fuse unit that, after circuit interruption, must be returned to the manufacturer to be restored for service. (SWG/PE) C37.40-1993, C37.100-1992

fade (A) The condition occurring during a braking cycle at low speed wherein the fundamental characteristics of the propulsion system utilized do not support the power requirement of the level of dynamic electric brake called for. Consequently, the level of dynamic electric brake actually generated decreases as a function of speed along an inherent characteristic. **(B)** In electric braking systems capable of supporting the level called for to zero speed, a deliberately created characteristic wherein the level of electric brake decreases as a function of speed to allow a smooth transition to friction brake for the purpose of the final stop. (VT) 1475-1999

fade depth The ratio, usually expressed in decibels, of a reference signal power to the signal power during a fade. (AP/PROP) 211-1997

fade duration The time interval during which a signal is below a reference value. (AP/PROP) 211-1997

fade in To increase signal strength gradually in a sound or television channel. (EEC/PE) [119]

fade out To decrease signal strength gradually in a sound or television channel. (EEC/PE) [119]

fade slope The time rate of change of the signal power during a fade, expressed in decibels per second. (AP/PROP) 211-1997

fading (1) (A) (data transmission) (Flat). That type of fading in which all frequency components of the received radio signal fluctuate in the same proportions simultaneously. *See also:* selective fading. **(B) (data transmission)** (Radio). The variation of radio field intensity caused by changes in the transmission medium, and transmission path, with time. *See also:* selective fading. (PE) 599-1985

(2) The temporal variation of received signal power caused by changes in the transmission medium or path(s). (AP/PROP) 211-1997

fading range The ratio of maximum signal to minimum signal during fading, usually expressed in dB. Often the fading range is specified over a range of percentages. For example, the 5–95% fading range is the ratio of the signal exceeded 5% of the time to that exceeded 95% of the time. (AP/PROP) 211-1997

fading rate The average number of fades occurring per unit time. (AP/PROP) 211-1997

fading spectrum The spatial or temporal frequency spectrum of a fading signal. (AP/PROP) 211-1990s

fail-off photocontrol A photocontrol that is designed so that the load remains off when the most likely failure occurs. (RL) C136.10-1996

fail-on photocontrol A photocontrol that is designed so that the load remains on when the most likely failure occurs. (RL) C136.10-1996

Fail-Operate system A system that can operate in the presence of faults. (C/BA) 896.9-1994w

failover (1) The process of reconfiguration after a fault or for planned maintenance. Failover may be manual or automatic. (C/BA) 896.2-1991w

(2) The transfer of a function or functions to a backup device. (SUB/PE) C37.1-1994

fail-safe (1) (software) Pertaining to a system or component that automatically places itself in a safe operating mode in the event of a failure; for example, a traffic light that reverts to blinking red in all directions when normal operation fails. *Contrast:* fail soft. *See also:* fault tolerance; fault secure. (C) 610.12-1990

(2) A designed property of an item that prevents its failures being critical failures. (R) [29]

(3) A design philosophy applied to safety-critical systems such that the result of a hardware failure or the effect of software error shall either prohibit the system from assuming or maintaining an unsafe state or shall cause the system to assume a state known to be safe. (VT/RT) 1475-1999, 1474.1-1999, 1483-2000

(4) A characteristic where, upon failure or malfunction of a component, subsystem, or system, the output automatically reverts to a predetermined design state of least critical consequence. Typical failsafe states are listed as follows:

Typical failsafe states

System or component	Preferred failsafe states
Cooling water valve	As is or open
Alarm system	Annunciate
Burner valve	Shutdown, limited, or as is an alarm
Propulsion speed control	As is
Feedwater valve	As is or open
Controllable pitch propeller	As is
Propulsion safety trip	As is and alarm

(IA/MT) 45-1998

fail-safe circuit A circuit in which the occurrence of a failure causes a specified set of outputs of the circuit to assume predetermined values. (C) 610.10-1994w

fail-safely The implementation of a function in a fail-safe manner. (VT/RT) 1483-2000

fail-safe sequential circuit A sequential circuit designed so that a failure in the internal logic causes the output to assume either a predetermined one or zero state. (C) 610.10-1994w

Fail-Safe system A system whose failures can only be, or are to an acceptable extent, benign failures. (C/BA) 896.9-1994w

fail soft Pertaining to a system or component that continues to provide partial operational capability in the event of certain failures; for example, a traffic light that continues to alternate between red and green if the yellow light fails. *Contrast:* fail-safe. *See also:* fault secure; fault tolerance. (C) 610.12-1990

Fail-Stop system A system whose failures can only be, or are to an acceptable extent, stopping failures. (C/BA) 896.9-1994w

failure (1) (A) The termination of the ability of an item to perform a required function. **(B)** (complete) Failure resulting from deviations in characteristic(s) beyond specified limits such as to cause complete lack of the required function. *Note:* The limits referred to in this category are special limits specified for this purpose. **(C)** (critical) Failure that is likely to cause injury to persons or significant damage to material. **(D)** (degradation) Failure that is both gradual and partial. *Note:* In time such a failure may develop into a complete failure. **(E)** (inherent weakness) Failure attributable to weakness inherent in the item when subjected to stresses within the stated capabilities of the item. **(F)** (intermittent) Failure of an item for a limited period of time, following which the item recovers its ability to perform its required function without being subjected to any external corrective action. *Note:* Such a failure is often recurrent. **(G)** (major) Failure, other than a critical failure, which is likely to reduce the ability of a more complex item to perform its required function. **(H)** (minor) Failure, other than a critical failure, which does not reduce the ability of a more complex item to perform its required function. **(I)** (nonrelevant) Failure to be excluded in interpreting test results or in calculating the value of a reliability characteristic. *Note:* The criteria for the exclusion should be stated. **(J)** (partial) Failure resulting from deviation in characteristic(s) beyond specified limits, but not such as to cause complete lack of the required function. *Note:* The limits referred to in this category are special limits specified for this purpose. **(K)** (primary) Failure of an item, not caused either directly or indirectly by the failure of another item. **(L)** (relevant) Failure to be included in interpreting test results or in calculating the value of a reliability characteristic. *Note:* The criteria for the inclusion should be stated. **(M)** (wear-out) Failure whose probability of occurrence increases with the passage of time and which occurs as a result of processes which are characteristic of the population. **(N) (software reliability)** The termination of the ability of a functional unit to perform its required function. **(O) (software reliability)** An event in which a system or system component does not perform a required function within specified limits. A failure may

be produced when a fault is encountered. **(P) (reliability data)** A subset of a fault and represents an irreversible state of a component such that it must be repaired in order for it to provide its design function. A component failure is generally defined in terms of the system in which it resides. For example, any leak might be considered a failure in a system where fission products are to be contained, and yet leaks may be considered as normal or even required states of other systems (for example, pump packing gland leakage). Failures are sometimes classified as either primary or secondary: A primary failure is the so-called "random failure" found in literature. *I* results from no external cause. A secondary failure results when the component is subject to conditions that exceed its design envelope (for example, excessive voltage, pressure, shock, vibration, temperature).
(R/SE/C/BA/PE/NP) [29], 982.2-1988, 982.1-1988, 896.9-1994, 500-1984

(2) (A) (test, measurement, and diagnostic equipment) (dependent) A failure that is caused by the failure of an associated item, distinguished from independent failure. **(B) (test, measurement, and diagnostic equipment)** (independent) A failure that occurs without being related to the failure of associated items, distinguished from dependent failure. (MIL) [2]

(3) (gradual) Failures that could be anticipated by prior examination or monitoring.

(4) (A) (major) (of a circuit breaker) Failure of a circuit breaker that causes the termination of one or more of its fundamental functions, which necessitates immediate action. *Note:* A major failure will result in an immediate change in system operation condition; e.g., the backup protective equipment being required to remove the fault, or will result in mandatory removal from service for non-scheduled maintenance (intervention required within 30 min). **(B)** (minor) (of a circuit breaker) Any failure of a part or a sub-assembly that does not cause a major failure of a circuit breaker. **(C)** Termination of the ability of an item to perform its required functions. *Note:* The occurrence of a failure does not necessarily imply the presence of a defect if the stress is beyond that originally specified. (PE) C37.10-1995

(5) (misuse) Failure attributable to the application of stresses beyond the stated capabilities of the item.

(6) (random) Any failure whose cause and/or mechanism make its time of occurrence unpredictable.

(7) (secondary) Failure of an item caused either directly or indirectly by the failure of another item.

(8) (sudden) Failure that could not be anticipated by prior examination or monitoring.

(9) (nuclear power generating station) The termination of the ability of an item to perform its required function. Failures may be unannounced and not detected until the next test (unannounced failure), or they may be announced and detected by any number of methods at the instant of occurrence (announced failure). 352-1975s

(10) (software) The inability of a system or component to perform its required functions within specified performance requirements. *Note:* The fault tolerance discipline distinguishes between a human action (a mistake), its manifestation (a hardware or software fault), the result of the fault (a failure), and the amount by which the result is incorrect (the error). *See also:* incipient failure; hard failure; soft failure; exception; failure mode; random failure; failure rate; crash. (C) 610.12-1990

(11) (raceway) (raceway systems for Class 1E circuits for nuclear power generating stations) The termination of the ability of the raceway system to perform its function. The level of damage done to the raceway system is such that either collapse is imminent or an electrical circuit is interrupted or degraded to an unacceptable level, or both. (PE/NP) 628-1987r

(12) (reliable industrial and commercial power systems planning and design) Any trouble with a power system component that causes any of the following to occur:

a) Partial or complete plant shutdown, or below-standard plant operation;
b) Unacceptable performance of user's equipment;
c) Operation of the electrical relaying or emergency operation of the plant electrical system;
d) Deenergization of any electric circuit or equipment. A failure on a public utility supply system may cause the user to have either of the following:
   - A power interruption or loss of service;
   - A deviation from normal voltage or frequency outside the normal utility profile

A failure on an in-plant component causes a forced outage of the component, that is, the component is unable to perform its intended function until it is repaired or replaced. *Synonym:* forced outage. (IA/PSE) 493-1997

(13) (safety systems equipment in nuclear power generating stations) The loss of ability of a component, equipment or system to perform a required function. (PE/NP) 627-1980r

(14) (supervisory control, data acquisition, and automatic control) An event that may limit the capability of an equipment or system to perform its function(s). (SUB/PE) C37.1-1994

(15) (nuclear power generating systems) The termination of the ability of an item to perform a required function.
(PE/NP) 338-1987r, 500-1984w, 308-1980s, 379-1994, 933-1999

(16) (outages) The inability of a component to perform its required function. (PE/PSE) 859-1987w

(17) (electric utility power systems) The termination of the ability of a transformer to perform its specified function. In the study of power transformer reliability, it is often difficult to distinguish between major and minor failures; therefore, the following failure definitions are given. (PE/TR) C57.117-1986r

(18) The inability of a system or component to perform its required functions within specified performance requirements. (C/BA) 896.3-1993w

(19) The loss of ability of a diagnostic unit, equipment, or system to perform a required function. The manifestation of a fault. Within the context of AI-ESTATE models, a manifestation is given by the outcome of a test unit. (ATLAS) 1232-1995

(20) The inability of a product to meet its operating specification. (EMC) C63.16-1993

(21) The termination of the capability of the subassembly to perform its required function. (SPD/PE) C62.38-1994r

(22) (supervisory) An event that may limit the capability of a piece of equipment or system to perform its function(s).

1) **critical.** Causes a false or undesired operation of apparatus under control.
2) **major.** Loss of control or apparatus that does not involve a false operation.
3) **minor.** Loss of data relative to power flow or equipment status.

(SWG/PE) C37.100-1992

(23) (A) (infant mortality) A characteristic pattern of failure, sometimes experienced with new equipment that may contain marginal components, wherein the number of failures per unit of time decrease rapidly as the number of operating hours increase. A burn-in period may be utilized to age (or mature) a piece of equipment to reduce the number of marginal components. **(B)** (random) The pattern of failures for equipment that has passed out of its infant mortality period and has not reached the wear-out phase of its operating lifetime. The reliability of equipment in this period may be computed by the equation

$$R = e^{-t}$$

where
R = failure rate
t = time period of interest

(C) (wear out) The pattern of failures experienced when equipment reaches its period of deterioration. Wear-out failure profiles may be approximated by a Gaussian (bell-curve) distribution centered on the nominal life of the equipment. (SWG/PE) C37.100-1992

(24) Any trouble with a power system component that causes any of the following to occur:

— Partial or complete shutdown, or below-standard plant operation
— Unacceptable performance of user's equipment
— Operation of the electrical protective relaying or emergency operation of the plant electrical system
— De-energization of any electric circuit or equipment

A failure on a public utility supply system can cause the user to have either of the following:

— A power interruption or loss of service
— A deviation from normal voltage or frequency outside the normal utility profile

A failure of an in-plant component causes a forced outage of the component, that is, the component is unable to perform its intended function until repaired or replaced. *Synonym:* forced outage. (IA) 399-1997

(25) Any deviation from specified post condition of a test case is considered a failure for that specific test case. Post conditions should require adherence to the specification, documentation, or functional base-line for the system. (C/PA) 2000.2-1999

failure analysis The logical, systematic examination of an item or its diagram(s) to identify and analyze the probability, causes, and consequences of potential and real failure. (SWG/MIL/PE) [2], C37.10-1995

failure category *See:* error category.

failure cause The circumstances during design, manufacture, or use which have led to failure. (SWG/R/PE) [29], C37.10-1995

failure commutation (thyristor converter) A failure to commutate the direct current from the conducting arm to the succeeding arm of a thyristor connection. *Note:* In inverter operation, a commutation failure results in a conduction-through. (IA/IPC) 444-1973w

failure criteria Rules for failure relevancy such as specified limits for the acceptability of an item. *See also:* reliability. (R) [29]

failure data *See:* error data.

failure, degradation *See:* degradation failure.

failure detection Examination to determine the position, evidence, and type of failure. (SWG/PE) C37.10-1995

failure distribution (supervisory control, data acquisition, and automatic control) The manner in which failures occur as a function of time; generally expressed in the form of a curve with the abscissa being time. (SWG/PE/SUB) C37.100-1992, C37.1-1994

failure management *See:* fault management.

failure mechanism (reliability data for pumps and drivers, valve actuators, and valves) (reliability data) The physical, chemical, or other process that results in failure. *Note:* The circumstance that induces or activates the process is termed the root cause of the failure. (PE/NP) 500-1984w

failure mode (1) (reliability data for pumps and drivers, valve actuators, and valves) (reliability data) The effect by which a failure is observed to occur. (PE/IA/NP/PSE) 500-1984w, 1100-1999

(2) (gas-insulated substations) A process of failure of equipment that causes a loss of its proper function. (SWG/PE/SUB) C37.122-1983s, C37.100-1992

(3) (software) The physical or functional manifestation of a failure. For example, a system in failure mode may be characterized by slow operation, incorrect outputs, or complete termination of execution. (C) 610.12-1990

(4) The manner in which failure occurs; generally categorized as electrical, mechanical, thermal, and contamination. (SWG/PE) C37.10-1995

failure modes and effects analysis (FMEA) (1) (Class 1E battery chargers and inverters) (FMEA) The identification of significant failures, irrespective of cause, and their consequences. This includes electrical and mechanical failures that could conceivably occur under specified service conditions and their effect, if any, on adjoining circuitry or mechanical interfaces displayed in a table, chart, fault tree or other format. (PE/NP) 650-1979s

(2) A systematic procedure for identifying the modes of failure and for evaluating their consequences. (PE/NP) 933-1999

failure mode types (A) (reliability data for pumps and drivers, valve actuators, and valves) (reliability data) Catastrophic. A failure mode which is both sudden and complete. *Note:* This failure causes cessation of one or more fundamental functions. This refers to system related failure modes. See Appendix A of IEEE Std 500-1984 P&V. **(B) (reliability data for pumps and drivers, valve actuators, and valves) (reliability data)** Degraded. A failure which is gradual, partial, or both. *Note:* Such a failure does not cease all function but compromises a function. The function may be compromised by any combination of reduced, increased, or erratic outputs. In time, such a failure may develop into catastrophic failure. **(C) (reliability data for pumps and drivers, valve actuators, and valves) (reliability data)** Incipient. An imperfection in the state or condition of an item or equipment so that a degraded or catastrophic failure can be expected to result if corrective action is not taken. (PE/NP) 500-1984

failure of continuously required function The inability of a component to perform a function that is continuously required. *Note:* Continuously required functions include carrying current, providing electrical isolation, and abstaining from tripping in the absence of a signal. Examples of inability to perform continuously required functions are: component short circuit, component open circuit, switching equipment opening without proper command, and switching equipment closing without proper command. (PE/PSE) 859-1987w

failure of response function The inability of a component to perform a function that is required as a response to system conditions or to a manually or automatically initiated command. *Note:* Response functions include responding to fault conditions (protective systems), to command (circuit breakers), and to manual operation (disconnect switches). Inabilities to perform a response function do not cause an immediate interruption of power flow, as they can be disclosed by subsequent inspection or by failure to respond to conditions as intended. This type of failure has been referred to as dormant failure, latent failure, and unrevealed failure. Examples are: switching equipment failing to open on command, switching equipment failing to close on command, and protection system tripped incorrectly (overreach during fault). (PE/PSE) 859-1987w

failure of thyristor-level A thyristor-level is deemed to have failed if it becomes short circuited or in any other way has degraded to the extent to make it functionally inoperative. (SUB/PE) 857-1996

failure rate (1) (general) (any point in the life of an item) The incremental change in the number of failures per associated incremental changes in time. (R)

(2) (reliability data for pumps and drivers, valve actuators, and valves) (reliability data) The expected number of failures of a given type, per item, in a given time interval (for example, valve failures per million valve hours). *Note:* For cyclic items or equipment insert "in a given number of operating cycles." (PE/NP) 500-1984w

(3) (nuclear power generating station) The expected number of failures of a given type, per item, in a given time interval (for example, capacitor short-circuit failures per million capacitor hours). (PE/NP) 380-1975w, 352-1975s, 933-1999

(4) (software) The ratio of the number of failures of a given category to a given unit of measure; for example, failures per unit of time, failures per number of transactions, failures per number of computer runs. (C) 610.12-1990

(5) (outages) The number of failures of a continuously required function per unit of time exposed to such failures = number of failures of a particular type/exposure time. *Notes:* 1. Rates for different failure modes can be calculated. The exposure time for each failure mode may be different. 2. Failure rates can be computed for a specific component, a class of components or units, or per unit of length in the case of lines, common structure, or common right-of-way exposure. (PE/PSE) 859-1987w

(6) (electric utility power systems) The ratio of the number of "failures with forced outages" of a given population over a given period of time, to the number of accumulated service years for all transformers in that population over the same period of time. *Note:* The failure rate defined here is composed of "failures with forced outages." This is used for statistical analysis in system mathematical studies. Other reports may be made using "failure with scheduled outages" and "defects." Tabulation of scheduled outages and defects needs to be aggressively pursued from the standpoint of reliability improvement. It should be recognized that reliability improvement is different from reliability measurement. Quantitative, mathematically correct, reliability measurement can only be accomplished by counting "failures with a forced outage." Reliability improvement, on the other hand, can be accomplished through tabulating and reporting a wide variety of problems. (PE/TR) C57.117-1986r

(7) The mean number of failures of a component per unit exposure time. Usually exposure time is expressed in years and failure rate is given in failures per year. (IA/PSE) 493-1997

(8) (forced outage rate) The mean number of failures of a component per unit of exposure time. Usually, expressed in failures per year. (IA) 399-1997

failure-rate acceleration factor The ratio of the accelerated testing failure rate to the failure rate under stated reference test conditions and time period. *See also:* reliability. (R) [29]

failure rate, assessed *See:* assessed failure rate.

failure rate, extrapolated *See:* extrapolated failure rate.

failure rate level For the assessed failure rate, a value chosen from a specific series of failure rate values and used for stating requirements or for the presentation of test results. *Note:* In a requirement, it denotes the highest permissible assessed failure rate. (R) [29]

failure rate, observed *See:* observed failure rate.

failure rate, predicted *See:* predicted failure rate.

failure ratio *See:* failure rate.

failure recovery (software) The return of a system to a reliable operating state after failure. *See also:* system; failure. (C/SE) 729-1983s

failures (A) (supervisory control, data acquisition, and automatic control) Infant mortality. A characteristic pattern of failure, sometimes experienced with new equipment which may contain marginal components, wherein the number of failures decreases rapidly as the number of operating hours increases. A burn-in period period may be utilized to age (mature) an equipment to reduce the number of marginal components. **(B) (supervisory control, data acquisition, and automatic control)** Random. The pattern of failures for equipment that has passed out of its infant mortality period and has not reached the wear-out phase of its operating lifetime. The reliability of an equipment in this period may be computed by the equation $R = e^{-\lambda t}$, where λ = failure rate, and t = time period of interest. **(C) (supervisory control, data acquisition, and automatic control)** Wear out. The pattern of failures experienced when equipment reaches its period of deterioration. Wear-out failure profiles may be approximated by a Gaussian (bell curve) distribution centered on the nominal life of the equipment. (SUB/PE) C37.1-1987

failures in time (FIT) A statistical measure of failure rate corresponding to one failure in 10^9 hours of device operation. (ED) 1005-1998

failure to trip In the performance of a relay or relay system, the lack of tripping that should have occurred considering the objectives of the relay system design. *See also:* relay. (SWG/PE/PSR) C37.100-1992, [103], [6], C37.90-1978s

failure with forced outage Failure of a transformer that requires its immediate removal from service. This is accomplished either automatically or as soon as switching operations can be performed. (PE/TR) C57.117-1986r

failure with scheduled outage Failure for which a transformer must be deliberately taken out of service at a selected time. (PE/TR) C57.117-1986r

fairlead (aircraft) A tube through which a trailing wire antenna is fed from an aircraft, with particular care in the design as to voltage breakdown and corona characteristics. *Note:* An antenna reel and counter are frequently a part of the assembly. (EEC/PE) [119]

fairness (1) (multiprocessor architecture) A bus request protocol in which a module refrains from acquiring the bus in order to allow other modules of the fairness class to use the bus. (C/MM) 896.1-1987s

(2) (NuBus) A property of some arbitration techniques that ensures all modules will get access to the bus on approximately the same terms. This prevents modules from being "starved." (C/MM) 1196-1987w

®NuBus is a registered trademark of Texas Instruments, Inc.

fairness interval (1) A group of back-to-back transfers during which each competing source using the fairness protocol gets a single transfer. The delimiters of the fairness interval are arbitration reset gaps. (C/MM) 1394-1995

(2) A time period delimited by arbitration reset gaps. Within a fairness interval, the total number of asynchronous packets that may be transmitted by a node is limited. Each node's limit may be explicitly established by the bus manager or it may be implicit. (C/MM) 1394a-2000

fair queuing In networking, a method for controlling congestion in a network node by restricting other nodes to an equal share of the node's bandwidth. *See also:* source quench. (C) 610.7-1995

fair weather The weather condition when the precipitation intensity is zero and the transmission line conductors are dry. *Note:* This should not be confused with the general connotation of fair weather as descriptive of pleasant weather conditions. Common usage is subject to misinterpretation, for it is a purely subjective description. Technically, when this term is used in weather forecasts, it is meant to imply no precipitation; less than 40% sky cover of low clouds; and no other extreme conditions of cloudiness, visibility, or wind. (T&D/PE) 539-1990

fair weather distribution A frequency or probability distribution of corona-effect data collected under fair weather conditions. (T&D/PE) 539-1990

fall arrester A device, such as a rope grab, which travels on a lifeline and will automatically engage the lifeline and lock so as to arrest an accidental fall of a worker. (T&D/PE) 1307-1996

fall arrest system The assemblage of equipment, such as a line-worker's body belt, aerial belt, or full body harness in conjunction with a connecting means, with or without an energy absorbing device, and an anchorage to limit the forces a worker can experience during a fall. *Note:* A fall arrest system is designed to prevent a worker, in the process of a fall, from falling more than the designed fall limit. After January 1, 1998 a line-worker's body belt is prohibited from use as part of a fall arrest system. (NESC/T&D/PE) C2-1997, 1307-1996

fall clearance distance The total fall distance plus the distances between the location of a line-worker's body belt or full body harness attachment point under load and the nearest possible point of contact, plus the dynamic elongation. (T&D/PE) 1307-1996

falling edge (1) A transition from a high to a low logic level. In positive logic, a change from logic 1 to logic 0. (C/TT) 1149.1-1990
(2) The transition from a logic one to logic zero. (TT/C) 1149.5-1995

fall prevention system (1) A system, which may include a positioning system, intended to prevent a worker from falling from an elevation. Such systems include positioning device systems, guardrail, barriers, and restraint systems. Fall prevention systems are used in an attempt to prevent workers from falling from an elevation. (T&D/PE) 1307-1996
(2) A system, which may include a positioning device system, intended to prevent a worker from falling from an elevation. (NESC) C2-1997

fall protection program A program intended to protect workers from injury due to falls from elevations. (NESC/T&D/PE) C2-1997, 1307-1996

fall protection system (hardware) Consists of either a fall prevention system or a fall arrest system. (NESC/T&D/PE) C2-1997, 1307-1996

fall time (1) (electric indicating instruments) The time, in seconds, for the pointer to reach 0.1 (plus or minus a specified tolerance) of the end scale from a steady end-scale deflection when the instrument is short-circuited. *See also:* moving element. (EEC/AII) [102]
(2) (pulse transformers) (last transition duration) The time interval of the pulse trailing edge between the instants at which the instantaneous value first reaches specified upper and lower limits of 90% and 10% of AT. (PEL/ET) 390-1987r
(3) (of a pulse) The interval between the 90% and 10% points (unless otherwise specified) on the last transition. (NPS) 300-1988r
(4) (A) The time required for a voltage or current pulse to decrease from 90% to 10% of its maximum value. *Synonym:* decay time. *Contrast:* rise time. **(B)** In digital logic, the time required to transition from a high state to a low state. (C) 610.10-1994
(5) (of a pulse) The time interval of the trailing edge of a pulse between stated limits. *See also:* pulse. (IM/HFIM) [40]

fall time t f (of a pulse) The interval on the last transition between the 90% and 10% points (unless other levels are specified) with respect to peak height. *See also:* transition. (NPS) 325-1996

false add To form a partial sum, that is, to add without recognizing a carry. (C) 610.10-1994w

false alarm (1) (test, measurement, and diagnostic equipment) An indicated fault where no fault exists. (MIL/ATLAS) [2], 1232-1995
(2) An indicated alarm where no danger, safeguards threat, or equipment failure condition exists. (PE/NP) 692-1997
(3) An erroneous radar target detection decision caused by noise or other interfering signals exceeding the detection threshold. (AES) 686-1997
(4) *See also:* false identification. (C) 610.4-1990w

false-alarm number The number of possible independent detection decisions during the false-alarm time. *Note:* When there is no pulse integration, it is equal to the reciprocal of the false-alarm probability. (AES) 686-1997

false-alarm probability The probability that noise or other interfering signals will erroneously cause a target detection decision. *See also:* detection probability. (AES) 686-1997

false-alarm time The average time between false alarms; that is, the average time between crossings of the target decision threshold by signals not representing targets. *Note:* In the early work of Marcum, false-alarm time is defined as the time in which the probability of one or more false alarms is one-half. Marcum's definition is no longer commonly used. (AES) 686-1997

false course (navigation systems normally providing one or more course lines) (navigation aid terms) A spurious additional course line indication due to undesired reflections or to a maladjustment of equipment. (AES/RS/GCS) 686-1982s, [42], 172-1983w

false identification In pattern classification, the assignment of a pattern to a pattern class other than its true pattern class. *Synonym:* false alarm. *Contrast:* misidentification. (C) 610.4-1990w

false operation probability False operation probability = number of unintended operations/exposure operations for which component should not respond. (PE/PSE) 859-1987w

false-proceed operation The creation or continuance of a condition of the vehicle apparatus in an automatic train control or cab signal installation that is less restrictive than is required by the condition of the track of the controlling section, when the vehicle is at a point where the apparatus, is or should be, in operative relation with the controlling track elements. *See also:* automatic train control. (EEC/PE) [119]

false-restrictive operation The creation or continuance of a condition of the automatic train control or cab signal vehicle apparatus that is more restrictive than is required by the condition of the track of the controlling section when the vehicle apparatus is in operative relation with the controlling track elements, or that is caused by failure or derangement of some part of the apparatus. *See also:* automatic train control. (EEC/PE) [119]

false start Occurs if the customer or distant office abandons without dialing any digits and before timeout. (COM/TA) 973-1990w

false tripping In the performance of a relay or relay system, the tripping that should not have occurred considering the objectives of the relay system design. *See also:* relay. (SWG/PE/PSR) C37.100-1992, C37.90-1978s, [6], [103]

FAMOS® Floating gate avalanche metal-oxide semiconductor (MOS) transistor used in erasable programmable read-only memory (EPROM). (ED) 1005-1998

fan (rotating machinery) (blower) The part that provides an air stream for ventilating the machine. (PE) [9]

fan-beam antenna An antenna producing a major lobe whose transverse cross section has a large ratio of major to minor dimensions. (AP/ANT) 145-1993

fan cover (rotating machinery) An enclosure for the fan that directs the flow of air. *See also:* fan. (PE) [9]

fan duty resistor A resistor for use in the armature or rotor circuit of a motor in which the current is approximately proportional to the speed of the motor. (IA/MT) 45-1998

fan fold paper *See:* continuous form.

fan housing (rotating machinery) The structure surrounding a fan and which forms the outer boundary of the coolant gas passing through the fan. (PE) [9]

fan-in network (power-system communication) A logic network whose output is a binary code in parallel form of n bits and having up to 2*n* inputs with each input producing one of the output codes. *See also:* digital. (PE) 599-1985w

fan marker (1) (electronic navigation) A marker having a vertically directed fan beam intersecting an airway to provide a position fix. *See also:* radio navigation. (AES) [42]
(2) (navigation aid terms) A vhf (very high frequency) radio facility having a vertically-directed fan beam intersecting an airway to provide a fix. (AES/GCS) 172-1983w

fan-marker beacon (navigation aid terms) A beacon that transmits vertical beam-horizontal cross section in the shape of a double convex lens. (AES/GCS) 172-1983w

fanout The pin count of a net (the number of pins connected to the net), minus one. This includes all input, output, and bidirectional pins on the net with the sole exception of one pin (assumed to be related to the particular timing arc currently of interest). Although less fundamental than pin count, fanout is frequently used in the definition of wireload models. (C/DA) 1481-1999

fan-out box A device that provides the capability to connect multiple devices to a single transceiver. *Synonyms:* multiport; multi-tap. *See also:* tap. (C) 610.7-1995

fan-out network (power-system communication) A logic network taking n input bits in parallel and producing a unique logic output on the one and only one of up to 2*n* outputs that corresponds to the input code. *See also:* digital. (PE) 599-1985w

fan shroud (rotating machinery) A structure, either stationary or rotating, that restricts leakage of gas past the blades of a fan. (PE) [9]

farad (metric practice) The capacitance of a capacitor between the plates of which there appears a difference of potential of one volt when it is charged by a quantity of electricity equal to one coulomb. (QUL) 268-1982s

faraday The number of coulombs (96 485) required for an electrochemical reaction involving one chemical equivalent. *See also:* electrochemistry. (EEC/PE) [119]

Faraday cage A conducting enclosure that is used to measure the net space charge per unit volume. (T&D/PE) 539-1990, 1227-1990r

Faraday cell (laser gyro) A biasing device consisting of an optical material with a Verdet constant, such as quartz, that is placed between two quarter-wave plates and surrounded by a magnetic field in such a fashion that a differential phase change is produced for oppositely directed plane polarized waves. (AES/GYAC) 528-1994

Faraday dark space (gas tube) The relatively nonluminous region in a glow-discharge cold-cathode tube between the negative flow and the positive column. *See also:* gas tube. (ED) [45]

Faraday effect *See:* Faraday rotation; magnetic rotation.

Faraday rotation (1) (communication satellite) The rotation of the plane of polarization of an electromagnetic wave when traveling through a magnetic field. In space communications this effect occurs when signals transverse the ionosphere. (COM) [25]

(2) (nonlinear, active, and nonreciprocal waveguide components) (nonreciprocal wave rotation) A nonreciprocal phenomenon in which the plane of polarization of a linearly polarized electromagnetic plane is rotated clockwise for one direction of propagation, and counterclockwise for the other direction (viewed from the source in each direction), when passing through a gyromagnetic material having a magnetostatic field component along the direction of propagation. (MTT) 457-1982w

(3) The rotation of the polarization ellipse of an electromagnetic wave as it propagates in a gyrotropic medium such as a plasma in the presence of a finite magnetic field, in a ferrite, or in some dielectric crystals. *Note:* A gyrotropic material is one in which the permittivity tensor, ε=, or the permeability tensor, π=, is antisymmetric such that $\varepsilon_{ij} = -\varepsilon_{ji}$ or $\mu_{ij} = -\mu_{ji}$, respectively. (AP/PROP) 211-1997

Faraday rotator (nonlinear, active, and nonreciprocal waveguide components) (nonreciprocal wave rotator) A nonreciprocal device providing Faraday rotation, usually in waveguide of circular or square cross section. (MTT) 457-1982w

Faraday's law (electromagnetic induction; circuit) The electromotive force induced is proportional to the time rate of change of magnetic flux linked with the circuit. (Std100) 270-1966w

faradic current (electrotherapy) An asymmetrical alternating current obtained from or similar to that obtained from the secondary winding of an induction coil operated by repeatedly interrupting a direct current in the primary. *See also:* electrotherapy. (EMB) [47]

faradization (faradism) (electrotherapy) The use of a faradic current to stimulate muscles and nerves. *See also:* electrotherapy; faradic current. (EMB) [47]

far-end crosstalk Crosstalk that is propagated in a disturbed channel in the same direction as the direction of propagation of the current in the disturbing channel. The terminal of the disturbed channel at which the far-end crosstalk is present and the energized terminals of the disturbing channel are ordinarily remote from each other. *See also:* coupling. (EEC/PE) [119]

far-field diffraction pattern (fiber optics) The diffraction pattern of a source (such as a light emitting diode (LED), injection laser diode (ILD), or the output end of an optical waveguide) observed at an infinite distance from the source. Theoretically, a far-field pattern exists at distances that are large compared with $(s^2)/\lambda$, where s is a characteristic dimension of the source and λ is the wavelength. Example: If the source is a uniformly illuminated circle, then s is the radius of the circle. *Note:* The far-field diffraction pattern of a source may be observed at infinity or (except for scale) in the focal plane of a well-corrected lens. The far-field pattern of a diffracting screen illuminated by a point source may be observed in the image plane of the source. *Synonym:* Fraunhofer diffraction pattern. *See also:* diffraction limited; diffraction. (Std100) 812-1984w

far-field radiation pattern Any radiation pattern obtained in the far-field of an antenna. *Note:* Far-field patterns are usually taken over paths on a spherical surface. *See also:* radiation sphere; radiation pattern cut. (AP/PE/T&D/ANT) 145-1993, 1260-1996

far-field region (1) (fiber optics) The region, far from a source, where the diffraction pattern is substantially the same as that at infinity. *See also:* far-field diffraction pattern. (Std100) 812-1984w

(2) (land-mobile communications transmitters) The region of the field of an antenna where the angular field distribution is essentially independent of the distance from the antenna. *Notes:* 1. If the antenna has a maximum overall dimension (D) that is large compared to the wavelength (λ), the far field region is commonly taken to exist at distances greater than $2D^2/\lambda$ from the antenna. 2. For an antenna focused at infinity, the far field region is sometimes referred to as the Fraunhofer region on the basis of analogy to optical terminology. (EMC) 377-1980r

(3) That region of the field of an antenna where the angular field distribution is essentially independent of the distance from a specified point in the antenna region. *Notes:* 1. In free space, if the antenna has a maximum overall dimension, D, that is large compared to the wavelength, the far-field region is commonly taken to exist at distances greater than $2D^2/\lambda$ from the antenna, λ being the wavelength. The far-field patterns of certain antennas, such as multi-beam reflector antennas, are sensitive to variations in phase over their apertures. For these antennas, $2D^2/\lambda$ may be inadequate. 2. In physical media, if the antenna has a maximum overall dimension, D, that is large compared to π/γ, the far-field region can be taken to begin approximately at a distance equal to $\gamma D^2/\pi$ from the antenna, γ being the propagation constant in the medium. (AP/ANT) 145-1993

(4) That region of the field of an antenna array where the angular field distribution is essentially independent of the distance from the center of the array. A general far field approximation is $2d^2/\lambda$, where d is the largest separation between elements in the array. (T&D/PE) 1260-1996

(5) That region of the field of an antenna where the angular field distribution is essentially independent of the distance from the antenna. In this region (also called the free space region), the field has a predominantly plane-wave character, i.e., locally uniform distributions of electric field strength and magnetic field strength in planes transverse to the direction of propagation. (NIR) C95.1-1999

(6) *See also:* Fraunhofer region. (AP/PROP) 211-1997

far-field region in physical media *See:* far-field region.

far field region, radiating *See:* radiating far field region.

far-side (of an SI or BI) That port of an SI or BI electrically farther from the originating master. (NID) 960-1993

fast approach Approach speeds that engender short, subnanosecond risetime ESD current waves. Fast-approach speed depends on the voltage difference between the intruder and receptor, e.g., for rounded electrodes of 8 mm diameter, greater than 0.05 m/s, 1 m/s, and 10 m/s at charge voltages of 4 kV, 8 kV, and 16 kV respectively. (SPD/PE) C62.47-1992r

FASTBUS The standard modular high-speed data acquisition and control system. (NID) 960-1993

FASTBUS protocol (FBP) (FASTBUS acquisition and control) The format and sequence of control and data messages in FASTBUS. Formats are specified by the FASTBUS signal line assignments. Sequences are specified by operations. (NID) 960-1993

fastener (lightning protection) A device used to secure the conductor to the structure that supports it. (NFPA) [114]

fast groove (disk recording) (fast spiral) An unmodulated spiral groove having a pitch that is much greater than that of the recorded grooves. *See also:* phonograph pickup. (SP) [32]

fast handshake A high-speed mode of operation that uses the same communication registers as the word-serial protocol and allows data transfer without the need for polling after each transfer. (C/MM) 1155-1992

Fast Link Pulse (FLP) Burst A group of no more than 33 and not less than 17 10BASE-T compatible link integrity test pulses. Each FLP Burst encodes 16 bits of data using an alternating clock and data pulse sequence. (C/LM) 802.3-1998

Fast Link Pulse (FLP) Burst Sequence The sequence of FLP Bursts transmitted by the local station. This term is intended to differentiate the spacing between FLP Bursts from the individual pulse spacings within an FLP Burst. (C/LM) 802.3-1998

fast-operate, fast-release relay A high-speed relay specifically designed for both short operate and short release time. (PE/EM) 43-1974s

fast-operate relay A high-speed relay specifically designed for short operate time but not necessarily short release time. (PE/EM) 43-1974s

fast-operate, slow-release relay A relay specifically designed for short operate time and long release time. (PE/EM) 43-1974s

fast packet switching A packet switching technique in which formats and procedures are designed to minimize packet processing time. *See also:* frame relay; cell relay. (C) 610.7-1995

fast spiral *See:* fast groove.

fast time (A) Simulated time with the property that a given period of actual time represents more than that period of time in the system being modeled; for example, in a simulation of plant growth, running the simulation for one second may result in the model advancing time by one full day; that is, simulated time advances faster than actual time. *Contrast:* slow time; real time. **(B)** The duration of activities within a simulation in which simulated time advances faster than actual time. *Contrast:* slow time; real time. (C) 610.3-1989

fast-time-constant circuit A circuit with short time-constant (such as a differentiator or high-pass filter) used to emphasize signals of short duration and reduce the receiver response to signals from extended clutter, long-pulse jamming, or noise. It is a form of pulsewidth discriminator (PWD). (AES) 686-1997

fast wave An electromagnetic wave propagating close to a boundary or within a bounded medium with a phase velocity greater than that of a free wave which would exist in an unbounded medium with the same electromagnetic properties. *See also:* slow wave. (AP/PROP) 211-1997

fast writing devices (metal-nitride-oxide field-effect transistor) Metal-nitride-oxide semiconductor (MNOS) memory transistors whose threshold window ΔvHL is sufficiently large after a writing pulse width of tw near 1 μs. A write cycle time of about 1 μs makes these devices useful for random access memory (RAM) applications. (ED) 581-1978w

fatal error An error that results in the complete inability of a system or component to function. (C) 610.12-1990

father *See:* parent node.

father file A file that contains data that have since been updated in another file, called the son file. *See also:* son file; grandfather file. (C) 610.5-1990w

fatigue The tendency for a metal to fracture in brittle manner under conditions of repeated cyclic stressing at stress levels below its tensile strength. (SWG/IA/PE) [59], [71], C37.100-1981s

fault (1) (wire or cable) A partial or total local failure in the insulation or continuity of a conductor. *See also:* center of distribution. (T&D/PE) [10]

(2) (components) A physical condition that causes a device, a component, or an element to fail to perform in a required manner, for example, a short-circuit, a broken wire, an intermittent connection. *See also:* pattern-sensitive fault; program-sensitive fault. (C/T&D/PE) [20], 1048-1990, [85]

(3) (surge arresters) A disturbance that impairs normal operation, for example, insulation failure or conductor breakage. (PE) [8], [84]

(4) (thyristor power converter) A condition existing when the conduction cycles of some semiconductors are abnormal. *Note:* This usually results in fault currents of substantial magnitude. (IA/IPC) 444-1973w

(5) *See also:* short circuit. (SWG/PE) C37.100-1981s

(6) (test, measurement, and diagnostic equipment) A degradation in performance due to detuning, maladjustment, misalignment, failure of parts, and so forth. (MIL) [2]

(7) (A) (software) An incorrect step, process, or data definition in a computer program. *Note:* This definition is used primarily by the fault tolerance discipline. In common usage, the terms "error" and "bug" are used to express this meaning. *See also:* intermittent fault; program-sensitive fault; data-sensitive fault; fault masking; equivalent faults. **(B) (protective grounding of power lines)** (current). A current that flows from one conductor to ground or to another conductor owing to an abnormal connection (including an arc) between the two. **(C) (software reliability)** An accidental condition that causes a functional unit to fail to perform its required function. **(D) (software reliability)** A manifestation of an error in software. A fault, if encountered, may cause a failure. *Synonym:* bug. (C) 610.12-1990

(8) (reliability data for pumps and drivers, valve actuators, and valves) Any undesired state of a component or system. A fault does not necessarily require failure (for example, a pump may not start when required because its feeder breaker was inadvertently left open—a "command block"). (PE/NP) 500-1984w

(9) A defect in a hardware device or component; for example, a short circuit or broken wire. *Synonym:* physical defect. (C/BA) 896.9-1994w, 610.10-1994w, 610.12-1990

(10) (components) A physical condition that causes a device, a component, or an element to fail to perform in a required manner, for example, a short-circuit, a broken wire, and an intermittent connection. (T&D/PE) 524a-1993r

(11) Erroneous hardware or software state resulting from component failure, operator error, physical interference from the environment, design error, program error, or data structure error. (C/BA) 896.3-1993w

(12) A physical condition that causes a device or a diagnostic unit to fail to perform nominally. (ATLAS) 1232-1995

(13) A defect or flaw in a hardware or software component. (SCC20) 1232.1-1997

fault bus A bus connected to normally grounded parts of electric equipment, so insulated that all of the ground current passes to ground through fault-detecting means. *Synonym:* fault ground bus. (SWG/PE) C37.100-1992

fault bus protection (relaying) A method of ground fault protection that makes use of a fault bus. (SWG/PE) C37.100-1992

fault category *See:* error category.

fault, circulating current *See:* circulating current fault.

fault-closure current rating The designated rms fault current that a load-break connector can close under specified conditions. (T&D/PE) 386-1995

fault coverage Quality metric used to measure diagnostic tests or automatic test equipment. (C/BA) 896.3-1993w

fault current (1) (health care facilities) A current in an accidental connection between an energized and a grounded or other conductive element resulting from a failure of insulation, spacing, or containment of conductors. (EMB) [47]
(2) (faulted circuit indicators) Any current through the sensor equal to or in excess of the trip current of the faulted circuit indicator (FCI). (T&D/PE) 495-1986w
(3) (general) A current that flows from one conductor to ground or to another conductor owing to an abnormal connection (including an arc) between the two. A fault current flowing to ground may be called a ground fault current. (T&D/PE) [10], [8], [84], 524a-1993r
(4) The current from the connected power system that flows in a short circuit. (SPD/PE) C62.11-1999, C62.62-2000

fault current division factor A factor representing the inverse of a ratio of the symmetrical fault current to that portion of the current that flows between the grounding grid and surrounding earth.

$$S_F = \frac{I_g}{3I_0}$$

where
S_f = the fault current division factor
I_g = the rms symmetrical grid current in A
I_0 = the zero-sequence fault current in A

Note: In reality, the current division factor would change during the fault duration, based on the varying decay rates of the fault contributions and the sequence of interrupting device operations. However, for the purposes of calculating the design value of maximum grid current and symmetrical grid current per definitions of symmetrical grid current and maximum grid current, the ratio is assumed constant during the entire duration of a given fault. (PE/SUB) 80-2000

fault current withstand (surge arresters) The maximum root-mean-square (rms) symmetrical fault current of a specified duration that a failed distribution class arrester will withstand without an explosive fracture of the arrester housing. (PE/SPD) C62.1-1981s

fault detection (test, measurement, and diagnostic equipment) One or more tests performed to determine if any malfunctions or faults are present in a unit. (MIL) [2]

fault-detector relay A monitoring relay whose function is to limit the operation of associated protective relays to specific system conditions. (SWG/PE) C37.100-1992

fault diagnosis The action of determining the cause of an error in location and nature. 896.9-1994w

fault dictionary A list of faults in a system or component, and the tests that have been designed to detect them. (C) 610.12-1990

faulted circuit indicator (FCI) A single or multiphase device designed to sense fault current and provide an indication that the fault current has passed through the power conductor(s) at the point where the FCI (faulted circuit indicator) sensor is installed. (T&D/PE) 495-1986w

fault electrode current (electron tube) The peak current that flows through an electrode under fault conditions, such as arc-backs and load short-circuits. *Synonym:* surge electrode current. *See also:* electrode current. (ED/EEC/ACO) 161-1971w, [84], [109]

fault ground bus *See:* fault bus.

fault hazard current (health care facilities) The hazard current of a given isolated power system with all devices connected except the line isolation monitor. *See also:* hazard current. (EMB) [47]

fault impedance An impedance, resistive or reactive, between the faulted power system phase conductor(s) or ground. (PE/PSR) C37.113-1999

fault-incidence angle The phase angle as measured between the instant of fault inception and a selected reference, such as the zero point on a current or voltage wave. (SWG/PE) C37.100-1992

fault indicator (test, measurement, and diagnostic equipment) A device that presents a visual display, audible alarm, and so forth, when a failure or marginal condition exists. (MIL) [2]

fault-initiating switch A mechanical switching device used in applied-fault protection to place a short circuit on an energized circuit and to carry the resulting current until the circuit has been de-energized by protective operation. *Notes:* 1. This switch is operated by a stored-energy mechanism capable of closing the switch within a specified rated closing time at its rated making current. The switch may be opened either manually or by a power-operated mechanism. 2. The applied short circuit may be intentionally limited to avoid excessive system disturbance. (SWG/PE) C37.100-1992

fault insertion *See:* fault seeding.

fault, intermittent *See:* intermittent fault.

fault interrupter A self-controlled mechanical switching device capable of making, carrying, and automatically interrupting an alternating current. It includes an assembly of control elements to detect overcurrents and control the fault interrupter. (SWG/PE) C37.100-1992

fault isolation (1) (test, measurement, and diagnostic equipment) Tests performed to isolate within the unit under test. (MIL) [2]
(2) Fault localization to a degree sufficient to undertake repair. (ATLAS) 1232-1995
(3) The process of reducing the number of anomalies that comprise a diagnosis. Identification of an anomaly or anomalies to a degree sufficient to undertake an appropriate corrective action. (SCC20) 1232.1-1997
(4) The process of reducing the number of anomalies that constitute a diagnosis; identification of an anomaly or anomalies to a degree sufficient to undertake an appropriate corrective action. (SCC20) 1232.2-1998

fault localization The reduction of ambiguity by the application of tests, observations, or other information. (ATLAS) 1232-1995

fault management In networking, a management function that is defined for detecting, isolating, and recovering from abnormal network behavior. *Synonym:* failure management. (C) 610.7-1995

fault masking The result of applying error compensation systematically, even in the absence of error. (C/BA) 896.9-1994w

fault, permanent *See:* permanent fault.

fault removal Methods and techniques aimed at reducing the presence (number, seriousness) of faults. (C/BA) 896.9-1994w

fault resistance (surge arresters) The resistance of that part of the fault path associated with the fault itself. (PE) [8], [84]

fault secure Pertaining to a system or component in which no failures are produced from a prescribed set of faults. *See also:* fail-safe; fail soft; fault tolerance. (C) 610.12-1990

fault seeding *See:* error seeding.

fault set A group of one or more faults with the same fault signature. (SCC20) 1445-1998

fault signature A set of unique primary output patterns in which the fault will produce a response different from the good machine response. (SCC20) 1445-1998

fault symptom (test, measurement, and diagnostic equipment) A measurable or visible abnormality in an equipment parameters. (MIL) [2]

fault time *See:* down time.

fault title A two-part description that includes a node name and a fault type [i.e, ⟨U5⟩6 SA1 (component: U5, pin: 6, fault type: Stuck at 1)]. (SCC20) 1445-1998

fault tolerance (1) (A) (software) The ability of a system or component to continue normal operation despite the presence of hardware or software faults. *See also:* error tolerance; robustness; fail-safe; fault secure; fail soft. **(B) (software)** The number of faults a system or component can withstand before normal operation is impaired. **(C) (software)** Pertaining to the study of errors, faults, and failures, and of methods for enabling systems to continue normal operation in the presence of faults. *See also:* redundancy; restart; recovery. (C) 610.12-1990
(2) The ability of a system or a component to continue normal operation despite the presence of hardware or software faults. (C/BA) 896.9-1994w, 896.3-1993w
(3) Methods and techniques aimed at providing a service complying with the specification in spite of faults. (C/BA) 896.9-1994w

fault tolerant (software) Pertaining to a system or component that is able to continue normal operation despite the presence of faults. (C) 610.12-1990, 610.10-1994w

fault-tolerant sequential circuit A sequential circuit designed so that a predetermined set of failures in internal state logic or output logic cause no error in the circuit output. (C) 610.10-1994w

fault, transient *See:* transient fault.

fault tree (1) An ordered arrangement of tests that are intended to lead to the localization of faults. (ATLAS) 1232-1995
(2) A graphical representation of an analytical technique whereby an undesired state of a system is specified and the patterns leading to that state can be evaluated to determine how the undesirable system failure can occur. (PE/NP) 1082-1997

fault tree analysis (FTA) (1) A technique by which failures that can contribute to an undesired event are organized deductively and represented pictorially. (PE/NP) 933-1999
(2) A structured analysis method used to comprehensively identify faults and combinations of faults of software and hardware components as they relate to a hazard. (VT/RT) 1483-2000

fault withstandability The ability of electrical apparatus to withstand the effects of prescribed electrical fault current conditions without exceeding specified damage criteria. (SWG/PE) C37.100-1981s

Faure plate (storage cell) (pasted plate) A plate consisting of electroconductive material, which usually consists of lead-antimony alloy covered with oxides or salts of lead, that is subsequently transformed into active material. *See also:* battery. (PE/EEC) [119]

fax *See:* facsimile.

FB+ *See:* Futurebus+.

f-bits *See:* frame bits.

FBP *See:* FASTBUS protocol.

FC assembly *See:* flat cable assembly.

FCA *See:* functional configuration audit.

FCC (Federal Communications Commission (ARCHIVE)) *See:* Federal Communications Commission; flow control character.

FCFS *See:* first-come, first-served.

FCI *See:* faulted circuit indicator.

FCode A computer programming language defined by this standard, which is semantically similar to the Forth programming language but is encoded as a sequence of binary byte codes representing a defined set of Forth definitions. (C/BA) 1275-1994

FCode driver A device driver, written in FCode, intended for use by Open Firmware and its client programs. (C/BA) 1275-1994

FCode evaluator The portion of Open Firmware that processes FCode programs by reading a sequence of bytes representing FCode numbers and executing or compiling the associated FCode functions. (C/BA) 1275-1994

FCode function A self-contained procedural unit of the FCode programming language to which an FCode number may be assigned. (C/BA) 1275-1994

FCode number A number from 0 to 4095 (conventionally written in hexadecimal as 0x00 to 0x0FFF) that denotes a particular FCode function. (C/BA) 1275-1994

FCode probing The process of locating and evaluating an FCode program. (C/BA) 1275-1994

FCode program A program encoded as a sequence of byte codes according to the rules of the FCode programming language. (C/BA) 1275-1994

FCode source An FCode program in text form. *See also:* tokenizer. (C/BA) 1275-1994

FCS *See:* frame check sequence.

FDDI *See:* fiber distributed data interface.

FDHM *See:* full width (duration) half maximum.

F-display A rectangular display in which a target appears as a centralized blip when the radar antenna is aimed at it. Horizontal and vertical aiming errors are respectively indicated by horizontal and vertical displacement of the blip.

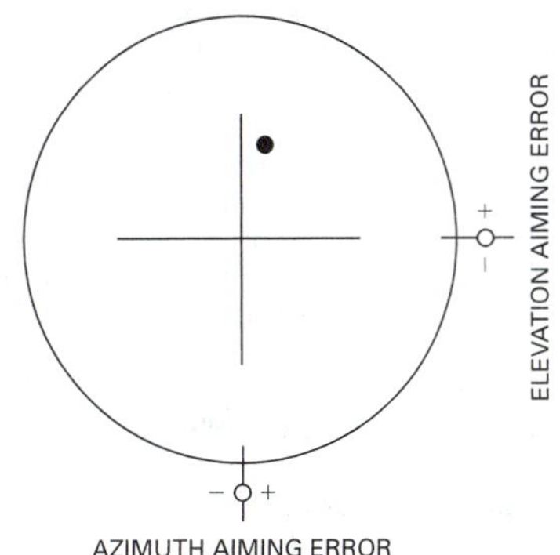

F-display

(AES) 686-1997

FDM *See:* frequency-division multiplexing.

FE *See:* format effector character.

feasibility The degree to which the requirements, design, or plans for a system or component can be implemented under existing constraints. (C) 610.12-1990

feature (1) A negotiable aspect of an interface. (C/PA) 1224.1-1993w
(2) An individual characteristic of a part, such as screw-thread, taper, or slot. (SCC14/QUL) SI 10-1997, 268-1982s
(3) (image processing and pattern recognition) In pattern recognition, an attribute of a pattern that may contribute to pattern classification; for example, size, texture, or shape. (QUL/C) 268-1982s, 610.4-1990w

feature extraction A step in pattern recognition, in which measurements or observations are processed to find attributes that can be used to assign patterns to pattern classes. (C) 610.4-1990w

feature reference An expression that unambiguously identifies a diagram feature in a diagram. (C/SE) 1320.1-1998

feature space In pattern recognition, a set of all possible n -tuples $(x_1, x_2, . . ., x_n)$ that can be used to represent n features of a pattern. *See also:* measurement space. (C) 610.4-1990w

feature test macro A `defined` symbol used to determine whether a particular set of features will be included from a header. (C/PA) 9945-1-1996, 9945-2-1993

FEC *See:* forward error correction.

Federal Communications Commission (FCC) A U.S. regulatory body operating under the Communications Act of 1934 to regulate all interstate telecommunications systems in the United States. (C/Std100) 610.7-1995, 610.10-1994w

federal information processing standard publication (FIPS PUB XXXX) Issued by the National Institute of Standards and Technology and available from the National Technical Information Service. (C/BA) 896.3-1993w

feed (1) (A) (machines) To supply the material to be operated upon to a machine. **(B) (machines)** A device capable of feeding as in (A). (C) 162-1963
(2) (A) To supply the material to be operated upon to a machine. *See also:* friction feed; tractor feed; single-sheet feed; continuous feed; cut-sheet feed. **(B)** A device capable of feeding as in (A). *See also:* card feed; hand-feed punch; automatic-feed punch; paper feed. **(C)** A command or signal sent to a printer to instruct it to perform a feed operation as in (A). *See also:* form feed; line feed. (C) 610.10-1994
(3) (A) For continuous aperture antennas, the feed is the primary radiator; for example, a horn feeding a reflector. **(B)** For array antennas, that portion of the antenna system which functions to produce the excitation coefficients. (AP/ANT) 145-1993

feedback (1) (transmission system or section thereof) The returning of a fraction of the output of the input. (AP/BT/ANT) 145-1983s, 182-1961w
(2) That portion of the output of a control system used as input for another phase of the system, particularly for self-correcting, self-regulating, or control purposes, as in closed-loop control. (C) 610.2-1987
(3) (A) A signal that is derived from the output of a circuit and applied to one or more inputs of the same circuit. **(B)** Pertaining to components or subcircuits that transform a portion of the output of a circuit into a form suitable for application to input of the same circuit. *See also:* servomechanism. (C) 610.10-1994
(4) *See also:* control loopback. (C/SE) 1320.1-1998

feedback admittance, short-circuit *See:* short-circuit feedback admittance.

feedback branch *See:* feedback node.

feedback control *See:* closed-loop control.

feedback control system (1) (hydraulic turbines) A control system in which the controlled quantity is measured and compared with a standard representing the desired value of the controlled quantity. In hydraulic governors, any deviation from the standard is fed back into the control system in such a sense that it will reduce the deviation between the controlled quantity and the standard providing negative feedback. (PE/EDPG) 125-1977s
(2) (general) A control system that operates to achieve prescribed relationships between selected system variables by comparing functions of these variables and using the comparison to effect control. See the diagram below.

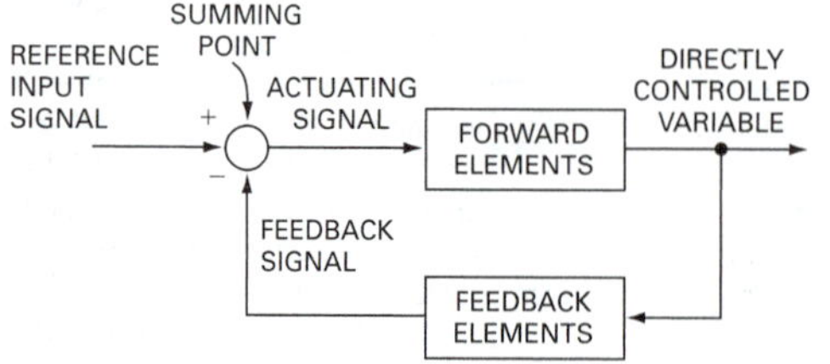

Simplified block diagram indicating essential elements of an automatic control system.

control system, feedback

feedback elements The elements in the controlling system that change the feedback signal in response to the directly controlled variable. *See also:* feedback control system. (PE/EDPG) 421-1972s

feedback impedance (analog computer) In an analog computer, a passive network connected between the output terminal of an operational amplifier and its summing junction. (C) 165-1977w

feedback limit *See:* limiter circuit.

feedback limiter A limiter circuit that limits the amount of positive or negative signal in an operational amplifier. *See also:* limiter circuit. (C) 610.10-1994w

feedback loop (numerically controlled machines) The part of a closed-loop system that provides controlled response information allowing comparison with a referenced command. (IA/EEC) [61], [74]

feedback node (network analysis) A node (branch) contained in a loop. *Synonym:* feedback branch. (CAS) 155-1960w

feedback oscillator An oscillating circuit, including an amplifier, in which the output is coupled in phase with the input, the oscillation being maintained at a frequency determined by the parameters of the amplifier and the feedback circuits such as inductance-capacitance, resistance-capacitance, and other frequency-selective element. *See also:* oscillatory circuit. (AP/ANT) 145-1983s

feedback signal (1) (general) A function of the directly controlled variable in such form as to be used at the summing point. *See also:* feedback control system. (IA/IAC) [60]
(2) (control system feedback) The return signal that results from the reference input signal. (See the corresponding figure.) *See also:* feedback control system.

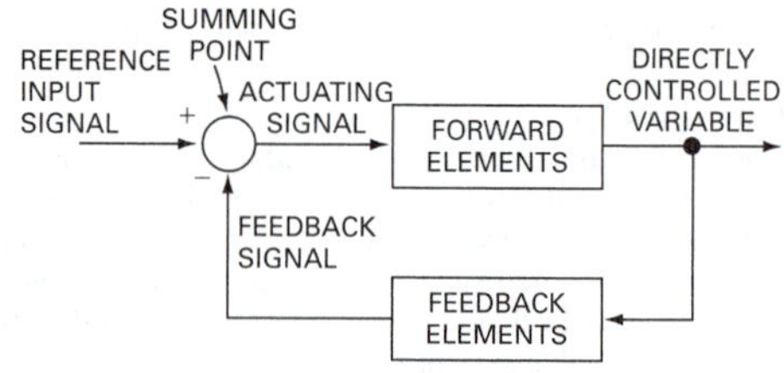

Simplified block diagram including essential elements of an automatic control system

signal, feedback

(PE/EDPG) 421-1972s, [3]

feedback winding (saturable reactor) A control winding to which a feedback connection is made. (PE/EEC) [119]

feed circuit (1) An arrangement for supplying dc power to a telephone set and an ac path between the telephone set and a terminating circuit. (COM/TA) 269-1992
(2) An electrical circuit for supplying dc power to a handsfree telephone set and an acdc path between the handsfree telephone and a terminating circuit. (COM/TA) 1329-1999

feed direction On most printers, the direction that the medium is moved through the marking engine. For a printer in which the medium is not moved, the feed direction may be considered as along the Y axis. The across feed direction is the direction orthogonal to the feed direction; it is also called the crossfeed or scan direction on some printers. (C/MM) 1284.1-1997

feeder (1) All circuit conductors between the service equipment, or the generator switchboard of an isolated plant, and the final branch-circuit overcurrent device. (NESC/NEC) [86]
(2) (packaging machinery) The circuit conductors between the service equipment, or the generator switchboard of an isolated plant, and the branch-circuit overcurrent device. (IA/PKG) 333-1980w
(3) (system) The portion of a broadband coaxial cable system that distributes signals to and receives signals from the user outlet ports. Characterized primarily by the presence of cable taps and distribution amplifiers. (LM/C) 802.7-1989r
(4) A cable or set of conductors that originates at a main distribution center (main switchboard) and supplying secondary distribution centers, transformers, or motor control centers. (Bus tie circuits between generator and distribution switchboards, including those between main and emergency switchboards, are not considered as feeders.) (IA/MT) 45-1998

feeder assembly The overhead or under-chassis feeder conductors, including the grounding conductor, together with the necessary fittings and equipment or a power-supply cord approved for mobile home use, designed for the purpose of delivering energy from the source of electrical supply to the distribution panelboard within the mobile home. (NESC/NEC) [86]

feeder cable (communication practice) A cable extending from the central office along a primary route (main feeder cable) or from a main feeder cable along a secondary route (branch feeder cable) and providing connections to one or more distribution cables. *See also:* cable. (EEC/PE) [119]

feeder distribution center A distribution center at which feeders or subfeeders are supplied. *See also:* distribution center. (EEC/PE) [119]

feeder maker A splitting device used to provide multiple line connections from trunk amplifiers. (LM/C) 802.7-1989r

feeder reactor (power and distribution transformers) A current-limiting reactor for connection in series with an alternating-current feeder circuit for the purpose of limiting and localizing the disturbance due to faults on the feeder. *See also:* reactor. (PE/TR) C57.12.80-1978r, C57.16-1996, [57]

feed function (numerically controlled machines) The relative velocity between the tool or instrument and the work due to motion of the programmed axis (axes). (IA/EEC) [61], [74]

feed groove (rotating machinery) A groove provided to direct the flow of oil in a bearing. *See also:* bearing. (PE) [9]

feed hole A hole punched in a data medium to enable it to be positioned or fed into a machine. *Synonym:* sprocket hole. (C) 610.10-1994w

feeding bridge A device to supply telephone lines with feeding current and signaling current. Feeding bridges are typically located in local exchanges, private branch exchanges, and remote terminals of subscriber line carrier systems. *Notes:* 1. Feeding current (for a telephone set) is the direct current primarily used to power the speech circuit in a telephone set. 2. Signaling current (in telephony) is the direct current supplied to a telephone set to enable the set to send supervision and addressing signals.

feeding coefficients *See:* excitation coefficients.

feeding point The point of junction of a distribution feeder with a distribution main or service connection. *See also:* center of distribution. (T&D/PE) [10]

feed line (1) A transmission line interconnecting an antenna and a transmitter or receiver or both. (AP/ANT) 145-1993 **(2) (rotating machinery)** A supply pipe line. *See also:* oil cup. (PE) [9]

feed pitch The distance between corresponding points of adjacent feed holes along the feed track. (C) 610.10-1994w

feed punch *See:* automatic-feed punch.

feed rate bypass (numerically controlled machines) A function directing the control system to ignore programmed feed rate and substitute a selected operational rate. (IA/EEC) [61], [74]

feed rate override (numerically controlled machines) A manual function directing the control system to modify the programmed feed rate by a selected multiplier. (IA/EEC) [61], [74]

feedthrough power meter (1) (bolometric power meters) A power-measuring system in which the detector structure is inserted or incorporated in a waveguide or coaxial transmission line to provide a means for measuring (monitoring) the power flow through or beyond the system. (IM) 470-1972w **(2) (measuring system) (electrothermic power meters)** A device which is inserted or incorporated in a waveguide or transmission line and provides a means for measuring (monitoring) the power flow through or beyond the system. (IM) 544-1975w

feedthrough signal The undelayed signal resulting from direct coupling between the input and output of the device. (UFFC) 1037-1992w, [22]

feed track A track of a data medium that contains the feed holes. *Synonym:* sprocket track. (C) 610.10-1994w

feed tube (cable plowing) A tube attached to the blade of a plow which guides and protects the cable as it enters the earth. *See also:* fixed feed tube; hinged removable feed tube; floating removable feed tube. (T&D/PE) 590-1977w

FEFO *See:* first-ended, first-out.

fence (A) A line or network of early-warning radars. **(B)** The locus of the positions of a surveillance radar beam that describes the search area covered by space-based radar. *See also:* clutter fence. (AES) 686-1997

fenestra method (illuminating engineering) A procedure for predicting the interior illuminance received from daylight through windows. (EEC/IE) [126]

fenestration (illuminating engineering) Any opening or arrangement of openings (normally filled with media for control) for the admission of daylight. (EEC/IE) [126]

FEO page *See:* For Exposition Only page.

FEP *See:* fuse-enclosure package.

ferreed relay Coined name (Bell Telephone Laboratories) for a special form of dry reed switch having a return magnetic path of high remanence material that provides a bistable, or latching, transfer contact. (PE/EM) 43-1974s

ferri-diode limiter (nonlinear, active, and nonreciprocal waveguide components) A hybrid power limiting device incorporating a ferrite power limiter in cascade with a p-i-n diode or varactor limiter. *See also:* ferrite limiter. (MTT) 457-1982w

ferrite An iron compound frequently used in the construction of magnetic cores components. (C) 610.10-1994w

ferrite devices figure of merit (nonlinear, active, and nonreciprocal waveguide components) A measure of performance of the device. It is usually expressed as the ratio of the quantity of interest to the insertion loss in decibels (dB).

ferrite limiter (nonlinear, active, and nonreciprocal waveguide components) A power limiter utilizing the nonlinear characteristics of ferrimagnetic material above a critical or threshold radio-frequency (rf) power level. *See also:* ferri-diode limiter. (MTT) 457-1982w

ferritic The body-centered cubic crystal structure of ferrous metals. (IA) [59], [71]

ferrodynamic instrument An electrodynamic instrument in which the forces are materially augmented by the presence of ferromagnetic material. *See also:* instrument. (EEC/PE) [119]

ferroelastic crystal (primary ferroelectric terms) One that has two or more orientation states in the absence of mechanical stress and electric field, and can be shifted from one to another of these states by a mechanical stress. (UFFC) 180-1986w

ferroelectric axis The crystallograph direction is parallel to the spontaneous polarization vector. *Note:* In some materials the ferroelectric axis may have several possible orientations with respect to the macroscopic crystal. *See also:* ferroelectric domain. (UFFC) 180w

ferroelectric ceramic (primary ferroelectric terms) Typically, a sintered polycrystalline material comprising an aggregate of ferroelectric single crystal grains (or crystallites). Each ceramic grain has properties similar to a ferroelectric single crystal, with the possible exception of grains with major dimensions $<< 1\ \mu m$. *Notes:* 1. A ceramic is, in general, any inorganic, nonmetallic, ordered or disordered material. A ceramic is commonly typified by polycrystallinity and the unique properties associated with grain boundaries. 2. Both single crystal and sintered materials are, strictly speaking, ferroelectric ceramics even though they are often separated in common usage of the terms. (UFFC) 180-1986w

ferroelectric Curie point (1) (primary ferroelectric terms) Temperature at which a ferroelectric material undergoes a

structural phase transition to a state where spontaneous polarization vanishes. *Note:* The Curie point is determined at zero applied field. (UFFC) 180-1986w
(2) The temperature T_C at which a ferroelectric material undergoes a structural phase transition to a state in which the spontaneous polarization vanishes in the absence of an applied electric field. *Note:* In a normal ferroelectric, the Curie point can be shifted by application of an external electric field, a mechanical stress or by doping with chemical impurities. *See also:* Curie-Weiss temperature; spontaneous polarization. (UFFC) [21]

ferroelectric Curie temperature The temperature above which ferroelectric materials do not exhibit reversible spontaneous polarization. *Note:* As the temperature is lowered from above the ferroelectric Curie temperature spontaneous polarization is detected by the onset of a hysteresis loop. The ferroelectric Curie temperature should be determined only with unstrained crystals, at atmospheric pressure, and with no externally applied direct-current fields. (In some ferroelectric multiple hysteresis-loop patterns may be observed at temperatures slightly higher than the ferroelectric Curie temperature under alternating-current fields.) *See also:* ferroelectric domain. (UFFC) 180w

ferroelectric domain (1) A region of a crystal exhibiting homogeneous and uniform spontaneous polarization. *Note:* An unpoled ferroelectric material may exhibit a complex domain structure consisting of many domains, each with a different polarization orientation. The direction of the spontaneous polarization within each domain is constrained to a small number of equivalent directions (see polar axis) dictated by the symmetry of the crystal structure above the ferroelectric Curie point. The transition region between two ferroelectric domains is called a domain wall. Domains can usually be detected by pyroelectric, optical, powder decoration or electrooptic means. *See also:* ferroelectric Curie point; spontaneous polarization; poling; polar axis. (UFFC) [21]
(2) A region of a ferroelectric crystal exhibiting homogeneous and uniform spontaneous polarization. In the close vicinity of domain walls, P_s is different from that in the bulk of the domain, due to the energy associated with the domain wall. The equilibrium domain structure is determined by minimization of the domain wall energy and the depolarizing energy. In a conducting ferroelectric crystal, the depolarizing fields can be neutralized by free charge so that the depolarizing energy vanishes and a single domain structure is energetically the most favorable in a perfect crystal. The formation of a domain wall is affected by the local electric field and mechanical stresses. In crystals with many small domains, the field and stress gradients can be large enough to change the measured apparent spontaneous polarization. in small domains P_s can vary considerably across a domain, and even at the domain center may differ from that measured in a large single domain. *Note:* An unpoled ferroelectric material may exhibit a complex domain structure consisting of many domains, each with a different polarization orientation. The direction of the spontaneous polarization within each domain is constrained to a small number of equivalent directions dictated by the symmetry of the prototype. The boundary region between two ferroelectric domains is called a domain wall. Domains can usually be observed by pyroelectric, optical, powder decoration, or electrooptic means. (UFFC) 180-1986w

ferroelectric glass-ceramics (primary ferroelectric terms) A multiphase solid containing ferroelectric crystal grains and other phases, one of which must be a glass. Typical grain sizes vary from $10^{-2} - 50$ μm, depending on chemical composition and nucleation-crystallization conditions. The principal fabrication process of a glass-ceramic has three steps:

1) A molten solution is formed of the ferroelectric and the glass-forming constituents.
2) The melt is rapidly quenched to form a vitreous body.
3) Controlled devitrification is accomplished by annealing. A chief advantage of a glass-ceramic is a wide range of formability, depending upon the viscosity-temperature behavior prior to devitrification.

(UFFC) 180-1986w

ferroelectric material (A) (primary ferroelectric terms) A material that exhibits, over some range of temperature, a spontaneous electric polarization that can be reversed or reoriented by application of an electric field. The requirement of a nonvanishing spontaneous polarization P_s is a necessary criterion, and the requirement of reversibility or reorientability of P_s is a suficient criterion for a ferroelectric phase. Materials belonging to nonpolar crystal classes at all temperatures, and in which a metastable polar state can be induced by an applied electric field, can also show reversible pyroelectric behavior, but are not included in the definition of ferroelectrics. The various possible stable orientations of P_s for a given ferroelectric phase are designated as orientation states. A ferroelectric crystal has two or more such orientation states in the absence of an electric field, and it can be switched from one to another of these states by a realizable electric field. Any two of the orientation states are identical (or enantiomorphous) in crystal structure, but different in their P_s orientation at zero electric field. **(B)** A crystalline material that exhibits, over some range of temperature, a remanent polarization that can be reversed or reoriented by application of an external electric field. *Note:* The saturation remanent polarization is equal to the spontaneous polarization in a single domain ferroelectric material. Since the spontaneous polarization in a ferroelectric material is strongly temperature dependent, poled ferroelectric materials exhibit a large pyroelectric effect near the Curie point. Ferroelectric materials also exhibit anomalies in small-signal dielectric permittivity, dielectric loss tangent, piezoelectric coefficients, and electrooptic coefficients near their ferroelectric Curie point. *See also:* polarization; Curie-Weiss temperature; ferroelectric domain; paraelectric region; poling; remanent polarization; antiferroelectric material; small-signal permittivity; polar axis; ferroelectric Curie point; spontaneous polarization. (UFFC) 180-1986, [21]

ferroelectric polymers (primary ferroelectric terms) Typically, a semicrystalline polymer with a large net dipole moment per unit volume. These materials exhibit a spontaneous electric polarization that can be reversed by the application of a strong electric field (0.1−0.5 MV/cm) and also exhibit piezoelectric and pyroelectric behavior. Examples of ferroelectric polymers include polyvinyl fluoride (PVF), polyvinylidene fluoride (PVF_2), and copolymers of vinylidene fluoride with vinyl trifluoroethylene or tetrafluoroethylene. The copolymers of vinylidene fluoride with trifluoroethylene exhibit Curie temperatures in the range of 50-160°C. The higher trifluoroethylene content materials (greater than 55 mol% VF_3) have a second-order transition at lower Curie points; while the materials with higher vinylidene fluoride content (greater than 65 mol% VF_2) have a first-order transition at higher Curie points. (UFFC) 180-1986w

ferromagnetic material (A) (electrical heating systems) A material that, in general, exhibits hysteresis phenomena and whose permeability is dependent on the magnetizing force. **(B)** Material whose relative permeability is greater than unity and depends upon the magnetizing force. A ferroemagnetic material usually has relatively high values of relative high values of relative permeability and exhibits hysteresis. (Std100/IA/PC) 844-1991, 270-1966

ferroresonance (1) (power and distribution transformers) A phenomenon usually characterized by overvoltages and very irregular wave shapes and associated with the excitation of one or more saturable inductors through capacitance in series with the inductor. (PE/TR) C57.12.80-1978r
(2) An electrical resonant condition associated with the saturation of a ferromagnetic device, such as a transformer, through capacitance. Ferroresonance can arise when (1) due

to dissimilar phase switching, the capacitance normally in shunt with the ferromagnetic device becomes energized in series with the device, (2) a weak source is isolated with a lightly loaded feeder containing power factor correction capacitors. For example, if the resulting voltage buildup produces saturation of the feeder transformers, there will be an interchange of energy between the system capacitance and the nonlinear magnetizing reactance of the transformers. (SWG/PE) C37.100-1992

(3) Occurs between the capacitance to ground of an ungrounded circuit and voltage transformers with primary windings that are grounded. This phenomenon is also possible in gas-insulated systems. (SPD/PE) C62.22-1997

(4) The steady-state mode of operation that exists when an alternating voltage of sufficient magnitude is applied to a circuit consisting of capacitance and ferromagnetic inductance causing changes in the ferromagnetic inductance that are repeated each half cycle. *Note:* When certain critical relations exist among circuit parameters, self-sustaining subharmonic or harmonic oscillations may also be excited in the circuit. (PEL) 449-1998

ferroresonant voltage regulation The effect obtained by the limiting action of the saturation characteristic of the magnetic material in a ferroresonant circuit, which regulates the output voltage over a specified range of input voltages and a specified frequency of excitation. *Note:* This effect regulates the half-cycle average value of the output voltage. (PEL) 449-1998

ferroresonant voltage regulator provided with a frequency-compensating network Output voltage of a ferroresonant voltage regulator changes considerably with the change of the input frequency. An LC network can be added to the regulator output, in series with the load, to compensate this voltage change. See corresponding figure. *Note:* Frequency compensating networks, of this series type, are effective in cases where regulators are operated with constant loads but they produce only limited improvement of regulation when loads are variable.

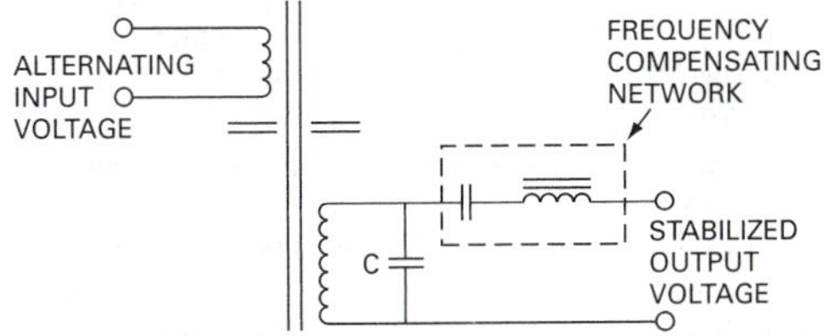

ferroresonant voltage regulator provided with a frequency-compensating network

(MAG/ET) 449-1984s

ferroresonant voltage regulator transformer A high-reactance transformer employing magnetic shunts that allow the magnetic functions of the basic series parallel ferroresonant regulator circuits to be combined into a single magnetic component. See the figures below.

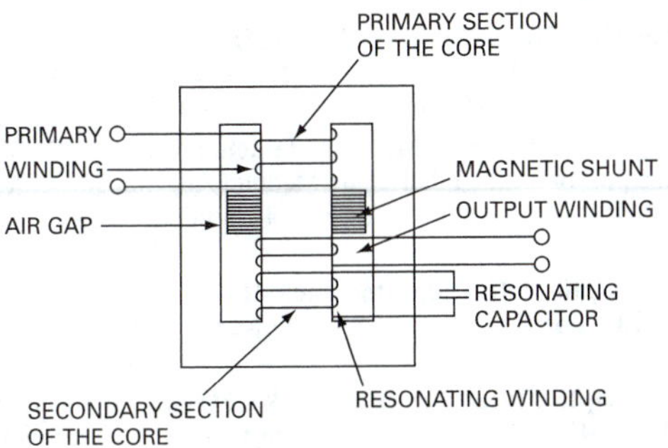

Common form of the ferroresonant transformer voltage regulator

Ferroresonant voltage regulator transformer

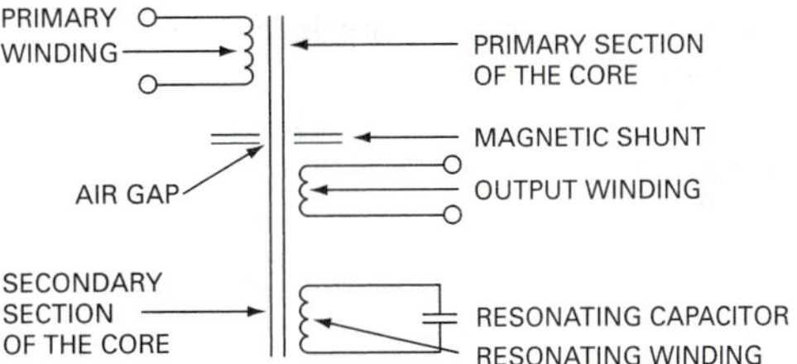

Schematic of a common form of the ferroresonant transformer voltage regulator

Ferroresonant voltage regulator transformer

(PEL) 449-1998

ferroresonant voltage regulator with compensating winding (ferroresonant voltage regulators) A ferroresonant voltage regulator having a compensating winding connected in series with the output winding to attain improved load and line regulation. See figures below.

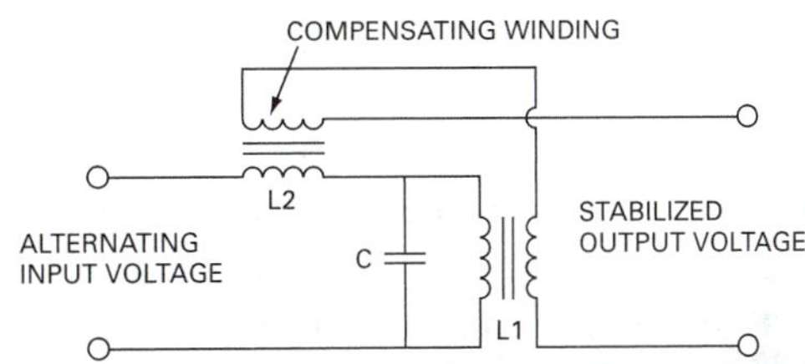

Two-core ferroresonant circuit with compensating winding

ferroresonant voltage regulator with compensating winding

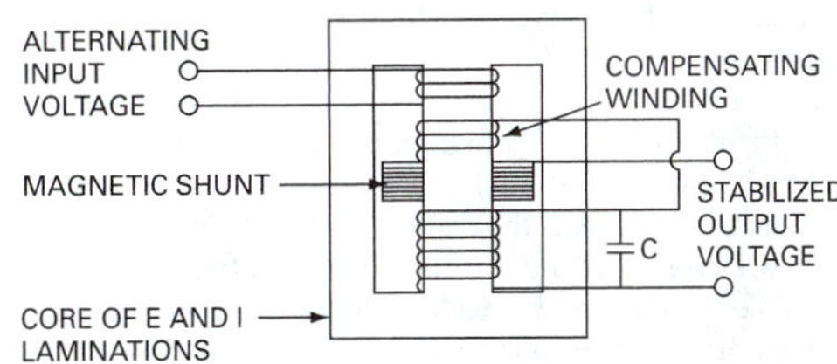

ferroresonant transformer circuit with compensating winding

(MAG/ET) 449-1984s

ferroresonant voltage regulator with compensation for varying load power factor (ferroresonant voltage regulators) Reduction of the amount of output voltage change caused by other than resistive loading and by large changes of load power factor is obtained by providing a capacitive impedance, inserted in series with the output, that essentially matches the output reactance of the regulator. The power factor compensation circuit is usually a capacitive reactance obtained by capacitors alone. (MAG/ET) 449-1984s

ferroresonant voltage regulator with harmonic filter (A) (harmonic neutralized) (ferroresonant voltage regulators) (magnetically coupled type). Reduction of output harmonics is obtained by effectively filtering the odd harmonics through use of a neutralizing winding that is magnetically coupled to the resonating winding as shown in the corresponding figure. **(B) (ferroresonant voltage regulators) (harmonic neutralized)** (electrically coupled tuned). Cancellation type reduction of output harmonics is obtained by effectively filtering the odd harmonics through use of an inductance in series with the resonating capacitor which effectively filters the major harmonic (the third harmonic) and a saturating inductor to produce odd harmonics which are induced back into the circuit of the regulator to cancel out the remaining odd harmonics. This type of filtering is shown in the corresponding figure. **(C) (ferroresonant voltage regulators) (harmonic neutralized)** (tuned type). Reduction of output harmonics is obtained by dividing the resonating capacitance into several sections and connecting them to filter the various odd harmonics that exist in the output of the basic regulator.

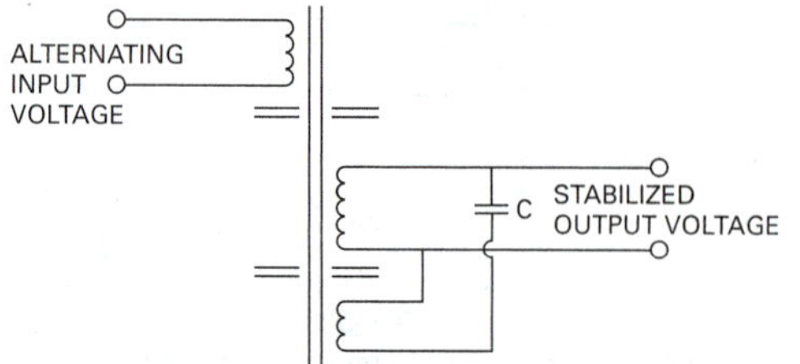

magnetically coupled tuned-cancellation type harmonic filter

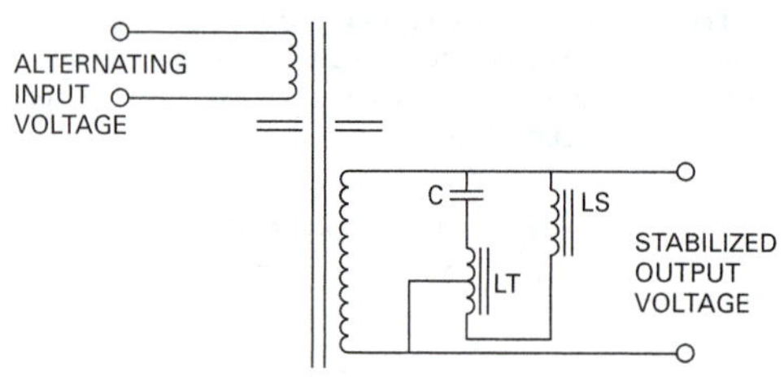

electrically connected tuned-cancellation type harmonic filter

(MAG/ET) 449-1984

ferrule (1) (fiber optics) A mechanical fixture, generally a rigid tube, used to confine the stripped end of a fiber bundle or a fiber. *Notes:* 1. Typically, individual fibers of a bundle are cemented together within a ferrule of a diameter designed to yield a maximum packing fraction. 2. Nonrigid materials such as shrink tubing may also be used for ferrules for special applications. *See also:* reference surface; packing fraction; fiber bundle. (Std100) 812-1984w
(2) (protection and coordination of industrial and commercial power systems) The cylindrical-shaped fuse terminal that also encloses the end of the fuse. In low-voltage fuses, the design is only used in fuses rated up to and including 60 A. The ferrule may be made of brass or copper, and may be plated with various materials. (IA/PSP) 242-1986r
(3) (of a cartridge fuse) A fuse terminal of cylindrical shape at the end of a cartridge fuse. (SWG/PE) C37.100-1992

fertilizer *See:* explosives.

festoon lighting A string of outdoor lights suspended between two points more than 15 feet apart. (NESC/NEC) [86]

FET *See:* field-effect transistor.

fetch To locate and load computer instructions or data from storage. *See also:* store; move. (C) 610.12-1990
(2) (A) That portion of an instruction cycle in which the next instruction is loaded from memory into the processor. **(B)** To obtain a data item from a storage location. (C) 610.10-1994

fetch cycle That portion of an instruction cycle during which a fetch takes place. (C) 610.10-1994w

FET photodetector (fiber optics) A photodetector employing photogeneration of carriers in the channel region of a field-effect transistor (FET) structure to provide photodetection with current gain. *See also:* photodiode; photocurrent. (Std100) 812-1984w

FF *See:* form feed character.

F Filter A 5 kHz to 245 kHz bandpass filter used for measuring the power of a High-bit-rate Digital Subscriber Line (HDSL) signal, noise, or impulse noise on an HDSL. (COM/TA) 743-1995

F format *See:* fixed format.

FGRAAL *See:* FORTRAN Extended GRAph Algorithmic Language.

fiber *See:* optical fiber.

fiber axis (fiber optics) The line connecting the centers of the circles that circumscribe the core, as defined under bold tolerance field. *Synonym:* optical axis. *See also:* tolerance field. (Std100) 812-1984w

fiber bandwidth (fiber optics) The lowest frequency at which the magnitude of the fiber transfer function decreases to a specified fraction of the zero frequency value. Often, the specified value is one-half the optical power at zero frequency. *See also:* transfer function. (Std100) 812-1984w

fiber buffer (fiber optics) A material that may be used to protect an optical fiber waveguide from physical damage, providing mechanical isolation or protection or both. *Note:* Cable fabrication techniques vary, some resulting in firm contact between fiber and protective buffering, others resulting in a loose fit, permitting the fiber to slide in the buffer tube. Multiple buffer layers may be used for added fiber protection. *See also:* fiber bundle. (Std100) 812-1984w

fiber bundle (1) (fiber optics) An assembly of unbuffered optical fibers. Usually used as a single transmission channel, as opposed to multifiber cables, which contain optically and mechanically isolated fibers, each of which provides a separate channel. *Notes:* 1. Bundles used only to transmit light, as in optical communications, are flexible and are typically unaligned. 2. Bundles used to transmit optical images may be either flexible or rigid, but must contain aligned fibers. *See also:* optical fiber; multifiber cable; optical cable; packing fraction; aligned bundle; ferrule; fiberoptics. (Std100) 812-1984w
(2) An assembly of unbuffered optical fibers, usually employed as a single transmission channel. (C) 610.7-1995

fiber distributed data interface An ANSI standard based on fiber optics configured in a dual, counter-rotating ring and operating at 125 million baud with a user data rate of 100 Mb/s. FDDI uses a token passing MAC so that it can operate on non-fiber media such as unshielded twisted pair. *Note:* With the physical layer protocol overhead removed, the net throughput is 100 000 000 b/s and with the MAC overhead removed, the net throughput is less than 100 000 000 b/s. (C) 610.7-1995

fiber laser (interferometric fiber optic gyro) A laser in which the lasing medium is an externally pumped optical fiber doped with low levels of rare-earth halides to make it capable of amplifying light. (AES/GYAC) 528-1994

fiber optic cable (1) A cable containing one or more of the optical fibers. (C) 610.7-1995
(2) A cable containing one or more optical fibers as specified in IEEE 802.3. (LM/C) 802.3u-1995s

fiber-optic cable—communication A fiber-optic cable meeting the requirements for a communication line and located in the communication space of overhead or underground facilities. (NESC) C2-1997

fiber-optic cable—supply A fiber-optic cable located in the supply space of overhead or underground facilities. (NESC) C2-1997

fiber-optic conductor *See:* fiber-optic cable—supply; fiber-optic cable—communication.

Fiber Optic Inter-Repeater Link (FOIRL) A Fiber Optic Inter-Repeater Link segment and its two attached Medium Attachment Units (MAUs). (C/LM) 802.3-1998

Fiber Optic Physical Medium Attachment For 10BASE-F, the portion of the Fiber Optic Medium Attachment Unit (FOMAU) that contains the functional circuitry. (C/LM) 802.3-1998

Fiber Optic Inter-Repeater Link (FOIRL) bit error rate (BER) For 10BASE-F, the mean bit error rate of the FOIRL. (C/LM) 802.3-1998

Fiber Optic Inter-Repeater Link (FOIRL) collision For 10BASE-F, the simultaneous transmission and reception of data in a Fiber Optic Medium Attachment Unit (FOMAU). (C/LM) 802.3-1998

Fiber Optic Inter-Repeater Link (FOIRL) Compatibility Interface For 10BASE-F, the FOMDI and Attachment Unit

Interface (AUI) (optional); the two points at which hardware compatibility is defined to allow connection of independently designed and manufactured components to the baseband optical fiber cable link segment. (C/LM) 802.3-1998

Fiber Optic Inter-Repeater Link (FOIRL) Segment A fiber optic link segment providing a point-to-point connection between two FOIRL Medium Attachment Units (MAUs) or between one FOIRL MAU and one 10BASE-FL MAU. *See also:* link segment. (C/LM) 802.3-1998

Fiber Optic Medium Attachment Unit A medium attachment unit (MAU) for fiber applications. (C/LM) 802.3-1998

Fiber Optic Medium Attachment Unit's (FOMAU's) Receive Optical Fiber For 10BASE-F, the optical fiber from which the local FOMAU receives signals. (C/LM) 802.3-1998

Fiber Optic Medium Attachment Unit's (FOMAU's) Transmit Optical Fiber For 10BASE-F, the optical fiber into which the local FOMAU transmits signals. (C/LM) 802.3-1998

Fiber Optic Medium Dependent Interface (FOMDI) For 10BASE-F, the mechanical and optical interface between the optical fiber cable link segment and the Fiber Optic Medium Attachment Unit (FOMAU). (C/LM) 802.3-1998

fiber optic physical medium attachment (FOPMA) For 10BASE-F, the portion of the FOMAU that contains the functional circuitry. (LM/C) 802.3u-1995s

fiber-optic plate (camera tubes) An array of fibers, individually clad with a lower index-of-refraction material, that transfers an optical image from one surface of the plate to the other. *See also:* camera tube. (ED) [45]

fiber-optic receiver operating range The range of optical power over which the fiber-optic receiver will meet the specified bit error rate (BER). (C/BA) 1393-1999

fiberoptics (1) (data transmission) The branch of optical technology concerned with the transmission of radiant power through fibers made of transparent materials, such as glass, fused silica plastic. *Notes:* 1. Communications applications of fiber optics employ flexible fibers. Either a single discrete fiber or a nonspatially aligned fiber bundle may be used for each information channel. Such fibers are generally referred to as "optical waveguides" to differentiate from fibers employed in noncommunications applications. 2. Various industrial and medical applications employ typically high-loss flexible fiber bundles in which individual fibers are spatially aligned, permitting optical relay of an image. An example is the endoscope. 3. Some specialized industrial applications employ rigid (fused) aligned fiber bundles for image transfer. An example is the fiber optics cathode-ray tube (CRT) faceplate used on some high-speed oscilloscopes. (PE) 599-1985w, 812-1984w

(2) A technology that uses light as a digital information carrier. (C) 610.7-1995, 610.10-1994w

fiber pair Optical fibers interconnected to provide two continuous light paths terminated at each end in an optical connector. (C/LM) 802.3-1998

Fibonacci number An integer in the Fibonacci series. (C) 1084-1986w

Fibonacci search A dichotomizing search in which, at each step in the search, the set of items is partitioned in accordance with the Fibonacci series. For example, a set of 8 items is partitioned to 5 and 3, the subset of 5 is partitioned to 3 and 2, and so on. If the number of items in the original set is other than a Fibonacci number, the next higher Fibonacci number is used to partition the set. *Contrast:* interpolation search; binary search. (C) 610.5-1990w

Fibonacci series A series of integers formulated by the Italian mathematician Leonardo Fibonacci, in which each integer is equal to the sum of the two preceding integers in the series, that is, 0, 1, 1, 2, 3, 5, 8, 13,.... Represented mathematically by

$$x_i = x_{i-1} + x_2$$

where

$x_0 = 0$

$x_1 = 1$

(C) 1084-1986w

fibre A filament-shaped optical waveguide made of dielectric materials. (LM/C) 11802-4-1994, 8802-3-1990s, 802.3u-1995s, 610.7-1995

Fibre Distributed Data Interface (FDDI) A 100 Mb/s, fiber optic-based, token-ring local area network standard (ISO/IEC 9314, formerly X3.237-1995). (C/LM) 802.3-1998

fibre-optic cable A cable containing one or more optical fibres. (C/LM) 8802-12-1998, 11802-4-1994

fibre optic channel The data path from any transmitting station's FMIC or transmitting concentrator's FMIC to the next receiving FMIC. (C/LM) 11802-4-1994

fibre optic concentrator lobe port *See:* fibre optic trunk coupling unit.

fibre optic interface (FOI) The interface between a station's PHY and the optical medium. It is bounded on one side by the MIC or PHY-layer I/O interface, and on the other side by the FMIC. (C/LM) 11802-4-1994

fibre-optic link A link segment configured from fibre optic cables and two attached Medium Dependent Interface (MDI) connectors.local area networks. (C) 8802-12-1998

fibre optic medium interface connection (FMIC) The mechanical and optical interface between the station or FOI and the fibre optic cable. This is a duplex optical port at which conformance testing is performed. (C/LM) 11802-4-1994

fibre optic station (FODTE) A compliant token ring station with an FOI as described in ISO/IEC TR 11802-4:1994. (C/LM) 11802-4-1994

fibre optic trunk coupling unit (FOTCU) A physical device that enables a fibre optic station to connect to a trunk cable. The FOTCU contains the means for inserting the fibre optic station into the ring, or conversely, bypassing the fibre optic station. (C/LM) 11802-4-1994

fibre pair Optical fibres interconnected to provide two continuous light paths terminated at both ends in an optical connector.local area networks. (C) 8802-12-1998

fibrillation (medical electronics) A continued, uncoordinated activity in the fibers of the heart, diaphragm, or other muscles consisting of rhythmical but asynchronous contraction and relaxation of individual fibers. (EMB) [47]

fictitious power (A) (polyphase circuit) At the terminals of entry, a vector equal to the (vector) sum of the ficititious powers for the individual terminals of entry. *Note:* The fictitious power for each terminal of entry is determined by considering each phase conductor and the common reference point as a single-phase circuit, as described for distortion power. The sign given to the distortion power in determining the fictitious power for each single-phase circuit shall be the same as that of the total active power. Fictitious power for a polyphase circuit has as its two rectangular components the reactive power and the distortion power. If the voltages have the same waveform as the corresponding currents, the magnitude of the fictitious power becomes the same as the reactive power. Fictitious power is expressed in volt-amperes when the voltages are in volts and the currents in amperes. **(B) (single-phase two-wire circuit)** At the two terminals of entry into a delimited region, a vector quantity having as its rectangular components the reactive power and the distortion power. *Note:* Its magnitude is equal to the square root of the difference of the squares of the apparent power and the amplitude of the active power. Its magnitude is also equal to the square root of the sum of the squares of the amplitudes of reactive power and distortion power. If voltage and current have the same waveform, the magnitude of the fictitious power is equal to the reactive power. The magnitude of the ficititious power is given by the equation

$$F = (U^2 - p^2)^{1/2}$$
$$= (Q^2 - D^2)^{1/2}$$
$$\left\{\sum_{r=1}^{r=\infty}\sum_{q=1}^{q=\infty}\left[E_r^2 I_q^2 - E_r E_q I_r I_q \cos(\alpha_r - \beta_r)\cos(\alpha_q - \beta_q)\right]\right\}^{1/2}$$

where the symbols are those of power, apparent (single-phase two-wire circuit). In determining the vector position of the fictitious power, the sign of the distortion power component must be assigned arbitrarily. Fictitious power is expressed in volt-amperes when the voltage is in volts and the current in amperes. *See also:* distortion power. (Std100) 270-1966

fiche *See:* microfiche.

fidelity (1) The degree with which a system, or a portion of a system, accurately reproduces at its output the essential characteristics of the signal that is impressed upon its input. (AP/PE/ANT) 145-1983s, 599-1985w
(2) (modeling and simulation) The degree of similarity between a model and the system properties being modeled. *Synonym:* correspondence. *See also:* model validation. (C) 610.3-1989w
(3) The degree to which the representation within a simulation is similar to a real-world object, feature, or condition in a measurable or perceivable manner. (DIS/C) 1278.1-1995

field (1) (television) One of the two (or more) equal parts into which a frame is divided in interlaced scanning. (BT/AV) [34]
(2) (computers) (record) A specified area used for a particular category of data, for example, a group of card columns used to represent a wage rate or a set of bit locations in a computer word used to express the address of the operand. (C) [20], [85]
(3) (diode-type camera tube) A single raster scan of the target. In the usual 2:1 interlace scan, two fields are required to completely scan the raster frame. (ED) 503-1978w
(4) (power cable systems) (in the field) The terms "field" or "in the field" refer generally to apparatus installed in the operating location. However, this may include material not yet installed or material that has been removed from the operating environment. (PE/IC) 400-1991
(5) (electric submersible pump cable) The term field or in the field may include cable not yet installed or cable that has been removed from its operating environment. (IA/PC) 1017-1985s
(6) (A) (data management) A specified area within a record, used for a particular data item; for example a group of card columns in which a telephone number is recorded. **(B) (data management)** The smallest unit of data that can be referred to in a database. *See also:* database segment. (C/Std100) 610.5-1990
(7) A group of any number of adjacent binary digits operated on as a unit. (SUB/PE) 999-1992w
(8) A series of contiguous bits treated as an instance of a particular data type that may be part of a higher level data structure. (DIS/C) 1278.1-1995
(9) In the shell command language, a unit of text that is the result of parameter expansion, arithmetic expansion, command substitution, or field splitting. During command processing, the resulting fields are used as the command name and its arguments. (C/PA) 9945-2-1993
(10) (A) A region near an electric charge, a source of electromagnetic radiation, or a magnet in which components or materials may be affected. **(B)** A portion of a computer instruction. *See also:* operation field; address field; operand field. **(C)** A portion of a data item such as the zone field of zoned decimal data. (C) 610.10-1994
(11) A defined subdivision of an ATLAS statement. (SCC20) 771-1998
(12) An area of the display screen that is reserved for the display of data or for user entry of a data item. In a database, it is a specified area used for a particular category of data (e.g., equipment operational status). *See also:* display. (PE/NP) 1289-1998

field accelerating relay A relay that functions automatically to maintain the armature current within limits, when accelerating to speeds above base speed, by controlling the excitation of the motor field. (IA/ICTL/IAC) [60]

field application relay (1) (power system device function numbers) A relay that automatically controls the application of the field excitation to an alternating-current (ac) motor at some predetermined point in the slip cycle. (SUB/PE) C37.2-1979s
(2) A relay that initiates the application of field excitation to a synchronous machine under specified conditions. *Note:* It is usually a polarized relay sensitive to the slip frequency of the induced field current. It may also remove excitation during an out-of-step condition. (SWG/PE) C37.100-1992

field assembled (1) Heating cable supplied in bulk form with terminating components and connections to be assembled by field personnel. (BT/IA/AV/PC) 152-1953s, 515.1-1995
(2) Heating cable or surface heating device supplied in bulk form with terminating components to be assembled in the field. (IA) 515-1997

field bar (line waveform distortion) A composite pulse, nominally of 8 ms duration, of reference-white amplitude. The field bar is composed of line bars as defined. This signal when displayed on a picture monitor has the form of the window signal shown in the corresponding figure.

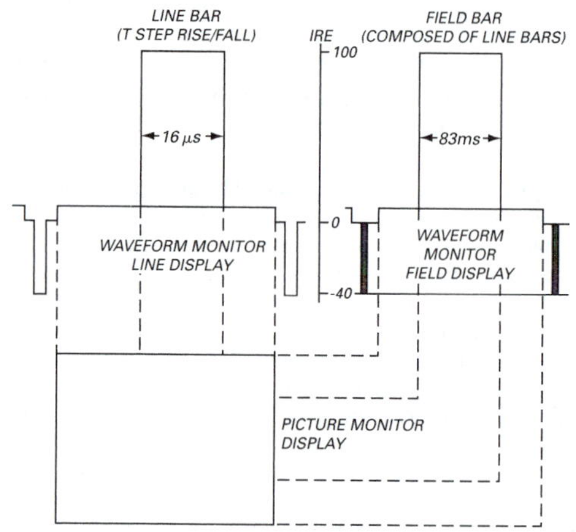

The window signal

field bar

(BT) 511-1979w

field-changing contactor (power system device function numbers) A contactor that functions to increase or decrease, in one step, the value of field excitation on a machine. (SUB/PE) C37.2-1979s

field charging Charging of aerosols by small ions moving under the influence of an electric field. (T&D/PE) 539-1990

field circuit breaker (power system device function numbers) A device that functions to apply or remove the field excitation of a machine. (SUB/PE) C37.2-1979s

field coil (A) (rotating machinery) (direct-current and salient-pole alternating-current machines). A suitably insulated winding to be mounted on a field pole to magnetize it. **(B) (rotating machinery)** (cylindrical-rotor synchronous machines). A group of turns in the field winding; occupying one pair of slots. *See also:* asynchronous machine; direct-current commutating machine. (PE) [9]

field-coil flange (rotating machinery) Insulation between the field coil and the pole shoe, and between the field coil and the member carrying the pole body, in a salient-pole machine. *See also:* stator; rotor. (PE) [9]

field contacts (sequential events recording systems) Electrical contacts that define the state of monitored equipment or a process. (PE/EDPG) [5], [1]

field contact voltage (sequential events recording systems) The voltage applied to field contacts for the purpose of sensing contact status. (PE/EDPG) [5], [1]

field control (motors) A method of controlling a motor by means of a change in the magnitude of the field current. *See also:* control. (IA/ICTL/APP/IAC) [69], [60]

field, critical *See:* critical field.

field, cutoff *See:* critical field.

field data Data from observations during field use. *Note:* The time stress conditions, and failure or success criteria should be stated in detail. (R) [29]

field decelerating relay A relay that functions automatically to maintain the armature current or voltage within limits, when decelerating from speeds above base speed, by controlling the excitation of the motor field. *See also:* relay. (IA/ICTL/IAC) [60]

field discharge (as applied to a switching device) A qualifying term indicating that the switching device has main contacts for energizing and de-energizing the field of a generator, motor, synchronous condenser or exciter; and has auxiliary contacts for short-circuiting the field through a discharge resistor at the instant preceding the opening of the main contacts. The auxiliary contacts also disconnect the field from the discharge resistor at the instant following the closing of the main contacts. *Note:* For dc generator operation, the auxiliary contacts may open before the main contacts close. (SWG/PE) C37.100-1992

field discharge circuit breaker (1) (rotating electric machinery) A circuit breaker having main contacts for energizing and deenergizing the field of a generator, motor, synchronous condenser, or rotating exciter, and having discharge contacts for short-circuiting the field through the discharge resistor at the instant preceding the opening of the circuit breaker main contacts. The discharge contacts also disconnect the field from the discharge resistor at the instant following the closing of the main contacts. For direct-current generator operation, the discharge contacts may open before the main contacts close. *Note:* When used in the main field circuit of an alternating or direct-current generator, motor, or synchronous condenser, the circuit breaker is designated as a main field discharge circuit breaker. When used in the field circuit of the rotating exciter of the main machine, the circuit breaker is designated as an exciter field discharge circuit breaker. (SWG/PE) C37.18-1979r

(2) (excitation systems for synchronous machines) A circuit breaker having main contacts for energizing and deenergizing the field of a synchronous machine or rotating exciter and having discharge contacts for short-circuiting the field through the discharge resistor prior to the opening of the circuit breaker main contacts. The discharge contacts also disconnect the field from the discharge resistor following the closing of the main contacts. *Notes:* 1. When used in the main field of a synchronous machine the circuit breaker is designated as a main field discharge circuit breaker. 2. When used in the field circuit of a rotating exciter of the main machine, the circuit breaker is designated as an exciter field discharge circuit breaker. (PE/EDPG) 421.1-1986r

field discharge protection A control function or device to limit the induced voltage in the field when the field current is disrupted or when an attempt is made to change the field current suddenly. *See also:* control. (IA/ICTL/APP/IAC) [69], [60]

field displacement (nonlinear, active, and nonreciprocal waveguide components) The condition, in a uniform single-mode waveguide, in which the presence of magnetized gyromagnetic material causes the transverse-plane field distributions to be significantly different in the two directions of propagation. (MTT) 457-1982w

field-disturbance sensor (measurement procedure for field-disturbance sensors) A device that employs a point source of radio-frequency (rf) energy to detect motion in the vicinity of the source, and in which the emitter and the receiver (or detector) are essentially at the same point, that is, a space-protected system. (EMC) 475-1983r

field-effect transistor A transistor in which the conduction is due entirely to the flow of majority carriers through a conduction channel controlled by an electric field arising from a voltage applied between the gate and source electrodes. (ED) 641-1987w

field emission Electron emission from a surface due directly to high-voltage gradients at the emitting surface. *See also:* electron emission. (ED) 161-1971w, [45]

field-enhanced photoelectric emission The increased photoelectric emission resulting from the action of a strong electric field on the emitter. *See also:* phototube. (ED) 161-1971w, [45]

field-enhanced secondary emission The increased secondary emission resulting from the action of a strong electric field on the emitter. *See also:* electron emission. (ED) [45], 161-1971w

field excitation current (Hall effect devices) The current producing the magnetic flux density in a Hall multiplier. (MAG) 296-1969w

field-failure protection The effect of a device, operative on the loss of field excitation, to cause and maintain the interruption of power in the motor armature circuit. (IA/ICTL/IAC) [60]

field-failure relay A relay that functions to disconnect the motor armature from the line in the event of loss of field excitation. *See also:* relay. (IA/ICTL/IAC) [60]

field flashing Short-time application of an external direct current source to the field of a synchronous generator to enable it to build up its voltage and become self-excited. (PE/EDPG) 1020-1988r

field forcing (1) (excitation systems for synchronous machines) A control function that rapidly drives the field current of a synchronous machine in the positive or in the negative direction. (PE/EDPG) 421.1-1986r

(2) A control function that temporarily overexcites or underexcites the field of a rotating machine to increase the rate of change of flux. *See also:* control. (IA/ICTL/IAC/APP) [60], [75]

field forcing relay A relay that functions to increase the rate of change of field flux by underexciting the field of a rotating machine. *See also:* relay. (IA/ICTL/IAC) [60]

field frame *See:* frame yoke.

field-free emission current (1) (general) The emission current from an emitter when the electric gradient at the surface is zero. (ED) [45], [84]

(2) (cathode) The electron current drawn from the cathode when the electric gradient at the surface of the cathode is zero. *See also:* electron emission. (ED) 161-1971w, [45]

field frequency (television) The product of frame frequency multiplied by the number of fields contained in one frequency. *See also:* television. (BT/AV) [34]

field intensity *See:* average detector.

field-intensity meter* A calibrated radio receiver for measuring field intensity. *See also:* interference measurement; interference. (IA) 54-1955w

* Deprecated.

field I^2R loss The product of the measured resistance, in ohms, of the field winding, corrected to a specified temperature, and the square of the field current in amperes. (PE) [9], [84]

field-lead insulation (rotating machinery) The dielectric material applied to insulate the enclosed conductor connecting the collector rings to the coil end windings. *Note:* Field leads also include the pole jumpers forming the series connection between the concentric windings on each pole. Where rectangular strap leads are employed, the insulation may consist of either taped mica and glass or moduled mica and glass or moduled mica and glass composites. Where circular rods are used, moulded laminate tubing is frequently employed as the primary insulation. *See also:* asynchronous machine; direct-current commutating machine. (PE) [9]

field length The number of words or characters in a field. (C) 610.5-1990w

field length type An indication of whether the field is fixed or variable in length. *Note:* If a field is a variable length type, the field length expresses the maximum length possible. (C) 610.5-1990w

field-limiting adjusting means The effect of a control function or device (such as a resistor) that limits the maximum or minimum field excitation of a motor or generator. *See also:* control. (IA/ICTL/IAC/APP) [60], [75]

field-locking *See:* lock.

field mark A mark that identifies the beginning or the end of a field. (C) 610.10-1994w

field measuring instrument A device used to sense and read out the electric or magnetic field intensities surrounding a VDT under test. (For this standard, this instrumentation consists of three parts: probe; readout detector, where the signal from the probe is processed and the data displayed; and any leads between the probe and readout detector.) (EMC) 1140-1994r

field mill A device in which a conductor is alternately exposed to the electric field to be measured and then shielded from it. *Note:* The resulting current induced in the conductor is a measure of the electric field strength at the conductor surface. *Synonym:* generating electric field meter. (T&D/PE) 539-1990, 1227-1990r

field molded joint (power cable joints) A joint in which the solid-dielectric joint insulation is fused and curved thermally at the job site. (PE/IC) 404-1986s

field pattern *See:* radiation pattern.

field pole (rotating machinery) A structure of magnetic material on which a field coil may be mounted. *Note:* There are two types of field poles: main and commutating. *See also:* direct-current commutating machine; asynchronous machine. (PE/EEC) [119]

field probe An electrically small field sensor or set of multiple field sensors with various electronics (for example, diodes, resistors, amplifiers, etc.). The output from a field probe cannot be theoretically determined from easily measured physical parameters. (EMC) 1309-1996

field programmable gate array (FPGA) A device containing many circuits whose interconnections and functions are programmable by the user. *Note:* Generally larger than a field programmable logic array. *See also:* dynamically programmable logic gate. (C) 610.10-1994w

field programmable logic array (FPLA) A logic array integrated circuit which can be programmed after manufacture, typically at the time of installation. *Note:* The programming is typically done by passing a high current through fusible links on the integrated circuit. *See also:* programmable logic array; field programmable gate array. (C) 610.10-1994w

field protection The effect of a control function or device to prevent overheating of the field excitation winding by reducing or interrupting the excitation of the shunt field while the machine is at rest. *See also:* control. (IA/ICTL/IAC/APP) [60], [75]

field protective relay A relay that functions to prevent overheating of the field excitation winding by reducing or interrupting the excitation of the shunt field. *See also:* relay. (IA/ICTL/IAC) [60]

field relay (power system device function numbers) A relay that functions on a given or abnormally low value or failure of machine field current, or on an excessive value of the reactive component of armature current in an alternating-current (ac) machine indicating abnormally low field excitation. (PE/SUB) C37.2-1979s

field-reliability test A reliability compliance or determination test made in the field where the operating and environmental conditions are recorded and the degree of control is stated. (R) [29]

field-renewable fuse *See:* renewable fuse.

field-renewable fuse unit (1) (high-voltage switchgear) A fuse unit that, after circuit interruption, may be readily restored for service by the replacement of the fuse link or refill unit. (SWG/PE) C37.40-1993

(2) *See also:* renewable fuse; fuse unit. (SWG/PE) C37.100-1981s

field-renewable fuse or fuse unit *See:* renewable fuse unit.

field replaceable unit (FRU) The smallest subassembly that can be swapped in the field to repair afault. (C/BA) 896.3-1993w

field-reversal permanent-magnet focusing (microwave tubes) Magnetic focusing by a limited series of field reversals, not periodic, whose location is usually related to breaks in the slow-wave circuit. *See also:* magnetron. (ED) [45]

field rheostat A rheostat designed to control the exciting current of an electric machine. (IA/IAC) [60]

field sensor An electrically small device without electronics(passive) that is used for measuring electric or magnetic fields, with a minimum of perturbation to field being measured. The field sensor transfer function (ratio of output signal-to-input electromagnetic field) can be theoretically determined from measured physical (geometrical) properties, such as length, radius, area, etc., as well as the electrical characteristics of the construction material. The measured physical properties must be traceable to internationally accepted standards via a national standards authority (for example, NIST in the USA). (EMC) 1309-1996

field separator A character or byte used to identify a boundary between two fields. (C) 610.5-1990w

field sequential (color television) Sampling of primary colors in sequence with successive television fields. (BT/AV) 201-1979w

field shunting control (shunted-field control) A system of regulating the tractive force of an electrically driven vehicle by shunting, and thus weakening, the traction motor series fields by means of a resistor. *See also:* multiple-unit control. (EEC/PE) [119]

field splice (nuclear power generating station) A permanent joining and reinsulating of conductors in the field to meet the service conditions required. (PE/NP) 380-1975w, 383-1974r

field spool (rotating machinery) A structure for the support of a field coil in a salient-pole machine, either constructed of insulating material or carrying field-spool insulation. *See also:* direct-current commutating machine; asynchronous machine. (PE) [9]

field-spool insulation (rotating machinery) Insulation between the field spool and the field coil in a salient-pole machine. *See also:* asynchronous machine; direct-current commutating machine. (PE) [9]

field strength (1) (electromagnetic wave) A general term that usually means the magnitude of the electric field vector, commonly expressed in volts per meter, but that may also mean the magnitude of the magnetic field vector, commonly expressed in amperes (or ampere-turns) per meter. *Note:* At frequencies above about 100 megahertz, and particularly above 1000 megahertz, field strength in the far zone is sometimes identified with power flux density P. For a linearly polarized wave in free space $P = E^2/(\mu_v \epsilon_v)$, where E is the electric field strength, and μ_v and ϵ_v are the magnetic and electric constants of free space, respectively. When P is expressed in watts per square meter and E in volts per meter, the denominator is often rounded off to 120π. *See also:* magnetic field strength; measurement system; electric field strength. (IM/COM) 284-1968w, [48]

(2) (overhead power lines) *See also:* voltage gradient. (T&D/PE) 539-1990

(3) The magnitude of the electric field vector. (T&D/PE) 1260-1996

(4) *See also:* radio field strength. (AP/PROP) 211-1997

field strength meter A calibrated radio receiver for measuring field strength. These meters employ a shielded loop antenna, which measures the magnetic component of the electromagnetic field, and then converts it to an electric field by multiplying the magnetic field strength by the impedance of free space for a plane wave. (PE/T&D/IA) 1260-1996, 169-1955w, 54-1955w

field system (rotating machinery) The portion of a direct-current or synchronous machine that produces the excitation flux. *See also:* asynchronous machine; direct-current commutating machine. (PE) [9]

field terminal (rotating machinery) A terminations for the field winding. *See also:* rotor; stator. (PE) [9]

field tests (1) (power cable joints) Tests that may be made on the cable and accessories after installation. (PE/IC) 404-1986s

(2) (metal-clad and station-type cubicle switchgear) (metal-enclosed interrupter switchgear) (metal-enclosed bus and calculating losses in isolated-phase bus) (metal-enclosed low-voltage power circuit-breaker switchgear) Tests made after the assembly has been installed at its place of utilization. (SWG/PE) C37.23-1987r, C37.20.2-1993, C37.20.4-1996, C37.20.1-1993r, C37.20.3-1996

(3) Tests that may be made on a cable system (including the high-voltage cable terminations) by the user after installation, as an acceptance or proof test. (PE/IC) 48-1996

(4) (A) Tests made on operating systems usually for the purpose of investigating the performance of switchgear or its component parts under conditions that cannot be duplicated in the factory. *Note:* Field tests are usually supplementary to factory tests and therefore may not provide a complete investigation of capabilities. **(B)** Tests made after the assembly has been installed at its place of utilization. (SWG/PE) C37.100-1992

field-time waveform distortion (video signal transmission measurement) The linear TV waveform distortion of time components from 64 μs to 16 ms, that is, time components of field-time domain. (BT) 511-1979w

field-turn insulation (rotating machinery) Insulation in the form of strip or tape separating the individual turns of a field winding. *See also:* asynchronous machine; direct-current commutating machine. (PE) [9]

field uniformity The extent to which the magnitude and direction of a field are uniform at any instant of time and at all points within a defined region. (T&D/PE) 539-1990

field voltage, base The synchronous machine field voltage required to produce rated voltage on the air-gap line of the synchronous machine at field temperatures of 75°C for field windings designed to operate at rating with a temperature rise of 60°C or less; or 100°C for field windings designed to operate at rating with a temperature rise greater than 60°C. *Note:* This defines one per unit excitation system voltage for use in computer representation of excitation systems. (PE/EDPG) 421-1972s

field voltage, no-load The voltage required across the terminals of the field winding of an electric machine under conditions of no load, rated speed and terminal voltage, and with the field winding at 25°C. (PE/EDPG) 421-1972s

field voltage, rated-load The voltage required across the terminals of the field winding of an electric machine under rated continuous-load conditions with the field winding at 75°C for field windings designed to operate at rating with a temperature rise of 60°C or less; or 100°C for field windings designed to operate at rating with a temperature rise greater than 60°C. (PE/EDPG) 421-1972s

field vulcanized A joint that is constructed in the field with externally applied heat and pressure to cross-link the joint dielectric. (PE/IC) 404-1993

field winding (rotating machinery) (excitation systems for synchronous machines) A winding on either the stationary or the rotating part of a synchronous machine whose sole purpose is the production of the main electromagnetic field of the machine. (PE/EDPG) [9], 421.1-1986r

field winding terminals (excitation systems for synchronous machines) The place of input to the field winding of the synchronous machine. If there are brushes and sliprings these are to be considered to be part of the field winding. (PE/EDPG) 421.1-1986r

FIFO *See:* first-in, first-out.

FIFO special file (FIFO) A type of file with the property that data written to such a file is read on a first-in-first-out basis. (PA/C) 9945-1-1996, 9945-2-1993, 1003.5-1999

fifo special file A type of file with the property that data written to such a file is read on a first-in-first-out basis. Other characteristics of FIFOs are described in packages POSIX_Files and POSIX_IO. *Synonym:* FIFO. (C/PA) 1003.5b-1995

fifth generation A period during the evolution of electronic computer in which very large scale integration is employed, along with approaches to computing that include artificial intelligence, knowledge engineering, and distributed processing. *Note:* Introduced in mid-1980's, this generation of computers has not yet reached maturity. *See also:* second generation; third generation; first generation; fourth generation. (C) 610.10-1994w

fifth generation language A computer language that incorporates the concepts of knowledge-based systems, expert systems, inference engines, and natural language processing. *Contrast:* machine language; assembly language; fourth generation language; high-order language. (C) 610.12-1990, 610.13-1993w

fifth normal form One of the forms used to characterize relations; a relation is said to be in fifth normal form if it is in fourth normal form and if every join dependency in the relation is a consequence only of the candidate keys of the relation. *Synonym:* projection/join normal form. (C) 610.5-1990w

fifth voltage range *See:* voltage range.

50 cm test point (50TP) The acoustic test point 50 cm from the front center of the handsfree telephone (HFT) and 30 cm above the test table. (COM/TA) 1329-1999

fifty percent disruptive discharge voltage (V_{50}) The prospective value of the test voltage that has a 50% probability of producing a disruptive discharge. (PE/PSIM) 4-1995

figurative constant A data name that is reserved for a specific constant in a programming language. For example, the data name THREE may be reserved to represent the value 3. *See also:* literal. (C) 610.12-1990

figure (metric practice) (numerical) An arithmetic value expressed by one or more digits. (SCC14/QUL) SI 10-1997, 268-1982s

figure of merit (1) (magnetic amplifier) The ratio of power amplification to time constant in seconds. (MAG) 107-1964w

(2) (thermoelectric couple)

$$\alpha^2 \left[(\rho_1\kappa_1)^{1/2} + (\rho_2\kappa_2)^{1/2} \right]^{-2}$$

where α is the Seebeck coefficient of the couple and ρ_1, ρ_2, κ_1, and κ_2 are the respective electric resistivities and thermal conductivities of materials 1 and 2. *Note:* This figure of merit applies to materials for the thermoelectric devices whose operation is based on the Seebeck effect or the Peltier effect. *See also:* thermoelectric device.

(3) (thermoelectric couple, ideal)

$$\bar{\alpha}^{-2} [(\overline{\rho_1\kappa_1})^{1/2} + (\overline{\rho_2\kappa_2})^{1/2}]^{-2}$$

where α is the average value of the Seebeck coefficient of the coupled and $\rho_1\kappa_1$ and $\rho_2\kappa_2$ are the average values of the products of the respective electric resistivities and thermal conductivities of materials 1 and 2, where the averages are found by integrating the parameters over the specified temperature range of the couple. *See also:* thermoelectric device.

(4) (thermoelectric material) The quotient of "the square of the absolute Seebeck coefficient α" by "the product of the electric resistivity ρ and the thermal conductivity κ"

$\alpha^2/\rho\kappa$

Note: This figure of merit applies to materials for thermoelectric devices whose operation is based on the Seebeck effect or the Peltier effect. *See also:* thermoelectric device. (ED) [46], 221-1962w

(5) (of an antenna) The ratio of the gain to the noise temperature of an antenna. *Notes:* 1. Usually the antenna-receiver system figure of merit is specified. For this case, the figure of merit is the gain of the antenna divided by the system noise temperature referred to the antenna terminals. 2. The system figure of merit at any reference plane in the RF system is the same as that taken at the antenna terminals since both the gain and system noise temperature are referred to the same reference plane. (AP/ANT) 145-1993

(6) (dynamically tuned gyro) A design constant that relates the rotor polar moment of inertia and the principal moments of inertia of the gimbal(s). A simplified expression for the figure of merit is:

$$\mathrm{FOM} = \frac{C}{\sum_{1}^{N} (A_n + B_n - C_n)}$$

where

C = rotor polar moment of inertia

A_n, B_n = transverse moments of inertia of the nth gimbal

C_n = polar moment of inertia of the nth gimbal

(AES/GYAC) 528-1994

(7) *See also:* pumped figure of merit; beta figure of merit; ferrite devices figure of merit; detector figure of merit.

filament (electron tube) A hot cathode, usually in the form of a wire or ribbon, to which heat may be supplied by passing current through it. *Note:* This is also known as a filamentary cathode. *See also:* electrode. (ED) 161-1971w, [45]

filamentary transistor A conductivity-modulation transistor with a length much greater than its transverse dimensions. *See also:* semiconductor; transistor. (ED) 216-1960w

filament current The current supplied to a filament to heat it. *See also:* heater current; electronic controller. (ED) [45]

filament power supply (electron tube) The means for supplying power to the filament. *See also:* power pack. (ED) [45]

filament voltage The voltage between the terminals of a filament. *See also:* electronic controller; electrode voltage. (EEC/PE) [119]

File An instance of the class `IEEE1451_File` or of a subclass thereof. (IM/ST) 1451.1-1999

file (1) (computers) A collection of related records treated as a unit. *Note:* Thus in inventory control, one line of an invoice forms an item, a complete invoice forms a record, and the complete set of such records forms a file. (C) [20], [85]

(2) (software) (data management) A set of related records treated as a unit. For example, in stock control, a file could consist of a set of invoice records. *See also:* logical file; data set; data file. (C) 610.5-1990w, 610.12-1990

(3) (information transfer) One named collection of data. (MM/C) 949-1985w

(4) An object that can be written to, or read from, or both. A file has certain attributes, including access permissions and type. File types include regular file, character special file, block special file, FIFO special file, and directory. Other types of files may be defined by the implementation. (C/PA) 9945-1-1996, 9945-2-1993

(5) An object that can be written to, or read from, or both. A file has certain attributes, including access permissions and type. File types include regular file, character special file, block special file, FIFO special file, socket, character special file for use with XTI calls, and directory. Other types of files may be defined by the implementation. (C) 1003.5-1999

(6) A set of related records usually treated as a named unit of storage. (C/MM) 855-1990

(7) An abstraction of the mechanism for the allocation, deallocation, initialization, and use of memory resources in a device. (IM/ST) 1451.1-1999

file access mode The type of access allowed for a given file and a given user. For example, the file access mode for a given file might be read-only access for one user, and read/write access for another. *Synonym:* access type. (C) 610.5-1990w

file access permissions A concept of the underlying system, as follows: The standard file access control mechanism uses the file permission bits, as described below. These bits are set at file creation by *open*(), *creat*(), *mkdir*(), and *mkfifo*() and are changed by *chmod*(). These bits are read by *stat*() or *fstat*(). Implementations may provide *additional* or *alternate* file access control mechanisms, or both. An additional access control mechanism shall only further restrict the access permissions defined by the file permission bits. An alternate access control mechanism shall:

1) Specify file permission bits for the file owner class, file group class, and file other class of the file, corresponding to the access permissions, to be returned by *stat*() or *fstat*().
2) Be enabled only by explicit user action, on a per-file basis, by the file owner or a user with the appropriate privilege.
3) Be disabled for a file after the file permission bits are changed for that file with *chmod*(). The disabling of the alternate mechanism need not disable any additional mechanisms defined by an implementation.

Whenever a process requests file access permission for read, write, or execute/search, if no additional mechanism denies access, access is determined as follows:

1) If a process has the appropriate privilege:
   a) If read, write, or directory search permission is requested, access is granted.
   b) If execute permission is requested, access is granted if execute permission is granted to at least one user by the file permission bits or by an alternate access control mechanism; otherwise, access is denied.
2) Otherwise:
   a) The file permission bits of a file contain read, write, and execute/search permissions for the file owner class, file group class, and file other class.
   b) Access is granted if an alternate access control mechanism is not enabled and the requested access permission bit is set for the class (file owner class, file group class, or file other class) to which the process belongs, or if an alternate access control mechanism is enabled and it allows the requested access; otherwise, access is denied.

(C/PA/C/PA) 9945-1-1996, 9945-2-1993

file attribute A property, feature, or characteristic of a file. (C) 610.5-1990w

file attributes The name and other identifiable properties of a file. (C/PA) 1238.1-1994w

file cleanup The removal of superfluous data from a file. *Synonym:* file tidying. (C) 610.5-1990w

file description *See:* open file description.

file descriptor (1) A value used to identify an open file for the purpose of file access. File descriptors are unique within a process. (C/PA) 1003.5-1992r

(2) A per-process unique nonnegative integer value used to identify an open file for the purpose of file access. (C/PA) 1003.5-1999, 9945-1-1996, 9945-2-1993

file directory (A) A list of files and their locations within a computer system. *See also:* catalog. **(B)** A list of the files and their locations on a particular storage device or volume. (C) 610.5-1990

file gap (1) An area on a storage medium, such as tape, used to indicate the end of a file. (C) [20], [85]

(2) **(data management)** An unused area on a data medium between the end of one file or group of data and the beginning of another file or group of data. (C) 610.5-1990w
(3) An area between two consecutive files used to indicate the end of the file. *Note:* Frequently used for other purposes such as to indicate the end or beginning of some other group of data. (C) 610.10-1994w

file group class A property of a file indicating access permissions for a process related to the group identification of the process. A process is in the file group class of a file if the process is not in the file owner class and if the effective group ID or one of the supplementary group IDs of the process matches the group ID associated with the file. Other members of the class may be implementation defined. (C/PA) 1003.5-1999, 9945-1-1996, 9945-2-1993

file hierarchy A concept of the underlying system, as follows. Files in the system are organized in a hierarchical structure in which all of the nonterminal nodes are directories and all of the terminal nodes are any other type of file. Because multiple directory entries may refer to the same file, the hierarchy is properly described as a "directed graph." (C/PA) 9945-1-1996, 9945-2-1993

file layout The arrangement and structure of data in a file. *Synonym:* file organization. (C) 610.5-1990w

file-locking *See:* lock.

file maintenance (1) (computers) The activity of keeping a file up to date by adding, changing, or deleting data. (C) [20], [85]
(2) **(data management)** The activity of adding, changing, or deleting data in a file as needed. (C) 610.5-1990w

file mark A mark that identifies the end of a file. (C) 610.10-1994w

file mode An object containing the file permission bits and other characteristics of a file. (C/PA) 9945-1-1996, 9945-2-1993

file mode bits The file permission bits, set-user-ID-on- execution bit (S_ISUID), and set-group-ID-on-execution bit (S_ISGID) of a file. (C/PA) 9945-2-1993

file name (A) One or more characters used to identify a file. **(B)** A name associated with a set of file data or output data. (C) 610.5-1990

filename (1) A name consisting of 1 to {NAME_MAX} bytes used to name a file. The characters composing the name may be selected from the set of all character values excluding the slash character and the null character. The filenames dot and dot-dot have special meaning. A filename is sometimes referred to as a pathname component. *See also:* pathname resolution. (C/PA) 9945-1-1996, 9945-2-1993
(2) A POSIX.1 filename with characters drawn from the POSIX.1 portable filename character set. (C/PA) 1387.2-1995
(3) A nonempty string that is used to name a file. A filename consists of, at most, `POSIX_Limits.Filename_Maxima'Last` components of type `POSIX.POSIX_Character`. The characters composing the name may be selected from the set of all the character values excluding the slash character and the null character. A filename is sometimes referred to as a pathname component. (C) 1003.5-1999

file offset The byte position in the file where the next I/O operation begins. Each open file description associated with a regular file, block special file, or directory has a file offset. A character special file that does not refer to a terminal device may have a file offset. There is no file offset specified for a pipe or FIFO. (C/PA) 9945-1-1996, 9945-2-1993, 1003.5-1999

filename character string A sequence of characters from the portable filename character set, not including the / (slash) character. Within software definition files of exported catalogs, all such strings shall be encoded using IRV. (C/PA) 1387.2-1995

file organization The order of physical records within a file that determines the access method to be implemented in order to use the file. *See also:* file layout. (C) 610.5-1990w

File-Oriented Interpretive Language (FOIL) A computer language, based on FORTRAN, used to provide conversational lesson-writing; used commonly in computer-aided instruction applications. (C) 610.13-1993w

file other class A property of a file indicating access permissions for a process related to the user and group information of the process. A process is in the file other class of a file if the process is not in the file owner class or file group class. (C/PA) 1003.5-1999, 9945-2-1993, 9945-1-1996

file owner class A property of a file indicating access permissions for a process related to the user identification of the process. A process is in the file owner class of a file if the effective user ID of the process matches the user ID of the file. (C/PA) 1003.5-1999, 9945-1-1996, 9945-2-1993

file permission Information about a file that is used, along with other information, to determine whether a process has read, write, or execute/search permission to a file. The file permission information is divided into three parts: owner, group, and other. Each part is used with the corresponding file class of processes. (C) 1003.5-1999

filename portability A concept of the underlying system, as follows: Filenames should be constructed from the portable filename character set because the use of other characters can be confusing or ambiguous in certain contexts. (C/PA) 9945-1-1996, 9945-2-1993

file permission bits Information about a file that is used, along with other information, to determine if a process has read, write, or execute/search permission to a file. The bits are divided into three parts: owner, group, and other. Each part is used with the corresponding file class of processes. These bits are contained in the file mode. (C/PA) 9945-1-1996, 9945-2-1993

file processing The periodic updating of one or more master files to reflect the effects of current data, often from a transaction file. For example, a monthly run updating the inventory file. (C) 610.2-1987

file-protection ring *See:* write ring.

file serial number (1) A per-file-system unique identifier for a file. File serial numbers are unique throughout a file system. (C/PA) 9945-1-1996, 9945-2-1993
(2) A per-file-system unique value for a file. File serial numbers are unique throughout a file system. (C) 1003.5-1999

file server On a network, a server that provides access to requesters at the file level; that is, an entire file or a file segment is sent to a requestor. *See also:* mail server; disk server; database server; print server; network server; terminal server. (C) 610.7-1995

file system (1) A collection of files and certain of their attributes. It provides a name space for file serial numbers referring to those files. (C/PA) 9945-1-1996, 9945-2-1993
(2) A collection of files, together with certain of their attributes. Each file system provides a separate binding of file serial numbers to files. A given file serial number is associated with at most one file in a file system, but it may refer to distinct files in distinct file systems. In other words, each file system defines a new *name space*, giving meaning to the *names* (file serial numbers) that designate files. (C) 1003.5-1999

fileset Defines the files that make up a software object, and is the lowest level of software object that can be specified as input to the software administration utilities. (C/PA) 1387.2-1995

file storage structure The storage directories in the software packaging layout under which the actual software files for each fileset are located. (C/PA) 1387.2-1995

filestore action One of the actions specified as part of the definition of the virtual filestore. (C/PA) 1238.1-1994w

file tidying *See:* file cleanup.

file times update A concept of the underlying system, as follows. Each file has three distinct associated time values: *st_atime*, *st_mtime*, and *st_ctime*. The *st_atime* field is associated with the times that the file data is accessed; *st_mtime* is associated with the times that the file data is modified; and *st_ ctime* is associated with the times that file status is changed. These values are returned in the file characteristics structure. Any function in this standard that is required to read or write file data or change the file status indicates which of the appropriate time-related fields are to be "marked for update." If an implementation of such a function marks for update a time- related field not specified by this standard, this shall be documented, except that any changes caused by pathname resolution need not be documented. For the other functions in this standard (those that are not explicitly required to read or write file data or change file status, but that in some implementations happen to do so), the effect is unspecified. An implementation may update fields that are marked for update immediately, or it may update such fields periodically. When the fields are updated, they are set to the current time and the update marks are cleared. All fields that are marked for update shall be updated when the file is no longer open by any process or when a *stat*() or *fstat*() is performed on the file. Other times at which updates are done are unspecified. Updates are not done for files on read-only file systems. (C/PA) 9945-1-1996, 9945-2-1993

file transfer protocol A protocol for transferring files between computers. (C) 610.7-1995

file type *See:* file.

filiform corrosion *See:* underfilm corrosion.

fill (1) (computer graphics) To insert a color, pattern, or hatch into a closed polygon or area bounded by lines or curves. *Synonyms:* area fill; polygon fill. (C) 610.6-1991w
(2) A sequence of data symbols of any combination of 0 and 1 data bits (as opposed to non-data-J and non-data-K bits) whose primary purpose is to maintain timing and spacing between frames and tokens. (C/LM) 8802-5-1998
(3) (data management) *See also:* filler character; zero fill; character fill. (C) 610.5-1990w

fill area A display element that consists of a closed polygon that is hollow or filled with a uniform color, pattern, or hatch.

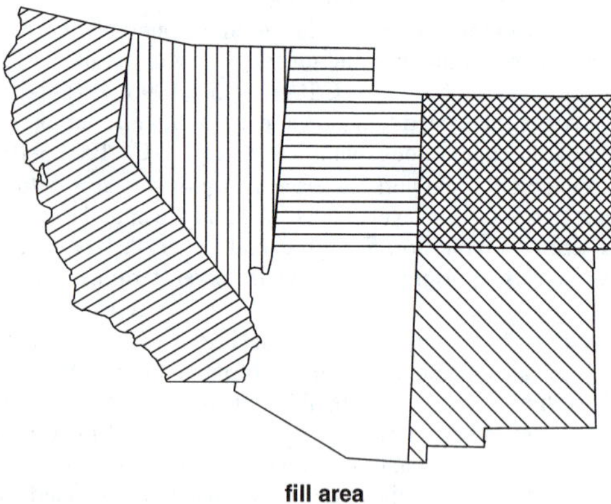

fill area

(C) 610.6-1991w

fill area attribute A characteristic of a filled region. For example, color index, interior style. (C) 610.6-1991w

filled A joint that consists of an outer shell that is filled with an insulating material to occupy the space around the individual insulated conductor(s). (PE/IC) 404-1993

filled-core annular conductor A conductor composed of a plurality of conducting elements disposed around a nonconducting supporting material that substantially fills the space enclosed by the conducting elements. *See also:* conductor. (T&D/PE) [10]

filled-system thermometer An all-metal assembly consisting of a bulb, capillary tube, and Bourdon tube (bellows and diaphragms are also used) containing a temperature-responsive fill. A mechanical device associated with the Bourdon is designed to provide an indication or record of temperature. *See also:* Bourdon. (PE/PSIM) 119-1974w

filled tape Fabric tape that has been thoroughly filled with a rubber or synthetic compound, but not necessarily finished on either side with this compound. *See also:* conductor. (T&D/PE) [10]

filler (1) (rotating machinery) Additional insulating material used to insure a tight depth-wise fit in the slot. *See also:* rotor; stator. (PE) [9]
(2) (mechanical recording) The inert material of a record compound as distinguished from the binder. *See also:* phonograph pickup. (SP) [32]
(3) (data management) One or more data items adjacent to an item of data that forces that item to take on a specified size; for example, in an 80-character output record in which a 30-character NAME, 20-character ADDRESS, and a 3-character AGE is to be placed, filler would be used to expand the data to be 80 characters. *Synonym:* filler strip. *See also:* padding; pad; character fill. (C) 610.5-1990w
(4) Three- or six-bit reserved code patterns that are appended to the end of individual MII channel data streams to equalize the stream lengths in all four channels.local area networks. (C) 8802-12-1998

filler character (A) A character used to occupy an area on a printed medium; for example, on a legal document, dashes or asterisks used to fill out a field to ensure that nothing is added to the field once the document has been issued. *See also:* filler. **(B)** A character that does not itself convey data but that may delete unwanted data, as in blanks used to fill out a field. *See also:* filler; character fill. (C) 610.5-1990

filler strip *See:* filler.

filling compound (power cable joints) A dielectric material poured or otherwise injected into the joint housing. Filling compounds may require heating or mixing prior to filling. Some filling compounds may also serve as the insulation. (PE/IC) 404-1986s

filling factor A factor that describes the fraction of energy flow confined in the substrate of a planar transmission line of inhomogeneous cross section.

a) *dielectric filling factor (single-layer microstrip).* The ratio of

$$\frac{\epsilon_r eff - 1}{\epsilon_r - 1}$$

where
ϵ_r = relative dielectric constant of the substrate of a microstrip line
$\epsilon_r eff$ = effective dielectric constant

b) *dielectric loss filling factor.* The ratio of

$$\frac{\tan \delta_{eff}}{\tan \delta}$$

where
$\tan \delta$ = dielectric loss tangent of the substrate of a microstrip line
$\tan \delta_{eff}$ = effective dielectric loss tangent of the line

c) *magnetic filling factor.* The ratio of

$$\frac{\dfrac{1}{\mu_{eff}} - 1}{\dfrac{1}{\mu} - 1}$$

where
μ = relative permeability of the substrate of a microstrip line
μ_{eff} = effective relative permeability

d) *magnetic loss filling factor.* The ratio of

$$\frac{\tan \delta_{m,\ eff}}{\tan \delta_m}$$

where
$\tan \delta_m$ = magnetic loss tangent of the substrate of a microstrip line
$\tan \delta_{m,\ eff}$ = effective magnetic loss tangent of the line

Synonym: filling fraction. (MTT) 1004-1987w

filling fraction *See:* filling factor.

fill light (illuminating engineering) Supplementary illumination used to reduce shadow or contrast. (EEC/IE) [126]

film (1) (rotating machinery) Sheeting having a nominal thickness not greater than 0.030 centimeters and being substantially homogeneous in nature. *See also:* electrochemical valve; direct-current commutating machine; asynchronous machine. (PE) [9]
(2) (electrochemical valve) The layer adjacent to the valve metal and in which is located the high-potential drop when current flows in the direction of high impedance. *See also:* electrochemical valve. (PE/EEC) [119]

film frame In micrographics, a line on microfilm, perpendicular to the document reference edge, on which binary characters may be written or read. (C) 610.2-1987

film integrated circuit An integrated circuit whose elements are films formed in situ upon an insulating substrate. *Note:* To further define the nature of a film integrated circuit, additional modifiers may be prefixed. Examples are: thin-film integrated circuit, and thick-film integrated circuit. *See also:* magnetic thin film; integrated circuit; electrochemical valve; thin film. (ED) 274-1966w, [46]

film storage *See:* magnetic thin film storage.

FILO *See:* first-in, last-out.

filter (1) (wave filter) A transducer for separating waves on the basis of their frequency. *Note:* A filter introduced relatively small insertion loss to waves in one or more frequency bands and relatively large insertion loss to waves of other frequencies. (SP) 151-1965w
(2) (A) A device or program that separates data, signals, or material in accordance with specified criteria. **(B)** A mask. (C) [20], [85]
(3) (illuminating engineering) A device for changing, by transmission or reflection, the magnitude or the spectral composition, or both, of the flux incident upon it. Filters are called selective (or colored) or neutral, according to whether or not they alter the spectral distribution of the incident flux. (EEC/IE) [126]
(4) (broadband local area networks) A circuit that selects or rejects one or more components of a signal related to frequency. (LM/C) 802.7-1989r
(5) A generic term used to describe those types of equipment whose purpose is to reduce the harmonic current or voltage flowing in or being impressed upon specific parts of an electrical power system, or both. (IA/SPC) 519-1992
(6) A command whose operation consists of reading data from standard input or a list of input files and writing data to standard output. Typically, the function of a filter is to perform some transformation on the data stream. (C/PA) 9945-2-1993
(7) (A) A circuit that eliminates certain portions of a signal, by frequency, voltage, or some other parameter. **(B)** A mathematical model which performs the same function on a sampled version of the signal. *Synonym:* mask. (C) 610.10-1994
(8) An assertion about the presence or value of certain attributes of an entry in order to limit the scope of a search. (C/PA) 1328.2-1993w, 1224.2-1993w, 1326.2-1993w, 1327.2-1993w
(9) *See also:* low-pass filter; band-pass filter; high-pass filter. (PE) 599-1985w

filter, active *See:* active filter.

filter, all-pass *See:* all-pass filter.

filter attenuation band (filter stop band) A continuous range of frequencies over which the filter introduces an insertion loss whose minimum value is greater than a specific value. (CAS) [13]

filter, band-elimination *See:* band-elimination filter.

filter bank A contiguous set of filters covering the Doppler frequency range of interest, used to separate moving targets. Commonly used in continuous wave (CW) and pulsed-Doppler radars and in the moving target detector (MTD) for detecting moving targets in clutter. (AES) 686-1997

filter, Butterworth *See:* Butterworth filter.

filter capacitor A capacitor used as an element of an electric wave filter. *See also:* electronic controller. (IA/IAC) [60]

filter capacitors Capacitors utilized with inductors and/or resistors for controlling harmonic problems in the power system, such as reducing voltage distortion due to large rectifier loads or arc furnaces.power systems relaying. (T&D/PE) 1036-1992, C37.99-2000

filter, Chebyshev *See:* Chebyshev filter.

filter, comb *See:* comb filter.

filter, damped *See:* damped filter.

filter effectiveness (shunt) Defined by the following two terms:

ρ_f = the impedance ratio that determines the per unit current that will flow into the shunt filter

ρ_s = the impedance ratio that determines the per unit current that will flow into the power source

ρ_f should approach unity and ρ_s should be very small at the tuned frequency. (IA/SPC) 519-1992

filter factor (illuminating engineering) The transmittance of "black light" by a filter. *Note:* The relationship among these terms is illustrated by the following formula for determining the luminance of fluorescent materials exposed to "black light":

candelas per square meter

$$= \frac{1}{\pi^*} \frac{\text{fluorens}}{\text{square meter}} \times \text{glow factor} \times \text{filter factor}$$

*π is omitted when luminance is in footlamberts and the area is in square feet. When integral-filter "black light" lamps are used, the filter factor is dropped from the formula because it already has been applied in assigning fluoren ratings to these lamps. (EEC/IE) [126]

filter, high-pass *See:* high-pass filter.

filter impedance compensator An impedance compensator that is connected across the common terminals of electric wave filters when the latter are used in parallel in order to compensate for the effects of the filters on each other. *See also:* network analysis; filter. (Std100) 270-1966w

filter inductor An inductor used as an element of an electric wave filter. *See also:* electronic controller. (IA/IAC) [60]

filter, low-pass *See:* low-pass filter.

filter matching loss The loss in output signal-to-noise ratio relative to a matched filter, caused by using a filter whose response is not matched to the transmitted signal. (AES/RS) 686-1990

filter mismatch loss The loss in output signal-to-noise ratio of a filter relative to the signal-to-noise ratio from a matched filter. *Note:* Filter mismatch loss is caused by using a filter whose response is not matched to the transmitted signal. (AES) 686-1997

filter pass band A frequency band of low attenuation (low relative to other regions termed stop bands). *See also:* filter transmission band. (CAS) [13]

filter, passive *See:* passive filter.

filter reactor (power and distribution transformers) A reactor used to reduce harmonic voltage in alternating-current or direct-current circuits. *See also:* reactor. (PE/TR) C57.12.80-1978r, [57]

filter, rejection *See:* rejection filter; filter.

filters (power supplies) Resistance-capacitance or inductance-capacitance networks arranged as low-pass devices to attenuate the varying component that remains when alternating-current voltage is rectified. *Note:* In power supplies without subsequent active series regulators, the filters determine the amount of ripple that will remain in the direct-current output. In supplies with active feedback series regulators, the regulator mainly controls the ripple, with output filtering serving chiefly for phase-gain control as a lag element. (AES/PE) [41], [78]

filter, series *See:* series filter.

filter, shunt *See:* shunt filter.

filter, sound effects *See:* sound-effects filter.

filter stop band A frequency band of high attenuation (high relative to other regions termed pass bands). *See also:* filter attenuation band. (CAS) [13]

filter transmission band A continuous range of frequencies over which the filter introduces an insertion loss whose maximum value does not exceed a specified value. *See also:* filter pass band. (CAS) [13]

filter, tuned A filter generally consisting of combinations of capacitors, inductors, and resistors that have been selected in such a way as to present a relative minimum (maximum) impedance to one or more specific frequencies. For a shunt (series) filter, the impedance is a minimum (maximum). Tuned filters generally have a relatively high Q (X/R). (IA/SPC) 519-1992

final approach path *See:* approach path.

final condition The values assumed by the variables in a system, model, or simulation at the completion of some specified duration of time. *Synonym:* equilibrium condition. *Contrast:* boundary condition; initial condition. (C) 610.3-1989w

final contact pressure The force exerted by one contact against the mating contact when the actuating member is in the final contact-closed position. *Note:* Final contact pressure is usually measured and expressed in terms of the force that must be exerted on the yielding contact while the actuating member is held in the final contact-closed position, and with the mating contact fixed in position, in order to separate the mating contact surfaces. *See also:* contactor. (IA/ICTL/IAC) [60]

final controlling element (1) (electric power system) The controlling element that directly changes the value of the manipulated variable. (PE/PSE) 94-1970w

(2) That forward controlling element that directly changes the value of the manipulated variable. (CS/PE/EDPG) [3]

final design (nuclear power quality assurance) Approved design output documents and approved changes thereto. (PE/NP) [124]

final emergency circuits All circuits (including temporary emergency circuits) that, after failure of a ship's service supply, may be supplied by the emergency generator. *See also:* emergency electric system. (EEC/PE) [119]

final emergency lighting Temporary emergency lighting plus manually controlled lighting of the boat deck and overside to facilitate lifeboat loading and launching. *See also:* emergency electric system. (EEC/PE) [119]

final relay actuation time The time of termination of chatter following contact actuation. (EEC/REE) [87]

final sag The sag of a conductor under specified conditions of loading and temperature applied, after it has been subjected for an appreciable period to the loading prescribed for the loading district in which it is situated, or equivalent loading, and the loading removed. Final sag shall include the effect of inelastic deformation (creep). (NESC/T&D) C2-1997, C2.2-1960

final state The values assumed by the state variables of a system, component, or simulation at the completion of some specified duration of time. *Contrast:* initial state. (C) 610.3-1989w

final-terminal stopping device (elevators) A device that automatically causes the power to be removed from an electric elevator or dumbwaiter driving-machine motor and brake or from a hydraulic elevator or dumbwaiter machine independent of the functioning of the normal-terminal stopping device, the operating device, or any emergency terminal stopping device, after the car has passed a terminal landing. *See also:* control. (EEC/PE) [119]

final test result code A test result code obtained from an assertion test that requires no further processing. (C/PA) 2003-1997

final trunk (data transmission) A group of trunks to the higher class office which has no alternate route, and in which the number of trunks provided results in a low probability of calls encountering "all trunks busy." (PE) 599-1985w

final unloaded conductor tension (electric systems) The longitudinal tension in a conductor after the conductor has been stretched by the application for an appreciable period, and subsequent release, of the loadings of ice and wind, and temperature decrease, assumed for the loading district in which the conductor is strung (or equivalent loading). *See also:* conductor. (BT/AV) [34]

final unloaded tension The longitudinal tension in a conductor after it has been subjected for an appreciable period to the loading prescribed for the loading district in which it is situated, or equivalent loading, and the loading removed. Final unloaded tension shall include the effect of inelastic deformation (creep). (NESC) C2-1997

final unloaded sag (1) (general) The sag of a conductor after it has been subjected for an appreciable period to the loading prescribed for the loading district in which it is situated, or equivalent loading, and the loading removed. *See also:* sag. (BT/AV) [34]

(2) The sag of a conductor after it has been subjected for an appreciable period to the loading prescribed for the loading district in which it is situated, or equivalent loading, and the loading removed. Final unloaded sag shall include the effect of inelastic deformation (creep). (NESC) C2-1997

final value The steady-state value of a specified variable. *See also:* control. (IA/ICTL/APP/IAC) [69], [60]

final voltage *See:* cutoff voltage.

finder switch An automatic switch for finding a calling subscriber line or trunk and connecting it to the switching apparatus. (EEC/PE) [119]

finding (telephone switching systems) Locating a circuit requesting service. (COM) 312-1977w

fine chrominance primary (national television system committee color television) An obsolete term. Use the preferred term, I chrominance signal. (BT/AV) 201-1979w

fine-grain parallel architecture Parallel architecture that uses between 1K and 256K processors. *Contrast:* coarse-grain parallel architecture; medium-grain parallel architecture. (C) 610.10-1994w

fines (cable plowing) Particles of earth or rock smaller than .125 in greatest dimension. (T&D/PE) 590-1977w

finger *See:* end finger.

finger line (conductor stringing equipment) A lightweight line, normally sisal, manila, or synthetic fiber rope, which is placed over the traveler when it is hung. It usually extends from the ground, passes through the traveler and back to the ground. It is used to thread the end of the pilot line or the pulling line over the traveler and eliminates the need for workers on the structure. These lines are not required if pilot lines are installed when the travelers are hung. (T&D/PE) 524a-1993r, 524-1992r

finger overlap (acoustic aperture) The length of a finger pair between which electromechanical interaction takes place. (UFFC) 1037-1992w

finishing (1) (electrotype) The operation of bringing all parts of the printing surface into the same plane, or, more strictly speaking, into positions having equal printing values. (EEC/PE) [119]

(2) An operation or group of operations performed on the printed media after it emerges from the printer output mechanism. Finishing includes operations such as stapling (stitching), punching, binding, folding, cutting, etc., which may or may not be considered part of the printing process. Note that operations of collating and sorting are normally considered printer output functions rather than finishing. (C/MM) 1284.1-1997

finishing rate (storage battery) (storage cell) The rate of charge expressed in amperes to which the charging current for some types of lead batteries is reduced near the end of charge to prevent excessive gassing and temperature rise. *See also:* charge. (PE/EEC) [119]

finite difference frequency domain (FDFD) A numerical technique for solving partial differential equations by first Fourier transforming the time variable of the equation from the time domain to the frequency domain. Then the resultant partial equation is discretized and solved using the finite difference method. (AP/PROP) 211-1997

finite difference time domain (FDTD) A numerical technique for solving a partial differential equation involving time and space variables. The solution is implemented sequentially in the time domain. (AP/PROP) 211-1997

finite element frequency domain (FEFD) A numerical technique for solving partial differential equations by first Fourier transforming the time variable of the equation to the frequency domain and then using the finite element method. (AP/PROP) 211-1997

finite element time domain (FETD) A numerical technique for solving a partial differential equation directly in the time domain. Discretization of the time variable can be accomplished by the finite difference scheme or by the Galerkin method. *Note:* This method differs from the finite difference time domain (FDTD) method in that the space variable is made discrete by the finite element method rather than the finite difference method. *See also:* finite difference time domain. (AP/PROP) 211-1997

finite state machine (software) A computational model consisting of a finite number of states and transitions between those states, possibly with accompanying actions. (C) 610.12-1990

finite-time stability (solutions) For all initial states that originate in a specified region R at time t_0, the resulting solutions remain in another specified region R_ε over the given time interval $t_0 \leq t \leq T$. *Notes:* 1. In the definition of finite-time stability the quantities R_π, R_ε, and T are prespecified. Obviously, R must be included in R_ε. 2. A system may be Lyapunov unstable and still be finite-time stable. For example, a system with dynamics x = ax, a > 0, is Lyapunov unstable, but if

R_δ: $|x| \leq \delta$

R_ε: $|x| \leq \varepsilon$

and $T < a^{-1} \ln(\varepsilon/\delta)$, the system is finite-time stable (relative to the given values of δ, ε, and T). *See also:* control system. (CS/IM) [120]

finline An E-plane line in which the planar conducting structure is affixed to a dielectric substrate. The thin conducting strips (fins) may be insulated or grounded. They can be arranged in various configurations. (MTT) 1004-1987w

finsen The recommended practical unit of erythermal flux or intensity of radiation. It is equal to one unit of erythermal flux per square centimeter. (EEC/PE) [119]

FIPS PUB XXXX *See:* federal information processing standard publication.

fire-alarm system An alarm system signaling the presence of fire. *See also:* protective signaling. (EEC/PE) [119]

fire-control radar (navigation aid terms) A radar whose prime function is to provide information for the manual or automatic control of artillery or other weapons. (AES/GCS) 686-1997, 172-1983w

fire detection and fire protection systems (nuclear power generating station) Definitions of terms relating to fire detection and protection systems and equipment may be found in the National Fire Protection Association (NFPA) Handbook. (PE/NP) 567-1980w

fire-door magnet An electromagnet for holding open a self-closing fire door. (EEC/PE) [119]

fire-door release system A system providing remotely controlled release of self-closing doors in fire-resisting bulkheads to check the spread of fire. *See also:* marine electric apparatus. (EEC/PE) [119]

fire endurance A measure of the elapsed time during which a material or assembly continues to exhibit fire resistance under specified conditions of test and performance. (DEI) 1221-1993w

fire exposure The heat flux of a fire, with or without direct flame impingement, to which a material, product, building element, or assembly is exposed. (DEI) 1221-1993w

fire gases The airborne products emitted by a material undergoing pyrolysis or combustion that at the relevant temperature exist in the gas phase. (DEI) 1221-1993w

fire hazard A fire risk greater than an acceptable level. (DEI) 1221-1993w

fire performance characteristic A response of a material, product, or assembly to a prescribed source of heat or flame under controlled fire conditions. Such characteristics include ease of ignition, flame spread, smoke generation, fire endurance of the material, corrosiveness, and toxicity of the smoke generated. (DEI) 1221-1993w

fire performance test A procedure that measures a response of a material, product, or assembly to heat or flame under controlled fire conditions. (DEI) 1221-1993w

fire point The lowest temperature at which a specimen will sustain burning for five seconds. (DEI) 1221-1993w

fire products Heat, smoke, and toxic and corrosive products. (DEI) 1221-1993w

fire propagation The movement of a flame front on the surface of materials and products beyond the ignition zone. (DEI) 1221-1993w

fire-protected cable systems Cable systems to which a fire-protective enclosure material has been applied either in direct contact with the cables or applied over the raceway to protect cables from fire. (PE/IC) 848-1996

fire-protective coatings A material applied to a completed cable or assembly of cables to prevent the propagation of flame. Fire-protective coatings include liquids, mastics, and tapes. (SUB/PE) 525-1992r

fire quenching Shock cooling by immersion of liquid or molten material in a cooling medium (crushed stones in collecting pits). (SUB/PE) 980-1994

fire rating (cable penetration fire stop qualification test) The term applied to cable penetration fire stops to indicate the endurance in time (hours and minutes to the standard time-temperature curve in ANSI/ASTM E119-76, while satisfying the acceptance criteria specified in this standard. (PE) 634-1978w

fire resistance The property of a material or assembly to withstand fire or give protection from it. As applied to elements of buildings, it is characterized by the ability to confine a fire or to continue to perform a given structural function, or both. (DEI) 1221-1993w

fire-resistance rating The measured time, in hours or fractions thereof, that the material or construction will withstand fire exposure as determined by fire tests conducted in conformity to recognize standards. (EEC/PE) [119]

fire-resistant So constructed or treated that it will not be injured readily by exposure to fire. (EEC/PE) [119]

fire-resistive barrier (A) (cable penetration fire stop qualification test) A wall, floor, or floor-ceiling assembly erected to prevent the spread of fire. (To be effective, fire barriers must have sufficient fire resistance to withstand the effects of the most severe fire that may be expected to occur in the area adjacent to the fire barrier and must provide a complete barrier to the spread of fire.) **(B) (nuclear power generating station) (cable-penetration fire stops, fire breaks, and system enclosures)** A wall, floor, or floor-ceiling assembly erected to prevent the spread of fire. (PE/SUB/EDPG) 634-1978, 525-1992, 690-1984

fire-resistive barrier rating This is expressed in time (hours and minutes) and indicates that the wall, floor, or floor-ceiling assembly can withstand, without failure, exposure to a standard fire for that period of time. (PE/SUB/EDPG) 634-1978w, 690-1984r, 525-1992r

fire-resistive construction A method of construction that prevents or retards the passage of hot gases or flames as defined by the fire-resistance rating. (EEC/PE) [119]

fire-retardant coatings Material applied along the length of cables or in localized areas, as deemed necessary, to retard the flame propagation properties of cables in trays. (PE/IC) 848-1996

fired tube (nonlinear, active, and nonreciprocal waveguide components) (microwave gas tubes) The condition of the tube during which a radio frequency glow exists at either the resonant gap, the resonant window, or both. *See also:* gas tube. (MTT/ED) 457-1982w, [45], 161-1971w

fireproofing (of cables) The application of a fire-resistant covering. (NESC) C2-1997

firing angle *See:* angle of retard.

firing power (nonlinear, active, and nonreciprocal waveguide components) The radio-frequency (rf) power level above which a gas tube becomes nonlinear. *See also:* gas tube. (MTT) 457-1982w

firm capacity (1) (electric power supply) (purchase or sales) That firm capacity that is purchased, or sold, in transactions with other systems and that is not from designated units, but is from the over-all system of the seller. *Note:* It is understood that the seller provides reserve capacity for this type of transaction. *See also:* generating station. (PE/PSE) [54]
(2) Capacity that is purchased, or sold, at the highest level of system generation availability. It is understood that the seller treats this type of transaction as a demand obligation. (PE/PSE) 858-1993w

firm power (1) (power operations) Power intended to be available at all times during the period covered by a commitment, even under adverse conditions. (PE/PSE) 858-1987s
(2) (emergency and standby power) Power intended to be always available, even under emergency conditions. *See also:* generating station. (IA/PE/T&D/PSE) 446-1987s, [10]

firm transfer capability (electric power supply) (transmission) The maximum amount of power that can be interchanged continuously, over an extended period of time. *See also:* generating station. (PE/PSE) [54]

firmware (1) (software) The combination of a hardware device and computer instructions and data that reside as read-only software on that device. *Notes:* 1. This term is sometimes used to refer only to the hardware device or only to the computer instructions or data, but these meanings are deprecated. 2. The confusion surrounding this term has led some to suggest that it be avoided altogether. (C) 610.12-1990
(2) (supervisory control, data acquisition, and automatic control) Hardware used for the nonvolatile storage of instructions or data that can be read only by the computer. Stored information is not alterable by any computer program. *See also:* station. (SWG/SUB/PE) C37.1-1987s, C37.100-1992
(3) (watthour meters) A register control program stored in read-only memory and considered to be an integral part of the register. (ELM) C12.13-1985s
(4) (electromechanical watthour meters) A program to control the solid-state demand register that is stored in read-only memory and considered to be an integral part of the register that cannot be changed in its operating environment. (ELM) C12.15-1990
(5) The combination of software and data that reside on read-only memory. (PE/NP) 7-4.3.2-1993
(6) A program, typically stored in read-only memory, that controls a computer from the time that it is turned on until the time that the primary operating system assumes control of the computer. (C/BA) 1275-1994
(7) The combination of a hardware device and computer instructions and data that reside as read-only software on that device. *Notes:* 1. This term is sometimes used to refer only to the hardware device or only to the computer instructions or data, but these meanings are deprecated. 2. The confusion surrounding this term has led some to suggest that it be avoided altogether. (C) 610.10-1994w
(8) The combination of a hardware device and computer instructions and/or computer data that reside as read-only software on the hardware device. (C/SE) J-STD-016-1995

firmware device driver A device driver intended for use by firmware. *Contrast:* operating system device driver. *See also:* device driver. (C/BA) 1275-1994

first-bit access time (of a BORAM) The time interval between the application of addressing and enabling signals and the availability at an output of the first bit from a block of data. (ED) 641-1987w

first-class object An object that can be the value of a variable or can be stored in a data structure. In Scheme, first-class objects have unlimited extent. *See also:* extent. (C/MM) 1178-1990r

first-come, first-served *See:* first-in, first-out.

first contingency incremental transfer capability (power operations) The amount of power, incremental above normal base power transfers, that can be transferred over the transmission network in a reliable manner, based on the following conditions:

a) With all transmission facilities in service, all facility loadings are within normal ratings, and all voltages are within normal limits;
b) The bulk power system is capable of absorbing the dynamic power swings and remaining stable following a disturbance resulting in the loss of any single generating unit, transmission circuit or transformer;
c) After the dynamic power swings following a disturbance resulting in the loss of any single generating unit, transmission circuit, or transformer, but before operator-directed system adjustments are made, all transmission facility loadings are within emergency ratings, and all voltages are within emergency limits.

(PE/PSE) 858-1987s

first dial (register) That graduated circle or cyclometer wheel, the reading on which changes most rapidly. The test dial or dials, if any, are not considered. *See also:* watthour meter. (EEC/PE) [119]

first-ended, first-out A queueing technique for concurrent processes in which items are retrieved from the queue based on the time at which the item is placed completely in the queue. That is, the item whose final segment is placed in the queue before those of all other items, will exit the queue before those other items. *Note:* Often used in message queueing applications. (C) 610.5-1990w

first Fresnel zone (data transmission) In optics and radio communication, the circular portion of a wave front transverse to the line between an emitter and a more distant point where the resultant disturbance is being observed, whose center is the intersection of the front with the direct ray and whose radius is such that shortest path from the emitter through the periphery to the receiving point is one-half wave longer than the ray. *Note:* A second zone, a third, etc., are defined by successive increases of the path by half-wave increments. (PE) 599-1985w

first generation A period during the evolution of electronic computers in which computers were designed around vacuum tubes. *Note:* Introduced in 1949, first generation computers were thought to have been the state of the art from 1951 to 1959, when the transistor was developed. *See also:* third generation; fourth generation; fifth generation; second generation. (C) 610.10-1994w

first-generation language *See:* machine language.

first-in, first-out (A) A technique for managing a set of items to which additions and deletions are to be made; items are appended to one end of a list and retrieved from the other end. *See also:* queue. **(B)** Pertaining to a system in which the next item to exit the system is the item that has been in the system for the longest time. *Synonym:* first-come, first-served. *Contrast:* last-in, first-out. (C) 610.5-1990

first-in-first-out (FIFO) queue A data structure from which entries are removed in the same order that they were added. In this document, entries are added at the "tail" of the queue and removed from the "head." (C/MM) 1212.1-1993

first-in, last-out *See:* last-in, first-out.

first (last) transition duration (pulse terminology) The transition duration of the first (last) transition waveform in a pulse waveform. *See also:* waveform epoch. (IM/WM&A) 194-1977w

first-level address *See:* direct address.

first-line release (telephone switching systems) Release under the control of the first line that goes-on-hook. (COM) 312-1977w

first normal form One of the forms used to characterize relations; a data structure or relation is said to be in first normal form if it has no repeating groups. For example:

First Normal Form

UNNORMALIZED
ORDER0 = {ORDER-NO} + DATE + CUSTOMER-NO
+ CUSTOMER-NAME + CUSTOMER-ADDRESS
+ ((SEQUENCE-NO + ITEM-NO + ITEM-DESCRIPTION
+ QUANTITY-ORDERED + UNIT-PRICE
+ EXTENDED-PRICE)) + TOTAL-ORDER-AMOUNT

FIRST NORMAL FORM
ORDER1 = {ORDER-NO} + DATE + CUSTOMER-NO
+ CUSTOMER-NAME + CUSTOMER-ADDRESS
+ TOTAL-ORDER-AMOUNT
ITEM1 = {ORDER-NO + SEQUENCE-NO} + ITEM-NO
+ ITEM-DESCRIPTION + QUANTITY-ORDERED
+ UNIT-PRICE + EXTENDED-PRICE

(C) 610.5-1990w

Note: repeating group enclosed in parenthesis. Keys in brackets.

first open (1) (of a file) When a process opens a file that is not currently an open file within any process. (C/PA) 9945-1-1996
(2) (of a file) The act when a process opens a file, message queue, or shared memory object that is not currently open within any process. (C) 1003.5-1999

first-order lag The change in phase due to a linear element of transfer function. 1/(1 + *Ts*). *Synonym:* linear lag. *See also:* lag. (ELM) C12.1-1982s

first-order lead (control system feedback) The change in phase due to a factor (1 + *Ts*) in the numerator of a transfer function. *See also:* feedback control system. (IM) [120]

first Townsend discharge (gas) A semi-self- maintained discharge in which the additional ions that appear are due solely to the ionization of the gas by electron collisions. *See also:* discharge. (ED) [45], [84]

first transition (pulse terminology) The major transition waveform of a pulse waveform between the base and the top. (IM/WM&A) 194-1977w

first voltage range *See:* voltage range.

fishbone antenna An end-fire, traveling wave antenna consisting of a balanced transmission line to which is coupled, usually through lumped circuit elements, an array of closely spaced, coplanar dipoles. (AP/ANT) 145-1993

fish tape (fishing wire) (snake) A tempered steel wire, usually of rectangular cross section, that is pushed through a run of conduit or through an inaccessible space, such as a partition, and that is used for drawing in the wires. (PE/EEC) [119]

fission (power operations) The splitting of a nucleus into parts (which are nuclei of lighter elements), accompanied by the release of a relatively large amount of energy (about 200 million electron volts per fission in the case of 235U fission) and frequently one or more neutrons. (PE/PSE) 858-1987s, 346-1973w

fitting An accessory such as a locknut, bushing, or other part of a wiring system that is intended primarily to perform a mechanical rather than an electrical function. (NESC/NEC) [86]

fittings (raceway systems for Class 1E circuits for nuclear power generating stations) (raceway) Raceway sections that are joined to other raceway sections for the purpose of coupling together or changing the size or direction of the raceway system. These include such items as couplings, elbows, tees, wyes, pulling sleeves, and pull boxes. (PE/NP) 628-1987r

five-bit byte *See:* quintet.

5B6B encoding A method whereby data quintets are mapped (encoded) as code sextets.local area networks. (C) 8802-12-1998

5GL *See:* fifth generation language.

fix (1) (navigation) A position determined without reference to any former position. (AES/RS/GCS) 686-1982s, 172-1983w
(2) A device or equipment modification to prevent interference or to reduce an equipment's susceptibility to interference. *See also:* electromagnetic compatibility. (EMC/INT) [53]
(3) (mathematics of computing) To convert a number from floating-point representation to fixed-point representation. *Contrast:* float. (C) 1084-1986w

fixed *See:* read-only access.

fixed addressing (64-bit) An address model implemented by bus standards supporting only 64-bit addresseses. The initial node space is large (258 Tbytes), is fixed in size, and extended spaces are not supported. (C/MM) 1212-1991s

fixed appliance (electric systems) An appliance that is fastened or otherwise secured at a specific location. *See also:* appliance. (NESC) [86]

fixed bank A capacitor bank that does not have a capacitor control and must be manually switched.power systems relaying. (T&D/PE) 1036-1992, C37.99-2000

fixed binary data *See:* fixed-point binary data.

fixed block format (numerically controlled machines) A format in which the number and sequence of words and characters appearing in successive blocks is constant. (IA/EEC) [61], [74]

fixed-called-address line (telephone switching systems) A line for originating calls to a fixed called address. (COM) 312-1977w

fixed cycle (numerically controlled machines) A preset series of operations that direct machine axis movement and/or cause spindle operation to complete such actions as boring, drilling, tapping, or combinations thereof. (IA) [61], [84]

fixed-cycle operation An operation that is completed in a specified number of regularly timed execution cycles. (C) [20], [85], 610.10-1994w

fixed decimal data *See:* fixed-point real data.

fixed disk A magnetic disk that is permanently mounted within a disk drive. *Synonym:* nonremovable disk. *Contrast:* removable disk. *See also:* hard disk. (C) 610.10-1994w

fixed error *See:* bias.

fixed error in sample time A nonrandom error in the instant of sampling. A fixed error in sample time may be fixed with respect to the data samples acquired or correlated with an event that is detected by the sampling process. Unless otherwise specified, usually taken to mean the maximum fixed error that may be observed. (IM/WM&A) 1057-1994w

fixed feed tube (cable plowing) A feed tube permanently attached to a blade. It may have removable back plate. *See also:* feed tube. (T&D/PE) 590-1977w

fixed format A file organization in which all logical records in the file are of fixed length. *Synonym:* F format. *Contrast:* variable format. (C) 610.5-1990w

fixed-frequency transmitter A transmitter designed for operation on a single carrier frequency. *See also:* radio transmitter. (AP/BT/PE/ANT) 145-1983s, 182-1961w, 599-1985w

fixed head A magnetic head that is in a fixed position, and which can access data only within a particular disk track. *Contrast:* floating head. *See also:* head-per-track disk drive. (C) 610.10-1994w

fixed impedance-type ballast A reference ballast designed for use with one specific type of lamp and, after adjustment during the original calibration, is expected to hold its established impedance throughout normal use. (EEC/LB) [97]

fixed-instruction computer A computer in which the instruction set cannot be changed. *Contrast:* user-programmable computer. (C) 610.10-1994w

fixed investment costs (power operations) Monies associated with investment in plant. (PE/PSE) 858-1987s

fixed length Pertaining to a record or field that has a constant length, regardless of the specific data contained in it. Filler characters may be used to maintain the fixed length. *Contrast:* variable length. *See also:* fixed format. (C) 610.5-1990w

fixed-length field A field whose length is constant. *Contrast:* variable-length field. *See also:* fixed format. (C) 610.5-1990w

fixed light (illuminating engineering) A light having a constant luminous intensity when observed from a fixed point. (EEC/IE) [126]

fixed motor connections A method of connecting electric traction motors wherein there is no change in the motor interconnections throughout the operating range. *Note:* This term is used to indicate that a transition from series to parallel relation is not provided. *See also:* traction motor. (EEC/PE) [119]

fixed operation cost Cost other than that associated with investment in plant, which does not vary or fluctuate with changes in operation or use of plant. (PE/PSE) 858-1993w

fixed point (mathematics of computing) (data management) Pertaining to a numeration system in which the position of the radix point is fixed with respect to one end of the numerals, according to some convention. *Contrast:* floating point; variable point. (MIL/C) [2], 162-1963w, [20], 1084-1986w, [85], 610.5-1990w

fixed-point arithmetic A method of arithmetic in which the numbers are expressed in the fixed-point representation system. *Contrast:* floating-point arithmetic. (C) 1084-1986w

fixed-point binary data Fixed-point data used to represent signed binary numbers.

decimal	75_{10}
fixed-point binary	$0100\ 1011_2$
decimal	-91_{10}
fixed point binary	$1010\ 0101_2$

Synonym: real fixed binary data. (C) 610.5-1990w

fixed-point data Integer data that can be expressed in a specific number of digits, with a radix point implicitly located at a predetermined position. *Synonym:* computational data. *Contrast:* floating-point data. *See also:* fixed-point real data; fixed-point binary data. (C) 610.5-1990w

fixed-point number A number expressed in fixed-point representation. (C) 1084-1986w

fixed-point part *See:* significand.

fixed-point real data Fixed-point data used to represent signed decimal numbers. For example, 75.6, 0, and -253. *Synonyms:* fixed real data; fixed decimal data; real fixed decimal data. (C) 610.5-1990w

fixed-point register A register used to manipulate fixed-point data. (C) 610.10-1994w

fixed-point representation system A numeration system in which the position of the radix point is fixed with respect to one end of the numerals, according to some convention. (C) 1084-1986w

fixed-point system *See:* point.

fixed-program read-only storage A form of read-only storage in which the data content of each storage cell is determined during manufacture and is thereafter unalterable. (C) 610.10-1994w

fixed rack An assembly enclosed at the top and sides, either open or with door(s) for access, with a top-to-bottom front panel opening for equipment mounting (for example, nominal 19-inch-wide chassis and subpanel assemblies). *Synonym:* cabinet. (SWG/PE) C37.100-1992, C37.21-1985r

fixed-radix notation A radix notation system in which all digit positions have the same radix. The weights of successive digit places are successive integral powers of a single radix. *Synonyms:* fixed-radix scale; numeration system. (C) 1084-1986w

fixed-radix numeration system *See:* fixed-radix notation.

fixed-radix scale *See:* fixed-radix notation.

fixed real data *See:* fixed-point real data.

fixed routing A routing strategy for store-and-forward network, in which the next path to each specific destination is always the same at each point in the network. (C) 610.7-1995

fixed sequential format A means of identifying a word by its location in the block. *Note:* Words must be presented in a specific order and all possible words preceding the last desired word must be present in the block. (IA/EEC) [61], [74]

fixed signal The signal of fixed location indicating a condition affecting the movement of a train or engine. (EEC/PE) [119]

fixed storage (computers) A storage device that stores data not alterable by computer instructions, for example, magnetic core storage with a lockout feature or punched paper tape. *See also:* nonerasable storage; permanent storage. (C) [20], [85]

fixed temperature heat detector (fire protection devices) A device that will respond when its operating element becomes heated to a predetermined level. (NFPA) [16]

fixed threshold transistor (metal-nitride-oxide field-effect transistor) Another name for a metal-oxide semiconductor (MOS) type transistor, used in contradistinction to the metal-nitride-oxide semiconductor (MNOS) transistor, which has a variable threshold voltage. (ED) 581-1978w

fixed transmitter A transmitter that is operated in a fixed or permanent location. *See also:* radio transmitter. (AP/BT/ANT) 145-1983s, 182-1961w

fixed word length (test, measurement, and diagnostic equipment) Property of a storage device in which the capacity for bits in each storage word is fixed. (MIL) [2]

fixing (electrostatography) The act of making a developed image permanent. *See also:* electrostatography. (ED) 224-1965w, [46]

fixnum A limited-precision computer representation for integers, where the limitation is imposed by machine-architecture constraints. *See also:* bignum. (C/MM) 1178-1990r

fixture *See:* luminaire.

fixture stud A threaded fitting used to mount a lighting fixture to an outlet box. *Synonym:* stud. *See also:* cabinet. (EEC/PE) [119]

flag (1) (microprocessor operating systems parameter types) A yes/no or true/false value. (MM/C) 855-1985s
(2) A character that signals the occurrence of some event. Usually a field of 1 b. (SUB/PE) 999-1992w
(3) A signal used to delimit packets in parallel signal transmission implementations. For example, in the 16bit parallel implementation the flag is a 17th signal. In some serial implementations special symbols could be used in place of flag transitions. (C/MM) 1596-1992
(4) A signal used to delimit packets in parallel-signal-transmission implementations. (C/MM) 1596.3-1996
(5) The first character of an ATLAS statement used to mark that statement as having a special purpose or capability. (SCC20) 771-1998

flag alarm An indicator in certain types of navigation instruments used to warn when the readings are unreliable. *See also:* navigation. (AES/RS/GCS) 686-1982s, [42], 172-1983w

flag register (A) A register used to hold one or more bit indicators called flags, for example: a register holding the negative, zero, and overflow bits. *See also:* condition code register. **(B)** A register used to hold a flag. (C) 610.10-1994

flame detector (1) (fire protection devices) A device which detects the infrared, or ultraviolet, or visible radiation produced by a fire. (NFPA) [16]
(2) (power system device function numbers) A device that monitors the presence of the pilot or main flame in such apparatus as a gas turbine or a steam boiler. (PE/SUB) C37.2-1979s

flame flicker detector (fire protection devices) A photoelectric flame detector including means to prevent response to visible light unless the observed light is modulated at a frequency characteristic of the flicker of a flame. (NFPA) [16]

flameproof apparatus Apparatus so treated that it will not maintain a flame or will not be injured readily when subjected to flame. (EEC/PE) [119]

flameproof terminal box A terminal box so designed that it may form part of a flameproof enclosure. (PE) [9]

flame protection of vapor openings Self-closing gauge hatches, vapor seals, pressure-vacuum breather valves, flame arresters, or other reasonably effective means to minimize the possibility of flame entering the vapor space of a tank. *Note:* Where such a device is used, the tank is said to be flameproofed. (NFPA) [114]

flame-resistant cable A portable cable that will meet the flame test requirements of the United States Bureau of Mines. *See also:* mine feeder circuit. (EEC/PE) [119]

flame resisting *See:* flame-retarding.

flame retardant (1) (Class 1E equipment and circuits) (nuclear power generating station) Capable of limiting the propagation of a fire beyond the area of influence of the energy source that initiated the fire. (PE/NP) 384-1992r
(2) So constructed or treated that it will not support flame. (SWG/PE) C37.100-1992

flame-retardant coatings A material applied to a completed cable or assembly of cables to prevent the propagation of flame when exposed to a flame source. Flame-retardant coatings include tapes, blankets, liquids, or mastics. (PE/IC) 817-1993w

flame-retarding (electric installations on shipboard) Flame-retarding materials and structures should have such fire-resisting properties that they will not convey flame nor continue to burn for longer times than specified in the appropriate flame test. Compliance with the requirements of the preceding paragraph should be determined with the apparatus and according to the methods described in the Underwriters' Laboratories Standards for the materials and structures unless specific applicable tests are invoked in these recommendations. (IA/MT) 45-1983s

flame spread index A number or classification indicating a comparative measure derived from observations made during the progress of the boundary of a zone of flame under defined test conditions. (DEI) 1221-1993w

flammable Subject to easy ignition and rapid flaming combustion. (DEI) 1221-1993w

flammable air-vapor mixtures When flammable vapors are mixed with air in certain proportions, the mixture will burn rapidly when ignited. *Note:* The combustion range for ordinary petroleum products, such as gasoline, is from 1 1/2 to 6% of vapor by volume, the remainder being air. (NFPA) [114]

flammable anesthetics (health care facilities) Gases or vapors such as fluroxene, cyclopropane, divinyl ether, ethyl chloride, ethyl ether, and ethylene, which may form flammable or explosive mixtures with air, oxygen, or reducing gases such as nitrous oxide. (NESC/NEC) [86]

flammable anesthetizing location (health care facilities) Any operating room, delivery room, anesthetizing room, corridor, utility room, or any other area if intended for the application of flammable anesthetics. (NESC/NEC) [86]

flammable vapors The vapors given off from a flammable liquid at and above is flash point. (NFPA) [114]

flange *See:* coupling flange.

flange, choke *See:* choke flange.

flange, contact *See:* contact flange.

FLAP A programming language used widely for manipulating formulas and performing symbolic mathematical calculations. (C) 610.13-1993w

flare-out (navigation aid terms) That portion of the approach path of an aircraft in which the slope is modified to provide the appropriate rate of descent at touchdown. *See also:* navigation. (AES/RS/GCS) 686-1982s, [42], 172-1983w

flarescan (navigation) A ground-based navigation system used in conjunction with an instrument approach system to provide flare-out vertical guidance to an aircraft by the use of a pulse-space-coded vertically scanning fan beam that provides elevation angle data. *See also:* navigation. (AES/RS/GCS) 686-1982s, [42], 172-1983w

flash An on-hook/off-hook signal with a duration between specified lower and upper bounds,from either the calling or called party to the SPCS indicating that some calling feature is desired. *See also:* hit, flash, and disconnect timing; hit; disconnect. (COM/TA) 973-1990w

flash barrier (rotating machinery) A screen of fire-resistant material to prevent the formation of an arc or to minimize the damage caused thereby. (PE) [9]

flash card In micrographics, a target printed with distinctive markings to be photographed to facilitate the indexing of microfilm. *Synonym:* flash target. *See also:* flash indexing. (C) 610.2-1987

flash current (primary cell) The maximum electric current indicated by an ammeter of the dead-beat type when connected directly to the terminals of the cell or battery by wires that together with the meter have a resistance of 0.01 ohm. *See also:* electrolytic cell. (EEC/PE) [119]

flash EEPROM An electrically erasable programmable read-only memory (EEPROM) in which clearing can be performed only on blocks or the entire array. (ED) 1005-1998

flasher A device for alternately and automatically lighting and extinguishing electric lamps. *See also:* appliance. (EEC/PE) [119]

flasher relay A relay that is so designed that when energized its contacts open and close at predetermined intervals. *See also:* appliance. (EEC/PE) [119]

flash indexing In micrographics, the process of dividing a roll of microfilm into batches of information using flash cards to identify each of the sections, thus providing a method of retrieval. *Synonym:* flash target coding. (C) 610.2-1987

flashing light (illuminating engineering) A rhythmic light in which the periods of light are of equal duration and are clearly shorter than the periods of darkness. (EEC/IE) [126]

flashing-light signal A railroad-highway crossing signal the indication of which is given by two red lights spaced horizontally and flashed alternately at predetermined intervals to give warning of the approach of trains, or a fixed signal in which the indications are given by color and flashing of one or more of the signal lights. (EEC/PE) [119]

flashing signal (telephone switching systems) A signal for indicating a change or series of changes of state, such as on-hook/off-hook, used for supervisory purposes. (COM) 312-1977w

flashlight battery A battery designed or employed to light a lamp of an electric hand lantern or flashlight. *See also:* battery. (EEC/PE) [119]

flashover (1) (general) A disruptive discharge through air around or over the surface of solid or liquid insulation, between parts of different potential or polarity, produced by the application of voltage wherein the breakdown path becomes sufficiently ionized to maintain an electric arc. (PE/T&D/PE/T&D/PE/T&D/PSIM) 28-1974, [8], 270-1966w, [10], [55], 1243-1997, 1410-1997

(2) A disruptive discharge over the surface of a solid insulation in a gas or liquid. (PE/PSIM) 4-1995
(3) A disruptive discharge around or over the surface of an insulating member, between parts of different potential or polarity, produced by the application of voltage wherein the breakdown path becomes sufficiently ionized to maintain an electric arc. (PE/IC) 48-1996
(4) The transition from a localized fire to the general conflagration within the compartment when all fuel surfaces are burning. (DEI) 1221-1993w
(5) A disruptive discharge around or over the surface of a solid or liquid insulator. (SPD/PE) C62.11-1999, C62.62-2000

flash plate A thin electrodeposited coating produced in a short time. (EEC/PE) [119]

flash point (1) The minimum temperature at which a liquid will give off vapor in sufficient amount to form a flammable air-vapor mixture that can be ignited under specified conditions. (NFPA) [114]
(2) The lowest temperature of a sample at which application of an ignition source causes the vapor of the sample to ignite momentarily under specified conditions of test. (DEI) 1221-1993w

flash target *See:* flash card.

flash target coding *See:* flash indexing.

flash timing *See:* hit, flash, and disconnect timing; hit; disconnect; flash.

flashtube (illuminating engineering) A tube of glass or quartz with electrodes at the ends and filled with a gas, usually xenon. It is designed to produce high intensity flashes of light of extremely short duration. (EEC/IE) [126]

flat-band voltage (metal-nitride-oxide field-effect transistor) Gate voltage that results in zero field at the surface of the silicon. It is related to the threshold voltage by a constant, generally small, voltage increment. (ED) 581-1978w

flatbed plotter A plotter that draws an image on a display surface mounted on a flat surface. (C) 610.10-1994w

flat cable assembly An assembly of parallel conductors formed integrally with an insulating material web specifically designed for field installation in metal surface raceway approved for the purpose. (NESC/NEC) [86]

flat-compound A qualifying term applied to a compound-wound generator to denote that the series winding is so proportioned that the terminal voltage at rated load is the same as at no load. (EEC/PE) [119]

flat file (A) A set of records that are identically formatted to contain no more than one occurrence of each data item. *Note:* records in such a file do not contain data aggregates or repeating groups. *See also:* relational file. **(B)** A two-dimensional array of data items that is stored as in definition (A). (C) 610.5-1990

flat flange *See:* cover flange.

flat leakage power (nonlinear, active, and nonreciprocal waveguide components) (microwave gas tubes) The peak radio-frequency power transmitted through the tube after the establishment of the steady-state radio-frequency discharge. *See also:* gas tube. (MTT/ED) 457-1982w, [45], 161-1971w

flat loss (1) (gain) The frequency independent contribution to the total transfer-function loss (or gain) or a four-terminal network. (CAS) [13]
(2) (broadband local area networks) Loss created by a component or set of components that maintains a constant attenuation across a specified bandwidth. (LM/C) 802.7-1989r

flat pack An integrated circuit package that has leads extending from the package in the same plane as the package so that leads can be spot welded to terminals on a substrate or soldered to a printed circuit board. (C) 610.10-1994w

flat-panel display device A display device whose physical depth (front-to-back) is relatively small. For example, a plasma panel or a liquid-crystal diode display device. (C) 610.10-1994w

flat-rate call (telephone switching systems) A call for which no billing is required. (COM) 312-1977w

flat-rate service (telephone switching systems) Service in which a fixed charge is made for all answered local calls during the billing interval. (COM) 312-1977w

flat-strip conductor *See:* strip-type transmission line.

flat-top antenna A short vertical monopole antenna with an end capacitor whose elements are all in the same horizontal plane. *See also:* top-loaded vertical antenna; end capacitor. (AP/ANT) 145-1993

flat-type relay armature An armature that rotates about an axis perpendicular to that of the core, with the pole face on a side surface of the core. (EEC/REE) [87]

flat weighting (data transmission) A noise measuring set measuring amplitude frequency characteristics which are flat over a specified frequency range. The frequency range must be stated. Flat noise power may be expressed in dBrn (F1-F2) or in dBm (F1-F2). The terms "3 kHz flat weighting" and "15 kHz flat weighting" from 30 Hz mean to the upper frequency indicated. (PE) 599-1985w

FLAVORS An object-oriented language originally developed as an extension of LISP. (C) 610.13-1993w

flection-point emission current That value of current on the diode characteristic for which the second derivative of the current with respect to the voltage has its maximum negative value. *Note:* This current corresponds to the upper flection point of the diode characteristic. *See also:* electron emission. (ED) 161-1971w, [45]

flexcircuit Flexible printed wiring board. A patterned arrangement of printed circuit and components utilizing flexible base materials with or without flexible cover layers. (C/BA) 14536-1995

flexibility The ease with which a system or component can be modified for use in applications or environments other than those for which it was specifically designed. *Synonym:* adaptability. *See also:* extendability; maintainability. (C) 610.12-1990

flexibility of the electric system The adaptability to development and expansion as well as to changes to meet varied requirements during the life of the building. (IA/PSE) 241-1990r

flexible connector (rotating machinery) An electric connection that permits expansion, contraction, or relative motion of the connected parts. (PE) [9]

flexible coupling (rotating machinery) A coupling having relatively high transverse or torsional compliance. *Notes:* 1. May be used to reduce or eliminate transverse loads or deflections of one shaft from being carried, or felt by the other coupled shaft. 2. May be used to reduce the torsional stiffness between two rotating masses in order to change torsional natural frequencies of the shaft system or to limit transient or pulsating torques carried by the shafts. *See also:* rotor. (PE) [9]

flexible disk *See:* floppy disk.

flexible equipment Equipment, structures, and components whose lowest resonant frequency is less than the cutoff frequency on the response spectrum. (PE/SUB/NP) 693-1997, 344-1987r

flexible manufacturing system A computer-integrated manufacturing system that can be reprogrammed to make a variety of parts or products. (C) 610.2-1987

flexible metal conduit A flexible raceway of circular cross section specially constructed for the purpose of the pulling in or the withdrawing of wires or cables after the conduit and its fittings are in place. *See also:* raceway. (EEC/PE) [119]

flexible mounting (rotating machinery) A flexible structure between the core and foundation used to reduce the transmission of vibration. (PE) [9]

flexible nonmetallic tubing (loom) A mechanical protection for electric conductors that consists of a flexible cylindrical tube having a smooth interior and a single or double wall of nonconducting fibrous material. *See also:* raceway. (PE/EEC) [119]

flexible tower (frame) A tower that is dependent on the line conductors for longitudinal stability but is designed to resist transverse and vertical loads. *See also:* tower. (T&D/PE) [10]

flexible waveguide A waveguide constructed to permit limited bending and twisting or stretching, or both, without appreciable change in its electrical properties. (MTT) 147-1979w, 146-1980w

flex model *See:* CODASYL model.

flexure (inertial sensors) (dynamically tuned gyro) An elastic element in a dynamically tuned gyro rotor suspension system, which permits limited angular freedom about axes perpendicular to the spin axis. (AES/GYAC) 528-1994

flicker (1) (A) (television) (general) Impression of fluctuating brightness or color, occurring when the frequency of the observed variation lies between a few hertz and the fusion frequencies of the images. **(B) (television)** A repetitive variation in luminance of a given area in a monochromatic or color display, the visibility of which is a function of repetition rate, duty cycle, luminance, and the decay characteristic. (BT/AV) 201-1979
(2) (computer graphics) The undesirable blinking of a graphical display image that occurs when the refresh rate is so low that regeneration of the display image is noticeable. (C) 610.6-1991w
(3) A perceptible change in electric light source intensity due to a fluctuation of input voltage. *Note:* The general meaning of this term could make it applicable to describe the pulsation of luminous flux from a low-inertia source (such as gas discharge lamps) caused by the zero crossings of the supply voltage at twice the power-system frequency. However, in the context of power supply disturbances, the term applies to perceptible, subjective, objectionable, and random or periodic variations of the light output. (T&D/PE) 1250-1995
(4) A variation of input voltage, either magnitude or frequency, sufficient in duration to allow visual observation of a change in electric light source intensity. (IA/PSE) 1100-1999

flicker effect (electron tube) The random variations of the output-current in a valve or tube with an oxide-coated cathode. *Note:* Its value varies inversely with the frequency. *See also:* electron tube. (ED) [45], [84]

flicker fusion frequency (illuminating engineering) The frequency of intermittent stimulation of the eye at which flicker disappears. (EEC/IE) [126]

flicker noise (nonlinear, active, and nonreciprocal waveguide components) One of the sources of noise associated with solid-state devices such as mixers or diode detectors, the amplitude of which varies inversely with frequency. It is also referred to as 1/f noise. In the audio-frequency region this noise becomes more significant than either thermal or shot noise. (MTT) 457-1982w

flicker threshold (television) The luminance at which flicker is just perceptible at a given repetition rate, with other variables held constant. (BT/AV) 201-1979w

flight instrument (navigation aid terms) A vehicle instrument used in the control of the direction of flight, attitude, altitude, or speed of a vehicle. (AES/GCS) 172-1983w

flight path (navigation) A proposed route in three dimensions. *See also:* navigation; course line. (AES/RS/GCS) 686-1982s, 172-1983w, [42]

flight-path computer (electronic navigation) Equipment providing outputs for the control of the motion of a vehicle by along a flight path. *See also:* navigation. (AES/RS/GCS) 686-1982s, [42], 172-1983w

flight-path deviation (electronic navigation) The amount by which the flight track of a vehicle differs from its flight path expressed in terms of either angular or linear measurement. *See also:* navigation. (AES/RS) 686-1982s, [42]

flight-path-deviation indicator (electronic navigation) A device providing a visual display of flight-path deviation. *See also:* navigation. (AES/RS) 686-1982s, [42]

flight track (electronic navigation) The path in space actually traced by a vehicle. *See also:* track. (AES/RS) 686-1982s, [42]

flip-flop (1) (A) (electronic computation) A circuit or device, containing active elements, capable of assuming either one or two stable states at a given time, the particular state being dependent upon the nature of an input signal, for example, its polarity, amplitude, and duration, and which of two input terminals last received the signal. *Note:* The input and output coupling networks, and indicators, may be considered as an integral part of the flip-flop. **(B) (electronic computation)** A device, as in definition (A), that is capable of counting modulo 2, in which case it might have only one input terminal. **(C) (electronic computation)** A sequential logic element having properties similar to definition (A) or (B) above. *See also:* toggle; feedback control system. (C) 162-1963
(2) A circuit or device capable of assuming either of two stable states, and which can be made to switch states by applying the proper signal or combination of signals to its inputs. *See also:* trigger circuit; R-S flip-flop; latch. (C) 610.10-1994w

flip-flop circuit A trigger circuit having two conditions of permanent stability, with means for passing from one to the other by an external stimulus. *See also:* trigger circuit. (PE) 599-1985w

flip-out The action of a worker or test torso being unintentionally separated from the body support component during or after fall arrest. (T&D/PE) 1307-1996

flippy *See:* floppy disk.

float (1) (mathematics of computing) To convert a number from fixed-point representation to floating-point representation. *Contrast:* fix. (C) 1084-1986w
(2) (gyros) An enclosed gimbal assembly housing the spin motor and other components, such as the pickoff and torquer. This assembly is immersed in a fluid, usually at the condition of neutral buoyancy. (AES/GYAC) 528-1994

float charge A constant potential normally applied to a battery to maintain it in a charged condition. (PE/EDPG) 1106-1995

float-displacement hysteresis (accelerometer) (gyros) The difference in rebalance torque or equivalent input after displacing the float about the output axis from its null position in successive clockwise and counterclockwise directions by equal amounts (up to its full range of angular freedom, unless otherwise specified). The float may be displaced by applying torques to the float through a torquer or through gyroscopic or acceleration torques in either open or closed-loop mode. The amount of float-displacement hysteresis may depend on the methods of applying torques, on the mode of operation (open or closed loop), and on the amount and duration of float displacement. (AES/GYAC) 528-1994

floating A method of operation for storage batteries in which a constant voltage is applied to the battery terminals sufficient to maintain an approximately constant stage of charge. *See also:* charge; trickle charge. (EEC/PE) [119]

floating battery A storage battery that is kept in operating condition by a continuous charge at a low rate. (EEC/PE) [119]

floating carrier *See:* controlled carrier.

floating character A character placed in the position that is one place more significant than the otherwise most significant position. (C) 1084-1986w

floating control *See:* floating control system.

floating control system (automatic control) A control system in which the rate of change of the manipulated variable is a continuous (or at least a piecewise continuous) function of the actuating signal. *Note:* The manipulated variable can remain at any value in its operating range when the actuating signal is zero and constant. Hence the manipulated variable is said to "float." When the forward elements in a control loop have integral control action only, the mode of control has been called "proportional-speed floating." The use of the term integral control action is recommended as a replacement for "proportional-speed floating control." *Synonym:* floating control. *See also:* integral control action; neutral zone; single-speed floating control system; multiple-speed floating control system. (PE/EDPG) [3]

floating decimal* *See:* floating point.

* Deprecated.

floating grid (electron tube) An insulated gird, the potential of which is not fixed. (ED) [45], [84]

floating head A magnetic head that is suspended on a layer of air at a small distance away from the surface of the recording medium and which can move from track to track. *Synonyms:* air-floating head; movable head; flying head. *Contrast:* fixed head. *See also:* head positioner. (C) 610.10-1994w

floating network or component A network or component having no terminal at ground potential. (CAS) [13]

floating neutral One whose voltage to ground is free to vary when circuit conditions change. *See also:* center of distribution. (PE/NP) 338-1977s

floating point (1) (mathematics of computing) (data management) Pertaining to a numeration system in which each number is represented as a fractional quantity multiplied by an integral power of the radix. *Contrast:* fixed point; variable point. (C) 610.5-1990w, 1084-1986w
(2) Pertaining to a system in which the location of the point does not remain fixed with respect to one end of numerical expressions, but is regularly recalculated. The location of the point is usually given by expressing a power of the base. *See also:* variable point; fixed point. (C) 162-1963w

floating-point arithmetic A method of arithmetic in which the numbers are expressed in the floating-point representation system. *Contrast:* fixed-point arithmetic. (C) 1084-1986w

floating-point coding compaction A method of numerical data compaction that uses the floating-point representation system. (C) 1084-1986w

floating-point coefficient *See:* exponent.

floating-point data Real data in which numbers are represented using only an exponent, *y*, and a mantissa, *x*, where *x* and *y* are integers. *Note:* The number is expressed in the form $x \cdot 10^y$, and only *x* and *y* are stored in fixed-point binary format.

decimal	$12.3 = .123 \cdot 10^2$
floating-point	$0111\ 1011\ 0000\ 0010_2 = 7B02_{16}$
----x---- ----y---	

Synonym: floating-point real data. *Contrast:* fixed-point data. (C) 610.5-1990w

floating-point exception An *fp_exception* of one of the following floating-point trap types, caused by execution of an FPop: *unfinished_FPop, unimplemented_FPop, sequence_error, hardware_error, invalid_fp_register,* or floating-point *IEEE_754_ exception* that occurs while the corresponding bit in FSR.TEM is set to 1. (C/MM) 1754-1994

floating-point IEEE-754 exception A floating-point exception, as specified by IEEE Std 754-1985. (C/MM) 1754-1994

floating-point number (radix-independent floating-point arithmetic) A digit string characterized by three components: a sign, a signed exponent, and a significand. Its numerical value, if any, is the signed product of its significand and the radix raised to the power of its exponent. (C/MM) 854-1987r, 1084-1986w

floating-point operate (FPop) instructions Instructions that perform floating-point calculations, as defined by the FPop1 and FPop2 opcodes. FPop instructions do not include FBfcc instructions, nor loads and stores between memory and the FPU. (C/MM) 1754-1994

floating-point real data *See:* floating-point data.

floating-point register A register used to manipulate floating point data. (C) 610.10-1994w

floating-point representation system A numeration system in which each number is represented as a sign, a signed exponent, and a significand, where the numerical value, if any, is the signed product of its significand and the radix raised to the power of the exponent. (C) 1084-1986w

floating-point system *See:* point.

floating-point trap type The specific type of floating-point exception that has occurred, encoded in the FSR.*ftt* field. (C/MM) 1754-1994

floating-point unit (FPU) A processing unit that contains a set of floating-point registers and performs floating-point operations, as defined by this standard. (C/MM) 1754-1994

floating removable feed tube (cable plowing) A feed tube removably attached to a blade so relative motion may occur between the feed tube and the blade around axis that are essentially vertical and horizontal (perpendicular to direction of travel). *See also:* feed tube. (T&D/PE) 590-1977w

floating speed (process control) In single-speed or multiple-speed floating control systems, the rate of change of the manipulated variable. (PE/EDPG) [3]

floating zero (numerically controlled machines) A characteristic of a numerical machine control permitting the zero reference point on an axis to be established readily at any point in the travel. *Note:* The control retains no information on the location of any previously established zeros. *See also:* zero offset. (IA/EEC) [61], [74]

float service applications Storage batteries applied for reserve use and maintained at a continuous "float" voltage point selected to just exceed the batteries' internal (self-discharge) losses. (PE/EDPG) 1184-1994

float-state A logic value that indicates the lack of an active drive condition, generally used in an environment with multiple drivers connected to a single signal, and commonly referenced in digital simulation as a "Z" state. (C/TT) 1450-1999

float storage (gyros) The sum of attitude storage and torque command storage in a rate-integrating gyro. *See also:* torque-command storage; attitude storage. (AES/GYAC) 528-1994

float switch (liquid-level switch) A switch in which actuation of the contacts is effected when a a float reaches a predetermined level. *See also:* switch. (IA/ICTL/IAC) [60], [84]

float voltage (1) The voltage applied to a battery to maintain it in a fully charged condition during normal operation. (PE/EDPG) 450-1995
(2) The voltage maintained across the battery by the charger in order to keep the battery at its best operational condition with minimum water loss. Float voltage is expressed in volts/cell. (IA/PSE) 602-1996

flonum A floating-point number. (C/MM) 1178-1990r

flood (charge-storage tubes) To direct a large-area flow of electrons, containing no spatially distributed information, toward a storage assembly. *Note:* A large-area flow of electrons with spatially distributed information is used in image-converter tubes. *See also:* charge-storage tube. (ED) 158-1962w, [45]

flooded cable A special coaxial cable containing a corrosion-resistant material between the aluminum sheath and the outer jacket. The corrosion inhibitor flows into imperfections in the jacket to prevent sheath corrosion in high moisture environments. (LM/C) 802.7-1989r

flooded cell A liquid electrolyte filled vented cell. (PE/EDPG) 1184-1994

flood gun A device in a cathode ray tube that emits a stream of electrons that uniformly covers the entire screen, used to maintain the energy level of the phosphors that have been energized by the electron gun. (C) 610.6-1991w, 610.10-1994w

floodlight (illuminating engineering) A projector designed for lighting a scene or object to a brightness considerably greater than its surroundings. It usually is capable of being pointed in any direction and is of weatherproof construction. *Note:* The beam spread of floodlights may range from relatively narrow (10 degrees) to wide (more than 100 degrees). (EEC/IE) [126]

floodlighting (illuminating engineering) A system designed for lighting a scene or object to a brightness greater than its surroundings. It may be for utility, advertising, or decorative purposes. (EEC/IE) [126]

flood-lubricated bearing (rotating machinery) A bearing in which a continuous flow of lubricant is poured over the top of the bearing or journal at about normal atmospheric pressure. *See also:* bearing. (PE) [9]

flood projection (facsimile) The optical method of scanning in which the subject copy is floodlighted and the scanning spot is defined in the path of the reflected or transmitted light. *See also:* scanning. (COM) 168-1956w

floor The result obtained by rounding a number down to the nearest integer. For example, the floor of 5.3 is 5. *Contrast:* ceiling. (C) 1084-1986w

floor acceleration (1) (seismic qualification of Class 1E equipment for nuclear power generating stations) The acceleration of a particular building floor (or equipment mounting) resulting from the motion of a given earthquake. The maximum floor acceleration is the zero period acceleration (ZPA) of the floor response spectrum. (PE/NP) 344-1987r
(2) (nuclear power generating station) The acceleration of a particular building floor (or equipment mounting) resulting from a given earthquake's motion. The maximum floor acceleration can be obtained from the floor response spectrum as the acceleration at high frequencies (in excess of 33 Hz) and is sometimes referred to as the ZPA (zero period acceleration). (PE/NP) 380-1975w

floor bushing A bushing intended primarily to be operated entirely indoors in a substantially vertical position to carry a circuit through a floor or horizontal grounded barrier. Both ends must be suitable for operating in air. *See also:* bushing 49-1948w

floor cavity ratio (illuminating engineering) For a cavity formed by the work-plane, the floor, and the wall surfaces between these two planes, the FCR is computed by using the distance from the floor to the work plane (hf) as the cavity height in the equations given in the definition for cavity ratio. (EEC/IE) [126]

floor lamp (illuminating engineering) A portable luminaire on a high stand suitable for standing on the floor. (EEC/IE) [126]

floor-standing equipment Equipment designed to be used directly in contact with the floor, or supported above the floor on a surface designed to support both the equipment and the operator (e.g., a raised computer floor). (EMC) C63.4-1991

floor trap (burglar-alarm system) A device designed to indicate an alarm condition in an electric protective circuit whenever an intruder breaks or moves a thread or conductor extending across a floor space. *See also:* protective signaling. (EEC/PE) [119]

floppy disk A magnetic disk made of flexible plastic material that is coated with magnetic material and encased in a protective plastic cover. *Note:* Although the name implies that the disk itself is flexible, this term is also used to refer to magnetic disks with rigid plastic covers. *Synonyms:* flexible disk; floppy diskette. *Contrast:* hard disk. *See also:* minifloppy disk; double-sided disk; single-sided disk; high-density disk; double-density disk; microfloppy disk. (C) 610.10-1994w

floppy diskette *See:* floppy disk.

flotation fluid (accelerometer) (inertial sensors) (gyros) The fluid that suspends the float inside the instrument case. The float may be fully or partially floated within the fluid. The degree of flotation varies with temperature because the specific gravity of the fluid varies with temperature. In addition, the fluid provides damping. *See also:* damping fluid. (AES/GYAC) 528-1994

flow Water movement in a stream or conduit. It is generally measured in cubic feet per second or cubic meters per second. (PE/EDPG) 1020-1988r

flow angle (gas tube) That portion, expressed as an angle, of the cycle of an alternating voltage during which current flows. *See also:* gas tube. (ED) [45]

flowchart (1) (computers) A graphical representation for the definition, analysis, or solution of a problem, in which symbols are used to represent operations, data, flow, and equipment. *See also:* logic diagram. (MIL/C) [2], [20], [85]
(2) (software) A control flow diagram in which suitably annotated geometrical figures are used to represent operations, data, or equipment, and arrows are used to indicate the sequential flow from one to another. *Synonym:* flow diagram. *See also:* structure chart; box diagram; graph; block diagram; input-process-output chart; bubble chart.

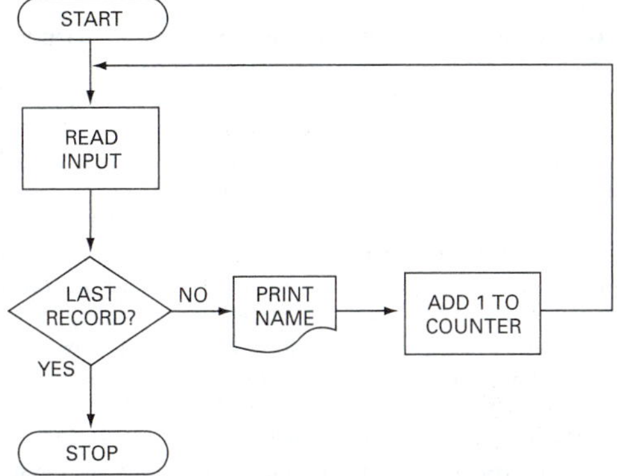

flowchart

(C) 610.12-1990

flowcharter A software tool that accepts as input a design or code representation of a program and produces as output a flowchart of the program. (C) 610.12-1990

flow control (1) A mechanism used in open systems to regulate data communications to ensure that no data is lost in the case of insufficient buffer size, or other limited resources. Flow control is done using receive not ready (RNR), receive ready (RR), and the flow control primitives. Flow control in devices or systems entails operations from the Data Link through the Application layer. (EMB/MIB) 1073.3.1-1994
(2) A mechanism for signaling the producer when messages may or must not be sent. It is used to avoid overrunning the limits of the consumer, memory, queue-depth, or message-passing facilities. (C/MM) 1212.1-1993
(3) (local area networks) An operational capability that allows a peer MAC client entity to cause a temporary delay in further packet transmission from the MAC to that entity. (C) 802.12c-1998
(4) The mechanism employed by a communications provider that constrains a sending entity to wait until the receiving entities can safely receive additional data without loss. (C) 1003.5-1999
(5) The function performed by a receiving entity to limit the amount of data that is sent by a transmitting entity. (C/MM) 1284.4-2000

flow control character (FCC) A control character transmitted on a link in the opposite direction to data flow for each direction of data flow, i.e., to the transmitter of data from the receiver, indicating that the receiver has space reserved to receive a further F N_chars. The value of F is specified separately for each technology in this standard. (C/BA) 1355-1995

flow diagram (1) Graphic representation of a program or a routine. (C) 270-1966w
(2) (software) *See also:* flowchart. (C) 610.12-1990

flow down Passing to a lower level or tier contractual requirements related to administrative, technical or quality performance. (NI) N42.23-1995

flow-duration curve Graphical representation that shows how stream flow has varied historically. (PE/EDPG) 1020-1988r

Flowmatic The first automatic programming language, developed specifically for the UNIVAC II computer. (C) 610.13-1993w

flow of control (software) *See also:* control flow. (C) 610.12-1990

flow relay A relay that responds to a rate of fluid flow. (SWG/PE) C37.100-1992

flow soldering *See:* dip soldering.

flow switch (power system device function numbers) A switch which operates on given values, or on a given rate of change, of flow. (SUB/PE) C37.2-1979s

FLP Burst *See:* Fast Link Pulse (FLP) Burst.

FLP Burst Sequence *See:* Fast Link Pulse (FLP) Burst Sequence.

fluctuating power (rotating machinery) A phasor quantity of which the vector represents the alternating part of the power, and that rotates at a speed equal to double the angular velocity of the current. *See also:* asynchronous machine. (PE) [9]

fluctuating target A radar target whose echo amplitude varies as a function of time. *See also:* target fluctuation. (AES) 686-1997

fluctuation (1) (pulse terminology) Dispersion of the pulse amplitude or other magnitude parameter of the pulse waveforms in a pulse train with respect to a reference pulse amplitude or a reference magnitude. Unless otherwise specified by a mathematical adjective, peak-to-peak fluctuation is assumed. *See also:* mathematical adjectives. (IM/WM&A) 194-1977w
(2) (radar) *See also:* target fluctuation. (AES) 686-1997

fluctuation loss The change in radar detectability or measurement accuracy for a target of given average echo return power due to target fluctuation. *Note:* It may be measured as the change in required average echo return power of a fluctuating target as compared to a target of constant echo return, to achieve the same detection probability or measurement accuracy. (AES) 686-1997

fluctuation noise *See:* random noise.

fluence (solar cells) The total time-integrated number of particles that cross a plane unit area from either side. (AES/SS) 307-1969w

fluence-to-dose-equivalent conversion factor The numerical quantity that relates the neutron fluence to the dose equivalent at a specified depth in tissue. (NI) N42.20-1995

fluid-filled joints (power cable joints) Joints in which the joint housing is filled with an insulating material that is fluid at all operating temperatures. (PE/IC) 404-1986s

fluid loss (rotating machinery) That part of the mechanical losses ina machine having liquid in its air gap that is caused by fluid friction. *See also:* asynchronous machine; direct-current commutating machine. (PE) [9]

fluidly delayed overcurrent trip *See:* overcurrent release; fluidly delayed release.

fluidly delayed release (trip) A release delayed by fluid displacement or adhesion. (SWG/PE) C37.100-1992

fluid pressure supply system (hydraulic turbines) The pumps, means for driving them, pressure and sump tanks, valves and piping connecting the various parts of the governing system and associated and accessory devices. (PE/EDPG) 125-1977s

fluids from essential freeze protection (electric pipe heating systems) The use of electric pipe heating systems to prevent the temperature of fluids from dropping below the freezing point of the fluid in desirably available or essential outdoor (usually) piping systems at fossil fueled generating stations. An example of an essential freeze protection system is the heating for the feedwater system. (PE/EDPG) 622A-1984r

fluorescence (illuminating engineering) The emission of light as the result of, and only during, the absorption of radiation of shorter wavelengths. (EEC/IE) [126]

fluorescent lamp (illuminating engineering) A low-pressure mercury electric-discharge lamp in which a fluorescing coating (phosphor) transforms some of the ultraviolet energy generated by the discharge into light. (EEC/IE) [126]

flush (A) To empty one or more storage locations of their contents; for example, to clear the contents of a buffer after saving its contents on disk. **(B)** To ensure that a buffer has been written to the permanent storage location. (C) 610.10-1994

flush antenna (aircraft) An antenna having no projections outside the streamlined surface of the aircraft. In general, flush antennas may be considered as slot antennas. (PE/EEC) [119]

flush instruction A flush (cache-control) instruction changes a line to the uncached state. If the data are dirty, they are copied back to memory before the old cache line is invalidated. (C/MM) 1596-1992

flush left In text formatting, justification of text such that it is aligned on the left and has a ragged right margin. *Contrast:* flush right. *See also:* left justification. (C) 610.2-1987

flush-mounted antenna An antenna constructed into the surface of a mechanism, or of a vehicle, without affecting the shape of that surface. *Contrast:* conformal antenna. (AP/ANT) 145-1993

flush-mounted device (power and distribution transformers) A device in which the body projects only a small specified distance in front of the mounting surface. (SWG/PE/TR) C37.100-1992, C57.12.80-1978r

flush-mounted or recessed (illuminating engineering) A luminaire that is mounted above the ceiling (or behind a wall or other surface) with the opening of the luminaire level with the surface. (EEC/IE) [126]

flush mounting So designed as to have a minimal front projection when set into and secured to a flat surface. (PE/TR) C57.12.80-1978r

flush right In text formatting, justification of text such that it is aligned on the right and has a ragged left margin. *Contrast:* flush left. *See also:* right justification. (C) 610.2-1987

flutter (sound recording and reproducing equipment) Frequency modulation of the signal in the range of approximately 6 Hz to 100 Hz resulting in distortion which may be perceived as a roughening of the sound quality of a tone or program. (SP) 193-1971w

flutter echo A rapid succession of reflected pulses resulting from a single initial pulse. (SP) [32]

flutter rate (sound recording and reproducing) The number of frequency excursions in hertz, in a tone that is frequency-modulated by flutter. *Notes:* 1. Each cyclical variation is a complete cycle of deviation, for example, from maximum-frequency to minimum-frequency and back to maximum-frequency at the rate indicated. 2. If the over-all flutter is the resultant of several components having different repetition rates, the rates and magnitudes of the individual components are of primary importance.

flux (1) (photovoltaic power system) The rate of flow of energy through a surface. *See also:* photovoltaic power system. (AES) [41]
(2) (soldering) (connections) A liquid or solid which when heated exercises a cleaning and protective action upon the surfaces to which it is applied. (SWG/PE) [103]
(3) (fiber optics) (solar cells) The number of particles that cross a plane unit area per unit time from either side. *Synonym:* radiant power. (AES/SS) 307-1969w, 812-1984w

fluxgate magnetometer An instrument for measuring magnetic fields by making use of the nonlinear magnetic characteristics of a probe or sensing element that has a ferromagnetic core. (T&D/PE) 1308-1994

flux guide (induction heating usage) Magnetic material used to guide electromagnetic flux in desired channels. *Note:* The guides may be used either to direct flux to preferred locations or to prevent the flux from spreading beyond definite regions. *See also:* induction heater. (IA) 54-1955w, 169-1955w

flux linkages The sum of the fluxes linking the turns forming the coil, that is, in a coil having N turns, the flux linkage is
$\lambda = \phi_1 + \phi_2 + \phi_3 \ldots \phi_N$
where ϕ_1 = flux linking turn 1, ϕ_2 = flux linking turn 2, etc., and ϕ_N = flux linking the Nth turn. (CHM) [51]

fluxmeter An instrument for use with a test coil to measure magnetic flux. It usually consists of a moving-coil galvanometer in which the torsional control is either negligible or compensated. *See also:* magnetometer. (EEC/PE) [119]

flux method *See:* lumen method.

flux transfer theory (illuminating engineering) A method of calculating the illuminance in a room by taking into account the interreflection of the light flux from the room surfaces based on the average flux transfer between surfaces. (EEC/IE) [126]

fly ash The finely divided particles of ash entrained in flue gases arising from the combustion of fuel. The particles of ash may contain incompletely burned fuel. The term has been applied predominantly to the gas-borne ash from boilers with spreader stoker, underfeed stoker, and pulverized fuel (coal) firing. *Note:* The above definition is consistent with the generic concept of the word ash. However, all the particulates (including unburned carbon) in suspension in the flue gases are generally called fly ash and the term herein is used in this sense. (PE/EDPG) 548-1984w

flyback (television) The rapid return of the beam in a cathode-ray tube in the direction opposite to that used for scanning. (BT/AV) 201-1979w

flying head *See:* floating head.

flying spot scanner (optical character recognition) A device employing a moving spot of light to scan a sample space, the intensity of the transmitted or reflected light being sensed by a photoelectric transducer. (C) [20], [85]

flywheel ring (rotating machinery) A heavy ring mounted on the spider for the purpose of increasing the rotor moment of inertia. *See also:* rotor. (PE) [9]

FM *See:* frequency modulation.

FM-CW radar *See:* frequency-modulated continuous wave radar.

FM-FM *See:* frequency modulation-frequency modulation.

FM-FM telemetry *See:* frequency modulation-frequency modulation telemetry.

FMIC *See:* fibre optic medium interface connection.

FM radio broadcast band (overhead-power-line corona and radio noise) A band of frequencies assigned for frequency-modulated broadcasting to the general public. *Note:* In the United States and Canada, the frequency band is 88 MHz−108 MHz. (T&D/PE) 539-1990

FMS *See:* flexible manufacturing system.

FN *See:* function key.

FOA *See:* oil-immersed transformer.

focal length (laser maser) The distance from the secondary nodal point of a lens to the primary focal point. In a thin lens, the focal length is the distance between the lens and the focal point. (LEO) 586-1980w

focal point (laser maser) The point toward which radiation converges or from which radiation diverges or appears to diverge. (LEO) 586-1980w

FOCUS A fourth-generation language used to develop information systems, characterized by its integrated database manipulation language and its ability to be used on a wide range of computer platforms. (C) 610.13-1993w

focus (oscillograph) Maximum convergence of the electron beam manifested by minimum spot size on the phosphor screen. *See also:* oscillograph; astigmatism. (IM/HFIM) [40]

focused tests Tests performed to identify a particular area of failure. (SWG/PE) C37.10-1995

focus emphasis A type of emphasis that indicates the current location for entering text. (C) 1295-1993w

focusing (1) (electron tube) The process of controlling the convergence of the electron beam. (ED/BT/AV) 161-1971w, [34], [84]
(2) The concentration of electromagnetic energy into a smaller region of space. *See also:* defocusing. (AP/PROP) 211-1997

focusing and switching grille (color picture tubes) A color-selecting-electrode system in the form of an array of wires including at least two mutually-insulated sets of conductors in which the switching function is performed by varying the potential difference between them, and focusing is accomplished by maintaining the proper average potentials on the array and on the phosphor screen. (ED) 161-1971w

focusing coil *See:* focusing magnet.

focusing device An instrument used to locate the filament of an electric lamp at the proper focal point of lens or reflector optical systems. (EEC/PE) [119]

focusing, dynamic *See:* dynamic focusing.

focusing electrode (beam tube) An electrode the potential of which is adjusted to focus an electron beam. *See also:* electrode. (ED/NPS/BT/AV) 161-1971w, [84], 398-1972r, [34]

focusing, electrostatic *See:* electrostatic focusing.

focusing grid *See:* focusing electrode.

focusing magnet An assembly producing a magnetic field for focusing an electron beam. (ED) 161-1971w, [45]

focusing, magnetic *See:* magnetic focusing.

focusing transducer An interdigital transducer with curved electrodes to focus the launched acoustic wave to a narrower beamwidth. (UFFC) 1037-1992w

FODTE *See:* fibre optic station.

fog Visible aggregate of minute water droplets suspended in the atmosphere near the earth's surface. According to international definition, fog reduces visibility below 1 km. Fog differs from clouds only in that the base of fog is at the earth's surface while clouds are above its surface. When composed of ice crystals, it is termed ice fog. Fog is easily distinguished from haze by its appreciable dampness and gray color. Mist may be considered as intermediate between fog and haze. Mist particles are microscopic in size. Mist is less damp than fog and does not restrict visibility to the same extent. There is no distinct division, however, between any of these categories. Near industrial and heavy traffic areas, fog often is mixed with smoke and vehicle exhaust, and this combination is known as smog. *Note:* Under fog or other dew formation conditions, conductors can become wet or dry depending upon the level of the load current in the conductors. Medium to high load currents produce enough heat through I^2R (resistance) losses to discourage dew formation. Load current also speeds up the drying process after rain, fog, wet snow, etc. (T&D/PE) 539-1990

fog-bell operator A device to provide automatically the periodic bell signals required when a ship is anchored in fog. (EEC/PE) [119]

fog lamps (A) Lamps that may be used in lieu of headlamps to provide road illumination under conditions of rain, snow, dust, or fog. *Synonym:* adverse-weather lamps. *See also:* headlamp. **(B) (illuminating engineering)** Units that may be used in lieu of headlamps or in connection with the lower beam headlights to provide road illumination under conditions of rain, snow, dust, or fog. *Synonym:* adverse-weather lamps. (EEC/IE) [126]

FOI *See:* fibre optic interface.

FOIL *See:* File-Oriented Interpretive Language.

foil (burglar-alarm system) (foil tape) A fragile strip of conducting material suitable for fastening with an adhesive to glass, wood, or other insulating material in order to carry the alarm circuit and to initiate an alarm when severed. *See also:* protective signaling. (EEC/PE) [119]

foil shield A thin, self-supported, metallic tape wrapped longitudinally or spirally around the cable core, and intended to act as a shield against EMI. (PE/IC) 1143-1994r

FOIRL *See:* Fiber Optic Inter-Repeater Link.

FOIRL BER For 10BASE-F, the mean bit error rate of the FOIRL. (LM/C) 802.3u-1995s

FOIRL Compatibility Interface For 10BASE-F, the FOMDI and AUI (optional); the two points at which hardware compatibility is defined to allow connection of independently designed and manufactured components to the baseband optical fiber cable link segment. (LM/C) 802.3u-1995s

FOIRL Segment *See:* Fiber Optic Inter-Repeater Link (FOIRL) Segment.

FOIRL collision For 10BASE-F, the simultaneous transmission and reception of data in a FOMAU. (LM/C) 802.3u-1995s

FOIRL compatibility interfaces The FOMDI (fiber-optic medium dependent interface) and the AUI (optional); the two points at which hardware compatibility is defined to allow connection of independently designed and manufactured components to the baseband optical fiber cable link segment. (LM/C) 802.3b-1989s, 802.3d-1989s, 802.3c-1989s, 802.3e-1989s

folded backplane A backplane in which the electrical bus runs the length of the backplane twice, connecting to alternating modules on each pass. (C/BA) 14536-1995

folded dipole (antenna) An antenna composed of two or more parallel, closely-spaced dipole antennas connected together at their ends with one of the dipole antennas fed at its center and the others short-circuited at their centers. (AP/ANT) 145-1993

folded monopole antenna A monopole antenna formed from half of a folded dipole with the unfed element(s) directly connected to the imaging plane. (AP/ANT) 145-1993

foldover convolution (self convolution) The undesired spurious convolution response that occurs when a portion of an input waveform is reflected into the convolution region and interacts with the input signals. (UFFC) 1037-1992w

Foldy's approximation The approximate solution for the propagation constant of the mean field in a random medium based on the scattering properties of a single particle. (AP/PROP) 211-1997

Foldy-Twersky theory *See:* Foldy's approximation.

follow (surge arresters) (power) The current from the connected power source that flows through an arrester during and following the passage of discharge current. (PE/SPD) 28-1974, C62.1-1981s

follow current (1) The current from the connected power source that flows through a surge-protective device during and following the passage of discharge current. (SPD/PE) C62.62-2000
(2) The current from the connected power source that flows through an arrester during and following the passage of discharge or surge current. (RL) C136.10-1996
(3) (surge arresters) (power) (gas tube surge-protective device) The current from the connected power source that flows through an arrester during and following the passage of discharge current.
(SPD/PE) C62.31-1987r, C62.32-1981s, [8], C62.1-1981s

follower drive (slave drive) A drive in which the reference input and operation are direct functions of another drive, called the master drive. *See also:* feedback control system. (IA/ICTL/APP/IAC) [69], [60]

follow-up potentiometer A servo potentiometer that generates the signal for comparison with the input signal. *See also:* electronic analog computer. (C) 165-1977w

FOMAU *See:* Fiber Optic Medium Attachment Unit.

FOMAU's receive optical fiber For 10BASE-F, the optical fiber from which the local FOMAU receives signals. (LM/C) 802.3u-1995s

FOMAU's transmit optical fiber For 10BASE-F, the optical fiber into which the local FOMAU transmits signals. (LM/C) 802.3u-1995s

FOMDI (Fiber Optic Medium-Dependent Interface) *See:* Fiber Optic Medium Dependent Interface.

F1A line weighting (data transmission) A noise weighting used in a noise measuring set to measure noise on a line that would be terminated by a 302 type, or similar, subset. The meter scale readings are in dBa (F1A). (PE) 599-1985w

F1 layer The lower of the two ionized layers normally existing in the F region in the day hemisphere. *See also:* F region. (AP/PROP) 211-1997

font (1) (computers) A family or assortment of characters of a given size and style. *See also:* type font. (C) [20], [85]
(2) (mathematics of computing) A family or related set of characters and symbols of a particular style of type. (C) 610.6-1991w
(3) A family or related set of characters and symbols of a particular style of type face; for example, 10-point Times Roman. *See also:* optical font; outline font; bit map font; downloadable font; character font. (C) 610.10-1994w

font cartridge A removable storage medium that is used with an output device such as a printer to store on-line fonts. *Note:* By changing the font cartridge, the user can access new fonts. *See also:* cartridge font; font disk. (C) 610.10-1994w

font disk (1) In phototypesetting, a glass disk, imprinted with a specific character font, used by a phototypesetter to generate characters in that character font. (C) 610.2-1987
(2) A disk that is used to store one or more fonts. *See also:* font cartridge. (C) 610.10-1994w

foot (rotating machinery) The part of the stator structure, end shield, or base, that provides means for mounting and fastening a machine to its foundation. *See also:* stator. (PE) [9]

footcandle (fc) (1) (A) (illuminating engineering) A unit of illuminance. One footcandle is one lumen per square foot (lm/ft^2). **(B)** A unit of illuminance (light incident upon a surface) that is equal to 1 lm/ft^2. In the international system, the unit of illuminance is lux (1 fc = 10.76 lux). (EEC/IE/IA/PSE) [126], 241-1990
(2) (television) *See also:* illumination. (BT/AV) 201-1979w

footer *See:* running footer.

footings (foundations) Structures set in the ground to support the bases of towers, poles, or other overhead structures. *Note:* Footings are usually skeleton steel pyramids, grills, or piers of concrete. *See also:* tower. (T&D/PE) [10]

footlambert* (1) (light-emitting diodes) (television) (illuminating engineering) A lambertian unit of luminance equal to ($1/\pi$) candela per square foot. This term is obsolete. (IE/EEC/BT/ED/AV) [126], 201-1979w, [127]
(2) A unit of luminance (photometric brightness) equal to $1/\pi$ candela per square foot ($10.7639/\pi$ candelas per square meter), or to the uniform luminance of a perfectly diffusing surface emitting or reflecting light at the rate of 1 lumen per square foot (10.7639 lumens per square meter), or to the average luminance of any surface emitting or reflecting light at that rate. *Notes:* 1. A footcandle is a unit of incident light, and a footlambert is a unit of emitted or reflected light. For a perfectly reflecting or perfectly diffusing surface, the numbers of footcandles is equal to the number of footlamberts. 2. The average luminance of any reflecting surface in footlamberts is, therefore, the product of the illumination in footcandles by the luminous reflectance of the surface. (BT/AV) 201-1979w
(3) A unit of luminance (photometric brightness) equal to $1/\pi$ candela per square foot, or to the uniform luminance of a perfectly diffusing surface emitting or reflecting light at the rate of one lumen per square foot. (ED) [127]
(4) (electric power systems in commercial buildings) The unit of illuminance that is defined as 1 lm uniformly emitted by an area of 1 ft^2. In the international system, the unit of luminance is candela per square meter (cd/m^2). (IA/PSE) 241-1990r

* Deprecated.

footprint (1) (of an antenna beam on a specified surface) An area bounded by a contour on a specified surface formed by the intersection of the surface and that portion of the beam of an antenna above a specified minimum gain level, the orientation of the beam with respect to the surface being specified. (AP/ANT) 145-1993

(2) The physical space that a device occupies on a desk or other work surface. *Synonym:* real estate. (C) 610.10-1994w

foot switch A switch that is suitable for operation by an operator's foot. *See also:* switch. (IA/ICTL/APP/IAC) [69], [60]

FOPMA (Fiber Optic Physical Medium Attachment) *See:* fiber optic physical medium attachment.

forbidden character *See:* illegal character.

forbidden combination A code expression that is defined to be nonpermissible and whose occurrence indicates a mistake or malfunction. (ED) 161-1971w

forbidden-combination check (1) (data management) A check in which a combination of bits or other representations is not valid according to some criteria. *Contrast:* illegal character. (C) 610.5-1990w

(2) (electronic computation) *See also:* forbidden combination; check.

force Any physical cause that is capable of modifying the motion of a body. The vector sum of the forces acting on a body at rest or in uniform rectilinear motion is zero. 270-1966w

forced-air cooling system (1) (rectifier) An air cooling system in which heat is removed from the cooling surfaces of the rectifier by means of a flow of air produced by a fan or blower. *See also:* rectification. (IA/EEC/PCON) [62], [110]

(2) (thyristor controller) A cooling system in which the heat is removed from the cooling surfaces of the thyristor controller components by means of a flow of air produced by a fan or blower. (IA/IPC) 428-1981w

forced collision A collision that occurs when a packet is transmitted even when traffic is detected on the network and, therefore, the packet will collide with other packets already on the network. (C) 610.7-1995

forced derated hours (electric generating unit reliability, availability, and productivity) The available hours during which a Class 1, 2, or 3 unplanned derating was in effect. (PE/PSE) 762-1987w

forced drainage (underground metallic structures) A method of controlling electrolytic corrosion whereby an external source of direct-current potential is employed to force current to flow to the structure through the earth, thereby maintaining it in a cathode condition. *See also:* inductive coordination. (EEC/PE) [119]

forced interruption (electric power system) An interruption caused by a forced outage. *See also:* outage. (PE/T&D/PSE) [54], 346-1973w, 1366-1998

forced-lubricated bearing (rotating machinery) A bearing in which a continuous flow of lubricant is forced between the bearing and journal. (PE) [9]

forced oscillation (linear constant-parameter system) The response to an applied driving force. *See also:* network analysis. (Std100) 270-1966w

forced outage (1) (emergency and standby power) A power outage that results from the failure of a system component, requiring that it be taken out of service immediately, either automatically or by manual switching operations, or an outage caused by improper operation of equipment or human error. This type of power outage is not directly controllable and is usually unexpected. (IA/PSE) 446-1995

(2) (electric power system) An outage that results from conditions directly associated with a component requiring that it be taken out of service immediately, either automatically or as soon as switching operations can be performed, or an outage caused by improper operation of equipment or human error. *Notes:* 1. This definition derives from transmission and distribution applications and does not necessarily apply to generation outages. 2. The key test to determine if an outage should be classified as forced or scheduled is as follows. If it is possible to defer the outage when such deferment is desirable, the outage is a scheduled outage; otherwise, the outage is a forced outage. Deferring an outage may be desirable, for example, to prevent overload of facilities or an interruption of service to consumers. (PE/PSE) 346-1973w

(3) (outages of electrical transmission facilities) An automatic outage, or a manual outage that cannot be deferred. (PE/PSE) 859-1987w

(4) An outage (failure) that cannot be deferred. (IA/PSE) 493-1997

(5) *See also:* failure.

forced outage duration *See:* repair time.

forced outage hours (electric generating unit reliability, availability, and productivity) The number of hours a unit was in a Class 1, 2, or 3 unplanned outage state. *See also:* unplanned outage. (PE/PSE) 762-1987w

forced release (telephone switching systems) Release initiated from sources other than the calling or called line. (COM) 312-1977w

forced response (1) The response of a system resulting from the application of an energy source with the system initially free of stored energy. (CAS) [13]

(2) (automatic control) A time response which is produced by a stimulus external to the system or element under consideration. *Note:* The response may be described in terms of the causal variable. *See also:* acceleration-forced response; impulse-forced response. (PE/EDPG) [3]

forced response [test] method A test that determines the power dissipation characteristics of a damper by the measurement of the force and velocity imparted to a damper that is mounted directly on the shaker. (T&D/PE) 664-1993

forced-triggered gap A bypass gap that is designed to operate on external command on quantities such as varistor energy, current magnitude, or rate of change of such quantities. The sparkover of the gap is initiated by a trigger circuit. After initiation, an arc is established in the power gap. Forced-triggered gaps typically sparkover only during internal faults. (T&D/PE) 824-1994

forced unavailability The long-term average fraction of time that a component or system is out of service due to a forced outage (failure). (IA/PSE) 493-1997, 399-1997

forced-ventilated machine *See:* open pipe-ventilated machine.

force factor (A) (electroacoustic transducer) The complex quotient of the pressure required to block the acoustic system divided by the corresponding current in the electric system. **(B) (electroacoustic transducer)** The complex quotient of the resulting open-circuit voltage in the acoustic system. *Note:* Force factors (A) and (B) have the same magnitude when consistent units are used and the transducer satisfies the principle of reciprocity. (SP) [32]

(2) (A) (electromechanical transducer) The complex quotient of the force required to block the mechanical system divided by the corresponding current in the electric system. **(B) (electromechanical transducer)** The complex quotient of the resulting open-circuit voltage in the electric system divided by the velocity in the mechanical system. *Notes:* 1. Force factors (A) and (B) have the same magnitude when consistent units are used and the transducer satisfies the principle of reciprocity. 2. It is sometimes convenient in an electrostatic or piezoelectric transducer to use the ratios between force and charge or electric displacement, or between voltage and mechanical displacement. (EEC/PE) [119]

force test A test used to assure that the fall arrest system itself does not severely injure a worker during a fall arrest. A test to ensure that the worker is not severely injured by limiting forces on the worker's body to 4.0 kN (900 pounds) with a line-worker's body belt or 8.0 kN (1800 pounds) with a full body harness. (T&D/PE) 1307-1996

forcing The application of control impulses to initiate a speed adjustment, the magnitude of which is greater than warranted by the desired controlled speed in order to bring about a greater rate of speed change. *Note:* Forcing may be obtained by directing the control impulse so as to effect a change in the field or armature circuit of the motor, or both. *See also:* electric drive. (IA/ICTL/IAC) [60]

foreground In job scheduling, the computing environment in which highpriority processes or those requiring user interaction are executed. *Contrast:* background. *See also:* foreground processing. (C) 610.12-1990

foreground image The part of a display image that can be modified. *Contrast:* background image. (C) 610.6-1991w

foreground job *See:* foreground process group.

foreground process A process that is a member of a foreground process group. (C/PA) 9945-1-1996, 9945-2-1993, 1003.5-1999

foreground process group [foreground job] A process group whose member processes have certain privileges, denied to processes in background process groups, when accessing their controlling terminal. Each session that has established a connection with a controlling terminal has exactly one process group of the session as the foreground process group of that controlling terminal. *Synonym:* foreground process group. (C/PA) 9945-1-1996, 9945-2-1993

foreground process group (1) A process group whose member processes have certain privileges, denied to processes in background process groups, when accessing their controlling terminal. Each session that has established a connection with a controlling terminal has exactly one process group of the session as the foreground process group of that controlling terminal. *Synonym:* foreground process group. (C/PA) 9945-1-1996, 9945-2-1993
(2) A group of processes that have certain privileges, denied to processes in background process groups, when accessing their controlling terminal. Each session that has established a connection with a controlling terminal has exactly one process group of the session as the foreground process group of that controlling terminal. (C) 1003.5-1999

foreground process group ID The process group ID of the foreground process group. (C/PA) 1003.5-1999, 9945-1-1996

foreground processing The execution of a high-priority process while lower-priority processes await the availability of computer resources, or the execution of processes that require user interaction. *Contrast:* background processing. (C) 610.12-1990

foreign area (telephone switching systems) A numbering plan area other than the one in which the calling customer is located. (COM) 312-1977w

foreign data dictionary A data dictionary developed by a non-Intelligent Transportation Systems (ITS) community. (SCC32) 1489-1999

foreign data source A data dictionary or message set developed by a non-ITS community. (SCC32) 1488-2000

foreign exchange An exchange that connects a customer's location to a remote customer. (C) 610.7-1995

foreign exchange circuit A circuit that provides foreign exchange service. *See also:* dial-up circuit; simplex circuit; four-wire circuit; two-wire circuit; leased circuit. (C) 610.7-1995

foreign exchange line (1) (data transmission) A subscriber line by means of which service is furnished to a subscriber at his request from an exchange other than the one from which service would normally be furnished. (PE) 599-1985w
(2) (telephone switching systems) A loop form an exchange other than the one from which service would normally be furnished. (COM) 312-1977w

foreign exchange service A service that provides a connection between a customer and a central office other than the one that serves the exchange area in which the customer is located. (C) 610.7-1995

foreign key (1) (A) An attribute that is a primary key, not to the record it is in, but to some related record. **(B)** In a relational data model, nonprime attributes of some relation that is defined on the same domain as a prime attribute of another relation. (C) 610.5-1990
(2) An attribute, or combination of attributes, of a child or category entity instance whose values match those in the primary key of a related parent or generic entity instance. A foreign key results from the migration of the parent or generic entity's primary key through a generalization structure or a relationship. (C/SE) 1320.2-1998

foreign potential Any voltage and resultant current imposed on telecommunications plant or equipment that is not supplied from the central office or from telecommunications equipment. (IA/PSE) 1100-1999

forensic engineering The application of engineering knowledge to questions of law affecting life and property. (SWG/PE) C37.10-1995

forest A set of disjoint trees. (C) 610.5-1990w

forestalling switch *See:* acknowledger.

For Exposition Only page A model page that contains pictorial and graphical information (in contrast to text) about a specific diagram. Unlike a diagram, the contents of a For Exposition Only page (FEO page) need not comply with IDEF0 rules. (C/SE) 1320.1-1998

fork *See:* branch.

fork beat *See:* carrier beat.

form (1) Any article, such as a printing plate, that is used as a pattern to be reproduced. (EEC/PE) [119]
(2) A medium, sometimes preprinted, on which information is to be printed or plotted. *See also:* form feed; printed card form; index hole; continuous form. (C) 610.10-1994w

FORMAC *See:* FORmula Manipulation Compiler; FORmula Manipulation Language.

formalization The precise description of the semantics of a language in terms of a formal language such as first order logic. (C/SE) 1320.2-1998

formal language (software) A language whose rules are explicitly established prior to its use. Examples include programming languages and mathematical languages. *Contrast:* natural language. (C) 610.12-1990, 610.13-1993w

formal logic The study of the structure and form of valid argument without regard to the meaning of the terms in the argument. (C) [20], [85]

formal parameter (software) A variable used in a software module to represent data or program elements that are to be passed to the module by a calling module. *Contrast:* argument. (C) 610.12-1990

formal qualification review (FQR) The test, inspection, or analytical process by which a group of configuration items comprising a system are verified to have met specific contractual performance requirements. *Contrast:* requirements review; test readiness review; design review; code review. (C) 610.12-1990

formal specification (A) (software) A specification written and approved in accordance with established standards. **(B) (software)** A specification written in a formal notation, often for use in proof of correctness. (C) 610.12-1990

formal testing (software) Testing conducted in accordance with test plans and procedures that have been reviewed and approved by a customer, user, or designated level of management. *Contrast:* informal testing. (C) 610.12-1990

formal test specification A specification of the assertion test using a formal method specified by the test method specification. The test method specification shall specify whether the formal test specification is normative or informative. (C/PA) 2003-1997

format (1) (computers) The general order in which information appears on the input medium.
(2) (data transmission) Arrangement of code characters within a group, such as a block or message. (COM) [49]
(3) Physical arrangement of possible locations of holes or magnetized areas. *See also:* address format. (MAG/EEC) 296-1969w, [74]
(4) (data management) The arrangement, order, or layout of data in or on a data medium. *See also:* variable format; fixed format. (C) 610.5-1990w

(5) **(A)** The structure or appearance of an object such as a storage medium, file, field, or page of text. **(B)** To establish or change the structure or appearance of an object as in definition (A). *See also:* high-level format; low-level format. (C) 610.10-1994

format character A control character used to control a printer. (C) 610.5-1990w

format classification (numerically controlled machines) A means, usually in an abbreviated notation, by which the motions, dimensional data, type of control system, number of digits, auxiliary functions, etc., for a particular system can be denoted. (MAG/EEC) 296-1969w, [74]

format detail (numerically controlled machines) Describes specifically which words and of what length are used by a specific system in the format classification. (IA) [61], [84]

format effector character Any control character used to control the positioning of printed, displayed, or recorded data. *Synonym:* layout character. *See also:* backspace character. (C) 610.5-1990w

formation lights (illuminating engineering) A navigation light especially provided to facilitate formation flying. (EEC/IE) [126]

formation voltage The final impressed voltage at which the film is formed on the valve metal in an electrochemical valve. *See also:* electrochemical valve. (EEC/PE) [119]

format status line A line displayed by many word processing systems that shows the current setting of text formatting parameters such as tabulation stops and margin positions. (C) 610.2-1987

formatted **(A)** Pertaining to magnetic media, such as tapes or diskettes, that have been initialized and prepared to accept and store data. **(B)** Pertaining to text that has been organized into a particular arrangement for output or display. (C) 610.2-1987

formatted information Information that has been arranged into discrete units and structures in a manner that facilitates its access and processing. *Contrast:* narrative information. (C) 610.5-1990w

form C converter A single converter unit in which the direct current can flow in one direction only and which is capable of inverting energy from the load to the ac supply. (IA/ID) 995-1987w

form designation (watthour meter) An alphanumeric designation denoting the circuit arrangement for which the meter is applicable and its specific terminal arrangement. The same designation is applicable to equivalent meters of all manufacturers. (ELM) C12.1-1982s

formed character printer A printer in which each character is a fully formed entity on a slug, drum, mask or other medium. *Contrast:* dot matrix printer. (C) 610.10-1994w

formette *See:* form-wound motorette.

form factor (1) (electric process heating) Coil ratio of conductor width to turn to turn space. *See also:* coil shape factor. (IA) 54-1955w

(2) (illuminating engineering) (f_{1-2}) The ratio of the flux directly received by surface 2 (and due to lambertian surface 1) to the total flux emitted by surface 1. It is used in flux transfer theory. (EEC/IE) [126]

(3) (overhead power lines) (dc electric-field strength and ion-related quantities) An empirical parameter representing the increased electric field at the surface of a dc field meter that is mounted above the ground plane. The increased field is due to field perturbation by the instrument. In a uniform field, the unperturbed electric field is given by the measured field divided by the form factor for the instrument. (T&D/PE) 539-1990, 1227-1990r

(4) (of a periodic function) (*ff*)The ratio of the rms value to the average absolute value $ff = y_{rms}/y_{AAV}$. (PE/PSIM) 120-1989r

(5) (periodic function) The ratio of the root square value to the average absolute value, averaged over a full period of the function. (IA/PSE) 1100-1999

⟨`form-feed`⟩ A character that in the output stream shall indicate that printing should start on the next page of an output device. The ⟨`form-feed`⟩ shall be the character designated by `'\f'` in the C-language binding. If ⟨`form-feed`⟩ is not the first character of an output line, the result is unspecified. It is unspecified whether this character is the exact sequence transmitted to an output device by the system to accomplish the movement to the next page. (C/PA) 9945-2-1993

form feed A command or signal sent to a printer to instruct it to eject the current page and go to the top of the next page. *See also:* tractor feed. (C) 610.10-1994w

form feed character (1) A format effector character that causes the print or display position to move to the next predetermined first line on the next form, the next page, or the equivalent. *Synonyms:* paper throw character; page eject character. (C) 610.5-1990w

(2) A format effector character that instructs a device to move to the top of the next page or screen. (C) 610.10-1994w

form, fit, and function In configuration management, that configuration comprising the physical and functional characteristics of an item as an entity, but not including any characteristics of the elements making up the item. *See also:* configuration identification. (C) 610.12-1990

forming (1) (electrical) (semiconductor devices) The process of applying electric energy to a semiconductor device in order to modify permanently the electric characteristics. *See also:* semiconductor. (IA) [12]

(2) (semiconductor rectifiers) The electrical or thermal treatment, or both, of a semiconductor rectifier cell for the purpose of increasing the effectiveness of the rectifier junction. *See also:* rectification. (IA) 59-1962w, [12]

(3) (electrochemical) The process that results in a change in impedance at the surface of a valve metal to the passage of current from metal to electrolyte, when the voltage is first applied. *See also:* electrochemical valve. (EEC/PE) [119]

forming shell A metal structure designed to support a wet-niche lighting fixture assembly and intended for mounting in a swimming pool structure. (NESC/NEC) [86]

form letter *See:* iterative document.

form overlay A pattern used as a background image. For example, drawing format, report form, title block. (C) 610.6-1991w

FORmula Manipulation Compiler (FORMAC) An extension of PL/1 used to perform symbolic manipulation of mathematical expressions. (C) 610.13-1993w

FORmula Manipulation Language (FORMAC) An extension of FORTRAN used to perform formal algebraic manipulations. (C) 610.13-1993w

FORmula TRANslator (FORTRAN (Fortran)) A high-order programming language used widely for solving scientific, mathematical and numerical problems. *Note:* At the time that this standard was written, FORTRAN 77 and Fortran 90 were both accepted IEEE language standards. *See also:* FGRAAL; common language; FOIL; algebraic language; DYNAMO; GASP IV. (C) 610.13-1993w

form-wound (rotating machinery) (performed winding) Applied to a winding whose coils are formed essentially to their final shape prior to assembly into the machine. *See also:* stator; rotor. (PE) [9]

form-wound motorette (rotating machinery) (formette) A motorette for form-wound coils. *See also:* asynchronous machine; direct-current commutating machine. (PE) [9]

FORTH A high-order programming language that can be used for a wide range of applications due to its ability to be used as an interpreter, command language, and even an operating system. *Note:* FORTH is not an acronym. *See also:* extensible language; Polyforth. (C) 610.13-1993w

Forth word *See:* command.

FORTRAN 66 A dialect of FORTRAN developed as a standard language in 1966. (C) 610.13-1993w

FORTRAN 77 A dialect of FORTRAN developed as a standard language in 1977. (C) 610.13-1993w

Fortran 90 A dialect of FORTRAN developed as a standard language in 1990. (C) 610.13-1993w

FORTRAN IV A dialect of FORTRAN developed as a standard language in 1962. (C) 610.13-1993w

FORTRAN Extended GRAph Algorithmic Language (FGRAAL) An extension of FORTRAN used widely to solve graph problems. *Note:* Includes facilities for manipulating sets and graphs. (C) 610.13-1993w

FORTRAN (Fortran) *See:* FORmula TRANslator.

fortuitous distortion (data transmission) A random distortion of telegraph signals such as that commonly produced by interference. (PE) 599-1985w

fortuitous telegraph distortion Distortion that includes those effects that cannot be classified as bias or characteristic distortion and is defined as the departure, for one occurrence of a particular signal pulse, from the average combined effects of bias and characteristic distortion. *Note:* Fortuitous distortion varies from one signal to another and is measured by a process of elimination over a long period. It is expressed in percent of unit pulse. *See also:* distortion. (AP/ANT) 145-1983s

forward The direction of motion of the train corresponding to the direction of vision of an operator or attendant when occupying his or her normal position in a normal orientation. *Note:* For an unattended vehicle, forward may be defined by the prevailing direction of operation on the guideway segment being utilized. (VT) 1475-1999

forward-acting regulator A transmission regulator in which the adjustment made by the regulator does not affect the quantity that caused the adjustment. *See also:* transmission regulator. (EEC/PE) [119]

forward admittance, short-circuit *See:* admittance, short-circuit forward.

forward annotation The annotation of information from further upstream (earlier in the design flow) in the design process. *See also:* forward annotation file. (C/DA) 1481-1999

forward annotation file A file containing information to be read by a tool for the purpose of forward annotation, for example a Standard Delay Format (SDF) file containing PATHCONSTRAINTS. *See also:* forward annotation. (C/DA) 1481-1999

forward bias (VF) (light-emitting diodes) (forward voltage) The bias voltage which tends to produce current flow in the forward direction. (ED) [127]

forward breakover (thyristor) The failure of the forward blocking action of the thyristor during a normal OFF-state period. (IA/IPC) 428-1981w

forward channel **(1)** Data path from the host to the peripheral. (C/MM) 1284-1994
(2) A channel used to transmit data in which the direction of transmission coincides with that in which information is being transferred. *Contrast:* backward channel. (C) 610.10-1994w

forward controlling elements The elements in the controlling system that change a variable in response to the actuating signal. *See also:* feedback control system. (IM/PE/EDPG) [120], [3]

forward current **(1) (metallic rectifier)** The current that flows through a metallic rectifier cell in the forward direction. *See also:* rectification. (EEC/PE) [119]
(2) (semiconductor rectifier device) The current that flows through a semiconductor rectifier device in the forward direction. *See also:* rectification. (IA) 59-1962w, [12]
(3) (reverse-blocking or reverse-conducting thyristor) The principal current for a positive anode-to-cathode voltage. *See also:* principal current. (IA/ED/IA) 223-1966w, [12], [46], [62]
(4) (light-emitting diodes) The current that flows through a semiconductor junction in the forward direction. (SP) 347-1972w

forward current, average, rating *See:* average forward current rating.

forward direction **(1) (metallic rectifier)** The direction of lesser resistance to current flow through the cell; that is, from the negative electrode to the positive electrode. *See also:* rectification. (PE/EEC) [119]
(2) (semiconductor rectifier device) The direction of lesser resistance to steady direct-current flow through the device; for example, from the anode to the cathode. *See also:* semiconductor rectifier stack; semiconductor.
(3) (semiconductor rectifier diode) The direction of lower resistance to steady-state direct-current; that is, from the anode to the cathode. (IA) [12]

forward (reverse) direction in isolator (nonlinear, active, and nonreciprocal waveguide components) That direction of propagation between two ports of an isolator for which attenuation of waves is lower (higher) than in the opposite direction. (MTT) 457-1982w

forward elements (automatic control) Those elements situated between the actuating signal and the controlled variable in the closed loop being considered. *See also:* feedback control system. (PE/EDPG) 421-1972s

forward error-correcting system A system employing an error-correcting code and so arranged that some or all signals detected as being in error are automatically corrected at the receiving terminal before delivery to the data sink or to the telegraph receiver. (COM) [49]

forward error correction A technique that identifies errors incurred in transmission and allows corrections to be done at the receiving station without retransmission of the message. *See also:* hamming code. (C) 610.7-1995

forward gate current (thyristor) The gate current when the junction between the gate region and the adjacent anode or cathode region is forward biased. *See also:* principal current. (IA/ED) 223-1966w, [45], [62], [12]

forward gate voltage (thyristor) The voltage between the gate terminal and the terminal of the adjacent anode or cathode region resulting from forward gate current. *See also:* principal voltage-current characteristic. (IA/ED) 223-1966w, [62], [45], [12]

forward offset mho characteristic A variant of a mho characteristic in which the reach does not encompass the intersection of the *R*-*X* axes. See figure below.

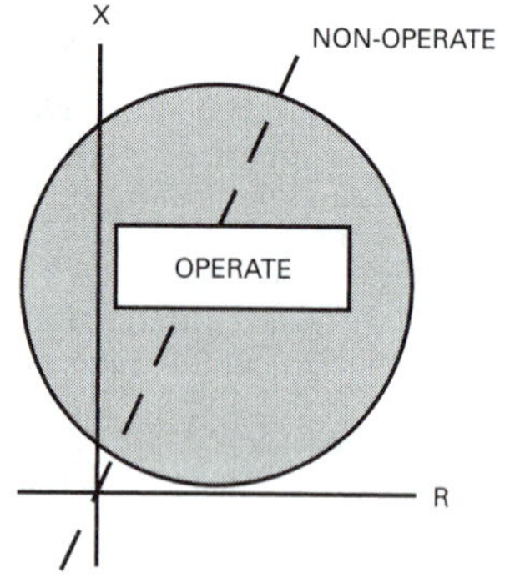

FORWARD OFFSET MHO

(SWG/PE) C37.100-1992

forward path (feedback-control loop) (signal-transmission system) The transmission path from the loop-error signal to the loop-output signal. *See also:* feedback. (IE) [43]

forward period (rectifier circuits) (rectifier circuit element) The part of an alternating-voltage cycle during which forward voltage appears across the rectifier circuit element. *Note:* The forward period is not necessarily the same as the conducting period because of the effect of circuit parameters and semiconductor rectifier cell characteristics. *See also:* rectifier circuit element. (IA) 59-1962w, [12]

forward power dissipation (semiconductor) The power dissipation resulting from forward current. (IA) [12]

forward power loss (semiconductor devices) The power loss within a semiconductor rectifier device resulting from the flow of forward current. *See also:* rectification; semiconductor rectifier stack. (IA) 59-1962w, [12], [62]

forward progress A situation in which a module is not blocked from performing the tasks necessary to achieve its goal. Forward process is guaranteed only in the absence of deadlock or starvation.
(C/BA) 1014.1-1994w, 896.4-1993w, 896.3-1993w, 10857-1994

forward recovery (A) The reconstruction of a file to a given state by updating an earlier version, using data recorded in a chronological record of changes made to the file. *Contrast:* inline recovery; backward recovery. **(B)** A type of recovery in which a system, program, database, or other system resource is restored to a new, not previously occupied state in which it can perform required functions.
(C) 610.5-1990, 610.12-1990

forward recovery time (semiconductor diode) The time required for the current or voltage to recover to a specified value after instantaneous switching from a stated reverse voltage condition to a stated forward current or voltage condition in a given circuit. *See also:* rectification.
(IA) 59-1962w, [12]

forward resistance (metallic rectifier) The resistance measured at a specified forward voltage drop or a specified forward current. *See also:* rectification. (EEC/PE) [119]

forward scattering Scattering of an electromagnetic wave into directions that are at acute angles to the average direction of propagation of the original wave. (AP/PROP) 211-1997

forward-scattering cross section *See:* radar cross section.

forward supervision The use of supervisory sequences sent from a primary station or node to a secondary station or node. *Contrast:* backward supervision. (C) 610.7-1995

forward transadmittance (electron tube) The complex quotient of: the fundamental component of the short-circuit current induced in the second of any two gaps; and the fundamental component of the voltage across the first.
(ED) 161-1971w

forward transfer impedance An attribute similar to internal impedance of a power source, but at frequencies other than the nominal (e.g., 60 Hz power frequency). Knowledge of the forward transfer impedance allows the designer to assess the capability of the power source to provide load current (at the harmonic frequencies) needed to preserve a good output voltage waveform. Generally, the frequency range of interest is 60 Hz to 3 kHz for 50 to 60 Hz power systems, and 20 to 25 kHz for 380 to 480 Hz power systems.
(IA/PSE) 1100-1999

forward voltage (1) (rectifiers) Voltage of the polarity that produces the larger current, hence, the voltage across a semiconductor rectifier diode resulting from forward current. *See also:* ON-state voltage; forward voltage drop.
(2) (reverse-blocking or reverse-conducting thyristor) A positive anode-to-cathode voltage. *See also:* principal characteristics. (IA/ED) 223-1966w, [12], [46], [62]

forward voltage drop (1) (metallic rectifier) The voltage drop in the metallic rectifying cell resulting from the flow of current through a metallic rectifier cell in the forward direction.
(2) (semiconductor rectifiers) *See also:* forward voltage.
(IA) 59-1962w, [12], 332-1972w

forward voltage overshoot (thyristor) The difference between the maximum forward OFF-state voltage following turn-off and the instantaneous ac voltage. (IA/IPC) 428-1981w

forward wave (traveling-wave tubes) A wave whose group velocity is in the same direction as the electron steam motion.
(ED) [45]

forward-wave structure (microwave tubes) A slow-wave structure whose propagation is characterized on a ω/β diagram (ω versus phase shift/section) by a positive slope in the region $O < \beta < \pi$ (in which the group and phase velocity therefore have the same sign). (ED) [45]

Foster's reactance theorem States that the driving-point impedance of a network composed of purely capacitive and inductive reactances is an odd rational function of frequency (ω) that has the following characteristics: a positive slope, and the poles and zeros of the function are on the *j*ω axis, they are simple, they occur in complex conjugate pairs, and they alternate. (CAS) [13]

FOT *See:* Frequence Optimum de Travail.

FOTCU *See:* fibre optic trunk coupling unit.

foul electrolyte An electrolyte in which the amount of impurities is sufficient to cause an undesirable effect on the operation of the electrolytic cells in which it is employed.

fouling The accumulation and growth of marine organisms on a submerged metal surface. (IA) [59], [71]

fouling point (railway practice) The location in a turnout back of a frog at or beyond the clearance point at which insulated joints or details are placed. (EEC/PE) [119]

foul weather The weather condition when there is precipitation or that can cause the transmission line conductors to be wet. Fog is not a form of precipitation, but it causes conductors to be wet. Dry snow is a form of precipitation, but it may not cause the conductors to be wet. (T&D/PE) 539-1990

foul weather distribution A frequency or probability distribution of corona-effect data collected under foul weather conditions. Other distributions can also be defined for more specific foul weather conditions, such as rain, snow, fog, sleet, frost, etc. (T&D/PE) 539-1990

foundation (rotating machinery) The structure on which the feet or base of a machine rest and are fastened. (PE) [9]

foundation bolt (rotating machinery) A bolt used to fasten a machine to a foundation. (PE) [9]

foundation-bolt cone (rotating machinery) A cone placed around a foundation bolt when imbedded in a concrete foundation to provide clearance for adjustment during erection.
(PE) [9]

four-address Pertaining to an instruction code in which each instruction has four address parts. *Note:* In a typical four-address instruction the address specify the location of two operands, the destination of the result, and the location of the next instruction to be interpreted. *See also:* three-plus-one address. (ED) 161-1971w

four-address instruction (1) (software) A computer instruction that contains four address fields. For example, an instruction to add the contents of locations A, B, and C, and place the result in location D. *Contrast:* three-address instruction; one-address instruction; two-address instruction; zero-address instruction. (C) 610.12-1990
(2) An instruction containing four addresses. *Synonym:* quadruple-address instruction. *See also:* address format.
(C) 610.10-1994w

four-bit byte *See:* quartet.

four bolt *See:* conductor grip.

four conductor bundle *See:* bundle.

4GL *See:* fourth generation language.

488-VXIbus interface device A message-based device that provides communication between an IEEE 488 interface and the VXIbus instruments. (C/MM) 1155-1992

Fourier series A single-valued periodic function (that fulfills certain mathematical conditions) may be represented by a Fourier series as follows

$$f(x) = 0.5A_0 + \sum_{n=1}^{n=\infty} [A_n \cos nx + B_n \sin nx]$$

$$= 0.5A_0 + \sum_{n=1}^{n=\infty} C_n \sin (nx + \theta_n)$$

where

$$A_n = \frac{1}{\pi} \int_0^{2\pi} f(x) \cos nx \, dx$$

$n = 0, 1, 2, 3, . . .$

$$B_n = \frac{1}{\pi} \int_0^{2} f(x) \sin nx \, dx$$

$$C_n = +(A_n^2 + B_n^2)^{1/2}$$

$$\theta_n = \text{arc tan } A_n/B_n$$

Note: $0.5A_0$ is the average of a periodic function $f(x)$ over one primitive period. (Std100) 270-1966w

Fourier spectrum (seismic qualification of Class 1E equipment for nuclear power generating stations) A complex valued function that provides amplitude and phase information as a function of frequency for a time domain waveform. (PE/NP) 344-1987r

four-plus-one address (computers) Pertaining to an instruction that contains four operand addresses and a control address. (C) [20], [85]

four-plus-one address format *See:* address format.

four-plus-one address instruction A computer instruction that contains five address fields, the fifth containing the address of the instruction to be executed next. For example, an instruction to add the contents of locations A, B, and C, place the results in location D, then execute the instruction at location E. *Contrast:* three-plus-one address instruction; two-plus-one address instruction; one-plus-one address instruction. (C) 610.12-1990

four-pole *See:* two-terminal pair network.

four quadrant DAM *See:* four quadrant digital-to-analog multiplier.

four quadrant digital-to-analog multiplier (hybrid computer linkage components) A digital-to-analog multiplier (DAM) that accepts both signs of the digital value, giving correct sign output in all four quadrants. (C) 166-1977w

four-quadrant multiplier (1) (analog computer) A multiplier in which operation is unrestricted as to the sign of both of the input variables. (C) 165-1977w, 166-1977w
(2) A multiplier in which the multiplication operation is unrestricted as to the sign of both of the input variables. *Contrast:* two-quadrant multiplier; one-quadrant multiplier. (C) 610.10-1994w

four-terminal network A network with four accessible terminals. *Note:* See two-terminal-pair network for an important special case. *See also:* two-terminal pair network; quadri pole. (Std100) 270-1966w

fourth generation A period during the evolution of electronic computers in which large scale integration is employed, enabling thousands of circuits to be incorporated on one chip, known as an integrated circuit. *Note:* Appearing in the mid-1970's, this generation is thought to be the state of the art at this time. *See also:* third generation; first generation; fifth generation; second generation. (C) 610.10-1994w

fourth generation language A computer language designed to improve the productivity achieved by high-order (third generation) languages and, often, to make computing power available to non-programmers. Features typically include an integrated database management system, query language, report generator, and screen definition facility. Additional features may include a graphics generator, decision support function, financial modeling, spreadsheet capability, and statistical analysis functions. *Contrast:* high-order language; machine language; assembly language; fifth generation language. (C) 610.12-1990, 610.13-1993w

fourth normal form (data management) One of the forms used to characterize relations; a relation R is said to be in fourth normal form if it is in Boyce/Codd normal form and if, when there exists a non-trivial multivalued dependency A →→ B, then all attributes in R are also functionally dependent on A. (C) 610.5-1990w

4-UTP (local area networks) Four-pair 100 Ω balanced cable meeting or exceeding the Category 3 specifications in ISO/IEC 11801:1995. (C) 8802-12-1998

fourth voltage range *See:* voltage range.

fourth-wire control (telephone switching systems) The wire (in addition to the tip, ring, and sleeve wires) used for transmission of special signals necessary in the establishment or supervision of a call. (COM) 312-1977w

four-wire channel (1) (telephone loop performance) Consists of two unidirectional channels carrying signals in opposite directions. (COM/TA) 820-1984r
(2) (data transmission) *See also:* four-wire circuit. (PE) 599-1985w

four-wire circuit (1) (data transmission) A two-way circuit using two paths so arranged that the electric waves are transmitted in one direction only by one path and in the other direction only by the other path. *Note:* The transmission paths may or may not employ four wires. (PE) 599-1985w
(2) A leased circuit in which two pairs of conductors are set up for a two-way transmission path. *See also:* dial-up circuit; simplex circuit; two-wire circuit; foreign exchange circuit. (C) 610.7-1995

four-wire device A handset or headset having separate transmitting and receiving leads, each a pair of wires. (COM/TA) 1206-1994

four-wire repeater (data transmission) A telephone repeater for use in a four-wire circuit and in which there are two currents in one side of the four-wire circuit and the other serving to amplify the telephone currents in the other side of the four-wire circuit. (PE) 599-1985w

four-wire system A three-phase system consisting of three phase conductors and a neutral conductor. (PE/EDPG) 665-1995

four-wire switching (telephone switching systems) Switching using a separate path, frequency, or time interval for each direction of transmission. (COM) 312-1977w

four-wire terminating set (data transmission) A hybrid set for interconnecting a four-wire and two-wire circuit. (PE) 599-1985w
(2) (A) An arrangement in which four-wire circuits are terminated on a two-wire basis for interconnection with two-wire circuits. **(B)** An arrangement by which a four-wire equivalent circuit is converted to a four-wire circuit. (C) 610.7-1995

fovea (illuminating engineering) A small region at the center of the retina, subtending about 2 degrees, that contains cones but no rods, and forms the site of most distinct vision. (EEC/IE) [126]

foveal vision *See:* central vision.

FOW *See:* oil-immersed transformer.

Fowler-Nordheim (F-N) tunneling A quantum-mechanical effect in which electrons penetrate through a barrier region in which they have no allowed states and emerge in the conduction band of the barrier as a result of an externally applied electric field. (ED) 1005-1998

FPGA *See:* field programmable gate array.

FPLA *See:* field programmable logic array.

FPU *See:* floating-point unit.

FQR *See:* formal qualification review.

fractal surface A mathematically generated, irregular shape that can be used to model natural three-dimensional shapes such as coastlines or terrain on a graphical display device. (C) 610.6-1991w

fraction (1) (binary floating-point arithmetic) The field of the significand that lies to the right of its implied binary point. (C/MM) 754-1985r
(2) (mathematics of computing) In floating point arithmetic, the component of the significand that lies to the right of its implied radix point. (C/MM) 854-1987r, 1084-1986w

fractional binary Pertaining to a binary numeral with the binary point (expressed or implied) at the left end, representing a fraction. (C) 1084-1986w

fractional error (measurement) The magnitude of the ratio of the error to the true value. (IM/HFIM) 314-1971w

fractional fixed point Pertaining to fixed-point numeration system in which each number is represented by a numeral with the radix point (expressed or implied) at the left end. All numbers greater than or equal to one must be scaled accordingly. (C) 1084-1986w

fractional-horsepower brush (rotating machinery) A brush with a cross-sectional area of 1/4 square inch (thickness x width) or less and not exceeding 1 1/2 inches in length, but larger than a miniature brush and smaller than an industrial brush. *See also:* brush. (PE) [9]

fractional-horsepower motor (rotating machinery) A motor built in a frame smaller than that of a motor of open construction having a continuous rating of 1 horsepower at 1700-1800 revolutions per minute. *See also:* direct-current commutating machine; asynchronous machine. (PE) [9]

fractional-slot winding (rotating machinery) A distributed winding in which the average number of slots per pole per phase is not integral, for example 3 2/7 slots per pole per phase. *See also:* direct-current commutating machine; asynchronous machine. (PE) [9]

fragility (nuclear power generating station) (seismic qualification of Class 1E equipment) (seismic testing of relays) Susceptibility of equipment to malfunction as the result of structural or operational limitations, or both.
(SWG/PE/NP/PSR) 380-1975w, C37.98-1977s, 344-1975s, C37.100-1992

fragility level (nuclear power generating station) (seismic qualification of Class 1E equipment) (seismic testing of relays) The highest level of input excitation, expressed as a function of input frequency, that a piece of equipment can withstand and still perform the required Class 1E functions.
(SWG/PE/PSR/NP) C37.98-1977s, C37.100-1992, C37.81-1989r, 344-1975s

fragility response spectrum (FRS) (nuclear power generating station) (seismic qualification of Class 1E equipment) (seismic testing of relays) A TRS (test response spectrum) obtained from tests to determine the fragility level of equipment. *See also:* test response spectrum.
(SWG/PE/PSR/NP) C37.98-1977s, C37.81-1989r, 344-1975s, C37.100-1992

Frame A unit of data transmission on an IEEE 802 LAN MAC that conveys a protocol data unit (PDU) between MAC Service users. There are three types of frame: *untagged*, *VLAN-tagged*, and *priority-tagged*. (C/LM) 802.1Q-1998

frame (1) (television) The total area, occupied by the picture, that is scanned while the picture signal is not blanked.
(2) (facsimile) A rectangular area, the width of which is the available line and the length of which is determined by the service requirements. (BT/COM/AV) [34], 168-1956w
(3) (test, measurement, and diagnostic equipment) A cross section of tape containing one bit in each channel and possibly a parity bit. *Synonym:* tape line. (MIL) [2]
(4) (data) (data transmission) A set of consecutive digit time slots in which the position of each digit time slot can be identified by reference to a framing signal. (PE) 599-1985w
(5) (telecommunications circuits and systems) A cyclic set of consecutive timeslots in which the relative position of each timeslot can be identified. (COM/TA) 1007-1991r
(6) A component of the module that provides structural support and enhanced thermal performance.
(C/BA) 1101.7-1995
(7) (A) A group of digits transmitted as a unit that carries a protocol data unit on a network. **(B)** A unit of transmission at the data link layer or, sometimes, the physical layer.
(C) 610.7-1995
(8) A set of consecutive time slots in which the position of each time slot can be identified by reference to a framing signal. (C/BA) 1393-1999
(9) A transmission unit that carries a protocol data unit (PDU) on the ring. (C/LM) 8802-5-1998
(10) (local area networks) The logical organization of control and data fields (e.g., addresses, data, error check sequences) defined for a MAC sublayer. (C) 8802-12-1998
(11) A continuous transmission of octets from one station [bedside communications controller (BCC) or device communications controller (DCC)] to the other station. A Physical layer frame is also referred to as a Physical layer protocol data unit (PhPDU). The Physical layer service data unit (PhSDU) passed between the Data Link layer and the Physical layer consists of the data octets portion of the frame. The PhSDU consists of an integral number of binary octets. The frame consists of these octets, plus other encoded symbols that are added by the Physical layer. For low-speed operation, each octet consists of a binary-encoded start bit, eight data bits, and a stop bit. The first octet of a low-speed frame consists of a flag octet. The last octet of a low-speed frame is either a flag octet or an abort octet. For high-speed operation, each octet consists of eight Manchester biphase-encoded data bits. For high-speed operation, each octet also has a start delimiter and either an end delimiter or an abort delimiter, indicating the beginning and end of individual frames, respectively. The operation of concatenating the delimiters to the data octets is performed by the Physical layer.
(EMB/MIB) 1073.4.1-2000
(12) *See also:* display frame. (C) 610.6-1991w
(13) *See also:* MAC frame. (C/LM) 802.1G-1996

frame alignment The state in which the frame of the receiving equipment is synchronized with respect to that of the received signal. (COM/TA) 1007-1991r

frame alignment signal The distinctive signal(s) inserted in every frame or once in *n* frames, always occupying the same relative position(s) within the frame, and used to establish and maintain frame alignment. (COM/TA) 1007-1991r

frame bits (f-bits) *See also:* frame alignment signal.
(COM/TA) 1007-1991r

frame buffer *See:* bit map.

frame check sequence (1) The field immediately preceding the closing delimiter of a frame. The FCS used is the 16 b polynomial defined by the cyclic redundancy check sequence specified by ITU-T (RC-ITU-T). This field allows the detection of errors by the receiving station.
(EMB/MIB) 1073.3.1-1994
(2) A field in a bit-oriented protocol frame containing the remainder of the cyclic redundancy check calculation on the contents of the frame. (C) 610.7-1995
(3) (local area networks) A Cyclic Redundancy Check (CRC) used by the transmit and receive algorithms to detect errors in the bit sequence of a MAC frame.
(C) 8802-12-1998

frame check sequence error An error in which the frame check sequence value contained in a received frame does not match the frame check sequence value calculated by the receiver. *See also:* cyclic redundancy check. (C) 610.7-1995

framed plate (storage cell) A plate consisting of a frame supporting active material. *See also:* battery. (EEC/PE) [119]

frame, DS1 *See:* DS1 frame.

frame frequency (television) The number of times per second that the frame is scanned. *See also:* television.
(EEC/PE) [119]

frame grabber An input device for digitizing, transferring and storing video frames, such as TV signals, in a computer. *See also:* frame store. (C) 610.10-1994w

frame, intermediate distributing *See:* intermediate distributing frame.

frame, main distributing *See:* main distributing frame.

frame rate (data transmission) The repetition rate of the frame. (PE) 599-1985w

frame relay A fast packet switching technology that provides a virtual circuit service relaying variable-size frames but only employing physical layer and data link layer protocols. *See also:* cell relay. (C) 610.7-1995

Frame relay The function of the Forwarding Process that forwards frames between the Ports of a Bridge.
(C/LM) 802.1Q-1998

frame ring (rotating machinery) A plate or assembly of flat plates forming an annulus in a radial plane and serving as a part of the frame to stiffen it. (PE) [9]

frame size (as applied to a low-voltage circuit breaker) The maximum continuous current rating in amperes for all parts except the coils of the direct-acting trip device. (SWG/PE) C37.100-1992

frame split (rotating machinery) A joint at which a frame may be separated into parts. (PE) [9]

frame store (A) Storage used for data to be sent to a display device. **(B)** Storage used to store data received from a frame grabber. (C) 610.10-1994

frame synchronization (data transmission) The process whereby a given channel at the receiving end is aligned with the corresponding channel at the transmitting end. (PE) 599-1985w

frame validity checking Verification, by a receiving station, of correct frame transmission by the transmitting station. Frame checking entails verifying for correct encoding of all transmitted delimiters and start and stop bits, data bits, and octet encoding. For the medical information bus (MIB), frame validity checking is performed by both the Physical layer and the Data Link layer. (EMB/MIB) 1073.4.1-2000

framework (1) (rotating machinery) A stationary supporting structure. (PE) [9]
(2) A conceptual system of tasks or activities used in a specified type of analysis. (PE/NP) 1082-1997
(3) A collection of classes created specifically to serve the needs of an application area. (SCC20) 1226-1998
(4) A reusable design (models and/or code) that can be refined (specialized) and extended to provide some portion of the overall functionality of many applications. (C/SE) 1320.2-1998

frame yoke (rotating machinery) (field frame) The annular support for the poles of a direct-current machine. *Note:* It may be laminated or of solid metal and forms part of the magnetic circuit. (PE) [9]

framing (facsimile) The adjustment of the picture to a desired position in the direction of line progression. *See also:* recording. (COM) 168-1956w

framing bit errors Frame bits that are in error. (COM/TA) 1007-1991r

framing signal (facsimile) A signal used for adjustment of the picture to a desired position in the direction of line progression. *See also:* facsimile signal. (COM) 168-1956w

Francis turbine Reaction-type turbine in which the water enters radially and leaves axially. (PE/EDPG) 1020-1988r

Fraunhofer diffraction pattern *See:* far-field diffraction pattern.

Fraunhofer pattern A radiation pattern obtained in the Fraunhofer region of an antenna. *Note:* For an antenna focused at infinity, a Fraunhofer pattern is a far-field pattern. (AP/ANT) 145-1993

Fraunhofer region (1) (data transmission) That region of the field in which the energy flow from an antenna proceeds essentially as though coming from a point source located in the vicinity of the antenna. *Note:* If the antenna has a well-defined aperture D in a given aspect, the Fraunhofer region in that aspect is commonly taken to exist at distances greater than $2D^2/$ from the aperture, being the wavelength. (PE) 599-1985w
(2) The region in which the field of an antenna is focused. *Note:* In the Fraunhofer region of an antenna focused at infinity, the values of the fields, when calculated from knowledge of the source distribution of an antenna, are sufficiently accurate when the quadratic phase terms (and higher order terms) are neglected. *See also:* far-field region. (AP/ANT) 145-1993
(3) That region around an electromagnetic radiator or scatterer (maximum dimension D) where the fields can be described in terms of a radial distance and azimuthal and polar angles. *Note:* In this region, the distances of all points to the source's center are larger than $2D^2/\lambda$. *Synonym:* far-field region. (AP/PROP) 211-1997

FRE Conduit fabricated from fiberglass reinforced epoxy. (SUB/PE) 525-1992r

free-body meter A meter that measures the electric field strength at a point above the ground and that is supported in space without conductive contact to earth. *Note:* Free-body meters are commonly constructed to measure the induced current between two isolated parts of a conductive body. Since the induced current is proportional to the time derivative of the electric field strength, the meter's detector circuit often contains an integrating stage in order to recover the waveform of the electric field. The integrated current waveform also coincides with that of the induced charge. The integrating stage is also desirable, particularly for measurements of electric fields with harmonic content because this stage (i.e., its integrating property) eliminates the excessive weighting of the harmonic components in the induced current signal. (T&D/PE) 539-1990, 1308-1994

free bystander A free bystander can be a participating slave that is no longer an entrant, or a potential master that has no current need to acquire the bus and is not fairness inhibited. (C/MM) 896.1-1987s

free capacitance (1) (conductor) The limiting value of its self-capacitance when all other conductors, including isolated ones, are infinitely removed.
(2) (between two conductors) The limiting value of the plenary capacitance as all other, including isolated, conductors are infinitely removed. 270-1966w

free-code call (telephone switching systems) A call to a service or office code for which no charge is made. (COM) 312-1977w

free cyanide (electroplating) (electrodepositing solution) The excess of alkali cyanide above the minimum required to give a clear solution, or above that required to form specified soluble double cyanides. *See also:* electroplating. (PE/EEC) [119]

free fall distance The vertical displacement of a fall arrest attachment point on the line-worker's body belt, aerial belt, or full body harness between onset of the fall and just before the system begins to apply force to arrest the fall. This distance excludes deceleration distance, lifeline and lanyard elongation, but includes any energy absorbing device slide distance or self-retracting lifeline/lanyard extension before they operate and fall arrest forces occur.The component slack (D-ring slide) distance should be included in the free fall distance. (T&D/PE) 1307-1996

free field (1) A field (wave or potential) in a homogeneous, isotropic medium free from boundaries. In practice, a field in which the effects of the boundaries are negligible over the region of interest. *Note:* The actual pressure impinging on an object (for example, electroacoustic transducer) placed in an otherwise free sound field will differ from the pressure that would exist at that point with the object removed, unless the acoustic impedance of the object matches the acoustic impedance of the medium. (SP) [32]
(2) Also known as a free space field. The electromagnetic field in a volume far removed from physical objects, conductive or non conductive; it is usually thought of, but not restricted to, a plane wave. For the case of a plane wave, the electrical and magnetic vectors are transverse to the propagation vector and to each other (TEM), and their ratio yields the intrinsic impedance of free space. (EMC) 1309-1996

free-field current response (receiving current sensitivity) (electroacoustic transducer used for sound reception) The ratio of the current in the output circuit of the transducer when the output terminals are short-circuited to the free-field sound pressure existing at the transducer location prior to the introduction of the transducer in the sound field. *Notes:* 1. The free-field response is defined for a plane progressive sound wave whose direction of propagation has a specified orientation with respect to the principal axis of the transducer. 2. The free-field current response is usually expressed in decibels, namely, 20 times the logarithm to the base 10 of the quotient of the observed ratio divided by the reference ratio, usually 1 ampere per newton per square meter. (SP) [32]

free-field microphone (audible noise measurements) A microphone that has been designed to have a flat frequency response to sound waves arriving with perpendicular incidence (i.e., straight at the microphone). (T&D/PE) 539-1990, 656-1992

free-field voltage response (receiving voltage sensitivity) (electroacoustic transducer used for sound reception) The ratio of the voltage appearing at the output terminals of the transducer when the output terminals are open-circuited to the free-field sound pressure existing at the transducer location prior to the introduction of the transducer in the sound field. *Notes:* 1. The free-field response is determined for a plane progresive sound wave whose direction of propagation has a specified orientation with respect to the principal axis of the transducer. 2. The free-field voltage response is usually expressed in decibels, namely, 20 times the logarithm to the base 10 of the quotient of the observed ratio divided by the reference ratio, usually 1 volt per newton per square meter. (SP) [32]

free-form typing In word processing, the process of entering text that does not include text formatting commands. (C) 610.2-1987

free gyro A two-degree-of-freedom gyro in which the spin axis may be oriented in any specified attitude. In this gyro, output signals are produced by an angular displacement of the case about an axis other than the spin axis. (AES/GYAC) 528-1994

free impedance (transducer) The impedance at the input of the transducer when the impedance of its load is made zero. *Note:* The approximation is often made that the free electric impedance of an electroacoustic transducer designed for use in water is that measured with the transducer in air. *See also:* self-impedance. (SP) [32]

free-line call (telephone switching systems) A call to a directory number for which no charge is made. (COM) 312-1977w

free motion (automatic control) One whose nature is determined only by parameters and initial conditions for the system itself, and not by external stimuli. *Note:* For a linear system, this motion is described by the complementary function of the associated homogeneous differential equation. *Synonym:* free oscillation. (PE/EDPG) [3]

free motional impedance (electroacoustics) (transducer) The complex remainder after the blocked impedance has been subtracted from the free impedance. *See also:* self-impedance. (SP) [32]

free oscillation The response of a system when no external driving force is applied and energy previously stored in the system produces the response. *Note:* The frequency of such oscillations is determined by the parameters in the system or circuit. The term shock-excited oscillation is commonly used. *See also:* oscillatory circuit. (AP/ANT) 145-1983s, 270-1966w

free progressive wave (free wave) A wave in a medium free from boundary effect. A free wave in a steady state can only be approximated in practice. (SP/ACO) [32]

free-radiation frequencies for industrial, scientific, or medical (ISM) apparatus Center of a band of frequencies assigned to industrial, scientific, or medical equipment either nationally or internationally for which no power limit is specified. *See also:* ISM apparatus; electromagnetic compatibility. (EMC/EEC/IE/INT) [70], [126]

free-running frequency The frequency at which a normally synchronized oscillator operates in the absence of a synchronizing signal. (BT/AV) [34]

free-running sweep (non-real time spectrum analyzer) (oscilloscopes) (spectrum analyzer) A sweep that recycles without being triggered and is not synchronized by any applied signal. *See also:* oscillograph. (IM/AES) [14], 748-1979w, [41]

free space Space that is free of obstructions and characterized by the constitutive parameters of a vacuum. (AP/PROP) 211-1997

free-space field intensity The radio field intensity that would exist at a point in a uniform medium in the absence of waves reflected from the earth or other objects. *See also:* radiation. (EEC/PE) [119]

free-space loss The loss between two isotropic radiators in free space, expressed as a power ratio. *Note:* The free-space loss is not due to dissipation, but rather due to the fact that the power flux density decreases with the square of the separation distance. It is usually expressed in decibels and is given by the formula $20\log(4\pi R/\lambda)$, where R is the separation of the two antennas and λ is the wavelength. (AP/ANT) 145-1993

free space permeability (μ_0) A scalar constant such that, in vacuum, its product with the magnetic field $\vec{H}$ is equal to the magnetic flux density:

$\vec{B} = \mu_0\vec{H}$

The numerical value of μ_0 is $4\pi \times 10^{-7}$ H/m. (AP/PROP) 211-1997

free space permittivity (ε_0) A scalar constant such that in vacuum, the product of ε_0 and the electric field, $\vec{E}$, is equal to the electric flux density:

$\vec{D} = \varepsilon_0\vec{E}$

The numerical value for ε_0 is $8.854*10^{-12}$ F/m. (AP/PROP) 211-1997

free-space transmission (mobile communication) Electromagnetic radiation that propagates unhindered by the presence of obstructions, and whose power or field intensity decreases as a function of distance squared. *See also:* mobile communication system. (VT) [37]

free time (availability) The period of time during which an item is in a condition to perform its required function but is not required to do so. (R) [29]

free wave *See:* free progressive wave.

freeze-out (telephone circuit) A short-time denial to a subscriber by a speech-interpolation system. (EEC/PE) [119]

freeze protection (1) (electric pipe heating systems) The use of electric pipe heating systems to prevent the temperature of fluids from dropping below the freezing point of the fluid. Freeze protection is usually associated with piping, pumps, valves, tanks, instrumentation, etc., such as water lines, that are located outdoors, or in unheated buildings. (PE/EDPG) 622A-1984r

(2) (electric heat tracing systems) The use of electric heat tracing systems to prevent the temperature of fluids from dropping below the freezing point of the fluid. Freeze protection is usually associated with piping, pumps, valves, tanks, instrumentation, etc, such as water lines, that are located outdoors or in unheated buildings. (PE/EDPG) 622B-1988r

freezing fog A fog whose droplets freeze upon contact with exposed objects and form a coating of hoarfrost and/or glaze. (T&D/PE) 539-1990

freezing rain Rain that falls in liquid form but freezes on impact to form a coating of glaze upon the ground and on exposed objects. (PE/T&D) 539-1990

F region The region of the terrestrial ionosphere from about 150–1000 km altitude. *Notes:* 1. The daytime F region is characterized by an F1 layer and an F2 layer, and at night the lower (F1) layer merges with the upper (F2) layer. 2. The maximum (or peak) of the F2 layer normally occurs in the 300–600 km altitude range. (AP/PROP) 211-1997

***f* register** One of the floating-point registers. (C/MM) 1754-1994

freight elevator An elevator primarily used for carrying freight on which only the operator and the persons necessary for loading and unloading the freight are permitted to ride. *See also:* elevator. (EEC/PE) [119]

Frenkel defect (solar cells) A defect consisting of the displacement of a single atom from its place in the atomic lattice of a crystal, the atom then occupying an interstitial position. (AES/SS) 307-1969w

Frequence Optimum de Travail (FOT) (radio-wave propagation) The French phrase for Optimum Working Frequency (OWF) applies to ionospheric propagation. *Note:* The FOT is estimated as 0.85 of the predicted monthly median maximum useable frequency (MUF). *See also:* optimum working frequency. (AP/PROP) 211-1990s

frequency (1) (automatic control) The number of periods, or specified fractions of periods, per unit time. *Notes:* 1. The frequency may be stated in cycles per second, or in radians per second, where 1 cycle = 2 pi radians. (PE/EDPG) [3]
(2) (periodic function) (data transmission) (Wherein time is the independent variable)

a) (general). The number of periods per unit time.
b) (automatic control). The number of periods, or specified fractions of periods, per unit time. Note: The frequency may be stated in cycles per second, or in radians per second, where 1 cycle = two radians.
c) (transformer). The number of periods occurring per unit time.
d) (pulse terms). The reciprocal of period.

(PE) 599-1985w
(3) (pulse terminology) A pulse radar in which the transmitter carrier frequency is changed between pulses in a random or pseudo-random way by an amount comparable to the reciprocal of the pulsewidth, or a multiple thereof. (AES/RS) 686-1982s
(4) (power and distribution transformers) The number of periods occurring per unit time. (PE/TR) C57.12.80-1978r
(5) (overhead power lines) The number of complete cycles of sinusoidal variation per unit time. *Notes:* 1. Typically, for ac power lines, the power frequency is 60 Hz in North America and certain other parts of the world and 50 Hz in Europe and many other areas of the world. 2. Electric and magnetic field strength components produced by power lines have frequencies equal to that of power line voltages and currents. 3. The term "power frequency" is often used to avoid specifying whether the power line in question operates at 50 Hz or 60 Hz. (PE/T&D) 539-1990
(6) (broadband local area networks) The number of times a periodic signal repeats itself in a unit of time, usually one second. One hertz (Hz) is one cycle per second. One kilohertz (kHz) is 1000 cycles per second. One megahertz (MHz) is 1 000 000 cycles per second. (LM/C) 802.7-1989r
(7) The number of complete cycles of sinusoidal variation per unit time. *Note:* 1) Electric and magnetic field components have a fundamental frequency equal to that of the power line voltages and currents. 2) For ac power lines, the most widely used frequencies are 60 Hz and 50 Hz. (T&D/PE) 644-1994
(8) The number of times per second that a wave cycle (one peak and one trough) repeats at a given amplitude. (C) 610.7-1995
(9) (of a periodic oscillation or wave) The number of identical cycles per second, measured in Hertz. (AP/PROP) 211-1997
(10) The number of periods occurring in unit time of a periodic quantity, in which time is the independent variable. (IA/MT) 45-1998

frequency-agile radar A pulse radar in which the transmitter-carrier frequency is changed between pulses or between groups of pulses by an amount comparable to or greater than the pulse bandwidth. (AES) 686-1997

frequency allocation The process of designating radio-frequency bands for use by specific radio services. *See also:* frequency allocation table; electromagnetic compatibility. (EMC) [53]

frequency allocation table The table of frequency allocations resulting from the process of designating radio-frequency bands for use by specific radio services. *See also:* electromagnetic compatibility; frequency allocation. (EMC) [53]

frequency allotment The process of designating radio frequencies within an allocated band for use within specific geographic areas. *See also:* frequency allotment plan; electromagnetic compatibility. (EMC) [53]

frequency allotment plan The plan (of frequency allotment) resulting from the process of designating radio frequencies within an allocated band for use within specific geographic areas. *See also:* frequency allotment; electromagnetic compatibility. (EMC) [53]

frequency assignment The process of designating radio frequency for use by a specific station under specified conditions of operations. *See also:* electromagnetic compatibility; frequency assignment list. (EMC) [53]

frequency assignment list The list of frequency assignments resultig from the process of designating radio frequency for use by a specific station under specified conditions of operations. *See also:* frequency assignment; electromagnetic compatibility. (EMC) [53]

frequency band (1) A continuous range of frequencies extending between two limiting frequencies. *Note:* The term frequency band or band is also used in the sense of the term bandwidth. *See also:* channel; signal; signal wave. (IM/IE/BT/AP/ANT) [14], [43], 270-1966w, 182-1961w, 145-1983s
(2) (overhead-power-line corona and radio noise) A continuous range of frequencies extending between two limiting frequencies. *Note:* Some bands of frequencies that are defined by agreement are called "channels." A band used in a particular communication link is also called a channel. (T&D/PE) 539-1990
(3) (spectrum analyzer) A continuous range of frequencies extending between two limiting frequencies. (IM) 748-1979w

frequency-band number The number N in the expression 0.3 ×10N that defines the range of band N. Frequency band N extends from 0.3 × 10N hertz, the lower limit exclusive, the upper limit inclusive. (Std100) 270-1966w

frequency band of emission (communication band) The band of frequencies effectively occupied by that emission, or the type of transmission and the speed of signaling used. *See also:* radio transmission. (AP/ANT) 145-1983s

frequency bands (mobile communication) The frequency allocations that have been made available for land mobile communications by the Federal Communications Commission, including the spectral bands: 25.0 to 50.0 megahertz, 150.8 to 173.4 megahertz, and 450.0 to 470.0 megahertz. (VT) [37]

frequency bias (1) (electric power system) An offset of the scheduled net interchange that varies with frequency error. (PE/PSE) 858-1993w, 94-1991w
(2) (electric power system) An offset in the scheduled net interchange power of a control area that varies in proportion to the frequency deviation. *Note:* This offset is in a direction to assist in restoring the frequency to schedule. *See also:* power system. (PE/PSE) [54]

frequency bias setting A coefficient that, when multiplied by frequency error, yields the frequency bias component of the area control error. (PE/PSE) 94-1991w

frequency changer (1) (general) A motor-generator set that changes power of an alternating-current system from one frequency to one or more different frequencies, with or without a change in the number of phases, or in voltage. *See also:* converter. (EEC/PE) [119]
(2) (rotating machinery) A motor-generator set or other equipment which changes power of an alternating-current system from one frequency to another. (PE) [9]
(3) (self-commutated converters) An alternating current (ac) converter for changing frequency. (IA/SPC) 936-1987w

frequency-changer set (rotating machinery) A motor-generator set that changes the power of an alternating-current system from one frequency to another. (PE) [9]

frequency-change signaling (telecommunications) A method in which one or more particular frequencies correspond to each desired signaling condition. *Note:* The transition from one set of frequencies to the other may be either a continuous or a discontinuous change in frequency or in phase. *See also:* frequency modulation. (COM) [49]

frequency characteristic (telephone sets) Electrical and acoustical properties as functions of frequency. *Note:* Examples include an amplitude-frequency characteristic and an impedance-frequency characteristic. (COM/TA) 269-1971w

frequency, chopped *See:* chopping rate.

frequency control The regulation of frequency within a narrow range. *See also:* generating station. (T&D/PE) [10]

frequency-conversion transducer *See:* conversion transducer.

frequency converter (1) A machine, device, or system for changing ac at one frequency to ac at a different frequency. (PEL/ET) 388-1992r

(2) *See also:* frequency changer.

frequency converter, commutator type (rotating machinery) A polyphase machine the rotor of which has one or two windings connected to slip rings and to a commutator. *Note:* By feeding one set of terminals with a voltage of given frequency, a voltage of another frequency may be obtained from the other set of terminals. *See also:* asynchronous machine. (PE) [9]

frequency, corner *See:* corner frequency.

frequency, cyclotron *See:* cyclotron frequency.

frequency, damped *See:* damped frequency.

frequency departure (telecommunications) The amount of variation of a carrier frequency or center frequency from its assigned value. *Note:* The term frequency deviation, which has been used for this meaning, is in conflict with this essential term as applied to phase and frequency modulation and is therefore deprecated for future use in the above sense. *See also:* radio transmission. (AP/ANT) 145-1983s

frequency-dependent negative resistor An impedance of the form $1/(Ks^2)$, where K is a real positive constant and s is the complex frequency variable. (CAS) [13]

frequency-derived channel (1) A channel obtained from multiplexing a channel by frequency division. (C) 610.7-1995
(2) A channel obtained from multiplexing a channel by frequency-division. (C) 610.10-1994w

frequency deviation (1) (power system) System frequency minus the scheduled frequency. *See also:* frequency modulation; frequency departure. (PE/PSE) 94-1970w, 858-1993w
(2) (telecommunication; frequency modulation) The peak difference between the instantaneous frequency of the modulated wave and the carrier frequency.
(AP/PE/IM/ANT/HFIM) 145-1983s, 599-1985w, [40], 270-1964w
(3) (frequency modulation broadcast receivers) The difference between the instantaneous frequency of the modulated wave and the carrier frequency. (BT) 185-1975w
(4) Instantaneous, normalized, or fractional frequency departure from a nominal frequency. (SCC27) 1139-1999
(5) An increase or decrease in the power frequency from nominal. The duration of a frequency deviation can be from several cycles to several hours. (IA/PSE) 1100-1999

frequency distortion A term commonly used for that form of distortion in which the relative magnitude of the different frequency components of a complex wave are changed in transmission. *Note:* When referring to the distortion of the phase-versus-frequency characteristic, it is recommended that a more specific term such as phase-frequency distortion or delay distortion be used. *See also:* amplitude distortion; distortion; distortion, amplitude-frequency.
(AP/ANT) 145-1983s

frequency diversity *See:* frequency diversity reception.

frequency diversity radar A radar that operates at more than one frequency, using either parallel channels or sequential groups of pulses. *Note:* Parallel channels may have complete duplicate transmitters and receivers, or may divide the transmitted pulse into subpulses at different frequencies, to which parallel receiver channels are tuned. (AES) 686-1997

frequency diversity reception (data transmission) That form of diversity reception that utilizes transmission at different frequencies. (PE) 599-1985w

frequency divider (1) A device for delivering an output wave whose frequency is a proper function, usually a submultiple, of the input frequency. *Note:* Usually the output frequency is an integral submultiple or an integral proper fraction of the input frequency. *See also:* harmonic conversion transducer.
(AP/PE/ANT) 145-1983s, 599-1985w
(2) (nonlinear, active, and nonreciprocal waveguide components) A device for delivering output power at a frequency that is usually an integral proper fraction or integral submultiple of the input frequency. (MTT) 457-1982w

frequency division multiple access (communication satellite) A method of providing multiple access to a communication satellite in which the transmissions from a particular earth station occupy a particular assigned frequency band. In the satellite the signals are simultaneously amplified and transposed to a different frequency band and retransmitted. The earth station identifies its receiving channel according to its assigned frequency band in the satellite signal.
(COM) [19]

frequency-division multiplex (data transmission) (telecommunications) The process or device in which each modulating wave modulates a separate subcarrier and the subcarriers are spaced in frequency. *Note:* Frequency division permits the transmission of two or more signals over a common path by using different frequency bands for the transmission of the intelligence of each message signal.
(AP/PE/ANT) 145-1983s, 599-1985w, 270-1964w

frequency-division multiplexing (1) Dividing a communication channel's bandwidth among several sub-channels with different carrier frequencies. Each sub-channel can carry separate data signals. (LM/C) 802.7-1989r
(2) A multiplexing technique for sharing a transmission channel wherein carrier signals of different frequencies are transmitted simultaneously. (C) 610.7-1995

frequency-division switching (telephone switching systems) A method of switching that provides a common path with a separate frequency band for each of the simultaneous calls.
(COM) 312-1977w

frequency domain A function in which frequency is the independent variable. (EMC) 1128-1998

frequency domain calibration A result which is the transfer function of the sensor or probe. A continuous wave calibration is a transfer function at a single frequency.
(EMC) 1309-1996

frequency doubler (nonlinear, active, and nonreciprocal waveguide components) A device for delivering output power at a frequency that is twice the input frequency.
(MTT/AP/ANT) 457-1982w, 145-1983s

frequency drift (1) (nonreal time spectrum analyzer) Gradual shift or change in displayed frequency over a period of time due to change in components (Hz/sec), (Hz/°C), etc.
(IM) [14]
(2) (spectrum analyzer) Gradual shift or change in displayed frequency over a period of time due to internal changes in the spectrum analyzer (Hz/s, Hz/°C, etc.). (IM) 748-1979w

frequency droop (electric power system) The absolute change in frequency between steady-state no load and steady-state full load. (IA/PSE) 446-1995

frequency error (1) System frequency minus the scheduled frequency. (PE/PSE) 94-1991w
(2) (power system) System frequency minus the scheduled frequency. *See also:* frequency modulation; frequency departure. (PE/PSE) 94-1970w, 858-1993w

frequency hopping (communication satellite) A modulation technique used for multiple access; frequency-hopping systems employ switching of the transmitted frequencies at a rate equal to or lower than the sampling rate of the information transmitted. Selection of the particular frequency to be transmitted can be made from a fixed sequence or can be selected in pseudo-random manner from a set of frequencies covering a wide bandwidth. The intended receiver would frequency-hop in the same manner as the transmitter in order to retrieve the desired information. (COM) [19]

frequency, image *See:* image frequency.

frequency influence (instruments other than frequency meters) (electric instruments) The percentage change (of full-scale value) in the indication of an instrument that is caused solely by a frequency departure from a specified reference frequency. *Note:* Because of the dominance of 60 hertz as the common frequency standard in the United States, alternating-current (power-frequency) instruments are always supplied for that frequency unless otherwise specified. *See also:* accuracy rating. (EEC/ERI/AII) [111], [102]

frequency instability ($S_y(f)$) One-sided spectral density of the fractional frequency deviation. (SCC27) 1139-1999

frequency, instantaneous *See:* instantaneous frequency.

frequency interlace (color television) The effect of intermeshing of the frequency spectrum of a modulated color subcarrier and the harmonics of the horizontal scanning frequency for the purpose of minimizing the visibility of the modulated color subcarrier. (BT/AV) 201-1979w

frequency linearity (non-real time spectrum analyzer) The linearity of the relationship between the input frequency and the displayed frequency. *See also:* linearity. (IM) [14], 748-1979w

frequency lock (1) (power-system communication) A means of recovering in a single-sideband suppressed-carrier receiver the exact modulating frequency that is applied to a single-sideband transmitter. *See also:* power-line carrier. (PE) 599-1985w
(2) For a vibrating beam accelerometer (VBA), the phenomenon where, in a certain band of acceleration around the crossover point of the dual resonator frequencies, the resonator frequencies lock together and do not normally respond to changes in acceleration. (AES/GYAC) 1293-1998

frequency locus For a nonlinear system or element whose describing function is both frequency-dependent and amplitude-dependent, a plot of the describing function, in any convenient coordinate system. (CS/PE/EDPG) [3]

frequency meter An instrument for measuring the frequency of an alternating current. *See also:* instrument. (EEC/PE) [119]

frequency meter, cavity resonator (waveguide components) A cavity resonator used to determine frequency. *See also:* cavity resonator. (MTT) 147-1979w

frequency-modulated continuous wave radar A radar transmitting a continuous carrier modulated by a periodic function such as a sinusoid or sawtooth wave to provide range data. (AES) 686-1997

frequency-modulated cyclotron A cyclotron in which the frequency of the accelerating electric field is modulated in order to hold the positively charged particles in synchronism with the accelerating field despite their increase in mass at very high energies. (ED) [45]

frequency-modulated radar A form of radar in which the radiated wave is frequency modulated and the returning echo beats with the wave being radiated, thus enabling the range to be measured. *See also:* radar. (EEC/PE) [119]

frequency-modulated ranging A technique in which a continuous carrier is frequency modulated by a sinusoidal or triangular waveform, permitting the echo time delay to be measured as the phase shift of the sinusoid or the difference between transmitted and received frequencies. (AES) 686-1997

frequency-modulated transmitter A transmitter that transmits a frequency-modulated wave. *See also:* radio transmitter. (AP/BT/ANT) 145-1983s, 182-1961w

frequency modulation (FM) (1) (electrical conversion) The cyclic or random dynamic variation, or both, of instantaneous frequency about a mean frequency during steady-state electric system operation. (AES) [41]
(2) (telecommunications) (data transmission) Angle modulation in which the instantaneous frequency of a sine-wave carrier is caused to depart from the carrier frequency by an amount proportional to the instantaneous value of the modulating wave. *Note:* Combinations of phase and frequency modulation are commonly referred to as frequency modulation. (COM/AP/PE/ANT) [49], 270-1964w, 145-1983s, 599-1985w
(3) (overhead-power-line corona and radio noise) Modulation in which the instantaneous frequency of a sine wave carrier is caused to depart from the carrier frequency by an amount proportional to the instantaneous value of the modulating signal. *Note:* Combinations of phase and frequency modulation are commonly referred to as frequency modulation. (T&D/PE) 539-1990
(4) A modulation technique in which a data signal is sent onto a carrier by modifying the transmitted frequency. (C) 610.7-1995

frequency modulation-frequency modulation (FM-FM) *See:* frequency modulation-frequency modulation telemetry.

frequency modulation-frequency modulation telemetry (communication satellite) A method of multiplexing many telemetry channels by first frequency modulating subcarriers, combining the modulated subcarriers and finally frequency modulating the radio carrier. This method is widely used for satellite transmissions and follows standards set by Inter Range Instrumentation Group (IRIG). *Synonym:* FM-FM telemetry. (COM) [19]

frequency-modulation (friction) noise ("scrape flutter") Frequency modulation of the signal in the range above approximately 100 Hz resulting in distortion which may be perceived as a noise added to the signal (that is, a noise not present in the absence of a signal). (SP) 193-1971w

frequency monitor An instrument for indicating the amount of deviation of a frequency from its assigned value. *See also:* instrument. (EEC/PE) [119]

frequency multiplier (1) (nonlinear, active, and nonreciprocal waveguide components) A device for delivering output power at a frequency that is an exact positive integer (except for 0 and 1) multiple of an input frequency. Frequency doublers, triplers, quadruplers, etc., are all special cases of frequency multipliers. (MTT) 457-1982w
(2) A device for delivering an output wave whose frequency is an exact integral multiple of the input frequency. *Note:* Frequency doublers and triplers are common special cases of frequency multipliers. *See also:* harmonic conversion transducer. (PE/AP/ANT) 599-1985w, 145-1983s

frequency of charging, resonance The frequency at which resonance occurs in the charging circuit of a pulse-forming network. (MAG) 306-1969w

frequency of occurrence If a process is repeated n times, during which an event occurs m times, the frequency of occurrence of the event, h, is defined as $h = m/n$. For large values of n, the frequency approaches the asymptotic value, called probability of occurrence. (T&D/PE) 539-1990

frequency pulling (oscillators) A change of the generated frequency of an oscillator caused by a change in load impedance. *See also:* oscillatory circuit; waveguide. (ED) 161-1971w, [45]

frequency, pulse repetition *See:* pulse-repetition frequency.

frequency quadrupler (nonlinear, active, and nonreciprocal waveguide components) A device for delivering output power at a frequency that is four times the input frequency. (MTT) 457-1982w

frequency range (1) (general) A specifically designated part of the frequency spectrum.
(2) (transmission system) The frequency band in which the system is able to transmit power without attenuating or distorting it more than a specified amount.
(3) (device) The range of frequencies over which the device may be considered useful with various circuit and operating conditions. *Note:* Frequency range should be distinguished from bandwidth, which is a measure of useful range with fixed circuits and operating conditions. *See also:* signal wave. (ED) 161-1971w, 270-1966w, [45]
(4) (acousto-optic deflector) The frequency range, Δf, over which the diffraction efficiency is greater than some specified minimum. (UFFC) [17]
(5) (spectrum analyzer) That range of frequency over which the instrument performance is specified (hertz to hertz). (IM) [14], 748-1979w

frequency record (electroacoustics) A recording a various known frequencies at known amplitudes, usually for the purpose of testing or measuring. *See also:* phonograph pickup. (SP) [32]

frequency regulation (1) (emergency and standby power) The percentage change in emergency or standby power frequency from steady-state no load to steady-state full load.

$$\%R = \frac{F_{\mathrm{nl}} - F_{f1}}{F_{f1}} \cdot 100$$

(IA/PSE) 446-1995
(2) (ferroresonant voltage regulators) The maximum amount that the output voltage or current will change as the result of a specified change in line frequency. *See also:* overall regulation. (PEL/ET) 449-1990s

frequency relay (1) (power system device function numbers) A relay that responds to the frequency of an electrical quantity, operating when the frequency of an electrical quantity, operating when the frequency or rate of change of frequency exceeds or is less than a predetermined value. (PE/SUB) C37.2-1979s
(2) A relay that responds to the frequency of an alternating electrical input quantity. (SWG/PE) C37.100-1992

frequency resolution The ability of a receiver or signal processing system to detect or measure separately two or more signals that differ only in frequency. *Note:* The classic measure of frequency resolution is the minimum frequency separation of two otherwise identical signals that permits the given system to distinguish that two frequencies are present and to extract the desired information from each of them. When the separation is done by means of a tunable bandpass filter system, the resolution is often specified as the width of the frequency-response lobe measured at a specific value (such as three decibels) below the peak response. *See also:* angular resolution. (AES) 686-1997

frequency response (1) (power supplies) The measure of an amplifier or power supply's ability to respond to a sinusoidal program. *Notes:* 1. The frequency response measures the maximum frequency for full-output voltage excursion. 2. Frequency response connotes amplitude-frequency response, which should be used in full, particularly if phase-frequency response is significant. This frequency is a function of the slewing rate and unity-gain bandwidth. *See also:* amplitude-frequency response. (AES/PE) [41], [78], 599-1985w
(2) (spectrum analyzer) The peak-to-peak variation of the displayed amplitude over a specified center frequency range, measured at the center frequency, (dB). *Note:* Frequency response is closely related to display flatness. The main difference is that the tuning control of the spectrum analyzer is readjusted so as to center the display. (IM) 748-1979w
(3) (speed governing of hydraulic turbines) A characteristic, expressed by formula or graph, which describes the dynamic and steady-state response of a physical system in terms of the magnitude ratio and the phase displacement between a sinusoidally varying input quantity and the fundamental of the corresponding output quantity as a function of the fundamental frequency. (PE/EDPG) 125-1977s
(4) (broadband local area networks) The change of a parameter (usually signal amplitude) with frequency. (LM/C) 802.7-1989r
(5) Electrical, acoustic, or electroacoustic sensitivity as a function of frequency. (COM/TA) 269-1992, 1206-1994
(6) (bandwidth) The change in response (reading) of a field meter to a field of constant amplitude but different frequencies. *Note:* The range of frequencies over which the field meter response is constant to within 3 dB is often referred to as the bandwidth of the field meter. (T&D/PE) 1308-1994
(7) The ratio of the magnitude of the system output to the magnitude of the stimulus over a specified frequency range. (COM/TA) 1027-1996
(8) Electrical, acoustic, or electroacoustic sensitivity (output/input) or gain as a function of frequency. (COM/TA) 1329-1999
(9) *See also:* transfer function. (PAS) 812-1984w
(10) (telecommunications) *See also:* attenuation distortion. (COM/TA) 1007-1991r

frequency-response characteristic (1) (signal-transmission system, industrial control) The frequency-dependent relation, in both gain and phase difference, between steady-state sinusoidal outputs. *Notes:* 1. With nonlinearity, as evidenced by distortion of a sinusoidal input of specified amplitudes, the relation is based on that sinusoidal component of the output having the frequency of the input. 2. Mathematically, the frequency-response characteristic is the complex function of $S = j\omega$:

$$A_0(j\omega)/A_i(j\omega) \exp \{j[\theta_0(j\omega) - \theta_i(j\omega)]\}$$

See also: signal; feedback control system. (IE/PE/IA/EDPG/IAC/APP) [43], [3], [60], [69]
(2) (linear system) In a linear system, the frequency-dependent relation, in both gain and phase difference, between steady-state sinusoidal inputs and the resultant steady-state sinusoidal outputs. (PE/EDPG) 421.2-1990
(3) (automatic generation control on electric power systems) The sum of an area's generation-frequency and load-frequency characteristics. (PE/PSE) 858-1993w, 94-1991w

frequency-response equalization (1) The effect of all frequency discriminative means employed in a transmission system to obtain a desired over-all frequency response. (SP/ACO) [32]
(2) The process of modifying a frequency response of one network by introducing a frequency response of another network so that, within the band of interest, the combined response follows a specified characteristic. (CAS) [13]

frequency selective bandwidth The inverse of the product $2\pi\sigma_t$, where σ_t is the time delay spread. *Synonyms:* coherent bandwidth; dispersive bandwidth. (AP/PROP) 211-1997

frequency selective fading Fading which alffects unequally the different spectral components of a radio signal. (AP/PROP) 211-1997

frequency-selective ringing (telephone switching systems) Selective ringing that employs currents of several frequencies to activate ringers, each of which is tuned mechanically or electrically, or both, to one of the frequencies so that only the desired ringer responds. (COM) 312-1977w

frequency-selective voltmeter A selective radio receiver, with provisions for output indication. (EMC) [53], 263-1965w

frequency selectivity (1) (A) (selectivity) A characteristic of an electric circuit or apparatus in virtue of which electric currents or voltages of different frequencies are transmitted with different attenuation. **(B) (selectivity)** The degree to which a transducer is capable of differentiating between the desired signal and signals or interference at other frequencies. *See also:* transducer. (EEC/PE) [119]
(2) (attenuator) (characteristic insertion loss) Peak-to-peak variation in decibels through the specified frequency range. (IM/HFIM) 474-1973w

frequency-sensitive relay A relay that operates when energized with voltage, current, or power within specific frequency bands. *See also:* relay. (EEC/REE) [87]

frequency shift A condition on a connection where all the frequencies in a signal are shifted by the same amount. (COM/TA) 743-1995

frequency shifter, optical *See:* optical frequency shifter.

frequency shift keying (FSK) (1) (data transmission) That form of frequency modulation in which the modulating signal shifts the output frequency between predetermined values, and the output wave has no phase discontinuity. (PE) 599-1985w
(2) (telecommunications) The form of frequency modulation in which the modulating wave shifts the output frequency between or among predetermined values, and the output wave has no phase discontinuity. *Note:* Commonly, the instantaneous frequency is shifted between two discrete values termed the mark and space frequencies. (IT) [7]
(3) A modulation technique in which binary 0 and 1 are represented by two different frequencies. *See also:* binary phase shift keying; amplitude shift keying. (C) 610.7-1995

frequency-shift pulsing (telephone switching systems) A means of transmitting digital information in which a sequence of two frequencies is used. (COM) 312-1977w

frequency span (nonreal time spectrum analyzer) (spectrum analyzer) The magnitude of the frequency segment displayed (Hz, Hz/div). (IM) [14], 748-1979w

frequency spectrum The distribution of the amplitude (and sometimes the phase) of the frequency components of a signal, as a function of frequency. (T&D/PE) 539-1990, 656-1992

frequency stability (1) (network analyzers) A measure of the amount that a signal source can be expected to vary from its nominal value in a specified time. *Notes:* 1. This can be separated into a short-term stability of limited excursion such as phase-jitter and noise, and a long-term stability such as drift. 2. May cause inaccuracies in measuring narrow band networks; may cause errors in stored corrections. (IM/HFIM) 378-1986w
(2) (data transmission) The measure of the ability to remain on its assigned channel as determined on both a short term (1-second) and a long term (24-hour) basis. (PE) 599-1985w

frequency stabilization The process of controlling the center or carrier frequency so that it differs from that of a reference source by not more than a prescribed amount. *See also:* frequency modulation. (AP/ANT) 145-1983s

frequency standard (1) (A) (electric power system) A device that produces a standard frequency. *See also:* standard frequency; speed-governing system. **(B)** A device that produces a standard frequency. (PE/PSE) 94-1970, 94-1991
(2) (facsimile) A local precision source supplying a stable frequency which is used, among other things, for control of synchronous scanning and recording devices. (COM) 167-1966w

frequency swing (data transmission) In frequency modulation, the peak difference betwee n the maximum and the minimum values of the instantaneous frequency. *Note:* The term "frequency swing" is sometimes used to describe the maximum swing permissible under specified conditions. Such usage should preferably include a specific statement of the conditions. (PE) 599-1985w

frequency time matrix (communication satellite) A modulation technique used for multiple access: frequency-time matrix systems require the simultaneous presence of energy in more than one time and frequency assignment to produce an output signal. The requirement for presence in several time and.or frequency slots reduces the probability of mutual interference when a number of users are simultaneously transmitting. (COM) [19]

frequency tolerance (radio transmitters) The extent to which a characteristic frequency of the emission, for example, the carrier frequency itself or a particular frequency in the sideband, may be permitted to depart from a specified reference frequency within the assigned band. *Note:* The frequency tolerance may be expressed in hertz or as a percentage of the reference frequency. *See also:* radio transmitter. (EEC/PE) [119]

frequency transformation The replacing of the frequency variable s in a function f (s) with a new variable z implicitly defined by s, g (z). This may be done, as examples, to convert a low-pass function into a band-pass function or to make calculations less affected by rounding errors. (CAS) [13]

frequency translation (1) (data transmission) The amount of frequency difference between the received audio signals and the original audio signals after passing through a communication channel. (PE) 599-1985w
(2) (broadband local area networks) Shifting the spectral location of a RF signal frequency from one location to another. (LM/C) 802.7-1989r

frequency translator *See:* translator.

frequency tripler (1) (nonlinear, active, and nonreciprocal waveguide components) A device for delivering output power at a frequency that is three times the input frequency. (MTT) 457-1982w
(2) A device delivering output voltage at a frequency that is three times the input frequency. (AP/ANT) 145-1983s

frequency-type telemeter A telemeter that employs the frequency of a periodically recurring electric signal as the translating means. (SWG/PE/SUB) C37.100-1992, C37.1-1994

frequency, undamped *See:* undamped frequency.

frequently-repeated overload rating The maximum direct current that can be supplied by the converter on a repetitive basis under normal operating conditions. *See also:* power rectifier. (IA/PCON) [62]

freshening charge (1) (lead storage batteries) (nuclear power generating station) The charge given to a storage battery following nonuse or storage. (PE/NP/EDPG) 380-1975w, 484-1987s
(2) The charging of batteries to assure that they are maintained "fresh" in a near-maximum state of charge, and to assure that there is no deterioration of the battery plates due to self-discharge and resulting sulfation. Freshening charges are usually performed using the manufacturer's recommended equalization or cycle-service charging voltage. (SCC21) 937-2000

Fresnel coefficients (for reflection and transmission) The ratio of the phasor value of the parallel or perpendicular polarization component of the electric field of a reflected or transmitted plane wave to that of the corresponding component of the incident plane wave, evaluated at an infinite planar interface separating two homogeneous media. *Synonyms:* reflection coefficient; reflection factor. (AP/PROP) 211-1997

Fresnel contour The locus of points on a surface for which the sum of the distances to a source point and an observation point is a constant, differing by a multiple of a half-wavelength from the minimum value of the sum of the distances. *Note:* This definition applies to media which are isotropic and homogeneous. For the general case, the distances along optical paths must be employed. (AP/ANT) 145-1993

Fresnel diffraction pattern *See:* near-field diffraction pattern.

Fresnel ellipse For a ground-reflected ray, the Fresnel ellipse is the locus of points in the ground plane for which the sum of the distances from the two antennas is an integral number of half wavelengths greater than the length of the specularly reflected ray. (AP/PROP) 211-1997

Fresnel ellipsoid The locus of points for which the sum of distances from two antennas is an integral number of half wavelengths greater than the length of the direct ray between the two antennas. The antennas are at the focal points of the set of ellipsoids. (AP/PROP) 211-1997

Fresnel emissivity The emissivity of an infinite planar interface between two homogeneous media. (AP/PROP) 211-1997

Fresnel lens antenna An antenna consisting of a feed and a lens, usually planar, that transmits the radiated power from the feed through the central zone and alternate Fresnel zones of the illuminating field on the lens. *Synonym:* zone-plate lens antenna. (AP/ANT) 145-1993

Fresnel pattern A radiation pattern obtained in the Fresnel region. (AP/ANT) 145-1993

Fresnel reflection (fiber optics) The reflection of a portion of the light incident on a planar interface between two homogeneous media having different refractive indices. *Notes:* 1. Fresnel reflection occurs at the air-glass interfaces at entrance and exit ends of an optical waveguide. Resultant transmission losses (on the order of 4 percent per interface) can be virtually eliminated by use of antireflection coatings or index matching materials. 2. Fresnel reflection depends upon the index difference and the angle of incidence; it is zero at Brewster's angle for one polarization. In optical elements, a thin transparent film is sometimes used to give an additional Fresnel reflection that cancels the original one by interference. This is called an antireflection coating. *See also:* reflectance; Brewster's angle; index matching material; refractive index; antireflection coating; reflection. (Std100) 812-1984w

Fresnel reflection method (fiber optics) The method for measuring the index profile of an optical fiber by measuring the reflectance as a function of position on the end face. *See also:* reflectance; index profile; Fresnel reflection. (Std100) 812-1984w

Fresnel region (1) (data transmission) The region between the antenna and the Fraunhofer region. *Note:* If the antenna has a well-defined aperture D in a given aspect, the Fresnel region in that respect is commonly taken to extend a distance of 2D^2 / in that aspect, being the wavelength. (PE/PSCC) 599-1985w

(2) The region (or regions) adjacent to the region in which the field of an antenna is focused (that is, just outside the Fraunhofer region). *Note:* In the Fresnel region in space, the values of the fields, when calculated from knowledge of the source distribution of an antenna, are insufficiently accurate unless the quadratic phase terms are taken into account, but are sufficiently accurate if the quadratic phase terms are included. *See also:* radiating near-field region. (AP/PE/ANT) 145-1993, 599-1985w

(3) The region around an electromagnetic radiator or scatterer (maximum dimension D) up to a distance of $2D^2$ divided by the wavelength. Outside of this region the dominant part of the fields decay as $1/r$, while inside the region the distance dependence is more complicated. (AP/PROP) 211-1997

Fresnel zone (1) The region on a surface between successive Fresnel contours. *Note:* Fresnel zones are usually numbered consecutively, with the first zone containing the minimum path length. (AP/ANT) 145-1993

(2) In general, any surface or region bounded by adjacent Fresnel ellipses or ellipsoids. For instance, any plane through both antennas will intersect Fresnel ellipses and define Fresnel zones in that plane. Any plane normal to the ray path between antennas will define a series of circular (annular) Fresnel zones. (AP/PROP) 211-1997

fretting Deterioration resulting from repetitive slip at the interface between two surfaces. *Note:* When deterioration is further increased by corrosion, the term fretting-corrosion is used. (IA) [59], [71]

friction and windage loss (rotating machinery) The power required to drive the unexcited machine at rated speed with the brushes in contact, deducting that portion of the loss that results from: A) forcing the gas through any part of the ventilating system that is external to the machine and cooler (if used); B) the driving of direct-connected flywheels or other direct-connected apparatus. *See also:* asynchronous machine; direct-current commutating machine. (PE) [9], [84]

friction brake The system of pneumatic, electropneumatic, hydraulic, electrohydraulic, or electric valves, controls, actuators and associated components which, in combination, provide the capability of braking the vehicle to a stop purely by the action of friction devices upon the wheel tread, disc rotors, or other surfaces. (VT) 1475-1999

friction electrification *See:* triboelectrification.

friction feed A method for feeding paper into a printer in which friction is used to hold the paper in place. *Contrast:* tractor feed. *See also:* single-sheet feed; continuous feed. (C) 610.10-1994w

friction tape A fibrous tape impregnated with a sticky moisture-resistant compound that provides a protective covering for insulation.

Friis transmission formula The relationship defining the power transfer ratio between two antennas:

$$\frac{P_r}{P_t} = \frac{A_r A_t}{(\lambda d)^2}$$

where

A_r = the effective area of the receiving antenna
A_t = the effective area of the transmitting antenna
λ = the wavelength
d = the separation between the antennas
P_r = the received power
P_t = the transmitted power

Notes: 1. The antennas must be in free space and separated by at least $2D^2/\lambda$, where D is the largest dimension of the larger antenna. 2. The formula accounts for free space propagation loss and antenna gains. (AP/PROP) 211-1997

fringing capacitance (nonlinear, active, and nonreciprocal waveguide components) (semiconductor) The fixed capacitance between the connecting devices (wires and straps) and the pedestal of a diode enclosure. (MTT) 457-1982w

fritting, relay *See:* relay fritting.

frogging (measuring longitudinal balance of telephone equipment operating in the voice band) A switching technique whereby the tip and ring leads of the test specimen are reversed relative to the driving or terminating test circuits, or both. (COM/TA) 455-1985w

frog-leg winding (rotating machinery) A composite winding consisting of one lap winding and one wave winding placed on the same armature and connected to the same commutator. *See also:* direct-current commutating machine. (PE) [9]

front (motor or generator) The front of a normal motor or generator is the end opposite the largest coupling or driving pulley. *See also:* asynchronous machine; direct-current commutating machine. (PE) [9]

front- and back-connected device A device in which one or more current-carrying conductors are connected directly to the fixed terminals located at the front of the mounting base, with the remaining conductors connected to the studs on the back of the mounting base. (SWG/PE) C37.100-1992

front- and back-connected fuse A fuse in which one or more current-carrying conductors are connected directly to the fixed terminals located at the front of the mounting base, with the remaining conductors connected to the studs on the back of the mounting base. (SWG/PE) C37.40-1993

front-and-back connected switch (high-voltage switchgear) A switch having provisions for some of the circuit connections to be made in front of, and others in back of, the mounting base. (SWG/PE) C37.30-1971s

front-connected device A device in which the current-carrying conductors are connected to the fixed terminals in front of the mounting base. (SWG/PE) C37.100-1992

front-connected fuse (high-voltage switchgear) A fuse in which the current-carrying conductors are fastened to the fixed terminals in front of the mounting base. (SWG/PE) C37.40-1993

front-connected switch (high-voltage switchgear) A switch in which the current-carrying conductors are connected to the fixed terminal blocks in front of the mounting base. (SWG/PE) C37.30-1971s

front contact (1) (general) A part of a relay against which, when the relay is energized, the current-carrying portion of the movable neutral member is held so as to form a continuous path for current. *See also: a* contact.
(2) (relay systems) A contact that is closed when the relay is picked up. *Synonym: a* contact. (PE/PSR) C37.90-1978s
(3) (utility-consumer interconnections) A contact that is open when the relay is deenergized. (PE/PSR) C37.95-1973s

front end (communication satellite) The first stage of amplification or frequency conversion immediately following the antenna in a receiving system. (COM) [24]

frontend Pertaining to one part of a process which has two parts, the frontend and the backend; the frontend usually denotes what the user sees and the backend denotes some special process. *Contrast:* backend; frontend. *See also:* backend computer. (C) 610.10-1994w

front end computer A communications computer associated with a host computer. It may perform line control, message handling, code conversion, error control and applications functions such as control and operation of special purpose terminals. *See also:* communications computer. (LM/COM) 168-1956w

front-end computer A computer that interfaces between a group of terminals, communication links, and a host computer and performs communications, error checking code conversion, and other special purpose functions. *Synonym:* front-end processor. *Contrast:* backend computer. *See also:* communications computer. (C) 610.7-1995, 610.10-1994w

front-end processor *See:* front-end computer.

front-of-wave impulse sparkover voltage The impulse sparkover voltage with a wavefront that rises at a uniform rate and causes sparkover on the wavefront. (SPD/PE) C62.11-1999

front-of-wave lightning impulse test (power and distribution transformers) A voltage impulse, with a specified rate-of-rise, that is terminated intentionally by sparkover of a gap that occurs on the rising front of the voltage wave with a specified time to sparkover, and a specified minimum crest voltage. Complete front-of-wave tests (transformer) involve application of the following sequence of impulse waves: one reduced full wave; two front-of-waves; two chopped waves; one full wave. (PE/TR) C57.12.80-1978r

front-of-wave lightning impulse voltage shape A voltage impulse, with a specified rate-of-rise, that is terminated intentionally by sparkover of a gap that occurs on the rising front of the voltage wave with a specified time to sparkover, and a specified minimum crest voltage. (PE/C) 1313.1-1996

front pitch (rotating machinery) The coil pitch at the connection end of a winding (usually in reference to a wave winding). (PE) [9]

front porch (television) The portion of a composite picture signal that lies between the leading edge of the horizontal blanking pulse and the leading edge of the corresponding synchronizing pulse. *See also:* television. (BT/AV) [34]

front relay contacts Sometimes used for relay contacts, normally open. (EEC/REE) [87]

front-to-back ratio (1) (general) The ratio of the directivity of an antenna to directive gain in a specified direction toward the back. [35]
(2) (data transmission) For a directional antenna, the ratio of its effectiveness toward the front to its effectiveness toward the back. (PE) 599-1985w
(3) The ratio of the maximum directivity of an antenna to its directivity in a specified rearward direction. *Notes:* 1. This definition is usually applied to beamtype patterns. 2. If the rearward direction is not specified, it shall be taken to be that of the maximum directivity in the rearward hemisphere relative to the antenna's orientation. (AP/ANT) 145-1993

FRS *See:* fragility response spectrum.

FRU *See:* field replaceable unit.

fruit *See:* fruit pulse.

fruit pulse A pulse reply received as the result of interrogation of a transponder by interrogators not associated with the responsor in question. *Synonym:* fruit. (IM/WM&A) 194-1977w

F scan *See:* F-display.

F-scope A cathode-ray oscilloscope arranged to present an F-display. (AES/RS) 686-1990

FSK *See:* frequency shift keying.

fsm *See:* finite state machine.

FS to AM converter *See:* facsimilie receiving converter.

FTAM regime The initial regime negotiated between a pair of FTAM ASEs, within which a series of file selection regimes may be established to select or create individual files. (C/PA) 1238.1-1994w

FTC circuit *See:* fast-time-constant circuit.

FTP *See:* file transfer protocol.

F2 layer The single ionized layer normally existing in the F region in the night hemisphere and the higher of the two layers normally existing in the F region in the day hemisphere. *See also:* F region. (AP/PROP) 211-1997

F2V A transaction that is originated at the Futurebus+ (the Bridge acts as a Futurebus+ slave), and has its destination at the VME64 (the Bridge acts as a VME64 master). (C/BA) 1014.1-1994w

F-type connector A 75-Ω connector used to connect coaxial cable to equipment. (LM/C) 802.7-1989r

fuel (fuel cells) A chemical element or compound that is capable of being oxidized. *See also:* electrochemical cell. (AES/IA/APP) [41], [73]

fuel adjustment clause (electric power utilization) (power operations) A clause in a rate schedule that provides for adjustment of the amount of the bill as the cost of fuel varies from a specified base amount per unit. (PE/PSE) 858-1987s, 346-1973w

fuel-and-oil quantity electric gauge A device that measures, by means of bridge circuits and an indicator with separate pointers and scales, the quantity of fuel and oil in the aircraft tanks. (EEC/PE) [119]

fuel battery An energy-conversion device consisting of more than one fuel cell connected in series, parallel, or both. *See also:* fuel cell. (AES/IA/APP) [41], [73]

fuel-battery power-to-volume ratio The kilowatt output per envelope volume of the fuel battery (exclusive of the fuel, oxidant, storage, and auxiliaries). *See also:* fuel cell. (AES/IA/APP) [41], [73]

fuel-battery power-to-weight ratio The kilowatt output per unit weight of the fuel battery (exclusive of the fuel, oxidant, storage, and auxiliaries). *See also:* fuel cell. (AES/IA/APP) [41], [73]

fuel cell An electrochemical cell that can continuously change the chemical energy of a fuel and oxidant to electric energy by an isothermal process involving an essentially invariant electrode-electrolyte system. (AES/IA/APP) [41], [73]

fuel-cell Coulomb efficiency The ratio of the number of electrons obtained from the consumption of a mole of the fuel to the electrons theoretically available from the stated reaction.

$$\text{Coulomb efficiency} = \frac{\int_0^{t_m} i\,dt}{nF} \times 100$$

where

t_m = time required to consume a mole of fuel
i = instantaneous current
n = number of electrons furnished in the stated reaction by the fuel molecule
F = Faraday's constant = 96485.3 ± 10.0 absolute joules per absolute volt gram equivalent

See also: fuel cell. (AES/IA/APP) [41], [73]

fuel-cell standard voltage (at 25°C) The voltage associated with the stated reaction and determined from the equation

$$E^0 = \frac{-J\Delta G^0}{nF}$$

where

E^0 = fuel-cell standard voltage
J = Joule's equivalent = 4.1840 absolute joules per calorie
ΔG^0 = standard free energy changes in kilocalories/mole of fuel
n = number of electrons furnished in the stated reaction by the fuel molecule
F = Faraday's constant = 96485.3 ± 10.0 absolute joules per absolute volt gram equivalent

See also: fuel cell. (AES/IA/APP) [41], [73]

fuel-cell system An energy conversion device consisting of one or more fuel cells and necessary auxiliaries. *See also:* fuel cell. (AES/IA/APP) [41], [73]

fuel-cell-system energy-to-volume ratio The kilowatt-hour output per displaced volume of the fuel-cell system (including the fuel, oxidant, and storage). *See also:* fuel cell. (AES/IA/APP) [41], [73]

fuel-cell-system energy-to-weight ratio The kilowatt-hour output per unit weight of the fuel-cell system (including the fuel, oxidant, and storage). *See also:* fuel cell. (AES/IA/APP) [41], [73]

fuel-cell-system power-to-volume ratio The kilowatt output per displaced volume of the fuel-cell system (exclusive of the fuel, oxidant, and storage). *See also:* fuel cell. (AES/IA/APP) [41], [73]

fuel-cell-system power-to-weight ratio The kilowatt output per unit weight of the fuel-cell system (exclusive of the fuel, oxidant, and storage). *See also:* fuel cell. (AES/IA/APP) [41], [73]

fuel-cell-system standard thermal efficiency The efficiency of a system made up of a fuel cell and auxiliary equipment. *Note:* This efficiency is expressed as the ratio of 1) the electric energy delivered to the load circuit to 2) the enthalpy change for the stated cell reaction.

$$\text{thermal efficiency} = \frac{\int_0^{t_m} (E_{IL} \times i_L)dt}{\Delta H^0}$$

where

t_m = time required to consume a mole of fuel
E_{IL} = fuel-cell-system working voltage
i_L = instantaneous current into the load
ΔH^0 = enthalpy change for the stated cell reaction at standard conditions

See also: fuel cell. (AES/IA/APP) [41], [73]

fuel-cell working voltage The voltage at the terminals of a single fuel-cell delivering current into system auxiliaries and load. *See also:* fuel cell. (AES/IA/APP) [41], [73]

fuel-control mechanism (gas turbines) All devices, such as power-amplifying relays, servomotors, and interconnections required between the speed governor and the fuel-control valve. *See also:* speed-governing system. (PE/EDPG) 282-1968w, [5]

fuel-control system (gas turbines) Devices that include the fuel-control valve and all supplementary fuel-control devices and interconnections necessary for adequate control of the fuel entering the combustion system of the gas turbine. *Note:* The supplementary fuel-control devices may or may not be directly actuated by the fuel-control mechanism. *See also:* speed-governing system. (PE/EDPG) 282-1968w, [5]

fuel-control valve (gas turbines) A valve or any other device operating as a final fuel-metering element controlling fuel input to the gas turbine. *Notes:* 1. This valve or device may be directly or indirectly controlled by the fuel-control mechanism. 2. Variable-displacement pumps, or other devices that operate as the final fuel-control element in the fuel-control system, and that control fuel entering the combustion system are fuel-control valves. *See also:* speed-governing system. (PE/EDPG) [5], 282-1968Dw

fuel economy The ratio of the chemical energy input to a generating station to its net electric output. *Note:* Fuel economy is usually expressed in British thermal units per kilowatthour. *See also:* generating station. (T&D/PE) [10]

fuel elements, nuclear *See:* nuclear fuel elements.

fuel-pressure electric gauge A device that measures the fuel pressure (usually in pounds per square inch) at the carburetor of an aircraft engine. *Note:* It provides remote indication by means of a self-synchronous generator and motor. (EEC/PE) [119]

fuel replacement energy (1) (power operations) Energy generated to substitute for energy which would otherwise have been generated by a different fuel source. (PE/PSE) 858-1987s
(2) Energy generated at a hydroelectric plant as a substitute for energy which would otherwise have been generated by a thermal-electric plant. (T&D/PE/PSE) 346-1973w

fuel reprocessing, nuclear *See:* nuclear fuel reprocessing.

fuel stop valve (gas turbines) A device that, when actuated, shuts off all fuel flow to the combustion system, including that provided by the minimum fuel limiter. *See also:* speed-governing system. (PE/EDPG) 282-1968w, [5]

fulguration *See:* electrodesiccation.

full adder An adder that accepts three inputs (two operands and a carry digit), producing a sum and a carry as outputs according to the table below. *Synonym:* three-input adder. *Contrast:* quarter adder; half adder. *See also:* full subtracter.

input #1	0	0	1	1	0	0	1	1
input #2	0	1	0	1	0	1	0	1
input carry	0	0	0	0	1	1	1	1
output sum	0	1	1	0	1	0	0	1
output carry	0	0	0	1	0	1	1	1

full adder

(C) 610.10-1994w

full automatic plating Mechanical plating in which the cathodes are automatically conveyed through successive cleaning and plating tanks. (EEC/PE) [119]

full availability (telephone switching systems) Availability that is equal to the number of outlets in the desired group. (COM) 312-1977w

full backup To perform a backup in which all data within a system is stored on the backup copy. *Contrast:* incremental backup. (C) 610.5-1990w

full binary tree *See:* complete binary tree.

full data transport address The combination of a *functional address* and a *channel address*, that specifies whether data is being read or written, to which function, and to which channel. (IM/ST) 1451.2-1997

full-direct trunk group (telephone switching systems) A full trunk group between end offices. (COM) 312-1977w

full duplex (1) (communication circuits) (telecommunications) (data transmission) Method of operation where each end can simultaneously transmit and receive. *Note:* Refers to a communications system or equipment capable of transmission simultaneously in two directions. (PE) 599-1985w
(2) A mode of operation of a network, DTE, or Medium Attachment Unit (MAU) that supports duplex transmission as defined in IEEE Std 100-1996. This mode of operation allows for simultaneous communication between a pair of stations, provided that the Physical Layer is capable of supporting simultaneous transmission and reception without interference. (C/LM) 802.3-1998
(3) (local area networks) A link segment capable of transferring signals in both directions simultaneously. (C) 8802-12-1998

(4) An operating condition which allows simultaneous communication in both send and receive directions with 3 dB or less switched loss in either direction. (COM/TA) 1329-1999

full-duplex operation (local area networks) A mode of operation in a network link that supports duplex transmission as defined in IEEE Std 610.7-1995. (C) 802.12c-1998

full duplex transmission *See:* duplex transmission.

full-energy peak (1) A peak in the spectrum resulting from the complete (total) absorption of a photon of a given energy in the active volume of the Ge crystal and the collection of all of the resulting charge. (NI) N42.14-1991
(2) The peak in a pulse height spectrum that corresponds to total absorption of a gamma photon in the NaI(Tl) detector. (NI) N42.12-1994
(3) (for a monoenergetic photon spectrum for a semiconductor spectrometer system) The distribution of events within the peak of the pulse-height distribution spectrum representing response to the monoenergetic photon source. *Note:* Notwithstanding other definitions or procedures for subtracting background and other distortions, the full energy peak intensity is defined as not including any events that exceed a Gaussian distribution by more than a factor ot two σ. (NPS/NID) 759-1984r

full energy peak efficiency (1) (x-ray energy spectrometers) (of a semiconductor radiation detector) The ratio of the number of events in the full energy peak of the spectral distribution to the total number of photons incident on the active detector volume during the same time interval. (NPS/NID) 759-1984r
(2) The ratio between the number of counts in the net area of the full-energy peak to the number of photons of that energy emitted by a source with specified characteristics for a specified source-to-detector distance. (NI) N42.14-1991

full field In a propulsion system, the motor connection in which, for series motors, the exciting field current is the same as the armature current, or, for separately excited motors, the exciting field current is at its maximum value relative to the armature current. (VT) 1475-1999

full-field relay A relay that functions to maintain full field excitation of a motor while accelerating on reduced armature voltage. *See also:* relay. (IA/ICTL/IAC) [60]

full float (constant potential) operation Operation of a dc system with the battery, battery charger, and load all connected in parallel and with the battery charger supplying the normal dc load plus any self-discharge or charging current, or both, required by the battery. (The battery will deliver current only when the load exceeds the charger output.) (PE/EDPG) 1115-1992

full float operation Operation of a dc system with the battery, battery charger, and load all connected in parallel and with the battery charger supplying the normal dc load plus any charging current required by the battery. (The battery will deliver current only when the load exceeds the charger output.) (SCC29) 485-1997

full functional dependency A functional dependency in which no attribute of the determinant can be omitted without voiding the dependent condition. (C) 610.5-1990w

full-height disk drive A disk drive that requires the whole height of the front panel of a standard computer cabinet bay. *Note:* Approximately 3.5 in, a full-height disk drive is twice the size of a half-height disk drive. (C) 610.10-1994w

Full Implementation Refers to the implementation of an operation on an Object. A Full Implementation means that all the referenced visible functionality of the operation is implemented as specified for the operation. (IM/ST) 1451.1-1999

full impulse voltage An aperiodic transient voltage that rises rapidly to a maximum value and falls, usually less rapidly, to zero. See the corresponding figure. *See also:* full-wave voltage impulse.

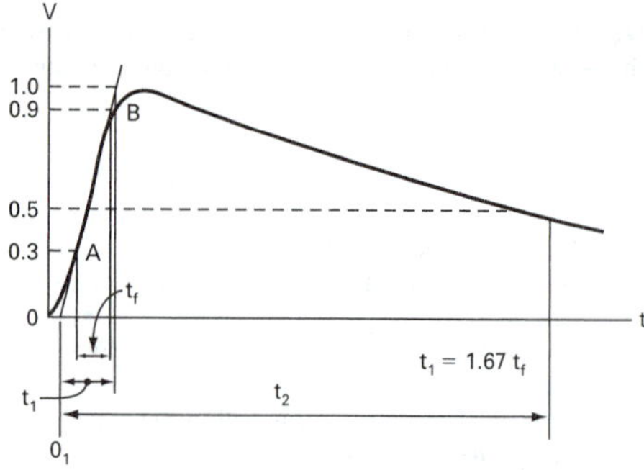

full impulse voltage

(PE/PSIM) 4-1978s

full-impulse wave (surge arresters) An impulse wave in which there is no sudden collapse. (PE) [8], [84]

full justification In text formatting, justification resulting in even margins on both the left and right margins. (C) 610.2-1987

full lightning impulse A lightning impulse not interrupted by any type of discharge. (PE/PSIM) 4-1995

full load (test, measurement, and diagnostic equipment) The greatest load that a circuit is designed to carry under specific conditions: any additional load is overload. (MIL/PE) [2], 599-1985w

full-load speed (electric drive) The speed that the output shaft of the drive attains with rated load connected and with the drive adjusted to deliver rated output at rated speed. *Note:* In referring to the speed with full load connected and with the drive adjusted for a specified condition other than for rated output at rated speed, it is customary to speak of the full-load speed under the (stated) conditions. *See also:* electric drive. (IA/ICTL/IAC) [60]

full magnetic controller An electric controller having all of its basic functions performed by devices that are operated by electromagnets. (IA/MT) 45-1998

full-page display device A display device that can display the contents of a full 8.5 × 11 in page at one time. (C) 610.10-1994w

full-pitch winding (rotating machinery) A winding in which the coil pitch is 100%; that is, equal to the pole pitch. *See also:* direct-current commutating machine; asynchronous machine. (PE) [9]

full range (accelerometer) (gyros) The algebraic difference between the upper and lower values of the input range. (AES/GYAC) 528-1994

full-range current-limiting fuse A fuse capable of interrupting all currents from the rated interrupting current down to the minimum continuous current that causes melting of the fusible element(s), with the fuse applied at the maximum ambient temperature specified by the fuse manufacturer. (SWG/PE) C37.40-1993

full scale In an analog computer, the nominal maximum value of a computer variable or the nominal maximum value at the output of a computing element. *Note:* Sometimes used to indicate the entire computing voltage range, for example, 20 V is full scale for a computer whose voltages ranges from +10 V to −10 V. (C) 610.10-1994w, 165-1977w

full scale input (accelerometer) (gyros) The maximum magnitude of the two input limits. (AES/GYAC) 528-1994

full-scale range The difference between the maximum and the minimum recordable input values as specified by the manufacturer. (IM/WM&A) 1057-1994w

full-scale signal A signal that spans the entire manufacturer's specified amplitude range of the instrument. (IM/WM&A) 1057-1994w

full-scale value (1) The largest value of the actuating electrical quantity that can be indicated on the scale or, in the case of instruments having their zero between the ends of the scale, the full-scale value is the arithmetic sum of the values of the

actuating electrical quantity corresponding to the two ends of the scale. *Note:* Certain instruments, such as power-factor meters, are necessarily excepted from this definition. *See also:* instrument; accuracy rating. (EEC/ERI/AII) [111], [102]
(2) (mechanical demand registers) The maximum scale capacity of the register. If a multiplier exists, the full-scale value will be the product of the maximum scale marking and the multiplying constant. (ELM) C12.4-1984

full-screen editing A method of text editing that allows the user to view a full display screen of text at one time and to enter or alter text by using either commands or cursor control. Scrolling functions allow the user to move up and down within the document. *Contrast:* line editing. (C) 610.2-1987

full-screen editor A text editor that allows the user to view a full display screen of data at one time and to enter or alter text by using either commands or cursor control. Scrolling functions allow the user to move up and down within the document. *Synonym:* screen editor. *Contrast:* line editor. (C) 610.2-1987

full span−max span (spectrum analyzer) A mode of operation in which the spectrum analyzer scans an entire selected frequency band. (IM) 748-1979w

full speed (data transmission) Referring to transmission of data in teleprinter systems at the full rated speed of the equipment. (PE) 599-1985w

full subtracter A subtracter that accepts three inputs (two operands and a borrow digit), producing a difference and a borrow as outputs according to the table below. *Contrast:* half subtracter. *See also:* full adder.

input #1	0	0	1	1	0	0	1	1
input #2	0	1	0	1	0	1	0	1
input carry	0	0	0	0	1	1	1	1
output differences	0	1	1	0	1	0	0	1
output borrow	0	1	0	0	1	1	0	1

full subtracter

(C) 610.10-1994w

full tree *See:* complete tree.

full-trunk group (telephone switching systems) A trunk group, other than a final trunk group, that does not overflow calls to another trunk group. (COM) 312-1977w

full-voltage starter A starter that connects the motor to the power supply without reducing the voltage applied to the motor. *Note:* Full-voltage starters are also designated as across-the-line starters. *See also:* starter. (IA/ICTL/IAC) [60]

full-wave lightning impulse test (power and distribution transformers) Application of the "standard lightning impulse" wave, a full wave having a front time of 1.2 microseconds and a time to half value of 50 microseconds, described as a 1.2/50 impulse. (PE/TR) C57.12.80-1978r

full-wave rectification (rectifying process) (power supplies) Full-wave rectification inverts the negative half-cycle of the input sinusoid so that the output contains two half-sine pulses for each input cycle. A pair of rectifiers arranged as shown with a center-tapped transformer or a bridge arrangement of four rectifiers and no center tap are both methods of obtaining full-wave rectification. *See also:* rectifier; rectification. (AES/IA/PE) [41], [62], [78]

full-wave rectifier *See:* full-wave rectification.

full-wave rectifier circuit A circuit that changes single-phase alternating current into pulsating unidirectional current, utilizing both halves of each cycle. *See also:* rectification. (IA/NPS) 59-1962w, 325-1971w, [12]

full-wave voltage impulse (surge arresters) A voltage impulse that is not interrupted by sparkover, flashover, or puncture. *See also:* full impulse voltage. (PE) [8]

full width at fiftieth maximum (FWFM) (x-ray energy spectrometers) Same as full width at half maximum (FWHM), except that measurement is made at one fiftieth of the maximum ordinate rather than one half. (NPS/NID) 759-1984r

full width at half maximum (FWHM) (1) (A) (germanium gamma-ray detectors) (x-ray energy spectrometers) (charged-particle detectors) The full width of a distribution measured at half the maximum ordinate. For a normal distribution, it is equal to $2(2 \ln 2)^{1/2}$ times the standard deviation σ. **(B)** The full width of a distribution measured at 50% of its peak height. If the distribution is a spectral line due to radiation, it is assumed that the background level was averaged over the base of that line and the average subtracted from the ordinates of the distribution before the FWHM was determined. For a normal distribution,

$$\text{FWHM} = 2(2 \ln 2)^{0.5} = 2.355\,\sigma$$

where σ = standard deviation and also the root-mean-square value of the distribution. (NPS/NID) 759-1984, 300-1988
(2) (germanium detectors) The full width of a gamma-ray peak distribution measured at half the maximum ordinate above the continuum. (PE/EDPG) 485-1983s
(3) (scintillation counters) The full width of a distribution measured at half the maximum ordinate. For a normal distribution, it is equal to $2(2 \ln 2)^{1/2}$ times the standard deviation (σ). *Note:* The expression full width at half maximum, given either as an absolute value or as a percentage of the value of the argument at the maximum of the distribution curve, is frequently used in nuclear physics as an approximate description of a distribution curve. Its significance can best be made clear by reference to a typical distribution curve, shown in the figure, of the measurement of the energy of the gamma rays from Cs^{137} with a scintillation counter spectrometer. The measurement is made by determining the number of gamma-ray photons detected in a prescribed interval of time, having measured energies falling within a fixed energy interval (channel width) about the values of energy (channel position) taken as argument of the distribution function. The abscissa of the curve shown is energy in megaelectronvolts (MeV) units and the ordinate is counts per given time interval per megaelectronvolt energy interval. The maximum of the distribution curve shown has an energy E_1 megaelectronvolts. The height of the peak is A_1 counts/100 seconds/megaelectronvolts. The full width at half maximum ΔE is measured at a value of the ordinate equal to $A_1/2$. The percentage full width at half maximum is $100\ \Delta E/E_1$. It is an indication of the width of the distribution curve, and where (as in the example cited) the gamma-ray photons are monoenergetic, it is a measure of the resolution of the detecting instrument. When the distribution curve is a Gaussian curve, the percentage full width at half maximum is related to the standard deviation σ by

$$100\frac{\Delta E}{E_1} = 100 \times 2(2 \ln 2)^{1/2} \times \sigma$$

See also: scintillation counter.

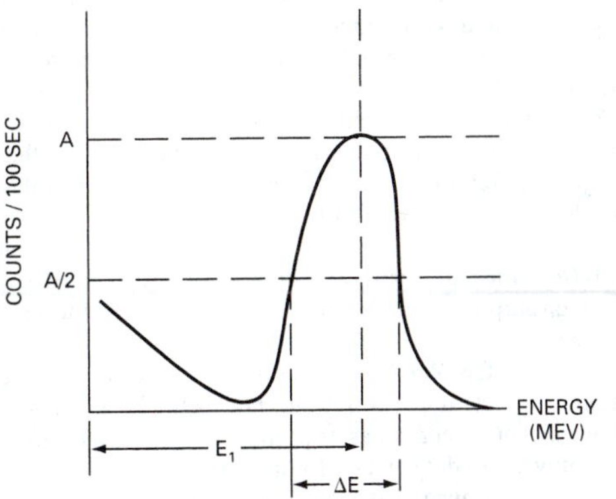

full width at half maximum

(NPS) 398-1972r
(4) The full width of a gamma photon peak distribution measured at an ordinate half way between the maximum ordinate of the peak and the background. (NI) N42.12-1994

(5) The full width of a distribution measured at half the maximum ordinate. For a normal distribution, FWHM = 2 · (2 ln 2)1/2 = 2.355 times the standard deviation, s. (NPS) 325-1996

full width at one-tenth maximum (FW.1M) The full width of a distribution measured at one tenth of the maximum ordinate measured above the background. (NPS) 300-1988r

full width at tenth maximum (FWTM) (x-ray energy spectrometers) Same as full width at half maximum (FWHM), except measurement is made at one tenth of the maximum ordinate rather than one half. (NPS/NID) 759-1984r

full width (duration) half maximum (fiber optics) A measure of the extent of a function. Given by the difference between the two extreme values of the independent variable at which the dependent variable is equal to half of its maximum value. The term "duration" is preferred when the independent variable is time. *Note:* Commonly applied to the duration of pulse waveforms, the spectral extent of emission or absorption lines, and the angular or spatial extent of radiation patterns. (Std100) 812-1984w

fullword *See:* computer word; word.

fully concatenated key *See:* concatenated key.

fully connected network A network in which each node is directly connected with every other node. (LM/COM) 168-1956w

fully decoded EAROM organization An EAROM organization in which rows and columns of memory cells are addressable through on-chip decoding circuitry. (ED) 641-1987w

fully inverted file An file that has been inverted on all secondary keys in the file. *Contrast:* partially inverted file. (C) 610.5-1990w

fully qualified software_spec A `software_spec` that always identifies a software object unambiguously. (C/PA) 1387.2-1995

fully relational Pertaining to a database management system that supports a relational database and a language that provides the functionality of the relational algebra. (C) 610.5-1990w

fume-resistant So constructed that it will not be injured readily by exposure to the specified fume. (SWG/PE/IA/ICTL/IAC) C37.100-1981s, [56], [60]

function (1) (general) When a mathematical quantity u depends on a variable quantity x so that to each value of x (within the interval of definition) there correspond one or more values of u, then u is a function of x written $u = f(x)$. The variable x is known as the independent variable or the argument of the function. When a quantity u depends on two or more variables $x_1, x_2, \ldots x_n$ so that for every set of values of $x_1, x_2, \ldots x_n$ (within given intervals for each of the variables) there correspond one or more values of u, then u is a function of $x_1, x_2, \ldots x_n$ and is written $u = f(x_1, x_2, \ldots x_n)$. The variables $x_1, x_2, \ldots x_n$ are the independent variables or arguments of the function. (Std100) 270-1966w

(2) (vector) When a scalar or vector quantity u depends upon a variable vector V so that if for each value of V (within the region of definition) there correspond one or more values of u, then u is a function of the vector V. (Std100) 270-1966w

(3) (test, measurement, and diagnostic equipment) The action or purpose which a specific item is intended to perform or serve. (MIL) [2]

(4) (A) (software) A defined objective or characteristic action of a system or component. For example, a system may have inventory control as its primary function. **(B) (software)** A software module that performs a specific action, is invoked by the appearance of its name in an expression, may receive input values, and returns a single value. *See also:* subroutine. (C) 610.12-1990

(5) (mathematics of computing) A mathematical entity whose value is uniquely determined by the value of one or more independent variables. (C) 610.2-1987

(6) Normal or characteristic action of a component or the system of which it is a part. (PE/EDPG) 803.1-1992

(7) A task, action, or activity expressed as a verb-noun combination (e.g., Brake Function: stop vehicle) to achieve a defined outcome. (C/SE) 1220-1994s

(8) (software user documentation) A specific purpose of an entity or its characteristic action. (SE/C) 1063-1987r, 1074-1995s

(9) A programming language construct, modeled after the mathematical concept. A function encapsulates some behavior. It is given zero or more arguments as input, performs some processing, and returns some results. Functions are also known as procedures, subprograms, or subroutines. (C/PA) 1327.2-1993w, 1224.2-1993w, 1326.2-1993w, 1328.2-1993w

(10) A logical component of a unit that operates mostly independently. A function is independent of bus interface and node topology: a unit has registers that are externally accessible in the bus address space, a function may or may not. For example, a multi-function node may contain SCSI controller, LAN interface, and terminal interface Functions that are accessed via a shared DMA multiplex Unit. A function may also be a software entity (e.g., a Function driver), especially when messages are used for Processor-to-Processor communication. (C/MM) 1212.1-1993

(11) A primitive operation on system-controlled resources. This standard defines a collection of functions together with suitable input and output parameters. (C/MM) 855-1990

(12) A task, action, or activity that must be accomplished to achieve a desired outcome. (C/SE) 1233-1998

(13) A single-valued mapping. The mapping `M` from `D` to `R` is a *function* if for any `X` in `D` and `Y` in `R`, there is at most one pair [`X`, `Y`] in `M`. *Synonym:* single-valued. *Contrast:* multi-valued. (C/SE) 1320.2-1998

(14) A transformation of inputs to outputs, by means of some mechanisms, and subject to certain controls, that is identified by a function name and modeled by a box. *Synonyms:* activity; task; process; operation. (C/SE) 1320.1-1998

(15) (scheme programming language) *See also:* procedure. (C/MM) 1178-1990r

functional A link interface becomes functional when the start-up procedure has successfully completed and the link interface is ready to transmit data. (C/BA) 1355-1995

functional address (FA) (1) The portion of a full data transport address that specifies the read or write function that is to be performed. (IM/ST) 1451.2-1997

(2) (local area networks) A bit-significant address used in the ISO/IEC 8802-5 MAC format to identify well-known functional groups. (C) 8802-12-1998

functional address instruction An instruction whose format contains no operation field because the operation is implicitly specified by its address fields. (C) 610.10-1994w

functional adjectives (A) (pulse terminology) Linear. Pertaining to a feature whose magnitude varies as a function of time in accordance with the following relation or its equivalent:

$$m = a + bt$$

(B) (pulse terminology) Exponential. Pertaining to a feature whose magnitude varies as a function of time in accordance with either of the following relations or their equivalents:

$$m = ae^{-bt}$$

$$m = a(1 - e^{-bt})$$

(C) (pulse terminology) Gaussian. Pertaining to a waveform or feature whose magnitude varies as a function of time in accordance with the following relation or its equivalent:

$$m = ae^{-b\,(t-c)^2},\ b > 0$$

(D) (pulse terminology) Trigonometric. Pertaining to a waveform or feature whose magnitude varies as a function of time in accordance with a specified trigonometric function or by a specified relationship based on trigonometric functions (for example, cosine squared). (IM/WM&A) 194-1977

functional architecture An arrangement of functions and their subfunctions and interfaces (internal and external) that defines the execution sequencing, conditions for control or data flow, and the performance requirements to satisfy the requirements baseline. (C/SE) 1220-1998

functional area (nuclear power generating station) (or areas) Location(s) designated within the control room to which displays and controls relating to specific function(s) are assigned. (PE/NP) 566-1977w

functional-area data dictionary A data dictionary that standardizes data element syntax and semantics, within and among application areas within the same functional area. *Note:* Functional-area data dictionaries contain among their contents refined or synthesized composites of the contents of application-specific data dictionaries, primarily in the form of logical application data elements. (SCC32) 1489-1999

functional baseline In configuration management, the initial approved technical documentation for a configuration item. *Contrast:* allocated baseline; product baseline; developmental configuration. (C) 610.12-1990

functional character *See:* control character.

functional cohesion A type of cohesion in which the tasks performed by a software module all contribute to the performance of a single function. *Contrast:* procedural cohesion; communicational cohesion; sequential cohesion; temporal cohesion; logical cohesion; coincidental cohesion. (C) 610.12-1990

functional component A device that performs a necessary function for the proper operation and application of a unit of equipment. (SWG/PE) C37.100-1992

functional configuration audit An audit conducted to verify that the development of a configuration item has been completed satisfactorily, that the item has achieved the performance and functional characteristics specified in the functional or allocated configuration identification, and that its operational and support documents are complete and satisfactory. *See also:* physical configuration audit; configuration management. (C) 610.12-1990

functional configuration identification In configuration management, the current approved technical documentation for a configuration item. It prescribes all necessary functional characteristics, the tests required to demonstrate achievement of specified functional characteristics, the necessary interface characteristics with associated configuration items, the configuration item's key functional characteristics and its key lower level configuration items, if any, and design constraints. *Contrast:* product configuration identification; allocated configuration identification. *See also:* functional baseline. (C) 610.12-1990

functional decomposition (software) A type of modular decomposition in which a system is broken down into components that correspond to system functions and subfunctions. *See also:* hierarchical decomposition; stepwise refinement. (C) 610.12-1990

functional dependency A type of dependency between two attributes A and B in a relation, in which B is functionally dependent on A if, and only if, at every instant in time, each value of A is associated with no more than one value of B. *Note:* A is said to "identify" or "functionally determine" B. Written A - B. *See also:* full functional dependency; join dependency. (C) 610.5-1990w

functional design (1) (A) (software) The process of defining the working relationships among the components of a system. **(B) (software)** The result of the process in definition (A). (C) 610.12-1990, 610.10-1994
(2) The result of the process in (A). (C) 610.10-1994w

functional designation (1) (general) Letters, numbers, words, or combinations thereof, used to indicate the function of an item or a circuit, or of the position or state of a control of adjustment. *See also:* symbol for a quantity; abbreviation; reference designation; letter combination. (GSD) 267-1966
(2) (electric and electronics parts and equipment) Words, abbreviations, or meaningful number or letter combinations, usually derived from the function of an item (for example: slew, yaw), used on drawings, instructional material, and equipment to identify an item in terms of its function. *Note:* A functional designation is not a reference designation nor a substitute for it. (GSD) 200-1975w

functional diagram (test, measurement, and diagnostic equipment) A diagram that represents the functional relationships among the parts of a system. (MIL) [2]

functional dynamic tests (1) Operation of the MNOS array at nominal speed to determine functional performance. (ED) 641-1987w
(2) Operation of the array at specified dynamic conditions to determine functional performance. (ED) 1005-1998

functional element (1) A set of one or more modules that perform a particular function. (C/BA) 14536-1995
(2) A component of the AI-ESTATE architectural concept that is expected to perform specific duties. These include reasoning system, human presentation system, unit under test (UUT), knowledge/data base management system, test system, maintenance data/knowledge collection system, and other system. (ATLAS) 1232-1995

functional fault tree (FFT) A structured analysis method used to identify vital functions at the system functional level by comprehensively examining system functional faults that could precipitate hazards. (VT/RT) 1483-2000

functional grouping A grouping of functions into sets such that all the functions within the same group are performed by a homogenous set of equipment. *Note:* Grouping is based upon equipment's function rather than upon the actual physical realization. One function may be spread over multiple physical boxes or one physical device may perform several functions. (C) 610.7-1995

functional level The level of verification activities at which vital system functions are identified from system functional and operational requirements. (VT/RT) 1483-2000

functionality The capabilities of the various computational, user interface, input, output, data management, and other features provided by a product. (C/SE) 1362-1998

functional language A programming language used to express programs as a sequence of functions and function calls. Examples include LISP and C. *See also:* algebraic language; algorithmic language. (C) 610.13-1993w, 610.12-1990

function-limiting control table The first table in each decade specifies the limits designed into an end device with respect to variables used within the decade. (AMR/SCC31) 1377-1997

functionally determined *See:* functional dependency.

functional model An OSI management model that provides a conceptual and terminological framework for specific management functional areas. (C) 610.7-1995

functional module (1) (VSB) A collection of electronic circuitry that resides on one board and works to accomplish a specific task. Functional modules are used as a vehicle for discussing bus protocols, and should not be considered to constrain the design of actual logic. (C/MM) 1096-1988w
(2) (VMEbus) A collection of electronic circuitry that resides on one printed-circuit board (pcb) and works together to accomplish a task. (BA/C) 1014-1987

functional nomenclature (generating stations electric power system) Words or terms which define the purpose, equipment, or system for which the component is required. (PE/EDPG) 505-1977r

functional partitioning The logical separation of system or unit elements along interfaces that define and isolate these elements on the basis of function or purpose. (SCC20) 1226-1998

functional performance test Tests of the steady-state performance of control functions and sequences. (PE/SUB) 1378-1997

functional quality (FQ) A measure of the service level and performance expected in the support of a BSR by the technology solution proposed. These FQs may be used as assessment criteria for performance and conformance testing as well as for influencing the choice of standards to populate the physical design and the choice of products to turn the physical design into a (operational) solution that can be implemented. (C/PA) 1003.23-1998

functional requirement (1) (software) A requirement that specifies a function that a system or system component must be able to perform. (C) 610.12-1990
(2) A statement that identifies what a product or process must accomplish to produce required behavior and/or results. (C/SE) 1220-1998

functional specification (1) (software) A document that specifies the functions that a system or component must perform. Often part of a requirements specification. (C) 610.12-1990
(2) A formal description of the essential requirements of a software product. It specifies the objectives of the software application, the functions that will meet those objectives, the information requirements, the internal data flows, and the external interfaces. (PE/EDPG) 1150-1991w

functional test (1) (evaluation of thermal capability) (thermal classification of electric equipment and electrical insulation) A means of evaluation in which an insulating material, insulation system, or electric equipment is exposed to factors of influence, which simulate or are characteristic of actual service conditions. (EI) 1-1986r
(2) (test pattern language) A test in which the cells of a memory are accessed in a specific order and at a specific rate, while data is being written into them, or read from them. (TT/C) 660-1986w
(3) (nuclear power generating station) A test to determine the ability of a component or system to perform an intended purpose. (PE/NP) 338-1987r
(4) A sequence of tests applied to a unit under test (UUT) to establish whether it is functioning correctly. (SCC20) 771-1998
(5) A test that is intended to verify that a test subject is behaving as specified. (SCC20) 1226-1998

functional testing (A) Testing that ignores the internal mechanism of a system or component and focuses solely on the outputs generated in response to selected inputs and execution conditions. *Synonym:* black-box testing. *Contrast:* structural testing. **(B)** Testing conducted to evaluate the compliance of a system or component with specified functional requirements. *See also:* performance testing. (C) 610.12-1990

functional test pattern *See:* pattern.

functional unit (1) A system element that performs a task required for the successful operation of the system. *See also:* system. (SMC) [63]
(2) An entity of hardware, software, or both, capable of accomplishing a specified purpose. (C/SE) 610.7-1995, 729-1983s, 610.10-1994w
(3) A group of related functions. (C/PA) 1224.1-1993w

functional vector A pattern generated to exercise a device's functional behavior. Generally defined to run the device at system speeds to verify system behavior of a design. *Contrast:* structural vector. (C/TT) 1450-1999

functional verification The process of evaluating whether or not the functional architecture satisfies the validated requirements baseline. (C/SE) 1220-1998

Function Block An instance of a subclass of `IEEE1451_FunctionBlock`. (IM/ST) 1451.1-1999

function check A check of master and remote station equipment by exercising a predefined component or capability.

a) *Analog.* Monitor a reference quantity.
b) *Control.* Control and indication from a control-check relay.
c) *Scan.* Accomplished when control function check has been performed with all remotes.
d) *Poll.* Accomplished when analog function is performed with all remotes.
e) *Logging.* Accomplished when results of the control function check are logged.

(SWG/PE/SUB) C37.100-1992, C37.1-1987s

function Class-A (back-up) current-limiting fuse A fuse capable of interrupting all currents from the rated maximum interrupting current down to the rated minimum interrupting current. *Note:* The rated minimum interrupting current for such fuses is higher than the minimum melting current that causes melting of the fusible element in one hour. (SWG/PE) C37.100-1981s

function Class-G (general purpose) current-limiting fuse (as applied to a high-voltage current-limiting fuse) A fuse capable of interrupting all currents from the rated maximum interrupting current down to the current that causes melting of the fusible element in one hour. (SWG/PE) C37.100-1981s

function code (subroutines for CAMAC) The symbol *f* represents an integer that is the function code for a CAMAC action. (NPS) 758-1979r

function codes (subroutines for CAMAC) The symbol *fa* represents an array of integers, each of which is the function code for a CAMAC action. The length of *fa* is given by the value of the first element of *cb* at the time the subroutine is executed. *See also:* control block. (NPS) 758-1979r

function, coupling *See:* coupling function.

function, describing *See:* describing function.

function driver A software function in the Processor that is a logical peer to a hardware or firmware function in an I/O Unit. For example, a SCSI Function driver is the target of an I/O completion message from a SCSI Controller Function on an I/O module. (C/MM) 1212.1-1993

function, error transfer *See:* error transfer function.

function field *See:* operation field.

function generator (1) (analog computer) A computing element whose output is a specified nonlinear function of its input or inputs. Normal usage excludes multipliers and resolvers. (C) 165-1977w
(2) (electric power system) A device in which a mathematical function such as $y, f(x)$ can be stored so that for any input equal to x, an output equal to $f(x)$ will be obtained. *See also:* speed-governing system. (PE/PSE) 94-1970w
(3) A device whose output analog variable is equal to some function of its input variables. *See also:* digitally-controlled function generator. (C) 610.10-1994w

function generator, bivariant *See:* bivariant function generator.

function generator, card set *See:* card set function generator.

function generator, curve-follower *See:* curve-follower function generator.

function generator, digitally controlled *See:* digitally-controlled function generator.

function generator, diode *See:* diode function generator.

function generator, map-reader *See:* map-reader function generator.

function generator, servo *See:* servo function generator.

function generator, switch-type *See:* switch-type function generator.

function key A control key used to initiate a desired functional operation. *Note:* A function key is distinguished from other control keys in that the functional operation can usually be programmed or defined dynamically. *Synonym:* user-definable key. *See also:* control key; alternate function key; command key. (C) 610.10-1994w

function, loop-transfer *See:* loop-transfer function.

function name An active verb or verb phrase that describes what is to be accomplished by a function. A box takes as its box name the function name of the function represented by the box. (C/SE) 1320.1-1998

function, output-transfer *See:* output-transfer function.

function pin An analog or digital pin on an integrated component that takes data into and/or out of the core circuit when operating in normal function mode. *Note:* All pins carrying signals that affect the behavior of the component, including reference supply pins but excluding power supply pins and compliance-enable pins, are regarded as function pins. *Synonym:* system pin. (C/TT) 1149.4-1999

function point A measure of the delivered software functionality. (C/SE) 1045-1992

function potentiometer (1) A potentiometer employed in analog computers in which the voltage at the moveable contact of the potentiometer follows a prescribed functional relationship to the displacement of the contact. (C) 610.10-1994w
(2) A multiplier potentiometer in which the voltage at the movable contact follows a prescribed functional relationship to the displacement of the contact. *See also:* linearity. (C) 165-1977w

function, probability density *See:* probability density function.

function, probability distribution *See:* probability density function.

function relay (analog computer) In an analog computer, a relay used as a computing element, generally driven by a comparator. (C) 165-1977w

function, return-transfer *See:* return-transfer function.

function switch In an analog computer, a manually operated switch used as a computing element. For example, a switch may be used to modify a circuit or to add or delete an input function or constant. (C/C) 610.10-1994w, 165-1977w

function, system-transfer *See:* system-transfer function.

function, transfer *See:* transfer function.

function, weighting *See:* weighting function.

function, work *See:* work function.

fundamental component The fundamental frequency component in the harmonic analysis of a wave. *See also:* signal wave. (Std100) 154-1953w

fundamental efficiency (thyristor) The ratio of the fundamental load power to the fundamental line power. (IA/IPC) 428-1981w

fundamental frequency (1) (A) (data transmission) (Signal-transmission system). The reciprocal of the period of a wave. **(B) (data transmission)** (Mathematically). The lowest frequency component in the Fourier representation of a periodic quantity. **(C) (data transmission)** (Data transmission) (periodic quantity). The frequency of a sinusoidal quantity having the same period as the periodic quantity. (PE) 599-1985
(2) The frequency of the primary power-producing component of a periodic waveform supplied by the generation system (component of order 1 of the waveform's Fourier series representation). (PEL) 1515-2000

fundamental matrix *See:* transition matrix.

fundamental mode (fiber optics) The lowest order mode of a waveguide. In fibers, the mode designated LP01 or HE11. *See also:* mode. (Std100) 812-1984w

fundamental mode of propagation (laser maser) The mode in a beamguide or beam resonator which has a single maximum for the transverse field intensity over the cross-section of the beam. (LEO) 586-1980w

fundamental power (thyristor) The product of the root-mean-square (rms) value of the fundamental current and the rms value of the fundamental voltage multiplied by the cosine of the phase angle by which the fundamental current lags the fundamental voltage. (IA/IPC) 428-1981w

fundamental-type piezoelectric crystal unit A unit designed to utilize the lowest frequency of resonance for a particular mode of vibration. *See also:* crystal. (EEC/PE) [119]

furnace transformer (power and distribution transformers) A transformer that is designed to be connected to an electric arc furnace. (PE/TR) C57.12.80-1978r

furniture ESD (1) An ESD in which the intruder is an inanimate object such as a cart or chair, with or without a human in physical contact with the object. (EMC) C63.16-1993
(2) An electrostatic discharge in which the intruder is an inanimate object such as a cart or chair, with or without a human in electrical contact with the object. (SPD/PE) C62.47-1992r

fuse (1) An overcurrent protective device with a circuit-opening fusible part that is heated and severed by the passage of the overcurrent through it. *Note:* A fuse comprises all the parts that form a unit capable of performing the prescribed functions. It may or may not be the complete device necessary to connect it into an electric circuit. (SWG/NESC/PE) C37.100-1992, [86]
(2) (protection and coordination of industrial and commercial power systems) A device that protects a circuit by fusing open its current-responsive element when an overcurrent or short-circuit current passes through it. (IA/PSP) 242-1986r
(3) (electric power systems in commercial buildings) An overcurrent protective device with a circuit opening, fusible element part that is heated and severed by the passage of overcurrent through it. (To re-energize the circuit, the fuse should be replaced.) (IA/PSE) 241-1990r
(4) A current-responsive protective device with a circuit-opening fusible part that is heated and severed by passage of current through it, creating an arc within the fuse. The interaction of the arc with certain other parts of the fuse results in current interruption. *Note:* A fuse comprises all the parts that form a unit capable of performing the prescribed functions. It may or may not be the complete device necessary to connect it into an electric circuit. (SWG/PE) C37.40-1993

fuse-arcing time *See:* arcing time.

fuse blade (of a cartridge fuse) A cartridge-fuse terminal having a substantially rectangular cross section. (SWG/PE) C37.100-1992

fuse carrier (of an oil cutout) An assembly of a cap that closes the top opening of an oil-cutout housing, an insulating member, and fuse contacts with means for making contact with the conducting element and for insertion into the fixed contacts of the fuse support. *Note:* The fuse carrier does not include the conducting element (fuse link). (SWG/PE) C37.100-1992, C37.40-1993

fuse clearing time *See:* clearing time.

fuse clips The current carrying parts of a fuse support that engage the fuse carrier, fuseholder, fuse unit, or blade. *Synonyms:* fuse contact; contact clips. (SWG/PE) C37.40-1993, C37.100-1992

fuse condenser A device that, added to a vented fuse, converts it to a nonvented fuse by providing a sealed chamber for condensation of gases developed during circuit interruption. (SWG/PE) C37.40-1993, C37.100-1992

fuse contact *See:* fuse clips; fuse terminal.

fuse cutout A cutout having a fuse link or fuse unit. *Note:* A fuse cutout is a fuse-disconnecting switch. (SWG/PE) C37.100-1992

fused capacitor (1) (series capacitor) A capacitor in combination with a fuse, either external or internal to the case. (T&D/PE) 824-1994
(2) (power systems relaying) A capacitor having fuses mounted on its terminals, inside a terminal enclosure, or inside the capacitor case, for the purpose of interrupting a failed capacitor. (T&D/PE) C37.99-2000, 18-1992

fused capacitor unit (series capacitor) A capacitor unit in combination with a fuse, either external or internal to the case, intended to isolate a failed unit from the associated units. (T&D/PE) [26]

fused electrolyte (bath) (fused salt) (electrolyte) A molten anhydrous electrolyte. (PE/EEC) [119]

fused-electrolyte cell A cell for the production of electric energy when the electrolyte is in a molten state. *See also:* electrochemistry. (EEC/PE) [119]

fuse-disconnecting switch (disconnecting fuse) A disconnecting switch in which a fuse unit or fuseholder and fuse link forms all or a part of the blade. (SWG/PE) C37.100-1992, C37.40-1993

fused junction *See:* alloy junction.

fused-loadbreak way A fused way incorporating an integral switching device operated by opening and closing a fuse assembly. (SWG/PE) C37.73-1998

fused quartz (fiber optics) Glass made by melting natural quartz crystals; not as pure as vitreous silica. *See also:* vitreous silica. (Std100) 812-1984w

fused salt *See:* fused electrolyte.

fused silica *See:* vitreous silica; fused quartz.

fused switch A switch intended to operate with fuses connected in series, directly attached to or in close proximity to the switch. (SWG/PE) C37.20.4-1996

fused-switched way A way connected to the bus through a three-phase group-operated switch or single-phase switch in series with high-voltage fuses. (SWG/PE) C37.73-1998

fused trolley tap A specially designed holder with enclosed fuse for connecting a conductor of a portable cable to the trolley system or other circuit supplying electric power to equipment in mines. *See also:* mine feeder circuit. (EEC/PE/MIN) [119]

fused-type voltage transformer (instrument transformers) One that is provided with means for mounting one or more fuses as integral parts of the transformer in series with the primary winding. (PE/TR) C57.13-1993, C57.12.80-1978r

fused way A way connected to the bus through a high-voltage fuse. (SWG/PE) C37.73-1998

fuse-enclosure package (FEP) (1) An enclosure supplied with one or more fuses as a package for which application data covering the specific fuse(s) and enclosure are supplied. (SWG/PE) C37.40-1993
(2) An enclosure supplied with one or more fuses as a package for which application data covering the specific fuse(s) and enclosure are supplied. (SWG/PE) C37.100-1992

fuse filler *See:* arc-extinguishing medium.

fuseholder (1) (cutout base) A device intended to support a fuse mechanically and connect it electrically in a circuit. *See also:* cabinet. (PE/EEC) [119]
(2) (of a high-voltage fuse) An assembly of a fuse tube or tubes together with parts necessary to enclose the conducting element and provide a means of making contact with the conducting element and the fuse clips. The fuseholder does not include the conducting element (fuse link or refill unit). (SWG/PE) C37.40-1993, C37.100-1992
(3) (of a low-voltage fuse) An assembly of base, fuse clips, and necessary insulation for mounting and connecting into the circuit the current-responsive element, with its holding means if used for making a complete device. *Notes:* 1. For low-voltage fuses, the current-responsive element and holding means are called a fuse. 2. For high-voltage fuses, the general type of assembly defined above is called a fuse support or fuse mounting. The holding means (fuseholder) and the current-responsive or conducting element are called a fuse unit. (SWG/PE) C37.100-1992

fuse hook A hook provided with an insulating handle for opening and closing fuses or switches and for inserting the fuseholder, fuse unit, or disconnecting blade into, and for removing it from, the fuse support. *Synonym:* switch hook. (SWG/PE) C37.40-1993, C37.100-1992

fuselage lights (illuminating engineering) Aircraft aeronautical lights, mounted on the top and bottom of the fuselage, used to supplement the navigation lights. (EEC/IE) [126]

fuseless capacitor bank (power systems relaying) A capacitor bank without any fuses, internal or external, which is constructed of parallel strings of series-connected capacitor units between line and neutral (wye connection) or between line terminals (delta or single- phase). *See also:* unfused capacitor bank. (PE) C37.99-2000

fuse link (1) (protection and coordination of industrial and commercial power systems) In British terminology only, a complete enclosed cartridge fuse; in such cases the addition of the carrier, or holder, completes the fuse. In the USA, a renewable, fusible element for fuse cutouts. (IA/PSP) 242-1986r
(2) A replaceable part or assembly, comprised entirely or principally of the conducting element, required to be replaced after each circuit interruption to restore the fuse to operating conditions. (SWG/PE) C37.100-1992, C37.40-1993

fuse melting time *See:* melting time.

fuse mounting *See:* fuse support.

fuse muffler (1) An attachment for the vent of a fuse, or a vented fuse, that confines the arc and substantially reduces the venting from the fuse. (SWG/PE) C37.40-1993
(2) *See also:* muffler. (SWG/PE) C37.100-1981s

fuse support (1) (fuse mounting) (high-voltage switchgear) An assembly of base or mounting support or oil cutout housing, insulator(s) or insulator unit(s), and fuse clips for mounting a fuse carrier, fuse holder, fuse unit, or blade and connecting it into the circuit. *Synonym:* fuse mounting. (SWG/PE) C37.40-1993
(2) (of a high-voltage fuse) An assembly of base, mounting support or oil-cutout housing, fuse clips, and necessary insulation for mounting and connecting into the circuit the current-responsive element with its holding means if such means are used for making a complete device. *Notes:* 1. For high-voltage fuses, the holding means is called a fuse carrier or fuseholder, and in combination with the current-responsive or conducting element is called a fuse unit. 2. For low-voltage fuses, the general type of assembly defined above is called a fuseholder. *Synonym:* fuse mounting. (SWG/PE) C37.100-1992

fuse terminal The means for connecting the current-responsive element or its holding means, if such means is used for making a complete device, to the fuse clips. *Synonym:* fuse contact. (SWG/PE) C37.100-1992

fuse time-current characteristic The correlated values of time and current that designate the performance of all or a stated portion of the functions of the fuse. *Note:* The time-current characteristics of a fuse are usually shown as a curve. (SWG/PE) C37.100-1992, C37.40-1993

fuse time-current tests Tests that consist of the application of current to determine the relation between the root-mean-square (rms) alternating current or direct current and the time for the fuse to perform the whole or some specified part of its interrupting function. (SWG/PE) C37.40-1981s, C37.100-1992

fuse tongs Tongs provided with an insulating handle and jaws. Fuse tongs are used to insert the fuseholder or fuse unit into the fuse support or to remove it from the support. (SWG/PE) C37.40-1981s, C37.100-1992

fuse tube (1) (high-voltage switchgear) A tube of insulating material that encloses the conducting element. (SWG/PE) C37.40-1993
(2) A tube of insulating material that surrounds the current-responsive element, the conducting element, or the fuse link. (SWG/PE) C37.100-1992

fuse unit An assembly comprising a conducting element mounted in a fuseholder with parts and materials in the fuseholder essential to the operation of the fuse. (SWG/PE) C37.40-1993, C37.100-1992

fusible element (of a fuse) That part, having predetermined current-responsive melting characteristics, which may be all or part of the current-responsive element. (SWG/PE) C37.100-1992, C37.40-1993

fusible enclosed (safety) switch A switch complete with fuse holders and either with or without meter-testing equipment or accommodation for meters, having all current-carrying parts completely enclosed in metal, and operable without opening the enclosure. *See also:* switch. (IA/ICTL/IAC) [60]

fusible link A programmable integrated circuit in which circuits form bit patterns by being "blasted" open (that is, by use of a heavy destructive current) or by being left closed. *Note:* This "blasting" is also called "burning" a PROM.
(C) 610.10-1994w

fusion (power operations) The formation of a heavier nucleus from two lighter ones with the attendant release of energy.
(T&D/PE/PSE) 858-1987s, 346-1973w

fusion frequency (television) Frequency of succession of retinal images above which their differences of luminosity or color are no longer perceptible. *Note:* The fusion frequency is a function of the decay characteristic of the display.
(BT/AV) 201-1979w

fusion splice (fiber optics) A splice accomplished by the application of localized heat sufficient to fuse or melt the ends of two lengths of optical fiber, forming a continuous, single fiber. (Std100) 812-1984w

Futurebus+ **(1)** Refers to IEEE Std 896.1-1991 and IEEE Std 896.2-1991 which refine the earlier IEEE Std 896.1-1987. Those standards are intended for use with (or as an upgrade path from) MULTIBUSII (IEEE Std 1296-1987) systems, VME (IEEE Std 1014-1987) systems, and U.S.Navy next-generation hardware systems. They support cache-coherent multiprocessing with physical buses on the backplane. SCI may be used to interconnect Futurebus+ systems, since they share the same coherence line size and CSR Architecture.
(C/MM) 1596-1992
(2) A name that refers to IEEE 896.1-1991 and companion 896 series standards. Futurebus+ defines a physically bussed backplane bus standard, which supports 32-bit and 64-bit physical addresses. IEEE Std 896.3-1993 discusses bridge and other architectural considerations.
(C/MM) 1212.1-1993, 1212-1991s
(3) Refers to IEEE Std 896.1-1991 and IEEE Std 896.2-1991. Futurebus+ supports cache-coherent multiprocessing with physical buses on the backplane. SCI may be used to interconnect Futurebus+ systems, since they share the same coherence line size and CSR Architecture.
(C/MM) 1596.5-1993

Futurebus+ 1 (FB+ 1) Primary Futurebus+.
(C/BA) 14536-1995

Futurebus+ 2 (FB+ 2) Primary Futurebus+ 2.
(C/BA) 14536-1995

future point (for supervisory control or indication or telemeter selection) Provision for the future installation of equipment required for a point. *Note:* A future point may be provided with space only; drilling, or other mounting provisions only; or drilling, or other mounting provisions, and wiring only.
(SWG/PE) C37.100-1992

future regression Regression testing provides a quick way to test broad areas of a system's functionality. Future regression testing expands the normal regression process to include future data and advanced system dates. (C/PA) 2000.2-1999

FW *See:* forward wave.

FWFM *See:* full width at fiftieth maximum.

FWHM *See:* spectral width, full-width half maximum; full width at half maximum.

FW.1M *See:* full width at half maximum.

FWTM *See:* full width at tenth maximum.

FW.02M (germanium gamma-ray detectors) Same as FWHM (full width at half maximum) except that the width measurement is made at one fiftieth the maximum ordinate rather than at one half. (NPS) 325-1986s

FX *See:* foreign exchange.

G

g **(1)** Acceleration due to gravity; approximately 9.80 m/s^2 (32.2 ft/s^2). (C/BA) 1101.7-1995
(2) Acceleration due to gravity, that is 9.81 m/s^2 (32.2 ft/s^2). (PE/SUB) 693-1997

G Acceleration normalized with respect to acceleration of gravity at the surface of the Earth (9.81 ms^{-2}) (sea level). (C/BA) 1156.4-1997

GA *See:* geographical address.

GAC *See:* geographical address control.

gaff Tool used to handle and roll poles. *Synonym:* gaff hook. (T&D/PE) 751-1990

gaff hook *See:* gaff.

gain **(1)** **(waveguide)** The power increase in a transmission path in the mode or form under consideration. It is usually expressed as a positive ratio, in decibels. (MTT) 146-1980w
(2) **(A)** **(wood structures used in power transmission and distribution)** A flat surface cut into the side of a pole to facilitate connections. **(B)** **(wood structures used in power transmission and distribution)** A connection device to accomplish the same purpose without cutting the pole. (PE/T&D) 751-1990
(3) (in a given direction) The ratio of the radiation intensity, in a given direction, to the radiation intensity that would be obtained if the power accepted by the antenna were radiated isotropically. *Notes:* 1. Gain does not include losses arising from impedance and polarization mismatches. 2. The radiation intensity corresponding to the isotropically radiated power is equal to the power accepted by the antenna divided by 4π. 3. If an antenna is without dissipative loss, then in any given direction, its gain is equal to its directivity. 4. If the direction is not specified, the direction of maximum radiation intensity is implied. 5. The term **absolute gain** is used in those instances where added emphasis is required to distinguish gain from relative gain; for example, absolute gain measurements. *Synonym:* absolute gain. (AP/ANT) 145-1993
(4) In a circuit or device, the ratio between the input and output signals. (C) 610.10-1994w
(5) *See also:* realized gain; partial gain. (AP/ANT) 145-1993

gain and offset (A) (independently based). Gain and offset are the values by which the input values are multiplied and then to which the input values are added, respectively, to minimize the mean squared deviation from the output values. (B) (terminal-based). Gain and offset are the values by which the input values are multiplied and then to which the input values are added, respectively, to cause the deviations from the output values to be zero at the terminal points, that is, at the first and last codes. (IM/WM&A) 1057-1994w

gain-crossover frequency (hydraulic turbines) The frequency at which the gain becomes unity and its decibel value zero. (PE/EDPG) 125-1977s

gain hit or change A sudden increase or decrease in amplitude, usually not exceeding 12 dB of the received signal. Gain hits last at least 4 ms but may continue for hours. (Example: Modems that use amplitude modulation carry the information by the level of the signal, and a gain hit may look like data to these modems.) *Contrast:* impulse noise. (PE/IC) 1143-1994r

gain integrator **(1)** **(analog computer)** For each input, the ratio of the input to the corresponding time rate of change of the output. For fixed input resistors, the "time constant" is determined by the integrating feedback capacitor. (C) 165-1977w
(2) In an analog computer, a device which provides the ratio of the input to the corresponding time rate of change of the output, for each input. (C) 610.10-1994w

gain/loss The logarithmic ratio (expressed in decibels) of output power to input power. (COM/TA) 1007-1991r

gain margin **(1)** **(hydraulic turbines)** The reciprocal of the gain at the frequency at which the open-loop phase angle reaches 180 degrees. (PE/EDPG) 125-1977s
(2) The reciprocal of the gain of a control loop at the frequency for which there is 180 degrees of phase shift around the control loop. (PEL) 1515-2000

gain medium (laser gyro) A medium that, when energized, provides amplification of coherent light waves to maintain lasing within a closed optical path. (AES/GYAC) 528-1994

gain, partial *See:* partial gain.

gain, photomultiplier tube *See:* photomultiplier tube gain.

gain stability The ability of a device to maintain its gain within specified limits, under specified environmental and power variations. (COM/TA) 1007-1991r

gain time control *See:* sensitivity time control.

gain, transmission *See:* transmission gain.

galactic radio waves Radio waves originating in our galaxy. (AP/PROP) 211-1997

gallon One U.S. gallon equals 3.785 liters. (SUB/PE) 980-1994

galloping The sudden surging and stopping action of cables during high-tension pulls when excessive stretching occurs in the pull rope. (PE/IC) 1185-1994

gal/min Gallons per minute. (T&D/PE) 957-1987s

gal/s Gallons per second. (PE/T&D) 957-1987s

galvanic action Noise currents due to the junction or thermal potentials resulting from a combination of different metals. This current can be a part of the disturbing current passing through the conductors or shield of the signal cable. (PE/IC) 1143-1994r

galvanic isolation A method of electrical isolation where neither the signal nor the common of the output of the isolator is dc coupled to the signal or common of the input of the isolator, except for low-level leakage associated with nonideal components. (VT) 1482.1-1999, 1476-2000

game A physical or mental competition in which the participants, called players, seek to achieve some objective within a given set of rules. *See also:* game theory. (C) 610.3-1989w

game theory **(A)** The study of situations involving competing interests, modeled in terms of the strategies, probabilities, actions, gains, and losses of opposing players in a game. *See also:* war game; management game. **(B)** The study of games to determine the probability of winning given various strategies. *See also:* war game; management game. (C) 610.3-1989

gaming simulation *See:* simulation game.

gamma (television) The exponent of that power law that is used to approximate the curve of output magnitude versus input magnitude over the region of interest. (BT/AV) 201-1979w

gamma correction (television) The insertion of a nonlinear output-input characteristic for the purpose of changing the system transfer characteristic. (BT/AV) 201-1979w

gamma key The connector keying pin located at the center of the module connector, next to the guide pin. (C/BA) 1101.3-1993

gamma-ray branching ratio For a given excited state, the ratio of the emission rate of a particular gamma ray to the total transition rate from the level (not to be confused with emission probability per decay). (NI) N42.14-1991

gamma-ray emission rate The rate at which a gamma ray of a given energy from the decay of a particular radionuclide is emitted from a given source. The gamma-ray emission rate is the activity times the gamma-ray emission probability. (NI) N42.14-1991

gamma-ray resolution (1) (germanium detectors) The measured full width at half maximum (FWHM), after background subtraction, of a gamma-ray peak distribution, expressed in units of energy. (PE/EDPG) 485-1983s
(2) (sodium iodide detector) The measured full width at half maximum (FWHM), after background subtraction, of a gamma-ray peak distribution, expressed as a percentage of the energy corresponding to the centroid of the distribution. (NI) N42.12-1980s

GAMMA 3 A programming language used for generating matrices and reports in conjunction with a mathematical programming system. (C) 610.13-1993w

gang punch To punch identical hole patterns into each punch card of a card deck. (C) 610.10-1994w

gap (1) In Physical Design Exchange Format (PDEF), the spacing between rows and/or columns in a datapath. (C/DA) 1481-1999
(2) A period of idle bus. (C/MM) 1394a-2000

gap character A character that is included in a computer word for technical reasons but that does not represent data. (C) 610.5-1990w

gap discharge *See:* microspark.

gapless Not possessing gaps, series or parallel, as in "gapless arrester." (SPD/PE) C62.11-1999

gap loss (fiber optics) That optical power loss caused by a space between axially aligned fibers. *Note:* For waveguide-to-waveguide coupling, it is commonly called "longitudinal offset loss." *See also:* coupling loss. (Std100) 812-1984w

gap width The dimension of the air gap in a read/write head, measured along the radius of the disk. (C) 610.10-1994w

garage A building or portion of a building in which one or more self-propelled vehicles carrying volatile flammable liquid for fuel power are kept for use, sale, storage, rental, repair, exhibition, or demonstrating purposes, and all that portion of a building which is on or below the floor or floors in which such in which such vehicles are kept and which is not separated therefrom by suitable cutoffs. (NESC/NEC) [86]

garbage Unwanted or meaningless data. (C) 610.5-1990w

garbage collection (1) (A) (data management) A space optimization technique in which superfluous data are eliminated. **(B) (data management)** A database reorganization technique in which the contents of a database are made more compact by physically deleting garbage such as records that have been deleted logically but remain physically in the database. (C) 610.5-1990
(2) (software) In computer resource management, a synonym for memory compaction *Synonym:* memory compaction. (C) 610.12-1990

gas-accumulator relay A relay so constructed that it accumulates all or a fixed proportion of gas released by the protected equipment and operates by measuring the volume of gas so accumulated. (SWG/PE) C37.100-1992

gas admixture ratio (nonlinear, active, and nonreciprocal waveguide components) The ratio of partial pressures of the separate constituent gases of the total gas composition used in gas tubes. (MTT) 457-1982w

gas amplification *See:* gas multiplication factor.

gas-barrier insulator (1) An insulating support specifically designed to prevent passage of gas from one gas compartment to another. (SUB/PE) C37.122-1993, C37.122.1-1993
(2) A spacer insulator specifically designed to prevent passage of gas from one gas compartment to another. (SWG/PE) C37.100-1992, C37.122.1-1993

gas cleanup (nonlinear, active, and nonreciprocal waveguide components) The phenomenon that causes gas atoms or molecules to be absorbed into a solid medium during a gas discharge. (MTT) 457-1982w

gas density, minimum *See:* minimum gas density.

gas density, nominal *See:* nominal gas density.

gas density, normal *See:* normal gas density.

gas-discharge display device *See:* plasma display device.

gaseous discharge (illuminating engineering) The emission of light from gas atoms excited by an electric current. (EEC/IE) [126]

gas fill (nonlinear, active, and nonreciprocal waveguide components) The process by which a plasma limiter or gas tube is evacuated and an admixture of gases is inserted. (MTT) 457-1982w

gas filled joint (power cable joints) Joints in which the fluid filling the joint housing is in the form of a gas. (PE/IC) 404-1986s

gas-filled protector A discharge gap between two or more electrodes hermetically sealed in a ceramic or glass envelope. (PE/PSC) 487-1992

gas-filled transformer (power and distribution transformers) A sealed transformer, except that the windings are immersed in a dry gas which is other than air or nitrogen. (PE/TR) C57.12.80-1978r

gas-flow counter tube A radiation-counter tube in which an appropriate atmosphere is maintained by a flow of gas through the tube. (ED) [45]

gas flow error (laser gyro) The error resulting from the flow of gas in dc discharge tubes. (AES/GYAC) 528-1994

gas focusing (electron-beam tubes) A method of concentrating an electron beam by gas ionization within the beam. *See also:* gas tube. (ED) [45]

gas grooves (electrometallurgy) The hills and valleys in metallic deposits caused by streams of hydrogen or other gas rising continuously along the surface of the deposit while it is forming. *See also:* electrowinning. (EEC/PE) [119]

gas-insulated substation (1) A compact, multicomponent assembly, enclosed in a grounded metallic housing in which the primary insulating medium is a compressed gas and that normally consists of buses, switch-gear, and associated equipment (subassemblies). (SWG/PE/SUB) C37.100-1992, C37.122-1993, C37.122.1-1993
(2) A compact, multicomponent assembly, enclosed in a grounded metallic housing in which the primary insulating medium is a gas, and that normally consists of buses, switchgear, and associated equipment (subassemblies). (PE/SUB) 80-2000

gas-insulated-substation (GIS) surge arrester A surge arrester specifically designed for use in a gas-insulated substation. (SWG/PE/SUB) C37.122-1983s, C37.100-1992

gas-insulated surge arrester A metal-enclosed surge arrester specifically designed for use in a gas-insulated substation. (SUB/PE) C37.122-1993, C37.122.1-1993

gasket-sealed relay A relay in an enclosure sealed with a gasket. *See also:* relay. (EEC/REE) [87]

gasket, waveguide *See:* waveguide gasket.

gas leakage (1) Loss of insulating gas from the pressurized compartment. (SUB/PE) C37.122-1993, C37.122.1-1993
(2) Loss of insulating gas from the pressurized system. (SWG/PE) C37.100-1992

gas multiplication factor The ratio of the charge collected from the sensitive volume to the charge produced in that volume by the initial ionizing event. (NI/NPS) 309-1999

gas-oil sealed system (1) (power and distribution transformers) A system in which the interior of the tank is sealed from the atmosphere, over the temperature range specified, by means of an auxiliary tank or tanks to form a gas-oil seal operating on the manometer principle. (PE/TR) C57.12.80-1978r
(2) An oil preservation system in which the interior of the tank is sealed from the atmosphere, over the temperature range specified, by means of an ancillary tank or tanks to form a gas-oil seal operating on the manometer principle. (PE/TR) C57.15-1999

gasoline dispensing and service station A location where gasoline or other volatile flammable liquids or liquified flammable gases are transferred to the fuel tanks (including auxiliary fuel tanks) of self-propelled vehicles. (NESC/NEC) [86]

GASP *See:* gas plasma display device; plasma display device.

GASP IV A simulation language designed to be used within a FORTRAN program to facilitate the representation of discrete, continuous, and combined models. (C) 610.13-1993w

gas panel *See:* plasma panel.

gas plasma display device (GASP) *See:* plasma display device.

gas-pressure relay A relay so constructed that it operates by the gas pressure in the protected equipment. (SWG/PE) C37.100-1992

gasproof So constructed or protected that the specified gas will not interfere with successful operation. (SWG/PE/IA/APP/IAC) C37.100-1981s, [56], [75], [60]

gasproof or vaporproof (rotating machinery) So constructed that the entry of a specified gas or vapor under prescribed conditions cannot interfere with satisfactory operating of the machine. *See also:* asynchronous machine; direct-current commutating machine. (PE) [9]

gas ratio The ratio of the ion current in a tube to the electron current that produces it. *See also:* electrode current. (ED) [45]

gas seal (rotating machinery) A sealing arrangement intended to minimize the leakage of gas to or from a machine along a shaft. *Note:* It may be incorporated into a ball or roller bearing assembly. (PE) [9]

gassing The evolution of gases from one or more of the electrodes during electrolysis. *See also:* electrolytic cell. (EEC/PE) [119]

gas system (rotating machinery) The combination of parts used to ventilate a machine with any gas other than air, including facilities for charging and purging the gas in the machine. (PE) [9]

gastight (1) (lightning protection) So constructed that gas or air can neither enter nor leave the structure except through vents or piping provided for the purpose. (NFPA) [114]
(2) So constructed that the specified gas will not enter the enclosing case under specified pressure conditions. (SWG/PE) C37.100-1981s

gas tube An electron tube in which the pressure of the contained gas or vapor is such as to affect substantially the electrical characteristics of the tube. (ED) [45]

gas-tube relaxation oscillator (arc-tube relaxation oscillator) A relaxation oscillator in which the abrupt discharge is provided by the breakdown of a gas tube. *See also:* oscillatory circuit. (EEC/PE) [119]

gas-tube surge arrester (gas tube surge-protective device) A gap or series of gaps in an enclosed discharge medium, other than air at atmospheric pressure, designed to protect apparatus or personnel or both from high transient voltages. *Note:* The arrester in its mounting, with optional fuses and fail-safe devices, shall be called a "protector" to differentiate between the complete protection assembly and the arrester. (PE/SPD) [8], C62.31-1987r

gas-turbine-electric drive A self-contained system of power generation and application in which the power generated by a gas turbine is transmitted electrically by means of a generator and a motor (or multiples of these) for propulsion purposes. *Note:* The prefix gas-turbine-electric is applied to ships, locomotives, cars, buses, etc., that are equipped with this drive. *See also:* electric locomotive. (EEC/PE) [119]

gate (1) A device or element that, depending upon one or more specified inputs, has the ability to permit or inhibit the passage of a signal. *See also:* control. (IA/NPS/ICTL/APP/NID/IAC) [69], 759-1984r, [60]
(2) (A) A device having one output channel and one or more input channels, such that the output channel state is completely determined by the contemporaneous input channel states, except during switching transients. **(B)** A combinational logic element having at least one input channel. **(C)** An AND gate. **(D)** An OR gate. (C) 162-1963, [85], [20], 270-1966
(3) (cryotron) An output element of a cryotron. *See also:* superconductivity. (ED) [46]
(4) (thyristor) An electrode connected to one of the semiconductor regions for introducing control current. *See also:* anode. (IA/ED) 223-1966w, [62], [46], [12]
(5) (A) (navigation systems) An interval of time during which some portion of the circuit or display is allowed to be operative, or *See also:* navigation. **(B) (navigation systems)** the circuit which provides gating. *See also:* navigation. (AES/RS) [42]
(6) (metal-nitride-oxide field-effect transistor) This structural element of an insulated-gate field-effect transistor (IGFET) controls the current between source and drain by a voltage applied to its terminal. (ED) 581-1978w
(7) (nonlinear, active, and nonreciprocal waveguide components) (microwave) In elementary form, a two-port switch having a single-pole, single-throw function. *See also:* bang snuffer. (MTT) 457-1982w
(8) A combinational circuit that performs an elementary logic operation. *Note:* Usually involves at least one input and one output. *Synonym:* logic gate; logic element. (C) 610.10-1994w
(9) (A) An interval of time during which some portion of a circuit or display is allowed to be operative. **(B)** The circuit that provides gating. (AES) 686-1997
(10) In Physical Design Exchange Format (PDEF), the physical abstraction of a library primitive. (C/DA) 1481-1999

gate-controlled delay time (thyristor) The time interval, between a specified point at the beginning of the gate pulse and the instant when the principal voltage (current) has dropped (risen) to a specified value near its initial value during switching of a thyristor from the OFF state to the ON state by a gate pulse. *See also:* principal voltage-current characteristic. (IA/ED) 223-1966w, [62], [12], [46]

gate-controlled rise time (thyristor) The time interval between the instants at which the principal voltage (current) has dropped (risen) from a specified value near its initial value to a specified low (high) value, during switching of a thyristor from the OFF state to the ON state by a gate pulse. *Note:* This time interval will be equal to the rise time of the ON state current only for pure resistive loads. *See also:* principal voltage-current characteristic. (IA/ED) 223-1966w, [62], [12], [46]

gate-controlled turn-off time (turn-off thyristor) The time interval, between a specified point at the beginning of the gate pulse and the instant when the principal current has decreased to a specified value, during switching from the ON state to the OFF state by a gate pulse. *See also:* principal voltage-current characteristic. (IA/ED) 223-1966w, [46], [12], [62]

gate-controlled turn-on time (thyristor) The time interval, between a specified point at the beginning of the gate pulse and the instant when the principal voltage (current) has dropped (risen) to a specified low (high) value during switching of a thyristor from the OFF state to the ON state by a gate pulse. Turn-on time is the sum of delay time and rise time. *See also:* rise time; delay time; principal voltage-current characteristic. (IA/ED/CEM) 223-1966w, [58], [62], [46]

gate current (semiconductor) The current that results from the gate voltage. *Notes:* 1. Positive gate current refers to conventional current entering the gate terminal. 2. Negative gate current refers to conventional current leaving the gate terminal. (IA) [12]

gated integrator A circuit for obtaining an output pulse with an amplitude proportional to the integral of the input signal over a definite time interval. (NPS) 325-1996

gated sweep (oscilloscopes) A sweep controlled by a gate waveform. Also, a sweep that will operate recurrently (free-running, synchronized, or triggered) during the application of a gating signal. *See also:* oscillograph. (IM/HFIM) [40]

gate electric contact *See:* car-door contact.

gate limit (speed governing system, hydraulic turbines) A device which acts on the governor system to prevent the turbine-control servomotor from opening beyond the position for which the device is set. (PE/EDPG) [5]

gate nontrigger current (thyristor) The maximum gate current that will not cause the thyristor to switch from the OFF state to the ON state. *See also:* principal current; gate trigger current. (IA/ED/CEM) 223-1966w, [62], [58], [46]

gate nontrigger voltage (thyristor) The maximum gate voltage that will not cause the thyristor to switch from the OFF state to the ON state. *See also:* gate trigger voltage; principal voltage-current characteristic.
(IA/ED/CEM) 223-1966w, [46], [58], [62]

gate power closer *See:* car-door closer.

gate protective action (thyristor converter) Protective action that takes advantage of the switching property in the converter protection network. (IA/IPC) 444-1973w

gate suppression (thyristor power converter) Removal of gating pulses. (IA/IPC) 444-1973w

gate terminal (thyristor) A terminal that is connected to a gate. *See also:* anode. (IA/ED/CEM) 223-1966w, [46], [58]

gate trigger current (thyristor) The minimum gate current required to switch a thyristor from the OFF state to the ON state. *See also:* principal current.
(IA/ED/CEM) 223-1966w, [62], [58], [46]

gate trigger voltage (thyristor) The gate voltage required to produce the gate-trigger current. *See also:* principal voltage-current characteristic.
(IA/ED/CEM) 223-1966w, [58], [46], [62]

gate turn-off current (gate turn-off thyristor) The minimum gate current required to switch a thyristor from the ON state to the OFF state. *See also:* principal current.
(IA/ED/CEM) 223-1966w, [46], [58], [62]

gate turn-off voltage (gate turn-off thyristor) The gate voltage required to produce the gate turn-off current. *See also:* principal voltage-current characteristic.
(IA/ED/CEM) 223-1966w, [58], [46], [62]

gate voltage (thyristor) The voltage between a gate terminal and a specified main terminal. *See also:* principal voltage-current characteristic.
(IA/ED/CEM) 223-1966w, [62], [46], [58]

gateway A functional unit that interconnects a local area network (LAN) with another network having different higher layer protocols. (LM/C) 8802-6-1994
(2) (A) A dedicated computer that attaches to two or more networks and that routes packets from one to the other. **(B)** In networking, a device that connects two systems that use different protocols. *Contrast:* bridge. *See also:* router; mail gateway. (C) 610.7-1995

gather write A write operation in which information from non-adjacent storage areas is placed into a single physical record. *Contrast:* scatter read. (C) 610.10-1994w

gating (1) The process of selecting those portions of a wave that exist during one or more selected time intervals or that have magnitudes between selected limits. *See also:* wavefront; modulation. (AP/ANT) 145-1983s
(2) The application of enabling or inhibiting pulses during part of a cycle of equipment operation. (AES) 686-1997

gating signal (keying signal) A signal that activates or deactivates a circuit during selected time intervals.
(PE/EEC) [119]

gating techniques (thyristor) Those techniques employed to provide controller (thyristor) gating signals.
(IA/IPC) 428-1981w

gauss (centimeter-gram-second electromagnetic-unit system) The gauss is 10^{-4} webers per square meter or one maxwell per square centimeter. (Std100) 270-1966w

Gaussian beam (1) (fiber optics) A beam of light whose electric field amplitude distribution is gaussian. When such a beam is circular in cross section, the amplitude is $E(r) = E(0) \exp[-(r/w)^2]$ where r is the distance from beam center and w is the radius at which the amplitude is $1/e$ of its value on the axis; w is called the beamwidth. *See also:* beam diameter. (Std100) 812-1984w
(2) (laser maser) A beam of radiation having an approximately spherical wave front at any point along the beam and having transverse field intensity over any wave front that is a Gaussian function of the distance from the axis of the beam.
(LEO) 586-1980w

Gaussian density function (radar) Sometimes referred to as normal probability distribution, the Gaussian probability-density function is given by

$$f(X) = \frac{1}{\sigma\sqrt{2\pi}} \exp - \left(\frac{x^2}{2\sigma^2}\right)$$

Often used to describe statistical nature of random noise, where σ = standard deviation. (AES/RS) 686-1982s

Gaussian distribution A probability distribution characterized by the probability density function

$$f(x) = \frac{1}{\sqrt{2\pi}\,\sigma} \exp\left[-\frac{(x-m)^2}{2\sigma^2}\right]$$

where
x = the random variable
m = the mean
σ = the standard deviation

The Gaussian distribution is often used for analytical modeling of radar noise and various measurement errors. *Synonym:* normal distribution. (AES) 686-1997

Gaussian filter A polynomial filter whose magnitude-frequency response approximates the ideal Gaussian response, the degree of approximation depending on the complexity of the filter. The ideal Gaussian response is given by

$$\left|H(j\omega)\right| = \exp[-0.3466(\omega/\omega_c)^2]$$

where ω_c 3 dB frequency. Gaussian filters, because of their good transient characteristics (small overshoot and ringing), find applications in pulse systems. (CAS) [13]

Gaussian frequency shift keying (GFSK) A modulation scheme in which the data is first filtered by a Gaussian filter in the baseband and then modulated with a simple frequency modulation. (C/LM) 8802-11-1999

Gaussian noise Noise characterized by a wide frequency range with regard to the desired signal of communication channel, statistical randomness, and other stochastic properties.
(C) 610.7-1995

Gaussian pulse (1) (fiber optics) A pulse that has the waveform of a gaussian distribution. In the time domain, the waveform is

$$f(t) = A \exp[-(t/a)^2]$$

where A is a constant, and a is the pulse half duration at the $1/e$ points. *See also:* full width (duration) half maximum.
(Std100) 812-1984w
(2) A pulse shape tending to follow the Gaussian curve corresponding to $A(t) = e-a(b-t)^2$. *See also:* pulse.
(IM/HFIM) [40]

Gaussian random noise *See:* random noise.

Gaussian response (1) (amplifiers) A particular frequency-response characteristic following the curve $y(f) = e-af^2$. *Note:* Typically, the frequency response approached by an amplifier having good transient response characteristics. *See also:* amplifier. (IM/HFIM) [40]
(2) (oscilloscopes) (amplifiers) A particular frequency response characteristic following the curve

$$y(f) = e^{-af^2}$$

Typically, the frequency response approached by an amplifier having good transient response characteristics.
(IM) 311-1970w

Gaussian system (units) A system in which centimeter-gram-second electrostatic units are used for electric quantities and centimeter-gram-second electromagnetic units are used for magnetic quantities. *Note:* When this system is used, the factor *c* (the speed of light) must be inserted at appropriate places in the electromagnetic equations. (Std100) 270-1966w

Gauss' law (electrostatics) States that the integral over any closed surface of the normal component of the electric flux density is equal in a rationalized system to the electric charge Q_0 within the surface. Thus,

$$\int_{\text{closed surface}} (\mathrm{D} \cdot \mathrm{n})dA = \int_{\text{volume enclosed}} \rho_0 dV = Q_0$$

Here, D is the electric flux density, n is a unit normal to the surface, dA the element of area, ρ_0 is the space charge density in the volume V enclosed by the surface. (Std100) 270-1966w

gaussmeter A magnetometer provided with a scale graduated in gauss or kilogauss. *See also:* magnetometer. (EEC/PE) [119]

GB *See:* gigabyte.

GCA *See:* ground-controlled approach.

GCI *See:* ground-controlled intercept.

GCI radar *See:* ground-controlled intercept radar.

GCR *See:* constant-linear-velocity recording; group code recording.

GDG *See:* generation data group.

G-display A modified F-display in which wings appear to grow on the blip, the width of the wings being inversely proportional to target range.

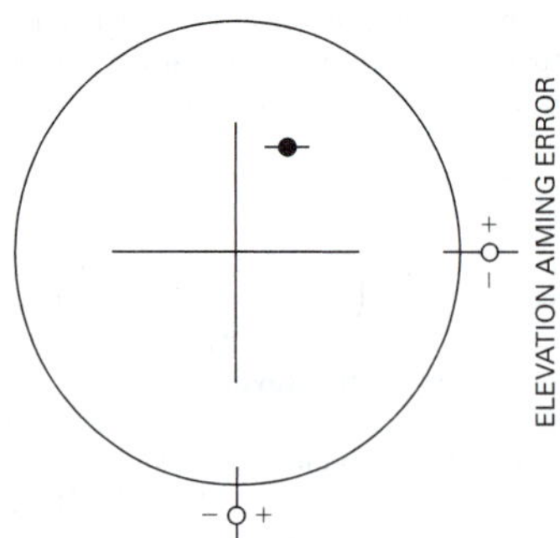

G-display

(AES) 686-1997

GDOP *See:* geometric dilution of precision.

G drift (electronic navigation) A drift component in gyros (sometimes in accelerometers) proportional to the nongravitational acceleration and caused by torques resulting from mass unbalance. Jargon. *See also:* navigation. (AES/RS) 686-1982s, [42]

geared-drive machine A direct-drive machine in which the energy is transmitted from the motor to the driving sheave, drum, or shaft through gearing. *See also:* driving machine. (EEC/PE) [119]

geared traction machine (elevators) A geared-drive traction machine. *See also:* driving machine. (EEC/PE) [119]

gearless motor A traction motor in which the armature is mounted concentrically on the driving axle, or is carried by a sleeve or quill that surrounds the axle, and drives the axle directly without gearing. *See also:* traction motor. (EEC/PE) [119]

gearless traction machine (elevators) A traction machine, without intermediate gearing, that has the traction sheave and the brake drum mounted directly on the motor shaft. *See also:* driving machine. (PE/EEC) [119]

gear pattern *See:* drive pattern.

gear ratio (watthour meter) The number of revolutions of the rotor of the first dial pointer, commonly denoted by the symbol R_g. (ELM) C12.1-1982s

Geiger-Mueller counter tube A radiation-counter tube designed to operate in the Geiger-Mueller region. (NI/NPS) 309-1999

Geiger-Mueller region The range of applied voltage in which the charge collected per isolated count is independent of the charge liberated by the initial ionizing event. (NI/NPS) 309-1999

Geiger-Mueller threshold The lowest applied voltage at which the charge collected per isolated tube count is substantially independent of the nature of the initial ionizing event. (NI/NPS) 309-1999

Geissler tube A special form of gas-filled tube for showing the luminous effects of discharges through rarefied gases. *Note:* The density of the gas is roughly one-thousandth of that of the atmosphere. *See also:* gas tube. (ED) [45], [84]

gel cell *See:* gelled electrolyte cell.

gelled electrolyte cell A valve-regulated lead-acid (VRLA) cell whose electrolyte has been immobilized by the addition of a gelling agent. *Synonym:* gel cell. (IA/PSE) 446-1995

gelled electrolyte Electrolyte in a VRLA cell that has been immobilized by the addition of a gelling agent. (SB) 1189-1996

general color rendering index (illuminating engineering) Measure of the average shift of eight standardized colors chosen to be of intermediate saturation and spread throughout the range of hues. If the color rendering index is not qualified as to the color samples used, R_a is assumed. (EEC/IE) [126]

general coordinated methods (general application to electric supply or communication systems) Those methods reasonably available that contribute to inductive coordination without specific consideration of the requirements for individual inductive exposures. *See also:* inductive coordination. (EEC/PE) [119]

general diffuse lighting (illuminating engineering) Lighting involving luminaires that distribute 40 to 60 percent of the emitted light downward and the balance upward, sometimes with a strong component at 90 degrees (horizontal). (EEC/IE) [126]

general insertion gain (waveguide) A gain resulting from placing two ports of a network between arbitrary generator and load impedances. It is the ratio of the power absorbed in the load when connected to the generator (reference power) to that when the network is inserted. *See also:* general insertion loss. (MTT) 146-1980w

general insertion loss (waveguide) A loss resulting from placing two ports of a network between arbitrary generator and load impedances. It is the ratio of the power absorbed in the load when connected to the generator (reference power) to that when the network is inserted. *See also:* general insertion gain. (MTT) 146-1980w

generality The degree to which a system or component performs a broad range of functions. *See also:* reusability. (C) 610.12-1990

generalization (A) Saying that a subclass S generalizes to a superclass C means that every instance of class S is also an instance of class C. Generalization is fundamentally different from a *relationship*, which <u>may</u> associate distinct instances. **(B)** A taxonomy in which instances of both entities represent the same real or abstract thing. One entity (the generic entity) represents the complete set of things and the other (category entity) represents a subtype or sub-classification of those things. The category entity may have one or more attributes, or relationships with instances of another entity, not shared by all generic entity instances. Each instance of the category entity is simultaneously an instance of the generic entity. (C/SE) 1320.2-1998

generalization hierarchy *See:* generalization taxonomy.

generalization network *See:* generalization taxonomy.

generalization structure A connection between a superclass and one of its more specific, immediate subclasses. (C/SE) 1320.2-1998

generalization taxonomy A set of generalization structures with a common generic ancestor. In a generalization taxonomy every instance is fully described by one or more of the classes in the taxonomy. The structuring of classes as a generalization taxonomy determines the inheritance of responsibilities among classes. (C/SE) 1320.2-1998

generalized entity *See:* generalized property.

generalized impedance converter A two-port active network characterized by the conversion factor *f* (*s*) of the complex frequency variable *s* and satisfying the following property: when port B is terminated with impedance Z (*s*) the impedance at port A is given by Z (*s*)*f* (*s*); when port A is terminated with impedance Z (*s*) the impedance at port B is given by Z (*s*)/*f* (*s*). (CAS) [13]

Generalized Information Retrieval Language A query language used by the United States Defense Nuclear Agency for information retrieval. (C) 610.13-1993w

Generalized Markup Language (GML) A page description language used to provide simplified tags in DCF for formatting documents. *See also:* Bookmaster. (C) 610.13-1993w

generalized property Any of the physical concepts in terms of examples of which observable physical systems and phenomena are described quantitatively. *Notes:* 1. Examples are the abstract concepts of length, electric current, energy, etc. 2. A generalized property is characterized by the qualitative attribute of physical nature, or dimensionality, but not by a quantitative magnitude. *Synonyms:* generalized quantity; generalized entity. (Std100) 270-1966w

generalized quantity *See:* generalized property.

general lighting (illuminating engineering) Lighting designed to provide a substantially uniform level of illuminance throughout an area, exclusive of any provision for special local requirements. (EEC/IE) [126]

general-purpose branch circuit A branch circuit that supplies a number of outlets for lighting and appliances. (NESC/NEC) [86]

general purpose circuit breaker A circuit breaker that has been designed, tested, and rated in accordance with general purpose circuit breaker requirements of applicable standards. (SWG/PE) C37.100-1992

general-purpose circuit breaker (alternating current high voltage circuit breakers) A circuit breaker that is not specifically designed for capacitance current switching. (SWG) 341-1972w

general-purpose computer A computer that is designed to solve a wide variety of problems. *Contrast:* special-purpose computer. (C) [20], [85], 610.10-1994w

general-purpose controller Any controller having ratings, characteristics, and mechanical construction for use under usual service conditions. *See also:* electric controller. (IA/C/ICTL/IAC) [60], [85]

general-purpose current-limiting fuse A fuse capable of interrupting all currents from the rated interrupting current down to the current that causes melting of the fusible element in no less than 1 h. (SWG/PE/SWG-OLD) C37.40-1993, C37.100-1992

general-purpose digital computer *See:* digital computer.

general-purpose enclosure (1) An enclosure used for usual service applications where special types of enclosures are not required. (SWG/PE) C37.100-1992
(2) An enclosure that primarily protects against accidental contact and slight indirect splashing but is neither dripproof nor splashproof. (IA/MT) 45-1998

general-purpose floodlight (illuminating engineering) A weatherproof unit so constructed that the housing forms the reflecting surface. The assembly is enclosed by a cover glass. (EEC/IE) [126]

general-purpose induction motor (rotating machinery) Any open motor having a continuous rating of 50 degrees Celsius rise by resistance for Class A insulation, or of 80 degrees Celsius rise for Class B, a service factor as listed in the following tabulation, and designed, listed, and offered in standard ratings with standard operating characteristics and mechanical construction, for use under usual service conditions without restrictions to a particular application or type of application.

Service Factor

	Synchronous Speed, revolutions per minute			
Horsepower	3600	1800	1200	900
1/20	1.4	1.4	1.4	1.4
1/12	1.4	1.4	1.4	1.4
1/8	1.4	1.4	1.4	1.4
1/6	1.35	1.35	1.35	1.35
1/4	1.35	1.35	1.35	1.35
1/3	1.35	1.35	1.35	1.35
1/2	1.25	1.25	1.25	
3/8	1.25	1.25		
1	1.25			

See also: asynchronous machine. (PE) [9]

general-purpose low-voltage dc power circuit breaker *See:* circuit breaker, general purpose low-voltage dc power.

general-purpose low-voltage power circuit breaker (low voltage dc power circuit breakers used in enclosures) A circuit breaker that during interruption does not usually prevent the fault current from rising to its sustained value. (SWG/PE) C37.14-1979s

general-purpose motor (rotating machinery) Any motor designed, listed and offered in standard ratings with operating characteristics and mechanical construction suitable for use under usual service conditions without restrictions to a particular application or type of application. (PE) [9]

general-purpose programming language A programming language that provides a set of processing capabilities applicable to most information processing problems and that can be used on many kinds of computers. For example, Ada, COBOL, FORTRAN, and PL/1. *See also:* CAL; JOSEF; SIMULA; common language; Pascal. (C) 610.13-1993w

general-purpose relay A relay that is adaptable to a variety of applications. *See also:* relay. (EEC/REE) [87]

general-purpose register A register, usually explicitly addressable, within a set of registers, that can be used for different purposes, for example, as an accumulator, as an index register, or as a special handler of data. *Synonym:* general register. (C) 610.10-1994w

General Purpose Systems Simulation (GPSS) A problem-oriented language used in performing discrete simulation problems, based on a block diagram approach, where each block represents a physical process and transactions move from one block to another. *See also:* CSS/II. (C) 610.13-1993w

general-purpose test equipment (test, measurement, and diagnostic equipment) Test equipment that is used for the measurement of a range of parameters common to two or more equipments or systems of basically different design. (MIL) [2]

general-purpose transformers (power and distribution transformers) Step-up or step-down transformers or autotransformers generally used in secondary distribution circuits of 600 V or less in connection with power and lighting service. (PE/TR) C57.12.80-1978r

general register *See:* general-purpose register.

general ROM format A format for the node-provided ROM. The general ROM format provides bus-dependent information and a root_directory; the root_directory directly provides additional ROM entries. (C/MM) 1212-1991s

General Space Planner A programming language based on FORTRAN that provides an interactive system for solving space planning problems. (C) 610.13-1993w

general statistical terms Terms applied to the procedures of data collection, classification, and presentation. (T&D/PE) 539-1990

general support maintenance *See:* depot maintenance.

general-use snap switch A form of general-use switch so constructed that it can be installed in flush device boxes or on outlet box covers, or otherwise used in conjunction with wiring systems recognized by this Code. (NESC/NEC) [86]

general-use switch A switch intended for use in general distribution and branch circuits. It is rated in amperes and it is capable of interrupting its rated current at its rated voltage. (NESC/NEC) [86]

generate (computers) To produce a program by selection of subsets from a set of skeletal coding under the control of parameters. (C) [20], [85]

generated address An address that has been calculated during the execution of a computer program. *Synonym:* synthetic address. *See also:* indirect address; effective address; relative address; absolute address. (C) 610.12-1990

generated error The total error resulting from the combined effects of using imprecise arguments in an inexact formula. For example, using a rounded number in a truncated series. (C) 1084-1986w

generated voltage (rotating machinery) A voltage produced in a closed path or circuit by the relative motion of the circuit or its parts with respect to magnetic flux. *See also:* Faraday's law; asynchronous machine; induced voltage; synchronous machine. (PE) [9]

generating availability data system (GADS) Reliability information available from the North American Electric Reliability Council. (PE/NP) 933-1999

generating electric field meter (gradient meter) A device in which a flat conductor is alternately exposed to the electric field to be measured and then shielded from it. *Note:* The resulting current to the conductor is rectified and used as a measure of the potential gradient at the conductor surface. *See also:* instrument. (EEC/PE) [119]

generating magnetometer (earth inductor) A magnetometer that depends for its operation on the electromotive force generated in a coil that is rotated in the field to be measured. *See also:* magnetometer. (EEC/PE) [119]

generating station (1) (power operations) A plant wherein electric energy is produced from some other form of energy (for example, chemical, mechanical, or hydraulic) by means of suitable apparatus. (PE/PSE) 858-1987s
(2) A plant wherein electric energy is produced by conversion from some other form of energy (for example, chemical, nuclear, solar, mechanical, or hydraulic) by means of suitable apparatus. This includes all generating station auxiliaries and other associated equipment required for the operation of the plant. Not included are stations producing power exclusively for use with communications systems. (PE/NESC/EDPG) 665-1995, C2-1997

generating-station auxiliaries Accessory units of equipment necessary for the operation of the plant. Example: Pumps, stokers, fans, etc. *Note:* Auxiliaries may be classified as essential auxiliaries or those that must not sustain service interruptions of more than 15 s to 1 min, such as boiler feed pumps, forced draft fans, pulverized fuel feeders, etc., and nonessential auxiliaries that may, without serious effect, sustain service interruptions of one to three minutes or more, such as air pumps, clinker grinders, coal crushers, etc. *See also:* generating station. (T&D/PE) [10]

generating-station auxiliary power The power required for operation of the generating station auxiliaries. *See also:* generating station. (T&D/PE) [10]

generating-station efficiency *See:* efficiency.

generating-station reserve *See:* reserve equipment.

generating unit (unique identification in power plants) The generator, or generators, associated prime mover or movers, auxiliaries and energy supply or supplies that are normally operated together as a single source of electric power. (PE/EDPG) 803-1983r

generation The production or storage, or both, of electric energy with the intent of enabling practical use of commercial sale of the available energy. This includes photovoltaic, windfarm, hydro, etc., as well as normal commercial and industrial thermal sources. (SUB/PE) 1109-1990w

generation data group (GDG) A collection of data files that are kept in chronological order and referenced by its generation number. *Note:* Each file is called a generation data set. (C) 610.5-1990w

generation data set One data file within a generation data group. (C) 610.5-1990w

generation-frequency characteristic The change in area generation of a utility or of a control area through governor action that results from a change in system frequency without supplementary control action. (PE/PSE) 858-1993w, 94-1991w

generation rate (semiconductor) The time rate of creation of electron-hole pairs. *See also:* semiconductor device. (ED) 216-1960w

generator (1) (rotating machinery) A machine that converts mechanical power into electric power. *See also:* direct-current commutating machine; asynchronous machine. (PE/TR) [9], C57.116-1989r
(2) (computers) A controlling routine that performs a generate function, for example, report generator, input-output generator. *See also:* function generator; noise generator. (C) [20], [85]
(3) A module or device that initiates a bus request (such as an interrupt request) as the master to that request. (C/BA) 1014.1-1994w

generator, alternating-current *See:* alternating-current generator.

generator, arc welder *See:* arc welder generator.

generator efficiency (thermoelectric couple) The ratio of the electric power output of a thermoelectric couple to its thermal power input. *Note:* This is an idealized efficiency assuming perfect thermal insulation of the thermoelectric arms. *See also:* thermoelectric device. (ED) [46]

generator-field accelerating relay A relay that functions automatically to maintain the armature current within prescribed limits when a motor supplied by a generator is accelerated to any speed, up to base speed, by controlling the generator field current. *Note:* This definition applies to adjustable-voltage direct-current drives. *See also:* relay. (IA/ICTL/IAC/APP) [60], [75]

generator-field control A system of control that is accomplished by the use of an individual generator for each elevator or dumbwaiter wherein the voltage applied to the driving-machine motor is adjusted by varying the strength and direction of the generator field. *See also:* control.

generator field decelerating relay A relay that functions automatically to maintain the armature current within prescribed limits when a motor, supplied by a generator, is decelerated from base speed, or less, by controlling the generator field-current. *Note:* This definition applies to adjustable-voltage direct-current drives. *See also:* relay. (IA/IAC/APP) [60], [75]

generator/motor A machine that may be used as either a generator or a motor, usually by changing rotational direction. *Notes:* 1. This type of machine has particular application in a pumped-storage operation, in which water is pumped into a reservoir during off-peak periods and released to provide generation for peaking loads. 2. This definition eliminates the confusion of terminology for this type of machine. A slant is used between the terms to indicate their equality, and also the machine serves one function or the other and not both at the same time. The word generator is placed first to provide a distinction in speech between this term and the commonly used term motor-generator, which has an entirely different meaning. *See also:* asynchronous machine. (PE) [9]

generator set A unit consisting of one or more generators driven by a prime mover. *See also:* direct-current commutating machine; asynchronous machine. (PE) [9]

generator-source short-circuit current The short-circuit current when the source is entirely from a generator through no transformation. (SWG/PE) C37.013-1997

generette (rotating machinery) A test jig designed on the principle of a motorette, for endurance tests on sample lengths of coils or bars for large generators. *See also:* asynchronous machine; direct-current commutating machine. (PE) [9]

generic actuator group (valve actuators) An actuator or family of actuators within a range of sizes with similar design principles, materials, manufacturing processes, limiting stresses, operating principles, and design margins. (PE/NP) 382-1985

generic ancestor (of a class) A superclass that is either an immediate superclass of the class or a generic ancestor of one of the superclasses of the class. *Contrast:* ancestor. *See also:* reflexive ancestor. (C/SE) 1320.2-1998

generic connection assembly (Class 1E connection assemblies) A connection assembly that represents a family of connection assemblies having similar materials, manufacturing processes, assembly techniques, limiting stresses, design, and operating principles. (PE/NP) 572-1985r

generic data element A data element related to or drawn from a large class of like data elements. (C) 610.5-1990w

generic design (electric penetration assemblies) A family of equipment units having similar materials, manufacturing processes, limiting stresses, design, and operating principles, that can be represented for qualification purposes by a representative unit(s). (PE/NP) 317-1983r

generic entity An entity whose instances are classified into one or more subtypes or subclassifications (category entities). *Synonyms:* superclass; supertype. (C/SE) 1320.2-1998

generic environment A set of environmental conditions intended to envelop the range of expected environments. (SWG/PE/NP) C37.100-1992, 649-1980s

generic equipment (1) (nuclear power generating station) A family of equipment units having similar materials, manufacturing processes, limiting stresses, design, and operating principles that can be represented for qualification purposes by representative units. (PE/NP) 649-1980s
(2) A family of equipment units having similar materials, manufacturing processes, limiting stresses, and design and operating principles that can be represented for qualification purposes by a representative unit(s). (SWG/PE) C37.100-1992

generic interface (1) The interface, defined at a level that is independent of any particular programming language. (C/PA) 1328-1993w, 1327-1993w, 1224-1993w
(2) A version of an interface that is independent of any particular programming language. (C/PA/C/PA) 1224.1-1993w, 1326.1-1993w

generic program unit A software module that is defined in a general manner and that requires substitution of specific data, instructions, or both in order to be used in a computer program. *See also:* instantiation. (C) 610.12-1990

generic property domain An expression of a pairing of a property and a value domain, without regard to any entity type with which it may be associated. *Note:* An example of a generic property domain is "Stop_location". (SCC32) 1489-1999

generic qualification (Class 1E connection assemblies) Qualification to a set of requirements designed to envelop the service conditions plus margin of a number of specific applications. (PE/NP) 572-1985r

generic response spectra (GRS) The response spectra that define the seismic ratings of metal-enclosed power switchgear. (SWG/PE) C37.100-1992, C37.81-1989r

genetic effect An alteration in DNA material within the cell. If germ cells (sperm, egg) are involved, mutations in offspring can result. If somatic (all other) cells are involved, effects such as premature aging or cancer can result. (T&D/PE) 539-1990

geocentric latitude (navigation) The acute angle between A) a line joining a point with the earth's geometric center and B) the earth's equatorial plane. (AES/RS) 686-1982s, [42]

geocentric vertical *See:* geometric vertical.

geodesic The shortest line between two points measured on any mathematically derived surface that includes the points. *See also:* navigation. (AES/RS) 686-1982s, [42]

geodesic lens antenna A lens antenna having a two-dimensional lens, with uniform index of refraction, disposed on a surface such that the rays in the lens follow geodesic (minimal) paths of the surface. (AP/ANT) 145-1993

geodetic latitude (navigation) The angle between the normal to the spheroid and the earth's equatorial plane: the latitude generally used in maps and charts. Also called geographic latitude. *See also:* navigation. (AES/RS) 686-1982s, [42]

geographical address (1) A unique identifier assigned to each physical module slot on the bus and assumed by any module connected to that slot. (C/BA) 10857-1994, 896.3-1993w, 896.4-1993w
(2) A unique identifier statically assigned to each slot by the backplane. (C/BA) 896.2-1991w, 896.10-1997
(3) The primary address of a device based on the physical (geographical) location of the module, and determined by coded backplane pins, or (on a cable segment) by switches. For a crate segment geographical address zero is for the rightmost position when the crate is viewed from the front and the address increases by one for each module position moved to the left. (NID) 960-1993

geographical address control (GAC) (FASTBUS acquisition and control) Logic associated with each segment for supervising and generating signals for geographical addressing. (NID) 960-1993

geographical addressing A scheme wherein each slot in the backplane is assigned a unique address. This address can be read by the board that is installed in the slot. The VSB specification defines the use of the geographical address for two purposes: (A) It forms part of the interrupt ID used during an interrupt-acknowledge cycle and, (B) it forms part of the arbitration ID used during a parallel arbitration cycle. The geographical address can also be used to set global board variables such as the base address of a memory board. (C/MM) 1096-1988w

geographic latitude *See:* geodetic latitude.

geographic vertical The direction of a line normal to the surface of the geoid. *See also:* navigation. (AES/RS) 686-1982s, [42]

geoid The shape of the earth as defined by the hypothetical extension of mean sea level continuously through all land masses. *See also:* navigation. (AES/RS) 686-1982s, [42]

geomagnetically induced currents (GIC) *See:* solar induced currents.

geomagnetic induced currents (GIC) Spurious, quasidirect currents flowing in grounded systems due to a difference in the earth surface potential caused by geomagnetic storms resulting from the particle emission of solar flares erupting from the surface of the sun. (PE/PSC) 367-1996

geometrical adjectives (A) (pulse terminology) *Trapezoidal.* Having or approaching the shape of a trapezoid. **(B) (pulse terminology)** *Rectangular.* Having or approaching the shape of a rectangle. **(C) (pulse terminology)** *Triangular.* Having or approaching the shape of a triangle. **(D) (pulse terminology)** *Sawtooth.* Having or approaching the shape of a right angle. *See also:* composite waveform. **(E) (pulse terminology)** *Rounded.* Having a curved shape characterized by a relatively gradual change in slope. (IM/WM&A) 194-1977

geometrical factor (navigation) The ratio of the change in a navigational coordinate to the change in distance, taken in the direction of maximum navigational coordinate change: the magnitude of the gradient of the navigational coordinate. *See also:* navigation. (AES/RS) 686-1982s, [42]

geometric capacitive current (I_C) A reversible current of comparatively high magnitude and short duration, which decays exponentially with time of voltage application, and which depends on the internal resistance of the measuring instrument and the geometric capacitance of the winding. (PE/EM) 43-2000

geometric correction An image restoration technique in which a geometrical transformation is performed on an image to compensate for geometrical distortions. (C) 610.4-1990w

geometric dilution of position (GDOP) (radar) An expression which refers to increased measurement errors in certain regions of coverage of the measurement system. It applies to systems which combine several surface of position measurements such as range only, angle only, or hyperbolic (range difference) to locate the object of interest. When two lines of position cross at a small acute angle, the measurement accuracy is reduced along the axis of the acute angle. (AES/RS) 686-1982s

geometric dilution of precision (GDOP) An increase in measurement errors in certain regions of coverage of a measurement system that combines several surface-of-position measurements, such as range only, angle only, or range difference (hyperbolic) to locate the object of interest. *Note:* When two lines of position cross at a small acute angle, the measurement accuracy is reduced along the axis of the acute angle. (AES) 686-1997

geometric distortion (television) The displacement of elements in the reproduced picture from the correct relative positions in the perspective plane projection of the original scene. (BT/AV) 201-1979w

geometric factor (cable calculations) (power distribution, underground cables) A parameter used and determined solely by the relative dimensions and geometric configuration of the conductors and insulation of a cable. (PE) [4]

geometric inertial navigation equipment The class of inertial navigation equipment in which the geographic navigational quantities are obtained by computations (generally automatic) based upon the outputs of accelerometers whose vertical axes are maintained parallel to the local vertical, and whose azimuthal orientations are maintained in alignment with a predetermined geographic direction (for example, north). *See also:* navigation. (AES/RS) 686-1982s, [42]

geometric mean The numerical result obtained by taking the *n* th root of the product of *n* quantities, *n* being equal to or greater than two. *Note:* In radio noise measurements, geometric means have been used to determine the long-line frequency spectrum from the short-line frequency spectrum by taking the geometric mean of the maximum and minimum values in microvolts per meter across the spectrum (or the arithmetic mean of values in decibels). (T&D/PE) 539-1990

geometric optics (1) (fiber optics) The treatment of propagation of light as rays. *Note:* Rays are bent at the interface between two dissimilar media or may be curved in a medium in which refractive index is a function of position. *See also:* physical optics; optical axis; skew ray; paraxial ray; meridional ray; axial ray. (Std100) 812-1984w
(2) The infinitesimal-wavelength limit of processes involved in scattering or propagation, in which case ray-optics apply. (AP/PROP) 211-1997

geometric perturbation of the electric field in the interelectrode space A change in the electric field caused by the presence of either a conducting object or one with a dielectric constant different from that of the medium in the interelectrode space. It is assumed that the introduced object does not change the distribution of charges on the energized electrodes. *Note:* The amount of perturbation depends on the geometry of the object, its location and electric potential, and, when applicable, its electrical parameters (i.e., dielectric constant, conductivity). (T&D/PE) 539-1990

geometric rectification error (accelerometer) The error caused by an angular motion of a linear accelerometer input reference axis when this angular motion is coherent with a vibratory cross acceleration input. This error occurs in the application of a linear accelerometer and is not caused by imperfections in the accelerometer. The error is proportional to the square of the cross acceleration and varies with the frequency. (AES/GYAC) 528-1994

geometric theory of diffraction (GTD) The theory of geometric optics modified to allow for rays propagating into shadow regions. Also includes the development of ray constructs for scattering from edges and removal of "infinities" in optical focusing predictions in inhomogeneous media. (AP/PROP) 211-1997

geometric vertical (navigation) The direction of the radius vector drawn from the center of the earth through the location of the observer. *See also:* navigation. (AES) [42]

geometry (oscilloscopes) The degree to which a cathode-ray tube can accurately display a rectilinear pattern. *Note:* Generally associated with properties of a cathode-ray tube: the name may be given to a cathode-ray-tube electrode or its associated control. *See also:* oscillograph. (IM/HFIM) [40]

geometry, detector *See:* detector geometry.

geometry, detector element *See:* detector element geometry.

geometry engine A hardware accelerator in some workstations that performs scaling, clipping, and other graphical translations between the display list and the display bit map. (C) 610.10-1994w

Georgia Tech Language An extension to ALGOL that contains access to LISP and other facilities. (C) 610.13-1993w

geosynchronous earth orbit (GEO) Circular orbit in the equatorial plane of the Earth—6.62 Earth radii from the center of the Earth. (C/BA) 1156.4-1997

geotropism (radiation protection) A change in instrument response with a change in instrument orientation as a result of gravitational effects. (NI) N323-1978r

germanium gamma-ray detector A complete assembly, including the detector element, cryostat, integral preamplifier, and high-voltage filter. *See also:* detector element. (NPS) 325-1996

germ cells *See:* genetic effect.

germicidal effectiveness *See:* bactericidal effectiveness.

germicidal efficiency of radiant flux *See:* bactericidal efficiency of radiant flux.

germicidal exposure *See:* bactericidal exposure.

germicidal flux *See:* bactericidal flux.

germicidal flux density *See:* bactericidal flux density.

get (A) To retrieve an item from a set of items as in retrieving a record from a file, or in obtaining a numerical value from a series of decimal digits. *Contrast:* put. **(B)** To select and retrieve a group of specified records from a database. (C) 610.5-1990

get next To select and retrieve the next record from a database that meets some specified criteria. *Note:* Used in conjunction with a placeholder point. *Contrast:* get unique. (C) 610.5-1990w

getter (electron tube) A substance introduced into an electron tube to increase the degree of vacuum by chemical or physical action on the residual gases. *See also:* electrode. (ED) [45], [84]

get unique To select and retrieve the first record from a database that meets some selection criteria. *Contrast:* get next. (C) 610.5-1990w

GFD *See:* ground flash density.

G Filter A 20 kHz to 1100 kHz band pass filter used for measuring the power of an Asymmetric Digital Subscriber Line (ADSL) signal, noise, or impulse noise on an ADSL. (COM/TA) 743-1995

ghost (1) (television) A spurious image resulting from an echo. *See also:* television. (EEC/PE) [119]
(2) (computer graphics) The residue of an old image, displayed at the same time as a new image, that occurs when the persistence is longer than the refresh rate. (C) 610.6-1991w

ghost hyphen *See:* discretionary hyphen.

ghost pulse *See:* ghost signals.

ghost signals (A) (loran) Identification pulses that appear on the display at less than the desired loran station full pulse repetition frequency. *See also:* navigation. **(B) (loran)** Signals appearing on the display that have a basic repetition frequency other than that desired. *See also:* navigation. (AES) [42]

ghost target An apparent target in a radar that does not correspond in position or frequency or both to any real target, but which results from distortion or misinterpretation by the radar circuitry of other real target signals that are present. *Note:* It may result from range-Doppler ambiguities in the radar waveform used, from intermodulation distortion due to circuit amplitude nonlinearities, or from combining data from two antenna systems or waveforms. (AES) 686-1997

GHz *See:* gigahertz.

Gibb's phenomenon Overshoot phenomenon obtained near a discontinuity point of a signal when the spectrum of that signal is truncated abruptly. (CAS) [13]

GIC *See:* generalized impedance converter; geomagnetically induced currents; geomagnetic induced currents.

gig Colloqual reference for gigabyte. (C) 610.10-1994w

giga (G) (mathematics of computing) A prefix indicating one billion (10^9). (C) 1084-1986w

Gigabit Media Independent Interface (GMII) The interface between the Reconciliation sublayer and the physical coding sublayer (PCS) for 1000 Mb/s operation. (C/LM) 802.3-1998

gigabyte Either 1 000 000 000 bytes or 2^{30} bytes. *Notes:* 1. The user of these terms shall specify the applicable usage. If the usage is 2^{10} or 1024 bytes, or multiples thereof, then note 2 below shall also be included with the definition. 2. As used in IEEE Std 610.10-1994, the terms kilobyte (kB) means 2^{10} or 1024 bytes, megabyte (MB) means 1024 kilobytes, and gigabyte (GB) means 1024 megabytes. *See also:* megabyte; kilobyte. (C) 610.10-1994w

gigahertz A unit of frequency equal to 1 000 000 000 Hz, that is, 10^9 Hz. (C) 610.7-1995

gigahertz transverse electromagnetic (GTEM) cell A tapered transverse electromagnetic (TEM) cell with an absorber-lined end wall terminated with an absorber load. (EMC) 1128-1998

gilbert (centimeter-gram-second electromagnetic-unit system) The unit of magnetomotive force. The gilbert is one oersted-centimeter. (Std100) 270-1966w

Gill-Morrell oscillator An oscillator of the retarding-field type in which the frequency of oscillation is dependent not only on electron transit time within the tube, but also on associated circuit parameters. *See also:* oscillatory circuit. (AP/ANT) 145-1983s

gimbal (gyros) A device that permits the spin axis to have one or two angular degrees of freedom. (AES/GYAC) 528-1994

gimbal error (gyros) The error resulting from angular displacements of gimbals from their reference positions such that gimbal pickoffs do not measure the true angular motion of the case about the input reference axis. (AES/GYAC) 528-1994

gimbal freedom (gyros) The maximum angular displacement of a gimbal about its axis. (AES/GYAC) 528-1994

gimbal lock (gyros) A condition of a two-degree-of-freedom gyro wherein the alignment of the spin axis with an axis-of-freedom deprives the gyro of a degree-of-freedom and, therefore, of its useful properties. (AES/GYAC) 528-1994

gimbal retardation (gyros) A measure of output axis friction torque when the gimbal is rotated about the output axis. It is expressed as an equivalent input. (AES/GYAC) 528-1994

gimbal-unbalance torque (dynamically tuned gyro) The acceleration-sensitive torque caused by gimbal unbalance along the spin axis due to non-intersection of the flexure axes. Under constant acceleration, it appears as a second harmonic of the rotor spin frequency because of the single-degree-of-freedom of the gimbal relative to the support shaft. When the gyro is subjected to vibratory acceleration, applied normal to the spin axis at twice the rotor spin frequency, this torque results in a rectified unbalance drift rate. *See also:* two-N (2N) translational sensitivity. (AES/GYAC) 528-1994

gin An assembly, which when attached to a support or assembled on a structure, provides a rigging point for rope blocks, blocks, etc., so as to manipulate various pieces of apparatus. The gin, unlike the davit, is not rigid since its boom swivels, affording greater maneuverability. (T&D/PE) 516-1995

GIRL *See:* Graphical Information Retrieval Language.

GIS *See:* compartment; gas-insulated substation; assembly.

GIS conductor end The end of the GIS high-voltage conductor inside the cable connection enclosure. (PE/IC) 1300-1996

GKS *See:* Graphical Kernel System.

gland seal (rotating machinery) A seal used to prevent leakage between a moveable and a fixed part. (PE) [9]

glare (1) (illuminating engineering) The sensation produced by luminances within the visual field that are sufficiently greater than the luminance to which the eyes are adapted to cause annoyance, discomfort, or loss in visual performance, or visibility. *Note:* The magnitude of the sensation of glare depends upon such factors as the size, position, and luminance of a source, the number of sources and the luminance to which the eyes are adapted. (EEC/IE) [126]
(2) (electric power systems in commercial buildings) The undesirable sensation produced by luminance within the visual field. (IA/PSE) 241-1990r

glass box (A) A system or component whose internal contents or implementation are known. *Synonym:* white box. *Contrast:* black box. **(B)** Pertaining to an approach that treats a system or component as in definition (A). (C) 610.12-1990

glass box model A model whose internal implementation is known and fully visible; for example, a model of a computerized change-return mechanism in a vending machine, in the form of a diagram of the circuits and gears that make the change. *Synonym:* white-box model. *Contrast:* black box model. (C) 610.3-1989w

glass-box testing *See:* structural testing.

glass half cell (glass electrode) A half cell in which the potential measurements are made through a glass membrane. *See also:* electrolytic cell. (EEC/PE) [119]

GLC circuit *See:* simple parallel circuit.

glide path (electronic navigation) The path used by an aircraft in approach procedures as defined by an instrument landing facility. *See also:* navigation. (AES/RS) 686-1982s, [42]

glide-path receiver An airborne radio receiver used to detect the transmissions of a ground-installed glide-path transmitter. *Note:* It furnishes a visual, audible, or electric signal for the purpose of vertically guiding an aircraft using an instrument landing system. (EEC/PE) [119]

glide slope (electronic navigation) An inclined surface generated by the radiation of electromagnetic waves and used with a localizer in an instrument landing system to create a glide path. *See also:* navigation. (AES/RS) 686-1982s, [41]

glide-slope angle (electronic navigation) The angle in the vertical plane between the glide slope and the horizontal. *See also:* navigation. (AES/RS) 686-1982s, [41]

glide-slope deviation (electronic navigation) The vertical location of an aircraft relative to a glide slope, expressed in terms of the angle measured at the intersection of the glide slope with the runway: or the linear distance above or below the glide slope. *See also:* navigation. (AES/RS) 686-1982s, [41]

glide-slope facility (navigation) The ground station of an ILS (instrument landing system) which generates the glide slope. (AES/RS) 686-1982s

glide-slope sector (instrument landing systems) A vertical sector containing the glide slope and within which the pilot's indicator gives a quantitative measure of the deviation above and below the glide slope: the sector is bounded above and below by a specified difference in depth of modulation, usually that which gives full-scale deflection of the glide-slope deviation indicator. *See also:* navigation. (AES/RS) 686-1982s, [41]

glint The inherent component of error in measurement of position and/or Doppler frequency of a complex target due to interference of the reflections from different elements of the target. *Notes:* 1. Glint may have peak values beyond the target extent in the measured coordinate. 2. Not to be confused with scintillation error. (AES) 686-1997

glitch A perturbation of the pulse waveform of relatively short duration and of uncertain origin. *See also:* pulse distortion. (IM/HFIM) [40]

glitch filter Filters out a fundamental transmission line effect found in bused backplane implementations. The effect is commonly called the *wire-OR glitch.* (C/BA) 896.4-1993w, 896.2-1991w, 896.10-1997

global Relating to the whole of an ATLAS test requirement. (SCC20) 771-1998

global broadcast A broadcast to slaves on all segments of a multi-segment system that can be reached from the originating segment. (NID) 960-1993

global compaction In microprogramming, compaction in which microoperations may be moved beyond the boundaries of the single entry, single exit sequential blocks in which they occur. *Contrast:* local compaction. (C) 610.12-1990

global data Data that can be accessed by two or more non-nested modules of computer program without being explicitly passed as parameters between the modules. *Synonym:* common data. *Contrast:* local data. (C) 610.12-1990

global (broadcast) DSAP address The predefined LLC DSAP address (all ones) used as a broadcast (all parties) address. It can never be the address of a single LLC on the data link. (C/LM/CC) 8802-2-1998

global identification A unique identifier assigned to each physical module slot in a system. This identifier would typically be both a bus identifier and a slot identifier. IEEE Std 1212-1991 specifies the format for such a global identifier. (C/BA) 10857-1994, 896.3-1993w, 896.4-1993w

global replace In text editing, an operation that substitutes a given textual pattern for all, or a given number of, occurrences of some other textual pattern found in the text. *See also:* global search. (C) 610.2-1987

global route, route In Physical Design Exchange Format (PDEF), the physical description of interconnect routing between logical and physical pins of cell, spare_cell, and/or cluster instances. (C/DA) 1481-1999

global search In text editing, an operation that identifies all, or a given number of, appearances of a given textual pattern in the text. *See also:* global replace. (C) 610.2-1987

global stability (solution $\phi(x(t_0);t)$ Stable for all initial perturbations, no matter how large they may be. *See also:* control system. (CS/IM) [120]

global system time SCI nodes may maintain time-of-day clocks as described in the CSR Architecture. Software may adjust each of these clocks in order to make them consistent to high accuracy. If this is done, the system is said to implement global system time. Otherwise each clock runs independently, which is sufficient for local timeout purposes but is not sufficient to implement the optional packet "time of death" feature. (C/MM) 1596-1992

global variable A variable that can be accessed by two or more non-nested modules of a computer program without being explicitly passed as a parameter between the modules. *Contrast:* local variable. (C) 610.12-1990

globe (illuminating engineering) A transparent or diffusing enclosure intended to protect a lamp, to diffuse and redirect its light, or to change the color of the light. (EEC/IE) [126]

glossary (1) The collection of the names and narrative descriptions of all terms that may be used for defined concepts (views, classes, subject domains, relationships, responsibilities, properties, and constraints) within an environment. (C/SE) 1320.2-1998

(2) A set of definitions that includes arrow labels and box names used in an IDEF0 model. (C/SE) 1320.1-1998

glossary page A model page that contains definitions for the arrow labels and box names in a specific diagram. (C/SE) 1320.1-1998

glossmeter (illuminating engineering) An instrument for measuring gloss as a function of the directionally selective reflecting properties of a material in angles near to and including the direction giving specular reflection. (EEC/IE) [126]

gloving A method of performing live-line maintenance on energized electrical conductors and equipment whereby a worker or workers, wearing specially-made and tested insulating gloves, with or without sleeves, and using cover-up equipment while supported by the structure or insulated aerial lift equipment, work(s) directly on the energized electrical conductor or equipment. (T&D/PE) 516-1995

glow corona (overhead-power-line corona and radio noise) Glow corona is a stable, essentially steady discharge of constant luminosity occurring at either positive or negative electrodes. (T&D/PE) 539-1979s

glow (mode) current (gas tube surge arresters) The current that flows after breakdown when circuit impedance limits the follow current to a value less than the glow-to-arc transition current. (PE) [8]

glow current (gas-tube surge protective devices) The current that flows after breakdown when circuit impedance limits the follow current to a value less than the glow-to-arc transition current. It is sometimes called the glow mode current. (PE/SPD) C62.31-1987r

glow discharge (1) (electron tube) A discharge of electricity through gas characterized by: A change of space potential, in the immediate vicinity of the cathode, that is much higher than the ionization potential of the gas; a low, approximately constant, current density at the cathode, and a low cathode temperature; the presence of a cathode glow. *See also:* gas tube; lamp. (ED) 161-1971w

(2) (illuminating engineering) An electric discharge characterized by a low, approximately constant, current density at the cathode, low cathode temperature, and a high, approximately constant, voltage drop. (EEC/IE) [126]

glow-discharge tube A gas tube that depends for its operation on the properties of a glow discharge. (ED) 161-1971w

glow factor (illuminating engineering) A measure of the visible light response of a fluorescent material to "black light." It is equal to ν times the luminance in candelas per square meter produced on the material divided by the incident "black light" flux density in milliwatts per square meter. ν is omitted when luminance is in footlamberts and the area is in square feet. It may be measured in lumens per milliwatt. *See also:* filter factor. (EEC/IE) [126]

glow lamp (illuminating engineering) An electric-discharge lamp whose mode of operation is that of a glow discharge, and in which light is generated in the space close to the electrodes. (EEC/IE) [126]

glow mode A stable, essentially steady discharge of constant luminosity occurring at either positive or negative electrodes. (T&D/PE) 539-1990

glow-mode current *See:* glow current.

glow, negative *See:* negative glow.

glow-switch An electron tube containing contacts operated thermally by means of a glow discharge. (ED) [45]

glow-to-arc transition current (gas-tube surge protective devices) The current required for the arrester to pass from the glow mode into the arc mode. (PE/SPD) C62.31-1987r

glow-tube *See:* glow-discharge tube.

glow voltage (gas-tube surge protective devices) The voltage drop across the arrester during glow-current flow. It is sometimes called the glow mode voltage. (SPD/PE) C62.31-1987r

glue-line heating (dielectric heating) An arrangement of electrodes designed to give preferential heating to a thin film of material of relatively high loss factor between alternate layers of relatively low loss factor material. (IA) 54-1955w, 169-1955w

glue logic A family of circuit logic consisting of various gates and simple logic elements, each of which serve as an interface between various parts of a computer such as processors, memory units and input-output devices. (C) 610.10-1994w

glyph A picture, logo, or symbol; used instead of text. (C/BA) 896.2-1991w

GLYPNIR A programming language with syntax similar to that of ALGOL, but with facilities to allow the programmer to specify the parallelism of an algorithm. (C) 610.13-1993w

GML *See:* Generalized Markup Language.

GO Availability analysis method similar to reliability block diagram with operators and event actions included. (PE/NP) 933-1999

go *See:* go/no-go.

Go-Back-N A transmission scheme where the transmitter may send multiple PDUs without waiting for an acknowledgment. If the receiver indicates that an error occurred in a given PDU, the sender will retransmit the errored PDU and all subsequently transmitted PDUs. *Note:* In this scheme, the receiver will only accept PDUs in sequential order. *Contrast:* selective retransmission. (C) 610.7-1995

go list In automatic indexing, a list of terms, words, or roots of words that are considered significant for purposes of information retrieval, and are to be used as keywords in an index. *Synonym:* inclusion list. *Contrast:* stop list. (C) 610.2-1987

goniometer (electronic navigation) A combining device used with a plurality of antennas so that the direction of maximum radiation or of greatest response may be rotated in azimuth without physically moving the antenna array. (AES/RS) 686-1982s, [42]

goniophotometer (illuminating engineering) A photometer for measuring the directional light distribution characteristics of sources, luminaires, media, and surfaces. (EEC/IE) [126]

go/no-go A set of terms (in colloquial usage) referring to the condition or state of operability of a unit that can only have two parameters: go, functioning properly, or no-go, not functioning properly. (PE/MIL/NP) 338-1987r, [2]

go/no-go test *See:* end-to-end test.

good neighbor A term used to describe "well-behaved" devices operating on a broadband medium that do not cause interference to any other service operating on the cable plant. (LM/C) 802.7-1989r

gooey Colloquial pronunciation for GUI, graphical user interface. (C) 610.10-1994w

go symbol An idle symbol that has been marked with the pertinent go bit (*idle.lg = 1* or *idle.hg = 1*) to give permission to a waiting node to transmit. (C/MM) 1596-1992

go to A computer program statement that causes a jump. *Contrast:* call; case; if-then-else. *See also:* branch. (C) 610.12-1990

Gouraud shading A technique for shading a three-dimensional solid object by interpolating the light intensities at the vertices of each polygon face, resulting in smooth shading. *See also:* Phong shading. (C) 610.6-1991w

governing system (hydraulic turbines) The combination of devices and mechanisms that detects speed deviation and converts it into a change in servomotor position. It includes the speed sensing elements, the governor control actuator, the hydraulic pressure supply system, and the turbine control servomotor. The terms "governor" and "governor equipment" are commonly used in the industry to describe the governing system and will be used interchangeably with the term "governing system." (PE/EDPG) [5], 125-1988r

governor (1) (power system device function numbers) The assembly of fluid, electrical, or mechanical control equipment used for regulating the flow of water, steam, or other medium to the prime mover for such purposes as starting, holding speed or load, or stopping. (SUB/PE) C37.2-1979s **(2) (hydroelectric power plants)** A system that controls speed and power output of a turbine. (PE/EDPG) 1020-1988r

governor actuator rating (speed governing systems, hydraulic turbines) The governor actuator rating is the flow rate in volume per unit time which the governor actuator can deliver at a specified pressure drop. The pressure drop shall be measured across the terminating pipe connections to the turbine control servomotors at the actuator. This pressure drop is measured with the specified minimum normal working pressure of the pressure supply system delivered to the supply port of the actuator distributing valve. (PE/EDPG) [5]

governor control actuator (hydraulic turbines) The combination of devices and mechanisms that detects a speed error and develops a corresponding hydraulic control output to the turbine control servomotors, but does not include the turbine control servomotors. Includes gate, blade, deflector, or needle control, or all equipment as appropriate. (PE/EDPG) 125-1988r

governor control actuator rating (hydraulic turbines) The flow rate in volume per unit time that the governor actuator can deliver at a specified pressure drop. The pressure drop shall be measured across the terminating pipe connections to the turbine control servomotors at the actuator. This pressure drop is measured with the specified minimum normal working pressure of the hydraulic pressure supply system delivered to the supply port of the actuator distributing valve. (PE/EDPG) 125-1988r

governor-controlled gates (on a hydro turbine) Gates that control the power input to the turbine and that are actuated by the speed governor directly or through the medium of the speed-control mechanism. (PE/PSE) 94-1991w

governor-controlled valves (on a steam turbine) Valves that control the power input to the turbine and that are actuated by the speed governor directly or through the medium of the speed-control mechanism. (PE/PSE) 94-1991w

governor dead band (automatic generation control) The magnitude of the total change in steady-rate speed within which there is no resulting measurable change in the position of the governor-controlled valves. *Note:* Dead band is the measure of the insensitivity of the speed-governing system and is expressed in percent of rated speed. (PE/PSE) 94-1970w

governor dead time (hydraulic turbines) Dead time is the time interval between the initiation of a specified change in steady-state speed and the first detectable movement of the turbine control servomotor. (PE/EDPG) 125-1977s

governor pin *See:* centrifugal-mechanism pin.

governor speed changer A device that adjusts the speed or power output of the turbine during operation. (PE/PSE) 94-1991w

governor speed-changer position The position of the speed changer indicated by the fraction of its travel from the position corresponding to minimum turbine speed to the position corresponding to maximum speed and energy input. It is usually expressed in percent. *See also:* speed-governing system. (PE/PSE) 94-1970w

governor spring *See:* centrifugal-mechanism spring.

governor weights *See:* centrifugal-mechanism weights.

GP *See:* group address.

GPR *See:* ground potential rise.

GPSS *See:* General Purpose Systems Simulation.

graded index optical waveguide (fiber optics) A waveguide having a graded index profile in the core. *See also:* graded index profile; step index optical waveguide. (Std100) 812-1984w

graded index profile (fiber optics) Any refractive index profile that varies with radius in the core. Distinguished from a step index profile. *See also:* profile dispersion; refractive index;

parabolic profile; multimode optical waveguide; normalized frequency; dispersion; mode volume; optical waveguide; step index profile; power-law index profile; profile parameter. (Std100) 812-1984w

graded insulation (electronic power transformer) The selective arrangement of the insulation components of a composite insulation system to more nearly equalize the voltage stresses throughout the insulation system. (PEL/ET) 295-1969r

graded junction (nonlinear, active, and nonreciprocal waveguide components) (semiconductor) A specially designed p-n junction with a p+ -type of region and an n-type of region whose doping levels increase linearly with distance from the junction. (MTT) 457-1982w

graded-time step-voltage test (rotating machinery) A controlled overvoltage test in which calculated voltage increments are applied at calculated time intervals. *Note:* Usually, a direct-voltage test with the increments and intervals so calculated that dielectric absorption appears as a constant shunt-conductance: to simplify interpretation. *See also:* asynchronous machine; direct-current commutating machine. (PE) [9]

grade-of-service (telephone switching systems) The proportion of total calls, usually during the busy hour, that cannot be completed immediately or served within a prescribed time. (COM) 312-1977w

gradient (1) (scalar field) At a point, a vector (denoted by ∇u) equal to, and in the direction of, the maximum space rate of change of the field. It is obtained as a vector field by applying the operator nabl to a scalar function. Thus, if $u = f(x, y, z)$

$$\nabla u = \text{grad } u = \mathbf{i}\frac{\partial u}{\partial x} + \mathbf{j}\frac{\partial u}{\partial y} + \mathbf{k}\frac{\partial u}{\partial z}$$

(Std100) 270-1966w

(2) (overhead power lines) *See also:* voltage gradient. (T&D/PE) 539-1990

gradient meter *See:* generating electric field meter.

gradient microphone A microphone the output of which corresponds to a gradient of the sound pressure. *Note:* Gradient microphones may be of any order as, for example, zero, first, second, etc. A pressure microphone is a gradient microphone of zero order. A velocity microphone is a gradient microphone of order one. Mathematically, from a directivity standpoint for plane waves, the root-mean-square response is proportional to $\cos^n$ q, where θ is the angle of incidence and n is the order of the microphone. *See also:* microphone. (EEC/PE) [119]

grading (telephone switching systems) Partial commoning or multipling of the outlets of connecting networks where there is limited availability to the outgoing group or subgroup of outlets. (COM) 312-1977w

grading device (composite insulators) A device for controlling the potential gradient at the end fittings, such as a grading ring or various semiconductive polymeric devices. (T&D/PE) 987-1985w

grading group (telephone switching systems) That part of a grading in which all inlets have access to the same outlets. (COM) 312-1977w

grading ring (surge arresters) (metal-oxide surge arresters for ac power circuits) A metal part, usually circular or oval in shape, mounted to modify electrostatically the voltage gradient or distribution. *Synonym:* control ring. (PE/SPD) 28-1974, C62.1-1981s, [8], C62.11-1999

gradual failure *See:* failure.

graduated (control) Marked to indicate a number of operating positions. (IA/ICTL/IAC) [60]

grain (photographic material) A small particle of metallic silver remaining in a photographic emulsion after development and fixing. *Note:* In the agglomerate, these grains form the dark area of a photographic image. (SP) [32]

graininess (photographic material) The visible coarseness under specified conditions due to silver grains in a developed photographic film. (SP) [32]

grandfather file A file that contains data that have since been updated in another file, called the father file, and further updated in a third file, called the son file. *See also:* son file; father file. (C) 610.5-1990w

grant A link control signal or link condition indicating that the receiving entity has been given permission to send a packet.local area networks. (C) 8802-12-1998

granular-filled fuse unit A fuse unit in which the arc is drawn through powdered, granular, or fibrous material. (SWG/PE) C37.40-1993, C37.100-1992

granularity (1) The depth or level of detail at which data is collected. (C/SE) 1045-1992
(2) Pertaining to the size of the standard meaningful unit with respect to a particular mode of operation; for example, in reference to computer processes, this term could be used to describe screen resolution, levels of manipulation of data, or the amount of time given to a background printing process. (C) 610.10-1994w

granularity period As defined in ISO/IECDIS 10164-11, the time between observations. For this standard, it is the time between two successive scans and is denoted by the symbol "GP." (LM/C) 802.1F-1993r

graph (A) (software) (data management) A diagram that represents the variation of a variable in comparison with that of one or more other variables; for example, a graph showing a bathtub curve. **(B) (data management) (software)** A diagram or other representation consisting of a finite set of nodes and internode connections called edges or arcs. *See also:* block diagram; input-process-output chart; box diagram; bubble chart; directed graph; undirected graph; structure chart.

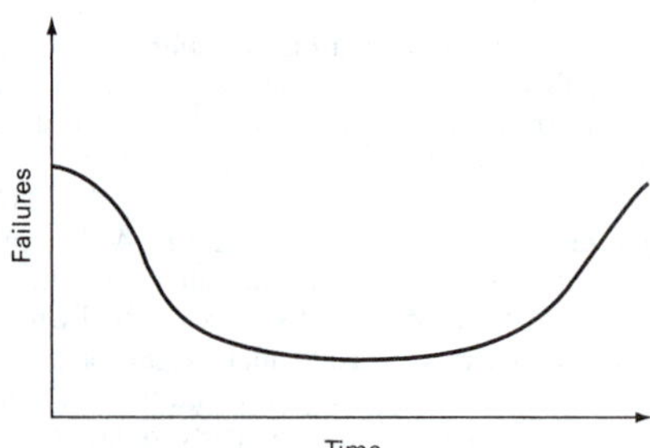

Variation of a variable in comparison with one or more other variables

graph A

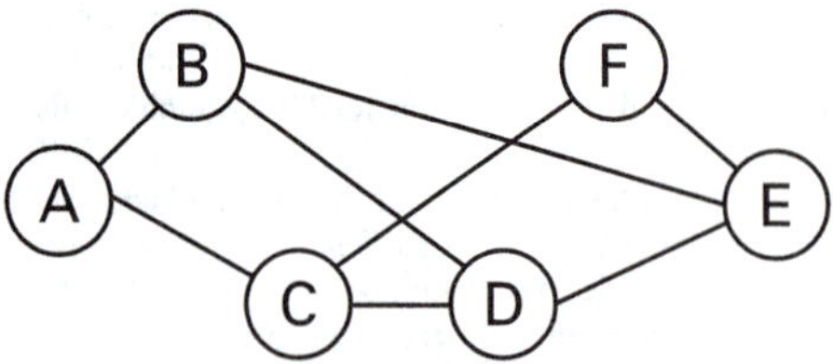

finite set of nodes and internode connections

graph B

(C) 610.12-1990, 610.5-1990

graph determinant (network analysis) One plus the sum of the loop set transmittances of all nontouching loop sets contained in the graph. *Notes:* 1. The graph determinant is conveniently expressed in the form

$$\Delta = (1 - \Sigma L_i + \Sigma L_i L_j - \Sigma L_i L_j L_k + \ldots)$$

where L_i is the loop transmittance of the i th loop of the graph, the second is over all of the different pairs of nontouching loops, and the third is over all the different triplets of nontouching loops, etc. 2. The graph determinant may be written alternatively as

$$\Delta = [(1 - L_i)(1 - L_2) \ldots (1 - L_n)]$$

where $L_1, L_2 \ldots, L_n$, are the loop transmittances of the n different loops in the graph, and where the dagger indicates that, after carrying out the multiplications within the brackets, a term will be dropped if it contains the transmittance product of two touching loops. 3. The graph determinant reduces to the return difference for a graph having only one loop. 4. The graph determinant is equal to the determinant of the coefficient equations. (CAS) 155-1960w

graphic A symbol produced by a process such as handwriting, drawing, or printing. *Synonym:* graphic symbol. (C) 610.2-1987, 610.10-1994w

graphical Pertaining to the pictorial representation of data. (C) 610.6-1991w

graphical display device A display device that can display graphical output. *Note:* Graphical display devices can display characters but they are in the form of graphical images. *See also:* display space; display surface. (C) 610.6-1991w

Graphical Information Retrieval Language (GIRL) A programming language used to manipulate information in arbitrary directed-graph structures, including facilities for insertion, retrieval, deletion, and comparison. (C) 610.13-1993w

graphical input device (A) An input device employed in the interactive process of identifying a location on a display surface; for example, a joystick, a data tablet, a control ball, a mouse, or a thumbwheel. **(B)** An input device employed in the entry of graphical images. (C) 610.6-1991

Graphical Kernel System (GKS) A computer graphics standard that provides a set of basic functions for producing computer generated pictures. It was developed by the International Standards Organization (ISO) and adopted by the American National Standards Institute (ANSI). (C) 610.6-1991w

graphical model A symbolic model whose properties are expressed in diagrams; for example, a decision tree used to express a complex procedure. *Contrast:* mathematical model; software model; narrative model. (C) 610.3-1989w

graphical user interface (GUI) (1) A user interface that is graphical in nature; that is, the user can enter commands by using a mouse, icons and windows. *Note:* Sometimes pronounced "gooey." *Contrast:* character-based user interface. (C) 610.10-1994w

(2) A means of presenting function to a user through the use of graphics. All such interfaces are outside the scope of this standard. (C/PA) 1387.2-1995

graphical user interface font *See:* screen font.

graphic character (1) A character, other than a control character, that is normally represented by a graphic. *Synonym:* optical character. (C) 610.2-1987

(2) A sequence of one or more *POSIX.POSIX_Characters* representing a single graphic symbol. (C) 1003.5-1999

graphic display (supervisory control, data acquisition, and automatic control) (station control and data acquisition) A hardware device [e.g., CRT, VDT, liquid crystal display (LCD), mapboard, plasma panel, arrays of lamps, or light emitting diodes] used to present pictorial information. (PE/SUB) C37.1-1994

graphic display device A display device that can display graphical output. *Note:* Graphic display devices can display characters but they are in the form of graphic images. *Contrast:* character display device. (C) 610.10-1994w

graphic printer A printer that can display both text and graphical output. *Contrast:* character printer. (C) 610.10-1994w

graphic input device An input device employed in the entry of graphic images. Examples include a joystick, a mouse, or a track ball. *See also:* digitizer. (C) 610.10-1994w

graphics adapter An expansion board that enhances the computer's ability to control the display device; for example, a graphics adapter that allows color output, or non-interlacing. *Synonym:* video board. (C) 610.10-1994w

graphics data *See:* display data.

graphics field* *See:* viewport.

* Deprecated.

graphics input The interactive process of entering data on a graphics system. (C) 610.6-1991w

graphics language A programming language that produces display data. (C) 610.6-1991w

graphics processor *See:* display processor.

GraphicString A value of the ASN.1 GraphicString restricted character string type. (C/PA) 1238.1-1994w

graphic symbol (1) (abbreviation) A geometric representation used to depict graphically the generic function of an item as it normally is used in a circuit. *See also:* abbreviation. (GSD) 267-1966

(2) A shorthand used to show graphically the functioning or interconnections of a circuit. A graphic symbol represents the functions of a part in the circuit. For example, when a lamp is employed as a nonlinear resistor, the nonlinear resistor symbol is used. Graphic symbols are used on single-line (one-line) diagrams, on schematic or elementary diagrams, or, as applicable, on connection or wiring diagrams. Graphic symbols are correlated with parts lists, descriptions, or instructions by means of designations. (GSD) 315-1975r

graphics system A collection of hardware or software allowing the use of graphical input or output in computer programs. (C) 610.6-1991w

graphic tablet A data tablet or digitizer that can be used with a stylus to trace existing graphic images, or for entering new images. (C) 610.10-1994w

graphic user terminal A terminal used to display and manipulate both alphanumeric symbols as well as graphic images. (C) 610.10-1994w

graphite brush A brush composed principally of graphite. *Note:* This type of brush is soft. Grades of brushes of this type differ greatly in current-carrying capacity and in operating speed from low to high. *See also:* brush. (PE/EEC/LB) [9], [101]

graph transmittance (network analysis) The ratio of signal at some specified dependent node, to the signal applied at some specified source node. *Note:* The graph transmittance is the weighted sum of the path transmittances of the different open paths from the designated source node to the designated dependent node, where the weight for each path is the path factor divided bt the graph determinant. (CAS) 155-1960w

grass A descriptive colloquialism referring to the appearance of noise on certain displays, such as an A-display. (AES) 686-1997

graticule (oscilloscopes) A scale for measurement of quantities displayed on the cathode-ray tube of an oscilloscope. *See also:* oscilloscope. (IM/HFIM) [40]

graticule area (oscilloscopes) The area enclosed by the continuous outer graticule lines. *Note:* Unless otherwise stated the graticule area shall be equal to or less than the viewing area. *See also:* quality area; oscillograph; viewing area. (IM/HFIM) [40]

graticule, internal *See:* internal graticule.

grating *See:* ultrasonic space grating.

grating lobe A lobe, other than the main lobe, produced by an array antenna when the interelement spacing is sufficiently large to permit the in-phase addition of radiated fields in more than one direction. (AP/ANT) 145-1993

gravitational acceleration unit (g, *g*) (1) A unit of acceleration that is approximately 32.2 ft/s^2 [9.8 m/s^2]. (C/BA) 1101.4-1993, 1101.3-1993

(2) The symbol *g* denotes a unit of acceleration equal in magnitude to the local value of gravity, unless otherwise specified. *Notes:* 1. In some applications, a standard value of *g* may be specified. 2. For an earthbound accelerometer, the attractive force of gravity acting on the proof mass must be treated as an applied upward acceleration of 1 *g*. (AES/GYAC) 528-1994

gravity gradient stabilization (communication satellite) The use of the gravity gradient along a satellite structure for controlling its attitude. This method usually requires long booms to create the necessary mass distribution. (COM) [19]

gravity vertical *See:* mass-attraction vertical.

gravity wave *See:* acoustic-gravity wave.

gray (metric practice) The absorbed dose when the energy per unit mass imparted to matter by ionizing radiation is one joule per kilogram. *Note:* The gray is also used for the ionizing radiation quantities: specific energy imparted, kerma, and absorbed dose index, which have the SI unit joule per kilogram. (QUL) 268-1982s

graybody (illuminating engineering) A temperature radiator whose spectral emissivity is less than unity and the same at all wavelengths. (EEC/IE) [126]

Gray code (mathematics of computing) A binary code in which sequential numbers are represented by binary expressions, each of which differs from the preceding expression in one place only.

Gray Code

DECIMAL DIGIT:	0	1	2	3	4	5
GRAY CODE:	000	001	011	010	110	111

Synonyms: reflected code; cyclic binary code; cyclic code; reflected binary unit-distance code; reflected binary code. (C) 1084-1986w

gray level A value associated with a pixel in a digital image, representing the brightness of the original scene in the vicinity of the point represented by the pixel. *Synonyms:* gray shade; gray tone. (C) 610.4-1990w

gray scale (1) (television) An optical pattern in discrete steps between light and dark. *Note:* A gray scale with ten steps is usually included in resolution test charts. (BT/AV) 201-1979w

(2) (image processing and pattern recognition) The range of gray levels that occur in an image. (C) 610.4-1990w

gray scale display device A monochrome display device that can display multiple shades of a single color in addition to the background color. (C) 610.10-1994w

gray scale manipulation An image enhancement technique in which the appearance of a digital image is improved by applying a point operator to each pixel in the image, adjusting its gray level. (C) 610.4-1990w

gray shade *See:* gray level.

grays in silicon A unit of absorbed dose as measured by its ionizing effect in silicon; 1 gray = 1 joule of energy deposited in a kilogram of irradiated silicon. *Notes:* 1. This number can be translated into the density of electron-hole pairs in silicon by the equation $n_{eh}\,4 \times 10^{-5} \times \propto$ grays. 2. One gray in silicon = 100 rd in silicon. However, the unit rads is not an SI unit and therefore is deprecated. 1 rd = 0.01 Gy. (ED) 641-1987w

gray tone *See:* gray level.

grazing angle The complement of the angle of incidence for large angles of incidence. *See also:* elevation angle. (AP/PROP) 211-1997

Green's function The response of a medium to an incident impulse function. (AP/PROP) 211-1997

Gregorian reflector antenna A paraboloidal reflector antenna with a concave subreflector, usually ellipsoidal in shape, located at a distance from the vertex of the main reflector that is greater than the prime focal length of the main reflector. *Note:* To improve the aperture efficiency of the antenna, the shapes of the main reflector and subreflector are sometimes modified. (AP/ANT) 145-1993

grid (1) In optical character recognition, two perpendicular sets of parallel lines used for specifying or measuring character images. (C) 610.2-1987

(2) (hydroelectric power plants) Network, usually of a power company, for transmitting and distributing electric power. (PE/EDPG) 1020-1988r

(3) (computer graphics) A two-dimensional array of points or lines used to determine a position in a graphics image space. For example, rectangular, radial.

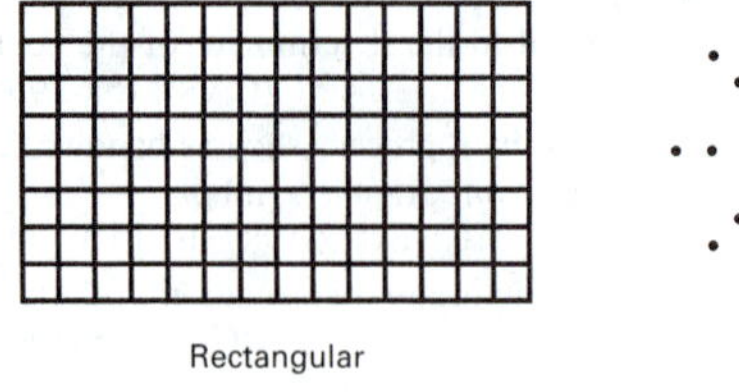

grid

(C) 610.6-1991w

grid constraint A process by which a point entered into a display image is automatically moved to the nearest grid point to achieve a neat appearance. *Synonyms:* modular constraint; constrained painting. (C) 610.6-1991w

grid control Control of anode current of an electron tube by means of proper variation (control) of the control-grid potential with respect to the cathode of the tube. *See also:* electronic controller. (IA/ICTL/IAC) [60]

grid-controlled mercury-arc rectifier A mercury-arc rectifier in which one or more electrodes are employed exclusively to control the starting of the discharge. *See also:* rectifier. (AP/ANT) 145-1983s

grid course (navigation) Course relative to grid north. *See also:* navigation. (AES/RS) 686-1982s, [42]

grid current (analog computer) The current flowing between the summing junction and the grid of the first amplifying stage of an operational amplifier. *Note:* Grid current results in an error voltage at the amplifier output. *See also:* electronic analog computer; electronic controller. (C) 165-1977w

grid-drive characteristic (electron tube) A relation, usually shown by a graph, between electric or light output and control-electrode voltage measured from cutoff. (ED) 161-1971w

grid driving power (electron tube) The average of the product of the instantaneous values of the alternating components of the grid current and the grid voltage over a complete cycle. *Note:* This power comprises the power supplied to the biasing device and to the grid. *See also:* electrode dissipation. (ED) 161-1971w

grid emission Electron or ion emission from a grid. *See also:* electron emission. (ED) 161-1971w

grid emission, primary *See:* primary grid emission.

grid emission, secondary *See:* secondary grid emission.

grid-glow tube A glow-discharge cold-cathode tube in which one or more control electrodes initiate but do not limit the anode current, except under certain operating conditions. *Note:* This term is used chiefly in the industrial field. (ED) [45]

grid, ground *See:* ground grid.

grid heading (navigation) Heading relative to grid north. *See also:* navigation. (AES/RS) 686-1982s, [42]

grid-leak detector A triode or multielectrode tube in which rectification occurs because of electron current to the grid. *Note:* The voltage associated with this flow through a high resistance in the grid circuit appears in amplified form in the plate circuit. (EEC/PE) [119]

grid mesh Any one of the open spaces enclosed by the grounding grid conductors. (PE/EDPG) 665-1995

grid modulation (electron tube) Modulation produced by the application of the modulating voltage to the control grid of any tube in which the carrier is present. *Note:* Modulation in which the grid voltage contains externally generated pulses is called grid pulse modulation. (AP/ANT) 145-1983s

grid neutralization (electron tube) The method of neutralizing an amplifier in which a portion of the grid-cathode alternating-current voltage is shifted 180 degrees and applied to the plate-cathode circuit through a neutralizing capacitor. *See also:* feedback; amplifier. (AP/ANT) 145-1983s

grid north (navigation) An arbitrary reference direction used in connection with a system of rectangular coordinates superimposed over a chart. *See also:* navigation. (AES/RS) 686-1982s, [42]

grid number *n* (electron tube) A grid occupying the *n* th position counting from the cathode. *See also:* electron tube. (ED) [45], [84]

grid pitch (electron tube) The pitch of the helix of a helical grid. *See also:* electron tube. (ED) [45], [84]

grid pulse modulation Modulation produced in an amplifier or oscillator by application of one or more pulses to a grid circuit. (AP/ANT) 145-1983s

grid (circuit) resistor A resistor used to limit grid current. *See also:* electronic controller. (IA/ICTL/IAC) [60]

grids (high-power rectifier) Electrodes that are placed in the arc stream and to which a control voltage may be applied. *See also:* rectification. (EEC/PE) [119]

grid system (substation grounding) A system consisting of interconnected bare conductors buried in the earth or in concrete to provide a common ground for electrical devices and metallic structures. (SUB/PE) 837-1989r

grid transformer Supplies an alternating voltage to a grid circuit or circuits. (IA/ICTL/IAC) [60]

grid voltage *See:* electrode voltage; electronic controller.

grid voltage supply (electron tube) The means for supplying to the grid of the tube a potential that is usually negative with respect to the cathode. *See also:* power pack. (EEC/PE) [119]

grip *See:* conductor grip.

grip, Chicago *See:* conductor grip.

grip, conductor *See:* conductor grip.

grip, vise *See:* strand restraining clamp.

grip, wire mesh *See:* woven wire grip.

grip, woven wire *See:* woven wire grip.

groove (mechanical recording) The track inscribed in the record by the cutting or embossing stylus. *See also:* phonograph pickup. (SP) [32]

groove angle (disk recording) The angle between the two walls of an unmodulated groove in a radial plane perpendicular to the surface of the recording medium. *See also:* phonograph pickup. (SP) [32]

groove diameter *See:* tape-wound core.

groove shape (disk recording) The contour of the groove in a radial plane perpendicular to the surface of the recording medium. *See also:* phonograph pickup. (SP) [32]

groove speed (disk recording) The linear speed of the groove with respect to the stylus. *See also:* phonograph pickup. (SP) [32]

groove width *See:* tape-wound core.

Grosch's law A guideline formulated by H. R. J. Grosch, stating that the computing power of a computer increases proportionally to the square of the cost of the computer. *See also:* computer performance evaluation. (C) 610.12-1990

gross actual generation (power system measurement) The energy that was generated by a unit in a given period. (PE/PSE) 762-1980s

gross available capacity (power system measurement) The gross dependable capacity, modified for equipment limitation at any time. (PE/PSE) 762-1980s

gross available generation (power system measurement) The gross energy that could have been generated in a given period if operated continuously at its gross available capacity. (PE/PSE) 762-1980s

gross demand load The summation of the demands for each of the several group loads. (IA/PSE) 241-1990r

gross demonstrated capacity The gross steady output that a generating unit or station has produced while demonstrating its maximum performance under stipulated conditions. *See also:* generating station. (PE/T&D) [10]

gross dependable capacity (power system measurement) The gross maximum capacity, modified for ambient limitations for a specified period of time, such as a month or a season. (PE/PSE) 762-1980s

gross generation (electric power system) The generated output power at the terminals of the generator. (PE/PSE) 858-1993w, 94-1991w

gross head (power operations) The difference of elevations between water surfaces of the forebay and tailrace under specified conditions. (PE/PSE) 858-1987s

gross heat rate (power operations) A measure of generating station thermal efficiency, generally expressed as British thermal unit per kilowatt-hour (Btu/kWh). *Note:* It is computed by dividing the total Btu content of the fuel burned (or of heat released from a nuclear reactor) by the resulting kilowatt-hours (kWh) generated. (PE/PSE) 858-1987s

gross information content A measure of the total information, redundant or otherwise, contained in a message. *Note:* It is expressed as the number of bits or hartleys required to transmit the message with specified accuracy over a noiseless medium without coding. *See also:* bit. (EEC/PE) [119]

gross maximum capacity (power system measurement) The maximum capacity that a unit can sustain over a specified period of time. To establish this capacity, formal demonstration is required. The test should be repeated periodically. This demonstrated capacity level shall be corrected to generating conditions for which there would be minimum ambient restriction. When a demonstration test has not been conducted, the estimated maximum capacity of the unit shall be used. (PE/PSE) 762-1980s

gross maximum generation (power system measurement) The energy that could have been produced by a unit in a given period of time if operated continuously at gross maximum capacity. (PE/PSE) 762-1980s

gross rated capacity The gross steady output that a generating unit or station can produce for at least two hours under specified operating conditions. *See also:* generating station. (T&D/PE) [10]

gross reserve generation (power system measurement) The energy that a unit could have produced in a given period but did not, because it was not required by the system. This is the difference between gross available generation and gross actual generation:

GRG = GAG + GAAG

(PE/PSE) 762-1980s

gross seasonal unavailable generation (power system measurement) The difference between the energy that would have been generated if operating continuously at gross maximum capacity and the energy that would have been generated if operating continuously at gross dependable capacity, calculated only during the time the unit was in the available state. (PE/PSE) 762-1980s

gross unit unavailable generation (power system measurement) The difference between the energy that would have been generated if operating continuously at gross dependable capacity and the energy that would have been generated if operating continuously at available capacity. This is the energy that could not be generated by a unit due to planned and unplanned outages and unit deratings. (PE/PSE) 762-1980s

ground (1) (A) (transmission path) A direct conducting connection to the earth or body of water that is a part thereof. **(B) (transmission path)** A conducting connection to a structure that serves a function similar to that of an earth ground (that is, a structure such as a frame of an air, space, or land vehicle that is not conductively connected to earth). (GSD) 315-1975

(2) (hydroelectric power plants) Connection to earth or to a common conducting body that serves in place of the earth. (PE/EDPG) 1020-1988r

(3) (A) A conducting connection, whether intentional or accidental, by which an electric circuit or equipment is connected to the earth, or to some conducting body of relatively large extent that serves in place of the earth. **(B)** High-frequency reference. *Note:* Grounds are used for establishing and maintaining the potential of the earth (or of the conducting body), or approximately that potential, on conductors connected to it and for conducting ground currents to and from earth (or the conducting body). *See also:* signal reference structure. (IA/PSE) 1100-1999
(4) A conducting connection, whether intentional or accidental, by which an electric circuit or equipment is connected to the earth or to some conducting body of relatively large extent that serves in place of the earth. (PE/SUB) 80-2000

groundable parts Those parts that may be connected to ground without affecting operation of the device. (SWG/PE) C37.40-1993, C37.100-1992

ground absorption (data transmission) The loss of energy in transmission of radio waves, due to dissipation in the ground. (PE) 599-1985w

ground acceleration (1) The acceleration of the ground resulting from a given earthquake's motion. The maximum ground acceleration can be obtained from the ground response spectrum as the acceleration at high frequencies (in excess of 33 Hz). (SWG/PE) C37.100-1992, C37.81-1989r
(2) The acceleration of the ground resulting from the motion of a given earthquake. The maximum or peak ground acceleration is the zero period acceleration (ZPA) of the ground response spectrum. (PE/SUB/NP) 693-1997, 344-1987r

ground and test device A term applied to a switchgear assembly accessory device that can be inserted in place of a drawout circuit breaker for the purpose of grounding the main bus and/or external circuits connected to the switchgear assembly and/or primary circuit testing. (SWG/PE) C37.20.6-1997, C37.100-1992, C37.20.2-1993

ground-area open floodlight (illuminating engineering) A unit providing a weatherproof enclosure for the lamp socket and housing. No cover glass is required. (EEC/IE) [126]

ground-area open floodlight with reflector insert (illuminating engineering) A weatherproof unit so constructed that the housing forms only part of the reflecting surface. An auxiliary reflector is used to modify the distribution of light. No cover glass is required. (EEC/IE) [126]

ground bar (lightning) A conductor forming a common junction for a number of ground conductors. (PE) [8], [84]

ground-based navigation aid An aid that requires facilities located upon land or sea. *See also:* navigation. (AES/RS) 686-1982s, [42]

ground, block *See:* traveler ground.

ground bus A bus to which the grounds from individual pieces of equipment are connected, and that, in turn, is connected to ground at one or more points. (SWG/PE) C37.100-1992

ground bushing (separable insulated connectors) An accessory device designed to electrically ground and mechanically seal a de-energized power cable terminated with an elbow. (T&D/PE) 386-1995

ground, butt *See:* structure base ground.

ground cable bond A cable bond used for grounding the armor or sheaths of cables or both. *See also:* ground. (T&D/PE) [10]

ground chain *See:* structure base ground.

ground clamp A clamp used in connecting a grounding conductor to a grounding electrode or to a thing grounded. *Synonym:* grounding clamp. *See also:* ground. (T&D/PE) [10]

ground clutter Clutter resulting from the ground or objects on the ground. *Synonym:* ground return. (AES) 686-1997

ground conductivity A property of the ground, expressed as the ratio of electric current density to electric field strength. (T&D/PE) 1260-1996

ground conductor (lightning) A conductor providing an electric connection between part of a system, or the frame of a machine or piece of apparatus, and a ground electrode or a ground bar. *See also:* grounded conductor. (PE) [8], [84]

ground conduit A conduit used solely to contain one or more grounding conductors. *See also:* ground. (T&D/PE) [10]

ground connection *See:* grounding connection.

ground contact (of a switchgear assembly) A self-coupling separable contact provided to connect and disconnect the ground connection between the removable element and the ground bus of the housing and so constructed that it remains in contact at all times except when the primary disconnecting devices are separated by a safe distance. *Note:* Safe distance, as used here, is a distance at which the equipment will meet its withstand-voltage ratings, both low-frequency and impulse, between line and load terminals with the switching device in the closed position. (SWG/PE) C37.100-1992

ground contact indicator *See:* line isolation monitor.

ground-controlled approach (GCA) A ground radar system providing information by which aircraft approaches to landing may be directed via radio communications; the system consists of a precision-approach radar (PAR) and an airport-surveillance radar (ASR). (AES/RS) 686-1990

ground-controlled approach radar A ground radar system providing information by which aircraft approaches to landing may be directed via radio communications. The system consists of a precision-approach radar (PAR) and an airport-surveillance radar (ASR). (AES) 686-1997

ground-controlled intercept (GCI) A radar system by means of which a controller on the ground may direct an aircraft to make an interception of another aircraft. (AES/RS) 686-1990

ground-controlled intercept radar A military radar system by which a controller on the ground may direct an aircraft to make an interception of another aircraft. (AES) 686-1997

ground current (1) (ground systems) Current flowing in the earth or in a grounding connection. (PE/PSIM) 81-1983
(2) A current flowing into or out of the earth or its equivalent serving as a ground. (PE/SUB) 80-2000

ground-derived navigation data (air navigation) Data obtained from measurements made on land or sea at locations external to the vehicle. *See also:* navigation. (AES/RS) 686-1982s, [42]

ground detection rings (rotating machinery) Collector rings connected to a winding and its core to facilitate the measurement of insulation resistance on a rotor winding. *See also:* rotor. (PE) [9]

ground detector An instrument or an equipment used for indicating the presence of a ground on an ungrounded system. *See also:* ground. (T&D/PE) [10]

ground detector relay (power system device function numbers) A relay that operates on failure of machine or other apparatus insulation to ground. *Note:* This function is not applied to a device connected in the secondary circuit of current transformers in a normally grounded power system, where other device numbers with a suffix G or N should be used, that is, 51N for an ac time overcurrent relay connected in the secondary neutral of the current transformers. (SUB/PE) C37.2-1979s

ground distance relay A distance relay designed to detect phase-to-ground faults. (PE/PSR) C37.113-1999

grounded (1) (ground systems) A system, circuit, or apparatus referred to is provided with a ground. (PE/PSIM) 81-1983
(2) (conductor stringing equipment) (power line maintenance) (electric systems) Connected to earth or to some extended conducting body that serves instead of the earth, whether the connection is intentional or accidental. (SPD/PE/T&D/TR) 32-1972r, C2.2-1960, 524a-1993r, 516-1995, C57.12.80-1978r, 524-1992r
(3) (safety in ac substation grounding) A system, circuit, or apparatus referred to is provided with ground for the purposes of establishing a ground return circuit and for maintaining its potential at approximately the potential of earth.

(4) (effectively grounded communication system) Permanently connected to earth through a ground connection of sufficiently low impedance and having sufficient ampacity to prevent the building up of voltages that may result in undue hazard to connected equipment or to persons. [86]
(5) Connected to or in contact with earth or connected to some extended conductive body that serves instead of the earth. (NESC) C2-1997
(6) A system, circuit, or apparatus provided with a ground(s) for the purposes of establishing a ground return circuit and for maintaining its potential at approximately the potential of earth. (PE/SUB) 80-2000

grounded capacitance *See:* ground.

grounded-cathode amplifier An electron-tube amplifier with the cathode at ground potential at the operating frequency, with input applied between the control grid and ground, and the output load connected between plate and ground. *Note:* This is the conventional amplifier circuit. *See also:* amplifier. (AP/ANT) 145-1983s

grounded circuit A circuit in which one conductor or point (usually the neutral conductor or neutral point of transformer or generator windings) is intentionally grounded, either solidly or through a noninterrupting current limiting grounding device. *See also:* grounded system; grounded conductor; ground. (SPD/PE/T&D) 32-1972r, [10]

grounded concentric wiring system A grounded system in which the external (outer) conductor is solidly grounded and completely surrounds the internal (inner) conductor through its length. The external conductor is usually uninsulated. *See also:* ground. (SPD/PE/T&D) 32-1972r, [10]

grounded conductor (1) (electric systems) A conductor that is intentionally grounded, either solidly or through a current limiting device. *See also:* ground. (SPD/PE/T&D) 32-1972r, C2.2-1960
(2) A system or circuit conductor that is intentionally grounded. (NESC/NEC) [86]
(3) A conductor that is intentionally grounded, either solidly or through a noninterrupting current-limiting device. (NESC) C2-1997

grounded, directly *See:* grounded solidly.

grounded, effectively *See:* effectively grounded.

grounded effectively *See:* effectively grounded.

grounded-grid amplifier An electron-tube amplifier circuit in which the control grid is at ground potential at the operating frequency, with input applied between cathode and ground, and output load connected between plate and ground. *Note:* The grid-to-plate impedance of the tube is in parallel with the load instead of acting as a feedback path. *See also:* amplifier. (AP/ANT) 145-1983s

grounded, impedance *See:* impedance grounded.

grounded impedance Grounded through impedance. *Note:* The components of the impedance need not be at the same location. (SPD/PE) 32-1972r

grounded member Any part in a substation that is normally connected to or in contact with earth. (SUB/PE) 1264-1993

grounded neutral system (surge arresters) A system in which the neutral is connected to ground, either solidly or through a resistance or reactance of low value. (PE) [8], [84]

grounded-neutral terminal type voltage transformer (1) One that has the neutral end of the high-voltage winding connected to the case or mounting base. (PE/TR) [57]
(2) One that has the neutral end of the primary winding connected to the case or mounting base in a manner not intended to facilitate disconnection. (PE/TR) C57.13-1993

grounded parts Parts that are intentionally connected to ground. (SWG/PE) C37.40-1993, C37.100-1992

grounded-plate amplifier (cathode-follower) An electron-tube amplifier circuit in which the plate is at ground potential at the operating frequency, with input applied between control grid and ground, and the output load connected between cathode and ground. *See also:* amplifier. (AP/ANT) 145-1983s

grounded potentiometer (analog computer) A potentiometer with one end terminal attached directly to ground. *See also:* electronic analog computer. (C) 165-1977w, 166-1977w

grounded solidly (system grounding) Connected directly through an adequate ground connection in which no impedance has been intentionally inserted. *Note:* This term, though commonly used, is somewhat confusing since a transformer may have its neutral solidly connected to ground, and yet the connection may be so small in capacity as to furnish only a very-high-impedance ground to the system to which it is connected. In order to define grounding positively and logically as to degree, the term effective grounding has come into use. The term solidly grounded will therefore be used in this standard only in referring to a solid metallic connection from system neutral to ground; that is, with no impedance intentionally added in the grounding circuit. (IA/PSE) 142-1982s

grounded system (1) A system of conductors in which at least one conductor or point is intentionally grounded, either solidly or through a noninterrupting current-limiting device. (NESC/T&D) C2-1997, C2.2-1960
(2) (system grounding) A system of conductors in which at least one conductor or point (usually the middle wire or neutral point of transformer or generator windings) is intentionally grounded, either solidly or through an impedance. *Note:* Various degrees of groundings are used, from solid or effective grounding to the high-impedance grounding obtained from a small grounding transformer used only to secure enough ground current for relaying, to the high-resistance grounding which secures control of transient overvoltage but may not furnish sufficient current for ground-fault relaying. In the figure on the next page, parts b and c show two points at which a system may be grounded and the corresponding voltage relationships. Note that according to NEMA SG 4-1975, there are system voltage limitations for corner grounding.
(3) (power and distribution transformers) A system of conductors in which at least one conductor or point (usually the middle wire or neutral point of transformer or generator windings) is intentionally grounded, either solidly or through a current-limiting device. (PE/TR) C57.12.80-1978r
(4) (surge arresters) An electric system in which at least one conductor or point (usually the neutral conductor or neutral point of transformer or generator windings) is intentionally grounded, either solidly or through a grounding device. (PE/SPD) C62.1-1981s, C62.62-2000

ground electrode (1) A conductor or group of conductors in intimate contact with the earth for the purpose of providing a connection with the ground. (IA/PSE) 1100-1999
(2) A conductor imbedded in the earth and used for collecting ground current from or dissipating ground current into the earth. (PE/SUB) 80-2000
(3) *See also:* ground rod. (T&D/PE) 524-1992r

ground electrode, concrete-encased *See:* concrete-encased ground electrode.

ground end (grounding device) The end or terminal of the device that is grounded directly or through another device. *See also:* grounding device. (PE/SPD) 32-1972r

ground equalizer inductors Coils of relatively low inductance, placed in the circuit connected to one or more of the grounding points of an antenna to distribute the current to the various points in any desired manner. *Note:* Broadcast usage only and now in disuse. *See also:* antenna. (AP/ANT) 145-1983s

ground fault (surge arresters) An insulation fault between a conductor and ground or frame. (PE) [8], [84]

ground-fault circuit-interrupter (1) (health care facilities) A device whose function is to interrupt the electric circuit to the load when a fault current to ground exceeds some predetermined value that is less than that required to operate the overcurrent protective device of the supply circuit. (EMB) [47]

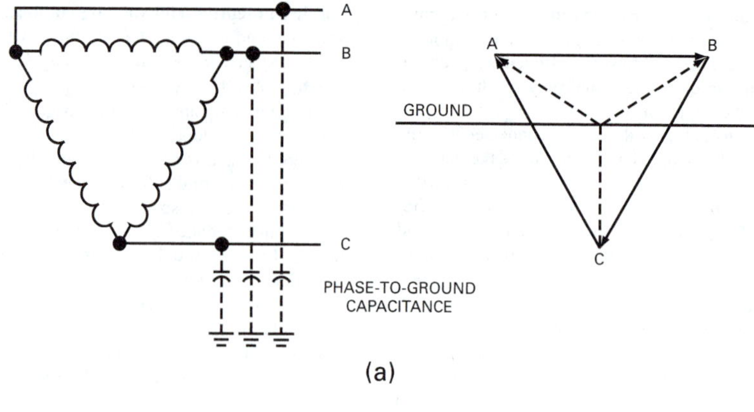

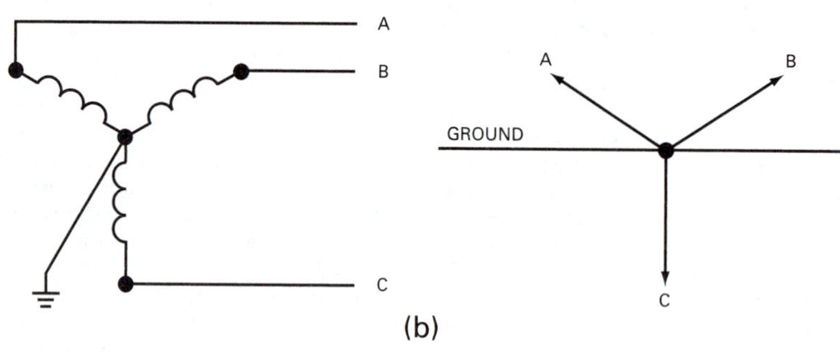

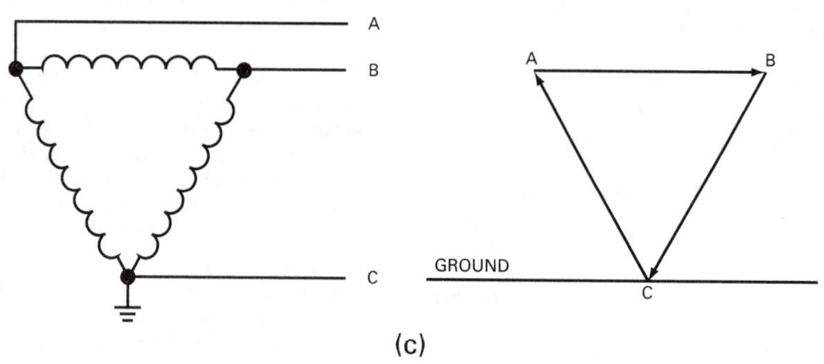

Voltages to ground under steady-state conditions: a) ungrounded system; b) grounded Wye-connected system; c) corner grounded Delta-connected system

grounded system

(2) A device intended for the protection of personnel that functions to interrupt the electric current to the load within an established period of time when a fault current to ground exceeds some predetermined value that is less than that required to operate the overcurrent protective device of the supply circuit. (NESC/NEC) [86]

ground fault factor The ratio of the highest power frequency voltage on an unfaulted phase during a line-to-ground fault to the phase-to-ground power-frequency voltage without the fault. *Notes:* 1. The ground-fault factor generally will be less than 1.3, if the zero-sequence reactance is less than three times the positive-sequence reactance, and the zero-sequence resistance does not exceed the positive-sequence reactance. 2. IEEE Std C62.1-1989 defines a "coefficient of grounding." This coefficient can be obtained by dividing the ground-fault factor by $\sqrt{3}$. (PE/C) 1313.1-1996

ground-fault neutralizer (neutral grounding devices) A grounding device that provides an inductive component of current in a ground fault that is substantially equal to and therefore neutralizes the rated-frequency capacitive component of the ground-fault current, thus rendering the system resonant grounded. (SPD/PE) 32-1972r

ground-fault neutralizer grounded (power and distribution transformers) (resonant grounded) Reactance grounded through such values of reactance that, during a fault between one of the conductors and earth, the rated-frequency current flowing in the grounding reactances and the rated-frequency capacitance current flowing between the unfaulted conductors and earth shall be substantially equal. *Notes:* 1. In the fault these two components of current will be substantially 180 degrees out of phase. 2. When a system is ground-fault neutralizer grounded, it is expected that the quadrature component of the rated-frequency single-phase-to-ground fault current will be so small that an arc fault in air will be self-extinguishing. (PE/TR) C57.12.80-1978r

ground-fault protection of equipment A system intended to provide protection of equipment from damaging line-to-ground arcing fault currents by operating to cause a disconnecting means to open all ungrounded conductors of the faulted circuit. This protection is provided at current levels less than that required to protect conductors from damage through the operation of a supply circuit overcurrent device. (NESC/NEC) [86]

ground flash density (GFD) (Ng) (1) The average number of lightning strokes per unit area per unit time at a particular location. (SUB/PE/T&D) 998-1996, 1410-1997
(2) The average number of lightning strokes to ground per unit area per unit time at a particular location. (PE/T&D) 1243-1997

ground gradient mat *See:* ground grid.

ground grid (1) (ground resistance) (ground systems) A system of grounding electrodes consisting of interconnected bare cables buried in the earth to provide a common ground for electrical devices and metallic structures. *Note:* It may be connected to auxiliary grounding electrodes to lower its resistance. *See also:* grounding device. (PE/PSIM) 81-1983
(2) (conductor string equipment) (temporary) A system of interconnected bare conductors arranged in a pattern over a specified area and on or buried below the surface of the earth. Normally, it is bonded to ground rods driven around and within its perimeter to increase its grounding capabilities and provide convenient connection points for grounding devices. The primary purpose of the grid is to provide safety for workers by limiting potential differences within its perimeter to safe levels in case of high currents that could flow if the circuit being worked became energized for any reason, or if an adjacent energized circuit faulted. Metallic surface mats and gratings are sometimes utilized for this same purpose. When used, these grids are employed at pull, tension, and midspan splice sites. *Synonyms:* ground mat; counterpoise; ground gradient mat. (T&D/PE/IA/PSE) 524a-1993r, 524-1992r, 1048-1990, 1100-1999
(3) A system of horizontal ground electrodes that consists of a number of interconnected bare conductors buried in the earth, providing a common ground for electrical devices or metallic structures, usually in one specific location. *Note:* Grids buried horizontally near the surface of the earth are also effective in controlling the surface potential gradients. A typical grid usually is supplemented by a number of ground rods and may be further connected to auxiliary ground electrodes to lower its resistance with respect to remote earth. (PE/EDPG) 665-1995

ground, high-frequency reference *See:* signal reference structure.

ground impedance tester A multifunctional instrument designed to detect certain types of wiring and grounding problems in low-voltage power distribution systems. (IA/PSE) 1100-1999

ground indication An indication of the presence of a ground on one or more of the normally ungrounded conductors of a system. *See also:* ground. (T&D/PE) [10]

grounding cable A cable used to make a connection to ground. *See also:* grounding conductor. (PE) [9]

grounding-cable connector The terminal mounted on the end of a grounding cable. (PE) [9]

grounding clamp A device used in making a connection between the electrical apparatus or conductors, and the ground bus, or grounding electrode. (T&D/PE) 1048-1990

grounding, coefficient of *See:* coefficient of grounding.

grounding conductor (1) The conductor that is used to establish a ground and that connects an equipment, device, wiring system, or another conductor (usually the neutral conductor) with the grounding electrode or electrodes. *Synonym:* ground system. (NESC/PE/PSIM) 81-1983, [86]
(2) A metallic conductor used to connect the metal frame or enclosure of an equipment, device, or wiring system with a mine track or other effective grounding medium. *See also:* mine feeder circuit. (PE/EEC/MIN) [119]
(3) A conductor used to connect equipment or the grounded circuit of a wiring system to a grounding electrode or electrodes. (NESC/NEC) [86]
(4) A conductor used to connect equipment or the grounded circuit of a wiring system to a grounding electrode or electrodes (i.e., ground grid). (PE/EDPG) 665-1995
(5) A conductor that is used to connect the equipment or the wiring system with a grounding electrode or electrodes. (T&D/NESC) C2.2-1960, C2-1997

grounding conductor, direct current equipment *See:* direct-current equipment grounding conductor.

grounding connection (ground systems) A connection used in establishing a ground and consists of a grounding conductor, a grounding electrode and the earth (soil) that surrounds the electrode or some conductive body which serves instead of the earth. (PE/PSIM) 81-1983

grounding device (electric power) An impedance device used to connect conductors of an electric system to ground for the purpose of controlling the ground current or voltages to ground or a nonimpedance device used to temporarily ground conductors for the purpose of the safety of workmen. *Note:* The grounding device may consist of a grounding transformer or a neutral grounding device, or a combination of these. Protective devices, such as surge arresters, may also be included as an integral part of the device. (SPD/PE) 32-1972r

grounding elbow An accessory device designed to electrically ground and mechanically seal a bushing insert or an integral bushing. (T&D/PE) 386-1995

grounding electrode A conductor used to establish a ground. *Synonyms:* ground system; ground electrode. (T&D/PE/PSIM) [10], 81-1983

grounding electrode conductor The conductor used to connect the grounding electrode to the equipment grounding conductor and/or to the grounded conductor of the circuit at the service equipment or at the source of a separately derived system. (NESC/NEC) [86]

grounding grid A system of horizontal ground electrodes that consists of a number of interconnected, bare conductors buried in the earth, providing a com-mon ground for electrical devices or metallic structures, usually in one specific location. *Note:* Grids buried horizontally near the earth's surface are also effective in controlling the surface potential gradients. A typical grid usually is supplemented by a number of ground rods and may be further connected to auxiliary ground electrodes to lower its resistance with respect to remote earth. (PE/SUB) 80-2000

grounding jumper (electric appliances) A strap or wire to connect the frame of the range to the neutral conductor of the supply circuit. *See also:* appliance outlet. (IA/APP) [90]

grounding outlet An outlet equipped with a receptacle of the polarity type having, in addition to the current-carrying contacts, one grounded contact that can be used for the connection of an equipment grounding conductor. *Note:* This type of outlet is used for connection of portable appliances. *Synonym:* safety outlet. *See also:* ground. (T&D/PE) [10]

grounding pad (rotating machinery) A contact area, usually on the stator frame, provided to permit the connection of a grounding terminal. *See also:* stator. (PE) [9]

grounding relays *See:* short-circuiting relays.

grounding switch A mechanical switching device by means of which a circuit or piece of apparatus may be electrically connected to ground. (SWG/PE) C37.100-1992, C37.30-1971s

grounding system (1) (health care facilities) A system of conductors which provides a low impedance return path for leakage and fault currents. It coordinates with, but may be locally more extensive than, the grounding system described in Article 250 of NFPA 70, National Electrical Code. (EMB) [47]
(2) (surge arresters) A complete installation comprising one or more ground electrodes, ground conductors, and ground bars as required. (PE) [8], [84]
(3) Comprises all interconnected grounding facilities in a specific area. (PE/SUB/T&D/PSIM) 80-2000, 524a-1993r, 81-1983

grounding terminal (rotating machinery) A terminal used to make a connection to a ground. *See also:* stator. (PE) [9]

grounding transformer (1) (power and distribution transformers) A transformer intended primarily to provide a neutral point for grounding purposes. *Note:* It may be provided with a d winding in which resistors or reactors are connected. *See also:* stabilizing winding; rated kilovolt-ampere; voltage rating. (PE/TR) C57.12.80-1978r
(2) A transformer(s), delta-wye or zig-zag connected, installed to establish a system ground and thus provide a source of zero-sequence current for ground fault detection. (PE/PSR) C37.113-1999

ground insulation (rotating machinery) Insulation used to insure the electric isolation of a winding from the core and mechanical parts of a machine. *See also:* asynchronous machine; coil insulation. (PE) [9]

ground isolation (sequential events recording systems) The disconnection of selected field contact circuits from the contact voltage supply to allow identification of the grounded field contact wires. *See also:* field contacts. (PE/EDPG) [1]

ground level (mobile communication) The elevation of the ground above mean sea level at the antenna site or other point of interest. *See also:* mobile communication system. (VT) [37]

ground light (illuminating engineering) Visible radiation from the sun and sky reflected by surfaces below the plane of the horizon. (EEC/IE) [126]

ground loop (1) A circuit in an analog computer when two or more points in the electrical system, that are nominally at ground potential, are connected by a conducting path such that either or both points are not at the same ground potential. (C) 610.10-1994w
(2) A potentially detrimental loop formed when two or more points in an electrical system that are nominally at ground potential are connected by a conducting path such that either or both points are not at the same ground potential. (IA/PSE) 1100-1999

ground, master *See:* master ground.

ground mat (1) (ground systems) A system of bare connectors, on or below the surface of the earth, connected to a ground or a ground grid to provide protection from dangerous touch voltages. *Note:* Plates and gratings of suitable area are common forms of ground mats. (PE/PSIM) 81-1983
(2) A solid metallic plate or a system of closely spaced bare conductors that are connected to and often placed in shallow depths above a ground grid or elsewhere at the earth's surface, in order to obtain an extra protective measure minimizing the danger of the exposure to high step or touch voltages in a critical operating area or places that are frequently used by people. Grounded metal gratings, placed on or above the soil surface, or wire mesh placed directly under the surface material, are common forms of a ground mat. (PE/SUB) 80-2000
(3) *See also:* ground grid. (T&D/PE) 524-1992r

ground, moving *See:* running ground.

ground overcurrent (1) A conducting or reflecting plane functioning to image a radiating structure. *Synonyms:* imaging plane; imaging plane. (AP/ANT) 145-1993, 145-1983s
(2) (adio-noise emission) A conducting surface or plate used as a common reference point for circuit returns and electric or signal potentials. (EMC) C63.4-1981, [53]
(3) An assumed plane of true ground or zero potential. *See also:* direct-stroke protection. (T&D/PE) [10]
(4) The net (phasor sum) current flowing in the phase and neutral conductors or the total current flowing in the normal neutral-to-ground connection that exceeds a predetermined value. (SWG/PE) C37.100-1992

ground, personal *See:* personal ground.

ground plane (1) A conducting surface or plate used as a common reference point for circuit returns and electric or signal potentials. (EMC/MTT) C63.5-1988, 1004-1987w, C63.4-1988s
(2) A conducting or reflecting plane functioning to image a radiating structure. *Synonyms:* imaging plane; imaging plane. (AP/ANT) 145-1983s, 145-1993

ground plane, effective *See:* effective ground plane.

ground plane field The electromagnetic field in near proximity to a conducting surface, with the boundary conditions that the tangential electric field approach zero and the normal magnetic remain continuous. The total normal electric field is related to the surface charge density by Gauss' law and the total tangential magnetic field to the surface current density by Ampere's law. (EMC) 1309-1996

ground plate (grounding plate) A plate of conducting material buried in the earth to serve as a grounding electrode. *See also:* ground. (T&D/PE) [10]

ground-position indicator (electronic navigation) A dead-reckoning tracer or computer similar to an air position indicator (API) with provision for taking account of drift. *See also:* radio navigation; navigation. (AES/RS) 686-1982s, [42]

ground potential difference voltage The voltage that results from current flow through the finite resistance and inductance between the receiver and driver circuit ground voltages. (C/MM) 1596.3-1996

ground potential rise (GPR) (1) The voltage that a station grounding grid may attain relative to a distant grounding point assumed to be at the potential of remote earth. (SPD/PE) C62.23-1995
(2) The product of a ground electrode impedance, referenced to remote earth, and the current that flows through that electrode impedance. (PE/PSC) 367-1996
(3) The difference in ground potential between a location in proximity to a point of large current injection into the ground and any remote ground point. GPR is usually caused by a short circuit of an energized power conductor to ground and is the result of the injected current flowing through the impedance of the ground circuit. (SWG/PE) C37.100-1992
(4) The maximum electrical potential that a substation grounding grid may attain relative to a distant grounding point assumed to be at the potential of remote earth. (PE/SUB) 1268-1997
(5) The maximum electrical potential that a substation grounding grid may attain relative to a distant grounding point assumed to be at the potential of remote earth. This voltage, GPR, is equal to the maximum grid current times the grid resistance. *Note:* Under normal conditions, the grounded electrical equipment operates at near zero ground potential. That is, the potential of a grounded neutral conductor is nearly identical to the potential of remote earth. During a ground fault the portion of fault current that is conducted by a substation grounding grid into the earth causes the rise of the grid potential with respect to remote earth. (PE/SUB) 80-2000

ground potential shift The difference in voltage between grounding or grounded (earthed) structures such as the opposite corners of a metal building. Generally, ground potential shift increases with distance of separation of ground locations and with the frequency or wave front rise time of the resulting current flow. Ground potential shift problems are generally exacerbated by surge events from lighting and utility power sources. (IA/PSE) 1100-1999

ground protection (1) (ground-fault protection) A method of protection in which faults to ground within the protected equipment are detected irrespective of system phase conditions. (SWG/PE/PSR) C37.90-1978s, [6], [56]
(2) A method of protection in which faults to ground within the protected equipment are detected. (SWG/PE) C37.100-1992

ground, radial *See:* radial ground.

ground range Distance along the ground between the points directly beneath the radar and the target. (AES) 686-1997

ground-referenced navigation data Data in terms of a coordinate system referenced to the earth or to some specified portion thereof. *See also:* navigation. (AES/RS) 686-1982s, [42]

ground reference meter A meter that measures the electric field at or close to the surface of the ground. Frequently implemented by measuring induced current or charge oscillating between an isolated electrode and ground. The isolated electrode is usually a plate located level with or slightly above the ground surface. *Note:* Ground reference meters measuring the induced current often contain an integrator circuit to compensate for the derivative relationship between the induced current and the electric field. (PE/T&D) 539-1990, 1308-1994

ground reference plane (GRP) A flat conductive surface whose potential is used as a common reference. Where applicable, the operating voltage of the EUT and the operator ground should also be referenced to the ground plane. (EMC) C63.16-1993

ground-reflected wave (data transmission) The component of the ground wave that is reflected from the ground. (PE) 599-1985w

ground relay A relay that by its design or application is intended to respond primarily to system ground faults. (SWG/PE) C37.100-1992

ground resistance (grounding electrode) The ohmic resistance between the grounding electrode and a remote grounding electrode of zero resistance. *Note:* By "remote" is meant "at a distance such that the mutual resistance of the two electrodes is essentially zero." (PE/PSIM) 81-1983

ground return *See:* ground clutter.

ground-return circuit (1) (ground systems) A circuit in which the earth is utilized to complete the circuit. (PE/PSIM) 81-1983
(2) (safety in ac substation grounding) A circuit in which the earth or an equivalent conducting body is utilized to complete the circuit and allow current circulation from or to its current source. (T&D/PE/SUB) 563-1978r, 80-2000
(3) (data transmission) A circuit which has a conductor (or two or more in parallel) between two points and which is completed through the ground or earth. (PE) 599-1985w
(4) A circuit in which the earth is utilized to complete the circuit. *See also:* transmission line; ground; telegraphy. (T&D/PE) [10]

ground-return current (line residual current) (electric supply line) The vector sum of the currents in all conductors on the electric supply line. *Note:* Actually the ground-return current in this sense may include components returning to the source in wires on other pole lines, but from the inductive coordination standpoint these components are substantially equivalent to components in the ground. *See also:* inductive coordination. (EEC/PE) [119]

ground-return system A system in which one of the conductors is replaced by ground. (PE) [8], [84]

ground rod (1) (protective grounding of power lines) (conductor stringing equipment) A rod that is driven into the ground to serve as a ground terminal, such as a copper-clad rod, solid copper rod, galvanized iron rod, or galvanized iron pipe. Copper-clad steel rods are commonly used during conductor stringing operations to provide a means of obtaining an electrical ground using portable grounding devices. *Synonym:* ground electrode. (T&D/PE) 524a-1993r, 1048-1990, 524-1992r
(2) A conducting rod serving as an electrical connection with the ground. (AP/ANT) 145-1993

ground roller *See:* running ground.

ground, rolling *See:* running ground; traveler ground.

ground, running *See:* running ground.

ground set *See:* master ground.

ground, sheave *See:* traveler ground.

ground source *See:* ground.

ground speed (navigation) The speed of a vehicle along its track. *See also:* navigation. (AES/RS) 686-1982s, [42]

ground-start signaling (telephone switching systems) A method of signaling using direct current in a ground return path to indicate a service request. (COM) 312-1977w

ground-state maser (laser maser) A maser in which the terminal level of the amplifying transition is appreciably populated at thermal equilibrium for the ambient temperature. (LEO) 586-1980w

ground stick *See:* master ground; personal ground.

ground stick, insulated *See:* insulated ground stick.

ground, structure *See:* structure base ground.

ground, structure base *See:* structure base ground.

ground support equipment (test, measurement, and diagnostic equipment) All equipment (implements, tools, test equipment devices--mobile or fixed--and so forth) required on the ground to make an aerospace system (aircraft, missile, and so forth) operational in its intended environment. (MIL) [2]

ground surveillance radar A radar set operated at a fixed point for observation and control of the position of aircraft or other vehicles in the vicinity. *See also:* navigation. (AES) [42]

ground system (1) That portion of an antenna consisting of a system of conductors or a conducting surface in or on the ground. (AP/ANT) 145-1993
(2) (ground systems) (general) Consists of all interconnected grounding connections in a specific area. *Synonym:* grounding system. (PE/T&D/PSIM) 81-1983, 524a-1993r
(3) That portion of an antenna closely associated with and including and extensive conducting surface, which may be the earth itself. *See also:* antenna. (AP/ANT) [35]

ground terminal (1) (lightning protection system) The portion extending into the ground, such as a ground rod, ground plate, or the conductor itself, serving to bring the lightning protection system into electric contact with the ground. (PE) [8], [84]
(2) The conducting part provided for connecting the arrester to ground. (SPD/PE) C62.11-1999

ground, tower *See:* structure base ground.

ground transformer *See:* grounding transformer.

ground, traveler *See:* traveler ground.

ground, traveling *See:* running ground.

groundwall insulation The main high voltage electrical insulation that separates the copper conductors from the grounded stator core in motor and generator stator windings. (DEI) 1043-1996

ground wave (1) (data transmission) A radio wave that is propagated over the earth and is ordinarily affected by the presence of the ground and troposphere. *Notes:* 1. The ground wave includes all components of a radio wave over the earth except ionospheric and tropospheric waves. 2. The ground wave is refracted because of variations in the dielectric constant of the troposphere including the condition known as a surface duct. *See also:* radiation; radio-wave propagation. (AP/PE/ANT) 149-1979r, 599-1985w
(2) From a source in the vicinity of the surface of the Earth, a wave that would exist in the vicinity of the surface in the absence of an ionosphere. *Note:* The ground wave can be decomposed into the Norton surface wave and a space wave consisting of the vector sum of a direct wave and a ground-reflected wave. (AP/PROP) 211-1997

ground well A hole with a diameter greater than an inserted ground rod, drilled to a specified depth, and backfilled with a highly conductive material. The backfill will be in intimate contact with the earth. (PE/EDPG) 665-1995

ground window The area through which all grounding conductors, including metallic raceways, enter a specific area. It is often used in communications systems through which the building grounding system is connected to an area that would otherwise have no grounding connection. (IA/PSE) 1100-1999

ground wire (1) (data transmission) (telecommunications) A conductor leading to an electric connection with the ground. (PE) 599-1985w

(2) (overhead power lines) A conductor having grounding connections at intervals, that is suspended usually above but not necessarily over the line conductor to provide a degree of protection against lightning discharges. *See also:* ground. (T&D/PE) [10]

ground, working *See:* personal ground.

group (1) (storage cell) An assembly of plates of the same polarity burned to a connecting strap. *See also:* battery. (PE/EEC) [119]

(2) (electric and electronics parts and equipment) A collection of units, assemblies, or subassemblies which is a subdivision of a set or system, but which is not capable of performing a complete operational function. Typical examples: antenna group, indicator group. (GSD) 200-1975w

(3) (data management) A set of items that are related to each other in some way; for example, a set of records that have the same value for a particular field, or a set of files in a generation data group. (C) 610.5-1990w

(4) A repeater port or a collection of repeater ports that can be related to the logical arrangement of ports within a repeater. (C/LM) 802.3-1998

(5) A Group associates

a) A group MAC address; and
b) A set of properties that define membership characteristics; and
c) A set of properties that define the forwarding/filtering behavior of a Bridge with respect to frames destined for members of that group MAC address;

with a set of end stations that all wish to receive information destined for that group MAC address. Members of such a set of end stations are said to be *Group members*. A Group is said to *exist* if the properties associated with that Group are visible in an entry in the Filtering Database of a Bridge, or in the GARP state machines that characterize the state of the Group; a Group is said to *have members* if the properties of the Group indicate that members of the Group can be reached through specific Ports of the Bridge. *Note:* An example of the information that Group members might wish to receive is a multicast video data stream. (C/LM) 802.1D-1998

(6) *See also:* channel group; Remote Bridge Group. (COM/C/LM) 802.1G-1996

group address (GP) (1) A predefined destination address that denotes a set of selected service access points (SAPs) from the medium access control (MAC) sublayer service offered by the DQDB layer to the logical link control (LLC) sublayer. (LM/C) 8802-6-1994

(2) The high order (left justified) bits assigned in the device address field of a FASTBUS address which are used to identify the segment on which a device is located; more than one group address may be assigned to a given segment. See base group address. (NID) 960-1993

group alerting (telephone switching systems) A central office feature for simultaneously signaling a group of customers from a control station providing an oral or recorded announcement. (COM) 312-1977w

group ambient temperature (cable or duct) (power distribution, underground cables) The no-load temperature in a group with all other cables or ducts in the group loaded. (PE) [4]

group-busy tone (telephone switching systems) A tone that indicates to operators that all trunks in a group are busy. (COM) 312-1977w

group code recording *See:* constant-linear-velocity recording.

group, commutating *See:* commutating group.

group delay (1) (network analyzers) In practice, $\Delta\omega$ must be sufficiently greater than zero to permit adequate measurement resolution. If $\Delta\omega$ is too large, however, the limit in the defining equation for group delay will not be reached, and the measured group delay will depend upon $\Delta\omega$. Therefore, the value of $\Delta\omega$ used in a measurement should be specified. (IM/HFIM) 378-1986w

(2) The derivative of radian phase with respect to radian frequency $\partial\phi/\partial\omega$. Group delay is equal to the phase delay for an ideal nondispersive delay device, but may differ greatly in actual devices where there is ripple in the phase vs. frequency characteristic (dispersive and nondispersive delay line). (UFFC) 1037-1992w

(3) In 10BROAD36, the rate of change of total phase shift, with respect to frequency, through a component or system. Group delay variation is the maximum difference in delay as a function of frequency over a band of frequencies. (C/LM) 802.3-1998

(4) *See also:* envelope delay. (AP/PROP) 211-1997

(5) (broadband local area networks) *See also:* time distortion. (LM/C) 802.7-1989r

group delay time The rate of change, with angular frequency, of the total phase shift through a network. *Notes:* 1. Group delay time is the time interval required for the crest of a group of interfering waves to travel through a 2-port network, where the component wave trains have slightly different individual frequencies. 2. Group delay time is usually very close in value to envelope delay and transmission time delay, and in the case of vanishing spectrum bandwidth of the signal these quantities become identical. *See also:* measurement system. (IM) 285-1968w

group (multicast) DSAP address A destination address assigned to a collection of LLCs to facilitate their being addressed collectively. The least significant bit shall be set equal to "1." (C/LM/CC) 8802-2-1998

group (multicast) destination service access point address (logical link control) A destination address assigned to a collection of LLCs to facilitate their being addressed collectively. The least significant bit shall be set equal to "1." (PE/TR) 799-1987w

group flashing light (illuminating engineering) A flashing light in which the flashes are combined in groups, each including the same number of flashes, and in which the groups are repeated at regular intervals. The duration of each flash is clearly less than the duration of the dark periods between flashes, and the duration of the dark periods between flashes is clearly less than the duration of the dark periods between groups. (EEC/IE) [126]

group ID (1) A nonnegative integer, which can be contained in an object of type *gid_t*, that is used to identify a group of system users. Each system user is a member of at least one group. When the identity of a group is associated with a process, a group ID value is referred to as a real group ID, an effective group ID, one of the (optional) supplementary group IDs, or an (optional) saved set-group-ID. (C/PA) 9945-1-1996, 9945-2-1993

(2) A value identifying a group of system users. Each system user is a member of at least one group. A group ID is defined in the package *POSIX_Process_Identification*. When the identity of a group is associated with a process, a group ID value is referred to as a real group ID, an effective group ID, one of the (optional) supplementary group IDs, or an (optional) saved set-group-ID. (C) 1003.5-1999

group index (fiber optics) (denoted N) For a given mode propagating in a medium of refractive index n, the velocity of light in vacuum, c, divided by the group velocity of the mode. For a plane wave of wavelength λ, it is related thus to the refractive index:

$$N = n - \lambda(dn/d\lambda)$$

See also: material dispersion parameter; group velocity. (Std100) 812-1984w

grouping (1) (facsimile) Periodic error in the spacing of recorded lines. *See also:* facsimile signal. (COM) 168-1956w
(2) (electroacoustics) Nonuniform spacing between the grooves of a disk recording. (SP) [32]

group item *See:* data aggregate.

group loop (analog computer) A potentially detrimental loop formed when two or more points in an electrical system that are nominally at group potential are connected by a conducting path such that either or both points are not at the same ground potential. (C) 165-1977w

group mark A mark that identifies the beginning or the end of a set of data; for example, a mark at the beginning of a block. (C) 610.10-1994w

group operation The operation of all poles of a multipole switch device by one operating mechanism. (SWG/PE) C37.100-1992

group path *See:* group path length.

group path length For a pulsed signal traveling between two points in a medium, the product of the speed of light in vacuum and the travel time of the pulse between the two points, provided the shape of the pulse is not significantly changed. (AP/PROP) 211-1997

groups, commutating, set of (thyristor converter) Two or more commutating groups that have simultaneous commutations. (IA/IPC) 444-1973w

group-series loop insulating transformer (power and distribution transformers) An insulating transformer whose secondary is arranged to operate a group of series lamps and/or a series group of individual-lamp transformers. *See also:* specialty transformer. (PE/TR) C57.12.80-1978r, [57]

group velocity (1) (A) (fiber optics) For a particular mode, the reciprocal of the rate of change of the phase constant with respect to angular frequency. *Note:* The group velocity equals the phase velocity if the phase constant is a linear function of the angular frequency. *See also:* group index; differential mode delay; phase velocity. **(B) (fiber optics)** Velocity of the signal modulating a propagating electromagnetic wave. *See also:* group index; phase velocity; differential mode delay. (Std100) 812-1984
(2) (waveguide) Of a traveling wave at a single frequency, and for a given mode, the velocity at which the energy is transported in the direction of propagation. (MTT) 146-1980w
(3) (of a traveling wave) The velocity of propagation of the envelope, provided that the envelope moves without significant change of shape. The magnitude of the group velocity is equal to the reciprocal of the rate of change of phase constant with angular frequency. (AP/PROP) 211-1997

grout (rotating machinery) A very rich concrete used to bond the feet, sole plates, bedplate, or rail of a machine to its foundation. (PE) [9]

grown junction (semiconductor) A junction produced during growth of a crystal from a melt. *See also:* semiconductor device. (ED) 216-1960w

GRP *See:* ground reference plane.

GRS *See:* generic response spectra.

G-scope A cathode-ray oscilloscope arranged to present a G-display. (AES/RS) 686-1990

G/T ratio (of an antenna) The ratio of the gain to the noise temperature of an antenna. *Notes:* 1. Usually the antenna-receiver system figure of merit is specified. For this case the figure of merit is the gain of the antenna divided by the system noise temperature referred to the antenna terminals. 2. The system figure of merit at any reference plane in the RF system is the same as that taken at the antenna terminals since both the gain and system noise temperature are referred to the same reference plane at the antenna terminals. *Synonym:* figure of merit. (AP/ANT) 145-1983s

G^2 drift (electronic navigation) A drift component in gyros (sometimes in accelerometers) proportional to the square of the nongravitational acceleration and caused by anisoelasticity of the rotor supports. Jargon. *See also:* navigation. (AES/RS) 686-1982s, [42]

guarantee *See:* work permit.

guard (1) (interference terminology) A conductor situated between a source of interference and a signal path in such a way that interference currents are conducted to the return terminal of the interference source without entering the signal path. *See also:* interference. (PE/PSR) [6]
(2) One or more conducting elements arranged and connected on an electrical instrument or measuring circuit so as to divert unwanted currents from the measuring means. (PE/TR) C57.12.90-1999

guardband (1) (data transmission) A frequency band between two channels which gives a margin of safety against mutual interference. (PE) 599-1985w
(2) (broadband local area networks) (channel) A designated unoccupied portion of the frequency spectrum that exists between two occupied portions of the spectrum. (LM/C) 802.7-1989r

guard channel One or more auxiliary parallel processing channels to control the main processing channel in order to reject interference that is partly in, but not centered on, the main channel. *Note:* Guard channels may be displaced in time (range), Doppler frequency, carrier frequency, or angle. Sometimes called "guard gates," "guard bands," or "sidelobe blanking" (not cancellation). Guard channel is used against range gate stealers, velocity gate stealers, sidelobe jamming, and to enhance apparent angle resolution in identification, friend or foe (IFF). May use auxiliary displays. *See also:* sidelobe blanker. (AES) 686-1997

guard circle (disk recording) An inner concentric groove inscribed, on disk records, to prevent the pickup from being damaged by being thrown to the center of the record. (SP) [32]

guarded (1) Covered, shielded, fenced, enclosed, or otherwise protected by means of suitable covers or casings, barrier rails or screens, mats, or platforms to remove the likelihood of the dangerous contact or approach by persons or objects to a point of danger. (NESC/IA/T&D/PC) 463-1993w, [86], C2.2-1960
(2) Covered, fenced, enclosed, or otherwise protected, by means of suitable covers or casings, barrier rails or screens, mats or platforms, designed to limit the likelihood under normal conditions, of dangerous approach or accidental contact by persons or objects. *Note:* Wires that are insulated but not otherwise protected are not normally considered to be guarded. (NESC) C2-1997

guarded enclosure An enclosure in which all openings giving direct access to live or rotating parts (except smooth rotating surfaces) are limited in size by the structural parts or by screens, baffles, grilles, expanded metal, or other means to prevent accidental contact with hazardous parts. The openings in the enclosure shall be such that they will not permit the passage of a rod larger than 12 mm (1/2 in) in diameter, except where the distance of exposed live parts from the guard is more than 102 mm (4 in); then the openings may be of such shape as not to permit the passage of a rod larger than 19 mm (3/4 in) in diameter. (IA/MT) 45-1998

guarded input (oscilloscopes) A shielded input where the shield is driven by a signal in phase with and equal in amplitude to the input signal. (IM) 311-1970w

guarded machine (rotating machinery) An open machine in which all openings giving direct access to live or rotating parts (except smooth shafts) are limited in size by the design of the structural parts, or by screens, grilles, expanded metal, etc., to prevent accidental contact with such parts. Such openings are of such size as not to permit the passage of a cylindrical rod 1/2 inch in diameter, except where the distance from the guard to the live or rotating parts is more than 4 inches: they are of such size as not to permit the passage of a cylindrical rod 3/4 inch in diameter. *See also:* asynchronous machine. (PE) [9]

guarded release (telephone switching systems) A technique for retaining a busy condition during the restoration of a circuit to its idle state. (COM) 312-1977w

guard electrode (testing of electric power system components) One or more electrically conducting elements, arranged and connected in an electric instrument or measuring circuit so as to divert unwanted conduction or displacement currents from, or confine wanted currents to, the measurement device. (PE/PSIM) [55]

guard frequency A reserved area within a range of frequencies that separates two channels in a carrier system or frequency-derived channel. (C) 610.10-1994w

guard-ground system (interference terminology) A combination of guard shields and ground connections that protects all or part of a signal transmission system from common-mode interference by eliminating ground loops in the protected part. *Note:* Ideally, the guard shield is connected to the source ground. The source is usually grounded also to the source ground by bonding of the transducer to the test body. The filter, signal receiver, etc., are floating with respect to their own grounded cases. This necessitates physically isolating the signal receiver and filter chassis from the cases and using isolation transformers in power supplies, or isolating input circuits from cases and using isolating input transformers. This arrangement in effect places the signal receiver and filter electrically at the source. By means of a similar guard, the load can be placed effectively at the source. See the figures below. *See also:* interference.

guard shield (interference terminology) A guard that is in the form of a shielding enclosure surrounding all or part of the signal path. *Note:* A guard shield is effective against both capacitively coupled and conductively coupled interference whereas a simple guard conductor is usually effective only against conductively coupled interference. *See also:* guard-ground system; interference

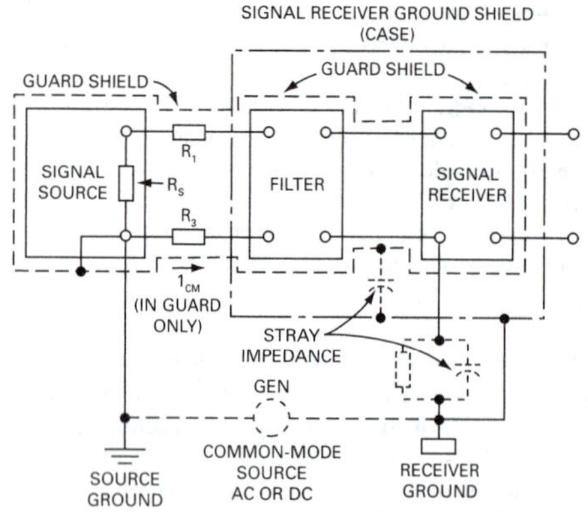

[The guard shield consists of the signal source shield (if present), the signal shield, the filter shield, and the signal receiver shield.]

ideal guard shield

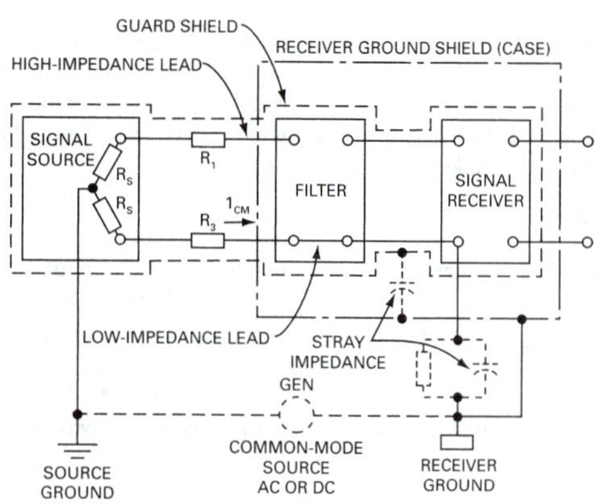

[Guard shield connections when connection at source is not convenient. When $(R_3 + R_S') \ll (R_1 + R_S)$, this arrangement causes common-mode current in the low-impedance lead, but protects the more critical high-impedance lead from current flow.]

guard shield

(PE/PSR) [6]

guard signal A signal sent over a communication channel to make the system secure against false information by preventing or guarding against the relay operation of a circuit breaker or other relay action until the signal is removed and replaced by a tripping or permissive signal. (SWG/PE) C37.100-1992

guard structure *See:* crossing structure.

guard wire A grounded wire erected near a lower-voltage circuit or public crossing in such a position that a high (or higher) voltage overhead conductor cannot come into accidental contact with the lower-voltage circuit, or with persons or objects on the crossing without first becoming grounded by contact with the guard wire. *See also:* ground. (T&D/PE) [10]

GUI *See:* graphical user interface.

guidance (missile) The process of controlling the flight path through space through the agency of a mechanism within the missile. *See also:* guided missile. (EEC/PE) [119]

guide (1) (high-voltage switchgear) An attachment used to secure proper alignment when operating a fuse or switch. (SWG/PE) C37.40-1993
(2) Document in which alternative approaches to good practice are suggested but no clear-cut recommendations are made. (C/SE) 730.1-1995

guide bearing (rotating machinery) A bearing arranged to limit the transverse movement of a vertical shaft. *See also:* bearing. (PE) [9]

guided missile An unmanned device whose flight path through space may be controlled by a self-contained mechanism. *See also:* beam rider guidance; preset guidance; homing guidance; command guidance; guidance. (EEC/PE) [119]

guided mode *See:* bound mode.

guided ray (fiber optics) In an optical waveguide, a ray that is completely confined to the core. Specifically, a ray at radial position r having direction such that

$$\sin \theta(r) = [n^2(r) - n^2(a)]^{1/2}$$

where $\theta(r)$ is the angle the ray makes with the waveguide axis, $n(r)$ is the refractive index, and $n(a)$ is the refractive index at the core radius. Guided rays correspond to bound (or guided) modes in the terminology of mode descriptors. *Synonyms:* bound ray; trapped ray. *See also:* bound mode; leaky ray. (Std100) 812-1984w

guided wave A propagating wave whose energy is concentrated within or near boundaries between media having different electromagnetic properties. (AP/PROP) 211-1997

guide flux *See:* shield; form factor.

guideline (nuclear power quality assurance) A suggested practice that is not mandatory in programs intended to comply with a standard. The word "should" denotes a guideline; the word "shall" denotes a requirement. (PE/NP) [124]

guide pin A pin used for guidance of the connector during module insertion and extraction. (C/BA) 1101.3-1993

guide rib (1) A rib provided for initial module alignment at the time of installation into the chassis module slot. For conduction-cooled modules only, guide ribs form the heat transfer paths to the chassis and the mounting for the module retainers at the alpha and beta ends of the module heatsink. (C/BA) 1101.3-1993
(2) The alpha and beta ends of the module. These form the heat transfer paths and the mounting for the module retainers and provide for module alignment. (C/BA) 1101.4-1993, 1101.7-1995

guide wavelength (1) The wavelength in a waveguide, measured in the longitudinal direction. *See also:* waveguide. (AP/ANT) [35]
(2) (planar transmission lines) For a travelling wave in a uniform transmission line at a given frequency and for a given mode, the distance along the axis of propagation between corresponding points at which a field component (or the voltage or current) differs in phase by 2π rad. (MTT) 1004-1987w

guise A function that provides the capability for an entity to be viewed with one appearance by one group of participants, and with another appearance by another group. (DIS/C) 1278.1-1995

gulp Slang for a group of bytes. (C) 610.10-1994w

gun-control switch A switch that closes an electric circuit, thereby actuating the gun-trigger-operating mechanism of an aircraft, usually by means of a solenoid. (EEC/PE) [119]

Gunn oscillator (nonlinear, active, and nonreciprocal waveguide components) A direct dc-to-rf (direct current to radio frequency) conversion device in which the active element of the oscillator is a bulk III-V semiconductor device having a negative dc resistance characteristic. Conduction can occur in either direction, although the substrate contact is considered to be the cathode. The Gunn diode is a transferred electron device with practical output frequencies ranging from approximately 4 GHz (gigahertz) to more than 60 GHz. (MTT) 457-1982w

guy A tension member having one end secured to a fixed object and the other end attached to a pole, crossarm, or other structural part that it supports. *See also:* guy wire; tower. (T&D/PE) [10]

guy anchor The buried element of a guy assembly that provides holding strength or resistance to guy wire pull. *Note:* The anchor may consist of a plate, a screw or expanding device, a log of timber, or a mass of concrete installed at sufficient depth and of such size as to develop strength proportionate to weight of earth or rock it tends to move. The anchor is designed to provide attachment for the anchor rod which extends above surface of ground for convenient guy connection. *See also:* tower; guy. (T&D/PE) [10]

guy insulator An insulating element, generally of elongated form with transverse holes or slots for the purpose of insulating two sections of a guy or provide insulation between structure and anchor and also to provide protection in case of broken wires. Porcelain guy insulators are generally designed to stress the porcelain in compression, but wooden insulators equipped with suitable hardware are generally used in tension. *See also:* tower. (PE/T&D) 1410-1997, [10]

guy wire A stranded cable used for a semiflexible tension support between a pole or structure and the anchor rod, or between structures. *See also:* tower; guy. (T&D/PE) [10], 1410-1997

GW BAsic A dialect of BASIC, designed for use with microprocessors and microcomputers. (C) 610.13-1993w

Gypsy A specification language used primarily for computer security applications; one of the two specification languages accepted for use by the US National Computer Security Center. (C) 610.13-1993w

gyration impedance A characteristic of a gyrator that may be expressed in terms of the impedance matrix elements as

$$\sqrt{z \times z_{21}}$$

See also: gyrator. (CAS) [13]

gyrator (nonlinear, active, and nonreciprocal waveguide components) A two-port nonreciprocal device that provides insertion phases differing by 180 degrees for the two opposite directions of propagation. (MTT) 457-1982w
(2) (A) A directional phase changer in which the phase changes in opposite directions differ by π radians or 180 degrees. *See also:* waveguide. **(B)** Any nonreciprocal passive element employing gyromagnetic properties. *See also:* waveguide. (AP/ANT) [35], [84]

gyro (gyroscope) A device using angular momentum (usually of a spinning rotor) to sense angular motion of its case with respect to inertial space about one or two axes orthogonal to the spin axis. *Notes:* 1. This definition does not include more complex systems, such as stable platforms, using gyros as components. 2. Certain devices, such as laser gyros, that perform similar functions but do not use angular momentum may also be classified as gyros. (AES/GYAC) 528-1994

gyrocompass A compass consisting of a continuously driven Foucault gyroscope whose supporting ring normally confines the spinning axis to a horizontal plane, so that the earth's rotation causes the spinning axis to assume a position in a plane passing through the earth's axis, and thus to point to true north. (EEC/PE) [119]

gyrocompass alignment (inertial systems) A process of self-alignment in azimuth based upon measurements of misalignment drift about the nominal east-west axis of the system. *See also:* navigation. (AES) [42]

gyrocompassing *See:* gyrocompass alignment.

gyro flux-gate compass A device that uses saturable reactors in conjunction with a vertical gyroscope, to sense the direction of the magnetic north with respect to the aircraft heading. *Synonym:* gyro flux-valve compass. (EEC/PE/GYAC) [119]

gyro flux-valve compass *See:* gyro flux-gate compass.

gyro-frequency (f_H) The lowest natural frequency at which charged particles spiral in a fixed magnetic field. It is given by:

$$f_H = q \frac{|\vec{B}|}{2\pi m}$$

where

q = the charge of the particles
$|\vec{B}|$ = the magnitude of magnetic flux density
m = the mass of the particles.

For a linear medium, the gyro frequency is the same as cyclotron frequency. (AP/PROP) 211-1997

gyro gain The ratio of the output angle of the gimbal to the input angle of a rate-integrating gyro at zero frequency. It is numerically equal to the ratio of the rotor angular momentum to the damping coefficient. (AES/GYAC) 528-1994

gyro horizon electric indicator An electrically driven device for use in aircraft to provide the pilot with a fixed artificial horizon. *Note:* It indicates deviation from level flight. (EEC/PE) [119]

gyromagnetic effect (nonlinear, active, and nonreciprocal waveguide components) The phenomenon by which the magnetization of a material or medium, subjected to a magnetostatic field, upon disturbance relaxes back to equilibrium by damped precessional motion about the direction of that field. (MTT) 457-1982w

gyromagnetic filter (nonlinear, active, and nonreciprocal waveguide components) (garnet, YIG) A filter whose operation depends on the gyromagnetic effect. (MTT) 457-1982w

gyromagnetic limiter (nonlinear, active, and nonreciprocal waveguide components) (ferrite, garnet, YIG) A power limiter whose operation depends on saturation effects in a gyromagnetic material. (MTT) 457-1982w

gyromagnetic material (nonlinear, active, and nonreciprocal waveguide components) (medium) A material (medium), such as ferrite or garnet, capable of exhibiting the gyromagnetic effect. (MTT) 457-1982w

gyromagnetic permeability tensor (nonlinear, active, and nonreciprocal waveguide components) A tensor used to describe the permeability properties exhibited by a gyromagnetic material, appropriate to electromagnetic wave propagation. (MTT) 457-1982w

gyromagnetic resonance absorption (nonlinear, active, and nonreciprocal waveguide components) That amount of power continuously absorbed in a gyromagnetic material subjected to a magnetostatic field when a disturbance causes a steady precession of the magnetization of that material at a rate near the gyromagnetic resonance frequency. (MTT) 457-1982w

gyromagnetic resonance field (nonlinear, active, and nonreciprocal waveguide components) The magnetostatic field that, when applied to a gyromagnetic material, causes gyromagnetic resonance to occur at a particular frequency. (MTT) 457-1982w

gyromagnetic resonance frequency (nonlinear, active, and nonreciprocal waveguide components) The damped natural frequency for precession of the magnetization of a gyromagnetic material subjected to a particular magnetostatic field. (MTT) 457-1982w

gyromagnetic resonance linewidth (nonlinear, active, and nonreciprocal waveguide components) The difference between magnetostatic field levels slightly above and slightly below gyromagnetic resonance for which the gyromagnetic resonance absorption falls to half the peak value. This resonance occurs in a gyromagnetic material with uniformly precessing magnetization at a fixed frequency. (MTT) 457-1982w

gyro operating null (dynamically tuned gyro) The condition where minimum change in drift rate occurs due to changes in wheel speed. (AES/GYAC) 528-1994

H

H *See:* H beacon.

HAL (A) A high-order problem-oriented language used in aerospace applications; characterized by its strong orientation toward mathematical computations and built-in vector/matrix arithmetic. **(B)** Abbreviation for High-order Ada Language; a computer language similar to Ada. (C) 610.13-1993

halation (cathode-ray tubes) An annular area surrounding a spot, that is due to the light emanating from the spot being reflected from the front and rear sides of the face plate. *See also:* cathode-ray tube. (Std100) [84]

half-adder A combinational logic element having two outputs, *S* and *C*, and two inputs, *A* and *B*, such that the outputs are related to the inputs according to the following equations:

$S = AORB$ (exclusive OR)

$C = A + B$

S denotes sum without carry, *C* denotes carry. Two half-adders may be used for performing binary addition. (C) [20]

half adder An adder that accepts two inputs, producing a sum and a carry as outputs according to the table below. *Synonym:* two-input adder. *Contrast:* quarter adder; full adder. *See also:* half subtracter.

input #1	0	0	1	1
input #2	0	1	0	1
output sum	0	1	1	0
output carry	0	0	0	1

half adder

(C) 610.10-1994w

half-adjust (mathematics of computing) To round a number by changing the least significant digit to zero and adding one to the next digit if the value of the least significant digit was half the radix or greater. (C) 1084-1986w

half-amplitude recovery time The time interval from the start of a full-amplitude pulse to the instant a succeeding pulse can attain an amplitude of 50% of the maximum amplitude of a full-amplitude pulse. (NI/NPS) 309-1999

half carry (mathematics of computing) A carry process in which a carry digit generated in the most significant digit place of the less-significant half of a sum is transferred to the least significant digit place of the more significant half. (C) 1084-1986w

half cell An electrode immersed in a suitable electrolyte. *See also:* electrolytic cell. (EEC/PE) [119]

half coil *See:* armature bar.

half duplex (1) (data transmission) Pertaining to a transmission over a circuit capable of transmitting in either direction, but only one direction at a time. (COM/PE) 599-1985w **(2)** Transmission over a circuit capable of transmitting in either direction, but only in one direction at a time. *Contrast:* duplex. (SUB/PE) 999-1992w **(3)** A mode of operation of a CSMA/CD local area network (LAN) in which DTEs contend for access to a shared medium. Multiple, simultaneous transmissions in a half duplex mode CSMA/CD LAN result in interference, requiring resolution by the CSMA/CD access control protocol. (C/LM) 802.3-1998 **(4) (local area networks)** A link segment capable of transferring signals in either direction along the link, but not in both directions simultaneously. Requires line turnaround to change signal direction. (4-UTP links are full duplex in control mode, but only half duplex in data mode.). (C) 8802-12-1998 **(5)** An operating condition that allows communication in either send and receive directions with more than 20 dB switched loss in either direction. (COM/TA) 1329-1999

half-duplex channel (data transmission) (half duplex operation) A channel of a duplex system arranged to permit operation in either direction but not in both directions simultaneously. (PE) 599-1985w

half-duplex operation (telegraph system) Operation of a duplex system arranged to permit operation in either direction but not in both directions simultaneously. *See also:* telegraphy. (EEC/PE) [119]

half-duplex repeater A duplex telegraph repeater provided with interlocking arrangements that restrict the transmission of signals to one direction at a time. *See also:* telegraphy. (EEC/PE) [119]

half-duplex transmission Transmission in which data may be sent in either direction but only in one direction at a time on a transmission medium. *Contrast:* simplex transmission; duplex transmission. (C) 610.7-1995

half-height disk drive A disk drive that uses the same width, but approximately one-half the front panel height as a standard disk drive, known as a full-height disk drive. (C) 610.10-1994w

half-power beamwidth In a radiation pattern cut containing the direction of the maximum of a lobe, the angle between the two directions in which the radiation intensity is one-half the maximum value. *See also:* principal half-power beamwidths. (AP/ANT) 145-1993

half section A bisected tee or pi section. A basic L-section building block of image-parameter filters. (CAS) [13]

half subtracter A subtracter that accepts two inputs, producing a sum and a borrow digit as output according to the table below. *Contrast:* full subtracter. *See also:* half adder.

input #1	0	0	1	1
input #2	0	1	0	1
output sum	0	1	1	0
output carry	0	1	0	0

half subtracter

(C) 610.10-1994w

halftone characteristic (facsimile) A relation between the density of the recorded copy and the density of the subject copy. *Note:* The term may also be used to relate the amplitude of the facsimile signal to the density of the subject copy or the record copy when only a portion of the system is under consideration. In a frequency-modulation system an appropriate parameter is to be used instead of the amplitude. *See also:* recording. (COM) 168-1956w

halftones *See:* level; storage tube.

halftoning (computer graphics) A technique for displaying an image with many gray levels on a monochrome display device in which the gray levels are approximated by variable-sized black and white dots. *See also:* dithering. (C) 610.6-1991w

half-wave dipole A wire antenna consisting of two straight collinear conductors of equal length, separated by a small feeding gap, with each conductor approximately a quarter-wavelength long. *Note:* This antenna gets its name from the fact that its overall length is approximately a half-wavelength. In practice, the length is usually slightly smaller than a half-wavelength—enough to cause the input impedance to be pure real ($jX = 0$). (AP/ANT) 145-1993

half-wave rectification (power supplies) In the rectifying process, half-wave rectification passes only one-half of each incoming sinusoid, and does not pass the opposite half-cycle. The output contains a single half-sine pulse for each input cycle. A single rectifier provides half-wave rectification. Because of its poorer efficiency and larger alternating-current component, half-wave rectification is usually employed in noncritical low-current circumstances. See the accompanying figure. *See also:* rectifier circuit element; rectification; rectifier.

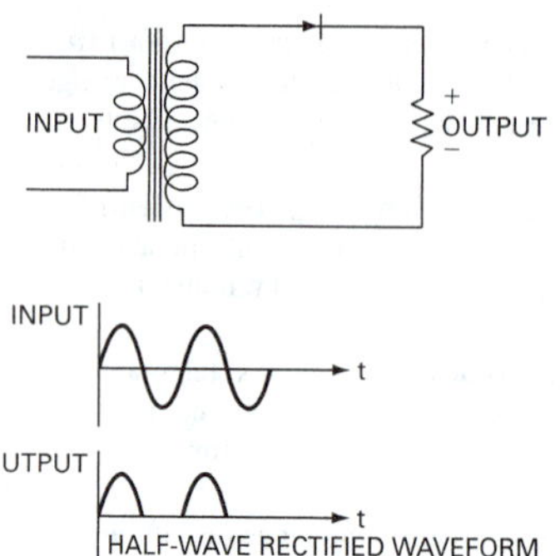

half-wave rectification

(AES) [41]

halfword (1) An aligned **doublet.** *Note:* The definition of this term is architecture-dependent, and so may differ from that used in other processor architectures. (C/MM) 1754-1994
(2) A contiguous sequence of bits or characters that comprises half of a computer word and which is capable of being addressed as a unit. (C) 610.10-1994w
(3) For the purpose of this standard, a half-word is a 16-bit data item taken as a unit. (C/MM) 1196-1987w
(4) Two bytes or 16 bits operated on as a unit. The most significant byte carries the index value 0 and the least significant byte carries the index value 1. (C/BA) 1496-1993w

half-word Slave An SBus Slave having a data path only through bits D[31:16] of the data bus. (C/BA) 1496-1993w

Hall analog multiplier A Hall multiplier specifically designed for analog multiplication purposes. (MAG) 296-1969w

Hall angle The angle between the electric field vector and the current density vector. (MAG) 296-1969w

Hall coefficient The coefficient of proportionality R in the relation

$$E_{\mathrm{H}} = R(\boldsymbol{J} \times \boldsymbol{B})$$

where
E_{H} is the resulting transverse electric field
$\boldsymbol{J}$ is the current density
$\boldsymbol{B}$ is the magnetic flux density

Note: The sign of the majority carrier charge can usually be inferred from the sign of the Hall coefficient.

(MAG) 296-1969w

Hall effect (A) (Hall effect devices) *(in conductors and semiconductors).* The change of the electric conduction caused by that component of the magnetic field vector normal to the current density vector, which, instead of being parallel to the electric field, forms an angle with it. *Note:* In conductors and semiconductors of noncubic single crystals, the current density and electric field vectors may not be parallel in the absence of an applied magnetic field. For such crystals the more general definition below should be used. **(B) (Hall effect devices)** *(in any material, including ferromagnetic and similar materials).* The change of the electric conduction caused by that component of the magnetic field vector applied normal to the current density vector, which causes the angle between the current density vector and the electric field to change from the magnitude that existed prior to the introduction of the magnetic field. *Note:* For ferromagnetic and similar materials there are two effects, the "ordinary" Hall effect due to the applied external magnetic flux as described for conductors and semiconductors and the "extraordinary" Hall effect due to the magnetization in the ferromagnetic or similar material. In the absence of the "extraordinary" Hall effect and the effects outlined in the preceding note, the current density vector and the electric field vector will be parallel when there is no external magnetic flux. (MAG) 296-1969

Hall effect device A device in which the Hall effect is utilized. (MAG) 296-1969w

Hall-effect probe A magnetic flux density sensor containing an element exhibiting the Hall-effect to produce a voltage proportional to the magnetic flux density. *Note:* Hall-effect probes respond to static as well as time varying magnetic flux densities. Due to saturation problems sometimes encountered when attempting to measure small power frequency flux densities in the presence of the substantial static geomagnetic flux of the earth, Hall-effect probes have seldom been used under ac power lines. (T&D/PE) 539-1990, 1308-1994

Hall generator A Hall plate, together with leads, and, where used, encapsulation and ferrous or nonferrous backing plate(s). (MAG) 296-1969w

Hall mobility (electric conductor) The quantity μH in the relation μH = Rσ, where R = Hall coefficient and σ = conductivity. *See also:* semiconductor. (ED) 216-1960w

Hall modulator A Hall effect device that is specifically designed for modulation purposes. (MAG) 296-1969w

Hall multiplier A Hall effect device that contains a Hall generator together with a source of magnetic flux density and that has an output that is a function of the product of the control current and the field excitation current.

(MAG) 296-1969w

Hall plate A three-dimensional configuration of any material in which the Hall effect is utilized. (MAG) 296-1969w

Hall probe A Hall effect device specifically designed for measurement of magnetic flux density. (MAG) 296-1969w

Hall terminals The terminals between which the Hall voltage appears. (MAG) 296-1969w

Hall voltage The voltage generated in a Hall plate due to the Hall effect. (MAG) 296-1969w

halogen-quenched counter tube A self-quenched counter tube in which the quenching agent is a halogen, usually bromine or chlorine. (ED) [45]

halt (A) Most commonly, a synonym for **stop. (B)** Less commonly, a synonym for **pause.** (C) 610.12-1990

halt instruction *See:* pause instruction.

halving interval (thermal classification of electric equipment and electrical insulation) (evaluation of thermal capability) The number corresponding to the interval in °C determined from the thermal endurance relationship expresses the halving of the time-to-end-point centered on the temperature of the TI or RTI. In case of graphical derivation the times corresponding to the TI or RTI (for example, 20 000 h) and one half that value (for example, 10 000 h) will usually produce an acceptable approximation. *See also:* thermal endurance graph. (EI) 1-1986r

hamming code (mathematics of computing) Any of several error-correcting codes invented by the mathematician Richard Hamming, which use redundant information bits to detect and correct any single error in a transmitted character. *See also:* error-correcting code. (C) 610.7-1995, 1084-1986w

hamming distance (1) (mathematics of computing) The number of digit positions in which two binary numerals, characters, or words of the same length are different. For example, the Hamming distance between 100101 and 101001 is two. *Synonyms:* signal distance; code distance.

(C) 1084-1986w

(2) The minimum number of incorrect bits that shall be received in order for a packet to be considered invalid. For example, the hamming distance 4 means that all one-, two-, and three-bit errors are detectable. (PE/SUB) 1379-1997

hand (head or butt) cable (mining) A flexible cable used principally in making electric connections between a mining machine and a truck carrying a reel of portable cable. *See also:* mine feeder circuit. (EEC/PE) [119]

hand burnishing (electroplating) Burnishing done by a hand tool, usually of steel or agate. *See also:* electroplating.

(EEC/PE) [119]

hand elevator An elevator utilizing manual energy to move the car. *See also:* elevator. (EEC/PE) [119]

hand-feed punch A card punch into which cards are manually entered and removed one at a time. *Synonym:* hand punch. *Contrast:* automatic-feed punch. (C) 610.10-1994w

hand-held computer A portable computer small enough to be held and operated while holding it in one hand. (C) 610.10-1994w

handhole (1) An opening in an underground system containing cable, equipment, or both into which workmen reach but do not enter. (T&D) C2.2-1960
(2) An access opening, provided in equipment or in a below-the-surface enclosure in connection with underground lines, into which personnel reach but do not enter, for the purpose of installing, operating, or maintaining equipment or cable or both. (NESC) C2-1997

handler (1) A module or device that responds to a bus request (such as an interrupt request) as the slave to that request. (C/BA) 1014.1-1994w
(2) A program or routine that performs or controls one task (e.g., error detection). (SCC20) 1226-1998

handling device (of metal-clad switchgear) That accessory used for the removal, replacement, or transportation of the removable element. (SWG/PE) C37.100-1992

handling zone The portion of a disk or other storage medium that may be touched by the gripping mechanism or actuator. *Contrast:* recording area. (C) 610.10-1994w

hand/metal discharge *See:* hand/metal ESD.

hand/metal ESD An ESD from an intruding human hand that occurs from an intervening metal object such as a ring, tool, key, etc. *Synonym:* hand/metal discharge. (EMC/PE/SPD) C63.16-1993, C62.47-1992r

hand operation Actuation of an apparatus by hand without auxiliary power. *See also:* switch. (IA/ICTL/IAC) [60], [84]

hand-printed character font An international standard optical font for use on hand-generated documents. *See also:* OCR-B; OCR-A. (C) 610.2-1987

hand-print recognition Optical character recognition of hand-printed characters. (C) 610.2-1987

hand punch *See:* hand-feed punch.

hand receiver An earphone designed to be held to the ear by the hand. (EEC/PE) [119]

hand-reset relay A relay so constructed that it remains in the picked-up condition even after the input quantity is removed; specific manual action is required to reset the relay. *Synonym:* mechanically reset relay. (SWG/PE) C37.100-1992

handset (1) (transmission performance of telephone sets) An assembly that includes a handle and a telephone set transmitter and receiver. Other components such as the speech network may also be located in the handset. (COM/TA) 269-1983s
(2) An assembly intended to be held in the hand of the user that includes a transmitter and receiver. (For the purposes of this standard, a handset is a four-wire device, that is, it does not include a built-in speech network.) (COM/TA) 1206-1994

handset telephone *See:* hand telephone set.

handsfree reference point (HFRP) The calibration point on the reference axis of the mouth simulator, 50 cm in front of the lip plane. (COM/TA) 1329-1999

handsfree telephone (HFT) A device for connection to a telephone network capable of two-way voice communication without close coupling to the user's mouth or ear. (COM/TA) 1329-1999

handsfree telephone test circuit An assembly consisting of a handsfree telephone set(s) and interface(s) as may be required to realize simulated partial telephone connections. (COM/TA) 1329-1999

handshake (1) (FASTBUS acquisition and control) An interlocked exchange of signals between a master and a slave, controlling the transfer of data. (NID) 960-1993
(2) (test, measurement, and diagnostic equipment) A hardware or software sequence of events requiring mutual consent of conditions prior to change. (MIL) [2]
(3) (STEbus) An interlocked sequence of signals between interconnected boards in which each board waits for an acknowledgement of its previous signal before proceeding. (C/MM) 1000-1987r

handshake cycle (digital interface for programmable instrumentation) The process whereby digital signals effect the transfer of each data byte across the interface by means of an interlocked sequence of status and control signals. (An interlocked sequence is a fixed sequence of events in which one event in the sequence must occur before the next event may occur.) *See also:* interlocked sequence. (IM/AIN) 488.1-1987r

handshake status A status transfer which indicates the exchange of data between bus owner and replying agent(s). (C/MM) 1296-1987s

handshaking The exchange of predetermined signals or control measures between two systems or system components upon initial exchanges. *Note:* When the connection is established, the two components acknowledge each other. (C) 610.7-1995, 610.10-1994w

hand telephone set (telephone) A telephone set having a handset and a mounting that serves to support the handset when the latter is not in use. *Note:* The prefix desk, wall, drawer, etc., may be applied to the term hand telephone set to indicate the type of mounting. *See also:* telephone station. (PE/EEC) [119]

hand-to-metal impedance The impedance between the human hand and the metal object with which it is associated in a hand/metal ESD. The metal object is usually the intruder discharge electrode. Examples of hand-to-metal impedance include resistance and capacitance between the fingers and a key, between the wrist and a metal watch or bracelet, and between the hand and a screwdriver. (SPD/PE) C62.47-1992r

handwheel A wheel the rim of which serves as a handle for manual operation of a rotary device. (IA/ICTL/IAC) [60]

hand winding (rotating machinery) A winding placed in slots or around poles by a human operator. *See also:* rotor; stator. (PE) [9]

hang-off (accelerometer) (gyros) The displacement of an inertial sensing element from its null position that occurs when an input is applied and that is due to the finite compliance of a capture loop or a restoring spring. (AES/GYAC) 528-1994

hangover *See:* tailing.

hang-over time (T_H) Time from the input signal going below the threshold level until 3 dB of switched loss is inserted in the output signal. *Synonyms:* decay time; release time. (COM/TA) 1329-1999

hang-up hand telephone set (bracket-type handset telephone) (suspended-type handset telephone) A hand telephone set in which the mounting is arranged for attachment to a vertical surface and is provided with a switch bracket from which the handset is suspended. *See also:* telephone station. (PE/EEC) [119]

hang-up signal (telephone switching systems) A signal transmitted over a line or trunk to indicate that the calling party has released. (COM) 312-1977w

HA1 receiver weighting (data transmission) A noise weighting used in a noise measuring set to measure noise across the HA1 receiver of a of a subset with a number 302 receiver or a similar subset. The meter scale readings are in the dBa (HA1). (PE) 599-1985w

hard copy (1) (computer graphics) A printed copy of computer output in a readable form; for example, a printed report, a listing. *Contrast:* soft copy. (C) 610.2-1987, 610.6-1991w
(2) A paper record of information (e.g., reports, listings, logs, and charts). (SUB/PE) C37.1-1994
(3) A permanent record of information in readable form for human use, for example, reports, listings, displays, logs, and charts. (SWG/PE) C37.100-1992

hard cover *See:* conductor cover.

hard disk A magnetic disk that consists of a rigid platter. *Synonym:* fixed disk. *Contrast:* floppy disk. *See also:* Winchester disk. (C) 610.10-1994w

hardened computer A computer that is physically designed to function reliably in harsh environments such as extremes of temperature, shock and vibration, humidity or radiation. *Note:* Often required for space and military applications. *See also:* hostile environment computer. (C) 610.10-1994w

hard error **(A)** An error caused by a hardware failure or by accessing incompatible hardware. **(B)** A storage error in which the data that is retrieved is wrong and the storage cell will no longer hold the data written to it. *Contrast:* transient error; soft error. (C) 610.10-1994

hard failure **(1)** A failure that results in complete shutdown of a system. *Contrast:* soft failure. (C) 610.12-1990
(2) A cessation of some system or system component from which there is no possible recovery. (C) 610.10-1994w

hard limiting A type of limiting characterized by very little variation in the output within the range where the output is subject to limiting. *Contrast:* soft limiting. (C) 610.10-1994w

hard line (test, measurement, and diagnostic equipment) Any direct electrical connection between the unit under test and the testing device. (MIL) [2]

hard link **(1)** The relationship between two directory entries that represent the same file; the result of an execution of the `ln` utility or the POSIX.1 *link*() function. (C/PA) 9945-2-1993
(2) A directory entry. (C/PA) 1387.2-1995

hard macro A cluster whose cell placements relative to each other are fixed. Often the interconnect routing between the cells is also fixed and a parasitics file describing the interconnect is available for the hard macro. The location of the hard macro in the floorplan may or may not be fixed. (C/DA) 1481-1999

hard region A cluster that has defined physical boundaries in a floorplan. All cells contained in the cluster shall be placed within the boundaries of the cluster. (C/DA) 1481-1999

hard-sector Pertaining to a magnetic disk that is segmented by physical, non-alterable means such as a hole, known as an index hole, in the disk. *Contrast:* soft-sector. (C) 610.10-1994w

hardware **(1)** **(software)** Physical equipment used to process, store, or transmit computer programs or data. *Contrast:* software. (C) 610.12-1990, 610.10-1994w
(2) Physical equipment used in data processing, as opposed to programs, procedures, rules, and associated documentation. (C/PA) 14252-1996

hardware accelerator **(A)** A circuit which performs operations normally done in software much faster than they can be done in software. **(B)** A circuit that performs hardware operations much faster than the original hardware. For example: an 80386 based accelerator for an 80286 based machine. (C) 610.10-1994

hardware check *See:* automatic check.

hardware configuration item (HWCI) An aggregation of hardware that is designated for configuration management and treated as a single entity in the configuration management process. *Contrast:* computer software configuration item. *See also:* configuration item. (C) 610.12-1990

hardware description language (HDL) A general-purpose computer language designed to serve as an interface to the design, documentation, and validation of computer hardware. *Synonym:* computer hardware description language. *See also:* hardware design language. (C) 610.10-1994w

hardware design language (HDL) **(1)** A specification language with special constructs and, sometime, verification protocols, used to develop, analyze, and document a hardware design. *Contrast:* program design language. *See also:* design language; CINEMA. (C) 610.13-1993w, 610.12-1990
(2) A design language with special constructs and, sometimes verification protocols, used to develop, analyze, and document, a hardware design or computer architecture. *See also:* hardware description language. (C) 610.10-1994w

hardware failure A change in the characteristics of a system hardware element beyond its design tolerances. (VT/RT) 1483-2000

hardware item An aggregation of hardware that is designated for purposes of specification, testing, interfacing, configuration management, or other purposes. (C/SE) J-STD-016-1995

hardware language *See:* hardware description language; hardware design language; machine language.

hardware monitor **(A)** A device that measures or records specified events or characteristics of a computer system; for example, a device that counts the occurrences of various electrical events or measures the time between such events. *See also:* monitor; software monitor. **(B)** A software tool that records or analyzes hardware events during the execution of a computer program. *See also:* monitor; software monitor. (C) 610.12-1990

hardwire (test, measurement, and diagnostic equipment) Circuitry with the absence of electrical elements, such as resistors, inductors, capacitors: circuits containing only wire and terminal connections with no intervening switching inherent. (MIL) [2]

hardwired **(1)** **(supervisory control, data acquisition, and automatic control) (station control and data acquisition)** The implementation of processing steps within a device by way of the placement of conductors between components within the device. The processing steps are not alterable except by modifying the conducting paths between components. (SWG/PE/SUB) C37.1-1987s, C37.100-1992
(2) **(hydroelectric power plants)** Wired interconnections of relays and other control devices. (PE/EDPG) 1020-1988r
(3) Pertaining to a circuit or device whose characteristics are permanently determined by the interconnections between components. *Contrast:* programmable. (C) 610.10-1994w

hardwired logic A group of logic circuits permanently interconnected to perform a specific function. (C) 610.10-1994w

harmful interference Any emission, radiation, or induction that endangers the functioning, or seriously degrades, obstructs, or repeatedly interrupts a radiocommunication service or any other equipment or system operating in accordance with regulations. *See also:* electromagnetic compatibility. (EMC) [53]

harmful quantity of oil A discharge of oil that violates applicable water quality standards, causes a film or sheen upon or discoloration of the surface of the water or adjoining shorelines, or causes a sludge or emulsion to be deposited beneath the surface of the water or upon adjoining shorelines. (SUB/PE) 980-1994

harmonic (harmonic control and reactive compensation of static power converters) (converter characteristics) (self-commutated converters) A sinusoidal component of a periodic wave or quantity having a frequency that is an integral multiple of the fundamental frequency. *Note:* For example, a component, the frequency of which is twice the fundamental frequency, is called a second harmonic. *See also:* noncharacteristic harmonic; characteristic harmonic; relative harmonic content; harmonic components; harmonic content. (IA/SPD/PE/T&D/SPC) 936-1987w, C62.48-1995, 599-1985w, 519-1992, 1250-1995

harmonic analyzer A mechanical device for measuring the amplitude and phase of the various harmonic components of a periodic function from its graph. *See also:* wave analyzer; signal wave; instrument. (EEC/PE) [119]

harmonic, characteristic *See:* characteristic harmonic.

harmonic components (converter characteristics) (self-commutated converters) The components of the harmonic content as expressed in terms of the order and rms (root-mean-square) values of the Fourier series terms describing the periodic function. (IA/SPC) 936-1987w

harmonic conjugate *See:* Hilbert transform.

harmonic content (1) (converter characteristics) (self-commutated converters) The function obtained by subtracting the dc (direct current) and fundamental components from a nonsinusoidal periodic function. (IA/SPC) 936-1987w
(2) (nonsinusoidal periodic wave) The deviation from the sinusoidal form, expressed in terms of the order and magnitude of the Fourier series terms describing the wave. *See also:* rectification; power rectifier. (IA/SPC) [62]
(3) Distortion of a sinusoidal waveform characterized by indication of the magnitude and order of the Fourier series terms describing the wave. *Note:* For power lines, the harmonic content is small and of little concern for the purpose of field measurements, except at points near large industrial loads (saturated power transformers, rectifiers, aluminum and chlorine plants, etc.) where certain harmonics may reach 10% of the line voltage. Laboratory installations also may have voltage or current sources with significant harmonic content. (T&D/PE) 644-1994, 539-1990
(4) A measure of the presence of harmonics in a voltage or current wave form expressed as a percentage of the amplitude of the fundamental frequency at each harmonic frequency. The total harmonic content is expressed as the square root of the sum of the squares of each of the harmonic amplitudes (expressed as a percentage of the fundamental). (IA/PSE) 446-1995

harmonic conversion transducer (frequency multiplier, frequency divider) A conversion transducer in which the output-signal frequency is a multiple or submultiple of the input frequency. *Notes:* 1. In general, the output-signal amplitude is a nonlinear function of the input-signal amplitude. 2. Either a frequency multiplier or a frequency divider is a special case of harmonic conversion transducer. *See also:* transducer; heterodyne conversion transducer. (ED) 161-1971w

harmonic distortion (1) (data transmission) Nonlinear distortion of a system or transducer characterized by the appearance in the output of harmonics other than the fundamental component when the input wave is sinusoidal. *Note:* Subharmonic distortion may also occur. (PE) 599-1985w
(2) (broadband local area networks) A form of interference caused by the generation of signals according to the relationship N_f, where N is an integer greater than one and f is the original signal's frequency. (LM/C) 802.7-1989r
(3) For a pure sine wave input, output components at frequencies that are an integer multiple of the applied sine wave frequency. (IM/WM&A) 1057-1994w
(4) Nonlinear distortion that appears as harmonics of a single-frequency input. (PE/IC) 1143-1994r
(5) The mathematical representation of the distortion of the pure sine waveform. *See also:* distortion factor. (IA/PSE) 1100-1999

harmonic factor The ratio of the root-sum-square (rss) value of all the harmonics to the root-mean-square (rms) value of the fundamental.

$$\text{harmonic factor (for voltage)} = \frac{\sqrt{E_3^2 + E_5^2 + E_7^2 \ldots}}{E_1}$$

$$\text{harmonic factor (for current)} = \frac{\sqrt{I_3^2 + I_5^2 + I_7^2 \ldots}}{I_1}$$

(IA/SPC) 519-1992

harmonic leakage power (TR and pre-TR tubes) The total radio-frequency power transmitted through the fired tube in its mount at frequencies other than the fundamental frequencies generated by the transmitter. (ED) 161-1971w

harmonic, noncharacteristic *See:* noncharacteristic harmonic.

harmonic-restraint relay A restraint relay so constructed that its operation is restrained by harmonic components of one or more separate input quantities. (SWG/PE) C37.100-1992

harmonics *See:* harmonic components.

harmonic series A series in which each component has a frequency that is an integral multiple of a fundamental frequency. (SP) [32]

harmonic telephone ringer A telephone ringer that responds only to alternating current within a very narrow frequency band. *Note:* A number of such ringers, each responding to a different frequency, are used in one type of selective ringing. *See also:* telephone station. (EEC/PE) [119]

harmonic test (rotating machinery) A test to determine directly the value of one or more harmonics of the waveform of a quantity associated with a machine, relative to the fundamental of that quantity. *See also:* asynchronous machine. (PE) [9]

harmonization The process of ensuring that profiles do not overlap or conflict. (C/PA) 14252-1996

harness A component with a design of straps that is fastened about the worker in a manner so as to contain the torso and distribute the fall arrest forces over at least the upper thighs, pelvis, chest, and shoulders with means for attaching it to other components and subsystems. (NESC/T&D/PE) C2-1997, 1307-1996

harsh environment (nuclear power generating station) An environment expected as a result of the postulated service conditions appropriate for the design basis and post-design basis accidents of the station. (A design basis accident is that subset of a design basis event which requires safety function performance). Harsh environments are the result of a loss of cooling accident (LOCA)/high energy line break (HELB) inside containment and post-LOCA or HELB outside containment. (PE/NP) 323-1974s

hartley A unit of information content, equal to one decadal decision, or the designation of one of ten possible and equally likely values or states of anything used to store or convey information. *Notes:* 1. A hartley may be conveyed by one decadal code element. One hartley equals (log of 10 to base 2) times one bits. 2. If, in the definition of information content, the logarithm is taken to the base ten, the result will be expressed in hartleys. *Synonym:* dit. *See also:* bit. (IT/PE) [123], 599-1985w

Hartley oscillator An electron tube or solid state circuit in which the parallel-tuned tank circuit is connected between grid and plate, the inductive element of the tank having an intermediate tap at cathode potential, and the necessary feedback voltage obtained across the grid-cathode portion of the inductor. *See also:* radio-frequency generator. (IA) 54-1955w

Harvard class architecture A computer architecture with separate paths to main storage for instructions and data, allowing for a high memory bandwidth. *Contrast:* Von Neumann architecture. (C) 610.10-1994w

hash To calculate the hash value for a given item. *See also:* hashing. (C) 610.5-1990w

hash address *See:* hash value.

hash addressing *See:* hashing.

hash clash *See:* collision.

hash coding *See:* hashing.

hash function In hashing, the function used to determine the position of a given item in a set of items. *Note:* The function operates on a selected field, called a key, in each item and the function is generally a many-to-one mapping. *Synonyms:* key transformation function; calc algorithm. *See also:* key folding function; division transformation function; algebraic coding function; key transformation; mid-square function; radix transformation function; multiplication transformation function; digit transformation function. (C) 610.5-1990w

hash index *See:* hash value.

hashing A technique for arranging a set of items, in which a hash function is applied to the key of each item to determine its hash value. The hash value identifies each item's primary position in a hash table, and if this position is already occupied, the item is inserted either in an overflow table or in another available position in the table. *Synonyms:* scatter storage; hash coding; randomizing. *See also:* open-address hashing; separate chaining; hash addressing; collision resolution. (C) 610.5-1990w

hash search The use of a hash function and collision resolution to locate an item in a hash table. (C) 610.5-1990w

hash table A two-dimensional table of items in which a hash function is applied to the key of each item to determine its hash value. The hash value identifies each item's primary position in the table, and if this position is already occupied, the item is inserted either in an overflow table or in another available position in the table. (C) 610.5-1990w

hash total The result of summing two or more values of a set for purposes of validation or error detection. *Synonym:* control total. (C) 610.5-1990w

hash value The number generated by a hash function to indicate the position of a given item in a hash table. *Synonyms:* hash index; hash address. (C) 610.5-1990w

hatch A series of one or more sets of evenly spaced parallel lines within a closed boundary on a display surface. *See also:* crosshatch.

hatch

(C) 610.6-1991w

HATS *See:* head and torso simulator.

hauptnutzzeit *See:* utilization time.

Hay bridge A 4-arm alternating-current bridge in which the arms adjacent to the unknown impedance are nonreactive resistors and the opposite arm comprises a capacitor in series with a resistor. *Note:* Normally used for the measurement of inductance in terms of capacitance, resistance, and frequency. Usually, the bridge is balanced by adjustment of the resistor that is in series with the capacitor, and of one of the nonreactive arms. The balance depends upon the frequency. It differs from the Maxwell bridge in that in the arm opposite the inductor, the capacitor is in series with the resistor. *See also:* bridge.

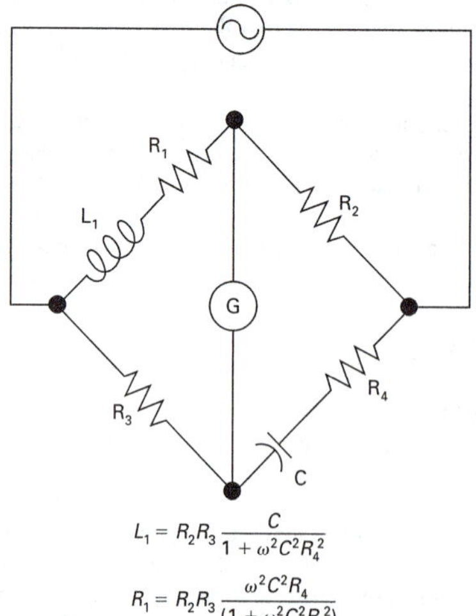

$$L_1 = R_2R_3\frac{C}{1+\omega^2C^2R_4^2}$$

$$R_1 = R_2R_3\frac{\omega^2C^2R_4}{(1+\omega^2C^2R_4^2)}$$

Hay bridge

(EEC/PE) [119]

hazard (1) (nuclear power generating station) A specified result of a design basis event that could cause unacceptable damage to systems or components important to safety. (PE/NP) 384-1981s

(2) (overhead power lines) A threat to the health, survival, or reproduction of an organism from some natural or artificial agent or event. (T&D/PE) 539-1990

(3) An intrinsic property or condition that has the potential to cause harm or damage. (DEI) 1221-1993w

(4) An existing or potential condition that can result in a mishap. (VT/RT) 1473-1999, 1483-2000

hazard analysis A systematic qualitative or quantitative evaluation of software for undesirable outcomes resulting from the development or operation of a system. These outcomes may include injury, illness, death, mission failure, economic loss, property loss, environmental loss, or adverse social impact. This evaluation may include screening or analysis methods to categorize, eliminate, reduce, or mitigate hazards. (C/SE) 1012-1998

hazard beacon (illuminating engineering) An aeronautical beacon used to designate a danger to air navigation. *Synonym:* obstruction beacon. (EEC/IE) [126]

hazard buoy *See:* danger buoy.

hazard current (1) (health care facilities) For a given set of connections in an isolated power system, the total current that would flow through a low impedance if it were connected between either isolated conductor and ground. The various hazard currents are: fault hazard current; monitor hazard current; total hazard current. *See also:* total hazard current; monitor hazard current; fault hazard current. (EMB) [47]

(2) (health care facilities) For a given set of connections in an isolated system, the total current that would flow through a low impedance if it were connected between either isolated conductor and ground. Fault Hazard Current: The hazard current of a given isolated system with all devices connected except the line isolation monitor. Monitor Hazard Current: The hazard current of the line isolation monitor alone. Total Hazard Current: The hazard current of a given isolated system with all devices, including the line isolation monitor, connected. (NESC/NEC) [86]

hazard-free logic A group of logic circuits that are not subject to failures due to logic failure conditions. (C) 610.10-1994w

hazardous (classified) locations Class I Locations. Class I locations are those in which flammable gases or vapors are or may be present in the air in quantities sufficient to produce explosive or ignitible mixtures. Class I locations shall include those specified in (a) and (b) below.

a) Class I, Division 1. A Class I, Division 1 location is a location:
   1) in which hazardous concentrations of flammable gases or vapors exist continuously intermittently, or periodically under normal operating conditions; or
   2) in which hazardous concentrations of such gases or vapors may exist frequently because of repair or maintenance operations or because of leakage; or
   3) in which breakdown or faulty operations of equipment or processes might release hazardous concentrations of flammable gases or vapors, and might also cause simultaneous failure of electric equipment. This classification usually includes locations where volatile flammable liquids or liquefied flammable gases are transferred from one container to another; interiors of spray booths and areas in the vicinity of spraying and painting operations where volatile flammable solvents are used; locations containing open tanks or vats of volatile flammable liquids; drying rooms or compartments for the evaporation of flammable solvents; portions of cleaning and dyeing plants where hazardous liquids are used; gas generator rooms and other portions of gas manufacturing plants where flammable gas may escape; inadequately ventilated pump rooms for flammable gas or for volatile flammable liquids; the interiors of refrigerators and freezers in which volatile flammable materials are stored in open, lightly stoppered, or easily ruptured containers; and all other locations where hazardous concentrations of flammable vapors or gases are likely to occur in the course of normal operations.

b) Class I, Division 2. A Class I, Division 3 location is a location
1) in which volatile flammable liquids or flammable gases are handled, processed, or used, but in which the hazardous liquids, vapors, or gases will normally be confined within closed containers or closed systems from which they can escape only in case of accidental rupture or breakdown of such containers or systems, or in case of abnormal operation of equipment; or
2) in which hazardous concentrations of gases or vapors are normally prevented by positive mechanical ventilation, and which might become hazardous though failure or abnormal operation of the ventilating equipment; or
3) that is adjacent to a Class I, Division I location, and to which hazardous concentrations of gases or vapors might occasionally be communicated unless such communication is prevented by adequate positive-pressure ventilation from a source of clean air, and effective safeguards against ventilation failure are provided. This classification usually includes locations where volatile flammable liquids or flammable gases or vapors are used, but which, in the judgment of the authority having jurisdiction, would become hazardous only in case of an accident or of some unusual operating condition, The quantity of hazardous material that might escape in case of accident, the adequacy of ventilating equipment, the total area involved, and the record of the industry or business with respect to explosions or fires are all factors that merit consideration in determining the classification and extent of each location. Piping without valves, checks, meters and similar devices would not ordinarily introduce a hazardous condition even even though used for hazardous liquids or gases. Locations used for the storage of hazardous liquids or of compressed gases in sealed containers would not normally be considered hazardous unless subject to other hazardous conditions also. Electrical conduits and their associated enclosures separated from process fluids by a single seal or barrier shall include those specified in (a) and (b) below.

c) Class II, Division 1. A Class II, Division I location is a location
1) in which combustible dust is or may be in suspension in the air continuously, intermittently, or periodically under normal operating conditions, in quantities sufficient to produce explosive or ignitible mixtures; or
2) where mechanical failure or abnormal operation of machinery or equipment might cause such explosive or ignitible mixtures to be produced and might also provide a source of ignition through simultaneous failure of electric equipment, operation of protection devices or from other causes; or
3) in which combustible dusts of an electrically conductive nature may be present. This classification usually includes the working areas of grain handling and storage plants; rooms containing grinders or pulverizers, cleaners, graders, scalpers, open conveyors or spouts, open bins or hoppers, mixers or blenders, automatic hopper scales, packing machinery, elevator heads and boots, stock distributors, dust and stock collectors (except all-metal collectors vented to the outside), and all similar dust-producing machinery and equipment in grain-processing plants, starch plants, sugar-pulverizing plants, malting plants, hay-grinding plants, and other occupancies of similar nature; coal-pulverizing plants (except where the pulverizing equipment is essentially dust-tight); all working areas where metal dusts and powders are produced, processed, handled, packed, or stored (except in tight containers); and all other similar locations where combustible dust may, under normal operating conditions be present in the air in quantities sufficient to produce explosive or ignitible mixtures. Combustible dusts which are electrically nonconductive include dusts produced in the handling and processing of grain and grain products, pulverized sugar and cocoa, dried egg and milk powders, pulverized spices, starch and pastes, potato and woodflour, oil meal from beans and seed, dried hay, and other oganic materials which may produce combustible dusts when processed or handled. Electrically conductive nonmetallic dusts include dusts from pulverized coal, coke, carbon black, and charcoal. Dusts containing magnesium or aluminum are particularly hazardous and the use of extreme precaution will be necessary to avoid ignition and explosion.

d) Class II, Division 2. A Class II, Division 2 location is a location in which combustible dust will not normally be in suspension in the air or will not be likely to be thrown into suspension by the normal operation of equipment or apparatus in quantities sufficient to produce explosive or ignitible mixtures, but;
1) where deposits or accumulations of such combustible dust may be sufficient to interfere with the safe dissipation of heat from electric equipment or apparatus; or
2) where such deposits or accumulations of combustible dust on, in, or in the vicinity of electric equipment might be ignited by arcs, sparks, or burning material from such equipment. Locations where dangerous concentrations of suspended dust would not be likely but where dust accumulations might form on, or in the vicinity of electric equipment, would include rooms and areas containing only closed spouting and conveyors, closed bins or hoppers, or machines and equipment from which appreciable quantities of dust would escape only under abnormal operating conditions; rooms or areas adjacent to a Class II, Division 1 location as described in (a) above.

(NESC) [86]

hazardous area class I The locations in which flammable gases or vapors are or may be present in the air in quantities sufficient to produce explosive or ignitible mixtures. *See also:* explosion-proof apparatus. (NESC) [86]

hazardous electrical condition (electrolytic cell line working zone) Exposure of personnel to surfaces, contact with which may result in the flow of injurious electrical current. (IA/PC) 463-1993w

hazardous levels of nonionizing electromagnetic radiation (radio frequency radiation hazard warning symbol) Incident electromagnetic energy that may be biologically detrimental or may directly or indirectly cause ignition of explosive materials or vapors. (NIR) C95.2-1982r

hazardous levels of radio-frequency energy Term used to describe incident RF energy that may be biologically detrimental or may directly or indirectly cause ignition of explosive materials or vapors. (NIR/SCC28) C95.2-1999

hazardous location (illuminating engineering) An area where ignitable vapors or dust may cause a fire or explosion created by energy emitted from lighting or other electrical equipment or by electrostatic generation. (EEC/IE) [126]

hazardous material (1) Any material that has been so designated by governmental agencies or adversely impacts human health or the environment. (PE/SUB) 1127-1998
(2) Those vapors, dusts, fibers or flyings which are explosive under certain conditions. (SWG/PE) C37.100-1981s

hazardous substance (liquid-filled power transformers) A quantity of material offered for transportation in one package or transport vehicle, when the material is not packaged that equals or exceeds the reportable quantity (RQ) specified for the material in Code of Federal Regulations (CFR), Title 40, Parts 116 and 117. (LM/C) 802.2-1985s

haze *See:* fog.

H beacon (electronic navigation) A designation applied to two types of facilities: A) A nondirectional radio beacon for homing by means of an airborne direction finder. B) A radar air navigation system using an airborne interrogator to measure the distances from two ground transponders. *See also:* navigation. (AES/RS) 686-1982s, [42]

H channel A wideband channel (i.e., a channel that contains multiples of 64 kbit/s). (C/COM/LM) 8802-9-1996, 802.9a-1995w

HCI *See:* human/computer interface.

HCL *See:* relay, high, common, low.

HCP Horizontal coupling plane. *See also:* coupling plane. (EMC) C63.16-1993

HC threshold voltage *See:* high conduction threshold voltage.

HDA *See:* head/disk assembly.

HDAM *See:* hierarchical direct access method.

HDBH *See:* high day busy-hour load.

HDBH load *See:* high day busy-hour load; time-consistent traffic measures.

H-display (1) (navigation aid terms) A type of radar display format. (AES/GCS) 172-1983w
(2) A B-display modified to include an indication of angle of elevation. The target appears as two closely spaced blips approximating a short bright line, the slope of which is in proportion to the tangent of the angle of target elevation.

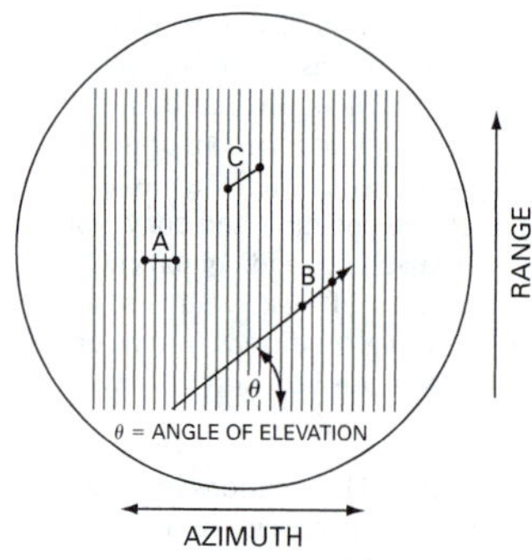

H-display

(AES) 686-1997, [42]

HDL *See:* hardware description language; hardware design language.

HDLC *See:* high-level data link control.

head (1) A device that reads, records, or erases data on a storage medium. *Note:* For example, a small electromagnet used to read, write, or erase data on a magnetic drum or tape, or the set of perforating, reading, or marking devices used for punching, reading, or printing on paper tape. (C) [20], [85]
(2) (hydroelectric power plants) The difference in hydraulic energy between two points, which includes the elevation head, pressure head, and velocity head. (PE/EDPG) 1020-1988r
(3) (data management) The first data item in a list. *Synonym:* header. (C) 610.5-1990w
(4) (A) A device that reads, writes, or erases data on a storage medium. For example, a small electromagnet (magnetic head) used to read, write, or erase data on a magnetic drum or magnetic tape, or a device such as a laser that reads and writes data on an optical storage medium. *See also:* write head; read/write head; read head; magnetic head. **(B)** A device within an output device, such as a printer or display device, that controls the creation of images on the device. *See also:* scan head; plotting head; display head; print head. (C) 610.10-1994
(5) *See also:* head water benefits; gross head; rated head; critical head; net head.

head and torso simulator (HATS) A device that accurately reproduces the sound transmission and pick-up characteristics of the median head and torso of adult humans. (COM/TA) 1206-1994, 1329-1999

head crash The sudden and complete failure of a disk drive caused by a physical collision between the read/write head and the surface of the recording medium. *Note:* Usually results in destruction of the head and part or all of the data on the medium. *See also:* disk crash. (C) 610.10-1994w

head/disk assembly In a magnetic disk device, an assembly that includes the magnetic disk, magnetic head, and an access mechanism. (C) 610.10-1994w

headed brush (rotating machinery) A brush having a top (cylindrical, conical, or rectangular) with a smaller cross section than the cross section of the body of the brush. *Note:* The length of the head shall not exceed 25 percent of the overall length. *See also:* brush. (EEC/LB) [101]

headend (1) (A) (broadband local area networks) The central location that has access to signals traveling in both inbound and outbound directions. The logical root of the broadband coaxial cable system. **(B) (broadband local area networks)** The physical location where the inbound and outbound paths are accessible. The headend is also called the central retransmission facility. (LM/C) 802.7-1989
(2) A point where two or more half-duplex data paths are joined on the communications network. (C) 610.7-1995
(3) In 10BROAD36, the location in a broadband system that serves as the root for the branching tree comprising the physical medium; the point to which all inbound signals converge and the point from which all outbound signals emanate. (C/LM) 802.3-1998

headend port (broadband local area networks) (or ports) An interface where connection(s) may be made to insert signals or remove signals from the cable plant. The headend ports are usually referred to with a direction (inbound or outbound). Headend ports provide central access to the cable system. (LM/C) 802.7-1989r

head-end system (railways) A system in which the electrical requirements of a train are supplied from a generator or generators, located on the locomotive or in one of the cars, customarily at the forward part of the train. *Note:* The generators may be driven by steam turbine, internal-combustion engine, or, if located in one of the cars, by a mechanical drive from a car axle. *See also:* axle-generator system. (EEC/PE) [119]

header (1) A transverse raceway for electric conductors, providing access to predetermined cells of a cellular metal floor, thereby permitting the intallation of electric conductors from a distribution center to the cells. (NESC/NEC) [86]
(2) (A) (software) A block of comments placed at the beginning of a computer program or routine. **(B) (software)** Identification or control information placed at the beginning of a file or message. *Contrast:* trailer. (C) 610.12-1990
(3) (A) (data management) Pertaining to data that describes and pertains to other data. For example, the header record for a file might describe the format for the remaining records in the file. **(B) (data management)** *See also:* head. (C) 610.5-1990
(4) *See also:* running header. (C) 610.2-1987
(5) A structure attached to or integral to the top of the heatsink used for structural performance and marking. (C/BA) 1101.3-1993
(6) The contiguous control bits preceding a frame, packet, block, or other data stream of bits that contain information about the message such as the address, type of frame, and/or sequencing. *Contrast:* trailer. (C) 610.7-1995

header card A punch card that contains information identifying data in the cards that follow. *Contrast:* trailer card. (C) 610.10-1994w

header hub (HH) The highest-level hub in a hierarchy of hubs. The HH broadcasts signals transmitted to it by lower level hubs or DTEs such that they can be received by all DTEs that may be connected to it either directly or through intermediate hubs. (LM/C) 802.3-1998, 610.7-1995

header label An internal label, immediately preceding the first record of a file, that identifies the file and contains data used in file control. (C) 610.5-1990w

HEADER packet A packet originating in the MTM-Bus Master that is the first packet of an MTM-Bus message. The HEADER packet includes an address and a command field. The address identifies which S-module(s) are to interpret and act upon the command contained within the command field. (TT/C) 1149.5-1995

head gap (1) (test, measurement, and diagnostic equipment) The space or gap intentionally inserted into the magnetic circuit of the head in order to force or direct the recording flux into or from the recording medium. (MIL) [2]
(2) The distance between a read/write head and the surface of a recording medium. (C) 610.10-1994w

heading (navigation) The horizontal direction in which a vehicle is pointed, expressed as an angle between a reference line and the line extending in the direction the vehicle is pointed, usually measured clockwise from the reference line. *See also:* navigation. (AES/RS) 686-1982s, [42]

heading-effect error (navigation) A manifestation of polarization error causing an error in indicated bearing that is dependent upon the heading of a vehicle with respect to the direction of signal propagation. *Note:* Heading-effect error is a special case of attitude-effect error where the vehicle is in a straight level flight: it is sometimes referred to as course push (or pull). *See also:* navigation. (AES/RS) 686-1982s, [42]

headlamp (illuminating engineering) A major lighting device mounted on a vehicle and used to provide illumination ahead of it. *Synonym:* headlight. (EEC/IE) [126]

headlight *See:* headlamp.

head loading zone The relative distance that a read/write head travels with respect to a rotating storage device in order to achieve the proper clearance between the head and the surface of the medium. (C) 610.10-1994w

headloss (hydroelectric power plants) Loss of potential energy mainly due to hydraulic friction. This loss is usually expressed in feet or meters of head. (PE/EDPG) 1020-1988r

head of bus function The function that generates *empty Queued Arbitrated (QA) slots, Pre-Arbitrated (PA) slots,* and *management information octets* at the point on each bus where data flow starts. The Head of Bus function also inserts the *virtual channel identifier* in the *PA segment header* of PA slots. (LM/C) 8802-6-1994

head or butt cable *See:* hand (head or butt) cable (mining).

head-per-track disk drive A disk drive in which one fixed head is located over each track on the drive. (C) 610.10-1994w

head positioner A component within a storage device that positions a floating head over a specific track on the storage medium. (C) 610.10-1994w

headquarters system (direct-connected system) A local system to which has been added means of transmitting system signals to and receiving them at an agency maintained by the local government, for example, in a police precinct house, or fire station. *See also:* protective signaling. (PE/EEC) [119]

head receiver An earphone designed to be held to the ear by a headband. *Note:* One or a pair (one for each ear) of head receivers with associated headband and connecting cord is known as a headset. (EEC/PE) [119]

headset An assembly, including a transmitter and receiver, intended to be worn on the head of the user. (COM/TA) 1206-1994

head space (test, measurement, and diagnostic equipment) The space between the reading or recording head and the recording medium, such as tape, drum or disc. (MIL) [2]

head switching (A) The use of two read/write heads, one to read from the medium and one to write on another medium. **(B)** The process of switching from one head to another, either on the same or on different storage media. (C) 610.10-1994

headwater (hydroelectric power plants) Source of energy for a hydraulic turbine. (PE/EDPG) 1020-1988r

head water benefits (power operations) The benefits brought about by the storage or release of water by a reservoir project upstream. Application of the term is usually in reference to benefits to a downstream hydroelectric power plant. (PE/PSE) 858-1987s

headway The time interval between the passing of the front ends of successive vehicles or trains moving along the same lane or track in the same direction. (VT/RT) 1474.1-1999

health Summary information regarding the current ability of a system or subsystem to perform its intended function. (VT) 1482.1-1999

health care facilities (health care facilities) Buildings, portions of buildings, and mobile facilities, that contain but are not necessarily limited solely to premises designed for use as hospitals, nursing homes, residential custodial care facilities, clinics, or medical and dental offices. (NESC/NEC) [86]

health information system (HIS) *See:* hospital information system.

heap (data management) A complete binary tree in which the key for each child node contains the key from its parent plus some additional value.

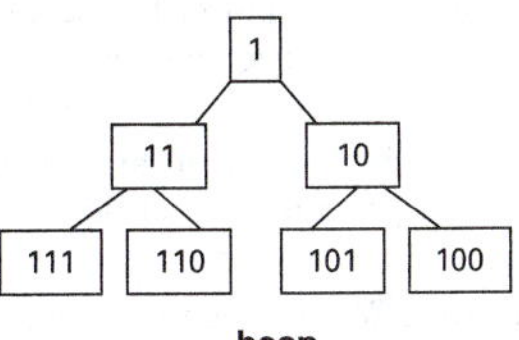

heap

(C) 610.5-1990w

heapsort A tree selection sort in which the items to be sorted are used to build a heap, and the items are then selected from the heap in the sorted order. (C) 610.5-1990w

hearing loss (1) (for speech) The difference in decibels between the speech levels at which the average normal ear and the defective ear, respectively, reach the same intelligibility, often arbitrarily set at 50 %.
(2) (hearing-threshold level) (ear at a specified frequency) The amount, in decibels, by which the threshold of audibility for that ear exceeds a standard audiometric threshold. *Notes:* 1. *See:*American Standard Specification for Audiometers for General Diagnostic Purposes. 2. This concept was at one time called deafness: such usage is now deprecated. 3. Hearing loss and deafness are both legitimate qualitative terms for the medical condition of a moderate or severe impairment of hearing, respectively. Hearing level, however, should only be used to designate a quantitative measure of the deviation of the hearing threshold from a prescribed standard. (SP) [32]

heartbeat A signal or a message passed between cooperating processes to indicate continuing proper operations. *See also:* signal quality error heartbeat. (C) 610.7-1995

heat capacity (1) The amount of heat necessary to raise the temperature of a given mass of a substance 1°—the mass multiplied by the specific heat. (IA/PSE) 241-1990r
(2) *(A) homogeneous conductors:* The specific heat of the conductor's material times the mass per unit length. *(B) nonhomogeneous conductors:* The sum of the heat capacities of the conductor's component materials. (T&D/PE) 738-1993
(3) The heat required to raise the temperature of a unit mass of material by one degree. (DEI) 1221-1993w

heat detector (1) (burglar-alarm system) A temperature-sensitive device mounted on the inside surface of a vault to initiate an alarm in the event of an attack by heat or burning. *See also:* protective signaling. (PE/EEC) [119]
(2) (fire alarm system) A device that detects abnormally high temperature or rate-of-temperature rise to initiate a fire alarm. (NFPA) [16]

heater (1) (electric pipe heating systems) A length of resistance material connected between terminals and used to generate heat electrically. Heaters, as used in this application, can take the form of cables with various sheath materials, blankets, and pads. *Synonym:* heating element. (PE/EDPG) 622-1979s
(2) (electron tube) An electric heating element for supplying heat to an indirectly heated cathode. *See also:* electrode. (ED) 161-1971w

heater coil *See:* load, work, or heater coil.

heater connector (heater plug) A cord connector designed to engage the male terminal pins of a heating or cooking appliance. (PE/EEC) [119]

heater current The current flowing through a heater. *See also:* filament current; electronic controller; preheating time cathode. (ED) [45]

heater de-energized maximum intermittent exposure temperature The maximum temperature of any surface adjacent to the heating device that the de-energized heating device can withstand for specified periods. (IA) 515-1997

heater energized maximum intermittent exposure temperature The maximum temperature of any surface adjacent to the heating device that the energized heating device can withstand for specified periods. (IA) 515-1997

heater transformer Supplies power for electron-tube filaments or heaters of indirectly heated cathodes. *See also:* electronic controller. (IA/ICTL/IAC) [60]

heater voltage The voltage between the terminals of a heater. *See also:* electronic controller; electrode voltage. (ED) [45]

heater warm-up time *See:* cathode heating time.

heat exchanger *See:* cooler.

heat-exchanger cooling system (1) (rectifier) A cooling system in which the coolant, after passing over the cooling surfaces of the rectifier, is cooled in a heat exchanger and recirculated. *Note:* The coolant is generally water of a suitable purity, or water that has been treated by a corrosion-inhibitive chemical. Antifreeze solutions may also be used where there is exposure to low temperatures. The heat exchanger is usually either:

1) Water-to-water where the heat is removed by raw water,
2) Water to-air where the heat is removed by air supplied by a blower,
3) Air-to-water,
4) Air-to-air,
5) Refrigeration cycle. The liquid in the closed system may be other than water, and the gas in the closed system may be other than air.

See also: rectifier; rectification. (IA) [62]
(2) (thyristor controller) A cooling system in which the coolant, after passing over the cooling surfaces of the thyristor controller components, is cooled in a heat exchanger and recirculated. *Note:* Heat may be removed from the thyristor controller component's cooling surfaces by liquid or air using the following types of heat exchangers:

a) Water-to-water,
b) Water-to-air,
c) Air-to-water,
d) Air-to-air,
e) Refrigeration cycle. The liquid in the closed system may be other than water, and the gas in the closed system may be other than air.

(IA/IPC) 428-1981w

heat flux The flow of heat per unit area; i.e., thermal energy incident upon a surface area per unit time [kW/m^2 (kJ/m^2/s)]. (DEI) 1221-1993w

heating cycle One complete operation of the thermostat from ON to ON or from OFF to OFF. (IA/APP) [90]

heating device Heating cable or surface heating unit. (IA) 515-1997

heating element A length of resistance material connected between terminals and used to generate heat electrically. *See also:* appliance outlet. (IA/APP) [90]

heating, glue line *See:* glue-line heating.

heating pattern The distribution of temperature in a load or charge. *See also:* induction heating. (IA) 54-1955w, 169-1955w

heating station (dielectric heating) The assembly of components, which includes the work coil or applicator and its associated production equipment. (IA) 54-1955w

heating system, radiant *See:* radiant heating system.

heating time, tube *See:* preheating time.

heating unit (1) (electrical appliances) An assembly containing one or more heating elements, electric terminals or leads, electrical insulation, and a frame, casing, or other suitable supporting means. *See also:* appliance outlet; appliance. (IA/APP) [90]
(2) A structure containing one or more heating elements, electrical terminals or leads, electric insulation and a frame or casing, all of which are assembled into one unit. (IA/PSE) 241-1990r

heating-up run (rotating electric machinery) A period of operation with current and ventilation designed to bring the machine to approximately its temperature-rise limit. (PE/EM) 11-1980r

heating ventilation and air conditioning (HVAC) system pull down The condition wherein the air conditioning system in the car is turned on and is required to cool a car that has been sitting in the heat of the day. *Note:* This condition usually presents the highest/longest sustained power demands on the auxiliary inverter. (VT) 1476-2000

heat loss (1) (electrical heating systems) A quantitative value of energy flow from a pipe, a vessel, or equipment to the surrounding ambient. (BT/IA/AV/PC) 152-1953s, 844-1991, 515-1997
(2) (waveguide terms) The part of the transmission loss due to the conversion of electric energy into heat. (MTT) 146-1980w
(3) A quantitative value of the rate of thermal energy flow from a pipe, vessel, or equipment to the surrounding ambient. (IA/PC) 515.1-1995

heat of combustion The thermal energy (chemical, convective, and radiative) per unit mass, i.e., MJ/kg (BTU/lb), released during burning. (DEI) 1221-1993w

heat of gasification The heat required to convert a unit mass of material to a vapor. (DEI) 1221-1993w

heat pump (electric power systems in commercial buildings) A refrigerating system employed to transfer heat into a space or substance. The condenser provides the heat, while the evaporator is arranged to pick up heat from the air, water, etc. By shifting the flow of air or other fluid, a heat pump system may also be used to cool a space. (IA/PSE) 241-1990r

heat rate (generating station) A measure of generating station thermal efficiency, generally expressed as BTU per kilowatt hour. *Note:* It is computed by dividing the total BTU content of the fuel burned (or of heat released from a nuclear reactor) by the resulting kilowatt hours generated. (PE/PSE) 346-1973w

heat rejection rate (nuclear power generating station) The rate at which a module emits heat energy to its environment (watts/hr or btu). (PE/NP) 381-1977w

heat release rate The rate of energy release associated with the combustion of a material, i.e., kW/m^2(kJ/m^2/s). (DEI) 1221-1993w

heat-run test A test that verifies equipment to be within thermal design when operated for an extended period at maximum current. *Synonym:* loading test. (PE/SUB) 1378-1997

heat-shield (cathode) (electron tube) A metallic surface surrounding a hot cathode, in order to reduce the radiation losses. *See also:* electron tube. (ED) [45], [84]

heat shrink A joint that consists of a tube or a series of tubes that are applied over the conductor and reduced in diameter over the cable with the use of externally applied heat. (PE/IC) 404-1993

heat sink (1) (electrical heat tracing for industrial applications) A part that conducts and dissipates heat away from the pipeline or vessel (the pipe or equipment). Heat sinks, as related to pipe heating systems, can be pipe supports, valve operators, etc. (BT/IA/AV/PC) 152-1953s, 844-1991

(2) (electric pipe heating systems) A part that absorbs heat. Heat sinks, as related to electric pipe heating systems, are those masses of materials that are directly connected to mechanical piping, valves, tanks, etc. that can absorb the heat generated by heaters, thus reducing the effect of the heater. Typical heat sinks can be pipe hangers, valve operators, etc. (PE/EDPG) 622-1979s

(3) (semiconductor rectifier diode) A mass of metal generally having much greater thermal capacity than the diode itself and intimately associated with it. It encompasses that part of the cooling system to which heat flows from the diode by thermal conduction only and from which heat may be removed by the cooling medium. (IA) [62]

(4) (photovoltaic power system) A material capable of absorbing heat: a device utilizing such material for the thermal protection of components or systems. (AES) [41]

(5) A part, also referred to as a frame, that serves both as a structural support and the principal thermal conduction medium. The PWB is attached to the heat sink. The heat sink spreads and conducts the heat from the components to the guide ribs (for conduction-cooled modules only). (C/BA) 1101.3-1993

(6) A part that conducts and dissipates heat away from the pipe or equipment. Heat sinks, as related to pipe heating systems, can be pipe supports, valve operators, etc. (IA) 515-1997

heat, specific *See:* specific heat.

heat tracing (1) (electrical heating systems) A heating system where the externally applied heat source follows (traces) the object to be heated. (IA/PC) 844-1985s

(2) (electrical heat tracing for industrial applications) The utilization of electric heating cables, other electric heating devices, and support components that are externally applied and used to maintain or raise the temperature of fluids in piping and associated equipment. (BT/AV) 152-1953s

heat-transfer aids (electrical heat tracing for industrial applications) Thermally conductive materials, such as metallic foils or heat-transfer cements, used to increase the heat-transfer rates from the heating cables to the process piping or equipment. (BT/IA/AV) 152-1953s, 515-1997

Heaviside-Campbell mutual-inductance bridge A mutual-inductance bridge of the Heaviside type in which one of the inductive arms contains a separate inductor that is included in the bridge arm during the first of a pair of measurements and is short-circuited during the second. *Note:* The balance is independent of the frequency. *See also:* Heaviside mutual-inductance bridge; bridge.

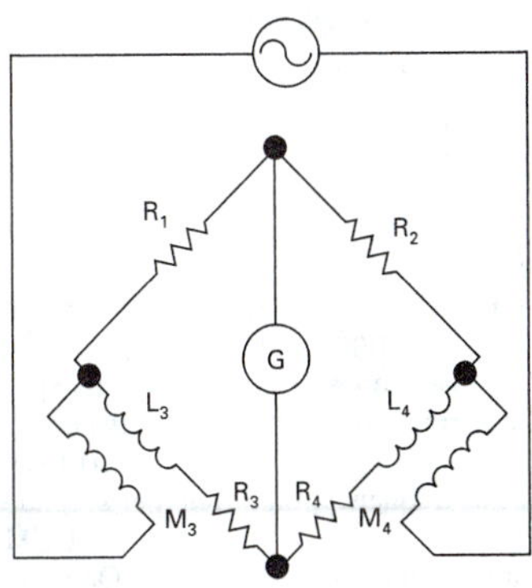

$$R_x = (R_3 - R'_3)\frac{R_2}{R_1}$$

$$L_x = (M - M')\left(1 + \frac{R_2}{R_1}\right)$$

Heaviside-Campbell mutual-inductance bridge

(EEC/PE) [119]

Heaviside-Lorentz system of units A rationalized system based on the centimeter, gram, and second and is similar to the Gaussian system but differs in that a factor 4ν is explicitly inserted to multiply r^2 in each of the formulations of the Coulomb Laws. (Std100) 270-1966w

Heaviside mutual-inductance bridge An alternating-current bridge in which two adjacent arms contain self-inductance, and one or both of these have mutual inductance to the supply circuit, the other two arms being normally nonreactive resistors. *Note:* Normally used for the comparison of self- and mutual inductances. The balance is independent of the frequency. *See also:* bridge.

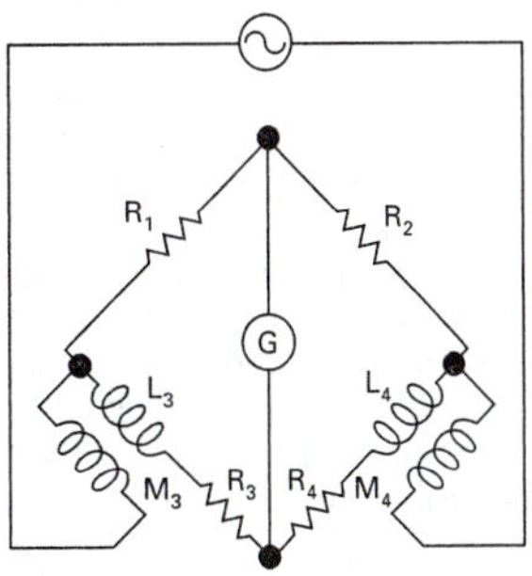

$$R_1R_4 = R_2R_3$$

$$L_3 - L_4\left(\frac{R_1}{R_2}\right) = -(M_3 - M_4)\left(1 + \frac{R_1}{R_2}\right)$$

Heaviside mutual-inductance bridge

(EEC/PE) [119]

heavy-duty floodlight (hd) (illuminating engineering) A weatherproof unit having a substantially constructed metal housing into which is placed a separate and removable reflector. A weatherproof hinged door with cover glass encloses the assembly but provides an unobstructed light opening at least equal to the effective diameter of the reflector. (IE/EEC) [126]

heavy load *See:* test current.

heavy rail transit A mode of rail rapid transit generally characterized by fully grade-separated construction, operation on exclusive rights-of-way, and station platforms at the floor level of the vehicles. (VT/RT) 1475-1999, 1474.1-1999, 1476-2000

heavy rail vehicle A vehicle operating on a heavy rail transit system. Typically, electrically propelled, bidirectional, capable of operating in multiple unit, and designed for rapid, high-level boarding and discharging of passengers. (VT) 1475-1999, 1476-2000

HE11mode (fiber optics) Designation for the fundamental mode of an optical fiber. *See also:* fundamental mode. (Std100) 812-1984w

height (1) (data management) In a tree, the maximum number of levels between the root node and a terminal node. *Synonym:* depth. *See also:* height-balanced tree. (C) 610.5-1990w

(2) By convention, the height axis is parallel to the connectors. (C/MM) 1101.2-1992

height balance In a tree, the maximum difference in height of any two subtrees of any node. *Note:* A height balance of *k* is written HB-*k*. (C) 610.5-1990w

height-balanced *k*-tree A tree whose height balance is *k*. (C) 610.5-1990w

height-balanced tree A tree whose height balance is 1. *Synonym:* balanced tree. *Contrast:* weight-balanced tree. *See also:* B-tree; Adel'son-Velskii and Landis tree; *n–m* tree. (C) 610.5-1990w

height-finding radar A radar whose function is to measure the range and elevation angle to a target, thus permitting computation of altitude or height. *Note:* Such a radar usually accompanies a surveillance radar that determines other target parameters. *See also:* nodding-beam height finder; three-dimensional radar. (AES) 686-1997

height gain The variation in electromagnetic field strength above a surface, expressed as gain relative to a fixed reference height. *Notes:* 1. This ratio is generally expressed in decibels and may have a negative value. 2. The reference height may be at the Earth's surface. (AP/PROP) 211-1997

height marker *See:* calibration marks.

helical antenna (data transmission) An antenna whose configuration is that of a helix. *Note:* The diameter, pitch, and number of turns in relation to the wavelength provide control of the polarization state and directivity of helical antennas. (PE/AP/ANT) 599-1985w, 145-1993

helical plate (storage cell) A plate of large area formed by helically wound ribbed strips of soft lead inserted in supporting pockets or cells of hard lead. *See also:* battery. (PE/EEC) [119]

helicopter work A technique of using a helicopter for performing live maintenance on energized wires and equipment, whereby one or more line workers work directly on an energized part using live tools after being raised and bonded to the energized wire or equipment. (T&D/PE) 516-1995

Helmholtz coils A pair of coils proportioned to provide a known uniform magnetic field that may be used to calibrate a probe coil. (COM/TA) 1027-1996

help *See:* help information.

help file A file containing help information. (C) 610.5-1990w

help information Information available for display to the user of a computer system, describing system features and use. *Synonym:* help. *See also:* help menu. (C) 610.2-1987

help menu A menu that gives the user a choice of topics for which help information is available on a given computer system. (C) 610.2-1987

hemispherical-conical reflectance (illuminating engineering) Ratio of reflected flux collected over a conical solid angle to the incident flux from the entire hemisphere. (See corresponding figure.) *Note:* The direction and extent of the cone must be specified.

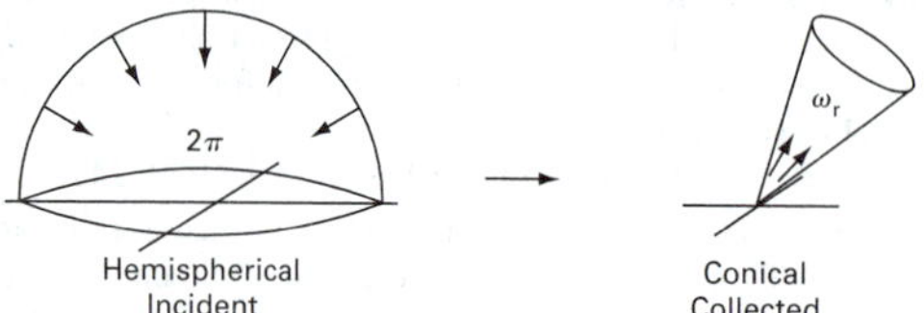

hemispherical-conical reflectance

(EEC/IE) [126]

hemispherical-conical transmittance (illuminating engineering) Ratio of transmitted flux collected over a conical solid angle to the incident flux from the entire hemisphere. (See corresponding figure.) *Note:* The direction and extent of the cone must be specified.

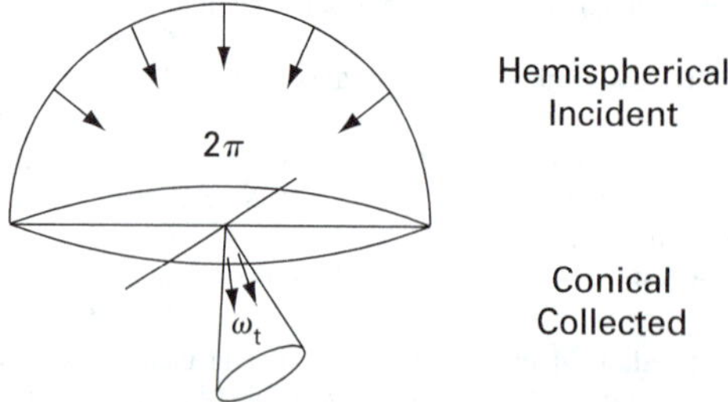

hemispherical-conical transmittance

(EEC/IE) [126]

hemispherical-directional reflectance (illuminating engineering) Ratio of reflected flux over an element of solid angle surrounding the given direction to the incident flux from the entire hemisphere. (See corresponding figure.) *Note:* The direction of collection and the size of the solid angle "element" must be specified.

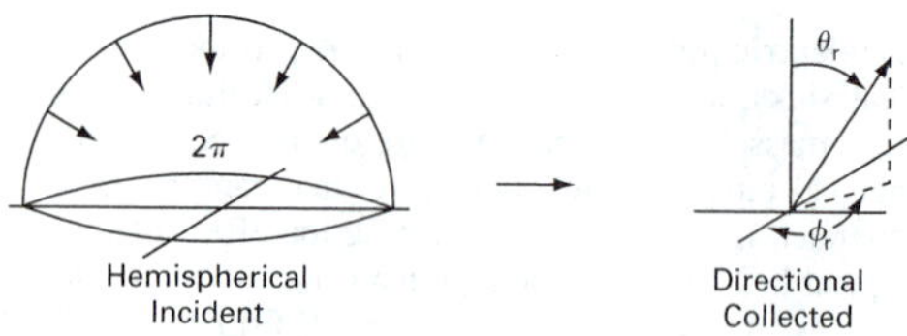

hemispherical-directional reflectance

(EEC/IE) [126]

hemispherical-directional transmittance (illuminating engineering) Ratio of transmitted flux collected over an element of solid angle surrounding the given direction to the incident flux from the entire hemisphere. *Note:* The direction of collection and size of the solid angle "element" must be specified. (EEC/IE) [126]

hemispherical reflectance† **(illuminating engineering)** The ratio of all of the flux leaving a surface or medium by reflection to the incident flux. Note: This term is obsolete, and retained for reference purposes only. (EEC/IE) [126]

† Obsolete.

hemispherical transmittance (illuminating engineering) The ratio of the transmitted flux leaving a surface or medium to the incident flux. *Note:* If transmittance is not preceded by an adjective descriptive of the angles of view, hemispherical transmittance is implied. *See also:* transmission. (EEC/IE) [126]

henry (metric practice) The inductance of a closed circuit in which an electromotive force of one volt is produced when the electric current in the circuit varies uniformly at a rate of one ampere per second. (QUL) 268-1982s

heptode A seven-electrode electron tube containing an anode, a cathode, a control electrode, and four additional electrodes that are ordinarily grids. (ED) 161-1971w

hermetically sealed relay A relay in a gastight enclosure that has been completely sealed by fusion or other comparable means to insure a low rate of gas leakage over a long period of time. *See also:* relay. (EEC/REE) [87]

hermetic motor A stator and rotor without shaft, end shields, or bearings for installation in refrigeration compressors of the hermetically sealed type. (PE) [9]

hermetic refrigerant motor-compressor (air-conditioning and refrigerating equipment) A combination consisting of a compressor and motor, both of which are enclosed in the same housing, with no external shaft or shaft seals, the motor operating in the refrigerant. (NESC/NEC) [86]

hermitian form (1) The nxn matrix is hermitian if its conjugate transpose is equal to itself. In terms of a set of complex variables; $x_1, x_2, \ldots x_n$; the quantity

$$[\overline{x_1}\overline{x_2}. \ . \ \overline{x_n}][A]\begin{bmatrix} \underline{x_1} \\ \underline{x_2} \\ \underline{\cdot\ \cdot\ \cdot} \\ x_n \end{bmatrix}$$

is the hermitian form of $[A]$. (CAS) [13]

(2) (metric practice) The frequency of a periodic phenomenon of which the period is one second. (QUL) 268-1982s

hertz (Hz) (1) (laser maser) The unit which expresses the frequency of a periodic oscillation in cycles per second. (LEO) 586-1980w

(2) The unit of frequency, one cycle per second. (IA/MT) 45-1998

(3) The unit for expressing frequency, f. One hertz equals one cycle per second. (NIR) C95.1-1999

Hertzian electric dipole An elementary source consisting of a time-harmonic electric current element of specified direction and infinitesimal length. *Notes:* 1. The continuity equation relating current to charge requires that opposite ends of the current element be terminated by equal and opposite amounts of electric charge, these amounts also varying harmonically with time. 2. As its length approaches zero, the current must approach infinity in such a manner that the product of current and length remains finite. (AP/ANT) 145-1993

Hertzian magnetic dipole A fictitious elementary source consisting of a time-harmonic magnetic current element of specified direction and infinitesimal length. *Notes:* 1. The continuity equation relating current to charge requires that opposite ends of the current element be terminated by equal and opposite amounts of magnetic charge, these amounts also varying harmonically with time. 2. As its length approaches zero, the current must approach infinity in such a manner that the product of current and length remains finite. 3. A magnetic dipole has the same radiation pattern as an infinitesimally small electric current loop. (AP/ANT) 145-1993

heterodyne (nonlinear, active, and nonreciprocal waveguide components) The process occurring in a frequency converter by which the signal input frequency is changed by superimposing a local oscillation to produce an output having the same modulation information as the original signal but at a frequency which is either the sum or the difference of the signal and local oscillator frequencies. (MTT) 457-1982w

heterodyne conversion transducer (converter) A conversion transducer in which the useful output frequency is the sum or difference of the input frequency and an integral multiple of the frequency of another wave usually derived from a local oscillator. *Note:* The frequency and voltage or power of the local oscillator are parameters of the conversion transducer. Ordinarily, the output-signal amplitude is a linear function of the input-signal amplitude over its useful operating range. (ED) 161-1971w

heterodyne frequency *See:* beats.

heterodyne reception (beat reception) The process of reception in which a received high-frequency wave is combined in a nonlinear device with a locally generated wave, with the result that in the output there are frequencies equal to the sum and difference of the combining frequencies. *Note:* If the received waves are continuous waves of constant amplitude, as in telegraphy, it is customary to adjust the locally generated frequency so that the difference frequency is audible. If the received waves are modulated the locally generated frequency is generally such that the difference frequency is superaudible and an additional operation is necessary to reproduce the original signal wave. *See also:* superheterodyne reception. (PE/EEC) [119]

heterogeneous computer network A computer network of different host computers, such as those of different manufacturers. *Contrast:* homogeneous computer network. (C) 610.7-1995

heterogeneous LAN A network of interconnected LANs of mixed media access control types. *Contrast:* homogeneous local area network. (C) 610.7-1995

heterojunction (fiber optics) A junction between semiconductors that differ in their doping level conductivities, and also in their atomic or alloy compositions. *See also:* homojunction. (Std100) 812-1984w

heteropolar machine (rotating machinery) A machine having an even number of magnetic poles with successive (effective) poles of opposite polarity. *See also:* asynchronous machine; direct-current commutating machine. (PE) [9]

heuristic (1) Pertaining to exploratory methods of problem solving in which solutions are discovered by evaluation of the progress made toward the final result. *See also:* algorithm. (MIL/C) [2], [85], [20]

(2) (modeling and simulation) Pertaining to experimental, especially trial-and-error, methods of problem-solving. *Note:* The resulting solution may not be the most desirable solution to the problem. (C) 610.3-1989w

Hevea rubber Rubber from the Hevea brasiliensis tree. *See also:* insulation.

Hewlett-Packard Graphics Language (HPGL) A page description language used by many laser printers. (C) 610.13-1993w

Hewlett-Packard Printer Control Language A page description language used in many laser printers. (C) 610.13-1993w

hex *See:* hexadecimal.

hexadecimal (A) (mathematics of computing) Pertaining to a selection in which there are sixteen possible outcomes. *Synonym:* sexadecimal. **(B) (mathematics of computing)** Pertaining to the numeration system with a radix of 16. *Synonym:* sexadecimal. (C) 1084-1986

hexadecimal character string A sequence of characters from the set of hexadecimal digits, preceded by the two characters `0x` (zero followed by a lowercase "x"). Hexadecimal character strings shall consist only of the following characters:

0 1 2 3 4 5 6 7 8 9 A B C D E F x

Within software definition files of exported catalogs, all such strings shall be encoded using IRV. (C/PA) 1387.2-1995

hexadecimal digit A numeral used to represent one of the 16 digits in the hexadecimal numeration system; 0, 1, 2, 3, 4, 5, 6, 7, 8, 9, A, B, C, D, E, or F. (C) 1084-1986w

hexadecimal notation Any notation that uses the hexadecimal digits and the radix 16. (C) 1084-1986w

hexadecimal number (A) A quantity that is expressed using the hexadecimal numeration system. **(B)** Loosely, a hexadecimal numeral. (C) 1084-1986

hexadecimal number system* *See:* hexadecimal numeration system.

* Deprecated.

hexadecimal numeral A numeral in the hexadecimal numeration system. For example, the hexadecimal numeral 17 is equivalent to the decimal numeral 23. (C) 1084-1986w

hexadecimal numeration system The numeration system that uses the hexadecimal digits and the radix 16. *Synonym:* hexadecimal system. (C) 1084-1986w

hexadecimal point The radix point in the hexadecimal numeration system. (C) 1084-1986w

hexadecimal system *See:* hexadecimal numeration system.

hexadecimal-to-decimal conversion The process of converting a hexadecimal numeral to an equivalent decimal numeral. For example, hexadecimal 8B.4 is converted to decimal 139.25. (C) 1084-1986w

hexlet (1) Sixteen bytes (128 bits) of data. (C/MM) 1754-1994

(2) A 16-byte data format or data type. The name hexadeclet would more accurately describe these 16-byte formats, but for notational convenience this abbreviated term is used throughout this standard. (C/MM) 1596.5-1993

hexode A six-electrode electron tube containing an anode, a cathode, a control electrode, and three additional electrodes that are ordinarily grids. (ED) 161-1971w

HF *See:* high frequency.

HFC *See:* horizontal footcandles.

HF radar *See:* high-frequency radar.

H-frame *See:* crossing structure.

HH *See:* header hub.

hickey (A) A fitting used to mount a lighting fixture in an outlet box or on a pipe or stud. *Note:* It has openings through which fixture wires may be brought out of the fixture stem. **(B)** A pipe-bending tool. (EEC/PE) [119]

HID Abbreviation for high-intensity discharge. *See also:* high-intensity discharge lamp; high-intensity discharge lamps.

HIDAM *See:* hierarchical indexed direct access method.

hidden A general term covering both private and protected. *Contrast:* public. *See also:* private; protected. (C/SE) 1320.2-1998

hidden line A line or line segment in a three-dimensional display image that is not visible because of the presence of surfaces closer to the viewer. *Note:* Such a line may be left invisible or may be displayed as a dashed or dotted line to enhance the realism of the image. (C) 610.6-1991w

hidden line/hidden surface removal A process of detecting hidden lines and hidden surfaces in an image and removing them from the rendering of that image before it is rendered. (C) 610.6-1991w

hidden surface A surface in a three-dimensional graphics display image that is not visible because of the presence of surfaces closer to the viewer. (C) 610.6-1991w

hierarchical Pertaining to a hierarchy, as in a hierarchical database or a hierarchical structure. (C) 610.5-1990w

hierarchical branching A tree-like menu structure that allows selection among alternatives without requiring the opening and closing of a series of menus. The alternatives are contained in one menu. (PE/NP) 1289-1998

hierarchical computer network A computer network in which processing and control functions are performed at several levels by computers suited for the functions performed. (C) 610.7-1995

hierarchical database (A) A database system that uses tree structures to represent the data. **(B)** A database in which data are organized into records, known as segments, that represent nodes in a hierarchy or tree structure. *Note:* Within the hierarchy, a subordinate to a given segment is known as its child segment and a superordinate is known as its parent segment. *Synonym:* sequential precedential database. *Contrast:* relational database; network database. (PE/EDPG) 1150-1991

hierarchical decomposition (software) A type of modular decomposition in which a system is broken down into a hierarchy of components through a series of top-down refinements. *See also:* functional decomposition; stepwise refinement. (C) 610.12-1990

hierarchical direct access method (HDAM) A database access method for hierarchical databases in which pointers maintain the structure itself as well as the control of the storage and retrieval functions of the database. All records are stored and retrieved using these pointers. *Contrast:* hierarchical sequential access method. *See also:* hierarchical indexed direct access method; hierarchical indexed sequential access method. (C) 610.5-1990w

hierarchical indexed direct access method (HIDAM) A database access method for hierarchical databases in which indices access root segments and pointers access dependent segments. *Contrast:* hierarchical indexed sequential access method. (C) 610.5-1990w

hierarchical indexed sequential access method (HISAM) A database access method for hierarchical databases in which indices control access to both root and dependent segments. *Contrast:* hierarchical indexed direct access method. (C) 610.5-1990w

hierarchical input-process-output *See:* input-process-output.

hierarchical instance The concrete appearance of a design unit at some hierarchical level. Because higher-level design units may be instantiated multiple times, a single such appearance may give rise to multiple instances of the lower-level design units within it. Where instances are referred to as "occurrences," hierarchical instances are referred to simply as instances. (C/DA) 1481-1999

hierarchical level A member of a linearly ordered set (i.e., hierarchy) of levels, e.g., a number in the range from 0 to 255. (C/LM) 802.10g-1995, 802.10-1998

hierarchically consecutive An unbroken unidirectional traversal of all nodes between two specified nodes in a tree. All nodes between the origin and destination nodes shall be visited during a traversal. All traversals from any node to its adjacent node shall be made in the same direction, either towards the root of the tree of towards the leaves of the tree. Typically, hierarchically consecutive is taken to imply from ancestral node (closer to the root) to descendent node (closer to the leaves). (C/SE) 1320.1-1998

hierarchical model (A) A data model whose pattern of organization is in the form of a tree structure. **(B)** A data model that provides a tree structure for relating data elements, where each node of the tree corresponds to a group of data elements or a record type, and has only one superior node or parent. (C) 610.5-1990

hierarchical modeling A technique used in computer performance evaluation, in which a computer system is represented as a hierarchy of subsystems, the subsystems are analyzed to determine their performance characteristics, and the results are used to evaluate the performance of the overall system. (C) 610.12-1990

hierarchical random-access memory (HRAM) A type of storage that consists of several layers of varying-speed storage in which information is stored in the fastest available storage. (C) 610.10-1994w

hierarchical routing A routing based on a hierarchical addressing scheme. *Note:* There are five classes of telephony offices in the North America:

- class 1 office: regional center
- class 2 office: sectional center
- class 3 office: primary center
- class 4 office: toll center
- class 5 office: end office

(C) 610.7-1995

hierarchical sequence In a hierarchical database, the sequence of root and dependent segments defined by traversing the database in some specified order. (C) 610.5-1990w

hierarchical sequential access method (HSAM) A database access method for hierarchical databases in which data items are stored and retrieved sequentially. *Contrast:* hierarchical direct access method. *See also:* hierarchical indexed sequential access method. (C) 610.5-1990w

hierarchical structure A collection of entities that are organized in a hierarchical fashion. *Contrast:* network structure. (C) 610.5-1990w

hierarchy (software) (data management) A structure in which components are ranked into levels of subordination; each component has zero, one, or more subordinates; and no component has more than one superordinate component. *See also:* link; hierarchical decomposition; tree; hierarchical modeling; network; data hierarchy.

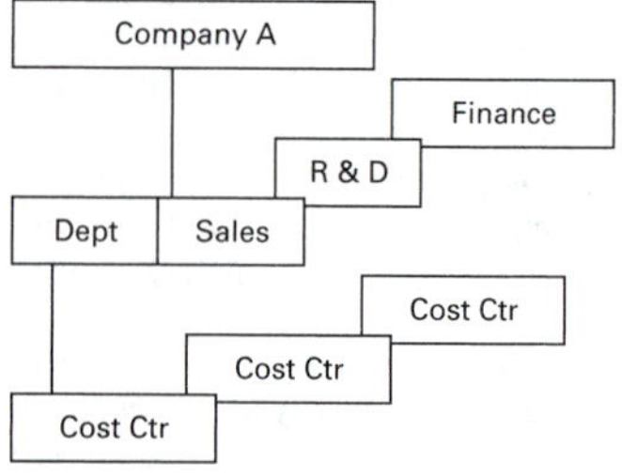

hierarchy

(C) 610.5-1990w, 610.12-1990

hierarchy chart *See:* structure chart.

high The higher of the two voltages used to convey a single bit of information. For positive logic, a logic 1. (TT/C) 1149.1-1990

high-availability computer A computer designed with various fault tolerant systems that enables it to function when one or more of its components fail. *Note:* A computer is so designated due to its high percentage of user availability. (C) 610.10-1994w

high conduction threshold voltage (metal-nitride-oxide field-effect transistor) The threshold voltage level resulting from a write-high pulse, which puts the transistor into the HC (high-conduction) state. (ED) 581-1978w

high day *See:* time-consistent traffic measures.

high day busy-hour load The one day among the same 10 days that has the highest traffic during the busy hour is designated the (annually recurring) "high day." The traffic level in the busy hour of the high day is termed the HDBH load. (There may be some other hour of the high day or another day of the year with a higher traffic level, but normally it would not be used in the engineering data base.) *See also:* time-consistent traffic measures. (COM/TA) 973-1990w

high-density disk A floppy disk that is capable of storing information at a higher density than that of the same size double-density disk. (C) 610.10-1994w

high dielectric cable Cable that provides high-voltage insulation between conductors, between conductors and shield, and between shield and earth. (PE/PSC) 487-1992

high direct voltage (power cable systems) A direct voltage above 5 000 V supplied by test equipment of limited capacity. (PE/EM/IC) 95-1977r, 400-1991

high-emphasis filtering In image processing, a sharpening technique in which rapid fluctuations in gray levels are emphasized. (C) 610.4-1990w

high-energy piping (nuclear power generating station) Piping serving as the pressure boundary for fluid systems that, during normal plant conditions, are either operating or maintaining temperature or pressure when the maximum operating temperature exceeds 200 F or the maximum operating pressure exceeds 275 pounds per square inch gauge (psig). (PE/NP) 567-1980w

higher layer The conceptual layer of control or processing logic existing in the hierarchical structure of a station that is above the data link layer and upon which the performance of data link layer functions are dependent; for example, device control, buffer allocation, LLC station management, etc. (C/LM/CC) 8802-2-1998

higher-order language (1) (software) A programming language that usually includes features such as nested expressions, user defined data types, and parameter passing not normally found in lower order languages, that does not reflect the structure of any one given computer or class of computers, and that can be used to write machine independent source programs. A single higher order language may represent multiple machine operations. *Synonym:* high-level language. *See also:* computer; assembly language; data type; machine language; source program; programming language. (C/SE) 729-1983s
(2) *See also:* high-order language. (C) 610.13-1993w

higher-order mode (waveguide or transmission line) Any mode of propagation characterized by a field configuration other than that of the fundamental or first-order mode with lowest cutoff frequency. *See also:* waveguide. (IM/HFIM) [40]

higher-order mode of propagation (1) (laser maser) A mode in a beamguide or beam resonator which has a plurality of maxima for the transverse field intensity over the cross-section of the beam. (LEO) 586-1980w
(2) (planar transmission lines) Any mode of propagation characterized by a field configuration other than that of the dominant or first order mode with the lowest cutoff frequency. (MTT) 1004-1987w

higher order service A service that provides a complex behavior of a diagnostic reasoner, possibly defined using a combination of primitive services. (SCC20) 1232.2-1998

high, false, 1 Unasserted state of a bus line. (C/MM) 1196-1987w

high-fidelity signal (speech quality measurements) A signal transmitted over a system comprised of a microphone, amplifier, and loudspeaker or earphones. A tape recorder may be part of the system. All components should be of the best quality the state of the art permits. 297-1969w

high-field-emission arc (gas) An electric arc in which the electron emission is due to the effect of a high electric field in the immediate neighborhood of the cathode, the thermionic emission being negligible. *See also:* discharge. (ED) [45], [84]

high frequencies Frequencies allocated for transmission in the outbound direction. In a mid-split broadband system, approximately 160–300 MHz or higher. (LM/C) 802.7-1989r

high frequency (HF) (1) A radar frequency band between 3 megahertz and 30 megahertz. (AES/RS) 686-1982s
(2) 3–30 MHz. *See also:* radio spectrum. (AP/PROP) 211-1997

high-frequency furnace (coreless-type induction furnace) An induction furnace in which the heat is generated within the charge, or within the walls of the containing crucible, or in both, by currents induced by high-frequency flux from a surrounding solenoid. (PE/EEC) [119]

high-frequency induction heater or furnace A device for causing electric current flow in a charge to be heated, the frequency of the current being higher than that customarily distributed over commercial networks. *See also:* induction heating. (IA) 54-1955w, 169-1955w

high-frequency radar A radar operating at frequencies between 3 to 30 megahertz. (AES/RS) 686-1982s

high frequency radar (radar) A radar operating at frequencies between 3 to 20 MHz. *Synonym:* HF radar. (AES/RS) 686-1982s

high-frequency stabilized arc welder A constant-current arc-welding power supply including a high-frequency arc stabilizer and suitable controls required to produce welding current primarily intended for tungsten-inert-gas arc welding. *See also:* constant-current arc-welding power supply. (EEC/AWM) [91]

high-gain dc amplifier (analog computer) An amplifier that is capable of amplification substantially greater than required for a specified operation throughout a frequency band extending from zero to some maximum. Also, an operational amplifier without feedback circuit elements. *See also:* operational amplifier. (C/Std100) 165-1977w, 610.10-1994w

high-impedance ac system An ac/dc system having low or very low SCR. (PE/T&D) 1204-1997

high-impedance rotor An induction-motor rotor having a high-impedance squirrel cage, used to limit starting current. *See also:* rotor. (PE) [9]

high initial response (excitation systems for synchronous machines) An excitation system capable of attaining 95% of the difference between ceiling voltage and rated-load field voltage in 0.1 s or less under specified condition. (PE/EDPG) 421.1-1986r

high-impedance value (1) The enumeration literal 'Z' of the type STD_ULOGIC defined by IEEE Std 1164-1993. (C/DA) 1076.3-1997
(2) The enumeration literal "Z" of the type STD_ULOGIC (or subtype STD_LOGIC) defined by IEEE Std 1164-1993. (For example, a latch.). (C/DA) 1076.6-1999

high-intensity discharge lamp (illuminating engineering) An electric discharge lamp in which the light producing arc is stabilized by wall temperature, and the arc tube has a bulb wall loading in excess 3W/cm^2. HID lamps include groups of lamps known as mercury, metal halide, and high-pressure sodium. *See also:* high-intensity discharge lamps. (EEC/IE) [126]

high-intensity discharge lamps A group of lamps filled with various gases that are generically known as mercury, metal halide, high-pressure sodium, and low-pressure sodium. *See also:* high-intensity discharge lamp. (IA/PSE) 241-1990r

high-key lighting (illuminating engineering) A type of lighting which, applied to a scene, results in a picture having graduations falling primarily between gray and white; dark grays or blacks are present, but in very limited areas. (EEC/IE) [126]

high level A level within the more positive (less negative) of the two ranges of the logic levels chosen to represent the logic states. (GSD/C/BA) 91-1984r, 1496-1993w

high-level data link control (HDLC) (1) A set of Data Link layer communication protocols defined by ISO/IEC 3309: 1993, ISO/IEC 4335: 1993, ISO/IEC 7809: 1993, and ISO/IEC 8885: 1993. These standards define a multiplicity of point-to-point and multidrop protocols. These include both master/slave and peer-to-peer types of data links, employing both half-duplex and full-duplex methodologies. (For the data link portion of this standard, a particular subset, known as TWANRM, is utilized. TWANRM defines a half-duplex master/slave variation of HDLC). (EMB/MIB) 1073.3.1-1994

(2) A standard protocol defined by ISO for bit-oriented, frame-delimited data communications. (C/EMB/MIB) 610.10-1994w, 1073.3.2-2000

high-level data link control protocol A standard protocol, defined by ISO, for bit-oriented, frame-delimited data communication protocol. (C) 610.7-1995

high-level firing time (microwave) (switching tubes) The time required to establish a radio-frequency discharge in the tube after the application of radio-frequency power. *See also:* gas tube. (ED) 161-1971w, [45]

high-level format To prepare a disk or a partition of a disk to be used by a particular operating system. *Note:* In most instances, this includes scanning the surface of the disk for defective areas. *Synonym:* logical format. *Contrast:* low-level format. (C) 610.10-1994w

high-level language (HLL) (1) (high-level microprocessor language) High-level language to be extended by IEEE trial use Std 755-1985. HLLs so extended are sometimes known as implementation languages. (C/MM) 755-1985w
(2) *See also:* high-order language. (C/SE) 729-1983s, 610.13-1993w

high-level modulation Modulation produced at a point in a system where the power level approximates that at the output of the system. (AP/BT/ANT) 145-1983s, 182-1961w

high-level radio-frequency signal (1) (microwave gas tubes) A radio-frequency signal of sufficient power to cause the tube to become fired. *See also:* gas tube. (ED) 161-1971w
(2) (nonlinear, active, and nonreciprocal waveguide components) (microwave gas tubes) A radio-frequency signal above the threshold power level necessary to cause the tube to become nonlinear (fired). *See also:* gas tube. (MTT) 457-1982w

high-level testing (mechanical) Testing performed to determine a damping of complete assemblies, subassemblies, or components. (SUB/PE) C37.122.1-1993

high-level voltage standing-wave ratio (nonlinear, active, and nonreciprocal waveguide components) (microwave switching tubes) The voltage standing-wave ratio caused by a fired tube located between a generator and matched termination in the waveguide. *See also:* gas tube. (ED/MTT) 161-1971w, 457-1982w

highlight (A) A technique in which a display element is emphasized through visual modification such as blinking, brightening, or intensity modulation. **(B)** To draw attention to a display element by visual modification as in definition (A). *See also:* blink. (C) 610.6-1991

high lights (any metal article) Those portions that are most exposed to buffing or polishing operations, and hence have the highest luster. (EEC/PE) [119]

high-limit temperature (1) (electrical heat tracing for industrial applications) The maximum allowable heat-tracing system temperature. (BT/AV) 152-1953s
(2) The maximum allowable temperature, including the piping, the fluid, and the heating system. (IA) 515-1997

high-low signaling (telephone switching systems) A method of loop signaling in which a high-resistance bridge is used to indicate an on-hook condition and a low resistance bridge is used to indicate an off-hook condition. (COM) 312-1977w

high media rate (HMR) Used to indicate a data rate of 100 Mbit/s or greater. (C/LM) 802.5t-2000

high-order Pertaining to the left-most digit or digits of a numeral. (C) 1084-1986w

high-order language (HOL) Any programming language that requires little knowledge of the computer hardware on which a program will run, can be translated into several different machine languages, allows symbolic naming of operations and addresses, provides features designed to facilitate expression of data structures and program logic, and usually results in several machine instructions for each program statement. Examples include Ada, ALGOL, COBOL, FORTRAN, Pascal. *Synonym:* third generation language. *Contrast:* machine language; assembly language; fifth generation language; fourth generation language. (C) 610.13-1993w, 610.12-1990

high-order position The leftmost position in a string; for example, the letter 'A' in 'APPLE' or the digit 9 in 965. *Contrast:* low-order position. *See also:* most significant digit; most significant character. (C) 610.5-1990w

high-pass filter (harmonic control and reactive compensation of static power converters) (data transmission) A filter having a single transmission band extending from some cutoff frequency (not zero) up to infinite frequency. (SP/IA/PE/SPC) 151-1965w, 519-1992, 599-1985w

high peaking The introduction of an amplitude-frequency characteristic having a higher relative response at the higher frequencies. *See also:* television. (BT/AV) [34]

high pot *See:* high-potential test.

high-potential test (power operations) A test that consists of the application of a voltage higher than the rated voltage for a specified time for the purpose of determining the adequacy against breakdown of insulating materials and spacings under normal conditions. *Note:* The test is used as a proof test of new apparatus, a maintenance test on older equipment, or as one method of evaluating developmental insulation systems. *Synonym:* high pot. (PE/PSE) 858-1987s

high-power-factor mercury-lamp ballast A multiple-supply type power-factor-corrected ballast, so designed that the input current is at a power factor of not less than 90 percent when the ballast is operated with center rated voltage impressed upon its input terminals and with a connected load, consisting of the appropriate reference lamp(s), operated in the position for which the ballast is designed. (EEC/LB) [97]

high-power−factor transformer (power and distribution transformers) A high-reactance transformer that has a power-factor-correcting device, such as a capacitor, so that the input current is at a power factor of not less than 90% when the transformer delivers rated current to its intended load device. *See also:* specialty transformer. (PE/TR) C57.12.80-1978r, [116]

high-pressure contact (as applied to high-voltage disconnecting switches) One in which the pressure is such that the stress in the material of either of the contact surfaces is near the elastic limit of the material so that conduction is a function of pressure. (SWG/PE) C37.100-1992

high-pressure sodium lamp (illuminating engineering) A high intensity discharge (HID) lamp in which light is produced by radiation from sodium vapor operating at a partial pressure about 1.33×10^4Pa (100Torr). Includes clear and diffuse-coated lamps. (EEC/IE) [126]

high-pressure vacuum pump A vacuum pump that discharges at atmospheric pressure. *See also:* rectification. (EEC/PE) [119]

high profile Terminations or connections designed for use outside of thermal insulation, or away from the surface being heated. (IA/PC) 515.1-1995

high-profile connection Terminations or connections designed for use outside of the thermal insulation, or away from the surface being heated. (IA) 515-1997

high-pulse-repetition frequency A pulsed-radar system whose pulse-repetition frequency is such that targets of interest are ambiguous with respect to range. *See also:* MPRF. (AES/RS) 686-1990

high-pulse-repetition-frequency waveform A waveform whose pulse-repetition frequency (PRF) is high enough to have no Doppler ambiguities for a given maximum-speed target. *See also:* low-pulse-repetition-frequency waveform; medium-pulse-repetition-frequency waveform. (AES) 686-1997

high-purity germanium (HPGe) Germanium with a low, net electrically active, uncompensated defect concentration usually less than $\approx 10^{10}$ cm^{-3}. (NPS) 325-1996

high-rate charge The application of a constant potential charge, at a higher level than the float charge, to a partially or fully discharged battery to recharge it. (PE/EDPG) 1106-1995

high-reactance rotor An induction-motor rotor having a high-reactance squirrel cage, used where low starting current is required and where low locked-rotor and breakdown torques are acceptable. *See also:* rotor. (PE) [9]

high-reactance transformer (power and distribution transformers) An energy-limiting transformer that has sufficient inherent reactance to limit the output current to a maximum value. *See also:* specialty transformer. (PE/TR) C57.12.80-1978r, [57]

(2) (A) (secondary short-circuit current rating) The current in the secondary winding when the primary winding is connected to a circuit of rated primary voltage and frequency and when the secondary terminals are short-circuited. **(B)** (kilovolt-ampere or voltampere short-circuit input rating) The input kilovolt-amperes or volt-amperes at rated primary voltage with the secondary terminals short-circuited. (PE/TR) [57]

high-resistance rotor (rotating machinery) An induction motor rotor having a high-resistance squirrel cage, used when reduced locked-rotor current and increased locked-rotor torque are required. (PE) [9]

high-resistance sheath A metallic covering with a characteristic resistance at a level high enough to prevent usage as an effective ground path. More specifically, it is a metallic covering that either does not have a conductance greater than or equal to that of the largest conductor under evaluation, based on the resistance of an equivalent sized copper conductor, or is incapable of passing an overcurrent test at levels of 1.10, 1.35, and 2.00 times the maximum branch circuit overcurrent protection for 7 h, 1 h, and 2 min, respectively. (IA/PC) 515.1-1995

high rupturing capacity (HRC) (protection and coordination of industrial and commercial power systems) In British and Canadian terminology, high rupturing capacity, equivalent to USA high interrupting capacity and generally indicating capability of interruption of at least 100 000 root-mean-square (rms) amperes (A) for low-voltage fuses. (IA/PSP) 242-1986r

high-speed buffer A cache or a set of logically partitioned blocks that provides significantly faster access to instructions and data than provided by main storage. (C) 610.10-1994w

high-speed carry (1) (electronic computation) A carry process such that if the current sum in a digit place is exactly one less than the base, the carry input is bypassed to the next place. *Note:* The processing necessary to allow the bypass occurs before the carry input arrives. Further processing required in the place as a result of the carry input, occurs after the carry has passed by. *Contrast:* cascaded carry. *See also:* standing-on-nines carry. (C) 162-1963w

(2) (mathematics of computing) A carry process in which, if the current sum in a given digit place is one less than the base, the sum is set to zero and the carry input is passed to the next place. *Contrast:* cascaded carry. *See also:* standing-on-nines carry. (C) 1084-1986w

high-speed excitation system An excitation system capable of changing its voltage rapidly in response to a change in the excited generator field circuit. *See also:* generating station. (PE/T&D) [10]

high-speed grounding switch *See:* fault-initiating switch.

high-speed limit (control systems for steam turbine-generator units) (speed/load reference) A device or input that limits the speed/load reference setting to a predetermined upper limit. This device may establish the upper limit of the synchronizing speed range. (PE/EDPG) 122-1985s

high-speed low-voltage dc power circuit breaker (1) A low-voltage dc power circuit breaker which, during interruption, limits the magnitude of the fault current so that its crest is passed not later than a specified time after the beginning of the fault current transient, where the system fault current, determined without the circuit breaker in the circuit, falls between specified limits of current at a specified time. *Note:* The specified time in present practice is 0.01 second. (SWG/PE) C37.100-1981s

(2) (low-voltage dc power circuit breakers used in enclosures) A circuit breaker which, when applied in a circuit with the parameter values specified in American National Standard C37.16-1979, Preferred Rating, Related Requirements and Application Recommendations for Low-Voltage Power Circuit Breakers and AC Power Circuit Protectors, Tables 12 and 12A, tests "b" (5 A/μS initial rate of rise of current), forces a current crest during interruption within 0.01 s after the current reaches the pickup setting of the instantaneous trip device. *Note:* For total performance characteristics at other than test circuit parameter values, consult the manufacturer. (SWG/PE) C37.14-1979s

high-speed metal-oxide semiconductor (HMOS) *See:* n-channel metal-oxide semiconductor.

high-speed printer (HSP) A printer that operates at a very high speed. (C) 610.10-1994w

high-speed regulator (power supplies) A power supply regulator circuit that, by the elimination of its output capacitor, has been made capable of much higher slewing rates than are normally possible. *Note:* High-speed regulators are used where rapid step-programming is needed: or as current regulators, for which they are ideally suited. *See also:* slewing rate. (AES) [41]

high-speed relay A relay that operates in less than a specified time. *Note:* The specified time in present practice is 50 ms (three cycles on a 60 Hz basis). (SWG/PE) C37.100-1992

high-speed short-circuiting switch *See:* fault-initiating switch.

high-split A frequency division scheme that allows two-way traffic on a single cable. Inbound path signals come to the headend from 5 to 174 MHz. Outbound path signals go from the headend from 234 MHz to the upper frequency limit. The guardband is located from 174 to 234 MHz. (LM/C) 802.7-1989r

high state (1) (programmable instrumentation) The relatively more-positive signal level used to assert a specific message content associated with one of two binary logic states. (IM/AIN) 488.1-1987r

(2) (signals and paths) (SBX bus) (microcomputer system bus) (STEbus) The more positive voltage level used to represent one of two logical binary states. (C/MM) 796-1983r, 959-1988r, 1000-1987rr

(3) (696 interface devices) (signals and paths) The electrically more positive signal level used to assert a specific message content associated with one of two binary logic states. (MM/C) 696-1983w

high symbol An idle symbol that has been marked for consumption by highest-priority nodes. Sometimes called high-idle symbol. (C/MM) 1596-1992

high-temperature Used to describe materials, insulation systems, and transformers that are designed to operate at a maximum hottest-spot temperature above 120C. (PE/TR) 1276-1997

high-temperature insulation system An insulation system composed of all high-temperature solid insulation materials, with or without high-temperature fluids. (PE/TR) 1276-1997

high-usage trunk (data transmission) A group of trunks for which an engineered alternate route is provided, and for which the number of trunks is determined on the basis of relative trunk efficiencies and economic considerations. (PE) 599-1985w

high-usage trunk group (telephone switching systems) A trunk group engineered on the basis of relative trunk efficiencies and economic considerations which will overflow traffic. (COM) 312-1977w

high-velocity camera tube (anode-voltage stabilized camera tube) A camera tube operating with a beam of electrons having velocities such that the average target voltage stabilizes at a value approximately equal to that of the anode. (ED) [45]

high-velocity scanning (electron tube) The scanning of a target with electrons of such velocity that the secondary-emission rate is greater than unity. *See also:* television. (ED) 161-1971w

high voltage (hv) (1) (electric power systems in commercial buildings) (system voltage ratings) A class of nominal system voltages equal to or greater than 100 000 V and equal to or less than 230 000 V. *See also:* nominal system voltage; medium voltage; low voltage. (IA/PSE/APP) 241-1990r, [80]
(2) A term applied to voltage levels that are greater than 1000 V. (T&D/PE) 516-1995

high-voltage aluminum-sheathed power cable (aluminum sheaths for power cables) Cable used in an electric system having a maximum phase-to-phase rms ac voltage above 72 500 V to 242 000 V, the cable having an aluminum sheath as a major component in its construction. (PE/IC) 635-1989r

high-voltage and low-voltage windings (power and distribution transformers) The terms high voltage and low voltage are used to distinguish the winding having the greater from that having the lesser voltage rating. (PE/TR) C57.12.80-1978r

high-voltage cable termination A device used for terminating alternating-current power cables having laminated or extruded insulation rated 2.5 kV and above, which are classified according to the following:

a) Class 1 termination: Provides electric stress control for the cable insulation shield terminus; provides complete external leakage insulation between the cable conductor(s) and ground; and provides a seal to the end of the cable against the entrance of the external environment and maintains the operating design pressure, if any, of the cable system. This class is divided into three types:
— Class 1A: For use on extruded dielectric cable
— Class 1B: For use on laminated dielectric cable
— Class 1C: Expressly for pressure-type cable systems

b) Class 2 termination: Provides electric stress control for the cable insulation shield terminus, and provides complete external leakage insulation between the cable conductor(s) and ground.

c) Class 3 termination: Provides electric stress control for the cable insulation shield terminus.

Note: Some cables below 15 kV do not have an insulation shield. Terminations for such cables would not be required to provide electric stress control. In such cases, this provision would not be part of the definition. (PE/IC) 48-1996

high-voltage disconnect jack A device used to disconnect cable pairs for testing purposes. Used to help safeguard personnel from remote ground potentials. (PE/PSC) 487-1992

high-voltage isolating relay A device that provides for the repeating of dc on/off signals while maintaining longitudinal isolation. High-voltage isolating relays may be used in conjunction with isolating transformers or may be used as standalone devices for dc tripping or dc telemetering. (PE/PSC) 487-1992

high-voltage power vacuum interrupter (X-radiation limits for ac high-voltage power vacuum interrupters used in power switchgear) An interrupter in which the separable contacts function within a single evacuated envelope and which is intended for use in power switchgear. 553-1980

high-voltage relay (A) A relay adjusted to sense and function in a circuit or system at a specific maximum voltage. **(B)** A relay designed to handle elevated voltages on its contacts, coil, or both. (SWG) 341-1980

high-voltage system An electric system having a maximum root-mean-square ac voltage above 72.5 kV. (PE/EDPG) 665-1995

high-voltage telephone repeater (wire-line communication facilities) A high-voltage telephone repeater provides high-voltage longitudinal isolation, while permitting voice and signalling to pass. This is accomplished by using a short span, carrier transmission system and high-voltage, isolation capacitors or transformers. The repeater is intended to provide ordinary telephone service in a power station environment without interference to other noninterruptible, critical circuits. (PE/PSC) 487-1980s

high-voltage time test An accelerated life test on a cable sample in which voltage is the factor increased. (PE/T&D) [10]

highway (CAMAC system) An interconnection between CAMAC crate assemblies or between one or more CAMAC crate assemblies and an external controller. (NPS) 583-1975s

highway crossing back light (railway practice) An auxiliary signal light used for indication in a direction opposite to that provided by the main unit of a highway crossing signal. (EEC/PE) [119]

highway crossing bell (railway practice) A bell located at a railroad-highway grade crossing and operated to give a characteristic and arrestive signal to give warning of the approach of trains. (EEC/PE) [119]

highway crossing signal An electrically operated signal used for the protection of highway traffic at railroad-highway grade crossings. (EEC/PE) [119]

high-Z A condition in which the driver on a pin is inactive. The state of a net attached to a pin that is at high-Z is determined by values applied to other parts of the net. *Contrast:* CD state. *See also:* net; core disconnect. (C/TT) 1149.4-1999

Hilbert transform (harmonic conjugate) (real functional $x(t)$ of the real variable t) The real function $x(t)$ that is the Cauchy principal value of

$$\frac{1}{\pi}\int_{-\infty}^{\infty}\frac{X(\tau)d\tau}{t-\tau}$$

This transformation shifts all Fourier components by 90°−cosω^t, for example, into sinω^t and sinωt into -cosω^t. *See also:* network analysis. (IT) [7]

hinge axis *See:* output axis.

hinge clip (of a switching device) The clip to which the blade is movably attached. (SWG/PE) C37.100-1992

hinged-iron ammeter A special form of moving-iron ammeter in which the fixed portion of the magnetic circuit is arranged so that it can be caused to encircle the conductor, the current in which is to be measured. This conductor then constitutes the fixed coil of the instrument. *Note:* The combination of a current transformer of the split-core type with an ammeter is often used similarly to measure alternating current, but should be distinguished from the hinged-iron ammeter. *See also:* instrument. (EEC/PE) [119]

hinged removable feed tube (cable plowing) A feed tube removably attached to a blade so relative motion may occur between the feed tube and the blade around an essentially vertical axis. (PE/T&D) 590-1977w

hipot (test, measurement, and diagnostic equipment) A colloquialism for high potential test: A testing technique whereby a high voltage source is applied to an insulating material to determine the condition of that material. (MIL) [2]

HIS *See:* health information system; hospital information system.

HISAM *See:* hierarchical indexed sequential access method.

hiss Noise in the audio-frequency range, having subjective characteristics analogous to prolonged sibilant sounds. (SP/ED) 151-1965w, [45]

historical data All relevant information available concerning the product, tests, and test equipment. This includes test observations (raw measurement data), derived test outcomes (i.e., LO, HI, GO), diagnostic conclusions derived from test results and the knowledge base, test subject mission and configuration history, test resources mission and history, etc. (SCC20) 1226-1998

hit **(1)** **(A)** A comparison of two items of data that satisfies specified conditions; for example, when comparing "X" with the string "ACDFXYN", a hit would be encountered in the fifth character position. **(B)** In disk caching, a condition where the target data is located within the cache storage, eliminating the need to reference secondary storage. *Synonym:* cache hit. *See also:* hit ratio; hit probability. (C) 610.10-1994
(2) **(A)** **(telecommunications)** A short on-hook signal that does not initiate any call-processing functions. *See also:* flash; hit, flash, and disconnect timing; disconnect. **(B)** An on-hook/off-hook sequence that is too short to be accepted by the system as a valid on-hook/off-hook signal. (COM/TA) 973-1990
(3) **(A)** **(data management)** In a search, the condition that occurs when the key value of an item is equal to the search argument; that is, a successful search results in a hit. **(B)** A record that produces the condition in definition (A); for example, a student record in which the student's home state matches the home state being searched for. *Contrast:* match. (C) 610.5-1990
(4) A target echo from one single pulse. (AES) 686-1997

hit file A file containing all records that resulted from a successful search. (C) 610.5-1990w

hit, flash, and disconnect timing The duration of an on-hook signal shorter than, between, or longer than certain bounds, which determines whether the signal is a hit, a flash, or a disconnect signal. During a stable call, the switching system may receive a disconnect signal from either the calling or called party. A disconnect signal is the change from an off-hook to on-hook state that persists beyond a prescribed time limit and that may last indefinitely thereafter. *See also:* flash; hit; disconnect. (COM/TA) 973-1990w

hit probability The probability that a cache storage contains the target data. *Note:* generally expressed in percent. (C) 610.10-1994w

hit ratio **(1)** In a search, the number of hits divided by the total number of items searched. (C) 610.5-1990w
(2) The proportion of cache hits to all accesses. (C) 610.10-1994w

hit timing *See:* hit, flash, and disconnect timing; hit; disconnect; flash.

HLL *See:* high-level language; high-order language.

HMI *See:* human-machine interface.

HMOS *See:* high-speed metal-oxide semiconductor.

H network A network composed of five branches, two connected in series between an input terminal and an output terminal, two connected in series between another input terminal and output terminal, and the fifth connected from the junction point of the first two branches to the junction point of the second two branches. *See also:* network analysis.

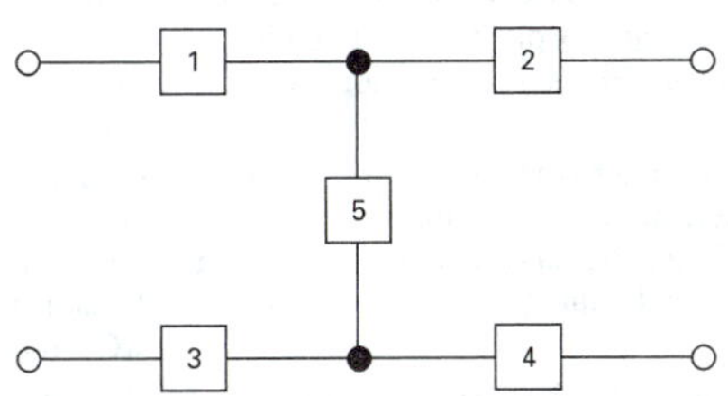

Branches 1 and 2 are the first two branches between an input and an output terminal; branches 3 and 4 are the second two branches; and branch 5 is the branch between the junction points.

H network

(Std100) 270-1966w

hoarfrost A deposit of interlocking ice crystals (hoar crystals) formed by direct sublimation on objects, usually those of small diameter freely exposed to the air such as tree branches, plant stems and leaf edges, wires, poles, etc. The deposition of hoarfrost on an object is similar to the process by which dew is formed, except that the temperature of the object must be below freezing. It forms when air with a dew point below freezing is brought to saturation by cooling. (T&D/PE) 539-1990

hodoscope An apparatus for tracing the path of a charged particle in a magnetic field. *See also:* electron optics. (ED) [45], [84]

Hoeppner connection (power and distribution transformers) A three-phase transformer connection involving transformation from a wye winding to the combination of a delta winding and a zigzag winding which are connected permanently in parallel. *Note:* This connection is used when a wye-delta connection is needed, with ground connections on both primary and secondary windings. (PE/TR) C57.12.80-1978r

hoghorn antenna A reflector antenna consisting of a sectoral horn that physically intersects a reflector in the form of a parabolic cylinder, a part of one of the nonparallel sides of the horn being removed to form the antenna aperture. (AP/ANT) 145-1993

hoist An apparatus for moving a load by the application of pulling force (not including a car or platform running in guides). These devices are normally designed using roller or link chain and built-in leverage to enable heavy loads to be lifted or pulled. They are often used to dead-end a conductor during sagging and clipping-in operations and when tensioning guys *Synonyms:* Coffing; chain tugger; coffin hoist; chain hoist; Coffing hoist; puller; drum. (T&D/PE) 524a-1993r, 524-1992r, 516-1995

hoist back-out switch A switch that permits operation of the hoist only in the reverse direction in case of overwind. *See also:* mine hoist. (EEC/PE/MIN) [119]

hoist, double drum *See:* puller, two drum, three drum.

hoisting-rope equalizer A device installed on an elevator car or counterweight to equalize automatically the tensions in the hoisting wire ropes. *See also:* elevator. (EEC/PE) [119]

hoist overspeed device A device that can be set to prevent the operation of a mine hoist at speeds greater than predetermined values and usually causes an emergency brake application when the predetermined speed is exceeded. *See also:* mine hoist. (EEC/PE/MIN) [119]

hoist overwind device A device that can be set to cause an emergency break application when a cage or skip travels beyond a predetermined point into a danger zone. *See also:* mine hoist. (EEC/PE/MIN) [119]

hoist signal code Consists of prescribed signals for indicating to the hoist operator the desired direction of travel and whether men or materials are to be hoisted or lowered in mines. *See also:* mine hoist. (EEC/PE/MIN) [119]

hoist signal system A system whereby signals can be transmitted to the hoist operator (and in some instances by him to the cager) for control of mine hoisting operations. *See also:* mine hoist. (EEC/PE/MIN) [119]

hoist, single drum *See:* drum puller.

hoist slack-brake switch A device for automatically cutting off the power from the hoist motor and causing the brake to be set in case the links in the brake rigging require tightening or the brakes require relining. *See also:* mine hoist. (PE/EEC/MIN) [119]

hoist, triple drum *See:* puller, two drum, three drum.

hoist trip recorder A device that graphically records information such as the time and number of hoists made as well as the delays or idle periods between hoists. *See also:* mine hoist. (EEC/PE/MIN) [119]

hoistway Any shaftway, hatchway, well hole, or other vertical opening or space in which an elevator or dumbwaiter is designed to operate. (NESC/NEC) [86]

hoistway access switch (elevators) A switch, located at a landing, the function of which is to permit operation of the car with the hoistway door at this landing and the car door or gate open, in order to permit access at the top of the car or to the pit. *See also:* control. (PE/EEC) [119]

hoistway-door combination mechanical lock and electric contact (elevators) A combination mechanical and electric device, the two related, but entirely independent, functions of which are: To prevent operation of the driving machine by the normal operating device unless the hoistway door is in the closed position, and to lock the hoistway door in the closed position and prevent it from being opened from the landing side unless the car is within the landing zone. *Note:* As there is no positive mechanical connection between the electric contact and the door-locking mechanism, this device insures only that the door will be closed, but not necessarily locked, when the car leaves the landing. Should the lock mechanism fail to operate as intended when released by a stationary or retiring car-cam device, the door can be opened from the landing side even though the car is not at the landing. If operated by a stationary car-cam device, it does not prevent opening the door from the landing side as the car passes the floor. *See also:* hoistway. (EEC/PE) [119]

hoistway-door electric contact (elevators) An electric device, the function of which is to prevent operation of the driving machine by the normal operating device unless the hoistway door is in the closed position. *See also:* hoistway. (EEC/PE) [119]

hoistway-door interlock (elevators) A device having two related and interdependent functions that are (1) to prevent the operation of the driving machine by the normal operating device unless the hoistway door is locked in the closed position: and (2) to prevent the opening of the hoistway door from the landing side unless the car is within the landing zone and is either stopped or being stopped. *See also:* hoistway. (EEC/PE) [119]

hoistway-door or gate locking device (elevators) A device that secures a hoistway or gate in the closed position and prevents it from being opened from the landing side except under specified conditions. *See also:* hoistway. (PE/EEC) [119]

hoistway enclosure The fixed structure, consisting of vertical walls or partitions, that isolates the hoistway from all other parts of the building or from an adjacent hoistway and in which the hoistway doors and door assemblies are installed. *See also:* hoistway. (EEC/PE) [119]

hoistway-gate separate mechanical lock (elevators) A mechanical device, the function of which is to lock a hoistway gate in the closed position after the car leaves a landing and prevent the gate from being opened from the landing side unless the car is within the landing zone. *See also:* hoistway. (PE/EEC) [119]

hoistway-unit system (elevators) A series of hoistway-door interlocks, hoistway-door electric contacts, or hoistway-door combination mechanical locks and electric contacts, or a combination thereof, the function of which is to prevent operation of the driving machine by the normal operating device unless all hoistway dobrs are in the closed position and, where so required, are locked in the closed position. *See also:* hoistway. (PE/EEC) [119]

HOL *See:* high-order language.

hold (1) An untimed delay in the program, terminated by an operator or interlock action. (C/IA) [61]
(2) (analog computer) In an analog computer, the computer control state in which the problem solution is stopped and held at its last values usually by automatic disconnect of integrator input signals. (C) 165-1977w
(3) A control function that arrests the further speed change of a drive during the acceleration or deceleration portion of the operating cycle. *See also:* feedback control system. (IA/ICTL/IAC) [60]
(4) To maintain storage elements at an equilibrium voltage by electron bombardment. *See also:* charge-storage tube; data processing. (ED) 161-1971w, 158-1962w
(5) (A) (test, measurement, and diagnostic equipment) The function of retaining information in one storage device after transferring it to another device; and. **(B) (test, measurement, and diagnostic equipment)** A designed stop in testing. (MIL) [2]
(6) (data management) When performing a get operation on a record, to lock the record for update by the requesting process at the same time the get operation is performed. (C) 610.5-1990w
(7) (A) An untimed delay in a computer program, terminated by an operator or an interlock operation. **(B)** In an analog computer, the computer control state in which the problem solution is stopped and held at its last values, usually by automatic disconnect of integrator input signals. (C) 610.10-1994

hold card *See:* hold out.

hold-closed mechanism (automatic circuit recloser) A device that holds the contacts in the closed position following the completion of a predetermined sequence of operations as long as current flows in excess of a predetermined value. (SWG/PE) C37.100-1981s, [56]

hold-closed operation (automatic circuit recloser) An opening followed by the number of closing and opening operations that the hold-closed mechanism will permit before holding the contacts in the closed position. (SWG/PE) C37.100-1981s

hold-down bail (separable insulated connectors) An externally mounted device designed to prevent separation at the operating interface of an elbow and an apparatus bushing. (T&D/PE) 386-1995

hold-down block (conductor stringing equipment) A device designed with one or more single groove sheaves to be placed on the conductor and used as a means of holding it down. This device functions essentially as a traveler used in an inverted position. It is normally used in midspan to control conductor uplift caused by stringing tensions, or at splicing locations to control the conductor as it is allowed to rise after splicing is completed. *Synonyms:* roller, hold-down; traveler, hold-down; splice release block. (T&D/PE) 524-1992r

holding amplifier A receiver circuit incorporating feedback that maintains the present input logic level in the absence of any other drive signals on the signal line. (C/BA) 1496-1993w

holding current (thyristor) The minimum principal current required to maintain the thyristor in the ON-state, after latching current has been reached and after removal of gate signal. *See also:* latching current. (IA/IPC) 428-1981w

holding-down bolt A bolt that fastens a machine to its bedplate, rails, or foundation. (PE) [9]

holding frequency (take the swings) A condition of operating a generator or station to maintain substantially constant frequency irrespective of variations in load. *Note:* A plant so operated is said to be regulating frequency. *See also:* generating station. (PE/T&D/PSE) 94-1970w, [10]

holding load A condition of operating a generator or station at substantially constant load irrespective of variations in frequency. *Note:* A plant so operated is said to be operating on base load. *See also:* base load; generating station. (T&D/PE) [10]

holding register (hybrid computer linkage components) The register, in a double-buffered digital-to-analog converter (DAC) or a digital-to-analog multiplier (DAM), that holds the next digital value to be transferred into the dynamic register. (C) 166-1977w

holding time (1) (data transmission) The length of time a communication channel is in use for each transmission. Includes both message time and operating time. (PE) 599-1985w
(2) The interval of time within which the decrease of the test voltage due to leakage, prior to the discharge, is not greater than 10% when measured with an instrument that has a dc resistance greater than 10^{16} Ω and a capacitance less than 10 pF. (EMC) C63.16-1993

holding tone (1) (telecommunications) A test tone in the range of 1002 to 1020 Hz, having specific requirements as specified in IEEE Std 743-1984. The level of the holding tone is specified as part of the test requirement. The tone is used to measure analog circuit impairments. (COM/TA) 1007-1991r

(2) The tone, near 1 kHz, transmitted over a telecommunication circuit for performing noise-with-tone, jitter, and transient impairment measurements. (COM/TA) 743-1995

hold off *See:* suspension of reclosing.

hold-off diode (charging inductors) A diode that is placed in series with the charging inductor and connected to the common junction of the switching element and the pulse-forming network in a pulse generator. *Note:* The use of a hold-off diode in the charging circuit of a pulse-forming network allows the capacitors of the network to charge to full voltage and remain at this voltage until the switch conducts. This permits the use of pulse-repetition frequencies of equal to or less than twice the frequency of resonance charging. (MAG) 306-1969w

hold order *See:* suspension of reclosing.

hold out Operating order, operating-order identification tag, or marker *Synonym:* hold card. (T&D/PE) 516-1995

hold time (1) The total time that a trunk, channel, or circuit is occupied by a call. (C) 610.7-1995
(2) (A) The amount of time information may be retained in dynamic storage before needing to be refreshed or the information lost. *Synonym:* output hold time. **(B)** The elapsed time during which a program is on hold. **(C)** The amount of time during which data presented to a flip-flop must be maintained after the clock transition in order for the data to be accurately stored. (C) 610.10-1994
(3) *See also:* setup/hold timing check; nochange timing check. (C/DA) 1481-1999

hold timing check A timing check that establishes only the end of the stable interval for a setup/hold timing check. If no setup timing check is provided for the same arc, transitions, and state, the stable interval is assumed to begin at the reference signal transition and a negative value for the hold time is not meaningful. *See also:* setup/hold timing check. (C/DA) 1481-1999

holdup-alarm attachment A general term for the various alarm-initiating devices used with holdup-alarm systems, including holdup buttons, footrails, and others of a secret or unpublished nature. *See also:* protective signaling. (EEC/PE) [119]

holdup-alarm system An alarm system signaling a robbery or attempted robbery. *See also:* protective signaling. (EEC/PE) [119]

hole (1) (semiconductor) A mobile vacancy in the electronic valence structure of a semiconductor that acts like a positive electron charge with a positive mass. *See also:* semiconductor. (ED) 216-1960w
(2) (image processing and pattern recognition) In image processing, a connected component of the complement of a region, that is surrounded by the region.

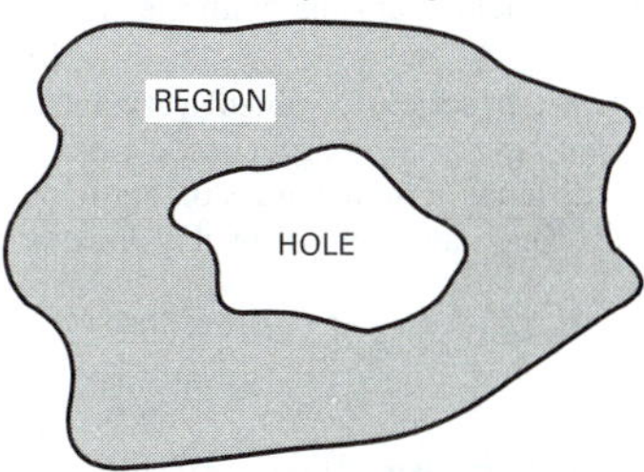

hole

(C) 610.4-1990w
(3) In a semiconductor, a conceptual unit of charge opposite to that of an electron. *Note:* A hole occurs when an electron is lost from an atom, and "moves" when an electron is lost from an adjacent atom. (C) 610.10-1994w

hole burning (laser maser) (of an absorption or an emission line) The frequency dependent saturation of attenuation or gain that occurs in an inhomogeneously broadened transition when the saturating power is confined to a frequency range small compared with the inhomogeneous linewidth. (LEO) 586-1980w

hole pattern A punching configuration or an array of holes that represent a single character in a data medium such as paper tape, or punch cards. (C) 610.10-1994w

Hollerith card *See:* punch card.

hollow-core annular conductor (hollow-core conductor) A conductor composed of a plurality of conducting elements disposed around a supporting member that does not fill the space enclosed by the elements: alternatively, a plurality of such conducting elements disposed around a central channel and interlocked one with the other or so shaped that they are self-supporting. *See also:* conductor. (T&D/PE) [10]

hollow-core conductor *See:* hollow-core annular conductor.

home address The information written on every track of a magnetic disk, identifying the relative track number of that track. (C) 610.10-1994w

home area (telephone switching systems) The numbering plan area in which the calling customer is located. (COM) 312-1977w

home computer A personal computer designed to be used in the home. (C) 610.2-1987, 610.10-1994w

home directory The current directory associated with a user at the time of login. (C/PA) 9945-2-1993

home key (A) A cursor control key that moves the cursor to the starting point of the screen, usually the upper left-hand corner. **(B)** A cursor control key that moves the cursor to the starting point of a file. (C) 610.10-1994

home signal (railway practice) A fixed signal at the entrance of a route or block to govern trains or engines entering or using that route or block. (EEC/PE) [119]

homing (1) (navigation) Following a course directed toward a point by maintaining constant some navigational coordinate (other than altitude). *See also:* radio navigation. (AES/RS) 686-1982s, [42]
(2) (telephone switching systems) Resetting of a sequential switching operation to a fixed starting point. (COM) 312-1977w

homing beacon (navigation aid terms) A beacon that provides homing guidance. (AES/GCS) 172-1983w

homing guidance That form of missile guidance wherein the missile steers itself toward a target by means of a mechanism actuated by some distinguishing characteristic of the target. *See also:* guided missile. (EEC/PE) [119]

homing relay A stepping relay that returns to a specified starting position prior to each operating cycle. *See also:* relay. (EEC/REE) [87]

homochromatic gain (optoelectronic device) The radiant gain or luminous gain for specified identical spectral characteristics of both incident and emitted flux. *See also:* optoelectronic device. (ED) [46]

homodyne reception (zero-beat reception) A system of reception by the aid of a locally generated voltage of carrier frequency. (EEC/PE) [119]

homogeneous cladding (fiber optics) That part of the cladding wherein the refractive index is constant within a specified tolerance, as a function of radius. *See also:* cladding; tolerance field. (Std100) 812-1984w

homogeneous computer network A computer network of similar host computers, such as those of one model by the same manufacturer. *Contrast:* heterogeneous computer network. (C) 610.7-1995

homogeneous dense medium Medium in which the refractive index is significantly different from that of a vacuum. *See also:* sparse medium. (AP/PROP) 211-1997

homogeneous Helmholtz equation The wave equation for the electromagnetic potential, Φ, given by:

$$(\nabla^2 + k^2)\Phi = 0$$

where k is the wavenumber in the medium. The homogeneous Helmholtz equation is also the scalar wave equation for a scalar component of the electric field represented by Φ. *Note:* Sometimes k^2 is replaced by $-\gamma^2$, where γ is the propagation constant. (AP/PROP) 211-1997

homogeneous LAN *See:* homogeneous local area network.

homogeneous line-broadening (laser maser) An increase of the width of an absorption or emission line, beyond the natural linewidth, produced by a disturbance (for example, collisions, lattice vibrations, etc.) which is the same for each of the emitters. (LEO) 586-1980w

homogeneous local area network (homogeneous LAN) A network of interconnected LANs, all of which use the same media access control type. *Contrast:* heterogeneous LAN. (C) 610.7-1995

homogeneous medium A medium whose properties are spatially invariant. (AP/PROP) 211-1997

homogeneous redundancy In fault tolerance, realization of the same function with identical means, for example, use of two identical processors. *Contrast:* diversity. (C) 610.12-1990

homogeneous plane wave A wave in which the planes of constant magnitude and constant phase are parallel. *Note:* Homogeneous plane waves are sometimes called uniform plane waves. (AP/PROP) 211-1997

homogeneous series (of current-limiting fuse units) A series of fuse units, deviating from each other only in such characteristics that, for a given test, the testing of one or a reduced number of particular fuse units of the series may be taken as representative of all the fuse units of the series. (SWG/PE) C37.100-1992, C37.40-1993

homojunction (fiber optics) A junction between semiconductors that differ in their doping level conductivities but not in their atomic or alloy compositions. *See also:* heterojunction. (Std100) 812-1984w

homopolar generator A dc generator in which the magnetic field flux passes in the same direction from one member to the other over the whole of a single air gap area. Characteristically the machines are high-current, low-voltage generators. (IA/MT) 45-1998

homopolar machine* (rotating machinery) A machine in which the magnetic flux passes in the same direction from one member to the other over the whole of a single air-gap area. Preferred term is acyclic machine. (PE) [9]

* Deprecated.

honeycomb coil A coil in which the turns are wound in crisscross fashion to form a self-supporting structure or to reduce distributed capacitance. *Synonym:* duolater coil. (PE/EM) 43-1974s

hood *See:* insulator cover.

hook, conductor lifting A device resembling an open boxing glove designed to permit the lifting of conductors from a position above them. It is normally used during clipping-in operations. Suspension clamps are sometimes used for this purpose. *Synonyms:* lip; lifting shoe; boxing glove; conductor hook. (T&D/PE) 524-1992r

hook ladder *See:* tower ladder.

hook operation *See:* stick operation.

hook ring (air switch) A ring provided on the switch blade for operation of the switch with a switch stick. (SWG/PE) C37.100-1992

hook stick *See:* switch stick.

hopper *See:* card hopper.

horizontal amplifier (oscilloscopes) An amplifier for signals intended to produce horizontal deflection. *See also:* oscillograph. (IM/HFIM) [40]

horizontal bushing A bushing intended to be mounted horizontally at an angle 70° to 90° from the vertical. (PE/TR) C57.19.03-1996

horizontal component of the electric field strength The rms value of the component of the electric field strength in a horizontal plane passing through the point of measurement. (T&D/PE) 539-1990

horizontal deflection axis (oscilloscopes) The horizontal trace obtained when there is a horizontal deflection signal but no vertical deflection signal. (IM) 311-1970w

horizontal feed Pertaining to the motion of a punch card along a card feed path with the long edge first. *Contrast:* vertical feed. (C) 610.10-1994w

horizontal footcandles (HFC) Illuminance measured in a horizontal plane. (RL) C136.10-1996

horizontal hold control (television) A synchronizing control that adjusts the free-running period of the horizontal deflection oscillator. (BT/AV) 201-1979w

horizontally integrated microprocessor A microprocessor in which horizontal microinstructions can be performed. *Contrast:* vertically integrated microprocessor. (C) 610.10-1994w

horizontally polarized field vector A linearly polarized field vector whose direction is horizontal. (AP/ANT) 145-1993

horizontally polarized plane wave A plane wave whose electric field vector is horizontally polarized. (AP/ANT) 145-1993

horizontally polarized wave (1) (general) A linearly polarized wave whose direction of polarization is horizontal. *See also:* radiation. (EEC/PE) [119]
(2) A linearly polarized wave whose electric field vector is perpendicular to the plane of incidence or parallel to the Earth's surface in radio propagation. Same as S-polarization in optics; perpendicular polarization in physics. *See also:* transverse electric wave. (AP/PROP) 211-1997

horizontal machine A machine whose axis of rotation is approximately horizontal. (PE) [9]

horizontal microinstruction A microinstruction that specifies a set of simultaneous operations needed to carry out a given machine language instruction. *Contrast:* diagonal microinstruction; vertical microinstruction. (C) 610.10-1994w, 610.12-1990

horizontal plane (illuminating engineering) (of a searchlight) The plane that is perpendicular to the vertical plane through the axis of the searchlight drum and in which the train lies. (EEC/IE) [126]

horizontal tabulation (A) On an impact printer or a typewriter, movement of the imprint position a predetermined number of character spaces along the writing line. **(B)** On a display device, movement of the cursor a predetermined number of display positions along a display line. *Contrast:* vertical tabulation. (C) 610.10-1994

horizontal tabulation character (HT) A format effector character that causes the print or display position to move forward to the next of a series of predetermined positions along the same horizontal line. (C) 610.5-1990w

horizontal ring induction furnace A device for melting metal, comprising an annular horizontal placed open trough or melting channel, a primary inductor winding, and a magnetic core which links the melting channel with the primary winding. (IA) 54-1955w

horn (1) (acoustic practice) A tube of varying cross-sectional area for radiating or receiving acoustic waves. *Note:* Normally it has different terminal areas that provide a change of acoustic impedance and control of the directional response pattern. (SP) [32]
(2) An antenna consisting of a waveguide section in which the cross sectional area increases towards an open end that is the aperture. (AP/ANT) 145-1993

horn antenna (data transmission) A radiating element having the shape of a horn. (PE) 599-1985w

horn gap An air-gap metal electrode device, consisting of a straight vertical round electrode and an angularly shaped round electrode. In the case of a telephone pair, there is one common grounded central straight vertical electrode and two angular electrodes, one for each side of the pair. The gaps are usually adjustable. Horn gaps are used usually outdoors on open-wire lines exposed to high-voltage power transmission lines and in conjunction with isolating or drainage transformers. They are also frequently used alone out along the open-wire pair. They provide protection against both lightning and power contacts. (PE/PSC) 487-1992

horn-gap switch A switch provided with arcing horns. (SWG/PE) C37.36b-1990r, C37.100-1992

horn mouth Normally the opening, at the end of a horn, with larger cross-sectional area. (SP) [32]

horn reflector antenna (1) (communication satellite) A form of reflector antenna, where the energy coming from the throat of a horn is reflected by a segment of a paraboloid. This type of antenna has a very low backlobe. *See also:* Cassegrainian feed. (COM) [24]
(2) An antenna consisting of a portion of a paraboloidal reflector fed with an offset horn that physically intersects the reflector, part of the wall of the horn being removed to form the antenna aperture. *Note:* The horn is usually either pyramidal or conical, with an axis perpendicular to that of the paraboloid. (AP/ANT) 145-1993

horn throat (audio and electroacoustics) Normally the opening, at the end of a horn, with the smaller cross-sectional area. (SP) [32]

horsepower rating, basis for single-phase motor A system of rating for single-phase motors, whereby horsepower values are determined, for various synchronous speeds, from the minimum value of breakdown torque that the motor design will provide. (PE) [9]

hose (1) (liquid cooling) (rotating machinery) The flexible insulated or insulating hydraulic connections applied between the conductors and either a central manifold or coolant passage. (PE) [9]
(2) *See also:* conductor cover. (PE/T&D) 516-1987s

hose clamp *See:* strand restraining clamp.

hoseproof *See:* waterproof machine.

hospital (health care facilities) A building or part thereof used for the medical, psychiatric, obstetrical or surgical care, on a 24-hour basis, of 4 or more inpatients. Hospital, wherever used in this Code, shall include general hospitals, mental hospitals, tuberculosis hospitals, children's hospitals, and any such facilities providing inpatient care. (NESC/NEC) [86]

hospital information system (HIS) An automated system used in hospitals and other health care facilities to perform such tasks as communication between staff members, statistical analysis, inventory planning, and scheduling of medication, blood analysis, and patient testing. *Note:* Hospital information systems typically use interactive operations on a hierarchical file structure based on a patient-oriented record. *Synonyms:* medical information system; health information system. (C) 610.2-1987

Hospital Operating System-Structured Programming Language An application-oriented language with some FORTRAN features, used to manipulate string variables and to support structured programming. (C) 610.13-1993w

host (1) A device to which other devices (peripherals) are connected and that generally controls those devices. (EMC/EMB/MIB) C63.4-1991, 1073.3.1-1994
(2) A device, typically a personal computer, that will control the communications with attached peripherals. (C/MM) 1284-1994
(3) In general, an electronic circuit assembly (e.g., a plug-in unit, a back-plane, a mother board, or another mezzanine card), that provides electrical and/or mechanical connections to a subordinate assembly. (C/BA) 1301.4-1996
(4) Whatever is driving (i.e., providing) commands or data to the printer (e.g., a workstation, a print server, or spooler.) (C/MM) 1284.1-1997
(5) The client or host station/computer, with which the RTU equipment communicates. *Synonym:* master. (PE/SUB) 1379-1997
(6) *See also:* host computer. (C/DIS) 610.7-1995, 1278.2-1995

host byte order The native representation of an integer: unsigned integer m is the representation in *host byte order* of bit string $b_n\, b_{n-1} \ldots b_0$ (where b_n is the *most significant*, or *highest order* bit, and b_0 is the *least significant* or *lowest order* bit) if $m = 2^n * b_n + 2^{n-1} * b_{n-1} + \ldots + 2^0 * b_0$. (C) 1003.5-1999

host character string A sequence of characters describing a host. Within software definition files of an exported catalog, all data that can be encoded using IRV, shall be. Any such data that cannot be so encoded shall be transformed using UTF-8. (C/PA) 1387.2-1995

host computer (1) A computer, attached to a network, providing primary services such as computation, data base access or special programs or programming languages. *See also:* communications computer. (LM/COM/EMB/MIB) 168-1956w, 1073.3.1-1994
(2) A computer that supports one or more simulation applications. All host computers participating in a simulation exercise are connected by network(s) including local area networks, wide area networks, radio frequency links, etc. (DIS/C) 1278.1-1995, 1278.2-1995r
(3) (A) The primary or controlling computer in a multiple computer installation. *Synonyms:* host machine; host. *See also:* bifunctional machine. **(B)** The primary or controlling processor in a multiprocessor computer or a computer with multiple processing elements, some of which may be dedicated to specific functions. For example, intelligent adapters; math coprocessors. *Synonym:* host processor. (C) 610.7-1995, 610.10-1994
(4) A computer used to prepare computer programs for use on another computer or on another computer system; for example, a computer used to compile, link, or test programs to be used on another system. *Synonym:* host processor. (C) 610.10-1994w

host computer system A computational-based system used for clinical support of patient care, limited to medical device data interchange. *Synonym:* patient care information system. (EMB/MIB) 1073.4.1-2000

hostile environment computer A computer designed for use in an environment not conducive to safe operation, such as one with many dust particles in the air; precautions must be taken to ensure that dust particles do not enter the disk drives or settle on the heads. *See also:* hardened computer. (C) 610.10-1994w

host interface The interface between a communications network and a host computer. (C) 610.7-1995, 610.10-1994w

host language A programming language such as COBOL or PL/I into which data manipulation language statements are embedded. *See also:* data sublanguage. (C) 610.5-1990w

host machine (1) (A) (software) A computer used to develop software intended for another computer. *Contrast:* target machine. **(B) (software)** A computer used to emulate another computer. *Contrast:* target machine. **(C) (software)** The computer on which a program or file is installed. **(D) (software)** In a computer network, a computer that provides processing capabilities to users of the network. (C) 610.12-1990
(2) *See also:* host computer. (C) 610.7-1995

host processor (1) (FASTBUS acquisition and control) The data processing and control processor assigned to exercise overall supervision over a FASTBUS system. Contains detailed knowledge of the system topology. (NID) 960-1993
(2) A processor that controls all or part of a user application network. *See also:* host computer. (C) 610.10-1994w

hot *See:* energized.

hot cathode (thermionic cathode) A cathode that functions primarily by the process of thermionic emission. (ED) 161-1971w

hot-cathode lamp (illuminating engineering) An electric-discharge lamp whose mode of operation is that of an arc discharge. The cathodes may be heated by the discharge or by external means. (EEC/IE) [126]

hot-cathode tube (thermionic tube) An electron tube containing a hot cathode. (ED) 161-1971w, [45]

hot electron programming The injection of energetic "hot" electrons onto the floating gate. (ED) 1005-1998

hot-end termination (electrical heat tracing for industrial applications) The termination applied to the end of a heating cable, opposite where the power is supplied. (BT/AV) 152-1953s

hot-line protectors *See:* open-wire protectors.

hot plate (including portable) An appliance fitted with heating elements and arranged to support a flat-bottomed utensil containing the material to be heated. (PE/EEC) [119]

hot reserve The thermal reserve generating capacity maintained at a temperature and in a condition to permit it to be placed into service promptly. *See also:* generating station. (T&D/PE) [10]

hot-spot A non-recommended abbreviated term frequently used as a synonym for the maximum or hot-test-spot temperature rise of a winding. (PE/TR) C57.134-2000

hot stick *See:* stick.

hot swap (1) The act of connecting or disconnecting a Smart Transducer Interface Module and a Network Capable Application Processor without first turning off the power that the Network Capable Application Processor supplies to the Smart Transducer Interface Module over the Transducer Independent Interface. (IM/ST) 1451.2-1997
(2) A disconnection of a subcomponent and a reconnection of the same or a replacement component without first turning off the power at the connection interface. (IM/ST) 1451.1-1999

hottest-spot differential temperature The temperature difference between the hottest spot of the conductors in contact with insulation and the average winding temperature. (PE/TR) 1276-1997

hottest-spot temperature (1) (electric equipment) (thermal classification of electric equipment and electrical insulation) The highest temperature attained in any part of the insulation of electric equipment. (Difficulties in its determination are encountered. See IEEE Std 1-1986, Section 4). (EI) 1-1986r
(2) (power and distribution transformers) The highest temperature inside the transformer winding. It is greater than the measured average temperature (using the resistance change method) of the coil conductors. (PE/TR) C57.12.80-1978r

hottest-spot temperature allowance (1) (thermal classification of electric equipment and electrical insulation) (electrical equipment) The designated difference between the hottest-spot temperature and the observable insulation temperature. (The value is arbitrary, difficult to determine, and depends on many factors, such as size and design of the equipment). (EI) 1-1986r
(2) (equipment rating) A conventional value selected to approximate the degrees of temperature by which the limiting insulation temperature rise exceeds the limiting observable temperature rise. (Std100) [83]

hot-wire instrument An electrothermic instrument that depends for its operation on the expansion by heat of a wire carrying a current. *See also:* instrument. (EEC/PE) [119]

hot-wire microphone A microphone that depends for its operation on the change in resistance of a hot wire produced by the cooling or heating effects of a sound wave. *See also:* microphone. (EEC/PE) [119]

hot-wire relay A relay in which the operating current flows directly through a tension member whose thermal expansion actuates the relay. *See also:* relay. (EEC/REE) [87]

hot zone In text formatting, a predefined region at the right end of each line of text, having the characteristic that any word that begins in the region and extends beyond it is automatically moved to the next line, and any word that begins before the region and extends beyond it must be hyphenated. *Synonyms:* line-end zone; line-ending zone; margin-adjust zone. (C) 610.2-1987

hot zone hyphenation In text formatting, semi-manual hyphenation in which any word that extends beyond the hot zone must be either hyphenated or moved to the next line. (C) 610.2-1987

Houlding measurement (nonlinear, active, and nonreciprocal waveguide components) A method used in the determination of a varactor diode figure of merit (cutoff frequency). The measurement involves matching the device under test at a fixed bias level in a tunable cavity and interpreting the reflection coefficient data when the bias level is changed. (MTT) 457-1982w

hour Sixty contiguous minutes starting at a clock hour or half-hour. *See also:* time-consistent traffic measures. (COM/TA) 973-1990w

hour rate (1) The discharge rate of a battery expressed in terms of the length of time a fully-charged battery can be discharged at a specific current before reaching a specified end-of-discharge voltage:

Hour rate = C/I

where C = rated capacity of the battery at the specified discharge current I. For example, if a fully-charged battery rated at 100 Ah can be discharged at 5 A for a period of 20 h before reaching the end-of-discharge voltage, discharge of the battery at 5 A is referred to as the 20 h rate (C/I = 100 Ah/5A). (PV) 1013-1990
(2) The discharge rate of a battery expressed in terms of the length of time a fully charged battery can be discharged at a specific current before reaching a specified end-of-discharge voltage: hour rate = C/I, where C = rated capacity of the battery at the specified discharge current I. (PV) 1144-1996

house cable (communication practice) A distribution cable within the confines of a single building or a series of related buildings but excluding cable run from the point of entrance to a cross-connecting box, terminal frame, or point of connection to a block cable. *See also:* cable. (PE/EEC) [119]

housekeeping operation A computer operation that establishes or reestablishes a set of initial conditions to facilitate the execution of a computer program; for example, initializing storage areas, clearing flags, rewinding tapes, opening and closing files. *Synonym:* overhead operation. (C) 610.12-1990

house turbine A turbine installed to provide a source of auxiliary power. *See also:* generating station. (PE/T&D) [10]

housing (1) (rotating machinery) Enclosing structure, used to confine the internal flow of air or to protect a machine from dirt and other harmful material. (PE) [9]
(2) (power cable joints) A metallic or other enclosure for the insulated splice. (PE/IC) 404-1986s
(3) (of an oil cutout) A part of the fuse support that contains the oil and provides means for mounting the fuse carrier, entrance terminals, and fixed contacts. The housing includes the means for mounting the cutout on a supporting structure and openings for attaching accessories such as a vent or an expansion chamber. *Synonym:* body. (SWG/PE) C37.100-1992, C37.40-1993

HPCL *See:* Hewlett-Packard Printer Control Language.

HPGe *See:* high-purity germanium.

HPGL *See:* Hewlett-Packard Graphics Language.

H-plane, principal For a linearly polarized antenna, the plane containing the magnetic field vector and the direction of maximum radiation. (AP/ANT) 145-1993

H-plane tee junction (shunt tee) (waveguide components) A waveguide tee junction in which the magnetic field vectors of the dominant mode in all arms are parallel to the plane containing the longitudinal axes of the arms. (MTT) 147-1979w

HPPCL *See:* Hewlett-Packard Printer Control Language.

HRAM *See:* hierarchical random-access memory.

HRC *See:* high rupturing capacity.

HSAM *See:* hierarchical sequential access method.

H-scope A cathode-ray oscilloscope arranged to present an H-display. (AES/RS) 686-1990

HSP *See:* high-speed printer.

HT *See:* horizontal tabulation character.

hub (1) (broadband local area networks) A central location of a network that connects network nodes through spokes. Similar to a headend for bi-directional networks except that it more often associated with a star architecture. A hub is usually a site that is responsible for providing services to headends located at remote sights. Microwaves or other communication methods may be used to connect the hub to a headend. (LM/C) 802.7-1989r
(2) A reference point established through a land survey. A hub or point on tangent (POT) is a reference point for use during construction of a line. The number of such points that are established will vary with the job requirements. Monuments, however, are usually associated with state or federal surveys and are intended to be permanent reference points. Any of these points may be used as a reference point for transit sagging operations, provided that all necessary data pertaining to them is known. It is quite common to establish additional temporary hubs as required for this purpose. *Synonyms:* monument; point on tangent. (T&D/PE) 524-1992r
(3) (A) A device to which multiple LAN station lobes are connected. *Note:* Multiple hubs may be interconnected to create a single LAN. In some circumstances the hub may implement a part of the LAN protocols. *See also:* gateway; bridge; router. **(B)** A socket on a control panel or plugboard into which an electrical lead or plug wire may be connected in order to carry signals—particularly to distribute the signals over many other wires. (C) 610.7-1995
(4) A device used to provide connectivity between DTEs. Hubs perform the basic functions of restoring signal amplitude and timing, collision detection, and notification and signal broadcast to lower-level hubs and DTEs. (LM/C) 802.3-1998, 610.7-1995

hue (1) (television) The attribute of visual sensation designated by blue, green, yellow, red, purple, etc. *Note:* This attribute is the psychosensorial correlate (or nearly so) of the colorimetric quantity dominant wavelength. (BT/AV) 201-1979w
(2) (illuminating engineering) (of a perceived color) The attribute which determines whether it is red, yellow, green, blue, or the like. (EEC/IE) [126]

hum A component of transmission-line audible noise consisting of pure tones of the power frequency and its harmonics. *Note:* For ac transmission lines, this is caused by ion motion in the air surrounding the conductors. (PE/T&D) 539-1990

human action The observable result (often a bodily movement) of a person's intention. (PE/NP) 1082-1997

human-centered simulation (human-centred simulation) A simulation carried out by people; for example, a simulation in which a human participant operates a mock-up of an instrument to establish a good ergonomic design of the instrument console. *Contrast:* computer-based simulation. *See also:* human-machine simulation. (C) 610.3-1989w

human/computer interface (HCI) The boundary across which physical interaction between a human being and the application platform takes place. (C/PA) 14252-1996

human factors engineering An interdisciplinary science and technology concerned with the process of designing for human use. (PE/NP) 1023-1988r

human interaction A human action or set of actions that affects equipment, response of systems, or other human actions. (PE/NP) 1082-1997

human interface *See:* user interface.

human-machine interface (HMI) Includes keyboards, displays, keypads, touch screens, and similar devices to allow human interaction with a system. (IM/ST) 1451.1-1999

human-machine simulation A simulation carried out by both human participants and computers, typically with the human participants asked to make decisions and a computer performing processing based on those decisions; for example, a simulation in which humans make automotive design decisions and a computer determines and displays the results of those decisions. (C) 610.3-1989w

human presentation system One system within the AI-ESTATE architectural concept. This system supports all users by providing a user interface for data entry and display to an AI-ESTATE implementation and knowledge acquisition support. (ATLAS) 1232-1995

human-system interface (HSI) The interaction between workers and their equipment. This interaction requires information to flow in two directions. The system provides status information to the user, and the user provides control information to the system. (PE/NP) 845-1999

human systems engineering The activities involved throughout the system life cycle that addresses the human element of system design (including usability, measures of effectiveness, measures of performance, and total ownership cost). These activities include the definition and synthesis of manpower, personnel, training, human engineering, health hazards, and safety issues. (C/SE) 1220-1998

human/touch ESD The ESD that occurs directly from a human fingertip, without the presence of any metallic structure in the ESD path. (EMC) C63.16-1993

humidity Water vapor within a given space. (IA/PSE) 241-1990r

humidity, relative The ratio of the mole fraction of water vapor that is present in the air to the mole fraction of water vapor that is present in saturated air. (IA/PSE) 241-1990r

hum sidebands (spectrum analyzer) Undesired responses created within the spectrum analyzer, appearing on the display, that are separated from the desired response by the fundamental or harmonics of the power line frequency. (IM) 748-1979w

hum test Measures low-frequency shield effectiveness against electric field coupling (dc to 100 kHz). (PE/IC) 1143-1994r

hunt group A series of telephone numbers in sequence that allows a calling party to connect with the first available line. (C) 610.7-1995

Huygen's principle Principle proposed by Christian Huygen in 1678 that states that:

— Each point on the wavefront of a light disturbance can be considered to be a new source of secondary spherical waves, and
— The wavefront at any other point in space can be found by constructing the envelopes of the secondary wavelets.

(AP/PROP) 211-1997

Huygens' source radiator An elementary radiator having the radiation properties of an infinitesimal area of a propagating electromagnetic wavefront. (AP/ANT) 145-1993

Huygens' sources Electric and magnetic sources that, if properly distributed on a closed surface *S* in substitution for the actual sources inside *S*, will ensure the result that the electromagnetic field at all points outside *S* is unchanged. *Synonym:* equivalent sources. (AP/ANT) 145-1993

HVdc converter station filter system (high-voltage direct-current systems) The harmonic filter system is designed to suppress, at their source, one or more predominant harmonic frequency currents and voltages which appear on the ac and dc transmission lines because of the ac-dc and dc-ac conversion processes. Harmonic filter system components consist of resistors and reactors (capacitive and inductive) of fixed or variable (controlled) values which make up discrete tuned filters (band-pass-to-ground) designed to limit the magnitude of a specific harmonic current and voltage, or high-pass (broad-band-to-ground) filters which can be effective over a wide frequency range. Carrier frequency noise can be caused by high-frequency current oscillations occurring during solid-state or mercury arc valve commutation. One method of noise suppression is by placing series inductors in the ac supply to the conversion equipment. In addition, precautions should be taken to minimize coupling though the interwinding capacitance of the converter transformers and also through the dc neutral circuit. Methods of reducing power system influence levels in this range (5 k Hz and above) need to be carefully analyzed at the design state to provide adequate filtering in order that the interference to carrier frequency communications systems can be avoided. A high-pass (broad-band-to-ground) filter is used to suppress harmonics over a wide frequency range. It consists of a parallel RL network in series with a capacitor and is not sharply tuned. It can be designed to be effective above any of the harmonics from about the 11th to the 20th harmonic, and can also serve to suppress carrier frequency noise. A discrete frequency (band-pass-to-ground) filter can be tuned to one or more specific lower order harmonics up to the 17th or 18th harmonic (that is, odd multiples of a fundamental on the ac bus, and even multiples on the dc side). On the dc side, series smoothing reactors and surge capacitors should be considered, in addition to band-pass and high-pass filters, as means of suppressing harmonics. (COM/TA) 368-1977w

HVdc converter station noise (high-voltage direct-current systems) The processes of rectification and inversion create undesirable harmonic voltages and currents on both the ac and dc portions of the power system. Audio-frequency harmonics are low-order multiples of the fundamental ac frequencies which exist at the HVdc converter station line terminals. The audio-frequency range of major concern is up to approximately 5 kHz. The voltage and current wave distortion which result from the ac-dc conversion process produce frequencies in the carrier range of approximately 5 to 100 kHz and above. Currents at these frequencies appear on the ac or dc, or both, systems at excessive levels if they are unfiltered or inadequately reduced by other means at the converter terminals. Harmonic and carrier frequency currents can propagate for up to 160 km (99.4 mi) or more on transmission lines. The actual distance of propagation varies and is dependent on the frequency, wavelength, and attenuation by the power system impedance. The order of characteristic voltage harmonics found on a dc transmission line is given by KP, and the order of characteristic current harmonics found on the ac lines at the HVdc line terminals is given by KP ± 1, where K, the rectifier phase number, is the total number of rectifier conduction pulses per cycle based on the ac system frequency, and P = 1, 2, 3,..., any positive integer. The predominant audio- frequency range harmonics and carrier range frequencies, if unfiltered or improperly filtered, will tend to create undesirable longitudinal voltages in communications circuits located in proximity to the HVdc line or associated ac lines, by means of electromagnetic or elctrostatic, or both (also dc ionic drift) coupling. Filters for these noise sources are usually installed at the converter station, but because of design limitations, may not be completely effective. These longitudinally induced harmonics can, if sufficient in amplitude, be manifested as audible noise in communications systems because they act on inherent communication circuit unbalances (the conversion of common-mode potentials to differential-mode potentials). Data transmission and pulse-type signals can also be adversely affected depending on the situation. (COM/TA) 368-1977w

HVdc transmission facility (high-voltage direct-current systems) A Facility consisting of converters located at terminal stations connected by a transmission line, bus, or cable systems, which operate at elevated potentials and currents, and transmit electrical energy between ac systems. The converters. when functioning as a rectifier, change alternating current to direct (unidirectional) current; the transmission line transfers the power between terminal stations where the converters, functioning as an inverter, change the direct current back into alternating current. HVdc transmission facilities can also serve as asynchronous ties between ac systems. (COM/TA) 368-1977w

H.V. power flow control reactor A transmission class reactor connected in series with the transmission system in order to optimize power flow by altering the line reactance. (PE/TR) C57.16-1996

HWCI *See:* hardware configuration item.

hybrid balance (data transmission) A measure of the degree of balance between two impedances connected to two conjugate sides of a hybrid set, and is given by the formula for return loss. (PE) 599-1985w

hybrid circuit (A) A circuit, usually in the form of a module or substrate, that is made up of discrete components and integrated circuits. *Contrast:* monolithic integrated circuit. **(B)** A circuit that uses a combination of digital and analog components, modes of operation, or techniques. (C) 610.10-1994

hybrid coil (bridge transformer) (data transmission) A single transformer having effectively three windings, which is designed to be connected to four branches of a circuit so as to render these branches conjugate in pairs. (PE) 599-1985w

hybrid computer (1) (analog computer) A computer which consists of two main computers, one a dc analog computer, and the other a digital computer, with appropriate control and signal interface, such that they may simultaneously operate or solve, or both, upon different portions of a single problem. (C) 165-1977w
(2) A computer consisting of both a DC analog computer and a digital computer that can process both analog and digital data. (C) 610.10-1994w

hybrid coupling A type of coupling in which different subsets of the range of values that a data item can assume are used for different and unrelated purposes in different software module. *Contrast:* pathological coupling; data coupling; common-environment coupling; content coupling; control coupling. (C) 610.12-1990

hybrid device A VMEbus compatible device that includes application-specific subsets of VXIbus protocols. (C/MM) 1155-1992

hybrid high-temperature insulation system An insulation system usually composed of high-temperature solid insulation material adjacent to winding conductors and cellulose materials in the areas where the maximum temperature at rated load does not exceed 120°C. This system typically uses conventional mineral oil as the insulating liquid. (PE/TR) 1276-1997

hybrid junction (waveguide components) A waveguide or transmission-line arrangement with four ports which, when the ports have reflectionless terminations, has the property that energy entering ay one port is transferred (usually equally) to two of the remaining three. (MTT) 147-1979w

hybrid mode (1) (fiber optics) A mode possessing components of both electric and magnetic field vectors in the direction of propagation. *Note:* Such modes correspond to skew (non-meridional) rays. *See also:* skew ray; mode; transverse electric mode; transverse magnetic mode. (Std100) 812-1984w
(2) (waveguide) A waveguide mode such that both the electric and magnetic field vectors have components in the direction of propagation of the mode as well as transverse components. (MTT) 146-1980w

hybrid-mode horn (antenna) A horn antenna excited by one or more hybrid waveguide modes in order to produce a specified aperture illumination. (AP/ANT) 145-1993

hybrid network A local area network or wide area network that contains a mixture of topologies and access methods. (C) 610.7-1995

hybrid scheme A relay scheme (usually a pilot scheme) combining the logic of two or more conventional schemes. (PE/PSR) C37.113-1999

hybrid set (data transmission) Two or more transformers interconnected to form a network having four pairs of accessible terminals to which may be connected four impedances so that the branches containing them may be made conjugate in pairs when the impedances have the proper values but not otherwise. (PE) 599-1985w

hybrid simulation A simulation, portions of which are designed to be executed on an analog system and portions on a digital system. Interaction between the two portions may take place during execution. *See also:* analog simulation; digital simulation. (C) 610.3-1989w

hybrid tee (waveguide components) (magic tee) A hybrid junction composed of an E-H tee with internal matching elements, which is reflectionless for a wave propagating into the junction from one pot when the other three ports have reflectionless terminations. (MTT) 147-1979w

hybrid wave (radio-wave propagation) An electromagnetic wave in which either the electric or magnetic field vector is linearly polarized normal to the plane of propagation and the other vector is elliptically polarized in this plane. *See also:* transverse-magnetic hybrid wave; transverse-electric hybrid wave. (AP) 211-1977s

hydraulically-release-free (trip-free) (as applied to a hydraulically operated switching device) A term indicating that by hydraulic control the switching device is free to open at any position in the closing stroke if the release is energized. *Note:* This release-free feature is operative even though the closing control switch is held closed. (SWG/PE) C37.100-1992

hydraulic operation Power operation by movement of a liquid under pressure. (SWG/PE) C37.100-1992

hydraulic pressure supply system The pumps, means for driving them, pressure and sump tanks, valves and piping connecting the various parts of the governing system, and associated and accessory devices. (PE/EDPG) 125-1988r

hydro capability (power operations) The capability supplied by hydroelectric sources under specified water conditions. (PE/PSE) 858-1987s

hydroelectric station (power operations) An electric generating station utilizing hydroenergy for the motive force of its prime movers. (PE/PSE) 858-1987s

hydrolysis (composite insulators) The chemical reaction between the ions of water and polymer materials resulting in depolymerization and a change of electrical and mechanical properties. (PE/T&D) 987-1985w

hydromagnetic wave* *See:* magneto-hydrodynamic wave.

* Deprecated.

hydro-thermal coordination Coordinated operation of hydroelectric, pumped-storage hydro, and steam electric stations so as to obtain minimum costs for the system over a predetermined period. (PE/PSE) 858-1993w

hyperabrupt junction (nonlinear, active, and nonreciprocal waveguide components) (semiconductor) A specially designed p-n junction that provides a greater capacitance change over a given voltage range than does an abrupt junction. These devices offer a linear frequency versus voltage characteristic over a limited voltage range when used in a voltage controlled oscillator. The slope of a log-log plot of abrupt-junction capacitance versus voltage is 0.5, whereas a hyperabrupt junction has a slope between 0.5 and 2.0 in the hyperabrupt voltage region. (MTT) 457-1982w

hypercube architecture A computer architecture in which processors are arranged as nodes in multiple dimensions with direct channel communication among neighboring nodes. *Note:* An n-dimensional hypercube has 2^n nodes. (C) 610.10-1994w

hypermedia The integration of text, graphics, sound, or video into an associative, user-controllable system of information storage and retrieval; for example, a hypermedia presentation on exercise might include dynamic links to additional topics such as health, anatomy, and sporting events. *Note:* If the information is primarily presented in text form, the product is known as hypertext. *See also:* multimedia. (C) 610.10-1994w

hypertape drive A high-speed tape drive that uses tape cartridges that can be automatically loaded. (C) 610.10-1994w

hypertext *See:* hypermedia.

hyphenation In text formatting, a manual, semi-manual, or fully automatic process of selecting appropriate word breaks at the end of a line of text and inserting a hyphen at one of those breaks. *Contrast:* hyphenless justification. *See also:* automatic hyphenation; semi-manual hyphenation; manual hyphenation. (C) 610.2-1987

hyphen drop In word processing, the automatic omission of a discretionary hyphen from formatted text when the hyphen is not needed to achieve justification. (C) 610.2-1987

hyphenless justification In text formatting, justification in which any word that will not fit entirely on one line is moved to the next line, and intercharacter or interword spacing is used to justify the text. *Contrast:* hyphenation. (C) 610.2-1987

hypothesis (overhead power lines) Statement of a concept that attempts to explain or predict some phenomenon in such a way that the hypothesis is testable. (T&D/PE) 539-1990

hysteresis The maximum difference in values for a digitizer code transition level when the transition level is approached from either side of the transition. (IM/WM&A) 1057-1994w

hysteresis coupling (1) (electric coupling) An electric coupling in which torque is transmitted by forces arising from the resistance to reorientation of established magnetic fields within a ferromagnetic material. (EM/PE) 290-1980w
(2) An electric coupling in which torque is transmitted from the driving to the driven member by magnetic forces arising from the resistance to reorientation of established magnetic flux fields within ferromagnetic material usually of high coercivity. *Note:* The magnetic flux field is normally produced by current in the excitation winding, provided by an external source. (PE) [9]

hysteresis error (accelerometer) (gyros) The maximum separation due to hysteresis between up-scale-going and down-scale-going indications of the measured variable (during a full-range traverse, unless otherwise specified) after transients have decayed. It is generally expressed as an equivalent input. (AES/GYAC) 528-1994

hysteresis loss (power and distribution transformers) (magnetic) The energy loss in magnetic material that results from an alternating magnetic field as the elementary magnets within the material seek to align themselves with the reversing magnetic field. (PE/TR) C57.12.80-1978r

Hz *See:* hertz.

I

I *See:* invalid; in-phase video.

***I^2t* (1) (protection and coordination of industrial and commercial power systems)** The measure of heat energy developed within a circuit during the fuses melting or clearing. Generally stated as melting I^2t or clearing I^2t. (IA/PSP) 242-1986r

(2) The integral of the square of the current during a given time interval in A^2-s:

$$I^2t = \int_{t_0}^{t_1} i^2 dt \ (A^2\text{-s})$$

Notes: 1. The melting I^2t is equal to the integral of the square of the current during the melting time of the fuse. 2. The clearing I^2t is equal to the integral of the square of the current during the clearing time of the fuse. The clearing time is equal to the sum of melting time and arcing time. 3. The I^2t (A^2-s) multiplied by the resistance (ohms) through which the current flows is equal to the energy (Joules) that will be produced in the resistance. (SWG/PE) C37.40b-1996

***I^2t* characteristic** (of a fuse) The amount of ampere-squared seconds passed by the fuse during a specified period and under specified conditions. *Notes:* 1. The specified period may be the melting, arcing, or total clearing time. The sum of melting and arcing I^2t is the clearing I^2t. 2. The melting characteristic is related to a specified current wave shape, and the arcing I^2t to specified voltage and circuit-impedance conditions. (SWG/PE) C37.100-1992

IA *See:* laser gyro axes; internal address field; input axis.

IACK daisy-chain driver A functional module that activates the interrupt-acknowledge daisy-chain whenever an interrupt handler acknowledges an interrupt request. This daisy-chain ensures that only one interrupter responds with its status/ID when more than one has generated an interrupt request. (C/BA) 1014-1987

IAGC *See:* instantaneous automatic gain control.

IAM Abbreviation for initial address message. (COM/TA) 973-1990w

I_{avg} output The average output load current at which to test the unit under test, (Imin + Imax)/2. (PEL) 1515-2000

IC *See:* information center; interexchange carrier; instruction counter; integrated circuit.

ICEA The Insulated Cable Engineers Association. Founded in 1925, the ICEA is a professional society of insulated cable engineers to promote the reliability of covered and insulated conductors for the transmission and distribution of electric energy, control, and instrumentation of equipment and communications. (T&D/PE) 524a-1993r

ice detection light (illuminating engineering) An inspection light designed to illuminate the leading edge of the wing to check for ice formation. (EEC/IE) [126]

ice proof (1) (high voltage air switches, insulators, and bus supports) So constructed or protected that ice will not interfere with successful operation. (SWG/PE) C37.30-1971s

(2) So constructed or protected that ice of a specified composition and thickness will not interfere with successful operation. (SWG/PE) C37.100-1992

ICES *See:* Integrated Civil Engineering System.

ice tests Design tests made to determine the rated ice-breaking ability of the switching equipment. (SWG/PE) C37.100-1992

I chrominance signal (national television system committee color television) The sidebands resulting from suppressed-carrier modulation of the chrominance subcarrier by the I video signal. *Note:* The signal is transmitted in vestigial form, the upper sideband being limited to a frequency within the top of the picture transmission channel (approximately 0.6 MHz above the chrominance subcarrier), and the lower sideband extending to approximately 1.5 MHz below the subcarrier. The phase of the signal, for positive I video signals, is 123 deg with respect to the (B-Y) axis. (BT/AV) 201-1979w

ICOM code An expression in one diagram that unambiguously identifies an arrow segment in another diagram. An ICOM code is used to associate a boundary arrow of a child diagram with an arrow attached to an ancestral box. *Synonym:* arrow reference. (C/SE) 1320.1-1998

ICOM label An arrow label attached without a squiggle directly to the arrowhead of an output boundary arrow or to the arrowtail of an input, control, or mechanism boundary arrow. An ICOM label associates a boundary arrow of a child diagram with an arrow label of an arrow attached to an ancestral box. (C/SE) 1320.1-1998

Icon A high-order programming language designed primarily to process non-numerical data, as in the applications such as analyzing natural language, transforming or generating computer programs, and formatting documents. (C) 610.13-1993w

icon A symbol that is a pictorial indication of a command or object and is located on a graphics tablet or an on-screen menu.

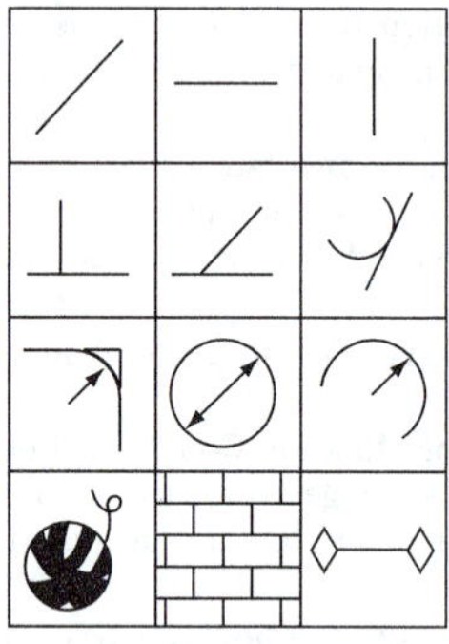

icon

(C) 610.6-1991w

iconic model A physical model that looks like the system being modeled; for example, a non-functional replica of a computer tape drive used for display purposes. *See also:* scale model. (C) 610.3-1989w

ICV *See:* integrity check value.

ICW *See:* interrupted continuous wave.

IDA *See:* independent disk array.

IDCODE (identity code) A defined instruction for the test logic defined by 1149.1-1990. (TT/C) 1149.1-1990

I_{DDQ} Current measurement taken at the ground rail during quiescent operation. (C/TT) 1450-1999

IDE *See:* integrated device electronics.

ideal capacitor (nonlinear capacitor) A capacitor whose transferred charge characteristic is single-valued. *See also:* nonlinear capacitor. (ED) [46]

ideal code bin width *Q* The full-scale range divided by the total number of code states. (IM/WM&A) 1057-1994w

ideal codec A codec that has theoretically optimum characteristics. (COM/TA) 269-1992

ideal conductor *See:* perfect conductor.

ideal dielectric *See:* perfect dielectric.

ideal filter (A) (frequency domain). A filter that passes, without attenuation, all frequencies inside specified frequency limits while rejecting all other frequencies. **(B)** (time domain). A filter with a time domain response identical to the excitation except for a constant delay. (CAS) [13]

idealized system (automatic control) An imaginary system whose ultimately controlled variable has a stipulated relationship to specified commands. *Note:* It is a basis for performance standards. *See also:* feedback control system. (IM/PE/EDPG) [120], [3]

ideally conducting medium *See:* perfect conductor.

ideal noise diode A diode that has an infinite internal impedance and in which the current exhibits full shot noise fluctuations. *See also:* signal-to-noise ratio. (ED) [45]

ideal paralleling (rotating machinery) Paralleling by adjusting the voltage, and frequency and phase angle for alternating-current machines, such that the conditions of the incoming machine are identical with those of the system with which it is being paralleled. *See also:* asynchronous machine. (PE) [9]

ideal site A test site on which the reflective surface is flat and has infinite conductivity and size. (EMC) C63.5-1988, C63.4-1988s

ideal synchronous machine A hypothetical synchronous machine that has certain idealized characteristics that facilitate analysis. *Note:* The results of the analysis of ideal machines may be applied to similar actual machines by making, when necessary, approximate corrections for the deviations of the actual machine from the ideal machine. The ideal machine has, in general, the following properties:

- — the resistance of each winding is constant throughout the analysis, independent of current magnitude or its rate of change;
- — the permeance of each portion of the magnetic circuit is constant throughout the analysis, regardless of the flux density;
- — the armature circuits are symmetrical with respect to each other;
- — the electric and magnetic circuits of the field structure are symmetrical about the direct axis or the quadrature axis;
- — the self-inductance of the field, and every circuit on the field structure, is constant;
- — the self-inductance of each armature circuit is a constant or a constant plus a second-harmonic sinusoidal function of the angular position of the rotor relative to the stator;
- — the mutual inductance between any circuit on the field structure and any armature circuit is a fundamental sinusoidal function of the angular position of the rotor relative to the stator;
- — the mutual inductance between any two armature circuits is a constant or a constant plus a second-harmonic sinusoidal function of the angular position of the rotor relative to the stator;
- — the amplitude of the second-harmonic component of variation of the self-inductance of the armature circuits and of the mutual inductances between any two armature circuits is the same;
- — effects of hysteresis are negligible;
- — effects of eddy currents are negligible or, in the case of solid-rotor machines, may be represented by hypothetical circuits on the field structure symmetrical about the direct axis and the quadrature axis.

(PE) [9]

ideal transducer (for connecting a specified source to a specified load) A hypothetical passive transducer that transfers the maximum available power from the source to the load. *Note:* In linear transducers having only one input and one output, and for which the impedance concept applies, this is equivalent to a transducer that:

a) dissipates no energy; and
b) when connected to the specified source and load presents to each its conjugate impedance.

See also: transducer. (Std100) 270-1966w

ideal transformer A hypothetical transformer that neither stores nor dissipates energy. *Note:* An ideal transformer has the following properties: Its self and mutual impedances are equal and are pure inductances of infinitely great value; Its self-inductances have a finite ratio; Its coefficient of coupling is unity; Its leakage inductance is zero; The ratio of the primary to secondary voltage is equal to the ratio of secondary to primary current. (CHM) [51]

ideal value (1) (automatic control) (control) The value of a selected variable that would result from a perfect system operating from the same command as the actual system under consideration. *See also:* feedback control system. (CS/IA/ICTL/IAC) [60]

(2) (automatic control) The value of the ultimately controlled variable of an idealized system under consideration. *Synonym:* desired value. *See also:* ideal value. (PE/EDPG) [3]

(3) (synchronous-machine regulating system) The value of a controlled variable (for example, generator terminal voltage) that results from a desired or agreed-upon relationship between it and the commands (commands such as voltage regulator setting, limits, and reactive compensators). (PE) [9]

IDEF1X model A set of one or more IDEF1X views, often represented as view diagrams that depict the underlying semantics of the views, along with definitions of the concepts used in the views. (C/SE) 1320.2-1998

IDEF0 model Abstractly, a hierarchical set of IDEF0 diagrams that depict, for a specific purpose and from a specific viewpoint, the functions of a system or subject area, along with supporting glossary, text, and For Exposition Only (FEO) information. Concretely, a set of model pages that include at least an A-0 context diagram and an A0 decomposition diagram, a glossary or specific glossary pages, one or more text pages to accompany each diagram, and FEO pages and model pages of other types as needed. (C/SE) 1320.1-1998

identified (1) (as applied to equipment) Recognized as suitable for the specific service, purpose, function, use, environment, application, etc., where described in a particular Code requirement. (FPN) Suitability of equipment for a specific purpose, environment or application may be determined by a qualified testing laboratory, inspection agency, or other organization concerned with product evaluation. Such identification may include labeling or listing. *Note:* For more information, refer to Section 90-6 of the NEC. *See also:* equipment; listed; labeled. (NESC/NEC/DEI) 1221-1993w, [86]

(2) (data management) *See also:* functional dependency. (C) 610.5-1990w

identification (of a target) The knowledge that a particular radar return signal is from a specific target. This knowledge may be obtained by determining size, shape, timing, position, maneuvers, rate of change of any of these parameters, signal modulation characteristics, or by means of coded responses through secondary radar. *Note:* Distinguished from target classification, in which only the type of target is determined (and which is often the intended meaning), and from target recognition, a more general term that encompasses identification and classification. (AES) 686-1997

identification beacon (navigation aid terms) A beacon that transmits coded signals to identify a geographic position. (AES/GCS) 172-1983w

identifier (1) (software) The name, address, label, or distinguishing index of an object in a computer program. (C) 610.12-1990

(2) Within an IDEF0 model, a model name, a box name, or an arrow label. (C/SE) 1320.1-1998

(3) A means of designating or referring to a specific entity instance. (SCC32) 1489-1999

identifier dependency A kind of constraint between two related entities requiring the primary key in one (child entity) to contain the entire primary key of the other (parent entity). Identifying relationships and categorization structures represent identifier dependencies. (C/SE) 1320.2-1998

identifier-dependent entity *See:* dependent entity.

identifier-independent entity *See:* independent entity.

identifiers Names assigned to each of the unique parametric data elements in the data file. They cannot be the same as any Parametric Data Log (PDL) reserved words. These names can be created from any combination of characters with a few exceptions:

a) The space character (i.e., ASCII code 32) may not be used unless enclosed in double quotes.
b) They shall be unique within the file.

(SCC20) 1545-1999

identifying relationship A kind of specific (not many-to-many) relationship in which every attribute in the primary key of the parent entity is contained in the primary key of the child entity. *Contrast:* nonidentifying relationship. (C/SE) 1320.2-1998

identity (1) The Boolean operation whose result has the value 1 if and only if all the operands have the same value. *Note:* An identity operation on two operands is called an equivalence operation. *Contrast:* nonidentity. *See also:* identity gate. (C) 610.10-1994w
(2) The inherent property of an instance that distinguishes it from all other instances. Identity is intrinsic to the instance and independent of the instance's property values or the classes to which the instance belongs. (C/SE) 1320.2-1998

identity element *See:* identity gate.

identity friend or foe (IFF) Equipment used for transmitting radio signals between two stations located on ships, aircraft, or ground, for automatic identification. *Notes:* 1. The usual basic parts of equipment are interrogators, transpondors, and respondors. 2. Usually the initial letters of the name (IFF) are used instead of the full name. *See also:* radio transmission. (EEC/PE) [119]

identity gate A gate that performs the Boolean operation of an identity operation. *Synonym:* identity element. (C) 610.10-1994w

identity operation A Boolean operation whose result is true if and only if the operands are all true or all false. *Note:* An identity operation on two operands is the same as an equivalence operation. (C) 1084-1986w

identity simulation A simulation in which the roles of the participants are investigated or defined; for example, a simulation that identifies aircraft based on their physical profiles, speed, altitude, and acoustic characteristics. (C) 610.3-1989w

identity-style view A view produced using the identity-style modeling constructs. (C/SE) 1320.2-1998

I-display A display used in a conical-scan radar, in which a target appears as a complete circle when the radar antenna is pointed at it and in which the radius of the circle is proportional to target range. The incorrect aiming of the antenna changes the circle to a segment whose arc length is inversely proportional to the magnitude of the pointing error, and the position of the segment indicates the direction in which the antenna should be moved to restore correct aiming.

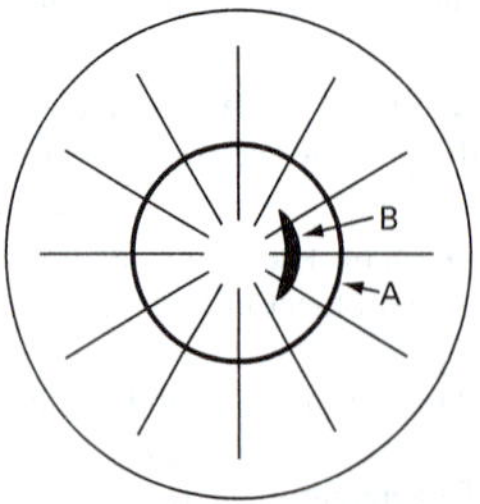

I-display

(AES) 686-1997

IDL *See:* idle.

idle (1) A signal condition where no transition occurs on the transmission line. It is used to define the time between packets. (LM/C) 610.7-1995
(2) (software) Pertaining to a system or component that is operational and in service, but not in use. *See also:* down; busy; up. (C) 610.12-1990
(3) A signal condition where no transition occurs on the transmission line, that is used to define the end of a frame and ceases to exist after the next LO or HI transition on the Attachment Unit Interface (AUI) or Media Independent Interface (MII) circuits. An IDL always begins with a HI signal level. A driver is required to send the IDL signal for at least 2 bit times and a receiver is required to detect IDL within 1.6 bit times. (C/LM) 802.3-1998
(4) (Idle_Up, Idle_Down) A link control signal indicating that the sending entity currently has no traffic pending for the entity connected to the other end of the link.local area networks. (C) 8802-12-1998

idle bar (rotating machinery) An open circuited conductor bar in the rotor of a squirrel-cage motor, used to give low starting current in a moderate torque motor. *See also:* rotor. (PE) [9]

idle channel code A repetitive pattern (code) to identify an idle channel. In some situations this code can produce a signal on a channel. (COM/TA) 1007-1991r

idle channel noise (1) The short-term average noise level as measured according to IEEE Std 743-1984. The measurement for PTS may be made with flat or C-message weighting. The measurements are made at any analog or digital interface with the far end terminated in the appropriate code or impedance(s). (COM/TA) 973-1990w
(2) Noise present on a channel with the distant end terminated and no input signal. (COM/TA) 1007-1991r

idle character A control character that is sent when there is no information to be sent. (C) 610.5-1990w

idle circuit (telephone loop performance) The condition of a transmission channel in the talk state when no signal is present. (COM/TA) 820-1984r

Idle Interrupts Enabled IIE bit A bit in the Slave Status register of every S-module that is set to indicate that the S-module may generate an interrupt during S-idle states. (TT/C) 1149.5-1995

idle packet Four consecutive null bytes, which usually contain no information. Idle packets are used to fill the space between *RamLink* packets. They may be used to convey an interrupt request. (C/MM) 1596.4-1996

idle period (gas tube) That part of an alternating-voltage cycle during which a certain arc path is not carrying current. (ED) [45], [84]

idler circuit (nonlinear, active, and nonreciprocal waveguide components) (parametric device) A portion of a parametric device that chiefly determines the behavior of the device at an idler frequency. *See also:* parametric device. (MTT) 457-1982w

idler frequency (nonlinear, active, and nonreciprocal waveguide components) (parametric device) A sum frequency (or difference frequency) generated within the parametric device other than the input, output, or pump frequencies that requires specific circuit consideration to achieve the desired device performance. *See also:* parametric device. (MTT) 457-1982w

idle slave *See:* slave.

idle state (1) Any Link Layer Controller state the name of which begins with the uppercase letters IDLE. Such states in the MTM-Bus Master Link Layer Controller are called M-idle states and in the MTM-Bus Slave Link Layer Controller are called S-idle states. (TT/C) 1149.5-1995
(2) The inactive (nontransmitting) state of a serial link transmitter. The output of an inactive transmitter shall present a high impedance to a bidirectional serial link. (EMB/MIB) 1073.4.1-2000

idle symbol (1) A symbol that is not inside a packet, and is therefore not protected by a CRC. Idle symbols serve to keep links running and synchronized when no other data are being transmitted. The idle symbol also contains flow-control information. (C/MM) 1596-1992

(2) A symbol that is not inside a packet and is therefore not protected by a CRC. Idle symbols serve to keep links running and synchronized when no other data are being transmitted. The idle symbol also contains flow-control information. (C/MM) 1596.3-1996, 1596-1992

idle time **(1)** **(A)** The period of time during which a system or component is operational and in service, but not in use. *Synonym:* standby time. *See also:* down time; busy time; setup time. **(B)** **(electric drive)** The portion of the available time during which a system is believed to be in good operating condition but is not in productive use. *See also:* electric drive. (C/IA/IAC) 610.12-1990, [60]

(2) That part of up time during which a functional unit is not performing useful operations. *Contrast:* operating time. (C) 610.10-1994w

IDP *See:* integrated data processing.

IDS/1 An extension to COBOL that permits data to be represented in ring type lists. (C) 610.13-1993w

IDT *See:* interdigital transducer.

IEC **(1)** **(television)** International Electrotechnical Commission. *Note:* The French name is Commission Electrotechnique Internationale (CEI). (BT/AV) 201-1979w

(2) *See also:* interexchange carrier. (C) 610.7-1995

IED *See:* intelligent electronic device.

IEEE *See:* Institute of Electrical and Electronics Engineers.

IEEE 802 *See:* LAN/MAN Standards Committee.

IEEE 802 LAN **(1)** A local area network used to carry LLC frames. (LM/C) 802.1H-1995

(2) LAN technologies that provide a MAC Service equivalent to the MAC Service defined in ISO/IEC 15802-1. IEEE 802 LANs include IEEE Std 802.3 (CSMA/CD), ISO/IEC 8802-4 (Token Bus), ISO/IEC 8802-5 (Token Ring), ISO/IEC 8802-6 (DQDB), ISO/IEC 8802-9 (IS-LAN), IEEE Std 802.11 (Wireless), ISO/IEC 8802-12 (Demand Priority), and ISO 9314-2 (FDDI) LANs. *Note:* The connectionless service part of ISO/IEC 8802-6 provides an equivalent MAC Service. (C/LM) 802.1D-1998

IEEE 1284-compatible device *See:* compatible device.

IEEE 1284-compliant device *See:* compliant device.

IEEE 1284-A connector A plug or receptacle 25-pin subminiature D-shell connector. This is the type of connector used on the MS-DOS compatible PC printer port adapter. (C/MM) 1284-1994

IEEE 1284-B connector A plug or receptacle 36-pin ribbon type connector. This type of connector is also known as a "Centronics Connector." (C/MM) 1284-1994

IEEE 1284-C connector A miniature plug or receptacle 36-pin ribbon type connector. (C/MM) 1284-1994

IEEE Standard A standard published by the Institute of Electrical and Electronics Engineers. (C) 610.7-1995

IF *See:* intermediate frequency.

IF-AND-ONLY-IF *See:* equivalence.

IF-AND-ONLY-IF element *See:* IF-AND-ONLY-IF gate.

IF-AND-ONLY-IF gate A gate that performs the Boolean operation of equivalence. *Synonyms:* exclusive-NOR element; IF-AND-ONLY-IF element. (C) 610.10-1994w

IF-A-THEN-NOT-B gate *See:* NAND gate.

IFIP *See:* International Federation of Information Processing.

IFF *See:* identity friend or foe.

IF-THEN *See:* implication.

IF-THEN element *See:* IF-THEN gate.

if-then-else A single-entry, single-exit two-way branch that defines a condition, specifies the processing to be performed if the condition is met and, optionally, if it is not, and returns control in both instances to the statement immediately following the overall construct. *Contrast:* go to; jump; case. *See also:* dyadic selective construct; monadic selective construct.

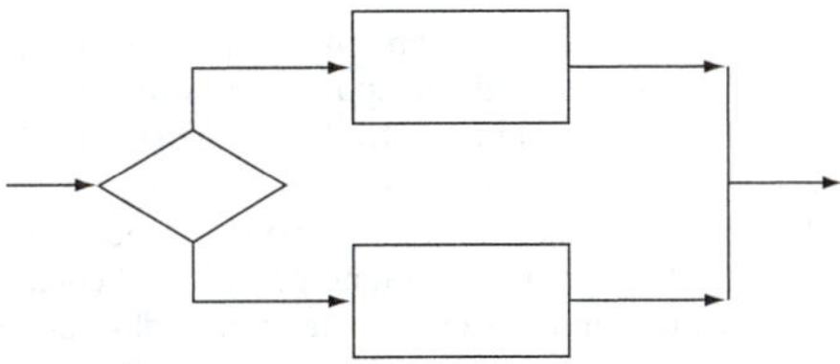

if-then-else construct

(C) 610.12-1990

IF-THEN gate A gate that performs the Boolean operation of implication. *Synonym:* IF-THEN element. (C) 610.10-1994w

IGES *See:* Initial Graphics Exchange Specification.

IGFET *See:* insulated-gate field-effect transistor.

ignition *See:* breakdown; sparkover.

ignition control Control of the starting instant of current flow in the anode circuit of a gas tube. *See also:* electronic controller. (IA/ICTL/IAC) [60]

ignition switch A manual or automatic switch for closing or interrupting the electric ignition circuit of an internal-combustion engine at the option of the machine operator, or by an automatic function calling for unattended operation of the engine. *Note:* Provisions for checking individual circuits of the ignition system for relative performance may be incorporated in such switches. *See also:* switch. (IA/ICTL/IAC) [60]

ignition temperature The lowest temperature at which sustained combustion of a material can be initiated under specified conditions. (DEI) 1221-1993w

ignition transformer (power and distribution transformers) Step-up transformer generally used for electrically igniting oil, gas, or gasoline in domestic, commercial, or industrial heating equipment. (PE/TR) C57.12.80-1978r

ignitor A stationary electrode that is partly immersed in the cathode pool and has the function of initiating a cathode spot. *See also:* electrode; electronic controller. (ED) [45]

ignitor-current temperature drift (microwave gas tubes) The variation in ignitor electrode current caused by a change in ambient temperature of the tube. *See also:* gas tube. (ED) 161-1971w

ignitor discharge (microwave switching tubes) (nonlinear, active, and nonreciprocal waveguide components) A direct-current glow discharge between the ignitor electrode and a suitably located electrode, used to facilitate radio-frequency ionization. *See also:* gas tube. (MTT) 457-1982w

ignitor electrode (microwave switching tubes) An electrode used to initiate and sustain the ignitor discharge. *See also:* gas tube. (ED) 161-1971w

ignitor firing time (microwave switching tubes) (nonlinear, active, and nonreciprocal waveguide components) The time interval between the application of a direct voltage to the ignitor electrode and the establishment of the ignitor discharge. *See also:* gas tube. (MTT) 457-1982w

ignitor interaction (microwave gas tubes) The difference between the insertion loss measured at a specified ignitor current and that measured at zero ignitor current. *See also:* gas tube. (ED) 161-1971w

ignitor leakage resistance (microwave switching tubes) The insulation resistance, measured in the absence of an ignitor discharge, between the ignitor electrode terminal and the adjacent radio-frequency electrode. *See also:* gas tube. (ED) 161-1971w

ignitor oscillations (microwave gas tubes) Relaxation oscillations in the ignitor circuit. *Note:* If present, these oscillations may limit the characteristics of the tube. *See also:* gas tube. (ED) 161-1971w

ignitor voltage drop (microwave switching tubes) The direct voltage between the cathode and the anode of the ignitor discharge at a specified ignitor current. *See also:* gas tube. (ED) 161-1971w

ignitron A single-anode pool tube in which an ignitor is employed to initiate the cathode spot before each conducting period. *See also:* electronic controller. (ED) 161-1971w

ignored Used to describe an instruction field, the contents of which are arbitrary and have no effect on the execution of the instruction. The contents of an ignored field will continue to be ignored in future versions of the architecture. *See also:* unused; reserved. (C/MM) 1754-1994

ignored conductor *See:* isolated conductor.

IH *See:* intermediate hub.

ihandle A cell-sized datum identifying a particular package instance. (C/BA) 1275-1994

IIL *See:* integrated injection logic.

IITRAN A programming language similar to PL/1; designed for use as an educational tool. (C) 610.13-1993w

ILD *See:* injection laser diode.

illegal character A character or combination of bits that is not valid according to some criteria; for example, a character that is not a member of some specified alphabet. *Synonyms:* forbidden character; improper character. *Contrast:* forbidden combination. (C) 610.5-1990w

Illegal Command (ILC) bit A bit in the Bus Error register of all S-modules. An S-module sets this bit to indicate that the module has received an illegal command. (TT/C) 1149.5-1995

Illegal Port Selected (IPS) bit A bit in the Bus Error register of all S-modules. An S-module sets this bit to indicate that the module has received a command addressed to an unsupported port. (TT/C) 1149.5-1995

illuminance The unit density of light flux (lm/unit area) that is incident on a surface. (IA/PSE) 241-1990r

illuminance, E = $d\Phi/dA$ (illuminating engineering) The density of the luminous flux incident at a point on a surface. Average illuminance is the quotient of the luminous flux incident on a surface by the area of the surface. (EEC/IE) [126]

illuminance (footcandle or lux) meter (1) (illuminating engineering) An instrument for measuring illuminance on a plane. Instruments that accurately respond to more than one spectral distribution are color corrected; that is, the spectral response is balanced to V(λ) or V′(λ). Instruments that accurately respond to more than one spatial distribution of incident flux are cosine corrected; that is, the response to a source of unit luminous intensity, illuminating the detector from a fixed distance and from different directions decreases as the cosine of the angle between the incident direction and the normal to the detector surface. The instrument is comprised of some form of photodetector with or without a filter driving a digital or analog readout through appropriate circuitry. (EEC/IE) [126]

(2) (television) *See also:* illumination. (BT/AV) 201-1979w

illumination* (1) (illuminating engineering) An alternate, but deprecated, term for illuminance. It is frequently used since illuminance is subject to confusion with luminance and illuminants, especially when not clearly pronounced. *Note:* The term illumination also is commonly used in a qualitative or general sense to designate the act of illuminating or the state of being illuminated. Usually the context will indicate which meaning is intended, but occasionally it is desirable to use the expression level of illumination to indicate that the quantitative meaning is intended. (EEC/IE) [126]

(2) (A) (television) (general) The density of the luminous flux incident on a surface; it is the quotient of the luminous flux by the area of the surface when the latter is uniformly illuminated. **(B) (television)** (at a point of a surface) The quotient of the luminous flux incident on an infinitesimal element of surface containing the point under consideration by the area of that element. *Notes:* 1. The term illumination also is commonly used in a qualitative or general sense to designate the act of illuminating or the state of being illuminated. Usually the context will indicate which meaning is intended, but occasionally it is desirable to use the expression level of illumination to indicate that the quantitative meaning is intended. The term illuminance, which sometimes is used in place of illumination, is subject to confusion with luminance and illuminates, especially when not clearly pronounced. 2. The units of measurements are: footcandle (lumen per square foot, lm/ft^2 lux (lumen per square meter, lx or lm/m^2). This unit of illumination is recommended by the IEC phot (lumen per square centimeter, lm/cm^2). (BT/ED/AV) 201-1979, [127]

(3) *See also:* aperture illumination. (ANT)

* Deprecated.

illumination (footcandle) meter An instrument for measuring the illumination on a surface. *Note:* Most such instruments consist of barrier-layer cells connected to a meter calibrated in footcandles. *See also:* photometry. (EEC/IE) [126]

illumination sensitivity (camera tubes or phototubes) The quotient of signal output current by the incident illumination, under specified conditions of illumination. *Notes:* 1. Since illumination sensitivity is not an absolute characteristic but depends on the spectral distribution of the incident flux, the term is commonly used to designate the sensitivity to radiation from a tungsten-filament lamp operating at a color temperature of 2870 K. 2. Illumination sensitivity is usually measured with a collimated beam at normal incidence. *See also:* transfer characteristic. (ED) 161-1971w

illuminator That part of a semiactive guidance missile weapon system that radiates electromagnetic waves in the direction of a designated target so that echo signals reflected from the illuminated target can be used by another sensor (the missile seeker) for purposes of homing. (AES) 686-1997

illustration Material that is labeled, numbered, set apart from the main body of text, and, normally, cited within the main text. (C/SE) 1063-1987r

illustrative diagram A diagram whose principal purpose is to show the operating principle of a device or group of devices without necessarily showing actual connections or circuits. Illustrative diagrams may use pictures or symbols to illustrate or represent devices or their elements. Illustrative diagrams may be made of electric, hydraulic, pneumatic, and combination systems. They are applicable chiefly to instruction books, descriptive folders, or other media whose purpose is to explain or instruct. *See also:* control. (IA/ICTL/IAC) 270-1966w, [60]

ILS *See:* instrument landing system.

ILS reference point A point on the centerline of the ILS runway designated as the optimum point of contact for landing: in International Civil Aviation Organization standards this point is from 150 to 300 meters (500 to 1000 feet) from the approach end of the runway. (AES/RS) 686-1982s

image (1) (optoelectronic device) A spatial distribution of a physical property, such as radiation, electric charge, conductivity, or reflectivity, mapped from another distribution of either the same or another physical property. *Note:* The mapping process may be carried out by a flux of photons, electric charges, or other means. *See also:* optoelectronic device. (ED) [46]

(2) (computer graphics) A displayed or drawn representation. (C) 610.6-1991w

(3) (image processing and pattern recognition) A two-dimensional representation of a scene. *Synonym:* picture. *See also:* digital image. (C) 610.4-1990w

(4) (A) In image processing, a two-dimensional representation of a scene. **(B)** In graphics, a displayed or drawn representation. (C) 610.10-1994

(5) *See also:* card image.

Image The data structure contained in the Load Server that the Loadable Device wishes to load. (C) 15802-4-1994

image analysis The process of describing or evaluating an image in terms of its parts, properties, and relationships. (C) 610.4-1990w

image antenna The imaginary counterpart of an actual antenna, assumed for mathematical purposes to be located below the surface of the ground and symmetrical with the actual antenna above ground. *See also:* antenna. (AP/ANT) [35]

image area In micrographics, that part of the film frame reserved for an image. (C) 610.2-1987

image attenuation The real part of the image transfer constant. *See also:* image transfer constant. (CAS) [13]

image burn *See:* retained image.

image camera tube *See:* image tube.

image compression The process of eliminating redundancy or approximating an image in order to represent the image in a more compact manner. *See also:* lossless encoding; adaptive coding; run length encoding; interframe coding; predictive coding; contour encoding. (C) 610.4-1990w

image converter (solid state) An optoelectronic device capable of changing the spectral characteristics of a radiant image. *Note:* Examples of such changes are infrared to visible and x ray to visible. *See also:* optoelectronic device. (ED) [46]

image-converter panel A thin, usually flat, multicell image converter. *See also:* optoelectronic device. (ED) [46]

image-converter tube (camera tubes) An image tube in which an infrared or ultraviolet image input is converted to a visible image output. *See also:* camera tube. (ED) [45]

image dissector (1) (optical character recognition) (computers) A mechanical or electronic transducer that sequentially detects the level of light in different areas of a completely illuminated sample space. (C) [20], [85]
(2) In optical character recognition, a mechanical or electronic device that sequentially detects the level of light intensity in different areas of a completely illuminated sample space. (C) 610.2-1987

image dissector tube (dissector tube) A camera tube in which an electron image produced by a photoemitting surface is focused in the plane of a defining aperture and is scanned past that aperture. *See also:* television. (ED) 161-1971w

image element (optoelectronic device) The smallest portion of an image having a specified correlation with the corresponding portion of the original. *Note:* In some imaging systems the size of the image elements is determined by the structure of the image space, in others by the carrier employed for the mapping process. *See also:* optoelectronic device. (ED) [46]

image enhancement The process of improving the appearance of an image by using techniques such as contrast stretching, edge enhancement, gray scale manipulation, smoothing, and sharpening. (C) 610.4-1990w

image frequency (heterodyne frequency converters in which one of the two sidebands produced by beating is selected) An undesired input frequency capable of producing the selected frequency by the same process. *Note:* The word image implies the mirrorlike symmetry of signal and image frequencies about the beating-oscillator frequency or the intermediate frequency, whichever is the higher. *See also:* radio receiver. (EEC/PE) [119]

image guide A planar dielectric waveguide composed of one or more dielectric strips of finite width affixed to one side of a single extended conducting ground plane. (MTT) 1004-1987w

image iconoscope A camera tube in which an electron image is produced by a photoemitting surface and focused on one side of a separate storage target that is scanned on the same side by an electron beam, usually of high-velocity electrons. *See also:* television. (ED) 161-1971w

image impedances (transducer) The impedances that will simultaneously terminate all of its inputs and outputs in such a way that at each of its inputs and outputs the impedances in both directions are equal. *Note:* The image impedances of a four-terminal transducer are in general not equal to each other, but for any symmetrical transducer the image impedances are equal and are the same as the iterative impedances. *See also:* self-impedance; transducer. (IM/HFIM) [40]

image intensifier (solid state) An optoelectronic amplifier capable of increasing the intensity of a radiant image. *See also:* optoelectronic device. (ED) [46]

image-intensifier panel A thin, usually flat, multicell image intensifier. *See also:* optoelectronic device. (ED) [46]

image-intensifier tube An image tube in which the output radiance is in approximately the same spectral region as, and substantially greater than, the photocathode irradiance. *See also:* image tube; camera tube. (ED) [45]

image line A planar dielectric waveguide composed of one or more dielectric strips of finite width affixed to one side of a single extended conducting ground plane. (MTT) 1004-1987w

image matching An image processing technique in which similar patterns are detected by comparing corresponding points of two images. *See also:* template matching. (C) 610.4-1990w

image memory A discrete portion of memory used to hold a representation of an image. *Note:* Sometimes known as a display buffer, or a bit map. *See also:* video RAM. (C) 610.10-1994w

image operator A function that transforms an input image into an output image. *Synonyms:* image transform operator; image transform. *See also:* point operator; neighborhood operator. (C) 610.4-1990w

image orthicon A camera tube in which an electron image is produced by a photoemitting surface and focused on a separate storage target, which is scanned on its opposite side by a low-velocity electron beam. (ED) 161-1971w

image parameters Fundamental network functions, namely image impedances and the image transfer function, that are used to design or describe a filter. *See also:* image transfer constant; image impedances. (CAS) [13]

image phase The imaginary part of the image transfer constant. *See also:* image transfer constant. (CAS) [13]

image phase constant The imaginary part of the image transfer constant. *See also:* transfer constant; transducer. (Std100) 270-1966w

image processing The manipulation of images by computer. *Synonym:* picture processing. *See also:* change detection; edge linking; congruencing; thinning; sampling. (C) 610.4-1990w

image ratio (heterodyne receiver) The ratio of the field strength at the image frequency to the field strength at the desired frequency, each field being applied in turn, under specified conditions, to produce equal outputs. *See also:* radio receiver. 188-1952w

image reconstruction The process of recovering an image from integrals of its gray levels taken along thin strips or slices of the image. (C) 610.4-1990w

image registration The process of positioning two images of the same scene with respect to one another so that corresponding points in the images represent the same point in the scene. *See also:* registered images. (C) 610.4-1990w

image response Response of a superheterodyne receiver to the image frequency, as compared to the response to the desired frequency. *See also:* radio receiver. (EEC/PE) [119]

image restoration The process of returning an image to its original condition by reversing the effects of known or estimated degradations. *See also:* geometric correction. (C) 610.4-1990w

image segmentation The process of dividing an image into regions. *Synonym:* object extraction. *See also:* line detection; region growing; edge detection; border detection; tracking. (C) 610.4-1990w

image space *See:* display space.

image storage (diode-type camera tube) The ability of the diode array target to integrate an image for times longer than the conventional frame time. (ED) 503-1978w

image-storage device An optoelectronic device capable of retaining an image for a selected length of time. *See also:* optoelectronic device. (ED) [46]

image-storage panel (optoelectronic device) A thin, usually flat, multicell image-storage device. *See also:* optoelectronic device. (ED) [46]

image-storage tube A storage tube into which the information is introduced by means of radiation, usually light, and read at a later time as a visible output. *See also:* storage tube. (ED) 158-1962w

image transfer constant (electric transducer) (transfer constant) The arithmetic mean of the natural logarithm of the ratio of input to output phasor voltages and the natural logarithm of the ratio of the input to output phasor currents when the transducer is terminated in its image impedances. *Note:* For a symmetrical transducer the transfer constant is the same as the propagation constant. *See also:* transducer. (Std100) 270-1966w

image transform *See:* image operator.

image transformation A transformation composed from the translation, rotation, or scaling of a graphical image. (C) 610.6-1991w

image transform operator *See:* image operator.

image tube An electron tube that reproduces on its fluorescent screen an image of an irradiation pattern incident on its photosensitive surface. *See also:* camera tube. (ED) 161-1971w

imaginary part If a complex quantity is represented by two components $A + jB$, B is called the imaginary part. (CAS) [13]

imaging plane *See:* ground plane.

imaging radar A high-resolution radar whose output is a representation of the radar cross section within the resolution cell (backscatter coefficient) from the object or scene resolved in two or three spatial dimensions. *Note:* The radar may use real aperture (such as a sidelooking airborne radar), synthetic-aperture radar (SAR), inverse synthetic aperture radar (ISAR), interferometric SAR, or tomographic techniques. (AES) 686-1997

I_{max} output The maximum allowable output load current over which the unit under test output voltage is required to be maintained within the specified operational limits. Imax is also known as the rated current. (PEL) 1515-2000

imbedded temperature-detector insulation (rotating machinery) The insulation surrounding a temperature detector, taking the place of a coil separator in its area. (PE) [9]

IMD *See:* intermodulation distortion.

I_{min} output The minimum allowable output load current over which the unit-under-test output voltage is required to be maintained within the specified operational limits. (PEL) 1515-2000

immediate access Access to a storage device or register in which access time is virtually equal to zero. *Note:* Access time measured in nanoseconds is considered to be virtually equal to zero. *Synonyms:* simultaneous access; instantaneous access. (C) 610.5-1990w

immediate access storage A type of storage whose access time is extremely small, relative to those of alternative types of storage. *Synonym:* instantaneous storage. (C) 610.10-1994w

immediate address (1) (computers) An instruction in which an address part contains the value of an operand rather than its address. *See also:* zero-level address. (C) [20], [85]
(2) An instruction address in which the address field is the operand itself. *Synonym:* zero-level address. (C) 610.10-1994w
(3) (software) *See also:* immediate data. (C) 610.12-1990

immediate addressing An addressing mode in which instructions contain the operand itself and not the address of the operand. (C) 610.10-1994w

immediate control *See:* bit steering.

immediate data Data or operands that are contained in the address field of a computer instruction. *Contrast:* n-level address; direct address; indirect address. (C) 610.10-1994w, 610.12-1990

immediate effect An effect of a transaction, which appears to occur between the time the request subaction is accepted and the response subaction is returned. If a bus standard allows CSR transactions to be split, and sufficient time is allowed between the acceptance of a request subaction and the return of a response subaction, an immediate effect can be emulated by a processor on the node. (C/MM) 1212-1991s

immediate entry A read-only memory (ROM) entry that provides a 24-bit immediate data value. (C/BA/MM) 896.10-1997, 896.2-1991w, 1212-1991s

immediate instruction (1) A computer instruction whose address fields contain the values of the operands rather than the operands' addresses. *Contrast:* indirect instruction; direct instruction. *See also:* absolute instruction; immediate data; effective instruction. (C) 610.12-1990
(2) A computer instruction whose address fields contain immediate data, or the values of the operands rather than the operands' addresses. (C) 610.10-1994w

immediate-nonsynchronized ringing (telephone switching systems) An arrangement whereby a pulse of ringing is sent to the called line when the connection is completed, irrespective of the state of the ringing cycle. (COM) 312-1977w

immediate restoration of service (health care facilities) Automatic restoration of operation with an interruption of not more than 10 seconds as applied to those areas and functions served by the Emergency System, except for areas and functions for which Article 700 [of the National Electrical Code] otherwise makes specific provisions. (NESC/NEC) [86]

immediate subclass A subclass, of a class C, having no superclasses that are themselves subclasses of C. (C/PA) 1328-1993w, 1327-1993w, 1224-1993w

immediate subobject One object that is a value of an attribute of another. (C/PA) 1328-1993w, 1224-1993w, 1327-1993w

immediate subordinate An entry in the DIT that is immediately below another entry in the tree. The distinguished name of the immediate subordinate is formed by appending its relative distinguished name to the distinguished name of the other entry. (C/PA) 1328.2-1993w, 1224.2-1993w, 1326.2-1993w, 1327.2-1993w

immediate superclass The superclass, of a class C, having no subclasses that are themselves superclasses of C. (C/PA) 1328-1993w, 1327-1993w, 1224-1993w

immediate superior An entry in the DIT that is immediately above another entry in the tree. The distinguished name of the immediate superior, followed by the relative distinguished name of the other entry, forms the distinguished name of the other entry. Each entry (except the root) has exactly one immediate superior. (C/PA) 1328.2-1993w, 1326.2-1993w, 1327.2-1993w, 1224.2-1993w

immediate superobject One object that contains another among its attribute values. (C/PA) 1328-1993w, 1224-1993w, 1327-1993w

immediate-synchronized ringing (telephone switching systems) An arrangement whereby the ringing cycle starts with a complete interval of ringing sent to the called line when the connection is completed. (COM) 312-1977w

immersed gun (microwave tubes) A gun in which essentially all the flux of the confining magnetic field passes perpendicularly through the emitting surface of the cathode. (ED) [45]

immersion plating The deposition, without application of an external electromotive force, of a thin metal coating upon a less noble metal by immersing the latter in a solution containing a compound of the metal to be deposited. *Synonym:* dip plating. *See also:* electroplating. (EEC/PE) [119]

immittance (linear passive networks) A response function for which one variable is a voltage and the other a current. *Note:* Immittance is a general term for both impedance and admittance, used where the distinction is irrelevant. (CAS) 156-1960w

immittance comparator An instrument for comparing the impedance or admittance of the two circuits, components, etc. *See also:* auxiliary device to an instrument. (IM/HFIM) [40]

immittance converter A two-port circuit capable of making the input immittance of one port (H_{in}) the product of the immittance connected to the other port (H_1) a positive or negative real constant ($\pm 1k$) and some internal immittance (H_i) i.e., $H_{in} = \pm kH_1H_i$. (CAS) [13]

immittance matrix A two-dimensional array of immittance quantities that relate currents to voltages at the ports of a network. (CAS) [13]

immobilized electrolyte (1) Electrolyte in a cell that is retained by either using gelled or absorbed electrolyte technology. (IA/PSE) 446-1995
(2) Electrolyte in a VRLA cell that is retained by using either gelled or absorbed electrolyte technology. (SB) 1189-1996

immunity (to a disturbance) The ability of a device, equipment, or system to perform without degradation in the presence of an electromagnetic disturbance. (EMC) C63.12-1987

immunity to interference The property of a receiver or any other equipment or system enabling it to reject a radio disturbance. *See also:* electromagnetic compatibility. (EMC) [53]

immunological effect Effect pertaining to the immune system. (T&D/PE) 539-1990

immutable class A class for which the set of instances is fixed; its instances do not come and go over time. *Contrast:* mutable class. *See also:* value class. (C/SE) 1320.2-1998

IMPact Avalanche Transit Time oscillator (nonlinear, active, and nonreciprocal waveguide components) A direct dc-rf (direct current to radio frequency) conversion device in which the active element of the oscillator is a p-n junction diode biased into the avalanche breakdown mode. *Synonym:* IMPATT oscillator. (MTT) 457-1982w

impact ionization gain (diode-type camera tube) The dimensionless ratio of the target signal current to the photocathode current which produced this signal, both averaged over a frame time or over a time long compared to the frame time. (ED) 503-1978w

impact printer A printer in which printing results from mechanical impacts with the paper. *Contrast:* nonimpact printer. *See also:* on-the-fly printer; element printer; dot matrix printer; formed character printer. (C) 610.10-1994w

impaired insulation (insulation systems of synchronous machines) The word "impaired" is here used in the sense of causing any change that could disqualify the insulating material for continuously performing its intended function whether creepage spacing, mechanical support, or dielectric barrier action. The electrical and mechanical properties of the insulation must not be impaired by the prolonged application of the hottest spot or limiting observable temperature permitted for the specific insulation class. (REM) [115]

impairment (of signal quality) An undesired change of a signal reducing the telephone transmission performance. Not closed. Impairment may be caused by noise, distortion, or other phenomena. (COM/TA) 823-1989w

IMPATT oscillator *See:* IMPact Avalanche Transit Time oscillator.

IMPDU *See:* initial MAC protocol data unit.

impedance (1) (A) (general) (linear constant-parameter system). The corresponding impedance function with *p* replaced by *jw* in which *w* is real. *Note:* Definitions (A) and (B) are equivalent. **(B) (general)** (linear constant-parameter system). The ratio of the phasor equivalent of a steady-state sine- wave voltage or voltage-like quantity (driving force) to the phasor equivalent of a steady-state sine-wave current or current-like quantity (response). *Note:* Definitions (A) and (B) are equivalent. **(C) (general)** A physical device or combination of devices whose impedance as defined in definition (A) or (B) can be determined. *Note:* This sentence illustrates the double use of the word impedance, namely for a physical characteristic of a device or system [definitions (A) and (B) and for a device definition (C)]. In the latter case, the word impedor may be used to reduce confusion. Definition (C) is a second use of "impedance" and is independent of definitions (A) and (B). *See also:* resistance; feedback impedance; reactance; input impedance; impedance function; network analysis. (Std100) 270-1966
(2) Linear operator expressing the relation between voltage (increments) and current (increments). Its inverse is called the admittance of an electric machine. *Notes:* 1. If a matrix has as its elements impedances, it is usually referred to as impedance matrix. Frequently the impedance matrix is called impedance for short. 2. Usually such impedances are defined with the mechanical angular velocity of the machine at steady state. *See also:* asynchronous machine. (PE/EM) [9]
(3) (two-conductor transmission line) The ratio of the complex voltage between the conductors to the complex current on one conductor in the same transverse plane. (AP/ANT) [35]
(4) (circular or rectangular waveguide) A nonuniquely defined complex ratio of the voltage and current at a given transverse plane in the waveguide, which depends on the choice of representation of the characteristic impedance. *See also:* characteristic impedance; waveguide. (AP/ANT) [35]
(5) (automatic control) (linear system under sinusoidal stimulus) The complex-number ratio of a force-like variable to the resulting velocity-like steady-state variable: a type of transfer function expressed as voltage per unit current, force per unit velocity, pressure difference per unit volume or mass flux, temperature difference per unit heat flux. *See also:* transfer function. (PE/EDPG) [3]
(6) (of a waveguide) A value relating any two of the three quantities, power (*P*), complex voltage (*V*), and complex current (*I*), in a given mode at a specified transverse plane in a waveguide; the value is nonunique, depending on how the voltage and current quantities are defined and on the selected ratio ($V^{2/P}$, P/I^2, or V/I). (MTT) 146-1980w
(7) (broadband local area networks) A measure of the complex resistive and reactive attributes of a component in an alternating-current circuit. (LM/C) 802.7-1989r
(8) The resistance to the flow of alternating current in a circuit. (C) 610.7-1995
(9) (of a series reactor) The phasor sum of the reactance and effective resistance, expressed in ohms. (PE/TR) C57.16-1996
(10) (shunt reactors over 500 kVA) (of a shunt reactor) The phasor sum of the reactance and resistance, expressed in ohms per phase, it may be derived from the rated kilovoltampere (kVA) and rated voltage. (PE/TR) C57.21-1981s
(11) *See also:* mutual impedance; intrinsic impedance; input impedance; self-impedance. (ANT)

impedance bond (railway practice) An iron-core coil of low resistance and relatively high reactance used on electrified railroads to provide a continuous direct-current path for the return propulsion current around insulated joints and to confine the alternating-current signaling energy to its own track circuit. (EEC/PE) [119]

impedance characteristic A nondirectional relay characteristic in which the threshold of operation for the basic form plots as a circle on an *R-X* diagram, with the reach a constant impedance in all four quadrants. See following figure.

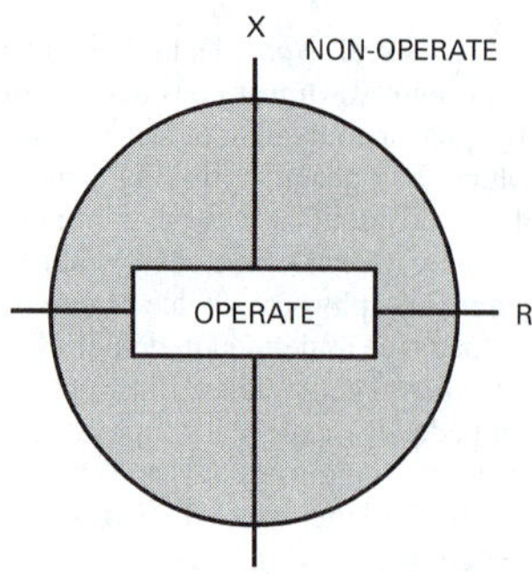

impedance characteristic

(SWG/PE) C37.100-1992

impedance, characteristic wave *See:* characteristic wave impedance.

impedance compensator An electric network designed to be associated with another network or a line with the purpose of giving the impedance of the combination a desired characteristic with frequency over a desired frequency range. *See also:* network analysis. (Std100) 270-1966w

impedance, conjugate *See:* conjugate impedance.

impedance drop (power and distribution transformers) The phasor sum of the resistance voltage drop and the reactance voltage drop. *Note:* For transformers, the resistance drop, the reactance drop, and the impedance drop are, respectively, the sum of the primary and secondary drops reduced to the same terms. They are determined from the load-loss measurements and are usually expressed in per unit, or in percent. (PE/TR) C57.12.80-1978r

impedance, effective input *See:* effective input impedance.

impedance, essentially zero source (transformer electrical tests) Source impedance low enough so that the test currents under consideration would cause less than five percent (5%) distortion (instantaneous) in the voltage amplitude or waveshape at the load terminals. (PEL/ET) 295-1969r

impedance feedback (analog computer) A passive network connected between the output terminal of an operational amplifier and its summing junction. (C) [20], 165-1977w

impedance function (defined for linear constant-parameter systems or parts of such systems) That mathematical function of p that is the ratio of a voltage or voltage-like quantity (driving force) to the corresponding current or current-like quantity (response) in the hypothetical case in which the former is e^{pt} (e is the natural log base, p is arbitrary but independent of t, t is an independent variable that physically is usually time) and the latter is a steady-state response of the form $e^{pt}/Z(p)$. *Note:* In electric circuits voltage is always the driving force and current is the response even though as in nodal analysis the current may be the independent variable: in electromagnetic radiation electric field strength is always considered the driving force and magnetic field strength the response, and in mechanical systems mechanical force is always considered as a driving force and velocity as a response. In a general sense the dimension (and unit) of impedance in a given application may be whatever results from the ratio of the dimensions of the quantity chosen as the driving force to the dimensions of the quantity chosen as the response. However, in the types of systems cited above any deviation from the usual convention should be noted. *See also:* network analysis. (Std100) 270-1966w

impedance grounded (power and distribution transformers) Grounded through impedance. *Note:* The components of impedance and the device to be grounded need not be at the same location.
(SPD/PE/T&D/TR) 32-1972r, C57.12.80-1978r, [8], [10]

impedance grounded neutral system A system whose neutral point(s) are grounded through an impedance (to limit ground fault currents). (PE/C) 1313.1-1996

impedance heating An electric heating system where the object to be heated generates heat as a result of an ac current passing through it. (IA/PC) 844-1991

impedance, image *See:* image impedances.

impedance, input *See:* input impedance.

impedance, intrinsic *See:* intrinsic impedance.

impedance inverter (A) network possessing an input (output) impedance that is proportional to the reciprocal of the load (source) impedance. **(B)** A symmetrical four-terminal network having the impedance inverting and phase characteristics of a quarter-wave length transmission line at its specified frequency or a chain matrix where A, D, O, B, jK and C, j/K (K is a constant relating the input impedance Z to the load impedance Z_L by the relationship Z, K^2/Z_L). (CAS) [13]

impedance irregularity (data transmission) A term used to denote impedance mismatch in a transmission medium. For example, a section of cable in an open-wire line constitutes an impedance irregularity. (PE) 599-1985w

impedance, iterative *See:* iterative impedance.

impedance kilovolt-amperes (1) (regulator) The kilovolt-amperes (kVA) measured in the shunt winding with the series winding short-circuited and with sufficient voltage applied to the shunt winding to cause rated current to flow in the windings. *See also:* voltage regulator. (PE/TR) C57.15-1968s **(2) (rated) (power and distribution transformers)** The kilovolt-amperes (kVA) measured in the excited winding with the other winding short-circuited and with sufficient voltage applied to the excited winding to cause rated current to flow in the winding. (PE/TR) C57.12.80-1978r

impedance, load *See:* load impedance.

impedance, loaded applicator *See:* loaded applicator impedance.

impedance, matching *See:* load matching.

impedance matching (glass industry) (electrical heating applications to melting furnaces and forehearths in the glass industry). The use of a transformer to match line-supply voltage levels to the voltage levels required by the molten-glass load. (IA) 668-1987w

impedance matrix (multiport network) A matrix operator that interrelates the voltages at the various ports to the currents at the same and other ports. (IM/HFIM) [40]

impedance mismatch factor The ratio of the power accepted by an antenna to the power incident at the antenna terminals from the transmitter. *Note:* The impedance mismatch factor is equal to one minus the magnitude squared of the input reflection coefficient of the antenna. (AP/ANT) 145-1993

impedance, normalized *See:* normalized impedance.

impedance, output *See:* output impedance.

impedance permeability (magnetic core testing) An ac permeability related to the total rms exciting current, including harmonics.

$$\mu_z = \frac{B_i}{H_z \mu_0}$$

where

$$H_z = \frac{\sqrt{2}NI}{1}$$

= equivalent peak field strength, amperes/meters
B_i = maximum intrinsic flux density tesla
I = rms exciting current, amperes
N = exciting coil turns

(MAG) 393-1977s

impedance ratio (divider) The ratio of the impedance of the two arms connected in series to the impedance of the low-voltage arm. *Note:* In determining the ratio, account should be taken of the impedance of the measuring cable and the instrument. The impedance ratio is usually given for the frequency range within which it is approximately independent of frequency. For resistive dividers the impedance ratio is generally derived from a direct-current measurement such as by means of a Wheatstone bridge. (PE/PSIM) 4-1978s, [55]

impedance ratio factor The ratio of the source impedance, at the point in the system under consideration, to the equivalent total impedance from the source to the converter circuit elements that commutate simultaneously. (IA/SPC) 519-1992

impedance relay (1) A distance relay in which the threshold value of operation depends only on the magnitude of the ratio of voltage to current applied to the relay, and is substantially independent of the phase angle of the impedance. (SWG/PE) C37.100-1992
(2) A distance relay in which the threshold value of operation depends only on the magnitude of the ratio of voltage to current applied to the relay, and is substantially independent of the phase angle between the applied voltage and current. (PE/PSR) C37.113-1999

impedance, source *See:* source impedance.

impedance, unloaded applicator *See:* unloaded applicator impedance.

impedance voltage (1) (of a regulator) The voltage required to circulate rated current through one of two specified windings of a transformer when the other winding is short-circuited, with the windings connected as for rated voltage operation. *Note:* It is usually expressed in per unit, or percent, of the rated voltage of the winding in which the voltage is measured. (PE/TR) C57.12.80-1978r
(2) (constant-current transformer) The measured primary voltage required to circulate rated secondary current through the short-circuited secondary coil for a particular coil separation. *Note:* It is usually expressed in per unit, or percent, of the rated primary voltage. *See also:* constant-current transformer. (PE/TR) [57], C57.12.80-1978r
(3) (current-limiting reactor) The product of its rated ohms impedance and rated current. *See also:* reactor. C57.16-1958w
(4) (neutral grounding devices) An effective resistance component corresponding to the impedance losses, and a reactance component corresponding to the flux linkages of the winding. (PE/SPD) 32-1972r
(5) The voltage required to circulate rated current through one winding of the regulator when another winding is short-circuited, with the respective windings connected as for a rated voltage operation. Impedance voltage is usually referred to the series winding, and then that voltage is expressed in per unit, or percent, of the rated voltage of the regulator. (PE/TR) C57.15-1999

impedance voltage drop (1) (of a series reactor) The product of its rated ohms impedance and its rated current. (PE/TR) C57.16-1996
(2) The phasor sum of the resistance voltage drop and the reactance voltage drop. For regulators, the resistance drop, the reactance drop, and the impedance drop are, respectively, the sum of the primary and secondary drops reduced to the same terms. They are usually expressed in per unit or percent of the rated voltage of the regulator. Since they differ at different operating positions of the regulator, two values of impedance shall be considered, in practice, to be the tap positions that result in the minimum and the maximum impedance. Neutral position has the minimum amount of impedance. (PE/TR) C57.15-1999

impedance, wave *See:* wave impedance.

impedor A device, the purpose of which is to introduce impedance into an electric circuit. *See also:* network analysis. (Std100) 270-1966w

imperative construct A sequence of one or more steps not involving branching or iteration. (C) 610.12-1990

imperative statement *See:* instruction.

imperfect debugging (software) In reliability modeling, the assumption that attempts to correct or remove a detected fault are not always successful. *See also:* fault; reliability. (C/SE) 729-1983s

imperfection (crystalline solid) Any deviation in structure from that of an ideal crystal. *Note:* An ideal crystal is perfectly periodic in structure and contains no foreign atoms. *See also:* semiconductor. (ED) 216-1960w

impingement attack Localized erosion-corrosion resulting from turbulent or impinging flow of liquids. (IA) [59]

impingement plume *See:* positive prebreakdown streamers.

implementation (1) (A) (software) The process of translating a design into hardware components, software components, or both. *See also:* coding. **(B) (software)** The result of the process in definition (A). (C) 610.12-1990
(2) Hardware and/or software that conforms to all the specifications of an ISA. (C/MM) 1754-1994
(3) An object providing to applications and users the services defined by this standard. The word *implementation* is to be interpreted to mean that object, after it has been modified in accordance with the manufacturer's instructions to:

— Configure it for conformance with IEEE Std 2003.2-1996;
— Select some of the various optional facilities described by this standard through customization by local system administrators or operators.

An exception to this meaning occurs when discussing conformance documentation or using the term implementation defined. (C/PA) 2003.2-1996
(4) That which implements the requirements of a base standard, or a profile. Test method specifications shall define specifically what an implementation is within the meaning of that specification. Implementation, as used here, is not to be confused with implementation-defined. (C/PA) 2003-1997

implementation architecture The logic structure of a computer system that describes how the functions described by the architecture are carried out. (C) 610.10-1994w

implementation defined (1) An indication that the implementation provider shall define and document the requirements for correct program constructs and correct data of a value or behavior. When the value or behavior in the implementation is designed to be variable or customizable on each instantiation of the system, the implementation provider shall document the nature and permissible ranges of this variation. (C/PA) 2003.2-1996
(2) (A) Possibly differing between processors, but defined for any particular processor. **(B)** A value or behavior is implementation defined if the implementation defines and documents the requirements for correct program construct and correct data. (C/PA/PAS) 1003.1-1988
(3) An indication that the implementation shall define and document the requirements for correct program constructs and correct data of a value or behavior. (C) 1003.5-1999

implementation-dependent (1) (pascal computer programming language) Possibly differing between processors and not necessarily defined for any particular processor. (Std100) 812-1984w
(2) Describes an aspect of the architecture that may legitimately vary among implementations of the architecture. In many cases, the permitted range of variation is specified in the standard. When a range is specified, compliant implementations shall not deviate from that range. Compliant implementations shall not add to or deviate from this standard except in aspects described as **implementation-dependent.** (C/MM) 1754-1994

implementation design *See:* database design.

implementation level The level of verification activities at which system components implementing vital functions are comprehensively identified and analyzed to verify that all functions identified as vital are implemented fail-safely. (VT/RT) 1483-2000

implementation phase (software verification and validation) (software) The period of time in the software life cycle during which a software product is created from design documentation and debugged. *See also:* test phase; software life cycle; software product. (C/SE) 1012-1986s, 610.12-1990

implementation requirement (software) A requirement that specifies or constrains the coding or construction of a system or system component. *Contrast:* physical requirement; performance requirement; functional requirement; design requirement; interface requirement. (C) 610.12-1990

implementation under test (IUT) **(1)** The Futurebus+ module that is being tested for conformance. (C/BA) 896.4-1993w
(2) That which implements the standard(s) being tested. An IUT may consist of hardware and software located on different systems. Test method specifications shall define specifically what an implementation is composed of within the meaning of that specification. (C/PA) 2003-1997

implication **(1) (mathematics of computing)** A dyadic Boolean operator having the property that if P is a statement and Q is a statement, then the expression "P implies Q" is true in all cases except when P is true and Q is false. *Note:* P implies Q is often represented as P → Q.

Implication Truth Table

P	Q	P → Q
0	0	1
0	1	1
1	0	0
1	1	1

Synonyms: IF-THEN; conditional implication. (C) 1084-1986w
(2) The dyadic Boolean operation whose result yields the value 0 if and only if the first operand has the value 1 and the second has the value 0. *See also:* IF-THEN gate. (C) 610.10-1994w

implicit address instruction *See:* zero-address instruction.

implicit computation (analog computer) Computation using a self-nulling principle in which, for example, the variable sought is first assumed to exist, after which a synthetic variable is produced according to an equation and compared with a corresponding known variable and the difference between the synthetic and the known variable driven to zero by adjusting the assumed variable. Although the term applies to most analog circuits, even to a single operational amplifier, it is restricted usually to computation performed by (A) circuits in which a function is generated at the output of a single high-gain dc amplifier by inserting an element generating the inverse function in the feedback path, (B) circuits in which combinations of computing elements are interconnected in closed loops to satisfy implicit equations, or (C) circuits in which linear or nonlinear differential equations yield the solutions to a system of algebraic or transcental equations in the steady-state. (C) 165-1977w

implied addressing **(1)** A method of addressing in which the operation field of an computer instruction implies the address of the operands. For example, if a computer has only one accumulator, an instruction that refers to the accumulator needs no address information describing it. Types include one-ahead addressing, repetitive addressing. *See also:* relative address; indirect address; direct address. (C) 610.12-1990
(2) An addressing mode in which the operation field of an instruction implicitly addresses operands. *See also:* repetitive addressing. (C) 610.10-1994w

implied binary point *See:* assumed binary point.

implied decimal point *See:* assumed decimal point.

implied radix point *See:* assumed radix point.

importance measures A quantitative analysis to determine the importance of variations in equipment reliability to system risk and/or reliability. (PE/NP) 933-1999

imprecision *See:* precision.

impregnant (rotating machinery) A solid, liquid, or semiliquid material that, under conditions of application, is sufficiently fluid to penetrate and completely or partially fill or coat interstices and elements of porous or semiporous substances or composites. (PE) [9]

impregnate (rotating machinery) The act of adding impregnant (bond or binder material) to insulation or a winding. *Note:* The impregnant, if thermosetting, is usually cured in the process. (PE) [9]

impregnated (fibrous insulation) A suitable substance replaces the air between the fibers, even though this substance does not fill completely the spaces between the insulated conductors. *Note:* To be considered suitable, the impregnating substance must have good insulating properties and must cover the fibers and render them adherent to each other and to the conductor. (EEC/PE) [119]

impregnated insulation (insulation systems of synchronous machines) Insulating is considered to be "impregnated" when a suitable substance provides a bond between components of the structure and also a suitable degree of filling and surface coverage sufficient to give adequate performance under the extremes of temperature, surface contamination (moisture, dirt, etc.), and electrical and mechanical stress expected in service. The impregnant must not flow or deteriorate enough at operating temperature so as to seriously affect performance in service. (REM) [115]

impregnated tape *See:* magnetic-powder-impregnated tape.

impregnation, winding *See:* winding impregnation.

impression The process of marking the media. A single-sided, one-color printer requiring one pass per sheet would produce one impression per sheet. A similar printer printing duplex would produce two impressions per sheet. A two-pass printer providing a base color and a highlight color would produce two impressions per side, etc. (C/MM) 1284.1-1997

improper character *See:* illegal character.

improper ferroelectric (primary ferroelectric terms) A ferroelectric in which the polarization is not the primary order parameter. (UFFC) 180-1986w

improper mode Refers to a mode of propagation that cannot be excited by a physical source in the absence of other modes, (e.g., Zenneck surface wave). (AP/PROP) 211-1997

improvement threshold (angle modulation) The condition of unity for the ratio of peak carrier voltage to peak noise voltage after selection and before any nonlinear process such as amplitude limiting and detection. *See also:* amplitude modulation. (AP/ANT) 145-1983s

impulse **(1) (automatic control) (mathematics)** A pulse that begins and ends within a time so short that it may be regarded mathematically as infinitesimal although the area remains finite. (Std100) 270-1966w
(2) An intentionally applied transient voltage or current that usually rises rapidly to a peak value and then falls more slowly to zero. (PE/PSIM) 4-1995
(3) A surge of unidirectional polarity, for example, a 1.2/50 μs voltage surge. (T&D/PE) 1250-1995
(4) A surge of unidirectional polarity. (SPD/PE) C62.22-1997, C62.62-2000, C62.11-1999
(5) *See also:* transient. (IA/PSE) 1100-1999

impulse bandwidth **(1) (electromagnetic site survey)** The ratio of the maximum value of the voltage at the output of a network (when properly corrected for network sinewave gain at the stated reference frequency) to the spectrum amplitude of the pulse applied at the input. In networks with a single-humped response the reference frequency is taken as that at which the gain is maximum. (EMC) 473-1985r
(2) (general) When an inpulse is passed through a network with a restricted passband, the output generally consists of a wave train, the envelope of which builds up to a maximum value and then decays approximately exponentially. The impulse bandwidth of such a network is defined as the ratio of that maximum value (when properly corrected for network sine wave gain at a stated reference frequency) to the spectrum amplitude of the pulse applied at the input. In networks with a single humped response, the reference frequency is taken as that at which the gain is maximum. (Overcoupled or stagger-tuned networks should not be used for measurement of spectrum amplitude of impulses.) *See also:* impulse strength. (EMC) 376-1975r
(3) (radio noise from overhead power lines) The peak value of the response envelope divided by the spectrum amplitude

of an applied impulse. *See also:* electromagnetic compatibility. (T&D/PE) 430-1986w, 539-1990

(4) (spectrum analyzer) (non-real time spectrum analyzer) The peak value of the time response envelope divided by the spectrum amplitude (assumed flat within the bandpass) of an applied pulse. (IM) 748-1979w

impulse circuitry (nonlinear, active, and nonreciprocal waveguide components) A term given to the circuitry associated with either a step recovery or a dual mode varactor frequency multiplier. As charge is stored in the multiplier diode during each positive cycle of the input frequency and released during each negative cycle, a magnetic field is built up in an impulse inductor that stores all the circuit energy as charge approaches zero. Multiplication occurs when the inductor releases its energy in the form of an impulse voltage across the diode at the time of switching or high-impedance recovery. (MTT) 457-1982w

impulse current (current testing) Ideally, an aperiodic transient current that rises rapidly to a maximum value and falls usually less rapidly to zero. A rectangular impulse current rises rapidly to a maximum value, remains substantially constant for a specified time and then falls rapidly to zero. (PE/PSIM) 4-1978s

impulse currents (high voltage testing) Two types of impulse currents are dealt with. The first type has a shape which increases from zero to a crest value in a short time, and thereafter decreases to zero, either approximately exponentially or in the manner of a heavily damped sine curve. This type is defined by the virtual front time T1 and the virtual time to half-value T2. The second type has an approximately rectangular shape and is defined by the virtual duration of the peak and the virtual total duration. (PE/PSIM) 4-1978s

impulse current tests *See:* virtual time to half-value; impulse currents; value of the test current; virtual origin; virtual total duration of a rectangular impulse current; virtual duration.

impulse (shock) excitation A method of producing oscillator current in a circuit in which the duration of the impressed voltage is relatively short compared with the duration of the current produced. *See also:* oscillatory circuit. (AP/BT/ANT) 145-1983s, 182-1961w

impulse flashover voltage (1) (insulators) The crest value of the impulse wave that, under specified conditions, causes flashover through the surrounding medium. (EEC/IEPL) [89]

(2) (surge arresters) The crest voltage of an impulse causing a complete disruptive discharge through the air between electrodes of a test specimen. *See also:* insulator; critical impulse flashover voltage. (PE) [8], 64

impulse flashover volt-time characteristic A curve plotted between flashover voltage for an impulse and time to impulse flashover, or time lag of impulse flashover. *See also:* insulator. (PE/T&D) [10]

impulse-forced response (automatic control) The total (transient plus steady-state) time response resulting from an impulse at the input. *Synonym:* impulse response. (PE/EDPG) [3]

impulse generator A standard reference source of broadband impulse energy. (EMC) 263-1965w

impulse inertia (surge arresters) The property of insulation whereby more voltage must be applied to produce disruptive discharge, the shorter the time of voltage application. (PE) [8], 64

impulse insulation level An insulation strength expressed in terms of the crest value of an impulse withstand voltage. *See also:* basic impulse insulation level. (EEC/LB) [100]

impulse noise (1) (A) (data transmission) (overhead-power-line corona and radio noise) Noise characterized by transient disturbances separated in time by quiescent intervals. *Notes:* 1. The frequency spectrum of these disturbances are substantially uniform over the useful pass band of the transmission system. 2. The same source may produce impulse noise in one system and random noise in a different system. *See also:* signal-to-noise ratio. **(B)** Any burst of noise that produces a voltage exceeding the root-mean-square (rms) noise voltage (i.e., the mean noise as measured with a standard noise measuring set using C-message weighting) by a given magnitude. Impulse noise is a spike that exceeds the rms value of the background or quantizing noise by at least 12 dB. The impulse noise counter registers the number of instances in which the measured noise exceeds a preset threshold. (T&D/PE/AP/COM/TA/ANT) 539-1990, 599-1985, 145-1983, 973-1990

(2) A component of the received noise signal that is much greater in amplitude than the normal peaks of the message circuit noise, and that occurs as short-duration spikes or energy bursts. (PE/IC) 1143-1994r

(3) Noise characterized by electrical pulses of high amplitude and narrow width, often originating from switching devices or electrical storms. (C) 610.7-1995

(4) Any burst of noise that produces a voltage exceeding the rms value of the background or quantizing noise by more than 12 dB. (COM/TA) 743-1995

(5) Noise characterized by transient disturbances. Impulses exceeding a specified threshold are counted for a specific duration with an impulse noise counter designed in accordance with IEEE Std 743-1984. (COM/TA) 1007-1991r

impulse-noise selectivity (receiver) A measure of the ability to discriminate against impulse noise. (VT) [37]

impulse protective level For a defined wave shape, the higher of the maximum sparkover value or the corresponding discharge-voltage value. (SPD/PE) C62.11-1999

impulse protective volt-time characteristic The discharge-voltage-time response of the device to impulses of a designated wave shape and polarity, but of varying magnitudes. (SPD/PE) C62.11-1999

impulse radar A radar whose transmitted pulse consists of one or a few cycles of carrier, usually generated by application of a short video pulse to a wideband radio frequency (RF) amplifier (e.g., a traveling-wave tube) or directly to a wideband antenna (e.g., a dipole). (AES) 686-1997

impulse ratio (surge arresters) The ratio of the flashover, sparkover, or breakdown voltage of an impulse to the crest value of the power-frequency, sparkover, or breakdown voltage. (T&D/PE) [10], [8]

impulse relay (A) A relay that follows and repeats current pulses, as from a telephone dial. **(B)** A relay that operates on stored energy of a short pulse after the pulse ends. **(C)** A relay that discriminates between length and strength of pulses, operating on long or strong pulses and not operating on short or weak ones. **(D)** A relay that alternately assumes one of two positions as pulsed. **(E)** Erroneously used to describe an integrating relay. *See also:* relay. (PE/EM) 43-1974

impulse response (1) (linear network) The response, as a function of time, of a network when the excitation is a unit impulse. Hence, the impulse response of a network is the inverse Laplace transform of the network function in the frequency domain. (CAS) [13]

(2) (fiber optics) The function h(t) describing the response of an initially relaxed system to an impulse (Dirac-delta) function applied at time $t = 0$. The root-mean-square (rms) duration, o_{rms}, of the impulse response is often used to characterize a component or system through a single parameter rather than a function:

$$o_{rms} = [1/M_0 \int_{-\infty}^{\infty} (T-t)^2 h(t)dt]^{1/2}$$

where

$$M_0 = \int_{-\infty}^{\infty} h^{(t)dt}$$

$$T = 1/M_0 \int_{-\infty}^{\infty} th^{(t)dt}$$

Note: The impulse response may be obtained by deconvolving the input waveform from the output waveform, or as the

inverse Fourier transform of the transfer function. *See also:* transfer function; root-mean-square pulse duration. (Std100) 812-1984w
(3) (automatic control) *See also:* impulse-forced response.

impulse rms sound level *See:* impulse root-mean-square sound level.

impulse root-mean-square sound level (measurement of sound pressure levels of ac power circuit breakers) The maximum rms value reached by a sound wave, with the mean (or average) taken over a short, specified time interval. Unit: decibel (dB A, B, or C). For the purposes of IEEE Std C37.082-1982 and IEEE Std C37.100-1992, the averaging time is that given by a resistance-capacitance charging circuit with a 35 millisecond (ms) time constant. Syn: impulse rms sound level. *Synonym:* impulse rms sound level. (SWG/PE) C37.100-1992, C37.082-1982r

impulses (high voltage testing) An intentionally applied aperiodic transient voltage or current which usually rises rapidly to a peak value and then falls more slowly to zero. Such an impulse is in general well represented by the sum of two exponentials. For special purposes, impulses having approximately linearly rising fronts or of oscillating or approximately rectangular form are used. The term "impulse" is to be distinguished from the term "surge," which refers to transients occurring in electrical equipment or networks in service. (PE/PSIM) 4-1978s

impulse sparkover voltage (1) (gas-tube surge protective devices) (low-voltage air-gap surge-protective devices) The highest value of voltage attained by an impulse of a designated wave shape and polarity applied across the terminals of an arrester prior to the flow of discharge current. Sometimes referred to as surge or impulse breakdown voltage. (PE/SPD) C62.31-1987r, C62.32-1981s
(2) The highest value of voltage attained by an impulse of a designated wave shape and polarity applied across the terminals of an arrester that will cause gap sparkover prior to the flow of discharge current. (SPD/PE) C62.11-1999
(3) The highest value of voltage attained by an impulse of a designated wave shape and polarity applied across the terminals of a gap-type surge-protective device prior to the flow of discharge current. (SPD/PE) C62.62-2000

impulse sparkover voltage-time curve (gas-tube surge protective devices) (arresters) A curve that relates the impulse sparkover voltage to the time to sparkover. (SPD/PE) C62.31-1987r

impulse sparkover volt-time characteristic (1) The gap sparkover response of the device to impulses of a designated wave shape and polarity, but of varying magnitudes. *Note:* For an arrester, this characteristic is shown by a graph of crest voltage values plotted against time-to-sparkover. (SPD/PE) C62.11-1999
(2) The sparkover response of a gap-type surge-protective device, when subjected to voltage impulses that have varying magnitudes and specified wave shape and polarity. (SPD/PE) C62.62-2000

impulse strength The area under the amplitude-time relation for the impulse. *Note:* This definition can be clarified with the aid of the figure below. Let $A(t)$ be some function of time having a value other than zero only between the times t_1 and $t_1 + \delta$. Then let the area under the curve $A(t)$ be designated by σ:

$$\sigma = \int_{\infty}^{\infty} A(t)dt = \int_{t_1}^{t_1+\delta} + A(t)dt$$

To define the theoretical or ideal impulse, let $A(t)$ vary in a reciprocal manner with δ such that the value σ remains constant, so that

$$\sigma = \lim \delta \rightarrow 0 \int_{t_1}^{t_1+\delta} + A(t)dt$$

In the limit the function $A(t)$ becomes an ideal "impulse" of "strength" σ. As an example, consider the function shown in the second part of the figure. Here a rectangular pulse of finite duration Δt and height A is shown. Now let $A = \sigma/\Delta t$ where σ is (for the present argument) an arbitrary constant, and let $\Delta t \rightarrow 0$. In the limit we have an impulse of strength σ. When $\sigma = 1$, one has a "unit impulse." In many conventional applications the amplitude $A(t)$ has the dimension volts and σ then has the dimension volt-seconds.

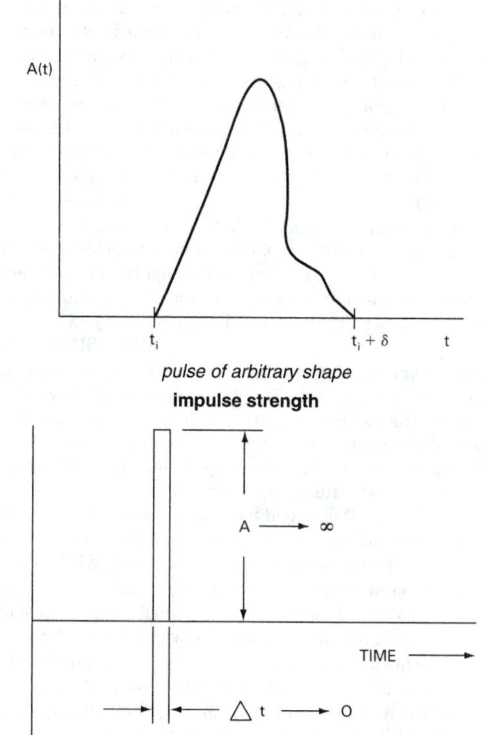

impulse strength

impulse test (1) (rotating machinery) A test for applying to an insulated component an aperiodic transient voltage having predetermined polarity, amplitude, and wave-form. *See also:* asynchronous machine. (SPD/PE) 32-1972r
(2) (surge arresters) (power and distribution transformers) An insulation test in which the voltage applied is an impulse voltage of specified wave shape. (PE/TR) C57.12.80-1978r
(3) (neutral grounding devices) Dielectric test in which the voltage applied is an impulse voltage of specified wave shape. The wave shape of an impulse test wave is the graph of the wave as a function of time or distance. *Note:* It is customary in practice to express the wave shape by a combination of two numbers, the first part of which represents the wave front and the second the time between the beginning of the impulse and the instant at which one-half crest value is reached on the wave tail, both values being expressed in microseconds, such as a 1.2×50 μs wave. (PE/SPD) 32-1972r

impulse time margin (in the operation of a relay) The difference between characteristic operating times and critical impulse times. (SWG/EEC/PE/CON/PSR) [28], C37.90-1978s, C37.100-1992

impulse transmission That form of signaling, used principally to reduce the effects of low-frequency interference, that employs impulses of either or both polarities for transmission to indicate the occurrence of transitions in the signals. *Note:* The impulses are generally formed by suppressing the low-frequency components, including direct current, of the signals. *See also:* telegraphy. (EEC/PE) [119]

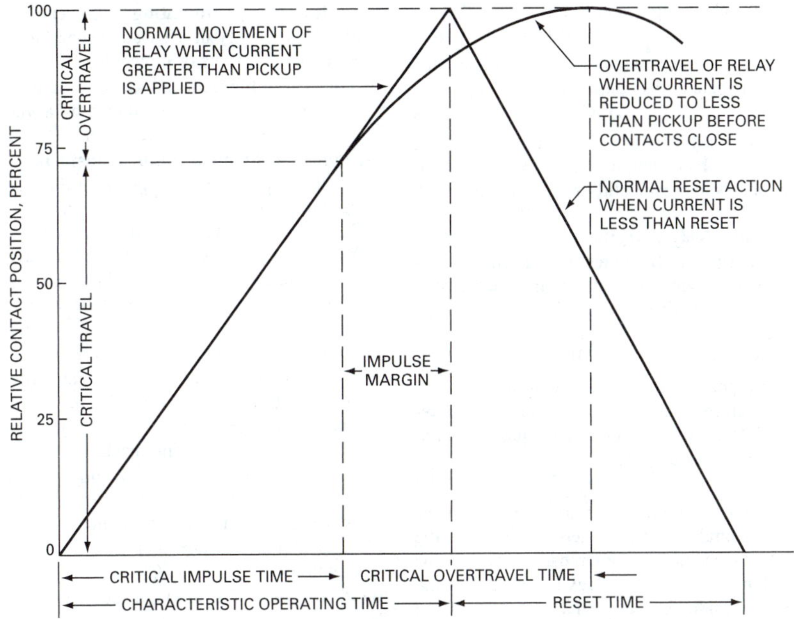

relationship of relay operating time for electromechanical relays (PSRC)

impulse transmitting relay A relay that closes a set of contacts briefly while going from the energized to the de-energized position or vice versa. *See also:* relay. (EEC/REE) [87]

impulse trip device (low-voltage dc power circuit breakers used in enclosures) A trip device that is designed to operate only by the discharge of a capacitor into its release (trip) coil and is utilized on high-speed circuit breakers to produce the tripping times that are independent of *di*/*dt*. (SWG/PE) C37.14-1999

impulse turbine A turbine that uses nozzles to convert water pressure into kinetic energy at atmospheric pressure to develop power. (PE/EDPG) 1020-1988r

impulse-type telemeter A telemeter that employs characteristics of intermittent electric signals, other than their frequency, as the translating means. *See also:* telemetering. (EEC/PE) [119]

impulse voltage (surge arresters) (current) Synonymous with voltage of an impulse wave (current of an impulse wave). (PE) [8], [84]

impulse wave (surge arresters) A unidirectional wave of current or voltage of very short duration containing no appreciable oscillatory components. *See also:* insulator. (PE) [8], [84]

impulse withstand voltage (1) (general) The crest value of an applied impulse voltage that, under specified conditions, does not cause a flashover, puncture, or disruptive discharge on the test specimen. *See also:* surge arrester; insulator. (SPD/PE) 32-1972r

(2) The crest voltage of an impulse that, under specified conditions, can be applied without causing flashover or puncture. (SWG/PE) C37.40-1993, C37.100-1992

(3) The crest value of an impulse that, under specified conditions, can be applied without causing a disruptive discharge. (SPD/PE) C62.11-1999, C62.62-2000

impulsive noise (1) (A) (control of system electromagnetic compatibility) Noise characterized by transient disturbances separated in time by quiescent intervals. *Notes:* 1. The frequency spectrum of these disturbances must be substantially uniform over the useful pass band of the transmission system. 2. The same source may produce an output characteristic of impulsive noise in one system and of random noise in a different system. **(B)** Electromagnetic noise that, when incident on a particular device or equipment, manifests itself as a succession of distinct pulses or transients. *Note:* The frequency spectrum of these disturbances must be substantially uniform over the useful pass band of the transmission system. 2. The same source may produce an output characteristic of impulsive noise in one system and of random noise in a different system. (EMC) C63.12-1984, C63.12-1987

(2) (measurement of sound pressure levels of ac power circuit breakers) A noise characterized by brief excursions of sound pressure (acoustic impulses) which significantly exceed the ambient noise. The duration of a single impulse is usually less than one second. For the purpose of IEEE Std C37.082-1982 and IEEE Std C37.100-1992, the noise produced by the closing or opening of a circuit breaker, or their combination, is classified as impulsive noise. Other components, such as compressor unloader exhausts, may be sources of impulsive noise. (SWG/PE) C37.100-1992, C37.082-1982r

(3) Noise, the effect of which is resolvable into a succession of discrete impulses in the normal operation of the particular system concerned. *See also:* electromagnetic compatibility. (EMC/INT) [53], [70]

impurity (1) (crystalline solid) An imperfection that is chemically foreign to the perfect crystal. *See also:* semiconductor. (ED) 216-1960w

(2) (chemical) (semiconductor) An atom within a crystal, that is foreign to the crystal. *See also:* dopant; semiconductor device. (IA) [12]

impurity, stoichiometric A crystalline imperfection arising from a deviation from stoichiometric composition. (ED) 216-1960w

inaccessible object An object for which the client does not possess a valid designator or handle. (C/PA) 1328-1993w, 1224-1993w, 1327-1993w

inactive (1) (A) Pertaining to a record or file that has not been accessed by an update transaction during a given processing cycle. **(B)** Pertaining to a record that will not be processed by future transactions. *See also:* active; logically deleted; purged. **(C) (696 interface devices) (signals and paths)** A signal in its logically false state. (C/MM) 610.2-1987, 696-1983

(2) When referring to an output driver (e.g., in the phrase *an inactive driver*), this term describes the mode in which the driver is not capable of determining the voltage of the network to which it is connected. (TT/C) 1149.5-1995, 1149.1-1990

inactive mode When the alarm input is in an unmonitored mode for testing. (PE/NP) 692-1997

inactive modules This category contains all the modules that have no need to participate in the control acquisition process in any way. (C/MM) 896.1-1987s

inactive region (1) (of a semiconductor radiation detector) A region of a detector in which charge created by ionizing radiation does not contribute significantly to the output signal. (NPS) 325-1996
(2) (germanium gamma-ray detectors) (charged-particle detectors) (x-ray energy spectrometers) (of a semiconductor radiation detector) A region of a detector in which charge created by ionizing radiation does not contribute significantly to the signal charge. (NPS/NID) 759-1984r, 301-1976s, 300-1988r

inadequate inertia systems An ac system having limited local generation, and therefore rotational inertia, so that the required voltage and frequency cannot be adequately maintained during transient ac or dc faults. (PE/T&D) 1204-1997

inadvertent interchange (1) (control area) (electric power system) The time integral of the net interchange minus the time integral of the scheduled net interchange. *Note:* This includes the intentional interchange energy resulting from the use of frequency and.or other bias as well as the unscheduled interchange energy resulting from human or equipment errors. (PE/PSE) 94-1970w, [54]
(2) The difference between the control area's net actual interchange and net scheduled interchange. (PE/PSE) 858-1993w

inadvertent interchange energy The time integral of a control area's net interchange error. (PE/PSE) 94-1991w

inadvertent write protection The circuitry that is used to prevent writing from occurring when the control signals of the memory enter an uncontrolled state during power on, power off, or noise transients. (ED) 1005-1998

INA JO A computer language used for proving or verifying program correctness. (C) 610.13-1993w

in-band signaling (1) The transmission of a signal using a frequency that is within the bandwidth of the information channel. *Synonym:* in-channel signaling. *Contrast:* out-of-band signaling. (LM/C) 610.7-1995, 802.3-1998
(2) Signaling applications in which the signaling information is transmitted in the same information flow as the data. (C/LM/COM) 802.9a-1995w, 8802-9-1996
(3) Signaling which utilizes frequences within the voice or intelligence band of a channel. (PE) 599-1985w
(4) Analog generated signaling that uses the same path as a message and in which the signaling frequencies are in the same band used for the message. (COM) 312-1977w

in-band tones Typically, a signal on the communication path in the range of 400–3300 Hz. (AMR/SCC31) 1390-1995, 1390.3-1999, 1390.2-1999

in-basket simulation A simulation in which a set of issues is presented to a participant in the form of documents on which action must be taken; for example, a simulation of an unfolding international crisis as a sequence of memos describing relevant events and outcomes of the participant's actions on previous memos. (C) 610.3-1989w

inbound (A) The direction of RF signal flow toward the headend. Referred to in the CATV industry as "upstream" or "reverse." **(B)** The direction of RF signal flow toward the headend location from the user outlet ports. (LM/C) 802.7-1989

inbound queue A queue carrying messages from one or more I/O Unit Functions to a Processor. (C/MM) 1212.1-1993

inbound telemetry Communication initiated by a telemetry interface unit (TIU) toward a utility or enhanced service provider (ESP). (AMR/SCC31) 1390-1995, 1390.2-1999, 1390.3-1999

incandescence (illuminating engineering) The self-emission of radiant energy in the visible spectrum, due to the thermal excitation of atoms or molecules. (EEC/IE) [126]

incandescent filament lamp (illuminating engineering) A lamp in which light is produced by a filament heated to incandescence by an electric current. (EEC/IE) [126]

incandescent-filament-lamp transformer (series type) A transformer that receives power from a current-regulated series circuit and that transforms the power to another circuit at the same or different current from that in the primary circuit. *Note:* If of the insulating type, it also provides protection to the secondary circuit, casing, lamp, and associated luminaire from the high voltage of the primary circuit. (EEC/LB) [97]

in-channel signaling *See:* in-band signaling.

inching (rotating machinery) Electrically actuated angular movement or slow rotation of a machine, usually for maintenance or inspection. *See also:* asynchronous machine; direct-current commutating machine. (PE) [9]

inch-pound units Units based upon the yard and the pound commonly used in the United States of America and defined by the National Institute of Standards and Technology. *Note:* Units having the same names in other countries may differ in magnitude. (SCC14/QUL) SI 10-1997, 268-1982s

incidence angle *See:* angle of incidence.

incident *See:* software test incident.

incidental amplitude-modulation factor (signal generators) That modulation factor resulting unintentionally from the process of frequency modulation and.or phase modulation. *See also:* signal generator. (IM/HFIM) [40]

incidental and restricted radiation Radiation in the radio-frequency spectrum from all devices excluding licensed devices. *See also:* mobile communication system. (VT) [37]

incidental frequency modulation (signal generators) The ratio of the peak frequency deviation to the carrier frequency, resulting unintentionally from the process of amplitude modulation. *See also:* signal generator. (IM/HFIM) [40]

incidental phase modulation (signal generators) The peak phase deviation of the carrier, in radians, resulting unintentionally from the process of amplitude modulation. *See also:* signal generator. (IM/HFIM) [40]

incidental radiation of conducted power (frequency-modulated mobile communications receivers) Radio-frequency energy generated or amplified by the receiver, which is detectable outside the receiver. (VT) 184-1969w

incidental radiator A device that produces radio-frequency energy during the course of its operation, although the device is not intentionally designed to generate or emit radio-frequency energy. Examples of incidental radiators are dc motors, mechanical light switches, etc. (EMC) C63.4-1991

incidental time *See:* miscellaneous time.

incident field That component of the exciting field identical to the field that would have been present in the absence of all particles, surfaces, and volumes. *See also:* exciting field. (AP/PROP) 211-1997

incident wave (1) (forward wave) (uniform guiding systems) A wave traveling along a waveguide or transmission line in a specified direction toward a discontinuity, terminal plane, or reference plane. *See also:* waveguide; reflected wave. (IM/HFIM) [40]
(2) (surge arresters) A traveling wave before it reaches a transition point. (PE) [8], [84]
(3) (waveguide) At a transverse plane in a transmission line or waveguide, a wave traveling in a reference direction. *See also:* reflected wave; transmitted wave. (MTT) 146-1980w
(4) A wave that impinges on a discontinuity in refractive index or a medium of different propagation characteristics. The incident wave is the total field in the absence of the discontinuity. (AP/PROP) 211-1997

incipient An imperfection in the state or condition of equipment that could result in a degraded or immediate failure if corrective action is not taken. (PE/NP) 933-1999

incipient failure A failure that is about to occur. (C) 610.12-1990

incipient failure detection (nuclear power generating station) Tests designed to monitor performance characteristics and detect degradation prior to failure(s) which would prevent performance of the Class 1E functions. *Note:* Incipient failure testing requires module test checks, inspection, etc., on a sufficient time basis to establish performance trends. At the outset, the test cycle and corresponding limits of deviation of module performance or status must be established. Specific parameter trend patterns, exceeding of performance limits shall require that the module be removed and adjusted, replaced or serviced. As used here "sufficient time basis" would never be less than periodic surveillance test interval. In any event the internal must be justified technically based upon such things as manufacturer recommended, periodic preventive maintenance procedures, past operating experience, etc. These module tests require testing on line and/or removal from service. (PE/NP) 381-1977w

in-circuit testing A method of testing a printed circuit assembly by making direct physical contact between automatic test equipment and nets that are connected to the pins of individual components. *See also:* net. (C/TT) 1149.4-1999

inclined-blade blower (rotating machinery) A fan made with flat blades mounted so that the plane of the blades is parallel to and displaced from the axis of rotation of the rotor. *See also:* fan. (PE) [9]

inclined orbit (communication satellite) An orbit of a satellite which is not equatorial, and not polar. (COM) [19]

inclosed *See:* enclosed.

inclusion (1) (fiber optics) Denoting the presence of extraneous or foreign material. (Std100) 812-1984w
(2) (mathematics of computing) *See also:* implication. (C) 1084-1986w

inclusion list *See:* go list.

inclusive NOR gate *See:* NOR gate.

inclusive OR *See:* OR.

inclusive OR gate *See:* OR gate.

incoherent (fiber optics) Characterized by a degree of coherence significantly less than 0.88. *See also:* coherent; degree of coherence. (Std100) 812-1984w

incoherent field *See:* diffuse field.

incoherent sampling Sampling of a waveform such that the relationship between the input frequency, sampling frequency, number of cycles in the data record, and the number of samples in the data record does not meet the definition of coherent sampling. (IM/WM&A) 1057-1994w

incoherent scattering Scattering produced when the wave of an exciting field encounters random fluctuations of complex permittivity or permeability. The fluctuations may be either discrete or continuous (turbid or turbulent in the case of scattering from atmospheric refractive index fluctuations). The scattered fields exhibit random variations in phase and magnitude and thus constitute a zero mean process. (AP/PROP) 211-1997

incoming (local area networks) A link control signal indicating that a packet may soon be sent to the receiving entity. (C) 8802-12-1998

incoming calling line identification (ICLID) Provides calling party data on calls originated from both analog and digital lines. (SCC31) 1390.3-1999

incoming first failure to match *See:* incoming matching loss.

incoming matching loss Incoming matching loss is the matching loss on trunk-to-line connections. It includes the connection of a trunk to an individual line or to any idle line in a multiline hunting group. If only first-trial connections are considered, the loss is termed "incoming first failure to match." (COM/TA) 973-1990w

incoming traffic (telephone switching systems) Traffic received directly from trunks by a switching entity. (COM) 312-1977w

incomplete cluster *See:* partial cluster.

incomplete diffusion (illuminating engineering) That in which the diffusing medium partially redirects the incident flux by scattering while the remaining fraction of incident flux is redirected without scattering, that is, a fraction of the incident flux can remain in an image-forming state. (EEC/IE) [126]

incomplete line A sequence of text consisting of one or more non-⟨`newline`⟩ characters at the end of the file. (C/PA) 9945-2-1993

incomplete sequence relay (power system device function numbers) A relay that generally returns the equipment to the normal, of off, position and locks it out if the normal starting, operating, or stopping sequence is not properly completed within a predetermined time. If the device is used for alarm purposes only, it should preferably be designated as 48A (alarm). (SUB/PE) C37.2-1979s

in-core sort *See:* internal sort.

incorrect call progress signals Any signal that misleads or confuses a knowledgeable caller. (COM/TA) 973-1990w

Incorrect Packet Count (IPC) bit A bit in the Bus Error register of all S-modules. An S-module sets this bit to indicate that, with respect to a just completed message transfer, either the S-module has received a request for an ACKNOWLEDGE packet and was not given the opportunity to send it or, in the case of an S-module in which the packet counting option is implemented, that it received a different number of packets than was specified in the PACKET COUNT packet. (TT/C) 1149.5-1995

incorrect relaying-system performance Any operation or lack of operation of the relays or associated equipment that, under existing conditions, does not conform to correct relaying-systems performance. (SWG/PE/PE) C37.100-1992, C37.90-1978s

incorrect relay operation Any output response or lack of output response by the relay that, for the applied input quantities, is not correct. (SWG/PE/PSR) C37.100-1992, C37.90-1978s

increment (test pattern language) The action of increasing the arithmetic value of a counter by one. (TT/C) 660-1986w
(2) (A) (mathematics of computing) The quantity by which a variable is increased. **(B) (mathematics of computing)** To increase the value of a variable. *Contrast:* decrement. **(C) (mathematics of computing)** To increase the value of a variable by one. *Contrast:* decrement. (C) 610.10-1994, 1084-1986

incremental backup To perform a backup of a system in which the only data that is stored on the backup is data that has been modified since the last full backup was performed. *Contrast:* full backup. (C) 610.5-1990w

incremental binary representation *See:* binary incremental representation.

incremental compiler A compiler that completes as much of the translation of each source statement as possible during the input or scanning of the source statement. Typically used for on-line computer program development and checkout. *Synonyms:* conversational compiler; online compiler; interactive compiler. (C) 610.12-1990

incremental computer A special-purpose computer that is specifically designed to process changes in the variables as well as the absolute value of the variables themselves, for example, digital differential analyzer. (C/IA/APP) [20], [75], 610.10-1994w

incremental cost of delivered power (source) The additional per-unit cost incurred when supplying another increment of power. (PE/PSE) 94-1991w

incremental cost of reference power (source) The additional per-unit cost incurred when supplying another increment of power to a designated reference point on a transmission system. (PE/PSE) 94-1991w

incremental delivered power (electric power system) The percent of an increment of power delivered from a source to any specific point, such as the system load. (PE/PSE) 94-1991w

incremental development A software development technique in which requirements definition, design, implementation, and testing occur in an overlapping, iterative (rather than sequential) manner, resulting in incremental completion of the overall software product. *Contrast:* waterfall model. *See also:* structured design; stepwise refinement; rapid prototyping; transform analysis; spiral model; object-oriented design; input-process-output; modular decomposition; transaction analysis; data structure-centered design. (C) 610.12-1990

incremental dimension (numerically controlled machines) A dimension expressed with respect to the preceding point in a sequence of points. *See also:* long dimension; normal dimension; dimension; short dimension. (IA/EEC) [61], [74]

incremental energy cost (1) The cost incurred by increasing the production of electric energy above some base level. (PE/PSE) 858-1993w
(2) (electric power supply) The additional cost of producing or transmitting electric energy above some base cost. (PE/PSE) 346-1973w

incremental feed (numerically controlled machines) A manual or automatic input of preset motion command for a machine axis. (IA) [61], [84]

incremental fuel cost of generation (any particular source) The cost, usually expressed in mill.kilowatt-hour, that would be expended for fuel in order to produce an additional increment of generation at any particular source. (PE/PSE) 94-1970w

incremental generating cost (A) (electric power system) (source at any particular value of generation) The ratio of the additional cost incurred in producing an increment of generation to the magnitude of that increment of generation. *Note:* All variable costs should be taken into account including maintenance. **(B)** The additional cost of producing an increment of generation divided by the increment of generation. (PE/PSE) 94-1970, 94-1991

incremental heat rate (A) (steam turbo-generator unit at any particular output) The ratio of a small change in heat input per unit time to the corresponding change in power output. *Note:* Usually, it is expressed in British thermal unit/kilowatt-hour. **(B)** For a steam turbine-generator unit, the rate of a change in heat input per unit of time to the corresponding change in power output. (PE/PSE) 94-1970, 94-1991

incremental hysteresis loss (magnetic material) The hysteresis loss in a magnetic material when it is subjected simultaneously to a biasing and an incremental magnetizing force. (Std100) 270-1966w

incremental inductance The incremental inductance of a smoothing reactor is the inductance of the smoothing reactor, in Henries, determined on the basis of a small current increase (or decrease) at a predefined dc current. The incremental inductance is, therefore, defined as a function of dc current from the minimum current up to the maximum peak short-circuit current. (PE/TR) 1277-2000

incremental induction At a point in a material that is subjected simultaneously to a polarizing magnetizing force and a symmetrical cyclically varying magnetizing force, one-half the algebraic difference of the maximum and minimum values of the magnetic induction at that point. (Std100) 270-1966w

incremental integrator In an analog computer, a digital integrator so modified that the output signal is maximum negative, zero, or maximum positive when the value of the input is negative, zero, or positive. (C) 610.10-1994w

incremental justification In text formatting, the use of extra intercharacter spacing to form even margins. *Contrast:* line filling. (C) 610.2-1987

incremental loading (electric power system) The assignment of loads to generators so that the additional cost of producing a small increment of additional generation is identical for all generators in the variable range. (PE/PSE) 94-1970w

incremental magnetizing force (magnetic material) At a point that is subjected simultaneously to a biasing magnetizing force and a symmetrical cyclic magnetizing force, one-half of the algebraic difference of the maximum and minimum values of the magnetizing force at the point. (Std100) 270-1966w

incremental maintenance cost (A) (electric power system) (any particular source) The additional cost for maintenance that will ultimately be incurred as a result of increasing generation by an additional increment. **(B)** The additional cost for maintenance per increment of power incurred by changing generation level. (PE/PSE) 94-1970, 94-1991

incremental permeability (1) (general) (magnetic induction) The ratio of the cyclic change in the magnetic induction to the corresponding cyclic change in magnetizing force when the mean induction differs from zero. *Note:* In anisotropic media, incremental permeability becomes a matrix. (Std100) 270-1966w
(2) (magnetic core testing) The permeability with stated alternating magnetic field conditions in the presence of a stated static magnetic field.

$$\mu_\Delta = \frac{1}{\mu_0}\frac{\Delta B}{\Delta H}$$

$\mu\Delta$ = relative incremental permeability, ΔH = total cyclic variation of the magnetic field strength, ΔB = corresponding total cyclic change in induction. (MAG) 393-1977s

incremental productivity The productivity computed periodically during development. (C/SE) 1045-1992

incremental representation Any number representation system in which the numerals express changes in the variables rather than the variables themselves. *See also:* binary incremental representation; ternary incremental representation. (C) 1084-1986w

incremental resistance (semiconductor) (forward or reverse of a semiconductor rectifier diode) The quotient of a small incremental voltage by a small incremental current at a stated point on the static characteristic curve. (IA) [12]

incremental sensitivity (instrument) (nuclear techniques) A measure of the smallest change in stimulus that produces a statistically significant change in response. Quantitatively it is usually expressed as the change in the stimulus that produces a change in response equal to the standard deviation of the response. *See also:* ionizing radiation. (NPS) 175-1960w

incremental sweep (oscilloscopes) A sweep that is not a continuous function, but that represents the independent variable in discrete steps. *See also:* oscillograph; stairstep sweep. (IM/HFIM) [40]

incremental tape drive A tape drive capable of handling one character at a time, creating interrecord gaps only when explicitly directed. (C) 610.10-1994w

incremental ternary representation *See:* ternary incremental representation.

incremental time constant (electric coupling) The time constant applicable for a small incremental change of excitation voltage about a specified operating value. (EM/PE) 290-1980w

incremental total dose test A test of the permanent changes induced by radiation obtained by a comparison of characteristics before and after exposure to successively higher increments of integrated total dose levels. (ED) 641-1987w

incremental total test dosage Test of permanent changes induced by radiation that are obtained by a comparison of characteristics before and after exposure to a given integrated total dose level. (ED) 1005-1998

incremental transmission loss (electric power system) The change in power loss incurred when power flow within the transmission network is changed and/or redistributed. (PE/PSE) 94-1991w

incremental vector A representation of test vectors containing only the changing signals and new signal values in each vector. Parallel vectors can be generated from incremental vectors by maintaining test-specified state information for signals that did not change. (C/TT) 1450-1999

incremental water rate For a hydro turbine-generator unit, the ratio of a change in water input at a constant per unit of time (at a constant net head) to the corresponding change in power output. (PE/PSE) 94-1991w

incremental worth of power (designated point on a transmission system) The additional per-unit cost that would be incurred in supplying another increment of power from any variable source of a system in economic balance to such designated point. *Note:* When the designated point is the composite system load, the incremental worth of power is commonly called lambda or Lagrangian multiplier. (PE/PSE) 94-1970w

increment size The minimum distance between two points or parallel lines of a display surface. *See also:* plotter step size. (C) 610.6-1991w, 610.10-1994w

increment (network) starter A starter that applies starting current to a motor in a series of increments of predetermined value and at predetermined time intervals in closed-circuit transition for the purpose of minimizing line disturbance. One or more increments may be applied before the motor starts. *See also:* starter. (IA/ICTL/IAC/APP) [60], [75]

indefinite admittance matrix (network analysis) A matrix associated with an n-node network whose elements have the dimension of admittance and, when multiplied into the vector of node voltages, gives the vector of currents entering the n nodes. (CAS) [13]

independence (Class 1E equipment and circuits) (Class 1E power systems for nuclear power generating stations) The state in which there is no mechanism by which any single design basis event, such as a flood, can cause redundant equipment to be inoperable. (PE/NP) 384-1992r, 308-1991

independent auxiliary (generating stations electric power system) An item capable of performing its function without dependence on a similar item or the component it serves. (PE/EDPG) [5], 505-1977r

independent ballast (mercury lamp) A ballast that can be mounted separately outside a lighting fitting or fixture. (EEC/LB) [95]

independent basic service set (IBSS) A BSS that forms a self-contained network, and in which no access to a distribution system (DS) is available. (C/LM) 8802-11-1999

independent biaxial test (seismic testing of relays) The horizontal and the vertical acceleration components are derived from two different input signals, which are phase incoherent. (SWG/PE/PSR) C37.98-1977s, C37.100-1992

independent conformity *See:* conformity.

independent contact A contacting member designed to close one circuit only. (EEC/PE) [119]

independent copy A copy of the object plus independent copies of all its subobjects (applied recursively). (C/PA) 1328-1993w, 1327-1993w, 1224-1993w

independent disk array A form of RAID storage, known as levels 4 and 5, in which the individual drives within the array may be accessed. *Note:* With level 4, all data drives use a common parity drive and with level 5, parity is performed across all drives. (C) 610.10-1994w

independent entity An entity for which each instance can be uniquely identified without determining its relationship to another entity. *Synonym:* identifier-independent entity. *Contrast:* dependent entity. (C/SE) 1320.2-1998

independent firing The method of initiating conduction of an ignitron by obtaining power for the firing pulse in the ignitor from a circuit independent of the anode circuit of the ignitron. *See also:* electronic controller. (IA/ICTL/IAC) [60]

independent ground electrode (surge arresters) A ground electrode or system such that its voltage to ground is not appreciably affected by currents flowing to ground in other electrodes or systems. (PE) [8], [84]

independent linearity *See:* linearity of a signal.

independently powered An adjective used to describe a node on a bus, when the node's power supply may fail while other nodes remain powered and operational. (C/MM) 1212-1991s

independent manual operation (of a switching device) A stored-energy operation where manual energy is stored and released, such that the speed and force of this operation are independent of the action of the attendant. (SWG/PE) C37.100-1992

independent operation The ability, when supplied with appropriate energy, and with control signals from internal sources or through one or more coupler interfaces, to perform all of the functions of which the installed equipment is intended to be capable. (VT) 1473-1999

independent pole tripping The application of multipole circuit breakers in such a manner that a malfunction of one or more poles or associated control circuits will not prevent successful tripping of the remaining pole(s). *Notes:* 1. Circuit breakers used for independent pole tripping must inherently be capable of individual pole opening. 2. Independent pole tripping is applied on ac power systems to enhance system stability by maximizing the probability of clearing at least some phases of a multiphase fault. (SWG/PE) C37.100-1992

independent power operation An operation by means of energy other than manual where the completion of the operation is independent of the continuity of the power supply. (SWG/PE) C37.100-1992

independent state class A state class that is not a dependent state class. *Contrast:* dependent state class. (C/SE) 1320.2-1998

independent telephone company A company not associated with a regional Bell operating company (non-Bell operating company). (AMR/SCC31) 1390-1995, 1390.2-1999, 1390.3-1999

independent transformer A transformer that can be mounted separately outside a luminaire. (EEC/LB) [98]

independent variable A variable whose value is not dependent on the values of other variables. *Contrast:* dependent variable. (C) 610.3-1989w

independent verification and validation (IV&V) **(1) (software)** Verification and validation performed by an organization that is technically, managerially, and financially independent of the development organization. (C) 610.12-1990

(2) Systematic evaluation of software products and activities by an organization that is not responsible for developing the product or performing the activity being evaluated. (C/SE) J-STD-016-1995

(3) V&V processes performed by an organization with a specified degree of technical, managerial, and financial independence from the development organization. (C/SE) 1012-1998

Independent Virtual Local Area Network (VLAN) Learning (IVL) Configuration and operation of the Learning Process and the Filtering Database such that, for a given set of VLANs, if a given individual MAC Address is learned in one VLAN, that learned information is not used in forwarding decisions taken for that address relative to any other VLAN in the given set. *Note:* In a Bridge that supports only IVL operation, the "given set of VLANs" is the set of all VLANs. (C/LM) 802.1Q-1998

Independent Virtual Local Area Network (VLAN) Learning (IVL) Bridge A type of Bridge that supports only Independent VLAN Learning. (C/LM) 802.1Q-1998

index (A) (electronic computation) An ordered reference list of the contents of a file or document, together with keys or reference notations for identification or location of those contents. **(B) (electronic computation)** A symbol or a number used to identify a particular quantity in an array of similar quantities. For example, the terms of an array represented by X1, X2, . . . , X100 have the indexes 1, 2, . . . , 100, respec-

tively. **(C) (electronic computation)** Pertaining to an index register. (C) [20], [85]
(2) (A) (data management) A data item that identifies a particular element in a set of items such as an array. **(B) (data management)** A list or table used to locate records within an indexed file that contains the location and unique key value of each record. *Synonym:* directory. *See also:* alternate index; cross-index. **(C) (data management)** To prepare a table as in definition (B). (C) 610.5-1990

indexed access The process of accessing stored data in such a way that indices are used to locate records within data storage. *Synonym:* keyed access. *See also:* indexed sequential access. (C) 610.5-1990w

indexed address An address that must be added to the contents of an index register to obtain the address of the storage location to be accessed. *Synonym:* variable address. *See also:* self-relative address; offset; relative address. (C) 610.12-1990, 610.10-1994w

indexed addressing An addressing mode in which an index register or index word is used to permit automatic modification of the referred address without altering the instruction. *Note:* Particularly useful when programming repetitive instruction sequences on many sets of data. (C) 610.10-1994w

index dip (fiber optics) A decrease in the refractive index at the center of the core, caused by certain fabrication techniques. Sometimes called profile dip. *See also:* refractive index profile. (Std100) 812-1984w

indexed file A file that may be accessed using an index. *Contrast:* partitioned data set; sequential file. (C) 610.5-1990w

indexed segment In a database, a segment that is located by an indexing segment. *Synonym:* index target segment. (C) 610.5-1990w

indexed sequential access The process of accessing stored data using the indexed sequential access mode. *Contrast:* direct access; sequential access. *See also:* indexed access. (C) 610.5-1990w

indexed sequential access method (ISAM) An access method by which data records may be stored and retrieved using either the sequential access method or the direct access method. *See also:* basic sequential access method; virtual sequential access method. (C) 610.5-1990w

indexed sequential access mode An access mode in which data records may be stored and retrieved using either direct access mode or sequential access mode. *Note*: The records are actually stored in a sequential fashion, but an index is maintained to allow direct access. *Contrast:* direct access mode; sequential access mode. (C) 610.5-1990w

INDEX file The file within an exported catalog containing the metadata describing the software objects and attributes for all bundles, products, subproducts and filesets. (C/PA) 1387.2-1995

index hole A hole found in hard-sectored media, such as magnetic disks, or paper tape, in which the hole indicates the start of the first sector, the first record, or the top of the form. *Contrast:* index mark. (C) 610.10-1994w

indexing segment In a database, a segment that contains a pointer to another segment, called the indexed segment, containing data. *Synonym:* index pointer segment. (C) 610.5-1990w

index mark A mark found on soft-sectored media, such as magnetic disks, in which a magnetic indicator is placed on the disk to indicate the beginning of each track within the sector. *Synonym:* address mark. (C) 610.10-1994w

index matching material (fiber optics) A material, often a liquid or cement, whose refractive index is nearly equal to the core index, used to reduce Fresnel reflections from a fiber end face. *See also:* mechanical splice; refractive index; Fresnel reflection. (Std100) 812-1984w

index of cooperation, international (facsimile in rectilinear scanning) The product of the total length of a scanning or recording line by the number of scanning or recording lines per unit length divided by pi. *Notes:* 1. For rotating devices the index of cooperation is the product of the drum diameter times the number of lines per unit length. 2. The prior IEEE index of cooperation was defined for rectilinear scanning or recording as the product of the total line length by the number of lines per unit length. This has been changed to agree with international standards. (COM) 168-1956w

index of illuminant metamerism (illuminating engineering) (of two objects that are metameric when illuminated by a reference source) Measure of the degree of color difference between the two objects when a specified test source is substituted for the reference source. (EEC/IE) [126]

index of observer metamerism (illuminating engineering) (of two objects that are metameric when viewed by a reference observer) Measure of the degree of color difference between the two objects when a specfied test observer is substituted for the reference observer. (EEC/IE) [126]

index of refraction *See:* refractive index.

index of sensation (illuminating engineering) (of a source) A number that expresses the effects of source luminance (L_s), solid angle factor (Q), position index (P), and the field luminance (F) on discomfort glare rating.

$$M = \frac{L_{sQ}}{PF^{0.44}}$$

(See solid angle factor for an equation defining Q). *Note:* A restatement of this formula lends itself more directly to computer applications. *See also:* discomfort glare rating. (EEC/IE) [126]

index pointer segment *See:* indexing segment.

index profile (fiber optics) In an optical waveguide, the refractive index as a function of radius. *See also:* step index profile; profile parameter; power-law index profile; profile dispersion parameter; parabolic profile; graded index profile; profile dispersion. (Std100) 812-1984w

index register (1) (computers) A register whose content is added to or subtracted from the operand address prior to or during the execution of an instruction. (MIL/C) [2], [85]
(2) A register whose contents can be used to modify an operand address during the execution of computer instructions; it can also be used as a counter. *Note:* may be used to control the execution of a loop, to control the use of an array, for table lookup or as a pointer. *Synonyms:* cycle counter; B-box; B-line. (C) 610.10-1994w

index target segment *See:* indexed segment.

index word In indexed addressing, a word containing an index modifier that is applied to the address field of a computer instruction. (C) 610.10-1994w

indicated bearing (direction finding systems) A bearing from a direction-finder site to a target transmitter obtained by averaging several readings: the indicated bearing is compared to the apparent bearing to determine accuracy of the equipment. *See also:* navigation. (AES/RS) 686-1982s, [42]

indicated bearing offset (navigation aid terms) (direction finder [DF] installations) The mean different between the indicated and apparent bearings of a number of signal sources, the sources being, for the most part, uniformly distributed in azimuth. (AES/GCS) 172-1983w

indicated value (A) (power meters) The uncorrected value determined by observing the indicating display of the instrument. **(B)** A scale reading or displayed value. (IM/NI) 470-1972, 544-1975, N42.17B-1989

indicating circuit That portion of the control circuit of a control apparatus or system that carries the results of logic functions to visual or audible devices that indicate the state of the apparatus controlled. (IA/MT) 45-1998

indicating control switch A switch that indicates its last control operation. (SWG/PE) C37.100-1992

indicating demand meter (metering) A demand meter equipped with a readout that indicates demand, maximum demand, or both. (ELM) C12.1-1982s

indicating fuse A fuse that automatically indicates that the fuse has interrupted the circuit. (SWG/PE) C37.40-1993, C37.100-1992

indicating instrument (glass industry) (electrical heating applications to melting furnaces and forehearths in the glass industry) An instrument in which only the present value of the quantity measured is visually indicated. (IA) 668-1987w

indicating or recording mechanism (demand meter) That mechanism that indicates or records the measurement of the electrical quantity as related to the demand interval. *Note:* This mechanism may be operated directly by and be a component part of the electric mechanism, or may be structurally separate from it. The demand may be indicated or recorded in kilowatts, kilovolt-amperes, amperes, kilovars, or other suitable units. This mechanism may be of an indicating type, indicating by means of a pointer related to its position on a scale or by means of the cumulative reading of a number of dial or cyclometer indicators: or a graphic type, recording on a circular or strip chart: or of a printing type, recording on a tape. It may record the demand for each demand interval or may indicate only the maximum demand. *See also:* demand meter. (EEC/PE) [119]

indicating scale (recording instrument) A scale attached to the recording instrument for the purpose of affording an easily readable value of the recorded quantity at the time of observation. *Note:* For recording instruments in which the production of the graphic record is the primary function, the chart scale should be considered the primary basis for accuracy ratings. For instruments in which the graphic record is secondary to a control function the indicating scale may be more accurate and more closely related to the control than is the chart scale. *See also:* moving element. (EEC/PE) [119]

indication (1) (supervisory control, data acquisition, and automatic control) (station control and data acquisition) An audio or visual signal that signifies a particular condition. (PE/SUB) C37.1-1994

(2) A light or other signal (audio or visual) provided by the man/machine interface that signifies a particular condition. (SWG/PE) C37.100-1992

(3) A mechanism informing an entity of the occurrence of an event in a lower layer entity. Alternatively, an indication may provide evidence of a request by a remote station entity. (EMB/MIB) 1073.4.1-2000

indication (status) function The capability of a supervisory system to accept, record, or display, or do all of these, the status of a device. The status of a device may be derived from one or more inputs giving the following two or more states of indication: *Two-state indication.* Only one of the two possible positions of the supervised device is displayed at one time. Such display may be derived from a single set of contacts.; *Three-state indication.* One in which the transitional state or security indication as well as the terminal positions of the supervised device is displayed. Such a display is derived from at least two sets of initiating contacts; *Multistate indication.* Only one of the predefined states (transitional or discrete, or both) is indicated at a time. Such a display is derived from multiple inputs; *Indication with memory.* An indication function with the additional capability of storing single or multiple changes of status that occur between scans. (SUB/PE) C37.1-1994

indication point (railway practice) The point at which the train control or cab signal impulse is transmitted to the locomotive or vehicle apparatus from the roadway element. (EEC/PE) [119]

indication (status) point interfaces Master Station or RTU (or both) element(s) that accept(s) a digital input signal for the function of indication. The input/output elements of a SCADA system provide the physical interface to external devices. It is preferred that a point serve one of the functions described below. In some earlier applications, a Control and Indication (C and I) point has been used to specify a combination control and indication point for a specific device (e.g., circuit breaker). The functions are: *Two-state indication.* Only one of the two possible positions of the supervised device is displayed at one time. Such display may be derived from a single set of contacts. *Tree-state indication* One in which the transitional state or security indication as well as the terminal positions of the supervised device is displayed. Such a display is derived from at least two sets of initiating contacts; *Multistate indication.* Only one of the predefined states (transitional or discrete, or both) is indicated at a time. Such a display is derived from multiple inputs; *Indication with memory* An indication function with the additional capability of storing single or multiple changes of status that occur between scans. (SUB/PE) C37.1-1994

indication with memory *See:* supervisory control functions.

indicator (1) (faulted circuit indicators) That portion of the FCI (faulted circuit indicator) which indicates that fault current has been sensed. (T&D/PE) 495-1986w

(2) (software) A device or variable that can be set to a prescribed state based on the results of a process or the occurrence of a specified condition. For example, a flag or semaphore. (C) 610.12-1990

(3) *See also:* display.

indicator light A light that indicates whether or not a circuit is energized. *See also:* appliance outlet. (IA/APP) [90]

indicators (Class 1E power systems for nuclear power generating stations) Devices that display information to the operator. (PE/NP) 380-1975w, 308-1980s

indicator symbol (logic diagrams) A symbol that identifies the state or level of an input or output of a logic symbol with respect to the logic symbol definition. (GSD) 91-1973s

indicator travel The length of the path described by the indicating means or the tip of the pointer in moving from one end of the scale to the other. *Notes:* 1. The path may be an arc or a straight line. 2. In the case of knife-edge pointers and others extending beyond the scale division marks, the pointer shall be considered as ending at the outer end of the shortest scale division marks. *See also:* moving element. (EEC/EMI) [112]

indicator tube An electron-beam tube in which useful information is conveyed by the variation in cross section of the beam at a luminescent target. (ED) 161-1971w, [45]

indices Plural form of index. (C) 610.5-1990w

indicial admittance The instantaneous response to unit step driving force. *Note:* This is a time function that is not an admittance of the type defined under admittance. *See also:* network analysis. (Std100) 270-1966w

indicial response (process control) The output of a system or element, expressed as a function of time, when forced from initial equilibrium by a unit-step input. *Note:* In the time domain, it is the graphic statement of the characteristic of a system or element analogous to the frequency-response characteristic of the transfer function. (PE/EDPG) [3]

indigenous error A computer program error that has not been purposely inserted as part of an error-seeding process. (C) 610.12-1990

indigenous fault (software) A fault existing in a computer program that has not been inserted as part of a fault seeding process. *See also:* fault seeding; fault; computer program. (C/SE) 729-1983s

indirect-acting machine voltage regulator A machine voltage regulator having a voltage-sensitive element that acts indirectly, through the medium of an interposing device such as contractors or a motor, to control the excitation of an electric machine. *Note:* A regulator is called a generator voltage regulator when it acts in the field circuit of a generator and is called an exciter voltage regulator when it acts in the field circuit of the main exciter. (SWG/PE) C37.100-1992

indirect-acting recording instrument A recording instrument in which the level of measurement energy of the primary detector is raised through intermediate means to actuate the marking device. *Note:* The intermediate means are commonly either mechanical, electric, electronic, or photoelectric. *See also:* instrument. (EEC/PE) [119]

indirect address (1) An address that specifies a storage location containing either a direct address or another indirect address. (C) [20], 610.10-1994w, [85]
(2) (software) An address that identifies the storage location of another address. The designated storage location may contain the address of the desired operand or another indirect address; the chain of addresses eventually leads to the operand. *Synonym:* multilevel address. *Contrast:* direct address; immediate data. *See also:* indirect instruction; *n*–level address. (C) 610.12-1990

indirect addressing An addressing mode in which the address field of an instruction contains an indirect address. *Contrast:* direct addressing. *See also:* n-level address. (C) 610.10-1994w

indirect-arc furnace An arc furnace in which the arc is formed between two or more electrodes. (EEC/PE) [119]

indirect commutation (auxiliary commutation) (circuit properties) (self-commutated converters) A commutation between a principal switching branch and an auxiliary switching branch succeeded by a commutation to the next principal switching branch. Indirect commutation is employed in some types of converters using circuit-commutated thyristors, where the auxiliary branch includes a commutating capacitor(s) to turn off the outgoing principal switch when the auxiliary switch is turned on. *Note:* In some converter circuits, several auxiliary branches may be involved consecutively. (IA/SPC) 936-1987w

indirect component (illuminating engineering) That portion of the luminous flux from a luminaire which arrives at the work-plane after being reflected by room surfaces. (EEC/IE) [126]

indirect-drive machine (elevators) An electric driving machine, the motor of which is connected indirectly to the driving sheave, drum, or shaft by means of a belt or chain through intermediate gears. *See also:* driving machine. (EEC/PE) [119]

indirect ESD event An ESD event taking place between an intruder and a receptor in proximity to equipment that is the victim. (SPD/PE) C62.47-1992r

indirect ESD test A test in which ESD is applied to a coupling plane in the vicinity of the EUT. (EMC) C63.16-1993

indirect instruction A computer instruction that contains indirect addresses for its operands. *Contrast:* direct instruction; immediate instruction. *See also:* effective instruction; absolute instruction. (C) 610.12-1990, 610.10-1994w

indirect lighting (illuminating engineering) Lighting involving luminaires which distribute 90% to 100% of the emitted light upward. (EEC/IE) [126]

indirectly controlled variable (control) (automatic control) A variable that is not directly measured for control but that is related to, and influenced by, the directly controlled variable. *See also:* feedback control system. (IA/ICTL/IAC) [60]

indirectly heated cathode (unipotential cathode) (equipotential cathode) A hot cathode to which heat is supplied by an independent heater. *See also:* electrode. (ED) 161-1971w

indirect manual operation (of a switching device) Operation by hand through an operating handle mounted at a distance from, and connected to the switching device by, mechanical linkage. (SWG/PE) C37.100-1992

indirect operation (of a switching device) Operating by means of an operating mechanism connected to the main operating shaft or an extension of it, through offset linkages and bearings. (SWG/PE) C37.100-1992

indirect release (trip) (of a switching device) A release energized by the current in the main circuit through a current transformer, shunt, or other transducing device. (SWG/PE) C37.100-1992

indirect stroke (surge arresters) A lightning stroke that does not directly strike any part of a network but induces an overvoltage in it. (PE/T&D) 1410-1997, [84], [8]

indirect-stroke protection (lightning) Lightning protection designed to protect a network or electric installation against indirect strokes. *See also:* direct-stroke protection. (T&D/PE) [10]

individual address (1) An address that identifies a single source or destination service access point. (LM/C) 8802-6-1994
(2) (local area networks) The unique address identifying an individual end node. (C) 8802-12-1998

individual branch circuit A branch circuit that supplies only one utilization equipment. (NESC/NEC) [86]

individual capacitor fuse A fuse applied to disconnect an individual faulted capacitor from its bank. *Synonyms:* individual capacitor fuse; capacitor fusepower systems relaying. (SWG/PE/T&D) C37.40b-1996, 1036-1992, C37.99-2000

individual-equipment test requirements The set of explicit requirements specifying the test conditions, instrumentation, equipment under test (EUT) operation, etc, to be used in testing a specific EUT for conducted and radiated radio noise. Such requirements should take precedence over the requirements of this standard. (EMC) C63.4-1988s

individual-lamp autotransformer (power and distribution transformers) A series autotransformer that transforms the primary current to a higher or lower current as required for the operation of an individual street light. (PE/TR) C57.12.80-1978r

individual-lamp insulating transformer (power and distribution transformers) An insulating transformer used to protect the secondary circuit, casing, lamp, and associated luminaire of an individual street light from the high-voltage hazard of the primary circuit. *See also:* specialty transformer. (PE/TR) C57.12.80-1978r, [57]

individual line (1) (telephone switching systems) A line arranged to serve one main station. (COM) 312-1977w
(2) (data transmission) A subscriber line arranged to serve only one main station although additional stations may be connected to the line as extensions. An individual line is not arranged for discriminatory ringing with respect to the stations on that line. (PE) 599-1985w

individual line downtime (switching system) The time during which the customer is out of service as a result of system failures. This does not include the time out of service due to congestion, unless the congestion is due to a switching system failure. *See also:* out of service. (COM/TA) 973-1990w

individual pole operation (of a multiple circuit breaker or switching device) A descriptive term indicating that any pole(s) of the device can be caused to change state (open or close) without changing the state of the remaining pole(s). Devices may have capability for individual pole opening, individual pole closing, or both. (SWG/PE) C37.100-1992

individual trunk (telephone switching systems) A trunk, link, or junctor that serves only one input group of a grading. (COM) 312-1977w

Individual Virtual Port A Subgroup Port that represents the capability for communication with one other Remote Bridge in the Remote Bridge Group to which the Port attaches. *Note:* An Individual Virtual Port always has another Individual Virtual Port as its peer Port; the two Ports connect their respective Remote Bridges in a two-member Subgroup. (C/LM) 802.1G-1996

indivisible access A data access for which the entire datum is read or written as a whole, with no possibility of partial interleaved access by another processor. (C/MM) 1596.5-1993

indivisible-access cycle A DTB cycle that is used to access slave locations indivisibly and without permitting any other master to access these locations until the operation is complete. (C/MM) 1096-1988w

indoor (1) Not suitable for exposure to the weather. *Note:* For example, an indoor capacitor unit is designed for indoor service or for use in a weatherproof housing. *See also:* outdoor. (T&D/PE/TR) 18-1992, C57.12.80-1978r

(2) Designed for use inside buildings or weatherproof (weather-resistant) enclosures. *Note:* Because of the wide variety of enclosures available, when a fuse that is designed for indoor application is installed inside an outdoor enclosure, such installations should be verified with the fuse manufacturer. (SWG/PE) C37.40-1993

(3) Designed for use only inside buildings, or weather-resistant enclosures. (SWG/PE) C37.100-1992

indoor arrester An arrester that, because of its construction, must be protected from the weather. (SPD/PE) C62.11-1999

indoor bushing A bushing in which both ends are in ambient air but are not exposed to external atmospheric conditions. *Note:* An outdoor bushing may be used indoors but an indoor bushing may not be used outdoors. (PE/TR) C57.19.03-1996

indoor current transformer One that, because of its construction, must be protected from the weather. (PE/TR) C57.13-1993

indoor enclosure (power system communication equipment) An enclosure for use where another housing provides protection against exposure to the weather. (PE/PSC) 281-1984w

indoor-immersed bushing A bushing in which one end is in ambient air but not exposed to external atmospheric conditions and the other end is immersed in an insulating medium such as oil or gas. (PE/TR) C57.19.03-1996

indoor reactor A reactor that, because of its construction, must be protected from the weather. (PE/TR) C57.16-1996

indoor regulator A regulator that, because of its construction, must be protected from the weather. (PE/TR) C57.15-1999

indoor shunt reactor (shunt reactors over 500 kVA) One which, because of its construction, must be protected from the weather. (PE/TR) C57.21-1981s

indoor surge-protective device A surge-protective device that, because of its construction, shall be protected from the weather. (SPD/PE) C62.62-2000

indoor termination A termination intended for use where it is protected from direct exposure to both solar radiation and precipitation. Terminations designed for use in sealed enclosures where the external dielectric strength is dependent upon liquid or special gaseous dielectrics are also included in this category. (PE/IC) 48-1990s

indoor termination—dry A termination intended for use where it is protected from solar radiation and precipitation and not subject to periodic condensation, or other excessive humidity (90% RH or more). May be installed in air conditioned or heated areas. (PE/IC) 48-1996

indoor termination—wet A termination intended for use where it is protected from direct exposure to both solar radiation and precipitation, but is subjected to climatic conditions that can cause condensation onto the termination surfaces. (PE/IC) 48-1996

indoor transformer (power and distribution transformers) A transformer which, because of its construction, must be protected from the weather. (PE/TR) C57.12.80-1978r

indoor wall bushing A wall bushing, of which both ends are suitable for operating only where protection from the weather is provided. *See also:* bushing. 49-1948w

induced charge The charge that flows when the condition of the device is changed from that of zero applied voltage (after having previously been saturated with either a positive or negative voltage) to at least that voltage necessary to saturate in the same sense. *Note:* The induced charge is dependent on the magnitude of the applied voltage, which should be specified in describing this characteristic of ferroelectric devices. *See also:* ferroelectric domain. (UFFC) 180w

induced control voltage (Hall effect devices) The electromotive force induced in the loop formed by the control current leads and the current path through the Hall plate by a varying magnetic flux density, when there is no control current. (MAG) 296-1969w

induced current (1) (general) Current in a conductor due to the application of a time-varying electromagnetic field. *See also:* induction heating. (IA) 54-1955w

(2) (interference terminology) The interference current flowing in a signal path as a result of coupling of the signal path with an interference field, that is, a field produced by an interference source. *See also:* interference. (IE) [43]

(3) (lightning strokes) The current induced in a network or electric installation by an indirect stroke. *See also:* direct-stroke protection. (T&D/PE) [10]

induced electrification The separation of charges of opposite sign onto parts of a conductor as a result of the proximity of charges on other objects. *Note:* The charge on a portion of such a conductor is often called an induced charge or a bound charge. (Std100) 270-1966w

induced emission *See:* stimulated emission.

induced field current (synchronous machines) The current that will circulate in the field winding (assuming the circuit is closed) due to transformer action when an alternating voltage is applied to the armature winding, for example, during starting of a synchronous motor. (PE) [9]

induced-potential tests (electric power) Dielectric tests in which the test voltages are suitable-frequency alternating voltages, applied or induced between the terminals. (PE/SPD) 32-1972r

induced voltage (1) (general) A voltage produced around a closed path or circuit by a change in magnetic flux linking that path. *Notes:* 1. Sometimes more narrowly interpreted as a voltage produced around a closed path or circuit by a time rate of change in magnetic flux linking that path when there is no relative motion between the path or circuit and the magnetic flux. 2. A single-phase stator winding energized with alternating current, produces a pulsating magnetic field which causes a voltage to be induced in a blocked rotor circuit, and the same magnetic field may be interpreted in terms of two magnetic fields of constant amplitude traveling in opposite directions around the air gap, causing two voltages to be generated in a blocked rotor circuit. 3. Whether a voltage is defined as being induced or generated is often simply a matter of point of view. *See also:* induction motor; generated voltage; Faraday's law. (PE) [9]

(2) (lightning strokes) The voltage induced on a network or electric installation by an indirect stroke. *See also:* direct-stroke protection. (T&D/PE) [10], 1410-1997

(3) (corona measurement) Voltage that is induced in a winding. Induced voltage also includes voltage applied across a winding. (MAG/ET) 436-1977s

induced voltage tests (power and distribution transformers) Induced voltage tests are dielectric tests on transformer windings in which the appropriate test voltages are developed in the windings by magnetic induction. *Note:* Power for induced voltage tests is usually supplied at higher-than-rated frequency to avoid core saturation and excessive excitation current. (PE/TR) C57.12.80-1978r

inducing current The current that flows in a single conductor of an electric supply line with ground return to give the same value of induced voltage in a telecommunication line (at a particular separation) as the vectorial sum of all voltages induced by the various currents in the inductive exposure as a result of ground fault. (PE/PSC) 367-1996

inductance (1) The property of an electric circuit by virtue of which a varying current induces an electromotive force in that circuit or in a neighboring circuit. (CHM) [51]

(2) A force that resists the sudden buildup of electric current. *Note:* Inductance can cause errors during transmission. (C) 610.7-1995

inductance coil *See:* inductor.

inductance coupling (interference terminology) The type of coupling in which the mechanism is mutual inductance between the interference is induced in the signal system by a magnetic field produced by the interference source. *See also:* interference. (IE) [43]

inductance, effective *See:* effective inductance.

inductance grounded (system grounding) Grounded through impedance, the principal element of which is inductance. *Note:* The conditions of an inductance-grounded system are that X_0/X_1 lie within the range of 3–10 and $R_0/X_0 \leq 1$. The ground-fault current becomes 25% or more of the three-phase fault current. Inductance grounding becomes "effective grounding" if X_0/X_1 is reduced to 3 or less. (IA/PSE) 142-1982s

induction (A) The process of generating time-varying voltages and/or currents in conductive objects or electric circuits by the influence of the time-varying electric, magnetic, or electromagnetic fields. **(B) (coupling)** The process of generating time-varying voltages and/or currents in otherwise unenergized conductive objects or electric circuits by the influence of the time-varying electric and/or magnetic fields. (T&D/PE) 539-1990, 1048-1990

induction coil (A) A transformer used in a telephone set for interconnecting the transmitter, receiver, and line terminals. **(B)** A transformer for converting interrupted direct current into high-voltage alternating current. *See also:* telephone station. (PE/EM) 43-1974

induction compass A device that indicates an aircraft's heading, in azimuth. Its indications depend on the current generated in a coil revolving in the earth's magnetic field. (EEC/PE) [119]

induction-conduction heater A heating device in which electric current is conducted through but is restricted by induction to a preferred path in a charge. *See also:* induction heating. (IA) 54-1955w

induction coupling (electric coupling) An electric coupling in which torque is transmitted by the interaction of the magnetic field produced by magnetic poles on one rotating member and induced currents in the other rotating member. *Note:* The magnetic poles may be produced by direct current excitation, permanent magnet excitation, or alternating current excitation. The induced currents may be carried in a wound armature or squirrel cage, or may appear as eddy currents. (PE/EM) 290-1980w, [9]

induction cup (of a relay) A form of relay armature in the shape of a cylinder with a closed end that develops operating torque by its location within the fields of electromagnets that are excited by the input quantities. (SWG/PE/PSR) C37.100-1992, C37.90-1978s

induction cylinder (of a relay) A form of relay armature in the shape of an open-ended cylinder that develops operating torque by its location within the fields of electromagnets that are excited by the input quantities. (SWG/PE/PSR) C37.100-1992, C37.90-1978s

induction disk (1) (of a relay) A form of relay armature in the shape of a disk that usually serves the combined function of providing an operating torque by its location within the fields of an electromagnet excited by the input quantities and a restraining force by motion within the field of a permanent magnet. (SWG/PE/PSR) C37.100-1992, C37.90-1978s
(2) (utility consumer interconnections) A thin circular (or spiraled) disk of nonmagnetic conducting material in which eddy currents are produced to create torque about an axis of rotation. (PE/PSR) C37.95-1973s

induction, electrostatic *See:* electrostatic induction.

induction factor (magnetic core testing) Under stated conditions, the self inductance that a coil of specified shape and dimensions placed on the core in a given position should have, if it consisted of one turn.

$$A_L = \frac{L}{N^2}$$

A_L = Induction factor (henrys/turns^2), L = Self inductance of the coil on the core, in henrys, N = Number of turns of the coil. (MAG) 393-1977s

induction frequency converter A wound-rotor induction machine in which the frequency conversion is obtained by induction between a primary winding and a secondary winding rotating with respect to each other. *Notes:* 1. The secondary winding delivers power at a frequency proportional to the relative speed of the primary magnetic field and the secondary member. 2. In case the machine is separately driven, this relative speed is maintained by an external source of mechanical power. 3. In case the machine is self-driven, this relative speed is maintained by motor action within the machine obtained by means of additional primary and secondary windings with number of poles differing from the number of poles of the frequency-conversion windings. In special cases one secondary winding performs the function of two windings, being short-circuited with respect to the poles of the driving primary winding and open-circuited with respect to the poles of the primary-conversion winding. *See also:* converter. (PE) [9]

induction furnace A transformer of electric energy to heat by electromagnetic induction. (EEC/PE) [119]

induction generator (1) (A) (rotating machinery) An induction machine, when driven above synchronous speed by an external source of mechanical power, used to convert mechanical power to electric power. *See also:* asynchronous machine. **(B)** A generator that produces power with rotor speeds slightly higher than synchronous speed. It does not have the rotor field excitation requirement of synchronous generators. (PE/EDPG) [9], 1020-1988
(2) An induction machine driven above synchronous speed by an external source of mechanical power. (IA/MT) 45-1998

induction heater (interference terminology) A device for causing electric current to flow in a charge of material to be heated. Types of induction heaters can be classified on the basis of frequency of the induced current, for example, a low-frequency induction heater usually induces power-frequency current in the charge: a medium-frequency induction heater induces currents of frequencies between 180 and 540 hertz: a high-frequency induction heater induces currents having frequencies from 1000 hertz upward. (PE/PSR) [6]

induction heating (electrical heating systems) The generation of heat in any conducting material by means of magnetic flux-induced currents. (IA/PC) 844-1991

induction instrument An instrument that depends for its operation on the reaction between a magnetic flux (or fluxes) set up by one or more currents in fixed windings and electric currents set up by electromagnetic induction in movable conducting parts. *See also:* instrument. (EEC/PE) [119]

induction loop (of a relay) A form of relay armature consisting of a single turn or loop that develops operating torque by its location within the fields of electromagnets that are excited by the input quantities. (SWG/PE/PSR) C37.100-1992, C37.90-1978s

induction loudspeaker A loudspeaker in which the current that reacts with the steady magnetic field is induced in the moving member. (SP) [32]

induction machine An asynchronous ac machine that comprises a magnetic circuit interlinked with two electric circuits, or sets of circuits, rotating with respect to each other and in which power is transferred from one circuit to another by electromagnetic induction. Examples of induction machines are induction generators, induction motors, and certain types of frequency converters and phase converters. (IA/MT) 45-1998

induction motor A polyphase ac motor in which the secondary field current is created solely by induction. The motor operates at less than synchronous speed and less than unity power factor. The operating speed is dependent on the frequency of the power source. It is generally the motor of choice for auxiliary drives. (IA/MT) 45-1998

induction-motor meter A motor-type meter in which the rotor moves under the reaction between the currents induced in it and a magnetic field. *See also:* electricity meter. (ELM) C12.1-1982s

induction regulator (electrical heating applications to melting furnaces and forehearths in the glass industry) A regulating transformer, having a primary winding in shunt and a secondary winding in series with a circuit for gradually adjusting the voltage, phase relation, or both, of the circuit by changing the relative magnetic coupling of the existing (primary) and series (secondary) windings. (IA) 668-1987w

induction ring heater A form of core-type induction heater adapted principally for heating electrically conducting charges of ring or loop form, the core being open or separable to facilitate linking the charge. *See also:* induction heater. (IA) 54-1955w, 169-1955w

induction vibrator A device momentarily connected between the airplane direct-current supply and the primary winding of the magneto, thus converting the magneto to an induction coil. *Note:* It provides energy to the spark plugs of an aircraft engine during its starting period. (EEC/PE) [119]

induction voltage regulator (power and distribution transformers) A regulating transformer having a primary winding in shunt and a secondary winding in series with a circuit, for gradually adjusting the voltage or the phase relation, or both, of the circuit by changing the relative magnetic coupling of the exciting (primary) and series (secondary) windings. (PE/TR) C57.12.80-1978r

induction watthour meter A motor-type meter in which currents induced in the rotor interact with a magnetic field to produce the driving torque. (ELM) C12.1-1982s

induction zone (of EMI) The area where the distance to the source of electromagnetic interference is less than the wavelength of the interference. In the induction zone the circuit or system will be affected by transverse or longitudinal fields. *Contrast:* radiation zone. (PE/IC) 1143-1994r

inductive assertion method (software) A proof of correctness technique in which assertions are written describing program inputs, outputs, and intermediate conditions, a set of theorems is developed relating satisfaction of the input assertions to satisfaction of the output assertions, and the theorems are proved or disproved using proof by induction. (C) 610.12-1990

inductive coordination (electric supply and communication systems) The location, design, construction, operation, and maintenance in conformity with harmoniously adjusted methods that will prevent inductive interference. (EEC/PE) [119]

inductive coupling (ground system) (1) (communication circuits) The association of two or more circuits with one another by means of inductance mutual to the circuits or the mutual inductance that associates the circuits. *Note:* This term, when used without modifying words, is commonly used for coupling by means of mutual inductance, whereas coupling by means of self-inductance common to the circuits is called direct inductive coupling.
(2) (inductive coordination practice) The interrelation of neighboring electric supply and communication circuits by electric or magnetic induction, or both. (PE/PSIM) 81-1983

inductive exposure A situation of proximity between electric supply and communication circuits under such conditions that inductive interference must be considered. *See also:* inductive coordination. (EEC/PE) [119]

inductive gap (microwave receiver protectors) (nonlinear, active, and nonreciprocal waveguide components) In cell-type waveguide receiver protectors, this is the slot width or distance between iris plates. *See also:* resonant gap. (MTT) 457-1982w

inductive influence (electric supply circuit with its associated apparatus) Those characteristics that determine the character and the intensity of the inductive field that it produces. Inductive influence is a measure of the interfering effect of the power system. (COM/TA) 469-1988w

inductive interference (electric supply and communication systems) An effect, arising from the characteristics and inductive relations of electric supply and communication systems, of such character and magnitude as would prevent the communication circuits from rendering service satisfactorily and economically if methods of inductive coordination were not applied. *See also:* inductive coordination. (EEC/PE) [119]

inductively coupled circuit A coupled circuit in which the common element is mutual inductance. *See also:* network analysis. (Std100) 270-1966w

inductive microphone *See:* inductor microphone.

inductive neutralization (coil neutralization) (shunt neutralization) A method of neutralizing an amplifier whereby the feedback susceptance due to an interelement capacitance is canceled by the equal and opposite susceptance of an inductor. *See also:* amplifier; feedback. (AP/ANT) 145-1983s

inductive residual voltage (Hall effect devices) The electromotive force induced in the loop formed by the Hall voltage leads and the conductive path through the Hall plate by a varying magnetic flux density, when there is no control current. (MAG) 296-1969w

inductive susceptiveness (communication circuits) Those characteristics that determine, so far as such characteristics are able to determine, the extent to which the service rendered by the circuit can be adversely affected by a given inductive field. *See also:* inductive coordination. (PE/EEC) [119]

inductor (1) (general) A device consisting of one or more associated windings, with or without a magnetic core, for introducing inductance into an electric circuit.
(2) (railway practice) A roadway element consisting of a mass of iron with or without a winding, that acts inductively on the vehicle apparatus of the train control, train stop, or cab signal system. (PE/EM) 43-1974s

inductor alternator An inductor machine for use as a generator, the voltage being produced by a variation of flux linking the armature winding without relative displacement of field magnet or winding and armature winding. (PE) [9]

inductor, charging *See:* charging inductor.

inductor circuit (railway practice) A circuit including the inductor coil and the two lead wires leading therefrom taken through roadway signal apparatus as required. (EEC/PE) [119]

inductor dynamotor (rotating machinery) A dynamotor inverter having toothed field poles and an associated stationary secondary winding for conversion of direct current to high-frequency alternating current by inductor-generator action. (PE) [9]

inductor frequency-converter (rotating machinery) An inductor machine having a stationary input alternating-current winding, which supplies the excitation, and a stationary output winding of a different number of poles in which the output frequency is induced through change in field reluctance by means of a toothed rotor. *Note:* If the machine is separately driven, the rotor speed is maintained by an external source of mechanical power. If the machine is self-driven, the primary winding and rotor function as in a squirrel-cage induction motor or a reluctance motor. (PE) [9]

inductor machine (rotating machine) A synchronous machine in which one member, usually stationary, carries main and exciting windings effectively disposed relative to each other, and in which the other member, usually rotating, is without windings but carries a number of regular projections. (Permanent magnets may be used instead of the exciting winding). (PE) [9]

inductor microphone (inductive microphone) A moving-conductor microphone in which the moving element is in the form of a straight-line conductor. *See also:* microphone. (EEC/PE) [119]

inductor type synchronous generator A generator in which the field coils are fixed in magnetic position relative to the armature conductors. The electromotive forces are produced by the movement of masses of magnetic material. (IA/MT) 45-1998

inductor-type synchronous motor An inductor machine for use as a motor, the torques being produced by forces between armature magnetomotive force and salient rotor teeth. *Note:* Such motors usually have permanent-magnet field excitation, are built in fractional-horsepower ratings, frames and operate at low speeds, 300 revolutions per minute or less. (PE) [9]

industrial brush (rotating machinery) A brush having a cross-sectional area (width × thickness) of more than 1.4 square inch or a length of more than 1 1.2 inches, but larger than a fractional-horsepower brush. *See also:* brush. (PE) [9]

industrial control Broadly, the methods and means of governing the performance of an electric device, apparatus, equipment, or system used in industry. (IA/IAC) [60]

industrial electric locomotive An electric locomotive, used for industrial purposes, that does not necessarily conform to government safety regulations as applied to railroads. *Note:* This term is generally applied to locomotives operating in surface transportation and does not include mining locomotives. A prefix diesel-electric, gas-electric, etc., may replace the word electric. *See also:* electric locomotive. (EEC/PE) [119]

industrial process supervisory system A supervisory system that initiates signal transmission automatically upon the occurrence of an abnormal or hazardous condition in the elements supervised, which include heating, air-conditioning, and ventilating systems, and machinery associated with industrial processes. *See also:* protective signaling. (EEC/PE) [119]

industrial zone A zone that includes manufacturing plants where fabrication or original manufacturing is done, as defined by local ordinances. (PE/SUB) 1127-1998

ineffective attempts A switch-related ineffective attempt within a stored program control switching system (SPCS) is any valid bid for service that is not connected to the correct termination as defined by the received digits and busy/idle state of the equipment. A valid bid for service is an originating or incoming call attempt for which the switching system receives the expected number of digits. In any switching system, some ineffective attempts must naturally exist because of competition for shared resources. Other ineffective attempts may occur because equipment that normally performs a call setup function is malfunctioning (this includes impaired digital signals in a time-division system), operating at reduced capacity, or failing totally. Failures of hardware or software or errors in procedural or office data may cause such equipment problems. The ineffective attempt rate, as defined, applies for all causes, including congestion. (COM/TA) 973-1990w

ineffective machine attempts Any valid bid for service that does not complete due to a switch failure. The failure can be due to hardware, software, or errors in procedure or office data. A valid bid for service is defined as any originating or incoming call attempt for which the expected number of digits are delivered to the switching system. Misdialings or incomplete dialings caused by customers are not included in this definition. Calls that cannot be completed due to traffic congestion are also not included in this definition unless the congestion is caused by a system or subsystem fault or error. (COM/TA) 973-1990w

inequality relation A VHDL relational expression in which the relational operator is /=. (C/DA) 1076.3-1997

inequivalence *See:* exclusive OR.

inertance (automatic control) A property expressible by the quotient of a potential difference (temperature, sound pressure, liquid level) divided by the related rate of change of flow: the thermal or fluid equivalent of electrical inductance or mechanical moment of inertia. *See also:* feedback control system. (PE/EDPG) [3]

inert gas-pressure system (power and distribution transformers) A system in which the interior of the tank is sealed from the atmosphere, over the temperature range specified, by means of a positive pressure of inert gas maintained from a separate inert gas source and reducing valve system. (PE/TR) C57.12.80-1978r

inertia compensation The effect of a control function during acceleration or deceleration to cause a change in motor torque to compensate for the driven-load inertia. *See also:* feedback control system. (IA/ICTL/IAC/APP) [60], [75]

inertia constant (machine) The energy stored in the rotor when operating at rated speed expressed as kilowatt-seconds per kilovolt-ampere rating of the machine. *Note:* The inertia constant is

$$h = \frac{0.231 \times WK^2 \times n^2 \times 10^{-6}}{\text{kVA}}$$

where h = inertia constant in kilowatt-seconds per kilovolt-ampere, Wk^2 = moment of inertia in pound-feet^2, n = speed in revolutions per minute, kVA = rating of machine in kilovolt-amperes. *See also:* asynchronous machine. (PE) [9]

inertialess scanning *See:* electronic scanning.

inertial navigation equipment A type of dead-reckoning navigation equipment whose operation is based upon the measurement of accelerations: accelerations are sensed dynamically by devices stabilized with respect to inertial space, and the navigational quantities (such as vehicle velocity, angular orientation, or positional information) are determined by computers and/or other instrumentation. *See also:* navigation. (AES/RS) 686-1982s, [42]

inertial navigator A self-contained, dead-reckoning navigation aid using inertial sensors, a reference direction, and initial or subsequent fixes to determine direction, distance, and speed; single integration of acceleration provides speed information and a double integration provides distance information. *See also:* navigation. (AES/RS) 686-1982s, [42]

inertial sensor A position, attitude, or motion sensor whose references are completely internal, except possibly for initialization. (AES/GYAC) 528-1994

inertial space (navigation) A frame of reference defined with respect to the fixed stars. *See also:* navigation. (AES/RS) 686-1982s, [42]

inertia relay A relay with added weights or other modifications that increase its moment of inertia in order either to slow it or to cause it to continue in motion after the energizing force ends. *See also:* relay. (EEC/REE) [87]

infant mortality The set of failures that occur during the early-failure period of a system or component. (C) 610.12-1990

infant mortality failures (station control and data acquisition) A characteristic pattern of failure wherein the number of failures per unit of time decreases rapidly as the number of operating hours increase. (SUB/PE) C37.1-1994

infeed A source of fault current between a relay location and a fault location. (PE/PSR) C37.113-1999

inference engine A software engine within an expert system that draws conclusions from rules and situational facts. *See also:* parallel inference machine. (C) 610.10-1994w

infiltration Leakage of outside air into a building. (IA/PSE) 241-1990r

infinite multiplication factor (power operations) The ratio of the average number of neutrons produced in each generation of nuclear fissions to the average number of corresponding neutrons absorbed. Since neutron leakage out of the system is ignored, k is the effective multiplication factor for an infinitely large assembly. (PE/PSE) 858-1987s

infix notation (mathematics of computing) A method of forming mathematical expressions in which each operator is written between its operands and the expression is interpreted subject to rules of operator precedence and grouping symbols. For example, A added to B and the result multiplied by C is represented as (A + B) · C. *Contrast:* postfix notation. (C) 1084-1986w

inflection point (tunnel-diode characteristic) The point on the forward current-voltage characteristic at which the slope of the characteristic reaches its most negative value. *See also:* peak point. (ED) 253-1963w, [46]

inflection-point current (tunnel-diode characteristic) The current at the inflection point. *See also:* peak point. (ED) 253-1963w, [46]

inflection-point emission current (electron tube) That value of current on the diode characteristic for which the second derivative of the current with respect to the voltage is zero. *Note:* This current corresponds to the inflection point of the diode characteristic and is, under suitable conditions, an approximate measure of the maximum space-charge-limited emission current. (ED) 161-1971w

inflection-point voltage (tunnel-diode characteristic) The voltage at which the inflection point occurs. *See also:* peak point. (ED) 253-1963w, [46]

influence (1) (specified variable or condition) (upon an instrument) The change in the indication of the instrument caused solely by a departure of the specified variable or condition from its reference value, all other variables being held constant. (EEC/AII) [102]
(2) (upon a recording instrument) The change in the recorded value caused solely by a departure of the specified variable or condition from its reference value, all other variables being held constant. *Note:* If the influences in any direction from reference conditions are not equal, the greater value applies. (EEC/ERI) [111]

influence quantity A radiation field, electromechanical condition, or environmental condition that may provoke a response. (NI) N42.17B-1989r

INFO file For each product and fileset, the file within an exported catalog containing the metadata describing the software_file objects and attributes. (C/PA) 1387.2-1995

informal testing Testing conducted in accordance with test plans and procedures that have not been reviewed and approved by a customer, user, or designated level of management. *Contrast:* formal testing. (C) 610.12-1990

informatics *See:* information science.

information (1) (general) The meaning assigned to data by known conventions. (C) [20], [85]
(2) (nuclear power generating station) Data describing the status and performance of the plant. (PE/NP) 566-1977w
(3) (data management) The meaning that humans assign to data by means of known conventions that are applied to the data. *See also:* narrative information; information traffic; formatted information. (C) 610.2-1987, 610.5-1990w

information access service (IAS) A component of infrared link management protocol (IrLMP). (EMB/MIB) 1073.3.2-2000

information center (IC) (A) A user-oriented computer system that provides non-technical users direct access to data and software for information processing tasks such as report generation, data modeling and manipulation, and word processing. *Synonym:* information resource center. *See also:* decision support services. **(B)** Support personnel for a computer system as in definition (A). (C) 610.2-1987

information content (message or a symbol from a source) The negative of the logarithm of the probability that this particular message or symbol will be emitted by the source. *Notes:* 1. The choice of logarithmic base determines the unit of information content. 2. The probability of a given message or symbol's being emitted may depend on one or more preceding messages or symbols. 3. The quantity has been called self-information. *See also:* information theory; hartley; bit. (Std100) 171-1958w

information display channel (accident monitoring instrumentation) An arrangement of electrical and mechanical components or modules, or both, from measured process variable to display device as required to sense and display conditions within the generating stations. (PE/NP) 497-1981w

information display channel failure (accident monitoring instrumentation) A situation where the display disagrees, in a substantive manner, (that is, the maximum error within which the information must be conveyed to the operator has been exceeded), with the conditions or status of the plant. (PE/NP) 497-1981w

information efficiency The efficiency with which information is handled by an organization. *See also:* information traffic. (C) 610.2-1987

information field The sequence of octets occurring between the control field and the end of the LLC PDU. The information field contents of I, TEST, and UI PDUs are not interpreted at the LLC sublayer. (C/LM/CC) 8802-2-1998

information graphics The use of a computer to produce low quality, low cost graphical output for peer group presentations. *Synonym:* peer graphics. *Contrast:* presentation graphics. (C) 610.2-1987

information hiding (software) A software development technique in which each module's interfaces reveal as little as possible about the module's inner workings and other modules are prevented from using information about the module that is not in the module's interface specification. *See also:* encapsulation. (C) 610.12-1990

information interchange The process of sending and receiving data in such a manner that the information content or meaning associated with the data is not altered during the transmission. *See also:* data interchange. (C) 610.5-1990w

information, mutual *See:* mutual information.

information object A well-defined piece of information, definition, or specification that requires a name to identify its use in an instance of communication. (C/LM) 802.10g-1995

information overload A condition resulting from presentation of too much data to be assimilated and acted upon without further organization. *See also:* exception reporting. (C) 610.2-1987

information processing *See:* data processing.

Information Processing Language (IPL) A high-order language used for performing list processing. (C) 610.13-1993w

information processing system *See:* information system.

information resource center (IRC) *See:* information center.

information retrieval The techniques used to recover information from an organized body of knowledge. *See also:* information storage. (C) 610.5-1990w

information science A branch of technology concerned with the way in which data are processed and transmitted through digital equipment. (C) 610.2-1987

information separator (IS) Any control character used to delimit like units of data in a hierarchical arrangement of data. The name of the separator does not necessarily indicate the units of data that it separates. *Synonym:* separating character. (C) 610.5-1990w

information services interface (ISI) The boundary across which external, persistent storage service is provided. (C/PA) 14252-1996

information source *See:* message source.

information storage The theory and techniques for the organization, storage, and searching of an organized body of knowledge. *Note:* Generally refers to a large body of data. *See also:* information system; information retrieval. (C) 610.5-1990w

information storage and retrieval *See:* information retrieval; information storage.

information storage and retrieval system *See:* information system.

information system (1) A mechanism used for acquiring, filing, storing, and retrieving an organized body of knowledge. *Synonym:* information storage and retrieval system. *See also:* information storage and retrieval. (C) 610.5-1990w, 610.10-1994w
(2) A data processing system integrated with such other processes as office automation and data communication. *Synonym:* information processing system. *See also:* data processing system. (C) 610.10-1994w

information systems service A high-level description of the services used to support a BSR. IS Services are cross-referenced to the BSRs they support, the IT Services that deliver them, and the technology components that house them. (C/PA) 1003.23-1998

information technology center *See:* information center.

information technology equipment Unintentional radiator equipment designed for one or more of the following purposes:

- Receiving data from an external source (such as a data input line or via a keyboard)
- Performing some processing functions of the received data (such as computation, data transformation or recording, filing, sorting, storage, transfer of data)
- Providing a data output (either to other equipment or by the reproduction of data or images).

Note: This definition includes electrical/electronic units or systems that predominantly generate a multiplicity of periodic binary pulsed electrical/electronic waveforms and are designed to perform data processing functions such as word processing, electronic computation, data transformation, recording, filing, sorting, storage, retrieval and transfer, and reproduction of data as images. (EMC) C63.4-1991

information technology service The most atomic level of technology. A group of IT services will interoperate to deliver an IS service in support of a BSR. IT services are described in terms of protocols, APIs, and service components. *See also:* application program interface. (C/PA) 1003.23-1998

information technology service model A textual and graphical representation of an IT service where all the low-level service components and interfaces are identified. (C/PA) 1003.23-1998

information theory (A) In the narrowest sense, is used to describe a body of work, largely about communciation problems but not entirely about electrical communication, in which the information measures are central. **(B)** In a broader sense it is taken to include all statistical aspects of communication problems, including the theory of noise, statistical decision theory as applied to detection problems, and so forth. *Note:* This broader field is sometimes called "statistical communication theory." **(C)** In a still broader sense its use includes theories of measurement and observation that use other measures of information, or none at all, and indeed work on any problem in which information, in one of its colloquial senses, is important. (Std100) [123]

information traffic The flow of information through an organization. Typically included are origination; production, consolidation, and presentation; reproduction; recording and storage; and distribution. *Synonyms:* document traffic; paper traffic. *See also:* document cycle. (C) 610.2-1987

information transfer (data transmission) The final result of data transmission from a data source to a data sink. The information transfer rate may or may not be equal to the transmission modulation rate. (PE) 599-1985w

informative annex An annex in a standard that is for information only and that is not an official part of the standard itself. (C/MM) 1754-1994

infrared (IR) (fiber optics) The region of the electromagnetic spectrum between the long-wavelength extreme of the visible spectrum (about 0.7 μm) and the shortest microwaves (about 1 mm). (Std100) 812-1984w

infrared radiation (A) (illuminating engineering) For practical purposes any radiant energy within the wavelength range 770 to 106 nm (nanometers) is considered infrared energy. *See also:* regions of electromagnetic spectrum. **(B) (laser maser)** Electromagnetic radiation with wavelengths that lie within the range 0.7 μm to 1 mm. (EEC/IE/LEO) [126], 586-1980

infrastructure The infrastructure includes the distribution system medium (DSM), access point (AP), and portal entities. It is also the logical location of distribution and integration service functions of an extended service set (ESS). An infrastructure contains one or more APs and zero or more portals in addition to the distribution system (DS). (C/LM) 8802-11-1999

ingress The process whereby unwanted signals enter the cable system to occupy spectrum that would otherwise remain free of signal energy. (LM/C) 802.7-1989r

inherent availability (IA) A measure of availability for a system operating in an ideal support environment in which schedule maintenance, standby, and logistic time are ignored. (PE/NP) 933-1999

inherent delay angle (thyristor converter) The delay angle that occurs in some connections (for example, 12-pulse connections) in certain operating conditions even where no phase control is applied. (IA/IPC) 444-1973w

inherent error *See:* inherited error.

inherent reliability The potential reliability of an item present in its design. *See also:* reliability. (R) [29]

inherent transient recovery voltage (transient recovery voltage) The TRV (transient recovery voltage) produced by the circuit with no modifying effect of the switching device. *Note:* The magnitude of the TRV for a given circuit and voltage is affected by the degree of current asymmetry. Symmetrical current usually produces the highest TRV magnitudes and is used as the basis for TRV-rated values. An asymmetrical current normally reduces the TRV magnitude from the symmetrical current case. (SWG/PE) C37.04E-1985w, C37.100-1992, C37.100B-1986w, C37.4D-1985w

inheritance (1) Using the same method to implement an operation as the immediate superclass in the class hierarchy. Inheritance greatly simplifies the definition of new object classes and is one of the main reasons for organizing classes into a hierarchy. (C) 1295-1993w
(2) The way in which the attribute definitions of a common object class are used as a part of the definition of other object classes. The definition of the new object class includes the definition of the common class plus the additional definitions specific to the new object class. (C/PA) 1387.2-1995
(3) A semantic notion by which the responsibilities (properties and constraints) of a subclass are considered to include the responsibilities of a superclass, in addition to its own, specifically declared responsibilities. (C/SE) 1320.2-1998

inherited attribute (A) An attribute that is a characteristic of a class by virtue of being an attribute of a generic ancestor. **(B)** An attribute that is a characteristic of a category entity by virtue of being an attribute in its generic entity or a generic ancestor entity. (C/SE) 1320.2-1998

inherited error (1) (mathematics of computing) An error that is input to a given operation, either from a previous operation or from the initial condition of a variable. *Synonym:* inherent error. *Contrast:* propagated error. (C) 1084-1986w
(2) (software) An error carried forward from a previous step in a sequential process. (C) 610.12-1990

inhibit (supervisory control, data acquisition, and automatic control) (station control and data acquisition) To prevent a specific event from occurring (e.g., alarm inhibit). (SWG/PE/SUB) C37.100-1992, C37.1-1994

inhibited bystander An inhibited bystander is a potential master that has no current need to acquire the bus and is fairness inhibited. (C/MM) 896.1-1987s

inhibited oil (power and distribution transformers) Mineral transformer oil to which a synthetic oxidation inhibitor has been added. (PE/TR) C57.12.80-1978r

inhibitor (insulating oil) Any substance that when added to an electrical insulating fluid retards or prevents undesirable reactions. (PE/TR) 637-1985r

inhomogeneous dense medium A medium having discrete or continuous spatial variations in its permittivity or permeability, such that multiple scattering must be considered. *See also:* sparse medium. (AP/PROP) 211-1997

inhomogeneous line-broadening (laser maser) An increase of the width of an absorption or emission line, beyond the natural linewidth, produced by a disturbance (for example, strain, imperfections, etc.) which is not the same for all of the source emitters. (LEO) 586-1980w

inhomogeneous medium A medium whose properties are not spatially invariant. (AP/PROP) 211-1997

inhomogeneous plane wave A wave for which the planes of constant magnitude and planes of constant phase are not parallel. Sometimes called a heterogeneous plane wave, but this use is deprecated. (AP/PROP) 211-1997

in-house system *See:* in-plant system.

initial condition (1) (analog computer) The value of a variable at the start of computation. A more restricted definition refers solely to the initial value of an integrator. Also used as a synonym for the computer-control state "reset." *See also:* reset. (C) 165-1977w
(2) (modeling and simulation) The values assumed by the variables in a system, model, or simulation at the beginning of some specified duration of time. *Contrast:* boundary condition; final condition. (C) 610.3-1989w

initial contact pressure The force exerted by one contact against the mating contact when the actuating member is in the initial contact-touch position. *Note:* The initial contact pressure is usually measured and expressed in terms of the force that must be exerted on the yielding contact while the actuating member is held in the initial contact-touch position in order to separate the mating contact surface against the action of the spring or other contact pressure device. *See also:* electric controller. (IA/ICTL/IAC) [60]

initial current pulse The subnanosecond risetime, and greater than 1 ns to perhaps 3 ns duration pulses that can occur at the start of the current wave from an ESD. Also called initial pulse, initial spike, and fast discharge mode. Its leading edge is the initial slope. (SPD/PE) C62.47-1992r

initial element *See:* primary detector.

initial erection (gyros) The mode of operation of a vertical gyro in which the gyro is being erected or slaved initially. (AES/GYAC) 528-1994

Initial Graphics Exchange Specification (IGES) (1) A computer graphics standard that provides a method for exchanging geometry and associated data among computer graphics systems, intended for human interpretation. It was publicly developed, and sponsored by the National Institute of Standards and Technology (formerly National Bureau of Standards) then adopted as an American National Standards Institute (ANSI) standard. *See also:* Product Data Exchange Specification. (C) 610.6-1991w
(2) A specification for representing product data, defined by ANSI/ASME Y14.26M-1989. (ATLAS) 1226-1993s

initial ionizing event An ionizing radiation interaction event that initiates a tube count. (NI/NPS) 309-1999

initialization A process of setting initial values of variables, constants, state, and other artifacts to establish the startup conditions for an object. (IM/ST) 1451.1-1999

initialization packet Special packets that are only generated by the controller during the RamLink initialization process. There are three types of initialization packets: *sync, wait,* and *wake.* (C/MM) 1596.4-1996

initialization test A test or collection of tests that does not require cooperation of any other node(s) on the bus. The default and vendor-dependent initialization tests are invoked by writing to the TEST_START register. (C/MM) 1212-1991s

initialization vector (IV) A binary vector used at the beginning of a cryptographic operation to allow cryptographic chaining. (LM/C) 802.10-1992

initialize To set a variable, register, or other storage location to a starting value. *See also:* reset; clear. (C) 610.12-1990, 610.10-1994w

initializing state A node state that is reflected by the value of 1 in the STATE_CLEAR.*state* field. The initializing state is an optional transient state which is entered immediately after a power_reset or command_reset event. (C/MM) 1212-1991s

initial luminous exitance (illuminating engineering) The density of luminous flux leaving a surface within an enclosure before interreflections occur. *Note:* For light sources this is the luminous exitance as defined herein. For nonself-luminous surfaces it is the reflected luminous exitance of the flux received directly from sources within the enclosure or from daylight. (EEC/IE) [126]

initial MAC protocol data unit (IMPDU) A protocol data unit (PDU) formed in the DQDB Layer by the addition of protocol control information (including address information) to a MAC Service Data Unit received from the Logical Link Control (LLC) Sublayer. The IMPDU is segmented into 44-octet segmentation units for transfer in Derived MAC Protocol Data Units (DMPDUs). (LM/C) 8802-6-1994

initial memory space A portion of the initial node space, which provides a RAM-access window for a memory-controller unit architecture. Unit architectures can also be mapped to nonconflicting portions of the initial memory space. The initial memory space is only relevant to bus standards implementing 64-bit fixed addressing. (C/MM) 1212-1991s

initial node space (1) The address space that is initially mapped to a node. The initial node space contains the initial register space (2 Kbytes) and initial units space. On buses implementing 32-bit or 64-bit extended addressing, the initial units space is 4 Kbytes in size. On buses implementing 64-bit fixed addressing, the initial node space is 256 Tbytes in size and also includes the node's initial memory and private spaces. (C/MM) 1212-1991s
(2) The 256 terabytes of Serial Bus address space that is available to each node. Addresses within initial node space are 48 bits and are based at zero. The initial node space includes initial memory space, private space, initial register space, and initial units space. (C/MM) 1394a-2000

initial program load *See:* bootstrap.

initial program loader A bootstrap loader used to load that part of an operating system needed to load the remainder of the operating system. (C) 610.12-1990

initial register space (1) A 2-Kbyte portion of the initial node space that is adjacent to the initial units space. The registers defined by the CSR Architecture are located within the initial register space. The initial register space also provides addresses for defining bus-dependent registers. (C/MM) 1212-1991s
(2) The address space reserved for resources accessible immediately after a reset. This includes the registers defined by the CSR Architecture as well as those defined by this standard. (C/MM) 1394-1995
(3) A 2 kilobyte portion of initial node space with a base address of FFFF F000 0000_{16}. This address space is reserved for resources accessible immediately after a bus reset. Core registers defined by ISO/IEC 13213:1994 are located within initial register space, as are Serial Bus-dependent registers defined by this standard. (C/MM) 1394a-2000

initial relay actuation time The time of the first closing of a previously open contact or the first opening of a previously closed contact. (EEC/REE) [87]

initial slope The slope, in amperes per nanosecond (A/ns), that occurs at the start of the ESD current wave. *Synonym:* rising slope. (SPD/PE) C62.47-1992r

initial state The values assumed by the state variables of a system, component, or simulation at the beginning of some specified duration of time. *Contrast:* final state. (C) 610.3-1989w

initial symmetrical ground fault current (safety in ac substation grounding) The maximum root-mean-square (rms) value of symmetrical fault current after the instant of a ground fault initiation. As such, it represents the rms value of the symmetrical component in the first half-cycle of a current wave that develops after the instant of fault at time zero. Generally, $I_{f(0+)} = 3I_0''$ where $I_{f(0+)}$ = initial symmetrical ground fault current I_0'' = rms value of zero-sequenced symmetrical current that develops immediately after the instant of fault initiation; that is, reflecting the subtransient reactances of rotating machines contributing to the fault. *Note:* Elsewhere in the guide, this initial symmetrical fault current is shown in an abbreviated notation, as If, or is referred to only as 3i0. The underlying reason for this latter notation is that, for purposes of this guide, the initial symmetrical fault current is assumed to remain constant for the entire duration of the fault. (T&D/PE) 563-1978r

initial transient recovery voltage (ITRV) A component of the transient recovery voltage that appears in the very short time immediately after current interruption. The initial transient recovery voltage is a result of traveling waves on the substation bus adjacent to the circuit-switching device. (SWG/PE) C37.100-1992

initial units space **(1)** The portion of the initial node space that is adjacent to but above the initial register space. When its size is sufficient, unit architectures are expected to be located within this space. (C/MM) 1212-1991s
(2) A portion of initial node space with a base address of FFFF F000 0800_{16}. This places initial units space adjacent to and above initial register space. The CSR's and other facilities defined by unit architectures are expected to lie within this space. (C/MM) 1394a-2000

initial unloaded sag The sag of a conductor prior to the application of any external load. (NESC) C2-1997

initial unloaded tension The longitudinal tension in a conductor prior to the application of any external load. (NESC) C2-1997

initiating cause A cause that directly leads to the failure. (SWG/PE) C37.10-1995

initiating relay A programming relay whose function is to constrain the action of dependent relays until after it has operated. (SWG/PE) C37.100-1992

initiation queue A DMA queue that is used primarily to pass I/O transaction-initiation messages. (C/MM) 1212.1-1993

initiator **(1)** The function that starts an I/O transaction-initiation/transaction-completion exchange by sending an initiation message to the responder. (C/MM) 1212.1-1993
(2) The file service user that requests FTAM regime establishment. (C/PA) 1238.1-1994w

injected current The current that flows through the test circuit breaker from the voltage source of a current injection circuit when this circuit is applied to the test circuit breaker. (SWG/PE) C37.100-1992, C37.081-1981r, C37.083-1999

injected-current frequency (ac high-voltage circuit breakers) The frequency of the injected current. (SWG/PE) C37.100-1992, C37.083-1999, C37.081-1981r

injection fiber *See:* launching fiber.

injection laser diode (ILD) (fiber optics) A laser employing a forward-biased semiconductor junction as the active medium. *Synonyms:* diode laser; semiconductor laser. *See also:* chirping; superradiance; active laser medium; laser. (Std100) 812-1984w

injection time (ac high-voltage circuit breakers) The time with respect to the power frequency current zero when the voltage circuit is applied. (SWG/PE) C37.081-1981r, C37.083-1999, C37.100-1992

ink jet printer A nonimpact printer in which the characters are formed by projecting a jet of ink droplets onto paper. (C) 610.10-1994w

in-line (monitoring radioactivity in effluents) A system where the detector assembly is adjacent to or immersed in the total effluent stream. (NI) N42.18-1980r

in-line code A sequence of computer instructions that is physically contiguous with the instructions that logically precede and follow it. (C) 610.12-1990

in-line connection Connection of two heater cables together electrically in series or parallel on the same pipe. (IA) 515-1997

inline recovery Recovery performed by resuming a process at a point preceding the occurrence of a failure. *Contrast:* backward recovery; forward recovery. (C) 610.5-1990w

inner jacket A jacket that is extruded over the cable core covering to provide additional dielectric strength when it is needed between the conductors and the shield. An inner jacket may be used in cables that are used for direct burial and also where high ground potential rise is to be withstood. *See also:* cable jacket. (PE/PSC) 789-1988w

inner storage *See:* internal storage.

I_{nom} output The nominal load current for which the unit-under-test output voltage is required to be maintained within the specified operational limits; the value should be between the minimum and maximum values. (PEL) 1515-2000

inoperable time The part of down-time in which all environmental conditions are satisfied, during which a device would not yield correct results if it were operated. (C) 610.10-1994w

inorder traversal The process of traversing a binary tree in a recursive fashion as follows: the left subtree is traversed in order, then the root is visited, then the right subtree is traversed in order. *Synonym:* symmetric traversal. *Contrast:* preorder traversal; postorder traversal. *See also:* converse inorder traversal. (C) 610.5-1990w

in-phase spring rate (inertial sensors) (dynamically tuned gyro) The residual difference, in a dynamically tuned gyro, between the dynamically induced spring rate and the flexure spring rate. (PE/AES/PSE/GYAC) 762-1987w, 528-1994

in-phase video One of a pair of coherent, bipolar video signals derived from the RF or IF signal by a pair of synchronous detectors with a 90° phase difference between the coherent oscillator (coho) reference inputs used for each. *Note:* The in-phase component is often identified as *I* and the other of the pair as quadrature video or *Q*. *See also:* quadrature video. (AES) 686-1997

in-plant system A communications system whose parts, including remote terminals, may be all situated in one building or several buildings. *Synonym:* in-house system. (C) 610.7-1995

input **(1)** **(A)** **(data transmission)** The data to be processed. **(B)** **(data transmission)** The state or sequence of states occurring on a specified input channel. **(C)** **(data transmission)** The device or collective set of devices used for bringing data into another device. **(D)** **(data transmission)** A channel for impressing a state on a device or logic element. **(E)** **(data transmission)** The process of transferring data from an external storage to an internal storage. (PE) 599-1985
(2) **(A)** **(software)** Pertaining to data received from an external source. *Contrast:* output. **(B)** **(software)** Pertaining to a device, process, or channel involved in receiving data from an external source. *Contrast:* output. **(C)** **(software)** To receive data from an external source. *Contrast:* output. **(D)** **(software)** To provide data from an external source. *Contrast:* output. **(E)** **(software)** Loosely, input data. *Contrast:* output. (C) 610.12-1990
(3) Pertaining to a device, process, or channel involved in the reception of data. (C) 610.10-1994w
(4) (to a relay) A physical quantity or quantities to which the relay is designed to respond. *Notes:* 1. A physical quantity that is not directly related to the prescribed response of a relay

(though necessary, to or in some way affecting the relay operation), is not considered part of input. 2. Time is not considered a relay input, but it is a factor in performance. (SWG/PE) C37.100-1992
(5) In an IDEF0 model, that which is transformed by a function into output. (C/SE) 1320.1-1998
(6) A pin or port that shall only receive logic signals from a connected net or interconnect structure. (C/DA) 1481-1999

input angle (gyros) The angular displacement of the case about an input axis. (AES/GYAC) 528-1994

input area An area of storage reserved for input data. (C) 610.10-1994w

input argument The designation given to an operation argument that will always have a value at the invocation of the operation. *Contrast:* output argument. (C/SE) 1320.2-1998

input arrow An arrow or arrow segment that expresses IDEF0 input, i.e., an object type set whose instances are transformed by a function into output. The arrowhead of an input arrow is attached to the left side of a box. (C/SE) 1320.1-1998

input assertion (software) A logical expression specifying one or more conditions that program inputs must satisfy in order to be valid. *Contrast:* output assertion; loop assertion. *See also:* inductive assertion method. (C) 610.12-1990

input axis (IA) (1) (accelerometer) The axis along or about which an input causes a maximum output. (AES/GYAC) 528-1994
(2) (gyros) The axis about which a rotation of the case causes a maximum output. For a conventional gyro, the input axis is normal to the spin axis. For an optical gyro, the input axis is perpendicular to a plane established by the light beams. (AES/GYAC) 528-1994

input-axis misalignment (accelerometer) (gyros) The angle between an input axis and its associated input reference axis when the device is at a null condition. (The magnitude of this angle is unambiguous, but when components are reported, the convention should always be identified. IEEE standards use both direction cosines and right-handed Euler angles, depending on the principal field of application. Other conventions, differing both in signs and designation of axes, are sometimes used.) (AES/GYAC) 528-1994

input block (A) (test, measurement, and diagnostic equipment) A section of internal storage of a computer, reserved for the receiving and processing of input information. *Synonym:* input area. **(B) (test, measurement, and diagnostic equipment)** A block used as an input buffer. **(C) (test, measurement, and diagnostic equipment)** A block of machine words, considered as a unit and intended to be transferred from an external source or storage or storage medium to the internal storage of the computer. (MIL) [2]

input buffer *See:* buffer.

input buffer register A data buffer register that accepts data from an input unit such as a magnetic tape drive or magnetic disk and which then transfers this data to internal storage. (C) 610.10-1994w

input capacitance (*n*-terminal electron tubes) The short-circuit transfer capacitance between the input terminal and all other terminals, except the output terminal, connected together. *Note:* This quantity is equivalent to the sum of the interelectrode capacitances between the input electrode and all other electrodes except the output electrode. *See also:* electron-tube admittances. (ED) 161-1971w

input channel A channel employed only for data input; for example, to impress a state on a device or logic element; or to transfer data from an external storage unit to an internal storage unit. *See also:* input-output channel; output channel. (C) 610.10-1994w

input data (test pattern language) The binary data that is written into a memory array. It is identified by the symbol "D." (TT/C) 660-1986w

input device A device used to enter data into a computer system. *Note:* Commonly used input devices include light pens and keyboards. *Synonym:* input unit. *Contrast:* output device. *See also:* string device; cursor control device; logical input device; graphical input device; graphic input device; pick device; input-output device. (C) 610.10-1994w, 610.6-1991w, 1084-1986w

input impedance (1) (analog computer) In an analog computer, a passive network connected between the input terminal or terminals of an operational amplifier and its summing junction. (C) 165-1977w
(2) (at a transmission line port) (waveguide) The impedance at the transverse plane of the port. *Note:* This impedance is independent of the generator impedance. (MTT) 146-1980w
(3) The impedance between the signal input of the waveform recorder and ground. (IM/WM&A) 1057-1994w
(4) (of an antenna) The impedance presented by an antenna at its terminals. (AP/ANT) 145-1993

input jitter tolerance The maximum level of input jitter, specified in terms of unit intervals peak to peak, that does not result in an onset of errors. (COM/TA) 1007-1991r

input leakage current (amplifiers) A direct current (of either polarity) that would flow in a short circuit connecting the input terminals of an amplifier. (IM) 311-1970w

input limiter A limiter circuit employing biased diodes in the amplifier input channel, that operates by limiting current entering the summing junction. (C) 610.10-1994w

input limits (accelerometer) (gyros) The extreme values of the input, generally plus or minus, within which performance is of the specified accuracy. (AES/GYAC) 528-1994

input loopback Loopback of output from one function to be input for another function in the same diagram. (C/SE) 1320.1-1998

input media Media that are employed as input; for example, punched cards; magnetic disks. *Contrast:* output media. (C) 610.10-1994w

input-output Pertaining to input, output, or both. (C) 610.10-1994w

input-output area *See:* buffer.

input-output bound (io bound) Pertaining to any process that performs input-output operations which take a long time relative to the time of CPU operations performed. *Contrast:* compute-bound. (C) 610.10-1994w

input-output channel A channel that handles the transfer of data between internal storage and peripheral equipment. *Synonyms:* data channel; computer channel. *See also:* input-output controller; selector channel; output channel; input channel. (C) 610.7-1995, 610.10-1994w

input-output characteristic (1) (transmission performance of telephone sets) Electric or acoustic output level as a function of the input level. (COM/TA) 269-1983s
(2) The accompanying diagram shows the relationship between the input-output characteristics of an accelerometer or gyro. (AES/GYAC) 528-1994

input-output circuit A circuit that connects a computer to another device. (C) 610.10-1994w

input-output control electronics The electronics required to interface an input-output device to a central processing unit. (C) 610.10-1994w

input-output controller (IOC) A controller that controls one or more input-output channels. *Synonym:* peripheral controller. *See also:* selector channel; input-output channel. (C) 610.7-1995, 610.10-1994w

input-output coupling *See:* data coupling.

input-output device A device through which data may be entered into a computer system, received from the system, or both. *Synonym:* input-output unit. *See also:* output device; input device. (C) 610.10-1994w

input/output model *See:* black box model.

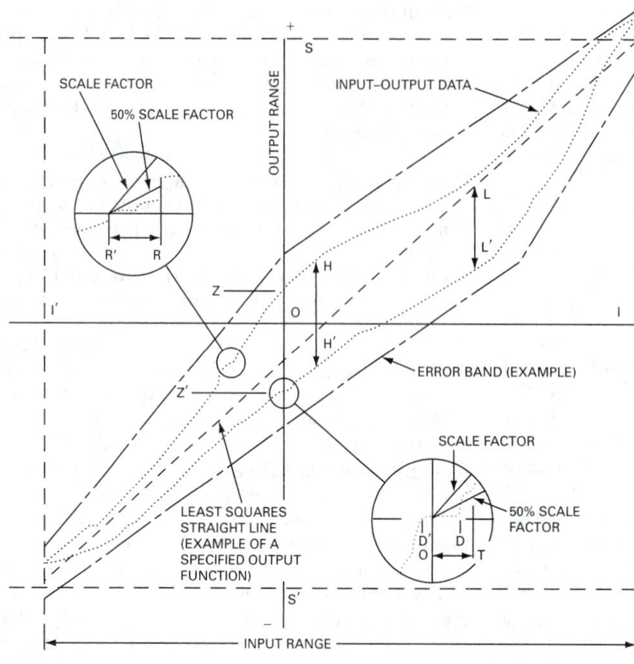

Composite error $= \frac{100LL'}{OS}$ or $\frac{100LL'}{OS'}$

Dead band $= DD'$

Dynamic range $= \frac{OI}{RR'}$ or $\frac{OI'}{RR'}$

Hysteresis error $= HH' \cdot \frac{II'}{SS'}$

Input limits $= I, I'$

Input range $= I'$ to I

Full range $= I - I'$

Output range $= S'$ to S

Output span $= S - S'$

Resolution $= RR'$

Scale factor $= \frac{SS'}{II'}$

Threshold $= OT$

Zero offset $= \frac{Z + Z'}{2} \cdot \frac{II'}{SS'}$

input-output characteristics

input-output port A port that is configured or programmed to provide a data path between the central processing unit and its peripheral devices. (C) 610.10-1994w

input-output processor (IOP) A processor dedicated to controlling input and output transfers. (C) 610.10-1994w

input-output unit *See:* input-output device.

input pin A component pin that receives signals from the external connections. (TT/C) 1149.1-1990

input power (total) (converters having dc input) (self-commutated converters) The mean value of the instantaneous power into the input terminals, taken over one period of the ripple component. *Note:* If either the voltage or the current is ripple-free, the dc power is the total power. (IA/SPC) 936-1987w

input power factor (of a system) The ratio at the input of active power (measured in watts or kilowatts) to input apparent power (measured in volt-amperes or kilovolt-amperes) at rated or specified voltage and load. *See also:* power factor, displacement; total power factor. (IA/PSE) 1100-1999

input primitive (1) (computer graphics) An element of data obtained from a logical input device. (C) 610.6-1991w **(2)** The effort to develop software products, expressed in units of staff-hours. (C/SE) 1045-1992

input-process-output (software) A software design technique that consists of identifying the steps involved in each process to be performed and identifying the inputs to and outputs from each step. *Note:* A refinement called hierarchical input-process-output identifies the steps, inputs, and outputs at both general and detailed levels of detail. *See also:* transaction analysis; modular decomposition; structured design; transform analysis; input-process-output chart; stepwise refinement; rapid prototyping. (C) 610.12-1990

input-process-output chart (software) A diagram of a software system or module, consisting of a rectangle on the left listing inputs, a rectangle in the center listing processing steps, a rectangle on the right listing outputs, and arrows connecting inputs to processing steps and processing steps to outputs. *See also:* block diagram; graph; box diagram; bubble chart; structure chart.

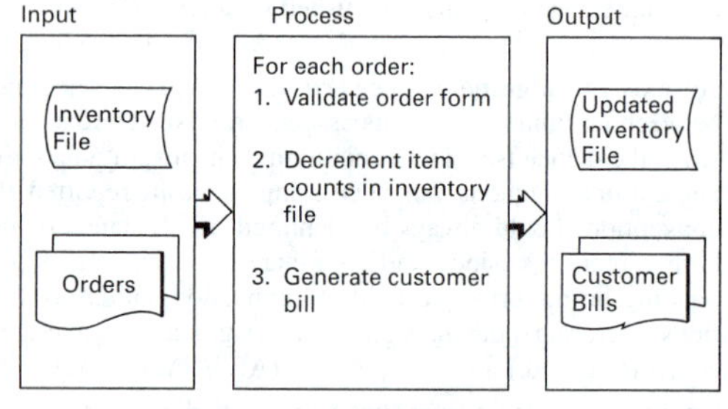

input-process-output chart

(C) 610.12-1990

input pulse shape (pulse transformers) Current pulse or source voltage pulse applied through associated impedance. The shape of the input pulse is described by a current- or voltage-time relationship and is defined with the aid of the corresponding input pulse shape figure. *Note:* A general amplitude quantity is designated by *A*, which may be current *I* or voltage *V*. *See also:* return swing; tilt; pulse duration; trailing edge amplitude; voltage-time product; leading edge linearity; backswing; fall time; output pulse shape; pulse top; ringing; trailing edge; overshoot; voltage-time product rating; rise time; quiescent value; leading edge.

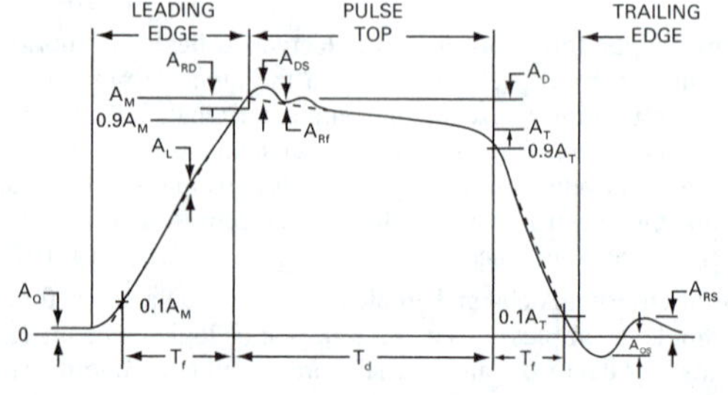

input pulse shape

(PEL/ET) 390-1987r

input queue The database that the service uses to convey objects to the client of the MT interface. (C/PA) 1224.1-1993w

input range (accelerometer) (gyros) The region between the input limits within which a quantity is measured, expressed by stating the lower- and upper-range value. For example: an angular displacement input range of $-5°$ to $+6°$. (AES/GYAC) 528-1994

input rate (gyros) The angular displacement per unit time of the case about an input axis. (AES/GYAC) 528-1994

input reference axis (accelerometer) (gyros) The direction of an axis (nominally parallel to an input axis) as defined by the case mounting surfaces or external case markings, or both. (AES/GYAC) 528-1994

input signal (1) (hydraulic turbines) A control sign injected at any point into a control system. (PE/EDPG) 125-1977s
(2) (control system feedback) A signal applied to a system or element. See the figure attached to definition "3" of error signal. *See also:* feedback control system. (PE/EDPG) 421-1972s, [3]

input signal level sensitivity (spectrum analyzer) The input signal level that produces an output equal to twice the value of the average noise alone. This may be power or voltage relationship, but must be stated so. *See also:* sensitivity; equivalent input noise sensitivity. (IM) 748-1979w

input span *See:* full range.

input station *See:* data input station.

input terminal (A) A terminal used to accept input. **(B)** Any point in a system or communication network at which data can enter the system. *Contrast:* output terminal. (C) 610.10-1994

input unit *See:* input device.

input variable A variable applied to a system or element. *See also:* feedback control system. (IM/PE/EDPG) [120], [3]

input voltage range (of a power system) The range of input voltage over which the system can operate properly. (IA/PSE) 1100-1999

input workstation *See:* data input station.

INQUIRE A nonprocedural database manipulation language used to access data stored in VSAM databases; characterized by its suitability for storage and retrieval of textual data. (C) 610.13-1993w

inrush The amount of current that a load or device draws when first energized. (IA/PSE) 1100-1999

inrush current (1) (electronic power transformer) The maximum root-mean-square or average current value, determined for a specified interval, resulting from the excitation of the transformer with no connected load, and with essentially zero source impedance, and using the minimum primary turns tap available and its rated voltage. (PEL/ET) 295-1969r
(2) (packaging machinery) Of a solenoid or coil, the steady-state current taken from the line with the armature blocked in the rated maximum open position. (IA/PKG) 333-1980w
(3) The rapid change of current with respect to time upon motor energization. Inrush current is dependent upon the voltage impressed across the motor terminals and the motor inductance by the relationship V = L di/dt. (PE/NP) 1290-1996

insert The station transmit signals will be routed to the next active downstream station, and input signals will be routed from the next active upstream station. (C/LM) 11802-4-1994

insertion (1) The opening of the capacitor bypass device to place the series capacitor in service with or without load current flowing. (T&D/PE) 824-1994
(2) A signalling pattern sent by a fibre optic station to request to join the ring (INSERT). This pattern consists of an alternating pattern of normal data signals and low light-level signals. (C/LM) 11802-4-1994

insertion character A character within a picture specification that represents a character that is inserted into the representation only under certain circumstances; for example, the value 1234, when represented using the picture specification 9,999 (the comma is the insertion character), is "1,234." (C) 610.5-1990w

insertion current The steady state root-mean-square current that flows through the series capacitor after the bypass device has opened. (T&D/PE) 824-1994

insertion/extraction levers (1) Mechanical levers used to provide a mechanical advantage during the insertion and extraction of a module to allow operation without tools. The insertion/extraction levers shall be part of each module. (C/BA) 1101.3-1993
(2) A generic term meaning levers that insert and extract the module. (C/BA) 1101.4-1993

insertion levers A device used to insert and extract the module. (C/BA) 1101.7-1995

insertion gain *See:* matched insertion gain; matched load insertion gain; matched generator insertion gain; general insertion gain.

insertion key echo The return of the Insertion Key to the station by the FOTCU. (C/LM) 11802-4-1994

insertion loss (1) (audible noise measurements) The insertion loss of a component (for example, a microphone windscreen) is the difference in decibels between the sound-pressure level measured before the insertion of the component and the sound-pressure level measured after the insertion of the component, provided that the source of the sound and all other conditions remain unchanged. The effect of the added component on the frequency response of a sound-measurement system should be considered and recorded. (T&D/PE) 656-1985s
(2) (data transmission) Resulting from the insertion of a transducer in a transmission system, the ratio of (A) the power delivered to that part of the system following the transducer, before insertion of the transducer, to (B) the power delivered to that same part of the system after insertion of the transducer. *Note:* If the input or output power, or both, consist of more than one component, such as multifrequency signal or noise, then the particular components used and their weighting are specified. (PE) 599-1985w
(3) (fiber optics) The total optical power loss caused by the insertion of an optical component such as a connector, splice, or coupler. (Std100) 812-1984w
(4) (overhead power lines) The difference, in decibels, between the sound pressure level of a component (e.g., windscreen) measured before the insertion of the component and the sound pressure level measured after the insertion of the component (provided that the source of the noise remains unchanged). (T&D/PE) 539-1990
(5) (broadband local area networks) The loss of signal level in a cable path caused by insertion of a passive device. *Synonym:* through loss. (LM/C) 802.7-1989r
(6) The signal loss that results when a channel is inserted between a transmitter and a receiver, which is the ratio of the signal level delivered to a receiver before a channel is inserted, to the signal level after the channel is inserted. (C/LM) 8802-5-1998, 610.7-1995
(7) *See also:* insertion loss. (COM/TA) 1007-1991r
(8) *See also:* matched insertion loss; matched generator insertion loss; matched load insertion loss; general insertion loss. (MTT) 146-1980w

insertion loss ripple The peak-to-peak variation of the insertion loss; i.e., the difference between the maximum and minimum insertion loss over a specified frequency range of the device. (UFFC) 1037-1992w

insertion reactor A reactor that is connected momentarily across the open contacts of a circuit-interrupting device for synchronizing and/or switching transient suppression purposes. (PE/TR) C57.16-1996

insertion sort A sort in which each item in the set to be sorted is inserted into its proper position among those items already considered. *See also:* linear sort; radix insertion sort; tree insertion sort; diminishing increment sort; binary insertion sort; list insertion sort; distribution counting sort; two-way insertion sort; address calculation sort. (C) 610.5-1990w

insertion voltage The steady state root-mean-square voltage appearing across the series capacitor upon the interruption of the bypass current with the opening of the bypass device. (T&D/PE) 824-1994

insert receiver A receiver that uses an ear mold or ear tip inserted into the ear canal. (COM/TA) 1206-1994

in service (1) (power system measurement) (electric generating unit reliability, availability, and productivity) The state in which a unit is electrically connected to the system. (PE/PSE) 762-1987w

(2) Lines and equipment are considered in service when connected to the system and intended to be capable of delivering energy or communication signals, regardless of whether electric loads or signaling apparatus are presently being served from such facilities. (NESC) C2-1997

in-service forced derated hours (electric generating unit reliability, availability, and productivity) The in-service hours during which a Class 1, 2, or 3 unplanned derating was in effect. (PE/PSE) 762-1987w

in-service inspection Visual periodic investigation of the principal features of the circuit breaker in service, without dismantling. This investigation is generally directed toward pressures and/or levels of fluids, tightness, position of relays, pollution of insulating parts; but actions such as lubricating, cleaning, washing, etc., that can be carried out with the circuit breaker in service, are included. *Note:* The observations resulting from inspection can lead to the decision of carrying out overhaul. (SWG/PE) C37.10-1995

in-service maintenance derated hours (electric generating unit reliability, availability, and productivity) The in-service hours during which a Class 4 unplanned derating was in effect. (PE/PSE) 762-1987w

in-service planned derated hours (electric generating unit reliability, availability, and productivity) The in-service hours during which a basic or extended planned derating was in effect. (PE/PSE) 762-1987w

in-service state The component or unit is energized and fully connected to the system. (PE/PSE) 859-1987w

in-service test (metering) A test made during the period that the meter is in service. It may be made on the customer's premises without removing the meter from its mounting, or by removing the meter for test, either on the premises or in a laboratory or meter shop. (ELM) C12.1-1982s

in-service unit derated hours (electric generating unit reliability, availability, and productivity) The in-service hours during which a unit derating was in effect. (PE/PSE) 762-1987w

in-service unplanned derated hours (electric generating unit reliability, availability, and productivity) The in-service hours during which an unplanned derating was in effect. (PE/PSE) 762-1987w

inside air temperature *See:* average inside air temperature.

inside communications cables A telephone-type cable intended for indoor use that is not designed to withstand solar radiation or precipitation and may or may not be shielded. (PE/PSC) 789-1988w

inside top air temperature (power and distribution transformers) The temperature of the air inside a dry-type transformer enclosure, measured in the space above the core and coils. (PE/TR) C57.12.80-1978r

in sight from (within sight from, within sight) Where this Code specifies that one equipment shall be "in sight from," "within sight from," or "within sight", etc., of another equipment, one of the equipments specified shall be visible and not more than 50 ft distant from the other. (NESC/NEC) [86]

***in situ* study** Referring to studies involving organisms in their natural condition or environment. (T&D/PE) 539-1990

in-situ total dose test A test of the instantaneous effects of an environment containing radiation obtained by monitoring the device characteristics during exposure to a relatively low, constant dose rate. (ED) 641-1987w

inspection (1) (nuclear power quality assurance) Examination or measurement to verify whether an item or activity conforms to specified requirements. (PE/NP) [124]

(2) (software) A static analysis technique that relies on visual examination of development products to detect errors, violations of development standards, and other problems. Types include code inspection; design inspection. (C) 610.12-1990

(3) A visual examination of a software product to detect and identify software anomalies, including errors and deviations from standards and specifications. Inspections are peer examinations led by impartial facilitators who are trained in inspection techniques. Determination of remedial or investigative action for an anomaly is a mandatory element of a software inspection, although the solution should not be determined in the inspection meeting. (C/SE) 1028-1997

inspection and test plan (ITP) A summary of prerequisites, system configurations, step-by-step procedures, and evaluation criteria of the tests in one place for permanent record. (SUB/PE) 1303-1994

inspection, meter installation *See:* meter installation inspection.

inspector (nuclear power quality assurance) A person who performs inspection activities to verify conformance to specific requirements. (PE/NP) [124]

installation The period of time in the software life cycle during which a software product is integrated into its operational environment and tested in this environment to ensure that it performs as required. (C/SE) 1074-1995s, 610.12-1990

installation manual A document that provides the information necessary to install a system or component, set initial parameters, and prepare the system or component for operational use. *See also:* user manual; programmer manual; operator manual; diagnostic manual; support manual. (C) 610.12-1990

installed incremental transfer capability (power operations) The amount of power, incremental above normal base power transfers, that can be transferred over the transmission network without giving consideration to the effect of transmission facility outages. All facility loadings are within normal ratings and all voltages are within normal limits. (PE/PSE) 858-1987s

installed life (1) (electric penetration assemblies) The interval of time from installation to permanent removal from service, during which the electric penetration assembly is expected to perform its required function(s). *Note:* Components of the assembly may require periodic replacement; thus, the installed life of such components is less than the installed life of the assembly. (PE/NP) 317-1983r

(2) (safety systems equipment in nuclear power generating stations) (valve actuators) The interval from installation to removal, during which the equipment or component thereof may be subject to design service conditions and system demands. *Note:* Equipment may have an installed life of 40 years with certain components changed periodically; thus, the installed life of the changed components would be less than 40 years. (SWG/PE/NP) 382-1985, 627-1980r, C37.100-1992, 323-1974s, 649-19802s

(3) (Class 1E battery chargers and inverters) The interval from installation to removal, during which the equipment or component thereof may be subject to design service conditions and system demands. *Note:* Equipment may have an installed life of 20 years with certain components changed periodically; thus, the installed life of the components would be less than 20 years. (PE/NP) 650-1979s

installed nameplate capacity (electric generating unit reliability, availability, and productivity) The full-load continuous gross capacity of a unit under specified conditions, as calculated from the electric generator nameplate based on the rated power factor. *Note:* The nameplate rating of the electric generator may not be indicative of the unit maximum or dependable capacity, since some other item or equipment (such as the turbine) may limit unit output. (PE/PSE) 762-1987w

installed reserve (power operations) (electric power supply) The reserve capacity installed on a system. (PE/PSE) 858-1987s, 346-1973w

installed software Any software object created by the use of the `swinstall` utility. (C/PA) 1387.2-1995

installed_software A software_collection containing installed software. This software is in a state ready for use, or ready to be shared by client systems. A directory path on a system and an installed_software catalog together identify a unique installed_software object. (C/PA) 1387.2-1995

installed_software catalog The catalog of metadata for an installed_software software_collection. Unlike a catalog for a distribution object, the storage and format of an installed_software catalog is undefined within this standard. The ability to dump and restore all or part of an installed_software catalog into an exported catalog structure is included in this standard. (C/PA) 1387.2-1995

installed_software path The root directory of an installed_software object; the pathname below which all software for that object shall be installed. (C/PA) 1387.2-1995

installer *See:* constructor.

Instance (1) The mapping of an Activity that processes all of its Input Information and generates all of its Output Information. *Contrast:* Invocation; Iteration. *See also:* mapping. (C/SE) 1074-1997
(2) *See also:* package instance.

instance (1) A discrete, bounded thing with an intrinsic, immutable, and unique identity. Anything that is classified into a class is said to be an instance of the class. All the instances of a given class have the same responsibilities, i.e., they possess the same kinds of knowledge, exhibit the same kinds of behavior, participate in the same kinds of relationships, and obey the same rules. (C/SE) 1320.2-1998
(2) The specific object that results from allocating resources to implement a single consistent set of internal variables required for the operations and behavior defined for a class definition. (IM/ST) 1451.1-1999
(3) An individual occurrence of an entity that belongs to a particular type of entity. (SCC32) 1489-1999

instance-level attribute A mapping from the instances of a class to the instances of a value class. (C/SE) 1320.2-1998

instance-level operation A mapping from the (cross product of the) instances of the class and the instances of the input argument types to the (cross product of the) instances of the other (output) argument types. (C/SE) 1320.2-1998

instance-level responsibility A kind of responsibility that applies to each instance of the class individually. *Contrast:* class-level responsibility. (C/SE) 1320.2-1998

instance net An abstraction expressing the idea of an electrical connection between various points in a design. In a hierarchical representation of the design, nets can occur at all levels and may connect to pins of lower hierarchical levels (including cell instances), ports of the current hierarchical level, and each other. In a flattened (unfolded and elaborated) design, electrically connected nets are collapsed and each net instance corresponds to an unique interconnect structure in the implementation. (C/DA) 1481-1999

instantaneous A qualifying term indicating that no delay is purposely introduced in the action of the device. (SWG/PE) C37.100-1992

instantaneous access *See:* immediate access.

instantaneous automatic gain control (IAGC) (nonlinear, active, and nonreciprocal waveguide components) (radar) A fast-acting automatic gain control that responds to variations of received signal, avoiding receiver saturation. (MTT) 457-1982w
(2) (A) That part of a receiver system that automatically adjusts the gain of an amplifier within the time duration of each pulse so that a substantially constant output pulse peak amplitude is obtained when the input pulse peak amplitudes are varying, the adjustment being sufficiently fast to operate during the time a pulse is passing through the amplifier. **(B)** A quick-acting automatic gain control that responds to variations of mean clutter or jamming level over different range or angular regions, avoiding receiver saturation. (AES) 686-1997

instantaneous demand (power operations) The load at any instant. (PE/PSE) 858-1987s, 346-1973w

instantaneous frequency (1) (data transmission) The time rate of change of the angle of an angle-modulated wave. *Note:* If the angle is measured in radians, the frequency in hertz is the time rate of change of the angle divided by 2. (PE) 599-1985w
(2) $1/(2\pi)$ times the time rate of change of phase of a wave. (AP/PROP) 211-1997

instantaneous overcurrent or rate-of-rise relay (power system device function numbers) A relay that functions instantaneously on an excessive value of current or on an excessive rate of current rise. (PE/SUB) C37.2-1979s

instantaneous peak power (waveguide) The maximum instantaneous power passing through the transverse section of a waveguide during the interval of interest. (MTT) 146-1980w

instantaneous phase or ground trip element *See:* instantaneous; direct-acting overcurrent trip device.

instantaneous power (1) (circuit) At the terminals of entry into a delimited region the rate at which electric energy is being transmitted by the circuit into or out of the region. *Note:* Whether power into the region or out of the region is positive is a matter of convention and depends upon the selected reference direction of energy flow. (Std100) 270-1966w
(2) (polyphase circuit) At the terminals of entry into a delimited region, the algebraic sum of the products obtained by multiplying the voltage between each terminal of entry and some arbitrarily selected common point in the boundary surface (which may be neutral terminal of entry) by the current through the corresponding terminal of entry. *Notes:* 1. The reference direction of each current must be the same, either into or out of the delimited region. The reference polarity for each voltage must be consistently chosen, either with all the positive terminals at the terminals of entry and all negative terminals at the common reference point, or vice-versa. If the reference direction for currents is into the delimited region and the positive reference terminals for voltage are at the phase terminals of entry, the power will be positive when the energy flow is into the delimited region and negative when the flow is out of the delimited region. Reversal of either the reference direction or the reference polarity will reverse the relation between the sign of the power and the direction of energy flow. 2. When the circuit has a neutral terminal of entry, it is usual to select the neutral terminal as the common point for voltage measurement, because one of the voltages is then always zero, and, when both the currents and voltages form symmetrical polyphase sets of the same phase sequence, the average power for each single-phase circuit consisting of one phase conductor and the neutral conductor, will be the same. When the voltages and currents are sinusoidal and the voltages are measured to the neutral terminal of entry as the common point, the instantaneous power at the four points of entry of a three-phase circuit with neutral is given by

$$p = E_aI_a\,[\cos(\alpha_a - \beta_a) + \cos(2\omega t + \alpha_a + \beta_a)]$$
$$+ E_bI_b\,[\cos(\alpha_b - \beta_b) + \cos(2\omega t + \alpha_b + \beta_b)]$$
$$+ E_cI_c\,[\cos(\alpha_c - \beta_c) + \cos(2\omega t + \alpha_c + \beta_c)]$$

where E_a, E_b, E_c are the root-mean-square amplitudes of the voltages from the phase conductors, *a*, *b*, and *c*, respectively, to the neutral conductor at the terminals of entry; I_a, I_b, I_c are the root-mean-square amplitudes of the currents in the phase conductors *a*, *b*, and *c*. a_a, a_b, a_c are the phase angles of the voltages E_a, E_b, E_c with respect to a common reference; β_a, β_b, β_c are the phase angles of the currents I_a, I_b, I_c with respect to the same reference as the voltages. 3. If there is no neutral conductor, so that there are only three terminals of entry, the point of entry of one of the phase conductors may be chosen as the common voltage point, and the voltages from that conductor to the common point become zero. If, in the preceding, the terminal of entry of phase conductor b is chosen as the common point, E_a is replaced by E_{ab} in the first line, E_c is replaced by E_{cb} in the third line, and the second line, being zero, is omitted. 4. If both the voltages and currents in the preceding equations constitute symmetrical polyphase sets of the same phase sequence, then $p = 3E_a I_a \cos(\alpha_a - \beta_a)$. Because this expression and similar expressions for m phases are independent of time, it follows that the instantaneous power is constant when the voltages and currents constitute polyphase symmetrical sets of the same phase sequence. 5. However, if the polyphase sets have single-phase symmetry or zero-phase symmetry rather than polyphase symmetry, the higher frequency terms do not cancel, and the instantaneous power is not a constant. 6. In general, the instantaneous power p at the $(m + 1)$ terminals of entry of a polyphase circuit of m phases to a delimited area, when one of the terminals is that of the neutral conductor, is expressed by the equation

$$p = \sum_{s=1}^{s=m} e_s i_s$$

$$= \sum_{s=1}^{s=m} \sum_{r=1}^{r=\infty} \sum_{q=1}^{q=\infty} E_{sr} I_{sq}$$

$$(\cos\,[(r - q)\omega t + \alpha_{sr} - \beta_{sq}]$$

$$+ \cos\,[(r + q)\omega t + \alpha_{sr} + \beta_{sq}])$$

where e_s is the instantaneous alternating voltage between the sth terminal entry and the terminal of voltage reference, which may be the true neutral point, the neutral conductor, or another point in the boundary surface. i_s is the instantaneous alternating current through the sth terminal of entry. E_{sr} is the root-mean-square amplitude of the rth harmonic of voltage e_s. I_{sq} is the root-mean-square amplitude of the qth harmonic of current i_s. αi_{sr} is the phase angle of the rth harmonic of e_s with respect to a common reference. β_{sq} is the phase angle of the qth harmonic of i_s with respect to the same reference as the voltages. The index s runs through the phase letters identifying the m-phase conductor of an m-phase system, a, b, c, etc., and then concludes with the neutral conductor n, if one exists. The indexes r and q identify the order of the harmonic term in each e_s and i_s, respectively, and run through all the harmonics present in the Fourier series representation of each alternating voltage and current. If the terminal voltage reference is that of the neutral conductor, the terms for $s = n$ will vanish. If the voltages and current are quasi-periodic, of the form given in "power, instantaneous (two-wire circuit)," this expression is still valid but E_{sr} and I_{sq} become periodic functions of time. 7. Instantaneous power is expressed in watts when the voltages are in volts and the currents in amperes. *See also:* zero-phase symmetrical set; single-phase symmetrical set. (Std100) 270-1966w

(3) (two-wire circuit) At the two terminals of entry into a delimited region, the product of the instantaneous voltage between one terminal of entry and the second terminal of entry, considered as the reference terminal, and the current through the first terminal. *Notes:* 1. The entire path selected for the determination of each voltage must lie in the boundary surface of the delimited region or be so selected that the voltage is the same as that analog such a path. 2. Mathematically the instantaneous power p is given by $p = ei$ in which e is the voltage between the first terminal of entry and the second (reference) terminal of entry and i is the current through the first terminal of entry in the reference direction. 3. If both the voltage and current are sinusoidal and of the same period, the instantaneous power at any instant t is given by the equation

$$p = ei = [(2)^{1/2} E \cos(\omega t + \alpha)]$$

$$\times [(2)^{1/2} I \cos(\omega t + \beta)]$$

$$= 2EI\cos(\omega t + \alpha)\cos(\omega t + \beta)$$

$$= EI[\cos(\alpha - \beta) + \cos(2\omega t + \alpha + \beta)]$$

in which E and I are the root-mean-square amplitudes of voltage and current, respectively, and α and β are the phase angles of the voltage and current, respectively, from the same reference. 4. If the voltage is an alternating voltage and the current is an alternating current of the same primitive period [see alternating voltage; alternating current; and period (primitive period) of a function], the instantaneous power is given by the equation

$$p = ei$$

$$= E_1 I_1[\cos(\alpha_1 - \beta_1) + \cos(2\omega t + \alpha_1 + \beta_1)]$$

$$+ E_2 I_2[\cos(\omega t + \alpha_2 - \beta_1) + \cos(3\omega t + \alpha_2 + \beta_1)]$$

$$= E_1 I_2[\cos(\omega t - \alpha_1 + \beta_2) + \cos(3\omega t + \alpha_1 + \beta_2)]$$

$$+ E_2 I_2[\cos(\alpha_2 - \beta_2) + \cos(4\omega t + \alpha_2 + \beta_2)]$$

$$+ \;\ldots$$

This equation can be written conveniently as a double summation

$$p = \sum_{r=1}^{r=\infty} \sum_{q=1}^{q=\infty} E_r I_q \cos[(r - q)\omega t + \alpha_r - \beta_q]$$

$$+ \cos[(r + q)\omega t + \alpha_r - \beta_q]$$

in which r is the order of the harmonic component of the voltage and q is the order of the harmonic component of the current (see "harmonic components (harmonics)"), and E, I, α, and β apply to the harmonic denoted by the subscript. 5. If the voltage and current are quasi-periodic functions of the form

$$e = (2)^{1/2} \sum_{r=1}^{r=\infty} E_r(t)\cos(r\omega t + \alpha_r)$$

$$i = (2)^{1/2} \sum_{q=1}^{q=\infty} I_q(t)\cos(q\omega + \beta_r)$$

where $E_r(t)$, $I_q(t)$ are aperiodic functions of t, the instantaneous power is given by the equation

$$p = ei$$

$$= E_1(t)I_1(t)\,[\cos(\alpha_1 - \beta_1)$$

$$+ \cos(2\omega t + \alpha_1 + \beta_1)]$$

$$+ E_2(t)I_1(t)\,[\cos(\omega t + \alpha_2 - \beta_1)$$

$$+ \cos(3\omega t + \alpha_2 + \beta_1)]$$

$$+ E_1(t)I_2(t)\,[\cos(\omega t - \alpha_1 + \beta_2)$$

$$+ \cos(3\omega t + \alpha_1 + \beta_2)]$$

$$+ E_2(t)I_2(t)\,[\cos(\alpha_2 - \beta_2)$$

$$+ \cos(4\omega t + \alpha_2 + \beta_2)]$$

$$+ \;\ldots$$

6. Instantaneous power is expressed in watts when the voltage is in volts and the current in amperes. 7. See "reference direction of energy." The sign of the energy will be positive if the flow of power is in the reference direction and negative if the flow is in the opposite direction. (Std100) 270-1966w

instantaneous power output The rate at which energy is delivered to a load at a particular instant. *See also:* radio transmitter. (AP/ANT) 145-1983s

instantaneous Poynting vector ($\vec{\mathbf{P}}(\mathbf{t},\vec{\mathbf{r}})$) (of an electromagnetic wave) The vector product of the instantaneous electric and magnetic field vectors. The integral of $\vec{P}(t, \vec{r})$ over a surface is the instantaneous electromagnetic power flow through the surface. (AP/PROP) 211-1997

instantaneous recording (mechanical recording) A phonograph recording that is intended for direct reproduction without further processing. *See also:* phonograph pickup. (SP) [32]

instantaneous relay recovery time Recovery time of a thermal relay measured when the heater is de-energized at the instant of contact operation. (EEC/REE) [87]

instantaneous relay reoperate time Reoperate time of a thermal relay measured when the heater is de-energized at the instant of contact operation. (EEC/REE) [87]

instantaneous sampling The process for obtaining a sequence of instantaneous values of a wave. *Note:* These values are called instantaneous samples. (AP/ANT) 145-1983s

instantaneous sound pressure (at a point) The total instantaneous pressure at that point minus the static pressure at that point. *Note:* The commonly used unit is the newton per square meter. (SP) [32]

instantaneous storage *See:* immediate access storage.

instantaneous suppression with automatic current regulation (thyristor) A combination of instantaneous trip or suppression and current regulation in which suppression is followed immediately by a regulated current. (IA/IPC) 428-1981w

instantaneous trip (1) (as applied to Circuit Breakers) A qualifying term indicating that no delay is purposely introduced in the tripping action of the circuit breaker. (NESC) [86]
(2) The means to sense an overload and reduce the output current to zero, as fast as practicable. (IA/IPC) 428-1981w

instantiation (software) The process of substituting specific data, instructions, or both into a generic program unit to make it usable in a computer program. (C) 610.12-1990

instant of chopping The instant when the initial discontinuity appears. (PE/PSIM) 4-1995

instant start fluorescent lamp (illuminating engineering) A fluorescent lamp designed for starting by a high voltage without preheating of the electrodes. (EEC/IE) [126]

Institute of Electrical and Electronics Engineers (1) An organization that, among other functions, sponsors standards development. (C/BA) 14536-1995
(2) An international professional organization that is accredited by American National Standards Institute to develop standards for them. (C) 610.7-1995, 610.10-1994w

institutional design Emphasizes reliability, resistance to wear and use, safety to public, and special aesthetic considerations, such as the "agelessness" of the structure. (IA/PSE) 241-1990r

instruction (1) (programmable digital computer systems in safety systems of nuclear power generating stations) A meaningful expression in a computer programming language that specifies an operation to a digital computer. 554-1990
(2) (btl interface circuits) A binary data word shifted serially into the test logic defined by this standard in order to define its subsequent operation. (TT/C) 1149.1-1990
(3) (software) *See also:* computer instruction. (C) 610.12-1990
(4) A statement or expression consisting of an operation and its operands (if any), which can be interpreted by a computer in order to perform some function or operation. *See also:* computer instruction; microinstruction; macroinstruction. (C) 610.10-1994w

instruction address (A) The address of an instruction. **(B)** The address that must be used to fetch an instruction. (C) 610.10-1994

instruction address register An address register used to hold the address of an instruction. *Synonyms:* instruction pointer register; program register. *See also:* P register. (C) 610.10-1994w

instruction address stop An instruction address that, when it is fetched, causes execution to stop. *See also:* address stop. (C) 610.10-1994w

instructional character *See:* control character.

instructional game An instruction method employed by some computer-assisted instruction systems, in which a game is used to instruct the student on some subject. *Contrast:* simulation. (C) 610.2-1987

instructional simulation (modeling and simulation) A simulation intended to provide an opportunity for learning or to evaluate learning or educational potential; for example, a simulation in which a mock-up of an airplane cockpit is used to train student pilots. *Synonyms:* academic simulation; tutorial simulation. (C) 610.3-1989w

instruction cache A cache that stores instructions for fast access by the processor. *Contrast:* data cache. (C) 610.10-1994w

instruction code *See:* computer instruction code.

instruction control unit In a processor, the part that retrieves instructions in proper sequence, interprets each instruction, and applies the proper signals to the arithmetic and logic unit and other parts in accordance with this interpretation. *Synonym:* computer control unit. (C) 610.10-1994w

instruction counter (IC) (software) A register that indicates the location of the next computer instruction to be executed. *Synonym:* program counter. (C) 610.12-1990

instruction cycle (software) The process of fetching a computer instruction from memory and executing it. *See also:* instruction time. (C) 610.12-1990, 610.10-1994w

instruction decoder (A) The portion of the computer that determines which functions of the execution unit and the operand handler must be performed to execute the instruction. *Note:* Often implemented as part of the instruction fetch unit. **(B)** A functional component that analyzes the operation to be performed, as indicated in an instruction. *See also:* instruction processor. (C) 610.10-1994

instruction fetch unit The portion of a computer that reads the next instruction word from memory and converts the commands to the internal format used by the instruction decoder. (C) 610.10-1994w

instruction field A bit field within an instruction word. (C/MM) 1754-1994

instruction format The number and arrangement of the fields (operand, operation, and address) in a computer instruction. *See also:* address format. (C) 610.10-1994w, 610.12-1990

instruction length (software) The number of words, bytes, or bits needed to store a computer instruction. (C) 610.12-1990

instruction modifier (software) A word or part of a word used to alter a computer instruction. (C) 610.12-1990

instruction pointer register *See:* instruction address register.

instruction processor A functional component that carries out the action indicated by the instruction decoder, resulting in a possible change of machine or data state; for example, instruction decision and execution. (C) 610.10-1994w

instruction register A register that is used to hold an instruction for interpretation. (C) 610.10-1994w

instruction repertoire *See:* instruction set.

instruction set The complete set of instructions recognized by a given computer or provided by a given programming language. *Note:* In computer hardware, this term is considered to be synonymous with a computer's architecture. *Synonym:* instruction repertoire. *See also:* computer instruction set. (C) 610.10-1994w, 610.12-1990

instruction set architecture (1) (software) An abstract machine characterized by an instruction set. *See also:* abstract machine; instruction set. (C/SE) 729-1983s
(2) An ISA defines instructions, registers, instruction and data memory, the effect of executed instructions on the registers and memory, and an algorithm for controlling instruction execution. An ISA does not define clock-cycle times, cycles per instruction, data paths, etc. This standard defines an ISA. (C/MM) 1754-1994

instruction time (1) (software) The time it takes a computer to fetch an instruction from memory and execute it. *See also:* instruction cycle. (C) 610.12-1990, 610.10-1994w
(2) The time it takes to perform one instruction cycle. (C) 610.10-1994w

instruction trace *See:* trace.

instruction word A word that represents an instruction. *See also:* very long instruction word. (C) 610.10-1994w

instrument (1) (plutonium monitoring) A complete system designed to quantify a particular type of ionizing radiation. (NI) N317-1980r

(2) (radiation protection) A complete system designed to quantify one or more particular ionizing radiation or radiations. (NI) N323-1978r
(3) (software) In software and system testing, to install or insert devices or instructions into hardware or software to monitor the operation of a system or component. (C) 610.12-1990
(4) (airborne radioactivity monitoring) A complete system designed to quantify one or more characteristics of ionizing radiation or radioactive material. (NI) N42.17B-1989r
(5) A device whose purpose is usually the generation or measurement of a class of signal. (SCC20) 1226-1998

instrumentation (software) Devices or instructions installed or inserted into hardware or software to monitor the operation of a system or component. (C) 610.12-1990

instrumentation cable A cable that carries low level electric energy from a transducer to a measuring or controlling device. It may be used in environments such as high temperature, high radiation levels, and high electromagnetic fields. An instrument cable may consist of a group of two or more paired or unpaired, shielded or unshielded, solid or stranded insulated conductors. (PE/PSC) 789-1988w

instrumentation tool (software) A software tool that generates and inserts counters or other probes at strategic points in another program to provide statistics about program execution such as how thoroughly the program's code is exercised. *See also:* program; execution; code. (C/SE) 729-1983s

instrument cable Cable used for instrument applications where the cable construction is generally 300 V, twisted pairs or triads, in wire sizes 16 AWG (1.31 mm^2) or 18 AWG (0.823 mm^2). For the purposes of this document, coaxial, triaxial, and fiber optic cables are not considered instrument cable because of differences in cable installation limits. (PE/IC) 1185-1994

instrument landing system (ILS) (1) (general) A generic term for a system which provides the necessary lateral, longitudinal and vertical guidance in an aircraft for a low approach or landing. *Synonym:* ILS.
(2) (navigation) An internationally adopted instrument landing system for aircraft, consisting of a vhf localizer, a uhf glide slope and 75 MHz markers. *See also:* instrument landing system reference point. (AES/RS) 686-1982s, [42]

instrument landing system marker beacon *See:* navigation; boundary marker.

instrument landing system reference point (electronic navigation) A point on the centerline of the instrument landing system runway designated as the optimum point of contact for landing: in standards of the International Civil Aviation Organization this point is from 500 to 1000 feet from the approach end of the runway. *See also:* navigation. (AES) [42]

instrument multiplier A particular type of series resistor that is used to extend the voltage range beyond some particular value for which the instrument is already complete. *See also:* auxiliary device to an instrument; voltage range multiplier. (EEC/AII) [102]

instrument quality control chart A chart developed to evaluate an instrument's response to predetermined, statistically-based limits as determined by an appropriate QC source. The predetermined statistical limits are not typically developed with the overall quality performance (bias and precision) parameters for an analytical technique in mind. (NI) N42.23-1995

instrument relay A relay whose operation depends upon principles employed in measuring instruments such as the electrodynamometer, iron vane, D'Arsonval galvanometer, and moving magnet. *See also:* relay. (EEC/REE) [87]

instrument response (1) (dynamic) The behavior of the instrument output as a function of the measured signal, both with respect to time. *See also:* frequency response; accuracy rating; ramp response; step response. (EEC/EMI) [112]
(2) (forced) The total steady-state plus transient time response resulting from an external input. (IM) [120]

instrument shunt (direct-current instrument shunts) A particular type of resistor designed to be connected in parallel with the measuring device to extend the current range beyond some particular value for which the instrument is already complete. (PE/PSIM) 316-1971w

instrument switch A switch used to connect or disconnect an instrument, or to transfer it from one circuit or phase to another. *Examples:* Ammeter switch, voltmeter switch. (SWG/PE) C37.100-1992

instrument terminals (direct-current instrument shunts) Those terminals which provide a voltage drop proportional to the current in the shunt and to which the instrument or other measuring device is connected. (PE/PSIM) 316-1971w

instrument tolerance chart A chart developed to evaluate an instrument's response to a predetermined, tolerance level as determined by an appropriate quality control source. The predetermined tolerance level, typically expressed as a percent, is set with the overall quality performance (bias and precision) parameters for an analytical technique in mind. For practical reasons, the response of most instruments is held in control to a tolerance not to exceed 3 and 5 percent as related to the initial instrument calibration. (NI) N42.23-1995

instrument transformer (1) (power and distribution transformers) A transformer that is intended to reproduce in its secondary circuit, in a definite and known proportion, the current or voltage of its primary circuit, with the phase relations and waveform substantially preserved. *See also:* voltage transformer; transformer correction factor; phase angle correction factor; marked ratio; true ratio; window-type current transformer; three-wire type current transformer; rated secondary voltage; percent ratio; ratio correction factor; double-secondary current transformer; phase angle of an instrument transformer; excitation losses for an instrument transformer; thermal burden rating of a voltage transformer; polarity; multi-ratio current transformer; cascade-type voltage transformer; insulated-neutral terminal type voltage transformer; rated current; double-secondary voltage transformer; current transformer; rated secondary current; fused-type voltage transformer; multiple-secondary current transformer; rated voltage. (PE/PSR/TR) C37.110-1996, C57.12.80-1978r
(2) One that is intended to reproduce in its secondary circuit, in a definite and known proportion, the current or voltage of its primary circuit with the phase relations substantially preserved. (PE/TR) C57.13-1993

instrument transformer—accuracy class The limits of transformer correction factor, in terms of percent error, that have been established to cover specific performance ranges for line power factor conditions between 1.0 and 0.6 lag. (ELM) C12.1-1988

instrument tranformer—accuracy rating for metering The accuracy class, together with a standard burden for which the accuracy class applies. (ELM) C12.1-1988

instrument-transformer correction factor *See:* transformer correction factor.

insulated (1) Separated from other conducting surfaces by a dielectric substance or air space permanently offering a high resistance to the passage of current and to disruptive discharge through the substance or space. *Note:* When any object is said to be insulated, it is understood to be insulated in a manner suitable for the conditions to which it is subjected. Otherwise, within the purpose of this definition, it is uninsulated. Insulating covering of conductors is one means for making the conductors insulated. (IA/PC) 463-1993w
(2) Separated from other conducting surfaces by a dielectric (including air space) offering a high resistance to the passage of current. *Note:* When any object is said to be insulated, it is understood to be insulated for the conditions to which it is normally subjected. Otherwise, it is, within the purpose of these rules, uninsulated. (NESC) C2-1997

insulated bearing (rotating machinery) A bearing that is insulated to prevent the circulation of stray currents. *See also:* bearing. (PE) [9]

insulated bearing housing (rotating machinery) A bearing housing that is electrically insulated from its supporting structure to prevent the circulation of stray currents. *See also:* bearing. (PE) [9]

insulated bearing pedestal (rotating machinery) A bearing pedestal that is electrically insulated from its supporting structure to prevent the circulation of stray currents. *See also:* bearing. (PE) [9]

insulated bolt A bolt provided with insulation. (EEC/PE) [119]

insulated cap (separable insulated connectors) An accessory device designed to electrically insulate and shield and mechanically seal a bushing insert or integral bushing. (T&D/PE) 386-1995

insulated-case circuit breaker (ICCB) A circuit breaker that is assembled as an integral unit in a supporting and enclosing housing of insulating material and with a stored energy mechanism. (IA/PSP) 1015-1997

insulated conductor (1) A conductor covered with a dielectric (other than air) having a rated insulating strength equal to or greater than the voltage of the circuit in which it is used. (NESC/T&D) C2-1997, C2.2-1960
(2) A conductor encased within material of composition and thickness that is recognized by this Code as electrical insulation. (NESC/NEC) [86]

insulated coupling (rotating machinery) A coupling whose halves are insulated from each other to prevent the circulation of stray current between shafts. *See also:* rotor. (PE) [9]

insulated flange (piping) Element of a flange-type coupling, insulated to interrupt the electrically conducting path normally provided by metallic piping. *See also:* rotor. (PE) [9]

insulated-gate field-effect transistor (IGFET) (1) (metal-nitride-oxide field-effect transistor) A four-terminal device consisting of two separate areas of one conductivity type called source and drain with a terminal each, separated from each other by a substrate of opposite conductivity type with its terminal and straddled by an electrode with terminal called gate, which is insulated from the silicon by a layer of insulator material, frequently silicon dioxide, called gate. (ED) 581-1978w
(2) A field-effect transistor that has one or more gate electrodes that are electrically insulated from the channel. (ED) 1005-1998, 641-1987w

insulated-gate field-effect transistor symbols (metal-nitride-oxide field-effect transistor) IGFET types may be categorized as memory—nonmemory, enhancement mode—depletion mode, and *n*-channel−p-channel. Standard symbols for memory transistors do not exist. The diagram below presents the standard electrical symbols for the nonmemory transistors and the symbols used in this standard for memory transistors. The symbols used for the memory transistors must be considered provisional until specific standards have been finalized. (ED) 581-1978w

insulated ground stick An insulated rod, fabricated from fiberglass reinforced plastic, with specialized connection hardware, operating mechanism, and of sufficient length to allow for safe gripping and installation of grounding clamps. (T&D/PE) 524a-1993r

insulated image guide A planar dielectric waveguide composed of one or more dielectric strips of finite width affixed to an extended dielectric layer of lower dielectric constant and finite thickness, attached in turn to an extended conducting ground plane. (MTT) 1004-1987w

insulated joint (1) A coupling or joint used to insulate adjacent pieces of conduits, pipes, rods, or bars. (VT/CON/LT) 16-1955w
(2) (cable) A device that mechanically couples and electrically insulates the sheath and armor of contiguous lengths of cable. *See also:* tower. (T&D/PE) [10]

insulated-neutral terminal type voltage transformer One that has the neutral end of the primary winding insulated from the case or base and connected to a terminal that provides insulation for a lower voltage than required for the line terminal. (The neutral may be connected to the case or mounting base in a manner intended to facilitate temporary disconnection for dielectric testing.) (PE/TR) C57.13-1993

insulated parking bushing An accessory device designed to electrically insulate and shield and mechanically seal a power cable terminated with an elbow. (T&D/PE) 386-1995

insulated rail joint A joint used to insulate abutting rail ends electrically from one another. (EEC/PE) [119]

insulated splice (power cable joints) A splice with a dielectric medium applied over the connected conductors and adjacent cable insulation. (PE/IC) 404-1986s

insulated static wire An insulated conductor on a power transmission line whose primary function is protection of the transmission line from lightning and one of whose secondary function is communications. (PE) 599-1985w

insulated supply system *See:* ungrounded system.

insulated tool or device (A) (power line maintenance) A tool or device that has conductive parts and is either coated or covered with a dielectric material. **(B)** A tool or device designed primarily to provide insulation from an energized part or conductor. It can be composed entirely of insulating materials. *Examples:* conductor cover; stick; insulating tape. (T&D/PE) 458-1985, 516-1995

insulated turnbuckle An insulated turnbuckle is one so constructed as to constitute an insulator as well as a turnbuckle. *See also:* tower. (T&D/PE) [10]

insulating (covering of a conductor, or clothing, guards, rods, and other safety devices) A device that, when interposed between a person and current-carrying parts, protects the person making use of it against electric shock from the current-carrying parts with which the device is intended to be used: the opposite of conducting. *See also:* insulation; insulated. (T&D) C2.2-1960

insulating cell (rotating machinery) An insulating liner, usually to separate a coil-side from the grounded surface at a slot. *See also:* stator; rotor. (PE) [9]

insulating clothing Clothing made of natural or synthetic material that is designed primarily to provide insulation from an energized part or conductor. (T&D/PE) 516-1995

insulating envelope An envelope of inorganic or organic material such as a ceramic or cast resin placed around the energized conductor and insulating material. (PE/TR) C57.19.03-1996

insulating (isolating) joint (power cable joints) A cable joint which mechanically couples and electrically separates the sheath, shield, and armor on contiguous lengths of cable. (PE/IC) 404-1986s

insulating material (1) (rotating machinery) (insulant) A substance or body, the conductivity of which is zero or, in practice, very small. *See also:* asynchronous machine; direct-current commutating machine. (PE) [9], [84]
(2) A material that cannot conduct electricity under normal conditions. *Contrast:* conducting material; semiconducting material. (C) 610.10-1994w

insulating-material classifications For the purpose of establishing temperature limits, insulating materials shall be classified as follows:

Class 90. Materials or combinations of materials such as cotton, silk, and paper without impregnation. Other materials, or combinations of materials may be included in this class if, by experience or accepted tests, they can be shown to be capable of operation at 90°C.

Class 105. Materials or combinations of materials such as cotton, silk and paper when suitably impregnated or coated or when immersed in a dielectric liquid such as oil. Other materials or combinations of materials may be included in this class if, by experience or accepted tests, they can be shown to be capable of operation at 105°C.

Class 130. Materials or combinations of materials such as mica, glass fiber, asbestos, etc., with suitable bonding substances. Other materials or combinations of materials, not necessarily inorganic, may be included in this class if, by experience or accepted tests, they can be shown to be capable of operation at 130°C.

Class 155. Materials or combinations of materials such as mica, glass fiber, asbestos, etc., with suitable bonding substances. Other materials or combinations of materials, not necessarily inorganic, may be included in this class if, by experience or accepted tests, they can be shown to be capable of operation at 155°C.

Class 180. Materials or combinations of materials such as silicone elastomer, mica, glass fiber, asbestos, etc., with suitable bonding substances such as appropriate silicone resins. Other materials or combinations of materials may be included in this class if, by experience or accepted tests, they can be shown to be capable of operation at 180°C.

Class 220. Materials or combinations of materials that by experience or accepted tests can be shown to be capable of operation at 220°C.

Over Class 220. Insulation that consists entirely of mica, porcelain, glass, quartz, and similar inorganic materials. Other materials or combinations of materials may be included in this class if, by experience or accepted tests, they can be shown to be capable of operation at temperatures over 220°C.

Notes: 1. Insulation is considered to be *impregnated* when a suitable substance provides a bond between components of the structure and also a degree of filling and surface coverage sufficient to give adequate performance under the extremes of temperature, surface contamination (moisture, dirt, etc.), and mechanical stress expected in service. The impregnant shall not flow or deteriorate enough at operating temperature so as to seriously affect performance in service. 2. The electrical and mechanical properties of the insulation shall not be *impaired* by the prolonged application of the limiting insulation temperature permitted for the specific insulation class. The word impaired is used here in the sense of causing any change that could disqualify the insulating material for continuously performing its intended function, whether it is creepage, spacing, mechanical support, or dielectric barrier action. 3. In the above descriptions of insulating materials classifications, the words *accepted tests* refer to recognized test procedures established for the thermal evaluation of materials by themselves or in simple combinations. Experience or test data, used in classifying insulating materials, are distinct from the experience or test data derived for the use of materials in complete insulation systems. The thermal endurance of complete systems may be determined by test procedures specified by the responsible technical committees. A material that is classified as suitable for a given temperature in the above tabulation may be found suitable for a different temperature other than the given one, either higher or lower, by an insulation system test procedure. For example, it has been found that some materials suitable for operation at one temperature in air may be suitable for a higher temperature when used in a system operated in an inert gas atmosphere. 4. It is important to recognize that other characteristics, in addition to thermal endurance, such as mechanical strength, moisture resistance, and corona endurance, are required in varying degrees in different applications for the successful use of insulating materials. (SWG/PE) C37.40-1993

insulating member The part of the substation that isolates the energized conductor from a grounded member or other energized conductors. (SUB/PE) 1264-1993

insulating spacer Insulating material used to separate parts. *See also:* stator; rotor. (PE) [9]

insulating tape *See:* insulating tool.

insulating tool (A) (power line maintenance) A tool or device designed primarily to provide insulation from an energized part or conductor. It can be composed entirely of insulating materials. Examples: conductor cover; stick; insulating tape. **(B) (power line maintenance)** A tool or device that has conductive parts separated by dielectric parts. (T&D/PE) 516-1995

insulating transformer A transformer used to insulate one circuit from another. *See also:* specialty transformer. (PE/IA/TR/PC) C57.12.80-1978r, 463-1993w, [57]

insulation (1) (rotating machinery) (electric systems) Material or a combination of suitable nonconducting materials that provide electric isolation of two parts at different voltages. (PE) [9]

(2) (power cable joints) A material of suitable dielectric properties, capable of being field-applied, and used to provide and maintain continuity of insulation across the splice. The material need not be identical to the cable insulation, but should be electrically and physically compatible, including factor-molded insulating components that are field-installed. (PE/IC) 404-1986s

(3) (high-voltage switchgear) A material having the property of an insulator used to separate parts of the same or different potential. (SWG/PE) C37.40-1993, C37.100-1992

(4) (as applied to cable) That which is relied upon to insulate the conductor from other conductors or conducting parts or from ground. (NESC/T&D/PE) C2-1997, [10]

insulation breakdown (electrical insulation tests) A rupture of insulation that results in a substantial transient or steady increase in leakage current at the specified test voltage. (AES/ENSY) 135-1969w

insulation breakdown current (electrical insulation tests) The current delivered from the test apparatus when a dielectric breakdown occurs. (AES/ENSY) 135-1969w

insulation class (1) (grounding device) A number that defines the insulation levels of the device. (PE/SPD) 32-1972r

(2) Divided into classes according to the thermal endurance of the system for temperature rating purposes. NEMA classes of insulation systems used in motors include Classes A, B, F and H. These classes have been established in accordance with IEEE Std 1-1986 (Reaff 1992). Other classes of insulation are constantly being developed for such use. Insulation systems shall be classified as follows:

1) *NEMA Class A*. An insulation system (105°C temperature limit including a 40°C ambient or 65°C rise) that by experience or accepted test can be shown to have suitable thermal endurance when operating at the limiting Class A temperature specified in the temperature rise standard for the machine under consideration.
2) *NEMA Class B*. An insulation system (130°C temperature limit including a 40°C ambient or 90°C rise) that by experience or accepted test can be shown to have suitable thermal endurance when operating at the limiting Class B temperature specified in the temperature rise standard for the machine under consideration.
3) *NEMA Class F*. An insulation system (155°C temperature limit including a 40°C ambient or 115°C rise) that by experience or accepted test can be shown to have suitable thermal endurance when operating at the limiting Class F temperature specified in the temperature rise standard for the machine under consideration.
4) *NEMA Class H*. An insulation system (180 °C temperature limit including a 40°C ambient or 140°C rise) that by experience or accepted test can be shown to have suitable thermal endurance when operating at the limiting Class H temperature specified in the temperature rise standard for the machine under consideration.

(IA/PC) 1068-1996

(3) *See also:* insulation level. (PE/TR) C57.12.80-1978r

(4) (outdoor apparatus bushings) The voltage by which the bushing is identified and which designates the level on which the electrical performance requirements are based. (PE/TR) 21-1976, C57.12.80-19782r

insulation configuration The complete geometric configuration of the insulation, including all elements (insulating and conducting) that influence its dielectric behavior. *See also:* phase-to-ground insulation configuration; longitudinal insulation configuration; phase-to-phase insulation configuration. (PE/C) 1313.1-1996

insulation coordination *See:* coordination of insulation.

insulation failure (1) The state of the insulation in which relevant physical, chemical, or electrical properties are altered sufficiently to cause operational failure or to jeopardize future failure-free operation under postulated conditions. (DEI/RE) 775-1993w
(2) (thyristor) The failure of a semiconductor or an insulator to support its rated voltage. *Synonym:* device breakdown. (IA/IPC) 428-1981w

insulation fault (surge arresters) Accidental reduction or disappearance of the insulation resistance between conductor and ground or between conductors. (PE) [8], [84]

insulation, graded *See:* graded insulation.

insulation level (1) (power and distribution transformers) An insulation strength expressed in terms of a withstand voltage. (PE/TR) C57.12.80-1978r
(2) The withstand values of the impulse and power frequency test voltages to ground, and where appropriate, between the phases, and between those parts where insulation is required. (PE/TR) C57.131-1995
(3) The combination of power frequency and impulse test voltage values that characterize the insulation of the capacitor bank with regard to its capability of withstanding the electric stresses between platform and earth, or between platform-mounted equipment and the platform. (T&D/PE) 824-1994
(4) A combination of voltage values (both power frequency and impulse) that characterize the insulation of an equipment with regard to its capability of withstanding dielectric stresses. (SPD/PE) C62.22-1997

insulation power factor (1) (rotating machinery) The ratio of dielectric loss in an insulation system to the applied apparent power, when measured at power frequency under designated conditions of voltage, temperature, and humidity. *Note:* Being the sine of an angle normally small, it is practically equal to loss tangent or dissipation factor, the tangent of the same angle. The angle is the complement of the angle whose cosine is the power factor. *See also:* loss tangent; asynchronous machine. (PE/TR) 21-1976, [9]
(2) (insulation) (outdoor apparatus bushings) The ratio of the power dissipated in the insulation, in watts, to the product of the effective voltage and current in voltamperes, when tested under a sinusoidal voltage and prescribed conditions. *Note:* The insulation power factor is equal to the cosine of the phase angle between the voltage and the resulting current when both the voltage and current are sinusoidal. (PE/TR) 21-1976, C57.19.03-1996, C57.12.80-1978r

insulation resistance (1) (aircraft, missile, and space equipment) The electrical resistance measured at specified direct-current potentials between any electrically insulated parts, such as a winding and other parts of the machine. (AES/ENSY) 135-1969w
(2) The capability of the electrical insulation of a winding to resist direct current. The quotient of applied direct voltage of negative polarity divided by current across machine insulation, corrected to 40°C, and taken at a specified time (t) from start of voltage application. The voltage application time is usually 1 min (IR_1) or 10 min (IR_{10}), however, other values can be used. Unit conventions: values of 1 through 10 are assumed to be in minutes, values of 15 and greater are assumed to be in seconds. (PE/EM) 43-2000
(3) The resistance, measured at a specified dc voltage, between any specified exposed conductive surface or line terminal (including ground) and one or more of the other line terminals of the device. (SPD/PE) C62.62-2000

insulation-resistance test A test for measuring the resistance of insulation under specified conditions. (PE) [9]

insulation-resistance versus voltage test (rotating machinery) A series of insulation-resistance measurements, made at increasing direct voltages applied at successive intervals and maintained for designated periods of time, with the object of detecting insulation system defects by departures of the measured characteristic from a typical form. Usually, this is a controlled overvoltage test. *See also:* asynchronous machine. (PE) [9]

insulation shielding (1) (power distribution) Conducting and/or semiconducting elements applied directly over and in intimate contact with the outer surface of the insulation. Its function is to eliminate ionizable voids at the surface of the insulation and confine the dielectric stress to the underlying insulation. (PE) [4]
(2) (cable systems) A nonmagnetic, metallic material applied over the insulation of the conductor or conductors, to confine the electric field of the cable to the insulation of the conductor or conductors. (PE/EDPG) 422-1977
(3) An envelope that encloses the insulation of a cable and provides an equipotential surface in contact with the cable insulation. (NESC) C2-1997

insulation sleeving (tubing) A varnish-treated or resin-coated flexible braided tube providing insulation when placed over conductors, usually at connections or crossovers. (PE) [9]

insulation slot separator (rotating machinery) Insulation member placed in a slot between individual coils, such as between main and auxiliary windings. *See also:* rotor; stator. (IA/APP) [90]

insulation system (1) (power and distribution transformers) An assembly of insulating materials in a particular type, and sometimes size, of equipment. (PE/TR) C57.12.80-1978r
(2) An assembly of insulation materials. For definition purposes, the insulation systems of synchronous machine windings (either field or armature) are divided into three components. These components are the coil insulation with its accessories, the connection and winding support insulation, and the associated structural parts. (REM) [115]
(3) *Class A.* A system utilizing materials having a preferred temperature index of 105 and operating at such temperature rises above stated ambient temperature as the equipment standard specifies based on experience or accepted test data. This system may alternatively contain materials of any class, provided that experience or a recognized system test procedure for the equipment has demonstrated equivalent life expectancy. The preferred temperature classification for a Class A insulation system is 105°C. *Class B.* A system utilizing materials having a preferred temperature index of 130 and operating at such temperature rises above stated ambient temperature as the equipment standard specifies based on experience or accepted test data. This system may alternatively contain materials of any class, provided that experience or a recognized system test procedure for the equipment has demonstrated equivalent life expectancy. The preferred temperature classification for a Class B insulation system is 130°C. *Class C.* A system utilizing materials having a preferred temperature index of over 240 and operating at such temperatures above stated ambient temperatures as the equipment standard specifies based on experience or accepted test data. This system may alternatively contain materials of any class, provided that experience or a recognized test procedure for the equipment has demonstrated equivalent life expectancy. The preferred temperature classification for a Class C insulation system is over 240°C. *Class F.* A system utilizing materials having a preferred temperature index of 155 and operating at such temperature rises above stated ambient temperatures as the equipment standard specifies based on experience or accepted test data. This system may alternatively contain materials of any class, provided that experience or a recognized test procedure for the equipment has demonstrated equivalent life expectancy. The preferred temperature classification for a Class F insulation system is 155°C. *Class H.* A system utilizing materials having a preferred temperature index of 180 and operating at such temperature rises above

stated ambient temperature as the equipment standard specifies based on experience or accepted test data. This system may alternatively contain materials of any class, provided that experience or a recognized test procedure for the equipment has demonstrated equivalent life expectancy. The preferred temperature classification for a Class H insulation system is 180°C. *Class N.* A system utilizing materials having a preferred temperature index of 200 and operating at such temperature rises above stated ambient temperatures as the equipment standard specifies based on experience or accepted test data. This system may alternatively contain materials of any class, provided that experience or a recognized test procedure for the equipment has demonstrated equivalent li fe expectancy. The preferred temperature classification for a Class N insulation system is 200°C. *Class R.* A system utilizing materials have a preferred temperature index of 220 and operating at such temperatures above stated ambient temperatures as the equipment standard specifies based on experience or accepted test data. This system may alternatively contain materials of any class, provided that experience or a recognized test procedure for the equipment has demonstrated equivalent life expectancy. The preferred temperature classification for a Class R insulation system is 220°C. *Class S.* A system utilizing materials having a preferred temperature index of 240 and operating at such temperatures above stated ambient temperatures as the equipment standard specifies based on experience or accepted test data. This system may alternatively contain materials of any class, provided that experience or a recognized test procedure for the equipment has demonstrated equivalent life expectancy. The preferred temperature classification for a Class S insulation system is 240°C. (PE/EM) 117-1974r

(4) An assembly of one or more dielectric materials used in a functional mode in the design of an electrical component or equipment, including adjacent components that may influence aging. (DEI/RE) 775-1993w

(5) An assembly of insulating materials in association with the conductors and the supporting structural parts. All of the components described below that are associated with the stationary winding constitute one insulation system, and all of the components that are associated with the rotating winding constitute another insulation system.

— *Coil insulation with its accessories.* All of the insulating materials that envelop and separate the current-carrying conductors and their component turns and strands, and form the insulation between them and the machine structure; includes wire coating, varnish, encapsulants, slot insulation, slot fillers, tapes, phase insulation, polebody insulation, and retaining ring insulation when present.

— *Connection and winding support insulation.* All of the insulation materials that envelop the connections that carry current from coil to coil, and from stationary or rotating coil terminals to the points of external circuit attachment; and the insulation of any metallic supports for the winding.

— *Associated structural parts (insulation system).* Items such as slot wedges, space blocks, and ties that are used to position the coil ends and connections; any nonmetallic supports for the winding; and field-coil flanges.

— *Insulation class.* Divided into classes according to the thermal endurance of the system for temperature rating purposes. NEMA classes of insulation systems used in motors include Classes A, B, F, and H. These classes have been established in accordance with IEEE Std 1-1986. Other classes of insulation are constantly being developed for such use.

Insulation systems shall be classified as follows:

1) *NEMA Class A.* An insulation system (105°C temperature limit including a 40°C ambient or 65°C rise) that by experience or accepted test can be shown to have suitable thermal endurance when operating at the limiting Class A temperature specified in the temperature rise standard for the machine under consideration.

2) *NEMA Class B.* An insulation system (130°C temperature limit including a 40°C ambient or 90°C rise) that by experience or accepted test can be shown to have suitable thermal endurance when operating at the limiting Class B temperature specified in the temperature rise standard for the machine under consideration.

3) *NEMA Class F.* An insulation system (155°C temperature limit including a 40°C ambient or 115°C rise) that by experience or accepted test can be shown to have suitable thermal endurance when operating at the limiting Class F temperature specified in the temperature rise standard for the machine under consideration.

4) *NEMA Class H.* An insulation system (180°C temperature limit including a 40°C ambient or 140°C rise) that by experience or accepted test can be shown to have suitable thermal endurance when operating at the limiting Class H temperature specified in the temperature rise standard for the machine under consideration.

(IA/PC) 1068-1996

(6) A system composed of solid insulating materials and insulating fluid. *Synonym:* system. (PE/TR) 1276-1997

(7) An assembly of insulating materials in a particular type of equipment. The class of the insulation system may be designated by letters, numbers, or other symbols. An insulation system class utilizes material having an appropriate temperature index and operates at such temperatures above stated ambient temperatures as the equipment standard specifies based on experience or accepted test data. The system may alternatively contain materials of any class, provided that experience or a recognized test procedure for the equipment has demonstrated equivalent life expectancy. (IA/MT) 45-1998

insulation, thermal *See:* thermal insulation.

insulator (1) A device intended to give flexible or rigid support to electrical conductors or equipment and to insulate these conductors or equipment from ground or from other conductors or equipment. An insulator comprises one or more insulating parts to which connecting devices (metal fittings) are often permanently attached. (SWG/PE) C37.100-1981s

(2) Insulating material in a form designed to support a conductor physically and electrically separate it from another conductor or object. (NESC/T&D) C2-1997, C2.2-1960

(3) A device made of a material in which electrons or ions cannot be moved easily, hence preventing the flow of electric current. (C) 610.10-1994w

insulator arcing horn A metal part, usually shaped like a horn, placed at one or both ends of an insulator or of a string of insulators to establish an arcover path, thereby reducing or eliminating damage by arcover to the insulator or conductor or both. *See also:* tower. (T&D/PE) [10]

insulator arcing ring A metal part, usually circular or oval in shape, placed at one or both ends of an insulator or of a string of insulators to establish an arcover path, thereby reducing or eliminating damage by arcover to the insulator or conductor or both. *See also:* tower. (T&D/PE) [10]

insulator arcing shield (insulator grading shield) An arcing ring so shaped and located as to improve the voltage distribution across or along the insulator or insulator string. *See also:* tower. (T&D/PE) [10]

insulator arcover A discharge of power current in the form of an arc, following a surface discharge over an insulator. *See also:* tower. (T&D/PE) [10]

insulator cover Electrical protection equipment designed specifically to cover insulators. Examples: dead-end cover, poletop cover, ridge-pin cover *Synonyms:* hood; pocketbook. *See also:* cover-up equipment. (T&D/PE) 516-1995

insulator grading shield *See:* insulator arcing shield.

insulator lifter A device designed to permit insulators to be lifted in a *string* to their intended position on a structure. *Synonyms:* insulator saddle; potty seat. (T&D/PE) 524-1992r

insulator saddle *See:* insulator lifter.

insulator stack A rigid assembly of two or more switch and bus insulating units. *See also:* tower. (PE/T&D/EEC/IEPL) [10], [89]

insulator string Two or more suspension insulators connected in series. *See also:* tower. (T&D/PE) [10]

insulator unit An insulator assembled with such metal parts as may be necessary for attaching it to other insulating units or device parts. (SWG/PE) C37.40-1993, C37.100-1992

inta *See:* integer array.

intake opening (rotating machinery) A port for the entrance of ventilation air. (PE) [9]

intake port (rotating machinery) An opening provided for the entrance of a fluid. (PE) [9]

integer (1) (microprocessor operating systems parameter types) A whole number that may be positive, negative, or zero and has a range of at least -32767 to $+32767$. The qualifier "long" is used to qualify the size the size of an integer. A long integer has a range of at least $(-2^{**}31 + 1)$ to $(+2^{**}31 - 1)$. (C/MM) 855-1985s
(2) (mathematics of computing) A positive or negative whole number, including zero. (C) 1084-1986w

integer adjectives (pulse terminology) The ordinal integers (that is, first, second, . . . *n*th, last) or the cardinal integers (that is, 1, 2, . . . *n*) may be used as adjectives to identify or distinguish between similar or identical features. The assignment of integer modifiers should be sequential as a function of time within a waveform epoch or within features thereof. (IM/WM&A) 194-1977w

integer arithmetic Fixed-point arithmetic in which the radix point is assumed to lie immediately to the right of the least significant digit in each numeral; that is, all numbers are assumed to be integers. (C) 1084-1986w

integer array (inta) (subroutines for CAMAC) The symbol inta represents an integer array, the length and contents of which are not defined in this standard. It is intended to contain system-dependent or implementation-dependent information associated with the definition of a LAM. If no such information is required, the array need not be used. This information can include parameters necessary for interrupt linkage, event specification, etc. The documentation for an implementation must describe the requirements for any parameters contained in this array. *Synonym:* inta. (NPS) 758-1979r

integer character string A decimal character string, an octal character string, or a hexadecimal character string. (C/PA) 1387.2-1995

integer data Numeric data used to represent whole numbers; that is, numeric values without fractional parts. For example, 0, +1, −1, +2, −2, ... *See also:* zoned decimal data; fixed-point data; packed decimal data; unsigned packed decimal data. (C) 610.5-1990w

integer programming In operations research, a class of procedures for locating the maximum or minimum of a function under given constraints, one of which is that some or all variables must have integer values. *Synonym:* discrete programming. (C) 610.2-1987

integer type A data type whose members can assume only integer values and can be operated on only by integer arithmetic operations, such as addition, subtraction, and multiplication. *Contrast:* character type; enumeration type; real type; logical type. (C) 610.12-1990

integer unit A processing unit that performs integer and control-flow operations and contains general-purpose integer registers and processor state registers, as defined by this standard. (C/MM) 1754-1994

integer variable A variable that may assume only integer (non-fractional) values. (C) 1084-1986w

integral action rate (proportional plus integral control action devices) (process control) For a step input, the ratio of the initial rate of change of output due to integral control action to the change in steady-state output due to proportional control action. *Note:* Integral action rate is often expressed as the number of repeats per minute because it is equal to the number of times per minute that the proportional response to a step input is repeated by the initial integral response. (PE/EDPG) [3]

Integral Activity Group An Activity Group that is needed to complete project Activities, but is outside the management and development Activity Groups. (C/SE) 1074-1997

integral bushing An apparatus bushing designed for use with another connector component, such as an elbow. (T&D/PE) 386-1995

integral control action (1) Control action in which the output is proportional to the time integral of the input. *See also:* control system, feedback. (PE/IA/ICTL/PSE/IAC) 94-1970w, [60]
(2) Control action in which the output is proportional to the time integral of the input, that is the rate of change of output is proportional to the input. *Note:* In the practical embodiment of integral control action the relation between output and input, neglecting high frequency terms, is given by

$$\frac{Y}{X} = \pm \frac{I/s}{\frac{bI}{s} + 1} \quad 0 \le b \ll 1$$

where

b = reciprocal of static gain
$I/2\pi$ = gain cross-over frequency in cycles per unit time
s = complex variable
X = input transform
Y = output transform

(CS/PE/EDPG) [3]
(3) A control mode is designated to provide a controlled output proportional to the time integral of the input. (PE/PSE) 94-1991w

integral coupling (rotating machinery) A coupling flange that is a part of a shaft and not a separate piece. *See also:* rotor. (PE) [9]

integral-horsepower motor A motor built in a frame as large as or larger than that of a motor of open construction having a continuous rating of one horsepower at 1700−1800 revolutions per minute. *See also:* asynchronous machine. (EEC/PE) [119]

integral nonlinearity (1) (A) (x-ray energy spectrometers) (of a pulse amplifying system) The maximum nonlinearity (deviation) over the specified operating range of a system, usually expressed as a percentage of the maximum of the specified range. **(B) (percent) (semiconductor radiation detectors)** The departure from linear response expressed as a percentage of the maximum rated output pulse amplitude. (NPS/NID) 759-1984, 301-1976
(2) The maximum difference between the ideal and actual code transition levels after correcting for gain and offset. (IM/WM&A) 1057-1994w

integral process A process that is needed to successfully complete project Activities, but is outside the management and development processes. The integral processes are verification and validation, software configuration management, documentation development, and training. (C/SE) 1074.1-1995

integral-slot winding (rotating machinery) A distributed winding in which the number of slots per pole per phase is an integer and is the same for all poles. *See also:* stator; rotor. (PE) [9]

integral switch and fuse A switch and fuse assembly mounted on the same frame. (SWG/PE) C37.20.4-1996

integral test equipment *See:* self-test.

integral time (speed governing of hydraulic turbines) The integral time, T_X, of an integrating element is the time required for the element's percent output to be equal in magnitude to the element's percent input where that input is a step function. The integral gain of an element is the reciprocal of its integral time.

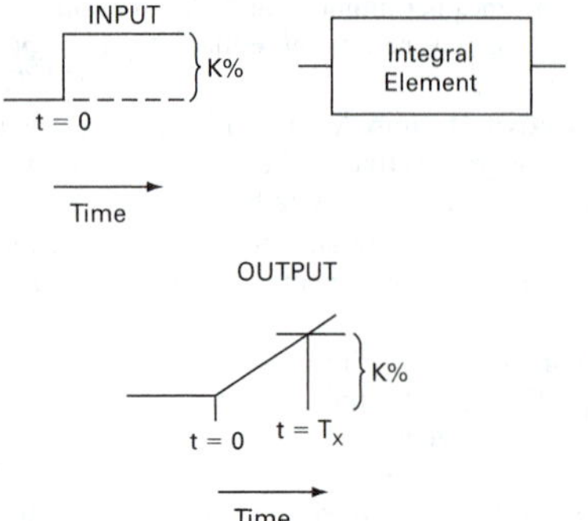

integral time

(PE/EDPG) 125-1977s

integral unit substation A unit substation in which the incoming, transforming, and outgoing sections are manufactured as a single compact unit. (SWG/PE/TR) C37.100-1992, C57.12.80-1978r

integrated (A) (germanium gamma-ray detectors) (pulse amplifier) (x-ray energy spectrometers) (charged-particle detectors) (pulse) A pulse is integrated when it is passed through a low-pass network, such as a single resistance-capacitance (RC) network or a cascaded RC network. **(B) (pulse) (pulse amplifier)** A pulse that is passed through a low-pass network, such as a single *RC* network or a cascaded *RC* network. (NPS/NID) 759-1984, 325-1986, 300-1988

integrated alarm system (alarm monitoring and reporting systems for fossil-fueled power generating stations) An alarm display system consisting of window annunciators combined with cathode-ray tube (CRT), printer, or mimic display. (PE/EDPG) 676-1986w

integrated antenna system A radiator with an active or nonlinear circuit element or network incorporated physically within the structure of the radiator. (AP/ANT) 145-1993

integrated circuit (IC) (solid state) A combination of interconnected circuit elements inseparably associated on or within a continuous substrate. *Note:* To further define the nature of an integrated circuit, additional modifiers may be prefixed. Examples are: 1) dielectric-isolated monolithic integrated circuit, 2) beamlead monolithic integrated circuit, 3) silicon-chip tantalum thin-film hybrid integrated circuit. *See also:* chip. (ED) 274-1966w, [46], 1005-1998 **(2) (A)** A combination of connected circuit elements (such as transistors, diodes, resistors, and capacitors) inseparably associated on or within a continuous substrate. **(B)** A solid-state circuit consisting of interconnected active and passive semiconductor devices diffused into a single silicon chip. *Synonyms:* chip; microcircuit. *See also:* monolithic integrated circuit; very-high-speed integrated circuit. (ED/C) [46], 610.10-1994

Integrated Civil Engineering System (ICES) A general-purpose software system including several programming languages, such as COGO and STRUDL, and subsystems that are designed for use in civil engineering and engineering management. (C) 610.13-1993w

integrated data dictionary A data dictionary that is functionally involved in data accesses, performing required checks for value limits and data types and disallowing illegal modifications to data elements within the system that is described. (C) 610.5-1990w

integrated database A repository for storing all information pertinent to the systems engineering process to include all data, schema, models, tools, technical management decisions, process analysis information, requirement changes, process and product metrics, and trade-offs. (C/SE) 1220-1998

integrated data package The evolving output of the systems engineering process that documents hardware, software, life cycle processes, and human engineering designs. (C/SE) 1220-1998

integrated data processing (IDP) The use of computers to coordinate a number of processes and improve overall efficiency by reducing or eliminating redundant data entry or processing operations. (C) 610.2-1987

integrated demand (1) The demand integrated over a specified period divided by that period. (PE/PSE) 858-1993w **(2) (electric power utilization)** The demand integrated over a specified period. (PE/PSE) 346-1973w

integrated-demand meter (block-interval demand meter) A meter that indicates or records the demand obtained through integration. *See also:* demand meter; electricity meter. (ELM) C12.1-1982s

integrated device electronics A data-transfer interface in which the control electronics for the disk drive are physically located on the drive itself rather than on an expansion board or drive adapter. *Synonym:* integrated drive electronics. (C) 610.10-1994w

integrated diagnostics A process that covers the entire spectrum of diagnostic activities in all phases of weapon system acquisition. (ATLAS) 1226-1993s

integrated drive electronics *See:* integrated device electronics.

integrated electronics The portion of electronic art and technology in which the interdependence of material, device, circuit, and system-design consideration is especially significant: more specifically, that portion of the art dealing with integrated circuits. *See also:* integrated circuit. (ED) 274-1966w, [46]

integrated energy curve (power operations) A curve of demand versus energy showing the amount of energy represented under a load curve, or a load duration curve, above any point of demand. *Synonym:* peak percent curve. *See also:* generating station. (PE/PSE) 858-1987s, 346-1973w

integrated heating system A complete system consisting of components such as pipelines, vessels, heating elements, heat transfer medium, thermal insulation, moisture barrier, non-heating leads, temperature controller, safety signs, junction boxes, conduit and fittings. (NESC/NEC) [86]

integrated injection logic A family of circuit logic in which the logic state is defined by current flow rather than by voltage level. (C) 610.10-1994w

integrated mica *See:* mica paper.

integrated microprocessor One or more large scale integration devices so interconnected as to provide all of the functions of a central processing unit within a single LSI circuit. *Note:* This use of the term is deprecated typically "microprocessor" is used. *See also:* horizontally integrated microprocessor; diagonally integrated microprocessor; vertically integrated microprocessor. (C) 610.10-1994w

integrated-numbering plan (telephone switching systems) In the world-numbering plan, arrangements for identifying telephone stations within a geographical area identified by a world-zone number which is also used as a country code. *See also:* world-zone number. (COM) 312-1977w

integrated numbering-plan area (telephone switching systems) A geographical area of the world that is identified by a world-zone number which is also used as a country code. *See also:* world-zone number. (COM) 312-1977w

integrated optical circuit (IOC) (fiber optics) An optical circuit, either monolithic or hybrid, composed of active and passive components, used for coupling between optoelectronic devices and providing signal processing functions. (Std100) 812-1984w

integrated plow (static or vibratory plows) (cable plowing) A self-contained or integral plow-prime mover unit.

integrated precipitable water vapor The equivalent liquid water height (in centimeters) of a vertical column of water vapor in the atmosphere with 1 cm^2 horizontal cross-section. (AP/PROP) 211-1997

integrated pulse In the absence of qualifiers (such as "true integration"), a pulse that has passed through one or more low-pass networks ("integrators"). (NPS) 325-1996

integrated radiation (laser maser) The integral of the radiance over the exposure duration. (LEO) 586-1980w

integrated services terminal equipment (ISTE) A device that serves as an information source and/or an information sink for the provision of voice, facsimile, video, data, and other information. (C/LM/COM) 802.9a-1995w, 8802-9-1996

integrating (block interval) demand meter (metering) A meter that integrates power or a related quantity over a fixed time interval, and indicates or records the average. (ELM) C12.1-1982s

integrating accelerometer A device that produces an output that is proportional to the time integral of an input acceleration. (AES/GYAC) 528-1994

integrating amplifier (1) (analog computer) An operational amplifier that produces an output signal equal to the time integral of a weighted sum of the input signals. *Note:* In an analog computer, the term integrator is synonymous with integrating amplifier. (C) 165-1977w, 166-1977w
(2) In an analog computer, an operational amplifier that produces an output signal equal to the time integral of a weighted sum of the input signals. (C) 610.10-1994w

integrating circuit *See:* integrator.

integrating motor In an analog computer, a motor designed to give a constant ratio of output shaft rotational speed to input signal. *Note:* The angle of rotation of the shaft with respect to a datum is thus proportional to the time integral of the applied signal. (C) 610.10-1994w

integrating network *See:* integrator.

integrating photometer (illuminating engineering) One which enables total luminous flux to be determined by a single measurement. The usual type is the Ulbricht sphere with associated photometric equipment for measuring the indirect illuminance of the inner surface of the sphere. (The measuring device is shielded from the source under measurement.) (EEC/IE) [126]

integrating preamplifier (germanium gamma-ray detectors) (x-ray energy spectrometers) A pulse preamplifier in which individual pulses are intentionally integrated by passive or active circuits. (NPS/NID) 325-1986s, 759-1984r

integrating relay A relay that operates on the energy stored from a long pulse or a series of pulses of the same or varying magnitude, for example, a thermal relay. *See also:* relay.

integration (1) (software) The process of combining software components, hardware components, or both into an overall system. (C) 610.12-1990
(2) The merger or combining of two or more lower-level elements into a functioning and unified higher-level element with the functional and physical interfaces satisfied. (C/SE) 1220-1994s
(3) Providing a single set of data for use between multiple activities or departments (multiple application data base). (PE/EDPG) 1150-1991w
(4) (of radar signals) The combination by addition (or the logical equivalent) of echo pulses or signal samples obtained by a radar as it illuminates a target so as to increase the output signal-to-noise ratio beyond that available from a single pulse or sample. (AES) 686-1997
(5) The service that enables delivery of medium access control (MAC) service data units (MSDUs) between the distribution system (DS) and an existing, non-IEEE 802.11 local area network (via a portal). (C/LM) 8802-11-1999

integration loss The loss incurred by integrating a signal noncoherently (postdetection) instead of coherently. (AES) 686-1997

integration test (1) The testing of several subsystems that perform in combination. (PE/SUB) 1378-1997
(2) (programmable digital computer systems in safety systems of nuclear power generating stations) Test(s) performed during the hardware-software integration process prior to computer system validation to verify compatibility of the software and the computer system hardware. 554-1990

integration testing (1) (software) Testing in which software components, hardware components, or both are combined and tested to evaluate the interaction between them. *See also:* component testing; unit testing; system testing; interface testing. (C) 610.12-1990
(2) An orderly progression of testing of incremental pieces of the software program in which software elements, hardware elements, or both are combined and tested until the entire system has been integrated to show compliance with the program design, and capabilities and requirements of the system. (C/SE) 1012-1998

integrator (1) (analog computer) A device producing an output proportional to the integral of one variable or of a sum of variables, with respect to another variable, usually time. *See also:* integrating amplifier. (C) 165-1977w
(2) (digital differential analyzer) A device using an accumulator for numerically accomplishing an approximation to the mathematical process of integration. (C) 162-1963w
(3) A functional unit whose output analog variable is the integral of an input analog variable with respect to time, or a variable other than time. *See also:* incremental integrator; storage integrator. (C) 610.10-1994w
(4) A device having an output proportional to the time integral of the input signal. (COM/TA) 1027-1996
(5) (as applied to relaying) A transducer whose output wave form is substantially the time integral of its input wave form. (SWG/PE) C37.100-1992

integrator, gain *See:* gain integrator.

integrity (1) (software) (data management) The degree to which a system or component prevents unauthorized access to, or modification of, computer programs or data. *See also:* database integrity; data integrity. (C) 610.5-1990w, 610.12-1990
(2) The condition of being unimpaired. (C/BA) 896.9-1994w

integrity check value (ICV) A value that is derived by performing an algorithmic transformation on the data unit for which data integrity services are provided. The ICV is sent with the protected data unit and is recalculated and compared by the receiver to detect data modification. (C/LM) 802.10-1998

INTELLECT A natural language front-end processor for an SQL database manipulation language. (C) 610.13-1993w

intellectual property An output of creative human thought process that has some intellectual or informational value. Intellectual property can be protected by patents, copyrights, trademarks, or trade secrets. (C/SE) 1420.1b-1999

intelligence bandwidth The sum of the audio- (or video-) frequency bandwidths of the one or more channels. (AP/ANT) 145-1983s

intelligent data model A data model that describes the logic, controls, and constraints that should be applied whenever the data are accessed. (C) 610.5-1990w

intelligent electronic device (IED) Any device incorporating one or more processors with the capability to receive or send data/control from or to an external source (e.g., electronic multifunction meters, digital relays, controllers). (SUB/PE) C37.1-1994

intelligent terminal A terminal that can send and receive information as well as perform some processing, such as making decisions or performing calculations, independent of the computer. *Synonym:* programmable terminal. *Contrast:* dumb terminal. *See also:* smart terminal. (C) 610.10-1994w

intended polarization The polarization of the radio wave for which an antenna system is designed. (AES) 686-1997

intensifier (non-real time spectrum analyzer) (baseline clipper) A means of changing the relative brightness between the signal and baseline portion of the display. (IM) [14]

intensifier electrode An electrode causing post acceleration. *See also:* electrode; post-accelerating (deflection) electrode. (ED) 161-1971w

intensity (1) (fiber optics) The square of the electric field amplitude of a light wave. Intensity is proportional to irradiance and may be used in place of the term "irradiance" when only relative values are important. *See also:* irradiance; radiant intensity; radiometry. (Std100) 812-1984w
(2) (oscilloscopes) A term used to designate brightness or luminance of the spot. *See also:* oscillograph. (IM/HFIM) [40]

intensity amplifier (oscilloscopes) An amplifier for signals controlling the intensity of the spot. *See also:* oscillograph. (IM/HFIM) [40]

intensity cueing *See:* depth cueing.

intensity (candlepower) distribution curve (illuminating engineering) A curve, often polar, that represents the variation of luminous intensity of a lamp or luminaire in a plane through the light center. *Note:* A vertical candlepower distribution curve is obtained by taking measurements at various angles of elevation in a vertical plane through the light center; unless the plane is specified, the vertical curve is assumed to represent an average such as would be obtained by rotating the lamp or luminaire about its vertical axis. A horizontal intensity distribution curve represents measurements made at various angles of azimuth in a horizontal plane through the light center. (EEC/IE) [126]

intensity level (specific sound-energy flux level) (sound-energy flux density level) In decibels, of a sound is 10 times the logarithm to the base 10 of the ratio of the intensity of this sound to the reference intensity. The reference intensity shall be stated explicitly. *Note:* In discussing sound measurements made with pressure or velocity microphones, especially in enclosures involving normal modes of vibration or in sound fields containing standing waves, caution must be observed in using the terms intensity and intensity level. Under such conditions it is more desirable to use the terms pressure level or velocity level since the relationship between the intensity and the pressure or velocity is generally unknown. (SP/ACO) [32]

intensity modulation (1) (general) The process, or effect, of varying the electron-beam current in a cathode-ray tube resulting in varying brightness or luminance of the trace. *See also:* television; oscillograph. (IM/HFIM) [40]
(2) (radar) A process used in certain displays whereby the luminance of the signal indication is a function of the received signal strength. (AES) 686-1997, [42]

intensity of magnetization *See:* magnetization.

intentional disconnect A primitive passed to the upper layers of both BCCs and DCCs to allow the device or system to recognize that an upcoming network disconnection is intentional. The method to trigger this primitive is left to the device or system designer. (EMB/MIB) 1073.3.1-1994

intentional radiator A device that intentionally generates and emits radio-frequency energy by radiation or induction. (EMC) C63.4-1991

interaction (nuclear power generating station) A direct or indirect effect of one device or system upon another. (PE/NP) 577-1976r

interaction-circuit phase velocity (traveling-wave tubes) The phase velocity of a wave traveling on the circuit in the absence of electron flow. *See also:* magnetron; electron device. (ED) [45]

interaction crosstalk coupling (between a disturbing and a disturbed circuit in any given section) The vector summation of all possible combinations of crosstalk coupling, within one arbitrary short length, between the disturbing circuit and all circuits other than the disturbed circuit (including phantom and ground-return circuits) with crosstalk coupling, within another arbitrary short length, between the disturbed circuit and all circuits other than the disturbing circuit. *See also:* coupling. (EEC/PE) [119]

interaction factor (1) (transducer) The factor in the equation for the received current that takes into consideration the effect of multiple reflections at its terminals. *Note:* For a transducer having a transfer constant θ, image impedances Z_{I_1} and Z_{I_2}, and terminating impedances Z_S and Z_R, this factor is

$$\frac{1}{1-\dfrac{Z_{I_2}-Z_R}{Z_{I_2}+Z_R}\times\dfrac{Z_{I_1}-Z_S}{Z_{I_1}+Z_S}\times e^{-2\theta}}$$

(2) (electrothermic power meters) The ratio of power incident from an rf source to the power delivered by the source to a nonreflecting load: mathematically, $|1 - \Gamma_g|^2$ where Γ_g is the complex reflection coefficient of the source. (IM) 544-1975w
(3) A factor in the equation for the insertion voltage ratio that takes into account the impedance mismatch variation at one end of the network due to an impedance mismatch at the other end. The factor is written in terms of the source and load impedance, the image impedances and the image transfer function of the four-terminal network. (CAS) [13]

interaction gap An interaction space between electrodes. (ED) 161-1971w

interaction impedance (traveling-wave tubes) A measure of the radio-frequency field strength at the electron stream for a given power in the interaction circuit. It may be expressed by the following equation

$$K=\frac{E^2}{2(\omega/v)^2P}$$

where E is the peak value of the electric field at the position of electron flow, ω is the angular frequency, v is the interaction-circuit phase velocity, and P is the propagating power. If the field strength is not uniform over the beam, an effective interaction impedance may be defined. (ED) 161-1971w

interaction loss (transducer) The interaction loss expressed in decibels is 20 times the logarithm to base 10 of the scalar value of the reciprocal of the interaction factor. *See also:* attenuation.

interaction space (traveling-wave tubes) A region of an electron tube in which electrons interact with an alternating electromagnetic field. (ED) 161-1971w

interactive (1) (software) Pertaining to a system or mode of operation in which each user entry causes a response from or action by the system. *Contrast:* batch. *See also:* real time; conversational; online. (C) 610.2-1987, 610.12-1990, 610.10-1994w
(2) The behavior of a utility or control_script which requires input from the user during its execution. (C/PA) 1387.2-1995

interactive compiler *See:* incremental compiler.

interactive electrical systems Two or more interconnected and compatible electrical systems with appropriate protection and measuring provisions at their interconnection point(s). (SUB/PE) 1109-1990w

interactive graphics A method of operation of a computer graphics system where the graphics system requests and accepts input from a user then allows the user to dynamically control the processing operation. *Contrast:* passive graphics. (C) 610.6-1991w

interactive language A nonprocedural language in which a program is created as a result of interactive dialog between the user and the computer system. The system provides questions, forms, and so on, to aid the user in expressing the results to be achieved. *See also:* command language; declarative language; rule-based language. (C) 610.12-1990, 610.13-1993w

interactive plotting The use of a display device to view the output of a graphic or computational process. Applications include computer-assisted instruction, computer-aided design, and control operations. (C) 610.2-1987

interactive shell A processing mode of the shell that is suitable for direct user interaction. (C/PA) 9945-2-1993

interblock gap An area between two consecutive blocks. *Synonyms:* record gap; block gap. (C) 610.5-1990w, 610.10-1994w

intercalated tapes (insulation) Two or more tapes, generally of different composition, applied simultaneously in such a manner that a portion of each tape overlies a portion of the other tape. (T&D/PE) [10]

intercardinal plane Any plane that contains the intersection of two successive cardinal planes and is at an intermediate angular position. *Note:* In practice, the intercardinal planes are located by dividing the angle between successive cardinal planes into equal parts. Often, it is sufficient to bisect the angle so that there is only one intercardinal plane between successive cardinal planes. (AP/ANT) 145-1993

intercarrier sound system A television receiving system in which use of the picture carrier and the associated sound-channel carrier produces an intermediate frequency equal to the difference between the two carrier frequencies. *Note:* This intermediate frequency is frequency modulated in accordance with the sound signal. *See also:* television. (EEC/PE) [119]

intercept call (telephone switching systems) A call to a line or an unassigned code that reaches an operator, a recorded announcement, or a vacant-code tone. (COM) 312-1977w

intercept trunk (telephone switching systems) A central office termination that may be reached by a call to a vacant number, changed number, or line out-of-order. (COM) 312-1977w

interchange Energy transferred from one power system to another. (PE/PSE) 858-1993w

interchangeable Said of two modules that, although possibly of different design, perform identical functions and have identical interface characteristics. (TT/C) 1149.5-1995

interchangeable bushing (outdoor apparatus bushings) A bushing designed for use in both power transformers and circuit breakers. (PE/TR) 21-1976

interchange circuit (data transmission) The length of cable used for signaling between the digital subset and the customer's equipment. (PE) 599-1985w

interchange energy (power operations) Energy delivered to or received by one electric system from another. (PE/PSE) 858-1987s, 346-1973w

interchannel interference (modulation systems) In a given channel, the interference resulting from signals in one or more other channels. (Std100) 270-1964w

intercharacter gap The space between the last bar of one character and the first bar of the next character which separates the two adjacent characters. (Also called intercharacter space). (PE/TR) C57.12.35-1996

intercharacter spacing In text formatting, the amount of space left between characters on a line. *Synonym:* letter spacing. *Contrast:* interword spacing. *See also:* kerning; incremental justification. (C) 610.2-1987

interclutter visibility The ability of a radar to detect moving targets that occur in resolution cells among patches of strong clutter; usually applied to moving-target indication (MTI) or pulsed-Doppler radars. *Note:* The higher the radar range and/or angle resolution, the better the interclutter visibility. (AES) 686-1997

intercom (interphone) The interference resulting from signals in one or more other channels. *See also:* intercommunicating system. (Std100) 270-1964w

intercommunicating system A privately owned two-way communication system without a central switchboard, usually limited to a single vehicle, building, or plant area. Stations may or may not be equipped to originate a call but can answer any call. *Synonyms:* intercom; interphone. (EEC/PE) [119]

interconnect **(1)** A collective term for structures (in an integrated circuit) that propagate a signal between the pins of cell instances with as little change as possible. These structures include metal and polysilicon segments, vias, fuses, antifuses, etc. Interconnect shall not include such structures if they occur as part of the fixed layout of a cell. (C/DA) 1481-1999
(2) The system of wiring that carries data and control signals between the different components mounted on a printed circuit assembly. *See also:* simple interconnect; extended interconnect; differential interconnect. (C/TT) 1149.4-1999

interconnected delta connection (power and distribution transformers) A three-phase connection using six windings (two per phase) connected in a six-sided circuit with six bushings to provide a fixed phase-shift between two three-phase circuits without change in voltage magnitude. *Note:* The interconnected delta connection is sometimes described as a "hexagon autotransformer," or a "squashed delta." (PE/TR) C57.12.80-1978r

interconnected star connection of polyphase circuits *See:* zigzag connection of polyphase circuits.

interconnected system (A) (electric power system) A system consisting of two or more individual power systems normally operating with connecting tie lines. *See also:* power system. **(B)** Two or more power systems connected by transmission facilities. (PE/PSE) 94-1970, [54], 94-1991

interconnecting channel (of a supervisory system) The transmission link, such as the direct wire, carrier, or microwave channel (including the direct current, tones, etc.) by which supervisory control or indication signals or selected telemeter readings are transmitted between the master station and the remote station or stations, in a single supervisory system. (SWG/PE) C37.100-1992

interconnection **(1)** The physical plant and equipment required to facilitate the transfer of electric energy between two or more entities. It can consist of a substation and an associated transmission line and communications facilities or only a simple electric power feeder. (SUB/PE) 1109-1990w
(2) The facilities that connect two power systems or control areas. (PE/PSE) 858-1993w

interconnection device *See:* adapter.

interconnection diagram (packaging machinery) A diagram showing the connections between the terminals in the control panel and outside points, such as connections to motors and auxiliary devices. (IA/PKG) 333-1980w

interconnection tie A feeder interconnecting two electric supply systems. *Note:* The normal flow of energy in such a feeder may be in either direction. *See also:* center of distribution. (T&D/PE) [10]

interconnect space The address space used for board identification, system configuration, and board specific functions such as testing and diagnostics. (C/MM) 1296-1987s

interconnect template A definition of the contents of the interconnect space of an agent. (C/MM) 1296-1987s

interdendritic corrosion Corrosive attack that progresses preferentially along interdendritic paths. *Note:* This type of attack results from local differences in composition, that is, coring, commonly encountered in alloy castings. (IA) [59]

interdigital magnetron A magnetron having axial anode segments around the cathode, alternate segments being connected together at one end, remaining segments connected together at the opposite end. (ED) 161-1971w, [45]

interdigital transducer (IDT) A comb-like conductive structure that is fabricated on the surface of a substrate and consists of interleaved metal electrodes (fingers) whose function is to transform electrical energy into acoustic energy or vice versa by means of the piezoelectric effect. (UFFC) 1037-1992w

interdigit interval (telephony) (dial-pulse address signaling systems) In dial-pulse signaling, an extended make interval used to separate and distinguish successive dial-pulse address digits. (COM/TA) 753-1983w

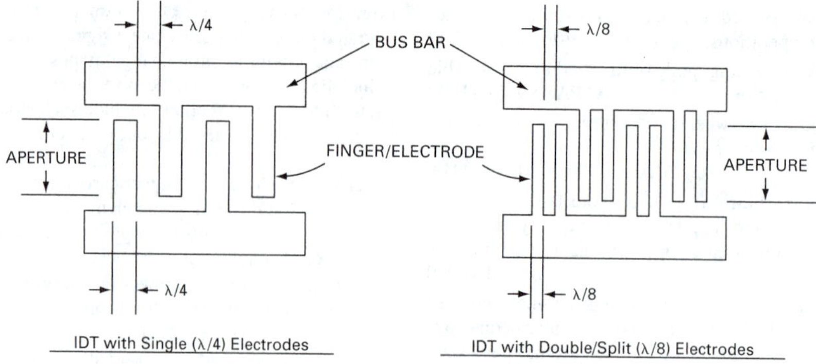

illustration of transducer parameters

interdigit (interdigital) time (measuring the performance of tone address signaling systems) The time interval between successive signal present intervals during which no signal present condition exists. This time includes the signal off condition and transition intervals between signal off condition and signal present condition on both state transitions. (COM/TA) 752-1986w

interelectrode capacitance (*j–l* interelectrode capacitance c_{jl} of an *n*-terminal electrode tube) The capacitance determined from the short-circuit transfer admittance between the *j*th and the *l*th terminals. *Note:* This quantity is often referred to as direct interelectrode capacitance. *See also:* electron-tube admittances. (ED) 161-1971w

interelectrode transadmittance (*j–l* interelectrode transadmittance of an *n*-electrode electron tube) The short-circuit transfer admittance from the *j*th electrode to the *l*th electrode. *See also:* electron-tube admittances. (ED) 161-1971w

interelectrode transconductance (*j–l* interelectrode transconductance) The real part of the *j–l* interelectrode transadmittance. *See also:* electron-tube admittances. (ED) 161-1971w

interelement influences (polyphase wattmeters) The percentage change in the recorded value that is caused solely by the action of the stray field of one element upon the other element. *Note:* This influence is determined at the specified frequency of calibration with rated current and rated voltage in phase on both elements or such lesser value of equal currents in both elements as gives end-scale deflection. Both current and voltage in one element shall then be reversed, and, for rating purposes, one-half the difference in the readings in percent is the interelement influence. *See also:* accuracy rating. (EEC/ERI/AII) [111], [102]

interexchange carrier In the United States, a common carrier limited by law to carry telephone traffic between local exchange and transport areas. (C) 610.7-1995

interexchange channel A direct channel or circuit between exchanges. (C) 610.7-1995

interface (1) (696 interface devices) A shared electrical boundary between parts of a computer system, through which information is conveyed. (C/MM) 696-1983w
(2) (microprocessor operating systems) A shared boundary between two layers or modules of software. (C/MM) 855-1985s
(3) (watthour meters) The means for transmitting information between the register and peripheral equipment. (ELM) C12.13-1985s
(4) (general) A shared boundary. (C) [20], [85]
(5) (Class 1E equipment and circuits) (nuclear power generating station) A junction or junctions between a Class 1E equipment and another equipment or device. (Examples: connection boxes, splices, terminal boards, electrical connections, grommets, gaskets, cables, conduits, enclosures, etc.) (PE/NP) 380-1975w, 323-1974s
(6) (programmable instrumentation) A common boundary between a considered system and another system, or between parts of a system, through which information is conveyed. (IM/AIN) 488.1-1987r
(7) (test, measurement, and diagnostic equipment) A shared boundary involving the specification of the interconnection between two equipments or systems. The specification includes the type, quantity and function of the interconnection circuits and the type and form of signals to be interchanged via those circuits. *See also:* adapter. (MIL) [2]
(8) (A) (data transmission) A common boundary; for example, a physical connection between two systems or two devices. The boundary may be mechanical such as the physical surfaces and spacings in mating parts, modules, components, or subsystems, or electrical, such as matching signal levels, impedances, or power levels of two or more subsystems. **(B) (data transmission)** A concept involving the specification of the interconnection between two equipments or systems. The specification includes the type, quantity, and function of the interconnection circuits and the type and form of signals to be interchanged by these circuits. (PE) 599-1985
(9) (A) (software) A shared boundary across which information is passed. **(B) (software)** A hardware or software component that connects two or more other components for the purpose of passing information from one to the other. **(C) (software)** To connect two or more components for the purpose of passing information from one to the other. **(D) (software)** To serve as a connecting or connected component as in definition (B). (C) 610.12-1990
(10) (STEbus) A shared boundary between two or more systems, or between two or more elements within a system, through which information is conveyed. (MM/C) 796-1983r, 1000-1987r
(11) (SBX bus) A shared boundary, between two systems or parts of systems, through which information is transferred. (C/MM) 959-1988r
(12) (electromechanical watthour meters) The means for communications between devices. (ELM) C12.15-1990
(13) A device placed between the line output of a digital telephone set and test equipment. The device performs at least one of the following functions: simulation of a normal network connection, control of the telephone set under test, or access for the reference codec to the digital voice signal. (COM/TA) 269-1992
(14) A junction or junctions between a Class 1E equipment and another equipment or device. (For motors, typical interfaces include, as applicable: mechanical mounting connection to the driven equipment and the motor mounting to its base, and force transmitted to the motor, electrical connection, cooling system connections, and lubrication system connection. *See also:* user interface; data-transfer interface. (PE/NP) 334-1994r

(15) (MULTIBUS) A shared boundary between modules or agents of a computer system, through which information is conveyed. (C/MM) 1296-1987s
(16) A shared boundary between two functional entities. A standard specifies the services in terms of the functional characteristics and behavior observed at the interface. The standard is a contract in the sense that it documents a mutual obligation between the service user and provider and assures a stable definition of that obligation. (C/PA) 14252-1996
(17) Hardware or software that provides a point of communication between two or more processes, persons, or other physical entities. (C) 610.7-1995, 610.10-1994w
(18) A shared boundary between two objects such as devices, systems, or networks, across which information is passed. *See also:* user interface; data-transfer interface. (SUB/PE/C) 999-1992w, 610.10-1994w
(19) Either the MA interface or the MT interface without distinction, or one of the two in particular. (C/PA) 1224.1-1993w
(20) A junction or junctions between a Class 1E equipment and another equipment or device. (For motors, typical interfaces include, as applicable: mechanical mounting connection to the driven equipment and the motor mounting to its base, any force transmitted to the motor, electrical connection, cooling system connections, and lubrication system connection.) (PE/NP) 334-1994r
(21) In software development, a relationship among two or more entities (such as software item - software item, software item - hardware item, software item - user, or software unit - software unit) in which the entities share, provide, or exchange data. An interface is not a software item, software unit, or other system component; it is a relationship among them. (C/SE) J-STD-016-1995
(22) A shared boundary between two layers or modules of software. (C/MM) 855-1990
(23) A shared boundary that specifies the interconnection between two units or systems, hardware or software. In hardware, the specification includes the type, quantity, and function of the interconnection circuits and the type and form of signals to be interchanged via those circuits. In software, the specification includes the object type and, where necessary, the name or instance handle of specific objects copied or shared between the two systems. (SCC20) 1226-1998
(24) The declaration of the meaning and the signature for a property or constraint. The interface states "what" a property (responsibility) knows or does or what a constraint (responsibility) must adhere to. The interface specification consists of the meaning (semantics) and the signature (syntax) of a property or constraint. (C/SE) 1320.2-1998

interface-CCITT (data transmission) The present European, and possible world standard, for interface requirements between data processing terminal equipment and data communication equipment. The CCITT standard resembles very closely the American EIA, Standard RS-232-C. This standard is considered mandatory in Europe and on the other continents. (PE) 599-1985w

interface control (1) (software, configuration management) The process of identifying all functional and physical characteristics relevant to the interfacing of two or more configuration items provided by one or more organizations, and ensuring that proposed changes to these characteristics are evaluated and approved prior to implementation. *See also:* configuration status accounting; configuration identification; configuration control; configuration audit; configuration management; configuration item; software library; baseline; configuration control board. (C/SE) 828-1983s, 610.12-1990
(2) (software) (DoD usage) In configuration management, the administrative and technical procedures and documentation necessary to identify functional and physical characteristics between and within configuration items provided by different developers, and to resolve problems concerning the specified interfaces. *See also:* configuration control. (C) 610.12-1990

Interface Definition Language (IDL) (1) A machine-compilable language used to describe interfaces that clients call and implementations provide. IDL provides a neutral way to define an interface. [IDL is an Object Management Group (OMG) product.] (SCC20) 1226-1998
(2) A programming language-independent method of specifying operation syntax. (IM/ST) 1451.1-1999

interface design document (IDD) Documentation that describes the architecture and design of interfaces between system and components. These descriptions include control algorithms, protocols, data contents and formats, and performance. (C/SE) 1012-1998

interface device *See:* adapter.

interface error An error condition caused by hardware incompatibility, software incompatibility or other incompabilities between any two items of equipment. (C) 610.10-1994w

Interface Only Implementation Refers to the implementation of an operation on an Object. This means that only the interface properties of the operation are implemented. The operation will return an appropriate error code but otherwise have no other effect on the Object. (IM/ST) 1451.1-1999

interface, operating *See:* operating interface.

interface operation *See:* operation.

interface plane An assigned plane on the bottom surface of the module connector from which the connector's electrical pins protrude, thus forming the mating surface. This surface is used as a reference for module dimensions. (C/BA) 1101.3-1993

interface requirement (software) A requirement that specifies an external item with which a system or system component must interact, or that sets forth constraints on formats, timing, or other factors caused by such an interaction. *Contrast:* physical requirement; performance requirement; functional requirement; design requirement; implementation requirement. (C) 610.12-1990

interface requirement specification (IRS) Documentation that specifies requirements for interfaces between systems or components. These requirements include constraints on formats and timing. (C/SE) 1012-1998

interface specification (1) (software) A document that specifies the interface characteristics of an existing or planned system or component. (C) 610.12-1990
(2) The description of essential functional, performance, and design requirements and constraints at a common boundary between two or more system elements. This includes interfaces between humans and hardware or software, as well as interfaces between humans themselves. (C/SE) 1220-1998

interface system (1) (696 interface devices) (general system term) The device independent functional, electrical, and mechanical elements of an interface necessary to effect unambiguous communication among a set of devices. Driver and receiver circuits, signal line descriptions, timing and control conventions, data transfer protocols, and functional logic circuits are typical system elements. (MM/C) 696-1983w
(2) (general system) (microcomputer system bus) The device-dependent electrical and functional interface elements necessary for communication between devices. Typical elements are: driver and receiver circuits and functional logic circuits. (MM/C) 796-1983r
(3) (STEbus) The device-independent electrical, mechanical, and functional interface elements required for unambiguous communication between two or more devices. Typical elements include

- driver and receiver circuitry
- signal line descriptions
- timing and control conventions
- communication protocols
- functional logic circuits.

(MM/C) 1000-1987r
(4) (programmable instrumentation) The device-independent mechanical, electrical, and functional elements of an

interface necessary to effect communication among a set of devices. Cables, connector, driver and receiver circuits, signal line descriptions, timing and control conventions, and functional logic circuits are typical interface system elements. (IM/AIN) 488.1-1987r

interface test A test to check interaction among equipment through permanent interconnections. (SUB/PE) 1303-1994

interface test adapter (1) A device or series of devices designed to provide a compatible connection between the unit under test (UUT) and the test equipment. May include proper stimuli or loads not contained in the test equipment. (ATLAS) 1232-1995
(2) *See also:* adapter. (SCC20) 1226-1998

interface testing (software) Testing conducted to evaluate whether systems or components pass data and control correctly to one another. *See also:* system testing; component testing; unit testing; integration testing. (C) 610.12-1990

interfacial connection (soldered connections) A conductor that connects conductive patterns on opposite sides of the base material. (EEC/AWM) [105]

interference (1) (electric-power-system measurements) Any spurious voltage or current appearing in the circuits of the instrument. *Note:* The source of each type of interference may be within the instrument case or external. The instrument design should be such that the effects of interference arising internally are negligible. (EEC/EMI) [112]
(2) (induction or dielectric-heating usage) The disturbance of any electric circuit carrying intelligence caused by the transfer of energy from induction- or dielectric-heating equipment. (IA) 54-1955w, 169-1955w
(3) (fiber optics) In optics, the interaction of two or more beams of coherent or partially coherent light. *See also:* coherent; diffraction; degree of coherence. (Std100) 812-1984w
(4) (overhead-power-line corona and radio noise) Impairment to a useful signal produced by natural or man-made sources. *Note:* Distortions caused by reflections, shielding, or extraneous power in a signal's frequency range are all examples of interference. *Synonym:* radio interference. (T&D/PE) 539-1979s
(5) Field strength produced by a radio disturbance, such as signals from other stations. (T&D/PE) 1260-1996
(6) (data transmission) In a signal transmission path, either extraneous power which tends to interfere with the reception of the desired signals or the disturbance of signal which results. (PE) 599-1985w

interference, common-mode *See:* common-mode interference.

interference coupling ratio (signal-transmission system) The ratio of the interference produced in a signal circuit to the actual strength of the interfering source (in the same units). *See also:* interference. (IE) [43]

interference, differential-mode *See:* differential-mode interference.

interference field strength Field strength produced by a radio disturbance. *Note:* Such a field strength has only a precise value when measured under specified conditions. Normally, it should be measured according to publications of the International Special Committee on Radio Interference. *See also:* electromagnetic compatibility. (EMC/INT) [53], [70]

interference filter (fiber optics) An optical filter consisting of one or more thin layers of dielectric or metallic material. *See also:* optical filter; interference; dichroic filter. (Std100) 812-1984w

interference guard bands The two bands of frequencies additional to, and on either side of, the communication band and frequency tolerance, which may be provided in order to minimize the possibility of interference. *See also:* channel. (AP/ANT) 145-1983s

interference, longitudinal *See:* common-mode interference.

interference measurement (induction or dielectric-heating usage) A measurement usually of field intensity to evaluate the probability of interference with sensitive receiving apparatus. *See also:* induction heating. (IA) 54-1955w

interference, normal-mode *See:* normal-mode interference.

interference pattern The resulting space distribution when progressive waves of the same frequency and kind are superposed. *See also:* wavefront. (EEC/PE) [119]

interference power Power produced by a radio disturbance. *Note:* Such a power has only a precise value when measured under specified conditions. *See also:* electromagnetic compatibility. (EMC/INT) [53], [70]

interference, series-mode *See:* differential-mode interference.

interference susceptibility (mobile communication) A measure of the capability of a system to withstand the effects of spurious signals and noise that tend to interfere with reception of the desired intelligence. *See also:* mobile communication system. (VT) [37]

interference testing (test, measurement, and diagnostic equipment) A type of on-line testing that requires disruption of the normal operation of the unit under test. *See also:* noninterference testing. (MIL) [2]

interference-test input (amplitude-modulation broadcast receivers) The least interfering-signal or signal field, of specified carrier frequency, which results in interference test output. It is expressed in decibels below one volt, or in microvolts, or in the case of loop measurements in decibels below one volt per meter or microvolts per meter. (CE) 186-1948w

interference-test output (amplitude-modulation broadcast receivers) Output that is 30 dB less than, or 0.001 of the power of, the normal test output. (CE) 186-1948w

interference, transverse *See:* differential-mode interference.

interference voltage Voltage produced by a radio disturbance. *Note:* Such a voltage has a precise value only when measured under specified conditions. Normally, it should be measured according to recommendations of the International Special Committee on Radio Interference. *See also:* electromagnetic compatibility. (EMC/INT) [53], [70]

interferometer (1) (fiber optics) An instrument that employs the interference of light waves for purposes of measurement. *See also:* interference. (Std100) 812-1984w
(2) An antenna and receiving system that determines the angle of arrival of a wave by phase comparison of the signals received at widely separated antennas. *Note:* In radar, the angle measurement made by an interferometer is generally ambiguous, and means must be used to resolve the ambiguities. (AES) 686-1997

interferometer antenna An array antenna in which the interelement spacings are large compared to wavelength and element size so as to produce grating lobes. (AP/ANT) 145-1993

interflectance method† **(illuminating engineering)** A lighting design procedure for predetermining the luminances of walls, ceiling, and floor and the average illuminance on the work-plane based on integral equations. It takes into account both direct and reflected flux. (This term is retained for reference and literature searches.) (EEC/IE) [126]

† Obsolete.

interflected component (illuminating engineering) That portion of luminous flux from a luminaire which arrives at the work-plane after being reflected one or more times from room surfaces, as determined by the flux transfer theory. *See also:* flux transfer theory. (EEC/IE) [126]

interflection (illuminating engineering) The multiple reflection of light by the various room surfaces before it reaches the work-plane or other specified surface of a room. *Synonym:* interreflection. (EEC/IE) [126]

interframe coding An image compression technique in which a sequence of images is compressed by taking advantage of redundancies between successive images. (C) 610.4-1990w

intergranular corrosion Corrosion that occurs preferentially at grain boundaries. (IA) [59]

interior The set of pixels in a region of a digital image that are not adjacent to pixels in the region's complement. *Contrast:* border. (C) 610.4-1990w

interior communication systems (marine) Those systems providing audible or visual signals or transmission of information within or on a vessel. (PE/EEC) [119]

interior wiring system ground A ground connection to one of the current-carrying conductors of an interior wiring system. *See also:* ground. (T&D/PE) [10]

interlaboratory standards Those standards that are used for comparing reference standards of one laboratory with those of another, when the reference standards are of such nature that they should not be shipped. *See also:* measurement system. (IM) 285-1968w, [38]

interlace (A) To arrange, access, select, or display in an alternating fashion. **(B)** To refresh a display device using two passes of the writing beam to complete the full display; the first pass draws every other line and the second fills in those skipped. (C) 610.10-1994

interlaced Pertaining to a display device in which every other line of pixels is refreshed on each pass. *Contrast:* noninterlaced. (C) 610.10-1994w

interlace factor (television) A measure of the degree of interlace of nominally interlaced fields. *Note:* In a two-to-one interlaced raster, the interlace factor is the ratio of the smaller of two distances between the centers of adjacent scanned lines to one-half the distance between the centers of sequentially scanned lines at a specified point. (BT/AV) 201-1979w

interlace scan A raster scan technique in which the electron beam alternately refreshes all even, then all odd, scan lines of a display surface. (C) 610.6-1991w

interlaced scanning (television) A scanning process in which the distance from center to center of successively scanned lines is two or more times the nominal line width, and in which the adjacent lines belong to different fields. *See also:* television. (BT/AV) [34]

interlacing impedance voltage of a Scott-connected transformer (power and distribution transformers) The interlacing impedance voltage of Scott-connected transformers is the single-phase voltage applied from the midtap of the main transformer winding to both ends, connected together, which is sufficient to circulate in the supply lines a current equal to the rated three-phase line current. The current in each half of the winding is 50% of this value. The per-unit or percent interlacing resistance is the measured watts expressed on the base of the rated kVA of the teaser winding. The per-unit or percent interlacing impedance is the measured voltage expressed on the base of the teaser-voltage. (PE/TR) C57.12.80-1978r

interLATA In the United States, a collection of circuits that cross local access and transport area boundaries and are passed onto an interexchange carrier. *See also:* intraLATA. (C) 610.7-1995

interleave (1) To arrange parts of one sequence of things or events so that they alternate with parts of one or more other sequences of things or events and so that each sequence retains its identity. (C/C) [20], [85]
(2) (software) To alternate the elements of one sequence with the elements of one or more other sequences so that each sequence retains its identity; for example, to alternately perform the steps of two different tasks in order to achieve concurrent operation of the tasks. (C) 610.12-1990
(3) To arrange parts of one sequence of things or events so that they alternate with parts of one or more other sequences of the same nature such that each sequence retains its identity; For example, to assign successive addresses to physically separated storage locations in such a way as to reduce access time. (C) 610.10-1994w

interleaved array In PL/1, an array whose name refers to noncontiguous storage. (C) 610.5-1990w

interleaved memory A type of memory in which two or more separate arrays are used to fill alternate accesses in such a way as to speed the average access time of the memory. For example, the odd addresses are all in one memory array and the even addresses are in a second. (C) 610.10-1994w

interleaved windings (power and distribution transformers) (of a transformer) An arrangement of transformer windings where the the primary and secondary windings, and the tertiary windings, if any, are subdivided into disks (or pancakes) or layers and interleaved on the same core. (PE/TR) C57.12.80-1978r

interleaving The process of alternating two or more operations or functions through the overlapped use of a computer facility. *See also:* interleaved memory. (C) 610.10-1994w

interlock (1) A device actuated by the operation of some other device with which it is directly associated, to govern succeeding operations of the same or allied devices. *Note:* An interlock system is a series of interlocks applied to associated equipment in such a manner as to prevent or allow operation of the equipment only in a prearranged sequence. Interlocks are classified into three main divisions: mechanical interlocks, electrical interlocks, and key interlocks, based on the type of interconnection between the associated devices. (SWG/PE/TR) C37.100-1992, C57.12.80-1978r
(2) To prevent one device from interfering with another. For example, to lock the switches to prevent manual movement of the switches while a program is executing. (C) 610.10-1994w
(3) Device that permits equipment or controls to operate only after other conditions have been fulfilled. (PE/EDPG) 1020-1988r

interlock bypass A command to temporarily circumvent a normally provided interlock. (IA/EEC) [61], [74]

interlocked sequence A fixed sequence of events in which one event in the sequence must occur before the next event may occur. (IM/AIN) 488.1-1987r

interlocking (1) (interlocking plant) (railways) An arrangement of apparatus in which various devices for controlling track switches, signals, and related appliances are so interconnected that their movements must succeed one another in a predetermined order, and for which interlocking rules are in effect. *Note:* It may be operated manually or automatically. (PE/EEC) [119]
(2) An arrangement of switch, lock, and signal devices that is located where rail tracks cross, join, separate, and so on. The devices are interconnected in such a way that their movements must succeed each other in a predefined order, thereby preventing opposing or conflicting train movements. (VT/RT) 1474.1-1999

interlocking deactivating means (defeater) A manually actuated provision for temporarily rendering an interlocking device ineffective, thus permitting an operation that would otherwise be prevented. For example, when applied to apparatus such as combination controllers or control centers, it refers to voiding of the mechanical interlocking mechanism between the externally operable disconnect device and the enclosure doors to permit entry into the enclosure while the disconnect device is closed. *See also:* electric controller. (IA/ICTL/IAC) [60]

interlocking limits (interlocking territory) (railways) An expression used to designate the trackage between the opposing home signals of an interlocking. *See also:* interlocking. (PE/EEC) [119]

interlocking machine (railways) An assemblage of manually operated levers or equivalent devices, for the control of signals, switches, or other units, and including mechanical or circuit locking, or both, to establish proper sequence of movements. *See also:* interlocking. (PE/EEC) [119]

interlocking plant *See:* interlocking.

interlocking relay (railways) A relay that has two independent magnetic circuits with their respective armatures so arranged that the dropping away of either armature prevents the other armature from dropping away to its full stroke. (EEC/PE) [119]

interlocking signals (railways) The fixed signals of an interlocking. (PE/EEC) [119]

interlocking station (railways) A place from which an interlocking is operated. (EEC/PE) [119]

interlocking territory *See:* interlocking limits.

interlock relay A relay with two or more armatures having a mechanical linkage, or an electric interconnection, or both, whereby the position of one armature permits, prevents, or causes motion of another armature. *See also:* relay. (EEC/REE) [87]

intermateability The capability for units of equipment to fit together mechanically but not necessarily work together electrically. (C/BA) 14536-1995

intermateable Mechanical compatibility between modules with the backplane, card cage, and system into which they are inserted. Sometimes includes compatibility in the assignment of power and ground pins. (C/BA) 896.2-1991w, 896.10-1997

intermediate contacts (of a switching device) Contacts in the main circuit that part after the main contacts and before the arcing contacts have parted. (SWG/PE/TR) C37.100-1992, C57.12.44-1994

intermediate current ratings (of distribution fuse links) A series of distribution fuse-link ratings chosen from a series of preferred numbers that are spaced between the preferred current ratings, but may not provide coordination therewith. Coordination between adjacent intermediate ratings may be secured to the same degree as between adjacent preferred current ratings. (SWG/PE) C37.40-1993

intermediate datatype Any of the basic datatypes in terms of which the other, substantive datatypes of the interface are defined. (C/PA) 1328-1993w, 1327-1993w, 1224-1993w

intermediate distributing frame (telephone switching systems) A frame where crossconnections are made only between units of central office equipment. (COM) 312-1977w

intermediate frequency (IF) (1) (A) (nonlinear, active, and nonreciprocal waveguide components) (general). A frequency to which a signal wave is shifted locally as an intermediate step in transmission or reception. **(B) (nonlinear, active, and nonreciprocal waveguide components)** (superheterodyne reception). The difference frequency resulting from a frequency conversion before demodulation. (MTT/PE) 457-1982, 599-1985, 188-1952
(2) (overhead-power-line corona and radio noise) The frequency resulting from a frequency conversion that is amplified locally in the receiver before demodulation. (T&D/PE) 539-1990

intermediate-frequency-harmonic interference (super-heterodyne receivers) Interference due to radio-frequency-circuit acceptance of harmonics of an intermediate-frequency signal. 188-1952w

intermediate-frequency interference ratio *See:* radio receiver; intermediate-frequency response ratio.

intermediate-frequency response ratio (superheterodyne receivers) The ratio of the field strength at a specified frequency in the intermediate frequency band to the field strength at the desired frequency, each field being applied in turn, under specified conditions, to produce equal outputs. *See also:* radio receiver. 188-1952w

intermediate-frequency transformer A transformer used in the intermediate-frequency portion of a heterodyne system. *Note:* Intermediate-frequency transformers are frequently narrow-band devices. (CHM) [51]

intermediate hub (IH) A hub that occupies any level below the header hub in a hierarchy of hubs. *Note:* This term is contextually specific to IEEE Std 802.3, clause 12. (LM/C) 610.7-1995, 802.3-1998

intermediate layer (solar cells) The material on the solar cell surface that provides improved spectral match between the cell and the medium in contact with this surface. (AES/SS) 307-1969w

intermediate macro Any of the basic macros in terms of which the other, substantive macros used to realize the dispatcher are defined. (C/PA) 1328-1993w, 1327-1993w

intermediate maintenance (test, measurement, and diagnostic equipment) Maintenance which is the responsibility of and performed by designated maintenance activities for direct support of using organizations. Its phases normally consist of calibration, repair or replacement of damaged or unserviceable parts, components or assemblies: the emergency manufacture of nonavailable parts and providing technical assistance to using organizations. (MIL) [2]

intermediate means (measurement sequence) All system elements that are used to perform necessary and distinct operations in the measurement sequence between the primary detector and the end device. *Note:* The intermediate means, where necessary, adapts the operational results of the primary detector to the input requirements of the end device. *See also:* measurement system. (EEC/EMI) [112]

intermediate metal conduit A metal raceway of circular cross section with integral or associated couplings, connectors and fittings approved for the installation of electrical conductors. (NESC/NEC) [86]

intermediate office (telephone switching systems) A switching entity where trunks are terminated for purposes of interconnection to other offices. (COM) 312-1977w

intermediate product *See:* partial product.

intermediate repeater (data transmission) A repeater for use in a trunk of line at a point other than an end. (PE) 599-1985w

intermediate subcarrier A carrier that may be modulated by one or more subcarriers and that is used as a modulating wave to modulate a carrier or another intermediate subcarrier. *See also:* carrier; subcarrier. (AP/ANT) 145-1983s

intermediate system In OSI context, an open system that performs a relay function that is neither the data source nor the data sink for a given instance of communication. (C) 610.7-1995

intermediate test result code A test result code, obtained from an assertion test, that requires further processing to determine the final test result code. (C/PA) 2003-1997

intermediate-voltage power supply (IVPS) Power supply that converts the third rail or catenary high-voltage dc or ac into an intermediate-voltage ac or dc to feed other power supplies. (VT) 1476-2000

intermittent duty (1) (rotating machinery) A duty in which the load changes regularly or irregularly with time. *See also:* asynchronous machine; voltage regulator; direct-current commutating machine. (PE) [9]
(2) Operation for alternate intervals of load and no load; or load and rest; or load, no load and rest. (NESC/NEC) [86]
(3) A requirement of service that demands operation for alternate periods of current loading and rest, such alternate intervals being definitely specified. (PE/TR) C57.16-1996
(4) A requirement of service that demands operation for alternate periods (1) load and no load; or (2) load and rest; or (3) load, no load and rest, as specified. (IA/MT) 45-1998

intermittent-duty rating The specified output rating of a device when operated for specified intervals of time other than continuous duty. (AP/ANT) 145-1983s

intermittent failure *See:* failure.

intermittent fault (1) (surge arresters) A fault that recurs in the same place and due to the same cause within a short period of time. (PE) [8], [84]
(2) (software) A temporary or unpredictable fault in a component. *See also:* transient error; random failure. (C/BA) 896.9-1994w, 610.12-1990

(3) A recurring temporary error caused by component degradation or inadequate design (e.g. noise margin). (C/BA) 896.3-1993w

intermittent inductive train control Intermittent train control in which the impulses are communicated to the vehicle-carried apparatus inductively. *See also:* automatic train control. (EEC/PE) [119]

intermittent rating *See:* periodic rating.

intermittent test (batteries) A service test in which the battery is subjected to alternate discharges and periods of recuperation according to a specified program until the cutoff voltage is reached. *See also:* battery. (PE/EEC) [119]

intermittent train control A system of automatic train control in which impulses are communicated to the locomotive or vehicle at fixed points only. *See also:* intermittent inductive train control; automatic train control. (EEC/PE) [119]

intermodal distortion *See:* multimode distortion.

intermodulation (nonlinear transducer element) The modulation of the components of a complex wave by each other, as a result of which waves are produced that have frequencies equal to the sums and differences of integral multiples of those of the components of the original complex wave. *See also:* modulation. (AP/ANT) [53], 145-1983s

intermodulation distortion (IMD) (1) (data transmission) Nonlinear distortion of a system or transducer, characterized by the appearance in the output of frequencies equal to the sums and differences of integral multiples of the two or more component frequencies present in the input wave. Harmonic components also present in the output are usually not included as part of the intermodulation distortion. When harmonics are included, a statement to that effect should be made. (PE/COM/TA) 599-1985w, 1007-1991r

(2) (nonlinear, active, and nonreciprocal waveguide components) Distortion produced by undesired intermodulation. (MTT) 457-1982w

(3) Nonlinear distortion of multiple-frequency inputs that shows up as harmonics of the individual inputs plus the sum and difference products of the inputs and their harmonics. (PE/IC) 1143-1994r

(4) An analog line impairment in which modulation on one channel or at one frequency distorts the modulation on another channel or frequency. (C) 610.7-1995

(5) Intermodulation distortion refers to the family of system performance impairments caused by the nonlinear transfer characteristic of a broadband system, which produces spurious output signals (called "intermodulation products") at frequencies that are linear combinations of those of the input. The system output (S_o) can generally be related to the system input by the transfer equation:

$$S_o = AS_o + BS_i^2 + CS_i^3$$

AS_i = the fundamental signal term

The terms BS_i^2 and CS_i^2 represent the second-order and third-order distortion terms, respectively. The second-order term produces a second harmonic frequency component for every input signal frequency and intermodulation frequency components of the form $f_1 \pm f_2$. The third-order term produces a third harmonic frequency component for every input signal frequency and intermodulation frequency components of the form $f_1 \pm f_2 \pm f_3$ and $2f_1 \pm f_2$. Third-order distortion also produces cross-modulation where modulation of one carrier can appear on another carrier on the system even when the second carrier is unmodulated when input into the system. In CATV systems where the video carriers are spaced at 6 MHz intervals, the summation of the third-order intermodulation distortion signals is called "composite triple-beat distortion." CTB can become significant when all distortion components fall near a video carrier. In CATV systems, it is common practice to specify composite triple beat and cross-modulation distortion, and design the cable system to meet these specifications. The amplifier distortion levels are specified by the manufacturer for a full channel load condition (a single video carrier in each 6 MHz channel at a given level). In broadband systems, different types of carriers and modulation techniques may be operating on a cable system so that the composite triple beat and cross-modulation distortions are difficult to determine. In broadband systems it is common practice to design the system to composite triple beat specifications based on CATV practices. In addition, the carrier to discrete second-order beat and third-order beat ratios are specified. The distortion ratios are specified independently under referenced conditions for the inbound and outbound paths. Inbound distortion is specified at the headend with signals injected prior to the most remote amplifier of the worst-case inbound path. Outbound distortion is specified following the most remote (from the headend) amplifier of the worst-case outbound path. *Second-order distortion.* This parameter describes the spurious signals that are produced as a result of the second-order curvature of the transfer characteristic of the system components, when two discrete input signals are applied. The dominant members lie at frequencies given by

$$F_{21} = \left| F_a + F_b \right| \text{ and}$$

$$F_{22} = \left| F_a - F_b \right|$$

where

F_a and F_b = the frequencies of the input signals.

Third-order distortion. This parameter describes the spurious output signals that are produced as a result of the third-order curvature of the transfer characteristic of the system components, when three discrete input signals are applied. The dominant member lies at frequencies given by

$$F_3 = \left| F_a \pm F_b \pm F_c \right|$$

where F_a, F_b, and F_c = the frequencies of the input signals. (LM/C) 802.7-1989r

(6) Nonlinear distortion of a system or transducer characterized by the appearance of frequencies at the output equal to the sums and differences of integral multiples of two or more of the input frequencies. IMD is the ratio, in dB, of the test signal to specific spurious output signals generated by the nonlinearity. (COM/TA) 743-1995

(7) Nonlinear distortion of a system or transducer characterized by the appearance in the output of frequencies equal to the sums and differences of integral multiples of the two or more component frequencies present in the input wave. *Note:* Harmonic components also present in the output are usually not included as part of the intermodulation distortion. When harmonics are included, a statement to that effect should be made. *See also:* distortion. (SP) 151-1965w

intermodulation interference (mobile communication) The modulation products attributable to the components of a complex wave that on injection into a nonlinear circuit produces interference on the desired signal. *See also:* mobile communication system. (VT) [37]

intermodulation noise Noise characterized by the intermodulation of signals from two independent lines or separate components of the desired signal causing interference. (C) 610.7-1995

intermodulation product intercept point (nonlinear, active, and nonreciprocal waveguide components) Intermodulation products have an output-versus-input characteristic which, when graphically displayed, would theoretically intercept the plot of the desired output-versus-input if the nonlinear device continued to operate linearly without compression. The signal input level at which this theoretical point would occur is called the intercept point and is usually defined in dBm (decibel referred to one milliwatt). The corresponding figure is a graphical representation of the intercept points for a single-tone second order and a two-tone third-order intermodulation product.

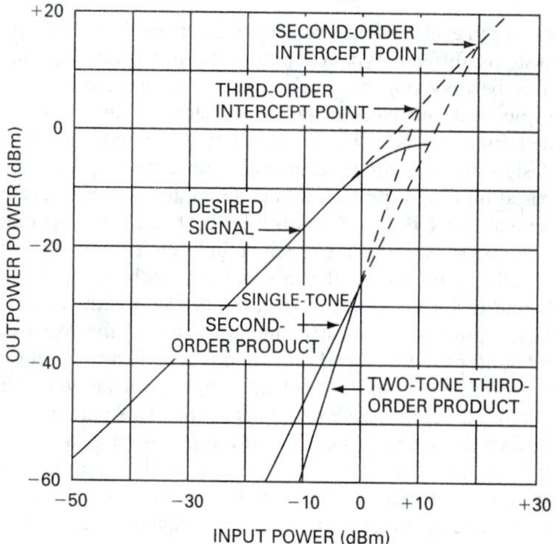

intermodulation product intercept point

(MTT) 457-1982w

intermodulation products (1) (nonlinear, active, and nonreciprocal waveguide components) The undesired responses in a nonlinear device that result from harmonics of two or more signals. (MTT) 457-1982w
(2) *See also:* intermodulation distortion.

intermodulation rejection (spectrum analyzer) The ratio of the sensitivity level and the level of either of two equal amplitude signals which produce any intermodulation product at the sensitivity level. (IM) 748-1979w

intermodulation spurious emission (land-mobile communications transmitters) External radio frequency (RF) emission of a transmitter which is a product of the nonlinear mixing process in the final stage of the transmitter which occurs when external RF power is coupled through the antenna output. (EMC) 377-1980r

intermodulation spurious response (1) (receiver performance) The receiver audio output resulting from the mixing of *n*th-order frequencies, in the nonlinear elements of the receiver, in which the resultant carrier frequency is equivalent to the assigned frequency. *See also:* spurious response. (VT) [37]
(2) (nonreal time spectrum analyzer) The spectrum analyzer response resulting from the mixing of the nth order frequencies, in the nonlinear elements of the spectrum analyzer, in which the resultant response is equivalent to the tuned frequency. *Synonym:* intermodulation distortion. (IM) [14]
(3) (frequency-modulated mobile communications receivers) The response resulting from the mixing of two or more undesired frequencies in the nonlinear elements of the receiver in which a resultant frequency is generated that falls within the receiver passband. (VT) 184-1969w
(4) (spectrum analyzer) The spectrum analyzer response resulting from the mixing of the nth order frequencies of the input signal in the nonlinear elements of the spectrum analyzer, in which the resultant response is equivalent to the tuned frequency. *Synonym:* intermodulation distortion. (IM) 748-1979w

internal address field (IA) The group of low-order bits (right justified and contiguous to the device address field on the left) assigned in the address of a FASTBUS device which is used to identify internal locations within a FASTBUS device. Secondary address cycles allow the number of different locations accessed to exceed that available in the internal address field. (NID) 960-1993

internal arrow An arrow connected at both ends (source and use) to a box in a diagram. *Contrast:* boundary arrow. (C/SE) 1320.1-1998

internal audit (nuclear power quality assurance) An audit of those portions of an organization's quality assurance program retained under its direct control and within its organizational structure. (PE/NP) [124]

internal bias (teletypewriter) Bias, either marking or spacing, that may occur within a start-stop printer receiving mechanism and that will have an effect on the margins of operation. (COM) [49]

internal blocking (telephone switching systems) The unavailability of paths in a switching network between a given inlet and any suitable idle outlet. (COM) 312-1977w

internal connector (pothead) A connector that joins the end of the cable to the other current-carrying parts of a pothead. *See also:* pothead; transformer. (PE/TR) [107], 48-1975s, [108]

internal correction voltage (electron tube) The voltage that is added to the composite controlling voltage and is the voltage equivalent of such effects as those produced by initial electron velocity and contact potential. *See also:* composite controlling voltage. (ED) 161-1971w

internal font A font that is permanently loaded in a printer's memory. *Synonyms:* permanent font; built-in font. (C) 610.10-1994w

internal fuse (of a capacitor) A fuse connected inside a capacitor unit, in series with an element or a group of elements.power systems relaying. (PE) C37.99-2000

internal graticule (oscilloscopes) A graticule whose rulings are a permanent part of the inner surface of the cathode-ray tube faceplate. (IM) 311-1970w

internal heating (electrolysis) The electrolysis of fused electrolytes is the method of maintaining the electrolyte in a molten condition by the heat generated by the passage of current through the electrolyte. *See also:* fused electrolyte. (EEC/PE) [119]

internal impedance (rotating machinery) The total self-impedance of the primary winding under steady conditions. *Note:* For a three-phase machine, the primary current is considered to have only a positive-sequence component when evaluating this quantity. *See also:* asynchronous machine. (PE) [9]

internal impedance drop (rotating machinery) The product of the current and the internal impedance. *Note:* This is the phasor difference between the generated internal voltage and the terminal voltage of a machine. *See also:* asynchronous machine. (PE) [9]

internal insulation (1) (surge arresters) (apparatus) The insulation that is not directly exposed to atmospheric conditions. (PE/TR) C57.12.80-1978r, [84], [8]
(2) (high voltage testing) Internal insulation comprises the internal solid, liquid, or gaseous elements of the insulation of equipment, which are protected from the effects of atmospheric and other external conditions such as contamination, humidity, vermin, etc. (C) 1313.1-1996
(3) Insulation comprising solid, liquid, or gaseous elements, which are protected from the effects of atmospheric and other external conditions such as contamination, humidity, vermin, etc. (PE/PSIM) 4-1995
(4) Insulating material provided in a radial direction around the energized conductor in order to insulate it from the ground potential. (PE/TR) C57.19.03-1996

internal label (1) A machine-readable label recorded on a data medium that provides information about the data recorded on the medium. *Contrast:* external label. *See also:* header label; end-of-volume label; end of file. (C) 610.5-1990w
(2) A label contained within a data medium, used to mark something such as the beginning or end of a file. *Contrast:* external label. *See also:* end-of-volume label; beginning-of-file label; beginning-of-volume label; end-of-file label. (C) 610.10-1994w
(3) A record in a known format in a known position within a volume that identifies the volume. (C/SS) 1244.1-2000

internal line fault A fault that occurs on the transmission line section that includes the series capacitor installation. (T&D/PE) 824-1994

internal load (power operations) Equal to customer load plus the station service load plus the transmission losses. (PE/PSE) 858-1987s

internal logic state A logic state assumed to exist inside a symbol outline at an input or an output. See figure below.

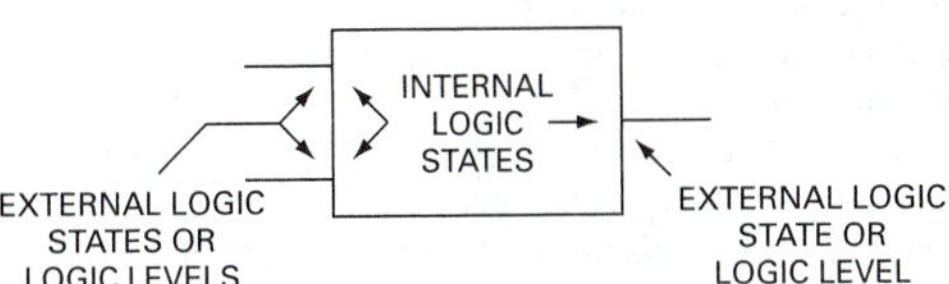

internal logic state

(GSD) 91-1984r

internally fused capacitor unit A capacitor unit that includes internal fuses.power systems relaying. (PE) C37.99-2000

internally-programmed automatic test equipment (test, measurement, and diagnostic equipment) An automated tester using any programming technique in which a substantial amount of programming information is stored within the equipment, although it may originate from external media. (MIL) [2]

internal memory *See:* internal storage.

internal merge sort A merge sort performed within main storage. *Contrast:* external merge sort. *See also:* two-way merge sort; Batcher's parallel sort; list merge sort. (C) 610.5-1990w

internal model A data model depicting entities within the conceptual schema of a database for a specific application. (C) 610.5-1990w

internal node *See:* nonterminal node.

internal ohmic measurements (battery) The measurement of either the internal impedance, conductance, or resistance of battery cells/units. (SB) 1188-1996

internal oxidation *See:* subsurface corrosion.

internal record A record within an internal view. *Synonym:* stored record. (C) 610.5-1990w

internal reference voltage An internally developed bias voltage used as an applied gate voltage in the read mode. (ED) 641-1987w

internal remanent residual voltage (Hall effect devices) That portion of the zero field residual voltage which is due to the remanent magnetic flux density in the ferromagnetic encapsulation of the Hall generator. (MAG) 296-1969w

internal resistance (batteries) The resistance to the flow of an electric current within a cell or battery. *See also:* battery. (EEC/PE) [119]

internal schema (A) A description of the format and layout of the entire contents of a database including the data as well as overhead portions such as indices. *Note:* Written using data definition language. *Contrast:* conceptual schema; external schema. **(B)** A description of the data as it is physically stored in a database, including a description of the environment in which the database is to reside. (C) 610.5-1990

internal short circuit (thyristor power converter) A short circuit caused by converter faults. *Note:* An internal short circuit may be fed from both ac and dc circuits: for example, in the cases of: converters with battery or motor loads; converters in a double converter; converters operating as inverters. (IA/IPC) 444-1973w

internal sort A sort performed within main storage. *Synonym:* in-core sort. *Contrast:* external sort. (C) 610.5-1990w

internal storage (1) (test, measurement, and diagnostic equipment) Storage facilities forming an integral part of the machine. (MIL) [2]
(2) Storage that is accessible by a processor without the use of input-output channels. *Note:* Includes main storage, and may include other kinds of storage such as cache memory and special-purpose registers. *Synonyms:* processor storage; internal memory. (C) 610.10-1994w

internal stress An aging stress that arises from the operational use of the equipment, e.g., voltage, temperature rises or gradients from losses, or self-induced mechanical vibration. (DEI/RE) 775-1993w

internal test (INTEST) A defined instruction for the test logic defined by IEEE Std 1149.1-1990. *Synonym:* INTEST. (TT/C) 1149.1-1990

internal test bus The system of wiring that carries analog test signals around the interior of an integrated circuit. *See also:* test bus interface circuit. (C/TT) 1149.4-1999

internal traffic (telephone switching systems) Traffic originating and terminating within the network being considered. (COM) 312-1977w

Internal Translator (IT) A programming language developed to handle numerical applications, scientific applications, or expressions evaluated from left to right ignoring operator precedence. (C) 610.13-1993w

internal triggering (1) (oscilloscopes) The use of a portion of a deflection signal (usually the vertical deflection signal) as a triggering-signal source. *See also:* oscillograph. (IM/HFIM) [40]
(2) (non-real time spectrum analyzer) The use of a deflection signal (usually the vertical deflection signal) as a triggering source. (IM) [14], 748-1979w

internal variable *See:* endogenous variable.

internal view The format, layout, and contents of the entire data content and overhead content of a database, as described in an internal schema. *Note:* There may be many external views of a database, but only one internal view. (C) 610.5-1990w

internal voltage surge (thyristor converter) Voltage surge caused by sources within a converter. *Note:* It may originate from blowing fuses, hole storage recovery phenomena, etc. Internal voltage surges are substantially under control of the circuit designer. (IA/IPC) 444-1973w

International Algebraic Language A forerunner of the ALGOL language. (C) 610.13-1993w

international call (telephone switching systems) A call to a destination outside of the national boundaries of the calling customer. (COM) 312-1977w

International Commission on Illumination *See:* CIE.

international direct distance dialing (telephone switching systems) The automatic establishing of international calls by means of signals from the calling device of a customer. (COM) 312-1977w

international distance dialing (telephone switching systems) The automatic establishing of international calls by means of signals from the calling device of either a customer or an operator. (COM) 312-1977w

International Federation of Information Processing An international organization of societies that serves information-processing professionals. *See also:* AFIPS. (C) 610.10-1994w

international interzone call (telephone switching systems) A call to a destination outside of a national or integrated numbering-plan area. (COM) 312-1977w

international intrazone call (telephone switching systems) A call to a destination within the boundaries of an integrated numbering-plan area, but outside the national boundaries of the calling customer. (COM) 312-1977w

internationalization The process of designing and developing an implementation with a set of features, functions, and options intended to satisfy a variety of cultural environments. *See also:* localization. (C/PA) 14252-1996

International Morse code (Continental code) A system of dot and dash signals, differing from the American Morse code only in certain code combinations, used chiefly in international radio and wire telegraphy. *See also:* telegraphy. (PE/EEC) [119]

international number (telephone switching systems) The combination of digits representing a country code plus a national number. (COM) 312-1977w

international operating center (telephone switching systems) In World Zone 1, a center where telephone operators handle originating and terminating international interzone calls and may also handle international intrazone calls. *See also:* world-zone number. (COM) 312-1977w

International Organization for Standardization An international organization that establishes and maintains standards for many different industries. (C) 610.7-1995, 610.10-1994w

international originating toll center (telephone switching systems) In World Zone 1, a toll center where telephone operators handle originating international interzone calls. *See also:* world-zone number. (COM) 312-1977w

international switching center (telephone switching systems) A toll office that normally serves as a point of entry or exit for international interzone calls. (COM) 312-1977w

International System of Electrical Units A system that uses the international ampere and the international ohm. *Notes:* 1. The international ampere was defined as the current that will deposit silver at the rate of 0.00111800 gram per second: and the international ohm was defined as the resistance at 0 degrees Celsius of a column of mercury having a length of 106.300 centimeters and a mass of 14.4521 grams. 2. The International System of Electrical Units was in use between 1893 and 1947 inclusive. By international agreement it was discarded, effective January 1, 1948 in favor of the MKSA system. 3. Experiments have shown that as these units were maintained in the United States of America, 1 international ohm equalled 1.000495 ohm and that 1 international ampere equalled 0.999835 ampere. *See also:* International System of Units. (Std100) 270-1966w

International System of Units (SI) A universal coherent system of units in which the following six units are considered basic: meter, kilogram, second, ampere, Kelvin degree, and candela. *Notes:* 1. The MKSA system of electrical units (MKSA System of Units) is a constituent part of this system adequate for mechanics and electromagnetism. 2. The electrical units of this system should not be confused with the units of the earlier International System of Electrical Units which was discarded January 1, 1948. 3. The International System of Units (abbreviated SI) was promulgated in 1960 by the Eleventh General Conference on Weights and Measures. *See also:* units and letter symbols. (Std100) 270-1966w

International Telecommunication Union-Telecommunications Standardization Sector (ITU-TSS) An international organization formerly known as Consultative Committee on International Telegraphy and Telephony (CCITT). *Note:* In March 1993 the CCITT was reorganized and renamed as ITU-TSS. (C) 610.7-1995

International Telegraph and Telephone Consultative Committee An international organization that studies and issues recommendations on issues related to communication technology. *Note:* Also know as CCITT, acronym for Comité Consultatif International de Télégraphique et Téléphonique (French). (C) 610.10-1994w

internetworking The network communication that occurs among devices across multiple networks. (C) 610.7-1995

interoffice call (telephone switching systems) A call between lines connected to different central offices. (COM) 312-1977w

interoffice trunk (1) (telephony) A direct trunk between local central offices in the same exchange. (EEC/PE) [119]
(2) A trunk connecting two telephone offices. (C) 610.7-1995

interoperability (1) (software) The ability of two or more systems or elements to exchange information and to use the information that has been exchanged. *See also:* compatibility (ATLAS/C/PA/SCC20) 1232-1995, 610.12-1990, 14252-1996, 1232.1-1997
(2) The capability for units of equipment to work together to do useful functions. (C/BA) 14536-1995
(3) The capability, promoted but not guaranteed by joint conformance with a given set of standards, that enables heterogeneous equipment, generally built by various vendors, to work together in a network environment. (DIS/C) 1278.2-1995
(4) The ability of two or more systems or components to exchange information in a heterogeneous network and use that information. (C/SE) 1430-1996

interoperability interface A concept of embodying NIRL policies in software objects, interfaces, and services used during interoperation. This allows for enforcement of policy which provides assurance of transaction integrity for interoperation of reuse libraries. (C/SE) 1430-1996

interoperability testing Testing conducted to ensure that a modified system retains the capability of exchanging information with systems of different types, and of using that information. (C/SE) 1219-1998

interoperable Said of two modules indicating that they may both be placed on the same physical MTM-Bus without causing errors of operation. (TT/C) 1149.5-1995

interoperable modules A set of modules that have the properties necessary to allow them to work together in a system to do useful functions. The necessary parameters include items such as electrical, protocol, mechanical, thermal, and I/O interfaces. (C/BA) 14536-1995

interoperate, interoperability Compatibility of modules with each other or with the system/back-plane into which they are inserted. When a pair of such units is said to *interoperate,* (a) they cannot suffer damage as a consequence of being powered and functioning in the same system; (b) the modules and the system will each be able to perform the basic function for which they were designed; (c) the modules will be able to communicate with each other using specified Futurebus+ transactions. Some modules or backplanes will have optional features not shared by others in the same system, so they need to be able to default to the smaller common set in order to communicate with each other. A module or system working in such a reduced-capability mode may suffer performance degradation that may be severe, but will still be interoperable in that system. (C/BA) 896.10-1997, 896.2-1991w

Inter-Packet Gap (IPG) A delay or time gap between CSMA/CD packets intended to provide interframe recovery time for other CSMA/CD sublayers and for the Physical Medium. For example, for 10BASE-T, the IPG is 9.6 μs (96 bit times); for 100BASE-T, the IPG is 0.96 μs (96 bit times). (C/LM) 802.3-1998

interphase rod or shaft A component of a switch-operating mechanism designed to connect two or more poles of a multipole switch for group operation. (SWG/PE) C37.100-1992, C37.30-1971s

interphase transformer (1) An autotransformer, or a set of mutually coupled reactors, used to obtain parallel operation between two or more simple rectifiers that have ripple voltages that are out of phase. *See also:* rectifier transformer. (PE/TR) [57], C57.12.80-1978r
(2) (thyristor power converter) An autotransformer, or a set of mutually coupled inductive reactors, used to obtain multiple operation between two or more simple converters that have ripple voltages that are out of phase. (IA/IPC) 444-1973w

interphase-transformer loss (rectifier transformer) The losses in the interphase transformer that are incident to the carrying of rectifier load. *Note:* They include both magnetic core loss and conductor loss. *See also:* rectifier transformer. (Std100) C57.18-1964w

interphase-transformer rating Consists of the root-mean-square current, root-mean-square voltage, and frequency, at the terminals of each winding, for the rated load of the rectifier unit, and a designated amount of phase control, as assigned to it by the manufacturer. *See also:* rectifier transformer; duty. (PE/TR) [57]

interphone *See:* intercommunicating system.

interphone equipment (aircraft) Equipment used to provide telephone communications between personnel at various locations in an aircraft. (EEC/PE) [119]

interpolation (signal interpolation) (submarine cable telegraphy) A method of reception characterized by synchronous restoration of unit-length signal elements which are weak or missing in the received signals as a result of one or more of such factors as suppression at the transmitter, attenuation in transmission, or discrimination in the receiving networks. *Note:* This is sometimes referred to as local correction. *See also:* telegraphy. (PE/EEC) [119]

interpolation function (burst measurements) A function that may be used to obtain additional values between sampled values. *See also:* burst. (SP) 265-1966w

interpolation search A searching technique in which, at each step of the search, an estimate is made of where the desired record is apt to be. *Synonyms:* external entry search; estimated entry search. *Contrast:* Fibonacci search; dichotomizing search; binary search. (C) 610.5-1990w

interpole *See:* commutating pole.

interposing relay (1) (supervisory control, data acquisition, and automatic control) (station control and data acquisition) A device that enables the energy in a high-power circuit to be switched by a low-power control signal.
(SWG/SUB/PE) C37.1-1987s, C37.100-1992
(2) (of a supervisory system) An auxiliary relay at the master or remote station, the contacts of which serve: (1) to energize a circuit (for closing, opening, or other purpose) of an element of remote station equipment when the selection of a desired point has been completed and when suitable operating signals are received through the supervisory equipment from the master station; or (2) to connect in the circuit the telemeter transmitting and receiving equipment, respectively, at the remote and master stations. *Note:* The interposing relays are considered part of a supervisory system.
(SWG/PE) C37.100-1992

interpret (software) To translate and execute each statement or construct of a computer program before translating and executing the next. *Contrast:* compile; assemble.
(C) 610.12-1990

interpretation line A human-readable interpretation of the bar code that is clearly identifiable with the bar code symbol that shall represent the encoded characters.
(PE/TR) C57.12.35-1996

interpreted card A punch card whose information content is made readable to the human eye by being printed across the top portion of the card. (C) 610.10-1994w

interpreter (1) (software) A computer program that translates and executes each statement or construct of a computer program before translating and executing the next. *Contrast:* assembler; compiler. (C) 610.12-1990
(2) A device that prints on a punch card the characters corresponding to hole patterns punched in the card. *Note:* The result is called an "interpreted card." *See also:* transfer interpreter. (C) 610.10-1994w
(3) A functional entity that translates one or more printer control or page description languages into a form suitable for the marking engine. Since printers sometimes emulate original implementations of these languages, interpreters are sometimes called emulations. (C/MM) 1284.1-1997

interpreter language The printer machine language, or page description language, by which information to be imaged or to be used in imaging is coded. (C/MM) 1284.1-1997

interpretive code Computer instructions and data definitions expressed in a form that can be recognized and processed by an interpreter. *Contrast:* machine code; assembly code; compiler code. (C) 610.12-1990

inter-record gap (test, measurement, and diagnostic equipment) An interval of space or time deliberately left between recording portions of data or records. Such spacing is used to prevent errors through loss of data or overwriting and permits tape stop-start operations. (MIL) [2]

interrecord gap The space between two consecutive records on a data medium. (C) 610.10-1994w

interreflection *See:* interflection.

inter-repeater link (IRL) A mechanism for connecting two and only two repeater sets. (C/LM) 802.3-1998

interrogation In a transponder system, the signal or combination of signals intended to trigger a response.
(AES) 686-1997

interrogative supervisory system A system whereby the master station controls all operations of the system, and whereby all indications are obtained on a master station request or interrogation basis. *Note:* The normal state is usually one of continuous interrogation or polling of the remote stations for changes in status. (SWG/PE) C37.100-1992

interrogator The transmitter of a secondary radar system.
(AES) 686-1997

interrogator-transponder A combined interrogator and transponder. (AES) 686-1997

interrogator-respondor (IR) (radar) A combined radio transmitter and receiver for interrogating a transponder and reporting the resulting replies independently of a radar echo display. *See also:* interrogator. (AES/RS) 686-1990

interrupt (1) (A) (software) The suspension of a process to handle an event external to the process. *Synonym:* interruption. *See also:* interrupt service routine; interrupt mask; priority interrupt; interrupt latency; interrupt priority. **(B) (software)** To cause the suspension of a process. **(C) (software)** Loosely, an interrupt request.
(C) 610.12-1990
(2) A signal for a processor to suspend one process and begin another. As implemented in IEEE Std 1451.2-1997, an interrupt is a signal from the Smart Transducer Interface Module that enables it to request service from the Network Capable Application Processor. (IM/ST) 1451.2-1997
(3) A mechanism provided to allow a device communications controller (DCC) to request service from a bedside communications controller (BCC) prior to the next scheduled polling time. Interrupt is also used to establish initial connection for high-speed DCCs. (EMB/MIB) 1073.4.1-2000

interrupt-acknowledge cycle (1) A data transfer bus (DTB) cycle, initiated by an interrupt handler, that reads a status/ID from an interrupter. An interrupt handler generates this cycle whenever it detects an interrupt request from an interrupter and it has control of the DTB. (C/BA) 1014-1987
(2) A DTB cycle that is initiated by a master in response to an interrupt request from a slave. An interrupt-acknowledge cycle involves two types of slaves. "Contending slaves" have an interrupt request pending and participate in the cycle. The "responding slave" is the one that transfers its status/ID information to the master. During the interrupt-acknowledge cycle, all contending slaves drive an interrupt ID on the bus. This ID is a combination of the geographical address of the board that is supplied by the backplane slot, and a priority code that is supplied by user-defined on-board logic. The interrupt ID is used to determine which of the contending slaves will respond to the cycle. *See also:* contending slave; responding slave. (C/MM) 1096-1988w

interrupted continuous wave (ICW) A continuous wave that is interrupted at a constant audio-frequency rate. *See also:* radio transmission.
(AP/BT/ANT) 145-1983s, 182A-1964w

interrupted quick-flashing light (illuminating engineering) A quick flashing light in which the rapid alternations are interrupted by periods of darkness at regular intervals.
(EEC/IE) [126]

interrupter (1) An element designed to interrupt specified currents under specified conditions. (SWG/PE) C37.100-1992
(2) A functional module that generates an interrupt request on the priority interrupt bus, and then provides status/ID information when the interrupt handier requests it.
(C/BA) 1014-1987

interrupter blade (of an interrupter switch) A blade used in the interrupter for breaking the circuit. (SWG/PE) C37.100-1992

interrupter relay contacts An additional set of contacts on a stepping relay, operated directly by the armature. (EEC/REE) [87]

interrupter switch (1) An air switch, equipped with an interrupter, for making or breaking specified currents, or both. *Note:* The nature of the current made or broken or both may be indicated by suitable prefix; that is load-interrupter switch, fault-interrupter switch, capacitor-current interrupter switch, etc. (SWG/PE/IA/PSE) C37.100-1992, 241-1990r
(2) A switching device, designed for making specified currents and breaking specified steady state currents. (HVS, Swg). *Note:* The nature of the current made or broken, or both, may be indicated by suitable prefix; that is, load interrupter switch, loop interrupter switch, unloaded line interrupter switch, etc. (SWG/PE) 1247-1998

interrupt handler A functional module that detects interrupt requests generated by interrupters and responds to those requests by asking for status/ID information. (C/BA) 1014-1987

interruptible load (1) Demand that can be interrupted by the supplying system in accordance with contractual provisions. (PE/PSE) 858-1993w
(2) (electric power utilization) A load which can be interrupted as defined by contract. *See also:* generating station. (PE/PSE) 346-1973w

interruptible load reserve (power operations) The operating reserve available through disconnection of interruptible loads. (PE/PSE) 858-1987s

interruptible power (power operations) Power which can be interrupted as defined by contract. (PE/PSE) 858-1987s

interrupting aid An arc-interrupting device that can be attached to an air switch to improve its interrupting capability. (SWG/PE) C37.100-1992, C37.36b-1990r

interrupting capacity (packaging machinery) The highest current at rated voltage that the device can interrupt. (IA/PKG) 333-1980w

interrupting current The current in a pole of a switching device at the instant of the initiation of the arc. *Synonym:* breaking current. (SWG/PE) C37.100-1992

interrupting device A device capable of being reclosed whose purpose is to interrupt faults and restore service or disconnect loads. These devices can be manual, automatic, or motor-operated. Examples may include transmission breakers, feeder breakers, line reclosers, motor-operated switches, or others. (PE/T&D) 1366-1998

interrupting device event The operation associated with the interrupting device for cases where a reclosing device operates but does not lockout and where a switch is opened only temporarily. (PE/T&D) 1366-1998

interrupting device operation The operation associated with a reclosing device for cases where the switch opens and closes once but does not lockout. (PE/T&D) 1366-1998

interrupting rating (protection and coordination of industrial and commercial power systems) A rating based on the highest root-mean-square (rms) alternating current that the fuse is required to interrupt under the conditions specified. The interrupting rating, in itself, has no direct bearing on any current-limiting effect of the fuse. (IA/PSP) 242-1986r

interrupting tests Tests that are made to determine or check the interrupting performance of a switching device. (SWG/PE) C37.100-1992, C37.40-1981s

interrupting time (of a mechanical switching device) The interval between the time when the actuating quantity of the release circuit reaches the operating value, the switching device being in a closed position, and the instant of arc extinction on the primary arcing contacts. *Notes:* 1. Interrupting time is numerically equal to the sum of opening time and arcing time. 2. In multipole devices, interrupting time may be measured for each pole or for the device as a whole, in which latter case, the interval is measured to the instant of arc extinction in the last pole to clear. *Synonym:* total break time. (SWG/PE) C37.100-1992

interruption (1) The complete loss of voltage for a time period. The time-base of the interruption is characterized as follows:
— Instantaneous: 0.5 to 30 cycles
— Momentary: 30 cycles to 2 s
— Temporary: 2 s to 2 min
— Sustained: greater than 2 min
(T&D/PE) 1250-1995
(2) The loss of service to one or more customers. *Note:* It is the result of one or more component outages, depending on system configuration. (EEC/PE/T&D/ACO) [109], 1366-1998
(3) The loss of electric power supply to one or more loads. (IA/PSE) 493-1997, 399-1997
(4) The complete loss of voltage for a time period. (IA/PSE) 1100-1999
(5) The suspension of a process to handle an event external to the process. (C) 610.12-1990

interruption duration (1) (electron power systems) The period from the initiation of an interruption to a consumer or other facility until service has been restored to that consumer or facility. *See also:* outage. (PE/PSE) [54]
(2) The period (measured in seconds, or minutes, or hours, or days) from the initiation of an interruption to a customer or other facility until service has been restored to that customer or facility. An interruption may require step-restoration tracking to provide reliable index calculation. It may be desirable to record the duration of each interruption. (PE/T&D) 1366-1998

interruption duration index (electric power system) The average interruption duration for consumers interrupted during a specified time period. It is estimated from operating history by dividing the sum of all consumer interruption durations during the specified period by the number of consumer interruptions during that period. *See also:* outage. (PE/PSE) [54]

interruption, forced *See:* forced interruption.

interruption frequency The expected (average) number of power interruptions to a load per unit time, usually expressed as interruptions per year. (IA/PSE) 493-1997, 399-1997

interruption frequency index (electric power system) The average number of interruptions per consumer served per time unit. *Note:* It is estimated from operating history by dividing the number of consumer interruptions observed in a time unit by the number of consumers served. A consumer interruption is considered to be one interruption of one consumer. *See also:* outage. (PE/PSE) [54]

interruption, momentary event *See:* momentary event interruption.

interruptions caused by events outside of distribution For most utilities, this type of interruption is a small percentage of the total interruptions. It will be defined here to account for the cases where outside influences are a major occurrence. Three categories that may be helpful to monitor are: transmission, generation, and substations. (PE/T&D) 1366-1998

interruption, scheduled *See:* scheduled interruption.

interruption to service The isolation of an electrical load from the system supplying that load, resulting from an abnormality in the system. (SWG/PE) C37.100-1992

interrupt latency The delay between a computer system's receipt of an interrupt request and its handling of the request. *See also:* interrupt priority. (C) 610.12-1990

interrupt mask A mask used to enable or disable interrupts by retaining or suppressing bits that represent interrupt requests. (C) 610.12-1990

interrupt operation (FASTBUS acquisition and control) A FASTBUS write operation to an interrupt service device, notifying it that the sender requires attention. (NID) 960-1993

interrupt priority The importance assigned to a given interrupt request. This importance determines whether the request will cause suspension of the current process and, if there are several outstanding interrupt requests, which will be handled first. (C) 610.12-1990

interrupt register A special-purpose register that holds data necessary for handling interrupts. (C) 610.10-1994w

interrupt request A signal or other input requesting that the currently executing process be suspended to permit performance of another process. (C) 610.12-1990

interrupt service device (ISD) (FASTBUS acquisition and control) A processor or other device that can respond to interrupt operations. (NID) 960-1993

interrupt service routine A routine that responds to interrupt requests by storing the contents of critical registers, performing the processing required by the interrupt request, restoring the register contents, and restarting the interrupted process. (C) 610.12-1990

interrupt vector A value provided by an input-output device, used to distinguish between different input-output functions which can generate the same interrupt. *Note:* A table of interrupt vectors is often provided to allow a processor to look up the address of a service routine. (C) 610.10-1994w

intersection (1) A relational operator that combines two relations having the same degree and results in a relation containing all of the tuples that are in both of the original relations. *See also:* selection; projection; product; join; union; difference.

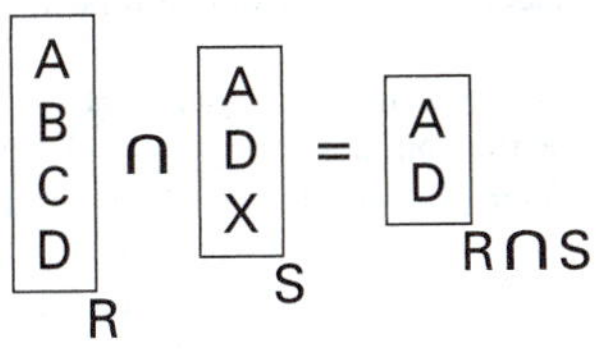

A B C D R ∩ A D X S = A D R∩S

intersection

(C) 610.5-1990w

(2) *See also:* AND. (C) 1084-1986w

interspersing (rotating machinery) Interchanging the coils at the edges of adjacent phase belts. *Note:* The purpose of interspersing depends on the type of machine in which it is done. In asynchronous motors it is used to reduce harmonics that can cause crawling. *See also:* asynchronous machine. (PE) [9]

interstage punching A mode of card punching such that either the odd- or even-numbered card columns are used. (C) 610.10-1994w

intersymbol interference (modulation systems) (transmission system) Extraneous energy from the signal in one or more keying intervals that tends to interfere with the reception of the signal in another keying interval, or the disturbance that results. (Std100) 270-1964w

intersymbol interference penalty The power penalty due to the finite bandwidth of the link. (C/LM) 802.3-1998

intersystem electromagnetic compatibility (control of system electromagnetic compatibility) The condition that enables a system to function without perceptible degradation due to electromagnetic sources in another system. C63.12-1984

intertoll dialing (telephony) Dialing over intertoll trunks. (PE/EEC) [119]

intertoll trunk (telephone switching systems) A trunk between two toll offices. (COM) 312-1977w

interturn insulation (rotating machinery) The insulation between adjacent turns, often in the form of strips. (PE) [9]

interturn test *See:* turn-to-turn test.

interval (1) The spacing between two sounds in pitch or frequency, whichever is indicated by the context. *Note:* The frequency interval is expressed by the ratio of the frequencies or by a logarithm of this ratio. (SP/ACO) [32]

(2) (pulse terminology) The algebraic time difference calculated by subtracting the time of a first specified instant from the time of a second specified instant. (IM/WM&A) 194-1977w

interval-oriented simulation A continuous simulation in which simulated time is advanced in increments of a size suitable to make implementation possible on a digital system. (C) 610.3-1989w

interval, sweep holdoff *See:* sweep holdoff interval.

interval timer A timer, sometimes programmable, which generates a periodic interrupt to a processor, used as a time reference. (C) 610.10-1994w

intervening slave The participating slave that, although not the repository of last resort of the requested data, finds it necessary to prevent the repository of last resort from providing the requested data. Having done so, the intervening slave provides the data instead. (C/BA) 10857-1994, 896.4-1993w, 896.3-1993w

interword spacing In text formatting, the amount of space left between words on a line. *Contrast:* intercharacter spacing. *See also:* line filling. (C) 610.2-1987

interworkability The capability for units of equipment to coexist in the same system, or on the same backplane, and accomplish useful work. (However, units that are interworkable only are not required to work together to accomplish a task. They should only not interfere with one another.) (C/BA) 14536-1995

interworking unit (IWU) A unit that provides the functions needed to allow interworking between a PSN and another network, e.g., interworking between a PSN and an ITU-T X.25 packet-switched public data network (PSPDN). (C/LM/COM) 802.9a-1995w, 8802-9-1996

INTEST *See:* internal test.

intrabeam viewing (laser maser) The viewing condition whereby the eye is exposed to all or part of a laser beam. (LEO) 586-1980w

intra-concha receiver A receiver that essentially fills the ear concha but does not enter the ear canal. (COM/TA) 1206-1994

intraLATA In the United States, a collection of circuits that are totally within a single local access and transport area and are the sole responsibility of the local telephone company. *See also:* interLATA. (C) 610.7-1995

intramodal distortion (fiber optics) That distortion resulting from dispersion of group velocity of a propagating mode. It is the only distortion occurring in single mode waveguides. *See also:* distortion; dispersion. (Std100) 812-1984w

intranet A managed network operating strictly within a single legal entity. More than one intranet may exist within the legal entity, and may be isolated for security reasons. (C) 2001-1999

intra-office blocking Matching loss averaged over all line-to-line connection classes on an ABSBH basis. A connection class is a group of network terminals distinguished by their relative locations in the switching network. Examples include all lines on a single switch, all lines on a single concentrator, all lines on a single switching frame, etc. Matching losses differ among connection classes. *See also:* matching loss. (COM/TA) 973-1990w

intra-office call (telephone switching systems) A call between lines connected to the same central office. (COM) 312-1977w

intrasystem electromagnetic compatibility (control of system electromagnetic compatibility) The condition that enables the various portions of a system to function without perceptible degradation due to electromagnetic sources in other portions of the same system. (Std100) C63.12-1984

intrinsic The specification that a property is total (i.e., mandatory), single-valued, and constant. (C/SE) 1320.2-1998

intrinsically safe circuit A circuit in which any spark or thermal effect is incapable of causing ignition of a mixture of flammable or combustible material in air under prescribed test conditions. Test conditions generally consider opening, shorting, grounding, or field wiring, along with failures in the circuit. (IA/MT) 45-1998

intrinsically safe equipment and wiring Equipment and wiring that are incapable of releasing sufficient electrical or thermal energy under normal or abnormal conditions to cause ignition of a specific hazardous atmospheric mixture in its most easily ignited concentration. This equipment is suitable for use in division 1 locations. Division 2 Equipment and Wiring are equipment and wiring which in normal operation would not ignite a specific hazardous atmosphere in its most easily ignited concentration. The circuits may include sliding or make-and-break contacts releasing insufficient energy to cause ignition. Circuits not containing sliding or make-and-break contacts may have higher energy levels potentially capable of causing ignition under fault conditions. (NEC/IA/PC) [11]

intrinsic coercive force The magnetizing force at which the intrinsic induction is zero when the material is in a symmetrically cyclically magnetized condition. (Std100) 270-1966w

intrinsic date An "instant" in time that is valid without dependence on any specific calendar system or date format. *Note:* An abstract date to and from which all other calendar systems and date representations can be transformed. (C/PA) 2000.1-1999

intrinsic font *See:* bit map font.

intrinsic impedance For a monochromatic (time harmonic) electromagnetic wave propagating in a homogeneous isotropic medium, the ratio of the complex amplitude of the electric field to that of the magnetic field. *Note:* The intrinsic impedance of a medium is sometimes referred to as the characteristic impedance of the medium. (AP/PROP) 211-1997

intrinsic induction (magnetic polarization) At a point in a magnetized body, the vector difference between the magnetic induction at that point and the magnetic induction that would exist in a vacuum under the influence of the same magnetizing force. This is expressed by the equation $\mathbf{B}_i = \mathbf{B} - \mu_0\mathbf{H}$. *Note:* In the centimeter-gram-second electromagnetic-unit system, $\mathbf{B}_i/4\pi$ is often called magnetic polarization. (Std100) 270-1966w

intrinsic joint loss (fiber optics) That loss, intrinsic to the fiber, caused by fiber parameter (for example, core dimensions, profile parameter) mismatches when two nonidentical fibers are joined. *See also:* gap loss; extrinsic joint loss; lateral offset loss; angular misalignment loss. (Std100) 812-1984w

intrinsic loss (or gain) (waveguide) A loss (or gain) resulting from placing two ports of a network between generator and load impedance whose values are adjusted for maximum power absorbed in the load. It is the ratio of the available power from the generator (without generator adjustment) to the power delivered to the load with the network present. (MTT) 146-1980w

intrinsic permeability The ratio of intrinsic normal induction to the corresponding magnetizing force. *Note:* In anisotropic media, intrinsic permeability becomes a matrix. (Std100) 270-1966w

intrinsic properties (semiconductor) The properties of a semiconductor that are characteristic of the pure, ideal crystal. *See also:* semiconductor. (AES/IA/ED) [41], 270-1966w, [12], 216-1960w

intrinsic relationship A kind of relationship that is total, single-valued, and constant from the perspective of (at least) one of the participating classes, referred to as a *dependent class*. Such a relationship is considered to be an integral part of the essence of the dependent class. For example, a transaction has an intrinsic relationship to its related account because it makes no sense for an instance of a transaction to "switch" to a different account since that would change the very nature of the transaction. *Contrast:* nonintrinsic relationship. (C/SE) 1320.2-1998

intrinsic semiconductor (1) (germanium gamma-ray detectors) (charged-particle detectors) (x-ray energy spectrometers) A semiconductor containing an equal number of free holes and electrons throughout its volume. The term "intrinsic germanium" is often used incorrectly for "high purity germanium". *See also:* intrinsic semiconductor. (NPS/NID) 759-1984r, 300-1988r, 325-1996

(2) A semiconductor whose charge-carrier concentration is substantially the same as that of the ideal crystal. *See also:* semiconductor. (ED) 216-1960w

(3) (power semiconductor) A semiconductor in which holes and electrons are created solely by thermal excitation across the energy gap. In an intrinsic semiconductor the concentration of holes and electrons must always be the same. (PE/EDPG) [93]

intrinsic temperature range (semiconductor) The temperature range in which the charge-carrier concentration of a semiconductor is substantially the same as that of an ideal crystal. *See also:* semiconductor. (IA/CEM) 270-1966w, [12], [64]

introspective testing *See:* self-test.

intruder A body that is in motion in an ESD event. The intruder is usually, but not necessarily, charged relative to its surroundings. It is always at a potential different from that of the receptor. (EMC/PE/SPD) C63.16-1993, C62.47-1992r

intruder electrode geometry The size and shape of that surface of the intruder, termed the intruder electrode, at which the ESD takes place. (SPD/PE) C62.47-1992r

intrusion Unauthorized human access to the substation property through physical presence or external influence. (PE/SUB) 1402-2000

intrusion detection Sensing the presence of an intruder or object within specific confines. (PE/NP) 692-1997

***in utero* study** Referring to studies involving the unborn animal. (T&D/PE) 539-1990

invalid (I) An attribute assigned to a cache line if there is not an up-to-date copy in the module's cache. (C/BA) 896.4-1993w

invalid character* *See:* illegal character.

* Deprecated.

invalid date-component value A conditionally invalid date-component value or an unconditionally invalid date-component value. (C/PA) 2000.1-1999

invalid date-specifier A date representation that contains one or more invalid date-component values. (C/PA) 2000.1-1999

invalid frame (1) (local area networks) A frame that is marked with an Invalid Packet Marker (IPM) or that has been identified by the MAC, Repeater Medium Access Control (RMAC), or lower sublayers as containing errors. (C) 8802-12-1998

(2) A PDU that either

1) Does not contain an integral number of octets,
2) Does not contain at least two address octets and a control octet, or
3) Is identified by the physical layer or MAC sublayer as containing data bit errors.

(C/LM/CC) 8802-2-1998

invalid packet marker (IPM) (local area networks) A pattern used by a repeater that is substituted for the end of stream delimiter (esd) in the MII channel transmission frames to identify a packet that was received with transmission errors. (C) 8802-12-1998

invariant An assertion that should always be true for a specified segment or at a specified point of a computer program. (C) 610.12-1990

invariant imbedding A mathematical technique that can be employed to treat radiative transfer problems in the presence of inhomogeneous profiles of absorption and temperature. In it a boundary-value problem is converted to an initial value problem that incorporates the boundary conditions in the equations themselves. The equations are in the form of first-order ordinary differential equations and can be solved by standard methods of initial value problems. (AP/PROP) 211-1997

inverse *See:* ones complement.

inverse binary state *See:* ones complement.

inverse electrode current The current flowing through an electrode in the direction opposite to that for which the tube is designed. *See also:* electrode current. (ED) 161-1971w, [45]

inverse magnitude contours *See:* magnitude contours.

inverse networks Two two-terminal networks are said to be inverse when the product of their impedances is independent of frequency within the range of interest. *See also:* network analysis. (Std100) 270-1966w

inverse neutral telegraph transmission That form of transmission employing zero current during marking intervals and current during spacing intervals. *See also:* telegraphy. (COM) [49]

inverse Nyquist diagram *See:* Nyquist diagram.

inverse-parallel connection An electric connection of two rectifying elements such that the cathode of the first is connected to the anode of the second, and the anode of the first is connected to the cathode of the second. (IA/ICTL/IAC) [60]

inverse period (rectifier element) The nonconducting part of an alternating-voltage cycle during which the anode has a negative potential with respect to the cathode. *See also:* rectification. (PE/EEC) [119]

inverse ratio photocontrol A photocontrol with the turn-off at a lower value than turn-on. (RL) C136.10-1996

inverse standing wave ratio [test] method A test that determines the power dissipation characteristics of a damper by the measurement of antinodal and nodal amplitudes on the span at each tunable harmonic. (T&D/PE) 664-1993

inverse-square law (illuminating engineering) A law stating that the illuminance E at a point on a surface varies directly with the intensity I of a point source, and inversely as the square of the distance d between the source and the point. If the surface at the point is normal to the direction of the incident light, the law is expressed by $E = I/d^2$. *Note:* For sources of finite size having uniform luminance this gives results which are accurate within one percent when d is at least five times the maximum dimension of the source as viewed from the point on the surface. Even though practical interior luminaires do not have uniform luminance, this distance, d, is frequently used as the minimum for photometry of such luminaires, when the magnitude of the measurement error is not critical. (EEC/IE) [126]

inverse-synthetic-aperture radar (ISAR) An imaging radar in which cross-range resolution (angular resolution) of a target (such as a ship, aircraft, or other reflecting object) is obtained by resolving in the Doppler domain the different Doppler frequencies produced by echoes from the individual parts of the object, when these different Doppler frequencies are caused by the object's own angular rotation relative to the radar. (AES) 686-1997

inverse time (1) (as applied to circuit breakers) A qualifying term indicating there is purposely introduced a delay in the tripping action of the circuit breaker, which delay decreases as the magnitude of the current increases. (NESC/NEC) [86]

(2) *See also:* inverse-time relay. (ICTL)

inverse-time delay A qualifying term indicating that there is purposely introduced a delaying action, the delay decreasing as the operating force increases. (PE) C37.100-1992

inverse-time overcurrent relay A current operated relay that produces an inverse time-current characteristic by integrating a function of current *F(I)* with respect to time. The function *F(I)* is positive above and negative below a predetermined input current called the pickup current. Pickup current is therefore the current at which integration starts positively and the relay produces an output when the integral reaches a predetermined positive set value. For the induction relay, it is the disk velocity that is the function of current *F(I)* that is integrated to produce the inverse-time characteristic. The velocity is positive for current above and negative for current below a predetermined pickup current. The predetermined set value of the integral represents the disk travel required to actuate the trip output. (PE/PSR) C37.112-1996

inverse-time relay A relay in which the input quantity and operating time are inversely related throughout at least a substantial portion of the performance range. *Note:* Types of inverse-time relays are frequently identified by such modifying adjectives as "definite minimum time," "moderately," "very," and "extremely" to identify relative degree of inverseness of the operating characteristics of a given manufacturer's line of such relays. (SWG/PE) C37.100-1992

inverse transfer function The reciprocal of a transfer function. (CS/PE/EDPG) [3]

inverse transfer locus The locus of the inverse transfer function. (CS/PE/EDPG) [3]

inverse voltage (rectifier) The voltage applied between the two terminals in the direction opposite to the forward direction. This direction is called the backward direction. *See also:* rectification of an alternating current. (Std100) [84]

inversion (1) (A) (metal-nitride-oxide field-effect transistor) The state of the silicon surface in the insulated-gate field-effect transistor (IGFET) structure when the voltage leading to depletion has been further increased such that a thin layer of minority carriers becomes stable at the surface. **(B) (mathematics of computing)** In Boolean algebra, the same as NOT. **(C) (mathematics of computing)** The process of taking the reciprocal of a number. (ED/C) 581-1978, 1084-1986

(2) (data management) The process of constructing an inverted list to be used to access a set of records. (C) 610.5-1990w

inversion efficiency The ratio of output fundamental power to input direct power expressed in percent. *See also:* self-commutated inverters. (IA) [62]

inversion ratio (laser maser) In a maser medium, the negative of the ratio of the population difference between two non-degenerate energy states under a condition of population inversion to the population difference at equilibrium. (LEO) 586-1980w

invert (A) To change a binary variable to its opposite logic state. **(B)** To take the reciprocal of a number. (C) 1084-1986

inverted (rotating machinery) Applied to a machine in which the usual functions of the stationary and revolving members are interchanged. Example: an induction motor in which the primary winding is on the rotor and is connected to the supply through sliprings, and the secondary is on the stator. *See also:* asynchronous machine. (PE) [9]

inverted file (A) A file whose elements may be retrieved by searching either the primary key or secondary key of each record. *Note:* An inverted file is distinguished from other files by the logical relationship and organization of items and records. In an inverted file, each value of each data item in the records appears exactly once, instead of once in each record. *See also:* fully inverted file; partially inverted file; secondary index. **(B)** A file whose initial sequence has been reversed or whose contents may be searched in reverse order. (C) 610.5-1990

inverted input (oscilloscopes) An input such that the applied polarity causes a deflection polarity opposite from conventional deflection polarity. (IM) 311-1970w

inverted list (A) A list whose contents may be retrieved by searching either the primary key or the secondary key of each element. **(B)** A technique for organizing records in which the primary keys for records that have equivalent values for a given secondary key are stored in a secondary index. *Contrast:* multilist. *See also:* inversion. (C) 610.5-1990

inverted microstrip A compound planar transmission line consisting of one or more thin conducting strips of finite width affixed to an insulating substrate of finite thickness and suspended above a single extended conducting ground plane with the strips facing the ground plane and separated from it by free space. The semi-infinite space above the substrate is also free space. (MTT) 1004-1987w

inverted strip dielectric waveguide A planar dielectric waveguide consisting of one or more dielectric strips of finite width affixed to an extended conducting ground plane on one side and to an extended dielectric layer of finite thickness and higher dielectric constant on the other side. (MTT) 1004-1987w

inverted-turn transposition (rotating machinery) A form of transposition used on multiturn coils in which one or more turns are given a 180-degree twist in the end winding or at the coil nose or series loop. *See also:* rotor; stator. (PE) [9]

inverter (1) A circuit or device whose output analog variable is equal in magnitude to its input analog variable, but is of opposite sign or polarity. (C) 610.10-1994w
(2) (electric power) A machine, device, or system that changes direct-current power to alternating-current power. *See also:* electronic analog computer; inverting amplifier. (IA/PEL/C/PE/ID/ET/EDPG/SPC) 10-1977, 995-1987w, 388-1992r, 165-1977w, 1020-1988r, 936-1987w
(3) Equipment that converts direct current (dc) to alternating current (ac). Any static power converter (SPC) with control, protection, and filtering functions used to interface an electric energy source with an electric utility system. *Notes:* 1. The term "inverter" is popularly used for the converter that serves as the interface device between the PV system dc output and the utility system. However, the definition for SPC more accurately describes this interface device. Because of popular usage, the term "inverter" is used throughout this recommended practice. It should be born in mind that this inverter includes the control, protection, and filtering functions as described in the definition for SPC. 2. Because of its integrated nature, the inverter is only required to be totally disconnected from the utility for service or maintenance. At all other times, whether the inverter is transferring PV energy to the utility or not, the control circuits remain connected to the utility to monitor utility conditions. The phrase "cease to energize the utility line" is used throughout this document to acknowledge that the inverter does not become totally disconnected from the utility when a trip function occurs, such as an overvoltage trip. The inverter can be completely disconnected from the utility for inverter maintenance by opening the ac-disconnect switch required by the NEC®. *Synonym:* static power converter. (SCC21) 929-2000

inverting amplifier (analog computer) An operational amplifier that produces an output signal of nominally equal magnitude and opposite algebraic sign to the input signal. *Note:* In an analog computer, the term inverter is synonymous with inverting amplifier. (C) 165-1977w

inverting parametric device A parametric device whose operation depends essentially upon three frequencies, a harmonic of the pump frequency and two signal frequencies, of which the higher signal-frequency is the difference between the pump harmonic and the lower signal frequency. *Note:* Such a device can exhibit gain at either of the signal frequencies provided power is suitably dissipated at the other signal frequency. It is said to be inverting because if one of the two signals is moved upward in frequency, the other will move downward in frequency. *See also:* parametric device. (ED) [46]

invisible range *See:* visible range.

Invocation The mapping of a parallel initiation of Activities of an Integral Activity Group that perform a distinct function and return to the initiating Activity. *Contrast:* Instance; Iteration. *See also:* mapping. (C/SE) 1074-1997

***in vitro* study** Referring to studies and/or effects that occur outside the living organism; e.g., within a test tube or Petri dish. (T&D/PE) 539-1990

***in vivo* study** Referring to studies and/or effects that occur within the body of living organisms. (T&D/PE) 539-1990

invoke To perform the actions described in 3.9.1.1of IEEE Std 1003.2-1992, except that searching for shell functions and special built-ins is suppressed. *See also:* execute. (C/PA) 9945-2-1993

invoked process That portion of an Integral Process that is called like a subroutine. (C/SE) 1074.1-1995

invoke ID An identifier used to distinguish one directory operation from all other outstanding operations. (C/PA) 1328.2-1993w, 1224.2-1993w, 1327.2-1993w, 1326.2-1993w

invoker identifier (IID) The specific element identifier (EID) of a responding transponder. (SCC32) 1455-1999

inward-wats service (telephone switching systems) A reverse-charge, flat-rate, or measured-time direct distance dialing service to a specific directory number. (COM) 312-1977w

I/O (1) (input/output) Input or output or both. (C) [20], [85]
(2) Input/output points. (SUB/PE) C37.1-1994

IOC *See:* input-output controller; integrated optical circuit.

I/O channel *See:* input-output channel.

I/O circuit *See:* input-output circuit.

I/O controller *See:* input-output controller.

ion (1) An electrically charged atom or radical. (IA) [59]
(2) The isolated atom, molecule, molecular cluster, or aerosol that by loss or gain of one or more electrons has acquired a net electric charge. *Note:* The inclusion of aerosols (particles) under this definition is consistent with historical usage. The use of the terms "small ion" and "charged aerosol" is encouraged. (T&D/PE) 539-1990, 1227-1990r

ion activity (ion species) The thermodynamic concentration, that is, the ion concentration corrected for the deviation from the law of ideal solutions. *Note:* The activity of a single ion species cannot, however, be measured thermodynamically. *See also:* ion. (EEC/PE) [119]

ion burn *See:* ion spot.

ion charge The resultant positive or negative charge of an ion, expressed as a multiple of the electron charge. (T&D/PE) 539-1990

ion charging (charge-storage tubes) Dynamic decay caused by ions striking the storage surface. *See also:* charge-storage tube. (ED) 158-1962w

ion concentration (species of ion) The concentration equal to the number of those ions, or of moles or equivalent of those ions, contained in a unit volume of an electrolyte. (EEC/PE) [119]

ion conduction current The portion of ion current resulting from ion transport due to the electric field. (T&D/PE) 539-1990

ion convection current The portion of ion current resulting from ion transport by fluid dynamic forces, such as wind. (T&D/PE) 539-1990

ion counter An instrument that determines monopolar space-charge density by measuring the charge collected from a known volume of air. (T&D/PE) 539-1990, 1227-1990r

ion current The flow of electric charge resulting from the motion of ions. (T&D/PE) 539-1990

ion exchange technique (fiber optics) A method of fabricating a graded index optical waveguide by an ion exchange process. *See also:* double crucible method; graded index profile; chemical vapor deposition technique. (Std100) 812-1984w

ion gun A device similar to an electron gun but in which the charged particles are ions. Example: proton gun. *See also:* electron optics. (ED) [45], [84]

ionic-heated cathode (electron tube) A hot cathode that is heated primarily by ionic bombardment of the emitting surface. (ED) 161-1971w, [45]

ionic-heated-cathode tube An electron tube containing an ionic-heated cathode. (ED) 161-1971w

ion implantation (A) (germanium gamma-ray detectors) (charged-particle detectors) A process in which a beam of energetic ions incident upon a solid results in the imbedding of those ions into the material. **(B)** A process in which a beam of energetic ions incident upon a solid results in the implantation of those ions into the material. (NPS) 325-1996, 300-1988

ion-implanted contact A detector contact consisting of a junction produced by the process of ion implantation. *See also:* ion implantation. (NPS) 325-1996, 300-1988r

ionization (1) (A) A breakdown that occurs in parts of a dielectric when the electric stress in those parts exceeds a critical value without initiating a complete breakdown of the insulation system. *Note:* Ionization can occur both on internal and external parts of a device. It is a source of radio noise and can damage insulation. **(B)** The process by which an atom or molecule receives enough energy (by collision with electrons, photons, etc.) to split it into one or more free electrons and a positive ion. Ionization is a special case of charging. (PE/IA/T&D/PL/APP) [8], [79], 539-1990

(2) (A) (outdoor apparatus bushings) The formation of limited avalanches of electrons developed in insulation due to an electric field. **(B) (outdoor apparatus bushings)** Ionization current is the result of capacitive discharges in an insulating medium due to electron avalanches under the influence of an electric field. *Note:* The occurrence of such currents may cause radio noise and/or damage to insulation. (PE/TR) 21-1976

(3) (corona measurement) Any process by which neutral molecules or atoms dissociate to form positively and negatively charged particles. (MAG/ET) 436-1977s

ionization current The electric current resulting from the movement of electric charges in an ionized medium, under the influence of an applied electric field. (SPD/PE) C62.11-1999

ionization extinction voltage (cable) (corona level) The minimum value of falling root-mean-square voltage that sustains electric discharge within the vacuous or gas-filled spaces in the cable construction or insulation. (PE) [4]

ionization factor (power distribution, underground cables) (dielectric) The difference between percent power factors at two specified values of electric stress. The lower of the two stresses is usually so selected that the effect of the ionization on power factor at this stress is negligible. (PE) [4]

ionization-gauge tube An electron tube designed for the measurement of low gas pressure and utilizing the relationship between gas pressure and ionization current. (ED) [45]

ionization measurement The measurement of the electric current resulting from the movement of electric charges in an ionized medium under the influence of the prescribed electric field. (PE/TR) 21-1976

ionization or corona detector *See:* discharge detector.

ionization or corona inception voltage *See:* discharge inception voltage.

ionization or corona probe *See:* discharge probe.

ionization smoke detector (fire protection devices) A device which has a small amount of radioactive material which ionizes the air in the sensing chamber, thus rendering it conductive and permitting a current flow through the air between two charged electrodes. This gives the sensing chamber an effective electrical conductance. When smoke particles enter the ionization area, they decrease the conductance of the air by attaching themselves to the ions, causing a reduction in mobility. When the conductance is less than the predetermined level, the detector circuit responds. (NFPA) [16]

ionization time (gas tube) The time interval between the initiation of conditions for and the establishment of conduction at some stated value of tube voltage drop. *Note:* To be exact the ionization time of a gas tube should be presented as a family of curves relating such factors as condensed-mercury temperature, anode and grid currents, anode and grid voltages, and regulation of the grid current. (ED) 161-1971w

ionization vacuum gauge A vacuum gauge that depends for its operation on the current of positive ions produced in the gas by electrons that are accelerated between a hot cathode and another electrode in the evacuated space. *Note:* It is ordinarily used to cover a pressure range of 10^{-4} to 10^{-10} conventional millimeters of mercury. *See also:* instrument. (EEC/PE) [119]

ionization voltage A high-frequency voltage appearing at the terminals of an arrester, generated by all sources, but particularly by ionization current within the arrester, when a power-frequency voltage is applied across the terminals. (SPD/PE) C62.11-1999

ionizing event (gas-filled radiation counter tube) Any interaction by which one or more ions are produced. (ED) 161-1971w

ionizing radiation (1) (A) (air) Particles or photons of sufficient energy to produce ionization in their passage through air. **(B) (air)** Particles that are capable of nuclear interactions with the release of sufficient energy to produce ionization in air. (NPS) 175-1960

(2) Particles or photons of sufficient energy to produce ionization in interactions with matter. (NI/NPS) 309-1999

ion migration A movement of ions in an electrolyte as a result of the application of an electric potential. *See also:* ion. (EEC/PE) [119]

ion mobility (1) The theoretical drift speed of a single, isolated ion in a liquid or gas, per unit electric field strength. The preferred unit is m^2/Vs; another commonly used unit is cm^2/Vs. Ion mobility depends on the ionic species. In air, several ionic species can exist simultaneously. (T&D/PE) 539-1990

(2) The drift speed of an ion in a liquid or gas per unit electric-field strength. The preferred unit is m^2/Vs; another commonly used unit is cm^2/Vs. (T&D/PE) 1227-1990r

ionogram A record showing the group path delay of ionospheric echoes as a function of frequency. (AP/PROP) 211-1997

ionosonde A swept-frequency or stepped frequency instrument that transmits radio waves vertically or obliquely to the ionosphere and uses the echoes to form an ionogram. (AP/PROP) 211-1997

ionosphere (1) (data transmission) That part of the earth's outer atmosphere where ions and free electrons are normally present in quantities sufficient to affect propagation of radio waves. (PE) 599-1985w

(2) That part of a planetary atmosphere where ions and free electrons are present in quantities sufficient to affect the propagation of radio waves. (AP/PROP) 211-1997

ionosphere disturbance A variation in the state of ionization of the ionosphere beyond the normally observed random day-to-day variation from average values for the location, date, and time of day under consideration. *Note:* Since it is difficult to draw the line between normal and abnormal viations, this definition must be understood in a qualitative sense. *See also:* radiation. (EEC/PE) [119]

ionosphere-height error (electronic navigation) The systematic component of the total ionospheric error due to the difference in geometrical configuration between ground paths and ionospheric paths. *See also:* navigation. (AES/RS) 686-1982s, [42]

ionospheric error (electronic navigation) The total systematic and random error resulting from the reception of the navigational signal via ionospheric reflections: this error may be due to variations in transmission paths, nonuniform height of the ionosphere, and nonuniform propagation within the ionosphere. *See also:* navigation. (AES/RS) 686-1982s, [42]

ionospheric absorption The loss of energy from an electromagnetic wave caused by collisions, primarily between electrons and neutral species and ions in the ionosphere. (AP/PROP) 211-1997

ionospheric mode of propagation Representation of a transmission path by the number of hops between the end points of the path, the ionospheric layers producing the ionospheric reflections being indicated for each hop. For example, 1F + 1E represents a hop with an ionospheric reflection in the F region followed by a reflection at the ground, followed, in turn, by a hop with a reflection from the E region. *Synonym:* mechanism of propagation. (AP/PROP) 211-1997

ionospheric storm An ionospheric disturbance characterized by wide variations from normal in the state of the ionosphere, including effects such as turbulence in the F region, increases in absorption, and often decreases in ionization density and increases in virtual height. *Note:* The effects are most marked in high magnetic latitudes and are associated with abnormal solar activity. *See also:* radiation. (EEC/PE) [119]

ionospheric tilt error (electronic navigation) The component of the ionospheric error due to nonuniform height of the ionosphere. *See also:* navigation. (AES/RS) 686-1982s, [42]

ionospheric wave *See:* sky wave.

ion repeller (charge-storage tubes) An electrode that produces a potential barrier against ions. *See also:* charge-storage tube. (ED) 158-1962w

ion sheath *See:* electron sheath.

ion size Physical dimensions and mass of an ion. Ions are usually classified as small, medium, and large. *Note:* The radius and mass of an ion depend on the number and type of molecules in the cluster forming the ion. The diameter of an ion comprised of a *single* molecule is about 3×10^{-10} m. (T&D/PE) 539-1990

ion spot (A) (camera tubes or image tubes) The spurious signal resulting from the bombardment or alteration of the target or photocathode by ions. *See also:* television. **(B)** (cathode-ray-tube screen) An area of localized deterioration of luminescence caused by bombardment with negative ions. (ED/BT/AV) 161-1971, [34]

ion transfer (ionotherapy) (electrotherapy) (ion therapy) (iontophoresis) (ionic medication) (medical ionization) The forcing of ions through biological interfaces by means of an electric field. *See also:* electrotherapy. (EMB) [47]

ion trap (cathode-ray tubes) A device to prevent ion burn by removing the ions from the beam. *See also:* isolating transformer.

IOP *See:* input-output processor.

I/O port *See:* input-output port.

I/O processor *See:* input-output processor.

I/O space The address space used for accessing peripheral devices such as communication controllers and mass storage devices. (C/MM) 1296-1987s

I/O transaction An instance of activity between Functions, usually composed of an Initiation and a Completion, although not necessarily bound one to one. A disk read and network data delivery are examples of I/O transactions. (C/MM) 1212.1-1993

I/O Unit memory Memory that is located in the I/O Unit. For most purposes in this document, the externally visible locations may interchangeably be in either the initial node space (4 kbytes) or unit-extension (or even the memory-extension) areas defined in IEEE Std 1212-1991. (C/MM) 1212.1-1993

Ip *See:* peak current.

IPG *See:* Inter-Packet Gap.

IPL *See:* Information Processing Language.

IPM *See:* invalid packet marker.

IPO chart *See:* input-process-output chart.

IPSE *See:* programming support environment.

$I_p \cdot T$ product (as applicable to transformers) The $I \cdot T$ product measured in the primary of a transformer when the secondary is open-circuited. It is abbreviated as $I_p \cdot T$. (COM/TA) 469-1988w

IR *See:* interrogator-respondor; infrared.

IRA *See:* laser gyro axes.

IRC *See:* information center; information resource center.

IR drop (electrolytic cells) The drop equal to the product of the current passing through the cell and the resistance of the cell. (EEC/PE) [119]

IR-drop compensation transformer (power and distribution transformers) A provision in the transformer by which the voltage drop due to transformer load current and internal transformer resistance is partially or completely neutralized. Such transformers are suitable only for one-way transformation, that is, not interchangeable for step-up and step-down transformations. (PE/TR) C57.12.80-1978r

Iref *See:* reference current.

iris (1) (waveguide technique) A metallic plate, usually of small thickness compared with the wavelength, perpendicular to the axis of a waveguide and partially blocking it. *Notes:* 1. An iris acts like a shunt element in a transmission line: it may be inductive, capacitive, or resonant. 2. When only a single mode can be supported an iris acts substantially as a shunt admittance. (AP/ANT) [35]
(2) (waveguide components) A partial obstruction at a transverse cross-section formed by one or more metal plates of small thickness compared with the wavelength. (MTT) 147-1979w
(3) (laser maser) The circular pigmented membrane which lies behind the cornea of the human eye. The iris is perforated by the pupil. (LEO) 586-1980w

IRL *See:* inter-repeater link.

IRM *See:* isochronous resource manager.

ironclad plate (storage cell) A plate consisting of an assembly of perforated tubes of insulating material and of a centrally placed conductor. *Note:* "Ironclad" is a registered trademark of ESB Incorporated. *See also:* battery. (PE/EEC) [119]

irradiance (1) (laser maser) (at a point of a surface) (E) Quotient of the radiant flux incident on an element of the surface containing the point by the area of that element. Unit: W · cm^{-2}. (LEO) 586-1980w
(2) (fiber optics) Radiant power incident per unit area upon a surface, expressed in watts per square meter. "Power density" is colloquially used as a synonym. *See also:* radiometry. (Std100) 812-1984w

irrational number A real number that is not a rational number. *Contrast:* rational number. (C) 610.5-1990w

irreversible dark current increase (diode-type camera tube) That dark current increase which results from irradiation of the target by soft x rays. (ED) 503-1978w

irreversible process An electrochemical reaction in which polarization occurs. *See also:* electrochemistry. (EEC/PE) [119]

irreversible target dark current increase (diode-type camera tube) That dark current increase which is permanent and increases with hours of operation. (ED) 503-1978w

irrigation machines An irrigation machine is an electrically driven or controlled machine, with one or more motors, not hand portable, and used primarily to transport and distribute water for agricultural purposes. (NESC/NEC) [86]

IS *See:* information separator.

ISA instruction set architecture. (C/BA) 14536-1995

ISAM *See:* indexed sequential access method.

ISAR *See:* inverse-synthetic-aperture radar.

I-scope A cathode-ray oscilloscope arranged to present an I-display. *See also:* I-display. (AES/RS) 686-1990

ISD *See:* interrupt service device.

I-series A series of ISDN standards reccommended by CCITT. (C) 610.10-1994w

ISI *See:* information services interface.

island (1) An operating part of a DQDB *subnetwork* that is isolated from the node containing the *default slot generator function*. (LM/C) 8802-6-1994
(2) That part of a power system consisting of one or more power sources and load that is, for some period of time, separated from the rest of the system. (SWG/PE) C37.100-1992

island effect (electron tube) The restriction of the emission from the cathode to certain small areas of it (islands) when the grid voltage is lower than a certain value. (Std100) [84]

islanding (1) (utility-interconnected static power converters) Operation of the power converter and part of the utility load while isolated from the remainder of the electric utility system. (DESG) 1035-1989w
(2) (windfarm generating stations) Operation of non-utility electric generation equipment, with or without a portion of an electric utility system, isolated from the remainder of the utility system. (DESG) 1094-1991w
(3) A condition in which a portion of the utility system that contains both load and distributed resources remains energized while isolated from the remainder of the utility system. (SCC21) 929-2000

ISM apparatus (industrial, scientific, and medical apparatus; electromagnetic compatibility) Apparatus intended for generating radio-frequency energy for industrial, scientific or medical purposes. *See also:* electromagnetic compatibility. (INT) [53], [70]

ISO *See:* International Organization for Standardization.

isocandela line (illuminating engineering) A line plotted on any appropriate set of coordinates to show directions in space, about a source of light, in which the intensity is the same. A series of such curves, usually for equal increments of intensity, is called an isocandela diagram. (EEC/IE) [126]

isoceraunic map *See:* isokeraunic map.

isochronous (1) The time characteristic of an event or signal recurring at known, periodic time intervals. (LM/C) 8802-6-1994
(2) A communication stream transport that is uniform in time. The delivery of the physical stream of information is recurring at regular intervals. (C/LM/COM) 802.9a-1995w, 8802-9-1996
(3) The essential characteristic of a time-scale or a signal such that the time intervals between consecutive significant instances either have the same duration or durations that are integral multiples of the shortest duration. (C/MM) 1394-1995
(4) Uniform in time (i.e., having equal duration) and recurring at regular intervals. (C/MM) 1394a-2000

isochronous channel A relationship between a talker and one or more listeners, identified by a channel number. One packet for each channel is sent during each isochronous cycle. Channel numbers are assigned using the isochronous resource management facilities. (C/MM) 1394-1995

isochronous cycle An operating mode of the bus that begins after a cycle start is sent, and ends when a subaction gap is detected. During an isochronous cycle, only isochronous subactions may occur. An isochronous cycle begins every 125 μs, on average. (C/MM) 1394-1995

isochronous gap (1) The period of idle bus before the start of arbitration for an isochronous subaction. (C/MM) 1394-1995
(2) For an isochronous subaction, the period of idle bus that precedes arbitration. (C/MM) 1394a-2000

isochronous period A period that begins after a cycle start packet is sent and ends when a subaction gap is detected. During an isochronous period, only isochronous subactions may occur. An isochronous period begins, on average, every 125 μs. (C/MM) 1394a-2000

isochronous resource manager (1) The node that contains the facilities needed to manage isochronous resources. In particular, the isochronous resource manager includes the BUS_ MANAGER_ID, BANDWIDTH_AVAILABLE, and CHANNELS_AVAILABLE registers. In addition, if there is no bus manager on the local bus, the isochronous resource manager may also perform limited power management and select a node to be the cycle master. (C/MM) 1394-1995
(2) A node that implements the BUS_MANAGER_ID, BANDWIDTH_AVAILABLE, CHANNELS_AVAILABLE and BROADCAST_CHANNEL registers (some of which permit the cooperative allocation of isochronous resources). Subsequent to each bus reset, one isochronous resource manager is selected from all nodes capable of this function. (C/MM) 1394a-2000

isochronous service octet A single octet of data passed isochronously between the DQDB layer and the isochronous service user (ISU). (LM/C) 8802-6-1994

isochronous service user (ISU) The entity that uses the isochronous service provided by the DQDB layer to transfer isochronous service octets over an established isochronous connection. (LM/C) 8802-6-1994

isochronous speed governing (gas turbines) Governing with steady-state speed regulation of essentially zero magnitude. (PE/EDPG) [5], 282-1968w

isochronous subaction (1) A complete link layer operation (arbitration and isochronous packet) that is sent only during an isochronous cycle. (C/MM) 1394-1995
(2) Within the isochronous period, either a concatenated packet or a packet and the gap that preceded it. (C/MM) 1394a-2000

isocon mode (camera tubes) A low-noise return-beam mode of operation utilizing only back-scattered electrons from the target to derive the signal, with the beam electrons specularly reflected by the electrostatic field near the target being separated and rejected. *See also:* camera tube. (ED) [45]

isoelectric point A condition of net electric neutrality of a colloid, with respect to its surrounding medium. *See also:* ion. (EEC/PE) [119]

isokeraunic level (lightning) The average annual number of thunderstorm days. *See also:* direct-stroke protection. (T&D/PE) [10]

isokeraunic lines Lines on a map connecting points having the same keraunic level. (SUB/PE) 998-1996

isokeraunic map A map showing equal levels of thunderstorm activity. Usually shown in mean annual days of thunderstorm activity. *Synonym:* isoceraunic map. *See also:* keraunic level. (T&D/PE) 751-1990

isolated (A) Physically separated, electrically and mechanically, from all sources of electrical energy. Such separation may not eliminate the effects of electrical induction. **(B)** Not readily accessible to persons unless special means for access are used. (PE/T&D/IA/NESC/PC) 516-1987, 524-1992, 1048-1990, 458-1990, [86], 463-1977, C2.2-1960, C2-1997

isolated bonding network (IBN) (A) A bonding network that has a single point of connection (single-point ground) to either the common bonding network (CBN) or another isolated bonding network. **(B)** Typically a system-level grounding topology used by the original equipment manufacturer (OEM) to desensitize its equipment to suspected or known site environmental issues such as power fault and surge, lightning, and grounding potential rise. The IBN requires the use of a single-point connection location (also known in the telephone industry as a *ground window*) to interface the rest of the building metallics (the CBN). *Note:* The IBN may also be known in the public telephone network as an *isolated ground plane*. (IA/PSE) 1100-1999

isolated by elevation Elevated sufficiently so that persons may safely walk underneath. (NESC) C2-1997

isolated capacitor bank A capacitor bank that is not in parallel with other capacitor banks. (T&D/PE) 1036-1992

isolated conductor (ignored conductor) In a multiple-conductor system, a conductor either accessible or inaccessible, the charge of which is not changed and to which no connection is made in the course of the determination of any one of the capacitances of the remaining conductors of the system. (Std100) 270-1966w

isolated equipment ground An isolated equipment grounding conductor runs in the same conduit or raceway as the supply conductors. This conductor is insulated from the metallic raceway and all ground points throughout its length. It originates at an isolated-ground-type receptacle or equipment input terminal block and terminates at the point where neutral and ground are bonded at the power source. (IA/PSE) 1100-1999

isolated impedance (of an array element) The input impedance of a radiating element of an array antenna with all other elements of the array absent. (AP/ANT) 145-1993

isolated-neutral system A system that has no intentional connection to ground except through indicating, measuring, or protective devices of very-high impedance. *See also:* grounded system. (PE) [8], [84]

isolated node A node without active ports; the node's ports may be disabled, disconnected, or suspended in any combination. (C/MM) 1394a-2000

isolated patient lead (health care facilities) A patient lead whose impedance to ground or the power line is sufficiently high that connecting the lead to ground, or to either conductor of the power line, results in current flow in the lead which is below a hazardous limit. (EMB) [47]

isolated-phase bus (1) (generating station grounding) A metal-enclosed bus in which each phase conductor is enclosed by an individual metal housing separated from adjacent conductor housings by an air space. (PE/EDPG) 665-1995 **(2)** A bus in which each phase conductor is enclosed by an individual metal housing separated from adjacent conductor housings by an air space. *Note:* The bus may be self-cooled or may be forced-cooled by means of circulating a gas or liquid. (SWG/PE) C37.100-1992

isolated plant (electric power) An electric installation deriving energy from its own generator driven by a prime mover and not serving the purpose of a public utility. (EEC/PE) [119]

isolated power system A system comprising an isolating transformer or its equivalent, a line isolation monitor, and its ungrounded circuit conductors. (NEC/NESC/EMB) [86], [47]

isolated redundant UPS configuration Uses a combination of automatic transfer switches and a reserve system to serve as the bypass source for any of the active systems. (IA/PSE) 241-1990r

isolating amplifier (signal-transmission system) An amplifier employed to minimize the effects of a following circuit on the preceding circuit. *Example:* An amplifier having effective direct-current resistance and/or alternating-current impedance between any part of its input circuit and any other of its circuits that is high compared to some critical resistance or impedance value in the input circuit. *See also:* signal. (IE) [43]

isolating contactor (power system device function numbers) A device that is used expressly for disconnecting one circuit from another for the purposes of emergency operation, maintenance, or test. (PE/SUB) C37.2-1979s

isolating device A device in a circuit that prevents malfunction in one section of a circuit from causing acceptable influences in other sections of the circuit or other circuits. (PE/NP) 308-1991

isolating switch (A) A switch intended for isolating an electric circuit from the source of power. It has no interrupting rating, and it is intended to be operated only after the circuit has been opened by some other means. **(B)** Device used to isolate plant electrical equipment from the rest of the circuit. (NESC/NEC/PE/EDPG) [86], 1020-1988

isolating time (of a sectionalizer) The time between the cessation of a current above the minimum actuating current value that caused the final counting and opening operation and the maximum separation of the contacts. (SWG/PE) C37.100-1992

isolating transformer (electroacoustics) A transformer inserted in a system to separate one section of the system from undesired influences of other sections. *Note:* Isolating transformers are commonly used to isolate system grounds and prevent the transmission of undesired currents. (SP/IE) 151-1965w, [43]

isolating (insulating) transformers Transformers that provide longitudinal (common mode) isolation of the telecommunication facility. They can be designed for use in a combined isolating-drainage transformer configuration and also can be designed for a low longitudinal to metallic conversion. *Synonym:* insulating transformer. (PE/PSC) 487-1992

isolating transformers with high-voltage isolating relays An assembly that provides protection for standard telephone service and consists basically of an isolating transformer and a high-voltage isolating relay. The transformer provides a path for voice and ringing frequencies while the relay provides a means for repeating dc signals around the transformer. A locally supplied battery or dc power supply is required for operation of the telephone and relay. (PE/PSC) 487-1992

isolation (1) (A) (nonlinear, active, and nonreciprocal waveguide components) *circulator.* The ratio of insertion loss to an isolated port relative to insertion loss to the coupled port in a circulator. **(B) (nonlinear, active, and nonreciprocal waveguide components)** *ferrite isolator.* The ratio of insertion loss in the reverse direction to insertion loss in the forward direction in an isolator. **(C) (nonlinear, active, and nonreciprocal waveguide components)** *mixer.*The degree to which the amplitude of an undesired wave is suppressed relative to the amplitude of the desired wave. **(D) (nonlinear, active, and nonreciprocal waveguide components)** *switch.* The ratio of insertion loss in the OFF (open) state to the insertion loss in the ON (closed) state in a switch. (ELM) C12.9-1982 **(2)** A measure of power transfer from one antenna to another. *Note:* The isolation between antennas is the ratio of power input to one antenna to the power received by the other, usually expressed in decibels. *See also:* radiation. (AP/ANT) 145-1993 **(3)** Separation of one section of a system from undesired influences of other sections. (IA/PSE) 1100-1999

isolation amplifier (buffer) An amplifier employed to minimize the effects of a following circuit on the preceding circuit. *See also:* amplifier. (BT/AV) [34]

isolation between antennas A measure of power transfer from one antenna to another. *Note:* The isolation between antennas is the ratio of power input to one antenna to the power received by the other, usually expressed in decibels. *See also:* radiation. (AP/ANT) 145-1993

isolation boundary (periodic testing of diesel-generator units applied as standby power supplies in nuclear power generating stations) A supporting system, subsystem, or device (valve, control power circuit breaker, switch, etc.) which provides a boundary with the diesel-generator unit. Failures of the device or the supporting system, or subsystem are not considered diesel-generator unit failures. (PE/NP) 749-1983w

isolation by elevation *See:* isolated by elevation.

isolation device (Class 1E equipment and circuits) (nuclear power generating station) A device in a circuit that prevents malfunctions in one section of a circuit from causing unacceptable influences in other sections of the circuit or other circuits. (PE/NP) 380-1975w, 384-1992r

isolation transformer (1) (health care facilities) A transformer of the multiple-winding type, with the primary and secondary windings physically separated, which inductively couples its ungrounded secondary winding to the grounded feeder system that energizes its primary winding. (EMB) [47] **(2) (health care facilities)** A transformer of the multiple-winding type, with the primary and secondary windings physically separated, that inductively couples its secondary winding to the grounded feeder systems that energize its primary winding, thereby preventing primary circuit potential from being impressed on the secondary circuits. (NESC/NEC) [86]

isolation voltage (power supplies) A rating for a power supply that specifies the amount of external voltage that can be connected between any output terminal and ground (the chassis). This rating is important when power supplies are connected in series. (AES) [41]

isolation zone Any area, adjacent to a perimeter physical barrier, cleared of objects that could conceal or shield an individual. The inner isolation zone is inside of the perimeter physical barrier; the outer isolation zone is outside of the perimeter physical barrier. (PE/NP) 692-1997

isolator (1) (SWG)
(2) (waveguide) A passive attenuator in which the loss in one direction is much greater than that in the opposite direction. *See also:* waveguide. [84]
(3) (fiber optics) A device intended to prevent return reflections along a transmission path. *Note:* The Faraday isolator uses the magneto-optic effect. (Std100) 812-1984w

isolator, optical *See:* optical isolator.

isolux (isofootcandle) line (illuminating engineering) A line plotted on any appropriate set of coordinates to show all the points on a surface where the illuminance is the same. A series of such lines for various illuminance values is called an isolux (isofootcandle) diagram. (EEC/IE) [126]

isophase *See:* equal interval (isophase) light.

isopreference (speech quality measurements) Two speech signals are isopreferent when the votes averaged over all listeners show an equal preference for the speech test and speech reference signals. 297-1969w

ISO Standard A standard approved and published by International Organization for Standardization. (C) 610.7-1995

isothermal (electric power systems in commercial buildings) A process that occurs at a constant temperature. (IA/PSE) 241-1990r

isotropic (fiber optics) Pertaining to a material whose electrical or optical properties are independent of direction of propagation and of polarization of a traveling wave. *See also:* birefringent medium; anisotropic. (Std100) 812-1984w

isotropic radiator A hypothetical, lossless antenna having equal radiation intensity in all directions. *Note:* An isotropic radiator represents a convenient reference for expressing the directive properties of actual antennas. (AP/ANT) 145-1993

isotropic scatterer A non-physical scatterer that scatters equally in all directions. (AP/PROP) 211-1997

ISTE *See:* integrated services terminal equipment.

$I_s \cdot T$ product (as applicable to transformers) The $I \cdot T$ product measured in the secondary (low voltage) of a transformer when the primary (high voltage) is open-circuited. It is abbreviated as $I_s \cdot T$. (COM/TA) 469-1988w

ISU *See:* isochronous service user.

IT *See:* Internal Translator.

ITA *See:* interface test adapter.

item (1) (nuclear power generating station) Any level of unit assembly, including structure, system, subsystem, subassembly, module, component, part, equipment or material. *Note:* This term applies specifically to the subject matter of IEEE Std 467-1980. (PE/NP) 467-1980w
(2) (nuclear power quality assurance) An all-inclusive term used in place of any of the following: appurtenance, assembly, component, equipment, material, module, part, structure, subassembly, subsystem, system, or unit. (PE/NP) [124]
(3) (computers) A collection of related characters, treated as a unit. *See also:* file. (C) [85]
(4) An all-inclusive term to denote any level of hardware assembly: that is, system, segment of a system, subsystem, equipment, component, part, etc. *Note:* Item includes items, population of items, sample, etc., where the context of its use so justifies. *See also:* reliability. (R) [29]
(5) (data management) One member of a group; for example, a field in a record or a record in a file. *See also:* data item. (C) 610.5-1990w

item condition A disjunction of two or more atomic conditions such that the name of the data item is the same in each atomic condition. For example, "LASTNAME = 'JONES' or LASTNAME = 'SMITH' or LASTNAME = 'GREEN.' ". *Note:* The disjunction may be implied, as in the example "LASTNAME = ('JONES','SMITH','GREEN' "). *See also:* record condition. (C) 610.5-1990w

item, nonrepaired *See:* nonrepaired item.

item or equipment hazard rate (reliability data for pumps and drivers, valve actuators, and valves) The instantaneous failure rate of an item or equipment or its conditional probability of failure versus time. (PE/NP) 500-1984w

item, repaired *See:* repaired item.

Iteration The mapping of any execution of an Activity where at least some Input Information is processed and some Output Information is created. One or more Iterations comprise an Instance. *Contrast:* Instance; Invocation. *See also:* mapping. (C/SE) 1074-1997

iteration (A) (software) The process of performing a sequence of steps repeatedly. *See also:* loop; recursion. **(B) (software)** A single execution of the sequence of steps in definition (A). (C) 610.12-1990

iterative (test, measurement, and diagnostic equipment) Describing a procedure or process which repeatedly executes a series of operations until some condition is satisfied. An iterative procedure may be implemented by a loop in a routine. (MIL) [2]

iterative construct *See:* loop.

iterative document A document that will be produced multiple times with relatively few changes in the text. For example, a letter meant to be prepared 100 times, each with a different name and address in the salutation. *Synonym:* form letter. (C) 610.2-1987

iterative impedance (transducer or a 2-port network) The impedance that, when connected to one pair of terminals, produces a like impedance at the other pair of terminals. *Notes:* 1. It follows that the iterative impedance of a transducer or a network is the same as the impedance measured at the input terminals when an infinite number of identically similar units are formed into an iterative or recurrent structure by connecting the output terminals of the first unit to the input terminals of the second, the output terminals of the second to the input terminals of the third, etc. 2. The iterative impedances of a four-terminal transducer or network are, in general, not equal to each other but for any symmetrical unit the iterative impedances are equal and are the same as the image impedances. The iterative impedance of a uniform line is the same as its characteristic impedance. (SP/IM/HFIM) 151-1965w, [40]

iterative operation (analog computer) Similar in many respects to repetitive operation, except that the automatic recycling of the computer is controlled by programmed logic circuits, which generally include a program change for a parameter(s), variable(s), or combinations of these between successive solutions, resulting in an iterative process which tends to converge on desired values of the parameter(s) or variables(s) that have been changed. *See also:* repetitive operation. (C) 165-1977w

ITP *See:* inspection and test plan.

$I \cdot T$ product The inductive influence expressed in terms of the product of its root-mean-square magnitude (I), in amperes, times its telephone influence factor (TIF). (COM/IA/SUB/PE/TA/SPC) 469-1988w, 519-1992, 1303-1994

I · T product kV · T product (voice-frequency electrical-noise test) Inductive influence usually expressed in terms of the product of its root-mean-square magnitude in kilovolts times its telephone influence factor (TIF), abbreviated as kV · T product. (COM/TA) 469-1977s

ITRV *See:* initial transient recovery voltage.

ITU-TSS *See:* X.200; X.25; V-series; X-series; International Telecommunication Union-Telecommunications Standardization Sector; X.400; X.75.

I-unit *See:* instruction fetch unit.

IV *See:* initialization vector.

***I* Video Signal (National Television System Committee color television)** One of the two video signal (E'_I and E'_Q) controlling the chrominance in the NTSC system. *Note:* It is a linear combination of gamma-corrected primary color signals, E'_R, E'_G, and E'_B as follows:

$$E'_I = -0.27(E'_B - E'_Y) + 0.74(E'_R - E'_Y)$$
$$= 0.60E'_R - 0.28E'_G - 0.32E'_B$$

(BT/AV) 201-1979w

IVV *See:* independent verification and validation.

IWU *See:* interworking unit.

IXC *See:* interexchange carrier; interexchange channel.

J

J *See:* joule.

jabber A condition wherein a station transmits for a period of time longer than the maximum permissible packet length, usually due to a fault condition. (LM/C) 610.7-1995, 802.3-1998

jabber control The ability of a station to interrupt automatically the transmission of data and inhibit an abnormally-long output data stream. *Note:* This term is contextually specific to IEEE Std 802.3. (C) 610.7-1995

jabber function A mechanism for controlling abnormally long transmissions (i.e., jabber). (C/LM) 802.3-1998

jack (1) (electric circuits) A connecting device, ordinarily designed for use in a fixed location, to which a wire or wires of a circuit may be attached and that is arranged for the insertion of a plug. (PE/EM) 43-1974s
(2) A connecting device within a circuit to which one or more wires may be attached and which is arranged so that a plug may be attached. *See also:* RJ-45; RJ-11. (C) 610.7-1995

jack bolt (rotating machinery) A bolt used to position or load an object. (PE) [9]

jacket (1) (electrical heat tracing for industrial applications) (cable) A thermoplastic or thermosetting plastic covering, sometimes fabric reinforced, applied over the insulation, core, metallic sheath, or armor of a cable. (BT/PE/AV) 152-1953s, [4]
(2) (primary dry cell) An external covering of insulating material, closed at the bottom. *See also:* electrolytic cell. (PE/EEC) [119]
(3) A polymeric sheath, sometimes fabric reinforced, applied over the insulation or core of a cable. (IA/PC) 515.1-1995
(4) A protective covering over the insulation, core, or sheath of a cable. (NESC) C2-1997

jack shaft (rotating machinery) A separate shaft carried on its own bearings and connected to the shaft of a machine. *See also:* rotor. (PE) [9]

jack system (rotating machinery) A system design to raise the rotor of a machine. *See also:* rotor. (PE) [9]

Jacob's ladder *See:* rope ladder.

jaggies *See:* stairstepping.

jam (1) (A) An external signal introduced deliberately into a transmission to prevent successful transmission. **(B)** A signal that carries a message that informs other stations that they must not transmit. (C) 610.7-1995
(2) A mis-feed in the feed mechanism of a printer or card reader. (C) 610.10-1994w

jamming A form of electronic countermeasures (ECM) in which interfering signals are transmitted at frequencies in the receiving band of a radar for the purpose of obscuring the radar signal (as in noise jamming) or causing confusion in interpreting the radar signal (as in repeater jamming). (AES) 686-1997

jam strobe Indication of jammer azimuth bearing, one form being a marker on the radar plan-position indicator (PPI) display. It can also show the jammer signal strength and severity of main and sidelobe jamming. (AES) 686-1997

jam transfer (hybrid computer linkage components) The transfer operation, in a double-buffered digital-to-analog converter (DAC) or digital-to-analog multiplier (DAM), in which the digital value is simultaneously loaded into both the holding and dynamic registers. (C) 166-1977w

Jansky A unit of spectral power flux density: 10^{-26} times one watt per square meter per Hertz. (AP/PROP) 211-1997

jar (storage cell) The container for the element and electrolyte of a lead-acid storage cell and unattacked by the electrolyte. *See also:* battery. (EEC/PE) [119]

JCL *See:* job control language.

J-display A modified A-display in which the time base is a circle and targets appear as radial deflections from the time base.

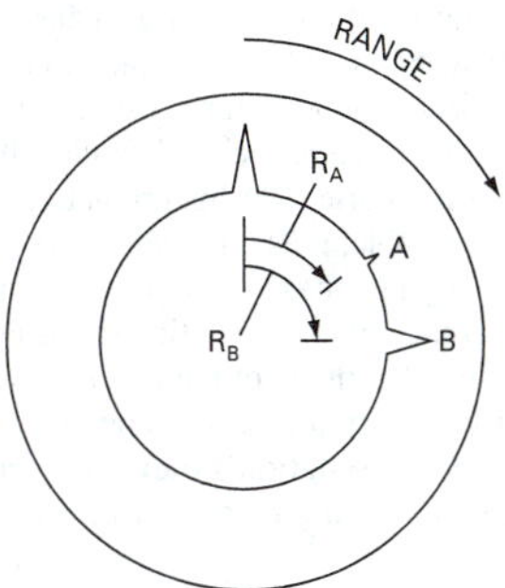

Note: Two targets, A and B, at different ranges

J-display

(AES) 686-1997

jet-engine modulation (JEM) Amplitude and/or frequency modulation of the radar echo from a jet-powered aircraft by motion of the compressor or turbine blades. *Note:* These modulations may cause errors in measuring the Doppler frequency of the target, but they also provide information useful in non-cooperative target recognition. *See also:* target recognition. (AES) 686-1997

jerk (inertial sensors) A vector that specifies the time rate of change of the acceleration; the third derivative of displacement with respect to time. (AES/GYAC) 528-1994

jitter (1) (A) (data transmission) (repetitive wave) Time-related, abrupt, spurious variations in the duration of any specified, related interval. **(B) (data transmission) (repetitive wave)** Amplitude-related, abrupt, spurious variations in the magnitude of successive cycles. **(C) (data transmission) (repetitive wave)** Frequency-related, abrupt, spurious variations in the frequency of successive pulses. **(D) (data transmission) (repetitive wave)** Phase-related, abrupt, spurious variations in the phase of the frequency modulation of successive pulses referenced to the phase of a continuous oscillator. *Note:* Qualitative use of jitter requires the use of a generic derivation of one of the categories to identify whether the jitter is time, amplitude, frequency, or phase related and to specify which form within the category, for example, pulse delay-time jitter, pulse-duration jitter, pulse-separation jitter. Quantitative use of jitter requires that a specified measure of the time or amplitude related variation, (for example, average, root-mean-square, or peak-to-peak) be included in addition to the generic term that specifies whether the jitter is time-, amplitude-, frequency-, or phase-related. (PE) 599-1985
(2) (oscilloscopes, electronic navigation, and television) Small, rapid aberrations in the size or position of a repetitive display, indicating spurious deviations of the signal or instability of the display circuit. *Note:* Frequently caused by mechanical or electronic switching systems or faulty components. It is generally continuous, but may be random or periodic. (BT/PE/AV) 201-1979w, 599-1985w
(3) (facsimile) Raggedness in the received copy caused by erroneous displacement of recorded spots in the direction of scanning. *See also:* recording. (COM) 168-1956w
(4) (pulse terminology) Dispersion of a time parameter of the pulse waveforms in a pulse train with respect to a reference time, interval, or duration. Unless otherwise specified by a mathematical adjective, peak-to-peak jitter is assumed. *See also:* mathematical adjectives. (IM/WM&A) 194-1977w
(5) The time varying phase of a pulse train relative to the phase of the reference pulse train. *See also:* phase jitter; amplitude jitter. (C) 610.7-1995

(6) Refers to the time-uncertainty of a transitioning edge recurring in a repetitive signal. This uncertainty is only with respect to other edges in that signal. Jitter is commonly measured using random bit patterns and accumulating an eye pattern to show the worst-case difference in transitions. (C/MM) 1596.3-1996
(7) **(A)** Small, rapid variations in the size, shape, or position of observable information, frequently caused by mechanical and electronic switching systems or faulty components. It also refers to zero-mean random errors in successive target position measurements due to target echo characteristics, propagation, or receiver thermal noise. **(B)** Intentional variation of a radar parameter, for example, pulse interval. (AES) 686-1997
(8) The time varying phase of a pulse train relative to the phase of a reference pulse train. Jitter is usually measured as the difference in edge times of the receiver's recovered clock or transmitter data output to a reference clock or data signal, typically the preceding station's transmitter clock or data output. The specifications are measured in nanoseconds. (C/LM) 8802-5-1998

jitter, maximum output *See:* maximum output jitter.

jitter, timing *See:* timing jitter.

jitter tolerance, input *See:* input jitter tolerance.

jitter transfer function The ratio between input jitter and output jitter in specified frequency bands throughout the applicable jitter mask. The jitter transfer is controlled by the gain and cutoff frequency of the jitter transfer characteristic. (COM/TA) 1007-1991r

jnd *See:* just noticeable difference.

job (1) A user-defined unit of work that is to be accomplished by a computer. For example, the compilation, loading, and execution of a computer program. *See also:* job stream; job step; job control language. (C) 610.12-1990
(2) A set of processes comprising a shell pipeline, and any processes descended from it, that are all in the same process group. (C/PA) 9945-2-1993
(3) That entity originated or initiated by a user which is handled by the printer interface control unit. A job need not result in the imaging of information on media. (C/MM) 1284.1-1997
(4) *See also:* batch job. (PA/C) 1003.2d-1994

job control (1) A facility that allows users to selectively stop (suspend) the execution of processes and continue (resume) their execution at a later point. The user typically employs this facility via the interactive interface jointly supplied by the terminal I/O driver and a command interpreter. Conforming implementations may optionally support job control facilities; the presence of this option is indicated to the application at compile time or run time by the definition of the {_POSIX_JOB_CONTROL} symbol. (C/PA) 9945-1-1996, 9945-2-1993
(2) A facility that allows users to stop (suspend) selectively the execution of processes and continue (resume) their execution at a later time. The user typically employs this facility via the interactive interface jointly supplied by the terminal I/O driver and a command interpreter. Conforming implementations may optionally support job control facilities. The presence of this option is indicated to the application at compile time by the subtype `Job_Control_Support` in package `POSIX_Options` or at run time by the value returned by the function `Job_Control_Is_Supported` in package `POSIX_Configurable_System_Limits;`. (C) 1003.5-1999

job control job ID A handle that is used to refer to a job. The job control job ID can be any of the forms shown in the table.

Job Control Job ID Formats

Job Control Job ID	Meaning
%%	Current job
%+	Current job
%−	Previous job
%*n*	Job number *n*
%*string*	Job whose command begins with *string*
%?*string*	Job whose command contains *string*

(C/PA) 9945-2-1993

job control language (JCL) A command language used to identify a sequence of jobs, describe their requirements to an operating system, and control their execution. *Note:* Commonly used in batch-oriented environments such as IBM's 370 Computer. (C) 610.13-1993w, 610.12-1990

job function A group of engineering processes that is identified as a unit for the purposes of work organization, assignment, or evaluation. Examples are design, testing, or configuration management. (C) 610.12-1990

job identifier A unique name for a job. A name that is unique among all other job identifiers in a batch system and that identifies the server to which the job was originally submitted. (C/PA) 1003.2d-1994

job name A label that is an attribute of a job. The job name is not necessarily unique. (C/PA) 1003.2d-1994

job-oriented terminal A terminal that is designed for a particular application, for example, a terminal used for airline checking or for point of sale. (C) 610.10-1994w

job owner The *username@hostname* of the user submitting the job, where *username* is a user name defined by Section 2.2.2.88 of POSIX.1 and *hostname* is a *network host name.* (C/PA) 1003.2d-1994

job priority An attribute used in selecting a job for execution. A value specified by the user that may be used by an implementation to determine the order in which jobs will be selected to be executed. *Job priority* has a numeric value in the range −1024 to 1023. *Note:* The *job priority* is not the execution priority (*nice value*) of the job. (C/PA) 1003.2d-1994

jobsite The assembly point at the structure or equipment where the workers, tools, and vehicles are assembled to perform the climbing to the worksite. (T&D/PE) 1307-1996

job state An attribute of a batch job. The state of a job determines the types of requests that the batch server that manages the job can accept for the job. Valid states include `QUEUED`, `RUNNING`, `HELD`, `WAITING`, `EXITING`, and `TRANSITING`. (C/PA) 1003.2d-1994

job step A user-defined portion of a job, explicitly identified by a job control statement. A job consists of one or more job steps. (C) 610.12-1990

job stream A sequence of programs or jobs set up so that a computer can proceed from one to the next without the need for operator intervention. *Synonym:* run stream. (C) 610.12-1990

jog (inch) (control) A control function that provides for the momentary operation of a drive for the purpose of accomplishing a small movement of the driven machine. *See also:* electric drive. (IA/ICTL/IAC/APP) [60], [75]

jogging (packaging machinery) The quickly repeated closure of the circuit to start a motor from rest for the purpose of accomplishing small movements of the driven machine. *Synonym:* inching. (IA/PKG) 333-1980w

jogging speed The steady-state speed that would be attained if the jogging pilot device contacts were maintained closed. *Note:* It may be expressed either as an absolute magnitude of speed or a percentage of maximum rated speed. *See also:* feedback control system. (IA/ICTL/IAC) [60]

JOHNNIAC Open Shop System (JOSS) A procedural language used for performing numerical computations and mathematics. (C) 610.13-1993w

Johnson noise (1) (interference terminology) The noise caused by thermal agitation (of electron charge) in a dissipative body. *Notes:* 1. The available thermal (Johnson) noise power N from a resistor at temperature T is $N = kT\Delta f$, where k is Boltzmann's constant and Δf is the frequency increment. 2. The noise power distribution is equal throughout the radio frequency spectrum, that is, the noise power is equal in all equal frequency increments. *See also:* signal. (IE) [43]
(2) (broadband local area networks) *See also:* noise figure; noise. (LM/C) 802.7-1989r

join (1) A relational operator that combines two relations having a common attribute and which results in a relation containing all of the attributes from both of the original relations. *See also:* intersection; projection; product; selection; union; difference.

P		Q		P join Q
A 50	join	50 3	=	A 50 3
B 50		60 4		B 50 3
C 60		70 5		C 60 4

join

(C) 610.5-1990w

(2) A junction at which an arrow segment (going from source to use) merges with one or more other arrow segments to form a root arrow segment. May denote bundling of arrow meanings, i.e., the inclusion of multiple object types within an object type set. (C/SE) 1320.1-1998

join dependency A type of dependency within a relation *R*, in which *R* is join dependent on *X*, *Y*, . . ., *Z* (subsets of attributes in *R*) if and only if *R* is equal to the join of its projections on *X*, *Y*, . . ., *Z*. *See also:* functional dependency. (C) 610.5-1990w

joint (interior wiring) A connection between two or more conductors. (EEC/PE) [119]

joint, compression *See:* compression joint.

joint insulation *See:* connection insulation.

jointly owned generation (power operations) Generation facility owned jointly by several electric utilities each entitled to a share of the capability. *Note:* One of the participating utilities operates the facility. (PE/PSE) 858-1987s

joint, protector *See:* protector joint.

joint review A process or meeting involving representatives of both the acquirer and the developer, during which project status, software products, and/or project issues are examined and discussed. (C/SE) J-STD-016-1995

joint use Simultaneous use by two or more kinds of utilities. (NESC/T&D) C2-1997, C2.2-1960

Jordan bearing A sleeve bearing and thrust bearing combined in a single unit. *See also:* bearing. (PE) [9]

JOSEF A general-purpose programming language similar to Pascal. (C) 610.13-1993w

JOSS *See:* JOHNNIAC Open Shop System.

joule (1) (metric practice) The work done when the point of application of a force of one newton is displaced a distance of one meter in the direction of the force. (QUL) 268-1982s

(2) (laser maser) A unit of energy: one (1) joule = 1 watt · second. (LEO) 586-1980w

Joule effect The evolution of thermal energy produced by an electric current in a conductor as a consequence of the electric resistance of the conductor. *See also:* Joule's law; thermoelectric device. (ED) [46]

Joule heat The thermal energy resulting from the Joule effect. *See also:* thermoelectric device. (ED) [46]

Joule's law (heating effect of a current) The rate at which heat is produced in an electric circuit of constant resistance is equal to the product of the resistance and the square of the current. (Std100) 270-1966w

journal (1) (shaft) A cylindrical section of a shaft that is intended to rotate inside a bearing. *See also:* armature; bearing. (PE) [9]

(2) (data management) A chronological record of the changes made to a set of data. *Note:* This record may be used as an audit trail to reconstruct a previous version of the data. *Synonym:* log. (C) 610.2-1987, 610.5-1990w

journal bearing (rotating machinery) A bearing that supports the cylindrical journal of a shaft. *See also:* bearing. (PE) [9]

joystick A cursor control device consisting of a lever having at least two degrees of freedom and that can be used as a locator. (C) 610.6-1991w, 1084-1986w, 610.10-1994w

jpd *See:* just perceptible difference.

***J* scan** *See:* J-display.

J-scope A cathode-ray oscilloscope arranged to present a J-display. (AES/RS) 686-1990

JTAG (1) Test access port and boundary scan architecture. (C/BA) 896.3-1993w

(2) An abbreviation for Joint Test Activity Group that is used to describe the serial diagnostic signals that have been defined by this group. (C/MM) 1596.4-1996

jukebox A storage device that holds multiple disks and which has one or more disk drives that can mount the disks in the disk drive as they are needed. *Synonym:* autochanger. (C) 610.10-1994w

Jules' Own Version of International Algorithmic Language A high-order programming language used primarily for solving scientific and control problems. *Note:* Based on ALGOL 58. *See also:* TINT. (C) 610.13-1993w

jump (A) (electronic computation) To (conditionally or unconditionally) cause the next instruction to be obtained from a storage location specified by an address part of the current instruction when otherwise it would be specified by some convention. **(B) (electronic computation)** An instruction that specifies a jump. *Note:* If every instruction in the instruction code specifies the location of the next instruction (for example, in a three-plus-one-address code), then each one is not called a jump instruction unless it has two or more address parts that are conditionally selected for the jump. *See also:* transfer; conditional jump; unconditional jump. (C) 162-1963

(2) (A) To depart from the implicit or declared order in which computer program statements are being executed. *Synonym:* transfer. **(B)** A program statement that causes a departure as in definition (A). *Contrast:* if-then-else; case. *See also:* go to; branch. **(C)** The departure described in definition (A). *See also:* unconditional jump; conditional jump. (C) 610.12-1990

jumper (1) (telephone switching systems) Crossconnection wire(s). (COM) 312-1977w

(2) (A) (protective grounding of power lines) A metallic wire connecting the conductors on opposite sides of a dead-end structure so that continuity is maintained. **(B) (protective grounding of power lines)** A conductor placed across the clear space between the ends of two conductors or metal pulling lines that are being spliced together. Its purpose then is to act as a shunt to prevent workers from accidentally placing themselves in series between the two conductors. (T&D/PE) 1048-1990

(3) A conductive tool used to maintain electrical continuity across equipment, or a conductor that shall be opened mechanically to enable various operations of live-line work to be performed *Synonym:* bypass. (T&D/PE) 516-1995

(4) (A) (conductor stringing equipment) The conductor that connects the conductors on opposite sides of a dead-end structure. **(B)** A conductor placed across the clear space between the ends of two conductors or metal pulling lines that are being spliced together. Its purpose, then, is to act as a shunt to prevent workmen from accidentally placing themselves in series between the two conductors. (T&D/PE) 524a-1993, 524-1992

jumper cable assembly An electrical or optical assembly, used for the bidirectional transmission and reception of information, consisting of a pair of transmission lines terminated at their ends with plug connectors. This assembly may or may not contain additional components, located between the plug connectors, to perform equalization. (C/LM) 802.3-1998

jump instruction (A) A computer instruction that specifies a jump. *Contrast:* unconditional jump instruction. *See also:* conditional jump instruction. **(B)** An instruction that changes the sequence in which computer instructions are performed.

Note: A jump instruction generally specifies the next instruction in terms of a real address. *See also:* branch instruction. (C) 610.10-1994

jump resonance A phenomenon associated with ferroresonant regulators where the output voltage suddenly changes to the regulating mode of operation at some value of the ascending input voltage (see the figures below), or suddenly drops out of the regulating mode of operation with descending input voltage.

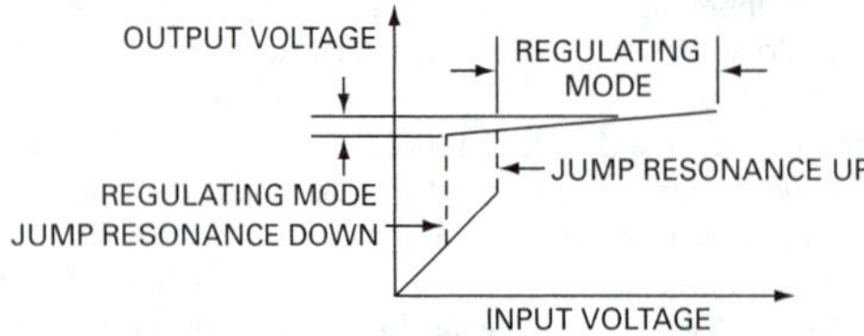

Output versus input voltage with jump resonance

Jump resonance

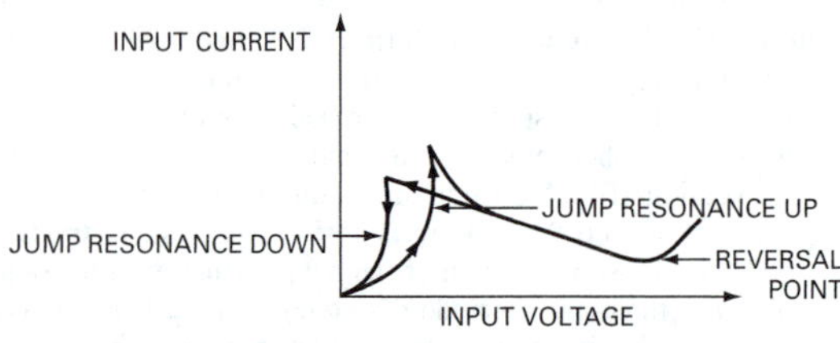

Reversal point with jump resonance

Jump resonance

(PEL) 449-1998

junction (1) (germanium gamma-ray detectors) (charged-particle detectors) (x-ray energy spectrometers) (of a semiconductor radiation detector) A region of transition between semiconductor regions of different electrical properties (for example, n-n$^+$, p-n, p-p$^+$ conductors) or between a metal and a semiconductor. (NPS/NID) 759-1984r, 300-1988r **(2)** The transition boundary between semiconductor regions of different electrical properties (for example: n-n$^+$, p−n, p−p$^+$ semi-conductors, or between a metal and a semiconductor). (NPS) 325-1996 **(3)** A point at which either a root arrow segment divides into branching arrow segments or arrow segments join into a root arrow segment. (C/SE) 1320.1-1998

junction, alloy *See:* alloy junction.

junction box An enclosed distribution panel for connecting or branching one or more corresponding electric circuits without the use of permanent splices. *See also:* cabinet. (T&D/PE) [10]

junction circuit A circuit that connects two other circuits. (C) 610.10-1994w

junction, collector *See:* collector junction.

junction depth (1) (germanium gamma-ray detectors) (charged-particle detectors) (x-ray energy spectrometers) (of a p-n semiconductor radiation detector) The distance below the crystal surface at which the conductivity type changes. (NPS/NID) 325-1986s, 759-1984r, 300-1988r **(2) (solar cells)** The distance from the illuminated surface to the center line of the junction in a solar cell. (AES/SS) 307-1969w

junction, diffused *See:* diffused junction.

junction, doped *See:* doped junction.

junction, emitter *See:* emitter junction.

junction frequency (JF) The frequency at which the traces seen on an oblique-incidence ionogram corresponding to the low-angle ray and to the high-angle ray respectively, for a given mode, merge together. *Note:* The high-angle ray is also called the Pederson ray. (AP/PROP) 211-1997

junction, fused *See:* alloy junction.

junction, grown *See:* grown junction.

junction loss (wire communication) That part of the repetition equivalent assignable to interaction effects arising at trunk terminals. *See also:* transmission loss. (EEC/PE) [119]

junction, *n-n* *See:* *n-n* junction.

junction, *p-n* *See:* *p-n* junction.

junction point *See:* node.

junction pole (wire communication) A pole at the end of a transposition section of an open wire line or the pole common to two adjacent transposition sections. (EEC/PE) [119]

junction, *p-p* *See:* *p-p* junction.

junction, rate-grown *See:* rate-grown junction.

junction, rectifying *See:* rectifying junction.

junction resistance (thermoelectric device) The difference between the resistance of two joined materials and the sum of the resistances of the unjoined materials. *See also:* thermoelectric device. (ED) [46]

junction temperature (light-emitting diodes) The temperature of the semiconductor junction. (IE/EEC) [126]

junction transistor A transistor having a base electrode and two or more junction electrodes. *See also:* transistor. (ED) 216-1960w

junction transposition (wire communication) A transposition located at the junction pole (s pole) between two transposition sections of an open wire line. (EEC/PE) [119]

junctor (1) (wire communication) (crossbar systems) A circuit extending between frames of a switching unit and terminating in a switching device on each frame. (PE/EEC) [119] **(2) (telephone switching systems)** Within a switching system, a connection or circuit between inlets and outlets of the same or different switching networks. (COM) 312-1977w

justification (A) In text formatting, the process of aligning text to form even margins or to achieve desired vertical spacing. *See also:* right justification; incremental justification; full justification; left justification; vertical justification; hyphenless justification. **(B)** The result of the process in (A). (C) 610.2-1987

justification range In text formatting, the permitted minimum and maximum space that can be inserted between words or characters by a justification routine. (C) 610.2-1987

justification routine In text formatting, a routine that produces justified text by using interword or intercharacter spacing. (C) 610.2-1987

justify (1) To shift a numeral so that the most significant digit, the least significant digit, or the radix point is placed at a specific position in a register. (C) 1084-1986w **(2)** Align text to a margin. Right justification adjusts the text so that its end touches the right margin or end of the text field. Left justification adjusts the text so that its beginning touches the left margin or beginning of the text field. (C) 1295-1993w

just noticeable difference (jnd) (visual) (television) The smallest difference between luminances or colors, occurring either alone or together, of (usually) adjacent areas that is easily discernible or obvious in the course of ordinary observation. (BT/AV) 201-1979w

just operate value, relay *See:* relay just-operate value.

just perceptible difference (television) (visual) The smallest difference between luminances or colors, occurring either alone or together, of (usually) adjacent areas that is discernible in the course of careful observation under the most favorable conditions. (BT/AV) 201-1979w

just scale A musical scale formed by taking three consecutive triads each having the ratio 4:5:6, or 10:12:15. *Note:* Consecutive triads are triads such that the highest note of one is the lowest note of the next. (SP/ACO) [32]

K

K *See:* kilo.

k *See:* kilo.

Ka-band A range of high frequencies (20–30 GHz) allotted for satellite transmission. (C) 610.7-1995

K_a-band A radar-frequency band between 27 GHz and 40 GHz, usually in the International Telecommunication Union (ITU) allocated band 33.4–36 GHz. (AES) 686-1997

karabiner *See:* carabiner.

Karnaugh map (mathematics of computing) A rectangular diagram of a logical expression drawn with overlapping rectangles representing a unique combination of the logic variables such that an intersection is shown for all combinations. The rows and columns are headed with combinations of the variables in a Gray code sequence. *See also:* Mahoney map; logic map. (C) 1084-1986w

Karn's algorithm An algorithm that allows transport protocols to distinguish between good and bad round trip time samples, thus improving round trip estimation. (C) 610.7-1995

kB *See:* kilobyte.

K-band A radar frequency band between 18 GHz and 27 GHz, usually in the International Telecommunication Union (ITU) allocated band 23–24.2 GHz. (AES) 686-1997

Kbyte Kilobyte. Indicates 2^{10} bytes. (C/MM) 1212-1991s

K-display A modified A-display used with a lobe-switching antenna, in which a target appears as a pair of vertical deflections. When the radar antenna is correctly pointed at the target, the deflections (blips) are of equal height, and when not so pointed, the difference in blip height is an indication of the direction and magnitude of pointing error.

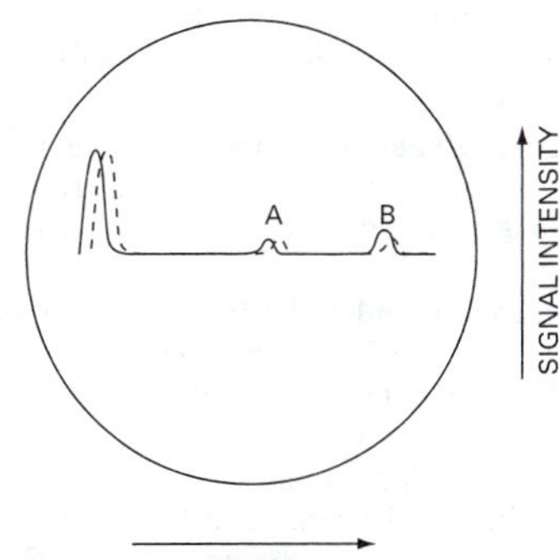

Note: Two targets, A and B, at different ranges; radar aimed at target A.

K-display

(AES) 686-1997

keep-alive circuit (1) (nonlinear, active, and nonreciprocal waveguide components) (transmit-receive, receiver protector) A circuit for providing residual ionization for the purpose of reducing the initiation time of the main discharge. *See also:* duplexer; ignitor discharge. (MTT) 457-1982w
(2) In a transmit-receive switch (TR switch), a circuit for producing residual ionization for the purpose of reducing the initiation (breakdown) time of the main discharge. (AES) 686-1997

Kellem *See:* woven wire grip; conductor grip.

Kellem grip *See:* conductor grip.

kelvin (metric practice) Unit of thermodynamic temperature, it is the fraction 1/273.16 of the thermodynamic temperature of the triple point of water (adopted by 13th General Conference on Weights and Measures). (QUL) 268-1982s

Kelvin bridge (double bridge) (thomson bridge) A 7-arm bridge intended for comparing the 4-terminal resistances of two 4-terminal resistors or networks, and characterized by the use of a pair of auxiliary resistance arms of known ratio that span the adjacent potential terminals of the two 4-terminal resistors that are connected in series by a conductor joining their adjacent current terminals.

$$R_x = R_s \frac{R_2}{R_1} - \frac{R_c R_3 (R_4/R_3 - R_2/R_1)}{R_3 + R_4 + R_c}$$

See also: bridge.

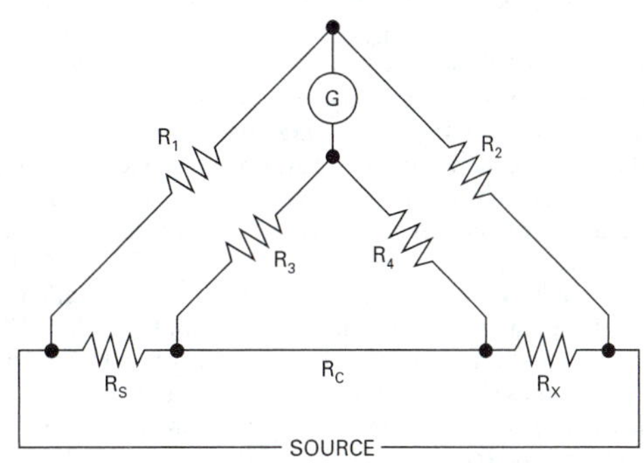

Kelvin bridge

(QUL/PE/EEC) 268-1982s, [119]

Kendall effect (facsimile) A spurious pattern or other distortion in a facsimile record, caused by unwanted modulation products arising from the transmission of a carrier signal, appearing in the form of a rectified baseband that interferes with the lower sideband of the carrier. *Note:* This occurs principally when the single sideband width is greater than half the facsimile carrier frequency. *See also:* recording. (COM) 168-1956w

kenotron (tube or valve) A hot-cathode vacuum diode. *Note:* This term is used primarily in the industrial and x-ray fields. (ED) [45]

keraunic level (1) Number of thunderstorm days in a given area for a specified period of time. *Synonym:* ceraunic level. *See also:* isokeraunic map. (T&D/PE) 751-1990
(2) The average annual number of thunderstorm days or hours for a given locality (1) A daily keraunic level is called a thunderstorm-day and is the average number of days per year in which thunder is heard during a 24 h period. (2) An hourly keraunic level is called a thunderstorm-hour and is the average number of hours per year that thunder is heard during a 60 min period. (SUB/PE) 998-1996

kernel (1) (A) (software) That portion of an operating system that is kept in main memory at all times. *Synonyms:* resident control program; nucleus. **(B) (software)** A software module that encapsulates an elementary function or functions of a system. *See also:* security kernel. (C) 610.12-1990
(2) The nucleus of the operating system. (C/PA) 1387.2-1995

kernel benchmark program A benchmark program consisting of portions of actual computer programs such that the portion chosen is believed to be responsible for most of the execution time. (C) 610.10-1994w

kernel element (KE) An abstraction of data processing and data communication resources. An application KE represents a logical component of application layer functionality. (SCC32) 1455-1999

kernel fileset A fileset in which one or more of the referenced files forms part of the kernel, and denoted by having the value of its *is_kernel* attribute set to `true`. (C/PA) 1387.2-1995

kerning In text formatting, the use of inter-character spacing to expand or compact a word or a line of text. *Synonyms:* white space reduction; mortising; white space expansion. (C) 610.2-1987

Kerr electrostatic effect *See:* electro-optic effect.

key (1) (telephone switching systems) A hand-operated switching device ordinarily comprising concealed spring contacts with an exposed handle or pushbutton, capable of closing or opening one or more parts of a circuit. (COM) 312-1977w
(2) (rotating machinery) A bar that by being recessed partly in each of two adjacent members serves to transmit a force from one to the other. *See also:* rotor. (PE) [9]
(3) (software) One or more characters, within a set of data, that contains information about the set, including its identification. *See also:* data. (C/SE) 729-1983s
(4) (A) (data management) In data management, a data element or concatenation of data elements that identifies an item within a set of items. *Note:* Such a data element is also known as a key field. *Synonyms:* sequence field; key field. *See also:* key value; primary key; secondary key; concatenation; sort key. **(B) (data management)** In a relational data model, one or more attributes that, when taken together, identify the relation to which the attributes belong. **(C) (data management)** In a tree, the portion of each node that identifies that node. (C) 610.5-1990
(5) A sequence of symbols that controls the operations of encipherment and decipherment. (LM/C) 802.10-1992
(6) When used in the context of a ROM entry, refers to an 8-bit field whose value identifies a ROM location as an immediate entry, offset entry, leaf entry, or subdirectory entry. This is a term used (but not defined) in ISO/IEC 13213: 1994. (C/BA) 896.2-1991w
(7) (A) A manually activated mechanism on a keyboard, used for entering a character or command into a computer system. *See also:* typing key; control key. **(B)** To press a lever or button. (C) 610.10-1994
(8) (relational data base) A field or group of fields in a relational data-base table that uniquely defines each row within that table. A composite key is made up of more than one field in the table. (PE/EDPG) 1150-1991w
(9) When used in the context of a read-only memory (ROM) entry, a key is an 8-bit field whose value identifies a ROM location as an immediate entry, offset entry, leaf entry, or subdirectory entry. (C/BA) 896.10-1997

keyboard (1) (test, measurement, and diagnostic equipment) A device for the encoding of data by key depression that causes the generation of the selected code element. (MIL) [2]
(2) An input device consisting of a systematic arrangement or layout of keys, used to encode data. *See also:* live keyboard; QWERTY keyboard; Dvorak keyboard; keypad; membrane keyboard. (C) 610.10-1994w

keyboard punch *See:* keypunch.

keyboard scanner A unit within a keyboard that detects the depression of a key and generates an encoded signal indicating the identity of that key. (C) 610.10-1994w

keyboard send/receive (KSR) A teletypewriter unit with keyboard and printer. *Contrast:* automatic send/receive. (C) 610.10-1994w

keyboard-to-disk *See:* key-to-disk converter.

key code An alpha or alphanumeric designator used to identify the style and angular position of the keying pins. (C/BA) 1101.4-1993, 1101.7-1995, 1101.3-1993

key compression The elimination of data from the beginning and the end of a key in which these characters are not needed to distinguish the key from other keys in the set. (C) 610.5-1990w

key distribution system The manual or automated means by which cryptographic keys are communicated between nodes of a computer or communications system. (C) 610.7-1995

keyed access *See:* indexed access.

keyer A device that changes the output of a transmitter from one value of amplitude or frequency to another in accordance with the intelligence to be transmitted. *Note:* This applies generally to telegraph keying. *See also:* radio transmission. (AP/ANT) 145-1983s

key field *See:* key.

key folding function A hash function in which the original key is split into two or more parts and some portion of their sum is returned as the hash value. For example, in the function below, the key is divided into three parts and the sum of the three parts is returned as the hash value.

Original key	Calculation	Hash value
96472135	964 + 721 + 35 = 1738	1738
90007810	900 + 078 + 10 = 988	988

(C) 610.5-1990w

key gases Gases generated in oil-filled transformers that can be used for qualitative determination of fault types, based on which gases are typical or predominant at various temperatures. (PE/TR) C57.104-1991

key generation The process of generating the key values for the items in a set according to some algorithm. (C) 610.5-1990w

keying (1) (modulating systems) Modulation involving a sequence of selections from a finite set of discrete states. *See also:* telegraphy. (IT) [7]
(2) (telegraph) The forming of signals, such as those employed in telegraph transmission, by an abrupt modulation of the output of a direct-current or an alternating-current source as, for example, by interrupting it or by suddenly changing its amplitude or frequency or some other characteristic. *See also:* telegraphy. (AP/ANT) 145-1983s
(3) (television) A signal that enables or disables a network during selected time intervals. *See also:* television. (BT/AV) [34]

keying interval (periodically keyed transmission system) (modulation systems) One of the set of intervals starting from a change in state and equal in length to the shortest time between changes of state. *Note:* The keying interval equals the symbol duration. (IT) [7]

keying rate (modulation systems) The reciprocal of the duration of the keying interval. (Std100) 270-1964w

keying wave *See:* marking wave.

keyless ringing (telephony) A form of machine ringing on manual switchboards that is started automatically by the insertion of the calling plug into the jack of the called line. (EEC/PE) [119]

key letter in context index (KLIC) A variation of a keyword in context (KWIC) index in which letters are used as the fundamental indexing units instead of keywords. *See also:* key phrase in context index. (C) 610.2-1987

key light (illuminating engineering) The apparent principal source of directional illumination falling upon a subject or area. (EEC/IE) [126]

key management The generation, storage, distribution, deletion, archiving, and application of keys in accordance with a security policy. (LM/C) 802.10-1992

Key Management Stack The protocols residing above SDE that request services via an SDE SAP that is supported by the use of a bootstrap SAID with either of the two values reserved for key management. (C/LM) 802.10-1998

key migration The modeling process of placing the primary key of a parent or generic entity in its child or category entity as a foreign key. (C/SE) 1320.2-1998

keypad A small group of keys that are set up for convenience and greater flexibility such that they are grouped together physically on a keyboard. For example, a numeric keypad or a cursor control keypad. (C) 610.10-1994w

key performance indicator (KPI) A measurement of the performance of a particular business system in terms of the aims and goals of an enterprise. (C/PA) 1003.23-1998

key phrase in context index (KPIC) A variation of a keyword in context (KWIC) index in which phrases are used as the fundamental indexing units instead of keywords. *See also:* key letter in context index. (C) 610.2-1987

key pin A hardware implementation that prevents mating of incompatible modules. (C/BA) 1101.4-1993, 1101.7-1995

key pulsing (telephone switching systems) A switchboard arrangement using a nonlocking keyset and providing for the transmission of a signal corresponding to each of the keys depressed. (COM) 312-1977w

key-pulsing signal (telephone switching systems) In multifrequency and key pulsing, a signal used to prepare the equipment for receiving digits. (COM) 312-1977w

keypunch A keyboard-activated card punch that punches holes in a card, according to input received from the keyboard. *Synonym:* keyboard punch. *See also:* numeric keypunch. (C) 610.10-1994w

keypunching The process of using a card punch to generate punch cards. (C) 610.10-1994w

key range A particular range of values of the keys found in some set of data. *Note:* Key ranges may be used to partition the set into subsets. (C) 610.5-1990w

key sequence Pertaining to a set of data that has been sequenced according to the value of some key. (C) 610.5-1990w

keyshelf (telephone switching systems) The shelf on which are mounted control keys for use by operators or other personnel. (COM) 312-1977w

key sorting A sorting technique in which a table of sort keys and corresponding addresses that point to the items to be stored are manipulated instead of moving the items themselves. *See also:* address table sorting. (C) 610.5-1990w

keystone distortion (1) (television) A form of geometric distortion that results in a trapezoidal display of a nominally rectangular raster or picture. *See also:* television. (BT/AV) [34]
(2) (camera tubes) A distortion such that the slope or the length of a horizontal line trace or scan line is linearly related to its vertical displacement. *Note:* A system having keystone distortion distorts a rectangular pattern into a trapezoidal pattern. (ED) 161-1971w

keystroke The action of pressing one of the keys on a keyboard. (C) 610.10-1994w

keystroke counter A counter that records the number of key depressions made on a given unit within some period of time. (C) 610.10-1994w

key-style view A view that represents the structure and semantics of data within an enterprise, i.e., data (information) models. The key-style view is backward-compatible with FIPS PUB 184. (C/SE) 1320.2-1998

key-to-disk converter An input device that converts data from a keyboard to disk storage. *Synonym:* keyboard-to-disk. *See also:* key-to-tape converter; card-to-disk converter. (C) 610.10-1994w

key-to-tape converter An input device that converts data from a keyboard to magnetic tape. *See also:* card-to-tape converter; key-to-disk converter. (C) 610.10-1994w

key transformation In searching, the process of mapping a set of keys into a set of integers, using a hash function. (C) 610.5-1990w

key transformation function *See:* hash function.

key value The contents of a key. (C) 610.5-1990w

keyway (rotating machinery) A recess provided for a key. *See also:* key. (PE) [9]

keyword In automatic indexing, a significant word in a title or document that characterizes the content of the document. *Synonyms:* descriptor; lead term. (C) 610.2-1987

keyword and context (KWAC) *See:* keyword and context index.

keyword and context index (KWAC) A type of keyword in context index in which items are presented in the form of a permutation index. *See also:* permutation on subject headings index. (C) 610.2-1987

keyword in context (KWIC) *See:* keyword in context index.

keyword in context index (KWIC) An automatic index in which keywords are placed in alphabetical order in a central column and the remainder of the information is given to the right and left, preserving the context. *Contrast:* keyword out of context index. *See also:* keyword and context index; key letter in context index; key phrase in context index; author and keyword in context index. (C) 610.2-1987

keyword out of context (KWOC) *See:* keyword out of context index.

keyword out of context index (KWOC) An automatic index in which the keywords are extracted from their normal context and are displayed in a left hand column with the full context following on the right in its normal order. *Contrast:* keyword in context index. *See also:* keyword out of title index; word and author index. (C) 610.2-1987

keyword out of title *See:* keyword out of title index.

keyword out of title index A type of keyword out of context index in which the items being indexed are document titles. (C) 610.2-1987

kHz *See:* kilohertz.

kick-sorter *See:* pulse; pulse-height analyzer.

kilo (A) (mathematics of computing) A prefix indicating one thousand (10^3). **(B) (mathematics of computing)** In statements involving size of computer storage, a prefix indicating 2^{10}, or 1024. (C) 1084-1986

kilobyte (1) Either 1000 bytes or 2^{10} or 1024 bytes. *Notes:* 1. The user of these terms shall specify the applicable usage. If the usage is 2^{10} or 1024 bytes, or multiples thereof, then note 2 below shall also be included with the definition. 2. In IEEE Std 610.10-1994, the terms kilobyte (kB) means 2^{10} or 1024 bytes, megabyte (MB) means 1024 kilobytes, and gigabyte (GB) means 1024 megabytes. *See also:* megabyte; gigabyte. (C) 610.10-1994w
(2) Equivalent to 1024 bytes. *Note:* There is a plan to deprecate this base-2 usage because of the potential for confusion with the base-10 definition. An alternative notation for base-2 is under development. Because the above mentioned base-2 usage is widespread in some areas of application, however, it may be employed in IEEE publications for a limited time. (C/MM) 1284.1-1997
(3) A quantity of data equal to 2^{10} bytes, or 1024 bytes. (C/MM) 1394a-2000

kilogram (metric practice) The unit of mass; it is equal to the mass of the international prototype of the kilogram (adopted by 1st and 3rd General Conference on Weights and Measures 1889 and 1901). (QUL) 268-1982s

kilohertz (1) A unit of frequency equal to 1000 cycles per second. (LM/C) 802.7-1989r
(2) A unit of frequency equal to 1000-Hz, that is, 10^3 Hz. (C) 610.7-1995

kilopascal (kilo pascals) Metric unit for water or air pressure. (T&D/PE) 957-1987s, 957-1995

kilovar (1000 vars) The practical unit of reactive power, equal to the product of the root-mean-square (rms) voltage in kilovolts (kV), the rms current in amperes (A), and the sine of the angle between them.power systems relaying. (PE/T&D) C37.99-2000, 18-1992

kilovolt-ampere rating (voltage regulators) The product of the rated load amperes and the rated range of regulation in kilovolts. *Note:* The kilovolt-ampere rating of a three-phase voltage regulator is the product of the rated load amperes and the rated range of regulation in kilovolts multiplied by 1.732. *See also:* voltage regulator. (PE/TR) C57.15-1968s

kilowatt A measure of the instantaneous power requirement. (IA/PSE) 241-1990r

kinematic drift *See:* misalignment drift.

kinescope *See:* picture tube.

kinetic depth effect The effect achieved in computer graphics when a three-dimensional object is rotated about an axis and lines nearer the viewer appear to move more rapidly than those farther away. (C) 610.6-1991w

kinetic energy The energy that a mechanical system possesses by virtue of its motion. *Note:* The kinetic energy of a particle at any instant is $(1.2)mv^2$, where m is the mass of the particle and v is its velocity at that instant. The kinetic energy of a body at any instant is the sum of the kinetic energies of its several particles. (Std100) 270-1966w

Kingsbury bearing *See:* tilting-pad bearing.

kino gun *See:* end injection.

Kirchhoff approximation When used in computing electromagnetic scattering from surfaces, this approximation assumes that the surface is locally planar and the field on the surface is equal to the field that would have existed had the surface been a plane tangent to the actual surface at that point. *Notes:* 1. Valid when the local radius of curvature is much greater than a wavelength and multiple scattering can be neglected. 2. Not valid in shadow regions. (AP/PROP) 211-1997

Kirchhoff's laws (A) (electric networks) The algebraic sum of the currents toward any point in a network is zero. **(B) (electric networks)** The algebraic sum of the products of the current and resistance in each of the conductors in any closed path in a network is equal to the algebraic sum of the electromotive forces in that path. *Note:* These laws apply to the instantaneous values of currents and electromotive forces, but may be extended to the phasor equivalents of sinusoidal currents and electromotive forces by replacing algebraic sum by phasor sum and by replacing resistance by impedance. *See also:* network analysis. (Std100) 270-1966

Kirchhoff's theory of diffraction A scalar wave theory, valid when the diffracting aperture has dimensions that are large relative to the wavelength. The diffracted field at any point is found by evaluating an integral involving the field and its derivative over the aperture only, and these quantities are assumed to be the same as they would be were the aperture infinitely large in both directions. (AP/PROP) 211-1997

KLIC *See:* key letter in context index.

Klein *See:* conductor grip.

klydonograph *See:* Lichtenberg figure camera.

klystron A velocity-modulated tube comprising, in principle, an input resonator, a drift space, and an output resonator. *See also:* reflex klystron; power klystron. (Std100) [84]

knee-point voltage (1) (Class C transformers) The point on the excitation curve where the tangent is at 45° to the abscissa. The excitation curve shall be plotted on log-log paper with square decades. *Note:* The above definition is for non-gapped CT. When the current transformer has a gapped core, the knee-point voltage is the point where the tangent to the curve makes an angle of 30° with the abscissa. (PE/PSR) C37.110-1996
(2) That sinusoidal voltage of rated frequency applied to the secondary terminals of the transformer, all other windings being open circuited, which, when increased by 10% causes the exciting current to increase by 50%. (PE/PSR) C37.110-1996

knife-edge diffraction *See:* edge diffraction.

knife switch A form of switch in which the moving element, usually a hinged blade, enters or embraces the stationary contact clips. *Note:* In some cases, however, the blade is not hinged and is removable. (SWG/PE) C37.100-1992

knockout (power and distribution transformers) A portion of the wall of a box or cabinet so fashioned that it may be readily removed by a hammer, screwdriver, and pliers at the time of installation in order to provide a hole for the attachment of a raceway cable or fitting. (PE/TR) C57.12.80-1978r

knot (pulse terminology) A point, $t_j k\ m_k$ ($k = 1, 2, 3, \ldots, n$), in a sequence of points wherein $t_k \leq t_{k\pi \mathrm{A}}$ 1 through which a spline function passes. (See corresponding figure.) (IM/WM&A) 194-1977w

knowledge (1) Information structure to facilitate derivation of new information. (SCC20) 1232.1-1997
(2) The aspect of an instance's specification that is determined by the values of its attributes, participant properties, and constant, read-only operations. (C/SE) 1320.2-1998

knowledge base (1) A collection of interrelated information, facts, or statements. (C) 610.5-1990w, 610.12-1990
(2) A combination of structure, data, and function used by reasoning systems. (ATLAS) 1232-1995

knowledge/data base management system One system within the AI-ESTATE architectural concept. This system supports

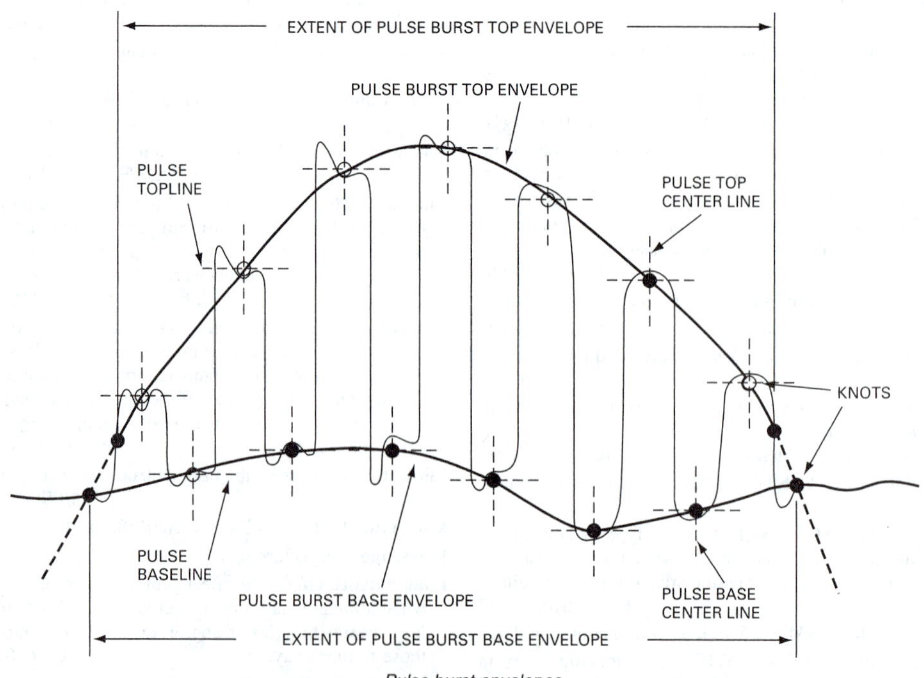

Pulse burst envelopes

knot

access to data and knowledge in external data stores and knowledge bases which will be accessed by services external to AI-ESTATE. (ATLAS) 1232-1995

knowledge reference Knowledge that associates, either directly or indirectly, a DIT entry with the DSA in which it is located. (C/PA) 1328.2-1993w, 1326.2-1993w, 1224.2-1993w, 1327.2-1993w

KOPS A measure of computer processing speed. *See also:* MIPS; MFLOPS. (C) 610.12-1990

Kordic algorithm A widely used algorithm that calculates the sine and cosine of an angle using only addition, subtraction, and shifting operations in scaled arithmetic. (C) 1084-1986w

KP Key pulse. (COM/TA) 973-1990w

kPa (kilo pascals) Metric unit for water or air pressure. (T&D/PE) 957-1987s, 957-1995

KPIC *See:* key phrase in context index.

K-scope A cathode-ray oscilloscope arranged to present a K-display. (AES/RS) 686-1990

KSR *See:* keyboard send/receive.

kth code transition level ***T[k]*** The input value corresponding to the transition between codes $k-1$ and k. (IM/WM&A) 1057-1994w

K28.5 The 8B/10B control (K) symbol used as one of the space-borne fiber-optic data bus (SFODB) Frame Sync characters. (C/BA) 1393-1999

K28.7 The 8B/10B control (K) symbol used as the other space-borne fiber-optic data bus (SFODB) Frame Sync character. (C/BA) 1393-1999

K_u-band (1) A range of microwave frequencies (15–17 GHz) whose uses include satellite transmission. (C) 610.7-1995 **(2)** A radar-frequency band between 12 GHz and 18 GHz, usually in one of the International Telecommunication Union (ITU) allocated bands 13.4–14.4 GHz or 15.7–17.7 GHz. (AES) 686-1997

kV Phase-to-phase voltage of the circuit(s). When phase-to-ground voltage is the intention, it should be so noted. (T&D/PE) 957-1995, 957-1987s

kVA or volt-ampere short-circuit input rating of a high-reactance transformer (power and distribution transformers) One that designates the input kVA or volt-amperes at rated primary voltage with the secondary terminals short-circuited. (PE/TR) C57.12.80-1978r

kV·T product Inductive influence expressed in terms of the product of its root-mean-square magnitude, in kilovolts, times its telephone influence factor (TIF). (COM/IA/TA/SPC) 469-1988w, 519-1992

KWAC *See:* keyword and context; keyword and context index.

KWIC *See:* keyword in context; keyword in context index.

KWOC *See:* keyword out of context; keyword out of context index.

L

L *See:* Bell Laboratories' Low-level Linked List Language.

LA *See:* laser gyro axes.

label (1) (A) (software) A name or identifier assigned to a computer program statement to enable other statements to refer to that statement. *Synonym:* identifier. **(B) (software)** One or more characters within or attached to a set of data, that identify or describe the data. *Synonym:* identifier. (C) 610.12-1990
(2) A visual user interface control that is noneditable text that can be used as a title, a message, to identify a field, or to provide additional information about a field. (C) 1295-1993w
(3) A word or phrase that is attached to or part of a model graphic. A label typically consists of a model construct's name (or one of the aliases) and may contain additional textual annotations (such as a note identifier). (C/SE) 1320.2-1998

labeled (1) Equipment or materials to which has been attached a label, symbol, or other identifying mark of an organization acceptable to the authority having jurisdiction and concerned with product evaluation, that maintains periodic inspection of production of labeled equipment or materials and by whose labeling the manufacturer indicates compliance with appropriate standards or performance in a specified manner. (NESC/NEC) [86]
(2) Conductors, equipment, or materials that meet appropriate standards or that have been tested and found suitable for use in a specified manner. (NEC/DEI) 1221-1993w

laboratory (A) (meter) A laboratory responsible for maintaining reference standards and assigning values to the working standards used for the testing of electricity meters and auxiliary devices. **(B)** (independent standards). A standards laboratory maintained by, and responsible to, a company or authority that is not under the same administrative control as the laboratories or companies submitting instruments for calibration. (ELM) C12.1-1988

laboratory—independent standards A standards laboratory maintained by, and responsible to, a company or authority that is not under the same administrative control as the laboratories or companies submitting instruments for calibration. (ELM) C12.1-1988

laboratory—meter A laboratory responsible for maintaining reference standards and assigning values to the working standards used for the testing of electricity meters and auxiliary devices. (ELM) C12.1-1988

laboratory reference standards (metering) Standards that are used to assign and check the values of laboratory secondary standards. (ELM) C12.1-1982s, C12.1-1988

laboratory resistivity Of a fly ash deposit is the ratio of the applied electric potential across the layer to the induced current density. The value of resistivity is specific for a given ash sample and depends upon the test variables or conditions: temperature, composition of the gaseous environment (especially water and sulfuric acid vapor content), the magnitude of the applied electric field strength, as well as the porosity of the ash layer. Measured resistivity also is a function of test procedure details, including items such as initial test temperature, the rate of heating or cooling, and the time that the voltage is applied prior to reading current. (PE/EDPG) 548-1984w

laboratory secondary standards (metering) Standards that are used in the routine calibration tasks of the laboratory. (ELM) C12.1-1982s, C12.1-1988

laboratory simulation A simulation developed and used under highly controlled conditions; for example, a simulation of a medical technique implemented in the controlled environment of a laboratory. (C) 610.3-1989w

laboratory test *See:* shop test.

laboratory working standards Those standards that are used for the ordinary calibration work of the standardizing laboratory. *Note:* Laboratory working standards are calibrated by comparison with the reference standards of that laboratory. *See also:* measurement system. (IM) [38]

labyrinth seal ring (rotating machinery) Multiple oil catcher ring surrounding a shaft with small clearance. *See also:* bearing. (PE) [9]

laced card A punch card punched accidentally or intentionally with holes in excess of the hole patterns required by the character set in use. (C) 610.10-1994w

lacing, stator-winding end-wire (rotating machinery) Cord or other lacing material used to bind the stator-winding end wire, to hold in place labels and devices placed on or in the end wire, and to position lead cables at their take-off points on the end wire. *See also:* stator. (PE) [9]

lacquer-film capacitor A capacitor in which the dielectric is primarily a solid lacquer film and the electrodes are thin metallic coatings deposited thereon. (PE/EM) 43-1974s

lacquer master* *See:* lacquer original.

* Deprecated.

lacquer original An original recording on a lacquer surface for the purpose of making a master. *See also:* phonograph pickup. (SP) [32]

lacquer recording (electroacoustics) Any recording made on a lacquer recording medium. *See also:* phonograph pickup. (SP) [32]

ladder, hook *See:* tower ladder.

ladder network (1) (general) A network composed of a sequence of H, L, T, or pi networks connected in tandem. *See also:* network analysis. (Std100) 270-1966w
(2) A cascade or tandem connection of alternating series and shunt arms. (CAS) [13]

ladder, rope *See:* rope ladder.

ladder, tower *See:* tower ladder.

ladder-winding insulation (rotating machinery) An element of winding insulation in the form of a single sheet precut to fit into one or more slots and with a broad area at each end to provide end-wire insulation. *See also:* rotor; stator. (PE) [9]

lag (1) (general) The delay between two events. (C) [20], [85]
(2) (control circuits) Any retardation of an output with respect to the casual input. *See also:* feedback; telegraphy; control system. (PE/IA/EDPG/IAC) [3], [60]
(3) (telegraph system) The time elapsing between the operation of the transmitting device and the response of the receiving device. *See also:* telegraphy. (PE/EEC) [119]
(4) (automatic control) (distance/velocity) A delay attributable to the transport of material or the finite rate of propagation of a signal or condition. *See also:* control system; feedback. (PE/EDPG) [3]
(5) (camera tubes) A persistence of the electrical-charge image for a small number of frames. (ED) 161-1971w
(6) (electric power systems in commercial buildings) The delay in action of a sensing element of a control element. (IA/PSE) 241-1990r

lag, first-order *See:* first-order lag.

lagged-demand meter A meter in which the indication of the maximum demand is subject to a characteristic time lag by either mechanical or thermal means. *See also:* electricity meter. (ELM) C12.1-1982s

lagged variable *See:* lag variable.

lagging operation Inductive megavars absorption of the static var compensator (SVC), similar to a shunt reactor. (PE/SUB) 1031-2000

lag, multi-order *See:* multiorder lag.

lag networks (power supplies) Resistance-reactance components, arranged to control phase-gain roll-off versus frequency. *Note:* Used to assure the dynamic stability of a power-supply's comparison amplifier. The main effect of a lag network is a reduction of gain at relatively low frequencies so that the slope of the remaining rolloff can be relatively more gentle. (AES) [41]

Lagrange stability (system) Every solution that is generated by a finite initial state is bounded. *Note:* An example of a system that is Lagrange stable is a second-order system with a single stable limit cycle. Although this system must contain a point inside the limit cycle that is Lyapunov unstable, the system is still Lagrange stable because every solution remains bounded. *See also:* control system. (CS/IM) [120]

lag, second-order *See:* second-order lag.

lag, transfer *See:* first-order lag; multiorder lag.

lag variable (A) In a discrete simulation, a variable that is an output of one period and an input for some future period. *Synonym:* lagged variable. **(B)** In an analog simulation, a variable that is a function of an output variable and that is used as input to the simulation to provide a time delay response or feedback. *Synonym:* lagged variable. (C) 610.3-1989

LAM access specifier (*m*) (subroutines for CAMAC) The symbol *m* represents an integer which is used to indicate the mode of access of a LAM and the lowest-order address component for LAM addressing. If *m* is zero or positive, it is interpreted as subaddress and the LAM is assumed to be accessed via dataless functions at this subaddress. If *m* is negative, it is interpreted as the negative of a bit position for a LAM which is accessed via reading, setting, or clearing bits in the group 2 registers at subaddresses 12, 13, or 14. (NPS) 758-1979r

lambda (λ) (1) System incremental cost. The additional cost of delivering another one MW of power to the load center. (PE/PSE) 858-1993w
(2) *See also:* system incremental cost. (PE/PSE) 94-1991w

lambert* (television) A unit of luminance (photometric brightness) equal to $1/\pi$ candela per square centimeter and, therefore, equal to the uniform luminance of a perfectly diffusing surface emitting or reflecting light at the rate of one lumen per square centimeter. *Note:* The lambert is also the average luminance of any surface emitting or reflecting light at the rate of one lumen per square centimeter. For the general case, the average must take account of variation of luminance with angle of observation, and also of its variation from point to point on the surface considered. (BT/AV) 201-1979w
* Deprecated.

Lambert-Beer Law *See:* Beer-Lambert Law.

Lambertian radiator *See:* Lambert's cosine law.

Lambertian surface (1) (illuminating engineering) A surface that emits or reflects light in accordance with Lambert's cosine law. A lambertian surface has the same luminance regardless of viewing angle. (EEC/IE) [126]
(2) (laser maser) An ideal surface whose emitted or reflected radiance is independent of the viewing angle. (LEO) 586-1980w
(3) A surface with a cosine dependence of the scattered power on both the incident and scattering angles and, hence, a cosine squared dependence on angle in the backscatter direction. (AP/PROP) 211-1997

lambertian units of luminance (illuminating engineering) The luminance of a surface in a specified direction also has been expressed in terms of the luminous exitance the surface would have if the luminances in all direction, within the hemisphere on the side of the surface being considered, were the same as the luminance in the specified direction. In other words, luminance has been expressed as the luminous exitance of a lambertian surface whose luminance equals the luminance of the surface in question and in the specified direction.

Lambertian Luminance Unit	Lambertian Luminous Exitance	Equivalent Luminance
1 asb	= 1 lm/m^2	= $(1/\pi)$ cd/m^2
1 fL	= 1 lm/ft^2	= $(1/\pi)$ cd/ft^2
		$(1/144\pi)$ cd/in^2
1 L	= 1 lm/cm^2	= $(1/\pi)$ cd/m^2

Note: The lambertian units of luminance are numerically equal to the corresponding units of luminous exitance. Thus, the luminance (in lambertian units) of a surface could be determined, and the numerical value directly used in the equation for a lambertian reflecting surface relating illuminance and luminous exitance, that is, $M = \rho E$ where ρ is the reflectance. In practice no surface follows exactly the cosine formula of emission or reflection and many do not even approximate it; hence the luminance is not uniform but varies with the angle from which it is viewed. Since the raison d'etre for this system was the use of a relation of generally unknown and variable accuracy when applied to real surfaces, the use of lambertian units of luminance has not been acceptable since 1967. *See also:* Lambert's cosine law. (EEC/IE) [126]

Lambert's cosine law (1) (fiber optics) The statement that the radiance of certain idealized surfaces, known as Lambertian radiators, Lambertian sources, or Lambertian reflectors, is independent of the angle from which the surface is viewed. *Note:* The radiant intensity of such a surface is maximum normal to the surface and decreases in proportion to the cosine of the angle from the normal. *Synonym:* cosine emission law. (Std100) 812-1984w
(2) (illuminating engineering) The luminous intensity in any direction from a plane perfectly diffusing surface element varies as the cosine of the angle between that direction and the perpendicular to the surface element. (EEC/IE) [126]

LAM identifier lam (subroutines for CAMAC) The symbol *lam* represents an integer that is used as the identifier of a CAMAC LAM signal. The information associated with the identifier must include not only the CAMAC address but also information about the means of accessing and controlling the LAM, that is, whether it is accessed via dataless functions at a subaddress or via read/write functions in group 2 registers. The value of lam is explicitly defined to be a non-zero integer; whether it is an encoded representation of the information required to describe the LAM or simply provides a key for accessing the information in a system-data structure is an implementation decision. The value 0 is used to indicate, where appropriate, that no LAM is being specified. (NPS) 758-1979r

laminar insulations (cable-insulation materials) Dielectric materials, either fibrous or film, or composite, comprising two or more layers of insulation arranged in series, and normally impregnated or flooded with an insulating liquid, or both. (PE) 402-1974w

laminated A joint in which both cables have a dielectric that consists of fluid-impregnated paper or paper/synthetic laminated tape, or varnished cloth. (PE/IC) 404-1993

lamp (1) (illuminating engineering) A generic term for a man-made source of light. By extension, the term is also used to denote sources that radiate in regions of the spectrum adjacent to the visible. *Note:* A lighting unit consisting of a lamp with shade, reflector, enclosing globe, housing, or other accessories is also called a "lamp." In such cases, in order to distinguish between the assembled unit and the light source within it, the latter is often called a "bulb" or "tube," if it is electrically powered. *See also:* luminaire. (EEC/IE) [126]
(2) (electric power systems in commercial buildings) Generic term for a manmade source of light. (IA/PSE) 241-1990r

lamp burnout factor (illuminating engineering) The fractional loss of task illuminance due to burned out lamps left in place for long periods. (EEC/IE) [126]

lamppost (illuminating engineering) A standard support provided with the necessary internal attachments for wiring and the external attachments for the bracket and luminaire. (EEC/IE) [126]

lamp shielding angle (illuminating engineering) The angle between the plane of the baffles or louver grid and the plane most nearly horizontal that is tangent to both the lamps and the louver blades. *Note:* The lamp shielding angle frequently is larger than the louver shielding angle, but never smaller. (EEC/IE) [126]

LAN *See:* local area network.

land (electroacoustics) The record surface between two adjacent grooves of a mechanical recording. *See also:* phonograph pickup. (SP) [32]

landing beacon (navigation aid terms) A beacon used to guide aircraft in landing. (AES/GCS) 172-1983w

landing direction indicator (illuminating engineering) A device to indicate visually the direction currently designated for landing and takeoff. (EEC/IE) [126]

landing light (illuminating engineering) An aircraft aeronautical light designed to illuminate a ground area from the aircraft. (EEC/IE) [126]

landing zone (elevators) A zone extending from a point eighteen inches below a landing to a point eighteen inches above the landing. *See also:* elevator landing. (PE/EEC) [119]

landmark beacon (illuminating engineering) An aeronautical beacon used to indicate the location of a landmark used by pilots as an aid to enroute navigation. (EEC/IE) [126]

landscape image *See:* comic-strip oriented image.

landscape orientation A page orientation of a display surface having greater width than height. *Note:* Derived from pictures of landscapes, which are normally horizontal in format. *Contrast:* portrait orientation. (C) 610.10-1994w

lane (navigation systems) (electronic navigation) The projection of a corridor of airspace on a navigation chart, the right and left sides of the corridor being defined by the same (ambiguous) values of the navigation coordinate (phase or amplitude), but within which lateral position information is provided (for example, a Decca lane in which there is a 360-degree change of phase). *See also:* navigation. (AES/RS) 686-1982s, [42]

Langevin ions *See:* charged aerosol.

language (1) (software requirements specifications) A means of communication, with syntax and semantics, consisting of a set of representations, conventions and associated rules used to convey information. (C/SE) 830-1984s

(2) (A) A system consisting of:

1) a well defined, usually finite, set of characters:
2) rules for combining characters with one another to form words or other expressions:
3) a specific assignment of meaning to some of the words or expressions, usually for communicating information or data among a group of people, machines, etc.

See also: machine language; code. **(B)** A system similar to that in definition (A) without any specific assignment of meanings. Such systems may be distinguished from those in definition (A), when necessary, by referring to them as formal or uninterpreted languages. Although it is sometimes convenient to study a language independently of any meanings, in all practical cases at least one set of meanings is eventually assigned. *See also:* code; machine language. (C/MIL) 162-1963, [2]

(3) (A) (software) A systematic means of communicating ideas by the use of conventionalized signs, sounds, gestures, or marks and rules for the formation of admissible expressions. *See also:* computer language. **(B) (software)** A means of communication, with syntax and semantics, consisting of a set of representations, conventions, and associated rules used to convey information. *See also:* computer language. (C) 610.12-1990, 610.13-1993

(4) A systematic means of communicating ideas by the use of conventional signs, sounds, gestures, or marks, and rules for the formation of admissible expressions. (C) 610.13-1993w

language-binding API specification A specification that documents the source code method, consistent with a specific programming language, used by an application to access services provided by an application platform. (C/PA) 14252-1996

language code (telephone switching systems) On an international call, an address digit that permits an originating operator to obtain assistance in a desired language. (COM) 312-1977w

language-description language *See:* metalanguage.

Language for Conversational Computing An interactive programming language combining ALGOL-like syntax with many of the features of JOSS. (C) 610.13-1993w

Language for the Expression of Associative Procedures A programming language based on ALGOL 60, containing set-theoretic and associative operations and data types. (C) 610.13-1993w

Language for Your Remote Instruction by Computer (LYRIC) An application-oriented language used primarily for computer-assisted instruction. (C) 610.13-1993w

language-independent service specification A specification that defines a set of required functional semantics independent of the syntax and semantics of a programming language. (C/PA) 14252-1996

language-independent specification The format for describing services that is not tied to any specific computer language. (SCC20) 1226-1998

Language of Temporal Ordering Specifications (LOTOS) A specification language used for telecommunications and distributed systems. (C) 610.13-1993w

language printout (dedicated-type sequential events recording systems) A word description composed of alphanumeric characters used to further identify inputs and their status. *See also:* language. (PE/EDPG) [1]

language processor (software) A computer program that translates, interprets, or performs other tasks required to process statements expressed in a given language. *See also:* compiler; assembler; translator; interpreter. (C) 610.12-1990

language standard A standard that describes the characteristics of a language used to describe a requirements specification, a design, or test data. *See also:* standard language. (C) 610.12-1990, 610.10-1994w, 610.13-1993w

LAN/MAN Management The management functionality specific to the management of IEEE 802 Local or Metropolitan Area subnetworks. (LM/C) 15802-2-1995

LAN/MAN Standards Committee (LMSC) The IEEE standards committee that develops LAN and MAN standards. *Synonym:* IEEE 802®. (C) 610.7-1995

LAN/WAN communications architecture A communications architecture that supports LAN and WAN networks. (C) 610.7-1995

lanyard (1) A flexible line of webbing, rope, wire rope, or strap which generally has a connector at each end for connecting the line-worker's body belt, aerial belt, or full body harness to a energy absorbing device, lifeline, or anchorage. Special use only. Required in operations where the lanyard is subject to being cut. Prohibited in the vicinity of energized facilities. (T&D/PE) 1307-1996

(2) A flexible line or webbing, rope, wire rope, or strap that generally has a connector at each end for connecting the line-worker's body belt, aerial belt, or full body harness to an energy absorbing device, lifeline, or anchorage. (NESC) C2-1997

lapel microphone A microphone adapted to positioning on the clothing of the user. *See also:* microphone. (EEC/PE) [119]

Laplace's equation The special form taken by Poisson's equation when the volume density of charge is zero throughout the isotropic medium. It is $\Delta^2 v = 0$. (Std100) 270-1966w

Laplace transform (function) (unilateral) The quantity obtained by performing the operation

$$F(s) = \int_0^{\infty} f(t)e^{-st}dt$$

where $s = \sigma + j\omega$. *See also:* feedback control system. (PE/EDPG) [3]

laptop computer A portable computer designed for use on one's lap. (C) 610.2-1987, 610.10-1994w

lap winding A winding that completes all its turns under a given pair of main poles before proceeding to the next pair of poles. *Note:* In commutator machines the ends of the individual coils of a simplex lap winding are connected to adjacent commutator bars; those of a duplex lap winding are connected to alternate commutator bars etc. *See also:* asynchronous machine. (EEC/PE) [119]

large ion (1) (overhead power) Ion comprised of charged particles, liquid or solid, suspended in air. Typical radius is in the range of 2×10^{-8} m to 2×10^{-7} m. Mobility is in the range of 10^{-9} m^2/Vs to 10^{-7} m^2/Vs. *Note:* Historically, these have been referred to as large or Langevin ions. The use of the term "charged aerosols" in encouraged. *Synonym:* charged aerosol. (T&D/PE) 539-1990
(2) *See also:* charged aerosol. (T&D/PE) 1227-1990r

large ions *See:* charged aerosol.

large scale integration (LSI) (A) Pertaining to an integrated circuit containing between 500 and 2×104 transistors in its design. *Contrast:* ultra-large scale integration; medium scale integration; small scale integration; very large scale integration. **(B)** Pertaining to an integrated circuit containing between 100 and 5,000 elements. (C) 610.10-1994

large signal One whose peak-to-peak amplitude is as large as practical but is recorded by the instrument within, but not including, the maximum and minimum amplitude data codes. As a minimum, the signal must span at least 90% of the full-scale range of the waveform recorder. (IM/WM&A) 1057-1994w

large signal performance (1) (excitation systems for synchronous machines) Response of an excitation control system, excitation system, or elements of an excitation system to signals which are large enough that nonlinearities must be included in the analysis of the response to obtain realistic results. (PE/EDPG) 421.1-1986r
(2) (dynamic performance of excitation systems) The response to signals that are large enough so that nonlinearities are significant. (PE/EDPG) 421.2-1990

larry A motor-driven burden-bearing track-mounted car designed for side or end dumping and used for hauling material such as coal, coke, or mine refuse. (PE/EEC/MIN) [119]

laser (1) (fiber optics) A device that produces optical radiation using a population inversion to provide Light Amplification by Stimulated Emission of Radiation and (generally) an optical resonant cavity to provide positive feedback. Laser radiation may be highly coherent temporally, or spatially, or both. *See also:* injection laser diode; active laser medium; optical cavity. (Std100) 812-1984w
(2) (laser maser) A device that produces an intense, coherent, directional beam of light by stimulating electronic, ionic, or molecular transitions to lower energy levels. Also, an acronym for light amplification by stimulated emission of radiation. (LEO) 586-1980w
(3) The laser produces a highly monochromatic and coherent (spatial and temporal) beam of radiation. A steady oscillation of nearly a single electromagnetic mode is maintained in a volume of an active material bounded by highly reflecting surfaces called a resonator. The frequency of oscillation varies according to the material used and by the methods of initially exciting or pumping the material. (EEC/IE) [126]
(4) (A) A device that can generate a laser beam. *Note:* "laser" is an acronym for light amplification by stimulated emissions of radiation. **(B)** Loosely, pertaining to a device that uses a laser beam, as in a laser printer. (C) 610.10-1994

laser beam An extremely narrow, coherent beam of electromagnetic energy in the form of light. (C) 610.10-1994w

laser beam printer *See:* laser printer.

laser diode *See:* injection laser diode.

laser disk An optical disk, typically 12 inches in diameter. *See also:* compact disc. (C) 610.10-1994w

laser gyro A device that measures angular rotation by optical heterodyning of internally generated, counter-propagating, optical beams. (AES/GYAC) 528-1994

laser gyro axes LA and NA are two perpendicular axes in the plane of the laser beams, and are normal to the IA. The IRA, LRA, and NRA are reference axes defined with respect to the mounting provisions. These axes are nominally parallel to IA, LA, and NA, respectively. Generally, the laser gyro will have two electrodes of one sign and one of the other. Plane axis LA is the center line of the laser leg containing the single electrode. Plane axis NA is the line in the laser beam plane perpendicular to LA and bisecting the leg containing LA. Generally, plane axis NA can be thought of as the axis of symmetry. (AES/GYAC) 528-1994

laser medium *See:* active laser medium.

laser printer A nonimpact, xerographic printer that uses a laser beam to create a latent image which is then made visible by a toner and transferred and fixed on paper. *Synonym:* laser beam printer. (C) 610.10-1994w

laser radar A radar whose carrier frequency is produced by a laser, usually in the infrared or visible light region. (AES) 686-1997

laser safety officer (laser maser) One who is knowledgeable in the evaluation and control of laser hazards and has authority for supervision of the control of laser hazards. (LEO) 586-1980w

laser system (laser maser) An assembly of electrical, mechanical, and optical components which includes a laser. (LEO) 586-1980w

lasing medium (laser maser) A material emitting coherent radiation by virtue of stimulated electronic or molecular transitions to lower energy levels. (LEO) 586-1980w

lasing threshold (1) (fiber optics) The lowest excitation level at which a laser's output is dominated by stimulated emission rather than spontaneous emission. *See also:* spontaneous emission; laser; stimulated emission. (Std100) 812-1984w
(2) (laser gyro) The discharge current at which the gain of the laser just equals the losses. (AES/GYAC) 528-1994

last close (1) (of a file) When a process closes a file, resulting in the file not being an open file within any process. (C/PA) 9945-1-1996
(2) (of a file) Occurs when a process closes a file, message queue, or shared memory object and this results in it not being open within any process. (C/PA) 1003.5b-1995
(3) The act of a process closing a file, message queue, or shared memory object that results in the file, message queue, or shared memory object no longer being open within any process. (C) 1003.5-1999

last-in, first-out (A) A technique for managing a set of items to which additions and deletions are to be made; items are appended to one end and retrieved from that same end. *See also:* stack. **(B)** Pertaining to a system in which the next item to exit the system is the item that has been in the system for the shortest time. *Synonym:* first-in, last-out. *Contrast:* first-in, first-out. (C) 610.5-1990

last-line release (telephone switching systems) Release under control of the last line that goes on-hook. (COM) 312-1977w

last transition (pulse terminology) The major transition waveform of a pulse waveform between the top and the base. (IM/WM&A) 194-1977w

LATA *See:* local area transport area.

latch An attachment used to hold a fuse or switch in the closed position. (SWG/PE) C37.40-1993, C37.100-1992

(2) (A) A circuit that can be used to hold data in a ready position until required; usually controlled by another circuit. *See also:* latching; transparent latch. **(B)** A circuit consisting of one or more latches as in (A) that is used to store digital data. *See also:* register. (C) 610.10-1994

latching The process of holding data in a circuit until other circuits are ready to change the latch circuit. (C) 610.10-1994w

latching current (1) (thyristor) The minimum principal current required to maintain the thyristor in the ON-state immediately after switching from the OFF-state to the ON-state has occurred and the triggering signal has been removed. *See also:* holding current. (IA/IPC) 428-1981w
(2) (of a switching device) The making current during a closing operation in which the device latches or the equivalent. (SWG/PE) C37.100-1992

latching relay A relay that is so constructed that it maintains a given position by means of a mechanical latch until released mechanically or electrically. (SWG/PE) C37.100-1992

latch-in relay A relay that maintains its contacts in the last position assumed without the need of maintaining coil energization. (PE/EM) 43-1974s

latch-up A state in which a low-impedance path results from an input, output, or supply excessive operating condition that triggers a parasitic structure and persists after the removal or cessation of the triggering condition. (ED) 1005-1998

late mode The very last edge that propagates through a given cone of logic. (C/DA) 1481-1999

latency (1) (biological electronics) The condition in an excitable tissue during the interval between the application of a stimulus and the first indication of a response. (EMB) [47]
(2) (electronic computation) The time between the completion of the interpretation of an address and the start of the actual transfer from the addressed location. (C) [20], [85]
(3) (software) The time interval between the instant at which an instruction control unit issues a call for data and the instant at which the transfer of data is started. (C) 610.12-1990
(4) The time, expressed in number of symbols, it takes for a signal to pass through a ring component. *See also:* ring latency; cumulative latency. (C/LM) 8802-5-1998

latency adjust buffer (LAB) A first-in first-out (FIFO) function within the CFBIU with the ability to store one full frame of Rx Data. The LAB function insures an integral number of frames are rotating on the SFODB ring. (C/BA) 1393-1999

latent period (electrobiology) The time elapsing between the application of a stimulus and the first indication of a response. *See also:* excitability. (EMB) [47]

lateral compression (software) In software design, a form of demodularization in which two or more modules that execute one after the other are combined into a single module. *Contrast:* downward compression; upward compression. (C) 610.12-1990

lateral conductor (1) A wire or cable extending in a general horizontal direction at an angle to the general direction of the line conductors. (T&D) C2.2-1960
(2) A wire or cable extending in a general horizontal direction at an angle to the general direction of the line conductors, and entirely supported on one structure. (NESC) C2-1997

lateral critical speeds (rotating machinery) The speeds at which the amplitudes of the lateral vibrations of a machine rotor due to shaft rotation reach their maximum values. *See also:* rotor. (PE) [9]

lateral-cut recording *See:* lateral recording.

lateral insulator (storage cell) An insulator placed between the plates and the side wall of the container in which the element is housed. *See also:* battery. (PE/EEC) [119]

lateral offset loss (fiber optics) A power loss caused by transverse or lateral deviation from optimum alignment of source to optical waveguide, waveguide to waveguide, or waveguide to detector. *Synonym:* transverse offset loss. (Std100) 812-1984w

lateral profile (1) (radio noise from overhead power lines and substations) The radio noise field strength at ground level plotted as a function of the horizontal distance from, and at a right angle to, the power line conductors. (T&D/PE) 430-1986w
(2) (overhead-power-line corona and radio noise) The profile of a parameter, usually near ground level, plotted as a function of the horizontal distance from and at a right angle to the line conductors. For example, a lateral profile of the vertical component of the electric field strength, of the radio noise field strength, etc. (T&D/PE) 539-1990

lateral recording (lateral-cut recording) A mechanical recording in which the groove modulation is perpendicular to the motion of the recording medium and parallel to the surface of the recording medium. (SP) [32]

lateral wave A wave, not predicted by geometrical optics, excited at and propagated along the interface of two (possibly lossy) dielectric media. For sufficiently large distances from the source, the magnitude of the field varies as the inverse square of the distance measured along the interface. *Note:* Lateral wave is similar to the component of the radio ground wave when the geometrical-optical component is separated out. *See also:* Norton surface wave. (AP/PROP) 211-1997

lateral width (light distribution) The lateral angle between the reference line and the width line, measured in the cone of maximum candlepower. *Note:* This angular width includes the line of maximum candlepower. *See also:* streetlighting luminaire. (EEC/IE) [126]

lateral working space (electric power distribution) The space reserved for working between conductor levels outside the climbing space, and to its right and left. *See also:* tower. (BT/AV) [34]

late relay contacts Contacts that open or close after other contacts when the relay is operated. (EEC/REE) [87]

LaTeX A text-formatting language based on TeX. (C) 610.13-1993w

lattice (1) (navigation) A pattern of identifiable intersecting lines of position, which lines are laid down by a navigation aid. *See also:* navigation. (AES/RS) 686-1982s, [42]
(2) A partial ordering imposed on a structure of model entities. (SCC20) 1232.1-1997

lattice channeling (germanium gamma-ray detectors) (charged-particle detectors) (in a semiconductor radiation conductor) A phenomenon that results in a crystallographic directional dependence of the rate of energy loss of ionizing particles. (NPS) 325-1986s, 300-1982s

lattice network A network composed of four branches connected in series to form a mesh, two nonadjacent junction points serving as input terminals, while the remaining two junction points serve as output terminals. *See also:* network analysis.

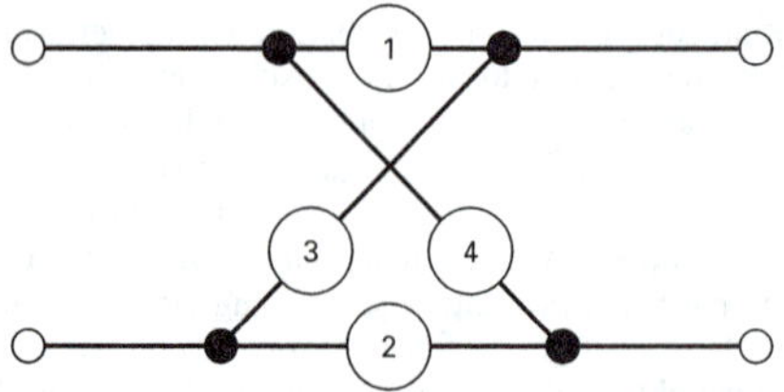

The junction points between branches 4 and 1 and between 3 and 2 are the input terminals; the junction points between branches 1 and 3 and between branches 2 and 4 are the output terminals.

lattice network

(Std100) 270-1966w

launch angle (fiber optics) The angle between the light input propagation vector and the optical axis of an optical fiber or fiber bundle. *See also:* launch numerical aperture. (Std100) 812-1984w

launcher (waveguide components) An adapter used to provide a waveguide or transmission line port for a wave propagating structure. (MTT) 147-1979w

launching fiber (fiber optics) A fiber used in conjunction with a source to excite the modes of another fiber in a particular fashion. *Note:* Launching fibers are most often used in test systems to improve the precision of measurements. *Synonym:* injection fiber. *See also:* pigtail; mode. (Std100) 812-1984w

launch numerical aperture (LNA) (fiber optics) The numerical aperture of an optical system used to couple (launch) power into an optical waveguide. *Notes:* 1. LNA may differ from the stated NA of a final focusing element if, for example, that element is underfilled or the focus is other than that for which the element is specified. 2. LNA is one of the parameters that determine the initial distribution of power among the modes of an optical waveguide. *See also:* launch angle; acceptance angle. (Std100) 812-1984w

laundry area An area containing or designed to contain either a laundry tray, clothes washer, and/or a clothes dryer. (NESC/NEC) [86]

Lawrence Radiation Laboratory TRANslator A FORTRAN-based language used to perform vector arithmetic, bit and byte manipulation, and pointer manipulation. (C) 610.13-1993w

lay (1) (cable) The helical arrangement formed by twisting together the individual elements of a cable. (PE/T&D) [4], [10]

(2) (helical element of a cable) The axial length of a turn of the helix of that element. *Notes:* 1. Among the helical elements of a cable may be each strand in a concentric-lay cable, or each insulated conductor in a multiple-conductor cable. 2. Also termed "pitch." 30-1937w

layer (1) In the OSI model, one of a collection of network-processing functions representing a level of a hierarchy of functions. *See also:* medium access control sublayer; application layer; presentation layer; transport layer; session layer; entity layer; physical layer; sublayer; network layer; data link layer; logical link control sublayer; client layer. (C) 610.7-1995

(2) In Physical Design Exchange Format (PDEF), a particular level of interconnect on which a logical or physical pin is located. (C/DA) 1481-1999

(3) A subdivision of the Open Systems Interconnection (OSI) architecture, constituted by subsystems of the same rank. (EMB/MIB) 1073.4.1-2000

layered protocol A protocol to follow the layering principle so that layer *n* at the destination machine receives exactly what layer *n* at the source machine sends. (C) 610.7-1995

layer management (1) Functions related to the management of the (N)-layer partly performed in the (N)-layer itself according to the (N)-protocol of the layer, and partly performed as a subset of systems management. (LM/C) 802.10-1992

(2) Functions related to the administration of a given Open Systems Interconnection (OSI) *layer*. These functions are performed in the layer itself according to the protocol of the layer and partly performed as a subset of *network management or systems management*. (LM/C) 8802-6-1994

layer management entity (LME) (1) The entity in a *layer* that performs local management of a *layer*. The LME provides information about the layer, effects control over it, and indicates the occurrence of certain events within it. (LM/C) 8802-6-1994

(2) The logical portion of the repeater responsible for collection of operational and error statistics and for management of the network.local area networks. (C) 8802-12-1998

layer management interface (LMI) The service interface provided by the *Layer Management Entity (LME)* to the *Network Management Process (NMP)*. (LM/C) 8802-6-1994

layer manager (LV) A systems management service application for which a particular exchange of systems management information has taken a manager role of the (N)-layer. (LM/C) 802.10-1992

layer service The service that a layer provides to the next higher layer. (C/LM/COM) 802.9a-1995w, 8802-9-1996

layout character *See:* format effector character.

layover The projection of the signature of an object having greater vertical extent than another object having less vertical extent but in the same range-Doppler (or range-angle) resolution cell. The higher object's signature in the image is superimposed or "overlaid" onto the signature of other vertically lower, but horizontally displaced, scene content. (AES/RS) 686-1990

layup The preparation of wood strips for gluing together. (T&D/PE) 751-1990

lay-up (nuclear power generating station) Idle condition of equipment and systems during and after installation, with protective measures applied as appropriate. (PE/NP) 380-1975w, 336-1980s

lay-up condition Idle condition of the equipment and systems between service operations. (VT) 1476-2000

L-band A radar-frequency band between 1 GHz and 2 GHz, usually in the International Telecommunication Union (ITU) allocated band 1.215–1.4 GHz. (AES) 686-1997

L–band radar (-band radar) A radar operating at frequencies between 1 and 2 GHz, usually in the ITU assigned band 1.215 to 1.4 GHz; may refer also to the 0.89 to 0.94 GHz ITU assignment. (AES/RS) 686-1982s

LC *See:* low conduction threshold voltage LC_v.

LC auxiliary switch *See:* LC contact; auxiliary switch.

L_char *See:* link character.

LC contact A latch-checking contact that is closed when the operating mechanism linkage is relatched after an opening operation of the switching device. (SWG/PE) C37.100-1992

LCD *See:* liquid crystal device; liquid crystal display.

LCFO *See:* loop current feed open.

LCMARC *See:* MAchine-Readable Cataloging.

LC-MARC *See:* MAchine-Readable Cataloging.

LD *See:* Loadable Device.

L_d The maximum average area control error allowed for each of the six ten-minute periods during an hour. It is calculated by the following formula:

$L_d = 0.025\ \Delta L + 5\ \text{MW}$

where ΔL is either the greatest hourly change (either increasing or decreasing) in the control area's net energy for load that occurred on the day of the control area's winter or summer peak demand, or the average of any ten hourly changes (either increasing or decreasing) in net energy for load that occurred during the year. (PE/PSE) 858-1993w

Ld(s) *See:* direct-axis operational inductance.

L-display Deflections are of equal amplitude when the radar antenna is pointed directly at the target, any inequality representing relative pointing error. The time base (range scale) can be vertical, as in the L-display illustration, or horizontal. The L-display is also known as a bearing-deviation indicator.

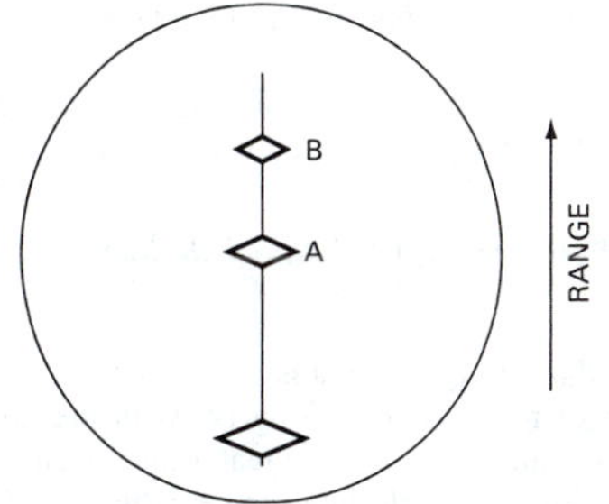

Note: Two targets, A and B at different ranges: radar aimed at target A.

L-display

(AES) 686-1997

lead (power and distribution transformers) A conductor that connects a winding to its termination (that is, terminal, bushing, terminal board, or connection to another winding). (PE/TR) C57.12.80-1978r

lead box (rotating machinery) (terminal housing) A box through which the leads are passed in emerging from the housing. (PE) [9]

lead cable (rotating machinery) A cable type of conductor connected to the stator winding, used for making connections to the supply line or among circuits of the stator winding. *See also:* stator. (PE) [9]

lead clamp (rotating machinery) (salient-pole construction) A device used to retain and support the field leads between the hub and rotor rim along the rotor spider arms. *See also:* rotor. (PE) [9]

lead collar (rotating machinery) (salient-pole construction) A bushing used to insulate field leads at a point of support between a collector ring and the rotor rim. *See also:* rotor. (PE) [9]

lead-covered cable (lead-sheathed cable) A cable provided with a sheath of lead for the purpose of excluding moisture and affording mechanical protection. *See also:* armored cable; underground cable. (T&D/PE) [10]

leader (1) (computers) The blank section of tape at the beginning and end of a reel of tape. *Note:* The otherwise blank section may, however, include a parity check. (IA) [61], [84]

(2) In text formatting, a character used to lead the reader's eye across the page and to indicate logical connection between two items on a line. For example, one of a series of periods used in a table of contents to associate a chapter title with its corresponding page number. (C) 610.2-1987

(3) (data management) The blank section of magnetic tape at the beginning of a reel. *Contrast:* trailer. (C) 610.5-1990w, 610.10-1994w

(4) *See also:* pilot line. (T&D/PE) 524-1992r

leader cable A navigational aid consisting of a cable around which a magnetic field is established, marking the path to be followed. *See also:* radio navigation. (EEC/PE) [119]

leader-cable system A navigational aid in which a path to be followed is defined by the detection and comparison of magnetic fields emanating from a cable system that is installed on the ground or under water. *See also:* navigation. (AES/RS) 686-1982s, [42]

leader cone A tapered cone made of rubber, neoprene, or polyurethane that is used to lead a conductor splice through the travelers, thus making a smooth transition from the smaller diameter conductor to the larger diameter splice. It is also used at the connection point of the pulling line and running board to assist in a smooth transition of the running board over the travelers, thus significantly reducing the shock loads. *Synonyms:* nose cone; tapered hose. (T&D/PE) 524-1992r

lead, first-order *See:* first-order lead.

lead-in That portion of an antenna system that connects the elevated conductor portion to the radio equipment. *See also:* antenna. (AP/ANT) [35]

leading In photocomposition, the use of white space between lines and paragraphs of a document. *See also:* reverse leading. (C) 610.2-1987

leading decision A loop control that is executed before the loop body. *Contrast:* trailing decision. *See also:* WHILE. (C) 610.12-1990

leading edge (1) (pulse transformers) (first transition) That portion of the pulse occurring between the time the instantaneous value first becomes greater than A_Q to the time of the intersection of straight-line segments used to determine A_M. (PEL/MAG/ET) 390-1987r, 391-1976w

(2) (radar) A radar range tracking technique in which the range error signal is based on the range delay of the leading edge of the received echo. This provides ability to reject delayed interference, chaff, and more distant sources. (AES/RS) 686-1982s

(3) (television) The major portion of the rise of a pulse. *See also:* television.

(4) The end of a perforated tape that first enters a perforated-tape reader. (C) 610.10-1994w

leading edge linearity (a_l) (power pulse transformers) (first transition)The maximum amount by which the instantaneous pulse value deviates during the rise time interval from a straight line intersecting the 10% and 90% A M amplitude points used in determining rise time. It is expressed in amplitude units or in percent of 0.8 A M. *See also:* input pulse shape. (PEL/MAG/ET) 390-1987r, 391-1976w

leading-edge pulse The first major transition away from the pulse baseline occurring after a reference time. (IM/HFIM) [40]

leading-edge pulse time The time at which the instantaneous amplitude first reaches a stated fraction of the peak pulse amplitude. *See also:* television. (IM/IE/EEC/WM&A) 194-1977w, 270-1966w, [126]

leading-edge tracking A radar range tracking technique in which the range error signal is based on the range delay of the leading edge of the received echo. *Note:* Leading-edge tracking provides the ability to reject delayed interference, chaff, and more distant sources. (AES) 686-1997

lead-in groove (disk recording) (lead-in spiral) (electroacoustics) A blank spiral groove at the beginning of a recording generally having a pitch that is much greater than that of the recorded grooves. *See also:* phonograph pickup. 157-1951w

leading operation Capacitive megavars generation of the static var compensator (SVC), similar to a shunt capacitor. (PE/SUB) 1031-2000

leading wire An insulated wire strung separately or as a twisted pair, used for connecting the two free ends of the circuit of the blasting caps to the blasting unit. *See also:* blasting unit. (EEC/PE/MIN) [119]

leading zero A zero that precedes the first non-zero digit in a numeric representation; for example, the two zeros in "00324.6." *Contrast:* trailing zero. (C) 610.5-1990w

lead-in spiral *See:* lead-in groove.

lead-in wire A conductor connecting an electrode to an external circuit. *See also:* electron tube. (ED) [45], [84]

lead line *See:* pilot line.

lead networks (power supplies) Resistance-reactance components arranged to control phase-gain roll-off versus frequency. *Note:* Used to assure the dynamic stability of a power-supply's comparison amplifier. The main effect of a lead network is to introduce a phase lead at the higher frequencies, near the unity-gain frequency. (AES/PE) [41], [78]

lead-out groove (disk recording) (throw-out spiral) A blank spiral groove at the end of a recording generally of a pitch that is much greater than that of the recorded grooves and that terminates in either a locked or an eccentric groove. *See also:* phonograph pickup. (SP) [32]

lead-over groove (crossover spiral) In disk recording, a groove cut between successive short-duration recordings on the same disk, to enable the pickup stylus to travel from one cut to the next. *See also:* phonograph pickup. (SP) [32]

lead polarity (power and distribution transformers) A designation of the relative instantaneous direction of the currents in the leads of a transformer. Primary and secondary leads are said to have the same polarity when, at a given instant, the current enters the primary lead in question and leaves the secondary lead in question in the same direction as though the two leads formed a continuous circuit. The lead polarity of a single-phase distribution or power transformer may be either additive or subtractive. If adjacent leads from each of the two windings in question are connected together and voltage applied to one of the windings: (A) the lead polarity is additive if the voltage across the other two leads of the windings in question is greater than that of the higher voltage winding alone; (B) the lead polarity is subtractive if the voltage across the other two leads of the windings in question is less than that of the higher voltage winding alone. The polarity of a polyphase transformer is fixed by the internal connections between phases; it is usually designated by means of a phasor diagram showing the angular displacements of the voltages in the windings and a sketch showing the marking of the leads. The phasors of the phasor diagrams represent induced voltages. The standard rotation of phasors is counterclockwise. (PE/TR) C57.12.80-1978r

leads, load *See:* load leads.

lead storage battery A storage battery the electrodes of which are made of lead and the electrolyte consists of a solution of sulfuric acid. *See also:* battery. (EEC/PE) [119]

lead term *See:* keyword.

lead variable (A) In a discrete simulation, a variable that is an output of one period and that predicts what the output of some future period will be. **(B)** In an analog simulation, a variable that is a function of an output variable and that is used as input to the simulation to provide advanced time response or feedback. (C) 610.3-1989

leaf (1) (A) A terminal node in a search tree. **(B)** In a tree, a node that has no children. (C) 610.5-1990

(2) A contiguous information field that is pointed to by a ROM-directory entry. A leaf contains a header (length and CRC specification) and other information fields. (C/MM) 1212-1991s

leaf diagram A diagram that has no descendent diagrams, i.e., a diagram that does not contain any function that has been decomposed. (C/SE) 1320.1-1998

leaf entry (1) A ROM entry that specifies the address of a leaf. (C/MM) 1212-1991s

(2) A ROM entry that specifies the address of a variable-length data block. (C/BA) 896.2-1991w

(3) A directory entry that has no subordinates. It can be an alias entry or an object entry. (C/PA) 1327.2-1993w, 1326.2-1993w, 1224.2-1993w, 1328.2-1993w

(4) A read-only memory (ROM) entry that specifies the address of a variable-length data block. (C/BA) 896.10-1997

leaf node (1) A device node that has no children. (C/BA) 1275-1994

(2) A function that is not decomposed. A box that represents a leaf node does not have a box detail reference. (C/SE) 1320.1-1998

leak (handling and disposal of transformer grade insulating liquids containing PCBs) Any instance in which a polychlorinated biphenyl (PCB) unit (PCB article, PCB container, PCB equipment) has any PCBs on any portion of its external surface. (LM/C) 802.2-1985s

leakage (1) (A) (analog computer) Undesirable conductive paths in certain components, specifically, in capacitors, a path through which a slow discharge may take effect in problem boards, interaction effects between electrical signals through insufficient insulation between patch bay terminals. **(B) (analog computer)** Current flowing through such paths. (C) 165-1977

(2) (signal-transmission system) Undesired current flow between parts of a signal-transmission system or between the signal-transmission system and point(s) outside the system. *See also:* signal. (IE) [43]

(3) (health care facilities) Any current, including capacitively coupled current, not intended to be applied to a patient but which may be conveyed from exposed metal parts of an appliance to ground or to other accessible parts of an appliance. (EMB) [47]

(4) (transmission lines and waveguides) Radiation or conduction of signal power out of or into an imperfectly closed and shielded system. The leakage is usually expressed in decibels below a specified reference power. (IM/HFIM) [40]

(5) (insulation) The current that flows through or across the surface of insulation and defines the insulation resistance at the specified direct-current potential. (AES/ENSY) 135-1969w

(6) (semiconductor radiation detectors) The total detector current flowing at the operating bias in the absence of radiation. (NID) 301-1976s

leakage current (1) (germanium gamma-ray detectors) (charged-particle detectors) (x-ray energy spectrometers) (of a semiconductor radiation conductor) The total detector current flowing at the operating bias in the absence of radiation. (NPS/NID) 759-1984r, 300-1988r

(2) (maintenance of energized power lines) A component of the measured current that flows along the surface of the tool or equipment, due to the properties of the tool or equipment surface, including any surface deposits. (T&D/PE) 516-1995

(3) In the absence of external ionizing radiation and at the operating bias, the total current flowing through or across the surface of the detector element. (NPS) 325-1996

(4) Current that is not functional, including current in earth conductors and enclosures. (EMB/MIB) 1073.4.1-2000

(5) The current flowing in the equipment-grounding conductor (including a conductive case) when the device is connected as intended to the energized power system at rated voltage. (SPD/PE) C62.62-2000

leakage (conduction) current (1) Current resulting from the resistance of the dielectric insulation and surface leakage. (PE/TR) C57.19.03-1996

(2) (rotating machinery) The nonreversible constant current component of measured current that remains after the capacitive current and absorption current have disappeared. *Note:* Leakage current passes through the insulation volume, through any defects in the insulation, and across the insulation surface. (PE/EM) 95-1977r

leakage current, input *See:* input leakage current.

leakage distance (insulators) The sum of the shortest distances measured along the insulating surfaces between the conductive parts, as arranged for dry flashover test. *Note:* Surfaces coated with semiconducting glaze shall be considered as effective leakage surfaces, and leakage distance over such surfaces shall be included in the leakage distance. *See also:* insulator. (EEC/IEPL) [89]

leakage distance of external insulation *See:* creepage distance.

leakage flux Any magnetic flux, produced by current in an instrument transformer winding, which does not link all turns of all windings. (PE/TR) [57], C57.13-1993

leakage flux, relay *See:* relay leakage flux.

leakage inductance (one winding with respect to a second winding) A portion of the inductance of a winding that is related to a difference in flux linkages in the two windings; quantitatively, the leakage inductance of winding 1 with respect to winding 2

$$L_{\text{if}} = \frac{\partial\left(\lambda_{11} - \frac{N_1}{N_2}\lambda_{21}\right)}{\partial i_1}$$

where λ_{11} and λ_{21} are the flux linkages of windings 1 and 2, respectively, resulting from current i_1 in winding 1, and N_1 and N_2 are the number of turns of windings 1 and 2, respectively. (CHM) [51]

leakage power (nonlinear, active, and nonreciprocal waveguide components) (transmit-receive tubes and power limiters) The radio-frequency power transmitted through a fired tube. *See also:* flat leakage power; harmonic leakage power. (SWG/MTT/PE) 457-1982w, C37.100-1981s

leakage radiation (radio transmitting system) Radiation from anything other than the intended radiating system. *See also:* radio transmission. (AP/ANT) 145-1983s

leaky mode (fiber optics) In an optical waveguide, a mode whose field decays monotonically for a finite distance in the transverse direction, but which becomes oscillatory everywhere beyond that finite distance. Specifically, a mode for which

$[n^2(a)k^2 - (l/a)^2]^{1/2} = \beta = n(a)k$

where β is the imaginary part (phase term) of the axial propagation constant, l is the azimuthal index of the mode, $n(a)$ is the refractive index at $r = a$, the core radius, and k is the free-space wavenumber, $2\pi/\lambda$, and λ is the wavelength. Leaky modes correspond to leaky rays in the terminology of geometric optics. *Note:* Leaky modes experience attenuation, even if the waveguide is perfect in every respect. *Synonym:* tunneling mode. *See also:* bound mode; cladding mode; unbound mode; leaky ray; mode. (Std100) 812-1984w

leaky ray (fiber optics) In an optical waveguide, a ray for which geometric optics would predict total internal reflection at the core boundary, but which suffers loss by virtue of the curved core boundary. Specifically, a ray at radial position r having direction such that

$n^2(r) - n^2(\mathrm{a}) = \sin^2\theta(r)$

and

$\sin^2\theta(r) = [n^2(r) - n^2(a)]/[1 - (r/a)^2\cos^2\phi(r)]$

where θ (r) is the angle the ray makes with the waveguide axis, $n(r)$ is the refractive index, a is the core radius, and ϕ (r) is the azimuthal angle of the projection of the ray on the transverse plane. Leaky rays correspond to leaky (or tunnelling) modes in the terminology of mode descriptors. *Synonym:* tunneling ray. *See also:* leaky mode; cladding ray; guided ray; bound mode. (Std100) 812-1984w

leaky wave An electromagnetic wave associated with a fast wave guided along a surface. The wave radiates (or "leaks") energy continuously as it travels along the surface and thus decreases exponentially in the direction of propagation. Over a limited region, it may increase with height above the surface. *Note:* Leaky waves may be created by periodic as well as uniform, open guiding structures. (AP/PROP) 211-1997

leaky-wave antenna An antenna that couples power in small increments per unit length, either continuously or discretely, from a traveling wave structure to free space. (AP/ANT) 145-1993

learning bridge A bridge that learns the location of attached stations by examining the address in the frames that it processes. *See also:* mail bridge. (C) 610.7-1995

learning system An adaptive system with memory. *See also:* system science. (SMC) [63]

leased channel (data transmission) A point-to-point channel reserved for sole use of a single leasing customer. (PE) 599-1985w

leased circuit A telecommunication circuit that provides a clear unbroken communications path from one station to another and that is always available for use. *Synonym:* dedicated circuit. *See also:* two-wire circuit; foreign exchange circuit; four-wire circuit; dial-up circuit; simplex circuit. (C) 610.7-1995

leased line A line contracted out for exclusive use from a common carrier. *Synonyms:* dedicated line; private line. *Contrast:* dial-up line. (C) 610.7-1995

leased line or private wire network (data transmission) A series of points interconnected by telegraph or telephone channels, and reserved for the exclusive use of one customer. (PE) 599-1985w

least significant Within a group of data items (e.g., bits or bytes) that, taken as a whole, represents a numerical value, the item within the group with the smallest numerical weighting. (C/BA) 1275-1994

least significant bit (1) (mathematics of computing) The bit having the smallest effect on the value of a binary numeral; usually the rightmost bit. (TT/C) 1149.5-1995, 1084-1986w

(2) (station control and data acquisition) In an n bit binary word its contribution is (0 or 1) toward the maximum word value of $2^n - 1$. (SWG/PE/SUB) C37.100-1992, C37.1-1987s

(3) (test access port and boundary-scan architecture) The digit in a binary number representing the lowest numerical value. For shift-registers, the bit located nearest to the serial output, or the first bit to be shifted out. The least significant bit of a binary word or shift-register is numbered 0. (TT/C) 1149.1-1990

(4) The bit in the binary notation of a number that is the coefficient of the lowest exponent possible. (IM/ST) 1451.2-1997

least significant character The character in the rightmost position in a character string. *Contrast:* most significant character. (C) 610.5-1990w, 1084-1986w

least significant digit The digit having the smallest effect on the value of a numeral; usually the right-most digit; for example, the 4 in 756.4. (C) 610.5-1990w, 1084-1986w

least significant word (lsw) In a multiword representation of a binary number, the word containing the lsb of that number. (TT/C) 1149.5-1995

leave-word call (telephone switching systems) A person-to-person call on which the designated called person was not available and the operator left instructions for its later establishment. (COM) 312-1977w

LEC *See:* link error counter.

Lecher wires (data transmission) Two parallel wires on which standing waves are set up, frequently for the measurement of wavelength. (PE) 599-1985w

LED *See:* light-emitting diode.

Leduc current (electrotherapy) A pulsed direct current commonly having a duty cycle of 1:10. *See also:* electrotherapy. (EMB) [47]

left-hand polarization of a field vector *See:* sense of polarization.

left-hand polarization of a plane wave *See:* sense of polarization.

left-hand polarized wave A circularly or an elliptically polarized electromagnetic wave for which the electric field vector, when viewed with the wave approaching the observer, rotates clockwise in space. *Notes:* 1. This definition is consistent with observing a counterclockwise rotation when the electric field vector is viewed in the direction of propagation. 2. A left-handed helical antenna radiates a left-hand polarized wave. (AP/PROP) 211-1997

left justification In text formatting, justification of text such that the left margin is aligned. *Contrast:* ragged left margin. (C) 610.2-1987

left-threaded tree A tree in which the left link field in each terminal node is made to point to its predecessor with respect to a particular order of traversal. *Contrast:* right-threaded tree. (C) 610.5-1990w

leg Any one of the conductors of an electric supply circuit between which is maintained the maximum supply voltage. *See also:* center of distribution. (CAS/T&D/PE) [10]

legacy region A set of LAN segments interconnected such that there is physical connectivity between any pair of segments using only ISO/IEC 15802-3-conformant, VLAN-unaware MAC Bridges. *Note:* In other words, if, in a Bridged LAN containing both ISO/IEC 15802-3 and IEEE 802.1Q Bridges, all the IEEE 802.1Q Bridges were to be removed, the result would be a set of one or more Bridged LANs, each with its own distinct Spanning Tree. Each of those Bridged LANs is a legacy region. (C/LM) 802.1Q-1998

legibility The quality of a display that allows groups of characters and symbols to be easily discriminated and recognized. *See also:* display. (PE/NP) 1289-1998

legitimate access The proper and correct access authorization. (PE/NP) 692-1997

leg wire One of the two wires attached to and forming a part of an electric blasting cap or squib. *See also:* blasting unit. (EEC/PE/MIN) [119]

Lenard tube An electron beam tube in which the beam can be taken through a section of the wall of the evacuated enclosure. (ED) [45]

length (1) (A) A measure of the magnitude of a unit of data, usually expressed as a number of subunits, for example, the length of a record is 32 blocks, the length of a word is 40 binary digits, etc. *See also:* word length; storage capacity; double length. **(B)** The number of subunits of data, usually digits or characters, that can be simultaneously stored linearly in a given device, for example, the length of the register is 12 decimal digits or the length of the counter is 40 binary digits. *See also:* word length; double length; storage capacity. **(C)** A measure of the amount of time that data are delayed when being transmitted from point to point, for example, the length of the delay line is 384 microseconds. *See also:* word length; storage capacity; double length. (C) 162-1963
(2) The number of bytes required to represent any string. (C/PA/C/PA/C/PA) 1328-1993w, 1327-1993w, 1224-1993w
(3) *See also:* block length; record length. (C) 610.5-1990w

length of lay (cable) (power distribution, underground cables) The axial length of one turn of the helix of any helical element. (PE) [4]

lens and aperture (phototube housing) The cooperating arrangement of a light-refracting member and an opening in an opaque diaphragm through which all light reaching the phototube cathode must pass. *See also:* electronic controller. (IA/ICTL/IAC) [60]

lens antenna An antenna consisting of an electromagnetic lens and a feed that illuminates it. (AP/ANT) 145-1993

lens distance relay A distance relay that has an operating characteristic comprising the common area of two intersecting circular relay characteristics. (SWG/PE) C37.100-1992

lens, electromagnetic *See:* electromagnetic lens.

lens multiplication factor (phototube housing) The maximum ratio of the light flux reaching the phototube cathode with the lens and aperture in place to the light flux with the lens and aperture removed. *See also:* photoelectric control. (IA/ICTL/IAC) [60]

lenticular characteristic A distance relay characteristic having a lens shape on a resistance-reactance (R-X) diagram. (PE/PSR) C37.113-1999

Lenz's law (induced current) The current in a conductor as a result of an induced voltage is such that the change in magnetic flux due to it is opposite to the change in flux that caused the induced voltage. (Std100) 270-1966w

letter An alphabetic character used for the representation of sounds in a spoken language. (C) [20], [85]

letter combination One or more letters that form a part of the graphic symbol for an item and denote its distinguishing characteristic. *Compare with:* functional designation; reference designation; symbol for a quantity. *See also:* functional designation; symbol for a quantity; abbreviation; reference designation. (GSD) 267-1966

letter-quality (LQ) Pertaining to printed output that is suitable for high quality correspondence. *Note:* This term implies that "letter quality" output matches that of a standard typewriter. *See also:* draft quality; near-letter quality. (C) 610.10-1994w

letter spacing *See:* intercharacter spacing.

let-through sparkover (surge arresters) A measure of the highest lightning surge an arrester is likely to withstand without sparkover in 3 μs or less. The value determined by a 1.2 × 50-μs impulse sparkover test. *See also:* IEEE. (PE) [8]

level (1) (data transmission) The magnitude of a quantity, especially when considered in relation to an arbitrary reference value. Level may be stated in the units in which the quantity itself is measured (for example, dB) expressing the ratio to a reference value. (PE) 599-1985w
(2) (A) (software) The degree of subordination of an item in a hierarchical arrangement. *See also:* hierarchy. **(B) (software)** A rank within a hierarchy. An item is of the lowest level if it has no subordinates and of the highest level if it has no superiors. *See also:* hierarchy. (C/SE) 729-1983
(3) (quantity) (general) Magnitude, especially when considered in relation to an arbitrary reference value. Level may be stated in the units in which the quantity itself is measured (for example, volts, ohms, etc.) or in units (for example, decibels) expressing the ratio to a reference value. *Notes:* 1. Examples of kinds of levels in common use are electric power level, sound-pressure level, voltage level. 2. The level as here defined is measured in two common units: in decibels when the logarithmic base is 10, or in nepers when the logarithmic base is e. The decibel requires that k be 10 for ratios of power, or 20 for quantities proportional to the square root of power. The neper is used to represent ratios of voltage, current, sound pressure, and particle velocity. The neper requires that k be 1. 3. In symbols,

$$L = k\log_r(q/q_0)$$

where
L = level of kind determined by the kind of quantity under consideration
r = base of the logarithm of the reference ratio
q = the quantity under consideration
q_0 = reference quantity of the same kind
k = a multiplier that depends upon the base of the logarithm and the nature of the reference quantity

See also: reference white level; transmission level; signal level; reference black level; blanking level. (T&D/PE/SP) 539-1990, [32]
(4) (charge-storage tubes) A charge value which can be stored in a given storage element and distinguished in the output from other charge values. *See also:* channel. (ED) 161-1971w
(5) The level in dBm of a signal or tone into a resistive load equal to the designated impedance at the point of measurement. Levels measured into complex impedances, as specified by ITU-T Recommendation O.41, clause 3.13, are not recognized by this standard. (COM/TA) 743-1995
(6) A designation of the coverage and detail of a view. There are multiple levels of view; each is intended to be distinct, specified in terms of the modeling constructs to be used. (C/SE) 1320.2-1998
(7) *See also:* transit. (T&D/PE) 524-1992r

level above threshold (sound) (sensation level) The pressure level of the sound in decibels above its threshold of audibility for the individual observer. (SP/ACO) [32]

level band pressure (electroacoustics) (for a specified frequency band) The sound-pressure level for the sound contained within the restricted band. *Notes:* 1. The reference pressure must be specified. 2. The band may be specified by its lower and upper cutoff frequencies or by its geometric or arithmetic center frequency and bandwidth. The width of the band may be indicated by a prefatory modifier, for example, octave band (sound pressure) level, half-octave band level, third-octave band level, 50-hertz band level. (SP) [32]

level compensator (signal transmission) An automatic transmission-regulating feature or device to minimize the effects of variations in amplitude of the received signal. *See also:* telegraphy. (EEC/PE) [119]

level detector (as applied to relaying) A device that produces a change in output at a prescribed input level. (SWG/PE) C37.100-1992

leveling *See:* platform erection.

leveling block *See:* leveling plate.

leveling error (accelerometer) (inertial sensors) The angle between the local horizontal and the input reference axis when the accelerometer output is zero. (AES/GYAC) 528-1984s

leveling plate (rotating machinery) (leveling block) A heavy pad built into the foundation and used to support and align the bed plate or rails using shims for adjustment before grouting. (PE) [9]

leveling zone (elevators) The limited distance above or below an elevator landing within which the leveling device may cause movement of the car toward the landing. *See also:* elevator car-leveling device. (PE/EEC) [119]

level n repeater A repeater that is (n−1) link segments below the root repeater in a cascade. (C) 8802-12-1998

level of documentation (software) A description of required documentation indicating its scope, content, format, and quality. Selection of the level may be based on project cost, intended usage, extent of effort, or other factors. *See also:* quality; documentation. (C/SE) 729-1983s

level of maintenance A level at which diagnostics can operate (e.g., maintenance depot, factory, in the field). (ATLAS) 1232-1995

Level 1 device A device that supports the Level 1 electrical interface. (C/MM) 1284-1994

level, power dBm *See:* power dBm level.

level, relay *See:* relay level.

level sensitive Pertaining to a circuit that can be held in one state as long as an input signal maintains a certain value. *Contrast:* edge sensitive. *See also:* transparent latch. (C) 610.10-1994w

level-sensitive scan design A variant of the scan design technique that results in race-free, testable digital electronic circuits. (TT/C) 1149.1-1990

level-sensitive signal Signals whose high or low state is sampled based on the leading and/or trailing edges of edge-sensitive strobe signals. (C/BA) 896.9-1994w

level-sensitive storage element A storage element mapped to by a synthesis tool that

a) Propagates the value at the data input whenever an appropriate value is detected on a clock control input, and
b) Preserves the last value propagated at all other times, except when any asynchronous control inputs become active.

(C/DA) 1076.6-1999

levels, usable *See:* usable levels.

level switch (power system device function numbers) A switch which operates on given values, or on a given rate of change, of level. (SUB/PE) C37.2-1979s

level, tracking *See:* tracking level.

Level 2 device A device that supports the Level 2 electrical interface. (C/MM) 1284-1994

lever blocking device (railway signaling) A device for blocking a lever so that it cannot be operated. (EEC/PE) [119]

lever indication (railway signaling) The information conveyed by means of an indication lock that the movement of an operated unit has been completed. (EEC/PE) [119]

LEX A compiler specification language in which the input is (1) a specification of a set of regular expressions and (2) actions to be taken upon recognizing each of these. The output of Lex is a lexical analysis program that can process the specified language. *Note:* Used in writing portions of compilers, as well as in textual pattern matching. (C) 610.13-1993w

Lex *See:* LEX.

LF *See:* low frequency.

LFC *See:* load-frequency control.

LHN *See:* long haul network.

liability The state of being responsible or answerable under a legal obligation. (C/SE) 1420.1b-1999

liberator tank (electrorefining) Sometimes known as a depositing-out tank, an electrolytic cell equipped with insoluble anodes for the purpose of either decreasing, or totally removing the metal content of the electrolyte by plating it out on cathodes. *See also:* electrorefining. (EEC/PE) [119]

librarian *See:* software librarian.

library (1) (integrated circuit) A collection of circuit functions, implemented in a particular integrated circuit technology, that an integrated circuit designer or electric design automation (EDA) synthesis application can select in order to implement a design. *See also:* cell. (C/DA) 1481-1999
(2) An automated or manual cartridge-storing facility, e.g., an automated library or a vault. A library may contain zero or more drives, input/output ports and attachments to other libraries. (C/SS) 1244.1-2000

library automation The application of automated techniques to library operations such as processing of documents, interlibrary communication, and on-line catalogue access. *See also:* MAchine-Readable Cataloging. (C) 610.2-1987

library control statements These statements control the logical organization and loading of subrules in a technology library. *See also:* technology library. (C/DA) 1481-1999

library data model The organizing principles and concepts underlying structured data in a reuse library and the means of representing that structure. (C/SE) 1420.1-1995

library routine (high-level microprocessor language) A function (which returns a value) or a procedure (which does not return a value) supplied with the implementation of the high-level language (HLL). (MM/C) 755-1985w

license A legal agreement between two parties, the licensor and the licensee, as to the terms and conditions for the use or transfer of an intellectual property right from the licensor to the licensee. (C/SE) 1420.1b-1999

licensee event report (LER) Reports submitted by the licensee to Nuclear Regulatory Commission (NRC) under Regulatory Guide 1.16. (PE/NP) 933-1999

licensing standard A standard that describes the characteristics of an authorization given by an official or a legal authority to an individual or organization to do or own a specific thing. (C) 610.12-1990

Lichtenberg figure camera (surge-voltage recorder) (klydonograph) A device for indicating the polarity and approximate crest value of the voltage surge by the appearance and dimensions of the Lichtenberg figure produced on a photographic plate or film, the emulsion coating of which is in contact with a small electrode coupled to the circuit in which the surge occurs. *Note:* The film is backed by an extended plane electrode. *See also:* instrument. (EEC/PE) [119]

life (1) The period during which a fully charged battery is capable of delivering at least a specified percentage of its rated capacity. (PV) 1145-1999
(2) The period during which a fully charged battery is capable of delivering at least a specified percentage of its capacity, generally 80%. (SCC21) 937-2000

life cycle The system or product evolution initiated by a perceived customer need through the disposal of the products. (C/SE) 1220-1998

life-cycle cost The total investment in product development, manufacturing, test, distribution, operation, support, training, and disposal. (C/SE) 1220-1998

life-cycle phase (software verification and validation plans) Any period of time during software development or operation that may be characterized by a primary type of activity (such as design or testing) that is being conducted. These phases may overlap one another; for verification and validation (V&V) purposes, no phase is concluded until its development products are fully verified. (C/SE) 1012-1986s

life-cycle processes (1) The following eight essential functional processes that may be necessary to provide total consumer satisfaction and meet public acceptance. Once the need for a life cycle process is identified, the life cycle process is treated as a system, and the systems engineering process is applied to define, design, and establish the life cycle process, and the supporting products and processes to maintain the life cycle process in an operational condition.

a) *Development.* The planning and execution of system and subsystem definition tasks required to evolve the system from customer needs to product solutions and their life cycle processes.
b) *Manufacturing.* The tasks, actions, and activities for fabrication and assembly of engineering test models and

brassboards, prototypes, and production of product solutions and their life cycle process products.

c) *Test.*

1) The tasks, actions, and activities for planning for evaluation and conducting evaluation of synthesis products against the functional architecture or requirements baseline, or the functional architecture against the requirements baseline.

2) The tasks, actions, and activities for evaluating the product solutions and their life cycle processes to measure specification conformance or customer satisfaction.

d) *Distribution.* The tasks, actions, and activities to initially transport, deliver, assemble, install, test, and checkout products to effect proper transition to users, operators, or consumers.

e) *Operations.* The tasks, actions, and activities that are associated with the use of the product or a life cycle process.

f) *Support.* The tasks, actions, and activities to provide supply, maintenance, and support material and facility management for sustaining operations.

g) *Training.* The measurable tasks, actions, and activities (including instruction and applied exercises) required to achieve and maintain the knowledge, skills, and abilities necessary to efficiently and effectively perform operations, support, and disposal throughout the system life cycle. Training is inclusive of the tools, devices, techniques, procedures, and materials developed and employed to provide training for all required tasks.

h) *Disposal.* The tasks, actions, and activities to ensure that disposal or recycling of destroyed or irreparable consumer and life cycle processes and by-products comply with applicable environmental regulations and directives. (C/SE) 1220-1998

(2) A set of interrelated activities that result in the development or assessment of software products. Each activity consists of tasks. The life cycle processes may overlap one another. For V&V purposes, no process is concluded until its development products are verified and validated according to the defined tasks in the SVVP. (C/SE) 1012-1998

life-cycle process product An end item required to perform a life-cycle process in support of a consumer product. This end item may be a product, process, or service. (C/SE) 1220-1994s

lifeline A component consisting of a flexible line for connection to an anchorage or anchorage connector at one end to hang vertically (vertical lifeline), or for connection to anchorages or anchorage connectors at both ends to span horizontally (horizontal lifeline). Serves as a means for connecting other components of a personal fall arrest system to the anchorage. A lifeline serves to extend the range of the user through the slidable connection of a fall arrester in the case of a vertical lifeline or a connector or other device in the case of a horizontal lifeline. (T&D/PE) 1307-1996

life performance curve (illuminating engineering) A curve that represents the variation of a particular characteristic of a light source (luminous flux, intensity, etc.) throughout the life of the source. *Note:* Life performance curves sometimes are called maintenance curves as, for example, lumen maintenance curves. (EEC/IE) [126]

life safety branch (health care facilities) A subsystem of the Emergency System consisting of feeders and branch circuits, meeting the requirements of Article 700 of NFPA 70-1978, National Electrical Code, and intended to provide adequate power needs to ensure safety to patients and personnel, and which can be automatically connected to alternate power sources during interruption of the normal power source. (EMB) [47]

life support equipment (nuclear power generating station) The breathing apparatus, medical supplies, sanitary facilities, and food and water supplies required to sustain operators for an extended period of time during abnormal operating conditions. (PE/NP) 567-1980w

life test *See:* accelerated test.

life test of lamps (illuminating engineering) A test in which lamps are operated under specified conditions for a specified length of time, for the purpose of obtaining information on lamp life. Measurements of photometric and electric characteristics may be made at specified intervals of time during this test. (EEC/IE) [126]

lifetime rated pulse currents (low voltage varistor surge arresters) Derated values of rated peak single pulse transient current for impulse durations exceeding that of an 8×20 μs waveshape, and for multiple pulses that may be applied over the device rated lifetime. (PE) [8]

lifetime, volume *See:* volume lifetime.

lifter, insulator *See:* insulator lifter.

lifting eye (of a fuseholder, fuse unit, or disconnecting blade) An eye provided for receiving a fuse hook or switch hook for inserting the fuse or disconnecting blade into, and for removing it from, the fuse support. (SWG/PE) C37.40-1993, C37.100-1992

lifting-insulator switch A switch in which one or more insulators remain attached to the blade, move with it, and lift it to the open position. (SWG/PE) C37.100-1992

lifting shoe *See:* conductor lifting hook.

light (1) (A) (fiber optics) In a strict sense, the region of the electromagnetic spectrum that can be perceived by human vision, designated the visible spectrum and nominally covering the wavelength range of 0.4 μm to 0.7 μm. *See also:* optical spectrum; ultraviolet; infrared. **(B) (fiber optics)** In the laser and optical communication fields, custom and practice have extended usage of the term to include the much broader portion of the electromagnetic spectrum that can be handled by the basic optical techniques used for the visible spectrum. This region has not been clearly defined but, as employed by most workers in the field, may be considered to extend from the near-ultraviolet region of approximately 0.3 μm, through the visible region, and into the mid-infrared region to 30 μm. *See also:* ultraviolet; optical spectrum; infrared. (Std100) 812-1984

(2) (illuminating engineering) Radiant energy that is capable of exciting the retina and producing a visual sensation. The visible portion of the electromagnetic spectrum extends from about 380 to 770 nm. *Note:* The subjective impression produced by stimulating the retina is sometimes designated as light. Visual sensations are sometimes arbitrarily defined as sensations of light, and in line with this concept it is sometimes said that light cannot exist until an eye has been stimulated. Electrical stimulation of the retina or the visual cortex is described as producing flashes of light. In illuminating engineering, however, light is a physical entity—radiant energy weighted by the luminous efficiency function. It is a physical stimulus which can be applied to the retina. *See also:* values of spectral luminous efficiency for photopic vision; spectral luminous efficiency. (EEC/IE) [126]

light adaptation (illuminating engineering) The process by which the retina becomes adapted to a luminance greater than about 3.4 cd/m^2 (2.2×10^{-3} cd/in^2) (1.0 fL). (EEC/IE) [126]

light amplification by stimulated emission of radiation *See:* laser.

light center (illuminating engineering) The center of the smallest sphere that would completely contain the light-emitting element of the lamp. (EEC/IE) [126]

light center length (illuminating engineering) The distance from the light center to a specified reference point on the lamp. (EEC/IE) [126]

light current *See:* photocurrent.

lighted beacon (navigation aid terms) A beacon that transmits signals by light waves (for example, light house). (AES/GCS) 172-1983w

lighted buoy (navigation aid terms) A buoy with a light that has characteristics for detection and identification. *See also:* buoy. (AES/GCS) 172-1983w

light-emitting diode (1) (fiber optics) A p-n junction semiconductor device that emits incoherent optical radiation when biased in the forward direction. *See also:* incoherent. (Std100) 812-1984w
(2) (illuminating engineering) A p-n junction solid-state diode whose radiated output is a function of its physical construction, material used, and exciting current. The output may be in the infrared or in the visible region. (EEC/IE) [126]
(3) A p-n junction semiconductor device that emits low-coherence optical radiation by spontaneous emission when biased in the forward direction. (AES/GYAC) 528-1994
(4) A p-n junction solid-state diode, whose radiated output is a function of its physical construction, material used, and exciting current. The output may be in the infrared or in the visible region. When used as a visible element, the output shall be limited to the visible region. (VT) 1477-1998

light-emitting diode display device A display device in which a single light-emitting diode is used for each segment of a character to be displayed. *Note:* Usually used in small control panels and displays such as clocks and appliances. (C) 610.10-1994w

lightguide *See:* optical waveguide.

lighting branch circuit A circuit that supplies energy to lighting outlets. A lighting branch circuit may also supply portable desk or bracket fans, small heating appliances, motors of 190 W (1/4 hp) and less, and other portable apparatus of not over 600 W each. (IA/MT) 45-1998

lighting effectiveness factor (illuminating engineering) The ratio of equivalent sphere illumination to measured or calculated task illuminance. (EEC/IE) [126]

lighting outlet (1) An outlet intended for the direct connection of a lampholder, a lighting fixture, or a pendant cord terminating in a lampholder. (NESC/NEC) [86]
(2) An outlet intended for the direct connection of a lamp holder or a lighting fixture. (IA/MT) 45-1998

light line *See:* pulling line.

light load (watthour meter) The current at which the meter is adjusted to bring its response near the lower end of the load range to the desired value. It is usually 10 percent of the test current for a revenue meter and 25 percent for a standard meter. (ELM) C12.1-1982s

light loss factor (illuminating engineering) The ratio of the illuminance on a given area after a period of time to the initial illuminance on the same area. *Note:* The light loss factor is used in lighting calculations as an allowance for the depreciation of lamps, light control elements, and room surfaces to values below the initial or design conditions, so that a minimum desired level of illuminance may be maintained in service. The light loss factor had formerly been widely interpreted as the ratio of average to initial illuminance. This term was formerly called "maintenance factor." (EEC/IE) [126]

lightness (illuminating engineering) (of a perceived patch of surface color) The brightness of an area judged relative to the brightness of a similarly illuminated area that appears to be white or highly transmitting. (EEC/IE) [126]

lightning An electric discharge that occurs in the atmosphere between clouds or between clouds and ground. (SPD/PE) C62.11-1999, C62.62-2000

lightning flash The complete lightning discharge, most often composed of leaders from a cloud followed by one or more return strokes. (SUB/PE/T&D) 998-1996, 1410-1997, 1243-1997

lightning first stroke A lightning discharge to ground initiated when the tip of a downward stepped leader meets an upward leader from the earth. (PE/T&D) 1243-1997, 1410-1997

lightning impulse An impulse with front duration up to a few tens of microseconds. (PE/PSIM) 4-1995

lightning impulse dry withstand voltage The crest value of a voltage impulse with a front duration from less than one to a few tens of microseconds that, under specified conditions, can be applied without causing flashover or puncture. (SWG/PE) C37.34-1994

lightning impulse insulation level (power and distribution transformers) An insulation level expressed in kilovolts of the crest value of a lightning impulse withstand voltage. (PE/TR) C57.12.80-1978r

lightning impulse protective level of a surge protective device The maximum lightning impulse voltage expected at the terminals of a surge-protective device under specified conditions of operation. *Note:* The lightning impulse protective levels are given by following: a) Front-of-wave impulse sparkover or discharge voltage, and b) the higher of either a 1.2/50 impulse sparkover voltage or the discharge voltage for a specified current magnitude and waveshape. (C/PE/TR) 1313.1-1996, C57.12.80-1978r

lightning impulse test (power and distribution transformers) Application of the following sequence of impulse waves:

a) one reduced full wave,
b) two chopped waves,
c) one full wave.

(PE/TR) C57.12.80-1978r

lightning mast A column or narrow-base structure containing a vertical conductor from its tip to earth, or that is itself a suitable conductor to earth. Its purpose is to intercept lightning strokes so that they do not terminate on objects located within its zone of protection. (SUB/PE) 998-1996

lightning outage A power outage following a lightning flashover that results in system fault current, thereby necessitating the operation of a switching device to clear the fault. (PE/T&D) 1243-1997, 1410-1997

lightning overvoltage (1) A type of transient overvoltage in which a fast front voltage is produced by lightning or fault. Such overvoltage is usually unidirectional and of very short duration. (PE/C) 1313.1-1996
(2) The crest voltage appearing across an arrester or insulation caused by a lightning surge. (SPD/PE) C62.22-1997

lightning subsequent stroke A lightning discharge that may follow a path already established by a first stroke. (PE/T&D) 1243-1997, 1410-1997

lightning surge A transient electric disturbance in an electric circuit caused by lightning. (SPD/PE) C62.22-1997, C62.11-1999

light pattern (buchmann-meyer pattern) (mechanical recording) (optical pattern) A pattern that is observed when the surface of the record is illuminated by a light beam of essentially parallel rays. *Note:* The width of the observed pattern is approximately proportional to the signal velocity of the recorded groove. (SP) [32]

light pen (A) A light-sensitive pick device, resembling a fountain pen, that detects light that emanates from a CRT beam when the end of the pen is held against the CRT device. *Note:* Often used as a locator. *Synonyms:* optical wand; selector pen. *See also:* sonic pen; electronic pen. **(B)** An input device that reads bar-codes by transmitting a beam of light onto the code and that receives and interprets the reflections from the barcode. *See also:* bar code scanner. (C) 610.10-1994

light pipe An optical transmission element that utilizes unfocused transmission and reflection to reduce photon losses. *Note:* Light pipes have been used to distribute the light more uniformly over a photocathode. *See also:* phototube. (NPS) 175-1960w

light rail transit A mode of rail transit characterized by its ability to operate on exclusive rights-of-way, street running, center reservation running, and grade crossings, and its ability to board and discharge passengers at track or vehicle floor level. (VT/RT) 1475-1999, 1474.1-1999, 1476-2000

light rail vehicle A vehicle that operates on a light rail transit system, capable of boarding and discharging passengers at track or vehicle floor level. (VT) 1475-1999, 1476-2000

light ray (fiber optics) The path of a point on a wavefront. The direction of a light ray is generally normal to the wavefront. *See also:* geometric optics. (Std100) 812-1984w

light-sensitive Pertaining to an input device that can detect energy in the form of light. *See also:* light pen; touch-sensitive. (C) 610.10-1994w

light source A device to supply radiant energy capable of exciting a phototube or photocell. *See also:* photoelectric control. (IA/ICTL/IAC) [60]

light-source color (illuminating engineering) The color of the light emitted by the source. *Note:* The color of a point source may be defined by its luminous intensity and chromaticity coordinates; the color of an extended source may be defined by its luminance and chromaticity coordinates. (EEC/IE) [126]

light transition load (rectifier circuits) The transition load that occurs at light load, usually at less than 5 percent of rated load. *Note:* Light transition load is important in multiple rectifier circuits. A similar effect occurs in rectifier units using saturable-reactor control. *See also:* rectifier circuit element; rectification.

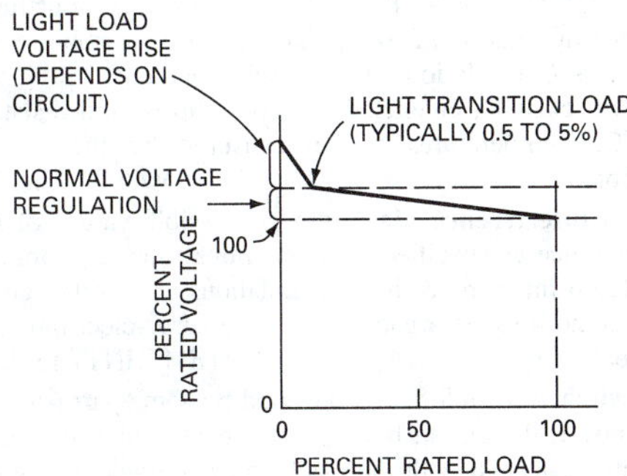

Voltage regulation characteristic showing light transition load.

light transition load

(IA) 59-1962w, [12]

light valve (electroacoustics) A device whose light transmission can be made to vary in accordance with an externally applied electrical quantity, such as voltage, current, electric field, magnetic field, or an electron beam. *See also:* phonograph pickup. (BT/SP/AV) [34], [32]

lightwatt *See:* spectral luminous efficiency.

limb sounding A technique for making observations of planetary atmospheres wherein satellite-borne sensors are oriented in a direction tangent to the spherical atmospheric layer of interest, to afford the longest pathlength and highest vertical resolution through the atmosphere. (AP/PROP) 211-1997

limit (1) (mathematical) A boundary of a controlled variable. **(2) (synchronous machine regulating systems)** The boundary at or beyond which a limiter functions. (PE) [9] **(3)** The designated quantity is controlled so as not to exceed a prescribed boundary condition. *See also:* feedback control system. (IA/ICTL/IAC) [60] **(4) (test, measurement, and diagnostic equipment)** The extreme of the designated range through which the measured value of a characteristic may vary and still be considered acceptable. (MIL) [2]

limit check A consistency check that ensures that a certain item limit is not exceeded. For example, if a record can hold four transactions, a limit check will reveal an error situation if an attempt is made to add a fifth transaction to a record. (C) 610.5-1990w

limit cycle A closed curve in the state space of a particular control system, from which state trajectories may recede, or which they may approach, for all initial states sufficiently close to the curve. (CS/PE/EDPG) [3]

limit cycle, stable *See:* stable limit cycle.

limit cycle, unstable *See:* unstable limit cycle.

limited access highways Limited access highways include both fully controlled highways and partially controlled highways where access is controlled by a governmental authority for purposes of improving traffic flow and safety. Fully controlled access highways have no grade crossings and have carefully designed access connections. (NESC) C2-1997

limited availability (telephone switching systems) Availability that is less than the number of outlets in the desired group. (COM) 312-1977w

limited-domain data element A data element whose domain is bounded. For example, a data element SEX with a domain of [M,F]. (C) 610.5-1990w

limited proportionality, region of *See:* region of limited proportionality.

limited signal (radar) A signal that is limited in amplitude by the dynamic range of the system. (AES/RS) 686-1990, [42]

limited stability A property of a system characterized by stability when the input signal falls within a particular range and by instability when the signal falls outside this range. (Std100) 154-1953w

limiter (1) (excitation systems for synchronous machines) An element of the excitation system which acts to limit a variable by modifying or replacing the functions of the primary detector element when predetermined conditions have been reached. *Note:* Examples:

a) An under excitation limiter prevents the voltage regulator from lowering the excitation of the synchronous machine below a prescribed level.
b) An over excitation limiter prevents the voltage regulator from raising the excitation of the synchronous machine above a level that would cause a thermal overload in the machine field; refer to ANSI C50.13-1977.
c) A volts per hertz limiter acts, through the voltage regulator to correct for a machine terminal voltage to frequency ratio that is considered abnormal.
d) Other types of limiters may be used to control various quantities, such as, rotor angle, excitation output, etc.

See also: ferri-diode limiter; multipactor limiter; p-i-n diode limiter; ferrite limiter; passive limiter; gyromagnetic limiter. (PE/EDPG) 421.1-1986r

(2) (A) (data transmission) A device in which some characteristic of the output is automatically prevented from exceeding a predetermined value. **(B) (data transmission)** More specifically, a transducer in which the output amplitude is substantially linear with regard to the input up to a predetermined value and substantially constant thereafter. *Note:* For waves having both positive and negative values, the predetermined value is usually independent of sign. (PE) 599-1985

(3) (rotating machinery) An element or group of elements that acts to limit by modifying or replacing the functioning of a regulator when predetermined conditions have been reached. *Note:* Examples are minimum excitation limiter, maximum excitation limiter, maximum armature-current limiter. (PE) [9]

(4) (radio receivers) A transducer whose output is constant for all inputs above a critical value. *Note:* A limiter may be used to remove amplitude modulation while transmitting angle modulation. *See also:* transducer; radio receiver. (AP/ANT) 145-1983s

(5) A device that is used to prevent an analog variable from exceeding specified limits. *See also:* limiter circuit; feedback limit. (C) 610.10-1994w

limiter circuit A circuit that limits the amplitude of a signal so that interfering noise can be kept to a minimum, or to protect components from excessive stress. *Synonyms:* clamping circuit; circuit limiter. *See also:* feedback limiter; input limiter; bridge limiter. (C) 610.10-1994w

limiter circuits (analog computer) A circuit of nonlinear elements that restrict the electrical excursion of a variable in accordance with some specified criteria. "Hard limiting" is a limiting action with negligible variation in output in the range where the output is limited. "Soft limiting" is a limiting action with appreciable variation in output in the range where the output is limited. A "bridge limiter" is a bridge circuit used as a limiter circuit. In an analog computer, a "feedback limiter" is a limiter circuit usually employing biased diodes shunting the feedback component of an operational amplifier; an "input limiter" is a limiter circuit usually employing biased diodes in the amplifier input channel that operates by limiting the current entering the summing junction. "Linear system or element"—a system with the properties: if y_1 is the response to x_1 and y_2 is the response to x_2, then (i) $(y_1 + y_2)$ is the response to $(x_1 + x_2)$ and (ii) ky_1 is the response to kx_1. *See also:* stop; analog computer. (C) 165-1977w

limiting (1) (automatic control) The intentional imposition or inherent existence of a boundary on the range of a variable, for example, on the speed of a motor. (PE/EDPG) 421-1972s, [3]

(2) The process of preventing a data value or variable from exceeding a specified limit. *See also:* limiter; soft limiting; hard limiting. (C) 610.10-1994w

limiting ambient temperature (1) (electric equipment) (thermal classification of electric equipment and electrical insulation) The highest (or lowest) ambient temperature at which electric equipment is expected to give specified performance under specified conditions, for example, rated load. (EI) 1-1986r

(2) (equipment) An upper or lower limit of a range of ambient temperatures within which equipment is suitable for operation at its rating. Where the term is used without an adjective the upper limit is meant. *See also:* limiting insulation system temperature. (Std100) [83]

limiting angular subtense (laser maser) (α_{min}) The apparent visual angle which divides intrabeam viewing from extended source viewing. (LEO) 586-1980w

limiting aperture (laser maser) The maximum circular area over which radiance and radiant exposure can be averaged. (LEO) 586-1980w

limiting condition for operation The lowest functional capability or performance level of equipment required for safe operation of the facility. (PE/NP) 338-1987r

limiting hottest-spot temperature (thermal classification of electric equipment and electrical insulation) (electric equipment) The highest temperature attained in any part of the insulation of electric equipment, which is operating under specified conditions, usually at maximum rating and the upper limiting ambient temperature. (EI) 1-1986r

limiting insulation system temperature (power and distribution transformers) (limiting hottest-spot temperature) The maximum temperature selected for correlation with a specified test condition of the equipment with the object of attaining a desired service life of the insulation system. (PE/TR) C57.12.80-1978r

limiting insulation temperature rise (equipment) The difference between the limiting insulation temperature and the limiting ambient temperature. *Synonym:* limiting hottest-spot temperature. *See also:* limiting insulation system temperature. (Std100) [83]

limiting observable temperature rise (equipment) The limit of observable temperature rise specified in equipment standards. *See also:* limiting insulation system temperature. (Std100) [83]

limiting oxygen index (LOI) *See:* oxygen index.

limiting polarization The resultant polarization of a wave after it has emerged from a magneto-ionic medium. (AP/PROP) 211-1997

limiting resolution (1) (diode-type camera tube) The high frequency bar pattern which can be visually distinguished as separate bars on a display. Units: Television lines per raster height (TVL/RH). (ED) 503-1978w

(2) (television) A measure of overall system resolution usually expressed in terms of the maximum number of lines per picture height discriminated on a television test chart. *Note:* For a number of lines N (alternate black and white lines), the width of each line is $1/N$ times the picture height. (BT/AV) 201-1979w, 208-1995

limiting temperature (power and distribution transformers) The maximum temperature at which a component or material may be operated continuously with no sacrifice in normal life expectancy. (IA/MT) 45-1983s

limit, lower *See:* lower limit.

limit of error *See:* uncertainty.

limit of temperature rise (1) (contacts) The temperature rise of contacts, above the temperature of the cooling air, when tested in accordance with the rating shall not exceed the following values. All temperatures shall be measured by the thermometer method. Laminated contacts: 50°C; solid contacts: 75°C.

(2) (resistors) The temperature rise of resistors above the temperature of the cooling air, when test is made in accordance with the rating, shall not exceed the following temperatures for the several classes of resistors: Class A, cast resistors, 450°C; Class B, imbedded resistors, outside of imbedding material, 250°C; Class C, strap or ribbon wound on Class C insulation, 600°C continuous and 800°C intermittent; class D, enameled wire or strap wound resistance, 350°C. Temperatures to be measured by thermocouple method. (VT/LT) 16-1955w

limits of interference Maximum permissible values of radio interference as specified in International Special Committee on Radio Interference recommendations or by other competent authorities or organizations. *See also:* electromagnetic compatibility. (EMC/INT) [53], [70]

limit switch A switch that is operated by some part or motion of a power-driven machine or equipment to alter the electric circuit associated with the machine or equipment. *See also:* switch. (IA/IAC) [60]

limit, upper *See:* upper limit.

linc The interface circuitry that attaches to an SCI ringlet. A linc typically contains control/status registers (including identification ROM and reset command registers) that are initially defined in a 4 Kbyte (minimum) ringlet-visible initial node address space. (C/MM) 1596-1992

line (1) (electric power) A component part of a system extending between adjacent stations or from a station to an adjacent interconnection point. A line may consist of one or more circuits. *See also:* system. (SPD/PE) 32-1972r

(2) (cathode-ray tubes) (trace) The path of a moving spot. (ED) [45]

(3) The block of memory (sometimes called a sector) that is managed as a unit for coherence purposes; i.e., cache tags are maintained on a per-line basis. SCI directly supports only one line size, 64 bytes. (C/MM) 1596-1992

(4) An aligned block of data on which coherence checks are performed. Several bus standards use 64-byte lines, so it is the expected line size for the CSR Architecture (although other line sizes are not prohibited). For example, the 64-byte line size is reflected in the size of bus transactions used for message passing. (C/MM) 1212-1991s

(5) (A) A circuit connecting two or more devices. **(B)** The portion of a data circuit that is external to data circuitry terminating equipment, and that either connects the data circuit-terminating equipment to one or more other data circuit-terminating equipment or connects the data switching exchange to other data switching exchanges. *See also:* link; channel. **(C)** A wire or set of wires over which a current is propagated. **(D)** A sequence of characters. *See also:* channel. (C) 610.7-1995, 610.10-1994

(6) A sequence of text consisting of zero or more non-⟨`newline`⟩ characters plus a terminating ⟨`newline`⟩ character. (C/PA) 9945-2-1993

(7) *See also:* cache line. (C/MM) 1596.5-1993

(8) (in a spectrum) A sharply peaked portion of a spectrum that represents a specific feature of the incident radiation, usually the full energy of a monoenergetic X ray, gamma ray, or charged particle. (NPS) 325-1996

lineal electric current element *See:* Hertzian electric dipole.

lineal magnetic current element *See:* Hertzian magnetic dipole.

line amplifier An amplifier that supplies a transmission line or system with a signal at a stipulated level. *See also:* amplifier. (BT/AV) [34]

linear accelerometer A device that measures translational acceleration along an input axis. In this accelerometer, an output signal is produced by the reaction of the proof mass to a translatory acceleration input. The output is usually an electrical signal proportional to applied translational acceleration. (AES/GYAC) 528-1994

linear amplifier (magnetic) An amplifier in which the output quantity is essentially proportional to the input quantity. *Note:* This may be interpreted as an amplifier that has no intentional discontinuities in the output characteristic over the useful input range of the amplifier. (MAG) 107-1964w

linear antenna An antenna consisting of one or more segments of straight conducting cylinders. *Note:* This term has restricted usage, and applies to straight cylindrical wire antennas. This term should not be confused with the conventional usage of "linear" in circuit theory. *Contrast:* linear array antenna. (AP/ANT) 145-1993

linear array (1) (data management) A one-dimensional array. (C) 610.5-1990w

(2) *See also:* linear array antenna.

linear array antenna A one-dimensional array of elements whose corresponding points lie along a straight line. (AP/ANT) 145-1993

linear Bayliss distribution *See:* Bayliss distribution.

linear broadcast (FASTBUS acquisition and control) A broadcast to a subset of the segments affected by a global broadcast. The subset can be either a specified segment or up to a specified segment or beyond a specified segment. (NID) 960-1993

linear charging (direct current) (charging inductors) A special case of resonance charging of the capacitance in an oscillatory series resistance-inductance-capacitance (RLC) circuit where the capacitor is repetitively discharged at a predetermined voltage at a rate much greater than twice the natural resonance of the RLC circuit. *Note:* The inductance of the charging inductor for linear charging is much greater than that for resonance charging for a given pulse-repetition frequency. Under the above conditions, the current through the charging inductor at the time the capacitance is discharged is not zero and the voltage across the capacitance is still rising. (MAG) 306-1969w

line conductor (Overhead supply or communication lines.) A wire or cable intended to carry electric currents, extending along the route of the line, supported by poles, towers, or other structures, but not including vertical or lateral conductors. (NESC) C2-1997

line counter A counter that records the number of lines processed or listed by some unit within some timer period. (C) 610.10-1994w

linear cross-field amplifier (microwave tubes) A crossed-field amplifier in which a nonre-entrant beam interacts with a forward wave. (ED) [45]

linear data structure A nonprimitive data structure that can represent data that is one-dimensional in nature. For example, a vector. *Contrast:* nonlinear data structure. (C) 610.5-1990w

linear detection The form of detection of an amplitude-modulated signal in which the output voltage is a linear function of the envelope of the input wave. (IT) [7]

linear distortion (linear waveform distortion) That distortion of an electric signal which is independent of the signal amplitude. *Note:* A small-signal nonuniform frequency response is an example of linear distortion. By contrast, nonlinear distortions of an electrical signal are those distortions that are dependent on the signal amplitude, for example, compression, expansion, and harmonic distortion, etc. (BT) 511-1979w

linear electrical parameters (uniform line) The series resistance, series inductance, shunt conductance, and shunt capacitance per unit length of line. *Note:* The term constant is frequently used instead of parameter. *See also:* transmission line. (AP/ANT) [35]

linear electric current element *See:* Hertzian electric dipole.

linear electron accelerator An evacuated metal tube in which electrons are accelerated through a series of small gaps (usually in the form of cavity resonators in the high-frequency range) so arranged and spaced that at a specific excitation frequency, the stream of electrons on passing through successive gaps gains additional energy from the electric field in each gap. (ED) [45]

linear element (fiber optics) A device for which the output electric field is linearly proportional to the input electric field and no new wavelengths or modulation frequencies are generated. A linear element can be described in terms of a transfer function or an impulse response function. (Std100) 812-1984w

linear energy transfer (LET) (1) The amount of energy that is transferred from an ionized particle to the medium through which it passes per normalized unit of distance traveled. *Note:* Usually expressed in units of MeV $\times$ cm^2/mg. (ED) 1005-1998

(2) The ratio $\Delta E/\Delta X$ of the average energy loss ΔE over the path length ΔX of a charged particle of specified energy in a medium. (C/BA) 1156.4-1997

linear filtering Pulse shaping in which the filter response does not change with time. [CR − (RC) n shaping is an example of linear filtering.] (NPS) 325-1996

linear FM (chirp) The modulation in a waveform whose phase varies quadratically with frequency, and, consequently, whose frequency varies linearly with time (inside the specified invelope of the impulse response). (UFFC) 1037-1992w

linear FM chirp filter A chirp filter that manifests a linear delay variation with frequency. (UFFC) 1037-1992w

linear gate (x-ray energy spectrometers) A gate whose presence does not affect the linearity of the gated signal. (NPS/NID) 759-1984r

linear-impedance relay A distance relay for which the operating characteristic on an *R-X* diagram is a straight line. *Note:* It may be described by the equation $Z = K/\cos(\theta - \alpha)$ where K and α are constants and θ is the phase angle by which the input voltage leads the input current. (SWG/PE) C37.100-1992

linear interpolation (numerically controlled machines) A mode of contouring control that uses the information contained in a block to produce velocities proportioned to the distance moved in two or more axes simultaneously. (IA/EEC) [61], [74]

linearity (1) (analog computer) A property of a component describing a constant ratio of incremental cause and effect. *Proportionality* is a special case of linearity in which the straight line passes through the origin. *Zero-error reference* of a linear transducer is a selected straight-line function of the input from which output errors are measured. *Zero-based linearity* is transducer linear defined in terms of a zero-error reference where zero input coincides with zero output. (C) 165-1977w

(2) (nuclear power generating station) The closeness with which a curve of a function approximates a straight line. *See also:* feedback control system. (IA/PE/ICTL/APP/NP/IAC) [69], 381-1977w, [60]

(3) (test, measurement, and diagnostic equipment) The condition wherein the change in the value of one quantity is directly proportional to the change in the value of another quantity. (MIL) [2]

linearity control (television) A control to adjust the variation of scanning speed during the trace interval to minimize geometric distortion. (BT/AV) 201-1979w

linearity error (1) (analog computer) An error which is the deviation of the output quantity, from a specified linear reference curve. (C) 165-1977w
(2) (Hall effect devices) The deviation of the actual characteristic curve of a Hall generator from the linear approximation to this curve. (MAG) 296-1969w
(3) (accelerometer) (gyros) The deviation of the output from a least-squares linear fit of the input-output data. It is generally expressed as a percentage of full scale, or percent of output, or both. (AES/GYAC) 528-1994

linearity error, percent of full scale (Hall effect devices) The maximum deviation, expressed as a percent of full scale, of the actual characteristic curve of a Hall effect device from the straight-line approximation to the characteristic curve derived by minimizing and equalizing the positive and negative deviations of the curve from the straight line. (MAG) 296-1969w

linearity error, percent of reading (Hall effect devices) The maximum percent deviation of the actual characteristic curve of a Hall effect device from the straight-line approximation to the curve derived by minimizing and equalizing the positive and negative percent deviations of the characteristic curve from the straight line. (MAG) 296-1969w

linearity of a multiplier (A) (analog computer) The ability of an electromechanical or electronic multiplier to generate an output voltage that varies linearly with either one of its two inputs, provided the other input is held constant. **(B) (analog computer)** The accuracy with which the above requirement is met. Linearity of a potentiometer is the accuracy with which a potentiometer yields a linear but not necessarily a proportional relationship between the angle of rotation of its shaft and the voltage appearing at the output arm, in the absence of loading errors. (C) 165-1977

linearity of a potentiometer The accuracy with which a potentiometer yields a linear but not necessarily a proportional relationship between the angle of rotation of its shaft and the voltage appearing at the output arm, in the absence of loading errors. *See also:* normal linearity. (C) 165-1977w

linearity of a signal (automatic control) The closeness with which its plot against the variable it represents approximates a straight line. *Note:* The property is usually expressed as a "non-linearity," for example, a maximum deviation. The straight line should be specified as drawn to give limited absolute deviation (independent linearity), to give minimum rms deviation (dependent linearity), to pass through the zero point, or to pass through both end points. (PE/EDPG) [3]

linearity, programming *See:* programming linearity.

linearity region (instrument approach system and similar guidance systems) The region in which the deviation sensitivity remains constant within specified values. *See also:* navigation. (AES) [42]

linearity, sweep *See:* sweep linearity.

linearity, vertical *See:* vertical linearity.

linear lag *See:* lag.

linear light (illuminating engineering) A luminous signal having a perceptible physical length. (EEC/IE) [126]

linear linked list *See:* linked linear list.

linear list A list that preserves the relationship of adjacency between data items in the list. (C) 610.5-1990w

linear load A load that draws a sinusoidal current wave when supplied by a sinusoidal voltage source. (IA/PSE) 1100-1999

linear magnetic current element *See:* Hertzian magnetic dipole.

linear object A synthetic environment object that is geometrically anchored to the terrain with one point and has a segment size and orientation. (C/DIS) 1278.1a-1998

linearly polarized field vector At a point in space, a field vector whose extremity describes a straight line segment as a function of time. *Note:* Linear polarization may be viewed as a special case of elliptical polarization where the axial ratio has become infinite. (AP/ANT) 145-1993

linearly polarized mode (LP) (fiber optics) A mode for which the field components in the direction of propagation are small compared to components perpendicular to that direction. *Note:* The LP description is an approximation which is valid for weakly guiding waveguides, including typical telecommunication grade fibers. *See also:* weakly guiding fiber; mode. (Std100) 812-1984w

linearly polarized plane wave A plane wave whose electric field vector is linearly polarized. (AP/ANT) 145-1993

linearly polarized wave An electromagnetic wave for which the locus of the tip of the electric field vector is a straight line in a plane orthogonal to the wave normal. (AP/PROP) 211-1997

linearly rising front-chopped impulse (high voltage testing) (chopped impulses) A voltage rising with approximately constant steepness, until it is chopped by a disruptive discharge, as described as a linearly rising front-chopped impulse. To define such an impulse, the best fitting straight line is drawn through the part of the front at the impulse between 50- and 90-percent of the amplitude at the instant of chopping; the intersections of this line with the 50- and 90-percent amplitudes then being designated E and F, respectively (see figure below). The impulse is defined by: (1) the voltage at the instant of chopping; (2) the rise time T_r (this is the time interval between E and F multiplied by 2.5); and (3) the virtual steepness S (this is the slope of the straight line E-F, usually expressed in kilovolts per microsecond). The impulse is considered to be approximately linear if the front, from 50 percent amplitude up to the instant of chopping, is entirely enclosed between two lines parallel to the line E-F, but displaced from it in time by $\pm$ 0.05 T_r.

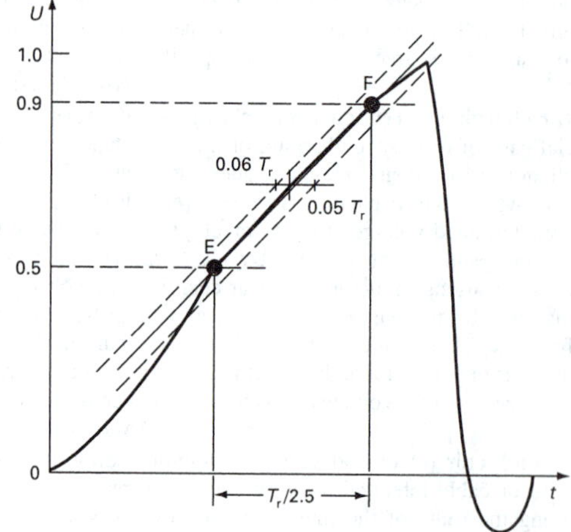

linearly rising front-chopped impulse

(PE/PSIM) 4-1978s

linearly rising switching impulse An impulse in which the impulse voltage rises at an approximately constant rate. Its amplitude is limited by the occurrence of a disruptive discharge that chops the impulse. 332-1972w

linear modulator A modulator in which, for a given magnitude of carrier, the modulated characteristic of the output wave bears a substantially linear relation to the modulating wave. *See also:* network analysis. (EEC/PE) [119]

linear network (or system) A network (or system) that has both the proportionality and the superposition properties. For example $H(\alpha x_1 + \beta x_2) = \alpha H(x_1) + \beta H(x_2)$. (CAS) [13]

linear operation A winding arrangement in which the tap winding is directly connected to the main winding and where the taps can be used only once when travelling through the tapping range. (PE/TR) C57.131-1995

linear optimization *See:* linear programming.

linear polarization (illuminating engineering) The process by which the transverse vibrations of light waves are oriented in or parallel to the same plane. Polarization may be obtained by using either transmitting or reflecting media. (EEC/IE) [126]

linear potentiometer (analog computer) A potentiometer in which the voltage at a movable contact is a linear function of the displacement of the contact. (C) 165-1977w

linear power amplifier A power amplifier in which the signal output voltage is directly proportional to the signal input voltage. *See also:* amplifier. (AP/ANT) 145-1983s

linear probing Open-address hashing in which collision resolution is handled by inserting an item that has a duplicate hash value into the next available position in the hash table. *Synonym:* consecutive spill method. *Contrast:* uniform probing; random probing. (C) 610.5-1990w

linear programming (LP) (1) (mathematics of computing) In operations research, a procedure for locating the maximum or minimum of a linear function of variables that are subject to linear constraints. *Contrast:* nonlinear programming. (C) 610.2-1987, 1084-1986w
(2) (general) Optimization problem characterization in which a set of parameter values are to be determined, subject to given linear constraints, optimizing a cost function that is linear in the parameter. *See also:* system. (SMC) [63]
(3) (computers) The analysis or solution of problems in which linear function of a number of variables is to be maximized or minimized when those variables are subject to a number of constraints in the form of linear inequalities. (C) [20], [85]

linear pulse amplifier (pulse techniques) A pulse amplifier in which the peak amplitude of the output pulses is directly proportional to the peak amplitude of the corresponding input pulses, if the input pulses are alike in shape. *See also:* pulse. (NPS) 175-1960w

linear quantizing *See:* equal interval quantizing.

linear rectifier A rectifier, the output current or voltage of which contains a wave having a form identical with that of the envelope of an impressed signal wave. *See also:* rectifier. (AP/BT/ANT) 145-1983s, 182-1961w

linear search *See:* sequential search.

linear sort (A) A insertion sort in which each item in the set to be sorted is inserted into the sorted set by scanning the sorted set sequentially to locate the proper place. *Synonym:* straight line sort. **(B)** A sort in which the items in the set to be sorted exist in a linear list. *Synonym:* straight line sort. (C) 610.5-1990

linear system (hydraulic turbines) A system with the properties that if y_1 is the response to x_1, and y_2 is the response to x_2, then $(y_1 + y_2)$ is the response to $(x_1 + x_2)$, and ky_1 is the response to kx_1. (PE/EDPG) 125-1977s

linear system or element (1) (analog computer) A system or element with the properties that if y_1 is the response to x_1, and y_2 is the response to x_2, then $(y_1 + y_2)$ is the response to $(x_1 + x_2)$, and ky_1 is the response to kx_1. *See also:* feedback control system. (C) 165-1977w
(2) An electrical element whose cause and effect relationship follows the proportionality rule. (CAS) [13]
(3) (automatic control) One whose time response to several simultaneous inputs is the sum of their independent time responses. *Note:* It is representable by a linear differential equation, and has a transfer function which is constant for any value of input within a specified range. A system or element not meeting these conditions is described as "nonlinear." (PE/EDPG) [3]

linear system with one degree of freedom *See:* damped harmonic system.

linear Taylor distribution *See:* linear Taylor distribution.

linear transducer A transducer for which the pertinent measures of all the waves concerned are linearly related. *Notes:* 1. By linearly related is meant any relation of linear character whether by linear algebraic equation, by linear differential equation, by linear integral equation, or by other linear connection. 2. The term waves concerned connotes actuating waves and related output waves, the relation of which is of primary interest in the problem at hand. *See also:* transducer. (Std100) 196-1952w, 270-1966w

linear TV waveform distortion (linear waveform distortion) The distortion of the shape of a waveform signal where this distortion is independent of the amplitude of the signal. *Notes:* 1. A TV video signal may contain time components with duration from as long as a TV field to as short as a picture element. The shapes of all these time components are subject to distortions. For ease of measurement it is convenient to group these distortions in three separate time domains: short-time waveform distortion, line-time waveform distortion, and field-time waveform distortion. 2. The waveform distortions for times from one field to tens of seconds is not within the scope of this standard. (BT) 511-1979w

linear varying parameter (varying parameter) A parameter that varies with time or position or both, but not with any dependent variable. *Note:* Unless otherwise specified, varying parameter refers to a linear-varying parameter, not to a nonlinear parameter. (Std100) 270-1966w

linear-varying-parameter network A linear network in which one or more parameters vary with time. (Std100) 154-1953w

line bar (linear waveform distortion) A pulse, nominally of 18 microseconds duration, of reference-white amplitude. The rise and fall portions of the line bar are T steps as defined. *See also:* field bar. (BT) 511-1979w

line breaker (electrically driven vehicles) (line switch) A device that combines the functions of a contactor and of a circuit breaker. *Note:* This term is also used for circuit breakers that function to interrupt circuit faults and do not combine the function of a contactor. (EEC/PE) [119]

line, bull *See:* bull line.

line-busy tone (telephone switching systems) A tone that indicates that a station termination is not available. (COM) 312-1977w

line-charging capacity (synchronous machines) The reactive power when the machine is operating synchronously at zero power factor, rated voltage, and with the field current reduced to zero. *Note:* This quantity has no inherent relationship to the thermal capability of the machine. (PE/EM) 115-1983s

line charging current The current supplied to an unloaded line or cable. (SWG/PE) C37.100-1981s

line circuit (1) (railway signaling) A signal circuit on an overhead or underground line. (EEC/PE) [119]
(2) (telephone switching systems) An interface circuit between a line and a switching system. (COM) 312-1977w

line-closing switching-surge factor The ratio of the line-closing switching-surge maximum voltage to the crest of the normal-frequency line-to-ground voltage at the source side of the closing switching device immediately prior to closing. (SWG/PE) C37.100-1992

line-closing switching-surge maximum voltage The maximum transient crest voltage to ground measured on a transmission line during a switching surge that results from energizing that line. (SWG/PE) C37.100-1992

line code An encoding technique used in digital networks for transmission on digital facilities. (COM/TA) 1007-1991r

line concentrator (telephone switching systems) A concentrator in which the inlets are lines. (COM) 312-1977w

line conditioning *See:* conditioning.

line conductor (overhead supply or communication lines) A wire or cable intended to carry electric currents, extending along the route of the line, supported by poles, towers or other structures, but not including vertical or lateral conductors. (NESC/T&D/PE/BT/AV) C2-1984s, [10], [34]

line coordination (data transmission) The process of ensuring that equipment at both ends of a circuit are set up for a specific transmission. (PE) 599-1985w

line current (thyristor) The current in the lines of the supplying power system. (IA/IPC) 428-1981w

line-deletion character A character within a line of terminal input specifying that it and all previous characters on the line are to be removed from the line; for example, if "*" is the line-deletion character in the string "ABCD*APPLE," the following would appear on the terminal: "APPLE." *See also:* character-deletion character. (C) 610.5-1990w

line density *See:* line width.

line designer A party who develops structure loading criteria, structure types, and structure locations based on line routing, maintenance and construction requirements. The line designer should establish design criteria for construction and maintenance that will affect the structure designer and constructor. The line designer could be an owner or an agent acting for the owner. (T&D/PE) 1025-1993r, 951-1996

line detection An image segmentation technique in which line pixels are identified by examining their neighborhoods. (C) 610.4-1990w

line discipline *See:* control procedure.

line display (spectrum analyzer) The display produced on a spectrum analyzer when the resolution bandwidth is less than the spacing of the individual frequency components. (IM) 748-1979w

line-drop compensator (1) (voltage regulators) A device which causes the voltage-regulating relay to increase the output voltage by an amount that compensates for the impedance drop in the circuit between the regulator and a predetermined location on the circuit (sometimes referred to as the load center). *See also:* voltage regulator. (PE/TR) C57.12.80-1978r **(2)** A device that causes the voltage regulating device to vary the output voltage an amount that compensates for the impedance voltage drop in the circuit between the regulator and a predetermined location on the circuit (sometimes referred to as the load center). (PE/TR) C57.15-1999

line-drop signal (manual switchboard) A drop signal associated with a subscriber line. (PE/EEC) [119]

line-drop voltmeter compensator A device used in connection with a voltmeter that causes the latter to indicate the voltage at some distant point of the circuit. *See also:* auxiliary device to an instrument. (EEC/PE) [119]

line editing A method of text editing that allows the user to change text, with cursor control, on only one line at a time. Multiple lines may be viewed or changed through editing commands. *Synonym:* line-oriented editing. *Contrast:* full-screen editing. *See also:* context editing. (C) 610.2-1987

line editor A text editor that allows the user to change text, with cursor control, on only one line at a time. Multiple lines may be viewed or changed through editing commands. *Contrast:* full-screen editor. (C) 610.2-1987

line end and ground end (A) (electric power) Line end is that end of a neutral grounding device that is connected to the line circuit directly or through another device. *See also:* grounding device. **(B) (electric power)** Ground end is that end that is connected to ground directly or through another device. *See also:* grounding device. (PE/SPD/T&D) 32-1972, [10]

line-ending zone *See:* hot zone.

line-end zone *See:* hot zone.

line extender (amplifiers) An RF amplifier required to compensate for losses in the feeder system. (LM/C) 802.7-1989r

line failure rate The expected frequency of outages a subscriber line can experience due to switching system and subsystem malfunctions. Problems that may occur in the station or wiring, outside plant, or loop electronics are excluded. The line failure rate may be given for hardware faults alone. (COM/TA) 973-1990w

line feed A command or signal sent to a printer to instruct it to move to the next writing line. *Note:* Often used in conjunction with a carriage return to move to the beginning of the next writing line. (C) 610.10-1994w

line fill The ratio of the number of connected main telephone stations on a line to the nominal main-station capacity of that line. *See also:* cable. (EEC/PE) [119]

line filling In text formatting, the use of extra interword spacing to form even margins. *Contrast:* incremental justification. (C) 610.2-1987

line-focus tube (x-ray tubes) An X-ray tube in which the focal spot is roughly a line. (ED) [45]

line frequency (television) The number of times per second that a fixed vertical line in the picture is crossed in one direction by the scanning spot. *Note:* Scanning during vertical return intervals is counted. *See also:* television. (BT/AV) [34]

line-frequency line current (thyristor) The root-mean-square (rms) value of the fundamental component of the line current, the frequency of which is the line frequency. (IA/IPC) 428-1981w

line-frequency line voltage (thyristor) That sine wave component of the line voltage, whose frequency is the line frequency. The root-mean-square (rms) value of that component. (IA/IPC) 428-1981w

line group (telephone switching systems) A multiplicity of lines served by a common set of links. (COM) 312-1977w

line guy Tensional support for poles or structures by attachment to adjacent poles or structures. *See also:* tower. (T&D/PE) [10]

line, hard *See:* hard line.

line hit (data transmission) An electric interference causing the introduction of spurious signals on a circuit. (PE) 599-1985w

line-impedance stabilization network (LISN) (1) (radio-noise emissions) A network inserted in the supply mains lead of apparatus to be tested that provides, in a given frequency range, a specified load impedance for the measurement of disturbance voltages and which may isolate the apparatus from the supply mains in that frequency range. *Note:* A LISN unit may contain one or more individual LISN circuits. *Synonym:* artificial mains network. (EMC) C63.4-1991 **(2)** A LISN is used to provide a standard connection from the power source to the UUT for EMI measurements. A LISN has input (power source), output (to unit under test), signal output port, and a chassis ground connection. The signal output port must be terminated with a 50 ohm load when no measurement instrumentation is attached. (PEL) 1515-2000 **(3)** *See also:* artificial mains network. (EMC)

line-insulation resistance (telephone switching systems) The resistance between the loop conductors and ground or between each other. (COM) 312-1977w

line insulator (pin, post) An assembly of one or more shells, having means for semirigidly supporting line conductors. *See also:* insulator. (EEC/IEPL) [89]

line integral (dot-product line integral) The line integral between two points on a given path in the region occupied by a vector field is the definite integral of the dot product of a path element and the vector. Thus,

$$I = \int_a^b \mathrm{V}\cos\theta \mathrm{d}s$$

$$= \int_a^b \mathrm{V} \cdot \mathrm{d}s$$

$$= \int_a^b (V_x \mathrm{d}x + V_y \mathrm{d}y + V_z \mathrm{d}z)$$

where V is the vector having a magnitude V, ds the vector element of the path, θ the angle between V and ds. Example: The magnetomotive force between two points on a line connecting two points in a magnetic field is the line integral of the magnetic field strength, that is, the definite integral between the two points of the dot product of a vector element of the length of the line and the magnetic strength at the element. (Std100) 270-1966w

line isolation monitor (1) (health care facilities) An instrument which continually checks the hazard current from an isolated circuit to ground. (EMB) [47]

(2) (health care facilities) A test instrument designed to continually check the balanced and unbalanced impedance from each line of an isolated circuit to ground and equipped with a built-in test circuit to exercise the alarm without adding to the leakage current hazard. Line isolation monitor was formerly known as ground contact indicator. (NESC/NEC) [86]

line lamp (wire communication) A switchboard lamp for indicating an incoming line signal. (PE/EEC) [119]

line, lead *See:* pilot line.

line, life *See:* safety life line.

life line *See:* safety life line.

linelet A 64-byte collection of other aligned data formats and/or field containers. (C/MM) 1596.5-1993

line, light *See:* pulling line.

line lightning performance The performance of a line expressed as the annual number of lightning flashovers on a circuit mile or tower-line mile basis. *See also:* direct-stroke protection. (T&D/PE) [10], 1243-1997, 1410-1997

line load A measure of the maximum capacity that can be handled by a circuit or line. (C) 610.7-1995

line-load control (telephone switching systems) A means of selectively restricting call attempts during emergencies so as to permit the handling of essential traffic. (COM) 312-1977w

line lockout (telephone switching systems) A means for the handling of permanent line signals to prevent further recognition as a call attempt. (COM) 312-1977w

line loss Energy loss on a transmission or distribution line. (PE/PSE) 858-1993w

lineman (1) (power line maintenance) A person qualified to perform various line-work operations including aerial and ground work. (T&D/PE) 516-1987s
(2) (cleaning insulators) A person qualified to perform various work operations in electric transmission or distribution, including on the ground or aerial. (T&D/PE) 957-1987s

lineman's body belt *See:* line-worker's body belt.

lineman's platform *See:* lineperson's platform.

line microphone A directional microphone consisting of a single straight line element, or an array of contiguous or spaced electroacoustic transducing elements disposed on a straight line, or the acoustical equivalent of such an array. *See also:* microphone; line transducer. (EEC/PE) [119]

line number, television *See:* television line number.

line of nodes (communication satellite) The line which is common to both the orbital plane and the equatorial plane and which passes through the ascending and descending nodes. (COM) [19]

line of position (LOP) (navigation) The intersection of two surfaces of position, normally plotted as lines on the earth's surface, each line representing the locus of constant indication of the navigational information. *See also:* navigation. (AES/RS) 686-1982s, [42]

line-oriented editing *See:* line editing.

line or input regulation (electrical conversion) Static regulation caused by a change in input. (AES) [41]

line or trace (cathode-ray tubes) The path of the moving spot on the screen or target. (ED) 161-1971w

line pairs per raster height (diode-type camera tube) The spatial frequency of a uniform periodic array referred to a unit length equal to the raster height. The array may be sinusoidal or comprised of equal width alternating light and dark bars (lines). Each period or cycle includes one light and one dark region or line, hence the term line pairs. (ED) 503-1978w

line parameters A sufficient set of parameters to specify the transmission characteristics of the line. (Std100) 270-1966w

lineperson's platform A device designed to be attached to a wood pole or metal structure, or both, to serve as a supporting surface for workers engaged in deadending operations, clipping-in, insulator work, etc. The designs of these devices vary considerably. Some resemble short cantilever beams, others resemble swimming pool diving boards, and still others as long as 40 ft (12 m) are truss structures resembling bridges. Materials commonly used for fabrication are wood, fiberglass, and metal. *Synonyms:* baker board; dead-end board; diving board; dead-end platform; D-board. (T&D/PE) 524-1992r

line, pilot *See:* pilot line.

line pixel A pixel contained in an arc that approximates a straight line. (C) 610.4-1990w

line power (thyristor) The total power delivered from the line to the controller. (IA/IPC) 428-1981w

line printer A printer that prints a line of characters as a unit. *Contrast:* page printer; character-at-a-time printer. (C) 610.10-1994w

line printing The printing of an entire line of characters as a unit. (C) [20]

line protocol A protocol that uses a control program to perform data communication functions, such as moving the data between transmit and receive locations over network lines. (C) 610.7-1995

line, pulling *See:* pulling line.

liner (1) (A) (rotating machinery) A separate insulating member that is placed against a grounded surface. *See also:* slot liner. **(B) (rotating machinery)** A layer of insulating material that is deposited on a grounded surface. *See also:* slot liner; pole-cell insulation. (PE) [9]
(2) (dry cell) (primary cell) Usually a paper or pulpboard sheet covering the inner surface of the negative electrode and serving to separate it from the depolarizing mix. *See also:* electrolytic cell. (EEC/PE) [119]

line regulation (1) (power supplies) The maximum steady-state amount that the output voltage or current will change as the result of a specified change in input line voltage (usually for a step change between 105–125 or 210–250 volts, unless otherwise specified). Regulation is given either as a percentage of the output voltage or current, or as an absolute change Δ E or Δ I. (PE) [78]
(2) (ferroresonant voltage regulators) The maximum amount that the output voltage or current will change as the result of a specified change in line voltage. *See also:* overall regulation. (PEL/ET) 449-1990s

line relay (railway practice) A relay that receives its operating energy over a circuit that does not include the track rails. (EEC/PE) [119]

line replaceable unit (LRU) (1) (test, measurement, and diagnostic equipment) A unit which is designated by the plan for maintenance to be removed upon failure from a larger entity (equipment, system) in the latter's operational environment. (MIL) [2]
(2) The smallest or lowest piece of equipment or subassembly that is replaceable onboard the vessel. Systems and subsystems may contain multiple numbers of LRUs. Typical LRUs include circuit cards, sealed bearings, and drive motors. (IA/MT) 45-1998

line residual current *See:* ground-return current.

lines (1) (A) communication lines. The conductors and their supporting or containing structures which are used for public or private signal or communication service, and which operate at potentials not exceeding 400 V to ground or 750 V between any two points of the circuit, and the transmitted power of which does not exceed 150 watts. When operating at less than 150 V, no limit is placed on the transmitted power of the system. Under specified conditions, communication cables may include communication circuits exceeding the preceding limitation where such circuits are also used to supply power solely to communication equipment. *Note:* Telephone, telegraph, railroad-signal, data, clock, fire, police-alarms, cable television and other systems conforming with the above are included. Lines used for signaling purposes, but not included under the above definition, are considered as supply lines of the same voltage and are to be insulated. **(B)** electric

supply lines. Those conductors used to transmit electric energy and their necessary supporting or containing structures. Signal lines of more than 400 V are always supply lines within the meaning of the rules, and those of less than 400 V may be considered as supply lines, if so run and operated throughout. (NESC) C2-1977
(2) (telecommunications switching systems) Local subscriber terminations. (COM/TA) 973-1990w
(3) *See also:* communication lines; electric supply lines.

line, safety *See:* safety life line.

line, safety life *See:* safety life line.

line section A portion of an overhead line or a cable bounded by two terminations, a termination and a tap point, or two tap points. (PE/PSE) 859-1987w

line segment A portion of a line section that has a particular type of construction or is exposed to a particular type of failure, and therefore which may be regarded as a single entity for the purpose of reporting and analyzing failure and exposure data. *Note:* A line segment is a subcomponent of a line section. (PE/PSE) 859-1987w

line side (data transmission) Data terminal connections to a communications circuit between two data terminals. (PE) 599-1985w

line-side converter ac voltage The rms power-frequency voltage from line to line at the ac terminals of the line-side converter. *Note:* This voltage is the same as the rated system voltage times the turns ratio of the input interface equipment. (IA/ID) 995-1987w

line-side converter alternating current The magnitude of the current expressed in amperes of the rms value of current at the output terminals of the line-side interface and the input terminals of line-side converter. (IA/ID) 995-1987w

line, sock *See:* pulling line.

line source A continuous distribution of sources of electromagnetic radiation, lying along a line segment. *Note:* Most often in practice the line segment is straight. (AP/ANT) 145-1993
(2) (A) (fiber optics) In the spectral sense, an optical source that emits one or more spectrally narrow lines as opposed to a continuous spectrum. *See also:* monochromatic. **(B) (fiber optics)** In the geometric sense, an optical source whose active (emitting) area forms a spatially narrow line. *See also:* monochromatic. (Std100) 812-1984

line source corrector A linear array antenna feed with radiating element locations and excitations chosen to correct for aberrations present in the focal region fields of a reflector. (AP/ANT) 145-1993

line spectrum (1) (fiber optics) An emission or absorption spectrum consisting of one or more narrow spectral lines, as opposed to a continuous spectrum. *See also:* spectral width; spectral line; monochromatic. (Std100) 812-1984w
(2) (spectrum analyzer) A spectrum composed of discrete frequency components. (IM) 748-1979w

line speed (data transmission) The maximum rate at which signals may be transmitted over a given channel; usually in baud or b/s (bits/sec). *See also:* baud. (PE/C) 599-1985w, 610.10-1994w

line spread function (diode-type camera tube) The spatial distribution of the signal, or the time distribution of the signal resulting from the scanning process, produced when an image of an extremely narrow line is formed on the photosurface. (ED) 503-1978w

line, straw *See:* pilot line.

line stretcher (waveguide components) A section of waveguide or transmission line having an adjustable physical length. (MTT) 147-1979w

line style *See:* line type.

line surge arrester A protective device for limiting surge voltages on transmission-line insulation by discharging or bypassing surge current; it prevents continued flow of follow-current to ground and is capable of repeating these functions. (PE/T&D) 1243-1997

line switch (1) (power system device function numbers) A switch used as a disconnecting, load-interrupter, or isolating switch in an ac or dc power circuit, when this device is electrically operated or has electrical accessories, such as an auxiliary switch, magnetic lock, etc. (PE/SUB) C37.2-1979s
(2) *See also:* step-by-step switch. (C) 610.7-1995

line switching (1) (data transmission) The switching technique of temporarily connecting two lines together so that the stations directly exchange information. (PE) 599-1985w
(2) *See also:* circuit switching. (C) 610.7-1995

line switching system *See:* step-by-step system.

line, tag *See:* tag line.

line tap (in respect to system protection) A connection to a line with equipment that does not feed energy into a fault on the line in sufficient magnitude to require consideration in the relay plan. (SWG/PE) C37.100-1992

line terminal (1) (surge arresters) The conducting part provided for connecting the arrester to the circuit conductor. *Note:* When a line terminal is not supplied as an integral part of the arrester, and the series gap is obtained by providing a specified air clearance between the line end of the arrester and a conductor, or arcing electrode, etc., the words line terminal used in the definition refer to the conducting part that is at line potential and that is used as the line electrode of the series gap. (SPD/PE) C62.1-1981s
(2) (rotating machinery) A termination of the primary winding for connection to a line (not neutral or ground) of the power supply or load. *See also:* asynchronous machine; rotor; stator. (PE) [9]
(3) (in respect to system protection) A connection to a line with equipment that can feed energy into a fault on the line in sufficient magnitude to require consideration in the relay plan and that has means for automatic disconnection. (SWG/PE) C37.100-1992
(4) The conducting part of the arrestor provided for connecting the arrester to the circuit conductor. *Note:* When a line terminal is not supplied as an integral part of the arrester, and the series gap is obtained by providing a specified air clearance between the line end of the arrester and a conductor, or arcing electrode, etc., the words "line terminal" used in the definition refer to the conducting part that is at line potential and that is used as the line electrode of the series gap. (SPD/PE) C62.11-1999

line-time waveform distortion (linear waveform distortion) The linear TV waveform distortion of time components from 1 microsecond to 64 microseconds, that is, time components of the line-time domain. (BT) 511-1979w

line-to-background ratio (x-ray energy spectrometers) (of a spectral line) The ratio of the intensity of a monoenergetic line to the intensity of the background immediately adjacent to the line. (NPS/NID) 759-1984r

line-to-line voltage *See:* voltage sets.

line-to-neutral voltage *See:* voltage sets.

line transducer A directional transducer consisting of a single straight-line element, or an array of contiguous or spaced electroacoustic transducing elements disposed on a straight line, or the acoustical equivalent of such an array. *See also:* line microphone. (SP) [32]

line transformer A transformer for interconnecting a transmission line and terminal equipment for such purposes as isolation, line balance, impedance matching, or for additional circuit connections. (SP) 151-1965w

line triggering (oscilloscopes) Triggering from the power line frequency. *See also:* oscillograph. (IM/HFIM) [14], [40], 311-1970w, 748-1979w

line type An attribute specifying the manner in which a line is displayed on a graphical display device. For example: solid, dashed, dotted, dot-dash. *Synonym:* line style. (C) 610.6-1991w

line-type fire detector (fire protection devices) A device in which detection is continuous along a path. (NFPA) [16]

line voltage (thyristor) The voltage between the lines of the supplying power system. (IA/IPC) 428-1981w

line voltage notch The dip in the supply voltage to a converter due to the momentary short-circuit of the ac lines during a commutation interval. Alternatively, the momentary dip in supply voltage caused by the reactive drops in the supply circuit during the high rates of change in currents occurring in the ac lines during commutation. (IA/SPC) 519-1992

line weighting (data transmission) A noise weighting used in a noise measuring set to measure noise on a line that is terminated by a subset with a number 144 receiver or a similar subset. The meter-scale readings are in dBrn (144 line). *See also:* noise definitions (data transmission). (PE) 599-1985w

line width (computer graphics) An attribute specifying the thickness of a line displayed on a graphical display device. *Synonym:* line density. (C) 610.6-1991w

linewidth (1) (laser maser) The interval in frequency or wavelength units between the points at which the absorbed power (or emitted power) of an absorption (or emission) line is half of its maximum value when measured under specified conditions. (LEO) 586-1980w
(2) (fiber optics) *See also:* spectral width. 812-1984w

line-work (power line maintenance) Various operations performed by a person on electrical facilities, including ground work, aerial work, and associated maintenance. (T&D/PE) 516-1995

line worker (1) A person qualified to perform various line-work operations, including aerial and ground work *Synonym:* lineman. (T&D/PE) 516-1995
(2) A person qualified to perform various work operations on electric transmission or distribution, including on the ground or from aerial devices. (T&D/PE) 957-1995

line-worker's body belt A belt that consists of a belt strap and D-rings, and may include a cushion section or a tool saddle. (NESC/T&D/PE) C2-1997, 1307-1996

linger To wait for a period of time before terminating a connection to allow outstanding data to be transferred. (C) 1003.5-1999

LINK *See:* link layer.

link (1) (protection and coordination of industrial and commercial power systems) The current-responsive element in a fuse that is designed to melt under overcurrent conditions and so interrupt the circuit. A renewal link is one intended for use in Class H low-voltage renewable fuses. (IA/PSP) 242-1986r
(2) (data transmission) A channel or circuit designed to be connected in tandem with other channels or circuits. In automatic switching, a link is a path between two units of switching apparatus within a central office. (PE) 599-1985w
(3) (communication satellite) A complete facility over which a certain type of information is transmitted, including all elements from source transducer to output transducer. (COM) [24]
(4) (telephone switching systems) A connection between switching stages within the same switching system. (COM) 312-1977w
(5) (A) (data management) To establish a pointer; for example, to link two items in a hierarchy. *See also:* pointer. **(B) (data management)** In relation theory, a relationship between two or more entities or records. **(C) (data management)** To append an item to a linked list. *See also:* push; link field; pointer. (C) 610.5-1990
(6) (A) (software) To create a load module from two or more independently translated object modules or load modules by resolving cross-references among them. *See also:* linkage editor. **(B) (software)** A part of a computer program, often a single instruction or address, that passes control and parameters between separate modules of the program. *Synonym:* linkage. **(C) (software)** To provide a link as in definition (B). (C) 610.12-1990
(7) (token ring access method and physical layer specifications) A unidirectional physical and media connection between two stations. (LM/C) 802.5c-1991r
(8) The physical connection and electronic hardware by which data is transferred between the host and the peripheral. This standard is concerned only with the link layer; the only information transfer defined or required as part of this standard is that necessary to control and synchronize the communication of peripheral data. The defined interface is transparent to the peripheral data communicated at data or higher layers. (C/MM) 1284-1994
(9) A means of communicating digital information bidirectionally in serial format between two devices or subsystems. A link comprises two link interfaces connected by an appropriate medium (or media, for connections between boards or cabinets), such that the link output of each interface is connected to the link input of the other. (C/BA) 1355-1995
(10) A channel or a point-to-point line. *See also:* circuit; line. (C) 610.7-1995, 610.10-1994w
(11) The transmission path between any two interfaces of generic cabling. (LM/C) 802.3u-1995s
(12) The physical or logical connection between a host and a printer. (C/MM) 1284.1-1997
(13) *See also:* directory entry. (C/PA) 1003.1b-1993s, 1003.2-1992s, 1003.5-1999
(14) *See also:* connector link. (T&D/PE) 524-1992r

linkage (1) (programming) Coding that connects two separately coded routines. (C) [85]
(2) (software) *See also:* link. (C) 610.12-1990

linkage editor (software) A computer program that creates a single load module from two or more independently translated object modules or load modules by resolving cross-references among the modules and, possibly, by relocating elements. May be part of a loader. *Synonym:* linker. *See also:* linking loader. (C) 610.12-1990

linkage product A device that provides an interface between network segments such as gateways or bridges. (C) 610.7-1995

linkage voltage test, direct-current test (rotating machinery) A series of current measurements, made at increasing direct voltages, applied at successive intervals, and maintained for designated periods of time. *Note:* This may be a controlled overvoltage test. *See also:* asynchronous machine. (PE) [9]

Link Aggregation Group A group of links that appear to a MAC Client as if they were a single link. All links in a Link Aggregation Group connect between the same pair of Aggregation Systems. One or more conversations may be associated with each link that is part of a Link Aggregation Group. (C/LM) 802.3ad-2000

link-attached terminal A terminal that is connected to the computer by telecommunication lines or by a data link. *Contrast:* channel-attached terminal. (C) 610.10-1994w

link-break cutout A load-break fuse cutout that is operated by breaking the fuse link to interrupt the load current. (SWG/PE) C37.40-1993, C37.100-1992

link cable The physical medium connecting two link interfaces, comprising of two or more electrical or optical cables. (C/BA) 1355-1995

Link Code Word The 16 bits of data encoded into a Fast Link Pulse (FLP) Burst. (C/LM) 802.3-1998

link communication A physical means of connecting one location to another for the purpose of transmitting and receiving information. (C) 610.7-1995

link character (L_char) Control characters which are used on a link in order to ensure flow control and the proper functioning of the link. *See also:* normal character. (C/BA) 1355-1995

link, connector *See:* connector link.

link count The number of directory entries that refer to a particular file. (C/PA) 9945-1-1996, 9945-2-1993, 1003.5-1999

linked linear list A linear list in which each item contains a pointer to the next item in the list, making it unnecessary for the items to be physically sequential. *Note:* the items are still logically adjacent. *Synonym:* linear linked list. (C) 610.5-1990w

linked list (1) A list in which each item contains a pointer to the next or preceding item in the list, making it unnecessary for the items to be physically sequential. *Note:* Unless the list is circular, the last item in the list contains a null link field. *Synonyms:* singly linked list; chain. *See also:* chained list; circularly-linked list; linked linear list. (C) 610.5-1990w **(2) (software)** *See also:* chained list. (C/SE) 729-1983s

Linked List A software data structure composed of individually allocated memory objects, in which each item points to the next. It is used as a FIFO queue in this document. (C/MM) 1212.1-1993

linker *See:* linkage editor.

link error counter (LEC) A counter maintained by a network station to track all data link errors. The maximum permissible value is a system parameter. (EMB/MIB) 1073.3.1-1994

link field (A) A field in each item of a linked list, containing a pointer to the next or preceding item in the list. *Synonym:* chain field. **(B)** In a tree, that portion of each node that contains a pointer to other nodes in the tree. (C) 610.5-1990

link identifier (LID) A string identifier generated by the lower layer service on the onboard equipment (OBE) transponder each time the transponder initiates a new session with a lower layer service on the roadside equipment (RSE). (SCC32) 1455-1999

linking loader A computer program that reads one or more object modules into main memory in preparation for execution, creates a single load module by resolving cross-references among the separate modules, and, in some cases, adjusts the addresses to reflect the storage locations into which the code has been loaded. *See also:* absolute loader; relocating loader; linkage editor. (C) 610.12-1990

link input A connection point for receiving signals. *See also:* link interface. (C/BA) 1355-1995

link interface (port) A connection point comprising a link input and a link output and implementing one of the relevant conformance subsets defined in IEEE Std 1355-1995. *See also:* link. (BA/C) 1355-1995

link is clear A condition in which there is no incoming energy on the link. *See also:* clearlocal area networks. (C) 8802-12-1998

link layer (1) (Layer 2) The layer of the ISO Reference Model that provides the functional and procedural means to transfer data between stations, and to detect and correct errors that can occur in the Physical layer. (DIS/C) 1278.2-1995 **(2)** The layer, in a stack of three protocol layers defined for the Serial Bus, that provides the service to the transaction layer of one-way data transfer with confirmation of reception. The link layer also provides addressing, data checking, and data framing. The link layer also provides an isochronous data transfer service directly to the application. See figure 34 for the relation of the link layer to the Serial Bus protocol stack. (C/MM) 1394-1995 **(3)** The Serial Bus protocol layer that provides confirmed and unconfirmed transmission or reception of primary packets. (C/MM) 1394a-2000

link output A connection point for transmitting signals. *See also:* link interface. (C/BA) 1355-1995

link pair A pair of links going in opposite directions between two stations. (LM/C) 802.5c-1991r

link partner The device at the opposite end of a link segment from the local station. The link partner device may be either a DTE or a repeater. (C/LM) 802.3-1998

link penalties For fiber optic links, the power penalties of a link not attributed to link attenuation. These power penalties include modal noise, relative intensity noise (RIN), intersymbol interference (ISI), mode partition noise, extinction ratio, and eye-opening penalties. (C/LM) 802.3-1998

link protocol In OSI, a protocol that ensures that the transmission of bits received are the same as the bits sent. (C) 610.7-1995

link pulse Communication mechanism used in 10BASE-T and 100BASE-T networks to indicate link status and (in Auto-Negotiation-equipped devices) to communicate information about abilities and negotiate communication methods. 10BASE-T uses Normal Link Pulses (NLPs), which indicate link status only. 10BASE-T and 100BASE-T nodes equipped with Auto-Negotiation exchange information using a Fast Link Pulse (FLP) mechanism that is compatible with NLP. (C/LM) 802.3-1998

link segment (1) The point-to-point full-duplex medium connection between two and only two Medium Dependent Interfaces (MDIs). (C/LM) 802.3-1998 **(2)** The physical interconnection between two repeaters or between a repeater and an end node. A link segment includes the link medium (twisted pairs or optical fibres) and its two attached Medium Dependent Interface (MDI) connectors.local area networks. (C) 8802-12-1998

Link Segment Delay Value (LSDV) A number associated with a given segment that represents the delay on that segment used to assess path delays for 100 Mb/s CSMA/CD networks. LSDV is similar to SDV; however, LSDV values do not include the delays associated with attached end stations and/or repeaters. (C/LM) 802.3-1998

link, swivel *See:* swivel link.

lin-log receiver A receiver having a linear amplitude response for small-amplitude signals and a logarithmic response for large-amplitude signals. *Note:* The lin-log reciever is a practical method of approximating a logarithmic receiver. (AES) 686-1997

lip *See:* conductor lifting hook.

lip microphone A microphone adapted for use in contact with the lip. *See also:* microphone. (EEC/PE) [119]

liquid Refers to both synthetic fluids and mineral transformer oil. *Note:* Some synthetic fluids may be unsuitable for use in the arcing environment of a step-voltage regulator. (PE/TR) C57.15-1999

liquid controller An electric controller in which the resistor is a liquid. *See also:* electric controller. (IA/ICTL/IAC) [60]

liquid cooling *See:* manifold insulation.

liquid counter tube (radiation counters) A counter tube suitable for the assay of liquid samples. It often consists of a thin glass-walled Geiger-Mueller tube sealed into a test tube providing an annular space for the sample. *See also:* anticoincidence. (ED) [45]

liquid crystal device A display device in which light is directed through a liquid crystal, a device possessing many of the optical properties of solid crystals but one whose molecular order is not as firmly fixed. *Note:* The liquid crystal becomes opaque when it is energized by electrodes attached to its glass casing, enabling it to be employed in forming characters or graphic representations. *See also:* liquid crystal display. (C) 610.10-1994w

liquid crystal display A display made of material whose reflectance or transmittance changes when an electric field is applied. (VT) 1477-1998

liquid crystal display device *See:* liquid crystal device.

liquid development (electrostatography) Development in which the image-forming material is carried to the field of the electrostatic image by means of a liquid. *See also:* electrostatography. (ED) 224-1965w, [46]

liquid-filled fuse unit A fuse unit in which the arc is drawn through a liquid. (SWG/PE) C37.40-1993, C37.100-1992

liquid-filled, or liquid-cooled transformer (electrical heating applications to melting furnaces and forehearths in the glass industry) A transformer that is immersed in a tank of insulating fluid that removes heat from the windings by conduction. The liquid is then cooled by running through external radiators. (IA) 668-1987w

liquid-flow counter tube (radiation counters) A counter tube specially constructed for measuring the radioactivity of a flowing liquid. *See also:* anticoincidence. (ED) [45]

liquid-function potential The contact potential between two electrolytes. It is not susceptible of direct measurement. (EEC/PE) [119]

liquid-immersed regulator A regulator in which the core and coils are immersed in an insulating liquid. (PE/TR) C57.15-1999

liquid-immersed self-cooled/forced-air-cooled/forced-liquid-cooled regulator A regulator having its core and coils immersed in liquid and having a self-cooled rating with cooling obtained by the natural circulation of air over the cooling surface; a forced-air-cooled rating with cooling obtained by the forced circulation of air over this same air cooling surface; and a forced-liquid-cooled rating with cooling obtained by the forced circulation of liquid over the core and coils and adjacent to this same cooling surface over which the cooling air is being forced-circulated. (PE/TR) C57.15-1999

liquid-immersed self-cooled/forced-air-cooled regulator A regulator having its core and coils immersed in liquid and having a self-cooled rating with cooling obtained by the natural circulation of air over the cooling surface and a forced-air-cooled rating with cooling obtained by the forced circulation of air over this same cooling surface. (PE/TR) C57.15-1999

liquid-immersed self-cooled regulator A regulator having its core and coil immersed in a liquid and cooled by the natural circulation of air over the cooling surfaces. (PE/TR) C57.15-1999

liquid-immersed transformer (power and distribution transformers) A transformer in which the core and coils are immersed in an insulating liquid. (PE/TR) C57.12.80-1978r

liquid-immersed type arrester An arrester designed for use immersed in an insulating liquid. (SPD/PE) C62.22-1997, C62.11-1999

liquid-immersed water-cooled A regulator having its core and coils immersed in a liquid and cooled by the natural circulation of the liquid over the water-cooled surface. (PE/TR) C57.15-1999

liquid-immersed water-cooled/self-cooled A regulator having its core and coils immersed in liquid and having a water-cooled rating with cooling obtained by the natural circulation of liquid over the water-cooled surface, and a self-cooled rating with cooling obtained by the natural circulation of air over the air-cooled surface. (PE/TR) C57.15-1999

liquid-in-glass thermometer A thin-walled glass bulb attached to a glass capillary stem closed at the opposite end, with the bulb and a portion of the stem filled with an expansive liquid, the remaining part of the stem being filled with the vapor of the liquid or a mixture of this vapor and an inert gas. Associated with the stem is a scale in temperature degrees so arranged that when calibrated the reading corresponding to the end of the liquid column indicates the temperature of the bulb. (PE/PSIM) 119-1974w

liquid-level switch *See:* float switch.

liquid resistor A resistor comprising electrodes immersed in a liquid. (IA/ICTL/IAC) [60], [84]

liquid-scintillation solution A solution consisting of an organic solvent (or mixture of solvents) and one or more organic scintillator solutes. (NI) N42.15-1990, N42.16-1986

liquidtight flexible metal conduit A raceway of circular cross section having an outer liquidtight, nonmetallic, sunlight-resistant jacket over an inner flexible metal core with associated couplings, connectors, and fittings and approved for the installation of electric conductors. (NESC/NEC) [86]

LISP *See:* LISt Processing.

Lissajous figure (1) (oscilloscopes) A special case of an X–Y plot in which the signals applied to both axes are sinusoidal functions. For a stable display the signals must be harmonics. Lissajous figures are useful for determining phase and harmonic relationships. (IM) 311-1970w

(2) (modeling and simulation) A pattern of two mutually orthogonal sets of curve forms, used to generate characters or curves, or for testing purposes.

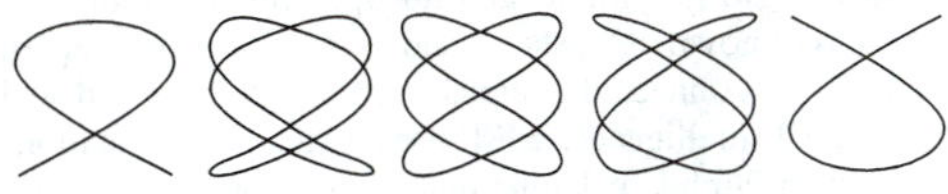

Lissajous figures

(C) 610.6-1991w

list (1) An ordered single-dimensioned set of data. (ATLAS) 771-1980s

(2) (A) (software) (data management) A set of data items, each of which has the same data definition. *See also:* linked list; unordered list; stack; linear list; queue; ordered list. **(B) (data management) (software)** To print or otherwise display a set of data items. (C) 610.5-1990, 610.12-1990

(3) The language-independent syntax for a family of datatypes constructed from a base datatype. A value of list datatype contains a sequence of zero or more values of the base datatype. Applying an operation to all members of a list may be supported through either of two programming paradigms. In the first, the sequencing control is provided by the application. In the second, it is provided by the implementation. (C/PA) 1351-1994w

(4) A datatype constructed from a base datatype. A list value contains a sequence of zero or more values of the base datatype. Applying an operation to all members of a list may be supported through either of two programming paradigms. In the first, the sequencing control is provided by the application; in the second, it is provided by the implementation. (C/PA) 1224.1-1993w

(5) A kind of collection class that contains no duplicates and whose members are ordered. *Contrast:* bag; set. (C/SE) 1320.2-1998

(6) (computers) *See also:* pushup list; pushdown list.

list box A visual user interface control that contains a list of items that a user can select. (C) 1295-1993w

listed (1) Equipment or materials included in a list published by an organization acceptable to the authority having jurisdiction and concerned with product evaluation, that maintains periodic inspection of production of listed equipment or materials and whose listing states either that the equipment or material meets appropriate standards or has been tested and found suitable for use in a specified manner. The means for identifying listed equipment may vary for each organization concerned with product evaluation, some of which do not recognize equipment as listed unless it is also labeled. The authority having jurisdiction should utilize the system employed by the listing organization to identify a listed product. (NESC/NEC) [86]

(2) *See also:* labeled. (DEI) 1221-1993w

listener (1) A node that receives an isochronous subaction for an isochronous channel. There may be zero, one, or more than one listeners for any given isochronous channel. (C/MM) 1394-1995

(2) An application at a node that receives a stream packet. (C/MM) 1394a-2000

listener echo Echo that reaches the ear of the listener. (EEC/PE) [119]

listener sidetone The sidetone acoustic output of the telephone caused by a diffuse noise field. (COM/TA) 269-1992

listening group (speech quality measurements) A group of persons assembled for the purpose of speech quality testing. Number, selection, characteristics, and training of the listeners depend upon the purpose of the test. 297-1969w

listing (software) An ordered display or printout of data items, program statements, or other information. (C) 610.12-1990

list insertion sort An insertion sort implemented using the list sorting technique. For example, a linear sort. (C) 610.5-1990w

list merge sort A merge sort implemented using the list sorting technique. (C) 610.5-1990w

LISt Processing (LISP) A list processing language designed for manipulating symbols and for operating on strings of information known as "lists"; handles recursive and repetitive handling of connected character strings. *Note:* Used widely in artificial intelligence, LISP uses a functional notation derived from lambda calculus that permits programs and data to have the same structure. *See also:* Common LISP; SCHEME; FLAVORS; EL1; functional language. (C) 610.13-1993w

list processing (1) (software) A method of processing data in the form of lists. Usually, chained lists are used so that the logical order of items can be changed without altering their physical locations. *See also:* chained list; list; data. (C/SE) 729-1983s
(2) (data management) The manipulation of data that is or is going to be stored in list structures. (C) 610.5-1990w

list processing language (1) A programming language designed to facilitate the manipulation of data expressed in the form of lists. Examples are LISP and IPL. *See also:* logic programming language; algorithmic language; algebraic language. (C) 610.12-1990
(2) A programming language designed to manipulate data expressed in the form of lists or character strings. Examples are LISP, LOGO, LPL, PROLOG, and SAM76. *Synonyms:* symbol manipulation language; symbolic language. (C) 610.13-1993w

List Processing Language (LPL) An extension of PL/1 used to provide list processing facilities in which the user can define and operate on cells of varying characteristics. (C) 610.13-1993w

list sorting A sorting technique in which the items to be sorted form a linked list and the links between the items in the list are manipulated in such a way that, in the final list, the items form a linked list in sorted order. *See also:* key sorting; address table sorting. (C) 610.5-1990w

list structure (A) A list, each item of which is either a single data item or a list structure itself. **(B)** A data structure that contains one or more lists. *Synonym:* compound list. (C) 610.5-1990

literal (1) (data management) Composed of characters, as in a literal variable name used to contain a customer's name. (C) 610.5-1990w
(2) (software) In a source program, an explicit representation of the value of an item; for example, the word FAIL in the instruction: If $x = 0$ then print "FAIL". *See also:* immediate data; figurative constant. (C) 610.12-1990
(3) The denotation of a specific instance of a value class. (C/SE) 1320.2-1998

lithium drifting (germanium gamma-ray detectors) (charged-particle detectors) (semiconductor radiation detectors) (x-ray energy spectrometers) A technique for compensating p-type material by causing lithium ions to move through a crystal under an applied electric field in such a way as to compensate the charge of the bound acceptor impurities. (NPS/NID) 325-1986s, 300-1988r, 301-1976s, 759-1984r

lithium-drifted detector (germanium gamma-ray detectors) (charged-particle detectors) (x-ray energy spectrometers) (semiconductor radiation detectors) A detector made by the lithium compensation process. (NPS/NID) 325-1986s, 759-1984r, 300-1988r, 301-1976s

littleAdd A bus transaction that adds an integer *next* argument to a specified data address and returns the previous data value from that address. All values in this transaction are assumed to be little-endian integers. In the CSR Architecture this is called a little_add transaction. (C/MM) 1596.5-1993

little addressian The physical location of data-byte addresses on a multiplexed address/data bus. On a little addressian bus, the data byte with the smallest address is multiplexed (in time or space) with the least-significant byte of the address. (C/MM/BA) 1212-1991s, 896.2-1991w, 896.10-1997

little addressan Bus that multiplexes the least significant byte of the address with the data byte that has the lowest address. (C/BA) 896.3-1993w

little endian (1) The least significant byte of a data item, it has the lowest relative address. Correspondingly, more significant bytes have higher relative memory addresses. (C/BA) 896.3-1993w
(2) A representation of multibyte numerical values in which bytes with lesser numerical significance appear at lower memory addresses. (C/BA) 1275-1994
(3) The arithmetic significance of data-byte addresses within a multibyte register. Within a little endian register or register set, the data byte with the smallest address is the least significant. (C/MM) 1212-1991s
(4) A multibyte data value that is stored in memory with the most significant data byte through least significant data byte in the highest through lowest memory addresses, respectively. (C/MM) 1212.1-1993
(5) A method of storing multibyte data in a byte addressable memory such that the least significant byte of the data is stored at the lowest address. (C/BA) 896.10-1997

little-endian A specified ordering of bytes within a data structure where the low-order byte (byte 0) is placed in the least significant byte lane of that data structure. (C/BA) 1014.1-1994w

little-endian processor A Processor architecture that is optimized for the processing of little endian data values, as opposed to big endian data values. (C/MM) 1212.1-1993

litz wire A conductor composed of a number of fine, separately insulated strands, usually fabricated in such a way that each strand assumes, to substantially the same extent, the different possible positions in the cross section of the conductor. (EEC/PE) [119]

live State of a part that, when connection is made to that part, can cause a current exceeding the allowable leakage current for the part connected to flow from that part to an accessible part of the same equipment. (EMB/MIB) 1073.4.1-2000

live cable test cap A protective structure at the end of a cable that insulates the conductors and seals the cable sheath. (T&D/PE) [10]

live-front So constructed that there are exposed live parts on the front of the assembly. *See also:* live-front switchboard. (PE/IA/ICTL/TR/IAC) C57.12.80-1978r, [60]

live-front switchboard A switchboard that has exposed live parts on the front. (SWG/PE) C37.100-1992

live insertion (1) The process of inserting boards into or withdrawing boards from a backplane when power is on. Insertion of boards into a backplane can be performed when power is off or when power is on. (C/BA) 896.3-1993w, 896.4-1993w, 896.2-1991w
(2) The process of inserting boards into a backplane when power is on. Insertion of boards into a backplane can be performed when power is off or when power is on. (C/BA) 896.10-1997

live insertion state A state in which a module may be removed from a powered, operational system without interfering with the operation of the remainder of the system or causing safety hazard to the maintenance operator. (C/BA) 896.2-1991w

live keyboard A keybord that lets users interact with the system while a program is running, allowing the examination or modification of program variables. (C) 610.10-1994w

live-line permit *See:* suspension of reclosing.

live-line tool *See:* stick.

livelock (1) A metastable situation in which some modules acquire and release resources in a way that none of them make forward progress. (C/BA) 10857-1994, 896.4-1993w, 896.3-1993w
(2) A system condition that occurs when two or more modules acquire and release resources without any of them doing useful work for an indefinite period of time. (C/BA) 1014.1-1994w

live-metal part (power and distribution transformers) A part consisting of electrically conductive material which can be energized under conditions of normal use of the equipment. (PE/TR) C57.12.80-1978r

live parts Those parts that are designed to operate at voltage different from that of the earth. (SWG/PE/SUB/PSIM) C37.40-1993, C37.100-1992, 119-1974w, 1268-1997

live removal The process of removing boards from a backplane when power is on. (C/BA) 896.2-1991w

live room (audio and electroacoustics) A room that has an unusually small amount of sound absorption. (SP) [32]

live tank switching device A switching device in which the vessel(s) housing the interrupter(s) is at a potential above ground. (SWG/PE) C37.100-1992

live time (1) The time interval of a count during which a counting system is capable of processing input pulses. Elapsed (or real) time equals live time plus dead time. (NI) N42.14-1991

(2) The live time, in seconds and fraction thereof, of acquisition of the spectrum. It is expressed as 14 characters including decimal point with leading zeros interpreted as zeros. (NPS/NID) 1214-1992r

(3) The total time of the measurement minus the total dead time. (NI) N42.12-1994

live work Work on or near (e.g., part of tools being used or worker's body less than minimum approach distance) energized or potentially energized lines (i.e., grounding, live tool work, hot stick work, gloving and barehand work). (T&D/PE) 516-1995

live zone The period(s) in the operating cycle of a machine during which corrective functions can be initiated. (IA/ICTL/IAC) [60]

LLC *See:* logical link control.

LLC encoding *See:* logical link control encoding.

LLC frame *See:* logical link control frame.

LLC procedure *See:* logical link control procedure.

LLC sublayer *See:* logical link control sublayer.

LME *See:* layer management entity.

LMI *See:* layer management interface.

LMSC *See:* LAN/MAN Standards Committee.

LNA *See:* launch numerical aperture.

L network (1) (general) A network composed of two branches in series, the free ends being connected to one pair of terminals and the junction point and one free end being connected to another pair of terminals. *See also:* network analysis. (Std100) 270-1966w

(2) An unbalanced ladder network composed of a series arm and a shunt arm.

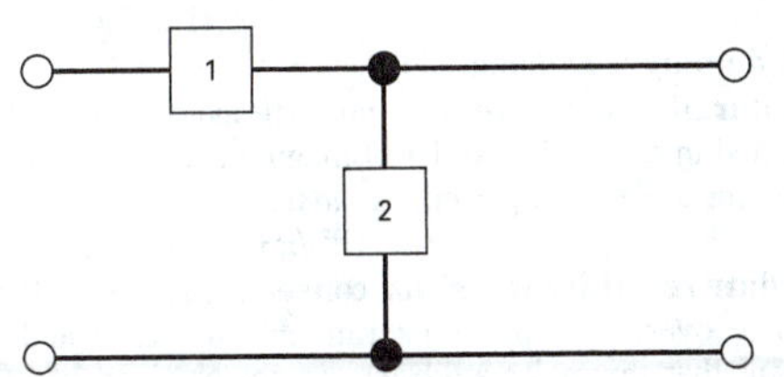

The free ends are left-hand terminal pair; the junction point and one free end are the right-hand terminal pair.

L network

(CAS) [13]

load (1) (charge) (induction and dielectric heating usage) The material to be heated. *See also:* induction heating. (IA) 54-1955w

(2) (power and distribution transformers) (output) The apparent power in megavolt-amperes, kilovolt-amperes, or volt-amperes that may be transferred by the transformer. (PE/TR) C57.12.80-1978r

(3) (rotating machinery) All the numerical values of the electrical and mechanical quantities that signify the demand to be made on a rotating machine by an electric circuit or a mechanism at a given instant. *See also:* direct-current commutating machine. (PE) [9]

(4) (programming) To place data into internal storage. (C) [20], [85]

(5) (electric) (electric utilization) The electric power used by devices connected to an electrical generating system. *See also:* generating station. (PE/PSE) [54]

(6) (A) (automatic control) An energy-absorbing device. **(B) (automatic control)** The material, force, torque, energy, or power applied to or removed from a system or element. (PE/EDPG) [3]

(7) (data transmission) A power-consuming device connected to a circuit. One use of the word "load" is to denote a resistor or impedance which replaces some circuit element temporarily or permanently removed. For example, if a filter is disconnected from a line, the line may be artificially terminated in an impedance which simulates the filter that was removed. The artificial termination is then called a load or a dummy load. (PE) 599-1985w

(8) (A) (test, measurement, and diagnostic equipment) To read information from cards or tape into memory. **(B) (test, measurement, and diagnostic equipment)** Building block or adapter providing a simulation of the normal termination characteristics of a unit under test. **(C) (test, measurement, and diagnostic equipment)** The effect that the test equipment has on the unit under test or vice versa. (MIL) [2]

(9) (A) (software) To read machine code into main memory in preparation for execution and, in some cases, to perform address adjustment and linking of modules. *See also:* loader. **(B) (software)** To copy computer instructions or data from external storage to internal storage or from internal storage to registers. *Contrast:* store. *See also:* fetch; move. (C) 610.12-1990

(10) (data management) To insert data values into a database that previously contained no data. *Synonym:* populate. *See also:* download; upload. (C) 610.5-1990w

(11) To move the image of a client program from a long-term storage medium (such as a disk) into memory where it may be executed. (C/BA) 1275-1994

(12) (A) In computer operations, the amount of scheduled work to be performed on a computer system. *See also:* line load. **(B)** In electronics, the amount of current drawn by a device. *Note:* This determines the "drive strength" of the circuit. *See also:* loading. (C) 610.7-1995, 610.10-1994

(13) To enter data or programs into storage or working registers. *See also:* mount. (C) 610.10-1994w

(14) Demand or energy. (PE/PSE) 858-1993w

(15) The true or apparent power consumed by power utilization equipment performing its normal function. (SWG/PE) C37.100-1992

loadability (of an air switch) The ratio of allowable continuous current at 25°C ambient temperature to rated current. *Note:* Loadability is a measure of the average allowable continuous current over a range of ambient temperatures from 10°C to 40°C for the air surrounding air switches. (SWG/PE) C37.30-1992s, C37.34-1994, C37.37-1996

loadability factor (of an air switch) The ratio of allowable continuous current at a given ambient temperature to rated current. (SWG/PE) C37.37-1996

Loadable Device (LD) A station on the network that is capable of accepting a load from a Load Service. (LM/C) 15802-4-1994

load&add A data-access operation that adds a *next* value to a specified data type and returns the previous data value. (C/MM) 1596.5-1993

load-and-go (1) An operating technique in which there are no stops between the loading and execution phases of a program, and which may include assembling or compiling. (C) [20], [85]

(2) (software) An operating technique in which there are no stops between the loading and execution phases of a computer program. (C) 610.12-1990

load angle (synchronous machines) The angular displacement, at a specified load, of the center line of a field pole from the axis of the armature magnetomotive force pattern. (PE) [9]

load-angle curve (synchronous machines) (load-angle characteristic) A characteristic curve giving the relationship between the rotor displacement angle and the load, for constant values of armature voltage, field current, and power factor. (PE) [9]

load balancing reactor A series connected reactor used to correct the division of current between parallel-connected transformers or circuits which have unequal impedance voltages under steady-state and short-circuit conditions. (PE/TR) C57.16-1996

load-band of regulated voltage (rotating machinery) The band or zone, expressed in percent of the rated value of the regulated voltage, within which the synchronous-machine regulating system will hold the regulated voltage of a synchronous machine during steady or gradually changing conditions over a specified range of load. (PE) [9]

load, base *See:* base load.

load, binder *See:* binder load.

load-break connector (1) (separable insulated connectors) A connector designed to close and interrupt current on energized circuits. (T&D/PE) 386-1995

(2) (power and distribution transformers) A separable insulated connector designed to close and interrupt current on energized circuits. (PE/TR) C57.12.80-1978r

load-break cutout A cutout with means for interrupting load currents. (SWG/PE) C37.40-1993, C37.100-1992

load-break tests (load-interrupting tests) (high-voltage switchgear) Tests that consist of manual or remote control opening of a device, which is provided with a means for breaking load, while the device is carrying a prescribed current under specified conditions. (SWG/PE) C37.40-1981s

load capacity *See:* peak code.

load cell *See:* dynamometer.

load center (1) (electric power utilization) A point at which the load of a given area is assumed to be concentrated. *See also:* generating station. (PE/PSE) [54]

(2) A point at which the load of a given area is assumed to be concentrated for technical evaluation. (PE/PSE) 858-1993w

load characteristic (electron tube) For an electron tube connected in a specified operating circuit at a specified frequency, a relation, usually represented by a graph, between the instantaneous values of a pair of variables such as an electrode voltage and an electrode current, when all direct electrode supply voltages are maintained constant. *Synonym:* dynamic characteristic. (ED) 161-1971w

load circuit (induction and dielectric heating usage) The network including leads connected to the output terminals of the generator. *Note:* The load circuit consists of the coupling network and the load material at the proper position for heating. (IA) 54-1955w

load-circuit efficiency (induction and dielectric heating usage) The ratio of the power absorbed by the load to the power delivered at the generator output terminals. *See also:* network analysis; induction heating. (IA) 54-1955w

load-circuit power input The power delivered to the load circuit. *Note:* It is the product of the alternating component of the voltage across the load circuit, the alternating component of the current passing through it (both root-mean-square values), and the power factor associated with these two quantities. *See also:* network analysis. (AP/ANT) 145-1983s

load coil (induction heating usage) An electric conductor that, when energized with alternating current, is adapted to deliver energy by induction to a charge to be heated. (IA) 169-1955w

load commutated inverter A converter in which the commutation voltages are supplied by the ac load. *Note:* For the purpose of this recommended practice the ac load is a synchronous machine operating as a motor. (IA/ID) 995-1987w

load control (watthour meters) A switching output for external load management. (ELM) C12.15-1990, C12.13-1985s

load controller (1) (control systems for steam turbine-generator units) The load controller includes only those components and control elements that are responsive to energy output and load reference, that furnish an input signal to the control mechanism for the purpose of controlling the load. (PE/EDPG) 122-1985s

(2) (hydroelectric power plants) Auxiliary control device that adjusts turbine flow in response to plant output requirements. (PE/EDPG) 1020-1988r

load current (1) (thyristor) The current in the load. (IA/IPC) 428-1981w

(2) (electron tube) (tube) The current output utilized in an external load circuit. *See also:* electron tube. (Std100) [84]

(3) (watthour meter) *See also:* test current. (ELM) C12.1-1982s

load current interrupting rating (of an interrupter switch) The highest rms current in amperes between unity and 0.8 power factor lagging that a device is required to interrupt, without requiring maintenance, at its rated maximum voltage and at rated frequency, for a number of operations equal to the operating life expectancy of the switch. (SWG/PE) C37.30-1992s

load current output (arc-welding apparatus) The current in the welding circuit under load conditions. (EEC/AWM) [91]

load curve A curve of power versus time showing the value of a specific load for each unit of the period covered. *See also:* generating station. (T&D/PE) [10]

load curves, daily *See:* daily load curves.

load-dependent delay That part of a delay through a cell instance attributed to the admittance (load) presented to the arc sink pin and the internal impedance of the output. (C/DA) 1481-1999

load diversity The difference between the sum of the maxima of two or more individual loads and the coincident of combined maximum load usually measured over a specified period. (PE/PSE) 858-1993w

load diversity power The rate of transfer of energy necessary for the realization of a saving of system capacity brought about by load diversity. *See also:* generating station. (T&D/PE) [10]

load division A control function that divides the load in a prescribed manner between two or more power devices connected to the same load. *See also:* feedback control system. (IA/ICTL/IAC) [60]

load, dummy *See:* dummy load.

load duration curve (power operations) A curve of loads, plotted in descending order of magnitude, against time intervals for a specified period. *See also:* generating station. (PE/PSE) 858-1987s, 28-1974

load duty repetitive (thyristor converter) A type of load duty where overloads appear intermittent but cyclic and so frequent that thermal equilibrium is not obtained between all overloads. (IA/IPC) 444-1973w

loaded applicator impedance (dielectric heating) The complex impedance measured at the point of application, with the load material at the proper position for heating, at a specified frequency. (IA) 54-1955w, 169-1955w

loaded impedance (transducer) The impedance at the input of the transducer when the output is connected to its normal load. *See also:* self-impedance. (SP) [32]

loaded linear antenna *See:* sectionalized linear antenna.

loaded loop (telephone loop performance) A loop into which lumped inductance (loading coil) is introduced at fixed intervals to compensate for the distributed cable capacitance. The addition of loading coils, properly placed, reduces mid-voiceband loss, flattens the frequency response over most of the voiceband, but creates a sharp cut-off at the high-frequency band edge. (COM/TA) 820-1984r

loaded origin The address of the initial storage location of a computer program at the time the program is loaded into main memory. *Contrast:* assembled origin. *See also:* starting address; offset. (C) 610.12-1990

loaded Q (1) (electric impedance) (working Q) The value of Q of such impedance when coupled or connected under working conditions. (PE/EEC) [119]
(2) (switching tubes) The unloaded Q of the tube modified by the coupled impedances. *Note:* As here used, Q is equal to 2π times the energy stored at the resonance frequency divided by the energy dissipated per cycle in the tube, or for cell-type tubes, in the tube and its external resonant circuit. (ED) 161-1971w

loaded voltage ratio (electronic power transformer) Equal to the secondary voltage divided by the primary voltage. For linear loads the ratio is stated for a specified load current and power factor. For rectifier loads, the ratio should be given for the specified circuit configuration, including the filters, and the rated direct-current load. Unless otherwise stated, the ratio is given for rated conditions, line voltage, frequency, load, and stabilized temperature. Primary voltages is given as line to line and secondary voltages as leg values (terminal to neutral or center tap if used) unless otherwise indicated. *See also:* turns ratio. (PEL/ET) 295-1969r

loader (A) (software) A computer program that reads machine code into main memory in preparation for execution and, in some cases, adjusts the addresses and links the modules. Types include absolute loader, linking loader, relocating loader. *See also:* bootstrap; linkage editor. **(B) (software)** Any program that reads programs or data into main memory. (C) 610.12-1990

load, estimated maximum *See:* estimated maximum load.

load factor The ratio of the average load over a designated period of time to the peak load occurring in that period. *Note:* Although not part of the official definition, the term *load factor* is used by some utilities and others to describe the equivalent number of hours per period the peak or average demand must prevail in order to produce the total energy consumption for the period. *See also:* generating station. (IA/PE/T&D/PSE) 141-1993r, 241-1990r, [10], [54]

load-frequency characteristic (1) The change in power requirements of a control-area load that is caused by a change in system frequency. (PE/PSE) 94-1991w
(2) The change in power requirements of a control area load that is caused by a change in system frequency. (PE/PSE) 858-1993w

load-frequency control (LFC) (1) The regulation of the power output of electric generators within a prescribed area in response to changes in system frequency, tie line loading, or the relation of these to each other, so as to maintain the scheduled system frequency or the established interchange with other areas within predetermined limits or both. *See also:* generating station. (T&D/PE) [10]
(2) (automatic generation control on electric power systems) An automatic generation control subsystem that changes control area generation in response to an area control error. (PE/PSE) 94-1991w

load ground (signal-transmission system) The potential reference plane at the physical location of the load. *See also:* signal. (IE) [43]

load group (nuclear power generating station) An arrangement of buses, transformers, switching equipment, and loads fed from a common power supply within a division. (PE/NP) 308-1991

load impedance (1) (germanium gamma-ray detectors) (charged-particle detectors) (x-ray energy spectrometers) (semiconductor radiation detectors) The impedance shunting the detector, and across which the detector output voltage signal is developed. (NPS/NID) 325-1986s, 759-1984r, 300-1982s, 301-1976s
(2) (general) The impedance presented by the load to a source or network. (IM/HFIM) [40]

load-impedance diagram (oscillators) A chart showing performance of the oscillator with respect to variations in the load impedance. *Note:* Ordinarily, contours of constant power and of constant frequency are drawn on a chart whose coordinates are the components of either the complex load impedance or of the reflection coefficient. *See also:* Rieke diagram; oscillatory circuit. (ED) 161-1971w

load-indicating automatic reclosing equipment Automatic reclosing equipment that provides for reclosing the circuit interrupter automatically in response to sensing of predetermined conditions of the load circuit. *Note:* This type of automatic reclosing equipment is generally used for direct-current load circuits. (SWG/PE) C37.100-1992

load-indicating resistor A resistor used, in conjunction with suitable relays or instruments, in an electric circuit, for the purpose of determining or indicating the magnitude of the connected load. (SWG/PE) C37.100-1992

loading (1) The modification of a basic antenna such as a dipole or monopole caused by the addition of conductors or circuit elements that change the input impedance or current distribution or both. (AP/ANT) 145-1993
(2) (communication practice) (data transmission) The insertion of reactance in a circuit for the purpose of improving its transmission characteristics in a given frequency band. *Note:* The term is commonly applied, in wire communication practice, to the insertion of loading coils in series in a transmission line at uniform intervals, and in radio practice, to the insertion of one or more loading coils anywhere in a transmission circuit. (PE) 599-1985w
(3) (data transmission) A commonly used type of reactive load used on leased lines to produce a specific bandwidth characteristic. (PE) 599-1985w
(4) (automatic control) Act of transferring energy into or out of a system. (PE/EDPG) [3]
(5) To add reactance in a circuit in order to minimize amplitude distortion. (C) 610.7-1995, 610.10-1994w

loading coil An inductor inserted in a circuit to increase its inductance for the purpose of improving its transmission characteristics in a given frequency band. *See also:* loading. (PE/EM) 43-1974s

loading-coil spacing The line distance between the successive loading coils of a coil-loaded line. *See also:* loading. (EEC/PE) [119]

loading error (1) (analog computer) An error due to the effect of a load impedance upon the transducer or signal source driving it. (C) 165-1977w
(2) (test, measurement, and diagnostic equipment) The error introduced when data are incorrectly transferred from one medium to another. (MIL) [2]

loading factor (A) The maximum amount of usable space in a physical block after accounting for block overhead. **(B)** The ratio of the number of stored entities in a file to the maximum number of entries that can be stored in a unit of data medium. (C) 610.5-1990

loading machine A machine for loading materials such as coal, ore, or rock into cars or other means of conveyance for transportation to the surface of the mine. (PE/EEC/MIN) [119]

loading test *See:* heat-run test.

load, internal *See:* internal load.

load-interrupter switch An interrupter switch designed to interrupt currents not in excess of the continuous-current rating of the switch. *Notes:* 1. It may be designed to close and carry abnormal or short-circuit currents as specified. 2. In international (IEC) practice a device with such performance characteristics is called a switch. (SWG/PE) C37.100-1992

load, interruptible *See:* interruptible load.

load leads (induction and dielectric heating usage) The connections or transmission line between the power source or generator and load, load coil or applicator. *See also:* induction heating. (IA) 54-1955w

load-limit changer (1) (speed governing systems) A device that acts on the speed-governing system to prevent the governor-controlled fuel valves from opening beyond the position for which the device is set. *See also:* speed-governing system. (PE/EDPG) 282-1968w, [5]
(2) (automatic generation control on electric power systems) A device on the speed-governing system that prevents the governor-controlled valves from opening beyond a preset limit. (PE/PSE) 94-1991w

load losses (1) (power and distribution transformers) Those losses which are incident to the carrying of a specified load. Load losses include I^2 R loss in the windings due to load and eddy currents; stray loss due to leakage fluxes in the windings, core clamps, and other parts, and the loss due to circulating currents (if any) in parallel windings, or in parallel winding strands. (PE/TR) C57.12.80-1978r
(2) (series transformer) (copper losses) The load losses of a series transformer are the I^2R losses, computed from the rated currents for the windings and the measured direct-current resistances of the windings corrected to 75 degrees Celsius. (EEC/LB) [98]
(3) (of a regulator) Those losses that are incident to the carrying of the load. Load losses include I^2R loss in the windings due to load current, stray loss due to stray fluxes in the windings, core clamps, and so forth. (PE/TR) C57.15-1999

load management A means of achieving a reduction of demand and energy by interrupting the customer's electric supply to specific devices or by modifying the devices' use characteristics. (PE/PSE) 858-1993w

load map (software) A computer-generated list that identifies the location or size of all or selected parts of memory-resident code or data. (C) 610.12-1990

load matching (1) (induction and dielectric heating) The process of adjustment of the load-circuit impedance to produce the desired energy transfer from the power source to the load. (IA) 54-1955w
(2) The technique of either adjusting the load-circuit impedance or inserting a network between two parts of a system to produce the desired power transfer or signal transmission. (CAS) [13]

load-matching network An electric network for accomplishing load matching. *See also:* induction heating. (IA) 54-1955w

load-matching switch (induction and dielectric heating) A switch in the load-matching network to alter its characteristics to compensate for some sudden change in the load characteristics, such as passing through the Curie point. *See also:* induction heating. (IA) 54-1955w

load module (software) A computer program or subprogram in a form suitable for loading into main storage for execution by a computer; usually the output of a linkage editor. *See also:* object module. (C) 610.12-1990

load point The position on a magnetic tape that is indicated by the beginning-of-tape marker. (C) 610.10-1994w

load power (thyristor) The total power delivered from the controller to the load. (IA/IPC) 428-1981w

load power factor (converters having ac output) (self-commutated converters) Characteristic of an ac (alternating current) load in terms expressed by the ratio of active power to apparent power assuming an ideal sinusoidal voltage. (IA/SPC) 936-1987w

load profile (1) (electric power systems in commercial buildings) The graphic representation of the demand load, usually on an hourly basis, for a particular day. (IA/PSE) 241-1990r
(2) The magnitude and duration of loads (kW and kvar) applied in a prescribed time sequence, including the transient and steady-state characteristics of the individual loads. (PE/NP) 387-1995
(3) The recording, storage, and analysis of consumption data over a period of time for a particular installation. (AMR/SCC31) 1377-1997

load range (watthour meter) The range in amperes over which the meter is designed to operate continuously with specified accuracy. (ELM) C12.1-1982s

load regulation (1) (ferroresonant voltage regulators) The maximum amount that the output voltage will change as the result of a specified change in load current. *See also:* overall regulation. (PEL/ET) 449-1990s
(2) (excitation systems) The magnitude of voltage change resulting from a load change. (PE/EDPG) 421.4-1990
(3) (synchronous machines) The steady-state decrease of the value of the specified variable resulting from a specified increase in load, generally from no-load to full-load unless otherwise specified. (PE/EDPG) 421-1972s

load resistance (germanium gamma-ray detectors) (semiconductor radiation detectors) (x-ray energy spectrometers) (charged-particle detectors) (of a semiconductor radiation detector) The resistive component of the load impedance. (NPS/NID) 325-1986s, 759-1984r, 301-1976s, 300-1982s

load-resistor contactor (power system device function numbers) A contactor that is used to shunt or insert a step of load limiting, shifting, or indicating resistance in a power circuit, or to switch a space heater in circuit, or to switch a light or regenerative load resistor of a power rectifier or other machine in and out of circuit. (PE/SUB) C37.2-1979s

load restoration The process of scheduled load restoration when the abnormality causing load shedding has been corrected. (SWG/PE) C37.100-1992

load rheostat A rheostat whose sole purpose is to dissipate electric energy. *Note:* Frequently used for load tests of generators. *See also:* control. (IA/IAC) [60], [84]

load saturation curve (synchronous machines) (load characteristic) The saturation curve of a machine on a specified constant load current. (PE) [9]

Load Server (LS) A station on the network that is capable of providing a load for a Loadable Device. (LM/C) 15802-4-1994

load sharing Distributing a given load among two or more computers on a network. (C) 610.7-1995

load shedding (1) (emergency and standby power) The process of deliberately removing preselected loads from a power system in response to an abnormal condition in order to maintain the integrity of the system. (SWG/IA/PE/PSE) 446-1995, C37.100-1992
(2) Disconnecting or interrupting the electrical supply to a customer load by the utility, usually to mitigate the effects of generating capacity deficiencies or transmission limitations. (PE/PSE) 858-1993w

load-shifting resistor A resistor used in an electric circuit to shift load from one circuit to another. (SWG/PE) C37.100-1992

load-side converter ac voltage The rms value of the power-frequency sinusoidal envelope of voltage at the ac terminals of the load-side converter. This is the commutating voltage of the load-side converter. *Note:* This is the voltage that appears at the terminals of the synchronous machine and is not the rms fundamental sine-wave voltage of the machine. (IA/ID) 995-1987w

load, sliding *See:* sliding load.

load-store computer *See:* reduced instruction set computer.

load switch or contactor (induction heating) The switch or contactor in an induction heating circuit that connects the high-frequency generator or power source to the heater coil or load circuit. *See also:* induction heating. (IA) 54-1955w

load, system *See:* system load.

load, system maximum hourly *See:* system maximum hourly load.

load tap changer (LTC) A selector switch device, which may include current interrupting contactors, used to change transformer taps with the transformer energized and carrying full load. (PE/TR) C57.131-1995, C57.12.80-1978r

load-tap-changing transformer (power and distribution transformers) A transformer used to vary the voltage, or the phase angle, or both, of a regulated circuit in steps by means of a device that connects different taps of tapped winding(s) without interrupting the load. (PE/TR) C57.12.80-1978r

load time (hybrid computer linkage components) The time required to read in the digital value to a register of the digital-to-analog converter, measured from the instant that the digital computer commands a digital-to-analog converter or a digital-to-analog multiplier "load." (C) 166-1977w

load transfer switch A switch used to connect a generator or power source optionally to one or another load circuit. *See also:* induction heating. (IA) 54-1955w

load, transition *See:* transition load.

load variation (power operations) Maximum system load over a time interval minus the integrated load over the same time interval. (PE/PSE) 858-1987s

load variation within the hour (electric power utilization) The short-time (three minutes) net peak demand minus the net 60-minute clock-hour integrated peak demand of a supplying system. *See also:* generating station. (PE/PSE) [54]

load voltage (1) (arc-welding apparatus) The voltage between the output terminals of the welding power supply when current is flowing in the welding circuit. (EEC/AWM) [91]
(2) (thyristor) The voltage across the load. (IA/IPC) 428-1981w

load voltage unbalance (thyristor) If the voltages measured between the pairs of load terminals are not equal, a voltage unbalance exists. (IA/IPC) 428-1981w

load weighing A function incorporated in the propulsion or friction brake system that measures changes in sprung vehicle weight. Its purpose is to permit control of tractive effort in order to achieve a constant effort-to-weight ratio for a given master control command. (VT) 1475-1999

load, work, or heater coil (induction heating) An electric conductor that, when energized with alternating current, is adapted to deliver energy by induction to a charge to be heated. *See also:* induction heating. (IA) 54-1955w

lobe (1) (directional lobe) (antenna lobe) (radiation lobe) (data transmission) A lobe is a portion of the directional pattern bounded by one or two cones of nulls. (PE) 599-1985w
(2) Those elements in the data path between a station's transmitter and its own receiver when it is in Bypass State. (C/LM) 11802-4-1994
(3) *See also:* minor lobe; side lobe; back lobe; shoulder lobe; vestigial lobe; major lobe. (AP/ANT) 145-1993

lobe cabling The cabling used to interconnect the station MICs to the TCUs. This cabling includes all work area cabling, horizontal cabling, and patch cables. The lobe cable only carries ring signals when the station is actively connected on the ring (inserted). When the station is not inserted in the ring, the lobe cable may contain local test signals. (C/LM) 8802-5-1998

lobe switching (1) A form of scanning in which the direction of maximum radiation is discretely changed by switching. *See also:* sequential lobing. (AP/ANT) 145-1993
(2) A means of direction finding in which a directive radiation pattern is periodically shifted in position so as to produce a variation of the signal at the target. *Note:* The signal variation provides information on the amount and direction of displacement of the target from the pattern mean position. *Synonym:* sequential lobing. (AES) 686-1997

Local As an adjective shall indicate that the modified term's extent is a single process space. Thus, a Local operation may be invoked only from within the process space of the object. (IM/ST) 1451.1-1999

local Relating to a bounded part of an ATLAS test requirement. (SCC20) 771-1998

local ability *See:* ability.

local access and transport area (LATA) (1) In the United States, a local geographic area in which a local telephone company is allowed to offer communications services. *See also:* intraLATA; interLATA. (C) 610.7-1995
(2) An area typically served by a regional Bell operating company (RBOC). (SCC31) 1390.2-1999
(3) A local telephone exchange area that serves to distinguish local phone service from long-distance phone service. Customers are served by a regional Bell operating company (RBOC) or other telephone company (OTC). (SCC31) 1390.3-1999

local adaptation The process of modifying a product that is specific to one culture to make it specific to another culture. (C/PA) 14252-1996

local area network (LAN) (1) A non-public data network in which serial transmission is used without store and forward techniques for direct data communication among data stations located on the user's premises. (LM/C) 8802-6-1994
(2) A communications network designed for a moderate size geographic area and characterized by moderate to high data transmission rates, low delay, and low bit error rates. (DIS/C) 1278.2-1995, 1278.3-1996
(3) A communication network to interconnect a variety of intelligent devices (e.g., personal computers, workstations, printers, file storage devices) that can transmit data over a limited area, typically within a facility. (PE/SUB/EMB/MIB) C37.1-1994, 1073.3.2-2000
(4) A computer network in which communication is limited to a geographic span of a few kilometers. *Note:* The medium is generally of a wider bandwidth than that provided by commercial carrier, the packet length frequently is much larger than network propagation time, and the bit error rate in the physical medium is on the order of one bit in 10^8 or better. *Synonym:* local network. *Contrast:* long haul network. *See also:* metropolitan area network; wide area network. (C) 610.7-1995

LAN Port A MAC-sublayer point of attachment of a Bridge to a Local Area Network. *Note:* This definition is equivalent to that of Bridge Port in ISO/IEC 10038, as applied to Local MAC Bridges. (C/LM) 802.1G-1996

local area transport area (LATA) An area typically served by a regional Bell operating company (RBOC). (AMR) 1390-1995

local backup A form of backup protection in which the backup protective relays are at the same station as the primary protective relays. (SWG/PE) C37.100-1992

local benchmark program A benchmark program that is specific to a particular site, application or environment. (C) 610.10-1994w

local broadcast A broadcast that is effective only on the originating segment. (NID) 960-1993

local_bus_ID A bus_ID with a value of 11111111112. (C/MM) 1394-1995

local central office *See:* central office.

local channel (data transmission) A channel connecting a communications subscriber to a central office. (PE) 599-1985w

local compaction In microprogramming, compaction in which microoperations are not moved beyond the boundaries of the single entry, single exit sequential blocks in which they occur. *Contrast:* global compaction. (C) 610.12-1990

local control (programmable instrumentation) A method whereby a device is programmable by means of its local (front or rear panel) controls in order to enable the device to perform different tasks. (Also referred to as manual control.). (IM/AIN) 488.1-1987r

local control/alarm (1) (electric pipe heating systems) The locations where control, alarm, or both signal or function take place. With respect to electric pipe heating systems, these are usually mounted in close proximity to the individual heating circuits that they operate. (PE/EDPG) 622A-1984r
(2) (electric heat tracing systems) The locations where control, alarm, or both signal or function take place. With respect to electric heat tracing systems, these are usually mounted in close proximity to the individual heating circuits that they operate. (PE/EDPG) 622B-1988r

local data Data that can be accessed by only one module or set of nested modules in a computer program. *Contrast:* global data. (C) 610.12-1990

local device The local device that may attempt to perform Auto-Negotiation with a link partner. The local device may be either a DTE or repeater. (C/LM) 802.3-1998

locale (1) The definition of the user environment that depends on language and cultural conventions. (C/PA) 14252-1996
(2) The definition of the subset of the environment of a user that depends on language and cultural conventions. (C/PA) 9945-2-1993

local environment The environment of the local MTA. (C/PA) 1224.1-1993w

local exchange A switching entity in the public telephone network to which subscribers are connected. Evolving practice in North America defines this entity as "end office." *See also:* end office; subscriber's line. (COM/TA) 823-1989w

local exchange access line *See:* subscriber's line; subscriber loop.

local interconnect A special interconnect among a few boards. This interconnect might be used to create a multi-board functional element. (C/BA) 14536-1995

local interprocess communication (local IPC]) The transfer of data between processes in the same system. (C) 1003.5-1999

localization The process of utilizing internationalization features to adapt an internationalized product to a specific cultural environment. *See also:* internationalization. (C/PA) 14252-1996

localized general lighting (illuminating engineering) Lighting utilizing luminaires above the visual task and also contributing to the illuminance of the surround. (EEC/IE) [126]

local lighting (illuminating engineering) Lighting providing illuminance over a relatively small area or confined space without providing any significant general surrounding lighting. (EEC/IE) [126]

local line *See:* local loop.

local loop (1) (data transmission) That part of a communication circuit between the subscriber's equipment and the line terminating equipment in the exchange (either 2-wire or 4-wire). (PE) 599-1985w
(2) A line connecting a subscriber's instrument or private branch exchange directly to the local end office. *Synonyms:* subscriber's loop; local line. (C) 610.7-1995
(3) The communication path between the telephone company's switching office and the end user. (AMR/SCC31) 1390-1995, 1390.2-1999, 1390.3-1999

locally-attached terminal *See:* channel-attached terminal.

Locally Bridged Local Area Network A Bridged Local Area Network interconnected by Local Bridges only; a single Local Area Network is also considered as (a degenerate case of) a Locally Bridged Local Area Network. (C/LM) 802.1G-1996

locally homogeneous medium A medium in which the changes in electrical properties are small when measured over a distance of a wavelength. (AP/PROP) 211-1997

Local MAC Bridge, Local Bridge A MAC Bridge that interconnects only Local Area Networks to which it attaches directly as specified in the relevant MAC Standards. (C/LM) 802.1G-1996

local MD The MD of which the local MTA is a part. (C/PA) 1224.1-1993w

local MTA The MTA that comprises the client and the service in the context of the MT interface. (C/PA) 1224.1-1993w

local network *See:* local area network.

local oscillator (data transmission) (beating oscillator) An oscillator in a superheterodyne circuit whose output is mixed with the received signal to produce a sum or difference frequency equal to the intermediate frequency of the receiver. (PE) 599-1985w

local port (local area networks) The repeater port that allows connection to a lower- level entity (e.g., end node, repeater, bridge, LAN analyzer). (C) 8802-12-1998

local receiving system (of a complete telephone connection) That part of a complete connection that consists of a telephone set used for receiving, a feeding bridge, and the line (physical cable) connecting the telephone set to the feeding bridge. (COM/TA) 823-1989w

local sending system (of a complete telephone connection) That part of a complete connection that consists of a telephone set used for sending, a feeding bridge, and the line (physical cable) connecting the telephone set to the feeding bridge. (COM/TA) 823-1989w

local source An emitter located close enough to a shielding enclosure for its electromagnetic energy to illuminate only a localized portion of a shielding face. The effect is assessed by choosing the poorest performance in the set of measured locations. (EMC/STCOORD) 299-1997

local systems environment An environment in which information processing systems or resources are not conforming to the services and protocols of OSI. *Contrast:* OSI environment. (C) 610.7-1995

local terminal A terminal that is directly connected and relatively close to the computer with which it communicates. *Contrast:* remote terminal. (C) 610.10-1994w

local variable A variable that can be accessed by only one module or set of nested modules in a computer program. *Contrast:* global variable. (C) 610.12-1990

Local visibility Operations or objects are locally visible if they can not be accessed or invoked except from within the process in which the object or operation executes. In IEEE 1451.1, objects not specifically designated as Network Visible are only Locally visible. (IM/ST) 1451.1-1999

locatable fileset A fileset for which permission is granted to install the files in a different location as specified by the user, and denoted by having the value of its *is_locatable* attribute set to `true`. (C/PA) 1387.2-1995

locatable software Software that contains locatable filesets. (C/PA) 1387.2-1995

location Any place in which data may be stored. *See also:* storage location; protected location. (C) 610.10-1994w

location cursor A visual cue that indicates where the interaction of the user with the keyboard will occur. (C) 1295-1993w

location monitor (1) A functional module that monitors data transfers over the data transfer bus (DTB) to detect accesses to the locations it has been assigned to watch. When an access to one of these assigned locations occurs, the location monitor generates an on-board signal. (C/BA) 1014-1987
(2) A function that monitors data transfers over the VMEbus in order to detect accesses to the locations it has been assigned to watch. When an access occurs to one of these assigned locations, the location monitor generates a local signal. (C/MM) 1155-1992

locator (1) A logical input device used to specify position information on a display surface. Typical physical devices include data tablets, joysticks, and thumbwheels. (C) 610.6-1991w
(2) An input device that can provide the coordinance of a cursor within some context or coordinate system. For example, a mouse, or a data tablet. *See also:* puck; joystick; track ball. (C) 610.10-1994w

lock (A) To exclude users from updating data that is being updated by another user. *Note:* Depending on the implementation, locking may occur on a field, record or an entire file. *See also:* exclusive lock; deadlock. **(B)** To exclude users from accessing data. *Synonyms:* file-locking; field-locking. (C) 610.5-1990

locked A condition of the bus which guarantees exclusive access to the parallel system bus and to resources on the replying agent(s). This inhibits transfer operations between the replying agent and any other bus interface. (C/MM) 1296-1987s

locked-rotor current (packaging machinery) The steady-state current taken from the line with the rotor locked and with rated voltage (and rated frequency in the case of alternating-current motors) applied to the motor. (IA/PKG) 333-1980w

locked-rotor torque (1) The torque output capability of the VAM at zero speed and rated voltage and frequency. (PE/NP) 1290-1996
(2) The minimum torque of a motor developed for all angular positions of the rotor, when at rest, and with rated voltage and frequency applied. (IA/MT) 45-1998

lock-in (laser gyro) The phenomenon exhibited by laser gyros that is characterized by frequency locking of the clockwise and counterclockwise beams at low input rates. It is caused by the coupling of energy between the laser beams and results in a threshold error near zero input angular rate unless a corrective means, such as biasing, is used. (AES/GYAC) 528-1994

locking (1) (data management) In code extension characters, having the characteristic that a change in interpretation applies to all coded representations following, or to all coded representations of a given class, until the next appropriate code extension character occurs. *Contrast:* nonlocking. (C) 610.5-1990w
(2) A facility whereby a module is requested to guarantee exclusive access to addressed data, blocking other modules from accessing that data. This allows indivisible operations to be performed on addressed resources. (C/BA) 10857-1994, 896.4-1993w, 896.3-1993w, 1014.1-1994w

locking carabiner Has a self-closing, self-locking gate that remains closed and locked until intentionally unlocked and opened for connection or disconnection. (PE/T&D) 1307-1996

locking-in (data transmission) The shifting and automatic holding of one or both of the frequencies of two oscillating systems which are coupled together, so that the two frequencies have the ratio of two integral numbers. (PE) 599-1985w

locking ring (rotating machinery) A ring used to prevent motion of a second part. (PE) [9]

locking shift character A shift-out character that causes all characters that follow to be interpreted as members of a different character set from the original one until the shift-in character of the original character set is encountered. *Contrast:* nonlocking shift character. (C) 610.5-1990w

lock-in rate (laser gyro) One half of the absolute value of the algebraic difference between the two rates defining the region over which lock-in occurs. (AES/GYAC) 528-1994

lockout (1) (telephone circuit controlled by two voice-operated devices) The inability of one or both subscribers to get through, either because of excessive local circuit noise or continuous speech from either or both subscribers. (IM/PE/EEC/HFIM) [40], [119]
(2) (software) A computer resource allocation technique in which shared resources (especially data) are protected by permitting access by only one device or process at a time. (C) 610.12-1990
(3) The final operation of a recloser or circuit breaker in an attempt to clear a persistent fault. The overcurrent protective device locks open their contacts under these conditions. (PE/T&D) 1366-1998

lockout-free (as applied to a recloser or sectionalizer) A general term denoting that the lockout mechanism can operate even though the manual operating level is held in the closed position. *Note:* When used as an adjective modifying a device, the device has this operating capability. (SWG/PE) C37.100-1992

lockout mechanism (of an automatic circuit recloser) A device that locks the contacts in the open position following the completion of a predetermined sequence of operations. (SWG/PE) C37.100-1992

lockout operation (of a recloser) An opening operation followed by the number of closing and opening operations that the mechanism will permit before locking the contacts in the open position. (SWG/PE) C37.100-1992

lockout protection device (series capacitor) A device to block the opening of the bypass device and insertion of the switching step following the closure of the bypass device from a cause that warrants inspection or maintenance. (T&D/PE) 824-1985s

lockout protection function A function that blocks the opening of the bypass switch and prevents insertion of the switching steps, following the closure of the bypass switch from a cause that warrants inspection or maintenance. (T&D/PE) 824-1994

lockout relay (1) An electrically reset or hand-reset auxiliary relay whose function is to hold associated devices inoperative until it is reset. (SWG/PE/EDPG) 1020-1988r, C37.100-1992
(2) (power system device function numbers) A hand or electrically reset auxiliary relay that is operated upon the occurrence of abnormal conditions to maintain associated equipment or devices inoperative until it is reset. (SUB/PE) C37.2-1979s

lockplate (rotating machinery) (locking plate) A plate used to prevent motion of a second part (for example, to prevent a bolt or nut from turning). (PE) [9]

lock-request packet The packet transmitted during the request subaction portion of a lock transaction. (C/MM) 1394-1995

lock-response packet The packet transmitted during the response subaction portion of a lock transaction. (C/MM) 1394-1995

lock transaction A transaction that passes an address, subcommand, and data parameter(s) from the requester to the responder and returns a data value from the responder to the requester. The subcommand specifies which indivisible update is performed at the responder; the returned data value is the previous value of the updated data. (C/MM) 1212-1991s, 1394-1995

lock-up relay A relay that locks in the energized position by means of permanent magnetic bias (requiring a reverse pulse for releasing) or by means of a set of auxiliary contacts that keep its coil energized until the circuit is interrupted. *Note:* Differs from a latching relay in that locking is accomplished magnetically or electrically rather than mechanically. (2) Sometimes used for latching relay. *See also:* relay. (EEC/REE) [87]

LOF *See:* lowest observed frequency.

log (1) (supervisory control) (data acquisition) (automatic control) A printed record of data. (SWG/PE/SUB) C37.100-1992, C37.1-1994
(2) (data management) *See also:* journal. (C) 610.2-1987, 610.5-1990w

logarithmic decrement (1) (underdamped harmonic system) The natural logarithm of the ratio of a maximum of the free oscillation to the next following maximum. The logarithmic decrement of an under-damped harmonic system is

$$\ln\left(\frac{X_1}{X_2}\right) = \frac{2\pi F}{(4MS - F^2)^{1/2}}$$

where X_1 and X_2 are the two maxima. (Std100) 270-1966w

(2) (automatic control) A measure of damping of a second-order linear system, expressed as the Napierian logarithm (with negative sign) of the ratio of the greater to the lesser of a pair of consecutive excursions of the variable (in opposite directions) about an ultimate steady-state value. *Note:* For the system characterized by a quadratic factor

$1/(s^2 + z\omega_n s + \omega_n^2)$

its value is

$-\pi z/(1 - z^2)^{1/2}$

Twice this value defines the envelopes of the damping, but C85 prefers the above definition for reasons of convenience noted under damping factor. *See also:* subsidence ratio; damping. (PE/EDPG) [3]

logarithmic receiver A receiver whose output is proportional to the log of the input. *Note:* A logarithmic receiver is often used to prevent receiver saturation by large signals. Sometimes called log receiver or logarithmic detector. *See also:* lin-log receiver; log fast-time-constant receiver. (AES) 686-1997

logarithmic search *See:* binary search.

log fast-time-constant receiver A receiver whose video amplifier has a logarithmic input-output characteristic and which is followed by a high-pass filter, or FTC. Its purpose is to suppress distributed clutter or to help achieve a constant-false-alarm-rate (CFAR) for clutter whose statistics can be described by a Rayleigh distribution. (AES) 686-1997

logger (1) A functional unit that records events and physical conditions, usually with respect to time, along with the time of the occurrence. (C) 610.10-1994w
(2) A device that enables a user to login to a computer system and to logout. (C) 610.10-1994w

logging function check Accomplished when results of the control function check are logged. A check of master and remote station equipment by exercising a predefined component or capability. (SUB/PE) C37.1-1994

logic (1) (A) The result of planning a data-processing system or of synthesizing a network of logic elements to perform a specified function. **(B)** Pertaining to the type or physical realization of logic elements used, for example, diode logic, and logic. *See also:* logic design; formal logic; symbolic logic. (C/MIL) 162-1963, [2]
(2) (control or relay logic) Predetermined sequence of operation of relays and other control devices. (PE/EDPG) 1020-1988r

logic add *See:* OR.

logical (A) Pertaining to a view or description of data that does not depend on the characteristics of the computer system or the physical storage. **(B)** Pertaining to the form of data organization, hardware or system that is processed by an application program; it may be different from the real (physical) form. *Contrast:* physical. (C) 610.5-1990

logical add *See:* OR; disjunction.

logical address (1) (FASTBUS acquisition and control) A primary address of 32 bits consisting of the device address and internal address. It is independent of the location of the device on a segment. (NID) 960-1993
(2) An 8-bit number that uniquely identifies each VXIbus device in a system. It defines a device's A16 register addresses, and indicates commander/servant relationships. (C/MM) 1155-1992

logical channel One connection over a single physical communication link. There may be multiple logical channels over a single physical communication link. (C/MM) 1284.4-2000

logical child segment (A) In a hierarchical database, a child segment in a logical database. **(B)** A pointer segment that establishes a child/parent relationship between a physical segment and a logical parent segment. *See also:* physical child segment. (C) 610.5-1990

logical cohesion A type of cohesion in which the tasks performed by a software module perform logically similar functions; for example, processing of different types of input data. *Contrast:* procedural cohesion; functional cohesion; communicational cohesion; temporal cohesion; sequential cohesion; coincidental cohesion. (C) 610.12-1990

logical comparison The examination of two binary variables to determine whether they have the same value. *Synonyms:* logic decision; logical decision; logic comparison. (C) 1084-1986w

logical connective *See:* logical operator.

logical data Data used to represent the result of some logical operation. (C) 610.5-1990w

logical database (A) A database as it is perceived by its users. *Synonyms:* logical view; application view. **(B)** A database containing a collection of related segments that may reside in one or more physical databases. *Note:* A logical database is sometimes referred to as a logical view or application view of a physical database. *See also:* view integration; logical segment. **(C)** A database containing a subset of the segments in a physical database. *Note:* The root segment in the logical database must be the root segment in the physical database. *Contrast:* physical database. (C) 610.5-1990

logical data model A data model that represents the meaning of the data contained in a data structure. *Contrast:* physical data model. (C) 610.5-1990w

logical decision *See:* logical comparison.

logical device In networking, an abstract specification of the operation of a physical device. (C) 610.7-1995

logical diagram *See:* logic diagram.

logical difference A set consisting of all elements belonging to set A but not to set B, when two sets of elements, A and B, are given. *Synonym:* logic difference. (C) 1084-1986w

logical error An error in the binary content (payload) of a digital signal; e.g., bit error. (COM/TA) 1007-1991r

logical expression (1) A combination of symbols and variables representing a logical relationship. (C) 1084-1986w
(2) An expression of boolean terms that may be combined using logical *and* (&&) and *or* (||) operators that evaluate to either **TRUE** or **FALSE**. The resulting value is used to determine which branch of an **'If . . . Else'** condition to take. (C/PA) 2003-1997

logical file (data management) A file independent of its physical environment. Portions of the same logical file may be located in different physical files, and several logical files or parts of logical files may be located in one physical file. (C/SE) 610.5-1990w, 729-1983s

logical format *See:* high-level format.

logical input device An abstract input device, defined by its function rather than by a specific implementation in hardware. For example, choice device, locator, pick device, valuator. (C) 610.6-1991w

logical input value A value provided by a logical input device. (C) 610.6-1991w

logical link control (LLC) That part of a data station that supports the logical link control functions of one or more logical links. The LLC generates command PDUs and response PDUs for sending and interprets received command PDUs and response PDUs. Specific responsibilities assigned to an LLC include

1) Initiation of control signal interchange,
2) Organization of data flow,
3) Interpretation of received command PDUs and generation of appropriate response PDUs, and
4) Actions regarding error control and error recovery functions in the LLC sublayer.

(C/LM/CC) 8802-2-1998

logical link control encoding The use of LLC addressing information in a frame as a protocol identifier associated with the MAC Service user data carried in the frame. (C/LM) 802.1Q-1998

logical link control frame A token ring frame containing an LLC PDU exchanged between peer entities using the MAC services. (C/LM) 8802-5-1998

logical link control procedure In a local area network (LAN) or a metropolitan area network (MAN), the part of the protocol that governs the assembling of data link layer frames and their exchange between data stations independently of how the transmission medium is shared. (LM/C) 8802-6-1994

logical link control sublayer (LLC) **(1)** In a local area network (LAN) or metropolitan area network (MAN), that part of the data link layer that supports medium-independent data link functions, and uses the medium access control (MAC) sublayer service to provide services to the network layer. (LM/C) 8802-6-1994
(2) (local area networks) That part of the data link layer that supports medium independent data link functions, and uses the services of the MAC to provide services to the network layer. (LM/C) 8802-5-1998, 8802-12-1998
(3) The upper sublayer of the data link layer of the seven-layer OSI model; provides media-independent functions and the logical connection between the stations within the local area network. *See also:* entity layer; medium access control sublayer; session layer; presentation layer; sublayer; transport layer; network layer; physical layer; data link layer; application layer; client layer. (C) 610.7-1995

logically connected The state following a successful negotiation between a DCC and a BCC for network connection. The trigger for this state is a set normal response mode (SNRM) being sent from a BCC to a DCC. (EMB/MIB) 1073.3.1-1994

logically deleted Pertaining to a record that no longer appears available to the user but is physically present in the file. *See also:* purged; active; inactive. (C) 610.2-1987

logical multiply *See:* AND.

logical operation **(1)** **(A)** An operation involving logical variables and operators. **(B)** Loosely, any nonarithmetic computer operation. *Synonym:* logic operation. (C) 1084-1986
(2) An operation for which the VHDL operator is **and**, **or**, **nand**, **nor**, **xor**, **xnor**, or **not**. (C/DA) 1076.3-1997
(3) An operation for which the VHDL operator is **and**, **or**, **nand**, **nor**, **xor**, or **not**. (C/DA) 1076.6-1999

logical operations with pulses (pulse terminology) This section considers the pulse as a logical operator. Some operations defined in operations on a pulse, operations by a pulse, and operations involving the interaction of pulses, frequently are logical operations in the sense of this section.

a) *General.* AND, NAND, OR, NOR, EXCLUSIVE OR, INVERSION, inhibiting, enabling, disabling, counting, or other logical operations may be performed.
b) *Slivering.* A process in which a (typically, unwanted) pulse of relatively short duration is produced by a logical operation. Typically, slivering is a result of partial pulse coincidence.
c) *Gating.* A process in which a first pulse enables or diables portions of a second pulse or other event for the duration of the first pulse.
d) *Shifting.* A process in which logical states in a specified sequence are transferred without alteration of the sequence from one storage element to another by the action of a pulse.

(IM/WM&A) 194-1977w

logical operator A symbol that represents a logical operation to be performed on the associated operands. *Synonyms:* logical connective; logic operator. (C) 1084-1986w

logical parent segment **(A)** In a hierarchical database, a parent segment in a logical database. *See also:* physical parent segment. **(B)** A segment that is pointed to by a logical child segment, establishing a parent/child relationship between the logical parent segment and some physical segment. *Note:* A logical parent segment may also be a physical parent segment. (C) 610.5-1990

logical product The result obtained from the AND operation. *Synonym:* logic product. (C) 1084-1986w

logical record (software) A record independent of its physical environment. *Note:* Portions of the same logical record may be located in different physical records, and several logical records or parts of logical records may be located in one physical record. (C) 610.5-1990w

logical schema A schema that defines a data model. (C) 610.5-1990w

logical segment A segment in a logical database. *See also:* logical twin segment; logical parent segment; logical child segment. (C) 610.5-1990w

logical shift (mathematics of computing) A shift that affects all positions in a register, word, or numeral, including the sign position. For example, +231.702 shifted two places to the left becomes 23170.200. *Note:* A logical shift may be applied to the multiple-precision representation of a number. *Synonyms:* logic shift; nonarithmetic shift. *Contrast:* arithmetic shift. (C) 1084-1986w

logical source statements (LSS) Source statements that measure software instructions independently of the physical format in which they appear. (C/SE) 1045-1992

logical structure *See:* data structure.

logical sum **(1)** The result obtained from the OR operation. *Synonym:* logic sum. (C) 1084-1986w
(2) The answer arrived at when adding two operands using the logical OR operation. For example: 0110 + 0101 = 0111 (all numbers binary). *Contrast:* algebraic sum. *See also:* disjunction. (C) 610.10-1994w

logical symbol *See:* logic symbol.

logical terminal A terminal addressable by its logical function rather than its physical address. (C) 610.10-1994w

logical trace An execution trace that records only branch or jump instructions. *See also:* subroutine trace; variable trace; execution trace; retrospective trace; symbolic trace. (C) 610.12-1990

logical truth value (subroutines for CAMAC) (l) The symbol l represents a logical truth value which can be either true or false. (NPS) 758-1979r

logical twin segment A twin segment in a logical database. *Contrast:* physical twin segment. (C) 610.5-1990w

logical type A data type whose members can assume only logical values (usually TRUE and FALSE) and can be operated on only by logical operators, such as AND, OR, and NOT. *Contrast:* character type; enumeration type; real type; integer type. (C) 610.12-1990

logical unit (LU) An addressable, functional group. In the case of a printer, scanner, or facsimile it is a functional group concerned with the storage, acquisition and/or processing of a textual and/or pictorial image. A printer may have one or more interpreters. The design of a particular printer determines if these interpreters are capable of concurrent operation. (C/MM) 1284.1-1997

logical variable *See:* switching variable.

logical view *See:* logical database; external schema.

logic board (power-system communication) An assembly of decision-making circuits on a printed-circuit mounting board. *See also:* digital. (PE) 599-1985w

logic circuit A circuit that is designed to perform one or more logic operations or to represent logic functions. *See also:* sequential circuit; combinational circuit; voter; asynchronous circuit; hardwired logic; power-fail circuit; hazard-free logic. (C) 610.10-1994w

logic comparison *See:* logical comparison.

logic decision *See:* logical comparison.

logic design **(A)** **(electronic computation)** The planning of a computer or data-processing system prior to its detailed engineering design. **(B)** **(electronic computation)** The synthesizing of a network of logic elements to perform a specified function. **(C)** **(electronic computation)** The result of (A) and

(B) above, frequently called the logic of the system, machine, or network. (C) 162-1963

logic diagram (1) (digital computers) A diagram representing the logic elements and their interconnections without necessarily expressing construction or engineering details. (C/MIL) 162-1963w, [2]
(2) A diagram that depicts the two-state device implementation of logic functions with logic symbols and supplementary notations, showing details of signal flow and control, but not necessarily the point-to-point wiring. (GSD) 91-1973s
(3) (mathematics of computing) A graphical representation of a system's logic elements and their interconnections. *Synonym:* logical diagram. (C) 1084-1986w

logic difference *See:* logical difference.

logic element A combinational logic element or sequential logic element. (C) 162-1963w

logic errors *See:* logical error.

logic function (1) A definition of the relationships that hold among a set of input and output logic variables. (GSD) 91-1984r
(2) (mathematics of computing) *See also:* switching function. (C) 1084-1986w

logic gate *See:* gate.

logic instruction (computers) An instruction that executes an operation that is defined in symbolic logic, such as AND, OR, NOR. (C) [85]
(2) (A) An instruction in which the operation field specifies a logic operation; for example, a conditional branch instruction. *Contrast:* arithmetic instruction. **(B)** An instruction that specifies an operation defined in symbolic logic, such as NOT, OR, AND. (C) 610.10-1994

logic level (1) Any level within one or two overlapping ranges of values of a physical quantity used to represent the logic states. *Note:* A logic variable may be equated to any physical quantity for which two distinct ranges of values can be defined. In IEEE Std 91-1984, these distinct ranges of values are referred to as logic levels and are denoted H and L. H is used to denote the logic level with the more positive algebraic value, and L is used to denote the logic level with the less positive algebraic value. In the case of systems in which logic states are equated with other physical properties (for example, positive or negative pulses, presence or absence of a pulse), H and L may be used to represent these properties or may be replaced by more suitable designations. (GSD) 91-1984r
(2) Any level within one of two non-overlapping ranges of values of voltage used to represent the logic states. *See also:* high; low. (C/BA) 1496-1993w

logic map A worksheet used by logic designers in the process of logic development, simplification, or optimization. *See also:* Karnaugh map; Mahoney map. (C) 1084-1986w

logic multiply *See:* AND.

logic 1 The highest voltage value of the two logic levels on an active-high signal and the lowest voltage value of the two logic levels on an active-low signal. (TT/C) 1149.5-1995

logic operation (1) (A) (electronic computation) (general) Any nonarithmetical operation. *Note:* Examples are: extract, logical (bit-wise) multiplication, jump, data transfer, shift, compare, etc. **(B) (sometimes)** Only those nonarithmetical operations that are expressible bit-wise in terms of the propositional calculus or two-valued Boolean algebra. (C) 162-1963
(2) An operation that follows the rules of symbolic logic. (C) 610.10-1994w
(3) (mathematics of computing) *See also:* logical operation. (C) 1084-1986w

logic operator *See:* AND; NAND; exclusive OR; logical operator; NOT.

logic product *See:* logical product.

logic programming language A programming language used to express programs in terms of control constructs and a restricted predicate calculus; for example, PARLOG; PROLOG or STRAND. *See also:* algebraic language; LISt Processing. (C) 610.13-1993w

logic-seeking printer A printer that is able to detect and skip over blank spaces, resulting in faster printing. (C) 610.10-1994w

logic shift (1) (computers) A shift that affects all positions. (C) [85]
(2) (mathematics of computing) *See also:* logical shift. (C) 1084-1986w

logic state (1) (mathematics of computing) One of the two possible values a binary variable may assume. (C) 1084-1986w
(2) One of two possible abstract states that may be taken on by a binary logic variable. (GSD/C/BA) 91-1984r, 1496-1993w
(3) The representation a simulator uses to describe the state of a circuit during digital logic simulation. There are four types of logic states that exist in a typical simulator: 0, 1, Z, and X. (SCC20) 1445-1998

logic sum *See:* logical sum.

logic symbol (A) (electronic computation) A symbol used to represent a logic element graphically. **(B) (electronic computation)** A symbol used to represent a logic connective. (C) 162-1963, [85]
(2) (A) (mathematics of computing) A symbol used to denote a logical operator. **(B) (mathematics of computing)** A symbol used to graphically represent a logic element. *Synonym:* logical symbol. (C) 1084-1986

logic unit A part of a computer that performs logic operations and related operations. *See also:* arithmetic and logic unit. (C) 610.10-1994w

logic variable* *See:* switching variable.
* Deprecated.

logic 0 The lowest voltage value of the two logic levels on an active-high signal and the highest voltage value of the two logic levels on an active-low signal. (TT/C) 1149.5-1995

logic, 0–1 The representation of information by two states termed 0 and 1. *See also:* bit; dot cycle; digital. (PE) 599-1985w

logic 0, logic 1 The two logic voltage levels for digital signals. In positive logic systems, the more positive of the two logic voltage levels is taken to be logic 1. *Notes:* 1. The voltage levels representing logic 0 and logic 1 are not necessarily the same at every function pin. 2. Logic 0 and logic 1 are often represented as LO and HI, respectively. (C/TT) 1149.4-1999

login (1) The process of establishing communication with and verifying the authority to use a network or computer. *Synonyms:* sign-on; logon. *Contrast:* logoff. *See also:* remote login. (C) 610.7-1995, 610.10-1994w
(2) The unspecified activity by which a user gains access to the system. Each login shall be associated with exactly one login name. (C/PA) 9945-1-1996, 9945-2-1993
(3) The activity by which a user gains access to the system. Each login shall be associated with exactly one login name. (C) 1003.5-1999

login name A user name that is associated with a login. (C/PA) 9945-1-1996, 9945-2-1993, 1003.5-1999

logistics time (LT) The downtime occasioned by the unavailability of spares, replacement parts, test equipment, maintenance facilities, or personnel. (PE/NP) 933-1999

log-normal distribution A probability distribution characterized by the probability density function:

$$f(x) = \frac{1}{x\sigma\sqrt{2\pi}} \exp\left[-\frac{(\ln x - \ln x_m)^2}{2\sigma^2}\right], x \geq 0$$
$$= 0, \qquad x < 0$$

where

x = the random variable
σ = the standard deviation of $\ln x$
x_m = the median value of x

Note: This function is often used for statistical modeling of the radar cross section of certain types of radar targets and clutter. (AES) 686-1997

LOGO A high-order list processing language designed for interactive educational applications; characterized by its simple vocabulary and built-in graphics capability, known as "turtle graphics." *See also:* extensible language. (C) 610.13-1993w

logoff The process of terminating communication with a computer. *Synonyms:* sign-off; logout. *Contrast:* login. (C) 610.7-1995, 610.10-1994w

logon *See:* login.

logout *See:* logoff.

log periodic antenna Any one of a class of antennas having a structural geometry such that its impedance and radiation characteristics repeat periodically as the logarithm of frequency. (PE/T&D/AP/ANT) 539-1990, 145-1993

LOI *See:* limiting oxygen index.

long dimension (numerically controlled machines) Incremental dimensions whose number of digits is one more to the left of the decimal point than for a normal dimension, and the last digit shall be zero, that is, XX.XXX0 for the example under normal dimension. (IA) [61]

long-distance navigation Navigation utilizing self-contained or external reference aids or methods usable at comparatively great distances. *Note:* Examples of long-distance aids are loran, Doppler, inertial, and celestial navigation. *See also:* approach navigation; navigation; short-distance navigation. (AES/RS) 686-1982s, [42]

long-distance trunk (data transmission) That type of trunk which permits trunk-to-trunk connection and which interconnects local, secondary, primary, and zone centers. (PE) 599-1985w

long haul network (LHN) **(1)** A computer network most frequently used to transfer data over distances from several thousand feet to several thousand miles. *Contrast:* local area network. *See also:* wide area network; metropolitan area network. (C) 610.7-1995

(2) A communications network designed for large geographical areas. *Synonym:* wide area network. (DIS/C) 1278.3-1996

longitudinal attenuation (overhead-power-line corona and radio noise) The decrease in electromagnetic noise field strength caused by dissipation of energy as a result of propagation along an overhead power line and through the earth. *Note:*

1) In North American practice, units are decibels per mile (dB/mile) or decibels per km (dB/km).
2) For multiconductor systems, such as those normally found in electric power systems, it is convenient to describe wave propagation as made up of a set of noninteracting modes, each with its own attenuation constant.
3) In the context of this standard, the electromagnetic noise energy is the result of corona and gap discharges.

See also: propagation mode. (T&D/PE) 539-1990

longitudinal balance **(1) (analog voice frequency circuits)** The electrical symmetry of the two wires comprising a pair with respect to ground. *See also:* longitudinal circuit. (COM/TA) 743-1995

(2) (data transmission) A measure of the similarity of impedance to ground (or common) for the two or more conductors of a balanced circuit. This term is used to express the degree of susceptibility to common mode interference. (PE) 599-1985w

(3) (communication and control cables) The ratio of the disturbing longitudinal rms voltage (V_s) to ground and the resulting metallic rms voltage (V_m) of the network under test, expressed in decibels as follows: longitudinal balance $= 20 \log_{10} V_s/V_m$ (in dB), where V_s and V_m are at the same frequency. (PE/PSC) 789-1988w

(4) (telecommunications) The ratio of the disturbing longitudinal voltage V_s and the resulting metallic voltage V_m of the network under test, expressed in decibels. (COM/TA) 1007-1991r

longitudinal balance, degree of *See:* degree of longitudinal balance.

longitudinal circuit (measuring longitudinal balance of telephone equipment operating in the voice band) A circuit formed by one communication conductor (or by two or more communication conductors in parallel) with a return through ground or through any other conductors except those which are taken with the original conductor or conductors to form a metallic circuit. (COM/TA) 455-1985w

longitudinal circuit port (measuring longitudinal balance of telephone equipment operating in the voice band) A place of access in the longitudinal transmission path of a device or network where energy may be supplied or withdrawn, or where the device or network variables may be measured. (COM/TA) 455-1985w

longitudinal electric field* *See:* longitudinal electromotive force.

* Deprecated.

longitudinal electromotive force* Voltage per unit length of a circuit, induced by the magnetic field, when the circuit is in the vicinity of a power line. *Synonym:* longitudinal electric field. (T&D/PE) 539-1990

* Deprecated.

longitudinal impedance (measuring longitudinal balance of telephone equipment operating in the voice band) Impedance presented by a longitudinal circuit at any given single frequency. (COM/TA) 455-1985w

longitudinal insulation configuration An insulation configuration between terminals belonging to the same phase, but which are temporarily separated into two independently energized parts (e.g., open switching device). (PE/C) 1313.1-1996

longitudinal interference *See:* signal; common-mode interference.

longitudinal magnetic recording A type of magnetic recording in which magnetic polarities representing data are aligned along the length of the recording track. *Contrast:* perpendicular magnetic recording. (C) 610.10-1994w

longitudinal magnetization Magnetization of the recording medium in a direction essentially parallel to the line of travel. (SP/MR) [32]

longitudinal mode (laser maser) Refers to modes that have the same field distributions transverse to the beam, but a different number of half period field variations along the axis of the beam. *See also:* longitudinal resonances. (LEO) 586-1980w

longitudinal (common) mode voltage (low voltage surge protective devices [gas-tube surge-protective devices]) The voltage common to all conductors of a group as measured between that group at a given location and an arbitrary reference (usually earth). (SPD/PE) C62.31-1987r, [8]

longitudinal noise (data transmission) In telephone practice, the 1/1000th part of the total longitudinal-circuit noise current at any given point in one or more telephone wires. (PE) 599-1985w

longitudinal offset loss *See:* gap loss.

longitudinal overvoltage An overvoltage that appears between the open contact of a switch. (PE/C) 1313.1-1996

longitudinal profile **(1) (overhead-power-line corona and radio noise)** The profile of a parameter, usually near ground level, measured at a constant lateral distance from the power line and plotted as a function of distance along the line. For example, a longitudinal profile of the vertical component of the electric field strength, of the radio noise field strength, etc. (T&D/PE) 539-1990

(2) (radio noise from overhead power lines and substations) The radio noise field strength at ground level measured at constant horizontal distance from the power line and plotted as a function of distance along the line. (T&D/PE) 430-1986w

longitudinal redundancy check (LRC) (1) (data transmission) A system of error control based on the formation of a block check following preset rules. *Note:* The check formation rule is applied in the same manner to each character. (COM) [49]
(2) A parity check performed bit-wise on the rows of a string of characters represented in matrix form, with each bit for each character representing one column in the matrix. *Note:* The LRC is the comparison of the parity of the rows before and after an operation such as a magnetic tape read or transmission through a data communication channel. *See also:* vertical redundancy check. (C) 610.7-1995

longitudinal resonances (laser maser) (in a beam resonator) Resonances corresponding to modes having the same field distribution transverse to the beam, but differing in the number of half period field variations along the axis of the beam. *Note:* Such resonances are separated in frequency by approximately $v/2L$ where v is the speed of light in the resonator and $2L$ is the round trip length of the beam in the resonator. (LEO) 586-1980w

longitudinal (common mode) signal (telephone loop performance) The longitudinal voltage is half the algebraic sum of the voltages to ground in the two conductors (tip and ring). The longitudinal current is the algebraic sum of the current in these conductors. (COM/TA) 820-1984r

longitudinal voltage (1) (power fault effects) A voltage acting in series with the longitudinal circuit. (PE/PSC) 367-1996
(2) *See also:* common-mode voltage. (LM/C) 802.3i-1990s

longitudinal wave (1) A wave in which the direction of displacement at each point of the medium is the same as the direction of the propagation. (Std100) 270-1966w
(2) In a plasma, the type of wave whose restoring force is electrostatic. The associated electric field and particle velocity is in the direction of propagation with accompanying charge density fluctuations. (AP/PROP) 211-1997

long-lever relay armature An armature with an armature ratio greater than 1 : 1. (EEC/REE) [87]

long-line adapter (telephone switching systems) Equipment inserted between a line circuit and the associated station(s) to allow conductor loop resistances greater than the maximum for which a system is designed. (COM) 312-1977w

long-line current Current (positive electricity) flowing through the earth from an anodic to a cathodic area that returns along an underground metallic structure. *Note:* Usually used only where the areas are separated by considerable distance and where the current results from concentration cell action. *See also:* stray-current corrosion. (IA) [59]

long packet A packet with a length of over 1518 B. *Synonym:* over-sized packet. *Contrast:* short packet. (C) 610.7-1995, 610.10-1994w

long-pitch winding (rotating machinery) A winding in which the coil pitch is greater than the pole pitch. *See also:* direct-current commutating machine. (PE) [9]

long-term settling error The absolute difference between the final value specified for short-term settling time, and the value 1 s after the beginning of the step, expressed as a percentage of the step amplitude. (IM/WM&A) 1057-1994w

long-term stability (LTS) (power supplies) (ferroresonant voltage regulators) The change in output voltage or current as a function of time, at constant line voltage, load, and ambient temperature (sometimes referred to as "drift"). *See also:* overall regulation. (AES/PEL/ET) [41], 449-1990s

long-term timebase stability The change in time base frequency (usually given in parts per million) over a specified period of time at a specified sampling rate. (IM/WM&A) 1057-1994w

long-time-delay phase trip element A direct-acting trip device element that functions with a purposely delayed action (seconds). (SWG/PE) C37.100-1992

long-time test current (thyristor converter) The specified value of direct current that a converter unit or section shall be capable of carrying for a sustained period (minutes or hours) following continuous operation at a specified lower dc value under specific conditions. (IA/IPC) 444-1973w

long-time rating A rating based on an operating interval of five minutes or longer. (NESC/NEC) [86]

longwall machine A power-driven machine used for undercutting coal on relatively long faces. (PE/EEC/MIN) [119]

long-wire antenna A wire antenna that, by virtue of its considerable length in comparison with the operating wavelength, provides a directional radiation pattern. (AP/ANT) 145-1993

longword serial A form of word-serial communication that allows 32-bit data transfers between commanders and servants. (C/MM) 1155-1992

look A colloquial expression for a single attempt at detection of a target. (AES) 686-1997

look up To use a code-decode table or look-up table to obtain data values or other information. (C) 610.5-1990w

look-up table A table of values used in obtaining the value of a function using a table look-up procedure. *See also:* code-decode table. (C) 610.5-1990w, 1084-1986w

loom *See:* flexible nonmetallic tubing.

loop (1) (telephone loop performance) The transmission and signaling channel, with or without gain, between the center of the end office switch and the network interface. It also extends direct current (dc) power to the network interface. (COM/TA) 820-1984r
(2) (signal-transmission system and network analysis) A set of branches forming a closed current path, provided that the omission of any branch eliminates the closed path. *See also:* mesh; signal; ground loop. (CAS/IE) 155-1960w, [43]
(3) (A) (software) A sequence of computer program statements that is executed repeatedly until a given condition is met or while a given condition is true. *Synonym:* iterative construct. *See also:* UNTIL; WHILE; loop control; loop body. **(B) (software)** To execute a sequence of computer program statements as in definition (A). (C) 610.12-1990
(4) (data transmission) (telephone circuit) In communications, loop signifies a type of facility, normally the circuit between the subscriber and central office. (Usually a metallic circuit). (PE) 599-1985w

loop antenna (1) (data transmission) An antenna consisting of one or more complete turns of conductor, excited so as to provide an essentially uniform circulatory current, and a radiation pattern approximating that of an elementary magnetic dipole. (PE) 599-1985w
(2) (overhead-power-line corona and radio noise) An antenna consisting of one or more turns of a conductor. If the circulatory current is essentially uniform, the antenna will have a radiation pattern approximating that of an elementary magnetic dipole. *Note:* The loop antenna responds to the magnetic field component of the electromagnetic wave, in the direction of the loop axis. (T&D/PE) 539-1990
(3) An antenna whose configuration is that of a loop. *Note:* If the electric current in the loop, or in multiple parallel turns of the loop, is essentially uniform and the loop circumference is small compared with the wavelength, the radiated pattern approximates that of a Hertzian magnetic dipole. (AP/ANT) 145-1993

loop assertion A logical expression specifying one or more conditions that must be met each time a particular point in a program loop is executed. *Synonym:* loop invariant. *Contrast:* input assertion; output assertion. *See also:* inductive assertion method. (C) 610.12-1990

loopback An internal arrow that is the output of a box whose box number is greater than the box number of the box that uses that arrow as input, control, or mechanism. These uses are referred to as input loopback, control loopback, and mechanism loopback, respectively. (C/SE) 1320.1-1998

loopback test A test for faults over a transmission medium in which received data is returned to the sending point and compared with the data sent, thus completing a loop. (C) 610.7-1995

loopback testing Testing in which signals or data from a test device are input to a system or component, and results are returned to the test device for measurement or comparison. (C) 610.12-1990

loop body The part of a loop that accomplishes the loop's primary purpose. *Contrast:* loop control. (C) 610.12-1990

loop circuit (railway signaling) A circuit that includes a source of electric energy, a line wire that conducts current in one direction, and connections to the track rails at both ends of the line to complete the circuit through the two rails in parallel in the other direction. (EEC/PE) [119]

loop control (1) The effect of a control function or a device to maintain a specified loop of material between two machine sections by automatic speed adjustment of at least one of the driven sections. *See also:* feedback control system. (IA/ICTL/IAC/APP) [60], [75]
(2) (software) The part of a loop that determines whether to exit from the loop. *Contrast:* loop body. *See also:* leading decision; trailing decision. (C) 610.12-1990

loop-control variable A program variable used to determine whether to exit from a loop. (C) 610.12-1990

loop converter (data transmission) A device used for conversion of dc (direct-current) loop current pulses to relay contact closures and thereby provide circuit isolation. (PE) 599-1985w

loop current (power supplies) A direct current flowing in the feedback loop (voltage control) independent of the control current generated by the reference Zener diode source and reference resistor. *Note:* The loop (leakage) current remains when the reference current is made zero. It may be compensated for, or nulled, in special applications to achieve a very-high impedance (zero current) at the feedback (voltage control) terminals. *See also:* leakage current. (AES) [41]

loop current feed open (LCFO) A Bellcore-defined, switch generated, fixed open (no voltage) on the line within the range of 150–350 ms as sent by the switch. Its purpose is to signal certain digital loop carrier (DLC) devices to assign a time slot (transmission path) for a 15 s interval. (AMR/SCC31) 1390-1995, 1390.2-1999, 1390.3-1999

looped dual bus A DQDB subnetwork with the head of bus functions for both Bus A and Bus B collocated. (LM/C) 8802-6-1994

loop elements (control system feedback) (a closed loop) (a closed loop) All elements in the signal path that begins with the loop error signal and ends with the loop return signal. *See also:* feedback control system. (PE/EDPG) [3]

loop equations *See:* mesh equations.

loop factor *See:* path factor.

loop feeder (power distribution) A number of tie feeders in series, forming a closed loop. *Note:* There are two routes by which any point on a loop feeder can receive electric energy, so that the flow can be in either direction. *See also:* center of distribution. (T&D/PE) [10]

loop gain (data transmission) The sum of the gains which are given to a signal of a particular frequency in passing around a closed loop. The loop may be a repeater, carrier terminal, or a complete system. The loop gain may be less than the sum of the individual amplifier gain because singing may occur if full amplification is used. The maximum usable gain is determined by, and may not exceed, the losses in the closed path. (COM/PE) 599-1985w

loop graph (network analysis) A signal flow graph each of whose branches is contained in at least one loop. *Note:* Any loop graph embedded in a general graph can be found by removing the cascade branches. (CAS) 155-1960w

looping-in (interior wiring) A method of avoiding splices by carrying the conductor or cable to and from the outlet to be supplied. (EEC/PE) [119]

loop invariant *See:* loop assertion.

loop loss The difference in signal level for the inbound and outbound paths at a user outlet. (LM/C) 802.7-1989r

loop noise bandwidth (communication satellite) One of the fundamental parameters of a phase lock loop. It is the equivalent bandwidth of a square cut-off lowpass filter, which, when multiplied by a flat input noise spectral density, produces the loop noise variance. (COM) [25]

loop phase angle (automatic control) (closed loop) The value of the loop phase characteristic at a specified frequency. *See also:* phase characteristic; feedback control system. (PE/EDPG) [3]

loop pulsing (telephone switching systems) Dial pulsing using loop signaling. (COM) 312-1977w

LOOPS An object-oriented language designed as an expert system shell. (C) 610.13-1993w

loop sensitivity-test input (amplitude-modulation broadcast receivers) The least signal field of a specified carrier frequency, modulated 30 percent at 400 cycles, and applied as induced pickup in the loop of the receiver, which results in normal test output when all controls are adjusted for greatest sensitivity. It is expressed in decibels below 1 volt per meter, or in microvolts per meter. (CE) 186-1948w

loop service (power distribution) Two services of substantially the same capacity and characteristics supplied from adjacent sections of a loop feeder. *Note:* The two sections of the loop feeder are normally tied together on the consumer's bus through switching devices. *See also:* service. (T&D/PE) [10]

loop-service feeder A feeder that supplies a number of separate loads distributed along its length and that terminates at the same bus from which it originated. *Synonym:* ring feeder. (SWG/PE) C37.100-1992

loop-set transmittance (network analysis) The product of the negatives of the loop transmittances of the loops in a set. (CAS) 155-1960w

loop signaling (telephone switching systems) A method of signaling over direct current circuit paths that utilize the metallic loop formed by the trunk conductors and terminating bridges. (COM) 312-1977w

loop stability The stability of a control loop as measured against some criteria, e.g., phase margin and gain margin. (PEL) 1515-2000

loop stick antenna A loop receiving antenna with a ferrite rod core used for increasing its radiation efficiency. (AP/ANT) 145-1993

loop test A method of testing employed to locate a fault in the insulation of a conductor when the conductor can be arranged to form part of a closed circuit or loop. (EEC/PE) [119]

loop timing Transmit timing derived from the received signal. (COM/TA) 1007-1991r

loop timing transfer function The difference between the bit rates of the incoming and outgoing signals. The difference should be reported in parts per million. (COM/TA) 1007-1991r

loop topology *See:* ring topology.

loop-transfer function (control system feedback) (closed loop) The transfer function obtained by taking the ratio of the Laplace transform of the return signal to the Laplace transform of its corresponding error signal. *See also:* feedback control system. (IM/PE/EDPG) [120], [3]

loop transmittance (1) (network analysis) The product of the branch transmittances in a loop. (CAS) 155-1960w
(2) (branch) The loop transmittance of an interior node inserted in that branch. *Note:* A branch may always be replaced by an equivalent sequence of branches, thereby creating interior nodes. (CAS) 155-1960w
(3) (node) The graph transmittance from the source node to the sink node created by splitting the designated node. (CAS) 155-1960w

loose coupling Any degree of coupling less than the critical coupling. *See also:* coupling. (EEC/PE) [119]

loose leads (rotating machinery) A form of termination in which the machine terminals are loose cable leads. (PE) [9]

loosely coupled A condition that exists when simulation entities are not involved in very close interaction such that every action of an entity does not need to be immediately accounted for by the other entities. Two tanks moving over terrain five miles apart from each other is an example of a loosely coupled situation. (DIS/C) 1278.2-1995

LOP *See:* line of position.

loran A long-range radio navigational aid of the hyperbolic type whose position lines are determined by the measurement of the difference in the time of arrival of synchronized pulses. (These devices are rapidly being replaced by satellite-based global positioning systems.) (IA/MT) 45-1998

loran repetition rate *See:* pulse-repetition frequency.

Lorenz-Mie scattering *See:* Mie scattering.

Lorenz number The quotient of the electronic thermal conductivity by the product of the absolute temperature and the component of the electric conductivity due to electrons and holes. *See also:* thermoelectric device. (ED) [46], 221-1962w

lorhumb line (navigation system chart, such as a loran chart with its overlapping families of hyperbolic lines) A line drawn so that it represents a path along which the change in values of one of the families of lines retains a constant relation to the change in values of another of the families of lines. *See also:* navigation. (AES/RS) 686-1982s, [42]

loss (1) (power) *(power).* Power expended without accomplishing useful work. Such loss is usually expressed in watts. *(communications).* The ratio of the signal power that could be delivered to the load under specified reference conditions to the signal power delivered to the load under actual operating conditions. Such loss is usually expressed in decibels. *Note:* Loss is generally due to dissipation or reflection due to an impedance mismatch or both. *See also:* transmission loss. (CHM) [51]

(2) (waveguide) The power reduction in a transmission path in the mode or modes under consideration. It is usually expressed as a positive ratio, in decibels. (MTT) 146-1980w

(3) (broadband local area networks) (fiber optics) (network analysis) *See also:* insertion loss; microbend loss; macrobend loss; differential mode attenuation; backscattering; gap loss; nonlinear scattering; reflection; angular misalignment loss; transmission loss; related transmission terms; extrinsic joint loss; lateral offset loss; absorption; Rayleigh scattering; material scattering; attenuation; waveguide scattering; intrinsic joint loss. (LM/C) 802.7-1989r, 812-1984w

loss angle (1) (biological) The complement φ of the phase angle θ (between the electrode potential vector and the current vector). (EMB) [47]

(2) (A) (magnetic core testing) *Dissipation factor.* The angle by which the fundamental component of the magnetizing current lags the fundamental component of the exciting current in a coil with a ferromagnetic core. The tangent of this angle is defined as the ratio of the in-phase and quadrature components of the impedance of the coil.

$$\tan \delta_n = \frac{R_s}{\omega L_s} = \frac{\omega L_p}{R_p} = \frac{\mu''_s}{\mu'_s} = \frac{\mu'_p}{\mu''_p}$$

where δ_n = loss angle. **(B) (magnetic core testing)** *Relative dissipation factor.* Defined as:

$$\frac{\tan \delta_n}{\mu_i} = \frac{\mu''_s}{\mu'_s)^2} = \frac{R_s}{\mu_i \omega L_s} = \frac{\omega L_p}{\mu_i R_p} = \frac{\omega L_O}{R_p}$$

(C) (magnetic core testing) *Quality factor Q. See also:* quality factor.

(3) (network, structure, or material) For inductive devices it is defined as the inverse of the tangent of the loss angle.

$Q = 1/(\tan \delta_n)$

Q = quality factor

δ_n = loss angle.

(MAG) 393-1977s

loss, electric system *See:* electric system loss.

losses (1) (grounding device) (electric power) I^2R loss in the windings, core loss, dielectric loss (for capacitors), losses due to stray magnetic fluxes in the windings and other metallic parts of the device, and in cases involving parallel windings, losses due to circulating currents. *Notes:* 1. The losses as here defined do not include any losses produced by the grounding device in adjacent apparatus or materials not part of the device. Losses will normally be considered at the maximum rated neutral current but may in some cases be required at other current ratings, if more than one rating is specified, or at no load, as for grounding transformers. 2. The losses may be given at 25°C or at 75°C. *See also:* grounding device. (PE/SPD) 32-1972r

(2) (of a series reactor) Losses that are incident to the carrying of current. They include:

— The resistance and eddy-current loss in the winding due to load current
— Losses caused by circulating current in parallel windings
— Stray losses caused by magnetic flux in other metallic parts of the reactor, in the reactor support structure and in the reactor enclosure when the support structure and the enclosure are supplied as an integral part of the reactor installation.

Note: The losses produced by magnetic flux in adjacent apparatus or material not an integral part of the reactor or it's enclosure (if supplied) are not included. (PE/TR) C57.16-1996

loss factor (1) (electric power generation) The ratio of the average power loss to the peak-load power loss during a specified period of time. *See also:* generating station. (T&D/PE) [10]

(2) (dielectric heating) The product of its dielectric constant and the tangent of its dielectric loss angle. *See also:* depth of current penetration; electric constant. (IA) 54-1955w

loss function An instantaneous measure of the cost of being in state x and of using control u at time t. *See also:* performance index. (CS/IM) [120]

loss, insertion *See:* insertion loss.

lossless encoding Any image compression technique that represents gray levels compactly but permits exact reconstruction of the image; for example, contour encoding; run length encoding. (C) 610.4-1990w

loss of control Control of the switching system status, such as "active copy" or "out-of-service," is important for proper operation or maintenance of the switch. The loss of such a capability may be expressed as a time duration or in minutes per year. (COM/TA) 973-1990w

loss of control power protection (series capacitor) A means to initiate the closing of the bypass device upon the loss of normal control power. (T&D/PE) 824-1994

loss of diagnostic capability *See:* loss of visibility.

loss-of-excitation relay A relay that compares the alternating voltages and currents at the terminal of a synchronous machine and operates to produce an output if the relationship between these quantities indicates that the machine has substantially lost its field excitation. (SWG/PE) C37.100-1992

loss of forming (semiconductor rectifiers) A partial loss in the effectiveness of the rectifier junction. *See also:* rectification. (IA) 59-1962w, [12]

loss of frame alignment Frame alignment will be considered lost when a significant ratio of frame alignment signal bits are received in error. (COM/TA) 1007-1991r

loss of service The loss of electrical power, a complete loss of voltage, to one or more customers or meters. This does not include any of the power quality issues (sags, swells, impulses, or harmonics). (PE/T&D) 1366-1998

loss of signal A loss of signal shall be declared when n consecutive zeros are detected on an incoming signal. (COM/TA) 1007-1991r

loss of visibility The switching system should be designed such that the status indicators are visible to the craft. This capability is called visibility. The craft must also be able to initiate diagnostic tests. The loss of visibility or diagnostic capability may be expressed as a time duration or in minutes per year. *Synonym:* loss of diagnostic capability. (COM/TA) 973-1990w

loss of voltage condition A voltage reduction to a level that results in the immediate loss of equipment capability to perform an intended function. (SWG/PE/NP) 741-1997, C37.40-1981s

loss on ignition (fly ash resistivity) In a fly ash sample a measure of the completeness of the combustion process. It is calculated as the ratio of weight loss upon ignition at 750C of a previously dried (100C) sample to the weight of the dry sample, expressed in a percentage. (PE/EDPG) 548-1984w

loss, return *See:* return loss.

loss tangent (1) (general) The ratio of the imaginary part of the complex dielectric constant of a material to its real part. (IM/HFIM) [40]

(2) (rotating machinery) (tan δ) The ratio of dielectric loss in an insulation system, to the apparent power required to establish an alternating voltage across it of a specified amplitude and frequency, the insulation being at a specified temperature. *Note:* It is the cotangent of the power-factor angle. *See also:* asynchronous machine. (PE) [9]

(3) (of a material) The ratio of the imaginary part of the complex permittivity to the real part. (AP/PROP) 211-1997

loss-tangent test (rotating machinery) (dissipation-factor test) A test for measuring the dielectric loss of insulation at predetermined values of temperature, frequency, and voltage or dielectric stress, in which the dielectric loss is expressed in terms of the tangent of the complement of the insulation power-factor angle. *See also:* asynchronous machine. (PE) [9]

loss, total *See:* total loss.

loss, transmission *See:* transmission loss.

lossy medium (laser maser) A medium which absorbs or scatters radiation passing through it. (LEO) 586-1980w

lost call (telephone switching systems) A call that cannot be completed due to blocking. (COM) 312-1977w

lot (1) The quantity of any one type of insulator manufactured with an identical process and materials not exceeding 5000 units. (T&D/PE) 1024-1988w

(2) A quantity of line hardware selected and agreed upon by the manufacturer and customer as being representative of a homogeneous population. Each lot, as far as is practicable, shall consist of units of product of a single type, grade, class, size, and composition that are manufactured at essentially the same time and under essentially the same conditions. Consideration should be given to limit the lot size, where applicable, to each heat-treating and/or annealing process of a group of units. As necessary, the manufacturer shall provide adequate and suitable storage space for each lot and means for proper identification. Any lot shall not exceed 35 000 units. (PE/T&D) C135.61-1997

LOTOS *See:* Language of Temporal Ordering Specifications.

loudness equation (loudness ratings of telephone connections) Loudness voltages (in millivolts) and pressures (in pascals) are determined in accordance with

$$S_E, S_M, V_W, \text{ or } V_T = \left\{ \frac{\sum_{j=2}^{N} \left(\log_{10} \frac{f_j}{f_{j-1}}\right) \left[\frac{\left(10^{\frac{x_j}{20}}\right)^{\frac{1}{2.2}} + \left(10^{\frac{x_{j-1}}{20}}\right)^{\frac{1}{2.2}}}{2} \right]}{\log_{10}(f_N/f_1)} \right\}^{2.2}$$

where

f_j = specific frequencies of the N frequencies selected for analysis

x_j = the signal level (in dBPa or dBmV) at frequency f_j

Loudness voltages and pressure are expressed in decibel-like form using the following equation

$$S_E, S_M, V_W, \text{ or } V_T = 20 \log_{10} \times \left\{ \frac{\sum_{j=2}^{N} \left(\log_{10} \frac{f_j}{f_{j-1}}\right) \left[\frac{\left(10^{\frac{x_j}{20}}\right)^{\frac{1}{2.2}} + \left(10^{\frac{x_{j-1}}{20}}\right)^{\frac{1}{2.2}}}{2} \right]}{\log_{10}(f_N/f_1)} \right\}^{2.2}$$

(COM/TA) 661-1979r

loudness rating (loudness ratings of telephone connections) The amount of frequency independent gain that must be inserted into a system under test so that speech sounds from the system under test and a reference system are equal in loudness. (COM/TA) 661-1979r

loudness rating guard-ring position (LRGP) The position a handset assumes when it is placed on an artificial test head as described in CCITT Recommendation P. 76. (COM/TA) 269-1992, 1206-1994

loudspeaking telephone A telephone with handsfree receive but not handsfree send capability. (COM/TA) 1329-1999

louver (illuminating engineering) A series of baffles used to shield a source from view at certain angles or to absorb unwanted light. The baffles usually are arranged in a geometric pattern. *Synonym:* louver grid. (EEC/IE) [126]

louvered ceiling (illuminating engineering) A ceiling area lighting system comprising a wall-to-wall installation of multicell louvers shielding the light sources mounted above it. (EEC/IE) [126]

louver grid *See:* louver.

louver shielding angle, θ (illuminating engineering) The angle between the horizontal plane of the baffles or louver grid and the plane at which the louver conceals all objects above. *Note:* The planes usually are so chosen that their intersection is parallel with the louvered blade. (EEC/IE) [126]

low The lower of the two voltages used to convey a single bit of information. For positive logic, a logic 0. (TT/C) 1149.1-1990

low-capacitance relay contacts A type of contact construction providing low intercontact capacitance. (EEC/REE) [87]

lowclass If an instance is in a class `S` and not in any subclass of `S`, then `S` is the lowclass for the instance. (C/SE) 1320.2-1998

low conduction threshold voltage LCv (LC) (metal-nitride-oxide field-effect transistor). The threshold voltage level resulting from a write-low pulse, which puts the transistor into the LC state. (ED) 581-1978w

low earth orbit (LEO) An orbit below 1500 km to 1800 km altitude. (C/BA) 1156.4-1997

low-energy power circuit A circuit that is not a remote-control or signaling circuit but has its power supply limited in accordance with the requirements of Class 2 and Class 3 circuits. (NESC/NEC) [86]

lower beams (illuminating engineering) (passing beams) One or more beams directed low enough on the left to avoid glare in the eyes of oncoming drivers, and intended for use in congested areas and on highways when meeting other vehicles within a distance of 300 m (1000 ft). Formerly "traffic beam." (EEC/IE) [126]

lower bracket (rotating machinery) A bearing bracket mounted below the level of the core of a vertical machine. (PE) [9]

lower burst reference (audio and electroacoustics) A selected multiple of the long-time average magnitude of a quantity, smaller than the upper burst reference. *See also:* burst duration; burst. (SP) [32]

lower coil support (rotating machinery) A support to restrain field-coil motion in the direction away from the air gap. *See also:* stator; rotor. (PE) [9]

lower curtate The adjacent card rows at the bottom of a punch card. (C) 610.10-1994w

lower guide bearing (rotating machinery) A guide bearing mounted below the level of the core of a vertical machine. (PE) [9]

lower-half bearing bracket (rotating machinery) The bottom half of a bracket that can be separated into halves for mounting or removal without access to a shaft end. (PE) [9]

lower layer service A generic service required by this standard that provides the minimum subset of the functionality defined by Layers 4 through 1 of the open systems interconnection (OSI) model. (SCC32) 1455-1999

lower limit (test, measurement, and diagnostic equipment) The minimum acceptable value of the characteristic being measured. (MIL) [2]

lower-range value The lowest quantity that a device is adjusted to measure. *Note:* The following compound terms are used with suitable modifications in the units: measured-variable lower-range value, measured signal lower-range value, etc. *See also:* instrument. (EEC/EMI) [112]

lower-sideband parametric down-converter An inverting parametric device used as a parametric down-converter. *See also:* parametric device. (ED) [46]

lowest observed frequency (LOF) In ionospheric sounding, LOF is the lowest frequency for which signals transmitted from a sounder and propagated via the ionosphere are observed on the ionogram, regardless of the precise propagation path involved. *Note:* The LOF is a function of the ionosonde's transmit power, antenna gain, and receiver noise environment. (AP/PROP) 211-1997

lowest usable frequency (LUF) The lowest frequency that would permit acceptable performance of a radio circuit by signal propagation via the ionosphere between given terminals at a given time under specified working conditions. *Notes:* 1. The LUF is a system-dependent parameter and is determined by factors such as ionospheric absorption, transmitter power, antenna gain, receiver characteristics, type of service, and noise conditions. 2. LUF is sometimes referred to as lowest usable high frequency. 3. The use of lowest useful frequency for LUF is deprecated. (AP/PROP) 211-1997

lowest useful frequency (LUF) (radio-wave propagation) For sky-wave signals in the MF/HF spectrum, the lowest frequency effective under specified conditions for ionospheric propagation of radio waves between two points. *Note:* The lowest useful frequency is a system-dependent parameter and is determined by factors such as ionospheric absorption, transmitter power, antenna gain, receiver characteristics, type of service, and noise conditions. (AP/PROP) 211-1990s

lowest useful high frequency (radio-wave propagation) The lowest high frequency effective for ionospheric propagation of radio waves between two specified points, under specified ionospheric conditions, and under specified factors such as absorption, transmitter power, antenna gain, receiver characteristics, type of service, and noise conditions. *See also:* radiation; radio-wave propagation. (AP) 211-1977s

low frequencies Frequencies allocated for transmission in the inbound direction in a mid-split broadband system, approximately 5–108 MHz. (LM/C) 802.7-1989r

low frequency (LF) 30–300 kHz. *See also:* radio spectrum. (AP/PROP) 211-1997

low-frequency dry-flashover voltage The root-mean-square voltage causing a sustained disruptive discharge through the air between electrodes of a clean dry test specimen under specified conditions. (T&D/PE) [10]

low-frequency flashover voltage (insulators) The root-mean-square value of the low-frequency voltage that, under specified conditions, causes a sustained disruptive discharge through the surrounding medium. *See also:* insulator. (EEC/IEPL) [89]

low-frequency furnace (core-type induction furnace) An induction furnace that includes a primary winding, a core of magnetic material, and a secondary winding of one short-circuited turn of the material to be heated. (EEC/PE) [119]

low-frequency high-potential test (rotating machinery) A high-potential test that applies a low-frequency voltage, between 0.1 hertz and 1.0 hertz, to a winding. (PE) [9]

low-frequency impedance corrector An electric network designed to be connected to a basic network, or to a basic network and a building-out network, so that the combination will simulate at low frequencies the sending-end impedance, including dissipation, of a line. *See also:* network analysis. (EEC/PE) [119]

low-frequency induction heater or furnace A device for inducing current flow of commercial power-line frequency in a charge to be heated. *See also:* induction heater. (IA) 54-1955w, 169-1955w

low-frequency puncture voltage (insulators) The root-mean-square value of the low-frequency voltage that, under specified conditions, causes disruptive discharge through any part of the insulator. *See also:* insulator. (EEC/IEPL) [89]

low-frequency wet-flashover voltage The root-mean-square voltage causing a sustained disruptive discharge through the air between electrodes of a clean test specimen on which water of specified resistivity is being sprayed at a specified rate. (T&D/PE) [10]

low-frequency withstand voltage (insulators) The root-mean-square value of the low-frequency voltage that, under specified conditions, can be applied without causing flashover or puncture. *See also:* insulator. (EEC/IEPL) [89]

low-key lighting (illuminating engineering) A type of lighting which, applied to a scene, results in a picture having graduations falling primarily between middle gray and black, with comparatively limited areas of light grays and whites. (EEC/IE) [126]

low level (1) A level with the more negative (less positive) of the two ranges of logic levels chosen to represent the logic states. (GSD) 91-1984r
(2) A signal voltage within the more negative (less positive) of the two ranges of logic levels chosen to represent the logic states. (C/BA) 1496-1993w

low-level analog signal cable (cable systems in power generating stations) Cable used for transmitting variable current or voltage signals for the control or instrumentation of plant equipment and systems, or both. (PE/EDPG) 422-1977

low-level digital signal circuit cable (cable systems in power generating stations) Cable used for transmitting coded information signals, such as those derived from the output of an analog-to-digital converter, or the coded output from a digital computer or other digital transmission terminals. (PE/EDPG) 422-1977

low-level format To format a blank storage medium in order to establish tracks and sectors on the medium. *Synonym:* physical format. *Contrast:* high-level format. (C) 610.10-1994w

low-level language *See:* assembly language.

low-level modulation Modulation produced at a point in a system where the power level is low compared with the power level at the output of the system. (COM/AP/BT/ANT) 145-1983s, 182-1961w

low-level radio-frequency signal (transmit-receive, anti-transmit receive, and pre-transmit-receive tubes) A radio-frequency signal with insufficient power to cause the tube to become fired. (ED) 161-1971w

low-level relay contacts Contacts that control only the flow of relatively small currents in relatively low-voltage circuits: for example, alternating currents and voltages encountered in voice or tone circuits, direct currents and voltages of the order of microamperes and microvolts, etc. (EEC/REE) [87]

low-level testing (1) (mechanical) Testing performed to determine natural frequencies of complete assemblies, subassemblies, or components. (SUB/PE) C37.122.1-1993

(2) Mechanical testing performed to determine natural frequencies and dampings of complete assemblies, subassemblies, or components. (SWG/PE/SUB) C37.100-1992, C37.122-1983s

low-order Pertaining to the right-most digit or digits of a numeral. (C) 1084-1986w

low-order position The rightmost position in a string; for example, the letter "E" in "APPLE" or the "5" in "965." *Contrast:* high-order position. *See also:* least significant character; least significant digit. (C) 610.5-1990w

low-pass filter (data transmission) (power line filters) A filter having a single transmission band extending from zero to some cutoff frequency, not infinite. *See also:* filter. (PE/EMC/SP) 599-1985w, C63.13-1991, 151-1965w

low (normal) power-factor mercury lamp ballast A ballast of the multiple-supply type that does not have a means for power-factor correction. (EEC/LB) [97]

low-power factor transformer (power and distribution transformers) A high-reactance transformer that does not have means for power-factor correction. *See also:* specialty transformer. (PE/TR) C57.12.80-1978r, [116]

low-pressure contact (area contact) A contact in which the pressure is such that stress in the material is well below the elastic limit of both contact surface materials, such that conduction is a function of area. (SWG/PE) C37.100-1992

low-pressure sodium lamp (illuminating engineering) A discharge lamp in which light is produced by radiation from sodium vapor operating at a partial pressure of 0.13 to 1.3 Pa (10^{-3} to 10^{-2} Torr). (EEC/IE) [126]

low-pressure vacuum pump A vacuum pump that compresses the gases received directly from the evacuated system. *See also:* rectification. (EEC/PE) [119]

low-priority effort (electric generating unit reliability, availability, and productivity) Repairs were carried out with less than a normal effort. *See also:* repair urgency. (PE/PSE) 762-1987w

low profile Terminations or connections designed to be an integral part of the heating cable. (IA/PC) 515.1-1995

low-profile connection Terminations or connections designed to be an integral part of the heating cable and installed under the insulation. (IA) 515-1997

low-pulse-repetition-frequency waveform A pulsed-radar waveform whose pulse-repetition frequency is such that targets of interest are unambiguously resolved with respect to range. *See also:* high-pulse-repetition-frequency waveform; medium-pulse-repetition-frequency waveform. (AES) 686-1997

low remanence current transformer One with a remanence not exceeding 10 percent of maximum flux. (PE/TR) [57], C57.13-1993

low-speed limit (control systems for steam turbine-generator units) (speed/load reference) A device or input that limits the speed/load reference setting to a predetermined lower limit. This device may establish the lower limit of the synchronizing speed range. (PE/EDPG) 122-1985s

low state (1) (programmable instrumentation) The relatively less-positive signal level used to assert a specific message content associated with one of two binary logic states. (IM/AIN) 488.1-1987r

(2) (signals and paths) (SBX bus) (STEbus) (microcomputer system bus) The more negative voltage level; used to represent one of two logical binary states. (MM/C) 796-1983r, 1000-1987r, 959-1988r

(3) (696 interface devices) (signals and paths) The electrically less positive signal level used to assert a specific message content associated with one of two binary logis states. (MM/C) 696-1983w

low symbol An idle symbol that has been marked for consumption by lower-priority nodes. Sometimes called low-idle symbol. May also be consumed by a highest-priority node when it is taking its fair share of lower-priority bandwidth. (C/MM) 1596-1992

low, true, 0 Asserted state of a bus line. (C/MM) 1196-1987w

low-velocity camera tube (cathode-voltage-stabilized camera tube) A camera tube operating with a beam of electrons having velocities such that the average target voltage stabilizes at a value approximately equal to that of the electron-gun cathode. (ED) [45]

low-velocity scanning (electron tube) The scanning of a target with electrons of velocity less than the minimum velocity to give a secondary-emission ratio of unity. *See also:* television. (ED/BT/AV) 161-1971w, [34], [45]

low voltage (1) An electromotive force rated nominal 24 volts, nominal or less, supplied from a transformer, converter, or battery. (NEC/NESC/IA/PSE) 241-1990r, [86]

(2) (system voltage ratings) A class of nominal system voltages 1000 or less. *See also:* high voltage; medium voltage; nominal system voltage. (IA/APP) [80]

low-voltage ac power circuit breaker *See:* circuit breaker.

low-voltage aluminum-sheathed power cable (aluminum sheaths for power cables) Cable used in an electric system having a maximum phase-to-phase rms ac voltage of 1000 V or less, the cable having an aluminum sheath as a major component in its construction. (PE/IC) 635-1989r

low-voltage electrical and electronic equipment (radio-noise emissions) Electrical and electronic equipment with operating input voltages of up to 600 V dc or rms ac. (EMC) C63.4-1991

low-voltage integrally fused power circuit breaker An assembly of a general-purpose ac low-voltage power circuit breaker and integrally mounted current-limiting fuses that together function as a coordinated protective device. (SWG/PE) C37.100-1992

low-voltage power cable (s) (1) (cable systems in power generating stations) Cable designed to supply power to utilization devices of the plant auxiliary system, operated at 600 V or less. (PE/EDPG) 422-1977

(2) Cable designed to supply power to utilization devices of the plant auxiliary system, operated at 600 V or 2000 V in sizes ranging from 14 AWG (2.08 mm^2) to 2000 kcmil (1010.0 mm^2). (PE/IC) 1185-1994

(3) Those cables used on systems operating at 1000 V or less. They are designed to supply operating power to utilization devices. (SUB/PE) 525-1992r

low-voltage power circuit breaker (LVPCB) A mechanical switching device, capable of making, carrying, and breaking currents under normal circuit conditions and also, making and carrying for a specified time and breaking currents under specified abnormal circuit conditions such as those of short-circuit. Rated 1000 V ac or below, or 3000 V dc and below, but not including molded-case circuit breakers. (IA/PSP) 1015-1997

low-voltage power supply (LVPS) A power supply that provides dc power to the low-voltage devices or circuits contained on the vehicle. It can also be configured to charge the vehicle battery. (VT) 1476-2000

low-voltage protection (1) The effect of a device operative on the reduction or failure of voltage so as to cause and maintain the interruption of power supply to the equipment protected. (NESC) C2-1984s

(2) *See also:* undervoltage protection. (SWG/PE) C37.100-1992

low-voltage release The effect of a device, operative on the reduction or failure of voltage, to cause the interruption of power supply to the equipment, but not preventing the reestablishment of the power supply on return of voltage. (T&D) C2.2-1960

low-voltage system (electric power) An electric system having a maximum root-mean-square alternating-current voltage of 1000 volts or less. *See also:* voltage classes. (IA/PSE) 570-1975w

low-voltage winding instrument transformer Winding that is intended to be connected to the measuring or control devices. (PE/TR) [57]

LP *See:* linear programming; linearly polarized mode.

LPL *See:* List Processing Language.

LP01 mode (fiber optics) Designation of the fundamental linearly polarized (LP) mode. *See also:* fundamental mode. (Std100) 812-1984w

LQ *See:* letter-quality.

L_q(s) *See:* quadrature-axis operational inductance.

LRA *See:* laser gyro axes.

LRC *See:* longitudinal redundancy check.

LRFD Acronym for load and resistance factor design. (T&D/PE) 751-1990

LRGP *See:* loudness rating guard-ring position.

LRU *See:* line replaceable unit.

L/s Abbreviation for liters per second. (T&D/PE) 957-1987s

L(s) *See:* operational inductance.

LS *See:* Load Server.

LS dividing network *See:* dividing network.

LSB *See:* least significant bit.

***L*-scope (radar)** A cathode-ray oscillosope arranged to present an L-display. (AES/RS) 686-1982s

L-scope A cathode-ray oscilloscope arranged to present an L-display. (AES/RS) 686-1990

LSDV *See:* Link Segment Delay Value.

LSI *See:* large scale integration.

LSS *See:* logical source statements.

L_{stub} Stub length, measured from the connector via to the center of the transceiver pin surface-mount pad (include etch length from dispersion via, if present). (C/BA) 896.2-1991w

lsw *See:* least significant word.

LTC *See:* load tap changer.

LTS *See:* long-term stability.

lubricant Any material applied on the cable or into a conduit to reduce friction and hence tension during cable pulling operations. (PE/IC) 1185-1994

LUF *See:* lowest usable frequency.

luff Pulling additional cable out of the conduit, using a split grip or mare's tail, to be used to facilitate terminating, racking, etc. (PE/IC) 1185-1994

lug A wire connector device to which the electrical conductor is attached by mechanical pressure or solder. (IA/MT) 45-1998

lug, stator mounting *See:* stator mounting lug.

Lukasiewicz notation *See:* prefix notation.

lumen (lm) (1) (television) (color terms) The unit of luminous flux. The luminous flux emitted within unit solid angle (one steradian) by a point source having a uniform intensity of one candela. (BT/AV) 201-1979w
(2) (illuminating engineering) SI unit of luminous flux. Radio-metrically, it is determined from the radiant power. Photometrically, it is the luminous flux emitted within a unit solid angle (one steradian) by a point source having a uniform luminous intensity of one candela. *See also:* luminous flux. (EEC/IE) [126]
(3) (electric power systems in commercial buildings) The international unit of luminous flux or the time rate of the flow of light. (IA/PSE) 241-1990r

lumen hour A unit of quantity of light (luminous energy). It is the quantity of light delivered in one hour by a flux of one lumen. *See also:* light. (EEC/IE) [126]

lumen method (illuminating engineering) A lighting design procedure used for determining the relation between the number and types of lamps or luminaires, the room characteristics, and the average level of illuminance on the work-plane. It takes into account both direct and reflected flux. *Synonym:* flux method. *See also:* inverse-square law. (EEC/IE) [126]

lumen-second (illuminating engineering) A unit of quantity of light, the SI unit of luminous energy (also called a talbot). It is the quantity of light delivered in one second by a luminous flux of one lumen. (EEC/IE) [126]

lumens per watt The ratio of lumens generated by a lamp to the watts consumed by the lamp. *See also:* efficacy. (IA/PSE) 241-1990r

luminaire (electric power systems in commercial buildings) A complete lighting unit that consists of parts designed to position a lamp (or lamps) in order to connect it to the power supply and to distribute its light. (IA/EEC/IE/PSE) 241-1990r, [126]

luminaire ambient temperature factor (illuminating engineering) The fractional loss of task illuminance due to improper operating temperature of a gas discharge lamp. (EEC/IE) [126]

luminaire dirt depreciation (illuminating engineering) The fractional loss of task illuminance due to luminaire dirt accumulation. (EEC/IE) [126]

luminaire efficiency (illuminating engineering) The ratio of luminous flux (lumens) emitted by a luminaire to that emitted by the lamp or lamps used therein. (IE/EEC/IA/PSE) [126], 241-1990r

luminaire surface depreciation factor (illuminating engineering) The loss of task illuminance due to permanent deterioration of luminaire surfaces. (EEC/IE) [126]

luminance (1) ($L = d^2\varphi/(d\omega da\cos\theta)$) (illuminating engineering) (in a direction and at a point of a real or imaginary surface) The quotient of the luminous flux at an element of the surface surrounding the point and propagated in directions defined by an elementary cone containing the given direction, by the product of the solid angle of the cone and the area of the orthogonal projection of the element of the surface on a plane perpendicular to the given direction. The luminous flux may be leaving, passing through, and arriving at the surface or both. Formerly, photometric brightness. By introducing the concept of luminous intensity, luminance may be expressed as $L = dE/(d\omega\cos\theta)$. Here, luminance at a point of a surface in a direction, is interpreted as the quotient of luminous intensity in the given direction, produced by an element of the surface surrounding the point, by the area of the orthogonal projection of the element of surface on a plane, perpendicular to the given direction. (Luminance may be measured at a receiving surface by using $L = dE/(d\omega\cos\theta)$. This value may be less than the the luminance of the emitting surface due to the attenuation of the transmitting media.) *Note:* In common usage the term brightness usually refers to the strength of sensation which results from viewing surfaces or spaces from which light comes to the eye. In much of the literature the term brightness, used alone, refers to both luminance and sensation. The context usually indicates which meaning is intended. Previous usage notwithstanding, neither the term brightness, not the term photometric brightness should be used to denote the concept of luminance. (IE/EEC) [126]
(2) (average photometric brightness) (average luminance) The total lumens actually leaving the surface per unit area. *Note:*

1) Average luminance specified in this way is identical in magnitude with luminous exitance, which is the preferred term.
2) In general, the concept of average luminance is useful only when the luminance is reasonably uniform throughout a very wide angle of observation and over a large area of the surface considered. It has the advantage that it can be computed readily for reflecting surfaces by multiplying the incident luminous flux density (illumination) by the luminous reflectance of the surface. For a transmitting body it can be computed by multiplying the incident luminous flux density by the luminous transmittance of the body.

(BT/AV) 201-1979w

(3) (electric power systems in commercial buildings) The light emanating from a light source or the light reflected from a surface (the metric unit of measurement is cd/cm^2). (IA/PSE) 241-1990r

(4) The luminous intensity per unit projected area of a given surface as viewed from a given direction. Measured in candelas per square meter or footlamberts. (PE/NP) 1289-1998

luminance channel (color television) Any path that is intended to carry the luminance signal. (BT/AV) 201-1979w

luminance channel bandwidth (color television) The bandwidth of the path intended to carry the luminance signal. (BT/AV) 201-1979w

luminance coefficient (illuminating engineering) The ratio of average initial wall or ceiling cavity luminance times π to the total lamp flux (lumens) divided by the floor area. *Note:*

1) If the luminance is in cd/in^2, the floor area must be in square inches.
2) If the luminance is in cd/ft^2 (or in cd/m^2), the floor area must be in square feet (or in square meters).
3) If the luminance is in footlamberts, the "π" is omitted and the floor area must be in square feet.

(EEC/IE) [126]

luminance contrast (1) (illuminating engineering) The relationship between the luminances of an object and its immediate background. It is equal to $(L_1 - L_2)/L_1$ or $(L_2 - L_1)/L_1 = \Delta L/L_1$, where L_1 and L_2 are the luminances of the background and object, respectively. The form of the equation must be specified. The ratio $\Delta L/L$ is known as Weber's fraction. Because of the relationship among luminance, illumination, and reflectance, contrast often is expressed in terms of reflectance when only reflecting surfaces are involved. Thus, contrast is equal to $(\rho_1 - \rho_2)/\rho_1$ or $(\rho_2 - \rho_1)/\rho_1$ where ρ_1 and ρ_2 are the reflectances of the background and object, respectively. This method of computing contrast holds only for perfectly diffusing surfaces; for other surfaces it is only an approximation unless the angles of incidence and view are taken into consideration. *See also:* luminance; reflectance. (EEC/IE) [126]

(2) The ratio of the luminance of the features of the object being viewed, in particular of the feature to be discriminated, to the luminance of the background. *See also:* luminance. (PE/NP) 1289-1998

luminance difference threshold (illuminating engineering) This can apply to the difference between two separated objects on a common background, or the difference between two juxtaposed patches separated by a contrast border, or the difference between a uniform small object and its background. In the latter case, a contrast border separates the object from its background. (EEC/IE) [126]

luminance factor The ratio of the luminance (photometric brightness) of a surface or medium under specified conditions of incidence, observation, and light source, to the luminance (photometric brightness) of a perfectly reflecting or transmitting, perfectly diffusing surface or medium under the same conditions. *Note:* Reflectance or transmittance cannot exceed unity, but luminance factor may have any value from zero to values approaching infinity. *See also:* lamp. (EEC/IE) [126]

luminance factor of room surfaces (illuminating engineering) Factors by which the average work-plane illuminance is multiplied to obtain the average luminances of walls, ceilings and floors. (EEC/IE) [126]

luminance flicker (color television) The flicker that results from fluctuation of luminance only. (BT/AV) 201-1979w

luminance primary (color television) One of a set of three transmission primaries whose amount determines the luminance of a color. *Note:* This is an obsolete term because it is useful only in a linear system. (BT/AV) 201-1979w

luminance ratio (illuminating engineering) The ratio between the luminances of any two areas in the visual field. *See also:* luminance. (EEC/IE) [126]

luminance signal (ntsc color television) A signal that has major control of the luminance. *Note:* It is a linear combination of gamma-corrected primary color signals, E'_R, E'_G, and E'_B as follows:

$$E'_Y = 0.30E'_R + 0.59E'_G + 0.11E'_B$$

The proportions expressed are strictly true only for television systems using the NTSC original standard receiver primaries having the CIE color points listed below, when they are mixed to produce white light having the same appearance as standard illuminant C.

Color	*x*	*y*
Red (R)	0.67	0.33
Green (G)	0.21	0.71
Blue (B)	0.14	0.08

(BT/AV) 201-1979w

luminance threshold (photometric brightness) The minimum perceptible difference in luminance for a given state of adaptation of the eye. *See also:* luminance; visual field. (IE/EEC) [126]

luminescence (illuminating engineering) Any emission of light not ascribable directly to incandescence. (EEC/IE) [126]

luminescent-screen tube A cathode-ray tube in which the image on the screen is more luminous than the background. *See also:* cathode-ray tube. (ED) [45], [84]

luminosity (television) Ratio of luminous flux to the corresponding radiant flux at a particular wavelength. It is expressed in lumens per watt. (BT/AV) 201-1979w

luminosity coefficients (television) The constant multipliers for the respective tristimulus values of any color, such that the sum of the three products is the luminance of the color. (BT/AV) 201-1979w

luminous ceiling (illuminating engineering) A ceiling area lighting system comprising a continuous surface of transmitting material of a diffusing or light controlling character with light sources mounted above it. (EEC/IE) [126]

luminous density, w = dQ/dV (illuminating engineering) Quantity of light (luminous energy) per unit volume. (EEC/IE) [126]

luminous efficacy of a source of light (illuminating engineering) The quotient of the total luminous flux emitted by the total lamp power input. It is expressed in lm/W. *Note:* The term luminous efficiency has in the past been extensively used for this concept. (EEC/IE) [126]

luminous efficacy of radiant flux (illuminating engineering) The quotient of the total luminous flux by the total radiant flux. It is expressed in lm/W. (EEC/IE) [126]

luminous efficiency (television) The ratio of the luminous flux to the radiant flux. *Note:* Luminous efficiency is usually expressed in lumens per watt of radiant flux. It should not be confused with the term efficiency as applied to a practical source of light, since the latter is based on the power supplied to the source instead of the radiant flux from the source. For energy radiated at a single wavelength, luminous efficiency is synonymous with luminosity. (BT/AV) 201-1979w

luminous exitance, M = $d\varphi/d$A (illuminating engineering) The density of luminous flux leaving a surface at a point. Formerly luminous emittance. *Note:* This is the total luminous flux emitted, reflected, and transmitted from the surface and is independent of direction. (EEC/IE) [126]

luminous flux (1) (television) The time rate of flow of light. (BT/AV) 201-1979w

(2) (illuminating engineering) Radiant flux (radiant power), the time rate of flow of radiant energy, evaluated in terms of a standardized visual response.

$$\Phi_V = K_m \int \Phi_e,\lambda V(\lambda) d\lambda$$

where Φ_V is in lumens, Φ_e,λ is in watts per nanometer, λ is in nanometers, $V(\lambda)$ is the spectral luminous efficiency, in lm/W. Unless otherwise indicated, the luminous flux is defined for photopic vision. For scotopic vision, the corresponding spectral luminous efficiency $V'(\lambda)$ and the corresponding maximum spectral luminous efficiency K'_m are submitted in the above equation. K_m and K'_m are derived from the basic SI definition of luminous intensity and have the values 683 lm/W and 1754 lm/W respectively. *Note:* The value of K_m 683 lm/W was recommended by the International Committee for Weights and Measures in 1977. *See also:* spectral luminous efficiency; candela; values of spectral luminous efficiency for photopic vision; values of spectral luminous efficiency for scotopic vision. (EEC/IE) [126]

luminous flux density at a surface, $d\Phi/dA$ (illuminating engineering) The luminous flux per unit area at a point on a surface. *Note:* This need not be a physical surface; it may also be a mathematical plane. *See also:* illuminance; luminous exitance, M = $d\varphi/dA$. (EEC/IE) [126]

luminous gain (optoelectronic device) The ratio of the emitted luminous flux to the incident luminous flux. *Note:* The emitted and incident luminous flux are both determined at specified ports. *See also:* optoelectronic device. (ED) [46]

luminous intensity (A) (television) (of a source of light in a given direction) The luminous flux per unit solid angle in the direction in question. Hence, it is the luminous flux on a small surface normal to that direction, divided by the solid angle (in steradians) that the surface subtends at the source. *Note:* Mathematically, a solid angle must have a point as its apex; the definition of luminous intensity, therefore, applies strictly only to a point source. In practice, however, light emanating from a source whose dimensions are negligible in comparison with the distance over which it is observed may be considered as coming from a point. **(B)** (of a point source of light in a given direction) The luminous flux per unit solid angle in the direction in question. Hence, it is the luminous flux on a small surface centered on and normal to that direction divided by the solid angle (in steradians) which the surface subtends at the source. Luminous intensity may be expressed in candelas or in lumens per steradian (lm/sr). *Note:* Mathematically a solid angle must have a point as its apex; the definition of luminous intensity, therefore, applies strictly only to a point source. In practice, however, light emanating from a source whose dimensions are negligible in comparison with the distance from which it is observed may be considered as coming from a point. Specifically, this implies that with change of distance (1) the variation in solid angle subtended by the source at the receiving point approaches $1/(\text{distance})^2$ and that (2) the average luminance of the projected source area as seen from the receiving point does not vary appreciably. For extended sources see equivalent luminous intensity of an extended source at a specified distance. The word intensity as defined above is used to designate luminous intensity (or candlepower). It is also widely used in other ways either informally or formally in other disciplines. Stimulus intensity may be used to designate the retinal illuminance of a proximal stimulus or the luminance of a distal stimulus. Intensity is used in the same sense with other modalities such as audition. Intensity has been used to designate the level of illuminance on a surface or the flux density in the cross section of a beam of light. In physical optics, intensity usually refers to the square of the wave amplitude. *See also:* distal stimuli; proximal stimuli. (BT/EEC/IE/AV) 201-1979, [126]

luminous reflectance (illuminating engineering) Any of the geometric aspects of reflectance in which both the incident and transmitted flux are weighed by the luminous efficiency of radiant flux [V (λ)]. *Note:* Unless otherwise qualified, the term "luminous reflectance" is meant by the term "reflectance." (EEC/IE) [126]

luminous sensitivity (camera tubes or phototubes) The quotient of signal output current by incident luminous flux, under specified conditions of illumination. *Notes:* 1. Since luminous sensitivity is not an absolute characteristic but depends on the spectral distribution of the incident flux, the term is commonly used to designate the sensitivity to radiation from a tungsten-filament lamp operating at a color temperature of 2870 K. 2. Luminous sensitivity is usually measured with a collimated beam at normal incidence. (ED) 161-1971w

luminous transmittance (illuminating engineering) Any of the geometric aspects of transmittance in which the incident and transmitted flux are weighed by the luminous efficiency of radiant flux [V(λ)]. *Note:* unless otherwise qualified, the term luminous transmittance is meant by the term transmittance. (EEC/IE) [126]

luminous-tube transformer (power and distribution transformers) Transformers, autotransformers, or reactors (having a secondary open-circuit rms of 1000 V or more) for operation of cold-cathode and hot-cathode luminous tubing generally used for signs, illumination, and decoration purposes. (PE/TR) C57.12.80-1978r

lumped Effectively concentrated at a single point. (EEC/PE) [119]

lumped capacitive load A lumped capacitance that is switched as a unit. (SWG/PE) C37.100-1992

lumped element circuit (microwave tubes) A circuit consisting of discrete inductors and capacitors. (ED) [45]

Luneburg lens antenna A lens antenna with a circular cross section having an index of refraction varying only in the radial direction such that a feed located on or near a surface or edge of the lens produces a major lobe diametrically opposite the feed. (AP/ANT) 145-1993

lux (1) (lx) (illuminating engineering) The SI unit of illuminance. One lux is one lumen per square meter (lm/m^2). (EEC/IE) [126]
(2) (metric practice) The illuminance produced by a luminous flux of one lumen uniformly distributed over a surface of one square meter. (QUL) 268-1982s
(3) (electric power systems in commercial buildings) The metric measure of illuminance that is equal to 1 lm uniformly incident on 1 m^2 (1 lux = 0.0929 fc). (IA/PSE) 241-1990r

Luxemburg effect A nonlinear effect in the ionosphere, as a result of which, the modulation on a strong carrier wave is transferred to another carrier passing through the same region. *See also:* radiation. (EEC/PE) [119]

LV *See:* layer manager; metal-enclosed low-voltage power circuit-breaker switchgear.

Lyapunov function (equilibrium point $\mathbf{x}_e$ of a system) A scalar differentiable function V(x) defined in some open region including $\mathbf{x}_e$ such that in that region

$V(\mathbf{x}) > 0$ for $\mathbf{x} \pm \mathbf{x}_e$ (1)

$V(\mathbf{x}_e) = 0$ (2)

$V'(\mathbf{x}) \le 0$ (3)

Notes: 1. The open region may be defined by (norm of $\mathbf{x} - \mathbf{x}_e$) < constant. 2. For the system $\mathbf{x}' = \mathbf{f}(\mathbf{x})$, $V'(\mathbf{x}) \equiv [\text{grad } V(\mathbf{x}) \cdot \mathbf{f}(\mathbf{x})]$. *See also:* feedback control system. (CS/IM) [120]

Lyapunov stability (of a solution $\phi(x(t_0);t)$) For every given $\varepsilon > 0$ there exists a $\delta > 0$ (which, in general, may depend on ε and on t_0) such that $\|\Delta\mathbf{x}(t_0)\| \le \delta$ implies $\|\Delta\phi\| \le \varepsilon$ for $t \ge t_0$. *Notes:* 1. The solution x = 0 of the system x = ax is Lyapunov stable if a < 0, and is Lyapunov unstable if a > 0. 2. For a linear system with an irreducible transfer function $T(s)$, Lyapunov stability implies that all the poles of $T(s)$ are in the left-half s plane and that those on the $j\omega$ axis are simple. *See also:* control system. (CS/IM) [120]

LYRIC *See:* Language for Your Remote Instruction by Computer.

M

m *See:* milli.

M *See:* mega.

MA *See:* module address.

MAC *See:* medium access control.

MAC address An address that identifies a particular medium access control (MAC) sublayer service access point (SAP). (LM/C) 8802-6-1994

MAC frame **(1)** Except where explicitly stated otherwise, a MAC frame conveying MAC user data across a LAN, or a representation of such a MAC frame within a MAC Bridge or on the non-LAN communications equipment of a Remote Bridge Group. (C/LM) 802.1G-1996
(2) A token ring frame containing a MAC PDU exchanged between MAC entities used to convey information that is used by the MAC protocol or management of the MAC sublayer. (C/LM) 8802-5-1998
(3) The logical organization of control and data fields (e.g., addresses, data, error check sequences) defined for the MAC sublayer. The MAC frame may be constructed in either ISO/IEC 8802-3 or ISO/IEC 8802-5 format. *See also:* framelocal area networks. (C) 8802-12-1998

machine **(1)** **(general)** An article of equipment consisting of two or more resistant, relatively constrained parts that, by a certain predetermined intermotion, may serve to transmit and modify force, motion, or electricity so as to produce some given effect or transformation or to do some desired kind of work. *Notes:* 1. If a matrix has as its elements impedances, it is usually referred to as impedance matrix. Frequently the impedance matrix is called impedance for short. 2. Usually such impedances are defined with the mechanical angular velocity of the machine at steady state. *See also:* asynchronous machine. (PE) [9], 270-1966w
(2) **(sound measurements)** Any rotating electrical device of which the acoustical characteristics are to be measured. *Notes:* 1. A small machine has a maximum linear dimension of 250 mm. This dimension is over major surfaces, excluding minor surface protuberances as well as shaft extension, and is measured either parallel to the shaft, or at right angles to it, according to which dimension gives the greater measurement. 2. A medium machine has a maximum linear dimension from 250 mm to 1 m as measured for small machine. 3. A large machine has a maximum linear dimension in excess of 1 m as measured for small machine.
(3) A generic term for a device such as a processor or computer. (C) 610.10-1994w
(4) **(computers)** *See also:* universal Turing machine; Turing machine.

machine address *See:* absolute address.

machine-aided Pertaining to a process or function performed with the assistance of one or more computers. (C) 610.2-1987

machine-aided translation Translation from one natural language to another, with the assistance of computer based aids such as automated lexicons and automated thesauri. *Synonym:* machine translation. *See also:* mechanical translation. (C) 610.2-1987

machine, aircraft electric *See:* aircraft electric machine.

machine-centered simulation *See:* computer simulation.

machine check **(A)** An automatic check. **(B)** A programmed check of machine functions. *See also:* automatic; check. (C) 162-1963

machine code **(1)** **(computers)** An operation code that a machine is designed to recognize. (MIL/C) [2], [20], [85]
(2) **(software)** Computer instructions and data definitions expressed in a form that can be recognized by the processing unit of a computer. *Contrast:* compiler code; interpretive code; assembly code. *See also:* computer instruction code. (C) 610.12-1990, 610.10-1994w

machine current The rms magnitude of the fundamental sinusoidal current at the terminals of the machine. (IA/ID) 995-1987w

machine cycle The time required for a processor to perform one internal operation, excluding those which may be accomplished in parallel. *Synonym:* microcycle. (C) 610.10-1994w

machine-dependent Pertaining to software that relies on features unique to a particular type of computer and therefore executes only on computers of that type. *Contrast:* machine independent. (C) 610.12-1990

machine, electric *See:* electric machine.

machine equation *See:* computer equation.

machine final-terminal stopping device (elevators) A final-terminal stopping device operated directly by the driving machine. *Synonym:* stop-motion switch. *See also:* control. (EEC/PE) [119]

machine independent Pertaining to software that does not rely on features unique to a particular type of computer, and therefore executes on computers of more than one type. *Contrast:* machine-dependent. *See also:* portability. (C) 610.12-1990

machine instruction **(1)** **(computers)** An instruction that a machine can recognize and execute. (MIL/C) [2], [20], [85]
(2) An instruction in the machine language of a particular processing unit of a computer. *See also:* computer instruction; machine code. (C) 610.10-1994w

machine instruction set *See:* computer instruction set.

machine language **(1)** **(A)** A language, occurring within a machine, ordinarily not perceptible or intelligible to persons without special equipment or training. **(B)** A translation or transliteration of definition (A) above into more conventional characters but frequently still not intelligible to persons without special training. (C) 162-1963
(2) **(software)** A language that can be recognized by the processing unit of a computer. Such a language usually consists of patterns of 1's and 0's, with no symbolic naming of operations or addresses. *Synonyms:* machine-oriented language; first-generation language. *Contrast:* symbolic language; assembly language; fifth generation language; fourth generation language; high-order language. (C) 610.12-1990
(3) A programming language that is directly executed by the central processing unit (CPU) portion of a computer. *Note:* No further translation, mapping, or decoding is required. *Synonyms:* machine-oriented language; first-generation language. *Contrast:* symbolic language; assembly language; fifth generation language; fourth generation language; high-order language. (C) 610.13-1993w
(4) A programming language that is directly executed by the ALU portion of the processor in a computer. *Synonym:* hardware language. (C) 610.10-1994w

machine operation *See:* computer operation.

machine-oriented language **(1)** **(A)** **(test, measurement, and diagnostic equipment)** A language designed for interpretation and use by a machine without translation. **(B)** **(test, measurement, and diagnostic equipment)** A system for expressing information which is intelligible to a specific machine; for example, a computer or class of computers. Such a language may include instructions which define and direct machine operations, and information to be recorded by or acted upon by these machine operations. **(C)** **(test, measurement, and diagnostic equipment)** The set of instructions expressed in the number system basic to a computer, together with symbolic operation codes with absolute addresses, relative addresses, or symbolic addresses. *Synonym:* assembly language; machine language. (MIL) [2]
(2) **(software)** *See also:* machine language. (C) 610.12-1990

machine or transformer thermal relay (power system device function numbers) A relay that functions when the temperature of a machine armature or other load-carrying winding or element of a machine or the temperature of a power rectifier or power transformer (including a power rectifier transformer) exceeds a predetermined value. (SUB/PE) C37.2-1979s

machine positioning accuracy, precision, or reproducibility (numerically controlled machines) Accuracy, precision, or reproducibility of position sensor or transducer and interpeting system, the machine elements, and the machine positioning servo. *Note:* Cutter, spindle, and work deflection, and cutter wear are not included. (May be the same as control positioning accuracy, precision, or reproducibility in some systems.) (IA/EEC) [61], [74]

machine readable Pertaining to data in a form that can be automatically input to a computer; for example, data encoded on a diskette. (C) 610.12-1990

machine-readable (A) Pertaining to a medium that can record information and convey it to a machine or sensing device. *Synonym:* machine-sensible. **(B)** Pertaining to information that can be read and processed by a machine. (C) 610.10-1994

MAchine-Readable Cataloging In library automation, an internationally-accepted standard for systems used to create catalogs of machine-readable bibliographic records. *Synonyms:* LCMARC; LC-MARC. (C) 610.2-1987

machine-readable medium (1) A data medium that is machine-readable. *Synonym:* automated data medium. (C) 610.5-1990w
(2) A data medium that can be used to convey data. (C) 610.10-1994w

machine recognition *See:* pattern recognition.

machine ringing (telephone switching systems) Ringing that once started continues automatically, rhythmically until the call is answered or abandoned. (COM) 312-1977w

machinery control room An enclosed or separated space generally located within the machinery spaces that functions as a central control station. (IA/MT) 45-1998

machinery spaces Spaces that are primarily used for machinery of any type, or equipment for the control of such machinery, such as boiler, engine, generator, motor, pump and evaporator rooms. (IA/MT) 45-1998

machine-sensible *See:* machine-readable.

machine simulation A simulation that is executed on a machine. *See also:* computer simulation. (C) 610.3-1989w

machine time *See:* time.

machine-tool control transformers (power and distribution transformers) Step-down transformers which may be equipped with fuse or other overcurrent protection device, generally used for the operation of solenoids, contactors, relays, portable tools, and localized lighting. (PE/TR) C57.12.80-1978r

machine translation (MT) *See:* machine-aided translation.

machine voltage The rms value of the fundamental sinusoidal voltage at the terminals of the machine. (IA/ID) 995-1987w

machine winding (rotating machinery) A winding placed in slots or around poles directly by a machine. *See also:* rotor; stator. (PE) [9]

machine word *See:* word; computer word.

machining accuracy, precision, or reproducibility Accuracy, precision, or reproducibility obtainable on completed parts under normal operating conditions. (IA/EEC) [61], [74]

MACRO A macro language used in Digital's VAX/VMS environment. (C) 610.13-1993w

macro (1) (software) In software engineering, a predefined sequence of computer instructions that is inserted into a program, usually during assembly or compilation, at each place that its corresponding macroinstruction appears in the program. (C) 610.12-1990
(2) In word processing, a predefined sequence of text and text formatting commands collected under a single user-defined name. Each time the name is entered, it is automatically replaced by the sequence of text and commands. (C) 610.2-1987
(3) A defined procedure or sequence of operations or characters that is inserted in the procedure each time its name is invoked. (SCC20) 771-1998

macroassembler An assembler that includes, or performs the functions of, a macrogenerator. (C) 610.12-1990

macrobending (fiber optics) In an optical waveguide, all macroscopic deviations of the axis from a straight line; distinguished from microbending. *See also:* microbend loss; macrobend loss; microbending. (Std100) 812-1984w

macrobend loss (fiber optics) In an optical waveguide, that loss attributable to macrobending. Macrobending usually causes little or no radiative loss. *Synonym:* curvature loss. *See also:* macrobending; microbend loss. (Std100) 812-1984w

macro definition *See:* macro.

macro generating program *See:* macrogenerator.

macrogenerator A routine, often part of an assembler or compiler, that replaces each macroinstruction in a source program with the predefined sequence of instructions that the macroinstruction represents. *Synonym:* macro generating program. (C) 610.12-1990

macroinstruction (1) (software) A source code instruction that is replaced by a predefined sequence of source instructions, usually in the same language as the rest of the program and usually during assembly or compilation. *See also:* macro; macrogenerator. (C) 610.12-1990, 610.10-1994w
(2) (computers) An instruction in a source language that is equivalent to a specified sequence of machine instructions. (C) [20], [85]

macro language A language used to define macros or macroinstructions. *Note:* IEEE Std 610.12-1990 defines terminology relating to macroinstructions. (C) 610.13-1993w

macro library A collection of macros available for use by a macrogenerator. *See also:* system library. (C) 610.12-1990

macroprocessor (software) A routine or set of routines provided in some assemblers and compilers to support the definition and use of macros. (C) 610.12-1990

macroprogramming Computer programming using macros and macroinstructions. (C) 610.12-1990

MAC service The unconfirmed connectionless-mode MAC service defined in ISO/IEC 10039, as an abstraction of the features common to a number of specific MAC services for Local Area Networks. (C/LM) 802.1G-1996

MAC service data unit (MSDU) The user data unit received in an MA-UNITDATA request for transfer by the medium access control (MAC) sublayer. (LM/C) 8802-6-1994

MAC sublayer *See:* medium access control sublayer.

MACSYMA An interactive programming system used to perform formal algebraic manipulation and symbolic mathematics. (C) 610.13-1993w

MAD *See:* minimum approach distance; Michigan Algorithmic Decoder.

MADCAP An extensible language used to perform numerical computation and set-theoretic operations, using input and output devices that permit two-dimensional input and output. (C) 610.13-1993w

mag card A colloquial reference to magnetic card. (C) 610.10-1994w

magic tee *See:* hybrid tee.

magner (polyphase circuit) At the terminals of entry into a delimited region, the algebraic sum of the reactive power for the individual terminals of entry when the the voltages are all determined with respect to the same arbitrarily selected common reference point in the boundary surface (which may be the neutral terminal of entry). The reference direction for the currents and the reference polarity for the voltages must be the same as for the instantaneous power and the active power.

The reactive power for each terminal entry is determined by considering each conductor and the common reference point as a single-phase two-wire circuit and finding the reactive power for each in accordance with the definition of **magner (single-phase two-wire circuit)**. If the voltages and currents are sinusoidal and of the same period, the reactive power Q for a three-phase circuit is given by

$$Q = E_aI_a\sin(\alpha_a - \beta_a) + E_bI_b\sin(\alpha_b - \beta_b) + E_cI_c\sin(\alpha_c - \beta_c)$$

where the symbols have the same meaning as in **power, instantaneous (polyphase circuit)**. If there is no neutral conductor and the common point for voltage measurement is selected as one of the phase terminals of entry, the expression will be changed in the same way as that for **power, instantaneous (polyphase circuit)**. If both the voltages and currents in the preceeding equations constitute symmetrical polyphase set of the same phase sequence

$$Q = 3E_aI_a\sin(\alpha_a - \beta_a)$$

In general the reactive power Q at the $(m + 1)$ terminals of entry of a polyphase circuit of m phases to a delimited region, when one of the terminals is the neutral terminal of entry, is expressed by the equation

$$Q = \sum_{s=1}^{s=m} \sum_{r=k}^{r=\infty} E_{sr}I_{sr}\sin(\alpha_{sr} - \beta_{sr})$$

where the symbols have the same meaning as in **power, active (polyphase circuit)**. The reactive power can also be stated in terms of the root-mean-square amplitude of the symmetrical components of the voltages and currents as

$$Q = \mathrm{m} \sum_{k=0}^{k=m-1} \sum_{r=k}^{r=\infty} E_{kr}I_{kr}\sin(\alpha_{kr} - \beta_{kr})$$

where the symbols have the same meaning as in **power, active (polyphase circuit)**. When the voltages and currents are quasi- periodic and the amplitudes of the voltages and currents are slowly varying, the reactive power for the circuit of each conductor may be determined for this condition as in **power, reactive (magner) (single-phase two-wire circuit)**. The reactive power for the polyphase circuit is the sum of the reactive power values for the individual conductors. Reactive power is expressed in vars when the voltages are in volts and the currents in amperes. *Note:* The sign of reactive power resulting from the above definition is the opposite of that given by the definition in the 1941 edition of the American Standard Definitions of Electrical Terms. The change has been made in accordance with a recommendation approved by the Standards Committee of the Institute of Electrical and Electronics Engineers, by the American National Standard Institute, and by the International Electrotechnical Commission. (Std100) 270-1966w

(2) (single-phase two-wire circuit) At the two terminals of entry of a single-phase two-wire circuit into a delimited region, for the special case of a sinusoidal voltage and a sinusoidal current of the same period, is equal to the product obtained by multiplying the root-mean-square value of the voltage between one terminal of entry and the second terminal of entry, considered as the reference terminal, by the root-mean-square value of the current through the first terminal and by the sine of the angular phase difference by which the voltage leads the current. The reference direction for the current and the reference polarity for the voltage must be the same as for active power at the same two terminals. Mathematically, the reactive power Q, for the case of sinusoidal voltage and current, is given by

$$Q = EI\sin(\alpha - \beta)$$

in which the symbols have the same meaning as in **power, instantaneous (two-wire circuit).** For the same conditions, the reactive power Q is also equal to the imaginary part of the product of the phasor voltage and the conjugate of the phasor current, or to the negative of the imaginary part of the product of the conjugate of the phasor voltage and the phasor current. Thus,

$$Q = \mathrm{Im}\mathbf{EI}^* = -\mathrm{Im}\mathbf{E}^*\mathbf{I} = \frac{1}{2j}[\mathbf{EI}^* - \mathbf{E}^*\mathbf{I}$$

in which **E** and **I** are the phasor voltage and phasor current, respectively, and * denotes the conjugate of the phasor to which it is applied. If the voltage is an alternating voltage and the current is an alternating current, the reactive power for each harmonic component is equal to the product obtained by multiplying the root-mean-square amplitude of that harmonic component of the voltage by the root-mean-square amplitude of the same harmonic component of the current and by the sine of the angular phase difference by which that harmonic component of the voltage leads the same harmonic component of the current. Mathematically, the reactive power of the r th harmonic component of Q_r is given by

$$Q_r = E_rI_r\sin(\alpha_r - \beta_r) = Im\mathbf{E}_r\mathbf{I}_r^* = \frac{1}{2j}[\mathbf{E}_r\mathbf{I}_r^* - \mathbf{E}_r^*\mathbf{I}_r] = -\mathrm{Im}\mathbf{E}_r^*\mathbf{I}_r$$

in which the symbols have the same meaning as in **power, instantaneous (two-wire circuit)** and **power, active (single-phase two-wire circuit) (average power) (power)**. The reactive power at the two terminals of entry of a single-phase two- wire circuit into a delimited region, for an alternating voltage and current, is equal to the sum of the values of reactive power for every harmonic component. Mathematically, the reactive power Q for an alternating voltage and current, is given by

$$Q = Q_1 + Q_2 + Q_3 + Q_4 + \ldots + Q_r + \ldots = E_1I_1\sin(\alpha_1 - \beta_1) + E_2I_2\sin(\alpha_2 - \beta_2) + \ldots = \sum_{r=1}^{r=\infty} Q_r = \sum_{r=1}^{r=\infty} E_rI_r\sin(\alpha_r - \beta_r)$$

in which the symbols have the same meaning as in **power, instantaneous (two-wire circuit)**. If the voltage and current are quasi-periodic functions of the form given in **power, instantaneous (two-wire circuit)** and the amplitudes are slowly varying, so that each may be considered to be constant during any one period, but to have slightly different values in successive periods, the reactive power at any time t may be taken as

$$Q = \sum_{r=1}^{r=\infty} E_r(t)I_r(t)\sin(\alpha_r - \beta_r)$$

by analogy with the expression for active power. When the reactive power is positive, the direction of flow of quadergy is in the reference direction of energy flow. Because the reactive power for each harmonic may have either sign, the direction of the reactive power for a harmonic component may be the same as or opposite to the direction of the total reactive power. The value of reactive power is expressed in vars when the voltage is in volts and the current in amperes. *Notes:* 1. The sign of reactive power resulting from the above definition is the opposite of that given by the definition in the 1941 edition of the American Standard Definitions of Electrical Terms. The change has been made in accordance with a recommendation approved by the Standards Committee of the Institute of Electrical and Electronics Engineers, by the American National Standards Institute, and by the Electrotechnical Commission. 2. Any designation of positive reactive power as inductive reactive power is deprecated. If the reference direction is from the generator toward the load, reactive power is positive if the load is predominantly inductive and negative if the load is predominantly capacitive. Thus a capacitor is a source of quadergy and an inductor is a consumer of quadergy. Designations of two kinds of reactive power are unnecessary and undesirable. (Std100) 270-1966w

(3) The product of voltage and the component of alternating current that is 90° out of phase with it. In a passive network reactive power represents the energy that is exchanged alternatively between a capacitive and an inductive storage medium. (CAS) [13]

magnesium cell A primary cell with the negative electrode made of magnesium or its alloy. *See also:* electrochemistry. (EEC/PE) [119]

magnet A body that produces a magnetic field external to itself. (Std100) 270-1966w

magnet, focusing *See:* focusing magnet.

magnetic (1) Pertaining to any form of storage medium in which patterns of magnetization are used to store or represent information; for example, magnetic storage, or a magnetic delay line. (C) 610.10-1994w

(2) (as applied to a switching device) A term indicating that interruption of the circuit takes place between contacts separable in an intense magnetic field. *Note:* With respect to contactors, this term indicates the means of operation. (SWG/PE) C37.100-1992

magnetic air circuit breaker *See:* circuit breaker; magnetic; air circuit breaker.

magnetically shielded type instrument An instrument in which the effect of external magnetic fields is limited to a stated value. The protection against this influence may be obtained either through the use of a physical magnetic shield or through the instrument's inherent construction. *See also:* instrument. (EEC/AII) [102]

magnetic amplifier A device using one or more saturable reactors, either alone or in combination with other circuit elements, to secure power gain. Frequency conversion may or may not be included. (PE/EM) 43-1974s

magnetic area moment *See:* magnetic moment.

magnetic-armature loudspeaker A magnetic loudspeaker whose operation involves the vibration of a ferromagnetic armature. (EEC/PE) [119]

magnetic axis (rotating machinery) (coil or winding) The line of symmetry of the magnetic-flux density produced by current in a coil or winding, this being the location of approximately maximum flux density, with the air gap assumed to be uniform. *See also:* stator; rotor. (PE) [9]

magnetic bearing (navigation) Bearing relative to magnetic north. *See also:* navigation. (AES/RS) 686-1982s, [42]

magnetic biasing The simultaneous conditioning of the magnetic recording medium during recording by the superposing of an additional magnetic field upon the signal magnetic field. *Note:* In general, magnetic biasing is used to obtain a substantially linear relationship between the amplitude of the signal and the remanent flux density in the recording medium. *See also:* phonograph pickup; direct-current magnetic biasing; alternating-current magnetic biasing. (SP/MR) [32]

magnetic bias, relay *See:* relay magnetic bias.

magnetic blowout A magnet, often electrically excited, whose field is used to aid the interruption of an arc drawn between contacts. *See also:* contactor. (IA/ICTL/IAC) [60]

magnetic brake A friction brake controlled by electromagnetic means. (IA/ICTL/IAC) [60]

magnetic-brush development (electrostatography) Development in which the image-forming material is carried to the field of the electrostatic image by means of ferromagnetic particles acting as carriers under the influence of a magnetic field. *See also:* electrostatography. (ED) [46]

magnetic bubble memory *See:* bubble memory.

magnetic card (1) (computers) A card with a magnetic surface on which data can be stored by selective magnetization of portions of the flat surface. (C) [20], [85]

(2) A card with a magnetic surface that can be used for data storage. (C) 610.10-1994w

magnetic cell A storage cell in which patterns of magnetization are used to represent information. *Synonym:* static magnetic cell. (C) 610.10-1994w

magnetic character A character that is formed on paper using a special magnetic ink. (C) 610.10-1994w

magnetic circuit A region at whose surface the magnetic induction is tangential. *Note:* The term is also applied to the minimal region containing essentially all the flux, such as the core of a transformer. (Std100) 270-1966w

magnetic compass A device for indicating the direction of the horizontal component of a magnetic field. *See also:* magnetometer. (EEC/PE) [119]

magnetic-compass repeater indicator A device that repeats the reading of a master direction indicator, through a self-synchronous coupling means. (EEC/PE) [119]

magnetic constant (1) (permeability of free space) (pertinent to any system of units) The magnetic constant is the scalar dimensional factor that in that system relates the mechanical force between two currents to their magnitudes and geometrical configurations. More specifically, μ_0 is the magnetic constant when the element of force $\mathbf{dF}$ of a current element $I_1\mathbf{dI}_1$ on another current element $I_2\mathbf{dI}_2$ at a distance r is given by

$$\mathbf{dF} = \mu_0 I_1 I_2 \mathbf{dI}_1 \times (\mathbf{dI}_2 \times \mathbf{r}_1)/nr^2$$

where $\mathbf{r}_1$ is a unit vector in the direction from $\mathbf{dI}_1$ to $\mathbf{dI}_2$, and n is a dimensionless factor which is unity in unrationalized systems and 4π in a rationalized system. *Note:* In the centimeter-gram-second (cgs) electromagnetic system, μ_0 is assigned the magnitude unity and the dimension numeric. In the centimeter-gram-second (cgs) electrostatic system, the magnitude of μ_0 is that of $1/c^2$ and the dimension is $[L^{-2}T^2]$. In the International System of Units (SI) μ_0 is assigned the magnitude $4\pi \cdot 10^{-7}$ and has the dimension $[LMT^{-2}I_{-2}]$. (Std100) 270-1966w

(2) (radio-wave propagation) *See also:* permeability. (AP/PROP) 211-1990s

magnetic contactor A contactor actuated by electromagnetic means. *See also:* contactor. (IA/ICTL/IAC) [60]

magnetic control relay A relay that is actuated by electromagnetic means. *Note:* When not otherwise qualified, the term refers to a relay intended to be operated by the opening and closing of its coil circuit and having contacts designed for energizing and/or de-energizing the coils of magnetic contactors or other magnetically operated device. *See also:* relay. (IA/IAC) [60]

magnetic core (1) A configuration of magnetic material that is, or is intended to be, placed in a rigid spatial relationship to current-carrying conductors and whose magnetic properties are essential to its use. *Note:* For example, it may be used to concentrate an induced magnetic field as in a transformer, induction coil, or armature; to retain a magnetic polarization for the purpose of storing data; or for its nonlinear properties as in a logic element. It may be made of iron wires, iron oxide, coils of magnetic tape, ferrite, thin film, etc. (C/MIL) 162-1963w, [20], [85], [2]

(2) A tiny doughnut-shaped piece of magnetic material used for its non-linear properties to store data in main storage. *Synonym:* memory core. (C) 610.10-1994w

magnetic coupling *See:* coupling; electrical coupling.

magnetic course (navigation) Course relative to magnetic north. *See also:* navigation. (AES/RS) 686-1982s, [42]

magnetic deflection (cathode-ray tubes) Deflecting an electron beam by the action of a magnetic field. *See also:* cathode-ray tube. (ED) [45], [84]

magnetic delay line A delay line whose operation is based on the time of propagation of magnetic waves. (C) [20], 610.10-1994w, [85]

magnetic deviation Angular difference between compass north and magnetic north caused by magnetic effects in the vehicle. *See also:* navigation. (AES) [42]

magnetic device (packaging machinery) A device actuated by electromagnetic means. (IA/PKG) 333-1980w

magnetic dipole *See:* Hertzian magnetic dipole.

magnetic dipole moment (centimeter-gram-second electromagnetic-unit system) The volume integral of magnetic polarization is often called magnetic dipole moment. *See also:* magnetic polarization. (Std100) 270-1966w

magnetic direction indicator (MDI) An instrument providing compass indication obtained electrically from a remote gyro-stabilized magnetic compass or equivalent. *See also:* radio navigation. (EEC/PE) [119]

magnetic disk (1) A flat circular plate with a magnetic surface on which data can be stored by selective polarization of portions of the flat surface.
(C/MIL) 162-1963w, [2], [20], [85]
(2) A disk made of plastic or metal that is coated with a magnetizable surface on one or both sides, on which information can be stored. *Contrast:* optical disk. *See also:* magneto-optical disk; diskette; hard disk; platter; floppy disk.
(C) 610.10-1994w

magnetic disk drive A disk drive that can access a magnetic disk. (C) 610.10-1994w

magnetic dissipation factor (magnetic material) The cotangent of its loss angle or the tangent of its hysteretic angle.
(Std100) 270-1966w

magnetic drum (1) A right circular cylinder with a magnetic surface on which data can be stored by selective polarization of portions of the curved surface.
(C/MIL) 162-1963w, [20], [2], [85]
(2) A cylinder whose entire surface is coated with a magnetic material on which information can be stored in tracks running the circumference of the cylinder. (C) 610.10-1994w

magnetic field ($\vec{H}$) (1) The field surrounding any current-carrying conductor. *See also:* electric field.
(PE/IC) 1143-1994r
(2) For time harmonic fields in a medium with linear and isotropic magnetic properties, the magnetic flux density divided by the permeability of the medium.
(AP/PROP) 211-1997

magnetic field induction (1) (inductive coupling) The process of generating voltages and/or currents in a conductive object or electric circuit by means of time-varying magnetic fields. *Notes:* 1. "Magnetic field induction" was formerly called "electromagnetic induction." This usage is now deprecated because electromagnetic induction refers to combined electric and magnetic field effects. 2. "Magnetic field induction" is preferred over "magnetic induction" because the latter is reserved to mean magnetic flux density.
(T&D/PE) 1048-1990
(2) (overhead power lines) The induction process that results from time-varying quasi-static magnetic fields. *Notes:* 1. "Magnetic field induction" was formerly called "electromagnetic induction". This usage is now deprecated because electromagnetic induction refers to combined electric and magnetic field effects. 2. The term "magnetic field induction" is preferred over "magnetic induction" because the latter may be taken to mean magnetic flux density. *Synonym:* inductive coupling (ground system). *See also:* electromagnetic field induction. (T&D/PE) 539-1990
(3) (inductive coupling) The process of generating voltages and/or currents in conductive objects or electric circuits by the induction process that results from time-varying quasi-static magnetic fields. *Notes:* 1. "Magnetic field induction" was formerly called "electromagnetic induction." This usage is now depreciated because electromagnetic induction refers to the combined electric and magnetic field effects. 2. The term "magnetic field induction" is preferred over "magnetic induction" because the latter may be taken to mean magnetic flux density. 3. The fields in the vicinity of a transmission line can be adequately described as an electric field and a magnetic field. Electromagnetic may imply one or both of these fields. There should be no questions as to what is meant when the electric field or the magnetic field is discussed.
(T&D/PE) 524a-1993r

magnetic field integral equation A Fredholm integral equation of the second kind for the electric current density induced on the surface S of a perfect electric conductor. *Note:* The tangential component of the incident magnetic field acts as the source for the current, hence, the name. The equation is as follows:

$$\vec{J}_S = 2\hat{n} \times \vec{H}^i + 2\hat{n} \times \int_{S_0} (\vec{J}_S \times \nabla_0 g) dS_0$$

where
$\hat{n}$ = unit normal to S
$\vec{H}^i$ = incident magnetic field
$g = \exp[-jk_0|\vec{r} - \vec{r}_0|]/4\pi|\vec{r} - \vec{r}_0|$
$\vec{J}_s$ = surface current density
(AP/PROP) 211-1997

magnetic field intensity *See:* magnetic field strength.

magnetic field interference A form of interference induced in the circuits of a device due to the presence of a magnetic field. *Note:* It may appear as common-mode or normal-mode interference in the measuring circuit. *See also:* accuracy rating. (EEC/ERI) [111]

magnetic field strength (*H*) (1) (magnetizing force) That vector point function whose curl is the current density and that is proportional to magnetic flux density in regions free of magnetized matter. *Note:* A consequence of this definition is that the familiar formula

$$\mathbf{H} = \frac{1}{4\pi} \int \mathbf{J} \times \nabla(1/r)dv - \frac{1}{4\pi} \nabla \int \mathbf{M} \cdot \nabla (1/r)dv$$

(where **H** is the magnetizing force, **J** is current density, and **M** is magnetization) is a mathematical identity.
(Std100) 270-1966w
(2) (overhead power lines) A vector quantity, often denoted as $\vec{H}$, related to the magnetic flux density, $\vec{B}$, by:

$\vec{H} = (\vec{B}/\mu_0) - \vec{M}$

where μ_0 = the magnetic permeability of free space, $\vec{M}$ = the magnetization of the magnetic medium. In free space, $\vec{M}$ vanishes and the relationship between $\vec{H}$ and $\vec{B}$ becomes

$\vec{H} = \vec{B}/\mu_0$

The preferred unit for $\vec{H}$ is amperes per meter (A/m).
(T&D/PE) 539-1990
(3) The magnitude of a time-varying magnetic field, which, when coupled to a coil, induces a voltage. The unit of magnetic field strength is ampere per meter (A/m).
(COM/TA) 1027-1984s
(4) The magnitude of the magnetic field vector $\vec{H}$. The units of magnetic field strength are in amperes per meter. *Synonym:* magnetizing force. (AP/PROP) 211-1997
(5) A field vector that is equal to the magnetic flux density divided by the permeability of the medium. Magnetic field strength is expressed in units of amperes per meter (A/m).
(NIR) C95.1-1999

magnetic field strength produced by an electric current (Biot-Savart law and/or Ampere's law) The magnetic field strength, at any point in the neighborhood of a circuit in which there is an electric current i, can be computed on the assumption that every infinitesimal length of circuit produces at the point an infinitesimal magnetizing force and the resulting magnetizing force at the point is the vector sum of the contributions of all the elements of the circuit. The contribution, dH, to the magnetizing force at a point P caused by the current i in an element ds of a circuit that is at a distance r from P, has a direction t at is perpendicular to both ds and r and a magnitude equal to

$$\frac{i \text{ ds} \sin \theta}{r^2}$$

where θ is the angle between the element ds and the line r. In vector notation

$$dH = \frac{i[r \times \text{ds}]}{r^2}$$

This law is sometimes attributed to Biot and Savart, sometimes to Ampere, and sometimes to Laplace, but no one of them gave it in its differential form. (Std100) 270-1966w

magnetic field vector (A) (radio-wave propagation) *(any point in a magnetic field).* The magnetic induction divided by the permeability of the medium. **(B) (radio-wave propagation)** In a medium with linear and isotropic magnetic property, the magnetic induction divided by the permeability of the medium. *Synonym:* magnetic vector. *See also:* radio-wave propagation. (AP) 211-1977

magnetic figure of merit The ratio of the real part of complex apparent permeability to magnetic dissipation factor. *Note:* The magnetic figure of merit is a useful index of the magnetic efficiency of a material in various electromagnetic devices. (Std100) 270-1966w

magnetic filling factor *See:* filling factor.

magnetic flux (φ) (1) (through an area) The surface integral of the normal component of the magnetic induction over the area. Thus,

$$\phi_A = \int_A (\mathrm{B} \cdot dA)$$

where ϕ_A is the flux through the area A, and **B** is the magnetic induction at the element dA of this area. *Note:* The net magnetic flux through any closed surface is zero. (Std100) 270-1966w

(2) A condition in a medium produced by a magnetomotive force such that, when altered in magnitude, a voltage is induced in an electric circuit linked with the flux. (IA/PC) 844-1991

magnetic flux density (*B*) (1) (β) (induction motors) Flux per unit area through an element of area normal to the direction of flux. (IA/PC) 844-1991

(2) (overhead power lines) The vector quantity, often denoted as $\vec{\mathrm{B}}$, of zero divergence at all points which determines the component of the Coulomb-Lorentz force that is proportional to the velocity of a moving charge. *Notes:* 1. In a zero electric field, the force, $\vec{\mathrm{F}}$, is given by

$$\vec{\mathrm{F}} = q\vec{v} \times \vec{\mathrm{B}}$$

where $\vec{v}$ = the velocity of the electric charge q. The vector properties of the field produced by currents in power lines are the same as those given above for the electric field. The preferred unit for the magnitude of the field components is the tesla (T) (1 T = 10^4 Gauss). 2. For time-varying (ac) fields, values are expressed as their rms values unless stated otherwise. *Synonym:* magnetic field. (T&D/PE) 539-1990

(3) (magnetic field) The vector quantity (*B*-field) of divergence zero at all points, which determines the component of the Coulomb-Lorentz force, that is proportional to the velocity of the charge carrier. *Note:* In a zero electric field, the force F is given by $\vec{\mathrm{F}} = q\vec{\mathrm{v}} \times \vec{\mathrm{B}}$, where $\vec{\mathrm{v}}$ is the velocity of the electric charge q. The vector properties of the field produced by currents in power lines are the same as those given above for the electric field. The magnitudes of the field components are expressed by their rms values in tesla (1T= 10^4G). (T&D/PE) 644-1994

(4) A vector field that acts on moving charges (q) such that the force per unit charge ($\vec{\mathrm{F}}$) is equal to the vector (cross) product of the velocity ($\vec{\mathrm{v}}$) of the particle and $\vec{\mathrm{B}}$, the magnetic flux density:

$$\frac{\vec{\mathrm{F}}}{q} = \vec{\mathrm{v}} \times \vec{\mathrm{B}}$$

The units of magnetic flux density are in volt seconds per meter squared. (AP/PROP) 211-1997

(5) A field vector quantity that results in a force (F) that acts on a moving charge or charges. The vector product of the velocity (v) at which an infinitesimal unit test charge, q, is moving with B, is the force that acts on the test charge divided by q.

$$\frac{F}{q} = (v \times B)$$

Magnetic flux density is expressed in units of tesla (T). One tesla is equal to 10^4 gauss (G). (NIR) C95.1-1999

magnetic flux density meter (1) (overhead power lines) A meter designed to measure magnetic flux density. These meters may use any of several types of flux density sensors or probes. *Note:* For measurement of the magnetic flux density from ac power systems, the meter shall conform to IEEE Std 644-1987. (T&D/PE) 539-1990

(2) A meter designed to measure the magnetic flux density. *Note:* Several types of meters are in common use, e.g., field meters with air core coil probes, meters with Hall-effect probes, and meters that combine two coils with a ferromagnetic core as in a fluxgate magnetometer. (T&D/PE) 1308-1994

magnetic flux leakage That portion of the total magnetic flux in a circuit that does not intercept the material that contains the magnetic flux that is heating. 844-1991

magnetic focusing (electron beams) A method of focusing an electron beam by the action of a magnetic field. (BT/ED/AV) [34], [45]

magnetic friction clutch (coupling) (electric coupling) A friction clutch (coupling) in which the pressure between the friction surfaces is produced by magnetic attraction. (EM/PE) 290-1980w

magnetic friction coupling An electric coupling in which torque is transmitted by means of mechanical friction. Pressure normal to the rubbing surfaces is controlled by means of an electromagnet and a return spring. *Note:* Couplings may be either magnetically engaged or magnetically released depending upon application. (PE) [9]

magnetic head (1) A transducer for converting electric variations into magnetic variations for storage on magnetic media, or for reconverting energy so stored into electric energy, or for erasing such stored energy. (SP/MR) [32]

(2) A head that can read, write, or erase on a magnetic storage medium. *See also:* erase head; floating head; fixed head; cylinder; access arm. (C) 610.10-1994w

magnetic heading (navigation) Heading relative to magnetic north. *See also:* navigation. (AES/RS) 686-1982s, [42]

magnetic hysteresis (electrical heating systems) The property of a magnetic material to convert electric energy to heat by virtue of the fact that the magnetic induction for a given magnetizing force depends upon the previous conditions of magnetization. (IA/PC) 844-1991

magnetic hysteresis loss (A) (magnetic material) The power expended as a result of magnetic hysteresis when the magnetic induction is periodic. **(B) (magnetic material)** The energy loss per cycle in a magnetic material as a result of magnetic hysteresis when the induction is cyclic (not necessarily periodic). *Note:* Definitions (A) and (B) are not equivalent; both are in common use. (Std100) 270-1966

magnetic hysteretic angle The mean angle by which the exciting current leads the magnetizing current. *Note:* Because of hysteresis, the instantaneous value of the hysteretic angle will vary during the cycle; the hysteretic angle is taken to be the mean value. (Std100) 270-1966w

magnetic induction* (1) (signal-transmission system) The process of generating currents or voltages in a conductor by means of a magnetic field. *See also:* magnetic flux density; signal. (IE) [43]

(2) *See also:* magnetic flux density. (AP/PROP) 211-1997

* Deprecated.

magnetic ink (1) An ink that contains particles of a magnetic substance whose presence can be detected by magnetic sensors. (C) [85], [20]

(2) A special ink containing magnetic particles that can be detected and traced by input devices designed specifically for that purpose. (C) 610.2-1987

(3) Special ink that can be read by a magnetic scanner, such as is used on bank checks. *See also:* magnetic character. (C) 610.10-1994w

magnetic ink character A character imprinted on a document using magnetic ink. *Synonym:* magnetic character. (C) 610.2-1987

magnetic ink character reader (MICR) A character reader that recognizes characters using magnetic ink character recognition. *Contrast:* optical character reader. (C) 610.10-1994w

magnetic ink character recognition (MICR) The automatic recognition of magnetic ink characters. *Contrast:* optical character recognition. *See also:* code for magnetic characters. (C) 610.2-1987, 610.10-1994w

magnetic ink scanner A scanner that can read magnetic ink characters. (C) 610.10-1994w

magnetic latching relay (A) A relay that remains operated from remanent magnetism until reset electrically. **(B)** A bistable polarized (magnetically latched) relay. (PE/EM) 43-1974

magnetic loading (rotating machinery) The average flux per unit area of the air-gap surface. *See also:* asynchronous machine. (PE) [9]

magnetic loss That contribution to the attenuation constant of a propagating mode on a planar transmission line that represents losses associated with the magnetic properties of the substrates (and overlays) materials involved, which may also include conduction mechanisms. (MTT) 1004-1987w

magnetic loss angle (1) (core) The angle by which the fundamental component of the core-loss current leads the fundamental component of the exciting current in an inductor having a ferromagnetic core. *Note:* The loss angle is the complement of the hysteretic angle. (Std100) 270-1966w
(2) For a pure sinusoidal wave in a medium with complex permeability μ, the angle defined by the equation

$$f_m = \tan^{-1}\left(\frac{\mu''}{\mu'}\right)$$

where
μ'' = the imaginary part of the complex permeability
μ' = the real part of the complex permeability
(AP/PROP) 211-1997

magnetic loss factor, initial (material) The product of the real component of its complex permeability and the tangent of its magnetic loss angle, both measured when the magnetizing force and the induction are vanishingly small. *Note:* In anisotropic media, magnetic loss factor becomes a matrix. (Std100) 270-1966w

magnetic loss filling factor *See:* filling factor.

magnetic loudspeaker A loudspeaker in which acoustic waves are produced by mechanical forces resulting from magnetic reactions. (EEC/PE) [119]

magnetic microphone *See:* variable-reluctance microphone.

magnetic microscope An electron microscope with magnetic lenses. *See also:* electron optics. (ED) [45], [84]

magnetic mine A submersible explosive device with a detonator actuated by the distortion of the earth's magnetic field caused by the approach of a mass of magnetic material such as the hull of a ship. (EEC/PE) [119]

magnetic moment (1) (magnetized body) The volume integral of the magnetization

$$\mathrm{m} = \int \mathrm{M}dv$$

(2) (current loop)

$$\mathrm{m} = I\int \mathrm{n}da = (I/2)\int r \times dr$$

where n is the positive normal to a surface spanning the loop, and r is the radius vector from an arbitrary origin to a point on the loop. *Notes:* 1. The numerical value of the moment of a plane current loop is *IA*, where *A* is the area of the loop. 2. The reference direction for the current in the loop indicates a clockwise rotation, when the observer is looking through the loop in the direction of the positive normal. 270-1966w

magnetic north The direction of the horizontal component of the earth's magnetic field toward the north magnetic pole. *See also:* navigation. (AES/RS) 686-1982s, [42]

magnetic overload relay An overcurrent relay the electric contacts of which are actuated by the electromagnetic force produced by the load current or a measure of it. *See also:* relay. (IA/IAC) [60]

magnetic-particle coupling (1) (electric coupling) An electric coupling that transmits torque through the medium of magnetic particles in a magnetic field between coupling members. (EM/PE) 290-1980w
(2) A type of electric coupling in which torque is transmitted by means of a fluid whose viscosity is adjustable by virtue of suspended magnetic particles. *Note:* The coupling fluid is incorporated in a magnetic circuit in which the flux path includes the two rotating members, the fluid, and a magnetic yoke. Flux density, and hence the fluid viscosity, are controlled through adjustment of current in a magnetic coil linking the flux path. (PE) [9]

magnetic pickup *See:* variable-reluctance pickup.

magnetic-plated wire A magnetic wire having a core of nonmagnetic material and a plated surface of ferromagnetic material. (SP) [32]

magnetic-platform influence (electric instruments) The change in indication caused solely by the presence of a magnetic platform on which the instrument is placed. *See also:* accuracy rating. (EEC/AII) [102]

magnetic polarization In the centimeter-gram-second electromagnetic-unit system, the intrinsic induction divided by 4π is sometimes called magnetic polarization or magnetic dipole moment per unit volume. *See also:* intrinsic induction. (Std100) 270-1966w

magnetic poles (magnet) Those portions of the magnet toward which or from which the external magnetic induction appears to converge or diverge, respectively. *Notes:* 1. By convention, the north-seeking pole is marked with N, or plus, or is colored red. 2. The term is also sometimes applied to a fictitious magnetic charge. (Std100) 270-1966w

magnetic pole strength (magnet) The magnetic moment divided by the distance between its poles. *Note:* Many authors use the above quantity multiplied by the magnetic constant; the two choices are numerically equal in the centimeter-gram-second electromagnetic-unit system. (Std100) 270-1966w

magnetic-powder-impregnated tape A magnetic tape that consists of magnetic particles uniformly dispersed in a nonmagnetic material. *Synonym:* impregnated tape. (SP) [32]

magnetic power factor The cosine of the magnetic hysteretic angle (the sine of the magnetic loss angle). (Std100) 270-1966w

magnetic recorder Equipment incorporating an electromagnetic transducer and means for moving a magnetic recording medium relative to the transducer for recording electric signals as magnetic variations in the medium. *Note:* The generic term magnetic recorder can also be applied to an instrument that has not only facilities for recording electric signals as magnetic variations, but also for converting such magnetic variations back into electric variations. *See also:* phonograph pickup. (SP) [32]

magnetic recording (1) (facsimile) Recording by means of a signal-controlled magnetic field. *See also:* recording. (COM) 168-1956w
(2) A method for storing data by selectively magnetizing portions of a magnetizable material. *See also:* nonreturn-to-reference recording; longitudinal magnetic recording; phase-modulation recording; perpendicular magnetic recording; return-to-reference recording. (C) 610.10-1994w

magnetic recording head In magnetic recording, a transducer for converting electric currents into magnetic fields, in order to store the electric signal as a magnetic polarization of the magnetic medium. (SP) [32]

magnetic recording medium A material usually in the form of a wire, tape, cylinder, disk, etc., on which a magnetic signal may be recorded in the form of a pattern of magnetic polarization. (SP) [32]

magnetic relay freezing Sticking of the relay armature to the core as a result of residual magnetism. (EEC/REE) [87]

magnetic reproducer Equipment incorporating an electromagnetic transducer and means for moving a magnetic recording medium relative to the transducer, for reproducing magnetic signals as electric signals. (SP) [32]

magnetic reproducing head In magnetic recording, a transducer for collecting the flux due to stored magnetic polarization (the recorded signal) and converting it into an electric voltage. (SP) [32]

magnetic rotation (polarized light) When a plane polarized beam of light passes through certain transparent substances along the lines of a strong magnetic field, the plane of polarization of the emergent light is different from that of the incident light. *Synonym:* Faraday effect. (Std100) 270-1966w

magnetic sensitivity (Hall effect devices) The ratio of the voltage across the Hall terminals to the magnetic flux density for a given magnitude of control current. (MAG) 296-1969w

magnetic shunt The section of the core of the ferroresonant transformer that provides the major path for flux generated by the primary winding current that does not link the secondary winding. In addition, the shunts provide a major path for the flux resulting from the output and resonating winding currents that do not link the primary winding. (PEL) 449-1998

magnetic spectrograph An electronic device based on the action of a constant magnetic field on the paths of electrons, and used to separate electrons with different velocities. *See also:* electron device. (ED) [84], [46]

magnetic starter (packaging machinery) A starter actuated by electromagnetic means. (IA/PKG) 333-1980w

magnetic storage (1) A method of storage that uses the magnetic properties of matter to store data by magnetization of materials such as cores, films, or plates, or of material located on the surfaces of tapes, discs, or drums, etc. *See also:* magnetic core; magnetic drum; magnetic tape. (C) 162-1963w
(2) Any storage medium that stores data using magnetic properties such as magnetic cores, disks, or tapes. *Contrast:* semiconductor storage. (C) 610.10-1994w

magnetic storm (1) A disturbance in the Earth's magnetic field, associated with abnormal solar activity, and capable of seriously affecting both radio and wire transmission. *See also:* radio transmitter. (EEC/PE) [119]
(2) A disturbance of the Earth's magnetic field, generally lasting one or more days and characterized by significant changes in the strength of this field. (AP/PROP) 211-1997

magnetic susceptibility (isotropic medium) In rationalized systems, the relative permeability minus unity.

$$\kappa = \mu_r - 1 = B_i/\mu_0 H$$

Notes: 1. In unrationalized systems, $k = (\mu_r - 1)4\pi$. 2. The susceptibility divided by the density of a body is called the susceptibility per unit mass, or simply the mass susceptibility. The symbol is χ. Thus, $\chi = \kappa/\rho$ where ρ is the density. χ multiplied by the atomic weight is called the atomic susceptibility. The symbol is χ_A. 3. In anisotropic media, susceptibility becomes a matrix. (Std100) 270-1966w

magnetic tape (1) (A) (homogeneous or coated) A tape with a magnetic surface on which data can be stored by selective polarization of portions of the surface. **(B) (homogeneous or coated)** A tape of magnetic material used as the constituent in some forms of magnetic cores. *See also:* coated magnetic tape. (C/MIL) 162-1963, [85], [2]
(2) A storage medium made of a flexible plastic ribbon that is coated with magnetic material (such as an iron oxide compound) on which information can be stored. (C) 610.10-1994w

magnetic tape cartridge A cartridge holding magnetic tape, on which information can be stored. (C) 610.10-1994w

magnetic tape cassette A cassette holding magnetic tape on which information can be stored. *See also:* magnetic tape cartridge. (C) 610.10-1994w

magnetic tape drive *See:* tape drive.

magnetic tape handler (test, measurement, and diagnostic equipment) A device that handles magnetic tape and usually consists of a tape transport and magnetic tape reader with associated electrical and electronic equipments. Most units provide for tape to be wound and stored on reels; however, some units provide for the tape to be stored loosely in closed bins. (MIL) [2]

magnetic tape reader (1) (test, measurement, and diagnostic equipment) A device capable of converting information from magnetic tape where it has been stored as variations in magnetizations into a series of electrical impulses. (MIL) [2]
(2) A reader capable of reading information on magnetic tape. (C) 610.10-1994w

magnetic tape storage A type of sequential access storage in which information is stored by magnetic recording on the surface of a magnetic tape. (C) 610.10-1994w

magnetic test coil A coil that, when connected to a suitable device, can be used to measure a change in the value of magnetic flux linked with it. *Note:* The change in the flux linkage may be produced by a movement of the coil or by a variation in the magnitude of the flux. Test coils used to measure magnetic induction B are often called B coils; those used to determine magnetizing force H may be called H coils. A coil arranged to rotate through an angle of 180 degrees about an axis of symmetry perpendicular to its magnetic axis is sometimes called a flip coil. *See also:* magnetometer. (EEC/PE) [119]

magnetic thin film (1) A layer of magnetic material, usually less than 10 000 angstroms thick. *Note:* In electronic computers, magnetic thin films may be used for logic or storage elements. *See also:* coated magnetic tape; magnetic core; magnetic tape. (C) 162-1963w
(2) A layer of magnetic material, usually less than one micron thick, applied to a carrier or base for use as storage cells. (C) 610.10-1994w

magnetic thin film storage A type of magnetic storage in which information is stored by magnetic recording on a magnetic thin film. (C) 610.10-1994w

magnetic track braking A system of braking in which a shoe or slipper is applied to the running rails by magnetic means. *See also:* electric braking. (EEC/PE) [119]

magnetic variometer An instrument for measuring differences in a magnetic field with respect to space or time. *Note:* The use of variometer to designate a continuously adjustable inductor is deprecated. *See also:* magnetometer. (EEC/PE) [119]

magnetic vector (radio-wave propagation) *See also:* magnetic field vector. (AP) 211-1977s

magnetic vector potential An auxiliary solenoidal vector point function characterized by the relation that its curl is equal to the magnetic induction and its divergence vanishes. Curl **A** = **B**, Divergence **A** = **0**. *Note:* These relations are satisfied identically by

$$A = (\mu_0/4\pi)\left[\int \mathbf{M} \times \nabla(1/r)dv + \int (\mathbf{J}/r)dv\right]$$

where v is the volume. (Std100) 270-1966w

magnetization (intensity of magnetization) (at a point of a body) The intrinsic induction at that point divided by the magnetic constant of the system of units employed:

$$M = B_i/\mu_0 = (B - \mu_0 H)/\mu_0$$

Note: The magnetization can be interpreted as the volume density of magnetic moment. (Std100) 270-1966w

magnetizing current (1) A hypothetical current assumed to flow through the magnetizing inductance of a transformer. (CHM) [51]

(2) **(rotating machinery)** The quadrature (leading) component (with respect to the induced voltage) of the exciting current supplied to a coil. (PE) [9], 270-1966w

magnetizing force *See:* magnetic field strength.

magnetizing inductance A hypothetical inductance, assumed to be in parallel with the core-loss resistance, that would store the same amount of energy as that stored in the core for a specified value of excitation. (CHM) [51]

magnet meter (magnet tester) An instrument for measuring the magnetic flux produced by a permanent magnet under specified conditions of use. It usually comprises a torque-coil or a moving-magnet magnetometer with a particular arrangement of pole-pieces. *See also:* magnetometer. (PE/EEC) [119]

magneto *See:* magnetoelectric generator.

magneto central office A telephone central office for serving magneto telephone sets. (COM) [48]

magnetoelectric generator (electric installations on shipboard) An electric generator, in which the magnetic flux is provided by one or more permanent magnets. (IA/MT) 45-1983s

magnetographic printer A nonimpact printer that creates, by means of magnetic heads operating on a metallic drum, a latent image which is made visible by a toner and transferred and fixed on paper. (C) 610.10-1994w

magneto-hydrodynamic wave (radio-wave propagation) A low-frequency wave in an electrically highly conducting fluid (such as a plasma) permeated by a static magnetic field. The restoring forces of the waves are, in general, the combination of a magnetic tensile stress along the magnetic field lines and the compressive stress between the field lines and the fluid pressure (for example, an Alfvén wave). (AP/PROP) 211-1990s

magneto-ionic medium An ionized gas that is permeated by a fixed magnetic field. (AP/PROP) 211-1997

magneto-ionic wave At a given frequency, either of the two characteristic plane electromagnetic waves that can travel in a homogeneous magneto-ionic medium without change of polarization. *Note:* These characteristic waves are also called the ordinary and extraordinary waves. *See also:* extraordinary wave; ordinary wave. (AP/PROP) 211-1997

magnetometer An instrument for measuring the intensity or direction (or both) of a magnetic field or of a component of a magnetic field in a particular direction. *Note:* The term is more usually applied to instruments that measure the intensity of a component of a magnetic field, such as horizontal-intensity magnetometers, vertical-intensity magnetometers, and total-intensity magnetometers. (EEC/PE) [119]

magnetomotive force (acting in any closed path in a magnetic field) The line integral of the magnetizing force around the path. (Std100) 270-1966w

magneto-optic (fiber optics) Pertaining to a change in a material's refractive index under the influence of a magnetic field. Magneto-optic materials generally are used to rotate the plane of polarization. (Std100) 812-1984w

magneto-optical disk A disk that uses optical methods, such as a laser, to record information on a magnetic storage medium. *Synonym:* optically assisted magnetic storage. *See also:* magnetic disk; optical disk. (C) 610.10-1994w

magnetopause The transition region between the planetary and the interplanetary magnetic fields. (AP/PROP) 211-1997

magnetoresistive coefficient (Hall generator) The ratio at a specified magnetic flux density B of the rate of change of resistance with magnetic flux density to the resistance R_B at the specified magnetic flux density B, defined by the equation

$$\alpha_B = \frac{1}{R_B}\frac{dR_B}{dB}$$

(MAG) 296-1969w

magnetoresistive effect The change in the resistance of a current-carrying Hall plate when acted upon by a magnetic field. *Notes:* 1. An increase in magnetic field may cause either an increase or a decrease in ferromagnetic and similar Hall plates, whereas there is usually an increase with Hall plates made of other material. 2. There are two factors affecting the changes in resistance: first, a bulk effect due to the characteristics of the Hall plate, and second, a geometric effect due to the shape of the Hall plate and to the presence or absence of shorting bars made of conducting material deliberately, as in the shorting bars plated on some magnetoresistors or the microconductors dispersed in other magnetoresistors or inadvertently, as in the case of the control current electrodes in a Hall generator, added to the current-carrying Hall plate. (MAG) 296-1969w

magnetoresistive ratio (Hall generator) The ratio of the resistance R_B, at a magnetic flux density B, to the resistance R_0, at zero magnetic flux density, defined by the equation

$$\alpha_M = \frac{R_B}{R_0}$$

(MAG) 296-1969w

magnetosphere The region of a planetary atmosphere where the planetary magnetic field, as modified by the solar wind and the interplanetary magnetic field, controls the motions of charged particles. *Note:* The Earth's magnetosphere includes part of the F region of the terrestrial ionosphere up to the magnetopause. (AP/PROP) 211-1997

magnetostriction The phenomenon of elastic deformation that accompanies magnetization. (Std100) 270-1966w

magnetostriction loudspeaker A loudspeaker in which the mechanical displacement is derived from the deformation of a material having magnetostrictive properties. (EEC/PE) [119]

magnetostriction microphone A microphone that depends for its operation on the generation of an electromotive force by the deformation of a material having magnetostrictive properties. *See also:* microphone. (EEC/PE) [119]

magnetostriction oscillator An oscillator with the plate circuit inductively coupled to the grid circuit through a magnetostrictive element, the frequency of oscillation being determined by the magnetomechanical characteristics of the coupling element. *See also:* oscillatory circuit. (AP/ANT) 145-1983s

magnetostrictive relay A relay in which operation depends upon dimensional changes of a magnetic material in a magnetic field. *See also:* relay. (EEC/REE) [87]

magneto switchboard (telephone switching systems) A telecommunication switchboard for serving magneto telephone sets. (COM) 312-1977w

magneto telephone set A local-battery telephone set in which current for signaling by the telephone station is supplied from a local hand generator, usually called a magneto. *See also:* telephone station. (EEC/PE) [119]

magneto-telluric (M-T) An adjective denoting natural magnetic and electric fields, and effects produced by them. (COM) 365-1974w

magneto-telluric current A current in the Earth associated with time-varying geomagnetic fields. (AP/PROP) 211-1997

magneto-telluric fields Electric and magnetic fields induced in the Earth by external time-varying sources that are usually of ionospheric origin. (AP/PROP) 211-1997

magnetron (induction and dielectric heating) An electron tube characterized by the interaction of electrons with the electric field of a circuit element in crossed steady electric and magnetic fields to produce alternating-current power output. (IA) 54-1955w

magnetron injection gun (microwave tubes) A gun that produces a hollow beam of high total permeance that flows parallel to the axis of a magnetic field. *See also:* magnetron. (ED) [45]

magnetron oscillator An electron tube in which electrons are accelerated by a radial electric field between the cathode and one or more anodes and by an axial magnetic field that provides a high-energy electron stream to excite the tank circuits. *See also:* magnetron. (AP/ANT) 145-1983s

magnet valve (electric controller) A valve controlling a fluid, usually air, operated by an electromagnet. *See also:* multiple-unit control. (VT/LT) 16-1955w

magnet wire (rotating machinery) Single-strand wire with a thin flexible insulation, suitable for winding coils. *See also:* rotor; stator. (PE) [9]

magnified sweep (oscilloscopes) A sweep whose time per division has been decreased by amplification of the sweep waveform rather than by changing the time constants used to generate it. *See also:* oscillograph. (IM/HFIM) [40]

magnitude (1) The quantitative attribute of size, intensity, extent, etc., that allows a particular entity to be placed in order with other entities having the same attribute. *Notes:* 1. The magnitude of the length of a given bar is the same whether the length is measured in feet or in centimeters. 2. The word magnitude is used in other senses. The definition given here is the basic one needed for the logical buildup of later definitions. (Std100) 270-1966w
(2) The real number indicating the maximum or peak value of a periodically varying quantity. *See also:* amplitude. (AP/PROP) 211-1997

magnitude characteristic (linear passive networks) The absolute value of a response function evaluated on the imaginary axis of the complex-frequency plane. (CAS) 156-1960w

magnitude contours (control system feedback) Loci of selected constant values of the magnitude of the return transfer function drawn on a plot of the loop transfer function for real frequencies. *Note:* Such loci may be drawn on the Nyquist or inverse Nyquist diagrams, or Nichols chart. *See also:* feedback control system. (PE/EDPG) [3]

magnitude origin line (pulse terminology) A line of specified magnitude which, unless otherwise specified, has a magnitude equal to zero and extends through the waveform epoch. (IM/WM&A) 194-1977w

magnitude parameters and references (pulse terminology) (Unless otherwise specified, derived from data within the waveform epoch.) *See also:* magnitude reference lines; top magnitude; base magnitude; magnitude reference points. (IM/WM&A) 194-1977w

magnitude ratio (hydraulic turbines) The ratio of the peak magnitude of the output signal to the peak magnitude of a constant-frequency constant-amplitude sinusoidal input signal. (PE/EDPG) 125-1977s

magnitude reference points *See:* proximal (distal) point; mesial point.

magnitude reference line (pulse terminology) A line parallel to the magnitude origin line at a specified magnitude. (IM/WM&A) 194-1977w

magnitude reference lines (1) (pulse terminology) The magnitude reference line at the base (top) magnitude. *Synonym:* baseline. *See also:* waveform epoch. (IM/WM&A) 194-1977w
(2) (percent reference magnitude) A reference magnitude specified by:

$$(x)\%M_r = M_b + \frac{x}{100}(M_t - M_b)$$

where
$0 < x < 100$
$(x)\%M_r$ = percent reference magnitude
M_b = base magnitude
M_r = top magnitude

M_b, M_t, and $(x)\%M_r$ are all in the same unit of measurement.
(3) (proximal line) (distal) A magnitude reference line at a specified magnitude in the proximal (distal) region of a pulse waveform. Unless otherwise specified, the proximal (distal) line is at the 10 (90) percent reference magnitude. *See also:* waveform epoch.
(4) (mesial line) A magnitude reference line at a specified magnitude in the mesial region of a pulse waveform. Unless otherwise specified, the mesial line is at the 50 percent reference magnitude. *See also:* waveform epoch.

magnitude-referenced point (pulse terminology) A point at the intersection of a magnitude reference line and a waveform. (IM) 194-1977w

magnitude-related adjectives (A) (pulse terminology) Proximal (distal). Of or pertaining to a region near to (remote from) a first state or region of origin. **(B) (pulse terminology)** Mesial. Of or pertaining to region between the proximal and distal regions. (IM/WM&A) 194-1977

mag tape A colloquial reference to magnetic tape. (C) 610.10-1994w

Mahoney map (mathematics of computing) A diagram used in logic design, simplification, or optimization; invented by Matthew V. Mahoney. *See also:* Karnaugh map; logic map. (C) 1084-1986w

MAID *See:* minimum air insulation distance.

mailbox (1) A register that maps write data to unit-internal message storage locations. The mapping between registers and message-storage locations is transparent to the one or more producers. A message is sent to a mailbox by performing an atomic block copy of the message bytes to the mailbox register. Following this message write, the mailbox hardware moves the mapping to the next message storage location to prepare for the next message write. (C/MM) 1212.1-1993
(2) The mechanism for storing and retrieving intelligent transportation systems (ITS) messages between the back office equipment (BOE) and onboard equipment (OBE). The specific processing mechanisms are defined by the resource manager. (SCC32) 1455-1999

mailbox service *See:* electronic mail.

mail bridge A bridge that screens mail that is passing between two networks to ensure that the mail items meet administrative constraints. *See also:* learning bridge. (C) 610.7-1995

mail exploder The part of an electronic mail delivery system that accepts a piece of mail and a list of addressees as input and sends a copy of the message to each addressee on the list. (C) 610.7-1995

mail gateway A device that connects to two or more electronic mail systems, especially dissimilar mail systems on two different networks, and transfers mail messages between them. (C) 610.7-1995

mail server On a network, a server that allows users to exchange mail messages. *See also:* file server; disk server; network server; terminal server; database server; print server. (C) 610.7-1995

mail system gateway Software that uses the MT interface (the client). (C/PA) 1224.1-1993w

main (interior wiring) A feeder extending from the service switch, generator bus, or converter bus to the main distribution center. (EEC/PE) [119]

main amplifier (shaping amplifier) The section of amplifier following the preamplifier that contains the pulse-shaping networks (filter networks). These networks optimize the signal-to-noise ratio in the amplifying chain. (NPS) 325-1996

main anode (pool-cathode tube) An anode that conducts load current. *Note:* The word main is used only when it is desired to distinguish the anode to which it is applied from an auxiliary electrode such as an excitation anode. It is used only in connection with pool-tube terms. *See also:* electrode. (ED) [45]

main bang (1) (radar) A transmitted pulse. (IM/WM&A) 194-1977w
(2) The transmitted pulse of a radar, especially its amplitude trace as viewed on an A-display at the start of transmission. (AES) 686-1997

main bonding jumper The connection between the grounded circuit conductor and the equipment grounding conductor at the service. (NESC/NEC) [86]

main capacitance (capacitance potential devices) The capacitance between the network connection and line. *See also:* outdoor coupling capacitor. 31-1944w

main circuit All the conducting parts of the gas-insulated substation assembly included in or connected to the circuits that its switching devices are designed to close or open. (SWG/SUB/PE) C37.122-1983s, C37.122.1-1993, C37.100-1992

main console *See:* master console.

main contacts (1) For resistance-type LTCs, a set of through current-carrying contacts that have no transition impedance between the transformer winding and the contacts and commutates the current to the main switching contacts without any arc. (PE/TR) C57.131-1995
(2) (of a switching device) Contacts that carry all or most of the main current. (SWG/PE/TR) C37.100-1992, C57.12.44-1994

main control unit In a processor with more than one instruction control unit, that unit to which, for a given interval of time, the other units are subordinated. (C) 610.10-1994w

main discharge current wave The relatively long portion of the ESD current wave that follows the initial current pulse, or that occurs by itself when the initial current pulse does not exist. It may be unidirectional or oscillatory; its initial slope may be fast or slow. (SPD/PE) C62.47-1992r

main distributing frame (telephone switching systems) A frame where crossconnections are made between the outside plant and central office equipment. (COM) 312-1977w

main distribution center A distribution center supplied directly by mains. *See also:* distribution center. (EEC/PE) [119]

main distribution frame *See:* wiring closet.

main distribution function *See:* wiring closet.

main exciter (1) (rotating machinery) An exciter that supplies all or part of the power required for the excitation of the principal electric machine or machines. *See also:* asynchronous machine. (PE) [9]
(2) (synchronous machines) The source of all or part of the field current for the excitation of an electric machine, exclusive of another exciter. (PE/EDPG) 421-1972s

main exciter response ratio (nominal exciter response) The numerical value obtained when the response, in volts per second, is divided by the rated-load field voltage, which response, if maintained constant, would develop, in one-half second, the same excitation voltage-time area as attained by the actual exciter. *Note:* The response is determined with no load on the exciter, with the exciter voltage initially equal to the rate-load field voltage, and then suddenly establishing circuit conditions that would be used to obtain nominal exciter ceiling voltage. For a rotating exciter, the response should be determined at the rated speed. This definition does not apply to main exciters having one or more series fields, except a light differential series field, or to electronic exciters. (PE/EEC) [119]

main file *See:* master file.

mainframe (1) A rigid framework that provides mechanical support for modules inserted into the backplane, ensuring that connectors mate properly and that adjacent modules do not contact each other. It also provides cooling airflow, and ensures that modules do not disengage from the backplane due to vibration or shock. (C/MM) 1155-1992
(2) The cabinet that houses the central processor and main storage of a computer system. *Note:* This term is sometimes used as an abbreviation for mainframe computer. (C) 610.10-1994w

mainframe computer A computer employing one or more mainframes. *Note:* The distinction between a microcomputer, minicomputer, and mainframe is not yet standardized, however, in 1991 a typical mainframe is IBM's 3090, a typical minicomputer is Digital's VAX, and a typical microcomputer is IBM's PS/2. *See also:* mainframe. (C) 610.10-1994w

main gap (glow-discharge tubes) The conduction path between a principal cathode and a principal anode. (ED) 161-1971w

main ground bus (1) A conductor or system of conductors that provides for connecting all designated metallic components of the gas-insulated substation to station ground (ground grid). (SWG/PE/SUB) C37.100-1992, C37.122-1983s
(2) A conductor or system of conductors provided for connecting all designated metallic components of the gas-insulation substation (GIS) to a substation grounding system. (PE/SUB) 80-2000

main lead (rotating machinery) A conductor joining a main terminal to the primary winding. *See also:* asynchronous machine. (PE) [9]

main lobe *See:* major lobe.

main memory *See:* main storage.

main model The top-level unit under test (UUT) model description that includes a list of component packages and a netlist. (SCC20) 1445-1998

main program A software component that is called by the operating system of a computer and that usually calls other software components. *See also:* subprogram; routine. (C) 610.12-1990

main protection *See:* primary protection.

main reflector The largest reflector of a multiple reflector antenna. (AP/ANT) 145-1993

main ring path Principal transmission path in the trunk cabling. The main ring path carries the data in the primary direction. *Contrast:* backup path. (C/LM) 8802-5-1998

mains (1) The ac power source available at the point of use in a facility. It consists of the set of electrical conductors (referred to by terms including "service entrance," "feeder," or "branch circuit") for delivering power to connected loads at the utilization voltage level. (SPD/PE) C62.48-1995, C62.41-1991r
(2) *See also:* primary distribution mains; center of distribution; secondary distribution mains.

mains coupling coefficient *See also:* mains decoupling factor.

mains decoupling factor (mains coupling coefficient) The ratio of the radio-frequency voltage at the mains terminal to the interfering apparatus to the radio-frequency voltage at the aerial terminals of the receiver. *Note:* Generally expressed in logarithmic units. *See also:* electromagnetic compatibility. (INT) [53], [70]

main secondary terminals The main secondary terminals provide the connections to the main secondary winding. *See also:* main secondary winding. 31-1944w

main secondary winding (capacitance potential devices) Provides the secondary voltage or voltages on which the potential device ratings are based. *See also:* main secondary terminals. 31-1944w

mains-interference immunity (mains-interference ratio) The degree of protection against interference conducted by its supply mains as measured under specified conditions. *Note:* See International Special Committee on Radio Interference recommendation 25.1 and International Electrotechnical Commission publication 69 or subsequent publications where the term "mains-interference ratio" is used. *See also:* electromagnetic compatibility. (INT) [53], [70]

main station A telephone station with a distinct call number designation, directly connected to a central office. *See also:* telephone station. (EEC/PE) [119]

main-station code (telephone switching systems) The digits designating a main station; these usually follow an office code. (COM) 312-1977w

main storage That part of internal storage into which instructions and other data must be loaded for subsequent execution or processing. *Synonyms:* primary storage; main memory. *Contrast:* auxiliary storage. *See also:* real storage; common storage; random-access memory. (C) 610.10-1994w

main switchgear connections Those that electrically connect together devices in the main circuit, or connect them to the bus, or both. *Synonym:* primary switchgear connections. (SWG/PE) C37.100-1992

main switching contacts For resistance-type LTCs, a set of contacts that has no transition impedance between the transformer winding and the contacts and makes and breaks current. (PE/TR) C57.131-1995

maintainability (1) (A) (software) The ease with which a software system or component can be modified to correct faults, improve performance or other attributes, or adapt to a changed environment. *See also:* flexibility; extendability. **(B) (software)** The ease with which a hardware system or component can be retained in, or restored to, a state in which it can perform its required functions. (C) 610.12-1990
(2) Ability of an item, under stated conditions of use, to be retained in or restored to a state in which it can perform its required functions, when maintenance is performed under stated conditions and using prescribed procedures and resources. *Note:* Maintainability can, depending on the particular analysis situation, be stated by one or several maintainability characteristics, such as discrete probability distribution, mean active maintenance time, etc. 2. The value of the maintainability characteristic may differ for different maintenance situations. 3. When the term maintainability is used as a maintainability characteristic, it always denotes the probability that the active maintenance is carried out within a given period of time. 4. The required function may be defined as a stated condition. (R) [29]
(3) The measure of the ability of an item to be retained in, or restored to, a specified condition when maintenance is performed by personnel having specified skill levels, using prescribed procedures and resources, at each prescribed level of maintenance and repair. (C/BA) 896.3-1993w

Maintaining, Preparing & Producing Executive Reports (MAPPER) A nonprocedural programming language for UNIVAC computers, designed for novice users. (C) 610.13-1993w

maintaining voltage (operating voltage) (glow lamp) The voltage measured across the lamp electrodes when the lamp is operating. (EEC/EL) [104]

maintain temperature (1) Specified temperature of the fluid or process material that the heating system is designed to hold at equilibrium under specified design conditions. (BT/IA/AV/PC) 152-1953s, 844-1991
(2) Specified temperature of the fluid or process material that the heat tracing is designed to hold at equilibrium under specified design conditions. (IA) 515-1997

maintenance (1) (computers) Any activity intended to keep equipment, programs or a data base in satisfactory working condition, including tests, measurements, replacements, adjustments, and repairs. *See also:* software maintenance; file maintenance. (C) [20], [85]
(2) (test, measurement, and diagnostic equipment) Activity intended to keep equipment (hardware) or programs (software) in satisfactory working condition, including tests, measurements, replacements, adjustments, repairs, program copying, and program improvement. Maintenance is either preventive or corrective. (MIL) [2]
(3) The combination of all technical and corresponding administrative actions intended to retain an item in, or restore it to, a state in which it can perform its required function. *Note:* The required function may be defined as a stated condition. (R) [29]
(4) (A) (software) The process of modifying a software system or component after delivery to correct faults, improve performance or other attributes, or adapt to a changed environment. **(B) (software)** The process of retaining a hardware system or component in, or restoring it to, a state in which it can perform its required functions. *See also:* preventive maintenance. (C) 610.12-1990
(5) *See also:* software maintenance. (C/SE) J-STD-016-1995
(6) The act of preserving or keeping in existence those conditions that are necessary in order for equipment to operate as it was originally intended. (IA/PSE) 902-1998

maintenance bypass Removal of the capability of a channel, component, or piece of equipment to perform a protective action due to a requirement for replacement, repair, test, or calibration. *Note:* A maintenance bypass is not the same as an operating bypass. A maintenance bypass may reduce the degree of redundancy of equipment, but it does not result in the loss of a safety function. (PE/NP) 603-1998

maintenance concept (test, measurement, and diagnostic equipment) A description of the general scheme for maintenance and support of an item in the operational environment. (MIL) [2]

maintenance data/knowledge collection system One system within the AI-ESTATE architectural concept. This system supports collection of data and knowledge necessary for a maintenance function. It is a special form of the knowledge/data base management system of the AI-ESTATE architectural concept. (ATLAS) 1232-1995

maintenance, depot *See:* depot maintenance.

maintenance derated hours (electric generating unit reliability, availability, and productivity) The available hours during which a Class 4 unplanned derating was in effect. (PE/PSE) 762-1987w

maintenance engineering analysis (test, measurement, and diagnostic equipment) A process performed during the development stage to derive the required maintenance resources such as personnel, technical data, support equipment, repair parts, and facilities. (MIL) [2]

maintenance factor *See:* light loss factor.

maintenance, intermediate *See:* intermediate maintenance.

maintenance interval (1) (Class 1E battery chargers and inverters) The period, defined in terms of real time, operating time, number of operating cycles, or a combination of these, during which satisfactory performance is required without maintenance or adjustments. (PE/NP) 650-1979s
(2) (switchgear assemblies for Class 1E applications in nuclear power generating stations) The period, defined in terms of real time, operating time, number of operating cycles, or a combination of these, during which satisfactory performance is expected without maintenance or adjustments. (SWG/PE) C37.100-1992, C37.82-1971s

maintenance level (test, measurement, and diagnostic equipment) The level at which maintenance is to be accomplished, that is, organizational, intermediate, and depot. (MIL) [2]

maintenance manual *See:* support manual.

maintenance measurement accuracy The ratio of good measurements to total measurements. Total measurements include missed measurements or peg counts, as well as errored measurements or peg counts. Maintenance peg counts should register all detected events under all conceivable operating conditions. If, under certain infrequent trouble conditions, registration of detected events requires unreasonable expense or switching system action that interferes with normal call processing, then the events may not be counted. The switching system should tag such defective measurements on the maintenance measurement report. Maintenance usage counts are the ratios of bad measurements to total measurements. (COM/TA) 973-1990w

maintenance operation device A removable device for use with power-operated circuit breakers that is used for manual operation of a de-energized circuit breaker during maintenance only. *Note:* This device is not to be used for closing the circuit breaker on an energized circuit. (SWG/PE) C37.100-1992

maintenance, organizational *See:* organizational maintenance.

maintenance outage hours (electric generating unit reliability, availability, and productivity) The number of hours a unit was in a Class 4 unplanned outage state. (PE/PSE) 762-1987w

maintenance panel A part of a unit of equipment used to display information or provide access to test points for maintenance. (C) 610.10-1994w

maintenance phase *See:* operation and maintenance phase.

maintenance plan (software) A document that identifies the management and technical approach that will be used to maintain software products. Typically included are topics such as tools, resources, facilities, and schedules. *See also:* document. (C/SE) 729-1983s

maintenance proof test (rotating machinery) A test applied to an armatue winding after being in service that is suitable for continued service. It is usually made at a lower voltage than the acceptance proof test. (PE/EM) 95-1977r

maintenance, scheduled *See:* scheduled maintenance.

maintenance temperature *See:* maintain temperature.

maintenance test (1) (electric submersible pump cable) Test made after removal of the cable from the well. It is intended to detect deterioration of the cable to determine suitability for reuse. (IA/PC) 1017-1985s
(2) (power cable systems) A test made during the operating life of a cable system. It is intended to detect deterioration of the system and to check the entire workmanship so that suitable maintenance procedures can be initiated. (PE/IC) 400-1991

MA interface The X.400 Application API. (C/PA) 1224.1-1993w

main-terminal (1) (A) (bidirectional thyristor) The main terminal that is named 1 by the device manufacturer. **(B) (bidirectional thyristor)** The main terminal that is named 2 by the device manufacturer. *See also:* anode. (IA/ED) 223-1966, [46], [12]
(2) (rotating machinery) A termination for the primary winding. *See also:* asynchronous machine. (PE) [9]

main terminals (thyristor) The terminals through which the principal current flows. *See also:* anode. (IA/ED) 223-1966w, [46], [12]

main transformer (power and distribution transformers) The term "main transformer" as applied to two single-phase Scott-connected units for three-phase to two-phase or two-phase to three-phase operation, designates the transformer that is connected directly between two of the phase wires of the three-phase lines. *Note:* A tap is provided at the midpoint for connection to the teaser transformer. (PE/TR) C57.12.80-1978r

main trunk *See:* trunk line.

main unit (power and distribution transformers) The core and coil unit that furnishes excitation to the series unit. (PE/TR) C57.12.80-1978r

main winding, single-phase induction motor A system of coils acting together, connected to the supply line, that determines the poles of the primary winding, and that serves as the principal winding for transfer of energy from the primary to the secondary of the motor. *Note:* In some multispeed motors, the same main winding will not be used for both starting operation and running operation. *See also:* asynchronous machine. (PE) [9]

main window The primary window for an application. (C) 1295-1993w

major alarm (telephone switching systems) An alarm indicating trouble or the presence of hazardous conditions needing immediate attention in order to restore or maintain the system capability. (COM) 312-1977w

major cycle (electronic computation) In a storage device that provides serial access to storage positions, the time interval between successive appearances of a given storage position. (C) 162-1963w

major defect A unit of product that, when tested, falls below 85% of its specified rated ultimate strength. (PE/T&D) C135.61-1997

major event A catastrophic event that exceeds design limits of the electric power system and that is characterized by the following (as defined by the utility):

a) Extensive damage to the electric power system;
b) More than a specified percentage of customers simultaneously out of service;
c) Service restoration times longer than specified.

Some examples are extreme weather, such as a one in five year event, or earthquakes. (PE/T&D) 1366-1998

major failure *See:* failure.

major insulation (1) (outdoor apparatus bushings) Insulating material internal to the bushing between the line potential conductor and ground. (PE/TR) 21-1976
(2) The insulating material providing the dielectric, which is necessary to maintain proper isolation between the energized conductor and ground potential. It consists of internal insulation and the insulating envelope(s). (PE/TR) C57.19.03-1996

majority (1) (computers) A logic operator having the property that if P is a statement, Q is a statement, R is a statement..., then the majority of P, Q, R, ..., is true if more than half the statements are true, false if half or less are true. (C) [20], [85]
(2) (mathematics of computing) A Boolean operator having the property that if P is a statement, Q is a statement, R is a statement, . . . then the majority of P,Q,R, . . . is true if more than half the statements are true, false if half or less are true. (C) 1084-1986w

majority carrier (semiconductor) The type of charge carrier constituting more than one half the total charge-carrier concentration. *See also:* semiconductor; semiconductor device. (ED) 216-1960w

majority circuit A circuit with multiple inputs whose output is related to the state of the majority of its inputs. *Note:* Majority circuits are typically used in fault tolerant computers. *See also:* voting computer; majority gate. (C) 610.10-1994w

majority element *See:* majority gate.

majority emitter (transistor) An electrode from which a flow of majority carriers enters the interelectrode region. *See also:* transistor. (IA) [12]

majority gate A gate that performs a majority operation. *Synonym:* majority element. (C) 610.10-1994w

majority operation A threshold operation in which each of the operands may take only the values 0 and 1; it takes the value 1 if an only if the number of operands having the value 1 is greater than the number of operands that have the value 0. *See also:* majority gate. (C) 610.10-1994w

major key *See:* primary key.

major lobe The radiation lobe containing the direction of maximum radiation. *Note:* In certain antennas, such as multilobed or split-beam antennas, there may exist more than one major lobe. *Synonym:* main lobe. *See also:* antenna. (AP/PE/T&D/ANT) [35], 145-1993, 1260-1996

major loop (control) A continuous network consisting of all of the forward elements and the primary feedback elements of the feedback control system. *See also:* feedback control system. (IA/IAC) [60]

major modification Includes conversion from one type of machine to another type of machine, conversion from one type of enclosure to another type of enclosure, or conversion from one rating to another rating or both. (IA/PC) 1068-1996

major pulse waveform features *See:* base; first transition; top; last transition.

major scheduled generation station shutdown Periodic shutdowns of the generating station for an extended time scheduled for major reconditioning of the station, for example, fuel reloading. (PE/NP) 380-1975w

major storm disaster Designates weather that exceeds design limits of facilities, and that satisfies all of the following: Extensive mechanical damage to facilities; More than a specified percentage of customers out of service; Service restoration longer than a specified time. *Notes:* 1. Typical industry criteria are 10% of customers out of service and 24 hours or more restoration time. Percentage of customers out of service

may be related to a company operating area rather than to an entire company. Examples of major storm disasters are hurricanes and major ice storms. 2. It is suggested that the specified percentage of customers out of service and restoration times be 10% and 24 hours. Percentage of customers out of service may be related to a company operating area rather than to an entire company. Examples of major storm disasters are hurricanes and major ice storms.
(PE/PSE) 859-1987w, 346-1973w

make-break operation (pulse operation) (data transmission) Used to describe a method of data transmission by means of opening and closing a circuit to produce a series of current pulses. (PE) 599-1985w

make-break relay contacts A contact form in which one contact closes connection to another contact and then opens its prior connection to a third contact. (EEC/REE) [87]

make busy (telephone switching systems) Conditioning a circuit to be unavailable for service. (COM) 312-1977w

make-busy signal (telephone switching systems) A signal transmitted from the terminating end of a trunk to prevent the seizure of the originating end. (COM) 312-1977w

make, % make (dial-pulse address signaling systems) (telephony) In dial-pulse signaling, make is that portion of the signal in which the dialing contacts are closed (make). % is the ratio of make time to the total pulse period (make + break) time. (COM/TA) 753-1983w

makeup time That part of available time needed for reruns due to faults or mistakes in operations. (C) 610.10-1994w

making capacity The maximum current or power that a contact is able to make under specified conditions. *See also:* contactor. (IA/ICTL/IAC) [60], [84]

making current (of a switching device) The value of the available current at the time the device closes. *Notes:* 1. Its rms value is measured from the envelope of the current wave at the time of the first major current peak. 2. The making current may also be expressed in terms of instantaneous value of current, in which case it is measured at the first major peak of the current wave. This is designated peak making current.
(SWG/PE) C37.100-1992, C37.30-1971s

making current, rated (switching device) The maximum root-mean-square current against which the recloser is required to close under specified conditions. *Notes:* 1. The root-mean-square value is measured from the envelope of the current wave at the time of the first major current peak. 2. See ANSI C37.05-1964 (R1969), Methods for Determining the Values of a Sinusoidal Current Wave and Normal-Frequency Recovery Voltage for AC High-Voltage Circuit Breakers.
(SWG/PE) C37.60-1981r

making-current tests (high-voltage switchgear) Tests that consist of manual or remote control closing of the device against a prescribed current. (SWG/PE) C37.40-1981s

malfunction (1) An error that results from failure in the hardware. *See also:* error; fault; mistake.
(C/MIL) 162-1963w, 165-1977w, [2]

(2) (seismic qualification of Class 1E power equipment) The loss of capability of Class 1E equipment to initiate or sustain a required function, or the initiation of an undesired spurious action that can result in consequences adverse to safety.
(SWG/PE/NP) 649-1980s, C37.81-1989r, 650-1979s, 344-1975s

(3) The loss of capability to initiate or sustain a required function, often a protective action, or the initiation of undesired spurious action. *Note:* A certain degree of equipment degradation may be acceptable in one system and not in another. In such cases, an evaluation of the equipment or device application should include a determination that the degree of relay contact bounce, changes in device calibration, or degradation of pressure-retaining boundaries are within acceptable limits.
(SWG/SUB/PE/PE) C37.122-1983s, C37.100-1992, C37.122.1-1993

(4) (analog computer) *See also:* error. (C) 166-1977w

(5) (test, measurement, and diagnostic equipment) *See also:* fault. (MIL) [2]

malicious call (telephone switching systems) A call of an harassing, abusive, obscene, or threatening nature.
(COM) 312-1977w

managed network A network or set of networks established and controlled by one or more organizations to meet specific organizational or business needs. (C) 2001-1999

MAN *See:* metropolitan area network.

managed object The OSI structure of management information (ISO/IEC 10165-2:1992) term used as an abstract representation of a resource. This managed object has a set of attributes. These attributes are equivalent to data objects.
(LM/C) 802.10-1992

management A process that consists of functions such as planning, organizing, controlling and supervising, and is performed to set and meet the stated objectives.
(C) 610.7-1995

management game A simulation game in which participants seek to achieve a specified management objective given pre-established resources and constraints; for example, a simulation in which participants make decisions designed to maximize profit in a given business situation and a computer determines the results of those decisions. *See also:* war game.
(C) 610.3-1989w

Management Information Base (MIB) A repository of information to describe the operation of a specific network device.
(C/LM) 802.3-1998

management information base (MIB) (1) A conceptual database of information contained in the collection of all the managed object classes and their instances.
(LM/C) 802.10-1992

(2) A simple network management protocol (SNMP) compatible data structure that defines the functional groups and management objects of a unit or system.
(C/MM) 1284.1-1997

management information format (MIF) A desktop management interface (DMI) compatible data structure that defines the functional groups and management objects of a unit or system. (C/MM) 1284.1-1997

management information model A model that identifies the entities and their relationships that participate in managing an OSI environment. (C) 610.7-1995

management information octets DQDB Layer Protocol Data Units (PDUs) used to carry DQDB Layer Management Protocol information between peer DQDB Layer Management Entities (LMEs). (LM/C) 8802-6-1994

management information system (MIS) An automated system designed to provide managers with the information required to make basic decisions. *Synonyms:* executive information system; business information system. *See also:* decision support system; computer-aided management. (C) 610.2-1987

management interface An interface provided by both the Media Independent Interface (MII) or Gigabit Media Independent Interface (GMII) that provides access to management parameters and services. (C/LM) 802.3-1998

management review A systematic evaluation of a software acquisition, supply, development, operation, or maintenance process performed by or on behalf of management that monitors progress, determines the status of plans and schedules, confirms requirements and their system allocation, or evaluates the effectiveness of management approaches used to achieve fitness for purpose. (C/SE) 1028-1997

manager role Where each task is initiated. The *manager role* is concerned with taking appropriate action at the completion or failure of a task. (C/PA) 1387.2-1995

Manchester biphase-L encoding A signal transmission method defined for the representation of binary data bits. Manchester biphase-L encoding specifies two "half-bits," so that a guaranteed mid-bit transition occurs in the transmitted signal. The transition is defined to be positive for encoding a logic "0" and negative for encoding a logic "1." (EMB/MIB) 1073.4.1-2000

Manchester encoding A method of encoding data in which separate data and clock signals can be combined into a single, self-synchronizable data stream, suitable for transmission on a serial channel. (C) 610.7-1995

mandatory A syntax keyword used to specify a total mapping. *Contrast:* optional. *See also:* total. (C/SE) 1320.2-1998

mandatory category A category that is essential to establish a common definition and to provide common terminology and concepts for communication among projects, business environments, and personnel. (C/SE) 1044-1993

mandatory nonidentifying relationship A kind of nonidentifying relationship in which an instance of the child entity must be related to an instance of the parent entity. *Contrast:* optional nonidentifying relationship. *See also:* nonidentifying relationship. (C/SE) 1320.2-1998

manhole (1) (More accurately termed splicing chamber or cable vault) A subsurface chamber, large enough for a man to enter, in the route of one or more conduit runs, and affording facilities for placing and maintaining in the runs, conductors, cables, and any associated apparatus. *See also:* splicing chamber. (T&D/PE) [10]
(2) An opening in an underground system that workmen or others may enter for the purpose of installing cables, transformers, junction boxes, and other devices, and for making connections and tests. *See also:* splicing chamber; distribution center. (BT/AV) [34]
(3) A subsurface enclosure that personnel may enter used for the purpose of installing, operating, and maintaining submersible equipment and cable. (NESC/T&D) C2-1997, C2.2-1960

man-centered simulation (man-centred simulation) *See also:* human-centered simulation. (C) 610.3-1989w

manhole chimney A vertical passageway for workmen and equipment between the roof of the manhole and the street level. (T&D/PE) [10]

manhole cover A removable lid that closes the opening to a manhole or similar subsurface enclosure. (NESC) C2-1997

manhole cover frame The structure that caps the manhole chimney at ground level and supports the cover. (T&D/PE) [10]

manhole grating A grid that provides ventilation and a protective cover for a manhole opening. (NESC) C2-1997

manifold insulation (rotating machinery) (liquid cooling) The insulation applied between ground and a manifold connecting several parallel liquid-cooling paths in a winding. *See also:* stator. (PE) [9]

manifold-pressure electric gauge A device that measures the pressure of fuel vapors entering the cylinders of an aircraft engine. *Note:* The gauge is provided with a scale, usually graduated in inches of mercury, absolute. It provides remote indication by means of a self-synchronous generator and motor. (EEC/PE) [119]

manipulated variable (control) A quantity or condition that is varied as a function of the actuating signal so as to change the value of the directly controlled variable. *Note:* In any practical control system, there may be more than one manipulated variable. Accordingly, when using the term it is necessary to state which manipulated variable is being discussed. In process control work, the one immediately preceding the directly controlled system is usually intended. *See also:* feedback control system. (IA/ICTL/IAC) [60]

manipulation detection A mechanism used to detect whether a data unit has been modified (either accidentally or intentionally). (LM/C) 802.10-1992

man-machine interface (1) (man-machine performance in nuclear power generating stations) The devices through which personnel receive information from the system or process and the devices through which personnel exercise their control of the system or process. (PE/NP) 845-1988s, 1023-1988r
(2) (station control and data acquisition) The operator contact with equipments governed by IEEE Std C37.1-1979, Mil-Std-1472 is recommended as a reference for use in the design and evaluation of the man/machine interface to equipments governed by this standard. Alternative human engineering data may be specified by the user. The man/machine interface for operation concerns standards and recommendations for information displays, control capabilities, colors and man/machine interaction of equipments governed by IEEE Std C37.1-1979. (SUB/PE) C37.1-1979s
(3) (software) *See also:* user interface. (C) 610.12-1990

man-machine simulation *See:* human-machine simulation.

man-made noise Noise generated in machines or other technical devices. *See also:* electromagnetic compatibility. (EMC) [53]

manned space flight network (MSFN) (communication satellite) A network of ground communication and tracking facilities maintained for the support of manned space flight programs. (COM) [24]

mantissa (A) (mathematics of computing) The fractional part of a logarithm. *Contrast:* characteristic. **(B) (mathematics of computing)** For floating-point arithmetic, *See also:* significand. (C) 1084-1986

manual (1) (electric systems) Operated by mechanical force, applied directly by personal intervention. *See also:* distribution center. (IA/IAC) [60]
(2) Capable of being operated by personal intervention. (NESC) C2-1997

manual block-signal system A block or a series of consecutive blocks governed by block signals operated manually upon information by telegraph, telephone, or other means of communication. *See also:* block-signal system. (EEC/PE) [119]

manual central office A central office of a manual telephone system. (COM) [48]

manual checkout (test, measurement, and diagnostic equipment) A checkout system which relies completely on manual operation, operator decision and evaluation of results. (MIL) [2]

manual control (1) (excitation systems for synchronous machines) In excitation control system usage, manual control refers to maintaining synchronous machine terminal voltage by operator action. *Note:* Manual control means may include an exciter field rheostat, controlled rectifiers, or a direct-current (dc) regulator controlling either exciter field current or exciter output voltage, or other means that do not include regulation of synchronous machine terminal voltage. (PE/EDPG) 421.1-1986r
(2) Those elements in the excitation control system which provide for manual adjustment of the synchronous machine terminal voltage by open-loop control. (PE/EDPG) 421-1972s
(3) Control in which the main devices under control, whether manually or power operated, are controlled by an attendant. (SWG/PE/SUB) C37.100-1992, C37.1-1994
(4) (programmable instrumentation) *See also:* local control. (IM/AIN) 488.1-1987r

manual controller An electric controller having all of its basic functions performed by devices that are operated by hand. (IA/MT) 45-1998

manual data input (numerically controlled machines) A means for the manual insertion of numerical control commands. (IA) [61]

manual fire-alarm system A fire-alarm system in which the signal transmission is initiated by manipulation of a device provided for the purpose. *See also:* protective signaling. (EEC/PE) [119]

manual holdup-alarm system An alarm system in which the signal transmission is initiated by the direct action of the person attacked or of an observer of the attack. *See also:* protective signaling. (EEC/PE) [119]

manual hyphenation In text formatting, hyphenation in which all line-ending and word break decisions are made by the user. *See also:* semi-manual hyphenation; automatic hyphenation. (C) 610.2-1987

manual input (A) (computers) The entry of data by hand into a device at the time of processing. **(B) (computers)** The data entered as in definition (A). (C) [20], [85]

manual load (armature current) division The effect of a manually operated device to adjust the division of armature currents between two or more motors or two or more generators connected to the same load. *See also:* feedback control system. (IA/ICTL/IAC) [60]

manual locking carabiner Has a self-closing gate which remains closed but not locked (unless purposely locked by the user) until intentionally opened by the user for connection or disconnection. (T&D/PE) 1307-1996

manual lockout device A device that holds the associated device inoperative unless a predetermined manual function is performed to release the locking feature. (SWG/PE) C37.100-1992

manually operated door or gate A door or gate that is opened and closed by hand. *See also:* hoistway. (EEC/PE) [119]

manually release-free *See:* mechanically release-free.

manually trip-free *See:* mechanically release-free.

manual mobile telephone system A mobile communication system manually interconnected with any telephone network, or a mobile communication system manually interconnected with a telephone network. (VT) [37]

manual operation Operation by hand without the use of any other source of power. (SWG/PE) C37.100-1992

manual outage An outage occurrence that results from intentional or inadvertent operator controlled opening of switching devices. (PE/PSE) 859-1987w

manual potentiometer (analog computer) A potentiometer that is set by hand, also known as a "hand-set potentiometer." (C) 165-1977w

manual release (electromagnetic brake) A device by which the braking surfaces may be manually disengaged without disturbing the torque adjustment. *See also:* electric drive. (IA/ICTL/IAC/APP) [60], [75]

manual reset A function that requires a manual operation to re-establish specific conditions. (SWG/PE) C37.100-1981s

manual-reset relay A relay that may be reset manually after an operation. (EEC/REE) [87]

manual-reset thermal protector (rotating machinery) A thermal protector designed to perform the function by opening the circuit to or within the protected machine, but requiring manual resetting to close the circuit. *See also:* starting-switch assembly. (PE) [9]

manual ringing (telephone switching systems) Ringing that is started by the manual operation of a key and continues only while the key is held operated. (COM) 312-1977w

manual speed adjustment A speed adjustment accomplished manually. *See also:* electric drive. (IA/ICTL/IAC) [60]

manual switchboard (telephone switching systems) A telecommunication switchboard for making interconnections manually by plugs and jacks or keys. (COM) 312-1977w

manual telecommunications exchange (telephone switching systems) A telecommunications exchange in which connections between stations are manually set by means of plugs and jacks or keys. (COM) 312-1977w

manual telecommunication system (telephone switching systems) A telecommunications system in which connections between customers are ordinarily established manually by operators in accordance with orders given orally by the calling parties. (COM) 312-1977w

manual test A test or collection of tests that requires an operator. The tests may involve power failure testing, on-line replacement testing, media testing, or cable connection tests (when loop-back cables are required). A manual test is invoked by writing to the TEST_START register. (C/MM) 1212-1991s

manual test equipment (test, measurement, and diagnostic equipment) Test equipment that requires separate manipulations for each task (for example, connection to signal to be measured, selection of suitable range, and insertion of stimuli). (MIL) [2]

manual testing Testing that requires a human to execute some or all of a test procedure. (SCC20) 1226-1998

manual transfer or selector device (power system device function numbers) A manually operated device that transfers the control circuits in order to modify the plan of operation of the switching equipment or of some of the devices. (PE/SUB) C37.2-1979s

manual trip device A device that is connected to the tripping linkage and that can be operated manually to trip a switching device. (SWG/PE) C37.100-1992

manufacture (software) In software engineering, the process of copying software to disks, chips, or other devices for distribution to customers or users. (C) 610.12-1990

manufactured building Any building that is of closed construction and that is made or assembled in manufacturing facilities on or off the building site, other than mobile homes or recreational vehicles. (NESC/NEC) [86]

manufacturer *See:* builder **(rotating electric machinery)** The organization supplying the electric machinery to the purchaser. (PE/EM) 11-1980r

manufacturing phase (software) The period of time in the software life cycle during which the basic version of a software product is adapted to a specified set of operational environments and is distributed to a customer base. (C) 610.12-1990

manufacturer A The supplier of the initial GIS. *See also:* gas-insulated substations. (PE) 1416-1998

manufacturer B The supplier of the extension GIS. *See also:* gas-insulated substations. (PE) 1416-1998

manuscript (numerically controlled machines) An ordered list of numerical control instructions. *See also:* programming. (IA) [61]

manuscript editor *See:* document editor.

many-to-many relationship (1) (data management) A relationship between two entities A and B such that any instance of A may be more associated with than one instance of B, and vice-versa. *Note:* The use of "m:n relationship" as a synonym for this term is deprecated. *Synonym:* m:n relationship. (C) 610.5-1990w

(2) A kind of relationship between two state classes (not necessarily distinct) in which each instance of one class may be associated with any number of instances of a second class (possibly none), and each instance of the second class may be related to any number of instances of the first class (possibly none). (C/SE) 1320.2-1998

many-to-one decoder *See:* decoder.

MAP *See:* memory allocation and protection.

map (1) (A) (data management) To establish a correspondence between the elements of one set and the elements of another set. *Synonym:* map over. **(B) (data management)** To establish a correspondence between the logical structure of a database and the physical structure of that database. (C/IA/APP) [20], 610.5-1990, [75]

(2) To create an association between a page-aligned range of the address space of a process and a range of physical memory or some memory object, such that a reference to an address in that range of the address space results in a reference to the associated physical memory or memory object. The mapped memory or memory object is not necessarily memory-resident. (C/PA) 9945-1-1996

map a range of addresses To create an association process's address space and a range of physical memory or some memory object, such that a reference to an address in that range of the address space results in a reference to the associated physical memory or memory object. The mapped memory or memory object is not necessarily memory-resident. (C) 1003.5-1999

map over *See:* map.

MAPPER *See:* Maintaining, Preparing & Producing Executive Reports.

mapping (1) Establishing a sequence of the Activities in this standard according to a selected software life cycle model (SLCM). *See also:* Instance; Invocation; Iteration. (C/SE) 1074-1997
(2) Process of correspondence between the elements of one set and the elements of another set. (SCC20) 1226-1998
(3) An assigned correspondence between two things that is represented as a set of ordered pairs. Specifically, a mapping from a class to a value class is an attribute. A mapping from a state class to a state class is a participant property. A mapping from the (cross product of the) instances of the class and the instances of the input argument types to the (cross product of the) instances of the other (output) argument types is an operation. (C/SE) 1320.2-1998

mapping completeness A designation of whether a mapping is complete (totally mapped) or incomplete (partial). *See also:* partial; total. (C/SE) 1320.2-1998

mapping function (computer graphics) A transformation that converts display elements from one coordinate system to another. (C) 610.6-1991w

mapping onto network protocol The embodiment of hardware and software as it relates to supporting IEEE 1451.1 communications on a specific bus standard. (IM/ST) 1451.1-1999

map program (software) A software tool, often part of a compiler or assembler, that generates a load map. (C) 610.12-1990

map-reader function generator A variant function generator using a probe to detect the voltage at a point on a conducting surface and having coordinates proportional to the inputs. *See also:* electronic analog computer. (C) 165-1977w

map vertical *See:* geographic vertical.

MAR *See:* memory address register.

MARC *See:* MAchine-Readable Cataloging.

margin (1) (electric penetration assemblies) The difference between the most severe design service conditions and the conditions used in the design qualification to account for normal variations in commercial production of equipment and reasonable errors in defining satisfactory performance. (PE/NP) 317-1983r
(2) (nuclear power generating station) (valve actuators) (safety systems equipment in nuclear power generating stations) The difference between service conditions and the conditions used for equipment qualification. (PE/NP) 382-1985, 323-1974s, 649-1980s, 627-1980r
(3) (switchgear assemblies for Class 1E applications in nuclear power generating stations) The difference between the demonstrated capability of the equipment and that required in service for specific conditions. (SWG/PE) C37.100-1992, C37.82-1971s
(4) (A) (data transmission) (Digital) Of a receiving equipment, the maximum degree of distortion of the received signal which is compatible with the correct translation of all of the signals which it may possibly receive. *Note:* This maximum degree of distortion applies without reference to the form of distortion effecting the signals. In other words, it is the maximum degree of the most unfavorable distortion acceptable, beyond which incorrect translation occurs, which determines the value of the margin. The condition of the measurements of the margin are to be specified in accordance with the requirements of the system. **(B) (data transmission)** (Analog) The excess of receive level beyond that needed for proper operation. (PE) 599-1985
(5) (teletypewriter) (orientation margin) (printing telegraphy) That fraction of a perfect signal element through which the time of selection may be varied in one direction from the normal time of selection, without causing errors while signals are being received. *Note:* There are two distinct margins, determined by varying the time of selection in either direction from normal. *See also:* telegraphy. (COM) [49]
(6) (nickel-cadmium storage batteries) The combination of design margin and aging factor originally used in determining the battery's initial capacity requirements. (PE/EDPG) 1106-1987s

margin-adjust hyphenation *See:* hot zone hyphenation.

margin-adjust zone *See:* hot zone.

marginal check (electronic computation) A preventive maintenance procedure in which certain operating conditions (for example, supply voltage or frequency) are varied about their nominal values in order to detect and locate incipient defective parts. *See also:* check. (C) 162-1963w

marginal checking (test, measurement, and diagnostic equipment) A system or method of determining circuit weaknesses and incipient malfunctions by varying the operating conditions of the circuitry. (MIL) [2]

marginal relay A relay that functions in response to predetermined changes in the value of the coil current or voltage. *See also:* relay. (EEC/REE) [87]

marginal testing (test, measurement, and diagnostic equipment) Testing that presents results on an indicator that has tolerance bands for evaluating the signal or characteristic being tested. (For example: a green band might indicate an acceptable tolerance range; a yellow band, a tolerance range representing marginal operation; and a red band, a tolerance that is unsatisfactory for operation of the item). (MIL) [2]

margin of commutation γ (margin angle) The time, expressed in degrees (one cycle of the ac waveform, 360°) from the termination of commutation in inverter operation to the next point of intersection between the two halfwaves of the voltage phases which have just commutated. *Note:* At this point of intersection, the converter circuit element which has just terminated conduction changes from reverse blocking state to OFF state. (IA/IPC) 444-1973w

marine distribution panel A panel receiving energy from a distribution or subdistribution switchboard and distributing energy to energy-consuming devices or other distribution panels or panelboards of a ship. *See also:* marine electric apparatus. (EEC/PE) [119]

marine electric apparatus Electric apparatus designed especially for use on shipboard to withstand the conditions peculiar to such application. (EEC/PE) [119]

marine generator and distribution switchboard Receives energy from the generating plant and distributes directly or indirectly to all equipment of a ship supplied by the generating plant. *See also:* marine electric apparatus. (EEC/PE) [119]

Marinelli beaker (1) (germanium semiconductor detector) Reentrant (inverted well) beaker. It is available in a variety of sizes for use in large volume, low level, measurements. The beaker specified herein is shown in a corresponding figure. A schematic of a typical sample-detector geometry is also illustrated in a corresponding figure. The specified beaker is considered to be of 450 mL capacity. The actual volume is greater than this, but, for purposes of this standard, the beaker is to be filled to 450 mL ± 2 mL. The beaker specified was selected because of: high counting efficiency for the sample material used; commercial availability at low cost; common usage in many laboratories; physical convenience. (NPS) 680-1978w
(2) A reentrant (inverted well) beaker that can be fitted over a detector endcap for the purpose of holding a radioactive sample in a configuration that surrounds a major portion of the detector. (NPS) 325-1996

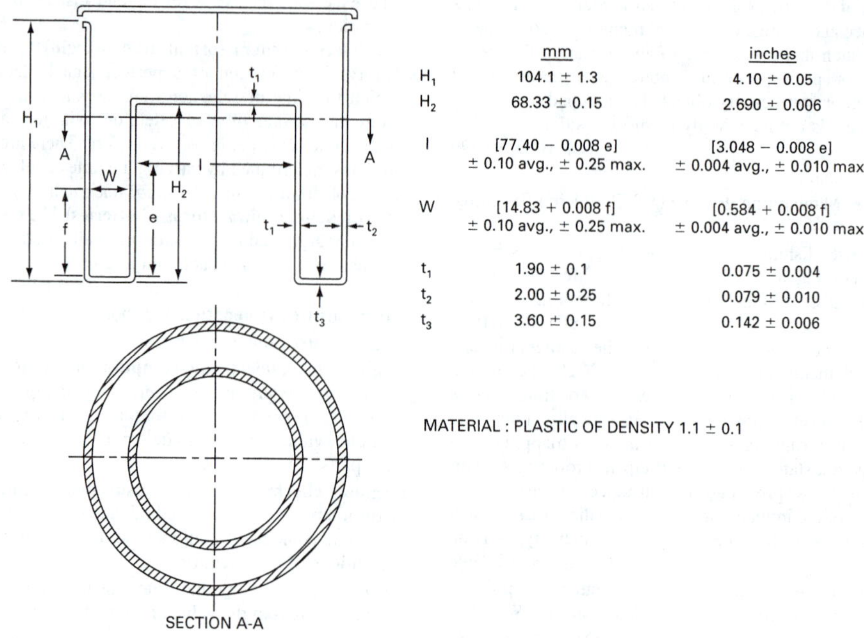

standard Marinelli beaker

Marinelli beaker standard source (germanium semiconductor detector) A standard Marinelli beaker containing a carrier with radioactive material. An MBSS may be a certified MBSS, a calibrated MBSS, a certified solution MBSS, or a calibrated solution MBSS. The calibration uncertainty of the photon emission rate for the filled beaker shall be not more than three percent unless otherwise stated. *Note:* The photon emission rate as used in IEEE Std 680-1978w is the number of photons per second resulting from the decay of radionuclides in the source, and is thus higher than the detected rate at the surface. (NPS) 680-1978w

marine panelboard A single panel or a group of panel units assembled as a single panel, usually with automatic overcurrent circuit breakers or fused switches, in a cabinet for flush or surface mounting in or on a bulkhead and accessible only from the front, serving lighting branch circuits or small power branch circuits of a ship. *See also:* marine electric apparatus. (EEC/PE) [119]

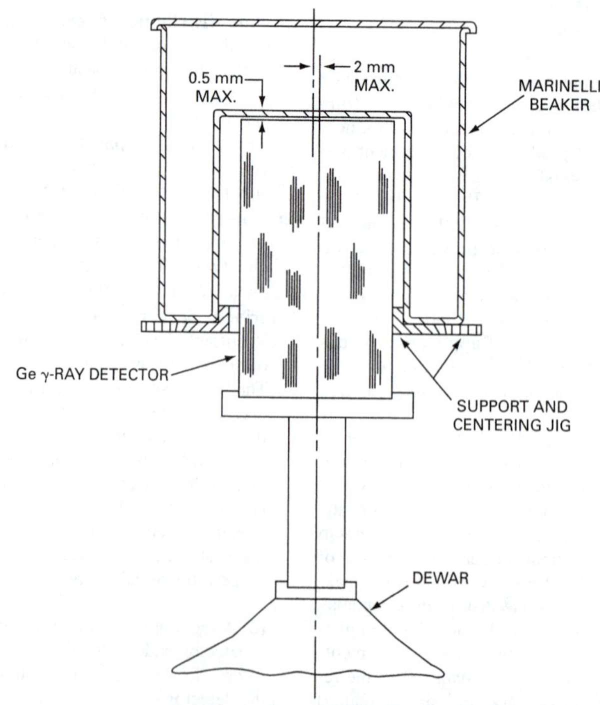

Marinelli beaker with solid state detector

marine subdistribution switchboard Essentially a section of the marine generator and distribution switchboard (connected thereto by a bus feeder and remotely located) that distributes energy in a certain section of a vessel. *See also:* marine electric apparatus. (EEC/PE) [119]

mariner's compass A magnetic compass used in navigation consisting of two or more parallel polarized needles secured to a circular compass card that is delicately pivoted and enclosed in a glass-covered bowl filled with alcohol to support by flotation the weight of the moving parts. *Note:* The compass bowl is supported in gimbals mounted in the binnacle. The compass card is graduated to show the 32 points of the compass in addition to degrees. (EEC/PE) [119]

mark (1) (liquid-filled power transformers) The descriptive name, instructions, cautions, or other information applied to polychlorinated biphenyls (PCBs) and PCB items or other objects subject to these regulations. (LM/C) 802.2-1985s
(2) (computers) *See also:* flag.
(3) (data management) A symbol or group of symbols that indicates the beginning or end of a field, a word, an item of data, or a set of data such as a file, a record, or a block. (C) 610.5-1990w
(4) A symbol or symbols that indicate the beginning or the end of a field, of a word, or of a data item in a file, record, or block. *Synonym:* marker. *See also:* beginning-of-tape marker; field mark; file mark; end mark; address mark; word mark; group mark; end-of-tape marker; index mark. (C) 610.10-1994w

mark detection *See:* mark sensing.

marked (liquid-filled power transformers) The marking of polychlorinated biphenyl (PCB) items and PCB storage areas and transport vehicles by means of applying a legible mark by painting, fixation of an adhesive label, or by any other method that meets the requirement of these regulations. (LM/C) 802.2-1985s

marked ratio The ratio of the rated primary value to the rated secondary value as stated on the nameplate. (PE/TR/PSR) C57.13-1993, C37.110-1996, C57.12.80-1978r

marker (1) (telephone switching systems) A wired-logic control circuit that, among other functions, tests, selects, and establishes paths through a switching stage or stages. (COM) 312-1977w
(2) (navigation aid terms) A radio beacon to designate a small area. (AES/GCS) 172-1983w
(3) (computer graphics) A symbol with a specific appearance that identifies a particular location. *See also:* polymarker. (C) 610.6-1991w
(4) *See also:* plumb marker pole. (T&D/PE) 524-1992r

marker beacon (navigation aid terms) A radio beacon to designate a small area. (AES/GCS) 172-1983w

marker-beacon receiver A receiver used in aircraft to receive marker-beacon signals that identify the position of the aircraft when over the marker-beacon station. (EEC/PE) [119]

marker lamp (railway practice) A signal lamp placed at the side of the rear end of a train or vehicle, displaying light of a particular color to indicate the rear end and to serve for identification purposes. (EEC/PE) [119]

marker light (railway practice) A light that by its color or position, or both, is used to qualify the signal aspect. (EEC/PE) [119]

marker radio beacon (navigation aid terms) A beacon that indicates a specific location. (AES/GCS) 172-1983w

marker signal (oscilloscopes) A signal introduced into the presentation for the purpose of identification, calibration, or comparison. (IM/HFIM) [40]

marker size An attribute specifying the size of a marker, relative to the standard size of a marker on the display device. (C) 610.6-1991w

marker type An attribute specifying the geometric shape of a marker; for example, dot, asterisk, circle. (C) 610.6-1991w

market Demand and supply of goods and services. (C/SE) 1430-1996

marketplace An infrastructure that supports the exchange of goods and services. (C/SE) 1430-1996

MARK IV A procedural language used for report writing and data manipulation. (C) 610.13-1993w

marking (marking and spacing intervals) (data transmission) (telegraph communication) Intervals that correspond, according to convention, to one condition or position of the originating transmitting contacts, usually a closed condition; spacing intervals are the intervals that correspond to another condition of the originating transmitting contacts, usually an open condition. *Note:* The terms "mark" and "space" frequently used for the corresponding conditions. The waves corresponding to the marking and spacing intervals are frequently designated as marking and spacing waves, respectively. (PE) 599-1985w

marking and spacing intervals (telegraph communication) Intervals that correspond, according to convention, to one condition or position of the originating transmitting contacts, usually a closed condition; spacing intervals are the intervals that correspond to another condition of the originating transmitting contacts, usually an open condition. *Note:* The terms mark and space are frequently used for the corresponding conditions. The waves corresponding to the marking and spacing intervals are frequently designated as marking and spacing waves, respectively. *See also:* telegraphy. (PE/EEC) [119]

marking engine A set of electrical and mechanical components that moves the print media and marks that media. In some implementations, a facsimile transmission function is considered to be a marking engine. (C/MM) 1284.1-1997

marking pulse (teletypewriter) The signal pulse that, in direct current, neutral, operation, corresponds to a circuit-closed or current-on condition. (COM) [49]

marking wave (telegraph communication) (keying wave) The emission that takes place while the active portions of the code characters are being transmitted. *See also:* radio transmitter. (AP/ANT) 145-1983s

Markov chain A discrete Markov process. (C) 610.3-1989w

Markov chain model A discrete, stochastic model in which the probability that the model is in a given state at a certain time depends only on the value of the immediately preceding state. *Synonym:* Markov model. *See also:* semi-Markov model. (C) 610.3-1989w, 1084-1986w

Markov model *See:* Markov chain model.

Markov process A stochastic process which assumes that in a series of random events, the probability for occurrence of each event depends only on the immediately preceding outcome. *See also:* semi-Markov process. (C) 610.3-1989w

mark scanning Optical sensing of marks recorded manually on a data medium. *Contrast:* mark sensing. (C) 610.2-1987

mark sensing Electrical sensing of conductive marks recorded manually on a nonconductive data medium. For example, graphite marks on paper. *Synonym:* mark detection. *Contrast:* mark scanning. (C) 610.2-1987

mark-sensing card A card that can be marked with a special electrographic pencil, then read directly into a computer. (C) 610.10-1994w

mark sensing column A vertical line of positions on a data medium, capable of being detected by mark sensing. (C) 610.2-1987

markup language *See:* page description language.

M-array glide slope (instrument landing systems) A modified null-reference glide-slope antenna system in which the modification is primarily an additional antenna used to obtain a high degree of energy cancellation at the low elevation angles. *Note:* Called M because it was 13th in a series of designs. This system is used at locations where higher terrain exists in front of the approach end of the runway, in order to reduce unwanted reflections of energy into the glide-slope sector. *See also:* navigation. (AES/RS) 686-1982s, [42]

marshaling/demarshaling The mapping of information typed as given in the signatures of `Perform-`, `Execute-`, and `Publish-`like operations into the network-specific, on-the-wire formats, including any required endian issues. Demarshaling is the reverse process. (IM/ST) 1451.1-1999

MARshall System for Aerospace Simulation (MARSYAS) A simulation language used for simulating large physical systems, designed for use by people inexperienced in simulation or programming. Allows equations and FORTRAN subroutines to be written along with the statements describing a block diagram model. (C) 610.13-1993w

MARSYAS *See:* MARshall System for Aerospace Simulation.

maser (1) (data transmission) (microwave amplification by stimulated emission of radiation) The general class of microwave amplifiers based on molecular interaction with electromagnetic radiation. The nonelectronic nature of the maser principle results in very low noise. (PE) 599-1985w
(2) (laser maser) A device for amplifying or generating radiation by induced transitions of electrons, atoms, molecules, or ions between two energy levels having a population inversion; microwave amplification by stimulated emission of radiation. (LEO) 586-1980w

mask (1) (A) (computers) A pattern of characters that is used to control the retention or elimination of portions of another pattern of characters. **(B) (computers)** A filter. (C) [20], [85]
(2) (software) A pattern of bits or characters designed to be logically combined with an unknown data item to retain or suppress portions of the data item; for example, the bit string "00000011" when logically ANDed with an eight-bit data item, gives a result that retains the last two bits of the data item and has zero in all the other bit positions. *See also:* interrupt mask. (C) 610.12-1990

mask document In word processing, a form displayed on a display screen with blank areas for the user to complete. (C) 610.2-1987

masking (1) (A) The process by which the threshold of audibility for one sound is raised by the presence of another (masking) sound. **(B)** The amount by which the threshold of audibility of a sound is raised by the presence of another (masking) sound. The unit customarily used is the decibel. (ACO)
(2) (color television) A process to alter color rendition in which the appropriate color signals are used to modify each other. *Note:* The modification is usually accomplished by suitable cross coupling between primary color-signal channels. *See also:* television. (BT/SP/AV) [34], [32]

masking audiogram A graphic presentation of the masking due to a stated noise. *Note:* This is plotted in decibels as a function of the frequency of the masked tone. (SP) [32]

masking, fault *See:* fault masking.

mask&swap A data-access operation that stores a *next* value to the *test* specified bits within a specified data type and returns the previous data value. (C/MM) 1596.5-1993

maskSwap A bus transaction that stores bits of a *next* argument to a specified data address and returns the previous data value from that address. The affected bits are specified by a *test* argument. In the CSR Architecture this is called a mask_swap transaction. (C/MM) 1596.5-1993

masquerade The pretense by an entity to be a different entity. (LM/C) 802.10-1992

mass (International System of Units (SI)) The SI unit of mass is the kilogram. This unit, or one of the multiples formed by attaching an SI prefix to gram, is preferred for all applications. Among the base and derived units of SI, the unit of mass is the only one whose name, for historical reasons, contains a prefix. Names of decimal multiples and submultiples of the unit of mass are formed by attaching prefixes to the word gram. The megagram (Mg) is the appropriate unit for measuring large masses such as have been expressed in tons. However, the name ton has been given to several large mass units that are widely used in commerce and technology: the long ton of 2240 lb, the short ton of 2000 lb, and metric ton of 1000 kilograms (also called the tonne). None of these terms are SI. The term metric ton should be restricted to commercial usage, and no prefixes should be used with it. Use of the term tonne is deprecated. *See also:* units and letter symbols. (QUL) 268-1982s

Massachusetts General Hospital Utility Multi-Programming System (MUMPS) An ANSI standard programming system containing its own operating system, command language, and interactive programming language; designed specifically for medical applications and is particularly adaptable to string handling functions and management of hierarchical data. (C) 610.13-1993w

mass-attraction vertical The normal to any surface of constant geopotential; it is the direction that would be indicated by a plumb bob if the earth were not rotating. *See also:* navigation. (AES/RS) 686-1982s, [42]

mass burning rate Mass loss per unit time by materials burning under specified conditions. (DEI) 1221-1993w

mass loading The change in phase velocity of a surface acoustic wave produced by a thin layer on the substrate of higher density than that of the substrate; perturbations in reflections, velocity, and dispersion that occur due to loading effects of thin films on the substrate surface. (UFFC) 1037-1992w

mass spectrograph An electronic device based on the action of a constant magnetic field on the paths of ions, used to separate ions of different masses. *See also:* electron device. (Std100) [84]

mass storage An area of storage, or a storage device, having a very large storage capacity. *Note:* Sometimes referred to as secondary storage in order to differentiate from main storage. *Synonym:* bulk storage. (C) 610.10-1994w

mass unbalance (gyros) The characteristic of a gyro resulting from lack of coincidence of the center of supporting forces and the center of mass. It gives rise to torques caused by linear accelerations that lead to acceleration-sensitive drift rates. (AES/GYAC) 528-1994

mast (power transmission and distribution) A column or narrow-base structure of wood, steel, or other material, supporting overhead conductors, usually by means of arms or brackets, span wires, or bridges. *Note:* Broad-base lattice steel supports are often known as towers; narrow-base steel supports are often known as masts. *See also:* pole; tower. (T&D/PE) [10]

mast arm *See:* bracket.

Master *See:* SBus Master.

master (1) (FASTBUS acquisition and control) A device that is capable of asserting or controlling an operation on a segment according to the FASTBUS protocol. A master may, in addition, contain slave logic. (NID) 960-1993
(2) (STD bus) A card controlling a bus transaction. The master that is currently controlling the bus is the current master. The card that is host to all other masters is the permanent master. All masters that are not the permanent master are temporary masters. (C/MM) 961-1987r
(3) (VMEbus) A functional module that initiates data transfer bus (DTB) cycles to transfer data between itself and a slave module. (BA/C) 1014-1987
(4) (VSB) A functional module that initiates bus cycles in order to transfer data between itself and VSB slaves. The master that is currently in control of the DTB is referred to as the *active* master. (MM/C) 1096-1988w
(5) (NuBus) A bus device that initiates a transaction. (C/MM) 1196-1987w
(6) (NuBus) A module that has acquired control of the bus through the control acquisition procedure. (C/BA) 1014.1-1994w, 896.3-1993w, 896.4-1993w, 10857-1994
(7) A device that initiates communications requests to gather data or perform controls. (PE/SUB) 1379-1997

master antenna television system (MATV) A small television antenna distribution system usually restricted to one or two buildings. (LM/C) 802.7-1989r

master-capable Said of an MTM-Bus module that is an S-module at a given time, but contains appropriate circuitry so that it may be converted by system control to an M-module if required. (TT/C) 1149.5-1995

master clock *See:* clock.

master clock node The node managing the overall synchronization process, which initiates clock synchronization cycles. *Synonym:* reference clock node. (C/BA) 896.2-1991w

master compass A magnetic or gyro compass arranged to actuate repeaters, course recorders, automatic pilots, or other devices. (EEC/PE) [119]

master console In a computer system with more than one console, the primary console that is used to control the computer. *Synonym:* main console. *Contrast:* auxiliary console. *See also:* remote console. (C) 610.10-1994w

master contactor (power system device function numbers) A device, generally controlled by [a master element] device function 1 or the equivalent and the required permissive and protective devices, that serves to make and break the necessary control circuits to place an equipment into operation under the desired conditions and to take it out of operation under other abnormal conditions. (SUB/PE) C37.2-1979s

master control The train-borne device or system directly providing the control signals to the train. (VT/RT) 1475-1999, 1474.1-1999

master controller (1) (load-frequency control) (electric power generators) The central device that develops corrective action for execution at one or more generating units. (PE/PSE) 94-1991w

(2) (car retarders) A controller that governs the operation of one or more magnetic or electropneumatic controllers. *Note:* It is designed to coordinate the movement or the pressure of the retarder with the movement of the retarder level. *See also:* car retarder; multiple-unit control. (VT/LT) 16-1955w

(3) (land transportation vehicles) A device that generates local and trainlike control signals to the propulsion and/or brake systems. (VT/LT) 16-1955w

(4) A physical device utilized by a human operator to provide the master control of a train. (VT) 1475-1999

master controller state A state of the finite state machine (fsm) required of M-modules that controls M-module Link Layer behavior with regard to message transmission. (TT/C) 1149.5-1995

master direction indicator A device that provides a remote reading of magnetic heading. It receives a signal from a magnetic sensing element. (PE/NP) 344-1975s

master drive A drive that sets the reference input for one or more follower drives. *See also:* feedback control system. (IA/ICTL/IAC) [60]

master elect A module that has won the most recent arbitration competition. (C/BA) 10857-1994, 896.4-1993w, 896.3-1993w

master element (power system device function numbers) The initiating device, such as a control switch, etc. which serves either directly or through such permissive devices as protective and time-delay relays to place an equipment in or out of operation. *Note:* This number is normally used for a hand-operated device, although it may also be used for an electrical or mechanical device for which no other function number is suitable. (SUB/PE) C37.2-1979s

master file (data management) An organized collection of records that is relatively permanent; for example, a file containing employee names, addresses, and salary information. *Synonym:* main file. *Contrast:* transaction file. (C) 610.2-1987, 610.5-1990w

master form An original form from which, directly or indirectly, other forms may be prepared. (EEC/PE) [119]

master ground A portable device designed to short circuit and connect (bond) a de-energized circuit or piece of equipment, or both, to an electrical ground. Normally located remote from, and on both sides of, the immediate work site. Primarily used to provide safety for personnel during construction, reconstruction, or maintenance operations *Synonyms:* ground set; ground stick. (T&D/PE) 524a-1993r, 516-1995, 524-1992r

master library (software) A software library containing master copies of software and documentation from which working copies can be made for distribution and use. *Contrast:* system library; software repository; software development library; production library. (C) 610.12-1990

master oscillator (data transmission) An oscillator so arranged as to establish the carrier frequency of the output of an amplifier. (PE) 599-1985w

master physical layer In a 100BASE-T2 link containing a pair of PHYs, the PHY that uses an external clock for generating its clock signals to determine the timing of transmitter and receiver operations. It also uses the master transmit scrambler generator polynomial for side-stream scrambling. Master and slave PHY status is determined during the Auto-Negotiation process that takes place prior to establishing the transmission link. *See also:* slave Physical Layer. (C/LM) 802.3-1998

master reference system for telephone transmission Adopted by the International Advisory Committee for Long Distance Telephony (CCIF), a primary reference telephone system for determining, by comparison, the performance of other telephone systems and components with respect to the loudness, articulation, or other transmission qualities of received speech. *Note:* The determination is made by adjusting the loss of a distortionless trunk in the master reference system for equal performance with respect to the quality under consideration. (EEC/PE) [119]

master remote unit (MRU) An intelligent electronic device that acts as a data concentrator or master to other intelligent electronic devices. (That is, an MRU acquires data from and sends data to other intelligent electronic devices). *Synonyms:* submaster; remote master. (PE/SUB) 1379-1997

master routine *See:* subroutine.

master sequence device (power system device function numbers) A device such as a motor-operated multicontact switch, or the equivalent, or a programming device, such as a computer, that establishes or determines the operating sequence of the major devices in an equipment during starting and stopping or during other sequential switching operations. (SUB/PE) C37.2-1979s

mastership (FASTBUS acquisition and control) A master is asserting mastership when it has control of the segment to which it is attached and is asserting grant acknowledge (GK) or address sync (AS). (NID) 960-1993

master/slave operation (power supplies) A system of interconnection of two regulated power supplies in which one (the master) operates to control the other (the slave). *Note:* Specialized forms of the master.slave configuration are used in: 1) complementary tracking (plus and minus tracking around a common point); 2) parallel operation to obtain increased current output for voltage regulation; 3) compliance extension to obtain increased voltage output for current regulation. (AES) [41]

master state *See:* supervisor state.

master station (1) (A) (data transmission) (supervisory system). The station from which remotely located units of switchgear or other equipment are controlled by supervisory control or that receives supervisory indications or selected telemeter readings. **(B) (data transmission)** (electronic navigation). One station of a group of stations, as in LORAN, that is used to control or synchronize the emission of the other stations. (SWG/SWG/PE) C37.100-1992, 599-1985

(2) (station control and data acquisition) (of a supervisory system) The entire complement of devices, functional modules, and assemblies that are electrically interconnected to effect the master station supervisory functions. The equipment includes the interface with the communication channel but does not include the interconnecting channel. During communication with one or more remote stations, the master station is the superior in the communication hierarchy. (PE/SUB) C37.100-1992, C37.1-1994

(3) (electronic navigation) One station of a group of stations, as in loran, that is used to control or synchronize the emission of the other stations. *See also:* radio navigation. (AES/RS) 686-1982s, [42]

(4) A station that controls other terminals sharing multiple-access transmission medium on a multipoint circuit. (C) 610.7-1995

master-station supervisory equipment (data transmission) That part of a (single) supervisory system that includes all necessary supervisory control relays, keys, lamps, and associated devices located at the master station for selection, control, indication, and other functions to be performed. (SWG/PE) C37.100-1992, 599-1985w

master switch A switch that dominates the operation of contactors, relays, or other remotely operated devices. (IA/MT) 45-1998

master terminal (1) A dedicated terminal that is reserved for the operator of the system or other authorized persons that are privileged to initiate conversations, and to control system-wide processes and operations. *Synonyms:* control terminal; operator console. (C) 610.10-1994w

(2) The entire complement of devices, functional modules, and assemblies that are electrically interconnected to effect the master terminal supervisory functions (of a supervisory system). The equipment includes the interface with the communication channel, but does not include the interconnecting channel. (SUB/PE) 999-1992w

master terminal unit (station control and data acquisition) The master station of a supervisory control system. *See also:* station; master station. (SWG/PE/SUB) C37.100-1992, C37.1-1994

mast-type antenna for aircraft A rigid antenna of streamlined cross section consisting essentially of a formed conductor or conductor and supporting body. (EEC/PE) [119]

MAT *See:* machine-aided translation.

mat (rotating machinery) A randomly distributed unwoven felt of fibers in a sheetlike configuration having relatively uniform density and thickness. *See also:* stator; rotor. (PE) [9]

match (A) A condition in which the values of corresponding components of two or more data items are equal. *See also:* hit. **(B)** To compare two or more data items to determine whether their corresponding components are equal as in definition "A." (C) 610.5-1990

matched A state applying to a sequence of zero or more characters when the characters in the sequence correspond to a sequence of characters defined by a BRE or ERE pattern. (C/PA) 9945-2-1993

matched condition *See:* matched termination.

matched filter A filter that maximizes the output ratio of peak signal power to mean noise power. *Note:* For white noise, a matched filter has a frequency response function that is the complex conjugate of the transmitted spectrum. Its impulse response is the time inverse of the transmitted waveform. (AES) 686-1997

matched generator insertion gain (waveguide) A gain resulting from placing two ports of a network between a load having an arbitrary impedance and a matched generator. It is the ratio of the power absorbed in the load when connected to the generator (reference power) to that when the network is inserted. *Contrast:* matched generator insertion loss. (MTT) 146-1980w

matched generator insertion loss (waveguide) A loss resulting from placing two ports of a network between a load having an arbitrary impedance and a matched generator. It is the ratio of the power absorbed in the load when connected to the generator (reference power) to that when the network is inserted. *Contrast:* matched generator insertion gain. (MTT) 146-1980w

matched impedances Two impedances are matched when they are equal. *Note:* Two impedances associated with an electric network are matched when their resistance components are equal and when their reactance components are equal. *See also:* network analysis. (Std100) 270-1966w

matched insertion gain (waveguide) A gain resulting from placing two ports of a network between a matched generator and a matched load. It is the ratio of the power absorbed in the load when connected to the generator (reference power) to that when the network is inserted. *Contrast:* matched insertion loss. (MTT) 146-1980w

matched insertion loss (waveguide) A loss resulting from placing two ports of a network between a matched generator and a matched load. It is the ratio of the power absorbed in the load when connected to the generator (reference power) to that when the network is inserted. *Contrast:* matched insertion gain. (MTT) 146-1980w

matched load insertion gain (waveguide) A gain resulting from placing two ports of a network between a generator having an arbitrary impedance and a matched load. It is the ratio of the power absorbed in the load when connected to the generator (reference power) to that when the network is inserted. *Contrast:* matched load insertion loss. (MTT) 146-1980w

matched load insertion loss (waveguide) A loss resulting from placing two ports of a network between a generator having an arbitrary impedance and a matched load. It is the ratio of the power absorbed in the load when connected to the generator (reference power) to that when the network is inserted. *Contrast:* matched load insertion gain. (MTT) 146-1980w

matched terminated line (waveguide) A transmission line having no reflected wave at any transverse section. (MTT) 146-1980w

matched termination (waveguide components) A termination matched with regard to the impedance in a prescribed way; for example, a reflectionless termination or a conjugate termination. *See also:* transmission line; reflectionless termination. (MTT) 147-1979w

matched transmission line (data transmission) A transmission line is said to be matched at any transverse section if there is no wave reflection at that section. (PE) 599-1985w

matched waveguide *See:* matched terminated line.

matching *See:* image matching.

matching error (analog computer) An error resulting from inaccuracy in matching (two resistors) or mating (a resistor and a capacitor) passive elements. *See also:* electronic analog computer. (EEC/IE) [126]

matching, impedance *See:* load matching.

matching interaction An instruction method employed by some computer-assisted instruction systems, in which the student is asked to match answers to questions. (C) 610.2-1987

matching, load *See:* load matching.

matching loss (1) (radar) The loss in S/N (signal-to-noise) output relative to a matched filter, caused by using a filter of other than matched response to the transmitted signal. *Synonym:* mismatch loss. (AES/RS) 686-1982s

(2) (telecommunications) The net probability of not being able to establish a network path between an originating line or incoming trunk and a terminating line or trunk when the terminating line or trunk is idle. *Synonym:* overflow loss. (COM/TA) 973-1990w

matching section (waveguide) (waveguide transformer) (transforming section) A length of waveguide of modified cross section, or with a metal or dielectric insert, used for impedance transformation. *See also:* waveguide. (AP/ANT) [35]

matching transformer (induction heater) A transformer for matching the impedance of the load to the optimum output characteristic of the power source. (IA) 54-1955w

material (nuclear power generating station) A substance or combination of substances used as constituents in the manufacture of components, modules, or items. *Note:* This term applies specifically to the subject matter of IEEE Std 467-1980. (PE/NP) 467-1980w

material absorption *See:* absorption.

material dispersion (fiber optics) That dispersion attributable to the wavelength dependence of the refractive index of material used to form the waveguide. Material dispersion is characterized by the material dispersion parameter M. *See also:* dispersion; material dispersion parameter; distortion; waveguide dispersion; profile dispersion parameter. (Std100) 812-1984w

material dispersion parameter (M) (fiber optics)

$$M(\lambda) = -1/c(dN/d\lambda) = \lambda/c(d^2n/d\lambda^2)$$

where n is the refractive index, N is the group index: $N = n - \lambda(dn/d\lambda)$, λ is the wavelength, and c is the velocity of light in vacuum. *Notes:* 1. For many optical waveguide materials, M is zero at a specific wavelength λ_0, usually found in the 1.2 to 1.5 μm range. The sign convention is such that M is positive for wavelengths shorter than λ_0 and negative for wavelengths longer than λ_0. 2. Pulse broadening caused by material dispersion in a unit length of optical fiber is given by M times spectral linewidth ($\Delta\lambda$), except at $\lambda = \lambda_0$, where terms proportional to $(\Delta\lambda)^2$ are important. (See Note 1). *See also:* group index; material dispersion. 812-1984w

material scattering (fiber optics) In an optical waveguide, that part of the total scattering attributable to the properties of the materials used for waveguide fabrication. *See also:* scattering; waveguide scattering; Rayleigh scattering. (Std100) 812-1984w

material temperature class (thermal classification of electric equipment and electrical insulation) (evaluation of thermal capability) The lowest value of a range of temperature indices for insulating materials. (EI) 1-1986r

Mathematica A programming language designed to manipulate equations symbolically. (C) 610.13-1993w

mathematical adjectives All definitions are stated in terms of time (the independent variable) and magnitude (the dependent variable). Unless otherwise specified, the following terms apply only to waveform data within a waveform epoch. These adjectives may be used to describe the relation(s) between other specified variable pairs (for example, time and power, time and voltage, etc.).

a) *Instantaneous.* Pertaining to the magnitude at a specified time.

b) *Positive (negative) peak.* Pertaining to the maximum (minimum) magnitude.

c) *Peak-to-peak.* Pertaining to the absolute value of the algebraic difference between the positive peak magnitude and the negative peak magnitude.

d) *Root-mean-square (rms).* Pertaining to the square root of the average of the square of the magnitude. If the magnitude takes on n discrete values m_j, the root-mean-square magnitude is

$$M_{rss} = \left[\sum_{j=1}^{j=n} m_j^2\right]^{1/2}$$

If the magnitude is a continuous function of time m(t),

$$M_{rss} = \left[\int_{t_1}^{t_2} m^2(t)dt\right]^{1/2}$$

The summation or the integral extends over the interval of time for which the rms magnitude is desired or, if the function is periodic, over any integral number of periodic repetitions of the function.

e) *Average.* Pertaining to the mean of the magnitude. If the magnitude takes on n discrete values m_j, the average magnitude is

$$M_{rms} = \left[\left(\frac{1}{n}\right)\sum_{j=1}^{j=n} m_j^2\right]^{1/2}$$

If the magnitude is a continuous function of time $m(t)$

$$M_{rms} = \left[\left(\frac{1}{t_2 - t_1}\right)\int_{t_1}^{t_2} m^2(t)dt\right]^{1/2}$$

The summation or the integral extends over the interval of time for which the average magnitude is desired or, if the function is periodic, over any integral number of periodice repetitions of the function.

f) *Average absolute* Pertaining to the mean of the absolute magnitude. If the magnitude takes on n discrete values m_j, the average absolute magnitude is

$$\overline{M} = \left(\frac{1}{n}\right)\sum_{j'=1}^{j'=n} m_j$$

If the magnitude is a continuous function of time $m(t)$

$$\overline{M} = \left(\frac{1}{t_2 - t_1}\right)\int_{t_1}^{t_2} m(t)dt$$

The summation or the integral extends over the interval of time for which the average absolute magnitude is desired or, if the function is periodic, over any integral number of periodic repetitions of the function.

g) *Root sum of squares (rss).*Pertaining to the square root of the arithmetic sum of the squares of the magnitude. If the magnitude takes on n discrete values m_j, the root sum of squares magnitude is

$$|\overline{M}| = \left(\frac{1}{n}\right)\sum_{j'=1}^{j'=n} |m_j|$$

If the magnitude is a continuous function of time $m(t)$,

$$|\overline{M}| = \left(\frac{1}{t_2 - t_1}\right)\int_{t_1}^{t_2} |m(t)|dt$$

The summation or the integral extends over the interval of time for which the root sum of squares magnitude is desired *r*, if the function is periodic, over any integral number of periodic repetitions of the function.

(IM/WM&A) 194-1977w

mathematical check (1) (graphic symbols for electrical and electronics diagrams) A programmed check of a sequence of operations that makes use of the mathematical properties of the sequence. Sometimes called a control. *See also:* programmed check. (GSD) 315-1975r
(2) (mathematics of computing) A check of the accuracy of a calculation by performing additional calculations. For example, verification of multiplication results by dividing the product by the multiplier to obtain the multiplicand. *Synonym:* arithmetic check. (C) 1084-1986w

mathematical model (1) (analog computer) A set of equations used to represent a physical system. (C) 165-1977w
(2) (modeling and simulation) A symbolic model whose properties are expressed in mathematical symbols and relationships; for example, a model of a nation's economy expressed as a set of equations. *Contrast:* graphical model; narrative model; software model. (C) 610.3-1989w

mathematical programming In operations research, a procedure for locating the maximum or minimum of a function subject to constraints. (C) 610.2-1987

Mathematical Programming System Extended (MPSX) A programming language used widely for controlling the solution strategy for mathematical programming problems. (C) 610.13-1993w

mathematical quantity *See:* mathematico-physical quantity.

mathematical simulation (analog computer) The use of a model of mathematical equations generally solved by computers to represent an actual or proposed system. (C) 165-1977w

mathematical symbol (abbreviation) A graphic sign, a letter or letters (which may have letters or numbers, or both, as subscripts or superscripts, or both), used to denote the performance of a specific mathematical operation, or the result of such operation, or to indicate a mathematical relationship.

See also: abbreviation; symbol for a unit; symbol for a quantity. (GSD) 267-1966

mathematico-physical quantity (symbolic quantity) (mathematical quantity) (abstract quantity) A concept, amenable to the operations of mathematics, that is directly related on one (or more) physical quantity and is represented by a letter symbol in equations that are statements about that quantity. *Note:* Each mathematical quantity used in physics is related to a corresponding physical quantity in a way that depends on its defining equation. It is characterized by both a qualitative and a quantitative attribute (that is, dimensionality and magnitude). (Std100) 270-1966w

matrix (1) (A) (color television) An array of coefficients symbolic of a color coordinate transformation. *Note:* This definition is consistent with mathematical usage. **(B) (color television)** To perform a color coordinate transformation by computation or by electrical, optical, or other means. (BT/AV) 201-1979

(2) (A) (mathematics) A two-dimensional rectangular array of quantities. Matrices are manipulated in accordance with the rules of matrix algebra. **(B) (mathematics)** By extension, an array of any number of dimensions. (C) [20], [85]

(3) A logic network whose configuration is an array of intersections of its input-output leads, with elements connected at some of these intersections. The network usually functions as an encoder or decoder. *Note:* A translating matrix develops several output signals in response to several input signals; a decoder develops a single output signal in response to several input signals (therefore sometimes called an and matrix); an encoder develops several output signals in response to a single input signal and a given output signal may be generated by a number of different input signals (therefore sometimes called an OR matrix). *See also:* encode; translate; decode. (C) 162-1963w

(4) (general) Loosely, any encoder, decoder, or translator. (Std100) 270-1966w

(5) (electrochemistry) A form used as a cathode in electroforming. (EEC/PE) [119]

(6) (data management) A two-dimensional array conceptually arranged in rows and columns. *Note:* A matrix with m rows and n columns is said to be of size $m \times n$ (m-by-n). *See also:* row-major order; table; column-major order. (C) 610.5-1990w

matrix-addressed storage display device A raster display device that does not require refresh. For example, a plasma panel. (C) 610.10-1994w

matrix character generator A character generator that creates characters composed of selected dots. *Contrast:* stroke character generator. (C) 610.6-1991w

matrix circuit *See:* matrix unit.

matrix, fundamental *See:* transition matrix.

matrix of controls In networking, a two-dimensional matrix that shows the relationship between all the controls in the communications network and the specific threats they mitigate. (C) 610.7-1995

matrix printer *See:* dot matrix printer.

matrix spike An aliquot of a sample which is spiked with a known concentration of the analyte of interest. (NI) N42.23-1995

matrix storage A type of storage whose elements are arranged in such a manner that access to any location requires the use of two or more coordinates; for example, cathode ray storage. (C) 610.10-1994w

matrix, system *See:* system matrix.

matrix, transition *See:* transition matrix.

matrix unit (color television) A device that performs a color coordinate transformation by electrical, optical, or other means. *Synonym:* matrix circuit. (BT/AV) 201-1979w

matte dip (electroplating) A dip used to produce a matte surface on a metal. *See also:* electroplating. (PE/EEC) [119]

matte surface (illuminating engineering) A surface from which the reflection is predominantly diffuse, with or without a negligible specular component. *See also:* diffuse reflection. (EEC/IE) [126]

MATV *See:* master antenna television system.

MAU *See:* medium attachment unit.

maximum allowable conductor temperature The maximum temperature limit that is selected in order to minimize loss of strength, sag, line losses, or a combination of the above. (T&D/PE) 738-1993

maximum arrest force The peak force measured by the test instrumentation during arrest of the test weight in the dynamic test. (T&D/PE) 1307-1996

maximum asymmetric short-circuit current (rotating machinery) The instantaneous peak value reached by the current in the armature winding within a half of a cycle after the winding has been suddenly short-circuited, when conditions are such that the initial value of any aperiodic component of current is the maximum possible. (PE) [9]

maximum available power (MAP) The maximum power that can be obtained by increasing dc current while not controlling the ac voltage. (PE/T&D) 1204-1997

maximum average power (attenuator) That maximum specified input power applied for a minimum of one hour (unless specified for a longer period) at the maximum operating temperature with output terminated in the characteristic impedance which will not permanently change the specified properties of the attenuator after return to ambient temperature at a power level 20 dB below maximum specified input power. (IM/HFIM) 474-1973w

maximum average power output (television) The maximum radio-frequency output power that can occur under any combination of signals transmitted, averaged over the longest repetitive modulation cycle. *See also:* television. (EEC/PE) [119]

maximum bundle gradient (overhead-power-line corona and radio noise) For a bundle of two or more subconductors, the highest value among the maximum gradients of the individual subconductors. For example, for a three-conductor bundle with individual maximum subconductor gradients of 16.5, 16.9, and 17.0 kV/cm, the maximum bundle gradient would be 17.0 kV/cm. (T&D/PE) 539-1990

maximum capability (power operations) The maximum generation expressed in kilowatt-hours per hour (kWh/h) which a generating unit, station, power source, or system can be expected to supply under optimum operating conditions. (PE/PSE) 858-1987s

maximum capacity (electric generating unit reliability, availability, and productivity) The maximum capacity that a unit can sustain over a specified period of time. The maximum capacity can be expressed as gross maximum capacity (GMC) or net maximum capacity (NMC). To establish this capacity, formal demonstration is required.The test is repeated periodically. This demonstrated capacity level is corrected to generating conditions for which there is minimum ambient restriction. When a demonstration test has not been conducted, the estimated maximum capacity of the unit is used. (PE/PSE) 762-1987w

maximum common-mode signal level The maximum level of the common-mode signal at which the common mode rejection ratio is still valid. (IM/WM&A) 1057-1994w

maximum continuous exposure temperature The highest temperature to which a component of the heat-tracing system may be continuously exposed (heater de-energized). (IA) 515-1997

maximum continuous operating voltage (MCOV) (1) The maximum rms value of power-frequency voltage that may be applied continuously between the terminals of the arrester without degradation or deleterious effects. (PE) C62.34-1996

(2) The maximum designated root-mean-square (rms) value of power-frequency voltage that may be applied continuously between the terminals of the arrester. (SPD/PE) C62.11-1999

(3) The maximum designated root-mean-square value of power frequency voltage that may be applied continuously between the terminals of the overvoltage protective device. (SPD/PE) C62.62-2000

maximum continuous rating (rotating machinery) The maximum values of electric and mechanical loads at which a machine will operate successfully and continuously. *Note:* An overload may be implied, along with temperature rises higher than normal standards for the machine. *See also:* asynchronous machine. (PE) [9]

maximum control current (magnetic amplifier) The maximum current permissible in each control winding either continuously or for designated operating intervals as specified by the manufacturer and shall be specified as either root-mean-square or average. (MAG) 107-1964w

maximum credible voltage or current transient That voltage or current transient that may exist in circuits, as determined by test or analysis, taking into consideration the circuit location, routing, and interconnections combined with failures that the circuits may credibly experience. (PE/NP) 384-1992r

maximum current (wattmeter or power-factor meter) (instrument) A stated current that, if applied continuously at maximum stated operating temperature and with any other circuits in the instrument energized at rated values, will not cause electric breakdown or any observable physical degradation. *See also:* instrument. (EEC/AII) [102]

maximum-deflection angle The maximum plane angle subtended at the deflection center by the usable screen area. *Note:* In this term, the hyphen is frequently omitted. (ED) 161-1971w

maximum demand (1) (power operations) The largest of a particular type of demand occurring within a specified period. (PE/PSE) 858-1987s
(2) (electric power systems in commercial buildings) The greatest of all the demands that have occurred during a specified period of time; determined by measurement over a prescribed time interval. (IA/PSE) 241-1990r
(3) The greatest of all demands that have occurred during a specified period of time such as one-quarter, one-half, or one hour. *Note:* For utility billing purposes the period of time is generally one month. (IA/PSE) 141-1993r
(4) The highest demand measured over a selected period of time, e.g., one month. (AMR/SCC31) 1377-1997
(5) *See also:* demand—maximum. (ELM) C12.1-1981

maximum-demand pointer (friction pointer of a demand meter) (demand meter) A means used to indicate the maximum demand that has occurred since its previous resetting. The maximum-demand pointer is advanced up the scale of an indicating demand meter by the pointer pusher. When not being advanced, it is held stationary, usually by friction, and it is reset manually when the meter is read for billing purposes. *See also:* demand meter. (EEC/PE) [119]

maximum design cantilever load-static (MDCL-static) The maximum cantilever load the surge arrester is designed to continuously carry. (SPD/PE) C62.11-1999

maximum design rating (composite insulators) The maximum mechanical load that the insulator is designed to withstand continuously for the life of the insulator. (T&D/PE) 987-1985w

maximum design voltage (1) (device) The highest voltage at which the device is designed to operate. *Note:* When expressed as a rating this voltage is termed rated maximum voltage. (PE/PSR) C37.90-1978s
(2) (outdoor electric apparatus) (to ground) The maximum voltage at which the bushing is designed to operate continuously. (PE/TR) 21-1976
(3) (power and distribution transformers) The highest rms phase-to-phase voltage that equipment components are designed to withstand continuously, and to operate in a satisfactory manner without derating of any kind. (PE/TR) C57.12.80-1978r
(4) (A) (of a device) The highest voltage at which the device is designed to operate. **(B)** (of a relay) The highest root-mean-square (rms) or dc voltage at which a relay is designed to be energized continuously. (SWG/SWG/PE) C37.100-1992

maximum-deviation sensitivity (in frequency-modulation receivers) Under maximum system deviation, the least signal input for which the output distortion does not exceed a specified limit. *See also:* frequency modulation. 188-1952w

maximum differential input The largest value of peak-to-peak differential (ppd) amplitude at which a receiver is expected to operate, under worst-case conditions, without exceeding the objective bit error ratio. (C/LM) 802.3-1998

maximum discharge current The maximum surge current that the surge protective device withstands without damage. The maximum discharge current is a peak impulse current, with a wave shape of 8/20. (PE) C62.34-1996

maximum effort (electric generating unit reliability, availability, and productivity) Repairs were accomplished in the shortest possible time. *See also:* repair urgency. (PE/PSE) 762-1987w

maximum excursion (electric conversion) The maximum positive or negative deviation from the initial or steady value caused by a transient condition. (AES) [41]

maximum exposure temperature (1) (electrical heating systems) The highest temperature to which an object may be exposed continuously. (IA/PC) 844-1991
(2) (electrical heat tracing for industrial applications) The highest temperature to which a device in the heat-tracing system may be exposed for a given period of time. (BT/AV) 152-1953s
(3) The highest temperature to which a component of the heat-tracing system may be exposed either continuously or for a specified period of time. (IA/PC) 515.1-1995

maximum frequency of interest (A) For switching power supplies: 10 times the maximum power switch switching frequency. **(B)** For filter products: 10 times the 3dB point. (PEL) 1515-2000

maximum generation (electric generating unit reliability, availability, and productivity) The energy that could have been produced by a unit in a given period of time if operated continuously at maximum capacity. Maximum generation can be expressed as gross maximum generation (GMG) or net maximum generation (NMG).

$$MG = \text{period hours} \cdot \text{maximum capacity} = PH \cdot MC$$
$$GMG = PH \cdot GMC$$
$$NMG = PH \cdot NMC$$

(PE/PSE) 762-1987w

maximum grid current A design value of the maximum grid current, defined as follows:

$$I_G = D_f \times I_g$$

where
I_G = the maximum grid current in A
D_f = the decrement factor for the entire duration of fault t_f, given in s
I_g = the rms symmetrical grid current in A

(PE/SUB) 80-2000

maximum ground acceleration (seismic design of substations) The maximum value of acceleration input to the equipment during a given earthquake for a particular site. (PE/SUB) 693-1984s, C37.122.1-1993

maximum hottest conductor temperature Used in discussions involving the life testing of materials, in lieu of the phrase winding hottest-spot temperature (θ_h). (PE/TR) 1276-1997

maximum instantaneous fuel change (gas turbines) The fuel change allowable for an instantaneous or sudden increased or decreased load or speed demand. *Note:* It is expressed in terms of equivalent load change in percent of rated load. (PE/EDPG) 282-1968w, [5]

maximum keying frequency (facsimile) (fundamental scanning frequency) The frequency in hertz numerically equal to the spot speed divided by twice the scanning spot X dimension. *See also:* scanning. (COM) 168-1956w

maximum limiting resolution (diode-type camera tube) The highest value of limiting resolution obtained under optimum irradiance conditions using a stationary bar pattern. Units: LP/RH. (ED) 503-1978w

maximum maintain temperature (1) Specified maximum temperature of a surface or process which the heat tracing is capable of maintaining continuously. (IA/PC) 515.1-1995
(2) Specified maximum temperature of a surface or process that the heat-tracing cable or surface heating device is capable of maintaining continuously. (IA) 515-1997

maximum mechanical load The largest service load allowed on a composite insulator or bushing. The maximum mechanical load (MML) is within the reversible elastic range and is supplied by the manufacturer. (PE/SUB) 693-1997

maximum modulating frequency (facsimile) The highest picture frequency required for the facsimile transmission system. *Note:* The maximum modulating frequency and the maximum keying frequency are not necessarily equal. *See also:* facsimile transmission. (COM) 168-1956w

maximum momentary speed variation (hydraulic turbines) The maximum momentary change of speed when the load is suddenly changed a specified amount. (PE/EDPG) 125-1977s

maximum observed frequency (MOF) In oblique-incidence ionospheric sounding, the MOF is the highest frequency for which the signals transmitted from a sounder are observed on the ionogram, regardless of the propagation path involved. (AP/PROP) 211-1997

maximum OFF voltage (magnetic amplifier) The maximum output voltage existing before trip ON control signal is reached as the control signal is varied from trip OFF to trip ON. (MAG) 107-1964w

maximum operating common-mode signal The largest common-mode signal for which the waveform recorder will meet its effective bits specifications in recording a simultaneously-applied, normal-mode signal. (IM/WM&A) 1057-1994w

maximum operating voltage (*Vm*) (1) (household electric ranges) The maximum voltage to which the electric parts of the range may be subjected in normal operation. *See also:* appliance outlet. (IA/APP) [90]
(2) The maximum system operating rms phase-to-phase (or phase-to-ground for single phase, or pole-to-ground for dc) voltage, which is also equal to the 1 per unit (p.u.) base. For clearance calculation, the maximum operating crest phase-to-ground voltage is equal to 1 per unit (p.u.). (T&D/PE) 516-1995

maximum output (receivers) The greatest average output power into the rated load regardless of distortion. *See also:* radio receiver. 188-1952w

maximum output jitter The peak-to-peak jitter acceptable to enable satisfactory interconnection of digital networks and equipment. (COM/TA) 1007-1991r

maximum output voltage (magnetic amplifier) The voltage across the rated load impedance with maximum control current flowing through each winding simultaneously in a direction that increases the output voltage. *Notes:* 1. Maximum output voltage shall be specified either as root-mean-square or average. 2. While specification may be either root-mean-square or average, it remains fixed for a given amplifier. (MAG) 107-1964w

maximum peak power (attenuator) That maximum peak power at the maximum specified pulse-length and average power which, when applied for a minimum of one hour (unless specified for a longer period) at the maximum operating temperature, while the output is terminated in the characteristic impedance, will not permanently change the specified properties of the attenuator when returned to ambient temperature at a power level 20 dB below the maximum specified input power or lower. (IM/HFIM) 474-1973w

maximum permissible exposure (MPE) The rms and peak electric and magnetic field strengths, their squares, or the plane-wave equivalent power densities associated with these fields and the induced and contact currents to which a person may be exposed without harmful effect and with an acceptable safety factor. (NIR) C95.1-1999

maximum power output (hydraulic turbines) The maximum output which the turbine-generator unit is capable of developing at rated speed with maximum head and maximum gate. (PE/EDPG) 125-1977s

maximum pulse rate (metering) The number of pulses per second at which a pulse device is nominally rated. (ELM) C12.1-1982s

maximum pulse repetition rate (digital delay line) The maximum pulse repetition rate shall be equal to 1/2 Dτ, where Dτ is the time spacing between the peaks of the output doublet. (UFFC) [22]

maximum rated step voltage The highest value of rated step voltage for which the LTC is designed. (PE/TR) C57.131-1995

maximum rated through current The rated through current for which both the temperature rise of the contacts and the service duty test apply. (PE/TR) C57.131-1995

maximum rate of fuel change (gas turbines) The rate of fuel change that is allowable after the maximum instantaneous fuel change, when an instantaneous speed or load demand upon the turbine is greater than that corresponding to the maximum instantaneous fuel change. *Note:* It is expressed in percent of equivalent load change per second. (PE/EDPG) 282-1968w, [5]

maximum relative side lobe level The maximum relative directivity of the highest side lobe with respect to the maximum directivity of the antenna. (AP/ANT) 145-1993

maximum retention time (storage tubes) The maximum time between writing into a storage tube and obtaining an acceptable output by reading. *See also:* storage tube. (ED) 161-1971w

maximum safe input power (1) (spectrum analyzer) The power applied at the input which will not cause degradation of the instrument characteristics. *Note:* Input signal conditions, for example, peak or average power, should be specified. (IM) 748-1979w
(2) (non-real time spectrum analyzer) The power applied at the input which will not cause degradation of the instrument characteristics. (IM) [14]
(3) (electrothermic unit) The maximum peak pulse or cw input power that will cause no permanent change in the calibration or characteristics of the electrothermic unit. Specify in watt-microseconds the maximum (safe) input energy per pulse and the applicable pulse repetition frequency in hertz or in kilohertz. Specify in watts or milliwatts the maximum (safe) input peak pulse power. (IM) 544-1975w

maximum sensitivity (frequency-modulation systems) The least signal input that produces a specified output power. 188-1952w

maximum sine-current differential permeability (toroidal magnetic amplifier cores) The maximum value of sine-current differential permeability obtained with a specified sine-current magnetizing force. (Std100) 106-1972

maximum single-conductor gradient (overhead-power-line corona and radio noise) The maximum value attained by the gradient $E(\theta)$ as θ varies over the range 0 to 2π, where $E(\theta)$ is the gradient on the surface of the power line conductor expressed as a function of angular position (θ). Unless otherwise stated, the gradient is a nominal gradient. *See also:* nominal conductor gradient; maximum single-subconductor gradient. (T&D/PE) 539-1990

maximum single-subconductor gradient (overhead-power-line corona and radio noise) The maximum value attained by the gradient $E(\theta)$ as θ varies over the range 0 to 2π, where $E(\theta)$ is the gradient on the surface of the power line subconductor expressed as a function of angular position (θ). Unless otherwise stated, the gradient is a nominal gradient.

See also: maximum single-conductor gradient; nominal conductor gradient. (T&D/PE) 539-1990

maximum sound pressure (for any given cycle of a periodic wave) The maximum absolute value of the instantaneous sound pressure occurring during that cycle. *Note:* In the case of a sinusoidal sound wave this maximum sound pressure is also called the pressure amplitude. (SP) [32]

maximum speed The highest speed within the operating speed range of the drive. *See also:* electric drive. (IA/ICTL/IAC) [60]

maximum static error (MSE) The maximum difference between any code transition level and its ideal value. (IM/WM&A) 1057-1994w

maximum surge current rating (semiconductor rectifiers) (rectifier circuits) (nonrepetitive) The maximum forward current having a specified waveform and short specified time interval permitted by the manufacturer under stated conditions. *See also:* average forward current rating. (IA) [62]

maximum surge energy absorbed (for rating purposes only) The maximum allowable surge energy that the photocontrol can absorb without changing its operating characteristics. This is based on a single 10/1000 μs current waveform. (RL) C136.10-1996

maximum system deviation (frequency-modulation systems) The greatest frequency deviation specified in the operation of the system. *Note:* Maximum system deviation is expressed in kilohertz. In the case of FCC authorized frequency modulation broadcast systems in the range from 88 to 108 MHz, the maximum system deviation is ±75 kHz. (BT) 185-1975w

maximum system voltage (1) (electrical systems in commercial buildings) The highest system voltage that occurs under normal operating conditions, and the highest system voltage for which equipment and other components are designed for satisfactory continuous operation without derating of any kind. (IA/PSE) 241-1990r
(2) The highest voltage at which a system is operated. (This voltage excludes voltage transients and temporary overvoltages caused by abnormal system conditions such as faults, load rejection, etc.). *Note:* This is generally considered to be the maximum system voltage. (SPD/PE) C62.22-1991s
(3) The highest rms phase-to-phase voltage that occurs on the system under normal operating conditions, and the highest rms phase-to-phase voltage for which equipment and other system components are designed for satisfactory continuous operation without deterioration of any kind. (C/PE/TR) 1313.1-1996, C57.12.80-1978r
(4) The highest voltage at which a system is operated. *Note:* This is generally considered to be the maximum system voltage as prescribed in ANSI C84.1-1995. (SPD/PE) C62.11-1999, C62.62-2000

maximum test output voltage (magnetic amplifier) (nonreversible output) The output voltage equivalent to the summation of the minimum output voltage plus 66 2/3 percent of the difference between the rated and minimum output voltages. (MAG) 107-1964w
(2) (A) (reversible output) Positive maximum test output voltage is the output voltage equivalent to 66 2/3 percent of the rated output voltage in the positive direction. **(B) (reversible output)** Negative maximum test output voltage is the output voltage equivalent to 66 2/3 percent of the rated output voltage in the negative direction. (MAG) 107-1964

maximum theoretical deviation from a sine wave (self-commutated converters) (converter characteristics) For a nonsinusoidal wave, the ratio of the arithmetic sum of the amplitudes (rms) of all harmonics in the wave to the amplitude (rms) of the fundamental. (IA/SPC) 936-1987w, 519-1992

maximum total sag The total sag at the midpoint of the straight line joining the two points of support of the conductor. *See also:* sag. (NESC/BT/T&D/AV) C2-1997, [34], C2.2-1960

maximum transfer unit The largest amount of data that can be transferred across a given physical network. (C) 610.7-1995

maximum undistorted output (amplitude-modulation broadcast receivers) The so-called maximum undistorted output is arbitrarily taken as the least power output which contains, under given operating conditions, a total power at harmonic frequencies equal to one percent of the apparent power at the fundamental frequency. This corresponds to a root-sum-square total voltage at harmonic frequencies equal to 10 percent of the root-sum-square voltage at the fundamental frequency, if measured across a pure resistance. (The root-sum-square voltage of a complex wave is the square root of the sum of the squares of the component voltages.) (CE) 186-1948w

maximum usable frequency (MUF) The highest frequency by which a radio wave can propagate between given terminals, on a specified occasion, by ionospheric refraction alone. *Notes:* 1. Where the MUF is restricted to a particular ionospheric propagation mode, the values may be quoted together with an indication of that mode (e.g., 1E MUF, 2F2 MUF). 2. If the extraordinary component of the wave is involved, then this is noted [e.g., 1F2 MUF (X)]. Absence of a specific response to the magneto-ionic component implies that the quoted value relates to the ordinary wave. 3. It is sometimes useful to quote the ground range for which the MUF applies. This is indicated in kilometers following the indication of the mode type [e.g., 1F2 (4000) MUF (X)]. (AP/PROP) 211-1997

maximum usable reading time (storage tubes) The length of time a storage element, line, or area can be read before a specified degree of decay occurs. *Notes:* 1. This time may be limited by static decay, dynamic decay, or a combination of the two. 2. It is assumed that rewriting is not done. 3. The qualifying adjectives maximum usable are frequently omitted in general usage when it is clear that the maximum usable reading time is implied. *See also:* storage tube. (ED) 158-1962w

maximum usable read number (storage tubes) The number of times a storage element, line, or area can be read without rewriting before a specified degree of decay results. *Note:* The qualifying adjectives maximum usable are frequently omitted in general usage when it is clear that the maximum usable read number is implied. *See also:* storage tube. (ED) 158-1962w, [45]

maximum usable viewing time (storage tubes) The length of time during which the visible output of a storage tube can be viewed, without rewriting, before a specified decay occurs. *Note:* The qualifying adjectives maximum usable are frequently omitted in general usage when it is clear that maximum usable viewing time implied. *See also:* storage tube. (ED) 158-1962w

maximum usable writing speed (storage tubes) The maximum speed at which information can be written under stated conditions of operation. Note the qualifying adjectives maximum usable are frequently omitted in general usage when it is clear that the maximum usable writing speed is implied. *See also:* storage tube. (ED) 158-1962w

maximum useful output *See:* maximum undistorted output.

maximum value of magnetic field (measurement of power frequency electric and magnetic fields from ac power lines). At a given point, the root-mean-square (rms) value of the semimajor axis of the magnetic field ellipse. (T&D/PE) 1308-1994

maximum value of the electric field strength At a given point, the rms value of the semimajor axis magnitude of the electric field ellipse. *See also:* electric field strength. (T&D/PE) 644-1994

maximum value of the field (overhead power lines) At a given point, the rms value of the major semi-axis magnitude of the field ellipse; i.e., the largest value of the field that would be measured at that point. (T&D/PE) 539-1990

maximum value of the magnetic field At a given point, the rms value of the semimajor axis magnitude of the magnetic field ellipse. (T&D/PE) 644-1994

maximum voltage (instrument) (wattmeter, power-factor meter or frequency meter) A stated voltage that, if applied continuously at the maximum stated operating temperature and with any other circuits in the instrument energized at rated values, will not cause electric breakdown or any observable physical degradation. *See also:* accuracy rating; instrument. (EEC/AII) [102]

maximum voltage rating The highest phase-to-ground or phase-to-ground and phase-to-phase voltage (rms) at which a connector is designed to operate. (T&D/PE) 386-1995

maximum (hottest-spot) winding temperature The maximum or hottest temperature of the current carrying components of a transformer winding in contact with insulation or insulating fluid. The hottest spot temperature is a naturally occurring phenomena due to the generation of losses and the heat transfer phenomena. It is the highest temperature inside the transformer winding and is greater than the measured average winding temperature of the coil conductors. All transformers have a maximum (hottest-spot) winding temperature. (PE/TR) C57.134-2000

maximum (hottest-spot) winding temperature rise The arithmetic difference between maximum (hottest-spot) winding temperature and the ambient temperature. (PE/TR) C57.134-2000

maxwell (line) The unit of magnetic flux in the centimeter-gram-second electromagnetic system. *Note:* The maxwell is 10^{-8} weber. (Std100) 270-1966w

Maxwell bridge (general) A 4-arm alternating-current bridge characterized by having in one arm an inductor in series with a resistor and in the opposite arm a capacitor in parallel with a resistor, the other two arms being normally nonreactive resistors. *Note:* Normally used for the measurement of inductance (or capacitance) in terms of resistance and capacitance (or inductance). The balance is independent of the frequency, and at balance the ratio of the inductance to the capacitance is equal to the product of the resistances of either pair of opposite arms. It differs from the Hay bridge in that in the arm opposite the inductor, the capacitor is shunted by the resistor. (See the corresponding figure.) *See also:* bridge.

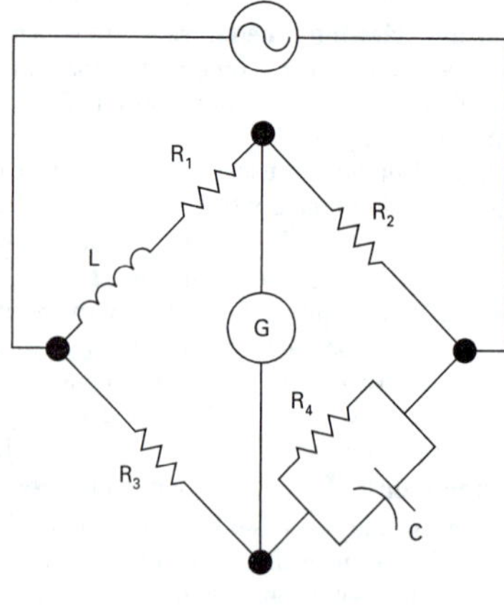

$R_1R_4 = R_2R_3 = L/C$

Maxwell bridge

(PE/EEC) [119]

Maxwell direct-current commutator bridge A 4-arm bridge characterized by the presence in one arm of a commutator, or 2-way contactor, that, with a known periodicity, alternately connects the unknown capacitor in series with the bridge arm and then opens the bridge arm while short-circuiting the capacitor, the other three arms being nonreactive resistors. (See the corresponding figure.) *Note:* Normally used for the measurement of capacitance in terms of resistance and time. The bridge is normally supplied from a battery and the detector is a direct-current galvanometer.

$$C = \frac{R_1}{nR_2R_3} \times \frac{\left[1 - \dfrac{R_1^2}{(R_1 + R_2 + R_B)(R_1 + R_3 + R_G)}\right]}{\left[1 + \dfrac{R_1R_B}{R_3(R_1 + R_2 + R_B)}\right]\left[1 + \dfrac{R_1R_G}{R_2(R_1 + R_3R_G)}\right]}$$

See also: bridge.

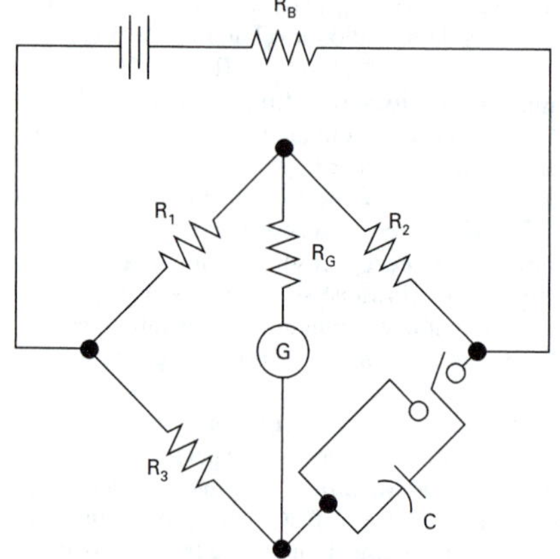

Maxwell direct-current commutator bridge

(EEC/PE) [119]

Maxwell-Garnett mixing formula Gives the approximate dielectric constant of a medium containing a few small spherical dielectric inclusions, ε_r, as:

$$\varepsilon_r = \varepsilon_1 \frac{1 + 2fy}{1 - fy} \quad \text{with} \quad y = \frac{\varepsilon_2 - \varepsilon_1}{\varepsilon_2 + 2\varepsilon_1}$$

where

ε_1 = the relative dielectric constant of the background medium

ε_2 = the relative dielectric constant of the inclusions

f = the volume fraction of the inclusions

The Maxwell-Garnett mixing formula is valid when f is less than a few percent and the inclusions are small compared to the wavelength. (AP/PROP) 211-1997

Maxwell inductance bridge A 4-arm alternating-current bridge characterized by having inductors in two adjacent arms and usually, nonreactive resistors in the other two arms. (See the corresponding figure.) *Note:* Normally used for the comparison of inductances. The balance is independent of the frequency. *See also:* bridge.

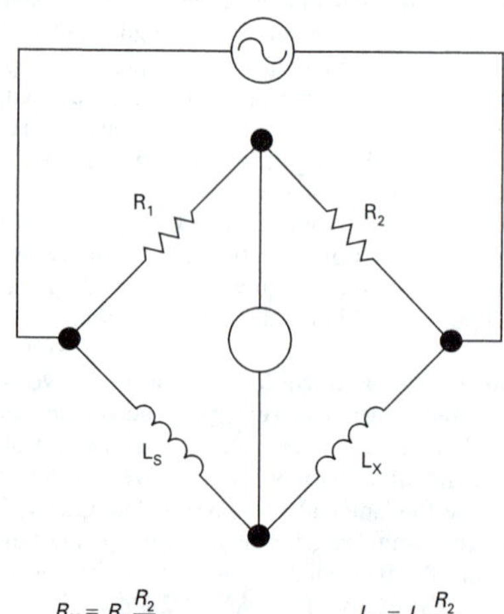

$R_X = R_S\frac{R_2}{R_1}$ $L_X = L_S\frac{R_2}{R_1}$

Maxwell inductance bridge

(EEC/PE) [119]

Maxwell mutual-inductance bridge An alternating-current bridge characterized by the presence of mutual inductance between the supply circuit and that arm of the network that includes one coil of the mutual inductor, the other three arms being normally non-reactive resistors. (See the corresponding figure.) *Note:* Normally used for the measurement of mutual inductance in terms of self-inductance. The balance is independent of the frequency. *See also:* bridge.

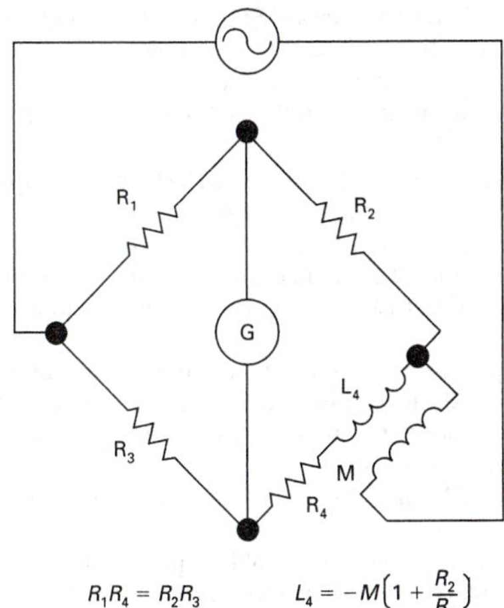

Maxwell mutual-inductance bridge

(EEC/PE) [119]

Maxwell's equations (Maxwell's laws) The fundamental equations of macroscopic electrmagnetic field theory. All real (physical) electric and magnetic fields satisfy Maxwell's equations, namely

$$\nabla \times \mathbf{E} = \frac{\partial \mathbf{B}}{\partial t}$$

$$\nabla \times \mathbf{H} = \frac{\partial \mathbf{D}}{\partial t} + \mathbf{J}$$

$$\nabla \cdot \mathbf{B} = 0$$

$$\nabla \cdot \mathbf{D} = q_r$$

where **E** is electric field strength, **D** is electric flux density, **H** is magnetic field strength, **B** is magnetic flux density, **J** is current density, and q_r is volume charge density. (Std100) 270-1966w

may (1) With respect to implementations, the word may is to be interpreted as an optional feature that is not required in this standard but can be provided. With respect to strictly conforming POSIX applications, the word may means that the optional feature shall not be used. (C/PA) 1003.1-1988s

(2) An indication of an optional feature or behavior of the implementation that is not required by this standard, although there is no prohibition against providing it. A Strictly Conforming POSIX.2 Application is permitted to use such features, but shall not rely on the implementation's actions in such cases. To avoid ambiguity, the reverse sense of *may* is not expressed as *may not,* but as *need not*. (C/PA) 2003.2-1996

(3) An indication of an optional feature.

1) With respect to implementations, the word *may* is to be interpreted as an optional feature that is not required in a standard but can be provided.
2) With respect to Strictly Conforming POSIX.5 Applications, the word *may* means that the optional feature shall not be used.

(C) 1003.5-1999

May Day *See:* radio distress signal.

MB *See:* megabyte.

MBASIC A dialect of the BASIC programming language. (C) 610.13-1993w

Mbit Indicates 2^{20} bits. (C/MM) 1596.4-1996

MBWO *See:* microwave backward-wave oscillator.

Mbyte Megabyte. Indicates 2^{20} bytes. (C/MM) 1212-1991s, 1596.4-1996

MC (metal-cladswitchgear) *See:* message code.

MCA *See:* multichannel analyzer; multichannel pulse-height analyzer.

MC cable (metal-clad cable) A factory assembly of one or more conductors, each individually insulated and enclosed in a metallic sheath of interlocking tape, or a smooth or corrugated tube. (NEC/NESC) [86]

McCulloh circuit A supervised, metallic loop circuit having manually or automatically operated switching equipment at the receiving end, that, in the event of a break, a ground, or a combination of a break and a ground at any point in the metallic circuit, conditions the circuit, by utilizing a ground return, for the receipt of signals from suitable signal transmitters on both sides of the point of trouble. (EEC/PE) [119]

M channel A 96 kbit/s maintenance channel that is used to convey Physical Layer status and control information to the far end of the link. (C/LM) 802.9a-1995w

MCOV *See:* maximum continuous operating voltage.

MDA *See:* minimum detectable amount; mirrored disk array.

MDC *See:* minimum detectable concentration.

MDI (Medium Dependent Interface) *See:* medium dependent interface; magnetic direction indicator.

M-display A type of A-display in which one target range is determined by moving an adjustable pedestal, notch, or step along the baseline until it coincides with the horizontal position of the target-signal deflection; the control that moves the pedestal is calibrated in range. *Note:* This display is usually identified as a variant of an A-display.

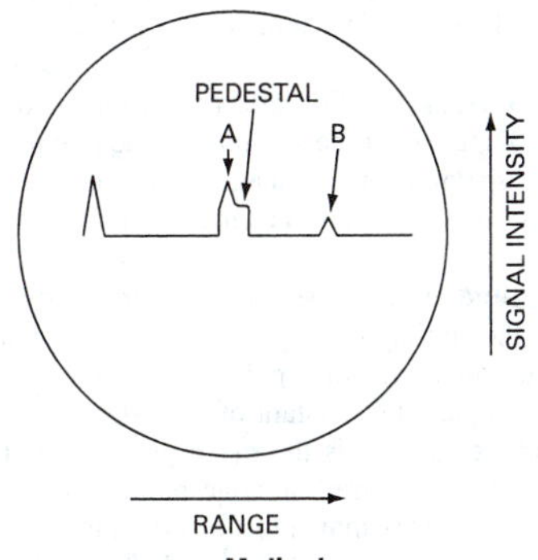

M-display

(AES) 686-1997

MDR *See:* memory data register; memory buffer register.

MDS *See:* minimum detectable signal.

MDV *See:* minimum detectable velocity.

ME *See:* metal-enclosed power switchgear.

MEA *See:* minimum en-route altitude.

mean access time The average access time identified with the normal operation of a device. (C) 610.10-1994w

mean charge (nonlinear capacitor) The arithmetic mean of the transferred charges corresponding to a particular capacitor voltage, as determined from a specified alternating charge characteristic. *See also:* nonlinear capacitor. (ED) [46]

mean-charge characteristic (nonlinear capacitor) The function relating mean charge to capacitor voltage. *Note:* Mean-charge characteristic is always single-valued. *See also:* nonlinear capacitor. (ED) [46]

mean first slip time (communication satellite) The mean time for a phase-lock loop starting to lock to a slip one or more cycles. (COM) [25]

mean free path For sound waves in an enclosure, the average distance sound travels between successive reflections in the enclosure. (ED/SP/ACO) [46], [32]

mean Hall plate temperature (Hall effect devices) The value of the temperature averaged over the volume of the Hall plate. (MAG) 296-1969w

mean horizontal intensity (illuminating engineering) (candlepower) The average intensity (candelas) of a lamp in a plane perpendicular to the axis of the lamp and which passes through the luminous center of the lamp. (EEC/IE) [126]

meaning (of a responsibility) A statement of what the responsibility means. The statement of responsibility is written from the point of view of the requester, not the implementer. The statement of responsibility states what the requester needs to know to make intelligent use of the property or constraint. That statement should be complete enough to let a requester decide whether to make the request, but it should stop short of explaining how a behavior or value is accomplished or derived. Meaning is initially captured using freeform natural language text in a glossary definition. It may be more formally refined into a statement of *pre-conditions* and *post-conditions* using the *specification language*. (C/SE) 1320.2-1998

mean instrument reading The arithmetic average of a series of readings. (NI) N42.17B-1989r

mean life The arithmetic mean of the times to failure of a group of nominally identical items. *See also:* reliability. (R) [29]

mean life, assessed *See:* assessed mean life.

mean life, extrapolated *See:* extrapolated mean life.

mean life, observed *See:* observed mean life.

mean life, predicted *See:* predicted mean life.

mean logistics time (MLT) The mean downtime occasioned by the unavailability of spares, replacement parts, test equipment, maintenance facilities, or personnel. (PE/NP) 933-1999

mean of reversed direct-current values (alternating-current instruments) The simple average of the indications when direct current is applied in one direction and then reversed and applied in the other direction. *See also:* accuracy rating. (EEC/AII) [102]

mean outage duration The mean duration of outage occurrences of a specified type = outage time due to outages of a specified type/number of outage occurrences of a specified type. *Note:* Also referred to as mean time to restoration. (PE/PSE) 859-1987w

mean pulse time The arithmetic mean of the leading-edge pulse time and the trailing-edge pulse time. *Note:* For some purposes, the importance of a pulse is that it exists (or is large enough) at a particular instant of time. For such applications the important quantity is the mean pulse time. The leading-edge pulse time and trailing-edge pulse time are significant primarily in that they may allow a certain tolerance in timing. (IM/WM&A) 194-1977w

mean radiating temperature For a non-isothermal body or medium:

— The temperature that would give rise to the same total brightness as the actual medium, or

— The temperature that would give rise to the same spectral brightness as the actual medium.

(AP/PROP) 211-1997

mean side lobe level The average value of the relative power pattern of an antenna taken over a specified angular region, which excludes the main beam, the power pattern being relative to the peak of the main beam. (AP/ANT) 145-1993

means of grounding (neutral grounding in electrical utility systems) The generic agent by which various degrees of grounding are achieved; for example, inductance grounding, resistance grounding, and resonant grounding. (SPD/PE) C62.92-1987r

mean spherical luminous intensity (illuminating engineering) Average value of the luminous intensity in all directions for a source. Also, the quotient of the total emitted luminous flux of the source by 4π.

$$I_{ms} = \left(\frac{1}{4\pi}\right) \int_0^{\pi} I \, d\omega = \frac{\Phi_{total}}{4\pi}$$

(EEC/IE) [126]

mean temperature coefficient of output voltage (Hall effect devices) The arithmetic average of the percentage changes in output voltage per degree Celsius taken over a given temperature range for a given control current magnitude and a given magnetic flux density. (MAG) 296-1969w

mean time before failures (MTBF) (power supplies) A measure of reliability giving the time before first failure. Mean time before failures may be approximated or predicted by summing the reciprocal failure rates of individual components in an assembly. (AES) [41]

mean time between failures (MTBF) (1) (power supplies) For repairable equipment, a measure of reliability giving the average time between repairs. Mean time between failures may be approximated or predicted by summing the reciprocal failure rates of individual components in an assembly. (AES) [41]

(2) (supervisory control, data acquisition, and automatic control) (station control and data acquisition) The time interval (hours) that may be expected between failures of an operating equipment. (SWG/PE/SPD/SUB) C37.100-1992, C62.1-1981s, C37.1-1994

(3) The average time (preferably expressed in hours) between failures of a continuously operating device, circuit, or system. (PE/PSC) 599-1985w

(4) (nuclear power generating station) The arithmetic average of operating times between failures of an item. (PE/NP) 352-1975s, 933-1999

(5) (repairable items) The product of the number of items and their operating time divided by the total number of failures. (R) [29]

(6) The expected or observed time between consecutive failures in a system or component. (C) 610.12-1990

(7) For a stated period in the life of a device, the mean value of the lengths of time between consecutive failures under stated conditions. *Note:* Used to measure equipment reliability—the higher the MTBF, the more reliable the equipment. (C) 610.10-1994w

(8) The mean exposure time between consecutive failures of a component. It can be estimated by dividing the exposure time by the number of failures in that period, provided that a sufficient number of failures has occurred in that period. (IA/PSE) 493-1997

mean time between failures, assessed (repaired items) The mean time between failures of an item determined by a limiting value or values of the confidence interval associated with a stated confidence level, based on the same data as the observed mean time between failures of nominally identical items. *Notes:* 1. The source of the data shall be stated. 2. Results can be accumulated (combined) only when all conditions are similar. 3. The assumed underlying distribution of failures against time shall be stated. 4. It should be stated whether a one-sided or a two-sided interval is being used. 5. Where one limiting value is given, this is usually the lower limit. (R) [29]

mean time between failures, extrapolated (repaired items) Extension by a defined extrapolation or interpolation of the observed or assessed mean time between failures for duration and.or conditions different from those applying to the observed or assessed mean time between failures. *Note:* The validity of the extrapolation shall be justified. (R) [29]

mean time between failures, observed (repaired items) For a stated period in the life of an item, the mean value of the length of time between consecutive failures, computed as the ratio of the cumulative observed time to the number of failures under stated conditions. *Notes:* 1. The criteria for what constitutes a failure shall be stated. 2. Cumulative time is the sum of the times during which each individual item has been performing its required function under stated conditions.

3. This is the reciprocal of the observed failure rate during the period. (R) [29]

mean time between failures, predicted (repaired items) For the stated conditions of use, and taking into account the design of an item, the mean time between failures computed from the observed, assessed, or extrapolated failure rates of its parts. *Note:* Engineering and statistical assumptions shall be stated, as well as the bases used for the computation (observed or assessed). (R) [29]

mean time between hazardous events (MTBHE) The average time between occurrences of events, where hazardous events and the equipment that may precipitate them are defined at the system level. The hazardous events included in MTBHE are those whose consequences are of a given severity, as determined by the organization generating the safety goals. (VT/RT) 1483-2000

mean time to diagnosis (MTTD) The average length of time taken to isolate and diagnose the failure of a system or system component. (C) 610.10-1994w

mean time to failure (nonrepaired items) The total operating time of a number of items divided by the total number of failures. *See also:* reliability. (R) [29]

mean time to failure, assessed (non-repaired items) The mean time to failure of an item determined by a limiting value or values of the confidence interval associated with a stated confidence level, based on the same data as the observed mean time to failure of nominally identical items. *Notes:* 1. The source of the data shall be stated. 2. Results can be accumulated (combined) only when all conditions are similar. 3. The assumed underlying distribution of failures against time shall be stated. 4. It should be stated whether a one-sided or a two-sided interval is being used. 5. Where one limiting value is given this is usually the lower limit. (R) [29]

mean time to failure, extrapolated (non-repaired items) Extension by a defined extrapolation or interpolation of the observed or assessed mean time to failure for durations and/or conditions different from those applying to the observed or assessed mean time to failure. *Note:* The validity of the extrapolation shall be justified. (R) [29]

mean time to failure, observed (non-repaired items) For a stated period in the life of an item, the ratio of the cumulative time for a sample to the total number of failures in the sample during the period, under stated conditions. *Notes:* 1. The criteria for what constitutes a failure shall be stated. 2. Cumulative time is the sum of the times during which each individual item has been performing its required function under stated conditions. 3. This is the reciprocal of the observed failure rate during the period. (R) [29]

mean time to failure, predicted (non-repaired items) For the stated conditions of use, and taking into account the design of an item the mean time to failure computed from the observed, assessed or extrapolated mean times to failure of its parts. *Note:* Engineering and statistical assumptions shall be stated, as well as the bases used for the computation (observed or assessed). (R) [29]

mean time to fix (MTTF) *See:* mean time to repair.

mean time to outage The mean time to outage occurrence of a specified type = service time/number of outage occurrences of the specified type. *Note:* There are other indices such as mean time between failure. (PE/PSE) 859-1987w

mean time to repair (MTTR) (1) (supervisory control, data acquisition, and automatic control) (station control and data acquisition) The time interval (hours) that may be expected to return failed equipment to proper operation. (SWG/PE/SUB) C37.100-1992, C37.1-1994

(2) (nuclear power generating station) The arithmetic average of time required to complete a repair activity. (PE/NP) 380-1975w, 352-1975s, 933-1999

(3) (software) The expected or observed time required to repair a system or component and return it to normal operations. (C) 610.12-1990

(4) For a stated period in the life of a device, the average time required for corrective maintenance to be performed. *Note:* Used to measure the complexity and modularity of equipment—the higher the MTTR, the more complex the equipment. (C) 610.10-1994w

(5) The mean time to repair or replace a failed component. It can be estimated by dividing the summation of repair times by the number of repairs, and, therefore, it is practically the average repair time. (IA/PSE) 493-1997

mean zonal candlepower (illuminating engineering) The average intensity (candelas) of a symmetrical luminaire or lamp at an angle to the luminaire or lamp axis that is in the middle of the zone under consideration. (EEC/IE) [126]

measurand A physical or electrical quantity, property, or condition that is to be measured. (SWG/PE) C37.100-1992

measure (1) The number (real, complex, vector, etc.) that expresses the ratio of the quantity to the unit used in measuring it. (Std100) 270-1966w

(2) (software reliability) A quantitative assessment of the degree to which a software product or process possesses a given attribute. (SE/C) 982.2-1988, 982.1-1988

(3) (A) A way to ascertain or appraise value by comparing it to a norm. **(B)** To apply a metric. (C/SE) 1061-1998

measured current (rotating machinery) The total direct current resulting from the application of direct voltage to insulation and including the leakage current, the absorption current, and, theoretically, the capacitive current. Measured current is the value read on the microammeter during a direct high voltage test of insulation. (PE/EM) 95-1977r

measured limiting voltage The maximum magnitude of voltage that is measured across the terminals of the surge protective device during the application of a series of impulses of specified wave shape and amplitude. (PE) C62.34-1996

measured service (telephone switching systems) Service in which charges are assessed in terms of the number of message units during the billing interval. (COM) 312-1977w

measured signal (automatic null-balancing electric instrument) The electrical quantity applied to the measuring-circuit terminals of the instrument. *Note:* It is the electrical analog of the measured variable. *See also:* measurement system. (EEC/EMI) [112]

measured value (power meters) An estimate of the value of a quantity obtained as a result of a measurement. *Note:* Indicated values may be corrected to give measured values. (IM) 470-1972w, 544-1975w

measured variable (automatic null-balancing electric instrument) (measurand) The physical quantity, property, or condition that is to be measured. *Note:* Common measured variables are temperature, pressure, thickness, speed, etc. *See also:* measurement system. (EEC/EMI) [112]

measure equations Equations in which the quantity symbols represent pure numbers, the measures of the physical quantities corresponding to the symbols. (Std100) 270-1966w

measurement (1) The determination of the magnitude or amount of a quantity by comparison (direct or indirect) with the prototype standards of the system of units employed. (Std100) 270-1966w

(2) The act or process of assigning a number or category to an entity to describe an attribute of that entity. A figure, extent, or amount obtained by measuring. (C/SE) 1061-1998

measurement component A general term applied to parts or subassemblies that are primarily used for the construction of measurement apparatus. *Note:* It is used to denote those parts made or selected specifically for measurement purposes and does not include standard screws, nuts, insulated wire, or other standard materials. *See also:* measurement system. (EEC/PE) [119]

measurement device An assembly of one or more basic elements with other components and necessary parts to form a separate self-contained unit for performing one or more measurement operations. *Note:* It includes the protecting, supporting, and connecting, as well as the functioning, parts, all of which are necessary to fulfill the application requirements of the device. It should be noted that end devices (which see) are frequently but not always complete measurement devices in themselves, since they often are built-in with all or part of the intermediate means or primary detectors to form separate self-contained units. *See also:* measurement system. (EEC/PE) [119]

measurement devices For an automatic control system, a device that measures physical and electrical quantities. (PE/PSE) 94-1991w

measurement energy The energy required to operate a measurement device or system. *Note:* Measurement energy is normally obtained from the measurand or from the primary detector. *See also:* measurement system. (EEC/PE) [119]

measurement equipment A general term applied to any assemblage of measurement components, devices, apparatus, or systems. *See also:* measurement system. (EEC/PE) [119]

measurement inverter *See:* measuring modulator.

measurement mechanism An assembly of basic elements and intermediate supporting parts for performing a mechanical operation in the sequence of measurement. *Note:* For example, it may be a group of components required to effect the proper motion of an indicating or recording means and does not include such parts as bases, covers, scales, and accessories. It may also be applied to a specific group of elements by substituting a suitable qualifying term, such as time-switch mechanism or chart-drive mechanism. *See also:* measurement system. (EEC/PE) [119]

measurement range (instrument) That part of the total range within which the requirements for accuracy are to be met. *See also:* instrument. (EEC/PE) [119]

Measurements Assurance Program A program that allows manufacturers to verify the accuracy of their measurements through exchange and measurement of samples with NIST. This involves the analysis of blind test samples sent to the manufacturers by NIST, and NIST measurement of sources certified and provided by the manufacturers. (NI) N42.22-1995

measurement space In pattern recognition, a set of all possible n-tuples $(x_1, x_2, \ldots, x_n)$ that can be used to represent n measurements of a pattern. *See also:* feature space. (C) 610.4-1990w

measurement standard A standard that describes the characteristics of evaluating a process of product. (C) 610.12-1990

measurement system One or more measurement devices and any other necessary system elements interconnected to perform a complete measurement from the first operation to the end result. *Note:* A measurement system can be divided into general functional groupings, each of which consists of one or more specific functional steps or basic elements. (EEC/PE) [119]

measurement uncertainty (1) (test, measurement, and diagnostic equipment). The limits of error about a measured value between which the true value will lie with the confidence stated. *Note:* Uncertainty of measurement comprises, in general, many components. Some of these components may be estimated on the basis of the statistical distribution of the results of series of measurements and can be characterized by experimental standard deviations. Estimates of other components can be based on experience or other information. (MIL/T&D/PE) [2], 1308-1994
(2) The limits of error for a measured value, between which the true value will lie with the confidence stated. (Measurement uncertainty is associated with measuring the absolute value of the electric and magnetic fields.) (EMC) 1140-1994r

measurement voltage divider (voltage ratio box) (volt box) A combination of two or more resistors, capacitors, or other circuit elements so arranged in series that the voltage across one of them is a definite and known fraction of the voltage applied to the combination, provided the current drain at the tap point is negligible or taken into account. *Note:* The term volt box is usually limited to resistance voltage dividers intended to extend the range of direct-current potentiometers. *See also:* auxiliary device to an instrument. (PE/EEC) [119]

measure of effectiveness (MOE) (1) Measure of how the system/individual performs its functions in a given environment. Used to evaluate alternative approaches' ability to meet functional objectives and mission needs. Examples of such measures include loss exchange results, face effectiveness contributions, and tons delivered per day. (DIS/C) 1278.3-1996
(2) The metrics by which a customer will measure satisfaction with products produced by the technical effort. (C/SE) 1220-1998
(3) Measure of how the system/individual performs its functions in a given environment. Used to evaluate whether alternative approaches meet functional objectives and mission needs. (C/DIS) 1278.4-1997

measure of performance (MOP) (1) Measure of how the system/individual performs its functions in a given environment (e.g., number of targets detected, reaction time, number of targets nominated, susceptibility of deception, task completion time). It is closely related to inherent parameters (physical and structural), but measures attributes of system/individual behavior. (DIS/C) 1278.3-1996
(2) An engineering performance measure that provides design requirements that are necessary to satisfy a measure of effectiveness. There are generally several measures of performance for each measure of effectiveness. (C/SE) 1220-1998
(3) Measure of how the system/individual performs its functions in a given environment. It is closely related to inherent parameters (physical and structural), but measures system/individual behavior. (C/DIS) 1278.4-1997

measuring accuracy, precision, or reproducibility (numerically controlled machines) Accuracy, precision, or reproducibility of position sensor or transducer and interpreting system. (IA/EEC) [61], [74]

measuring and test equipment (1) (nuclear power generating station) Devices or systems used to calibrate, measure, gage, test, inspect or control to acquire research, development, test or operational data or to determine compliance with design, specifications or other technical requirements. (PE/NP) 498-1985s
(2) *See also:* test, measurement, and diagnostic equipment. (MIL) [2]

measuring element (automatic control) That portion of the feedback elements which converts the signal from the primary detecting element to a form compatible with the reference input. (PE/EDPG) [3]

measuring mechanism (recording instrument) The parts that produce and control the motion of the marking device. *See also:* moving element. (EEC/ERI) [111]

measuring modulator (measurement inverter) (chopper) An intermediate means in a measurement system by which a direct-current or low-frequency alternating-current input is modulated to give a quantitatively related alternating-current output usually as a preliminary to amplification. *Note:* The modulator may be of any suitable type such as mechanical, magnetic, or varistor. The mechanical types, which are actuated by vibrating or rotating members, may be classified as contacting, microphonic, or generating (capacitive or inductive). *See also:* auxiliary device to an instrument. (EEC/PE) [119]

measuring or test equipment (1) (nuclear power quality assurance) Devices or systems used to calibrate, measure, gage, test, or inspect in order to control or to acquire data to verify conformance to specified requirements. (PE/NP) [124]
(2) *See also:* test, measurement, and diagnostic equipment. (MIL) [2]

measuring unit Any analog or digital device that analyzes input currents or voltages or both to produce an output to the relay logic. (PE/PSR) C37.90.1-1989r

measuring units and relay logic (surge withstand capability tests) Analog or digital devices which analyze the input currents and voltages to determine the immediate status of that part of the power system that they were installed to protect and to provide the control signal to trip circuit breakers. (PE/PSR) C37.90-1978s

mechanical back-to-back test (rotating machinery) A test in which two identical machines are mechanically coupled together and the total losses of both machines are calculated from the difference between the electrical input to one machine and the electrical output of the other machine. *See also:* efficiency. (PE) [9]

mechanical bias *See:* relay mechanical bias.

mechanical braking The kinetic energy of the drive motor and the driven machinery is dissipated by the friction of a mechanical brake. *See also:* electric drive. (IA/ICTL/IAC) [60]

mechanical condition monitor (power system device function numbers) A device that functions upon the occurrence of an abnormal mechanical condition (except that associated with bearings as covered under device function) [bearing protective device], such as excessive vibration, eccentricity, expansion, shock, tilting, or seal failure. (SUB/PE) C37.2-1979s

mechanical current rating (neutral grounding devices) (electric power) The symmetrical root-mean-square alternating-current component of the completely offset current wave that the device can withstand without mechanical failure. *Note:* The mechanical forces depend upon the maximum crest value of the current wave. However, for convenience, the mechanical current rating is expressed in terms of the root-mean-square value of the alternating-current component only of a completely displaced current wave. Specifically, the crest value of the completely offset wave will then be 2.82 times the mechanical current rating. *See also:* grounding device. (PE/SPD) 32-1972r

mechanical cutter (mechanical recorder) An equipment for transforming electric or acoustic signals into mechanical motion of approximately like form and inscribing such motion in an appropriate medium by cutting or embossing. (SP) [32]

mechanical data processing A method of data processing that involves the use of small, simple, mechanical machines. (C) 610.2-1987

mechanical failure (of a circuit breaker) Failure other than an electrical failure. (SWG/PE) C37.10-1995

mechanical fatigue test (rotating machinery) A test designed to determine the effect of a specific repeated mechanical load on the life of a component. *See also:* asynchronous machine; direct-current commutating machine. (PE) [9]

mechanical filter *See:* mechanical wave filter.

mechanical freedom (accelerometer) The maximum linear or angular displacement of the accelerometer's proof mass, relative to its case. (AES/GYAC) 528-1994

mechanical-hydraulic governor (hydraulic turbines) A governor in which the control signal proportional to speed error and necessary stabilizing signals are developed mechanically, summed by a mechanical system, and are then hydraulically amplified. (PE/EDPG) 125-1988r

mechanical-impact strength (insulators) The impact that, under specified conditions, the insulator can withstand without damage. *See also:* insulator. (EEC/IEPL) [89]

mechanical impedance The complex quotient of the effective alternating force applied to the system, by the resulting effective velocity. (SP) [32]

mechanical inertia time (hydraulic turbines) A characteristic of the machine due to the inertia of the rotating components of the machine defined as:

$$T_M = \frac{(Wk^2)(N^2)(10^{-6})}{(1.61)(\mathrm{HP})}$$

where

W = weight of machine rotating parts, pounds

k = radius of gyration, feet

N = rated speed, rev/min

HP = rated output of turbine, horsepower

Notes: 1. T_M is also approximately equal to $2H$ where H is the inertia constant. 2. To calculate T_M using International SI units:

$T_M = J\omega_0^2 / P_0$

where J = the polar moment of inertia in kgm^2 calculated by dividing Wk^2 in newton-meters by acceleration of gravity 9.81 m/second^2.

$= Mk^2 = GD^2$

$\omega_0 = \pi 2N/60$ rad/second

P_0 = rated output of turbine, watts.

(PE/EDPG) 125-1977s

mechanical inspection *See:* computer-aided inspection.

mechanical interchangeability (of fuse links) The characteristic that permits the designs of various manufacturers to be interchanged physically so they fit into and withstand the tensile stresses imposed by various types of prescribed cutouts made by different manufacturers. (SWG/PE) C37.100-1992, C37.40-1993

mechanical interlocking machine An interlocking machine designed to operate the units mechanically. *See also:* interlocking. (EEC/PE) [119]

mechanical latching relay A relay in which the armature or contacts may be latched mechanically in the operated or unoperated position until reset manually or electrically. (PE/EM) 43-1974s

mechanical limit (neutral grounding devices) The rated maximum instantaneous value of current, in amperes, that the device will withstand without mechanical failure. (PE/SPD) 32-1972r

mechanically delayed overcurrent trip *See:* overcurrent release; mechanically delayed release.

mechanically delayed release (trip) A release delayed by a mechanical device. (SWG/PE) C37.100-1992

mechanically de-spun antenna (communication satellite) A rotating directional antenna, mounted to a rotating object (namely spin stabilized communication satellite); the rotation of the antenna is counter to the rotation of the body it is mounted to, such that the antenna beam points into the same direction of space. (COM) [24]

mechanically release-free (trip-free) (as applied to a switching device) A term indicating that the release can open the device even though in a manually operated switching device the operating lever is being moved toward the closed position; or in a power-operated switching device, such as solenoid- or spring-actuated types, the operating mechanism is being moved toward the closed position either by continued application of closing power or by means of a maintenance closing lever. (SWG/PE/TR) C37.100-1992, C57.12.44-1994

mechanically reset relay *See:* hand-reset relay.

mechanically switched capacitor (MSC) A shunt-connected circuit containing a mechanical power-switching device in series with a capacitor bank and sometimes also a damping reactor. (PE/SUB) 1031-2000

mechanically switched reactor (MSR) A shunt-connected circuit containing a mechanical power switching device in series with a reactor. (PE/SUB) 1031-2000

mechanically timed relays Relays that are timed mechanically by such features as clockwork, escapement, bellows, or dashpot. *See also:* relay. (EEC/REE) [87]

mechanical modulator (A) (electronic navigation) A device that varies some characteristic of a carrier wave so as to transmit information, the variation being accomplished by physically moving or changing a circuit element. **(B) (electronic**

navigation) In instrument landing systems, a particular arrangement of radio-frequency transmission lines and bridges with resonant sections coupled to the lines and motor-driven capacitor plates that alter the resonance so as to produce 90- and 150-hertz modulations. *See also:* navigation. (AES/RS) 686-1982, [42]

mechanical mouse A mouse whose motion-sensing component is mechanical in nature such as a control ball or a pair of wheels. *Contrast:* optical mouse. (C) 610.10-1994w

mechanical operation (of a switch) Operation by means of an operating mechanism connected to the switch by mechanical linkage. *Note:* Mechanically operated switches may be actuated either by manual, electrical, or other suitable means. (SWG/PE) C37.100-1992

mechanical part Any part having no electric or magnetic function. (PE/EM) [9]

mechanical plating Any plating operation in which the cathodes are moved mechanically during the deposition. *See also:* electroplating. (EEC/PE) [119]

mechanical recorder *See:* mechanical cutter.

mechanical rectifier A rectifier in which rectification is accomplished by mechanical action. *See also:* rectification. (EEC/PE) [119]

mechanical reference plane (standard connector) A transverse plane of the waveguide or transmission line to which all critical, longitudinal dimensions are referenced to assure nondestructive mating; it is the only plane where a mated standard connector pair butt against one another. *Note:* Usually a stable, rugged metal surface. (IM/HFIM) 474-1973w

mechanical register (pulse techniques) An electromechanical indicating pulse counter. *See also:* pulse. (NPS) 175-1960w

mechanical relay damping ring A loose member mounted on a contact spring to reduce contact chatter. (EEC/REE) [87]

mechanical reproducer *See:* phonograph pickup.

mechanical shock A significant change in the position of a system in a nonperiodic manner in a relatively short time. *Note:* It is characterized by suddenness and large displacements and develops significant internal forces in the system. *See also:* shock motion. (SP) [32]

mechanical short-circuit rating The maximum asymmetrical (peak) fault current that the reactor is capable of withstanding with no loss of electrical or mechanical integrity. (PE/TR) C57.16-1996

mechanical short-time current rating (current transformer) The root-mean-square value of the alternating-current component of a completely displaced primary current wave that the transformer is capable of withstanding, with secondary short-circuited. *Note:* Capable of withstanding means that after a test the current transformer shows no visible sign of distortion and is capable of meeting the other specified applicable requirements. *See also:* instrument transformer. (PE/TR) [57]

mechanical splice (fiber optics) A fiber splice accomplished by fixtures or materials, rather than by thermal fusion. Index matching material may be applied between the two fiber ends. *See also:* optical waveguide splice; index matching material; fusion splice. (Std100) 812-1984w

mechanical switching device A switching device designed to close and open one or more electric circuits by means of guided separate contacts. *Note:* The medium in which the contacts separate may be designated by suitable prefix; that is, air, gas, oil, etc. (SWG/PE) C37.100-1992

mechanical terminal load (high voltage air switches, insulators, and bus supports) The external mechanical load at each terminal equivalent to the combined mechanical forces to which the air switch may be subjected. (SWG/PE) C37.30-1992s

mechanical time constant (critically damped indicating instrument) The period of free oscillation divided by 2ν. *See also:* electromagnetic compatibility. (INT) [53], [70]

mechanical translation Translation from one natural language to another by a computer or through some other mechanical means. *See also:* machine-aided translation. (C) 610.2-1987

mechanical transmission system An assembly of elements adapted for the transmission of mechanical power. *See also:* phonograph pickup. (SP) [32]

mechanical trip (railway practice) (trip arm) A roadway element consisting in part of a movable arm that in operative position engages apparatus on the vehicle to effect an application of the brakes by the train-control system. (EEC/PE) [119]

mechanical wave filter (mechanical filter) A filter designed to separate mechanical waves of different frequencies. *Note:* Through electromechanical transducers, such a filter may be associated with electric circuits. *See also:* filter. (PE/EEC) [119]

mechanical wrap or connection (soldered connections) The securing of a wire or lead prior to soldering. (EEC/AWM) [105]

mechanism (1) (indicating instrument) The arrangement of parts for producing and controlling the motion of the indicating means. *Note:* It includes all the essential parts necessary to produce these results but does not include the base, cover, dial, or any parts, such as series resistors or shunts, whose function is to adapt the instrument to the quantity to be measured. *See also:* moving element; instrument. (EEC/AII) [102]
(2) (recording instrument) Includes the arrangement for producing and controlling the motion of the marking device; the marking device; the device (clockwork, constant-speed motor, or equivalent) for driving the chart; the parts necessary to carry the chart. *Note:* It includes all the essential parts necessary to produce these results but does not include the base, cover, indicating scale, chart, or any parts, such as series resistors or shunts, whose function is to make the recorded value of the measured quantity correspond to the actual value. *See also:* moving element. (EEC/PE) [119]
(3) (overhead power lines) In the context of biological effects, the process(es) by which an agent (physical or chemical) causes the effect; e.g., causing a change in hormone production or in the function of cell membranes. (T&D/PE) 539-1990
(4) (of a switching device) The complete assembly of levers and other parts that actuates the moving contacts of a switching device. (SWG/PE) C37.100-1992
(5) In an IDEF0 model, the means used by a function to transform input into output. (C/SE) 1320.1-1998

mechanism arrow An arrow or arrow segment that expresses IDEF0 mechanism, i.e., an object type set whose instances are used by a function to transform input into output. The arrowhead of an mechanism arrow is attached to the bottom side of a box. (C/SE) 1320.1-1998

mechanism loopback Loopback of output from one function to be mechanism for another function in the same diagram. (C/SE) 1320.1-1998

mechanism of propagation *See:* ionospheric mode of propagation.

media (1) (A) A means of communication. *See also:* hypermedia. **(B)** Material on which information can be stored or transported. *Note:* Media is the plural form of medium. *See also:* input media; output media. (C) 610.7-1995, 610.10-1994
(2) Any readable or writable data storage area. (C/SS) 1244.1-2000

Media Access Control (MAC) The data link sublayer that is responsible for transferring data to and from the Physical Layer. (C/LM) 802.3-1998

media-independent information transfer (information transfer) Used as a general term to refer to any volume conforming to IEEE Trial-Use Std 949-1985w. (MM/C) 949-1985w

Media Independent Interface (MII) (1) A transparent signal interface at the bottom of the Reconciliation sublayer. (C/LM) 802.3-1998
(2) A set of signals with electrical, logical, and physical definitions that provide the complete interface between the medium access control (MAC) layer and Physical Layer (PHY) (via the RS). (C/LM) 802.5t-2000

Medium Dependent Interface The mechanical and electrical interface between the transmission medium and the Medium Attachment Unit (MAU) (10BASE-T) or PHY (100BASE-T). (C/LM) 802.3-1998

median (L_{50}) The level exceeded 50% of the time over a specified time period with a specified weather condition. (T&D/PE) 539-1990

median water conditions (power operations) Precipitation and runoff conditions which provide water for hydroelectric energy development approximating the median amount and distribution available over a long time period, usually the period of record. (PE/PSE) 858-1987s

medical devices Any product that interfaces to the patient directly or operates in the patient environment. (EMB/MIB) 1073.4.1-2000

medical information bus (MIB) The informal name for the IEEE 1073 family of standards. (EMB/MIB) 1073.3.1-1994

medical device system (MDS) A bedside medical device which is actively connected to a 1073-type communications link. (EMB/MIB) 1073-1996

medical information system (MIS) *See:* hospital information system.

medium (1) (computers) The material, or configuration thereof, on which data are recorded; for example, paper tape, cards, magnetic tape. (C) [20], [85]
(2) (information transfer) A vehicle capable of transferring data. (MM/C) 949-1985w
(3) (broadband local area networks) The physical layer utilized to allow transmission of signals to communicate to various devices connected to it. For example, the medium of a CATV system is a broadband coaxial cable. (LM/C) 802.7-1989r
(4) The physical material from which the link is constructed.local area networks. (C) 8802-12-1998
(5) In data communications, a path over which communication flows, such as coaxial cable; optical fiber. *Note:* Medium is the singular form of media. (C) 610.7-1995
(6) The singular form of the term media. (C) 610.10-1994w
(7) The material on which the data may be transmitted. STP, UTP, and optical fibers are examples of media. (C/LM) 8802-5-1998
(8) (data management) *See also:* empty medium; virgin medium; machine-readable medium; data medium. (C) 610.5-1990w

medium access control (1) That part of a data station that supports the medium access control functions that reside just below the LLC sublayer. The MAC procedures include framing/deframing data units, performing error checking, and acquiring the right to use the underlying physical medium. (C/LM/PE/CC/TR) 8802-2-1998, 799-1987w
(2) *See also:* medium access control sublayer. (LM/C/CC) 8802-6-1994

medium access control management protocol data unit (MMPDU) The unit of data exchanged between two peer MAC entities to implement the MAC management protocol. (C/LM) 8802-11-1999

medium access control procedure (MAC procedure) In a local area network (LAN) or metropolitan area network (MAN), the part of the protocol that governs access to the transmission medium independently of the physical characteristics of the medium, but taking into account the topological aspects of the subnetwork, in order to enable the exchange of data between nodes. The MAC procedures include framing, error protection, and acquiring the right to use the underlying transmission medium. (LM/C) 8802-6-1994

medium access control protocol data unit (MPDU) The unit of data exchanged between two peer MAC entities using the services of the physical layer (PHY). (C/LM) 8802-11-1999

medium access control service data unit (MSDU) Information that is delivered as a unit between MAC service access points (SAPs). (C/LM) 8802-11-1999

medium access control sublayer (MAC) (1) The portion of the data station that controls and mediates the access to the ring. (C/LM) 8802-5-1998
(2) In a local area network (LAN), the part of the data link layer that supports topology-dependent functions and uses the services of the physical layer to provide service to the logical link control (LLC) sublayer. In ISO/IEC 8802, the combined set of functions in the DQDB Layer that support the MAC Sublayer service to the logical link control (LLC) sublayer. (LM/C) 8802-6-1994
(3) The portion of the data link layer that controls access to the medium. The MAC sublayer is required in end nodes.local area networks. (LM/C) 8802-12-1998
(4) The lower sublayer of the data link layer of seven-layer OSI model; provides topology-dependent functions between the physical layer and the logical link control sublayer. *See also:* network layer; transport layer; sublayer; application layer; client layer; data link layer; presentation layer; physical layer; entity layer; session layer; logical link control sublayer. (C) 610.7-1995

medium attachment unit (MAU) (1) The device that interfaces the communications system to the medium. The MAU incorporates the circuitry from the PLS (physical layer signaling interface) to the medium interface. (LM/C) 802.7-1989r
(2) In a local area network (LAN), a device used in a data station to couple the data terminal equipment (DTE) to the transmission medium. (C/LM) 802.9a-1995w
(3) In a local area network, a device used in a data station to couple the data terminal equipment (DTE) to the transmission medium. *Note:* This term is contextually specific to IEEE Std 802.3. (C) 610.7-1995
(4) A device containing an attachment unit interface (AUI), physical medium attachment (PMA), and medium dependent interface (MDI) that is used to connect a repeater or data terminal equipment (DTE) to a transmission medium. (C/LM) 802.3-1998

medium dependent interface (MDI) (1) (medium attachment units and repeater units) The mechanical and electrical interface between the trunk cable medium and the medium attachment unit (MAU). *See also:* coaxial cable interface. (LM/C) 8802-3-1990s
(2) The mechanical and electrical interface between the transmission medium and the MAU (10BASE-T) or PHY (100BASE-T). (LM/C) 802.3u-1995s
(3) The physically exposed interface between the link segment medium and the PMD of the end node or repeater, for which all mechanical, electrical or optical, and transmitted signal requirements are specified.local area networks. (C) 8802-12-1998

medium frequency (MF) 300 kHz to 3 MHz. *See also:* radio spectrum. (AP/PROP) 211-1997

medium-grain parallel architecture Parallel architecture that uses between 32 and 1024 processors. *Contrast:* coarse-grain parallel architecture; fine-grain parallel architecture. (C) 610.10-1994w

medium independent interface (MII) The logical interface between the Physical Medium Independent (PMI) and PMD in an end node or repeater. Optionally, the MII may be implemented as a physically exposed interface with specified signaling timing and electrical characteristics.local area networks. (C) 8802-12-1998

medium interface connector (MIC) A connector interface at which signal transmit and receive characteristics are specified for attaching stations and concentrators. One class of MICs is the connection between the attaching stations and the lobe cabling. A second set is the attachment interface between the concentrator and its lobes. A third set is the interface between the concentrator and the trunk cabling. Two types of connectors are specified: one for connecting to STP media and one for connecting to UTP media. (C/LM) 8802-5-1998

medium ion (dc electric-field strength and ion-related quantities) Ion comprised of several molecules or molecular clusters bound together by charge that is larger and less mobile than a small ion due to more mass or a greater number of molecular clusters. Typical radius is in the range of 10^{-9} m to 2×10^{-8} m. Mobility is in the range of 10^{-7} m^2/Vs to 10^{-5} m^2/Vs. (T&D/PE) 539-1990, 1227-1990r

medium noise (sound recording and reproducing system) The noise that can be specifically ascribed to the medium. *See also:* noise. 191-1953w

medium-pulse-repetition-frequency waveform A pulsed-radar waveform whose pulse-repetition frequency (PRF) is such that targets of interest are ambiguous with respect to both range and Doppler shift. *See also:* high-pulse-repetition-frequency waveform; low-pulse-repetition-frequency waveform. (AES) 686-1997

medium scale integration (MSI) **(A)** Pertaining to an integrated circuit containing between 100 and 500 transistors in its design. *Contrast:* large scale integration; very large scale integration; small scale integration; ultra-large scale integration. **(B)** Pertaining to an integrated circuit containing between 10 and 100 elements. (C) 610.10-1994

medium voltage (1) (cable systems in power generating stations) 601 to 15 000 V. (PE/EDPG) 422-1977
(2) (system voltage ratings) A class of nominal system voltages greater than 1000 V and less than 100 000 V. *See also:* low voltage; nominal system voltage; high voltage. (IA/PSE/APP) 241-1990r, [80]

medium-voltage aluminum-sheathed power cable (aluminum sheaths for power cables) Cable used in an electric system having a maximum phase-to-phase rms ac voltage above 1 000 V to 72 500 V, the cable having an aluminum sheath as a major component in its construction. (PE/IC) 635-1989r

medium-voltage power cable (1) (cable systems in power generating stations) Cables designed to supply power to utilization devices of the plant auxiliary system, operated at 601 to 15 000 V. (PE/EDPG) 422-1977
(2) Cable designed to supply power to utilization devices of the plant auxiliary system, operated at 5000–46 000 V in sizes ranging from 8 AWG (8.37 mm^2) to 2000 kcmil (1010.0 mm^2). (PE/IC) 1185-1994

medium-voltage system (electric power for industrial and commercial systems only) An electric system having a maximum root-mean-square alternating-current voltage above 1000 volts to 72 500 volts. *See also:* voltage classes. (IA/PSE) 570-1975w

meet *See:* AND.

meg Colloquial reference for megabyte. (C) 610.10-1994w

mega (M) (A) (mathematics of computing) A prefix indicating one million (10^6). **(B) (mathematics of computing)** In statements involving size of computer storage, a prefix indicating 2^{20}, or 1 048 576. (C) 1084-1986

megabyte Either 1 000 000 bytes or 2^{20} bytes. *Notes:* 1. The user of these terms shall specify the applicable usage. If the usage is 2^{10} or 1024 bytes, or multiples thereof, then note 2 below shall also be included with the definition. 2. As used in IEEE Std 610.10-1994, the terms kilobyte (kB) means 2^{10} or 1024 bytes, megabyte (MB) means 1024 kilobytes, and gigabyte (GB) means 1024 megabytes. *See also:* gigabyte. (C) 610.10-1994w

megacycle One million cycles. (C) 610.10-1994w

megahertz (MHz) (1) A unit of frequency equal to 1 000 000 cycles per second. (LM/C) 802.7-1989r
(2) A unit of frequency equal to 1 000 000 Hz, that is, 10^6 Hz. (C) 610.7-1995

Meissner oscillator An oscillator that includes an isolated tank circuit inductively coupled to the input and output circuits of an amplifying device to obtain the proper feedback and frequency. *See also:* oscillatory circuit. (AP/ANT) 145-1983s

mel A unit of pitch. By definition, a simple tone of frequency 1000 hertz, 40 decibels above a listener's threshold, produces a pitch of 1000 mels. *Note:* The pitch of any sound that is judged by the listener to be *n* times that of the 1-mel tone is *n* mels. (SP) [32]

melting channel The restricted portion of the charge in a submerged resistor or horizontal-ring induction furnace in which the induced currents are concentrated to effect high energy absorption and melting of the charge. *See also:* induction heating. (IA) 54-1955w, 169-1955w

melting-speed ratio (1) The ratio between between 0.1 s and 300 s or 600 s minimum melting currents, whichever is specified, which designates the relative speed of the fuse link. (SWG/PE) C37.40-1993
(2) (of a fuse) A ratio of the current magnitudes required to melt the current-responsive element at two specified melting times. *Notes:* 1. Specification of the current wave shape is required for time less than one-tenth of a second. 2. The lower melting time in present use is 0.1 s, and the higher minimum melting current times are 100 a for low-voltage fuses and 300 s or 600 s, whichever specified, for high-voltage fuses. (SWG/PE) C37.100-1992

melting time (1) (protection and coordination of industrial and commercial power systems) The time required to melt the current-responsive element on a specified overcurrent. Where the fuse is current limiting in less than half-cycle, the melting time may be approximately half or less of the clearing time. (IA/PSP) 242-1986r
(2) (of a fuse) The time required for overcurrent to sever the current-responsive element. (SWG/PE/SWG-OLD) C37.100-1992, C37.40-1993, C37.40b-1996

member In data management, a subunit contained in a partitioned data set. (C) 610.5-1990w

membrane keyboard A type of keyboard in which the keys are not raised, rather it is composed of a semi-flexible plastic sheet with a conductive surface below. *Synonym:* pressure-sensitive keyboard. (C) 610.10-1994w

membrane potential The potential difference, of whatever origin, between the two sides of a membrane. *See also:* electrobiology. (EMB) [47]

memory (1) All of the addressable storage in a processing unit and other internal storage that is used to execute instructions. *See also:* main storage. (C) 610.10-1994w
(2) *See also:* storage medium; storage.

memory action (of a relay) A method of retaining an effect of an input after the input ceases or is greatly reduced, so that this input can still be used in producing the typical response of the relay. *Note:* For example, memory action in a high-speed directional relay permits correct response for a brief period after the source of voltage input necessary to such response is short-circuited. (PE) C37.100-1992

memory address An address of a particular storage location in memory. (C) 610.10-1994w

memory address register A register containing the address of the memory location to be accessed. (C) 610.10-1994w

memory agent A module that uses split transactions to assume all the rights and responsibilities of some number of remote memory modules. (C/BA) 896.4-1993w

memory allocation and protection **(A)** To allocate physical sections of memory into logical partitions with read/write protection provided within each partition. **(B)** Pertaining to the hardware components that perform the allocation as in (A). (C) 610.10-1994

memory array (1) A matrix of memory locations arranged in a rectangular geometric pattern on an integrated circuit. (C) 610.10-1994w
(2) *See also:* array. (ED) 1005-1998

memory bank *See:* bank.

memory board A circuit board that provides random-access memory to a system. (C) 610.10-1994w

memory boundary The last address of an aligned data block. The maximum data block size that can be transferred by an IUT Master is the product of data width and data length. (C/BA) 896.4-1993w

memory buffer register A register in which a word is stored as it is read from memory or as it is written to memory. *Synonym:* memory data register. (C) 610.10-1994w

memory bus A bus connecting memory to the devices which can access it, including the processor and peripheral devices. (C) 610.10-1994w

memory capacity (1) The maximum number of bits that a memory is capable of storing. (ED) 641-1987w
(2) (software) The maximum number of items that can be held in a given computer memory; usually measured in words or bytes. (C) 610.12-1990
(3) *See also:* capacity. (ED) 1005-1998
(4) (electronic computation) *See also:* channel capacity; storage capacity.

memory cell (1) The smallest subdivision of a memory into which a unit of data has been or can be entered, in which it is or can be stored, and from which it can be retrieved. (ED) 641-1987w
(2) The combination of one or more single or merged transistors formed to provide a means of accessing, changing, and storing data. (ED) 1005-1998

memory compaction (A) A storage allocation technique in which the contents of all allocated storage areas are moved to the beginning of the storage space and the remaining storage blocks are combined into a single block. *Synonym:* garbage collection. **(B)** A storage allocation technique in which contiguous blocks of nonallocated storage are combined to form single blocks. (C) 610.12-1990

memory core *See:* magnetic core.

memory cycle (1) (test, measurement, and diagnostic equipment) The time required to read information from memory and replace it. (MIL) [2]
(2) A single complete access (read or write) of memory. (C) 610.10-1994w

memory data register *See:* memory buffer register.

memory device A device that contains only memory and implements configuration registers. (C/MM) 1155-1992

memory dump A display of the contents of all or part of a computer's internal storage, usually in binary, octal, or hexadecimal form. *See also:* static dump; selective dump; snapshot dump; dynamic dump; change dump. (C) 610.12-1990

memory image A series of bits that can be stored within a contiguous portion of transponder memory and that may be passed as a parameter within commands initiated by the roadside equipment (RSE). (SCC32) 1455-1999

memory integrated circuit An integrated circuit consisting of memory cells and usually including associated circuits such as signal amplification and address selection. (ED) 1005-1998

memory location A subdivision of a memory, including one or several memory cells, that is the smallest part of the memory that can be addressed. *Note:* The content of a memory location is usually called a bit, a byte, or a word, as appropriate. (ED) 1005-1998

memory management unit (MMU) A device that performs address translation between a CPU's virtual addresses and the physical addresses of some bus; typically, the bus represented by the root node. (C/BA) 1275-1994

memory map (1) A diagram that shows where programs and data are stored in a computer's memory. (C) 610.12-1990
(2) A list of all the current addresses in a computer. *Note:* This may indicate what is currently allocated, who is using it and where it is located. *Synonym:* memory map list. (C) 610.10-1994w

memory map list *See:* memory map.

memory mapping (A) The manner in which an address is translated into a physical address of a storage location. *See also:* biasing; segmenting; paging. **(B)** The process of translating addresses as in definition (A). (C) 610.10-1994

memory-mode agent An agent that communicates with others by using memory and/or I/O space on the parallel system bus. (C/MM) 1296-1987s

memory-mode system A system in which the agents communicate with one another with data structures in memory and/ or I/O space. (C/MM) 1296-1987s

memory object (1) Either a file or shared memory object. When used in conjunction with *mmap*(), a memory object will appear in the address space of the calling process. (C/PA) 9945-1-1996
(2) Either a file or shared memory object. When used in conjunction with `Map_Memory`, `Open_And_Map_Shared_Memory`, or `Open_Or_Create_And_Map_Shared_Memory`, a memory object will appear in the address space of the calling process. (C) 1003.5-1999

memory organization The arrangement of memory cells, either by geometrical arrangement in rows and columns or by organization of the data to be stored. (ED) 1005-1998, 641-1987w

memory page A segment of transponder memory that is assigned a unique location by which it may be referenced. (SCC32) 1455-1999

memory relay (A) A relay having two or more coils, each of which may operate independent sets of contacts, and another set of contacts that remain in a position determined by the coil last energized. **(B)** Sometimes erroneously used for polarized relay. *See also:* relay. (EEC/REE) [87]

memory-resident Managed by the implementation in such a way as to provide an upper bound on memory access times. (C/PA) 9945-1-1996, 1003.5-1999

memory space The address space used for accessing physical memory devices for storage and retrieval of code and data. (C/MM) 1296-1987s

memory window The difference in threshold voltage between the low- and high-conductance logic states of a memory cell. (ED) 641-1987w

MENTOR A block-structured language used widely in computer-aided instruction; characterized by its ability to model a student's knowledge. (C) 610.13-1993w

menu (1) A list of options available for selection by the user of a computer system. *Synonyms:* display menu; help menu; menu selection. (C) 610.2-1987, 610.6-1991w
(2) A rectangular visual user interface control containing a group of controls used to select an action from a group of choices. (C) 1295-1993w

menu bar A visual user interface control that is the bounded area near the top of a window, below the title bar, and above the rest of the window that contains cascade buttons that provide access to other menus. (C) 1295-1993w

menu by-pass In a menu-driven system, a feature that permits advanced users to perform functions in a command-driven mode without selecting options from the menus. (C) 610.12-1990

menu-driven Pertaining to a system or mode of operation in which the user directs the system through menu selections. *Contrast:* command-driven. *See also:* menu by-pass. (C) 610.12-1990

menu selection (A) The process of choosing an item from a menu. **(B)** The item chosen from a menu. (C) 610.2-1987

mercury-arc converter, pool-cathode *See:* pool-cathode mercury-arc converter; oscillatory circuit.

mercury-arc rectifier A gas-filled rectifier tube in which the gas is mercury vapor. *See also:* rectification. (ED) [45], [84]

mercury cells Electrolytic cells having mercury cathodes with which deposited metals form amalgams. (EEC/PE) [119]

mercury-contact relays (A) (mercury plunger relay) A relay in which the magnetic attraction of a floating plunger by a field surrounding a sealed capsule displaces mercury in a pool to effect contacting between fixed electrodes. **(B) (mercury-wetted-contact relay)** A form of reed relay in which the reeds and contacts are glass enclosed and are wetted by a film of mercury obtained by capillary action from a mercury pool in the base of a glass capsule vertically mounted. **(C) (mercury-contact relay)** A relay mechanism in which mercury establishes contact between electrodes in a sealed capsule as a result of the capsule's being tilted by an electromagnetically actuated armature, either on pick-up or dropout or both. *See also:* mercury relay.

mercury fluorescent lamp (illuminating engineering) An electric discharge lamp having a high-pressure mercury arc in an arc tube, and an outer envelope coated with a fluorescing substance (phosphor) which transforms some of the ultraviolet energy generated by the arc into light. (EEC/IE) [126]

mercury-hydrogen spark-gap converter (dielectric heating) A spark-gap generator or power source which utilizes the oscillatory discharge of a capacitor through an inductor and a spark gap as a source of radio-frequency power. The spark gap comprises a solid electrode and a pool of mercury in a hydrogen atmosphere. *See also:* induction heating. (IA) 54-1955w, 169-1955w

mercury lamp (illuminating engineering) A high intensity discharge (HID) lamp in which the major portion of the light is produced by radiation from mercury operating at a partial pressure in excess of 1.013×10^5 Pa (one atmosphere). Includes clear, phosphor-coated (mercury-fluorescent), and self-ballasted lamps. (EEC/IE) [126]

mercury-lamp ballast *See:* ballast.

mercury-lamp transformer *See:* constant-current (series) mercury-lamp transformer.

mercury motor meter A motor-type meter in which a portion of the rotor is immersed in mercury, which serves to direct the current through conducting portions of the rotor. *See also:* electricity meter.

mercury oxide cell A primary cell in which depolarization is accomplished by oxide of mercury. *See also:* electrochemistry. 341

mercury-pool cathode (gas tube) A pool cathode consisting of mercury. (ED) [45], [84]

mercury relay A relay in which the movement of mercury opens and closes contacts. (EEC/REE) [87]

mercury storage A type of storage that utilizes the acoustic wave propagation properties of mercury to store data. *See also:* acoustic delay line. (C) 610.10-1994w

mercury vapor lamp transformers (power and distribution transformers) (multiple-supply type) Transformers, autotransformers, or reactors for operating mercury or metallic iodide vapor lamps for all types of lighting applications, including indoor, outdoor area, roadway, uviarc, and other process and specialized lighting. (PE/TR) C57.12.80-1978r

mercury-vapor tube A gas tube in which the active gas is mercury vapor. (ED) 161-1971w

merge (1) (computers) To combine two or more sets of items into one, usually in a specified sequence. (C/C) [20], [85]
(2) (data management) To combine the items of two or more sets, all in the same order, into one set in that order. *See also:* unbalanced merge; collate; bitonic merge; coalesce; order-by-merging; balanced merge; merge sort. (C) 610.5-1990w

merge exchange sort *See:* Batcher's parallel sort.

merge search A sequential search in which the set of search arguments is ordered in the same sequence as the set to be searched; the set is searched sequentially, using the first search argument, until an equal or greater search key is found, the former case signifying a successful search, the latter, an unsuccessful search; the search for the next search argument begins where the last search left off. (C) 610.5-1990w

merge sort A sort in which the set to be sorted is divided into subsets, the items in each subset are sorted, and the sorted subsets are merged. *Synonym:* merging sort. *See also:* internal merge sort; external merge sort. (C) 610.5-1990w

merging Reconfiguration function that involves dual ring stations ceasing to use contra-rotating links in favor of a restored link or station. (LM/C) 802.5c-1991r

merging sort *See:* merge sort.

meridional ray (fiber optics) A ray that passes through the optical axis of an optical waveguide (in contrast with a skew ray, which does not). *See also:* optical axis; skew ray; paraxial ray; axial ray; geometric optics; numerical aperture. (Std100) 812-1984w

Merritt and Miller's Own Block Structured Simulation Language (MOBSSL-UAF) A simulation language used to model continuous systems using an augmented block structure. (C) 610.13-1993w

Mesa An application development language used by Xerox to program Viewpoint applications. (C) 610.13-1993w

mesh (1) A set of branches forming a closed path in a network, provided that if any one branch is omitted from the set, the remaining branches of the set do not form a closed path. *Note:* The term loop is sometimes used in the sense of mesh. *See also:* network analysis. (Std100) 270-1966w
(2) (computer graphics) A group of polygons that, when placed on the surface of a three-dimensional object, visually describes the shape of the exterior surface. (See the corresponding figure.)

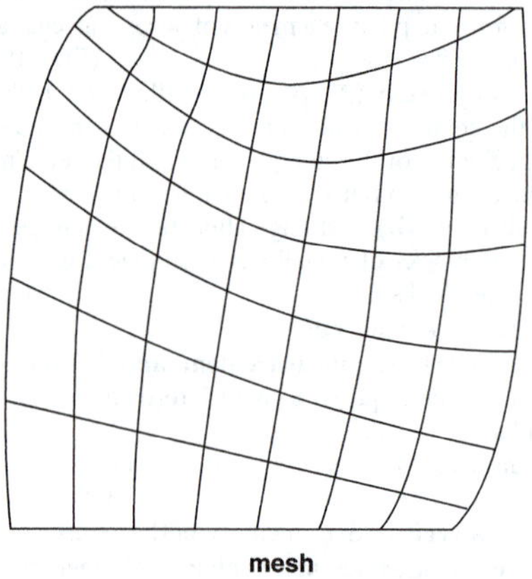

mesh

(C) 610.6-1991w

mesh-connected circuit A polyphase circuit in which all the current paths of the circuit extend directly from the terminal of entry of one phase conductor to the terminal of entry of another phase conductor, without any intermediate interconnections among such paths and without any connection to the neutral conductor, if one exists. *Note:* In a three-phase system this is called the delta (or D) connection. *See also:* network analysis. (Std100) 270-1966w

mesh current A current assumed to exist over all cross sections of a given closed path in a network. *Note:* A mesh current may be the total current in a branch included in the path, or it may be a partial current such that when combined with others the total current is obtained. *See also:* network analysis. (Std100) 270-1966w

mesh equations Any set of equations (of minimum number) such that the independent mesh or loop currents of a specified network may be determined from the impressed voltages. *Notes:* 1. For a given network, different sets of equations, equivalent to one another, may be obtained by different choices of mesh or loop currents. 2. The equations may be differential equations, or algebraic equations when impedances and phasor equivalents of steady-state single-frequency sine-wave quantities are used. *Synonym:* loop equations. *See also:* network analysis. (Std100) 270-1966w

mesh table A multidimension table that defines every type of delay model in terms of discrete points. Each point represents a delay value in terms of several cell parameters or interconnect parameters. The delay calculation module is expected to interpolate between these points based on a mathematical expression defined by the technology file. (C/DA) 1481-1999

mesh voltage The maximum touch voltage within a mesh of a ground grid. (PE/SUB) 80-2000

mesial point (pulse terminology) A magnitude referenced point at the intersection of a waveform and a mesial line. *See also:* waveform epoch. (IM/WM&A) 194-1977w

mesopause The upper boundary of the mesosphere. (AP/PROP) 211-1997

mesopic vision (illuminating engineering) Vision with fully adapted eyes at luminance conditions between those of photopic and scotopic vision, that is, between about 3.4 cd/m^2 (2.2 × 10^{-3} cd/in^2) (1.0 fL) and 0.034 cd/m^2 (2.2 × 10^{-5} cd/in^2) (0.01 fL). (EEC/IE) [126]

mesosphere That part of the Earth's atmosphere, located above the stratosphere, in which the temperature decreases with increasing height. The mesosphere extends to an altitude of around 85 km, where the temperature reaches a minimum value. (AP/PROP) 211-1997

message (1) (telephone switching systems) An answered call or the information content thereof. (COM) 312-1977w
(2) **(A)** In telecommunications, a combination of characters and symbols transferred from one point to another. **(B)** For bisync-type devices, the data unit from the beginning of a transmission to the first end-of-text (ETX) characters. **(C)** In information theory, an ordered series of characters or bits intended to convey information. **(D)** A group of characters and control bit sequences transferred as an entity from a data source to a data sink, where the arrangement of characters is determined by the data source. **(E)** An arbitrary amount of information whose beginning and end are defined or implied. (PE/SUB) 999-1992
(3) A value or set of values representing an interface event between functions. The term as used here is intended to be very primitive, not implying a particular structure or interface protocol unless modified by an appropriate adjective (like transaction-initiation message). A message can be arbitrarily simple (a signal) or complicated. (C/MM) 1212.1-1993
(4) A set of packets starting with a HEADER, consisting of that HEADER and all (ACKNOWLEDGE and DATA) packets transmitted as the immediate consequence of the command in that HEADER, and terminating when the M-module returns to the IDLE Master Controller state. (TT/C) 1149.5-1995
(5) An ordered series of characters used to convey information. (C) 610.7-1995
(6) Information that can be transferred among processes or threads by being added to and removed from a message queue. A message consists of a fixed-size message buffer. (C/PA) 9945-1-1996
(7) A logical grouping of one or more packets sent either from host to printer (a command message) or from printer to host (a response message). (C/MM) 1284.1-1997
(8) A package of information meeting a standard format that is sent to or from a transponder's memory. (SCC32) 1455-1999
(9) A communication sent from one object to another. *Message* encompasses requests to meet responsibilities as well as simple informative communications. *See also:* request. (C/SE) 1320.2-1998
(10) Information that can be transferred among tasks (possibly in different processes) by being added to and removed from a message queue. A message queue consists of a fixed-size buffer. (C) 1003.5-1999
(11) A grouping of data elements and/or data frames, as well as associated message metadata, that is used to convey a complete unit of information. For the purposes of this document, a message is an abstract description using a message set template (MST); it is not a specific instance. (SCC32) 1488-2000
(12) A set of ordered data (possibly empty) that includes a message boundary indication. Message data may span multiple packets. A packet shall not hold data from more than one message. (C/MM) 1284.4-2000

message attribute Information that describes a message and which may specify, at the logical level, relevant associated requirements for data exchange, interpretation, and handling. (SCC32) 1488-2000

message-based device An intelligent device that implements the defined VXIbus registers and communication protocols. (C/MM) 1155-1992

message body That portion of a message specification that describes the data elements and/or data frames contained within the message. (SCC32) 1488-2000

message box A visual user interface control used to display information not requested by the user but displayed in a secondary window by an application in response to an unexpected event or a possibility of something undesirable happening. (C) 1295-1993w

message code (MC) The predefined 12-bit code contained in an Auto-Negotiation Message Page. (LM/C) 802.3-1998

message_extension An allocated buffer in System Memory containing items that either would not fit in the primary_message or that are only needed for unusually large messages. (C/MM) 1212.1-1993

message group A collection of related messages. (SCC32) 1488-2000

message identifier An identifier used to identify derived MAC protocol data units (DMPDUs) derived from the same initial MAC protocol data unit (IMPDU). (LM/C) 8802-6-1994

message instance An occurrence of a message containing the actual values for the data elements and, in some cases, data about the message. (SCC32) 1488-2000

message length Although messages can be of any length up to 65 539 bytes, the packet size should be selected for effective transmissions over the physical link without requiring disassembly and reassembly. For connections through a network, the packet size of that network would generally be the most efficient. (C/MM) 1284.1-1997

message-mode agent An agent that exclusively uses message space for communication with other agents. (C/MM) 1296-1987s

message-mode system A system in which communication between agents is via blocks of data transmitted in the message space. (C/MM) 1296-1987s

message page (MP) An Auto-Negotiation Next Page encoding that contains a predefined 12-bit Message Code. (C/LM) 802.3-1998

message queue (1) A data structure and related procedures for passing a sequence of primary_messages from one or more producers to a consumer. (C/MM) 1212.1-1993
(2) An object to which messages can be added and removed. Messages may be removed in the order in which they were added or in priority order. (C/PA) 9945-1-1996, 1003.5-1999

message queue descriptor A per-process unique value used to identify an open message queue. (C) 1003.5-1999

message set A collection of messages based on the ITS class names. (SCC32) 1488-2000

message set template (MST) An abstract structure addressing the message attributes and syntax used to specify ITS messages, as well as rules for producing message standards using the MST (e.g., conformance statements). (SCC32) 1488-2000

message sink The part of a communications system that is the final destination of a message. *Contrast:* message source. (C) 610.7-1995

message source (1) That part of a communication system where messages are assumed to originate. *See also:* information theory. (Std100) 171-1958w

(2) The part of a communications system from which a message originates. *Synonym:* information source. *Contrast:* message sink. (C) 610.7-1995

message space The address space used for packet based communications ranging from interrupts to negotiated data movement. *See also:* packet. (C/MM) 1296-1987s

messages, species of A group of messages having in the Command fields of their respective HEADER packets a common command code. The name, S, of a message species is the same as the name of the command that defines the message species. (TT/C) 1149.5-1995

message stream modification Attempts to modify, delete, reorder, duplicate, or insert information while the message stream is being transmitted over a communication channel. Message stream modification attacks may be perpetrated at any point in the communication architecture (e.g., data link, network, transport, application), and could result in unauthorized modification of information or unauthorized receipt of services. (C/BA) 896.3-1993w

message switch (data transmission) A technique whereby messages are routed to the appropriate receiver by way of message address codes rather than by switching of the communication channel itself. (PE) 599-1985w

message switching (1) A method of handling messages over communications networks. The entire message is transmitted to an intermediate point (that is, a switching computer), stored for a period of time, perhaps very short, and then transmitted again towards its destination. The destination of each message is indicated by an address integral to the message. *See also:* circuit switching. (LM/COM) 168-1956w
(2) In data communications, a method of transporting messages by receiving, storing, and forwarding complete messages over communications networks. *See also:* time multiplexed switching; circuit switching; space-division switching. (C) 610.7-1995

message telecommunication network (telephone switching systems) An arrangement of switching and transmission facilities to provide telecommunication services to the public. (COM) 312-1977w

message-timed release (telephone switching systems) Release effected automatically after a measured interval of communication. (COM) 312-1977w

message unit (telephone switching systems) A basic chargeable unit based on the duration and destination of a call. (COM) 312-1977w

message-unit call (telephone switching systems) A call for which billing is in terms of accumulated message units. (COM) 312-1977w

meta (1) A word denoting a description that is one level of abstraction above the entity being described. (SCC32) 1489-1999
(2) A Greek prefix meaning that which pertains to the whole or overall entity or that which is in common or shared with all member entities comprising the whole. (IM/ST) 1451.2-1997

meta-attribute A documenting characteristic of a data concept. (SCC32) 1489-1999

metacomment A VHDL comment (--) that is used to provide synthesis-specific interpretation by a synthesis tool. (C/DA) 1076.6-1999

metacompiler *See:* compiler generator.

metadata (1) Data that describes other data; for example, a data dictionary contains a collection of metadata. (C) 610.5-1990w
(2) The information kept about software. It consists of the values of the various attributes of each of the objects. (C/PA) 1387.2-1995
(3) Information about the way asset description data is stored and organized within a library. (C/SE) 1430-1996
(4) Data that defines and describes other data. (SCC32) 1489-1999

metafile A file of device-independent commands, typically used to store graphical information to be displayed at a later time or on a different system or device. (C) 610.6-1991w

META 5 (A) A programming language used for symbolic data manipulation and for syntax-directed computing. **(B)** An assembly language for CDC computers. (C) 610.13-1993

metalanguage (1) A language used to specify some or all aspects of itself or of another language; for example, Backus-Naur form. (C) 610.13-1993w, 610.12-1990
(2) A form of notation used to rigorously define the syntax, and sometimes the semantics, of another language. (SCC20) 771-1998

metal clad The conducting parts are entirely enclosed in a metal casing. (EEC/PE) [119]

metal-clad switchgear (1) (electric power distribution for industrial plants) Metal-enclosed power switchgear characterized by the following necessary features.

a) The main circuit switching and interrupting device is of the removable type arranged with a mechanism for moving it physically between connected and disconnected positions and equipped with self-aligning and self-coupling primary and secondary disconnecting devices.
b) Major parts of the primary circuit, such as the circuit switching or interrupting devices, buses, potential transformers, and control power transformers, are enclosed by grounded metal barriers. Specifically included is an inner barrier in front of or a part of the circuit interrupting device to ensure that no energized primary circuit components are exposed when the unit door is opened.
c) All live parts are enclosed within grounded metal compartments. Automatic shutters prevent exposure of primary circuit elements when the removable element is in the test, disconnected, or fully withdrawn position.
d) Primary bus conductors and connections are covered with insulating material throughout. For special configurations, insulated barriers between phases and between phase and ground may be specified.
e) Mechanical interlocks are provided to ensure a proper and safe operating sequence.
f) Instruments, meters, relays, secondary control devices, and their wiring are isolated by grounded metal barriers from all primary circuit elements with the exception of short lengths of wire, such as at instrument transformer terminals.
g) The door through which the circuit interrupting device is inserted into the housing may serve as an instrument or relay panel and may also provide access to a secondary or control compartment within the housing.

Notes: 1. Auxiliary frames may be required for mounting associated auxiliary equipment, such as potential transformers, control power transformers, etc. 2. The term metal-clad switchgear can be properly used only if metal-enclosed switchgear conforms to the foregoing definition. All metal-clad switchgear is metal-enclosed, but not all metal-enclosed switchgear can be correctly designated as metal-clad. The most prevalent type of switching and interrupting device used in metal-clad switchgear is the air-magnetic power circuit breaker over 1000 volts (V). (IA/PSE) 141-1986s
(2) Switchgear that is characterized by the following necessary features:

a) The main switching and interrupting device is of the removable (drawout) type arranged with a mechanism for moving it physically between connected and disconnected positions and equipped with self-aligning and self-coupling primary disconnecting devices and disconnectable control wiring connections.
b) Major parts of the primary circuit, that is, the circuit switching or interrupting devices, buses, voltage transformers, and control power transformers, are completely enclosed by grounded metal barriers, that have no intentional openings between compartments. Specifically included is a metal barrier in front of, or a part of, the circuit

interrupting device to ensure that, when in the connected position, no primary circuit components are exposed by the opening of a door.

c) All live parts are enclosed within grounded metal compartments.
d) Automatic shutters that cover primary circuit elements when the removable element is in the disconnected, test, or removed position.
e) Primary bus conductors and connections are covered with insulating material throughout.
f) Mechanical interlocks are provided for proper operating sequence under normal operating conditions.
g) Instruments, meters, relays, secondary control devices and their wiring are isolated by grounded metal barriers from all primary circuit elements with the exception of short lengths of wire such as at instrument transformer terminals.
h) The door through which the circuit-interrupting device is inserted into the housing may serve as an instrument or relay panel, and may also provide access to a secondary or control compartment within the housing.

Notes: 1. Auxiliary vertical sections may be required for mounting devices or for use as a bus transition. 2. The term metal- clad (as applied to switchgear assemblies) is correctly used only in connection with switchgear conforming fully to this definition for metal-clad switchgear. Metal-clad switchgear is metal-enclosed, but not all metal-enclosed switchgear can be correctly designated as metal-clad.
(SWG/PE) C37.100-1992, C37.20.2-1993

metal distribution ratio (electroplating) The ratio of the thicknesses (weights per unit areas) of metal upon two specified parts of a cathode. *See also:* electroplating.
(PE/EEC) [119]

metal-enclosed (1) (metal-enclosed bus and calculating losses in isolated-phase bus) (as applied to metal-enclosed bus) Surrounded by a metal case or housing, with provisions for grounding. (SWG/PE) C37.23-1987r
(2) (as applied to a switchgear assembly or components thereof) Surrounded by a metal case or housing, usually grounded. (SWG/PE) C37.100-1992

metal-enclosed bus (1) (electric power distribution for industrial plants) An assembly of rigid electrical buses with associated connections, joints, and insulating supports, all housed within a grounded metal enclosure. Three basic types of metal-enclosed bus construction are recognized: nonsegregated phase, segregated phase, and isolated phase. The most prevalent type used in industrial power systems is the nonsegregated phase, which is defined as one in which all phase conductors are in a common metal enclosure without barriers between the phases. When metal-enclosed buses over 100 V are used with metal-clad switchgear, the bus conductors and connections are covered with insulating material throughout. When metal-enclosed buses are associated with metal-enclosed 1000 V and below power circuit breaker switchgear or metal-enclosed interrupter switchgear, the primary bus conductors and connections are usually bare.
(IA/PSE) 141-1986s
(2) (metal-enclosed bus and calculating losses in isolated-phase bus) An assembly of conductors with associated connections, joints, and insulating supports within a grounded metal enclosure. The conductors may be either rigid or flexible. (SWG/PE) C37.23-1987r
(3) An assembly of conductors with associated connections, joints, and insulating supports within a grounded metal enclosure. (PE/EDPG) 665-1995
(4) An assembly of conductors with associated connections, joints, and insulating supports within a grounded metal enclosure. The conductors may be either rigid or flexible. *Note:* In general, three basic types of construction are used: nonsegregated-phase, segregated-phase, and isolated-phase.

- *nonsegregated-phase bus.* A bus in which all phase conductors are in a common metal enclosure without barriers between phases. When associated with metal-clad switchgear, the primary bus and connections shall be covered with insulating material equivalent to the switchgear insulation system.
- *segregated-phase bus.* A bus in which all phase conductors are in a common metal enclosure but are segregated by metal barriers between phases.
- *isolated-phase bus.*A bus in which each phase conductor is enclosed by an individual metal housing separated from adjacent conductor housing by an air space. The bus may be self-cooled or may be forced-cooled by means of circulating a gas or liquid.

(SWG/PE) C37.100-1992

metal-enclosed equipment A capacitor equipment assembly enclosed in a metal enclosure or metal house, usually grounded, to prevent accidental contact with live parts *Synonym:* metal-housed equipment. (T&D/PE) 18-1992

metal-enclosed interrupter switchgear (1) (electric power distribution for industrial plants) Metal-enclosed power switchgear including the following equipment as required: interrupter switches; power fuses; bare bus and connections; instrument and control power transformers; control wiring and accessory devices. The interrupter switches and power fuses may be of the stationary or removable type. For the removable type, mechanical interlocks are provided to ensure a proper and safe operating sequence.
(PE/SWG/IA/PSE) 141-1986s
(2) Metal-enclosed power switchgear including the following equipment as required: Interrupter switches; Power fuses (current limiting or noncurrent limiting); Bare bus and connections; Instrument transformers; Control wiring and accessory devices. The interrupter switches and power fuses may be stationary or removable (drawout) type. When removable type, automatic shutters that cover primary circuit elements when the removable element is in the disconnected, test, or removed position, and mechanical interlocks are to be provided for proper operating sequence.
(SWG/PE) C37.20.3-1996
(3) Metal-enclosed power switchgear that includes the following equipment as required: (1) interrupter switches, (2) power fuses, (3) bare bus and connections, (4) instrument transformers, and (5) control wiring and accessory devices. The interrupter switches and power fuses may be of the stationary or removable type. When of the removable type, mechanical interlocks are provided to ensure a proper and safe operating sequence. (SWG/PE) C37.100-1992

metal-enclosed low-voltage power circuit-breaker switchgear (LV) (A) (metal-enclosed low-voltage power circuit-breaker switchgear) Low-voltage (LV) switchgear of multiple or individual enclosures, including the following equipment as required: low-voltage power circuit breakers (fused or unfused); bare bus and connections; instrument and control power transformers; instruments, meters, and relays; control wiring and accessory devices. The low-voltage power circuit breakers are contained in individual grounded metal compartments and controlled either remotely or from the front of the enclosure. The circuit breakers may be stationary or removable (drawout) type; when of removable type, mechanical interlocks are provided for proper operating sequence. **(B)** Metal-enclosed power switchgear, including the following equipment as required: 1000 V and below power circuit breakers (fused or unfused); bare bus and connections; instrument and control power transformers; instruments, meters, relays; control wiring and accessory devices; cable and busway termination facilities. The 1000 V and below power circuit breakers are contained in individual grounded metal compartments and controlled either remotely or from the front of the panels. The circuit breakers are usually of the drawout type, but may be nondrawout. When drawout-type circuit breakers are used, mechanical interlocks must be provided to ensure a proper and safe operating sequence.
(SWG/PE/IA/PSE) C37.20.1-1993, C37.100-1992, 141-1986

metal-enclosed power switchgear (ME) (metal-clad and station-type cubicle switchgear) (metal-enclosed low-voltage power circuit-breaker switchgear) A switchgear assembly completely enclosed on all sides and top with sheet metal (except for ventilating openings and inspection windows) containing primary power-circuit switching or interrupting devices or both, with buses and connections. The assembly may include control and auxiliary devices. Access to the interior of the enclosure is provided by doors or removable covers or both. *Note:* Metal-clad switchgear, station-type cubicle switchgear, metal-enclosed interrupter switchgear, and low-voltage power circuit-breaker switchgear are specific types of metal-enclosed power switchgear.
(SWG/PE) C37.20.1-1993r, C37.20.3-1996, C37.100-1992, C37.81-1989r, C37.20.2-1993

metal fog (electrolysis) A fine dispersion of metal in a fused electrolyte. *Synonym:* metal mist. *See also:* fused electrolyte. (PE/EEC) [119]

metal-graphite brush (rotating machinery) A brush composed of varying percentages of metal and graphite, copper or silver being the metal generally used. *Note:* This type of brush is soft. Grades of brushes of this type have extremely high current-carrying capacities, but differ greatly in operating speed from low to high. *See also:* brush. (PE/EEC/LB) [9], [101]

metal halide lamp (illuminating engineering) A high intensity discharge (HID) lamp in which the major portion of the light is produced by radiation of metal halides and their products of dissociation possibly in combination with metallic vapors such as mercury. Includes clear and phosphor-coated lamps. (EEC/IE) [126]

metal-housed equipment *See:* metal-enclosed equipment.

metallic circuit (measuring longitudinal balance of telephone equipment operating in the voice band) A circuit of which the ground (earth) forms no part.
(COM/PE/TA) 455-1985w, 599-1985w

metallic covering A metal sheath or braid used to provide physical protection for heating cable and, in some cases, to provide an electrical ground path. (IA/PC) 515-1997, 515.1-1995

metallic enclosure A grounded, leak-tight enclosure that contains the compressed insulating gas and associated electrical equipment.
(SWG/PE/SUB) C37.100-1992, C37.122-1993, C37.122.1-1993

metallic E-plane line An E-plane line in which there is no insulating substrate. (MTT) 1004-1987w

metallic impedance (measuring longitudinal balance of telephone equipment operating in the voice band) (telephone equipment) Impedance presented by a metallic circuit at any given single frequency, at or across the terminals of one of its transmission ports. (COM/TA) 455-1985w

metallic longitudinal induction ratio The ratio of the metallic-circuit current or noise-metallic arising in an exposed section of open wire telephone line, to the longitudinal-circuit current or noise-longitudinal in sigma. It is expressed in microamperes per milliampere or the equivalent. *Synonym:* M-L ratio. *See also:* inductive coordination; metallic noise. (EEC/PE) [119]

metallic noise (data transmission) The weighted noise current in a metallic circuit at a given point when the circuit is terminated at that point in the nominal characteristic impedance of the circuit. (PE) 599-1985w

metallic outer covering (electrical heat tracing for industrial applications) A metal sheath or braid used to provide physical protection for heating cable and. in some cases, to provide an electrical ground path. (BT/AV) 152-1953s

metallic rectifier (electric installations on shipboard) A metallic rectifier cell is a device consisting of a conductor and semiconductor forming a junction. The junction exhibits a difference in resistance to current flow in the two directions through the junction. This results in effective current flow in one direction only. A metallic rectifier stack is a single columnar structure of one or more metallic rectifier cells.
(IA/MT) 45-1983s

metallic rectifier cell A device consisting of a conductor and a semiconductor forming a junction. *Notes:* 1. Synonymous with metallic rectifying cells. 2. Such cells conduct current in each direction but provide a rectifying action because of the large difference in resistance to current flow in the two directions. 3. A metallic rectifier stack is a single columnar structure of one or more metallic rectifier cells. *See also:* rectification. (IA/MT) 45-1983s

metallic rectifier stack assembly The combination of one or more stacks consisting of all the rectifying elements used in one rectifying circuit. *See also:* rectification.
(EEC/PE) [119]

metallic rectifier unit An operative assembly of a metallic rectifier, or rectifiers, together with the rectifier auxiliaries, the rectifier transformers, and the essential switchgear. *See also:* rectification. (EEC/PE) [119]

metallic signal (telephone loop performance) (differential) The metallic voltage is the algebraic difference between the voltages to ground in the two conductors (tip and ring). The metallic current is half the algebraic difference between the current in these conductors. (COM/TA) 820-1984r

metallic transmission port (measuring longitudinal balance of telephone equipment operating in the voice band) (telephone equipment) A place of access in the metallic transmission path of a device or network where energy may be supplied or withdrawn, or where the device or network variables may be measured. The terminals of such a port are sometimes referred to as the tip and ring terminals. *Note:* In any particular case, the transmission ports are determined by the way the device is used, and not by its structure alone.
(COM/TA) 455-1985w

metallic voltage (measuring longitudinal balance of telephone equipment operating in the voice band) (telephone equipment) The voltage across a metallic circuit.
(COM/TA) 455-1985w

metallized brush *See:* metal-graphite brush.

metallized paper capacitor A capacitor in which the dielectric is primarily paper and the electrodes are thin metallic coatings deposited thereon. (PE/EM) 43-1974s

metallized screen (cathode-ray tubes) A screen covered on its rear side (with respect to the electron gun) with metallic film, usually aluminized, transparent to electrons and with a high optical reflection factor, which passes on to the viewer a large part of the light emitted by the screen on the electron-gun side. *See also:* cathode-ray tube. (ED) [45], [84]

metal master *See:* original master.

metal mist *See:* metal fog.

metal negative *See:* original master.

metal-nitride-oxide-memory section The portion of an integrated circuit memory built using a MNOS technology.
(ED) 641-1987w

metal-nitride-oxide-semiconductor transistor In analogy with the metal-oxide-semiconductor (MOS) transistor, this acronym derives from the layer sequence in the gate region of the IGFET, namely, Metal-Nitride-Oxide-Semiconductor: MNOS Memory Transistor. Usually it has a variable threshold voltage. Some devices with this layer sequence have fixed threshold voltages. *Synonym:* MNOS transistor.
(ED) 581-1978w, 641-1987w

metalogical value (1) One of the enumeration literals 'U', 'X', 'W', or '-' of the type STD_ULOGIC defined by IEEE Std 1164-1993. (C/DA) 1076.3-1997
(2) One of the enumeration literals, 'U', 'X', 'W', or '-', of the type STD_ULOGIC (or subtype STD_LOGIC) defined by IEEE Std 1164-1993. (C/DA) 1076.6-1999

metal-oxide semiconductor (MOS) A semiconductor technology using field-effect transistors in which the metal gate electrode is isolated from the channel by an oxide film. *Contrast:* bipolar. *See also:* complementary metal-oxide semiconductor
(C) 610.10-1994w

metal-oxide-semiconductor transistor A type of IGFET, referring specifically to the layer sequence in the gate region of the IGFET. (ED) 581-1978w, 641-1987w

metal-oxide surge arrester (MOSA) A surge arrester utilizing valve elements fabricated from nonlinear resistance metal-oxide materials. (SPD/PE) C62.11-1999

metal-plastic laminate A tape made of aluminum, copper, lead, or other metal substrate that is laminated on one or both sides with a tightly adhering plastic film. The film used may consist of either an adhesive polyolefin copolymer that self-bonds to the metal substrate during the laminating process or another polymeric compound that is adhered through the use of a supplemental adhesive. (PE/IC) 1142-1995

metal-to-metal touch voltage (1) The voltage between metallic objects or structures within the substation site that may be bridged by direct hand-to-hand or hand-to-feet contact. (SUB/PE) 1268-1997

(2) The difference in potential between metallic objects or structures within the substation site that may be bridged by direct hand-to-hand or hand-to-feet contact. *Note:* The metal-to-metal touch voltage between metallic objects or structures bonded to the ground grid is assumed to be negligible in conventional substations. However, the metal-to-metal touch voltage between metallic objects or structures bonded to the ground grid and metallic objects internal to the substation site, such as an isolated fence, but not bonded to the ground grid may be substantial. In the case of gas-insulated substations (GIS), the metal-to-metal touch voltage between metallic objects or structures bonded to the ground grid may be substantial because of internal faults or induced currents in the enclosures. In a conventional substation, the worst touch voltage is usually found to be the potential difference between a hand and the feet at a point of maximum reach distance. However, in the case of a metal-to-metal contact from hand-to-hand or from hand-to-feet, both situations should be investigated for the possible worst reach conditions. (PE/SUB) 80-2000

metalworking machine tool A power-driven machine not portable by hand, used to shape or form metal by cutting, impact, pressure, electrical techniques, or a combination of these processes. (NESC/NEC) [86]

metamers (illuminating engineering) Lights of the same color but of different spectral energy distribution. *Note:* The term "metamers" is also used to denote objects that, when illuminated by a given source and viewed by a given observer, produce metameric lights. (EEC/IE) [126]

metamodel A metamodel `Vm` for a subset of $\text{IDEF}_{\text{object}}$ is a view of the constructs in the subset that is expressed using those constructs such that there exists a valid instance of `Vm` that is a description of `Vm` itself. (C/SE) 1320.2-1998

Meta-TEDS The collection of those Transducer Electronic Data Sheet data fields that pertain to the whole or overall entity or those that are in common or shared with all member entities (channels) comprising the whole transducer product. (IM/ST) 1451.2-1997

metatype A collection of defined linguistic entities that share some common features. (C/TT) 1450-1999

meter (1) (laser maser) A unit of length in the international systems of units: currently defined as a fixed number of wavelengths, in vacuum, of the orange-red line of the spectrum of krypton 86. Typically, the meter is sub-divided into the following units: centimeter (cm) 10^{-2} m, millimeter (mm) 10^{-3} m, micrometer (μm) 10^{-6} m, nanometer (nm) 10^{-9} m. (LEO/QUL) 586-1980w, 268-1982s

(2) A device that measures and records the consumption or usage of the product/service. (AMR/SCC31) 1377-1997

(3) *See also:* watthour meter; electricity meter; demand meter. (ELM) C12.1-1981

meter installation inspection (metering) Examination of the meter, auxiliary devices, connections, and surrounding conditions, for the purpose of discovering mechanical defects or conditions that are likely to be detrimental to the accuracy of the installation. Such an examination may or may not include an approximate determination of the percentage registration of the meter. (ELM) C12.1-1982s

meter laboratory *See:* laboratory.

meter relay Sometimes used for instrument relay. *See also:* relay. (EEC/REE) [87]

meter shop *See:* shop—meter.

meter socket An enclosure that has matching jaws to accommodate the bayonet-type (blade) terminals of a detachable watthour meter and has a means of connections for the termination of the circuit conductors. It may be a single-position socket for one meter or a multiposition trough socket for two or more meters. (ELM) C12.7-1993

meter support That part of a ringless-type meter socket that positions and supports a detachable watthour meter. (ELM) C12.7-1993

method (1) A software procedure associated with a package. (C/BA) 1275-1994

(2) A procedure implementing one of the operations supported by an object class. (C) 1295-1993w

(3) A formal, well-documented approach for accomplishing a task, activity, or process step governed by decision rules to provide a description of the form or representation of the outputs. (C/SE) 1220-1994s

(4) A property of a class that defines a specific behavior. (SCC20) 1226-1998

(5) A statement of how property values are combined to yield a result. (C/SE) 1320.2-1998

method of pulse measurement A method of making a pulse measurement comprises: the complete specification of the functional characteristics of the devices, apparatus, instruments, and auxiliary equipment to be used; the essential adjustments required; the procedure to be used in making essential adjustments; the operations to be performed and their sequence; the corrections that will ordinarily need to be made; the procedures for making such corrections; the conditions under which all operations are to be carried out. *See also:* pulse measurement. (IM/WM&A) 181-1977w

methodology (1) A comprehensive, integrated series of techniques or methods creating a general systems theory of how a class of thought-intensive work ought to be performed. (C/SE) 730.1-1995

(2) A body of methods, rules, and postulates employed by a discipline. (C/SE) 1074-1995s

methods or types of grounding (neutral grounding in electrical utility systems) The equipment, procedure, or scheme used for attaining the particular means. (SPD/PE) C62.92-1987r

method standard A standard that describes the characteristics of the orderly process or procedure used in the engineering of a product or performing a service. (C) 610.12-1990

metric (1) A quantitative measure of the degree to which a system, component, or process possesses a given attribute. (C) 610.12-1990

(2) A value calculated from observed attribute values. (LM/C) 802.1F-1993r

metric algorithm The behaviour of a metric managed object that models a formalized process to calculate specified results. (LM/C) 802.1F-1993r

metric attribute An attribute of a metric managed object whose value is either used as a parameter of one or more metric algorithms or whose value represents the output of such an algorithm. (LM/C) 802.1F-1993r

metric managed object A managed object that contains at least one metric attribute whose value is calculated from values of attributes observed in managed objects. (LM/C) 802.1F-1993r

metrics framework A decision aid used for organizing, selecting, communicating, and evaluating the required quality attributes for a software system. A hierarchical breakdown of quality factors, quality subfactors, and metrics for a software system. (C/SE) 1061-1998

metrics sample A set of metric values that is drawn from the metrics database and used in metrics validation. (C/SE) 1061-1998

metric validation The act or process of ensuring that a metric reliably predicts or assesses a quality factor. (C/SE) 1061-1998

metric value A metric output or an element that is from the range of a metric. (C/SE) 1061-1998

metrology (test, measurement, and diagnostic equipment) The science of measurement for determination of conformance to technical requirements including the development of standards and systems for absolute and relative measurements. (MIL) [2]

metropolitan area network (MAN) (1) A network for connecting a group of individual stations and networks [for example, local area networks (LANS)] located in the same urban area. *Note:* A MAN generally operates at a higher speed than the networks interconnected, crosses network administrative boundaries, may be subject to some form of regulation, and supports several access methods. (LM/C) 8802-6-1994
(2) A computer network in which the geographic span is generally 5–50 km and operates at speeds greater than 1 Mb/s with physical layer data error ratio comparable to a LAN. *See also:* local area network; wide area network; long haul network. (C) 610.7-1995

MEW *See:* microwave early warning.

mezzanine card An electronic card assembly mounted parallel to, and with electrical and/or mechanical connections to, its host. (C/BA) 1301.4-1996

mezzanine card boundary envelope A three dimensional set of separation planes within which a mezzanine card and all its components reside. It will be treated as a single unit. (C/BA) 1301.4-1996

MF *See:* medium frequency.

MFLOPS Millions of floating point operations per second. A measure of computer processing speed. *See also:* MIPS; KOPS. (C) 610.12-1990

mho (siemens) The unit of conductance (and of admittance) in the International System of Units (SI). The mho is the conductance of a conductor such that a constant voltage of one volt between its ends produces a current of one ampere in it. (Std100) 270-1966w

mho characteristic An inherently directional distance relay characteristic in which the threshold of operation for the basic form plots as a circle on an *R-X* diagram, with the circle passing through the origin. See figure below. *Note:* For a self-polarized relay, the plot of the characteristic passes through the intersection of the *R-X* axes; for a cross-polarized relay it includes this intersection, but the relay retains its full directional characteristic.

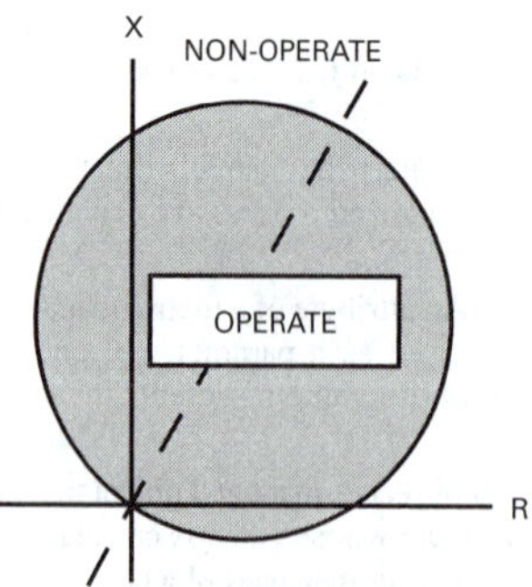

mho characteristic

(SWG/PE) C37.100-1992

mho relay A distance relay for which the inherent operating characteristic on an *R-X* diagram is a circle that passes through the origin. *Note:* The operating characteristic may be described by the equation $Z = K\cos(\theta - \alpha)$ where K and α are constants and θ is the phase angle by which the input voltage leads the input current. (SWG/PE) C37.100-1992

mho unit A distance relaying unit having a circular impedance tripping locus that passes through the origin on an R-X diagram. (PE/PSR) C37.113-1999

MHz *See:* megahertz.

MIB *See:* Management Information Base; medical information bus; management information base.

MIC *See:* medium interface connector.

mica flake (rotating machinery) Mica lamina in thickness not over approximately 0.0028 centimeter having a surface area parallel to the cleavage plane under 1.0 centimeter square. *See also:* stator; rotor. (PE) [9]

mica folium (rotating machinery) A relatively thin flexible bonded sheet material composed of overlapping mica splittings with or without backing or facing. *See also:* rotor; stator. (PE) [9]

mica paper (rotating machinery) (integrated mica) (reconstituted mica) Mica flakes having an area under approximately 0.200 centimeter square combined laminarly into a substantial sheet-like configuration with or without binder, backing, or facing. *See also:* stator; rotor. (PE) [9]

mica sheet (rotating machinery) A composite of overlapping mica splittings bonded into a planar structure with or without backing or facing. *See also:* rotor; stator. (PE) [9]

mica splitting (rotating machinery) Mica lamina in thickness approximately 0.0015 centimeter to 0.0028 centimeter having a surface area parallel to the basal cleavage plane of at least 1.0 centimeter square. *See also:* rotor; stator. (PE) [9]

mica tape (rotating machinery) A composite tape composed of overlapping mica splittings bonded together with or without backing or facing. *See also:* stator; rotor. (PE) [9]

Michigan Algorithmic Decoder (MAD) A programming language used widely for doing numerical computations. *Note:* MAD was designed to permit the development of a very fast compiler. *See also:* NOMAD. (C) 610.13-1993w

MICR *See:* magnetic ink character reader; magnetic ink character recognition.

micro (μ) (mathematics of computing) A prefix indicating one millionth (10^{-6}). (C) 1084-1986w

microarchitecture (1) The microword definition, data flow, timing constraints, and precedence constraints that characterize a given microprogrammed computer. (C) 610.12-1990
(2) The architecture of a microprogrammed computer. (C) 610.10-1994w

microbar A unit of pressure formerly in common usage in acoustics. One microbar is equal to one dyne per square centimeter and equals 0.1 newton per square meter. The newton per square meter is now the preferred unit. *Note:* The term bar properly denotes a pressure of 10^6 dynes per square centimeter. Unfortunately, the bar was once used in acoustics to mean one dyne per square centimeter, but this is no longer correct. (SP) [32]

microbend loss (fiber optics) In an optical waveguide, that loss attributable to microbending. (Std100) 812-1984w

microbending (fiber optics) In an optical waveguide, sharp curvatures involving local axial displacements of a few micrometers and spatial wavelengths of a few millimeters. Such bends may result from waveguide coating, cabling, packaging, installation, etc. *Note:* Microbending can cause significant radiative losses and mode coupling. *See also:* macrobending. (Std100) 812-1984w

microchannel plate (electron image tube) An array of small aligned channel multipliers usually used for intensification. *See also:* camera tube; amplifier. (ED) [45]

microcircuit *See:* integrated circuit.

microcode (1) (software) A collection of microinstructions, comprising part of, all of, or a set of microprograms. (C) 610.12-1990
(2) A collection of microinstructions comprising part of, or all of a microprogram. (C) 610.10-1994w

microcode assembler A computer program that translates microprograms from symbolic form to binary form. (C) 610.12-1990

microcomputer (micro) A computer that contains at least one microprocessor as its main computing element. *Note:* The distinction between a microcomputer, minicomputer, and mainframe is not yet standardized, however, in 1991 a typical mainframe is IBM's 3090, a typical minicomputer is Digital's VAX, and a typical microcomputer is IBM's PS/2. (C) 610.10-1994w

microcopy A copy of an image or document so reduced in size from its original that it cannot be read by the unaided human eye. For example, microform, microfiche, microfilm, microimage. (C) 610.2-1987

microcycle *See:* machine cycle.

microelectronic device (electric and electronics parts and equipment) An item of inseparable parts and hybrid circuits, usually produced by integrated circuit techniques. Typical examples are microcircuit, integrated circuit package, micromodule. (GSD) 200-1975w

microfacsimile Transmission and reception of microimages via facsimile communication. (C) 610.2-1987

microfiche A sheet of microfilm capable of containing microimages in a grid pattern. The sheet usually contains a title that can be read without magnification. *Synonym:* fiche. (C) 610.2-1987

microfilm (A) A high resolution film for recording microimages. **(B)** To record microimages on film. *See also:* computer output microfilm. (C) 610.2-1987, 610.10-1994

microfilmer *See:* computer output microfilmer.

microfloppy disk A floppy disk that is 3.5 inches wide. *Contrast:* minifloppy disk. (C) 610.10-1994w

microfont* *See:* optical character recognition-B.

* Deprecated.

microform A medium that contains micro-images. For example, microfiche, microfilm. (C) 610.2-1987

micrographics That branch of science and technology concerned with methods and techniques for converting information to or from microform. *Synonym:* microphotographics. *See also:* office automation. (C) 610.2-1987

microimage An image that is too small to be read by the human eye without magnification. (C/C) 610.2-1987, 610.10-1994w

microinstruction In microprogramming, an instruction that specifies one or more of the basic operations needed to carry out a machine language instruction. Types include diagonal microinstruction, horizontal microinstruction, and vertical microinstruction. *See also:* nanoinstruction; microcode; microprogram. (C) 610.10-1994w, 610.12-1990

micrometer (laser maser) A unit of length equal to 10^{-6} m. In common practice a micrometer is a micron. (LEO) 586-1980w

micron (metric system) The millionth part of a meter. *Note:* According to the set of submultiple prefixes now established in the International System of Units, the preferred term would be micrometer. However, use of the same word to denote a small length, and also to denote an instrument for measuring a small length, could occasionally invite confusion. Therefore it seems unwise to deprecate, at this time, the continued use of the word micron. (Std100) [123]

microoperation In microprogramming, one of the basic operations needed to carry out a machine language instruction. *See also:* microinstruction. (C) 610.12-1990

microphone An electroacoustic transducer that responds to sound waves and delivers essentially equivalent electric waves. (T&D/PE/SP) 539-1990, [32]

microphonics (1) (general) The noise caused by mechanical shock or vibration of elements in a system. (SP) 151-1965w
(2) (interference terminology) Electrical interference caused by mechanical vibration of elements in a signal transmission system. *See also:* signal. (IE) [43]
(3) (electron tube) (microphonic effect) (microphonism) The undesired modulation of one or more of the electrode currents resulting from the mechanical vibration of one or more of the valve or tube elements. *See also:* electron tube. (ED) 161-1971w

microphonics, microphonic noise Electrical noise caused by mechanical or audio induced vibration of the detector assembly. (NPS) 325-1996

microphonism *See:* microphonics.

microphotographics *See:* micrographics.

microprint A positive microcopy photographically printed onto paper. (C) 610.2-1987

microprocessor An integrated circuit that contains the logic elements for manipulating data and for making decisions. *See also:* microcomputer; processor. (C) 610.10-1994w

microprogram (software) A sequence of instructions, called microinstructions, specifying the basic operations needed to carry out a machine language instruction. *See also:* control store; microcode. (C) 610.12-1990, 610.10-1994w

microprogrammable computer A microprogrammed computer in which microprograms can be created or altered by the user. *Contrast:* fixed-instruction computer. (C) 610.12-1990, 610.10-1994w

microprogrammed computer A computer in which machine language instructions are implemented by microprograms rather than by hard-wired logic. *Note:* A microprogrammed computer may or may not be a microcomputer; the concepts are not related despite the similarity of the terms. *See also:* microarchitecture; microprogrammable computer. (C) 610.12-1990, 610.10-1994w

microprogramming (1) The process of designing and implementing the control logic of a computer by identifying the basic operations needed to carry out each machine language instruction and representing these operations as sequences of instructions in a special memory called control store. This method is an alternative to hard wiring the control signals necessary to carry out each machine language instruction. Techniques include bit steering, compaction, residual control, single-level encoding, two-level encoding. *See also:* microinstruction; microcode; microprogram. (C) 610.12-1990
(2) The process of designing and implementing the control logic of a computer by identifying the basic operations needed to carry out each machine language instruction and then representing these operations in appropriate sequence in a special memory, called a control store. (C) 610.10-1994w

micropublishing The production and distribution of information via microform. The information may be original or may have been previously published in another form. (C) 610.2-1987

micropulsation Small magnitude fluctuations (usually much less than 10^{-6} of the Earth's magnetic field) with periods on the order of seconds or minutes (f<1 Hz). *Note:* Micropulsations usually result from current fluctuations in the E region. (AP/PROP) 211-1997

microradiometer (radio-micrometer) A thermosensitive detector of radiant power in which a thermopile is supported on and connected directly to the moving coil of a galvanometer. *Note:* This construction minimizes lead losses and stray electric pickup. *See also:* electric thermometer. (EEC/PE) [119]

micrositing Of, or related to, the characteristics of a particular wind-turbine site, as contrasted to those characteristics that prevail over the entire windfarm. (DESG) 1094-1991w

microspark (overhead-power-line corona and radio noise) A spark breakdown occurring in the miniature air gap formed by two conducting or insulating surfaces. (This is sometimes called a "gap discharge.") (T&D/PE) 539-1990

microstrip (1) A class of planar transmission lines consisting of one or more thin conducting strips of finite width parallel to a single extended conducting ground plane. In its common form, the strips are affixed to an insulating substrate attached

to the ground plane. The semi-infinite space above the strips is filled with a medium of relative permittivity and permeability equal or less than the substrate. (MTT) 1004-1987w
(2) *See also:* strip-type transmission line.

microstrip antenna An antenna that consists of a thin metallic conductor bonded to a thin grounded dielectric substrate. *Note:* The metallic conductor typically has some regular shape; for example, rectangular, circular, or elliptical. Feeding is often by means of a coaxial probe or a microstrip transmission line. (AP/ANT) 145-1993

microstrip array An array of microstrip antennas. (AP/ANT) 145-1993

microstrip dipole A microstrip antenna of rectangular shape with its width much smaller than its length. (AP/ANT) 145-1993

microsyn (accelerometer) (gyros) An electromagnetic device often used as a pickoff in single-degree-of-freedom gyros and accelerometers. It has a stator, fastened to the sensor case, containing primary and secondary sets of windings, and a rotor, without windings, that is attached to the float. (AES/GYAC) 528-1994

microwave Pertaining to the portion of the radio frequency spectrum above 1 GHz. *See also:* microwave link. (C) 610.7-1995

microwave amplification by stimulated emission of radiation *See:* maser.

microwave backward-wave oscillator *See:* carcinotron.

microwave early warning (MEW) A U.S. high-power, long-distance radar of the WWII era with a number of separate displays giving high resolution and large traffic handling capacity in detecting and tracking targets. (AES/RS) 686-1990

microwave landing system (MLS) (radar) An airfield approach radar generating a guideline for target landing. (AES/RS) 686-1982s

microwave link A communications system in which information is conveyed by microwave transmissions. (C) 610.7-1995

microwave-pilot protection A form of pilot protection in which the communication means between relays is a beamed microwave radio channel. (SWG/PE) C37.100-1992

microwave plumbing *See:* plumbing.

microwaves (data transmission) A term used rather loosely to signify radio waves in the frequency range from about 1000 megahertz (mHz) upwards. (PE) 599-1985w

microwave therapy The therapeutic use of electromagnetic energy to generate heat within the body, the frequency being greater than 100 megahertz. *See also:* electrotherapy. (EMB) [47]

microword An addressable element in the control store of a microprogrammed computer. (C) 610.12-1990, 610.10-1994w

midband The part of the electromagnetic frequency spectrum that is located between television Channels 6 and 7. (LM/C) 802.7-1989r

middle marker A marker facility in an ILS (instrument landing system) that is installed approximately 1000 m (3500 ft) from the approach end of the runway on the localizer course line to provide a fix. (AES/RS) 686-1982s

middle-square function *See:* mid-square function.

MID page A set of one message identifier value. (LM/C) 8802-6-1994

mid-peak period (watthour meters) The period of time during which the specified mid-peak rate applies. (ELM) C12.13-1985s

midrange computer *See:* minicomputer.

mid-split A frequency division scheme that allows two-way traffic on a single cable. Inbound path signals come to the headend from 5 to 108 MHz. Outbound path signals go from the headend from 162 MHz to the upper frequency limit. The guardband is located from 108 to 162 MHz. (LM/C) 802.7-1989r

mid-square function In hashing, a hash function that returns the middle digits of the square of the original key. For example, in the function below, the middle three digits are returned.

Original key	Calculation	Hash Value
2964	2964 × 2964 = 8,785,296	852
1119	1119 × 1119 = 110,781	78

Synonym: middle-square function. (C) 610.5-1990w

Mie scattering Scattering by spherical particles whose diameters are comparable to or greater than a wavelength. *Synonym:* Lorenz-Mie scattering. (AP/PROP) 211-1997

migrated attribute A foreign key attribute of a child entity. (C/SE) 1320.2-1998

migrated key *See:* foreign key.

MII *See:* Media Independent Interface; medium independent interface.

MIIT *See:* media-independent information transfer.

mild environment (nuclear power generating station) An environment expected as a result of normal service conditions and extremes (abnormal) in service conditions where seismic is the only design basis event (DBE) of consequence. (PE/NP) 323-1974s

mile of standard cable (MSC) Two units, both loosely designated as a mile of standard cable, were formerly used as measures of transmission efficiency. One, correctly known as an 800-hertz mile of standard cable, signified an attenuation constant, independent of frequency, of 0.109. The other signified the effect upon speech volume of an actual mile of standard cable, equivalent to an attenuation constant of approximately 0.122. Both units are now obsolete, having been replaced by the decibel. One 800-hertz mile of standard cable is equal to approximately 0.95 decibel. One standard cable mile is equivalent in effect on speech volume to approximately 1.06 decibels. (EEC/PE) [119]

milestone (1) (software) A scheduled event for which some project member or manager is held accountable and which is used to measure progress, for example, a formal review, issuance of a specification, product delivery. (C/SE) 729-1983s
(2) A scheduled event used to measure progress. Examples of major milestones for software projects may include an acquirer or managerial sign-off, baselining of a specification, completion of system integration, and product delivery. Minor milestones might include baselining of a software module or completion of a chapter of the user's manual. (C/SE) 1058-1998

milli (m) (mathematics of computing) A prefix indicating one thousandth (10^{-3}). (C) 1084-1986w

millimeter-wave radar A radar whose carrier frequency is from 30–300 GHz (wavelength is 1–10 mm). (AES) 686-1997

milliroentgen (mR) The amount of X-radiation that produces $2.58 \cdot 10^{-7}$ coulomb per kilogram of air. (SWG/PE) C37.100-1992, 553-1983

mill scale The heavy oxide layer formed during hot fabrication or heat treatment of metals. Especially applied to iron and steel. (IA) [59]

Mills cross antenna system A multiplicative array antenna system consisting of two linear receiving arrays positioned at right angles to one another and connected together by a phase modulator or switch such that the effective angular response of the output is related to the product of the radiation patterns of the two arrays. (AP/ANT) 145-1993

MIMD *See:* multiple instruction, multiple data.

MIMIC A problem-oriented programming language for solving engineering problems, particularly those involving differential equations. (C) 610.13-1993w

mimic bus A single-line diagram of the main connections of a system constructed on the face of a switchgear or control panel, or assembly. (SWG/PE) C37.100-1992

MI mineral-insulated, metal sheathed cable A factory assembly of one or more conductors insulated with a highly compressed refractory mineral insulation and enclosed in a liquidtight and gastight continuous copper sheath. (NESC/NEC) [86]

mine-fan signal system A system that indicates by electric light or electric audible signal, or both, the slowing down or stopping of a mine ventilating fan. *See also:* dispatching system. (EEC/PE) [119]

mine feeder circuit A conductor or group of conductors, including feeder and sectionalizing switches or circuit breakers, installed in mine entries or gangways and extending to the limits set for permanent mine wiring beyond which limits portable cables are used. (EEC/PE) [119]

mine hoist A device for raising or lowering ore, rock, or coal from a mine and for lowering and raising men and supplies.

mine jeep A special electrically driven car for underground transportation of officials, inspectors, repair, maintenance, surveying crews, and rescue workers. (EEC/PE) [119]

mine radio telephone system A means to provide communication between the dispatcher and the operators on the locomotives where the radio impulses pass along the trolley wire and down the trolley pole to the radio telephone set. *See also:* dispatching system. (EEC/PE) [119]

miner's electric cap lamp A lamp for mounting on the miner's cap and receiving electric energy through a cord that connects the lamp with a small battery. (EEC/PE) [119]

miner's hand lamp A self-contained mine lamp with handle for convenience in carrying. (EEC/PE) [119]

mine tractor A trackless, self-propelled vehicle used to transport equipment and supplies and for general service work. (EEC/PE) [119]

mine ventilating fan A motor-driven disk, propeller, or wheel for blowing (or exhausting) air to provide ventilation of a mine. (EEC/PE) [119]

mini *See:* minicomputer.

miniature brush A brush having a cross-sectional area of less than 1/64 square inch with the thickness and width thereof less than 1/8 inch or, in the case of a cylindrical brush, a diameter less than 1/8 inch. *See also:* brush. (EEC/EM/LB) [101]

minicartridge *See:* quarter-inch cartridge.

minicomputer (mini) A computer of smaller size relative to a mainframe, but generally larger and more powerful than a microcomputer. *Note:* The distinction between a microcomputer, minicomputer, and mainframe is not yet standardized, however, in 1991 a typical mainframe is IBM's 3090, a typical minicomputer is Digital's VAX, and a typical microcomputer is IBM's PS/2. (C) 610.10-1994w

minifloppy disk A floppy disk that is 5.25 inches wide. *Contrast:* microfloppy disk. (C) 610.10-1994w

minimally conformant network An IEEE 802.11 network in which two stations in a single basic service area (BSA) are conformant with ISO/IEC 8802-11: 1999. (C/LM) 8802-11-1999

minimally consistent object An object that satisfies various conditions set forth in the definition of its class. (C/PA) 1328-1993w, 1224-1993w, 1327-1993w

minimal perceptible erythema The erythemal threshold. *See also:* ultraviolet radiation. (EEC/IE) [126]

minimal ROM format A format for the node-provided ROM. The minimal ROM format provides a 24-bit company_id value; although additional ROM parameters can be provided, their format and meaning are vendor-dependent. (C/MM) 1212-1991s

minimum access code (test, measurement, and diagnostic equipment) A system of coding which minimizes the effect of delays for transfer of data or instructions between storage and other machine units. (MIL) [2]

minimum air insulation distance (MAID) The shortest distance in air between electrical apparatus and/or a line worker's body at different potential. This minimum air insulation distance, with a floating electrode in the gap, is equal to or greater than the sum of the individual minimum approach distances. This is the electrical component and does not include any factor for inadvertent movement. (T&D/PE) 516-1995

minimum approach distance (MAD) **(1)** The minimum air insulation distance plus a modifier for inadvertent movement. (T&D/PE) 516-1995
(2) The closest distance a qualified employee is permitted to approach either an energized or a grounded object, as applicable for the work method being used. (NESC) C2-1997

minimum bend radius The curvature to which a cable can be bent without sustaining damage or significant degradation of performance. (C) 610.7-1995

minimum clearance between poles (phases) The shortest distance between any live parts of adjacent poles (phases). *Note:* Cautionary differentiation should be made between clearance and spacing or center-to-center distance. (SWG/PE) C37.40-1993, C37.100-1992

minimum clearance to ground The shortest distance between any live part and adjacent grounded parts. (SWG/PE) C37.40-1993, C37.100-1992

minimum conductance function *See:* minimum-resistance (conductance) function.

minimum cut-set A set of components that, if removed from the system, results in loss of continuity to the load point being investigated and that does not contain as a subset any set of components that is itself a cut-set of the system. (IA/PSE) 493-1997

minimum delay programming A programming technique in which storage locations for computer instructions and data are chosen so that access time is minimized. (C) 610.12-1990

minimum delta (power supplies) A qualifier, often appended to a percentage specification to describe that specification when the parameter in question is a variable, and particularly when that variable may approach zero. The qualifier is often known as the minimum delta *V*, or minimum delta *I*, as the case may be. (AES) [41]

minimum demand The lowest demand measured over a selected period of time, such as one month. (AMR/SCC31) 1377-1997

minimum detectable activity That activity giving an indication corresponding to 4.65 times the standard deviation of the indication given by a specific background, divided by the appropriate conversion actor to result in units of activity. (NI) N42.17B-1989r

minimum detectable amount The amount of a radionuclide, which if present in a sample, would be detected with a β probability of non-detection while accepting a probability α of erroneously detecting that radionuclide in an appropriate blank sample. For IEEE Std N42.23-1995, the α and β probabilities are both set at 0.05. (NI) N42.23-1995

minimum detectable concentration The minimum detectable amount expressed in concentration units. (NI) N42.23-1995

minimum detectable signal (MDS) The minimum signal level that gives reliable detection in the presence of white Gaussian noise. *Note:* MDS must be described in terms of a probability of detection and a probability of false alarm, due to its statistical nature. (AES) 686-1997

minimum detectable velocity (MDV) In a Doppler processing radar for detection of moving targets, the minimum target velocity that can be detected. (AES) 686-1997

minimum differential sensitivity The smallest value of peak-to-peak differential (ppd) amplitude at which a receiver is expected to operate, under worst-case conditions, without exceeding the objective bit error ratio. (C/LM) 802.3-1998

minimum discernible signal The minimum detectable signal for a system using an operator and display or aural device for detection. (AES) 686-1997

minimum-distance code (1) (computers) A binary code in which the signal distance does not fall below a specified minimum value. (C) [20], [85]
(2) (mathematics of computing) A BCD code in which the Hamming distance between consecutive numerals does not fall below a specified minimum value. (C) 1084-1986w

minimum-driving-point function (linear passive networks) A driving-point function that is a minimum-resistance, minimum-conductance, minimum-reactance, and minimum-susceptance function. (CAS) 156-1960w

minimum en-route altitude (electronic navigation) The lowest altitude between radio fixes that assures acceptable navigational signal coverage and meets obstruction clearance requirements for instrument flight. *See also:* navigation. (AES/RS) 686-1982s, [42]

minimum firing power (microwave switching tubes) The minimum radio-frequency power required to initiate a radio-frequency discharge in the tube at a specified ignitor current. *See also:* gas tube. (ED) 161-1971w

minimum flashover voltage (impulse) The crest value of the lowest voltage impulse, of a given wave shape and polarity that causes flashover. (PE) [8]

minimum fuel limiter (gas turbines) A device by means of which the speed-governing system can be prevented from reducing the fuel flow below the minimum for which the device is set as required to prevent unstable combustion or blowout of the flame. (PE/EDPG) [5]

minimum gas density The minimum operating gas density at which the gas-insulated substation and its components are certified to meet their assigned electrical ratings. (SWG/PE/SUB) C37.100-1992, C37.122.1-1993, C37.122-1993

minimum illumination (sensitivity) The minimum level, in footcandles of a photoelectric lighting control, at which it will operate. *See also:* photoelectric control. (IA/ICTL/IAC) [60]

minimum impulse flashover voltage (neutral grounding devices) The crest value of the lowest voltage impulse at a given wave shape and polarity that causes flashover. (SPD/PE) 32-1972r

minimum input shaft torque (electric coupling) The minimum input torque required to drive an electric coupling with zero output torque load, either with or without rated excitation as specified. (EM/PE) 290-1980w

minimum melting current The smallest current at which a current responsive fuse element will melt at any specified time. (SWG/PE) C37.40-1993, C37.100-1992

minimum ON-state voltage (thyristor) The minimum positive principal voltage for which the differential resistance is zero with the gate open-circuited. *See also:* principal voltage-current characteristic. (ED) [46]

minimum ON voltage (magnetic amplifier) The minimum output voltage existing before the trip OFF control signal is reached as the control signal is varied from trip ON to trip OFF. (MAG) 107-1964w

minimum output voltage (magnetic amplifier) The minimum voltage attained across the rated load impedance as the control ampere-turns are varied between the limits established by positive maximum control currents flowing through all the corresponding control windings simultaneously and negative maximum control currents flowing through all the corresponding control windings simultaneously. (MAG) 107-1964w

minimum perceptible erythema (illuminating engineering) The erythemal threshold. (EEC/IE) [126]

minimum-phase function (linear passive networks) A transmittance from which a nontrivial realizable allpass function cannot be factored without leaving a nonrealizable remainder. *Note:* For lumped-parameter networks, this is equivalent to specifying that the function has no zeros in the interior of the right half of the complex-frequency plane. (CAS) 156-1960w

minimum phase network (1) (data transmission) A network for which the phase shift at each frequency equals the minimum value which is determined uniquely by the attenuation-frequency characteristic in accordance with the following equation:

$$B_c = \frac{1}{\pi}\int_{-\infty}^{+\infty} \frac{dA}{du} \log \coth \frac{|u|}{2}\, du$$

where B_c is phase shift (radians) at a particular frequency f_c, A is attenuation (nepers) as a function of frequency f, and u is $\log(f/f_c)$. *Note:* A ladder network employing lumped impedances, with no coupling between the branches, is an example of a minimum phase network. A bridged T or lattice network of the all-pass type is a nonminimum phase network. (PE) 599-1985w
(2) (excitation systems) See definition above. *Notes:* 1. A network for which the transfer function expressed as a function of s has neither poles nor zeros in the right-hand s plane. Networks (elements) or systems having either poles or zeros in the right half s-plane do not have minimum phase characteristics, assuming there is no right-hand S-plane pole-zero cancellation. 2. Elements whose response is described by transfer functions having transport lags also exhibit non-minimum phase characteristics. The frequency response characteristics of a typical element with transport lag is given in the corresponding figure.

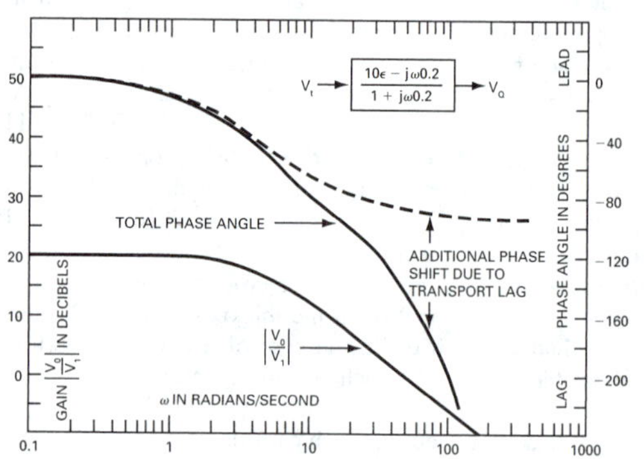

Element with one time constant and a transport lag.

minimum phase network

(PE/EDPG) 421A-1978s
(3) (two specified terminals or two branches) A network for which the transfer admittance expressed as a function of p has neither poles nor zeros in the right-hand p plane. *Note:* A simple T section of real lumped constant parameters without coupling between branches is a minimum-phase network, whereas a bridged-T or lattice section of all-pass type may not be. *See also:* network analysis; impedance function. (Std100) 270-1966w

minimum pulse down time In order for all devices to detect the zero state of a signal between the occurrence of two successive one states of the signal, the zero state must last for at least a bus dependent minimum pulse down time. (NID) 960-1993

minimum-reactance function (linear passive networks) A driving-point impedance from which a reactance function cannot be subtracted without leaving a nonrealizable remainder. *Notes:* 1. For lumped-parameter networks, this is equivalent to specifying that the impedance function has no poles on the imaginary axis of the complex-frequency plane, including the point at infinity. 2. A driving-point impedance (admittance) having neither poles nor zeros on the imaginary axis is both a minimum-reactance and a minimum-susceptance function. (CAS) 156-1960w

minimum reception altitude (MRA) (electronic navigation) The lowest en-route altitude at which adequate signals can be received to determine specific radio-navigation fixes. *See also:* navigation. (AES/RS) 686-1982s, [42]

minimum-resistance (conductance) function (linear passive networks) A driving-point impedance (admittance) from which a positive constant cannot be subtracted without leaving a nonrealizable remainder. (CAS) 156-1960w

minimum single-conductor gradient (overhead-power-line corona and radio noise) The minimum value attained by the gradient $E(\theta)$ as given in the definition of "maximum single-conductor (or subconductor) gradient" as θ varies over the range 0 to 2π. *See also:* maximum single-conductor gradient; maximum single-subconductor gradient; minimum single-subconductor gradient. (T&D/PE) 539-1990

minimum single-subconductor gradient (overhead-power-line corona and radio noise) The minimum value attained by the gradient $E(\theta)$ as given in the definition of "maximum single-subconductor gradient" as θ varies over the range 0 to 2π. *See also:* minimum single-conductor gradient; maximum single-conductor gradient; maximum single-subconductor gradient. (T&D/PE) 539-1990

minimum speed (adjustable-speed drive) The lowest speed within the operating speed range of the drive. *See also:* electric drive. (IA/ICTL/IAC) [60]

minimum-susceptance function (linear passive networks) A driving-point admittance from which a susceptance function cannot be subtracted without leaving a nonrealizable remainder. *Notes:* 1. For lumped-parameter networks, this is equivalent to specifying that the admittance function has no poles on the imaginary axis of the complex-frequency plane, including the point at infinity. 2. A driving-point immitance having neither poles nor zeros on the imaginary axis is both a minimum-susceptance and a minimum-reactance function. (CAS) 156-1960w

minimum tasks Those V&V tasks required for the software integrity level assigned to the software to be verified and validated. (C/SE) 1012-1998

minimum test output voltage (magnetic amplifier) (nonreversible output) The output voltage equivalent to the summation of the minimum output voltage plus 33 1/3 percent of the difference between the rated and minimum output voltages. (MAG) 107-1964w

minimum tool distance The minimum distance that must be maintained between tools and energized lines or devices. (T&D/PE) 516-1995

minimum tool-insulation distance The shortest permissible distance between energized electrical apparatus and any part of a worker's body or conducting object while performing live work with an insulating tool in the air gap. (T&D/PE) 516-1995

minimum usable reading speed (storage tubes) The slowest scanning rate under stated operating conditions before a specified degree of decay occurs. *Note:* The qualifying adjectives minimum usable are frequently omitted in general usage when it is clear that the minimum usable reading speed is implied. *See also:* storage tube. (ED) 158-1962w

minimum usable writing time (storage tubes) The time required to write stored information from one specified level to another under stated conditions of operation. *Note:* The qualifying adjectives minimum usable are frequently omitted in general usage when it is clear that the minimum usable writing time is implied. *See also:* storage tube. (ED) 158-1962w, [45]

minitrack (communication satellite) A ground based tracking system for satellites using interferometers. It requires a minimum satellite instrumentation, hence the name. (COM) [24]

minor alarm (telephone switching systems) An alarm indicating trouble which does not seriously impair the system capability. (COM) 312-1977w

minor cycle (electronic computation) In a storage device that provides serial access to storage positions, the time interval between the appearance of corresponding parts of successive words. (C) 162-1963w, 270-1966w

minor failure *See:* failure.

minority carrier (semiconductor) The type of charge carrier constituting less than one half the total charge-carrier concentration. *See also:* semiconductor. (ED) 216-1960w

minority emitter (transistor) An electrode from which a flow of minority carriers enters the interelectrode region. *See also:* transistor; semiconductor. (PE/EEC) [119]

minor key *See:* secondary key.

minor lobe Any radiation lobe except a major lobe. *See also:* side lobe; back lobe. (AP/T&D/PE/ANT) 145-1993, 1260-1996

minor loop A continuous network consisting of both forward elements and feedback elements and is only a portion of the feedback control system. *See also:* feedback control system. (IA/ICTL/IAC) [60]

minor railway tracks Railway tracks included in the following list:

a) Spurs less than 2000 feet long and not exceeding two tracks in the same span.
b) Branches on which no regular service is maintained or which are not operated during the winter season.
c) Narrow-gauge tracks or other tracks on which standard rolling stock cannot, for physical reasons, be operated.
d) Tracks used only temporarily for a period not exceeding one year.
e) Tracks not operated as a common carrier, such as industrial railways used in logging, mining, etc.

(T&D) C2.2-1960

minuend A number from which another number (the subtrahend) is subtracted to produce a result (the difference). (C) 1084-1986w

minus *See:* difference.

minus input *See:* inverted input.

MIPS A measure of computer processing speed. *See also:* KOPS; MFLOPS. (C) 610.12-1990

mirrored disk array A form of RAID storage, known as level 1, in which each block of data is duplicated on a mirror drive. (C) 610.10-1994w

mirroring The rotation of one or more display elements one hundred and eighty degrees about an axis in the plane of the display surface. (C) 610.6-1991w

MIS *See:* medical information system; management information system.

misaligned frame A frame that erroneously includes a fragmentary byte and that contains a frame check sequence error. *Note:* This term is contextually specific to IEEE Std 802.3. (C) 610.7-1995

misalignment *See:* input-axis misalignment.

misalignment drift (gyros) The part of the total apparent drift component due to uncertainty of orientation of the gyro input axis with respect to the coordinate system in which the gyro is being used. *See also:* navigation. (AES/RS) 686-1982s, [42]

misalignment loss *See:* angular misalignment loss; gap loss; lateral offset loss.

miscellaneous function (numerically controlled machines) An on-off function of a machine such as spindle stop, coolant on, clamp. (IA/EEC) [61], [74]

miscellaneous time The part of up time that is not rerun time, system production time, or system test time, but is time typically used for demonstrations or operator training. *Synonym:* incidental time. (C) 610.10-1994w

MISD *See:* multiple instruction, single data.

misdetection (image processing and pattern recognition) In pattern recognition, the failure to detect the existence of a pattern. (C) 610.4-1990w

misfire (1) (gas tube) A failure to establish an arc between the main anode and cathode during a scheduled conducting period. *See also:* rectification; gas tube. (ED) [45]
(2) The failure of a blasting charge to explode when expected. *Note:* In electric firing, this usually is the result of a broken blasting circuit or insufficient current through the electric blasting cap. *See also:* blasting unit. (EEC/PE/MIN) [119]

mishap An unplanned event or series of events resulting in death, injury, occupational illness, or damage to or loss of equipment or property, or damage to the environment; an accident. (VT/RT) 1483-2000

misidentification In pattern classification, the failure to assign a pattern to its true pattern class. *Synonym:* type I error. *Contrast:* false identification. (C) 610.4-1990w

mismatch The condition in which the impedance of a load does not match the impedance of the source to which it is connected. *See also:* self-impedance. (AP/ANT) [35]

mismatch factors (power meters) Resulting from a combination of interaction factor and reflection factor resulting from reflective source and load impedances which relate incident, absorbed, and delivered power to a nonreflection load. (IM) 544-1975w

mismatch loss *See:* matching loss.

mismatch uncertainty (power meters) Uncertainty in an assigned value that is caused by uncorrected or uncertain values for one or both of the mismatch factors. (IM) 470-1972w, 544-1975w

misordering data A form of unauthorized data modification in which the reception sequence of data units is altered from the original transmission sequence in an unauthorized manner. This can be attempted by a combination of techniques involving deleting, delaying, and reinserting data; or modifying sequence control information; or both. (C/LM) 802.10-1998

misrouted calls A category of call setup irregularities caused by switching system errors. These errors result in a call attempt becoming ineffective after a subscriber has initiated the call correctly (i.e., has gone off-hook, received dial tone, and signaled correctly in the allotted time). This category includes, but is not limited to, calls to the wrong office or wrong number or calls receiving an incorrect tone or announcement or no tone or announcement. Not included in the misrouted calls category are misroutings caused by either erroneous inputs to the switching system, unavailability of service circuits, or failure to establish a network path. (COM/TA) 973-1990w

missing (thyristor converter) (misgating) A condition where the onset of conduction of an arm is substantially delayed from its correct instant of time. *Note:* If an arm fails to turn on during inverter service, there is a commutation failure resulting in a conduction-through. (IA/IPC) 444-1973w

mission (1) The operating objective for which the system was intended. *See also:* system. (SMC) [63]
(2) (nuclear power generating station) The singular objective, task, or purpose of an item or system. (PE/NP) 352-1975s, 933-1999

mission time (nuclear power generating station) The time during which the mission should be performed without interruption. (PE/NP) 933-1999, 352-1975s

mist *See:* fog.

mistake (1) (electronic computation) A human action that produces an unintended result. *Note:* Common mistakes include incorrect programming, coding, manual operation, etc. (C) 162-1963w, [20], [85]
(2) (analog computer) *See also:* error. (C) 165-1977w
(3) (software) A human action that produces an incorrect result. *Note:* The fault tolerance discipline distinguishes between the human action (a mistake), its manifestation (a hardware or software fault), the result of the fault (a failure), and the amount by which the result is incorrect (the error). (C) 610.12-1990

mistrigger (thyristor) (misfire) The failure of a thyristor to conduct at the correct instant of time. (IA/IPC) 428-1981w

misuse Use of processing or communication services for other than official or authorized purposes (e.g., personal gain, espionage). Misuse includes the threats of inadvertent or intentional execution of malicious functions (e.g., computer virus, Trojan horse), performance of undesirable functions (e.g., erasing the file system), and general perpetration of errors of commission, omission, and oversight. Misuse could result in unauthorized disclosure or modification of information, unauthorized receipt of services, or denial of service to legitimate users or critical functions. (C/BA) 896.3-1993w

mixed-base notation *See:* mixed-radix notation.

mixed-base numeration system *See:* mixed-radix notation.

Mixed-Configuration Group A Group to which the member Remote Bridges attach by a mixture of Individual Virtual Ports and Multipeer Virtual Ports. *Note:* It is possible for some members of a Mixed-Configuration Group to attach only by Individual Virtual Ports. (C/LM) 802.1G-1996

mixed frequency fields The superposition of two or more electromagnetic fields of differing frequency. (NIR) C95.1-1999

mixed highs (color television) Those high-frequency components of the picture signal that are intended to be reproduced achromatically in a color picture. (BT/AV) 201-1979w

mixed logic (logic diagrams) The defining of the 1-state of the variables as the more positive or less positive of the two possible levels, depending upon the absence or presence of the polarity indicator symbol. *Synonym:* direct polarity indication. (GSD) 91-1973s

mixed-loop series street-lighting system A street-lighting system that comprises both open loops and closed loops. *See also:* alternating-current distribution; direct-current distribution. (T&D/PE) [10]

mixed mode Pertaining to an expression that contains two or more different data types. For example, $Y := X + N$, where X and Y are floating point variables and N is an integer variable. *Synonym:* mixed type. (C) 610.12-1990

mixed-mode interference Interference that consists of components from both common- and differential-mode interference. (EMC) C63.13-1991

mixed-pressure turbine, condensing or noncondensing (control systems for steam turbine-generator units) Steam enters the turbine at two or more pressures through separate inlet openings with means for controlling the inlet steam pressures or turbine power output. (PE/EDPG) 122-1985s

mixed radix Pertaining to a numeration system that uses more than one radix, such as the biquinary system. (C) [20], [85]

mixed-radix notation A radix notation system in which all digit positions do not have the same radix. For example, biquinary notation in which the digit positions have the radix 2 or 5, alternately. *Synonyms:* mixed-radix numeration system; mixed-base numeration system; mixed-base notation. *Contrast:* fixed-radix notation. (C) 1084-1986w

mixed-radix numeration system *See:* mixed-radix notation.

mixed rain and snow Precipitation consisting of a mixture of rain and wet snow. It usually occurs when the temperature of the air layer near the ground is slightly above freezing. (T&D/PE) 539-1990

mixed-signal circuit A circuit in which some variables are represented by analog (continuously variable) quantities, and some variables are represented by digital (discrete) quantities. (C/TT) 1149.4-1999

mixed sweep (oscilloscopes) In a system having both a delaying sweep and a delayed sweep, a means of displaying the delaying sweep to the point of delay pickoff and displaying the delayed sweep beyond that point. *See also:* oscillograph. (IM/HFIM) [40]

mixed transaction An address beat followed by any number or combination of data write and data read transfers to a single location using the single address transfer mode. This is terminated by the appropriate style of end beat. (C/MM) 896.1-1987s

mixed type *See:* mixed mode.

mixer (A) (data transmission) In a sound transmission, recording or reproducing system, a device having two or more inputs, usually adjustable, and a common output, which operates to combine linearly in a desired proportion the separate input signals to produce an output signal. **(B) (data transmission)** The stage in a heterodyne receiver in which the incoming signal is modulated with the signal from the local oscillator to produce the intermediate-frequency signal. **(C) (data transmission)** A process of intermingling of data traffic flowing between concentration and expansion stages. (PE) 599-1985

mixer tube An electron tube that performs only the frequency-conversion function of a heterodyne conversion transducer when it is supplied with voltage or power from an external oscillator. (ED) 161-1971w

mixing ratio (of water vapor) The ratio of the mass of water vapor to the mass of dry air in a given volume of air. This ratio is generally expressed in grams per kilogram. (AP/PROP) 211-1997

mixing rules Various theoretical and often empirical models to predict the effective medium constitutive parameters. (AP/PROP) 211-1997

mixing segment A medium that may be connected to more than two Medium Dependent Interfaces (MDIs). (C/LM) 802.3-1998

MKSA system of units A system in which the basic units are the meter, kilogram, and second, and the ampere is a derived unit defined by assigning the magnitude $4\pi \times 10^{-7}$ to the rationalized magnetic constant (sometimes called the permeability of space). *Notes:* 1. At its meeting in 1950 the International Electrotechnical Commission recommended that the MKSA system be used only in the rationalized form. 2. The electrical units of this system were formerly called the practical electrical units. 3. If the MKSA system is used in the unrationalized form the magnetic constant is 10^{-7} henry/meter and the electric constant is $10^7/c^2$ farads/meter. Here c, the speed of light, is approximately 3×10^8 meters/second. 4. In this system, dimensional analysis is customarily used with the four independent (basic) dimensions: mass, length, time, current. (Std100) 270-1966w

MLHG *See:* multiline hunt group.

M-L ratio *See:* metallic longitudinal induction ratio.

MLS *See:* microwave landing system.

MMI *See:* man-machine interface; user interface.

MMU *See:* memory management unit.

mnemonic (1) (test, measurement, and diagnostic equipment) Assisting or intending to assist a human memory and understanding. Thus a mnemonic term is usually an abbreviation, that is easy to remember; for example, mpy for multiply and acc for accumulator. (MIL) [2]
(2) An abbreviation or other shortened keyboard notation that is used to substitute for a more complicated action, such as selecting an object or performing an operation on it. (C) 1295-1993w

mnemonic code (test, measurement, and diagnostic equipment) A pseudo code in which information, usually instructions, is represented by symbols or characters which are readily identified with the information. (MIL) [2]

mnemonic symbol (software) A symbol chosen to assist the human memory, for example, an abbreviation such as "mpy" for "multiply." (C/SE) 729-1983s

MNOS Acronym for metal nitride oxide semiconductor. (ED) 641-1987w

MNOS transistor *See:* metal-nitride-oxide-semiconductor transistor.

m:n relationship* *See:* many-to-many relationship.
* Deprecated.

mobile (x-ray) Equipment mounted on a permanent base with wheels and/or casters for moving while completely assembled. (NEC/NESC) [86]

mobile communication system Combinations of interrelated devices capable of transmitting intelligence between two or more spatially separated radio stations, one or more of which shall be mobile. (VT) [37]

mobile home A factory-assembled structure or structures equipped with the necessary service connections and made so as to be readily movable as a unit or unit(s) without a permanent foundation. The phrase "without a permanent foundation" indicates that the support system is constructed with the intent that the mobile home placed thereon will be moved from time to time at the convenience of the owner. (NESC/NEC) [86]

mobile home accessory building or structure Any awning, cabana, ramada, storage cabinet, carport, fence, windbreak or porch established for the use of the occupant of the mobile home upon a mobile home lot. (NESC/NEC) [86]

mobile home lot A designated portion of a mobile home park designed for the accommodation of one mobile home and its accessory buildings or structures for the exclusive use of its occupants. (NESC/NEC) [86]

mobile home park A contiguous parcel of land which is used for the accommodation of occupied mobile homes. (NESC/NEC) [86]

mobile home service equipment The equipment containing the disconnecting means, overcurrent protective devices, and receptacles or other means for connecting a mobile home feeder assembly. (NESC/NEC) [86]

mobile radio service Radio service between a radio station at a fixed location and one or more mobile stations, or between mobile stations. *See also:* radio transmission. (EEC/PE) [119]

mobile station (1) A radio station designed for installation in a vehicle and normally operated when in motion. *See also:* mobile communication system. (COM/VT) [37]
(2) A type of station that uses network communications while in motion. (C/LM) 8802-11-1999

mobile substation equipment Substation equipment mounted and readily movable as a system of transportable devices. (PE/SUB) 1268-1997

mobile telemetering Electric telemetering between points that may have relative motion, where the use of interconnecting wires is precluded. *Note:* Space radio is usually employed as an intermediate means for mobile telemetering, but radio may also be used for telemetering between fixed points. *See also:* telemetering. (EEC/PE) [119]

mobile telephone system (automatic channel access) A mobile telephone system capable of operation on a plurality of frequency channels with automatic selection at either the base station or any mobile station of an idle channel when communication is desired. *See also:* mobile communication system. (VT) [37]

mobile transformer Transformers that are usually mounted on trailers for easy transport to temporarily replace stationary transformers taken out of service because of failure or maintenance. (PE/TR) 1276-1997

mobile transmitter A radio transmitter designed for installation in a vessel, vehicle, or aircraft, and normally operated while in motion. *See also:* radio transmitter. (AP/ANT) 145-1983s

mobile unit substation A unit substation mounted and readily movable as a unit on a transportable device. (SWG/PE) C37.100-1992

mobility *See:* drift mobility.

mobility, Hall *See:* Hall mobility.

mobility spectrum The distribution of ions as a function of mobility. Historically, ions have been classified by mobility as small (10^{-5} m²/Vs to 2×10^{-4} m²/Vs), medium (10^{-7} m²/Vs to 10^{-5} m²/Vs), and large (10^{-9} m²/Vs to 10^{-7} m²/Vs). (T&D/PE) 539-1990

MOBSSL-UAF *See:* Merritt and Miller's Own Block Structured Simulation Language.

mock-up A full-sized structural, but not necessarily functional, model built accurately to scale, used chiefly for study, testing, or display; for example, a full-sized model of an airplane displayed in a museum. *See also:* physical model. (C) 610.3-1989w

mod *See:* modulo.

modal A state in which the user has to complete the request of the mode before continuing. (C) 1295-1993w

modal analysis (power-system communication) A method of computing the propagation of a wave on a multiconductor power line. (PE) 599-1985w

modal channel (x-ray energy spectrometers) That channel in the distribution containing the largest number of counts. (NPS/NID) 759-1984r

modal direction (transmission performance of telephone sets) The assumed direction of speech transmission on a modal head. Also, the axis of an artificial mouth,. (COM/TA) 269-1983s

modal distance (telephony) The distance between the center of the grid of a telephone-handset transmitter cap and the center of the lips of a human talker (or the reference point of an artificial mouth), when the handset is in the modal position. (COM/TA) 269-1971w

modal head (transmission performance of telephone sets) Head dimensions that are modal for a human population. The modal head is the same as that adopted by the Comité Consultatif International Télégraphique et Téléphonique (CCITT) for the measurement of Affaiblissement équivalent pour netteté (equivalent articulation loss). The applicable dimensions are shown in the correpsonding figure.

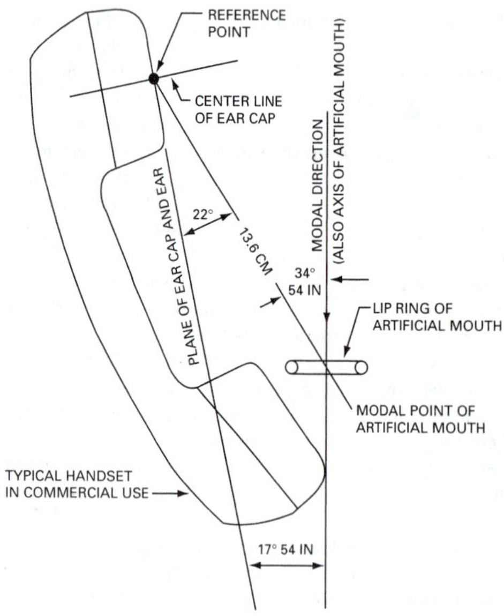

modal head

(COM/TA) 269-1983s

modal noise (fiber optics) Noise generated in an optical fiber system by the combination of mode dependent optical losses and fluctuation in the distribution of optical energy among the guided modes or in the relative phases of the guided modes. *Synonym:* speckle noise. *See also:* mode. (Std100) 812-1984w

modal participation factor The magnitude of each structural mode (natural frequency) that participates to compose the final dynamic response of the system. Each participation factor is a function of the system mass distribution and the generalized mode shape at each natural frequency. (SWG/PE) C37.100-1992, C37.81-1989r

modal point (transmission performance of telephone sets) The position of the center of the lips of a modal head. Also, the corresponding reference point of an artificial mouth, the center of the external plane of the lip ring. (COM/TA) 269-1983s

modal position (transmission performance of telephone sets) The position a telephone-set handset assumes when the earcap of the handset is held in close contact with the ear of a modal head and the modal direction is in the plane defined by the axes of the transmitter cap and ear-cap. (COM/TA) 269-1983s

mode (1) (radix-independent floating-point arithmetic) (binary floating-point arithmetic) A variable that a user may set, sense, save, and restore to control the execution of subsequent arithmetic operations. The default mode is the mode that a program can assume to be in effect unless an explicitly contrary statement is included in either the program or its specification. The following mode is implemented: 1) Rounding, to control the direction of rounding errors. 2) In certain implementations, rounding precision, to shorten the precision of results. 3) The implementor may, at his option, implement the following modes: traps disabled or enabled, to handle exceptions. (MM/C) 854-1987r, 754-1985r
(2) (electron tube) A state of a vibrating system to which corresponds one of the possible resonance frequencies (or propagation constants). *Note:* Not all dissipative systems have modes. *See also:* oscillatory circuit. (ED) 161-1971w, [45]
(3) (fiber optics) In any cavity or transmission line, one of those electromagnetic field distributions that satisfies Maxwell's equations and the boundary conditions. The field pattern of a mode depends on the wavelength, refractive index, and cavity or waveguide geometry. *See also:* bound mode; linearly polarized mode; single mode optical waveguide; differential mode delay; unbound mode; multimode distortion; transverse magnetic mode; multimode optical waveguide; transverse electric mode; hybrid mode; multimode laser; mode volume; equilibrium mode distribution; leaky mode; cladding mode; fundamental mode; equilibrium mode simulator; differential mode attenuation. (Std100) 812-1984w
(4) (radio-wave propagation) A characteristic solution to the wave equation for specified boundary conditions. Other uses of the term mode are common. *See also:* mode of propagation, ionospheric. (AP/PROP) 211-1990s
(5) (mathematics of computing) A variable that a user may set, sense, save, and restore to control the execution of subsequent arithmetic operations. (C) 1084-1986w
(6) An operating condition of a function, subfunction, or physical element of the system. (C/SE) 1220-1994s
(7) A collection of attributes that specifies a file's type and its access permissions. *See also:* file access permissions. (C/PA) 9945-1-1996, 9945-2-1993, 1003.5-1999
(8) A set of related features or functional capabilities of a product, (e.g., on-line, off-line, and maintenance modes). (C/SE) 1362-1998

mode conversion (waveguide) The transformation of an electromagnetic wave from one mode of propagation to one or more other modes. (MTT) 146-1980w

mode conversion gain (waveguide) The gain due to the conversion of power from one waveguide mode to another. (MTT) 146-1980w

mode conversion loss (waveguide) The loss due to the conversion of power from one waveguide mode to another. (MTT) 146-1980w

mode coupler (waveguides) A coupler that provides preferential coupling to a specific wave mode. *See also:* waveguide. (IM/HFIM) [40]

mode coupling (fiber optics) In an optical waveguide, the exchange of power among modes. The exchange of power may reach statistical equilibrium after propagation over a finite distance that is designated the equilibrium length. *See also:* equilibrium length; equilibrium mode distribution; mode; mode scrambler. (Std100) 812-1984w

mode dispersion *See:* multimode distortion.

modal distortion *See:* multimode distortion.

mode distortion *See:* multimode distortion.

mode filter (1) (fiber optics) A device used to select, reject, or attenuate a certain mode or modes. (Std100) 812-1984w
(2) (waveguide components) A device designed to pass energy along a waveguide in one or more selected modes of propagation, and substantially to reject energy carried in other modes. (MTT) 147-1979w

mode, higher-order *See:* higher-order mode.

MODEL An application-oriented language used widely for simulating digital circuits. (C) 610.13-1993w

model (1) A mathematical or physical representation of the system relationships. (SMC) [63]
(2) (A) (modeling and simulation) An approximation, representation, or idealization of selected aspects of the structure, behavior, operation, or other characteristics of a real-world process, concept, or system. *Note:* Models may have other models as components. **(B) (modeling and simulation)** To serve as a model as in definition (A). **(C) (modeling and simulation)** To develop or use a model as in definition (A). (C) 610.3-1989
(3) (computer graphics) An accurate and complete graphical representation of an object. *See also:* modeling system. (C) 610.6-1991w
(4) A representation of one or more aspects of a system. (C/SE) 1016.1-1993w
(5) An analog representation, which may be conceptual, qualitative, or quantitative. (PE/NP) 1082-1997
(6) A representation of a real world process, device, or concept. (C/SE) 1233-1998
(7) (A) A representation of something that suppresses certain aspects of the modeled subject. This suppression is done in order to make the model easier to deal with and more economical to manipulate and to focus attention on aspects of the modeled subject that are important for the intended purpose of the model. For instance, an accurate model of the solar system could be used to predict when planetary conjunctions will take place and the phases of the moon at a particular time. Such a model would generally not attempt to represent the internal workings of the sun or the surface composition of each planet. **(B)** An interpretation of a theory for which all the axioms of the theory are true. [logic sense]. (C/SE) 1320.2-1998

model a cell The creation of a specific elaboration of a model using `modelSearch`. (C/DA) 1481-1999

model accreditation (or simulation accreditation) The official certification that a model or simulation is acceptable for use for a specific purpose. (DIS/C) 1278.3-1996

mode-less A state that does not interfere with the user performing any other action. (C) 1295-1993w

model glossary The collection of the names and definitions of all defined concepts that appear within the views of a model. (C/SE) 1320.2-1998

model hierarchy The diagrams that correspond to the nodes of the hierarchical graph structure of an IDEF0 model. (C/SE) 1320.1-1998

modeling Technique of system analysis and design using mathematical or physical idealizations of all or a portion of the system. Completeness and reality of the model are dependent on the questions to be answered, the state of knowledge of the system, and its environment. *See also:* system. (SMC) [63]

modeling procedures The action of a circuit with respect to timing and power. These actions include creating segments and nodes, determining the propagation properties, and setting the delay and slew equations to use. (C/DA) 1481-1999

modeling statements Delay calculation language (DCL) statements that map cell configurations to modeling procedures. (C/DA) 1481-1999

modeling system A system in which a computer graphics model can be defined or transformed using world coordinates. (C) 610.6-1991w

model name A unique, descriptive name that distinguishes one IDEF0 model from other IDEF0 models with which it may be associated. An IDEF0 model's model name and model name abbreviation are placed in the A-0 context diagram along with the model's purpose statement and viewpoint statement. (C/SE) 1320.1-1998

model name abbreviation A unique short form of a model name that is used to construct diagram references. (C/SE) 1320.1-1998

model note A textual and/or graphical component of a diagram that records a fact not otherwise depicted by a diagram's boxes and arrows. (C/SE) 1320.1-1998

model note number An integer number, placed inside a small square, that unambiguously identifies a model note in a specific diagram. (C/SE) 1320.1-1998

model page A logical component of an IDEF0 model that can be presented on a single sheet of paper. Model pages include diagram, text, FEO, and glossary pages. (C/SE) 1320.1-1998

model space The coordinate system used by a computer graphics model. (C) 610.6-1991w

MODEL 204 (M204) A database manipulation language with English-like syntax. (C) 610.13-1993w

model validation (1) The process of determining the degree to which the requirements, design, or implementation of a model are a realization of selected aspects of the system being modeled. *Contrast:* model verification. *See also:* fidelity. (C) 610.3-1989w
(2) The process of determining the degree to which a model is an accurate representation of the real world from the perspective of the intended use(s) of the model. (DIS/C) 1278.3-1996

model verification (1) The process of determining the degree of similarity between the realization steps of a model; for example, between the requirements and the design, or between the design and its implementation. *Contrast:* model validation. (C) 610.3-1989w
(2) The process of determining that a model implementation accurately represents the developer's conceptual description and specifications. (DIS/C) 1278.3-1996

modem (1) (data transmission) A contraction of MOdulator-DEModulator, an equipment that connects data terminal equipment to a communication line. (PE) 599-1985w
(2) (supervisory control, data acquisition, and automatic control) A modulator/demodulator device that converts serial binary digital data to and from the signal form appropriate for the respective communication channel. (SWG/PE/SUB) C37.100-1992, C37.1-1994
(3) (broadband local area networks) A modulator-demodulator device. The modulator encodes digital information onto an analog carrier signal by varying the amplitude, frequency, or phase of that carrier. The demodulator extracts digital information from a similarly modified carrier. A modem transforms digital signals into a form suitable for transmission over an analog medium. (LM/C) 802.7-1989r
(4) A device that modulates and demodulates signals transmitted over data communication facilities. One of the functions of a modem is to enable digital data to be transmitted over analog transmission facilities. (SUB/PE) 999-1992w
(5) (A) A device that performs modulation and demodulation functions necessary to transmit signals over communication lines. *Note:* This term originated as an abbreviation for modulator-demodulator. *Synonyms:* modulator-demodulator; demodulator-modulator; data set. *See also:* acoustic coupler. **(B)** A device that transforms a digital signal received into an analog signal and vice versa. (C) 610.7-1995

modem control The monitoring of modem status lines. (C/PA) 2003.1-1992

mode mixer *See:* mode scrambler.

mode of operation (rectifier circuits) The characteristic pattern of operation determined by the sequence and duration of commutation and conduction. *Note:* Most thyristor converters and rectifier circuits have several modes of operation, which may be identified by the shape of the current wave. The particular mode obtained at a given load depends upon the circuit constants. *See also:* rectifier circuit element; rectification. (IA/IPC) 444-1973w

mode of propagation (1) (A) (waveguides) A form of propagation of guided waves that is characterized by a particular field pattern in a plane transverse to the direction of propagation, which field pattern is independent of position along the axis of the waveguide. *Note:* In the case of uniconductor waveguides the field pattern of a particular mode of propagation is also independent of frequency. *See also:* waveguide. **(B)** A form of electromagnetic wave than can advance and can transport energy along the axis of a transmission line without change in the form of the electromagnetic field pattern in successive transverse sections (except for a monotonic decrease in amplitude along the direction of propagation, due to energy dissipation, which is present to some degree in every transmission line). (MTT) 148-1959, 146-1980, 1004-1987 **(2)** A form of propagation of guided waves where the transverse field pattern is invariant with range (i.e., as in a waveguide). *Note:* Inappropriate uses of the term mode are common. *Synonym:* mechanism of propagation. *See also:* ionospheric mode of propagation. (AP/PROP) 211-1997

mode of propagation, ionospheric Representation of a transmission path by the number of hops between the end points of the path, the ionospheric layers producing the ionospheric reflections being indicated for each hop. Example: 1F + 1E represents a hop with an ionospheric reflection in the F region followed by a reflection at the ground, followed, in turn, by a hop with a reflection from the E region. (AP/PROP) 211-1990s

mode of resonance (waveguide) A form of natural electromagnetic oscillation in a resonator, characterized by a particular field pattern. (MTT) 146-1980w

mode of vibration (vibratory body, such as a piezoelectric crystal unit) A pattern of motion of the individual particles due to stresses applied to the body, its properties, and the boundary conditions. Three common modes of vibration are flexural, extensional, and shear. *See also:* crystal. (EEC/PE) [119]

mode scrambler (A) (fiber optics) A device for inducing mode coupling in an optical fiber. *Synonym:* mode mixer. *See also:* mode coupling. **(B) (fiber optics)** A device composed of one or more optical fibers in which strong mode coupling occurs. *Note:* Frequently used to provide a mode distribution that is independent of source characteristics or that meets other specifications. *Synonym:* mode mixer. *See also:* mode coupling. (Std100) 812-1984

mode shape (mechanical) A plot that shows displacements of various points in the vibrating structure at a particular instant in time. There is a characteristic mode shape associated with each natural frequency of a vibrating structure. (SWG/SUB/PE) C37.122-1983s, C37.122.1-1993, C37.100-1992

mode stripper *See:* cladding mode stripper.

mode transducer (waveguide components) A device for transforming an electromagnetic wave from one mode of propagation to another. (MTT) 147-1979w

mode transformer *See:* mode transducer.

mode voltage *See:* glow voltage.

mode volume (fiber optics) The number of bound modes that an optical waveguide is capable of supporting; for *V*5, approximately given by $V^2/2$ and $(V^2/2[g/(g+2)]$, respectively, for step index and power-law profile waveguides, where g is the profile parameter, and V is normalized frequency. *See also:* step index profile; power-law index profile; normalized frequency; mode; V number; effective mode volume. (Std100) 812-1984w

modification (A) (software) A change made to software. *See also:* software. **(B) (software)** The process of changing software. *See also:* software. (C/SE) 729-1983

modification request (MR) A generic term that includes the forms associated with the various trouble/problem-reporting documents (e.g., incident report, trouble report) and the configuration change control documents [e.g., software change request (SCR)]. (C/SE) 1219-1998

modified calling line disconnect A telephone network feature that, if the end user is off-hook, requires the end user's line to go on-hook before receiving a dial tone from the telephone network. (SCC31) 1390.3-1999, 1390.2-1999

modified circuit transient recovery voltage The circuit transient recovery voltage modified in accordance with the normal-frequency recovery voltage and the asymmetry of the current wave obtained on a particular interruption. *Note:* This voltage indicates the severity of the particular interruption with respect to recovery-voltage phenomena. (SWG/PE) C37.100-1992

modified cosecant-squared antenna pattern A cosecant-squared antenna pattern modified to obtain increased antenna gain at the higher elevation angles so as to provide larger echo signals from targets at high altitude and short range that would normally be too weak to be detected when sensitivity time control (STC) is used along with the conventional cosecant-squared antenna pattern. *Note:* Sometimes called a thumb pattern. Commonly used in 2-D air-traffic control radars. (AES) 686-1997

modified impedance relay An impedance form of distance relay for which the operating characteristic of the distance unit on an *R-X* diagram is a circle having its center displaced from the origin. *Note:* It may be described by the equation

$$Z^2 = 2K_1Z_{\cos}(\theta - \alpha) = K_2{}^2 - K_1{}^2$$

here K_1, K_2, and α are constants and θ is the phase angle by which the input voltage leads the input current. (SWG/PE) C37.100-1992

modified index of refraction In the troposphere, the sum of the refractive index at a given height *h* above the mean local surface and the ratio of this height to the geometrical mean radius of the Earth. (AP/PROP) 211-1997

modified inherent transient recovery voltage (transient recovery voltage) The TRV (transient recovery voltage) that results from the interaction of a circuit (that produces the inherent transient recovery voltage) and the impedance (capacitors, resistors, etc.) of an interrupting device without the modifying effects of an arc and its voltage. Modifying impedances, such as capacitors and resistors, are sometimes included as part of a switching device to modify the TRV. (SWG/PE) C37.04E-1985w, C37.4D-1985w, C37.100B-1986w, C37.100-1992

modified-off-the-shelf (MOTS) A software product that is already developed and available, usable either "as is" or with modification, and provided by the supplier, acquirer, or a third party. (C/SE) 1062-1998

modified performance test A test, in the "as found" condition, of a battery's capacity and its ability to provide a high-rate, short-duration load (usually the highest rate of the duty cycle) that will confirm the battery's ability to meet the critical period of the load duty cycle, in addition to determining its percentage of rated capacity. (PE/EDPG) 450-1995

modified source statements Original source statements that have been changed. (C/SE) 1045-1992

modified z transform (data processing) The modified *z* transform of $f(t)$, denoted $F(z,m)$, is the delayed *z* transform of f(t) with the substitution $\Delta = 1 - m$; that is,

$$F(z, m) = \sum_{n=0}^{\infty} f[nT - (1 - m)T]u[nT - (1 - m)T]z^{-n}$$

$$0 < m < 1$$

(IM) [52]

modify (A) To change the contents of a database. **(B)** To change the logical structure of a database. *See also:* alter. (C) 610.5-1990

Modula 2 *See:* MODUlar LAnguage II.

MODULA II *See:* MODUlar LAnguage II.

modular (software) Composed of discrete parts. *See also:* modular decomposition; modular programming. (C) 610.12-1990

modular assembly A circuit breaker element consisting of sealed interrupters, mechanism, and connecting terminals. (SWG/PE) C37.59-1996

modular constraint *See:* grid constraint.

modular decomposition (software) The process of breaking a system into components to facilitate design and development; an element of modular programming. *Synonym:* modularization. *See also:* factoring; hierarchical decomposition; demodularization; cohesion; coupling; packaging; functional decomposition. (C) 610.12-1990

modularity (software) The degree to which a system or computer program is composed of discrete components such that a change to one component has minimal impact on other components. *See also:* cohesion; coupling. (C) 610.12-1990

modularization *See:* modular decomposition.

MODUlar LAnguage II (MODULA II) A programming language developed, as an expanded version of Pascal, to support modular design, structured programs, and mathematical calculations. *See also:* block-structured language. (C) 610.13-1993w

modular programming (software) A software development technique in which software is developed as a collection of modules. *See also:* stepwise refinement; data structure-centered design; transaction analysis; rapid prototyping; modular decomposition; input-process-output; structured design; transform analysis; object-oriented design. (C) 610.12-1990

MODULAR II *See:* MODUlar LAnguage II.

modulate (A) To convert voice or data signal for transmission over a communications network. *Contrast:* demodulate. **(B)** To vary one or more attributes of a carrier (amplitude, frequency, phase) such that the frequency information in the modulating signal can be recovered by its inverse process. (C) 610.7-1995

modulated 12.5T pulse (linear waveform distortion) A burst of color subcarrier frequency of nominally 3.58 MHz. The envelope of the burst is $\sin^2$ shaped with a HAD of nominally 1.56 μs. The MOD 12.5T pulse consists of a luminance and a chrominance component. The envelope of the frequency spectrum consists of two parts, namely signal energy concentrated in the luminance region below 0.6 MHz and in the chrominance region from roughly 3 MHz to 4.2 MHz.

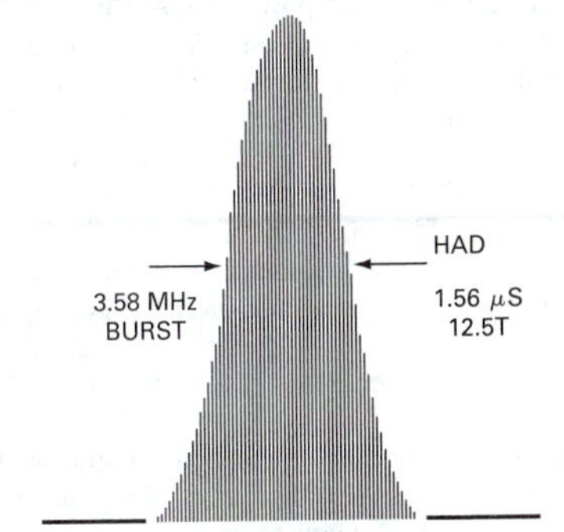

Envelope of frequency spectrum of modulated 12.5T pulse

modulated 12.5T pulse

(BT) 511-1979w

modulation (1) (A) (data transmission) *(Carrier).* (i) The process by which some characteristic of a carrier is varied in accordance with a modulating wave. (ii) The variation of some characteristic of a carrier. *See also:* angle modulation; modulation index. **(B) (data transmission)** *(Signal transmission system).* (i) A process whereby certain characteristics of a wave, often called the carrier, are varied or selected in accordance with a modulating function. (ii) The result of such a process. *See also:* angle modulation; modulation index. (PE) 599-1985

(2) (diode-type camera tube) The ratio of the difference between the maximum and minimum signal currents divided by the sum. To avoid ambiguity, the optical input image intensity shall be assumed to be sinusoidal in the direction of scan. (ED) 503-1978w

(3) (fiber optics) A controlled variation with time of any property of a wave for the purpose of transferring information. (Std100) 812-1984w

(4) (overhead-power-line corona and radio noise) The process by which some characteristic of a carrier is varied in accordance with a modulating signal. (T&D/PE) 539-1990

(5) (broadband local area networks) The method whereby information is superimposed onto a RF carrier to transport signals through a communications channel. (LM/C) 802.7-1989r

(6) The process of changing or regulating the characteristics of a carrier that is vibrating at a certain amplitude and frequency so that the variations represent meaningful information. *Contrast:* demodulation. (C) 610.7-1995

modulation contrast (diode-type camera tube) The ratio of the difference between the peak and the minimum values of irradiance to the sum of the peak and the minimum value of irradiance of an image or specified portion of an image. (ED) 503-1978w

modulation index (angle modulation with a sinusoidal modulating function) (data transmission) The ratio of the frequency deviation of the modulated wave to the frequency of the modulating function. *Note:* The modulation index is numerically equal to the phase deviation expressed in radians. (PE) 599-1985w

modulation threshold (illuminating engineering) In the case of a square wave or sine wave grating, manipulation of luminance differences can be specified in terms of modulation and the threshold may be called the modulation threshold.

$$\text{modulation} = \frac{L_{max} - L_{min}}{L_{max} + L_{min}}$$

Periodic patterns that are not sine wave can be specified in terms of the modulation of the fundamental sine wave component. The number of periods or cycles per degree of visual angle represents the spatial frequency. (EEC/IE) [126]

modulator A device that converts a signal into a modulated signal that is suitable for transmission. (C) 610.7-1995

modulation transfer function (diode-type camera tube) $R_o(N)$, the modulus of the optical transfer function (OTF), is synonymous with the sine amplitude response. That is, the response of the imaging sensor to sinewave images. When the modulation transfer functions or MTFs of a linear sensor's components are known, the overall system MTF can be found by multiplying the individual component MTFs together. (ED) 503-1978w

modulator-demodulator *See:* modem.

module (1) (cable penetration fire stop qualification test) An opening in a fire resistive barrier so located and spaced from adjacent modules (openings) that its respective cable penetration fire stop's performance will not affect the performance of cable penetration fire stops in any adjacent module. A module may take on any shape to permit the passage of cables from one or any number of raceways. (ED) 581-1978w

(2) (A) (software) A program unit that is discrete and identifiable with respect to compiling, combining with other units, and loading; for example, the input to, or output from, an

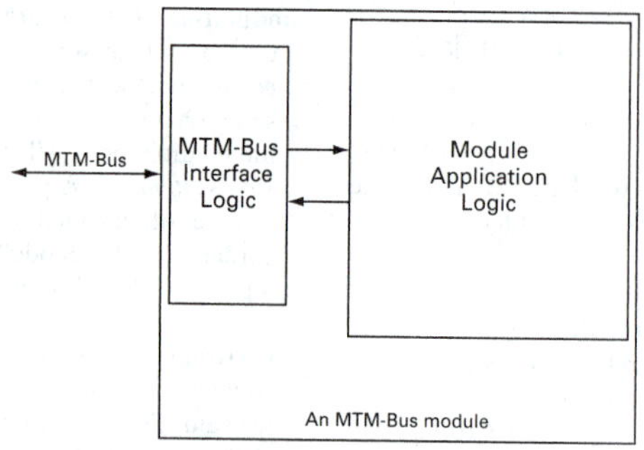

NOTE–An MTM-Bus module consists of MTM-Bus interface logic and module application logic.

MTM-Bus module

assembler, compiler, linkage editor, or executive routine. **(B) (software)** A logically separable part of a program. *Note:* The terms "module," "component," and "unit" are often used interchangeably or defined to be sub-elements of one another in different ways depending upon the context. The relationship of these terms is not yet standardized. (C) 610.12-1990

(3) (STEbus) A plug-in unit consisting of one or more boards that contains at least one bus interface conforming to IEEE Std 1000-1987, which plugs into the backplane. (C/MM) 1000-1987r

(4) (MULTIBUS) A basic functional unit within an agent. (C/MM) 1296-1987s

(5) Collection of circuitry designed to perform specific functions that includes an interface to Futurebus+. (C/BA) 10857-1994, 896.4-1993w, 896.3-1993w

(6) (NuBus) *See also:* board. (C/MM) 1196-1987w

(7) A board or board set that comprises a single physical unit. It provides mechanical mounting and protection of electronic components, thermal transfer of heat away from the components to an external heat sink, and electrical and fiber-optic connections. A module is removable and replaceable. (BA/C) 14536-1995

(8) A plug-in unit per IEC 50. (C/BA) 1101.4-1993, 1101.3-1993

(9) A board, or board set, consisting of one or more nodes, that share a physical interface to SCI. If a module has multiple boards with backplane-mating connectors, it only uses one for the logical connection to the node. The others may provide additional power or I/O for their associated boards, but otherwise merely pass the input link signals through to the output link to provide continuity in case the module is plugged into a ring-connected backplane. (C/MM) 1596-1992

(10) Typically a board assembly and its associated mechanical parts, front panel, optional shields, etc., which contains everything required to occupy a slot in a mainframe. A module may occupy one or more slots. (C/MM) 1155-1992

(11) A collection of circuitry that is designed to perform a specific operation. This is standard terminology for Futurebus+, while VME64 uses board synonymously. (C/BA) 1014.1-1994w

(12) A board, or board set, consisting of one or more nodes that share a physical interface. Although only one board in a module connects to bus signals, each board connector could provide power from the bus. (C/MM) 1212-1991s

(13) An electronic circuit assembly that connects to one or more slots on the backplane. It is removable from and replaceable in a backplane assembly via connectors. (C/BA) 896.2-1991w

(14) An addressable unit or interconnected set of units attached to the MTM-Bus and fully supporting the MTM-Bus protocols. The boundary of an MTM-Bus module may correspond to the physical partitioning of the system, but is not required to do so. For the purposes of this document, a module is comprised of an MTM-Bus interface and module application logic, as shown in the figure below. (TT/C) 1149.5-1995

(15) (FASTBUS module) Any FASTBUS Device that can be housed in a FASTBUS crate, that can connect to a crate segment and that conforms with the mandatory specifications for a FASTBUS module. (NID) 960-1993

(16) A packaged functional hardware unit designed for use with other components. (C) 610.10-1994w

(17) The smallest component of physical management; i.e., a replaceable device. (C/MM) 1394-1995

(18) Multiple cells/units in a single assembly. (SB) 1188-1996

(19) A board or board set consisting of one or more nodes that share a physical interface, although only one board in a module connects to bus signals. Each board connection could provide power from the bus. (C/BA) 1156.4-1997

(20) Any assembly of interconnected components that constitutes an identifiable device, instrument, or piece of equipment. A module can be disconnected, removed as a unit, and replaced with a spare. It has definable performance characteristics that permit it to be tested as a unit. A module could be a card, a drawout circuit breaker, or other subassembly of a larger device, provided it meets the requirements of this definition. (PE/NP) 603-1998

module accelerated aging (nuclear power generating station) (advanced life conditioning) The acceleration process designed to achieve an advanced life condition in a short period of time. It is the process of subjecting a module or component to stress conditions in accordance with known measurable physical or chemical laws of degradation in order to render its physical and electrical properties similar to those it would have at an advanced age operating under expected service conditions. In addition, when operations of a device are cyclical, acceleration is achieved by subjecting the device to the number of cycles anticipated during its qualified life. (PE/NP) 381-1977w

module accuracy (nuclear power generating station) Conformity of a measurement value to an accepted standard value or true value. *Note:* For further information, see Process Measurement and Control Terminology SAMA PMC-20.1-1973. (PE/NP) 381-1977w

module address (MA) (1) (FASTBUS acquisition and control) The group of bits assigned in the device address field of a FASTBUS address which identifies the module on its segment. The module address may partially overlap the group address. (NID) 960-1993

(2) An eight-bit value uniquely identifying an MTM-Bus module. (TT/C) 1149.5-1995

module aging (nuclear power generating station) (natural) The change with passage of time of physical chemical, or electrical properties of a component or module under design range operating conditions which may result in degradation of significant performance characteristics. (PE/NP) 381-1977w

module auxiliary connector (FASTBUS acquisition and control) The standard connector that mounts above the module segment connector on a module circuit board. (NID) 960-1993

module calibration (nuclear power generating station) Adjustment of a device, to bring the module's output to a desired value or series of values, within a specified tolerance, for a particular value or series of values of the input or measurements used to establish the input-output function of the module. (PE/NP) 381-1977w

module circuit board The printed board that is the circuit part of a FASTBUS module on which the module segment connector is mounted for mating with the crate segment connector. (NID) 960-1993

module common-mode rejection (nuclear power generating station) The ability of a module with a differential input stage to cancel or reject a signal applied equally to both inputs. (PE/NP) 381-1977w

module components (nuclear power generating station) Items from which the module is assembled (for example, resistors, capacitors, wires, connectors, transistors, springs, etc.). (PE/NP) 381-1977w

module conformity (nuclear power generating station) The closeness with which the curve of a function approximates a specified curve. (PE/NP) 381-1977w

module contact rating (nuclear power generating station) The electrical power-handling capability of relay or switch contacts. This should be specified as continuous or interrupting, resistive or inductive, ac or dc. (PE/NP) 381-1977w

module counter A counter that reverts to zero in the counting sequence after reaching a value of $n - 1$. (C) 1084-1986w

module design range operating conditions (nuclear power generating station) The range of environmental and energy supply operating conditions within which a module is designed to operate. (PE/NP) 381-1977w

module drift (nuclear power generating station) A change in output-input relationship over a period of time, normally determined as the change in output over a specified period of time for one or more input values which are held constant under specified reference operating conditions. (PE/NP) 381-1977w

module electromagnetic interference (nuclear power generating station) Any unwanted electromagnetically transmitted energy appearing in the circuitry of a module. (PE/NP) 381-1977w

module energy supply (nuclear power generating station) Electrical energy, compressed fluid, manual force or other such input to the module that will establish the power for its operation. (PE/NP) 381-1977w

Module Fail Status (MFS) bit A bit in the Slave Status register of every S-module that is set by the S-module when the module's built-in test has failed or is currently executing. (TT/C) 1149.5-1995

module failure trending (nuclear power generating station) Systematic documentation and analysis of the frequency of a particular failure mode. (PE/NP) 381-1977w

module frequency response (nuclear power generating station) The frequency-dependent relation, in both amplitude and phase, between steady-state sinusoidal inputs and the resulting fundamental sinusoidal outputs. (PE/NP) 381-1977w

module header A structure attached to or integral to the top of a module's frame that is used for structural performance, marking, and optional component-mounting. (C/BA) 1101.4-1993, 1101.7-1995

module input overrange constraints (nuclear power generating station) The upper and/or lower values of the input signal which may be applied to a module without causing damage or otherwise altering permanent characteristics of the module or causing undesired saturation effects. (PE/NP) 381-1977w

module input signal range (nuclear power generating station) The region between the limits within which a quantity is measured or received, expressed by stating the lower and upper values of the input signal. (PE/NP) 381-1977w

module interface All aspects of the electrical, fiber optic, protocol, mechanical, and thermal interfaces of a module to associated modules, the backplane, I/O connections, module cage and cabinet mounting, conduction and/or convection cooling, and power and ground. (C/BA) 14536-1995

module interface plane An assigned plane on the bottom surface of the connector, from which the connector's electrical pins protrude, forming the mating surface. This surface is used as a reference for module dimensions. (C/BA) 1101.4-1993

module interference plane An assigned plane on the bottom surface of the connector from which the connector's electrical pins protrude, forming the mating surface. (C/BA) 1101.7-1995

module isolation characteristics (nuclear power generating station) provisions for electrical isolation of particular sections of a module from each other; such as input and output circuitry, control and protection circuitry, and redundant protection circuitry. (PE/NP) 381-1977w

module load capability (nuclear power generating station) The range of load values within which a module will perform to its specified performance characteristics. (PE/NP) 381-1977w

module output impedance (nuclear power generating station) The internal impedance presented by a module at its output terminals to a load. (PE/NP) 381-1977w

module output ripple (nuclear power generating station) The ac component of a dc output signal harmonically related in frequency to either the supply voltage or a voltage generated within the module (for example, carrier demodulation). (PE/NP) 381-1977w

module output signal range (nuclear power generating station) The region between the limits within which a quantity is transmitted, expressed by stating the lower and upper values of the output signal. (PE/NP) 381-1977w

module pulse characteristics (nuclear power generating station) Information such as pulse duration, amplitude, rise time, decay time, separation, and shape. (PE/NP) 381-1977w

module qualified life (nuclear power generating station) The life expectancy in years (or cycles of operation, if applicable) over which the module has been demonstrated to be qualified for use, as established by type tests, analysis or other qualification method. (PE/NP) 381-1977w

module range and characteristics of adjustments (nuclear power generating station) Such information as upper and lower range-limits of calibration capability and where applicable, their relationship to the calibrated range of the module. (PE/NP) 381-1977w

module reference operating conditions (nuclear power generating station) The range of environmental operating conditions of a module within which environmental influences are negligible. (PE/NP) 381-1977w

module reproducibility (nuclear power generating station) The closeness of agreement among repeated measurements of the output for the same value of input made under the same operating conditions over a period of time, approaching from both directions. (PE/NP) 381-1977w

module response time (nuclear power generating station.) The time required for an output change from an initial value to a specified percentage of the final steady-state value, resulting from the application of a specified input change under specified conditions. For digital equipment (that is, relays, solid state logic, delay networks, etc.). Response time is the time required for a change from an initial state to a specified final state resulting from application of specified input under specified conditions. (PE/NP) 381-1977w

module retainer The device used to secure the module in the chassis and to hold module guide ribs against the chassis webs to attain a good thermal interface. *Synonym:* expansion element. (C/BA) 1101.4-1993, 1101.3-1993, 1101.7-1995

modules (electric pipe heating systems) Any assembly of interconnected components that constitutes an identifiable device, instrument, or piece of equipment that can be disconnected, removed as a unit and replaced with a spare, and has definable performance characteristics which permit it to be tested as a unit. A module can be a card or other subassembly. (PE/EDPG) 622A-1984r

module segment connector (FASTBUS acquisition and control) The standard connector that mounts on a FASTBUS module and mates with the crate segment connector for connection of the module to the segment. (NID) 960-1993

module signal to noise rato (nuclear power generating station) The output signal with input signal applied minus the output signal with no input signal applied divided by the output signal with no input signal applied. (PE/NP) 381-1977w

Module Status register A status register that is recommended to be implemented in the MTM-Bus interface circuitry of every S-module. The bits in this register are defined by the manufacturer of the MTM-Bus interface circuitry of an S-bits. The bits of such a register may serve to record error-condition detection or the module's application-related status. (TT/C) 1149.5-1995

module strength *See:* cohesion.

module supplementary board Any board in a FASTBUS module that does not make direct connection with the crate segment. (NID) 960-1993

module testing *See:* component testing.

module-type tests (nuclear power generating station) Tests made on one or more production units to demonstrate that the performance characteristics of the module(s) conform to the module's specifications. (PE/NP) 381-1977w

modulo An arithmetic operation that yields the remainder of an integer division problem. For example 39-3 modulo 6. (C) 1084-1986w

modulo *N* check (data transmission) A form of check digits, such that the number of ones in each number *A* operated upon is compared with a check number *B*, carried along with *A* and equal to the remainder of *A* when divided by *N*; for example, in a modulo 4 check, the check number will be 0, 1, 2, or 3 and the remainder of *A* when divided by 4 must equal the reported check number *B*, or else an error or malfunction has occurred; a method of verification by congruences; for example, casting out nines. *See also:* residue check. (COM) [49]

modulo-n counter A counter in which the state represented reverts to zero after reaching a maximum value of n-1. (C) 610.10-1994w

modulo-*n* residue The remainder obtained by dividing a number by *n*. (C) 1084-1986w

modulo-two sum *See:* exclusive OR.

modulus (1) (mathematics of computing) The number of integers that can be represented in a numeration system. For example, in a system with a modulus of five, the only integers that can be represented are 0, 1, 2, 3, and 4. (C) 1084-1986w
(2) (phasor) Its absolute value. The modulus of a phasor is sometimes called its amplitude. (Std100) 270-1966w

MOE *See:* measure of effectiveness.

MOF *See:* maximum observed frequency.

Moho (mohorovicic discontinuity) Seismic discontinuity situated about 35 km below the continents and about 10 km below the oceans. Crudely speaking, it separates the earth's crust and mantle. (COM) 365-1974w

moiré (television) The spurious pattern in the reproduced picture resulting from interference beats between two sets of periodic structures in the image. *Note:* The most common cause of moiré is the interference between scanning lines and some other periodic structure such as a line pattern in the original scene, a mesh or dot pattern in the camera sensor (for example, the target mesh in an image orthicon), or the phosphor dots or other structure in a shadow-mask picture tube. Moiré may result from the interference between the subcarrier elements of the chrominance signal and another periodic structure. In systems using an fm carrier, such as magnetic or video-disc record-playback systems, moire may also be caused by interference between the upper sidebands of the fm carrier and lower sidebands of harmonics of the fm carrier. In general, moiré may be caused by interference beats between any two periodic structures that are not perfectly aligned and not of the same frequency. (BT/AV) 201-1979w

moisture barrier A metal barrier that prevents moisture from permeating radially into the cable core. (PE/IC) 1142-1995

moisture block A means for preventing moisture from migrating longitudinally along the cable core, either through the conductor or within the space allowable between the extruded insulation shield and the jacket. (PE/IC) 1142-1995

moisture content The amount of water in parts per million by volume (ppmv) that is in the gaseous state and mixed with the insulating gas. *Synonym:* moisture content. (SWG/PE/SUB) C37.100-1992, C37.122.1-1993, C37.122-1993

moisture-resistant (1) (packaging machinery) So constructed or treated that exposure to a moist atmosphere will not readily cause damage. (IA/PKG) 333-1980w
(2) Not readily injured by exposure to a moist atmosphere. (SWG/PE) C37.100-1992

molded-case circuit breaker (MCCB) (1) One that is assembled as an integral unit in a supporting and enclosing housing of molded insulating material. (SWG/PE) C37.100-1992
(2) A circuit breaker assembled as an integral unit in a supporting and enclosing housing of insulating material; the overcurrent and tripping means being of the thermal type, the magnetic type, the electronic type, or a combination thereof. (IA/MT) 45-1998
(3) A circuit breaker that is assembled as an integral unit in a supporting and enclosing housing of insulating material. (IA/PSP) 1015-1997

mole (metric practice) The amount of substance of a system which contains as many elementary entities as there are atoms in 0.012 kilogram of carbon-12 (adopted by 14th General Conference on Weights and Measures). *Note:* When the mole is used, the elementary entities must be specified and may be atoms, molecules, ions, electrons, other particles, or specified groups of such particles. (QUL) 268-1982s

molecular data element *See:* composite data element.

momentary When used as a modifier to quantify the duration of a short duration variation, refers to a time range at the power frequency from 30 cycles to 3 s. (SCC22) 1346-1998

momentary current The current flowing in a device, an assembly, or a bus at the major peak of the maximum cycle as determined from the envelope of the current wave. *Note:* The current is expressed as the rms value, including the dc component, and may be determined by the method shown in IEEE Std C37.09-1979. (SWG/PE) C37.100-1992

momentary disturbance A variation in the level of the steady-state supply voltage that results from surges, sags, faults, circuit and equipment switching, or from the operation of circuit breakers or reclosers resulting from their response to abnormal circuit conditions. *See also:* transient. (T&D/PE) 1250-1995

momentary event interruption An interruption of duration limited to the period required to restore service by an interrupting device. *Note:* Such switching operations must be completed in a specified time not to exceed 5 min. This definition includes all reclosing operations that occur within 5 min of the first interruption. For example, if a recloser or breaker operates two, three, or four times and then holds, the event shall be considered one momentary interruption event. (PE/T&D) 1366-1998

momentary interruption (1) (electric power system) An interruption of duration limited to the period required to restore service by automatic or supervisory-controlled switching operations or by manual switching at locations where an operator is immediately available. *Note:* Such switching operations must be completed in a specified time not to exceed 5 minutes. (PE/PSE) 346-1973w
(2) Single operation of an interrupting device which results in a voltage zero. For example, two breaker or recloser operations equals two momentary interruptions. (PE/T&D) 1366-1998
(3) A type of short duration variation. The complete loss of voltage (<0.1 pu) on one or more phase conductors for a time period between 0.5 cycles and 3 s. (SCC22) 1346-1998
(4) (A) (power quality monitoring) A type of short duration variation. **(B) (power quality monitoring)** The complete loss of voltage (< 0.1 pu) on one or more phase conductors for a time period between 0.5 cycles and 3 s. (IA/PSE) 1100-1999

momentary rating (x-ray) A rating based on an operating interval that does not exceed five seconds. (NESC/NEC) [86]

monadic (mathematics of computing) Pertaining to an operation involving a single operand. *Contrast:* dyadic. (C) 1084-1986w

monadic Boolean operation A logical operation involving one operand. For example, the NOT operation. *Contrast:* dyadic Boolean operation. (C) 1084-1986w

monadic operation An operation involving one operand. For example, the square root operation. *Synonym:* unary operation. *Contrast:* dyadic operation. (C) 1084-1986w

monadic operator An operator that specifies an operation on one operand. For example, the square root operator. *Synonym:* unary operator. *Contrast:* dyadic operator. (C) 1084-1986w

monadic selective construct An if-then-else construct in which processing is specified for only one outcome of the branch, the other outcome resulting in skipping this processing. *Contrast:* dyadic selective construct. (C) 610.12-1990

monarch A processor that has the responsibility for initializing a part of the system, such as a ringlet. If a system has multiple monarchs, they eventually defer to an emperor that coordinates the initialization process. (C/MM) 1596-1992

monarch processor (1) The processor selected to manage the configuration and initialization of all modules on one logical bus. *See also:* monarch; emperor processor. (C/BA) 896.3-1993w, 896.4-1993w, 10857-1994
(2) The processor that is selected to partially initialize the local-bus resources and fetch the initial boot code. (C/BA/MM) 896.2-1991w, 1212-1991s, 896.10-1997

monitor (1) (token ring access method) That function that recovers from various error situations. It is contained in each ring station; however, only the monitor in one of the stations on a ring is the active monitor at any point in time. The monitor function in all other stations on the ring is in standby mode. (LM/C) 802.5-1985s
(2) (radioactivity monitoring instrumentation) An instrument that provides a continual measurement of one or more parameters and generates a signal to record or transmit that measurement. (NI) N42.17B-1989r
(3) (software) A software tool or hardware device that operates concurrently with a system or component and supervises, records, analyzes, or verifies the operation of the system or component. *Synonym:* execution monitor. *See also:* hardware monitor; software monitor. (C) 610.12-1990, 610.10-1994w
(4) (A) A device that observes and records selected activities with a computer system for analysis. **(B)** A generic term refering to any kind of display device. (C) 610.10-1994
(5) Continual or periodic testing with comparison to observe or determine trends. (SWG/PE) C37.10-1995

monitor direction In T101, refers to transmission from the controlled station (RTU/IED) to the controlling station (master/RTU). (PE/SUB) 1379-1997

monitor functions The functions that recover from various error situations and are contained in each ring station. In normal operation only one of the stations on a ring is the active monitor at any point in time. The monitor functions in all other stations on the ring ensures that the active monitor function is being performed. (C/LM) 8802-5-1998

monitor hazard current (health care facilities) The hazard current of the line isolation monitor alone. *See also:* hazard current. (EMB) [47]

monitoring (1) (data transmission) In communication, an observation of the characteristics of transmitted signals. (PE) 599-1985w
(2) (electric pipe heating systems) To check the operation and performance of an equipment or system by sampling the results of the operation. Monitoring with respect to electric pipe heating systems usually consists of checking system temperatures or operation of the heater circuits; voltage, current, etc. (PE/EDPG) 622A-1984r
(3) (electric heat tracing systems) To check the operation and performance of an equipment or system by sampling the results of the operation. Monitoring with respect to electric heat tracing systems usually consists of checking system temperatures or operation of the heater circuits; voltage, current, etc. (PE/EDPG) 622B-1988r
(4) The process of observing a system to verify that its parameters are within prescribed limits. (PE/EM) 1129-1992r
(5) That aspect of performance management concerned with tracking the system activities in order to gather the appropriate data for determining performance. (LM/C) 802.1F-1993r

monitoring amplifier (electroacoustics) An amplifier used primarily for evaluation and supervision of a program. *See also:* amplifier. (BT/AV) [34]

monitoring laboratory An accredited laboratory that prepares and distributes test materials to a service laboratory for the purpose of monitoring the day-to-day operation of the service laboratory. The service laboratory may also be a monitoring laboratory providing third party testing materials to another service laboratory or a special QA function within the service laboratory not involved in the routine processing of samples. (NI) N42.23-1995

monitoring relay A relay that has as its function to verify that system or control-circuit conditions conform to prescribed limits. (SWG/PE) C37.100-1992

monkey tail *See:* running board.

monochromatic (1) (color) (television) Having spectral emission over an extremely small region of the visible spectrum. (BT/AV) 201-1979w
(2) (fiber optics) Consisting of a single wavelength or color. In practice, radiation is never perfectly monochromatic but, at best, displays a narrow band of wavelengths. *See also:* spectral width; line source; coherent. (Std100) 812-1984w

monochromator (fiber optics) An instrument for isolating narrow portions of the spectrum. (Std100) 812-1984w

monochrome (television) Having only one chromaticity, usually achromatic. (BT/AV) 201-1979w

monochrome channel (television) Any path that is intended to carry the monochrome signal. (BT/AV) 201-1979w

monochrome channel bandwidth (television) The bandwidth of the path intended to carry the monochrome signal. (BT/AV) 201-1979w

monochrome display device A display device that can display only one color, or shades of that color, in addition to the background color. *Contrast:* color display device. *See also:* gray scale display device. (C) 610.10-1994w, 610.6-1991w

monochrome signal* (television) (monochrome television) A signal wave for controlling the luminance values in the picture. *See also:* luminance signal.

* Deprecated.

monochrome television The electric transmission and reception of transient visual images in only one chromaticity, usually achromatic. *Note:* Also termed black-and-white television. (BT/AV) 201-1979w

monochrome transmission (television) The transmission of a signal wave for controlling the luminance values in the picture, but not the chromaticity values. *Note:* Also termed black-and-white transmission. (BT/AV) 201-1979w

monoclinic system (piezoelectricity) A monoclinic crystal has either a single axis of twofold symmetry or a single plane of reflection symmetry, or both. Either the twofold axis or the normal to the plane of symmetry (they are the same if both exist, and this direction is called the unique axis in any case) is taken as the *b* or *Y* axis. Of the two remaining axes, the smaller is the *c* axis. In class 2, $+Y$ is chosen so that d_{22} is positive; $+Z$ is chosen parallel to c (sense trivial), and $+X$ such that it forms a right-handed system with $+Z$ and $+Y$. In class m, + Z is chosen so that d_{33} is positive, and $+X$ so that d_{11} is positive, and $+Y$ to form a right–handed system. *Note:* "Positive" and "negative" may be checked using a carbon-zinc flashlight battery. The carbon anode connection will have the same effect on meter deflection as the + end of the crystal axis upon release of compression. *See also:* crystal systems. (UFFC) 176-1978s

monocular visual field (illuminating engineering) The field for a single eye. (EEC/IE) [126]

monolithic integrated circuit An integrated circuit formed in a single piece of the substrate material. *Contrast:* hybrid circuit. (C) 610.10-1994w

monomode optical waveguide *See:* single mode optical waveguide.

monopinch Single-axis monompulse used in search radars to provide effective beam narrowing by displacing the displayed azimuth by the target angle off axis. Also called "ECS (electrical correction system)" and "EBS (electrical beam sharpening)." (AES/RS) 686-1990

monopolar ion density (dc electric-field strength and ion-related quantities) The number of ions of a given polarity per unit volume. The preferred unit is m^{-3}; another commonly used unit is cm^{-3}. (T&D/PE) 539-1990, 1227-1990r

monopolar space-charge density (overhead power lines) The space charge density of one polarity. The preferred unit is C/m^3. (T&D/PE) 539-1990, 1227-1990r

monopole An antenna, constructed above an imaging plane, that produces a radiation pattern approximating that of an electric dipole in the half-space above the imaging plane. (AP/ANT) 145-1993

monoprocessor architecture *See:* single processor architecture.

monopulse (1) Simultaneous lobing whereby direction-finding information is obtainable from a single pulse. (AP/ANT) 145-1993
(2) A radar technique in which information concerning the angular location of a target is obtained by comparison of signals received in two or more simultaneous antenna beams. *Notes:* 1. The simultaneity of the beams makes it possible to obtain a 2-D angle estimate from a single pulse (hence the term "monopulse"), although multiple pulses are usually employed to improve the accuracy of the estimate or to provide Doppler resolution. 2. The simultaneous lobe technique used in continuous wave (CW) radars is also referred to as monopulse, although pulses are not used. (AES) 686-1997

monostable Pertaining to a circuit or device that is capable of assuming one of two states, one of which is stable. *Synonym:* one-shot. *See also:* monostable circuit; bistable. (C) 610.10-1994w

monostable circuit A trigger circuit that has one stable and one quasistable state. (C) 610.10-1994w

monostatic cross section (1) The scattering cross section in the direction toward the source. *Synonym:* backscattering cross section. *Contrast:* bistatic cross section. (AP/ANT) 145-1993
(2) The scattering cross-section of a target in the retro-direction. (AP/PROP) 211-1997

monostatic radar (1) A radar where the transmit and receive antennas are collocated. (AP/PROP) 211-1997
(2) A radar system that transmits and receives through either a common antenna or through collocated antennas. (AES) 686-1997

monostatic reflectivity Reflectivity in which the reflected and incident waves follow the same path but in opposite directions. The transmit and receive antennas are in the same location, and normal incidence occurs at the reflecting surface. (EMC) 1128-1998

monotonic recorder A recorder that has output codes that do not decrease (increase) for a uniformly increasing (decreasing) input signal, disregarding random noise. (IM/WM&A) 1057-1994w

Monte Carlo method (1) In modeling and simulation, any method that employs Monte Carlo simulation to determine estimates for unknown values in a deterministic problem. (C) 610.3-1989w, 1084-1986w
(2) A numerical statistical technique that simulates random propagation and scattering processes by repeatedly calculating outcomes using parameters chosen at random from the processes' parameter space. (AP/PROP) 211-1997

Monte Carlo simulation A deterministic simulation in which random statistical sampling techniques are employed such that the result determines estimates for unknown values. (C) 610.3-1989w

month A service observing month, which is not generally a calendar month. *See also:* time-consistent traffic measures. (COM/TA) 973-1990w

month-end processing The operations required to complete a monthly cycle. For example, monthly ledger processing. (C) 610.2-1987

monthly cycle One complete execution of a data processing function that must be performed once a month. *See also:* daily cycle; annual cycle; weekly cycle. (C) 610.2-1987

monthly peak duration curve (power operations) A curve showing the total number of days within the month during which the net 60 min clock-hour integrated peak demand equals or exceeds the percent of monthly peak values shown. *See also:* generating station. (PE/PSE) 858-1987s, 346-1973w

monument *See:* hub.

MOP *See:* measure of performance.

mortising *See:* kerning.

MOS *See:* metal-oxide semiconductor.

MOSA *See:* metal-oxide surge arrester.

MOS transistor *See:* metal-oxide-semiconductor transistor.

most significant Within a group of data items (e.g., bits or bytes) that, taken as a whole, represents a numerical value, the item within the group with the greatest numerical weighting. (C/BA) 1275-1994

most significant bit (1) (mathematics of computing) The bit having the greatest effect on the value of a binary numeral; usually the leftmost bit. (TT/C) 1149.5-1995, 1084-1986w
(2) (test access port and boundary-scan architecture) The digit in a binary number representing the greatest numerical value. For shift-registers, the bit furthest from the serial output, or the last bit to be shifted out. Logic values expressed in binary form are shown with their most significant bit on the left. (TT/C) 1149.1-1990

(3) In an n bit binary word its contribution is (0 or 1 times 2^{n-1}) toward the maximum word value of 2^{n-1}. (SWG/PE/SUB) C37.100-1992, C37.1-1987s
(4) The bit in the binary notation of a number that is the coefficient of the highest exponent possible. (IM/ST) 1451.2-1997

most significant character The character in the leftmost position in a character string. *Contrast:* least significant character. (C) 610.5-1990w

most significant digit The digit having the greatest effect on the value of a numeral; usually the leftmost digit; for example, the 7 in 756.4. *Contrast:* least significant digit. (C) 610.5-1990w, 1084-1986w

most significant word (msw) In a multiword representation of a binary number, the word containing the msb of that number. (TT/C) 1149.5-1995

motherboard The printed circuit board on which an SBus Card is mounted through the connectors specified by this standard. (C/BA) 1496-1993w
(2) **(A)** The main circuit board within a computer, bearing the primary components of a computer system, including the processor, main storage, support circuitry, bus controller and bus connector. *See also:* backplane; daughter board. **(B)** A standard size printed circuit board to which are attached one or more daughterboards that add functionality and provide a selection of interface buffering. (C) 610.10-1994

motion To move the pointer while a mouse button is pressed. (C) 1295-1993w

motion picture display *See:* cine-oriented image.

MOTIS The set of ISO standards for Message-Oriented Text Interchange Systems. (C/PA) 1224.1-1993w

MOTS *See:* modified-off-the-shelf.

motive power (valve actuators) The electric, fluid, air, nitrogen, or mechanical energy required to operate the actuator. (PE/NP) 382-1985

motor A rotating machine that converts electrical energy into mechanical energy. As used in IEEE Std 1068-1996, the term can also be used to mean a generator. (IA/PC) 1068-1996

motor branch circuit A branch circuit that supplies energy to one or more motors and associated motor controllers. (IA/MT) 45-1998

motor-circuit switch (1) A switch, rated in horsepower, capable of interrupting the maximum operating overload current of a motor of the same horsepower rating as the switch at the rated voltage. (NESC/NEC) [86]
(2) (packaging machinery) A switch intended for use in a motor branch circuit. It is rated in horsepower and is capable of interrupting the maximum operating overload current of a motor of the same rating at the rated voltage. (IA/PKG) 333-1980w

motor conduit box (packaging machinery) An enclosure on a motor for the purpose of terminating a conduit run and joining motor to power conductors. (IA/PKG) 333-1980w

motor control center (MCC) (1) (nuclear power generating station) A floor mounted assembly of one or more enclosed vertical sections having a common horizontal power bus and principally containing combination motor starting units. These units are mounted one above the other in the vertical sections. The sections may incorporate vertical buses connected to the common power bus, thus extending the common power supply to the individual units. Units may also connect directly to the common power bus by suitable connections. (PE/NP) 649-1980s
(2) A group of devices assembled for the purpose of switching and protecting a number of load circuits. The control center may contain transformers, contactors, circuit breakers, protective devices, and other devices intended primarily for energizing or de-energizing load circuits. (IA/MT) 45-1998

motor control circuit The circuit of a control apparatus or system that carries the electric signals directing the performance of the controller but does not carry the main power current. Motor control circuits tapped from the load side of the motor branch-circuits, short-circuit protective devices shall not be considered to be branch circuits and shall be permitted to be protected by either supplementary or branch-circuit overcurrent protective devices. (NESC/NEC) [86]

motor-generator set A machine that consists of one or more motors mechanically coupled to one or more generators to convert electric power from one frequency to another, or to create an isolated power source. (IA/MT) 45-1998

motor home A vehicular unit designed to provide temporary living quarters for recreational, camping or travel use built on or permanently attached to a self-propelled vehicle chassis or on a chassis cab or van which is an integral part of the completed vehicle. *See also:* recreational vehicle. (NESC/NEC) [86]

motoring An induction or synchronous generator operating as a motor and drawing power from the grid. (PE/EDPG) 1020-1988r

motor lead extension cable (electric submersible pump cable) Three-conductor cable running from above the pump to the motor including motor connecting plug. (IA/PC) 1017-1985s

motor meter A meter comprising a rotor, one or more stators, and a retarding element by which the resultant speed of the rotor is made proportional to the quantity being integrated (for example, power or current) and a register connected to the rotor by suitable gearing so as to count the revolutions of the rotor in terms of the accumulated integral (for example, energy or charge). *See also:* electricity meter. (ELM) C12.1-1982s

motor parts (electric) A term applied to a set of parts of an electric motor. Rotor shaft, conventional stator-frame (or shell), end shields, or bearings may not be included, depending on the requirements of the end product into which the motor parts are to be assembled. (PE) [9]

motor reduction unit A motor with an integral mechanical means of obtaining a speed different from the speed of the motor. *Note:* Motor reduction units are usually designed to obtain a speed lower than that of the motor, but may also be built to obtain a speed higher than that of the motor. (IA/MT) 45-1998

motor starting reactor A current limiting reactor used to limit the starting current of a machine. (PE/TR) C57.16-1996

motor supply line (rotating machinery) The source of electric power to which the windings of a motor are connected. *See also:* direct-current commutating machine; asynchronous machine. (PE) [9]

motor synchronizing Synchronizing by means of applying excitation to a machine running at slightly below synchronous speed. *See also:* asynchronous machine. (PE) [9]

motor-type watthour meter A motor in which the speed of the rotor is proportional to the power, with a readout device that counts the revolutions of the rotor. (ELM) C12.1-1982s

mount (1) (switching tubes) The flange or other means by which the tube, or tube and cavity, are connected to a waveguide. *See also:* gas tube. (ED) 161-1971w, [45]
(2) **(A)** To place a data medium in a position and condition so that it can be accessed; for example, to mount a magnetic tape on a tape drive and connect the tape drive to an application. **(B)** To insert a removable storage medium into place so that it can be accessed. (C) 610.10-1994
(3) The action of making a cartridge, side, partition, or volume accessible.

a) Mounting a *volume* is a logical action that implies mounting the one or more partitions that make up the volume.
b) Mounting a *partition* is a logical action that implies mounting the underlying side of a cartridge and engaging a software interface to gain access to that partition.
c) Mounting a *side* is the physical action of mounting a cartridge in a drive in a particular orientation.

d) Mounting a *cartridge* is the physical action of moving a cartridge to a drive and loading it into the drive.

(C/SS) 1244.1-2000

mounted plow (static or vibratory plows) (cable plowing) A unit which, to be operable, is semipermanently attached to and dependent upon a prime mover.

mounting lug, stator *See:* stator mounting lug.

mounting pitch (mp) (1) The interval of distance between repeated features, parts, or assemblies in a given space. (C/BA) 1301.2-1993
(2) The interval of distance between parts or assemblies in a given space. (C/MM) 1301.3-1992r
(3) The pitch used to arrange parts or assemblies in a given space. (C/MM) 1301.1-1991

mounting position (of a switch or fuse support) A position determined by, and corresponding to, the position of the base of the device. *Note:* The usual positions are

a) Vertical
b) Horizontal upright (when the fuse holder or fuse unit is mounted above the supporting insulators)
c) Horizontal underhung (when the fuse holder or fuse unit is mounted below the supporting insulators)
d) Angle (from vertical)

(SWG/PE) C37.40-1993, C37.100-1992

mounting rabbet (1) (packaging machinery) Any channel for holding wires, cables, or bus bars; designed expressly for, and used solely for, this purpose. (IA/PKG) 333-1980w
(2) (rotating machinery) A male or female pilot on a face or flange type of end shield of a machine, used for mounting the machine with a mating rabbet. The rabbet may be circular, or of other configuration and need not be continuous. (PE) [9]

mounting ring (rotating machinery) A ring of resilient or nonresilient material used for mounting an electric machine into a base at the end shield hub. (PE) [9]

mounting structure A structure for mounting an insulating support. (PE/SUB) 605-1998

mount point Either the root directory or a directory for which the *st_dev* field of the POSIX.1 *struct stat* differs from that of its parent directory. (C/PA) 9945-2-1993

mouse (1) A cursor control device, used as a locator, consisting of a hand-held control box within some sort of motion-sensing component such that the position or movement of the mouse on a surface controls the position of a cursor on a display. It is used to provide coordinate input data to a display device. *See also:* puck. (C) 610.6-1991w
(2) An electronic or mechanical input device that allows manipulation of the screen pointer. Mice also generally have one or more buttons that allow selection and manipulation of objects. (C) 1295-1993w
(3) A cursor control device used as a locator, consisting of a hand-held control box with some sort of motion-sensing component such that the position or movement of the mouse on a surface controls the motion of a cursor on a display device. *Note:* A mouse usually includes one or more buttons which provide additional input information. *See also:* mechanical mouse; shaft recorder; track ball; puck; bus mouse; optical mouse; serial mouse. (C) 610.10-1994w

mouse port A port used to interface with a mouse.

(C) 610.10-1994w

mouth reference point (MRP) (1) A point on the axis of the artificial mouth, 25 mm in front of the center of the external plane of the lip ring. (COM/TA) 269-1992
(2) The point on the reference axis, 25 mm in front of the lip plane. (COM/TA) 1206-1994
(3) The point on the reference axis of the mouth simulator, 25 mm in front of the lip plane. (COM/TA) 1329-1999

mouth simulator A device consisting of a loudspeaker mounted in an enclosure and having a directivity and radiation pattern similar to those of the average human mouth.

(COM/TA) 1329-1999

***m*-out-of-*n* code** A binary code in which *m* of the *n* digits that represent a word, character, or digit are in one state, and the other digits are in the opposite state. *See also:* two-out-of-five code. (C) 1084-1986w

movable bridge coupler (drawbridge coupler) A device for engaging and disengaging signal or interlocking connections between the shore and a movable bridge span.

(EEC/PE) [119]

movable bridge rail lock (drawbridge)A mechanical device used to insure that the movable bridge rails are in proper position for the movement of trains. *See also:* interlocking.

(EEC/PE) [119]

movable head *See:* floating head.

movable relay contact The member of a contact pair that is moved directly by the actuating system. (EEC/REE) [87]

move (1) (A) To read data from a source, altering the contents of the source location, and to write the same data elsewhere in a physical form that may differ from that of the source. For example, to move data from one file to another. *Contrast:* copy. **(B)** Sometimes, a synonym for **copy.** *See also:* store; load; fetch. (C) 610.12-1990
(2) The physical action of moving a cartridge from one location to another, where a location is a slot, a drive, or a port. (C/SS) 1244.1-2000

movement authority The authority for a train to enter and travel through a specific section of track, in a given travel direction. Movement authorities are assigned, supervised, and enforced by a communications-based train control system to maintain safe train separation, and to provide protection through interlockings. (VT/RT) 1474.1-1999

moving average An average calculated on a selected, changing subset of a time series of data. For example, a four-point moving average would be the average of the last four data points in the time series. (C) 1084-1986w

moving-base navigation aid An aid that requires cooperative facilities located upon a moving vehicle other than the one being navigated. *Notes:* 1. The cooperative facilities may move along a predictable path that is referenced to a specified coordinate system such as in the case of a nongeostationary navigation satellite. 2. Such an aid may also be designed solely to permit one moving vehicle to home upon another. *See also:* navigation. (AES/RS) 686-1982s, [42]

moving-base-derived navigation data Data obtained from measurements made at moving cooperative facilities located external to the navigated vehicle. *See also:* navigation.

(AES/RS) 686-1982s, [42]

moving-base-referenced navigation data Data in terms of a coordinate system referenced to a moving vehicle other than the one being navigated. *See also:* navigation.

(AES/RS) 686-1982s, [42]

moving-coil loudspeaker (dynamic loudspeaker) A moving-conductor loudspeaker in which the moving conductor is in the form of a coil conductively connected to the source of electric energy. (PE/EEC) [119]

moving-coil microphone (dynamic microphone) A moving-conductor microphone in which the movable conductor is in the form of a coil. *See also:* microphone. (EEC/PE) [119]

moving-conductor loudspeaker (moving conductor) A loudspeaker in which the mechanical forces result from magnetic reactions between the field of the current and a steady magnetic field. (EEC/PE) [119]

moving-conductor microphone A microphone the electric output of which results from the motion of a conductor in a magnetic field. *See also:* microphone. (EEC/PE) [119]

moving contact A conducting part that bears a contact surface arranged for movement to and from the stationary contact.

(SWG/PE) C37.100-1992

moving-contact assembly (rotating machinery) That part of the starting switch assembly that is actuated by the centrifugal mechanism. *See also:* centrifugal starting switch. (EEC/PE) [119]

moving element (instrument) Those parts that move as a direct result of a variation in the quantity that the instrument is measuring. *Notes:* 1. The weight of the moving element includes one-half the weight of the springs, if any. 2. The use of the term movement is deprecated. (EEC/ERI/AII) [111], [102]

moving-iron instrument An instrument that depends for its operation on the reactions resulting from the current in one or more fixed coils acting upon one or more pieces of soft iron or magnetically similar material at least one of which is movable. *Note:* Various forms of this instrument (plunger, vane, repulsion, attraction, repulsion-attraction) are distinguished chiefly by mechanical features of construction. *See also:* instrument. (EEC/PE) [119]

moving-magnet instrument An instrument that depends for its operation on the action of a movable permanent magnet in aligning itself in the resultant field produced either by another permanent magnet and by an adjacent coil or coils carrying current, or by two or more current-carrying coils, the axes of which are displaced by a fixed angle. *See also:* instrument. (EEC/PE) [119]

moving-magnet magnetometer A magnetometer that depends for its operation on the torques acting on a system of one or more permanent magnets that can turn in the field to be measured. *Note:* Some types involve the use of auxiliary magnets (Gaussian magnetometer), others electric coils (sine or tangent galvanometer). *See also:* magnetometer. (EEC/PE) [119]

moving target detector (MTD) A low pulse-repetition frequency (PRF) pulsed Doppler system usually characterized by employing a filter bank, adaptive thresholds, clutter map, and more than one coherent processing interval at different PRFs. (AES) 686-1997

moving-target indicator improvement factor *See:* moving-target indication improvement factor.

moving-target indication (MTI) A technique that enhances the detection and display of moving radar targets by suppressing fixed targets. *Note:* Doppler processing is one method of implementation. (AES) 686-1997

moving-target indication improvement factor The signal-to-clutter power ratio at the output of the clutter filter divided by the signal-to-clutter power ratio at the input of the clutter filter, averaged uniformly over all target radial velocities of interest. *Synonym:* clutter improvement factor. (AES) 686-1997

MP A dialect of ALGOL 60 having extensible language features; used largely as a programming language for system software. (C) 610.13-1993w

mp *See:* mounting pitch.

MPE *See:* minimum perceptible erythema.

M peak (closed loop) **(1) (control system feedback)** The maximum value of the magnitude of the return transfer function for real frequencies, the value at zero frequency being normalized to unity. *See also:* feedback control system. (PE/EDPG) [3]

(2) (excitation systems) Mp is the maximum value of the closed-loop amplitude response. *See also:* bandwidth. (PE/EDPG) 421A-1978s

m-phase circuit A polyphase circuit consisting of m distinct phase conductors, with or without the addition of a neutral conductor. *Note:* In this definition it is understood that m may be assigned the integral value of three or more. For a two-phase circuit, see: two-phase circuit; two-phase, three-wire circuit; two-phase, four-wire circuit; two-phase, five-wire circuit. *See also:* network analysis. (Std100) 270-1966w

MPRF *See:* medium-pulse-repetition-frequency waveform.

MPSX *See:* Mathematical Programming System Extended.

MR *See:* modification request.

MRA *See:* minimum reception altitude.

MRP *See:* mouth reference point.

msb *See:* most significant bit.

MSB *See:* most significant bit.

MSC *See:* mile of standard cable; multistrip coupler.

M scan *See:* M-display.

M-scope A cathode-ray oscilloscope arranged to present an M-display. (AES/RS) 686-1990

MSDU *See:* MAC service data unit.

MSE *See:* maximum static error.

MSFN *See:* manned space flight network.

MSI *See:* medium scale integration.

msw *See:* most significant word.

MT *See:* machine translation.

M-T *See:* magneto-telluric.

MTBF (meantime between failure, mean time between failure) *See:* mean time before failures; mean time between failures.

MTE *See:* manual test equipment.

MTF *See:* mean time between failures.

m^3/s (cubic meters per second) Volume of water or liquid discharged per second under standard conditions. (T&D/PE) 957-1987s, 957-1995

MTI *See:* moving-target indication.

MTI improvement factor *See:* moving-target indication improvement factor.

MT interface The X.400 Gateway API. (C/PA) 1224.1-1993w

MTM-Bus A serial, backplane, test and maintenance bus, consisting of one or more logic boards, that can be used to integrate modules from different design teams or vendors into testable and maintainable subsystems, as specified in this Standard. (TT/C) 1149.5-1995

MTM-Bus interface logic The portion of a module that is designed for the purpose of MTM-Bus-compliant communication and through which takes place all the communication between the given module and any other on a given MTM-Bus implementation. MTM-Bus interface logic need not be defined on the basis of physical package boundaries. (TT/C) 1149.5-1995

MTM-Bus Master The module in control of the MTM-Bus. This is the module that, at a given time, is sourcing MCTL and MMD. (TT/C) 1149.5-1995

MTM-Bus mastership Property of being the current MTM-Bus Master module. (TT/C) 1149.5-1995

MTM-Bus Slave An MTM-Bus module that cannot command actions of other modules on the bus, but that may be selected by the MTM-Bus Master module to participate in a message. (TT/C) 1149.5-1995

MTTD *See:* mean time to diagnosis.

MTTF *See:* mean time to fix; mean time to repair.

MTTR *See:* mean time to repair.

MTU *See:* maximum transfer unit.

M204 *See:* MODEL 204.

M-type backward-wave oscillator *See:* carcinotron.

mu circuit (feedback amplifier) (μ circuit) That part that amplifies the vector sum of the input signal and the fed-back portion of the output signal in order to generate the output signal. *See also:* feedback. (EEC/PE) [119]

Mueller matrix The matrix that relates the Stokes vector of the scattered wave to the Stokes vector of the incident wave. *Synonym:* Stokes matrix. (AP/PROP) 211-1997

MUF *See:* maximum usable frequency.

mu factor (*n*−terminal electron tubes) (μ factor) The ratio of the magnitude of infinitesimal change in the voltage at the *j*th electrode to the magnitude of an infinitesimal change in the voltage at the l th electrode under the conditions that the current to the m th electrode remain unchanged and the voltages of all other electrodes be maintained constant. *See also:* electron-tube admittances. (ED) 161-1971w, [45]

muffler (of a fuse) An attachment for the vent of a fuse, or a vented fuse, that confines the arc and substantially reduces the venting from the fuse. (SWG/PE) C37.100-1992

multi-access contention protocol The protocol used on the Aloha network by which a station transmits a message at will and then listens for an acknowledgment. If no ACK is received within a randomly selected timeout interval, the message is re-transmitted. (C) 610.7-1995

multiaddress *See:* multiple-address.

multiaddress format An address format that contains more than one address field; for example, a three-address instruction. (C) 610.10-1994w

multiaddress instruction (1) A computer instruction that contains more than one address field. *Synonym:* multiple-address instruction. *Contrast:* one-address instruction. (C) 610.12-1990
(2) A computer instruction that contains more than one address. *Synonym:* multiple instruction. (C) 610.10-1994w

multianode tank (multianode tube) An electron tube having two or more main anodes and a single cathode. *Note:* This term is used chiefly for pool-cathode tubes. (ED) [45]

multiaperture core A magnetic core, usually used for non-destructive reading, with two or more holes through which wires may be passed in order to create more than one magnetic path. *Synonym:* multiple aperture core. (C) 610.10-1994w

multiband image A set of images of the same scene, each formed by radiation from a different segment of the spectrum. (C) 610.4-1990w

multibeam antenna An antenna capable of creating a family of major lobes from a single non-moving aperture, through use of a multiport feed, with one-to-one correspondence between input ports and member lobes, the latter characterized by having unique main beam pointing directions. *Note:* Often, the multiple main beam angular positions are arranged to provide complete coverage of a solid angle region of space. (AP/ANT) 145-1993

multibeam oscilloscope An oscilloscope in which the cathode-ray tube produces two or more separate electron beams that may be individually or jointly controlled. *See also:* oscillograph; dual-beam oscilloscope. (IM/HFIM) [40], 311-1970w

multibit point interface Multibit (e.g., BCD, gray code) point. Master Station or RTU (or both) element(s) that inputs a series of multibit quantities in parallel. (SUB/PE) C37.1-1994

multicable penetrator A device consisting of multiple nonmetallic cable seals assembled in a surrounding metal frame, for insertion in openings in decks, bulkheads, or equipment enclosures and through which cables may be passed to penetrate decks or bulkheads or to enter equipment without impairing their original fire or watertight integrity. (IA/MT) 45-1998

multicast (1) A transmission mode in which a single message is sent to multiple network destinations, (i.e., one-to-many). (DIS/C) 1278.1-1995, 1278.2-1995
(2) A mode of operation in which the M-module transmits data simultaneously (i.e., during a single message) to a predefined subset of the S-modules currently connected to the bus. Also, a message transmitted in this mode. (TT/C) 1149.5-1995
(3) A technique that allows copies of a single packet to be passed to a selected subset of all possible destinations. *Contrast:* broadcast. (C) 610.7-1995
(4) A medium access control (MAC) address that has the group bit set. A multicast MAC service data unit (MSDU) is one with a multicast destination address. A multicast MAC protocol data unit (MPDU) or control frame is one with a multicast receiver address. (C/LM) 8802-11-1999

multicast address A special address indicating a specific group of end nodes.local area networks. (C) 8802-12-1998

multicast select bit 0; multicast select bit 1 Those bits in the Slave Status register of every S-module by means of which the S-module is programmed to be a member of one of the four possible multicast select groups. (TT/C) 1149.5-1995

multicast select group A group of S-modules that may be addressed simultaneously in a multicast. Four such groups are possible. Each has an address defined by IEEE 1149.5-1995. The multicast select group of an S-module is programmable. (TT/C) 1149.5-1995

multicavity magnetron A magnetron in which the circuit includes a plurality of cavities. *See also:* magnetron. (ED) 161-1971w

multicellular horn (electroacoustics) A cluster of horns with juxtaposed mouths that lie in a common surface. *Note:* The purpose of the cluster is to control the directional pattern of the radiated energy. (SP) [32]

multichannel analyzer (MCA) (x-ray energy spectrometers) An instrument which digitizes analog amplitude signal pulses and stores them in a memory as a function of their analog amplitude. (NPS/NID) 759-1984r

multichannel pulse-height analyzer (MCA) An electronic device that records and stores pulses according to their height. It consists of three function segments: an ADC to provide a means of measuring pulse amplitude; memory registers (one for each channel of the spectrum) to tally the number of pulses having an amplitude within a given voltage increment; an input/output section that permits transfer of the spectral information to other devices, such as a computer, oscilloscope display, or permanent storage media (disk or magnetic tape storage). *Synonym:* multichannel analyzer. (NI) N42.14-1991

multichannel radio transmitter A radio transmitter having two or more complete radio-frequency portions capable of operating on different frequencies, either individually or simultaneously. *See also:* radio transmitter. (AP/ANT) 145-1983s

multicharacter collating element A sequence of two or more haracters that collate as an entity. For example, in some coded character sets, an accented character is represented by a (non-spacing) accent, followed by the letter. Another example is the Spanish elements "ch" and "ll." (C/PA) 9945-2-1993

multichip integrated circuit An integrated circuit whose elements are formed on or within two or more semiconductor chips that are separately attached to a substrate. *See also:* integrated circuit. (ED) 274-1966w

multicomputer *See:* multiprocessor.

multiconductor bundle *See:* bundle.

multiconstant speed motor (rotating machinery) A multi-speed motor whose two or more definite speeds are constant or substantially constant over its normal range of loads; for example A synchronous or an induction motor with windings capable of various pole groupings. (PE) [9]

multidimensional system A system whose state vector has more than one element. *See also:* control system. (PE/EDPG) [3]

multidrop (1) Said of the configuration of a bus with a single shared medium segment that allows one or more of its module connectors to be unoccupied without disturbing bus operation. (TT/C) 1149.5-1995
(2) Pertaining to a communication arrangement where several devices share a common transmission channel. *Contrast:* multipoint. *See also:* point-to-point. (C) 610.7-1995

multielectrode tube An electron tube containing more than three electrodes associated with a single electron stream. (ED) 161-1971w

multi-element conduction interval (thyristor) That part of the conduction interval when ON-state current flows in more than one basic control element simultaneously. (IA/IPC) 428-1981w

multifamily dwelling A building containing three or more dwelling units. (NESC/NEC) [86]

multifiber cable (fiber optics) An optical cable that contains two or more fibers, each of which provides a separate information channel. *See also:* optical cable assembly; fiber bundle. (Std100) 812-1984w

multifiber joint (fiber optics) An optical splice or connector designed to mate two multifiber cables, providing simultaneous optical alignment of all individual waveguides. *Note:* Optical coupling between aligned waveguides may be achieved by various techniques including proximity butting (with or without index matching materials), and the use of lenses. (Std100) 812-1984w

multifield key *See:* concatenated key.

multiframe A cyclic set of consecutive frames in which the relative position of each frame can be identified. (COM/TA) 1007-1991r

multiframe alignment *See:* frame alignment.

multiframe alignment signal *See:* frame alignment signal.

multifrequency test A broad band test motion, simulating a typical seismic motion, that can produce a simultaneous response from all applicable modes of a multidegree-of-freedom system. (SWG/PE) C37.100-1992, C37.81-1989r

multifrequency transmitter A radio transmitter capable of operating on two or more selectable frequencies, one at a time, using present adjustments of a single radio-frequency portion. *See also:* radio transmitter. (AP/BT/ANT) 145-1983s, 182-1961w

multigrounded neutral system (power and distribution transformers) A distribution system of the four-wire type where all transformer neutrals are grounded, and neutral conductors are directly grounded at frequent points along the circuit. (PE/TR) C57.12.80-1978r

multilateration The location of an object by means of two or more range measurements from different reference points. It is a useful technique with radar because of the inherent accuracy of radar range measurement. *Note:* The use of three reference points to obtain target location is known as trilateration. (AES) 686-1997

multilayer Pertaining to a printed circuit board with several layers of printed circuit etched or patterned, one over the other and interconnected by electroplated holes which can also receive component leads. (C) 610.10-1994w

multilayer filter *See:* interference filter.

multilevel address *See:* indirect address.

multilevel network subject A network subject that causes information to flow through a network at two or more security levels without risk of compromise by transmitting sensitivity labels along with the data. *Contrast:* single-level network subject. (C) 610.7-1995

multilevel security (1) (software) A mode of operation permitting data at various security levels to be concurrently stored and processed in a computer system when at least some users have neither the clearance nor the need-to-know for all data contained in the system. *See also:* data; security; computer system. (C/SE) 729-1983s
(2) The capability of simultaneously separating and protecting information of two or more security levels during processing. (C/BA) 896.3-1993w

multilevel storage *See:* virtual storage.

multiline hunt group (MLHG) A group of lines that have a fixed alternate routing should one or more of the lines in the group be busy. (SCC31/AMR) 1390.3-1999, 1390.2-1999, 1390-1995

multilist A technique for organizing records in which records that have equivalent values for a given secondary key form a linked list. *Synonym:* multiple threaded list.

Student	Name	Homeroom	Link
1	MARY	25	4
2	JOE	27	15
3	JOHN	10	6
4	ANNE	25	5
5	SUSAN	25	–
6	KIM	10	21
7	BOB	26	16
•	•	•	•

multilist

(C) 610.5-1990w

multimedia A form of hypermedia consisting of a combination of two or more forms of the following: text, audio, graphics, animation, and full-motion video. (C) 610.10-1994w

multimode distortion (fiber optics) In an optical waveguide, that distortion resulting from differential mode delay. *Note:* The term "multimode dispersion" is often used as a synonym; such usage, however, is erroneous since the mechanism is not dispersive in nature. *Synonyms:* mode distortion; intermodal distortion. *See also:* distortion. (Std100) 812-1984w

multimode fiber (interferometric fiber optic gyro) An optical fiber waveguide that will allow more than one bound mode to propagate. (AES/GYAC) 528-1994

multimode group delay *See:* differential mode delay.

multimode laser (fiber optics) A laser that produces emission in two or more transverse or longitudinal modes. *See also:* mode; laser. (Std100) 812-1984w

multi-mode optical fiber An optical fiber that has a relatively large core in which the light bounces off the walls of the core. This results in multiple signal paths through the fiber which limits the maximum signaling rate more and more as the fiber length increases. *See also:* single-mode optical fiber. (C) 610.7-1995

multimode optical waveguide (fiber optics) An optical waveguide that will allow more than one bound mode to propagate. *Note:* May be either a graded index or step index waveguide. *See also:* mode; bound mode; single mode optical waveguide; power-law index profile; multimode distortion; mode volume; step index optical waveguide; normalized frequency. (Std100) 812-1984w

multimode SAW oscillator A surface acoustic wave oscillator in which more than one frequency satisfies the oscillation conditions. (UFFC) 1037-1992w

multimode waveguide A waveguide used to propagate power in more than one mode at a frequency of interest. (MTT) 146-1980w

multinomial A linear sum of terms involving powers of more than one variable.

$$\sum_{i_1=0}^{N_1}\sum_{i_2=0}^{N_2}\cdots\sum_{i_m=0}^{N_m} A(i_1, i_2, \ldots i_m)X_1^{i_2}X_2^{i_2}\ldots X_m^{i_m}$$

(IM/ST) 1451.2-1997

multioffice exchange (telephone switching systems) A telecommunications exchange served by more than one local central office. (COM) 312-1977w

multiorder lag (automatic control) In a linear system or element, lag of energy storage in two or more separate elements of the system. *Note:* It is evidenced by a differential equation of order higher than one, or by more than one time-constant. It may sometimes be approximated by a delay followed by a first-order or second-order lag. *See also:* lag. (PE/EDPG) [3]

multioutlet assembly A type of surface or flush raceway designed to hold conductors and receptacles, assembled in the field or at the factory. (NESC/NEC) [86]

multipactor limiter (nonlinear, active, and nonreciprocal waveguide components) A high-vacuum device that uses the multipacting phenomenon to limit high microwave power levels. (MTT) 457-1982w

multiparty ringing (telephone switching systems) By custom, any arrangement that provides for the individual ringing of more than four parties. (COM) 312-1977w

multipath **(1)** The propagation of a wave from one point to another by more than one path. When multipath occurs in radar, it usually consists of a direct path and one or more indirect paths by reflection from the surface of the earth or sea or from large man-made structures. At frequencies below approximately 40 MHz, it may also include more than one path through the ionosphere. (AES) 686-1997
(2) (facsimile) *See also:* multipath transmission.

multipath error The error in a radar-observed parameter caused by multipath. (AES) 686-1997

multipath transmission The propagation phenomenon that results in signals reaching the receiving antenna by two or more paths. When two or more signals arrive simultaneously, wave interference results. The received signal fades if the wave interference is time varying or if one of the terminals is in motion. (AP/PROP) 211-1997

multiphase transducer An interdigital transducer having more than two connections and that is driven in different phases; usually used for unidirectional transducers. (UFFC) 1037-1992w

Multipeer Virtual Port A Subgroup Port that represents the capability for communication with more than one other Remote Bridge in the Remote Bridge Group to which the Port attaches. *Note:* A Multipeer Virtual Port always has Multipeer Virtual Ports as its peer Ports. (C/LM) 802.1G-1996

multiple **(1)** A group of terminals arranged to make a circuit or group of circuits accessible at a number of points at any one of which connection can be made.
(2) To connect in parallel, or to render a circuit accessible at a number of points at any one of which connection can be made. (EEC/PE) [119]
(3) (analog computer) A junction into which patch cords may be plugged to form a common connection. (C) 165-1977w

multiple access (communication satellite) The capability of having simultaneous access to one communication satellite from a number of ground stations. (COM) [19]

multiple-address (multiaddress) (computers) Pertaining to an instruction that has more than one address part. (C) 162-1963w

multiple-address instruction *See:* multiaddress instruction.

multiple aperture core *See:* multiaperture core.

multiple arithmetic A system or method of performing ordinary arithmetic with a digital computer where several parts of one or more numbers are utilized in an arithmetic operation, yielding several results. (C) 1084-1986w

multiple-beam headlamp (illuminating engineering) A headlamp so designed as to permit the driver of a vehicle to use any one of two or more distributions of light on the road. (EEC/IE) [126]

multiple-beam klystron (microwave tubes) An O-type tube having more than one electron beam, and resonators coupled laterally but not axially. (ED) [45]

multiple-break relay contacts Contacts that open a circuit in two or more places. (EEC/REE) [87]

multiple-choice interaction An instruction method employed by some computer-assisted instruction systems, in which the student is asked to choose one of a set of multiple choice answers in response to a question. (C) 610.2-1987

multiple circuit Two or more circuits connected in parallel. *See also:* center of distribution. (PE/T&D) [10]

multiple-conductor cable A combination of two or more conductors cabled together and insulated from one another and from sheath or armor where used. *Note:* Specific cables are referred to as 3-conductor cable, 7-conductor cable, 50-conductor cable, etc. (T&D/PE) [10]

multiple-conductor concentric cable A cable composed of an insulated central conductor with one or more tubular stranded conductors laid over it concentrically and insulated from one another. *Note:* This cable usually has only two or three conductors. Specific cables are referred to as 2-conductor concentric cable, 3-conductor concentric cable, etc. (T&D/PE) [10]

multiple-current generator A generator capable of producing simultaneously currents or voltages of different values, either alternating-current or direct-current. (PE) [9]

multiple data stream *See:* multiple instruction, single data; multiple instruction, multiple data.

multiple ESD event An ESD event in which more than one discharge occurs. The time interval between successive discharges may be several microseconds to several tens of milliseconds. Related terms include multiple ESD, multiple discharge, and multiple. (SPD/PE) C62.47-1992r

multiple exclusive selective construct *See:* case.

multiple feeder A feeder that is connected to a common load in multiple with one or more feeders from independent sources. (SWG/PE) C37.100-1992

multiple frame transmission A transmission where more than one frame is transmitted when a token is captured. (C/LM) 8802-5-1998

multiple-gun cathode-ray tube A cathode-ray tube containing two or more separate electron-gun systems. (ED) [45]

multiple hoistway (elevators) A hoistway for more than one elevator or dumbwaiter. *See also:* hoistway. (PE/EEC) [119]

multiple inclusive selective construct A special instance of the case construct in which two or more different values of the control expression result in the same processing. For example, values 1 and 2 cause one branch, 3 and 4 cause another, and so on. (C) 610.12-1990

multiple independent outages Outage occurrences, each having distinct and separate initiating incidents, where no outage occurrence is the consequence of any other, but the outage states overlap. (PE/PSE) 859-1987w

multiple inheritance The ability of a subclass to inherit responsibilities from more than one superclass. (C/SE) 1320.2-1998

multiple instruction *See:* multiaddress instruction.

multiple instruction, multiple data Pertaining to a computer architecture in which the processors receive both instructions and data from separate sources. *See also:* multiple instruction, single data; single instruction, multiple data. (C) 610.10-1994w

multiple instruction, single data Pertaining to a computer architecture in which all processors receive instructions from separate (multiple) sources but receive data from a single (common) source. *See also:* single instruction, single data; multiple instruction, multiple data. (C) 610.10-1994w

multiple lampholder (current tap) A device that by insertion in a lampholder, serves as more than one lampholder. (EEC/PE) [119]

multiple lightning stroke A lightning stroke having two or more components. *See also:* direct-stroke protection. (T&D/PE) [10]

multiple list insertion sort *See:* address calculation sort.

multiple metallic rectifying cell An elementary metallic rectifier having one common electrode and two or more separate electrodes of the opposite polarity. *See also:* rectification. (EEC/PE) [119]

multiple modulation A succession of processes of modulation in which the modulated wave from one process becomes the modulating wave for the next. *Note:* In designating multiple-modulation systems by their letter symbols, the processes are listed in the order in which the modulating function encounters them. For example, PPM-AM means a system in which one or more signals are used to position-modulate their respective pulse subcarriers which are spaced in time and are used to amplitude-modulate a carrier. (AP/ANT) 145-1983s

multiple outage event An outage event involving two or more components, or two or more units. (PE/PSE) 859-1987w

multiple outstanding transactions A state where more than one transaction has been issued and is pending completion. (C/MM) 1284.4-2000

multiple-packet error rejection Error handling and rejection notification occurs on a message-by-message basis. (C/MM) 1284.1-1997

multiple plug (cube tap) (plural tap) A device that, by insertion in a receptacle, serves as more than one receptacle. (PE/EEC) [119]

multiple-pointer form demand register (metering) An indicating demand register from which the demand is obtained by reading the position of the multiple pointers relative to their scale markings. The multiple pointers are resettable to zero. (ELM) C12.1-1982s

multiple precision (data management) (mathematics of computing) Pertaining to the use of two or more computer words to represent a number in order to preserve or gain precision. *Synonyms:* multiprecision; extended precision. *Contrast:* single precision. *See also:* triple precision; double precision. (C) 610.5-1990w, 1084-1986w

multiple-precision arithmetic Computer arithmetic performed with operands that are expressed in multiple-precision representation. (C) 1084-1986w

multiple punching Punching more than one hole in the same card column by several keystrokes, usually in order to extend the character set of the keypunch. (C) 610.10-1994w

multiple query error A word-serial protocol error that occurs when a servant receives a command requiring it to output a response to its data low register, and is unable to respond because of an unread response to a previous command. (C/MM) 1155-1992

multiple rectifier circuit A rectifier circuit in which two or more simple rectifier circuits are connected in such a way that their direct currents add, but their commutations do not coincide. *See also:* rectifier circuit element; rectification. (IA) [62]

multiple rho (electronic navigation) A generic term referring to navigation systems based on two or more distance measurements for determination of position. *See also:* navigation. (AES/RS) 686-1982s, [42]

multiple scatter A calculation of wave scattering from a surface or a collection of particles for which the field exciting each surface element or particle consists of the incident field plus the fields scattered by all the other surface elements or particles in their many interactions. A full multiple scattering solution is an exact solution to the problem. (AP/PROP) 211-1997

multiple-secondary current transformer (1) One that has three or more secondary windings, each on a separate magnetic circuit, with all magnetic circuits excited by the same primary winding. (PE/TR) C57.13-1993, C57.12.80-1978r **(2)** One that has two or more secondary coils each on a separate magnetic circuit with all magnetic circuits excited by the same primary winding. (PE/PSR) C37.110-1996

multiple-shot blasting unit A unit designed for firing a number of explosive charges simultaneously in mines, quarries, and tunnels. *See also:* blasting unit. (EEC/PE) [119]

multiple sound track Consists of a group of sound tracks, printed adjacently on a common base, independent in character but in a common time relationship, for example, two or more have been used for stereophonic sound recording. *See also:* phonograph pickup; multitrack recording system. (SP) [32]

multiple speed floating *See:* control system, multiple-speed floating.

multiple-speed floating control system (automatic control) A form of floating control system in which the manipulated variable may change at two or more rates each corresponding to a definite range of values of the actuating signal. (PE/EDPG) [3]

multiple spot scanning (facsimile) The method in which scanning is carried on simultaneously by two or more scanning spots, each one analyzing its fraction of the total scanned area of the subject copy. *See also:* scanning. (COM) 168-1956w

multiple street-lighting system A street-lighting system in which street lights, connected in multiple, are supplied from a low-voltage distribution system. *See also:* direct-current distribution; alternating-current distribution. (T&D/PE) [10]

multiple-supply-type ballast A ballast designed specifically to receive its power from an approximately constant-voltage supply circuit and that may be operated in multiple (parallel) with other loads supplied from the same source. *Note:* The deviation in source voltage ordinarily does not exceed plus or minus 5%, but in the case of ballasts designed for a stated input voltage range, the deviation may by greater as long as it is within the stated range. (EEC/LB) [97]

multiple switchboard (telephone switching systems) A telecommunications switchboard having each line connected to two or more jacks so that the line is within the reach of several operators. (COM) 312-1977w

multiple system (electrochemistry) The arrangement in a multielectrode electrolytic cell whereby in each cell all of the anodes are connected to the positive bus bar and all of the cathodes to the negative bus bar. *See also:* electrorefining. (EEC/PE) [119]

multiple threaded list *See:* multilist.

multiple thyristor converter A thyristor converter in which two or more simple thyristor converters are connected in such a way that their direct currents add, but their commutations do not coincide. (IA/IPC) 444-1973w

multiple-time-around echo *See:* second-time-around echo.

multiple transit signals Spurious signals having delay time related to the main signal delay by small odd integers. Specific multiple transit signals may be labeled the third transit (triple transit), fifth transit, etc. There is often a tradeoff available between multiple transit signal levels and bandwidth, delay time, insertion loss, and VSWR (dispersive and nondispersive delay lines). (UFFC) 1037-1992w, [22]

multiple transmission line A planar transmission-line configuration employing more than one parallel guiding structure, each of which could form a single planar transmission line. (MTT) 1004-1987w

multiple tube A space-charge-controlled tube or valve containing within one envelope two or more units or groups of electrodes associated with independent electron streams, through sometimes with one or more common electrodes. Examples: Double diode, double triode, triode-heptode, etc. *Synonym:* multiple valve. *See also:* multiple-unit tube. (ED) 161-1971w

multiple tube counts Spurious counts induced by previous tube counts. (NI/NPS) 309-1999

multiple-tuned antenna An antenna designed to operate, without modification, in any of a number of pre-set frequency bands. (AP/ANT) 145-1993

multiple twin quad (telephony) A quad in which the four conductors are arranged in two twisted pairs, and the two pairs twisted together. *See also:* cable. (EEC/PE) [119]

multiple unit A system of simultaneous control of all vehicles in a consist from one master control through the means of trainlines. (VT) 1475-1999

multiple-unit control (electric traction) A control system in which each motive-power unit is provided with its own controlling apparatus and arranged so that all such units operating together may be controlled from any one of a number of points on the units by means of a master controller. (EEC/PE) [119]

multiple-unit electric car An electric car arranged either for independent operation or for simultaneous operation with other similar cars (when connected to form a train of such cars) from a single control station. *Note:* A prefix diesel-electric, gas-electric, etc., may replace the word electric. *See also:* electric motor car. (EEC/PE) [119]

multiple-unit electric locomotive A locomotive composed of two or more multiple-unit electric motive-power units connected for simultaneous operation of all such units from a single control station. *Note:* A prefix diesel-electric, turbine-electric, etc., may replace the word electric. *See also:* electric locomotive. (EEC/PE) [119]

multiple-unit electric motive-power unit An electric motive-power unit arranged either for independent operation or for simultaneous operation with other similar units (when connected to form a single locomotive) from a single control station. *Note:* A prefix diesel-electric, gas-electric, turbine-electric, etc., may replace the word electric. *See also:* electric locomotive. (EEC/PE) [119]

multiple-unit electric train A train composed of multiple-unit electric cars. *See also:* electric motor car. (EEC/PE) [119]

multiple-unit tube *See:* multiple tube.

multiple valve *See:* multiple tube.

multiple-valve unit (MVU) A single structure comprising more than one valve. (SUB/PE) 857-1996

multiplex To interleave or simultaneously transmit two or more messages on a signal channel. (C/PE) 610.10-1994w, 599-1985w

multiplex equipment, asynchronous A transmission interconnection device that interleaves nonsynchronous low bit-rate digital signals to form a single high bit-rate digital signal. It also performs the reverse function of dividing a high bit-rate digital signal into multiple nonsynchronous low bit-rate signals. The two processes are referred to in this document as multiplexing (combining signals) and demultiplexing (separating signals). Similarly, the mechanisms used to perform these functions are referred to as multiplex equipment. (COM/TA) 1007-1991r

multiplex equipment, digital The equipment for combining digital signals from one digital level to a higher digital level. (COM/TA) 1007-1991r

multiplex equipment, primary The equipment for combining analog (vf) signals, or digital data signals, to a primary rate digital signal and vice versa. (COM/TA) 1007-1991r

multiplexer (A) (supervisory control, data acquisition, and automatic control) A device that allows the interleaving of two or more signals to a single line or terminal. **(B) (supervisory control, data acquisition, and automatic control)** A device for selecting one of a number of inputs and switching its information to the output. (SWG/PE/SUB) C37.1-1987, C37.100-1992

(2) (A) A device that allows the transmission of a number of different signals simultaneously over a single channel or transmission facility. *Synonym:* multiplexor. **(B)** A device capable of interleaving the events of two or more activities or of distributing the events of an interleaved sequence to their respective activities. *Contrast:* demultiplexer. (C) 610.7-1995

multiplexing (1) (modulation systems) (data transmission) The combining of two or more signals into a single wave (the multiplex wave) from which the signals can be individually recovered. (PE) 599-1985w

(2) The division of a transmission facility into two or more channels, either by splitting the frequency band transmitted by the channel into narrower bands, each of which is used to constitute a distinct channel (frequency division multiplexing) or by allotting this common channel to several different information channels one at a time (time-division multiplexing). (SUB/PE) 999-1992w

(3) Subdivision of a common channel to make two or more channels by splitting the frequency band transmitted by the common channel into narrower bands, by allotting this common channel to several different information channels, or by other means, one at a time. *Contrast:* demultiplexing. *See also:* time compression multiplexing; frequency-division multiplexing; time-division multiplexing; synchronous time division multiplexing; time multiplexed switching. (C) 610.7-1995

multiplex lap winding (rotating machinery) A lap winding in which the number of parallel circuits is equal to a multiple of the number of poles. (PE) [9]

multiplexor (hybrid computer linkage components) An electronic multiposition switch under the control of a digital computer, generally used in conjunction with an analog-to-digital converter (ADC), that allows for the selection of any one of a number of analog signals (up to the maximum capacity of the multiplexor), as the input to the ADC. A device that allows the interleaving of two or more signals to a single line or terminus. (C) 166-1977w

multiplex printing telegraphy That form of printing telegraphy in which a line circuit is employed to transmit in turn one character (or one or more pulses of a character) for each of two or more independent channels. *See also:* time-division multiplexing; telegraphy; frequency-division multiplexing. (EEC/PE) [119]

multiplex radio transmission The simultaneous transmission of two or more signals using a common carrier wave. *See also:* radio transmission. (AP/ANT) 145-1983s

multiplex wave winding (rotating machinery) A wave winding in which the number of parallel circuits is equal to a multiple of two, whatever the number of poles. (BT) 204-1961w

multiplicand A number to be multiplied by another number (the multiplier) to produce a result (the product). (C) 1084-1986w

multiplication factor (1) (power operations) A measure of the change in the neutron population in a reactor core from one generation to the subsequent generation. *See also:* effective multiplication factor; infinite multiplication factor. (PE/PSE) 858-1987s

(2) (multiplier type of valve or tube) (thermionics) The ratio of the output current to the primary emission current. *See also:* electron emission. (ED) [45], [84]

multiplication time *See:* multiply time.

multiplication transformation function In hashing, a hash function that returns the original key multiplied by some value. For example, in the function below, the original key is multiplied by the length of the record in which it is found.

Original Record	Calculation	Hash Value
35 Bob White	35 × 13 = 448	448
41 Richard Doe	41 × 17 = 697	697

See also: mid-square function. (C) 610.5-1990w

multiplicative array antenna system A signal-processing antenna system consisting of two or more receiving antennas and circuitry in which the effective angular response of the output of the system is related to the product of the radiation patterns of the separate antennas. (AP/ANT) 145-1993

multiplier (1) (general) A device that has two or more inputs and whose output is a representation of the product of the quantities represented by the input signals. (Std100) 270-1966w

(2) (analog computer) In an analog computer, a device capable of multiplying one variable by another. (C) 165-1977w

(3) (mathematics of computing) A number by which another number (the multiplicand) is multiplied to produce a result (the product). (C) 1084-1986w

(4) A device capable of multiplying one variable by another. *Contrast:* divider. *See also:* two-quadrant multiplier; one-quadrant multiplier; analog multiplier; four-quadrant multiplier. (C) 610.10-1994w

(5) *See also:* normal linearity; servo multiplier; constant multiplier.

multiplier, constant *See:* constant multiplier.

multiplier, electronic *See:* electronic multiplier.

multiplier, four-quadrant *See:* four-quadrant multiplier.

multiplier, one-quadrant *See:* one-quadrant multiplier.

multiplier phototube A phototube with one or more dynodes between its photocathode and output electrode. *See also:* amplifier; photocathode. (ED/NPS) 161-1971w, 398-1972r

multiplier potentiometer (analog computer) Any of the ganged potentiometers of a servo multiplier that permit the multiplication of one variable by a second variable. (C) 165-1977w, 166-1977w

multiplier section, electron *See:* electron multiplier.

multiplier servo An electromechanical multiplier in which one variable is used to position one or more ganged potentiometers across which the other variable voltages are applied. *See also:* electronic analog computer; multiplier. (C) 165-1977w, 166-1977w, 610.10-1994w

multiplier, two-quadrant *See:* two-quadrant multiplier.

multiplying-digital-to-analog converter *See:* digital-to-analog multiplier.

multiplying punch *See:* calculating punch.

multiply time The elapsed time required to perform one multiplication operation, not including the time required to obtain the operands or to return the result to storage. *Synonym:* multiplication time. *Contrast:* subtract time; add time. (C) 610.10-1994w

multipoint Pertaining to a circuit or a communication arrangement where one line connects several stations. *Contrast:* multidrop. *See also:* point-to-point. (C) 610.7-1995

multipoint circuit (data transmission) A circuit interconnecting several stations. (PE) 599-1985w

multipoint connection (1) A configuration in which more than two stations are connected to a shared communications channel. (LM/COM) 168-1956w

(2) A connection between two data stations using one or more intermediate stations. (C) 610.7-1995

multipole fuse (1) (high-voltage switchgear) An assembly of two or more single-pole fuses. (SWG/PE) C37.40-1993

(2) *See also:* pole. (SWG/PE) C37.100-1981s

multipole operation (of a circuit breaker or switching device) A description term indicating that all poles of the device are linked mechanically, electrically, or by other means such that they change state (open or close) substantially simultaneously. Devices may have capability for multipole opening, multipole closing, or both. (SWG/PE) C37.100-1992

multi-port *See:* fan-out box.

multiport bridge A bridge that interconnects two or more DQDB subnetworks. (LM/C) 8802-6-1994

multiport memory A type of memory that can be simultaneously read or written to by two or more devices through the use of separate address and data buses. (C) 610.10-1994w

multi-port module A module that has a special pin assignment to connect to multiple other modules. Examples of multi-port modules are switch modules and multi-port memory modules. (C/BA) 14536-1995

multiposition relay A relay that has more than one operate or nonoperate position, for example, a stepping relay. *See also:* relay. (EEC/REE) [87]

multiposition switches (A) (self-returning switch) A switch that returns to a stated position when it is released from any one of a stated set of other positions. **(B)** (spring return switch) A switch in which the self-returning function is effected by the action of a spring. **(C)** (gravity-return switch) A switch in which the self-returning function is effected by the action of weight. **(D)** (self-positioning switch) A switch that assumes a certain operating position when it is placed in the neighborhood of the position. *See also:* switch. (IA/ICTL/IAC) [60]

multiprecision *See:* multiple precision.

multipressure-zone pothead (electric power distribution) A pressure-type pothead intended to be operated with two or more separate pressure zones that may be at different pressures. *See also:* pressure-type pothead; single-pressure-zone potheads. (PE) 48-1975s

multipressure zone termination *See:* pressure-type termination.

multiprocessing (1) (computers) Pertaining to the simultaneous execution of two or more programs or sequences of instructions by a computer network consisting of two or more processors. *See also:* parallel processing; multiprogramming. (C) [20]

(2) (software) A mode of operation in which two or more processes are executed concurrently by separate processing units that have access (usually) to a common main storage. *Contrast:* multiprogramming. *See also:* time sharing; multitasking. (C) 610.12-1990

multiprocessor (computers) A computer capable of multiprocessing. (C) [20], [85]

(2) (A) A computer or network of computers that can execute two or more programs concurrently under integrated control. **(B)** A computer that has more than one processor. *Contrast:* uniprocessor. (C) 610.10-1994

multiprocessor architecture An architecture employing two or more stand-alone processors whose activities are coordinated under a central control. *Contrast:* single processor architecture. (C) 610.10-1994w

multiprogramming (1) (computers) Pertaining to the interleaved execution of two or more programs by a computer. *See also:* parallel processing. (C) [20], [85]

(2) (software) A mode of operation in which two or more computer programs are executed in an interleaved manner by a single processing unit. *Contrast:* multiprocessing. *See also:* multitasking; time sharing. (C) 610.12-1990

multi-radio-frequency-channel transmitter *See:* multichannel radio transmitter.

multirange amplifier An amplifier that has a switchable, programmable, or automatically set amplification factor in order to adapt different analog signal ranges to a specified output range. (C) 610.10-1994w

multirate meter A meter that registers at different rates or on different dials at different hours of the day. *See also:* electricity meter. (EEC/PE) [119]

multi-ratio current transformer (1) (instrument transformers) (power and distribution transformers) One from which more than one ratio can be obtained by the use of taps on the secondary winding. (PE/PSR) C37.110-1996, C57.12.80-1978r

(2) One with three or more ratios obtained by the use of taps on the secondary winding. (PE/TR) C57.13-1993

multirestraint relay A restraint relay so constructed that its operation may be restrained by more than one input quantity. (SWG/PE) C37.100-1992

multisection coil (rotating machinery) A coil consisting of two or more coil sections or a group of turns, each section or group being individually insulated. (PE) [9]

multisegment magnetron A magnetron with an anode divided into more than two segments, usually by slots parallel to its axis. *See also:* magnetron. (ED) 161-1971w

multispeed motor A motor that can be operated at any one of two or more definite speeds, each being practically independent of the load. For example, a dc motor with two armature windings, or an induction motor with windings capable of various pole groupings. (IA/MT) 45-1998

multistage tube (x-ray tubes) An x-ray tube in which the cathode rays are accelerated by multiple ring-shaped anodes at progressively higher potential. (ED) [45]

multistate indication *See:* supervisory control functions.

multistatic radar A radar system having two or more transmitting or receiving antennas with all antennas separated by large distances when compared to the antenna sizes. *See also:* bistatic radar. (AES) 686-1997

multistator electromechanical watthour meter A multistator meter (polyphase electromechanical watthour meter or multielement watthour meter) is a watthour meter containing more than one stator. (ELM) C12.10-1987

multi-step control *See:* step control system.

multi-step control system *See:* step control system.

multistream computer A computer that is capable of executing multiple streams of instructions simultaneously. *See also:* multiple instruction, multiple data; multiple instruction, single data. (C) 610.10-1994w

multistrip coupler (MSC) An array of metal strips deposited on a piezoelectric substrate with lengths transverse to the propagation direction, which transfers acoustic power from one acoustic path to an adjacent parallel path. (UFFC) 1037-1992w

multi-tap (1) A passive distribution component composed of a directional coupler and a splitter with two or more output connections. *See also:* tap. (LM/C) 802.7-1989r
(2) *See also:* fan-out box. (C) 610.7-1995

multitasking (1) A mode of operation in which two or more tasks are executed in an interleaved manner. *See also:* time sharing; multiprogramming; multiprocessing. (C) 610.12-1990
(2) A mode of operation that provides for concurrent performance or interleaved execution of two or more tasks. (C) 610.10-1994w

multiterminal A transmission line with more than two terminals having a source of power. (PE/PSR) C37.113-1999

multiterminal surge-protective device A protective device that has three or more terminals, usually containing both series and parallel elements between the terminals. (SPD/PE) C62.62-2000

multitrace (oscilloscopes) A mode of operation in which a single beam in a cathode-ray tube is shared by two or more signal channels. *See also:* chopped display; alternate display; oscillograph. (IM/HFIM) [40]

multitrack recording system A system that provides two or more recording tracks on a medium, resulting in either related or unrelated recordings in common time relationship. *See also:* multiple sound track; phonograph pickup. (SP) [32]

multivalent function If to any value of u there corresponds more than one value of x (or more than one set of values of $x_1, x_2, \ldots, x_n$) then u is a multivalent function. Thus $u = \sin x$, $u = x^2$ are multivalent. (Std100) 270-1966w

multivalued A mapping that is not a function. *Contrast:* function; single-valued. (C/SE) 1320.2-1998

multivalued dependency A type of dependency among three attributes A, B, and C in relation R, in which B is multivalued dependent on A if, and only if, the set of values of B that match a given pair of values for A and C depends only on the value for A and is independent of the value for C. *See also:* fourth normal form. (C) 610.5-1990w

multivalued function If to any value of x (or any set of values of $x_1, x_2, \ldots, x_n$) there corresponds more than one value of u, then u is a multivalued function. Thus $u = \cos^{-1} x$ is multivalued. (Std100) 270-1966w

multivalued property A property with a multi-valued mapping. *Contrast:* single-valued property. (C/SE) 1320.2-1998

multivariable function generator (analog computer) A function generator with more than one input. (C) 165-1977w

multivariable system A system whose input vector and.or output vector has more than one element. (CS/PE/EDPG) [3]

multivibrator A relaxation oscillator employing two electron tubes to obtain the in-phase feedback voltage by coupling the output of each to the input of the other through, typically, resistance-capacitance elements. *Notes:* 1. The fundamental frequency is determined by the time constants of the coupling elements and may be further controlled by an external voltage. 2. A multivibrator is termed free-running or driven, according to whether its frequency is determined by its own circuit constants or by an external synchronizing voltage. The name multivibrator was originally given to the free-running multivibrator, having been suggested by the large number of harmonics produced. 3. When such circuits are normally in a nonoscillating state and a trigger signal is required to start a single cycle of operation, the circuit is commonly called a one-shot, a flip-flop, or a start-stop multivibrator. *See also:* oscillatory circuit. (AP/ANT) 145-1983s

multivoltage control (elevators) A system of control that is accomplished by impressing successively on the armature of the driving-machine motor a number of substantially fixed voltages such as may be obtained from multicommutator generators common to a group of elevators. *See also:* control. (EEC/PE) [119]

multiway merge sort A merge sort in which the set to be sorted is divided into two or more ordered subsets that are merged by comparing the smallest items of each subset, outputting the smallest of those, then repeating the process. *See also:* two-way merge sort. (C) 610.5-1990w

multiway radix trie search A radix trie search using a trie of order greater than 2, in which more than one digit is considered on each branch. *See also:* binary radix trie search. (C) 610.5-1990w

multiway tree A tree of order greater than 2. (C) 610.5-1990w

multiwire branch circuit A branch circuit consisting of two or more ungrounded conductors having a potential difference between them, and a grounded conductor having equal potential difference between it and each ungrounded conductor of the circuit and which is connected to the neutral conductor of the system. (NESC/NEC) [86]

multiwire element A radiating element composed of several wires connected in parallel, the assemblage being the electrical equivalent of a single conductor larger than any one of the individual wires. (AP/ANT) 145-1993

MUMPS *See:* Massachusetts General Hospital Utility Multi-Programming System.

municipal fire-alarm system A manual fire-alarm system in which the stations are accessibly located for operation by the public, and the signals of which register at a central station maintained and operated by the municipality. *See also:* protective signaling. (EEC/PE) [119]

municipal police report system A system of strategically located stations from any one of which a patrolling policeman may report his presence to a supervisor in a central office maintained and operated by the municipality. *See also:* protective signaling. (EEC/PE) [119]

M-unit *See:* refractive modulus.

Munsell chroma (1) (illuminating engineering) The index of perceived (Y) and chromaticity coordinates (x,y) for CIE Standard Illuminant C and the CIE 1931 Standard Observer. (EEC/IE) [126]
(2) (television) The dimension of the Munsell system of color that corresponds most closely to saturation. *Note:* Chroma is frequently used, particularly in English works, as the equivalent of saturation. (BT/AV) 201-1979w

Munsell color system (illuminating engineering) A system of surface-color specification based on perceptually uniform color scales for the three variables: Munsell hue, Munsell value, and Munsell chroma. For an observer of normal color vision, adapted to daylight, and viewing the specimen when illuminated by daylight and surrounded with a middle gray to white background, the Munsell hue, value, and chroma of the color correlate well with the hue, lightness, and perceived chroma. *Note:* A number of other color specification systems have been developed, usually for specific commercial purposes. (IE/EEC/BT/AV) [126], 201-1979w

Munsell hue (1) (H illuminating engineering) The index of the hue of the perceived object color defined in terms of the luminance factor (V) and chromaticity coordinates (x,y) for

CIE Standard Illuminant C and the CIE 1931 Standard Observer. (EEC/IE) [126]
(2) (television) The index of the hue of the perceived object color defined in terms of the Y value and chromaticity coordinates (x,y) of the color of the light reflected or transmitted by the object. (BT/AV) 201-1979w

Munsell value (1) (V illuminating engineering) The index of the lightness of the perceived object color defined in terms of the luminance factor (Y) for CIE Standard Illuminant C and the CIE 1931 Standard Observer. *Note:* The exact definition gives Y as a 5th power function of V so that tabular or iterative methods are needed to find V as a function of Y. However, V can be estimated within ±0.1 by V11.6 (Y/100)1/3-1.6 or within ±0.6 by VY1/2 where Y is the luminance factor expressed in percent. (IE/EEC) [126]
(2) (television) The index of the lightness of the perceived object color defined in terms of the Y value. *Note:* Munsell value is approximately equal to the square root of the reflectance expressed in percent. (BT/AV) 201-1979w

musa antenna *See:* antenna.

must operate value *See:* relay-must-operate value.

mutable class A class for which the set of instances is not fixed; its instances come and go over time. *Contrast:* immutable class. *See also:* state class. (C/SE) 1320.2-1998

mutation *See:* program mutation.

mutation testing A testing methodology in which two or more program mutations are executed using the same test cases to evaluate the ability of the test cases to detect differences in the mutations. (C) 610.12-1990

mutex (1) A synchronization object used to allow multiple threads to serialize their access to shared data. This term is derived from the capability it provides, namely, mutual exclusion. The thread that has locked a mutex becomes its owner and remains the owner until that same thread unlocks the mutex. (C/PA) 9945-1-1996
(2) A synchronization object used to allow multiple tasks (possibly in different processes) to serialize their access to shared data or other shared resources. The name derives from the capability it provides, namely, mutual exclusion. (C) 1003.5-1999
(3) A mechanism for implementing mutual exclusion. (IM/ST) 1451.1-1999

mutex owner The task that last locked a mutex, until that same task unlocks the mutex. (C) 1003.5-1999

Mutex Service An instance of the class `IEEE1451_MutexService` or of a subclass thereof. (IM/ST) 1451.1-1999

mutual capacitance The capacitance between two conductors in a pair when the rate of change of the charges on the two are equal in magnitude but opposite in sign, and the potentials of the remaining conductors in the cable are held constant. (PE/PSC) 789-1988w

mutual coherence function The normalized coherence function. *See also:* coherence function. (AP/PROP) 211-1997

mutual conductance The control-grid-to-anode transconductance. *See also:* ON period. (ED) [45], [84]

mutual coupling effect (A) (on input impedance of an array element) The change in input impedance of an array element from the case when all other elements are present but open circuited to the case when all other elements are present and excited. **(B)** (on the radiation pattern of an array antenna) The change in antenna pattern from the case when a particular feeding structure is attached to the array and mutual impedances among elements are ignored in deducing the excitation, to the case when the same feeding structure is attached to the array and mutual impedances among elements are included in deducing the excitation. (AP/ANT) 145-1993

mutual impedance (1) The mutual impedance between any two terminal pairs in a multielement array antenna is equal to the open-circuit voltage produced at the first terminal pair divided by the current supplied to the second when all other terminal pairs are open circuited. *See also:* antenna. (AP/ANT) [35], 149-1979r, 145-1993
(2) The ratio of the total induced open-circuit voltage on the disturbed circuit to the disturbing electric supply system phase current, with the effect of all conductors taken into account. (PE/PSC) 367-1996

mutual inductance The common property of two electric circuits whereby an electromotive force is induced in one circuit by a change of current in the other circuit. *Notes:* 1. The coefficient of mutual inductance M between two windings is given by the following equation

$$M = \frac{\partial i}{\partial \lambda}$$

where λ is the total flux linkage of one winding and i is the current in the other winding. 2. The voltage e induced in one winding by a current i in the other winding is given by the following equation

$$e = -\left[M\frac{\mathrm{d}i}{\mathrm{d}t} + i\frac{\mathrm{d}M}{\mathrm{d}t}\right]$$

If M is constant

$$e = -M\frac{\mathrm{d}i}{\mathrm{d}t}$$

(CHM) [51]

mutual inductor An inductor for changing the mutual inductance between two circuits. (Std100) 270-1966w

mutual information The amount of information about one event, say $x = a$, provided by another event, say $y = b$, for example, $I_{x,y}^{(a;b)} = \log P_{x\,y}^{(a\,b)}$ over $P_x^{(a)}$, where $P_x^{(a)}$ is the probability that $x = a$ and $P_{x\,y}^{(a\,b)}$ is the probability that $x = a$ and $P_{x\,y}^{(a\,b)}$ is the conditional probability that $x = a$ given that $y = b$. *Note:* This quantity is symmetric in its two arguments; that is, $I_{y,x}^{(b;a)} = I_{x,y}^{(a;b)}$. *See also:* information. (IT) [123]

mutual interference chart A plot or matrix, with ordinate and abscissa representing the tuned frequencies of a single transmitter-receiver combination, that indicates potential interference to normal receiver operation by reason of interaction of the two equipments under consideration at any combination of tuned transmit/receive frequencies. *Note:* This interaction includes transmitter harmonics and other spurious emissions, and receiver spurious responses and images. *See also:* electromagnetic compatibility. (EMC) [53]

mutually exclusive events (nuclear power generating station) (reliability analysis of nuclear power generating station safety systems) Events that cannot exist simultaneously. (PE/NP) 352-1987r

mutual resistance of grounding electrodes Equal to the voltage change in one of them produced by a change of one ampere of direct current in the other, and is expressed in ohms. (PE/PSIM) 81-1983

mutual surge impedance (surge arresters) The apparent mutual impedance between two lines, both of infinite length. *Note:* It determines the relationship between the surge voltage induced into one line by a surge current of short duration in the other. (PE) [8], [84]

MUX *See:* multiplexer.

MVU *See:* multiple-valve unit.

N

n (1) When preceding a capitalized signal name, denotes a signal having negative true logic. (C/MM) 1284-1994
(2) A multiplier having integer values (e.g., 0, 1, 2, 3. . .). (C/BA) 1301.4-1996
(3) *See also:* access; nano. (C) 610.10-1994w

n Multiplier having integer values of 1, 2, 3, . . . (C/BA/MM) 1301.2-1993, 1301.3-1992r

NA *See:* laser gyro axes; numerical aperture.

n–address instruction (-address instruction) A computer instruction that contains *n* address fields, where *n* may be any non-negative integer. *Contrast: n*–plus-one address instruction. *See also:* two-address instruction; one-address instruction. (C) 610.12-1990

n-address instruction format An instruction format that specifies n address fields, referencing n storage locations. For example, a "three-address instruction" contains three addresses. (C) 610.10-1994w

***n*–adic** *See:* N–ary.

n–adic Boolean operation (-adic boolean operation) A Boolean operation involving exactly *n* operands. *See also:* monadic Boolean operation; dyadic Boolean operation. (C) 1084-1986w

NAK *See:* negative acknowledge character.

name (1) (data management) An alphanumeric term that identifies a data item such as a field, record, or file. (C) 610.5-1990w
(2) The language-independent syntax for an identifier used to represent a specific value of a specific datatype. When a name is bound to its value is not specified: at compile time, link time, or load time. The use of a name as a parameter in a datatype specification does not imply that its value is defined at compile time. (C/PA) 1351-1994w
(3) In the shell command language, a word consisting solely of underscores, digits, and alphabetics from the portable character set. The first character of a name shall not be a digit. (C/PA) 9945-2-1993
(4) A construct that singles out a particular (directory) object from all other objects. A name shall be unambiguous (that is, denote just one object); however, it need not be unique (that is, be the only name that unambiguously denotes the object). (C/PA) 1328.2-1993w, 1224.2-1993w, 1327.2-1993w, 1326.2-1993w
(5) An identifier used to represent a specific value of a specific datatype. This Standard does not specify when a name is bound to its value: at compile time, link time, or load time. The use of a name as a parameter in a datatype specification does not imply that its value is defined at compile time. (C/PA) 1328-1993w, 1224-1993w, 1327-1993w, 1224.1-1993w
(6) A word or phrase that designates some model construct (such as a class, responsibility, subject domain, etc.). (C/SE) 1320.2-1998
(7) An indexical term used by humans as a means of identifying data elements and other data concepts. (SCC32) 1489-1999

named constraint A constraint that is specific to a particular model, rather than being inherent in some modeling construct (such as a cardinality constraint.). A named constraint is explicitly named, its meaning is stated in natural language, and its realization is written in the specification language. (C/SE) 1320.2-1998

Named Tag Set A field containing a Tag Set Name and its associated set of security tags. (C/LM) 802.10g-1995

nameplate (rotating machinery) (rating plate) A plaque giving the manufacturer's name and the rating of the machine. (PE) [9]

name service A service that assigns names that are unique within the name space, and that can translate a unique name into the location of the named entity. (C/PA) 1003.2d-1994

name space The environment within which a name is known to be unique. (C/PA) 1003.2d-1994

NaN *See:* not a number.

NAND (mathematics of computing) A Boolean operator having the property that if P is a statement, Q is a statement, R is a statement, . . ., then the NAND of P, Q, R, . . . is true if and only if at least one statement is false. *Note:* The NAND of P and Q is often represented by $\overline{PQ}$.

NAND Truth Table

P	Q	$\overline{PQ}$
0	0	1
0	1	1
1	0	1
1	1	0

Synonyms: Sheffer stroke; nonconjunction. (C) 1084-1986w

NAND element *See:* NAND gate.

NAND gate A gate that performs the Boolean operation of nonconjunction. *Note:* NAND is synonymous with NOT-AND. *Synonyms:* IF-A-THEN-NOT-B gate; NAND element. (C) 610.10-1994w

nano A prefix indicating 10^{-9}. (C) 1084-1986w

nanocode A collection of nanoinstructions. *Synonym:* nanoprogram. (C) 610.10-1994w, 610.12-1990

nanoinstruction (1) (software) In a two-level implementation of microprogramming, an instruction that specifies one or more of the basic operations needed to carry out a microinstruction. (C) 610.12-1990
(2) In a two-level implementation of multiprogramming, an instruction that specifies one or more of the basic operations needed to carry out a microinstruction. (C) 610.10-1994w

nanoprogram *See:* nanocode.

nanostore (software) In a two-level implementation of microprogramming, a secondary control store in which nanoinstructions reside. (C) 610.12-1990

narrative information (data management) Information that is presented according to the syntactic order of a natural language. *Contrast:* formatted information. (C) 610.5-1990w

narrative model A symbolic model whose properties are expressed in words; for example, a written specification for a computer system. *Synonym:* verbal-descriptive model. *Contrast:* mathematical model; software model; graphical model. (C) 610.3-1989w

narrow-angle diffusion (illuminating engineering) That in which flux is scattered at angles near the direction which the flux would take by regular reflection or transmission. (EEC/IE) [126]

narrow angle luminaire (illuminating engineering) A luminaire which concentrates the light within a cone of comparatively small solid angle. (EEC/IE) [126]

narrow-band axis (color television) (phasor representation of the chrominance signal) The direction of the phasor representing the coarse chrominance primary. (BT/AV) 201-1979w, [34]

narrow-band electrical noise A disturbance at a single frequency or in a narrowband of frequencies about a single frequency. (PE/IC) 1143-1994r

narrow-band interference For purposes of measurement, a disturbance of spectral energy lying within the bandpass of the measuring receiver in use. *See also:* electromagnetic compatibility. (EMC) [53]

narrow-band radio noise Radio noise having a spectrum exhibiting one or more sharp peaks, narrow in width compared to the nominal bandwidth of, and far enough apart to be resolvable by, the measuring instrument (or the communication receiver to be protected). (EMC) C63.4-1988s

narrow-band response spectrum (seismic qualification of Class 1E equipment for nuclear power generating stations) A response spectrum that describes the motion in which amplified response occurs over a limited (narrow) range of frequencies. (PE/NP) 344-1987r

narrow-band spurious emission (land-mobile communications transmitters) Any spurious output emitted from a radio transmitter, other than on its assigned frequency, which produces a disturbance of spectral energy lying within the bypass of the measuring receiver in use. (EMC) 377-1980r

N–ary (1) (A) (software) Characterized by a selection, choice or condition that has *n* possible different values or states. **(B) (software)** Of a fixed radix numeration system, having a radix of *n*. (C/SE) 729-1983
(2) A code whose output alphabet consists of N symbols. (IT) [123]

n–ary (A) (mathematics of computing) Pertaining to a selection in which there are *n* possible outcomes. **(B) (mathematics of computing)** Pertaining to a numeration system with a radix of *n*. (C) 1084-1986

n–ary Boolean operation* (–ary relation) A relation with *n* attributes in each tuple. (C) 610.5-1990w
* Deprecated.

n-ary tree (–ary tree) A tree of order *n*; for example, an 8-ary tree. *See also:* octary tree; binary tree; multiway tree. (C) 610.5-1990w

Nassi-Shneiderman chart *See:* box diagram.

nat *See:* nit.

national call (telephone switching systems) A toll call to a destination outside of the local service area of the calling customer but within the boundaries of the country in which he or she is located. (COM) 312-1977w

national character (data management) Deprecated term for the characters #,@, and $. (C) 610.5-1990w

national distance dialing (telephone switching systems) The automatic establishing of a national call by means of signals from the calling device of either a customer or an operator. (COM) 312-1977w

National Electrical Code dimensions (protection and coordination of industrial and commercial power systems) Dimensions once stated in the National Electrical Code (NEC), but now found in ANSI/UL 198B-1982 and in ANSI/UL 198D-1982. These dimensions are common to Class H and K fuses and provide interchangeability between manufacturers for fuses and fusible equipment of a given ampere and voltage range. (IA/PSP) 242-1986r

National Institute for Standards and Technology (NIST) The US federal agency with legally mandated authority to maintain national physical standards for the activity of radionuclides (formerly US National Bureau of Standards, NBS). (NI) N42.22-1995

national number (telephone switching systems) The combination of digits representing an area code and a directory number that, for the purpose of distance dialing, uniquely identifies each main station within each of the world's geographical areas that is identified by a country code. (COM) 312-1977w

national-numbering plan (telephone switching systems) Any plan for identifying telephone stations within a geographical area identified by a unique country code. (COM) 312-1977w

national numbering-plan area (telephone switching systems) A geographical area of the world where a country code and the national boundaries are related uniquely. (COM) 312-1977w

national radioactivity standard source (germanium detectors) A calibrated radioactive source prepared and distributed as a standard reference material by the U.S. National Bureau of Standards. (PE/EDPG) 485-1983s

national standard (plutonium monitoring) An instrument source or other system or device maintained and promulgated as such by the United States National Bureau of Standards. (NI) N317-1980r

national standards (metering) Those standards of electrical measurements that are maintained by the National Bureau of Standards. (ELM) C12.1-1982s

National Television Standards Committee (NTSC) A standards-setting body for the television and video industries in the United States. (C) 610.10-1994w

nationwide toll dialing (telephony) A system of automatic switching whereby an outward toll operator can complete calls to any basic-numbering-plan area in the country covered by the system. (EEC/PE) [119]

native data type (data management) A data type that is built into a software or hardware system. (C) 610.5-1990w

native system demand (power operations) The net 60 min clock-hour peak integrated demand within the system less interruptible loads. *See also:* alternating-current distribution. (PE/PSE) 858-1987s, 346-1973w

NATURAL A database manipulation language used to access data stored in an ADABAS database. (C) 610.13-1993w

natural air cooling system (thyristor controller) A cooling system in which the heat is removed from the cooling surfaces of the thyristor controller components only by the natural action of the ambient air. (IA/IPC) 428-1981w

natural bandwidth (laser maser) The linewidth of an absorption or emission line when spontaneous emission is the dominant process determining spectral distribution. (LEO) 586-1980w

natural binary *See:* binary.

natural frequency (1) (surge arresters) The frequency or frequencies at which the circuit will oscillate if it is free to do so. (PE) [8], [84]
(2) The frequency or frequencies at which a body vibrates due to its own physical characteristics (mass, shape, boundary condition, and elastic restoring forces brought into play) when the body is distorted in a specific direction and then released. (SWG/PE/NP) 344-1987r, C37.81-1989r
(3) (accelerometer) (gyros) That frequency at which the output lags the input by 90°. It generally applies only to inertial sensors with approximate second-order response. (AES/GYAC) 528-1994
(4) A frequency at which a body or system vibrates due to its own physical characteristics (mass and stiffness) when the body or system is distorted and then released. (PE/SUB) 693-1997

natural language (software) A language whose rules are based on usage rather than being pre-established prior to the language's use. Examples include German and English. *Contrast:* formal language. *See also:* INTELLECT. (C) 610.12-1990, 610.13-1993w

natural linewidth (laser maser) The linewidth of an absorption or emission line when spontaneous emission is the dominant process determining spectral distribution. (LEO) 586-1980w

natural model A model that represents a system by another system that already exists in the real world; for example, a model that uses one body of water to represent another. (C) 610.3-1989w

natural noise Noise having its source in natural phenomena and not generated in machines or other technical devices. *See also:* electromagnetic compatibility. (EMC/INT) [53], [70]

natural number (A) A non-negative integer: {0, 1, 2,. . . }. **(B)** A positive integer {1, 2, 3,. . . }. (C) 1084-1986, 610.5-1990

natural period The period of the periodic part of a free oscillation of the body or system. *Notes:* 1. When the period varies with amplitude, the natural period is the period as the amplitude approaches zero. 2. A body or system may have several modes of free oscillation, and the period may be different for each. (Std100) 270-1966w

natural priority The order of packet transmission of a node given that all nodes start arbitration at the same instant using the same priority level. For the cable environment, the closer a node is to the root, the higher its natural priority. For the backplane environment, the priority level and node_offset are concatenated to give its natural priority. (C/MM) 1394-1995

natural two-way merge sort (data management) A two-way merge sort in which the set to be sorted is repeatedly divided into two ordered subsets and merged, taking advantage of runs which occur naturally in the input set. *Contrast:* straight two-way merge sort. (C) 610.5-1990w

Naval Electronics Laboratory International Algorithmic Compiler (NELIAC) A programming language used primarily for solving scientific and real-time control problems. (C) 610.13-1993w

Naveam (navigation aid terms) A radio navigational warning of dangers to shipping in the Eastern, Atlantic, Mediterranean, and Red Seas. (AES/GCS) 172-1983w

navigation (navigation aid terms) The process of directing a vehicle so as to reach the intended destination. (AES/GCS) 172-1983w

navigational aid (navigation aid terms) An instrument, system, device, chart, or method intended to assist in the navigation of a vehicle. (AES/GCS) 172-1983w

navigational astronomy (navigation aid terms) That part of astronomy of direct use to a navigator, comprised principally of celestial coordinates, time, and apparent motions of celestial bodies. (AES/GCS) 172-1983w

navigational radar (surface search radar) A high-frequency radio transmitter-receiver for the detection, by means of transmitted and reflected signals, of any object (within range) projecting above the surface of the water and for visual indication of its bearing and distance. *See also:* radio direction-finder.

navigational satellite (navigation aid terms) An artificial earth orbiting satellite designed for navigational purposes. *See also:* satellite navigation. (AES/GCS) 172-1983w

navigation coordinate (navigation aid terms) Any one of a set of quantities; the set serving to define a position. (AES/GCS) 172-1983w

navigation light system (illuminating engineering) A set of aircraft aeronautical lights provided to indicate the position and direction of motion of an aircraft to pilots of other aircraft or to ground observers. (EEC/IE) [126]

navigation parameter (navigation aid terms) A measurable characteristic of motion or position used in the process of navigation. (AES/GCS) 172-1983w

navigation quantity (navigation aid terms) A measured value of a navigation parameter. (AES/GCS) 172-1983w

n–bit byte (1) A group of *n* adjacent binary digits operated upon as a unit. (C) 1084-1986w
(2) A string that consists of n bits; for example, an octet is an eight-bit byte. (C) 610.10-1994w

N/C *See:* NC.

NC *See:* not connected; numerical control.

NCAP Block An instance of the class `IEEE1451_NCAPBlock` or of a subclass thereof. (IM/ST) 1451.1-1999

n–channel device (metal-nitride-oxide field-effect transistor) Insulated–gate field–effect transistor (IGFET) where source and drain are regions of *n*–type conductivity. (ED) 581-1978w

n-channel metal-oxide semiconductor (NMOS) A type of semiconductor technology which employs a metal-oxide semiconductor device, using electrons to conduct current in the semiconductor channel. *Note:* The channel has a predominantly negative charge. *Synonym:* high-speed metal-oxide semiconductor. *Contrast:* p-channel metal-oxide semiconductor. *See also:* complementary metal-oxide semiconductor; V-channel metal-oxide semiconductor. (C) 610.10-1994w

N_char *See:* normal character.

***n*–conductor cable** *See:* multiple-conductor cable.

n-conductor concentric cable *See:* multiple-conductor concentric cable.

n-core-per-bit storage A type of storage in which each storage cell uses n magnetic cores per binary character, where n may be any positive integer. *See also:* one-core-per-bit storage. (C) 610.10-1994w

NCP *See:* network control program.

N–contours *See:* phase contours.

NDB *See:* nondirectional beacon.

N-display A K-display having an adjustable pedestal, notch, or step, as in the M-display, for the measurement of range. *Note:* This display is usually regarded as a variant of an A-display or a K-display rather than as a separate type.

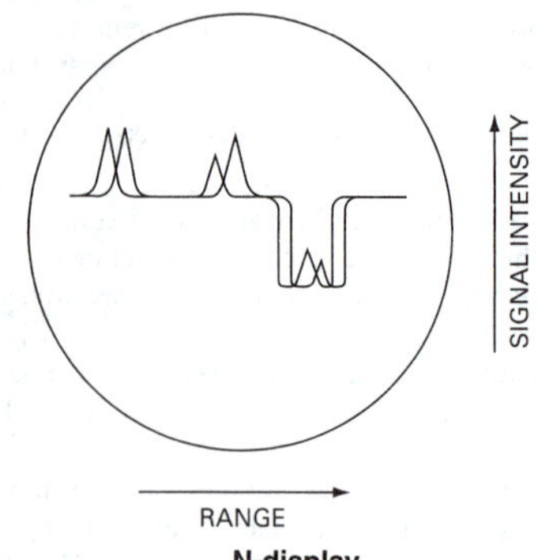

N-display

(AES) 686-1997

NDR *See:* nondestructive read.

near end crosstalk (NEXT) (1) Crosstalk that is propagated in a disturbed channel in the direction opposite to the direction of propagation of the current in the disturbing channel. *Note:* The terminal of the disturbed channel at which the near-end crosstalk is present is ordinarily near to or coincides with the energized terminal of the disturbing channel. *See also:* coupling. (EEC/PE) [119]
(2) Crosstalk that is propagated in a distributed channel in the direction opposite to the direction of the propagation of the signal in the disturbing channel and is measured at or near the source of the disturbing signal. (LM/C) 610.7-1995, 8802-5-1998

near field The field of a source at distances that are small compared to the wavelength. *Notes:* 1. The near field includes the quasi-static and induction fields varying as r^{-3} and r^{-2}, respectively, but does not include the radiation field varying as r^{-1}. 2. Often refers to points within the Fresnel region of a large antenna. (AP/PROP) 211-1997

near-field diffraction pattern (fiber optics) The diffraction pattern observed close to a source or aperture, as distinguished from far-field diffraction pattern. *Note:* The pattern in the output plane of a fiber is called the near-field radiation pattern. *Synonym:* Fresnel diffraction pattern. *See also:* diffraction; far-field diffraction pattern; far-field region. (Std100) 812-1984w

near-field pattern *See:* radiation pattern.

near-field radiation pattern (1) Any radiation pattern obtained in the near-field of an antenna. *Note:* Near-field patterns are usually taken over paths on planar, cylindrical, or spherical surfaces. *See also:* Fresnel pattern; radiation pattern cut. (AP/ANT) 145-1993
(2) *See also:* radiation pattern.

near-field region (1) That part of space between the antenna and far-field region. *Note:* In lossless media, the near-field may be further subdivided into reactive and radiating near-field regions. (AP/ANT) 145-1993
(2) (fiber optics) The region close to a source, or aperture. The diffraction pattern in this region typically differs significantly from that observed at infinity and varies with distance from the source. *See also:* far-field region; far-field diffraction pattern. (Std100) 812-1984w

(3) That part of space between the antenna array and the far-field region. Refers to the field of a source at distances small compared to the wavelength. *Note:* The near field includes the quasi-static and induction fields varying as r-3 and r-2, respectively, but does not include the radiation field varying as r-1. (T&D/PE) 1260-1996
(4) A region generally in proximity to an antenna or other radiating structure, in which the electric and magnetic fields do not have a substantially plane-wave character, but vary considerably from point to point. The near-field region is further subdivided into the reactive near-field region, which is closest to the radiating structure and that contains most or nearly all of the stored energy, and the radiating near-field region where the radiation field predominates over the reactive field, but lacks substantial plane-wave character and is complicated in structure. *Note:* For most antennas, the outer boundary of the reactive near-field region is commonly taken to exist at a distance of one-half wavelength from the antenna surface. (NIR) C95.1-1999

near-field region, radiating *See:* radiating near-field region.

near-field region, reactive That portion of the near-field region immediately surrounding the antenna, wherein the reactive field predominates. *Note:* For a very short dipole, or equivalent radiator, the outer boundary is commonly taken to exist at a distance $\lambda/2\pi$ from the antenna surface, where λ is the wavelength. (AP/ANT) 145-1993

near-field scanning (fiber optics) The technique for measuring the index profile of an optical fiber by illuminating the entrance face with an extended source and measuring the point-by-point radiance of the exit face. *See also:* refracted ray method. (Std100) 812-1984w

near-field test point (NFTP) The acoustical measurement point located 1 cm from the center of the handsfree telephones (HFTs) loudspeaker along the axis of the speaker. (COM/TA) 1329-1999

near-letter quality (NLQ) Pertaining to printed output that is of nearly as high quality as that of a standard typewriter. *Contrast:* letter-quality; draft quality. (C) 610.10-1994w

near-side (of an SI or BI) That port of an SI or BI that is electrically closer to the originating master. (NID) 960-1993

neck (cathode-ray tubes) The small tubular part of the envelope near the base. *See also:* cathode-ray tube. (ED) [45], [84]

negate (1) To perform the logic operation NOT. (C) [20], [85]
(2) (mathematics of computing) To perform the NOT operation. (C) 1084-1986w
(3) To drive a signal or a parallel set of signals to the zero logic state. (C/BA) 1496-1993w

negation The monadic Boolean operation whose result has a Boolean value opposite to that of the operand. *See also:* NOT gate. (C) 610.10-1994w

negative Pertaining to a voltage or charge that is associated with an excess of electrons. *Contrast:* positive. (C) 610.10-1994w

negative acknowledge character (NAK) A transmission control character transmitted by a station as a negative response to the station with which the connection has been set up. (C) 610.5-1990w

negative acknowledgment A reply transmitted by a receiving station to inform the sending station that an error in data has been detected. *Contrast:* acknowledgment. (C) 610.7-1995

negative after-potential (electrobiology) Relatively prolonged negativity that follows the action spike in a homogeneous fiber group. *See also:* contact potential. (EMB) [47]

negative conductance The conductance of a negative-resistance device. (CAS) [13]

negative conductor A conductor connected to the negative terminal of a source of supply. *Note:* A negative conductor is frequently used as an auxiliary return circuit in a system of electric traction. *See also:* center of distribution. (PE/T&D) [10]

negative-differential-resistance region (thyristor) Any portion of the principal voltage-current characteristic in the switching quadrant(s) within which the differential resistance is negative. *See also:* principal voltage-current characteristic. (IA/ED) 223-1966w, [46], [12]

negative electrode (1) (primary cell) The anode when the cell is discharging. *Note:* The negative terminal is connected to the negative electrode. *See also:* electrolytic cell.
(2) (metallic rectifier) The electrode from which the forward current flows within the cell. *See also:* rectification. (PE/EEC) [119]

negative feedback (1) The process by which part of the signal in the output circuit of an amplifying device reacts upon the input circuit in such a manner as to counteract the initial power, thereby decreasing the amplification. (CAS) [13]
(2) (control) A feedback signal in a direction to reduce the variable that the feedback represents. *See also:* feedback; feedback control system. (IA/ICTL/APP/IAC) [69], [60]
(3) (stabilized feedback) (data transmission) The process by which a part of the power in the output circuit of an amplifying device reacts upon the input circuit in such a manner as to reduce the initial power, thereby decreasing the amplification. *Synonym:* degeneration. (PE) 599-1985w

negative glow (1) (gas tube) The luminous glow in a glow-discharge cold-cathode tube between the cathode dark space and the Faraday dark space. *See also:* gas tube. (ED) [45]
(2) (overhead-power-line corona and radio noise) Corona mode that occurs at electric field strengths above those required for Trichel streamers. Negative glow is confined to a small portion of the electrode and appears as a small, stationary, luminous bluish fan. The corona current of negative glow is essentially pulse less. *See also:* Trichel streamers. (T&D/PE) 539-1990

negative logic An electronic logic system where the voltage representing one, active, or true has a more negative value than the voltage representing zero, inactive, or false. Also known as negative-true logic, it is normally used in electronic and computing data and communications switching systems for noise immunity reasons. (IM/ST) 1451.2-1997

negative logic convention The representation of the 1-state and the 0-state by the low (L) and high (H) levels, respectively. (GSD) 91-1984r

negative matrix A matrix the surface of which is the reverse of the surface to be ultimately produced by electroforming. (EEC/PE) [119]

negative modulation (amplitude-modulation television system) That form of modulation in which an increase in brightness corresponds to a decrease in transmitted power. *See also:* television. (PE/EEC) [119]

negative-phase-sequence impedance (rotating machinery) The quotient of the negative-sequence rated-frequency component of the voltage, assumed to be sinusoidal, at the terminals of a machine rotating at synchronous speed, and the negative-sequence component of the current at the same frequency. *Note:* It is equal to the asynchronous impedance for a slip equal to 2. *See also:* asynchronous machine. (PE) [9]

negative-phase-sequence (phase-reversal) relay A relay that responds to the negative-phase-sequence component of a polyphase input quantity. *Note:* Frequently employed in three-phase systems. *See also:* relay. (SWG/PE/PSR) C37.90-1978s, C37.100-1981s

negative-phase-sequence reactance (rotating machinery) The quotient of the reactive fundamental component of negative-sequence primary voltage due to sinusoidal negative-sequence primary current of rated frequency, and the value of this current, the machine running at rated speed. *See also:* asynchronous machine. (PE/EDPG) [5]

negative-phase-sequence relay A relay that responds to the negative-phase-sequence component of a polyphase input quantity. (SWG/PE) C37.100-1992

negative phase-sequence resistance (rotating machinery) The quotient of the in-phase fundamental component of negative-sequence primary voltage, due to sinusoidal negative-sequence primary current of rated frequency, and the value of this current, the machine running at rated speed. *See also:* asynchronous machine. (PE) [9]

negative-phase-sequence symmetrical components Of an unsymmetrical set of polyphase voltages or currents of M phases, that set of symmetrical components that have the $(m - 1)$st phase sequence. That is, the angular phase lag from the first member of the set to the second, from every other member of the set to the succeeding one, and from the last member to the first, is equal to $(m - 1)$ times the characteristic angular phase difference, or $(m - 1)2\pi/m$ radians. The members of this set will reach their positive maxima uniformly but in the reverse order of their designations. *Note:* The negative-phase-sequence symmetrical components for a three-phase set of unbalanced sinusoidal voltages ($m = 3$), having the primitive period, are represented by the equations

$$e_{a2} = (2)^{1/2}E_{a2}\cos(\omega t + \alpha_{a2})$$

$$e_{b2} = (2)^{1/2}E_{a2}\cos\left(\omega t + \alpha_{a2} - \frac{4\pi}{3}\right)$$

$$e_{c2} = (2)^{1/2}E_{a2}\cos\left(\omega t + \alpha_{a2} - \frac{2\pi}{3}\right)$$

derived from the equation of symmetrical components of a set of polyphase (alternating) voltages. Since in this case, $r = 1$ for every component (of first harmonic), the third subscript is omitted. Then k is 2 for $(m$ -1)st sequence, and s takes on the algebraic values 1, 2, and 3 corresponding to phases a, b, and c. The sequence of maxima occurs in the order, a, c, b, which is the reverse or negative of the order for $k = 1$. *See also:* symmetrical components. (Std100) 270-1966w

negative-polarity lightning stroke A stroke resulting from a negatively charged cloud that lowers negative charge to the earth. *See also:* direct-stroke protection. (T&D/PE) [10]

negative pre-breakdown streamers (overhead-power-line corona and radio noise) Streamers occurring at electric field strengths close to breakdown. The discharge appears as a bright filament with very little branching and extends far into the gap. The associated current pulse has high magnitude, long duration, and low repetition rate. (PE/T&D) 539-1990

negative-resistance device A resistance in which an increase in current is accompanied by a decrease in voltage over the working range. (Std100) 270-1966w

negative-resistance oscillator An oscillator produced by connecting a parallel-tuned resonant circuit to a two-terminal negative-resistance device. (One in which an increase in voltage results in a decrease in current.) *Note:* Dynatron and transitron oscillators, arc converters, and oscillators of the semiconductor type are examples. *See also:* oscillatory circuit. (AP/BT/ANT) 145-1983s, 182A-1964w

negative-resistance repeater (data transmission) A repeater in which gain is provided by a series or a shunt negative resistance, or both. (PE) 599-1985w

negative response An input string that matches one of the responses acceptable to the LC_MESSAGES category keyword `noexpr`, matching an ERE in the current locale. (C/PA) 9945-2-1993

negative-sequence reactance The ratio of the fundamental reactive component of negative-sequence armature voltage, resulting from the presence of fundamental negative-sequence armature current of rated frequency, to this current, the machine being operated at rated speed. *Notes:* 1. The rated current value of negative-sequence reactance is the value obtained from a test with a fundamental negative-sequence current equal to rated armature current. The rated voltage value of negative-sequence reactance is the value obtained from a line-to-line short-circuit test at two terminals of the machine at rated speed, applied from no load at rated voltage, the resulting value being corrected when necessary for the effect of harmonic components in the current. 2. For any unbalanced short-circuits, certain harmonic components of current, if present, may produce fundamental reactive components of negative-sequence voltage that modify the ratio of the total fundamental reactive component of negative-sequence voltage to the fundamental component of negative-sequence current. This effect can be included by multiplying the negative-sequence reactance, before it is used for short-circuit calculations, by a wave distortion factor, equal to or less than unity, that depends primarily upon the type of short-circuit, and upon the characteristics of the machine and the external circuit, if any, between the machine and the point of short-circuit. (EEC/PE) [119]

negative-sequence resistance The ratio of the fundamental component of in-phase armature voltage, due to the fundamental negative-sequence component of armature current to this component of current at rated frequency. *Note:* This resistance, which forms a part of the negative-sequence impedance for use in circuit calculations to establish relationships between voltages and currents, is not directly applicable in the calculations of the total loss in the machine caused by negative-sequence currents. This loss is the product of the square of the fundamental component of the negative-sequence current and the difference between twice the negative-sequence resistance and the positive-sequence resistance, that is, $I_2^2\,(2R_2 - R_1)$. (EEC/PE) [119]

negative shielding angle The shielding angle formed when the shield wire is located beyond the area occupied by the outermost conductors. *See also:* positive shielding angle; shielding angle. (SUB/PE) 998-1996

negative temperature (laser maser) An effective temperature used in the Boltzmann factor to describe a population inversion. *Note:* If n_2 particles populate the higher of two states and n_1 particles populate the lower state, their ratio is conventionally expressed by the Boltzmann factor

$$\frac{n_2}{n_1} = \exp\,(-\mathrm{kT})$$

where k is the Boltzmann constant and T the (effective) absolute temperature. (LEO) 586-1980w

negative terminal (of a battery) The terminal toward which positive electric charge flows in the external circuit from the positive terminal. *Note:* The flow of electrons in the external circuit is to the positive terminal and from the negative terminal. *See also:* battery. (EEC/PE) [119]

negative-transconductance oscillator Oscillator in which the output of the device is coupled back to the input without phase shift, the condition for oscillation being satisfied by the negative conductance of the device. *See also:* oscillatory circuit. (AP/ANT) 145-1983s

negative vectors Two vectors are mutually negative if their magnitudes are the same and their directions opposite. (Std100) 270-1966w

negentropy *See:* average information content, per symbol.

neighborhood (1) Of any point μ_0, in a three-dimensional space, the volume enclosed by a sphere drawn with μ_0 as center. Of any point μ_0, in a two-dimensional space, the area enclosed by a circle drawn with μ_0 as center. (Std100) 270-1966w

(2) (image processing and pattern recognition) A set of pixels located near a given pixel. (C) 610.4-1990w

neighborhood operator An image operator that assigns a gray level to each output pixel based on the gray levels in a neighborhood of the corresponding input pixel. *Contrast:* point operator. (C) 610.4-1990w

neighboring MTA An MTA with which the service is prepared to establish OSI connections in the context of the MT interface. (C/PA) 1224.1-1993w

NEITHER-NOR* *See:* NOR.

* Deprecated.

NELIAC *See:* Naval Electronics Laboratory International Algorithmic Compiler.

neon indicator A cold-cathode gas-filled tube containing neon, used as a visual indicator of a potential difference or a field. (ED) [45], [84]

neper (data transmission) A division of the logarithmic scale so that the number of nepers is equal to the natural logarithm of the scalar ratio of two currents or two voltages. *Notes:* 1. With I_1 and I_2 designating the scalar value of two currents, and n the number of nepers denoting their scalar ratio: $n\log_e (I_1/I_2)$. 2. One neper equals 0.8686 bel. 3. The neper is a dimensionless unit. (PE) 599-1985w

nerve-block (electrobiology) The application of a current to a nerve so as to prevent the passage of a propagated potential. *See also:* excitability. (EMB) [47]

NESC National Electrical Safety Code, ANSI C2-1993. (T&D/PE) 516-1995

nest (software) To incorporate a computer program construct into another construct of the same kind. For example, to nest one subroutine, block, or loop within another; to nest one data structure within another. (C) 610.12-1990

nest or section (multiple system) A group of electrolytic cells placed close together and electrically connected in series for convenience and economy of operation. *See also:* electrorefining. (EEC/PE) [119]

net (1) One complete circuit connecting at least one output to at least one input. *Note:* Must be some form of conductor such as a wire. *See also:* network. (C) 610.7-1995, 610.10-1994w
(2) An interconnection path between component pins, where the resistance of the path is not significantly different from zero. (C/TT) 1149.4-1999

net assured capability (electric power supply) The net dependable capability of all power sources available to a system, including firm power contracts and applicable emergency interchange agreements, less that reserve assigned to provide for scheduled maintenance outages, equipment and operating limitations, and forced outages of power sources. (PE/PSE) 346-1973w

net dependable capability (1) (electric power supply) The maximum generation, expressed in kilowatt hours per hour which a generating unit, station, power source, or system can be depended upon to supply on the basis of average operating conditions. (PE/PSE) 346-1973w
(2) The maximum system load, expressed in kilowatthours per hour that a generating unit, station, or power source can be depended upon to supply on the basis of average operating conditions. *Note:* This capability takes into account average conditions of weather, quality of fuel, degree of maintenance and other operating factors. It does not include provision for maintenance outages. *See also:* generating station. (T&D/PE/PSE) [10], [54]

net energy for load Net system generation plus net energy interchange. (PE/PSE) 858-1993w

net generation (electric power system) Gross generation minus station service or unit service power requirements. (PE/PSE) 858-1993w, 94-1991w

net head (power operations) The gross head less all hydraulic losses except those chargeable to the turbine. (PE/PSE) 858-1987s

net information content A measure of the essential information contained in a message. *Note:* It is expressed as the minimum number of bits or hartleys required to transmit the message with specified accuracy over a noiseless medium. *See also:* bit. (EEC/PE) [119]

net interchange (control area) (power and/or energy) The algebraic sum of the power on the tie lines of a control area. (PE/PSE) 94-1991w

net interchange deviation (electric power system) (control area) The net interchange minus the scheduled net interchange. (PE/PSE) 94-1970w

net interchange error The net actual interchange power minus the scheduled net interchange. (PE/PSE) 94-1991w

net interchange schedule programmer (speed governing systems) A means of automatically changing the net interchange schedule from one level to another at a predetermined time and during a predetermined period or at a predetermined rate. *See also:* speed-governing system. (PE/PSE) 94-1970w

netlist A point-to-point description of the interconnections between individual components in a circuit. (SCC20) 1445-1998

net load capability The maximum system load expressed in kilowatt hours per hour that a generating unit, station, or power source can be expected to supply under good operating conditions. *Notes:* 1. This capability provides for variations of load within the hour that it is assumed are to be spread among all of the power sources. 2. This is sometimes called net rated capability. *See also:* generating station. (T&D/PE/PE/PSE) [10], [54]

net loss (data transmission) (circuit equivalent) The sum of all the transmission losses occurring between the two ends of the circuit, minus the sum of all the transmission gains. (PE) 599-1985w

net rated capability *See:* net load capability.

net space charge The free, unbalanced charge in a given region, taking no account of the charges of both signs that balance each other. The preferred unit is the coulomb (C). *Note:* The term "space charge" is often used to refer to "net space charge." (T&D/PE) 539-1990, 1227-1990r

net space-charge density Net space charge (spacecharge) per unit volume. The preferred unit is C/m^3. This quantity provides no information about the monopolar space–charge density. (PE/T&D) 539-1990, 1227-1990r

net structure *See:* network structure.

net system energy (1) (power operations) Energy requirements of a system, including losses, defined as: A) net generation of the system, plus B) energy delivered to other systems. (PE/PSE) 858-1987s
(2) Energy requirements of a system, including losses, defined as A) net generation of the system, plus B) energy received from others, less C) energy delivered to other systems. (T&D/PE/PSE) 346-1973w

network (1) (measuring longitudinal balance of telephone equipment operating in the voice band) A combination of elements or devices. (COM/TA) 455-1985w
(2) (A) (data transmission) A series of points interconnected by communication channels. **(B) (data transmission)** The switched telephone network is the network of telephone lines normally used for dialed telephone calls. **(C) (data transmission)** A private network is a network of communications channels confined to the use of one customer. (COM/SUB/PE) 999-1992, 599-1985
(3) (A) (software) An interconnected or interrelated group of nodes. *See also:* documentation; node. **(B) (software)** In connection with a disciplinary or problem oriented qualifier, the combination of material, documentation, and human resources that are united by design to achieve certain objectives, for example, a social science network, a science information network. *See also:* node; documentation. (C/SE) 729-1983
(4) (distribution of electric energy) An aggregation of interconnected conductors consisting of feeders, mains, and services. *See also:* alternating-current distribution; network analysis. (T&D/PE) [10]
(5) (A) (data management) A data structure in which components are allowed to have more than one superordinate component. *See also:* hierarchy; graph. **(B) (data management)** A graph in which the edges connecting the nodes are assigned weights representing some characteristic, such as cost or quantity, related to the edge.

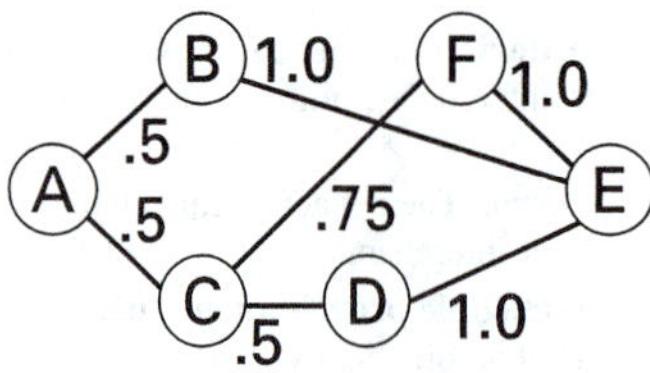

network

(C) 610.5-1990

(6) A collection of interconnected hosts. The term network in this document is used to refer to the network of hosts. The batch system is used to refer to the network of batch servers. (C/PA) 1003.2d-1994

(7) A network is any set of devices or subsystems connected by links joining (directly or indirectly) a set of terminal nodes. (C/BA) 1355-1995

(8) An arrangement of components, or nodes, and interconnecting branches. *See also:* computer network; circuit; store-and-forward switched network; public data network; public circuit-switched network; net; decentralized computer network; centralized computer network; packet switching network; distributed computer network; communications network; public switched network; switched network; value-added network; circuit-switched network; private network; public packet switching network; software defined network. (C) 610.7-1995, 610.10-1994w

(9) A communication mechanism for interconnecting multiple Smart Transducers. (IM/ST) 1451.1-1999

network address A network-visible identifier used to designate specific endpoints in a network. Specific endpoints on host systems shall have addresses, and host systems may also have addresses. (C) 1003.5-1999

network address attributes The Organizational-Unit-Name-1, Organizational-Unit-Name-2, Organizational-Unit-Name-3, and Organizational-Unit-Name-4 attributes specific to the class. (C/PA) 1224.1-1993w

network allocation vector (NAV) An indicator, maintained by each station, of time periods when transmission onto the wireless medium (WM) will not be initiated by the station whether or not the station's clear channel assessment (CCA) function senses that the WM is busy. (C/LM) 8802-11-1999

network analysis (networks) The derivation of the electrical properties, given its configuration and element values. (Std100) 270-1966w

network analyzer (network analyzers) A system that measures the two-port transmission and one-port reflection characteristics of a multiport in its linear range at a common input and output frequency. *Note:* This includes systems that: (1) Measure magnitude of reflection and transmission coefficient only as well as those systems that measure magnitude and phase; (2) operate manually or automatically; (3) cover a frequency range either continuously or in small enough steps to make coverage practically continuous; (4) vary frequency manually or automatically, stepped, or swept. (IM/HFIM) 378-1986w

network architecture *See:* computer network architecture.

network byte order An implementation-defined way of representing an integer so that, when transmitted over a network via a network endpoint, the integer shall be transmitted as an appropriate number of octets with the most significant octet first. This term has been used in historical systems and base documents to denote this representation. This representation is used in many networking protocols, including Internet and OSI. However, it should not be assumed this representation is useful with all network protocols. (C) 1003.5-1999

Network Capable Application Processor (NCAP, IEEE1451_NCAP) (1) A device between the Smart Transducer Interface Module and the network that performs network communications, Smart Transducer Interface Module communications, and data conversion functions. (IM/ST) 1451.2-1997

(2) A device that supports a network interface, application functionality, and generally access to the physical world via one or more transducers. An IEEE1451_NCAP is an NCAP conformant to this standard. (IM/ST) 1451.1-1999

network capacitance (charging inductors) The effective capacitance of the pulse-forming circuit. (MAG) 306-1969w

network control (telephone switching systems) The means of determining and establishing the required connections in response to information received from the system control. (COM) 312-1977w

network control host A network management central control center that is used to configure agents, communicate with agents, and display information collected from agents. (C/LM) 802.3-1998

network control program (NCP) A software program that deals with the operation of a front-end computer or communications controller. (C) 610.7-1995

network database (1) (data management) A database in which data are organized into segments which represent nodes within a network. *Contrast:* relational database. (C) 610.5-1990w

(2) A database system that uses directed graphics to represent the data. (PE/EDPG) 1150-1991w

network feeder A feeder that supplies energy to a network. *See also:* center of distribution. (T&D/PE) [10]

network function Any impedance function, admittance function, or other function of p that can be expressed in terms of or derived from the determinant of a network and its cofactors. *Notes:* 1. This includes not only impedance and admittance functions as previously defined, but also voltage ratios, current ratios, and numerous other quantities. 2. Certain network functions are sometimes classified together for a given purpose (for example, those with common zeros or common poles). These represent subgroups that should be specifically defined in each special case. 3. In the case of distributed-parameter networks, the determinant may be an infinite one: the definition still holds. *See also:* network analysis. (Std100) 270-1966w

network hierarchy In data communications, a network with a single computer having control over all the nodes connected to it. (C) 610.7-1995

network host name A string of characters in the portable character set that identifies a specific host in a network. (C/PA) 1003.2d-1994

network interface (1) (telephone loop performance) The interface between the public switched telephone network and the customer premises wiring. (COM/TA) 820-1984r

(2) (demarcation point) The point of connection between the local loop and the end user's (customer's) wiring. (AMR/SCC31/SCC31) 1390-1995, 1390.2-1999, 1390.3-1999

network interface controller A communication device which permits the connection of information processing devices to a network. (C) 610.7-1995

network interface definition language (NIDL) A model, proposed by the Institute of Electrical and Electronics Engineers, for parallel processing and logical process partitioning across a distributed computer network. (C) 610.7-1995

network layer (1) (Layer 3) The layer of the ISO Reference Model that performs those routing and relaying services necessary to support data transmission over interconnected networks. (DIS/C) 1278.2-1995

(2) In Open Systems Interconnection (OSI) architecture, the layer that provides services to establish a path between open systems with a predictable quality of service. (LM/C) 8802-6-1994

(3) The third layer of the seven-layer OSI model; responsible for establishing routes between the sending and receiving stations. *Note:* This layer appends and removes routine headers on the data packets, selects paths for them, and regulates their flow to prevent congestion. The network layer may use intermediate systems. *See also:* entity layer; application layer;

client layer; session layer; logical link control sublayer; medium access control sublayer; physical layer; sublayer; presentation layer; transport layer; data link layer.
(C) 610.7-1995

network leased line or private wire (data transmission) A series of points interconnected by telegraph or telephone channels, and reserved for the exclusive use of one customer.
(PE) 599-1985w

network limiter An enclosed fuse for disconnecting a faulted cable from a low-voltage network distribution system and for protecting the unfaulted portions of that cable against serious thermal damage.
(SWG/PE/TR) C37.100-1992, C57.12.44-1994

network loading unit (NLU) The time required to transmit one octet over the MIB Physical layer.
(EMB/MIB) 1073.3.1-1994

network, load matching *See:* load-matching network.

network management (1) The functions related to the management of Data Link Layer and Physical Layer resources and their status across an IEEE 802® local area network (LAN) or metropolitan area network (MAN).
(LM/C) 8802-6-1994
(2) The collection of administrative structures, policies, and procedures that collectively provide for the management of the organization and operation of the network as a whole.
(DIS/C) 1278.1-1995, 1278.2-1995
(3) In networking, a management function defined for monitoring, controlling, and coordinating the resources which allow communication to take place. *See also:* security management; performance management; configuration management; accounting management; fault management.
(C) 610.7-1995

network management environment An environment in which information processing system or resources communicate conforming to the services and protocols of network management. (C) 610.7-1995

network management process (NMP) The entity that provides access to network management functions on behalf of the user of the network management services. In order to perform this function, NMPs may intercommunicate in a peer-to-peer manner and may use the services of NMPs in other nodes via a network management protocol. The NMP at a node is the user of the service provided at the Layer Management Interface (LMI). (LM/C) 8802-6-1994

network management protocol A protocol in which objects can be accessed in a network management environment.
(C) 610.7-1995

network management system A system that monitors, controls, and coordinates the resources to facilitate communication. *See also:* security management; configuration management; accounting management; fault management; performance management. (C) 610.7-1995

network master relay A relay that functions as a protective relay by opening a network protector when power is backfed into the supply system and as a programming relay by closing the protector in conjunction with the network phasing relay when polyphase voltage phasors are within prescribed limits. (SWG/PE/TR) C37.100-1992, C57.12.44-1994

network model (data management) A data model in which entities are represented as nodes within a modified tree structure that permits all but the root to have multiple parents. *Note:* The network model was originally proposed as the CODASYL model by the Conference on Data Systems Languages (CODASYL) and described in their 1968 and 1971 publications. (C) 610.5-1990w

network object In computer security, a passive entity on a computer network that contains or receives information. For example: a data file, video display, clock, or printer. *Contrast:* network subject. (C) 610.7-1995

network operations center A center that is responsible for the operational aspects of a network. *Note:* Among these are monitoring and controlling, trouble-shooting, user-assistance.
(C) 610.7-1995

network pathname The pathname of a file that includes the name of the host on which it resides. This amendment allows two syntaxes for a network pathname:

pathname@hostname
hostname:pathname

(C/PA) 1003.2d-1994

network phasing relay A monitoring relay that has as its function to limit the operation of a network master relay so that the network protector may close only when the voltages on the two sides of the protector are in a predetermined phasor relationship.
(SWG/PE/TR) C37.100-1992, C57.12.44-1994

network primary distribution system A system of alternating-current distribution in which the primaries of the distribution transformers are connected to a common network supplied from the generating station or substation distribution buses. *See also:* alternating-current distribution. (T&D/PE) [10]

network, private line telegraph *See:* private line telegraph network.

network, private line telephone *See:* private line telephone network.

network protector (power and distribution transformers) An assembly comprising a circuit breaker and its complete control equipment for automatically disconnecting a transformer from a secondary network in response to predetermined electrical conditions on the primary feeder or transformer, and for connecting a transformer to a secondary network either through manual control or automatic control responsive to predetermined electrical conditions on the feeder and the secondary network. *Note:* The network protector is usually arranged to connect automatically its associated transformer to the network when conditions are such that the transformer, when connected, will supply power to the network and to automatically disconnect the transformer from the network when power flows from the network to the transformer.
(SWG/PE/TR) C37.100-1992, C57.12.80-1978r, C57.12.44-1994

network protector fuse A back-up device for the network protector. (PE/TR) C57.12.44-1994

network restraint mechanism A device that prevents opening of a network protector on transient power reversals that either do not exceed a predetermined value or persist for a predetermined time. (SWG/PE) C37.100-1992

network secondary distribution system A system of alternating-current distribution in which the secondaries of the distribution transformers are connected to a common network for supplying light and power directly to consumers' services. *See also:* alternating-current distribution.
(T&D/PE/TR) [10], C57.12.44-1994

network server On a network, a server that provides special network applications to all users at the network level; that is, it allows all users to share files more easily and have larger file storage areas available for their use. *See also:* disk server; file server; mail server; terminal server; database server; print server. (C) 610.7-1995

network service An application, such as file transfer protocol, available on a network. (C) 610.7-1995

network service access point (NSAP) The point at which an entity can request connection to the network. In IEEE 1073, both BCCs and DCCs have an NSAP.
(EMB/MIB) 1073.3.1-1994

network structure (data management) A collection of entities that are organized in a network fashion. *Synonym:* net structure. *Contrast:* hierarchical structure. (C) 610.5-1990w

network subject In computer security, an active entity on a computer network that causes information to flow through the network or changes the network state. For example, a processor or program operating in support of a process or user requirement. *Contrast:* network object. *See also:* single-level network subject; multilevel network subject.
(C) 610.7-1995

network synthesis (networks) The derivation of the configuration and element values, with given electrical properties. *See also:* network analysis. (Std100) 270-1966w

network theory The study of networks used to model processes such as communications, computer performance, routing problems, and project management. (C) 610.3-1989w

network transformer (power and distribution transformers) A transformer designed for use in a vault to feed a variable capacity system of interconnected secondaries. *Note:* A network transformer may be of the submersible or of the vault type. It usually, but not always, has provision for attaching a network protector. (PE/TR) C57.12.80-1978r

network tripping and reclosing equipment A piece of equipment that automatically connects its associated power transformer to an ac network when conditions are such that the transformer, when connected, will supply power to the network and that automatically disconnects the transformer from the network when power flows from the network to the transformer. (SWG/PE) C37.100-1992

Network Visible Refers to objects or operations that can be accessed or invoked from a different process than the process in which the object or operation executes, in particular if the processes are in different network capable application processors (NCAPs) communicating over a network. In IEEE 1451.1, Network Visible objects or operations are also locally visible. (IM/ST) 1451.1-1999

Neumann boundary condition A boundary condition applied to the solution of a partial differential equation in which the derivative of the function, applied in the direction normal to the boundary, is specified as a constant at the boundary. (AP/PROP) 211-1997

neuroelectricity Any electric potential maintained or current produced in the nervous system. *See also:* electrobiology. (EMB) [47]

neuromuscular effect Effect pertaining to the nervous system associated with muscle function. (T&D/PE) 539-1990

neurosensory effect Effect pertaining to the nervous system associated with sensory function. (PE/T&D) 539-1990

neutral (1) (rotating machinery) The point along an insulated winding where the voltage is the instantaneous average of the line terminal voltages during normal operation. *See also:* asynchronous machine. (PE) [9]
(2) (hydroelectric power plants) Common point of a star-connected generator or transformer winding. (PE/EDPG) 1020-1988r
(3) For use with the figures in this guide, the term neutral is understood to be the center tap of a grounded three-phase transformer winding. Since only single-phase loads are depicted, the other phases of the supply transformer have been omitted for clarity. (PE/EDPG) 1050-1996

neutral conductor (circuit consisting of three or more conductors). The conductor that is intended to be so energized, that, in the normal steady state, the voltages from every other conductor to the neutral conductor, at the terminals of entry of the circuit into a delimited region, are definitely related and usually equal in amplitude. *Note:* If the circuit is an alternating-current circuit, it is intended also that the voltages have the same period and the phase difference between any two successive voltages, from each of the other conductors to the neutral conductor, selected in a prescribed order, have a predetermined value usually equal to 2π radians divided by the number of phase conductors m. *See also:* network analysis; center of distribution. (Std100) 270-1966w

neutral direct-current telegraph system (single-current system) (single morse system) A telegraph system employing current during marking intervals and zero current during spacing intervals for transmission of signals over the line. *See also:* telegraphy. (EEC/PE) [119]

neutral ground (power and distribution transformers) An intentional ground applied to the neutral conductor or neutral point of a circuit, transformer, machine, apparatus, or system. (PE/EDPG/TR) 665-1995, C57.12.80-1978r

neutral grounding capacitor (electric power) A neutral grounding device the principal element of which is capacitance. *Note:* A neutral grounding capacitor is normally used in combination with other elements, such as reactors and resistors. *See also:* grounding device. (PE/SPD/T&D) 32-1972r, [10]

neutral grounding device (electric power) A grounding device used to connect the neutral point of a system of electric conductors to earth. *Note:* The device may consist of a resistance, inductance, or capacitance element, or a combination of them. *See also:* grounding device; neutral grounding impedor. (SPD/PE/T&D) 32-1972r, [10]

neutral grounding impedor (electric power) A neutral grounding device comprising an assembly of at least two of the elements resistance, inductance, or capacitance. *See also:* grounding device; neutral grounding device. (SPD/PE/T&D) 32-1972r, [10]

neutral grounding reactor (1) A current-limiting inductive reactor for connection in the neutral for the purpose of limiting and neutralizing disturbances due to ground faults. (PE/TR) C57.12.80-1978r
(2) (neutral grounding devices) One in which the principal element of which is inductive reactance. (SPD/PE) 32-1972r

neutral grounding resistor (electric power) A neutral grounding device, the principal element of which is resistance. *See also:* grounding device. (SPD/PE/T&D) 32-1972r, [10]

neutral grounding wave trap A neutral grounding device comprising a combination of inductance and capacitance designed to offer a very high impedance to a specified frequency or frequencies. *Note:* The inductances used in neutral grounding wave traps should meet the same requirements as neutral grounding reactors. (SPD/PE) 32-1972r

neutralization A method of nullifying the voltage feedback from the output to the input circuits of an amplifier through the tube interelectrode impedances. *Note:* Its principal use is in preventing oscillation in an amplifier by introducing a voltage into the input equal in magnitude but opposite in phase to the feedback through the interelectrode capacitance. *See also:* amplifier; feedback. (AP/BT/ANT) 145-1983s, 182A-1964w

neutralizing indicator An auxiliary device for indicating the degree of neutralization of an amplifier. (For example, a lamp or detector coupled to the plate tank circuit of an amplifier.) *See also:* amplifier. (AP/ANT) 145-1983s

neutralizing transformers (wire-line communication facilities) A device which introduces a voltage into a circuit pair to oppose an unwanted voltage. It neutralizes extraneous longitudinal voltages resulting from ground potential rise, or longitudinal induction, or both, which simultaneously allowing ac or dc metallic signals to pass. These transformers are primarily used to protect communication circuits at power stations, or along routes where exposure to power line induction is a problem, or both. (PE/PSC) 487-1980s

neutralizing transformers and reactors Devices that introduce a voltage into a circuit pair to oppose an unwanted voltage. These devices neutralize extraneous longitudinal voltages resulting from ground potential rise, or longitudinal induction, or both, while simultaneously allowing ac and dc metallic signals to pass. These transformers or reactors are primarily used to protect telecommunication or control circuits at power stations or along routes where exposure to power line induction is a problem, or both. (PE/PSC) 487-1992

neutralizing voltage The alternating-current voltage specifically fed from the grid circuit to the plate circuit (or vice versa), deliberately made 180 degrees out of phase with, and equal in amplitude to, the alternating-current voltage similarly transferred through undesired paths, usually the grid-to-plate tube capacitance. *See also:* amplifier. (AP/ANT) 145-1983s

neutral keying (data transmission) A form of telegraph signal which has current either on or off in the circuit with "on" as mark, "off" as space. (PE) 599-1985w

neutral lead (rotating machinery) A main lead connected to the common point of a star-connected winding. *See also:* asynchronous machine. (PE) [9]

neutral point (system) The point that has the same potential as the point of junction of a group of equal nonreactive resistances if connected at their free ends to the appropriate main terminals or lines of the system. *Note:* The number of such resistances is 2 for direct-current or single-phase alternating-current, 4 for two-phase (applicable to 4-wire systems only), and 3 for three-, six- or twelve-phase systems. *See also:* direct-current distribution; alternating-current distribution; center of distribution. (T&D/PE) [10]
(2) (A) (power and distribution transformers) The common point of a Y connection in a polyphase system. **(B) (power and distribution transformers)** The point of a symmetrical system which is normally at zero voltage. (PE/TR) C57.12.80-1978

neutral relay (1) A relay in which the movement of the armature does not depend upon the direction of the current in the circuit controlling the armature. (EEC/REE) [87]
(2) A relay that responds to quantities in the neutral of a power circuit. (SWG/PE) C37.100-1992

neutral terminal The terminal connected to the neutral of a machine or apparatus. *See also:* asynchronous machine. (PE) [9]

neutral wave trap (electric power) A neutral grounding device comprising a combination of inductance and capacitance designed to offer a very high impedance to a specified frequency or frequencies. *See also:* grounding device. (SPD/PE) 32-1972r

neutral zone (control element) The range of values of input for which no change in output occurs. *Note:* The neutral zone is an adjustable parameter in many two-step and floating control systems. *See also:* dead band.

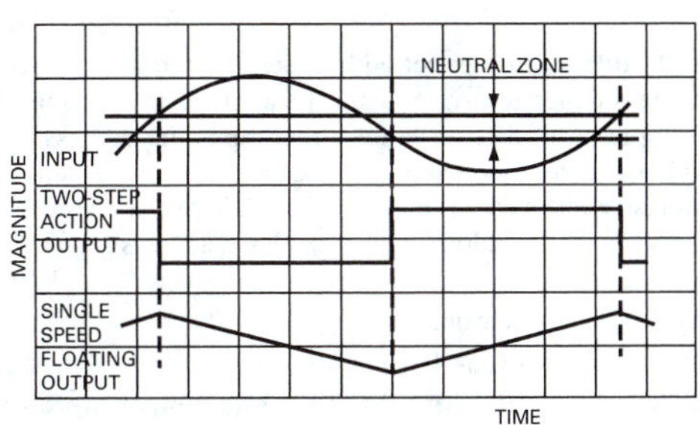

neutral zone

(PE/EDPG) [3]

new installation (elevators) Any installation not classified as an existing installation by definition, or an existing elevator, dumbwaiter, or escalator moved to a new location subsequent to the effective date of a code. *See also:* elevator. (EEC/PE) [119]

newline The character or characters necessary to generate the start of the next line of ASCII text. May also be known as a carriage-return (CR), linefeed (LF), or a CR-LF combination. (C/TT) 1450-1999

⟨newline⟩ A character that in the output stream shall indicate that printing should start at the beginning of the next line. The ⟨`newline`⟩ shall be the character designated by `'\n'` in the C-language binding. It is unspecified whether this character is the exact sequence transmitted to an output device by the system to accomplish the movement to the next line. (C/PA) 9945-2-1993

new-line character (NL) *See:* carriage return character.

newline string A white space string consisting only of the ⟨`newline`⟩ character. (C/PA) 1387.2-1995

new source statements The sum of the added and modified source statements. (C/SE) 1045-1992

new sync (data transmission) Allows for a rapid transition from one transmitter to another on multipoint private line data networks. (PE) 599-1985w

newton (metric practice) That force which, when applied to a body having a mass of one kilogram, gives it an acceleration of one meter per second squared. (QUL) 268-1982s

NEXT *See:* near end crosstalk.

Next Page General class of pages optionally transmitted by Auto-Negotiation able devices following the base Link Code Word negotiation. (C/LM) 802.3-1998

Next Page Algorithm (NPA) The algorithm that governs Next Page communication. (C/LM) 802.3-1998

Next Page Bit A bit in the Auto-Negotiation base Link Code Word or Next Page encoding(s) that indicates that further Link Code Word transfer is required. (C/LM) 802.3-1998

next program counter (nPC) Contains the address of the instruction to be executed next (if a trap does not occur). (C/MM) 1754-1994

next transfer address (NTA) A pointer in a slave to the module register that will be accessed during the next data transfer. The NTA register may be written during a primary address cycle and may be read or written during a secondary address cycle. During block and pipelined transfers the NTA may be automatically modified between data cycles. (NID) 960-1993

n gate thyristor (anode gate scr) A thyristor in which the gate terminal is connected to the n region adjacent to the region to which the anode terminal is connected and which is normally switched to the ON state by applying a negative signal between gate and anode terminals. (IA/IPC) 428-1981w

NIAM *See:* Nijssen's Information Analysis Method.

nibble (1) (MULTIBUS) A group of four adjacent bits operated on as a unit. (C/LM/MM) 802.9a-1995w, 1296-1987s
(2) Slang term for half a byte; or the first four or last four bits of an octet. *Synonym:* nybble. *See also:* quartet. (C) 610.10-1994w
(3) A group of four data bits. The unit of data exchange on the Media Independent Interface (MII). (C/LM) 802.3-1998
(4) Four bits of data. (C/MM) 1394a-2000
(5) (nybble) (mathematics of computing) *See also:* quartet. (C) 1084-1986w

Nibble Mode An asynchronous, reverse (peripheral-to-host) channel, under control of the host. Data bytes are transmitted as two sequential, 4-b nibbles using four peripheral-to-host status lines. Nibble Mode is used with Compatibility Mode to implement a bidirectional channel. These two modes cannot be active simultaneously. (C/MM) 1284-1994

Nichols chart (control system feedback) (Nichols diagram) A plot showing magnitude contours and phase contours of the return transfer function referred to ordinates of logarithmic loop gain and to abscissae of loop phase angle. *See also:* feedback control system. (PE/EDPG) [3]

nickel-cadmium storage battery An alkaline storage battery in which the positive active material is nickel oxide and the negative contains cadmium. *See also:* battery. (EEC/PE) [119]

NIDL *See:* network interface definition language.

night (illuminating engineering) The hours between the end of evening civil twilight and the beginning of morning civil twilight. *Note:* Civil twilight ends in the evening when the center of the sun's disk is six degrees below the horizon and begins in the morning when the center of the sun's disk is six degrees below the horizon. *See also:* civil twilight. (EEC/IE) [126]

night alarm An electric bell or buzzer for attracting the attention of an operator to a signal when the switchboard is partially attended. (EEC/PE) [119]

night effect (navigation aid terms) (radio navigation systems) A special case of error occurring predominantly at night when sky-wave propagation is at the maximum. (AES/GCS) 172-1983w

nighttime interference A radio disturbance caused by skywave signals from distant stations. Skywave propagation loss is lowest at night. (T&D/PE) 1260-1996

Nijssen's Information Analysis Method (NIAM) Named after Dr. G. M. Nijssen, leader of Control Data Corporation's research team on information systems development, NIAM is a method for performing information analysis based on a general framework for information systems in which terms like information, information flow, and information system are explained. In this framework, the conceptual grammar is an essential part that describes the static and dynamic aspects of the object system accurately and completely. Although formal, the conceptual grammar is perfectly understandable for users, enabling them to participate fully in the development of an information system. (ATLAS) 1232-1995

NIM **(1)** (A) A committee sponsored by the United States Department of Energy and associated with the United States National Institute of Standards and Technology. It produced the NIM instrumentation system specifications, endorsed the use of CAMAC, and collaborates with ESONE in the maintenance and extension of CAMAC. (B) A standardized modular instrumentation system consisting of NIM MODULES and NIM BINS as defined in DOE/ER-0457T, May 1990. (NID) 960-1993
(2) A standardized modular instrumentation system consisting of NIM Modules and NIM Bins as defined in DOE Report DOE/ER-0457T, May 1990. (NPS) 325-1996

NIM Bin A mounting unit or housing for NIM Modules that conforms to the requirements of DOE Report DOE/ER-0457T, May 1990 It includes bussed connectors at the rear that mate with connectors on the modules to provide the modules with power. (NPS) 325-1996

NIM Module, NIM Instrument A modular functional unit or instrument that mounts in a NIM bin And conforms to the requirements of DOE Report DOE/ER-0457T, May 1990. (NPS) 325-1996

911 call (telephone switching systems) (nine-one-one call) A call to an emergency service bureau. (COM) 312-1977w

nines check *See:* casting out nines.

nines complement **(1)** The radix-minus-one complement of a numeral whose radix is ten. (C) [20], [85]
(2) **(mathematics of computing)** The diminished-radix complement of a decimal numeral, which is formed by subtracting each digit from 9. For example, the nines complement of 4830 is 5169. *Synonym:* complement on nine. (C) 1084-1986w

1988 class A class for which all the specific attributes are for the 1988 version of the CCITT X.400 Recommendation alone. (C/PA) 1224.1-1993w

ninety-percent response time (electric instruments) (of a thermal converter) The time required for 90 percent of the change in output electromotive force to occur after an abrupt change in the input quantity to a new constant value. *See also:* thermal converter. (EEC/AII) [102]

nipple (rigid metal conduit) A straight piece of rigid metal conduit not more than two feet in length and threaded on each end. *See also:* raceway. (EEC/CON) [28]

NIST *See:* National Institute for Standards and Technology.

NIST source verification The process of verification of the capability of a manufacturer to calibrate a radionuclide source that is traceable to NIST. This is achieved through calibration by the manufacturer of a source distributed by NIST without disclosure of the known value. The reported value is compared to the NIST value and, if appropriate, a report of traceability is issued. (NI) N42.22-1995

NIST traceability The process of relating the measurement accuracy of radionuclide sources to national physical standards. Traceability is achieved by demonstrating the capability to produce accurate standardized sources by participation in a MAP with linkage to NIST and production of certified materials in accordance with a quality assurance program that meets the guidance provided in this standard. Traceability of sources requires demonstrated measurement traceability and applies only to products produced in accordance with this standard. (NI) N42.22-1995

nit **(1)** **(television)** The unit of luminance (photometric brightness) equal to one candela per square meter. *Note:* Candela per square meter is the unit of luminance in the International System of Units (SI). Nit is the name recommended by the International Commission on Illumination (CEI). (BT/AV) 201-1979w
(2) Also called "nat". (IT) [123]

Nixie tube display device A display device that employs a gas-filled digital indicator tube containing stacked metallic elements which, when energized, emit a glow in the shape of a number.

Nixie tube display device

(C) 610.10-1994w

NL *See:* new-line character.

N-layer A subdivision of the architecture, constituted by subsystems of the same rank (N). (C/LM/CC) 8802-2-1998

***n*–level address** **(1)** **(computers)** A multilevel address that specifies *n* levels of addressing. (C) [20], [85]
(2) **(software)** An indirect address that specifies the first of a chain of *n* storage locations, the first *n*–1 of which contain the address of the next location in the chain and the last of which contains the desired operand. For example, a two-level address. *Contrast:* direct address; immediate data. (C) 610.12-1990

n-level address In indirect addressing, the nth address sought in the attempt to arrive at the location of the operand; for example, a third-level address represents the address of the address of the address of the operand. *Note:* A zero-level address is the same as an immediate address; a one-level address is a direct address; and a two-level address is an indirect address. (C) 610.10-1994w

NLQ *See:* near-letter quality.

NLP *See:* Normal Link Pulse.

NLP Receive Link Integrity Test Function Auto-Negotiation's Link Integrity Test function that allows backward compatibility with the 10BASE-T Link Integrity Test function of IEEE 802.3. (LM/C) 802.3u-1995s

NLP sequence A Normal Link Pulse sequence, defined in IEEE 802.3 as TP_IDL. (LM/C) 802.3u-1995s

NLU *See:* network loading unit.

NMOS *See:* n-channel metal-oxide semiconductor.

NMP *See:* network management process.

N/M register set A generic term to indicate the microprocessor registers that deal with the retention of high and low memory addresses. *Note:* Usually used in scalable RISC computer discussions. *See also:* register set. (C) 610.10-1994w

***n*–*m* tree (data management)** A tree in which each node has at least *n* but no more than *m* subtrees; for example, 2-4 tree. *Note:* A 2-4 tree is sometimes written as 2-3-4 tree. (C) 610.5-1990w

***n-n* junction (semiconductor)** A region of transition between two regions having different properties in *n*-type semiconducting material. *See also:* semiconductor device. (EEC/PE) [119]

no-address instruction *See:* zero-address instruction.

no-backoff error A transmission state that results from a transceiver transmitting when there is no carrier, and without waiting for the necessary delay. (C) 610.7-1995

noble potential A potential substantially cathodic to the standard hydrogen potential. *See also:* stray-current corrosion. (IA) [59]

no-busy test call (telephone switching systems) A call in which busy testing is inhibited. (COM) 312-1977w

nochange timing check A timing check similar to a setup/hold timing check except the setup and hold times are referred to opposite transitions of the reference signal. The stable interval is extended to include the period between these transitions; that is, the time for which the reference signal stays in a specified state. This timing check is frequently applied to memory banks and latch banks to establish the stability of the address or to select inputs before, during, and after the write pulse. (C/DA) 1481-1999

nodalization (A) The set of nodes within a system being modeled. **(B)** The process of developing the nodes as in (A). (C) 610.3-1989

nodal point *See:* node.

nodding-beam height finder A height-finding radar with a fan beam oriented with its narrow beamwidth in elevation, and which mechanically sector scans (nods) in elevation to locate the target and determine its elevation angle. (AES) 686-1997

node (1) (network analysis) One of the set of discrete points in a flow graph. (CAS) 155-1960w
(2) (software) In a diagram, a point, circle, or other geometric figure used to represent a state, event, or other item of interest. *See also:* graph. (C) 610.12-1990
(3) (modeling and simulation) A single entity that is represented in a mathematical model; for example, in a model of a nuclear reactor, a water pump or section of pipe. (C) 610.3-1989w
(4) (data management) In a tree, an element that is used to contain information that describes some object; for which there is at least one key used to identify the node. *Note:* Nodes are connected to each other by link fields to form the tree. *Synonym:* vertex. *See also:* terminal node; child node; parent node; nonterminal node; root node. (C) 610.5-1990w
(5) (A) (broadband local area networks) A point of junction between two connectors. **(B) (broadband local area networks)** The location where a line has a defined position. **(C) (broadband local area networks)** The point where signals leave one system and enter another. (LM/C) 802.7-1989
(6) In the context of Open Firmware, node is a synonym for device node. *See also:* device node. (C/BA) 1275-1994
(7) A general term denoting either a switching element in a network or a host computer attached to a network. (DIS/C) 1278.1-1995, 1278.2-1995
(8) An entity associated with one or more interconnected lincs and optionally containing other functional units, such as cache and memory. In normal operation each node can be accessed independently (a control-register update on one node has no effect on the control registers of another node). (C/MM) 1596-1992
(9) A device or subsystem having one or more link interfaces. A node may be a terminal node (q.v.). A node may perform a routing function, routing packets between its node interfaces according to the information in the destination field of the packet. (C/BA) 1355-1995
(10) A device that consists of an access unit (AU) and a single point of attachment of the access unit to each bus of a DQDB subnetwork for the purpose of transmitting and receiving data on that subnetwork. Adjacent nodes are connected by a transmission link. (LM/C) 8802-6-1994
(11) (A) In networking, a point or a junction in a transmission system where lines or trunks from one or more systems meet. *Synonyms:* branch point; junction point; nodal point; vertex. **(B)** In data communications, a device or station that implements some part of the communication protocol. (C) 610.7-1995
(12) Within a circuit, a point of interconnection between two or more components such as input and output terminals. (C) 610.10-1994w
(13) An addressable device attached to the Serial Bus with at least the minimum set of control registers. Changing the control registers on one node does not affect the state of control registers on another node. (C/MM) 1394-1995
(14) The entity associated with a particular set of control register addresses (including identification ROM and reset command registers) that is initially defined in a 4 Kbyte (minimum) initial node address space. In normal operation each node can be accessed independently (a control register update on one node has no effect on the control registers of another node). (C/MM) 1596.5-1993
(15) The software visible station on a bus. (each node is allocated a set of control register addresses [including identification-ROM and reset command registers], which are initially defined in a 4 kbyte [minimum] initial node address space. Although multiple nodes may share one bus interface, each node can be reset independently [a reset of one node has no effect on any other node]. Each module consists of one or two nodes that are independently initialized and configured by operating system software.) (C/BA/MM) 14536-1995, 13213-1994
(16) A term used to describe a RamLink slave within the context of the CSR Architecture. The entity associated with a particular set of control-register addresses (including identification ROM and reset-command registers). In normal operation each node can be accessed independently (a control-register update on one node has no effect on the control registers of another node). (C/MM) 1596.4-1996
(17) A modeled function located within the hierarchical graph structure of an IDEF0 model by its designated node number; as a function, a node is represented in a diagram by a named box. (C/SE) 1320.1-1998
(18) A set of Control and Status Register (CSR) addresses (including identification read-only memory and reset command registers) that are initially defined in a 4 kB (minimum) initial node address space. Each node can be reset independently (a reset of one node has no effect on other nodes). (C/BA) 1014.1-1994w, 896.2-1991w, 896.3-1993w, 896.4-1993w, 896.10-1997
(19) A conceptual point (through which logic signals pass) that has been identified as an aid to modeling the timing properties of a cell but may not correspond to any physical structure. In Physical Design Exchange Format (PDEF), this is a physical pin that does not correspond to a logical structure. (C/DA) 1481-1999
(20) A Serial Bus device that may be addressed independently of other nodes. A minimal node consists of only a physical layer (PHY) without an enabled link. If the link and other layers are present and enabled they are considered part of the node. (C/MM) 1394a-2000
(21) *See also:* batch node.

node absorption (network analysis) A flow-graph transformation whereby one or more dependent nodes disappear and the resulting graph is equivalent with respect to the remaining node signals. *Note:* For example, a circuit analog of node absorption is the star-delta transformation. (CAS) 155-1960w

nodecast An adjective used to describe an interrupt or message transaction that is distributed to all units on a node. Also used as a verb; e.g., "transactions may be nodecast to all units on a node." (C/MM) 1212-1991s

node controller A component within a node that provides a coordination point for management functions exclusively local to a given node and involving the application, transaction, link, and physical elements located at that node. (C/MM) 1394-1995

node equations (networks) Any set of equations (of minimum number) such that the independent node voltages of a specified network may be determined from the impressed currents. *Notes:* 1. The number of node equations for a given network is not necessarily the same as the number of mesh or loop equations for that network. 2. Notes for mesh or loop equations, with appropriate changes, apply here. *See also:* network analysis. (Std100) 270-1966w

node_id **(1)** A 16 b value that determines the initial node address space. On some buses, the node_id value has two components: a *bus_id* field specifies one of 1024 bus address and an *offset_id* field specifies one of 64 node positions on the bus. During system initialization, software is expected to assign unique node_id values to nodes within a system. (C/MM) 1212-1991s
(2) A 16 b number that defines the initial node address space. During system initialization, systemwide unique node_id values may be assigned to nodes within a tightly coupled system. (C/BA) 896.9-1994w, 896.3-1993w
(3) A 16 b number that determines the node address space. After system initialization, unique node_id values have been assigned to all nodes within a tightly coupled system. The node_id is the part of the 64 b address that is used for routing packets. (C/MM) 1596-1992
(4) A unique 16 b number that distinguishes the node from other nodes in the system. The ten most significant bits of node_id are the same for all nodes on the same bus; this is the bus_id. The six least-significant bits of node_id are unique for each node on the same bus; this is called the physical_id. (C/MM) 1394-1995
(5) A 16-bit number that uniquely differentiates a node from all other nodes within a group of interconnected buses. The ten most significant bits of node ID are the same for all nodes on the same bus, i.e., the bus ID. The six least significant bits of node ID are unique for each node on the same bus; this is called the physical ID. The physical ID is assigned as a consequence of bus initialization. (C/MM) 1394a-2000

node index A text listing, often indented, of the nodes in an IDEF0 model, shown in outline order. Same meaning and node content as a node tree. (C/SE) 1320.1-1998

node interface A link interface on a switch. *See also:* link interface; switch. (C/BA) 1355-1995

node letter The letter that is the first character of a node number. (C/SE) 1320.1-1998

node name A text string of the form "`driver-name@unit-address:device-arguments`", which identifies a device node within the address space of its parent. (C/BA) 1275-1994

node number An expression that unambiguously identifies a function's position in a model hierarchy. A node number is constructed by concatenating a node letter, the diagram number of the diagram that contains the box that represents the function, and the box number of that box. (C/SE) 1320.1-1998

node ROM A range of register addresses that are mapped to address offsets 1024 to 2047. (C/BA) 896.10-1997, 896.2-1991w

node signal (network analysis) A variable X_k associated with node *k*. (CAS) 155-1960w

node tree A graphical listing of the nodes of an IDEF0 model, showing parent-child relationships as a graphical tree. Same meaning and node content as a node index. (C/SE) 1320.1-1998

node voltage (networks) The voltage from a reference point to any junction point (node) in a network. *Note:* The assumptions of lumped-network theory are such that the path of integration is immaterial. (Std100) 270-1966w

no-fill mode In text formatting, an operating mode in which no justification is performed. (C) 610.2-1987

no-go *See:* go/no-go.

noise (1) (analog computer) Unwanted disturbances superimposed upon a useful signal, which tend to obscure its information content. Random noise is the part of the noise that is unpredictable, except in a statistical sense. (C) 165-1977w
(2) (A) (general) (data transmission) An undesired disturbance within the useful frequency band. *Note:* Undesired disturbances within the useful frequency band produced by other services may be called interference. **(B)** Any unwanted disturbance in a system, such as random variations in voltage or current, or extra bits in data. **(C)** Any unwanted variation in a signal. *See also:* intermodulation noise; crosstalk; Gaussian noise; white noise; impulse noise. (PE) 599-1985
(3) (phototubes) The random output that limits the minimum observable signal from the phototube. *See also:* phototube. (NPS) 175-1960w, 398-1972r
(4) (facsimile) Any extraneous electric disturbance tending to interfere with the normal reception of a transmitted signal. (COM) 168-1956w
(5) (hybrid computer linkage components) Unwanted disturbances superimposed upon a useful signal that tends to obscure its information content expressed in millivolts peak and referred to the input voltage. (C) 166-1977w
(6) (electrical noise) Unwanted electrical signals that produce undesirable effects in the circuits of the control systems in which they occur. (IA/PE/T&D/ICTL/EDPG) 518-1982r, 1050-1996, 1250-1995
(7) (oscilloscopes) Any extraneous electric disturbance tending to interfere with the normal display. (IM) 311-1970w
(8) (broadband local area networks) Any unwanted signal in a communications system. White noise (or random noise) is random energy (e.g.,shot noise and thermal noise) that has a uniform distribution of energy across the band-pass. The analogy for white noise is white light. Johnson noise (thermal) is the noise generated by electron movement (current through a resistor) above absolute zero. The noise level is proportional to temperature. Shot noise is the type of unrandom noise generated when current flows across an abrupt junction. Shot noise is characteristic of semiconductor devices. (LM/C) 802.7-1989r
(9) (image processing and pattern recognition) Irrelevant data that hamper recognition and interpretation of data of interest. (C) 610.4-1990w
(10) Any deviation between the output signal (converted to input units) and the input signal, except deviations caused by linear time invariant system response (gain and phase shift), a dc level shift, or an error in the sample rate. For example, noise includes the effects of random errors, fixed pattern errors, nonlinearities and time base errors (fixed error in sample time and aperture uncertainty). (IM/WM&A) 1057-1994w
(11) Electrical noise is unwanted electrical signals that produce undesirable effects in the circuits of the control systems in which they occur. See figure below.

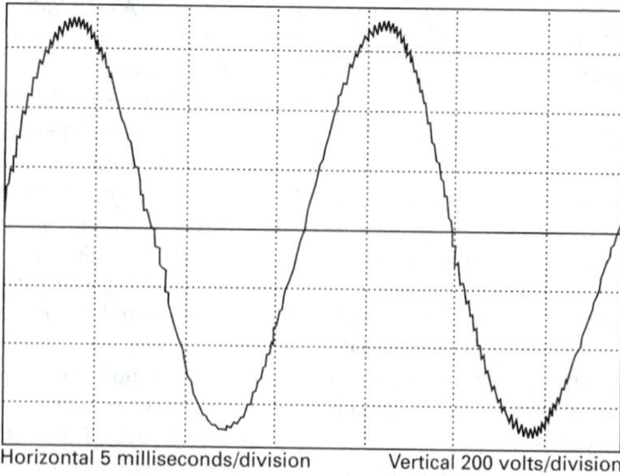

noise example

(IA/PE/SPD/PSE) 1100-1992s, C62.48-1995
(12) Undesirable sound emissions or undesirable electromagnetic signals/emissions. (PE/SUB) 1127-1998
(13) Unwanted electrical signals in the circuits of the control systems in which they occur. (SCC22) 1346-1998

noise amplitude distribution (control of system electromagnetic compatibility) A distribution showing the pulse amplitude that is equalled or exceeded as a function of pulse repetition rate. (EMC) C63.12-1987

noise, audio-frequency *See:* audio-frequency noise.

noise cleaning *See:* smoothing.

noise, common-mode *See:* common-mode noise.

noise-current generator A current generator, the output of which is described by a random function of time. *Note:* At a specified frequency, a noise-current generator can often be adequately characterized by its mean-square current within the frequency increment Δf or by its spectral density. If the circuit contains more than one noise-voltage generator or noise-current generator, the correlation coefficients among the generators must also be specified. *See also:* signal-to-noise ratio; network analysis; signal. (IE/ED) [43], [45]

noise, dBr*n* Decibels relative to one picowatt (0 dBr*n* equals −90 dBm) reference noise level. This is the customary North American unit for measurement of noise power on telecommunication circuits. (COM/TA) 1007-1991r

noise, dBr*n* C Decibels relative to one picowatt reference noise level, measured with C-message filter weighting. *Note:* there are several other filter weightings besides C-message. (COM/TA) 1007-1991r

noise definitions (data transmission) (1) dBa: For F1 A weighted noise measurement, usually obtained with a WECO 2B noise meter. 0 dBa equivalent to 1000 Hz tone with a power of -85 dBm. Or, a 3 kHz white noise band of -82 dBm. Filter produces a 3 dB loss over flat indication.
(2) dBmp: Filter produces a 2.5 dB loss compared to no weighting. For psophometrically weighted noise according to CCITT; 0 dBmp is equivalent to a 1000 Hz tone with a power of -1 dBm (or an 800 Hz tone with a power of 0 dBm). Or, a 3 kHz band of white noise with a power of +2.5 dBm. Filter produces a 2.5 dB loss compared to no weighting.
(3) dBrn 144: Obsolete unit for Type 144 weighted-noise measurement.
(4) dBrnC: For C-message weighted noise; 0 dBrnc is equivalent to a 1000 Hz tone with a power of -90 dBm (10^{-12} W (watt) or 90 dB below 1 mW (milliwatt). Or, 3 kHz white-noise band of a power of 88 dBm (actually 88.5 dBm). Filter produces a 2 dB loss compared to no weighting.
(5) dBrnCO, dBaO, pWpO, dBmpO: Noise units as measured in dBrnC, dBa, pWp, and dBmp at (or referred to) a 0 transmission level point.
(6) pWp: For psophometrically weighted noise; 1 pWp is equivalent to a 1000 Hz tone with a power of -91 dBm (or an 800 Hz tone of -90 dBm). Or, a 3 kHz band of white noise with a power of 88 dBm.

$$1 \text{ pWp} = \frac{1 \text{ pW}}{1.78} = 0.56 \text{ pW}$$

(PE) 599-1985w

noise, differential-mode *See:* transverse-mode noise.

noise diode, ideal *See:* ideal noise diode.

noise discriminator/noise guard/noise monitor A circuit or algorithm intended to discriminate between speech and noise. It can affect switching, transmission, and/or noise performance. (COM/TA) 1329-1999

noise, electrical *See:* electrical noise.

noise equivalent power (fiber optics) At a given modulation frequency, wavelength, and for a given effective noise bandwidth, the radiant power that produces a signal-to-noise ratio of 1 at the output of a given detector. *Notes:* 1. Some manufacturers and authors define NEP as the minimum detectable power per root unit bandwidth; when defined in this way, NEP has the units of watts/(hertz)$^{1/2}$. Therefore, the term is a misnomer, because the units of power are watts. 2. Some manufacturers define NEP as the radiant power that produces a signal-to-dark-current noise ratio of unity. This is misleading when dark-current noise does not dominate, as is often true in fiber systems. *See also:* detectivity; D-star. (Std100) 812-1984w

noise factor (1) (of a two-port transducer) At a specified input frequency the ratio of: a) the total noise power per unit bandwidth at a corresponding output frequency available at the output port when the noise temperature of its input termination is standard (290°K) at all frequencies, to b) that portion of *a* engendered at the input frequency by the input termination at the standard noise temperature (290°K). *Notes:* 1. For heterodyne systems there will be, in principle, more than one output frequency corresponding to a single input frequency, and vice versa: for each pair of corresponding frequencies a noise factor is defined. Definition b) includes only that noise from the input termination which appears in the output via the principal-frequency transformation of the system, that is, via the signal-frequency transformation(s), and does not include spurious contributions such as those from an unused image-frequency or an unused idler-frequency transformation. 2. The phrase "available at the output port" may be replaced by "delivered by system into an output termination." 3. To characterize a system by a noise factor is meaningful only when the admittance (or impedance) of the input termination is specified. *See also:* noise figure. (ED) 161-1971w
(2) The ratio noise energy of the input to the output of a device. The noise factor has an associated bandwidth of measurement. (LM/C) 802.7-1989r

noise figure (A) (interference terminology, linear system) (at a selected input frequency) The ratio of: a) the total noise power per unit bandwidth (at a corresponding output frequency) delivered by the system into an output termination, to b) the portion thereof engendered at the input frequency by the input termination, whose noise temperature is standard (290°K at all frequencies). *Notes:* 1. Numerically, the noise factor F at frequency f can be expressed as

$$\overline{F}(f) = \frac{P_{\text{noise out}}}{GP_{\text{noise in}}}$$

where $P_{\text{noise out}}$ and $P_{\text{noise in}}$ are taken at frequency f and 290 kelvins and G is the gain of the system at frequency f. 2. For heterodyne systems there will be, in principle, more than one output frequency corresponding to a single input frequency and vice versa; for each pair of corresponding frequencies a noise factor is defined. 3. The phrase "available at the output terminals" may be replaced by "delivered by the system into an output termination" without changing the sense of the definition. 4. The term "noise factor" is used where it is desired to emphasize that the noise figure is a function of input frequency. **(B) (interference terminology, linear system)** (average) The ratio of *a*, the total noise power delivered by the system into its output termination when the noise temperature of its input termination is standard (290 K) at all frequencies, to *b*, the portion thereof engendered by the input termination. *Notes:* 1. For heterodyne systems, portion *a* includes only that noise from the input termination that appears in the output via the principal frequency transformation of the system, and does not include spurious contributions such as those from image frequency transformations. 2. A quantitative relation between average noise factor F and spot noise factor $F\,(f)$ is

$$\overline{F} = \frac{\int_0^\infty F(f)G(f)\mathrm{d}f}{\int_0^\infty G(f)\mathrm{d}f}$$

where f is the input frequency and $F(f)$ is the ratio of *a*, the signal power delivered by the system into its output termination, to *b*, the corresponding signal power available from the input termination at the input frequency. For heterodyne systems, *a* comprises only power appearing in the output via the principal frequency transformation of the system. 3. To characterize a system by an average noise factor is meaningful only when the admittance (or impedance) of the input termination is specified. *Synonym:* noise factor. (PE) 599-1985

(2) **(A) (broadband local area networks)** The ratio of the total white noise energy at the output to the amount of Johnson noise at the output. The Johnson noise is due to the noise generated by the impedance of the signal source. **(B) (broadband local area networks)** The ratio noise energy of the input to the output of a device. The noise factor has an associated bandwidth of measurement. (LM/C) 802.7-1989

noise figure, average *See:* average noise figure.

noise floor The noise level at a referenced location and bandwidth. The video bandwidth is assumed to be a 4MHz bandwidth. A terminated 75Ω cable operating at 68°F or 20°C has a 4 MHz noise floor of − 59 dBmV. (LM/C) 802.7-1989r

noise-free equivalent amplifier (signal-transmission system) An ideal amplifier having no internally generated noise that has the same gain and input.output characteristics as the actual amplifier. *See also:* signal. (IE) [43]

noise generator (analog computer) In an analog computer, a computing element used purposely to introduce noise of specified amplitude distribution, spectral density, or root-mean square value, or appropriate combination therefore into other computing elements. (C) 165-1977w

noise generator diode A diode in which the noise is generated by shot effect and the noise power of which is a definite function of the direct current. (ED) [45]

noise, idle channel *See:* idle channel noise.

noise immunity The ability of a circuit to perform its function in the presence of noise. (C) 610.10-1994w

noise, impulse *See:* impulse noise.

noise jitter *See:* uncorrelated jitter.

noise killer (telegraph circuits) An electric network inserted usually at the sending end, for the purpose of reducing interference with other communication circuits. *See also:* telegraphy. (PE/EEC) [119]

noise level (1) (A) (audio and electroacoustics) The noise power density spectrum in the frequency range of interest, *See also:* signal-to-noise ratio. **(B) (audio and electroacoustics)** The average noise power in the frequency range of interest. *See also:* signal-to-noise ratio. **(C) (audio and electroacoustics)** The indication on a specified instrument. *Notes:* 1. In definition (C), the characteristics of the instrument are determined by the type of noise to be measured and the application of the results thereof. 2. Noise level is usually expressed in decibels relative to a reference value. *See also:* signal-to-noise ratio. (SP/PE) 151-1965, 599-1985
(2) (speech quality measurements) The A-weighted sound level of the noise. 297-1969w

noise-level test (rotating machinery) A test taken to determine the noise level produced by a machine under specified conditions of operation and measurement. *See also:* asynchronous machine. (PE) [9]

noise linewidth (1) (germanium gamma-ray detectors) (x-ray energy spectrometers) (charged-particle detectors) (semiconductor radiation detectors) The contribution of noise to the width of a spectral peak. (NPS/NID) 759-1984r, 301-1976s
(2) (charged-particle detectors) The contribution of electronic noise to the width of a spectral peak. (NPS) 300-1988r
(3) The contribution of electrical noise to the width of a spectral peak. (Electrical noise adds to the peak width in quadrature.) (NPS) 325-1996

noise, longitudinal *See:* longitudinal noise.

noise measurement units (data transmission) The following units are used to express weighted and unweighted circuit noise. They include terms used in American and International practice. *dBa-dBrn adjusted.* Weighted circuit noise power, in dB, referred to 3.16 pW (picowatts) (-85 dBm) which is 0 dBa. Use of F1 A line or HA1 receiver weighting is indicated in parenthesis as required. *dBrn (Describes above reference noise).* Weighted circuit noise power in dB referred to 1.0 pW (-90 dBm) which is 0 dBrn. Use of 144 line, 144 receiver, or C message weighting, parentheses, as required. With C-message weighting, as 1 mW (milliwatt), 1000 Hz tone will read +90 dBrn, but the same power as white noise distributed over a $3kH_3$ band (nominally 300 to 3300 Hz) will read approximately +88.5 dBrn (rounded off to +88 dBrn, due to frequency weighting). With 144 weighting, as 1 mW, 1000 Hz tone will also read +90 dBrn, but the same $3kH_3$ white-noise power would read only +82 dBrn, due to the different frequency weighting. (PE) 599-1985w

noise measuring set *See:* circuit noise meter.

noise, metallic *See:* metallic noise.

noise, normal-mode *See:* normal-mode noise.

noise power (Hall effect devices) The power generated by a random electromagnetic process. (MAG) 296-1969w

noise-power ratio (NPR) (data transmission) (expressed in decibels) This is a term commonly used in noise loading technique. Usually an uncalibrated receiver is used to measure the noise power in a channel of a system loaded with noise, first with full noise in the channel and the noise source. The ratio of these readings is the NPR. An NPR reading is independent of the noise bandwidth of the receiver; provided the same bandwidth is used in both noise measurements and the band stop filters are wide enough. (PE) 599-1985w

noise pressure equivalent (electroacoustic transducer or system used for sound reception) The root-mean-square sound pressure of a sinusoidal plane progressive wave that, if propagated parallel to the principal axis of the transducer, would produce an open-circuit signal voltage equal to the root-mean-square of the inherent open-circuit noise voltage of the transducer in a transmission band having a bandwidth of 1 hertz and centered on the frequency of the plane sound wave. *Note:* If the equivalent noise pressure of the transducer is a function of secondary variables, such as ambient temperature or pressure, the applicable values of these quantities should be stated explicitly. *See also:* phonograph pickup. (SP) [32]

noise quieting (receiver performance) (receiver) A measure of the quantity of radio-frequency energy, at a specified deviation from the receiver center frequency, required to reduce the noise output by a specified amount. (VT) [37]

noise reduction (photographic recording and reproducing) A process whereby the average transmission of the sound track of the print (averaged across the track) is decreased for signals of low level and increased for signals of high level. *Note:* Since the noise introduced by the sound track is less at low transmission, this process reduces film noise during soft passages. The effect is normally accomplished automatically. *See also:* phonograph pickup. (SP) [32]

noise referred to the input (of an amplifier) The electronic noise level at the input of a hypothetically noise-free amplifier that would produce the same noise as measured at the point of observation in the actual amplifier. This definition is implied when the term "amplifier noise" is used without qualifications. (NPS) 325-1996

noise sidebands (non-real time spectrum analyzer) Undesired response caused by noise internal to the spectrum analyzer appearing on the display around a desired response. (IM) 748-1979w

noise suppression *See:* smoothing.

noise temperature (1) (general) (at a pair of terminals and at a specific frequency) The temperature of a passive system having an available noise power per unit bandwidth equal to that of the actual terminals. *Note:* Thus, the noise temperature of a simple resistor is the actual temperature of the resistor, while the noise temperature of a diode may be many times the observed absolute temperature. *See also:* signal; signal-to-noise ratio. (IE) [43]
(2) (standard) The standard reference temperature T_0 for noise measurements is 290°K. *Note:* kT_0/e = 0.0250 volt, where e is the magnitude of the electronic charge and k is Boltzmann's constant. (ED) 161-1971w

(3) (of an antenna) The temperature of a resistor having an available thermal noise power per unit bandwidth equal to that at the antenna output at a specified frequency. *Note:* Noise temperature of an antenna depends on its coupling to all noise sources in its environment as well as noise generated within the antenna. (AP/ANT) 145-1993

(4) (at a port) The temperature of a passive system having an available noise power per unit bandwidth equal to that of the actual port, at a specified frequency. A uniform temperature throughout the passive system is implied. *See also:* thermal noise. (ED) 161-1971w

(5) (port) (selected frequency) (at a port and at a selected frequency) A temperature given by the exchangeable noise-power density divided by Boltzmann's constant, at a given port and at a stated frequency. *Notes:* 1. When expressed in units of kelvins, the noise temperature T is given by the relation

$$T = \frac{N}{k}$$

where N is the exchangeable noise-power density in watts per hertz at the port at the stated frequency and k is Boltzmann's constant expressed as joules per kelvin ($k \cong 1.38 \times 10^{-23}$ joules per kelvin). 2. Both N and T are negative for a port with an internal impedance having a negative real part. *See also:* signal-to-noise ratio; waveguide. (ED) [45]

(6) (average operating) Equivalent temperature of passive system having an available noise power equal to that of the operating system. In space communication systems the noise temperature is generally composed of contributions from the background, atmospheric, and receiver front end noise. (COM) [25]

noise-to-ground (data transmission) In telephone practice, the weighted noise current through the 100 000-Ω circuit of a circuit noise meter, connected between one or more telephone wires and ground. (PE) 599-1985w

noise transmission impairment (NTI) The noise transmission impairment that corresponds to a given amount of noise is the increase in distortionless transmission loss that would impair the telephone transmission over a substantially noise-free circuit by an amount equal to the impairment caused by the noise. Equal impairments are usually determined by judgment tests or intelligibility tests. (EEC/PE) [119]

noise, transverse-mode *See:* transverse-mode noise.

noise-voltage generator (interference terminology) A voltage generator the output of which is described by a random function of time. *Note:* At a specified frequency, a noise-voltage generator can often be adequately characterized by its mean-square voltage with the frequency increment Δf or by its spectral density. If the circuit contains more than one noise-current generator or noise-voltage generator, the correlation coefficients among the generators must be specified. *See also:* signal-to-noise ratio; signal; network analysis. (ED) 161-1971w

noise weighting (data transmission) In measurement of circuit noise, a specific amplitude frequency characteristic of a noise measuring set. It is designed to give numerical readings which approximate the amount of transmission impairment due to the noise, to an average listener, using a particular class of telephone subset. The noise weightings generally used were established by the agencies concerned with public telephone service and are based on characteristics of specific commercial telephone subsets, representing successive stages of technological development. The coding of commercial apparatus appears in the nomenclature of certain weightings. The same weighting nomenclature and units are used in the military versions and in the commercial noise measuring sets. (PE) 599-1985w

noise with tone, dBr*n* C-notched The measure of noise on a channel when a holding tone is present. A very narrow band elimination filter is used with a C-message filter to significantly reduce the level of the holding tone, and pass the noise in the remaining part of the channel. (COM/TA) 1007-1991r

no-load (rotating machinery) (adjective) The state of a machine rotating at normal speed under rated conditions, but when no output is required of it. *See also:* asynchronous machine; direct-current commutating machine. (PE) [9]

no-load (excitation) losses (power and distribution transformers) Those losses that are incident to the excitation of the transformer. No-load (excitation) losses include core loss, dielectric loss, conductor loss in the winding due to exciting current, and conductor loss due to circulating current in parallel windings. These losses change with the excitation voltage. (PE/TR) C57.12.80-1978r

no-load field current (excitation systems for synchronous machines) The direct current in the field winding of the synchronous machine required to produce rated voltage at no-load and rated speed. (PE/EDPG) 421.1-1986r

no-load field voltage (excitation systems for synchronous machines) The voltage required across the terminals of the field winding of the synchronous machine under conditions of no-load, rated speed, and rated terminal voltage, and with the field winding at 25°C. (PE/EDPG) 421.1-1986r

no-load losses (1) Those losses that are incident to the excitation of the regulator. No-load losses include core loss, dielectric loss, conductor loss in the winding due to exciting current, and conductor loss due to circulating current in parallel windings. These losses change with the excitation voltage. *Synonym:* excitation losses. (PE/TR) C57.15-1999

(2) (power operations) Power losses in an electric facility when energized at rated voltage and frequency, but not carrying load. (PE/PSE) 858-1987s

(3) (electronic power transformer) The input power, expressed in watts, to a completely assembled transformer that is excited at rated terminal voltage and frequency, but not supplying load current. (PEL/ET) 295-1969r

no-load saturation curve (no-load characteristic) (of a synchronous machine) The saturation curve of a machine on no-load. (PE) [9]

no-load speed (of an electric drive) The speed that the output shaft of the drive attains with no external load connected and with the drive adjusted to deliver rated output at rated speed. *Note:* In referring to the speed with no external load connected and with the drive adjusted for a specified condition other than for rated output at rated speed, it is customary to speak of the no-load speed under the (stated) conditions. *See also:* electric drive. (IA/ICTL/IAC) [60]

no-load test (synchronous machines) A test in which the machine is run as a motor providing no useful mechanical output from the shaft. (PE) [9]

NOMAD A fourth-generation programming language that permits a wide latitude of generality in algebraic expressions. *Note:* Adapted from MAD. (C) 610.13-1993w

nomenclature (generating stations electric power system) (electric power system) The words and terms used to identify electric power systems. (PE/EDPG) 505-1977r

nomenclature standard A standard that describes the characteristics of a system or set of names, or designations, or symbols. (C) 610.12-1990

nominal A term used to describe functional behavior as being within expected norms, or as designed. (ATLAS) 1232-1995

nominal band of regulated voltage (synchronous machines) The band of regulated voltage for a load range between any load requiring no-load field voltage and any load requiring rated-load field voltage with any compensating means used to produce a deliberate change in regulated voltage inoperative. *See also:* direct-current commutating machine. (EEC/PE/EDPG) [119], 421-1972s

nominal battery voltage The voltage computed on the basis of 2.0 volts per cell for the lead-acid type and 1.2 volts per cell for the alkali type. (NESC/NEC) [86]

nominal collector ring voltage *See:* rated-load field voltage.

nominal conductor gradient (overhead-power-line corona and radio noise) The gradient determined for a smooth cylindrical conductor whose diameter is equal to the outside diameter of the actual (stranded) conductor. (T&D/PE) 539-1990

nominal control current (Hall generator) That value of control current that, if exceeded, will cause the linearity error of the device to exceed a rated magnitude. (MAG) 296-1969w

nominal discharge current (1) (arresters) The discharge current having a designated peak value and waveshape, that is used to classify an arrester with respect to protective characteristics. *Note:* It is also the discharge current that is used to initiate follow current in the operating duty test. (PE) [8]
(2) The discharge current that can be applied to a surge protective device a specified number of times without causing damage to it. The nominal discharge current is a peak surge current, with a wave shape of 8/20. (PE) C62.34-1996

nominal gas density The manufacturer's recommended operating gas density (usually expressed as pressure at 20°C). (SWG/SUB/PE) C37.122-1983s, C37.122.1-1993, C37.100-1992

nominal input power (loudspeaker measurements) Nominal input power is equal to the square of the true rms voltage at the input terminals of the loudspeaker, divided by the rated impedance. *See also:* rated impedance. 219-1975w

nonfirm power (power operations) (electric power system) Power supplied or available under an arrangement that does not have the availability feature of firm power. (PE/PSE) 858-1987s, 346-1973w

nominal line pitch (television) The average separation between centers of adjacent lines forming a raster. (BT/AV) 201-1979w

nominal line width (facsimile) The average separation between centers of adjacent scanning or recording lines. *See also:* recording; scanning. (COM) 168-1956w

nominal metallic impedance (measuring longitudinal balance of telephone equipment operating in the voice band) Impedance based on lumped constants of a metallic circuit at a single given frequency. (COM/TA) 455-1985w

nominal pull-in torque (synchronous motor) The torque it develops as an induction motor when operating at 95% of synchronous speed with rated voltage applied at rated frequency. *Note:* This quantity is useful for comparative purposes when the inertia of the load is not known. (PE) [9]

nominal rate of rise (1) (of an impulse) For a wavefront, the slope of the line that determines the virtual zero. It is usually expressed in volts or amperes per microsecond. (SPD/PE) C62.22-1997, C62.11-1999
(2) (of an impulse wavefront) The slope of the line that determines the virtual zero. It is usually expressed in volts or amperes per microsecond. (SPD/PE) C62.62-2000

nominal ratio *See:* marked ratio.

nominal synchronous-machine excitation-system ceiling voltage The ceiling voltage of the excitation system with:

1) The exciter and all rotating elements at rated speed.
2) The auxiliary supply voltages and frequencies at rated values.
3) The excitation system loaded with a resistor having a value equal to the resistance of the field winding to be excited at a temperature of: (A) 75°C for field windings designed to operate at rating with a temperature rise of 60°C or less, and (B) 100°C for field windings designed to operate at rating with a temperature rise greater than 60°C.
4) The manual control means adjusted as it would be to produce the rated voltage of the excitation system, if this manual control means is not under the control of the voltage regulator when the regulator is in service, unless otherwise specified. (The means used for controlling the exciter voltage with the voltage regulator out of service is normally called the manual control means.)
5) The voltage sensed by the synchronous machine voltage regulator reduced to give the maximum output from the regulator. Note that the regulator action may be simulated in test by applying to the field of the exciter under regulator control the maximum output developed by the regulator.

(PE) [9]

nominal system voltage (1) (power and distribution transformers) The system voltage by which the system is designated and to which certain operating characteristics of the system are related. (The nominal voltage of a system is near the voltage level at which the system normally operates and provides a per-unit base voltage for system study purposes. To allow for operating contingencies, systems generally operate at voltage levels about 5 to 10 percent below the maximum system voltage for which system components are designed.) (PE/TR) C57.12.80-1978r
(2) (rotating electric machinery) A number used to denote the general level of voltage of the system described and, in the case of an externally powered system, selected from the list of preferred values below. (PE/EM) 11-1980r
(3) (electric power systems in commercial buildings) The voltage by which a portion of the system is designated, and to which certain operating characteristics of the system are related. Each nominal system voltage pertains to a portion of the system that is bounded by transformers or utilization equipment. (IA/PSE) 241-1990r
(4) The root-mean-square phase-to-phase voltage by which the system is designated and to which certain operating characteristics of the system are related. *Note:* The nominal system voltage is near the voltage level at which the system normally operates. To allow for operating contingencies, systems generally operate at voltage levels about five to ten percent below the maximum system voltage for which systems components are designed. (PE/C) 1313.1-1996
(5) A nominal value assigned to designate a system of a given voltage class.
(SWG/PE/SPD) C37.100-1992, C62.22-1997, C62.11-1999, C62.62-2000
(6) A nominal value assigned to a system or circuit of a given voltage for the purpose of convenient designation. The term nominal voltage designates the line-to-line voltage, as distinguished from the line-to-neutral voltage. It applies to all parts of the system or circuit. (PE/TR) C57.15-1999

nominal thickness (cable element) (power distribution, underground cables) The specified, indicated, or named thickness. *Note:* In general, measured thicknesses will approximate but will not necessarily be identical with nominal thicknesses. (PE) [4]

nominal utilization voltage The voltage rating of certain utilization equipment used on the system. (IA/PSE) 241-1990r

nominal value (1) An arbitrary reference value selected to establish equipment ratings. (IA/PSE) 602-1996
(2) (metric practice) A value assigned for convenient designation; existing in name only. (SCC14/QUL) SI 10-1997, 268-1982s

nominal voltage (Vn) (1) A nominal value assigned to a circuit or system for the purpose of conveniently designating its voltage class (as 120/240, 480Y/277, 600 etc.) The actual voltage at which a circuit operates can vary from the nominal within a range that permits satisfactory operation of equipment. (NESC/NEC/SPD/PE) C62.48-1995, [86]
(2) (of a system) A nominal value assigned to a system or circuit of a given voltage class for the purpose of convenient designation. *Note:* The term "nominal voltage" designates the "line to line" voltage, as distinguished from the "line to neutral" voltage. It applies to all parts of the system or circuit. (PE/TR) C57.16-1996
(3) A nominal value assigned to a circuit or system for the purpose of conveniently designating its voltage class (as 208/120, 480/277, 600). (SCC22) 1346-1998

no motion A safety-critical function utilized to indicate that the train is at zero speed or sufficiently close to zero speed. Used to inhibit the ability of the doors to open when the train is moving and can be used for functions, such as emergency brake reset, that may require an indication of the no motion condition. (VT) 1475-1999

nonabsorbing state In a Markov chain model, a state that can be left once it is entered. *Contrast:* absorbing state. (C) 610.3-1989w

nonadjustable (as applied to Circuit Breakers) A qualifying term indicating that the circuit breaker does not have any adjustment to alter the value of current at which it will trip or the time required for its operation. (NESC/NEC) [86]

nonarithmetic shift *See:* logical shift.

nonatmospheric paths (atmospheric correction factors to dielectric tests) Paths, such as through a gas or vacuum sealed from the atmosphere, through a liquid such as oil, or through a solid, or a combination thereof. 579-1975w

nonautomatic Action requiring personal intervention for its control. As applied to an electric controller, nonautomatic control does not necessarily imply a manual controller, but only that personal intervention is necessary. *See also:* automatic. (NESC/NEC) [86]

nonautomatic extraction turbine (control systems for steam turbine-generator units) (condensing or noncondensing) Steam is extracted from one or more stages, but without means of controlling the pressure(s) of the extracted steam. (PE/EDPG) 122-1985s

nonautomatic opening The opening of a switching device only in response to an act of an attendant. *Synonym:* nonautomatic tripping. (SWG/PE) C37.100-1992

nonautomatic operation Operation controlled by an attendant. (SWG/PE) C37.100-1992

nonautomatic tripping *See:* nonautomatic opening.

nonbilling error Occurs when a call that should be billed is not billed due to switching system error. (COM/TA) 973-1990w

non-blind testing Testing of capabilities when the service laboratory is aware that they are being tested for conformance. (NI) N42.23-1995

nonblocking Executing with `POSIX_IO.Non_Blocking` set. When executing with `POSIX_IO.Non_Blocking` set, functions do not wait for protocol events (*e.g.*, acknowledgments) to occur before returning control. *See also:* blocking. (C) 1003.5-1999

nonblocking switching network (telephone switching systems) A switching network in which any idle outlet can always be reached from any given inlet under all traffic conditions. (COM) 312-1977w

nonbridging relay contacts A contact arrangement in which the opening contact opens before the closing contact closes. (EEC/REE) [87]

non-bus-based architecture A computer architecture that is not designed on a bus connection; for example, a crosspoint switch. *Contrast:* bus-based architecture. (C) 610.10-1994w

noncanonical input processing The processing of terminal input as uninterpreted characters. (C) 1003.5-1999

nonce A random or pseudo-random value used within an authentication system. (SCC32) 1455-1999

nonceramic insulator Insulators made from polymer materials. (T&D/PE) 957-1995

noncharacteristic harmonic (1) (harmonic control and reactive compensation of static power converters) Those harmonics which are not produced by semiconductor converter equipment in the course of normal operation. These may be the result of beat frequencies, a demodulation of characteristic harmonics and the fundamental, or unbalance in the ac power system or unsymmetrical delay angle. (IA/SPC) 936-1987w

(2) Harmonics that are not produced by semiconductor converter equipment in the course of normal operation. These may be a result of beat frequencies; a demodulation of characteristic harmonics and the fundamental; or an imbalance in the ac power system, asymmetrical delay angle, or cycloconverter operation. (IA/SPC) 519-1992

nonclock (test access port and boundary-scan architecture) A signal where the transitions between the low and high logic levels do not themselves cause operation of stored-state devices. The logic level is important only at the time of a transition on a clock signal. (TT/C) 1149.1-1990

noncode fire-alarm system A local fire-alarm system in which the alarm signal is continuous and is usually sounded by vibrating bells. *See also:* protective signaling. (EEC/PE) [119]

noncoherent moving-target indication A form of MTI radar in which a moving target is detected by using the clutter echo as the reference signal. *Notes:* 1. No internal reference signal is required. 2. Sometimes called externally coherent MTI. (AES) 686-1997

noncoherent MTI *See:* noncoherent moving-target indication.

noncoherent transaction A transaction (typically read or write) that is completed without checking for consistency with other caches on the local bus. For example, noncoherent 4-byte read and write transactions are used to access CSRs. Note that noncoherent transactions may be converted into coherent transaction sequences when passing through bus bridges. (C/MM) 1212-1991s

noncoincident demand The sum of the individual maximum demands regardless of time occurrence within a specified period. *See also:* alternating-current distribution. (PE/PSE) 858-1993w, 346-1973w

noncomposite color picture signal (national television system committee color television) The electric signal that represents complete color picture information but excludes line and field sync signals. (BT/AV) 201-1979w

nonconcurrent bus operation In dual bus systems "concurrent bus operation capable" describes buses capable of conducting simultaneous unrelated bus transactions. "Nonconcurrent bus operation capable" describes buses capable of conducting a transaction on only one bus at a time. (C/BA) 14536-1995

nonconducting Made of a material of high dielectric strength. (T&D/PE) 957-1995

nonconductive *See:* nonconducting.

nonconformance (1) (nuclear power quality assurance) A deficiency in characteristic, documentation, or procedure that renders the quality of an item or activity unacceptable or indeterminate. (PE/NP) [124]

(2) A non-conformance is a deficiency in characteristic, documentation, or procedure which renders the quality of an item or service unacceptable or indeterminate. Any activity in the laboratory which adversely affects data quality can result in a nonconformance. (NI) N42.23-1995

nonconforming load (1) (electric power system) A customer load, the characteristics of which are such as to require special treatment in deriving incremental transmission losses. (PE/NP) 387-1984s

(2) (automatic generation control on electric power systems) A load that has characteristics dissimilar to those of the system load and that, depending on the modeling and/or control methodologies, may require special treatment. (PE/PSE) 94-1991w

nonconjunction The dyadic Boolean operation whose result has the Boolean value 0 if and only if each operand has the Boolean value 1. *Contrast:* conjunction. *See also:* NAND gate. (C) 610.10-1994w

noncontact plunger *See:* choke piston.

noncontinuous electrode A furnace electrode the residual length of which is discarded when too short for further effective use. *See also:* electrothermics. (EEC/PE) [119]

noncontinuous enclosure (1) A bus enclosure in which the consecutive sections of the enclosure for the same phase conductor are electrically insulated from each other, although each section is connected to ground. *Note:* This construction prevents longitudinal currents from flowing beyond each enclosure section. This design is no longer in common usage. (PE/SUB/SWG-OLD) C37.100-1992, C37.122-1993, C37.122.1-1993

(2) Refers to a type of isolated-phase bus in which the enclosure is sectionalized with insulation between sections to block the longitudinal flow of current in the enclosure. (PE/EDPG) 665-1987s

(3) A bus enclosure with the consecutive sections of the housing of the same phase conductor electrically isolated (or insulated from each other), so that no current can flow beyond each enclosure section. (PE/SUB) 80-2000

noncontinuously acting regulator (synchronous machines) A regulator that requires a sustained finite charge in the controlled variable to initiate corrective action. (PE/EDPG) 421.1-1986r, 421-1972s

noncooperative target recognition (NCTR) *See:* target recognition.

noncritical failure (test, measurement, and diagnostic equipment) Any failure that degrades performance or results in degraded operation requiring special operating techniques or alternative modes of operation which could be tolerated throughout a mission but should be corrected immediately upon completion of mission. (MIL) [2]

non-deliverable software product A software product that is not required by the contract to be delivered to the acquirer or other designated recipient. (C/SE) J-STD-016-1995

nondelivered source statements Source statements that are developed in support of the final product, but not delivered to the customer. (C/SE) 1045-1992

nondestructive addition (mathematics of computing) Addition performed on a computer in such a manner that the first operand placed in the arithmetic register is the augend. The addend is then added, and the sum becomes the new augend. *Contrast:* destructive addition. (C) 1084-1986w

nondestructive read (1) (accessed) A read operation that does not erase the data in the source location. *Contrast:* destructive read. *See also:* multiaperture core. (C) 610.10-1994w, 610.12-1990

(2) (magnetic cores) A method of reading the magnetic state of a core without changing its state. (Std100) 163-1959w

nondestructive reading (charge-storage tubes) Reading that does not erase the stored information. *See also:* charge-storage tube. (ED/MIL) 158-1962w, [2]

nondestructive testing (test, measurement, and diagnostic equipment) Testing of a nature which does not impair the usability of the item. (MIL) [2]

nondeveloped source statements Existing source statements that are reused or deleted. (C/SE) 1045-1992

non-deviative absorption Absorption that occurs in regions where the refractive index is close to unity (i.e., when radio ray bending is negligible). *Contrast:* deviative absorption. (AP/PROP) 211-1997

nondirected flow (oil-immersed forced-oil-cooled transformers) Indicates that the pumped oil from heat exchangers or radiators flows freely inside the tank, and is not forced to flow through the windings. (PE/TR) C57.91-1995

nondirectional beacon (NDB) (air navigation) (navigation aid terms) A radio facility which can be used with an airborne DF (direction finder) to provide a line of position. *Synonyms:* locator; H; H beacon; compass locator. (AES/GCS) 172-1983w

nondirectional microphone *See:* omnidirectional microphone.

nondisconnecting fuse An assembly consisting of a fuse unit or fuseholder and a fuse support having clips for directly receiving the associated fuse unit or fuseholder, which has no provision for guided operation as a disconnecting switch. (SWG/PE) C37.40-1993, C37.100-1992

nondisjunction The dyadic Boolean operation whose result has the Boolean value of 1 if and only if each operand has the Boolean value 0. *Synonym:* NEITHER-NOR. *Contrast:* disjunction. *See also:* NOR gate. (C) 610.10-1994w

nondispersive delay line A delay line that nominally has constant delay over a specified frequency band. The argument (phase) of the transfer function is a linear function of frequency. (UFFC) 1037-1992w, [22]

nondisruptive discharge A discharge between intermediate electrodes or conductors in which the voltage across the terminal electrodes is not reduced to practically zero. (PE/PSIM) 4-1995

nondisruptive test A test that is invoked through a write to the TEST_START register and does not disrupt the node's operation (the node does not enter the testing state). (C/MM) 1212-1991s

nonenclosed Not surrounded by a medium that will prevent a person accidentally contacting live parts. (PE/TR) C57.12.80-1978r

nonenclosed switches, indoor or outdoor Switches designed for service without a housing restricting heat transfer to the external medium. (SWG/PE) C37.100-1992

nonenergy-limiting transformer (power and distribution transformers) A constant-potential transformer that does not have sufficient inherent impedance to limit the output current to a thermally safe maximum value. *See also:* specialty transformer. (PE/TR) C57.12.80-1978r, [57]

nonequivalence The dyadic Boolean operation whose result has the Boolean value of 1 if and only if the operands have different Boolean values. *Contrast:* equivalence. *See also:* exclusive-OR gate; nonidentity. (C) 610.10-1994w

nonerasable storage *See:* fixed storage.

nonessential loads Those station auxiliary loads not immediately necessary to maintain full HVDC station output. (SUB/PE) 1158-1991r

no network response Occurs when a call attempt encounters a switching system trouble and the switching system does not notify the customer that the call cannot be completed. (COM/TA) 973-1990w

nonexposed installation (lightning) An installation in which the apparatus is not subject to overvoltages of atmospheric origin. *Note:* Such installations are usually connected to cable networks. (PE) [8], [84]

nongravitational acceleration (accelerometer) That component of the acceleration of an object that is caused by externally applied forces. In vector notation: $\vec{A} = \vec{G} + \vec{N}$ where $\vec{A}$ = kinematic acceleration; i.e., the second derivative of position with respect to time in inertial space; $\vec{G}$ = the acceleration due to gravity, the acceleration of an object that is allowed to fall freely in the local gravitational field; and $\vec{N}$ = nongravitational acceleration. Nongravitational acceleration is the force applied to an object, divided by its mass. For an accelerometer, it is the indicated acceleration, $\vec{A}_{ind} = \vec{N} = \vec{A} - \vec{G}$. A freely falling accelerometer ($\vec{A} = \vec{G}$) has no output, and a stationary accelerometer ($\vec{A} = 0$) indicates 1 g upward. (AES/GYAC) 528-1994

nongyroscopic angular sensor An inertial angular sensor whose functions do not depend on the angular momentum of a spinning rotor. (AES/GYAC) 528-1994

nonhoming (telephone switching systems) Resumption of a sequential switching operation from its last setting. (COM) 312-1977w

nonidentifying relationship A kind of specific (not many-to-many) relationship in which some or all of the attributes contained in the primary key of the parent entity do not participate in the primary key of the child entity. *Contrast:* identifying relationship. *See also:* mandatory nonidentifying relationship; optional nonidentifying relationship. (C/SE) 1320.2-1998

nonidentity The Boolean operation whose result has the Boolean value 1 if and only if all the operands do not have the same Boolean value. *Note:* A nonidentity operation on two operands is a nonequivalence operation. *Contrast:* identity. (C) 610.10-1994w

nonidentity operation (mathematics of computing) A Boolean operation whose result is true if and only if not all of the operands have the same Boolean value. *Note:* A nonidentity operation on two operands is the same as an exclusive-OR operation. (C) 1084-1986w

nonimpact printer A printer in which printing is not the result of a mechanical contact with the printing medium. *Contrast:* impact printer. *See also:* magnetographic printer; ink jet printer; electrostatic printer; electrosensitive printer; thermal printer; xerographic printer. (C) 610.10-1994w

noninjecting contact (1) (x-ray energy spectrometers) (charged-particle detectors) (of a semiconductor radiation detector) A contact at which the carrier density in the adjacent semiconductor material is not changed from its equilibrium value. (NPS/NID) 759-1984r, 300-1988r
(2) (germanium gamma-ray detectors) (of a semiconductor radiation detector) A purely resistive contact, that is, one that has a linear voltage-current characteristic throughout its entire operating range. (NPS) 325-1986s

noninterference testing (test, measurement, and diagnostic equipment) A type of on-line testing that may be carried out during normal operation of the unit under test without affecting the operation. *See also:* interference testing. (MIL) [2]

noninterlaced Pertaining to a display device in which every line of pixels is refreshed on each pass. *Contrast:* interlaced. (C) 610.10-1994w

nonintrinsic relationship A kind of relationship that is partial, is multi-valued, or may change. *Contrast:* intrinsic relationship. (C/SE) 1320.2-1998

noninverting parametric device A parametric device whose operation depends essentially upon three frequencies, a harmonic of the pump frequency and two signal frequencies, of which one is the sum of the other plus the pump harmonic. *Note:* Such a device can never provide gain at either of the signal frequencies. It is said to be noninverting because if either of the two signals is moved upward in frequency, the other will move upward in frequency. *See also:* parametric device. (ED) [46]

nonislanding inverter An inverter that will cease to energize the utility line in ten cycles or less when subjected to a typical islanded load in which either of the following is true:

a) There is at least a 50% mismatch in real power load to inverter output (that is, real power load is $<$ 50% or $>$ 150% of inverter power output).
b) The islanded-load power factor is $<$ 0.95 (lead or lag).

If the real-power-generation-to-load match is within 50% and the islanded-load power factor is $>$ 0.95, then a nonislanding inverter will cease to energize the utility line within 2 s whenever the connected line has a quality factor of 2.5 or less. *Note:* See Annex A for a test procedure that identifies an inverter as a nonislanding inverter. (SCC21) 929-2000

nonisolated amplifier An amplifier that has an electrical connection between the signal circuit and another circuit including ground. (C) 610.10-1994w

nonkey attribute An attribute that is not the primary or a part of a composite primary key of an entity. (C/SE) 1320.2-1998

nonlinear The amplitude of the output is not linearly proportional to the input. If a true sine wave were transmitted through a nonlinear device, its shape would be changed. (PE/IC) 1143-1994r

nonlinear capacitor A capacitor having a mean-charge characteristic or a peak-charge characteristic that is not linear, or a reversible capacitance that varies with bias voltage. (ED) [46]

nonlinear circuit *See:* nonlinear network.

nonlinear data structure (data management) A nonprimitive data structure that can represent data that is multidimensional in nature. For example, a tree. *Contrast:* linear data structure. (C) 610.5-1990w

nonlinear devices Nonlinear processing, including automatic gain control (AGC) circuits and noise guards. (COM/TA) 1329-1999

nonlinear distortion (1) Distortion caused by a deviation from a desired linear relationship between specified measures of the output and input of a system. *Note:* The related measures need not to be output and input values of the same quantity: for example, in a linear detector the desired relation is between the output signal voltage and the input modulation envelope: or the modulation of the input carrier and the resultant detected signal. *See also:* close-talking microphone; distortion. (SP) 151-1965w
(2) The generation of new signal components not present in the original transmitted signal. (PE/IC) 1143-1994r

nonlinear element capacitance (varactor measurements) The capacitance of the high frequency equivalent series RC circuit. (ED) 318-1971w

nonlinear exponent (low voltage varistor surge arresters) A measure of varistor voltage nonlinearity between two given operating currents, expressed as $I = KV^{\alpha}$; where I is any current between the two operating currents I_1 and I_2; K is a device constant; V is the varistor voltage; and a is the logarithm of the ratio of the two operating currents (I_1/I_2) divided by the logarithm of the ratio of the varistor voltages at the two operating currents (V_2/V_1). (PE) [8]

nonlinear FM The modulation in a waveform whose phase variation with frequency has terms of order greater than two and, consequently, whose frequency varies in a nonlinear fashion with time (inside the specified envelope of the impulse response). (UFFC) 1037-1992w

nonlinear ideal capacitor An ideal capacitor whose transferred-charge characteristic is not linear. *See also:* nonlinear capacitor. (ED) [46]

nonlinearity (1) (x-ray energy spectrometers) (of a pulse amplifying system) Distortion caused by a deviation from a desired linear relationship between specified measures of the output and input pulse amplitudes of a system or device. (NPS/NID) 759-1984r
(2) (accelerometer) (gyros) The systematic deviation from the straight line that defines the nominal input-output relationship. (AES/GYAC) 528-1994

nonlinear load A load that draws a nonsinusoidal current wave when supplied by a sinusoidal voltage source. (IA/SPC/PSE) 519-1992, 1100-1999

nonlinear load current Load current that is associated with a nonlinear load. *See also:* nonlinear load. (IA/PSE) 1100-1999

nonlinear network (A) (signal-transmission system) A network (circuit) not specifiable by linear differential equations with time and/or position coordinates as the independent variable. *Note:* It will not operate in accordance with the superposition theorem. *See also:* signal; network analysis. **(B) (signal-transmission system)** A network (circuit) in which the signal transmission characteristics depend on the input signal magnitude. *See also:* signal; network analysis. (IE) [43], 270-1966

nonlinear optimization *See:* nonlinear programming.

nonlinear programming (1) In operations research, a procedure for locating the maximum or minimum of a function of variables that are subject to constraints, when either the function or the constraints, or both, are nonlinear. *Synonym:* nonlinear optimization. *Contrast:* linear programming. *See also:* convex programming. (C) 610.2-1987
(2) Optimization problem in which any or all of the following are nonlinear in the variables:

a) The objective functions.
b) The defining interrelationships among the variables, the plant description.
c) The constraints.

See also: system. (SMC) [63]

nonlinear parameter A parameter dependent on the magnitude of one or more of the dependent variables or driving forces of the system. *Note:* Examples of dependent variables are current, voltage, and analogous quantities. (Std100) 270-1966w

nonlinear resistor-type arrester (valve type) An arrester having a single or a multiple spark gap connected in series with nonlinear resistance. *Note:* If the arrester has no series gap, the characteristic element limits the follow current to a magnitude that does not interfere with the operation of the system. *See also:* valve-type arrester. (PE) [8]

nonlinear scattering (fiber optics) Direct conversion of a photon from one wavelength to one or more other wavelengths. In an optical waveguide, nonlinear scattering is usually not important below the threshold irradiance for stimulated nonlinear scattering. *Note:* Examples are Raman and Brillouin scattering. *See also:* photon. (Std100) 812-1984w

nonlinear series resistor (arresters) The part of the lightning arrester that, by its nonlinear voltage-current characteristics, acts as a low resistance to the flow of high discharge currents thus limiting the voltage across the arrester terminals, and as a high resistance at normal power-frequency voltage thus limiting the magnitude of follow current. (PE) [8]

nonlinear system or element *See:* linear system or element.

nonlined construction (dry cell) (primary cell) A type of construction in which a layer of paste forms the only medium between the depolarizing mix and the negative electrode. *See also:* electrolytic cell. (PE/EEC) [119]

nonline frequency components (thyristor) Expressed by the frequency and the root-mean-square (rms) value of the components having a different frequency than the line frequency. (IA/IPC) 428-1981w

nonline frequency content (thyristor) The function obtained by subtracting the line frequency component from a nonsinusoidal periodic function. *Note:* For the case of asynchronous ON-OFF control, this function varies with time. (IA/IPC) 428-1981w

nonload-break connector A connector designed to be separated and engaged on de-energized circuits. (T&D/PE) 386-1977s

nonloaded (of an electric impedance) The value of Q of such impedance without external coupling or connection. (EEC/PE) [119]

nonlocking (data management) In code extension characters, having the characteristic that a change in interpretation applies only to a specified number of the coded representations following, commonly only one. *Contrast:* locking. (C) 610.5-1990w

nonlocking carabiner Has a self-closing gate which remains closed, but not locked, until intentionally opened by the user for connection or disconnection. (T&D/PE) 1307-1996

nonlocking gear train A power transmission gear train design that allows the power train to back-drive whenever the primary driving force is removed. (PE/NP) 1290-1996

nonlocking shift character (data management) A shift-out character that causes the character following it to be interpreted as a member of a different character set from the original set. *Contrast:* locking shift character. (C) 610.5-1990w

nonmagnetic relay armature stop A nonmagnetic member separating the pole faces of core and armature in the operated position, used to reduce and stabilize the pull from residual magnetism in release. (EEC/REE) [87]

nonmagnetic relay shim Sometimes used for relay armature stop, nonmagnetic. (EEC/REE) [87]

nonmagnetic ship A ship constructed with an amount of magnetic material so small that it causes negligible distortion of the earth's magnetic field. (EEC/PE) [119]

nonmechanical switching device A switching device designed to close or open, or both, one or more electric circuits by means other than by separable mechanical contacts. (SWG/PE) C37.100-1992

nonmetallic extensions An assembly of two insulated conductors within a nonmetallic jacket or an extruded thermoplastic covering. The classification includes both surface extensions, intended for mounting directly on the surface of walls or ceilings, and aerial cable, containing a supporting messenger cable as an integral part of the cable assembly. (NESC/NEC) [86]

nonmetallic-sheathed cable A factory assembly of two or more insulated conductors having an outer sheath of moisture-resistant, flame-retardant, nonmetallic material. (NESC/NEC) [86]

nonminimum phase function (linear networks) A network function that is not minimum-phase. *Contrast:* minimum-phase function. (CAS) [13]

nonmultiple switchboard (telephone switching systems) A telecommunications switchboard having each line connected to only one jack. (COM) 312-1977w

nonmultiple transit spurious signals Signals unrelated to the main signal delay by a simple integer and that may be labeled by their respective delay times (dispersive and nondispersive delay lines). (UFFC) 1037-1992w, [22]

non-NULL DATA packet A DATA packet other than a NULL packet. (TT/C) 1149.5-1995

nonnumerical action (switch) That action that does not depend on the called number (such as hunting an idle trunk). (PE/EEC) [119]

nonoperate value *See:* relay nonoperate value.

non-PCB transformer *See:* nonpolychlorinated biphenyl transformer.

non-permanent records Those records required to show evidence that an activity was performed in accordance with the applicable requirements but need not be retained for the life of the project. (NI) N42.23-1995

nonphantomed circuit A two-wire or four-wire circuit that is not arranged to form part of a phantom circuit. *See also:* transmission line. (EEC/PE) [119]

nonphysical primary color (color) (television) A primary represented by a point order outside the area of the chromaticity diagram enclosed by the spectrum locus and the purple boundary. *Note:* Nonphysical primaries cannot be produced because they require negative power at some wavelengths. However, they have properties that facilitate colorimetric calculation. (BT/AV) 201-1979w

nonplanar network A network that cannot be drawn on a plane without crossing of branches. *See also:* network analysis. (Std100) 270-1966w

nonpolarized electrolytic capacitor An electrolytic capacitor in which the dielectric film is formed adjacent to both metal electrodes and in which the impedance to the flow of current is substantially the same in both directions. (EEC/PE) [119]

non-polarized return-to-zero recording (RZ(NP)) Return-to-reference recording in which zeros are represented by the absence of magnetization, ones are represented by a specified condition of magnetization, and the reference condition is zero magnetization. The specified condition is usually saturation. Conversely, the absence of magnetization can be used to represent ones, and the magnetized condition to represent zeros. *Synonym:* dipole modulation. *Contrast:* polarized return-to-zero recording. (C) 610.10-1994w

nonpolychlorinated biphenyl transformer (liquid-filled power transformers) Transformers that contain less than 50 parts per million (ppm) PCB. No transformer may ever be considered to be a non-PCB transformer unless its dielectric fluid has been tested or otherwise verified to contain less than 50 ppm PCB. Examples: Transformers so identified. *Synonym:* non-PCB transformer.

nonpreferred ATLAS An ATLAS structure that has been rendered redundant by a later structure, but that is retained in the standard for upward compatibility. (ATLAS) 771-1989s

nonprime attribute (data management) An attribute that is not part of any candidate key of a relation. *Contrast:* prime attribute. (C) 610.5-1990w

nonprimitive data structure (data management) A structured set of primitive data structures. Structures may be linear, as in a vector, or nonlinear as in a tree. *Synonym:* complex data structure. *Contrast:* primitive data structure. *See also:* nonlinear data structure; linear data structure. (C) 610.5-1990w

nonprocedural language (software) A language in which the user states what is to be achieved without having to state specific instructions that the computer must execute in a given sequence. *Contrast:* procedural language. *See also:* interactive language; declarative language; rule-based language. (C) 610.12-1990, 610.13-1993w

nonprocedural programming language (software unit testing) A computer programming language used to express the parameters of a problem rather than the steps in an solution (for example, report writer or sort specification languages). *See also:* procedural programming language. (SE/C) 1008-1987r

nonprofessional projector Those types other than described under professional projector. (NESC/NEC) [86]

n-on-p solar cells (photovoltaic power system) Photovoltaic energy-conversion cells in which a base of p-type silicon (having fixed electrons in a silicon lattice and positive holes that are free to move) is overlaid with a surface layer of n-type silicon (having fixed positive holes in a silicon lattice with electrons that are free to move). *See also:* photovoltaic power system. (AES) [41]

nonquadded cable *See:* paired cable.

nonreactive power (A) (polyphase circuit) At the terminals of entry of a polyphase circuit, a vector equal to the (vector) sum of the nonreactive powers for the individual terminals of entry. *Note:* The nonreactive power for each terminal of entry is determined by considering each phase conductor and the common reference point as a single-phase circuit, as described for distortion power. The sign given to the distortion power in determining the nonreactive power for each single-phase circuit shall be the same as that of the total active power. Nonreactive power for a polyphase circuit has as its two rectangular components the active power and the distortion power. If the voltages have the same waveform as the corresponding currents, the magnitude of the nonreactive power becomes the same as the active power. Nonreactive power is expressed in volt-amperes when the voltages are in volts and the currents in amperes. **(B) (single-phase two-wire circuit)** At the two terminals of entry of a single-phase two-wire circuit into a delimited region, a vector quantity having as its rectangular components the active power and the distortion power. Its magnitude is equal to the square root of the difference of the squares of the apparent power and the amplitude of the reactive power. Its magnitude is also equal to the square root of the sum of the squares of the amplitudes of the active power and the distortion power. If voltage and current have the same waveform, the magnitude of the nonreactive power is equal to the active power. The amplitude of the nonreactive power is given by the equation

$$N = (U^2 - Q^2)^{1/2} = (P^2 + D^2)^{1/2}$$

$$= \left[\sum_{r=1}^{r=\infty} \sum_{q=1}^{q=\infty} [E_r^2 I_q^2 - E_r E_q I_r I_q \times \sin(\alpha_r - \beta_r) \sin(\alpha_q - \beta_q)] \right]^{1/2}$$

where the symbols are those in power, apparent (single-phase two-wire circuit). In determining the vector position of the nonreactive power, the sign of the distortion power component must be assigned arbitrarily. Nonreactive power is expressed in volt-amperes when the voltage is in volts and the current in amperes. *See also:* distortion power. (Std100) 270-1966

non-real-time service Any service function that does not require real time service. *See also:* real-time service. (DIS/C) 1278.2-1995

nonreciprocal A device or network that does not have the property of reciprocity. (CAS) [13]

nonreciprocal differential insertion phase (nonlinear, active, and nonreciprocal waveguide components) The difference between insertion phases in the two opposite directions of propagation between two ports of a junction. (MTT) 457-1982w

nonredundant UPS configuration *See:* nonredundant uninterruptible power supply configuration.

nonredundant uninterruptible power supply configuration Consists of one or more UPS modules operating in parallel with a bypass circuit transfer switch and a battery. (IA/PSE) 241-1990r

nonrelevant failure *See:* failure.

nonremovable disk *See:* fixed disk.

nonrenewable fuse (1) (one time) A fuse or fuse unit not intended to be restored for service after circuit interruption. *Synonyms:* nonrenewable fuse unit; one-time fuse. (SWG/PE) C37.100-1992
(2) A fuse unit that, after circuit interruption, cannot readily be restored for service. (SWG/PE) C37.40-1993

nonrenewable fuse unit *See:* nonrenewable fuse.

nonrepaired item An item that is not repaired after a failure. *See also:* reliability. (R) [29]

nonrepetitive peak line voltage (thyristor) The highest instantaneous value of any nonrepetitive transient line voltage. *Note:* The voltage V_{LSM} may originate from operating circuit breakers, atmospheric disturbances, etc. This kind of voltage may be minimized by the provision of surge suppression components. (IA/IPC) 428-1981w

nonrepetitive peak OFF-state voltage (thyristor) The maximum instantaneous value of any nonrepetitive transient OFF-state voltage that occurs across the thyristor. *See also:* principal voltage-current characteristic. (IA/ED) 223-1966w, [12], [46], [62]

nonrepetitive peak reverse voltage rating (rectifier circuits) (semiconductor rectifiers) The maximum value of non-repetitive peak reverse voltage permitted by the manufacturer under stated conditions. *See also:* average forward current rating.

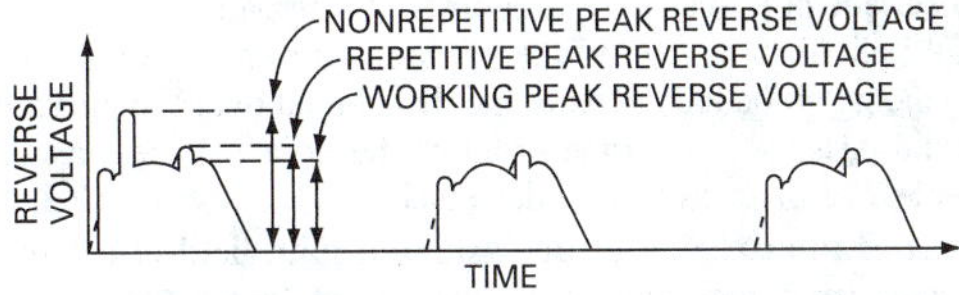

nonrepetitive peak reverse voltage rating

(IA) 59-1962w

nonrepetitive transient reverse voltage (1) (reverse-blocking thyristor) The maximum instantaneous value of any unrepetitive transient reverse voltage that occurs across a thyristor. *See also:* principal voltage-current characteristic. (IA) [62]
(2) (semiconductor rectifiers) The maximum instantaneous value of the reverse voltage, including all nonrepetitive transient voltages but excluding all repetitive transient voltages, that occurs across a semiconductor rectifier cell, rectifier diode, or rectifier stack. *See also:* semiconductor rectifier stack; rectification. (IA) 59-1962w

nonrequired time (availability) The period of time during which the user does not require the item to be in a condition to perform its required function. (R) [29]

nonrestorable fire detector (fire protection devices) A device whose sensing element is designed to be destroyed by the process of detecting a fire. (NFPA) [16]

nonreturn-to-reference recording The magnetic recording of binary characters such that patterns of magnetization used to represent zeros and ones occupy the whole storage cell, with no part of the cell magnetized to a reference condition. *Contrast:* return-to-reference recording. *See also:* nonreturn-to-zero (change) recording. (C) 610.10-1994w

nonreturn to zero (NRZ) (1) (magnetic tape pulse recorders for electricity meters) A method whereby a pulse is recorded on the magnetic tape by a polarity reversal of the recording head current. (ELM) C12.14-1982r
(2) A signaling technique in which a polarity level high represents a logical "1" (one) and a polarity level low represents a logical level "0" (zero). (C/MM) 1394-1995

nonreturn to zero code (power-system communication) A code form having two states termed zero and one, and no neutral or rest condition. *See also:* digital. (PE) 599-1985w

Non-Return-to-Zero, Invert on Ones (NRZI) An encoding technique used in FDDI where a polarity transition represents a logical ONE. The absence of a polarity transition denote a logical ZERO. (C/LM) 802.3-1998

Non-Return-to-Zero, Invert on Ones (NRZI)-bit A code-bit transferred in NRZI format. The unit of data passed across the Physical Medium Dependent (PMD) service interface in 100BASE-X. (C/LM) 802.3-1998

nonreturn-to-zero (change) recording (NRZ(C)) (1) Non-return-to-reference recording in which the reference condition is zero magnetization. (C) 610.10-1994w
(2) Non-return-to-reference recording in which zeros are represented by magnetization to a specified condition, and ones are represented by magnetization to a specified alternative condition. The two conditions may be saturation and zero magnetization but are more commonly saturation in opposite senses. This method is called "change recording" because the recorded magnetic condition is changed when, and only when, the recorded binary character changes from zero to one or from one to zero. (C) 610.10-1994w

nonreversible output *See:* maximum test output voltage (magnetic amplifier).

nonreversible power converter (semiconverter) An equipment containing assemblies of mixed power thyristor and diode devices that is capable of transferring energy in only one direction (that is, from the alternating-current side to the direct-current side). *See also:* power rectifier. (IA) [62]

nonreversing A control function that provides for operation in one direction only. *See also:* feedback control system. (IA/ICTL/IAC/APP) [60], [75]

nonsalient pole (rotating machinery) The part of a core, usually circular, that by virtue of direct-current excitation of a winding embedded in slots and distributed over the interpolar (and possibly over some or all of the polar) space, acts as a pole. *See also:* asynchronous machine. (PE) [9]

nonsaturation region of an insulated-gate field-effect transistor (metal-nitride-oxide field-effect transistor) (the acronym for insulated-gate field-effect transistor is IGFET) A portion of the I_{DS} versus V_{DS} characteristic where I_{DS} is strongly dependent on V_{DS}. This is true when $0 < V_{DS} < V_{GS} - V_T$. (ED) 581-1978w

nonsealed system A system in which a tank or compartment is vented to the atmosphere, usually with two breather openings to permit circulation of air across the gas space above the oil. Circulation can be made unidirectional when a pipe is extended up through the oil and is heated by the oil to induce movement of air drawn from the outside into the gas space and across the oil to a breather on top of the tank. (PE/TR) C57.15-1999

nonsegregated-phase bus (1) One in which all phase conductors are in a common metal enclosure without barriers between the phases. *Note:* When associated with metal-clad switchgear, the primary bus conductors and connections are covered with insulating material throughout. (SWG/PE/EDPG) 665-1995
(2) A bus in which all phase conductors are in a common metal enclosure without barriers between the phases. *Note:* When associated with metal-clad switchgear, the primary bus conductors and connections are covered with insulating material throughout. (SWG/PE) C37.100-1992

nonselective collective automatic operation (elevators) Automatic operation by means of one button in the car for each landing level served and one button at each landing, wherein all stops registered by the momentary actuation of landing or car buttons are made irrespective of the number of buttons actuated or of the sequence in which the buttons are actuated. *Note:* With this type of operation the car stops at all landings for which buttons have been actuated, making the stops in the order in which the landings are reached after the buttons have been actuated, but irrespective of its direction of travel. *See also:* control. (EEC/PE) [119]

nonself-maintained discharge (gas) A discharge characterized by the fact that the charged particles are produced solely by the action of an external ionizing agent. *See also:* discharge. (ED) [45], [84]

non-self-restoring insulation (1) Insulation that loses its insulating properties or does not recover them completely after a disruptive discharge. (PE/PSIM) 4-1995
(2) (power and distribution transformers) An insulation that loses its insulating properties or does not recover them completely after a disruptive discharge caused by the application of a test voltage; insulation of this kind is generally, but not necessarily, internal insulation. (SPD/PE/C/SWG-OLD/TR) C62.22-1997, 1313.1-1996, C37.100-1992, C57.12.80-1978r

nonsequential computer A computer that must be directed to the location of each instruction. *See also:* arbitrary sequence computer. (C) 610.10-1994w

nonshield insulated splice (power cable joints) An insulated splice in which no conducting material is employed over the insulation for electric stress control. (PE/IC) 404-1986s

nonshorting cap A device that provides an open circuit between line and load when a photocontrol is not used. (RL) C136.10-1996

nonsignificant code (data management) A code that identifies a particular item but does not yield further information about the properties or classification of the item. *Contrast:* significant code. (C) 610.5-1990w

non-source routed (NSR) Indicates that the frame does not make use of a routing information field (i.e., the RIF is null and the RII is not set). (C/LM/CC) 8802-2-1998

nonspecific subordinate reference A knowledge reference that holds information about a DSA that holds one or more unspecified subordinate entries. (C/PA) 1327.2-1993w, 1224.2-1993w, 1328.2-1993w, 1326.2-1993w

nonspinning reserve (power operations) (electric power supply) That operating reserve capable of being connected to the bus and loaded within a specified time. (PE/PSE) 858-1987s, 346-1973w

nonstop switch (elevators) A switch that, when operated, will prevent the elevator from making registered landing stops. *See also:* control. (PE/EEC) [119]

nonstorage display (display storage tubes) Display of nonstored information in the storage tube without appreciably affecting the stored information. *See also:* storage tube. (ED) 158-1962w

nonsustained disruptive discharge A momentary disruptive discharge. (PE/PSIM) 4-1995

nonsynchronous *See:* asynchronous.

nonsynchronous (interdigital) transducer An interdigital transducer that has nonuniform electrode center-to-center spacing. (UFFC) 1037-1992w

nonsynchronous transmission (data transmission) A transmission process so that between any two significant instants in the same group, there is always an integral number of unit intervals. Between two significant instants located in different groups, there is not always an integral number of unit intervals. *Note:* In data transmission, this group is a block or a character. In telegraphy, this group is a character. (PE) 599-1985w

nonsystematic jitter *See:* uncorrelated jitter.

nonterminal node (data management) In a tree, a node that can have one or more subtrees. *Synonyms:* branch node; internal node. *Contrast:* terminal node. *See also:* root node. (C) 610.5-1990w

no-test trunk access method Used to connect to an end user's line, the no-test trunk access method contains a direct metallic path and is primarily used for loop and customer premise equipment (CPE) testing purposes. (SCC31) 1390.3-1999

nonthermal fire hazard A hazard resulting from combustion products (such as smoke and toxic and corrosive fire products). (DEI) 1221-1993w

nontouching loop set (network analysis) A set of loops no two of which have a common node. (CAS) 155-1960w

nontransitive dependency A type of dependency among attributes in a relation, in which a nonprime attribute A is said to be nontransitively dependent on another attribute B if and only if A is dependent on B, and there is another attribute C that is functionally dependent on B but does not functionally determine A. *Contrast:* transitive dependency. (C) 610.5-1990w

nonuniformity (transmission lines and waveguides) The degree with which a characteristic quantity, for example, impedance, deviates from a constant value along a given path. *Note:* It may be defined as the maximum amount of deviation from a selected nominal value. For example, the nonuniformity of the characteristic impedance of a slotted coaxial line may be 0.05 ohm due to dimensional variations. (IM/HFIM) [40]

non-utility generator A facility for generating electricity that is not exclusively owned by an electric utility and that operates connected to an electric utility system. (PE/PSE) 858-1993w

nonvented fuse (or fuse unit) A fuse without intentional provision for the escape of arc gases, liquids, or solid particles to the atmosphere during circuit interruption. (SWG/PE) C37.40-1993, C37.100-1992

nonvented power fuse (installations and equipment operating at over 600 volts, nominal) A fuse without intentional provision for the escape of arc gases, liquids, or solid particles to the atmosphere during circuit interruption. (NESC/NEC) [86]

nonventilated (power and distribution transformers) So constructed as to provide no intentional circulation of external air through the enclosure. (PE/TR) C57.12.80-1978r

nonventilated dry-type transformer (power and distribution transformers) (dry-type general purpose distribution and power transformers) A dry-type transformer which is so constructed as to provide no intentional circulation of external air through the transformer, and operating at zero gauge pressure. (PE/TR) C57.94-1982r, C57.12.80-1978r

nonventilated enclosure An enclosure so constructed as to provide no intentional circulation of external air through the enclosure. *Note:* Doors or removable covers are usually gasketed and humidity control may be provided by filtered breathers. (SWG/PE) C37.100-1992, C37.23-1987r

nonvolatile memory (1) A memory in which the data content is retained when power is no longer supplied to it. (ED) 641-1987w, 1005-1998

(2) Memory whose contents are retained after power has been shut off. (C/BA) 14536-1995

(3) Computer memory whose contents are preserved when the system power is off. (C/BA) 1275-1994

(4) Memory that retains its contents even through power failures. (C/MM) 1596-1992

(5) Read/write storage that is preserved through losses of power. (C/MM) 1212-1991s

nonvolatile random-access memory (NVRAM) A semi-permanent type of data storage (memory) that is backed up by batteries to maintain stored data even if system power is lost. Can be both read and changed by the system. (PE/SUB) 1379-1997

nonvolatile storage (1) (test, measurement, and diagnostic equipment) A storage device which can retain information in the absence of power. Contrast to volatile storage. (MIL) [2]

(2) A type of storage whose contents are not lost when power is lost. *Contrast:* volatile storage. *See also:* erasable storage; bubble memory. (C) 610.10-1994w

no-op *See:* no-operation.

no-operation (no-op) (1) (computers) An instruction that specifically instructs the computer to do nothing, except to proceed to the next instruction in sequence. *Synonym:* no-op. (C) [20], [85]

(2) (software) A computer operation whose execution has no effect except to advance the instruction counter to the next instruction. Used to reserve space in a program or, if executed repeatedly, to wait for a given event. Often abbreviated no-op. *Synonyms:* no-op; do-nothing operation. (C) 610.12-1990

no-op instruction *See:* dummy instruction.

NOR (1) (mathematics of computing) A Boolean operator having the property that if P is a statement, Q is a statement, R is a statement, . . . , then the NOR of P,Q,R,. . . is true if and only if all statements are false. *Note:* P NOR Q is often represented by P ↓ Q. *Synonym:* nondisjunction.

P	Q	P↓Q
0	0	1
0	1	0
1	0	0
1	1	0

NOR truth table

(C) 1084-1986w

(2) (software) *See also:* notice of revision. (C) 610.12-1990

NOR element *See:* NOR gate.

NOR gate A gate that performs the Boolean operation of nondisjunction. *Synonyms:* inclusive NOR gate; NOR element; NOT-OR. *See also:* OR gate. (C) 610.10-1994w

norator A two-terminal ideal element the current through which and the voltage across which can each be arbitrary. (CAS) [13]

normal (1) (state of a superconductor) The state of a superconductor in which it does not exhibit superconductivity. Example: Lead is normal at temperatures above a critical temperature. *See also:* superconductivity; superconducting. (ED) [46]

(2) (power generation) *See also:* preferred.

(3) *See also:* interference, normal-mode; normal-mode interference. (SUB/PE) C37.1-1994

normal base current Rated current of a transformer corresponding to its rated voltage and rated base kilovoltamperes. (PE/TR) C57.109-1993

normal base power transfers (power operations) Those power transfers that are considered to be a part of normal base system loadings for the condition being analyzed. Other transfers, such as emergency power or opportunistic economy energy transfers, are excluded even though they may be provided for in contractual arrangements. (PE/PSE) 858-1987s

normal binary *See:* binary.

normal character (N_char) N_chars represent, at the minimum, the 256 values of a byte (i.e., all the data characters) plus a control character representing an end_of_packet marker. (C/BA) 1355-1995

normal clear A term used to express the normal indication of the signals in an automatic block system in which an indication to proceed is displayed except when the block is occupied. *See also:* centralized traffic-control system. (EEC/PE) [119]

normal clear system *See:* normal clear.

normal contact A contact that is closed when the operating unit is in the normal position. (EEC/PE) [119]

normal (through) dielectric heating applications The metallic electrodes are arranged on opposite sides of the material so that the electric field is established through it. *Note:* The electrodes may be classified as plate electrodes, roller electrodes, or concentric electrodes.

1) Plate electrodes may have plane surfaces or surfaces of any desired shape to meet a particular condition and the spacing between them may be uniform or varied.
2) Roller electrodes are rollers separated by the material which moves between them.
3) Concentric electrodes consist of an enclosed and a surrounding electrode, with the material placed between the two.

(IA) 54-1955w

normal dimension Incremental dimensions whose number of digits is specified in the format classification. For example, the format classification would be plus 14 for a normal dimension: X.XXXX. *See also:* incremental dimension; long dimension; dimension; short dimension. (IA) [61]

normal distribution *See:* Gaussian distribution.

normal effort (electric generating unit reliability, availability, and productivity) Repairs were carried out with normal repair crews working normal shifts. *See also:* repair urgency. (PE/PSE) 762-1987w

normal float A constant-potential charge application to a battery to maintain it in a charged condition. (PE/EDPG) 1106-1987s

normal form (data management) The form of a data structure, relation, or database that has been reduced to a simpler, more stable form than it was in its unnormalized form. *See also:* fourth normal form; Boyce/Codd Normal form; second normal form; first normal form; third normal form; projection/join normal form. (C) 610.5-1990w

normal frequency The frequency at which a device or system is designed to operate. (SWG/PE) C37.100-1992, C37.34-1994

normal-frequency dew withstand voltage (1) The normal-frequency withstand voltage applied to insulation completely covered with condensed moisture. *See also:* normal-frequency withstand voltage. (SWG/PE) C37.100-1992
(2) The root-mean-square (rms) voltage that can be applied to an insulator or a device, completely covered with condensed moisture, under specified conditions for a specified time without causing flashover or puncture. (SWG/PE) C37.40-1981s

normal-frequency dry withstand voltage (1) The normal-frequency withstand voltage applied to dry insulation. *See also:* normal-frequency withstand voltage. (SWG/PE) C37.100-1992
(2) The root-mean-square (rms) voltage that can be applied to a dry device under specified conditions for a specified time without causing flashover or puncture. (SWG/PE) C37.40-1981s

normal-frequency line-to-line recovery voltage The normal-frequency recovery voltage, stated on a line-to-line basis, that occurs on the source side of a three-phase circuit-interrupting device after interruption is complete in all three poles. (SWG/PE) C37.100-1992

normal-frequency pole-unit recovery voltage The normal-frequency recovery voltage that occurs across a pole unit of a circuit-interrupting device upon circuit interruption. (SWG/PE) C37.100-1992

normal-frequency wet withstand voltage (1) The normal-frequency withstand voltage applied to wetted insulation. *See also:* normal-frequency withstand voltage. (SWG/PE) C37.100-1992
(2) The root-mean-square (rms) voltage that can be applied to a wetted device under specified conditions for a specified time without causing flashover or puncture. (SWG/PE) C37.40-1981s

normal-frequency recovery voltage The normal-frequency root-mean-square (rms) voltage that occurs across the terminals of an ac circuit-interrupting device after the interruption of the current and after the high-frequency transients have subsided. *Note:* For determination of the normal-frequency recovery voltage, see IEEE C37.09-1979. (SWG/PE) C37.40-1981s, C37.100-1992

normal-frequency withstand voltage The normal-frequency voltage that can be applied to insulation under specified conditions for a specified time without causing flashover or puncture. *Note:* This value is usually expressed as a root-mean-square (rms) value. *See also:* normal-frequency wet withstand voltage; normal-frequency dry withstand voltage; normal-frequency dew withstand voltage. (SWG/PE) C37.100-1992

normal gas density The manufacturer's recommended operating gas density (usually expressed as a pressure at 20°C). (SUB/PE) C37.122-1993

normal-glow discharge (gas) The glow discharge characterized by the fact that the working voltage decreases or remains constant as the current increases. *See also:* discharge. (ED) [45], [84]

normal induction (magnetic material) The maximum induction in a magnetic material that is in a symmetrically cyclically magnetized condition. (Std100) 270-1966w

normalization (1) (A) (data management) The process of decomposing and restructuring a complex data structure in order to reduce the structure to a simpler, more stable form. *Note:* Such a data structure is said to be in "normal form." **(B) (data management)** The process of reducing a relation to its simplest form such that each attribute is derived from a single domain consisting of nondecomposable values. (C) 610.5-1990
(2) The decomposition of more complex data structures according to a set of rules designed to give simpler, more stable structures. (PE/EDPG) 1150-1991w

normalization transformation *See:* viewing transformation.

normalize (1) (A) (data management) In database design, to reduce a data structure, relation, or database to a simpler, more stable form. *Synonym:* standardize. *See also:* normal form; normalized form. **(B) (data management)** To alter or position data into a standard format, as in justification of text. **(C) (data management)** To adjust the exponent and mantissa of floating-point data such that the mantissa lies in a standard range. *See also:* normalized form. (C) 610.5-1990
(2) (mathematics of computing) To shift the fixed-point part of a floating-point number, and make the corresponding adjustment to the exponent, to ensure that the fixed-point part lies within some prescribed range. The number represented remains unchanged. *Synonym:* standardize. (C) 1084-1986w
(3) (A) (test, measurement, and diagnostic equipment) To adjust the characteristic and fraction of a floating decimal point number thus eliminating leading zeros in the fraction. **(B) (test, measurement, and diagnostic equipment)** To adjust a measured parameter to a value acceptable to an instrument or measurement technique. (MIL) [2]
(4) To divide an impedance or frequency by a reference quantity thereby making the result dimensionless. (CAS) [13]

normalized admittance (waveguide) The reciprocal of the normalized impedance. (MTT) 146-1980w

normalized amplitude The amplitude of a signal when driving its steady-state value; i.e., not under the influence of ringing or other dynamic influences. (C/LM) 802.3-1998

normalized device coordinate system A device-independent coordinate system in which all coordinate values are normalized to values between 0 and 1. *Note:* The world coordinate system is mapped to the normalized device coordinate system by a viewing transformation. (C) 610.6-1991w

normalized directivity *See:* antenna [aperture] illumination efficiency.

normalized form (A) (data management) In database design, the form assumed by data that have been normalized. *Contrast:* unnormalized form. *See also:* normal form. **(B) (data management)** The form taken by a floating-point representation when the fixed-point part lies within some standard range, so chosen that any given real number can be represented by a unique pair of numerals. Examples below illustrate real data and their corresponding normalized form such that the fixed-point portion is in the form x.xxx.

$0.123 * 10^4 = 1.23 * 10^3$

$0.999 * 10^{-1} = 9.99 * 10^{-2}$

(C) 610.5-1990

normalized frequency (fiber optics) A dimensionless quantity (denoted by V), given by

$$V = \frac{2\pi a}{\lambda}\sqrt{n_1^2 - n_2^2}$$

where a is waveguide core radius, λ is wavelength in vacuum, and n_1 and n_2 are the maximum refractive index in the core and refractive index of the homogeneous cladding, respectively. In a fiber having a power-law profile, the approximate number of bound modes is $(V^2/2)[g/(g + 2)]$, where g is the profile parameter. *Synonym:* V number. *See also:* single mode optical waveguide; bound mode; mode volume; power-law index profile; parabolic profile. (Std100) 812-1984w

normalized impedance (1) (waveguide) The ratio of an impedance and the corresponding characteristic impedance. *Note:* The normalized impedance is independent of the convention used to define the characteristic impedance, provided that the same convention is also taken for the impedance to be normalized. (MTT) 146-1980w
(2) The ratio of an impedance to a specified reference impedance. *Note:* For a transmission line or a waveguide, the reference impedance is usually a characteristic impedance. (IM/HFIM) [40]

normalized peak error Peak error divided by three times the standard deviation of the differences discussed in 4.5.1.1. *See also:* peak error. (IM/WM&A) 1057-1994w

normalized peak ESD current Ratio of the peak current to the charge voltage (e.g., 5A/kV). (SPD/PE) C62.47-1992r

normalized plateau slope (radiation counter tubes) The slope of the substantially straight portion of the counting rate versus voltage characteristic divided by the quotient of the counting rate by the voltage at the Geiger-Mueller threshold. (ED) 161-1971w

normalized relation *Contrast:* unnormalized relation. *See also:* relation; normalize. (C) 610.5-1990w

normalized response (automatic control) One obtained by dividing a measured value and dimension by some convenient reference value and dimension; usually the quotient is nondimensional. (PE/EDPG) [3]

normalized rising slope Ratio of the initial slope to the charge voltage (e.g., 3.75 A/ns/kV). (SPD/PE) C62.47-1992r

normalized site attenuation (NSA) Site attenuation divided by the antenna factors of the radiating and receiving antennas (all in linear units). (EMC) C63.5-1988, C63.4-1991

normalized transimpedance (magnetic amplifier) The ratio of differential output voltage to the product of differential control current and control winding turns. (MAG) 107-1964w

normal joint (power cable joints) A joint which is designed not to restrict movement of dielectric fluid between cables being joined. (PE/IC) 404-1986s

normal lightning current Lightning currents of 65 kA or less. (SPD/PE) C62.11-1999

normal linearity (computers) Transducer linearity defined in terms of a zero-error reference, that is chosen so as to minimize the linearity error. *Note:* In this case, the zero input does not have to yield zero output and full-scale input does not have to yield fullscale output. The specification of normal linearity, therefore, is less stringent than zero-based linearity. *See also:* linearity of a multiplier; electronic analog computer. (C) 165-1977w

Normal Link Pulse (NLP) An out-of-band communications mechanism used in 10BASE-T to indicate link status. (C/LM) 802.3-1998

Normal Link Pulse Receive Link Integrity Test function A test function associated with Auto-Negotiation that allows backward compatibility with the 10BASE-T Link Integrity Test function of IEEE 802.3. (C/LM) 802.3-1998

normally closed *See:* normally open and normally closed.

normally closed contact A contact, the current-carrying members of which are in engagement when the operating unit is in its normal position. (EEC/PE) [119]

normally closed relay contacts A contact pair that is closed when the coil is not energized. (EEC/REE) [87]

normally open and normally closed When applied to a magnetically operated switching device, such as a contactor or relay, or to the contacts thereof, these terms signify the position taken when the operating magnet is deenergized. Applicable only to nonlatching types of devices. *See also:* contactor. (IA/ICTL/IAC) [60]

normally open contact (open contact) A contact, the current-carrying members of which are not in engagement when the operating unit is in the normal position. *Synonym:* open contact. (EEC/PE) [119]

normally open relay contacts A contact pair that is open when the coil is not energized. (EEC/REE) [87]

normal-mode (1) (transverse or differential) The voltage that appears differentially between two signal wires and that acts on the circuit in the same manner as the desired signal. (PE/IC) 1143-1994r
(2) *See also:* normal-mode interference. (SUB/PE) C37.1-1994

normal-mode interference (signal-transmission system) A form of interference that appears between measuring circuit terminals. *See also:* differential-mode interference; accuracy rating. (EEC/SUB/PE/EMI) [112], C37.1-1994

normal-mode noise (1) (transverse or differential) The noise voltage which appears differentially between two signal wires and which acts on the signal sensing circuit in the same manner as the desired signal. (SUB/PE) 525-1992r
(2) (cable systems in power generating stations) (instrumentation and control equipment grounding in generating stations) (transverse or differential) The noise voltage that appears differentially between two signal wires and acts on the signal sensing circuit in the same manner as the desired signal. Normal mode noise maybe caused by one or more of the following:

a) electrostatic induction and differences in distributed capacitance between the signal wires and the surroundings;
b) electromagnetic induction and magnetic fields linking unequally with the signal wires;
c) junction or thermal potentials due to the use of dissimilar metals in the connection system;
d) common mode to normal mode noise conversion.

(PE/C/EDPG) 1050-1989s, 166-1977w
(3) *See also:* transverse-mode noise. (IA/PSE) 1100-1999

normal mode signal (or differential signal) The difference between the signal at the positive input and the negative input of a differential input waveform recorder. If the signal at the positive input is designated V+, and the signal at the negative input is designated V-, then the normal mode (or differential) signal (Vdm) is

$$V_{dm} = V+ - V-$$

(IM/WM&A) 1057-1994w

normal number (radix-independent floating-point arithmetic) (mathematics of computing) A non-zero number that is finite and not subnormal. (MM/C) 854-1987r, 1084-1986w

normal operating conditions The environmental stresses and operating stresses, such as voltage, current loading, or mechanical loading, encountered in normal service, including anticipated overloads and periodic testing but not including accidents or other extraordinary events. (DEI/RE) 775-1993w

normal operating load Any equipment operation that can reasonably be expected to occur during an earthquake, except short-circuit loads, that produces a force, stress, or load. (PE/SUB) 693-1997

normal operations area (nuclear power generating station) A functional area allocated for those displays and controls necessary for the tasks routinely performed during plant startup, shutdown and power operation modes. (PE/NP) 566-1977w

normal permeability The ratio of normal induction to the corresponding maximum magnetizing force. *Note:* In anisotropic media, normal permeability becomes a matrix. (Std100) 270-1966w

normal position (of a device) A predetermined position that serves as a starting point for all operations. (EEC/PE) [119]

normal random number (mathematics of computing) Any member of a random number sequence that has a normal, or Gaussian, distribution. (C) 1084-1986w

normal rating (1) The level of power flow that facilities can carry through a series of daily load cycles without loss of life to the facility involved. (PE/PSE) 858-1993w
(2) Capacity of installed equipment. (PE/PSE) 346-1973w

normal response mode (NRM) The secondary station (a DCC) initiates transmission only after receiving express permission to do so from the primary station (the BCC). (EMB/MIB) 1073.3.1-1994

normal stop system A term used to describe the normal indication of the signals in an automatic block signal system in which the indication to proceed is given only upon the approach of a train to an unoccupied block. *See also:* centralized traffic-control system. (EEC/PE) [119]

normal-terminal stopping device (elevators) A device, or devices, to slow down and stop an elevator or dumbwaiter car automatically at or near a terminal landing independently of the functioning of the operating device. *See also:* control. (PE/EEC) [119]

normal test output (A) (amplitude-modulation broadcast receivers) For receivers capable of delivering at least 1 watt maximum undistorted output, the normal test output is an audio-frequency power of 0.5 watt delivered to a standard dummy load. **(B) (amplitude-modulation broadcast receivers)** For receivers capable of delivering 0.1 but less than 1 watt maximum undistorted output, the normal test output 0.05 watt audio-frequency power delivered to a standard dummy load. When this value is used, it should be so specified. Otherwise, the 0.5-watt value is assumed. **(C) (amplitude-modulation broadcast receivers)** For receivers capable of delivering less than 0.1 watt maximum undistorted output, the normal test output is 0.005 watt audio-frequency power delivered to a standard dummy load. When this value is used, it should be so specified. **(D) (amplitude-modulation broadcast receivers)** For automobile receivers, normal test output is 1.0 watt audio-frequency power delivered to a standard dummy load. (CE) 186-1948

normal transfer capability (power operations) The maximum amount of power that can be transmitted continuously over the transmission network. (PE/PSE) 858-1987s

normal velocity storage (accelerometer) (digital accelerometer) The velocity information that is stored in the accelerometer during the application of an acceleration within its input range. (MTT/AES/GYAC) 457-1982w, 528-1994

normal voltage limit (power operations) The voltage range that is acceptable on a sustained basis. *See also:* emergency voltage limit. (PE/PSE) 858-1987s

normal weather (1) Includes all weather not designated as adverse or major storm disaster. (PE/PSE) [54], 859-1987w, 346-1973w
(2) (electric power system) All weather not designated as adverse. *See also:* outage. (PE/PSE) [54], 859-1987w

normal weather forced outage rate The number of forced outages per unit of service time in normal weather = number of forced outages during normal weather/service time during normal weather. *See also:* outage rate. (PE/PSE) 859-1987w

normal weather persistent-cause forced-outage rate (electric power system) (for a particular type of component) The mean number of outages per unit of normal weather time per component. *See also:* outage. (PE/PSE) [54]

normative A mandatory set of instructions or references. (C/PA) 1003.23-1998

normative annex An annex that is an official part of the standard but is placed after the clauses of the standard for convenience or to create a hierarchical distinction. (C/MM) 1754-1994

normative model A model that makes use of a familiar situation to represent a less familiar one; for example, a model that depicts the human cardiovascular system by using a mechanical pump, rubber hoses, and water. (C) 610.3-1989w

north (navigation aid terms) The primary reference direction relative to the earth. True north is the direction of the north geographical pole. Magnetic north is the direction north as determined by the earth's magnetic line of force. (AES/GCS) 172-1983w

North American digital hierarchy An ascending series of digital rates (1.544 Mb/s—DS1, 3.152 Mb/s—DS1C, 6.312 Mb/s—DS2, and 44.736 Mb/s—DS3) used for PCM transmission in North America. (COM/TA) 1007-1991r

north-stabilized plan-position indicator (radar) A special case of azimuth-stabilized PPI in which the reference direction is north. *See also:* navigation. (AES/RS) 686-1997, [42]

Norton's theorem States that a linear time-invariant one-port is equivalent to a circuit which consists of the driving-point admittance of the one-port shunted by the short-circuit current of the one-port. (CAS) [13]

Norton surface wave The propagating electromagnetic wave produced by a source over or on the ground. The Norton wave consists of the total field minus the geometrical-optics field. *See also:* ground wave; lateral wave. (AP/PROP) 211-1997

Norton transformation A four-terminal network transformation of a series (shunt) ladder element to an equivalent pi (tee) network in cascade with an ideal transformer. The pi (tee) network arms are identical to the series (shunt) arm except for multiplying factors related to the ideal transformer turns-ratio. (CAS) [13]

nose cone *See:* leader cone.

nose suspension A method of mounting an axle-hung motor or a generator to give three points of support, consisting of two axle bearings (or equivalent) and a lug or nose projecting from the opposite side of the motor frame, the latter supported by a truck or vehicle frame. *See also:* traction motor. (EEC/PE) [119]

NOT (mathematics of computing) A monadic Boolean operator having the property that if P is a statement, then the expression "NOT P" is true if P is false, and false if P is true. *Note:* NOT P is often represented by ~P, $\overline{P}$, or P′.

NOT Truth Table

P	$\overline{P}$
0	1
1	0

Synonyms: Boolean complementation; negation; inversion; complementary operator. (C) 1084-1986w

NOT-AND* *See:* NAND.

* Deprecated.

NOT AND gate *See:* NAND gate.

not a number (NaN) A symbolic entity encoded in floating-point format. There are two types of NaNs. Signaling NaNs signal the invalid (operation) exception whenever they appear as operands. Quiet NaNs propagate through almost every arithmetic operation without signaling exceptions. (C/MM/NI) 754-1985r, 1084-1986w, N13.4-1971w

notation (1) A system of symbols used to represent information, and the rules for their use. (C) 1084-1986w
(2) *See also:* positional notation.
(3) A set of symbols used to represent design entities and entity attributes. (C/SE) 1016.1-1993w

notation standard A standard that describes the characteristics of formal interfaces within a profession. (C) 610.12-1990

NOT-BOTH* *See:* NAND.
* Deprecated.

notch (1) A transient reduction in the magnitude (absolute value) of the quasi-sinusoidal mains voltage. (The duration is always less than a half-cycle and usually less than a few milliseconds.) *See also:* sag; undervoltage. (T&D/PE) 1250-1995
(2) A switching (or other) disturbance of the normal power voltage waveform, lasting less than a half cycle; which is initially of opposite polarity than the waveform, and is thus subtractive from the normal waveform in terms of the peak value of the disturbance voltage. This includes complete loss of voltage for up to a half cycle. *See also:* transient. (IA/PSE) 1100-1999

notch area The area of the line voltage notch. It is the product of the notch depth, in volts, times the width of the notch measured in microseconds. (IA/SPC) 519-1992

notch depth The average depth of the line voltage notch from the sine wave of voltage. (IA/SPC) 519-1992

notch filter A band-elimination filter, sometimes used to eliminate a single frequency for example, 60 Hz. (CAS) [13]

notching (relays) A qualifying term applied to a relay indicating that a predetermined number of separate impulses is required to complete operation. *See also:* relay. (PE/EEC) [119]

notching or jogging device (power system device function numbers) A device that functions to allow only a specified number of operations of a given device, or equipment, or a specified number of successive operations within a given time of each other. It is also a device that functions to energize a circuit periodically or for fractions of specified time intervals, or that is used to permit intermittent acceleration or jogging of a machine at low speeds for mechanical positioning. (PE/SUB) C37.2-1979s

notching relay A programming relay in which the response is dependent upon successive impulses of the input quantity. (SWG/PE) C37.100-1992

not connected (NC) (semiconductor memory) The inputs/outputs that are not connected to any active part of the circuit or any other pin or any conductive surface of the package. (TT/C) 662-1980s

note (1) A conventional sign used to indicate the pitch, or the duration, or both, of a tone. It is also the tone sensation itself or the oscillation causing the sensation. *Note:* The word serves when no distinction is desired among the symbol, the sensation, and the physical stimulus. (SP) [32]
(2) Helpful hint(s) and other material that may assist the user. (C/SE) 1063-1987r
(3) A body of free text that describes some general comment or specific constraint about a portion of a model. A note may be used in an early, high-level view prior to capturing constraints in the specification language; a note may further clarify a rule by providing explanations and examples. A note may also be used for "general interest" comments not involving rules. These notes may accompany the model graphics. (C/SE) 1320.2-1998

notebook computer A portable computer that is approximately the size of a standard 3-ring notebook binder; about 10 in × 11 in × 2 in. (C) 610.10-1994w

NOT element *See:* NOT gate.

no-test trunk (1) A specialized switch facility used for operator and service personnel metallic testing of lines. (AMR) 1390-1995
(2) A specialized switch facility used by the operator and service personnel for metallic testing of a subscriber's telephone lines (also known as a test trunk facility). (SCC31) 1390.3-1999, 1390.2-1999

NOT gate A combinational circuit that performs the Boolean operation of negation. *Synonym:* NOT element. (C) 610.10-1994w

notice of revision A form used in configuration management to propose revisions to a drawing or list, and, after approval, to notify users that the drawing or list has been, or will be, revised accordingly. *Synonym:* NOR. *See also:* specification change notice; configuration control; engineering change. (C) 610.12-1990

NOT-IF-THEN *See:* exclusion.

NOT-IF-THEN element *See:* NOT-IF-THEN gate.

NOT-IF-THEN gate A gate that performs the Boolean operation of exclusion. *Synonym:* NOT-IF-THEN element. (C) 610.10-1994w

NOT-OR* *See:* NOR.
* Deprecated.

NOT OR gate *See:* NOR gate.

not present When describing an argument to a function call, the case where a null pointer is supplied by the caller (as opposed to a pointer to an actual object of the type specified in the declaration). (C/PA) 1238.1-1994w

noughts complement *See:* radix complement.

novenary (A) (mathematics of computing) Pertaining to a selection in which there are nine possible out-comes. **(B) (mathematics of computing)** Pertaining to the numeration system with a radix of 9. (C) 1084-1986

novendenary (A) (mathematics of computing) Pertaining to a selection in which there are 19 possible outcomes. **(B) (mathematics of computing)** Pertaining to the numeration system with a radix of 19. (C) 1084-1986

NPA *See:* Next Page Algorithm; numbering-plan area code; numbering-plan area.

nPC *See:* next program counter.

n-plus-one address instruction format An n-address instruction format that also specifies the next instruction to be executed. *See also:* address format. (C) 610.10-1994w

***n*–plus-one address instruction** A computer instruction that contains $n + 1$ address fields, the last containing the address of the instruction to be executed next. *Contrast:* n–address instruction. *See also:* two-plus-one address instruction; one-plus-one address instruction. (C) 610.12-1990

NPR *See:* noise-power ratio.

NRA *See:* laser gyro axes.

NRC bulletins Publications titled NRC Information Notice and published by the Office of Nuclear Reactor Regulation (NRR) of the Nuclear Regulatory Commission (NRC). (PE/NP) 933-1999

NRM *See:* normal response mode.

nroff A text-formatting language. (C) 610.13-1993w

NRZ (non-return-to-zero recording) *See:* nonreturn to zero.

NRZ(C) *See:* nonreturn-to-zero (change) recording.

NRZI *See:* Non-Return-to-Zero, Invert on Ones.

NRZI-bit A code-bit transferred in NRZI format. The unit of data passed across the PMD service interface in 100BASE-X. (LM/C) 802.3u-1995s

NSA *See:* normalized site attenuation.

NSAP *See:* network service access point.

N-scope A cathode-ray oscilloscope arranged to present an N-display. (AES/RS) 686-1990

NSLR A country's National Standardizing Laboratory for radioactivity measurements. (NI) N42.14-1991

NSR *See:* non-source routed.

NTA *See:* next transfer address.

***n*–terminal electron tubes** *See:* input capacitance.

***N*–terminal network** A network with *N* accessible terminals. *See also:* network analysis. (Std100) 270-1966w

***N*–terminal pair network** A network with 2*N* accessible terminals grouped in pairs. *Note:* In such a network one terminal of each pair may coincide with a network node. *See also:* network analysis. (Std100) 270-1966w

N th field lag (th field lag) The fraction of the output signal which is read out in the *N* th field after the initial reading scan of an input signal which has been completely extinguished just before the scanning beam reaches that portion of the target irradiated by the input signal. This can be readily understood with reference to the decay characteristics diagram below. A similar definition can be made for the buildup of a signal as illustrated in the buildup characteristics diagram below. (ED) 503-1978w

***N* th harmonic** The harmonic of frequency *N* times that of the fundamental component. (Std100) 154-1953w

NTI *See:* noise transmission impairment.

NTSC *See:* National Television Standards Committee.

***n*–tuple (A)** An ordered set of values $(x_1, x_2, \ldots, x_n)$. **(B)** In a relational data model, a tuple. (C) 610.5-1990, 1084-1986

n-tuple length register A set of n registers that function as a single register. *Synonym:* n-tuple register. (C) 610.10-1994w

n-tuple register *See:* n-tuple length register.

N^2 diagram A system engineering or software tool for tabulating, defining, analyzing, and describing functional interfaces and interactions among system components. The N^2 diagram is a matrix structure that graphically displays the bidirectional interrelationships between functions and components in a given system or structure. (C/SE) 1362-1998

N-type A constant impedance connector compatible with certain coaxial cables. (C) 610.7-1995

***n*-type crystal rectifier** A crystal rectifier in which forward current flows when the semiconductor is negative with respect to the metal. *See also:* rectifier. (EEC/PE) [119]

***n*–type semiconductor** *See:* semiconductor, *n*-type.

NuBus® Refers to IEEE Std 1196-1987, IEEE Standard for a Simple 32-Bit Backplane Bus: Nubus. (C/MM) 1596-1992

nuclear burnup (A) (power operations) A measure of nuclear reactor fuel consumption, usually expressed as energy produced per unit weight of fuel exposed (megawatt-days per metric ton of fuel). **(B) (power operations)** Percentage of fuel atoms that have undergone fission (atom percent burnup). (PE/PSE) 858-1987, 346-1973

nuclear energy (power operations) The energy with which nucleons are bound together to form nuclei. When a nucleus is changed or rearranged in a nuclear reaction (fission, fusion, etc.) or by radioactive decay, nuclear energy may be released or absorbed in the form of kinetic energy of the reactants or products. *See also:* nuclear reactor; nuclear fuel elements; nuclear fuel reprocessing. (T&D/PE/PSE) 858-1987s, 346-1973w

nuclear fuel elements (power operations) (nuclear power generating station) An assembly of rods, tubes, plates, or other geometrical forms into which nuclear fuel is contained for use in a reactor. (PE/PSE) 858-1987s, 346-1973w

nuclear fuel reprocessing (power operations) (nuclear power generating station) The processing of irradiated reactor fuel to recover the unused fissionable material, or fission products, or both. (PE/PSE) 858-1987s, 346-1973w

nuclear plant reliability data system (NPRDS) A reliability database maintained by the Institute of Nuclear Power Operations (INPO) that receives failures reports from utilities within the United States. (PE/NP) 933-1999

nuclear power generating stations, class ratings (1) (class 1 structures and equipment) Structures and equipment that are essential to the safe shutdown and isolation of the reactor or whose failure or damage could result in significant release of radioactive material.
(2) (Class 1E) The safety classification of the electric equipment and systems that are essential to emergency reactor shutdown, containment isolation, reactor core cooling, and containment and reactor heat removal, or are otherwise essential in preventing significant release of radioactive material to the environment.
(3) (Class 1E control switchboard) A rack panel, switchboard, or similar type structure fitted with any Class 1E equipment.
(4) (Class 1E electric systems) The systems that provide the electric power used to shut down the reactor and limit the release of radioactive material following a design basis event.
(5) (class II structures and equipment) Structures and equipment that are important to reactor operation but are not essential to the safe shutdown and isolation of the reactor and whose failure cannot result in a significant release of radioactive material.
(6) (class III structures and equipment) Structures and equipment that are not essential to the operation, safe shutdown, or isolation of the reactor and whose failure cannot result in the release of radioactive material.
(PE/NP) 380-1975w, 323-1974s, 344-1975s, 336-1980s, 420-1982, 383-1974r, 382-1980s, 308-1980s

nuclear power generating station (station) A plant wherein electric energy is produced from nuclear energy by means of suitable apparatus. The station may consist of one or more generating units. (PE/NP) 308-1991

nuclear reactor (power operations) An apparatus by means of which a fission chain reaction can be initiated, maintained, and controlled. (PE/PSE) 858-1987s

nuclear safety related (nuclear power generating station) That term used to call attention to safety classifications incorporated in the body of the document so marked. *Note:* As used in IEEE Std 494-1974, the term calls attention to the safety classification Class 1E. (PE/NP) 494-1974w

nucleus That part of a control program that is resident in main storage. *Synonym:* resident control program. (C) 610.10-1994w

nugget A data structure used in the procedural interface (PI) for rapid switching between technologies. (C/DA) 1481-1999

nuisance alarm An alarm warning that does not represent danger, safeguards threat, or equipment failure conditions requiring an actual response. (PE/NP) 692-1997

nuisance failure Intermittent or sustained failure of equipment secondary to system safety or reliability. (PE/NP) 933-1999

NUL The null character ('\0'). (PE/SUB) 1379-1997

NULL A byte with all bits set to zero. (C/MM) 1284.1-1997

null (1) (direction finding systems) (navigation aid terms) The condition of minimum output as a function of the direction of arrival of the signal, or of the rotation of the response pattern of the DF (direction finder) antenna system. (AES/GCS) 172-1983w
(2) (microprocessor operating systems parameter types) A value whose definition is to be supplied within the context of a specific operating system. This value is a representation of the set of no numbers or no value for the operating system in use. (C/MM) 855-1985s
(3) (signal-transmission system) The condition of zero error-signal achieved by equality at a summing junction between an input signal and an automatically or manually adjusted balancing signal of phase or polarity opposite to the input signal. *See also:* signal. (IE) [43]
(4) The direction between radiation lobes where the signal drops to a minimum. In general, a null is any portion of the pattern where the signal level is less than 10% of the rms of the pattern. (T&D/PE) 1260-1996
(5) *See also:* null. (AES/GYAC) 528-1994

null address (local area networks) An all zero address that does not identify any network end node. The null address may be used as the destination address in training packets being sent from an end node or lower-level repeater to an upper-level repeater. The null address may be used in the Source Address (SA) field of training frames to verify link operation between an end node and the connected repeater. The null address may also be used in void frames. Packets with the null address are forwarded by the receiving repeater to all promiscuous ports. (C) 8802-12-1998

nullator An idealized one-port that is simultaneously an open and short circuit, that is, V-I-O. The nullator is a bilateral, loss-less one-port. (CAS) [13]

null balance (automatic null-balancing electric instrument) (instruments) The condition that exists in the circuits of an instrument when the difference between an opposing electrical quantity within the instrument and the measured signal does not exceed the dead band. *Note:* The value of the opposing electrical quantity produced within the instrument is related to the position of the end device. *See also:* measurement system; feedback control system. (EEC/EMI) [112]

null-balance system A system in which the input is measured by producing a null with a calibrated balancing voltage or current. *See also:* signal. (IE) [43]

null character (1) (data management) A control character that is used to accomplish media-fill or time-fill, and that may be inserted into or removed from, a sequence of characters without affecting the meaning of the sequence; however, the control of equipment or the format may be affected by this character. *See also:* space character. (C) 610.5-1990w
(2) The value `POSIX_Character'Value(0)`, if defined. The use of a null character in filenames or environment variable names or values produces undefined results. Implementations shall not return null characters to applications in filenames or environment variable names or values. (C) 1003.5-1999

null cycle A type of an attention cycle that is used to dismiss a resource lock and initiate a rearbitration. (C/MM) 1196-1987w

null data (data management) Data for which space is allocated but for which no value currently exists. (C) 610.5-1990w

null DATA packet *See:* NULL packet.

nullity (degrees of freedom on mesh basis) (networking) The number of independent meshes that can be selected in a network. The nullity N is equal to the number of branches B minus the number of nodes V plus the number of separate parts P. $N = B - V + P$. *See also:* network analysis. (Std100) 270-1966w

null junction (power supplies) The point on the Kepco bridge at which the reference resistor, the voltage-control resistance, and one side of the comparison amplifier coincide. *Note:* The null junction is maintained at almost zero potential and is a virtual ground. *See also:* summing point. (AES/PE) [41], [78]

null offsetting (accelerometer) (inertial sensors) (gyros) A calibration or test technique by which the electrical null position is intentionally shifted, resulting in a rotation of the input axis relative to the input reference axis. (AES/GYAC) 528-1994

null operation (FASTBUS acquisition and control) A primary address cycle followed by no data cycles. It determines if the system contains a device capable of responding to the primary address used. It can be used to reserve segment interconnects for an arbitration locked sequence. (NID) 960-1993

NULL packet (1) A special type of DATA packet containing a data field entirely filled with logic zeros and a parity bit equal to logic one. *Synonym:* null DATA packet. (TT/C) 1149.5-1995
(2) A packet in which no clocked data is transmitted on Serial Bus between DATA_PREFIX and DATA_END. (C/MM) 1394a-2000

null pointer (data management) A pointer that is empty; that is, a pointer that does not point to anything. (C) 610.5-1990w

null-reference glide slope (navigation aid terms) A glide-slope system using a two-element array in which the slope angle is defined by the first null above the horizontal in the field pattern of the upper antenna. (AES/GCS) 172-1983w

null steering To control, usually electronically, the direction at which a directional null appears in the radiation pattern of an operational antenna. (AP/ANT) 145-1993

null-steering antenna system An antenna having in its radiation pattern one or more directional nulls that can be steered, usually electronically. (AP/ANT) 145-1993

null string (1) (data management) A string containing no entries. *Note:* It is said that a null string has length zero. (C) 610.5-1990w
(2) *See also:* empty string. (C/PA) 1003.5-1999, 9945-2-1993

null transaction A transaction that has no effect on the master file that is being updated. It is usually used for documentation purposes only. *See also:* delete transaction; add transaction; change transaction; update transaction. (C) 610.2-1987

null tree (data management) A tree that has exactly one root and one descendant node. (C) 610.5-1990w

null triggering/zero-crossing triggering (thyristor) A method of triggering the controller circuit elements such that the associated angle of retard is zero. (IA/IPC) 428-1981w

number (A) (electronic computation) Formally, an abstract mathematical entity that is a generalization of a concept used to indicate quantity, direction, etc. In this sense a number is independent of the manner of its representation. **(B) (electronic computation)** Commonly: A representation of a number as defined above (for example, the binary number 10110, the decimal number 3695, or a sequence of pulses). **(C) (electronic computation)** An expression, composed wholly or partly of digits, that does not necessarily represent the abstract entity mentioned in the first meaning. *Note:* Whenever there is a possibility of confusion between meaning (A) and meaning (B) or (C), it is usually possible to make an unambiguous statement by using "number" for meaning (A) and "numerical expression" for meaning (B) or (C). *Synonym:* numerical expression. (C) 162-1963
(2) (A) (mathematics of computing) A mathematical abstraction indicating a quantity or amount. **(B) (mathematics of computing)** Loosely, a numeral. (C) 1084-1986

number-crunching Computer processing that relies heavily on the arithmetic and logical capabilities of the central processing unit, as contrasted with processing that entails extensive input/output or data movement. (C) 610.2-1987

number group (telephone switching systems) An arrangement for associating equipment numbers with mainstation codes. (COM) 312-1977w

numbering plan (telephone switching systems) A plan employing codes and directory numbers for identifying main stations and other terminations within a telecommunication system. (COM) 312-1977w

numbering-plan area (NPA) (telephone switching systems) A geographical subdivision of the territory covered by a national or integrated numbering plan. An NPA is identified by a distinctive area code (NPA code). (COM) 312-1977w

numbering-plan area code (NPA) (telephone switching systems) A one–, two–, or three-digit number that, for the purpose of distance dialing, designates one of the geographical areas within a country (and in some instances neighboring territories) that is covered by a separate numbering plan. (COM) 312-1977w

number of channels A six-character number (no decimal point) giving the number of channels included in this file. The last record contains blank data for the data in excess of the actual number of channels. Leading spaces are interpreted as zeros. (NPS/NID) 1214-1992r

number of inherent tap positions The highest number of tap positions for half a cycle of operation for which an LTC is designed. (PE/TR) C57.131-1995

number of loops (magnetically focused electron beam) The number of maxima in the beam diameter between the electron gun and the target, or between a point on the photocathode and the target. (ED) 161-1971w

number of rectifier phases (rectifier circuits) The total number of successive, nonsimultaneous commutations occuring within that rectifier circuit during each cycle when operating without phase control. *Note:* It is also equal to the order of the principal harmonic in the direct-current potential wave shape. The number of rectifier phases influences both alternating-current and direct-current waveforms. In a simple single-way rectifier the number of rectifier phases is equal to the number of rectifying elements. *See also:* rectifier circuit element; rectification. (IA) 59-1962w, [12]

number of scanning lines (numerically) (television) The total number of lines, both active and blanked, in a frame. *Note:* In any specified scanning system, this number is inherently the ratio of the line frequency to the frame frequency and is always a whole number. In a two-to-one odd-line interlaced system, it is always an odd whole number. (BT/AV) 201-1979w

number of service tap positions The number of tap positions for half a cycle of operation for which an LTC is used in a transformer. *Note:* The above two terms, 3.19 and 3.20, are generally given as the ± values of the relevant numbers, e.g., ± 16 positions. They are, in principle, valid also for the motor-drive mechanism. When the term *number of tap positions* is used in connection with a transformer, this always refers to the number of service tap positions of the LTC. (PE/TR) C57.131-1995

number 1 master *See:* original master.

number 1 mold (mother) (metal positive) A mold derived by electroforming from the original master. *See also:* phonograph pickup. (SP) [32]

number range (mathematics of computing) The set of values that a number may assume. (C) 1084-1986w

number representation (mathematics of computing) A representation of a number in a numeration system. *Synonym:* numeration. (C) 1084-1986w

number representation system (mathematics of computing) A system for the representation of numbers; for example, the decimal numeration system, the Roman numeral system, the binary numeration system. *Synonyms:* numeration system; numeral system. (C) 1084-1986w

number sign The character "#". (C/PA) 9945-2-1993

number system (electronic computation) Loosely, a numeration system. *See also:* positional notation. (C) [20], 1084-1986w, [85]

number 2, number 3, etc. master A master produced by electroforming from a number 1, number 2, etc. mold. *See also:* phonograph pickup. (SP) [32]

number 2, number 3, etc. mold A mold derived by electroforming from a number 2, number 3, etc. master. *See also:* phonograph pickup. (SP) [32]

numeral (mathematics of computing) A representation of a number. *See also:* binary numeral; decimal numeral; hexadecimal numeral; octal numeral. (C) [20], [85], 1084-1986w

numeral system *See:* number representation system.

numeration *See:* number representation.

numeration system (1) (numeral system) A system for the representation of numbers, for example, the decimal system, the Roman numeral system, the binary system. (C) [20], [85]
(2) (mathematics of computing) *See also:* number representation system. (C) 1084-1986w

numeric (data management) (mathematics of computing) Pertaining to data that can be expressed using only numbers and mathematical symbols, in contrast to characters or other special signs or symbols. *Synonym:* numerical. *See also:* pure numeric; arithmetic; numeric data. (C) 610.5-1990w, 1084-1986w

numerical *See:* numeric.

numerical action (switch) That action that depends on at least part of the called number. (EEC/PE) [119]

numerical analysis (mathematics of computing) The study of methods of obtaining useful quantitative solutions to problems that have been expressed mathematically, including the study of the errors and bounds on errors in obtaining such solutions. (C) [20], [85], 1084-1986w

numerical aperture (NA) (A) (fiber optics) The sine of the vertex angle of the largest cone of meridional rays that can enter or leave an optical system or element, multiplied by the refractive index of the medium in which the vertex of the cone is located. Generally measured with respect to an object or image point and will vary as that point is moved. **(B) (fiber optics)** For an optical fiber in which the refractive index decreases monotonically from n sub 1 on axis to n sub 2 in the cladding, the numerical aperture is given by

$$NA = \sqrt{n_1^2 - n_2^1}$$

(C) (fiber optics) Colloquially, the sine of the radiation or acceptance angle of an optical fiber, multiplied by the refractive index of the material in contact with the exit or entrance face. This usage is approximate and imprecise, but is often encountered. *See also:* meridional ray; radiation angle; launch numerical aperture; radiation pattern; acceptance angle. (Std100) 812-1984

numerical control (NC) (1) Pertaining to the automatic control of processes by the proper interpretation of numerical data. (C) [20], [85]
(2) Computer control of machines that produce manufactured parts. *See also:* distributed numerical control; direct numerical control; process control; computer numerical control. (C) 610.2-1987

numerical control machine (1) A machine that produces manufactured parts under automatic control. (C) 610.2-1987
(2) A machine that produces drilled, milled, and machined parts under automatic control. *Synonym:* numerical control tool. (C) 610.10-1994w

numerical control system (numerically controlled machines) A system in which actions are controlled by the direct insertion of numerical data at some point. *Note:* The system must automatically interpret at least some portion of these data. (IA/EEC) [61], [74]

numerical control tool *See:* numerical control machine.

numerical data Data in which information is expressed by a set of numbers or symbols that can only assume discrete values or configurations. (IA) [61]

numerical display (illuminating engineering) An electrically operated display of digits. Tungsten filaments, gas discharges, light-emitting diodes, liquid crystals, projected numerals, illuminated numbers and other principles of operation may be used. (EEC/IE) [126]

numerical expression *See:* number.

numerical model (A) A mathematical model in which a set of mathematical operations are reduced to a form suitable for solution by a simpler methods such as numerical analysis or automation; for example, a model in which a single equation representing a nation's economy is replaced by a large set of simple averages based on empirical observations of inflation rate, unemployment rate, gross national product, and other indicators. **(B)** A model whose properties are expressed by numbers. (C) 610.3-1989

numerical reliability (software) The probability that an item will perform a required function under stated conditions for a stated period of time. *See also:* function. (C/SE) 729-1983s

numerical shift *See:* arithmetic shift.

numeric attribute An attribute whose value may be either integer or real. (LM/C) 802.1F-1993r

numeric bit data *See:* binary picture data.

numeric character *See:* digit.

numeric character data *See:* decimal picture data.

numeric character set (data management) A character set that contains digits and may contain control characters, special characters, and the space character, but not letters. (C) 610.5-1990w

numeric code A code that uses numerals to represent data. (C) 610.5-1990w, 1084-1986w

numeric data (data management) Data used to represent numbers. *See also:* binary data; complex data; packed data; real data; integer data. (C) 610.5-1990w

numeric keypad A keypad comprising of the numeric keys and usually the dot, comma and return keys. *Note:* It is most often located near the full keyboard, the keys being arranged in a 3×4 or 4×4 array to facilitate numeric data entry. (C) 610.10-1994w

numeric keypunch A keypunch that processes only numeric data. (C) 610.10-1994w

numeric optical disk *See:* optical disk.

numeric printout (dedicated-type sequential events recording systems) A brief coded method of identifying inputs using numeric characters only. (PE/EDPG) [5], [1]

numeric punch A hole punched in one of the punch rows designated as zero through nine. *See also:* digit punch. (C) 610.10-1994w

numeric representation (data management) A discrete representation of data by numerals. (C) 610.5-1990w

numeric shift (data management) A control for selecting the numeric character set on an alphanumeric keyboard or printer. *Contrast:* alphabetic shift. *See also:* shift character. (C) 610.5-1990w

N-unit (N) A measurement of refractivity, usually in parts per million, where:

$N = (n - 1) \times 10^6$

where

n = the refractive index

See also: refractivity. (AP/PROP) 211-1997

nurses' stations (health care facilities) Areas intended to provide a center of nursing activity for a group of nurses working under one nurse supervisor and serving bed patients, where the patient's calls are received, nurses are dispatched, nurses' notes written, inpatient charts prepared, and medications prepared for distribution to patients. Where such activities are carried on in more than one location within a nursing unit, all such separate areas are considered a part of the nurses' station. (NESC/NEC) [86]

nursing home (health care facilities) A building or part thereof used for the lodging, boarding and nursing care, on a 24-hour basis, of 4 or more persons who, because of mental or physical incapacity, may be unable to provide for their own needs and safety without the assistance of another person. Nursing home, wherever used in the National Electrical Code shall include nursing and convalescent homes, skilled nursing facilities, intermediate care facilities, and infirmaries of homes for the aged. (NESC/NEC) [86]

N-user An N+1 entity that uses the services of the N-layer, and below, to communicate with another N+1 entity. (C/LM/CC) 8802-2-1998

nutating feed A technique of conical scanning in which the polarization remains unchanged. (AES) 686-1997

nutation (gyros) The oscillation of the spin axis of a gyro about two orthogonal axes normal to the mean position of the spin axis. (AES/GYAC) 528-1994

nutation frequency (inertial sensors) (gyros) The frequency of the coning or periodic wobbling motion of the rotor spin axis that results from a transient input. For an undamped rotor, the nutation frequency equals the product of the rotor spin frequency and the ratio of the rotor polar moment of inertia to the effective rotor transverse moment of inertia. (AES/GYAC) 528-1994

NVM *See:* nonvolatile memory.

nybble *See:* quartet; nibble.

Nyquist diagram (data transmission) The Nyquist diagram of a feedback amplifier is a plot, in rectangular coordinates, of the real and imaginary parts of the factor $\propto/\beta$ for frequencies from zero to infinity, where $\propto$ is the amplification in the absence of feedback, and β is the fraction of the output voltage that is superimposed on the amplifier input. *Note:* The criterion for stability of a feedback amplifier is that the curve of the Nyquist diagram shall not enclose the point $X = -1$, $Y = 0$, where μ/β equals $X + jY$. (PE) 599-1985w

Nyquist interval (data transmission) The maximum separation in time that can be given to regularly spaced instantaneous samples of a wave of bandwidth W for complete determination of the wave form of the signal. Numerically, it is equal to $1/2W$ seconds. (PE) 599-1985w

Nyquist rate (1) (channel) The reciprocal of the Nyquist interval. (IT) [123]

(2) The minimum rate that an analog signal must be sampled in order to be represented in digital form. This rate is twice the frequency of that signal. (PE/PSR) 1344-1995

O

OA *See:* output axis; oil-immersed transformer.

OBC *See:* bar code; optical bar code.

OBE *See:* operating basis earthquake.

OBI *See:* omnibearing indicator.

Object An instance of the class `IEEE1451_Entity` or of a subclass thereof. (IM/ST) 1451.1-1999

object **(1)** **(A)** Pertaining to the outcome of an assembly or compilation process. *See also:* object program; object module; object code. **(B)** A program constant or variable. **(C)** An encapsulation of data and services that manipulate that data. *See also:* object-oriented design. (C) 610.12-1990
(2) A passive entity in a system that contains or receives information. Typically, objects include files, directories, registers, buffers, cache, memories, bus lines, displays, and other input/output devices. (C/BA) 896.3-1993w
(3) A piece of data that can be defined by the operations performed on it. The Intrinsics represent an object internally as a pointer to a dynamically allocated data structure. (C) 1295-1993w
(4) A representation of a real-world entity. An object is an instance of a class and has values for the attributes and relationships defined for that class. (C/SE) 1420.1-1995
(5) An abstraction of a physical or logical resource. (C) 610.7-1995
(6) An instance in the software hierarchy that can be operated on using the software administration utilities. (C/PA) 1387.2-1995
(7) Any of the complex information objects created, examined, modified, or destroyed by means of the [OM] interface. (C/PA) 1238.1-1994w, 1328-1993w, 1224-1993w, 1327-1993w, 1224.1-1993w
(8) A member of an object set and an instance of an object type. An object represents something in the observable world that may be distinguished from other instances of its object type and may be uniquely identified. (C/SE) 1320.1-1998
(9) A data object that has an identifier (name) and a value. (C/LM) 802.10-1998
(10) A collection of data and operations. (IM/ST) 1451.1-1999
(11) *See also:* directory object; OM object. (C/PA) 1328.2-1993w, 1224.2-1993w
(12) *See also:* instance. (C/SE) 1320.2-1998

object class An identified family of objects (or conceivable objects) that share certain characteristics. *Synonym:* directory class. (C/PA) 1328.2-1993w, 1224.2-1993w, 1326.2-1993w, 1327.2-1993w

object code Computer instructions and data definitions in a form output by an assembler or compiler. An object program is made up of object code. *Contrast:* source code. (C) 610.12-1990

object color The color of the light reflected or transmitted by the object when illuminated by a standard light source, such as source A, B, or C of the Commission Internationale de l'Eclairage (CIE). *See also:* standard source; color. (EEC/IE) [126]

Object Dispatch Address A network-specific identifier for an endpoint of a client-server communication. Specifically, a value having datatype `ObjectDispatchAddress`. (IM/ST) 1451.1-1999

object entry *See:* entry.

object extraction *See:* image segmentation.

object file A regular file containing the output of a compiler, formatted as input to a linkage editor for linking with other object files into an executable form. The methods of linking are unspecified and may involve the dynamic linking of objects at run time. The internal format of an object file is unspecified, but a conforming application shall not assume an object file is a text file. (C/PA) 9945-2-1993

object identifier **(1)** A value (distinguishable from all other such values) that is associated with an information object. (C/PA) 1328.2-1993w, 1224.2-1993w, 1326.2-1993w, 1327.2-1993w
(2) In general, a unique representation (name) of a manageable object defined in a management information base (MIB). (C/MM) 1284.1-1997
(3) Some concrete representation for the identity of an object (instance). The object identifier (oid) is used to show examples of instances with identity, to formalize the notion of identity, and to support the notion in programming languages or database systems. (C/SE) 1320.2-1998

Objective C An object-oriented version of C. (C) 610.13-1993w

object configuration In an object, the configuration is the specification of its allowed communications and of the internal state or organization of the object. To configure an object means to make the necessary changes to the object to make real these specifications. (IM/ST) 1451.1-1999

objective evidence (nuclear power quality assurance) Any documented statement of fact, other information, or record, either quantitative or qualitative, pertaining to the quality of an item or activity, based on observations, measurements, or tests which can be verified. (PE/NP) [124]

objective loudness rating (loudness ratings of telephone connections) The rating of a connection or its components when measured according to this standard. (COM/TA) 661-1979r

objectives The desired goals and results of the evaluation/selection process in terms relevant to the organization(s) involved. (C/SE) 1209-1992w

object language *See:* target language.

Object Management Group (OMG) Organization of computer manufacturers, software developers, communications organizations, and computer users established to promote open object-oriented computer architectures and standards. (SCC20) 1226-1998

object model **(1)** An integrated abstraction that treats all activities as performed by collaborating objects and encompassing both the data and the operations that can be performed against that data. An object model captures both the meanings of the knowledge and actions of objects behind the abstraction of responsibility. (C/SE) 1320.2-1998
(2) A definition of data structures and operations organized in a formal specification. An object model provides applications with a common view and a common way of interfacing to an element of functionality. (IM/ST) 1451.1-1999

object module A computer program or subprogram that is the output of an assembler or compiler. *See also:* load module; object program. (C) 610.12-1990

Object Name A nonconfigurable name for an instance of an Object used to convey the purpose or function of the Object instance. For any Object, the operation `GetObjectName` returns a value, `object_name`, that has the same value as Object Name. (IM/ST) 1451.1-1999

object-oriented design A software development technique in which a system or component is expressed in terms of objects and connections between those objects. *See also:* transform analysis; rapid prototyping; stepwise refinement; structured design; data structure-centered design; transaction analysis; modular decomposition; input-process-output. (C) 610.12-1990

object-oriented language **(1)** A programming language that allows the user to express a program in terms of objects and messages between those objects. Examples include Smalltalk and LOGO. (C) 610.12-1990

(2) A computer language that allows the user to express a program in terms of objects and messages between those objects. Examples include SMALLTALK and LOGO. *See also:* LOOPS; FLAVORS; C++; EIFFEL. (C) 610.13-1993w

object program (software) A computer program that is the output of an assembler or compiler. *Synonym:* target program. *Contrast:* source program. *See also:* object module. (C) 610.12-1990

object repository An area for the storage of objects, like a library. (SCC20) 1226-1998

object set A subset of instantiations from the set of all possible instantiations of all object types within an object type set. An object set is a subset of the union of the members of an object type set; the set of object sets includes the empty set and the set of the union of the members of the object type set itself. An object set is modeled by an arrow segment. (C/SE) 1320.1-1998

Object Tag A configurable identifier for the endpoints of client-server communications. Specifically a value having datatype `ObjectTag`. For any Object, the operation `GetObjectTag` returns a value, `object_tag`, that has the same value as Object Tag. (IM/ST) 1451.1-1999

object type The set of all possible instantiations of a singular concept, either physical or data, within an IDEF0 model. An IDEF0 object type is generally analogous to an IDEF1X entity or an IDEF1 entity class. (C/SE) 1320.1-1998

object type set A named set of one or more object types. An object type set may include object types that are themselves grouped as object type sets. An object type set is designated by an arrow label. (C/SE) 1320.1-1998

oblique-incidence ionospheric sounding *See:* active sounding.

observable A property of a component of a state whereby its value at a given time can be computed from measurements on the output over a finite past interval. *See also:* control system. (CS/IM) [120]

observable, completely The property of a plant whereby all components of the state are observable. *See also:* control system; plant; observable. (CS/IM) [120]

observable insulation temperature (electric equipment) (thermal classification of electric equipment and electrical insulation) The temperature of the insulation in electric equipment, which is measured in a specified way; for example, with a thermometer, embedded thermocouple, resistance detector, or by winding resistance or other suitable procedure. (EI) 1-1986r

observable temperature (equipment) The temperature of equipment obtained on test or in operation. (EI) 1-1969s

observable temperature rise (thermal classification of electric equipment and electrical insulation) (electric equipment) The difference between the observable insulation temperature and the ambient temperature. (EI) 1-1986r

observation The raw data acquired by executing a test procedure. It represents the observed characteristics of a specific signal (e.g., the voltage peak of a sinusoid wave form), the observed characteristics of the environment (e.g., the ambient temperature), or the derived value of product characteristics (e.g., the measured value of gain). (SCC20) 1226-1998

observation time The time interval over which a radar echo signal may be integrated for detection or measurement. (AES) 686-1997

observed attribute An attribute of a managed object whose value is being observed by an EWMA metric managed object. (LM/C) 802.1F-1993r

observed failure rate For a stated period in the life of an item, the ratio of the total number of failures in a sample to the cumulative observed time on that sample. The observed failure rate is to be associated with particular, and stated time intervals (or summation of intervals) in the life of the items, and with stated conditions. *Notes:* 1. The criteria for what constitutes a failure shall be stated. 2. Cumulative time is the sum of the times during which each individual item has been performing its required function under stated conditions. (R) [29]

observed instantaneous availability At a stated instant of time the proportion of occasions when an item can perform a required function. *Notes:* 1. Occasions can refer to either a number of items at a single instant of time or to one or more items at instants repeated in time. 2. The run-up time is counted in down-time after repair and is counted in the up-time when the equipment is brought into use for the first time. 3. The observed instantaneous availability is to be associated with a period of time and with stated conditions of use and maintenance. (R) [29]

observed managed object A managed object with one or more observed attributes. (LM/C) 802.1F-1993r

observed mean active maintenance time The ratio of the sum of the active maintenance times to the total number of maintenance actions. *Note:* The maintenance conditions applied shall be stated. (R) [29]

observed mean availability The ratio of the cumulative time for which an item can perform a required function to the cumulative time under observation, or at instants of time (chosen by a sampling technique), the mean of the proportion of a number of nominally identical items which can perform their required function. *Notes:* 1. When one limiting value is given, this is usually the lower limit. 2. The observed mean availability is to be associated with a stated period of time and with stated conditions of use and maintenance. (R) [29]

observed mean life (non-repaired items) The mean value of the lengths of observed times to failure of all items in a sample under stated conditions. *Note:* The criteria for what constitutes a failure shall be stated. (R) [29]

observed reliability (A) (non-repaired items) For a stated period of time, the ratio of the number of items which performed their functions satisfactorily at the end of the period to the total number of items in the sample at the beginning of the period. **(B) (repaired item or items)** The ratio of the number of occasions on which an item or items performed their functions satisfactorily for a stated period of time to the total number of occasions the item or items were required to perform for the same period. *Note:* The criteria for what constitutes satisfactory function shall be stated. (R) [29]

obsolescent An indication that a certain feature may be considered for withdrawal in future revisions of a standard. (C/PA) 2003.2-1996, 1003.5-1999

obstacle gain The ratio, usually expressed in dB, of the electromagnetic field at a point in the vicinity of the geometrical shadow of an obstacle to the field which would occur in the absence of the obstacle. (AP/PROP) 211-1997

obstruction beacon *See:* hazard beacon.

obstruction lights (illuminating engineering) Aeronautical ground lights provided to indicate obstructions. (EEC/IE) [126]

Occam A general-purpose programming language designed in the early 1980's for use in parallel computer systems. (C) 610.13-1993w

occluded ear simulator Ear simulator that simulates the inner part of the ear canal, from the tip of an ear insert to the eardrum. (COM/TA) 1206-1994

occulting light (illuminating engineering) A rhythmic light in which the periods of light are clearly longer than the periods of darkness. (EEC/IE) [126]

occupational title standard A standard that describes the characteristics of the general areas of work or profession. (C) 610.12-1990

occupied bandwidth (radio-noise emissions) The frequency bandwidth such that, below its lower and above its upper frequency limits, the mean powers radiated are each equal to 0.5% of the total mean power radiated by a given emission. In some cases, for example multichannel frequency division systems, the percentage of 0.5% may lead to certain difficulties in the practical application of the definition of occupied bandwidth; in such cases a different percentage may be useful. (EMC) C63.4-1988s

occurrence An individual instance of an entity, record, or item, containing a specific set of values for its constituent parts. (C) 610.5-1990w

OCR *See:* optical character reader; optical character recognition.

OCR-A *See:* optical character recognition-A.

OCR-B *See:* optical character recognition-B.

octad (mathematics of computing) (octade) A group of three bits used to represent one octal digit. (C) 1084-1986w

octal (A) (mathematics of computing) Pertaining to a selection in which there are eight possible outcomes. **(B) (mathematics of computing)** Pertaining to the numeration system with a radix of eight. (C) 1084-1986

octal character string A sequence of characters from the set of octal digits the first of which shall be the digit zero. Octal character strings shall consist only of the following characters:

0 1 2 3 4 5 6 7

Within software definition files of exported catalogs, all such strings shall be encoded using IRV. (C/PA) 1387.2-1995

octal digit A numeral used to represent one of the eight digits in the octal numeration system; 0, 1, 2, 3, 4, 5, 6, or 7. (C) 1084-1986w

octal notation Any notation that uses the octal digits and the radix 8. (C) 1084-1986w

octal number (A) A quantity that is expressed using the octal numeration system. **(B)** Loosely, an octal numeral. (C) 1084-1986

octal number system* *See:* octal numeration system.

* Deprecated.

octal numeral A numeral in the octal numeration system. For example, the octal numeral 14 is equivalent to the decimal numeral 12. (C) 1084-1986w

octal numeration system The numeration system that uses the octal digits and the radix 8. *Synonym:* octal system. (C) 1084-1986w

octal point The radix point in the octal numeration system. (C) 1084-1986w

octal system *See:* octal numeration system.

octal-to-binary conversion The process of converting an octal numeral to an equivalent binary numeral. For example, octal 213.2 is converted to binary 10001011.01. (C) 1084-1986w

octal-to-decimal conversion The process of converting an octal numeral to an equivalent decimal numeral. For example, octal 213.2 is converted to decimal 139.25. (C) 1084-1986w

octant *See:* sextant.

octantal error (navigation) (navigation aid terms) An error in measured bearing caused by the finite spacing of the antenna elements in systems using spaced antennas to provide bearing information (such as VOR [very high-frequency omnidirectional range]): this error varies in a sinusoidal manner throughout the 360° and has four positive and four negative maximums. (AES/GCS) 172-1983w

octary tree A tree of order 8. *Note:* Such a tree is typically used to store three-dimensional data. *Synonyms:* octonary tree; octtree. (C) 610.5-1990w

octave (1) (data transmission) In electric communication, the interval between two frequencies having a ratio of 2 to 1. (PE) 599-1985w

(2) The interval between two frequencies that have a frequency ratio of 2 (e.g., 1 to 2 Hz, 2 to 4 Hz, 4 to 8 Hz, etc.). (SWG/PE/T&D/PSR) C37.98-1977s, 539-1990, C37.100-1992, C37.81-1989r

(3) (overhead power lines) The interval between two sounds having a fundamental frequency ratio of two. (T&D/PE) 539-1990

octave band, one-third octave band The integrated sound pressure level of all components in a frequency band corresponding to a specified octave. *Note:* The location of an octave band pressure level on a frequency scale, f_0, is usually specified as the geometric mean of the upper and lower frequencies of the octave. The lower frequency of the octave band is $f_0/\sqrt{2}$ and the upper frequency is $(/\sqrt{2})f_0$. A third-octave band extends from a lower frequency $f_0/^6/\sqrt{2}$ to an upper frequency of $(^6/\sqrt{2})f_0$. (T&D/PE) 656-1992

octave-band pressure level (1) (octave pressure level) (sound) The band pressure level for a frequency band corresponding to a specified octave. *Note:* The location of an octave-band pressure level on a frequency scale is usually specified as the geometric mean of the upper and lower frequencies of the octave. (SP/ACO) [32]

(2) (overhead power lines) The integrated sound pressure level of all components in a frequency band corresponding to a specified octave. *Note:* The location of an octave band pressure level on a frequency scale, f_0, is usually specified as the geometric mean of the upper and lower frequencies of the octave. The lower frequency of the octave band is $f_0/^6/\sqrt{2}$ and the upper frequency is $(^6/\sqrt{2})f_0$. A third-octave band extends from a lower frequency $f_0/^6/\sqrt{2}$ to an upper frequency of $(^6/\sqrt{2})f_0$. (T&D/PE) 539-1990

octet (1) A group of eight adjacent binary digits operated on as a unit. (SUB/PE/C) 999-1992w, 610.5-1990w, 1084-1986w

(2) A sequence of eight bits, usually operated upon as a unit. (DIS/C) 1278.1-1995

(3) A data unit composed of eight ordered binary bits. An octet is encoded as a pair of code symbols. (C/LM) 802.9a-1995w

(4) A byte composed of eight bits. (LM/C) 802.3u-1995s, 610.10-1994w

(5) An ordered sequence of 8 b. *Note:* Octets can be stored in larger objects if appropriate to a particular architecture. (C/PA) 1224-1993w, 1327-1993w

(6) An eight-bit data entity (byte). (C/MM) 1284.1-1997

(7) 8 b data object. *See also:* byte. (PE/SUB) 1379-1997

(8) A group of 8 bits, also known as a byte. (IM/ST) 1451.2-1997

(9) A sequence of eight bits. (AMR/SCC31) 1377-1997

(10) A bit-oriented element that consists of eight contiguous binary bits. (C/LM/CC) 8802-2-1998

(11) Unit of data representation that consists of eight contiguous bits. (C) 1003.5-1999

(12) A group of eight adjacent bits. (EMB/MIB) 1073.4.1-2000, 1073.3.2-2000

Octet Array A value of type `OctetArray`. (IM/ST) 1451.1-1999

octet string (1) A value of ASN.1 type octetstring. (C/PA) 1238.1-1994w

(2) A string composed of octets. (C/PA) 1328-1993w, 1327-1993w, 1224-1993w

octetstring type A simple type whose distinguished values are an ordered sequence of zero, one or more octets, each octet being an ordered sequence of 8 bits. (C/PA) 1238.1-1994w

octlet (1) A set of eight adjacent bytes. (C/BA) 10857-1994, 896.3-1993w, 896.4-1993w

(2) Eight bytes of data. (MM/C) 1394-1995, 1596-1992

(3) Eight bytes (64 bits) of data. Not to be confused with an "octet," which has been used to describe 8 bits of data. In this document, the term **byte,** rather than "octet," is used to describe 8 bits of data. (C/MM) 1754-1994

(4) An ordered set of eight adjacent bytes. (C/BA) 1014.1-1994w

(5) An 8-byte data format or data type. Not to be confused with an octet, which has been commonly used to describe 8 bits of data. In this document, the term byte, rather than octet, is used to describe 8 bits of data. (C/MM) 1596.5-1993

(6) Eight bytes of data. Not to be confused with an octet, which has been commonly used to describe eight bits of data.

In this document, the term byte, rather than octet, is used to describe eight bits of data. (C/MM) 1596.4-1996
(7) Eight bytes, or 64 bits, of data. (C/MM) 1394a-2000

octode An eight-electrode electron tube containing an anode, a cathode, a control electrode, and five additional electrodes that are ordinarily grids. (ED) 161-1971w

octodenary **(A)** Pertaining to a selection in which there are 18 possible outcomes. **(B)** Pertaining to the numeration system with a radix of 18. (C) 1084-1986

octonary* *See:* octal.
* Deprecated.

octonary tree *See:* octary tree.

octtree *See:* octary tree.

odd-even check *See:* parity check.

odd-even sort *See:* Batcher's parallel sort.

odd parity **(1)** An error detection method in which the number of ones in a binary word, byte, character, or message is maintained as an odd number. (C) 1084-1986w
(2) The property possessed by a binary word, byte, character, or message that has an odd number of ones. (C) 1084-1986w

O-display **(1)** **(navigation aid terms)** A type of radar display format. (AES/GCS) 172-1983w
(2) An A-display modified by the inclusion of an adjustable notch for measuring range. (AES) 686-1997

odolite (navigation aid terms) An optical instrument for accurately measuring horizontal and vertical angles. (AES/GCS) 172-1983w

odometer (navigation aid terms) A device attached to a vehicle for counting the number of revolutions of a drive shaft or wheel. (AES/GCS) 172-1983w

OEM *See:* original equipment manufacturer.

oersted The unit of magnetic field strength in the unrationalized centimeter-gram-second (cgs) electromagnetic system. The oersted is the magnetic field strength in the interior of an elongated uniformly wound solenoid that is excited with a linear current density in its winding of one abampere per 4π centimeters of axial length. (Std100) 270-1966w

off-axis mode (laser maser) An off-axis mode will incorporate one or more of the maxima which lie off the axis of a beam. *See also:* higher-order mode of propagation. (LEO) 586-1980w

off-center display A plan-position-indicator display, the center of which does not correspond to the position of the radar antenna. *See also:* radar. (EEC/PE) [119]

off-center PPI A plan-position indicator (PPI) that has the zero position of the time base at a point other than the center of the display, thus providing the equivalent of a larger display for a selected portion of the service area. (AES/RS) 686-1990

offered traffic (telephone switching systems) A measure of the calls requesting service during a given period of time. (COM) 312-1977w

off-hook **(1)** **(telephone switching systems)** A closed station line or any supervisory or pulsing condition is indicative of this state. (COM) 312-1977w
(2) In regard to a telephone set, activated÷ashthat is, a telephone set is in use. The off-hook condition indicates a (busy) condition to incoming calls. *Contrast:* on-hook. (C) 610.7-1995

office automation The automation of information traffic through the use of any or all of the following: voice processing; word and data processing; reprographics; records processing and micrographics; telecommunications. *See also:* paperless office; automatic calendar; electronic office; electronic mail. (C) 610.2-1987

office class (telephone switching systems) A designation (Class 1, 2, 3, 4, 5) given to each office in World Zone 1 involved in the completion of toll calls. The class is determined according to the office's switching function, its interrelation with other switching offices, and its transmission requirements. The class designation given to the switching points in the network determines the routing pattern for all calls. Class 1 is higher in rank than Class 2; Class 2 is higher than Class 3; and so on. *See also:* world-zone number. (COM) 312-1977w

office code (telephone switching systems) The digits that designate a block of main-station codes within a numbering-plan area. (COM) 312-1977w

office failure rate The expected frequency of entire outages because of malfunctions in the switching system. (COM/TA) 973-1990w

office of the future *See:* electronic office.

office test (meter) A test made at the request or suggestion of some department of the company to determine the cause of seemingly abnormal registration. *See also:* service test. (EEC/PE) [119]

Official Production System (OPS5) A nonprocedural programming language that uses precise rules, in the form of a rule-and-fact set model, to reach solutions to problem descriptions. *Note:* Used in artificial intelligence applications for building expert systems. (C) 610.13-1993w

OFF-impedance (thyristor) The differential impedance between the terminals through which the principal current flows, when the thyristor is in the OFF state at a stated operating point. *See also:* principal voltage-current characteristic. (IA/ED) 223-1966w, [12], [46]

offline **(1)** **(monitoring radioactivity in effluents)** A system where an aliquot is withdrawn from the effluent stream and conveyed to the detector assembly. (NI) N42.18-1980r
(2) **(A)** **(test, measurement, and diagnostic equipment)** Operation of input/output and other devices not under direct control of a device. **(B)** **(test, measurement, and diagnostic equipment)** Peripheral equipment operated outside of, and not under control of the system; for example, the off-line printer. (MIL) [2]
(3) **(software)** Pertaining to a device or process that is not under the direct control of the central processing unit of a computer. *Contrast:* online. *See also:* vary off-line. (C) 610.12-1990, 610.10-1994w
(4) Used to describe an MTM-Bus module when it is in a mode such that it will not respond to state transitions on MTM-Bus signals whether or not the module is connected to the bus. Also used to describe such a mode. (TT/C) 1149.5-1995
(5) In 1000BASE-X, a DTE in its nonfunctional state. (C/LM) 802.3-1998

offline operation **(A)** **(emergency and standby power)** Pertaining to computer systems not under direct control of the central processing unit. **(B)** **(emergency and standby power)** Pertaining to uninterruptible power supply systems whereby an inverter is off during normal operating conditions. (IA/PSE) 446-1987

offline storage Storage that is not under the control of a processing unit. *Contrast:* online storage. (C) 610.10-1994w

offline system A system that is dormant until it is called upon to operate, such as a diesel generator that is started up when a power failure occurs. (IA/PSE) 493-1997

off-line testing (test, measurement, and diagnostic equipment) Testing of the unit under test removed from its operational environment or its operational equipment. Shop testing. (MIL) [2]

off-net call (telephone switching systems) A call from a switched-service network to a station outside that network. (COM) 312-1977w

off-normal relay contacts Contacts on a multiple switch that are in one condition when the relay is in its normal position and in the reverse condition for any other position of the relay. (EEC/REE) [87]

off-peak energy (power operations) Energy supplied during designated periods of relatively low system demands. (T&D/PE/PSE) 858-1987s, 346-1973w

off-peak period (watthour meters) The period of time during which the specified off-peak rate applies. (ELM) C12.13-1985s

off-peak power Power supplied during designated periods of relatively low system demands. *See also:* generating station. (T&D/PE) [10]

OFF period (1) (electron tube) The time during an operating cycle in which the electronic tube is nonconducting. *See also:* ON period. (Std100) [84]
(2) (circuit switching element) (inverters) The part of an operating cycle during which essentially no current flows in the circuit switching element. *See also:* self-commutated inverters. (IA) [62]

off-road vehicle A vehicle specifically designed and equipped to traverse sand, swamps, muddy tundra, or rough mountainous terrain. Vehicles falling into this category are usually all wheel drive or tracked units. In some cases, units equipped with special air bag rollers having a soft footprint are utilized. *Synonyms:* all terrain vehicle; swamp buggy. (T&D/PE) 524-1992r

offset (1) (transducer) The component of error that is constant and independent of the inputs, often used to denote bias. (C) 166-1977w, 165-1977w
(2) (course computer) (electronic navigation) An automatic computer that translates reference navigational coordinates into those required for a predetermined course. *See also:* navigation. (AES/RS) 686-1982s, [42]
(3) (pulse terminology) The algebraic difference between two specified magnitude reference lines. Unless otherwise specified, the two magnitude reference lines are the waveform baseline and the magnitude origin line. *See also:* waveform epoch. (IM/WM&A) 194-1977w
(4) (A) (software) The difference between the loaded origin and the assembled origin of a computer program. *Synonym:* relocation factor. **(B) (software)** A number that must be added to a relative address to determine the address of the storage location to be accessed. This number may be the difference defined in (A) or another number defined in the program. *See also:* self-relative address; relative address; base address; indexed address. (C) 610.12-1990
(5) The *octet* position relative to the start of a *Pre-Arbitrated (PA) segment* used to carry an *isochronous service octet* for a particular *Isochronous Service User (ISU).* (LM/C) 8802-6-1994
(6) (A) The measure of unbalance between halves of a symmetrical circuit. **(B)** The change in input voltage needed to cause the output voltage of a linear amplifier to be zero. **(C)** The difference between the value or condition desired and that actually attained. **(D)** The difference between the address in a base register and the memory location of a datum. *See also:* relative address. (C) 610.10-1994
(7) (as used in data acquisition) A predetermined value modifying the actual value so as to improve the integrity of the system, for example, the use of a 4 mA signal to represent zero in a 4 mA to 20 mA system. (SWG/PE/SUB) C37.100-1992, C37.1-1994
(8) (as applied to a distance relay) The displacement of the operating characteristic on an *R-X* diagram from the position inherent to the basic performance class of the relay. *Note:* A relay with this characteristic is called an offset relay. (SWG/PE/PSR) C37.100-1992, C37.90-1978s
(9) *See also:* gain and offset. (IM/WM&A) 1057-1994w

offset angle (lateral disk reproduction) (electroacoustics) The offset angle is the smaller of the two angles between the projections into the plane of the disk of the vibration axis of the pickup stylus and the line connecting the vertical pivot (assuming a horizontal disk) of the pickup arm with the stylus point. *See also:* phonograph pickup. (SP) [32]

offset (outboard) bearing (air switch) A component of a switch-operating mechanism designed to provide support for a torsional operating member and a crank that provides reciprocating motion for switch operation. (SWG/PE) C37.100-1992

offset, clipping *See:* clipping offset.

offset course computer (navigation aid terms) An automatic computer which translates reference navigational coordinates into those required for a predetermined course. (AES/GCS) 172-1983w

offset entry A read-only memory (ROM) entry that provides a 24-bit offset value. The offset values specifies the location of a Control and Status Register (CSR) that provides a 32-bit parameter. (C/BA/MM) 896.10-1997, 896.2-1991w, 1212-1991s

offset marker pole *See:* plumb marker pole.

offset paraboloidal reflector *See:* paraboloidal reflector.

offset paraboloidal reflector antenna A reflector antenna whose main reflector is a portion of a paraboloid that is not symmetrical with respect to its focal axis, and does not include the vertex so that aperture blockage by the feed is reduced or eliminated. (AP/ANT) 145-1993

offset plan-position indicator (PPI) A PPI that has the zero position of the time base at a point other than the center of the display, thus providing the equivalent of a larger display for a selected portion of the coverage area. *Synonym:* off-center PPI. (AES) 686-1997

offset voltage (1) (power supplies) A direct-current potential remaining across the comparison amplifier's input terminals (from the null junction to the common terminal) when the output voltage is zero. The polarity of the offset voltage is such as to allow the output to pass through zero and the polarity to be reversed. It is often deliberately introduced into the design of power supplies to reach and even pass zero-output volts. (AES) [41]
(2) The driver offset voltage is the average dc voltage generated by the differential driver;

$$V_{os} = (V_{oa} + V_{ob})/2.$$

(C/MM) 1596.3-1996

offset waveform (pulse terminology) A waveform whose baseline is offset from, unless otherwise specified, the magnitude origin line. (IM/WM&A) 194-1977w

OFF state (thyristor) The condition of the thyristor corresponding to the high-resistance low-current portion of the principal voltage-current characteristic between the origin and the breakover point(s) in the switching quadrant(s). *See also:* principal voltage-current characteristic. (IA) [12]

OFF-state current (thyristor) The principal current when the thyristor is in the OFF state. *See also:* principal current. (IA) [12]

OFF-state power dissipation (thyristor) The power dissipation resulting from OFF-state current. (IA) [12]

OFF-state voltage (thyristor) The principal voltage when the thyristor is in the OFF state. *See also:* principal voltage-current characteristic. (IA) [12]

ohm (1) (general) The unit of resistance (and of impedance) in the International System of Units (SI). The ohm is the resistance of a conductor such that a constant current of one ampere in it produces a voltage of one volt between its ends. (Std100) 270-1966w
(2) (metric practice) The electric resistance between two points of a conductor when a constant difference of potential of one volt, applied between these two points, produces in this conductor a current of one ampere, this conductor not being the source of any electromotive force. (QUL) 268-1982s

ohmic contact (1) (semiconductor) A contact between two materials, possessing the property that the potential difference across it is proportional to the current passing through. *See also:* semiconductor. (AES) [41]
(2) (charged-particle detectors) (x-ray energy spectrometers) A purely resistive contact*f*one that has a linear voltage-current characteristic throughout its entire operating range. (NPS/ED/NID) 759-1984r, 325-1996, 216-1960w, 301-1976s, 300-1988r

ohmic resistance test (rotating machinery) A test to measure the ohmic resistance of a winding, using direct current. *See also:* asynchronous machine. (PE) [9]

ohmmeter A direct-reading instrument for measuring electric resistance. It is provided with a scale, usually graduated in either ohms, megohms, or both. If the scale is graduated in megohms, the instrument is usually called a megohmmeter. *See also:* instrument. (EEC/PE) [119]

Ohm's law The current in an electric circuit is inversely proportional to the resistance of the circuit and is directly proportional to the electromotive force in the circuit. *Note:* Ohm's law applies strictly only to linear constant-current circuits. (Std100) 270-1966w

OHR *See:* over-the-horizon radar.

oid *See:* object identifier.

oil (1) (packaging machinery) Used as a prefix and applied to a device that interrupts an electric circuit; indicates that the interruption occurs in oil. (IA/PKG) 333-1980w
(2) (power and distribution transformers) The term "oil" includes the following insulating and cooling liquids: Type I Mineral Oil (uninhibited oil), Type II Mineral Oil (inhibited oil), and Askarel. (PE/TR) C57.12.80-1978r
(3) (outdoor apparatus bushings) Mineral transformer oil. (PE/TR) 21-1976

oil buffer (elevators) A buffer using oil as a medium that absorbs and dissipates the kinetic energy of the descending car or counterweight. *See also:* elevator. (PE/EEC) [119]

oil-buffer stroke (elevators) (oil buffer) The oil-displacing movement of the buffer plunger or piston, excluding the travel of the buffer-plunger accelerating device. *See also:* elevator. (PE/EEC) [119]

oil catcher (rotating machinery) A recess to carry off oil. *See also:* oil cup. (PE) [9]

oil-containment system A system designed to collect and retain oil in order to prevent 1) its migration beyond the boundaries of the system and 2) the contamination of navigable waters. (SUB/PE) 980-1994

oil cup (rotating machinery) An attachment to the oil reservoir for adding oil and controlling its upper level. (PE) [9]

oil cutout (oil-filled cutout) (oil-filled cutout) A cutout in which all or part of the fuse support and its fuse link or disconnecting blade are mounted in oil with complete immersion of the contacts and the fusible portion of the conducting element (fuse link), so that arc interruption by severing of the fuse link or by opening of contacts will occur under oil. (SWG/PE) C37.40-1993, C37.100-1992

oil discharge Any leak or spillage of oil, regardless of volume and including those that do not reach navigable waters. A discharge includes but is not limited to any spilling, leaking, pumping, pouring, emitting, emptying, or dumping of oil. (SUB/PE) 980-1994

oil-electric drive *See:* diesel-electric drive.

oil feeding reservoirs Oil storage tanks situated at intervals along the route of an oil-filled cable or at oil-filled joints of solid cable for the purpose of keeping the cable constantly filled with oil under pressure. (T&D/PE) [10]

oil-filled (designated liquid-filled) (prefix) The prefix oil-filled or designated liquid-filled as applied to equipment indicates that oil or the designated liquid is the surrounding medium. (EEC/PE) [119]

oil-filled bushing (outdoor electric apparatus) A bushing in which the space between the inside surface of the weather casing and the major insulation (or conductor where no major insulation is used) is filled with oil. (PE/TR) 21-1976

oil-filled cable A self-contained pressure cable in which the pressure medium is low-viscosity oil having access to the insulation. *See also:* pressure cable; self-contained pressure cable. (T&D/PE) [10]

oil-filled pipe cable A pipe cable in which the pressure medium is oil having access to the insulation. *See also:* pressure cable; pipe cable. (T&D/PE) [10]

oil-fill stand pipe *See:* oil-overflow plug.

oil groove (rotating machinery) A groove cut in the surface of the bearing lining or sometimes in the journal to help to distribute the oil over the bearing surface. *See also:* oil cup. (PE) [9]

oil-immersed (1) Having the coils immersed in an insulating liquid. *Note:* The insulating liquid is usually (though not necessarily) oil. *See also:* oil-immersed transformer. (TRR/PE/TR) C57.15-1968s
(2) (grounding device) Means that the windings are immersed in an insulating oil. (SPD/PE) 32-1972r

oil-immersed forced-air-cooled shunt reactor (shunt reactors over 500 kVA) (class OFA) An oil-immersed shunt reactor which is cooled by forced circulation of the cooling air over the cooling surface. (PE/TR) C57.21-1981s

oil-immersed forced-oil-cooled transformer with forced-water cooler (class FOW) A transformer having its core and coils immersed in oil and cooled by the forced circulation of this oil through external oil-to-water heat-exchanger equipment utilizing forced circulation of water over its cooling surface. (PE/TR) C57.12.80-1978r

oil-immersed forced-oil-cooled with forced-air cooler shunt reactor (shunt reactors over 500 kVA) (class FOA) An oil-immersed shunt reactor cooled by the forced circulation of oil through external oil-to-air heat-exchanger equipment utilizing forced circulation of air over its cooling surface. (PE/TR) C57.21-1981s

oil-immersed forced-oil-cooled with forced-water cooler shunt reactor (class FOW) (shunt reactors over 500 kVA) An oil-immersed shunt reactor cooled by the forced circulation of the oil through external oil-to-water heat-exchanger equipment utilizing forced circulation od water over its cooling surface. (PE/TR) C57.21-1981s

oil-immersed self-cooled/forced-air-cooled/forced-air-cooled transformer (power and distribution transformers) (class OA/FA/FA) A transformer having its core and coils immersed in oil and having a self-cooled rating obtained by the natural circulation of air over the cooling surface, a forced-air-cooled rating obtained by the forced circulation of air over a portion of the cooling surface, and an increased forced-air-cooled rating obtained by the increased forced circulation of air over a portion of the cooling surface. (PE/TR) C57.12.80-1978r

oil-immersed self-cooled/forced-air-cooled/forced-oil-cooled transformer (power and distribution transformers) (class OA/FA/FOA) A transformer having its core and coils immersed in oil and having a self-cooled rating with cooling obtained by the natural circulation of air over the cooling surface, a forced-air-cooled rating with cooling obtained by the forced circulation of air over this same cooling surface, and a forced-oil-cooled rating with cooling surface, and a forced-oil-cooled rating with cooling obtained by the forced circulation of oil over the core and coils and adjacent to this same cooling surface over which the air is being forced circulated. (PE/TR) C57.12.80-1978r

oil-immersed self-cooled/forced-air-cooled transformer (power and distribution transformers) (Class OA/FA) A transformer having a self-cooled rating with cooling obtained by the natural circulation of air over the cooling surface, and a forced-air-cooled rating with cooling obtained by the forced circulation of air over this same cooling surface. (PE/TR) C57.12.80-1978r

oil-immersed self-cooled/forced-air, forced-oil-cooled/forced-air, forced-oil-cooled transformer (power and distribution transformers) (class OA/FOA/FOA) A transformer similar to class OA/FA/FOA transformer except that its auxiliary cooling controls are arranged to start a portion of the oil pumps and a portion of the fans for the first auxiliary rating and the remainder of the pumps and fans for the second auxiliary rating. (PE/TR) C57.12.80-1978r

oil-immersed self-cooled shunt reactor (class OA) (shunt reactors over 500 kVA) An oil-immersed shunt reactor which is cooled by natural circulation of the cooling air over the cooling surface. (PE/TR) C57.21-1981s

oil-immersed self-cooled transformer (power and distribution transformers) (class OA) A transformer having its core and coils immersed in oil, the cooling being effected by the natural circulation of air over the cooling surface. (PE/TR) C57.12.80-1978r

oil-immersed shunt reactor (shunt reactors over 500 kVA) One in which the coils and magnetic current are immersed in an insulating oil. (PE/TR) C57.21-1981s

oil-immersed transformer A transformer in which the core and coils are immersed in an insulating oil. (PE/TR) C57.12.80-1978r

oil-immersed water-cooled/self-cooled transformer (class OW/A) A transformer having its core and coils immersed in oil and having a water-cooled rating with cooling obtained by the natural circulation of oil over the water-cooled surface, and a self-cooled rating with cooling obtained by the natural circulation of air over the cooling surface. (PE/TR) C57.12.80-1978r

oil-immersed water-cooled shunt reactor (Class OW) (shunt reactors over 500 kVA) An oil-immersed shunt reactor which is cooled by the natural circulation of the cooling oil over the water-cooled surface. (PE/TR) C57.21-1981s

oil-immersed water-cooled transformer (power and distribution transformers) (Class OW) A transformer having its core and coils immersed in oil, the cooling being effected by the natural circulation of oil over the water-cooled surface. (PE/TR) C57.12.80-1978r

oil-immersible current-limiting fuse *See:* oil-immersible current-limiting fuse unit.

oil-immersible current-limiting fuse unit A current-limiting fuse unit suitable for application requiring total or partial immersion directly in oil or other dielectric liquid of a transformer or switchgear. *Synonym:* oil-immersible current-limiting fuse. (SWG/PE/SWG-OLD) C37.40-1993, C37.100-1992

oil-impregnated paper-insulated bushing (1) A bushing in which the major insulation is provided by paper impregnated with oil. (PE/TR) 21-1976
(2) A bushing in which the internal insulation consists of a condenser wound from paper and subsequently impregnated with oil. The condenser is contained in an insulating envelope, the space between the condenser and the insulating envelope being filled with oil. (PE/TR) C57.19.03-1996

oil leakage load The load applied to the top of the bushings at which oil leakage begins. (PE/SUB) 693-1997

oilless circuit breaker *See:* circuit breaker.

oil-level gauge (rotating machinery) An indicating device showing oil level in the oil reservoir. (PE) [9]

oil-lift bearing (rotating machinery) A journal bearing in which high-pressure oil is forced under the shaft journal or thrust runner to establish a lubricating film. *See also:* bearing. (PE) [9]

oil-lift system (rotating machinery) A system that lubricates a bearing before starting by forcing oil between the journal or thrust runner and bearing surfaces. *See also:* oil cup. (PE) [9]

oil-overflow plug (rotating machinery) (oil-fill stand-pipe) An attachment to the oil reservoir that can be opened to allow excess oil to escape, to inspect the oil level, or to add oil. (PE) [9]

oil pot (rotating machinery) (oil reservoir) A bearing reservoir for a vertical-shaft bearing. *See also:* oil cup. (PE) [9]

oil-pressure electric gauge A device that measures the pressure of oil in the line between the oil pump and the bearings of an aircraft engine. The gauge is provided with a scale, usually graduated in pounds per square inch. It provides remote indication by means of self-synchronous generator and motor. (EEC/PE) [119]

oil-proof enclosure An enclosure constructed so that oil vapors, or free oil not under pressure, that may accumulate within the enclosure will not prevent successful operation of, or cause damage to, the enclosed equipment. (IA/MT) 45-1998

oil reservoir *See:* oil pot.

oil-resistant gaskets (power and distribution transformers) Those made of material which is resistant to oil or oil fumes. (PE/TR) C57.12.80-1978r

oil ring (rotating machinery) A ring encircling the shaft in such a manner as to bring oil from the oil reservoir to the sleeve bearing and shaft. *See also:* bearing. (PE) [9]

oil-ring guide (rotating machinery) A part whose main purpose is the restriction of the motion of the oil ring. *See also:* oil cup. (PE) [9]

oil-ring lubricated bearing A bearing in which a ring, encircling the journal, and rotated by it, raises oil to lubricate the bearing from a reservoir into which the ring dips. (PE) [9]

oil-ring retainer (rotating machinery) A guard to keep the oil ring in position on the shaft. (PE) [9]

oil seal (rotating machinery) A part or combination of parts in a bearing assembly intended to prevent leakage of oil from the bearing. *Synonyms:* bearing seal; bearing oil seal. (PE) [9]

oil spill (spill event) A discharge of oil into or upon navigable waters or shorelines in harmful quantities. (SUB/PE) 980-1994

oil switch (high-voltage switchgear) A switch with contacts that separate in oil. (SWG/PE) C37.40-1993

oil thrower (rotating machinery) (oil slinger) A peripheral ring or ridge on a shaft adjacent to the journal and which is intended to prevent any flow of oil along the shaft. *See also:* oil pot. (PE) [9]

oiltight (power and distribution transformers) So constructed as to exclude oils, coolants, and similar liquids under specified test conditions. (PE/TR) C57.12.80-1978r

oil-tight enclosure An enclosure constructed so that oil vapors or free oil not under pressure, which may be present in the surrounding atmosphere, cannot enter the enclosure. (IA/MT) 45-1998

oiltight pilot devices Devices such as push-button switches, pilot lights, and selector switches that are so designed that, when properly installed, they will prevent oil and coolant from entering around the operating or mounting means. *See also:* switch. (IA/ICTL/IAC) [60]

oil, uninhibited *See:* uninhibited oil.

oil-well cover (rotating machinery) A cover for an oil reservoir. *See also:* oil cup. (PE) [9]

oil wick (rotating machinery) Wool, cotton, or similar material used to bring oil to the journal surface by capillary action. *See also:* oil cup. (PE) [9]

OL/2 A programming language designed to allow statement of mathematical problems, with emphasis on arrays and structures that exhibit the parallelism inherent in many algorithms. (C) 610.13-1993w

O/M *See:* engineering units.

OM attribute A component of an OM object, comprised of an integer denoting the type of the attribute and an ordered sequence of one or more attribute values, each accompanied by an integer denoting the syntax of the value. (C/PA) 1328.2-1993w, 1326.2-1993w, 1327.2-1993w, 1224.2-1993w

OM attribute type A category into which OM attribute values are placed on the basis of their purpose. (C/PA) 1328.2-1993w, 1326.2-1993w, 1327.2-1993w, 1224.2-1993w

OM attribute value An arbitrarily complex information item that can be viewed as a characteristic or property of an OM object. (C/PA) 1328.2-1993w, 1327.2-1993w, 1326.2-1993w, 1224.2-1993w

OM class A category into which OM objects are placed on the basis of both their purpose and their internal structure. (C/PA) 1328.2-1993w, 1326.2-1993w, 1224.2-1993w, 1327.2-1993w

omega (navigation aid terms) A very long distance navigation system operating at approximately 10 kHz (kilohertz), in which hyperbolic lines of position are determined by measurement of the difference in travel time of continuous wave signals from two transmitters separated by 5000 nmi (nautical miles) to 6000 nmi (9000 km [kilometers] to 11 000 km) or in which changes in distances from the transmitters are measured by counting rf (radio frequency) wavelengths in space of lanes as the vehicle moves from a known position, the lanes being counted by phase comparison with a stable oscillator aboard the vehicle. (AES/GCS) 172-1983w

OM interface The API to OSI Object Management. (C/PA) 1224.1-1993w

omnibearing (navigation aid terms) A magnetic bearing indicated by a navigational receiver on transmisssion from an omnirange. (AES/GCS) 172-1983w

omnibearing converter (navigation aid terms) A device which combines the omnibearing signal with vehicle heading information to furnish electrical signals for the operation of the pointer of a radio magnetic indicator. (AES/GCS) 172-1983w

omnibearing-distance facility (navigation aid terms) A combination of an omnirange and a distance measuring facility, so that both bearing and distance information may be obtained; tacan and VOR/DME are omnibearing distance facilities. (AES/GCS) 172-1983w

omnibearing-distance navigation (navigation aid terms) Radio navigation utilizing a polar coordinate system as a reference, making use of omnibearing-distance facilities. (AES/GCS) 172-1983w

omnibearing indicator (OBI) (navigation aid terms) An instrument that presents an automatic and continuous indication of an omnibearing. (AES/RS/GCS) 686-1982s, 172-1983w

omnibearing line *See:* radial.

omnibearing selector (navigation aid terms) A control used with an omnirange receiver so that any desired omnibearing may be selected; deviation from on-course for any selected bearing is displayed on the course line deviation indicator. (AES/GCS) 172-1983w

omnidirectional antenna An antenna having an essentially non-directional pattern in a given plane of the antenna and a directional pattern in any orthogonal plane. *Note:* For ground-based antennas, the omnidirectional plane is usually horizontal. (AP/ANT) 145-1993

omnidirectional microphone (nondirectional microphone) A microphone the response of which is essentially independent of the direction of sound incidence. *See also:* microphone. (EEC/PE) [119]

omnidirectional pattern A pattern with the same response in all azimuthal directions. *Note:* This radiation pattern results when only one tower is used to create the radiation pattern. (T&D/PE) 1260-1996

omnidirectional range (omnirange) (navigation aid terms) A radio facility providing bearing information at or from such facilities at all azimuths within its service area and providing direct indication of the magnetic bearing (omnibearing) of that station from any direction. (AES/GCS) 172-1983w

omni-font character recognition Character recognition of many or all character fonts. *Contrast:* single-font character recognition. (C) 610.2-1987

omnirange *See:* omnidirectional range.

OMNITAB II A programming language designed for nonprogrammers, to provide data, numerical, and statistical analysis; provides capability for performing calculations and statistical procedures such as regression and matrix inversion. (C) 610.13-1993w

OM object Any of the complex information objects created, examined, modified, or destroyed by means of the interface. (C/PA) 1328.2-1993w, 1326.2-1993w, 1327.2-1993w, 1224.2-1993w

OMR *See:* optical mark reading.

OM syntax A category into which an OM attribute value is placed on the basis of its form. *Synonym:* attribute value syntax. (C/PA) 1328.2-1993w, 1327.2-1993w, 1224.2-1993w, 1326.2-1993w

onboard equipment (OBE) Equipment located within a vehicle that supports the information exchange with roadside equipment (RSE). (SCC32) 1455-1999

on-chip interface An interface through which the computer communicates with outside devices and circuits. (C) 610.10-1994w

on-core type A moisture barrier applied directly over the cable core. (PE/IC) 1142-1995

on-course curvature (navigation) (navigation aid terms) The rate of change of the indicated course with respect to distance along the course line or path. (AES/GCS) 172-1983w

one A true logic state or a true condition of a variable. (C/BA) 1496-1993w

one-address Pertaining to an instruction code in which each instruction has one address part. Also called single address. In a typical one-address instruction the address may specify either the location of an operand to be taken from storage, the destination of a previously prepared result, or the location of the next instruction to be interpreted. In a one-address machine, the arithmetic unit usually contains at least two storage locations, one of which is an accumulator. For example, operations requiring two operands may obtain one operand from the main storage and the other from the storage location in the arithmetic unit that is specified by the operation part. *See also:* single-address. (C) 162-1963w

one-address instruction (1) A computer instruction that contains one address field. For example, an instruction to load the contents of location A. *Synonyms:* single-operand instruction; single-address instruction. *Contrast:* multiaddress instruction; zero-address instruction; two-address instruction; four-address instruction; three-address instruction. (C) 610.12-1990

(2) An instruction containing one address. *Synonyms:* single-operand instruction; single-address instruction. *See also:* address format. (C) 610.10-1994w

one-ahead addressing A method of implied addressing in which the operands for a computer instruction are understood to be in the storage locations following the locations of the operands used for the last instruction executed. *Contrast:* repetitive addressing. (C) 610.12-1990

1BASE5 IEEE 802.3 Physical Layer specification for a 1 Mb/s CSMA/CD local area network over two pairs of twisted-pair telephone wire. (C/LM) 802.3-1998

one-core-per-bit storage A type of storage in which each storage cell uses one magnetic core per binary character. (C) 610.10-1994w

one-family dwelling A building consisting solely of one dwelling unit. (NESC/NEC) [86]

one-fluid cell A cell having the same electrolyte in contact with both electrodes. *See also:* electrochemistry. (EEC/PE) [119]

1GL *See:* machine language.

one-hour rating (rotating electric machinery) The output that the machine can sustain for 1 hour starting cold under the conditions of Section 4 of IEEE Std 11-1980 without exceeding the limits of temperature rise of Section 5. (PE/EM) 11-1980r

100BASE-T2 IEEE 802.3 specification for a 100 Mb/s CSMA/CD local area network over two pairs of Category 3 or better balanced cabling. (C/LM) 802.3-1998

100BASE-FX IEEE 802.3 Physical Layer specification for a 100 Mb/s CSMA/CD local area network over two optical fibers. (C/LM) 802.3-1998

100BASE-T IEEE 802.3 Physical Layer specification for a 100 Mb/s CSMA/CD local area network. (C/LM) 802.3-1998

100BASE-T4 IEEE 802.3 Physical Layer specification for a 100 Mb/s CSMA/CD local area net-work over four pairs of Category 3, 4, and 5 unshielded twisted-pair (UTP) wire. (C/LM) 802.3-1998

100BASE-TX IEEE 802.3 Physical Layer specification for a 100 Mb/s CSMA/CD local area net-work over two pairs of Category 5 unshielded twisted-pair (UTP) or shielded twisted-pair (STP) wire. (C/LM) 802.3-1998

100BASE-X IEEE 802.3 Physical Layer specification for a 100 Mb/s CSMA/CD local area network that uses the Physical Medium Dependent (PMD) sublayer and Medium Dependent Interface (MDI) of the ISO/IEC 9314 group of standards developed by ASC X3T12 (FDDI). (C/LM) 802.3-1998

100 percent disruptive-discharge voltage (dielectric tests) A specified minimum voltage that is to be applied to a test object in a 100 percent disruptive-discharge test under specified conditions. The term applies mostly to impulse tests and has significance only in cases where the loss of dielectric strength resulting from a disruptive discharge is temporary. (PE/PSIM) 4-1978s

100 percent insulation level Cables in this category shall be permitted to be applied where the system is provided with relay protection such that ground faults will be cleared as rapidly as possible, but in any case within one minute. While these cables are applicable to the great majority of cable installations that are on grounded systems, they shall be permitted to be used also on other systems for which the application of cables is acceptable provided the above clearing requirements are met in completely de-energizing the faulted section. (NESC/NEC) [86]

133 percent insulation level This insulation level corresponds to that formerly designated for ungrounded systems. Cables in this category shall be permitted to be applied in situations where the clearing time requirements of the 100 percent level category cannot be met, and yet there is adequate assurance that the faulted section will be de-energized in a time not exceeding one hour. Also, they shall be permitted to be used when additional insulation strength over the 100 percent level category is desirable. (NESC/NEC) [86]

1 kHz envelope delay The envelope delay at a carrier frequency of 1020 Hz. (COM/TA) 743-1995

one-level address *See:* direct address; n-level address.

one-line diagram (single-line) A diagram which shows, by means of single lines and graphic symbols, the course of an electric circuit or system of circuits and the component devices or parts used therein. (GSD/ICTL) 315-1975r

one minus cosine (high voltage circuit breakers) The 1−cosine curve starting at zero and reaching a peak of E_2 at time T_2. The crest is denoted by *P*. *Note:* The 1-cosine curve is the standard envelope for rating circuit breaker transient recovery voltage performance for circuit breakers rated 72.5 kV and below. (SWG) 327-1972w

one minus cosine envelope (of a transient recovery voltage) A voltage-versus-time curve of the general form $e_2 E_2$ (1 − cos *Kt*) in which e_2 represents the transient voltage across a switching device pole unit, reaching its crest E_2 at a time T_2. (SWG/PE) C37.100-1992

one-N (1N) modulation (dynamically tuned gyro) The modulation of the pickoff output at spin frequency. (AES/GYAC) 528-1994

one-N (1N) translational sensitivity *See:* radial-unbalance torque.

ONE output (A) (magnetic cell) The voltage response obtained from a magnetic cell in a ONE state by a reading or resetting process. **(B) (magnetic cell)** The integrated voltage response obtained from a magnetic cell in a ONE state by a reading or resetting process. *See also:* ONE state. (Std100) 163-1959

one-plus call (telephone switching systems) A type of station-to-station call in which the digit one is dialed as an access code. (COM) 312-1977w

one-plus-one address Pertaining to an instruction that contains one operand address and a control address. (C) [20], [85]

one-plus-one address format *See:* address format.

one-plus-one address instruction A computer instruction that contains two address fields, the second containing the address of the instruction to be executed next. For example, an instruction to load the contents of location A, then execute the instruction at location B. *Contrast:* four-plus-one address instruction; three-plus-one address instruction; two-plus-one address instruction. (C) 610.12-1990

one-port surge protective device A surge protective device (SPD) with protective components connected in shunt with the circuit to be protected. A one-port SPD may have separate input and output terminals without a specified series impedance between these terminals. (PE) C62.34-1996

one-quadrant multiplier (1) A multiplier in which the multiplication operation is restricted to input variables of the same sign. *Contrast:* four-quadrant multiplier; two-quadrant multiplier. (C) 610.10-1994w
(2) A multiplier in which operation is restricted to a single sign of both input variables. *See also:* electronic analog computer. (C) 165-1977w

1RTT *See:* round trip time.

ones complement (mathematics of computing) The diminished-radix complement of a binary numeral, which is formed by subtracting each digit from 1. For example, the ones complement of 1101 is 0010. *Synonyms:* complement on one; inverse binary state. (C) 1084-1986w

one-shot *See:* monostable.

one-sided switching array (telephone switching systems) A switching array where each crosspoint interconnects multiples within one group. (COM) 312-1977w

1-state (logic) The logic state represented by the binary number 1 and usually standing for an active or true logic condition. (GSD) 91-1973s

ONE state A state of a magnetic cell wherein the magnetic flux through a specified cross-sectional area has a positive value, when determined from an arbitrarily specified direction of positive normal to that area. A state wherein the magnetic flux has a negative value, when similarly determined, is a ZERO state. A ONE output is (1) the voltage response obtained from a magnetic cell in a ONE state by a reading or resetting process, or (2) the integrated voltage response obtained from a magnetic cell in a ONE state by a reading or resetting process. A ZERO output is (1) the voltage response obtained from a magnetic cell in a ZERO state by a reading or resetting process, or (2) the integrated voltage response obtained from a magnetic cell in a ZERO state by a reading or resetting process. A ratio of a ONE output to a ZERO output is a ONE-to-ZERO ratio. A pulse, for example a drive pulse, is a write pulse if it causes information to be introduced into a magnetic cell or cells, or is a read pulse if it causes information to be acquired from a magnetic cell or cells. (Std100) 163-1959w

1T (linear waveform distortion) (video signal transmission measurement) Letter symbol for the duration of a half-period of the nominal upper cut-off frequency of a transmission system. Therefore

$$T = \frac{1}{2f_c}$$

Note: for the TV system M

$$T = \frac{1}{2 \times 4\ (\text{MHz})} = 125''\ (\text{ns})$$

The duration *T* is commonly referred to as the Nyquist interval. The concept of *T* is employed not only when the frequency cut-off is a physical property of a given system but also when the system is flat and there is no interest in the performance of the system beyond a given frequency. (BT) 511-1979w

one-third octave (1) (seismic testing of relays) The interval between two frequencies which have a frequency ratio of 2 1/3. For example, 1 to 1.26, 1.26 to 1.59, 159 to 2.0 Hz, etc. (PE/PSR) C37.98-1977s
(2) The interval between two frequencies that have a frequency ratio of the cube root of two. For example, 1 to 1.26, 1.26 to 1.59, 1.59 to 2.0 Hz, etc. (SWG/PE) C37.100-1992

1000BASE-CX 1000BASE-X over specialty shielded balanced copper jumper cable assemblies. (C/LM) 802.3-1998

1000BASE-LX 1000BASE-X using long wavelength laser devices over multimode and single-mode fiber. (C/LM) 802.3-1998

1000BASE-SX 1000BASE-X using short wavelength laser devices over multimode fiber. (C/LM) 802.3-1998

1000BASE-T IEEE 802.3 Physical Layer specification for a 1000 Mb/s CSMA/CD LAN using four pairs of Category 5 balanced copper cabling. (C/LM) 802.3-1998

1000BASE-X IEEE 802.3 Physical Layer specification for a 1000 Mb/s CSMA/CD LAN that uses a Physical Layer derived from ANSI X3.230-1994 (FC-PH). (C/LM) 802.3-1998

one-time fuse (1) (protection and coordination of industrial and commercial power systems) Strictly speaking, any nonrenewable fuse, but generally accepted and used to describe any Class H nonrenewable cartridge fuse, with a single (as opposed to dual) fusing element and intended to interrupt not over 10 000 amperes (A). (IA/PSP) 242-1986r
(2) (protection and coordination of industrial and commercial power systems)

O net loss (circuit equivalent) The net loss is the sum of all the transmission losses occurring between the two ends of the circuit, minus the sum of all the transmission gains. *See also:* transmission loss. (EEC/PE) [119]

one-sided z transform (data processing) Let T be a fixed positive number, and let $f(t)$ be defined for $t \geq 0$. The z transform of $f(t)$ is the function

$$[f(t)] = F(z) = \sum_{n=0}^{\infty} f(nT)z^{-n}$$

for

$$|z| > R = 1/\rho$$

where ρ is the radius of convergence of the series and z is a complex variable. If $f(t)$ is discontinuous at some instant $t = kT$, k an integer, the value used for $f(kT)$ in the z transform is $f(kT^{+})$. The z transform for the sequence $\{f_n\}$ is:

$$[\{f_n\}] = F(z) = \sum_{n=0}^{\infty} f_n z^{-n}$$

(IM) [52]

one-to-many relationship A kind of relationship between two state classes in which each instance of one class, referred to as the *child class*, is specifically constrained to relate to no more than one instance of a second class, referred to as the *parent* class. (C/SE) 1320.2-1998

ONE-to-partial-select ratio The ratio of a ONE output to a partial-select output. *See also:* coincident-current selection. (Std100) 163-1959w

ONE-to-ZERO ratio A ratio of a ONE output to a ZERO output. *See also:* ONE state. (Std100) 163-1959w

one-transistor cell A memory cell that is accessed within the physical confines of a single source and drain. (ED) 1005-1998

O network A network composed of four impedance branches connected in series to form a closed circuit, two adjacent junction points serving as input terminals while the remaining two junction points serve as output terminals. *See also:* network analysis. (EEC/PE) [119]

one-unit call (telephone switching systems) A call for which there is a single-unit charge for an initial minimum interval. (COM) 312-1977w

one-way *See:* linked list.

one-way automatic leveling device A device that corrects the car level only in case of under-run of the car, but will not maintain the level during loading and unloading. *See also:* elevator car-leveling device. (EEC/PE) [119]

one-way correction A method of register control that effects a correction in register in one direction only. (IA/ICTL/IAC) [60]

one-way-only operation A mode of operation of a data link in which data may be transmitted in a preassigned direction over one channel. *Synonym:* simplex operation. *See also:* two-way simultaneous operation; two-way alternate operation. (C) 610.7-1995

one-way trunk (telephone switching systems) A trunk between two switching entities accessible by calls from one end only. At the originating end, the one-way trunk is known as an outgoing trunk; at the terminating end, it is known as an incoming trunk. (COM) 312-1977w

one-wire circuit *See:* direct-wire circuit.

one-wire line *See:* open-wire pole line.

on-hook (1) (telephone switching systems) An open station line or any supervisory or pulsing condition is indicative of this state. (COM) 312-1977w
(2) In regard to a telephone set, deactivated—that is, a telephone set is not in use. *Contrast:* off-hook. (C) 610.7-1995

on-hook/off-hook Signaling conditions on a line in the form of dc impedance presented to the local loop by the telemetry interface unit (TIU). Off-hook implies that the TIU is in a low resistance state and is allowing significant current to flow. On-hook implies that the TIU is in a high resistance state and is not allowing significant current to flow. (SCC31/AMR) 1390.2-1999, 1390.3-1999, 1390-1995

ONI *See:* operator number identification.

ON impedance (thyristor) The differential impedance between the terminals through which the principal current flows, when the thyristor is in the ON state at a stated operating point. *See also:* principal voltage-current characteristic. (IA) [12]

online (A) Pertaining to a system or mode of operation in which input data enter the computer directly from the point of origin or output data are transmitted directly to the point where they are used. For example, an airline reservation system. *Contrast:* batch. *See also:* interactive; conversational; real time. **(B)** Pertaining to a device or process that is under the direct control of the central processing unit of a computer. *Contrast:* offline. *See also:* vary on-line. **(C)** Pertaining to equipment or devices under direct control of the central processing unit. **(D)** Pertaining to a user's ability to interact with a computer. (C/MIL) 610.12-1990, 610.10-1994, [20], [2]

online compiler *See:* incremental compiler.

online dialog *See:* dialog.

online font A font that may be reviewed and accessed automatically by a printer. *Contrast:* downloadable font. *See also:* internal font; printer font. (C) 610.10-1994w

online operation (A) (emergency and standby power) Pertaining to equipment or devices under direct control of the central processing unit. **(B) (emergency and standby power)** Pertaining to uninterruptible power supply systems whereby an inverter is on during normal operation conditions. (IA/PSE) 446-1987

online ordering *See:* teleordering.

online storage Storage under control of a processing unit. *Contrast:* offline storage. (C) 610.10-1994w

online system A system that is operating at all times, such as an inverter supplied by dc power via the primary power source through a battery charger. (IA/PSE) 493-1997

online testing (test, measurement, and diagnostic equipment) Testing of the unit under test in its operational environment.

See also: interference testing; noninterference testing. (MIL) [2]

on-load factor (thyristor) The ratio of the controller ON-state interval to the operating period in the ON-OFF control mode, often expressed as a percentage. (IA/IPC) 428-1981w

on-net call (telephone switching systems) A call within a switched-service network. (COM) 312-1977w

ON-OFF control (thyristor) The starting instant may be synchronous or asynchronous with respect to the line voltage. The controller ON-state interval is equal to or greater than half a line period. *See also:* operation modes. (IA/IPC) 428-1981w

on-off control system A two-step control system in which a supply of energy to the controlled system is either on or off. *See also:* feedback control system. (IM/PE/EDPG) [120], [3]

ON-OFF keying (modulation systems) A binary form of amplitude modulation in which one of the states of the modulated wave is the absence of energy in the keying interval. Note: The terms mark and space are often used to designate, respectively, the presence and absence of energy in the keying interval. *See also:* telegraphy. (Std100) 270-1964w

ON-OFF test (test, measurement, and diagnostic equipment) A test conducted by repeatedly switching on and off either the signal, power, or load connected to the unit under test while observing the reaction or performance of some parameter of that unit under test. A test frequently used to isolate offending equipment while conducting compatibility, interference, or system performance evaluations. (MIL) [2]

O noise unit An amount of noise judged to be equal in interfering effect to the one-millionth part of the current output of a particular type of standard generator of artificial noise, used under specified conditions. Note: This term was formerly used in connection with ear balance measurements, but has been largely superseded by dBa employed with indicating noise meter. Approximately seven noise units of noise on a telephone line are frequently taken as equivalent to reference noise. *See also:* signal-to-noise ratio. (EEC/PE) [119]

on-peak energy (power operations) Energy supplied during designated periods of relatively high system demands. (T&D/PE/PSE) 858-1987s, 346-1973w

on-peak period (watthour meters) The period of time during which the specified on-peak rate applies. (ELM) C12.13-1985s

on-peak power Power supplied during designated periods of relatively high system demands. *See also:* generating station. (T&D/PE) [10]

ON period (electron tube or valve) The time during an operating cycle in which the electron tube or valve is conducting. (ED) [45], [84]

on site (monitoring radioactivity in effluents) Location within a facility that is controlled with respect to access by the general public. (NI) N42.18-1980r

ON state (thyristor) The condition of the thyristor corresponding to the low-resistance low-voltage portion of the principal voltage-current characteristic in the switching quadrant(s). *Note:* In the case of reverse-conducting thyristors, this definition is applicable only for a positive anode-to-cathode voltage. *See also:* principal voltage-current characteristic. (IA) [12]

ON-state current (thyristor) The principal current when the thyristor is in the ON state. (IA) [12]

ON-state voltage (thyristor) The principal voltage when the thyristor is in the ON state. *See also:* principal voltage-current characteristic. (IA) [12]

on-the-fly printer An impact printer whose type slugs do not stop moving during the impression time. (C) 610.10-1994w

OOLR *See:* overall objective loudness rating.

op *See:* operation code.

opacity (electroacoustics) (optical path) The reciprocal of transmission. *See also:* transmission. (SP) [32]

opaque (1) The language-independent syntax for a family of datatypes with no order or other operations defined. An opaque datatype may have associated names that identify distinguished values. (C/PA) 1351-1994w
(2) A datatype with no order or other operations defined. An opaque datatype may have associated names that identify distinguished values. (C/PA) 1224.1-1993w

opcode A bit pattern that identifies a particular instruction. (C/MM) 1754-1994

open To create a package instance. (C/BA) 1275-1994

open-address hashing Hashing in which collision resolution is handled by inserting an item that has a duplicate hash value into another available position in the hash table. *Contrast:* separate chaining. *See also:* uniform probing; random probing; double hashing; linear probing. (C) 610.5-1990w

open amortisseur An amortisseur that has no connections between poles. (EEC/PE) [119]

open architecture (1) An architecture for which design parameters and specifications are made available to any and all vendors or manufacturing firms, thus encouraging development of compatible products and enhancements. *Contrast:* closed architecture. (C) 610.10-1994w
(2) An architecture from which a system can be assembled from multiple vendor-supplied interface components. The resulting system can execute applications written by arbitrary independent vendors and can be extended by users other than the original supplier. (SCC20) 1226-1998

open area *See:* test site.

open-area test site (OATS) A site that meets specified requirements for measuring radio-interference fields radiated by an equipment under test (EUT). (EMC) 1128-1998

open-center display A plan-position-indicator display on which zero range corresponds to a ring around the center of the display. *See also:* radar. (EEC/PE) [119]

open-center plan-position indicator A PPI in which the display of the initiation of the time base precedes that of the transmitted pulse. (AES) 686-1997

open-center PPI *See:* open-center plan-position indicator.

open-circuit characteristic *See:* open-circuit saturation curve.

open-circuit control A method of controlling motors employing the open-circuit method of transition from series to parallel connections of the motors. *See also:* multiple-unit control. (EEC/PE) [119]

open circuit cooling (rotating machinery) A method of cooling in which the coolant is drawn from the medium surrounding the machine, passes through the machine and then returns to the surrounding medium. (PE) [9]

open-circuit dc voltage The dc voltage on an ungrounded conductive object relative to ground, as a result of deposition of charge. (T&D/PE) 539-1990

open-circuit impedance (A) (general) An impedance of a network that has a specified pair or group of terminals open circuited. **(B) (general)** (four-terminal network or line). The input-output- or transfer-impedance parameters z_{11}, z_{22}, z_{12}, and z_{21} of a four-terminal network when the far-end is open circuited. (CAS) [13]

open-circuit induced voltage The rms power-frequency voltage on an ungrounded conductive object relative to ground or the voltage across the terminals of an open circuit loop, as a result of induction. (T&D/PE) 539-1990

open-circuit inductance The apparent inductance of a winding with all other windings open-circuited. (CHM) [51]

open-circuit potential The measured potential of a cell from which no current flows in the external circuit. 332-1972w

open-circuit saturation curve (synchronous machines) (open-circuit characteristic) The saturation curve of a machine with an open-circuited armature winding. (PE) [9]

open circuit signaling (data transmission) That type of signaling in which no current flows while the circuit is in the idle condition. (PE) 599-1985w

open-circuit test (synchronous machines) A test in which the machine is run as a generator with its terminals open-circuited. (PE) [9]

open-circuit transition (1) (multiple-unit control) A method of changing the connection of motors from series to parallel in which the circuits of all motors are open during the transfer. *See also:* multiple-unit control. (EEC/PE) [119]
(2) (reduced-voltage controllers, including star-delta controllers) A method of starting in which the power to the motor is interrupted during normal starting sequence. *See also:* electric controller. (IA/ICTL/IAC) [60]

open circuit transition auto-transformer starting (rotating machinery) The process of auto-transformer starting whereby the motor is disconnected from the supply during the transition from reduced to rated voltage. (PE) [9]

open-circuit voltage (1) (batteries) The voltage at its terminals when no appreciable current is flowing. (PE/EEC) [119]
(2) (arc-welding apparatus) The voltage between the output terminals of the welding power supply when no current is flowing in the welding circuit. (EEC/AWM) [91]
(3) (overhead power lines) A voltage on a conductive object or in an electric circuit as a result of induction or deposition of charge. (T&D/PE) 539-1990

open-collector A type of bus driver (only drives low or not at all). (C/MM) 1196-1987w

open conductor A type of electric supply or communication line construction in which the conductors are bare, covered, or insulated and without grounded shielding, individually supported at the structure either directly or with insulators. *Synonym:* open wire. (NESC) C2-1997

open contact *See:* normally open contact.

open cutout A cutout in which the fuse clips and fuseholder, fuse unit, or disconnecting blade are exposed.
(SWG/PE) C37.40-1993, C37.100-1992

open-delta connection (power and distribution transformers) A connection similar to a delta-delta connection utilizing three single-phase transformer, but with one single-phase transformer removed. Note: The two remaining transformers of an open-delta bank will carry 57.7 percent of the load carried by the bank using three identical transformers connected delta-delta. (PE/TR) C57.12.80-1978r

open dual bus A DQDB *subnetwork* with the *Head of Bus function* for Bus A and the Head of Bus function for Bus B at different *nodes*. (LM/C) 8802-6-1994

open-ended Pertaining to a process or system that can be augmented. (C) [20], [85]

open-ended coil (rotating machinery) A partly preformed coil the turns of which are left open at one end to facilitate their winding into the machine. *See also:* asynchronous machine. (PE) [9]

open file A file that is currently associated with a file descriptor.
(C/PA) 9945-1-1996, 9945-2-1993, 1003.5-1999

open file description A record of how a process or group of processes is accessing a file. Each file descriptor shall refer to exactly one open file description, but an open file description may be referred to by more than one file descriptor. A file offset, file status, and file access modes are attributes of an open file description.
(C/PA) 1003.5-1999, 9945-1-1996

Open Firmware The firmware architecture defined by IEEE Std 1275-1994 and its applicable supplements or, when used as an adjective, a software component compliant with such an architecture.
(C/BA) 1275.1-1994w, 1275.2-1994w, 1275.4-1995, 1275-1994

open-fuse trip device (1) (low-voltage ac power circuit protectors) (ac power circuit breakers) A device that operates to open (trip) all poles of a circuit breaker (protector) in response to the opening, or absence, of one or more fuses integral to the circuit protector on which the device is mounted. After operating, the device shall prevent closing of the circuit breaker (protector) until reset operation is performed. *Note:* Since some open-fuse trip devices may operate by sensing the voltage across the fuses, they may not prevent closing of the circuit breaker (protector) with an open or missing fuse, but in most cases will cause an immediate trip if such an operation is performed. There is a practical limit of load impedance above which the device (sensing voltage across an open or missing fuse) will not function as described.
(SWG/PE) C37.29-1981r, C37.13-1990r
(2) A device that operates to open (trip) all poles of a switching device in response to the opening, or absence, of one or more fuses integral to the switching device on which the device is mounted. After operating, the device prevents closing of the switching device until a reset operation is performed.
(SWG/PE) C37.100-1992

opening eye (of a fuse holder, fuse unit, or disconnecting blade) An eye provided for receiving a fuse hook or switch hook for opening and closing the fuse.
(SWG/PE) C37.100-1992, C37.40-1993

opening operating time (of a switch) The interval of time it takes during switch operation to move from the fully closed to the fully open position. (SWG/PE) C37.100-1992

opening operation *See:* open operation.

opening time (of a mechanical switching device) The interval of time between the time when the actuating quantity of the release circuit reaches the operating value, and the instant when the primary arcing contacts have parted. Any time delay device forming an integral part of the switching device is adjusted to its minimum setting or, if possible, is cut out entirely for the determination of opening time. *Note:* The opening time includes the operating time of an auxiliary relay in the release circuit when such a relay is required and supplied as part of the switching device. *See also:* isolating time.
(SWG/PE) C37.100-1992

open line wire charging current Current supplied to an unloaded open-wire line. *Note:* Current is expressed in rms amperes. (SWG/PE) C37.100-1992

open-link cutout A cutout that does not employ a fuseholder and in which the fuse support directly receives an open-link fuse link or a disconnecting blade.
(SWG/PE) C37.40-1993, C37.100-1992

open-link fuse link A replaceable part or assembly comprised of the conducting element and fuse tube, together with the parts necessary to confine and aid in extinguishing the arc and to connect it directly into the fuse clips of the open-link fuse support. (SWG/PE) C37.40-1993, C37.100-1992

open-link fuse support An assembly of base or mounting support, insulators or insulator unit, and fuse clips for directly mounting an open-link fuse link and for connecting it into the circuit.
(SWG/PE) C37.40-1993, C37.100-1992, C37.100B-1986w

open-line test A test that energizes a converter dc yard up to full voltage without energizing the remote station. The test can be configured to also test energize the transmission line.
(PE/SUB) 1378-1997

open listening A mode of telephone communication in which a telephone handset is used in the normal position for send. The incoming signal is received simultaneously by the handset and loudspeaker. (COM/TA) 1329-1999

open loop (automatic control) A signal path without feedback. *See also:* control system; feedback. (PE/EDPG) [3]

open-loop control (1) (station control and data acquisition) A form of control without feedback.
(SWG/PE/SUB) C37.100-1992, C37.1-1994
(2) Pertaining to a control system in which the output is permitted to vary in accordance with the inherent characteristics of the system, and no function of the output is used as feedback to the system. (C) 610.2-1987

open-loop control system (1) (general) A system in which the controlled quantity is permitted to vary in accordance with the inherent characteristics of the control system and the controlled power apparatus for any given adjustment of the con-

troller. *Note:* No function of the controlled variable is used for automatic control of the system. It is not a feedback control system. *See also:* network analysis; control; control system. (MAG/PEL/ET) 264-1977w, 111-1984w
(2) (hydraulic turbines) A control system that has no means for comparing the output with the input for control purposes. (PE/EDPG) 125-1977s

open-loop gain (power supplies) The gain, measured without feedback, is the ratio of the voltage appearing across the output terminal pair to the causative voltage required at the (input) null junction. The open-loop gain is denoted by the symbol A in diagrams and equations. *See also:* closed loop; loop gain. (AES/PE) [41], [78]

open loop measurement (data transmission) A measurement made in which a circuit has at least one of two hybrid sets disconnected and thereby opening the loop. (PE) 599-1985w

open-loop series street-lighting system A street-lighting system in which the circuits each consist of a single line wire that is connected from lamp to lamp and returned by a separate route to the source of supply. *See also:* direct-current distribution; alternating-current distribution. (T&D/PE) [10]

open-loop system A control system that has no means for comparing the output with input for control purposes. (EEC) [74]

open machine (1) (rotating machinery) A machine in which no mechanical protection as such is embodied and there is no restriction to ventilation other than that necessitated by good mechanical construction. *See also:* asynchronous machine; direct-current commutating machine. (PE) [9]
(2) A machine that has ventilating openings that permit passage of external cooling air over and around the windings. (IA/MT) 45-1998

open network A network that can be accessed from computers or terminals external to the network, using dial-up or dedicated lines, or other means. *Contrast:* closed network. (C) 610.7-1995

open-numbering plan (telephone switching systems) A numbering plan in which the number of digits dialed varies according to the requirements of the telecommunications message network. (COM) 312-1977w

open operation (of a switching device) The movement of the contacts from the normally closed to the normally open position. *Note:* The letter O signifies this operation: Open. (SWG/PE) C37.100-1992

open path (network analysis) A path along which no node appears more than once. (CAS) 155-1960w

open-phase protection A form of protection that operates to disconnect the protected equipment on the loss of current in one phase conductor of a polyphase circuit, or to prevent the application of power to the protected equipment on the absence of one or more phase voltages of a polyphase system. (SWG/PE) C37.100-1992

open-phase relay A polyphase relay designed to operate when one or more input phases of a polyphase circuit are open. (SWG/PE) C37.100-1992

open pipe-ventilated machine An open machine except that openings for the admission of the ventilating air are so arranged that inlet ducts or pipes can be connected to them. This air may be circulated by means integral with the machine or by means external to and not a part of the machine. In the latter case, this machine is sometimes known as a separately ventilated machine or a forced-ventilated machine. Mechanical protection may be defined as under dripproof machine, splashproof machine, guarded machine, or semiguarded machine. *See also:* direct-current commutating machine; asynchronous machine. (EEC/PE) [119]

open region (A) (three-dimensional space) A volume that satisfies the following conditions:

a) any point of the region has a neighborhood that lies within the region;

b) any two points of the region may be connected by a continuous space curve that lies entirely in the region.

(B) (two-dimensional space) An area that satisfies the conditions of definition (A). (Std100) 270-1966

open relay An unenclosed relay. *See also:* relay. (EEC/REE) [87]

open resonator (laser maser) An open resonator and a beam resonator are identical. (LEO) 586-1980w

open specifications Specifications that are maintained by an organization that uses an open, public consensus process to accommodate new technologies and user requirements over time. (C/PA) 14252-1996

open subroutine (1) (computers) A subroutine that must be relocated and inserted into a routine at each place it is used. *See also:* closed subroutine; subroutine; closed subroutine. (C/MIL) [20], [2], [85]
(2) (software) A subroutine that is copied into a computer program at each place that it is called. *Synonym:* direct insert subroutine. *Contrast:* closed subroutine. *See also:* macro; in-line code. (C) 610.12-1990

open switchgear assembly An assembly that does not have enclosures as part of the structure. (SWG/PE) C37.100-1992

open system A system that implements sufficient open specifications or standards for interfaces, services, and supporting formats to enable properly engineered application software

— To be ported with minimal changes across a wide range of systems from one or more suppliers
— To interoperate with other applications on local and remote systems
— To interact with people in a style that facilitates user portability

(C/PA) 14252-1996, 1003.23-1998

open system API A combination of standards-based interfaces specifying a complete interface between an application program and the underlying application platform. (C/PA) 14252-1996

open system environment (OSE) A comprehensive set of interfaces, services, and supporting formats, plus user aspects for interoperability or for portability of applications, data, or people, as specified by information technology standards and profiles. (C/PA) 14252-1996

open system hardware architecture An electronic system design that allows components, which are developed or built by multiple parties, to work together. (C/BA) 14536-1995

open systems interconnection (OSI) A model that provides a common basis for the coordination of standards development for the purpose of systems interconnection, while allowing existing standards to be placed into perspective within the overall reference model. The OSI model defines a seven-layer functional model, including descriptions of the functions defined for each layer. Refers to ISO 7498:84. (EMB/MIB) 1073.3.1-1994, 1073.4.1-2000

open systems interconnection model A computer network architecture model proposed as a standard model by the International Organization for Standardization. The model consists of seven layers, each consisting of entities, or sets of functions performed on bits, frames, packets, or messages. *Note:* Enables any OSI-compatible computer or device to communicate with any other OSI-compliant computer or device for a meaningful exchange of information. *Synonym:* OSI reference model. (C) 610.7-1995

open systems interconnection (OSI) (N)-service A capability of the (N)-layer, and the layers beneath it, that is provided to the (N)-entities at the boundary between the (N)-layer and the (N+1)-layer. (LM/C) 802.10-1992

open systems interconnection reference model (OSIRM) A model that organizes the data communication concept into seven layers and defines the services that each layer provides. (DIS/C) 1278.2-1995

open systems interconnect model A seven-layer network communications model developed by an International Organization for Standardization (ISO) subcommittee that governs communications interchange between systems. The model is an internationally accepted framework of standards for inter-system communications. (C/MM) 1284.4-2000

open terminal box (rotating machinery) A terminal box that is, normally, open only to the interior of the machine. (PE) [9]

open wire *See:* open conductor.

open-wire circuit A circuit made up of conductors separately supported on insulators. *Note:* The conductors are usually bare wire, but they may be insulated by some form of continuous insulation. The insulators are usually supported by crossarms or brackets on poles. (EEC/PE) [119]

open-wire lead *See:* open-wire pole line.

open-wire line charging current (high voltage circuit breakers) Current supplied to an unloaded open wire line. *Note:* Current is expressed in root-mean-square amperes. (SWG) 341-1972w

open-wire pole line (open-wire lead) (open-wire line) A pole line whose conductors are principally in the form of open wire. (PE/EEC) [119]

open-wire protectors Combined isolating and drainage transformer-type protectors used in conjunction with, but not limited to, horn gaps and grounding relays are used on open-wire lines to provide protection against lightning, power contacts, or high values of induced voltage. *Synonym:* hot-line protectors. (PE/PSC) 487-1992

open wiring (on insulators) An exposed wiring method using cleats, knobs, tubes, and flexible tubing for the protection and support of single insulated conductors run in or on buildings, and not concealed by the building structure. (NESC/NEC) [86]

operable equipment (test, measurement, and diagnostic equipment) An equipment which, from its most recent performance history and a cursory electrical and mechanical examination, displays an indication of operational performance for all required functions. (MIL) [2]

operand (1) (software) (mathematics of computing) A variable, constant, or function upon which an operation is to be performed. For example, in the expression A = B + 3, B and 3 are the operands. (C) 610.12-1990, 1084-1986w
(2) (microprocessor assembly language) Data which is to be operated on; also, an address denoting data which is to be operated on. (C/MM) 695-1985s
(3) An argument to a command that is generally used as an object supplying information to a utility necessary to complete its processing. Operands generally follow the options in a command line. (C/PA) 9945-2-1993
(4) An entity on which an operation is performed. (C) 610.10-1994w

operand field A field within a computer instruction that specifies an operand needed by the instruction. *See also:* operation field; address field. (C) 610.10-1994w

operand handler In a pipelined machine, the portion of the computer that fetches data from memory and stores results in memory. *Note:* It receives its instructions from the instruction decoder, and passes operands to or from the execution unit. (C) 610.10-1994w

operate (analog computer) In an analog computer, the computer-control state in which input signals are connected to all appropriate computing elements, including integrators, for the generation of the solution. (C) 165-1977w, 610.10-1994w

operated unit A switch, signal, lock, or other device that is operated by a lever or other operating means. (EEC/PE) [119]

operating basis earthquake (OBE) (1) (seismic qualification of Class 1E equipment for nuclear power generating stations) An earthquake that could reasonably be expected to occur at the plant site during the operating life of the plant considering the regional and local geology and seismology and specific characteristics of local subsurface material. It is that earthquake that produces the vibratory ground motion for which those features of the nuclear power plant, necessary for continued operation without undue risk to the health and safety of the public, are designed to remain functional. (PE/NP) 344-1987r
(2) (Class 1E battery chargers and inverters) (seismic testing of relays) (seismic qualification of Class 1E equipment) (nuclear power generating station) That earthquake which could reasonably be expected to affect the plant site during the operating life of the plant. It is that earthquake which produces the vibratory ground motion for which those features of the nuclear power plant necessary for continued operation without undue risk to the health and safety of the public are designed to remain functional.
(SWG/PE/NP/PSR) 649-1980s, C37.81-1989r, 650-1979s, C37.98-1977s, C37.100-1992, 382-1985

operating bypass (1) (nuclear power generating station) Normal and permissive removal of the capability to accomplish a protective function that could otherwise occur in response to a particular set of generating station conditions. *Note:* Typically, operating bypasses are used to permit a change to a different mode of generating station operation (for example, prevention of initiation of safety injection during cold shutdown conditions). (PE/NP) 279-1971w
(2) Inhibition of the capability to accomplish a safety function that could otherwise occur in response to a particular set of generating conditions. *Note:* An operating bypass is not the same as a maintenance bypass. Different modes of plant operation may necessitate an automatic or manual bypass of a safety function. Operating bypasses are used to permit mode changes (e.g., prevention of initiation of emergency core cooling during the cold shutdown mode). (PE/NP) 603-1998

operating characteristic (of a relay) The response of the relay to the input quantities that result in relay operation. (SWG/PE/PSR) C37.100-1992, C37.90-1978s

operating conditions (1) (reliability data for pumps and drivers, valve actuators, and valves) (reliability data) The loading or demand cyclic operation, or both, of an item between zero and 100% of its related capability(ies). (PE/NP) 500-1984w
(2) (general) The whole of the electrical and mechanical quantities that characterize the work of a machine, apparatus, or supply network, at a given time. (EI) 96-1969w

operating cycle (nuclear power generating station) The complete sequence of operations that occur during a response to a demand function. (PE/NP) 380-1975w, 382-1980s

operating device (elevators) The car switch, pushbutton, lever, or other manual device used to actuate the control. *See also:* control. (EEC/PE) [119]

operating duty (of a switching device) A specified number and kind of operations at stated intervals. (SWG/PE) C37.100-1992

operating duty cycle One or more unit operations, as specified. (SPD/PE) C62.11-1999, C62.62-2000

operating-duty test (surge arresters) A test in which working conditions are simulated by the application to the arrester of a specified number of impulses while it is connected to a power supply of rated frequency and specified voltage. (PE) [8], [84]

operating experience (1) (safety systems equipment in nuclear power generating stations) Verifiable service data for equipment. (PE/NP) 627-1980r
(2) (Class 1E battery chargers and inverters) Accumulation of verifiable service data for conditions equivalent to those for which particular equipment is to be qualified. (PE/NP) 650-1979s, 323-1974s

operating failure rate (reliability data for pumps and drivers, valve actuators, and valves) The probability (per hour) of failure for those operating components required to operate or function for a period of time. (PE/NP) 500-1984w

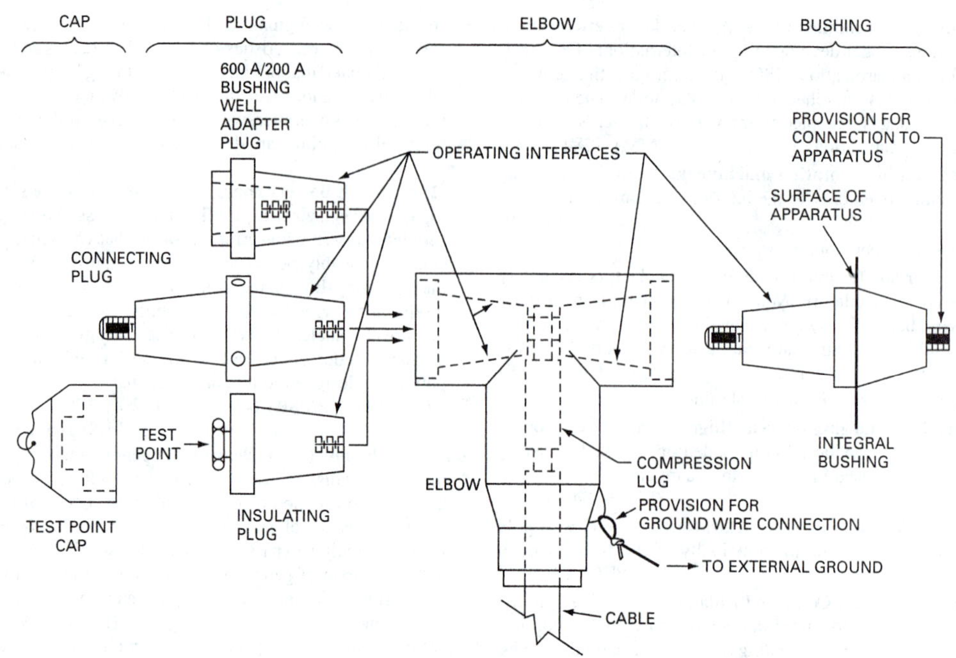

Typical components of 600 A separable insulated connector system
operating interface

operating floor (packaging machinery) A floor or platform used by the operator under normal operating conditions. (IA/PKG) 333-1980w

operating frequency (thyristor) The operating frequency is the reciprocal value of the operating period. (IA/IPC) 428-1981w

operating frequency line current (thyristor) The root-mean-square (rms) value of the fundamental component of the line current, whose frequency is the operating frequency. (IA/IPC) 428-1981w

operating frequency load voltage (thyristor) The root-mean-square (rms) value of the fundamental component of the load voltage, whose frequency is the operating frequency. (IA/IPC) 428-1981w

operating influence The change in a designated performance characteristic caused solely by a prescribed change in a specified operating variable from its reference operating condition to its extreme operating condition, all other operating variables being held within the limits of reference operating conditions. *Notes:* 1. It is usually expressed as a percentage of span. 2. If the magnitude of the influence is affected by direction, polarity, or phase, the greater value shall be taken. (EEC/EMI) [112]

operating interface (connector) The surfaces at which a connector is normally separated. (See the corresponding figure.) (PE/T&D) 386-1995

operating life (accelerometer) (gyros) The accumulated time of operation throughout which a gyro or accelerometer exhibits specified performance when maintained and calibrated in accordance with a specified schedule. (AES/GYAC) 528-1994

operating life expectancy (1) (of a fault-initiating switch) The number of closing operations at rated making current that a switch is capable of performing when it is new and tested at its rated making current. (SWG/PE) C37.30-1992s
(2) (of a load interrupter switch) The number of operations that a switch is capable of successfully performing when it is new and tested at its rated interrupting current. (SWG/PE) C37.30-1992s

operating line; operating curve The locus of all simultaneous values of total instantaneous electrode voltage and current for given external circuit conditions. (ED) [45]

operating mechanism (1) (power system device function numbers) The complete electrical mechanism or servomechanism, including the operating motor, solenoids, position switches, etc. for a tap changer, induction regulator, or any similar piece of apparatus that otherwise has no device function number. (SUB/PE) C37.2-1979s
(2) (of a switching device) The part of the mechanism that actuates all the main-circuit contacts of the switching device either directly or by the use of pole-unit mechanisms. (SWG/PE) C37.100-1992

operating modes (nuclear power generating station) The nuclear power plant modes as defined by the technical specifications for the plant. (PE/NP) 566-1977w

operating noise temperature The temperature in kelvins given by

$$T_{op} = \frac{N_0}{kG_s}$$

where N_0 is the output noise power per unit bandwidth at a specified output frequency flowing into the output circuit (under operating conditions), k is Boltzmann's constant, and G_s is the ratio of the signal power delivered at the specified output frequency into the output circuit (under operating conditions) to the signal power available at the corresponding input frequency or frequencies to the system (under operating conditions) at its accessible input terminations. *Notes:* 1. In a nonlinear system T_{op} may be a function of the signal level. 2. In a linear two-port transducer with a single input and a single output frequency, if the noise power originating in the output termination and reflected at the output port can be neglected, T_{op} is related to the noise temperature of the input termination T_i and the effective input noise temperature T_e by the equation

$$T_{op} = T_i + T_e$$

See also: transducer. (ED) 161-1971w

operating overload (packaging machinery) The overcurrent to which electric apparatus is subjected in the course of the normal operating conditions that it may encounter. *Notes:* 1. The maximum operating overload is to be considered six times normal full-load current for alternating-current industrial motors and control apparatus; four times normal full-load current for direct-current industrial motors and control apparatus used for reduced-voltage starting; and ten times normal full-load current for direct-current industrial motors and control apparatus used for full-voltage starting. 2. It should be understood that these overloads are currents that may persist for a very short time only, usually a matter of seconds. (IA/PKG) 333-1980w

operating period (thyristor) The time between starting instants of successive controller ON-state intervals in the ON-OFF control mode. (IA/IPC) 428-1981w

operating point (working point) The point on the family of characteristic curves corresponding to the average voltages or currents of the electrodes in the absence of a signal. *See also:* quiescent point. (ED) [45]

operating range (1) (navigation aid terms) The maximum distance at which reliable service is provided by an aid to navigation. (AES/GCS) 172-1983w
(2) (plutonium monitoring) The region between the limits within which a quantity is measured. (NI) N317-1980r

operating reserve (1) Generating capability above firm system demand available to provide for regulation, load forecasting error, equipment-forced and scheduled outages, and local area protection. It consists of spinning and non-spinning reserve. (PE/PSE) 858-1993w
(2) (electric power supply) That reserve above firm system load required to provide for: regulation within the hour to cover minute to minute variations; load forecasting error; loss of equipment; local area protection. The operating reserve consists of spinning or nonspinning reserve, or both. (PE/PSE) 346-1973w

operating speed range The range between the lowest and highest rated speeds at which the drive may perform at full load. *See also:* electric drive. (IA/IAC) [60]

operating system (software) A collection of software, firmware, and hardware elements that controls the execution of computer programs and provides such services as computer resource allocation, job control, input/output control, and file management in a computer system. (C) 610.12-1990

operating system device driver A device driver intended for use by a primary operating system. *Contrast:* firmware device driver. *See also:* device driver. (C/BA) 1275-1994

operating system software Application-independent software that supports the running of application software and manages the resources of the application platform. (C/PA) 14252-1996

Operating Systems Simulation Language (OSSL) A simulation language used to simulate hardware and software aspects of computer systems. (C) 610.13-1993w

operating tap voltage (capacitance potential devices) Indicates the root-mean-square voltage to ground at the point of connection (potential tap) of the device network to the coupling capacitor or bushing. This is the voltage on which certain insulation tests are based. *See also:* outdoor coupling capacitor. 31-1944w

operating temperature (T_{op}) (1) (power supplies) The range of environmental temperatures in which a power supply can be safely operated (typically, 20 to 50 degrees Celsius). (AES) [41]
(2) (accelerometer) (gyros) The temperature at one or more gyro or accelerometer elements when the device is in the specified operating environment. (AES/GYAC) 528-1994
(3) The temperature at which a unit under test operates. It is often specified over a range and can have various definitions. Definitions include ambient temperature, baseplate temperature, inlet air temperature, etc. (PEL) 1515-2000

operating temperature limits (attenuator) Maximum temperature in degrees Celsius at which attenuator will operate with full input power. *Note:* Derating function for maximum power versus temperature must be specified to show maximum temperature in degrees Celsius at which attenuator will operate 10 dB below full input power. (IM/HFIM) 474-1973w

operating temperature, maximum (electrical insulation tests) The stabilized temperature obtained from operation of the equipment at rated load and duty cycle in the maximum ambient temperature specified for the device under test. (AES/ENSY) 135-1969w

operating temperature range (Hall effect devices) The range of ambient temperature over which the Hall effect device may be operated with nominal control current and a specified maximum magnetic flux density. (MAG) 296-1969w

operating temperature, room (electrical insulation tests) The temperature of the equipment expected at rated load and duty cycle in an ambient temperature of 20°C ± 5° (68°F ± 9°). An equipment item that has been operated through its normal duty cycle or has stabilized to the approximate normal running temperature may be assumed to be at room ambient operating temperature. (AES/ENSY) 135-1969w

operating time (1) (reliability data for pumps and drivers, valve actuators, and valves) The period of time that an active item or equipment is functioning effectively. (PE/NP) 500-1984w
(2) The part of up time during which a functional unit is performing useful operations. *Contrast:* idle time. *See also:* miscellaneous time; rerun time. (C) 610.10-1994w
(3) (of a relay) The time interval from occurrence of specified input conditions to a specified operation. (SWG/PE/PSR) C37.100-1992, C37.98-1977s

operating voltage (1) The actual voltage applied to the heating cable when in service. (IA/PC) 515.1-1995, 515-1997
(2) The voltage of the system on which a device is operated. *Note:* This voltage, if alternating, is usually expressed as an rms value. (SWG/PE) C37.100-1992

operating voltage range (hybrid computer linkage components) The minimum and maximum values of the analog input voltage which can be represented by the output to within a given accuracy. (C) 166-1977w

operation (1) (FASTBUS acquisition and control) A primary address cycle followed by zero or more data cycles and a termination sequence. (NID) 960-1993
(2) (train control) The functioning of the automatic train-control or cab-signaling system that results from the movement of an equipped vehicle over a track element or elements for a block with the automatic train-control apparatus in service, or which results from the failure of some part of the apparatus. *See also:* automatic train control. (PE/EEC) [119]
(3) (elevators) The method of actuating the control. *See also:* control.
(4) (A) A defined action, namely, the act of obtaining a result from one or more operands in accordance with a rule that completely specifies the result for any permissible combination of operands. **(B)** The set of such acts specified by such a rule, or the rule itself. **(C)** The act specified by a single computer instruction. **(D)** A program step undertaken or executed by a computer, for example, addition, multiplication, extraction, comparison, shift, transfer. The operation is usually specified by the operator part of an instruction. **(E)** The event of specific action performed by a logic element. **(F)** Loosely: command. (C) 162-1963
(5) (A) (computers) The action specified by an operator on one or more operands. For example, in the expression $A = B + 3$, the process of adding B to 3 to obtain A. *Note:* Unlike the mathematical meaning, such an operation may not involve an operator or operands; for example, the operation Halt. **(B) (programming)** A defined action that can be performed by a computer system; for example, addition, comparison, branching. **(C)** The process of running a computer system in its intended environment to perform its intended functions. (C/Std100) 610.12-1990

(6) (SBX bus) The process whereby digital signals effect the transfer of data across the interface by means of a sequence of control signals. Operations may be either interlocked or full speed. (C/MM) 959-1988r
(7) The total activity associated with the exchange of data or events. (C/BA) 1014.1-1994w
(8) The language-independent syntax for a program abstraction with formal input and output parameters that are bound to objects or values in the application that invokes it. (C/PA) 1351-1994w
(9) A program step executed by a computer. *See also:* computer operation; logic operation. (C) 610.10-1994w
(10) A program abstraction with formal input and output parameters that are bound to objects or values in the application that invokes it.
(C/PA) 1328.2-1993w, 1328-1993w, 1224.2-1993w, 1327.2-1993w, 1224-1993w, 1327-1993w, 1224.1-1993w
(11) A high-level file transfer or management task performed by an FTAM API function. An operation is implemented by the underlying FTAM service provider, which performs a series of low-level file actions. (C/PA) 1238.1-1994w
(12) (of a switching device) Action of the parts of the device to perform its normal function. (SWG/PE) C37.100-1992
(13) An action defined by a procedure. (SCC20) 1226-1998
(14) A kind of property that is a mapping from the (cross product of the) instances of the class and the input argument types to the (cross product of the) instances of the other (output) argument types. The operations of a class specify the behavior of its instances. While an attribute or participant property is an abstraction of what an instance *knows*, an operation is an abstraction of what an instance *does*. Operations can perform input and output, and can change attribute and participant property values. Every operation is associated with one class and is thought of as a responsibility of that class. No operations are the joint responsibility of multiple classes. (C/SE) 1320.2-1998
(15) At the specification level, an operation is a service that a class knows how to carry out. (IM/ST) 1451.1-1999

operational (A) (software) Pertaining to a system or component that is ready for use in its intended environment. **(B) (software)** Pertaining to a system or component that is installed in its intended environment. **(C) (software)** Pertaining to the environment in which a system or component is intended to be used. (C) 610.12-1990

operational amplifier (1) (analog computer)

1) An amplifier, usually a high-gain dc amplifier, designed to be used with external circuit elements to perform a specified computing operation or to provide a specified transfer function.
2) An amplifier, usually a high-gain dc amplifier, with external circuit elements, used for performing a specified computing operation.

Notes: 1. The gain and phase characteristics are generally designed to permit large variations in the feedback circuit without instability. 2. The input terminal of an operational amplifier (1) is the summing junction of an operational amplifier (2) and is generally designed to draw current that is negligibly small relative to signal currents in the feedback impedance. *See also:* integrating amplifier; inverting amplifier; summing amplifier. (C) 165-1977w
(2) A two-input amplifier designed to perform control or mathematical operations by means of an external feedback circuit connecting the output to one input, and very high gain for voltage differences of either polarity at the inputs. *Note:* Operational amplifiers are used in analog computers to provide mathematical operations such as summing and integration. *See also:* integrating amplifier; summing amplifier. (C) 610.10-1994w

operational availability (A_O) (nuclear power generating station) The measured characteristic of an item expressed by the probability that it will be operable when needed as determined by periodic test and resultant analysis. (PE/NP) 933-1999, 338-1987r

operational character *See:* control character.

operational concept description (OCD) *See:* concept of operations document.

operational conditions The factors, including weather, human operations, external system interactions, etc. that contribute to defining operational scenarios or environments. (C/SE) 1220-1994s

operational environment The natural or induced environmental conditions, anticipated system interfaces, and user interactions within which the system is expected to be operated. (C/SE) 1220-1994s

operational gain *See:* closed-loop gain.

operational impedance (rotating machinery) Defined by the equation

$$Z(s) = \frac{V(s)}{I(s)}$$

where $V(s)$ is the Laplace transform of the voltage and $I(s)$ is the Laplace transform of the current. (PE) [9]

operational inductance (rotating machinery) Defined by the equation

$$L(s) = \frac{\Lambda(s)}{I(s)}$$

where $\Lambda(s)$ is the Laplace transform of the flux linkages and $I(s)$ is the Laplace transform of the current. (PE) [9]

operational maintenance *See:* noninterference testing.

operational maintenance influence (instruments) The effect of routine operations that involve opening the case, such as to inspect or mark records, change charts, add ink, alter control settings, etc. *See also:* accuracy rating. (EEC/EMI) [112]

operational maximum usable frequency (MUF) The highest frequency that would permit acceptable performance of a radio circuit by signal propagation via the ionosphere between given terminals at a given time under specified working conditions. *Notes:* 1. Acceptable performance may, for example, be quoted in terms of the maximum allowable bit error rate or required signal-to-noise ratio. 2. Specified working conditions may include such factors as antenna type, transmitter power, class of emission, and required information rate. *See also:* maximum usable frequency. (AP/PROP) 211-1997

operational power supply A power supply whose control amplifier has been optimized for signal-processing applications rather than the supply of steady-state power to a load. A self-contained combination of operational amplifier, power amplifier, and power supplies for higher-level operations. (AES) [41]

operational programming The process of controlling the output voltage of a regulated power supply by means of signals (which may be voltage, current, resistance, or conductance) that are operated on by the power supply in a predetermined fashion. Operations may include algebraic manipulations, multiplication, summing, integration, scaling, and differentiation. (AES) [41]

operational register Any register on a device that is not required for the system configuration process. (C/MM) 1155-1992

operational relay (analog computer) A relay that may be driven from one position or state to another by an operational amplifier or a relay amplifier. *See also:* function relay. (C) 165-1977w, 610.10-1994w

operational reliability (1) (software) The reliability of a system or software subsystem in its actual use environment. Operational reliability may differ considerably from reliability in the specified or test environment. *See also:* system; reliability. (C/SE) 729-1983s
(2) The assessed reliability of an item based on field data. *See also:* reliability. (R) [29]

(3) The assessed reliability of an item based on operational data. (PE/NP) 933-1999

operational stage The time following commissioning of the facility. (PE/SUB) 1402-2000

operational testing (1) (software) Testing conducted to evaluate a system or component in its operational environment. *Contrast:* development testing. *See also:* qualification testing; acceptance testing. (C) 610.12-1990
(2) All testing required to verify system operation in accordance with design requirements after the major component is energized or operated. (PE/EDPG) 1248-1998

operational tests (nuclear power generating station) Tests conducted in a qualification program to demonstrate operational capability. (SWG/PE/NP) 649-1980s, C37.100-1992

operation and maintenance phase (software) The period of time in the software life cycle during which a software product is employed in its operational environment, monitored for satisfactory performance, and modified as necessary to correct problems or to respond to changing requirements. (C/SE) 1012-1986s, 610.12-1990

operation code (op) (1) (A) The operations that a computing system is capable of executing, each correlated with its equivalent in another language; for example, the binary or alphanumeric codes in machine language along with their English equivalents; the English description of operations along with statements in a programming language such as Cobol, Algol, or Fortran. **(B)** The code that represents or describes a specific operation. The operation code is usually the operation part of the instruction. (C) 162-1963
(2) (computers) A character or set of characters that specifies a computer operation; for example, the code BNZ to designate the operation "branch if not zero." (C) 610.12-1990

operation, coordinated *See:* coordinated operation.

operation decoder A device that selects one or more control channels according to the operation field of a machine instruction. (C) 610.10-1994w

operation exception An exception that occurs when a program encounters an invalid operation code. *See also:* underflow exception; data exception; addressing exception; overflow exception; protection exception. (C) 610.12-1990

operation factor The ratio of the duration of actual service of a machine or equipment to the total duration of the period of time considered. *See also:* generating station. (T&D/PE) [10]

operation field The field of a computer instruction that specifies the function to be performed. *Synonyms:* function field; operation part. *See also:* address field; operand field. (C) 610.10-1994w, 610.12-1990

operation indicator *See:* target.

operation influence (electrical influence) The maximum variation in the reading of an instrument from the initial reading, when continuously energized at a prescribed point on the scale under reference conditions over a stated interval of time, expressed as a percentage of full-scale value. *See also:* accuracy rating. (EEC/AII) [102]

operation modes (thyristor) In thyristor ac power controllers different operation modes are possible. These operation modes may be periodic or nonperiodic. (IA/IPC) 428-1981w

Operation Name A name given to an operation defined for an IEEE 1451.1 class. (IM/ST) 1451.1-1999

operation part (1) (instruction) (electronic computation) The part that usually specifies the kind of operation to be performed, but not the location of the operands. (C) 162-1963w, 270-1966w
(2) *See also:* operation field. (C) 610.12-1990

operation, quantizing *See:* quantizing operation.

operations analysis *See:* operations research.

operations and maintenance Plant staff organized to perform these functions. (PE/NP) 933-1999

operations application A class of administrative application that provides an interface for the staff who operate a media library. Examples of operations applications include a program that directs operations staff to obtain a cartridge and load it into a drive, and a program that allows an operator to place a drive or robot in or out of service. (C/SS) 1244.1-2000

operations by a pulse (pulse terminology) (general) Activation, blanking, clearing, deactivation, deflection, reading, resetting, selection, sequencing, setting, starting, stopping, storing, switching, and writing may occur or be performed. (IM/WM&A) 194-1977w

operations involving the interaction of pulses (pulse terminology) Addition, chopping, coding, comparison, decoding, encoding, mixing, modulation, subtraction, summation, and superposition may occur or be performed. *See also:* complex waveforms. (IM/WM&A) 194-1977w

operations on a pulse (general) (pulse terminology) Amplification, attenuation, conditioning, conversion, coupling demodulation, detection, discrimination, filtering, inversion, reception, reflection, and transmission may occur or be performed. (IM/WM&A) 194-1977w

operations related outage A scheduled outage in which the unit or component is removed from service to improve system operating conditions. (PE/PSE) 859-1987w

operations research The design of models for complex problems concerning the optimal allocation of available resources, and the application of mathematical methods for the solution of these problems. *Synonym:* operations analysis. (C) 610.2-1987

operation, synchronized *See:* synchronized operation.

operation table A table that defines an operation by listing all appropriate combinations of values of the operands and indicates the result for each combination. *See also:* truth table. (C) 610.10-1994w

operation time (electron tube) The time after simultaneous application of all electrode voltages for a current to reach a stated fraction of its final value. Conventionally the final value is taken as that reached after a specified length of time. *Note:* All electrode voltages are to remain constant during measurement. The tube elements must all be at room temperature at the start of the test. (ED) 161-1971w

operation with minimum constant (γ) Operation of an inverter at minimum commutation margin angle γ in order to ensure transmission at the maximum dc voltage (possible only at powers below MAP; i.e., in the "stable" region of the ac voltage/dc power characteristic). (PE/T&D) 1204-1997

operation with variable (γ) Margin angle γ is varied around an average value in order to stabilize the ac voltage. This can be achieved either by direct control of the ac voltage or by indirectly controlling the dc voltage. Another way of stabilizing the receiving system ac voltage is to arrange for the inverter, and not the rectifier, to be the current-controlling station. These modes of control are normally used for operation beyond MAP; that is, in the "unstable" region of the ac voltage/dc power characteristic. (PE/T&D) 1204-1997

operator (1) (telephone switching systems) A person who handles switching and signaling operations needed to establish connections between stations or who performs various auxiliary functions associated therewith. (COM) 312-1977w
(2) (nuclear power generating station) A person licensed to operate the plant. (PE/NP) 566-1977w
(3) (A) (software) A mathematical or logical symbol that represents an action to be performed in an operation. For example, in the expression $A = B + 3$, $+$ is the operator, representing addition. **(B) (software)** A person who operates a computer system. (C) 610.12-1990
(4) (A) In symbol manipulation, a symbol that represents the action to be performed in an operation. **(B)** A person who operates a machine. (C/MM) 695-1985, 610.10-1994
(5) In the shell command language, either a control operator or a redirection operator. (C/PA) 9945-2-1993
(6) *See also:* batch operator.

operator code (telephone switching systems) The digits dialed by operators to reach other operators. (COM) 312-1977w

operator console *See:* operator control panel.

operator control panel A functional unit that allows an operator to control a computer system. (C) 610.10-1994w

operator field *See:* operation field.

operator-handled call (telephone switching systems) A call in which information necessary for its completion, other than the number of the calling station, is verbally given by or to an operator. (COM) 312-1977w

operator loss A loss in effective signal-to-noise ratio manifested by reduced detection probability or increased false-alarm rate, when detection is performed by a human operator rather than an ideal thresholding device. (AES) 686-1997

operator manual A document that provides the information necessary to initiate and operate a system or component. Typically described are procedures for preparation, operation, monitoring, and recovery. *Note:* An operator manual is distinguished from a user manual when a distinction is made between those who operate a computer system (mounting tapes, etc.) and those who use the system for its intended purpose. *See also:* support manual; installation manual; user manual; diagnostic manual; programmer manual. (C) 610.12-1990

operator number identification (ONI) (telephone switching systems) An arrangement in which the operator requests the identity of the calling station and enters it into the system for automatic message accounting. (COM) 312-1977w

operator's telephone set (operator's set) A set consisting of a telephone transmitter, a head receiver, and associated cord and plug, arranged to be worn so as to leave the operator's hands free. *See also:* telephone station. (EEC/PE) [119]

OPGW *See:* composite overhead groundwire with optical fibers.

O + I Originating plus incoming. (COM/TA) 973-1990w

opposition The relation between two periodic functions when the phase difference between them is one-half of a period. (Std100) 270-1966w

OPS5 *See:* Official Production System.

optical ammeter An electrothermic instrument in which the current in the filament of a suitable incandescent lamp is measured by comparing the resulting illumination with that produced when a current of known magnitude is used in the same filament. The comparison is commonly made by using a photoelectric cell and indicating instrument. *See also:* instrument. (EEC/PE) [119]

optical axis (fiber optics) In an optical waveguide, synonymous with "fiber axis." (Std100) 812-1984w

optical bandwidth (acoustically tunable optical filter) The width at the 50 percent (−3 decibel) points of the optical intensity versus optical wavelength response curve of the device, measured under the conditions of white light input and fixed acoustic frequency. (UFFC) [17]

optical bar code (OBC) *See:* bar code.

optical blank (fiber optics) A casting consisting of an optical material molded into the desired geometry for grinding, polishing, or (in the case of optical waveguides) drawing to the final optical/mechanical specifications. *See also:* preform. (Std100) 812-1984w

optical cable (1) (fiber optics) A fiber, multiple fibers, or fiber bundle in a structure fabricated to meet optical, mechanical, and environmental specifications. *Synonyms:* optical fiber cable. *See also:* fiber bundle; optical cable assembly. (Std100) 812-1984w
(2) A cable in which one or more of the conductors is an optical fiber, multiple fibers, or a fiber bundle fabricated to meet optical, mechanical, and environmental specifications. *Synonym:* optical fiber cable; optical fiber bundle. (C) 610.7-1995

optical cable assembly (fiber optics) An optical cable that is connector terminated. Generally, an optical cable that has been terminated by a manufacturer and is ready for installation. *See also:* fiber bundle; optical cable. (Std100) 812-1984w

optical cavity (fiber optics) A region bounded by two or more reflecting surfaces, referred to as mirrors, end mirrors, or cavity mirrors, whose elements are aligned to provide multiple reflections. The resonator in a laser is an optical cavity. *Synonym:* resonant cavity. *See also:* laser; active laser medium. (Std100) 812-1984w

optical character *See:* graphic character.

optical character reader (OCR) A character reader that recognizes characters by transmitting light onto a surface and interpreting its reflections. *Contrast:* magnetic ink character reader. *See also:* page reader; optical scanner. (C) 610.10-1994w

optical character recognition (OCR) The automatic recognition of graphic characters using light-sensitive devices such as optical mark readers. *Contrast:* magnetic ink character recognition. (SWG/C/PE) 610.2-1987, C37.20.1-1987s

optical character recognition-A An international standard optical font used on documents intended to be read by optical character recognition. *Note:* This font is generally considered to present a less natural appearance to the eye than OCR-B. (C) 610.2-1987

optical character recognition-B (OCR-B) An international standard optical font used on documents intended to be read by optical character recognition. *Note:* This font is generally considered to present a more natural appearance to the eye than OCR-A. (C) 610.2-1987

optical combiner (fiber optics) A passive device in which power from several input fibers is distributed among a smaller number (one or more) of input fibers. *See also:* star coupler. (Std100) 812-1984w

optical computer A computer in which light and optics replace some or all of the traditional wires and electronic circuits. (C) 610.10-1994w

optical conductor* *See:* optical waveguide.
* Deprecated.

optical connector *See:* optical waveguide connector.

optical coupler (wire-line communication facilities) An optical coupler provides isolation using a short length, optical path. (PE/PSC) 487-1980s

optical coupling coefficient (optoelectronic device) (between two designated ports) The fraction of the radiant or luminous flux leaving one port that enters the other port. *See also:* optoelectronic device. (ED) [46]

optical data bus (fiber optics) An optical fiber network, interconnecting terminals, in which any terminal can communicate with any other terminal. *See also:* optical link. (Std100) 812-1984w

optical density, Dλ (1) (laser maser) Logarithm to the base ten of the reciprocal of the transmittance: $D\lambda = -\log_{10\tau\mu}$, where τ is transmittance. *See also:* photographic transmission density. (LEO) 586-1980w
(2) (fiber optics) A measure of the transmittance of an optical element expressed by: $\log_{10}(1/T)$ or $-\log_{10}T$, where T is transmittance. The analogous term $\log_{10}(1/R)$ is called reflection density. *Note:* The higher the optical density, the lower the transmittance. Optical density times 10 is equal to transmission loss expressed in decibels (dB); for example, an optical density of 0.3 corresponds to a transmission loss of 3 dB. *See also:* transmission loss; transmittance. (Std100) 812-1984w

optical density of smoke A measure of the attenuation of a light beam passing through smoke, expressed as the common logarithm of the ratio of the incident flux, I_0, to the transmitted flux, I [$D = \log_{10}(I_0/I)$]. (DEI) 1221-1993w

optical depth The value of the integral of the extinction coefficient over a specified path. (AP/PROP) 211-1997

optical depolarizer (interferometric fiber optic gyro) A component placed in an optical path that results in depolarization of the input light, regardless of its state of polarization. *Notes:*

1. Depolarizers are usually composed of two or more birefringent sections (optical fiber or crystal material such as quartz), each of which introduces a relatively large and different retardation. 2. Depolarization depends on the bandwidth of the light, being more complete for wide bandwidth sources and not possible for purely monochromatic light. (AES/GYAC) 528-1994

optical detector (fiber optics) A transducer that generates an output signal when irradiated with optical power. *See also:* optoelectronic. (Std100) 812-1984w

optical directional coupler (interferometric fiber optic gyro) A device that combines or splits the optical wave(s) from one or more waveguides to produce one or more optical waves. (AES/GYAC) 528-1994

optical disk A disk on which information is stored and retrieved by optical means, using a laser. *Synonyms:* digital optical disk; numeric optical disk. *Contrast:* magnetic disk. *See also:* magneto-optical disk; laser disk; video disk; compact disc. (C) 610.10-1994w

optical fall time The time it takes for optical power to fall from 90% effective power to 10% effective power. When expressed as a percentage, the fall time is specified as a percentage of an encoded bit period. (C/BA) 1393-1999

optical fiber (1) (fiber optics) Any filament or fiber, made of dielectric materials, that guides light, whether or not it is used to transmit signals. *See also:* fiberoptics; optical waveguide; fiber bundle. (Std100) 812-1984w
(2) A filament-shaped optical waveguide made of dielectric materials. (LM/C) 11802-4-1994, 8802-3-1990s, 610.7-1995, 802.3-1998

optical fiber bundle *See:* optical cable.

optical fiber cable *See:* optical cable.

optical fiber cable interface *See:* Fiber Optic Medium Dependent Interface.

optical fiber cable link segment A length of optical fiber cable that contains two optical fibers and is comprised of one or more optical fiber cable sections and their means of interconnection, with each optical fiber terminated at each end in the optical connector plug. (C/LM) 802.3-1998

optical fiber waveguide *See:* optical waveguide.

optical field meter A meter that measures changes in the transmission of light through a fiber or crystal due to the influence of the electric field (for example, meters based on Pockel's effect). Optical field meters can be used to implement free-body or ground reference measurements. When optical fibers are used, the meter is inherently electrically isolated from ground. (T&D/PE) 539-1990

optical filter (fiber optics) An element that selectively transmits or blocks a range of wavelengths. (Std100) 812-1984w

optical font (1) A character font used in optical character recognition. For example, hand-printed character font, OCR-A, or OCR-B. (C) 610.2-1987
(2) A font that can be input by a special input device and translated into electronic form. (C) 610.10-1994w

optical frequency shifter (interferometric fiber optic gyro) A device that either increases or decreases the frequency of light passing through it by an amount equal to the frequency of an electrical control signal. *Note:* A commonly used optical frequency shifter is the acousto-optic Bragg cell. (AES/GYAC) 528-1994

optical glide path lights *See:* angle-of-approach lights.

optical idle signal The signal transmitted by the Fiber Optic Medium Attachment Unit (FOMAU) into its transmit optical fiber during the idle state of the DO circuit. (C/LM) 802.3-1998

optical image The result of projecting a scene onto a surface. For example, the image of a scene formed on film by a camera lens. (C) 610.4-1990w

optical interface The optical input and output connection interface to a 10BASE-FP Star. (C/LM) 802.3-1998

optical isolator (interferometric fiber optic gyro) A device intended to suppress return reflections along a transmission path. *Note:* The Faraday isolator uses the magneto-optic effect. (AES/GYAC) 528-1994

optical landing system (navigation aid terms) A shipboard gyro stabilized or shore-based device which indicates to the pilot his displacement from a preselected glide path. (AES/GCS) 172-1983w

optical link (fiber optics) Any optical transmission channel designed to connect two end terminals or to be connected in series with other channels. *Note:* Sometimes terminal hardware (for example, transmitter/receiver modules) is included in the definition. *See also:* optical data bus. (Std100) 812-1984w

optically active material (fiber optics) A material that can rotate the polarization of light that passes through it. *Note:* An optically active material exhibits different refractive indices for left and right circular polarizations (circular birefringence). *See also:* birefringent medium. (Std100) 812-1984w

optically assisted magnetic storage *See:* magneto-optical disk.

optically pumped laser (laser maser) A laser in which the electrons are excited into an upper energy state by the absorption of light from an auxiliary light source. (LEO) 586-1980w

optical mark reader A reader that can perform mark sensing of hand-written pencil marks, and pre-printed marks by detecting the presence or absence of reflected light. (C) 610.10-1994w

optical mark reading (OMR) The use of pattern recognition techniques to identify graphite marks by automatic means. (C) 610.2-1987

optical mouse A mouse in which motion is sensed by transmitting light onto a special surface and interpreting its reflections using an optical sensor. *Contrast:* mechanical mouse. (C) 610.10-1994w

optical path length (1) (fiber optics) In a medium of constant refractive index n, the product of the geometrical distance and the refractive index. If n is a function of position,

$$\text{optical path length} = \int n\,ds,$$

where *ds* is an element of length along the path. *Note:* Optical path length is proportional to the phase shift a light wave undergoes along a path. *See also:* optical thickness. (Std100) 812-1984w
(2) (interferometric fiber optic gyro) (laser gyro) The optical length of the path traversed in a single pass by an optical beam, taking into account the index of refraction of each medium supporting propagation. (AES/GYAC) 528-1994

optical pattern *See:* light pattern.

optical phase modulator (interferometric fiber optic gyro) A device that modulates the phase of a light wave as a function of an electrical control signal. *Note:* Commonly used phase modulators vary the optical path length by means of electro-optic or elasto-optic effects. (AES/GYAC) 528-1994

optical photons (scintillation counting) Photons with energies corresponding to wavelengths between approximately 120–1800 m. (NPS) 398-1972r

optical polarization controller (interferometric fiber optic gyro) A component, placed in the optical path, that can be adjusted to change the light from any state of polarization at the input to any desired polarization at the output. *Note:* In fiber optics, polarization controllers usually consist of two or three sections of birefringent fiber that can be rotated with respect to each other. (AES/GYAC) 528-1994

optical polarizer (interferometric fiber optic gyro) A device that selects a single, linear polarization state by suppressing the orthogonal state. (AES/GYAC) 528-1994

optical power *See:* radiant power.

optical printer *See:* electrostatic printer.

optical probe A flux density meter in which the transduction mechanism is optical. A number of physical effects (i.e., magnetostriction, change in birefringence) may be used to affect the light in a "witness crystal" or "sense fiber." (T&D/PE) 539-1990

optical pyrometer A temperature-measuring device comprising a standardized comparison source of illumination and source convenient arrangement for matching this source, either in brightness or in color, against the source whose temperature is to be measured. The comparison is usually made by the eye. *See also:* electric thermometer. (EEC/PE) [119]

optical reader *See:* optical character reader; optical mark reader; optical scanner.

optical recording A method for storing data by using optical means. (C) 610.10-1994w

optical repeater (fiber optics) In an optical waveguide communication system, an optoelectronic device or module that receives a signal, amplifies it (or, in the case of a digital signal, reshapes, retimes, or otherwise reconstructs it) and retransmits it. *See also:* modulation. (Std100) 812-1984w

optical rise time The time it takes for optical power to rise from 10% effective power to 90% effective power. When expressed as a percentage, the rise time is specified as a percentage of an encoded bit period. (C/BA) 1393-1999

optical scanner (A) (character recognition) A device that scans optically and usually generates an analog or digital signal. **(B) (character recognition)** A device that optically scans printed or written data and generates their digital representations. *See also:* electronic analog computer; visual scanner. (C) [20], [85]

(2) (A) A scanner that uses light for examining patterns. *See also:* bar code scanner; optical character reader. **(B)** A device that scans optically and generates a corresponding output signal. *See also:* digitizer. (C) 610.10-1994

optical sensor A device capable of detecting light and producing an analog or digital output signal. *See also:* optical mouse. (C) 610.10-1994w

optical sound recorder *See:* photographic sound recorder.

optical sound reproducer *See:* photographic sound reproducer.

optical spectrum (fiber optics) Generally, the electromagnetic spectrum within the wavelength region extending from the vacuum ultraviolet at 40 nanometers (nm) to the far infrared at 1 millimeter (mm). *See also:* infrared; light. (Std100) 812-1984w

optical storage Storage of information in which access to that information is obtained using optical signals. *Synonym:* photo-optic storage. *See also:* CD-ROM storage. (C) 610.10-1994w

optical switch (interferometric fiber optic gyro) A device in an optical path that can pass, stop, or redirect light, depending on control input. (AES/GYAC) 528-1994

optical thickness (fiber optics) The physical thickness of an isotropic optical element, times its refractive index. *See also:* optical path length. (Std100) 812-1984w

optical time domain reflectometry (fiber optics) A method for characterizing a fiber wherein an optical pulse is transmitted through the fiber and the resulting light scattered and reflected back to the input is measured as a function of time. Useful in estimating attenuation coefficient as a function of distance and identifying defects and other localized losses. *See also:* scattering; Rayleigh scattering. (Std100) 812-1984w

optical tracker (navigation aid terms) A device for determining the direction of a luminous body relative to a set of reference axes using visible light vice, infrared, or radio frequencies. (AES/GCS) 172-1983w

optical transfer function (diode-type camera tube) The spatial frequency response of an imaging sensor to a point source input. That is, the Fourier transform of the output image waveshape when the input image is a point, is known as the two-dimensional optical transfer function. In the one- dimensional case, the optical transfer function is the Fourier transform of the output image when the input image is a line. In the most common one-dimensional form, the OTF, designated $R_O(N)$, is written

$$R_O(N) = |R_O(N)| \exp[j\Phi(N)].$$

The OTF contains both amplitude and phase information. (ED) 503-1978w

optical wand *See:* light pen.

optical waveguide (A) (fiber optics) Any structure capable of guiding optical power. **(B) (fiber optics)** In optical communications, generally a fiber designed to transmit optical signals. *Note:* The use of "optical conductor" as a synonym for this term is deprecated. *Synonym:* optical fiber waveguide; lightguide. *See also:* optical fiber; cladding; tapered fiber waveguide; multimode optical waveguide; fiber bundle; fiberoptics; core. (Std100) 812-1984

optical waveguide connector (fiber optics) A device whose purpose is to transfer optical power between two optical waveguides or bundles, and that is designed to be connected and disconnected repeatedly. *See also:* optical waveguide coupler; multifiber joint. (Std100) 812-1984w

optical waveguide coupler (A) (fiber optics) A device whose purpose is to distribute optical power among two or more ports. *See also:* tee coupler; star coupler. **(B) (fiber optics)** A device whose purpose is to couple optical power between a waveguide and a source or detector. (Std100) 812-1984

optical waveguide preform *See:* preform.

optical waveguide splice (fiber optics) A permanent joint whose purpose is to couple optical power between two waveguides. (Std100) 812-1984w

optical waveguide termination (fiber optics) A configuration or a device mounted at the end of a fiber or cable which is intended to prevent reflection. *See also:* index matching material. (Std100) 812-1984w

optic amplifier An optoelectronic amplifier whose signal input and output ports are electric. *Note:* This is in accord with the accepted terminologies of other electric-signal input and output amplifiers such as dielectric, magnetic, and thermionic amplifiers. *See also:* optoelectronic device. (ED) [46]

optic axis (fiber optics) In an anisotropic medium, a direction of propagation in which orthogonal polarizations have the same phase velocity. Distinguished from "optical axis." *See also:* anisotropic. (Std100) 812-1984w

optic coupling device An isolation device using an optical link to provide the longitudinal isolation. Circuit arrangements on each side of the optical link convert the electrical signal into an optical signal for transmission through the optical link and back to an electrical signal. Various circuit arrangements provide one-way or two-way transmission and permit transmission to the various combinations of voice and/or dc signalling used by the power industry. The optical link may be either a quartz rod or a short length of optic fiber. Single-channel optic coupling devices may be used in conjunction with other isolation devices in protection systems. (PE/PSC) 487-1992

optic port A port where the energy is electromagnetic radiation, that is, photons. *See also:* optoelectronic device. (ED) [46]

optimal control An admissible control law that gives a performance index an extremal value. *See also:* control system. (IM) [120]

optimization The procedure used in the design of a system to maximize or minimize some performance index. May entail the selection of a component, a principle of operation, or a technique. *See also:* system. (SMC) [63]

optimum bunching (electron tube) (traveling-wave tube) The bunching condition that produces maximum power at the desired frequency in an output gap. *See also:* electron device. (ED) 161-1971w, [45]

optimum linearizing load resistance (Hall generator) The load resistance that produces the least linearity error. (MAG) 296-1969w

optimum working frequency (OWF, FOT) The frequency that is exceeded by the operational maximum usable frequency (MUF) during 90% of the specified period, usually a month. *Note:* The acronym FOT is the French abbreviation of "fréquence optimale de travail." (AP/PROP) 211-1997

option (1) An argument to a command that is generally used to specify changes in the default behavior of a utility. (C/PA) 9945-2-1993
(2) Any behavior or feature defined in the base standard that need not be present in all conforming implementations. (C/PA) 2003-1997

optional (1) The referenced item is not required to claim compliance with this standard. Implementation of an optional item shall be as defined in this standard. (C/BA) 1496-1993w
(2) A syntax keyword used to specify a partial mapping. *Contrast:* mandatory. *See also:* partial. (C/SE) 1320.2-1998

optional attribute An attribute that may have no value for an instance. (C/SE) 1320.2-1998

optional category A category that provides additional details that are not essential but may be useful in particular situations. (C/SE) 1044-1993

optional nonidentifying relationship A kind of nonidentifying relationship in which an instance of the child entity can exist without being related to an instance of the parent entity. *Contrast:* mandatory nonidentifying relationship. *See also:* nonidentifying relationship. (C/SE) 1320.2-1998

optional-pause instruction A pause instruction that allows manual suspension of a computer program. *Synonym:* optional stop instruction. (C) 610.10-1994w

optional stop (numerically controlled machines) A miscellaneous function command similar to a program stop except that the control ignores the command unless the operator has previously pushed a button to validate the command. (IA) [61]

optional stop instruction *See:* optional-pause instruction.

optional tasks Those V&V tasks that may be added to the minimum V&V tasks to address specific application requirements. (C/SE) 1012-1998

option-argument A parameter that follows certain options. In some cases an option-argument is included within the same argument string as the option; in most cases it is the next argument. (C/PA) 9945-2-1993

options file A file that can be specified with the -x option. This file contains extended option definitions that override default definitions. (C/PA) 1387.2-1995

optoelectronic (fiber optics) Pertaining to a device that responds to optical power, emits or modifies optical radiation, or utilizes optical radiation for its internal operation. Any device that functions as an electrical-to-optical or optical-to-electrical transducer. *Notes:* 1. Photodiodes, light emitting diodes (LED), injection lasers and integrated optical elements are examples of optoelectronic devices commonly used in optical waveguide communications. 2. "Electro-optical" is often erroneously used as a synonym. *See also:* optical detector; electro-optic effect. (Std100) 812-1984w

optoelectronic amplifier An optoelectronic device capable of power gain, in which the signal ports are either all electric ports or all optic ports. *See also:* optoelectronic device. (ED) [46]

optoelectronic cell The smallest portion of an optoelectronic device capable of independently performing all the specified input and output functions. Note: An optoelectronic cell may consist of one or more optoelectronic elements. *See also:* optoelectronic device. (ED) [46]

optoelectronic device An electronic device combining optic and electric ports. (ED) [46]

optoelectronic element A distinct constituent of an optoelectronic cell, such as an electroluminor, photoconductor, diode, optical filter, etc. *See also:* optoelectronic device. (ED) [46]

OR (mathematics of computing) A Boolean operator having the property that if P is a statement, Q is a statement, R is a statement, . . ., then the OR of P, Q, R, . . . is true if and only if at least one statement is true. *Note:* P OR Q is often represented by P$\vee$Q or P + Q.

OR Truth Table

P	Q	P$\vee$Q
0	0	0
0	1	1
1	0	1
1	1	1

Synonyms: OR-ELSE; inclusive OR; logical add; union; disjunction; logic add; Boolean add; false add. *Contrast:* exclusive OR. (C) 1084-1986w

ORA *See:* output reference axis.

orbit (navigation aid terms) The path of a celestial body relative to another body around which it revolves. (AES/GCS) 172-1983w

orbital inclination (communication satellite) The angle between the plane of the orbit and the plane of the equator measured at the ascending node. (COM) [19]

orbital plane (communication satellite) The plane containing the radius vector and the velocity vector of a satellite, the system of reference being that specified for defining the orbital elements. In the idealized case of the unperturbed orbit, the orbital plane is fixed relative to the equatorial plane of the primary body. (COM) [19]

orbital stability (1) (closed solution curve denoted Γ) Implies that for every given $\varepsilon > 0$ there exists a $\delta > 0$ (which, in general, may depend on ε and on t_0) such that $\rho(\Gamma, x(t_0)) \leq \delta$ implies $\rho(\Gamma, \phi(x(t_0); t)) \leq \varepsilon$ for $t \leq t_0$, where $\rho(\Gamma, a)$ denotes the minimum distance between the curve Γ and the point a. Here the point $x(t_0)$ is assumed to be off the curve Γ. *Notes:* 1. Orbital stability does not imply Lyapunov stability of a closed solution curve, since a point on the closed curve may not travel at the same speed as a neighboring point off the curve. 2. Only nonlinear systems can produce the type of solutions for which the concept of orbital stability is applicable. *See also:* control system. (CS/IM) [120]
(2) *See also:* stability of a limit cycle.

OR circuit *See:* OR gate.

OR element *See:* OR gate.

order (general) To put items in a given sequence. (C) [20], [85]
(2) (A) (in electronic computation) Synonym for instruction. **(B) (in electronic computation)** Synonym for command. **(C) (in electronic computation)** Loosely, synonym for operation part. *Note:* The use of order in the computer field as a synonym for terms similar to the above is losing favor owing to the ambiguity between these meanings and the more-common meanings in mathematics and business. *See also:* instruction. (C) 162-1963
(3) (A) (data management) To place items in an arrangement in accordance with a specified set of rules. *Note:* The arrangement need not be linear. *See also:* sort. **(B) (data management)** The result of an arrangement as in (A). **(C) (data management)** In a tree, the maximum number of subtrees of any node. *Note:* The use of "sequence" as a preferred term for "order" is deprecated. (C) 610.5-1990

order-by-merging To order the items of a set by splitting the set into subsets, ordering the subsets, and merging the subsets. *See also:* sort by merging; sequence by merging. (C) 610.5-1990w

order clash In software design, a type of structure clash in which a program must deal with two or more data sets that have been sorted in different orders. *See also:* data structure-centered design. (C) 610.12-1990

ordered list A list in which the data items are arranged in some specific order, either physically or logically by some key. *Contrast:* unordered list. (C) 610.5-1990w

ordered_set As used in the 1000BASE-X PCS, a single special code-group, or a combination of special and data code-groups, used for the delineation of a packet and synchronization between the transmitter and receiver circuits at opposite ends of a link. (C/LM) 802.3-1998

ordered tree A tree in which the left-to-right order of the subtrees of a given node is significant. *Contrast:* unordered tree. (C) 610.5-1990w

ordering bias The manner and degree by which the order of a set of items departs from the order of a randomly distributed set of items. The ordering bias of a set is inversely proportional to the effort required to sort the set. (C) 610.5-1990w, 1084-1986w

ordering relation A VHDL relational expression in which the relational operator is <, <=, >, or >=. (C/DA) 1076.3-1997

orderly release The graceful termination of a network connection with no loss of data. (C) 1003.5-1999

order parameter (primary ferroelectric terms) A parameter, or functionally related set of parameters, that can describe the reduction in symmetry occurring at a phase transition from a nonferroic phase to a ferroic phase. (UFFC) 180-1986w

order tone (telephone switching systems) A tone that indicates to an operator that verbal information can be transferred to another operator. (COM) 312-1977w

order wire (communication practice) An auxiliary circuit for use in the line-up and maintenance of communication facilities. (COM) [48]

ordinary binary *See:* binary.

ordinary wave (1) (radio-wave propagation) The magnetoionic wave component that, when viewed below the ionosphere in the direction of propagation, has counterclockwise or clockwise elliptical polarization, respectively, according as the earth's magnetic field has a positive or negative component in the same direction. *See also:* radiation. (EEC/PE) [119]
(2) That characteristic magneto-ionic wave component deviating the least, in most of its propagation characteristics, from those expected for a wave in a non-magnetized plasma of the same density. *Note:* For vertical incidence, the ordinary wave is reflected near the height at which the plasma frequency is equal to the wave frequency when the effects of collisions are negligible. *Synonym:* O wave. (AP/PROP) 211-1997

ordinary-wave component (radio-wave propagation) That magneto-ionic wave component deviating the least, in most of its propagation characteristics, relative to those expected for a wave in the absence of a fixed magnetic field. More exactly, if at fixed electron density, the direction of the fixed magnetic field were rotated until its direction was transverse to the direction of phase propagation, the wave component whose propagation would then be independent of the magnitude of the fixed magnetic field. *Synonym:* O wave. (AP) 211-1977s

OR-ELSE *See:* OR.

organic scintillator solute material (1) (liquid-scintillation counters) An organic compound that can absorb radiant energy and immediately (typically within 10^{-9} s) re-emit this energy as photons in the visible or ultraviolet range. This material is sometimes referred to as the scintillator or the fluor. (NI) N42.16-1986
(2) (liquid-scintillation counters) An organic compound that can absorb radiant energy and immediately (typically within 10^{-9} s) re-emit this energy as photons in the ultraviolet range. (NI) N42.15-1990

organizational maintenance (test, measurement, and diagnostic equipment) Maintenance which is the responsibility of and performed by using organizations on its assigned equipment. Its phases normally consist of inspecting, servicing, lubricating, adjusting and the replacing of parts, minor assemblies and subassemblies. (MIL) [2]

organizational model An OSI management model that describes the distribution of management controls in the OSI environment. (C) 610.7-1995

Organizational Process Asset (OPA) An artifact that defines some portion of an organization's software project environment. (C/SE) 1074-1997

organizational unit name attributes The Organizational-Unit-Name-1, Organizational-Unit-Name-2, Organizational-Unit-Name-3, and Organizational-Unit-Name-4 attributes specific to the class. (C/PA) 1224.1-1993w

organizing *See:* self-organizing.

Oregon State Conversational Aid to Research (OSCAR) An interactive programming system used for performing numerical calculations, string manipulations, vector and matrix operations, and complex arithmetic. (C) 610.13-1993w

OR gate (1) (electronic computation) (OR circuit) A gate whose output is energized when any one or more of the inputs is in its prescribed state. An OR gate performs the function of the logical OR. (Std100) 270-1966w
(2) A gate that performs the Boolean operation of disjunction. *Synonyms:* OR element; inclusive OR gate. *See also:* exclusive-OR gate; NOR gate. (C) 610.10-1994w

orientation (1) (illuminating engineering) The relation of a building with respect to compass directions. (EEC/IE) [126]
(2) The positioning of the dimensions of a spatial vector. For example, spatial vectors may be represented by a cartesian coordinate system oriented in space in some application-specific manner. (IM/ST) 1451.1-1999

orifice An opening or window in a side or end wall of a waveguide or cavity resonator, through which energy is transmitted. *See also:* waveguide. (EEC/PE) [119]

orifice plate (rotating machinery) A restrictive opening in a passage to limit flow. (PE) [9]

origin The address of the initial storage location assigned to a computer program in main memory. *Contrast:* starting address. *See also:* assembled origin; loaded origin. (C) 610.12-1990

originating port A transmitting port on a physical layer (PHY), which has no active receiving port. The source of the transmitted packet is either the PHY's local link or the PHY itself. (C/MM) 1394a-2000

originating traffic (telephone switching systems) Traffic received from lines. (COM) 312-1977w

origin attribute The classification of software as either developed or nondeveloped. (C/SE) 1045-1992

original equipment manufacturer The manufacturer of a component in a computer system such that the component is used in assembling a larger system or component by another manufacturer. Many peripheral devices are made by an OEM but sold as part of a complete computer system by another vendor. (C) 610.10-1994w

original master (number 1 master) (disk recording) (metal master) (metal negative) (electroacoustics) The master produced by electroforming from the face of a wax or lacquer recording. (SP) [32]

original source statements Source statements that are obtained from an external product. (C/SE) 1045-1992

original supplier (replacement parts for Class 1E equipment in nuclear power generating stations) Supplier of the original Class 1E equipment, as opposed to the original manufacturer of the part. (PE/NP) 934-1987w

originating domain The MD at which a communique or report originated. (C/PA) 1224.1-1993w

OR-parallelism Pertaining to the performance of multiple predicate operations concurrently, the successful completion of any results in a true response. *Note:* Successful termination of one predicate operation may cause the others to be immediately terminated. *Contrast:* AND-parallelism. *See also:* OR-tying. (C) 610.10-1994w

orphaned process group (1) A process group in which the parent of every member is either itself a member of the group or is not a member of the group's session. (C/PA) 9945-1-1996

(2) A process group in which the parent of every member is either itself a member of the group or is not a member of the session of the group. An orphaned process group is no longer a member of the session of the process that created it. A process group can become orphaned when all other members of the session exit. (C) 1003.5-1999

orphan prevention The ability of a text formatter to avoid placing the final one or two lines of a paragraph at the top of a page. *See also:* widow prevention. (C) 610.2-1987

orthicon A camera tube in which a beam of low-velocity electrons scans a photoemissive mosaic capable of storing an electric-charge pattern. *See also:* television. (ED) 161-1971w

ortho-axis An angle of 54.7 degrees to the edges and centerlines of each face of the DUT. This angle is the ortho-angle which is the angle that the diagonal of a cube makes to each side at the trihedral corners of the cube. (EMC) 1309-1996

orthogonality (oscilloscopes) The extent to which traces parallel to the vertical axis of a cathode-ray-tube display are at right angles to the horizontal axis. *See also:* oscillograph. (IM/HFIM) [40]

orthogonal polarization (1) (with respect to a specified polarization) In a common plane of polarization, the polarization for which the inner product of the corresponding polarization vector and that of the specified polarization is equal to zero. *Notes:* 1. The two orthogonal polarizations can be represented as two diametrical points on the Poincaré sphere. 2. Two elliptically polarized fields having the same plane of polarization have orthogonal polarizations if their polarization ellipses have the same axial ratio, major axes at right angles, and opposite senses of polarization. *See also:* polarization vector. (AP/ANT) 145-1993

(2) For a given wave, the unique polarization state containing no components of the given wave's polarization. *Note:* For linear polarization, the (linear) polarization perpendicular to the reference (linear) polarization. 2. For circular polarization, the (circular) polarization with the opposite sense of rotation. 3. For elliptical polarization, the polarization with the same axial ratio, opposite rotation sense and major axis perpendicular to that of the reference polarization. (AP/PROP) 211-1997

orthorhombic system (piezoelectricity) An orthorhombic crystal has three mutually perpendicular twofold axes or two mutually perpendicular planes of reflection symmetry, or both. The *a*, *b*, *c* axes are of unequal length. For classes 222 and 2/*m* 2/*m* 2/*m* unit distances are chosen such that $c_0 < a_0 < b_0$. For the remaining class, which is polar, *Z* will always be the polar axis regardless of whether it is *a*, *b*, or *c* in the crystallographer's notation. Axes *X* and *Y* will then be chosen so that *X* is parallel to the remaining axis that is smallest. This class therefore may be properly designated *mm*2, 2*mm*, or *m*2*m*, depending on whether *c*, *a*, or *b* is the polar axis. Axis sense is trivial except for the polar class for which +Z is chosen such that *d*33 is positive. *Note:* Positive and negative may be checked using a carbon-zinc flashlight battery. The carbon anode connection will have the same effect on meter deflection as the + end of the crystal axis upon release of compression. *See also:* crystal systems. (UFFC) 176-1978s

OR-tying The process of connecting together two or more logic gate outputs such that the common output is forced to ground when any of the individual gate outputs is low. *See also:* OR-parallelism. (C) 610.10-1994w

OSAM *See:* overflow sequential access method.

O scan *See:* O-display.

OSCAR *See:* Oregon State Conversational Aid to Research.

oscillating current A current that alternately increases and decreases but is not necessarily periodic. (Std100) 270-1966w

oscillating scan head A scan head that physically moves back and forth across the original page as it scans each line. (C) 610.10-1994w

oscillating sort An external merge sort in which sorts and merges are performed alternately; that is, the first two subsets are sorted and merged, the next subset is sorted and merged with the previously merged subsets, and so on, until all subsets are sorted and merged. (C) 610.5-1990w

oscillation (1) (general) The variation, usually with time, of the magnitude of a quantity with respect to a specified reference when the magnitude is alternately greater and smaller than the reference. *See also:* vibration. (SP) [32]

(2) (vibration) A generic term referring to any type of a response that may appear in a system or in part of a system. *Note:* Vibration is sometimes used synonymously with oscillation, but it is more properly applied to the motion of a mechanical system in which the motion is in part determined by the elastic properties of the body. (Std100) 270-1966w

(3) (gas turbines) The periodic variation of a function between limits above or below a mean value, for example, the periodic increase and decrease of position, speed, power output, temperature, rate of fuel input, etc. within finite limits. (PE/EDPG) 282-1968w, [5]

oscillator (1) (general) Apparatus intended to produce or capable of maintaining electric or mechanical oscillations. (Std100) 270-1966w, [84]

(2) (electronics) A nonrotating device for producing alternating current, the output frequency of which is determined by the characteristics of the device.

(3) A circuit that continuously alternates between two or more states. (C) 610.10-1994w

oscillator mode Frequency or frequencies for which the total phase shift around the oscillator loop is an integral multiple of 2π. (UFFC) 1037-1992w

oscillator starting time, pulsed *See:* pulsed-oscillator starting time.

oscillator tube, positive grid *See:* positive-grid oscillator tube.

oscillatory circuit A circuit containing inductance and capacitance so arranged that when shock excited it will produce a current or a voltage that reverses at least once. If the losses exceed a critical value, the oscillating properties will be lost. (EEC/PE) [119]

oscillatory surge A surge that includes both positive and negative polarity values. (SPD/PE) C62.11-1999, C62.62-2000

oscillatory transient A sudden, nonpower frequency change in the steady-state condition of voltage or current that includes both positive or negative polarity value. (SCC22) 1346-1998

oscillogram A record of the display presented by an oscillograph or an oscilloscope. *See also:* oscillograph. (IM/HFIM) [40]

oscillograph An instrument primarily for producing a record of the instantaneous values of one or more rapidly varying electrical quantities as a function of time or of another electrical or mechanical quantity. *Notes:* 1. Incidental to the recording of instantaneous values of electrical quantities, these values may become visible, in which case the oscillograph performs the function of an oscilloscope. 2. An oscilloscope does not have inherently associated means for producing records. 3. The term includes mechanical recorders.

oscillograph tube (oscilloscope tube) A cathode-ray tube used to produce a visible pattern that is the graphic representation of electric signals, by variations of the position of the focused spot or spots in accordance with these signals. (ED) 161-1971w

oscillography The art and practice of utilizing the oscillograph. *See also:* oscillograph. (IM/HFIM) [40]

oscilloscope (1) An instrument primarily for making visible the instantaneous value of one or more rapidly varying electrical quantities as a function of time or of another electrical or mechanical quantity. *See also:* oscillograph; cathode-ray oscilloscope. (EEC/PE) [119]

(2) An instrument for measuring and displaying an electrical quantity (typically voltage) as a function of time. The band-

width should be at least 10 times the maximum frequency of interest and include triggering capacity. (PEL) 1515-2000

oscilloscope, dual-beam *See:* dual-beam oscilloscope.

oscilloscope, multibeam *See:* multibeam oscilloscope.

O-scope A cathode-ray oscilloscope arranged to present an O-display. (AES/RS) 686-1990

OSE *See:* open system environment.

OSI environment An environment in which information processing system or resources communicate conforming to the services and protocols of OSI. *Contrast:* local systems environment. (C) 610.7-1995

OSI model *See:* open systems interconnection model.

OSI reference model *See:* open systems interconnection model.

OSSL *See:* Operating Systems Simulation Language.

OSI *See:* open systems interconnection.

O structure *See:* snub structure.

OSIRM *See:* open systems interconnection reference model.

Ostwald color system (illuminating engineering) A system of describing colors in terms of color content, white content, and black content. It is usually exemplified by color charts in triangular form with Full Color, White, and Black at the apexes providing a gray scale of White and Black mixtures, and parallel scales of Constant White Content as these grays are mixed with varying proportions of the Full Color. Each chart represents a constant dominant wavelength (called hue), and the colors lying on a line parallel to the gray scale represent constant purity (called Shadow Series). (EEC/IE) [126]

other insulation characteristics (insulation systems of synchronous machines) It is important to recognize that other characteristics, in addition to thermal endurance, such as mechanical strength, moisture resistance, and corona endurance are required in varying degrees in different applications for the successful use of insulating materials. (REM) [115]

other peripheries Other control room areas, such as remote emergency shutdown panels or remote emergency control rooms. (PE/NP) 845-1988s

other system One system within the AI-ESTATE architectural concept. This system covers functional capabilities not defined within the other systems of the AI-ESTATE conceptual architecture. Examples of other systems are operating systems, CAD/CAE design tools, and testability analysis tools. (ATLAS) 1232-1995

other tests (power and distribution transformers) Tests so identified in individual product standards which may be specified by the purchaser in addition to routine tests. (Examples: impulse, insulation power factor, audible sound.) *Note:* Transformer "General Requirements" Standards (such as ANSI C57.12.00-1973, IEEE Std 462-1973), General Requirements for distribution, Power and Regulating Transformers classify various tests as "routine," "design," or "other" depending on the size, voltage, and type of transformer involved. (PE/TR) C57.12.80-1978r

OTH radar *See:* over-the-horizon radar.

O-type tube or valve (microwave tubes) A microwave tube in which the beam, the circuit, and the focusing field have symmetry about a common axis. The interaction between the beam and the circuit is dependent upon velocity modulation, the suitably focused beam being launched by a gun structure outside one end of the microwave circuit and principally collected outside the other end of the microwave structure. (ED) [45]

Oudin current (resonator current) (desiccating current) (medical electronics) A brush discharge produced by a high-frequency generator that has an output range of 2 to 10 kilovolts and a current sufficient to evaporate tissue water without charring. It is usually applied through a small needlelike electrode with the reference or ground electrode being relatively large and diffuse. (EMB) [47]

Oudin resonator A coil of wire often with an adjustable number of turns, designed to be connected to a source of high-frequency current, such as a spark gap and induction coil, for the purpose of applying an effluve (convective discharge) to a patient. (EMB) [47]

outage (1) (electric power system) The state of a component when it is not available to perform its intended function due to some event directly associated with that component. *Notes:* 1. An outage may or may not cause an interruption of service to consumers, depending on system configuration. 2. This definition derives from transmission and distribution applications and does not necessarily apply to generation outages. (PE/T&D/PSE) [54], 346-1973w, 1366-1998

(2) The state of a component or system when it is not available to properly perform its intended function due to some event directly associated with that component or system. (IA/PSE) 493-1997, 399-1997

(3) *See also:* interruption. (IA/PSE) 1100-1992s

(4) *See also:* clearance. (T&D/PE) 524-1992r

outage duration (1) (electric power system) The period from the initiation of an outage until the affected component once again becomes available to perform its intended function. *Note:* Outage durations may be defined for specific types of outages; for example, permanent forced outage duration, transient forced outage duration, and scheduled outage duration. *See also:* outage. (PE/PSE) [54], 346-1973w

(2) (electric power plants) The period from the initiation of an outage occurrence until the component or unit is returned to the in-service state. *Notes:* 1. Outage durations are commonly summarized for specific types of outages as, for example, permanent forced outage duration, transient forced outage duration, and scheduled outage duration. 2. Outage duration is normally equal to the sum of switchingtime, repair time, and travel and material procurement time, but may be longer for reasons other than unavailability of manpower, equipment, or material. (PE/PSE) 859-1987w

outage duration, permanent forced *See:* permanent forced outage duration.

outage duration, scheduled *See:* scheduled outage duration.

outage duration, transient forced *See:* transient forced outage duration.

outage, equipment *See:* equipment outage.

outage event An event involving the outage occurrence of one or more units or components. (PE/PSE) 859-1987w

outage, forced *See:* forced outage.

outage initiation Outage occurrences are initiated either automatically or manually. (PE/PSE) 859-1987w

outage occurrence The change in the state of one component or one unit from the in-service state to the outage state. *Notes:* 1. The noun "outage" is commonly used to mean outage occurrence. When used as a predicate adjective, "outaged" is ambiguous (for example, "The unit is outaged") and more specific terminology is suggested (for example, "The unit is in an outage sence"). 2. If redundant components are used, a component outage occurrence may not imply a unit outage occurrence, depending on the failure mode or switching procedure. Examples of component outage occurrences not related to any unit outage occurrence include one circuit breaker in a ring bus configuration isolated for maintenance, and one of two line protection systems deactivated for testing. (PE/PSE) 859-1987w

outage, permanent forced *See:* permanent forced outage.

outage, power *See:* power outage.

outage rate (1) (electric power system) For a particular classification of outage and type of component, the mean number of outages per unit exposure time per component. *Note:* Outage rates may be defined for specific weather conditions and types of outages. For example, permanent forced outage rates may be separated into adverse weather permanent forced outage rate and normal weather permanent forced outage rate. (PE/PSE) 346-1973w

(2) (outage occurrences and outage states of electrical transmission facilities) The number of outage occurrences per unit of service time = number of outage occurrences/service time. *Notes:* 1. Usually the unit of service time is one year. 2. Outage rates can be subdivided by outage types, by the weather prevailing during the service time, or by season. For example:

a) scheduled outage rate: The number of scheduled outages per unit of service time = number of scheduled outages/service time. In some studies, scheduled outage rate may be defined as the number of outage occurrences per unit of exposure time (including both service time and outage time).
b) normal weather forced outage rate: The number of forced outages per unit of service time in normal weather = number of forced outages during normal weather/service time during normal weather.
c) summer outage rate: The number of outage occurrences per unit of service time during the summer. Summer outage rate = number of outages during the summer/service time during summer.

(PE/PSE) 859-1987w

outage rate, adverse weather permanent forced (electric power system) For a particular type of component, the mean number of outages per unit of adverse weather exposure time per component. (PE/PSE) 346-1973w

outage rate, normal weather permanent forced (electric power supplies) For a particular type of component, the mean number of outages per unit of normal weather exposure time per component. (PE/PSE) 346-1973w

outage, scheduled *See:* scheduled outage.

outage state The component or unit is not in the in-service state; that is, it is partially or fully isolated from the system. *Notes:* 1. A unit may be in the outage state due to a failure of a component within the unit or due to the outage occurrence of another unit or component. 2. A two-terminal transformer disconnected on one or both sides is in the outage state. 3. A three-terminal overhead transmission line disconnected from one terminal is, as a unit, in the outage stage. However, the two overhead line sections still carrying power are, as components, in the in-service state. A circuit breaker that is not energized on either side is in the outage state. Whether the circuit breaker is open or closed does not affect its in-service/outage state status. (PE/PSE) 859-1987w

outage time (1) The accumulated time one or more components or units are in the outage state during the reporting period. (PE/PSE) 859-1987w
(2) Mean time to repair plus time for logistics and approval. (PE/NP) 933-1999

outage times (reliability data for pumps and drivers, valve actuators, and valves) (reliability data) (1) Out of Service: The average time required to get the failure, analyze it, obtain spare parts, repair and return the item or equipment to service, including planned delays. (2) Restoration: The average time required to get to the failure, analyze it, obtain spare parts, repair, and return the item or equipment to service, excluding planned delays. (3) Repair: The average time required to analyze the failure, repair, and return the item or equipment to service. This excludes planned delays and waiting for spares or tools. (PE/NP) 500-1984w

outage, transient forced *See:* transient forced outage.

outage type Outage occurrences are classified by type according to the urgency with which the outage occurrence is initiated and by how the equipment is restored to service. (PE/PSE) 859-1987w

outbound The direction of RF signal flow away from the head-end and towards the user outlet ports. Referred to in the CATV industry as "downstream" or "forward." (LM/C) 802.7-1989r

outbound queue A queue carrying messages from a Processor to an I/O Unit, or from Processor to Processor. (C/MM) 1212.1-1993

outbound telemetry Communication initiated by a utility or an enhanced service provider (ESP) toward a telemetry interface unit (TIU). (SCC31/AMR) 1390.3-1999, 1390.2-1999, 1390-1995

outcome-oriented simulation A simulation in which the end result is considered more important than the process by which it is obtained; for example, a simulation of a radar system that uses methods far different from those used by the actual radar, but whose output is the same. *Contrast:* process-oriented simulation. (C) 610.3-1989w

outdoor (1) (power and distribution transformers) Suitable for installation where exposed to the weather. (PE/TR) C57.12.80-1978r
(2) Designed for use outside buildings. (SWG/PE) C37.100-1992
(3) (prefix) Designed for use outside buildings and to withstand exposure to the weather. (T&D/PE) 18-1992
(4) Designed for use outside buildings or enclosures. (SWG/PE) C37.40-1993

outdoor arrester An arrester that is designed for outdoor use. (SPD/PE) C62.11-1999

outdoor bushing A bushing in which both ends are in ambient air and exposed to external atmospheric conditions. (PE/TR) C57.19.03-1996

outdoor coupling capacitor A capacitor designed for outdoor service that provides, as its primary function, capacitance coupling to a high-voltage line. *Note:* It is used in this manner to provide a circuit for carrier-frequency energy to and from a high-voltage line and to provide a circuit for power-frequency energy from a high-voltage line to a capacitance potential device or other voltage-responsive device. 31-1944w

outdoor current (or voltage) transformer One of weather-resistant construction, suitable for service without additional protection from the weather. (PE/TR) C57.13-1993

outdoor enclosure (1) (power system communication equipment) An enclosure constructed to protect equipment therein from the weather and accidental contact that would interfere with the successful operation. (PE/PSC) 281-1984w
(2) An enclosure for outdoor application designed to protect against weather hazards such as rain, snow, or sleet. *Note:* Condensation is minimized by use of space heaters. (SWG/PE) C37.100-1992

outdoor-immersed bushing A bushing in which one end is in ambient air and exposed to external atmospheric condition and the other end is immersed in an insulating medium such as oil or gas. (PE/TR) C57.19.03-1996

outdoor-indoor bushing A bushing in which both ends are in ambient air but one end is intended to be exposed to external atmospheric conditions and the other end is intended not to be so exposed. (PE/TR) C57.19.03-1996

outdoor reactor A reactor of weatherproof construction. (PE/TR) C57.16-1996

outdoor regulator A regulator designed for use outside of buildings. (PE/TR) C57.15-1999

outdoor shunt reactor (shunt reactors over 500 kVA) One of weather-resistant construction. (PE/TR) C57.21-1981s

outdoor surge-protective device A surge-protective device that is designed for outdoor use. (SPD/PE) C62.62-2000

outdoor termination A termination intended for use where it is not protected from direct exposure to either solar radiation or precipitation. These are Class 1A, 1B, or 1C terminations. Class 2 terminations may also qualify. (PE/IC) 48-1996

outdoor termination—polluted A termination intended for use where it is not protected from direct exposure to either solar radiation or precipitation, and is exposed to nonstandard (unusual) service conditions such as extreme seacoast salt deposits, solid precipitates, etc. Often requires extra maintenance such as washing or extra creepage length. (PE/IC) 48-1996

outdoor transformer (power and distribution transformers) A transformer of weather-resistant construction suitable for service without additional protection from the weather. (PE/TR) C57.12.80-1978r

outdoor wall bushing A wall bushing on which one or both ends (as specified) are suitable for operating continuously outdoors. *See also:* bushing. 49-1948w

outdoor weatherproof enclosure (series capacitor) An enclosure so constructed or protected that exposure to the weather will not interfere with the successful operation of the equipment contained therein. (T&D/PE) [26]

outer frame (rotating machinery) The portion of a frame into which the inner frame with its assembled core and winding is installed. (PE) [9]

outer jacket A jacket that is extruded over the cable shield. It also may be extruded over both the shield and a supporting messenger cable. *See also:* cable jacket. (PE/PSC) 789-1988w

outer marker (navigation aid terms) A marker facility in an ILS (instrument landing system) which is installed at approximately 5 nmi (nautical miles) (9 km [kilometers]) from the approach end of the runway on the localizer course line to provide height, distance, and equipment functioning checks to aircraft on intermediate and final approach. (AES/GCS) 172-1983w

outfeed A current out of a terminal on a faulted line. Outfeed only occurs on multiterminal or series compensated lines. (PE/PSR) C37.113-1999

outgoing blocking The matching loss on line-to-trunk connections. Line-to-trunk connections include outgoing talking paths, line-to-service circuit connections, and the originating half of an intra-office call when such a connection is directly analogous to an outgoing connection. *See also:* matching loss. (COM/TA) 973-1990w

outgoing traffic (telephone switching systems) Traffic delivered directly to trunks from a switching entity. (COM) 312-1977w

outlet A point on the wiring system at which current is taken to supply utilization equipment. (IA/MT) 45-1998

outlet box A box used on a wiring system at an outlet. *See also:* cabinet. (EEC/PE) [119]

outline font A font defined in terms of mathematical curves that specify the visual representation of each character. *Note:* An outline font has no predetermined size, but rather is scaled to the desired size as needed. *Contrast:* vector font; bit map font. (C) 610.10-1994w

outline lighting An arrangement of incandescent lamps or electric discharge tubing to outline or call attention to certain features such as the shape of a building or the decoration of a window. (NESC/NEC) [86]

out-of-band packet A packet of exceptional data in the data stream. This type of packet may contain special information relating to the data at this position in the data stream, for example, "end of job." (C/MM) 1284.4-2000

out-of-band signaling (1) The transmission of a signal using a frequency that is within the pass band of the transmission facility but outside a frequency range normally used for data transmission. *Contrast:* in-band signaling. (LM/C) 610.7-1995, 802.3-1998

(2) Signaling applications in which the signaling information is outside of the user information channel, whether or not transmitted in a different physical or logical channel from the associated user data, e.g., over different physical paths, in different time-slots in a time division multiplex (TDM) stream. (C/LM/COM) 802.9a-1995w, 8802-9-1996

(3) Signaling which utilizes frequencies within the guard band between channels. This term is also used to indicate the use of a portion of a channel bandwidth provided by the medium such as a carrier channel, but denied to the speech or intelligence path by filters. It results in a reduction of the effective available bandwidth. (PE) 599-1985w

(4) Analog generated signaling that uses the same path as a message and in which the signaling frequencies are lower or higher than those used for the message. (COM) 312-1977w

out-of-frame condition A state that occurs when the receive frame alignment is not consistent with the transmit system frame alignment (for the same direction of transmission). *See also:* change-of-frame alignment. (COM/TA) 1007-1991r

out-of-phase (as prefix to a characteristic quantity) A qualifying term indicating that the characteristic quantity applies to operation of the circuit breaker in out-of-phase conditions. *See also:* out of step. (SWG) 417-1973w

out-of-phase conditions Abnormal circuit conditions of loss or lack of synchronism between the parts of an electrical system on either side of a circuit breaker in which, at the instant of operation of the circuit breaker, the phase angle between rotating phasors representing the generated voltages on either side exceeds the normal value and may be as much as 180° (phase opposition). (SWG) 417-1973w

out-of-roundness (conductor) The difference between the major and minor diameters at any one cross section. *See also:* waveguide. (EEC/REWS) [92]

out of service (1) Occurs when calls cannot be originated or completed due to system failure for an interval greater than a specified time. It is measured as expected long-term average time out of service in minutes per year. (COM/TA) 973-1990w

(2) Lines and equipment are considered out of service when disconnected from the system and not intended to be capable of delivering energy or communications signals. (NESC) C2-1997

out of step A system condition in which two or more synchronous machines have lost synchronism with respect to one another and are operating at different average frequencies. (SWG/PE/PSR) C37.100-1992, C37.90-1978s

out-of-step protection (power system) A form of protection that separates the appropriate parts of a power system, or prevents separation that might otherwise occur, in the event of loss of synchronism. (SWG/PE/PSR) C37.100-1992, C37.90-1978s

outpulsing (telephone switching systems) Pulsing from a sender. (COM) 312-1977w

output (1) (A) (data transmission) Data that have been processed. **(B) (data transmission)** The state or sequence of states occurring on a specified output channel. **(C) (data transmission)** The device or collective set of devices used for taking data out of a device. **(D) (data transmission)** A channel for expressing a state of a device or logic element. **(E) (data transmission)** The process of transferring data from an internal storage to an external storage device. **(F) (software)** Loosely, output data. *Contrast:* input. (C/PE) 599-1985, 610.12-1990

(2) (A) (rotating machinery) (generator). The power (active, reactive, or apparent) supplied from its terminals. **(B) (rotating machinery)** (motor). The power supplied by its shaft. *See also:* asynchronous machine. (PE) [9]

(3) (A) (software) Pertaining to data transmitted to an external destination. **(B) (software)** Pertaining to a device, process, or channel involved in transmitting data to an external destination. **(C) (software)** To transmit data to an external destination. (C) 610.12-1990, 610.10-1994

(4) In an IDEF0 model, that which is produced by a function. (C/SE) 1320.1-1998

(5) *See also:* buffer; display element. (C) 610.5-1990w, 610.6-1991w

output, acoustic *See:* acoustic output.

output angle (1) (gyros) The angular displacement of a gimbal about its output axis with respect to its support. (AES/GYAC) 528-1994

(2) *See also:* radiation angle. (PAS)

output area An area of storage reserved for output data. (C) 610.10-1994w

output argument An argument that has not been specified as an input argument. It is possible for an output argument to have no value at the time a request is made. *Contrast:* input argument. (C/SE) 1320.2-1998

output arrow An arrow or arrow segment that expresses IDEF0 output, i.e., an object type set whose instances are created by a function by transforming the function's input. The arrowtail of an output arrow is attached to the right side of a box. (C/SE) 1320.1-1998

output assertion (software) A logical expression specifying one or more conditions that program outputs must satisfy in order for the program to be correct. *Contrast:* input assertion; loop assertion. *See also:* inductive assertion method. (C) 610.12-1990

output attenuation (signal generators) The ratio, expressed in decibels (dB), of any selected output, relative to the output obtained when the generator is set to its calibration level. *Note:* It may be necessary to eliminate the effect of carrier distortion and/or modulation feedthrough by the use of suitable filters. *See also:* signal generator. (IM/HFIM) [40]

output axis (OA) (accelerometer) (gyros) An axis-of-freedom about which the output of the sensor is measured. A pickoff generates an output signal as a function of the output angle. In an accelerometer, it is sometimes referred to as the hinge axis. (AES/GYAC) 528-1994

output-axis-angular-acceleration drift rate (gyros) Drift rate that is proportional to angular acceleration of the gyro case about the output axis, with respect to inertial space. The relationship of this component of drift rate to angular acceleration can be stated by means of a coefficient having dimensions of angular displacement per unit time divided by angular displacement per unit time squared. (AES/GYAC) 528-1994

output buffer *See:* buffer.

output capacitance (*n*-terminal electron tube) The short-circuit transfer capacitance between the output terminal and all other terminals, except the input terminal, connected together. *See also:* electron-tube admittances. (ED) 161-1971w, [45]

output-capacitor discharge time (power supply) The interval between the time at which the input power is disconnected and the time when the output voltage of the unloaded regulated power supply has decreased to a specified safe value. *See also:* regulated power supply. 209-1950w

output channel A channel for transferring data from a device or logic gate to an external component. *See also:* input channel; input-output channel. (C) 610.10-1994w

output circuit (1) (protective relay system) An output from a relay system which exercises direct or indirect control of a power circuit breaker, such as trip or close. (PE/PSR) C37.90-1978s
(2) (protective relay system) A circuit from a relay system that exercises direct or indirect control of power apparatus such as tripping or closing of a power circuit breaker. (PE/PSR) C37.90.1-1989r

output control characteristics (thyristor) Output operating characteristics that can be deliberately selected or controlled, or both. (IA/IPC) 428-1981w

output control range (thyristor) The continuous range over which the output of a power controller can be changed by control signal input. (IA/IPC) 428-1981w

output current (self-commutated converters) (converters having ac output) The total rms (root-mean-square) current from the output terminals. (IA/SPC) 936-1987w

output-dependent overshoot and undershoot Dynamic regulation for load changes. (AES) [41]

output device A device in a computer system used for presenting information to the user. *Note:* Common output devices include printers, display devices and plotters. *Synonym:* output unit. *Contrast:* input device. *See also:* input-output device. (C) 610.10-1994w

output electrode (electron tube) The electrode from which is received the amplified, modulated, detected, etc., voltage. *See also:* electron tube. (ED) [45], [84]

output enable (semiconductor memory) The inputs that when false cause the output to be in the OFF or high impedance state. This pin must be true for the output to be in any other state. (TT/C) 662-1980s

output factor The ratio of the actual energy output, in the period of time considered, to the energy output that would have occurred if the machine or equipment had been operating at its full rating throughout its actual hours of service during the period. *See also:* generating station. (T&D/PE) [10]

output frequency stability (inverters) The deviation of the output frequency from a given set value. *See also:* self-commutated inverters. (IA) [62]

output gap (electron tube) (traveling-wave tubes) An interaction gap by means of which usable power can be abstracted from an electron stream. (ED) [45]

output hold time *See:* hold time.

output impedance (1) (analog computer) The impedance presented by the transducer to a load. (C) 165-1977w
(2) (self-commutated converters) (converters having ac output) The impedance presented by the converter to the load for specified frequencies. (IA/SPC) 936-1987w
(3) The electrical impedance at an output terminal of a circuit or device, as it appears to the circuit that uses the output signal. (C) 610.10-1994w
(4) (A) (device, transducer, or network) The impedance presented by the output terminals to a load. *Notes:* 1. Output impedance is sometimes incorrectly used to designate load impedance. 2. This is a frequency-dependent function, and is used to help describe the performance of the power supply and the degree of coupling between loads. *See also:* self-impedance. **(B) (power supplies)** The effective dynamic output impedance of a power supply is derived from the ratio of the measured peak-to-peak change in output voltage to a measured peak-to-peak change in load alternating current. Output impedance is usually specified throughout the frequency range from direct current to 100 kilohertz. *See also:* self-impedance. (AES) [41]
(5) The output electrode impedance at the output electrodes. *See also:* self-impedance. (ED) [45]
(6) (transformer-rectifier system) Internal impedance in ohms measured at the direct current terminals when the rectifier is continuously providing direct current to a load. This impedance is perferably expressed as a curve of impedance in ohms versus frequency, over the frequency range of interest to the application. (PEL/ET) 295-1969r
(7) (Hall generator) The impedance between the Hall terminals. (MAG) 296-1969w

output media Media that are generated as output; for example, paper reports, or magnetic tapes. *Contrast:* input media. (C) 610.10-1994w

output phase displacement (power inverters that have polyphase output) (inverters) The angular displacement between fundamental phasors. *See also:* self-commutated inverters. (IA) [62]

output pin A component pin that drives signals onto external connections. (TT/C) 1149.1-1990

output power (1) (general) The power delivered by a system or transducer to its load. (SP) 151-1965w
(2) (electron tube or valve) The power supplied to the load by the electron tube or valve at the output electrode. *See also:* ON period. (ED) [45], [84]

output primitives Primitives that include source statements, function points, and documents. (C/SE) 1045-1992

output pulse (accelerometer) A pulse that represents the minimum unit of velocity increment ($g \cdot$ s, m/s). (AES/GYAC) 528-1994, 530-1978r

output pulse amplitude (digital delay line) Peak amplitude of the output doublet which is obtained across the specified output load for a given amplitude of input step. (UFFC) [22]

output pulse duration (digital delay line) Time spacing between the 10 percent amplitude point of the rise of the first peak to the 10 percent amplitude point of the fall of the second peak. (UFFC) [22]

output pulse shape (pulse transformers) Load current pulse flowing in a winding or voltage pulse developed across a winding in response to application of an input pulse. The shape of the output pulse is described by a current- or voltage-time relationship. The following definitions for the input pulse shape apply to the output pulse shape: pulse amplitude; rise time; pulse duration; fall time; trailing edge; tilt; overshoot; backswing; return swing; rolloff; ringing; leading edge linearity; quiescent value; leading edge; pulse top; trailing edge. Typically, a prominent feature of the output pulse is an accentuated backswing (last transition overshoot), ABS. (PEL/MAG/ET) 390-1987r, 391-1976w

output queue The database that the client of the MT interface uses to convey objects to the service. (C/PA) 1224.1-1993w

output range (accelerometer) (gyros) The product of input range and scale factor. (AES/GYAC) 528-1994

output reference axis (ORA) (accelerometer) (gyros) The direction of an axis defined by the case mounting surfaces or external case markings or both. It is nominally parallel to the output axis. (AES/GYAC) 528-1994

output resonator (electron tube) (catcher) A resonant cavity, excited by density modulation of the electron beam, that supplies useful energy to an external circuit. *See also:* velocity-modulated tube. (ED) [45], [84]

output (reverse transfer) impedance (of a power source) Similar to forward transfer impedance, but it describes the characteristic impedance of the power source as seen from the load, looking back at the source. *See also:* forward transfer impedance. (IA/PSE) 1100-1999

output ripple voltage (regulated power supply) The portion of the output voltage harmonically related in frequency to the input voltage and arising solely from the input voltage. Note: Unless otherwise specified, percent ripple is the ratio of root-mean-square value of the ripple voltage to the average value of the total voltage expressed in percent. In television, ripple voltage is usually expressed explicitly in peak-to-peak volts to avoid ambiguity. *See also:* regulated power supply. 209-1950w

output signal (1) (hydraulic turbines) The physical reaction of of any element of a control system to an input signal. (PE/EDPG) 125-1977s
(2) (control system feedback) A signal delivered by a system or element. *See also:* feedback control system. (NESC/PE/EDPG) 421-1972s, [86]

output span (accelerometer) (gyros) The algebraic difference between the upper and lower values of the output range. (AES/GYAC) 528-1994

output-structure transit time That portion of the photomultiplier transit time occurring with the output structure. (NPS) 398-1972r

output terminal (A) A terminal used to display or generate output. **(B)** A point in a system or communication network at which data can leave the system *Contrast:* input terminal. (C) 610.10-1994

output torque without excitation (electric coupling) The torque an electric coupling will transmit or develop with zero excitation. (EM/PE) 290-1980w

output-transfer function (control system feedback) (closed loop) The transfer function obtained by taking the ratio of the Laplace transform of the output signal to the Laplace transform of the input signal. *See also:* feedback control system. (IM/PE/EDPG) [120], [3]

output unit *See:* output device.

output variable A variable delivered by a system or element. *See also:* feedback control system. (IM/PE/EDPG) [120], [3]

output voltage (converters having ac output) (self-commutated converters) The fundamental rms (root-mean-square) voltage (unless otherwise specified for a particular load) between the output terminals. (IA/SPC) 936-1987w

output voltage regulation (power supply) The change in output voltage, at a specified constant input voltage, resulting from a change of load current between two specified values. *See also:* regulated power supply. 209-1950w

output voltage stabilization (power supply) The change in output voltage, at a specified constant load current, resulting from a change of input voltage between two specified values. *See also:* regulated power supply. 209-1950w

output voltage versus input voltage characteristics Ferroresonant regulators may have output versus input characteristics as shown below and in the figure under jump resonance. (PEL) 449-1998

output winding (1) (secondary windings) The winding(s) from which the output is obtained. *See also:* magnetic amplifier. (Std100) [123]
(2) The winding of the ferroresonant transformer used to provide the regulated output voltage. *Note:* The output winding is wound on the secondary section of the core and separated from the primary by a magnetic shunt. (PEL) 449-1998

outrigger (of a switching-device terminal) An attachment that is fastened to or adjacent to the terminal pad of a switching device to maintain electrical clearance between the conductor and other parts or, when fastened adjacent, to relieve mechanical strain on the terminal, or both. (SWG/PE) C37.100-1992

outside plant (communication practice) That part of the plant extending from the line side of the main distributing frame to the line side of the station or private-branch-exchange protector or connecting block, or to the line side of the main distributing frame in another central office building. (PE/EEC) [119]

outside space block *See:* end finger.

outstanding directory operation A directory operation, invoked asynchronously (i.e., with **asynchronous** = **true** in the context), that has not yet been the subject of an invocation of the `ds_abandon` interface operation or the `ds_receive_results` interface operation. (C/PA) 1326.2-1993w, 1224.2-1993w, 1328.2-1993w, 1327.2-1993w

outward-wats service (telephone switching systems) A flat-rate or measured-time direct distance dialing service for defined geographical groups of numbering plan areas. (COM) 312-1977w

oven (analog computer) An enclosure and associated sensors and heaters for maintaining components at a controlled and usually constant temperature. (C) 165-1977w, 166-1977w

oven, wall-mounted *See:* wall-mounted oven.

overall The acoustic output level of a telephone set due to an acoustic input to another telephone set to which it is connected by a test circuit. (COM/TA) 269-1992

over-all electrical efficiency (dielectric and induction heater) The ratio of the power absorbed by the load material to the total power drawn from the supply lines. *See also:* induction heating; load-circuit efficiency. (IA) 54-1955w, 169-1955w

over-all generator efficiency (thermoelectric device) The ratio of (A) electric power output to (B) thermal power input. *See also:* thermoelectric device. (ED) [46]

overall objective loudness rating (OOLR) (loudness ratings of telephone connections)

$$\mathrm{OOLR} = -20 \log_{10} \frac{S_E}{S_M}$$

where
S_M = sound pressure at the mouth reference point (in pascals)
S_E = sound pressure at the ear reference point (in pascals)

(COM/TA) 661-1979r

overall power efficiency (laser maser) The ratio of the useful power output of the device to the total input power. (LEO) 586-1980w

overall regulation (power supplies) The maximum amount that the output will change as a result of the specified change in line voltage, output load, input frequency, temperature, or time. *Note:* Line regulation, load regulation, effect of frequency variation, stability, and temperature coefficient are defined and usually specified separately as follows:
— *Line regulation.* The maximum amount that the output voltage or current will change as the result of a specified change in line voltage. (Regulation is given either as a percentage of the rated output voltage or current, or as an absolute change, ΔE or ΔI.)
— *Load regulation.* The maximum amount that the output voltage will change as the result of a specified change in load current. (Regulation is given either as a percentage of the rated output voltage or as an absolute change, ΔE.)
— *Frequency regulation.* The maximum amount that the output voltage or current will change as the result of a specified change in line frequency. (Regulation is given either as a percentage of the rated output voltage or current, or as an absolute change, ΔE or ΔI.)
— *Temperature coefficient (power supplies).* The percent change in the output voltage or current as a result of a 1°C change in the ambient operating temperature (percent per degree Celsius).
— *Long-term stability (LTS) (power supplies).* The change in output voltage or current as a function of time, at constant line voltage, load, and ambient temperature (sometimes referred to as *drift*).

(PEL) 449-1998

overbilling error Occurs when a call is billed more that it should be. (COM/TA) 973-1990w

overbunching (electron tube) The bunching condition produced by the continuation of the bunching process beyond the optimum condition. (ED) 161-1971w

overburden (earth conductivity) The surface layers or regions of the earth that are water bearing and are subject to weathering. They comprise predominantly sand, gravel, clays, and poorly consolidated rocks. (COM) 365-1974w

overcast sky (illuminating engineering) A sky which has 100 percent cloud cover; the sun is not visible. (EEC/IE) [126]

overcharge The forcing of current through a battery after it has been fully recharged. (PV) 1013-1990, 1144-1996

overcompounded A qualifying term applied to a compound-wound generator to denote that the series winding is so proportioned that the terminal voltage at rated load is greater than at no load. (EEC/PE) [119]

overcurrent (1) Any current in excess of the rated current of equipment or the ampacity of a conductor. It may result from overload (see definition) short circuit, or ground fault. A current in excess of rating may be accommodated by certain equipment and conductors for a given set of conditions. Hence the rules for overcurrent protection are specific for particular situations. (NESC/NEC) [86]
(2) (packaging machinery) An abnormal current greater than the full-load value. (IA/PKG) 333-1980w

overcurrent protection (1) (thyristor) (power and distribution transformers) A form of protection(s) that operates when current exceeds a predetermined value.
(SWG/IA/PE/IPC/TR) 428-1981w, C57.12.80-1978r
(2) Protection of the battery charger against excessive current, including short circuit current. (IA/PSE) 602-1996
(3) A form of protection that operates when current exceeds a predetermined value. (SWG/PE) C37.100-1992
(4) (overload protection) The effect of a device, operative on excessive current (but not necessarily on short circuit), to cause and maintain the interruption of current flow to the device governed. (IA/MT) 45-1998

overcurrent protective device (packaging machinery) A device operative on excessive current that causes and maintains the interruption of power in the circuit.
(IA/PKG) 333-1980w

overcurrent relay A relay that operates when its input current exceeds a predetermined value.
(SWG/PE/PSR) C37.100-1992, C37.90-1978s

overcurrent release (trip) A release that operates when the current in the main circuit is equal to or exceeds the release setting. (SWG/PE) C37.100-1992

overcutting (disk recording) The effect of excessive level characterized by one groove cutting through into an adjacent one. *See also:* phonograph pickup. (SP) [32]

overdamped A degree of damping that is more than sufficient to prevent the oscillation of the output following an abrupt stimulus. *Note:* For a linear second order system the roots of the characteristic equation must then be real and unequal. *See also:* control system; feedback.
(CAS/IA/ICTL/IAC) [13], [60]

overdamping (aperiodic damping) The special case of damping in which the free oscillation does not change sign. A damped harmonic system is overdamped if $F^2 > 4MS$.
(Std100) 270-1966w

over-erased cell An unselected cell that has excessive source-drain leakage current resulting from an erase operation that has reduced the threshold of the cell below the applied control gate voltage. (ED) 1005-1998

over-erased device A device that cannot be read or programmed correctly because of excessive over-erase leakage.
(ED) 1005-1998

over-erase leakage The current on the bit-line caused by over-erased cell(s). (ED) 1005-1998

over-erase recover A custom programming algorithm to raise the threshold of an over-erased cell. (ED) 1005-1998

overfilled launch The overfilled launch condition that excites both radial and azimuthal modes. (C/LM) 802.3-1998

overflow (A) (mathematics of computing) The condition that arises when the result of an arithmetic operation exceeds the capacity of the number representation system used in a digital computer. **(B) (mathematics of computing)** The carry digit arising from this condition. *Synonym:* arithmetic overflow.
(C) 1084-1986

overflow area A physical location in which data are placed when there is no available space in the primary data area. Overflow areas may be allocated within stored record, physical blocks, disk tracks, or disk cylinders.
(C) 610.5-1990w

overflow error The error caused by an overflow condition in computer arithmetic. (C) 1084-1986w

overflow exception An exception that occurs when the result of an arithmetic operation exceeds the size of the storage location designated to receive it. *See also:* addressing exception; underflow exception; operation exception; protection exception; data exception. (C) 610.12-1990

overflow loss *See:* matching loss.

overflow sequential access method (OSAM) An access method for handling data overflow from ISAM.
(C) 610.5-1990w

overflow traffic (telephone switching systems) That part of the offered traffic that cannot be carried by a group of servers.
(COM) 312-1977w

overhang packing (rotating machinery) Insulation inserted in the end region of the winding to provide spacing and bracing. *See also:* stator; rotor. (PE) [9]

overhaul Work done with the objective of repairing or replacing parts that are found to be out of tolerance by inspection, or test, or examination, or as required by equipment maintenance manual, in order to restore the component and/or the circuit breaker to an acceptable condition. (SWG/PE) C37.10-1995

overhead bits (1) Bits assigned at the source that remain with the information payload until the payload reaches the sink. The overhead bits are used for functions associated with transporting the payload. (COM/TA) 1007-1991r
(2) In data communications, additional bits transmitted for control framing, synchronization, and error checking purposes. (C/Std100) 610.7-1995

overhead electric hoist A motor-driven hoist having one or more drums or sheaves for rope or chain, and supported overhead. It may be fixed or traveling. *See also:* hoist. (EEC/PE) [119]

overhead ground wire (OHGW) (1) (lightning protection) (conductor stringing equipment) Multiple grounded wire or wires placed above the phase conductor for the purpose of intercepting direct strokes in order to protect the phase conductors from the direct strokes. *Synonyms:* shield wire; static wire; sky wire; earth wire. (T&D/PE) 524a-1993r, 524-1992r, 1048-1990
(2) A grounded, bare conductor suspended horizontally between supporting rods or masts to provide protection from lightning strikes for structures, equipment, or suspended conductors within the zone of protection created by the combination of the masts and the overhead ground wire. (PE/EDPG) 665-1995
(3) Grounded wire or wires placed above phase conductors for the purpose of intercepting direct strokes in order to protect the phase conductors from the direct strokes. They may be grounded directly or indirectly through short gaps. *See also:* direct-stroke protection. (SPD/PE/T&D) C62.23-1995, 1243-1997, 1410-1997

overhead insulated ground (static or sky) wire-coupling protector (wire-line communication facilities) A device for protecting carrier terminals which are used in conjunction with overhead, insulated, ground wires (static wire) of a power transmission line. (PE/PSC) 487-1980s

overhead line charging current Current supplied to an unloaded overhead line. *Note:* Current is expressed in rms amperes. (SWG/PE) C37.100-1992

overhead operation *See:* housekeeping operation.

overhead power line, corona (overhead-power-line corona and radio noise) Corona occurring at the surfaces of electrodes during the positive or negative polarity of the power-line voltage. *Notes:* 1. Surfaces irregularities such as stranding, nicks, scratches, and semiconducting or insulating protrusions are usual corona sites. 2. Dry or wet airborne particles in proximity of electrodes may cause corona discharges. 3. Weather has a pronounced influence on the occurrence and characteristics of overhead-power-line corona. (T&D/PE) 539-1979s

overhead structure (elevators) All of the structural members, platforms, etc., supporting the elevator machinery, sheaves, and equipment at the top of the hoistway. *See also:* elevator. (PE/EEC) [119]

overhead system service-entrance conductors The service conductors between the terminals of the service equipment and a point usually outside the building, clear of building walls, where joined by tap or splice to the service drop. (NESC/NEC) [86]

overhead time The amount of time a computer system spends performing tasks that do not contribute directly to the progress of any user task; for example, time spent tabulating computer resource usage for billing purposes. (C) 610.12-1990

overjacket A polymeric sheath, sometimes fabric reinforced, applied over the metallic covering. (IA/PC) 515.1-1995

overlap The distance the control of one signal extends into the territory that is governed by another signal or other signals. *See also:* neutral zone. (EEC/PE) [119]

overlap angle (1) (gas tube) The time interval, in angular measure, during which two consecutive arc paths carry current simultaneously. (ED) [45], [84]
(2) (rectifier circuits) *See also:* commutating angle. (IA) [62]

overlap control *See:* two-step control system.

overlap interval (self-commutated converters) (circuit properties) The time interval during which two commutating converter branches are carrying principal current simultaneously. (IA/SPC) 936-1987w

overlapped execution A mode of operation in which the execution of one instruction overlaps the fetch and decode of the next to be executed. *See also:* pipelining. (C) 610.10-1994w

overlapping protection A situation in which the protected zone of one relay overlaps the protected zone of another relay (usually done to ensure protection of equipment at the border of a protected zone). This is often necessary due to the location of current transformers (CTs) on equipment. (PE/PSR) C37.113-1999

overlapping register set A set of registers, only part of which is available to an application at any given time. *Note:* A subset of the available registers is shared with the calling routine and a subset may be shared with any routines called by the current routine. (C) 610.10-1994w

overlap testing (nuclear power generating station) Overlap testing consists of channel, train, or load group verification by performing individual tests on the various components and subsystems of the channel, train, or load group. The individual component and subsystem tests check common parts of adjacent subsystems, such that the entire channel, train, or load group is verified by testing of individual components or subsystems and by repetitive testing of common parts of adjacent subsystems. (PE/NP) 338-1987r

overlap X (facsimile) The amount by which the recorded spot X dimension exceeds that necessary to form a most nearly constant density line. *Note:* This effect arises in that type of equipment which responds to a constant density in the subject copy by a succession of discrete recorded spots. *See also:* recording. (COM) 168-1956w

overlap Y (facsimile) The amount by which the recorded spot Y dimension exceeds the nominal line width. *See also:* recording. (COM) 168-1956w

overlay (1) (A) (software) A storage allocation technique in which computer program segments are loaded from auxiliary storage to main storage when needed, overwriting other segments not currently in use. **(B) (software)** A computer program segment that is maintained in auxiliary storage and loaded into main storage when needed, overwriting other segments not currently in use. **(C) (software)** To load a computer program segment from auxiliary storage to main storage in such a way that other segments of the program are overwritten. (C) 610.12-1990
(2) (transmission lines) A layer of dielectric material placed upon a single or coupled planar transmission line. It is often used to make the two modes of coupled transmission lines have phase velocities nearly the same. (MTT) 1004-1987w

overlay supervisor A routine that controls the sequencing and positioning of overlays. (C) 610.12-1990

overload (1) (power operations) Loading in excess of normal rating of equipment. (PE/PSE) 858-1987s
(2) (protection and coordination of industrial and commercial power systems) Generally used in reference to an overcurrent that is not of sufficient magnitude to be termed a short circuit. An overload is normally that overcurrent value from 100 percent of fuse rating up to ten times fuse rating. *See also:* short circuit. (IA/PSP) 242-1986r
(3) (power and distribution transformers) Output of current power, or torque, by a device, in excess of the rated

output of the device on a specified rating basis. (PE/TR) C57.12.80-1978r

(4) Operation of equipment in excess of normal, full load rating, or of a conductor in excess of rated ampacity which, when it persists for a sufficient length of time, would cause damage or dangerous overheating. A fault, such as a short circuit or ground fault, is not an overload. *See also:* overcurrent. (NEC/NESC/C/MM) [86], 1296-1987s

(5) (radiation protection) Response of less than full scale (that is, maximum scale reading) when exposed to radiation intensities greater than the upper detection limit. (NI) N323-1978r

(6) (test, measurement, and diagnostic equipment) To exceed the rated capacity of. (MIL) [2]

(7) (thyristor power computer) A condition existing when the load current exceeds the continuous rating of the converter unit in magnitude or time, or both, but the conduction cycles and waveforms remain essentially normal. (IA/IPC) 444-1973w

(8) (software) To assign an operator, identifier, or literal more than one meaning, depending upon the data types associated with it at any given time during program execution. (C) 610.12-1990

(9) A condition existing in an analog computer, within or at the output of a computing element, that causes a substantial computing error because of the voltage or current saturation of one or more of the parts of the computing element. *Note:* This condition is similar to an overflow of an accumulator in a digital computer. (C) 610.10-1994w

(10) A condition in which the maximum current of the power supply is exceeded. (PEL) 1515-2000

overload capacity (1) The current, voltage, or power level beyond which permanent damage occurs to the device considered. This is usually higher than the rated load capacity. *Note:* To carry load greater than the continuous rating, may be acceptable for limited use. (AP/ICTL/ANT) 145-1983s

(2) (accelerometer) The maximum acceleration to which an accelerometer may be subjected beyond the normal operating range without causing a permanent change in the specified performance characteristics. (AES/GYAC) 528-1994

overload capacity factor The number by which a maximum load is multiplied to assure that the system does not fail when loaded beyond the design load. (T&D/PE) 1307-1996

overload characteristic That portion of the output voltage versus output current characteristic of ferroresonant regulators from rated current to short-circuit current. See figures below.

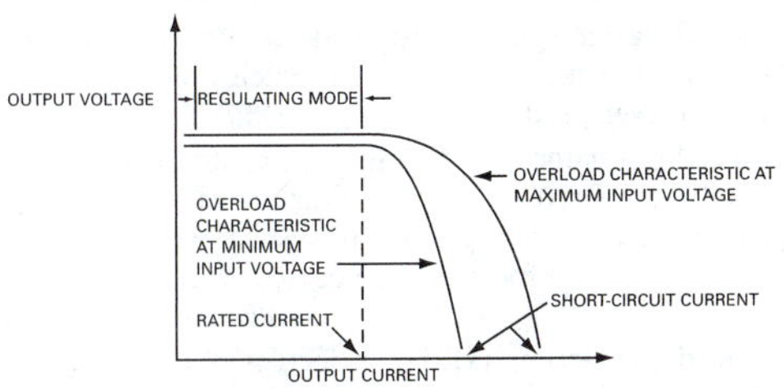

Overload characteristic with unsaturated series inductance

Overload characteristic

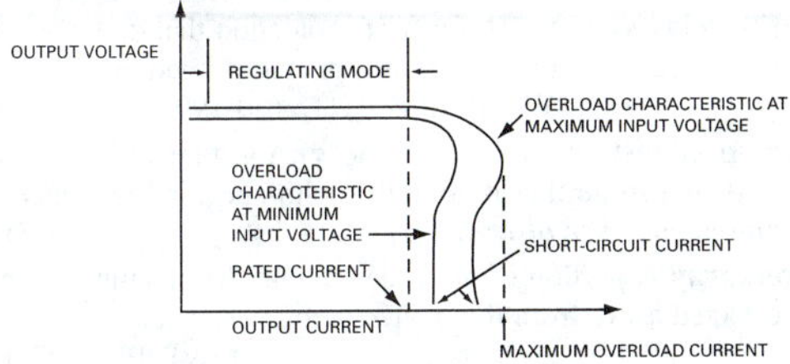

Overload characteristic with saturated series inductance

Overload characteristic

(PEL) 449-1998

overload detection A means to detect excessive overload of series capacitor bank components and to initiate an alarm signal, the closing of the associated bypass switch, or both. (T&D/PE) 824-1994

overload factor The ratio of the maximum value of a signal for which the operation of the predetector circuits of the receiver does not depart from linearity by more than one decibel, to the value corresponding to full-scale deflection of the indicating instrument. *See also:* electromagnetic compatibility. (EMC/INT) [53], [70]

overload level (system or component) That level above which operation ceases to be satisfactory as a result of signal distortion, overheating, or damage. *See also:* level. (SP) 151-1965w

overload ON-state current (thyristor) An ON-state current of substantially the same wave shape as the normal ON-state current and having a greater value than the normal ON-state current. *See also:* principal current. (IA) [62]

overload point, signal *See:* signal overload point.

overload protection The effect of a device operative on excessive current, but not necessarily on short circuit, to cause and maintain the interruption of current flow to the device governed. *See also:* overcurrent protection. (IA/ICTL/IAC) [60]

overload pulse (x-ray energy spectrometers) An signal that drives a section of the amplifying chain into saturation. (NPS/NID) 759-1984r

overload recovery time (diode-type camera tube) A measure of the ability of the camera tube to recover from a specified overload signal, defined as the increased time required for the readout process to reach its nonoverload third-field value. Units: seconds or numbers of fields. (ED) 503-1978w

overload relay (1) (general) A relay that responds to electric load and operates at a preset value of overload. *Note:* Overload relays are usually current relays but they may be power, temperature, or other relays. (SWG/PE) C37.100-1981s

(2) An overcurrent relay that functions at a predetermined value of overcurrent to cause the disconnection of the power supply. *Note:* An overload relay is intended to protect the motor or controller and does not necessarily protect itself. (IA/MT) 45-1998

overmoded waveguide A waveguide used to propagate a single mode, but capable of propagating more than one mode at the frequency of interest. *See also:* waveguide. (MTT) 146-1980w

overpotential *See:* overvoltage.

overprint In text formatting, to print the same or different characters at the same position on an output page. Used to create bold-face type, underlining, and special characters. (C) 610.2-1987

overpunch To punch holes into a column of a punch card that already contains one or more holes. *Note:* Often used to represent special characters. (C) 610.10-1994w

overrange (1) (noun) (system or element) Any excess value of the response above its nominal full-scale value, or deficiency below the nominal minimum value. (PE/EDPG) [3]

(2) (test, measurement, and diagnostic equipment) An input to a measuring device which exceeds in magnitude the capability of a given range. (MIL) [2]

overrange velocity storage (accelerometer) (digital accelerometer) The velocity information that can be stored in the accelerometer during the application of an acceleration exceeding its input range. (AES/GYAC) 530-1978r, 528-1994

overreaching protection A form of protection in which the relays at one terminal operate for faults beyond the next terminal. They may be constrained from tripping until an incoming signal from a remote terminal has indicated whether the fault is beyond the protected line section. (SWG/PE/PSR) C37.100-1992, C37.90-1978s

overreach (of a relay) The extension of the zone of protection beyond that indicated by the relay setting. (SWG/PE/PSR) C37.100-1992, C37.90-1978s

override (1) (general system) (microcomputer system bus) A bus master overrides the bus control logic when it is necessary to guarantee itself back-to-back bus cycles. This is called overriding the bus, temporarily preventing other masters from using the bus. (MM/C) 796-1983r
(2) The ability of a property in a subclass to respecify the realization of an inherited property of the same name while retaining the same meaning. (C/SE) 1320.2-1998

overriding property A property in a subclass that has the same meaning and signature as a similarly named property in one of its superclasses, but has a different realization. (C/SE) 1320.2-1998

overshoot (1) (pulse transformers) (first transition overshoot, a_{os}) The amount by which the first maximum occurring in the pulse-top region exceeds the straight-line segment fitted to the top of the pulse in determining A_M. It is expressed in amplitude units or as a percentage of A_M. (PEL/ET) 390-1987r
(2) (pulse terminology) A distortion that follows a major transition. *See also:* preshoot. (IM/WM&A) 194-1977w
(3) (oscilloscopes) In the display of a step function (usually of time), that portion of the waveform which, immediately following the step, exceeds its nominal or final amplitude. (IM) 311-1970w
(4) (A) (data transmission) *(Instrument).* The amount of the overtravel of the indicator beyond its final steady deflection when a new constant value of the measured quantity is suddenly applied to the instrument. The overtravel and deflection are determined in angular measure and the overshoot is expressed as a percentage of the change in steady deflection. *Notes:* 1. Since in some instruments the percentage depends on the magnitude of the deflection, a value corresponding to an initial swing from zero to end scale is used in determining the overshoot for rating purposes. 2. Overshoot and damping factor have a reciprocal relationship. The percentage overshoot may be obtained by dividing 100 by the damping factor. **(B) (data transmission)** *(Power supplies).* A transient rise beyond regulated output limits, occurring when the alternating current power input is turned on or off, and for line or load step changes.

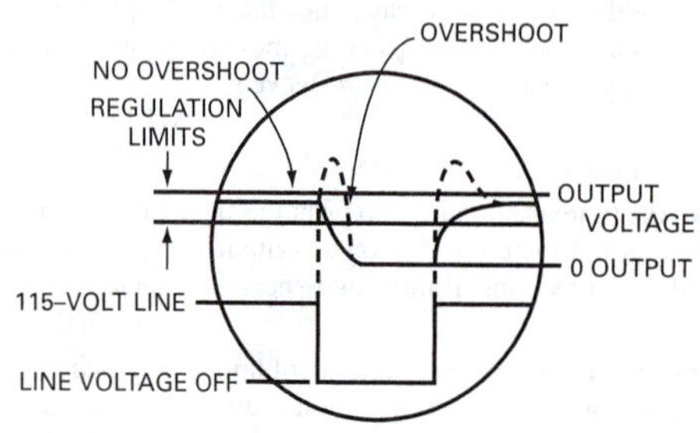

Scope view of turn-off/turn-on effects of a power supply, showing overshoot.

overshoot

(PE) 599-1985
(5) (television) That part of a distorted wave front characterized by a rise above (or a fall below) the final value, followed by a decaying return to that final value. *Note:* Generally overshoots are produced in transfer devices having excessive transient response. (BT/AV) 201-1979w
(6) The value by which a lightning impulse exceeds the defined crest value. (PE/PSIM) 4-1995
(7) The maximum amount by which the step response exceeds the topline, specified as a percentage of (recorded) pulse amplitude. (IM/WM&A) 1057-1994w

overshoot duration (low voltage varistor surge arresters) The time between the point at which the wave exceeds the clamping voltage and the point at which the voltage overshoot has decayed to 50 percent of its peak. For the purpose of this definition, clamping voltage is defined with an 8 × 29 μs current waveform of the same peak current amplitude as the waveform used for the overshoot duration. (PE) [8]

overshoot response time (low voltage varistor surge arresters) The time between the point at which the wave exceeds the clamping voltage and the peak of the voltage overshoot. For the purpose of this definition, clamping voltage is defined with an 8 × 20 μ current waveform of the same peak current amplitude as the waveform used for the response time. (PE) [8]

overshoot switch-off (transformer-rectifier system) The transient output voltage pulse occurring as the result of deenergization of the core on switch-off. (PEL/ET) 295-1969r

overshoot switch-on (transformer-rectifier system) The transient voltage on the output direct voltage following the completion of capacitor charging in the direct current circuit. It may be expressed as a percentage of excess over the steady-state direct voltage. (PEL/ET) 295-1969r

overshoot, system *See:* system overshoot.

overshoot transient *See:* transient deviation.

over-sized packet *See:* long packet.

oversized waveguide A waveguide operated in its dominant mode, but far above cutoff; sometimes termed quasioptical waveguide. (MTT) 146-1980w

over-size insulation (electrical heat tracing for industrial applications) A term applied to thermal insulation when the inside diameter of the thermal insulation must be larger than the nominal outside diameter of a particular pipe so as to accommodate the heating cable. (BT/IA/AV) 152-1953s, 515-1997

overslung car frame A car frame to which the hoisting-rope fastenings or hoisting-rope sheaves are attached to the crosshead or top member of the car frame. *See also:* hoistway. (EEC/PE) [119]

overspeed (1) (hydraulic turbines) Any speed in excess of rated speed expressed as a percent of rated speed. (PE/EDPG) 125-1977s
(2) (hydroelectric power plants) Any speed in excess of rated speed. (PE/EDPG) 1020-1988r

overspeed and overtemperature protection system (gas turbines) The overspeed governor, overtemperature detector, fuel stop valve(s), blow-off valve, other protective devices and their interconnections to the fuel stop valve, and to the blow-off valve, if used, that are required to shut off all fuel flow and shut down the gas turbine. (PE/EDPG) 282-1968w

overspeed device (power system device function numbers) Usually a direct-connected speed switch which functions on machine overspeed. (PE/SUB) C37.2-1979s

overspeed governor (gas turbines) A control element that is directly responsive to speed and that actuates the overspeed and overtemperature protection system when the turbine reaches the speed for which the device is set. (PE/EDPG) 282-1968w

overspeed protection (1) The effect of a device operative whenever the speed rises above a preset value to cause and maintain an interruption of power to the protected equipment or a reduction of its speed. *See also:* relay. (IA/ICTL/IAC) [60]
(2) (relay systems) A form of protection that operates when the speed of rotation exceeds a predetermined value. (SWG/PE/PSR) C37.100-1992, C37.90-1978s

overspeed test (rotating machinery) A test on a machine rotor to demonstrate that it complies with specified overspeed requirements. *See also:* rotor. (PE) [9]

overspray A portion of the water stream that is unintentionally directed away from the device being washed. (T&D/PE) 957-1995

overtemperature (rotating machinery) Unusually high temperature from causes such as overload, high ambient temperature, restricted ventilation, etc. (PE/EM) 432-1976s

overtemperature detector (gas turbines) The primary sensing element that is directly responsive to temperature and that actuates the overspeed and overtemperature protection system when the turbine temperature reaches the value for which the device is set. (PE/EDPG) 282-1968w

overtemperature protection (1) (power supplies) A thermal relay circuit that turns off the power automatically should an overtemperature condition occur. (AES) [41]

(2) A feature in a power supply that senses and responds to an over-temperature condition. (PEL) 1515-2000

overtesting Testing beyond requirements. (PE/SUB) 693-1997

over-the-horizon radar (OHR, OTH) (1) (navigation aid terms) Radar using sufficiently low carrier frequencies, usually in the high-frequency (hf) band, so that ground-wave or ionospherically refracted sky-wave propagation can allow detection far beyond the ranges allowed by line-of-sight propagation. (AES/GCS) 172-1983w

(2) Radar using sufficiently low carrier frequencies, usually in the high-frequency (HF) band typically from about 5–30 MHz, so that ionospherically refracted sky-wave propagation can allow detection at ranges (nominally from perhaps 1000–4000 km) far beyond the ranges allowed by line-of-sight propagation. At HF, the surface wave, or ground wave, mode of propagation can allow detection of low-altitude targets at ranges from perhaps 40–200 km, depending on the size of the target and the radar. (AES) 686-1997

overtone *See:* harmonic.

overtone-type piezoelectric-crystal unit *(A)* An overtone driven by the action of the piezoelectric effect; *(B) (crystal unit)* A resonator constructed from a piezoelectric crystal material and designed to operate in the vicinity of an overtone of that device. (CAS) [13]

overtravel (of a relay) The amount of continued movement of the responsive element after the input is changed to a value below pickup.

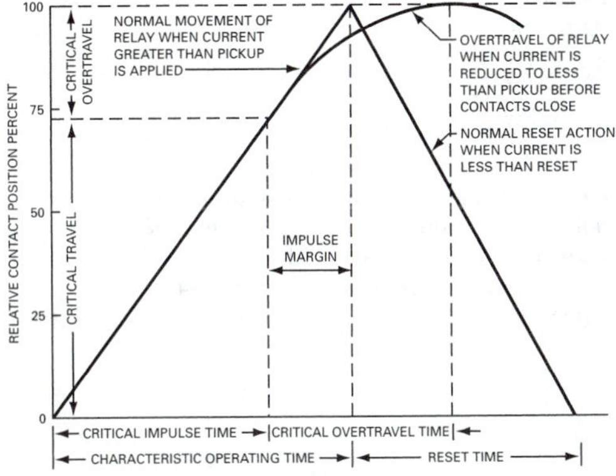

overtravel

(SWG/PE) C37.100-1992

overvoltage (1) (rotating machinery) An abnormal voltage higher than the normal service voltage, such as might be caused from switching or lightning surges. (PE/EM) 432-1976s

(2) (radiation counter tubes) The amount by which the applied voltage exceeds the Geiger-Mueller threshold. (ED) [45]

(3) (electrochemistry) The displacement of an electrode potential from its equilibrium (reversible) value because of flow of current. *Note:* This is the irreversible excess of potential required for an electrochemical reaction to proceed actively at a specified electrode, over and above the reversible potential characteristic of that reaction. (IA) [59]

(4) (rotating machinery) (overpotential) A voltage above the normal rated voltage or the maximum operating voltage of a device or circuit. A direct test overvoltage is a voltage above the peak of the alternating line voltage. (PE/EM) 95-1977r

(5) Any voltage whose magnitude is less than the maximum safe input voltage of the recorder but greater than the full-scale value for the selected range. (IM/WM&A) 1057-1994w

(6) An rms increase in the ac voltage, at the power frequency, for durations greater than a few seconds. (T&D/PE) 1250-1995

(7) Voltage, between one phase and ground or between two phases, having a crest value exceeding the corresponding crest of the maximum system voltage. Overvoltage may be classified by shape and duration as either *temporary* or *transient. Notes:* 1. Unless otherwise indicated, such as for surge arresters, overvoltage are expressed in per unit with reference to $V_m\sqrt{2}\sqrt{3}$. 2. A general distinction may be made between highly damped overvoltages of relatively short duration (*transient overvoltages*) and undamped or only slightly damped overvoltages of relatively long duration (*temporary overvoltages*). The transition between these two groups cannot be clearly defined. (PE/C) 1313.1-1996

(8) Abnormal voltage between two points of a system that is greater than the highest value appearing between the same two points under normal service conditions. Overvoltages may be low-frequency, temporary, and transient (surge). (SPD/PE) C62.22-1997

(9) When used to describe a specific type of long duration variation, refers to a measured voltage having a value greater than the nominal voltage for a period of time greater than 1 min. Typical values are 1.1 to 1.2 pu. (SCC22) 1346-1998

(10) When used to describe a specific type of long duration variation, refers to an RMS increase in the ac voltage, at the power frequency, for a period of time greater than 1 min. Typical values are 1.1-1.2 pu. *See also:* swell; transient. (IA/PSE) 1100-1999

overvoltage due to resonance (surge arresters) Overvoltage at the fundamental frequency of the installation, or of a harmonic frequency, resulting from oscillation of circuits. (PE) [8], [84]

overvoltage protection The effect of a device operative on excessive voltage to cause and maintain the interruption of power in the circuit or reduction of voltage to the equipment governed. (IA/IAC) [60]

overvoltage relay A relay that operates when its input voltage exceeds a predetermined value. (SWG/PE/SUB/PSR) C37.100-1992, C37.2-1979s, C37.90-1978s

overvoltage release (trip) A release that operates when the voltage of the main circuit is equal to or exceeds the release setting. (SWG/PE) C37.100-1992

overvoltage suppressors (thyristor) Devices used in the ac power controller to attenuate repetitive and nonrepetitive overvoltages of internal or external origin, for example, snubbers, surge arresters, limiters, etc. (IA/IPC) 428-1981w

overvoltage test (rotating machinery) A test at voltages above the rated operating voltage. (PE) [9]

overwriting (charge-storage tubes) Writing in excess of that which produces write saturation. *See also:* charge-storage tube. (ED) 158-1962w

OW *See:* oil-immersed water-cooled transformer.

O wave *See:* ordinary wave.

Owen bridge A 4-arm alternating-current bridge in which one arm, adjacent to the unknown inductor, comprises a capacitor and resistor in series: the arm opposite the unknown consists of a second capacitor, and the fourth arm of a resistor. *Note:* Normally used for the measurement of self-inductance in terms of capacitance and resistance. Usually, the bridge is balanced by adjustment of the resistor that is in series with the first capacitor and of another resistor that is inserted in series with the unknown inductor. The balance is independent of frequency. *See also:* bridge.

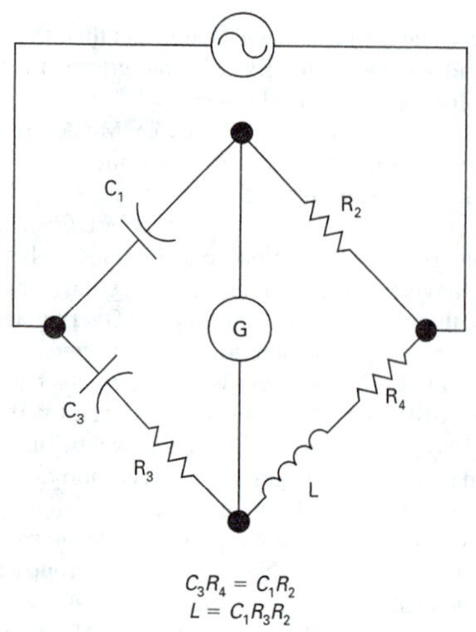

Owen bridge

(EEC/PE) [119]

owned attribute An attribute of an entity that has not migrated into the entity. (C/SE) 1320.2-1998

owner (1) (nuclear power quality assurance) The person, group, company, agency, or corporation who has or will have title to the nuclear power plant. (PE/NP) [124]
(2) A party who owns the transmission line during the construction phase of the line and may include a person who acts for, or on behalf of, an owner as the owner's agent or delegate. (T&D/PE) 1025-1993r, 951-1996
(3) A single point of contact, identified by organization position. (T&D/PE/C/SE) 1074-1995s
(4) The individual, corporation, or organization that intends to use the shield and that is the ultimate source of the shielding requirement. (EMC/STCOORD) 299-1997

ownership State of a master that has arbitrated and won the bus and has not yet lost a bus arbitration contest. (C/MM) 1196-1987w

oxidant A chemical element or compound that is capable of being reduced. *See also:* electrochemical cell. (IA/APP) [73]

oxidation (electrochemical cells and corrosion) Loss of electrons by a constituent of a chemical reaction. *See also:* electrochemical cell. (IM/HFIM) [40]

oxidation inhibitor (insulating oil) Any substance added to an insulating fluid to improve its resistance to deleterious attack in an oxidizing environment. For example, 2, 6-ditertiary-butyl para-cresol or 2,6-ditertiary-butyl phenol, or both, are sometimes added to petroleum insulating oil to improve its oxidation stability. (PE/TR) 637-1985r

oxide-cathode *See:* oxide-coated cathode.

oxide-coated cathode (oxide-cathode) (thermionics) A cathode whose active surface is a coating of oxides of alkaline earths on a metal. *See also:* electron emission. (ED) [45], [84]

oxidizing (electrotyping) The treatment of a graphited wax surface with copper sulfate and iron filings to produce a conducting copper coating. (PE/EEC) [119]

oxygen-concentration cell A galvanic cell resulting primarily from differences in oxygen concentration. *See also:* electrolytic cell. (IA) [59]

oxygen index The minimum concentration of oxygen, expressed as volume percent, in a mixture of oxygen and nitrogen that will just support flaming combustion of a material initially at room temperature, referred to in battery manufacturers' flammability designations for battery cases. *Synonym:* limiting oxygen index. (IA/PSE) 446-1995

oxygen recombination The process by which oxygen is generated at the positive plates and ultimately recombined with hydrogen ions at the negative plates and converted back to water. In this process, hydrogen gas formation and evolution are suppressed. (SB) 1189-1996

oxygen recombination efficiency The amount of oxygen ultimately converted to water at the negative plates expressed as a percentage of the total amount of oxygen produced at the positive plates:

$$O_{2\text{eff}} = \frac{O_2 \text{ converted to water at the negative plates}}{\text{total } O_2 \text{ produced at the positive plates}} \times 100$$

(SB) 1189-1996

ozone A colorless gas, Os_3, with a penetrating odor; an allotropic form of oxygen. *Note:* Corona and other electrical discharges dissociate the oxygen molecule, which can cause the following reactions:

$$O_2 \rightarrow O + O$$

$$O + O_2 \rightarrow O_3$$

(T&D/PE) 539-1990

ozone-producing radiation (illuminating engineering) Ultraviolet energy shorter than 220 nm (nanometers) that decomposes oxygen O_2, thereby producing ozone O_3. Some ultraviolet sources generate energy at 184.9 nm, which is particularly effective in producing ozone. (EEC/IE) [126]

P

p *See:* pico.

PA Pre-arbitrated. (LM/C) 802.6-1990r

PA access function *See:* pre-arbitrated (PA) access function.

PABX *See:* private automatic branch exchange.

pace voltage (surge arresters) A voltage generated by ground current between two points on the surface of the ground at a distance apart corresponding to the conventional length of an ordinary pace. (PE) [8], [84]

PACF *See:* phase angle correction factor.

pacing (1) (Class 1E connection assemblies) A method of ongoing qualification by parallel age conditioning. (PE/NP) 572-1985r

(2) A method to regulate the flow of bytes read from or written to a Smart Transducer Interface Module. (IM/ST) 1451.2-1997

pack (1) To compress several items of data in a storage medium in such a way that the individual items can later be recovered. (C) [20], [85]

(2) (data management) (software) To store data in a compact form in a storage medium, using known characteristics of the data and medium in such a way as to permit recovery of the data. *Contrast:* unpack. *See also:* packed data. (C) 610.5-1990w, 610.12-1990

package (1) (software) A separately compilable software component consisting of related data types, data objects, and subprograms. *See also:* encapsulation; information hiding; data abstraction. (C) 610.12-1990

(2) The combination of a node's properties, methods, and private data. (C/BA) 1275-1994

(3) An external container, substrate, or platform used to hold a semiconductor or circuit. *Note:* it may be made of plastic or ceramic with many interfacing pins. (C) 610.10-1994w

(4) A group of related OM classes. (C/PA) 1328.2-1993w, 1327.2-1993w, 1224.2-1993w, 1326.2-1993w

(5) A group of related classes. (C/PA) 1328-1993w, 1238.1-1994w, 1224-1993w, 1224.1-1993w, 1327-1993w

package closure The set of classes that need to be supported in order to be able to create all possible instances of all classes defined in the package. (C/PA) 1328-1993w, 1224-1993w, 1327-1993w

package, core *See:* core package.

package instance A data structure resulting from the opening of a particular package, consisting of a set of values for the package's private data. (C/BA) 1275-1994

packaged magnetron An integral structure comprising a magnetron, its ma gnetic circuit, and its output matching device. *See also:* magnetron. (ED) 161-1971w, [45]

packager role Where software that has been developed is organized in a form suitable for distribution. (C/PA) 1387.2-1995

packaging (1) (software) In software development, the assignment of modules to segments to be handled as distinct physical units for execution by a computer. (C) 610.12-1990

(2) The process of containing, connecting, protecting, and sealing circuits and components into enclosures such as devices, modules, or housings. (C) 610.10-1994w

packaging machine Any automatic, semiautomatic, or hand-operated apparatus that performs one or more packaging functions, such as, but not limited to, the fabrication, preparation, filling, closing, labeling, or preparing, or both, for final distribution of any type of package or container used to protect or display, or both, any product. (IA/PKG) 333-1980w

packed array An array in which all data elements in the set have non-trivial values. *Synonym:* dense list. (C) 610.5-1990w

packed binary data Binary data stored in a compact form in a storage medium, using known characteristics of the data and the medium to permit recovery of the data. (C) 610.5-1990w

packed data Data stored in a compact form in a storage medium, using known characteristics of the data and the medium to permit recovery of the data. *See also:* packed decimal data; packed binary data. (C) 610.5-1990w

packed decimal data Integer data stored in a compact form in a storage medium, using known characteristics of the data and the medium to permit recovery of the data. In the most common implementation, each decimal digit is represented in binary, occupying four bits, and the right-most decimal digit is followed by a four-bit sign digit (hexadecimal A,C,E, or F for positive; B or D for negative).

decimal	275_{10}
packed decimal	$0010\ 0111\ 0101\ 1111_2 = 275F_{16}$
decimal	-91_{10}
packed decimal	$0000\ 1001\ 0001\ 1011_2 = 091B_{16}$

See also: unsigned packed decimal data. (C) 610.5-1990w

packet (1) A group of binary digits including data and control elements which is switched and transmitted as a composite whole. The data and control elements and possibly error control information are arranged in a specified format. (LM/COM) 168-1956w

(2) (MULTIBUS II) A block of information that is transmitted within a single transfer operation in message space. *See also:* message space; transfer operation. (C/MM) 1296-1987s

(3) A collection of symbols that contains addressing information and is protected by a CRC. A subaction consists of two packets, a send packet and an echo packet. (C/MM) 1596-1992

(4) A 17-bit unit of data consisting of a 16-bit word plus 1 parity bit. (TT/C) 1149.5-1995

(5) A sequence of N_chars with a specific order and format. A packet consists of a destination followed by a payload. A packet is delimited by an end_of_packet marker. *See also:* destination; payload. (C/BA) 1355-1995

(6) A unit of data of some finite-size that is transmitted as a unit. *Note:* Usually consists of a header containing control information such as a sequence number, the network address of the station that originated the packet, and the network address of the packet's destination. *See also:* long packet; short packet. (C) 610.7-1995, 610.10-1994w

(7) A serial stream of clocked data bits. A packet is normally the PDU for the link layer, although the cable physical layer can also generate and receive special short packets for management purposes. (C/MM) 1394-1995

(8) A collection of symbols that contains addressing information and is protected by a CRC. A subaction consists of two packets: a send packet and an echo packet. (C/MM) 1596.3-1996

(9) A block of information that is transmitted within a single transfer operation. (C/MM) 1596.4-1996

(10) A structured field, having a start byte, a two-byte length field (the first two bytes), a flag byte, a command byte, followed by the subcommand and/or data fields. (C/MM) 1284.1-1997

(11) Consists of a data frame as defined previously, preceded by the Preamble and the Start Frame Delimiter, encoded, as appropriate, for the Physical Layer (PHY) type. (C/LM) 802.3-1998

(12) (local area networks) The total information transmitted over the link medium, including the preamble, the MAC frame, and the start of stream and end of stream delimiters. *See also:* frame. (C) 8802-12-1998

(13) A sequence of bits transmitted on Serial Bus and delimited by DATA_PREFIX and DATA_END. (C/MM) 1394a-2000

(14) A group of bytes, including address, data, and control elements. (C/MM) 1284.4-2000

packet assembler/disassembler (PAD) (1) A device that assembles and disassembles packets. (LM/C/COM) 8802-9-1996

(2) A protocol conversion device that performs packet assembly/disassembly. *Note:* Generally refers to a terminal multiplexer device that connects hosts and terminals on a network. (C) 610.7-1995

packet assembly/disassembly The process of dividing a message into packets for transmission over a packet switching network and then reassembling the packets in the original message. (C) 610.7-1995

PACKET COUNT packet A packet by means of which an M-module conveys to addressed S-modules the number of packets to follow in the current message. S-modules may or may not include the ability to use the data in this packet. (TT/C) 1149.5-1995

packet data network *See:* packet switching network.

packet data transfer A very fast, but technology-dependent, noncompelled transfer mechanism used to transfer data from a source to a destination. (C/BA) 1014.1-1994w

packet data transfer protocol A very fast but technology-dependent noncompelled transfer mechanism that uses a compelled protocol over the entire packet to provide flow control. (C/BA/C/BA) 10857-1994, 896.4-1993w

packet error An error that occurs when a packet is lost in the network. *See also:* type error; address error; abnormal preamble; alignment error. (C) 610.7-1995

packet handler (PH) A device for processing packets or frames in a manner so as to be able to route individual frames or packets out of one data stream into multiple different data streams. (LM/C/COM) 8802-9-1996

packetized data transfer Transfer of data through a network where data is conveyed in packets and/or frames in a statistical manner. Packets or frames are propagated through the network and delivered to destinations based on addressing information contained therein. (LM/C/COM) 8802-9-1996

packet latency During a send/receive message, the number of packet transfers by which the first non-NULL DATA packet returned by the addressed S-module lags the first non-null DATA packet transmitted by the M-module. (TT/C) 1149.5-1995

packet layer The layer of the protocol concerned with end-to-end transmission of information, possibly through a number of intermediate routers. It is at the packet layer that the routing decisions are taken. (C/BA) 1355-1995

packet mode protocol The access protocol based on the use of out-of-band control and Link Layer multiplexing. (LM/C/COM) 8802-9-1996

packet pair Two packets, one transmitted by the M-module and one by an S-module, such that the last 14 bits of the M-module-originated packet are transmitted simultaneously with the first 14 bits of the S-module-originated packet. (TT/C) 1149.5-1995

packet-switched data network (PSDN) A packet-switched subnetwork that can be a PPSN (private) or a PSPDN (public) packet-switched network. (LM/C/COM) 8802-9-1996

packet switched network *See:* packet switching network.

packet switched public data network (PSPDN) A public subnetwork that is accessed via the CCITT X.25 protocol and that provides virtual circuit service. (LM/C/COM) 8802-9-1996

packet switching (1) A data transmission process, utilizing addressed packets, whereby a channel is occupied only for the duration of transmission of the packet. *See also:* store-and-forward switching; message switching; circuit switching. (LM/COM) 168-1956w

(2) A technique used in data communications in which messages are broken into finite-size packets and are forwarded to the other party over the network. Packets may vary in size so long as the size does not exceed the maximum size convention for the local network or protocol in use. *Note:* The packets need not travel the same path. At the end of the circuit, the packets are reassembled into the messages and are then passed on to the receiving terminals. *Contrast:* cell switching. *See also:* fast packet switching; virtual circuit. (C) 610.7-1995

packet switching network A network that uses packet switching techniques for transmission of data. *Synonyms:* packet switched network; packet data network. (C) 610.7-1995

packet symbol A symbol contained within a packet and protected by the packet's CRC. (Exception: part of the second symbol in a packet contains flow control information that is not covered by the CRC, but the symbol as a whole is still considered to be within the packet.) (C/MM) 1596-1992

packing density (computers) The number of useful storage cells per unit of dimension, for example, the number of bits per inch stored on a magnetic tape or drum track. (MIL/C) [2], [85], [20]

packing fraction (fiber optics) In a fiber bundle, the ratio of the aggregate fiber cross-sectional core area to the total cross-sectional area (usually within the ferrule) including cladding and interstitial areas. *See also:* ferrule; fiber bundle. (Std100) 812-1984w

packing gland An explosion-proof entrance for conductors through the wall of an explosionproof enclosure, to provide compressed packing completely surrounding the wire or cable, for not less than 0.5 in measured along the length of the cable. *See also:* mine feeder circuit. (EEC/PE) [119]

PAD *See:* packet assembler/disassembler; packet assembly/disassembly.

pad (1) (data transmission) (attenuating pad) A nonadjustable passive network that reduces the power level of a signal without introducing appreciable distortion. *Note:* A pad may also provide impedance matching. (PE) 599-1985w

(2) (data management) To fill an item such as a record or block with one or more filler characters in order to satisfy some prescribed condition. For example, in order to right justify a seven-character string in a ten-position field, three blank characters are used to pad the data. *See also:* zero fill; character fill. (C) 610.5-1990w

(3) (local area networks) Any combination of octets used to extend the end of the data field of the ISO/IEC 8802-3 MAC frame so that it will meet minimum length requirements. (C) 8802-12-1998

padding (A) The technique of filling out a fixed-length block of data with dummy characters, words, or records. **(B)** Dummy characters, words, or records used to fill out a fixed-length block of data. (C) 610.12-1990

paddle A cursor control device consisting of a rotatable knob and potentiometer used to control the position of a cursor on a display device. (C) 610.10-1994w

pad electrode (dielectric heating) One of a pair of electrode plates between which a load is placed for dielectric heating. (IA) 54-1955w, 169-1955w

pad-mounted (1) A method of supporting equipment, generally at ground level. (NESC) C2-1977s

(2) A general term describing equipment positioned on a surface-mounted pad located outdoors. The equipment is usually enclosed with all exposed surfaces at ground potential. (SWG/PE) C37.100-1992

pad-mounted equipment A general term describing enclosed equipment, the exterior of which enclosure is at ground potential, positioned on a surface-mounted pad. (NESC) C2-1997

pad-mounted fused switchgear (PMFSG) Load-interrupter switch and fuse assemblies in which all energized parts are insulated and completely enclosed within an enclosure when the doors and covers are securely closed. The overall enclosure is of suitable environmental and tamper-resistant construction for out-door, above-ground installation. (SWG/PE) C37.73-1998

pad-mounted transformer (power and distribution transformers) An outdoor transformer utilized as part of an underground distribution system, with enclosed compartment(s) for high-voltage and low-voltage cables entering from below, and mounted on a foundation pad. (PE/TR) C57.12.80-1978r

pad-type bearing (rotating machinery) A journal or thrust-type bearing in which the bearing surface is not continuous but consists of separate pads. *See also:* bearing. (PE) [9]

PAGE A text-formatting language that uses two-character instruction codes to control typesetting. (C) 610.13-1993w

page (1) (A) A fixed-length segment of data or of a computer program treated as a unit in storage allocation. **(B)** In a virtual storage system, a fixed-length segment of data or of a computer program that has a virtual address and is transferred as a unit between main and auxiliary storage. (C/Std100) 610.12-1990, 610.5-1990
(2) A screenful of information on a video display terminal. (C) 610.12-1990
(3) (or segment) The smallest portion of memory that can be relocated by memory mappingtechniques. (BA/C) 896.3-1993w
(4) A unit of memory with a specific size defined as a power of 2. (C/BA) 1014.1-1994w
(5) (A) In virtual storage, a fixed length block of instructions or data that has a virtual address and that is transferred as a unit between real storage and auxiliary storage. **(B)** To transfer data between real and auxiliary storage as in definition (A). (C/Std100) 610.10-1994
(6) The granularity of process memory mapping or locking. Physical memory and memory objects can be mapped into the address space of a process on page boundaries and in integral multiples of pages. Process address space can be locked into memory—made memory-resident—on page boundaries and in integral multiples of pages. (C/PA) 9945-1-1996
(7) A section of the array that may be written to simultaneously. This usually refers to more than one byte. (ED) 1005-1998
(8) (A) A fixed-length block of data, especially that which fits into a single printed sheet. **(B)** To summon a particular page or the next logical page. *See also:* paging. (PE/NP) 1289-1998
(9) A logical representation of a single unit of printing media. It is a function of the document formatting rather than the printing process. There are one or more pages per impression. A "four-up" single-color printing will typically have four pages per impression. (C/MM) 1284.1-1997
(10) In Auto-Negotiation, the encoding for a Link Code Word. Auto-Negotiation can support an arbitrary number of Link Code Word encodings. Additional pages may have a predefined encoding or may be custom encoded. *See also:* message page; Unformatted Page. (C/LM) 802.3-1998
(11) The granularity of memory mapping and locking, *i.e.*, a fixed-length contiguous range of the address space of a process. Physical memory and memory objects can be mapped into the address space of a process on page boundaries and in integral multiples of pages. Process address space can be locked into memory (*i.e.*, made memory-resident) on page boundaries and in integral multiples of pages. *Note:* There is no implied requirement that usage of the term *page* in this interface for memory mapping necessarily be the same as the term might be used in a virtual memory implementation. (C) 1003.5-1999

page breakage A portion of main storage that is unused when the last page of data or of a computer program does not fill the entire block of storage allocated to it. *See also:* paging. (C) 610.12-1990

page description language (PDL) (1) A computer language in which commands from a text-formatting language are combined into higher-level instructions that can be used in other documents. Examples include GML, HPGL, Postscript, and TEX. *Synonym:* markup language. (C) 610.13-1993w
(2) A formal printer machine language, consisting of commands and data (or equivalently, operators and operands) used to specify and control the content and format of printed pages. A data stream encoded in a PDL is rendered into the printed page image by an interpreter. (C/MM) 1284.1-1997

page eject character *See:* form feed character.

page fault In demand paging, a condition that causes a program interrupt when a page must be read in from disk into main storage. (C) 610.10-1994w

page frame A block of main storage having the size of, and used to hold, a page. *See also:* paging. (C) 610.12-1990
(2) (A) In real storage, a storage location that has the size of a page. **(B)** An area of main storage used to hold a page. (C) 610.10-1994

page makeup *See:* computer-aided page makeup; photocomposition.

page mode A method of operation in which more than one memory location is written simultaneously. (ED) 1005-1998

page orientation The direction of print on a display device or page of paper; that is, left-to-right or top-to-bottom. *See also:* portrait orientation; landscape orientation. (C) 610.10-1994w

page printer A printer that prints one complete page of output at a time. For example, a computer-output microfilm printer or a laser printer. *Contrast:* character-at-a-time printer; line printer. (C) 610.10-1994w

pager A routine that initiates and controls the transfer of pages between main and auxiliary storage. *See also:* paging. (C) 610.12-1990

page reader A character reader whose input data are in the form of printed text. *See also:* optical character reader. (C) 610.10-1994w

page reference An expression that unambiguously identifies a model page. The page reference incorporates a diagram reference to the associated diagram, the type of page, and any sequencing data needed to distinguish different pages of the same type that are associated with the same diagram. (C/SE) 1320.1-1998

page size (1) The edge-to-edge dimensions of hard-copy documents, or the average characters per line and the number of lines per screen for electronically displayed documents. (C/SE) 1045-1992
(2) The number of storage units in a page. (C) 1003.5-1999

page swapping The exchange of pages between main storage and auxiliary storage. *See also:* paging. (C) 610.12-1990, 610.10-1994w

page table A table that identifies the location of pages in storage and gives significant attributes of those pages. *See also:* paging. (C) 610.12-1990

page turning *See:* paging.

page type letter The uppercase letter in a page reference that denotes a specific type of model page. (C/SE) 1320.1-1998

page zero In the paging method of storage allocation, the first page in a series of pages. (C) 610.12-1990

pagination *See:* automatic pagination.

paging (1) (A) A storage allocation technique in which programs or data are divided into fixed-length blocks called pages, main storage is divided into blocks of the same length called page frames, and pages are stored in page frames, not necessarily contiguously or in logical order. *Synonym:* block allocation. *Contrast:* contiguous allocation. **(B)** A storage allocation technique in which programs or data are divided into fixed-length blocks called pages, main storage is divided into

blocks of the same length called page frames, and pages are transferred between main and auxiliary storage as needed. *See also:* virtual storage; demand paging; anticipatory paging. **(C)** The transfer of pages as in definition (B). *Synonym:* page turning. *See also:* working set; page; page swapping; page frame; page zero; page breakage; page table; pager. (C) 610.12-1990
(2) **(A)** A storage allocation technique in which programs or data are divided into fixed-length blocks called pages, main stage is divided into blocks of the same called page frames, and pages are stored in page frames, not necessarily contiguously or in logical order. *See also:* segment. **(B)** The transfer of pages between main and auxiliary memory, as in definition (A). *Synonym:* page turning. (C) 610.10-1994
(3) A method of viewing and moving through data in which a user conceives of data as being grouped into pages and moves through it by discrete steps. *See also:* page. (PE/NP) 1289-1998

paging device An auxiliary storage device used primarily to hold pages. (C) 610.10-1994w

paging rate In a virtual memory system, the rate at which pages are being transferred between real storage and auxiliary storage. (C) 610.10-1994w

pair A term applied in electric transmission to two like conductors employed to form an electric circuit. *See also:* cable. (EEC/PE) [119]

paired brushes (rotating machinery) (pair of brushes) Two individual brushes that are joined together by a common shunt or terminal. *Note:* They are not to be confused with a split brush. *See also:* brush. (PE) [9]

paired cable (nonquadded cable) A cable in which all of the conductors are arranged in the form of twisted pairs, none of which is arranged with others to form quads. *See also:* cable. (EEC/PE) [119]

pairing (scanning) (television) The condition in which lines appear in groups of two instead of being equally spaced. (BT/AV) 201-1979w

pair of brushes *See:* paired brushes.

PAL *See:* programmable array logic.

PAM *See:* pulse amplitude modulation.

PAM5×5 A block coding technique utilizing a 5×5 matrix (representing two 5-level signals) to generate pairs of quinary codes representing data nibbles and control characters. In 100BASE-T2, PAM5×5 code pairs are sent in parallel across two wire pairs. (C/LM) 802.3-1998

pancake coil A coil having the shape of a pancake, usually with the turns arranged in the form of a flat spiral. (EEC/PE) [119]

pancake motor A motor that is specially designed to have an axial length that is shorter than normal. (PE) [9]

pane (1) The rectangular area within a window where an application displays text or graphics. (C) 1295-1993w
(2) A component of a window. (C) 610.10-1994w

paned window A window that allows a user to view information in different panes of a window, separated by sashes. Panes can be shrunk or enlarged by moving the sash, but the total screen area occupied by the pane window remains constant. (C) 1295-1993w

panel (1) (Class 1E control boards) A unit of one or more sections of flat material suitable for mounting electric devices. (SWG/PE/NP) 420-1982, C37.100-1992
(2) (packaging machinery) An element of an electric controller consisting of a slab or plate on which various component parts of the controller are mounted and wired. (IA/ICTL/PKG/IAC) 330-1980w, 333-1980w, [60]
(3) **(A)** **(photovoltaic converter)** Combination of shingles or subpanels as a mechanical and electric unit required to meet performance specifications. *See also:* semiconductor. **(B)** **(solar cells)** The largest unit combination of solar cells or subpanels that is mechanically designed to facilitate manufacture and handling and that will establish a basis for electrical performance by test. **(C)** A distinct portion of an equipment's surface, usually defined by or contained within a frame or border; for example, a maintenance panel or an operator control panel. *See also:* control panel; plasma panel; display panel. (AES/C/SS) 307-1969, 610.10-1994
(4) (computers) *See also:* problem board; control panel. (AES/SS) 307-1969w

panelboard A single panel or group of panel units designed for assembly in the form of a single panel; including buses, automatic overcurrent devices, and with or without switches for the control of light, heat, or power circuits; designed to be placed in a cabinet or cutout box placed in or against a wall or partition and accessible only from the front. *See also:* switchboard. (NESC/NEC) [86]

panel control An assembly of man-machine interface devices. (SWG/PE) C37.100-1992

panel efficiency (1) (photoelectric converter) The ratio of available power output to incident radiant power intercepted by a panel composed of photoelectric converters. *Note:* This is less than the efficiency of the individual photoelectric converters because of area not covered by photoelectric converters, Joule heating, and photoelectric-converter mismatches. *See also:* semiconductor. (AES) [41]
(2) (solar cells) The ratio of available electric power output to total incident radiant power intercepted by the area of a panel composed of solar cells. *Note:* This depends on the spectral distribution of the radiant power source and junction temperature(s), requires uniform normal illumination on the intercepting area, and results in an efficiency less than the efficiency of the individual solar cells because of area not covered by solar cells, incident energy heating, solar cell mismatch, optical losses, and wiring losses. (AES/SS) 307-1969w

panel-frame mounting (of a switching device) Mounting on a panel frame in the rear of a panel with the operating mechanism on the front of the panel. (SWG/PE) C37.100-1992

panel interface A screen-oriented user interface designed to permit interactive processing. (C) 610.10-1994w

panel system An automatic telephone switching system that is generally characterized by the following features:

a) The contacts of the multiple banks over which selection occurs are mounted vertically in flat rectangular panels;
b) The brushes of the selecting mechanism are raised and lowered by a motor that is common to a number of these selecting mechanisms;
c) The switching pulses are received and stored by controlling mechanisms that govern the subsequent operations necessary in establishing a telephone connection.

(EEC/PE) [119]

panic brake Using any available form of braking, whether or not fail-safe, to obtain the shortest possible stopping distance. (VT) 1475-1999

panning (1) (computer graphics) In computer graphics, the process of moving an entire display image in such a manner that new data appears within the viewport as old data disappears, to give a visual impression of horizontal movement of the image. *Note:* The term panning is sometimes used to mean horizontal or vertical movement. *Contrast:* scrolling. (C) 610.6-1991w
(2) *See also:* scrolling. (C) 610.2-1987

paper feed A mechanism that positions the printing medium as the paper is moved into a printing device. (C) 610.10-1994w

paperless office An office that has been automated so that no paper documents are needed. *See also:* office automation; electronic office. (C) 610.2-1987

paper-lined construction (dry cell) (primary cell) A type of construction in which a paper liner, wet with electrolyte, forms the principal medium between the negative electrode, usually zinc, and the depolarizing mix. (A layer of paste may lie between the paper liner and the negative electrode.) *See also:* electrolytic cell. (PE/EEC) [119]

paper tape *See:* punch tape.

paper tape reader A reader that senses hole patterns in punched paper tape and translates them into internal machine data representations. *Synonym:* perforated tape reader. (C) 610.10-1994w

paper throw character *See:* form feed character.

paper traffic *See:* information traffic.

PAR *See:* precision-approach radar.

parabolic approximation *See:* parabolic equation.

parabolic equation Results when the Helmholtz equation is approximated to emphasize preferred propagation in the axial direction, leading to a differential equation of parabolic form. *Synonym:* parabolic approximation. (AP/PROP) 211-1997

parabolic profile (fiber optics) A power-law index profile with the profile parameter, g, equal to 2. *Synonym:* quadratic profile. *See also:* multimode optical waveguide; profile parameter; power-law index profile; graded index profile. (Std100) 812-1984w

parabolic torus reflector A toroidal reflector formed by rotating a segment of a parabola about a non-intersecting co-planar line. (AP/ANT) 145-1993

paraboloidal reflector An axially symmetric reflector that is a portion of a paraboloid. *Note:* This term may be applied to any reflector that is a portion of a paraboloid, provided the term is appropriately qualified. For example, if the reflector is a portion of a paraboloid but does not include its vertex, then it may be called an off-set paraboloidal reflector. (AP/ANT) 145-1993

parachute harness An assembly of webbing, strapping, and attachments, that permits the attachment of a support device (e.g. D-ring), and fits the human such that the entire weight of same can be supported without injury. The lanyard is constructed so that it will support the normal weight of a human, but can be released by pulling on a disconnection means. Upon this action, there will be separation of the body support assembly from the lanyard that has been attached to the harness, thus permitting the movement of the body support means from the human body support assembly. A parachute harness allows separation of the lanyard supporting assembly from an object, commonly an aircraft or helicopter. As released, the force to complete the full separation of the harness is of the order of 1/3 of that required to produce separation resulting from the actuation of the master (main) release (disconnection) means. (T&D/PE) 1307-1996

paraelastic crystal (primary ferroelectric terms) By analogy with paraelectric crystals, a crystal in which mechanical strain *S* is a single-valued function of mechanical stress *T*, whose elastic compliance exhibits an obvious Curie-Weiss behavior with temperature over some given temperature range, and that at some critical temperature T_c undergoes a phase transition to a ferroelastic phase. Crystals that clearly have a paraelastic phase include the metallic alloys Nb_3 Sn, In-Th, Au-Zn-Sn, and lithium ammonium tartrate. (UFFC) 180-1986w

paraelastic phase (primary ferroelectric terms) A phase that encompasses the range of temperature in which the elastic compliance exhibits Curie-Weiss behavior. (UFFC) 180-1986w

paraelectric Curie temperature (of a ferroelectric material) The intercept of the linear portion of the plot of $1/\varepsilon$ versus *T*, where ε is the small signal dielectric permittivity measured at zero bias field and *T* is the absolute temperature in the region above the ferroelectric Curie temperature where ε generally follows the Curie-Weiss relation. *See also:* ferroelectric domain. (UFFC) 180w

paraelectric phase (primary ferroelectric terms) Encompasses the range of temperature or pressure over which the permittivity exhibits Curie-Weiss behavior. (UFFC) 180-1986w

paraelectric region The region above the Curie point where the small signal permittivity increases with decreasing temperature. *See also:* small-signal permittivity; ferroelectric Curie point. (UFFC) [21]

parallax (computer graphics) The apparent displacement of an object as seen from two different points. It is used to simulate a three-dimensional image on a graphical display device. (C) 610.6-1991w

parallel **(1) (A) (networks) (parallel elements)** Two-terminal elements are connected in parallel when they are connected between the same pair of nodes. **(B) (networks) (parallel elements)** Two-terminal elements are connected in parallel when any cut-set including one must include the others. *See also:* network analysis. (Std100) 270-1966

(2) (A) Pertaining to the simultaneity of two or more processes. **(B)** Pertaining to the simultaneity of two or more similar or identical processes. **(C)** Pertaining to simultaneous processing of the individual parts of a whole, such as the bits of a character and the characters of a word using separate facilities for the various parts. *See also:* serial-parallel. (C) 162-1963, [20], [85], 270-1966

(3) (radio-wave propagation) Of a propagating wave for which the electric field vector lies parallel to the plane of incidence. *Note:* Sometimes called vertical polarization. (AP) 211-1977s

(4) (software) Pertaining to the simultaneous transfer, occurrence, or processing of the individual parts of a whole, such as the bits of a character, using separate facilities for the various parts. *Contrast:* serial. *See also:* concurrent. (C) 610.12-1990

(5) Many bits transmitted over a single pathway simultaneously. *Contrast:* serial. *See also:* bit parallel. (C) 610.10-1994w

(6) In a propulsion system, the motor circuit in which the final parallel or series-parallel motor connection is achieved and the maximum available per-motor voltage is applied. (VT) 1475-1999

parallel adder An adder in which addition is performed concurrently on multiple digits of the operands. *Contrast:* serial adder. (C) 610.10-1994w

parallel addition Addition that is performed concurrently on all digit places of the operands. *Note:* This technique uses partial sums and partial carries to obtain its results. *Contrast:* serial addition. (C) 1084-1986w

parallel architecture A multiprocessor architecture in which parallel processing can be performed, that is, different parts of a single task can be executed concurrently on different processors. *See also:* fine-grain parallel architecture; medium-grain parallel architecture; coarse-grain parallel architecture. (C) 610.10-1994w

parallel classes A pair of classes that are distinct, are not mutually exclusive and have a common generic ancestor class and for which neither is a generic ancestor of the other. (C/SE) 1320.2-1998

parallel computer (A) A computer that has multiple arithmetic units or logic units that are used to accomplish parallel operations or parallel processing. *Contrast:* sequential computer; serial computer. **(B)** A computer design in which more than one operation can occur simultaneously. *See also:* simultaneous computer. (C) 610.10-1994

parallel-connected capacitance (as applied to interrupter switches) Capacitances are defined to be parallel-connected when the crest value of inrush current to the capacitance being switched exceeds the switch inrush current capability for single capacitance. (SWG/PE) C37.100-1992

parallel-connected capacitor unit A capacitor unit with the elements connected in parallel groups, with the parallel groups connected in series between the line terminals. A capacitor unit that has only one string of capacitor elements between the capacitor terminals is considered to be parallel-connected.power systems relaying. (PE) C37.99-2000

parallel connection The arrangement of cells in battery made by connecting all positive terminals together and all negative terminals together, the voltage of the group being only that of one cell and the current drain through the battery being divided among the several cells. *See also:* battery. (EEC/PE) [119]

parallel construct A program construct consisting of two or more procedures that can occur simultaneously. (C) 610.12-1990

parallel contention arbitration A process whereby modules assert their unique arbitration number on a parallel bus and release signals according to an algorithm such that after a period of time the winner's number appears on the bus. (C/BA) 10857-1994, 896.4-1993w

parallel detection In Auto-Negotiation, the ability to detect 100BASE-TX and 100BASE-T4 technology specific link signaling while also detecting the Normal Link Pulse (NLP) sequence or Fast Link Pulse (FLP) Burst sequence. (C/LM) 802.3-1998

parallel digital computer One in which the digits are handled in parallel. Mixed serial and parallel machines are frequently called serial or parallel according to the way arithmetic processes are performed. An example of a parallel digital computer is one that handles decimal digits in parallel although it might handle the bits that comprise a digit either serially or in parallel. *See also:* serial digital computer. (Std100) 270-1966w

parallel disk array (PDA) A form of RAID storage, known as level 3, in which an array of disk drives transfer data in parallel with one redundant drive that functions as a parity check disk. (C) 610.10-1994w

parallel feeder A feeder that operates in parallel with one or more feeders of the same type from the same source. *Note:* These feeders may be of the stub-, multiple-, or tie-feeder type. (SWG/PE) C37.100-1992

parallel heating cable Heating elements that are electrically connected in parallel, either continuously or in zones, so that watt density per lineal length is maintained irrespective of any change in length for the continuous type or for any number of discrete zones. (BT/IA/AV/PC) 152-1953s, 515.1-1995, 515-1997

parallel inference machine A computer containing an inference engine that can perform logic inference processing concurrently on two or more rules or goal clauses. (C) 610.10-1994w

paralleling (rotating machinery) The process by which a generator is adjusted and connected to run in parallel with another generator or system. *See also:* direct-current commutating machine; asynchronous machine. (PE) [9]

paralleling reactor (power and distribution transformers) A current-limiting reactor for correcting the division of load between parallel-connected transformers which have unequal impedance voltages. *See also:* reactor. (PE/TR) C57.12.80-1978r, [57]

parallelism (A) Concurrent operation of several parts of a computer system. *Note:* This could be simultaneous processing of multiple programs, or simultaneous operation of multiple computers. **(B)** Pertaining to specific techniques for implementing parallel operations. *See also:* AND-parallelism; OR-parallelism. (C) 610.10-1994

parallel-mode interference *See:* common-mode interference.

parallel noise (of a device) Electrical noise that can be attributed to a hypothetical white noise generator connected in parallel with the input of the device. (NPS) 325-1996

parallel operation (power supplies) Voltage regulators, connected together so that their individual output currents are added and flow in a common load. Several methods for parallel connection are used: spoiler resistors, master/slave connection, parallel programming, and parallel padding. Current regulators can be paralleled without special precaution. (AES) [41]

parallel padding (power supplies) A method of parallel operation for two or more power supplies in which their current limiting or automatic crossover output characteristic is employed so that each supply regulates a portion of the total current, each parallel supply adding to the total and padding the output only when the load current demand exceeds the capability or limit setting-of the first supply. (AES) [41]

parallel polarization (1) (facsimile) A linear polarization for which the field vector is parallel to some reference plane. *Note:* These terms are applied mainly to uniform plane waves incident upon a plane of discontinuity (surface of the earth, surface of a dielectric or a conductor). Then the convention is to take as reference the plane of incidence, that is, the plane containing the direction of propagation and the normal to the surface of discontinuity. If these two directions coincide, the reference plane must be specified by some other convention. (COM/AP/ANT) 167-1966w, 145-1993
(2) The polarization of a wave for which the electric field vector lies in the plane of incidence. *Note:* This is sometimes called vertical or transverse magnetic (TM) polarization. (EMC) 1128-1998
(3) The polarization of a wave for which the electric field vector lies parallel to the plane of incidence. *Note:* Sometimes called vertical or transverse magnetic (TM) polarization. In optics, it is called "p" polarization. In radio propagation, H is parallel to the ground. *Synonyms:* transverse-magnetic polarization; vertical polarization. (AP/PROP) 211-1997

parallel port A port that transfers data one byte at a time, each bit over its own line. *Contrast:* serial port. (C) 610.10-1994w

parallel printer A printer that receives its input data in the form of a parallel stream of data. *Contrast:* serial printer. (C) 610.10-1994w

parallel processing (1) (computers) Pertaining to the simultaneous execution of two or more sequences of instructions or one sequence of instructions operating on two or more sets of data, by a computer having multiple arithmetic and.or logic units. *See also:* multiprocessing; multiprogramming. (C) [20], [85]
(2) Pertaining to the concurrent or simultaneous execution of two or more processes in multiple devices, such as processing units or channels. *Contrast:* serial processing. *See also:* pipelining. (C) 610.10-1994w

parallel programming (power supplies) A method of parallel operation for two or more power supplies in which their feedback terminals (voltage-control terminals) are also paralleled. These terminals are often connected to a separate programming source. (AES) [41]

parallel projection (computer graphics) The projection of a three-dimensional image onto a two-dimensional surface such that objects that are farther from the viewer in three dimensions are rendered the same size as closer ones. *Note:* The resulting image is less realistic than that achieved by a perspective projection. *Contrast:* perspective projection.

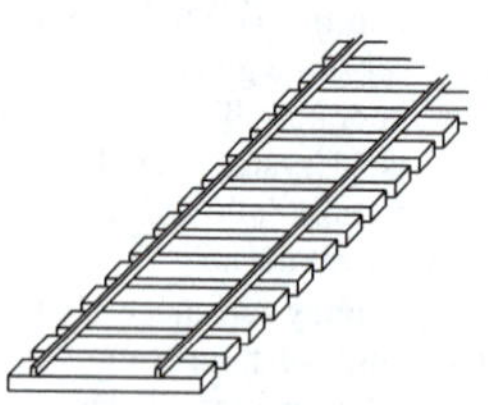

parallel projection

(C) 610.6-1991w

parallel rectifier A rectifier in which two or more similar rectifiers are connected in such a way that their direct currents add and their commutations coincide. *See also:* power rectifier. (IA) [62]

parallel rectifier circuit A rectifier circuit in which two or more simple rectifier circuits are connected in such a way that their direct currents add and their commutations coincide. *See also:* rectifier circuit element; rectification. (IA) [12]

parallel redundant UPS configuration Consists of two or more UPS modules with static inverter turn-off(s), a system control cabinet, and either individual module batteries or a common battery. (IA/PSE) 241-1990r

parallel resonance The sinusoidal steady state condition that exists in a circuit composed of an inductor and a capacitor connected in parallel when the applied frequency is such that: the driving-point impedance is a maximum; or the susceptance of the two parallel arms are equal in magnitude; or the phase-angle of the driving-point impedance is zero. Sometimes defined as above for more general RLC (resistance-inductance-capacitance) networks. (CAS) [13]

parallel search storage A type of storage in which one or more parts of all storage locations are queried simultaneously or concurrently. *See also:* associative storage. (C) 610.10-1994w

parallel-serial converter *See:* serializer.

parallel storage A storage device in which digits, characters, or words, are dealt with simultaneously or concurrently. (C) [20], [85], 610.10-1994w

parallel system bus The signals, media, and protocol used to interconnect agents in the IEEE Std 1296-1987 system). (C/MM) 1296-1987s

parallel thyristor converter A thyristor converter in which two or more simple converters are connected in such a way that their direct currents add and their commutations coincide. (IA/IPC) 444-1973w

parallel-T network (twin-T network) A network composed of separate T networks with their terminals connected in parallel. *See also:* network analysis. (EEC/PE) [119]

parallel transmission (1) (data transmission) Simultaneous transmission of the bits making up a character, either over separate channels or on different carrier frequencies on one channel. (COM) [49]
(2) In data communications, the simultaneous transmission of all the bits making up a character or byte where each bit travels on a different path. *Contrast:* serial transmission. (C) 610.7-1995

parallel two-terminal pair networks Two-terminal pair networks are connected in parallel at the input or at the output terminals when their respective input or output terminals are in parallel. *See also:* network analysis. (BT) 153-1950w

parallel vector A representation specifying a set of waveforms across all primary signals, to be applied to those signals in a parallel fashion (i.e., simultaneously). (C/TT) 1450-1999

paralyzable system (x-ray energy spectrometers) Any system or device whose response characteristics contain a region where the ratio of output-to-input count rate decreases with increasing input count rate. (NPS/NID) 759-1984r

paramagnetic material Material whose relative permeability is slightly greater than unity and practically independent of the magnetizing force. (Std100) 270-1966w

Parameter An instance of the class `IEEE1451_Parameter` or of a subclass thereof. A Parameter is a representation of a variable. Parameters may selectively be manipulated across a network, and can be an externally visible representation for data. (IM/ST) 1451.1-1999

parameter (1) (mathematical) A variable that is given a constant value for a specific purpose or process. (C) [20], [85]
(2) (A) (physical) One of the constants entering into a functional equation and corresponding to some characteristic property, dimension, or degree of freedom. **(B) (electrical)** One of the resistance, inductance, mutual inductance, capacitance, or other element values included in a circuit or network. Also called network constant. (Std100) 270-1966
(3) A quantity of property treated as a constant but which may sometimes vary or be adjusted. (CS/PE/EDPG) [3]
(4) (A) (test, measurement, and diagnostic equipment) Any specific quantity or value affecting or describing the theoretical or measurable characteristics of a unit being considered which behaves as an independent variable or which depends upon some functional interaction of other quantities in a theoretically determinable manner. **(B) (test, measurement, and diagnostic equipment)** In programming, a variable that is given a constant value for a specific purpose or process. (MIL) [2]
(5) (A) (software) A variable that is given a constant value for a specified application. *See also:* adaptation parameter. **(B) (software)** A constant, variable, or expression that is used to pass values between software modules. *See also:* argument; formal parameter. (C) 610.12-1990
(6) In the shell command language, an entity that stores values. There are three types of parameters: variables (named parameters), positional parameters, and special parameters. Parameter expansion is accomplished by introducing a parameter with the $ character. (C/PA) 9945-2-1993
(7) A data item required for the calculation of some result. (C/DA) 1481-1999
(8) An attribute usually representing some property subject to change, for example, a configuration parameter. (IM/ST) 1451.1-1999

parameter bunching One-half the product of the bunching angle in the absence of velocity modulation and the depth of velocity modulation. *Note:* In a reflex klystron the effective bunching angle must be used. *See also:* electron device. (ED) 161-1971w

parameterized collection class A kind of collection class restricted to hold only instances of a specified type (class). (C/SE) 1320.2-1998

parameter potentiometer A potentiometer employed in analog computers to represent a problem parameter such as a coefficient or a scale factor. *See also:* coefficient potentiometer. (C) 610.10-1994w, 166-1977w, 165-1977w

Parameter With Update An instance of the class `IEEE1451_ParameterWithUpdate` or of a subclass thereof. (IM/ST) 1451.1-1999

parameter word A word that directly or indirectly provides or designates one or more parameters. (C) 610.10-1994w

parametric amplifier An inverting parametric device used to amplify a signal without frequency translation from input to output. *Note:* In common usage, this term is a synonym for reactance amplifer. *See also:* parametric device. (ED) 254-1963w, [46]

parametric converter An inverting parametric device or noninverting parametric device used to convert an input signal at one frequency into an output signal at a different frequency. *See also:* parametric device. (ED) [46]

parametric data Data representing an internal state variable of an object that is represented by a Parameter. More specifically, physical parametric data generally represents some aspect of the physical world. (IM/ST) 1451.1-1999

parametric device A device whose operation depends essentially upon the time variation of a characteristic parameter usually understood to be a reactance. (ED) [46]

parametric down-converter A parametric converter in which the output signal is at a lower frequency than the input signal. *See also:* parametric device. (ED) [46]

parametric test A test that is performed to verify device behavior such as output drive current, input leakage current, or output voltage. (C/TT) 1450-1999

parametric test data The values of the observed or measured electrical or physical properties of a unit under test (UUT) that may be used to determine its characteristics or its conformance to requirements. (SCC20) 1545-1999

parametric up-converter A parametric converter in which the output signal is at a higher frequency than the input signal. *See also:* parametric device. (ED) [46]

parametric variation (automatic control) A change in those system properties generally regarded as constants which affect the dependent variables describing system operation. (PE/EDPG) [3]

parasitic element (1) (data transmission) A radiating element that is not coupled directly to the feed lines of an antenna and that materially affects the radiation pattern or impedance of an antenna, or both. (PE) 599-1985w
(2) An unwanted circuit element that is an unavoidable adjunct of a wanted circuit element. (CAS) [13]

(3) A radiating element that is not connected to the feed lines of an antenna and that materially affects the radiation pattern or impedance of an antenna, or both. *Contrast:* driven element. (AP/ANT) 145-1993

parasitic oscillations Unintended self-sustaining oscillations, or transient impulses. *See also:* oscillatory circuit. (AP/BT/ANT) 145-1983s, 182A-1964w

parasitics Electrical properties of a design (resistance, capacitance, and impedance) that arise due to the nature of the materials used to implement the design. (C/DA) 1481-1999

paraxial approximation An approximation in which the waves are constrained to travel predominantly in one direction. (AP/PROP) 211-1997

paraxial ray (fiber optics) A ray that is close to and nearly parallel with the optical axis. *Note:* For purposes of computation, the angle, θ, between the ray and the optical axis is small enough for sin θ or tan θ to be replaced by θ (radians). *See also:* light ray. (Std100) 812-1984w

parcel plating Electroplating upon only a part of the surface of a cathode. *See also:* electroplating. (EEC/PE) [119]

PARD *See:* PARD.

parent A widget that is the immediate superior of the current widget in the widget instance hierarchy. (C) 1295-1993w

parent box An ancestral box related to its child diagram by exactly one parent/child relationship, that is, a box detailed by a child diagram. The existence of this child diagram is indicated by a box detail reference. (C/SE) 1320.1-1998

parent diagram A diagram that contains a parent box. (C/SE) 1320.1-1998

parent directory (A) When discussing a given directory, the directory that contains a directory entry for the given directory. The parent directory is represented by the pathname dot-dot in the given directory. **(B)** When discussing other types of files, a directory containing a directory entry for the file under discussion. This concept does not apply to dot and dot-dot. (C/PA) 9945-1-1996, 9945-2-1993, 1003.5-1999

parent entity (1) An entity set that has other entities dependent on and related to it called subset entities. The subset entities are linked and related to other entity sets through the parent entity set. (PE/EDPG) 1150-1991w
(2) An entity in a specific relationship whose instances can be related to a number of instances of another entity (child entity). (C/SE) 1320.2-1998

parent function A function modeled by a parent box. (C/SE) 1320.1-1998

parenthesis-free notation *See:* prefix notation.

parent node In a tree, a node having a given node as a child node. *Synonym:* father. *Contrast:* dependent node; child node. *See also:* descendant node.

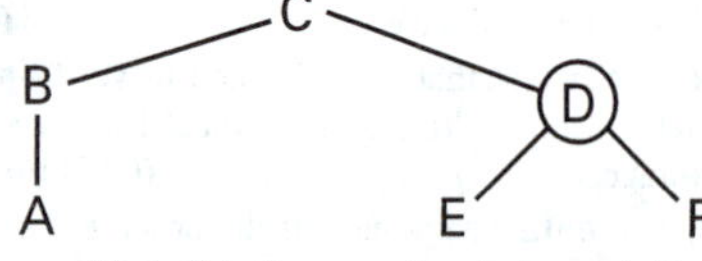

Node D is the parent node for node E

parent node

(C) 610.5-1990w
(2) The node to which a device node is attached. Each device node has exactly one parent node, except the root node, which has none. (A device node descends from its parent node. Traveling "up" the device tree takes one through parent nodes to the root node. Traveling "down" the device tree takes one through child nodes to the leaf nodes.) (C/BA) 1275-1994

parent process *See:* POSIX process.

parent process ID An attribute of a new process after it is created by a currently active process. The parent process ID of a process is the process ID of its creator, for the lifetime of the creator. After the lifetime of the creator has ended, the parent process ID is the process ID of an implementation-defined system process. (C/PA) 1003.5-1999, 9945-2-1993, 9945-1-1996

parent segment In a hierarchical database, a segment that has one or more dependent segments, called child segments, below it in a hierarchy. (C) 610.5-1990w

parity (1) (A) (mathematics of computing) An error detection method in which the total number of ones in a binary word, byte, character, or message is set to an odd or even number by appending a redundant bit. This number is subsequently checked to ensure that it remains odd or even. **(B) (mathematics of computing)** The property of oddness or evenness possessed by a word, byte, character, or message. This property is determined by the total number of ones. *See also:* even parity; odd parity. (C) 1084-1986
(2) (for FASTBUS) A bit, optionally appended to a FASTBUS word, whose value is chosen to make the total number of one bits (including the parity bit) odd. It is used for error checking since receipt of an even number of one bits implies a transmission error. (NID) 960-1993
(3) The value, even or odd, of the sum of a string of binary digits. For example, the parity of the string 0000111101001 is even. *Contrast:* cyclic redundancy check. *See also:* parity bit; parity error; parity check. (C) 610.7-1995

parity bit (1) (computers) A binary digit appended to an array of bits to make the sum of all the bits always odd or always even. (MIL/C) [2], [85], [20]
(2) (mathematics of computing) A binary digit appended to a binary word, byte, character, or message to make the total number of ones an odd or an even number. *See also:* parity check. (C) 1084-1986w
(3) An extra bit attached to a byte, character string, or word, used to enable detection of transmission errors. Based on system convention, the bit is set making the number of ones in a grouping of bits either always even or always odd. *Note:* This permits detection of bit groupings containing single errors. (C) 610.7-1995
(4) An extra bit attached to a byte, character string or word, used to detect transmission errors. (C) 610.10-1994w

parity check (1) (electronic computation) A summation check in which the bits in a character or block are added (modulo 2) and the sum checked against a single, previously computed parity digit; that is, a check that tests whether the number of ones is odd or even. (MIL/C/COM) [2], [85], [20], [49]
(2) (mathematics of computing) A check to determine whether the total number of ones in a binary word, byte, character, or message is odd or even. *Synonyms:* odd-even check; even-odd check. (C) 1084-1986w
(3) An error detecting code that uses the parity bit(s). *Contrast:* cyclic redundancy check. *See also:* vertical redundancy check; longitudinal redundancy check. (C) 610.7-1995

parity error (1) The failure of a binary word, byte, character, or message to pass a parity check. (C) 1084-1986w
(2) An error that occurs when the parity bit of a string is found to be incorrect. (C) 610.7-1995

Parity Error (PRE) bit A bit in the Bus Error register of all S-modules. An S-module sets this bit to indicate that the module has detected a Parity Error on a DATA packet received. (TT/C) 1149.5-1995

parity violation If the received parity and the parity calculation on the data do not agree, a parity violation has occurred. (COM/TA) 1007-1991r

park electrical wiring systems All of the electrical wiring fixtures, equipment and appurtenances related to electrical installations within a mobile home park, including the mobile home service equipment. (NESC/NEC) [86]

parking (1) (multiprocessor architecture) The process whereby the current master retains control of the bus when there are no other masters wishing to use it. (C/MM) 896.1-1987s
(2) (MULTIBUS II) The state of the bus owner where, after the completion of the current transfer operation, ownership is retained until there is a request by another agent for the use of the bus. (C/MM) 1296-1987s

parking brake A means that supplies static braking forces to maintain a vehicle or train in a no motion state. (VT) 1475-1999

parking lamp (illuminating engineering) A lighting device placed on a vehicle to indicate its presence when parked. (EEC/IE) [126]

parking stand A bracket, designed for installation on an apparatus, suitable for holding accessory devices, such as the insulated parking bushing and the grounding bushing. (T&D/PE) 386-1995

PARLOG A logic programming language used widely for parallel computing, supporting declarative programming. (C) 610.13-1993w

parse (1) (software) To determine the syntactic structure of a language unit by decomposing it into more elementary subunits and establishing the relationships among the subunits. For example, to decompose blocks into statements, statements into expressions, expressions into operators and operands. (C) 610.12-1990

(2) To resolve a request or response into component parts. In the context of messages, a device can break the message into pieces, each of which consists of a header and sometimes some corresponding data. If a device is able to parse a message, it can recognize each piece of a message. It does not necessarily make use of the data found in that message. However, it shall make any confirmation responses or other responses that the message requires. (PE/SUB) 1379-1997

PARSEC *See:* PARser and Extensible Compiler.

parsec (pc) The distance at which 1 astronomical unit subtends an angle of 1 second of arc; approximately, 1 pc = 206 265 AU = 30857 × 10^{12} m. (QUL) 268-1982s

parser A software tool that parses computer programs or other text, often as the first step of assembly, compilation, interpretation, or analysis. (C) 610.12-1990

PARser and Extensible Compiler (PARSEC) An extensible language using syntax similar to PL/1; PARSEC is derived from PROTEUS and is yused as the base language for writing PL/PROPHET. (C) 610.13-1993w

part (1) (unique identification in power plants and related facilities) An element of a component not amenable to further disassembly for maintenance purposes. (PE/EDPG) 804-1983r, 803-1983r

(2) The lowest element of a physical or system architecture, specification tree, or system breakdown structure that can not be partitioned (e.g., bolt, nut, bracket, semiconductor, computer software unit). (C/SE) 1220-1994s

partial An incomplete mapping, i.e., some instances map to no related instance. An attribute may be declared partial, meaning it may have no value. A participant property is declared optional as part of the relationship syntax. An operation is declared partial when it may have no meaning for some instances, i.e., it may not give an answer or produce a response. *Contrast:* total. *See also:* mapping completeness; optional. (C/SE) 1320.2-1998

(2) (A) A physical component of a complex tone. **(B) (audio and electroacoustics)** A component of a sound sensation that may be distinguished as a simple tone that cannot be further analyzed by the ear and that contributes to the timbre of the complex sound. *Notes:* 1. The frequency of a partial may be either higher or lower than the basic frequency and may or may not be an integral multiple or submultiple of the basic frequency. If the frequency is not a multiple or submultiple, the partial is inharmonic. 2. When a system is maintained in steady forced vibration at a basic frequency equal to one of the frequencies of the normal modes of vibration of the system, the partials in the resulting complex tone are not necessarily identical in frequency with those of the other normal modes of vibration. (SP) [32]

partial automatic control Control that is a combination of manual and automatic control. For example, to cause a voltage reduction, the local automatic load tap changing closed-loop control may be biased by way of a supervisory control command. (SWG/PE/SUB) C37.100-1992, C37.1-1994

partial-automatic station A station that includes protection against the usual operating emergencies, but in which some or all of the steps in the normal starting or stopping sequence, or in the maintenance of the required character of service, must be performed by a station attendant or by supervisory control. (SWG/PE) C37.100-1992

partial-automatic transfer equipment (or throwover) Equipment that automatically transfers load to another (emergency) source of power when the original (preferred) source to which it has been connected fails, but that will not automatically retransfer the load to the original source under any conditions. *Note:* The restoration of the load to the preferred source from the emergency source upon the reenergization of the preferred source after an outage may be of the continuous-circuit restoration type or the interrupted-circuit restoration type. (SWG/PE) C37.100-1992

partial-body exposure Exposure that results when RF fields are substantially nonuniform over the body. Fields that are nonuniform over volumes comparable to the human body may occur due to highly directional sources, standing-waves, reradiating sources, or in the near field region of a radiating structure. *See also:* radio-frequency hot spot. (NIR) C95.1-1999

partial-body irradiation (electrobiology) Pertains to the case in which part of the body is exposed to the incident electromagnetic energy. *See also:* electrobiology. (NIR) C95.1-1982s

partial carry (parallel addition) A technique in which some or all of the carries are stored temporarily instead of being allowed to propagate immediately. *See also:* carry. (C) [85], [20]

(2) (A) (mathematics of computing) A carry process in which the carry digits are stored temporarily, instead of being processed as they occur. *Contrast:* complete carry. *See also:* cascaded carry; partial sum. **(B) (mathematics of computing)** The numeral that represents the carry digits generated in definition "A". (C) 1084-1986

partial checkback message Message from the initiating end is mirrored by the receiving end back to the initiating end to verify error-free transmission of the message. (SWG/PE/SUB) C37.100-1992, C37.1-1987s

partial cluster A subclass cluster in which an instance of the superclass may exist without also being an instance of any of the subclasses. *Contrast:* total cluster. *See also:* superclass. (C/SE) 1320.2-1998

partial correctness (software) In proof of correctness, a designation indicating that a program's output assertions follow logically from its input assertions and processing steps. *Contrast:* total correctness. (C) 610.12-1990

partial dial abandon or a partial dial timeout Occurs if the call is abandoned or times out without sufficient digits dialed. (COM/TA) 973-1990w

partial-dialing timing The time interval following each dialed digit except the last that determines if the call shall be treated as if dialing had stopped prematurely. For nonimmediate start trunk types, the partial-dial timing interval may be shorter. Instead of timing each digit, an alternative for multifrequency trunks is an overall time limit from the beginning of the start signal until end of pulsing. (COM/TA) 973-1990w

Partial Differential Equation Language (PDEL) An application-oriented language used for solving partial differential equations in which the user does not have to program the numerical analysis algorithms. *Note:* Used as a preprocessor to PL/1. (C) 610.13-1993w

partial directivity (of an antenna for a given polarization) In a given direction, that part of the radiation intensity corresponding to a given polarization divided by the total radiation intensity averaged over all directions. *Note:* The (total) directivity of an antenna, in a specified direction, is the sum of the partial directivities for any two orthogonal polarizations. (AP/ANT) 145-1993

partial discharge (PD) (1) (power and distribution transformers) An electric discharge which only partially bridges the insulation between conductors, and which may or may not occur adjacent to a conductor. *Notes:* 1. Partial discharges occur when the local electric-field intensity exceeds the dielectric strength of the dielectric involved, resulting in local ionization and breakdown. Depending on intensity, partial discharges are often accompanied by emission of light, heat, sound, and radio influence voltage (with a wide frequency range). 2. The relative intensity of partial discharge can be observed at the transformer terminals by measurement of the apparent charge (coulombs). However, the apparent charge (terminal charge) should not be confused with the actual charge transferred across the discharging element in the dielectric which in most cases cannot be ascertained. Partial discharges tests using the radio influence voltage techniques which are responsive to the apparent terminal charges are generally used for measurement of relative discharge intensity. 3. Partial discharges can also be detected and located using sonic techniques. 4. "Corona" has also been used to describe partial discharges. This is a non-preferred term since it has other unrelated meanings. (PE/TR) C57.12.80-1978r **(2) (dry-type transformers)** An electric discharge that only partially bridges the insulation between conductors. The term "corona" has also been used frequently with this connotation. Such usage is imprecise and is gradually being discontinued in favor of the term "partial discharge." (PE/PSIM/TR) 454-1973w, C57.124-1991r **(3)** A localized electric discharge resulting from ionization in an insulation system when the voltage stress exceeds the critical value. This discharge partially bridges the insulation between electrodes. (SWG/PE) 1291-1993r, C37.100-1992 **(4) (liquid-filled power transformers)** An electric discharge that only partially bridges the insulation between conductors. (PE/TR/PSIM) C57.113-1988s, 62-1995 **(5)** A discharge that does not completely bridge the insulation between electrodes. (PE/PSIM) 4-1995 **(6)** An electric discharge that only partially bridges the insulation between conductors, and that may or may not occur adjacent to the conductor. (PE/TR) C57.104-1991 **(7)** A discharge that does not completely bridge the insulation between electrodes. *Note:* The term "corona" is preferably reserved for partial discharges in air around a conductor, but not within the bushing assembly. (PE/TR) C57.19.03-1996

partial discharge energy (dielectric tests) The energy dissipated by an individual discharge. The partial discharge energy is expressed in joules. (PE/PSIM) 454-1973w

partial discharge extinction voltage (1) (dry-type transformers) The voltage at which partial discharges exceeding a specified level cease under specified conditions when the voltage is gradually decreased from a value exceeding the inception voltage. This voltage is expressed as the peak value divided by the square root of two. (PE/TR) C57.124-1991r **(2)** The voltage at which partial discharge (corona) is no longer detectable on instrumentation adjusted to a specified sensitivity, following application of a specified higher voltage. (SWG/PE/IC) 1291-1993r, 48-1996

partial discharge-free test voltage (dry-type transformers) A specified voltage, applied in accordance with a specified test procedure, at which the test object should not exhibit partial discharges above the acceptable energized background noise level. (PE/TR) C57.124-1991r, C57.113-1991

partial discharge inception voltage (1) (dry-type transformers) The lowest voltage at which partial discharges exceeding a specified level are observed under specified conditions when the voltage applied to the test object is gradually increased from a lower value. This voltage is expressed as the peak value divided by the square root of two. (PE/TR) C57.124-1991r **(2)** The voltage that should be recorded on the device or system under test when raised to a point where the PD signal rises above the energized background noise level. (SWG/PE) 1291-1993r

partial discharge power (dielectric tests) The power fed into the terminals of the test object due to partial discharges. The average discharge power is expressed in watts. (PE/PSIM) 454-1973w

partial duplex An operating condition that allows simultaneous communication in both send and receive directions with 3-20 dB switched loss in either direction. (COM/TA) 1329-1999

partial effective area (of an antenna for a given polarization and direction) In a given direction, the ratio of the available power at the terminals of a receiving antenna to the power flux density of a plane wave incident on the antenna from that direction and with a specified polarization differing from the receiving polarization of the antenna. (AP/ANT) 145-1993

partial failure *See:* failure.

partial-fraction expansion A sum of fractions that is used to represent a function that is a ratio of polynomials. The denominators of the fractions are the poles of the function. (CAS) [13]

partial gain (of an antenna for a given polarization) In a given direction, that part of the radiation intensity corresponding to a given polarization divided by the radiation intensity that would be obtained if the power accepted by the antenna were radiated isotropically. *Note:* The (total) gain of an antenna, in a specified direction, is the sum of the partial gains for any two orthogonal polarizations. (AP/ANT) 145-1993

partially dead region or layer (x-ray energy spectrometers) (of a semiconductor detector) Any region or layer on or in the detector which contributes an output pulse which is less than the full energy peak for that incident radiation. (NPS/NID) 759-1984r

partially inverted file A file that has been inverted on some of its secondary keys. *Contrast:* fully inverted file. (C) 610.5-1990w

partially polarized wave A wave with some randomly polarized content. (AP/PROP) 211-1997

partially shielded insulated splice (power cable joints) An insulated splice in which a conducting material is employed over a portion of the insulation for electric stress control. (PE/IC) 404-1986s

partial outage state The component or unit is at least partially energized, or is not fully connected to all of its terminals, or both, so that it is not serving some of its functions within the power system. *Note:* A unit composed of a three-terminal line would be in the partial outage state if it were disconnected from one terminal with two line sections still carrying power. (PE/PSE) 859-1987w

partial product The result obtained by multiplying the multiplicand by one of the digits of the multiplier. *Synonym:* intermediate product. (C) 1084-1986w

partial-read pulse Any one of the currents applied that cause selection of a cell for reading. *See also:* coincident-current selection. (Std100) 163-1959w

partial realized gain (of an antenna for a given polarization) The partial gain of an antenna for a given polarization reduced by the loss due to the mismatch of the antenna input impedance to a specified impedance. (AP/ANT) 145-1993

partial reference designation (electric and electronics parts and equipment) A reference designation that consists of a basic reference designation and which may include, as prefixes, some but not all of the reference designations that apply to the subassemblies or assemblies within which the item is located. (GSD) 200-1975w

partial-select output (A) The voltage response of an unselected magnetic cell produced by the application of partial-read pulses or partial-write pulses. **(B)** The integrated voltage response of an unselected magnetic cell produced by the application of partial-read pulses or partial-write pulses. *See also:* coincident-current selection. (Std100) 163-1959

partial sum The result obtained from the addition of two or more numbers without regard to carries. *Note:* In the binary numeration system, the partial sum is the same result as is obtained from the exclusive-OR operation. *See also:* cascaded carry. (C) 1084-1986w

partial system downtime A weighted time out of service for switching system failures that put a number of lines out of service simultaneously. This measure is the expected probability that two randomly chosen lines will be out of service simultaneously due to the same fault cause. It may be expressed in minutes per year. It protects customers under circumstances that are not fully covered by individual line downtime or total system downtime. *Synonym:* simultaneous line downtime. (COM/TA) 973-1990w

partial-write pulse Any one of the currents applied that cause selection of a cell for writing. *See also:* coincident-current selection. (Std100) 163-1959w

participant property A kind of property of a state class that reflects that class' knowledge of a relationship in which instances of the class participate. When a relationship exists between two state classes, each class contains a participant property for that relationship. A participant property is a mapping from a state class to a related (not necessarily distinct) state class. The name of each participant property is the name of the role that the other class plays in the relationship, or it may simply be the name of the class at the other end of the relationship (as long as using the class name does not cause ambiguity). A value of a participant property is the identity of a related instance. (C/SE) 1320.2-1998

participate With regard to the action of an S-module during message transmission, to execute the command contained in the HEADER packet of the current message and return packets as required by that command and by the state of the Acknowledge bit in the HEADER packet. The handling of some errors may cause an S-module to cease to participate in a message (e.g., by ceasing to execute the current command, by returning NULL packets when data is expected, by driving a constant value on the MSD signal without regard to packet transmission timing, etc). (TT/C) 1149.5-1995

participating slave A slave involved in a transaction as either a selected slave, an intervening slave, a broadcast slave, or a slave involved in multiple packet mode. (C/BA) 10857-1994, 896.3-1993w, 896.4-1993w

particle accelerator Any device for accelerating charged particles to high energies, for example, cyclotron, betatron, Van der Graaff generator, linear accelerator, etc. (ED) [45]

particle size distribution The probability density function describing the size distribution of particles in a medium. (AP/PROP) 211-1997

particle velocity (sound field) The velocity of a given infinitesimal part of the medium, with reference to the medium as a whole, due to the sound wave. *Note:* The terms instantaneous particle velocity, effective particle velocity, maximum particle velocity, and peak particle velocity have meanings that correspond with those of the related terms used for sound pressure. (SP) [32]

particulate A small particle that is created by thermal decomposition of organic materials present inside the generator. (PE/EM) 1129-1992r

parting The selective corrosion of one or more components of a solid solution alloy. (IA) [59]

parting limit The maximum concentration of a more-noble component in an alloy, above which parting does not occur within a specific environment. (IA) [59]

Partition An addressable portion of a Partitioned File. (IM/ST) 1451.1-1999

partition **(1)** **(A)** A portion of a computer's main storage that is set aside to hold a single program. **(B)** A portion of a storage medium that is set aside for some special purpose; for example, the boot partition of a magnetic disk contains operating system files from which the computer can be booted. **(C)** A portion of a storage medium that is treated as if it were an individual medium; as in a partition of a hard disk. (C) 610.10-1994
(2) A partition is a program or part of a program that can be invoked from outside the Ada implementation. Each partition may run in a separate address space, possibly on a separate computer. An *active partition* is a partition that contains at least one task. Every active partition has an *environment task*, on which all the other tasks of that partition depend. An active partition corresponds to a POSIX process. (C) 1003.5-1999
(3) A subdivision of a file. (IM/ST) 1451.1-1999
(4) A portion of a side of a cartridge that is accessible as a unit. (C/SS) 1244.1-2000

partitioned access The process of storing and retrieving data from storage in such a way that the data is divided into subunits, called members, and the data may be processed as a whole or member by member. *Note:* The directory used to retrieve each member is stored along with the data. *See also:* partitioned data set; basic partitioned access method. (C) 610.5-1990w

partitioned data set A file that is divided into subunits, called members, each of which may be processed individually. *Contrast:* indexed file; sequential file. (C) 610.5-1990w

Partitioned File An instance of the class `IEEE1451_PartitionedFile` or of a subclass Thereof. (IM/ST) 1451.1-1999

partitioning (software) (software requirements specifications) Decomposition; the separation of the whole into its parts. (C/SE) 830-1984s, 610.12-1990

partition noise Noise caused by random fluctuations in the distribution of current between the various electrodes. (ED) [45]

partly cloudy sky (illuminating engineering) A sky that has 30% to 70% cloud cover. (EEC/IE) [126]

Partner The remote entity in a Link Aggregation Control Protocol exchange. (C/LM) 802.3ad-2000

part programming, computer (numerically controlled machines) The preparation of a manuscript in computer language and format required to accomplish a given task. The necessary calculations are to be performed by the computer. (IA/EEC) [61], [74]

part programming, manual (numerically controlled machines) The preparation of a manuscript in machine control language and format required to accomplish a given task. The necessary calculations are to be performed manually. (IA/EEC) [61], [74]

parts (replacement parts for Class 1E equipment in nuclear power generating stations) Items from which the equipment is assembled (for example, resistors, capacitors, wires, connectors, transistors, tubes, lubricants, O-rings, and springs). (PE/NP) 934-1987w

parts per million $x/10^6$ is x parts per million. (CAS) [13]

parts per million by volume (PPMV) **(1)** The volume of water vapor in the total SF6 system at the pressure the system is operated. (SUB/PE) C37.122-1983s
(2) One million times the ratio of the volume of water vapor present in the gas to the total volume of the gas (including water vapor). (PE/IC) 1125-1993

parts per million by weight (PPMW) **(1)** The weight of water vapor in the total SF6 system at the pressure the system is operated. (SUB/PE) C37.122-1983s
(2) One million times the ratio of the weight of water vapor present in the gas to the total weight of the gas (including water vapor). (PE/IC) 1125-1993

parts program A set of computer instructions used to control the action of a numerical control machine in producing a particular manufactured part. (C) 610.2-1987

part-winding starter A starter that applies voltage successively to the partial sections of the primary winding of an alternating-current motor. *See also:* starter. (IA/ICTL/IAC) [60]

part-winding starting (rotating machinery) A method of starting a polyphase induction or synchronous motor, by which certain specially designed circuits of each phase of the primary winding are initially connected to the supply line. The remaining circuit or circuits of each phase are connected to the supply in parallel with initially connected circuits, at a predetermined point in the starting operation. (PE) [9]

party line (1) (data transmission) A subscriber line arranged to serve more than one main station, with discriminatory ringing for each station. (PE) 599-1985w
(2) (telephone switching systems) A line arranged to serve more than one main station, with distinctive ringing for each station. (COM) 312-1977w
(3) A communication channel that services multiple terminals. (SUB/PE) 999-1992w

Pascal A general-purpose programming language standardized by IEEE & ASC X3 adapted for use on a variety of computers; characterized by its ability to handle algorithms, various data types, and block-structured. *Note:* Often used in teaching programming concepts, Pascal was named after Blaise Pascal, a French mathematician, and was developed by Niklaus Wirth. *See also:* block-structured language; MODULA II; JOSEF; Rascal; high-order language. (C) 610.13-1993w

Paschen's law (gas) The law stating that, at a constant temperature, the breakdown voltage is a function only of the product of the gas pressure by the distance between parallel plane electrodes. *See also:* discharge. (ED) [45], [84]

pass (1) A single cycle in the processing of a set of data, usually performing part of an overall process. For example, a pass of an assembler through a source program; a pass of a sort program through a set of data. (C) 610.12-1990
(2) The lack of any deviation from the expected post condition of a test case signifies a pass for that specific test case. Adherence to specification or documentation, or functional baseline indicates a pass. (C/PA) 2000.2-1999

pass band (1) (data transmission) A range of frequency spectrum that can pass at low attenuation. *See also:* band-pass filter. (T&D/PE) 1308-1994, 599-1985w
(2) A range of frequencies transmitted to a terminal at low attenuation. *See also:* bandwidth. (C/CAS) 610.7-1995, [13]
(3) A band of frequencies that pass through a filter with little attenuation (relative to other frequency bands such as a stop band). (CAS/T&D/PE) 1308-1994

pass-band ripple The difference between maxima and minima of loss in a filter passband. If the differences are of constant amplitude then the filter is said to be equiripple. (CAS) [13]

pass element (power supplies) A controlled variable-resistance device, either a vacuum tube or power transistor, in series with the source of direct-current power. The pass element is driven by the amplified error signal to increase its resistance when the output needs to be lowered or to decrease its resistance when the output must be raised. *See also:* series regulator. (AES) [41]

passenger elevator An elevator used primarily to carry persons other than the operator and persons necessary for loading and unloading. *See also:* elevator. (EEC/PE) [119]

passenger information sign A device that displays or annunciates transit trip information to passengers. (VT) 1477-1998

passenger vessel A vessel that carries more than 12 persons in addition to the crew. (IA/MT) 45-1998

pass/fail criteria Decision rules used to determine whether a software item or a software feature passes or fails a test. (C/SE) 829-1998

pass gate The transistor(s), controlled by the word-line, to enable the programming of bits along the selected word-line and the isolating of bits on unselected word-lines from the array source. (ED) 1005-1998

passivation The process or processes (physical or chemical) by means of which a metal becomes passive. (IA) [59]

passivator An inhibitor that changes the potential of a metal appreciably to a more cathodic or noble value. (IA) [59]

passive-active cell A cell composed of passive and active areas. *See also:* electrolytic cell. (IA) [59]

passive angle tracking (PAT) A tracking technique that uses a received signal other than the backscattered radar emissions with which to track an object, jammer, or other signal source. Passive homing, home-on-jam (HOJ), and track-on-jam (TOJ) are examples of PAT using a radar receiving system. (AES) 686-1997

passive concentrator A type of token ring concentrator that contains no active elements in the signal path of any lobe port. Embedded repeater functions may be provided by the ring in and ring out port. (C/LM) 8802-5-1998

passive data dictionary A data dictionary that is only a repository for data definitions. *Note:* No active measures are taken to ensure that the data dictionary is consistent with the data items actually used in the system. *Synonym:* stand-alone data dictionary. *Contrast:* active data dictionary. (C) 610.5-1990w

passive device A device that does not require power and contains no active components. The term encompasses taps, directional couplers, splitters, power inserters, and in-line equalizers. (LM/C) 802.7-1989r

passive electric network An electric network containing no source of energy. *See also:* network analysis. (EEC/PE) [119]

passive filter A filter network containing only passive elements, such as inductors, capacitors, resistors and transformers. (CAS) [13]

passive fire protection The selection of materials that resist ignition and fire propagation, and produce low levels of fire products. (DEI) 1221-1993w

passive graphics A method of operation of a computer graphics system in which the graphics system does not accept input from a user and does not allow the user to influence its processing while it is in progress. *Contrast:* interactive graphics. (C) 610.6-1991w

passive homing guidance (navigation aid terms) Guidance in which a craft or missile is directed toward a destination by means of natural radiation from the destination. (AES/GCS) 172-1983w

passive limiter (nonlinear, active, and nonreciprocal waveguide components) A nonlinear device that suppresses input radio-frequency (rf) power without the aid of an external bias. (MTT) 457-1982w

passive satellite (communication satellite) A communication satellite that is a reflector and performs no active signal processing. (COM) [24]

passive sensor (test, measurement, and diagnostic equipment) A sensor requiring no source of power other than the signal being measured. (MIL) [2]

Passive-Star Coupler A component of a 10BASE-FP fiber optic mixing segment that divides optical power received at any of N input ports among all N output ports. The division of optical power is approximately uniform. (C/LM) 802.3-1998

passive station (data transmission) All stations on a multipoint network, other than the master and slave(s), that, during the information message transfer state, monitor the line for supervisory sequences, ending characters, etc. (PE) 599-1985w

passive test (test, measurement, and diagnostic equipment) A test conducted upon an equipment or any part thereof when the equipment is not energized. *Synonym:* cold test. (MIL) [2]

passive transducer A transducer that has no source of power other than the input signal(s), and whose output signal-power cannot exceed that of the input. *Note:* The definition of a passive transducer is a restriction of the more general passive network, that is, one containing no impressed driving forces. *See also:* transducer. (Std100) 270-1966w

passivity (A) (chemical) The condition of a surface that retards a specified chemical reaction at that surface. *See also:* electrochemistry. **(B) (electrolytic or anodic)** Such a condition of an anode that the normal anodic reaction is retarded. *See also:* electrochemistry. (EEC/PE) [119]

paste (dry cell) (primary cell) A gelatinized layer containing electrolyte that lies adjacent to the negative electrode. *See also:* electrolytic cell. (EEC/PE) [119]

pasted sintered plate (alkaline storage battery) A plate consisting of fritted metal powder in which the active material is impregnated. *See also:* battery. (EEC/PE) [119]

patch (1) To connect circuits together temporarily by means of a cord, known as a patch cord. (EEC/PE) [119]
(2) (computers) To modify a routine in a rough or expedient way. (C) [20], [85]
(3) (A) (software) A modification made directly to an object program without reassembling or recompiling from the source program. **(B) (software)** A modification made to a source program as a last-minute fix or afterthought. **(C) (software)** Any modification to a source or object program. **(D) (software)** To perform a modification as in definitions (A), (B), or (C). (C) 610.12-1990
(4) (computer graphics) A portion of the surface of a three-dimensional object, as displayed on a graphical display device. (See the corresponding figure.)

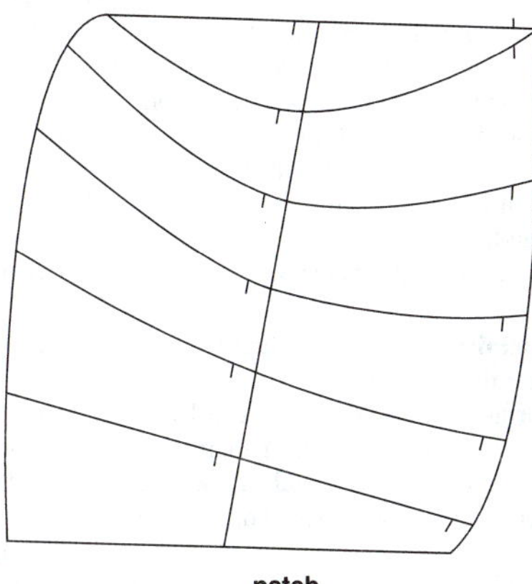

patch

(C) 610.6-1991w

patch bay (1) (analog computer) In an analog computer, a concentrated assembly of the inputs and outputs of computing elements, control elements, tie points, reference voltages, and ground points that offers a means of electrical connection. (C) 165-1977w
(2) A specially designed rewireable panel that allows its user to dynamically rewire or perform analog programming. *Synonym:* wiring panel. (C) 610.10-1994w

patch board A specially designed reconfigurable connection board used to prototype or test integrated circuits. *Synonym:* patch panel. *See also:* problem board. (C) 610.10-1994w

patchcord (test, measurement, and diagnostic equipment) An interconnecting cable for plugging or patching between terminals; commonly employed on patchboard, plugboard, and in maintenance operations. (MIL) [2]

patch cord Flexible cable unit or element with connectors(s) used to establish connections on a patch panel. (C/LM) 802.3-1998

patch panel *See:* electronic analog computer; problem board.

patent A document protecting an invention that grants an inventor the right to prevent others from making, using, or selling the invention for 20 years from the date of application for patent. (C/SE) 1420.1b-1999

path (1) (navigation) (navigation aid terms) A line connecting a series of points in space and constituting a proposed or traveled route. *See also:* course line; flight track; flight path. (AES/GCS) 172-1983w
(2) (network analysis) Any continuous succession of branches, traversed in the indicated branch directions. (CAS) 155-1960w
(3) (telephone switching systems) The set of links and junctors joined in series to establish a connection. (COM) 312-1977w
(4) (A) (telecommunications switching systems) A continuous physical connection. **(B) (telecommunications switching systems)** A time slot in a shared facility. (COM/TA) 973-1990
(5) (A) (software) In software engineering, a sequence of instructions that may be performed in the execution of a computer program. **(B) (software)** In file access, a hierarchical sequence of directory and subdirectory names specifying the storage location of a file. (C) 610.12-1990
(6) In the critical path method, any sequence of activities that goes from the beginning to the end of a project. (C) 610.2-1987
(7) (A) (data management) In a hierarchical database, a sequence of segments encountered in traversing from the root segment to an individual dependent segment. **(B) (data management)** With respect to a network or graph, some sequence of nodes such that each successive node is connected to its predecessor by an edge. *See also:* simple path. (C) 610.5-1990
(8) A bridged route between a source and a destination. (LM/C/CC) 8802-2-1998
(9) The concatenation of all the physical links between the link layers of two nodes. (C/MM) 1394-1995
(10) The sequence of segments and repeaters providing the connectivity between two DTEs in a single collision domain. In CSMA/CD networks there is one and only one path between any two DTEs. (LM/C/LM) 802.3-1998
(11) (data transmission) *See also:* channel. 599-1985w

path analysis (software) Analysis of a computer program to identify all possible paths through the program, to detect incomplete paths, or to discover portions of the program that are not on any path. (C) 610.12-1990

path assertion *See:* common ancestor constraint.

path cofactor *See:* path factor.

path condition (software) A set of conditions that must be met in order for a particular program path to be executed. (C) 610.12-1990

path delay value (PDV) The sum of all segment delay values for all segments along a given path. (C/LM) 802.3-1998, 802.9a-1995w

path expression A logical expression indicating the input conditions that must be met in order for a particular program path to be executed. (C) 610.12-1990

path factor (network analysis) The graph determinant of that part of the graph not touching the specified path (loop). *Notes:* 1. A path (loop) factor is obtainable from the graph determinant by striking out all terms containing transmittance products of loops that touch that path (loop). 2. For loop Lk, the loop factor is $-\partial\Delta/\partial \boldsymbol{L}_k$. (CAS) 155-1960w

path length (1) The length of a magnetic flux line in a core. *Note:* In a toroidal core with nearly equal inside and outside diameters, the value

$$l_m = \frac{\pi}{2}(\text{O.D.} + \text{I.D.})$$

where O.D. and I.D. are the outside and inside diameters of the core, is commonly used. (Std100) 163-1959w
(2) (interferometer fiber optic gyro) (laser gyro) The geometrical length of the path traversed in a single pass by an optical beam. *See also:* optical path length. (AES/GYAC) 528-1994

path loss The amount of attenuation between a headend port and a user outlet port. (LM/C) 802.7-1989r

pathname (1) A string that is used to identify a file. A pathname consists of, at most, {PATH_MAX} bytes, including the terminating null character. It has an optional beginning slash, followed by zero or more filenames separated by slashes. If the pathname refers to a directory, it may also have one or more trailing slashes. Multiple successive slashes are considered to be the same as one slash. A pathname that begins with two successive slashes may be interpreted in an implementation-defined manner, although more than two leading slashes shall be treated as a single slash. (C/PA) 9945-1-1996, 9945-2-1993
(2) A POSIX.1 pathname with characters drawn from the POSIX.1 portable character set. (C/PA) 1387.2-1995
(3) A nonempty string that is used to identify a file. A pathname consists of, at most, `POSIX_Limits.Pathname_Maxima'Last` components of type `POSIX.POSIX_Character`. It has an optional beginning slash followed by zero or more filenames separated by slashes. If the pathname refers to a directory, it may also have one or more trailing slashes. Multiple successive slashes are considered the same as one slash. A pathname that begins with two successive slashes may be interpreted in an implementation-defined manner, although more than two leading slashes shall be treated exactly the same as a single slash. (C) 1003.5-1999

pathname character string A sequence of characters from the portable filename character, including the / (slash) character. Within software definition files of exported catalogs, all such strings shall be encoded using IRV. (C/PA) 1387.2-1995

pathname component *See:* filename.

pathname resolution Pathname resolution is performed for a process to resolve a pathname to a particular file in a file hierarchy. There may be multiple pathnames that resolve to the same file. Each filename in the pathname is located in the directory specified by its predecessor (for example, in the pathname fragment "a/b", file "b" is located in directory "a"). Pathname resolution fails if this cannot be accomplished. If the pathname begins with a slash, the predecessor of the first filename in the pathname is taken to be the root directory of the process (such pathnames are referred to as absolute pathnames). If the pathname does not begin with a slash, the predecessor of the first filename of the pathname is taken to be the current working directory of the process (such pathnames are referred to as "relative pathnames"). The interpretation of a pathname component is dependent on the values of {NAME_MAX} and {_POSIX_NO_TRUNC} associated with the path prefix of that component. If any pathname component is longer than {NAME_MAX}, and {_POSIX_NO_TRUNC} is in effect for the path prefix of that component, the implementation shall consider this an error condition. Otherwise, the implementation shall use the first {NAME_MAX} bytes of the pathname component. The special filename, dot, refers to the directory specified by its predecessor. The special filename, dot-dot, refers to the parent directory of its predecessor directory. As a special case, in the root directory, dot-dot may refer to the root directory itself. A pathname consisting of a single slash resolves to the root directory of the process. A null pathname is invalid. (C/PA) 9945-1-1996, 9945-2-1993

pathocathode radiant sensitivity *See:* cathode radiant sensitivity.

pathological coupling A type of coupling in which one software module affects or depends upon the internal implementation of another. *Contrast:* control coupling; data coupling; content coupling; common-environment coupling; hybrid coupling. (C) 610.12-1990

path prefix A pathname, with an optional ending slash, that refers to a directory. (PA/C) 9945-1-1996, 9945-2-1993, 1003.5-1999

path testing Testing designed to execute all or selected paths through a computer program. *Contrast:* branch testing; statement testing. (C) 610.12-1990

path transmittance (network analysis) The product of the branch transmittances in that path. (CAS) 155-1960w

path variability value (PVV) (1) A value that is bounded by the sum of the segment variability values (SVVs) for all the segments along a given path. (C/LM) 802.9a-1995w
(2) The sum of all segment variability values for all the segments along a given path. (C/LM) 802.3-1998

pathway A facility for the placement of telecommunications. (IA/PSE) 1100-1999

patient care information system A broad classification of computational-based systems used for clinical support of patient care. For the purposes of this standard the scope of patient care information systems is further limited to medical device data exchange. (EMB/MIB) 1073-1996

patient care-related electrical appliance (health care facilities) An electrical appliance that is intended to be used for diagnostic, therapeutic or monitoring purposes in a patient care area. (EMB) [47]

patient care system (PCS) A patient care information system which is actively connected to a 1073-type communications link. (EMB/MIB) 1073-1996

patient data management system *See:* patient care information system.

patient equipment grounding point (health care facilities) A jack or terminal which serves as the collection point for redundant grounding of electric appliances serving a patient vicinity or for grounding other items in order to eliminate electromagnetic interference problems. (EMB) [47]

patient grounding point (health care facilities) A jack or terminal bus which serves as the collection point for redundant grounding of electric appliances serving a patient vicinity. (NESC/NEC) [86]

patient lead (health care facilities) Any deliberate electrical connection which may carry current between an appliance and a patient. This may be a surface contact (for example, an electrocardiogram (ECG) electrode); an invasive connection (for example, implanted wire or catheter); or an incidental long-term connection (for example, conductive tubing). It is not intended to include adventitious or casual contacts such as pushbutton, bed surface, lamp, hand-held appliance, etc. (EMB) [47]

patient vicinity (health care facilities) In an area which patients are normally cared for the patient vicinity is the space with surfaces likely to be contacted by the patient or an attendant who can touch the patient. This encloses a space within the room 6 ft beyond the perimeter of the bed in its nominal location, and extending vertically 7.5 ft above the floor. (NESC/NEC) [86]

patina A green coating consisting principally of basic sulfate and occasionally containing small amounts of carbonate or chloride, that forms on the surface of copper or copper alloys exposed to the atmosphere a long time. (IA) [59]

patrol tour An inspection by a member of the security organization along a predetermined route to observe the route area's security conditions. (PE/NP) 692-1997

patrol tour stations (nuclear security systems) Points along patrol tour routes where security force member progress is acknowledged. (PE/NP) 692-1986s

pattern (1) (image processing and pattern recognition) A meaningful regularity that can be used to classify objects or other items of interest. (C) 610.4-1990w
(2) (computer graphics) A series of repeated entities or a repetitious design within a closed boundary on a display surface. (See the corresponding figure.)

pattern

(C) 610.6-1991w

(3) (test pattern language) The sequence of addresses and data used to test a semiconductor memory. (TT/C) 660-1986w
(4) A sequence of characters used either with RE notation or for pathname expansion as a means of selecting various character strings or pathnames, respectively. The syntaxes of the two patterns are similar, but not identical; this standard always indicates the type of pattern being referred to in the immediate context of the use of the term. (C/PA) 9945-2-1993
(5) One or more vectors comprising a functionality test for a specific portion of a device under test (DUT). (C/TT) 1450-1999
(6) *See also:* radiation pattern.

pattern class One of a set of mutually exclusive categories into which a pattern can be classified. *Synonyms:* class; category. (C) 610.4-1990w

pattern classification The process of assigning patterns to pattern classes. *Synonym:* pattern identification. (C) 610.4-1990w

pattern distortion (oscilloscopes) Any deformation of the pattern from its intended form. (IEC 151-14.). *Notes:* 1. In an oscilloscope the intended form is rectilinear and rectangular. 2. An oscilloscope control that affects pattern distortion may be labeled "pattern" or "geometry." (IM) 311-1970w

pattern, heating *See:* heating pattern.

pattern identification *See:* pattern classification.

pattern insensitivity The ability to write, store, and read streams of data without regard to the sequence of the data and the pattern of distribution of ones and zeros across the memory locations. (ED) 641-1987w

pattern jitter *See:* correlated jitter.

pattern-propagation factor Ratio of the field strength that is actually present at a point in space to that which would have been present if free-space propagation had occurred with the antenna beam directed toward the point in question. This factor is used in the radar equation to modify the strength of the transmitted or received signal to account for the effect of multipath propagation, diffraction, refraction, and pattern of an antenna. (AES) 686-1997

pattern recognition (1) The identification of shapes, forms, or configurations by automatic means. (C) [20], [85]
(2) (image processing and pattern recognition) The analysis, description, identification, and classification of objects or other meaningful regularities by automatic or semiautomatic means. *Synonym:* machine recognition. (C) 610.4-1990w

patterns A set of unit under test (UUT) stimulus and expected response states. A pattern contains one unit of logic state (0, 1, X, Z) data for each UUT input and each UUT output pin. (SCC20) 1445-1998

pattern segmentation The process of determining which regions or areas of in-terest in an image or other set of data constitute patterns of interest for pattern recognition. (C) 610.4-1990w

pattern select A broadcast address specifying that all devices seeing the broadcast remain attached to the master only if their T pins are asserted during the immediately ensuing write data cycle. (NID) 960-1993

pattern-sensitive fault A fault that appears in response to some particular pattern of data. *See also:* data-sensitive fault. (C) [20], [85]

pause (1) To suspend the execution of a computer program. *Synonym:* halt. *Contrast:* stop. (C) 610.12-1990
(2) A mechanism for full duplex flow control. (C/LM) 802.3-1998

pause instruction A computer instruction that specifies suspension of the execution of a computer program. *Note:* A pause instruction does not cause an exit from the program. *Synonym:* halt instruction. *See also:* stop instruction; optional-pause instruction. (C) 610.10-1994w

Pause Interrupt Enabled (PIE) bit A bit in the Slave Status register of every S-module that is set to indicate that the S-module may generate an interrupt during the PAUSE Slave Controller state when the S-module is addressed. (TT/C) 1149.5-1995

PAX *See:* private automatic exchange.

payback A financial analysis technique where the cost to implement a project is compared to the annual savings due from the project. (SCC22) 1346-1998

payload (1) The information bits (within a frame). (COM/TA) 1007-1991r
(2) The data (a message, a memory access request, an acknowledgment, etc.) that is to be transferred from the source node to the destination node. It has a specific format, defined in the transaction layer. Note that a payload may be null. *See also:* packet. (C/BA) 1355-1995
(3) The portion of a primary packet that contains data defined by an application layer. (C/MM) 1394-1995
(4) The portion of a primary packet that contains data defined by an application. (C/MM) 1394a-2000

payout site *See:* tension site.

pay station *See:* public telephone station.

P-band A letter-band designation no longer applicable and which should not be used. Originally it denoted frequencies in the vicinity of 230 MHz, which are no longer allowed for radar usage. Later it was sometimes applied to denote the 420–450 MHz International Telecommunication Union (ITU) allocated radiolocation band that is now called UHF by IEEE Std 521-1984, IEEE Standard Letter Designations for Radar-Frequency Bands. (AES) 686-1997

PBX *See:* private branch exchange.

PBX trunk *See:* private-branch-exchange trunk.

PC *See:* personal computer; printed circuit.

PCA *See:* physical configuration audit.

PCB *See:* polychlorinated biphenyl; askarel.

PCC *See:* point of common coupling.

PCD *See:* POSIX Conformance Document.

P channel (1) (for the ISLAN16-T) A 10 Mbit/s packet channel, that provides a Physical Layer service to an ISO/IEC 8802-3 MAC. (C/LM) 802.9a-1995w
(2) A full duplex packet channel that provides IEEE 802 MAC Layer services. The P channel may optionally carry CCITT Q.93x (where x refers to the family of CCITT Q.930 protocols) call control in a logically out-of-band fashion, which may be highlighted by referring to it as P_D. Presently there is only one P channel defined per ISLAN interface. The minimum size of the P channel shall be limited by the application requirements of the topology that represents the connection of the AU to the backbone LAN application. The support of station management communication will be an important factor in the characterization of P channel bandwidth size. (LM/C/COM) 8802-9-1996

p-channel device (metal-nitride-oxide field-effect transistor) Insulated-gate field-effect transformer (IGFET) where source and drain are regions of p-type conductivity. (ED) 581-1978w

p-channel metal-oxide semiconductor (PMOS) A type of semiconductor technology which employs metal oxide field effect transistors, using holes to conduct current in the semiconductor channel. *Note:* The channel has a predominantly positive charge. *Contrast:* n-channel metal-oxide semiconductor. *See also:* complementary metal-oxide semiconductor. (C) 610.10-1994w

PCM *See:* punched card machine; pulse code modulation.

PC parallel port The parallel printer port used as the parallel interface for most printers and supported by most PCs. This interface has been variously defined by different PC and peripheral manufacturers. This standard describes the more prevalent variations of the interface and defines a family of signaling methods that are backward compatible with the typical PC parallel port. (C/MM) 1284-1994

PCS *See:* print contrast signal; power conditioning subsystem; physical coding sublayer; patient care system.

PCSN *See:* private circuit switching network.

PCTP *See:* POSIX Conformance Test Procedure.

PCTS *See:* POSIX Conformance Test Suite.

P.D *See:* control action, proportional plus derivative.

PD *See:* partial discharge.

PDA *See:* parallel disk array.

PDEL *See:* Partial Differential Equation Language.

PDES *See:* Product Data Exchange Specification.

P-display A name for the type of display commonly known as plan-position indicator (PPI). (AES) 686-1997

PDL *See:* program design language.

PDM *See:* pulse-duration modulation.

PDN *See:* public data network.

PDR *See:* preliminary design review.

PDS *See:* partitioned data set.

PDS/MaGen *See:* Problem Descriptor System.

PDU *See:* protocol data unit; command protocol data unit.

PDV *See:* path delay value.

PE Conduit fabricated from polyethylene. (SUB/PE) 525-1992r

PEAK Channel number corresponding to the peak of a distribution. (NPS) 398-1972r

peak *See:* line.

peak alternating gap voltage (electron tube) (traveling-wave tubes) The negative of the line integral of the peak alternating electric field taken along a specified path across the gap. *Note:* The path of integration must be stated. (ED) 161-1971w

peak anode current The maximum instantaneous value of the anode current. *See also:* electronic controller. (IA/IAC) [60]

peak burst magnitude (audio and electroacoustics) The maximum absolute peak value of voltage, current, or power for a burstlike excursion. *See also:* burst duration; burst. (SP) [32]

peak cathode current (steady-state) The maximum instantaneous value of a periodically recurring cathode current. (ED) 161-1971w

peak-charge characteristic (nonlinear capacitor) The function relating one-half the peak-to-peak value of transferred charge in the steady state to one-half the peak-to-peak value of a specified applied symmetrical alternating capacitor voltage. *Note:* Peak-charge characteristic is always single-valued. *See also:* nonlinear capacitor. (ED) [46]

peak code The codes that will produce the maximum decoder output level. (COM/TA) 1007-1991r

peak current (low-voltage dc power circuit breakers used in enclosures) The instantaneous value of current at the time of its maximum value. (SWG/PE) C37.14-1979s, C37.100-1992

peak demand *See:* maximum demand.

peak detector (overhead-power-line corona and radio noise) A detector, the output voltage of which is the true peak value of an applied signal or noise. (T&D/PE) 539-1990

peak discharge current The peak current occurring during a transient discharge. *Note:* Due to the very short discharge time, substantial peak currents (up to a few amperes) can be encountered in typical induction circumstances. (T&D/PE) 539-1990

peak distortion (data transmission) The largest total distortion of telegraph signals noted during a period of observation. (PE) 599-1985w

peak electrode current (electron tube) The maximum instantaneous current that flows through an electrode. *See also:* electrode current. (ED) [45]

peak error The residual with the largest absolute value. (IM/WM&A) 1057-1994w

peak flux density The maximum flux density in a magnetic material in a specified cyclically magnetized condition. (Std100) 163-1959w

peak forward anode voltage (electron tube) The maximum instantaneous anode voltage in the direction in which the tube is designed to pass current. *See also:* electronic controller; electrode voltage. (ED) [45]

peak forward current rating (rectifier circuit element) (repetitive) The maximum repetitive instantaneous forward current permitted by the manufacturer under stated conditions. *See also:* average forward current rating. (IA) [62]

peak forward voltage (of a rectifying element) The maximum instantaneous voltage between the anode and cathode during the positive nonconducting period. *See also:* rectification. (EEC/PE) [119]

peak full-width-half-maximum calibration coefficients The full-width-half-maximum (also called shape) (F) versus channel number (Ch) coefficients as

$$F = P + Q \cdot Ch^{I} + R \cdot Ch^{2I} + W \cdot Ch^{3I}$$

with the coefficients, P, Q, R, and W stored as four successive 14-character numbers including the decimal point, and I as a four-character number including the decimal point. Leading spaces are interpreted as zeros. Any values not used or calculated should be set to all spaces. The P term is usually called the offset or zero intercept. The I is the lowest exponent of the channel number. In most cases I will be 0.5 for a quadratic dependance of the FWHM with channel and I will be 1.0 for linear dependance of the FWHM with channel. The Q term is the multiplier for the lowest power dependance of the FWHM-channel curve. The R term is the multiplier of the second exponent term of the FWHM-channel curve. The W term is the multiplier of the third exponent term of the FWHM-channel curve. (NPS/NID) 1214-1992r

peak induction (of toroidal magnetic amplifier cores) The magnetic induction corresponding to the peak applied magnetizing force specified. *Note:* It will usually be slightly less than the true saturation induction. *Synonym:* peak flux density. (Std100) 106-1972

peaking circuit A circuit capable of converting an input wave into a peaked waveform. (EEC/PE) [119]

peaking network A type of interstage coupling network in which an inductance is effectively in series (series peaking network) or in shunt (shunt peaking network) with the parasitic capacitance to increase the amplification at the upper end of the frequency range. *See also:* network analysis. (EEC/PE) [119]

peaking station (power operations) A generating station that is normally operated to provide power only during maximum load periods. (PE/PSE) 858-1987s, 346-1973Tw

peaking time (1) (semiconductor radiation detectors) The time elapsed from the first zero crossing of the defined zero level to the departure from peak amplitude of a pulse equal to the maximum rated amplifier output. (NID) 301-1976s
(2) (germanium gamma-ray detectors) (charged-particle detectors) (of an amplifier output pulse) The time between the 1% amplitude point on the leading edge and the 100% amplitude point of a pulse (provided that the pulse does not have a flat top). For flat-top pulses, the peaking time is defined as the time between the 1% amplitude point and the midpoint of the flat top. (NPS) 325-1986s
(3) The time interval between the 1% point on the first transition and the top centerline. (NPS) 300-1988r

peaking time tp (of a pulse) The interval between the 1% point on the first transition, with respect to the peak height, and the top center line. *See also:* transition; top centerline. (NPS) 325-1996

peak inrush current (electronic power transformer) The peak instantaneous current value resulting from the excitation of the transformer with no connected load, and with essentially zero source impedance, and using the minimum turns primary tap and rated voltage. (PEL/ET) 295-1969r

peak instantaneous sound pressure level (measurement of sound pressure levels of ac power circuit breakers) Maximum unweighted positive or negative pressure peak value reached by an impulsive sound wave at any time during the period of observation. Unit: decibel (dB). Readings can be considered as peak instantaneous sound pressure level if the C-weighting is used and the response time of the instrument is 50 μs or less. Peak instantaneous sound pressure level is sometimes referred to as impact noise. (SWG/PE) C37.100-1992, C37.082-1982r

peak inverse anode voltage (electron tube) The maximum instantaneous anode voltage in the direction opposite to that in which the tube is designed to pass current. *See also:* electrode voltage; electronic controller. (ED) [45]

peak inverse voltage (PIV) (semiconductor diode) The maximum instantaneous anode-to-cathode voltage in the reverse direction that is actually applied to the diode in an operating circuit. *Notes:* 1. This is an applications term not to be confused with breakdown voltage, which is a property of the device. 2. In semiconductor work the preferred term is peak reverse voltage. *See also:* semiconductor; peak reverse voltage. (ED) 216-1960w

peak inverse voltage, maximum rated (semiconductor diode) The recommended maximum instantaneous anode-to-cathode voltage that may be applied in the reverse direction. *See also:* semiconductor. (ED) 216-1960w

peak jitter *See:* jitter.

peak let-through characteristic curve *See:* current-limiting characteristic curve.

peak let-through current (1) (protection and coordination of industrial and commercial power systems) The maximum instantaneous current through a current-limiting fuse during the total clearing time. Since this is an instantaneous value, it may well exceed the root-mean-square (rms) available current, but will be less than the peak current available without a fuse in the circuit if the fault level is high enough for it to operate in its current-limiting mode. (IA/PSP) 242-1986r
(2) The highest current flowing in the circuit following the inception of the fault that the circuit breaker and the protected system must withstand, expressed as an instantaneous rather than an rms value. (IA/PSP) 1015-1997

peak let-through cutoff current (of a current-limiting fuse) The highest instantaneous current passed by the fuse during the interruption of the circuit. (SWG/PE) C37.100-1992, C37.40-1993

peak limiter A device that automatically limits the magnitude of a signal to a predetermined maximum value by changing its amplification. *Notes:* 1. The term is frequently applied to a device whose gain is quickly reduced and slowly restored when the instantaneous magnitude of the input exceeds a predetermined value. 2. In this context, the terms instantaneous magnitude and instantaneous peak power are used interchangeably. (BT/AV) [34]

peak load (1) (A) The maximum load consumed or produced by a unit or group of units in a stated period of time. It may be the maximum instantaneous load or the maximum average load over a designated interval of time. *Note:* Maximum average load is ordinarily used. In commercial transactions involving peak load (peak power) it is taken as the average load (power) during a time interval of specified duration occurring within a given period of time, that time interval being selected during which the average power is greatest. *See also:* generating station. **(B)** The maximum load of a specified unit or group of units in a stated period of time. (T&D/PE/IA/PSE) [10], 241-1990, 141-1993
(2) (rotating machinery) (motors) The largest momentary or short-time load expected to be delivered by a motor. It is expressed in percent of normal power or normal torque. *See also:* asynchronous machine. (PE) [9]

peak load duty (thyristor converter) A type of duty where the rating of the converter is specified in terms of the magnitude and duration of the peak load together with the time of no-load between peaks. (IA/IPC) 444-1973w

peak load station (electric power supply) A generating station that is normally operated to provide power during maximum load periods. *See also:* generating station. (PE/PSE) [54]

peak magnetizing force (1) (toroidal magnetic amplifier cores) The maximum value of applied magnetomotive force per mean length of path of the core. (MAG) 393-1977s
(2) (peak field strength) The upper or lower limiting value of magnetizing force associated with a cyclically magnetized condition. (Std100) 163-1959w

peak nominal varistor voltage (low voltage varistor surge arresters) Voltage across the varistor measured at a specified pulsed direct-current (dc) current of specific duration coincident with a specified alternating-current (ac) current crest. (PE) [8]

peak overvoltages (for current-limiting fuses) The peak value of the voltage that can exist across a current-limiting fuse during its arcing interval. (SWG/PE) C37.100-1992, C37.40-1993

peak percent curve *See:* integrated energy curve.

peak point (tunnel-diode characteristic) The point on the forward current-voltage characteristic corresponding to the lowest positive (forward) voltage at which d/dV = 0.

peak-point current (tunnel-diode characteristic) The current at the peak point. *See also:* peak point. (ED) 253-1963w, [46]

peak-point voltage (tunnel-diode characteristic) The voltage at which the peak point occurs. *See also:* peak point. (ED) 253-1963w, [46]

peak power density The maximum instantaneous power density occurring when power is transmitted. (NIR) C95.1-1999

peak power, instantaneous *See:* instantaneous peak power.

peak power output (modulated carrier system) The output power, averaged over a carrier cycle, at the maximum amplitude that can occur with any combination of signals to be transmitted. *See also:* television; radio transmitter. (AP/ANT) 145-1983s

peak power pulse (waveguide) The root-mean-square (rms) value of rectangular pulse of radio frequency (RF) power passing through the transverse section of a waveguide. (MTT) 146-1980w

peak pulse amplitude (television) The maximum absolute peak value of the pulse excluding those portions considered to be unwanted, such as spikes. *Note:* Where such exclusions are made, it is desirable that the amplitude chosen be illustrated pictorially. *See also:* pulse. (PE) 599-1985w

peak pulse power, carrier-frequency The power averaged over that carrier-frequency cycle that occurs at the maximum of the pulse of power (usually one half the maximum instantaneous power). (IM/WM&A) 194-1977w

peak radiant responsivity (diode-type camera tube) The peak value of the spectral response of the tube usually specified together with the wavelength at which it occurs. Units: amperes watt^{-1} (AW^{-1}). (ED) 503-1978w

peak repetitive ON-state current (thyristor) The peak value of the on-state current including all repetitive transient currents. *See also:* principal current. (ED/IA) [46], [123], [12], [62]

peak responsibility The load of a customer, a group of customers, or a part of the system at the time of occurrence of the system peak load. *See also:* generating station. (T&D/PE) [10]

peak restriking voltage (surge arresters) The maximum instantaneous voltage that is attained by the re-striking voltage. (PE) [8], [84]

peak reverse voltage (semiconductor rectifiers) The maximum instantaneous value of the reverse voltage that occurs across a semiconductor rectifier device, or rectifier stack. *See also:* rectification. (IA) [12]

peak sound pressure (for any specified time interval) The maximum absolute value of the instantaneous sound pressure in that interval. *Note:* In the case of a periodic wave, if the time interval considered is a complete period, the peak sound pressure becomes identical with the maximum sound pressure. (SP) [32]

peak switching current (rotating machinery) The maximum peak transient current attained following a switching operation on a machine. *See also:* asynchronous machine. (PE) [9]

peak-to-Compton ratio for the 1332 keV ^{60}Co peak The ratio of the full-energy peak height, for ^{60}Co measured at 1332 keV, to the average height of the corresponding Compton plateau between 1040 keV and 1096 keV. (NI) N42.14-1991

peak torque (electric coupling) The maximum torque an electric coupling will transmit or develop for any speed relation on input and output members, with rated excitation and at specified operating conditions. (EM/PE) 290-1980w

peak value (1) (A) (of alternating voltage) The maximum value, disregarding small high-frequency oscillations (greater than 10 kHz) such as those arising from partial discharges. **(B)** (of impulse voltages) The maximum value of impulses that are smooth double exponential waves without overshoot. (PE/PSIM) 4-1995
(2) (surge arresters) (voltage or current) The maximum value of an impulse. If there are small oscillations superimposed at the peak, the peak value is defined by the maximum value and not the mean curve drawn through the oscillations. (PE) [8], [84]
(3) (electrical measurements in power circuits) The largest absolute value (y_p) of y. (PE/PSIM) 120-1989r
(4) *See also:* crest value. (SPD/PE) C62.22-1997

peak wavelength (1) (fiber optics) The wavelength at which the radiant intensity of a source is maximum. *See also:* spectral width; spectral line. (Std100) 812-1984w
(2) (light-emitting diodes) The wavelength at which the spectral radiant intensity is a maximum. (ED) [127]

peak withstand current (1) (of an air switch) The crest value of the total current during the maximum cycle that a switch is required to carry at rated frequency. (SWG/PE) C37.34-1994
(2) The maximum instantaneous current at the major peak of an offset power-frequency sinusoidal current that a switch is required to carry. (SWG/PE) 1247-1998

peak working voltage (1) (pulse transformers) The maximum instantaneous voltage stress that may appear under operation across the insulation being considered, including abnormal and transient conditions. (PEL/ET) 390-1987r
(2) (charging inductors) The algebraic sum of the maximum alternating crest voltage and the direct voltage of the same polarity appearing between the terminals of the inductor winding or between the inductor winding and the grounded elements. (MAG) 306-1969w
(3) (corona measurement) The maximum instantaneous voltage that may appear under normal rated conditions across the insulation being considered. This insulation may be within a winding, between windings, or between winding and grounds. (MAG/ET) 436-1977s

peanut *See:* rope connector.

PEARL *See:* Process and Experiment Automation Realtime Language.

PEC *See:* protocol error counter.

Pederson ray The upper ionospheric ray in oblique-incidence propagation. *See also:* junction frequency. (AP/PROP) 211-1997

pedestal A substantially flat-topped pulse that elevates the base level for another wave. *See also:* pulse. (EEC/PE) [119]

pedestal bearing (rotating machinery) A complete assembly of a bearing with its supporting pedestal. (PE) [9]

pedestal bearing insulation (rotating machinery) The insulation applied either below the bearing liner shell and the adjacent pedestal support or between the base of the pedestal and the machine bed plate, to break the current path that may be formed through the shaft to the outboard bearing to the frame to the drive-end bearing and thence back to the shaft. *Note:* The voltage is usually very low. However, very destructive bearing currents can flow in this path if some insulating break is not provided. High-pressure moulded laminates are usually employed for this type of insulation. (PE) [9]

pedestal delay time (amplitude, frequency, and pulse modulation) The time elapsed between the application of an electronic command signal to the electronic driver and the time the diffracted light reaches the 10% intensity point. (UFFC) [17]

peeling The unwanted detachment of a plated metal coating from the base metal. *See also:* electroplating. (EEC/PE) [119]

peer (1) Elements of a distributed system that communicate with each other using a common protocol. (DIS/C) 1278.2-1995
(2) Service layer on a remote node at the same level. For instance a "peer link layer" is the link layer on a different node. (C/MM) 1394-1995

peer-entity authentication The corroboration that a peer entity in an association is the one claimed. This service, when provided by the (N)-layer, provides corroboration to the (N+1)-entity that the peer entity is the claimed (N+1)-entity. This is primarily intended for, although not limited to, connection-oriented service and may be either unilateral or mutual. (LM/C) 802.10-1992

peer graphics *See:* information graphics.

peer Port (of a Virtual Port) A Virtual Port is a peer (Virtual) Port of a given Virtual Port when both Virtual Ports attach to the same Group and both represent the capability for communication between the Remote Bridges to which they belong. (C/LM) 802.1G-1996

peer protocol The sequence of message exchanges between two entities in the same layer that utilize the services of the underlying layers to effect the successful transfer of data and/or control information from one location to another location. (C/LM/CC) 8802-2-1998

peer-to-peer communication (A) Communication between two or more processes or programs by which both computers can exchange data freely. *Note:* Any physical differences between the computers are rendered transparent to the application. **(B)** Communication between two or more network nodes in which either node can initiate sessions, and is able to poll or answer to polls. (C) 610.7-1995

peg count (1) (telephone switching systems) The notation of the number of occurrences of an event. (COM) 312-1977w
(2) *See also:* traffic peg count. (COM/TA) 973-1990w

pel *See:* pixel.

Peltier coefficient, absolute The product of the absolute temperature and the absolute Seebeck coefficient of the material; the sign of the Peltier coefficient is the same as that of the Seebeck coefficient. *Note:* The opposite sign convention has also been used in the technical literature. *See also:* thermoelectric device. (ED) [46]

Peltier coefficient, quotient The quotient of: "A," the rate of Peltier heat absorption by the junction of the two conductors by "B," the electric current through the junction; the Peltier coefficient is positive if Peltier heat is absorbed by the junction when the electric current flows from the second-named conductor to the first conductor. *Notes:* 1. The opposite sign convention has also been used in the technical literature. 2. The Peltier coefficient of a couple is the algebraic difference of either the relative or absolute Peltier coefficients of the two conductors. *See also:* thermoelectric device. (ED) [46]

Peltier coefficient, relative The Peltier coefficient of a couple composed of the given material as the first-named conductor and a specified standard conductor. *Note:* Common standard conductors are platinum, lead, and copper. *See also:* thermoelectric device. (ED) [46]

Peltier effect The absorption or evolution of thermal energy, in addition to the Joule heat, at a junction through which an electric current flows; and in a nonhomogeneous, isothermal conductor, the absorption or evolution of thermal energy, in addition to the Joule heat, produced by an electric current. *Notes:* 1. For the case of a nonhomogeneous, nonisothermal conductor, the Peltier effect cannot be separated from the Thomson effect. 2. A current through the junction of two dissimilar materials causes either an absorption or liberation of heat, depending on the sense of the current, and at a rate directly proportional to it to a first approximation. *See also:* thermoelectric device. (ED) [46], 270-1966w

Peltier heat The thermal energy absorbed or evolved as a result of the Peltier effect. *See also:* thermoelectric device. (ED) [46]

penalty brake A function of the automatic train protection portion of the master control system, accomplished by a safety critical, full-service, or emergency brake application. *Note:* Although most commonly associated with an overspeed operating condition, penalty braking may be initiated for a variety of reasons, depending on the vehicle design and the requirements of the authority having jurisdiction. (VT) 1475-1999

penalty factor A factor that produces the incremental cost of delivered power from a source when multiplied by the incremental cost of power at the source. Mathematically, it is:

$$\frac{1}{(1 - \text{Incremental Transmission Loss})^*}$$

* Expressed as a decimal. (PE/PSE) 94-1991w

pencil-beam antenna **(1)** An antenna whose radiation pattern consists of a single main lobe with narrow principal half-power beamwidths and side lobes having relatively low levels. *Note:* The main lobe usually has approximately elliptical contours of equal radiation intensity in the angular region around the peak of the main lobe. This type of pattern is diffraction-limited in practice. It is often called a sum pattern in radar applications. (AP/ANT) 145-1993
(2) Antenna beam with a narrow radiation lobe with approximately equal azimuth and elevation beamwidths (i.e., one with circular or almost circular shape in the plane perpendicular to the direction of propagation). (AES) 686-1997

pendant A device or equipment that is suspended from overhead either by means of the flexible cord carrying the current or otherwise. (EEC/PE) [119]

pending master The master that participated in and won the most recent arbitration cycle. As a result it will assume bus mastership when the current master releases the bus. (NID) 960-1993

pendulosity (accelerometer) (gyros) The product of the inertial-sensing mass and the distance from the center of mass to the center of support measured along the pendulous axis. (AES/GYAC) 528-1994

pendulous accelerometer An accelerometer that employs a proof mass that is suspended to permit a rotation about an axis perpendicular to an input axis. (AES/GYAC) 528-1994

pendulous axis (accelerometer) In pendulous devices, an axis through the mass center of the proof mass, perpendicular to and intersecting the output axis. The positive direction is defined from the output axis to the proof mass. (AES/GYAC) 528-1994

pendulous integrating gyro accelerometer (PIGA) A device using a single-degree-of-freedom gyro having an intentional pendulosity along the spin axis that is servo-driven about the input axis at a rate that balances the torque induced by acceleration along the input axis. The angle through which the servoed axis rotates is proportional to the integral of applied acceleration. (AES/GYAC) 528-1994

pendulous reference axis (PRA) (accelerometer) The direction of an axis, as defined by the case mounting surfaces or external case markings or both. It is nominally parallel to the pendulous axis. (AES/GYAC) 528-1994

penetration Attacks by unauthorized persons in attempts to gain system access by defeating the system security perimeter (e.g., log-in controls, access controls, bypass controls). Penetration often is in conjunction with browsing or misuse, and could result in unauthorized disclosure or modification of information, unauthorized receipt of services, or denial of service to legitimate users or critical functions. (C/BA) 896.3-1993w

penetration CRT display device A CRT display device characterized by a display screen covered with several layers of phosphor that are selectively energized by varying the voltage of the electron beam, allowing the display of multiple colors. *Note:* Often used to add color to a random-scan display device. (C) 610.10-1994w

penetration depth **(1)** For a given frequency, the depth at which the electric field strength of an incident plane wave, penetrating into a lossy medium, is reduced to $1/e$ of its value just beneath the surface of the lossy medium. *Note:* The penetration depth, also called the skin depth, is equal to the reciprocal of the attenuation constant in the lossy medium. *Synonym:* skin depth. (AP/PROP) 211-1997
(2) For a plane electromagnetic wave incident on the boundary of a medium, the distance from the boundary into the medium along the direction of propagation in the medium, at which the field strengths of the wave have been reduced to $1/e$ (~36.8%) of the boundary values. (NIR) C95.1-1999

penetration, depth of *See:* depth of current penetration.

penetration frequency **(A)** For a given angle of incidence, the lowest frequency that just penetrates the ionosphere. **(B)** *See also:* critical frequency. (AP/PROP) 211-1997

penetration frequency, oblique incidence propagation (radio-wave propagation) For a given angle of incidence, the lowest frequency that just penetrates the ionosphere. (AP/PROP) 211-1990s

penetration frequency, vertical incidence propagation *See:* critical frequency.

pentode A five-electrode electron tube containing an anode, a cathode, a control electrode, and two additional electrodes that are ordinarily grids. (ED) 161-1971w

pen travel The length of the path described by the pen in moving from one end of the chart scale to the other. The path may be an arc or a straight line. *See also:* moving element. (EEC/EMI) [112]

perceived chroma of an area of surface color (illuminating engineering) The attribute according to which it appears to exhibit more or less chromatic color judged in proportion to the brightness of a similarly illuminated area that appears to be white or highly transmitting. In a given set of viewing conditions, and at luminance levels that result in photopic vision, a stimulus of a given chromaticity and luminance factor exhibits approximately constant perceived chroma for all levels of illumination; but for a stimulus of a given chromaticity viewed at a given level of illumination, the perceived chroma generally increases if the luminance factor is increased. (EEC/IE) [126]

perceived color (illuminating engineering) The proximal stimulus applied to the retina initiates color which is perceived as a substance occupying the space in front of the observer's eyes. Color may be perceived as self-luminous or as being reflected or transmitted light. It may be perceived as being confined to a point or line or arrayed as a surface or film or distributed in three dimensions as in the case of the perceived image of a patch of fog. A perceived image may be perceived as composed of volume color as in the case of fog or as covered by surface color as in the case of a piece of chalk. In the case of the sky or a patch of color seen through an aperture where it cannot be identified as belonging to a specific object, it is called aperture color and judgement is suspended as to whether the color is self-luminous or perceived by reflected or transmitted light. The color of a point source of light may be perceived and described as such. This is a special case of a self-luminous color. (EEC/IE) [126]

perceived light-source color (illuminating engineering) The color perceived to belong to a light source. (EEC/IE) [126]

perceived object color The color perceived to belong to an object, resulting from characteristics of the object, of the incident light, and of the surround, the viewing direction, and observer adaptation. *See also:* color. (EEC/IE) [126]

percentage differential relay A differential relay in which the designed response to the phasor difference between incoming and outgoing electrical quantities is modified by a restraining action of one or more of the input quantities. *Note:* The relay operates when the magnitude of the phasor difference exceeds the specified percentage of one or more of the input quantities. (SWG/PE) C37.100-1992

percentage error (watthour meter) The difference between a meter's percentage registration and 100%. A meter whose percentage registration is 95% is said to be 5% slow, or its error is −5%. A meter whose percentage registration is 105% is 5% fast, or its error is +5%. (ELM) C12.1-1982s

percentage immediate appreciation (telephone transmission system) The percentage of the total number of spoken sentences that are immediately understood without conscious deductive effort when each sentence conveys a simple and easily understandable idea. *See also:* volume equivalent. (PE/EEC) [119]

percentage modulation (A) In angle modulation, the fraction of a specified reference modulation, expressed in percent. **(B)** In amplitude modulation, the modulation factor expressed in percent. *Note:* It is sometimes convenient to express percentage modulation in decibels below 100% modulation. *See also:* radio transmission. (BT) 182A-1964

percentage modulation, effective (single, sinusoidal input component) The ratio of the peak value of the fundamental component of the envelope to the direct-current component in the modulated conditions, expressed in percent. *Note:* It is sometimes convenient to express percentage modulation in decibels below 100% modulation. (AP/ANT) 145-1983s

percentage registration (accuracy) (percentage accuracy) (watthour meter) The ratio of the actual registration of the meter to the true value of the quantity measured in a given time, expressed as a percentage. *See also:* electricity meter. (ELM) C12.1-1982s

percent energy resolution *See:* energy resolution.

percent flutter (reproduced tone) (sound recording and reproducing) The root-mean-square deviation from the average frequency, expressed as a percentage of average frequency. (SP) 193-1971w

percent harmonic distortion (electroacoustics) A measure of the harmonic distortion in a system or transducer, numerically equal to 100 times the ratio of the square root of the sum of the squares of the root-mean-square voltages (or currents) of each of the individual harmonic frequencies, to the root-mean-square voltage (or current) of the fundamental. *Note:* It is practical to measure the ratio of the root-mean-square amplitude of the residual harmonic voltages (or currents), after the elimination of the fundamental, to the root-mean-square amplitude of the fundamental and harmonic voltages (or currents) combined. This measurement will indicate percent harmonic distortion with an error of less than 5% if the magnitude of the distortion does not exceed 30%. *See also:* distortion. (SP) 151-1965w

percent impairment of hearing (percent hearing loss) An estimate of a person's ability to hear correctly. It is usually based, by means of an arbitrary rule, on the pure-tone audiogram. The specific rule for calculating this quantity from the audiogram now varies from state to state according to a rule or law. *Note:* The term disability of hearing is sometimes used for impairment of hearing. Impairment refers specifically to a person's illness or injury that affects his personal efficiency in the activities of daily living. Disability has the additional medicolegal connotation that an impairment reduces a person's ability to engage in gainful activity. Impairment is only a contributing factor to the disability. (SP) [32]

percent impedance (1) (rectifier transformer) The percent of rated alternating-current winding voltage required to circulate current equivalent to rated line kilovolt-amperes in the alternating-current winding with all direct-current winding terminals short-circuited. *See also:* rectifier transformer. (Std100) C57.18-1964w

(2) (rectifier transformer) The percent of rated primary winding voltage required to circulate current equivalent to rated kilovoltamperes in the primary winding with all secondary winding terminals short-circuited. (PE/TR) C57.18.10-1998

percent loss of life The equivalent aging in hours at the reference hottest-spot temperature over a time period (usually 24 h) times 100 divided by the total normal insulation life in hours at the reference hottest-spot temperature. The equivalent aging in hours at different hot-spot temperatures is obtained by multiplying the aging acceleration factors for the hottest-spot temperatures times the time periods of the various hottest-spot temperatures. (PE/TR) C57.91-1995

percent-make-and-break (telephone switching systems) The proportions of a dial pulse cycle during which the circuit is closed (make) and opened (break) respectively. (COM) 312-1977w

percent polarized The degree of polarization expressed in percent. *See also:* degree of polarization. (AP/PROP) 211-1997

percent pulse waveform distortion (pulse terminology) Pulse waveform distortion expressed as a percentage of, unless otherwise specified, the pulse amplitude of the reference pulse waveform. (IM/WM&A) 194-1977w

percent pulse waveform feature distortion (pulse terminology) Pulse waveform feature distortion expressed as a percentage of, unless otherwise specified, the pulse amplitude of the reference pulse waveform. (IM/WM&A) 194-1977w

percent ratio (instrument transformers) The true ratio expressed in percent of the marked ratio. (PE/TR) C57.13-1978s, C57.12.80-1978er

percent ratio correction The difference between the ratio correction factor and unity, expressed in percent [(RCF − 1) × 100]. *See also:* ratio correction factor. (PE/TR) C57.13-1993, [57]

percent ripple (1) (power system communication equipment) The ratio of the effective (root-mean-square) value of the ripple voltage or current to the average value of the total (direct current) voltage or current, expressed in percent. (PE/PSC) 281-1984w

(2) (electrical conversion) The ratio of the root-mean-square (RMS) value of the voltage pulsations (E_{max} to E_{min}) to the average value of the total voltage.

$$\text{percent ripple} = \frac{\text{RMS ripple}}{E_{nominal}}(100\%)$$

Note: In most applications, the definition has been revised to simplify the calculations by defining percent ripple as the ratio of the root-mean-square (RMS) value of the voltage pulsations to the nominal no-load output voltage of the converter $E_{nominal}$.

$$\text{percent ripple} = \frac{\text{RMS ripple}}{E_{av}}(100\%)$$

(AES) [41]

(3) The ratio of the value of the ripple voltage to the value of the total voltage multiplied by 100. (PEL/ET) 388-1992r

percent ripple voltage or current *See:* percent ripple.

percent steady-state deviation (control) The difference between the ideal value and the final value, expressed as a percentage of the maximum rated value of the directly controlled variable (or another variable if specified). *See also:* feedback control system. (IA/IAC) [60]

percent syllable articulation *See:* syllable articulation.

percent system deviation (control) At any given point on the time response, the difference between the ideal value and the instantaneous value, expressed as a percentage of the maximum rated value of the directly controlled variable (or another variable if specified). *See also:* deviation; feedback control system. (IA/IAC) [60]

percent total flutter (sound recording and reproducing) The value of flutter indicated by an instrument that responds uniformly to flutter of all rates from 0.5 up to 200 Hz. *Note:* Except for the most critical tests, instruments that respond uniformly to flutter of all rates up to 120 Hz are adequate, and their indications may be accepted as showing percent total flutter. (SP) 193-1971w

percent transformer correction-factor error Difference between the transformer correction factor and unity expressed in percent. *Note:* The percent transformer correction-factor error is positive if the transformer correction factor is greater than unity. If the percent transformer correction-factor error is positive, the measured watts or watthours will be less than the true value. (PE/TR) [57], C57.13-1978s

percent transient deviation (control) The difference between the instantaneous value and the final value, expressed as a percentage of the maximum rated value of the directly controlled variable (or another variable if specified). *See also:* feedback control system. (IA/IAC) [60]

percent unbalance of phase voltages (electrical conversion) The ratio of the maximum deviation of a phase voltage from the average of the total phases to the average of the phase voltages, expressed in percent.

$$\text{percent unbalance} = \frac{\text{RMS phase voltage} - \text{RMS average phase voltages}}{\text{RMS average phase voltage}} \times 100\%$$

(AES) [41]

perfect conductor A medium for which the conductivity is infinite. In a perfect conductor, the total electric and magnetic fields are identically zero regardless of the exciting source. *Synonyms:* ideal conductor; perfectly conducting medium. (AP/PROP) 211-1997

perfect dielectric (1) (ideal dielectric) A dielectric in which all of the energy required to establish an electric field in the dielectric is recoverable when the field or impressed voltage is removed. Therefore, a perfect dielectric has zero conductivity and all absorption phenomena are absent. A complete vacuum is the only known perfect dielectric. (Std100) 270-1966w
(2) A dielectric medium in which the conductive and dielectric losses are identically zero. *See also:* ideal dielectric. (AP/PROP) 211-1997

perfect diffusion (illuminating engineering) That in which flux is uniformly scattered in accordance with Lambert's cosine law. *See also:* Lambert's cosine law. (EEC/IE) [126]

perfective maintenance (1) (software) Software maintenance performed to improve the performance, maintainability, or other attributes of a computer program. *Contrast:* adaptive maintenance; corrective maintenance. (C) 610.12-1990
(2) Modification of a software product after delivery to improve performance or maintainability. (C/SE) 1219-1998

perfectly conducting medium *See:* perfect conductor.

perforated punch tape *See:* perforated tape.

perforated tape (1) Tape in which a code hole(s) and a tape-feed hole have been punched in a row. (IA) [61]
(2) A tape on which a pattern of holes is used to represent information. *Synonyms:* perforated punch tape; punched tape. *See also:* paper tape reader; chadless tape. (C) 610.10-1994w

perforated tape reader *See:* paper tape reader.

perforator (1) (telegraph practice) A device for punching code signals in paper tape for application to a tape transmitter. *Note:* A perforating device that is automatically controlled by incoming signals is called a reperforator. *See also:* telegraphy. (C/COM) [20], [49], [85]
(2) (test, measurement, and diagnostic equipment) A device for punching digital information into tape for application to a tape transmitter or tape reader. *Synonym:* tape punch. (MIL) [2]

performance (1) (software) The degree to which a system or component accomplishes its designated functions within given constraints, such as speed, accuracy, or memory usage. (C) 610.12-1990
(2) A measure of a computer system or subsystem to perform its functions; for example, response time, throughput, or number of transactions per second. The efficiency of a system in accomplishing pieces of work is an attribute of performance. (C/PA) 14252-1996

performance characteristic(s) (1) Those characteristics (such as impedance, losses, dielectric test levels, temperature rise, sound level, etc.) that describe the performance of the equipment under specified conditions of operation. (PE/TR) C57.12.80-1978r
(2) (of a device) An operating characteristic, the limit or limits of which are given in the design test specifications. (SWG/PE) C37.100-1992, [56], C37.40-1993
(3) The parameters that are essential to describe the behavior or applicability of the device under specified conditions of operation. (SPD/PE) C62.62-2000

performance chart (magnetron oscillators) A plot on coordinates of applied anode voltage and current showing contours of constant magnetic field, power output, and over-all efficiency. *See also:* magnetron. (ED) 161-1971w

performance criteria The established level of quality (bias, precision, detection sensitivity, etc.) and operational commitments (turnaround times, reporting protocol, etc.) that are:

1) Agreed upon between the Service Laboratory and the customer (or inter-governmental agencies or intra-company entities) within a formal contract.
2) Established by the Service Laboratory and documented within the Laboratory's Operational or Quality Assurance Program Manual or
3) Both 1 and 2 above.

(NI) N42.23-1995

performance criterion The criterion upon which the insulation strength or withstand voltages and clearances are selected. The performance criterion is based on an acceptable probability of insulation failure and is determined by the consequence of failure, required level of reliability, expected life of equipment, economics, and operational requirements. The criterion is usually expressed in terms of an acceptable failure rate (number of failures per year, years between failures, risk of failure, etc.) of the insulation configuration. (PE/C) 1313.1-1996

performance evaluation (1) (software) The technical assessment of a system or system component to determine how effectively operating objectives have been achieved. *See also:* component; system. (C/SE) 729-1983s
(2) The analysis, in terms of initial objectives and estimates, usually made on site to provide information on operating experience and to identify required corrective actions. (PE/NP) 933-1999

performance factor The ratio PL/RRS where PL, the performance level, is the level of ground shaking and RRS, the required response spectrum, is the test or analysis level. (PE/SUB) 693-1997

performance index (excitation systems) A scalar measure of the quality of system behavior. It is frequently a function of system output and control input over some specified time interval and/or frequency range. A quadratic performance index is a quadratic function of system states and this form finds wide applications to linear systems. (PE/EDPG) 421A-1978s

performance level (PL) A specified level of earthquake ground shaking that is used to define standardized seismic qualification levels (high, moderate, and low) for substation equipment. (PE/SUB) 693-1997

Performance Level 1 Environment primarily intended for aircraft applications subject to extreme vibration, shock, and temperature variations. (C/MM) 1156.1-1993

Performance Level 2 Environment primarily intended for shipboard, subsurface ship, and shore applications subject to substantial vibration, extreme shock, and temperature variations. (C/MM) 1156.1-1993

Performance Level 3 Environment primarily intended for shipboard, vehicular, and shore applications subject to vibration, shock, and temperature variations. (C/MM) 1156.1-1993

Performance Level 4 Environment primarily intended for sheltered applications subject to vibration, corrosion, shock, and temperature variations. (C/MM) 1156.1-1993

Performance Level 5 Environment primarily intended for sheltered applications subject to minimal vibration, shock, or temperature variations. (C/MM) 1156.1-1993

performance management In networking, a management function defined for controlling and analyzing the throughput and error rate of the network. (C) 610.7-1995

performance monitor (test, measurement, and diagnostic equipment) A device that continuously or periodically scans a selected number of test points to determine if the unit is operating within specified limits. The device may include provisions for insertion of stimuli. (MIL) [2]

performance monitoring Determining whether equipment is operating or capable of operating within specific limits. (PE/NP) 933-1999

performance requirement (1) (software) A requirement that imposes conditions on a functional requirement; for example a requirement that specifies the speed, accuracy, or memory usage with which a given function must be performed. (C) 610.12-1990

(2) A requirement that specifies a performance characteristic that a system or system component must possess; for example, speed, accuracy, frequency. (C/PA) 14252-1996

(3) The measurable criteria that identifies a quality attribute of a function, or how well a functional requirement must be accomplished. (C/SE) 1220-1998

performance specification (software) A document that specifies the performance characteristics that a system or component must possess. These characteristics typically include speed, accuracy, and memory usage. Often part of a requirements specification. (C) 610.12-1990

performance test A constant current or constant power capacity test, made on a battery after it has been in service, to detect any change in the capacity. (PE/SB/EDPG) 1106-1995, 450-1995, 1188-1996

performance testing Testing conducted to evaluate the compliance of a system or component with specified performance requirements. *See also:* functional testing. (C/PE/EDPG) 610.12-1990, 1248-1998

performance tests (rotating machinery) The tests required to determine the characteristics of a machine and to determine whether the machine complies with its specified performance. *See also:* direct-current commutating machine; asynchronous machine. (PE) [9]

performance verification (test, measurement, and diagnostic equipment) A short, precise check to verify that the unit under test is operational and performing its intended function. (MIL) [2]

performing activity The person(s) or organization that performs the tasks specified in this standard. (C/SE) 1220-1994s

periapsis (communication satellite) The least distant point from the center of a primary body (or planet) to an orbit around it. (COM) [19]

perigee (navigation aid terms) That orbital point nearest the earth when the earth is the center of attraction. (AES/GCS) 172-1983w

perimeter In image processing, the number of pixels in the border of a region. (C) 610.4-1990w

perimeter lights (illuminating engineering) Aeronautical ground lights provided to indicate the perimeter of a landing pad for helicopters. (EEC/IE) [126]

period (1) (pulse terminology) The absolute value of the minimum interval after which the same characteristics of a periodic waveform or a periodic feature recur. (IM/WM&A) 194-1977w

(2) (modeling and simulation) The time interval between successive events in a discrete simulation. (C) 610.3-1989w

(3) (NuBus) Time between two driving edges. (C/MM) 1196-1987w

(4) An interval of time in the battery duty cycle during which the load is assumed to be constant for purposes of cell sizing calculations. (PE/EDPG) 1115-1992

(5) The character ".". The term *period* is contrasted against *dot* which is used to describe a specific directory entry. (C/PA) 9945-2-1993

(6) An interval of time in the battery duty cycle during which the load is assumed to be constant for purposes of cell sizing calculations. (SCC29) 485-1997

period, critical hydro *See:* critical hydro period.

period hours (electric generating unit reliability, availability, and productivity) The number of hours a unit was in the active state. (PE/PSE) 762-1987w

periodically sampled equivalent time format (pulse measurement) A format that is identical to the periodically sampled real time format, below, except that the time coordinate is equivalent to and convertible to real time. Typically, each datum point is derived from a different measurement on a different wave in a sequence of waves. *See also:* sampled format. (IM/WM&A) 181-1977w

periodically sampled real time format (pulse measurement) A finite sequence of magnitudes m_0, m_1, m_2, . . .m_n each of which represents the magnitude of the wave at times t_0, $t_0 + \Delta t$, $t_0 + 2\Delta t$, . . .$t_0 + n\Delta t$, respectively, wherein . . . the data may exist in a pictorial format or as a list of numbers. *See also:* sampled format. (IM/WM&A) 181-1977w

periodic and random deviation The sum of all ripple and noise components measured over a specified bandwidth and state, unless otherwise specified, in peak-to-peak values. (PEL) 1515-2000

periodic-automatic-reclosing equipment A piece of equipment that provides for automatically reclosing a circuit-switching device a specified number of times at specified intervals between reclosures. *Note:* This type of automatic reclosing equipment is generally used for ac circuits. (SWG/PE) C37.100-1992

periodic check (test, measurement, and diagnostic equipment) A test or series of tests performed at designated intervals to determine if all elements of the unit under test are operating within their designated limits. (MIL) [2]

periodic damping *See:* underdamping.

periodic duty (packaging machinery) (power and distribution transformers) Intermittent operation in which the load conditions are regularly recurrent. (NEC/NESC/IA/PE/PKG/TR) 333-1980w, C57.16-1996, C57.12.80-1978r, [86]

periodic electromagnetic wave (radio-wave propagation) A wave in which the electric field vector is repeated in detail in either of two ways: at a fixed point, after the lapse of a time known as the period, or at a fixed time, after the addition of a distance known as the wavelength. (AP) 211-1977s

periodic frequency modulation (converters having ac output) (self-commutated converters) The periodic variation of the output frequency from its rated value. (IA/SPC) 936-1987w

periodic function A function that satisfies $f(x) = f(x + nk)$ for all x and for all integers n, k being a constant. For example, $\sin(x + a) = \sin(x + a + 2n\pi)$. (Std100) 270-1966w

periodic line (transmission lines) A line consisting of successive identically similar sections, similarly oriented, the electrical properties of each section not being uniform throughout. *Note:* The periodicity is in space and not in time. An example of a periodic line is the loaded line with loading coils uniformly spaced. *See also:* transmission line. (Std100) 270-1966w

periodic monitoring The process of sampling the state of some phenomenon at a sample interval greater than 1 s. (SWG/SUB/PE) C37.122-1983s, C37.122.1-1993, C37.100-1992

periodic output voltage modulation (self-commutated converters) (converters having ac output) The periodic variation of output voltage amplitude at frequencies less than the fundamental output frequency. (IA/SPC) 936-1987w

periodic permanent-magnet focusing (PPM) (microwave tubes) Magnetic focusing derived from a periodic array of permanent magnets. *See also:* magnetron. (ED) [45]

periodic pulse train (automatic control) A pulse train made up of identical groups of pulses, the groups repeating at regular intervals. (PE/EDPG) [3]

periodic rating (1) (electric power sources) The load that can be carried for the alternate periods of load and rest specified in the rating, the apparatus starting at approximately room temperature, and for the total time specified in the rating, without causing any of the specified limitations to be exceeded. *See also:* asynchronous machine. (IA/IAC) [60]
(2) (relay) A rating that defines the current or voltage that may be sustained by the relay during intermittent periods of energization as specified, starting cold and operating for the total time specified without causing any of the prescribed limitations to be exceeded. (SWG/PE) C37.100-1981s

periodic slow-wave circuit (microwave tubes) A circuit whose structure is periodically recurring in the direction of propagation. *See also:* microwaves. (ED) [45]

periodic tasks Tasks that have to be processed at regular intervals, and each instance shall be normally completed before the next instance of the same task arrives. (C/BA) 896.3-1993w

periodic test Test performed at scheduled intervals to detect failures and verify operability. (PE/NP) 379-1994, 308-1980Gs, 381-1977w, 338-1987r

periodic wave A wave in which the displacement at each point of the medium is a periodic function of the time. Periodic waves are classified in the same manner as periodic quantities. (Std100) 270-1966w

periodic waveguide A waveguide in which propagation is obtained by periodically arranged discontinuities or periodic modulations of the material boundaries. (MTT) 146-1980w

period timing check A timing check that specifies the allowable time between successive periods of a signal. (C/DA) 1481-1999

peripheral (1) A device, attached to a host via a communication link. (C/MM) 1284-1994
(2) Pertaining to a device that operates in combination or conjunction with the computer but is not physically part of the computer and is not essential to the basic operation of the system; for example, printers, keyboards, graphic digital converters, disks, and tape drives. *Note:* Such devices are often referred to as "peripherals" or "peripheral equipment." *See also:* input-output device. (C) 610.10-1994w

peripheral air-gap leakage flux (rotating machinery) The component of air-gap magnetic flux emanating from the rotor or stator, that flows from pole to pole without entering the radially opposite surface of the air gap. *See also:* rotor; stator. (PE) [9]

peripheral controller *See:* input-output controller.

peripheral control unit *See:* controller.

peripheral device A device connected to another device (host) that, in turn, controls its operation. (EMC) C63.4-1991

peripheral equipment (test, measurement, and diagnostic equipment) Equipment external to a basic unit. A tape unit, for example, is peripheral equipment to a computer. (MIL) [2]

peripheral personality The characteristics of a language processor or operating environment that a peripheral runs to interpret commands and data being sent. (C/MM) 1284-1994

peripheral stimulation Action by a chemical or physical agent at or near the surface of an organism. (T&D/PE) 539-1990

peripheral storage *See:* auxiliary storage.

peripheral transfer The process of transmitting data between two peripheral units. *See also:* radial transfer. (C) 610.10-1994w

peripheral unit With respect to a particular processing unit, any equipment that can communicate directly with that unit. (C) 610.10-1994w

peripheral vision (illuminating engineering) The seeing of objects displaced from the primary line of sight and outside the central visual field. (EEC/IE) [126]

peripheral visual field (illuminating engineering) That portion of the visual field which falls outside the region corresponding to the foveal portion of the retina. (EEC/IE) [126]

periphery The outer part of an integrated circuit where instances of cell types designed specifically to interface the internal circuitry to the "outside world" are placed. This part includes "pad" cells (which are input and output buffers) and power and ground pads; it may also include test circuitry, such as boundary scan cells. (C/DA) 1481-1999

periscope (navigation aid terms) An optical instrument which displaces the line of sight parallel to permit a view which otherwise may be obstructed itself. (AES/GCS) 172-1983w

periscope antenna An antenna consisting of a very directive feed located close to ground level and oriented so that its beam illuminates an elevated reflector that is oriented so as to produce a horizontal beam. (AP/ANT) 145-1993

periscopic sextant (navigation aid terms) A sextant designed to be mounted inside a vehicle, with a tube extending vertically upward through the skin of the vehicle. (AES/GCS) 172-1983w

permanent connection (substation grounding) A grounding connector that will retain its electrical and mechanical integrity for the design life of the conductor within limits established in IEEE Std 837-1984. (SUB/PE) 837-1989r

permanent echo A signal reflected from an object fixed with respect to the radar site. (AES) 686-1997

permanent fault (1) (surge arresters) A fault that can be cleared only by action taken at the point of fault. (PE) [8], [84]
(2) A continuous and stable failure or error. (C/BA) 896.3-1993w
(3) One that will persist regardless of how fast the system is de-energized or the number of times that the system is de-energized and re-energized. (T&D/PE) 1250-1995

permanent-field synchronous motor A synchronous motor similar in construction to an induction motor in which the member carrying the secondary laminations and windings carries also permanent-magnet field poles that are shielded from the alternating flux by the laminations. It starts as an induction motor but operates normally at synchronous speed. *See also:* permanent-magnet synchronous motor. (PE) [9]

permanent font *See:* internal font.

permanent forced outage (1) A forced outage where the component or unit is damaged and cannot be restored to service

until repair or replacement is completed. *Note:* Repairs can be further classified by urgency as high, normal, and low urgency repairs. (PE/PSE) 859-1987w
(2) (electric power system) An outage whose cause is not immediately self-clearing, but must be corrected by eliminating the hazard or by repairing or replacing the component before it can be returned to service. An example of a permanent forced outage is a lightning flashover which shatters an insulator thereby disabling the component until repair or replacement can be made. *Note:* This definition derives from transmission and distribution applications and does not necessarily apply to generation outages. (PE/PSE) 346-1973w

permanent forced outage duration (electric power system) The period from the initiation of the outage until the component is replaced or repaired. (PE/PSE) 346-1973w

permanently grounded device A grounding device designed to be permanently connected to ground, either solidly or through current transformers and/or another grounding device. (SPD/PE) 32-1972r

permanently installed decorative fountains and reflection pools Those that are constructed in the ground, on the ground, or in a building in such a manner that the pool cannot be readily disassembled for storage and are served by electrical circuits of any nature. These units are primarily constructed for their aesthetic value and not intended for swimming or wading. (NESC/NEC) [86]

permanently installed swimming, wading and therapeutic pools Those that are constructed in the ground, on the ground, or in a building in such a manner that the pool cannot be readily disassembled for storage, whether or not served by electrical circuits of any nature. (NESC/NEC) [86]

permanent magnet (PM) A ferromagnetic body that maintains a magnetic field without the aid of external electric current. (Std100) [84]

permanent-magnet erasing head (electroacoustics) A head that uses the fields of one or more permanent magnets for erasing. *See also:* phonograph pickup. (SP) [32]

permanent-magnet focusing (microwave tubes) Magnetic focusing derived from the use of a permanent magnet. *See also:* magnetron. (ED) [45]

permanent-magnet generator (1) (magneto) A generator in which the open-circuit magnetic flux field is provided by one or more permanent magnets. (PE) [9]
(2) An electric generator in which the magnetic flux is provided by one or more pairs of permanent magnets. (IA/MT) 45-1998

permanent-magnet loudspeaker A moving-conductor loudspeaker in which the steady field is produced by means of a permanent magnet. (EEC/PE) [119]

permanent-magnet moving-coil instrument (d'arsonval instrument) An instrument that depends for its operation on the reaction between the current in a movable coil or coils and the field of a fixed permanent magnet. *See also:* instrument. (PE/EEC) [119]

permanent-magnet moving-iron instrument (polarized-vane instrument) An instrument that depends for its operation on the action of an iron vane in aligning itself in the resultant magnetic field produced by a permanent magnet and the current in an adjacent coil of the instrument. *See also:* instrument. (EEC/PE) [119]

permanent-magnet, second-harmonic, self-synchronous system A remote-indicating arrangement consisting of a transmitter unit and one or more receiver units. All units have permanent-magnet rotors and toroidal stators using saturable ferromagnetic cores and excited with alternating current from a common external source. The coils are tapped at three or more equally spaced intervals, and the corresponding taps are connected together to transmit voltages that consist principally of the second harmonic of the excitation voltage. The rotors of the receiver units will assume the same angular position as that of the transmitter rotor. *See also:* synchro system. (EEC/PE) [119]

permanent-magnet synchronous motor (rotating machinery) A synchronous motor in which the field system consists of one or more permanent magnets. *See also:* permanent-field synchronous motor. (PE) [9]

permanent master *See:* master.

permanent record Records that shall be maintained for the lifetime of the project or as long as required by contract. These records may be forwarded to the customer for retention by the customer's record keeping system upon request or at the end of the contract period. (NI) N42.23-1995

permanent signal (A) (telephone switching systems) A sustained off-hook supervisory signal originating outside a switching system. **(B)** Occurs if no dialed (or outpulsed) digits are received and the system times out. (COM/TA) 312-1977, 973-1990

permanent-signal alarm (telephone switching systems) An alarm resulting from the simultaneous accumulation of a predetermined number of permanent signals. (COM) 312-1977w

permanent signal timing or no dial timing The time from the beginning of line or trunk seizure until the first character dialed is detected. This applies for individual lines and immediate start trunks. (COM/TA) 973-1990w

permanent-signal tone (telephone switching systems) A tone that indicates to an operator or other employee that a line is in a permanent-signal state. (COM) 312-1977w

permanent-split capacitor motor A capacitor motor with the same value of effective capacitance for both starting and running operations. *See also:* asynchronous machine. (PE) [9]

permanent storage A type of storage whose contents cannot be modified. *Synonym:* nonerasable storage. *Contrast:* erasable storage. *See also:* read-only storage. (C) 610.10-1994w

permanent virtual circuit A virtual circuit that is established at service subscription time and always connects the same two user end points. *Note:* Bandwidth on a PVC is always available but lacks the flexibility of dynamically connecting to different end users. *See also:* switched virtual circuit. (C) 610.7-1995

permeability (μ) (1) (A) (general) A general term used to express various relationships between magnetic induction and magnetizing force. These relationships are either:

a) absolute permeability, that in general is the quotient of a change in magnetic induction divided by the corresponding change in magnetizing force; or
b) specific (relative) permeability, which is the ratio of the absolute permeability to the magnetic constant.

Note: Relative permeability is a pure number that is the same in all unit systems; the value and dimension of absolute permeability depend upon the system of units employed. **(B) (general)** In anisotropic media, permeability becomes a matrix. (Std100) 270-1966
(2) (electrical heating systems) Ratio of the magnetic flux density to the corresponding magnetizing force. (IA/PC) 844-1991
(3) The drainage characteristic of soil that denotes its capacity to conduct or discharge fluids under a given hydraulic gradient. (SUB/PE) 980-1994
(4) A macroscopic material property of a medium that relates the magnetic flux density, $\vec{B}$, to the magnetic field, $\vec{H}$, in the medium. For a monochromatic wave in a linear medium, that relationship is described by the (phasor) equation:

$$\vec{B} = \pi = \cdot \vec{H}$$

where

$\pi =$ = a tensor that is generally frequency dependent

For an isotropic medium, the tensor reduces to a complex scalar:

$$\mu = \mu' - j\mu''$$

where
μ' = the real part of the permeability
μ'' accounts for losses

(AP/PROP) 211-1997

permeability, complex *See:* complex permeability.

permeability, relative complex *See:* relative complex permeability.

permeameter An apparatus for determining corresponding values of magnetizing force and flux density in a test specimen. From such values of magnetizing force and flux density, normal induction curves or hysteresis loops can be plotted and magnetic permeability can be computed. *See also:* magnetometer. (EEC/PE) [119]

permeance The reciprocal of reluctance. (Std100) 270-1966w

permissible mine equipment Equipment that complies with the requirements of and is formally approved by the United States Bureau of Mines after having passed the inspections and the explosion and.or other tests specified by that Bureau. *Note:* All equipment so approved must carry the official approval plate required as identification for permissible equipment. (EEC/PE) [119]

permissible response rate (A) (steam generating unit) The maximum assigned rate of change in generation for load-control purposes based on estimated and known limitations in the turbine, boiler, combustion control, or auxiliary equipment. The permissible response rate for a hydro-generating unit is the maximum assigned rate of change in generation for load-control purposes based on estimated and known limitations of the water column, associated piping, turbine, or auxiliary equipment. *See also:* speed-governing system. **(B)** steam-generating unit. The maximum allowable rate of change of generation under load control, which is based on the limitations of the turbine, boiler, combustion control, and/or auxiliary equipment. **(C)** hydro-generating unit. The maximum allowable rate of change of generation under load-control, which is based on the limitations of the water column, piping, turbine, and/or auxiliary equipment. (PE/PSE) 94-1970, 94-1991

permission *See:* file permission.

permissions *See:* file access permissions.

permissive (1) (as applied to a relay system) A general term indicating that functional cooperation of two or more relays is required before control action can become effective. (SWG/PE/PSR) C37.90-1978s, C37.100-1992 **(2)** A term that describes the constraints placed on accesses of a data format; the read, write, and lock accesses can be performed divisibly (i.e., using multiple transations) using a multiphase fetch and/or update process. (C/MM) 1596.5-1993 **(3)** Pertaining to a scheme requiring permission to trip from a remote terminal, usually in the form of a pilot signal. (PE/PSR) C37.113-1999

permissive block A block in manual or controlled manual territory, governed by the principle that a train other than a passenger train may be permitted to follow a train other than a passenger train in the block. *See also:* block-signal system; controlled manual block signal system. (EEC/PE) [119]

permissive connection A connection in which non-voice information can be sent over the voice communications network. *See also:* programmable connection; RJ-45. (C) 610.7-1995

permissive control (electric power system) An automatic generation control methodology that reduces generating unit control error only when unit change will reduce area control error. (PE/PSE) 94-1991w

permissive control device (power system device function numbers) Generally a two-position device that in one position permits the closing of a circuit breaker, or the placing of an equipment into operation, and in the other position prevents the circuit breaker or the equipment from being operated. (SUB/PE) C37.2-1979s

permissive [security] attribute A security attribute that identifies an active entity or a resource as member of a group. An entity is granted access to all resources in the groups of which it is a member. Permissive attributes could be used alone or in combination with restrictive attributes. Commonly, when used in combination with restrictive attributes, they are secondary in the determination of access privilege. (C/LM) 802.10-1998, 802.10g-1995

permit *See:* clearance.

permittivity (ε) (1) (primary ferroelectric terms) (small-signal, ferroelectric material) The incremental change in electric displacement per unit electric field when the magnitude of the measuring field is very small compared to the coercive electric field. The small signal relative permittivity, κ, is equal to the ratio of the absolute permittivity ε to the permittivity of free space ε_0, that is $\kappa = \varepsilon/\varepsilon_0$. Macroscopically, ε is found by measuring the capacitance. The units of permittivity are coulombs/volt-meter or farads/meter. In a ferroelectric, the measuring field or voltage must be sufficiently small in order to prevent ferroelectric domain reorientation from contributing to the permittivity. *Note:* The value of the small-signal permittivity may depend on the remanent polarization, electric field, mechanical stress, sample history, or frequency of the measuring field. (Measurements are usually made at a frequency of 1 kHz or higher.) (UFFC) 180-1986w **(2)** A macroscopic material property of the medium that relates the electric flux density, *D*, to the electric field, $\hat{E}$, in the medium. For a monochromatic wave in a linear medium, that relationship is described by the (phasor) equation:

$$\vec{D} = \varepsilon= \cdot \hat{E}$$

where ε=, the complex permittivity, is a tensor that is generally frequency dependent. For an isotropic medium, the tensor reduces to a complex scalar:

$$\varepsilon = \varepsilon' - j\varepsilon''$$

where
ε' = the real part of the permittivity
ε'' accounts for losses

(AP/PROP) 211-1997

permittivity, complex *See:* complex permittivity.

permittivity, free space *See:* free space permittivity.

permittivity in physical media The real part of the complex permittivity. *See also:* complex dielectric constant. (AP/ANT) 145-1983s

permittivity, relative complex *See:* relative complex permittivity.

permutation An ordered sequence of a given number of items chosen from a set. *Contrast:* combination. (C) 610.5-1990w

permutation index An automatic index in which each item appears repeatedly, each time with a different word of the item as the first word, followed by the subsequent words in the item, then by that part of the item that came before the word. *See also:* permutation on subject headings index; keyword and context index. (C) 610.2-1987

permutation on subject headings (POSH) *See:* permutation on subject headings index.

permutation on subject headings index A permutation index in which the item entries are subject headings. (C) 610.2-1987

perpendicular magnetic recording A type of magnetic recording in which magnetic polarities representing data are aligned perpendicularly to the plane of the recording surface. *Synonym:* vertical magnetic recording. *Contrast:* longitudinal magnetic recording. (C) 610.10-1994w

perpendicular magnetization Magnetization of the recording medium in a direction perpendicular to the line of travel and parallel to the smallest cross-sectional dimension of the medium. *Note:* In this type of magnetization, either single pole-piece or double pole-piece magnetic heads may be used. *See also:* phonograph pickup. (SP/MR) [32]

perpendicular polarization (1) (facsimile) A linear polarization for which the field vector is parallel to some reference plane. *Note:* These terms are applied mainly to uniform plane waves incident upon a plane of discontinuity (surface of the earth, surface of a dielectric or a conductor). Then the convention is to take as reference the plane of incidence, that is, the plane containing the direction of propagation and the normal to the surface of discontinuity. If these two directions coincide, the reference plane must be specified by some other convention. (COM/AP/ANT) 167-1966w, 145-1993
(2) The polarization of a wave for which the electric field is perpendicular to the plane of incidence. Sometimes called horizontal or transverse electric (TE) polarization. (EMC) 1128-1998
(3) The polarization of a wave for which the electric field vector is perpendicular to the plane of incidence. *Note:* Sometimes called horizontal or transverse electric (TE) polarization; in optics, such a wave is said to be "s" polarized. (AP/PROP) 211-1997

persistence (1) (oscilloscopes) The decaying luminosity of the luminescent screen (phosphor screen) after the stimulus has been reduced or removed. *See also:* phosphor decay. (IM) 311-1970w
(2) (A) (computer graphics) The length of time that a display image remains on a display surface without being refreshed. **(B) (computer graphics)** The tendency of a phosphor to continue to emit light when no longer energized by an electron beam. (C) 610.6-1991
(3) A mode for semaphores, shared memory, and message queues requiring that the object and its state (including data, if any) are preserved after the object is no longer referenced by any process. Persistence of an object does not imply that the state of the object is maintained across a system crash or a system reboot. (C/PA) 9945-1-1996
(4) A characteristic of semaphores, shared memory, and message queues requiring that the object and its state (including data, if any) are preserved after last close (the object is no longer referenced by any process). Persistence of an object does not necessarily imply that the state of the object is maintained across a system crash or a system reboot. (C) 1003.5-1999

persistence characteristic (1) (camera tubes) The temporal step response of a camera tube to illumination. (ED) 161-1971w
(2) (decay characteristic) (luminescent screen) A relation, usually shown by a graph, between luminance (or emitted radiant power) and time after excitation is removed. (ED) 161-1971w

persistent-cause forced outage (electric power system) A component outage whose cause is not immediately self-clearing but must be corrected by eliminating the hazard or by repairing or replacing the affected component before it can be returned to service. *Note:* An example of a persistent-cause forced outage is a lightning flashover that shatters an insulator thereby disabling the component until repair or replacement can be made. *See also:* outage. (PE/PSE) [54]

persistent-cause forced-outage duration (electric power system) The period from the initiation of a persistent-cause forced outage until the affected component is replaced or repaired and made available to perform its intended function. *See also:* outage. (PE/PSE) [54]

persistent current (superconducting material) A magnetically induced current that flows undiminished in a superconducting material or circuit. *See also:* superconductivity. (ED) [46]

persistent-image device An optoelectronic amplifier capable of retaining a radiation image for a length of time determined by the characteristics of the device. *See also:* optoelectronic device. (ED) [46]

persistent-image panel (optoelectronic device) A thin, usually flat, multicell persistent-image device. *See also:* optoelectronic device. (ED) [46]

persistent menu A menu that popped up and stayed visible for one round of use. Menus stay on the screen until the user chooses an item or dismisses the menu. (C) 1295-1993w

persistent URI A Uniform Resource Identifier (URI) is persistent if it is a reference that does not need to change at the link in a document, and can still reach the desired object even though that object may have changed locations. (C) 2001-1999

personal computer (1) (measurement of radio-noise emissions) A system, containing a host and a limited number of peripherals designed to be used in the home or in small offices, that enables individuals to perform a variety of computing or word-processing functions or both, and that typically is of a size permitting it and its peripherals to be located on a table surface. *Note:* Other definitions given in product standards or applicable regulations may take precedence. (EMC) C63.4-1991
(2) A single-user microcomputer designed for personally controllable applications. *See also:* workstation; laptop computer; desktop computer; home computer. (C) 610.2-1987, 610.10-1994w

personal computing (A) Computing performed using a personal computer. **(B)** Computing performed in an environment in which the user has complete control over the data and access to software with which the data may be manipulated. *Synonym:* personal processing. (C) 610.2-1987

personal ground (conductor stringing equipment) A portable device designed to connect (bond) a deenergized conductor or piece of equipment, or both, to an electrical ground. It is distinguished from a master ground in that it is utilized at the immediate site when work is to be performed on a conductor or piece of equipment that could accidentally become energized. *Synonyms:* ground stick; working ground; red head. (T&D/PE) 524a-1993r, 524-1992r, 1048-1990

personal name attributes The Organizational-Unit-Name-1, Organizational-Unit-Name-2, Organizational-Unit-Name-3, and Organizational-Unit-Name-4 attributes specific to the class. (C/PA) 1224.1-1993w

personal processing *See:* personal computing.

personnel security Procedures to ensure that personnel with access to sensitive information and critical services have the appropriate authorizations and training. (C/BA) 896.3-1993w

person-to-person call (telephone switching systems) A call intended for a designated person. (COM) 312-1977w

perspective projection (computer graphics) The projection of a three-dimensional image onto a two-dimensional surface such that objects that are farther from the viewer in three dimensions are rendered smaller than closer ones. *Contrast:* parallel projection.

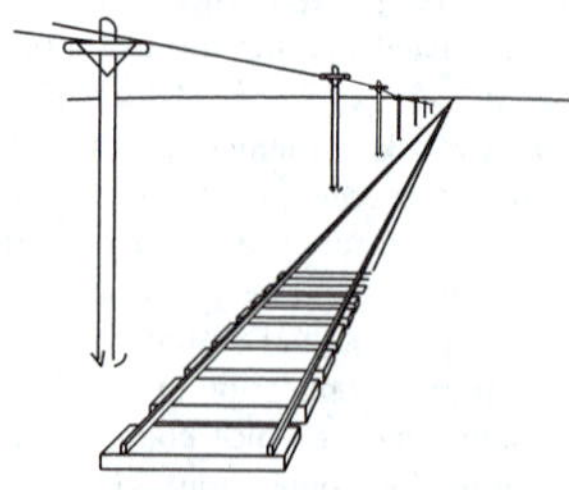

perspective projection

(C) 610.6-1991w

PERT *See:* program evaluation and review technique.

perturbation technique An approximate analytical method, the accuracy of which is based on the smallness of one or more characteristics of the medium or interface. (AP/PROP) 211-1997

perturbed electric or magnetic field (A) (electric and magnetic fields from ac power lines) (weakly perturbed field) The field at a point will be regarded as weakly perturbed if the magnitude does not change by more than 5% or the direction does not vary by more than 5 degrees, or both, when an object is introduced into the vicinity. The electric field at the surface of the object is in general strongly perturbed [see definition (C) below] by the presence of the object. At power frequencies the magnetic field is not in general perturbed by the presence of objects which are free of magnetic materials. Exceptions to this are regions near the surface of nonmagnetic electric conductors which develop eddy currents because of the B-field time variation. **(B) (electric and magnetic fields from ac power lines)** (moderately perturbed field) The field at a point will be regarded as moderately perturbed if the magnitude varies between 5% and 30% or the direction varies between 5 degrees and 30 degrees, or both, when an object is introduced into the vicinity. **(C) (electric and magnetic fields from ac power lines)** (strongly perturbed field) The field at a point will be regarded as strongly perturbed if the magnitude varies in excess of 30% or the direction varies in excess of 30 degrees, or both, when an object is introduced into the vicinity. (T&D/PE) 644-1979

perturbed field (1) (measurement of power frequency electric and magnetic fields from ac power lines) A field that is changed in magnitude or direction, or both, by the introduction of an object. *Note:* The electric field at the surface of the object is, in general, strongly perturbed by the presence of the object. At power frequencies the magnetic field is not, in general, greatly perturbed by the presence of objects that are free of magnetic materials. Exceptions to this are regions near the surface of thick electric conductors where eddy currents alter time-varying magnetic fields. (T&D/PE) 644-1994

(2) (overhead power lines) A field that is changed in magnitude and/or direction by the introduction of an object or by an electric charge in the region. *Note:* The electric field close to the object is, in general, strongly perturbed by the presence of the object. At power frequencies the magnetic field is not, in general, greatly perturbed by the presence of objects that are free of magnetic materials. Exceptions to this are regions near the surface of thick electric conductors where eddy currents alter time-varying magnetic fields. (T&D/PE) 539-1990

per-unit quantity (rotating machinery) The ratio of the actual value of a quantity to the base value of the same quantity. The base value is always a magnitude, or in mathematical terms, a positive, real number. The actual value of the quantity in question (current, voltage, power, torque, frequency, etc.) can be of any kind: root-mean-square, instantaneous, phasor, complex, vector, envelope, etc. *Note:* The base values, though arbitrary, are usually related to characteristic values, for example, in case of a machine, the base power is usually chosen to be the rated power (active or apparent), the base voltage to be the rated root-mean-square voltage, the base frequency, the rated frequency. Despite the fact that the choice of base values is rather arbitrary, it is of advantage to choose base values in a consistent manner. The use of a consistent per-unit system becomes a practical necessity when a complicated system is analyzed. *See also:* asynchronous machine. (PE) [9]

per unit reactance On a rated current base, a dimensionless quantity obtained by referencing the magnitude of the reactance to the rated system line-to-neutral voltage divided by the rated current of the reactor. *Note:* Per unit reactance can also be defined on an arbitrary MVA base. (PE/TR) C57.16-1996

per-unit resistance The measured watts expressed in per-unit on the base of the rated kilovolt-amperes of the teaser winding. *See also:* efficiency. (EEC/PE) [119]

per-unit system (1) (rotating machinery) The system of base values chosen in a consistent manner to facilitate analysis of a device or system, when per-unit quantities are used. Its importance becomes paramount when analog facilities (network analyzer, analog and hybrid computers) are utilized. *Note:* In electric network analysis and electromechanical system studies, usually four independent fundamental base values are chosen. The rest of the base values are derived from the fundamental ones. In most cases power, voltage, frequency, and time are chosen as fundamental base values. The base power must be the same for all types: apparent, active, reactive, instantaneous. The base time is usually one second. From the above, all other base values can be found, for example, base power times base time equals base energy, etc. The per-unit system can cover extensive networks because the base voltages of network sections connected by transformers can differ, in which case an ideal per-unit transformer is usually introduced having a turns ratio equal to the quotient of the effective turns ratio of the actual transformer and the ratio of base voltage values. By keeping the power, frequency, and time bases the same, only those base quantities will differ for different network sections that are directly or indirectly related to voltage (for example, current, impedance, reactance, inductance, capacitance, etc.) but those related to power, frequency, and time only (for example, energy, torque, etc.) will remain unchanged. *See also:* asynchronous machine. (PE) [9]

(2) The reference unit, established as a calculating convenience, for expressing all power system electrical parameters on a common reference base. One per unit (pu) is 100% of the base chosen. The pu system in power system engineering is used to obtain a better comparison of the performance of the power system elements of different ratings, similar to the decibel system used for equating the losses and levels of different telecommunications systems. (PE/PSC) 367-1996

per-unit value (ac rotating machinery) (basic per-unit quantities for ac rotating machines) The actual value divided by the base quantity when both actual and base values are expressed in the same units. (EM/PE) 86-1987w

perveance The quotient of the space-charge-limited cathode current by the three-halves power of the anode voltage in a diode. *Note:* Perveance is the constant G appearing in the Child-Langmuir-Schottky equation

$$i_k = \mathrm{G}e_b^{3/2}$$

When the term perveance is applied to triode or multigrid tube, the anode voltage e_b is replaced by the composite controlling voltage e of the equivalent diode. (ED) [45]

Petri net (software) An abstract, formal model of information flow, showing static and dynamic properties of a system. A Petri net is usually represented as a graph having two types of nodes (called places and transitions) connected by arcs, and markings (called tokens) indicating dynamic properties. (C) 610.3-1989w, 610.12-1990

PFM telemetry *See:* pulse frequency modulation telemetry.

p gate thyristor (cathode date SCR) A thyristor in which the gate terminal is connected to the p region adjacent to the region to which the cathode terminal is connected and that is normally switched to the ON state by applying a positive signal between gate and cathode terminals. (IA/IPC) 428-1981w

pH (of a solution A) The pH is obtained from the measurements of the potentials E of a galvanic cell of the form H_2: solution A: saturated potassium chloride (KCl); reference electrode with the aid of the equation

$$\mathrm{pH} = \frac{E - E_0}{(\mathrm{R}T/\mathrm{F})\ln 10} = \frac{E - E_0}{2.303\mathrm{R}T/\mathrm{F}}$$

in which E_0 is a constant depending upon the nature of the reference electrode, R is the gas constant in joules per mole per degree, T is the absolute temperature in kelvins, and F is the Faraday constant in coulombs per gram equivalent. Historically, pH was defined by

$$\mathrm{pH} = \log \frac{1}{[H+]}$$

in which $[H+]$ is the hydrogen ion concentration. According to present knowledge there is no simple relation between hydrogen ion concentration or activity and pH. Values of pH may be regarded as a convenient scale of acidities. *See also:* ion; ion activity. (EEC/PE) [119]

PH *See:* packet handler.

phandle A cell-sized datum identifying a particular package. (C/BA) 1275-1994

phanotron A hot-cathode gas diode. *Note:* This term is used primarily in the industrial field. (ED) [45]

phantom circuit (data transmission) A third circuit derived from two physical circuits by means of repeating coils installed at the terminals of the physical (side) circuits. A phantom circuit is a superimposed circuit derived from two suitably arranged pairs of wires, called side circuits, the two wires of each pair being effectively in parallel. (PE) 599-1985w

phantom-circuit loading coil A loading coil for introducing a desired amount of inductance in a phantom circuit and a minimum amount of inductance in the constituent side circuits. *See also:* loading. (EEC/PE) [119]

phantom-circuit repeating coil (phantom-circuit repeat coil) A repeating coil used at a terminal of a phantom circuit, in the terminal circuit extending from the midpoints of the associated side-circuit repeating coils. (EEC/PE) [119]

phantom group A group of four open wire conductors suitable for the derivation of a phantom circuit. (EEC/PE) [119]

phantom signaling A technique where a dc power source is superimposed on the transmit and receive signal pairs in a transparent or "phantom" fashion such that its application does not affect the data bearing signals on either pair. This dc power source is normally applied to request a concentrator to insert a lobe into the ring. (C/LM) 8802-5-1998

phantom target (A) (radar) An echo box, or other reflection device, that produces a particular blip on the radar indicator. **(B) (radar)** A condition, maladjustment, or phenomenon (such as a temperature inversion) that produces a blip on the radar indicator resembling blips of targets for which the system is being operated. *See also:* echo box; navigation. (AES/RS) 686-1982

phase (1) (of a periodic phenomenon $f(t)$, for a particular value of t) The fractional part t/P of the period P through which t has advanced relative to an arbitrary origin. *Note:* The origin is usually taken at the last previous passage through zero from the negative to the positive direction. (IM) [120]
(2) (A) A distinct part of a process in which related operations are performed, as in the shift phase of a shift-and-carry operation. **(B)** A relative measurement that describes the temporal relationship between two signals that have the same frequency. (C) 610.10-1994
(3) A major stage within the generating-plant life cycle. *See also:* plant life cycle. (PE/EDPG) 1150-1991w
(4) The time within a timing cycle when a primary input is in transition between logic states. (SCC20) 1445-1998

phase advancer A phase modifier that supplies leading reactive volt-amperes to the system to which it is connected. Phase advancers may be either synchronous or asynchronous. *See also:* converter. (EEC/PE) [119]

phase angle (1) (general) The measure of the progression of a periodic wave in time or space from a chosen instant or position. *Notes:* 1. The phase angle of a field quantity, or of voltage or current, at a given instant of time at any given plane in a waveguide is $(wt - \beta z + \theta)$, when the wave has a sinusoidal time variation. The term waveguide is used here in its most general sense and includes all transmission lines; for example, rectangular waveguide, coaxial line, strip line, etc. The symbol β is the imaginary part of the propagation constant for that waveguide, propagation is in the $+z$ direction, and θ is the phase angle when $z = t = 0$. At a reference time $t = 0$ and at the plane z, the phase angle $(-\beta z + \Theta)$ will be represented by Φ. 2. Phase angle is obtained by multiplying the phase by 360 degrees or by 2π radians. (IM) [38]
(2) (speed governing of hydraulic turbines) Referring to a simultaneous phasor diagram of the input and output, the angle by which the output signal lags or leads the input signal. (PE/EDPG) 125-1977s
(3) (electronics) The phase angle of a current transformer is the phase displacement between the primary and secondary currents. The phase angle is positive when the secondary current leads the primary current. (PEL/ET) 389-1990

phase angle correction factor (PACF) The ratio of the true power factor to the measured power factor. It is a function of both the phase angles of the instrument transformers and the power factor of the primary circuit being measured. *Note:* The phase angle correction factor corrects for the phase displacement of the secondary current or voltage, or both, due to the instrument transformer phase angle(s). For a current transformer, the phase angle correction factor:

$$\text{PACF} = \cos(\theta_2 + \beta)/\cos(\theta_2)$$

For a voltage transformer, the phase angle correction factor:

$$\text{PACF} = \cos(\theta_2 - \beta)/\cos(\theta_2)$$

When both voltage and current transformers are used, the combined phase angle correction:

$$\text{PACF} = \cos(\theta_2 + \beta - \gamma)/\cos(\theta_2)$$

θ_2 is the apparent power factor angle of the circuit being measured. (PE/TR) C57.13-1993

phase angle, dielectric *See:* dielectric phase angle.

phase angle, loop *See:* loop phase angle.

phase-angle measuring (power system device function numbers) A relay that functions at a predetermined phase angle between two voltages or between two currents or between voltage and current. (PE/SUB) C37.2-1979s

phase angle of an instrument transformer The phase displacement, in minutes or radians, between the primary and secondary values. The phase angle of a current transformer is designated by the Greek letter beta (β) and is positive when the current leaving the identified secondary terminal leads the current entering the identified primary terminal. The phase angle of a voltage transformer is designated by the Greek letter gamma (γ) and is positive when the secondary voltage from the identified to the unidentified terminal leads the corresponding primary voltage. (PE/TR) C57.13-1993, C57.12.80-1978r

phase back (electrical heating systems) The amount of retardation (expressed in percent or as an angle) during which the controlling element is prevented from conducting. (IA/PC) 844-1985s

phase-balance relay A relay that responds to differences between quantities of the same nature associated with different phases of a normally balanced polyphase circuit. (SWG/PE) C37.100-1992

phase belt (coil group) A group of adjacent coils in a distributed polyphase winding of an alternating-current machine that are ordinarily connected in series to form one section of a phase winding of the machine. Usually, there are as many such phase belts per phase as there are poles in the machine. *Note:* The adjacent coils of a phase belt do not necessarily occupy adjacent slots; the intervening slots may be occupied by coils of another winding on the same core. Such may be the case in a two-speed machine. *See also:* rotor; stator. (PE) [9]

phase center The location of a point associated with an antenna such that, if it is taken as the center of a sphere whose radius extends into the far-field, the phase of a given field component over the surface of the radiation sphere is essentially constant, at least over that portion of the surface where the radiation is significant. *Note:* Some antennas do not have a unique phase center. (AP/ANT) 145-1993

phase-change recording A method of recording information on optical storage in which the laser strikes the optical medium, causing it to crystallize in a controlled manner such that the change can be interpreted as a binary 0 or 1. (C) 610.10-1994w

phase characteristic (1) The variation with frequency of the phase angle of a phasor quantity. (Std100) 270-1966w

(2) (linear passive networks) The angle of a response function evaluated on the imaginary axis of the complex-frequency plane. (CAS) 156-1960w

phase characteristic, loop (automatic control) (closed loop) The phase angle of the loop transfer function for real frequencies. *See also:* feedback control system. (PE/EDPG) [3]

phase-coded transducer An interdigital transducer in which the electrodes do not strictly alternate in polarity, thus creating a coded time-phase relationship; the phase of the signal is determined by the polarity of the connections to the interdigital transducer bus bars. (UFFC) 1037-1992w

phase coherence *See:* coherent.

phase-coil insulation (rotating machinery) Additional insulation between adjacent coils that are in different phases. *See also:* asynchronous machine. (PE) [9]

phase-comparison monopulse A form of monopulse employing receiving beams with different phase centers as obtained, for example, from side-by-side antennas or separate portions of an array. *Note:* The information on target displacement from the antenna axis appears as a relative phase between signals received at the two phase centers. *See also:* amplitude-comparison monopulse; monopulse. (AES) 686-1997

phase-comparison protection A form of pilot protection that compares the relative phase-angle position of specified currents at the terminals of a circuit. (SWG/PE/PSR) C37.100-1992, C37.113-1999

phase conductor (alternating-current circuit) The conductors other than the neutral conductor. *Note:* If an alternating-current circuit does not have a neutral conductor, all the conductors are phase conductors. (Std100) 270-1966w

phase connections (rotating machinery) The insulated conductors (usually arranged in peripheral rings) that make the necessary connections between appropriate phase belts in an alternating-current winding. *See also:* stator; rotor. (PE) [9]

phase constant (β) (1) (A) (fiber optics) The imaginary part of the axial propagation constant for a particular mode, usually expressed in radians per unit length. *See also:* axial propagation constant. **(B)** The imaginary component of the propagation constant. This is the spatial rate of decrease of phase of a field component in the direction of propagation in radians per unit length. (Std100/MTT) 812-1984, 1004-1987

(2) (waveguide) Of a traveling wave, the space rate of change of phase of a field component (or of the voltage or current) in the direction of propagation, in radians per unit length. (MTT) 146-1980w

(3) The magnitude of the phase vector. *See also:* propagation vector. (AP/PROP) 211-1997

phase contours Loci of the return transfer function at constant values of the phase angle. *Note:* Such loci may be drawn on the Nyquist, inverse Nyquist, or Nichols diagrams for estimating performance of the closed loop with unity feedback. In the complex plane plot of $KG(j\omega)$, these loci are circles with centers at $-1/2, j/2N$ and radiuses such that each circle passes through the origin and the point $-1, j0$. In the inverse Nyquist diagram they are straight lines $\gamma = -N(x + 1)$ radiating from the point $-1,0$. *See also:* Nyquist diagram; Nichols chart; inverse Nyquist diagram. (IM) [120]

phase control (1) (rectifier circuits) The process of varying the point within the cycle at which forward conduction is permitted to begin through the rectifier circuit element. *Note:* The amount of phase control may be expressed in two ways: the reduction in direct-current voltage obtained by phase control or the angle of retard or advance. (IA) [62]

(2) (thyristor) The starting instant is synchronous with respect to the line voltage. The controller ON-state interval is equal to or less than half the line period. (IA/IPC) 428-1981w

phase control power (thyristor converter) The power used to synchronize the phase control of the thyristor converter to the ac supply input phases. (IA/IPC) 444-1973w

phase control range (thyristor) The range over which it is possible to adjust the angle of retard expressed in electrical degrees. (IA/IPC) 428-1981w

phase converter (rotating machinery) A converter that changes alternating-current power of one or more phases to alternating-current power of a different number of phases but of the same frequency. *See also:* converter. (PE) [9]

phase-corrected horn A horn designed to make the emergent electromagnetic wave front substantially plane at the mouth. *Note:* Usually this is achieved by means of a lens at the mouth. *See also:* waveguide; circular scanning. (AP/ANT) [35], [84]

phase correction (telegraph transmission) The process of keeping synchronous telegraph mechanisms in substantially correct phase relationship. *See also:* telegraphy. (EEC/PE) [119]

phase correction pattern A metallized area of varying width applied to the device substrate such that the resulting piezoelectric "shorting" effect adjusts the relative phase of the passing acoustic wavefront along its entire length. (UFFC) 1037-1992w

phase corrector A network that is designed to correct for phase distortion. *See also:* network analysis. (Std100) 270-1966w

phase-crossover frequency (hydraulic turbines) The frequency at which the phase angle reaches 180 degrees. (PE/EDPG) 125-1977s

phased-array antenna An array attenna whose beam directon or radiation pattern is controlled primarily by the relative phases of the excitation coefficients of the radiating elements. *See also:* antenna. (AP/ANT) 145-1983s

phase delay (1) (facsimile) (in the transfer of a single-frequency wave from one point to another in a system) The time delay of a part of the wave identifying its phase. *Note:* The phase delay is measured by the ratio of the total phase shift in cycles to the frequency in hertz. *See also:* facsimile transmission. (COM) 168-1956w

(2) (dispersive and nondispersive delay lines) The ratio of total radian phase shift, to the specified radian frequency, w. Phase delay is nominally constant over the frequency band of operation for nondispersive delay devices. *See also:* phase lag. (UFFC) [22]

(3) The ratio of total radian insertion phase shift ϕ to the specified radian frequency ω. Phase delay is nominally constant over the frequency band of operation for nondispersive delay devices (dispersive and nondispersive delay line). (UFFC) 1037-1992w

(4) (as applied to relaying) An equal delay of both the leading and trailing edges of a locally generated block. (SWG/PE) C37.100-1992

(5) The ratio of the total phase shift (radians) experienced by a sinusoidal signal in transmission through a system or transducer, to the frequency (radians/second) of the signal. *Note:* The unit of phase delay is the second. (SP) 151-1965w

phase delay distortion (system or transducer) The difference between the phase delay at one frequency and the phase delay at a reference frequency. (SP) 151-1965w

phase delay time In the transfer of a single-frequency wave from one point to another in a system, the time delay of a part of the wave identifying its phase. *Note:* The phase delay time is measured by the ratio of the total phase delay through the network, in cycles, to the frequency, in hertz. *See also:* measurement system. (IM) 285-1968w, [38]

phase deviation ($\phi(t)$) (1) (angle modulation) (phase modulation) The peak difference between the instantaneous angle of the modulated wave and the angle of the carrier. *Note:* In the case of a sinusoidal modulating function, the value of the phase deviation, expressed in radians, is equal to the modulation index. *See also:* angle or phase; phase modulation. (Std100) [123]

(2) (data transmission) The lack of direct proportionality of phase shift to frequency over the frequency range required

for transmission, or the effect of such departure on a transmitted signal. (PE) 599-1985w
(3) Instantaneous phase departure from a nominal phase. (SCC27) 1139-1999

phase difference (general) The difference in phase between two sinusoidal functions having the same periods. (Std100) 270-1966w
(2) (A) (automatic control) Between sinusoidal input and output of the same frequency, phase angle of the output minus phase angle of the input: it is called "phase lead" if the input angle is the smaller, "phase lag" if the larger. **(B) (automatic control)** Of two periodic phenomena (for example, in nonlinear systems) the difference between the phase angles of their two fundamental waveforms. *Note:* Regarded as part of the transfer function which relates output to input at a specified frequency, phase difference is simply the phase angle $\theta(j\omega)$ in $A(j\omega) \exp j\theta(j\omega)$. Measurement of phase difference in the complex case is sometimes made in terms of the angular interval between respective crossings of a mean reference line, but values so measured will generally differ from those made in terms of the fundamental waveforms. *See also:* phase shift. (PE/EDPG) [3]

phase distance relay A distance relay designed to detect phase-to-phase and three-phase faults. (PE/PSR) C37.113-1999

phase distortion (1) (data transmission) Either the lack of direct proportionality of phase shift to frequency over the frequency range required for transmission, or the effect of such departure on a transmitted signal. (PE) 599-1985w
(2) (facsimile) *See also:* delay distortion; phase-frequency distortion. (C) 610.7-1995

phased satellite (communication satellite) A satellite, the center of mass of which is maintained in a desired relation relative to other satellites, to a point on earth or to some other point of reference such as the sub-solar point. *Note:* If it is necessary to identify those satellites that are not phased satellites, the term "unphased satellites" may be used. (COM) [19]

phase-failure protection *See:* open-phase protection; phase-undervoltage protection.

phase-frequency distortion (facsimile) Distortion due to lack of direct proportionality of phase shift to frequency over the frequency range required for transmission. *Notes:* 1. Delay distortion is a special case. 2. This definition includes the case of a linear phase-frequency relation with the zero frequency intercept differing from an integral multiple of p. *See also:* phase delay distortion; phase distortion; facsimile transmission; distortion. (COM) 168-1956w

phase function matrix The matrix that results when the elements of the Mueller matrix are averaged over all scatterer orientations. The phase function matrix relates the average scattered Stokes vector to the incident Stokes vector. (AP/PROP) 211-1997

phase grouping The same phase of a number of circuit breaker' poles is grouped in an adjacent configuration along the line of the same row. (SWG/SUB/PE) C37.122-1983s, C37.100-1992

phase hit or change A sudden change in the received signal phase (or frequency) lasting longer than 4 ms. Since two common modulation techniques for high-speed data transmission are phase and frequency modulation, phase hits cause errors by looking like data. *See also:* gain hit or change; dropouts. (PE/IC) 1143-1994r

phase instability ($S_\phi(f)$) One-sided spectral density of the phase deviation. (SCC27) 1139-1999

phase-insulated terminal box (rotating machinery) A terminal box so designed that the protection of phase conductors against electric failure within the terminal box is by insulation only. (PE) [9]

phase jitter An instability in the phase of a transmission signal. *See also:* amplitude jitter. (C) 610.7-1995

phase lag (phase delay) (2-port network) The phase angle of the input wave relative to the output wave ($\phi_{in} - \phi_{out}$), or the initial phase angle of the output wave relative to the final phase angle of the output wave ($\phi_i - \phi_f$). *Note:* Under matched conditions, phase lag is the negative of the angle of the transmission coefficient of the scattering matrix for a 2-port network. *See also:* phase difference. (IM) 285-1968w, [38]

phase localizer (navigation aid terms) A localizer in which the on-course line is defined by the phase reversal of energy radiated by the sideband antenna system, a reference carrier signal being radiated and used for the detection of phase. (AES/GCS) 172-1983w

phase lock The state of synchronization between two ac signals in which they remain at the same frequency and with constant phase difference. This term is typically applied to a circuit that synchronizes a variable oscillator with an independent signal. (PE/PSR) 1344-1995

phase-locked Pertaining to two signals whose phases relative to each other are kept constant by a controlling device. (C) 610.10-1994w

phase lock loop (communication satellite) A circuit for synchronizing a variable local oscillator with the phase of a transmitted signal. Widely used in space communication for coherent carrier tracking, and threshold extension, bit synchronization and symbol synchronization. (COM) [24]

phase locus (for a loop transfer function, say G(s) H(s) A plot in the *s* plane of those points for which the phase angle, ang *GH*, has some specified constant value. *Note:* The phase loci for 180 degrees plus or minus *n* 360 degrees are also root loci. *See also:* feedback control system. (PE/EDPG) [3]

phase margin (1) (loop transfer function for a stable feedback control system) (excitation systems) 180 degrees minus the absolute value of the loop phase angle at a frequency where the loop gain is unity. *Note:* Phase margin is a convenient way of expressing relative stability of a linear system under parameter changes, in Nyquist, Bode, or Nichols diagrams. In a conditionally stable feedback control system where the loop gain becomes unity at several frequencies, the term is understood to apply to the value of phase margin at the highest of these frequencies. *See also:* feedback control system. (PE/EDPG) 421A-1978s
(2) (speed governing of hydraulic turbines) 180 degrees minus the absolute value of the open-loop phase angle at a frequency where the open-loop gain is unity. (PE/EDPG) 125-1977s
(3) The absolute value of loop phase angle subtracted from 180 degrees found in a feedback system at the frequency for which its gain reaches unity. The margin from 180 degrees represents a measure of dynamic stability. (PEL) 1515-2000

phase meter (phase-angle meter) An instrument for measuring the difference in phase between two alternating quantities of the same frequency. *See also:* instrument. (EEC/PE) [119]

phase modifier (rotating machinery) An electric machine, the chief purpose of which is to supply leading or lagging reactive power to the system to which it is connected. Phase modifiers may be either synchronous or asynchronous. *See also:* converter. (IA/PE/MT) 45-1983s, [9]

phase-modulated transmitter A transmitter that transmits a phase-modulated wave. (AP/BT/ANT) 145-1983s, 182-1961w

phase modulation (1) (data transmission) Angle modulation in which the angle of a carrier is caused to depart from its reference value by an amount proportional to the instantaneous value of the modulating function. *Notes:* 1. A wave phase modulated by a given function can be regarded as a wave frequency modulated by the time derivative of that function. 2. Combinations of phase and frequency modulation are commonly referred to as frequency modulation. *See also:* reactance modulator; angle or phase; pulse duration; phase deviation. (IT/AP/PE/ANT) 145-1983s, 599-1985w, [123]
(2) (overhead-power-line corona and radio noise) Modulation in which the angle of a carrier is caused to depart from

its reference value by an amount proportional to the instantaneous value of the modulating signal.
(T&D/PE) 539-1990

(3) A modulation technique in which a data signal is sent onto a fixed carrier frequency by modifying the phase of the carrier. (C) 610.7-1995

phase-modulation recording A type of magnetic recording in which each storage cell is divided into two regions that are each magnetized in opposite senses; the sequence of these senses indicates whether the binary character represented is zero or one. *See also:* double-pulse recording.
(C) 610.10-1994w

phase-modulation telemetering (electric power system) A type of telemetering in which the phase difference between the transmitted voltage and a reference voltage varies as a function of the magnitude of the measured quantity. *See also:* telemetering. (PE/PSE) 94-1970w

phase modulator, optical *See:* optical phase modulator.

phase nonlinearity The deviation in phase from a perfectly phase-linear response as a function of frequency. The phase response of a perfectly phase-linear system is directly proportional to frequency. (IM/WM&A) 1057-1989w

phase of a circularly polarized field vector In the plane of polarization, the angle that the field vector makes, at a time taken as the origin, with a reference direction and with the angle counted as positive if it is in the same direction as the sense of polarization and negative if it is in the opposite direction to the sense of polarization. (AP/ANT) 145-1993

phase overcurrent The current flowing in a phase conductor which exceeds a predetermined value.
(SWG/PE) C37.100-1981s

phase path (radio-wave propagation) For a monochromatic electromagnetic wave, the product of the phase constant and the physical path length. *Note:* In a slowly varying spatially inhomogeneous medium, the phase path length equals the line integral of the real part of the phase vector along the ray path.
(AP/PROP) 211-1990s

phase path length For a monochromatic electromagnetic wave, the product of the phase constant and the physical path length. *Note:* In a slowly varying spatially inhomogeneous medium, the path length equals the line integral of the real part of the phase constant along the ray path. *See also:* electrical length.
(AP/PROP) 211-1997

phase pattern (of an antenna) The spatial distribution of the relative phase of a field vector excited by an antenna. *Notes:* 1. The phase may be referred to any arbitrary reference. 2. The distribution of phase over any path, surface, or radiation pattern cut is also called a phase pattern.
(AP/ANT) 145-1993

phase recovery time (microwave gas tubes) The time required for a fired tube to deionize to such a level that a specified phase shift is produced in the low-level radio-frequency signal transmitted through the tube. *See also:* gas tube.
(ED) 161-1971w

phase, relative *See:* relative phase.

phase relay A relay that by its design or application is intended to respond primarily to phase conditions of the power system.
(SWG/PE) C37.100-1992

phase resolution The minimum change of phase that can be distinguished by a system. *See also:* measurement system.
(IM) [38]

phase-reversal protection *See:* phase-sequence reversal protection.

phase-reversals relay *See:* negative-phase-sequence relay.

phase-segregated terminal box A terminal box so designed that the protection of phase conductors against electric failure within the terminal box is by insulation, and additionally by grounded metallic barriers forming completely isolated individual phase compartments so as to restrict any electric breakdown to a ground fault. (PE) [9]

phase-selector relay A programming relay whose function is to select the faulted phase or phases, thereby controlling the operation of other relays or control devices.
(SWG/PE) C37.100-1992

phase-separated terminal box *See:* phase-segregated terminal box.

phase separator (rotating machinery) Additional insulation between adjacent coils that are in different phases. *See also:* stator; rotor. (PE) [9]

phase sequence (1) (set of polyphase voltages or currents) The order in which the successive members of the set reach their positive maximum values. *Note:* The phase sequence may be designated in several ways. If the set of polyphase voltages or currents is a symmetrical set, one method is to designate the phase sequence by specifying the integer that denotes the number of times that the angular phase lag between successive members of the set contains the characteristic angular phase difference for the number of phases m. If the integer is zero, the set is of zero phase sequence; if the integer is one, the set is of first phase sequence; and so on. Since angles of lag greater than 2*p* produce the same phase position for alternating quantities as the same angle decreased by the largest integral multiple of 2p contained in the angle of lag, it may be shown that there are only m distinct symmetrical sets normally designated from 0 to $m - 1$ phase sequence. It can be shown that only for the first phase sequence do all the members of the set reach their positive maximum in the order of identification at uniform intervals of time.
(PE) [9], 270-1966w

(2) (power and distribution transformers) The order in which the voltages successively reach their positive maximum values. *See also:* direction of rotation of phasors.
(PE/TR) C57.12.80-1978r

phase-sequence indicator A device designed to indicate the sequence in which the fundamental components of a polyphase set of potential differences, or currents, successively reach some particular value, such as their maximum positive value. *See also:* instrument. (EEC/PE) [119]

phase-sequence relay A relay that responds to the order in which the phase voltages or currents successively reach their maximum positive values. (SWG/PE) C37.100-1992

phase-sequence reversal A reversal of the normal phase sequence of the power supply. For example, the interchange of two lines on a three-phase system will give a phase reversal.
(IA/ICTL/IAC) [60]

phase-sequence reversal protection A form of protection that prevents energization of the protected equipment on the reversal of the phase sequence in a polyphase circuit.
(SWG/PE) C37.100-1992

phase-sequence test (rotating machinery) A test to determine the phase sequence of the generated voltage of a three-phase generator when rotating in its normal direction. *See also:* asynchronous machine. (PE) [9]

phase-sequence voltage relay (power system device function numbers) A relay that functions upon a predetermined value of polyphase voltage in the desired phase sequence.
(PE/SUB) C37.2-1979s

phase shift (1) The absolute magnitude of the difference between two phase angles. *Notes:* 1. The phase shift between two planes of a 2-port network is the absolute magnitude of the difference between the phase angles at those planes. The total phase shift, or absolute phase shift, is expressed as the total number of cycles, including any fractional number, between the two planes, where one complete cycle is 2π radians or 360 degrees. Relative phase shift is the total or absolute phase shift less the largest integral number of 2π radians or 360 degrees. The unit of phase shift is, therefore, the radian or the electrical degree. The term 2-port network is used in its most general sense to include structures of passive or active elements. This includes the case of a given length of waveguide but may also refer to any two ports of a multiport device, where it is understood that a signal is incident only at one port. 2. A phase shift can be either a phase lead (advance) or a phase lag (delay). *See also:* measurement system. (IM) 285-1968w, [38]

(2) (A) (transfer function) A change of phase angle with frequency, as between points on a loop phase characteristic. *See also:* feedback control system. **(B) (signal)** A change of phase angle with transmission. (IM) [120]

(3) The total number of degrees or radians that a continuous-wave signal experiences as it is transmitted through the delay device at a given frequency within the band of operation. The phase shift is nominally linearly proportional to frequency within the frequency band of operation for a nondispersive delay device (dispersive and nondispersive delay lines). (UFFC) 1037-1992w, [22]

(4) The displacement in time of one waveform relative to another of the same frequency and harmonic content. (SCC22) 1346-1998

(5) The displacement between corresponding points on similar wave shapes, and is expressed in degrees leading or lagging. (IA/AES/PSE) 1100-1999, [41]

phase-shift circuit Anetwork that provides a voltage component shifted in phase with respect to a reference voltage. *See also:* electronic controller. (IA/ICTL/IAC) [60]

phase shifter (data transmission) A device in which the output voltage (or current) may be adjusted, in use or in its design, to have some desired phase relation with the input voltage (or current). (PE) 599-1985w

phase shifter, waveguide *See:* waveguide phase shifter.

phase-shifting transformer (1) (metering) An assembly of one or more transformers intended to be connected across the phases of a polyphase circuit so as to provide voltages in the proper phase relations for energizing varmeters, varhour meters, or other measurement equipment. This type of transformer is sometimes referred to as a phasing transformer. (ELM) C12.1-1988

(2) A transformer that advances or retards the phase-angle relationship of one circuit with rspect to another. *Notes:* 1. The terms "advance" and "retard" describe the electrical angular position of the load voltage with respect to the source voltage. 2. If the load voltage reaches its positive maximum sooner than the source voltage, this is an "advance" position. 3. Conversely, if the load voltage reaches its positive maximum later than the source voltage, this is a "retard" position. *See also:* regulating winding; excitation-regulating winding; main unit; load-tap-changing transformer; excited winding; series winding; regulated circuit; excitation winding; voltage-regulating relay; series unit; phase-shifting transformer; regulating transformer; line-drop compensator; voltage winding for regulating equipment. (PE/TR) C57.12.80-1978r

phase-shift keying (PSK) (modulation systems) The form of phase modulation in which the modulating function shifts the instantaneous phase of the modulated wave among predetermined discrete values. (Std100) 270-1964w

phase-shift oscillator An oscillator produced by connecting any network having a phase shift of an odd multiple of 180 degrees (per stage) at the frequency of oscillation, between the output and the input of an amplifier. When the phase shift is obtained by resistance-capacitance elements, the circuit is an R-C phase-shift oscillator. *See also:* oscillatory circuit. (AP/ANT) 145-1983s

phase space (A) The state space augmented by the independent time variable. **(B)** One used synonymously with the state space, usually with the state variables being successive time derivatives of each other. *See also:* control system. (CS/IM) [120]

phase spacing (1) The distance between center-lines of adjacent devices of differing phases. (SWG/PE) C37.40-1993

(2) (of a fuse or switching device) The distance between center-lines of the current-carrying parts of the adjacent poles of the switching device. (SWG/PE) C37.100-1992

phase splitter (data transmission) (phase splitting circuit) A device which produces, from a single input wave, two or more output waves that differ in phase from one another. (PE) 599-1985w

phase-splitting circuit *See:* phase splitter.

phase swinging (rotating machinery) Periodic variations in the speed of a synchronous machine above or below the normal speed due to power pulsations in the prime mover or driven load, possibly recurring every revolution. (PE) [9]

phase-to-ground insulation configuration An insulation configuration between a terminal and the neutral or ground. (PE/C) 1313.1-1996

phase-to-ground per unit overvoltage (power and distribution transformers) The ratio of a phase-to-ground overvoltage to the phase-to-ground voltage corresponding to the maximum system voltage. (PE/TR) C57.12.80-1978r

phase-to-phase insulation configuration An insulation configuration between two-phase terminals. (PE/C) 1313.1-1996

phase-to-phase per unit overvoltage (power and distribution transformers) The ratio of a phase-to-phase overvoltage to the phase-to-phase voltage corresponding to the maximum system voltage. (PE/TR) C57.12.80-1978r

phase-to-phase voltage on an alternating-current electric system *See:* nominal system voltage; maximum system voltage; medium voltage; high voltage; low voltage; service voltage; utilization voltage.

phase transfer function The argument Φ (N) of the modulation transfer function is designated as the phase transfer function. (ED) 503-1978w

phase transition (primary ferroelectric terms) A change in the crystal structure, usually occurring at a well-defined temperature, which alters the orientation or magnitude, or both, of the electric polarization. (UFFC) 180-1986w

phase-tuned tube (microwave gas tubes) A fixed-tuned broadband transmit-receive tube, wherein the phase angle through and the reflection introduced by the tube are controlled within limits. *See also:* gas tube. (ED) 161-1971w

phase-undervoltage protection A form of protection that disconnects or inhibits connection of the protected equipment on deficient voltage in one or more phases of a polyphase circuit. (SWG/PE) C37.100-1992

phase-undervoltage relay A relay that operates when one or more phase voltages in a normally balanced polyphase circuit is less than a predetermined value. (SWG/PE/PSR) C37.100-1992, C37.90-1978s

phase vector ($\vec{\beta}$) The real part of the propagation vector, $\vec{k}$. *Note:* The phase vector points in the direction of maximum rate of change of the phase. *See also:* propagation vector. (AP/PROP) 211-1997

phase vector in physical media The imaginary part of the propagation vector. (AP/ANT) 145-1983s

phase velocity (1) (fiber optics) For a particular mode, the ratio of the angular frequency to the phase constant. *See also:* coherence time; axial propagation constant; group velocity. (Std100) 812-1984w

(2) (of a traveling plane wave at a single frequency) The velocity of an equiphase surface along the wave normal. *See also:* radio-wave propagation; waveguide. (AP/ANT/PROP) [35], [36]

(3) (waveguide) Of a traveling wave at a given frequency, and for a given mode, the velocity of an equiphase surface in the direction of propagation. (MTT) 146-1980w

(4) The velocity at which the equiphase planes of a propagating wave travel. *Note:* The minimum phase velocity is in the direction of the wave normal. (AP/PROP) 211-1997

phase-versus-frequency response characteristic A graph or tabulation of the phase shifts occurring in an electric transducer at several frequencies within a band. *See also:* transducer. (AP/ANT) 145-1983s

phase voltage (machine or apparatus) (of a winding) The potential difference across one phase of the machine or apparatus. *See also:* asynchronous machine. (PE) [9], [84]

phase weighting Response weighting by change in period of finger arrangement inside the interdigital transducer. (UFFC) 1037-1992w

phasing The adjustment of picture position along the scanning line. *See also:* scanning. (COM) 168-1956w

phasing signal A signal used for adjustment of the picture position along the scanning line. *See also:* facsimile signal. (COM) 168-1956w

phasing time (facsimile) The time interval during which the start positions of the scanning and recording strokes are aligned so as to ensure against a split image at the recorder. (COM) 167-1966w

phasing voltage (of a network protector) The voltage across the open contacts of a selected phase. *Note:* This voltage is equal to the phasor difference between the transformer voltage and the corresponding network voltage. (SWG/PE/TR) C37.100-1992, C57.12.44-1994

phasor (1) (metering) A complex number, associated with sinusoidally varying electrical quantities, such that the absolute value (modulus) of the complex number corresponds to either the peak amplitude or rms value of the quantity, and the phase (argument) to the phase angle at zero time. By extension, the term "phasor" can also be applied to impedance and related complex quantities that are not time-dependent. (ELM) C12.1-1988

(2) A complex number expressing the magnitude and phase of a time-varying quantity. Unless otherwise specified, it is used only within the context of steady-state alternating linear systems. In polar coordinates, it can be written as $Ae^{j\phi}$, where A is the amplitude or magnitude (usually rms, but sometimes indicated as peak value) and ϕ is the phase angle. The phase angle ϕ should not be confused with the space angle of a vector. *See also:* electric field strength. (T&D/PE) 644-1994

(3) A complex equivalent of a simple sine wave quantity such that the complex modulus is the sine wave amplitude and the complex angle (in polar form) is the sine wave phase angle. *See also:* vector. (PE/PSR) 1344-1995

phasor diagram (synchronous machines) A diagram showing the relationships of as many of the following phasor quantities as are necessary: armature current, armature voltages, the direct and quadrature axes, armature flux linkages due to armature and field winding currents, magnetomotive forces due to armature and field-winding currents, and the various components of air-gap flux. (PE) [9]

phasor difference *See:* phasor sum.

phasor function A functional relationship that results in a phasor. (Std100) 270-1966w

phasor notation For monochromatic fields, the complex notation used in the expressions for field quantities with the exponential time factor $\exp\{j\omega t\}$. For example, for plane waves

$$\vec{\epsilon}(\vec{r}, t) = \mathrm{Re}\{\vec{E}(\vec{r}, \omega)\exp(j\omega t)\}$$

where

$\vec{\epsilon}(\vec{r}, t)$ = the instantaneous electric field

Re indicates the real part

$\vec{E}(\vec{r}, \omega)$ = the phasor notation for the electric field

(AP/PROP) 211-1997

phasor power (rotating machinery) The phasor representing the complex power. *See also:* asynchronous machine. (PE) [9]

(2) (A) (polyphase circuit) At the terminals of entry of a polyphase circuit into a delimited region, a phasor (or plane vector) that is equal to the (phasor) sum of the phasor powers for the individual terminals of entry when the voltages are all determined with respect to the same arbitrarily selected common reference point in the boundary surface (which may be the neutral terminal of entry). The reference direction for the currents and the reference polarity for the voltages must be the same as for instantaneous power, active power, and reactive power. The phasor power for each terminal of entry is determined by considering each conductor and the common reference point as a single-phase, two-wire circuit and finding the phasor power for each in accordance with the definition of (B) below. The phasor power $\mathbf{S}$ is given by $\mathbf{S} = \mathbf{P} + \mathbf{jQ}$ where $\mathbf{P}$ is the active power for the polyphase circuit and $\mathbf{Q}$ is the reactive power for the same terminals of entry. If the voltages and currents are sinusoidal and of the same period, the phasor power $\mathbf{S}$ for a three-phase circuit is given by

$$\mathbf{S} = \mathbf{E}_a\mathbf{I}_a^* + \mathbf{E}_b\mathbf{I}_b^* + \mathbf{E}_c\mathbf{I}_c^*$$

where $\mathbf{E}_a$, $\mathbf{E}_b$, and $\mathbf{E}_c$ are the phasor voltages from the phase conductors a, b, and c, respectively, to the neutral conductor at the terminals of entry; $\mathbf{I}_a$, $\mathbf{I}_b$, and $\mathbf{I}_c$ are the conjugate of the phasor currents in the phase conductor, so that there are only three terminals of entry; the point of entry of one of the phase conductors may be chosen as the common voltage point; and the phasor from that conductor to the common voltage point becomes zero. If the terminal of entry of phase conductor b is chosen as the common point, the phasor power of a three- phase, three-wire circuit becomes

$$\mathbf{S} = \mathbf{E}_{ab}\mathbf{I}_a^* + \mathbf{E}_{cb}\mathbf{I}_c^*$$

where $\mathbf{E}_{ab}$, $\mathbf{E}_{cb}$ are the phasor voltages from phase conductor a to b and from c to b, respectively. If both the voltages and currents in the preceding equations constitute symmetrical sets of the same phase sequence $\mathbf{S} = 3\mathbf{E}_a\mathbf{I}_a^*$. In general, the phasor power at the $(m + 1)$ terminals of entry of a polyphase circuit of m phases to a delimited region, when one of the terminals is the neutral terminal of entry, and is expressed by the equation

$$\mathbf{S} = \sum_{s=1}^{s=m} \sum_{r=1}^{r=\infty} \mathbf{E}_{sr}\mathbf{I}_{sr}^*$$

where $\mathbf{E}_{sr}$ is the phasor representing the rth harmonic of the voltage from phase conductor s to neutral at the terminals of entry. $\mathbf{I}_{sr}^*$ is the conjugate of the phasor representing the rth harmonic of the current through the sth terminal of entry. The phasor power can also be stated in terms of the symmetrical components of the voltages and currents as

$$\mathbf{S} = m \sum_{k=0}^{k=m-1} \sum_{r=1}^{r=\infty} \mathbf{E}_{kr}\mathbf{I}_{kr}^*$$

where $\mathbf{E}_{kr}$ is the phasor representing the symmetrical component of kth sequence of the rth harmonic of the line-to-neutral set of polyphase voltages at the terminals of entry. $\mathbf{I}_{kr}^*$ is the conjugate of the phasor representing the symmetrical component of the kth sequence of the rth harmonic of the polyphase set of currents through the terminals of entry. Phasor power is expressed in voltamperes when the voltages are in volts and the currents in amperes. *Note:* This term was once defined as "vector power." With the introduction of the term "phasor quantity," the name of this term has been altered to correspond. The definition has also been altered to agree with the change in the sign of reactive power. *See also:* reactive power. **(B) (single-phase two-wire circuit)** At the two terminals of entry of a single-phase two-wire circuit into a delimited region, a phasor (or plane vector) of which the real component is the active power and the imaginary component is the reactive power at the same two terminals of entry. When either component of phasor power is positive, the direction of that component is in the reference direction. The phasor power $\mathbf{S}$ is given by $\mathbf{S} = P + jQ$ where P and Q are the active and reactive power, respectively. If both the voltage and current are sinusoidal, the phasor power is equal to the product of the phasor voltage and the conjugate of the phasor current.

$$\mathbf{E} = Ee^{j\alpha};\ \mathbf{I} = \mathbf{I}e^{j\beta};$$

the phasor power is

$$S = P + jQ = \mathbf{EI^*} = EIe^{j(\alpha-\beta)}$$
$$= EI[\cos(\alpha - \beta) + j\sin(\alpha - \beta)]$$

If the voltage is an alternating voltage and the current is an alternating current, the phasor power for each harmonic component is defined in the same way as for the sinusoidal voltage and sinsuoidal current. Mathematically, the phasor power of the rth harmonic component $\mathbf{S}_r$ is given by

$$\mathbf{S}_r = P_r + jQ_r = \mathbf{E}_r\mathbf{I}_r{}^* = E_rI_re^{j(\alpha r-\beta r)}$$
$$= E_rI_r[\cos(\alpha_r - \beta_r) + j\sin(\alpha_r - \beta_r)]$$

The phasor power at the two terminals of entry of a single-phase two-wire circuit into a delimited region, for an alternating voltage and current, is equal to the (phasor) sum of the values of the phasor power for every harmonic. Mathematically, this relation may be expressed

$$\mathbf{S} = \mathbf{S}_1 + \mathbf{S}_2 + \mathbf{S}_3 + \ldots = \Sigma \mathbf{S}_r$$
$$= \mathbf{E}_1\mathbf{I}_1{}^* + \mathbf{E}_2\mathbf{I}_2{}^* + \mathbf{E}_3\mathbf{I}_3{}^* \ldots = \Sigma\, \mathbf{E}_r\mathbf{I}_r{}^*$$
$$= (\mathbf{P}_1 + \mathbf{P}_2 + \mathbf{P}_3 + \ldots) + j(Q_1 + Q_2 + Q_3 + \ldots)$$
$$= \Sigma\, (P_r + jQ_r)$$

The amplitude of the phasor power is equal to the square root of the sum of the squares of the active power and the reactive power. Mathematically, if S is the amplitude of the phasor power and θ is the angle between the phasor power and the real-power axis,

$$\mathbf{S} + Se^{j\theta}$$
$$S = (P^2 + Q^2)^{1/2}$$
$$= [(P_1 + P_2 + P_3 + \ldots)^2 + (Q_1 + Q_2 + Q_3 + \ldots)^2]^{1/2}$$
$$\theta = \tan^{-1}\frac{Q}{P} = \tan^{-1}\frac{Q_1 + Q_2 + Q_3}{P_1 + P_2 + P_3 + \ldots}$$

If the voltage and current are quasi-periodic and the amplitude of the voltage and current components are slowly varying, the phasor power may still be taken as the phasor having P and Q as its components, the values of P and Q being determined for these conditions, as specified in "power, active (single-phase two-wire circuit) (average power) (power)" and "power, reactive (magner) (single-phase two-wire circuit)," respectively. For this condition the phasor power will be a function of time. If the voltage and current have the same waveform, the amplitude of the phasor power is equal to the apparent power, but they are not the same for all other cases. The phasor power is expressed in voltamperes when the voltage is in volts and the current in amperes. *Note:* This term was once defined as "vector power." With the introduction of the term "phasor quantity," the name of this term has been altered to agree with the change in the sign of reactive power. *See also:* alternating current; reactive power.
(Std100) 270-1966

phasor power factor (A) The power factor of the synchronous machine defined by the cosine of the phasor angle between the fundamental sinusoidal phase voltage and the fundamental sinusoidal phase current. *Note:* This is not the angle between the load converter commutating voltage and the machine current. **(B)** The ratio of the active power to the amplitude of the phasor power. The phasor power factor is expressed by the equation

$$F_{pp} = \frac{P}{S}$$

where F_{pp} is the phasor power factor, P is the active power, and S is the amplitude of phasor power. If the voltages and currents are sinusoidal and, for polyphase circuits, form symmetrical sets,

$$\mathrm{A} = |A|e^{j\theta A}$$
$$\mathrm{B} = |B|e^{j\theta B}$$

See also: power factor, displacement.
(Std100/IA/ID) 995-1987, 270-1966

phasor product (quotient) A phasor whose amplitude is the product (quotient) of the amplitudes of the two phasors and whose phase angle is the sum (difference) of the phase angles of the two phasors. If two phasors are

$$F_{pp} = \cos(\alpha - \beta)$$

the phasor product is

$$AB = |AB|e^{f(\theta A+\theta B)}$$

and the quotient is

$$\frac{A}{B} = \left|\frac{A}{B}\right| e^{j(\theta A-\theta B)}$$

(Std100) 270-1966w

phasor quantity (A) A complex equivalent of a simple sine-wave quantity such that the modulus of the former is the amplitude A of the latter, and the phase angle (in polar form) of the former is the phase angle of the latter. **(B)** Any quantity (such as impedance) that is expressed in complex form. *Note:* In definition "A," sinusoidal variation with t enters; in definition "B," no time variation (in constant-parameter circuits) enters. The term "phasor quantity" covers both cases.
(Std100) 270-1966

phasor quotient *See:* phasor product.

phasor reactive factor (A) The ratio of the reactive power to the amplitude of the phasor power. The phasor reactive factor is expressed by the equation

$$F_{qp} = \frac{Q}{S}$$

where F_{qp} is the phasor reactive factor, Q is the reactive power, and S is the amplitude of the phasor power. If the voltages and currents are sinusoidal and, for polyphase circuits, form symmetrical sets. **(B)** $F_{pp} = \sin(\alpha - \beta)$.
(Std100) 270-1966

phasor sum (difference) A phasor of which the real component is the sum (difference) of the real components of two phasors and the imaginary component is the sum (difference) of the imaginary components of two phasors. If two phasors are

$$A = a_1 + ja_2$$
$$B = b_1 + jb_2$$

phasor sum (difference) is

$$A \pm B = (a_1 \pm b_1) + j(a_2 \pm b_2)$$

(Std100) 270-1966w

PH_CHARACTERISTICS Physical layer characteristics indication primitive. A set of attributes used to delineate the optional capabilities of a BCC or DCC Physical layer. This set of capabilities is required to be labeled on a BCC or DCC. In addition, there is required to be a mechanism for providing this information to the BCC's or DCC's MIB upper layers. The list of attributes includes the following capabilities: low-speed (2400 Bd or 9600 Bd) operation, high-speed (1 Mb/s) operation, +12 V used from MIB connector, BCC capability, capability for DCC interrupt function, and capability for DCC sync function. (EMB/MIB) 1073.3.1-1994

PHIGS *See:* Programmer's Hierarchical Interactive Graphics System.

Philips gauge A vacuum gauge in which the gas pressure is determined by measuring the current in a glow discharge. *See also:* instrument. (EEC/PE) [119]

phi polarization (Φ polarization) The state of the wave in which the E vector is tangential to the lines of latitude of a given spherical frame of reference. *Note:* The usual frame of reference has the polar axis vertical and the origin at or near the antenna. Under these conditions, a vertical dipole will radiate only theta (θ) polarization, and a horizontal loop will radiate only phi (Φ) polarization. *See also:* antenna.
(AP/ANT) 149-1979r, 145-1983s

phon The unit of loudness level as specified in the definition of loudness level. (SP) [32]

Phong shading (computer graphics) A technique for shading a three-dimensional solid object by interpolating the normal vectors at the vertices of each polygon face, resulting in realistic highlights. *See also:* Gouraud shading.
(C) 610.6-1991w

phonograph pickup (mechanical reproducer) A mechanoelectrical transducer that is actuated by modulations present in the groove of the recording medium and that transforms this mechanical input into an electric output. *Notes:* 1. Where no confusion is likely the term phonograph pickup may be shortened to pickup. 2. A phonograph pickup generally includes a pivoted mounting arm and the transducer itself (the pickup cartridge). (SP) [32]

phosphene (A) (electrotherapy) (electrical) A visual sensation experienced by a human subject during the passage of current through the eye. *See also:* electrotherapy. **(B) (overhead power lines)** Visual sensations due to nonoptical stimulation of the visual system. (EMB/T&D/PE) [47], 539-1990

phosphor (1) A substance capable of luminescence. *See also:* television; radio navigation; fluorescent lamp; cathode-ray tube. (EEC/PE) [119]
(2) (computer graphics) A chemical coating, used on the inside face of a display surface, that emits light when energized by an electron beam. (C) 610.6-1991w

phosphor decay A phosphorescence curve describing energy emitted versus time. *See also:* oscillograph. (IM/HFIM) [40]

phosphorescence (illuminating engineering) The emission of light as the result of the absorption of radiation, and continuing for a noticeable length of time after excitation. (EEC/IE) [126]

phosphor screen All the visible area of the phosphor on the cathode-ray tube faceplate. *See also:* oscillograph. (IM/HFIM) [40]

phot* (illuminating engineering) A unit of illuminance equal to one lumen per square centimeter. (EEC/IE) [126]
* Deprecated.

photocathode An electrode used for obtaining a photoelectric emission when irradiated. *See also:* electrode; phototube. (NPS) 398-1972r

photocathode blue response The photoemission current produced by a specified luminous flux from a tungsten filament lamp at 2854 K color temperature when the flux is filtered by a CS 5-58 blue filter of half stock thickness (1.75–2.25 mm). This parameter is useful in characterizing response to scintillation counting sources. (NPS) 398-1972r

photocathode luminous sensitivity *See:* cathode luminous sensitivity.

photocathode response (diode-type camera tube) The response of a photocathode is the current emitted into vacuum per incident radiant power of specified spectral distribution. It is expressed in amperes watt^{-1} (AW^{-1}). (ED) 503-1978w

photocathode, semitransparent *See:* semitransparent photocathode.

photocathode spectral quantum efficiency (diode-type camera tube) The ratio of the average number of electrons emitted to the number of photons in the input signal irradiance on the photocathode face as a function of the photon energy, frequency, or wavelength. (ED) 503-1978w

photocathode transit time That portion of the photomultiplier transit time corresponding to the time for photoelectrons to travel from the photocathode to the first dynode. (NPS) 398-1972r

photocathode transit-time difference The difference in transit time between electrons leaving the center of the photocathode and electrons leaving the photocathode at some specified point on a designated diameter. (NPS) 398-1972r

photocell (1) (photoelectric cell) A solid-state photosensitive electron device in which use is made of the variation of the current-voltage characteristic as a function of incident radiation. *See also:* phototube. (NPS) 398-1972r
(2) (photoelectric cell) A device exhibiting photovoltaic or photoconductive effects. *See also:* phototube. (ED) [45], [84]
(3) A semiconductor device, the electrical properties of which are affected by illumination. *Note:* One common type of photocell is the photoelectric cell which generates electricity when exposed to light, and is used to power many portable devices. (C) 610.10-1994w

photochemical radiation (illuminating engineering) Energy in the ultraviolet, visible and infrared regions to produce chemical changes in materials. *Note:* Examples of photochemical processes are accelerated fading tests, photography, photoreproduction and chemical manufacturing. In many such applications a specific spectral region is of importance. (EEC/IE) [126]

photocomposer *See:* phototypesetter.

photocomposition The formation of text and graphics into discrete camera-ready pages. *Synonym:* page makeup. *See also:* computer-aided page makeup. (C) 610.2-1987

photoconductive cell A photocell in which the photoconductive effect is utilized. *See also:* phototube. (ED) [45], [84]

photoconductive effect (photoconductivity) A photoelectric effect manifested as a change in the electric conductivity of a solid or a liquid and in which the charge carriers are not in thermal equilibrium with the lattice. *Note:* Many semiconducting metals and their compounds (notably selenium, selenides, and tellurides) show a marked increase in electric conductance when electromagnetic radiation is incident on them. *See also:* photoemissive effect; phototube; photoelectric effect; photovoltaic effect. 270-1966w

photoconductivity (fiber optics) The conductivity increase exhibited by some nonmetallic materials, resulting from the free carriers generated when photon energy is absorbed in electronic transitions. The rate at which free carriers are generated, the mobility of the carriers, and the length of time they persist in conducting states (their lifetime) are some of the factors that determine the amount of conductivity change. *See also:* photoelectric effect. (Std100) 812-1984w

photocurrent (fiber optics) The current that flows through a photosensitive device (such as a photodiode) as the result of exposure to radiant power. Internal gain, such as that in an avalanche photodiode, may enhance or increase the current flow but is a distinct mechanism. *See also:* dark current; photodiode. (Std100) 812-1984w

photodetector A device that senses incident illumination. (C) 610.10-1994w

photodiode (fiber optics) A diode designed to produce photocurrent by absorbing light. Photodiodes are used for the detection of optical power and for the conversion of optical power to electrical power. *See also:* avalanche photodiode; photocurrent. (Std100) 812-1984w

photoelectric beam-type smoke detector (fire protection devices) A device which consists of a light source which is projected across the area to be protected into a photosensing cell. smoke between the light source and the receiving photosensing cell reduces the light reaching the cell, causing actuation. (NFPA) [16]

photoelectric cathode *See:* photocathode.

photoelectric color-register controller A photoelectric control system used as a longitudinal position regulator for a moving material or web to maintain a preset register relationship between repetitive register marks in the first color and reference positions of the printing cylinders of successive colors. *See also:* photoelectric control. (IA/IAC) [60]

photoelectric control Control by means of which a change in incident light effects a control function. (IA/ICTL/IAC) [60]

photoelectric counter A photoelectrically actuated device used to record the number of times a given light path is intercepted by an object. *See also:* photoelectric control. (IA/ICTL/IAC) [60]

photoelectric current The current due to a photoelectric effect. *See also:* photoelectric effect. (IA/IAC) [60], [84]

photoelectric cutoff register controller A photoelectric control system used as a longitudinal position regulator that maintains the position of the point of cutoff with respect to a repetitively referenced pattern on a moving material. *See also:* photoelectric control. (IA/ICTL/IAC) [60]

photoelectric directional counter A photoelectrically actuated device used to record the number of times a given light path is intercepted by an object moving in a given direction. *See also:* photoelectric control. (IA/CEM) [58]

photoelectric door opener A photoelectric control system used to effect the opening and closing of a power-operated door. *See also:* photoelectric control. (IA/CEM) [58]

photoelectric effect (A) (fiber optics) *External photoelectric effect.* The emission of electrons from the irradiated surface of a material. **(B) (fiber optics)** *Internal photoelectric effect.* Photoconductivity. (Std100) 812-1984

photoelectric emission (electron tube) The ejection of electrons from a solid or liquid by electromagnetic radiation. *See also:* field-enhanced photoelectric emission. (ED) 161-1971w

photoelectric flame detector (fire protection devices) A device whose sensing element is a photocell which either changes its electrical conductivity or produces an electrical potential when exposed to radiant energy. (NFPA) [16]

photoelectric lighting controller A photoelectric relay actuated by a change in illumination to control the illumination in a given area or at a given point. *See also:* photoelectric control. (IA/IAC) [60]

photoelectric loop control A photoelectric control system used as a position regulator for a strip processing line that matches the average linear speed in one section to the speed in an adjacent section to maintain the position of the loop located between the two sections. *See also:* photoelectric control. (IA/ICTL/IAC) [60]

photoelectric pinhole detector A photoelectric control system that detects the presence of minute holes in an opaque material. *See also:* photoelectric control. (IA/CEM) [58]

photoelectric power system *See:* photovoltaic power system.

photoelectric pyrometer An instrument that measures the temperature of a hot object by means of the intensity of radiant energy exciting a phototube. (IA/ICTL/IAC) [60]

photoelectric relay A relay that functions at predetermined values of incident light. *See also:* photoelectric control. (IA/IAC) [60]

photoelectric scanner A single-unit combination of a light source and one or more phototubes with a suitable optical system. *See also:* photoelectric control. (IA/ICTL/IAC) [60]

photoelectric side-register controller A photoelectric control system used as a lateral position regulator that maintains the edge of, or a line on, a moving material or web at a fixed position. *See also:* photoelectric control. (IA/ICTL/IAC) [60]

photoelectric smoke detector A photoelectric relay and light source arranged to detect the presence of more than a predetermined amount of smoke in air. *See also:* photoelectric control. (IA/ICTL/IAC) [60]

photoelectric smoke-density control A photoelectric control system used to measure, indicate, and control the density of smoke in a flue or stack. *See also:* photoelectric control. (IA/IAC) [60]

photoelectric spot-type smoke detector (fire protection devices) A device which contains a chamber with either overlapping or porous covers which prevent the entrance of outside sources of light but which allow the entry of smoke. The unit contains a light source and a special photosensitive cell in the darkened chamber. The cell is either placed in the darkened area of the chamber at an angle different from the light path or has the light blocked from it by a light stop or shield placed between the light source and the cell. With the admission of smoke particles, light strikes the particles and is scattered and reflected into the photosensitive cell. This causes the photosensing circuit to respond to the presence of smoke particles in the smoke chamber. (NFPA) [16]

photoelectric system (protective signaling) An assemblage of apparatus designed to project a beam of invisible light onto a photoelectric cell and to produce an alarm condition in the protection circuit when the beam is interrupted. *See also:* protective signaling. (EEC/PE) [119]

photoelectric tube An electron tube, the functioning of which is determined by the photoelectric effect. *See also:* phototube. (ED) [45]

photo-electron An electron liberated by the photoemissive effect. *See also:* photoelectric effect. (ED) [45]

photo-electron irradiation dark current increase (diode-type camera tube) That irreversible dark current increase which is caused by bombardment of the charge storage target by photo-electrons. (ED) 503-1978w

photo-electron irradiation deterioration (diode-type camera tube) That irreversible dark current increase which is associated with bombardment of the charge storage target by photo-electrons. (ED) 503-1978w

photoemissive effect *See:* photoelectric effect.

photoemission spectrum (scintillator material) The relative numbers of optical photons emitted per unit wavelength as a function of wavelength interval. The emission spectrum may also be given in alternative units such as wave number, photon energies, frequency, etc. *Note:* Optical photons are photons with energies corresponding to wavelengths between 2000 and 15 000 angstroms. (NPS) 175-1960w

photoflash lamp (illuminating engineering) A lamp in which combustible metal or other solid material is burned in an oxidizing atmosphere to produce light of high intensity and short duration for photographic purposes. (EEC/IE) [126]

photoformer A function generator that operates by means of a cathode-ray beam optically tracking the edge of a mask placed on a screen. *See also:* electronic analog computer. (C) 165-1977w

photographic emulsion The light-sensitive coating on photographic film consisting usually of a gelatin containing silver halide. (SP) [32]

photographic sound recorder (optical sound recorder) Equipment incorporating means for producing a modulated light beam and means for moving a light-sensitive medium relative to the beam for recording signals derived from sound signals. (SP) [32]

photographic sound reproducer (optical sound reproducer) A combination of light source, optical system, photoelectric cell, or other light-sensitive device such as a photoconductive cell, and a mechanism for moving a medium carrying an optical sound record (usually film), by means of which the recorded variations may be converted into electric signals of approximately like form. (SP) [32]

photographic transmission density (optical density) The common logarithm of opacity. Hence, film transmitting 100 percent of the light has a density of zero, transmitting 10 percent a density of 1, and so forth. Density may be diffuse, specular, or intermediate. Conditions must be specified. (SP) [32]

photo-ionization Ionization of atoms or molecules caused by infrared, visible, or ultraviolet photons. (AP/PROP) 211-1997

photometer (illuminating engineering) An instrument for measuring photometric quantities such as luminance, luminous intensity, luminous flux or illuminance. (EEC/IE) [126]

photometric brightness *See:* luminance.

photometry (illuminating engineering) The measurement of quantities associated with light. *Note:* Photometry may be visual in which the eye is used to make a comparison, or physical in which measurements are made by means of physical receptors. (EEC/IE) [126]
(2) (A) (television) (general) The measurement of quantities referring to radiation evaluated in accordance with the visual effect it produces, as based on certain conventions.

(B) (television) (visual) That branch of photometry in which the eye is used to make comparison. **(C) (television)** (physical) That branch of photometry in which the measurement is made by means of physical receptors. (BT/AV) 201-1979

photomultiplier *See:* multiplier phototube.

photomultiplier transit time (scintillation counting) The time difference between the incidence of a delta-function light pulse on the photocathode of the photomultiplier and the occurrence of the half-amplitude point on the output-pulse leading edge. (NPS) 398-1972r

photomultiplier tube *See:* multiplier phototube.

photomultiplier tube gain The ratio of the signal output current to the photoelectric signal current from the photocathode. (NI) N42.15-1990

photon (1) (A) (fiber optics) A quantum of electromagnetic energy. The energy of a photon is hν where h is Planck's constant and ν is the optical frequency. *See also:* nonlinear scattering; Planck's constant. **(B)** Ionizing electromagnetic radiation, irrespective of origin. (Std100/NI) 812-1984, N42.17B-1989

(2) (range protection) A quantum of electromagnetic radiation irrespective of origin. (NI) N323-1978r

photon emission spectrum, scintillator material (scintillation counting) The relative numbers of optical photons emitted per unit wavelength as a function of wavelength interval. The emission spectrum may also be given in alternative units such as wavenumber, photon energies, frequency, and so on. (NPS) 398-1972r

photon emitting diode (light-emitting diodes) A semiconductor device containing a semiconductor junction in which radiant flux is nonthermally produced when a current flows as a result of an applied voltage. (ED) [127]

photon noise *See:* quantum noise.

photo-optic storage *See:* optical storage.

photopic spectral luminous efficiency function (Vλ) (photometric standard observer for photopic vision) The photopic spectral luminous efficiency function gives the ratio of the radiant flux at wavelength λ_m to that at wavelength λ, when the two fluxes produce the same photopic luminous sensations under specified photometric conditions, λ_m being chosen so that the maximum value of this ratio is unity. Unless otherwise indicated, the values used for the spectral luminous efficiency function relate to photopic vision by the photometric standard observer having the characteristics laid down by the International Commission on Illumination (CIE). (ED) [127]

photopic vision (illuminating engineering) Vision mediated essentially or exclusively by the cones. It is generally associated with adaptation to a luminance of at least 3.4 cd/m^2 (2.2 $\times$ 10^{-3} cd/in^2) (1.0fL). (EEC/IE) [126]

photosensitive recording (facsimile) Recording by the exposure of a photosensitive surface to a signal-controlled light beam or spot. *See also:* recording. (COM) 168-1956w

photosensitive tube *See:* photoelectric tube.

photosensitizers (laser maser) Substances that increase the sensitivity of a material to irradiation by electromagnetic radiation. (LEO) 586-1980w

phototube (photoelectric tube) An electron tube that contains a photocathode and has an output depending at every instant on the total photoelectric emission from the irradiated area of the photocathode. *See also:* field-enhanced photoelectric emission. (NPS) 398-1972r

phototube gain (liquid-scintillation counting) The ratio of the signal output current to the photoelectric signal current from the photocathode. (NI) N42.15-1980s

phototube housing An enclosure containing a phototube and an optical system. *See also:* photoelectric control. (IA/ICTL/IAC) [60]

phototube, multiplier *See:* multiplier phototube.

phototypesetter A nonimpact printer that creates characters using photographic techniques. *Synonym:* photocomposer. (C) 610.10-1994w

phototypesetting The preparation of textual material for printing using an optical system with a light source, a type store, a lens system, and a light-sensitive recording medium. *See also:* computer-aided typesetting. (C) 610.2-1987

photovaristor A varistor in which the current-voltage relation may be modified by illumination, for example, cadmium sulphide or lead telluride. *See also:* semiconductor device. (Std100) 102-1957w

photovoltaic array (terrestrial photovoltaic power systems) The smallest installed assembly of photovoltaic (PV) panels, support structure, foundation, and other components as required, such as a tracker. *Synonym:* PV array. *See also:* array control. (PV) 928-1986r

photovoltaic array subfield (terrestrial photovoltaic power systems) One or more arrays associated by a distinguishing feature, such as field geometry or electrical interconnection. *Synonym:* PV array subfield. *See also:* array control. (PV) 928-1986r

photovoltaic cell (terrestrial photovoltaic power systems) The basic device that converts sunlight directly into dc electricity. *Synonym:* PV cell. *See also:* array control. (PV) 928-1986r

photovoltaic effect (fiber optics) The production of a voltage difference across a pn junction resulting from the absorption of photon energy. The voltage difference is caused by the internal drift of holes and electrons. *Synonym:* PV effect. *See also:* photon. (Std100) 812-1984w

photovoltaic module (terrestrial photovoltaic power systems) The smallest, complete, environmentally protected assembly of photovoltaic (PV) cells (flat plte-type), or receiver(s) and optics (concentrator-type), and related components, such as interconnects and mounting, that accepts unconcentrated sunlight. *Synonym:* PV module. *See also:* array control. (PV) 928-1986r

photovoltaic panel (terrestrial photovoltaic power systems) One or more photovoltaic (PV) modules assembled and wired and designed to provide a field-installable unit. *Synonym:* PV panel. *See also:* array control. (PV) 928-1986r

photovoltaic power system (terrestrial photovoltaic power systems) A system that converts sunlight directly into electric energy and processes it into a form suitable for use by the intended load. The system will include an array subsystem and may also include the following major subsystems: power conditioning, storage, thermal, and system monitor and control. A photovoltaic (PV) system-utility interface may also be included. *Synonym:* PV power system. *See also:* array control. (PV) 928-1986r

photovoltaic receiver (terrestrial photovoltaic power systems) An assembly of one or more photovoltaic (PV) cells that accepts concentrated sunlight and incorporates means for thermal and electric energy removal. *Synonym:* PV receiver. *See also:* array control. (PV) 928-1986r

photovoltaic system-utility interface (terrestrial photovoltaic power systems) The interconnection between the power conditioning subsystem, the on-site ac loads, and the utility. *Synonym:* PV system-utility interface. *See also:* array control. (PV) 928-1986r

PhPDU *See:* physical protocol data unit.

PHR *See:* physical record.

PhsDU *See:* physical interface data unit.

PhSDU *See:* physical protocol service unit; physical service data unit.

PhS_User *See:* physical service user.

PHY Abbreviation for physical. *See also:* physical layer. (LM/C) 802.5-1989s

PHY layer *See:* physical layer.

PHY packet (1) A packet either generated or received by the cable physical layer. These packets are always exactly 64 bits long where the last 32 bits are the bit complement of the first 32 bits. (C/MM) 1394-1995

(2) A 64-bit packet where the most significant 32 bits are the one's complement of the least significant 32 bits. (C/MM) 1394a-2000

physical (data management) Pertaining to the representation and storage of data on a data medium such as magnetic disk, or to characteristics of the data such as the length of data elements or records. *Contrast:* logical. (C) 610.5-1990w

physical address (1) A unique identifier that selects a particular device from the set of all devices connected to a particular bus. (C/BA) 1275-1994
(2) The address of a data item in physical memory. *See also:* virtual address. (C) 610.10-1994w

physical address space The set of possible physical addresses for a particular bus. (C/BA) 1275-1994

physical architecture An arrangement of physical elements that provides the design solution for a consumer product or life-cycle process intended to satisfy the requirements of the functional architecture and the requirements baseline. (C/SE) 1220-1994s

physical characteristics The physical *design* attributes or distinguishing features that pertain to a measurable description of a product or process. (C/SE) 1220-1994s

physical child segment In a hierarchical database, a child segment in a physical database. *See also:* logical child segment. (C) 610.5-1990w

physical circuit (data transmission) A two-wire metallic circuit that is not arranged for phantom use. (PE) 599-1985w

physical coding sublayer (PCS) A sublayer used in 100BASE-T and 1000BASE-X to couple the Media Independent Interface (MII) and the Physical Medium Attachment (PMA). The PCS contains the functions to encode data bits into code-groups that can be transmitted over the physical medium. Four PCS structures are defined—one for 100BASE-X, one for 100BASE-T4, one for 100BASE-T2, and one for 1000BASE-X. (C/LM) 802.3-1998

physical concept Anything that has existence or being in the ideas of man pertaining to the physical world. Examples are magnetic fields, electric currents, electrons. (Std100) 270-1966w

physical configuration audit (software) An audit conducted to verify that a configuration item, as built, conforms to the technical documentation that defines it. *See also:* functional configuration audit. (C) 610.12-1990

physical connection The full-duplex physical layer association between directly connected nodes. In the case of the cable physical layer, this is a pair of physical links running in opposite directions. (C/MM) 1394-1995

physical damage (rotating machinery) This contributes to electrical insulation failure by opening leakage paths through the insulation. Included here are: physical shock, vibration, overspeed, short-circuit forces, erosion by foreign matter, damage by foreign objects, and thermal cycling. (PE/EM) 432-1976s

physical database (A) (data management) A database as it is actually stored. **(B) (data management)** A database containing a collection of related segments or records that are physically stored together. *Note:* Segments within a physical database are known as physical segments. *Contrast:* logical database. (C) 610.5-1990

physical data model (data management) A data model that represents the implementation of the data contained in a data structure. *Contrast:* logical data model. (C) 610.5-1990w

physical defect *See:* fault.

physical element A product, subsystem, assembly, component, subcomponent, subassembly, or part of the physical architecture defined by its designs, interfaces (internal and external), and requirements (functional, performance, constraints, and physical characteristics). (C/SE) 1220-1994s

physical entity *See:* physical quantity.

physical format *See:* low-level format.

physical ID (1) The six least significant bits of the node ID. On a particular bus, each node's physical ID is unique. (C/MM) 1394a-2000
(2) The least-significant 6 bits of the node_ID. This number is unique on a particular bus and is chosen by the physical layer during initialization. (C/MM) 1394-1995

physical interface (1) The circuitry that interfaces the module, board(s), and node(s) to the bus signals. (C/MM) 1212-1991s
(2) The circuitry that interfaces a module's nodes to the input link, output link, and miscellaneous signals. (C/MM) 1596.3-1996, 1596-1992

physical interface data unit (PhsDU) An octet (8 data bits) that is communicated across the interface between a BCC or a DCC Physical layer and Data Link layer. (EMB/MIB) 1073.3.1-1994

physical layer (PHY) (1) (Layer 1) The layer of the ISO Reference Model that provides the mechanical, electrical, functional, and procedural characteristics access to the transmission medium. (DIS/C) 1278.2-1995
(2) In this part of ISO/IEC 8802, the subdivision that provides the protocol to allow transfer of *slot octets, management information octets,* and *DQDB Layer* timing information over the *transmission link* between *DQDB Layer subsystems* at adjacent *nodes.* The Physical Layer provides the service to the *DQDB Layer.* (LM/C) 8802-6-1994
(3) The first layer of the seven-layer OSI model; responsible for transporting bits between adjacent systems. *Note:* This layer accepts a bit stream, called a frame, from the data link layer and places it on the media. It also performs the inverse operation of extracting a bit stream from the physical media and passes it to the data link layer. This layer describes mechanical and electrical characteristics of the connection, as well as the required interchange circuits. *See also:* medium access control sublayer; session layer; logical link control sublayer; entity layer; client layer; application layer; presentation layer; data link layer; sublayer; transport layer; network layer. (C) 610.7-1995
(4) The layer, in a stack of three protocol layers defined for the Serial Bus, that translates the logical symbols used by the link layer into electrical signals on the different Serial Bus media. The physical layer guarantees that only one node at a time is sending data and defines the mechanical interfaces for the Serial Bus. There are different physical layers for the backplane and for the cable environment. See figure 34 for the relation of the physical layer to the Serial Bus protocol stack. (C/MM) 1394-1995
(5) The layer responsible for interfacing with the transmission medium. This includes conditioning signals received from the MAC for transmitting to the medium and processing signals received from the medium for sending to the MAC. (C/LM) 8802-5-1998
(6) The Serial Bus protocol layer that translates the logical symbols used by the link layer into electrical signals on Serial Bus media. The physical layer is self-initializing. Physical layer arbitration guarantees that only one node at a time is sending data. The mechanical interface is defined as part of the physical layer. There are different physical layers for the backplane and for the cable environment. (C/MM) 1394a-2000

physical layer convergence procedure (PLCP) The part of the *Physical Layer* that supports the transfer of *slot octets, management information octets,* and *DQDB Layer* timing information in a manner that adapts the capabilities of the *transmission system* to the service expected by the DQDB Layer. (LM/C) 8802-6-1994

physical layer entity (PHY) That portion of the physical layer between the medium dependent interface (MDI) and media independent interface (MII), or between the MDI and gigabit media independent interface (GMII), consisting of the physical coding sublayer (PCS), physical medium attachment (PMA), and, if present, physical medium dependent (PMD) sublayers. The PHY contains the functions that transmit, receive, and manage the encoded signals that are impressed on and recovered from the physical medium. (C/LM) 802.3-1998

physical layer protocol data unit (PhPDU) A frame, consisting of both nondata symbols (delimiters or start and stop bits) and data symbols (data bits), that is transmitted from a bedside communications controller (BCC) to a device communications controller (DCC) or from a DCC to a BCC. A PhPDU includes mechanisms indicating the start and end of transmission. PhPDUs may be preceded and followed by idle periods on the serial transmission medium. Conditions exist in which two consecutive PhPDUs may be transmitted contiguously by the same station. (EMB/MIB) 1073.4.1-2000

physical layer service access point (PhSAP) The interface between the Physical layer entity and the Physical layer user entity. It consists of the set of services performed by the Physical layer entity. (EMB/MIB) 1073.4.1-2000

physical layer service data unit (PhSDU) A set of octets, constituting a frame, that is transferred between the PhS_user layer entity and the Physical layer entity. A PhSDU transferred from the PhS_user layer entity is transmitted on the physical medium. Alternatively, a PhSDU is a frame of octets that is received from the physical medium and is transferred to the PhS_user layer entity. (EMB/MIB) 1073.4.1-2000

physical layer signaling sublayer The portion of the physical layer, contained within the DTE, that provides the logical and functional coupling between the MAU and the data link layer. (C) 610.7-1995

physical link In the cable physical layer, the simplex path from the transmit function of the port of one node to the receive function of a port of a directly connected node. (C/MM) 1394-1995

physically connected The state when an IEEE 1073 connector is connected at both ends of the cable. This is indicated to both the DCC and the BCC by voltage levels. (EMB/MIB) 1073.3.1-1994

physical media components (PMC) **(1)** The sublayer of the PHY responsible for interfacing with the transmission medium. The functions of the PMC include receive, transmit, clock recovery, and ring access control. (C/LM) 8802-5-1998
(2) The sublayer of the PHY responsible for interfacing with the transmission medium. The functions of the PMC include receive, transmit, clock recovery, and ring access control. The PMCs for different medium rates and types are different. (C/LM) 802.5t-2000

physical medium *See:* transmission medium.

physical medium attachment (PMA) (medium attachment units and repeater units) The portion of the medium attachment interface (MAU) that contains the functional circuitry. (LM/C) 802.3-1985s

physical medium attachment sublayer **(1)** The portion of the MAU that contains the functional circuitry. *Synonym:* PMA sublayer. (LM/C) 610.7-1995
(2) That portion of the physical layer that contains the functions for transmission, reception, and (depending on the PHY) collision detection, clock recovery, and skew alignment. (C/LM) 802.3-1998

physical medium dependent sublayer **(1)** In 100BASE-X, that portion of the physical layer responsible for interfacing to the transmission medium. The PMD is located just above the medium dependent interface (MDI). (C/LM) 802.3-1998
(2) (local area networks) The medium dependent portion of the physical layer. (C) 8802-12-1998

physical medium independent sublayer (local area networks) The medium independent portion of the Physical Layer. (C) 8802-12-1998

physical memory The main storage actually provided in a computer. *See also:* virtual storage. (C) 610.10-1994w

physical model A model whose physical characteristics resemble the physical characteristics of the system being modeled; for example, a plastic or wooden replica of an airplane. *Contrast:* symbolic model. *See also:* scale model; mock-up; iconic model. (C) 610.3-1989w

physical optics (fiber optics) The branch of optics that treats light propagation as a wave phenomenon rather than a ray phenomenon, as in geometric optics. (Std100) 812-1984w

physical optics approximation Estimates the field scattered by a body by considering only the interaction of the incident wave with the local geometry of the body at every point illuminated by the incident wave. The physical optics approximation is the Kirchhoff approximation in the illuminated part of the body and zero in its shadow. (AP/PROP) 211-1997

Physical Parameter An instance of a subclass of `IEEE1451_PhysicalParameter`. (IM/ST) 1451.1-1999

Physical Parameter Type The syntax and interpretation of the Physical Parameter's data and metadata. (IM/ST) 1451.1-1999

physical parent segment (data management) In a hierarchical database, a parent segment in a physical database. *See also:* logical parent segment. (C) 610.5-1990w

physical photometer (illuminating engineering) An instrument containing a physical receptor and associated filters, which is calibrated so as to read photometric quantities directly. *See also:* physical receptor. (EEC/IE) [126]

physical property Any one of the generally recognized characteristics of a physical system by which it can be described. (Std100) 270-1966w

physical protocol data unit (PhPDU) A frame, consisting of both nondata symbols (delimiters or start and stop bits) and data symbols (data bits), that is transmitted from a BCC to a DCC or from a DCC to a BCC. A PhPDU includes mechanisms indicating the start and end of transmission. PhPDUs may be preceded and followed by idle periods on the serial transmission medium. There are conditions in which two consecutive PhPDUs may be transmitted contiguously by the same station. (EMB/MIB) 1073.3.1-1994, 1073.4.1-1994s

physical protocol service unit (PhSDU) A symbol for a single data bit transmitted on the physical medium between a BCC and a DCC. (EMB/MIB) 1073.3.1-1994

physical quantity (concrete quantity) (physical entity) A particular example of a measurable physical property of a physical system. It is characterized by both a qualitative and a quantitative attribute (that is, kind and magnitude). It is independent of the system of units and equations by which it and its relation to other physical quantities are described quantitatively. (Std100) 270-1966w

physical receptor (illuminating engineering) A device which generates electric current or voltage or undergoes a change of resistance or generates a charge when radiation is incident on it. (EEC/IE) [126]

physical record (PHR) **(A) (data management)** A record whose characteristics depend on the manner or form in which it is stored, retained, or moved. *Note:* A physical record may consist of all or part of a logical record or several physical records. **(B) (data management)** That which is accessed by a single read or write operation. (C) 610.5-1990

physical requirement (software) A requirement that specifies a physical characteristic that a system or system component must possess; for example, material, shape, size, weight. *Contrast:* functional requirement; implementation requirement; design requirement; interface requirement; performance requirement. (C) 610.12-1990

physical security **(1)** Protection of system resources from physical access, tampering, and destruction, such as through the use of barriers, locks, seals, and intrusion detection systems. (C/BA) 896.3-1993w
(2) The application of methods for preventing malevolent acts against safeguards and security interests, detecting such acts as they occur, and responding to such acts. (PE/NP) 692-1997

physical segment (data management) In a hierarchical database, the smallest unit of accessible data. *See also:* physical child segment; physical twin segment; physical parent segment. (C) 610.5-1990w

physical sequential access *See:* sequential access.

physical service (PhS) The service performed by the Physical layer entity for the Physical layer user entity. (EMB/MIB) 1073.4.1-2000

physical service data unit (PhSDU) A set of octets, comprising a frame, that is transferred between the Ph_user layer entity and the Physical layer entity. A PhSDU transferred from the Ph_user layer entity is transmitted on the physical medium. Alternatively, a PhSDU is a frame of octets that is received from the physical medium and is transferred to the Ph_user layer entity. (EMB/MIB) 1073.4.1-1994s

physical service user (PhS_User) Refers to the Data Link layer entity, which is the user of a bedside communications controller (BCC) or a device communications controller (DCC) Physical layer entity. (EMB/MIB) 1073.4.1-2000

physical signaling components (PSC) (1) The sublayer of the PHY responsible for processing the signal elements received from the ring by the PMC for sending symbols to the MAC and for conditioning the symbols received from the MAC for inclusion as signal elements in the repeated data stream to the PMC. (C/LM) 8802-5-1998
(2) The sublayer of the PHY responsible for changing the signal elements received from the ring by the PMC into indicators and sending these indicators to the MAC. It is also responsible for conditioning the indicators received from the MAC for inclusion as signal elements by the PMC in the repeated data stream. At a particular medium rate, the PSC may be the same for different medium types. (C/LM) 802.5t-2000

physical signaling sublayer (PLS) In 10BASE-T, that portion of the physical layer contained within the data terminal equipment (DTE) that provides the logical and functional coupling between the medium attachment unit (MAU) and the data link layer. (C/LM) 802.3-1998

physical stimuli (illuminating engineering) May be either distal or proximal. (EEC/IE) [126]

physical structure (data management) The representation and storage of a database on a data medium. *See also:* reorganization; conceptual schema. (C) 610.5-1990w

physical source statements (PSS) Source statements that measure the quantity of software in lines of code. (C/SE) 1045-1992

physical system A part of the real physical world that is directly or indirectly observed or employed by mankind. (C) 270-1966w, 610.10-1994w

physical twin segment (data management) In a hierarchical database, a twin segment in a physical database. *Contrast:* logical twin segment. (C) 610.5-1990w

physical unit *See:* unit.

physical verification The process of evaluating whether or not the requirements of the physical architecture are traceable to the verified functional architecture and satisfy the validated requirements baseline. (C/SE) 1220-1994s

physical volume *See:* volume.

PI *See:* processor interface.

P.I *See:* proportional plus integral control action.

pick device (1) A logical input device used to select a display element on a display surface. A typical physical device is a light pen. *See also:* sonic pen. (C) 610.6-1991w
(2) An input device that is used to specify or detect a particular display element or segment. *Contrast:* pointing device. *See also:* electronic pen; light pen. (C) 610.10-1994w

pickle (electroplating) A solution or process used to loosen or remove corrosion products such as oxides, scale, and tarnish from a metal. *See also:* electroplating. (IA) [59]

pickling (A) (electroplating) (chemical) The removal of oxides or other compounds from a metal surface by means of a solution that acts chemically upon the compounds.
(B) (electrolytic) Pickling during which a current is passed through the pickling solution to the metal (cathodic pickling) or from the metal (anodic pickling). *See also:* electroplating. (PE/EEC) [119]

pickoff (1) (test, measurement, and diagnostic equipment) A sensing device that responds to movement to create a signal or to effect some type of control. (MIL) [2]
(2) (accelerometer) (gyros) A device that produces an output signal as a function of the relative linear or angular displacement between two elements. (AES/GYAC) 528-1994

pickoff axis (inertial sensors) (dynamically tuned gyro) The axis of angular displacement between the rotor and the case that results in the maximum signal per unit of rotation from the pickoff. (AES/GYAC) 528-1994

pickoff offset (dynamically tuned gyro) The difference in angular rotor position between operation at pickoff electrical null and at gyro operating null. (AES/GYAC) 528-1994

pickup (1) (electronics) A device that converts a sound, scene, or other form of intelligence into corresponding electric signals (for example, a microphone, a television camera, or a phonograph pickup). *See also:* television; phonograph pickup; microphone. (MIL/BT/AV) [2], [34]
(2) (of a relay) The action of a relay as it makes designated response to progressive increase of input. As a qualifying term, the state of a relay when all response to progressive increase of input has been completed. Also used to identify the minimum value of an input quantity reached by progressive increases that will cause the relay to reach the pickup state from reset. *Note:* In describing the performance of relays having multiple inputs, pickup has been used to denote contact operation, in which case pickup value of any input is meaningful only when related to all other inputs. (SWG/PE) C37.100-1992

pickup and seal voltage (magnetically operated device) The minimum voltage at which the device moves from its deenergized into its fully energized position. (IA/ICTL) 74-1958w

pickup current *See:* pickup value.

pickup factor, direction-finder antenna system An index of merit expressed as the voltage across the receiver input impedance divided by the signal field strength to which the antenna system is exposed, the direction of arrival and polarization of the wave being such as to give maximum response. *See also:* navigation. (AES/GCS/RS) 173-1959w, 686-1982s, [42], 172-1983w

pickup spectral characteristic (color television) The set of spectral responses of the device, including the optical parts, that converts radiation to electric signals, as measured at the output terminals of the pickup tubes. *Note:* Because of nonlinearity, the spectral characteristics of some kinds of pickup tubes depend upon the magnitude of radiance used in the measurement. (BT/AV) 201-1979w

pickup tube *See:* camera tube.

pickup value The minimum input that will cause a device to complete contact operation or similar designated action. *Note:* In describing the performance of devices having multiple inputs, the pickup value of an input is meaningful only when related to all other inputs. (SWG/PE) C37.100-1981s

pickup voltage (magnetically operated device) (or current) The voltage (or current) at which the device starts to operate when its operating coil is energized under conditions of normal operating temperature. *See also:* contactor. (VT/IA/LT/IAC) 16-1955w, [60]

pico (mathematics of computing) A prefix indicating 10^{-12}. (C) 1084-1986w

PICS *See:* protocol implementation conformance statement.

PICT A standard electronic format for exchanging graphical information. (ATLAS) 1232-1995

pictorial format (pulse measurement) A graph, plot, or display in which a waveform is presented for observation or analysis. Any of the waveform formats defined in the following subsections may be presented in the pictorial format. (IM/WM&A) 181-1977w

pictorial pattern recognition The recognition of patterns in visual or pictorial data. (C) 610.4-1990w

picture *See:* image.

picture data (data management) Data that are associated with a picture specification. *Synonym:* pictured data. *See also:* decimal picture data; binary picture data. (C) 610.5-1990w

pictured data *See:* picture data.

picture element (1) (pixel) The smallest area of a television picture capable of being delineated by an electric signal passed through the system or part thereof. *Note:* It has three important properties, namely: P_v, the vertical height of the picture element; P_h, the horizontal length of the picture element; and P_a, the aspect ratio of the picture element. In addition, N_p, the total number of picture elements in a complete picture, is of interest since this number provides a convenient way of comparing systems. For convenience, P_v and P_h are normalized for V, the vertical height of the picture; that is, P_v or P_h must be multiplied by V to obtain the actual dimension in a particular picture. P_v is defined as $P_v = 1/N$, where N is the number of active scanning lines in the raster. P_h is defined as $P_h = t_r A/t_e$, where t_r is the average value of the rise and delay times (10 percent to 90 percent) of the most rapid transition that can pass through the system or part thereof, t_e is the duration of the part of a scanning line that carries picture information, and A is the aspect ratio of the picture. (At present all broadcast television systems have a horizontal to vertical aspect ratio of 4/3.) P_a is defined as $P_a = P_h/P_v = t_r AN/te$ and N_p is defined as $N_p = (1/P_v) \times (A/P_h) = Nt_e/T_r$. *See also:* television. (BT) [33]
(2) (image processing and pattern recognition) (computer graphics) *See also:* pixel. (C) 610.4-1990w, 610.6-1991w

picture frequencies (facsimile) The frequencies which result solely from scanning subject copy. *Note:* This does not include frequencies that are part of a modulated carrier signal. *See also:* scanning. (COM) 168-1956w

picture inversion (facsimile) A process that causes reversal of the black and white shades of the recorded copy. *See also:* facsimile transmission. (COM) 168-1956w

picture processing *See:* image processing.

picture signal (television or facsimile) The signal resulting from the scanning process. *See also:* television. (BT/AV) [34]

picture specification A character-by-character description of the composition and characteristics of the representation of some data item; for example, the picture S99V999 (S = sign character; 9 = decimal digit character; V = radix point character) may be used to describe the following items, resulting in the picture data as indicated:

value	picture data
.06	+00.060
−10.342	−10.342
3	+03.000

(C) 610.5-1990w

picture transmission (telephotography) The electric transmission of a picture having a gradation of shade values. (EEC/PE) [119]

picture tube (kinescope) (television) A cathode-ray tube used to produce an image by variation of the beam current as the beam scans a raster. *See also:* television. (BT/AV) 201-1979w

P.I.D *See:* proportional plus integral plus derivative control action.

PIE *See:* Pause Interrupt Enabled (PIE) bit.

Pierce gun (microwave tubes) A gun that delivers an initially convergent electron beam. If a magnetic focusing scheme is used, the beam is made to enter the field at the minimum beam diameter or else, if the magnetic field threads through the cathode, the magnetic field must have a shape that is consistent with the desired beam that imparts certain flow characteristics to the electron beam. In Brillouin flow, angular electron velocity about the axis is imparted to the beam on entry into the magnetic field and the resulting inwardly directed force balances both the space charge and centrifugal forces. In practice, values of field up to twice the theoretical equilibrium value may be found necessary. In confined flow there is no overall angular velocity of the beam about the beam axis. Individual electron trajectories are tight helices (of radius small compared to beam radius) whose axis is along a magnetic-field line. The required magnetic field is several times greater than the Brillouin value and the flux must intersect the cathode surface. (ED) [45]

Pierce oscillator An oscillator that includes a piezoelectric crystal connected between the input and the output of a three-terminal amplifying element, the feedback being determined by the internal capacitances of the amplifying elements. *Note:* This is basically a Colpitts oscillator. *See also:* oscillatory circuit; Colpitts oscillator. (AP/BT/ANT) 145-1983s, 182A-1964w

piezoelectric accelerometer A device that employs a piezoelectric material as the principal restraint and pickoff. It is generally used as a vibration or shock sensor. (AES/GYAC) 528-1994

piezoelectric crystal cut, type *See:* type of piezoelectric crystal cut.

piezoelectric-crystal element A piece of piezoelectric material cut and finished to a specified geometrical shape and orientation with respect to the crystallographic axes of the material. *See also:* crystal. (PE/EM) 43-1974s

piezoelectric-crystal plate A piece of piezoelectric material cut and finished to specified dimensions and orientation with respect to the crystallographic axes of the material, and having two major surfaces that are essentially parallel. *See also:* crystal. (PE/EM) 43-1974s

piezoelectric-crystal unit A complete assembly, comprising a piezoelectric-crystal element mounted, housed, and adjusted to the desired frequency, with means provided for connecting it in an electric circuit. Such a device is commonly employed for purposes of frequency control, frequency measurement, electric wave filtering, or interconversion of electric waves and elastic waves. *Note:* Sometimes a piezoelectric-crystal unit may be an assembly having in it more than one piezoelectric-crystal plate. Such an assembly is called a mutliple-crystal unit. *See also:* crystal. (PE/EM) 43-1974s

piezoelectric effect Some materials become electrically polarized when they are mechanically strained. The direction and magnitude of the polarization depend upon the nature and amount of the strain, and upon the direction of the strain. In such materials the converse effect is observed, namely, that a strain results from the application of an electric field. (Std100) 270-1966w

piezoelectric loudspeaker *See:* crystal loudspeaker.

piezoelectric microphone *See:* crystal microphone.

piezoelectric pickup *See:* crystal pickup.

piezoelectric transducer A transducer that depends for its operation on the interaction between electric charge and the deformation of certain materials having piezoelectric properties. *Note:* Some crystals and specially processed ceramics have piezoelectric properties. (SP) [32]

PIGA *See:* pendulous integrating gyro accelerometer.

piggyback board *See:* daughter board.

piggybacking A technique in which an acknowledgment of a previously received protocol data unit is carried within an outgoing protocol data unit. (C) 610.7-1995

pigtail (fiber optics) A short length of optical fiber, permanently fixed to a component, used to couple power between it and the transmission fiber. *See also:* launching fiber. (Std100) 812-1984w

pileup *See:* relay pileup.

pile-up (x-ray energy spectrometers) (of signal pulses) Two pulses (signals) are said to be piled-up when a second pulse occurs before the transient response of the preceding pulse has decayed to a negligible value. (NPS/NID) 759-1984r

pile-up rejection (x-ray energy spectrometers) A technique used to identify and reject pulses (signals) that are piled up. (NPS/NID) 759-1984r

pillbox antenna A reflector antenna having a cylindrical reflector enclosed by two parallel conducting plates perpendicular to the cylinder, spaced less than one wavelength apart. *Contrast:* cheese antenna. (AP/ANT) 145-1993

PILOT *See:* Programmed Inquiry, Learning Or Teaching.

pilot (1) A signal transmitted either inbound or outbound through the system in order to provide a reference for automatic gain or automatic slope control (AGC or ASC circuits within the amplifier). (LM/C) 802.7-1989r
(2) A selected cell whose condition is assumed to indicate the condition of the entire battery. *See also:* battery. (EEC/PE) [119]

pilotage (navigation aid terms) The process of directing a vehicle by reference to recognizable landmarks or soundings, or to electronic or other aids to navigation. Observations may be by any means including optical, aural, mechanical, or electronic. (AES/GCS) 172-1983w

pilot cell (1) (storage battery) A selected cell whose condition is assumed to indicate the condition of the entire battery. *See also:* battery. (EEC/PE) [119]
(2) One or more cells chosen for monitoring the operating parameters, e.g., cell voltage, specific gravity and temperature, of the entire battery. (SCC21) 937-2000

pilot channel A channel over which a pilot is transmitted. (EEC/PE) [119]

pilot circuit The portion of a control apparatus or system that carries the controlling signal from the master switch to the controller. *See also:* control. (IA/ICTL/IAC) [60]

pilot communication scheme A protection scheme involving relays at two or more substations that share data or logic status via a communication channel to improve tripping speed and/or coordination. (PE/PSR) C37.113-1999

pilot director indicator A device that indicates to the pilot information as to whether or not the aircraft has departed from the target track during a bombing run. (EEC/PE) [119]

piloted ignition Initiation of combustion as a result of contact of a material or its vapors with an energy source such as a flame, spark, electrical arc, or glowing wire. (DEI) 1221-1993w

pilot exciter (1) (excitation systems for synchronous machines) The equipment providing the field current for the excitation of another exciter. (PE/EDPG) 421.1-1986r
(2) The source of all or part of the field current for the excitation of another exciter. (IA/MT) 45-1998

pilot fit (rotating machinery) (spigot fit) A clearance hole and mating projection used to guide parts during assembly. (PE) [9]

pilot house control (illuminating engineering) A mechanical means for controlling the elevation and train of a searchlight from a position on the other side of the bulkhead or deck on which it is mounted. (EEC/IE) [126]

pilot lamp A lamp that indicates the condition of an associated circuit. In telephone switching, a pilot lamp is a switchboard lamp that indicates a group of line lamps, one of which is or should be lit. (EEC/PE) [119]

pilot light A light, associated with a control, that by means of position or color indicates the functioning of the control. (EEC/PE) [119]

pilot line (conductor stringing equipment) A lightweight line, normally synthetic fiber rope, used to pull heavier pulling lines that, in turn, are used to pull the conductor. Pilot lines may be installed with the aid of finger lines or by helicopter when the insulators and travelers are hung. *Synonyms:* P-line; leader; lead line; P-line; pilot rope; leader; straw line. (T&D/PE) 524a-1993r, 524-1980s, 524-1992r

pilot line winder (1) (conductor stringing equipment) A device designed to payout and rewind pilot lines during stringing operations. It is normally equipped with its own engine which drives a drum or a supporting shaft for a reel mechanically, hydraulically or through a combination of both. These units are usually equipped with multiple drums or reels, depending upon the number of pilot lines required. The pilot line is payed out from the drum or reel, pulled through the travelers in the sag section, and attached to the pulling line on the reel stand or drum puller. It is then rewound to pull the pulling line through the travelers. (T&D/PE) 524-1980s
(2) (conductor stringing equipment) A device designed to payout and rewind pilot lines during stringing operations. It is normally equipped with its own engine, which drives a drum or a supporting shaft for a reel mechanically, hydraulically, or through a combination of both. These units are usually equipped with multiple drums or reels, depending upon the number of reels, pulled through the travelers in the sag section, and attached to the pulling line on the reel stand or drum puller. It is then rewound to pull the pulling line through the travelers. (T&D/PE) 524a-1993r

pilot protection A form of line protection that uses a communication channel as a means to compare electrical conditions at the terminals of a line. (SWG/PE) C37.100-1992

pilot rope *See:* pilot line.

pilot streamer (lightning) The initial low-current discharge that begins when the voltage gradient exceeds the breakdown voltage of air. *See also:* direct-stroke protection. (T&D/PE) [10]

pilot wire An auxiliary conductor used in connection with remote measuring devices or for operating apparatus at a distant point. *See also:* center of distribution. (T&D/PE) [10]

pilot-wire-controlled network A network whose switching devices are controlled by means of pilot wires. *See also:* alternating-current distribution. (T&D/PE) [10]

pilot wire protection Pilot protection in which a metallic circuit is used for the communicating means between relays at the circuit terminals. (SWG/PE) C37.100-1992

pilot-wire regulator An automatic device for controlling adjustable gains or losses associated with transmission circuits to compensate for transmission changes caused by temperature variations, the control usually depending upon the resistance of a conductor or pilot wire having substantially the same temperature conditions as the conductors of the circuits being regulated. *See also:* transmission regulator. (EEC/PE) [119]

pi mode (magnetrons) (π mode) The mode of operation for which the phases of the fields of successive anode openings facing the interaction space differ by p radians. *See also:* magnetron. (ED) 161-1971w

π-model A simplification of a general resistor/inductor/capacitor (RLC) network that represents the driving point admittance for an interconnect. (C/DA) 1481-1999

PIN *See:* plant information network.

pin (1) The point at which connection is made between the integrated circuit and the substrate on which it is mounted (e.g., the printed circuit board). For packaged components, this would be the package pin; for components mounted directly on the substrate, this would be the bonding pad. (TT/C) 1149.1-1990
(2) Any of the leads on a device that connect it to the system, each of which provides some function such as input, output, control, power, or ground. (C) 610.10-1994w
(3) A terminal point where an interconnect structure makes electrical contact with the fixed structures of a cell instance, or the conceptual point where a net connects to a lower level in the design hierarchy. (C/DA) 1481-1999

pinboard (1) A perforated board that accepts manually inserted pins to control the operation of equipment. (MIL/C) [2], [20], [85]
(2) *See also:* plugboard. (C) 610.7-1995

pinch (electron tube) The part of the envelope of an electron tube or valve carrying the electrodes and through which pass the connections to the electrodes. *See also:* electron tube. (ED) [45], [84]

pinch effect (1) (rheostriction) The phenomenon of transverse contraction and sometimes momentary rupture of a fluid conductor due to the mutual attraction of the different parts carrying currents. *See also:* induction heating; electrothermics. (Std100) 270-1966w
(2) (disk recording) A pinching of the reproducing stylus tip twice each cycle in the reproduction of lateral recordings due to a decrease of the groove angle cut by the recording stylus when it is moving across the record as it swings from a negative to a positive peak. (SP) [32]
(3) (induction heating) The result of an electromechanical force that constricts, and sometimes momentarily ruptures, a molten conductor carrying current at high density. *See also:* skin effect. (IA) 54-1955w

pin count The number of cell instance pins that an interconnect structure visits, including all input, output, and bidirectional pins. Pin count is the number of "places" the interconnect goes to on the chip. (C/DA) 1481-1999

pin-cushion distortion (1) A defect in a display surface that causes parallel lines to bow towards each other, causing a distorted image. *See also:* barrel distortion. (C) 610.6-1991w
(2) (camera tubes or image tubes) A distortion that results in a progressive increase in radial magnification in the reproduced image away from the axis of symmetry of the electron optical system. *Note:* For a camera tube, the reproducer is assumed to have no geometric distortion. *See also:* distortion, amplitude-frequency; hiss; distortion factor; percent harmonic distortion. (ED/BT) 161-1971w, 185-1975w

p-i-n detector (germanium gamma-ray detectors) (charged-particle detectors) (x-ray energy spectrometers) (semiconductor radiation detectors) A detector consisting of an intrinsic or nearly intrinsic region between a p- and n-region. (NPS/NID) 300-1988r, 301-1976s, 759-1984r, 325-1996

PIN diode (fiber optics) A diode with a large intrinsic region sandwiched between p- and n-doped semiconducting regions. Photons absorbed in this region create electron-hole pairs that are then separated by an electric field, thus generating an electric current in a load circuit. (Std100) 812-1984w

p-i-n diode attenuator (nonlinear, active, and nonreciprocal waveguide components) A device that provides a predetermined value of attenuation in a transmission line in response to a precise value of bias. (MTT) 457-1982w

p-i-n diode limiter (nonlinear, active, and nonreciprocal waveguide components) A passive microwave power limiter that utilizes the nonlinear conductivity of p-i-n diodes. (MTT) 457-1982w

pi network (π network) A network composed of three branches connected in series with each other to form a mesh, the three junction points forming an input terminal, an output terminal, and a common input and output terminal, respectively. *See also:* network analysis.

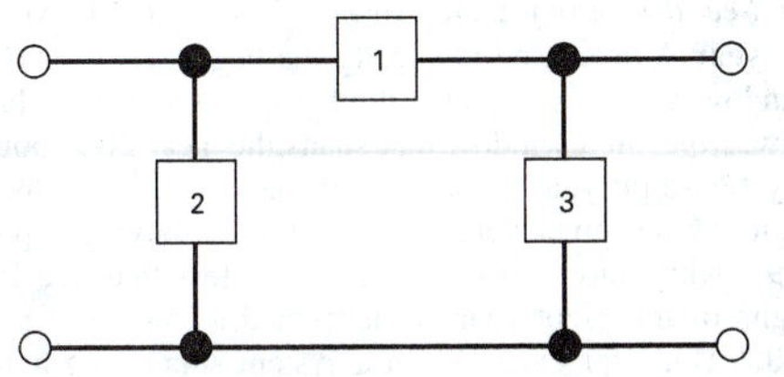

The junction point between branches 1 and 2 forms an input terminal, that between branches 1 and 3 forms an output terminal, and that between branches 2 and 3 forms a common input and output terminal.

pi (π) network

(Std100) 270-1966w

pin feed *See:* tractor feed.

ping Describes the transmission of a physical layer (PHY) packet to a particular node in order to time the response packet(s) provoked. (C/MM) 1394a-2000

ping-pong transmission technique *See:* time compression multiplexing.

pin insulator A complete insulator, consisting of one insulating member or an assembly of such members without tie wires, clamps, thimbles, or other accessories, the whole being of such construction that when mounted on an insulator pin it will afford insulation and mechanical support to a conductor that has been properly attached with suitable accessories. *See also:* insulator; tower. (T&D/PE) [10]

pin jack A single-conductor jack having an opening for the insertion of a plug of very small diameter. (PE/EM) 43-1974s

pink noise (speech quality measurements) A random noise whose spectrum level has a negative slope of 10 decibels per decade. 297-1969w

pin-on platform A platform attached by a pin to a boom to support a worker at an elevated worksite. A platform is a device used to support the worker in a standing position (generally without sides). (T&D/PE) 1307-1996

pins (electron tube or valve) Metal pins connected to the electrodes that plug into the holder. They ensure the electric connection between the electrodes and the external circuit and also mechanically fix the tube in its holder. *See also:* electron tube. (ED) [45], [84]

PIO *See:* programmed input-output.

pip A popular term for a sharp deflection in a visible trace. *See also:* radar. (EEC/PE) [119]

pipe (1) An object accessed by one of the pair of file descriptors created by the *pipe*() function. Once created, the file descriptors can be used to manipulate it, and it behaves identically to a FIFO special file when accessed in this way. It has no name in the file hierarchy. (C/PA) 9945-1-1996, 9945-2-1993
(2) The circuitry in a pipelined processor that implements the overlapping parallel functions. (C) 610.10-1994w
(3) An object accessed by one of the pair of file descriptors created by the `POSIX_IO.Create_Pipe` procedure. Once created, the file descriptors can be used to manipulate a pipe, and it behaves identically to a FIFO special file when accessed in this way. It has no name in the file hierarchy. (C) 1003.5-1999

pipe cable A pressure cable in which the container for the pressure medium is a loose-fitting rigid metal pipe. *See also:* oil-filled pipe cable; pressure cable. (T&D/PE) [10]

pipe guide A component of a switch-operating mechanism designed to maintain alignment of a vertical rod or shaft. (SWG/PE) C37.30-1992s

pipeline (1) (heating of pipelines and vessels) A length of pipe, including pumps, valves, flanges, control devices, strainers, or similar equipment for conveying fluids. (NESC/IA/PC) 844-1991, [86]
(2) (software) A software or hardware design technique in which the output of one process serves as input to a second, the output of the second process serves as input to a third, and so on, often with simultaneity within a single cycle time. (C) 610.12-1990
(3) In the shell command language, a sequence of one or more commands separated by the control operator "|". (C/PA) 9945-2-1993

pipelined transfer (FASTBUS acquisition and control) The portion of a FASTBUS operation in which a master either sends data to or causes data to be sent by an attached slave on every transition of data sync. The slave acknowledges receipt of or sends data with every transition of data acknowledge. The master does not wait for an acknowledge signal from the slave before causing another data sync transition. (NID) 960-1993

pipeline processing *See:* pipelining.

INSTRUCTION 1	instruction decode	operand fetch	instruction execution	operand store			
INSTRUCTION 2	instruction fetch	instruction decode	operand fetch	instruction execution	operand store		
INSTRUCTION 3		instruction fetch	instruction decode	operand fetch	instruction execution	operand store	
INSTRUCTION 4			instruction fetch	instruction decode	operand fetch	instruction execution	operand store
INSTRUCTION 5				instruction fetch	instruction decode	operand fetch	instruction execution
TIME:				CURRENT TIME:			

pipelining

pipeline processor A processor in which execution of instructions takes place as a series of units, arranged so that several units can be simultaneously processing appropriate parts of several instructions. (C) 610.10-1994w

pipelining The function of forwarding in sequence some or all of the Beginning of Message (BOM) and Continuation of Message (COM) *Derived MAC Protocol Data Units* (*DMPDUs*) before receipt of the End of Message (EOM) DMPDU. (LM/C) 8802-6-1994
(2) **(A)** Parallel processing in which instructions are executed in an assembly-line fashion: consecutive instructions are operated upon in sequence, but with several being initiated before the first is complete. *Synonym:* pipeline processing. **(B)** A technique for operation in which each instruction is broken into multiple steps, which are performed by different portions of the computer. A typical instruction stream allows a different instruction to be at each step in the pipeline at any point in time, allowing multiple instructions to overlap execution. See the corresponding figure. *Note:* In microprocessors, pipelining can make multiple cycle instructions appear to execute in a single clock cycle once the pipeline is full. (C) 610.10-1994

pipe-ventilated *See:* duct ventilated.

pip-matching display (navigation) (navigation aid terms) A display in which the received signal appears as a pair of blips, the comparison of the characteristics of which provides a measure of the desired quantity. (AES/GCS) 172-1983w

pi point (π point) A frequency at which the insertion phase shift of an electric structure is 180 degrees or an integral multiple thereof. (EEC/PE) [119]

Pirani gauge A bolometric vacuum gauge that depends for its operation on the thermal conduction of the gas present, pressure being measured as a function of the resistance of a heated filament ordinarily over a pressure range of 10^{-1} to 10^{-4} conventional millimeter of mercury. *See also:* instrument. (EEC/PE) [119]

piston (high-frequency communication practice) (plunger) A conducting plate movable along the inside of an enclosed transmission path and acting as a short-circuit for high-frequency currents. *See also:* waveguide. (PE/EEC) [119]

piston attenuator (waveguide) A variable cutoff attenuator in which one of the coupling devices is carried on a sliding member like a piston. *See also:* waveguide. (AP/ANT) [35], [84]

pistonphone A small chamber equipped witha reciprocating piston of measurable displacement that permits the establishment of a known sound pressure in the chamber. (SP) [32]

pit (rotating machinery) A depressed area in a foundation under a machine. (PE) [9]

pitch (1) (audio and electroacoustics) The attribute of auditory sensation in terms of which sounds may be ordered on a scale extending from low to high, such as a musical scale. *Notes:* 1. Pitch depends primarily upon the frequency of the sound stimulus, but it also depends upon the sound pressure and wave form of the stimulus. 2. The pitch of a sound may be described by the frequency of that simple tone, having a specified sound pressure or loudness level, that seems to the average normal ear to produce the same pitch. 3. The unit of pitch is the "mel." (SP/ACO) [32]
(2) (cable) *See also:* lay. 30-1937w
(3) The centerline to centerline spacing of modules or boards in a card cage or on a backplane. (C/BA) 14536-1995
(4) The center-to-center spacing between adjacent chassis module slots. (C/BA) 1101.3-1993
(5) The spacing from the side A surface of the guide rib on a module to the side A surface of the guide rib on an adjacent module. (C/BA) 1101.4-1993
(6) The nominal spacing from the centerline of a module to the centerline of an adjacent module. (C/BA) 1101.7-1995

pitch angle *See:* pitch attitude.

pitch attitude (navigation aid terms) The angle between the longitudinal axis of the vehicle and the horizontal. *Synonym:* pitch angle. (AES/GCS) 172-1983w

pitch factor (rotating machinery) The ratio of the resultant voltage induced in a coil to the arithmetic sum of the magnitudes of the voltages induced in the two coil sides. *See also:* armature. (PE) [9]

pits Depressions produced in metal surfaces by nonuniform electrodeposition or from electrodissolution; for example, corrosion. (EEC/PE) [119]

pitting Localized corrosion taking the form of cavities at the surface. (IA) [59]

pitting factor The depth of the deepest pit resulting from corrosion divided by the average penetration as calculated from weight loss. (IA) [59]

PIV *See:* peak inverse voltage; peak reverse voltage.

pivot-friction error Error caused by friction between the pivots and the jewels: it is greatest when the instrument is mounted with the pivot axis horizontal. *Note:* This error is included with other errors into a combined error defined in repeatability. *See also:* moving element. (EEC/AII) [102]

pivot year A year used to specify the beginning of a 100-year window and to interpret 2-digit year dates within that window. *Note:* In a window that spans the Year 2000 boundary, any two-digit year value greater than or equal to the last 2 digits of the pivot year is interpreted as having a prefix of "19," while any two-digit year value less than the last two digits of the pivot year is interpreted as having a prefix of "20." Thus, for example, in a system supporting a 1950 to 2049 window, a pivot year of 1950 causes two-digit year values between 50 and 99 to be interpreted as 1950 to 1999, and two-digit year values between 00 and 49 to be interpreted as 2000 to 2049. (C/PA) 2000.1-1999

pixel (1) (image processing and pattern recognition) In image processing, the smallest element of a digital image that can be assigned a gray level. *Note:* This term originated as a con-

traction for "picture element." *Synonyms:* pel; picture element; resolution cell. *See also:* edge pixel; line pixel. (C) 610.4-1990w
(2) (computer graphics) The smallest element of a display surface that can be assigned independent characteristics. *Note:* This term is derived from the term "picture element." *Synonyms:* pel; picture element. *See also:* voxel. (C) 610.6-1991w
(3) An abbreviation for picture element—the smallest unit of display on a video screen. (C) 1295-1993w
(4) The smallest element of a display surface whose characteristics are independent assigned. *Note:* This term is derived from the term "picture element." *Synonym:* pel; picture element. (C) 610.10-1994w

PL/1 *See:* Programming Language/1.

PL/I Deprecated for PL/1. (C) 610.13-1993w

PLA *See:* programmable logic array.

place (1) (elecronic digital computers) In positional notation, a position corresponding to a given power of the base, a given cumulated product, or a digit cycle of a given length. It can usually be specified as the nth character from one end of the numerical expression. (C) 162-1963w
(2) *See also:* digit place. (C) 1084-1986w

placeholder *See:* dummy.

place value In a positional notation system, the power of the radix that corresponds to a given place. For example, in a decimal integer the place values from right to left are 1, 100, etc. (C) 1084-1986w

plain conductor A conductor consisting of one metal only. *See also:* conductor. (T&D/PE) [10]

plain flange (waveguide) (plane flange) (plain connector) A coupling flange with a flat face. *See also:* waveguide. (AP/ANT) [35]

planar array A two-dimensional array of elements whose corresponding points lie in a plane. (AP/ANT) 145-1993

planar dielectric waveguide A planar transmission line consisting of one or more dielectric layers or dielectric strips of finite width, or both, located above a single or between a pair of extended conducting ground planes. (MTT) 1004-1987w

planar network A network that can be drawn on a plane without crossing of branches. *See also:* network analysis. (Std100) 270-1966w

planar transmission line A transmission line composed of one or more parallel plates, slabs, or sheets of conducting or insulating materials, including free space, and in which one or more layers are composed of materials of differing electromagnetic properties, arranged in strips of finite cross section and aligned with the axis of propagation to form the guiding structures. The line may be enclosed laterally by conducting walls aligned parallel to the axis of propagation. (MTT) 1004-1987w

Planckian locus (television) The locus of chromaticities of Planckian (blackbody or full) radiators having various temperatures. *See also:* chromaticity diagram. (BT/AV) 201-1979w

Planck radiation law (illuminating engineering) An expression representing the spectral radiance of a blackbody as a function of the wavelength and temperature. This law commonly is expressed by the formula

$$L_\lambda = I_\lambda/A' = c_{1L}\lambda^{-5}[e^{(c_2/\lambda T} - 1]^{-1}$$

in which L_λ is the spectral radiance, I_λ is the spectral radiant intensity, A' is the projected area ($A \cos \theta$) of the aperture of the blackbody, e is the base of natural logarithms (2.71828), T is the absolute temperature, and c_{1L} and $_2$ are constants designated as the first and second radiation constraints. *Notes:* 1. The designation c_{1L} is used to indicate that the equation in the form given here refers to the radiance L, or to the intensity I per unit projected area A', of the source. Numerical values are commonly given not for c_{1L} but for c_1, which applies to the total flux radiated from a blackbody aperture, that is, in a hemisphere (2π steradians), so that, with the Lambert cosine law taken into account, $c_1 = \pi c_{1L}$. The currently recommended value of c_1 is 3.7415×10^{-16} W · m^2 or 3.7415×10^{-12} W · cm^2. Then c_{1L} is 1.1910×10^{-16} W · m^2 · sr^{-1} or 1.1910×10^{-12} W · cm^2 · sr^{-1}. If, as is more convenient, wavelengths are expressed in micrometers and area in square centimeters, $c_{1L} = 1.1910 \times 101$W · μm^4 × cm^{-2} · sr − 1, L_λ being given in W · cm^{-2} · sr^{-1} · μ^{-1}. The presently recommended value of c^2 is 1.43879 cm kelvin. The Planck law in the following form gives the energy radiated from the blackbody in a given wavelength interval ($\lambda_1 - \lambda_2$):

$$(Q = \int_{\lambda_1}^{\lambda_2} Q_\lambda d\lambda) = Atc_1 \int_{\lambda_1}^{\lambda_2} \lambda^{-5}\,(e^{(c_2/\lambda T)-1}d\lambda$$

If A is the area of the radiation aperture or surface in square centimeters, t is the time in seconds, λ is the wavelength in micrometers, $c_1 = 3.7415 \times 10^4$ W · μm^4 · cm^{-2}, then Q is the total energy in watt seconds (joules) emitted from this area (that is, in the solid angle 2π) in time t, within the wavelength interval ($\lambda_1 - \lambda_2$). 2. It often is convenient, as is done here, to use different units of length in specifying wavelengths and areas, respectively. If both quantities are expressed in centimeters and the corresponding value for c_1(3.7415×10^{-5} erg · cm · sec^{-1}) is used, this equation gives the total emission of energy in ergs from area A (that is in the solid angle 2π), for time t, and for the interval $\lambda_1 - \lambda_2$ in centimeters. (EEC/IE) [126]

Planck's constant (fiber optics) The number h that relates the energy E of a photon with the frequency ν of the associated wave through the relation $E = h\nu/h = 6.626 \times 10^{-34}$ joule second. *See also:* photon. (Std100) 812-1984w

Planck's radiation law Defines the emission spectrum of a blackbody in terms of its physical temperature. (AP/PROP) 211-1997

plane angle (SI) (International System of Units (SI)) The SI unit for plane angle is the radian. Use of the degree and its decimal submultiples is permissible when the radian is not a convenient unit. Use of the minute and second is discouraged except for special fields such as cartography. *See also:* units and letter symbols. (QUL) 268-1982s

plane bend (waveguide components) (corner) A waveguide bend (corner) in which the longitudinal axis of the guide remains in a plane parallel to the plane of the magnetic field vectors throughout the bend (corner). (MTT) 147-1979w

plane-earth factor (radio-wave propagation) The ratio of the electric field strength that would result from propagation over an imperfectly conducting plane earth to that which would result from propagation over a perfectly conducting plane. (AP) 211-1977s

plane flange *See:* cover flange.

plane of contraflexure In an H-frame structure, locates the inflection points for each pole. An inflection point is a point in the pole that separates an outward pole curvature from an inward pole curvature. This is also a location of zero moment. (T&D/PE) 751-1990

plane of incidence The plane containing the normal to the surface of a boundary and the phase vector, $\vec{\beta}$, of the incident wave. (AP/PROP) 211-1997

plane of polarization (1) (data transmission) For a plane polarized wave, the plane containing the electric intensity and the direction of propagation. (PE) 599-1985w
(2) (radio-wave propagation) For a plane-polarized wave, the plane containing the electric and magnetic field vectors. (AP) 211-1977s
(3) A plane containing the polarization ellipse. *Notes:* 1. When the ellipse degenerates into a line segment, the plane of polarization is not uniquely defined. In general, any plane containing the segment is acceptable; however, for a plane wave in an isotropic medium, the plane of polarization is taken to be normal to the direction of propagation. 2. In optics, the expression *plane of polarization* is associated with a linearly polarized plane wave (sometimes called a *plane po-*

larized wave) and is defined as a plane containing the field vector of interest and the direction of propagation. This usage would contradict the above one and is deprecated. (AP/ANT) 145-1993

plane of propagation (radio-wave propagation) Of an electromagnetic wave, the plane containing the attenuation vector and the wave normal; in the common degenerate case where these vectors have the same direction, the plane containing the electric vector and the wave normal. (AP) 211-1977s

plane-parallel resonator (laser maser) A beam resonator comprising a pair of plane mirrors oriented perpendicular to the axis of the beam. (LEO) 586-1980w

plane-polarized wave (1) (radio-wave propagation) At a point in a homogeneous medium, an electromagnetic wave whose electric and magnetic field vectors at all times lie in a fixed plane. (AP) 211-1977s
(2) *See also:* plane of polarization. (AP/ANT) 145-1993

plane wave (1) (fiber optics) A wave whose surfaces of constant phase are infinite parallel planes normal to the direction of propagation. (Std100) 812-1984w
(2) A wave in which the only spatial dependence of the field vectors is through a common exponential factor whose exponent is a linear function of position. *Notes:* 1. In a linear, homogeneous, and isotropic space the electric field vector, magnetic field vector and the propagation vector are mutually perpendicular. The ratio of the magnitude of the electric field vector to the magnitude of the magnetic field vector is equal to the intrinsic impedance of the medium; for free space the intrinsic impedance is equal to 376.730Ω or approximately 120πΩ. 2. A plane wave can be resolved into two component waves corresponding to two orthogonal polarizations. The total power flux density of the plane wave at a given point in space is equal to the sum of the power flux densities in the orthogonal component waves. (AP/ANT) 145-1993
(3) A wave whose equiphase surfaces form a family of parallel planes. (AP/PROP) 211-1997

plane-wave equivalent power density A commonly used term associated with any electromagnetic wave, equal in magnitude to the power density of a plane wave having the same electric (*E*) or magnetic (*H*) field strength. (NIR) C95.1-1999

plane wave exponential factor (radio-wave propagation) The factor $\exp(-j\vec{k} \cdot \vec{r})$ in the phasor expression for plane wave fields, where $\vec{k}$ is the wave vector and $\vec{r}$ is the position vector. (AP/PROP) 211-1990s

plane wave propagation factor The factor $\exp(-j\vec{k} \cdot \vec{r})$ in the phasor expression for plane wave fields, where $\vec{k}$ is the propagation vector (a constant) and $\vec{r}$ is the position vector. (AP/PROP) 211-1997

plane wave, uniform *See:* uniform plane wave.

PLANIT *See:* Programming LANguage for Interactive Teaching.

planned derated hours (electric generating unit reliability, availability, and productivity) The available hours during which a basic or extended planned derating was in effect. (PE/PSE) 762-1987w

planned derating (electric generating unit reliability, availability, and productivity) That portion of the unit derating that is scheduled well in advance. *See also:* extended planned derating; basic planned derating. (PE/PSE) 762-1987w

planned outage (electric generating unit reliability, availability, and productivity) The state in which a unit is unavailable due to inspection, testing, nuclear refueling, or overhaul. A planned outage is scheduled well in advance. (PE/PSE) 762-1987w

planned outage hours (electric generating unit reliability, availability, and productivity) The number of hours a unit was in the basic or extended planned outage state. (PE/PSE) 762-1987w

planned stop *See:* optional stop.

PLANNER A computer language and reasoning model used commonly in artificial intelligence for proving theorems and for manipulating models in a robot. (C) 610.13-1993w

plan-position indication (PPI) (A) A type of radar display format. *See also:* display. **(B)** A display in which target blips are shown in plan position, thus forming a map-like display, with radial distance from the center representing range and with the angle of the radius vector representing azimuth angle.

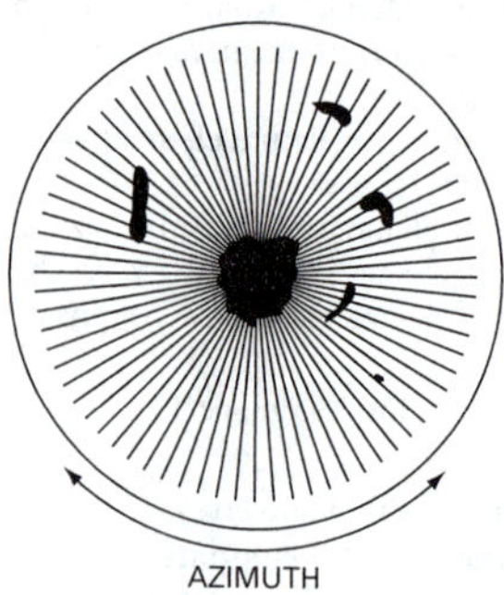

PPI display

(AES/RS) 686-1990

Plan-position indicator (PPI) A display in which target echoes (blips) are shown in plan position, thus forming a map-like display, with radial distance from the center representing range and with the angle of the radius vector representing azimuth angle.

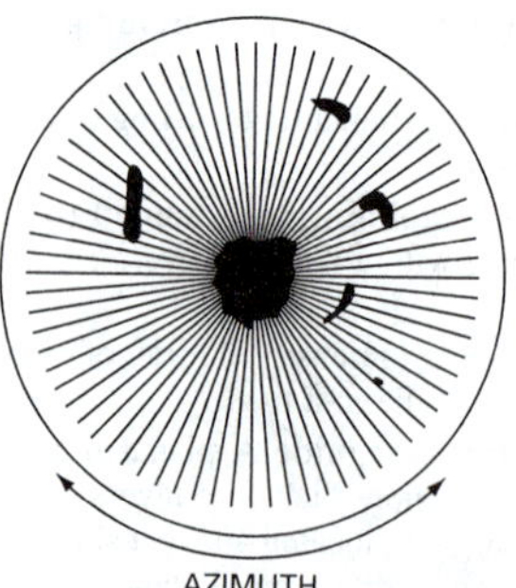

Plan-position indicator

(AES) 686-1997

plan-position indicator (PPI) (navigation aid terms) A type of radar display format. *See also:* display; display; display. (AES/GCS) 172-1983w, 686-1997

plan-position-indicator scope A cathode-ray oscilloscope arranged to present a plan-position-indicator display. (AES/RS) 686-1990

plan standard A standard that describes the characteristics of a scheme for accomplishing defined objectives or work within specified resources. (C) 610.12-1990

plant For a given system, that part which is to be controlled and whose parameters are unalterable. *See also:* process equipment. (CS/PE/EDPG) [3]

plant-capacity factor *See:* plant factor.

plant dynamics Equations which describe the behavior of the plant. *See also:* control system. (CS/IM) [120]

plant factor (plant-capacity factor) The ratio of the average load on the plant for the period of time considered to the aggregate rating of all the generating equipment installed in the plant. *See also:* generating station. (T&D/PE) [10]

plant information network (PIN) A representation of both the flow and the structure of the information that supports the discrete activities occurring during the life cycle of the plant. (PE/EDPG) 1150-1991w

plant life cycle The life of a power plant from conception to decommissioning. In this model, the cycle consists of five phases: site selection and plant concepts, design, construction, operation, and decommissioning. (PE/EDPG) 1150-1991w

plasma A macroscopically neutral assembly of charged and possibly also uncharged particles. *Note:* A plasma is said to be cold if the thermal effects of charged particles on dynamic processes in the plasma can be neglected for the particular problem involved. A plasma is said to be hot (or warm) if the thermal effects are not negligible. (AP/PROP) 211-1997

plasma display device A display device employing a plasma panel to display data on the display screen. *Note:* An image can persist for relatively long periods of time on such a device. *Synonyms:* gas-discharge display device; gas plasma display device. (C) 610.10-1994w

plasma frequency (f_N) A natural frequency of oscillation of charged particles in a plasma given by:

$$(f_N)^2 = (2\pi)^{-2} \frac{Nq^2}{m\varepsilon_0}$$

where

q = the charge per particle
m = the particle mass
N = the particle number density
ε_0 = the permittivity of free space

Note: For electrons, with f_N in Hertz and N in electrons per cubic meter:

$$(f_N)^2 = 80.6N$$

(AP/PROP) 211-1997

plasma panel A grid of electrodes encased within two flat glass plates separated by an ionizing gas in which the energizing of selected electrodes causes the gas to be ionized and light to be emitted at that point. *Note:* Also called a matrix-addressed storage display device. *Synonym:* gas panel. *See also:* plasma display device. (C) 610.10-1994w

plasmapause The outer boundary of the plasmasphere, characterized by a steep decrease of the plasma density. (AP/PROP) 211-1997

plasma sheath A layer of charged particles, of substantially one sign, that accumulates around a body in a plasma. (AP/PROP) 211-1997

plasmasphere That ionized region of the topside ionosphere that encircles the Earth around the equator and follows the rotation of the Earth. *Note:* In the equatorial plane, the plasmasphere extends to a distance of 3–7 Earth radii, depending on local time and geomagnetic activity. (AP/PROP) 211-1997

plasma waves Electrostatic waves associated with a "warm" plasma, giving rise to density and velocity fluctuations. (AP/PROP) 211-1997

plastic (rotating machinery) A material that contains as an essential ingredient an organic substance of large molecular weight, is solid in its finished state, and, at some stage in its manufacture or in its processing into finished articles, can be shaped by flow. (PE) [9]

plastic-clad silica fiber (fiber optics) An optical waveguide having silica core and plastic cladding. (Std100) 812-1984w

plate (electron tube) A common name for an anode in an electron tube. *See also:* electrode. (ED) 161-1971w

plateau The portion of the counting-rate-versus-voltage characteristic in which the counting rate is substantially independent of the applied voltage. (NI/NPS) 309-1999

plateau length The range of applied voltage over which the plateau of a radiation-counter tube extends. (NI/NPS) 309-1999

plateau slope The slope of the plateau of a gas-filled counter tube expressed as the percentage change in counting rate per 100 V change in applied voltage. Alternatively, the slope can be expressed in percent counting rate per volt change in applied voltage. (NI/NPS) 309-1999

plateau slope, normalized *See:* normalized plateau slope.

plateau slope, relative *See:* relative plateau slope.

plate-circuit detector A detector functioning by virtue of a nonlinearity in its plate-circuit characteristic. (EEC/PE) [119]

plated-through hole (electronics) Deposition of metal on the side of a hole and on both sides of a base to provide electric connection, and an enlarged portion of conductor material surrounding the hole on both sides of the base. (EEC/AWM) [105]

plated wire storage A type of magnetic storage in which information is stored by magnetically recording it on a plated wire surface. (C) 610.10-1994w

plate (anode) efficiency The ratio of load circuit power (alternating current) to the plate power input (direct current). *See also:* network analysis. (AP/ANT) 145-1983s

plate keying Keying effected by interrupting the plate-supply circuit. *See also:* telegraphy. (AP/ANT) 145-1983s

plate (anode) load impedance The total impedance between anode and cathode exclusive of the electron stream. *See also:* network analysis. (AP/ANT) 145-1983s

plate (anode) modulation Modulation produced by introducing the modulating signal into the plate circuit of any tube in which the carrier is present. (AP/BT/ANT) 145-1983s, 182A-1964w

platen In impact printers, the surface against which a print element strikes in order to make character imprints. (C) 610.10-1994w

plate (anode) neutralization A method of neutralizing an amplifier in which a portion of the plate-to-cathode alternating voltage is shifted 180 degrees and applied to the grid-cathode circuit through a neutralizing capacitor. (AP/BT/ANT) 145-1983s, 182A-1964w

plate out (monitoring radioactivity in effluents) A thermal, electrical, chemical, or mechanical action that results in a loss of material by deposition on surfaces between sampling point and detector. (NI/PE/NP) N42.18-1980r, 380-1975w

plate (anode) power input The power delivered to the plate (anode) of an electron tube by the source of supply. *Note:* The direct-current power delivered to the plate of an electron tube is the product of the mean plate voltage and the mean plate current. (AP/ANT) 145-1983s

plate (anode) pulse modulation Modulation produced in an amplifier or oscillator by application of externally generated pulses to the plate circuit. (AP/ANT) 145-1983s

platform *See:* aerial platform.

platform, aerial *See:* aerial platform.

platform control power Energy source(s) available at platform potential for performing protection and control functions. (T&D/PE) 824-1994

platform erection (navigation aid terms) In the alignment of inertial systems, the process of bringing the vertical axis of a stable platform system into agreement with the local vertical. (AES/GCS) 172-1983w

platform fault-detection A means used to detect insulation failure on the platform that results in current flowing from normal current-carrying circuit elements to the platform. (T&D/PE) 824-1994

platform fault-detection device (series capacitor) A device to detect insulation failure on the platform that results in current flowing from normal current carrying circuit elements to the platform. (T&D/PE) 824-1985s

platform internal interface (PII) The interface between application platform service components within that platform. (C/PA) 14252-1996

platform, lineperson's *See:* lineperson's platform.

platform profile A profile whose focus is on functionality and interfaces for a particular type of platform, which may be a single processor shared by a group of applications or a large distributed system with each application dedicated to a single processor. (C/PA) 14252-1996

platform-to-ground signaling devices Devices to transmit protection, control, and alarm functions to and from the platform. (T&D/PE) 824-1994

plating rack (electroplating) Any frame used for suspending one or more electrodes and conducting current to them during electrodeposition. *See also:* electroplating. (EEC/PE) [119]

platter One disk within a stack of disks, such that the disks are attached to a common spindle. *See also:* disk. (C) 610.10-1994w

playback (1) A term used to denote reproduction of a recording. (EEC/PE) [119]
(2) *See also:* reversible execution. (C) 610.12-1990
(3) To output data or text for review purposes. *Synonyms:* playout; printout. (C) 610.10-1994w

playback head *See:* read head.

playback loss *See:* translation loss.

playout *See:* playback.

PL/C A subset of PL/1 characterized by its enhanced debugging, quick compilation, code optimization and error checking facilities. *Note:* The C stands for Cornell University, where the language was developed. (C) 610.13-1993w

PLCP *See:* physical layer convergence procedure.

PL/DB *See:* Programming Language/Data Base.

plenary capacitance (between two conductors) The capacitance between two conductors when the changes in the charges on the two are equal in magnitude but opposite in sign and the other n − 2 conductors are isolated conductors. (Std100) 270-1966w

plenum An air compartment or chamber to which one or more ducts are connected and which forms part of an air-distribution system. (NESC/NEC) [86]

PLI Deprecated for PL/1. (C) 610.13-1993w

plier clamp *See:* strand restraining clamp.

P-line *See:* pilot line.

pliotron A hot-cathode vacuum tube having one or more grids. *Note:* This term is used primarily in the industrial field. (ED) [45]

PII *See:* platform internal interface.

PL/M A procedure-oriented programming language derived from PL/1, designed specifically for microcomputers. (C) 610.13-1993w

plotter An output device that presents data on paper in the form of a two-dimensional graphic representation. *See also:* drum plotter; analog plotter; digital plotter; flatbed plotter; electrostatic plotter; raster plotter. (C) 610.10-1994w

plotter step size (1) The minimum distance between two points or parallel lines on a plotter. *See also:* increment size. (C) 610.6-1991w, 610.10-1994w
(2) The distance between adjacent addressable points of a plotter. (C) 610.10-1994w

plotting board The flat surface of a flatbed plotter on which the output is displayed. *Synonym:* plotting table. (C) 610.10-1994w

plotting chart (navigation aid terms) A chart designed primarily for plotting dead reckoning lines of position. (AES/GCS) 172-1983w

plotting head A head within a plotter that is used to create marks on the display surface. (C) 610.10-1994w

plotting table *See:* plotting board.

plow (cable plowing) Equipment capable of laying cable, flexible conduit, etc., underground. (T&D/PE) 590-1977w

plow blade (cable plowing) A soil-cutting tool. (T&D/PE) 590-1977w

plow blade amplitude (cable plowing) Maximum displacement of plow blade tip from mean position induced by the vibrator (half the stroke). (T&D/PE) 590-1977w

plow blade frequency (cable plowing) Rate of blade tip vibration in hertz. (T&D/PE) 590-1977w

plowing (cable plowing) A process for installing cable, flexible conduit, etc., by cutting or separating the earth, permitting the cable or flexible conduit to be placed or pulled in behind the blade. (T&D/PE) 590-1977w

PLPC *See:* physical layer convergence procedure.

PL/PROPHET A language similar in style to PL/1, used in conjunction with the PROPHET system in pharmacology research. *See also:* PARSEC. (C) 610.13-1993w

plug (1) A device, usually associated with a cord, that by insertion in a jack or receptacle establishes connection between a conductor or conductors associated with the plug and a conductor or conductors connected to the jack or receptacle. (PE/EM) 43-1974s
(2) A cluster of blades fixed to a photocontrol, shorting cap, or nonshorting cap to establish an electrical and mechanical connection when inserted into a mating receptacle. (RL) C136.10-1996

plug adapter (plug body) A device that by insertion in a lampholder serves as a receptacle. (EEC/PE) [119]

plug adapter lampholder (current tap) A device that by insertion in a lampholder serves as one or more receptacles and a lampholder. (EEC/PE) [119]

plugboard (1) (general) A perforated board that accepts manually inserted plugs to control the operation of equipment. *See also:* control panel. (C) [20], [85]
(2) (test, measurement, and diagnostic equipment) Patchboard the use of which is restricted to punched card machines. *See also:* patch board. (MIL) [2]
(3) A printed circuit board into which plugs or pins may be placed to control the operation of equipment. *Synonym:* pinboard. *See also:* jack. (C) 610.7-1995, 610.10-1994w

plugboard chart A chart that shows, for a given job, where plugs may be inserted into a plugboard. *Synonym:* plugging chart. (C) 610.10-1994w

plug braking (rotating machinery) A form of electric braking of an induction motor obtained by reversing the phase sequence of its supply. *See also:* asynchronous machine. (PE) [9]

plug fuses (protection and coordination of industrial and commercial power systems) Plug fuses are rated 125 volts (V) and are available with current artings up to 30 amperes (A). Their use is limited to circuits rated 125 V or less. However, they may also be used in circuits supplied from a system having a grounded neutral, and in which no conductor operates at more than 150 V to ground. The National Electrical Code (NEC) requires type S fuses in all new installations of plug fuses because they are tamper resistant and size limiting, thus making it difficult to overfuse. (IA/PSP) 242-1986r

plugging A control function that provides braking by reversing the motor line voltage polarity or phase sequence so that the motor develops a counter-torque that exerts a retarding force. *See also:* electric drive. (IA/ICTL/IAC) [60]

plugging chart *See:* plugboard chart.

plug-in A communication device when it is so designed that connections to the device may be completed through pins, plugs, jacks, sockets, receptacles, or other forms of ready connectors. (EEC/PE) [119]

plug-in device A device that may be installed and removed at will, especially a device that is attached to a bus intended for system expansion. (C/BA) 1275-1994

plug-in driver A package, usually associated with a plug-in device and serving as the interface to that device, that is created by evaluating an FCode program resident on that device. (C/BA) 1275-1994

plug-in-type bearing (rotating machinery) A complete journal bearing assembly, consisting of a bearing liner and bearing housing and any supporting structure that is intended to be inserted into a machine end-shield. *See also:* bearing. (PE) [9]

plug-in unit *See:* CAMAC plug-in unit.

plug-receptacle interface temperature The temperature either at the top of the receptacle or at the bottom of the control base with the control mounted in the receptacle in an ambient temperature of 25°C. (RL) C136.10-1996

plumb-bob gravity At a site on the earth, the force per unit mass acting on a mass at rest relative to the earth, not including any reaction force from the suspension. The plumb bob gravity includes the gravitational attraction of the earth, the effect of the centripetal acceleration due to the earth rotation, and tidal effects. The time average of the plumb-bob-gravity acceleration in the earth-fixed frame is the vector difference of the earth-Newtonian-gravitational acceleration and the earth-rotation-centripetal acceleration. The earth-Newtonian-gravitational-acceleration vector points generally toward the center of mass of the earth, as modified by the effects of gravity anomalies and the earth's flattening due to the earth rotation. The earth-rotation-centripetal-acceleration vector points toward and is perpendicular to the earth-rotation axis. The magnitude and direction of the instantaneous plumb-bob-gravity acceleration vary about their averages by approximately ±0.15 μg and ±0.15 μrad, respectively, with an approximate 12.4 h period due to the effect of lunar-solar earth and ocean tides. The tidal acceleration is due to the difference between the lunar and solar attractions at the site and at the center of the earth, the variation in earth attraction caused by the variation in the distance of the site from the center of the earth due to tidal deformation and ocean tide loading, and the gravitational attraction of the solid earth and ocean tide deformations. The direction of the plumb-bob-gravity acceleration defines the local vertical down direction, and its magnitude defines the unit of acceleration *g* used in accelerometer scale factor calibration at a test site (with compensation for tidal effects if accuracy warrants). (AES/GYAC) 1293-1998

plumb-bob vertical (navigation aid terms) The direction indicated by a simple, ideal, frictionless pendulum that is motionless with respect to the earth; it indicates the direction of the vector sum of the gravitational and centrifugal accelerations of the earth at the location of the observer. (AES/GCS) 172-1983w

plumbing (1) (data transmission) A term employed in communication practice to designate coaxial lines or wave guides and accessory equipment for radio-frequency transmission. (COM/PE) 599-1985w
(2) A colloquial expression for pipe-like waveguide circuit elements and transmission lines. *Synonym:* microwave plumbing. (AES) 686-1997

plumb mark (conductor stringing equipment) A mark placed on the conductor located vertically below the insulator point of support for steel structures and vertically above the pole center line at ground level for wood pole structures used as a reference to locate the center of the suspension clamp. (T&D/PE) 524-1992r

plumb marker pole A small diameter, lightweight pole with a marking device attached to one end, having sufficient length to enable a worker to mark the conductor directly below him/her from a position on the bridge or arm of the structure. This device is utilized to mark the conductor immediately after completion of sagging. *Synonym:* marker. (T&D/PE) 524-1992r

plume *See:* positive prebreakdown streamers.

plunger relay A relay operated by a movable core or plunger through solenoid action. *See also:* relay. (EEC/REE) [87]

plural service *See:* dual service.

plural tap *See:* multiple plug.

plus input (oscilloscopes) An input such that the applied polarity causes a deflection polarity in agreement with conventional deflection polarity. (IM) 311-1970w

plus/minus operation A winding arrangement in which one or the other end of the tap winding is connected by a reversing change-over selector to the main winding, and allows use of the taps in a buck or boost mode when travelling through the tapping range. (PE/TR) C57.131-1995

PM *See:* permanent magnet; phase modulation; protective margin; permanent-magnet focusing.

PMA *See:* physical medium attachment.

PMA sublayer *See:* physical medium attachment sublayer.

PMOS *See:* p-channel metal-oxide semiconductor.

PMSG *See:* dead-front pad-mounted switchgear.

pnet A physical net that has no correspondence to the logical function of the design, such as a route segment, which is reserved for future routes across a hard macro, or a power net not described in the design netlist. (C/DA) 1481-1999

pneumatically release-free (trip-free) (as applied to a pneumatically operated switching device) A term indicating that by pneumatic control the switching device is free to open at any position in the closing stroke if the release is energized. *Note:* This release-free feature is operative even though the closing control switch is held closed. (SWG/PE) C37.100-1992

pneumatic bellows, relay *See:* relay pneumatic bellows.

pneumatic brake pipe A pressurized air line, continuous over the length of the train, used variously to indicate train integrity, provide indication of a emergency condition, equalize reservoir pressures, or propagate a brake application signal. (VT) 1475-1999

pneumatic controller A pneumatically supervised device or group of devices operating electric contacts in a predetermined sequence. *See also:* multiple-unit control. (VT/LT) 16-1955w

pneumatic loudspeaker A loudspeaker in which the acoustic output results from controlled variation of an air stream. (EEC/PE) [119]

pneumatic operation Power operation by means of compressed gas. (SWG/PE) C37.100-1992

pneumatic switch A pneumatically supervised device opening or closing electric contacts, and differs from a pneumatic controller in being purely an ON and OFF type device. *See also:* control switch. (VT/LT) 16-1955w

pneumatic transducer A unilateral transducer in which the sound output results from a controlled variation of an air stream. (SP) [32]

pneumatic tubing system (protective signaling) An automatic fire-alarm system in which the rise in pressure of air in a continuous closed tube, upon the application of heat, effects signal transmission. *Note:* Most pneumatic tubing systems contain means for venting slow pressure changes resulting from temperature fluctuations and therefore operate on the so-called rate-of-rise principle. *See also:* protective signaling. (EEC/PE) [119]

***p-n* junction (semiconductor)** A region of transition between *p*- and *n*-type semiconducting material. *See also:* semiconductor device. (PE/EEC) [119]

PN sequence *See:* pseudonoise sequence.

pocket A card stacker in a card sorter. *Synonym:* bin. (C) 610.10-1994w

pocketbook *See:* insulator cover; conductor grip.

pocket sort *See:* distribution sort.

pocket-type plate (of a storage cell) A plate of an alkaline storage battery consisting of an assembly of perforated oblong metal pockets containing active material. *See also:* battery. (EEC/PE) [119]

poid The curve traced by the center of a sphere when it rolls or slides over a surface having a sinusoidal profile. (SP) [32]

Poincaré sphere (1) A sphere whose points are associated in a one-to-one fashion with all possible polarization states of a plane wave [field vector] according to the following rules: The longitude equals twice the tilt angle and the latitude is twice the angle whose cotangent is the negative of the axial ratio of the polarization ellipse. *Notes:* 1. For this definition, the axial ratio carries a sign. 2. The points of the northern hemisphere of the Poincaré sphere represent polarizations with a left-hand sense and those of the southern hemisphere represent polarization with a right-hand sense. The north pole represents left-hand circular polarization and the south pole right-hand circular polarization. The points of the equator represent all possible linear polarizations. *See also:* axial ratio. (AP/ANT) 145-1993
(2) A tool for graphically displaying the polarization state of a monochromatic wave. For a fully polarized wave, each point on the sphere's surface defines a unique polarization state, with axial ratio and tilt angle mapping into latitude and longitude, respectively. (AP/PROP) 211-1997

point (1) (A) (positional notation) The character, or the location of an implied symbol, that separates the integral part of a numerical expression from its fractional part. For example, it is called the binary point in binary notation and the decimal point in decimal notation. If the location of the point is assumed to remain fixed with respect to one end of the numerical expressions, a fixed-point system is being used. If the location of the point does not remain fixed with respect to one end of the numerical expression, but is regularly recalculated, then a floating-point system is being used. *Note:* A fixed-point system usually locates the point by some convention, while a floating-point system usually locates the point by expressing a power of the base. *See also:* variable point; checkpoint; breakpoint; rerun point; floating point; fixed point; branch point. **(B) (positional notation)** The character, or implied location of such a character, that separates the integral part of a numerical expression from the fractional part. Since the place to the left of the point has unit weight in the most commonly used systems, the point is sometimes called the units point, although it is frequently called the binary point in binary notation and the decimal point in decimal notation. *See also:* floating point; breakpoint; fixed point. (C) [20], 270-1966, [85], 162-1963

(2) (lightning protection) The pointed piece of metal used at the upper end of the elevation rod to receive a lightning discharge. (EEC/PE) [119]

(3) The standard typographical unit of measurement, approximately equal to 1/72 in. (C) 1295-1993w

(4) (for supervisory control or indication or telemeter selection) All of the supervisory control or indication devices in a system, exclusive of the common devices, in the master station and in the remote station that are necessary for

1) Energizing the closing, opening, or other circuits of a unit, or set of units of switchgear or other equipment being controlled, or
2) Automatic indication of the closed or open or other positions of the unit, or set of units of switchgear or other equipment for which indications are being obtained, or
3) Connecting a telemeter transmitting equipment into the circuit to be measured and to transmit the telemeter reading over a channel to a telemeter receiving equipment.

Note: A point may serve for any two or all three of the purposes described above; for example, when a supervisory system is used for the combined control and indication of remotely operated equipment, point (for supervisory control) and point (for supervisory indication) are combined into a single control and indication point. (SWG/PE) C37.100-1992

(5) (lightning protection) *See also:* radix point. (C) 1084-1986w

point contact (semiconductors) A pressure contact between a semiconductor body and a metallic point. *See also:* semiconductor device; semiconductor. (ED) 216-1960w

point-contact transistor A transistor having a base electrode and two or more point-contact electrodes. *See also:* transistor; semiconductor. (ED) 216-1960w

point coordination function (PCF) A class of possible coordination functions in which the coordination function logic is active in only one station in a basic service set (BSS) at any given time that the network is in operation. (C/LM) 8802-11-1999

point detector A device that is a part of a switch-operating mechanism and is operated by a rod connected to a switch, derail, or movable-point frog to indicate that the point is within a specified distance of the stock rail. (EEC/PE) [119]

point equipment (point) Elements of a supervisory system, exclusive of the basic common equipment, that are peculiar to and required for the performance of a discrete supervisory function.

1) **alarm point.** Station (remote or master, or both) equipment that inputs a signal to the alarm function.
2) **accumulator point.** Station (remote or master, or both) equipment that accepts a pulsing digital input signal to accumulate a total of pulse counts.
3) **analog point.** Station (remote or master, or both) equipment that inputs an analog quantity to the analog function.
4) **control point.** Station (remote or master, or both) equipment that operates to perform the control function.
5) **indication (status) point.** Station (remote or master, or both) equipment that accepts a digital input signal for the function of indication.
6) **sequence of events point.** Station (remote or master, or both) equipment that accepts a digital input signal to perform the function of registering sequence of events.
7) **telemetering selection point.** Station (remote or master, or both) equipment for the selective connection of telemetering transmitting equipment to appropriate telemetering receiving equipment over an interconnecting communication channel. This type of point is more commonly used in electromechanical or stand-alone type of supervisory control.
8) **spare point.** Point equipment that is not being utilized, but is fully wired and equipped.
9) **wired point.** Point for which all common equipment, wiring, and space are provided. To activate the point requires only the addition of plug-in hardware.
10) **space-only point.** Point for which cabinet space only is provided for future addition or wiring and other necessary plug-in equipment.

Note: A point may serve for one or more of the purposes described above, for example, when a supervisory system is used for combined control and supervision of remotely operated equipment, a point for supervisory control and point for supervisory indication may be combined into a single control and indication point. *See also:* supervisory control functions. (SWG/PE/SUB) C37.100-1992, C37.1-1987s

pointer (1) (software) A data item that specifies the location of another data item; for example, a data item that specifies the address of the next employee record to be processed. *Synonym:* link. *See also:* pointer segment; stack pointer. (C) 610.5-1990w, 610.12-1990

(2) An identifier that indicates the address or storage location of an data item. (C) 610.10-1994w

pointer data Data used to represent the addresses of other data items. (C) 610.5-1990w

pointer optimization A database reorganization technique in which database access is made more efficient by reestablishing the pointers within the database so that fewer pointers are needed to represent the database structure. (C) 610.5-1990w

pointer pusher (demand meter) The element that advances the maximum demand pointer in accordance with the demand and in integrated-demand meters is reset automatically at the end of each demand interval. *See also:* demand meter. (EEC/PE) [119]

pointer segment (data management) A segment in a database that establishes a parent/child relationship between segments. *Note:* The segment contains only a pointer to the physical child segment for its parent segment. (C) 610.5-1990w

pointer shift due to tapping The displacement in the position of a moving element of an instrument that occurs when the instrument is tapped lightly. The displacement is observed by a change in the indication of the instrument. *See also:* moving element. (EEC/AII) [102]

point ID printout (sequential events recording systems) A brief coded method of identifying inputs using alphanumeric characters, usually used in computer based systems. (PE/EDPG) [1]

pointing accuracy (communication satellite) The angular difference between the direction in which the main beam of an antenna points and the required pointing direction. (COM) [25]

pointing device (1) A device, such as a mouse, trackball, or joystick, used to move a pointer on the screen. (C) 1295-1993w
(2) An input deice that is used to specify a particular addressable location. *Contrast:* pick device. *See also:* stylus; cursor control device. (C) 610.10-1994w, 610.6-1991w

point-junction transistor A transistor having a base electrode and both point-contact and junction electrodes. *See also:* transistor. (EEC/PE) [119]

point method (illuminating engineering) (formerly called "point-by-point method") A lighting design procedure for predetermining the illuminance at various locations in lighting installations, by use of luminaire photometric data. The direct component of illuminance due to the luminaires and the interreflected component of illuminance due to the room surfaces are calculated separately. The sum is the total illumination at a point. (EEC/IE) [126]

point object A synthetic environment object that is geometrically anchored to the terrain with a single point. (C/DIS) 1278.1a-1998

point of common coupling (PCC) (1) The connection point between the SVC and the power system at which performance requirements are defined. (SUB/PE) 1303-1994
(2) The busbar from which other loads sensitive to voltage may be connected as well as the static var compensator (SVC) and any disturbing load it is required to compensate. (PE/SUB) 1031-2000
(3) The point at which the electric utility and the customer interface occurs. Typically, this point is the customer side of the utility revenue meter. *Note:* In practice, for building-mounted photovoltaic (PV) systems (such as residential PV systems) the customer distribution panel may be considered the PCC for convenience in making measurements and performing testing. (SCC21) 929-2000

point of connection For a static var compensator (SVC) with a dedicated transformer, the high voltage (HV) bus to which the whole is connected. For an SVC connected to an existing transformer, or direct connected at low voltage, the busbar to which it is connected. (PE/SUB) 1031-2000

point of fixation (illuminating engineering) A point or object in the visual field at which the eyes look, and upon which they are focused. (EEC/IE) [126]

point of measurement Place at which the conventionally true values are determined and at which the reference point of the monitor is placed for test purposes. (NI) N42.20-1995

point of observation (illuminating engineering) For most purposes it may be assumed that the distribution of luminance in the field of view can be described as if there were a single point of observation located at the midpoint of the baseline connecting the centers of the entrance pupils of the two eyes. For many problems it is necessary, however, to regard the centers of the entrance pupils as separate points of observation for the two eyes. (EEC/IE) [126]

point of presence The point at which the local telephone company terminates subscribers' circuits for long distance dial-up or leased line communications. (C) 610.7-1995

point of regulation (POR) The location in the subsystem (unit under test) where voltage is sensed for voltage regulation. The point of regulation can be remote or local to the voltage regulation equipment. (PEL) 1515-2000

point-of-sale terminal (1) A device for recording sales data in machine readable form at the time each sale is made. (C) 610.2-1987
(2) A job-oriented terminal for recording sales data in machine-readable form at the time and place at which each sale is made. (C) 610.10-1994w

point on tangent (POT) *See:* hub.

point operator An image operator that assigns a gray level to each output pixel based on the gray level of the corresponding input pixel. *Contrast:* neighborhood operator. (C) 610.4-1990w

point source (laser maser) A source of radiation whose dimensions are small enough compared with the distance between source and receptor for them to be neglected in calculations. (LEO) 586-1980w

point test A predefined location within equipment or routines at which a known result should be present if the equipment or routine is operating properly. (SWG/PE/SUB) C37.100-1992, C37.1-1994

point-to-multipoint connection A connection with multiple endpoints, wherein one endpoint that is designated as the root (originator) is connected by a point-to-point connection with all other endpoints. (C/LM) 802.9a-1995w

point-to-point (1) Descriptive of a communication channel that services just two terminals. (SUB/PE) 999-1992w
(2) Pertaining to a channel, line, or a circuit that has only two end points. *See also:* multipoint; multidrop. (C) 610.7-1995

point-to-point communications Communications that take place between exactly two devices. (C/MM) 1284.4-2000

point-to-point configuration A network configuration in which two communicating stations are connected by a point-to-point channel. (C) 610.7-1995

point-to-point connection A connection with only two endpoints. (C/LM) 802.9a-1995w

point-to-point control system *See:* positioning control system.

point-to-point radio communication Radio communication between two fixed stations. *See also:* radio transmission. (EEC/PE) [119]

point transposition A transposition, usually in an open wire line, that is executed within a distance comparable to the wire separation, without material distortion of the normal wire configuration outside this distance. (EEC/PE) [119]

Poisson's equation In rationalized form:

$$\nabla^2 V = -\frac{\rho}{\varepsilon_0 \varepsilon}$$

where $\varepsilon_0\varepsilon$ is the absolute capacitivity of the medium, V the potential, and ρ the charge density at any point. (Std100) 270-1966w

polar axis (primary ferroelectric terms) A direction that is parallel to the spontaneous polarization vector. When a polar crystal is heated or cooled, the internal or external electrical conduction generally cannot provide enough current to compensate for the change in polarization with temperature, and the crystal develops an electric charge on its surface. For this reason, polar crystals are called pyroelectric. *Note:* For crystal class m, the polar axis is in an arbitrary direction in a plane (the mirror plane). The polar axis for crystal class 1 can be in any arbitrary direction. (UFFC) 180-1986w

polar cap Polar region bounded by the auroral zone. (AP/PROP) 211-1997

polar cap absorption (PCA) The intense absorption of radio waves in the polar regions caused by the arrival of high-energy solar protons concentrated in this region by the lines of force of the Earth's magnetic field. (AP/PROP) 211-1997

polar contact A part of a relay against which the current-carrying portion of the movable polar member is held so as to form a continuous path for current. (EEC/PE) [119]

polar direct-current telegraph system A system that employs positive and negative currents for transmission of signals over the line. *See also:* telegraphy. (EEC/PE) [119]

polar-duplex signaling (telephone switching systems) Any method of bidirectional signaling over a line using ground potential compensation and polarity sensing. (COM) 312-1977w

polarential telegraph system A direct-current telegraph system employing polar transmission in one direction and a form of differential duplex transmission in the other direction. *Note:* Two kinds of polarential systems, known as types A and B, are in use. In half-duplex operation of a type-A polarential system the direct-current balance is independent of line resistance. In half-duplex operation of a type-B polarential system the direct-current balance is substantially independent of the line leakage. *See also:* telegraphy. (COM) [49]

polarimetry The study of electromagnetic propagation, scattering, and emission that considers the complete polarization state of any arbitrarily polarized wave. (AP/PROP) 211-1997

Polaris correction (navigation aids) A correction to be applied to the corrected sextant altitude of Polaris to obtain latitude. (AES/GCS) 172-1983w

polarity (1) (instrument transformers) (power and distribution transformers) The designation of the relative instantaneous directions of the currents entering the primary terminals and leaving the secondary terminals during most of each half cycle. *Note:* Primary and secondary terminals are said to have the same polarity, when, at a given instant during most of each half cycle, the current enters the identified, similarity marked primary lead and leaves the identified, similarly marked secondary terminal in the same direction as though the two terminals formed a continuous circuit. (PE/PSR/TR) C37.110-1996, C57.12.80-1978r

(2) (batteries) An electrical condition determining the direction in which current tends to flow on discharge. By common usage the discharge current is said to flow from the positive electrode through the external circuit. *See also:* battery. (PE/EEC) [119]

(3) (television) (picture signal) The sense of the potential of a portion of the signal representing a dark area of a scene relative to the potential of a portion of the signal representing a light area. Polarity is stated as black negative or black positive. *See also:* television. (BT/AV) [34]

(4) The relative instantaneous directions of the currents entering the primary terminals and leaving the secondary terminals during most of each half cycle. *Note:* Primary and secondary terminals are said to have the same polarity when, at a given instant during most of each half cycle, current enters the primary terminal and leaves the secondary terminal in the same direction as though there was a continuous circuit between the two terminals. (PE/TR) C57.13-1993

(5) The orientation of any device that has poles or signed electrodes. (C) 610.10-1994w

(6) Polarity of the dc voltage with respect to ground, e.g., positive or negative. (PE/TR) C57.19.03-1996

(7) The polarity of a regulator is intrinsic in its design. Polarity is correct if the regulator boosts the voltage in the "raise" range and bucks the voltage in the "lower" range. The relative polarity of the shunt winding and the series windings of a step-voltage regulator will differ in the boost and buck modes between Type A and Type B regulators. (PE/TR) C57.15-1999

polarity and angular displacement (regulator) Relative lead polarity of a regulator or a transformer is a designation of the relative instantaneous direction of current in its leads. In addition to its main transformer windings, a regulator commonly has auxiliary transformers or auxiliary windings as an integral part of the regulator. The same principles apply to the polarity of all transformer windings. *Notes:* 1. Primary and secondary leads are said to have the same polarity when at a given instant the current enters an identified secondary lead in the same direction as though the two leads formed a continuous circuit. 2. The relative lead polarity of a single-phase transformer may be either additive or subtractive. If one pair of adjacent leads from the two windings is connected together and voltage applied to one of the windings, then:

a) The relative lead polarity is additive if the voltage across the other two leads of the windings is greater than that of the higher-voltage winding alone.
b) The relative lead polarity is subtractive if the voltage across the other two leads of the winding is less than that of the higher-voltage winding alone.

3. The polarity of a polyphase transformer is fixed by the internal connections between phases as well as by the relative locations of leads: it is usually designated by means of a vector line diagram showing the angular displacement of windings and a sketch showing the marking of leads. The vector lines of the diagram represent induced voltages and the recognized counterclockwise direction of rotation is used. The vector line representing any phase voltage of a given winding is drawn parallel to that representing the corresponding phase voltage of any other winding under consideration. *See also:* voltage regulator. (PE/TR) C57.15-1968s

polarity guard A device that guarantees proper tip and ring polarity. (SCC31) 1390.2-1999, 1390.3-1999

polarity marks (instrument transformers) The identifications used to indicate the relative instantaneous polarities of the primary and secondary current and voltages. *Notes:* 1. On voltage transformers during most of each half cycle in which the identified primary terminal is positive with respect to the unidentified primary terminal, the identified secondary terminal is also positive with respect to the unidentified secondary terminal. 2. The polarity marks are so placed on current transformers that during most of each half-cycle, when the direction of the instantaneous current is into the identified primary terminal, the direction of the instantaneous secondary current is out of the correspondingly identified secondary terminal. 3. This convention is in accord with that by which standard terminal markings H_1, X_1, etc., are correlated. *See also:* instrument transformer. (PE/TR) [57]

polarity or polarizing voltage device (power system device function numbers) A device that operates, or permits the operation of, another device on a predetermined polarity only, or verifies the presence of a polarizing voltage in the equipment. (PE/SUB) C37.2-1979s

polarity-related adjectives (A) (pulse terminology) Of, having, or pertaining to a single polarity. **(B) (bipolar)** Of, having, or pertaining to both polarities. (IM/WM&A) 194-1977

polarity reversal Change of voltage polarity from positive to negative or from negative to positive polarity. (PE/TR) C57.19.03-1996

polarity test (rotating machinery) A test taken on a machine to demonstrate that the relative polarities of the windings are correct. *See also:* asynchronous machine. (PE) [9]

polarizability The average electric dipole moment produced per molecule per unit of electric field strength. (Std100) 270-1966w

polarization (1) (primary ferroelectric terms) The electric dipole moment per unit volume. Polarization is related to electric displacement D through the linear expression

$$D_i = P_i + \varepsilon_0 E_i$$

where the derived constant ε_0 (usually called the permittivity of free space) equals $8.854 \cdot 10^{12}$ coulomb/volt-meter. In ferroelectric materials both D and P are nonlinear functions of E and may depend on previous history of the material. When the electric field is applied along a polar axis that is also a special axis of the prototype phase of the crystal, this expression may then be regarded as a scalar equation, since D, E, and P all point along the same direction. When the term $\varepsilon_0 E$ in the above expression is negligible compared to P (as in the case for most ferroelectric materials), D is nearly equal to P; therefore, the D versus E and P versus E plots of the hysteresis loop become, in practice, equivalent. *Note:* The polarization P may be expressed as the bound surface charge per unit area of a free surface normal to the direction of P. *See also:* desired polarization; polarization error. (UFFC) 180-1986w

(2) (primary ferroelectric terms) (of a waveguide mode) In some cases, the polarization of the electric field vector on the axis of symmetry of a waveguide. In general, however, the polarization of the mode is not identical to the polarization of the electric field vector in the mode, since the latter varies from point to point in the guide cross-section. *Notes:* 1. Polarization is that property of a degenerate waveguide mode which characterizes a particular mode within a set of degenerate modes. The main application of this concept is to waveguides of square or circular cross-section. 2. When two orthogonal modes can be identified in a square or circular waveguide, a polarization ellipse can be associated with the field vectors and considered in terms of axial ratio, etc. (MTT) 146-1980w

(3) (radiated wave) That property of a radiated electromagnetic wave describing the time-varying direction and amplitude of the electric field vector; specifically, the figure traced as a function of time by the extremity of the vector at a fixed location in space, as observed along the direction of propagation. *Note:* In general the figure is elliptical and it is traced in a clockwise or counterclockwise sense. The commonly referenced circular and linear polarizations are obtained when the ellipse becomes a circle or a straight line, respectively. Clockwise sense rotation of the electric vector is designated right-hand polarization and counterclockwise sense rotation is designated left-hand polarization. *See also:* radiation. (AP/ANT) [35]

(4) (electronic navigation) (desired) The polarization of the radio wave for which an antenna system is designed. *See also:* navigation. (AES/RS/GCS) 686-1982s, 173-1959w, [42]

(5) (batteries) The change in voltage at the terminals of the cell or battery when a specified current is flowing, and is equal to the difference between the actual and the equilibrium (constant open-circuit condition) potentials of the plates, exclusive of the IR drop. *See also:* polarization. (EEC/PE) [119]

(6) (waveguide) (of a field vector) For a field vector at a single frequency at a fixed point in space, the polarization is that property which describes the shape and orientation of the locus of the extremity of the field vector and the sense in which this locus is traversed. *Notes:* 1. For a time harmonic (or single frequency) vector, the locus is an ellipse with center at the origin. In some cases, this ellipse becomes a circle or a segment of a straight line. The polarization is then called "circular" and "linear, " respectively. 2. The orientation of the ellipse is defined by its plane, called the plane of polarization, and by the direction of its axes. (For a linearly polarized field, any plane containing the segment locus of the field vector is a plane of polarization.) (MTT) 146-1980w

(7) (A) (of an antenna) In a given direction from the antenna, the polarization of the wave transmitted by the antenna. *Note:* When the direction is not stated, the polarization is taken to be the polarization in the direction of maximum gain. **(B)** [of a wave (radiated by an antenna in a specified direction)] In a specified direction from an antenna and at a point in its far field, the polarization of the (locally) plane wave that is used to represent the radiated wave at that point. *Note:* At any point in the far field of an antenna, the radiated wave can be represented by a plane wave whose electric field strength is the same as that of the wave and whose direction of propagation is in the radial direction from the antenna. As the radial distance approaches infinity, the radius of curvature of the radiated wave's phase front also approaches infinity, and thus, in any specified direction, the wave appears locally as a plane wave. (AP/ANT) 145-1993

(8) (as applied to a relay) The input that provides a reference for establishing the direction of system phenomena such as direction of power or reactive flow, or direction to a fault or other disturbance on a power system. (SWG/PE) C37.100-1992

(9) (of an electromagnetic wave) The locus of the tip of the electric field vector observed in a plane orthogonal to the wave normal. *Notes:* 1. Elliptical polarization is the most general case. 2. The polarization of an electromagnetic wave is defined by the tilt angle, the axial ratio and the sign of the axial ratio, which expresses the sense of rotation of the polarization ellipse. *See also:* circularly polarized wave; elliptically polarized wave; linearly polarized wave; parallel polarization; perpendicular polarization. (AP/PROP) 211-1997

polarization capacitance (biological) The reciprocal of the product of electrode capacitive reactance and 2π times the frequency.

$$C_p = \frac{1}{2\pi f X_p}$$

See also: electrode impedance. (EMB) [47]

polarization controller, optical *See:* optical polarization controller.

polarization coupling loss That part of the transmission loss due to the mismatch between the polarization of the incoming wave and the polarization of the receiving antenna. (AP/PROP) 211-1997

polarization current Time-dependent, decaying current in the specimen, following the instant that a constant voltage is applied until steady-state conditions have been obtained. *Notes:* 1. Polarization current does not include the conductance current. The sum of the polarization and conductance currents in the specimen is that which is normally observed during measurements. 2. Polarization current includes both polarization absorption and capacitive-charge currents. (PE) 402-1974w

polarization, desired *See:* desired polarization.

polarization diversity reception (data transmission) That form of diversity reception that utilizes separate vertically and horizontally polarized receiving antennas. (PE) 599-1985w

polarization efficiency The ratio of the power received by an antenna from a given plane wave of arbitrary polarization to the power that would be received by the same antenna from a plane wave of the same power flux density and direction of propagation, whose state of polarization has been adjusted for a maximum received power. *Notes:* 1. The polarization efficiency is equal to the square of the magnitude of the inner product of the polarization vector describing the receiving polarization of the antenna and the polarization vector of the plane wave incident at the antenna. 2. If the receiving polarization of an antenna and the polarization of an incident plane wave are properly located as points on the Poincaré sphere, then the polarization efficiency is given by the square of the cosine of one-half the angular separation of the two points. *Synonym:* polarization mismatch factor. *See also:* polarization vector. (AP/ANT) 145-1993

polarization ellipse (waveguide) The locus of the extremity of a field vector at a fixed point in space. *See also:* polarization. (MTT) 146-1980w

polarization error (navigation aid terms) (navigation) The error arising from the transmission or reception of an electromagnetic wave having a polarization other than that intended for the system. (AES/GCS) 172-1983w

polarization index ($P.I._{t2/t1}$) (1) (rotating machinery) The ratio of the the insulation resistance of a machine winding measured at 1 min after voltage has been applied divided into the measurement at 10 min. (PE/EM) 95-1977r

(2) Variation in the value of insulation resistance with time. The quotient of the insulation resistance at time (t_2) divided by the insulation resistance at time (t_1). If times t_2 and t_1 are not specified, they are assumed to be 10 min and 1 min, respectively. Unit conventions: values of 1 through 10 are assumed to be in minutes, values of 15 and greater are assumed to be in seconds (e.g., $P.I._{60/15}$ refers to IR_{60s}/IR_{15s}). (PE/EM) 43-2000

polarization index test (rotating machinery) A test for measuring the ohmic resistance of insulation at specified time intervals for the purpose of determining the polarization index. *See also:* asynchronous machine; direct-current commutating machine. (PE) [9]

polarization maintaining fiber (interferometric fiber optic gyro) A single-mode fiber that preserves the plane of polarization of light coupled into it as the beam propagates through its length. (AES/GYAC) 528-1994

polarization match The condition that exists when a plane wave, incident upon an antenna from a given direction, has a polarization that is the same as the receiving polarization of the antenna in that direction. *See also:* receiving polarization. (AP/ANT) 145-1993

polarization mismatch factor *See:* polarization efficiency.

polarization mismatch loss The magnitude, expressed in decibels, of the polarization efficiency. (AP/ANT) 145-1993

polarization pattern (of an antenna) (A) The spatial distribution of the polarizations of a field vector excited by an antenna taken over its radiation sphere. (B) The response of a given antenna to a linearly polarized plane wave incident from a given direction and whose direction of polarization is rotating about an axis parallel to its propagation vector; the response being plotted as a function of the angle that the direction of polarization makes with a given reference direction. *Notes:* 1. When describing the polarizations over the radiation sphere [definition (A)], or a portion of it, reference lines shall be specified over the sphere, in order to measure the tilt angles of the polarization ellipses and the direction of polarization for linear polarizations. An obvious choice, though by no means the only one, is a family of lines tangent at each point on the sphere to either the θ or ϕ coordinate line associated with a spherical coordinate system of the radiation sphere. 2. At each point on the radiation sphere, the polarization is usually resolved into a pair of orthogonal polarizations, the co-polarization and the cross polarization. To accomplish this, the co-polarization must be specified at each point on the radiation sphere. For certain linearly polarized antennas, it is common practice to define the co-polarization in the following manner: First specify the orientation of the co-polar electric field vector at a pole of the radiation sphere. Then, for all other directions of interest (points on the radiation sphere), require that the angle that the co-polar electric field vector makes with each great circle line through the pole remain constant over that circle, the angle being that at the pole. In practice, the axis of the antenna's main beam should be directed along the polar axis of the radiation sphere. The antenna is then appropriately oriented about this axis to align the direction of its polarization with that of the defined co-polarization at the pole. This manner of defining co-polarization can be extended to the case of elliptical polarization by defining the constant angles using the major axes of the polarization ellipses rather than the co-polar electric field vector. The sense of polarization must also be specified. 3. The polarization pattern [definition (B)] generally has the shape of a dumbbell. The polarization ellipse of the antenna in the given direction is similar to one that can be inscribed in the dumbbell shape with points of tangency at the maxima and minima points; thus, the axial ratio and tilt angle can be obtained from the polarization pattern. *See also:* cross polarization; tilt angle; co-polarization. (AP/ANT) 145-1993

polarization-phase vector (for a field vector) The polarization vector, among all of those that define the same polarization, that carries the phase information of the field vector whose polarization it represents. *Note:* The polarization-phase vector of the field vector $\vec{E}$ is given by $\vec{E} == \vec{E}/E$ where E is magnitude of $\vec{E}$ that is, the positive square root of $\vec{E} * \vec{E}$. *See also:* polarization vector. (AP/ANT) 145-1993

polarization potential (biological) The boundary potential over an interface. *See also:* electrobiology. (EMB) [47]

polarization ratio The magnitude of a complex polarization ratio. (AP/ANT) 145-1993

polarization reactance (biological) The impedance multiplied by the sine of the angle between the potential vector and the current vector.

$X_p = Z_p \sin \theta$

See also: electrode impedance. (EMB) [47]

polarization, receiving *See:* receiving polarization.

polarization receiving factor The ratio of the power received by an antenna from a given plane wave of arbitrary polarization to the power received by the same antenna from a plane wave of the same power density and direction of propagation, whose state of polarization has been adjusted for the maximum received power. *Note:* It is equal to the square of the absolute value of the scalar product of the polarization unit vector of the given plane wave with that of the radiation field of the antenna along the direction opposite to the direction of propagation of the plane wave. *See also:* waveguide. (MTT) 146-1980w

polarization resistance (biological) The impedance multiplied by the cosine of the phase angle between the potential vector and the current vector. $R_p = Z_p \cos \theta$. *See also:* electrode impedance. (EMB) [47]

polarization state *See:* state of polarization.

polarization unit vector (field vector) (at a point) A complex field vector divided by its magnitude. *Notes:* 1. For a field vector of one frequency at a point, the polarization unit vector completely describes the state of polarization, that is, the axial ratio and orientation of the polarization ellipse and the sense of rotation on the ellipse. 2. A complex vector is one each of whose components is a complex number. The magnitude is the positive square root of the scalar product of the vector and its complex conjugate. *See also:* waveguide. (MTT) 146-1980w

polarization vector (for a field vector) A unitary vector which describes the state of polarization of a field vector at a given point in space. *Notes:* 1. Polarization vectors differing only by a unitary factor ($e^{j\alpha}$ where α is real) correspond to the same polarization state. 2. The appropriate inner product, $\langle \hat{e}_1, \hat{e}_2 \rangle$, for two polarization vectors in the same plane of polarization is given by $\langle \hat{e}_1,\hat{e}_2 \rangle = \hat{e}_1{}^* \cdot \hat{e}_2$ where $\hat{e}_1$ and $\hat{e}_2$ represent the polarization vectors corresponding to polarizations 1 and 2. 3. The inner product of polarization vectors representing the same polarization is equal to unity. The inner product of two polarization vectors representing two orthogonal polarizations is zero. 4. The inner product of a polarization vector corresponding to a specified polarization, $\hat{e}_1$, and a complex electric field vector $\vec{E}$, at a point in space will yield the component of the electric field vector corresponding to the specified polarization, $\vec{E}_1$; that is $\vec{E}_1 = (\hat{e}_1{}^* \cdot \vec{E})\hat{e}_1$. 5. The basis vectors for the components of the polarization vector may correspond to any two orthogonal polarizations, the most common being two orthogonal linear polarizations or right-hand and left-hand circular polarizations. (AP/ANT) 145-1993

polarized dipole magnetization *See:* polarized return-to-zero recording.

polarized electrolytic capacitor An electrolytic capacitor in which the dielectric film is formed adjacent to only one metal electrode and in which the impedance to the flow of current in one direction is greater than in the other direction. (EEC/PE) [119]

polarized plug (packaging machinery) A plug so arranged that it may be inserted in its counterpart only in a predetermined position. (IA/PKG) 333-1980w

polarized relay A relay that consists of two elements, one of which operates as a neutral relay and the other of which operates as a polar relay. *See also:* polar relay; neutral relay. (EEC/PE) [119]

polarized return-to-zero recording (RZ(P)) Return-to-zero recording in which zeros are represented by magnetization in one sense, ones are represented by magnetization in the opposite sense, and the reference condition is the absence of magnetization. *Synonym:* polarized dipole magnetization. *Contrast:* non-polarized return-to-zero recording. (C) 610.10-1994w

polarized snubber (converter circuit elements) (self-commutated converters) A snubber, including a diode, in which the limiting action depends on the direction of voltage or current. (IA/SPC) 936-1987w

polarizer A substance that when added to an electrolyte increases the polarization. *See also:* electrochemistry. (EEC/PE) [119]

polarizer, optical *See:* optical polarizer.

polarizing fiber (interferometric fiber optic gyro) A single-mode fiber that maintains one and only one polarization state as the beam propagates through its length by suppressing its orthogonal state. (AES/GYAC) 528-1994

polar mode *See:* resolver.

polar navigation (navigation aids) Navigation in polar regions where unique considerations and techniques are applied. (AES/GCS) 172-1983w

polar operation (data transmission) Circuit operation in which mark and space transitions are represented by a current reversal. (PE) 599-1985w

polar orbit (communication satellite) An inclined orbit with an inclination of 90°. The plane of a polar orbit contains the polar axis of the primary body. (COM) [19]

polar regions (navigation aid terms) The regions near the geographic poles. Definite limits for these regions are not recognized. (AES/GCS) 172-1983w

polar relay A relay in which the direction of movement of the armature depends upon the direction of the current in the circuit controlling the armature. *See also:* neutral relay; electromagnetic relay; polarized relay. (EEC/PE) [119]

pole (1) (illuminating engineering) A standard support generally used where overhead lighting distribution circuits are employed. (EEC/IE) [126]
(2) (electric power or communication) A column of wood or steel, or some other material, supporting overhead conductors, usually by means of arms or brackets. *See also:* pole shoe; field pole; tower. (T&D/PE) [10]
(3) (pole unit) (of a switching device or fuse) That portion of the device associated exclusively with one electrically separated conducting path of the main circuit of the device. *Notes:* 1. Those portions that provide a means for mounting and operating all poles together are excluded from the definition of a pole. 2. A switching device or fuse is called single-pole if it has only one pole. If it has more than one pole, it may be called multipole (two-pole, three-pole, etc.) and provided, in the case of a switching device, that the poles are or can be coupled in such a manner as to operate together. (SWG/PE) C37.100-1992
(4) The complex frequency where a Laplace transform is infinite. Combined with residues, this is a convenient mathematical notation for the impedance or transfer function of a passive circuit, such as a resistor/inductor/capacitor (RLC) circuit, since poles above this frequency can be ignored in calculations without significant loss of accuracy. (C/DA) 1481-1999
(5) *See also:* stick. (PE/T&D) 516-1995

pole body (rotating machinery) The part of a field pole around which the field winding is fitted. *See also:* asynchronous machine. (PE) [9]

pole-body insulation (rotating machinery) Insulation between the pole body and the field coil. *See also:* asynchronous machine. (PE) [9]

pole bolt (rotating machinery) A bolt used to fasten a pole to the spider. (PE) [9]

pole-cell insulation (rotating machinery) (salient pole) Insulation that constitutes the liner between the field pole coil and the salient pole body. *See also:* rotor. (PE) [9]

pole-changing winding (rotating machinery) A winding so designed that the number of poles can be changed by simple changes in the coil connections at the winding terminals. *See also:* rotor; stator. (PE) [9]

pole disagreement relay A protective relay designed to monitor currents in the three poles of a device, such as a circuit breaker, to verify the integrity of the electrical continuity of all its phases. (SWG/PE) C37.100-1992

pole end plate (rotating machinery) A plate or structure at each end of a laminated pole to maintain axial pressure on the laminations. *See also:* asynchronous machine. (PE) [9]

pole face (rotating machinery) The surface of the pole shoe or nonsalient pole forming one boundary of the air gap. *See also:* asynchronous machine; direct-current commutating machine. (PE) [9]

pole-face bevel (rotating machinery) The portion of the pole shoe that is beveled so as to increase the length of the radial air gap. *See also:* asynchronous machine. (PE) [9]

pole face, relay *See:* relay pole face.

pole-face shaping (rotating machinery) The contour of the pole shoe that is shaped other than by being beveled, so as to produce nonuniform radial length of the air gap. *See also:* rotor; stator. (PE) [9]

pole fixture A structure installed in lieu of a single pole to increase the strength of a pole line or to provide better support for attachments than would be provided by a single pole. Examples are A fixtures, H fixtures, etc. *See also:* open-wire pole line. (EEC/PE) [119]

pole guy A tension member having one end securely anchored and the other end attached to a pole or other structure that it supports against overturning. *See also:* tower. (T&D/PE) [10]

pole line A series of poles arranged to support conductors above the surface of the ground, and the structures and conductors supported thereon. *See also:* open-wire pole line. (EEC/PE) [119]

pole, offset marker *See:* offset marker pole.

pole piece A piece or an assembly of pieces of ferromagnetic material forming one end of a magnet and so shaped as to appreciably control the distribution of the magnetic flux in the adjacent medium. (Std100) 270-1966w

pole pitch (rotating machinery) The peripheral distance between corresponding points on two consecutive poles; also expressed as a number of slot positions. *See also:* stator; armature; rotor. (PE) [9]

pole, plumb marker *See:* plumb marker pole.

pole shoe The portion of a field pole facing the armature that serves to shape the air gap and control its reluctance. *Note:* For round-rotor fields, the effective pole shoe includes the teeth that hold the field coils and wedges in place. *See also:* stator; rotor; field pole. (PE) [9]

pole slipping (rotating machinery) The process of the secondary member of a synchronous machine slipping one pole pitch with respect to the primary magnetic flux. (PE) [9]

pole steps Devices attached to the side of a pole, conveniently spaced to provide a means for climbing the pole. *See also:* tower. (T&D/PE) [10]

pole strap *See:* positioning strap.

pole tip (rotating machinery) The leading or trailing extremity of the pole shoe. *See also:* stator; rotor. (PE) [9]

pole-top cover *See:* insulator cover.

pole-type regulator A regulator that is designed for mounting on a pole or similar structure. (PE/TR) C57.15-1999

pole-type transformer (power and distribution transformers) A transformer which is suitable for mounting on a pole or similar structure. (PE/TR) C57.12.80-1978r

pole-unit mechanism (of a switching device) That part of the mechanism that actuates the moving contacts of one pole. (SWG/PE) C37.100-1992

pole-zero cancellation (1) The pole-zero adjustment on the shaping amplifier adjusts the zero location of the pole-zero network to cancel exactly the preamplifier output pole and thus provide single-pole (i.e., no under or overshoot) response of the signal pulse at the amplifier output. This operation converts the long-tailed preamplifier pulse to a short-tailed pulse suitable for signal optimization and subsequent pulse-height analysis. Proper pole-zero cancellation is an absolute necessity to prevent spectral degradation at moderate (2000 s^{-1}) rates. (NI) N42.14-1991
(2) A technique used to cancel out the effects of a singularity in an amplifier's transfer function in order to effect a mono-

tonic return of signal pulses to the baseline. (NPS/NID) 759-1984r

policies (safety systems equipment in nuclear power generating stations) Management directives that describe the organization, principles, plans, or courses of action. (PE/NP) 600-1983w

poling (1) The process by which a direct-current electric field exceeding the coercive field is applied to a multidomain ferroelectric to produce a net remanent polarization. *See also:* remanent polarization; ferroelectric domain; polarization. (UFFC) [21], 180-1986w
(2) (general) The adjustment of polarity. Specifically, in wire line practice, it signifies the use of transpositions between transposition sections of open wire or between lengths of cable to cause the residual crosstalk couplings in individual sections or lengths to oppose one another. *See also:* open-wire pole line. (EEC/PE) [119]

polishing (electroplating) The smoothing of a metal surface by means of abrasive particles attached by adhesive to the surface of wheels or belts. *See also:* electroplating. (EEC/PE) [119]

Polish notation *See:* prefix notation.

poll (data transmission) A flexible, systematic method, centrally controlled for permitting stations on a multipoint circuit to transmit without contending for the line. (PE) 599-1985w

poll/final bit A bit set by the BCC to "1" to solicit a response from a DCC. The P/F bit is set to "1" by the DCC to indicate that the final frame of the message has been transmitted. (EMB/MIB) 1073.3.1-1994

poll function check Accomplished when analog function is performed with all remotes. A check of master and remote station equipment by exercising a predefined component or capability. *See also:* analog function check. (SUB/PE) C37.1-1994

polling (1) (supervisory control, data acquisition, and automatic control) (data request) The process by which a data acquisition system selectively requests data from one or more of its remote terminals. A remote terminal may be requested to respond with all, or a selected portion of, the data available. (SWG/PE/SUB) C37.100-1992, C37.1-1994
(2) A technique for sharing a multiple-access transmission medium, where devices cannot transmit until they have received implicit or explicit permission. *See also:* centralized polling; distributed polling. (C) 610.7-1995
(3) A scheduling scheme whereby the local process periodically checks until the prespecified events (*e.g.*, read, write) have occurred. (C) 1003.5-1999
(4) (supervisory control, data acquisition, and automatic control) (data request) *See also:* supervisory control.

polling supervisory system (station control and data acquisition) A system in which the master interrogates each remote to ascertain if there has been a change since the last interrogation. Upon detection of a change the master may request data immediately. (PE/SUB) C37.100-1992, C37.1-1994

polychlorinated biphenyl (liquid-filled power transformers) Any chemical substance that is limited to the biphenyl molecule that has been chlorinated to varying degrees or any combination of substances that contains such substance ≥50 parts per million (ppm) dry weight basis. Examples: PCB liquids and nonliquids. (LM/C) 802.2-1985s

polychlorinated biphenyl article (liquid-filled power transformers) Any manufactured article, other than a PCB container that contains PCBs and whose surface(s) has been in direct contact with PCBs. Examples: PCB large high- and low-voltage capacitors; PCB transformer; PCB cooler motor. (LM/C) 802.2-1985s

polychlorinated biphenyl article container (liquid-filled power transformers) Any package, can bottle, bag, barrel, drum, tank, or other device used to contain PCB articles or PCB equipment, and whose surface(s) has not been in direct contact with PCBs. Examples: Shipping or storage cartons for capacitors. (LM/C) 802.2-1985s

polychlorinated biphenyl container (liquid-filled power transformers) Any package, can, bottle, bag, barrel, drum, tank, or other device that contains PCBs or PCB articles and whose surface(s) has been in direct contact with PCBs. Examples: Bottle, barrel, drum, or box. (LM/C) 802.2-1985s

polychlorinated biphenyl contaminated electrical equipment (liquid-filled power transformers) Any electrical equipment, including but not limited to transformers (including those used in railway locomotives and self-propelled cars), capacitors, circuit breakers, reclosers, voltage regulators, switches (including sectionalizers and motor starters), bushings, electromagnets, and cable, that contain 50 parts per million (ppm) or greater PCBs but less than 500 ppm PCB. Oil-filled electrical equipment other than circuit breakers, reclosers, and cable whose PCB concentration is unknown must be assumed to be PCB-contaminated electrical equipment. Examples: Some oil-filled units; some retrofilled units. (LM/C) 802.2-1985s

polychlorinated biphenyl equipment (liquid-filled power transformers) Any manufactured item, other than a PCB container, that contains a PCB article or other PCB equipment. Examples: Microwave oven, power-factor-corrected lighting ballast. (LM/C) 802.2-1985s

polychlorinated biphenyl item (liquid-filled power transformers) Any PCB article, PCB article container, PCB container, or PCB equipment that deliberately or unintentionally contains or has a part of it any PCB or PCBs at a concentration of 50 parts per million (ppm) or greater. Examples: PCB askarel contaminated transformer (mineral) oil, or coolants retrofilled to transformer formerly cooled with askarel. (LM/C) 802.2-1985s

polychlorinated biphenyl storage for disposal (liquid-filled power transformers) The facilities meet the following criteria:

a) Adequate roof and walls to prevent rain water from reaching the stored PCBs and PCB items.
b) An adequate floor that has continuous curbing with a minimum 6 inches high curb. The floor and curbing provide a containment volume equal to at least two times the internal volume of the largest PCB article or PCB container stored therein or 25% of the total internal volume of all PCB articles or PCB containers stored therein, whichever is greater.
c) No drain valves, floor drains, expansion joints, sewer lines, or other openings that would permit liquids to flow from the curbed area.
d) Floors and curbing constructed of continuous smooth and impervious materials, such as Portland cement, concrete, or steel, to prevent or minimize penetration of PCBs.
e) Not located at a site that is below the 100-year flood water elevation.

(LM/C) 802.2-1985s

polychlorinated biphenyl transformer Any transformer that contains 500 parts per million (ppm) PCB or greater. Examples: PCB askarel-insulated units; some oil-filled units; some retrofilled units. (LM/C) 802.2-1985s

polycrystalline silicon A silicon layer that contains a multiplicity of crystals. This is the common form of silicon that is employed for transistor gates and interconnections on the surface of dielectric layers in MOS circuits. (ED) 1005-1998

Polyforth A dialect of FORTH. (C) 610.13-1993w

polygon A display element that consists of an area enclosed by a sequence of straight lines. (C) 610.6-1991w

polygon fill *See:* fill.

polyline A display element that consists of a set of connected lines.

polyline

(C) 610.6-1991w

polyline attribute A characteristic of the line segments that make up a polyline. For example, color index, line type, line width. (C) 610.6-1991w

polymarker A display element that consists of a set of locations, each of which is indicated by a marker.

polymarker

(C) 610.6-1991w

polymarker attribute A characteristic of the markers that make up a polymarker. For example, color index, marker size, marker type. (C) 610.6-1991w

Polymorphic Programming Language An interactive, extensible language containing facilities for defining new data types and operators. (C) 610.13-1993w

polyphase (as applied to a relay) A descriptive term indicating that the relay is responsive to polyphase alternating electrical input quantities. *Note:* A multiple-unit relay with individual units responsive to single-phase electrical inputs is not a polyphase relay even though the several single-phase units constitute a polyphase set. (SWG/PE) C37.100-1992

polyphase ac fields (1) Fields whose space components may not be in phase. These fields will be produced by polyphase power lines. The field at any point can be described by the field ellipse—that is, by the magnitude and direction of the semimajor axis and the magnitude and direction of its semiminor axis. *Note:* Such fields are sometimes referred to as being elliptically polarized. Certain power line geometries can produce circularly polarized fields. For polyphase power lines, the electric field at large distances (≥15 m) away from the outer phases (conductors) can frequently be considered a single-phase field because the minor axis of the electric field ellipse is only a fraction (<10%) of the major axis when measured at a height of 1 m above ground level. *See also:* electric field strength. (T&D/PE) 644-1994
(2) Fields whose space components may not be in time phase with each other. These fields will be the transversal fields produced by polyphase power lines. The field at any point can be described by the field ellipse; i.e., the magnitude and direction of the major semi-axis and the magnitude and direction of the minor semi-axis. The magnitude of the field strength is the magnitude of the major semi-axis. *Note:* For polyphase power lines, the electric field at a distance of 15 m or more away from the outer phases (conductors) can frequently be considered a single-phase field because the minor axis of the electric-field ellipse is only a fraction (less than 10%) of the major axis when measured at a height of 1 m. Similar remarks apply for the magnetic field. *See also:* ac electric field strength. (T&D/PE) 539-1990

polyphase circuit An alternating-current circuit consisting of more than two intentionally interrelated conductors that enter (or leave) a delimited region at more than two terminals of entry and that are intended to be so energized that in the steady state the alternating voltages between successive pairs of terminals of entry of the phase conductors, selected in a systematic chosen sequence, have:

a) the same period,
b) definitely related and usually equal amplitudes, and
c) definite and usually equal phase differences. If a neutral conductor exists, it is intended also that the voltages from the successive phase conductors to the neutral conductor be equal in amplitude and equally displaced in phase.

Note: For all polyphase circuits in common use except the two-phase three-wire circuit, it is intended that the voltage amplitudes and the phase differences of the systematically chosen voltages between phase conductors be equal. For a two-phase three-wire circuit it is intended that voltages between two successive pairs of terminals be equal and have a phase difference of p.2 radians, but that the voltage between the third pair of terminals have an amplitude $(2)^{1.2}$ times as great as the other two, and a phase difference from each of the other two of 3 p.4 radians. *See also:* zig-zag connection of polyphase circuits. (Std100) 270-1966w

polyphase code A pulse compression waveform in which a long pulse is subdivided into many subpulses and the phase of each subpulse is chosen with a quantization less than π radians. The Frank polyphase code is an example in which the phases are selected so as to obtain a discrete version of the continuous analog linear frequency-modulation waveform. (AES) 686-1997

polyphase machine (rotating machinery) A machine that generates or utilizes polyphase alternating-current power. These are usually three-phase machines with three voltages displaced 120 electrical degrees with respect to each other. *See also:* asynchronous machine. (PE) [9]

polyphase merge sort An unbalanced merge sort in which the distribution of the sorted subsets is based on a polynomial series such as the Fibonacci series. *See also:* cascade merge sort. (C) 610.5-1990w

polyphase symmetrical set (1) (polyphase voltages) A symmetrical set of polyphase voltages in which the angular phase difference between successive members of the set is not zero, π radians, or a multiple thereof. The equations of symmetrical set of polyphase voltages represent a polyphase symmetrical set of polyphase voltages if k/m is not zero, 1/2, or a multiple thereof. (The symmetrical set of voltages represented by the equations of symmetrical set of polyphase voltages may be said to have polyphase symmetry if k/m is not zero, 1/2, or a multiple of 1/2). *Note:* This definition may be applied to a two-phase four-wire or five-wire circuit if m is considered to be 4 instead of 2. It is not applicable to a two-phase three-wire circuit. (Std100) 270-1966w
(2) (polyphase currents) This definition is obtained from the corresponding definitions for voltage by substituting the word current for voltage, the symbol I for E, and β for α wherever they appear. The subscripts are unaltered. (Std100) 270-1966w

polyphase synchronous generator A generator whose field circuits are arranged so that two or more symmetrical alternating electromotive forces with definite phase relationships are produced at the terminals. Polyphase synchronous generators are usually two-phase, producing two electromotive forces displaced 90 electrical degrees apart, or three-phase, producing three electromotive forces displaced 120 electrical degrees apart. (Polyphase generators used for marine services are generally three phase.) (IA/MT) 45-1998

polyplastic A synonym for polyethylene-coated, nylon-reinforced hose, usually considered to be nonconductive. (In terms of this guide, the hose is used to carry water.) (T&D/PE) 957-1995

polyplexer Equipment combining the functions of duplexing and lobe switching. (AES/RS) 686-1990

poly-sol Plastic additive used in some washing applications to break down surface adhesion. (T&D/PE) 957-1987s

polyvinyl chloride An insulator in cable coatings and coaxial cable foam compositions. (C) 610.7-1995

pondage (power operations) Hydroreserve and limited storage capacity that provides only daily or weekly regulation of streamflow. (PE/PSE) 858-1987s

pondage station A hydroelectric generating station with storage sufficient only for daily or weekend regulation of flow. *See also:* generating station. (T&D/PE) [10]

p-on-n solar cells (photovoltaic power system) Photovoltaic energy-conversion cells in which a base of n-type silicon (having fixed positive holes in a silicon lattice and electrons that are free to move) is overlaid with a surface layer of p-type silicon (having fixed electrons in a silicon lattice and positive holes that are free to move). (AES) [41]

pool cathode A cathode at which the principal source of electron emission is a cathode spot on a metallic pool electrode. (ED) [45]

pool-cathode mercury-arc converter A frequency converter using a mercury-arc pool-type discharge device. (IA) 54-1955w, 169-1955w

pool rectifier A gas-filled rectifier with a pool cathode, usually mercury. (ED) [45], [84]

pool tube A gas tube with a pool cathode. *See also:* electronic controller. (ED) [45]

POP *See:* point of presence.

pop *See:* pull.

populate *See:* load.

population (1) (data management) The number of records in a file or database. (C) 610.5-1990w
(2) (utility power systems) Transformers that have given common specific characteristics. (PE/TR) C57.117-1986r

population, conceptual *See:* conceptual population.

population inversion (laser maser) A nonequilibrium condition of a system of weakly interacting particles (electronics, atoms, molecules, or ions) which exists when more than one-half of the particles occupy the higher of two energy states. (LEO) 586-1980w

pop-up menu (1) A menu that is brought into view as a result of a selection action other than choosing a menu-bar label. *Contrast:* pull-down menu. (PE/NP) 1289-1998
(2) A menu that appears outside of menu bar when requested, usually as the result of pressing BMenu or KMenu. (C) 1295-1993w

pores (electroplating) Micro discontinuities in a metal coating that extend through to the base metal or underlying coating. *See also:* electroplating. (LEO) 586-1980w

Port A Port Object. Context may indicate that the Port Object is of a specific class. (IM/ST) 1451.1-1999

port (1) (electronic devices or networks) A place of access to a device or network where energy may be supplied or withdrawn or where the device or network variables may be observed or measured. *Notes:* 1. In any particular case, the ports are determined by the way the device is used and not by its structure alone. 2. The terminal pair is a special case of a port. 3. In the case of a waveguide or transmission line, a port is characterized by a specified mode of propagation and a specified reference plane. 4. At each place of access, a separate port is assigned to each significant independent mode of propagation. 5. In frequency changing systems, a separate port is also assigned to each significant independent frequency response. *See also:* network analysis; optoelectronic device; waveguide. (ED/IM/HFIM) [46], [45], [40]
(2) (rotating machinery) An opening for the intake or discharge of ventilating air. (PE) [9]
(3) (rotating machinery) (for a waveguide component) A means of access characterized by a specified reference plane and a specified propagating mode in a waveguide which permits power to be coupled into or out of a waveguide component. *Note:* At low frequencies the port is synonymous with a terminal pair. 2. To each propagating mode at a specified reference plane there corresponds a distinct port. (MTT) 146-1980w
(4) (broadband local area networks) An electrical interface that has defined operating boundaries. The specific references within IEEE Std 802.7-1989 assume ports to be 75 Ω transmission line interfaces that have an associated connector to which the signals pass. (LM/C) 802.7-1989r
(5) A source or destination of data transferred by a Data Transfer class command into and/or out of an S-module. A port may be an on-module memory, on-module interface, a peripheral attached to a module, or some other mechanism to/from which data is passed.Within IEEE Std 1149.5-1995, a port is defined by a module address, a port ID meaningful to the MTM-Bus interface logic of that module, and the semantics and structure of packets by which data can be conveyed to and/or from that port. This latter often entails some description of the application to/from which data are passed. A port is selected/accessed/addressed via a Data Transfer class command. (TT/C) 1149.5-1995
(6) The physical interconnection point or an access point for a communication link. (C) 610.7-1995
(7) An input or output connection between a peripheral device and a computer. *See also:* parallel port; serial port; mouse port; input-output port. (C) 610.10-1994w
(8) A physical layer entity in a node that connects to either a cable or backplane and provides one end of a physical connection with another node. (C/MM) 1394-1995
(9) A signal interface provided by token ring stations, passive concentrator lobes, active concentrator lobes, or concentrator trunks that is generally terminated at a media interface connector (MIC). Ports may or may not provide physical containment of channels. *See also:* Bridge Port. (C/LM/C/LM) 802.1G-1996, 8802-5-1998
(10) An interface point connecting a communications channel and a device. (PE/SUB) 1379-1997
(11) A segment or Inter-Repeater Link (IRL) interface of a repeater unit. (C/LM) 802.3-1998
(12) A conceptual point at which a cell or a hierarchical design unit makes its interface available to higher levels in the design hierarchy. (C/DA) 1481-1999
(13) An abstraction of an access point to network communications. (IM/ST) 1451.1-1999
(14) The part of the physical layer (PHY) that allows connection to one other node. (C/MM) 1394a-2000
(15) A physical entity that allows import or export of one or more cartridges from a library. (C/SS) 1244.1-2000
(16) *See also:* link interface. (C/BA) 1355-1995
(17) *See also:* Bridge Port. (C/LM) 802.1G-1996

portability (1) (software) The ease with which a system or component can be transferred from one hardware or software environment to another. *Synonym:* transportability. *See also:* machine independent. (C) 610.12-1990
(2) (application software) The ease with which application software and data can be transferred from one application platform to another. (C/PA) 14252-1996
(3) The capability of being moved between differing environments without losing the ability to be applied or processed. (ATLAS) 1232-1995
(4) The capability of being read and/or interpreted by multiple systems. (SCC20) 1232.1-1997
(5) The ease with which software can be transferred from one system or environment to another. A relative measure of effort, inversely proportional to the level of modification required for software to be transferred from one system or environment to another. (SCC20) 1226-1998

portable (x-ray) X-ray equipment designed to be hand-carried. (NEC/NESC) [86]

portable appliance An appliance which is actually moved or can easily be moved from one place to another in normal use. For the purpose of this article, the following major appliances other than built-in are considered portable if cord-connected; refrigerators, gas range equipment, clothes washers, dishwashers without booster heaters, or other similar appliances. *See also:* appliance. (NESC/NEC) [86]

portable battery A storage battery designed for convenient transportation. *See also:* battery. (EEC/PE) [119]

portable character set The set of characters described in 2.4 that is supported on all conforming systems. This term is contrasted against the smaller *portable filename character set.* (C/PA) 9945-2-1993

portable character string A sequence of characters from the portable character set. Within software definition files of exported catalogs, all such strings shall be encoded using IRV. (C/PA) 1387.2-1995

portable computer A personal computer that is designed and configured to permit transportation as a piece of handheld luggage. *Note:* U.S. Federal regulations limit use of the term "portable" to objects weighing no more than 21 pounds. *See also:* notebook computer; transportable computer; hand-held computer; laptop computer. (C) 610.2-1987, 610.10-1994w

portable concentric mine cable A double-conductor cable with one conductor located at the center and with the other conductor strands located concentric to the center conductor with rubber or synthetic insulation between conductors and over the outer conductor. *See also:* mine feeder circuit. (EEC/PE) [119]

portable filename character set The set of characters from which portable filenames are constructed. For a filename to be portable across conforming implementations it shall consist only of the following characters:

A B C D E F G H I J K L M N O P Q R S T U V W X Y Z

a b c d e f g h i j k l m n o p q r s t u v w x y z

0 1 2 3 4 5 6 7 8 9 . _ -

The last three characters are the period, underscore, and hyphen characters, respectively. The hyphen shall not be used as the first character of a portable filename. Upper- and lowercase letters shall retain their unique identities between conforming implementations. In the case of a portable pathname, the slash character may also be used. (C/PA) 9945-1-1996, 9945-2-1993, 1003.5-1999

portable identifier character set The set of characters from which portable identifiers are constructed. This set shall consist only of the following characters:

A B C D E F G H I J K L M N O P Q R S T U V W X Y Z

a b c d e f g h i j k l m n o p q r s t u v w x y z

0 1 2 3 4 5 6 7 8 9 . _ ( ) -

(C/PA) 2003-1997

portable lighting (illuminating engineering) Lighting involving equipment designed for manual portability. (EEC/IE) [126]

portable luminaire (illuminating engineering) A lighting unit which is not permanently fixed in place. (EEC/IE) [126]

portable mine blower A motor-driven blower to provide secondary ventilation into spaces inadequately ventilated by the main ventilating system and with the air directed to such spaces through a duct. (EEC/PE) [119]

portable mine cable An extra-flexible cable, used for connecting mobile or stationary equipment in mines to a source of electric energy when permanent wiring is prohibited or impracticable. *See also:* mine feeder circuit. (EEC/PE) [119]

portable mining-type rectifier transformer A rectifier transformer that is suitable for transporting on skids or wheels in the restrictive areas of mines. *See also:* rectifier transformer. (Std100) C57.18-1964w

portable parallel duplex mine cable A double or triple-conductor cable with conductors laid side by side without twisting, with rubber or synthetic insulation between conductors and around the whole. The third conductor, when present, is a safety ground wire. *See also:* mine feeder circuit. (EEC/PE) [119]

portable pathname character set The set of characters from which portable pathnames are constructed. The set contains all the characters of the portable filename set, plus the character slash (/). (C) 1003.5-1999

portable platforms A platform temporarily installed on a pole or tower. The platforms are available in various lengths and materials. They may be fixed or may pivot. The platform may have an anchorage point for a positioning strap. (T&D/PE) 1307-1996

portable shunt (direct-current instrument shunts) An instrument shunt with insulating base which may be laid on, or fastened to, any flat surface. It may be used also for switchboard applications where the current is relatively low and connection bars are not used. (PE/PSIM) 316-1971w

portable standard watthour meter A portable meter, principally used as a standard for testing other meters. It is usually provided with several current and voltage ranges and with a readout indicating revolutions and fractions of a revolution of the rotor. *Note:* Electronic portable standards not using a rotor may have a readout indicating equivalent revolutions and fractions of revolutions, or other units such as percentage registration. (ELM) C12.1-1982s

portable station (1) (mobile communication) A mobile station designed to be carried by or on a person. Personal or pocket stations are special classes of portable stations. *See also:* mobile communication system. (VT) [37]
(2) A type of station that may be moved from location to location, but that only uses network communications while at a fixed location. (C/LM) 8802-11-1999

portable traffic control light (illuminating engineering) A signalling light designed for manual portability that produces a controllable distinctive signal for purposes of directing aircraft operations in the vicinity of an aerodrome. (EEC/IE) [126]

portable transmitter A transmitter that can be carried on a person and may or may not be operated while in motion. *Notes:* 1. This has been called a transportable transmitter, but the designation portable is preferred. 2. This includes the class of so-called walkie-talkies, handy-talkies, and personal transmitters. *See also:* radio transmission; transportable transmitter; radio transmitter. (AP/BT/ANT) 145-1983s, 182A-1964w

portable X- or gamma-radiation survey instrument (radiation survey instruments) An instrument with a self-contained energy source (for example, batteries) designed to measure exposure rate while being carried. Such instruments may also have the capability to measure integral exposure, but instruments with the capability of measuring integral exposure only are specifically are specifically excluded from this definition. (NI) N13.4-1971w

portal The logical point at which medium access control (MAC) service data units (MSDUs) from a non-IEEE 802.11 local area network (LAN) enter the distribution system (DS) of an extended service set (ESS). (C/LM) 8802-11-1999

port difference (hybrid) A port that yields an output proportional to the difference of the electric field quantities existing at two other ports of the hybrid. *See also:* waveguide. (IM/HFIM) [40]

PORT ID packet The first DATA packet transferred in a Data Transfer class message. This packet contains the identifier (Port Identifier) by which means a port is selected for the remainder of the message. (TT/C) 1149.5-1995

Port Object Any Object whose class is `IEEE1451_SubscriberPort` or a subclass thereof or a subclass of `IEEE1451_BasePort`. (IM/ST) 1451.1-1999

port protection system A computer security mechanism used to protect dial-up communication lines from unauthorized use, often requiring special passwords or using call-back procedures. *Synonym:* secure modem. (C) 610.7-1995

portrait image *See:* cine-oriented image.

portrait orientation A page orientation of a display surface having greater height than width. *Note:* Derived from portraits of people, which are usually vertical in format. *Contrast:* landscape orientation. (C) 610.10-1994w

port signal (data transmission) The signal used to telemeter the real time occurrence of polarity reversals of a power voltage or current. The signal may be a pulse train, a square voltage wave, a frequency shift keying (FSK) tone or an FSK carrier wave. Use is generally for frequency or phase-angle telemetering. (PE) 599-1985w

port sum (hybrid) A port that yields an output proportional to the sum of the electric-field quantities existing at two other ports of the hybrid. *See also:* waveguide. (IM/HFIM) [40]

port-to-port time The elapsed time between the application of a stimulus to an input interface and the appearance of the response at an output interface. *See also:* think time; response time; turnaround time. (C) 610.12-1990

Port Transfer Error A port-specific error indicating some failure with relation to transmission of command or data to/from a currently selected port. (TT/C) 1149.5-1995

Port Transfer Error (PTE) bit A bit in the Bus Error register of all S-modules. An S-module sets this bit to indicate that the port selected by the command in the current message has reported a Data Transfer Port Error. Such errors will be found defined in the port documentation of specific S-modules. The acronym stands for "Port Transfer Error". (TT/C) 1149.5-1995

POSH *See:* permutation on subject headings; permutation on subject headings index.

position (1) (FASTBUS acquisition and control) The location of a module in a crate. The position number corresponds to the geographical address. (NID) 960-1993
(2) (navigation) (navigation aids) The location of a point with respect to a specified or implied coordinate system. (AES/GCS) 172-1983w
(3) (A) (within a string) The ordinal position of one element of a string relative to another. **(B) (within an attribute)** The ordinal position of one value relative to another. (C/PA) 1328-1993, 1224-1993, 1327-1993
(4) (navigation aids) (navigation) *See also:* digit place. (C) 1084-1986w
(5) (navigation aids) (navigation) *See also:* punch position; sign position.

positional crosstalk (multibeam cathode-ray tubes) The variation in the path followed by an one electron beam as the result of a change impressed on any other beam in the tube. (ED) 161-1971w

positional notation (A) A number representation that makes use of an ordered set of digits, such that the value contributed by each digit depends on its position as well as on the digit value. *Note:* The Roman numeral system for example, does not use positional notation. *Synonym:* positional representation. *See also:* binary system; binary numeration system; decimal system; decimal numeration system; biquinary numeration system; Gray code. **(B)** One of the schemes for representing numbers, characterized by the arrangement of digits in sequence, with the understanding that successive digits are to be interpreted as coefficients of successive powers of an integer called the base (or radix) of the number system. *Notes:* 1. In the binary number system the successive digits are interpreted as coefficients of the successive powers of the base two, just as in the decimal number system they relate to successive powers of the base ten. 2. In the ordinary number systems each digit is a character that stands for zero or for a positive integer smaller than the base. 3. The names of the number systems with bases from 2 to 20 are: binary, ternary, quaternary, quinary, senary, septenary, octonary (also octal), novenary, decimal, unidecimal, duodecimal, terdenary, quaterdenary, quindenary, sexadecimal (also hexadecimal), septendecimal, octodenary, novendenary, and vicenary. The sexagenary number system has the base 60. The commonly used alternative of saying base-3, base-4, etc., in place of ternary, quaternary, etc., has the advantage of uniformity and clarity. 4. In the most common form of positional notation, the expression $\pm\ a_n a_{n-1} \ldots a_2 a_1 a_0 \cdot a_{-1} a_{-2} \ldots a_{-m}$ is an abbreviation for the sum

$$\pm \sum_{i=-m}^{n} a_i r^i$$

where the point separates the positive powers from the negative powers, the a_i are integers ($0 \le a_i < r$) called digits, and r is an integer, greater than one, called the base (or radix). *See also:* radix; base. **(C)** A number-representation system having the property that each number is represented by a sequence of characters such that successive characters of the sequence represent integral coefficients of accumulated products of a sequence of integers (or reciprocals of integers) and such that the sum of these products, each multiplied by its coefficient, equals the number. Each occurrence of a given character represents the same coefficient value. *Note:* The biquinary system is an example of (C). **(D)** A number-representation system such that if the representations are arranged vertically in order of magnitude with digits of like significance in the same column, then each column of digits consists of recurring identical cycles (for numbers sufficiently large in absolute value) whose length is an integral multiple of the cycle length in the column containing the next-less-significant digits. *Note:* (B), (C), and (D) are not mutually exclusive. The biquinary system is an example of (C) and (D); whereas the Gray code system is an example of (D) only. The binary and decimal systems are examples of (B), (C), and (D). (C) 162-1963, [20], 270-1966

positional parameter In the shell command language, a parameter denoted by a single digit or one or more digits in curly braces. (C/PA) 9945-2-1993

positional representation *See:* positional notation.

positional response (close-talking pressure-type microphone) The response-frequency measurements conducted with the principal axis of a microphone collinear with the axis of the artificial voice and the combination of microphone and artificial voice placed at various angles to the horizontal plane. *Note:* Variations in positional response of carbon microphones may be due to gravitational forces. (SP) 258-1965w

positional servomechanism (1) In an analog computer, a servomechanism in which a mechanical shaft is positioned, usually in the angle of rotation, in accordance with one or more input signals. *Note:* Frequently, the shaft is positioned (excluding transient motion) in a manner linearly related to the value of the input signal. *See also:* repeater servomechanism; servomechanism. (C) 610.10-1994w
(2) A servomechanism in which a mechanical shaft is positioned, usually in the angle of rotation, in accordance with one or more input signals. *Note:* Frequently, the shaft is positioned (excluding transient motion) in a manner linearly related to the value of the input signal. However, the term also applies to any servomechanism in which a loop input signal generated by a transmitting transducer can be compared to a loop feedback signal generated by a compatible or identical receiving transducer to produce a loop error signal that, when reduced to zero by movement of the receiving transducer, results in a shaft position related in a prescribed and repeatable manner to the position of the transmitting transducer. *See also:* repeater servomechanism; electronic analog computer. (C) 165-1977w

position changing mechanism (power system device function numbers) A mechanism that is used for moving a main device from one position to another in an equipment; as, for example, shifting a removable circuit breaker unit to and from the connected, disconnected, and test positions. (SUB/PE) C37.2-1979s

position-control system A control system that attempts to establish and.or maintain an exact correspondence between the reference input and the directly controlled variable, namely physical position. *See also:* feedback control system. (IA/ICTL/IAC) [60]

position index, P (illuminating engineering) A factor which represents the relative average luminance for a sensation at the borderline between comfort and discomfort (BCD), for a source located anywhere within the visual field. (EEC/IE) [126]

position indicating device (designator) A mechanical device that indicates, at the location the switch-operating mechanism, whether the contacts of the switch are in the open or closed position. (SWG/PE) C37.30-1992s

position indicator (elevators) A device that indicates the position of the elevator car in the hoistway. It is called a hall position indicator when placed at a landing or a car position indicator when placed in the car. *See also:* control. (PE/EEC) [119]

position influence (electric instruments) The change in the indication of an instrument that is caused solely by a position departure from the normal operating position. *Note:* Unless otherwise specified, the maximum change in the recorded value caused solely by an inclination in the most unfavorable direction from the normal operating position. *See also:* accuracy rating. (EEC/ERI/EMI/AII) [111], [112], [102]

positioning control system (1) (automatic control) A control system in which there is a predetermined relation between the actuating signal and the position of a final controlling element. *Note:* In a "proportional-position control system" there is a continuous linear relation between the value of the actuating signal and the position of a final controlling element. (PE/EDPG) [3]

(2) (numerically controlled machines) A system in which the controlled motion is required only to reach a given end point, with no path control during the transition from one end point to the next. (IA) [61]

positioning device system A system of equipment or hardware that, when used with its line-worker's body belt or full body harness, allows a worker to be supported on an elevated vertical surface, such as a pole or tower, and work with both hands free. (NESC) C2-1997

positioning strap A strap with snaphook(s) to connect to the D-rings of a line-worker's body belt or full body harness. Used as a positioning device (also known as pole strap or safety strap). (NESC/T&D/PE) C2-1997, 1307-1996

positioning time *See:* seek time.

position lights (illuminating engineering) The aircraft aeronautical lights which form the basic or internationally recognized navigation light system. *Note:* The system is composed of a red light showing from dead ahead to 110 degrees to the left, a green light showing from dead ahead to 110 degrees to the right, and a white light showing to the rear through 140 degrees. (EEC/IE) [126]

position light signal A fixed signal in which the indications are given by the position of two or more lights. (EEC/PE) [119]

position of the effective short (microwave switching tubes) The distance between a specified reference plane and the apparent short-circuit of the fired tube in its mount. *See also:* gas tube. (ED) 161-1971w

position readout (numerically controlled machines) Display of absolute position as derived from a position transducer. (IA) [61]

position seating A control scheme that uses the limit switch as the primary control for operation of a VAM. The limit switch controls the VAM by interrupting power to the motor contactor when the valve actuator has completed a predetermined number of revolutions. (PE/NP) 1290-1996

position-sensitive detector (1) A detector in which the centroid of the area of impact of ionizing radiation (at the surface of the detector) can be measured in one or two dimensions. (NPS) 300-1988r

(2) A detector in which the centroid of the area of impact of ionizing radiation can be determined from the signals issuing from its terminals. Depending on the design, position sensing can be in one or more dimensions. (NPS) 325-1996

position sensor or position transducer (numerically controlled machines) A device for measuring a position and converting this measurement into a form convenient for transmission. (IA) [61]

position stopping A control function that provides for stopping the driven equipment at a preselected position. *See also:* electric controller. (IA/ICTL/IAC) [60]

position switch (power system device function numbers) A switch that makes or breaks contact when the main device or piece of apparatus which has no device function number reaches a given position. (SUB/PE) C37.2-1979s

position-type telemeter *See:* ratio-type telemeter.

positive Pertaining to a voltage or charge that is associated with a deficiency of electrons. *Contrast:* negative. (C) 610.10-1994w

positive after-potential (electrobiology) Relatively prolonged positivity that follows the negative after-potential. *See also:* contact potential. (EMB) [47]

positive column (gas tube) The luminous glow, often striated, in a glow-discharge cold-cathode tube between the Faraday dark space and the anode. *See also:* gas tube. (ED) [45]

positive conductor A conductor connected to the positive terminal of a source of supply. *See also:* center of distribution. (T&D/PE) [10]

positive creep effect (semiconductor rectifiers) The gradual increase in reverse current with time, that may occur when a direct-current reverse voltage is applied to a semiconductor rectifier cell. *See also:* rectification. (IA) [12]

positive electrode (A) (primary cell) The cathode when the cell is discharging. The positive terminal is connected to the positive electrode. *See also:* electrolytic cell. **(B) (metallic rectifier)** The electrode to which the forward current flows within the metallic rectifying cell. *See also:* rectification. (PE/EEC) [119]

positive feedback (regeneration) (data transmission) The process by which a part of the power in the output circuit of an amplifying device reacts upon the input circuit in such a manner as to reinforce the initial power, thereby increasing the amplification. (PE) 599-1985w

positive glow A bright blue discharge appearing as a luminous sheet adhering closely and uniformly to the electrode. Positive glow appears at electric field strengths above those required for burst corona and onset streamers. The corona current of positive glow is essentially pulseless. *See also:* burst corona. (T&D/PE) 539-1990

positive grid *See:* retarding-field (positive-grid) oscillator.

positive-grid oscillator tube (Barkhausen tube) A triode operating under oscillating conditions such that the quiescent voltage of the grid is more positive than that of either of the other electrodes. (ED) [45]

positive logic An electronic logic system where the voltage representing one, active, or true has a more positive value than the voltage representing zero, inactive, or false. It is normally used in industrial and commercial control switching systems for safety reasons. (IM/ST) 1451.2-1997

positive logic convention The representation of the 1-state and the 0-state by the high (H) and low (L) levels, respectively. (GSD) 91-1984r

positive matrix (positive) A matrix with a surface like that which is to be ultimately produced by electroforming. (PE/EEC) [119]

positive modulation (in an amplitude-modulation television system) That form of modulation in which an increase in brightness corresponds to an increase in transmitted power. *See also:* television. (EEC/PE) [119]

positive nonconducting period (rectifier element) The nonconducting part of an alternating-voltage cycle during which the anode has a positive potential with respect to the cathode. *See also:* power rectifier; rectification. (IA) [62]

positive noninterfering and successive fire-alarm system A manual fire-alarm system employing stations and circuits such that, in the event of simultaneous operation of several stations, one of the operated stations will take control of the circuit, transmit its full signal, and then release the circuit for successive transmission by other stations that are held inoperative until they gain circuit control. *See also:* protective signaling. (EEC/PE) [119]

positive onset streamers Streamers occurring at electric field strengths at and slightly above the corona inception voltage gradient. These appear as bright blue "brushes" increasing in length to several inches as the voltage gradient is increased. The associated current pulses are of appreciable magnitude, short duration (in the range of hundreds of nanoseconds), and low repetition rate (less than 1 kHz). *Note:* Occurrence of burst corona and positive onset streamers requires the same range of electric field strength. (T&D/PE) 539-1990

positive-phase-sequence reactance (rotating machinery) The quotient of the reactive fundamental component of the positive-sequence primary voltage due to the sinusoidal positive-sequence primary current of rated frequency, and the value of this current, the machine running at rated speed. *See also:* asynchronous machine. (PE) [9]

positive-phase-sequence relay A relay that responds to the positive-phase-sequence component of a polyphase input quantity. (SWG/PE/PSR) C37.100-1992, C37.90-1978s

positive-phase-sequence resistance (rotating machinery) The quotient of the in-phase component of positive-sequence primary voltage corresponding to direct load losses in the primary winding and stray load losses due to sinusoidal positive-sequence primary current, and the value of this current, the machine running at rated speed. (PE) [9]

positive-phase-sequence symmetrical components (of an unsymmetrical set of polyphase voltages or currents of m phases) The set of symmetrical components that have the first phase sequence. That is, the angular phase lag from the first member of the set to the second, from every other member of the set to the succeeding one, and from the last member to the first, is equal to the characteristic angular phase difference, or $2\pi/m$ radians. The members of this set will reach their positive maxima uniformly in their designated order. The positive-phase-sequence symmetrical components for a three-phase set of unbalanced sinusoidal voltages ($m = 3$), having the primitive period, are represented by the equations

$$e_{a1} = (2)^{1/2}E_{a1}\cos\,(\omega t + \alpha_{a1})$$

$$e_{b1} = (2)^{1/2}E_{a1}\cos\left(\omega t + \alpha_{a1} - \frac{2\pi}{3}\right)$$

$$e_{c1} = (2)^{1/2}E_{a1}\cos\left(\omega t + \alpha_{a1} - \frac{4\pi}{3}\right)$$

derived from the equation of symmetrical components of a set of polyphase (alternating) voltages. Since in this case $r = 1$ for every component (of 1st harmonic) the third subscript is omitted. Then k is 1 for 1st sequence and s takes on the algebraic values 1, 2, and 3 corresponding to phases a, b, and c. The sequence of maxima occurs in the order a, b, c. *See also:* network analysis. (Std100) 270-1966w

positive plate (storage cell) The grid and active material from which current flows to the external circuit when the battery is discharging. *See also:* battery. (EEC/PE) [119]

positive-polarity lightning stroke A stroke resulting from a positively charged cloud that lowers positive charge to the earth. *See also:* direct-stroke protection. (T&D/PE) [10]

positive prebreakdown streamers Streamers occurring at electric field strengths above those required for onset streamers and positive glow. The discharge appears as a light blue filament with branching extending far into the gap. The associated current pulses have high magnitude, short duration (in the range of hundreds of nanoseconds), and low repetition rate (in the range of a few kilohertz). *Note:* When appearing as multiple discharges, these streamers are usually referred to as a "plume." When the plume occurs between an electrode and an airborne particle (snow, rain, aerosols, etc.) coming into near proximity or impacting on the electrode, it is referred to as an "impingement plume." When the plume occurs due to the disintegration of water drops resting on the electrode surface, it is referred to as a "spray plume." (T&D/PE) 539-1990

positive-sequence impedance The quotient of that component of positive-sequence rated-frequency voltage, assumed to be sinusoidal, that is due to the positive-sequence component of current, divided by the positive-sequence component of current. *See also:* asynchronous machine. (PE) [9]

positive-sequence resistance That value of resistance that, when multiplied by the square of the fundamental positive-sequence rated-frequency component of armature current and by the number of phases, is equal to the sum of the copper loss in the armature and the load loss resulting from that current, when the machine is operating at rated speed. Positive-sequence resistance is normally that corresponding to rated armature current. *Note:* Inasmuch as the load loss may not vary as the square of the current, the positive-sequence resistance applies accurately only near the current for which it was determined. (EEC/PE) [119]

positive shielding angle The shielding angle formed when the shield wire is located above and inside of the area occupied by the outermost conductors. *See also:* shielding angle; negative shielding angle. (SUB/PE) 998-1996

positive terminal (batteries) The terminal from which the positive electric charge flows through the external circuit to the negative terminal when the cell discharges. *Note:* The flow of electrons in the external circuit is to the positive terminal and from the negative terminal. *See also:* battery. (EEC/PE) [119]

POSIX Portable Operating Systems Interface. A family of standards, which define a standard operating system interface, plus the environment to support application portability at the source code level. (C/PA) 1003.23-1998

POSIX character A value of the type `POSIX_Character`. An array of POSIX characters, of type `POSIX_String` is called a ***POSIX** string*. (C) 1003.5-1999

POSIX Conformance Document (PCD) The conformance docuent required by a POSIX standard. (C/PA) 13210-1994, 2003.1-1992

POSIX Conformance Test Procedure (PCTP) The non-software procedures possibly used in conjunction with other test methods to measure conformance. (C/PA) 13210-1994, 2003.1-1992

POSIX Conformance Test Suite (PCTS) The collection of software possibly used in conjunction with other test methods to measure conformance. (C/PA) 13210-1994, 2003.1-1992

POSIX I/O The input/output operations defined by IEEE Std 1003.5-1992 and IEEE Std 1003.5b-1995. (C/PA) 1003.5-1992r, 1003.5b-1995

POSIX process A conceptual object, having an associated address space, one or more threads of control executing within that address space, a collection of system resources required for execution, and certain other attributes. A POSIX process is said to perform an action if any of the conceptual threads of control within it performs the action. A process is created by another process with procedures `POSIX_Process_Primitives.Start_Process`, `POSIX_Process_Primitives.Start_Process_Search`, or the function `POSIX_Unsafe_Process_Primitives.Fork`. The process that issues `Start_Process`, `Start_Process_Search`, or `Fork` is known as the parent process. The newly created process is the child process. (C) 1003.5-1999

POSIX SP *See:* POSIX Standardized Profile.

POSIX Standardized Profile (POSIX SP) A standardized profile that specifies the application of certain POSIX base standards in support of a class of applications and does not require any departure from the structure defined by the reference model for POSIX systems in this guide. (C/PA) 14252-1996

post (waveguide) A cylindrical rod placed in a transverse plane of the waveguide and behaving substantially as a shunt susceptance. *See also:* waveguide. (AP/ANT) [35]

post-accelerating (deflection) electrode (intensifier electrode) An electrode to which a potential is applied to produce post-acceleration. *See also:* electrode. (PE/PSR) C37.90-1978s

post acceleration (electron-beam tubes) Acceleration of the beam electrons after deflection. (ED) 161-1971w

postal telephone and telegraph Common carriers that are owned by the government and in which the government is the sole monopoly supplier of communication facilities. (C) 610.7-1995

postamble (1) In networking, a sequence of bits appended after the last bit of the frame check sequence. *See also:* preamble; abnormal preamble. (C) 610.7-1995
(2) A sequence of bits recorded at the end of each block on a magnetic medium for the purpose of synchronization when reading backward. *Contrast:* preamble. (C) 610.10-1994w
(3) In 10BROAD36, the bit pattern appended after the last bit of the Frame Check Sequence by the Medium Attachment Unit (MAU); the Broadband End-of-Frame Delimiter (BEOFD). (C/LM) 802.3-1998

postamble breakpoint *See:* epilog breakpoint.

post-arc current The current that flows through the arc gap of a circuit breaker immediately after current zero, and that has a substantially lower magnitude than the test current. (SWG/PE) C37.100-1992

post-condition A condition that is guaranteed to be true after a successful property request. (C/SE) 1320.2-1998

post-deflection acceleration *See:* post acceleration.

postdialing delay (1) The time interval between the end of dialing and a physical connection that ensures completion or correct call cisposition, insofar as these are under switch control. For example, the time from seizure of a ringing circuit until the actual start of ringing (or audible ringing) is not included in postdialing delay. This interval excludes the timing period sometimes required to detect the end of dialing. The originating system delay objective excludes the outpulsing interval. The terminating system delay objective excludes the interval between seizure of a ringing circuit and the start of called-party alerting and excludes the interval between the start of called-party alerting and the initiation of the audible ring signal to the caller. (COM/TA) 973-1990w
(2) In an automatic telecommunications system that time interval between the receipt of the last called address digit from the calling station and the application of ringing to the called station. (COM) 312-1977w

post disconnect timing A timing interval (normally about 12 s in length), initiated when the called party goes on-hook, in which the established connection remains in place as long as the calling party continues to remain off-hook. (AMR/SCC31) 1390-1995, 1390.3-1999, 1390.2-1999

posted transaction A transaction in which the request and response are performed within different transactions. (C/BA) 1014.1-1994w

post emphasis *See:* de-emphasis.

post equalization *See:* de-emphasis.

post-fault (event) A qualifying term that refers to an interval beginning with the clearing of a fault. (SWG/PE) C37.100-1992

postfix notation (mathematics of computing) A method of forming mathematical expressions in which each operator is preceded by its operands. For example, A added to B and the result multiplied by C is expressed as AB + CX. *Synonyms:* suffix notation; reverse Polish notation. *Contrast:* infix notation; prefix notation. (C) 1084-1986w

post insulator (composite insulators) Intended to be loaded in tension, bending, or compression. The most common types are a horizontal line post where the post projects nearly horizontally from a pole and is loaded in flexure by the conductor, and a station post insulator used as a bus support in an outdoor substation. (T&D/PE) 987-1985w

postmortem dump (1) A dump that is produced upon abnormal termination of a computer program. *See also:* static dump; selective dump; memory dump; dynamic dump; snapshot dump; change dump. (C) 610.12-1990
(2) A static dump used for debugging purposes that is performed at the end of a machine run. (C) [85]

postorder traversal The process of traversing a binary tree in a recursive fashion as follows: the left subtree is traversed, then the right tree is traversed, then the root is visited. *Synonym:* endorder traversal. *Contrast:* preorder traversal; inorder traversal. *See also:* converse postorder traversal. (C) 610.5-1990w

postprocessor (1) (software) A computer program or routine that carries out some final processing step after the completion of the primary process; for example, a routine that reformats data for output. *Contrast:* preprocessor. (C) 610.12-1990
(2) (numerically controlled machines) A set of computer instructions that transform tool centerline data into machine motion commands using the proper tape code and format required by a specific machine control system. Instructions such as feedrate calculations, spindle-speed calculations, and auxiliary-function commands may be included. (IA) [61]

post puller An electric vehicle having a powered drum handling wire rope used to pull mine props, after coal has been removed, for the recovery of the timber. (EEC/PE) [119]

Postscript A page description language used in many laser printers. (C) 610.13-1993w

post, waveguide *See:* waveguide post.

POT *See:* point on tangent.

potential diagram (electrode-optical system) A diagram showing the equipotential curves in a plane of symmetry of an electron-optical system. *See also:* electron optics. (ED) [45], [84]

potential energy The work required to bring the system from an arbitrarily chosen reference configuration to the given configuration without change in other energy of the system. (Std100) 270-1966w

potential false-proceed operation The existence of a condition of vehicle or roadway apparatus in an automatic train control or cab-signal installation under which a false-proceed operation would have occurred had a vehicle approached or entered a section where normally a restrictive operation would occur. (EEC/PE) [119]

potential gradient (1) A vector of which the direction is normal to the equipotential surface, in the direction of decreasing potential, and of which the magnitude gives the rate of variation of the potential. (Std100) [84]
(2) *See also:* voltage gradient. (T&D/PE) 539-1990

potential hydro energy (power operations) (electric power supply) The possible aggregate energy obtainable over a specified period by practical use of the available stream flow and river gradient. (PE/PSE) 858-1987s, 346-1973w

potentially blocking operation An operation that is not allowed within a protected action, because it may be required to block the calling task. Certain operations are defined by the Ada language to be potentially blocking. (C) 1003.5-1999

potential master (1) A potential master is a module that is capable of participating in the control acquisition process and taking full control of the bus. A potential master may be in any of these states:

a) entrant;
b) free bystander;

c) inhibited bystander;
d) competitor;
e) withholder;
f) master elect;
g) master;
h) recompeting master.
(C/MM) 896.1-1987s

(2) A module capable of acquiring control of the bus through the control acquisition process. (C/BA) 896.4-1993w

potential profile A plot of potential as a function of distance along a specified path. (PE/PSIM) 81-1983

potential slave A module that is capable of being addressed by and is able to carry out transactions with the master. (C/MM) 896.1-1987s

potential source-rectifier exciter (1) (excitation systems for synchronous machines) An exciter whose energy is derived from a stationary alternating-current (ac) potential source and converted to direct current by rectifiers. The exciter includes the power potential transformers and power rectifiers which may be either noncontrolled or controlled, including gate circuitry. It is exclusive of input control elements. The source of ac power may come from the machine terminals or from a station auxiliary bus or a separate winding within the synchronous machine. (PE/EDPG) 421.1-1986r

(2) (synchronous machines) An exciter whose energy is derived from a stationary alternating current potential source and converted to direct current by rectifiers. *Note:* (1) The exciter includes the power potential transformers, where used, and power rectifiers which may be either noncontrolled or controlled, including gate circuitry. (2) It is exclusive of input control elements. (PE/EDPG) 421-1972s

potential transformer (1) (voltage transformer) An instrument transformer that is intended to have its primary winding connected in shunt with a power-supply circuit, the voltage of which is to be measured or controlled. *See also:* instrument transformer. (ELM) C12.1-1982s

(2) (power and distribution transformers) *See also:* fused-type voltage transformer; cascade-type voltage transformer; double-secondary voltage transformer; insulated-neutral terminal type voltage transformer; voltage transformer; rated secondary voltage; thermal burden rating of a voltage transformer; rated voltage. (PE/TR) C57.12.80-1978r

potential transformer, cascade-type A single high-voltage line-terminal potential transformer with the primary winding distributed on several cores with the cores electromagnetically coupled by coupling windings and the secondary winding on the core at the neutral end of the high-voltage winding. Each core is insulated from the other cores and is maintained at a fixed potential with respect to ground and the line-to-ground voltage. *See also:* instrument transformer. (PE/TR) C57.13-1978s

potential transformer, double-secondary One that has two secondary windings on the same magnetic circuit insulated from each other and the primary. Either or both of the secondary windings may be used for measurement or control. *See also:* instrument transformer. (PE/TR) C57.13-1978s

potential transformer, fused-type One that is provided with the means for mounting a fuse, or fuses, as an integral part of the transformer in series with the primary winding. *See also:* instrument transformer. (PE/TR) C57.13-1978s

potential transformer, grounded-neutral terminal type One that has the neutral end of the high-voltage winding connected to the case or mounting base. *See also:* instrument transformer. (PE/TR) C57.13-1978s

potential transformer, insulated-neutral terminal type One that has the neutral end of the high-voltage winding insulated from the case or base and connected to a terminal that provides insulation for a lower-voltage insulation class than required for the rated insulation class of the transformer. *See also:* instrument transformer. (PE/TR) C57.13-1978s

potential transformer, single-high-voltage line terminal One that has the line end of the primary winding connected to a terminal insulated from ground for the rated insulation class. The neutral end of the winding may be (1) insulated from ground but for a lower insulation class than the line end (insulated neutral) or (2) connected to the case or base (grounded neutral). *See also:* instrument transformer. (PE/TR) C57.13-1978s

potential transformer, two-high-voltage line terminals One that has both ends of the high-voltage winding connected to separate terminals that are insulated from each other, and from other parts of the transformer, for the rated insulation class of the transformer. *See also:* instrument transformer. (PE/TR) C57.13-1978s

potentiometer (1) (measurement techniques) An instrument for measuring an unknown electromotive force or potential difference by balancing it, wholly or in part, by a known potential difference produced by the flow of known currents in a network of circuits of known electrical constants. *See also:* instrument. (EEC/PE) [119]

(2) (analog computer) A resistive element with two end terminals and a movable contact. *See also:* attenuator. (C) 165-1977w

(3) A resistor with an adjustable sliding contact that functions as an adjustable voltage divider. *See also:* function potentiometer; servo potentiometer; parameter potentiometer. (C) 610.10-1994w, 165-1977w

potentiometer, follow-up *See:* follow-up potentiometer.

potentiometer, function *See:* function potentiometer.

potentiometer granularity (analog computer) The physical inability of a potentiometer to produce an output voltage that varies in other than discrete steps, due either to contacting individual turns of wire in a wire-would potentiometer or to discrete irregularities of the resistance element of composition or film potentiometers. (C) 165-1977w, 166-1977w

potentiometer, grounded *See:* grounded potentiometer.

potentiometer, linear *See:* linear potentiometer.

potentiometer, manual *See:* manual potentiometer.

potentiometer, multiplier *See:* multiplier potentiometer.

potentiometer, parameter *See:* parameter potentiometer.

potentiometer, servo *See:* servo potentiometer.

potentiometer set In an analog computer, a computer-control state that supplies the same operating potentiometer loading as under computing conditions and thus allows correct potentiometer adjustment. (C) 610.10-1994w, 165-1977w, 166-1977w

potentiometer, sine-cosine *See:* sine-cosine potentiometer.

potentiometer, tapered *See:* tapered potentiometer.

potentiometer, tapped *See:* tapped potentiometer.

potentiometer, ungrounded *See:* ungrounded potentiometer.

pothead A device that seals the end of a cable and provides insulated egress for the conductor or conductors. (PE/TR) [107], [108], 48-1975s

pothead body The part of a pothead that joins the entrance fitting to the insulator or to the insulator lid. *See also:* transformer; pothead. (PE/TR) [107], 48-1975s, [108]

pothead bracket or mounting plate The part of the pothead used to attach the pothead to the supporting structure. *See also:* transformer; pothead. (PE) 48-1975s

pothead bracket or mounting-plate insulator An insulator used to insulate the pothead from the supporting structure for the purpose of controlling cable sheath currents. *See also:* pothead; transformer. (PE) 48-1975s

pothead entrance fitting A fitting used to seal or attach the cable sheath, armor, or other coverings to the pothead. *See also:* pothead; transformer. (PE) 48-1975s

pothead insulator An insulator used to insulate and protect each conductor passing through the pothead. *See also:* pothead; transformer. (PE/TR) [107], [108], 48-1975s

pothead insulator lid The part of a multi-conductor pothead used to join two or more insulators to the body. *See also:* transformer; pothead. (PE/TR) [107], [108], 48-1975s

pothead mounting plate The part of the pothead used to attach the pothead to the supporting structure. *See also:* transformer. (PE/TR) [107], [108]

pothead mounting-plate insulator An insulator used to insulate the pothead from the supporting structure for the purpose of controlling cable sheath currents. *See also:* transformer. (PE/TR) [107], [108]

pothead sheath insulator An insulator used to insulate an electrically conductive cable sheath or armor from the metallic parts of the pothead in contact with the supporting structure for the purpose of controlling cable sheath currents. *See also:* pothead. (PE) 48-1975s

Potier reactance (rotating machinery) An equivalent reactance used in place of the primary leakage reactance to calculate the excitation on load by means of the Potier method. *Note:* It takes into account the additional leakage of the excitation winding on load and in the overexcited region; it is greater than the real value of the primary leakage reactance. It is useful for the calculation of excitation of the machine at other loads and power factors. The height of a Potier reactance triangle determines the reactance drop, and the reactance X_P is equal to the reactance drop divided by the current. The value of Potier reactance is that obtained from the no-load normal-frequency saturation curve: and normally with the excitation for rated voltage and current at zero power factor (overexcited), and at rated frequency. Approximate values of Potier reactance may be obtained from test load excitations at loads differing from rated load, and at power factors other than zero.The excitation results in the range from zero power factor overexcited to unity power factor are close enough to the test values for most practical applications. (PE) [9]

Potter horn A circular horn with one or more abrupt changes in diameter that excites two or more waveguide modes in order to produce a specified aperture illumination. (AP/ANT) 145-1993

potting (encapsulation) The sealing of components and associated conductors in a filter assembly with an insulating, thermally conductive material to exclude contaminants. (EMC) C63.13-1991

potty seat *See:* insulator lifter.

Poulsen arc (also Poulsen singing arc or signing arc) A type of arc-gap transmitting circuit that uses a resistance-capacitance (rc) circuit to tune the arc. This technique substantially reduces the bandwidth used by the arc-gap transmitter. (EMC) 140-1990r

poured joint (power cable joints) A joint insulated by the means of a hot or cold poured insulating medium which solidifies. (PE/IC) 404-1986s

powder *See:* explosives.

power (Φ) (1) (laser maser) The time rate at which energy is emitted, transferred, or received; usually expressed in watts (or in joules per second). (LEO) 586-1980w

(2) (used as an adjective) A general term used by reason of specific physical or electrical characteristics to denote application or restriction, or both, to generating stations, switching stations, or substations. (PE/SWG-OLD) C37.100-1992, C37.40-1993

(3) The rate of generating, transferring, or using energy. (PE/PSE) 858-1993w

(4) (fiber optics) *See also:* irradiance; radiant intensity; radiant power. 812-1984w

power—active The time average of the instantaneous power over one period of the wave. *Note:* For sinusoidal quantities in a two-wire circuit, it is the product of the voltage, the current, and the cosine of the phase angle between them. For nonsinusoidal quantities, it is the sum of all the harmonic components, each determined as above. In a polyphase circuit, it is the sum of the active powers of the individual phases. (ELM) C12.1-1988

power, active *See:* active power.

power amplification (1) The ratio of the power level at the output terminals of an amplifier to that at the input terminals. Also called power gain. *See also:* power gain; amplifier. (AP/ANT) 145-1983s

(2) (magnetic amplifier) The product of the voltage amplification and the current amplification. (MAG) 107-1964w

power—apparent For sinusoidal quantities in either single-phase or polyphase circuits, apparent power is the square root of the sum of the squares of the active and reactive powers. *Note:* This is, in general, not true for nonsinusoidal quantities. (ELM) C12.1-1988

power, apparent *See:* apparent power.

power, auxiliary *See:* auxiliary power.

power, available *See:* available power.

power, average phasor *See:* average phasor power.

power budget The minimum optical power available to overcome the sum of attenuation plus power penalties of the optical path between the transmitter and receiver calculated as the difference between the transmitter launch power (min) and the receive power (min). (C/LM) 802.3-1998

power cable Cable used to supply power to plant auxiliary system devices. The classifications for power cable are: low voltage and medium voltage. (PE/IC) 1185-1994

power capacitor An assembly of dielectric and electrodes in a container (case), with terminals brought out, that is intended to introduce capacitance into an electric power circuit. (T&D/PE) 18-1992

power capacity (waveguide) The maximum power which can be carried by the waveguide under a specified set of environmental and circuit conditions with a desired safety factor. (MTT) 146-1980w

power, carrier-frequency, peak pulse *See:* peak pulse power, carrier-frequency.

power-circuit limit switch A limit switch the contacts of which are connected into the power circuit. *See also:* switch. (IA/ICTL/IAC) [60]

power circuit protector (low-voltage ac power circuit protectors) An assembly consisting of a modified low-voltage power circuit breaker, which has no direct-acting tripping devices, with a current-limiting fuse in series with the load terminals of each pole. (SWG/PE) C37.100-1992, C37.29-1981r

power-closed car door or gate (elevators) A door or gate that is closed by a car-door or gate power closer or by a door or gate power operator. *See also:* elevator. (PE/EEC) [119]

power coefficient (attenuator) (characteristic insertion loss) Temporary and reversible variation in decibels when input power is varied from 20 dB below full rated power or lower to full rated power after steady-state condition has been reached. (IM/HFIM) 474-1973w

power, commercial *See:* commercial power.

power conditioning subsystem (PCS) (terrestrial photovoltaic power systems) The subsystem that converts the dc power from the array subsystem to dc or ac power that is compatible with system requirements. *See also:* array control. (PV) 928-1986r

power connection The connection between the heating cable and incoming power. (IA/PC) 515.1-1995

power control center (power operations) The location where power system operators monitor, analyze, or control power systems using digital or analog teleprocessing systems. (PE/PSE) 858-1987s

power correction capacitor Device that provides a capacitive load to offset the demand for lagging reactive power. (PE/EDPG) 1020-1988r

power dBm level Decibels relative to one milliwatt. This is the customary unit worldwide for measurement of telecommunications signal power. Zero dBm equals one milliwatt. (COM/TA) 1007-1991r

power density (S) **(1) (control of system electromagnetic compatibility)** (of an electromagnetic wave) Emitted power per unit cross-sectional area normal to the direction of propagation. (EMC) C63.12-1987
(2) (of a traveling wave) The time average of the Poynting vector. *Synonym:* power flux density. *See also:* spectral power density; spectral power flux density. (AP/PROP) 211-1997
(3) Power per unit area normal to the direction of propagation, usually expressed in units of watts per square meter (W/m^2) or, for convenience, units such as milliwatts per square centimeter (mW/cm^2) or microwatts per square centimeter (μW/cm^2). For plane waves, power density, electric field strength (E), and magnetic field strength (H) are related by the impedance of free space, i.e., 377 Ω. In particular,

$$S = \frac{E^2}{377} = 377\ H^2$$

where E and H are expressed in units of V/m and A/m, respectively, and S in units of W/m^2. Although many survey instruments indicate power density units, the actual quantities measured are E or E^2 or H or H^2. (NIR) C95.1-1999
(4) (fiber optics) *See also:* irradiance. 812-1984w

power-density spectrum A plot of power density per unit frequency as a function of frequency. (EMC) [53]

power detection That form of detection in which the power output of the detecting device is used to supply a substantial amount of power directly to a device such as a loudspeaker or recorder. *See also:* detection. (EEC/PE) [119]

power dissipation (light-emitting diodes) The time average product of current times voltage of the device. (ED) [127]

power distribution, underground cables *See:* cable bedding; aluminum-covered steel wire; cable separator; base ambient temperature.

power disturbance Any deviation from the nominal value (or from some selected thresholds based on load tolerance) of the input ac power characteristics. (IA/PSE) 1100-1999

power disturbance monitor Instrumentation developed specifically to capture power disturbances for the analysis of voltage and current measurements. (IA/PSE) 1100-1999

power divider (waveguide) A device for producing a desired distribution of power at a branch point. *See also:* waveguide. (AP/ANT) [35], [84]

power, effective radiated *See:* effective radiated power.

power elevator An elevator utilizing energy other than gravitational or manual to move the car. *See also:* elevator. (EEC/PE) [119]

power, emergency *See:* emergency power.

power factor (1) (electrical heating systems) The ratio of the circuit power (watts) to the circuit voltamperes. (IA/PC) 844-1991
(2) (converter characteristics) (self-commutated converters) (total) The ratio of the total active power in watts to the total apparent power in voltamperes (the product of root-mean-square [rms] voltage and rms current) on the ac (alternating current) side of the converter. *Note:* This definition includes the effect of harmonic components of current and voltage, as well as the effect of phase displacement between current and voltage. (IA/SPC) 936-1987w
(3) (harmonic control and reactive compensation of static power converters) (total) The ratio of the total power input in watts to the total voltampere input to the converter. *Notes:* 1. This definition includes the effect of harmonic components of current and voltage, the effect of phase displacement between current and voltage, and the exciting current of the transformer. Voltamperes are the product of root-mean-square (rms) voltage and rms current. 2. The power factor is determined at the ac line terminals of the converter. (IA/SPC) 519-1981s
(4) The ratio of total watts to the total root-mean-square (RMS) voltamperes.

$$F_P = \frac{\Sigma \text{ Watts per Phase}}{\Sigma \text{ RMS Voltamperes per Phase}} = \frac{\text{Active Power}}{\text{Apparent Power}}$$

Note: If the voltages have the same waveform as the corresponding currents, power factor becomes the same as phasor power factor. If the voltages and currents are sinusoidal and, for polyphase circuits, form symmetrical sets, $F_P = \cos(\alpha - \beta)$. *See also:* asynchronous machine. (PE/AES) [9], 270-1966w, [84], [41]
(5) (rectifier or rectifier unit) (thyristor converter) The ratio of the total watts input (total power input in watts) to the total voltampere input to the rectifier, rectifier unit or converter. *Notes:* 1. This definition includes the effect of harmonic components of current and voltage, the effect of phase displacement between the current and voltage, and the exciting current of the transformer. Voltampere is the product of root- mean square volts and root-mean-square amperes. 2. It is determined at the alternating-current line terminals of the thyristor converter or rectifier unit. *See also:* power rectifier; rectification. (IA/IPC) 444-1973w
(6) (rotating machinery) (dielectric) The cosine of the dielectric phase angle or the sine of the dielectric loss angle. (PE) [9]
(7) (insulation) (outdoor apparatus bushings) The ratio of the power dissipated in the insulation, in watts, to the product of the effective voltage and current in voltamperes, when tested under a sinusoidal voltage and prescribed conditions. *Note:* The insulation power factor is equal to the cosine of the phase angle between the voltage and the resulting current when both the voltage and current are sinusoidal. (PE/TR) 21-1976, C57.19.03-1996, C57.12.80-1978r
(8) (metering) The ratio of the active power to the apparent power. (ELM) C12.1-1988
(9) (thyristor) The ratio of the total watts to the total voltamperes. *Note:* This definition includes the effect of harmonic components of current and voltage, and the effect of phase displacement between current and voltage. (IA/IPC) 428-1981w
(10) (hydroelectric power plants) Ratio of real to total apparent power (kW/kVA) expressed as a decimal or percent. (PE/EDPG) 1020-1988r
(11) (dielectric) The cosine of the phase angle between a sinusoidal voltage applied across a dielectric (or combinations of dielectrics) and the resulting current through the dielectric system. (PE/PSIM) 62-1995

power factor adjustment clause (power operations) A clause in a rate schedule that provides for an adjustment in the billing if the customer's power factor varies from a specified reference. (PE/PSE) 858-1987s

power-factor angle The angle whose cosine is the power factor. *See also:* asynchronous machine. (PE) [9]

power factor, arithmetic *See:* arithmetic power factor.

power factor, coil Q *See:* coil *Q*.

power-factor-corrected mercury-lamp ballast A ballast of the multiple-supply type that has a power-factor-correcting device, such as a capacitor, so that the input current is at a power factor in excess of that of an otherwise comparable low-power-factor ballast design, but less than 90%, when the ballast is operated with center rated voltage impressed upon its input terminals and with a connected load, consisting of the appropriate reference lamp(s), operated in the position for which the ballast is designed. The minimum input power factor of such a ballast should be specifically stated. (EEC/LB) [97]

power factor, dielectric *See:* dielectric power factor.

power factor, displacement (1) (thyristor) The ratio of the active power of the fundamental wave, in watts, to the apparent power of the fundamental wave in voltamperes. This is the cosine of the phase angle by which the fundamental current lags the fundamental voltage. This is the power factor as seen in utility metering by watthour and varhour meters assuming that the ac voltages are sinusoidal. (IA/IPC) 428-1981w

(2) (converter characteristics) (self-commutated converters) The displacement component of power factor; the ratio of the active power of the fundamental wave, in watts, to the apparent power of the fundamental wave, in volt-amperes. It is also equal to cosf1, the cosine of the phase displacement angle between the fundamental component of the voltage and current on the ac (alternating current) side of a converter. (IA/SPC) 936-1987w

(3) (A) The displacement component of power factor. **(B)** The ratio of the active power of the fundamental wave, in watts, to the apparent power of the fundamental wave, in volt-amperes. (IA/PSE) 1100-1999

power-factor influence (electric instruments) The change in the recorded value that is caused solely by a power-factor departure from a specified reference power factor maintaining constant power (or vars) at rated voltage, and not exceeding 120% of rated current. It is to be expressed as a percentage of the full-scale value. (EEC/ERI/AII) [111], [102]

power-factor meter A direct-reading instrument for measuring power factor. It is provided with a scale graduated in power factor. *See also:* instrument. (EEC/PE) [119]

power factor relay (power system device function numbers) A relay that operates when the power factor in an alternating-current (ac) circuit rises above or falls below a predetermined value. (SUB/PE) C37.2-1979s

power-factor tip-up (rotating-machinery stator-coil insulation) The difference between the power-factors measured at two different designated voltages applied to an insulation system, other conditions being constant. *Notes:* 1. Used mainly as a measure of discharges, and hence of voids, within the system at the higher voltage. 2. The incremental change in power factor divided by incremental change in voltage applied to an insulation system. 3. Tip-up tests may be made using dissipation factor (tan δ) instead of power factor. In this case the tip-up is often identified as Δ tan δ or delta tan delta. *See also:* asynchronous machine. (EM/PE) 286-1975w

power-factor tip-up test (rotating machinery) A test applied to insulation to determine the power-factor tip-up. *See also:* asynchronous machine. (PE) [9]

power factor, total *See:* total power factor.

power-factor-voltage characteristic (rotating machinery stator-coil insulation) The relation between the magnitude of the applied test voltage and the measured power factor of the insulation. *Note:* The characteristic is usually shown as a curve of power factor plotted against test voltage. *See also:* asynchronous machine. (EM/PE) 286-1975w

power-fail circuit A logic circuit that protects an operating program if primary power fails by informing the computer when power failure is imminent, initiating a routine that saves all volatile data. After power has been restored, the circuit initiates a routine that restores the data and restarts computer operations. (C) 610.10-1994w

power failure (emergency and standby power) Any variation in electric power supply that causes unacceptable performance of the user's equipment. (IA/PSE) 446-1995

power failure recovery A sequence of events that provides orderly control of system shutdown during a temporary power failure and start-up after power is restored. (C/MM) 1296-1987s

power feeder A feeder supplying principally a power or heating load. *See also:* feeder.

power-flow angle The angle (ϕ) between the direction of the power-flow vector and the direction of the propagation vector. (UFFC) 1037-1992w

power-flow vector Vector-characterizing energy propagation caused by a wave and giving magnitude and direction of power per unit-area propagating in the wave (i.e., analogous to Poynting vector). (UFFC) 1037-1992w

power flux density *See:* power density.

power frequency (1) The value of frequency used in the electrical power system, such as 50 Hz or 60 Hz. (EMC) C63.13-1991

(2) The frequency at which a device or system is designed to operate. (SWG/PE) C37.34-1994, C37.100-1992

power-frequency current-interrupting rating (surge arresters) (of an expulsion arrester) A designation of the range of the symmetrical root-mean-square fault currents of the system for which the arrester is designed to operate. An expulsion arrester is given a maximum current-interrupting rating and may also have a minimum current-interrupting rating. (PE/SPD) C62.1-1981s

power-frequency dew withstand voltage The rms voltage that can be applied to an insulator or a device, completely covered with condensed moisture, under specified conditions for a specified time without causing flashover or puncture. (SWG/PE) C37.40-1993

power-frequency dry withstand voltage The rms voltage that can be applied to a dry device under specified conditions for a specified time without causing flashover or puncture. (SWG/PE) C37.40-1993

power-frequency overvoltage A root-mean-square voltage in excess of the maximum (highest) system voltage that lasts longer than one cycle. (SPD/PE) C62.62-2000

power-frequency recovery voltage The power-frequency rms voltage that occurs across the terminals of an ac circuit-interrupting device after the interruption of the current and after the high-frequency transients have subsided. (SWG/PE) C37.40-1993

power-frequency sparkover voltage (1) The rms value of the lowest power-frequency sinusoidal voltage that will cause sparkover when applied across the terminals of an arrester. (SPD/PE) C62.11-1999

(2) The root-mean-square value of the lowest power-frequency sinusoidal voltage that will cause sparkover when applied across the terminals of a surge-protective device. (SPD/PE) C62.62-2000

power-frequency wet withstand voltage The rms voltage that can be applied to a wetted device under specified conditions for a specified time without causing flashover or puncture. (SWG/PE) C37.40-1993

power-frequency withstand voltage A specified root-mean-square test voltage, at a power frequency that will not cause a disruptive discharge. (SPD/PE) C62.62-2000, C62.11-1999

power fuse A fuse consisting of an assembly of a fuse support and a fuse unit or fuseholder that may or may not include the refill unit or fuse link. *Note:* The power fuse is identified by the following characteristics:

a) Dielectric withstand basic impulse insulation level (BIL) strengths at power levels
b) Application primarily in stations and substations
c) Mechanical construction basically adapted to station and substation mountings

(SWG/PE) C37.40-1993, C37.100-1992

power fuse unit (installations and equipment operating at over 600 volts, nominal) A vented, nonvented or controlled vented fuse unit in which the arc is extinguished by being drawn through solid material, granular material, or liquid, either alone or aided by a spring. (NEC/NESC) [86]

power gain (1) (data transmission) The ratio of the signal power that a transducer delivers to its load to the signal power absorbed by its input circuit. *Notes:* 1. Power gain is usually expressed in decibels. 2. If more than one component is involved in the input or output, the particular components used are specified. 3. If the output signal power is at a frequency other than the input signal power, the gain is a conversion gain. (PE) 599-1985w

(2) (two-port linear transducer) At a specified frequency, the ratio of:

a) the signal power that the transducer delivers to a specified load; to
b) the signal power delivered to its input port.

Note: The power gain is not defined unless the input impedance of the transducer has a positive real part. (ED) 161-1971w

(3) *See also:* partial gain. (AP/ANT) 145-1993

power gap The part of the bypass gap that carries the fault current after sparkover of the bypass gap. (T&D/PE) 824-1994

power influence (telephone loop performance) The power of a longitudinal signal induced in a telephone circuit by an electromagnetic field emanating from a conductor or conductors of a power system. In common usage, power influence is synonymous with longitudinal noise. (COM/TA) 820-1984r

power, instantaneous *See:* instantaneous power.

power inverter A component for converting dc power into ac power. (IA/MT) 45-1998

power klystron (microwave tubes) A klystron, usually an amplifier, with two or more cavities uncoupled except by the beam, designed primarily for power amplification or generation. (ED) [45]

power knock out A function, derived from friction brakes being applied above a low preset level on any truck, that removes propulsion power on every vehicle in the train. (VT) 1475-1999

power-law index profile (fiber optics) A class of graded index profiles characterized by the following equations:

$n(r) = n_1(1 - 2\Delta(r/a)^g)^{1/2}\ r = a$

$n(r) = n_2 = n_1(1 - 2\Delta)^{1/2}\ r = a$

where

$$\Delta = \frac{n_1^2 - n_2^2}{2n_1^2}$$

where $n(r)$ is the refractive index as a function of radius, n_1 is the refractive index on axis, n_2 is the refractive index of the homogeneous cladding, a is the core radius, and g is a parameter that defines the shape of the profile. *Notes:* 1. α is often used in place of g. Hence, this is sometimes called an alpha profile. 2. For this class of profiles, multimode distortion is smallest when g takes a particular value depending on the material used. For most materials, this optimum value is around 2. When g increases without limit, the profile tends to a step index profile. *See also:* graded index profile; step index profile; profile parameter; mode volume. (Std100) 812-1984w

power level (data transmission) The magnitude of power averaged over a specified interval of time. *Note:* Power level may be expressed in units in which the power itself is measured or in decibels indicating the ratio to a reference power. This ratio is usually expressed either in decibels, referred to one mW (milliwatt), abbreviated dBm, or in decibels referred to one W (watt), abbreviated dBW. (PE) 599-1985w

power level at DSX Power in dBm of an all-ones signal measured at a digital signal crossconnect (DSX). *See also:* d-BDSX. (COM/TA) 1007-1991r

power-line carrier (1) (overhead-power-line corona and radio noise) The use of RF energy, generally below 600 kHz, to transmit information, using power lines to guide the information transmission. (T&D/PE) 539-1990

(2) (protective relaying of utility-consumer interconnections) A high-frequency signal superimposed on the normal voltage on a power circuit. It is customarily coupled to the power line by means of a coupling capacitor. A tuning device provides series resonance at the carrier frequency. Prevention of shorting of the carrier signal by a fault external to the protected line is ordinarily provided by a line trap. (PE/PSR) C37.95-1973s

(3) The use of radio frequency energy to transmit information over transmission lines whose primary purpose is the transmission of power. (SUB/PE) 999-1992w

power-line-conducted radio noise Radio noise produced by equipment operation, which exists on the power line of the equipment and is measurable under specified conditions. *Note:* It may enter a receptor, such as ITE, by direct coupling or by subsequent radiation from some circuit element. (EMC) C63.4-1991

power loss (A) (data transmission) From a circuit, in the sense that it is converted to another form of power not useful for the purpose at hand (for example I^2R loss). A physical quantity measured in watts in the International System of Units (SI) and having the dimensions of power. For a given R, it will vary with the current in R. **(B) (data transmission)** Defined as the ratio of two powers. If P_o is the output power and P_i is the input power of a transducer or network under specified conditions, P_o/P_i is a dimensionless quantity that would be unity if $P_o = P_i$. **(C) (data transmission)** (Logarithmic). Loss may also be defined as the logarithm, or a quantity directly proportional to the logarithm of a power ratio, such as P_o/P_i. Thus if loss $= 10\log_{10} (P_o/P_i)$, the loss is zero when $P_o = P_i$. This is the standard for measuring loss in decibels. *Notes:* 1. In cases (B) and (C), the loss for a given linear system is the same whatever may be the power levels. Thus (B) and (C) give characteristics of the system, and do not depend, as (A) does, on the value of the current or other dependent quantity. 2. If more than one component is involved in the input or output, the particular components used must be specified. This ratio is usually expressed in decibels. 3. If the output signal power is at a frequency other than the input signal power, the loss is a conversion loss. **(D) (data transmission)** (Electric instrument) (watt loss). In the circuit of a current-or-voltage- measuring instrument, the active power at its terminals for end-scale indication. *Note:* For other than current or voltage- measuring instruments, for example, wattmeters, the power loss of any circuit is expressed at a stated value of current or of voltage. (PE) 599-1985

power monitor A functional module that monitors the status of the primary power source to the system, and signals when that power has strayed outside the limits required for reliable system operation. Since most systems are powered by an ac source, the power monitor is typically designed to detect drop-out or brown-out conditions on ac lines. (C/BA) 1014-1987

power, nonfirm *See:* nonfirm power.

power, nonreactive *See:* nonreactive power.

power of ten *See:* decade.

power-on retention time The retention time with the memory biased in the on condition in a nonoperating (that is, unclocked, deselected) mode. (ED) 641-1987w

power-operated door or gate (elevators) A hoistway door and/or a car door or gate that is opened and closed by a door or gate power operator. *See also:* hoistway. (EEC/PE) [119]

power operation Operation by other than hand power. (SWG/PE) C37.100-1992

power outage Complete absence of power at the point of use. (IA/PSE) 446-1995

power outlet An enclosed assembly which may include receptacles, circuit breakers, fuseholders, fused switches, buses and watt-hour meter mounting means; intended to supply and control power to mobile homes, recreational vehicles or boats, or to serve as a means for distributing power required to operate mobile or temporarily installed equipment. (NESC/NEC) [86]

power output (hydraulic turbines) The electrical output of the turbine generator unit as measured at the generator terminals. (PE/EDPG) 125-1977s

power output, instantaneous *See:* instantaneous power output.

power pack A unit for converting power from an alternating-current or direct-current supply into alternating-current or direct-current power at voltages suitable for supplying an electronic device. (EEC/PE) [119]

power, partial discharge *See:* partial discharge power.

power pattern *See:* radiation pattern.

power, phase control *See:* phase control power.

power, phasor *See:* phasor power.

power pool (power operations) Term referring to a group of power systems operating as an interconnected system and pooling their resources. (PE/PSE) 858-1987s

power primary detector (electric power system) A power-measuring device for producing an output proportional to power input. *See also:* speed-governing system. (PE/PSE) 94-1970w

power quality The concept of powering and grounding electronic equipment in a manner that is suitable to the operation of that equipment and compatible with the premise wiring system and other connected equipment. (IA/PSE) 1100-1999

power quantities (two-phase circuit) (single-phase three-wire circuit) The definitions of the power quantities for a single-phase circuit of more than two wires and of a two-phase circuit are essentially the same as those expressions involve m, the number of phases or phase conductors, the numeral 2 should be used for single-phase, three-wire systems, and the numeral 4 for two-phase, four-wire and five-wire systems. *See also:* polyphase symmetrical set. (Std100) 270-1966w

power rating (waveguide attenuator) The maximum power that, if applied under specified conditions of environment and duration, will not produce a permanent change that causes any performance characteristics to be outside of specifications. This includes characteristic insertion loss and standing-wave ratio. *See also:* waveguide. (IM/HFIM) [40]

power rating or voltage rating (coaxial transmission line) (line and connectors) That value of transmitted power or voltage that permits satisfactory operation of the line assembly and provides an adequate safety factor below the point where injury or appreciably shortened life will occur. *See also:* transmission line. (EEC/REWS) [92]

power—reactive For sinusoidal quantities in a two-wire circuit, reactive power is the product of the voltage, the current, and the sine of the phase angle between them. For nonsinusoidal quantities, it is the sum of all the harmonic components, each determined as above. In a polyphase circuit, it is the sum of the reactive powers of the individual phases. (ELM) C12.1-1988

power, reactive *See:* reactive power.

power rectifier A rectifier unit in which the direction of average energy flow is from the alternating-current circuit to the direct-current circuit. (IA) [62]

power rectifier transformer (power and distribution transformers) A rectifier transformer connected to mercury-arc or semiconductor rectifiers for electrochemical service, steel processing applications, electric furnace applications, mining applications, transportation applications, and direct-current transmissions. (PE/TR) C57.12.80-1978r

power reflectance* (1) (of a radome) At a given point on a radome, the ratio of the power flux density that is internally reflected from the radome to that incident on the radome from an internal radiating source. (AP/ANT) 145-1993
(2) *See also:* power reflection coefficient. (AP/PROP) 211-1997

* Deprecated.

power reflection coefficient The squared magnitude of the Fresnel reflection coefficient. *Synonym:* power reflection factor. (AP/PROP) 211-1997

power reflection factor *See:* power reflection coefficient.

power relay A relay that responds to a suitable product of voltage and current in an electric circuit. *See also:* active-power relay; reactive power relay. (SWG/PE/PSR) C37.100-1992, C37.90-1978s

power_reset An initialization event triggered by the restoration of primary power. On a backplane bus, a power_reset event is generally triggered by one or several specialized signals driven by the shared power supply. (C/MM) 1212-1991s

power response (close-talking pressure-type microphone) The ratio of the power delivered by a microphone to its load, to the applied sound pressure as measured by a Laboratory Standard Microphone placed at a stated distance from the plane of the opening of the artificial voice. *Note:* The power response is usually measured as a function of frequency in decibels (dB) above 1 milliwatt per newton per square centimeter [mW/(N/m^2)] or 1 milliwatt per 10 microbars [mW/10μbar]. *See also:* close-talking pressure-type microphones. (SP) 258-1965w

power selsyn (synchros or selsyns) An inductive type of positioning system having two or more similar mechanically independent slip-ring machines with corresponding slip rings of all machines connected together and the stators fed from a common power source. *See also:* synchro system. (PE) [9]

power sensitivity error The maximum deviation from linearity over each power range of either the electrothermic unit or the electrothermic power meter. Expressed in percent. (IM) 544-1975w

power service protector An assembly consisting of a modified low-voltage power circuit breaker, which has no direct-acting tripping devices, with a current-limiting fuse connected in series with the load terminals of each pole. (SWG/PE) C37.100-1992

power, signal electronics *See:* signal electronics power.

powers of units (International System of Units (SI)) An exponent attached to a symbol containing a prefix indicates that the multiple or submultiple of the unit (the unit with its prefix) is raised to the power expressed by the exponent. For example:

$$1\ \text{cm}^3 = (10^{-2}\ \text{m})^3 = 10^{-6}\ \text{m}^3$$

$$1\ \text{ns}^{-1} = (10^{-9}\ \text{s})^{-1} = 10^9\ \text{s}^{-1}$$

$$1\ \text{mm}^2/\text{s} = (10^{-3}\ \text{m})^2/\text{s} = 10^{-6}\ \text{m}^2/\text{s}$$

See also: prefixes and symbols; units and letter symbols. (QUL) 268-1982s

power source isolation Absence of a direct-current circuit (path) between the power source and the system power supply outputs. (PE/EDPG) [1]

power sources The electrical and mechanical equipment and their interconnections necessary to generate or convert power. (PE/NP) 603-1998

power spectral density (PSD) (1) (seismic qualification of Class 1E equipment for nuclear power generating stations) The mean squared amplitude per unit frequency of a waveform. PSD is expressed in g^{1}2/Hz versus frequency for acceleration waveforms. (PE/NP) 344-1987r
(2) The Fourier transform of the autocorrelation function of a time series of data. For a stationary ergodic time series, the PSD is equal to the expected value of the magnitude squared of the Fourier transform of the data. The PSD expresses the noise variance in an accelerometer's output as a function of positive and negative frequencies, usually in g^2/Hz, and is calculated as the ensemble or frequency average of the magnitude squared of the fast Fourier transform (FFT) of the data. The one-sided PSD plots twice the PSD value versus positive frequency f, since for a real time series the two-sided PSD values are equal for f and $-f$. The velocity random walk coefficient is calculated from the two-sided PSD acceleration white noise level (half the one-sided PSD level). (AES/GYAC) 1293-1998

power station battery (1) A battery that is a separate source of energy for communication equipment in power stations. (COM)
(2) (control) A battery that is a separate source of energy for the control of power apparatus in a power station. *See also:* battery. (PE) 599-1985w

power storage That portion of the water stored in a reservoir available for generating electric power. *See also:* generating station. (T&D/PE) [10]

power supply A unit that converts voltage from one level to another, usually regulating the output. *Note:* Typically used to convert an AC voltage to a DC voltage. *See also:* converter. (C) 610.10-1994w

power-supply assembly The conductors, including the grounding conductors, insulated from one another, the connectors, attachment plug caps, and all other fittings, grommets, or devices installed for the purpose of delivering energy from the source of electrical supply to the distribution panel within the recreational vehicle. (NESC/NEC) [86]

power supply circuit (relay system) An input circuit to a relay system that supplies power for the functioning of the relay system. (PE/PSR) C37.90.1-1989r

power supply, direct-current *See:* direct-current power supply.

power supply, direct-current regulated *See:* direct-current regulated power supply.

power supply, uninterruptible *See:* uninterruptible power supply.

power-supply voltage range (transmitter performance) The range of voltages over which there is not significant degradation in the transmitter or receiver performance. *See also:* audio-frequency distortion. (VT) [37]

power switchboard (1) A type of switchboard including primary power-circuit switching and interrupting devices together with their interconnections. *Note:* Knife switches, fuses, and air circuit breakers are the commonly used switching and interrupting devices. (SWG/PE) C37.100-1992 **(2)** A switchgear and control assembly that receives energy from the main generating plant and distributes directly or indirectly to all equipment supplied by the generating plant. (IA/MT) 45-1998

power system (1) (generating stations electric power system) The electric power sources, conductors, and equipment required to supply electric power. (PE/EDPG) 505-1977r **(2) (electric)** The generation resources and/or transmission facilities operated as an entity to meet load and/or interchange commitments. (PE/PSE) 94-1991w **(3)** The generation resources and/or transmission facilities operated under common management or supervision to meet load and interchange commitments. (PE/PSE) 858-1993w

power system, emergency *See:* emergency power system.

power system stabilizer (1) (excitation systems for synchronous machines) An element or group of elements that provide an additional input to the regulator to improve power system performance. *Note:* A number of different quantities may be used as input to the power system stabilizer, such as, shaft speed, frequency, synchronous machine electrical power, etc. (PE/EDPG) 421.1-1986r **(2) (excitation systems)** Used to provide damping at power system frequencies associated with local and intertie modes of oscillation. (PE/EDPG) 421.4-1990 **(3) (excitation systems) (synchronous machines)** An element or group of elements that provide an additional input to the regulator to improve the dynamic performance of the power system. *Note:* A number of different quantities may be used as input to the power system stabilizer, such as shaft speed, frequency, synchronous machine electrical power, and others. (PE/EDPG) 421.2-1990, 421-1972s

power system, standby *See:* standby power system.

power-temperature coefficient The change in power required to hold the bolometer element at the desired operating resistance per unit change in ambient temperature. *Note:* This quantity is expressed in microwatts per degree Celsius. (IM) 470-1972w

power termination connection The termination applied to the end of a heating cable where the power is supplied. (IA) 515-1997

power test method A test that determines the power dissipation characteristics of a damper by the measurement of the force and velocity imparted to the test span at the point of attachment to the shaker. (T&D/PE) 664-1993

power transfer relay A relay so connected to the normal power supply that the failure of such power supply causes the load to be transferred to another power supply. (EEC/PE) [119]

power transformer (power and distribution transformers) A transformer that transfers electric energy in any part of the circuit between the generator and the distribution primary circuits. (PE/TR) C57.12.80-1978r

power transmittance of a radome In a given direction, the ratio of the power flux density emerging from a radome with an internal source to the power flux density that would be obtained if the radome were removed. (AP/ANT) 145-1993

power type relay A term for a relay designed to have heavy-duty contacts usually rated 15 A or higher. Sometimes called a contactor. (PE/EM) 43-1974s

power up/down protection *See:* inadvertent write protection.

power, utility *See:* commercial power.

power vector (A) (single-phase two-wire circuit) At the two terminals of entry of a single-phase two-wire circuit into a delimited region, a vector whose magnitude is equal to the apparent power, and the three rectangular components of which are, respectively, the active power, the reactive power, and the distortion power at the same two terminals of entry. Mathematically, the vector power U is given by

$$\mathrm{U} = \mathrm{i}P + \mathrm{j}Q + \mathrm{k}D$$

where i, j, and k are unit vectors along the three perpendicular axes, respectively. P, Q, and D are the active power, reactive power, and distortion power, respectively. The direction cosines of the angles between the vector power U and the three rectangular axes are

$$\cos\phi = \frac{P}{\mathrm{U}}$$

$$\cos\Psi = \frac{Q}{\mathrm{U}}$$

$$\cos\theta = \frac{D}{\mathrm{U}}$$

The magnitude of the vector power is the apparent power, or

$$\mathrm{U} = (P^2 + Q^2 + D^2)^{\frac{1}{2}}$$

$$= \left(\sum_{r=1}^{r=\infty}\sum_{q=1}^{q=\infty} E_r{}^2 I_q{}^2\right)^{\frac{1}{2}}$$

where the symbols are those of the preceding definitions. The geometric power diagram shows the relationships among the different types of power. Active power, reactive power, and distortion power are represented in the directions of the three rectangular axes. The diagram corresponds to a case in which all three are positive. Since the sign of D is not definitely determined, D may be drawn in either direction along the axis. The position of U is thus also ambiguous, as it may occupy either of two positions, for D positive or negative. When the sign of D has been assumed, the vector positions of the fictitious power F and the nonreactive power N are determined. They have been shown in the corresponding figure, with the assumption that D has the same sign as P. Vector power is expressed in voltamperes when the voltage is in volts and the current in amperes. *Notes:* 1. The vector power becomes a plane vector having the same magnitude as the phasor power if the voltage and the current have the same wave form. This condition is fulfilled as a special case when the voltage and current are sinusoidal and of the same period. 2. The term vector power as defined in the 1941 edition of the American Standard Definitions of Electrical Terms has now been called phasor power and the present definition of vector power is new. *See also:* network analysis. **(B) (polyphase circuit)** At the terminals of entry of a polyphase circuit, a vector of which the three rectangular components are, respectively, the active power, the reactive power, and the distortion power at the same terminals of entry. In determining the components, the reference terminals for voltage measurement are taken as the neutral terminal of entry, if one exists, otherwise as the true neutral point. The vector power is also the (vector) sum of the vector powers for the individual terminals of entry. The vector power for each terminal of entry is determined by considering each phase conductor and the common reference point as a single-phase circuit, as described for distortion power. The sign given to the distortion power in determining the vector power for each single-phase circuit is the same as that of the total active power. The magnitude of the vector power is the apparent power. If the voltages have the same waveform as the corresponding currents, the magnitude of the vector power is equal to the amplitude of the phasor power. Vector power is expressed in voltamperes when the voltages are in volts and the currents in amperes. *See also:* network analysis.

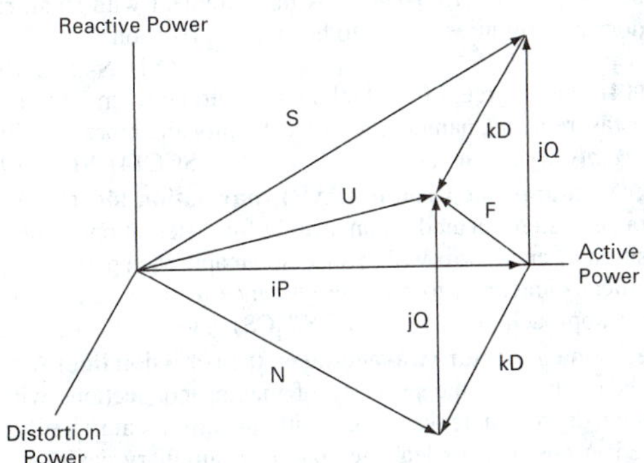

power vector

(Std100) 270-1966

power winding (saturable reactor) A winding to which is supplied the power to be controlled. Commonly the functions of the output and power windings are accomplished by the same winding, which is then termed the output winding. *See also:* magnetic amplifier. (PE/EEC) [119]

Poynting vector (1) If there is a flow of electromagnetic energy into or out of a closed region, the rate of flow of this energy is, at any instant, proportional to the surface integral of the vector product of the electric field strength and the magnetizing force. This vector product is called Poynting's vector. If the electric field strength is E and the magnetizing force is H, then Poynting's vector is given by

$$U = E \times H \text{ and } U = E \times H/4\pi$$

in rationalized and unrationalized systems, respectively. Poynting's vector is often assumed to be the local surface density of energy flow per unit time. (Std100) 270-1966w **(2)** *See also:* time-averaged Poynting vector; instantaneous Poynting vector. (AP/PROP) 211-1997

Poynting vector, instantaneous [$\vec{P}(t,\vec{r})$] *See:* instantaneous Poynting vector.

Poynting vector, time-averaged *See:* time-averaged Poynting vector.

PPCSN *See:* private packet/frame and circuit switching network.

***p*-percent disruptive discharge voltage (V_p)** The prospective value of the test voltage that has a *p*-percent probability of producing a disruptive discharge. (PE/PSIM) 4-1995

PPI *See:* plan-position indicator.

***p-p* junction (semiconductor)** A region of transition between two regions having different properties in *p*-type semiconducting material. *See also:* semiconductor device. (PE/EEC) [119]

ppm *See:* parts per million.

PPM *See:* periodic permanent-magnet focusing; pulse position modulation.

PPMV *See:* parts per million by volume.

PPMW *See:* parts per million by weight.

PPS *See:* preferred power supply.

PPSN *See:* private packet/frame switching network.

PR *See:* physical record.

PRA *See:* pendulous reference axis.

practical reference pulse waveform (pulse measurement) A reference pulse waveform which is derived from a pulse which is produced by a device or apparatus. (IM/WM&A) 181-1977w

practical stability *See:* finite-time stability.

practice Recommended approach, employed to prescribe a disciplined, uniform approach to the software life cycle. (C/SE) 730.1-1995

practices (software quality assurance) Requirements employed to prescribe a disciplined uniform approach to the software development process. *See also:* conventions; standards. (C) 610.12-1990

pragma A generic term used to define a construct with no predefined language semantics that influences how a synthesis tool will synthesize VHDL code into an equivalent hardware representation. (C/DA) 1076.6-1999

preallocation The reservation of resources in a system for a particular use. Preallocation does not imply that the resources are immediately allocated to that use, but merely indicates that they are guaranteed to be available in bounded time when needed. (C/PA) 9945-1-1996

preamble (1) In networking, a sequence of bits at the start of each new transmission to allow synchronization of clocks and other physical layer circuitry at other stations. *See also:* postamble; abnormal preamble. (C) 610.7-1995 **(2)** A sequence of bits recorded at the beginning of each block on a magnetic tape for the purpose of synchronization. *Contrast:* postamble. (C) 610.10-1994w

preamble breakpoint *See:* prolog breakpoint.

preamplifier (1) An amplifier connected to a low-level signal source to present suitable input and output impedances and provide gain so that the signal may be further processed without appreciable degradation in the signal-to-noise ratio. *Notes:* 1. A preamplifier may include provision for equalizing and.or mixing. 2. Further processing frequently includes further amplification in a main amplifier. *See also:* amplifier. (SP) 151-1965w

(2) The input section of an amplifier chain, usually located as close to the detector element as possible. (NPS) 325-1996

preamplifier, pulsed optical feedback *See:* pulsed optical feedback preamplifier.

pre-arbitrated (PA) access function The *access control function* in this part of ISO/IEC 8802 that uses assigned *offsets* in *Pre-Arbitrated (PA) slots* for the transfer of *isochronous service octets.* (LM/C) 8802-6-1994

pre-arbitrated (PA) segment A multiuser *segment* transferred using *Pre-Arbitrated Access (PA) functions*. The payload of the *PA segment* contains *isochronous service octets* from zero or more *Isochronous Service Users (ISUs).* (LM/C) 8802-6-1994

pre-arbitrated (PA) slot A *slot* that is dedicated by the *Head of Bus function* for transfer of *isochronous service octets* in the payload of a *PA segment.* (LM/C) 8802-6-1994

pre-arcing time *See:* melting time.

preassigned multiple access (communication satellite) A method of providing multiple access in which the satellite channels are preassigned at both ends of the path. (COM) [19]

precedence call (telephone switching systems) A call on which the calling party has elected to use one of several levels of priority available to him. (COM) 312-1977w

precedented system A system for which design examples exist within its class, so as to provide guidance for establishing the design architecture, engineering and technical plans, specifications, or low risk alternatives. (C/SE) 1220-1998

precession (1) (navigation aid terms) The change in the direction of the axis of rotation of a spinning body, as a gyroscope, when acted upon by a torque. (AES/GCS) 172-1983w
(2) (gyros) A rotation of the spin axis produced by a torque, T, applied about an axis mutually perpendicular to the spin axis and the axis of the resulting rotation. A constant precession rate, ω, is related to rotor angular momentum, H, and applied torque, T, by the equation $T = H\omega$.
(AES/GYAC) 528-1994

precipitation clutter Unwanted echoes from rain, snow, hail, sleet and other hydrometeorological particles.
(AES) 686-1997

precipitation intensity (overhead power lines) The rate of precipitation, usually expressed in millimeters per hour (mm/h). Since precipitation intensity in general is not constant, an average over some time period shorter than one h our is most useful, unless instantaneous intensity is measured. *Note:* Rain gauge measurements can provide an indication of intensity except for light rain, where the quantization error is large.
(T&D/PE) 539-1990

precipitation scatter Electromagnetic scattering caused by precipitating rain, hail, or snow particles.
(AP/PROP) 211-1997

precision (1) (mathematics of computing) (software) (data management) The degree of exactness or discrimination with which a quantity is stated; for example, a precision of 2 decimal places versus a precision of 5 decimal places. *Contrast:* accuracy. *See also:* multiple precision; triple precision; double precision; single precision.
(C) 610.5-1990w, 610.12-1990, 1084-1986w
(2) (A) (monitoring radioactivity in effluents) The degree of agreement of repeated measurements of the same property expressed in terms of dispersion of test results about the mean result obtained by repetitive testing of a homogeneous sample under specified conditions. The precision of a method is expressed quantitatively as the standard deviation computed from the results of a series of controlled determinations. **(B) (general)** The quality of being exactly or sharply defined or stated. A measure of the precision of a representation is the number of distinguishable alternatives from which it was selected, which is sometimes indicated by the number of significant digits it contains. *See also:* double precision; accuracy. (C) 162-1963, 165-1977
(3) (measurement process) The quality of coherence or repeatability of measurement data, customarily expressed in terms of the standard deviation of the extended set of measurement results from a well-defined (adequately specified) measurement process in a state of statistical control. The standard deviation of the conceptual population is approximated by the standard deviation of an extended set of actual measurements. *See also:* reproducibility; accuracy.
(AES/RS) 686-1982s, [42]
(4) (pulse measurement) The degree of mutual agreement between the results of independent measurements of a pulse characteristic, property, or attribute yielded by repeated application of a pulse measurement process.
(IM/WM&A) 181-1977w
(5) (analog computer) Exactly or sharply defined or stated. A measure of the precision of a representation is the number of distinguishable alternatives from which it was selected, which is sometimes indicated by the number of significant digits it contains. *See also:* accuracy. (C) 165-1977w
(6) (metric practice) The degree of mutual agreement between individual measurements, namely repeatability and reproducibility. *See also:* accuracy. (QUL) 268-1982s
(7) (measuring and test equipment) (nuclear power generating station) The quality of an instrument scale or readout being exactly or sharply defined or stated.
(PE/NP) 498-1985s
(8) (plutonium monitoring) (radiological monitoring instrumentation) The degree of agreement of repeated measurements of the same property, expressed quantitatively as the standard deviation computed from the results of the series of measurements.
(NI) N320-1979r, N317-1980r, N42.12-1994
(9) (airborne radioactivity monitoring) The degree of agreement of repeated measurements of the same parameter.
(NI) N42.17B-1989r, N42.20-1995
(10) The repeatability of measurement data, customarily expressed in terms of standard deviation. (ELM) C12.1-1988
(11) The discrepancy among individual measurements.
(PE/PSIM) 4-1995
(12) The degree of exactness or discrimination with which a quantity is stated. *Note:* The result of a calculation may have more precision than it has accuracy; for example, the true value of pi to six significant digits is 3.14159; the value of 3.14162 is precise to six digits but only five digits are accurate. *See also:* accuracy. (C) 610.10-1994w
(13) A concept employed to describe dispersion of measurements with respect to a measure of location or central tendency (ANSI N15.5-1982). A measurement with small random uncertainties is said to have high precision.
(NI) N42.23-1995
(14) The degree of mutual agreement between individual measurements, namely their repeatability and reproducibility. *See also:* accuracy. (SCC14) SI 10-1997

precision-approach radar (PAR) (navigation aid terms) A radar system located on an airfield for observation of the position of an aircraft with respect to an approach path and specifically intended to provide guidance to the aircraft during the approach. (AES/GCS) 686-1997, 172-1983w

precision connector (waveguide or transmission line) A connector that has the property of making connections with a high degree of repeatability without introducing significant reflections, loss or leakage. *See also:* auxiliary device to an instrument. (IM/HFIM) [40]

precision device (packaging machinery) A device that will operate within prescribed limits and will consistently repeat operations within those limits. (IA/PKG) 333-1980w

precision statistic An estimator of precision calculated from a finite sample of data using a specified formula.
(NI) N42.23-1995

precision wound (rotating machinery) A coil wound so that maximum nesting of the conductors occurs, usually with all crossovers at one end, with conductor aligned and positioned with respect to each adjacent conductor. *See also:* rotor; stator. (PE) [9]

precompiler (software) A computer program or routine that processes source code and generates equivalent code that is acceptable to a compiler. For example, a routine that converts structured FORTRAN to ANSI-standard FORTRAN. *See also:* preprocessor. (C) 610.12-1990

precondition A condition that is required to be true before making a property request. (C/SE) 1320.2-1998

preconditioning A control-function that provides for manually or automatically establishing a desired condition prior to normal operation of the system. *See also:* feedback control system. (IA/ICTL/IAC) [60]

preconditioning time The interval of time required by channel equipment (e.g., modems) to ready the channel for data transmission. (SUB/PE) 999-1992w

precursor *See:* undershoot.

predicted failure rate For the stated conditions of use, and taking into account the design of an item, the failure rate computed from the observed, assessed, or extrapolated failure rates of its parts. *Note:* Engineering and statistical assumptions shall be stated, as well as the bases used for the computation (observed or assessed). (R) [29]

predicted mean active maintenance time The mean active maintenance time of an item calculated by taking into account the reliability characteristics and the mean active maintenance time of all of its parts and other relevant factors according to the stated conditions. *Notes:* 1. Maintenance policy, statistical assumptions and computing methods shall be stated. 2. The source of the data shall be stated. (R) [29]

predicted mean life (non-repaired items) For the stated conditions of use, and taking into account the design of an item, the mean life computed from the observed, assessed or extrapolated mean life of its parts. *Note:* Engineering and statistical assumptions shall be stated, as well as the bases used for the computation (observed or assessed). (R) [29]

predicted reliability For the stated conditions of use, and taking into account the design of an item, the reliability computed from the observed, assessed, or extrapolated reliabilities of its parts. *Note:* Engineering and statistical assumptions shall be stated, as well as the bases used for the computation (observed or assessed). (R) [29], 1413-1998

predictive alarming (alarm monitoring and reporting systems for fossil-fueled power generating stations) A method of alerting the operator to a potential problem in time for him to respond and initiate corrective action to mitigate the problem. (PE/EDPG) 676-1986w

predictive assessment The process of using a predictive metric(s) to predict the value of another metric. (C/SE) 1061-1992s

predictive coding (image processing and pattern recognition) An image compression technique that uses the gray levels of preceding pixels to predict the gray level of the current pixel, so that only the difference between the predicted and measured value needs to be encoded. (C) 610.4-1990w

predictive maintenance The practice of conducting diagnostic tests and inspections during normal equipment operations in order to detect incipient weaknesses or impending failures. (IA/PSE) 902-1998

predictive metric A metric applied during development and used to predict the values of a software quality factor. (C/SE) 1061-1998

predictive metric value A numerical target related to a quality factor to be met during system development. This is an intermediate requirement that is an early indicator of final system performance. For example, design or code errors may be early predictors of final system reliability. (C/SE) 1061-1998

predictive model (modeling and simulation) A model in which the values of future states can be predicted or are hypothesized; for example, a model that predicts weather patterns based on the current value of temperature, humidity, wind speed, and so on at various locations. (C) 610.3-1989w

predissociation A process by which a molecule that has absorbed energy dissociates before it has had an opportunity to lose energy by radiation. (ED) 161-1971w

predistortion (system) (pre-emphasis) (transmitter performance) A process that is designed to emphasize or de-emphasize the magnitude of some frequency components with respect to the magnitude of others. *See also:* pre-emphasis. (VT) [37]

pre-emphasis (A) (pre-equalization) (General). A process in a system designed to emphasize the magnitude of some frequency components with respect to the magnitude of others, to reduce adverse effects, such as noise, in subsequent parts of the system. *Note:* After transmitting the pre-emphasized signal through the noisy part of the system, de-emphasis may be applied to restore the original signal with a minimum loss of signal-to-noise ratio. **(B) (pre-equalization)** (Modulating systems) (recording). An arbitrary change in the frequency response of a recording system from its basic response (such as constant velocity or amplitude) for the purpose of improvement in signal-to-noise ratio, or the reduction of distortion. (PE) 599-1985

pre-emphasis network A network inserted in a system in order to emphasize one range of frequencies with respect to another. *See also:* network analysis. (AP/ANT) 145-1983s

preempted thread A running thread whose execution is suspended due to another thread becoming runnable at a higher priority. (C/PA) 9945-1-1996

preempted process A running process whose execution is suspended due to another process becoming runnable at a higher priority. (C/PA) 1003.1b-1993s

preemption (1) (telephone switching systems) On a precedence call, the disconnection and subsequent reuse of part of an established connection of lower priority if all the relevant circuits are busy. (COM) 312-1977w
(2) Process in which the current bus master relinquishes the bus because another module has requested it. In some systems any module may cause preemption; in some systems only a module with a higher priority request may cause preemption. (C/BA) 896.3-1993w
(3) The release of the bus by the current bus master due to the request of another module. Note that in some systems, preemption occurs when the current bus master relinquishes the bus because another module has requested it. In some systems, any module may cause preemption; in some systems, only a module with a higher priority request may cause preemption. (C/BA) 896.4-1993w
(4) The release of the bus by the current bus master due to the request of another module. *Note:* In some systems, any module may cause preemption; in others, only a module with a higher priority request may cause preemption. (C/BA) 10857-1994

preemptive control (test, measurement, and diagnostic equipment) An action or function which, by reason of preestablished priority, is able to seize or interrupt the process in progress and cause to be performed a process of higher priority. (MIL) [2]

pre-fault (event) A qualifying term that refers to an interval ending with the inception of a fault. (SWG/PE) C37.100-1992

preference (channel supervisory control) (power-system communication) An assembly of devices arranged to prevent the transmission of any signals over a channel other than supervisory control signals when supervisory control signals are being transmitted. (Std100) [123]

preference level (speech quality measurements) The signal-to-noise ratio (S.N) of the speech reference signal when it is isopreferent to the speech test signal. 297-1969w

preferred (electric power system) (generating stations electric power system) That equipment and system configuration selected to supply the power system loads under normal conditions. (PE/EDPG) 505-1977r

preferred basic impulse insulation level (insulation strength) A basic impulse insulation level that has been adopted as a preferred American National Standard voltage value. *See also:* basic impulse insulation level. (EEC/LB) [100]

preferred current ratings (of distribution fuse links) A series of distribution fuse-link ratings so chosen from a series of preferred numbers that a specified degree of coordination may be obtained between adjacent sizes. (SWG/PE) C37.40-1993

preferred insulations system classification (electric installations on shipboard) The preferred insulation system classifications are classes A, B, F, H, C, or 105, 130, 155, 180, or greater than 220, and as designated by the equipment standard. (IA/MT) 45-1983s

preferred power supply (PPS) That power supply from the transmission system to the Class 1E distribution system that is preferred to furnish electric power under accident and post-accident conditions. (PE/NP) 308-1991, 765-1995

preferred values The preferred values for the parameters listed for various tests are preferred in the sense that their use promotes uniformity. However, specific applications may require values other than the listed preferred values. (PE) C62.34-1996

prefetching In a pipelined process, to fetch the next instruction, or instruction part, before the processing unit requires it, resulting in a performance improvement by eliminating the lag between completion of one instruction and the availability of the next. (C) 610.10-1994w

prefix code (telephone switching systems) One or more digits preceding the national or international number to implement direct distance dialing. (COM) 312-1977w

prefixes and symbols (International System of Units (SI)) Used to form names and symbols of the decimal mutiples and submultiples of the SI units:

Prefix	Abbreviation	Factor
tera- (megamega-	T (MM*)	10^{12}
giga- (kilomega-*)	G (kM*)	10^{9}
mega-	M	10^{6}
myria-		10^{4}
kilo-	k	10^{3}
hecto-	h	10^{2}
deka-		10
deci-	d	10^{-1}
centi-	c	10^{-2}
milli-	m	10^{-3}
decimilli-	dm	10^{-4}
micro-	μ	10^{-6}
nano- (millimicro-*)	n (mμ*)	10^{-9}
pico- (micromicro-*)	p (μμ*)	10^{-12}

*Deprecated.

These prefixes or their symbols are directly attached to names or symbols of units, forming multiples and submultiples of the units. In strict terms these must be called "multiples and submultiples of SI units," particularly in discussing the coherence of the system. In common parlance, the base units and derived units, along with their multiples and submultiples, are all called SI units. *See also:* units and letter symbols.

prefix multipliers The prefixes listed in the following table, when applied to the name of a unit, serve to form the designation of a unit greater or smaller than the original by the factor indicated.

Multiplication Factor	Prefix	Symbol
1 000 000 000 000 000 000 = 10^{18}	exa	E
1 000 000 000 000 000 = 10^{15}	peta	P
1 000 000 000 000 = 10^{12}	tera	T
1 000 000 000 = 10^{9}	giga	G
1 000 000 = 10^{6}	mega	M
1 000 = 10^{3}	kilo	k
100 = 10^{2}	hecto*	h
10 = 10^{1}	deka*	da
0.1 = 10^{-1}	deci*	d
0.01 = 10^{-2}	centi	c
0.001 = 10^{-3}	milli	m
0.000 001 = 10^{-6}	micro	μ
0.000 000 001 = 10^{-9}	nano	n
0.000 000 000 001 = 10^{-12}	pico	p
0.000 000 000 000 001 = 10^{-15}	femto	f
0.000 000 000 000 000 001 = 10^{-18}	atto	a

*Deprecated.

270-1966w

prefix notation (mathematics of computing) A parenthesis-free method of forming mathematical expressions devised by the Polish logician Jan Lukasiewicz, in which each operator is immediately followed by its operands. For example, A added to B and the result multiplied by C is expressed as X + ABC. *Synonyms:* parenthesis-free notation; Lukasiewicz notation; Polish notation. *Contrast:* postfix notation; infix notation. (C) 1084-1986w

preform (fiber optics) A glass structure from which an optical fiber waveguide may be drawn. *See also:* ion exchange technique; chemical vapor deposition technique; optical blank. (Std100) 812-1984w

preformed coil or coil side (rotating machinery) An element of a preformed winding, composed of conductor strands, usually insulated and sometimes transposed, cooling ducts in some designs, turn insulation where number of turns exceeds one, and coil insulation. *See also:* rotor; stator. (PE) [9]

preformed winding A winding consisting of coils which are given their shape before being assembled in the machine. (PE) [9]

P register A special-purpose instruction address register that holds the address of the next instruction to be fetched or executed. (C) 610.10-1994w

preheat (switch start) fluorescent lamp (illuminating engineering) A fluorescent lamp designed for operation in a circuit requiring a manual or automatic starting switch to preheat the electrodes in order to start the arc. (EEC/IE) [126]

preheating time (mercury-arc valve) The time required for all parts of the valve to attain operating temperature. (ED) [45], [84]

preheating time cathode (electron tube) The minimum period of time during which the heater voltage should be applied before the application of other electrod voltages. *See also:* heater current. (ED) [45]

preheat-starting (fluorescent lamps) (switch-starting systems) The designation given to those systems in which hot-cathode electric discharge lamps are started from preheated cathodes through the use of a starting switch, either manual or automatic in its operation. *Note:* The starting switch, when closed, connects the two cathodes in series in the ballast circuit so that current flows to heat the cathodes to emission temperature. When the switch is opened, a voltage surge is produced that initiates the discharge. Only the arc current flows through the cathodes after the lamp is in operation. *See also:* fluorescent lamp. (EEC/IE) [126]

preliminary design (A) (software) The process of analyzing design alternatives and defining the architecture, components, interfaces, and timing and sizing estimates for a system or component. *See also:* detailed design. **(B) (software)** The result of the process in definition "A." (C) 610.12-1990

preliminary design review (A) A review conducted to evaluate the progress, technical adequacy, and risk resolution of the selected design approach for one or more configuration items; to determine each design's compatibility with the requirements for the configuration item; to evaluate the degree of definition and assess the technical risk associated with the selected manufacturing methods and processes; to establish the existence and compatibility of the physical and functional interfaces among the configuration items and other items of equipment, facilities, software and personnel; and, as applicable, to evaluate the preliminary operational and support documents. *See also:* critical design review; system design review. **(B)** A review, as in definition "A" of any hardware or software component. (C) 610.12-1990

preliminary relay contacts Contacts that open or close in advance of other contacts when the relay is operating. (EEC/REE) [87]

premises wiring (system) That interior and exterior wiring, including power, lighting, control, and signal circuit wiring together with all of its associated hardware, fittings, and wiring devices, both permanently and temporarily installed, which extends from the load end of the service drop, or load end of the service lateral conductors to the outlet(s). Such wiring does not include wiring internal to appliances, fixtures, motors, controllers, motor control centers, and similar equipment. (NEC/NESC) [86]

premolded A joint that is factory molded in the shape which it will take when installed. Installation is performed by sliding the joint over the cable. The use of heat is not a part of the installation procedure. (PE/IC) 404-1993

pre-molded joint (power cable joints) A joint made of pre-molded components assembled in the field. (PE/IC) 404-1986s

preoperational system test (Class 1E power systems and equipment) A test to confirm that all individual component parts of a system function as a system and the system functions as designed. A preoperational test is performed following significant modifications or additions made to the facility at later dates. (PE/NP) 415-1986w

preoperational testing All testing required for system components prior to energizing or operating the major system component. (PE/EDPG) 1248-1998

preorder traversal The process of traversing a binary tree in a recursive fashion as follows: the root is visited, then the left subtree is traversed, then the right subtree is traversed. *Con-*

trast: inorder traversal; postorder traversal. *See also:* converse preorder traversal. (C) 610.5-1990w

PREP *See:* PRogrammed Electronics Patterns.

preparatory function (numerically controlled machines) A command changing the mode of operation of the control such as from positioning to contouring or calling for a fixed cycle of the machine.

prepatch panel *See:* problem board.

prepend To append to the beginning. For example, a Media Access Control (MAC) frame is prepended with a preamble and appended with a frame check sequence (FCS). (C/LM) 802.3-1998

preprocessing An operation performed before a primary process; for example, in pattern recognition, processing in which patterns are simplified to make classification easier. (C) 610.4-1990w

preprocessor (1) (software) A computer program or routine that carries out some processing step prior to the primary process; for example, a precompiler or other routine that reformats code or data for processing. *Contrast:* postprocessor. *See also:* D-TRAN; LYRIC; PDEL. (C) 610.12-1990, 610.13-1993w

(2) A device that effects preparatory computation or organization. (C) 610.10-1994w

pre-read head A read head that is placed before another read head and is used to read data before the same data are read by the other read head. (C) 610.10-1994w

prerecorded data medium A data medium on which certain preliminary items are preset; the remaining items are entered during subsequent operations. (C) 610.10-1994w

preregister operation (elevators) Operation in which signals to stop are registered in advance by buttons in the car and at the landings. At the proper point in the car travel, the operator in the car is notified by a signal, visual, audible, or otherwise, to initiate the stop, after which the landing stop is automatic. *See also:* control. (EEC/PE) [119]

prerequisite The specification in a software object that implies it shall not be installed until after some other software object is installed, and configured until after the other software object is configured. (C/PA) 1387.2-1995

pre-rip (cable plowing) A process using a plow blade to loosen the earth prior to plowing and installing the cable, flexible tube, etc. (T&D/PE) 590-1977w

prescribed surface (sound measurements) A hypothetical surface surrounding the machine on which sound measurements are made. (PE/EM) 85-1973w

prescriptive model A model used to convey the required behavior or properties of a proposed system; for example, a scale model or written specification used to convey to a computer supplier the physical and performance characteristics of a required computer. *Contrast:* descriptive model. (C) 610.3-1989w

preselector (A) A device placed ahead of a frequency converter or other device, that passes signals of desired frequencies and reduces others. **(B)** In automatic switching, a device that performs its selecting operation before seizing an idle trunk. (EEC/PE) [119]

presence tests (test, measurement, and diagnostic equipment) Actions which verify the presence or absence of signals or characteristics. Such signals or characteristics are those which are not tolerance critical to operation of the item. (MIL) [2]

present When describing an argument to a function call, the case where a pointer to an actual object of the type specified in the declaration is supplied by the caller (as opposed to a null pointer). (C/PA) 1238.1-1994w

presentation address (1) An unambiguous name that is used to identify a set of presentation service access points. Loosely, it is the network address of an OSI service. (C/PA) 1328.2-1993w, 1224.2-1993w, 1326.2-1993w, 1327.2-1993w

(2) A name unambiguous within the OSI environment that is used to identify a set of presentation service access points. (C/PA) 1238.1-1994w

present average demand An average value occurring during a current demand interval or subinterval (e.g., watts or voltamperes). (AMR/SCC31) 1377-1997

presentation graphics The use of a computer to produce high quality, high resolution graphical output. *Contrast:* information graphics. (C) 610.2-1987

presentation layer (1) (Layer 6) The layer of the ISO Reference Model that frees the application processes from concern with differences in data representation. (DIS/C) 1278.2-1995

(2) The sixth layer of the seven-layer OSI model; responsible for general user services related to the representation of user data. *Note:* This layer provides compression, encryption, character and file conversion on messages from the application layer. *See also:* application layer; data link layer; physical layer; network layer; sublayer; transport layer; session layer; client layer; entity layer; logical link control sublayer; medium access control sublayer. (C) 610.7-1995

preset To establish an initial condition, such as the control values of a loop. (C/C) [20], [85]

preset guidance (navigation aid terms) Guidance in which a predetermined path is set into the guidance mechanism of a craft and is not altered after launching. (AES/GCS) 172-1983w

preset speed A control function that establishes the desired operating speed of a drive before initiating the speed change. *See also:* electric drive. (IA/ICTL/IAC/APP) [60], [75]

preshoot (pulse terminology) A distortion which precedes a major transition. *Note:* Colloquial term which qualitatively describes a type of distortion. (IM/WM&A) 194-1977w

press To push and hold a mouse button. (C) 1295-1993w

pressing (disk recording) A pressing is a record produced in a record-molding press from a master or stamper. *See also:* phonograph pickup. (SP) [32]

pressure (solderless) connector A device that establishes a connection between two or more conductors or between one or more conductors and a terminal by means of mechanical pressure and without the use of solder. (NESC/NEC) [86]

pressure altimeter (navigation aid terms) An altimeter that measures and indicates altitude above a datum plane by means of an aneroid which responds to the change in atmospheric pressure with height. (AES/GCS) 172-1983w

pressure barrier seal (nuclear power generating station) Consists of an aperture seal and an electrical conductor seal. (PE/IM/NP) 380-1975w, [76]

pressure cable An oil-impregnated paper-insulated cable in which positive gauge pressure is maintained on the insulation under all operating conditions. (T&D/PE) [10]

pressure coefficient *See:* environmental coefficient.

pressure connector (packaging machinery) A conductor terminal applied with pressure so as to make the connection mechanically and electrically secure. (IA/PKG) 333-1980w

pressure-containing terminal box (rotating machinery) A terminal box so designed that the products of an electric breakdown within the box are completely contained inside the box. (PE) [9]

pressure controller (control systems for steam turbine-generator units) Includes only those components and control elements that generate one or more signal(s) for the control mechanism in response to pressure set point and pressure feedback signals for the purpose of controlling pressure. (PE/EDPG) 122-1985s

pressure control system (control systems for steam turbine-generator units) A system that controls the pressure at a sensing point in a designated location. Typically, it includes the pressure-sensing element, the controller, the control mechanism, and the controlled valve(s). (PE/EDPG) 122-1985s

pressure-gradient microphone A microphone in which the electric output substantially corresponds to a component of the gradient (space derivative) of the sound pressure. *Note:* Pressure-gradient microphones may be of any order; for example, zero, first, second, etc. Thus, a pressure microphone is a gradient microphone of zero order. The rms response to plane waves is proportional to $\cos^n\theta$, where θ is the angle of incidence and n is the order of the gradient. Because of the finite dimensions of all gradient microphones, however, the response characteristic and the directional characteristic, respectively, are only approximations of the derivative of the sound pressure and of the directional formula noted. (T&D/PE) 539-1990

pressure-lubricated bearing (rotating machinery) A bearing in which a continuous flow of lubricant is forced into the space between the journal and the bearing. *See also:* bearing. (PE) [9]

pressure microphone A microphone in which the electric output substantially corresponds to the instantaneous sound pressure of the impressed sound waves. *Note:* A pressure microphone is a gradient microphone of zero order and is nondirectional when its dimensions are small compared to a wavelength. *See also:* microphone. (T&D/PE/EEC) 539-1990, [119]

pressure reference changer (control systems for steam turbine-generator units) A device for producing the pressure reference signal to the pressure controller in response to a manual or automatic adjustment. (PE/EDPG) 122-1985s

pressure relay A relay that responds to liquid or gas pressure. (SWG/PE) C37.100-1992

pressure-relief device (arresters) A means for relieving internal pressure in an arrester and preventing explosive shattering of the housing, following prolonged passage of follow current or internal flashover of the arrester. (PE) [8]

pressure-relief terminal box (rotating machinery) A terminal box so designed that the products of an electric breakdown within the box are relieved through a pressure-relief diaphragm. (PE) [9]

pressure-relief test (arresters) A test made to ascertain that an arrester failure will not cause explosive shattering of the housing. (PE) [8]

pressure retaining boundary (nuclear power generating station) The pressure retaining boundary includes those surfaces of the aperture seal, the conductor feed-through plate, the conductor seal (or seals), and the conductor (or conductors) which are exposed to the containment environment. (PE/NP) 380-1975w

pressure-sensitive keyboard *See:* membrane keyboard.

pressure switch (1) A switch in which actuation of the contacts is effected at a predetermined liquid or gas pressure. (IA/ICTL/IAC) [60], [84]
(2) (power system device function numbers) A switch which operates on given values, or on a given rate of change, of pressure. (PE/SUB) C37.2-1979s

pressure system (protective signaling) A system for protecting a vault by maintaining a predetermined differential in air pressure between the inside and outside of the vault. Equalization of pressure resulting from opening the vault or cutting through the structure initiates an alarm condition in the protection circuit. *See also:* protective signaling. (EEC/PE) [119]

pressure-type pothead A pressure-type pothead is a pothead intended for use on positive-pressure cable systems. *See also:* single-pressure-zone potheads; multipressure-zone pothead. (PE) 48-1975s

pressure-type termination A Class 1C termination intended for use on positive pressure cable systems.

— Single pressure zone termination: A pressure-type termination intended to operate with one pressure zone.

— Multipressure zone termination: A pressure-type termination intended to operate with two or more pressure zones.

(PE/IC) 48-1996

pressure wire connector A device that establishes the connection between two or more conductors or between one or more conductors and a terminal by means of mechanical pressure and without the use of solder. (EEC/PE) [119]

pressurized (rotating machinery) Applied to a sealed machine in which the internal coolant is kept at a higher pressure than the surrounding medium. (PE) [9]

prestartup testing All testing required prior to rotating the generating unit under power (hydraulic or electrical) which is unique to the unit and not associated with system testing. (PE/EDPG) 1248-1998

prestore To store data that are required by a computer program or routine before the program or routine is entered. (C) 610.12-1990

prestressed-concrete structures Concrete structures that include metal tendons that are tensioned and anchored either before or after curing of the concrete. (NESC) C2-1997

prestrike The initiation of current between the contacts during a closing operation before the contacts have mechanically touched. (SWG/PE) C37.100-1992, C37.083-1999

prestrike current (lightning) The current that flows in a lightning stroke prior to the return stroke current. *See also:* direct-stroke protection. (T&D/PE) [10]

presumptive address *See:* base address.

presumptive instruction A computer instruction that is not an effective instruction until it has been modified in a prescribed manner. (C) 610.10-1994w

pretersonic Ultrasonic and with frequency higher than 500 megahertz. (UFFC) [122]

pre-tested iteration *See:* WHILE.

pretransmit-receive tube A gas-filled radio-frequency switching tube used to protect the transmit-receive tube from excessively high power and the receiver from frequencies other than the fundamental. *See also:* gas tube. (ED) 161-1971w

prettyprinting The use of indentation, blank lines, and other visual cues to show the logical structure of a program. (C) 610.12-1990

preventive autotransformer An autotransformer (or center-tapped reactor) used in load tap changing and regulating transformers, or step-voltage regulators to limit the circulating current when operating on a position in which two adjacent taps are bridged, or during the change of taps between adjacent position. (PE/TR) C57.131-1995, C57.12.80-1978r

preventive maintenance (1) (test, measurement, and diagnostic equipment) Tests, measurement, replacements, adjustments, repairs and similar activities, carried out with the intention of preventing faults or malfunctions from occurring during subsequent operation. Preventive maintenance is designed to keep equipment and programs in proper operating condition and is performed on a scheduled basis. (MIL) [2]
(2) The maintenance carried out at predetermined intervals or corresponding to prescribed criteria, and intended to reduce the probability of failure or the performance degradation of an item. (R) [29]
(3) (software) Maintenance performed for the purpose of preventing problems before they occur. (C) 610.12-1990
(4) Maintenance performed specifically to prevent faults from occurring. (C) 610.10-1994w
(5) The practice of conducting routine inspections, tests, and servicing so that impending troubles can be detected and reduced or eliminated. (IA/PSE) 902-1998
(6) A procedure in which the system is periodically checked and/or reconditioned to prevent or reduce the probability of failure or deterioration in subsequent service. (PE/NP) 933-1999

previously developed software Software that has been produced prior to or independent of the project for which the Plan is prepared, including software that is obtained or purchased from outside sources. (C/SE) 1228-1994

PRF *See:* pulse-repetition frequency.

PRI *See:* pulse-repetition interval.

Primacord *See:* explosives.

primaries (color) (television) The colors of constant chromaticity and variable amount that, when mixed in proper proportions, are used to produce or specify other colors. *Note:* Primaries need not be physically realizable. (BT/AV) 201-1979w

primary (1) (supervisory control, data acquisition, and automatic control) An equipment or subsystem that normally contributes to system operation. *See also:* backup. (SWG/PE/SUB) C37.100-1992, C37.1-1994
(2) (instrument transformers) The winding intended for connection to the circuit to be measured or controlled. (PE/TR) [57], C57.13-1993
(3) The part of a machine having windings that are connected to the power supply line (for a motor or transformer) or to the load (for a generator). (PE/EM) [9]
(4) (A) First to operate; for example, primary arcing contacts, primary detector. **(B)** First in preference; for example, primary protection. **(C)** Referring to the main circuit as contrasted to auxiliary or control circuits; for example, primary disconnecting devices. **(D)** Referring to the energy input side of transformers, or the conditions (voltages) usually encountered at this location; for example, primary unit substation. (SWG/PE) C37.100-1992

primary address (FASTBUS acquisition and control) An address assigned to a device by means of which a master is able to establish contact with the device or a subdivision of the device. Primary address types are logical, geographical and broadcast addresses. (NID) 960-1993

primary address cycle The portion of a FASTBUS operation in which a master addresses a slave on the address/data (AD) lines. The address type is specified by the EG and MS control lines. It begins with the master asserting the address sync (AS) line and terminates with the master receiving an address acknowledgment on the AK line. Logical, geographical or broadcast addresses are asserted during primary address cycles. (NID) 960-1993

primary arcing contacts (of a switching device) The contacts on which the initial arc is drawn and the final current, except for the arc-shunting-resistor current, is interrupted after the main contacts have parted. (SWG/PE) C37.100-1992

primary backplane bus A backplane bus that provides the main communications path among a cluster of modules. (C/BA) 14536-1995

primary battery *See:* primary cell; battery; electrochemistry.

primary bus The collection of signals that provides the system with the basic mechanism for exchanging data between boards. (C/BA) 896.9-1994w, 896.3-1993Gw

primary calibration (monitoring radioactivity in effluents) The determination of the electronic system accuracy when the detector is exposed in a known geometry to radiation from sources of known energies and activity levels traceable to the National Bureau of Standards (NBS). (NI) N42.18-1980r

primary cell A cell that produces electric current by electrochemical reactions without regard to the reversibility of those reactions. Some primary cells are reversible to a limited extent. *See also:* electrochemistry. (EEC/PE) [119]

primary center (1) (telephone switching systems) A toll office to which toll centers and toll points may be connected. Primary centers are classified as Class 3 offices. *See also:* office class. (COM) 312-1977w
(2) Class 3 office in the North American hierarchical routing plan; a control center connecting toll centers of the telephone system together. *See also:* regional center; sectional center; toll center; end office. (C) 610.7-1995

primary circuit The circuit on the input side of the regulator. (PE/TR) C57.15-1999

primary coating (fiber optics) The material in intimate contact with the cladding surface, applied to preserve the integrity of that surface. *See also:* cladding. (Std100) 812-1984w

primary-color unit (television) The area within a color cell occupied by one primary color. *See also:* television. (ED) 161-1971w

primary current ratio (electroplating) The ratio of the current densities produced on two specified parts of an electrode in the absence of polarization. It is equal to the reciprocal of the ratio of the effective resistances from the anode to the two specified parts of the cathode. *See also:* electroplating. (EEC/PE) [119]

primary data element A data element within a record that represents the subject of that record; for example, the data element "name" in a record containing "name," "city of birth," and "date of birth." *Contrast:* attribute data element. (C) 610.5-1990w

primary detecting element (automatic control) That portion of the feedback elements that first either utilizes or transforms energy from the controlled medium to produce a signal that is a function of the value of the directly controlled variable. (PE/EDPG) [3]

primary detector (1) (power systems) That portion of the measurement device which either utilizes or transforms energy from the controlled medium to produce a measurable effect which is a function of change in the value of the controlled variable. (PE/PSE) 94-1970w
(2) (or sensing element or initial element) The first system element or group of elements that responds quantitatively to the measurand and performs the initial measurement operation. A primary detector performs the initial conversion or control of measurement energy and does not include transformers, amplifiers, shunts, resistors, etc., when these are used as auxiliary means. *Synonyms:* sensing element; initial element. (SWG/PE) C37.100-1992

primary disconnecting devices (of a switchgear assembly) Self-coupling separable contacts provided to connect and disconnect the main circuits between the removable element and the housing. (SWG/PE) C37.100-1992

primary distribution feeder A feeder operating at primary voltage supplying a distribution circuit. *Note:* A primary feeder is usually considered as that portion of the primary conductors between the substation or point of supply and the center of distribution. (IA/CEM) [58]

primary distribution mains The conductors that feed from the center of distribution to direct primary loads or to transformers that feed secondary circuits. *See also:* center of distribution. (IA/CEM) [58]

primary distribution system A system of alternating-current distribution for supplying the primaries of distribution for supplying the primaries of distribution transformers from the generating station or substation distribution buses. *See also:* alternating-current distribution; center of distribution. (IA/CEM) [58]

primary distribution trunk line A line acting as a main source of supply to a distribution system. *See also:* center of distribution. (IA/CEM) [58]

primary electric shock An electric shock sufficiently severe to cause direct physiological harm. (T&D/PE) 539-1990

primary electron (thermionics) An electron in a primary emission. *See also:* electron emission. (ED) [45], [84]

primary fault The initial breakdown of the insulation of a conductor, usually followed by a flow of power current. *See also:* center of distribution. (IA/CEM) [58]

primary flow (carriers) A current flow that is responsible for the major properties of the device. (Std100) 102-1957w

primary grid emission (thermionic) Current produced by electrons or ions thermionically emitted from a grid. *See also:* electron emission. (ED) [45]

primary ground electrode A ground electrode specifically designed or adapted for discharging the ground fault current into the ground, often in a specific discharge pattern, as required (or implicitly called for) by the grounding system design. (PE/SUB) 80-2000

primary input (1) A node in a circuit in which the tester can apply stimulus. (SCC20) 1445-1998
(2) The point where a logic signal arrives at the boundary of the design as currently known to an electric design automation (EDA) application. For a complete integrated circuit design, for example, this point is the metal pad of an input or bidirectional pad cell. (C/DA) 1481-1999

primary key (1) (A) In sorting and searching, the key that is given the highest priority within a group of related keys. For example, after sorting, the values in the primary key will be in the given order, independent of the values of the other fields. *Synonyms:* major key; prime key. *Contrast:* secondary key. **(B)** In a relation, a specific minimal set of attributes that functionally determines all other attributes in the relation, and thus uniquely differentiates one entity from another. *Note:* More than one set of attributes with this property may exist. Each such set is known as a candidate key, but only one is chosen as the primary key. *See also:* alternate key; candidate key. (C) 610.5-1990
(2) The candidate key selected as the unique identifier of an entity. (C/SE) 1320.2-1998

primary line of sight (illuminating engineering) The line connecting the point of observation and the point of fixation. In terms of a single eye, it is the line connecting the point of fixation and the center of the entrance pupil. (EEC/IE) [126]

primary line-to-ground voltage (coupling capacitors and capacitance potential devices) Refers to the high-tension root-mean-square line-to-ground voltage of the phase to which the coupling capacitors or potential device, in combination with its coupling capacitor or bushing, is connected. *See also:* rated primary line-to-ground voltage. 31-1944w

primary_message The part of the interface event message carried by the DMA mechanism itself. The primary_message may point to message_extensions. (C/MM) 1212.1-1993

primary network A network supplying the primaries of transformers whose secondaries may be independent or connected to a secondary network. *See also:* center of distribution. (IA/CEM) [58]

primary oil containment A tank or enclosure designed for continuous containment of oil for operating or storage purposes. (SUB/PE) 980-1994

primary outage An outage occurrence within a related multiple outage event that occurs as a direct consequence of the initiating incident and is not dependent on any other outage occurrence. *Note:* A primary outage of a component or a unit may be caused by a fault on equipment within the unit or component or repair of a component within the unit. *See also:* related multiple outage event. (PE/PSE) 859-1987w

primary output (1) A node in a circuit in which the tester can observe a response. (SCC20) 1445-1998
(2) The point where a logic signal leaves the design as currently known to an electric design automation (EDA) application. For a complete integrated circuit design, for example, this point is the metal pad of an output or bidirectional pad cell. (C/DA) 1481-1999

primary output patterns (POPAT) A set of unique responses at the node in which a fault or a group of faults are detected. (SCC20) 1445-1998

primary overcurrent protective device of apparatus (nuclear power generating station) A device or apparatus which normally performs the function of circuit interruption. (PE/NP) 317-1976s

primary packet (1) A packet made up of whole quadlets that contains a transaction code in the first quadlet. Any packet that is not an acknowledge or a PHY packet. (C/MM) 1394-1995
(2) Any packet that is not an acknowledge or a physical layer (PHY) packet. A primary packet is an integral number of quadlets and contains a transaction code in the first quadlet. (C/MM) 1394a-2000

primary protection (as applied to a relay system) First-choice relay protection in contrast with backup relay protection. (SWG/PE/PSR) C37.100-1992, C37.90-1978s

primary radar (1) A radar system, subsystem, or mode of operation in which the return signals are the echoes obtained by reflection from the target. Since this is the normal method of radar operation, the word *primary* is omitted unless necessary to distinguish it from *secondary*. *See also:* secondary radar. (AES/GCS) 172-1983w
(2) A radar system in which the return signals are the echoes obtained by reflection from the target. Since this is the normal method of radar operation, the word *primary* is omitted unless necessary to distinguish it from *secondary*. *See also:* secondary radar. (AES) 686-1997

primary radiator The radiating element of a reflector or lens antenna that is coupled to the transmitter or receiver directly, or through a feed line. *Note:* For some applications, an array of radiating elements is employed. (AP/ANT) 145-1993

primary reactor starter A starter that includes a reactor connected in series with the primary winding of an induction motor to furnish reduced voltage for starting. It includes the necessary switching mechanism for cutting out the reactor and connecting the motor to the line. *See also:* starter. (IA/ICTL/IAC/APP) [60], [75]

primary representation The form in which the service supplies an attribute value to the client. (C/PA) 1328-1993w, 1327-1993w, 1224-19931w

primary resistor starter A starter that includes a resistor connected in series with the primary winding of an induction motor to furnish reduced voltage for staring. It includes the necessary switching mechanism for cutting out the resistor and connecting the motor to the line. *See also:* starter. (IA/ICTL/IAC) [60]

primary ring In a dual ring, this is the ring over which application data (LLC frames) are exchanged during normal operation. MAC1, which transmits the application data, is attached to the primary ring. When a dual ring station is in either WRAPA or WRAPB, MAC1 is switched to remain in the operational section of the wrapped ring. The primary ring uses the links ARx and BTx in each dual ring station. (LM/C) 802.5c-1991r

primary section of the core The section of the core of a ferroresonant transformer on which the primary winding is wound. (PEL) 449-1998

primary service area (radio broadcast transmitter) The area within which reception is not normally subject to objectionable interference or fading. *See also:* radio transmitter. (EEC/PE) [119]

primary shock A shock of such a magnitude that it may produce direct physiological harm. The results of primary shock are fibrillation, respiratory tetanus, and muscle contraction. (T&D/PE) 1048-1990, 524a-1993r

primary signal A signal at the interface between the physical device and the physical tester. Any and all information meant for test is defined on these signals; test translators need process these signals only. (C/TT) 1450-1999

primary socket identifier (PSID) A socket number identifying a particular endpoint on the primary device. (C/MM) 1284.4-2000

primary space allocation (data management) The amount of space that is reserved for a particular file when it is initially defined. *Contrast:* secondary space allocation. (C) 610.5-1990w

primary standard (luminous standards) (illuminating engineering) A light source by which the unit of light is established and from which the values of other standards are derived. This order of standard also is designated as the national standard. *Note:* A satisfactory primary (national) standard must be reproducible from specifications. Primary (national) standards usually are found in national physical laboratories, such as the National Bureau of Standards. *See also:* candela. (IE/EEC) [126]

primary state The state on a node that is initialized by a power_reset. For example, the CSRs defined by IEEE Std 1212-1991, CSR Architecture, are part of the node's primary state. (C/MM) 1212-1991s

primary station (1) The station that, at any given instant, has the right to select and to transmit information to a secondary station and the responsibility to insure information transfer. *See also:* secondary station. (C) 610.7-1995
(2) As defined by the infrared link access protocol (IrLAP), the station on the data link that assumes responsibility for the organization of data flow and for unrecoverable data link error conditions. It issues commands to the secondary stations and gives them permission to transmit. (EMB/MIB) 1073.3.2-2000

primary storage *See:* main storage.

primary supply voltage (mobile communication) The voltage range over which a radio transmitter, a radio receiver, or selective signaling equipment is designed to operate without degradation in performance. *See also:* mobile communication system. (VT) [37]

primary switchgear connections *See:* main switchgear connections.

primary transmission feeder A feeder connected to a primary transmission circuit. *See also:* center of distribution. (T&D/PE) [10]

primary unit substation (1) (power and distribution transformers) A substation in which the low-voltage section is rated above 1000 V. (PE/TR) C57.12.80-1978r
(2) *See also:* unit substation. (SWG/PE) C37.100-1981s

primary voltage rating of a general-purpose specialty transformer (power and distribution transformers) The input circuit voltage for which the primary winding is designed, and to which operating and performance characteristics are referred. (PE/TR) C57.12.80-1978r

primary winding (1) (power and distribution transformers) The winding on the energy input side. (PE/TR) C57.12.80-1978r
(2) (voltage regulators) The shunt winding. *See also:* voltage regulator. (PE/TR) C57.15-1968s
(3) (rotating machinery) (motor or generator) The winding carrying the current and voltage of incoming power (for a motor) or power output (for a generator). The choice of what constitutes a primary circuit is arbitrary for certain machines having bilateral power flow. In a synchronous or direct-current machine, this is more commonly called the armature winding. *See also:* armature. (PE) [9]
(4) (instrument transformers) The winding intended for connection to the circuit to be measured or controlled. (PE/TR) C57.13-1993, [57]
(5) The winding of the ferroresonant transformer to which the input voltage is applied. (PEL) 449-1998

primary window A window in which the main interaction between the user and an object or application takes place (also known as main window). (C) 1295-1993w

primary zone The part of a relay's protected zone where the relay operates with no intentional time delay. (PE/PSR) C37.113-1999

prime (charge-storage tubes) To charge storage elements to a potential suitable for writing. *Note:* This is a form of erasing. *See also:* television; charge-storage tube. (ED) 161-1971w

prime attribute An attribute that forms all or part of the primary key of a relation. *Contrast:* nonprime attribute. (C) 610.5-1990w

prime key *See:* primary key.

prime meridian (navigation aid terms) The meridiam of longitude 0°, almost universally considered as Greenwich, England. (AES/GCS) 172-1983w

prime mover (emergency and standby power) The machine used to develop mechanical horsepower to drive an emergency or standby generator to produce electrical power. (IA/PSE) 446-1995

prime power (1) The maximum potential power (chemical, mechanical, or hydraulic) constantly available for transformation into electric power. *See also:* generating station. (T&D/PE) [10]
(2) The source of supply of electrical energy that is normally available and used continuously day and night, usually supplied by an electric utility company, but sometimes supplied by base-loaded user-owned generation. (IA/PSE) 446-1995

prime source In the event that several vendors offer pin-for-pin compatible components, the prime source is the vendor who introduced the component type. (TT/C) 1149.1-1990

priming rate (charge-storage tubes) The time rate of priming a storage element, line, or area from one specified level to another. Note the distinction between this and "priming speed." *See also:* charge-storage tube. (ED) 158-1962w

priming speed (charge-storage tubes) The lineal scanning rate of the beam across the storage surface in priming. Note the distinction between this and "priming rate." *See also:* charge-storage tube. (ED) 158-1962w, 161-1971w

primitive (1) (software reliability) Data relating to the development or use of software that is used in developing measures or quantitative descriptions of software. Primitives are directly measurable or countable, or may be given a constant value or condition for a specific measure. Examples include: error, failure, fault, time, time interval, date, number of noncommentary source code statements, edges, and nodes. (C/SE) 982.2-1988, 982.1-1988
(2) The lowest level for which data is collected. (C/SE) 1045-1992
(3) A basic or fundamental unit, often referring to the lowest level of machine instruction or the lowest unit of a language. (C) 610.10-1994w
(4) (local area networks) A definition of a service provided by a sublayer to the sublayer immediately above. Primitives may be initiated by either the upper or lower sublayer. (C) 8802-12-1998
(5) A mechanism for communicating between the entities of different layers of the medical information bus (MIB) communications standard (e.g., between the Physical layer and Data Link layer entities). A primitive may request a particular service, provide indication of a request, provide a response to an indication, or provide confirmation of a response. (EMB/MIB) 1073.4.1-2000
(6) *See also:* display element. (C) 610.6-1991w

primitive attribute A characteristic that applies to a display element. For example, color, line style, character size. (C) 610.6-1991w

primitive Boolean function A Boolean expression having the property that all other Boolean expressions can be constructed using it alone. (C) 1084-1986w

primitive data structure A data structure that can be directly operated upon by machine-level instructions. Examples include integer, real, character, logical, and pointer. *Contrast:* nonprimitive data structure. (C) 610.5-1990w

primitive period *See:* period.

primitive service A service that accesses or manipulates defined elements of AI-ESTATE–conformant models (e.g., get, put, create, delete). (SCC20) 1232.2-1998

primitive type *See:* atomic type.

principal axis (1) (close-talking pressure-type microphone) The axis of a microphone normal to the plate of the principal acoustic entrance of a microphone, and that passes through the center of the entrance. *See also:* close-talking pressure-type microphones. (SP) 258-1965w
(2) (transducer used for sound emission or reception) A reference direction used in describing the directional characteristics of the transducer. It is usually an axis of structural

symmetry, or the direction of maximum response, but if one of these does not coincide with the reference direction, it must be described explicitly. (SP) [32]

principal axis of compliance (accelerometer) (gyros) An axis along which an applied force results in a displacement along that axis only. The acceleration-squared error due to anisoelasticity is zero when acceleration is along a principal axis of compliance. (AES/GYAC) 528-1994

principal branch (main branch) A branch involved in the major transfer of energy from one side of the converter to the other. (IA/ID/SPC) 995-1987w, 936-1987w

principal characteristics *See:* principal voltage-current characteristic.

principal-city office (telephone switching systems) An intermediate office that has the screening and routing capabilities to accept traffic to all end office within one or more numbering-plan areas. (COM) 312-1977w

principal current (1) (self-commutated converters) (circuit properties) (of a converter switching element or branch) The on-state current of the semiconductor devices in a switching element or branch flowing between its principal terminals. *Note:* The principal current is often referred to as the "current" of the switching element or branch. (IA/SPC) 936-1987w

(2) (thyristor) A generic term for the current through the collector junction. *Note:* It is the current though both main terminals. (IA/IPC) 428-1981w

principal E-plane (1) For a linearly polarized antenna, the plane containing the electric field vector and the direction of maximum radiation. (AP/ANT) 145-1993

(2) (linearly polarized antenna) The plane containing the electric field vector and the direction of maximum radiation. *See also:* radiation; antenna. (AP/ANT) [35]

principal half-power beamwidths For a pattern whose major lobe has a half-power contour that is essentially elliptical, the half-power beamwidths in the two pattern cuts that contain the major and minor axes of the ellipse, respectively. (AP/ANT) 145-1993

principal H-plane *See:* principal H-plane.

principal power (thyristor) The power which is consumed in the load circuit plus the losses in the power circuit elements including switching losses. (IA/IPC) 428-1981w

principal restraint (accelerometer) The means by which a measurable force or torque is generated to oppose the force or torque produced by an acceleration along or about an input axis. (AES/GYAC) 528-1984s

principal terminal (self-commutated converters) (converter circuit elements) A terminal (of a device or circuit element) through which passes the current transmitting the power that is controlled by the device or circuit element. The term is used for distinction from control terminals, monitoring signal terminals, etc. *Note:* Examples of principal terminals are the anode and cathode of thyristor or diode devices, the collector and emitter of bipolar transistor devices, and the source and drain of field-effect transistor devices. (IA/SPC) 936-1987w, 995-1987

principal voltage (thyristor) The voltage between the main terminals. *Notes:* 1. In the case of reverse blocking and reverse conducting thyristors, the principal voltage is called positive when the anode potential is higher than the cathode potential, and called negative when the anode potential is lower than the cathode potential. 2. For bidirectional thyristors, the principal voltage is called positive when the potential of main terminal 2 is higher than the potential of main terminal 1. (IA/IPC) 428-1981w

principal voltage-current characteristic (thyristor) The function, usually represented graphically, relating the principal voltage to the principal current with gate current, where applicable, as a parameter. *Synonym:* principal characteristics. (IA/ED) 223-1966w, [46], [12]

printable character (1) A character in the range 0x21 through 0x7E or the range 0xA1 through 0xFE. (C/BA) 1275-1994

(2) One of the characters included in the print character classification of the LC_CTYPE category in the current locale. (C/PA) 9945-2-1993

print bar *See:* type bar.

print chain In a chain printer, a revolving carrier on which the type slugs of an impact printer are mounted. *Synonym:* print train. (C) 610.10-1994w

print contrast ratio In optical character recognition, the ratio obtained by subtracting the reflectance at an inspection area from the maximum reflectance found within a specified distance from that area, and dividing the result by that maximum reflectance. *Contrast:* print contrast signal. (C) 610.2-1987

print contrast signal In optical character recognition, a measure of the contrast between a printed character and the paper on which the character is printed. *Contrast:* print contrast ratio. (C) 610.2-1987

print control character A control character for print operations such as line spacing, page ejection, or carriage return. (C) 610.5-1990w

print controller The parts within a printer that perform the processing required to generate an image. *Contrast:* print engine. (C) 610.10-1994w

print data set A data set in which data that is to be printed are stored. (C) 610.5-1990w

print drum In a drum printer, a rotating cylinder that presents characters at more than one printing position. (C) 610.10-1994w

printed card form The layout or format of the printed matter on a card; the printed matter usually describes the purpose of the card and designates the precise location of card fields. (C) 610.10-1994w

printed circuit (1) (soldered connections) A pattern comprising printed wiring formed in a predetermined design in, or attached to, the surface or surfaces of a common base. (EEC/AWM) [105]

(2) A circuit in which the conducting wires are "printed" as conductive strips on an insulating board. *Synonym:* etched circuit. *See also:* printed circuit board. (C) 610.10-1994w

printed circuit antenna An antenna of some desired shape bonded onto a dielectric substrate. *Note:* The microstrip antenna is a notable example. *See also:* microstrip antenna. (AP/ANT) 145-1993

printed-circuit assembly A printed-circuit board on which separately manufactured component parts have been added. (EEC/AWM) [105]

printed circuit board A circuit board onto which the pattern of copper foil connecting the components has been etched or printed. *Note:* The term "card" is often used synonymously with "printed circuit board." *Synonym:* printed circuit card. *Contrast:* wire-wrapped board. *See also:* plugboard. (C) 610.10-1994w

(2) (A) A board for mounting of components on which most connections are made by printed circuitry. **(B)** A board having printed circuits on both sides. **(C)** A board having printed circuits on one side only. (Std100) [123]

printed circuit card *See:* printed circuit board.

printed wiring (soldered connections) A portion of a printed circuit comprising a conductor pattern for the purpose of providing point-to-point electric connection only. (EEC/AWM) [105]

printed wiring board (PWB) (1) A generic form that includes other interconnection boards. (C/BA) 1101.3-1993

(2) A generic term that includes any interconnection board. (C/BA) 1101.4-1993

(3) Any interconnection board. (C/BA) 1101.7-1995

print element An interchangeable unit employed in element printers that contains a complete set of type slugs. By changing the print element, one can change the character font, size, and density. Examples include "daisy wheels," "golf balls," and "thimbles." *Synonym:* type element. (C) 610.10-1994w

print engine **(1)** The mechanism within a printer that actually transforms the desired image to the paper. *Contrast:* print controller. (C) 610.10-1994w
(2) That set of electrical and mechanical mechanisms that move the print media or paper and marks that paper. (C/MM) 1284.1-1997

printer **(1)** **(teletypewriter) (teleprinter)** A printing telegraph instrument having a signal-actuated mechanism for automatically printing received messages. It may have a keyboard similar to that of a typewriter for sending messages. The term receiving-only is applied to a printer having no keyboard. *See also:* telegraphy. (COM) [49]
(2) An output device that produces a hard copy record of data mainly in the form of discrete graphic characters belonging to one or more predetermined character sets. *See also:* graphic printer; character-at-a-time printer; teleprinter; line printer; color printer; high-speed printer; serial printer; page printer; continuous-stream printer; character printer; impact printer; bidirectional printer; nonimpact printer. (C) 610.10-1994w
(3) An intelligent device that includes, as a primary function, the ability to convert an electrically transmitted or stored image into a physical image formed by colorant on some medium (such as paper). (C/MM) 1284.1-1997

printer driver An application software component that allows the computer system to control and communicate with a particular printer without concern for the printer's hardware characteristics. (C) 610.10-1994w

printer font A font that resides in or is intended for a printer. *Note:* Can be internal, downloaded, or on a font cartridge. *Contrast:* screen font. *See also:* online font; internal font. (C) 610.10-1994w

printer interface control unit (PICU) The set of electronics that interfaces external communications ports, common peripheral interfaces (such as font cards or disk drives), the logical units, and the marking engine. It is the function of the printer interface control unit to coordinate and sequence all the functions and operations of the printer. (C/MM) 1284.1-1997

printer language The human language used for the American National Standard Code for Information Interchange (ASCII) strings within all command and response messages, other than those to be printed or those to be displayed on the local or remote consoles (e.g., English, French, German). (C/MM) 1284.1-1997

print formatter *See:* text formatter.

print formatting *See:* text formatting.

print head **(A)** A head within a printer that mechanically controls the creation of an image on paper. **(B)** A term commonly applied to that component in a dot-matrix printer that is responsible for forming characters using a pattern of dots. (C) 610.10-1994

printing That set of operations implemented by the printer that results in an image rendered as marks on the selected media. (C/MM) 1284.1-1997

printing demand meter An integrated demand meter that prints on a paper tape the demand for each demand interval and indicates the time during which the demand occurred. *See also:* electricity meter. (EEC/PE) [119]

printing line The writing line on a printer. *See also:* printing position. (C) 610.10-1994w

printing position **(A)** One character position in a printing line. **(B)** The location of the printer head. (C) 610.10-1994

printing recorder (protective signaling) An electromechanical recording device that accepts electric signal impulses from transmitting circuits and converts them to a printed record of the signal received. *See also:* protective signaling. (EEC/PE) [119]

print media That consumable upon which the marking engine marks so as to form a text and/or pictorial image, typically paper. (C/MM) 1284.1-1997

printout **(1)** **(test, measurement, and diagnostic equipment)** The output of a device which is printed on some type of printer. (MIL) [2]
(2) Computer output printed on paper. *See also:* playback. (C) 610.10-1994w

print record A record in a print data set. (C) 610.5-1990w

print server On a network, a server that is dedicated to queuing and sending printer output from the networked computers to a shared printer. *See also:* database server; file server; disk server; mail server; network server; terminal server. (C) 610.7-1995

print through An undesired transfer of a recorded signal from one part of a magnetic medium to another part when these parts are brought into close proximity. (C/SP) 610.10-1994w, [32]

print train In a train printer, a track in which the type slugs are engaged. *See also:* print chain. (C) 610.10-1994w

print wheel In a wheel printer, a rotating disk that presents the characters of a character set at a single printing position. *Synonym:* type wheel. (C) 610.10-1994w

print wire One of a set of wires that is used in a dot-matrix printer to transfer ink to the paper. (C) 610.10-1994w

priority **(1)** **(computer graphics)** A segment attribute that determines which of several overlapping segments is closer to the viewer. (C) 610.6-1991w
(2) **(software)** The level of importance assigned to an item. (C) 610.12-1990
(3) **(multiprocessor architecture)** A bus request protocol in which the module with the highest arbitration number acquires the bus. (C/MM) 896.1-1987s
(4) A nonnegative integer associated with processes or threads, whose value is constrained to a range defined by the applicable scheduling policy. Numerically higher values represent higher priorities. (C/PA) 9945-1-1996
(5) A rank order of status, activities, or tasks. Priority is particularly important when resources are limited. (C/SE) 1362-1998
(6) (use in primitives) A parameter used to convey the priority required or desired. (C/LM/CC) 8802-2-1998
(7) The general term for an integer-valued attribute of processes, tasks, messages, and asynchronous I/O operations, whose value is used in selecting among entities of the same kind. Numerically higher values represent higher priorities and are given preference for selection over lower priorities. The *priority of an Ada task* is an integer that indicates a degree of urgency and is the basis for resolving competing demands of tasks for resources. Unless otherwise specified, whenever tasks compete for processors or other implementation-defined resources, the resources are allocated to the task with the highest priority value. The *base priority* of a task is the priority with which it was created, or to which it was later set by `Dynamic_Priorities.Set_Priority`. At all times, a task also has an *active priority*, which generally reflects its base priority as well as any priority it inherits from other sources. *Priority inheritance* is the process by which the priority of a task or other entity (*e.g.*, a protected object is used in the evaluation of the active priority of another task. At any time, the active priority of a task is the maximum of all the priorities the task is inheriting at that instant. (C) 1003.5-1999

priority-based scheduling Scheduling in which the selection of a running thread is determined by the priority of the runnable threads. (C/PA) 9945-1-1996

priority inheritance This occurs when a lower priority message at the head of the queue uses the priority of a higher priority message that is blocked. (C/BA) 896.3-1993w

priority interrupt An interrupt performed to permit execution of a process that has a higher priority than the process currently executing. (C) 610.12-1990

priority interrupt bus One of the four buses provided by the backplane. The priority interrupt bus allows interrupter modules to send interrupt requests to interrupt-handler modules. (C/BA) 1014-1987

priority inversion Condition that can occur when a higher priority activity arrives while a lower priority activity is using the shared resource, and the higher priority activity cannot preempt the lower priority activity. (C/BA) 896.3-1993w

priority queue A list to which items may be appended to or retrieved from any position, depending on some property of the item being added or removed. *Note:* This data structure is misnamed in that it contradicts the definition of "queue." (C) 610.5-1990w

priority resolution A mechanism that allows a local device and its link partner to resolve to a single mode of operation given a set of prioritized rules governing resolution. (C/LM) 802.3-1998

Priority Resolution Table The look-up table used by Auto-Negotiation to select the network connection type where more than one common network ability exists (100BASE-TX, 100BASE-T4, 10BASE-T, etc.). The priority resolution table defines the relative hierarchy of connection types from the highest common denominator to the lowest common denominator. (C/LM) 802.3-1998

priority string (power-system communication) A series connection of logic circuits such that inputs are accommodated in accordance with their position in the string, one end of the string corresponding to the highest priority. *See also:* digital. (Std100) [123]

priority-tagged frame A tagged frame whose tag header carries priority information, but carries no VLAN identification information. *See also:* VLAN-tagged frame; untagged frame. (C/LM) 802.1Q-1998

privacy The service used to prevent the content of messages from being read by other than the intended recipients. (C/LM) 8802-11-1999

privacy mode (local area networks) A mode in which an end node receives only those packets specifically addressed to it. (C) 8802-12-1998

privacy protection The establishment of appropriate administrative, technical, and physical safeguards to ensure the security and confidentiality of data records and to protect both security and confidentiality against any anticipated threats or hazards that could result in substantial harm, embarrassment, inconvenience, or unfairness to any individual about whom such information is maintained. (C/SE) J-STD-016-1995

privacy system (radio transmission) A system designed to make unauthorized reception difficult. *See also:* radio transmission. (EEC/PE) [119]

private **(1)** A design feature intended solely for use by the component manufacturer. (TT/C) 1149.1-1990
(2) A responsibility that is visible only to the class or the receiving instance of the class (available only within methods of the class). *Contrast:* protected; public. *See also:* hidden. (C/SE) 1320.2-1998

private automatic branch exchange **(1) (telephone switching systems) (data transmission)** A private branch exchange that is automatic. (COM/PE) 312-1977w, 599-1985w
(2) *See also:* private branch exchange. (C) 610.7-1995

private automatic exchange **(1) (telephone switching systems) (data transmission)** A private nonbranch exchange that is automatic. (COM/PE) 312-1977w, 599-1985w
(2) A telephone exchange that provides private telephone service to an organization and does not allow calls to be transmitted to or from the public telephone network. (C) 610.7-1995

private branch exchange **(1) (telephone switching systems) (data transmission)** A private telecommunications exchange that includes access to a public telecommunications exchange. (COM/PE) 312-1977w, 599-1985w
(2) A telephone exchange on the user's premises, providing a switching facility for telephones on extension lines within the premises and access to the public telephone network. *Synonym:* private automatic branch exchange. (C) 610.7-1995

private branch exchange hunting (telephone switching systems) An arrangement for searching over a group of trunks at the central office, any one of which would provide a connection to the desired private branch exchange. (COM) 312-1977w

private-branch-exchange trunk (telephone switching systems) A line used as a trunk between a private branch exchange and the central office that serves it. (COM) 312-1977w

private circuit switching network (PCSN) A private network that only provides circuit switching functions, except that it may be able to transport packetized "user-to-user" information passed over the signalling channel. (LM/C/COM) 8802-9-1996

private data Data, associated with a package, that is used by the methods of that package but is not intended for use by other software. (C/BA) 1275-1994

private exchange A telephone exchange serving a single organization and having no means for connection with a public telephone exchange. (EEC/COM/CON) [28], [48]

private line **(1) (data transmission) (private wire)** A channel or circuit furnished to a subscriber for the subscriber's exclusive use. (PE) 599-1985w
(2) *See also:* leased line. (C) 610.7-1995

private line service A service in which the customer leases a circuit, not connected to the public switched telephone network, for the customer's exclusive use. (C) 610.7-1995

private line telegraph network (data transmission) A system of points interconnected by leased telegraph channels and providing hard-copy or five-track punched paper tape, or both, at both sending and receiving points. (PE) 599-1985w

private line telephone network (data transmission) A series of points interconnected by leased voice-grade telephone lines, with switching facilities or exchange operated by the customer. (PE) 599-1985w

private network A network established and operated by a private organization in which the customer leases circuits and, sometimes, switching capacity for the customer's use. *Contrast:* public data network. *See also:* software defined network. (C) 610.7-1995

private non-branch exchange (telephone switching systems) A series of points interconnected by leased voice-grade telephone lines, with switching facilities or exchange operated by the customer. (Std100) [123]

private object The internal representation of the language-independent model of an object in a service, and thus unspecified. (C/PA) 1328-1993w, 1238.1-1994w, 1327-1993w, 1224-1993w

private OM object The internal representation of the language-independent model of an OM object in a service, and thus unspecified. (C/PA) 1327.2-1993w, 1224.2-1993w, 1328.2-1993w, 1326.2-1993w

private packet/frame and circuit switching network (PPCSN) A private switching network that provides both circuit and packet/frame switching functions (i.e., all the functions of both PCSNs and PPSNs). (LM/C/COM) 8802-9-1996

private packet/frame switching network (PPSN) A private network that only provides packet/frame switching functions. (LM/C/COM) 8802-9-1996

private page A memory page that has been reserved for the exclusive use of a particular agency or function. Access credentials may be required to reference the page. (SCC32) 1455-1999

private residence A separate dwelling or a separate apartment in a multiple dwelling that is occupied only by the members of a single family unit. (EEC/PE) [119]

private residence elevator A power passenger electric elevator, installed in a private residence, and that has a rated load not in excess of 700 pounds, a rated speed not in excess of 50 feet per minute, a net inside platform area not in excess of 12 square feet, and a rise not in excess of 50 feet. *See also:* elevator. (EEC/PE) [119]

private-residence inclined lift A power passenger lift, installed on a stairway in a private residence, for raising and lowering persons from one floor to another. *See also:* elevator. (EEC/PE) [119]

private switching network (PSN) A private network that provides switching functions (circuit and/or packet/frame switching). It is operated by the user and located on user premises to cover the communications needs in the user's domain. The term private-switching network includes both the private circuit-switching network and the private packet-switching network. (LM/C/COM) 8802-9-1996

private telecommunication exchange (telephone switching systems) A telecommunications exchange for a single organization. (COM) 312-1977w

private telephone network A telephone network set up solely to meet the requirements of the particular organization. (COM/TA) 823-1989w

private type A data type whose structure and possible values are defined but are not revealed to the user of the type. *See also:* information hiding. (C) 610.12-1990

privileged An instruction (or register) that can only be executed (or accessed) when the processor is in supervisor mode (PSR.S = 1). *See also:* supervisor mode. (C/MM) 1754-1994

privileged instruction (software) A computer instruction that can be executed only by a supervisory program. (C) 610.12-1990

privileged state *See:* supervisor state.

probabilistic model *See:* stochastic model.

probabilistic risk assessment (PRA) A calculation of the probability and consequences of various known and postulated accidents. (PE/NP) 933-1999

probability density function (1) (A) (control of system electromagnetic compatibility) The first derivative of the probability distribution function; it represents the probability of obtaining a given value. **(B)** The derivative of the distribution function $P(x)$. (EMC) C63.12-1987
(2) (overhead power lines) The derivative of the probability distribution function $P(x)$. *Note:* An expression giving the probability of a discrete random variable x as a function of x or, for continuous random variables, the probability in an elemental range dx. The total probability is unity or 100%, so that the probability density function represents the proportion of probabilities for particular values of x. (T&D/PE) 539-1990
(3) Pertaining to a real random variable x, the derivative with respect to an arbitrary value X of the variable x, of the probability distribution function of X, if a derivative exists. *Note:* The mathematical expression for this function is

$$g(X) = \frac{d}{dX}[f(X)] = \frac{d}{dX}[P(x \le X)]$$

(CS/PE/EDPG) [3]

probability distribution (nuclear power generating station) The mathematical function that relates the probability of an event to an elapsed time or to a number of trials. (PE/NP) 352-1975s

probability distribution function (1) (control of system electromagnetic compatibility) ***[P(x)]*** The function of x whose value is the probability that the amplitude is greater than, or equal to, x. *Notes:* 1. The probability distribution function is a nondecreasing function ranging from zero to unity. C63.12-1984
(2) (reliability analysis of nuclear power generating station safety systems) The mathematical function that gives $(X < x)$ where X is a random variable and x is a particular value of X. (PE/NP) 352-1987r
(3) (overhead power lines) ***[P(x)]*** The probability that a parameter is less than or equal to a given value x. *Notes:* 1. The distribution function $P(x)$ for a random variable x is the total frequency of occurrence of members with particular random-variable values less than or equal to x. The total frequency of occurrence of all values of x is unity or 100%, so that the distribution function is the proportion of members bearing values less than or equal to x. Similarly, for n particular values of the random variable, $x_1, x_2, \ldots x_n$, the distribution function $P(x_1, x_2, \ldots x_n)$ is the frequency of values less than or equal to x_1 for the first values, x_2 for the second, and so on. 2. The terms "cumulative probability distribution" and, very often, simply "distribution" are used to denote the probability distribution function. (T&D/PE) 539-1990
(4) Pertaining to a real random variable x, the function of an arbitrary value X - X of this variable, whose value is the probability, P, that the random variable is less than or equal to X. *Note:* The mathematical expression for this function is $f(X) = P(x \le X)$. (CS/PE/EDPG) [3]

probability of failure indices The probability of a component failing to respond to a command, or responding when it should not = number of failures to respond as intended/exposure operations. (PE/PSE) 859-1987w

probability of failure to close on command Probability of failure to close on command = number of failures to close/number of commands to close. (PE/PSE) 859-1987w

probability of failure to open on command Probability of failure to open on command = number of failures to open/number of commands to open. *Note:* This index can be calculated separately for commands to open under fault and without fault. (PE/PSE) 859-1987w

probability of failure to operate on command Probability of failure to operate on command = number of failures to operate/number of commands to operate. (PE/PSE) 859-1987w

probability of occurrence The asymptotic value of the frequency of occurrence of the event. (T&D/PE) 539-1990

probability paper A graph paper with the grid along the ordinate specially ruled so that the distribution function of a specified distribution can be plotted as a straight line against the variable on the abscissa. These specially ruled grids are available for the normal, binomial, Poisson, lognormal, and Weibull distributions, respectively. (T&D/PE) 539-1990

probe (1) (gas) (potential) An auxiliary electrode of small dimensions compared with the gas volume, that is placed in a gas tube to determine the space potential. (ED) [45], [84]
(2) A tester instrument used to observe the state of a node. (SCC20) 1445-1998

probeable node Any node that is physically accessible to a tester probe. (SCC20) 1445-1998

probe address The address of a device that is known when the associated FCode program begins execution. (C/BA) 1275-1994

pro beam system Tunnel lighting system or luminaires having a light distribution that is greater in the direction of travel. (RL) C136.27-1996

probe coil A coil (air or magnetic material core) used to sense an alternating magnetic field. (COM/TA) 1027-1996

probe loading The effect of a probe on a network, for example, on a slotted line, the loading represented by a shunt admittance or a discontinuity described by a reflection coefficient. *See also:* measurement system. (IM/HFIM) [40]

probe pickup, residual *See:* residual probe pickup.

probe window The period of time during a pattern when a probe can capture activity on a node. (SCC20) 1445-1998

probing A fault diagnostic technique that incorporates the use of a portable device (hand-held or robotic) to monitor or capture unit under test (UUT) response data. The location of the probe placement is determined by the circuit response and the circuit topology. (SCC20) 1445-1998

problem *See:* benchmark problem.

problem board In an analog computer, a removable frame of receptacles for patch cords and plugs that offers a means for interconnecting the inputs and outputs of computing elements. *See also:* patch board; patch panel. (C) 610.10-1994w, 165-1977w

problem check (analog computer) One or more tests used to assist in obtaining the correct machine solution to a problem. Static check consists of one or more tests of computing elements, their interconnections, or both, performed under static conditions. Dynamic check consists of one or more tests of computing elements, their interconnections, or both, performed under dynamic conditions. Rate test is a test that verifies that the time constants of the integrators are those selected. This term also refers to the computer-control state that implements the rate test previously described. Dynamic problem check is any dynamic check used to ascertain the correct performance of some or all of the computer components. *See also:* computer-control state. (C) 165-1977w

Problem Descriptor System (PDS/MaGen) A programming language useful in a wide variety of operations research applications, and designed to facilitate the generation of matrices and reports for mathematical programming systems. (C) 610.13-1993w

problem domain A set of similar problems that occur in an environment and lend themselves to common solutions. (C/SE) 1362-1998

problem-oriented language (1) (computers) A programming language designed for the convenient expression of a given class of problems. (MIL/C) [2], [85], [20]
(2) (software) A programming language designed for the solution of a given class of problems. Examples are list processing languages, information retrieval languages, simulation languages. (C) 610.12-1990, 610.13-1993w

problem state In the operation of a computer system, a state in which programs other than the supervisory program can execute. *Synonyms:* user state; slave state. *Contrast:* supervisor state. (C) 610.12-1990

problem variable *See:* scale factor.

procedural cohesion (software) A type of cohesion in which the tasks performed by a software module all contribute to a given program procedure, such as an iteration or decision process. *Contrast:* temporal cohesion; logical cohesion; sequential cohesion; coincidental cohesion; functional cohesion; communicational cohesion. (C) 610.12-1990

procedural interface (PI) The set of C functions used by an application and a delay and power calculation module (DPCM) to exchange information and determine the timing calculation for a design. (C/DA) 1481-1999

procedural interface function One of the C functions that comprise the DPCS procedural interface. (C/DA) 1481-1999

procedural language (1) (software) A programming language in which the user states a specific set of instructions that the computer must perform in a given sequence. All widely-used programming languages are of this type. *Synonym:* procedure-oriented language. *Contrast:* nonprocedural language. *See also:* algebraic language; list processing language; logic programming language; algorithmic language. (C) 610.12-1990
(2) A computer language in which the user states a specific set of instructions that the computer must perform in a given sequence. Examples include BASIC, COBOL, FORTRAN, and Pascal. *Synonym:* procedure-oriented language. *Contrast:* nonprocedural language. (C) 610.13-1993w

procedural programming language (software unit testing) A computer programming language used to express the sequence of operations to be performed by a computer (for example, COBOL). *See also:* nonprocedural programming language. (C/SE) 1008-1987r

procedure (1) (computers) The course of action taken for the solution of a problem. (C) [20], [85]
(2) (nuclear power quality assurance) A document that specifies or describes how an activity is to be performed. (PE/NP) [124]
(3) (A) (software) A course of action to be taken to perform a given task. **(B) (software)** A written description of a course of action as in definition "A;" for example, a documented test procedure. **(C) (software)** A portion of a computer program that is named and that performs a specific action. (C) 610.12-1990
(4) (software user documentation) Ordered series of instructions that a user follows to do one or more tasks. (C/SE) 1063-1987r
(5) (scheme programming language) A parameterized program fragment, called a subroutine or function in some programming languages. (C/MM) 1178-1990r

procedure-oriented language (1) (computers) A programming language designed for the convenient expression of procedures used in the solution of a wide class of problems. (MIL/C) [2], [20], [85]
(2) (software) *See also:* procedural language. (C) 610.12-1990

process (1) (automatic control) The collective functions performed in and by the equipment in which a variable is to be controlled. *Synonym:* controlled system. (PE/EDPG) [3]
(2) (A) (software) A sequence of steps performed for a given purpose; for example, the software development process. **(B) (software)** An executable unit managed by an operating system scheduler. *See also:* job; task. **(C) (software)** To perform operations on data. (C) 610.12-1990
(3) An address space and one or more threads of control that execute within that address space, and their required system resources. (C/PA) 14252-1996
(4) A sequence of tasks, actions, or activities, including the transition criteria for progressing from one to the next, that bring about a result. (C/SE) 1220-1994s
(5) An address space and the single thread of control that executes within that address space, and its required system resources. A process is created by another process issuing the POSIX.1 *fork*() function. The process that issues *fork*() is known as the parent process, and the new process created by the *fork*() is known as the child process. The attributes of processes required by POSIX.2 form a subset of those in POSIX.1. (C/PA) 9945-2-1993
(6) An address space with one or more threads executing within that address space, and the required system resources for those threads. A process is created by another process issuing the *fork*() function. The process that issues *fork*() is known as the parent process, and the new process created by the *fork*() is known as the child process. Many of the system resources defined by this part of ISO/IEC 9945 are shared among all of the threads within a process. These include the process ID; the parent process ID; the process group ID; the session membership; the real, effective and saved-set user ID; the real, effective and saved-set group ID; the supplementary group IDs; the current working directory; the root directory; the file mode creation mask; and file descriptors. (C/PA) 9945-1-1996
(7) An address space, and the program (including any Ada tasks contained within the program) executing within that address space, and its required system resources. A process is created by another process with procedures `POSIX_Process_Primitives.Start_Process`, `POSIX_Process_Primitives.Start_Process_Search`, or the function `POSIX_Unsafe_Process_Primitives.Fork`. The process that issues `Start_Process`, `Start_Process_Search`, or `Fork` is known as the parent process, and the newly created process is the child process. (C/PA) 1003.5-1992r
(8) An address space, a single thread of control that executes within that address space, and its required system resources. On a system that implements threads, a process is redefined to consist of an address space with one or more threads ex-

ecuting within that address space and their required system resources. *Note:* The term process is used in contrast to "system process," or the OSI usage of the term "application process."
(C/PA) 1327.2-1993w, 1224.2-1993w, 1326.2-1993w, 1328.2-1993w
(9) An organized set of activities performed for a given purpose; for example, the software development process.
(C/SE) J-STD-016-1995
(10) A unit of activity characterized by a single sequential thread of execution, a current state, and an associated set of system resources. (C/MM) 855-1990
(11) Sequence of operations performed in and by the equipment in which a variable is to be controlled.
(SCC20) 1226-1998
(12) **(A)** A set of interrelated activities, which transforms inputs into outputs. *Note:* The term "activities" covers use of resources. **(B)** A series of actions bringing about a result.
(C/SE) 1490-1998
(13) Consists of all execution within a single distinct address space supported by the operating system of a computer.
(IM/ST) 1451.1-1999
(14) *See also:* POSIX process. (C) 1003.5-1999

Process A function that must be performed in the software life cycle. A Process is composed of Activities.
(C/SE) 1074-1995s

process architect The person or group that has primary responsibility for creating and maintaining the software life cycle process (SLCP). *See also:* software life cycle process.
(C/SE) 1074-1997

processable scored card A scored card including at least one separable part that can be processed after separation. *See also:* stub card. (C) 610.10-1994w

Process and Experiment Automation Realtime Language (PEARL) A general-purpose, high-order language designed to meet the requirements of real-time programming in process and experiment automation. (C) 610.13-1993w

Process Architect The person or group that manages the implementation of the Standard in an organization.
(C/SE) 1074.1-1995

process bound *See:* compute-bound.

process control (1) (electric pipe heating systems) The use of electric pipe heating systems to increase or maintain, or both, the temperature of fluids (or processes) in mechanical piping systems including pipes, pumps, tanks, instrumentation in nuclear power generating stations. (PE/EDPG) 622A-1984r
(2) (automatic control) Control imposed upon physical or chemical changes in a material. *See also:* feedback control system. (PE/EDPG) [3]
(3) (electric heat tracing systems) The use of electric heat tracing systems to increase or maintain, or both, the temperature of fluids (or processes) in mechanical piping systems including pipes, pumps, valves, tanks, instrumentation, etc, in power generating stations. (PE/EDPG) 622B-1988r
(4) Automatic control in which a computer is used to regulate continuous operations such as chemical processes, military operations, or manufacturing operations. *See also:* numerical control. (C) 610.2-1987

process equipment (automatic control) Apparatus with which physical or chemical changes in a material are produced. *Synonym:* plant. (PE/EDPG) [3]

process group (1) A collection of processes that permits the signaling of related processes. Each process in the system is a member of a process group that is identified by a process group ID. A newly created process joins the process group of its creator. (C/PA) 9945-1-1996, 9945-2-1993
(2) A collection of processes that permits the signaling of related processes. Each process in the system is a member of a process group that is identified by its process group ID. A newly created process joins the process group of its creator.
(C) 1003.5-1999

process group ID (1) The unique identifier representing a process group during its lifetime. A process group ID is a positive integer that can be contained in a *pid_t*. It shall not be reused by the system until the process group lifetime ends.
(C/PA) 9945-1-1996, 9945-2-1993
(2) A unique value identifying a process group during its lifetime. A process group ID shall not be reused by the system until the process group lifetime ends. (C) 1003.5-1999

process group leader (1) A process whose process ID is the same as its process group ID.
(C/PA) 9945-1-1996, 9945-2-1993
(2) The unique process, within a process group, that created the process group. (C) 1003.5-1999

process group lifetime (1) A period of time that begins when a process group is created and ends when the last remaining process in the group leaves the group, due either to the end of the last process's process lifetime or to the last remaining process calling the *setsid*() or *setpgid*() functions.
(C/PA) 9945-1-1996
(2) A period of time that begins when a process group is created and ends when the last remaining process in the group leaves the group, due either to the end of the process lifetime of the last process or to the last remaining process calling the `Set_Process_Group_ID` procedure. (C) 1003.5-1999

process ID (1) The unique identifier representing a process. A process ID is a positive integer that can be contained in a *pid_t*. A process ID shall not be reused by the system until the process lifetime ends. In addition, if there exists a process group whose process group ID is equal to that process ID, the process ID shall not be reused by the system until the process group lifetime ends. A process that is not a system process shall not have a process ID of 1.
(C/PA) 9945-1-1996, 9945-2-1993
(2) A unique value identifying a process during its lifetime. The process ID is a value of the type `Process_ID` defined in the package `POSIX_-Process_Identification`. A process ID shall not be reused by the system until the process lifetime ends. In addition, if a process group exists where the process ID of the process group leader is equal to that process ID, that process ID shall not be reused by the system until the process group lifetime ends. An implementation shall reserve a value of process ID for use by system processes. A process that is not a system process shall not have this process ID.
(C) 1003.5-1999

processing *See:* multiprocessing; parallel processing; data processing; information processing.

processing cycle A single, complete execution of data processing that is periodically repeated. *Synonym:* data processing cycle. *See also:* daily cycle; monthly cycle; weekly cycle; annual cycle. (C) 610.2-1987

processing unit A functional unit that consists of one or more processors and their storage. *See also:* central processing unit.
(C) 610.10-1994w

process lifetime (1) The period of time that begins when a process is created and ends when its process ID is returned to the system. After a process is created with a *fork*() function, it is considered active. At least one thread of control and the address space exist until it terminates. It then enters an inactive state where certain resources may be returned to the system, although some resources, such as the process ID, are still in use. When another process executes a *wait*() or *waitpid*() function for an inactive process, the remaining resources are returned to the system. The last resource to be returned to the system is the process ID. At this time, the lifetime of the process ends. (C/PA) 9945-1-1996
(2) A period of time that begins when a process is created and ends when its process ID is returned to the system. After a process is created, it is considered active. Its threads of control and address space exist until it terminates. It then enters an inactive state where certain resources may be returned to the system, although some resources, such as the process ID, are still in use. When another process executes a

`Wait_For_Child_Process` procedure for an inactive process, the remaining resources are returned to the system. The last resource to be returned to the system is the process ID. At this time, the lifetime of the process ends. (C) 1003.5-1999

process list An ordered set of runnable processes that all have the same ordinal value for their priority. The ordering of processes on the list is determined by a scheduling policy or policies. The set of process lists includes all runnable processes in the system. (C/PA) 1003.1b-1993s

process management The direction, control, and coordination or work performed to develop a product or perform a service. Example is quality assurance. (C) 610.12-1990

process metric A metric used to measure characteristics of the methods, techniques, and tools employed in developing, implementing, and maintaining the software system. (C/SE) 1061-1998

process model A model of the processes performed by a system; for example, a model that represents the software development process as a sequence of phases. *Contrast:* structural model. (C) 610.3-1989w

processor (1) (A) (computers) (hardware). A data processor. **(B) (computers)** (pascal computer programming language). A system or mechanism that accepts a program as input, prepares it for execution, and executes the process so defined with data to produce results. *Note:* A processor may consist of an interpreter, a compiler and run-time system, or other mechanism, together with an associated host computing machine and operating system, or other mechanism for achieving the same effect. A compiler in itself, for example, does not constitute a processor. (Std100/SUB/PE) 812-1984, C37.1-1994
(2) (software) A computer program that includes the compiling, assembling, translating, and related functions for a specific programming language, for example, Cobol processor, Fortran processor. *See also:* multiprocessor. (C) [20], [85]
(3) The combination of the IU, FPU, and CP (if present). (C/MM) 1754-1994
(4) (A) A device that interprets and executes instructions, consisting of at least an instruction control unit and an arithmetic unit. *See also:* coprocessor; preprocessor. **(B)** A device that contains a central processing unit. (C) 610.10-1994

Processor A main system processor unit that executes operating system code and manages system resources. It is usually constrained on the number of CSRs it can devote to the functions of a given I/O Unit. (C/MM) 1212.1-1993

processor architecture The system-visible interfaces to a central processor, including its instruction set, stack and register structures, and trap and interrupt-handling methods. (C) 610.10-1994w

process-oriented simulation A simulation in which the process is considered more important than the outcome; for example, a model of a radar system in which the objective is to replicate exactly the radar's operation, and duplication of its results is a lesser concern. *Contrast:* outcome-oriented simulation. (C) 610.3-1989w

processor interface (PI) (FASTBUS acquisition and control) The interface device between a processor and a FASTBUS segment. (NID) 960-1993

processor storage *See:* internal storage.

process standard A standard that deals with the series of actions or operations used in making or achieving a product. (C) 610.12-1990

process step Any task performed in the development, implementation, or maintenance of software (for example, identifying the software components of a system as part of the design). (C/SE) 1061-1992s

processtag (microprocessor operating systems parameter types) A "tag" returned by one function for use by another. Its contents may not be examined or changed. Its form is system dependent. A processtag is only valid within a given process and should not be passed between processes. (C/MM) 855-1985s

process-to-process communication The transfer of data between processes. (C) 1003.5-1999

procurement document (nuclear power quality assurance) Purchase requisitions, purchase orders, drawings, contracts, specifications, or instructions used to define requirements for purchase. (PE/NP) [124]

procurement documents (nuclear power generating station) Those documents such as specifications, contracts, letters of intent, work orders, purchase orders or proposals and their acceptance which authorize the seller to perform services or supply equipment, material or facilities to the purchaser. *Note:* This term applies specifically to the subject matter of IEEE Std 467-1980. (PE/NP) 467-1980w

producer (1) The node on a ringlet that transmits a send packet to the consumer and deletes the echo packet that is returned. (C/MM) 1596-1992
(2) A unit that adds a message to a DMA queue. (C/MM) 1212.1-1993

product (1) (mathematics of computing) The result of a multiplication operation. (C) 1084-1986w
(2) (data management) A relational operator that builds a relation from two specified relations consisting of all possible concatenated pairs of tuples, one from each of the two original relations. (See the corresponding figure.)

S		T		S × T	
A	×	X	=	A	X
B		Y		A	Y
		Z		A	Z
				B	X
				B	Y
				B	Z

product

(C) 610.5-1990w
(3) An element of the physical or system architecture, specification tree, or system breakdown structure that is a subordinate element to the system and is comprised of two or more subsystems. It represents a major consumer product (e.g., automobile, airplane) of a system or a major life-cycle process product (e.g., simulator, building, robot) related to a life-cycle process that supports a product or group of products. (C/SE) 1220-1994s
(4) A software object used to define a set of related software. Filesets are contained within products. (C/PA) 1387.2-1995
(5) Any output of the software development Activities (e.g., document, code, or model). *See also:* Activity. (C/SE) 1074-1997

product analysis The process of evaluating a product by manual or automated means to determine if the product has certain characteristics. (C) 610.12-1990

product and process data package The evolving output of the systems engineering process that documents hardware designs, software designs with their associated documentation, and life-cycle processes. (C/SE) 1220-1994s

product baseline In configuration management, the initial approved technical documentation (including, for software, the source code listing) defining a configuration item during the production, operation, maintenance, and logistic support of its life cycle. *Contrast:* developmental configuration; functional baseline; allocated baseline. *See also:* product configuration identification. (C) 610.12-1990

product certification *See:* certification.

product characteristic An observable attribute of a product. This includes functional, physical, and performance characteristics (e.g., gain and bandwidth of an amplifier). (SCC20) 1226-1998

product code *See:* bar code; universal product code.

product configuration identification The current approved or conditionally approved technical documentation defining a configuration item during the production, operation, maintenance, and logistic support phases of its life cycle. It prescribes all necessary physical or form, fit and function characteristics of a configuration item, the selected functional characteristics designated for production acceptance testing, and the production acceptance tests. *Contrast:* functional configuration identification; allocated configuration identification. *See also:* product baseline. (C) 610.12-1990

Product Data Exchange Specification A computer graphics standard that provides a method for representing and exchanging complete information among computer graphics systems, without requiring human interpretation. It was developed by an international team led by the National Institute of Standards and Technology (formerly National Bureau of Standards). *See also:* Initial Graphics Exchange Specification. (C) 610.6-1991w

Product Data Exchange using STEP A specification for representing product data using ISO/DIS 10303 STEP. (ATLAS) 1226-1993s

product engineering The technical processes to define, design, and construct or assemble a product. (C) 610.12-1990

production (routine) (power cable joints) Tests made on joint components or subassemblies during production for the purpose of quality control. (PE/IC) 404-1986s

production library (software) A software library containing software approved for current operational use. *Contrast:* software repository; system library; software development library; master library. (C) 610.12-1990

production tests (1) Those tests made to check the quality and uniformity of the workmanship and materials used in the manufacture of switchgear or its components. (SWG/PE) C37.100-1992, C37.81-1989r

(2) (routine) (power cable joints) Tests made on joint components or sub-assemblies during production for the purpose of quality control. (PE/IC) 404-1986s

(3) Tests made for quality control by the manufacturer on every device or representative samples, or on parts or materials as required to verify during production that the product meets the design specifications and applicable standards. *Note:* Certain quality assurance tests on identified critical parts of repetitive high-production devices may be tested on a planned statistical sampling basis. *Synonym:* routine tests (SWG/PE/SWG-OLD) C37.20.1-1993r, C37.20.4-1996, C37.21-1985r, C37.20.2-1993, C37.20.3-1996, C37.20.6-1997

(4) (circuit breakers) Those tests made to check the quality and uniformity of the workmanship and materials used in the manufacture of gas-insulated equipment. (SUB/PE) C37.122-1983s

(5) Those tests made to check the quality and uniformity of workmanship and materials used in the manufacturing of generator circuit breakers. (SWG/PE) C37.013-1997

(6) *See also:* routine tests. (SPD/PE) C62.11-1999

productivity ratio The relationship of an output primitive to its corresponding input primitive. (C/SE) 1045-1992

product line A collection of systems that are potentially derivable from a single domain architecture. (C/SE) 1517-1999

product management The definition, coordination, and control of the characteristics of a product during its development cycle. Example is configuration management. (C) 610.12-1990

product metric A metric used to measure the characteristics of any intermediate or final product of the software development process. (C/SE) 1061-1998

product modulator A modulator whose modulated output is substantially equal to the product of the carrier and the modulating wave. *Note:* The term implies a device in which intermodulation between components of the modulating wave does not occur. *See also:* modulation. (EEC/PE) [119]

product relay A relay that operates in response to a suitable product of two alternating electrical input quantities. (SWG/PE) C37.100-1992

product sensitivity The ratio of Hall voltage to the product of control current and magnetic flux density at any point on the product sensitivity characteristic curve of a Hall generator. (MAG) 296-1969w

product specification (A) A document that specifies the design that production copies of a system or component must implement. *Note:* For software, this document describes the as-built version of the software. *See also:* design description. **(B)** A document that describes the characteristics of a planned or existing product for consideration by potential customers or users. (C) 610.12-1990

product specification file (PSF) The input file used to define the structure and attributes of software objects and related files to be packaged by the `swpackage` utility. (C/PA) 1387.2-1995

product standard A standard that defines what constitutes completeness and acceptability of items that are used or produced, formally or informally, during the software engineering process. (C) 610.12-1990

product support The providing of information, assistance, and training to install and make software operational in its intended environment and to distribute improved capabilities to users. (C) 610.12-1990

product verification The process of verification of the traceability of a calibration standard of a manufacturer by NIST. This is achieved by the submission of a source from a manufacturer to NIST for verification of the value certified by the manufacturer. The reported value is compared to the NIST value, and if appropriate, a report of traceability is issued. (NI) N42.22-1995

professional projector The professional projector is a type using 35- or 70-millimeter film which has a minimum width of 1 3/8 inches and has on each edge 5.4 perforations per inch, or a type using carbon arc, Xenon, or other light source equipment which develops hazardous gases, dust or radiation. (NESC/NEC) [86]

professional standard (software) A standard that identifies a profession as a discipline and distinguishes it from other professions. (C) 610.12-1990

PROFILE A computer language used to match, score, and retrieve statistical data. (C) 610.13-1993w

profile (1) (overhead power lines) A diagram showing the variation of a quantity or parameter with location. (T&D/PE) 539-1990

(2) (overhead power lines) *See also:* step index profile; power-law index profile; graded index profile; index profile; parabolic profile. 812-1984w

(3) A set of one or more base standards and, where applicable, the identification of chosen classes, subsets, options, and parameters of those base standards that are necessary for accomplishing a particular function. (C/PA) 14252-1996

(4) The thickness of the module. (C/BA/C/BA) 1101.4-1993, 1101.3-1993

(5) The maximum thickness of the module. (C/BA) 1101.7-1995

(6) A set of one or more base standards, and, where applicable, the identification of chosen classes, subsets, options, and parameters of those base standards, necessary for accomplishing a particular function. In this standard that function is to provide communication services appropriate for DIS simulation applications. (DIS/C) 1278.2-1995

(7) *See also:* application environment profile. (C/BA) 896.2-1991w

(8) A unique numeric identifier that indicates a set of communications parameter values to be used in establishing a physical session. (SCC32) 1455-1999

Profile A module The assembled unit, containing a Futurebus+ interface and one or more nodes complying with Profile A, that is inserted into a compatible Profile A slot. Profile A modules may operate in systems complying with other profiles if the system meets Profile A physical requirements and if modules support a compatible transaction set when sharing data. (C/BA) 896.2-1991w

Profile A system The assembly of hardware elements made up of, at a minimum, the Profile A compliant backplane and subrack, power supply, fans, etc. Modules complying with other profiles may operate compatibly with Profile A systems and modules if they meet Profile A physical requirements and if their features constitute an identity or a superset of those implemented in a Profile A system as per Tables 59 and 60. (C/BA) 896.2-1991w

Profile B module The assembled plug-in unit, containing a Futurebus+ interface and one or two nodes complying with Profile B, which is inserted into a compatible Futurebus+ slot. Profile B modules may operate compatibly in systems complying with other profiles if the system meets Profile B mechanical requirements, and if non-Profile B nodes properly subset their transaction set when addressing Profile B modules, as specified in this profile. (C/BA) 896.2-1991w

Profile B system The assembly of hardware elements made up of, at a minimum, the Profile B compliant backplane and card cage, power supply, air mover, and a bridge to the rest of the system or to another bus. Modules complying with other profiles may operate compatibly with Profile B systems and modules if they meet Profile B mechanical requirements, and if their features constitute an identity or a superset of those mandated in this profile. (C/BA) 896.2-1991w

profile dispersion (A) (fiber optics) In an optical waveguide, that dispersion attributable to the variation of refractive index contrast with wavelength, where contrast refers to the difference between the maximum refractive index in the core and the refractive index of the homogeneous cladding. Profile dispersion is usually characterized by the profile dispersion parameter, defined by the following entry. **(B) (fiber optics)** In an optical waveguide, that dispersion attributable to the variation of refractive index profile with wavelength. The profile variation has two contributors:

1) variation in refractive index contrast, and
2) variation in profile parameter.

See also: distortion; refractive index profile; dispersion. (Std100) 812-1984

profile dispersion parameter (fiber optics)

$$P(\lambda) = \frac{n_1}{N_1}\frac{\lambda}{\Delta}\frac{d\Delta}{d\lambda}$$

where n_1, N_1 are, respectively, the refractive and group indices of the core, and

$$n_1\sqrt{1-2\Delta}$$

is the refractive index of the homogeneous cladding, $N_1 = n_1 - \lambda(dn_1/d\lambda)$, and Δ is the refractive index constant. Sometimes it is defined with the factor (-2) in the numerator. *See also:* dispersion. (Std100) 812-1984w

Profile F module The assembled unit, containing a Futurebus+ interface and one or two nodes complying with Profile F, which is inserted into a compatible Futurebus+ slot. Profile F modules may operate compatibly in systems complying with other profiles if the system meets Profile F mechanical requirements, and if non-Profile F nodes properly subset their transaction set when addressing Profile F modules, as specified in this profile. (C/BA) 896.2-1991w

Profile F system The assembly of hardware elements made up of, at a minimum, the Profile F compliant backplane and card cage, power supply, and an air mover. Modules complying with other profiles may operate compatibly with Profile F systems and modules if they meet Profile F mechanical requirements, and if their features constitute an identity or a superset of those mandated in this profile. (C/BA) 896.2-1991w

profile parameter (fiber optics) The shape-defining parameter, *g*, for a power-law index profile. *See also:* refractive index profile; power-law index profile. (Std100) 812-1984w

Profile S module The assembled unit, containing a Futurebus+ interface and one or two nodes complying with Profile S, which is inserted into a compatible Futurebus+ slot. (C/BA) 896.10-1997

Profile S system The assembly of hardware elements made up of, at a minimum, the Profile S compliant backplane, card cage, and a power supply. Modules complying with other profiles may operate compatibly with Profile S systems and modules if they meet Profile S mechanical requirements, and if their features constitute an identity or a superset of those mandated in this profile. (C/BA) 896.10-1997

prognosis (test, measurement, and diagnostic equipment) The use of test data in the evaluation of a system or equipment for potential or impending malfunctions. (MIL) [2]

program (1) (general) A sequence of signals transmitted for entertainment or information. (SP) 151-1965w
(2) (A) (electronic computation) A plan for solving a problem. **(B) (electronic computation)** Loosely, a routine. **(C) (electronic computation)** To devise a plan for solving a problem. **(D) (electronic computation)** Loosely, to write a routine. *See also:* communication; source program; object program; target program; computer program; programmed acceleration. (ED/ED/C) 581-1978, 641-1987, 162-1963
(3) (telephone switching systems) A set of instructions arranged in a predetermined sequence to direct the performance of a planned action or actions. (COM) 312-1977w
(4) (semiconductor memory) The inputs that when true enable programming, or writing into, a programmable read only memory (PROM). (TT/C) 662-1980s
(5) (software) To write a computer program. (C) 610.12-1990
(6) A prepared sequence of instructions to the system to accomplish a defined task. The term *program* in POSIX.2 encompasses applications written in the Shell Command Language, complex utility input languages (for example, awk, lex, sed, etc.), and high-level languages. (C/PA) 9945-2-1993
(7) The process of incorporating digital data onto an integrated circuit. (C) 610.10-1994w
(8) A collection of processes working together to accomplish a common task. (C/MM) 855-1990
(9) The operation of injecting electrons onto the floating gate of the memory cell. (ED) 1005-1998
(10) A set of partitions, which can execute in parallel with one another, possibly in a separate address space and possibly on a separate computer. (C) 1003.5-1999

program amplifier *See:* line amplifier.

program architecture (software) The structure and relationships among the components of a computer program. The program architecture may also include the program's interface with its operational environment. *See also:* computer program; component. (C/SE) 729-1983s

program attention key *See:* attention key.

program block (software) In problem-oriented languages, a computer program subdivision that serves to group related statements, delimit routines, specify storage allocation, delineate the applicability of labels, or segment paths of the computer program for other purposes. *See also:* computer program; label; segment; routine. (C/SE) 729-1983s

program correctness *See:* correctness.

program counter (1) A register in the processing unit that contains the address of the next instruction to be executed. *Synonym:* instruction address register. (C) 610.10-1994w
(2) *See also:* instruction counter. (C) 610.12-1990

program data set A data set in which user programs are stored. (C) 610.5-1990w

program definition language *See:* program design language.

program design language (1) (software) A specification language with special constructs and, sometimes, verification

protocols, used to develop, analyze, and document a program design. *See also:* hardware design language. (C) 610.12-1990
(2) (software) *See also:* design language.
(3) A specification language with special constructs and verification protocols, used to develop, analyze, and document a program design. *Contrast:* hardware design language. (C) 610.13-1993w

program disturb The corruption of data in one location caused by the programming of data at another location. (ED) 1005-1998

program editor A text editor user to enter, alter, and view source code for computer programs. Such an editor may have features that make it sensitive to the syntax of the source language on which it operates. *Contrast:* document editor. (C) 610.2-1987

program-erase cycle The event of writing a memory cell from the programmed state to the erased state and back to the programmed state. *Note:* This event may be used as a unit of measurement for endurance. Within a sequence, program-erase cycles are indistinguishable from erase-program cycles. *See also:* erase-program cycle. (ED) 1005-1998

program evaluation and review technique (PERT) (1) A variation of the critical path method in which minimum, maximum, and most likely times are used to estimate the mean and standard deviation of each activity item; these values are used to compute estimated path times and to find the critical path; and the critical path values are used to find the standard deviation of the completion time for the whole project. (C) 610.2-1987
(2) A diagrammatic method for establishing program goals and tracking. (PE/NP) 933-1999

program extension (software) An enhancement made to existing software to increase the scope of its capabilities. *See also:* software; enhancement. (C/SE) 729-1983s

program flowchart *See:* flowchart.

program instruction A computer instruction in a source program. *Note:* A program instruction is distinguished from a computer instruction that results from assembly, compilation, or other interpretation process. (C) 610.12-1990

program instrumentation (A) (software) Probes, such as instructions or assertions, inserted into a computer program to facilitate execution monitoring, proof of correctness, resource monitoring, or other activities. **(B) (software)** The process of preparing and inserting probes into a computer program. *See also:* computer program; execution; instruction; proof of correctness; assertion. (C/SE) 729-1983

program level The magnitude of program in an audio system expressed in volume units. (SP) 151-1965w

program library *See:* software library.

program listing A printout or other human readable display of the source and, sometimes, object statements that make up a computer program. (C) 610.12-1990

program loading Placing executable instructions into the memory of a computer where they can be executed. (C) 610.10-1994w

programmable (1) (programmable instrumentation) That characteristic of a device that makes it capable of accepting data to alter the state of its internal circuitry to perform a specific task(s). (IM/AIN) 488.1-1987r
(2) Pertaining to a device such as a circuit or a keyboard that can accept instructions that alter its basic functions. *Contrast:* hardwired. *See also:* user-programmable computer; microprogrammable computer. (C) 610.10-1994w

programmable array logic A programmable, two-level logic device in which the input decode (AND array) logic is programmable, but the output (OR array) is fixed. *Contrast:* programmable logic array. (C) 610.10-1994w

programmable breakpoint A breakpoint that automatically invokes a previously specified debugging process when initiated. *See also:* prolog breakpoint; dynamic breakpoint; data breakpoint; epilog breakpoint; code breakpoint; static breakpoint. (C) 610.12-1990

programmable connection A connection in which information is sent over data type circuits. *See also:* RJ-11; permissive connection. (C) 610.7-1995

programmable controller Solid state control system with programming capability that performs functions similar to a relay logic system. (PE/EDPG) 1020-1988r

programmable digital computer (programmable digital computer systems in safety systems of nuclear power generating stations) A device that can store instructions and is capable of the execution of a systematic sequence of operations performed on data that is controlled by internally stored instructions. 7432-1982w

programmable equipment (supervisory control, data acquisition, and automatic control) A remote or master station having one or more of its operations specified by a program contained in a memory device. (SWG/PE/SUB) C37.100-1992, C37.1-1994

programmable function key *See:* user-definable key.

programmable logic array A general-purpose integrated circuit that consists of an array of gates that can be programmed to perform various functions. *Contrast:* programmable array logic. *See also:* field programmable logic array. (C) 610.10-1994w

programmable measuring apparatus (programmable instrumentation) A measuring apparatus that performs specified operations on command from the system and may transmit the results of the measurement(s) to the system. (IM/AIN) 488.1-1987r

programmable read-only memory A type of read-only memory whose contents can be initialized, or burned, only once, and cannot thereafter be altered. *See also:* erasable programmable read-only memory; PROM burner; electrically erasable programmable read-only memory. (C) 610.10-1994w

programmable stimuli (test, measurement, and diagnostic equipment) Stimuli that can be controlled in accordance with instructions from a programming device. (MIL) [2]

programmable terminal *See:* intelligent terminal.

program margin The minimum measured difference between the programmed states and the sensing level for the array. (ED) 1005-1998

programmed acceleration A controlled velocity increase to the programmed rate. (IA/EEC) [61], [74]

programmed check A check procedure designed by the programmer and implemented specifically as a part of his program. *See also:* check problem; mathematical check. (C) [20], [85]

programmed control A control system in which the operations are determined by a predetermined input program from cards, tape, plug boards, cams, etc. *See also:* feedback control system. (IA/ICTL/IAC) [60]

programmed deceleration (numerically controlled machines) A controlled velocity decrease to a fixed percent of the programmed rate. (IA) [61]

PRogrammed Electronics Patterns (PREP) A programming language for use in designing integrated circuits. *Note:* PREP is conceptually similar to APT, except that it involves description of two-dimensional figures. (C) 610.13-1993w

Programmed Inquiry, Learning Or Teaching (PILOT) A programming language designed for writing computer-aided instruction applications; PILOT is simple and well-suited to support an interactive "question and answer" type of system. (C) 610.13-1993w

programmed input-output A method for transferring data between an interface and memory in which the program polls the input-output device to see if data is available. *Contrast:* direct memory access. *See also:* direct memory transfer. (C) 610.10-1994w

programmed instruction A self-instructional method using materials that lead the student through a systematic sequence of steps to a predetermined learning objective. (C) 610.2-1987

programmer **(1) (A) (test, measurement, and diagnostic equipment)** A device having the function of controlling the timing and sequencing operations. **(B) (test, measurement, and diagnostic equipment)** A person who prepares sequences of instructions for a programmable machine. (MIL) [2]
(2) An arrangement of operating elements or devices that initiates, and often controls, one or a series of operations in a given sequence. (SWG/PE) C37.100-1992

programmer-comparator **(A) (test, measurement, and diagnostic equipment)** A device that reads commands and data from a sequential program usually on tape or cards. **(B) (test, measurement, and diagnostic equipment)** A device that sets up delays, switching, and stimuli, and performs measurements as directed by the program. **(C) (test, measurement, and diagnostic equipment)** A device that compares the results of each measurement with fixed programmed tolerance limits to arrive at a decision. Often numerous other operations, such as branching on no-go or other conditions, are included. (MIL) [2]

programmer manual A document that provides the information necessary to develop or modify software for a given computer system. Typically described are the equipment configuration, operational characteristics, programming features, input/output features, and compilation or assembly features of the computer system. *See also:* diagnostic manual; installation manual; user manual; operator manual; support manual. (C) 610.12-1990

Programmer's Hierarchical Interactive Graphics System A computer graphics standard that provides a complete graphics system designed for interactive three-dimensional applications using complex data structures and modeling. It was developed by the American National Standards Institute (ANSI). (C) 610.6-1991w

programming **(1) (electronic computation)** The ordered listing of a sequence of events designed to accomplish a given task. *See also:* multiprogramming; automatic programming; linear programming. (MIL/IA) [2], [84], [61]
(2) (power supplies) The control of any power-supply functions, such as output voltage or current, by means of an external or remotely located variable control element. Control elements may be variable resistances, conductances, or variable voltage or current sources. (AES) [41]

programming (write) algorithm Typically for flash memories or ultraviolet-erasable programmable read-only memories (UV-EPROMs), the specified timed sequence of signals and levels necessary to cause the device to program. (ED) 1005-1998

programming delay D A relay whose function is to establish or detect electrical sequences. (SWG/PE) C37.100-1981s

PROgramming in LOGic (PROLOG) A declarative programming language that uses precise rules, in the form of a rule and fact set model, to reach solutions to problem descriptions. *Note:* Used in artificial intelligence applications for building expert systems. (C) 610.13-1993w

programming language A computer language used to express computer programs. *Contrast:* design language; query language; specification language; test language. *See also:* high-order language; machine language; common language; assembly language; general-purpose programming language. (C) 610.13-1993w, 610.12-1990

Programming Language/1 (PL/1) A programming language that is suitable for processing numerical, scientific, and business applications and that is standardized by ANSI. *See also:* LPL; ALPHA; FORMAC; block-structured language; IITRAN. (C) 610.13-1993w

programming language API specification The interface between applications and application platforms traditionally associated with programming language specifications, such as program control, math functions, string manipulation, etc. (C/PA) 14252-1996

programming language binding specification For a language-independent specification, a document that specifies, in terms of a particular programming language, the behavior that the language-independent specification specifies in language-independent terms. It may also specify additional behavior that is relevant to the usage of the particular programming language. (C/PA) 1328-1993w, 1224-1993w, 1326.1-1993w, 1224.1-1993w, 1327-1993w

Programming Language/Data Base (PL/DB) A dialect of PL/1 designed specifically for processing databases and including normal executable statements for arithmetic, conditional, and loop control and supports hierarchical data structures. (C) 610.13-1993w

Programming LANguage for Interactive Teaching (PLANIT) An instructional system consisting of a user language that supports the development of computer programs for preparing, editing, and presenting subject matters suitable for interactive presentations. (C) 610.13-1993w

programming, linear *See:* linear programming.

programming linearity (power supplies) The linearity of a programming function refers to the correspondence between incremental changes in the input signal (resistance, voltage, or current) and the consequent incremental changes in power-supply output. *Note:* Direct programming functions are inherently linear for the bridge regulator and are accurate to within a percentage equal to the supply's regulating ability. (AES) [41]

programming, nonlinear *See:* nonlinear programming.

programming, quadratic *See:* quadratic programming.

programming relay A relay whose function is to establish or detect electrical sequences. (SWG/PE/PSR) C37.100-1992, C37.90-1978s

programming speed (power supplies) Describes the time requires to change the output voltage of a power supply from one value to another. The output voltage must change across the load and because the supply's filter capacitor forms a resistance-capacitance network with the load and internal source resistance, programming speed can only be described as a function of load. Programming speed is the same as the recovery-time specification for current-regulated operation; it is not related to the recovery-time specification for voltage-regulated operation. (AES) [41]

programming support environment (software) An integrated collection of software tools accessed via a single command language to provide programming support capabilities throughout the software life cycle. The environment typically includes tools for specifying, designing, editing, compiling, loading, testing, configuration management, and project management. Sometimes called integrated programming support environment. *See also:* scaffolding. (C) 610.12-1990

programming system A set of programming languages and the support software (editors, compilers, linkers, etc.) necessary for using these languages with a given computer system. (C) 610.12-1990

program mutation **(A) (software)** A computer program that has been purposely altered from the intended version to evaluate the ability of test cases to detect the alteration. *See also:* mutation testing. **(B) (software)** The process of creating an altered program as in definition "A." (C) 610.12-1990

program network chart A diagram that shows the relationship between two or more computer programs. (C) 610.12-1990

program production time That part of system production time during which a computer program is successfully executed. (C) 610.10-1994w

program protection (software) The application of internal or external controls to preclude any unauthorized access or modification to a computer program. *See also:* modification; computer program. (C/SE) 729-1983s

program register *See:* instruction address register.

program-sensitive fault **(1)** **(computers)** A fault that appears in response to some particular sequence of program steps. (C) [20], [85]

(2) **(software)** A fault that causes a failure when some particular sequence of program steps is executed. *Contrast:* data-sensitive fault. (C) 610.12-1990

program specification (software) Any specification for a computer. program. *Synonym:* design specification. *See also:* requirements specification; performance specification; functional specification; design specification. (C/SE) 729-1983s

program status word (PSW) **(A)** A computer word that contains information specifying the current status of a computer program. The information may include error indicators, the address of the next instruction to be executed, currently enabled interrupts, and so on. **(B)** A special-purpose register that contains a program status word as in (A). (C) 610.12-1990

program stop (numerically controlled machines) A miscellaneous function command to stop the spindle, coolant, and feed after completion of other commands in the block. It is necessary for the operator to push a button in order to continue with the remainder of the program. (IA) [61], [84]

program support library *See:* software development library.

program structure diagram *See:* structure chart.

program synthesis (software) The use of software tools to aid in the transformation of a program specification into a program that realizes that specification. (C) 610.12-1990

program test time That part of system production time during which a computer program is tested. (C) 610.10-1994w

program tracking (communication satellite) A technique for tracking a satellite by pointing a high-gain antenna toward the satellite that employs a computer program for antenna pointing; known orbital parameters are used as an input to the computer program. (COM) [19]

program validation *See:* computer program validation; validation.

progressive grading (telephone switching systems) A grading in which the outlets of different grading groups are connected together in such a way that the number of grading groups connected to each outlet is larger for later choice outlets. (COM) 312-1977w

progressive scanning* *See:* sequential scanning.

* Deprecated.

project **(1)** **(unique identification in power plants)** A single- or multiple-unit power plant or major independent related facility. A project is composed of systems and structures and may be defined to include the design, construction, operation, and related activities associated with the project during its life cycle. (PE/EDPG) 803-1983r

(2) A subsystem that is subject to maintenance activity. (C/SE) 1219-1998

(3) A temporary endeavor undertaken to create a unique product or service. (C/SE) 1490-1998

(4) **(unique identification in power plants)** *See also:* projection. (C) 610.5-1990w

project agreement A document or set of documents baselined by the acquirer and the supplier that specifies the conditions under which the project will be conducted. A project agreement may include items such as the scope, objectives, assumptions, management interfaces, risks, staffing plan, resource requirements, price, schedule, resource and budget allocations, project deliverables, and acceptance criteria for the project deliverables. Documents in a project agreement may include some or all of the following: a contract, a statement of work, user requirements, system engineering specifications, software requirements specifications, a software project management plan, supporting process plans, a business plan, a project charter, or a memo of understanding. (C/SE) 1058-1998

project deliverable **(1)** A work product to be delivered to the acquirer. Quantities, delivery dates, and delivery locations are specified in a project agreement. Project deliverables may include the following: operational requirements, functional specifications, design documentation, source code, object code, test results, installation instructions, training aids, user's manuals, product development tools, and maintenance documentation. Project deliverables may be self-contained or may be part of a larger system's deliverables. (C/SE) 1058-1998

(2) The work product(s) to be delivered to the customer. The quantities, delivery dates, and delivery locations are specified in the project agreement. (C/SE) 1058.1-1987s

projected peak point (tunnel-diode characteristic) The point on the forward current-voltage characteristic where the current is equal to the peak-point current and where the voltage is greater than the valley-point voltage. *See also:* peak point. (ED) [46]

projected peak-point voltage (tunnel-diode characteristic) The voltage at which the projected peak point occurs. *See also:* peak point. (ED) [46]

project environment An environment that defines the objectives, success criteria, project milestones, and associated management priorities that govern the systems engineering activities in support of product development. (C/SE) 1220-1998

project evaluation and review technique* *See:* program evaluation and review technique.

* Deprecated.

project file (software) A central repository of material pertinent to a project. Contents typically include memos, plans, technical reports, and related items. *Synonym:* project notebook. (C) 610.12-1990

project function An activity that spans the entire duration of a software project. Examples of project functions include project management, configuration management, quality assurance, and verification and validation. (C/SE) 1058.1-1987s

projection **(1)** **(data management)** A relational operator that extracts specified attributes from a relation and results in a relation containing only those attributes. *Synonym:* project. *See also:* product; join; selection; intersection; union; difference.

Name	Homeroom
Mary	26A
Joe	43
Harry	27
Michael	25
Susan	25
Mickey	41

Projection of Relation *Students* in Fig to entity/attribute matrix on Attributes NAME and HOMEROOM

projection

(C) 610.5-1990w

(2) **(computer graphics)** The transformation of an *N*-dimensional image into an image in less than *N* dimensions. For example, a shadow cast by the sun is a two-dimensional projection of a person or object existing in three dimensions. *See also:* perspective projection; parallel projection; stereoscopic projection. (C) 610.6-1991w

projection/join normal form *See:* fifth normal form.

projection tube A cathode-ray tube specifically designed for use with an optical system to produce a projected image. (ED) [45]

project library *See:* software development library.

project life cycle A collection of generally sequential project phases whose name and number are determined by the control needs of the organization or organizations involved in the project. (C/SE) 1490-1998

project notebook *See:* project file.

projector (illuminating engineering) A lighting unit which, by means of mirrors and lenses, concentrates the light to a limited solid angle so as to obtain a high value of luminous intensity. (EEC/IE) [126]

project plan (software) A document that describes the technical and management approach to be followed for a project. The plan typically describes the work to be done, the resources required, the methods to be used, the procedures to be followed, the schedules to be met, and the way that the project will be organized. For example, a software development plan. (C) 610.12-1990

PROLOG *See:* PROgramming in LOGic.

prolog breakpoint (software) A breakpoint that is initiated upon entry into a program or routine. *Synonym:* preamble breakpoint. *Contrast:* epilog breakpoint. *See also:* dynamic breakpoint; code breakpoint; static breakpoint; programmable breakpoint; data breakpoint. (C) 610.12-1990

PROM Programmable read-only memory. A form of nonvolatile memory that is supplied with null contents and is loaded with its contents in the laboratory or in the field. Once programmed, its contents cannot be changed. (C/BA) 14536-1995

PROM burner *See:* PROM programmer.

promiscuous mode (local area networks) A mode in which a repeater port or an end node receives all message traffic transmitted on the network. (C) 8802-12-1998

PROM programmer A device used to program PROM devices and to reprogram EPROM, using electrical pulses. *Synonym:* PROM burner. (C) 610.10-1994w

prompt (A) (computer graphics) (software) A symbol or message displayed by a computer system requesting input from the user of the system. **(B) (software) (computer graphics)** To display a symbol or message as in definition "A." (C) 610.6-1991, 610.12-1990

proof (1) (packaging machinery) (used as a suffix) So constructed, protected, or treated that its successful operation is not interfered with when subjected to the specified material or condition. (SWG/IA/PKG) 333-1980w
(2) (suffix) An apparatus is designated as dustproof, splashproof, etc., when so constructed, protected, or treated that its successful operation is not interfered with when subjected to the specified material or condition. (T&D/PE/TR) 18-1992, C57.12.80-1978r
(3) (used as a suffix) So constructed, protected, or treated that successful operation is not interfered with when the device is subjected to the specified material or condition. *Note:* Explosionproof requires that the fuse shall not be injured and flame shall not be transmitted to the outside of the fuse for all current interruptions within the rating of the fuse. (SWG/PE/SWG-OLD) C37.40-1993, C37.100-1992

proof mass (accelerometer) The effective mass whose inertia transforms an acceleration along, or about, an input axis into a force or torque. The effective mass takes into consideration flotation and contributing parts of the suspension. (AES/GYAC) 528-1994

proof of correctness (A) (software) A formal technique used to prove mathematically that a computer program satisfies its specified requirements. *See also:* total correctness; formal specification; assertion; partial correctness; inductive assertion method. **(B) (software)** A proof that results from applying the technique in definition "A." (C) 610.12-1990

proof of performance The report submitted to the regulatory body which includes field strength measurements and other information to show that the measured radiation pattern meets the conditions specified in the station license. (T&D/PE) 1260-1996

proof test (1) (evaluation of thermal capability) (thermal classification of electric equipment and electrical insulation) A means of evaluation in which an arbitrary fixed level of a diagnostic factor is applied periodically. In this case, the number of failures among multiple test specimens (rather than the magnitude of the diagnostic factor) defines the end-point of the test. *See also:* diagnostic factor. (EI) 1-1986r
(2) (rotating machinery) (withstand test) A "fail" or "no fail" test of the insulation system of a rotating machine made to demonstrate whether the electrical strength of the insulation is above a predetermined minimum value. (PE/EM) 95-1977r

proof testing That test used to qualify equipment for a particular application or to a particular requirement. (SWG/PE) C37.100-1992, C37.81-1989r

proof-test load (composite insulators) The routine mechanical load that is applied to an insulator at the time of its manufacture. (T&D/PE) 987-1985w

propagated error An error that occurs in a GIVEN operation and is passed along to a later operation. *Contrast:* inherited error. (C) 1084-1986w

propagated potential (biological) A change of potential involving depolarization progressing along excitable tissue. (EMB) [47]

propagating mode (1) (waveguide) A waveguide mode such that the variation of phase along the direction of the guide is not negligible. (MTT) 146-1980w
(2) Refers to a mode where the imaginary part of the propagation constant is much greater than the real part, (i.e., the mode is not cut-off as in a metallic wave guide). (AP/PROP) 211-1997

propagation (data transmission) (electrical practice) The travel of waves through or along a medium. (PE) 599-1985w

propagation constant (γ) (1) (fiber optics) For an electromagnetic field mode varying sinusoidally with time at a given frequency, the logarithmic rate of change, with respect to distance in a given direction, of the complex amplitude of any field component. *Note:* The propagation constant is a complex quantity. (Std100) 812-1984w
(2) (A) (transmission lines and transducers) (per unit length of a uniform line). The natural logarithm of the ratio of the phasor current at a point of the line, to the phasor current at a second point, at unit distance from the first point along the line in the direction of transmission, when the line is infinite in length or is terminated in its characteristic impedance. **(B) (transmission lines and transducers)** (per section of a periodic line). The natural logarithm of the ratio of the phasor current entering a section, to the phasor current leaving the same section, when the periodic line is infinite in length or is terminated in its iterative impedance. **(C) (transmission lines and transducers)** (of an electric transducer). The natural logarithm of the ratio of the phasor current entering the transducer, to the phasor current leaving the transducer, when the transducer is terminated in its iterative impedance. (Std100) 270-1966
(3) (overhead power lines) The complex quantity of a traveling plane wave at a given frequency whose real part is the attenuation constant in nepers per unit length and whose imaginary part is the phase constant in radians per unit length. (T&D/PE/MTT) 539-1990, 146-1980w
(4) The image transfer constant for a symmetrical transducer. (CAS) [13]
(5) (planar transmission lines) A complex parameter having the dimension of inverse length, which characterizes the variation of the field magnitude and phase of a mode in the direction of propagation. The real part is denoted as the attenuation constant and the imaginary part as the phase constant. (MTT) 1004-1987w
(6) The complex scalar γ in expressions for one-dimensional wave propagation using the exponential factor $\exp(-\gamma z)$.

$$\gamma = jk = \alpha + j\beta$$

where scalar quantity α is the attenuation, scalar quantity β is the phase constant, and k is the wave number. *See also:* plane wave propagation factor; propagation vector. (AP/PROP) 211-1997

propagation delay (1) (power generation) (sequential events recording systems) The time interval between the appearance of a signal at any circuit input and the appearance of the associated signal at that circuit output. (PE/EDPG) [1] **(2)** In networking, the delay time between when a signal enters a channel and when it is received. *See also:* time delay. (C) 610.7-1995 **(3)** **(A)** The amount of time between when a signal is impressed on the input of a circuit and when it is received or detected at the output. **(B)** The time delay between when a signal is input to a device and a resultant action occurs on its output. *Synonym:* latency. (C) 610.10-1994 **(4)** The time required for energy to propagate between two specified points, determined by multiplying the group velocity of the wave by the distance between the two points projected onto the direction of propagation. (AP/PROP) 211-1997

propagation factor (radio-wave propagation) For a time-harmonic wave propagating from one point to another, the ratio of the complex electric field strength at the second point to that value which would exist at the second point if propagation took place in a vacuum. (AP) 211-1977s

propagation loss The total reduction in radiant power surface density. The propagation loss for any path traversed by a point on a wave front is the sum of the spreading loss and the attenuation loss for that path. *See also:* radio transmission. (Std100) 270-1966w

propagation mode (1) (laser maser) (in a periodic beamguide) A form of propagation characterized by identical field distributions over cross-sections of the beam at positions separated by one period of the guide. (LEO) 586-1980w **(2) (overhead-power-line corona and radio noise)** A concept for treating propagation of electromagnetic noise along a set of overhead power-line conductors. Modal waves form a complete set of noninteracting components into which the propagated wave may be separated. *Note:* For a three-phase horizontal single-circuit transmission line with one conductor per phase and without ground wires, the following modes are defined: Mode 1—The transmission path is between the center phase and the outside phases. It has the lowest attenuation and the lowest surge impedance. Mode 2—The transmission path is between outside phases. It has an intermediate attenuation and an intermediate surge impedance. Mode 3—The transmission path is along all three phases and returning through ground. It has the highest attenuation and the highest surge impedance. (T&D/PE) 539-1990

propagation model An empirical or mathematical expression used to compute propagation path loss. *See also:* electromagnetic compatibility. [53]

propagation ratio (radio-wave propagation) For a time-harmonic wave propagating from one point to another, the ratio of the complex field strength at the second point to that at the first point. (AP) 211-1977s

propagation sort *See:* bubble sort.

propagation vector (1) Vector-characterizing phase progression of a wave in direction normal to lines of constant phase with magnitude proportional to the reciprocal of the wavelength. (UFFC) 1037-1992w **(2)** The complex vector k in expressions for wave propagation using the exponential factor $\exp[-j(\vec{k}\cdot\vec{r})]$ is:

$$\vec{k} = \vec{\beta} - j\vec{\alpha}$$

where

$\vec{\beta}$ = the phase vector

$\vec{\alpha}$ = the attenuation vector

$\vec{r}$ = a position vector

(AP/PROP) 211-1997

propagation vector in physical media The complex vector $\bar{\gamma}$ in plane wave solutions of the form $e^{-\gamma\cdot r}$ for an e^{jwt} time variation and $\bar{r}$ the position vector. *See also:* attenuation vector in physical media; wave vector in physical media; phase vector in physical media. (AP/ANT) 145-1983s

propeller turbine A reaction turbine with fixed or variable pitch propeller-type blades. (PE/EDPG) 1020-1988r

propeller-type blower (rotating machinery) An axial-flow fan with air-foil-shaped blades. *See also:* fan. (PE) [9]

prop-encoded-array The primitive data type, consisting of a sequence of bytes, used to represent a property value. (C/BA) 1275-1994

proper ferroelectric (primary ferroelectric terms) A ferroelectric in which the polarization is the primary order parameter. (UFFC) 180-1986w

proper mode A mode of propagation that can be excited by a physical source. (AP/PROP) 211-1997

proper operation The functioning of the train control or cab signaling system to create or continue a condition of the vehicle apparatus that corresponds with the condition of the track of the controlling section when the vehicle apparatus is in operative relation with the track elements of the system. (EEC/PE) [119]

property (1) A kind of responsibility that is an inherent or distinctive characteristic or trait that manifests some aspect of an object's knowledge or behavior. Three kinds of property are defined: attributes, participant properties due to relationships, and operations. (C/SE) 1320.2-1998 **(2)** A documenting characteristic of an entity type used to group and differentiate individual entities. *Note:* An example of a property is "Stop." This property might be associated with the entity types "ROUTE.BUS" and "ROUTE.TRAVELER." (SCC32) 1489-1999

property encoding A specific data format, defined by this standard, that is used to represent various types of information within a prop-encoded-array. (C/BA) 1275-1994

property name A text string used to specify, or name, a particular property. (C/BA) 1275-1994

property value The data portion of a property, stored in property encoding format. (C/BA) 1275-1994

proportional amplifier An amplifier in which the output is a single value and an approximately linear function of the input over its operating range. *See also:* feedback control system. (IA/ICTL/IAC) [60]

proportional control action A control mode designed to provide a linear relationship between the controller output and input, assuming both are within normal range. *See also:* proportional control action. (PE/PSE) 94-1991w

proportional counter tube A radiation-counter tube designed to operate in the proportional region. (C) 165-1977w

proportional gain (hydraulic turbines) The proportional gain G_p of a proportional element is the ratio of the element's percent output to its percent input. A linear relationship is assumed.

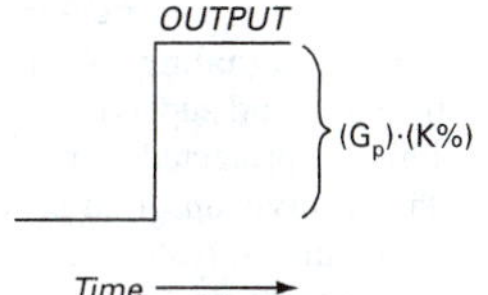

proportional gain

(PE/EDPG) 125-1977s

proportionally *See:* linearity.

proportional plus derivative control *See:* control action, proportional plus derivative.

proportional plus integral control *See:* proportional plus integral control action.

proportional plus integral control action Control action in which the output is proportional to a linear combination of the input and the time integral of the input. *Note:* In the practical embodiment of proportional plus integral action the relation between output and input, neglecting high frequency terms, is

$$\frac{Y}{X} = \pm P \frac{\frac{I}{s} + 1}{\frac{bI}{s} + 1} \quad 0 \leq b \ll 1$$

where
b = proportional gain/static gain
I = integral action rate
P = proportional gain
s = complex variable
X = input transform
Y = output transform

Synonym: P.I. (CS/PE/EDPG) [3]

proportional plus integral plus derivative control *See:* proportional plus integral plus derivative control action.

proportional plus integral plus derivative control action Control action in which the output is proportional to a linear combination of the input, the time integral of input and the time rate-of-change of input. *Note:* In the practical embodiment of proportional plus integral plus derivative control action the relationship of output and input, neglecting high frequency terms, is

$$\frac{Y}{X} = \pm P \frac{1 + sD}{1 + sD/a} \quad a > 1$$

where
a = derivative action gain
b = proportional gain/static gain
D = derivative action time constant
I = integral action rate
P = proportional gain
s = complex variable
X = input transform
Y = output transform

Synonym: P.I.D. (CS/PE/EDPG) [3]

proportional region (radiation counter tubes) The range of operating voltage for a counter tube in which the gas amplification is greater than unity and is independent of the amount of primary ionization. *Notes:* 1. In this region the pulse size from a counter tube is proportional to the number of ions produced as a result of the initial ionizing event. 2. The proportional region depends on the type and energy of the radiation. (ED) 161-1971w

proportional spacing Text formatting and output that takes into account the width of each character, rather than allocating the same amount of horizontal space to characters of all widths. (C) 610.2-1987

proportionate mortality ratio An index used in occupational epidemiological studies that expresses the proportion of deaths from a single cause. It is not a mortality rate and, therefore, does not necessarily indicate a risk value. Rather, it indicates within a group the relative importance of specific causes of death. (T&D/PE) 539-1990

proprietary system (protective signaling) A local system sounding and/or recording alarm and supervisory signals at a control center located within the protected premises, the control center being under the supervision of employees of the proprietor of the protected premises. *Note:* According to the United States Underwriters' rules, a proprietary system must be a recording system. *See also:* protective signaling. (EEC/PE) [119]

propulsion-control transfer switch Apparatus in the engine room for transfer of control from engine room to bridge and vice versa. *Note:* Engine-room control is provided on all ships. Bridge control with a transfer switch is optional and is used principally on small vessels such as tugs or ferries, usually with a direct-current propulsion system. (EEC/PE) [119]

propulsion set-up switch Apparatus providing ready means to set up for operation under varying conditions where practicable; for example, cutout of one or more generators when multiple units are provided. *See also:* electric propulsion system. (EEC/PE) [119]

propulsion system The system of motors, drive mechanisms, controls, and other devices that propels or retards a vehicle. (VT) 1475-1999

prorated section A complete, suitably housed part of an arrester, comprising all necessary components, including gaseous medium, in such a proportion as to accurately represent, for a particular test, the characteristics of a complete arrester. (SPD/PE) C62.11-1999

prorated unit (arresters) A completely housed prorated section of an arrester that may be connected in series with other prorated units to construct an arrester of higher voltage rating. (PE) [8]

prospective characteristics of a test voltage causing disruptive discharge The characteristics of a test voltage that would have been obtained if no disruptive discharge had occurred. (PE/PSIM) 4-1995

prospective current (1) (ac high-voltage circuit breakers) The current that would flow if it were not influenced by the circuit breaker. (SWG/PE) C37.081-1981r
(2) (surge arresters) (available current) The root-mean-square symmetrical short-circuit current that would flow at a given point in a circuit if the arrester(s) at that point were replaced by links of zero impedance. (PE) [8], [84]

prospective current of a circuit *See:* available current.

prospective crest value *See:* prospective peak value.

prospective peak value (surge arresters) (of a chopped impulse) The peak (crest) value of the full-wave impulse voltage from which a chopped impulse voltage is derived. *Synonym:* prospective crest value. (PE) [8]

prospective peak value of test voltage (switching impulse testing) The voltage that would be obtained if no disruptive discharge occurred before the crest. 332-1972w

prospective short-circuit current *See:* available short-circuit current.

prospective (available) short-circuit current (at a given point in a circuit) The maximum current that the power system can deliver through a given circuit point to any negligible impedance short circuit applied at the given point, or at any other point that will cause the highest current to flow through the given point. *Notes:* 1. This value can be in terms of either symmetrical or asymmetrical, peak, or rms current, as specified. 2. In some resonant circuits, the maximum available short-circuit current may occur when the short-circuit is placed at some other point than the given one where the available current is measured. (SWG/PE) C37.40-1993

prospective study An epidemiological study of a group exposed to some factor over time to determine if this factor is associated with the development of a particular disease, as compared to a nonexposed control group. (T&D/PE) 539-1990

protected A responsibility that is visible only to the class or the receiving instance of the class (available only within methods of the class or its subclasses). *Contrast:* private; public. *See also:* hidden. (C/SE) 1320.2-1998

protected area (PA) A controlled-access area encompassed by physical barriers. (PE/NP) 692-1997

protected enclosure (electric installations on shipboard) An enclosure in which all openings are protected with wire screen, expanded metal, or perforated covers. A common from of specifications for "protected enclosure" is: "The openings should not exceed 1/2 sq. in. in area and should be of such shape as not to permit the passage of a rod larger than 1/2 in. in diameter, except where the distance of exposed live parts from the guard is more than 4 in. the openings may be 3/4 sq. in. in area and must be of such shape as not to permit the passage of a rod larger than 3/4 in. in diameter." (IA/MT) 45-1983s

protected field On a display device, a display field in which a user cannot enter, modify or erase data. *Contrast:* unprotected field. (C) 610.10-1994w

protected location (1) (computers) A storage location reserved for special purposes in which data cannot be stored without undergoing a screening procedure to establish suitability for storage therein. (C) [20], [85]
(2) A location whose contents are protected against accidental alteration, improper alteration, or unathorized access. (C) 610.10-1994w

protected machine *See:* guarded machine.

protected outdoor transformer A transformer that is not of weatherproof construction but that is suitable for outdoor use if it is so installed as to be protected from rain or immersion in water. *See also:* transformer. (PE/TR) [57]

protected storage A type of storage in which data cannot be modified by an application program except under specified conditions. (C) 610.10-1994w

protected zone *See:* cone of protection.

protection (1) The process of observing a system, and automatically initiating an action to mitigate the consequences of an operating condition that has deviated from the established acceptable performance criteria. (PE/EM) 1129-1992r
(2) (software) An arrangement for restricting access to or use of all, or part, of a computer system. *See also:* storage protection; computer system. (C/SE) 729-1983s

protection character A character used to replace a suppressed zero in order to avoid error or false statements; for example, in the string "$********50.03" the asterisk is the protection character. (C) 610.5-1990w

protection exception (software) An exception that occurs when a program attempts to write into a protected area in storage. *See also:* overflow exception; addressing exception; underflow exception; operation exception; data exception. (C) 610.12-1990

protections Limitations imposed on the radiated signal for certain azimuths. *Note:* These limitations generally are set so that interference does not occur to stations in that direction. (T&D/PE) 1260-1996

protection system (1) (protection systems) The electrical and mechanical devices (from measured process variables to protective action system input terminals) involved in generating those signals associated with the protective functions. These signals include those that actuate reactor trip and actuate engineered safety features (for example, containment isolation, core spray, safety injection, pressure reduction, and air cleaning). (PE/NP) 279-1971w
(2) The part of the sense and command features involved in generating those signals used primarily for the reactor trip system and engineered safety features. (PE/NP) 603-1998

protective action (1) (protection systems) The initiation of a signal or operation of equipment within the safety system for the purpose of accomplishing a protective function in response to a generating station condition having reached a limit specified in the design basis. *Notes:* 1. Protective action at the channel level is the initiation of a signal by a single channel when the sensed variable(s) reaches a specified limit. 2. Protective action at the system level is the operation of sufficient actuated equipment including the appropriate auxiliary supporting features to accomplish a protective function. Examples of protective actions at the system level are: rapid insertion of control rods, closing of containment isolation valves, and operation of safety injection and core spray. (PE/NP) 279-1971w
(2) The initiation of a signal within the sense and command features or the operation of equipment within the execute features for the purpose of accomplishing a safety function. (PE/NP) 603-1998

protective action set point (nuclear power generating station) The reference value to which the measured variable is compared for the initiation of protective action. (PE/NP) 380-1975w

protective action system (1) (Class 1E power systems for nuclear power generating stations) The electrical and mechanical equipment (from the protection system output to and including the actuated equipment-to-process coupling) that performs a protective action when it receives a signal from the protective system. *Note:* Examples of protective action systems are: control rods and their trip mechanisms; isolation valves, their operators and their contactors; and emergency service water pumps and associated valves, their motors and circuit breakers. In some instances protective actions may be performed by protective action system equipment that responds directly to the process conditions (for example, check valves, self-actuating relief valves). (PE/NP) 603-1978, 308-1980s
(2) (nuclear power generating station) An arrangement of equipment that performs a protective action when it receives a signal from the protection system. (PE/NP) 279-1971w

protective buffer An optional single word buffer in a slave that always contains a copy of the most recent data asserted or received by the slave. (NID) 960-1993

protective covering (power cable joints) A field-applied material to provide environmental protection over the joint or housing, or both. (PE/IC) 404-1986s

protective device A bypass gap, varistor, or other device that limits the voltage on the capacitor segment to a predetermined level when overcurrent flows through the series capacitor (that is, during system faults, system swings, or other abnormal events), and that is capable of carrying capacitor discharge, system fault, and load current for the specified duration. (T&D/PE) 824-1994

protective function (1) (nuclear power generating station) Any one of the functions necessary to mitigate the consequences of a design basis event (for example, reduce power, isolate containment, or cool the core). *Note:* A protective function is a design basis objective that must be accomplished; a successfully completed protective action at the system level, including the sensing of one or more variables, will accomplish the protective function. However, the design may be such that a given protective function may be accomplished by any one of several protective actions at the system level. (PE/NP) 379-1977s
(2) (nuclear power generating station) The completion of those protective actions at the system level required to maintain plant conditions within the allowable limits established for a design basis event (for example, reduce power, isolate containment, or cool the core). (PE/NP) 603-1978, 308-1980s
(3) (nuclear power generating station) The sensing of one or more variables associated with a particular generating station condition, the signal processing and the initiation and completion of the protective action at values of the variables established in the design bases. (PE/NP) 279-1971w, 379-1977s

protective gap A gap placed between live parts and ground to limit the maximum overvoltage that may otherwise occur. (SWG/PE/T&D) C37.100-1992, 516-1995

protective level of the bypass gap The maximum instantaneous voltage (including tolerance) appearing across the capacitor immediately before or during operation of the bypass gap. (T&D/PE) 824-1994

protective level of the varistor The maximum instantaneous voltage appearing across the capacitor at a specified current through the varistor. (T&D/PE) 824-1994

protective lighting (illuminating engineering) A system intended to facilitate the nighttime policing of industrial and other properties. (EEC/IE) [126]

protective margin (PM) The value of the protective ratio (PR), minus one, expressed in percent. PM = (PR − 1) · 100. (C/PE) 1313.1-1996, [8]

protective power gap (series capacitor) A bypass gap that limits the voltage on the capacitor segment to a predetermined level when system fault occurs on the line, and that is capable of carrying capacitor discharge, system fault, and load currents for specified durations. (T&D/PE) [26]

protective ratio (1) (surge arresters) The ratio of the insulation withstand characteristics of the protected equipment to the arrester protective level, expressed as a multiple of the latter figure. (PE) [8]
(2) The ratio of the insulation strength of the protected equipment to the overvoltages appearing across the insulation. (PE/C) 1313.1-1996

protective relay (1) (power operations) A device whose function is to detect defective lines or apparatus or other power system conditions of an abnormal or dangerous nature and to initiate appropriate control action. (PE/PSE) 858-1987s
(2) A relay whose function is to detect defective lines or apparatus or other power system conditions of an abnormal or dangerous nature and to initiate appropriate control circuit action. *Note:* A protective relay may be classified according to its input quantities, operating principle, or performance characteristics. (SWG/PE/PSR) C37.98-1977s, C37.90-1978s, C37.100-1992

protective screen (burglar-alarm system) A lightweight barrier of either solid strip or lattice construction, carrying electric protection circuits, and barring access through a normal opening to protected premises. *See also:* protective signaling. (EEC/PE) [119]

protective signaling Protective signaling comprises the initiation, transmission, and reception of signals involved in the detection and prevention of property loss or damage due to fire, burglary, robbery, and other destructive conditions, and in the supervision of persons and of equipment concerned with such detection and prevention. (EEC/PE) [119]

protective switchgear (thyristor converter) The ac circuit devices and the dc circuit devices that may be used in the thyristor converter unit to clear fault conditions. (IA/IPC) 444-1973w

protective system (1) (Class 1E power systems for nuclear power generating stations) The electrical and mechanical devices (from measured process variables to protective action system input terminals) involved in generating those signals associated with the protective functions. These signals include those that initiate reactor trip, engineered safety features (for example, containment isolation, core spray, safety injection, pressure reduction, and air cleaning) and auxiliary supporting features. (PE/NP) 308-1980s
(2) (nuclear power generating station safety systems) The part of the sense and command features involved in generating those signals used primarily for the reactor trip system and engineered safety features. 379-1988s

protective system false operation rate Protective system false operation rate = number of false operations/exposure time. (PE/PSE) 859-1987w

protect notch *See:* write-protect notch.

protector joint A split sleeve that fits over a conductor compression joint used to protect the joint from bending or damage if the joint must pass through travelers. The joint protector usually has split rubber collars at each end to protect the conductor from damage where it exits at each end of the sleeve. (T&D/PE) 524-1992r

protector tube (1) (surge arresters) An expulsion arrester used primarily for the protection of line ard switch insulation. (PE) 28-1974, [8]
(2) (electron-tube type) A glow-discharge cold-cathode tube that employs a low-voltage breakdown between two or more electrodes to protect circuits against overvoltage. (ED) [45]

PROTEUS A computer language used in signal processing. (C) 610.13-1993w

protocol (1) (supervisory control, data acquisition, and automatic control) A strict procedure required to initiate and maintain communication. (SWG/SUB/PE) 999-1992w, C37.1-1994, C37.100-1992
(2) A formal set of conventions governing the format and relative timing of message exchange between two communications terminals. *See also:* control procedure. (LM/COM) 168-1956w
(3) (software) A set of conventions that govern the interaction of processes, devices, and other components within a system. (C) 610.12-1990
(4) (STEbus) The signaling rules used to convey information or commands between boards connected to the bus. (C/MM) 1000-1987r
(5) (MULTIBUS II) The set of signaling rules used to convey information between agents. (C/MM) 1296-1987s
(6) A set of semantic and syntactic rules that determine the behavior of entities that interact. (C/PA) 14252-1996
(7) A set of rules and formats (semantic and syntactic) that determines the communication behavior of simulation applications. (DIS/C) 1278.1-1995, 1278.2-1995
(8) A set of conventions or rules that govern the interactions of processes or applications within a computer system or network. (ATLAS) 1232-1995
(9) (A) A formal set of conventions governing the format and relative timing of message exchange in a computer system. **(B)** A set of semantic and syntactic rules that determine the behavior of functional units in achieving meaningful communication. (C) 610.7-1995, 610.10-1994
(10) A set of semantic and syntactic rules for exchanging information. (C) 1003.5-1999

protocol access A protocol that is adopted at a specified reference point between a user and a network to enable the user to employ the services and/or facilities of that network. (C) 610.7-1995

protocol control information (computer graphics) Information exchanged between entities to coordinate their joint operation. (LM/C) 8802-6-1994

protocol converter A dedicated device that translates the protocol native to an end-user device into a different protocol, allowing communication with another end-user device. *Note:* A protocol converter converts the message formats so both systems are compatible. (C) 610.7-1995

protocol data unit (PDU) (1) Information that is delivered as a unit between peer entities of a *local area network (LAN)* or a *metropolitan area network (MAN)* and that contains control information, address information, and may contain user data. (LM/C) 8802-6-1994
(2) A block of data that is exchanged between two devices using a protocol. (C) 610.7-1995
(3) Information delivered as a unit between peer entities that may contain control information, address information, and data. (C/MM) 1394-1995
(4) A unit of data specified in a protocol and consisting of protocol information and, possibly, user data. (C/LM) 802.10g-1995
(5) A Distributed Interactive Simulation (DIS) data message that is passed on a network between simulation applications according to a defined protocol. (DIS/C) 1278.1-1995, 1278.2-1995, 1278.4-1997
(6) Information delivered as a unit between peer entities that contains control information and, optionally, data. (C/LM) 8802-5-1998
(7) The sequence of contiguous octets delivered as a unit to the MAC sub-layer or received as a unit from the MAC sub-layer. A valid LLC PDU is at least 3 octets in length, and contains two address fields and a control field. A PDU may

or may not include an information field in addition. (C/LM/CC) 8802-2-1998

(8) Information delivered as a unit between peer entities that contains control information and, optionally, data. (EMB/MIB) 1073.3.2-2000

protocol engine A component of the DNI implementation model that is a conceptual machine that implements a particular communications protocol profile. (C) 1003.5-1999

protocol entity (1) An entity that provides one or more service access points for use by higher-level entities. (C) 610.7-1995

(2) An entity that follows a set of rules and formats (semantic and syntactic) that determines the communication behavior of other entities. (C/LM) 802.10g-1995

protocol error counter (PEC) A counter contained by a network station to keep track of protocol errors such as invalid frames. (EMB/MIB) 1073.3.1-1994

protocol event In the DNI implementation model, an event that is generated by a protocol engine and queued for attention by the event handler. (C) 1003.5-1999

protocol implementation conformance statement (PICS) **(1)** A statement of which capabilities and options have been implemented for a given Open Systems Interconnection (OSI) protocol. (C/LM) 8802-5-1998, 610.7-1995

(2) (local area networks) A statement of which capabilities and options have been implemented for a given interconnection protocol. (C) 8802-12-1998

protocol independent interface An interface that enables the application to be insulated from the specifics of the underlying protocol stack which provides the communication services. Protocol independent interfaces allow the application to be written so that it can be ported to various protocol stacks. (C) 1003.5-1999

protocol profile A set of one or more protocol definitions and, where applicable, the identification of chosen classes, subsets, option and parameters of those definitions, necessary for accomplishing a particular function. A protocol profile can be thought of as a vertical slice through a layered set of communications protocols. (C) 1003.5-1999

protocol stack (1) The hierarchy of protocols used in a computer network architecture. (C) 610.7-1995

(2) An instantiation of a set of protocols. (C/LM) 802.10a-1999

protocol suite A defined set of complementary protocols within the communication service profile. (DIS/C) 1278.2-1995

protocol type (PTYPE) A field in the RDE PDU information field that describes the protocol function of the PDU. (C/LM/CC) 8802-2-1998

proton microscope A device similar to the electron microscope but in which the charged particles are protons. *See also:* electron optics. (ED) [45], [84]

proton range (solar cells) The maximum distance traversed through a material by a proton of a given energy. (AES/SS) 307-1969w

prototype (1) (modeling and simulation) (software) A preliminary type, form, or instance of a system that serves as a model for later stages or for the final, complete version of the system. (C) 610.3-1989w, 610.12-1990

(2) An experimental model, either functional or nonfunctional, of the system or part of the system. A prototype is used to get feedback from users for improving and specifying a complex human interface, for feasibility studies, or for identifying requirements. (C/SE) 1233-1998

prototype standard A concrete embodiment of a physical quantity having arbitrarily assigned magnitude, or a replica of such embodiment. *Note:* As an illustration of the distinction between prototype standard and unit, the length of the United States Prototype Meter Bar is not exactly one meter. (Std100) 270-1966w

prototype transformer A transformer manufactured primarily to obtain engineering data or evaluate manufacturing or design feasibility. Prototypes may be pre-production units or units typical of current designs manufactured for test purposes to obtain data to comply with changes in industry standards or for other reasons. (PE/TR) C57.134-2000

prototyping (software) A hardware and software development technique in which a preliminary version of part or all of the hardware or software is developed to permit user feedback, determine feasibility, or investigate timing or other issues in support of the development process. *See also:* rapid prototyping. (C) 610.12-1990

proximal (distal) point (pulse terminology) A magnitude referenced point at the intersection of a waveform and a proximal (distal) line. *See also:* waveform epoch. (IM/WM&A) 194-1977w

proximal stimuli (illuminating engineering) The distribution of illuminance on the retina constitutes the proximal stimulus. (EEC/IE) [126]

proximity-coupled dipole array antenna An array antenna consisting of a series of coplanar dipoles, loosely coupled to the electromagnetic field of a balanced transmission line, the coupling being a function of the proximity and orientation of the dipole with respect to the transmission line. (AP/ANT) 145-1993

proximity discharge *See:* proximity ESD.

proximity effect (electric circuits and lines) The phenomenon of non-uniform current distribution over the cross section of a conductor caused by the time variation of the current in a neighboring conductor. *See also:* induction heating. (IA) 54-1955w, 270-1966w

proximity-effect error (navigation systems) (navigation aid terms) An error in determination of system performance caused by improper use of measurements made in the near field of the antenna system. (AES/GCS) 172-1983w

proximity-effect ratio (power distribution, underground cables) The quotient obtained by dividing the alternating-current resistance of a cable conductor subject to proximity effect, by the alternating-current resistance of an identical conductor free of proximity effect. (PE) [4]

proximity ESD *See:* indirect ESD event.

proximity influence The percentage change in indication caused solely by the fields produced from two edgewise instruments mounted in the closest possible proximity, one on each side (or above and below for horizontal-scale instruments). *Note:* Proximity influence of alternating-current instruments on either alternating-current or direct-current types is determined by energizing two instruments, one on each side of the test instrument (or above and below) at 90% of end-scale value (in phase with the current in the instrument under test, if the latter is alternating current). The current in the two outside instruments only shall be reversed. For rating purposes, the proximity influence shall be taken as one-half the difference in the readings in percentage of full scale. In direct-current permanent-magnet moving-coil instruments the field produced by the current in the instrument is small compared with the field from the permanent magnet. The proximity influence on either an alternating-current or direct-current test instrument will be the difference in reading, expressed as a percentage of full-scale value, of the instrument under test mounted alone on the panel, compared with the reading when two direct-current instruments are mounted in closest possible proximity, each with current applied to give 90% end-scale deflection. All three instruments shall be of the same manufacture and size. *See also:* accuracy rating. (EEC/AII) [102]

proximity switch A device that reacts to the proximity of an actuating means without physical contact or connection therewith. *See also:* switch. (IA/ICTL/IAC/APP) [60], [75]

proxy install A proxy install uses an alternate root directory as the target path. (C/PA) 1387.2-1995

PSC *See:* physical signaling components.

P scan *See:* plan-position indicator.

P-scope A cathode-ray oscilloscope arranged to present a P-display. (AES/RS) 686-1990

PSDN *See:* packet-switched data network.

pseudo code **(1)** **(software)** A combination of programming language constructs and natural language used to express a computer program design. For example: IF the data arrives faster than expected,THEN reject every third input.ELSE process all data received.ENDIF. (C) 610.12-1990
(2) **(test, measurement, and diagnostic equipment)** An arbitrary code, independent of the hardware of a computer, which has the same general form as actual computer code but which must be translated into actual computer code if it is to direct the computer. (MIL) [2]
(3) A combination of programming language constructs and natural language used to express a computer program design. For example:

IF the data arrives faster than expected

THEN reject every third input

ELSE process all data received

ENDIF

(C) 610.13-1993w

pseudo-coning (inertial sensors) (strapdown inertial system) A system error created when the system computer attempts to cancel a steady coning input term which in actuality does not exist. Because of certain coupling errors in the gyro, a rate input about only one axis can produce outputs on both axes of the gyro. If the coupling error, for example, is angular acceleration sensitivity, the two outputs produced will have the same form as if a true coning motion was applied to the gyro. (AES/GYAC) 528-1984s

pseudo-instruction **(1)** **(test, measurement, and diagnostic equipment)** An instruction which resembles the instructions acceptable to the computer but which must be translated into actual computer instructions in order to control the computer. (MIL) [2]
(2) **(software)** A source language instruction that provides information or direction to the assembler or compiler and is not translated into a target language instruction. For example, an instruction specifying the desired format of source code listings. *Synonyms:* pseudo operation; pragma; pseudo-op. (C) 610.12-1990

pseudolatitude (navigation aid terms) A latitude in a coordinate system which has been arbitrarily displaced from the earth's conventional latitude system so as to move the meridian convergence zone (polar region) away from the place of intended operation. (AES/GCS) 172-1983w

pseudolongitude (navigation aid terms) A longitude in a coordinate system which has been arbitrarily dosplace from the earth's conventional longitude system so as to move the meridian convergence zone (polar region) away from the place of intended operation. (AES/GCS) 172-1983w

pseudonoise sequence (communication satellite) A binary sequence with a very desirable transorthogonal auto-correlation property. In space communications commonly used for synchronization and ranging. Syn: PN sequence. (COM) [19]

pseudo-op *See:* pseudo-instruction.

pseudo operation *See:* pseudo-instruction.

pseudo-random Pertaining to the approximation of true, statistical randomness. (C) 1084-1986w

pseudo-random number Any member of a sequence of numbers sufficiently close to a random number sequence to permit its use in calculations formally requiring random numbers. (C) 1084-1986w

pseudo-random number sequence **(1)** A sequence of numbers, determined by some defined arithmetic process, that is satisfactorily random for a given purpose, such as by satisfying one or more of the standard statistical tests for randomness. Such a sequence may approximate any one of several statistical distributions, such as uniform distribution or normal Gaussian distribution. (C/C) [20], [85]
(2) **(mathematics of computing)** A sequence of numbers, determined by some defined arithmetic process, that is sufficiently close to a random number sequence to permit its use in calculations formally requiring a random number sequence. (C) 1084-1986w

pseudo-random test signal A signal consisting of a bit sequence that approximates a random signal. (COM/TA) 1007-1991r

pseudo signal A signal other than that at the interface between the device and the tester. This includes internal signals, derived signals, and any other signals that may be required by tools other than test translators to generate tests or test constructs. (C/TT) 1450-1999

pseudoternary coding A means of digital signaling in which three signal levels are used to encode binary data. (C) 610.10-1994w

PSF *See:* product specification file.

PSK *See:* phase-shift keying.

PSN *See:* public switched network; private switching network.

PSPDN *See:* packet switched public data network.

PSS *See:* physical source statements.

PSTN *See:* public switched telephone network.

PSW *See:* program status word.

psychometric chroma (illuminating engineering) A correlate of perceived chroma defined in terms of CIELUV or CIELAB. Equal scale intervals correspond approximately to equal differences in perceived chroma. (EEC/IE) [126]

psychometric hue-angle (illuminating engineering) A correlate of hue defined in terms of CIELUV or CIELAB. (EEC/IE) [126]

psychometric lightness (illuminating engineering) A correlate of lightness defined in terms of CIELUV or CIELAB. Equal scale intervals correspond approximately to equal differences in (perceived) lightness. (EEC/IE) [126]

psychometric saturation (illuminating engineering) A correlate of saturation defined in terms of CIELUV. Equal scale intervals correspond approximately to equal differences of (perceived) saturation. *Note:* Psychometric saturation cannot be calculated in terms of CIELAB. (EEC/IE) [126]

psychophysics Study of correlations between stimulus parameters and detection or perception of stimuli. (T&D/PE) 539-1990

PTE *See:* Port Transfer Error.

PTM *See:* pulse-time modulation.

PTT *See:* postal telephone and telegraph.

PTYPE *See:* protocol type.

p-type crystal rectifier A crystal rectifier in which forward current flows when the semiconductor is positive with respect to the metal. *See also:* rectifier. (EEC/PE) [119]

p-type semiconductor *See:* semiconductor, *p*-type.

public **(1)** A design feature, documented in the component data sheet, that may be used by purchasers of the component. (TT/C) 1149.1-1990
(2) A responsibility that is not hidden, i.e., visible to any requester (available to all without restriction). *Contrast:* private; protected. (C/SE) 1320.2-1998

Publication Contents The publisher-defined contents of a publication. (IM/ST) 1451.1-1999

Publication Domain A Domain for a specific publication. *See also:* Domain. (IM/ST) 1451.1-1999

Publication Key A publisher-defined identifier specifying the form and contents of a publication in a publish-subscribe communication. (IM/ST) 1451.1-1999

Publication Topic A configurable identifier for the name of a publication in a publish-subscribe communication. Specifically a value having datatype `PublicationTopic`. For a Publisher port, the operation `GetPublicationTopic` returns a value, `publication_topic`, that has the same value as Publication Topic. (IM/ST) 1451.1-1999

public-address system A system designed to pick up and amplify sounds for an assembly of people. (EEC/PE) [119]

public circuit-switched network A public data network using circuit-switching techniques. (C) 610.7-1995

public data network A network established and operated by communications common carriers or telecommunications administrations for the specific purpose of providing low error-rate data transmission services to the public. *Synonym:* public data transmission service. *Contrast:* private network. *See also:* public switched network; public packet switching network; public circuit-switched network. (C) 610.7-1995

public data transmission service *See:* public data network.

public key system An encryption system using a combination of a public encryption key and a private decryption key to provide message security or authentication. *See also:* electronic signatures. (C) 610.7-1995

public object The representation by the client of an object in a particular language binding.
(C/PA) 1328-1993w, 1224-1993w, 1327-1993w, 1238.1-1994w

public OM object The representation by the client of an object in a particular programming language binding.
(C/PA) 1327.2-1993w, 1326.2-1993w, 1328.2-1993w, 1224.2-1993'

public packet switching network A public data network that uses packet switching techniques. (C) 610.7-1995

public page A memory page that may be used by any agency or application. No access credentials are associated with such a page. (SCC32) 1455-1999

public specifications Specifications that are available, without restriction, to anyone for implementation, sublicensing, and distribution (i.e., sale) of that implementation.
(C/PA) 14252-1996

public switched network A public data network in which dedicated communications paths are established for customers.
(C) 610.7-1995

public switched telephone network (PSTN) **(1)** A telephone network in which connections are established as and when required and that is supplied, operated, and controlled by one or more telecommunications operating companies to provide telephone service that is available to the public.
(COM/TA) 823-1989w
(2) Commonly called the switched telephone network.
(AMR) 1390-1995
(3) A network of a complete public telephone system, including telephones, lines and exchanges. (C) 610.7-1995

public telecommunications exchange (telephone switching systems) A telecommunications exchange that serves the public. (COM) 312-1977w

public telephone network A network of public telephone system. *See also:* public switched telephone network.
(C) 610.7-1995

public telephone station (pay station) A station available for use by the public, generally on the payment of a fee that is deposited in a coin collector or is paid to an attendant. *See also:* telephone station. (EEC/PE) [119]

Public Transducer A Parameter or other class instance that is the public interface or abstraction of one or more physical transducers supported by an NCAP. (IM/ST) 1451.1-1999

Publisher object Any object that posts publications onto the network via an Object of class `IEEE1451_BasePublisherPort` or a subclass thereof.
(IM/ST) 1451.1-1999

Publisher Port An instance of the class `IEEE1451_PublisherPort` or of a subclass thereof.
(IM/ST) 1451.1-1999

publish-subscribe communication A communication pattern where one or more objects, publishers, communicate information on a specific topic to one or more subscriber objects interested in that topic without the necessity of any of these objects knowing the identity of any other object. In an IEEE 1451.1 system, the pattern is established via the Publication Topic, Publication Key, and Publication Domain of the publication. (IM/ST) 1451.1-1999

puck (computer graphics) An input device, used as a locator, consisting of a hand-held control box with crosshairs that can be moved over a data tablet containing an image such that the user can identify a point in the image. *Note:* This term is sometimes used to refer to a mouse.
(C) 610.6-1991w, 610.10-1994w

pull (data management) To retrieve data from a stack. *Synonym:* pop. *Contrast:* push. (C) 610.5-1990w

pull blade (cable plowing) A plow blade used to pull direct burial conductors into position by means of a suitable pulling grip attachment at the heel of the blade.
(T&D/PE) 590-1977w

pull box A box with a blank cover that is inserted in one or more runs of raceway to facilitate pulling in the conductors, and may also serve the purpose of distributing the conductors. *See also:* cabinet. (EEC/PE) [119]

pull-down menu A menu that is brought into view by selecting a menu-bar label. *Contrast:* pop-up menu.
(PE/NP) 1289-1998

puller *See:* bullwheel puller.

puller, bullwheel *See:* bullwheel puller.

puller, drum *See:* drum puller.

puller, reel *See:* reel puller.

puller, two drum, three drum *See:* two-drum, three-drum puller.

pulley (1) (rotating machinery) (sheave) A shaft-mounted wheel used to transmit power by means of a belt, chain, band, etc. *See also:* rotor. (PE) [9]
(2) *See also:* sheave. (T&D/PE) 524-1992r

pull function* *See:* dragging.
* Deprecated.

pulling eye A device that may be fastened to the conductor or conductors of a cable or formed by or fastened to the wire armor and to which a hook or rope may be directly attached in order to pull the cable into or from a duct. *Note:* Pulling eyes are sometimes equipped, like test caps, with facilities for oil feed or vacuum treatment. (T&D/PE) [10]

pulling figure (oscillators) The difference between the maximum and minimum values of the oscillator frequency when the phase angle of the load-impedance reflection coefficient varies through 360 degrees, while the absolute value of this coefficient is constant and equal to a specified value, usually 0.20. (Voltage standing-wave ratio 1.5.) *See also:* oscillatory circuit; waveguide. (ED) 161-1971w

pulling into synchronism (rotating machinery) The process of synchronizing by changing from asynchronous speed to synchronous. (PE) [9]

pulling iron An anchor secured in the wall, ceiling, or floor of a manhole or vault to attach rigging used to pull cable.
(NESC) C2-1997

pulling line (conductor stringing equipment) A high strength line, normally synthetic fiber rope or wire rope, used to pull the conductor. However, on reconstruction jobs where a conductor is being replaced, the old conductor often serves as the pulling line for the new conductor. In such cases, the old conductor must be closely examined for any damage prior to the pulling operation. *Synonyms:* hard line; light line; sock line; pulling rope; bull line.
(T&D/PE) 524a-1993r, 1048-1990, 524-1992r

pulling out of synchronism (rotating machines) The process of losing synchronism by changing from synchronous speed to a lower asynchronous speed (for a motor) or higher asynchronous speed (for a generator). (PE) [9]

pulling rope *See:* pulling line.

pulling tension The longitudinal force exerted on a cable during installation. (NESC) C2-1997

pulling vehicle (conductor stringing equipment) Any piece of mobile ground equipment capable of pulling pilot lines, pulling lines, or conductors. However, helicopters may be considered as a pulling vehicle when utilized for the same purpose. (T&D/PE) 524a-1993r, 524-1992r

pull-in test (synchronous machines) A test taken on a machine that is pulling into synchronism from a specified slip. (PE) [9]

pull-in time (acquisition time) (communication satellite) The time required for achieving synchronization in a phase-lock loop. (COM) [25]

pull-in torque (synchronous motor) The maximum constant torque under which the motor will pull its connected inertia load into synchronism, at rated voltage and frequency, when its field excitation is applied. *Note:* The speed to which a motor will bring its load depends on the power required to drive it and whether the motor can pull the load into step from this speed depends on the inertia of the revolving parts, so that the pull-in torque cannot be determined without having the Wk^2 as well as the torque of the load. (PE) [9]

pull or transfer box A box without a distribution panel, within which one or more corresponding electric circuits are connected or branched. (T&D/PE) [10]

pull-out test (rotating machinery) A test to determine the conditions under which an alternating-current machine develops maximum torque while running at specified voltage and frequency. *See also:* asynchronous machine. (PE) [9]

pull-out torque The maximum sustained torque that a synchronous motor will develop at synchronous speed with rated voltage applied at rated frequency and with normal excitation. (IA/MT) 45-1998

pull rope A rope, attached to the cable, that is used to pull the cable through the conduit system. *Synonym:* bull rope. (PE/IC) 1185-1994

pull setting *See:* sag section.

pull site (conductor stringing equipment) The location on the line where the puller, reel winder, and anchors (snubs) are located. This site may also serve as the pull or tension site (anchor) for the next sag section. *Synonyms:* reel setup; tugger setup. (T&D/PE) 524a-1993r, 524-1992r

pull-up torque (alternating-current motor) The minimum torque developed by the motor during the period of acceleration from rest to the speed at which breakdown torque occurs with rated voltage applied at rated frequency. (PE) [9]

pulsating function A periodic function whose average value over a period is not zero. For example, $f(t) = A + B \sin \omega t$ is a pulsating function where neither A nor B is zero. (Std100) 270-1966w

pulse (1) (impulse) (data transmission) A brief excursion of a quantity from normal. (PE) 599-1985w
(2) (automatic control) A variation of a signal whose value is normally constant; this variation is characterized by a rise and a decay, and has a finite duration. (PE/EDPG) [3]
(3) (pulse terminology) A wave that departs from a first nominal state, attains a second nominal state, and ultimately returns to the first nominal state. Throughout the remainder of this document the term pulse is included in the term wave. (IM/WM&A) 194-1977w
(4) A wave that departs from an initial level for a limited duration of time and ultimately returns to the original level. *Note:* In demand metering, the term "pulse" is also applied to a sudden change of voltage or current produced, for example, by the closing or opening of a contact. (ELM) C12.1-1988
(5) A variation in the value of a magnitude which is short in relation to the time schedule of interest, the final value being the same as the initial value. *Note:* In digital logic circuits, a pulse is usually a voltage. *See also:* recovery time; strobe. (C) 610.7-1995, 610.10-1994w
(6) (relaying) A brief excursion of a quantity from its initial level. (SWG/PE) C37.100-1992

pulse accumulator (or register) (of a telemeter system) A device that accepts and stores pulses and makes them available for readout on demand. (SWG/PE) C37.100-1992

pulse advance (pulse terminology) (delay) The occurrence in time of one pulse waveform before (after) another pulse waveform. (IM/WM&A) 194-1977w

pulse advance interval (pulse terminology) (delay interval) The interval by which, unless otherwise specified, the pulse start time of one pulse waveform preceded (follows), unless otherwise specified, the pulse start time of another pulse waveform. (IM/WM&A) 194-1977w

pulse amplifier (pulse techniques) An amplifier designed specifically for the amplification of electric pulses. *See also:* pulse. (NPS) 175-1960w

pulse amplifier or relay A device used to change the amplitude or waveform of a pulse for retransmission to another pulse device. (ELM) C12.1-1988

pulse amplitude, A_M (1) (pulse transformers) That quantity determined by the intersection of a line passing through the points on the leading edge where the instantaneous value reaches 10% and 90% of A_M and a straight line that is the best least-squares fit to the pulse in the pulse-top region (usually this is fitted visually rather than numerically). For pulses deviating greatly from the ideal trapezoidal pulse shape, a number of successive approximations may be necessary to determine A_M. *Note:* The pulse amplitude A_M may be arrived at by applying the following procedure.

1) Visually or numerically determine the best straight line fit to the pulse in the pulse-top region and extend this straight line into the leading-edge region.
2) An initial estimate of A_M is the first intersection of the pulse (in the late leading-edge or early pulse-top regions) with the straight line fitted to the pulse top.
3) Using the estimate of A_M calculate 0.1 A_M and 0.9 A_M and draw a straight line through these two points of the pulse-leading edge.
4) The intersection of the leading-edge straight line and the pulse-top straight line gives an improved estimate of A_M.
5) Repeat steps 3 and 4 until the estimate of A_M does not change. The converged estimate is the pulse amplitude A_M.

(PEL/ET) 390-1987r

(2) The algebraic difference between the top magnitude and the base magnitude. (IM/WM&A) 194-1977w
(3) A general term indicating the magnitude of a pulse measured with respect to the normally constant value unless otherwise stated. (ED) [127]

pulse amplitude modulation (PAM) (1) (data transmission) (pulse terminology) Modulation in which the modulating wave is caused to amplitude modulate a pulse carrier. (IM/AP/WM&A/ANT) 194-1977w, 145-1983s
(2) A form of pulse modulation in which the amplitude of a pulse carrier is varied. (C) 610.7-1995

pulse amplitude, peak *See:* peak pulse amplitude.

pulse amplitude, root-mean-square *See:* root-mean-square (effective) pulse amplitude.

pulse average time (light-emitting diodes) The time interval between the instants at which the instantaneous pulse amplitude first and last reaches a specified fraction of the peak pulse amplitude, namely, 50%. (ED) [127]

pulse bandwidth (pulse terminology) The smallest continuous frequency interval outside of which the amplitude (of the spectrum) does not exceed a prescribed fraction of the amplitude at a specified frequency. Caution: This definition permits the spectrum amplitude to be less than the prescribed amplitude within the interval. *Notes:* 1. Unless otherwise stated, the specified frequency is that at which the spectrum has its maximum amplitude. 2. This term should really be pulse spectrum bandwidth because it is the spectrum and not the pulse itself that has a bandwidth. However, usage has caused the contraction and for that reason the term has been accepted. *See also:* signal. (IM/WM&A) 194-1977w

pulse base (pulse waveform) (pulse techniques) That major segment having the lesser displacement in amplitude from the baseline, excluding major transitions. *See also:* pulse. (IM/HFIM) [40]

pulse, bidirectional *See:* bidirectional pulse.

pulse broadening (fiber optics) An increase in pulse duration. *Note:* Pulse broadening may be specified by the impulse response, the root-mean-square pulse broadening, or the full-duration-half-maximum pulse broadening. *See also:* root-mean-square pulse broadening; impulse response; full width (duration) half maximum. (Std100) 812-1984w

pulse burst (1) (pulse terminology) A finite sequence of pulse waveforms. (IM/WM&A) 194-1977w
(2) A sequence of closely spaced pulses. *Note:* Pulse bursts are usually generated coherently and batch-processed for Doppler resolution, and often have a total burst duration much less than the radar echo delay time. (AES) 686-1997

pulse burst base envelope (pulse terminology) Unless otherwise specified, the waveform defined by a cubic natural spline with knots at (A) that point of intersection of the pulse burst top envelope and the pulse burst waveform which precedes the first pulse waveform in a pulse burst, (B) each point of intersection of the base center line and the baseline between adjacent pulse waveforms in a pulse burst, and (C) that point of intersection of the pulse burst top envelope and the pulse burst waveform which follows the last pulse waveform in a pulse burst. *See also:* knot. (IM/WM&A) 194-1977w

pulse burst time-related definitions (A) (pulse terminology) (pulse burst duration) The interval between the pulse start time of the first pulse waveform and the pulse stop time of the last pulse waveform in a pulse burst. **(B) (pulse terminology)** (pulse burst separation) The interval between the pulse stop time of the last pulse waveform in a pulse burst and the pulse start time of the first pulse waveform of the immediately following pulse burst. **(C) (pulse terminology)** (pulse burst repetition period) The interval between the pulse start time of the first pulse waveform in a pulse burst and the pulse start time of the first pulse waveform in the immediately following pulse burst in a sequence of periodic pulse bursts. **(D) (pulse terminology)** (pulse burst reception frequency) The reciprocal of burst repetition period. (IM/WM&A) 194-1977

pulse burst top envelope (pulse terminology) Unless otherwise specified, the waveform defined by a cubic natural spline with knots at: the first transition mesial point of the first pulse waveform in a pulse burst; each point of intersection of the top centerline and the topline of each pulse waveform in a pulse burst; and the last transition mesial point of the last pulse waveform in a pulse burst. *See also:* knot. (IM/WM&A) 194-1977w

pulse capacity (metering) The number of pulses per demand interval that a pulse receiver can accept and register without loss. (ELM) C12.1-1988

pulse capture (accelerometer) (gyros) A technique that uses discrete quanta of torque-time (force-time) area to generate a restoring torque (force). (AES/GYAC) 528-1994

pulse carrier (1) A carrier consisting of a series of pulses. *Note:* Usually, pulse carriers are employed as subcarriers. *See also:* carrier. (IM/AP/WM&A/ANT) 194-1977w, 145-1983s
(2) A series of identical pulses intended for modulation. (C) 610.7-1995

pulse, carrier-frequency *See:* carrier-frequency pulse.

pulse code (A) A pulse train modulated so as to represent information. **(B)** Loosely, a code consisting of pulses, such as Morse code, Baudot code, binary code. *See also:* pulse. (IM/WM&A/HFIM) 194-1977, [40]

pulse code modulation (1) (data transmission) The type of pulse modulation where the magnitude of the signal is sampled and each sample is approximated to a nearest reference level (this process is called quantizing). Then a code, which represents the reference level, is transmitted to the distant location. The figure above is an example of one form of PCM which has eight reference levels. It can be seen that a straight binary code would require a group of three pulses to be transmitted for each sample. The main advantage of PCM is the fact that at the receiving end only the presence or absence of a pulse must be detected. (Std100) [123]
(2) (pulse terminology) A modulation process involving the conversion of a waveform from analog to digital form by means of coding. *Notes:* 1. This is a generic term, and additional specification is required for a specific purpose. 2. The term is commonly used to signify that form of pulse modulation in which a code is used to represent quantized values of instantaneous samples of the signal wave. (IM/WM&A) 194-1977w, 270-1964w
(3) A modulation technique in which an analog signal is converted to a bit stream for transmission. (C) 610.7-1995

pulse coder (navigation aid terms) A device for varying one or more of the characteristics of a pulse or of a pulse train so as to transmit information. (AES/RS/GCS) 686-1990, 172-1983w

pulse coincidence (noncoincidence) (pulse terminology) The occurrence (lack of occurrence) of two or more pulse waveforms in different waveforms either essentially simultaneously or for a specified interval. (IM/WM&A) 194-1977w

pulse coincidence (noncoincidence) duration (pulse terminology) The interval between specified points on two or more pulse waveforms in different waveforms during which pulse coincidence (noncoincidence) exists. (IM/WM&A) 194-1977w

pulse compression A method for obtaining the resolution of a short pulse with the energy of a long pulse of width T by internally modulating the phase or frequency of a long pulse so as to increase its bandwidth $B >> 1/T$, and using a

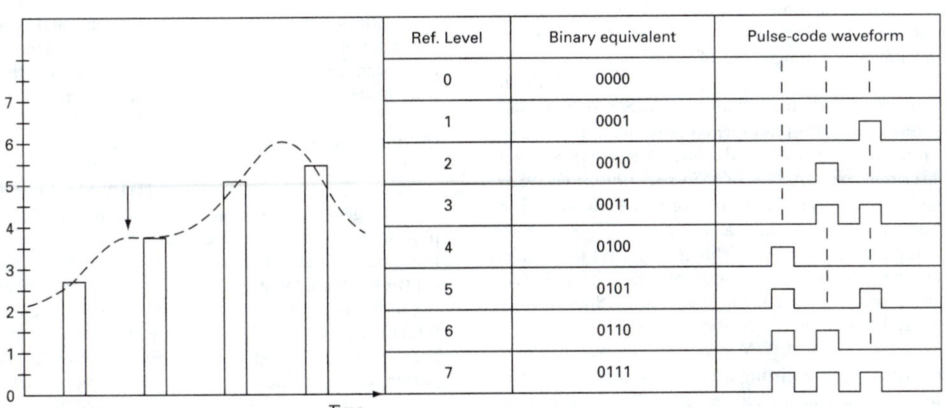

0
1
2
3
4
5
6
7
Time
Ref. Level
Binary equivalent
Pulse-code waveform
0 0000
1 0001
2 0010
3 0011
4 0100
5 0101
6 0110
7 0111

A form of pulse code modulation

pulse code modulation

matched filter (also called a pulse compression filter) on reception to compress the pulse of width T to a width of approximately $1/B$. Used to obtain high range resolution when peak-power limited. *See also:* chirp; coded pulse; Costas code. (AES) 686-1997

pulse control (rotating electric machinery) Means by which the voltage applied to a machine circuit departs from being essentially constant if unidirectional, or from being essentially sinusoidal if alternating. Pulse-control devices include, but are not limited to, choppers, inverters, and rectifiers. (PE/EM) 11-1980r

pulse corner (pulse waveform feature) A continuous pulse waveform feature of specified extent which includes a region of maximum curvature or a point of discontinuity in the waveform slope. (PE/PSR) [6]

pulse corrector (telephone switching systems) Equipment to reestablish, within predetermined limits, the make/break ratio of dial pulses. (COM) 312-1977w

pulse count (control of system electromagnetic compatibility) The number of pulses in some specified time interval. C63.12-1984

pulse-count deviation (metering) The difference between the number of recorded pulses and the number of pulses supplied to the input of terminals of a pulse recorder (true count), expressed as a percentage of the true count. Pulse-count deviation is applicable to each data channel of a pulse recorder. (ELM) C12.1-1988

pulse counter (pulse techniques) A device that indicates or records the total number of pulses that it has received during a time interval. *See also:* pulse. (NPS) 398-1972r

pulsed-Doppler radar A Doppler radar that uses pulsed transmissions. *Synonym:* pulse-Doppler radar. (AES/GCS) 686-1997, 172-1983w

pulse decay time (1) (germanium gamma-ray detectors) (x-ray energy spectrometers) The interval between the instants at which the instantaneous value last reaches specified upper and lower limits, namely, 90% and 10% of the peak pulse value unless otherwise stated. (In the case of a step function applied to an amplifier that has simple capacitance-resistance to resistance-capacitance (CR-RC) shaping, the decay time is given by $t_d = 3.36CR$.) (NPS/NID) 325-1986s, 301-1976s, 759-1984r

(2) (t_f) (pulse fall time) (light-emitting diodes) The interval between the instants at which the instantaneous amplitude last reaches specified upper and lower limits, namely, 90% and 10% of the peak pulse amplitude unless otherwise stated. *See also:* pulse. (NPS/BT/ED/AV/NID) 325-1971w, [34], 300-1982s, [127], 301-1976s

(3) (data transmission) The interval of time required for the trailing edge of a pulse to decay from 90% to 10% of the peak-pulse amplitude. (PE) 599-1985w

(4) (laser maser) The time duration of a laser pulse; usually measured as the time interval between the half-power points on the leading and trailing edges of the pulse. (LEO) 586-1980w

(5) (radar) *See also:* pulse width. (AES/RS) 686-1982s

pulse decoder (navigation aid terms) A device for extracting information from a pulse-coded signal. *Synonym:* constant-delay discriminator. (AES/RS/GCS) 686-1990, 172-1983w

pulse delay (transducer) The interval of time between a specified point on the input pulse and a specified point on the related output pulse. *Notes:* 1. This is a general term which applies to the pulse delay in any transducer, such as receiver, transmitter, amplifier, oscillator, etcetera. 2. Specifications may require illustrations. *See also:* transducer. (IM/PE/WM&A) 194-1977w, 599-1985w

pulse delay time (light-emitting diodes) The interval between the instants at which the instantaneous amplitudes of the input pulse and output pulses first reach a specified fraction of their peak pulse amplitudes, namely, 10%. (ED) [127]

pulse delay, transducer *See:* pulse delay.

pulse delay, transmitter *See:* pulse delay.

pulse density violation A violation that occurs if a signal contains more than a specified number of zeros, or the density of ones is less than specified. (COM/TA) 1007-1991r

pulse device (for electricity metering) The functional unit for initiating, transmitting, retransmitting, or receiving electric pulses, representing finite quantities, such as energy, normally transmitted from some form of electricity meter to a receiver unit. (ELM) C12.1-1988

pulse distortion (pulse techniques) The unwanted deviation of a pulse waveform from a reference waveform. *Note:* Some specific forms of pulse distortion have specific names. They include, but are not exclusive to, the following: overshoot, ringing, preshoot, tilt (droop), rounding (undershoot and dribble-up), glitch, bump, spike, and backswing. For further explanation of the forms of pulse distortion, see the following illustrations and IEEE Standard 194 (1977). *See also:* pulse. (IM/WM&A) 194-1977w

pulsed laser (laser maser) A laser that delivers its energy in the form of a single pulse or train of pulses. The duration of a pulse ≤ 0.25 s. (LEO) 586-1980w

pulse-Doppler radar *See:* pulsed-Doppler radar.

pulsed optical feedback preamplifier An integrating preamplifier in which the charge that accumulates on the feedback capacitor is periodically reset by a pulse of light incident on a photosensitive element, such as the gate of the n-p junction of the input FET. (NPS) 325-1996

pulsed oscillator An oscillator that is made to operate during recurrent intervals by self-generated or externally applied pulses. *See also:* oscillatory circuit. (AP/ANT) 145-1983s

pulsed-oscillator starting time The interval between the leading-edge pulse time of the pulse at the oscillator control terminals and the leading-edge pulse time of the related output pulse. (IM/WM&A) 194-1977w

pulse droop (television) A distortion of an otherwise essentially flat-topped rectangular pulse characterized by a decline of the pulse top. *See also:* television. (IM/WM&A) 194-1977w

pulse duration (1) (fiber optics) The time between a specified reference point on the first transition of a pulse waveform and a similarly specified point on the last transition. The time between the 10%, 50%, or 1/e points is commonly used, as is the root-mean-square (rms) pulse duration. *See also:* root-mean-square pulse duration. (Std100) 812-1984w

(2) (loosely) The duration of a rectangular pulse whose energy and peak power equal those of the pulse in question. *Note:* When determining the peak power, any transients of relatively short duration are frequently ignored. *See also:* phase modulation; pulse. (AP/PE/ANT) 145-1983s, 599-1985w

(3) (laser maser) The time duration of a laser pulse; usually measured as the time interval between the half-power points on the leading and trailing edges of the pulse. (LEO) 586-1980w

(4) (radiation counters) (telecommunications) (television) The time interval between the first and last instants at which the instantaneous amplitude reaches a stated fraction of the peak pulse amplitude. *See also:* signal; pulse. (BT/NPS/PE/AV) [34], 398-1972r, 599-1985w

(5) (pulse terminology) The duration between pulse start time and pulse stop time. *See also:* waveform epoch. (IM/WM&A) 194-1977w

(6) (charged-particle detectors) (of an amplifier output pulse) The width of a pulse measured between the first and last transitions at designated levels. (NPS) 300-1988r

(7) (pulse transformers) (90%) (t_p) The time interval between the instants at which the instantaneous value reaches 90% of A_M on the leading edge and 90% of A_T on the trailing edge. *Notes:* 1. Often the input pulse tilt (droop) is only a few percentages, and in those cases pulse duration may be considered as the time interval between the first and last instants at which the instantaneous value reaches 90% of A_M. 2. Pulse duration may be specified at a value other than 90% A_M and A_T in special cases. (PEL/ET) 390-1987r

(8) (at the baseline) The width of a pulse measured close to the baseline, typically at 1% of the peak height (t.01). (NPS) 325-1996

(9) *See also:* pulse width. (AES) 686-1997

pulse-duration discriminator A circuit in which the output is a function of the deviation of the input pulse duration from a reference. (AES/RS) 686-1990

pulse-duration distribution (electromagnetic site survey) The fraction of pulse duration at level vithat exceeds time *T*. *See also:* average crossing rate. (EMC) 473-1985r

pulse-duration modulation (1) (pulse terminology) Pulse-time modulation in which the value of each instantaneous sample of the modulating wave is caused to modulate the duration of a pulse. *Notes:* 1. The deprecated terms pulsew-

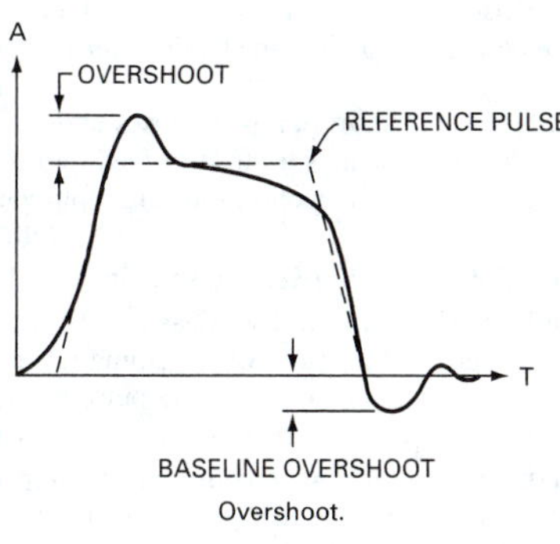

Overshoot.

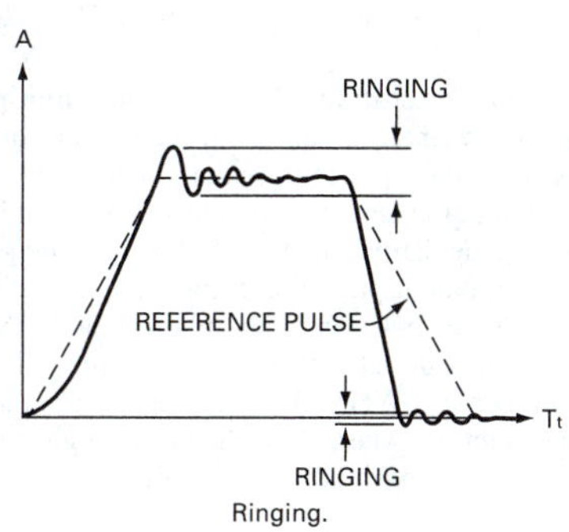

Ringing.

distortion, short-time waveform

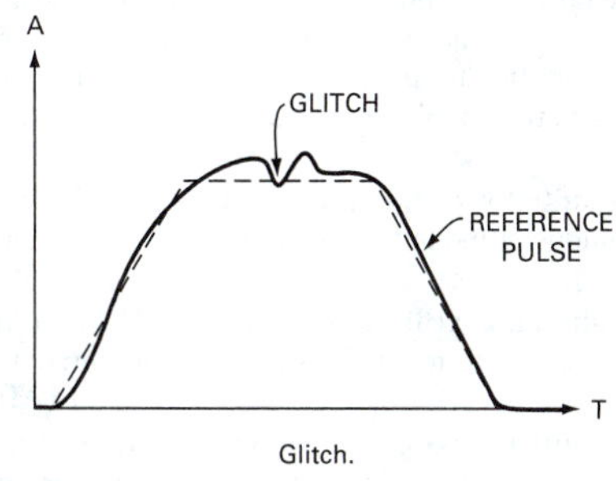

Glitch.

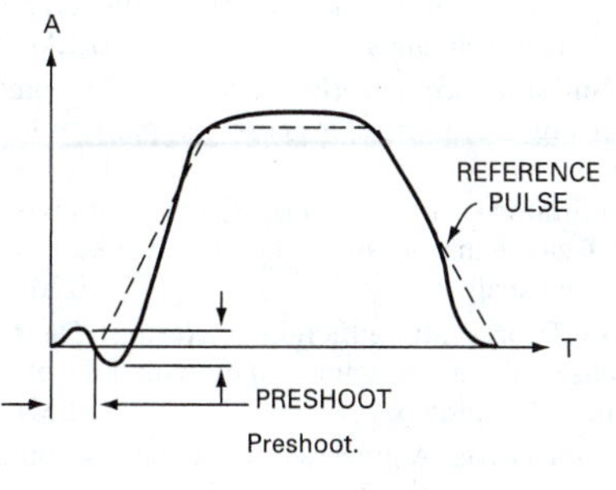

Preshoot.

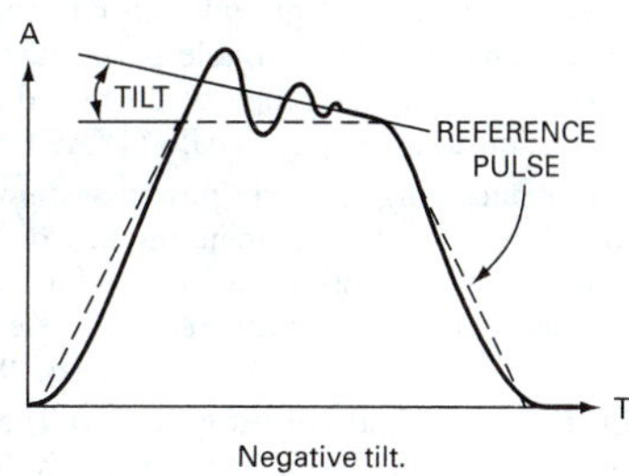

Negative tilt.

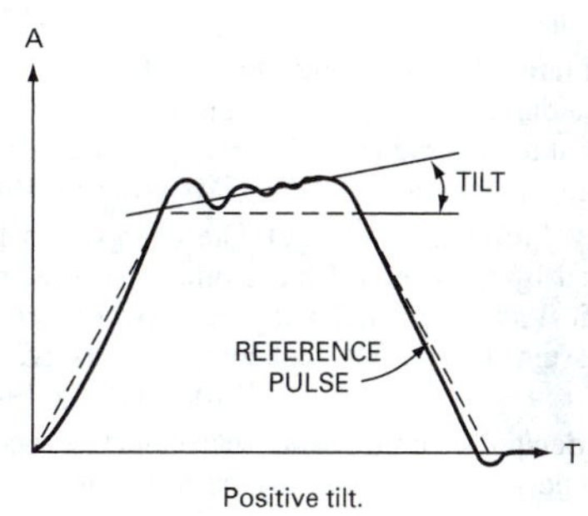

Positive tilt.

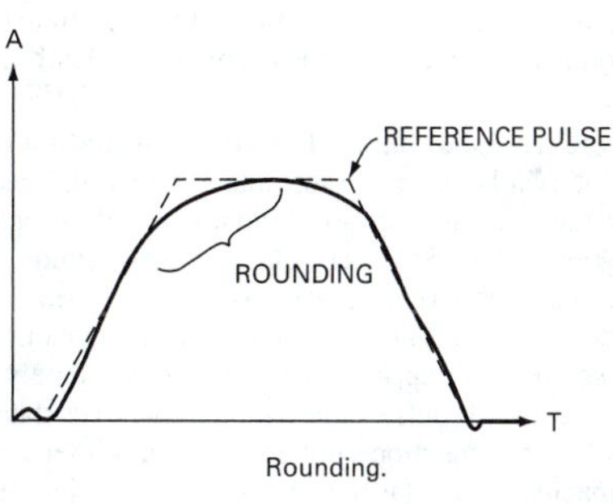

Rounding.

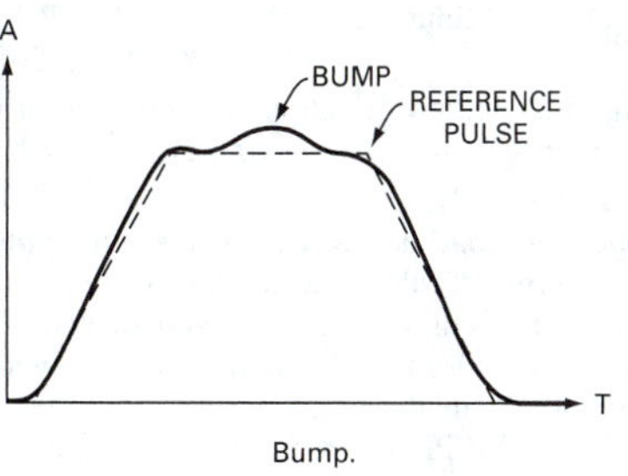

Bump.

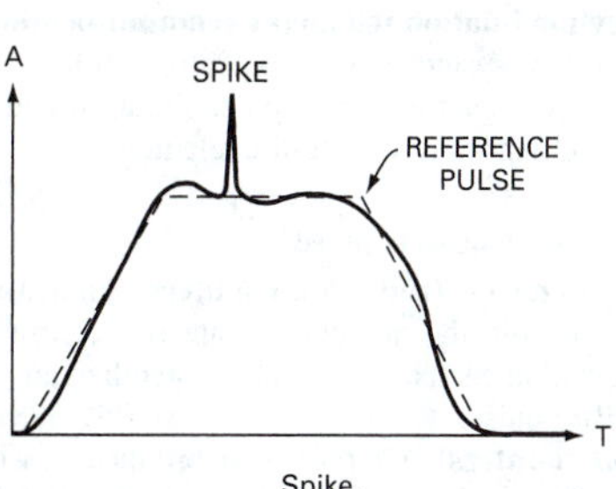

Spike.

distortion, pulse

idth modulation and pulse-length modulation also have been used to designate this system of modulation. 2. In pulse-duration modulation, the modulating wave may vary the time of occurrence of the leading edge, the trailing edge, or both edges of the pulse. (IM/AP/WM&A/ANT) 194-1977w, 145-1983s
(2) *See also:* pulse-width modulation.

pulse-duration-modulation torquing (accelerometer) (gyros) A torquing mechanization that provides current to a sensor torquer of fixed amplitude but variable pulse duration proportional to input. The duration may be quantized to enable digital interpretation and readout. (AES/GYAC) 528-1994

pulse-duration telemetering (electric power system) (pulse-width modulation) A type of telemetering in which the duration of each transmitted pulse is varied as a function of the magnitude of the measured quantity. *See also:* telemetering. (PE/PSE) 94-1970w

pulse-duration time (t_p) (light-emitting diodes) The time interval between the first and last instants at which the instantaneous amplitude reaches a stated fraction of the peak pulse amplitude, namely, 90%. (ED) [127]

pulse duty factor (light-emitting diodes) The ratio of the average pulse duration to the average pulse spacing. *Note:* This is equivalent to the product of the average pulse duration and the pulse-repetition rate. (IM/WM&A) 194-1977w

pulse energy (pulse terminology) The energy transferred or transformed by a pulse(s). Unless otherwise specified by a mathematical adjective, the total energy over a specified interval is assumed. *See also:* mathematical adjectives. (IM/WM&A) 194-1977w

pulse expander/pulse compressor An expander device spreads the input energy from a relatively narrow pulse over a relatively large interval of time. The compressor acts as a matched conjugate filter that recompresses the energy spread by the expander into a relatively small interval of time (mainlobe) accompanied by a certain level of time sidelobes. (UFFC) 1037-1992w

pulse fall time (A) (photomultipliers for scintillation counting) The interval between the instants at which the instantaneous amplitude last reaches specified upper and lower limits, namely, 90% and 10% of the peak pulse amplitude unless otherwise stated. **(B) (charged-particle detectors)** The interval between the instants at which the instantaneous value last reaches specified upper and lower limits, namely, 90% and 10% of the peak pulse value unless otherwise stated. (In the case of a step function applied to an amplifier that has simple capacitance-resistance to resistance-capacitance (CR-RC) shaping, the fall time is given by $t_f = 3.36$ CR.) (NPS) 398-1972, 300-1982

pulse-forming line A passive electric circuit in a radar modulator whose propagation delay determines the length of the modulation pulse. (AES) 686-1997

pulse frequency modulation (data transmission) (pulse terminology) A form of pulse-time modulation in which the pulse-repetition rate is the characteristic varied. *Note:* A more precise term for pulse-frequency modulation would be "pulse-repetition-rate modulation." (IM/PE/WM&A) 194-1977w, 599-1985w

pulse frequency modulation telemetry (communication satellite) A telemetry system where the information is coded according to subcarrier frequency, pulse duration and pulse repetition rate. Often used for satellite telemetry. (COM) [24]

pulse, Gaussian *See:* Gaussian pulse.

pulse-height analyzer (1) (radiation counters) An instrument capable of indicating the number or rate of occurrence of pulses falling within each of one or more specified amplitude ranges. *See also:* pulse. (NPS) 398-1972r
(2) (radiation counters) A circuit that produces an output signal if it receives an input pulse whose amplitude falls between upper and lower assigned values. (NI) N42.15-1990

pulse-height analyzer, multichannel A circuit that accepts all input pulses and assigns each pulse to a memory location corresponding to its amplitude. (NI) N42.15-1990

pulse-height discriminator (1) (liquid-scintillation counting) A circuit that produces an output signal if it receives an input pulse whose amplitude exceeds an assigned value. (NI) N42.15-1980s
(2) (pulse techniques) A circuit that produces a specified output pulse if and only if it receives an input pulse whose amplitude exceeds an assigned value. *See also:* pulse. (NPS) 175-1960w
(3) A circuit that responds to pulses above a designated amplitude and generates a pulse to feed into a counting circuit but that does not respond to lower amplitude pulses. (NI/NPS) 309-1999

pulse height radiation *See:* pulse-height resolution.

pulse-height resolution The measured FWHM, after ambient background subtraction, of a gamma-ray peak distribution, expressed as a percentage of the pulse height corresponding to the centroid of the distribution. (NI) N42.12-1994

pulse-height resolution constant, electron (photomultipliers) The product of the square of the electron (photomultiplier) pulse-height resolution expressed as the fractional full width at half maximum (FWHM/A_1), and the mean number of electrons per pulse from the photocathode. *See also:* phototube. (NPS) 175-1960w

pulse-height resolution, electron (photomultipliers) A measure of the smallest change in the number of electrons in a pulse from the photocathode that can be discerned as a change in height of the output pulse. Quantitatively, it is the fractional standard deviation (σ/A_1) of the pulse-height distribution curve for output pulses resulting from a specified number of electrons per pulse from the photocathode. *Note:* The fractional full width at half maximum of the pulse-height distribution curve (FWHM/A_1) is frequently used as a measure of this resolution, where A_1 is the pulse height corresponding to the maximum of the distribution curve. *See also:* pulse. (NPS) 398-1972r

pulse-height selector (pulse techniques) A circuit that produces a specified output pulse when and only when it receives an input pulse whose amplitude lies between two assigned values. *See also:* pulse. (NPS) 175-1960w

pulse initiator (metering) Any device, mechanical or electrical, used with a meter to initiate pulses, the number of which are proportional to the quantity being measured. It may include an external amplifier or auxiliary relay or both. (ELM) C12.1-1988

pulse-initiator coupling ratio (metering) The number of revolutions of the pulse-initiating shaft for each output pulse. (ELM) C12.1-1988

pulse-initiator gear ratio (metering) The ratio of meter rotor revolutions to revolutions of the pulse-initiating shaft. (ELM) C12.1-1988

pulse-initiator output constant (metering) The value of the measured quantity for each outgoing pulse of a pulse initiator, expressed in kilowatt hours per pulse, kilovarhours per pulse, or other suitable units. (ELM) C12.1-1988

pulse-initiator output ratio (metering) The number of revolutions of the meter rotor per output pulse of the pulse initiator. (ELM) C12.1-1988

pulse-initiator ratio (metering) The ratio of revolutions of the first gear of the pulse initiator to revolutions of the pulse-initiating shaft. (ELM) C12.1-1988

pulse-initiator shaft reduction (metering) The ratio of revolutions of the meter rotor to the revolutions of the first gear of the pulse initiator. (ELM) C12.1-1988

pulse interleaving A process in which pulses from two or more sources are combined in time-division multiplex for transmission over a common path. (IM/WM&A) 194-1977w

pulse interrogation The triggering of a transponder by a pulse or pulse mode. *Note:* Interrogations by means of pulse modes may be employed to trigger a particular transponder or group of transponders. (IM/WM&A) 194-1977w

pulse interval *See:* pulse spacing.

pulse-interval modulation A form of pulse-time modulation in which the pulse spacing is varied. (IM/WM&A) 194-1977w

pulse jitter A relatively small variation of the pulse spacing in a pulse train. *Note:* The jitter may be random or systematic, depending on its origin, and is generally not coherent with any pulse modulation imposed. (IM/WM&A) 194-1977w

pulse length (fiber optics) Often erroneously used as a synonym for pulse duration. (Std100) 812-1984w

pulse-length modulation *See:* pulse-duration modulation.

pulse measurement The assignment of a number and a unit of measurement to a characteristic, property, or attribute of a pulse wherein the number and unit assigned indicate the magnitude of the characteristic which is associated with the pulse. Typically, this assignment is accomplished by comparison of a transform of the pulse, its pulse waveform, with a scale or reference which is calibrated in the unit of measurement. *See also:* method of pulse measurement. (IM/WM&A) 181-1977w

pulse measurement process A realization of a method of pulse measurement in terms of specific devices, apparatus, instruments, auxiliary equipment, conditions, operators, and observers. *See also:* method of pulse measurement. (IM/WM&A) 181-1977w

pulse mode (A) A finite sequence of pulses in a prearranged pattern used for selecting and isolating a communication channel. **(B)** The prearranged pattern. (IM/WM&A) 194-1977

pulse mode, spurious *See:* spurious pulse mode.

pulse modulated field An electromagnetic field produced by the amplitude modulation of a continuous wave carrier by one or more pulses. (NIR) C95.1-1999

pulse modulation (1) (continuous-wave) Modulation of one or more characteristics of a pulse carrier. *Note:* In this sense, the term is used to describe methods of transmitting information on a pulse carrier. (IT) [7]
(2) The encoding of information by varying the basic characteristics of a sequence of pulses, such as width, duration, amplitude, phase or the number of pulses. *See also:* pulse amplitude modulation; pulse code modulation; pulse position modulation. (C) 610.7-1995, 610.10-1994w

pulse modulation, width *See:* pulse-duration modulation.

pulse modulator A device that applies pulses to the element in which modulation takes place. *See also:* modulation. (AP/ANT) 145-1983s

pulse number (1) (circuit properties) (self-commutated converters) (of a group of principal branches or of a complete converter) The number of nonsimultaneous commutations from one principal branch to another during one cycle of operation, considering the group or the complete converter, respectively. (IA/SPC) 936-1987w
(2) The total number of successive nonsimultaneous commutations occurring within the converter circuit during each cycle when operating without phase control. It is also equal to the order of the principal harmonic in the direct voltage, that is, the number of pulses present in the dc output voltage in one cycle of the supply voltage. (IA/SPC) 519-1992

pulse operation The method of operation in which the energy is delivered in pulses. *Note:* Pulse operation is usually described in terms of the pulse shape, the pulse duration, and the pulse-recurrence frequency. *See also:* pulse. (PE) 599-1985w

pulse packet The volume of space occupied by a single radar pulse. The dimensions of this volume are determined by the angular width of the beam, the duration of the pulse, and the distance from the antenna. (AES/RS) 686-1990

pulse-pair resolution (photomultipliers) The time interval between two equal-amplitude delta-function optical pulses such that the valley between the two corresponding anode pulses falls to fifty percent of the peak amplitude. (NPS) 398-1972r

pulse period (1) (measuring the performance of tone address signaling systems) When sending a sequence of signals, the time interval from the start of one signal present condition to the start of the next signal present condition. *Synonym:* cycle time. (COM/TA) 752-1986w
(2) (dial-pulse address signaling systems) (telephony) The time from the start of one break interval of a dial pulse in a train until the start of the next break interval. Milliseconds is the preferred unit of time to express the duration of the pulse period. (COM/TA) 753-1983w

pulse permeability (magnetic core testing) The value of amplitude permeability when the rate of change of induction (that is, the exciting voltage) is held substantially constant over a period of time during each cycle. The frequency, amplitude, duration of the exciting voltage, and the time interval for which the permeability is measured must be stated

$$\mu_{\pi} = \frac{1}{\mu_0}\frac{\Delta B}{\Delta H}$$

where

μ_{π} = pulse permeability, relative
ΔB = change in induction during the stated time interval
ΔH = associated change in magnetic field strength.

Note: When pulse permeability is to be related to a specific circuit condition, a second subscript may be used; for example, $\mu_{\pi a}$ would represent the relative amplitude permeability determined under pulsed excitation. (MAG) 393-1977s

pulse pileup Occurrence of two successive pulses closely associated in time but from separate decays such that they contribute to each other's pulse height and shape. Usually, the system processes the two inputs as a composite single pulse, which is stored in a spectral channel different from that at which either of the component pulses would have been stored. Pulse pileup is a function of the square of the counting rate and of the amplifier pulse width. *Synonym:* random summing. (NI) N42.14-1991

pulse position modulation (PPM) (1) (data transmission) (pulse-phase modulation) Pulse-time modulation in which the value of each instantaneous sample of a modulating wave is caused to modulate the position in time of a pulse. (AP/PE/ANT) 145-1983s, 599-1985w
(2) A form of pulse modulation in which the position in time of a pulse is varied, without modifying the pulse duration, to convey information. (C) 610.7-1995

pulse power (pulse terminology) The power transferred or transformed by a pulse(s). Unless otherwise specified by a mathematical adjective average power over a specified interval is assumed. *See also:* mathematical adjectives. (IM/WM&A) 194-1977w

pulse power, carrier-frequency peak *See:* peak pulse power, carrier-frequency.

pulse quadrant (pulse waveform feature) One of the four continuous and contiguous waveform features of specified extent that include a region of maximum curvature or a point of discontinuity in the waveform slope. (PE/PSR) [6]

pulse, radio-frequency *See:* radio-frequency pulse.

pulse rate (watthour meter) The number of pulses per demand interval at which a pulse device is nominally rated. *See also:* pulse-repetition frequency; auxiliary device to an instrument. (ELM) C12.1-1982s

pulse rate—maximum The number of pulses per second at which a pulse device is nominally rated. (ELM) C12.1-1988

pulse rate telemetering (electric power system) A type of telemetering in which the number of unidirectional pulses per unit time is varied as a function of the magnitude of the measured quantity. *See also:* telemetering. (PE/PSE) 94-1970w

pulse ratio (dial-pulse address signaling systems) (telephony) The percentage of the total dial-pulse period during which the circuit is in each of the two interval states. This ratio is customarily expressed as % break (preferred) of % make. (COM/TA) 753-1983w

pulse rebalance *See:* pulse capture.

pulse receiver (metering) The unit that receives and registers the pulses. It may include a periodic resetting mechanism, so that a reading proportional to demand may be obtained. (ELM) C12.1-1988

pulse recorder (metering) A device that receives and records pulses over a given demand interval. *Note:* It may record pulses in a machine-translatable form on magnetic tape, paper tape, or other suitable media. (ELM) C12.1-1988

pulse-recorder channel (metering) A means of conveying information. It consists of an individual input, output, and intervening circuitry required to record pulse data on the recording media. (ELM) C12.1-1988

pulse regeneration (data transmission) The process of restoring a series of pulses to their original timing, form, and relative magnitude. (PE) 599-1985w

pulse relay—totalizing A device used to receive and totalize pulses from two or more sources for proportional transmission to another totalizing relay or to a receiver. (ELM) C12.1-1988

pulse repeater (transponder) A device used for receiving pulses from one circuit and transmitting corresponding pulses into another circuit. It may also change the frequency and waveforms of the pulses and perform other functions. *See also:* pulse; repeater. (AP/ANT) 145-1983s

pulse-repetition frequency (PRF) **(1)** High prf = more than 1 Hz. (LEO) 586-1980w
(2) The number of pulses per unit time of a periodic pulse train or the reciprocal of the pulse period. *Note:* This term also includes the average number of pulses per unit time of aperiodic pulse trains where the periods are of random duration. *See also:* pulse. (IM/HFIM) [40]
(3) The number of pulses per unit of time, usually per second. *Synonym:* pulse-repetition rate. (AES) 686-1997

pulse-repetition frequency stagger The technique of varying the time between pulses of a pulse radar. This is useful in compensating for blind speeds in pulsed moving-target indication (MTI) radars. (AES) 686-1997

pulse-repetition interval (PRI) The time duration between successive pulses. *Note:* PRI is the reciprocal of the pulse-repetition frequency (PRF). *Synonym:* pulse-repetition period. (AES) 686-1997

pulse-repetition period *See:* pulse-repetition interval.

pulse-repetition rate *See:* pulse-repetition frequency.

pulse reply The transmission of a pulse or pulse mode by a transponder as the result of an interrogation. (IM/WM&A) 194-1977w

pulse response characteristics (pulse response curve) The relationship between the indication of a quasi-peak voltmeter and the repetition rate of regularly repeated pulses of constant amplitude. *See also:* electromagnetic compatibility. (EMC/INT) [53], [70]

pulse rise time **(1) (germanium gamma-ray detectors) (charged-particle detectors) (x-ray energy spectrometers)** The interval between the instants at which the instantaneous value first reaches specified lower and upper limits, namely, 10% and 90% of the peak pulse value unless otherwise specified. (In the case of a step function applied to an RC (resistance-capacitance) low-pass filter, the rise time is given by t_r = 2.2 RC. In the case of a step function applied to an amplifier that has simple CR-RC (capacitance-resistance to resistance-capacitance) shaping, that is, one high-pass and one low-pass RC filter of equal time constants, the rise time is given by t_r = 0.57 RC).
(NPS/NID) 325-1986s, 759-1984r, 301-1976s, 300-1982s
(2) (data transmission) The interval between the instants at which the instantaneous amplitude first reaches specified lower and upper limits, namely 10% and 90% of the peak-pulse amplitude unless otherwise stated. (PE) 599-1985w

pulse scaler (pulse techniques) A device that produces an output signal whenever a prescribed number of input pulses has been received. It frequently includes indicating devices for interpolation. *See also:* pulse. (NPS) 175-1960w

pulse separation (pulse terminology) The interval between the trailing-edge pulse time of one pulse and the leading-edge pulse time of the succeeding pulse. *See also:* leading-edge pulse time; trailing-edge pulse time; pulse. (IM/WM&A) 194-1977w

pulses, equalizing *See:* equalizing pulses.

pulse shape (A) (pulse terminology) For descriptive purposes a pulse waveform may be imprecisely described by any of the adjectives, or combinations thereof, in descriptive adjectives, major (minor); polarity related adjectives; geometrical adjectives; and functional adjectives, exponential. **(B) (pulse terminology)** For tutorial purposes a hypothetical pulse waveform may be precisely defined by the further addition of the adjective ideal. *See also:* descriptive adjectives. **(C) (pulse terminology)** For measurement or comparison purposes a pulse waveform may be precisely defined by the further addition of the adjective reference. *See also:* descriptive adjectives. (IM/WM&A) 194-1977

pulse shaper (pulse techniques) Any transducer used for changing one or more characteristics of a pulse. *Note:* This term includes pulse regenerators. *See also:* pulse. (IM/WM&A) 194-1977w

pulse shaping Intentionally changing the shape of a pulse. (IM/WM&A) 194-1977w

pulse spacing (pulse interval) The interval between the corresponding pulse times of two consecutive pulses. *Note:* The term pulse interval is deprecated because it may be taken to mean the duration of the pulse instead of the space or interval from one pulse to the next. Neither term means the space between pulses. (IM/WM&A) 194-1977w

pulse spacing distribution (electromagnetic site survey) The fraction of pulse spacing time at level v_1 that exceeds time T. *See also:* average crossing rate. (EMC) 473-1985r

pulse spectrum (signal-transmission system) The frequency distribution of the sinusoidal components of the pulse in relative amplitude and in relative phase. *See also:* signal. (IM/WM&A) 194-1977w

pulse speed (telephony) (dial-pulse address signaling systems) The number of dial pulses occurring per unit of time per unit of time. Pulses per second (pls/s) is the preferred unit to express pulse speed. (COM/TA) 753-1983w

pulse spike (automatic control) An unwanted pulse of relatively short duration, superimposed on the main pulse. *See also:* spike. (PE/EDPG) [3]

pulse start time (pulse terminology) The instant specified by a magnitude referenced point on the first transition of a pulse waveform. Unless otherwise specified, the pulse start time is at the mesial point on the first transition. *See also:* waveform epoch; pulse stop time. (IM/WM&A) 194-1977w

pulse stop time (pulse terminology) The instant specified by a magnitude referenced point on the last transition of a pulse waveform. Unless otherwise specified, the pulse stop time is at the mesial point on the last transition. *See also:* waveform epoch; pulse start time. (IM/WM&A) 194-1977w

pulse storage time (light-emitting diodes) The interval between the instants at which the instantaneous amplitudes of the input and output pulses last reach a specified fraction of their peak pulse amplitudes, namely, 90 percent. (ED) [127]

pulse stretcher (spectrum analyzer) A pulse shaper that produces an output whose duration is greater than that of the input pulse and whose amplitude is proportional to that of the peak amplitude of that input pulse. (NPS/IM) 398-1972r, 748-1979w

pulse string *See:* pulse train.

pulse stuffing A method in which pulses are inserted into a stream of pulses to achieve synchronization between two digital communications systems. (C) 610.7-1995

pulse techniques *See:* burst.

pulse tilt A distortion in an otherwise essentially flat-topped rectangular pulse characterized by either a decline or a rise of the pulse top. *See also:* television. (IM/BT/WM&A/AV) 194-1977w, [34]

pulse time, leading edge *See:* leading-edge pulse time.

pulse time, mean *See:* mean pulse time.

pulse-time modulation (PTM) (pulse terminology) (data transmission) Modulation in which the value of instantaneous samples of the modulating wave are caused to modulate the time of occurrence of some characteristic of a pulse carrier. *Note:* Pulse-duration modulation, pulse-position modulation, and pulse-interval modulation, are particular forms of pulse-time modulation. (IM/PE/AP/WM&A/ANT) 194-1977w, 599-1985w, 145-1983s

pulse-time reference points (A) (pulse terminology) (top center point). A specified time referenced point or magnitude referenced point on a pulse waveform top. If no point is specified, the top center point is the time referenced point at the intersection of a pulse waveform and the top center line. *See also:* waveform epoch. **(B) (pulse terminology)** [first (last) base point]. Unless otherwise specified, the first (last) datum point in a pulse epoch. *See also:* pulse train time-related definitions; waveform epoch. (IM/WM&A) 194-1977

pulse time symbology See figure below.

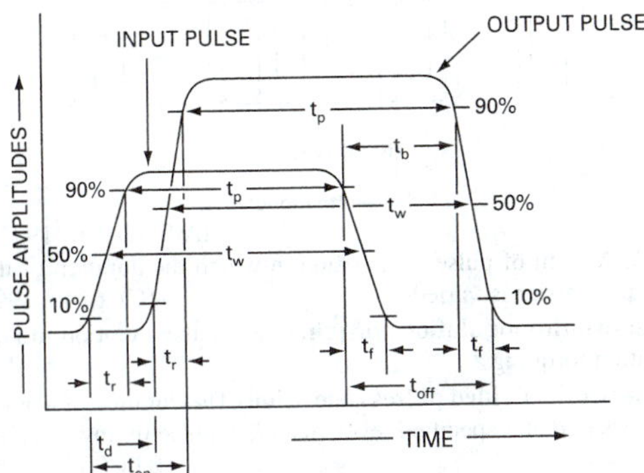

pulse-time symbology

pulse-time, trailing-edge *See:* trailing-edge pulse time.

pulse timing of video pulses (television) The determination of an occurrence of a pulse or a specified portion thereof at a particular time. *See also:* time of rise of video pulses; television; pulse width of video pulses. (BT) 207-1950w

pulse top (1) (pulse transformers) That portion of the pulse occurring between the time of intersection of straight-line segments used to determine A_T and A_T. (PEL/ET) 390-1987r
(2) That major segment of a pulse waveform having the greater displacement in amplitude from the baseline. *See also:* pulse. (IM/HFIM) [40]

pulse trailing edge The major transition towards the pulse baseline occuring before a reference time. *See also:* pulse. (IM/HFIM) [40]

pulse train (1) (dial-pulse address signaling systems) (telephony) In dial-pulse signaling, a series of contiguous pulses of undetermined length. (COM/TA) 753-1983w
(2) (pulse terminology) A continuous repetitive sequence of pulse waveforms. (IM/WM&A) 194-1977w
(3) (thyristor) A gate signal applied during the desired conducting interval, or parts thereof, made up of a train of pulses of predetermined duration, amplitude, and frequency. (IA/IPC) 428-1981w
(4) (signal-transmission system) A sequence of pulses. *See also:* signal; pulse. (BT/IE/PE/IA/ICTL/AV/EDPG/IAC) [34], 270-1966w, [43], [3], [60]
(5) A series of pulses with similar characteristics. *Synonym:* pulse string. (C) 610.7-1995
(6) A sequence of pulses at the pulse repetition frequency used to accomplish a function such as moving-target indication (MTI) or increased effective signal-to-noise ratio. *Note:* A pulse train of duration less than the radar echo delay time is usually referred to as a pulse burst. *See also:* pulse burst. (AES) 686-1997

pulse train, periodic *See:* periodic pulse train.

pulse-train spectrum (pulse-train frequency-spectrum) The frequency distribution of the sinusoidal components of the pulse train in amplitude and in phase angle. (IM/WM&A) 194-1977w

pulse train time-related definitions (A) (pulse terminology) (pulse repetition period) The interval between the pulse start time of a first pulse waveform and the pulse start time of the immediately following pulse waveform in a periodic pulse train. **(B) (pulse terminology)** (pulse repetition frequency) The reciprocal of pulse repetition period. **(C) (pulse terminology)** (pulse separation) The interval between the pulse stop time of a first pulse waveform and the pulse start time of the immediately following pulse waveform in a pulse train. **(D) (pulse terminology)** (duty factor) Unless otherwise specified, the ratio of the pulse waveform duration to the pulse repetition period of a periodic pulse train. **(E) (pulse terminology)** (on-off ratio) Unless otherwise specified, the ratio of the pulse waveform duration to the pulse separation of a periodic pulse train. **(F) (pulse terminology)** (base center line) The time reference line at the average of the pulse stop time of a first pulse waveform and the pulse start time of the immediately following pulse waveform in a pulse train. **(G) (pulse terminology)** (base center point) A specified time referenced point or magnitude referenced point on a pulse train waveform base. If no point is specified, the base center point is the time referenced point at the intersection of a pulse train waveform base and a base center line. **(H) (pulse terminology)** (pulse train epoch) The span of time in a pulse train for which waveform data are known or knowable and which extends from a first base center point to the immediately following base center point. (IM/WM&A) 194-1977

pulse-train top (base) envelope (pulse terminology) Unless otherwise specified, the waveform defined by a cubic natural spline with knots at each point of intersection of the top center line and topline (the base center line and the baseline) of each (between adjacent) pulse waveforms(s) in a pulse train. (IM/WM&A) 194-1977w

pulse train, unidirectional *See:* unidirectional pulse train.

pulse transmitter (1) A pulse-modulated transmitter whose peak power-output capabilities are usually large with respect to average power-output rating. *See also:* radio transmitter; pulse. (AP/ANT) 145-1983s
(2) (power system device function numbers) Used to generate and transmit pulses over a telemetering or pilot-wire circuit to the remote indicating or receiving device. (SUB/PE) C37.2-1979s

pulse turn-off time (light-emitting diodes) The arithmetic sum of the pulse storage time, and the pulse decay time of the output pulse. (ED) [127]

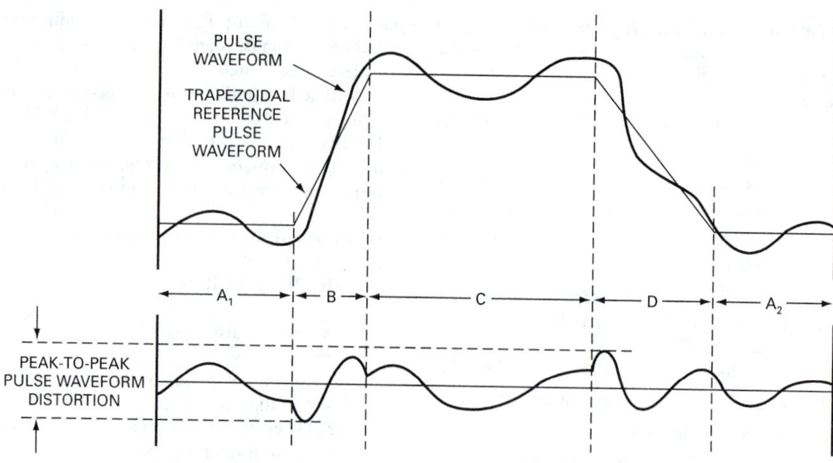

Pulse waveform distortion and pulse waveform feature distortion.

pulse waveform distortion

pulse turn-on time (light-emitting diodes) The arithmetic sum of the pulse delay time, and the pulse rise time of the output pulse. (ED) [127]

pulse-type telemeter A telemeter that employs characteristics of intermittent electric signals other than their frequency, as the translating means. *Note:* These pulses may be utilized in any desired manner to obtain the final indications, such as periodically counting the total number of pulses; or measuring their "on" time, their "off" time, or both. (SWG/PE/SUB) C37.100-1992, C37.1-1994

pulse waveform distortion (pulse terminology) The algebraic difference in magnitude between all corresponding points in time of a pulse waveform and a reference pulse waveform. Unless otherwise specified by a mathematical adjective, peak-to-peak pulse waveform distortion is assumed. *See also:* mathematical adjectives; pulse waveform feature distortion. (IM/WM&A) 194-1977w

pulse waveform feature distortion (pulse terminology) The algebraic difference in magnitude between all corresponding points in time of a pulse waveform and a reference pulse waveform feature. Unless otherwise specified by a mathematical adjective, peak-to-peak pulse waveform feature distortion is assumed. *See also:* mathematical adjectives. (IM/WM&A) 194-1977w

pulse width The time interval between the points on the leading and trailing edges at which the instantaneous value bears a specified relation to the maximum instantaneous value of the pulse, usually the time interval between the half-power points of the pulse. *Synonym:* pulse duration. (AES) 686-1997

pulse-width at half maximum, $T_{0.5}$ (germanium gamma-ray detectors) (of an amplifier output pulse) The time interval between the 50% of maximum amplitude points of a pulse. (NPS) 325-1986s

pulse width discriminator (PWD) A device that passes only those video pulses whose duration falls within specified limits. (AES) 686-1997

pulse-width modulation (self-commutated converters) (converter characteristics) Pulse-time modulation in which the value of each instantaneous sample of the modulating wave is caused to modulate the duration of a pulse. The modulating frequency may be fixed or variable. Examples of waveforms produced by PWM are shown in the corresponding figure. *See also:* pulse-duration modulation.

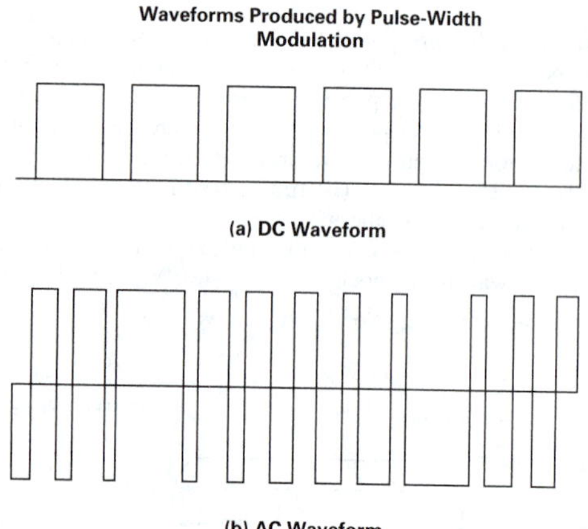

Pulse-Width Modulation

(IA/SPC) 936-1987w

(2) A form of pulse modulation in which the duration of the pulse carrier is varied. (C) 610.7-1995

pulse-width-modulation torquing *See:* pulse-duration-modulation torquing.

pulse width of video pulses (television) The duration of a pulse measured at a specified level. *See also:* pulse timing of video pulses. (BT) 207-1950w

pulse width timing check A timing check that specifies the minimum time a signal shall remain in a specified state once it has transitioned to that state. (C/DA) 1481-1999

pulsing (telephone switching systems) The signaling over the communication path of signals representing one or more digits required to set up a call. (COM) 312-1977w

pulsing circuit (peaking circuit) A circuit designed to provide abrupt changes in voltage or current of some characteristic pattern. *See also:* electronic controller. (IA/ICTL/IAC) [60]

pulsing transformer Supplies pulses of voltage or current. *See also:* electronic controller. (IA/ICTL/IAC) [60]

pump (parametric device) The source of alternating-current power that causes the nonlinear reactor to behave as a time-varying reactance. *See also:* parametric device. (ED) [46]

pump-back test (rotating machinery) (electrical back-to-back test) A test in which two identical machines are mechanically coupled together, and they are both connected electrically to a power system. The total losses of both machines are taken as the power input drawn from the system. *See also:* asynchronous machine; direct-current commutating machine. (PE) [9]

pumped figure of merit (parametric amplifier) (nonlinear, active, and nonreciprocal waveguide components) The ratio of the half-amplitude fundamental Fourier component of pumped elastance to the series resistance of the varactor diode. This quantity, designated as $m_1\omega_c$, has the dimensions of angular frequency and describes the noise, gain, and impedance characteristics of parametric amplifiers. The parameter m_1 is the normalized fundamental component of elastance and ω_c is the total angular cutoff frequency. (MTT) 457-1982w

pumped-storage hydro capability (power operations) The capability supplied by hydroelectric sources under specified water conditions using a reservoir that is alternately filled by pumping and depleted by generating. (PE/PSE) 858-1987s

pumped storage station (power operations) A hydroelectric generating station at which electrical energy is normally generated during periods of relatively high system demand by utilizing water that has been pumped into a storage reservoir usually during periods of relatively low system demand. (PE/PSE) 858-1987s, 346-1973w

pumped tube An electron tube that is continuously connected to evacuating equipment during operation. *Note:* This term is used chiefly for pool-cathode tubes. (ED) [45]

pump efficiency (laser maser) The ratio of the power or energy absorbed from the pump to the power or energy available from the pump source. (LEO) 586-1980w

pump-free control *See:* antipump device.

pump-free device *See:* antipump device.

pumping The unintentional cyclical tripping and closing of a network protector. (PE/TR) C57.12.44-1994

pumping load (power operations) Totals of loads caused by pumping in pumped-storage stations within the system. (PE/PSE) 858-1987s

pump runout A pump flow condition in which the pump is operating beyond its design point due to a reduction in the system head. As a result, the pump motor's brake-horsepower and full load current demand may be increased. (PE/NP) 741-1997

punch (A) A device for making holes in some data medium such as a card or paper tape. *See also:* card punch; card reproducing punch; spot punch; calculating punch. **(B)** A perforation created by a device as in definition (A). *See also:* zone punch; eleven punch; numeric punch; twelve punch; digit punch. **(C)** To make a perforation as in definition (B). *See also:* gang punch; overpunch. (C) 610.10-1994

punch card A card into which hole patterns can be punched such that the patterns can be used to store or represent information. *Note:* This term is often used in place of "punched card" a standard-sized punch card has twelve rows of 80 columns. *Synonym:* Hollerith card. *See also:* edge-punched card; control card; data card; scored card; binder-hole card; laced card; edge-notched card; interpreted card; trailer card; edge-coated card; binary card; header card; aperture card; check card; twelve-row punch card; short card. (C) 610.10-1994w

punched card (1) A card on which a pattern of holes or cuts is used to represent data. (MIL/C) [2], [20]
(2) A card punched with hole patterns such that the patterns store or represent information. *See also:* punch card. (C) 610.10-1994w

punched card holder *See:* card hopper.

punched card machine *See:* card reader.

punched card reader *See:* card reader.

punched paper tape *See:* perforated tape.

punched tape (computers) A tape on which a pattern of holes or cuts is used to represent data. (MIL/C) [2], [20], [85]

punched tape handler (test, measurement, and diagnostic equipment) A device that handles punched tape and usually consists of a tape transport and punched tape reader with associated electrical and electronic equipments. Most units provide for tape to be wound and stored in reels; however, some units provide for the tape to be stored loosely in closed bins. (MIL) [2]

punched tape reader (1) (test, measurement, and diagnostic equipment) A device capable of converting information from punched tape, where it has been stored in the form of a series of holes, into a series of electrical impulses. (MIL) [2]
(2) An input unit that senses the hole patterns in a perforated tape, transforming the hole patterns into electrical signals representing data. (C) 610.10-1994w

punching (rotating machinery) A lamination made from sheet material using a punch and die. *See also:* rotor; stator. (PE) [9]

punching station The place in a punch where a card or paper tape is punched. (C) 610.10-1994w

punch position (computers) A site on a punched tape or card where holes are to be punched. (MIL/C) [2], [20], [85]

punch tape A tape in which hole patterns can be punched such that the patterns can be used to store or represent information. (C) 610.10-1994w

punch-through voltage (x-ray energy spectrometers) (of a semiconductor radiation detector) The voltage at which a junction detector becomes fully depleted. *See also:* depletion voltage. (NPS/NID) 759-1984r

puncture (1) (voltage testing) A disruptive discharge through the body of a solid dielectric. (PE/PSIM) [55]
(2) A disruptive discharge through solid insulation. (PE/PSIM) 4-1995
(3) (A) A disruptive discharge through the body of a solid dielectric. **(B)** A disruptive discharge through solid insulation. **(C)** Term used to denote when a disruptive discharge occurs through a solid dielectric and produces permanent loss of dielectric strength; in a liquid or gaseous dielectric, the loss may be only temporary. (SPD/PE) C62.11-1999

puncture voltage (surge arresters) The voltage at which the test specimen is electrically punctured. (T&D/PE) [10], [8]

pupil (1) (laser maser) The variable aperture in the iris through which light travels toward the interior of the eye. (LEO) 586-1980w
(2) (pupillary aperture) (illuminating engineering) The opening in the iris which admits light into the eye. (EEC/IE) [126]

purchaser (1) (nuclear power quality assurance) The organization responsible for establishment of procurement requirements and for issuance, administration, or both, of procurement documents. (PE/NP) [124]
(2) (rotating electric machinery) The organization placing the contract for the machinery or its repair; often called the "user." (PE/EM) 11-1980r
(3) An entity that contractually acts as the customer. (VT) 1475-1999

pure alphabetic Pertaining to data that contains only the letters of the alphabet (AaBbCcDdEeFfGgHh. . .). *Contrast:* pure numeric; pure alphanumeric. (C) 610.5-1990w

pure alphanumeric (data management) Pertaining to data that contains only the letters of the alphabet (AaBbCcDdEeFfGgHh. . .) and the numerals (1234567890). *Contrast:* pure alphabetic; pure numeric. (C) 610.5-1990w

pure binary *See:* binary.

pure binary numeration system *See:* binary numeration system.

pure numeric Pertaining to data that contains only the numerals (1234567890). *Contrast:* pure alphanumeric; pure alphabetic. (C) 610.5-1990w

purged Pertaining to a record that has been physically deleted from a file. *See also:* active; logically deleted; inactive. (C) 610.2-1987

purge instruction A purge (cache-control) instruction changes a line to the uncached state, invalidating the old cache line without copying dirty data back to memory. (C/MM) 1596-1992

purging A ring state that occurs when the active monitor has detected a ring error and is returning the ring to an operational state by transmitting purge frames. (C/LM) 8802-5-1998

pure tone (1) (overhead power lines) A sound wave, the instantaneous sound pressure of which is a simple sinusoidal function of time. (T&D/PE) 539-1990
(2) (overhead power lines) *See also:* simple tone. (SP) [32]

purification of electrolyte The treatment of a suitable volume of the electrolyte by which the dissolved impurities are removed in order to keep their content in the electrolyte within desired limits. *See also:* electrorefining. (EEC/PE) [119]

purity *See:* excitation purity.

Purkinje phenomenon (illuminating engineering) The reduction in subjective brightness of a red light relative to that of a blue light when the luminances are reduced in the same proportion without changing the respective spectral distributions. In passing from photopic to scotopic vision, the curve of spectral luminous efficiency changes, the wavelength of maximum efficiency being displaced toward the shorter wavelengths. (EEC/IE) [126]

purple boundary (television) (illuminating engineering) The straight line drawn between the ends of the spectrum locus on a chromaticity diagram. (BT/IE/EEC/AV) 201-1979w, [126]

purported name A construct that is syntactically a name but that has not (yet) been shown to be a valid name. (C/PA) 1328.2-1993w, 1224.2-1993w, 1326.2-1993w, 1327.2-1993w

purpose statement A brief statement of the reason for an IDEF0 model's existence that is presented in the A-0 context diagram of the model. (C/SE) 1320.1-1998

push (data management) To append data onto a stack. *Contrast:* pull. (C) 610.5-1990w

push a button The act of moving the pointer to a button widget and then selecting the button. (C) 1295-1993w

push brace A supporting member, usually of timber, placed between a pole or other structural part of a line and the ground or a fixed object. *See also:* tower. (T&D/PE) [10]

push button A visual user interface control containing a button, labeled with text, graphics, or both, that represents the action initiated when a user selects it. (C) 1295-1993w

pushbutton Part of an electric device, consisting of a button that must be pressed to effect an operation. *See also:* switch. (IA/ICTL/IAC) [60], [84]

pushbutton dial (telephone switching systems) A type of calling device used in automatic switching that has an activator per digit that generates distinctive pulsing. (COM) 312-1977w

pushbutton station A unit assembly of one or more externally operable pushbutton switches, sometimes including other pilot devices such as indicating lights or selector switches, in a suitable enclosure. *See also:* switch. (IA/ICTL/IAC) [60]

pushbutton switch (pushbutton) A master switch, usually mounted behind an opening in a cover or panel, and having an operating plunger or button extending forward in the opening. Operation of the switch is normally obtained by pressure of the finger against the end of the button. *See also:* switch. (EEC/PE) [119]

pushbutton switching A reperforator switching system in which selection of the outgoing channel is initiated by an operator. (COM) [49]

pushdown list (1) (computers) A list that is constructed and maintained so that the next item to be retrieved is the most recently stored item in the list, that is, last in, first out. *See also:* pushup list. (C) [20], [85]
(2) (data management) *See also:* stack. (C) 610.5-1990w

push-down stack *See:* stack.

pushdown storage (1) (software) A storage device that handles data in such a way that the next item to be retrieved is the most recently stored item still in the storage device, that is, last-in-first-out (LIFO). *See also:* stack; data. (C/SE) 729-1983s
(2) A type of storage in which data are ordered in such a way that the next data item to be retrieved is the most recently stored item. *Note:* Commonly characterized as "last-in-first-out," or LIFO. *Synonym:* stack storage. (C) 610.10-1994w
(3) (software) *See also:* stack. (C) 610.5-1990w

pushing figure (oscillators) The change of oscillator frequency with a specified change in current, excluding thermal effects. *See also:* electronic tuning sensitivity; oscillatory circuit; television. (ED) 161-1971w

push-pull amplifier circuit *See:* balanced amplifier.

push-pull circuit A circuit containing two like elements that operate in 180-degree phase relationship to produce additive output components of the desired wave, with cancellation of certain unwanted products. *Note:* Push-pull amplifiers and push-pull oscillators are examples. *See also:* amplifier. (Std100) [123]

push-pull currents Balanced currents. *See also:* waveguide. (MTT) 146-1980w

push-pull microphone A microphone that makes use of two like microphone elements actuated by the same sound waves and operating 180 degrees out of phase. *See also:* microphone. (EEC/PE) [119]

push-pull operation The operation of two similar electron devices or of an equivalent double-unit device, in a circuit such that equal quantities in phase opposition are applied to the input electrodes, and the two outputs are combined in phase. (ED) [45]

push-pull oscillator A balanced oscillator employing two similar tubes in phase opposition. *See also:* oscillatory circuit; balanced oscillator. (AP/ANT) 145-1983s

push-pull voltages Balanced voltages. *See also:* waveguide. (MTT) 146-1980w

push-push circuit A circuit employing two similar tubes with grids connected in phase opposition and plates in parallel to a common load, and usually used as a frequency multiplier to emphasize even-order harmonics. (AP/ANT) 145-1983s

push-push currents Currents flowing in the two conductors of a balanced line that, at every point along the line, are equal in magnitude and in the same direction. *See also:* waveguide. (MTT) 146-1980w

push-push voltages Voltages (relative to ground) on the two conductors of a balanced line that, at every point along the line, are equal in magnitude and have the same polarity. *See also:* waveguide. (MTT) 146-1980w

push-to-type operation That form of telegraph operation, employing a one-way reversible circuit, in which the operator must keep a switch operated in order to send from his station. It is generally used in radio transmission where the same frequency is employed for transmission and reception. *See also:* telegraphy. (EEC/PE) [119]

pushup list (1) (computers) A list that is constructed and maintained so that the next item to be retrieved and removed is the oldest item still in the list, that is, first in, first out. *See also:* pushdown list. (C) [20], [85]
(2) (data management) *See also:* queue. (C) 610.5-1990w

pushup storage (1) A type of storage in which data are ordered in such a way that the next data item to be retrieved is the item that was stored first. *Note:* Commonly characterized as first-in-first-out, or FIFO order. *See also:* stack. (C) 610.10-1994w

(2) *See also:* queue. (C) 610.5-1990w

put To place an item into a set of items as in inserting a record into a file, or in representing a numerical value as a series of decimal digits. *Contrast:* get. (C) 610.5-1990w

PV array *See:* photovoltaic array.

PV array subfield *See:* photovoltaic array subfield.

PVC (1) (cable systems in power generating stations) Conduit fabricated from polyvinyl chloride. (PE/SUB/EDPG) 422-1977, 525-1992r

(2) *See also:* polyvinyl chloride; permanent virtual circuit. (C) 610.7-1995

PV cell *See:* photovoltaic cell.

PV effect *See:* photovoltaic effect.

PV module *See:* photovoltaic module.

PV panel *See:* photovoltaic panel.

PV power system *See:* photovoltaic power system.

PV receiver *See:* photovoltaic receiver.

PV system-utility interface *See:* photovoltaic system-utility interface.

PVV *See:* path variability value.

PWB *See:* printed wiring board.

PWM *See:* pulse-width modulation.

pyramidal horn antenna A horn antenna, the sides of which form a pyramid. (AP/ANT) [35], 145-1993

pyroconductivity Electric conductivity that develops with rising temperature, and notably upon fusion, in solids that are practically nonconductive at atmospheric temperatures. (EEC/PE) [119]

pyroelectric effect (primary ferroelectric terms) The appearance of an electric charge at the surface of a polar material when uniform heating or cooling changes the polarization. If the polar material is electroded and an external resistance is connected between the electrodes, the current that flows is a pyroelectric current. All pyroelectrics are polar; ferroelectrics are a subgroup of the polar materials and, therefore, they are both pyroelectric and piezoelectric. The differ from the more general pyroelectrics principally by the reversibility or reorientability of their spontaneous polarization P_s. *Note:* Due to the existence of free charge, the pyroelectric charge may be rapidly compensated. (UFFC) 180-1986w

pyrolysate A product of thermal decomposition. (PE/EM) 1129-1992r

pyrolysis Irreversible chemical decomposition caused by heat, usually without oxidation. (DEI) 1221-1993w

pyrometer A thermometer of any kind usable at relatively high temperatures (above 500 degrees Celsius). *See also:* electric thermometer. (EEC/PE) [119]

Q

Q In the context of a fiber optic communication system, one-half of the ratio of peak-to-peak signal to rms noise. (C/LM) 802.3-1998

QA *See:* quality assurance.

QA access function *See:* queued arbitrated access function.

QAM *See:* quadrature amplitude modulation.

QA segment *See:* queued arbitrated segment.

QA slot *See:* queued arbitrated slot.

QC *See:* quality control.

Q-channel *See:* quadrature video.

Q chrominance signal The sidebands resulting from suppressed-carrier modulation of the chrominance subcarrier by the Q video signal. *Note:* The signal is transmitted in double-sideband form, the sidebands extending approximately 0.6 MHz above and below the chrominance subcarrier. The phase of the signal, for positive Q video signals, is 33° with respect to the (B—Y) axis. (BT/AV) 201-1979w

Q coil *See:* coil *Q*.

Q energy (laser maser) The capacity for doing work. Energy content is commonly used to characterize the output from pulsed lasers, and is generally expressed in joules. (LEO) 586-1980w

Q-hour meter An electricity meter that measures the quantity obtained by effectively lagging the applied voltage to a watthour meter by 60°. This quantity is one of the quantities used in calculating quadergy (varhours). (ELM) C12.1-1988

QIC *See:* quarter-inch cartridge.

Q-meter (quality-factor meter) An instrument for measuring the quality factor Q of a circuit or circuit element. *See also:* instrument. (EEC/PE) [119]

Q of an electrically small tuned antenna An inverse measure of the bandwidth or an antenna as determined by its impedance. It is numerically equal to one half the magnitude of the ratio of the incremental change in impedance to the incremental change in frequency at resonance, divided by the ratio of the antenna resistance to the resonance frequency. *Note:* The Q of an antenna also is a measure of the energy stored to the energy radiated or dissipated per cycle. (AP/ANT) 145-1983s

Q of a resonant antenna The ratio of 2π times the energy stored in the fields excited by the antenna to the energy radiated and dissipated per cycle. *Note:* For an electrically small antenna, it is numerically equal to one-half the magnitude of the ratio of the incremental change in impedance to the corresponding incremental change in frequency at resonance, divided by the ratio of the antenna resistance to the resonant frequency. (AP/ANT) 145-1993

QOLB *See:* queue on lock bit.

QP detector *See:* quasi-peak detector.

Q-percentile life (1) (non-repaired items) (assessed) The Q-percentile life determined as limiting value or values of the confidence interval with a stated confidence level, based on the same data as the observed Q-percentile life of nominally identical items. *Notes:* 1. The source of the data should be stated. 2. Results can be accumulated (combined) only when all conditions are similar. 3. The assumed underlying distribution of failures against time should be stated. 4. It should be stated whether a one-sided or two-sided interval is being used. 5. Where one limiting value is given this is usually the lower limit. (R)

(2) (extrapolated) Extension by a defined extrapolation or interpolation of the observed or assessed Q-percentile life for stress conditions different from those applying to the assessed Q-percentile life and for different percentages. *Note:* The validity of the extrapolation should be justified.

(3) (observed) The length of observed time at which a stated proportion (Q%) of a sample of items has failed. *Notes:* 1. The criteria for what constitutes a failure should be stated. 2. The Q-percentile life is also that life at which (100-Q)% reliability is observed.

(4) (predicted) For the stated conditions of use, and taking into account the design of an item, assessed or extrapolated Q-percentile lives of its parts. *Note:* Engineering and statistical assumptions should be stated, as well as the bases used for the computation (observed or assessed).

Q response (subroutines for CAMAC) The symbol *q* represents a logical truth value which corresponds to the CAMAC Q response. It is set to true if the Q response is 1, to false if the Q response is 0. (NPS) 758-1979r

Q responses (subroutines for CAMAC) The symbol *qa* represents an array of Q response values. Each element of *qa* has the same form and can have the same values as the parameter *q*. The length of qa is given by the value of the first element of *cb* at the time the subroutine is executed. *See also:* control block; Q response. (NPS) 758-1979r

Q-switch (laser maser) A device for producing very short (Δ 30 ns), intense laser pulses by enhancing the storage and dumping of electronic energy in and out of the lasing medium, respectively. (LEO) 586-1980w

Q-switched laser (laser maser) A laser which emits short (Δ 30 ns), high-power pulses by utilizing a Q-switch. (LEO) 586-1980w

Q-switching (laser maser) Producing very short (Δ 30 ns), intense pulses by enhancing the storage and dumping of electronic energy in and out of the laser-maser medium, respectively. (LEO) 586-1980w

quad (1) A structural unit employed in cable, consisting of four separately insulated conductors twisted together. *See also:* cable. (EEC/PE) [119]

(2) *See also:* star quad. (LM/C/LM) 802.3-1998

quad-bundle *See:* bundle.

quadded cable A cable in which at least some of the conductors are arranged in the form of quads. *See also:* cable. (EEC/PE) [119]

quadded components The use of four identical components in a particular circuit configuration in order to reduce the probability of overall circuit failure due the possible occurrence of a fault in one or more of such components during circuit operation. (C) 610.10-1994w

quadded logic The quadruple replication of each individual gate in a logic circuit. (C) 610.10-1994w

quadergy (1) (general) Delivered by an electric circuit during a time interval when the voltages and currents are periodic, the product of the reactive power and the time interval, provided the time interval is one or more complete periods or is quite long in comparison with the time of one period. If the reference direction for energy flow is selected as into the region, the net delivery of quadergy will be into the region when the sign of the quadergy is positive and out of the region when the sign is negative. If the reference direction is selected as out of the region, the reverse will apply. The quadergy is expressed by

$$K = Qt$$

where Q is the reactive power and t is the time interval. If the voltages and currents form polyphase symmetrical sets, there is no restriction regarding the relation of the time interval to the period. If the voltages and currents are quasi-periodic and the amplitudes of the voltages and currents are slowly varying, the quadergy is the integral with respect to time of the reactive power, provided the integration is for a time that is one or more complete periods or that is quite long in comparison with the time of one period. Mathematically,

$$K = \int_{t_0}^{t_0+t} Q dt$$

where Q is the reactive power determined for the condition of voltages and current having slowly varying amplitudes. Quadergy is expressed in var-seconds or var-hours when the voltages are in volts and the currents in amperes, and the time is in seconds or hours, respectively. *See also:* network analysis. (Std100) 270-1966w
(2) (metering) The integral of reactive power with respect to time. (ELM) C12.1-1988

quadlet (1) A set of four adjacent bytes. (C/BA) 10857-1994, 896.3-1993w, 896.4-1993w, 1014.1-1994w
(2) A unit of computer data consisting of 32 bits. (C/BA) 1275-1994
(3) Four bytes (32 bits) of data. (C/MM) 1754-1994, 1596-1992, 1394-1995, 1212-1991s, 1394a-2000
(4) A 4-byte data format or data type. (C/MM) 1596.5-1993
(5) Four bytes of data. (C/MM) 1596.4-1996

quadlet aligned address An address with zeros in the least significant two bits. (C/MM) 1394-1995

quadrant This pertains to the investigation and study of tamper protection concepts and mechanisms. *See also:* tamper protection. (C/BA) 896.3-1993w

quadrantal error (navigation) (navigation aid terms) An angular error in measured bearing caused by characteristics of the vehicle or station which adversely affect the direction of signal propagation; the error varies in a sinusoidal manner throughout the 360° and has two positive and two negative maximums. (AES/GCS) 172-1983w

quadratic lag *See:* lag.

quadratic profile *See:* parabolic profile.

quadratic programming Optimization problem in which:

a) The objective function is a quadratic function of the variable.
b) The plant description is linear.

See also: system. (SMC) [63]

quadrature The relation between two periodic functions when the phase difference between them is one-fourth of a period. *See also:* network analysis. (Std100) 270-1966w

quadrature-acceleration drift rate (dynamically tuned gyro) A drift rate about an axis, normal to both the spin axis and the axis along which an acceleration is applied. This drift rate results from a torque about the axis of applied acceleration and is in quadrature with that due to mass unbalance. (AES/GYAC) 528-1994

quadrature amplitude modulation A modulation technique that uses variations in signal amplitude and phase to represent data-encoded symbols as a number of states. (C) 610.7-1995

quadrature axis (synchronous machines) The axis that represents the direction of the radial plane along which the main field winding produces no magnetization, normally coinciding with the radial plane midway between adjacent poles. *Notes:* 1. The positive direction of the quadrature axis is 90 degrees ahead of the positive direction of the direct axis, in the direction of rotation of the field relative to the armature. 2. The definitions of currents and voltages given in the terms listed below are applicable to balanced load conditions and for sinusoidal currents and voltages. They may also be applied under other conditions to the positive-sequence fundamental-frequency components of currents and voltages. More generalized definitions, applicable under all conditions, have not been agreed upon. (PE) [9]

quadrature-axis component (1) (armature voltage) That component of the armature voltage of any phase that is in time phase with the quadrature-axis component of current in the same phase. *Note:* A quadrature-axis component of voltage may be reproduced by:

a) rotation of the direct-axis component magnetic flux;
b) variation (if any) of the quadrature-axis component of magnetic flux;
c) resistance drop caused by flow of the quadrature-axis component of armature current. The quadrature-axis component of terminal voltage is related to the synchronous internal voltage by

$$\mathbf{E}_{aq} = \mathbf{E}_i - R\mathbf{I}_{aq} - jX_d\mathbf{I}_{ad}$$

See also: phasor diagram. (EEC/PE) [119]
(2) (armature current) That component of the armature current that produces a magnetomotive-force distribution that is symmetrical about the quadrature axis. (PE/EEC) [119]
(3) (rotating machinery) (magnetomotive force) The component of a magnetomotive force that is directed along an axis in quadrature with the axis of the poles. *See also:* asynchronous machine; direct-axis synchronous impedance. (PE) [9]

quadrature-axis current (rotating machinery) The current that produces quadrature-axis magnetomotive force. *See also:* direct-axis synchronous impedance. (PE) [9]

quadrature-axis magnetic flux (rotating machinery) The magnetic-flux component directed along the quadrature axis. *See also:* direct-axis synchronous impedance. (PE) [9]

quadrature-axis operational inductance (standstill frequency response testing) (synchronous machine parameters by standstill frequency testing) The ratio of the Laplace transform of the quadrature-axis armature flux linkages to the Laplace transform of the quadrature-axis current. (PE/EM) 115A-1987

quadrature-axis subtransient impedance (rotating machinery) The operator expressing the relation between the initial change in armature voltage and a sudden change in quadrature-axis armature current, with only the fundamental-frequency components considered for both voltaged and current, with no change in the voltage applied to the field winding, and with the rotor running at steady speed. In terms of network theory it corresponds to the quadrature-axis impedance the machine displays against disturbances (modulations) with infinite frequency. *Note:* If no rotor winding is along the quadrature axis and/or the rotor is not made out of solid steel, this impedance equals the quadrature-axis synchronous impedance. *See also:* asynchronous machine; direct-axis synchronous impedance. (PE) [9]

quadrature-axis subtransient open-circuit time constant The time in seconds required for the rapidly decreasing component (negative) present during the first few cycles in the direct-axis component of symmetrical short-circuit conditions with the machine running at rated speed, to decrease to 1/c Δ 0.368 of its initial value. (EEC/PE) [119]

quadrature-axis subtransient reactance The ratio of the fundamental component of reactive armature voltage due to the initial value of the fundamental quadrature-axis component of alternating-current component of the armature current, to this component of current under suddenly applied balanced load conditions and at rated frequency. Unless otherwise specified, the quadrature-axis subtransient reactance will be that corresponding to rated armature current. (EEC/PE) [119]

quadrature-axis subtransient short-circuit time constant The time in seconds required for the rapidly decreasing component present during the first few cycles in the quadrature-axis component of the alternating-current component of the armature current under suddenly applied symmetrical short-circuit conditions, with the machine running at rated speed to decrease to 1/e Δ 0.368 of its initial value. (EEC/PE) [119]

quadrature-axis subtransient voltage (rotating machinery) The quadrature-axis component of the terminal voltage that appears immediately after the sudden opening of the external circuit when the machine is running at a specified load, before any flux variation in the excitation and damping circuits has taken place. (PE) [9]

quadrature-axis synchronous impedance (rotating machinery) (synchronous machines) The impedance of the armature winding under steady-state conditions where the axis of the armature current and magnetomotive force coincides with the quadrature axis. In large machines where the armature resistance is negligibly small, the quadrature-axis synchronous impedance is equal to the quadrature-axis synchronous reactance. (PE) [9]

quadrature-axis synchronous reactance The ratio of the fundamental component of reactive armature voltage, due to the fundamental quadrature-axis component of armature current, to this component of current under steady-state conditions and at rated frequency. Unless otherwise specified, the value of quadrature-axis synchronous reactance will be that corresponding to rated armature current. (EEC/PE) [119]

quadrature-axis transient impedance (rotating machinery) The operator expressing the relation between the initial change in armature voltage and a sudden change in quadrature-axis armature current component, with only the fundamental frequency components considered for both voltage and current, with no change in the voltage applied to the field winding, with the rotor running at steady speed, and by considering only the slowest decaying component and the steady-state component of the voltage drop. In terms of network theory it corresponds to the quadrature-axis impedance the machine display against disturbances (modulation) with infinite frequency by considering only two poles pairs* (or poles*), namely those with smallest (including zero) real parts, of the impedance function. *Notes:* 1. If no rotor winding is along the quadrature axis and/or the rotor is not made out of solid steel, this impedance equals the quadrature-axis synchronous impedance. 2. The term pole refers here to the roots of the denominator of the impedance function. *See also:* direct-axis synchronous impedance; asynchronous machine. (PE) [9]

quadrature-axis transient open-circuit time constant The time in seconds required for the root-mean-square alternating-current value of the slowly decreasing component present in the direct-axis component of symmetrical armature voltage on open-circuit to decrease to 1/c Δ 0.368 of its initial value when the quadrature field winding (if any) is suddenly short-circuited with the machine running at rated speed. *Note:* This time constant is important only in turbine generators. (EEC/PE) [119]

quadrature-axis transient reactance The ratio of the fundamental component of reactive armature voltage, due to the fundamental quadrature-axis component of the alternating-current component of the armature current, to this component of current under suddenly applied load conditions and at rated frequency, the value of current to be determined by the extrapolation of the envelope of the alternating-current component of the current wave to the alternating-current component of the current wave to the instant of the sudden application of load, neglecting the high-decrement current during the first few cycles. *Note:* The quadrature-axis transient reactance usually equals the quadrature-axis synchronous reactance except in solid-rotor machines, since in general there is no really effective field current in the quadrature axis. (EEC/PE) [119]

quadrature-axis transient short-circuit time constant The time in seconds required for the root-mean-square alternating-current value of the slowly decreasing component present in the direct-axis component of the alternating-current component of the armature current under suddenly applied short-circuit conditions with the machine running at rated speed to decrease to 1/e Δ 0.368 of its initial value. (EEC/PE) [119]

quadrature-axis transient voltage (rotating machinery) The quadrature-axis component of the terminal voltage that appears immediately after the sudden opening of the external circuit when running at a specified load, neglecting the components with very rapid decay that may exist during the first few cycles. *See also:* asynchronous machine; direct-axis synchronous impedance. (PE) [9]

quadrature-axis voltage (rotating machinery) The component of voltage that would produce quadrature-axis current when resistance limited. (PE) [9]

quadrature hybrid (waveguide components) A hybrid junction that has the property that a wave leaving one output port is in phase quadrature with the wave leaving the other output port. (MTT) 147-1979w

quadrature modulation Modulation of two carrier components 90 degrees apart in phase by separate modulating functions. (Std100) 270-1964w

quadrature spring rate (dynamically tuned gyro) When the case of a dynamically tuned gyro is displaced with respect to the gyro rotor through an angle about an axis perpendicular to the spin axis, a torque proportional to and 90° away from the displacement acts in a direction to reduce this angle and to align the rotor with the case. The torque is usually due to windage, a squeeze-film force, or flexure hysteresis. This spring rate results in a drift rate coefficient having dimensions of angular displacement per unit time per unit angle of displacement about an input axis. (AES/GYAC) 528-1994

quadrature video One of a pair of coherent, bipolar video signals derived from the RF or IF signal by a pair of synchronous detectors with a 90° phase difference between the coherent oscillator (coho) reference inputs used for each. *Note:* The quadrature component is often identified as *Q* and the other of the pair as in-phase video or *I*. *Synonym:* Q-channel. *See also:* in-phase video. (AES) 686-1997

Quadrex *See:* explosives.

quadrilateral characteristic A distance relay characteristic on an R-X diagram created by a directional measurement, a reactance measurement, and two resistive measurements. (PE/PSR) C37.113-1999

quadri pole *See:* two-terminal pair network.

quadruple-address instruction *See:* four-address instruction.

quadruple-length register Four registers that function as a single register. *Synonym:* quadruple register. *See also:* n-tuple length register; double-length register; triple-length register. (C) 610.10-1994w

quadruple register *See:* quadruple-length register.

quadrupole parametric amplifier A beam parametric amplifier having transverse input and output couplers for the signal, separated by a quadrupole structure that is excited by a pump to obtain parametric amplification of a cyclotron wave. *See also:* parametric device. (ED) 254-1963w, [46]

quadword An aligned **hexlet.** *Note:* The definition of this term is architecture-dependent, and so may differ from that used in other processor architectures. (C/MM) 1754-1994

qualification (1) (nuclear power generating station) The generation and maintenance of evidence to ensure that the equipment will operate on demand to meet the system performance requirements. (PE/NP) 323-1974s
(2) (personnel) (nuclear power quality assurance) The characteristics or abilities gained through education, training, or experience, as measured against established requirements, such as standards or tests, that qualify an individual to perform a required function. (PE/NI/NP) [124], N42.23-1995
(3) (raceway systems for Class 1E circuits for nuclear power generating stations) (raceway system) Demonstration in the form of certificates of compliance, analysis reports, or testing reports that the raceway system meets the design requirements. (PE/NP) 628-1987r

qualification testing (1) (software) Formal testing, usually conducted by the developer for the consumer, to demonstrate that the software meets its specified requirements. *See also:* software; requirement; acceptance testing; system testing; formal testing. (C/SE) 729-1983s
(2) (mechanical) Testing of the complete assembly or subassemblies to determine acceptability by applying an actual input that has a test response spectrum (either ground- or

floor-response spectrum) equal to or larger than the design earthquake response spectrum.
(SWG/SUB/PE) C37.122-1983s, C37.122.1-1993G, C37.100-1992

(3) Testing performed to demonstrate to the acquirer that a software item or a system meets its specified requirements.
(C/SE) J-STD-016-1995

qualification tests (safety systems equipment in nuclear power generating stations) Tests conducted on safety systems equipment to demonstrate the capability to meet specified functional requirements under the action of specified test levels of environmental and operational parameters. *Note:* These tests subject a sample or samples of equipment to specified conditions designed to simulate normal, abnormal, containment test, design-basis-event, including loss-of-coolant, and post-design-basis-event conditions.
(PE/NP) 600-1983w

qualified Having adequate knowledge of the installation, construction, or operation of apparatus and the hazards involved.
(NESC) C2-1997

qualified climber A worker who, by reason of training and experience, understands the methods and has routinely demonstrated proficiency in climbing techniques and familiarity with the hazards associated with climbing.
(NESC/T&D/PE) C2-1997, 1307-1996

qualified diesel-generator unit A diesel-generator unit that meets the qualification of IEEE Std 387-1995.
(PE/NP) 387-1995

qualified life (1) (nuclear power generating station) (cable, field splice, and connection qualification) The period of time for which satisfactory performance can be demonstrated for a specific set of service conditions.
(PE/EDPG) 690-1984r

(2) (seismic qualification of Class 1E equipment for nuclear power generating stations) The period of time, prior to the start of a design basis event (DBE), for which the equipment was demonstrated to meet the design requirements for the specified service conditions.
(PE/NP) 344-1987r, 317-1983r, 627-1980ar, 382-1985, 323-1974ls

(3) (Class 1E battery chargers and inverters) (Class 1E motor control) The period of time for which satisfactory performance can be demonstrated for a specific set of service conditions. *Note:* The qualified life of a particular equipment item may be changed during its installed life where justified.
(SWG/PE/NP) 383-1974r, C37.100-1992, 649-1980s, 650-1979s, 323-1974as

qualified life test (electric penetration assemblies) Tests performed on preconditioned test specimens to verify that an electric penetration assembly will meet design requirements at the end of its qualified life.
(PE/NP) 317-1983r

qualified module (nuclear power generating station) Module that exhibits performance characteristics that are acceptable for Class 1E service in a nuclear power generating station and that satisfy the aging criteria and other requirements of this document.
(PE/NP) 381-1977w

qualified person One familiar with the construction and operation of the equipment and the hazardous involved.
(NESC/NEC) [86]

qualified procedures An approved and validated procedure that has been demonstrated to meet the specified requirement for its intended purpose.
(NI/PE/NP) N42.23-1995, [124]

qualified worker A worker who has received, as part of a training program, formal instruction and training in the techniques for climbing and working on structures and/or equipment. A qualified worker will have satisfactorily completed and demonstrated proficiency in the climbing portion and technical aspects of a formal training program. Climbing may be included in routine work assignments. Examples of a qualified worker are engineers and technicians doing inspections, testing, communications installations, etc.
(T&D/PE) 1307-1996

qualifying symbol A symbol added to the basic outline of an element to designate the physical or logic characteristics of an input or output of the element or the overall logic characteristics of the element.
(GSD) 91-1984r

qualitative adjectives *See:* geometrical adjectives; magnitude-related adjectives; time-related adjectives; polarity-related adjectives; descriptive adjectives.

qualitative distortion terms *See:* preshoot; valley; overshoot; ringing; rounding; spike; tilt.

qualitative techniques, results or data Judgmental scaling of the adequacy of the man-machine interface based, for example, on information from task analysis, or human engineering, training, or procedures review.
(PE/NP) 845-1988s

quality (software) The totality of features and characteristics of a product or service that bear on its ability to satisfy given needs. *See also:* software quality.
(C/SE) 729-1983s

quality area The area of the cathode-ray-tube phosphor screen that is limited by the cathode-ray tube and instrument specification. *Note:* If the quality area and the graticule area are not equal, this must be specified. *See also:* viewing area; graticule area; oscillograph.
(IM/HFIM) [40]

quality assurance (QA) (1) (software quality assurance) A planned and systematic pattern of all actions necessary to provide adequate confidence that the item or product conforms to established technical requirements.
(C/SE) 729-1983s, 730-1998

(2) All those planned and systematic actions necessary to provide adequate confidence that an analysis, measurement, or surveillance program will perform satisfactorily in service.
(NI) N42.23-1995

(3) (monitoring radioactivity in effluents) All planned and systematic actions necessary to provide adequate confidence that a system or component will perform satisfactorily in service.
(PE/NI/NP) 933-1999, N42.18-1980r

quality assurance program plan A document that contains or references the quality assurance elements established for an activity, group of activities, a scientific investigation or a project, and describes how conformance with such requirements is to be assured for the activities.
(NI) N42.23-1995

quality assurance record (nuclear power quality assurance) A completed document that furnishes evidence of the quality of items or activities, or both, affecting quality.
(PE/NP) [124]

quality attribute A characteristic of software, or a generic term applying to quality factors, quality subfactors, or metric values.
(C/SE) 1061-1998

quality control (1) (nuclear power generating station) Those quality assurance actions which provide a means to control and measure the characteristics of an item, process or facility to established requirements.
(PE/NP) 467-1980w

(2) Those actions that control and measure the attributes of the analytical process, standards, reagents, measurement equipment, components, system, or facility according to predetermined quality requirements.
(NI) N42.23-1995

quality control chart *See:* instrument quality control chart.

quality factor (1) (network, structure, or material) Two pi times the ratio of the maximum stored energy to the energy dissipated per cycle at a given frequency. *Notes:* 1. The Q of an inductor at any frequency is the magnitude of the ratio of its reactance to its effective series resistance at that frequency. 2. The Q of a capacitor at any frequency is the magnitude of the ratio of is susceptance to its effective shunt conductance at that frequency. 3. The Q of a simple resonant circuit comprising an inductor and a capacitor is given by

$$Q = \frac{Q_L Q_C}{Q_L + Q_C}$$

where Q_L and Q_C, are the Qs of the inductor and capacitor, respectively, at the resonance frequency. If the resonant circuit comprises an inductance L and a capacitance C in series with an effective R, the value of Q is

$$Q = \frac{1}{R}\left(\frac{L}{C}\right)^{1/2}$$

An approximate equivalent definition, which can be applied to other types of resonant structures, is that the Q is the ratio of the resonance frequency to the bandwidth between the frequencies on opposite sides of the resonance frequency (known as half-power points) where the response of the resonant structure differs by 3 decibels (dB) from that at resonance. 4. The Q of a magnetic or dielectic material at any frequency is equal to 2π times the ratio of the maximum stored energy to the energy dissipated in the material per cycle. 5. For networks that contain several elements, and distributed parameter systems, the Q is generally evaluated at a frequency of resonance. 6. The nonloaded Q of a system is the value of Q obtained when only the incidental dissipation of the system element is present. The loaded Q of a system is the value of Q obtained when the system is coupled to a device that dissipates energy. 7. The period in the expression for Q is that of the driving force, not that of energy storage, which is usually half that of the driving force. (Std100) 270-1966w

(2) In active filters the transfer functions are generally broken down into second-order sections expressed by biquadratic functions as follows:

$$T(s) = \frac{n_2 s^2 + n_1 s + n_0}{d_2 s^2 + d_1 s + d_0}$$

Such transfer functions are generally re-arranged in the following form where the zero Q-factor. QZ, and the pole Q-factor, QP, may be identified.

$$T(s) = K\frac{s^2 + s\dfrac{\omega_{0z}}{Q_z} + \omega_{0z}{}^2}{s^2 + s\dfrac{\omega_{0p}}{Q_p} + \omega_{0p}{}^2}$$

(CAS) [13]

(3) Two π times the ratio of the maximum stored energy to the energy dissipated per cycle at a given frequency. An approximate equivalent definition is that the Q is the ratio of the resonant frequency to the bandwidth between those frequencies on opposite sides of the resonant frequency, where the response of the resonant structure differs by 3 dB from that at resonance. If the resonant circuit comprises an inductance, L, and a capacitance, C, in series with an effective resistance, R, then the value of Q is

$$Q = \frac{1}{R}\sqrt{\frac{L}{C}}$$

(IA/SPC) 519-1992

(4) A management-oriented attribute of software that contributes to its quality. (C/SE) 1061-1998

(5) Two pi times the ratio of the maximum stored energy to the energy dissipated per cycle at a given frequency. *Note:* In a parallel resonant circuit, such as a load on a power system

$$Q = R\sqrt{\frac{C}{L}}$$

where
Q = quality factor
R = effective load resistance
C = effective load capacitance (including shunt capacitors)
L = effective load inductance

Or, on a power system, where real power, P, and reactive powers, P_{qL}, for inductive load, and P_{qC} for capacitive load, are known as

$$Q = (1/P)\sqrt{P_{qL} \times P_{qC}}$$

where
Q = quality factor
P = real power
P_{qL} = inductive load
P_{qC} = capacitive load

(SCC21) 929-2000

quality factor sample A set of quality factor values that is drawn from the metrics database and used in metrics validation. (C/SE) 1061-1998

quality factor value A value of the direct metric that represents a quality factor. *See also:* metric value. (C/SE) 1061-1998

quality management That aspect of the overall management function that determines and implements the quality policy. (C/SE) 1074-1995s

quality metric (software) A quantitative measure of the degree to which software possesses a given attribute which affects its quality. *See also:* software; quality. (C/SE) 729-1983s

quality of lighting (illuminating engineering) Pertains to the distribution of luminance in a visual environment. The term is used in a positive sense and implies that all luminances contribute favorably to visual performance, visual comfort, ease of seeing, safety, and esthetics for the specific visual tasks involved. (EEC/IE) [126]

quality of service (QoS) The four negotiated parameters for a link: signaling speed, maximum turnaround time, data size, and disconnect threshold. (EMB/MIB) 1073.3.2-2000

quality policy The overall quality intentions and direction of an organization as regards quality, as formally expressed by top management. (C/SE) 1074-1995s

quality requirement A requirement that a software attribute be present in software to satisfy a contract, standard, specification, or other formally imposed document. (C/SE) 1061-1998

quality subfactor A decomposition of a quality factor or quality subfactor to its technical components. (C/SE) 1061-1998

quantitative adjectives *See:* functional adjectives; integer adjectives; mathematical adjectives.

quantitative techniques, results or data Procedures for developing numerical information, or the numerical information itself, that is capable of being systematically combined or compared with a standard to assess man-machine performance. (PE/NP) 845-1988s

quantitative testing (test, measurement, and diagnostic equipment) Testing that monitors or measures the specific quantity, level, or amplitude of a characteristic to evaluate the operation of an item. The outputs of such tests are presented as finite or quantitative values of the associated characteristics. (MIL) [2]

quantity equations Equations in which the quantity symbols represent mathematico-physical quantities possessing both numerical values and dimensions. (Std100) 270-1966w

quantity of light (illuminating engineering) (luminous energy, $Q_v = \int \varphi_v dt$) The product of the luminous flux by the time it is maintained. It is the time integral of luminous flux. (EEC/IE) [126]

quantization (1) (telecommunications) A process in which the continuous range of values of an input signal is divided into non-overlapping subranges, and to each subrange a discrete value of the output is uniquely assigned. Whenever the signal value falls within a given subrange, the output has the corresponding discrete value. *Note:* "Quantized" may be used as an adjacent modifying various forms of modulation, for example, quantized pulse-amplitude modulation. *See also:* quantization distortion; quantization level. (C/COM) [20], [123], [49], [85]

(2) (data transmission) In communication, quantization is a process in which the range of values of a wave is divided into a finite number of smaller subranges, each of which is represented by an assigned (or quantized) value within the subrange. *Note:* "Quantized" may be used as an adjective modifying various forms of modulation, for example, quantized pulse amplitude modulation. (PE) 599-1985w

(3) (accelerometer) (gyros) The analog-to-digital conversion of a gyro or accelerometer output signal that gives an output that changes in discrete steps, as the input varies continuously. (AES/GYAC) 528-1994

quantization distortion (data transmission) (quantization noise) The inherent distortion introduced in the process of quantization. (PE) 599-1985w

quantization error (1) (supervisory control, data acquisition, and automatic control) The amount that the digital quantity differs from the analog quantity. (SWG/PE/SUB) C37.1-1987s, C37.100-1992
(2) *See also:* quantizing error. (AES) 686-1997

quantization level (1) (data transmission) A particular subrange of a symbol designating it. (PE) 599-1985w
(2) (telecommunications) The discrete value of the output designating a particular subrange of the input. *See also:* quantization. (Std100) 270-1964w

quantization noise *See:* quantization distortion.

quantize To subdivide the range of values of a variable into a finite number of non-overlapping subranges or intervals, each of which is represented by an assigned value within the subrange, for example, to represent a person's age as a number of whole years. (C) [20], [85]

quantized pulse modulation Pulse modulation that involves quantization. (EEC/PE) [119]

quantized system A system in which at least one quantizing operation is present. (PE/EDPG) [3]

quantizer A device that digitizes analog input; for example, a digitizing tablet, motion sensor, or a light pen. *See also:* digitizer. (C) 610.10-1994w

quantizing error An error caused by conversion of an analog variable having a continuous range of values to a quantized form having only discrete values, as in analog-to-digital conversion. The error is the difference between the original (analog) value and its quantized (digital) representation. *Synonym:* quantization error. (AES) 686-1997

quantizing loss (A) In phased arrays, a loss in peak gain that occurs when the beam is phase steered by digitally controlled phase shifters, due to the quantizing errors in the phase shifts applied to the various radiating elements. **(B)** In signal processing, a loss that occurs when elements of a composite signal (for example, complex amplitudes of pulses in a pulse train) are quantized (digitized) before being combined. *See also:* quantizing error. (AES) 686-1997

quantizing noise (1) (analog voice frequency circuits) The noise introduced during the process of digitally encoding an analog signal. (COM/TA) 743-1995
(2) (distortion) The impairment introduced during the process of digitally encoding and decoding an analog signal. (COM/TA) 1007-1991r
(3) (telecommunications) A random signal caused by the error of approximation in a quantizing process. It may be regarded as noise arising in the pulse-code modulation process when the code-derived facsimile does not exactly match the waveform of the original message. (COM/TA) 973-1990w

quantizing operation One which converts one signal into another having a finite number of predetermined magnitude values. (PE/EDPG) [3]

quantum efficiency (1) (photocathodes) The average number of electrons photoelectrically emitted from the photocathode per incident photon of a given wavelength. *Note:* The quantum efficiency varies with the wavelength, angle of incidence, and polarization of the incident radiation. *See also:* phototube; photocathode; semiconductor. (ED/NPS) 161-1971w, 398-1972r
(2) (laser, maser, laser material, or maser material) The ratio of the number of photons or electrons emitted by a material at a given transition to the number of absorbed particles. (LEO) 586-1980w
(3) (fiber optics) In an optical source or detector, the ratio of output quanta to input quanta. Input and output quanta need not both be photons. (Std100) 812-1984w

quantum noise (fiber optics) Noise attributable to the discrete or particle nature of light. *Synonym:* photon noise. (Std100) 812-1984w

quantum-noise-limited operation (fiber optics) Operation wherein the minimum detectable signal is limited by quantum noise. *See also:* quantum noise. (Std100) 812-1984w

quarter adder An adder that accepts two inputs, producing only a sum as output according to the table below. *Contrast:* full adder; half adder. *See also:* exclusive OR.

input #1	0	0	1	1
input #2	0	1	0	1
output sum	0	1	1	0

quarter adder

(C) 610.10-1994w

quarter-inch cartridge A type of storage medium for magnetic tapes in which each tape is encased in a small cartridge. *Synonym:* minicartridge. (C) 610.10-1994w

quarter-phase or two-phase circuit A combination of circuits energized by alternating electromotive forces that differ in phase by a quarter of a cycle, that is, 90 degrees. *Note:* In practice the phases may vary several degrees from the specified angle. *See also:* center of distribution. (T&D/PE) [10]

quarters (electric installations on shipboard) Where used in these recommendations, those spaces provided for passengers or crew, as specified, which are actually used for berthing, mess spaces, offices, private baths, toilets and showers, and lounging rooms, smoking rooms, and similar spaces. (IA/MT) 45-1983s

quarter-squares multiplier An analog multiplier incorporating inverters, analog adders, and square-law function generators, whose operation is based on the identity:

$$xy = \frac{(x + y)^2 - (x - y)^2}{4}$$

(C) 610.10-1994w

quarter-thermal-burden ambient-temperature rating The maximum ambient temperature at which the transformer can be safely operated when the transformer is energized at rated voltage and frequency and is carrying 25 % of its thermal-burden rating without exceeding the specified temperature limitations. (EEC/PE) [119]

quartet (1) (mathematics of computing) A group of four adjacent digits operated upon as a unit. (C) 1084-1986w
(2) A byte composed of four bits. *Synonym:* four-bit byte. *See also:* nibble. (C) 610.10-1994w

quasi-analog signal (data transmission) A digital signal after conversion to a form suitable for transmission over a specified analog channel. The specifications of an analog channel includes the frequency of range, frequency of bandwidth, signal-to-noise ratio (snr), and envelope delay distortion. When this form of signaling is used to convey message traffic over the public dial-up network telephone systems, it is often referred to as voice data. (PE) 599-1985w

quasi-asymptotic stability (solution $\phi x(x(t_0);t)$). Implies

$$\lim_{t \to \infty} \|\Delta\phi\| = 0$$

Notes: 1. Quasi-asymptotic stability is condition. 2. in the definition of asymptotic stability and, hence, need not imply Lyapunov stability. 3. An example of a solution that is quasi-asymptotically stable but not asymptotically stable is the solution $x(t) = 0$ of the system $x = x^2$. The solution of the above system for a perturbation $\Delta x(t_0)$ from 0 is

$$\rho\,(\Delta x(t_0);t) = \Delta x(t_0)/[1 - (t - t_0)\Delta x(t_0)]$$

Obviously, $\phi(\Delta x(t_0);t)$ approaches zero as t approaches ∞, yet is not Lyapunov stable since it is unbounded at $t = t_0 + (1/\Delta x(t_0))$. *See also:* control system. (CS/IM) [120]

quasi-crystalline approximation A formulation used to determine the coherent mean field of a wave propagating in a nontenuous (dense) random medium; a higher order approximation to Foldy's approximation. (AP/PROP) 211-1997

quasi-Gaussian pulse shape A pulse shape approximated by the curve of a normal or Poisson distribution. In this standard, four or more integrators shall be used in producing the pulse. *See also:* integrated pulse. (NPS) 325-1996

quasi-Gaussian shaping (1) (germanium gamma-ray detectors) (charged-particle detectors) Pulse shaping consisting of one differentiation followed by four or more integrations resulting in a pulse shape that is approximated by a Gaussian curve. For *n* integrations the shaping network is sometime denoted as $CR - (RC)^n$. (NPS) 325-1986s
(2) (charged-particle detectors) Pulse shaping in a linear amplifier in which the amplitude of the shaped pulse versus time approximates that of a normal distribution (in the case of unipolar shaping) or its mathematical derivative (in the case of bipolar shaping). Unless otherwise stated, the shaping network shall contain at least four cascaded low-pass sections (integrators). (NPS) 300-1988r

quasi-impulsive noise A superposition of impulsive and continuous noise. *See also:* electromagnetic compatibility. (EMC/INT) [53], [70]

quasi-peak detector A detector having specified electrical time constants that, when regularly repeated pulses of constant amplitude are applied to it, delivers an output voltage that is a fraction of the peak value of the pulses, the fraction increasing toward unity as the pulse repetition rate is increased. *Synonym:* QP detector. *See also:* electromagnetic compatibility. (EMC/T&D/PE/INT) [53], 539-1990, [70]

quasi-peak voltmeter A quasi-peak detector coupled to an indicating instrument having a specific mechanical time-constant. *See also:* electromagnetic compatibility. (EMC/INT) [53], [70]

quasi-random signal A pseudo-random test signal that has artificial constraints to limit the maximum number of zeros in the bit sequence. (COM/TA) 1007-1991r

quasi-square wave (converter characteristics) (self-commutated converters) The stepped waveform obtained from the difference of two phase-shifted square waves of equal amplitude.

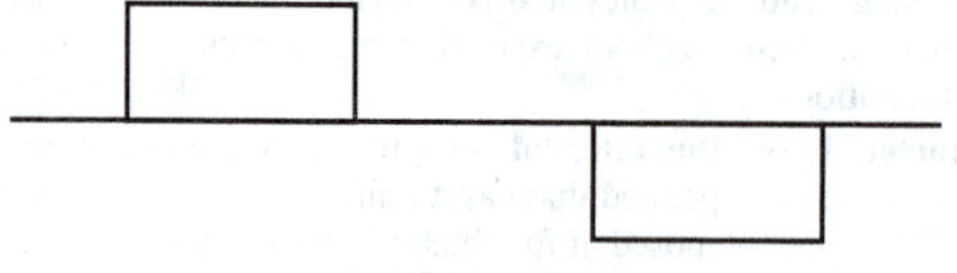

quasi-square wave

(IA/SPC) 936-1987w

quasi-static field A field that satisfies the condition $f << c/l\pi\sqrt{2}$, where f is the frequency of the field, c is the speed of light, and l is a characteristic dimension of the measurement geometry, e.g., the distance between the field source and the measurement point. *Note:* Power frequency magnetic and electric fields near power lines and appliances are examples of quasi-static fields. (T&D/PE) 1308-1994

quasi-triangular pulse shape (1) A pulse shape approximated by the curve of a normal or Poisson distribution. In this standard, four or more integrators shall be used in producing the pulse. *See also:* integrated pulse. (NPS) 325-1996
(2) A pulse shape approximated by a triangle with a truncated top. Such a shape can be obtained from a CR −(RC) n or sinen network by summing appropriate signal fractions from the n integrating sections. (NPS) 325-1996

quaternary code A code whose output alphabet consists of four symbols. *See also:* ternary code. (IT) [123]

quenched sample (1) (liquid-scintillation counters) A counting sample (material of interest plus liquid-scintillation solution) that contain adulterants that reduce the photon output from the vials. (NI) N42.16-1986
(2) (liquid-scintillation counters) A counting sample (material of interest plus liquid-scintillation solution) that contains chemical impurities that reduce the photon output to the photomultiplier tubes. (NI) N42.15-1990

quenched spark gap converter (dielectric heating) A spark-gap generator or power source that utilizes the oscillatory discharge of a capacitor through an inductor and a spark gap as a source of radio-frequency power. The spark gap comprises one or more closely spaced gap operating in series. *See also:* induction heating. (IA) 54-1955w, 169-1955w

quenching The process of terminating a discharge in a Geiger-Mueller radiation-counter tube by inhibiting a reignition. *Note:* This may be effected by self-quenching (internal quenching) by use of an appropriate gas or vapor filling, or externally (external quenching) by momentary reduction of the applied potential difference. (NI/NPS) 309-1999

quenching circuit (radiation counters) A circuit that reduces the voltage applied to a Geiger-Mueller tube after an ionizing event, thus preventing the occurrence of subsequent multiple discharges. Usually the original voltage level is restored after a period that is longer than the natural recovery time of the Geiger-Mueller tube. *See also:* anticoincidence. (ED) [45]

query (data transmission) The process by which a master station asks a slave station to identify itself and to give its status. *See also:* poll. (PE) 599-1985w

query language A database manipulation language used to access information stored in a database. *Contrast:* specification language; programming language. (C) 610.13-1993w

queue (1) (software) A list that is accessed in a first-in, first-out manner. *See also:* stack. (C/SE) 729-1983s
(2) *See also:* batch queue.

queue allocation protocols The protocols used to allocate queue space when several nodes are sending packets to a shared node. This involves rejecting packets (with a busy status), but reserving future queue space; the reserved queue space is eventually used during one of the packet's retransmissions. (C/MM) 1596-1992

queued arbitrated access function The *access control function* in this part of ISO/IEC 8802 that uses the *Distributed Queue* to access *empty Queued Arbitrated (QA) slots* for the transfer of *QA segments. Synonym:* QA access function. (LM/C) 8802-6-1994

queued arbitrated segment A *segment* transferred using *Queued Arbitrated (QA) Access functions. Synonym:* QA segment. (LM/C) 8802-6-1994

queued arbitrated slot A *slot* that is used for the transfer of a *QA segment. Synonym:* QA slot. (LM/C) 8802-6-1994

queue on lock bit (QOLB) A mechanism for efficiently sequencing the access to resources that are not to be used by more than one process at a time. (C/MM) 1596-1992

queue position The place a job occupies in a queue. This place is relative to other jobs in the queue and defined in part by submission time and its priority. *See also:* job priority. (C/PA) 1003.2d-1994

queue priority The maximum job priority allowed for any job in a given queue. The *queue priority* is set and may be changed by users with appropriate privilege. The priority shall be bounded in an implementation-defined manner. (C/PA) 1003.2d-1994

quick-break A term used to describe a device that has a high contact opening speed independent of the operator. (SWG/PE) C37.100-1992

quick-break switch (high-voltage switchgear) A switch that has a high contact opening speed independent of the operator. (SWG/PE) C37.40-1993

quick charge *See:* boost charge.

quick-flashing light (illuminating engineering) A rhythmic light exhibiting very rapid regular alternations of light and darkness. There is no restriction on the ratio of the durations of the light to the dark periods. (EEC/IE) [126]

Quick FORTRAN A subset of FORTRAN that is easy to learn due to its efficient debugging facility and interactive interface. (C) 610.13-1993w

quick-make A term used to describe a device that has a high contact closing speed independent of the operator. (SWG/PE) C37.100-1992

quick release (control brakes) The provision for effecting more rapid release than would inherently be obtained. *See also:* electric drive. (IA/IAC) [60]

quick set (control brakes) The provision for effecting more rapid setting than would inherently be obtained. *See also:* electric drive. (IA/IAC) [60]

quick startup reserve (power operations) The operating reserve available within a specified time through startup and synchronization of quick start internal combustion generation. (PE/PSE) 858-1987s

quiescent *See:* supervisory control.

quiescent-carrier telephony That form of carrier telephony in which the carrier is suppressed whenever there are no modulating signals to be transmitted. (EEC/PE) [119]

quiescent current (electron tube) The electrode current corresponding to the electrode bias voltage. (ED) [45], [84]

quiescent operating point (magnetic amplifier) The output obtained under any specified external conditions when the signal is non-time-varying and zero. (MAG) 107-1964w

quiescent point (amplifiers) That point on its characteristic that represents the conditions existing when the signal input is zero. *Note:* The quiescent values of the parameters are not in general equal to the average values existing in the presence of the signal unless, the characteristic is linear and the signal has no direct-current component. *See also:* operating point. (EEC/PE) [119]

quiescent state A state in which all modules on a Futurebus+ backplane bus have ceased all activity on all Futurebus+ defined signals and all user I/O field signals. (C/BA) 896.2-1991w

quiescent supervisory system (station control and data acquisition) A system that is normally alert but inactive, and transmits information only when a change in indication occurs at the remote station or when a command operation is initiated at the master station. *See also:* supervisory system. (SWG/PE/SUB/PE) C37.100-1992, C37.1-1994

quiescent value (pulse transformers) (base magnitude) The maximum value existing between pulses. (PEL/ET) 390-1987r

quiet automatic volume control Automatic volume control that is arranged to be operative only for signal strengths exceeding a certain value, so that noise or other weak signals encountered when tuning between strong signals are suppressed. *See also:* radio receiver. (EEC/PE) [119]

quiet ground (health care facilities) A system of grounding conductors, insulated from portions of the conventional grounding of the power system, that interconnects the grounds of electric appliances for the purpose of improving immunity to electromagnetic noise. (EMB) [47]

quieting sensitivity (1) (frequency-modulation receivers) The minimum unmodulated signal input for which the output signal-noise ratio does not exceed a specified limit, under specified conditions. *See also:* radio receiver. (VT) [37]
(2) (test, measurement, and diagnostic equipment) The level of a continuous wave (CW) input signal which will reduce the noise output level of a frequency-modulation (FM) receiver by a specified amount, usually 20 dB. (MIL) [2]

quiet sun (radio-wave propagation) The sun in the absence of any unusual electromagnetic activity. (AP/PROP) 211-1990s

quiet tuning A circuit arrangement for silencing the output of a radio receiver except when the receiver is accurately tuned to an incoming carrier wave. *See also:* radio receiver. (EEC/PE) [119]

quiet zone (1) The area immediately preceding the start character and following the stop character, which contains no markings. (PE/TR) C57.12.35-1996
(2) A described volume within an anechoic chamber where electromagnetic waves reflected from the walls, floor, and ceiling are stated to be below a certain specified minimum. The quiet zone may have a spherical, cylindrical, rectangular, etc., shape depending on the chamber characteristics. (EMC) 1128-1998

quill drive A form of drive in which a motor or generator is geared to a hollow cylindrical sleeve, or quill, or the armature is directly mounted on a quill, in either case, the quill being mounted substantially concentrically with the driving axle and flexibly connected to the driving wheels. *See also:* traction motor. (EEC/PE) [119]

quinary Five-level. (C/LM) 802.3-1998

quinhydrone electrode *See:* quinhydrone half cell.

quinhydrone half cell A half cell with an electrode of an inert metal (such as platinum gold) in contact with a solution saturated with quinhydrone. *Synonym:* quinhydrone electrode. *See also:* electrochemistry. (EEC/PE) [119]

quintet (1) A byte composed of five bits. *Synonym:* five-bit byte. (C) 610.10-1994w
(2) (local area networks) A contiguous string of five bits. (C) 8802-12-1998

quotation board A manually or automatically operated panel equipped to display visually the price quotations received by a ticker circuit. Such boards may provide displays of large size or may consist of small automatic units that ordinarily display one item at a time under control of the user. *See also:* telegraphy. (EEC/PE) [119]

quotient *See:* phasor product.

quotient relay A relay that operates in response to a suitable quotient of two alternating electrical input quantities. *See also:* relay. (SWG/PE) C37.100-1981s

Q video signal (National Television System Committee color television) One of the two video signals (E'_1 and E'_g) controlling the chrominance in the NTSC system. *Note:* It is a linear combination of gamma-corrected primary color signals, E'_R, E'_G, and E'_B, as follows:

$$E'_Q = 0.41(E'_B - E'_Y) + 0.48(E'_R - E'_Y)$$
$$= 0.21E'_R - 0.52E'_G + 0.31E'_B$$

(BT/AV) 201-1979w

QWERTY keyboard A standard keyboard layout, named for the first six keys of the third row from the bottom. That is, when the row is read across, the keys read Q W E R T Y U I O P. *Contrast:* Dvorak keyboard. (C) 610.10-1994w

R

rabbet, mounting *See:* mounting rabbet.

RAC *See:* reflective array compressor.

raceway (1) Any channel designed expressly and used solely for holding conductors. (NESC) C2-1997
(2) (raceway systems for Class 1E circuits for nuclear power generating stations) Any channel that is designed and used expressly for supporting wires, cables, or bus bars. Raceways consist primarily of, but are not restricted to, cable tray, conduits, and wireways.
(PE/NP/IC) 628-1987r, 848-1996, 384-1992r, 634-1978w
(3) (electric systems) Any channel for enclosing, and for loosely holding wires, cables, or busbars in interior work that is designed expressly for, and used solely for, this purpose. *Note:* Raceways may be of metal or insulating material and the term includes rigid metal conduit, rigid nonmetallic conduit, flexible metal conduit, electrical metallic tubing, underfloor raceways, cellular concrete-floor raceways, cellular metal-floor raceways, surface metal raceways, structural raceways, wireways and busways, and auxiliary gutters or moldings. (NESC) [86]
(4) An enclosed channel designed expressly for holding wires, cables or busbars with additional functions as permitted in this Code. (NESC/NEC) [86]

raceway penetration (raceway systems for Class 1E circuits for nuclear power generating stations) An opening for a raceway in a floor or wall to permit passage of cables from one side to the other. The raceway may or may not be continuous through the opening. (PE/NP) 628-1987r

raceway system (raceway systems for Class 1E circuits for nuclear power generating stations) An integrated assembly of raceways, fittings, supports, accessories, and anchorages. (PE/NP) 628-1987r

rachet demand (power operations) The maximum past or present demands that are taken into account to establish billings for previous or subsequent periods.
(PE/PSE) 858-1987s

rachet demand clause (power operations) A clause in a rate schedule that provides that maximum past or present demands be taken into account to establish billings for previous or subsequent periods. (PE/PSE) 858-1987s

rack (1) (control boards, panels, and racks) A framework, constructed of rails or steel members, for mounting an assembly of modules for monitoring, measuring, and controlling remotely operated systems. (PE/NP) 420-1982
(2) (electronic) A protective enclosure to house modules, backplane(s), I/O connector assemblies, internal cables, and other electronic, mechanical, and thermal devices. *Synonyms:* rack; cabinet; box; box; cabinet; rack. (C/BA) 14536-1995

rack, traveler *See:* traveler rack.

racon (navigation aid terms) A radar beacon that returns a coded signal providing identification of the beacon as well as range and bearing. *See also:* radar beacon.
(AES/GCS) 172-1983w

rad (photovoltaic power system) An absorbed radiation unit equivalent to 100 ergs/gram of absorber. *See also:* photovoltaic power system. (AES) [41]

radar (1) (navigation aid terms) A device for transmitting electromagnetic signals and receiving echoes from objects of interest (targets) within its volume of coverage. Presence of a target is revealed by its echo or its transponder reply. Additional information about a target provided by a radar includes one or more of the following: distance (range), by the elapsed time between transmissions of the signal and reception of the return signal; direction, by use of directive antenna patterns; rate of change of range, by measurement of Doppler shift; description or classification of target, by analysis of echoes and their variation with time. The name radar was originally an acronym for Radio Detection and Ranging. *Note:* Some radars can operate in a passive mode, in which the transmitter is turned off and information about targets is derived by receiving radiation emanating from the targets themselves or reflected by the targets from external sources.
(AES/GCS) 172-1983w
(2) An electromagnetic system for the detection and location of objects that operates by transmitting electromagnetic signals, receiving echoes from objects (targets) within its volume of coverage, and extracting location and other information from the echo signal. *Notes:* 1. Radar is an acronym for radio detection and ranging. 2. Radar equipment can be operated with the transmitter turned off, as a passive direction finder on sources radiating within the band of the receiving system. *See also:* passive angle tracking. (AES) 686-1997
(3) A device that radiates electromagnetic waves and utilizes the reflection of such waves from distant objects to determine their existence or position. (IA/MT) 45-1998

radar-absorbent material (RAM) Material used to reduce the radar cross section of an object. *Note:* Also used in anechoic chambers to reduce reflection from the walls.
(AES) 686-1997

radar altimeter *See:* radio altimeter.

radar astronomy That branch of astronomy that uses radar to study astronomical objects. (AP/PROP) 211-1997

radar backscattering cross-section Radar scattering cross-section as determined for coincident transmitter and receiver locations. *See also:* scattering cross section.
(AP/PROP) 211-1997

radar beacon (navigation aid terms) A transponder used for replying to interrogations from a radar. *See also:* secondary radar. (AES/GCS) 686-1997, 172-1983w

radar bearing (navigation aid terms) A bearing obtained by a radar. (AES/GCS) 172-1983w

radar camouflage The art, means, or result of concealing the presence of the nature of an object from radar detection by the use of coverings or surfaces that considerably reduce the radio energy reflected toward a radar. *See also:* radar.
(EEC/PE) [119]

radar cross section (RCS) (1) For a given scattering object, upon which a plane wave is incident, that portion of the scattering cross section corresponding to a specified polarization component of the scattered wave. *See also:* scattering cross section. (AP/ANT) 145-1993
(2) A measure of the reflective strength of a radar target; usually represented by the symbol σ and measured in square meters. RCS is defined as 4π times the ratio of the power per unit solid angle scattered in a specified direction of the power unit area in a plane wave incident on the scatterer from a specified direction. More precisely, it is the limit of that ratio as the distance from the scatterer to the point where the scattered power is measured approaches infinity. *Note:* Three cases are distinguished:

a) Monostatic or backscatter RCS when the incident and pertinent scattering directions are coincident but opposite in sense.
b) Forward-scatter RCS when the two directions and senses are the same.
c) Bistatic RCS when the two directions are different. If not identified, RCS is usually understood to refer to case a). In all three cases, radar cross section of a specified target is a function of frequency, transmitting and receiving polarizations, and target aspect angle (except for a sphere). For some applications, e.g., statistical detection analyses, it is described by its average value (or sometimes its median value) and statistical characteristics over an appropriate range of one or more of those parameters.

Synonyms: backscatter cross section; bistatic-scatter cross section; effective echoing area; forward-scattering cross section. (AES) 686-1997

radar display The visual representation of radar output data. See individual definitions and illustrations of various radar display formats in the table below. *Note:* The letter designations from A to P, plus R, for radar display formats were devised in the years during and following WWII in an effort to standardize nomenclature. Several of these letter designations are now rarely if ever used, as noted in the individual definitions, but they are still found in some technical literature. The additional designations of plan-position indicator (PPI) and range-height indicator (RHI) are also defined. The standardized type designations do not cover all possible display formats. (AES) 686-1997

radar duplexing assembly *See:* circulator; transmit-receive switch; duplexer.

radar equation A mathematical expression that relates the range of a radar at which specific performance is obtained to the parameters characterizing the radar, target, and environment. *Note:* The parameters in the radar equation can include the transmitter power, antenna gain and effective area, frequency, radar cross section of the target, range to the target, receiver noise figure, signal-to-noise ratio required for detection, losses in the radar system, and the effects of the propagation path. *Synonyms:* radar range equation; range equation. (AES) 686-1997

radar fix (navigation aid terms) A position fix established by means of radar data. (AES/GCS) 172-1983w

radar letter designations (radar frequency bands) The radar letter designations are consistent with the recommended nomenclature of the International Telecommunications Union (ITU), as shown in Table 2, below. Note that the high frequency (HF) and the very high frequency (VHF) definitions are identical in the two systems. The essence of the radar nomenclature is to subdivide the existing ITU bands, in accordance with radar practice, without conflict or ambiguity. The letter band designations are not to be construed as being a substitute for the specific frequency limits of the frequency bands. The specific frequency limits are to be used when appropriate, but when a letter designation of a radar-frequency band is called for, those of Table 1, below, are to be used. The letter designations described in IEEE Std 521-1984 are designed for radar usage and are used in current practice. They are not meant to be used for other radio or telecommunication purposes, unless they pertain to radar. The letter designations for Electronic Countermeasure operations as described in Air Force Regulation No 55-44, Army Regulation No 105-86, and Navy OPNAV Instruction 3430.9B are not consistent with radar practice and are not used to describe radar-frequency bands. (AES/RS) 521-1984r

radar performance figure The ratio of the pulse power of the radar transmitter to the power of the minimum signal detectable by the receiver. *Note:* Now seldom used as a measure of performance. (AES) 686-1997

radar range equation *See:* radar equation.

radar reflectivity A measure of backscattering from an inhomogeneous medium, defined as radar cross-section (RCS) per unit volume. Frequently used in radar measurements of meteorological phenomena. (AP/PROP) 211-1997

radar relay Equipment for relaying the radar video and appropriate synchronizing signals to a remote location. (AES) 686-1997

radar responder beacon *See:* racon.

radar shadow Absence of radar illumination because of an intervening reflecting, diffracting, or absorbing object. *Note:* The shadow is manifested on the display by the absence of blips from targets in the shadow area. (AES) 686-1997

radar transmitter The transmitter portion of a radio detecting and ranging system. (AP/VT/ANT) 145-1983s, [37]

radial (1) (navigation) (navigation aid terms) One of a number of lines of position defined by an azimuthal navigation facility; the radial is identified by its bearing (usually the magnetic bearing) from the facility. (AES/GCS) 172-1983w

(2) An azimuth where field strength measurements are taken, starting near the array and extending to well into the far field. Measurements along a radial can be used to establish the radiation in a certain azimuth after allowing for changes other than ground conductivity, such as near field effects, temperature changes, loss or gain due to elevation changes, shadow losses and absorption, and other effects. (T&D/PE) 1260-1996

radial air gap *See:* air gap.

radial-blade blower (rotating machinery) A fan made with flat blades mounted so that the plane of the blades passes through the axis of rotation of the rotor. *See also:* fan. (PE) [9]

radial distribution feeder *See:* radial feeder.

radial feeder A feeder supplying electric energy to a substation or a feeding point that receives energy by no other means. *Note:* The normal flow of energy in such a feeder is in one direction only. *See also:* center of distribution. (T&D/PE) [10]

radial ground A conductor connection by which separate electrical circuits or equipment are connected to earth at one point. Sometimes referred to as a *star ground.* (IA/PSE) 1100-1999

radially outer coil side *See:* bottom-coil slot.

radial magnetic pull (rotating machinery) The radial force acting between rotor and stator resulting from the radial displacement of the rotor from magnetic center. *Note:* Unless other conditions are specified, the value of radial magnetic pull will be for no load and rated voltage, and for rated no load field current and rated frequency as applicable. (PE) [9]

radial overfilled launch A launch condition created when a multimode optical fiber is illuminated by the coherent optical output of a source operating in its lowest-order transverse mode in a manner that excites predominantly the radial modes of the multimode fiber. (C/LM) 802.3-1998

radial power factor (paper-insulated power cable) The power factor of individual insulating tapes of a power cable as a function of the radial location of the insulating tapes through the insulation wall. (PE/IC) 83-1963w

radial probable error (RPE) *See:* circular probable error.

radial sensitivity The counting rate of a Geiger-Mueller counter as a function of radial position across the window of an end-window or pancake Geiger-Mueller counter. (NI/NPS) 309-1999

radial system A system in which independent feeders branch out radially from a common source of supply. *See also:* direct-current distribution; alternating-current distribution. (T&D/PE) [10]

radial-time-base display *See:* plan-position indicator.

radial transfer The transmission of information between a peripheral unit and a unit of equipment that is more central than that of the peripheral unit using a connection that is dedicated to that peripheral unit. *See also:* peripheral transfer. (C) 610.10-1994w

radial transmission feeder *See:* radial feeder.

radial transmission line (waveguide) A pair of parallel conducting planes used for propagating waves whose phase fronts are concentric coaxial circular cylinders having their common axis normal to the planes; sometimes applied to tapered versions, such as biconical lines. (MTT) 146-1980w

radial type A unit substation which has a single stepdown transformer and which has an outgoing section for the connection of one or more outgoing radial (stub end) feeders. (PE/TR) C57.12.80-1978r

radial-unbalance torque (1) (laser maser) A unit of angular measure equal to the angle subtended at the center of a circle by an arc whose length is equal to the radius of the circle. One (1) radian Δ 57.3 degrees; 2π radians = 360 degrees. (LEO) 586-1980w

Table 1
Standard Radar-Frequency Letter Band Nomenclature

Band Designation	Nominal Frequency Range	Specific Frequency Ranges for Radar Based on ITU Assignments for Region 2, see Note (1)
HF	3 MHz–30 MHz	Note (2)
VHF	30 MHz–300 MHz	138 MHz–144 MHz
		216 MHz–225 MHz
UHF	300 MHz–1000 MHz (Note 3)	420 MHz–450 MHz (Note 4)
		890 MHz–942 MHz (Note 5)
L	1000 MHz–2000 MHz	1215 MHz–1400 MHz
S	2000 MHz–4000 MHz	2300 MHz–2500 MHz
		2700 MHz–3700 MHz
C	4000 MHz–8000 MHz	5250 MHz–5925 MHz
X	8000 MHz–12000 MHz	8500 MHz–10680 MHz
K_u	12.0 GHz–18 GHz	13.4 GHz–14.0 GHz
		15.7 GHz–17.7 GHz
K	18 GHz–27 GHz	24.05 GHz–24.25 GHz
K_a	27 GHz–40 GHz	33.4 GHz–36.0 GHz
V	40 GHz–75 GHz	59 GHz–64 GHz
W	75 GHz–110 GHz	76 GHz–81 GHz
		92 GHz–100 GHz
mm (Note6)	110 GHz–300 GHz	126 GHz–142 GHz
		144 GHz–149 GHz
		231 GHz–235 GHz
		238 GHz–248 GHz (Note 7)

NOTES: (1) These frequency assignments are based on the results of the World Administrative Radio Conference of 1979. The ITU defines no specific service for radar, and the assignments are derived from those radio services that use radar: radiolocation, radionavigation, meteorlogical aids, earth exploration satellites, and space research.
(2) There are no official ITU radiolocation bands at HF. So-called HF radars might operate anywhere from just above the broadcast band (1.605 MHz) to 40 MHz or higher.
(3) The official ITU designation for the ultra-high-frequency band extends to 3000 MHz. In radar practice, however, the upper limit is usually taken as 1000 MHz. L and S bands being used to describe the higher UHF region.
(4) Sometimes called P band, but use is rare.
(5) Sometimes included in L band.
(6) The designation mm is derived from *millimeter* wave radar, and is also used to refer to V and W bands when general information relating to the region above 40 GHz is to be conveyed.
(7) The region from 300 GHz–3000 GHz is called the submillimeter band.

Table 2
Comparison of Radar-Frequency Letter Band Nomenclature with ITU Nomenclature

Radar Nomenclature		International Telecommunications Union Nomenclature			
Radar Letter Designation	Frequency Range	Frequency Range	Band No	Adjective Band Designation	Corresponding Metric Designation
HF	3 MHz–30 MHz	3 MHz–30 MHz	7	High-frequency (HF)	Dekametric waves
VHF	30 MHz–300 MHz	30 MHz–300 MHz	8	Very high frequency (VHF)	Metric Waves
UHF	300 MHz–1000 MHz				
L	1 GHz–2GHz	0.3 GHz–3 GHz	9	Ultra-high frequency (UHF)	Decimetric waves
S	2 GHz–4 GHz				
C	4 GHz–8 GHz				
X	8 GHz–12 GHz	3 GHz–30 GHz	10	Super-high frequency (SHF)	Centimetric waves
K_u	12 GHz–18 GHz				
K	18 GHz–27 GHz				
K_a	27 GHz–40 GHz				
V	40 GHz–75 GHz	30 GHz–300 GHz	11	Extremely high frequency (EHF)	Millimetric Waves
W	75 GHz–110 GHz				
mm	110 GHz–300 GHz				

(2) (dynamically tuned gyro) The acceleration-sensitive torque caused by radial unbalance due to noncoincidence of the flexure axis and the center of mass of the rotor. Under constant acceleration, it appears as a rotating torque at the rotor spin frequency. When the gyro is subjected to vibratory acceleration along the spin axis at the spin frequency, this torque results in a rectified drift rate. (AES/GYAC) 528-1994

radian (metric practice) The plane angle between two radii of a circle that cut off on the circumference an arc equal in length to the radius. (QUL) 268-1982s

radiance (1) (fiber optics) Radiant power, in a given direction, per unit solid angle per unit of projected area of the source, as viewed from that given direction. Radiance is expressed in watts per steradian per square meter. *See also:* brightness; radiometry; conservation of radiance. (Std100) 812-1984w
(2) (laser maser) Radiant flux or power output per unit solid angle unit area ($W \cdot sr^{-1} \cdot cm^{-2}$). (LEO) 586-1980w
(3) (television) (radiant intensity per unit area at a point on a surface and in a given direction) The quotient of the radiant intensity in the given direction of an infinitesimal element of the surface containing the point under consideration, by the area of the orthogonally projected area of the element on a plane perpendicular to the given direction. *Note:* The usual unit is the watt per steradian per square meter. This is the radiant analog of luminance. (BT/AV) 201-1979w
(4) (light-emitting diodes) [$L_e = d^2\phi_e/d\psi\ (dA \cos\theta)\ = dI_e/(dA \cos\theta)$). At a point of the surface of a source, of a receiver, or of any other real or virtual surface, the quotient of the radiant flux leaving, passing through or arriving at an element of the surface surrounding the point, and propagated in the direction defined by an elementary cone containing the given direction, by the product of the solid angle of the cone and the area of the orthogonal projection of the element of the surface on a plane perpendicular to the given direction. *Note:* In the defining equation θ (theta) is the angle between the normal to the element of the surface and the direction of observation. (ED) [127]
(5) (illuminating engineering) [$L = d^2\ \Phi/[d\omega\ (dA \cdot \cos\theta)] = dI/(dA \cdot \cos\theta)$] (in a direction at a point of the surface of a receiver, or of any other real or virtual surface) Properly, this should be a second partial derivative since area and solid angle are independent variables. However, the symbol "d" is used due to the convenience in printing and typing. For this

specific use, no possible errors or confusion are foreseen. This practice is in accord with the International Lighting Vocabulary (CIE No. 17 (E-1.1.) 1970) and the practice of the National Bureau of Standards (NBS Technical Note 910-1, 1976). The quotient of the radiant flux leaving, passing through, or arriving at an element of the surface surrounding the point, and propagated in directions defined by an elementary cone containing the given direction, by the product of the solid angle of the cone and the area of the orthogonal projection of the element of the surface on a plane perpendicular to the given direction. *Note:* In the defining equation fI theta fR is the angle between the normal to the element of the surface and the given direction. (EEC/IE) [126]

radian frequency (1) The number of radians per unit time. The unit of time is generally the second and the radian frequency omega is therefore $2\pi f$, where f is the frequency in hertz. (CAS) [13]
(2) *See also:* angular frequency. (AP/PROP) 211-1997

radiant density (light-emitting diodes) (we = dQe/dV) Radiant energy per unit volume; joules per m^3. (ED) [127]

radiant efficiency of a source of radiant flux (light-emitting diodes) The ratio of the total radiant flux to the forward power dissipation (total electrical lamp power input). (ED) [127]

radiant emittance (fiber optics) Radiant power emitted into a full sphere (4π steradians) by a unit area of a source; expressed in watts per square meter. *Synonym:* radiant exitance. *See also:* radiant flux density at a surface; radiometry. (Std100) 812-1984w

radiant energy (1) (fiber optics) Energy that is transferred via electromagnetic waves, that is, the time integral of radiant power; expressed in joules. *See also:* radiometry. (Std100) 812-1984w
(2) (light-emitting diodes) Energy traveling in the form of electromagnetic waves. It is measured in units of energy such as joules, ergs or kilowatt-hours. (IE/EEC/ED) [126], [127]
(3) (laser maser) Energy emitted, transferred, or received in the form of radiation. Unit: joule (J). (LEO) 586-1980w

radiant energy density (illuminating engineering) ($w = dQ/dV$) Radiant energy per unit volume; for example, joules per cubic meter. (EEC/IE) [126]

radiant exitance *See:* radiant emittance.

radiant exposure (laser maser) Surface density of the radiant energy received. Unit: $\mathbf{J} \cdot cm^{-2}$. (LEO) 586-1980w

radiant flux (1) (A) (light-emitting diodes) ($\theta_e = dQ_e/dt$) The time rate of flow of radiant energy. *Note:* It is expressed preferably in watts, or in ergs per second. **(B) (laser maser)** Power emitted, transferred, or received in the form of radiation. Unit: W. *Synonym:* radiant power. (BT/IE/EEC/LEO/AV) 201-1979, [34], [126], 586-1980
(2) (television) Power emitted, transferred, or received in the form of radiation. *Note:* It is expressed preferably in watts. (BT/AV) 201-1979w

radiant flux density at a surface (A) ($M_e = d\Phi_e/dA$, $E_e = d\Phi_e/dA$) The quotient of radiant flux at that element of surface of the area of that element: that is, watts per cm^2. When referring to radiant flux emitted from a surface, this has been called "radiant emittance" (symbol: M). The preferred term for radiant flux leaving a surface is "radiant exitance" (symbol: M). When referring to radiant flux incident on a surface, it is called "irradiance" (symbol: E). *Note:* "radiant emittance" is obsolete. **(B) (illuminating engineering)** ($d\Phi/dA$) The quotient of radiant flux at an element of surface to the area of that element; e.g., watts per square meter. When referring to radiant flux emitted from a surface, this has been called "radiant emittance" (a deprecated term) (symbol: M). The preferred term for radiant flux leaving a surface is "radiant exitance" (symbol: M). The radiant exitance per unit wavelength interval is called "spectral radiant exitance." When referring to radiant flux incident on a surface it is called "irradiance" (symbol: E). (ED/EEC/IE) [127], [126]

radiant gain (optoelectronic device) The ratio of the emitted radiant flux to the incident radiant flux. *Note:* The emitted and incident radiant flux are both determined at specified ports. *See also:* optoelectronic device. (ED) [46]

radiant heater A heater that dissipates an appreciable part of its heat by radiation rather than by conduction or convection. *See also:* appliance.

radiant heating system (electric power systems in commercial buildings) A heating system in which the heat radiated from panels is effective in providing heating requirements. The term "radiant heating" includes panel *and* radiant heating. (IA/PSE) 241-1990r

radiant incidence *See:* irradiance.

radiant intensity (1) (fiber optics) Radiant power per unit solid angle, expressed in watts per steradian. *See also:* intensity; radiometry. (Std100) 812-1984w
(2) (television) (of a source, in a given direction) The quotient of the radiant power emitted by a source, or by an element of source, in an infinitesimal cone containing the given direction, by the solid angle of that cone. *Note:* It is expressed preferably in watts per steradian. (BT/AV) 201-1979w
(3) (light-emitting diodes) ($I_e = d\Phi_e/d\omega$) The radiant flux proceeding from the source per unit solid angle in the direction considered; that is watts per steradian. (ED) [127]
(4) (laser maser) Quotient of the radiant flux leaving the source, propagated in an element of solid angle containing the given direction, by the element of solid angle. Unit: watt per steradian ($W \cdot sr^{-1}$). (LEO) 586-1980w
(5) (illuminating engineering) ($I = d\Phi/d\omega$) The radiant flux proceeding from a source per unit solid angle in the direction considered; for example, watts per steradian. *Note:* Mathematically a solid angle must have a point as its apex; the definition of radiant intensity, therefore, applies strictly only to a point source. In practice, however, radiant energy emanating from a source whose dimensions are negligible in comparison with the distance from which it is observed may be considered as coming from a point. Specifically, this implies that with change of distance, 1) the variation in solid angle subtended by the source at the receiving point approaches 1/(distance)^2 and that 2) the average radiance of the projected source area as seen from the receiving point does not vary appreciably. (EEC/IE) [126]

radiant power (fiber optics) The time rate of flow of radiant energy, expressed in watts. The prefix is often dropped and the term "power" is used. *Synonyms:* optical power; flux; radiant flux; power. *See also:* radiometry. (Std100) 812-1984w

radiant sensitivity (camera tubes or phototubes) The quotient of signal output current by incident radiant flux at a given wavelength, under specified conditions of irradiation. *Note:* Radiant sensitivity is usually measured with a collimated beam at normal incidence. *See also:* phototube; luminous flux; radiant flux. (ED) 161-1971w

radiated coupling Propagation of a wave through free space at radial distances greater than $\lambda/6$ from the power line carrying the disturbing energy. *See also:* coupling. (PE/PSC) 487-1992

radiated emission test site (1) (radio-noise emissions) A site meeting specified requirements suitable for measuring radio interference fields radiated by a device, equipment, or system under test. C63.5-1988
(2) (radio-noise emissions) A site meeting specified requirements suitable for measuring radio frequency fields radiated by an EUT. (EMC) C63.4-1991

radiated interference Radio interference resulting from radiated noise or unwanted signals. *See also:* electromagnetic compatibility. (EMC) [53]

radiated noise (1) (radio noise from overhead power lines and substations) Radio noise energy in the form of an electromagnetic field including both the radiation and induction components of the field. (T&D/PE) 430-1986w

(2) (power line filters) Electromagnetic interference that is radiated into the environment, either directly from equipment or from the power cord or any other cabling connected to it. (EMC) C63.13-1991

radiated power output (transmitter performance) The average power output available at the antenna terminals, less the losses of the antenna, for any combination of signals transmitted when averaged over the longest repetitive modulation cycle. *See also:* audio-frequency distortion. (VT) [37]

radiated radio noise (1) (overhead power lines) Radio noise that is propagated by radiation from a source into space in the form of electromagnetic waves; e.g., the undesired electromagnetic waves generated by corona sources on a transmission line. *Note:* Radiated radio noise includes both the radiation and the induction components of the electromagnetic fields generated by the noise source. (T&D/PE) 539-1990

(2) (measurement of radio-noise emissions) Radio-noise energy in the form of an electromagnetic field including both the radiation and induction components of the field. (EMC) C63.4-1991

radiated spurious emission power (land-mobile communications transmitters) Any part of the spurious emission power output radiated from the transmission enclosure, independent of any associated transmission lines or antenna, in the form of an electromagnetic field composed of variations of the intensity of electric and magnetic fields. (EMC) 377-1980r

radiating element A basic subdivision of an antenna that in itself is capable of radiating or receiving radio waves. *Note:* Typical examples of a radiating element are a slot, horn, or dipole antenna. (AP/ANT) 145-1993

radiating far field region (land-mobile communications transmitters) Measurement is performed at or beyond a distance of 3λ, but not less than 1 meter (m). *See also:* far-field region. (EMC) 377-1980r

radiating near-field region (1) (land-mobile communications transmitters) Measurement is limited to the region external to the induction field and extending to the outer boundary of the reactive field that is commonly taken to exist at a distance of $\lambda/2\pi$. Either the electric or magnetic component of the radiated energy may be used to determine the magnitude of power present. *See also:* near-field region. (EMC) 377-1980r

(2) That portion of the near-field region of an antenna between the farfield and the reactive portion of the near-field region, wherein the angular field distribution is dependent upon distance from the antenna. *Notes:* 1. If the antenna has a maximum overall dimension that is not large compared to the wavelength, this field region may not exist. 2. For an antenna focused at infinity, the radiating near-field region is sometimes referred to as the Fresnel region on the basis of analogy to optical terminology. (AP/ANT) 145-1993

radiation (1) (nuclear) (nuclear work) The usual meaning of radiation is extended to include moving nuclear particles, charged or uncharged. (ED) 161-1971w, [45]

(2) (data transmission) In radio communication, the emission of energy in the form of electromagnetic waves. The term is also used to describe the radiated energy. (PE) 599-1985w

(3) The electromagnetic waves or corpuscular emissions released as a result of atomic nuclear changes that penetrate into insulation to cause ionizing reactions that then change the chemical, electrical, or physical properties of the insulation. (DEI/RE) 775-1993w

(4) *See also:* ionizing radiation. (NI/NPS) 309-1999

radiation angle (fiber optics) Half the vertex angle of the cone of light emitted by a fiber. *Note:* The cone is usually defined by the angle at which the far-field irradiance has decreased to a specified fraction of its maximum value or as the cone within which can be found a specified fraction of the total radiated power at any point in the far field. *Synonym:* output angle. *See also:* numerical aperture; acceptance angle; far-field region. (Std100) 812-1984w

radiation condition A condition implying that at large distances from a source, only outgoing waves can exist. (AP/PROP) 211-1997

radiation counter An instrument used for detecting or measuring radiation by counting action. (ED) 161-1971w

radiation coupling (interference terminology) The type of coupling in which the interference is induced in the signal system by electromagnetic radiation produced by the interference source. *See also:* interference. (IE) [43]

radiation detector Any device whereby radiation produces some physical effect suitable for observation and/or measurement. *See also:* anticoincidence. (ED) [45]

radiation efficiency The ratio of the total power radiated by an antenna to the net power accepted by the antenna from the connected transmitter. *See also:* antenna. (AP/PE/ANT) 145-1993, 599-1985w

radiation, electromagnetic *See:* electromagnetic radiation.

radiation-induced data-loss characteristics (1) Characteristics that define stored data integrity as a function of total dose or dose rate after a given test pattern has been written into the device. (ED) 641-1987w

(2) The collection of threshold voltage data as a function of total dose or dose rate after initial high-conduction or low-conduction threshold voltage levels had been written into the device. (ED) 581-1978w

(3) Stored data integrity as a function of total dose or dose rate after a given test pattern has been written into the device. (ED) 1005-1998

radiation intensity (1) (data transmission) The radiation intensity in a given direction is the power radiated from an antenna per unit solid angle in that direction. (PE) 599-1985w

(2) In a given direction, the power radiated from an antenna per unit solid angle. (AP/ANT) 149-1979r, 145-1993

radiation lobe (antenna pattern) A portion of the radiation pattern bounded by regions of relatively weak radiation intensity. *See also:* antenna. (AP/ANT) 149-1979r, 145-1983s

radiation loss (1) (transmission system) That part of the transmission loss due to radiation of radio-frequency power. *See also:* waveguide. (MTT) 146-1980w

(2) (waveguide) A power loss due to electromagnetic radiation leaving a network. (MTT) 146-1980w

(3) (planar transmission lines) The loss associated with planar transmission lines having ideal conductors and lossless media, caused by energy leakage from the transmission line. (MTT) 1004-1987w

radiation mode (fiber optics) In an optical waveguide, a mode whose fields are transversely oscillatory everywhere external to the waveguide, and which exists even in the limit of zero wavelength. Specifically, a mode for which

$$\beta = [n^2(a)k^2 - (l/a)^2]^{1/2} \text{ [zc]}$$

where β is the imaginary part (phase term) of the axial propagation constant, l is the azimuthal index of the mode, $n(a)$ is the refractive index at $r = a$, the core radius, and k is the free-space wavenumber, $2\pi/\lambda$, where λ is the wavelength. Radiation modes correspond to refracted rays in the terminology of geometric optics. *Synonym:* unbound mode. *See also:* bound mode; leaky mode; refracted ray; mode. (Std100) 812-1984w

radiation pattern (1) (fiber optics) Relative power distribution as a function of position or angle. *Notes:* 1. Near-field radiation pattern describes the radiant emittance ($W \times m^{-2}$) as a function of position in the plane of the exit face of an optical fiber. 2. Far-field radiation pattern describes the irradiance as a function of angle in the far-field region of the exit face of an optical fiber. 3. Radiation pattern may be a function of the length of the waveguide, the manner in which it is excited, and the wavelength. *See also:* near-field region; far-field region. (Std100) 812-1984w

(2) The spatial distribution of a quantity that characterizes the electromagnetic field generated by an antenna. *Notes:* 1. The distribution can be expressed as a mathematical function or as a graphical representation. 2. The quantities that are most often used to characterize the radiation from an antenna are proportional to, or equal to, power flux density, radiation intensity, directivity, phase, polarization, and field strength. 3. The spatial distribution over any surface or path is also an antenna pattern. 4. When the amplitude or relative amplitude of a specified component of the electric field vector is plotted graphically, it is called an **amplitude pattern, field pattern,** or **voltage pattern.** When the square of the amplitude or relative amplitude is plotted, it is called a **power pattern.** 5. When the quantity is not specified, an amplitude or power pattern is implied. *Synonym:* antenna pattern. (AP/ANT) 145-1993

radiation pattern cut Any path on a surface over which a radiation pattern is obtained. *Note:* For far-field patterns, the surface is that of the radiation sphere. For this case, the path formed by the locus of points for which θ is a specified constant and ϕ is a variable is called a "conical cut." The path formed by the locus of points for which ϕ is a specified constant and θ is a variable is called a "great circle cut." The conical cut with θ equal to 90° is also a great circle cut. A spiral path that begins at the north pole ($\theta = 0°$) and ends at the south pole ($\theta = 180°$) is called a "spiral cut." (AP/ANT) 145-1993

radiation protection guide (electrobiology) Radiation level that should not be exceeded without careful consideration of the reasons for doing so. *See also:* electrobiology. (NIR) C95.1-1982s

radiation pyrometer (radiation thermometer) A pyrometer in which the radiant power from the object or source to be measured is utilized in the measurement of its temperature. The radiant power within wide or narrow wavelength bands filling a definite solid angle impinges upon a suitable detector. The detector is usually a thermocoupler or thermopile or a bolometer responsive to the heating effect of the radiant power, or a photosensitive device connected to a sensitive electric instrument. *See also:* electric thermometer. (EEC/PE) [119]

radiation resistance The ratio of the power radiated by an antenna to the square of the RMS antenna current referred to a specified point. *Notes:* 1. The total power radiated is equal to the power accepted by the antenna minus the power dissipated in the antenna. 2. This term is of limited utility for antennas in lossy media. (AP/ANT) 145-1993

radiation sphere (for a given antenna) A large sphere whose center lies within the volume of the antenna and whose surface lies in the far field of the antenna, over which quantities characterizing the radiation from the antenna are determined. *Notes:* 1. The location of points on the sphere are given in terms of the θ and ϕ coordinates of a standard spherical coordinate system whose origin coincides with the center of the radiation sphere. 2. If the antenna has a spherical coordinate system associated with it, then it is desirable that its coordinate system coincide with that of the radiation sphere. (AP/ANT) 145-1993

radiation thermometer *See:* radiation pyrometer.

radiation trapping (laser maser) The suppression or delay of fluorescence in an optically thick absorbing medium resulting from absorption and re-emission. (LEO) 586-1980w

radiation zone (of EMI) The area where the distance to the source of electromagnetic interference is greater than the wavelength of the interference. In the radiation zone, the circuit or system will be affected by plane waves. *Contrast:* induction zone. (PE/IC) 1143-1994r

radiative heat release The heat radiating from flames. (DEI) 1221-1993w

radiative relaxation time (laser maser) The relaxation time that would be observed if only processes involving the radiation of electromagnetic energy were effective in producing relaxation. (LEO) 586-1980w

radiative transfer theory A heuristic formulation for the calculation of the scattered specific intensity based on the conservation of energy. (AP/PROP) 211-1997

radiator (1) (illuminating engineering) An emitter of radiant energy. (EEC/IE) [126]
(2) (telecommunications) Any antenna or radiating element that is a discrete physical and functional entity. (AP/ANT) 145-1993
(3) (electric power systems in commercial buildings) A heating unit that provides heat transfer to objects within a visible range by radiation and by conduction to the surrounding air, which is circulated by natural convection. (IA/PSE) 241-1990r

radio-acoustic ranging (navigation aid terms) Determining distance by a combination of radio and sound. *Synonym:* echo ranging. (AES/GCS) 172-1983w

radioactive check source (liquid-scintillation counters) A radioactive sample used to monitor the operational status of an instrument. The approximate activity should be known. (NI) N42.16-1986

radioactive source A radionuclide prepared in a form convenient for use in testing a detector or spectrometer. (NPS) 325-1996

radioactivity standard source (1) Either a radioactivity standard that has been certified as to absolute radioactivity by a laboratory recognized as the National Standardizing Laboratory of a country for radioactivity measurements or a radioactivity standard that has been obtained from a supplier who participates in measurement assurance activities with the National Standardizing Laboratory when such standards are available. In such measurement assurance activities, the radioactivity calibration value of the suppliers should agree with the National Standardizing Laboratory value within the overall uncertainty stated by the supplier in its certification of the same batch of sources or in its certification of similar sources. (NI) N42.12-1994
(2) A radioactivity source that has been certified as to absolute radioactivity either by (a) the laboratory recognized as the National Standardizing Laboratory of the country for radioactivity measurements, NSLR, (NIST in the case of the U.S.) or (b) by a supplier who participates in measurement assurance activities with the National Standardizing Laboratory when such standards are available. In such measurement assurance activities, the radioactivity calibration value of the supplier shall agree with the National Standardizing Laboratory value within the overall uncertainty stated by the supplier in its verification of the same batch of sources or in its certification of similar sources. *Synonym:* calibrated source. (NI/NPS) 309-1999

radio altimeter (navigation aid terms) An altimeter using radar principles for height measurement. Height is determined by measurement of propagation time of a radio signal transmitted from the vehicle and reflected back to the vehicle from the terrain below. *Synonym:* radar altimeter. (AES/GCS) 686-1997, 172-1983w

radio astronomy The branch of astronomy dealing with the reception and analysis of radio waves from extraterrestrial sources. (AP/PROP) 211-1997

radio-autopilot coupler (navigation aid terms) Equipment providing means by which electrical signals from navigation receivers control the vehicle autopilot. (AES/GCS) 172-1983w

radio beacon (navigation aid terms) A facility, usually a nondirectional radio station, emitting identifiable signals intended for radio direction finding observations. *See also:* nondirectional beacon. (AES/GCS) 172-1983w

radio-beacon buoy (navigation aid terms) A buoy equipped with a marker-radio beacon. *See also:* buoy. (AES/GCS) 172-1983w

radio broadcasting Radio transmission intended for general reception. *See also:* radio transmission. (AP/BT/ANT) 145-1983s, 182-1961w

radio button A visual user interface control used to represent one of a group of mutually exclusive settings. When a radio button is selected, a visual indication is provided to indicate it is the selected button. (C) 1295-1993w

radio channel (data transmission) A band of frequencies of a width sufficient to permit its use for radio communication. *Note:* The width of the channel depends on the type of transmission and the tolerance for the frequency of emission. Normally allocated for radio transmission in a specified type of service or by a specified transmitter. (AP/PE/ANT) 145-1983s, 599-1985w

radio circuit A means for carrying out one radio communication at a time in either or both directions between two points. *See also:* radio transmission; radio channel. (EEC/PE) [119]

radio compass A direction-finder used for navigational purposes. *See also:* radio navigation. (EEC/PE) [119]

radio compass indicator A device that, by means of a radio receiver and rotatable loop antenna, provides a remote indication of the relationship between a radio bearing and the heading of the aircraft. (EEC/PE) [119]

radio compass magnetic indicator A device that provides a remote indication of the relationship between a magnetic bearing, radio bearing, and the aircraft's heading. (EEC/PE) [119]

radio control The control of mechanism or other apparatus by radio waves. *See also:* radio transmission. (EEC/PE) [119]

radio detection (radio warning) The detection of the presence of an object by radiolocation without precise determination of its position. *See also:* radio transmission. (EEC/PE) [119]

radio direction-finder (RDF) (navigation aid terms) A device used to determine the direction of arrival of radio signals. *Note:* At one time this term was used by the British to mean radio distance-finding—that is, radar. (AES/GCS) 172-1983w

radio direction finding (navigation aid terms) A procedure for determining the bearing, at a receiving point, of the source of a radio signal by observing the direction of arrival and other properties of the signal. (AES/GCS) 172-1983w

radio distress signal (SOS) Radiotelegraph distress signal consists of the group . . . --- . . . in Morse code, transmitted on prescribed frequencies. The radiotelephone distress signal consists of the spoken words May Day (*m'aidez* = help me). *Note:* By international agreement, the effect of the distress signal is to silence all radio traffic that may interfere with distress calls. (EEC/PE) [119]

radio disturbance An electromagnetic disturbance in the radio-frequency range. *See also:* radio interference; radio noise. (EMC) [53]

radio Doppler The direct determination of the radial component of the relative velocity of an object by an observed frequency change due to such velocity. *See also:* radio transmission. (EEC/PE) [119]

radio fadeout (Dellinger effect) A phenomenon in radio propagation during which substantially all radio waves that are normally reflected by ionospheric layers in or above the E region suffer partial or complete absorption. *See also:* radiation. (EEC/PE) [119]

radio field strength The electric or magnetic field strength at a radio frequency. *Synonym:* field strength. (AP/PROP) 211-1997

radio frequency (RF) (1) (A) (data transmission) (Loosely) The frequency in the portion of the electromagnetic spectrum that is between the audio-frequency portion and the infrared portion. **(B) (data transmission)** A frequency useful for radio transmission. *Note:* The present practicable limits of radio frequency are roughly 10 kHz (kilohertz) to 100 000 MHz (megahertz). Within this frequency range electromagnetic radiation may be detected and amplified as an electric current at the wave frequency. (PE) 599-1985

(2) (power line filters) A frequency in the portion of the electromagnetic spectrum that is between the audio frequency portion and the infrared portion. (EMC) C63.13-1991

(3) A frequency in the radio spectrum. *See also:* radio spectrum. (AP/PROP) 211-1997

(4) A frequency that is useful for radio transmission. (NIR) C95.1-1999

(5) (A) (Loosely) The frequency in the portion of the electromagnetic spectrum that is between the audio-frequency portion and the infrared portion. **(B)** A frequency useful for radio transmission. *Note:* The present practicable limits of radio frequency are roughly 10 kHz to 100 000 MHz. Within this frequency range, electromagnetic radiation may be detected and amplified as an electric current at the wave frequency. (EMB/MIB) 1073.3.2-2000

radio-frequency absorber A material designed to absorb electromagnetic energy. The material may have a flat face or may be formed into pyramids, wedges, or cones. Radar absorber material is commonly referred to as RAM. (EMC) 1128-1998

radio-frequency alternator A rotating-type generator for producing radio-frequency power. (AP/ANT) 145-1983s

radio-frequency attenuator (signal-transmission system) A low-pass filter that substantially reduces the radio-frequency power at its output relative to that at its input, but transmits lower-frequency signals with little or no power loss. *See also:* signal. (IE) [43]

radio-frequency converter A power source for producing electric power at a frequency of 10 kHz and above. (IE/IA) 169-1955w

radio-frequency electric current hazard advisory symbol Refers to the overall design and shape shown in the figure below.

RF electric current hazard advisory symbol

radio-frequency (RF) electric current hazard advisory symbol (NIR/SCC28) C95.2-1999

radio-frequency energy Includes radio frequency fields and radiation with frequencies between 3 kHz and 300 GHz, and includes microwave frequencies. (NIR/SCC28) C95.2-1999

radio-frequency energy advisory symbol Refers to the overall design, and shape shown in the figure below.

RF energy advisory symbol

radio-frequency (RF) energy advisory symbol (NIR/SCC28) C95.2-1999

radio-frequency generator (1) (signal-transmission system) A source of radio-frequency energy. (IE) [43]
(2) (induction heating) A power source for producing electric power at a frequency of 10 kHz and above.
(IA) 54-1955w

radio-frequency generator, electron tube type (induction and dielectric usage) A power source comprising an electron-tube oscillator, an amplifier if used, a power supply and associated control equipment. *See also:* magnetron; Colpitts oscillator; tuned grid-tuned plate oscillator; Hartley oscillator.
(IA) 54-1955w

radio-frequency hot spot A highly localized area of relatively more intense radio-frequency radiation that manifests itself in two principal ways:

a) The presence of intense electric or magnetic fields immediately adjacent to conductive objects that are immersed in lower intensity ambient fields (often referred to as re-radiation), and
b) Localized areas, not necessarily immediately close to conductive objects, in which there exists a concentration of radio-frequency fields caused by reflections and/or narrow beams produced by high-gain radiating antennas or other highly directional sources. In both cases, the fields are characterized by very rapid changes in field strength with distance.

RF hot spots are normally associated with very nonuniform exposure of the body (partial body exposure). This is not to be confused with an actual thermal hot spot within the absorbing body. (NIR) C95.1-1999

radio frequency interference (RFI) *See:* radio interference.

radio frequency link (test, measurement, and diagnostic equipment) A radio frequency channel or channels used to connect the unit under test with the testing device. *Synonym:* RF link. (MIL) [2]

radio frequency protection guides (radio frequency electromagnetic fields) The radio frequency field strengths or equivalent plane wave power densities which should not be exceeded without:

a) careful consideration of the reasons for doing so,
b) careful estimation of the increased energy deposition in the human body, and
c) careful consideration of the increased risk of unwanted biological effects.

(NIR) C95.1-1982s

radio-frequency pulse A radio-frequency carrier amplitude modulated by a pulse. The amplitude of the modulated carrier is zero before and after the pulse. *Note:* Coherence of the carrier (with itself) is not implied.
(IM/WM&A) 194-1977w

radio-frequency switching relay A relay designed to switch frequencies that are higher than commercial power frequencies with low loss. (PE/EM) 43-1974s

radio-frequency system loss (mobile communication) The ratio expressed in decibels of the power delivered by the transmitter to its transmission line to the power required at the receiver-input terminals that is just sufficient to provide a specified signal-to-noise ratio at the audio output of the receiver. *See also:* mobile communication system.
(VT) [37]

radio-frequency transformer A transformer for use with radio-frequency currents. *Note:* Radio-frequency transformers used in broadcast receivers are generally shunt-tuned devices that are tunable over a relatively broad range of frequencies. *See also:* radio transmission. (CHM) [51]

radio gain (radio-wave propagation) Of a radio system, the reciprocal of the system loss. (AP) 211-1977s

radio horizon (1) (data transmission) (of an antenna) The locus of the farthest points at which direct rays from the antenna become tangential to the planetary surface. *Note:* On a spherical surface the horizon is a circle. The distance to the horizon is affected by atmospheric reflection.
(AP/ANT) 145-1983s
(2) The locus of points at which the direct rays from a point source of radio waves are tangent to the surface of the Earth. *Note:* In general, the radio and geometric horizons differ because of atmospheric refraction. (AP/PROP) 211-1997

radio-influence field (RIF) Radio-influence field is the radio noise field emanating from an equipment or circuit, as measured using a radio noise meter in accordance with specified methods. *See also:* electromagnetic compatibility.
(EMC/CHM) [51]

radio-influence tests Tests that consist of the application of voltage and the measurement of the corresponding radio-influence voltage produced by the device being tested.
(SWG/PE) C37.40-1981s, C37.100-1992

radio-influence voltage (RIV) (1) (outdoor apparatus bushings) A high-frequency voltage generated as a result of ionization, which may be a propagated by conduction, induction, radiation or a combined effect of all three.
(PE/TR) 21-1976
(2) (high-voltage ac cable terminations) The radio noise appearing on conductors of electric equipment or circuits, as measured using a radio-noise meter as a two-terminal voltmeter in accordance with specified methods.
(PE/IC) 48-1996
(3) (overhead-power-line corona and radio noise) The radio frequency voltage appearing on conductors of electrical equipment or circuits, as measured using a radio noise meter as a two-terminal voltmeter in accordance with specified methods (generally termed conducted measurements). *Note:* The term *influence* was coined to avoid the general admission that power systems would generate and conduct interference. The term *influence* is used only in North America; the term *interference* is preferred elsewhere.
(T&D/PE) 539-1990
(4) (power and distribution transformers) A radio frequency voltage generally produced by partial discharge and measured at the equipment terminals for the purpose of determining the electromagnetic interference effect of the discharges. *Notes:* 1. "RIV" can be measured with a coupled radio interference measuring instrument and is commonly measured at approximately 1 MHz, although a wide frequency range is involved. 2. "RIV" values are often used as an "index" of "partial discharge" intensity. 3. The RIV of equipment was historically measured to determined the influence of energized equipment on radio broadcasting, hence—RIV. (PE/TR) C57.12.80-1978r
(5) A high-frequency voltage, generated by all sources of ionization current, that appears at the terminals of electric-power apparatus or on power circuits. (SPD/PE) C62.11-1999

radio interference Degradation of the reception of a wanted signal caused by radio frequency (RF) disturbance. *Notes:* 1. RF disturbance is an electromagnetic disturbance having components in the RF range. 2. The words "interference" and "disturbance" are often used indiscriminately. The expression "radio frequency interference" is also commonly applied to an RF disturbance or an unwanted signal. *Synonym:* radio frequency interference.
(EMB/T&D/PE/MIB) 1073.3.2-2000, 539-1990

radio interferometer A type of radio telescope that uses two or more physically separated collecting elements in order to achieve high angular resolution of the brightness temperature distribution of a radio source. (AP/PROP) 211-1997

radiolocation (navigation aid terms) Position determination by means of radio aids for purposes other than those of navigation. (AES/GCS) 172-1983w

radio magnetic indicator (RMI) (navigation aid terms) A combined indicating instrument which converts omnibearing indications to a display resembling an ADF (automatic direc-

RADIOMETRIC TERMS

TERM NAME	SYMBOL	QUANTITY	UNIT
Radiant energy	Q	Energy	joule (J)
Radiant power *Syn:* optical power	φ	Power	watt (W)
Irradiance	E	Power incident per unit area (irrespective of angle)	$W \cdot m^{-1}$
Spectral irradiance	E λ	Irradiance per unit wave Length interval at a given wavelength	$W \cdot m^{-2}$ $\cdot nm^{-2}$
Radiant emittance *Syn:* radiant excitance	W	Power emitted (into a full sphere) per unit area	$W \cdot m^{-1}$
Radiant intensity	I	Power per unit solid angle	$W \cdot sr^{-1}$
Radiance	L	Power per unit angle per unit projected area	$W \cdot sr^{-1}$ $\cdot m^{-2}$
Spectral radiance	L λ	Radiance per unit wavelength interval at a given wavelength	$W \cdot sr^{-1}$ $\cdot m^{-2}$ $\cdot nm^{-1}$

tion finder) display, one in which the indicator points toward the omnirange station; it combines omnibearing, vehicle heading, and relative bearing. (AES/GCS) 172-1983w

radiometric sextant (navigation aid terms) An instrument which measures the direction to a celestial body by detecting and tracking the nonvisible natural radiation of the body; such radiation includes radio, infrared, and ultraviolet. (AES/GCS) 172-1983w

radiometry (1) (fiber optics) The science of radiation measurement. The basic quantities of radiometry are listed below. (Std100) 812-1984w
(2) The measurement of quantities associated with radiant energy and power. (EEC/IE) [126]

radio navigation (navigation aids) Navigation based upon the reception of radio signals. (AES/GCS) 172-1983w

radio noise (1) (radio noise from overhead power lines and substations) Any unwanted disturbance within the radio frequency band, such as undesired electromagnetic waves in any transmission channel or device. (T&D/PE) 430-1986w
(2) (radio noise from overhead power lines and substations) An electromagnetic noise that may be superimposed upon a wanted signal and is within the radio-frequency range. (EMC) C63.5-1988, C63.4-1991
(3) (overhead-power-line corona and radio noise) Electromagnetic noise having components in the radio frequency range. (T&D/PE) 539-1990

radio noise field strength (overhead-power-line corona and radio noise) A measure of the field strength of the radiated radio noise at a given location. *Notes:* 1. In practice, the quantity measured is not the electromagnetic field strength of the interfering waves but some quantity that is proportional to, or bears a known relation to, the electromagnetic field strength. 2. The radio noise field strength is measured in average, rms, quasi-peak, or peak values, according to which detector function of the radio noise meter is used. 3. The radio noise field strength is expressed either in μV/m, or in dB above 1 μV/m, per unit bandwidth, or in a specified bandwidth. (T&D/PE) 539-1990

radiophare (navigation aid terms) A term often used in international terminology, meaning radio beacon. (AES/GCS) 172-1983w

radio propagation path (mobile communication) For a radio wave propagating from one point to another, the great-circle distance between the transmitter and receiver antenna sites. *See also:* mobile communication system. (VT) [37]

radio proximity fuse A radio device contained in a missile to detonate it within predetermined limits of distance from a target by means of electromagnetic interactions with the target. *See also:* radio transmission. (AP/ANT) 145-1983s

radio range (navigation aid terms) A radio facility that provides radial lines of position by having characteristics in its emission which are convertible to bearing information and useful in the lateral guidance of aircraft. (AES/GCS) 172-1983w

radio range-finding The determination of the range of an object by means of radio waves. *See also:* radio transmission. (EEC/PE) [119]

radio receiver A device for converting radio-frequency power into perceptible signals. (VT) [37]

radio relay system (radio relay) A point-to-point radio transmission system in which the signals are received and retransmitted by one or more intermediate radio stations. *See also:* radio transmission. (EEC/PE) [119]

radio shielding A metallic covering in the form of conduit and electrically continuous housings for airplane electric accessories, components, and wiring, to eliminate radio interference from aircraft electronic equipment. (EEC/PE) [119]

radio signal A carrier in the RF range that is modulated by an electromagnetic signal. (T&D/PE) 539-1990

radiosonde An automatic radio transmitter in the meteorological-aids service, usually carried on an aircraft, free balloon, kite, or parachute, that transmits meteorological data. *See also:* radio transmitter. (AP/ANT) 145-1983s

radio source In radio astronomy, a celestial object or region that emits radio waves. (AP/PROP) 211-1997

radio spectrum The radio frequency portion of the electromagnetic spectrum. The frequency ranges are shown in the following table:

Frequency designation	Frequency range
Ultra low frequency (ULF)	< 3 Hz
Extremely low frequency (ELF)	3 Hz to 3 kHz
Very low frequency (VLF)	3–30 kHz
Low frequency (LF)	30–300 kHz
Medium frequency (MF)	300 kHz to 3 MHz
High frequency (HF)	3–30 MHz
Very high frequency (VHF)	30–300 MHz
Ultra high frequency (UHF)	300 MHz to 3 GHz
Super high frequency (SHF)	3–30 GHz
Extremely high frequency (EHF)	30–300 GHz
Submillimeter	300 GHz to 3 THz

(AP/PROP) 211-1997

radio star (communication satellite) A discrete source in the celestial sphere emitting electrical random noise. *See also:* background noise. (COM) [25]

radio station A complete assemblage of equipment for radio transmission or reception, or both. *See also:* radio transmission. (EEC/PE) [119]

radio telescope An instrument used to detect and collect radio emissions from an object or region in space. (AP/PROP) 211-1997

radio transmission The transmission of signals by means of radiated electromagnetic waves other than light or heat waves. (EEC/PE) [119]

radio transmitter A device for producing radio-frequency power, for purposes of radio transmission. (AP/ANT) 145-1983s

radio warning *See:* radio detection.

radio wave An electromagnetic wave of radio frequency. Current usage includes frequencies up to 3 THz. *See also:* radio spectrum. (AP/PROP) 211-1997

radio-wave propagation The transfer of energy by electromagnetic radiation at radio frequencies. (AP/PROP) 211-1997

radix (1) (mathematics of computing) A quantity whose successive integer powers are the implicit multipliers of the sequence of digits that represent a number in some positional notation systems. For example, if the radix is 5, then 143.2 means 1 times 5 to the second power, plus 4 times 5 to the first power, plus 3 times 5 to the zero power, plus 2 times 5 to the minus-one power. *Synonyms:* base number; radix number; base. (C) 1084-1986w
(2) (radix-independent floating-point arithmetic) The base for the representation of floating point numbers. (MM/C) 854-1987r

radix alignment In text formatting, the formatting of numbers in a column such that their radix points, whether explicit or implicit, form a vertical line. *See also:* decimal alignment. (C) 610.2-1987

radix complement (mathematics of computing) The complement obtained by subtracting each digit of a given numeral from the largest digit in the numeration system, then adding 1 to the least significant digit of the result and executing any required carries. For example, twos complement in binary notation, tens complement in decimal notation. *Synonyms:* noughts complement; zero complement; true complement; base complement; complement on *n*. *Contrast:* diminished-radix complement. (C) 1084-1986w

radix exchange sort A radix sort in which items are compared and, if necessary, exchanged in multiple passes, using successive digits within the numeric representation of the sort key, starting with the most significant digit. *Synonym:* divide-and-conquer sort. (C) 610.5-1990w

radix insertion sort A radix sort in which each item is inserted into its proper position in the sorted set according to the digital properties of the numerical representation of the sort keys. (C) 610.5-1990w

radix list sort A radix sort implemented using the list sorting technique. (C) 610.5-1990w

radix-minus-one complement (1) A numeral in radix notation that can be derived from another by subtracting each digit from one less than the radix, for example, nines complement in decimal notation, ones complement in binary notation. (C) [20], [85]
(2) (mathematics of computing) *See also:* diminished-radix complement. (C) 1084-1986w

radix notation A positional representation system in which the ratio of the place values of adjacent digits is a positive integer (the radix). *Synonyms:* radix numeration system; radix scale. (C) 1084-1986w

radix number *See:* radix.

radix numeration system *See:* radix notation.

radix point (mathematics of computing) In positional notation, the character, expressed or implied, that separates the integral part of a numerical expression from the fractional part. For example, binary point, decimal point, hexadecimal point, or octal point. *Synonyms:* base point; arithmetic point; point. (C) 1084-1986w

radix point character A character within a picture specification that represents the radix point. *Synonym:* virtual point picture character. (C) 610.5-1990w

radix scale *See:* radix notation.

radix search A searching technique that takes advantage of the digital properties of the numerical representation of the search keys. *Contrast:* radix sort. *See also:* radix trie search; digital tree search; multiway radix trie search; binary radix trie search. (C) 610.5-1990w

radix sort A sort that takes advantage of the digital properties of the numerical representation of the sort keys; for example, sorting on keys with base 10 representation by first sorting on the hundreds place, then the tens place, then the ones place. *Contrast:* radix search. *See also:* radix insertion sort; radix list sort; digital sort; straight radix sort; radix exchange sort. (C) 610.5-1990w

radix transformation function In hashing, a hash function the result of which is the original key in a different numerical base from its original base. For example, in the function below, the original key (assumed to be in base 10) is expressed in base 16.

Original key	Calculation	Hash value
72	$72_{10} = 48_{16}$	48
157	$157_{10} = 9D_{16}$	9D

(C) 610.5-1990w

radix trie search A radix search in which the items in the set to be searched are placed in a trie. *Note:* The trie is traversed taking branches according to the search argument until a terminal node is encountered, and if the search is successful, the external node is equal to the search argument. *See also:* multiway radix trie search; binary radix trie search. (C) 610.5-1990w

radome A cover, usually intended for protecting an antenna from the effects of its physical environment without degrading its electrical performance. (AP/ANT) 145-1993

rads in Si (1) (metal-nitride-oxide field-effect transistor) Amount of radiation measured by its ionizing effect in silicon; 1 rad equals 100 erg of energy deposited in a gram of irradiated solid. This number can be translated into the density of electron-hole pairs (ehp/cm^3) by the following operation:

$$n_{eh}[\text{ehp/volume}] = \gamma[\text{energy/mass}] \times \rho[\text{mass/volume}] \times N_{eh}[\text{ehp/energy}]$$

where

n_{eh} = volume density of ehp,
γ = total radiation dose as energy dissipated per unit mass, typically expressed in rads,
d = density of solid in mass/volume,
N_{eh} = number of ehps created per energy dissipated.

In many solids, $N_{eh} \sim 1/b\text{Eg}$, and in silicon particularly, $b = 3.6[\text{ehp}]^{-1}$. *Note:* [15], and Eg = 1.0 eV. This means that one ehp is created per 3.6 electron-volts of energy dissipated. In order to permit the use of the total dose expressed in rads directly, use is made of the identity that 1 erg = 6.2×10^{11} eV. From this $N_{eh} \sim 6.2 \times 10^{11}$ [eV/erg] [ehp]/3.6[eV], and since 1 rad = 100 ergs/g, this makes $N_{eh} \sim 1.7 \times 10^{13}$ [ehp/g rads]. Thus, for silicon, n_{eh} [ehp/cm^3] = $1.7 \times 10^{13} \times \gamma$[rads] $\times \rho$[g/ cm^3]. (ED) 581-1978w
(2) Amount of radiation measured by its ionizing effect in silicon; one rad equals 100 erg of energy deposited in a gram of irradiated solid. (ED) 1005-1998

ragged left margin In text formatting, a left margin that is not aligned. *Contrast:* left justification. (C) 610.2-1987

ragged right margin In text formatting, a right margin that is not aligned. *Contrast:* right justification. (C) 610.2-1987

RAID *See:* redundant arrays of inexpensive disks.

RAID level 2 A form of RAID storage, known as level 2, in which Hamming codes are used for error correction. (C) 610.10-1994w

RAID storage Acronym for redundant arrays of inexpensive disks; a type of storage that uses several magnetic or optical disks, known as a disk array, working in tandem to increase disk capacity, improve data transfer rates, and provide higher system reliability. *Note:* Six basic architectures of RAID storage, referred to as levels 0 through 5, have been defined; see corresponding figure.

Level	Description
0	Data striping without parity
1	Mirrored disk array
2	
3	Parallel disk array
4	Independent disk array
5	Independent disk array

types of RAID storage

(C) 610.10-1994w

rail clamp A device for connecting a conductor or a portable cable to the track rails that serve as the return power circuit in mines. *See also:* mine feeder circuit. (EEC/PE/MIN) [119]

rain Precipitation in the form of liquid water drops with diameters greater than 0.5 mm, or, if widely scattered, smaller diameters. For observation purposes, the intensity of rainfall at any given time and place may be classified as Very light: scattered drops that do not completely wet an exposed surface regardless of duration; Light: the rate of fall being no more than 2.5 mm/h; Moderate: from 2.6 to 7.6 mm/h, the maximum rate of fall being no more than 0.76 mm in 6 min; Heavy: over 7.7 mm/h. When rain gauge measurements are not readily available to determine the rain intensity, estimates may be made according to a descriptive system set forth in observation manuals. *Notes:* 1. For corona studies, probability distributions for rain are produced from data obtained during "measurable rain"; i.e., rain intensities that can be measured with standard rain counters such as tipping buckets or instantaneous rate meters. 2. For ac lines, heavy rain levels are often considered representative of maximum or L_5 levels. Heavy rain data are often generated by artificial tests on conductors strung in high-voltage test setups. 3. The only other form of liquid precipitation, drizzle, is to be distinguished from rain in that drizzle drops are generally less than 0.5 mm in diameter, are very much more numerous, and reduce visibility much more than does light rain. (T&D/PE) 539-1990

rain clutter Radar echoes from rain that impair or obscure the echoes from desired targets. *See also:* precipitation clutter. (AES) 686-1997

rainproof (power and distribution transformers) So constructed, protected, or treated as to prevent rain from interfering with the successful operation of the apparatus under specified test conditions. (NEC/NESC/PE/TR) C57.12.80-1978r, [86]

rain rate A measure of the volume of water collected per unit area per unit time due to rain. The common unit is millimeters per hour. *Note:* Precipitation rate may refer to other hydrometeors such as snow, in which case the common units are either millimeters per hour or equivalent rainfall rate in millimeters per hour. *Synonym:* rainfall rate. (AP/PROP) 211-1997

rainfall rate *See:* rain rate.

raintight (1) So constructed or protected that exposure to a beating rain will not result in the entrance of water under specified test conditions. (NESC/NEC) [86]
(2) (power and distribution transformers) So constructed or protected as to exclude rain under specified test conditions. (PE/TR) C57.12.80-1978r

rake An inclination from the perpendicular. (T&D/PE) 751-1990

RAM (1) (random-access memory) A memory that permits access to any of its address locations in any desired sequence with similar access time to each location. *Note:* The term RAM, as commonly used, denotes a read/write memory. (ED) 641-1987w
(2) Reliability, availability, and maintainability of the plant. In economic analysis, the increase in RAM due to data integration or automation is sometimes quantified as a benefit. In computer applications, this refers to random access memory, which is not used in this text. (PE/EDPG) 1150-1991w
(3) High-speed read/write memory with an access time that is the same for all storage locations. *See also:* memory board; dynamic random-access memory; main storage; static random-access memory. (C) 610.10-1994w

RAM disk A simulated storage disk created and maintained by a special driver that stores data electronically (in RAM, or random-access storage) rather than magnetically. *Note:* Such storage is inherently dynamic. *Synonym:* virtual disk. (C) 610.10-1994w

Raman-Nath region (acousto-optic device) The region that occurs when the Bragg Region inequality is reversed, that is $L < n\Lambda^2/\lambda_0$. The angle of incidence is generally zero degrees, and light is diffracted into many diffraction orders. (UFFC) [23]

RAMIS *See:* Rapid Access Management Information System.

RamLink The packet-transfer architecture portion of this standard, which describes how packets are transferred between the master and slaves. Several signal-layer specifications can be used, including *RingLink* and *SyncLink.* (C/MM) 1596.4-1996

ramp (1) (thyristor) A controlled change in output at a predetermined linear rate, from one value to another. (IA/IPC) 428-1981w
(2) (railway control) A roadway element consisting of a metal bar of limited length, with sloping ends, fixed on the roadway, designed to make contact with and raise vertically a member supported on the vehicle. (PE/EEC) [119]
(3) (A) (pulse terminology) (single transition) A linear feature. **(B) (automatic control)** *See also:* unit-ramp signal. (IM/WM&A) 194-1977

ramp-forced automatic control response The total (transient plus steady-state) time response resulting from a sudden increase in the rate of change of input from zero to some finite value. *Synonym:* ramp response. (PE/EDPG) [3]

ramp-forced response time (automatic control) The time interval by which an output lags an input, when both are varying at a constant rate. (PE/EDPG) [3]

ramping The rate at which a generating unit increases or decreases its output, usually expressed in megawatts per minute. (PE/PSE) 858-1993w

ramp response (null-balancing electric instrument) A criterion of the dynamic response of an instrument when subjected to a measured signal that varies at a constant rate. *See also:* accuracy rating. (EEC/EMI) [112]

ramp response time (null-balancing electric instrument) The time lag, expressed in seconds, between the measured signal and the equivalent positioning of the end device when the measured signal is varying at constant rate. *See also:* accuracy rating. (EEC/EMI) [112]

ramp response-time rating (null-balancing electric instrument) The maximum ramp response time for all rates of change of measured signal not exceeding the average velocity corresponding to the span step-response-time-rating of the instrument when the instrument is used under rated operating conditions. Example: If the span step-response-time-rating is four seconds, the ramp response-time rating shall apply to any rate of change of measured signal not exceeding 25% of span per second. *See also:* accuracy rating. (EEC/EMI) [112]

ramp shoe *See:* shoe.

random (1) (data transmission) A condition not localized in time or frequency. (PE) 599-1985w
(2) (automatic control) Describing a variable whose value at a particular future instant cannot be predicted exactly, but can only be estimated by a probability distribution function. (PE/EDPG) [3]

(3) **(modeling and simulation)** Pertaining to a process or variable whose outcome or value depends on chance or on a process that simulates chance, often with the implication that all possible outcomes or values have an equal probability of occurrence; for example, the outcome of flipping a coin or executing a computer-programmed random number generator. (C) 610.3-1989w

random access (1) (A) (computers) Pertaining to the process of obtaining data from, or placing data into storage where the time required for such access is independent of the location of the data most recently obtained or placed in storage. **(B) (computers)** Pertaining to a storage device in which the access time is effectively independent of the location of the data. (MIL/C) [2], [20], [85]
(2) (data management) An access mode in which specific logical records are obtained from or placed into a file in a nonsequential manner. *Contrast:* direct access; sequential access. *See also:* direct access. (C) 610.5-1990w

random-access memory (RAM) (1) A memory that permits access to any of its address locations in any desired sequence with similar access time to each location (adapted from IEC 748-2). *Note:* The term RAM, as commonly used, denotes a read/write memory with unlimited data rewrite capability and equal read and write times. (ED) 1005-1998
(2) A type of temporary data storage (memory) that can be read and changed while the computer is in use. Data stored in random-access memory is lost if the system loses power. (PE/SUB) 1379-1997

random access method* *See:* direct access method.

* Deprecated.

random access programming (test, measurement, and diagnostic equipment) Programming without regard for the sequence required for access to the storage position called for in the program. (MIL) [2]

random array antenna *See:* array antenna.

random drift rate (gyros) The nonsystematic, time-varying component of drift rate under specified operating conditions. It is expressed as an rms value, or standard deviation of angular displacement per unit time. (AES/GYAC) 528-1994

random error (1) Errors that have unknown magnitudes and directions and that vary with each measurement. (PE/PSIM) 4-1995
(2) (measurement) A component of error whose magnitude and direction vary in a random manner in a sequence of measurements made under nominally identical conditions. (IM/HFIM) 314-1971w

random errors (1) (navigation aid terms) Those errors which cannot be predicted except on a statistical basis. (AES/GCS) 172-1983w
(2) Those errors that cannot be predicted except on a statistical basis. (AES/RS) 686-1990

random failure (1) (software) A failure whose occurrence is unpredictable except in a probabilistic or statistical sense. *See also:* transient error; intermittent fault. (C) 610.12-1990
(2) Any failure whose cause and/or mechanism makes its time of occurrence unpredictable. (SWG/PE/NP) C37.100-1992, 650-1979s

random failures The pattern of failures for equipment that has passed out of its infant-mortality period and has not reached the wear-out phase of its operating lifetime. The reliability of an equipment in this period may be computed by the equation:

$R = e^{-\lambda t}$

where
λ is failure rate
t is time period of interest

(SUB/PE) C37.1-1994

random-incidence microphone (audible noise measurements) (overhead power lines) A microphone that has been designed to have a flat frequency response in a diffuse sound field where sound waves are arriving equally from all directions. (T&D/PE) 539-1990, 656-1992

randomizing *See:* hashing.

randomly polarized Electromagnetic radiation in which the direction of the electric field vector changes randomly in time and/or space. (AP/PROP) 211-1997

random medium A medium in which the spatial variations of permittivity, discrete and/or continuous, are best described in terms of statistical measures. (AP/PROP) 211-1997

random noise (1) (overhead-power-line corona and radio noise) Noise that comprises transient disturbances occurring at random. *Notes:* 1. Random noise is the part of noise that is unpredictable except in a statistical sense. The term is most frequently applied to limiting cases where the number of transient disturbances per unit time is large, so that the spectral characteristics are the same as those of thermal noise. Thermal noise and shot noise are special cases of random noise. 2. A random noise whose instantaneous magnitudes occur according to the Gaussian distribution is called "Gaussian random noise." 3. In power line noise, "random noise" is a component of the total noise caused by discharges. *Synonym:* fluctuation noise.
(T&D/PE/C/EMC/PSR) 539-1990, C37.93-1976s, 165-1977w, C63.5-1988, C63.4-1988s
(2) Electromagnetic noise, the values of which at given instants are not predictable. *Note:* The part of the noise that is unpredictable except in a statistical sense. The term is most frequently applied to the limiting case in which the number of transient disturbances per unit time is large, so that the spectral characteristics are the same as those of thermal noise. Thermal noise and shot noise are special cases of random noise. (EMC) C63.12-1987
(3) A nondeterministic fluctuation in the output of a waveform recorder, described by its frequency spectrum and its amplitude statistical properties. (IM/WM&A) 1057-1994w
(4) (broadband local area networks) *See also:* noise. (LM/C) 802.7-1989r

random noise bandwidth (overhead-power-line corona and radio noise) The width in hertz of a rectangle having the same area and maximum amplitude as the square of the amplifier frequency response to a sinusoidal input. (T&D/PE) 539-1990

random number (mathematics of computing) A number selected by chance from a given set of numbers, and satisfying one or more of the standard tests for statistical randomness. (C) 1084-1986w

random number sequence (A) A sequence of random numbers, each of which is statistically independent of its predecessors. **(B)** Loosely, a pseudo-random number sequence. **(C)** A sequence of numbers in which no number can be predicted from knowledge of its predecessors. (C) 610.5-1990

random-ordered list *See:* unordered list.

random paralleling (rotating machinery) Paralleling of an alternating-current machine by adjusting its voltage to be equal to that of the system, but without adjusting the frequency and phase angle of the incoming machine to be sensibly equal to those of the system. *See also:* asynchronous machine. (PE) [9]

random photon summing (sodium iodide detector) The simultaneous detection of two or more photons originating from the disintegrations of more than one atom. (NI) N42.12-1994

random probing Open-address hashing in which collision resolution is handled by randomly selecting positions in the hash table until an available position is found. *Contrast:* uniform probing; linear probing. (C) 610.5-1990w

random separation Installed with no deliberate separation. (NESC) C2-1997

random-scan A technique employed in random-scan display devices in which the beam moves from point to point, creating an image composed of vectors. *See also:* vector graphics.

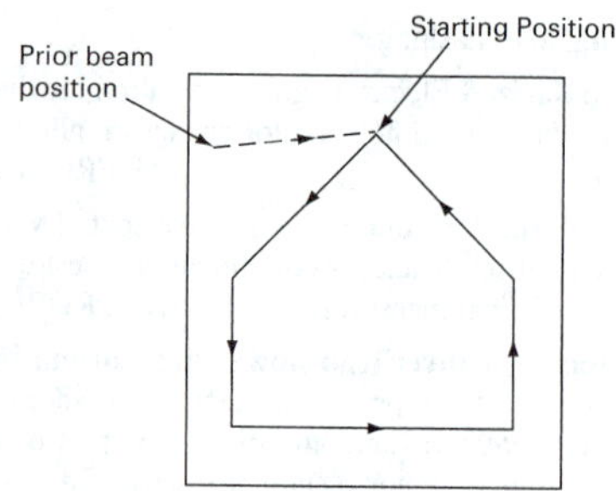

random-scan

(C) 610.6-1991w

random-scan display device A type of CRT display device in which the beam moves from point to point, creating an image composed of vectors. *Note:* This method is often called "vector graphics." *Synonyms:* stroker display; vector display device; refresh line-drawing display device. *Contrast:* raster display device. *See also:* refresh display device. (C) 610.10-1994w, 610.6-1991w

random summing (germanium detectors) The simultaneous detection of two or more photons originating from the disintegration of more than one atom. (PE/EDPG) 485-1983s

random surface A boundary surface, between two different but otherwise homogeneous media, whose height fluctuations are best described in terms of statistical measures. (AP/PROP) 211-1997

random walk (1) (angle) The angular error buildup with time, due to white noise in angular rate. This error is typically expressed in degrees per square root of hour $[°/\sqrt{h}]$. (AES/GYAC) 528-1994
(2) (rate) The drift rate error buildup with time, due to white noise in angular acceleration. This error is typically expressed in degrees per hour, per square root of hour $[(°/h)/\sqrt{h}]$. (AES/GYAC) 528-1994

random winding (rotating machinery) A winding in which the individual conductors of a coil side occupy random position in a slot. *See also:* rotor; stator. (PE) [9]

random-wound motorette A motorette for random-wound coils. (PE) [9]

range (1) (electric pipe heating systems) (electric heat tracing systems) The capability span of an instrument, the region between the lower and upper limits of a measured or generated function. With respect to electric pipe heating systems, range is usually defined as the difference between the lowest available set point and the highest available set point. (PE/EDPG) 622A-1984r, 622B-1988r
(2) (radiation protection) The set of values lying between the upper and lower detection limits. (NI) N320-1979r, N323-1978r
(3) (A) (computers) The set of values that a quantity or function may assume. **(B) (computers)** The difference between the highest and lowest value that a quantity or function may assume. *See also:* error range. (C) [20], [85]
(4) (A) (electronic navigation) An ambiguous term meaning either: a distance, as in artillery techniques and radar measurements or a line of position, located with respect to ground references, such as a very-high frequency omnidirectional radio range (VOR) station, or a pair of lighthouses, or an aural radio range (A–N) radio beacon. *Note:* In electronic navigation, the reader must be particularly wary, since the two meanings of the word range often occur in close proximity. *See also:* radio range. **(B) (health physics instrumentation)** All values lying between the upper and lower indicated limits. (NI) N42.17B-1989
(5) Distance between a radar and a target. (AES) 686-1997

range and elevation guidance for approach and landing (REGAL) (navigation aids) A ground-based navigation system used in conjunction with a localizer to compute vertical guidance for proper glide-slope and flare-out during an instrument approach and landing; it uses a digitally-coded vertically-scanning fan beam that provides data for both elevation angle and distance. *See also:* navigation. (AES/RS/GCS) 686-1982s, [42], 172-1983w

range check A consistency check that ensures that an item of data falls between pre-established maximum and minimum values. (C) 610.5-1990w

range curvature A term applied to a number of signal and image effects [for a synthetic-aperture radar (SAR)] that are due to the spherical nature of RF wavefronts. *Note:* These effects include quadratic phase used to synthesize an aperture, range walk, and image keystone distortions for spotlight SAR systems. (AES) 686-1997

range equation *See:* radar equation.

range extender (telephone switching systems) Equipment inserted in a switched connection to allow an increased loop resistance. (COM) 312-1977w

range-height indication (A) A type of radar display format. *See also:* display. **(B)** An intensity-modulated display in which horizontal and vertical distances of a blip from an origin in the lower-left part of the display represent target ground range and target height, respectively. The display is generated by successive range, sweeps starting at the origin and inclined at an angle that varies progressively in accordance with the elevation scan of the radar antenna at a selected azimuth. The height scale of the display is usually expanded relative to the range scale.

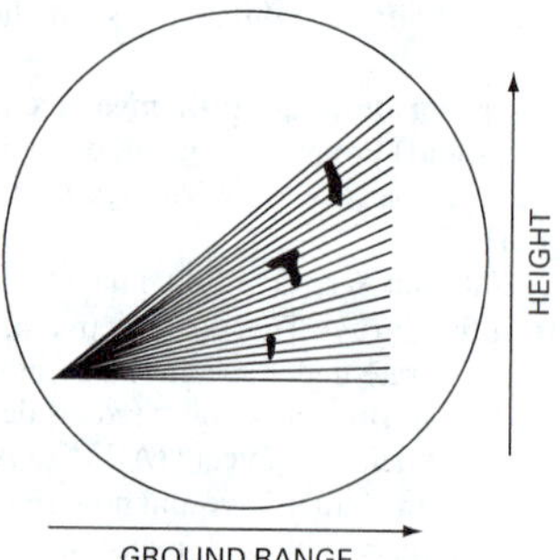

RHI display

(AES/RS) 686-1990

range-height indicator (RHI) (A) A type of radar display format. *See also:* display. **(B)** An intensity-modulated display in which horizontal and vertical distances of a blip from an origin in the lower-left part of the display represent target ground range and target height, respectively. *Note:* The display is generated by successive range weeps starting at the origin and inclined at an angle that varies progressively in accordance with the elevation scan of the radar antenna at a selected azimuth. The height scale of the display is usually expanded relative to the range scale.

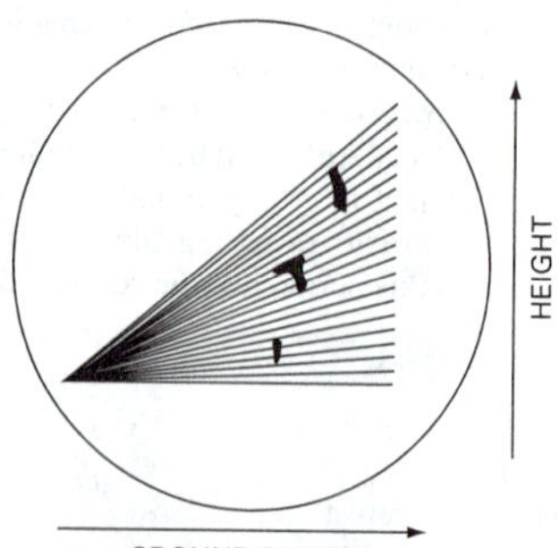

Range-height indicator

(AES) 686-1997

range lights (illuminating engineering) Groups of color-coded boundary lights provided to indicate the direction and limits of a preferred landing path normally on an aerodrome without runways but exceptionally on an aerodrome with runways. (EEC/IE) [126]

range mark A calibration marker used on a display to aid in measuring target range. *Synonym:* range marker. (AES) 686-1997

range marker *See:* range mark.

range noise The noise-like variation in the apparent distance of a target, caused by changes in phase and amplitude of the target-scattering sources, and including radial components of glint and scintillation error. (AES) 686-1997

range of a radio system The maximum distance for which a radiowave transmitting system, with specified installation and operating conditions, produces a usable signal strength at a specified radio receiver installation. (AP/PROP) 211-1997

range offset processing Synthetic-aperture processing in which the spectrum is translated from intermediate frequency (IF) to a carrier offset from zero frequency by approximately half the IF bandwidth. *See also:* synthetic-aperture radar. (AES) 686-1997

range resolution The ability to distinguish between two targets solely by the measurement of their ranges. Range resolution is usually expressed in terms of the minimum range separation at which two targets at the same azimuth and elevation angles can be distinguished. *Note:* The required separation should be specified for targets of given relative power level at the receiver. Equal powers are often assumed, but it may be necessary to specify the separation at two or more power ratios where resolution of targets of different powers is important. (AES) 686-1997

range walk The migration of a point scatterer from range cell to range cell during the signal integration period. *Note:* Range walk can occur in synthetic-aperture radar; typically caused by range curvature and/or target rotational or radar line-of-sight translational motion. Range walk also can occur in a very high resolution radar if the relative range rate between the target and the radar is high, relative to the ratio of the range cell to the integration period. (AES) 686-1997

ranging (communication satellite) The measurement of distance between two points and a precisely known reference point. A multiplicity of tones or a PN (pseudonoise) sequence ranging code is often used. (COM) [19]

rank (networks) (degrees of freedom on a node basis) The number of independent cut-sets that can be selected in a network. The rank R is equal to the number of nodes V minus the number of separate parts P. Thus $R = V - P$. *See also:* network analysis. (Std100) 270-1966w

rapid access loop (test, measurement, and diagnostic equipment) In internal memory machines, a small section of memory which has much faster accessibility than the remainder of the memory. (MIL) [2]

Rapid Access Management Information System (RAMIS) A nonprocedural database manipulation language that provides data management and decision support facilities. (C) 610.13-1993w

rapid prototyping A type of prototyping in which emphasis is placed on developing prototypes early in the development process to permit early feedback and analysis in support of the development process. *Contrast:* waterfall model. *See also:* transform analysis; structured design; modular decomposition; object-oriented design; input-process-output; spiral model; stepwise refinement; incremental development; data structure-centered design; transaction analysis. (C) 610.12-1990

rapid start fluorescent lamp (illuminating engineering) A fluorescent lamp designed for operation with a ballast that provides a low-voltage winding for preheating the electrodes and initiating the arc without a starting switch or the application of high voltage. (EEC/IE) [126]

rapid-starting systems (fluorescent lamps) The designation given to those systems in which hot-cathode electric discharge lamps are operated with cathodes continuously heated through low-voltage heater windings built as part of the ballast, or through separate low-voltage secondary transformers. Sufficient voltage is applied across the lamp and between the lamp and fixture to initiate the discharge when the cathodes reach a temperature high enough for adequate emission. The cathode-heating current is maintained even after the lamp is in full operation. *Note:* In Europe this system is sometimes referred to as an instant-start system. (EEC/LB) [94]

raptor A bird of prey. (T&D/PE) 751-1990

Rascal A dialect of Pascal. *See also:* Pascal. (C) 610.13-1993w

raster (1) (television) (cathode-ray tubes) A predetermined pattern of scanning lines that provides substantially uniform coverage of an area. (ED/BT/AV) 161-1971w, 201-1979w **(2)** *See also:* raster grid; pixel. (C) 610.6-1991w

raster burn A change in the characteristics of that area of the target that has been scanned, resulting in a spurious signal corresponding to that area when a larger or tilted raster is scanned. (ED) 161-1971w

raster CRT *See:* raster display device.

raster display device A cathode ray tube display device in which the electron beam makes a line by line sweep of the screen, called raster scanning, creating an image composed of dots by modifying the intensity of the beam. *Note:* This method is often called "raster graphics." *Synonym:* raster CRT. *Contrast:* random-scan display device. *See also:* raster scan; cell-organized raster display device; matrix-addressed storage display device. (C) 610.6-1991w, 610.10-1994w

raster font *See:* vector font.

raster graphics The representation of an image by an array of pixels arranged in rows and columns. *Contrast:* vector graphics. *See also:* raster display device. (C) 610.6-1991w

raster grid The grid of addressable coordinates on the display surface of a display device. (C) 610.6-1991w, 610.10-1994w

raster order (A) The order in which pixels are scanned in a raster display device; usually left-to-right; top-to-bottom. **(B)** The order in which pixel information is stored in memory such that it may be displayed on a raster display device. (C) 610.10-1994

raster plotter A plotter that generates a display image on a display surface using a line-by-line scanning technique. *Contrast:* digital plotter; analog plotter. *See also:* electrostatic plotter. (C) 610.10-1994w

raster scan (1) (computer graphics) A technique employed in raster display devices in which the electron beam "scans" the display surface line by line, illuminating the pixels, creating an image on the display surface. *See also:* raster graphics.

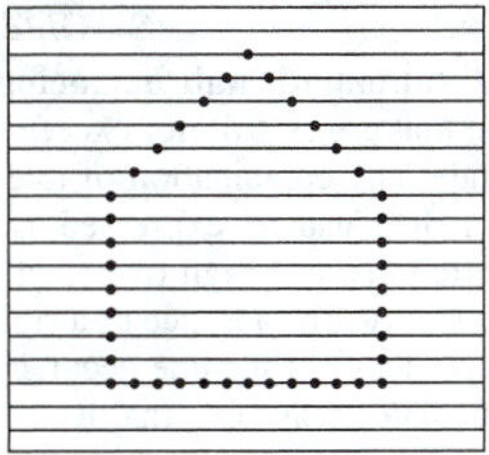

raster scan

(C) 610.6-1991w, 610.10-1994w

(2) A method of sweeping the electron beam of a cathode-ray tube screen or an antenna beam that is characterized by more than one sweep either from side to side or from top to bottom. (AES) 686-1997

raster unit The distance between two adjacent addressable locations on a cathode ray tube display device. (C) 610.6-1991w

ratchet demand (electric power utilization) The maximum past or present demands that are taken into account to establish billings for previous or subsequent periods. *See also:* alternating-current distribution. (PE/PSE) [54]

ratchet demand clause (electric power utilization) A clause in a rate schedule that provides that maximum past or present demands be taken into account to establish billings for previous or subsequent periods. *See also:* alternating-current distribution. (PE/PSE) [54], 346-1973w

ratchet relay A stepping relay actuated by an armature-driven ratchet. *See also:* relay. (EEC/REE) [87]

rate The change in a value over a specified period of time. *Note:* Instantaneous rate is the derivative of the value with respect to time and cannot generally be measured. The measured rate approaches the instantaneous rate as the specified period of time approaches zero. (LM/C) 802.1F-1993r

rate action (process control) That component of proportional plus rate control action or of proportional plus reset plus rate control action for which there is a continuous linear relation between the rate of change of the directly controlled variable and the position of a final control element. *See also:* control action. (PE/EDPG) [3]

rate base (power operations) The net plant investment or valuation bases specified by a regulatory authority, upon which a utility is permitted to earn a specified rate of return. (PE/PSE) 858-1987s

rate biasing (laser gyro) The action of intentionally rotating the laser gyro about the input axis to avoid the region in which lock-in occurs. *See also:* anti-lock means. (AES/GYAC) 528-1994

rate center In the United States, a defined geographic location used by telephone companies to determine distance measurements for interLATA and intraLATA mileage rates. (C) 610.7-1995

rate, chopping *See:* chopping rate.

rate compensation heat detector (fire protection devices) A device that will response when the temperature of the air surrounding the device reaches a predetermined level, regardless of the rate of temperature rise. (NFPA) [16]

rate control action (electric power system) Action in which the output of the controller is proportional to the input signal and the first derivative of the input signal. Rate time is the time interval by which the rate action advances the effect of the proportional control action. *Note:* Applies only to a controller with proportional control action plus derivative control action. *See also:* speed-governing system. (PE/PSE) 94-1970w

rated A qualifying term that, applied to an operating characteristic, indicates the designated limit or limits of the characteristic for application under specified conditions. *Note:* The specific limit or limits applicable to a given device is specified in the standard for that device, and included in the title of the rated characteristic, that is, rated *maximum* voltage, rated frequency *range,* etc. (SWG/PE) C37.100-1992

rated accuracy (1) (automatic null-balancing electric instrument) The limit that errors will not exceed when the instrument is used under any combination of rated operating conditions. *Notes:* 1. It is usually expressed as a percent of the span. It is preferred that a + sign or − sign or both precede the number or quantity. The absence of a sign infers a ± sign. 2. Rated accuracy does not include accuracy of sensing elements or intermediate means external to the instrument. *See also:* accuracy rating. (EEC/EMI) [112]
(2) (direct-current instrument shunts) The limit of error, expressed as a percentage of the rated output voltage, with two thirds rated current applied for one half hour to allow for self heating. It represents the expected accuracy of the shunt obtainable under normal conditions of use. (PE/PSIM) 316-1971w

rated accuracy of instrument shunts (electric power system) The limit of error, expressed as a percentage of rated voltage drop, with two-thirds rated current applied for one-half hour to allow for self-heating. *Note:* Practically, it represents the expected accuracy of the shunt obtainable over normal operating current ranges. *See also:* accuracy rating. (PE/PSIM) [55]

rated alternating voltage (rated alternating-current winding voltages) (rectifier unit) (rectifier) The root-mean-square voltages between the alternating-current line terminals that are specified as the basis for rating. *Note:* When the alternating-current winding of the rectifier transformer is provided with taps, the rated voltage shall refer to a specified tap that is designated as the rated-voltage tap. *See also:* rectification; rectifier transformer. (IA/EEC/PCON) [62], [110], C57.18-1964w

rated apparent efficiency (thyristor) Rated output volt-amperes divided by rated input power, generally expressed as percent. (IA/IPC) 428-1981w

rated asymmetrical making current The maximum rms current, at rated frequency, including the dc component, against which a device is required to close and latch under specified conditions. (SWG/PE) C37.100-1992

rated average tube current The current capacity of a tube, in average amperes, as assigned to it by the manufacturer for specified circuit conditions. *See also:* rectification. (EEC/PCON) [110]

rated burden (capacitance potential devices) The maximum unity-power-factor burden, specified in watts at rated secondary voltage, that can be carried for an unlimited period when energized at rated primary line-to-ground voltage, without causing the established limitations to be exceeded. *See also:* outdoor coupling capacitor. 31-1944w

rated capacity (C) (1) (nickel-cadmium cell) The capacity assigned to a nickel-cadmium cell by its manufacturer for a specific constant current charge, with a given discharge time, at a specified electrolyte temperature, to a given end-of-discharge voltage. *Note:* The conditions used to establish rated capacity are based on a constant current charge. (PE/PV/EDPG) 1115-1992, 1144-1996
(2) (battery) The manufacturer's statement of the number of ampere-hours or watt-hours that can be delivered by a fully charged battery at a specific discharge rate and electrolyte temperature, to a given end-of-discharge voltage. (IA/PSE) 446-1995
(3) The capacity assigned to a cell by its manufacturer for a given discharge rate, at a specified electrolyte temperature and specific gravity, to a given end-of-discharge voltage. (SCC29/PV) 485-1997, 1145-1999

rated capacitance switching transient overvoltage ratio The largest value of transient overvoltage ratio that a device will produce at either its source or load terminals when switching its rated capacitance switching current. (SWG/PE) C37.30-1992s

rated capacitive switching current The rms symmetrical value of the highest capacitive load current that a device is required to make and interrupt at a rated maximum voltage as part of its designated operation duty cycle. *Note:* The capacitive switching current rating should be at least 135% of the rated capacitor bank. The excess current can be caused by harmonics, overvoltage, or plus tolerance in the capacitor kvar. (SWG/PE) C37.100-1992

rated circuit voltage Used to designate the rated, root-mean-square, line-to-line, voltage of the circuit on which coupling capacitors or the capacitance potential device in combination with its coupling capacitor or bushings designed to operate. *See also:* outdoor coupling capacitor. 31-1944w

rated closing time (1) (of a fault-initiating switch) The specified interval in a closing operation between the energizing of the trip coil, at the minimum standard control voltage, and the making of the fault-initiating switch contacts. (SWG/PE) C37.30-1992s
(2) (of a generator circuit breaker) The interval between energizing of the close circuit at rated control voltage and rated fluid pressure of the operating mechanism and the closing of the main circuit. (SWG/PE) C37.013-1997

rated continuous controller current (thyristor) The rated root-mean-square (rms) value of the maximum controller current which can be carried continuously without exceeding established limitations under prescribed conditions of operation. (IA/IPC) 428-1981w

rated continuous current (1) (neutral grounding devices) The current expressed in amperes, root-mean-square, that the device can carry continuously under specified service conditions without exceeding the allowable temperature rise. (PE/SPD) 32-1972r

(2) (of a switching device or an assembly) The maximum rms current, in amperes at rated frequency, that a device or an assembly will carry continuously without exceeding the limit of observable temperature rise. (SWG/PE) C37.30-1992s

(3) The maximum rms current in amperes, at rated frequency, which a device will carry continuously without exceeding the allowable temperature rise and total temperature. (SWG/PE) C37.40-1993

(4) (of a generator circuit breaker) The designated limit of current in rms amperes at power frequency that a generator circuit breaker shall be required to carry continuously without exceeding any of its designated limitations. (SWG/PE) C37.013-1997

(5) The designed limit in rms amperes or dc amperes that a switch or circuit breaker will carry continuously without exceeding the limit of observable temperature rise. (IA/MT) 45-1998

rated continuous output current (converters having ac output) (self-commutated converters) The maximum output current that can be carried continuously without exceeding established limitations under prescribed conditions of operation. (IA/SPC) 936-1987w

rated controller current (thyristor) Rated root-mean-square (rms) value of the controller current which is specified by the manufacturer under the prescribed operation mode as a basis of declaring the duty cycles and overcurrent capability. (IA/IPC) 428-1981w

rated current (1) (power and distribution transformers) The primary current selected for the basis of performance specifications of a current transformer. (PE/TR) C57.13-1993, C57.12.80-1978r

(2) (shunt reactors over 500 kVA) (of a shunt reactor) Derived from the rated voltage and rated kilovoltamperes (kVA). (PE/TR) C57.21-1981s

(3) (neutral grounding devices) (electric power) The thermal current rating. The rated current of resistors whose rating is based on constant voltage is the initial root-mean-square symmetrical value of the current that will flow when rated voltage is applied. *See also:* grounding device. (PE/SPD) 32-1972r

(4) The rms power frequency current in amperes that can be carried for the duty specified, at rated frequency without exceeding the specified temperature limits, and within the limitations of established standards. (PE/TR) C57.16-1996

rated differential capacitance voltage (1) (maximum) The greatest value of differential capacitance voltage that may be impressed on the contacts of the interrupter unit at which the interrupter switch is required to make and interrupt all values of capacitance current up to its rated switching current. (SWG/PE) C37.30-1992s

(2) (minimum) The least value of differential capacitance voltage that may be impressed on the contacts of the interrupter unit at which the interrupter switch is required to make and interrupt all values of capacitance current up to its rated switching current. (SWG/PE) C37.30-1992s

rated direct current (thyristor converter) The current in terms of which all test and service current ratings are specified (for example, the per-unit base), except in the case of high-peak loads which are specified in erms of peak load duty. (IA/IPC) 444-1973w

rated direct current current The rated dc current of a smoothing reactor is the maximum continuous dc current at rated conditions. (PE/TR) 1277-2000

rated direct current voltage The rated dc voltage of a smoothing reactor is the maximum continuous dc voltage, pole to ground, that will be experienced by the smoothing reactor. (PE/TR) 1277-2000

rated direct-current winding voltage (rectifier) The root-mean-square voltage of the direct-current winding obtained by turns ratio from the rated alternating-current winding voltage of the rectifier transformer. *See also:* rectifier transformer. (Std100) C57.18-1964w

rated direct voltage (power inverter) The nominal direct input voltage. *See also:* self-commutated inverters. (IA) [62]

rated duty That duty that the particular machine or apparatus has been designed to comply with. (Std100) 270-1966w

rated dynamic short circuit load current (thyristor) The maximum permissible peak transient current which can be supplied into a short circuited load. This is stated in terms of I^2t, number of cycles and maximum peak value. In general this places a constraint on the minimum source of impedance. (IA/IPC) 428-1981w

rated efficiency (thyristor) Rated output power divided by rated input power, generally expressed as percent. (IA/IPC) 428-1981w

rated excitation-system voltage (rotating machinery) The main exciter rated voltage. (PE) [9]

rated fault-closing current The highest rms total current, including the dc component, that the device shall be required to close at rated maximum voltage and rated frequency and carry for a specified time under specified conditions. (SWG/PE) C37.100-1992

rated field current (excitation systems for synchronous machines) The direct current in the field winding of the synchronous machine when operating at rated voltage, current, power factor, and speed. (PE/EDPG) 421.1-1986r

rated field voltage (excitation systems for synchronous machines) The voltage required across the terminals of the field winding of the synchronous machine under rated continuous load conditions of the synchronous machine with its field winding at 75°C for field windings designed to operate at rating with a temperature rise of 60°C or less; or 100°C for field windings designed to operate at rating with a temperature rise greater than 60°C. (PE/EDPG) 421.1-1986r

rated 15-cycle current (of a disconnecting device or assembly) (15-cycle current rating) The rms symmetrical current of an asymmetrical wave produced by a circuit having a prescribed X/R ratio, which the device or assembly is required to carry for 15 cycles. *Note:* This rating is an index of the ability of the disconnecting device to withstand heat that may be generated under short-circuit conditions. (SWG/PE) C37.40-1993

rated frequency (1) (converters having ac output) (self-commutated converters) (frequency range) The rated value of the fundamental frequency of the output voltage or the range over which the fundamental frequency may be adjusted. (IA/SPC) 936-1987w

(2) (A) (power system or interconnected system) The frequency used in the specification of apparatus upon which test conditions and frequency limits are based. **(B) (power system or interconnected system)** The system frequency at which a power system normally operates. (PE/PSE) 94-1991

(3) (arresters) The frequency, or range of frequencies, of the power systems on which the arrester is designed to be used. (Std100) [84]

(4) (grounding device) The frequency of the alternating current for which it is designed. *Note:* Some devices, such as neutral wave traps, may have two or more rated frequencies; the rated frequency of the circuit and the frequencies of the harmonic or harmonics the devices are designed to control. *See also:* grounding device. (PE/SPD) 32-1972r

(5) (frequency rating) (of a fuse) The system frequency for which it is designed. (SWG/PE) C37.40-1993

(6) The frequency of the alternating current for which the LTC is designed. (PE/TR) C57.131-1995

(7) The power frequency at which a device is designed to operate. (SWG/PE) C37.100-1992

rated fundamental output current (converters having ac output) (self-commutated converters) The fundamental output current specified by the manufacturer as a basis for rating. (IA/SPC) 936-1987w

rated head (1) (hydraulic turbines) The value stated on the turbine nameplate. (PE/EDPG) 125-1977s
(2) (power operations) The head at which a turbine operating at rated speed will deliver rated capacity at specified gate and efficiency. (PE/PSE) 858-1987s

rated high-frequency transient making current The peak value of the high-frequency current, with specified damping, against which a device is required to close and latch under specified conditions. (SWG/PE) C37.100-1992

rated ice-breaking ability The maximum thickness of ice deposited on the device that will not interfere with the successful opening or closing of a device. (SWG/PE) C37.30-1992s

rated impedance (loudspeaker measurements) The rated impedance of a loudspeaker driver or system is that value of a pure resistance which is to be substituted for the driver of system when measuring the electric power delivered from the source. This should be specified by the manufacturer. 219-1975w

rated impulse protective level (arresters) The impulse protective level with the residual voltage referred to the nominal discharge current. (PE) [8]

rated impulse withstand voltage (apparatus) An assigned crest value of a specified impulse voltage wave that the apparatus must withstand without flashover, disruptive discharge, or other electric failure. (PE) [8]

rated inductance (of a series reactor) The total installed inductance at a specified frequency. It may consist of mutual as well as self inductance components. (PE/TR) C57.16-1996

rated input power (1) (thyristor) The total real power at the lines of the controller at rated line current and voltage. (IA/IPC) 428-1981w
(2) The input power to the ferroresonant regulator with the rated load and under stated operating conditions. (PEL) 449-1998

rated input voltamperes (1) The input voltamperes to the ferroresonant regulator with the rated load and under stated operating conditions. (PEL) 449-1998
(2) (thyristor) The product of rated line voltage and current. (IA/IPC) 428-1981w

rated insulation class (neutral grounding devices) (electric power) An insulation class expressed in root-mean-square kilovolts, that determines the dielectric tests that the device shall be capable of withstanding. *See also:* grounding device; outdoor coupling capacitor. (SWG/PE) C37.60-1981r

rated internal pressure (power cable joints) The rated internal pressure of a joint is the nominal internal operating pressure. This will depend on the types of cable being joined ad the service conditions. (PE/IC) 404-1986s

rated interrupting current (rated interrupting capacity) (current interrupting rating) (of a fuse) The designated value of the highest available rms short-circuit current that the fuse is required to interrupt successfully under stated conditions. (SWG/PE) C37.40-1993

rated interrupting time (of a generator circuit breaker) The maximum permissible interval between the energizing of the trip circuit at rated control voltage and rated fluid pressure of the operating mechanism and the interruption of the main circuit in all poles on an opening operation. (SWG/PE) C37.013-1997

rated kilovolt-ampere (1) (current-limiting reactor) The kilovolt-amperes that can be carried for the time specified at rated frequency without exceeding the specified temperature limitations, and within the limitations of established standards. *See also:* reactor. C57.16-1958w
(2) (shunt reactors over 500 kVA) (of a shunt reactor) The apparent power at rated voltage for which the shunt reactor is designed. (PE/TR) C57.21-1981s
(3) (power and distribution transformers) (of a transformer) The output that can be delivered for the time specified at rated secondary voltage and rated frequency without exceeding the specified temperature-rise limitations under prescribed conditions. (PE/TR) C57.12.80-1978r
(4) (power and distribution transformers) (of a grounding transformer) The short-time kilovolt-ampere rating is the product of the rated line-to-neutral voltage at rated frequency, and the maximum constant current that can flow in the neutral for the specified time without causing specified temperature-rise limitations to be exceeded, and within the limitations of established standards for such equipment. (PE/TR) C57.12.80-1978r

rated kilowatts (power and distribution transformers) (of a constant-current transformer) The kilowatt output at the secondary terminals with rated primary voltage and frequency, and with rated secondary current and power factor, and within the limitations of established standards. (PE/TR) C57.12.80-1978r

rated kVA tap (power and distribution transformers) (in a transformer) A tap through which the transformer can deliver its rated kVA output without exceeding the specified temperature rise. (PE/TR) C57.12.80-1978r

rated life (1) (glow lamp) The length of operating time, expressed in hours, that produces specified changes in characteristics. *Note:* In lamps for indicator use the characteristic usually is light output; the end of usual life is considered to be when light output reaches 50% of initial, or when the lamp becomes inoperative at line voltage. In lamps used as circuit components, the characteristic is usually voltage; life is determined as the length of time for a specified change from initial. (EEC/EL) [104]
(2) (of a ballast or a lamp) The number of burning hours at which 50% of the units have burned out and 50% have survived. (IA/PSE) 241-1990r

rated line current (thyristor) Rated root-mean-square (rms) value of the current in the lines at rated controller current for the specified controller connection. (IA/IPC) 428-1981w

rated line frequency (thyristor) The frequency or range of frequencies at which the controller can operate. *Note:* Some wide ranges may require a derating curve to express this rating meaningfully. (IA/IPC) 428-1981w

rated line kilovoltampere rating (rectifier transformer) The kilovoltampere rating assigned to it by the manufacturer corresponding to the kilovoltampere drawn from the alternating-current system at rated voltage and kilowatt load on the rectifier under the normal mode of operation. *See also:* rectifier transformer. (Std100) C57.18-1964w

rated line voltage (thyristor) Rated root-mean-square (rms) value of the line voltage. (IA/IPC) 428-1981w

rated load (1) (elevators) The load which the device is designed and installed to lift at the rated speed. *See also:* elevator. (EEC/PE) [119]
(2) (rectifier unit) The kilowatt power output that can be delivered continuously at the rated output voltage. It may also be designated as the one-hundred-percent-load or full-load rating of the unit. *Note:* Where the rating of a rectifier unit does not designate a continuous load it is considered special. *See also:* continuous rating; rectification. (IA) [62]

rated load-break current (load break current rating) The designated value of the maximum rms current that a device having operable means for interrupting load currents is required to interrupt successfully under stated conditions when opened by manual or remote control means. (SWG/PE) C37.40-1993

rated-load current (air-conditioning equipment) The rated-load current for a hermetic refrigerant motor-compressor is the current resulting when the motor-compressor is operated at the rated load, rated voltage and rated frequency of the equipment it serves. (NEC/NESC) [86]

rated-load field voltage (rotating machinery) (nominal collector ring voltage) The voltage required across the terminals of the field winding of an electric machine under rated continuous-load conditions with the field winding at:

a) 75°C for field windings designed to operate at rating with a temperature rise of 60°C or less;
b) 100°C for field windings designed to operate at rating with a temperature rise greater than 60°C.

(PE) [9]

rated load power factor (thyristor) A range of load power factors over which a controller may be operated. (IA/IPC) 428-1981w

rated-load torque (rotating machinery) (rated torque) The shaft torque necessary to produce rated power output at rated-load speed. *See also:* asynchronous machine. (PE) [9]

rated load voltage (thyristor) The root-mean-square (rms) voltage delivered at the controller load terminals with rated line voltage and rated continuous controller current. (IA/IPC) 428-1981w

rated locked rotor current The steady state current taken from the line with the rotor locked and with rated voltage and rated frequency applied to the motor. (PE/NP) 1290-1996

rated making current (1) The maximum rms current against which the device is required to close successfully when switched from the open to the closed position. (SWG/PE) C37.40-1993
(2) The maximum current that the switch shall be required to close (initiate) and carry under specified conditions. For transient currents, fault initiation, capacitive discharge, etc., the rated making current shall be the prospective current available from the circuit without the influence of the switching device. (SWG/PE) C37.30-1997

rated maximum interrupting of main contacts voltage (field discharge circuit breakers) The maximum direct-current (dc) voltage, including voltage induced in the machine field by current in the machine armature, at which the field discharge circuit breaker main contacts are required to interrupt the excitation source current. The magnitude of the dc component of the total voltage across the main contacts is equal to the displacement of the axis. (SWG/PE) C37.100-1992, C37.18-1979r

rated maximum voltage (1) (maximum voltage rating) (high-voltage switchgear) The highest root-mean-square (rms) voltage at which the device is designed to operate. *Note:* This voltage corresponds to the maximum tolerable zone primary voltage at distribution transformers for distribution cutouts and single-pole air switches, and at substations and on transmission systems for power fuses given in ANSI C84.1-1977. (SWG/PE) C37.40-1981s, C37.60-1981r
(2) (of a generator circuit breaker) The highest rms voltage for which the circuit breaker is designed, and the upper limit for operation. The rated maximum voltage is equal to the maximum operating voltage of the generator to which the circuit breaker is applied. (SWG/PE) C37.013-1997

rated mechanical operations (high voltage air switches, insulators, and bus supports) The minimum number of operating cycles that an air switch can perform without requiring replacement of parts. (SWG/PE) C37.30-1971s

rated mechanical terminal load (high voltage air switches, insulators, and bus supports) The static force of conductors equivalent to the external mechanical load, applied at each terminal in specified directions, than an air switch can withstand. (SWG/PE) C37.30-1971s

rated mechanism fluid operating pressure (of a generator circuit breaker) The pressure at which a gas- or liquid-operated mechanism is designed to operate. (SWG/PE) C37.013-1997

rated minimum displacement factor (thyristor) The minimum ratio of input power to the input volt-amperes (at fundamental line frequency) at which a controller may be operated. (IA/IPC) 428-1981w

rated minimum interrupting current (high-voltage switchgear) The designated value of the smallest current that a fuse is required to interrupt at a designated voltage under prescribed conditions. (SWG/PE) C37.40-1981s

rated minimum tripping current (automatic circuit reclosers) The minimum rms current which causes a device to operate. (SWG/PE) C37.60-1981r

rated momentary current (1) (maximum voltage rating) (high-voltage switchgear) The maximum current measured at the major peak of the maximum cycle, which the device or assembly is required to carry. *Notes:* 1. The current is expressed as the root-mean-square (rms) value including the direct-current component, as determined from the envelope of the current wave by the method shown in Appendix A of IEEE Std C37.41-1981. 2. This rating is an index of the ability of the disconnecting device to withstand electromagnetic forces under short-circuit conditions. (SWG/PE) C37.40-1981s
(2) (of an air switch) The rms total current that a switch is required to carry for at least one cycle at rated frequency. (SWG/PE) C37.30-1992s

rated nominal voltage class (field discharge circuit breakers) The voltage to which operating and performance characteristics are referred. (SWG/PE) C37.100-1992, C37.18-1979r

rated nonrepetitive peak line voltage (thyristor) The maximum value of the transient peak instantaneous voltage, ULSM, appearing across the lines with the controller disconnected. (IA/IPC) 428-1981w

rated nonrepetitive peak OFF-state voltage (thyristor) The maximum instantaneous value of any nonrepetitive transient off-state voltage which may occur across the thyristor without damage. (IA/IPC) 428-1981w

rated OFF voltage (magnetic amplifier) The output voltage existing with trip off control signal applied. (MAG) 107-1964w

rated ON voltage (magnetic amplifier) The output voltage existing with trip on control signal applied. Rated on voltage shall be specified either as root-mean-square or average. *Note:* While specification may be either root-mean-square or average it remains fixed for a given amplifier. (MAG) 107-1964w

rated operating conditions (automatic null-balancing electric instrument) The limits of specified variables or conditions within which the performance ratings apply. *See also:* measurement system. (EEC/EMI) [112]

rated output (1) (self-commutated converters) (converters having ac output) The apparent output power for specified load conditions. (IA/SPC) 936-1987w
(2) (electrical heat tracing for industrial applications) Total wattage or watt/unit length of heating cable, at rated voltage, temperature and length. (BT/AV) 152-1953s
(3) Total power or power/unit length of heating cable, at rated voltage or current, maintain temperature, and length, normally expressed as W/m or W/ft. (IA/PC) 515.1-1995
(4) The total power or power/unit length of heating cable or surface heating device, at rated voltage, temperature, and length normally expressed as W/m (W/ft) or kW. (IA) 515-1997

rated output capacity (inverters) The kilovoltampere output at specified load power-factor conditions. *See also:* self-commutated inverters. (IA) [62]

rated output current (1) (converters having ac output) (self-commutated converters) The total rms (root-mean-square) output current specified by the manufacturer as a basis of declaring the duty cycles and overcurrent capability, and of selecting the conductor to the load. (IA/SPC) 936-1987w
(2) (magnetic amplifier) Rated output current that the amplifier is capable of supplying to the rated load impedance, either continuously or for designated operating intervals, under nominal conditions of supply voltage, supply frequency, and ambient temperature such that the intended life of the

amplifier is not reduced or a specified temperature rise is not exceeded. Rated output current shall be specified either as root-mean-square or average. *Notes:* 1. When other than rated load impedance is used, the root-mean-square value of the rated output current should not be exceeded. 2. While specification may be either root-mean-square or average, it remains fixed for a given amplifier. (MAG) 107-1964w

rated output frequency (inverters) The fundamental frequency or the frequency range over which the output fundamental frequency may be adjusted. *See also:* self-commutated inverters. (IA) [62]

rated output power (thyristor) The total real power available to the controller load at rated controller current and rated load voltage. (IA/IPC) 428-1981w

rated output voltage (magnetic amplifier) The voltage across the rated load impedance when rated output current flows. Rated output voltage shall be specified by the same measure as rated output current (that is, both shall be stated as root-mean-square or average). *Note:* While specification may be either root-mean-square or average, it remains fixed for a given amplifier. (MAG) 107-1964w

rated output voltamperes (magnetic amplifier) (thyristor) The product of the rated output voltage and the rated output current. (MAG/IA/IPC) 107-1964w, 428-1981w

rated output voltamperes of the ferroresonant regulator The sum of the rated output winding voltamperes under stated operating conditions. (PEL) 449-1998

rated output winding voltamperes The product of the output voltage and output current (root-mean-square values) at the rated load and under stated operating conditions. (PEL) 449-1998

rated peak single pulse transient current (low voltage varistor surge arresters) Maximum peak current which may be applied for a single 8 × 20-μs impulse, with rated line voltage also applied, without causing device failure. (PE) [8]

rated peak single-surge transient current The maximum peak current that may be applied for a single impulse (with rated line voltage also applied) without causing device failure. (SPD/PE) C62.62-2000

rated performance (automatic null-balancing electric instrument) The limits of the values of certain operating characteristics of the instrument that will not be exceeded under an combination of rated operating conditions. (EEC/EMI) [112]

rated permissible tripping delay Y (of a generator circuit breaker) The maximum time the circuit breaker is required to carry rated short-circuit current after closing on this current and before interrupting. (SWG/PE) C37.013-1993s

rated power output (hydraulic turbines) The value stated on the generator nameplate. (PE/EDPG) 125-1977s

rated primary current (current transformer) Current selected for the basis of performance specifications. *See also:* instrument transformer. (PE/PSR/TR) C37.110-1996, C57.13-1978s

rated primary line-to-ground voltage The root-mean-square line-to-ground voltage for which the potential device, in combination with its coupling capacitor or bushing, is designed to deliver rated burden at rated secondary voltage. The rated primary line-to-ground voltage is equal to the rated circuit voltage (line-to-line) divided by $(3)^{1/2}$. *See also:* primary line-to-ground voltage. 31-1944w

rated primary voltage (power and distribution transformers) (constant-voltage transformer) The voltage calculated from the rated secondary voltage by turn ratio. *Notes:* 1. See turn ratio of a transformer and its note, for the definition of the turn ratio to be used. 2. In the case of a multiwinding transformer, the rated voltage of any other winding is obtained in a similar manner. (PE/TR) C57.12.80-1978r
(2) (A) (instrument transformers) The rated primary voltage (of a potential (voltage) transformer) is the voltage selected for the basis of performance guarantees.
(B) (instrument transformers) The rated primary voltage (of a current transformer) designates the insulation class of the primary winding. *Note:* A current transformer can be applied on a circuit having a nominal system voltage corresponding to or less than the rated primary voltage of the current transformer. *See also:* instrument transformer. (PE/TR) C57.13-1978

rated primary voltage of a constant current transformer (power and distribution transformers) The primary voltage for which the transformer is designed, and to which operation and performance characteristics are referred. (PE/TR) C57.12.80-1978r

rated range of regulation of a voltage regulator The amount that the regulator will raise or lower its rated voltage. The rated range may be expressed in per unit, or in percent, of rated voltage, or it may be expressed in kilovolts. (PE/TR) C57.15-1999

rated reactance (of a series reactor) The product of rated inductance and rated angular frequency that provides the required reduction in fault current or other desired modification to power circuit characteristics. (PE/TR) C57.16-1996

rated recurrent peak voltage (low voltage varistor surge arresters) Maximum recurrent peak voltage which may be applied for a specified duty cycle and waveform. (PE) [8]

rated secondary current (1) (power and distribution transformers) (constant-voltage transformer) The secondary current obtained by dividing the rated kVA by the rated secondary voltage, kV. *Note:* The relationship above applies directly for single-phase transformers, but requires additional consideration of the connections involved in three-phase transformers. (PE/TR) C57.12.80-1978r
(2) (power and distribution transformers) The rated current divided by the marked ratio. (PE/TR/PSR) C57.13-1993, C57.12.80-1978r, C37.110-1996

rated secondary current of a constant-current transformer (power and distribution transformers) The secondary current for which the transformer is designed and to which operation and performance characteristics are referred. (PE/TR) C57.12.80-1978r

rated secondary voltage (1) (power and distribution transformers) (constant-voltage transformer) The voltage at which the transformer is designed to deliver rated kVA and to which operating and performance characteristics are referred. (PE/TR) C57.12.80-1978r
(2) (power and distribution transformers) (voltage transformer) The voltage divided by the marked ratio. (PE/TR) C57.12.80-1978r
(3) (capacitance potential devices) This is the root-mean-square secondary voltage for which the potential device, in combination with its coupling capacitor or bushing, is designed to deliver its rated burden when energized at rated primary line-to-ground voltage. *See also:* outdoor coupling capacitor; secondary voltage. 31-1944w
(4) The rated voltage divided by the marked ratio. (PE/TR) C57.13-1993

rated short-circuit withstand current The maximum rms total current that it can carry momentarily without electrical, thermal, or mechanical damage or permanent deformation. The current shall be the rms value, including the dc component, at the major peak of the maximum cycle as determined from the envelope of the current wave during a given test time interval. (SWG/PE) C37.100-1992, C37.23-1987r

rated short-term output current (converters having ac output) (self-commutated converters) The maximum output current that can be carried for a specified time without exceeding the established limitations under prescribed conditions of operation. (IA/SPC) 936-1987w

rated short-time current (1) (metal-enclosed bus and calculating losses in isolated-phase bus) (rated for isolated-phase bus) The maximum symmetrical current that the bus must carry without exceeding a specified total temperature in a given time interval. (SWG/PE) C37.23-1987r

(2) **(high-voltage switchgear) (short-time current rating)** (of a disconnecting device) The maximum root-mean-square (rms) total current (including the direct-current component) which the device is required to carry successfully for a specified short-time interval. *Note:* The ratings recognized the limitations imposed by both thermal and electromagnetic effects. (SWG/PE) C37.40-1981s

rated short-time of main contacts voltage (field discharge circuit breakers) The highest direct-current (dc) voltage at which the circuit breaker main contacts shall be required to interrupt exciter short-circuit current. (SWG/PE) C37.100-1992, C37.18-1979r

rated short-time overcurrent The rms power frequency current, of magnitude greater than the continuous current rating, that can be carried for a specified period of time and depending on the ambient temperature may result in defined loss of the reactor's service life. (PE/TR) C57.16-1996

rated short-time withstand current The maximum rms total current that a circuit breaker can carry momentarily without electrical, thermal, or mechanical damage or permanent deformation. The current shall be the rms value, including the dc component, at the major peak of the maximum cycle as determined from the envelope of the current wave during a given test time interval. *Synonyms:* withstand rating; short-time rating. (IA/PSP) 1015-1997

rated single pulse transient energy (low voltage varistor surge arresters) Energy which may be dissipated for a single impulse of maximum rated current at a specified waveshape, with rated root-mean-square (rms) voltage or rated direct-current (dc) voltage also applied, without causing device failure. (PE) [8]

rated single-surge transient energy Energy that may be dissipated in a surge-protective device for a single impulse of maximum rated current at a specified waveshape, with rated root-mean-square voltage or rated dc voltage also applied, without causing device failure. (SPD/PE) C62.62-2000

rated source impedance (thyristor) The equivalent impedance of the line voltage source, including the connections to the terminals of the converter. (IA/IPC) 428-1981w

rated speed (hydraulic turbines) The value stated on the unit nameplate. (PE/EDPG) 125-1977s

rated standby power dissipation The power dissipated in a protective device while connected to an ac line that has a voltage and frequency equal to the rating of the device and that has no load current flowing and no surges applied. (SPD/PE) C62.62-2000

rated start torque A conservative value established by the manufacturer for motor torque at the locked rotor (zero speed) condition taking into account uncertainties in the manufacturing process. It can be more than 20% less than locked rotor torque. (PE/NP) 1290-1996

rated supply current (magnetic amplifier) The root-mean-square current drawn from the supply when the amplifier delivers rated output current. (MAG) 107-1964w

rated supply voltage (converters having dc input) (self-commutated converters) The supply voltage specified by the manufacturer as a basis for rating. (IA/SPC) 936-1987w

rated step voltage For each value of rated through current, the highest permissible voltage between successive tap positions. *Note:* Step voltage of resistance-type LTCs means tap to tap voltage (no bridging position). (PE/TR) C57.131-1995

rated switching current—parallel-connected capacitance The rms symmetrical value of the highest parallel-connected capacitance load current in amperes that a device is required to make and interrupt a number of times equal to its operating life expectancy. (SWG/PE) C37.30-1992s

rated switching current—single capacitance The rms symmetrical value of the highest single capacitance load current in amperes that a device is required to make and interrupt a number of times equal to its operating life expectancy. (SWG/PE) C37.30-1992s

rated symmetrical interrupting current (accelerometer) The root-mean-square value of the symmetrical component of the highest current which a device is required to interrupt under the operating duty, rated maximum voltage, and circuit constants specified. (SWG/PE) C37.60-1981r

rated system deviation The specified maximum permissible carrier frequency deviation. Nominal values for mobile communications systems are ± 15 kHz or ± 5 kHz. (VT) 184-1969w

rated system frequency The frequency expressed in hertz, of the power system alternating voltage. (IA/ID) 995-1987w

rated system voltage (A) (current-limiting reactor) The voltage to which operations and performance characteristics are referred. It corresponds to the nominal system voltage of the circuit on which the reactor is intended to be used. *See also:* reactor. **(B) (synchronous motor drives)** The rms power-frequency voltage from line to line that has been designated as the basis for the system rating. The line-side interface equipments may or may not have the same rated voltage. (PE/IA/TR/ID) C57.16-1996, 995-1987

rated thermal current (neutral grounding devices) The root-mean-square neutral current in amperes which the device is rated to carry under standard operating conditions for rated time without exceeding temperature limits. (SPD/PE) 32-1972r

rated three-second current (high-voltage switchgear) (three-second current rating) The root-mean-square (rms) total current, including the direct-current component which the device, or assembly, is required to carry for three seconds. *Note:* For practical purposes, this current is measured at the end of the first second. This rating is an index of the ability of the disconnecting device to withstand heat that may be generated under short-circuit conditions. (SWG/PE) C37.40-1981s

rated through current The current flowing through the LTC towards the external circuit, which the apparatus is capable of transferring from one tap to another at the relevant rated step voltage, and which can be carried continuously while meeting the requirements of this standard. *Note:* Concerning the relationship between rated through current and the relevant step voltage. (PE/TR) C57.131-1995

rated time (grounding device) (electric power) The time during which it will carry its rated current, or withstand its rated voltage, or both, under standard conditions without exceeding standard limitations, unless otherwise specified. *See also:* grounding device. (PE/SPD) 32-1972r

rated-time temperature rise (grounding device) The maximum temperature rise above ambient attained by the winding of a device as the result of the flow of rated thermal current (or, for certain resistors, the maintenance of rated voltage across the terminals) under standard operating conditions, for rated time and with a starting temperature equal to the steady-state temperature. It may be expressed as an average or a hot-spot winding rise. (SPD/PE) 32-1972r

rated torque *See:* rated-load torque.

rated transient average power dissipation (low voltage varistor surge arresters) Maximum average power that may be dissipated due to a group of pulses occurring within a specified isolated time period, without causing device failure. (PE) [8]

rated transient inrush frequency The highest frequency of the transient inrush current of a designated operating duty. (SWG/PE) C37.100-1992

rated values (thyristor) A specified value for the electrical, thermal, mechanical, and environmental quantities assigned by the manufacturer to define the operating conditions under which a controller is expected to give satisfactory service. The rated values may change if the operating mode is different from that specified. *Note:* For calculating or measuring the root-mean-square (rms) value, several integration intervals are possible depending upon the operation mode. The interval used should be exactly specified. (IA/IPC/SPC) 428-1981w, 936-1987w

rated voltage (1) (electric submersible pump cable) The rated voltage is expressed in terms of phase-to-phase voltage of a three-phase system. (IA/PC) 1017-1985s
(2) (power and distribution transformers) The voltage to which operating and performance characteristics of apparatus and equipment are referred. *Note:* Deviation from rated voltage may not impair operation of equipment, but specified performance characteristics are based on operation under rated conditions. However, in many cases apparatus standards specify a range of voltage within which successful performance may be expected.
(BT/PE/AV/TR) 152-1953s, C57.12.80-1978r
(3) (power cable joints) The rated voltage of a joint is the voltage at which it is designed to operate under usual service conditions. Unless otherwise specified, the voltage rating is assigned with the understanding that the joint will be applied on the three-phase circuits whose nominal phase-to-phase voltage rating does not exceed that of the joint.
(PE/IC) 404-1986s
(4) (power cable systems) For cables, either single-conductor or multiple-conductor, the rated voltage is expressed in terms of phase-to-phase voltage of a three-phase system. For single-phase systems, a rated voltage of the square root of three times the voltage to ground should be assumed.
(PE/IC) 400-1991
(5) (rotating electric machinery) The voltage specified at the terminals of a machine. (PE/EM) 11-1980r
(6) (arresters) The designated maximum permissible root-mean-square value of power-frequency voltage between its line and earth terminals at which it is designed to operate correctly. (PE) [8]
(7) (grounding device) (electric power) The root-mean-square voltage, at rated frequency, that may be impressed between its terminals under standard conditions for its rated time without exceeding standard limitations, unless otherwise specified. (SPD/PE) 32-1972r
(8) The primary voltage upon which the performance specifications of a voltage transformer are based.
(PE/TR) C57.13-1993
(9) (shunt reactors over 500 kVA) (of a shunt reactor) The voltage to which operating and performance characteristics are referred. (PE/TR) C57.21-1981s
(10) (instrument transformers) (power and distribution transformers) (of a voltage transformer) The primary voltage selected for the basis of performance specifications of a voltage transformer. (PE/TR) C57.12.80-1978r, C57.13-1978s
(11) (of equipment, of a winding) The voltage to which operating and performance characteristics are referred.
(PE/TR) C57.15-1999, C57.12.80-1978r
(12) The voltage to which operating and performance characteristics of heating cables are referred. (IA) 515-1997

rated voltage adjustment range (thyristor) The range over which the steady state load voltage can be varied.
(IA/IPC) 428-1981w

rated voltage of a step-voltage regulator The voltage for which the regulator is designed and on which performance characteristics are based. (PE/TR) C57.15-1999

rated voltage of a winding The rated voltage of a winding is the voltage to which operating and performance characteristics are referred. (PE/TR) C57.15-1999

rated voltage of the series winding of a step-voltage regulator The voltage between terminals of the series winding, with rated voltage applied to the regulator, when the regulator is in the position that results in maximum voltage change and is delivering rated output at 80% lagging power factor.
(PE/TR) C57.15-1999

rated watts input (household electric ranges) The power input in watts (or kilowatts) that is marked on the range nameplate, heating units, etc. *See also:* appliance outlet.
(IA/APP) [90]

rated withstand current (surge arresters) (surge current) The crest value of a surge, of given wave shape and polarity, to be applied under specified conditions without causing disruptive discharge on the test specimen. (PE) [8]

rated withstand voltage (insulation strength) The voltage that electric equipment is required to withstand without failure or disruptive discharge when tested under specified conditions and within the limitations of established standards. *See also:* basic impulse insulation level. (EEC/LB) [100]

rate-grown junction (semiconductor) A grown junction produced by varying the rate of crystal growth. *See also:* semiconductor device. (ED) 216-1960w

rate gyro Generally, a single-degree-of-freedom gyro having a primarily elastic restraint of the spin axis about the output axis. In this gyro, an output signal is produced by precession of the gimbal, the precession angle being proportional to the angular rate of the case about the input axis.
(AES/GYAC) 528-1994

rate-integrating gyro A single-degree-of-freedom gyro having a primarily viscous restraint of the spin axis about the output axis. In this gyro, an output signal is produced by precession of the gimbal, the precession angle being proportional to the integral of the angular rate of the case about the input axis.
(AES/GYAC) 528-1994

rate-of-change protection A form of protection in which an abnormal condition causes disconnection or inhibits connection of the protected equipment in accordance with the rate of change of current, voltage, power, frequency, pressure, etc.
(SWG/PE/PSR) C37.100-1992, C37.90-1978s

rate-of-change relay A relay that responds to the rate of change of current, voltage, power, frequency, pressure, etc.
(SWG/PE/PSR) C37.100-1992, C37.90-1978s

rate of decay (audio and electroacoustics) The time rate at which the sound pressure level (or other stated characteristic) decreases at a given point and at a given time. *Note:* Rate of decay is frequently expressed in decibels per second.
(SP) [32]

rate of punching (test, measurement, and diagnostic equipment) Number of characters, blocks, words, or frames of information placed in the form of holes distributed on cards or tape per unit time. The number of cards punched per unit time.
(MIL) [2]

rate of reading (test, measurement, and diagnostic equipment) Number of cards, characters, blocks, words, or frames sensed by a sensing device per unit time. (MIL) [2]

rate-of-rise current tripping *See:* rate-of-rise release.

rate-of-rise detector (fire protection devices) A device that will response when the temperature rises at a rate exceeding a predetermined amount. (NFPA) [16]

rate-of-rise limiters (thyristor) Devices used to control the rate of rise of current or voltage to the semiconductor device, or both. *Note:* Current rate of rise limiters may include linear or nonlinear devices. (IA/IPC) 428-1981w

rate-of-rise of reapplied forward voltage (thyristor) The average slope of the reapplied voltage measured as the slope, of a line from the intersection of the voltage waveform and the axis, to the point where the waveform achieves 63% of the maximum forward OFF-state voltage.
(IA/IPC) 428-1981w

rate-of-rise of restriking voltage (surge arresters) (transient recovery voltage rate) The rate, expressed in volts per microsecond, that is representative of the increase of the restriking voltage. *Synonym:* rrrv. (PE) [8], [84]

rate-of-rise release (trip) A release that operates when the rate of rise of the actuating quantity in the main circuit exceeds the release setting. *Synonym:* rate-of-rise trip.
(SWG/PE) C37.100-1992

rate-of-rise suppressors (semiconductor rectifiers) Devices used to control the rate of rise of current and/or voltage to the semiconductor devices in a semiconductor power converter. *See also:* semiconductor rectifier stack. (IA) [62]

rate-of-rise trip *See:* rate-of-rise release.

rate ramp (gyros) A stochastic process whose sample time functions exhibit linear time growth, and whose growth rate is a random variable. (AES/GYAC) 528-1994

rate random walk *See:* random walk.

rate servomechanism (1) In an analog computer, a servomechanism in which a mechanical shaft is translated or rotated at a rate proportional to an input signal amplitude. (C) 610.10-1994w

(2) A servomechanism in which a mechanical shaft is translated or rotated at a rate proportional to an input signal amplitude. *See also:* electronic analog computer. (C) 165-1977w

rate signal (1) A signal that is the time derivative of a specified variable. *See also:* feedback control system. (IA/ICTL/IAC) [60]

(2) A signal that is responsive to the rate of change of an input signal. (PE/EDPG) 421-1972s

rate-squared sensitivity (nongyroscopic angular sensors) (angular accelerometers) An error torque about the input axis proportional to the product of input rates on the other two axes. This is analogous to anisoinertia torque. (AES/GYAC) 528-1994

rate, sweep *See:* sweep time division.

rate test *See:* problem check.

rating (1) (rating of electric equipment) (general) The whole of the electrical and mechanical quantities assigned to the machine, apparatus, etc., by the designer, to define its working in specified conditions indicates on the rating plate. *Note:* The rating of electric apparatus in general is expressed in voltamperes, kilowatts, or other appropriate units. Resistors are generally rated in ohms, amperes, and class of service. (Std100/EI) 96-1969w, [123]

(2) (rotating machinery) The numerical values of electrical quantities (frequency, voltage, current, apparent and active power, power factor) and mechanical quantities (power, torque), with their duration and sequences, that express the capability and limitations of a machine. The rated values are usually associated with a limiting temperature rise of insulation and metallic parts. *See also:* asynchronous machine. (PE) [9]

(3) The designated limit(s) of the rated operating characteristic(s) of a device. *Note:* Such operating characteristics as current, voltage, frequency, etc., may be given in the rating. (SWG/PE/PSR) C37.40-1981s, C37.100-1992, C37.90-1978s

(4) (current-limiting reactor) The voltamperes that it can carry, together with any other characteristics, such as system voltage, current and frequency assigned to it by the manufacturer. *Note:* It is regarded as a test rating that defines an output that can be carried under prescribed conditions of test, and within the limitations of established standards. *See also:* reactor. C57.16-1958w

(5) (A) (interphase transformer) The root-mean-square current, root-mean-square voltage, and frequency at the terminals of each winding, when the rectifier unit is operating at rated load and with a designated amount of phase control. *See also:* rectifier transformer; duty. **(B) (rectifier transformer)** The kilovoltampere output, voltage, current, frequency, and number of phases at the terminals of the alternating-current winding; the voltage (based on turn ratio of the transformer), root-mean-square current, and number of phases at the terminals of the direct-current winding, to correspond to the rated load of the rectifier unit. *Notes:* 1. Because of the current wave shapes in the alternating- and direct-current windings of the rectifier transformer, these windings may have individual ratings different from each other and from those of power transformers in other types of service. The ratings are regarded as test ratings that define the output that can be taken from the transformer under prescribed conditions of test without exceeding any of the limitations of the standards. 2. For rectifier transformers covered by established standards, the root-mean-square current ratings and kilovoltampere ratings of the windings are based on values derived from rectangular rectifier circuit element currents without overlap. *See also:* rectifier transformer. **(C) (power and distribution transformers)** The rating of a transformer consists of a volt-ampere output together with any other characteristics, such as voltage, current, frequency, power factor, and temperature rise, assigned to it by the manufacturer. It is regarded as a rating associated with an output which can be taken from the transformer under prescribed conditions and limitations of established standards. (PE/TR) C57.12.80-1978

(6) A voltampere output together with any other characteristics, such as voltage, current, frequency, powerfactor, and temperature rise, assigned to it by the manufacturer. (PE/TR) C57.12.80-1978r

(7) (rotating electric machinery) The output at the shaft if a motor, or at the terminals if a generator, assigned to a machine under specified conditions of speed, voltage, temperature rise, etc. *See also:* relay rating. (PE/EM) 11-1980r

(8) The designation of an operating limit for a device. (SPD/PE) C62.11-1999, C62.62-2000

(9) (A) (in kVA of a voltage regulator) The rating that is the product of the rated load amperes and the rated "raise" or "lower" range of regulation in kilovolts (kV). If the rated raise and lower range of regulation are unequal, the larger shall be used in determining the rating in kVA. **(B)** (in kVA of a voltage regulator) The rating in kVA of a three-phase voltage regulator is the product of the rated load amperes and the rated range of regulation in kilovolts multiplied by 1.732. (PE/TR) C57.15-1999

rating, emergency *See:* emergency rating.

rating, normal *See:* normal rating.

rating of a series reactor The rating of a series reactor consists of the current that it can carry at its specified reactance together with any other defining characteristics, such as system voltage, BIL, duty and frequency. (PE/TR) C57.16-1996

rating of diesel-generator unit (A) (nuclear power generating station) Continuous rating. The electric power output capability that the diesel-generator unit can maintain in the service environment for 8760 h of operation per (common) year with only scheduled outages for maintenance. **(B) (nuclear power generating station)** Short time rating. The electric power output capability that the diesel-generator unit can maintain in the service environment for 2 h in any 24 h period, without exceeding the manufacturer's design limits and without reducing the maintenance interval established for the continuous rating. *Note:* Operation at this higher rating does not limit the use of the diesel-generator unit at its continuous rating. (PE/NP) 387-1984

rating of interphase transformer (power and distribution transformers) The root-mean-square current, root-mean-square voltage, and frequency at the terminals of each winding, when the rectifier unit is operating at rated load and with a designated amount of phase control. (PE/TR) C57.12.80-1978r

rating plate *See:* nameplate.

rating plug An interchangeable module of an electronic trip unit that, together with the sensor, sets the current rating range of the circuit breaker. For example, a 1200 A frame may contain an 800 A sensor, fixing the maximum rating that can be configured for the unit at 800 A adjustable by the following kind of settings. By installing a 600 A rating plug, the adjustable rating is correspondingly 600 A multiplied by the long-time pickup adjustment [i.e., the long-time pickup may be adjusted to "0.9" and the ampere rating or setting is $(0.9 \times 600\ \text{A}) = 540\ \text{A}$]. (IA/PSP) 1015-1997

ratio *See:* squareness ratio.

ratio control system (automatic control) A system that maintains two or more variables at a predetermined ratio. *Note:* Frequently some function of the value of an uncontrolled variable is the command to a system controlling another variable. (PE/EDPG) [3]

ratio correction factor (RCF) The ratio of the true ratio to the marked ratio. The primary current or voltage is equal to the secondary current or voltage multiplied by the marked ratio times the ratio correction factor. (PE/TR) C57.13-1993, C57.12.80-1978r

ratio meter An instrument that measures electrically the quotient or two quantities. A ratio meter generally has no mechanical control means, such as springs, but operates by the balancing of electromagnetic forces that are a function of the position of the moving element. *See also:* instrument. (EEC/PE) [119]

Rational FORTRAN An extension of FORTRAN that provides free format coding (as opposed to FORTRAN's strict column format), source file inclusion, and block structures. (C) 610.13-1993w

rationalized system of equations A rationalized system of electrical equations is one in which the proportionality factors in the equations that relate (A) the surface integral of electric flux density to the enclosed charge, and (B) the line integral of magnetizing force to the linked current, are each unity. *Notes:* 1. By these choices, some formulas applicable to configuration having spherical or circular symmetry contain an explicit factor of 4π or 2π; for example, Coulomb's law is $f = q_1q_2/(4\pi\ \epsilon_0 r^2)$. 2. The differences between the equations of a rationalized system and those of an unrationalized system may be considered to result from either (a) the use of a different set of units to measure the same quantities or (b) the use of the same set of units to measure quantities that are quantitatively different (though of the same physical nature) in the two systems. The latter consideration, which represents a changed relation between certain mathematicophysical quantities and the associated physical quantities, is sometimes called total rationalization. (Std100) 270-1966w

rational number (data management) A real number that can be expressed as a fraction x/y where x and y are integers and y is not equal to zero. *Contrast:* irrational number. (C) 610.5-1990w

ratio set (valve actuators) A set of performance parameter values described by a range of numerical values whoseboundaries have been established by doubling and halving the numerical mean value of a selected physical performance parameter. (PE/NP) 382-1985

ratio-type telemeter A telemeter that employs the relative phase position between, or the magnitude relation between, two or more electrical quantities as the translating means. *Note:* Examples of ratio-type telemeters include ac or dc position matching systems. *Synonym:* position-type telemeter. (SWG/PE/SUB) C37.100-1992, C37.1-1994

ratproof electric installing Apparatus and wiring designed and arranged to eliminate harborage and runways for rats. *See also:* marine electric apparatus. (EEC/PE) [119]

raw data Data that has not been processed or reduced from its original form. (C) 610.5-1990w

raw requirement An environmental or customer requirement that has not been analyzed and formulated as a well-formed requirement. (C/SE) 1233-1998

ray The path of a wave packet or energy flow in a homogeneous or a slowly varying medium. *Notes:* 1. Energy transport (per unit area) is generally associated with bundles of rays. 2. In isotropic but slowly varying media, the ray path is identical to the path of the wave normal, but this may not be the case in anisotropic media. (AP/PROP) 211-1997

Ray Dist (navigation aid terms) A radio navigation system used in hydrographic and geophysical surveying. (AES/GCS) 172-1983w

Rayleigh criterion A criterion that characterizes the roughness of a surface with respect to the reflection of an electromagnetic wave. The degree of roughness is expressed in terms of the quantity:

$$\frac{h\cos\theta}{\lambda}$$

where
h = the rms height of the surface irregularities
θ = the angle of incidence with respect to the mean surface
λ = the wavelength

The surface is considered specular (smooth) if:

$$\frac{h\cos\theta}{\lambda} < \frac{1}{100}$$

The surface is considered rough if:

$$\frac{h\cos\theta}{\lambda} > \frac{1}{10}$$

(AP/PROP) 211-1997

Rayleigh density function (radar) A probability density function describing the behavior of some variable, given by

$$f(X) = \frac{1}{\sigma_{AVG}} \exp - \left(\frac{X}{\sigma_{AVG}}\right)$$

Often used to describe the signal statistics after envelope detection. (AES/RS) 686-1982s

Rayleigh disk A special form of acoustic radiometer that is used for the fundamental measurement of particle velocity. (SP) [32]

Rayleigh distribution A probability distribution characterized by the probability density function:

$$\begin{aligned} f(x) &= \frac{x}{\sigma^2} \exp\left(-\frac{x^2}{2\sigma^2}\right), \quad x \geq 0 \\ &= 0, \qquad\qquad\qquad\qquad x < 0 \end{aligned}$$

where
x = the random variable
σ^2 = the average value of x^2

Notes: 1. Named after Lord Rayleigh. 2. This function is often used to model the statistics of the amplitude of noise at intermediate frequency (IF) or at video after linear envelope detection. 3. The Rayleigh distribution function also describes the distribution of the signal voltage of fluctuating signals consisting of four or more components having similar amplitudes and random phases. (AES) 686-1997

Rayleigh fading Signal level variations when the received wave is composed of numerous scattered waves with uniform relative phase distribution. (AP/PROP) 211-1997

Rayleigh hypothesis The Rayleigh hypothesis is an assumption that only outgoing waves exist everywhere above a rough interface, including the trough regions. (AP/PROP) 211-1997

Rayleigh scattering (1) (fiber optics) Light scattering by refractive index fluctuations (inhomogeneities in material density or composition) that are small with respect to wavelength. The scattered field is inversely proportional to the fourth power of the wavelength. *See also:* waveguide scattering; material scattering; scattering. (Std100) 812-1984w
(2) (laser maser) Scattering of radiation in the course of its passage through a medium containing particles, the sizes of which are small compared with the wavelength of the radiation. (LEO) 586-1980w
(3) Scattering by dielectric particles much smaller than a wavelength. For the special case of spherical particles in the Rayleigh scattering limit, the scattering cross-section is inversely proportional to the fourth power of the wavelength and directly proportional to the sixth power of the particle diameter. (AP/PROP) 211-1997

ray tracing A technique for displaying a three-dimensional object with shading and shadows, by tracing light rays backward from the viewing position to the light source, on a two-dimensional display surface. (C) 610.6-1991w

RBOC *See:* regional Bell operating company.

RCF *See:* ratio correction factor.

RCL *See:* Rule and Constraint Language.

RC integrator A low-pass electrical filter section consisting of a resistor in series with the signal path followed by a capacitor across it. (NPS) 325-1996

RDA *See:* reflective dot array.

RCS *See:* radar cross section.

RDF *See:* radio direction-finder.

RDL *See:* Resource Description Language.

R-display (1) (navigation aid terms) A type of radar display format. *See also:* display; display; display. (AES/GCS) 172-1983w
(2) An A-display with a segment of the time base expanded near the blip for greater precision in range measurement and visibility of pulse shape. *Note:* Usually regarded as an optional feature of an A-display rather than being identified by the term "R-display." (AES) 686-1997

reach (1) (protective relaying) The maximum distance from the relay location to a fault for which a particular relay will operate. The reach may be stated in terms of miles, primary ohms, or secondary ohms. (PE/PSR) C37.95-1973s
(2) (of a relay) The extent of the protection afforded by a relay in terms of the impedance or circuit length as measured from the relay location. *Note:* The measurement is usually to a point of fault, but excessive loading or system swings may also come within reach or operating range of the relay. (SWG/PE) C37.100-1992

reactance (1) (general) The imaginary part of impedance. *See also:* reactor. (IM/HFIM) 270-1966w, [40]
(2) (of a series reactor) The product of the inductance in Henries and the angular frequency of the system. (PE/TR) C57.16-1996
(3) The imaginary component ($\pm$ j, or $\pm$ 90° phase angle) of impedence, where resistance is the real component with a zero phase angle. Reactance appears in two forms, one as a capacitive reactance (Xc) for a capacitor (C), and the other as inductive reactance (X_l) for an inductor (L). The former has the negative 90° phase angle, and the latter a positive 90° angle. Their values can be expressed in the following relationships:

$$Xc = -j\frac{1}{2\pi fC}, \text{ and } X_1 = j2\pi fL,$$

where f is the excitation frequency. (IA/MT) 45-1998

reactance amplifier *See:* parametric amplifier.

reactance characteristic A nondirectional distance relay characteristic in which the threshold of operation for the basic form plots as a straight line on an *R-X* diagram, with the reach a constant reactance for all values of resistance. See figure below. *Note:* A small variation in the reactance reach for different values of resistance, as required in some applications, may also be referred to as a reactance characteristic.

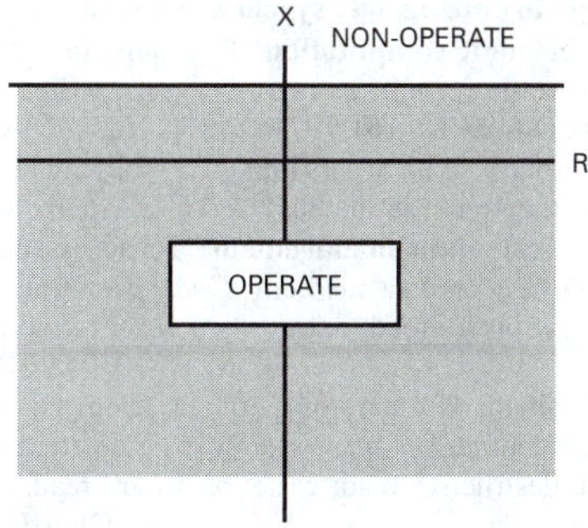

reactance characteristic

(SWG/PE) C37.100-1992

reactance drop (1) (power and distribution transformers) The component of the impedance voltage drop in quadrature with the current. (PE/TR) C57.12.80-1978r
(2) (general) The voltage drop in quadrature with the current. (PE/TR) C57.15-1968s

reactance, effective synchronous *See:* effective synchronous reactance.

reactance frequency multiplier A frequency multiplier whose essential element is a nonlinear reactor. *Note:* The nonlinearity of the reactor is utilized to generate harmonics of a sinusoidal source. *See also:* parametric device. (ED) 254-1963w, [46]

reactance function (1) (linear passive networks) The driving-point impedance of a lossless network. *Note:* This is an odd function of the complex frequency. (CAS) 156-1960w
(2) A function that is realizable as a driving-point impedance with ideal inductors and capacitors. It must meet the conditions described in Foster's reactance theorem. (CAS) [13]

reactance grounded (power and distribution transformers) Grounded through impedance, the principal element of which is reactance. *Note:* The reactance may be inserted either directly, in the connection to ground, or indirectly, by increasing the reactance of the ground return circuit. The latter may be done by intentionally increasing the zero-sequence reactance of apparatus connected to ground, or by omitting some of the possible connections from apparatus neutrals to ground. (SPD/PE/TR) 32-1972r, C57.12.80-1978r

reactance modulator A device, used for the purpose of modulation, whose reactance may be varied in accordance with the instantaneous amplitude of the modulating electromotive force applied thereto. *Note:* Such a device is normally an electron-tube circuit and is commonly used to effect phase or frequency modulation. *See also:* phase modulation; frequency modulation; modulation. (AP/BT/ANT) 145-1983s, 182A-1964w

reactance relay A linear-impedance form of distance relay for which the operating characteristic of the distance unit on an *R-X* diagram is a straight line on constant reactance. *Note:* The operating characteristic may be described by either equation $X = K$, or $Z\sin\theta = K$, where K is a constant, and θ is the angle by which the input voltage leads the input current. (SWG/PE) C37.100-1992

reactance voltage drop (1) The component of voltage drop in quadrature with the current. (PE/TR) C57.16-1996
(2) The component of the impedance voltage in quadrature with the current. (PE/TR) C57.15-1999

reaction curve (process control) The plot of a time response. (PE/EDPG) [3]

reaction frequency meter *See:* absorption frequency meter.

reaction time (illuminating engineering) The interval between the beginning of a stimulus and the beginning of the response of an observer. (EEC/IE) [126]

reaction torque A torque (or force) exerted on a gimbal, gyro rotor, or accelerometer proof mass, usually as a result of applied electrical excitations exclusive of torquer (or forcer) command signals. (AES/GYAC) 528-1994

reaction turbine A turbine that uses the velocity and pressure of the water flowing through the runner to develop power. (PE/EDPG) 1020-1988r

reaction wheels (communication satellite) A set of gyro wheels used for controlling the attitude of a satellite. (COM) [19]

reactivation date (electric generating unit reliability, availability, and productivity) The date a unit was returned to the active state from the deactivated shutdown state. (PE/PSE) 762-1987w

reactive Process by which a material undergoes a chemical or physical change. (PE/IC) 848-1996

reactive attenuator (waveguide) An attenuator that absorbs no energy. *See also:* waveguide. (AP/ANT) [35], [84]

reactive current (rotating machinery) The component of a current in quadrature with the voltage. *See also:* asynchronous machine. (PE) [9], [84]

reactive-current compensator (rotating machinery) A compensator that acts to modify the functioning of a voltage regulator in accordance with reactive current. (PE) [9]

reactive factor The ratio of the reactive power to the apparent power. The reactive factor is expressed by the equation

$$F_q = \frac{Q}{U}$$

where
F_q = reactive factor
Q = reactive power
U = apparent power

If the voltages have the same waveform as the corresponding currents, reactive factor becomes the same as phasor reactive factor. If the voltages and currents are sinusoidal and for polyphase circuits form symmetrical sets

$F_q = \sin(\alpha - \beta)$

See also: network analysis. (Std100) 270-1966w

reactive-factor meter An instrument for measuring reactive factor. It is provided with a scale graduated in reactive factor. *See also:* instrument. (EEC/PE) [119]

reactive field (of an antenna) Electric and magnetic fields surrounding an antenna and resulting in the storage of electromagnetic energy rather than in the radiation of electromagnetic energy. (AP/ANT) 145-1993

reactive ignition cable High-tension ignition cable, the core of which is so constructed to give a high reactive impedance at radio frequencies. *See also:* electromagnetic compatibility. (EMC/INT) [53], [70]

reactive near-field region That portion of the near-field region immediately surrounding the antenna, wherein the reactive field predominates. *Note:* For a very short dipole, or equivalent radiator, the outer boundary is commonly taken to exist at a distance $\lambda / 2\pi$ from the antenna surface, where λ is the wavelength. (AP/ANT) 145-1983s

reactive power (1) (metering) For sinusoidal quantities in a two-wire circuit, reactive power is the product of the voltage, the current, and the sine of the phase angle between them. For nonsinusoidal quantities, it is the sum of all harmonic components, each determined as above. In a polyphase circuit, it is the sum of the reactive powers of the individual phases. *See also:* magner. (ELM) C12.1-1982s

(2) The product of voltage and out-of-phase component of alternating current. In a passive network, reactive power represents the alternating exchange of stored energy (inductive or capacitive) between two areas. *See also:* magner. (PE) [9]

(3) (control of small hydroelectric plants) Power that is in quadrature with real power, such as used by capacitive or inductive loads, expressed in kvar. (PE/EDPG) 1020-1988r

(4) (electrical measurements in power circuits) The square root of the square of the apparent power S minus the square of the active power P.

$Q = (S^2 - P^2)^{1/2}$

Reactive power is developed when there are inductive, capacitive, or nonlinear elements in the system. It does not represent useful energy that can be extracted from the system but it can cause increased losses and excessive voltage peaks. (PE/PSIM) 120-1989r

(5) The vector difference between apparent and real power:

$Q = \sqrt{S^2 - p^2}$

(PEL) 1515-2000

reactive power relay A power relay that responds to reactive power. (SWG/PE) C37.100-1992

reactive reflector antenna *See:* reflective array antenna.

reactive voltampere-hour meter *See:* varhour meter.

reactive voltampere meter *See:* varmeter.

reactivity (power operations) A measure of the departure of a nuclear reactor from criticality. Mathematically,

$\rho = (k_{\text{eff}} - 1) + K_{\text{eff}}$

If ρ is positive (excess reactivity), the reactor is supercritical and its power level is increasing. If ρ is negative (negative reactivity), the power level of the reactor decreases. For a reactor at criticality, for instance, constant power level, the reactivity is zero. *Note:* Other measures are also used to express reactivity. *See also:* excess reactivity. (PE/PSE) 858-1987s

reactor (1) (power and distribution transformers) An electromagnetic device, the primary purpose of which is to introduce inductive reactance into a circuit. (PE/TR) C57.12.80-1978r

(2) (radiological monitoring instrumentation) A nuclear reactor designed for and capable of operation at a steady state reactor power level of ≥ R $1MW_{th}$. (NI) N320-1979r

(3) (shunt reactors over 500 kVA) A device used for introducing impedance into an electric circuit, the principal element of which is inductive reactance. (PE/TR) C57.21-1981s

(4) A device with the primary purpose of introducing reactance into an electric circuit for purposes such as motor starting, paralleling transformers, and control of current. (IA/MT) 45-1998

reactor, ac *See:* ac reactor.

reactor, amplistat *See:* amplistat reactor.

reactor, bus *See:* bus reactor.

reactor, current-balancing *See:* current-balancing reactor.

reactor, current-limiting *See:* current-limiting reactor.

reactor, dc *See:* direct-current reactor.

reactor, diode-current-balancing *See:* diode-current-balancing reactor.

reactor facility (radiological monitoring instrumentation) The structures, systems and components used for the operation of a nuclear reactor. If a site contains more than one nuclear reactor, reactor facility means all structures, systems and components used for operation of the nuclear reactors at the site. (NI) N320-1979r

reactor, feeder *See:* feeder reactor.

reactor, filter *See:* filter reactor.

reactor, paralleling *See:* paralleling reactor.

reactor starting (rotating machinery) The process of starting a motor at reduced voltage by connecting it initially in series with a reactor (inductor) which is short-circuited for the running condition. (PE) [9]

reactor, starting *See:* starting reactor.

reactor-start motor A single-phase induction motor of the split-phase type with a main winding connected in series with a reactor for starting operation and an auxiliary winding with no added impedance external to it. For running operation, the reactor is short-circuited or otherwise made ineffective, and the auxiliary winding circuit is opened. *See also:* asynchronous machine. (PE) [9]

reactor, synchronizing *See:* synchronizing reactor.

read (1) (electronic computation) To acquire information usually from some form of storage. *See also:* write; destructive read. (ED/C/MIL) 161-1971w, [85], 162-1963w, [20], [2]

(2) The process of an access unit (AU) copying bits of a data stream as they pass on the bus. (LM/C) 8802-6-1994

(3) (software) (data management) To access data from a storage device or data medium. *Contrast:* write. *See also:* update; destructive read; delete; dirty read; retrieve. (C) 610.5-1990w, 610.12-1990

(4) To obtain data from a storage device, from a data medium, or another source. *See also:* scatter read; delete; nondestructive read; destructive read; write; backward read; read cycle; read/write. (C) 610.10-1994w

readability The ease with which words and text can be read. Refers specifically to the functional relationships that exist between the properties of words and text and the observer's accuracy and speed of understanding words or text. (PE/NP) 1289-1998

read-around number (storage tubes) The number of times reading operations are performed on storage elements adjacent to any given storage element without more than a specified loss of information from that element. *Note:* The sequence of operations (including priming, writing, or erasing), and the storage elements on which the operations are performed, should be specified. *See also:* storage tube. (ED) 158-1962w, 161-1971w

read-around ratio* (1) (FASTBUS acquisition and control) A cycle in which the direction of data flow is from slave(s) toward a master. *Synonym:* read. (NID) 960-1986s
(2) (VMEbus) A data transfer bus (DTB) cycle that is used to transfer 1, 2, 3, or 4 bytes from a slave to a master. The cycle begins when the master broadcasts an address and an address modifier. Each slave captures the address and the address modifier, and verifies if it will respond to the cycle. If it is intended to respond, it retrieves the data from its internal storage, places it on the data bus and acknowledges the transfer. The master then terminates the cycle. (BA/C) 1014-1987
(3) *See also:* read-around number.

* Deprecated.

read-back check *See:* echo check.

read circuitry That part of the memory that is used in transferring the stored data from the memory section to external circuitry. (ED) 1005-1998

read cycle (1) A data transfer bus (DTB) cycle that is used to transfer 1, 2, 3, or 4 bytes from a slave to a master. The cycle begins when the master broadcasts an address and an address modifier. Each slave captures the address and the address modifier, and verifies if it will respond to the cycle. If it is intended to respond, it retrieves the data from its internal storage, places it on the data bus, and acknowledges the transfer. The master then terminates the cycle. (C/BA) 1014-1987
(2) (read) A cycle in which the direction of data flow is from slave(s) toward a master. (NID) 960-1993
(3) A cycle in which data are transferred from some storage location to the device that requested the read. *Contrast:* write cycle. (C) 610.10-1994w

read cycle time The minimum time interval between the starts of successive read cycles in a storage device that has separate read and write cycles. *Contrast:* write cycle time. (C) 610.10-1994w

read data transfer One or more data transfers from a replying agent to a bus owner, with uninterrupted bus ownership. (C/MM) 1296-1987s

read delay trd (metal-nitride-oxide field-effect transistor) Time period between the end of the writing pulse and the start of the read condition. (ED) 581-1978w

read disturb (1) (metal-nitride-oxide field-effect transistor) A change in the instantaneous threshold voltage of a metal-nitride-oxide-semiconductor (MNOS) transistor due to the very act of measuring it. (ED) 581-1978w
(2) The corruption of data that is caused by reading the memory. (ED) 1005-1998

read disturb cycles The number of consecutive read cycles that occur before a memory state becomes indistinguishable, due solely to reading. (ED) 641-1987w

reader (1) (A) An input device that is capable of sensing stored information, and of conveying that information into on-line storage. **(B)** Any device which can sense, detect, or convert data from one medium to another. *See also:* character reader; paper tape reader; optical mark reader; badge reader; card reader. (C) 610.10-1994
(2) A component of a roadside beacon that provides the capabilities for radio wave communications with a transponder. (SCC32) 1455-1999

reader note A comment made by a reader about an diagram and placed on the diagram page. A reader note is not part of the diagram itself, but rather is used for communication about a diagram during model development. (C/SE) 1320.1-1998

read frame The transfer of data from a Smart Transducer Interface Module to a Network Capable Application Processor. (IM/ST) 1451.2-1997

read head (1) (test, measurement, and diagnostic equipment) A sensor that converts information stored on punched tape, magnetic tape, magnetic drum, and so forth into electrical signals. (MIL) [2]
(2) A head capable only of reading information from the storage medium. *Synonym:* playback head. *Contrast:* write head; read/write head. *See also:* pre-read head. (C) 610.10-1994w

readily accessible (packaging machinery) (power and distribution transformers) Capable of being reached quickly for operation, renewal, or inspections, without requiring those to whom ready access is requisite to climb over or remove obstacles or to resort to portable ladders, chairs, etc. (NEC/NESC/IA/PE/PKG/TR) 333-1980w, C57.12.80-1978r, [86]

readily climbable Having sufficient handholds and footholds to permit an average person to climb easily without using a ladder or other special equipment. (NESC) C2-1997

readiness test (test, measurement, and diagnostic equipment) A test specifically designed to determine whether an equipment or system is operationally suitable for a mission. (MIL) [2]

reading (1) (recording instrument) The value indicated by the position of the index that moves over the indicating scale. *See also:* accuracy rating. (EEC/PE) [119]
(2) (radiation instrumentation) The indicated value of the readout. (NI) N42.17B-1989r

reading rate (storage tubes) The rate of reading successive storage elements. *See also:* storage tube. (ED) 158-1962w

reading speed (storage tubes) *See also:* storage tube; data processing. (ED) 158-1962w, 161-1971w

reading speed, minimum usable *See:* minimum usable reading speed.

reading time (storage tubes) The time during which stored information is being read. *See also:* storage tube. (ED) 158-1962w

reading time, maximum usable *See:* maximum usable reading time.

read-in lag (diode-type camera tube) The fraction of the steady-state ON signal that is read out in any field after initiation of irradiance. (ED) 503-1978w

read-modify-write cycle A data transfer bus (DTB) cycle that is used to both read from, and write to, a slave location without permitting any other master to access that location. This cycle is most useful in multiprocessing systems where certain memory locations are used to provide semaphore functions. (C/BA) 1014-1987

read-modify-write (RMW) cycle A cycle in which an item is read, its contents are modified, and then is written back to storage in a single operation. *See also:* write cycle; read cycle. (C) 610.10-1994w

read-mostly devices (metal-nitride-oxide field-effect transistor) Metal-nitride-oxide semiconductor (MNOS) memory transistors whose retention under constant read condition is in excess of one year. This makes these devices applicable in electrically-alterable read-only memories (EAROMs). A typical writing pulse width is one ms. (ED) 581-1978w

read number, maximum usable *See:* maximum usable read number.

read-only (1) Pertaining to a storage medium which can only be read from. *Contrast:* write-once/read-many; read/write. (C) 610.10-1994w
(2) A property that causes no state changes, i.e., it does no updates. (C/SE) 1320.2-1998

read-only access A type of access to data in which the data may be read but not changed or deleted. *Synonym:* fixed. *Contrast:* read/write access. *See also:* update access; delete access; write access. (C) 610.5-1990w

read-only file system A file system that has implementation-defined characteristics restricting modifications. (C/PA) 9945-1-1996, 9945-2-1993, 1003.5-1999

read-only memory (ROM) **(1)** The memory on a node that provides storage locations for normally read-only data or code. The ROM data are maintained across losses of primary and secondary power. In some implementations ROM may be writable, using (normally disabled) vendor-specific protocols. (C/MM) 1596-1992

(2) A form of nonvolatile memory whose contents are generally supplied during manufacture and cannot be altered. (C/BA) 14536-1995

(3) Memory that can only be read from. (C) 610.10-1994w

(4) A memory in which the contents are intended to be read only and not altered during normal operation. (ED) 1005-1998

(5) A type of permanent data storage (memory) that can be read but not altered by the system. Data stored in read-only memory is not affected by power loss to the system. (PE/SUB) 1379-1997

read-only storage A type of storage which can be read, but not modified except by a particular user, or when operating under particular conditions; for example, punched paper tape, or a storage device in which writing is prevented by a lock-out. *Synonym:* nonerasable storage. *See also:* fixed-program read-only storage; control read-only memory; fixed storage; protected storage. (C) 610.10-1994w

readout (radiation instrumentation) The device that conveys visual information regarding the measurement to the user. (NI) N42.17B-1989r, N323-1978r, N320-1979r, N317-1980r

(2) **(A)** **(test, measurement, and diagnostic equipment)** The device used to present output information to the operator, either in real time or as an output of a storage medium. **(B)** **(test, measurement, and diagnostic equipment)** The act of reading, transmitting, displaying information either in real time or from an internal storage medium of an operator or an external storage medium or peripheral equipment. (MIL) [2]

readout, command *See:* command readout.

readout device *See:* character display device.

read-out lag (diode-type camera tube) The fraction of the initial signal which is read out in any field after the image illumination is interrupted. (ED) 503-1978w

readout, position *See:* position readout.

read path In a reader, a path that has a read station. (C) 610.10-1994w

read pulse A pulse that causes information to be acquired from a magnetic cell or cells. *See also:* ONE state. (Std100) 163-1959w

read ready violation A word-serial protocol error that occurs when data is read from a servant while its read ready bit is zero (0). (C/MM) 1155-1992

read station The location in a reader where the data on a medium are read. *Synonym:* sensing station. (C) 610.10-1994w

read transaction A transaction that passes an address and size parameter from the requester to the responder and returns data values from the responder to the requester. The size parameter specifies the number of bytes that are transferred. (C/MM) 1212-1991s

read/write Pertaining to an operation, process, or object that is involved in both reading and writing. For example, a read/write head is a head that can perform both read and write operations. *Contrast:* write-once/read-many; read-only. *See also:* write; read. (C) 610.10-1994w

read/write access A type of access to data in which the data may be both retrieved, changed, and stored. *Contrast:* read-only access. *See also:* update access; delete access; write access. (C) 610.5-1990w

read/write cycle A cycle in which one read operation and one write (or rewrite) operation are performed. (C) 610.10-1994w

read/write head A head capable of both reading from or writing on the medium. *Synonyms:* record head; combined head. *Contrast:* write head; read head. (C) 610.10-1994w

read/write memory (RWM) Memory into which information may be stored (or written) and from which information may be retrieved (or read) for example, digital tape recorders and random-access memory. *Contrast:* read-only memory. (C) 610.10-1994w

read/write opening *See:* read/write slot.

read/write slot An opening in the jacket of a floppy disk allowing access to the storage medium by the read/write heads. *Synonym:* read/write opening. (C) 610.10-1994w

ready/busy An end-of-write indicator. (ED) 1005-1998

ready light An indicator light on a system or system component that indicates that the system is on and ready for operation. (C) 610.10-1994w

ready task A task that is not blocked. The ready tasks include those that are running as well as those that are waiting for a processor. *See also:* blocked task. (C) 1003.5-1999

ready-to-receive signal (facsimile) A signal sent back to the facsimile transmitter indicating that a facsimilee receiver is ready to accept the transmission. *See also:* facsimile signal. (COM) 168-1956w

reagent blank A volume of demineralized water for liquid samples carried through the entire analytical procedure. The volume or weight of the blank shall be approximately equal to the volume or weight of the sample processed. (NI) N42.23-1995

real address **(1)** The address of a storage location in the main storage part of a virtual storage system. *Contrast:* virtual address. (C) 610.12-1990

(2) The address of a storage location in real storage. *See also:* address translator. (C) 610.10-1994w

real data Data used to represent real numbers. *See also:* binary coded decimal real data; floating-point real data. (C) 610.5-1990w

real estate *See:* footprint.

real fixed binary data *See:* fixed-point binary data.

real fixed decimal data *See:* fixed-point real data.

real float binary data *See:* floating-point data.

real float decimal data *See:* floating-point data.

real group ID The attribute of a process that, at the time of process creation, identifies the group of the user who created the process. This value is subject to change during the process lifetime. *See also:* group ID. (C/PA) 9945-1-1996, 9945-2-1993, 1003.5-1999

realizable function (linear passive networks) A response function that can be realized by a network containing only positive resistance, inductance, capacitance, and ideal transformers. *Note:* This is the sense of realizability in the theory of linear, passive, reciprocal, time-invariant networks. (CAS) 156-1960w

realization The representation of interface responsibilities through specified algorithms and any needed representation properties. The realization states "how" a responsibility is met; it is the statement of the responsibility's method. Realization consists of any necessary representation properties together with the algorithm (if any). A realization may involve representation properties or an algorithm, or both. For example, an attribute typically has only a representation and no algorithm. An algorithm that is a "pure algorithm" (i.e., without any representation properties) uses only literals; it does not "get" any values as its inputs. Finally, a derived attribute or operation typically has both an algorithm and representation properties. (C/SE) 1320.2-1998

realized gain The gain of an antenna reduced by the losses due to the mismatch of the antenna input impedance to a specified impedance. *Note:* The realized gain does not include losses due to polarization mismatch between two antennas in a complete system. (AP/ANT) 145-1993

realized gain, partial *See:* partial realized gain.

realm *See:* area.

real number A member of the set of all positive and negative numbers, including integers, zero, mixed, fractional, rational, and irrational numbers. (C) 610.5-1990w, 1084-1986w

real storage (1) The main storage portion of a virtual storage system. *Contrast:* virtual storage. (C) 610.12-1990
(2) The main storage in a virtual storage system. *Note:* Although real storage and main storage are physically identical, conceptually real storage represents only parts of the range of addresses available to the user of a virtual storage system, whereas, the main storage includes the total range of addresses available to the user. (C) 610.10-1994w

real time (1) (processing) (emergency and standby power) Pertaining to the actual time during which a physical process transpires or pertaining to the performance of a computation during the actual time of related physical processing in order that results of the computation can be used in guiding the physical process. (IA/PSE) 446-1987s
(2) The actual time in the real world during which an event takes place. *Synonyms:* true time; actual time. (C) 610.10-1994w
(3) (analog computer) Using an ordinary clock as a time standard, the number of seconds measured between two events occurring in a physical system. By contrast, computer time is the number of seconds measured, with the same clock, between corresponding events in the simulated system. The ratio of the time interval between two events in a simulated system to the time interval between the corresponding events in the physical system is the time scale. Computer time is equal to the product of real time and the time scale. Real-time computation is computer operation in which the time scale is unity. Machine time is synonymous with computer time. *See also:* scale factor. (C) 165-1977w
(4) (software) Pertaining to a system or mode of operation in which computation is performed during the actual time that an external process occurs, in order that the computation results can be used to control, monitor, or respond in a timely manner to the external process. *Contrast:* batch. *See also:* conversational; interactive; online; interrupt. (C) 610.12-1990, 610.10-1994w
(5) (modeling and simulation) In modeling and simulation, simulated time with the property that a given period of actual time represents the same period of time in the system being modeled; for example, in a simulation of a radar system, running the simulation for one second may result in the model advancing time by one second; that is, simulated time advances at the same rate as actual time. *Contrast:* slow time; fast time. (C) 610.3-1989w
(6) An event or data transfer in which, unlessaccomplished within an allotted amount of time, the accomplishment of the action has either no value or diminishing value. (DIS/C) 1278.2-1995
(7) The real time, in seconds and fraction thereof, of acquisition of the spectrum. It is expressed as 14 characters including decimal point with leading zeros interpreted as zeros. (NPS/NID) 1214-1992r

real-time clock (RTC) (1) A device that signals the computer at regular intervals in order that it may keep up with some external event. *See also:* time-of-day clock. (C) 610.10-1994w
(2) A hardware system element that provides the system with a reference to real-world time. A common implementation would be a circuit containing a set of registers holding the current month, day, year, day-of-week number, and other time-related values, along with a circuit that continuously updates these register values. The circuit is normally provided with an alternate power source, such as a battery, which allows the RTC to continue to function when main system power is not available. (C/PA) 2000.2-1999

real-time environment profile A profile designed to support applications requiring bounded response. (C/PA) 1003.13-1998

real-time printout (sequential events recording systems) The recording of actual time that an input signal was received as correlated to a time standard. (PE/EDPG) [5], [1]

real-time service A service that satisfies timing constraints imposed by the service user. The timing constraints are user specific and should be such that the user will not be adversely affected by delays within the constraints. (DIS/C) 1278.2-1995

real-time system A system in which the correctness of a computation depends not only upon the results of the computations but also upon the time at which the outputs are generated. (C/BA) 896.3-1993w

real-time testing (test, measurement, and diagnostic equipment) The testing of a system or its components at its normal operating frequency or timing. (MIL) [2]

real type A data type whose members can assume real numbers as values and can be operated on by real number arithmetic operations, such as addition, subtraction, multiplication, division, and square root. *Contrast:* integer type; logical type; character type; enumeration type. (C) 610.12-1990

real user ID (1) The attribute of a process that, at the time of process creation, identifies the user who created the process. This value is subject to change during the process lifetime. *See also:* user ID. (C/PA) 9945-1-1996, 9945-2-1993
(2) The attribute of a process that, at the time of process creation, identifies the user who created the process. This value is subject to change during the process lifetime. *See also:* user ID. (C) 1003.5-1999

real variable A variable that may assume only real-number values. (C) 1084-1986w

real-world time The actual time in the real world, expressed as Universal Coordinated Time (UTC). (DIS/C) 1278.1-1995

reasoning system In the context of AI-ESTATE, a system that can combine elements of knowledge to draw conclusions. (ATLAS) 1232-1995

reassembly The function in the DQDB layer that provides for the reconstruction of an initial MAC protocol data unit (IMPDU). Reassembly is performed by concatenating the segmentation units received in derived MAC protocol data units (DMPDUs). This is the inverse process to segmentation. (LM/C) 8802-6-1994

reassociation The service that enables an established association [between access point (AP) and station (STA)] to be transferred from one AP to another (or the same) AP. (C/LM) 8802-11-1999

reboot fileset A fileset which, if installed, requires reboot of the operating system to complete its installation, and denoted by having the value of its *is_reboot* attribute set to `true`. (C/PA) 1387.2-1995

rebooting An implementation-defined procedure generally used to terminate and then restart operations on the target system. (C/PA) 1387.2-1995

recalescent point The temperature at which there is a sudden liberation of heat when metals are lowered in temperature. *See also:* coupling; induction heating. (IA/MET) 54-1955w, 169-1955w

receipt of a CCS message *See:* receipt of a CCS signal.

receipt of a CCS signal Occurs when the signal or complete message becomes available for acceptance by the processor (that is, stored in the input buffer). *Synonym:* receipt of a CCS message. (COM/TA) 973-1990w

receipt of a per-trunk-signaling supervisory signal Occurs when the state transition that begins the signal is received (that is, E-lead signal or loop open or closure). All times noted are exclusive of hit timing. (COM/TA) 973-1990w

receive (1) The acoustic output of a telephone set due to an electrical input to the telephone set or connecting test circuit. (COM/TA) 269-1992
(2) The acoustic output of a handset or headset due to an electrical input to the device or connecting test circuit. (COM/TA) 1206-1994

(3) The acoustic output of a handsfree telephone due to an electrical input. (COM/TA) 1329-1999

receive attenuation during double talk (A_{RDT}) Attenuation in the receive path, seen at the 50 cm test point (50TP), inserted during double talk. The send talker initiates the double talk. (COM/TA) 1329-1999

receive channel A channel used within a data circuit to receive data. *Contrast:* transmit channel. (C) 610.10-1994w

receive characteristic (telephony) The acoustic output level of a telephone set as a function of the electrical input level. The output is measured in an artificial ear, and the input signal is obtained from an available constant-power source of specified impedance. (IA) 169-1955w, [123]

received power (mobile communication) The root-mean-square value of radio-frequency power that is delivered to a load that correctly terminates an isotropic reference antenna. The reference antenna most commonly used is the half-wave dipole. *See also:* mobile communication system. (VT) [37]

receive electrical test point (RETP) The point in a battery feed circuit, reference codec, or wireless reference base station at which signals are applied to the handsfree telephone in the receive direction. (COM/TA) 1329-1999

receive loudness rating directionality (RLRD) Receive loudness rating versus angle around the handsfree telephone (HFT), normalized to the loudness rating at the 50 cm test point (50TP). (COM/TA) 1329-1999

receive-only equipment Data communication equipment capable of receiving signals, but not arranged to transmit signals. (COM) [49]

receiver (1) (facsimile) The apparatus employed to translate the signal from the communications channel into a facsimile record of the subject copy. *See also:* facsimile. (COM) 168-1956w

(2) (telephone switching systems) A part of an automatic switching system that receives signals from a calling device or other source for interpretation and action. (COM) 312-1977w

(3) (MULTIBUS) An agent that is the recipient of the data during a solicited message. *See also:* solicited messages. (C/MM) 1296-1987s

(4) A pin of a cell instance that is receiving or can receive a signal from an interconnect structure. (C/DA) 1481-1999

MULTIBUS® is a registered trademark of Intel Corporation.

receiver common-mode voltage The combination of three components: the driver-receiver ground potential difference (V_{gpd}); the longitudinally coupled peak noise voltage measured between the receiver circuit ground and the signal transmission media with the driver end shorted to ground (V_{noise}); 3 the driver offset voltage. (C/MM) 1596.3-1996

receiver differential noise margin high The tolerable signal voltage variation from any source that still results in the receiver producing a logic high output state when the driver is stimulated by a logic high input. Differential noise margin high is calculated by subtracting the receiver's minimum differential high input voltage from the driver's minimum high differential output voltage;

$V_{odh}(\text{min}) - V_{idh}(\text{min})$.

(C/MM) 1596.3-1996

receiver differential noise margin low Tolerable voltage variation to guarantee that the receiver produces a logic low output when the driver is stimulated by a logic low input;

$V_{idl}(\text{max}) - V_{odl}(\text{max})$.

(C/MM) 1596.3-1996

receiver gating The application of enabling or inhibiting pulses to one or more stages of a receiver only during the part of a cycle of operation when reception is either desired or undesired, respectively. *See also:* gating. (AES) 686-1997

receiver ground (signal-transmission system) The potential reference at the physical location of the signal receiver. *See also:* signal. (IE) [43]

receiver linear dynamic range (electromagnetic site survey) The interval between the minimum detectable signal and the 1 dB gain compression point within which the receiver gain deviates from a constant value by less than 1 dB. (EMC) 473-1985r

receiver-off-hook tone (telephone switching systems) A tone on a line to indicate an abnormal off-hook condition. (COM) 312-1977w

receiver 1 dB gain compression point (electromagnetic site survey) The input signal level to an otherwise linear receiver for which the gain has been decreased 1 dB (decibel) below the value measured for the mimimum detectable input signal (within the linear response range). (EMC) 473-1985r

receiver operating characteristic curves Plots of probability of detection versus probability of false alarm for various input signal-to-noise power ratios and detection threshold settings. (AES) 686-1997

receiver primaries *See:* display primaries.

receiver relay An auxiliary relay whose function is to respond to the output of a communications set such as an audio, carrier, radio, or microwave receiver. (SWG/PE/PSR) C37.100-1992, C37.90-1978s

receiver, telephone *See:* telephone receiver.

receiver training A startup routine in 100BASE-T2 used to acquire receiver parameters and synchronize the scramblers of two connected Physical Layers (PHYs). (C/LM) 802.3-1998

receive speech front end clipping during double talk (T_{RFDT}) The length of time that speech undergoes syllabic clipping, as seen at the 50 cm test point (50TP), just after the onset of double talk. The receive talker initiates the double talk. (COM/TA) 1329-1999

receiving (1) (nuclear power quality assurance) Taking delivery of an item at a designated location. (PE/NP) [124]

(2) (transmission performance of telephone sets) The acoustic output level of a telephone set due to an electric input to the telephone set or connecting test circuit. The electric input may be varied either in frequency or level. The output is measured in an artificial ear and the input is measured as the open-circuit voltage from a source of constant available power. (COM/TA) 269-1983s

receiving converter, facsimile *See:* facsimilie receiving converter.

receiving-end crossfire (telegraph channel) The crossfire from one or more adjacent telegraph channels at the end remote from the transmitting end. *See also:* telegraphy. (EEC/PE) [119]

receiving loop loss That part of the repetition equivalent assignable to the station set, subscriber line, and battery supply circuit that are on the receiving end. *See also:* transmission loss. (EEC/PE) [119]

receiving objective loudness rating (loudness ratings of telephone connections)

$$\text{ROLR} = -20\log_{10}\frac{S_E}{1/2\ V_W}$$

where

V_W = open-circuit voltage of the electric source (in millivolts)

S_E = sound pressure at the ear reference point (in pascals)

Note: Normally occurring ROLRs will be in the 40 to 55 decibel (dB) range. These numbers are a result of the units chosen and have no physical significance. (COM/TA) 661-1979r

receiving polarization (of an antenna) The polarization of a plane wave, incident from a given direction and having a given power flux density, that results in maximum available power at the antenna terminals. *Notes:* 1. The receiving polarization of an antenna is related to the antenna's polarization on transmit (see definition above) in the following way. In the same plane of polarization, the polarization ellipses have the same axial ratio, the same sense of polarization, and the same spatial orientation. Since their senses of polarization and spatial orientation are specified by viewing their polarization ellipses in the respective directions into which they are propagating, one should note that (a) although their senses of polarization are the same, they would appear to be opposite if both waves were viewed in the same direction; and (b) their tilt angles are such that they are the negative of one another with respect to a common reference. 2. The receiving polarization may be used to specify the polarization characteristic of a non-reciprocal antenna that may transmit and receive arbitrarily different polarizations. (AP/ANT) 145-1993

receiving voltage sensitivity *See:* free-field voltage response.

receptacle **(1)** A receptacle is a contact device installed at the outlet for the connection of a single attachment plug. A single receptacle is a single contact device with no other contact device on the same yoke. A multiple receptacle is a single device containing two or more receptacles. (NESC/NEC) [86]
(2) An outlet that is intended to be equipped electrically and mechanically to receive the plug. (RL) C136.10-1996
(3) A device installed in a receptacle outlet to accommodate an attachment plug. (IA/MT) 45-1998

receptacle circuit A branch circuit to which only receptacle outlets are connected. *See also:* branch circuit. (EEC/PE) [119]

receptacle circuit tester A device that, by a pattern of lights, is intended to indicate wiring errors in receptacles. Receptacle circuit testers have some limitations. They may indicate incorrect wiring, but cannot be relied upon to indicate correct wiring. (IA/PSE) 1100-1992s

receptacle outlet An outlet where one or more receptacles is installed. (IA/MT) 45-1998

reception diversity (data transmission) That method of radio reception whereby, in order to minimize the effects of fading, a resultant signal is obtained by combination or selection, or both, of two or more sources of received-signal energy that carry the same modulation or intelligence, but that may differ in strength or signal-to-noise ratio at any given instant. (PE) 599-1985w

receptive field (medical electronics) The region in which activity is observed by means of the pickup electrode. (EMB) [47]

receptor The body that is at rest in an ESD event. The receptor is usually but not necessarily at the same potential as its surroundings. It is always at a potential different from that of the intruder. (SPD/PE) C62.47-1992r

receptor electrode geometry The size and shape of that surface of the receptor, termed the receptor electrode, at which the ESD takes place. (SPD/PE) C62.47-1992r

reciprocal bearing (navigation aid terms) The opposite direction to a bearing. (AES/GCS) 172-1983w

reciprocal color temperature (illuminating engineering) Color temperature T_c expressed on a reciprocal scale ($1/T_c$). An important use stems from the fact that a given small increment in reciprocal color temperature is approximately equally perceptible regardless of color temperature. Also, color temperature conversion filters for sources approximating graybody sources change the reciprocal color temperature by nearly the same amount anywhere on the color temperature scale. *Note:* The unit is the reciprocal megakelvin (MK^{-1}). The reciprocal color temperature expressed in this unit has the numerical value $10^6/T_c$ when T_c is expressed in kelvins. The acronym "mirek" (for micro-reciprocal-kelvin) occasionally has been used in the literature. The acronym "mired" (for micro-reciprocal-degree) is now considered obsolete. (EEC/IE) [126]

reciprocal transducer A transducer in which the principle of reciprocity is satisfied. *Note:* The use of the term reversible transducer as a synonym for reciprocal transducer is deprecated. *See also:* transducer. (Std100) 270-1966w

reciprocating mechanism (high voltage air switches, insulators, and bus supports) An operating mechanism which produces longitudinal motion of the operating means to open or close the switching device. (SWG/PE) C37.30-1971s

reciprocity In wave propagation, the invariance of the complex amplitudes of the received signals to the interchange in location of transmitter and receiver, but not the antennas. *Note:* Reciprocity applies provided that the transmission medium is isotropic and that the antennas remain in place with only their transmit and receive functions interchanged. Reciprocity may not hold when the antennas are in different media. (AP/PROP) 211-1997

reciprocity theorem States that if an electromagnetic force E at one point in a network produces a current I at a second point in the network, then the same voltage E acting at the second point will produce the same current I at the first point. (EEC/PE) [119]

recirculated air (electric power systems in commercial buildings) Return air passed through the air conditioner before being supplied again to the conditioned space. (IA/PSE) 241-1990r

reclamation (insulating oil) The restoration to usefulness by the removal of contaminants and products of degradation such as polar, acidic, or colloidal materials from used electrical insulating liquids by chemical or adsorbent means. *Note:* The methods listed under reconditioning are usually performed in conjunction with reclaiming. Reclaiming typically includes treatment with clay or other absorbents. (PE/TR) 637-1985r

reclamation of oil The removal of harmful chemical contaminants. This is usually done with absorptive agents or alkali salts. (PE/TR) C57.106-1991w

reclosing device (power operations) A control device which initiates the reclosing of a circuit after it has been opened by a protective relay. (PE/PSE) 858-1987s

reclosing fuse A combination of two or more fuseholders, fuse units, or fuse links mounted on a fuse support or supports, mechanically or electrically interlocked, so that one fuse at a time can be connected into the circuit and the functioning of that fuse automatically connects the next fuse into the circuit, with or without intentionally added time delay, thereby permitting one or more service restorations without replacement of fuse links, refill units, or fuse units. (SWG/PE) C37.40-1993, C37.100-1992

reclosing interval (of an automatic circuit recloser) The open-circuit time between an automatic opening and the succeeding automatic reclosure. (SWG/PE) C37.100-1992

reclosing relay A programming relay whose function is to initiate the automatic reclosing of a circuit breaker. (SWG/PE/PSR) C37.100-1992, C37.90-1978s

reclosing time (of a circuit breaker) The interval between the time when the actuating quantity of the release (trip) circuit reaches the operating value (the breaker being in the closed position) and the reestablishment of the circuit on the primary arcing contacts on the reclosing stroke. (SWG/PE) C37.100-1992

reclosure (relay) The automatic closing of a circuit-interrupting device following automatic tripping. Reclosing may be programmed for any combination of instantaneous, time-delay, single-shot, multiple-shot, synchronism-check, dead-line-live-bus, or dead-bus-live-line operation. (PE/PSR) C37.95-1973s

recognition *See:* magnetic ink character recognition; pattern recognition; optical character recognition; character recognition.

recognition time The time elapsed between the change of the value of a digital input signal and its recognition by the digital input unit. (C) 610.10-1994w

recombinant Pseudonym for oxygen recombination. (SB) 1187-1996

recombination (overhead power lines) The process by which positive and negative ions recombine to neutralize each other. (T&D/PE) 539-1990

recombination center (solar cells) A defect having electrical properties so as to facilitate the recombination of mobile charge carriers (electrons or holes) with one each of the opposite polarity. (MAG) 306-1969w

recombination rate (1) (volume) The time rate at which free electrons and holes recombine within the volume of a semiconductor. *See also:* semiconductor device. (AES/SS) 307-1969w
(2) (overhead power lines) The rate at which positive and negative ions recombine in a given gas or liquid. (T&D/PE) 539-1990

recombination velocity (semiconductor surface) The quotient of the normal component of the electron (hole) current density at surface by the excess electron (hole) charge density at the surface. *See also:* semiconductor device. (ED) 216-1960w

recommended test position (RTP) An acoustic test point, other than the 50 cm test point (50TP), that corresponds to the most appropriate user position for nonstandard desktop and non-desktop applications. This may be specified by the handsfree telephone (HFT) manufacturer. (COM/TA) 1329-1999

recommended wearing position (RWP) A test position (corresponding to the manufacturer's instructions) for the transmitter of a headset that does not have a fixed spatial relationship between the location of its transmitter and receiver sound ports. (COM/TA) 1206-1994

recompeting master (multiprocessor architecture) The module that is in control of the bus and has initiated a control acquisition procedure in order to unlock slave interfaces. (C/MM) 896.1-1987s

recomplementation (mathematics of computing) The process of taking the complement of a complement. *Note:* The complement of a complement is the original numeral. (C) 1084-1986w

Reconciliation Sublayer (RS) A 100BASE-T mapping function that reconciles the signals at the Media Independent Interface (MII) to the Media Access Control (MAC)-Physical Signaling Sublayer (PLS) service definitions. (C/LM) 802.3-1998

reconditioned carrier reception (exalted-carrier reception) The method of reception in which the carrier is separated from the sidebands for the purpose of eliminating amplitude variations and noise, and then added at increased level to the sideband for the purpose of obtaining a relatively undistorted output. This method is frequently employed, for example, when a reduced-carrier single-sideband transmitter is used. *See also:* radio receiver. (EEC/PE) [119]

reconditioning (1) (insulating oil) The removal of insoluble contaminants, moisture, and dissolved gases from used, electrical insulating liquids by mechanical means. *Note:* The typical means employed are settling, filtering, centrifuging, and vacuum drying or degassing. (PE/TR) 637-1985r
(2) A general term covering the process of maintaining existing power switchgear equipment in operating condition as recommended by the manufacturer's instructions, using only the original manufacturer's recommended replacement parts, without altering the original design. (SWG/PE) C37.100-1992
(3) The process of maintaining existing power switchgear equipment in operating condition as recommended by the manufacturer's instructions, using only the original manufacturers' designed parts. *Note:* Reverse engineered parts (designs copied from existing parts by other manufacturers) are not considered to be the original manufacturer's design or recommended replacement parts. (SWG/PE) C37.59-1996

reconditioning of oil The mechanical removal of moisture and insoluble contaminants. (PE/TR) C57.106-1991w

reconfiguration (1) (dual ring operation with wrapback reconfiguration) A change of the path around which the token that is used for normal data transfer circulates. (LM/C) 802.5c-1991r
(2) (DQDB subnetwork of a metropolitan area network) The process by which the configuration control function activates and deactivates resources of a DQDB subnetwork to take account of a change in the operational status of a cluster, node, or transmission link in the subnetwork. (LM/C) 8802-6-1994
(3) A strategy for repairing components in which failing components are switched out of operation and replaced by failure-free components. (C) 610.10-1994w

reconfiguration management The management functions responsible for reconfiguration. This includes both dual ring management and any other management required for reconfiguration. (LM/C) 802.5c-1991r

reconstituted mica *See:* mica paper.

reconstruction (1) Replacement of any portion of an existing installation by new equipment or construction. Does not include ordinary maintenance replacements. (NESC) C2-1977s
(2) (image processing and pattern recognition) *See also:* image reconstruction. (C) 610.4-1990w

record (1) (data management) (software) A set of data items, called fields, treated as a unit. For example, in stock control, the data for each invoice could constitute one record. *Synonym:* data record. *See also:* entity; database record. (C) 610.5-1990w, 610.12-1990
(2) The language-independent syntax for a family of datatypes constructed from a sequence of base datatypes, each associated with a name. A value of record datatype contains, for each name, a value of the corresponding base datatypes. (C/PA) 1351-1994w
(3) A set of related data items treated as a unit. For example, in stock control, the data for each invoice could constitute one record. (C) 610.7-1995
(4) To put data into a storage device. (C) 610.10-1994w
(5) A datatype constructed from a sequence of base datatypes, each associated with a name. A record value contains, for each name, a value of the corresponding base datatype. (C/PA) 1224.1-1993w
(6) A collection of related data or words treated as a unit and saved in a position-dependent fashion within a file of other such units. (C/MM) 855-1990
(7) A collection of related data units or words that itself is treated as a unit. (C) 1003.5-1999

record condition (data management) A conjunction of two or more item conditions such that the name of the data item in each condition is distinct. For example, "LASTNAME = 'JONES' and SEX = 'FEMALE.'" (C) 610.5-1990w

recorded announcement (telephone switching systems) A prerecorded oral message received on a call. (COM) 312-1977w

recorded spot, X dimension (facsimile) The effective recorded-spot dimension measured in the direction of the recorded line. *Notes:* 1. By effective dimension is meant the largest center-to-center spacing between recorded spots which gives minimum peak-to-peak variation of density of the recorded line. 2. This term applies to that type of equipment which responds to a constant density in the subject copy by a succession of discrete recorded spots. *See also:* recording. (COM) 168-1956w

recorded spot, *Y* dimension (facsimile) The effective recorded-spot dimension measured perpendicularly to the recorded line. *Note:* By effective dimension is meant the largest center-to-center distance between recorded lines which gives minimum peak-to-peak variation of density across the recorded lines. *See also:* recording. (COM) 168-1956w

recorded value The value recorded by the marking device on the chart, with reference to the division lines marked on the chart. *See also:* accuracy rating. (EEC/PE) [119]

recorder (1) (analog computer) A device that makes a permanent record, usually graphic, of varying signals. *Synonym:* strip-chart recorder. (C) 165-1977w, 610.10-1994w
(2) (facsimile) That part of the facsimile receiver which performs the final conversion of electric picture signal to an

image of the subject copy on the record medium. *See also:* recording; facsimile. (COM) 168-1956w

recorder, strip-chart *See:* strip-chart recorder.

recorder-warning tone (telephone switching systems) A tone that indicates periodically that the conversion is being electrically recorded. (COM) 312-1977w

record gap (1) (computers) (storage medium) An area used to indicate the end of a record. (C) [20], [85]
(2) (test, measurement, and diagnostic equipment) An interval of space or time associated with a record to indicate or signal the end of the record. (MIL) [2]
(3) *See also:* interblock gap. (C) 610.5-1990w
(4) *See also:* interblock gap. (C) 610.10-1994w

record head *See:* read/write head.

recording (1) (facsimile) The process of converting the electrical signal to an image on the record medium. *See also:* electromechanical recording; electrothermal recording; photosensitive recording; magnetic recording; electrochemical recording; electrolytic recording; electrostatic recording. (COM) 168-1956w
(2) The process of storing information on some storage medium for later retrieval. *See also:* magnetic recording; optical recording. (C) 610.10-1994w

recording area (1) In micrographics, the maximum useful area of a microfilm or other medium that can record information, including the image as well as the document marks. (C) 610.2-1987
(2) The area on a disk or storage medium on which information can be recorded. *Synonym:* recording zone. *Contrast:* handling zone. (C) 610.10-1994w

recording channel (electroacoustics) The term refers to one of a number of independent recorders in a recording system or to independent recording tracks on a recording medium. *Note:* One or more channels may be used at the same time for covering different ranges of the transmitted frequency band, for multichannel recording, or for control purposes. *See also:* phonograph pickup. (SP) [32]

recording-completing trunk (telephone switching systems) A one-way trunk for operator recording, extending, and automatic completing of toll calls. (COM) 312-1977w

recording demand meter A demand meter that records on a chart the demand for each demand interval. *See also:* electricity meter. (EEC/PE) [119]

recording density The number of bits in a single linear track, measured in bits per unit of length or area of the recording medium. *Synonyms:* packing density; surface density; bit density. *See also:* track density. (C) 610.10-1994w

recording, instantaneous *See:* instantaneous recording.

recording instrument (electrical heating applications to melting furnaces and forehearths in the glass industry) An instrument that makes a graphic record of the value of one or more quantities as a function of another variable, usually time. (IA) 668-1987w

recording loss (mechanical recording) The loss in recorded level whereby the amplitude of the wave in the recording medium differs from the amplitude executed by the recording stylus. *See also:* phonograph pickup. (SP) [32]

recording medium The material on which program instructions and text are recorded; for example, magnetic tape. (C) 610.10-1994w

recording spot (facsimile) The image area found at the record medium by the facsimile recorder. *See also:* recording. (COM) 168-1956w

recording stylus (mechanical recording) A total that inscribes the groove into the recording medium. *See also:* phonograph pickup. (SP) [32]

recording trunk A trunk extending from a local central office or private branch exchange to a toll office, that is used only for communication with toll operators and not for completing toll connections. (COM/PE/EEC) [119]

recording zone *See:* recording area.

record layout (data management) The arrangement and structure of data in a record. (C) 610.5-1990w

record length (data management) The number of words or characters in a record. (C) 610.5-1990w

record length type (data management) The category to which a record belongs by virtue of having fixed or variable length. (C) 610.5-1990w

record-locking *See:* lock.

record medium (facsimile) A physical medium on which the facsimile recorder forms an image of the subject copy. *See also:* recording. (COM) 168-1956w

record of data A sequential collection of samples acquired by the waveform recorder. (IM/WM&A) 1057-1994w

record segmentation (data management) The allocation of individual data items in a record to separate physical storage areas or to different physical devices. (C) 610.5-1990w

record sheet (facsimile) The medium which is used to produce a visible image of the subject copy in record form. The record medium and the record sheet may be identical. *See also:* recording. (COM) 168-1956w

record spot (facsimile) The image of the recording spot on the record sheet. *See also:* recording. (COM) 168-1956w

records processing The process of manipulating, storing, and retrieving records in electronic form. *See also:* office automation. (C) 610.2-1987

record type (data management) The category to which a record belongs by virtue its format, content, or characteristics. (C) 610.5-1990w

recoverable light loss factors (illuminating engineering) Factors that give the fractional light loss that can be recovered by cleaning or lamp replacement. (EEC/IE) [126]

recovered charge (semiconductor) The charge recovered from a semiconductor device after switching from a forward current condition to a reverse condition. (IA) [12]

recovery (1) (data management) The restoration of a system, program, database, or other system resource to a prior state following a failure or externally caused disaster; for example, the restoration of a database to a point at which processing can be resumed following a system failure. (C) 610.5-1990w
(2) (software) The restoration of a system, program, database, or other system resource to a state in which it can perform required functions. (C) 610.12-1990
(3) The ability of the `swinstall` utility, for a failed software installe, to return the system to the state that it was in before the failure, including restoring the files. (C/PA) 1387.2-1995
(4) A set of interactions intended to restore failed equipment or to find alternatives to achieve its function. (PE/NP) 1082-1997
(5) The process of restoring the ring to normal operation. When the ring is beaconing, claiming, or purging, the ring is in a state of recovery. (C/LM) 8802-5-1998

recovery current (semiconductor rectifiers) The transient component of reverse current associated with a change from forward conduction to reverse voltage. *See also:* rectification. (IA) [12]

recovery cycle (electrobiology) The sequence of states of varying excitability following a conditioning stimulus. The sequence may include periods such as absolute refractoriness, relative refractoriness, supernormality, and subnormality. *See also:* excitability. (EMB) [47]

recovery, error *See:* error recovery.

recovery-of-frame alignment (telecommunications) The establishment of proper frame alignment of the receiver after an out-of-frame condition. (COM/TA) 1007-1991r

recovery phase (MULTIBUS II) The final phase of an exception operation in which the parallel system bus is allowed to

sit idle for a defined amount of time. *See also:* exception operation. (C/MM) 1296-1987s
MULTIBUS II® is a registered trademark of Intel Corporation.

recovery/removal timing check A timing check that establishes an interval with respect to a reference signal transition, during which an asynchronous control signal may not change from the active to inactive state. This timing check is frequently applied to flip-flops and latches to establish a stable interval for the set and reset inputs with respect to the active edge of the clock or the active-to-inactive transition of the gate. Two limit values are necessary to define the stable interval. The recovery time is the time before the reference signal transition when the stable interval begins. The removal time is the time after the reference signal transition when the stable interval ends. If the asynchronous control signal goes inactive during the stable interval, it is unknown whether the flip-flop or latch takes on the state of the data input, remains set, or is reset. (C/DA) 1481-1999

recovery timing check A timing check that establishes only the beginning of the stable interval for a recovery/removal timing check. If no removal timing check is provided for the same arc, transitions, and state, the stable interval is assumed to end at the reference signal transition, and a negative value for the recovery time is not meaningful. *See also:* recovery/removal timing check. (C/DA) 1481-1999

recovery time (1) (power supplies) Specifies the time needed for the output voltage or current to return to a value within the regulation specification after a step load or line change. *Notes:* 1. Recovery time, rather than response time, is the more meaningful and therefore preferred way of specifying power-supply performance, since it relates to the regulation specification. 2. For load change, current will recover at a rate governed by the rate-of-change of the compliance voltage across the load. This is governed by the resistance-capacitance time constant of the output filter capacitance, internal source resistance, and load resistance. *See also:* programming speed; radar.

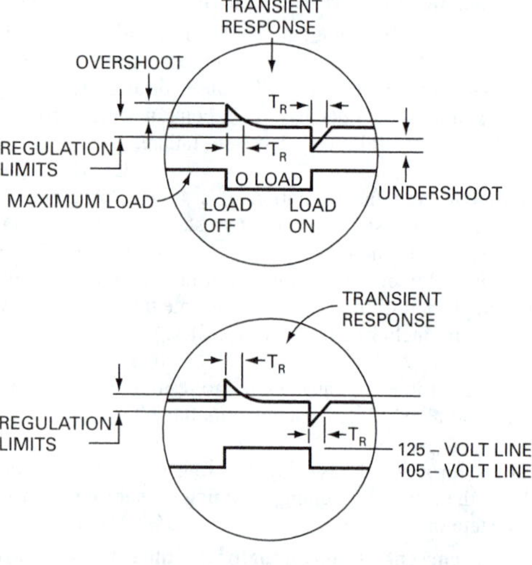

Recovery time. Oscilloscope views showing (top) the effects of a step load change, and (bottom) the effects of a step line change. T_R = recovery time.

recovery time

(AES/IA/PSE) [41], 1100-1999

(2) (A) (anti-transmit-receive tube) The time required for a fired tube to deionize to such a level that the normalized conductance and susceptance of the tube in its mount are within specified ranges. *Note:* Normalization is with respect to the characteristic admittance of the transmission line at its junction with the tube mount. *See also:* radar; gas tube. **(B) (gas tube)** The time required for the control electrode to regain control after anode current interruption. *Note:* To be exact, the deionization and recovery time of a gas tube should be presented as families of curves relating such factors as condensed-mercury temperature, anode current, anode and control electrode voltages, and control-circuit impedance. *See also:* radar. **(C) (TR and pre-TR tubes)** The time required for a fired tube to deionize to such a level that the attenuation of a low-level radio-frequency signal transmitted through the tube is decreased to a specified value. *See also:* relay recovery time. (ED) 161-1971

(3) (gas turbines) The interval between two conditions of speed occurring with a specified sudden change in the steady-state electric load on the gas-turbine-generator unit. It is the time in seconds from the instant of change from the initial load condition to the instant when the decreasing oscillation of speed finally enters a specified speed band. *Note:* The specified speed band is taken with respect to the midspeed of the steady-state speed band occurring at the subsequent steady-state load condition. The recovery time for a specified load increase and the same specified load decrease may not be identical and will vary with the magnitude of the load change. (PE/EDPG) [5]

(4) (A) When sending or receiving pulses, the time required between the end of a pulse and the beginning of the next pulse. **(B)** The time required by some peripheral devices between one access and another. (C) 610.10-1994

(5) The minimum time from the start of a counted pulse to the instant a succeeding pulse can attain a specified percentage of the maximum amplitude of the counted pulse. *See also:* half-amplitude recovery time. (NI/NPS) 309-1999

(6) (reverse-blocking thyristor or semiconductor diode) *See also:* reverse recovery time.

(7) *See also:* recovery/removal timing check. (C/DA) 1481-1999

recovery voltage (1) The voltage that occurs across the terminals of a pole of a circuit-interrupting device upon interruption of the current.
(SWG/PE/IA/SPD/PSE) C37.40-1993, C37.100-1992, 1100-1992s, C62.1-1981s, C62.62-2000

(2) The power frequency voltage that appears across each set of main switching, transition, or transfer contacts of the arcing switch or arcing tap switch after these contacts have broken the switched current. (PE/TR) C57.131-1995

recreational vehicle A vehicular type unit primarily designed as temporary living quarters for recreational, camping, or travel use, which either has its own motive power or is mounted on or drawn by another vehicle, The basic entities are: travel trailer, camping trailer, truck camper and motor home. (NESC/NEC) [86]

recreational vehicle park A plot of land upon which two or more recreational vehicle sites are located, established or maintained for occupancy by recreational vehicles of the general public as temporary living quarters for recreation or vacation purposes. (NESC/NEC) [86]

recreational vehicle site A plot of ground within a recreational vehicle park intended for the accommodation of either a recreational vehicle, tent, or other individual camping unit on a temporary basis. (NESC/NEC) [86]

recreational vehicle site feeder circuit conductors The conductors from the park service equipment to the recreational vehicle site supply equipment. (NESC/NEC) [86]

recreational vehicle site supply equipment The necessary equipment, usually a power outlet, consisting of a circuit breaker or switch and fuse and their accessories, located near the point of entrance of supply conductors to a recreational vehicle site and intended to constitute the disconnecting means for the supply to that site. (NESC/NEC) [86]

recreational vehicle stand That area of a recreational vehicle site intended for the placement of a recreational vehicle. (NESC/NEC) [86]

rectangular array *See:* rectangular grid array.

rectangular grid array A regular arrangement of array elements, in a plane, such that lines connecting corresponding points of adjacent elements form rectangles. (AP/ANT) 145-1993

rectangular impulse (surge arresters) An impulse that rises rapidly to a maximum value, remains substantially constant for a specified period, and then falls rapidly to zero. The parameters that define a rectangular impulse wave are polarity, peak value, duration of the peak, total duration. (PE) [8]

rectangular mode *See:* resolver.

rectangular-shape logic symbol A logic symbol in which the logic function is indicated by a qualifying symbol in its interior. (GSD) 91-1973s

rectangular wave (data transmission) A periodic wave which alternately assumes one of two fixed values, the time of transition being negligible in comparison with the duration of each fixed value. (PE) 599-1985w

rectification The term used to designate the process by which electric energy is transferred from an alternating-current circuit to a direct-current circuit. (IA) 59-1962w, [12]

rectification error (accelerometer) A steady-state error in the output while vibratory disturbances are acting on an accelerometer. Anisoelasticity, for example, is one source of rectification error. (AES/GYAC) 528-1994

rectification factor The quotient of the change in average current of an electrode by the change in amplitude of the alternating sinusoidal voltage applied to the same electrode, the direct voltages of this and other electrodes being maintained constant. *See also:* transrectification factor; conductance for rectification. (ED) 161-1971w

rectification failure (power system device function numbers) A device that functions if one or more anodes of a power rectifier fail to fire, or to detect an arc bac, or on failure of a diode to conduct or block properly. (SUB/PE) C37.2-1979s

rectification of an alternating current Process of converting an alternating current to a unidirectional current. *See also:* semiconductor device; electronic rectifier; inverse voltage. (ED) [45], [84]

rectified unbalance *See:* gimbal-unbalance torque.

rectified value (alternating quantity) The average of all the positive values of the quantity during an integral number of periods. Since the positive values of a quantity y are represented by the expression

$$\frac{1}{2}[y + |y|],$$

$$y_r = \frac{1}{T}\int_0^T \frac{1}{2}[y + |y|]dt$$

Note: The word positive and the sign + may be replaced by the word negative and the sign −. (Std100) 270-1966w

rectifier (1) (self-commutated converters) A converter for conversion from alternating current (ac) to direct current (dc). (IA/PEL/C/SPC/ET) 936-1987w, 388-1992r, 610.10-1994w **(2) (generating stations electric power system)** A device for converting ac to dc. (PE/EDPG) 505-1977r **(3) (ac adjustable-speed drives)** A converter for conversion from ac to dc. (IA/ID) 995-1987w **(4)** A component for converting ac to dc by inversion or suppression of alternate half cycles. (IA/MT) 45-1998

rectifier anode An electrode of the rectifier from which the current flows into the arc. *Note:* The direction of current flow is considered in the conventional sense from positive to negative. The cathode is the positive direct-current terminal of the apparatus and is usually a pool of mercury. The neutral of the transformer secondary system is the negative direct-current terminal of the rectifier unit. *See also:* rectification. (EEC/PE) [119]

rectifier assembly A complete unit containing rectifying components, wiring, and mounting structure capable of converting alternating-current power to direct-current power. *See also:* converter. (PE) [9]

rectifier cathode The electrode of the rectifier into which the current flows from the arc. *Note:* The direction of current flow is considered in the conventional sense from positive to negative. The cathode is the positive direct-current terminal of the rectifier unit and is usually a pool of mercury. The neutral of the transformer secondary system is the negative direct-current terminal of the rectifier unit. *See also:* rectification. (EEC/PE) [119]

rectifier circuit element A circuit element bounded by two circuit terminals that has the characteristic of conducting current substantially in one direction only. *Note:* The rectifier circuit element may consist of more than one semiconductor rectifier cell, rectifier diode, or rectifier stack connected in series or parallel or both, to operate as a unit. (IA) [12]

rectifier electric locomotive An electric locomotive that collects propulsion power from an alternating-current distribution system and converts this to direct current for application to direct-current traction motors by means of rectifying equipment carried by the locomotive. *Note:* A rectifier electric locomotive may be defined by the type of rectifier used on the locomotive, such as ignitron electric locomotive. *See also:* electric locomotive. (EEC/PE) [119]

rectifier electric motor car An electric motor car that collects propulsion power from an alternating-current distribution system and converts this to direct current for application to direct-current traction motors by means of rectifying equipment carried by the motor car. *Note:* A rectifier electric motor car may be defined by the type of rectifier used on the motor car, such as ignitron electric motor car. *See also:* electric motor car. (EEC/PE) [119]

rectifier instrument The combination of an instrument sensitive to direct current and a rectifying device whereby alternating currents or voltages may be measured. *See also:* instrument. (EEC/PE) [119]

rectifier junction (semiconductor rectifier cell or diode) The junction in a semiconductor rectifier cell that exhibits asymmetrical conductivity. *See also:* semiconductor rectifier stack; semiconductor. (IA) [12]

rectifier stack (semiconductor) An integral assembly of one or more rectifier diodes, including its associated mounting and cooling attachments if integral with it. *See also:* semiconductor rectifier stack. (IA) [12]

rectifier transformer (power and distribution transformers) A transformer that operates at the fundamental frequency of an alternating-current system and designated to have one or more output windings conductively connected to the main electrodes of a rectifier. *See also:* alternating-current winding; interphase transformer; rating of interphase transformer; anode paralleling reactor; power rectifier transformer; commutating reactor; direct-current winding. (PE/TR) C57.12.80-1978r

rectifier tube An electronic tube or valve designed to rectify alternating current. (ED) [45]

rectifier unit An operative assembly consisting of the rectifier, or rectifiers, together with the rectifier auxiliaries, the rectifier transformer equipment, and the essential switchgear. *See also:* rectification; rectifier transformer. (IA) [62]

rectifier valve *See:* rectifier tube.

rectifying device An elementary device, consisting of one anode and its cathode, that has the characteristic of conducting current effectively in only one direction. *See also:* rectification. (EEC/PE) [119]

rectifying element A circuit element that has the property of conducting current effectively in only one direction. *Note:* When a group of rectifying devices is connected, either in parallel or series arrangement, to operate as one circuit element, the group of rectifying devices should be considered as a rectifying element. *See also:* rectifying device; rectifier circuit element; rectification; rectifying junction. (EEC/PE) [119]

rectifying junction (1) (barrier layer) (blocking layer) The region in a metallic rectifier cell that exhibits the asymmetrical conductivity. *See also:* rectification. (EEC/PE) [119]

(2) A region between two materials, typically n-type or p-type semiconductors, or between a metal and a semiconductor, arranged to provide a very low resistance to current flow in one direction and a very high resistance to current flow in the opposite direction. (NPS) 325-1996

rectilinear scanning (television) The process of scanning an area in a predetermined sequence of straight parallel scanning lines. (BT/AV) 201-1979w

rector, shunt *See:* shunt reactor.

recurrence rate *See:* pulse-repetition frequency.

recurrent sweep A sweep that repeats or recurs regularly. It may be free-running or synchronized. *See also:* oscillograph. (IM/HFIM) [40]

recursion (A) (data management) A process in which a software module calls itself. *See also:* simultaneous recursion. **(B) (data management)** The process of defining or generating a process or data structure in terms of itself. (C) 610.12-1990

recursive (1) (A) (software) Pertaining to a software module that calls itself. **(B) (software)** Pertaining to a process or data structure that is defined or generated in terms of itself. (C) 610.12-1990

(2) (scheme programming language) Self-referential. In common usage, a recursive procedure is one that calls itself; similarly a recursion is a call by a procedure to itself. A set of procedures is mutually recursive if they refer to one another. (C/MM) 1178-1990r

recursive data structure (data management) A data structure that is defined in terms of itself. (C) 610.5-1990w

recursively defined sequence (data management) A sequence in which each item after the first is determined using a given operation for which one or more of the operands include one or more of the preceding items. (C) 610.5-1990w

recursive routine (software) A routine that may be used as a subroutine of itself, calling itself directly or being called by another subroutine, one that it itself has called. The use of a recursive routine usually requires the keeping of records of the status of its unfinished uses in, for example, a pushdown list. *See also:* subroutine; list; routine. (C/SE) 729-1983s

red Pertains to the parts of a computer or communications system in which data being transmitted or manipulated is not encrypted. *Contrast:* black. (C) 610.7-1995

red alarm A locally detected failure in a sink device; e.g., a primary multiplex equipment. Examples of locally detected failures could be loss of synchronization, incoming signal failure, a blown fuse, etc. (COM/TA) 1007-1991r

redefinition (A) (data management) The process of changing a database schema by adding, removing, or renaming attributes or relations. **(B) (data management)** In a relation, the process of changing the data type or size of an attribute, or altering the characteristics of a domain. (C) 610.5-1990

red, green, blue display device A color display device characterized by its ability to provide three different color responses independently to the screen lined with multi-colored phosphor. (C) 610.10-1994w

red head *See:* personal ground.

redirecting surfaces or media (illuminating engineering) Those which change the direction of the flux without scattering the redirected flux. (EEC/IE) [126]

redirection In the shell command language, a method of associating files with the input/output of commands. (C/PA) 9945-2-1993

redirection operator In the shell command language, a token that performs a redirection function; it is one of the following symbols:

< > >| << >> <& >& <<− <>

(C/PA) 9945-2-1993

redistribution (storage or camera tubes) The alteration of the charge pattern on an area of a storage surface by secondary electrons from any other part of the storage surface. *See also:* charge-storage tube. (ED) 158-1962w, 161-1971w

REDUCE A list processing language written in LISP, used primarily for performing symbolic operations and simplification of arrays and matrices. (C) 610.13-1993w

reduced full-wave test (power and distribution transformers) A wave similar in shape and duration to that involved in a "full-wave lightning impulse test," but reduced in magnitude. *Note:* The reduced full wave normally has a crest value between 50% and 70% of the full-wave value involved, and is used for comparison of oscillograms in failure detection. (PE/TR) C57.12.80-1978r

reduced generator efficiency (thermoelectric device) The ratio of a specified generator efficiency to the corresponding Carnot efficiency. *See also:* thermoelectric device. (ED) [46]

reduced instruction set computer (RISC) A computer characterized by a small instruction set and large collection of registers. *Note:* All or most instructions can be executed in a single clock cycle. *Synonym:* load-store computer. (C) 610.10-1994w

reduced kilovoltampere tap (power and distribution transformers) (in a transformer) A tap through which the transformer can deliver only an output less than rated kVA without exceeding the specified temperature rise. The current is usually that of the rated kVA tap. (PE/TR) C57.12.80-1978r

reduced-voltage starter A starter, the operation of which is based on the application of a reduced voltage to the motor. *See also:* starter. (IA/ICTL/IAC) [60], [84]

reducing joint A joint between two lengths of cable the conductors of which are not the same size. *See also:* branch joint; straight joint; cable joint. (T&D/PE) [10]

redundance The introduction of auxiliary elements and components into a circuit, module, or system unit to perform the same functions as similar elements in such units for the purpose of improving their overall reliability in performance and safety. Active redundance is that redundance wherein all redundant items are operating simultaneously, rather than being switched on when needed. Standby redundance is that redundance wherein the alternative means of performing the function is inoperative until needed and is switched in upon failure of the primary means of performing the function. (C) 610.10-1994w

redundancy (1) The provision of extra memory cells, usually rows or columns, that can be mapped into the memory array to replace defective cells. *Note:* In nonvolatile memory, the mapping may be controlled through EEPROM or other fuse techniques. (ED) 1005-1998

(2) The existence in a system of more than one means of accomplishing a given function. (VT/RT) 1475-1999, 1474.1-1999

redundancy, active *See:* active redundancy.

redundancy check *See:* redundant check.

redundancy factor The ratio of the total number of series thyristor-levels in the valve, Nt, to the same number minus the total number of redundant series thyristor-levels in the valve, Nr. The redundancy factor, f_r, is defined by:

$$fr = \frac{Nt}{Nt - Nr}$$

(SUB/PE) 857-1996

redundancy, standby *See:* standby redundancy.

redundant (1) Pertaining to characters that do not contribute to the information content. Redundant characters are often used for checking purposes or to improve reliability. *See also:* check digit; self-checking code; error-detecting code; parity. (C) 162-1963w

(2) (cable systems in power generating stations) Applied to two or more systems serving the same objective, where they are also either:

a) Systems where personnel or public safety is involved, such as fire pumps; or

b) Systems provided with redundancy because of the severity of economic consequences of equipment damage. (Turbine-generator ac and dc bearing oil pumps are examples of redundant equipment under this definition.) (PE/EDPG) 422-1977

(3) (electric pipe heating systems) The introduction of auxiliary elements and components to a system to perform the same function as other elements in the system for the purpose of improving reliability. Redundant electric pipe heating systems consist of two heaters and two controllers, each with its own sensor, supplied from two power systems, all independent of each other but all applied to the same mechanical piping, valves, tanks, etc. Redundant electric pipe heating systems are referred to as primary and backup in this recommended practice. *Synonym:* redundancy. (PE/EDPG) 622-1979s

redundant arrays of inexpensive disks (RAID) *See:* RAID storage.

redundant check (data transmission) (checking code) A check that uses extra digits (check bits) short of complete duplication, to help detect the absence of error within the character or block. (PE) 599-1985w

redundant equipment A piece of equipment or a system that duplicates the essential function of another piece of equipment or system to the extent that either may perform the required function, regardless of the state of operation or failure of the other. *Notes:* 1. Duplication of essential functions can be accomplished by the use of identical equipment, equipment diversity, or functional diversity. 2. Redundancy can be accomplished by use of identical equipment, equipment diversity, or functional diversity. *Synonym:* redundant system. (PE/NP) 603-1998, 308-1980s, 497-1981w, 384-1992r, 379-1994, 387-1995

redundant link (local area networks) A second link from an end node or from the cascade port of a repeater that provides an alternative path to maintain network connectivity in case of a repeater or link failure. (C) 8802-12-1998

redundant system *See:* redundant equipment.

redundant systems (cable systems) Two or more systems serving the same objective, where they are also either systems where personnel or public safety is involved, such as fire pumps, or systems provided with redundancy because of the severity of economic consequences of equipment damage. *Note:* Turbine-generator alternating-current and direct-current bearing oil pumps are examples of redundant equipment under this definition. (PE/EDPG) 422-1977

redundant thyristor-levels The maximum number of levels in the series string of thyristors in a valve that may be short circuited externally or internally during service without affecting the safe operation of the valve as demonstrated by type tests, and which, if and when exceeded, would require shut down of the valve to replace the failed thyristors or acceptance of increased risk of failure of the valve. (SUB/PE) 857-1996

reed relay A relay using glass-enclosed, magnetically closed reeds as the contact members. Some forms are mercury wetted. (PE/EM) 43-1974s

reel A cylinder with flanges on which tape or film may be wound. *Contrast:* spool. *See also:* write ring; leader. (C) 610.10-1994w

reel puller A device designed to pull a conductor during stringing operations. It is normally equipped with its own engine, which drives the supporting shaft for the reel mechanically, hydraulically, or through a combination of both. The shaft, in turn, drives the reel. The application of this unit is essentially the same as that for the drum puller. Some of these devices function as either a puller or tensioner. *See also:* drum puller. (T&D/PE) 524-1992r

reel setup *See:* pull site; tension site.

reel stand A device designed to support one or more conductor or groundwire reels having the possibility of being skid, trailer, or truck mounted. These devices may accommodate rope or conductor reels of varying sizes and are usually equipped with reel brakes to prevent the reels from turning when pulling is stopped. They are used for either slack or tension stringing. The designation of reel trailer or reel truck implies that the trailer or truck has been equipped with a reel stand (jacks) and may serve as a reel transport or *payout* unit, or both, for stringing operations. Depending upon the sizes of the reels to be carried, the transporting vehicles may range from single-axle trailers to semitrucks with trailers having multiple axles. *Synonyms:* reel truck; reel transporter; reel trailer. (T&D/PE) 524a-1993r, 524-1992r

reel tensioner (conductor stringing equipment) A device designed to generate tension against a pulling line or conductor during the stringing phase. Some are equipped with their own engines, which retard the supporting shaft for the reel mechanically, hydraulically, or through a combination of both. The shaft, in turn, retards the reel. Some of these devices function as either a puller or tensioner. Other tensioners are equipped only with friction type retardation. *Synonyms:* tensioner; retarder. (T&D/PE) 524a-1993r, 524-1980s

reel trailer *See:* reel stand.

reel transporter *See:* reel stand.

reel truck *See:* reel stand.

reel winder A device designed to serve as a recovery unit for a pulling line. It is normally equipped with its own engine, which drives a supporting shaft for a reel mechanically, hydraulically, or through a combination of both. The shaft, in turn, drives the reel. It is normally used to rewind a pulling line as it leaves the bullwheel puller during stringing operations. This unit is not intended to serve as a puller, but sometimes serves this function where only low tensions are involved. *Synonym:* takeup reel. (T&D/PE) 524-1992r

re-encoded checkback message Message from the initiating end that is re-encoded by the receiving end. A new message is sent to the initiating end to verify error-free receipt and proper interpretation of the message. In typical applications the initiating end is the master station and the receiving end is the RTU. Preferred usage is re-encoded, which allows the master station to verify not only that the communication was error free, but also that the RTU's I/O hardware and software acted correctly in interpreting the selection. (SUB/PE) C37.1-1994

re-engineering (1) The process of examining and altering an existing system to reconstitute it in a new form. May include reverse engineering (analyzing a system and producing a representation at a higher level of abstraction, such as design from code), restructuring (transforming a system from one representation to another at the same level of abstraction), redocumentation (analyzing a system and producing user or maintenance documentation), forward engineering (using software products derived from an existing system, together with new requirements, to produce a new system), retargeting (transforming a system to install it on a different target system), and translation (transforming source code from one language to another or from one version of a language to another). (C/SE) J-STD-016-1995

(2) The process of improving a system after production through modification to correct a design deficiency or to make an incremental improvement. (C/SE) 1220-1998

reenterable *See:* reentrant.

reentrant Pertaining to a software module that can be entered as part of one process while also in execution as part of another process and still achieve the desired results. *Synonym:* reenterable. (C) 610.12-1990

reentrant-beam crossed-field amplifier (amplitron) (microwave tubes) A crossed-field amplifier in which the beam is reentrant and interacts with either a forward or a backward wave. (ED) [45]

reentrant beam (microwave tubes) An undeterminated recirculating electron beam. (ED) [45]

reentrant circuit (microwave tubes) A slow-wave structure that closes upon itself. (ED) [45]

reentrant function A function whose effect, when called by two or more threads, is guaranteed to be as if the threads each executed the function one after another in an undefined order, even if the actual execution is interleaved. (C/PA) 9945-1-1996

reentrant switching network (telephone switching systems) A switching network in which outlets (usually last choice) from a given connecting stage are connected to inlets of the same or previous stage. (COM) 312-1977w

reentry communication (communication satellite) Communication during re-entry of a space vehicle into the atmosphere. Usually the ionization requires a special system of modulation to overcome the communication blackout. (COM) [19]

reentry point The place in a software module at which the module is reentered following a call to another module. (C) 610.12-1990

referee test (metering) A test made by or in the presence of one or more representatives of a regulatory body or other impartial agency. (ELM) C12.1-1982s

reference *See:* linearity.

reference accuracy (automatic null-balancing electric instrument) A number or quantity that defines the limit of error under reference operating conditions. *Notes:* 1. It is usually expressed as a percent of the span. It is preferred that a + sign or − sign or both precede the number of quantity. The absence of a sign infers a ± sign. 2. Reference accuracy does not include accuracy of sensing elements or intermediate means external to the instrument. *See also:* error and correction; accuracy rating. (EEC/EMI) [112]

reference address *See:* base address.

reference air line A uniform section of air-dielectric transmission line of accurately calculable characteristic impedance used as a standard immittance. *See also:* transmission line. (IM/HFIM) [40]

reference atmosphere for refraction *See:* standard atmosphere for refraction.

reference audio noise power output (mobile communications receivers) The average audio noise power present at the output of an unsquelched receiver having no radio-frequency signal input in which the audio gain has been adjusted for the reference audio power output. (VT) 184-1969w

reference audio power output (mobile communications receivers) The manufacturer's rated audio-frequency power available at the output of a properly terminated receiver, when responding to a standard test modulated radio-frequency input signal at a −80 dBW level. (VT) 184-1969w

reference ballast (illuminating engineering) A ballast which is specially constructed, having certain prescribed characteristics and which is used for testing electric-discharge lamps and other ballasts. (EEC/IE) [126]

reference ballasts Specially constructed series ballasts having certain prescribed characteristics. *Note:* They serve as comparison standards for use in testing ballasts or lamps and are used also in selecting the reference lamps that are necessary for the testing of ballasts. Reference ballasts are characterized by a constant impedance over a wide range of operating current. They also have constant characteristics that are relatively uninfluenced by time and temperature. *See also:* primary standard; fixed impedance-type ballast. (EEC/IE) [126]

reference black level (television) The picture-signal level corresponding to a specified maximum limit for black peaks. (BT/AV) [34]

reference block (numerically controlled machines) A block within the program identified by an *o* (letter o) in place of the word address *n* and containing sufficient data to enable resumption of the program following an interruption. This block should be located at a convenient point in the program that enables the operator to reset and resume operation. (IA) [61]

reference boresight A direction established as a reference for the alignment of an antenna. *Note:* The direction can be established by optical, electrical or mechanical means. *See also:* electrical boresight. (AP/ANT) 145-1993

reference clock A clock of very high stability and accuracy that may be completely autonomous and whose frequency serves as a basis of comparison for the frequency of other clocks. (COM/TA) 1007-1991r

reference clock node *See:* master clock node.

reference codec (1) A codec that approaches the performance of an ideal codec and has superior, well-defined characteristics used for testing digital telephone sets. (COM/TA) 269-1992
(2) A well-defined analog-to-digital and digital-to-analog converter for testing digital telephones using analog test equipment. (COM/TA) 1329-1999

reference conditions The values assigned for the different influence quantities at which or within which the instrument complies with the requirements concerning errors in indication. *See also:* accuracy rating. (EEC/AII) [102]

reference current (Iref) (1) (fluorescent lamps) The value of current specified in a specific lamp standard. *Note:* It is normally the same as the value of current for which the corresponding lamp is rated. Since the reference ballast is a standard that is representative of the impedance of lamp power sources installed, it is not necessary to change this current value unless major changes in lamp standards require modification of the ballast impedance. For this reason, reference ballast characteristics are specified in terms of, and with reference to, reference current. (EEC/LB) [96]
(2) The peak value of the resistive component of a power-frequency current high enough to make the effects of stray capacitance of the arrester negligible. This current level shall be specified by the manufacturer. *Note:* Depending on the arrester design, the Iref will typically be in the range of 0.05–1.0 mA per sq cm of disk area. (SPD/PE) C62.11-1999

reference deflection (volume measurements of electrical speech and program waves) The deflection to the meter-scale point marked 0 vu, 100, or both. *Note:* This is the deflection at which the meter should be used. *See also:* vu. (BT/AV) 152-1953s

reference designation (1) (abbreviation) (symbols) Numbers, or letters and numbers, used to identify and locate units, portions thereof, and basic parts of a specific set. Compare with: functional designation and symbol for a quantity. *See also:* abbreviation. (GSD) 267-1966
(2) (electric and electronics parts and equipment) Letters or numbers, or both, used to identify and locate discrete units, portions thereof, and basic parts of a specific set. *Note:* A reference designation is not a letter symbol, abbreviation, or functional designation for an item. (GSD) 200-1975w

reference direction (1) (navigation aid terms) A direction from which other directions are reckoned; for example, true north, grid north, and so on. (AES/GCS) 172-1983w
(2) (specified circuit) With reference to the boundary of a delimited region, the arbitrarily selected direction in which electric energy is assumed to be transmitted past the boundary, into or out of the region. *Notes:* 1. When the actual direction of energy flow is the same as the reference direction, the sign is negative. 2. Unless specifically stated to the contrary, it shall be assumed that the reference direction for all power, energy, and quadergy quantities associated with the circuit is the same as the reference direction of the energy flow. 3. In these definitions it will be assumed that the reference direction of the current in each conductor of the circuit is the same as the reference direction of energy flow. (Std100/EDPG) 270-1966w

reference directivity *See:* standard directivity.

reference distance (sound measurement) A standard 1 m distance from the major machine surfaces at which mean sound level data shall be reported. (PE/EM) 85-1973w

reference document A standard that shall be on hand and available to the user in order to help in the implementation of another standard. (ATLAS) 1226-1993s

referenced shared memory object (1) A shared memory object that is open or has one or more mappings defined on it. (C/PA) 9945-1-1996
(2) A shared memory object that is open or has one or more mappings defined on it. (C) 1003.5-1999

reference edge *See:* document reference edge.

reference excursion (analog computer) The range from zero voltage to nominal full-scale operating voltage. *See also:* electronic analog computer. (C) 165-1977w, 610.10-1994w

reference expression An expression that uniquely identifies a box, a node or function, a diagram, or a model page within an IDEF0 model. (C/SE) 1320.1-1998

reference frequency The frequency upon which a phasor or amplitude-or-phase representation of signals is based. (IT) [123]

reference frequency, upper and lower *See:* bandwidth.

reference grounding point (health care facilities) A terminal bus which is the equipment grounding bus or an extension of the equipment grounding bus and is a convenient collection point for grounding of electrical appliances and equipment, and, when necessary and appropriate, exposed conductive surfaces in a patient vicinity. (NESC/EMB) [47], [86]

reference-input elements (automatic control) The portion of the controlling system that changes the reference input signal in response to the command. *See also:* feedback control system. (IA/IAC) [60]

reference input signal (1) The command expressed in a form directly usable by the system. The reference input signal is in the terms appropriate to the form in which the signal is used, that is, voltage, current, ampere-turns, etc. *See also:* feedback control system. (IA/IAC) [60]
(2) (control system feedback) A signal external to a control loop that serves as the standard of comparison for the directly controlled variable. See the figure attached to the definition of **signal, feedback.** *See also:* feedback control system. (PE/EDPG) 421-1972s

reference laboratory A laboratory responsible for the national testing program for a sector of the radioassay community. A reference laboratory is authorized to prepare testing media by adding known amounts of radioactive material for distribution to service laboratories. The reference laboratory is responsible for evaluating the performance of the service laboratories in terms of accuracy and precision. (NI) N42.23-1995

reference lamp (1) (mercury) A seasoned lamp that under stable burning conditions, in the specified operating position (usually vertically, base up), and in conjunction with the reference ballast rated input voltage, operates at values of lamp volts, watts, and amperes, each within ± 2% of the nominal values. (EEC/LB) [98], [97]
(2) (fluorescent) Seasoned lamps that under stable burning conditions, in conjunction with the reference ballast specified for the lamp size and rating, and at the rated reference ballast supply voltage, operate at values of lamp volts, watts, and amperes each within ± 2 1/2% of the values, and under conditions established by present standards. *See also:* reference ballasts. (EEC/LB) [94]

reference line (1) (navigation aid terms) A line from which angular or linear measurements are reckoned. (AES/GCS) 172-1983w
(2) (illuminating engineering) Either of two radial lines where the surface of the cone of maximum candlepower is intersected by a vertical plane parallel to the curb line and passing through the light-center of the luminaire. (EEC/IE) [126]

reference lines and points (pulse terminology) Constructs which are (either actually or figuratively) superimposed on waveforms for descriptive or analytical purposes. Unless otherwise specified, all defined lines and points lie within a waveform epoch. *See also:* knot; magnitude origin line; cubic natural spline; time origin line; magnitude-referenced point; time referenced point; time reference line; magnitude reference line. (IM/WM&A) 194-1977w

reference material (standard) A material or substance of one or more properties that are sufficiently well established to be used for the calibration of an apparatus, the assessment of a measurement method, or for assigning values to materials. (NI) N42.23-1995, N42.22-1995

reference model A structured collection of concepts and their relationships that scope a subject and enable the partitioning of the relationships into topics relevant to the overall subject and that can be expressed by a common means of description. (C/PA) 14252-1996

reference modulation (navigation aids) (very high-frequency omnidirectional range) That modulation of the ground-station radiation which produces a signal in the air-borne receiver whose phase is independent of the bearing of the receiver; the reference signal derived from this modulation is used for comparison with the variable signal. (AES/GCS) 172-1983w

reference noise (data transmission) The magnitude of circuit noise that will produce a circuit-noise-meter reading equal to that produced by 10^{-12} watt of electric power at 1000 Hz (hertz). (PE) 599-1985w

reference operating conditions (automatic null-balancing electric instrument) The conditions under which reference performance is stated and the base from which the values of operating influences are determined. *See also:* measurement system. (EEC/EMI) [112]

reference orientation (radioactivity monitoring instrumentation) The orientation in which the instrument is normally intended to be operated as stated by the manufacturer. (NI) N42.17B-1989r

reference performance (1) (watthour meter) Performance at specified reference conditions for each test, used as a basis for comparison with performance under other conditions of the test. (ELM) C12.1-1982s
(2) (automatic null-balancing electric instrument) The limits of the values of certain operating characteristics of the instrument that will not be exceeded under any combination of reference operating conditions. *See also:* test; electricity meter. (EEC/EMI) [112]

reference plane (1) A plane perpendicular to the direction of propagation in a waveguide or transmission line, to which measurement or immittance, electric length, reflection coefficients, scattering coefficients, and other parameters may be referred. *See also:* waveguide. (IM/HFIM) [40]
(2) A theoretical plane, having neither thickness nor tolerance, used to separate space. (C/BA/MM) 1301.2-1993, 1301.1-1991, 1301.3-1992r
(3) A theoretical plane having neither thickness nor tolerance. (C/BA) 1301.4-1996

reference phantom A 30 cm × 30 cm × 15 cm block of polymethyl methacrylate (PMMA). This phantom follows the recommendations of ICRU and ISO for a simplified practical phantom for the calibration of dosimeters. *Note:* The conversion factors for the same sized phantom constructed of ICRU tissue should be used for evaluating the dose equivalent. (NI) N42.20-1995

reference plane, electrical *See:* electrical reference plane.

reference plane, mechanical *See:* mechanical reference plane.

reference point of an instrument A physical mark, or marks, on the outside of an instrument used to position it at a point where the conventionally true value of a quantity to be measured is known. (NI) N42.20-1995

reference power supply A regulated, electronic power supply furnishing the reference voltage. *See also:* electronic analog computer. (C) 165-1977w

reference quality voltage DC source that is capable of sourcing and sinking current over a defined range without appreciable change of voltage and whose stability over a given time interval is guaranteed. (C/TT) 1149.4-1999

reference radius (sound measurement) The sum of the reference distance and one half the maximum linear dimension as defined for small-, medium-, or large machines. *See also:* machine. (PE/EM) 85-1973w

reference sensitivity (mobile communications receivers) The level of a radio-frequency signal with standard test modulation which provides a 12-decibel sinad with at least 50% reference audio power output. (VT) 184-1969w

reference standards (nuclear power generating station) Standards (that is, primary, secondary and working standards, where appropriate) used in a calibration program. These standards establish the basic accuracy limits for that program. (PE/NP) 498-1985s

reference standard watthour meter A meter used to maintain the unit of electric energy. It is usually designed and operated to obtain the highest accuracy and stability in a controlled laboratory environment. (ELM) C12.1-1982s

reference surface (fiber optics) That surface of an optical fiber which is used to contact the transverse-alignment elements of a component such as a connector. For various fiber types, the reference might be the fiber core, cladding, or buffer layer surface. *Note:* In certain cases the reference surface may not be an integral part of the fiber. *See also:* ferrule; optical waveguide connector. (Std100) 812-1984w

reference system (loudness ratings of telephone connections) A system that provides 0 dB acoustic gain between a mouth reference point at 25 mm in front of a talker's lips and an ear reference point at the entrance to the ear canal of a listener, when the listener is using an earphone. This system is assigned a loudness rating of 0 dB. The frequency characteristic of the system must be flat over the range 300-3300 Hz and show infinite attenuation outside of this range. *Note:* If an actual reference system is constructed for subjective comparison purposes, the system response at 300 and 3300 Hz shall be down 3 ± 1 dB relative to the midband response. The gain of the system shall be adjusted to compensate for the finite slope of the filter skirts and deviation from flatness of the pass band. The amount of this adjustment can be determined by first calculating the objective loudness rating (OLR) over a frequency range that includes at least the 50 dB down points of the real response, and next calculating the OLR of the ideal response, over the same frequency range. The difference between the OLRs is the required gain adjustment. (COM/TA) 661-1979r

reference test field (direction-finder testing) (navigation aid terms) That field strength, in microvolts per meter, numerically equal to the DF (direction finder) sensitivity. (AES/GCS) 172-1983w

reference threshold squelch adjustment (mobile communications receivers) The minimum adjustment position of the squelch control required to reduce the reference audio noise power output by at least 40 dB. (VT) 184-1969w

reference time (magnetic storage) An instant near the beginning of switching chosen as an origin for time measurements. It is variously taken as the first instant at which the instantaneous value of the drive pulse, the voltage response of the magnetic cell, or the integrated voltage response reaches a specified fraction of its peak pulse amplitude. (C) [20]

reference voltage (1) (analog computer) In an analog computer, a voltage used as a standard of reference, usually the nominal full scale of the computer. (C) 165-1977w
(2) The lowest peak value independent of polarity of power-frequency voltage, divided by the square root of 2, required to produce a resistive component of current equal to the reference current of the arrester or arrester element. The reference voltage of a multiunit arrester is the sum of the reference voltages of the series units. The voltage level shall be specified by the manufacturer. (SPD/PE) C62.11-1999
(3) The point on the voltage/current (V/I) characteristic where the static var compensator (SVC) is at zero output (i.e., where no vars are absorbed from, or supplied to, the transmission system at the point of connection). (PE/SUB) 1031-2000

reference volume (volume measurements of electrical speech and program waves) The level which gives a reading of 0 vu on a standard volume indicator. *Notes:* 1. The methods of reading and calibration are described in Section 3 of IEEE Std 152-1953w. 2. The "reading of 0 vu" is the algebraic sum of the meter and attenuator readings on the standard volume indicator. *See also:* standard volume indicator; vu. (BT/AV) 152-1953s

reference volume control setting The volume control position resulting in a specified nominal receive loudness rating. (COM/TA) 1329-1999

reference waveguide A uniform section of waveguide with accurately fabricated internal cross-sectional dimensions used as a standard of immittance. *See also:* waveguide. (IM/HFIM) [40]

reference white (A) (television) (original scene) The light from a nonselective diffuse reflector that is lighted by the normal illumination of the scene. *Notes:* 1. Normal illumination is not intended to include lighting for special effects. 2. In the reproduction of recorded material, the word scene refers to the original scene. **(B) (television)** (color television display) That white with which the display device simulates reference white of the original scene. *Note:* In general, the reference whites of the original scene and of the display device are not colorimetrically identical. (BT/AV) 201-1979

reference white level (television) The picture-signal level corresponding to a specified maximum limit for white peaks. *See also:* television. (BT/AV) [34]

referential integrity (A) A guarantee that a reference refers to an object that exists. **(B)** A guarantee that all specified conditions for a relationship hold true. For example, if a class is declared to require at least one instance of a related state class, it would be invalid to allow an instance that does not have such a relationship. (C/SE) 1320.2-1998

referral An outcome that can be returned by a DSA that cannot perform a directory operation itself, and that identifies one or more other DSAs more able to perform the directory operation. (C/PA) 1328.2-1993w, 1224.2-1993w, 1327.2-1993w, 1326.2-1993w

refill unit (of a high-voltage fuse unit) An assembly comprised of a conducting element, the complete arc-extinguishing medium, and parts normally required to be replaced after each circuit interruption to restore the fuse unit to its original operating condition. (SWG/PE) C37.100-1992, C37.40-1993

reflectance (1) (A) (electrical power systems in commercial buildings) The ratio of the light reflected by a surface to the light incident. **(B) (fiber optics)** The ratio of reflected power to incident power. *Note:* In optics, frequently expressed as optical density or as a percent; in communication applications, generally expressed in decibels (dB). Reflectance may be defined as specular or diffuse, depending on the nature of the reflecting surface. Formerly: "reflection." *See also:* reflection. **(C) (illuminating engineering)** ($\rho=\Phi_r/\Phi_i$) (of a surface or medium) The ratio of the reflected flux to the incident flux. Reflectance is a function of:

a) Geometry (i) of the incident flux (ii) of collection for the reflected flux;
b) spectral distribution (i) characteristic of the incident flux (ii) weighting function for the collected flux; and
c) polarization (i) of the incident flux (ii) component defined for the collected flux.

Notes: 1. Unless the state of polarization for the incident flux and the polarized component of the reflected flux are stated, it shall be considered that the incident flux is unpolarized and that the total reflected flux (including all polarizations) is evaluated. 2. Unless qualified by the term "spectral" (see spectral reflectance) or other modifying adjectives, luminous reflectance (see luminous reflectance) is meant. 3. If no qualifying geometric adjective is used, reflectance for hemispherical collection is meant. 4. Certain of the reflectance terms are theoretically imperfect and are recognized only as practical concepts to be used when applicable. Physical measurements of the incident and reflected flux are always biconical in nature. Directional reflectances (see above) cannot exist since one component is finite while the other is infinitesimal; here the reflectance-distribution function is required. However, the concepts of directional and hemispherical reflectances have practical application in instrumentation, measurements, and calculations when including the aspect of the nearly zero or nearly 2π conical angle would increase complexity without appreciably affecting the immediate results. 5. In each case of conical incidence or collection, the solid angle is not restricted to a right circular cone, but may be of any cross section including rectangular, a ring, or a combination of two or more solid angles. *See also:* hemispherical-directional reflectance; conical-hemispherical reflectance, $\rho(\omega_i; 2\pi)$; bidirectional reflectance; biconical reflectance; bihemispherical reflectance; hemispherical-conical reflectance; bidirectional reflectance–distribution function. (Std100/IA/EEC/IE/PSE) 241-1990, 812-1984, [126]

(2) *See also:* power reflection coefficient. (AP/PROP) 211-1997

(3) (laser maser) (reflectivity, ρ) The ratio of total reflected radiant power to total incident power. (LEO) 586-1980w

(4) (computer graphics) The amount of light reflected from the surface of a three-dimensional object. This quality is used in rendering three-dimensional objects in computer graphics. (C) 610.6-1991w

reflectance factor (illuminating engineering) The ratio of the flux actually reflected by a sample surface to that which would be reflected into the same reflected-beam geometry by an ideal (lossless), perfectly diffuse (lambertian) standard surface irradiated in exactly the same way as the sample. Note the analogies to reflectance in the fact that nine canonical forms are possible paralleling bihemispherical reflectance, hemispherical-conical reflectance, hemispherical-directional reflectance, conical-hemispherical reflectance, biconical reflectance, conical-directional reflectance, directional-hemispherical reflectance, directional-conical reflectance, and bidirectional reflectance, that spectral may be applied as a modifier, that it may be luminous or radiant reflectance factor, etc. (EEC/IE) [126]

reflected binary code *See:* Gray code.

reflected binary unit-distance code *See:* Gray code.

reflected code *See:* Gray code.

reflected glare (illuminating engineering) Glare resulting from reflections of high luminance in polished or glossy surfaces in the field of view. It usually is associated with reflections from within a visual task or areas in close proximity to the region being viewed. *See also:* veiling reflections. (EEC/IE) [126]

reflected harmonics (electric conversion) Harmonics produced in the prime source by operation of the conversion equipment. These harmonics are produced by current-impedance (IZ) drop due to nonsinusoidal load currents, and by switching or commutating voltages produced in the conversion equipment. (AES) [41]

reflected-light scanning The scanning of changes in the magnitude of reflected light from the surface of an illuminated web. (IA/ICTL/IAC) [60]

reflected wave (1) (data transmission) When a wave in one medium is incident upon a discontinuity or a different medium, the reflected wave is the wave component that results in the first medium in addition to the incident wave. *Note:* The reflected wave includes both the reflected rays of geometrical optics and the diffracted wave. (PE) 599-1985w

(2) (waveguide) At a transverse plane in a transmission line or waveguide, a wave returned from a reflecting discontinuity in a direction opposite to the incident wave. *See also:* incident wave. (MTT) 146-1980w

(3) (surge arresters) A wave, produced by an incident wave, that returns in the opposite direction to the incident wave after reflection at the point of transition. (PE) [9], [84]

(4) (overhead power lines) When a wave in one medium is incident upon a discontinuity or a different medium, the reflected wave includes the wave component traveling in a different direction to the incident wave in the first medium, as well as the incident wave. If the wave is in a unidimensional medium, i.e., a transmission line, then the reflected wave travels in the opposite direction to the incident wave. (T&D/PE) 539-1990

(5) For two media, separated by a planar interface, that part of the incident wave that is returned to the first medium. The direction of propagation of the reflected wave is given by Snell's law of reflection. (AP/PROP) 211-1997

reflecting slave An unselected slave that forces the selected slave into a write-only mode. In read transfers the reflecting slave substitutes itself for the selected slave in providing data while causing the selected slave to store the data that appears on the bus. In write transfers the reflecting slave copies the data into itself as well as the selected slave. Reflecting slaves can operate only during single-slave mode transactions. (C/MM) 896.1-1987s

reflection (1) (fiber optics) The abrupt change in direction of a light beam at an interface between two dissimilar media so that the light beam returns into the medium from which it originated. Reflection from a smooth surface is termed specular, whereas reflection from a rough surface is termed diffuse. *See also:* total internal reflection; critical angle; reflectivity; reflectance. (Std100) 812-1984w

(2) (illuminating engineering) A general term for the process by which the incident flux leaves a (stationary) surface or medium from the incident side, without change in frequency. *Note:* Reflection is usually a combination or regular and diffuse reflection. *See also:* regular reflection; diffuse reflection. (EEC/IE) [126]

(3) (laser maser) Deviation of radiation following incidence on a surface. (LEO) 586-1980w

(4) A form of data modification in which PDUs sent by an entity are returned in an unauthorized manner. This can be attempted by a combination of techniques involving deleting, delaying, and reinserting data; and/or modifying address or sequence control information. (C/LM) 802.10-1998

reflection coefficient (1) (waveguide) At a given frequency, at a given point, and for a given mode of propagation, the ratio of some quantity associated with the reflected wave to the corresponding quantity in the incident wave. *Note:* The reflection coefficient may be different for different associated quantities, and the chosen quantity must be specified. The voltage reflection coefficient is most commonly used and is defined as the ratio of the complex electrical field strength (or voltage) of the reflected wave to that of the incident wave. Examples of other quantities are power or current. (MTT) 146-1980w

(2) (overhead power lines) At a given frequency, at a given point, and for a given mode of propagation, the ratio of some quantity associated with the reflected wave to the corresponding quantity in the incident wave. (T&D/PE) 539-1990

(3) *See also:* Fresnel coefficients. (AP/PROP) 211-1997

reflection color tube A color-picture tube that produces an image by means of electron reflection techniques in the screen region. *See also:* television. (ED) 161-1971w, [45]

reflection error (navigation aids) The error due to the fact that some of the total received signal arrives from a reflection rather than all by way of the direct path. (AES/GCS) 172-1983w

reflection factor (1) (data transmission) The reflection factor between two impedances Z_1 and Z_2 is:

$$\frac{(4Z_1Z_2)^{1/2}}{Z_1 + Z_2}$$

Physically, the reflection factor is the ratio of the current delivered to a load, whose impedance is not matched to the source, to the current that would be delivered to a load of matched impedance. (PE) 599-1985w

(2) (reflex klystrons) The ratio of the number of electrons of the reflected beam to the total number of electrons that enter the reflector space in a given time. *See also:* velocity-modulated tube. (ED) [45], [84]

(3) (electrothermic power meters) The ratio of the power absorbed in, to the power incident upon, a load; mathematically, $1 - |\Gamma_t|^2$, where $|\Gamma_t|$ is the magnitude of the reflection coefficient of the load. (IM) 544-1975w

(4) (electrothermic power meters) *See also:* reflection coefficient.

(5) *See also:* Fresnel coefficients. (AP/PROP) 211-1997

reflection loss (1) (data transmission) The reflection loss for a given frequency at the junction of a source of power and a load is given by the formula

$$20\log_{10} \frac{Z_1 + Z_2}{(4Z_1Z_2)^{1/2}} \text{ dB}$$

where Z_1 is the impedance of the source of power and Z_2 is the impedance of the load. Physically, the reflection loss is the ratio, expressed in decibels (dB), of the scalar values of the volt-amperes delivered to a load of the same impedance as the source. The reflection loss is equal to the number of decibels which corresponds to the scalar value of the reciprocal of the reflection factor. *Note:* When the two impedances have opposite phases and appropriate magnitudes, a reflection gain may be obtained. (PE) 599-1985w

(2) (waveguide) (or gain) The ratio of incident to transmitted power at a reference plane of a network. (MTT) 146-1980w

reflectionless termination A termination that terminates a waveguide or transmission line without causing a reflected wave at any transverse section. *See also:* waveguide; transmission line. (MTT) 146-1980w

reflectionless transmission line A transmission line having no reflected wave at any transverse section. *See also:* transmission line. (IM/HFIM) [40]

reflectionless waveguide A waveguide having no reflected wave at any transverse section. *See also:* waveguide. (IM/HFIM) [40]

reflection mode photocathode (photomultipliers for scintillation counting) A photocathode wherein photoelectrons are emitted from the same surface as that on which the photons are incident. (NPS) 398-1972r

reflection modulation (storage tubes) A change in character of the reflected reading beam as a result of the electrostatic fields associated with the stored signal. A suitable system for collecting electrons is used to extract the information from the reflected beam. *Note:* Typically the beam approaches the target closely at low velocity and is then selectively reflected toward the collection system. *See also:* charge-storage tube. (ED) 158-1962w

reflections (broadband local area networks) Echoes created in a cable system by impedance mismatches and cable discontinuities or irregularities. *See also:* echo. (LM/C) 802.7-1989r

reflection, specular *See:* specular reflection.

reflective array antenna An antenna consisting of a feed and an array of reflecting elements arranged on a surface and adjusted so that the reflected waves from the individual elements combine to produce a prescribed secondary pattern. *Note:* The reflecting elements are usually waveguides containing electrical phase shifters and are terminated by short circuits. *Synonym:* reactive reflector antenna. (AP/ANT) 145-1993

reflective array compressor (RAC) A device that uses reflections of the surface acoustic wave from an array of oblique grooves, metal stripes, or dots to achieve the desired dispersive delay function that provides energy-spreading or compression. (UFFC) 1037-1992w

reflective dot array (RDA) A type of device that uses reflections of the surface acoustic wave from oblique rows of metallic dots to achieve desired filter function. (UFFC) 1037-1992w

reflective memory Memory that may physically reside in more than one location (on multiple modules) but that contains the identical information within that memory at all the physical locations. (C/BA) 1014.1-1994w

reflectivity (1) (fiber optics) The reflectance of the surface of a material so thick that the reflectance does not change with increasing thickness; the intrinsic reflectance of the surface, irrespective of other parameters such as the reflectance of the rear surface. No longer in common usage. (Std100) 812-1984w

(2) (photovoltaic power system) The reflectance of an opaque, optically smooth, clean portion of material. (AES) [41]

(3) For a radio-frequency (RF) absorber, the ratio of the plane wave reflected power density (P_r) to the plane wave incident power density (P_i) at a reference point in space. It is expressed in dB as $R = 10\log_{10}(P_r/P_i)$. (EMC) 1128-1998

(4) The ratio of reflected to incident power densities of a plane wave incident on a surface and equal to the square of the magnitude of the reflection coefficient. (AP/PROP) 211-1997

reflectometer (1) (illuminating engineering) A photometer for measuring reflectance. *Note:* Reflectometers may be visual or physical instruments. (EEC/IE) [126]

(2) (illuminating engineering) An instrument for the measurement of the ratio of reflected-wave to incident-wave amplitudes in a transmission system. *Note:* Many instruments yield only the magnitude of this ratio. *See also:* instrument. (IM/HFIM) [40]

reflector (1) (data transmission) One or more conductors or conducting surfaces for reflecting radiant energy. (PE) 599-1985w

(2) (illuminating engineering) A device used to redirect the flux from a source by the process of reflection. (EEC/IE) [126]

(3) (wave propagation) A reflector comprises one or more conductors or conducting surfaces for reflecting radiant energy. *See also:* reflector element; antenna. (PE/PSIM) 81-1983

(4) A surface acoustic wave reflecting component that normally makes use of the periodic discontinuity provided by an array of metal strips, dots, or grooves. (UFFC) 1037-1992w

(5) *See also:* subreflector; horn reflector antenna; Gregorian reflector antenna; corner reflector; Cassegrain reflector antenna; offset paraboloidal reflector; paraboloidal reflector; main reflector; reflector element; spherical reflector; cylindrical reflector; reflector antenna; umbrella reflector antenna; toroidal reflector. (AP/ANT) 145-1993

reflector antenna An antenna consisting of one or more reflecting surfaces and a radiating [receiving] feed system. *Note:* Specific reflector antennas often carry the name of the reflector used as part of the term used to specify it; for example, paraboloidal reflector antenna. (AP/ANT) 145-1993

reflector element A parasitic element located in a direction other than forward of the driven element of an antenna intended to increase the directivity of the antenna in the forward direction. (AP/ANT) 145-1993

reflector space (reflex klystrons) The part of the tube following the buncher space, and terminated by the reflector. *See also:* velocity-modulated tube. (ED) [45], [84]

reflex baffle (audio and electroacoustics) A loudspeaker enclosure in which a portion of the radiation from the rear of the diaphragm is propaged outward after controlled shift of phase or other modification, the purpose being to increase the useful radiation in some portion of the frequency spectrum. (SP) [32]

reflex bunching The bunching that occurs in an electron stream that has been made to reverse its direction in the drift space. (ED) 161-1971w

reflex circuit A circuit through which the signal passes for amplification both before and after a change in its frequency. (EEC/PE) [119]

reflexive ancestor (of a class) The class itself or any of its generic ancestors. *Contrast:* ancestor. *See also:* generic ancestor. (C/SE) 1320.2-1998

reflex klystron (microwave tubes) A single-resonator oscillator klystron in which the electron beam is reversed by a negative electrode so that it passes twice through the resonator, thus providing feedback. (ED) [45]

reformatting *See:* reorganization.

reforming (semiconductor rectifiers) The operation of restoring by an electric or thermal treatment, or both, the effectiveness of the rectifier junction after loss of forming. *See also:* rectification. (IA) 59-1962w, [12]

refractance The amount that light rays are bent upon intersecting the surface of a three-dimensional object. This property is used in rendering three-dimensional objects in computer graphics. (C) 610.6-1991w

refracted near-field scanning method *See:* refracted ray method.

refracted ray (fiber optics) In an optical waveguide, a ray that is refracted from the core into the cladding. Specifically, a ray at radial position r having direction such that

$$\frac{n^2(r) - n^2(a)}{1 - (r/a)^2\cos^2\phi(r)} \leq \sin^2\theta(r)$$

where $\phi(r)$ is the azimuthal angle of projection of the ray on the transverse plane, $\theta(r)$ is the angle the ray makes with the waveguide axis, $n(r)$ is the refractive index at the core radius, and a is the core radius. Refracted rays correspond to radiation modes in the terminology of mode descriptors. *See also:* radiation mode; cladding ray; leaky ray; guided ray. (Std100) 812-1984w

refracted ray method (fiber optics) The technique for measuring the index profile of an optical fiber by scanning the entrance face with the vertex of a high numerical aperture cone and measuring the change in power of refracted (unguided) rays. *Synonym:* refracted near-field scanning method. *See also:* refracted ray; refraction. (Std100) 812-1984w

refracted wave (1) (data transmission) That part of an incident wave that travels from one medium into a second medium. *Synonym:* transmitted wave. (PE) 599-1985w
(2) For two media, that part of the incident wave that travels from the first medium into the second medium. *Note:* For planar interfaces, the direction of propagation of the refracted wave is given by Snell's law. *Synonym:* transmitted wave. (AP/PROP) 211-1997

refraction (1) (fiber optics) The bending of a beam of light in transmission through an interface between two dissimilar media or in a medium whose refractive index is a continuous function of position (graded index medium). *See also:* angle of deviation; refractive index. (Std100) 812-1984w
(2) (radio-wave propagation) Of a traveling wave, the change in direction of propagation resulting from the spatial variation of refractive index of the medium. (AP) 211-1977s

refraction error (1) (navigation aids) Error due to the bending of one or more wave paths by the propagation medium. (AES/GCS) 172-1983w
(2) Error in angle and/or range due to the bending of one or more wave paths by changes in the refractive index of the propagation medium. (AES) 686-1997

refraction loss That part of the transmission loss due to refraction resulting from nonuniformity of the medium. (SP) [32]

refractive index (1) (A) (data transmission) (wave transmission medium). The ratio of the phase velocity in free space to that in the medium. **(B) (data transmission)** (dielectric for electromagnetic wave). The ratio of the sine of the angle of incidence to the sine of the angle of refraction as the wave passes from a vacuum into the dielectric. The angle of incidence θ_i is the angle between the direction of travel of the wave in vacuum and the normal to the surface of the dielectric. The angle of refraction θ_r is the angle between the direction of travel of the wave after it has entered the dielectric and the normal to the surface. Refractive index is related to the dielectric constant through the following relation:

$$n = \frac{\sin\theta_i}{\sin\theta_r} = (\varepsilon')^{1/2}$$

where ε' is the real dielectric constant. Since ε' and n vary with frequency, the above relation is strictly correct only if all quantities are measured at the same frequency. The refractive index is also equal to the ratio of the velocity of the wave in the vacuum to the velocity in the dielectric medium. (PE) 599-1985
(2) (fiber optics) (of a medium) Denoted by n, the ratio of the velocity of light in vacuum to the phase velocity in the medium. *Synonym:* index of refraction. *See also:* index profile; scattering; core; linearly polarized mode; optical path length; numerical aperture; step index optical waveguide; normalized frequency; group index; index matching material; fused silica; cladding; profile dispersion; weakly guiding fiber; critical angle; graded index optical waveguide; dispersion; mode; power-law index profile; material dispersion; Fresnel reflection. (Std100) 812-1984w
(3) A dimensionless complex quantity, characteristic of a medium and so defined that its real part, called the refractive index, n, is the ratio of the phase velocity in free space to the phase velocity in the medium. The product of the imaginary part of the refractive index and the free space propagation constant is the attenuation constant in the medium. (AP/PROP) 211-1997

refractive index, complex *See:* complex refractive index.

refractive index contrast (fiber optics) Denoted by Δ, a measure of the relative difference in refractive index of the core and cladding of a fiber, given by $\Delta = (n_1^2 - n_2^2)/2n_1^2$ where n_1 and n_2 are, respectively, the maximum refractive index in the core and the refractive index of the homogeneous cladding. (Std100) 812-1984w

refractive index gradient The change of the atmospheric refractive index with height. Refraction may be included in propagation calculations by using an effective Earth radius of K times the geometrical radius of the Earth (6375 km) and straight line propagation. The refraction types of the atmosphere and their corresponding refractive index gradients are shown in the following table:

Refraction types	Refractive index gradients		
	dN/dh (N-Units/km)	dM/dh (M-Units/km)	K-Factor
Homogeneous	0	157	1
Adiabatic	−23	134	1.2
Standard	−39.2	118	4/3
Subrefractive	> −39.2	> 118	< 4/3
Extreme subrefractive	> 0	> 157	< 1
Superrefractive	< −39.2	< 118	> 4/3
Ducting threshold	−157	0	∞
Ducting	< −157	< 0	∞

(AP/PROP) 211-1997

refractive index profile (fiber optics) The description of the refractive index along a fiber diameter. *See also:* step index profile; power-law index profile; profile parameter; graded index profile; profile dispersion; profile dispersion parameter; parabolic profile. (Std100) 812-1984w

refractive modulus (M) (1) (data transmission) (excess modified index of refraction) The excess over unity of the modified index of refraction, expressed in millionths. It is represented by M and is given by the equation:

$$M = (n + h/a - 1)10^6$$

where n is the index of refraction at a height h above sea level, and a is the radius of the earth. (PE) 599-1985w
(2) In the troposphere, the excess over unity of the modified index of refraction, expressed in millionths:

$$M = (n + h/a - 1)10^6$$

where

a = the mean geometrical radius of the Earth

n = the refractive index at a height, h, above the local surface and $h/a << 1$

(AP/PROP) 211-1997

refractivity The amount by which the real part of the refractive index, n, exceeds unity. Refractivity is often measured in parts per million, called N-units, where $N = (n-1) \times 10^6$. (AP/PROP) 211-1997

refractivity profile The height dependence of refractivity in the atmosphere. *See also:* refractive index gradient. (AP/PROP) 211-1997

refractometer An instrument used to measure the refractive index of the atmosphere. (AP/PROP) 211-1997

refractor (illuminating engineering) A device used to redirect the flux from a source, primarily by the process of refraction. (EEC/IE) [126]

refractory A nonmetallic material highly resistant to fusion and suitable for furnace roofs and linings. (EEC/PE) [119]

refresh (1) To redraw an image on a non-permanent display surface. *Synonyms:* repaint; regenerate. (C) 610.6-1991w
(2) To ensure that the information on the terminal screen of the user is up-to-date. (C/PA) 9945-2-1993
(3) (A) The process of repeatedly producing a display image on a display surface so that the image remains visible. **(B)** To write data periodically to dynamic storage so that it is not lost. (C) 610.10-1994
(4) A periodic referencing of all storage locations in a DRAM, which typically recharges data-storage capacitors in order to maintain data integrity. (C/MM) 1596.4-1996

refresh buffer *See:* bit map.

refresh cycle *See:* display cycle.

refresh display device A display device whose screen surface does not retain an image for a long period of time, requiring the image to be continuously refreshed to remain visible and avoid flicker. *Note:* The refresh method can be either random-scan or raster scan. *Synonym:* refresh tube. *Contrast:* direct-view storage tube. (C) 610.6-1991w, 610.10-1994w

refresh line-drawing display device *See:* random-scan display device.

refresh period Applicable to RAM that is in the autorefresh mode. The maximum elapsed time between refresh commands that is sufficient to ensure that RAM contents remain defined. (C/MM) 1596.4-1996

refresh rate (1) (supervisory control, data acquisition, and automatic control) The number of times in each second that the information displayed on a nonpermanent display, for example, a crt, is rewritten or reenergized.
(SWG/PE/SUB) C37.1-1987s, C37.100-1992
(2) (computer graphics) The frequency with which an image is regenerated on a display surface. (C) 610.6-1991w

refresh tube *See:* refresh display device.

REGAL *See:* range and elevation guidance for approach and landing.

regenerate (1) (electronic storage devices) To bring something into existence again after decay of its own accord or after intentional destruction. (C) 162-1963w
(2) (storage devices in which physical states used to represent data deteriorate) To restore the device to its latest undeteriorated state. *See also:* rewrite. (C) 162-1963w
(3) (storage devices in which physical states used to represent data deteriorate) *See also:* refresh. (C) 610.6-1991w
(4) *See also:* refresh. (C) 610.10-1994w

regenerated leach liquor (electrometallurgy) The solution that has regained its ability to dissolve desired constituents from the ore by the removal of those constituents in the process of electrowinning. *See also:* electrowinning. (PE/EEC) [119]

regeneration (1) (storage tubes) The replacing of stored information lost through static decay and dynamic decay. *See also:* storage tube. (ED) 158-1962w, 161-1971w
(2) (telecommunications) The process of receiving and reconstructing a digital signal so that the amplitudes, waveforms, and timing of its signal elements are re-established within specified limits. (COM/TA) 1007-1991r

regeneration of electrolyte The treatment of a depleted electrolyte to make it again fit for use in an electrolyte cell. *See also:* electrorefining. (EEC/PE) [119]

regenerative brake A form of dynamic brake in which the electrical energy generated by braking is returned to the power supply line, provided to on-board loads, or a combination thereof during the braking cycle instead of being dissipated in resistors. (VT) 1475-1999

regenerative braking A form of dynamic braking in which the kinetic energy of the motor and driven machinery is returned to the power supply system. *See also:* dynamic braking; electric drive; asynchronous machine. (PE/IA/ICTL/IAC) [9], [60]

regenerative branch (self-commutated converters) (converter circuit elements) An auxiliary branch intended to transfer energy from the load to the supply side of the converter. (IA/SPC) 936-1987w

regenerative divider (regenerative modulator) A frequency divider that employs modulation, amplification, and selective feedback to produce the output wave. (PE/EEC) [119]

regenerative fuel-cell system A system in which the reactance may be regenerated using an external energy source. *See also:* fuel cell. (AES) [41]

regenerative repeater (1) (data transmission) A repeater that performs pulse regeneration. *Note:* The retransmitted signals are practically free from distortion. (PE) 599-1985w
(2) (fiber optics) A repeater that is designed for digital transmission. *Synonym:* regenerator. *See also:* optical repeater. (Std100) 812-1984w
(3) A repeater whose function is to re-time and re-transmit the received signal impulses that have been restored to their original strength. (C) 610.7-1995

regenerative track That part of a track on a magnetic drum or magnetic disk, used in conjunction with a read/write head, such that the heads are connected to function as circulating storage. *Synonym:* revolver track. (C) 610.10-1994w

regenerator *See:* regenerative repeater.

region (1) A connected subset of an image.

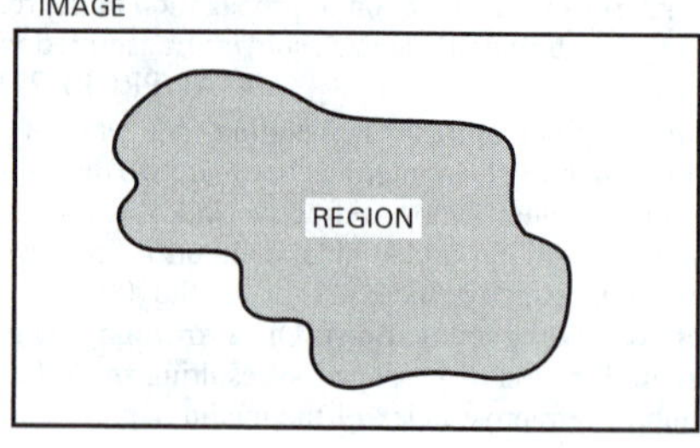

region

(C) 610.4-1990w
(2) (A) As relates to the address space of a process, a sequence of addresses. **(B)** As relates to a file, a sequence of offsets. (C/PA) 9945-1-1996, 1003.5-1999
(3) A region pertains to a particular physical section or block of a floorplan. *See also:* cluster. (C/DA) 1481-1999

regional Bell operating company (RBOC) A regional telephone company that may or may not be made up of individual operating companies.
(SCC31/AMR) 1390.2-1999, 1390.3-1999, 1390-1995

regional center (1) (telephone switching systems) A toll office to which a number of sectional enters are connected. Regional centers are classified as Class 1 offices. *See also:* office class. (COM) 312-1977w
(2) Class 1 office in the North American hierarchical routing plan; a control center connecting sectional centers of the telephone system. *See also:* toll center; sectional center; end office; primary center. (C) 610.7-1995

region, Geiger-Mueller *See:* Geiger-Mueller region.

region growing (image processing and pattern recognition) An image segmentation technique in which regions are formed by repeatedly taking the union of subregions that are similar in gray levels or textures. *See also:* region partitioning. (C) 610.4-1990w

region of limited proportionality (radiation counter tubes) The range of applied voltage below the Geiger-Mueller threshold, in which the gas amplification depends upon the charge liberated by the initial ionizing event. (ED) 161-1971w

region partitioning (image processing and pattern recognition) An image segmentation technique in which regions are formed by repeatedly taking the union of sub-regions that are similar in gray levels or textures and by repeatedly splitting apart subregions that are dissimilar. *See also:* region growing. (C) 610.4-1990w

region, proportional *See:* proportional region.

regions of electromagnetic spectrum (1) (illuminating engineering) For convenience of reference, the electromagnetic spectrum is arbitrarily divided as follows:

Vacuum ultraviolet
Extreme ultraviolet 10–100 nm
Far ultraviolet 100–200 nm
Middle ultraviolet 200–300 nm
Near ultraviolet 300–380 nm
Visible 380–770 nm
Near (short wavelength) 770–1400 nm infrared
Intermediate infrared 1400–5000 nm
Far (long wavelength) 5000–1 000 000 nm infrared

Note: The spectral limits indicated above have been chosen as a matter of practical convenience. There is a gradual transition from region to region without sharp delineation. Also, the division of the spectrum is not unique. In various fields of science, the classifications may differ due to the phenomena of interest. Another division of the ultraviolet spectrum often used by photobiologists is given by the International Commission on Illumination (CIE):

- UV-A 315 to 400 nm
- UV-B 280 to 315 nm
- UV-C 100 to 280 nm

(EEC/IE) [126]

(2) (light-emitting diodes) For convenience of reference the electromagnetic spectrum near the visible spectrum is divided as follows.

Spectrum	Wavelength in Nanometers
far ultraviolet	10–280
middle ultraviolet	280–315
near ultraviolet	315–380
visible	380–780
infrared	790–10^5

Note: The spectral limits indicated above should not be construed to represent sharp delineations between the various regions. There is a gradual transition from region to region. The above ranges have been established for practical purposes. *See also:* radiant energy. (EEC/IE) [126]

register (1) (electronic computation) A device capable of retaining information, often that contained in a small subset (for example, one word), of the aggregate information in a digital computer. *See also:* address register; index register; circulating register; shift register. (C) 162-1963w
(2) (telephone switching systems) A part of an automatic switching system that receives and stores signals from a calling device or other source for interpretation and action. (COM) 312-1977w
(3) A term used to describe quadlet addresses that can be read or written by software. In the context of this document, a register does not imply a specific hardware implementation. If a bus standard allows transactions to be split, and sufficient time is allowed between the request and response subactions, the functionality of the register can be emulated by a processor on the module. (C/MM) 1212-1991s
(4) A storage device or storage location having a specified storage capacity. *See also:* strobe. (C) 610.10-1994w
(5) A set of records (paper, electronic, or a combination) maintained by a Registration Authority containing assigned names and the associated information. (C/LM) 802.10g-1995

register architecture A computer architecture whose design is based on the maintenance of data items in registers. *Contrast:* stack architecture. (C) 610.10-1994w

register-arithmetic and logic unit An arithmetic and logic unit which also contains a register array. (C) 610.10-1994w

register array *See:* register file.

register-based device A servant-only device that supports VXI-bus configuration registers. Register-based devices are typically controlled by message-based devices via device-dependent register reads and writes. (C/MM) 1155-1992

register constant (meter) The factor by which the register reading must be multiplied in order to provide proper consideration of the register, or gear, ratio and of the instrument transformer ratios to obtain the registration in the desired unit. *Note:* It is commonly denoted by the symbol *Kr*. *See also:* electricity meter; moving element. (ELM) C12.1-1982s

registered images Two or more images of the same scene that have been positioned with respect to one another so that corresponding points in the images represent the same point in the scene. (C) 610.4-1990w

register file A set of registers which may be addressed by their number in the set. *Synonym:* register array. (C) 610.10-1994w

register length (1) (electronic computation) The number of characters that a register can store. (Std100) 270-1966w
(2) The storage capacity of a register. (C) 610.10-1994w

register marks Any mark or line printed or otherwise impressed on a web of material and which is used as a reference to maintain register. *See also:* photoelectric control. (IA/ICTL/IAC) [60]

register, mechanical *See:* mechanical register.

register memory (A) Use of high-speed general purpose registers as one would use memory, as in using registers to hold frequently-used data items. **(B)** Registers specifically included in the machine design for use as high-speed storage. (C) 610.10-1994

register ratio (watthour meter) The number of revolutions of the first gear of the register, for one revolution of the first dial pointer. *Note:* This is commonly denoted by the symbol R_r. (ELM) C12.1-1982s

register reading The numerical value indicated by the register. Neither the register constant nor the test dial (or dials), if any exist, is considered. *See also:* electricity meter. (EEC/PE) [119]

register set A subset of the full array of registers in a machine which the processing unit is currently allowed to use. *Note:* Machines may have N registers of which the processor may be able to address only M at a time; this divides the register array into N/M register sets. (C) 610.10-1994w

register transfer language (RTL) A computer language used to represent the flow of information on a system level; for example, to show data at the level of computer devices such as registers, gates, and ALUs. (C) 610.10-1994w

register-transfer level (RTL) (1) A description of computer operations where data transfers from register to register, latch to latch and through logic gates are described. *Note:* This may be an abstract description or microcoding. (C) 610.10-1994w
(2) A level of description of a digital design in which the clocked behavior of the design is expressly described in terms of data transfers between storage elements, which may be implied, and combinational logic, which may represent any computing or arithmetic-logic-unit logic. RTL modeling al-

lows design hierarchy that represents a structural description of other RTL models. (C/DA) 1076.6-1999

registration (1) Accurate positioning relative to a reference. (C) [20], [85]

(2) (display device) The condition in which corresponding elements of the primary-color images are in geometric coincidence. *See also:* registration. (PE/EEC) [119]

(3) (camera device) The condition in which corresponding elements of the primary-color images are scanned in time sequence. (PE/EEC) [119]

(4) Alignment of coordinate systems and phenomenological agreement between environment models. (DIS/C) 1278.3-1996

(5) (watthour meter) *See also:* watthhour meter—percentage registration. (ELM) C12.1-1988

(6) (watthour meter) *See also:* image registration. (C) 610.4-1990w

regressed (illuminating engineering) A luminaire which is mounted above the ceiling with the opening of the luminaire above the ceiling line. (EEC/IE) [126]

regression test Retesting to detect faults introduced by modification. (C/SE) 1219-1998

regression testing (software) Selective retesting of a system or component to verify that modifications have not caused unintended effects and that the system or component still complies with its specified requirements. (C) 610.12-1990

regular binary *See:* binary.

regular expression A pattern (sequence of characters or symbols) constructed according to the rules defined in 2.8. (C/PA) 9945-2-1993

regular file A file that is a randomly accessible sequence of bytes, with no further structure imposed by the system. (C/PA) 9945-1-1996, 9945-2-1993, 1003.5-1999

regular (specular) reflectance (illuminating engineering) The ratio of the flux leaving a surface or medium by regular (specular) reflection to the incident flux. (EEC/IE) [126]

regular reflection (illuminating engineering) That process by which incident flux is redirected at the specular angle. *See also:* specular angle. (EEC/IE) [126]

regular transmission (illuminating engineering) That process by which incident flux passes through a surface or medium without scattering. (EEC/IE) [126]

regular transmittance (illuminating engineering) The ratio of the regularly transmitted flux leaving a surface or medium to the incident flux. (EEC/IE) [126]

regulated circuit The circuit on the output side of the regulator, where it is desired to control the voltage, or the phase relation, or both. The voltage may be held constant at any selected point on the regulated circuit. (PE/TR) C57.15-1999

regulated frequency Frequency so adjusted that the average value does not differ from a predetermined value by an appreciable amount. *See also:* generating station. (T&D/PE) [10]

regulated power supply A power supply that maintains a constant output voltage (or current) for changes in the line voltage, output load, ambient temperature, or time. (AES) [41]

regulated-power-supply efficiency The ratio of the regulated output power to the input power. *See also:* regulated power supply. 209-1950w

regulated voltage, band of *See:* band of regulated voltage.

regulated voltage, nominal band of *See:* nominal band of regulated voltage.

regulating autotransformer (rectifier) A transformer used to vary the voltage applied to the alternating-current winding of rectifier transformer by means of de-energized autotransformer taps, and with load-tap-changing equipment to vary the voltage over a specified range on any of the autotransformer taps. *See also:* rectifier transformer. (Std100) C57.18-1964w

regulating circuit (thyristor) A circuit that together with the power controller and the thyristor trigger equipment forms a system for automatic control of the desired variable. (IA/IPC) 428-1981w

regulating device (power system device function numbers) A device that functions to regulate a quantity, or quantities, such as voltage, current, power, speed, frequency, temperature, and load at a certain value or between certain (generally close) limits for machines, tie lines, or other apparatus. (PE/SUB) C37.2-1979s

regulating limit setter (speed governing systems) A device in the load-frequency-control system for limiting the regulating range on a station or unit. *See also:* speed-governing system. (PE/PSE) 94-1970w

regulating range (load-frequency control) That range of power output within which a generating unit is permitted to respond to load frequency control. (PE/PSE) 94-1991w

regulating relay A relay whose function is to detect a departure from specified system operating conditions and to restore normal conditions by acting through supplementary equipment. (SWG/PE) C37.100-1992

regulating system, synchronous-machine *See:* synchronous-machine regulating system.

regulating transformer A transformer used to vary the voltage, or the phase angle, or both, of an output circuit (referred to as the regulated circuit) controlling the output within specified limits, and compensating for fluctuations of load and input voltage (and phase angle, when involved within specified limits. *See also:* voltage-regulating relay; line-drop compensator; series unit; regulating winding; primary circuit; main unit; excitation-regulating winding; excited winding; regulated circuit; series winding; excitation winding. (PE/TR) C57.12.80-1978r

regulating winding (power and distribution transformers) The winding of the main unit in which taps are changed to control the voltage or phase angle of the regulated circuit through the series unit. (PE/TR) C57.12.80-1978r

regulation (1) (rotating machinery) The amount of change in voltage or speed resulting from a load change. *See also:* asynchronous machine. (PE) [9]

(2) (overall) (power supplies) The maximum amount that the output will change as a result of the specified change in line voltage, output load, temperature, or time. *Note:* Line regulation, load regulation, stability, and temperature coefficient are defined and usually specified separately. *See also:* stability; temperature coefficient; load regulation; line regulation. (AES) [41]

(3) (electrical conversion) The change of one of the controlled or regulated output parameters resulting from a change of one or more of the unit's variables within specificaton limits. (AES) [41]

(4) (transformer-rectifier system) The change in output voltage as the load current is varied. It is usually expressed as a percentage of the rated load voltage when the load current is changed by its rated value.

$$\text{percent regulation} = 100\,\frac{(E_1 - E_2)}{E_2}$$

where E_1 is the no-load voltage and E_2 is the voltage at rated load current and the line voltage is held constant at rated value. (PEL/ET) 295-1969r

(5) (power supplies) *See also:* load regulation; overall regulation. (PEL/ET) 449-1990s

regulation changer (speed governing systems, hydraulic turbines) A device by means of which the speed regulation may be adjusted while the turbine is operating. (PE/EDPG) [5]

regulation curve (generator) A characteristic curve between voltage and load. The speed is either held constant, or varied according to the speed characteristics of the prime mover. The excitation is held constant for separately excited fields, and the rheostat setting is held constant for self-excited machines. (EEC/PE) [119]

regulation, frequency *See:* frequency regulation.

regulation, load *See:* load regulation.

regulation pull-out (power supply) (regulation drop-out) The load currents at which the power supply fails to regulate when the load current is gradually increased or decreased. *See also:* regulated power supply. 209-1950w

regulator, continuously acting *See:* continuously acting regulator.

regulator, noncontinuously acting *See:* noncontinuously acting regulator.

regulator, rheostatic-type *See:* rheostatic-type regulator.

regulator, synchronous machine *See:* synchronous machine regulator.

reguline A word descriptive of electrodeposits that are firm and coherent. (EEC/PE) [119]

rehashing *See:* collision resolution.

reheat turbine, condensing or noncondensing (control systems for steam turbine-generator units) Steam enters the turbine initially at one pressure, then is extracted at a lower pressure and temperature, and reheated. The steam is then readmitted to the turbine. (PE/EDPG) 122-1985s

reignition (1) A resumption of current between the contacts of a switching device during an opening operation after an interval of zero current of less than 1/4 cycle at normal frequency. (SWG/PE) C37.100-1992

(2) A process by which multiple counts are generated within a counter tube by atoms or molecules excited or ionized in the discharge accompanying a tube count. (NI/NPS) 309-1999

reinforced plastic (rotating machinery) A plastic with some strength properties greatly superior to those of the base resin, resulting from the presence of high-strength fillers imbedded in the composition. *Note:* The reinforcing fillers are usually fibers, fabrics, or mats made of fibers. The plastic laminates are the most common and strongest. (PE) [9]

reinsertion The restoration of load current to the series capacitor from the bypass path (see the corresponding figure). (T&D/PE) 824-1994

reinsertion current (series capacitor) The transient current, load current, or both, flowing through the series capacitor after the opening of the bypass path. (T&D/PE) 824-1994

reinsertion voltage (1) (series capacitor) The transient voltage, steady-state voltage, or both, appearing across the series capacitor after the opening of the bypass path. (T&D/PE) 824-1994

(2) (uniconductor waveguide) The frequency range below the cutoff frequency. *See also:* waveguide. (MTT) 146-1980w

(3) *See also:* stop band.

rejection filter (signal-transmission system) A filter that attenuates alternating currents between given upper and lower cutoff frequencies and transmits substantially all others. Also, a filter placed in the signal transmission path to attenuate interference. *See also:* signal. (IE) [43]

related multiple outage event A multiple outage event in which one outage occurrence is the consequence of another outage occurrence, or in which multiple outage occurrences were initiated by a single incident, or both. Each outage occurrence in a related multiple outage event is classified as either a primary outage or a secondary outage depending on the relationship between that outage occurrence and its initiating incident. (A) (primary outage) An outage occurrence within a related multiple outage event that occurs as a direct consequence of the initiating incident and is not dependent on any other outage occurrence. *Notes:* 1. A primary outage of a component or a unit may be caused by a fault on equipment within the unit or component or repair of a component within the unit. (B) (secondary outage) An outage occurrence that is the result of another outage occurrence. 2. Secondary outages of components or units may be caused by repair of other components or units requiring physical clearance, failure of a circuit breaker to clear a fault, or a protective relay system operating incorrectly and overreaching into the normal tripping zone of another unit. 3. Some secondary outages are solely the result of system configuration; for example, two components connected in series will always go out of service together. These secondary outages may be given special treatment when compiling outage data. 4. At present, primary outages have been referred to in the industry as independent outage occurrences, and secondary outages as dependent or related outage occurrences. (C) (common-mode outage event) A related multiple outage event consisting of two or more primary outage occurrences initiated by a single incident or underlying cause where the outage occurrences are not consequences of each other. 5. Primary outage occurrences in a common-mode outage event are referred to as common-mode outage event are referred to as common-mode outages. Examples of common-mode outage events are a single lightning stroke causing tripouts of both circuits on a common tower, and an external object causing the outage of two circuits on the same right-of-way. (PE/PSE) 859-1987w

related transmission terms (loss and gain) The term loss used with different modifiers has different meanings, even when applied to one physical quantity such as power. In view of definitions containing the word loss (as well as others containing the word gain), the following brief explanation is presented. (A) Power loss from a circuit, in the sense that it is converted to another form of power not useful for the purpose at hand (for example, I^2R loss) is a physical quantity measured in watts in the International System of Units (SI) and having the dimensions of power. For a given R, it will vary with the square of the current in R. (B) Loss may be defined as the ratio of two powers; for example: if P_o is the output power and P_i the input power of a network under specified conditions, P_i/P_o is a dimensionless quantity that would be unity if P_o, P_i. Thus, no power loss in the sense of "A" means a loss, defined as the ratio P_i/P_o, of unity. The concept is closely allied to that of efficiency. (C) Loss may also be defined as the logarithm, or as directly proportional to the logarithm, of a power ratio such as P_i/P_o. Thus if loss $= 10 \log_{10} P_i/P_o$ the loss is zero when P_i, P_o. This is the standard for measuring loss in decibels. It should be noted that in cases "B" and "C" the loss (for a given linear system) is the same whatever may be the power levels. Thus (B) and (C) give characteristics of the system and do not depend (as (A) does) on the value of the current or other dependent quantity. Power refers to average power, not instantaneous power. *See also:* network analysis. (Std100) 270-1966w

relation In a relational data model or relational database, a set of tuples, each of which has the same attributes. *Note:* Often thought of as a table of data. (C) 610.5-1990w

relational algebra An algebra that includes a set of relational operators, such as join and projection, to manipulate relations and the axioms of those operators. (C) 610.5-1990w

relational database (1) A database in which data are organized into one or more relations that may be manipulated using a relational algebra. *Contrast:* network database. (C) 610.5-1990w

(2) A database that represents data as a collection of tables linked through common entries. (PE/EDPG) 1150-1991w

relational database model An external data model that represents a relational database. (C) 610.5-1990w

relational database schema A collection of relation schemas that define the structural properties of a relational database. (C) 610.5-1990w

relational data model (A) A data model whose pattern or organization is based on a set of relations, each of which consists of an unordered set of tuples. **(B)** A data model that provides for the expression of relationships among data elements as formal mathematical relations. (C) 610.5-1990

relational engine A database engine for relational databases. *See also:* SQL engine. (C) 610.10-1994w

relational file (A) A file, consisting of tuples, in which all data items are associated via the same relationship. *Note:* Also called a flat file. **(B)** Any file resulting from relational algebra. (C) 610.5-1990

relational language A query language that may be used to access and retrieve data from a relational database. (C) 610.5-1990w

relationally complete Pertaining to a query language or system that can be used to form expressions from a relational algebra. (C) 610.5-1990w

relational model *See:* relational database model; relational data model.

relational operator An operator that performs an operation on relations; for example, the join or projection operators. *See also:* relational algebra. (C) 610.5-1990w

relation schema The set of all attribute names for a relation. (C) 610.5-1990w

relationship (1) A directed connection between two or more data items or attributes. (C) 610.5-1990w
(2) An association between two classes. (C/SE) 1420.1-1995
(3) A specific association that exists between entity sets or entities of a set that can be described by a single word or phrase. (PE/EDPG) 1150-1991w
(4) A kind of association between two (not necessarily distinct) classes that is deemed relevant within a particular scope and purpose. The association is named for the sense in which the instances are related. A relationship can be represented as a time-varying binary relation between the instances of the current extents of two state classes. (C/SE) 1320.2-1998

relationship instance An association of specific instances of the related classes. (C/SE) 1320.2-1998

relationship name A verb or verb phrase that reflects the meaning of the relationship expressed between the two entities shown on the diagram on which the name appears. (C/SE) 1320.2-1998

relative address (1) (computers) The number that specifies the difference between the absolute address and the base address. (MIL/C) [2], [20], [85]
(2) (software) An address that must be adjusted by the addition of an offset to determine the address of the storage location to be accessed. *Contrast:* absolute address. *See also:* self-relative address; indexed address; base address. (C) 610.12-1990
(3) An address to which a base address must be added in order to form an absolute address of a particular storage location. *See also:* symbolic address; absolute address; base address; relocatable address. (C) 610.10-1994w

relative addressing (1) (microprocessor assembly language) An addressing mode in which the effective address is formed by adding an offset to the program counter (or a portion thereof) during execution. (MM/C) 695-1985s
(2) An addressing mode in which a base address is used to store the beginning address of some area in storage, and all locations within that are expressed in terms of their displacement from the beginning, or the relative address. (C) 610.10-1994w

relative bearing (navigation aids) Bearing relative to heading. (AES/GCS) 172-1983w

relative bias The relative bias statistic for the ith measurement in a category with respect to the "true or expected" value (value of the spike known by comparison with or derivation from a standard reference material) is defined as:

$$B_{ri} = (A_i - A_{ai}) / A_{ai}$$

where

A_i = the ith measurement in a category being tested, not necessarily a replicate, but possibly a different quantity of spike for each measurement.

A_{ai} = the stated quantity in the test sample, as defined by the spike.

In order to avoid the expense of a large number of replicates at each radioactivity level in each category, the relative bias (which may be obtained at differing quantity levels) for that test category is calculated from N individual relative bias B_{ri} values and is defined as:

$$B_r = \overline{B}_{ri} = \sum_{i-1}^{N} B_{ri} / N$$

(NI) N42.23-1995

relative capacitivity *See:* relative dielectric constant.

relative chrominance level (linear waveform distortion) The difference between the level of the luminance and chrominance signal components. An inaccuracy in RCL will cause saturation inaccuracy of all colors in a color TV picture. (BT) 511-1979w

relative chrominance time (linear waveform distortion) The difference in absolute time between the luminance and chrominance signal components. An inaccuracy in RCT will cause registration inaccuracy of all colors relative to their luminance components in a color TV picture. (BT) 511-1979w

relative coding (computers) Coding that uses machine instructions with relative addresses. (C) [20], [85]

relative complex dielectric constant (complex capacitivity) (complex permittivity) (homogeneous isotropic material) The ratio of the admittance between two electrodes of a given configuration of electrodes with the material as a dielectric to the admittance between the same two electrodes of the configuration with vacuum as dielectric or

$$\varepsilon^* \equiv \varepsilon' - j\varepsilon'' = Y/(j\omega C_v)$$

where Y is the admittance with the material and $j\omega C_v$ is the admittance with vacuum. Experimentally, vacuum must be replaced by the material at all points where it makes a significant change in the admittance. *Note:* The word relative is frequently dropped. *See also:* relative dielectric constant. (Std100) 270-1966w

relative complex permeability (μ_r) The complex permeability of a medium normalized to the permeability of free space μ_0. (AP/PROP) 211-1997

relative complex permittivity (ε_r) The complex permittivity, of a medium normalized to the free space permittivity ε_0. (AP/PROP) 211-1997

relative contrast sensitivity (illuminating engineering) The relation between the reciprocal of the luminous contrast of a task at visibility threshold and the background luminance expressed as a percentage of the value obtained under a very high level of diffuse task illumination. (EEC/IE) [126]

relative cross-polar side lobe level *See:* cross-polar side lobe level, relative.

relative damping (1) (instrument) (specific damping) Under given conditions, the ratio of the damping torque at a given angular velocity of the moving element to the damping torque that, if present at this angular velocity, would produce the condition of critical damping. *See also:* accuracy rating. (EEC/PE) [119]
(2) (automatic control) (under damped system). A number expressing the quotient of the actual damping of a second-order linear system or element by its critical damping. *Note:* For any system whose transfer function includes a quadratic factor $s^2 + 2z\omega_n s + \omega_n^2$, relative damping is the value of z, since $z = 1$ for critical damping. Such a factor has a root $-\sigma + j\omega$ in the complex s-plane, from which $z = \sigma/\omega_n = \sigma(\sigma^2 + \omega^2)^{1/2}$. (PE/EDPG) [3]

relative dielectric constant (relative permittivity) (relative capacitivity) (homogeneous isotropic material) The ratio of the capacitance of a given configuration of electrodes with the material as a dielectric to the capacitance of the same electrode configuration with a vacuum (or air for most practical purposes) as the dielectric or

$$\varepsilon' = C_x/C_v$$

where C_x is the capacitance with the material and C_v is the capacitance with vacuum. Experimentally, vacuum must be replaced by the material at all points where it makes a significant change in the capacitance. *See also:* electric flux density; relative capacitivity. (Std100) 270-1966w

relative directive gain (physical media) In a given direction and at a given point in the far field, the ratio of the power flux per unit area from an antenna to the power flux per unit area from a reference antenna at a specified location and delivering the same power from the antenna to the medium. *Note:* All or part of the reference antenna must be within the smallest sphere containing the subject antenna. (AP/ANT) 145-1983s

relative distinguished name A set of AVAs, each of which is true, concerning the distinguished values of a particular entry. (C/PA) 1328.2-1993w, 1327.2-1993w, 1224.2-1993w, 1326.2-1993w

relative error The ratio of an error to the correct value. *Contrast:* absolute error. (C) 1084-1986w

relative error of indication The quotient (I) of the error of indication of a measured quantity by the conventionally true value of that measured quantity. It may be expressed as a percentage, for example:

$$I = \left(\frac{H_i - H_t}{H_t}\right) \times 100\%$$

(NI) N42.20-1995

relative full-energy peak efficiency for detector specification The ratio of full-energy peak detection efficiency for a point source of ^{60}Co(1332 keV photons) to that of a NaI(Tl) crystal, 7.6 cm diameter × 7.6 cm high, for a source-to-detector distance of 25 cm. (NI) N42.14-1991

relative fundamental content (converter characteristics) (self-commutated converters) The ratio of the rms (root-mean-square) value of the fundamental component to the rms value of the total nonsinusoidal periodic function. (IA/SPC) 936-1987w

relative gain (of an antenna) The ratio of the gain of an antenna in a given direction to the gain of a reference antenna. *Note:* Unless otherwise specified, the maximum gains of the antennas are implied. (AP/ANT) 145-1993

relative grid (navigation aid terms) Navigation in a relative grid as opposed to an absolute coordinate system (for example, geo-referenced). A relative grid, arbitrarily constructed by designating a point as the origin and constructing a set of axes U, V, W enables members to navigate in this relative grid by virtue of their U, V, W coordinates. (AES/GCS) 172-1983w

relative harmonic content (converter characteristics) (self-commutated converters) The ratio of the rms (root-mean-square) value of the harmonic content to the rms value of the total nonsinusoidal periodic function. (IA/SPC) 936-1987w

relative humidity (RH) (1) The ratio between the amount of water vapor in the gas at the time of measurement and the amount of water vapor that could be in the gas when condensation begins, at a given temperature. (PE/IC) 1125-1993 **(2)** (with respect to water or ice) The ratio, expressed as a percentage, of the water vapor pressure in moist air to the saturation vapor pressure with respect to a plane pure water (ice) surface at the same temperature. (AP/PROP) 211-1997

relative integer date A date defined as the number of days starting from any given Gregorian calendar date determined at the time of initial implementation of a given integer date system, and extending in a contiguous sequence greater than one leap year (366 days). *Note:* The freedom of choice of a start date is the principal characteristic of the relative integer date. Knowledge of the start date is not essential other than to confirm that the interval or range of dates used follows the adopted start date. Synchronization between two disparate relative integer date systems is readily achieved by converting a given Gregorian date to its relative integer date representation in both systems. The difference between the resulting relative integer dates then becomes the required offset to be added or subtracted for each system to correctly interpret the other's relative integer date. (C/PA) 2000.1-1999

relative intensity noise The ratio of the variance in the optical power to the average optical power. (C/LM) 802.3-1998

relative interfering effect (single-frequency electric wave in an electroacoustic system) The ratio, usually expressed in decibels, of the amplitude of a wave of specified reference frequency to that of the wave in question when the two waves are equal in interfering effect. The frequency of maximum interfering effect is usually taken as the reference frequency. Equal interfering effects are usually determined by judgment tests or intelligibility tests. *Note:* When applied to complex waves, the relative interfering effect is the ratio, usually expressed in decibels, of the power of the reference wave to the power of the wave in question when the two waves are equal in interfering effect. (EEC/PE) [119]

relative ionospheric opacity meter (radio-wave propagation) A radio-frequency receiving device that measures the ionospheric absorption experienced by cosmic radio noise passing through the ionosphere. *Synonym:* riometer. (AP) 211-1977s

relative lead polarity A designation of the relative instantaneous directions of current in its leads. *Notes:* 1. Primary and secondary leads are said to have the same polarity when at a given instant during most of each half cycle, the current enters an identified, or marked, primary lead and leaves the similarly identified, or marked, secondary lead in the same direction as though the two leads formed a continuous circuit. 2. The relative lead polarity of a single-phase transformer may be either additive or subtractive. If one pair of adjacent leads from the two windings is connected together and voltage applied to one of the windings, then:

a) The relative lead polarity is additive if the voltage across the other two leads of the windings is greater than that of the higher-voltage winding alone.

b) The relative lead polarity is subtractive if the voltage across the other two leads of the windings is less than that of the higher-voltage winding alone.

3. The relative lead polarity is indicated by identification marks on primary and secondary leads of like polarity, or by other appropriate identification. *See also:* routine test; constant-current transformer. (PE/TR) [57]

relative loader *See:* relocating loader.

relative luminosity (television) The ratio of the value of the luminosity at a particular wavelength to the value at the wavelength of maximum luminosity. (BT/AV) 201-1979w

relative plateau slope (radiation counter tubes) The average percentage change in the counting rate near the midpoint of the plateau per increment of applied voltage. *Note:* Relative plateau slope is usually expressed as the percentage change in counting rate per 100 V change in applied voltage.

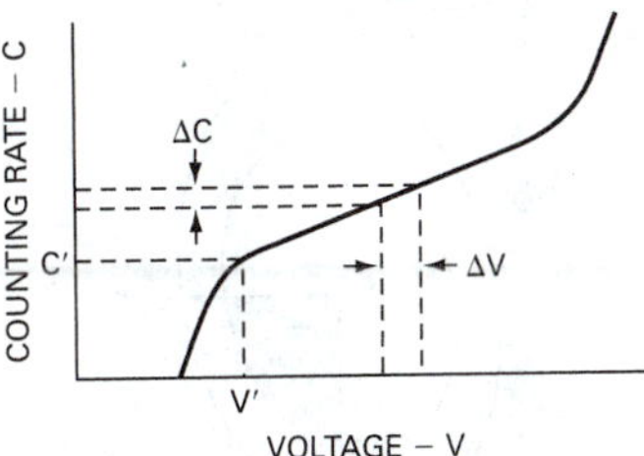

Counting rate-voltage characteristic in which

$$\text{relative plateau slope} = 100\,\frac{\Delta C/C}{\Delta V}$$

$$\text{normalized plateau slope} = \frac{\Delta C/\Delta V}{C'/V'} = \frac{\Delta C/C'}{\Delta V/V'}$$

plateau slope, relative

(ED/NPS/NID) 161-1971w, 309-1970s

relatively prime Describes integers whose greatest common divisor is 1. (IM/WM&A) 1057-1994w

relatively refractory state (electrobiology) The portion of the electric recovery cycle during which the excitability is less than normal. *See also:* excitability. (EMB) [47]

relative nonline frequency content (thyristor) The ratio of the root-mean-square (rms) value of the nonline frequency content to the total rms value of the nonsinusoidal periodic function. 445 subline frequency components (thyristor). Expressed by the frequency and the root-mean-square (rms) value of the components having a lower frequency than the line frequency (dc included). (IA/IPC) 428-1981w

relative partial gain (of an antenna with respect to a reference antenna of a given polarization) In a given direction, the ratio of the partial gain of an antenna, corresponding to the polarization of the reference antenna, to the maximum gain of the reference antenna. (AP/ANT) 145-1993

relative permeability The ratio of normal permeability to the magnetic constant. *Note:* In anisotropic media, relative permeability becomes a matrix. (Std100) 270-1966w

relative permittivity in physical media The ratio of the complex permittivity to the permittivity of free space. *See also:* relative dielectric constant. (AP/ANT) 145-1983s

relative phase (of an elliptically polarized field vector) The phase angle of the unitary factor by which the polarization-phase vector for the given field vector differs from that of a reference field vector with the same polarization. *Notes:* 1. The relative phase of an elliptically polarized field $\vec{E}_1$ can be defined with respect to that of another field $\vec{E}_0$ having the same polarization. In that case, the polarization vectors $\hat{e}_1$ and $\hat{e}_0$ have the same direction and, being of unit magnitudes, they differ only by a unitary factor: $e_1 = e^{j\alpha}e_0$. The angle α is the phase difference between $\vec{E}_1$ and $\vec{E}_0$. 2. The field vectors $\vec{E}_1(t) = \text{Re}\ \vec{E}_1 e^{j\omega t}$ and $\vec{E}_0(t) = \text{Re}\ \vec{E}_0 \hat{e}^{j\omega t}$ describe similar ellipses as t varies. The angle α is 2π times the area of the sector shown on the figure divided by the area of the ellipse described by the extremity of $\vec{E}_0(t)$. For circular polarization, α is the angle between $\vec{E}_0$ and $\vec{E}_1$ at any instant of time. 3. The phase of an elliptically polarized field vector can be expressed relative to a spatial direction in its plane of polarization. For example, the phase angle is given by 2π times the area of the sector shown on the figure, which is bounded by $\vec{E}(0)$ and the reference, divided by the area of the ellipse described by $\vec{E}(t)$. The angle is positive if it is in the same direction as the sense of polarization and negative if it is in the direction opposite to the sense of polarization.

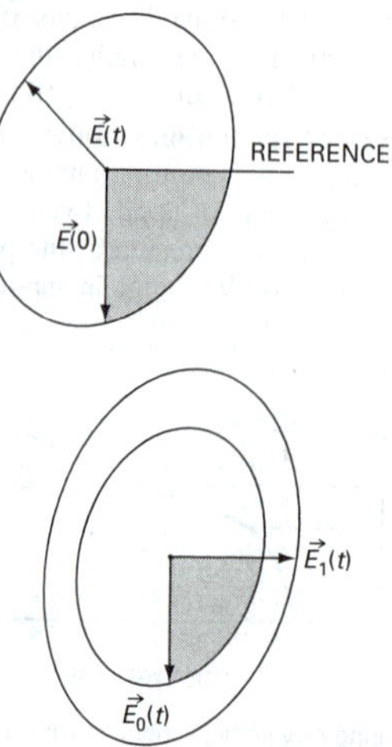

(AP/ANT) 145-1993

relative power gain The ratio of the power gain in a given direction to the power gain of a reference antenna in its reference direction. *Note:* Common reference antennas are half-wave dipoles, electric dipoles, magnetic dipoles, monopoles, and calibrated horn antennas. (AP/ANT) 145-1983s

relative power gain in physical media In a given direction and at a given point in the far field, the ratio of the power flux per unit area from an antenna to the power flux per unit area from a reference antenna at a specified location with the same power input to both antennas. *Note:* All or part of the reference antenna must be within the smallest sphere containing the subject antenna. (AP/ANT) 145-1983s

relative precision The relative precision of the measurement process is selected for the purposes of this standard to be the relative dispersion of the values of B_{ri} from its mean B_r, and is defined to be:

$$S_A = \sqrt{\frac{\sum_{i=1}^{N} (B_{ri} - B_r)^2}{(N - 1)}}$$

where N is the number of test samples measured by an individual service laboratory in a given test category. The sample size N should be at least five. (NI) N42.23-1995

relative redundancy (of a source) The ratio of the redundancy of the source to the logarithm of the number of symbols available at the source. *See also:* information theory. (IT) 171-1958w

relative refractive index (radio-wave propagation) Of two media, the ratio of their refractive indices. (AP) 211-1977s

relative response (audio and electroacoustics) The ratio, usually expressed in decibels, of the response under some particular conditions to the response under reference conditions. Both conditions should be stated explicitly. (SP) [32]

relative Seebeck coefficient The Seebeck coefficient of a couple composed of the given material as the first-named conductor and a specified standard conductor. *Note:* Common standards are platinum, lead, and copper. *See also:* thermoelectric device. (ED) [46]

relative side lobe level The maximum relative directivity of a side lobe with respect to the maximum directivity of an antenna, usually expressed in decibels. (AP/ANT) 145-1993

relative stability (stable underdamped system) (automatic control) The property measured by the relative setting times when parameters are changed. *See also:* feedback control system. (IM) [120]

relative temperature index (1) (evaluation of thermal capability) (thermal classification of electric equipment and electrical insulation) The temperature index of a new or candidate insulating material, which corresponds to the accepted temperature index of a reference material for which considerable test and service experience has been obtained. Both new and reference material are subjected to the same aging and diagnostic procedure in a comparative test. *See also:* thermal endurance graph. (EI) 1-1986r
(2) (solid electrical insulating materials) Derived at an arbitrary time by comparing the life values from thermal endurance graphs from a new and a referenced material with considerable service experience. (EI) 98-1984r

relative visual performance The potential task performance based upon the illuminance and contrast of the lighting system performance. (IA/PSE) 241-1990r

relaxation (laser maser) The spontaneous return of a system towards its equilibrium condition. (LEO) 586-1980w

relaxation oscillator Any oscillator whose fundamental frequency is determined by the time of charging or discharging of a capacitor or inductor through a resistor, producing waveforms that may be rectangular or sawtooth. *Note:* The frequency of a relaxation oscillator may be self-determined or determined by a synchronizing voltage derived from an external source. *See also:* electronic controller; oscillatory circuit. (AP/ANT) 145-1983s

relaxation time (laser maser) The time required for the deviation from equilibrium of some system parameter to diminish to $1/e$ of its initial value. (LEO) 586-1980w

relay **(1) (general)** An electric device designed to respond to input conditions in a prescribed manner and, after specified conditions are met, to cause contact operation or similar abrupt change in associated electric control circuits. *Notes:* 1. Inputs are usually electrical, but may be mechanical, thermal, or other quantities, or a combination of quantities. Limit switches and similar simple devices are not relays. 2. A relay may consist of several relay units, each responsive to a specified input, with the combination of units providing the desired overall performance characteristic(s) of the relay. (SWG/PE/IM/HFIM/PSR) C37.100-1992, 474-1973w, C37.90-1989r

(2) (electric and electronics parts and equipment) An electrically controlled, usually two-state, device that opens and closes electrical contacts to effect the operation of other devices in the same or another electric circuit. *Notes:* 1. A relay is a device in which a portion of one or more sets of electrical contacts is moved by an armature and its associated operating coil. 2. This concept is extended to include assembled reed relays in which the armature may act as a contact. *See also:* switch. (GSD) 200-1975w

(3) (packaging machinery) A device that is operative by a variation in the conditions of one electric circuit to affect the operation of other devices in the same or another electric circuit. *Note:* Where relays operate in response to changes in more than one condition, all functions should be mentioned. (IA/PKG) 333-1980w

(4) A special-purpose switch that is activated by an electrical signal. *See also:* operational relay. (C) 610.10-1994w

relay actuation time The time at which a specified contact functions. (EEC/REE) [87]

relay actuation time, effective *See:* effective relay actuation time.

relay actuation time, final *See:* final relay actuation time.

relay actuation time, initial *See:* initial relay actuation time.

relay actuator The part of the relay that converts electric energy into mechanical work. (EEC/REE) [87]

relay adjustment The modification of the shape or position of relay parts to affect one or more of the operating characteristics, that is, armature gap, restoring spring, contact gap. (EEC/REE) [87]

relay air gap Air space between the armature and the pole piece. This is used in some relays instead of a nonmagnetic separator to provide a break in the magnetic circuit. (EEC/REE) [87]

relay, alternating-current *See:* alternating-current relay.

relay amplifier (1) An amplifier that drives an electromechanical relay. *See also:* electronic analog computer. (C) 165-1977w, 166-1977w

(2) In an analog computer, an amplifier that drives an electromechanical relay. (C) 610.10-1994w

relay antifreeze pin Sometimes used for relay armature stop, nonmagnetic. (EEC/REE) [87]

relay armature (electromechanical relay) The moving element that contributes to the designed response of the relay and that usually has associated with it a part of the relay contact assembly. (SWG/PE) C37.100-1981s

relay armature, balanced *See:* balanced relay armature.

relay armature bounce *See:* relay armature rebound.

relay armature card An insulating member used to link the movable springs to the armature. (EEC/REE) [87]

relay armature contact (A) A contact mounted directly on the armature. **(B)** Sometimes used for relay contact, movable. (EEC/REE) [87]

relay armature, end-on *See:* end-on relay armature.

relay armature, flat-type *See:* flat-type relay armature.

relay armature gap The distance between armature and pole face. (EEC/REE) [87]

relay armature hesitation Delay or momentary reversal of armature motion in either the operate or release stroke. (EEC/REE) [87]

relay armature lifter *See:* relay armature stud.

relay armature, long-lever *See:* long-lever relay armature.

relay armature overtravel The portion of the available stroke occurring after the contacts have touched. (EEC/REE) [87]

relay armature ratio The ratio of the distance through which the armature stud or card moves to the armature travel. (EEC/REE) [87]

relay armature rebound Return motion of the armature following impact on the backstop. (EEC/REE) [87]

relay armature, short-lever *See:* short-lever relay armature.

relay armature, side *See:* side relay armature.

relay armature stop Sometimes used for relay backstop. (EEC/REE) [87]

relay armature stop, nonmagnetic *See:* nonmagnetic relay armature stop.

relay armature stud An insulating member that transmits the motion of the armature to an adjacent contact member. (EEC/REE) [87]

relay armature travel The distance traveled during operation by a specified point on the armature. (EEC/REE) [87]

relay back contacts Sometimes used for relay contacts, normally closed. (EEC/REE) [87]

relay backstop The part of the relay that limits the movement of the armature away from the pole face or core. In some relays a normally closed contact may serve as backstop. (EEC/REE) [87]

relay backup That part of the backup protection that operates in the event of failure of the primary relays. (SWG/PE) C37.100-1992, [56]

relay bank *See:* relay level.

relay bias winding An auxiliary winding used to produce an electric bias. (EEC/REE) [87]

relay blades Sometimes used for relay contact springs. (EEC/REE) [87]

relay bracer spring A supporting member used in conjunction with a contact spring. (EEC/REE) [87]

relay bridging (A) A result of contact erosion, wherein a metallic protrusion or bridge is built up between opposite contact faces to cause an electric path between them. **(B)** A form of contact erosion occurring on the break of a low-voltage, low-inductance circuit, at the instant of separation, that results in melting and resolidifying of contact metal in the form of a metallic protrusion or bridge. **(C)** Make-before-break contact action, as when a wiper touches two successive contacts simultaneously while moving from one to the other. (EEC/REE) [87]

relay brush *See:* relay wiper.

relay bunching time The time during which all three contacts of a bridging contact combination are electrically connected during the armature stroke. (EEC/REE) [87]

relay bushing Sometimes used for relay spring stud. (EEC/REE) [87]

relay chatter time The time interval from initial actuation of a contact to the end of chatter. (EEC/REE) [87]

relay coil One or more windings on a common form. (EEC/REE) [87]

relay coil, concentric-wound *See:* concentric-wound relay coil.

relay-coil dissipation The amount of electric power consumed by a winding. For the most practical purposes, this equals the I^2R loss. (EEC/REE) [87]

relay-coil resistance The total terminal-to-terminal resistance of a coil at a specified temperature. (EEC/REE) [87]

relay-coil serving A covering, such as thread or tape, that protects the winding from mechanical damage. (EEC/REE) [87]

relay-coil temperature rise The increase in temperature of a winding above the ambient temperature when energized under specified conditions for a given period of time, usually the time required to reach a stable temperature. (EEC/REE) [87]

relay-coil terminal A device, such as a solder lug, binding post, or similar fitting, to which the coil power supply is connected. (EEC/REE) [87]

relay-coil tube An insulated tube upon which a coil is wound. (EEC/REE) [87]

relay comb An insulating member used to position a group of contact springs. (EEC/REE) [87]

relay contact actuation time The time required for any specified contact on the relay to function according to the following subdivisions. When not otherwise specified contact actuation time is relay initial actuation time. For some purposes, it is preferable to state the actuation time in terms of final actuation time or effective actuation time. (EEC/REE) [87]

relay contact arrangement The combination of contact forms that make up the entire relay switching structure. (EEC/REE) [87]

relay contact bounce Sometimes used for relay contact chatter, when internally caused. (EEC/REE) [87]

relay contact chatter The undesired intermittent closure of open contacts or opening of closed contacts. It may occur either when the relay is operated or released or when the relay is subjected to external shock or vibration. (EEC/REE) [87]

relay contact chatter, armature hesitation Chatter ascribed to delay or momentary reversal in direction of the armature motion during either the operate or the release stroke. (EEC/REE) [87]

relay contact chatter, armature impact Chatter ascribed to vibration of the relay structure caused by impact of the armature on the pole piece in operation, or on the backstop in release. (EEC/REE) [87]

relay contact chatter, armature rebound Chatter ascribed to the partial return of the armature to its operated position as a result of rebound from the backstop in release. (EEC/REE) [87]

relay contact chatter, externally caused Chatter resulting from shock or vibration imposed on the relay by external action. (EEC/REE) [87]

relay contact chatter, external shock Chatter ascribed to impact experienced by the relay or by the apparatus of which it forms a part. (EEC/REE) [87]

relay contact chatter, initial Chatter ascribed to vibration produced by opening or closing the contacts themselves, as by contact impact in closure. (EEC/REE) [87]

relay contact chatter, internally caused Chatter resulting from the operation or release of the relay. (EEC/REE) [87]

relay contact chatter, transmitted vibration Chatter ascribed to vibration originating outside the relay and transmitted to it through its mounting. (EEC/REE) [87]

relay contact combination (A) The total assembly of contacts on a relay. **(B)** Sometimes used for contact form. (EEC/REE) [87]

relay contact, fixed *See:* stationary relay contact.

relay contact follow The displacement of a stated point on the contact-actuating member following initial closure of a contact. (EEC/REE) [87]

relay contact follow, stiffness The rate of change of contact force per unit contact follow. (EEC/REE) [87]

relay contact form A single-pole contact assembly. (EEC/REE) [87]

relay contact functioning The establishment of the specified electrical state of the contacts as a continuous condition. (EEC/REE) [87]

relay contact gap *See:* relay contact separation.

relay contact, movable *See:* movable relay contact.

relay contact pole Sometimes used for relay contact, movable. (EEC/REE) [87]

relay contact rating A statement of the conditions under which a contact will perform satisfactorily. (EEC/REE) [87]

relay contacts The current-carrying parts of a relay that engage or disengage to open or close electric circuits. (EEC/REE) [87]

relay contacts, auxiliary *See:* auxiliary relay contacts.

relay contacts, back *See:* back relay contacts.

relay contacts, break *See:* normally closed relay contacts.

relay contacts, break-make *See:* break-make relay contacts.

relay contacts, bridging *See:* bridging relay contacts.

relay contacts, continuity transfer *See:* continuity transfer relay contacts.

relay contacts, double break *See:* double break relay contacts.

relay contacts, double make *See:* double make relay contacts.

relay contacts, dry *See:* dry relay contacts.

relay contacts, early *See:* early relay contacts.

relay contact separation The distance between mating contacts when the contacts are open. (EEC/REE) [87]

relay contacts, front *See:* front relay contacts.

relay contacts, interrupter *See:* interrupter relay contacts.

relay contacts, late *See:* late relay contacts.

relay contacts, low-capacitance *See:* low-capacitance relay contacts.

relay contacts, low-level *See:* low-level relay contacts.

relay contacts, make *See:* normally open relay contacts.

relay contacts, make-break *See:* make-break relay contacts.

relay contacts, multiple-break *See:* multiple-break relay contacts.

relay contacts, nonbridging *See:* nonbridging relay contacts.

relay contacts, normally closed *See:* normally closed relay contacts.

relay contacts, normally open *See:* normally open relay contacts.

relay contacts, off-normal *See:* off-normal relay contacts.

relay contacts, preliminary *See:* preliminary relay contacts.

relay contact spring (A) A current-carrying spring to which the contacts are fastened. **(B)** A non-current-carrying spring that positions and tensions a contact-carrying member. (EEC/REE) [87]

relay contacts, sealed *See:* sealed relay contacts.

relay contacts, snap-action *See:* snap-action relay contacts.

relay contact, stationary *See:* stationary relay contact.

relay contact wipe The sliding or tangential motion between two contact surfaces when they are touching. (EEC/REE) [87]

relay core The magnetic member about which the coil is wound. (EEC/REE) [87]

relay critical current *See:* relay critical voltage.

relay critical voltage That voltage (current) that will just maintain thermal relay contacts operated. *Synonym:* relay critical current. (EEC/REE) [87]

relay cycle timer A controlling mechanism that opens or closes contacts according to a preset cycle. (EEC/REE) [87]

relay damping ring, mechanical *See:* mechanical relay damping ring.

relay, direct-current *See:* direct-current relay.

relay, double-pole *See:* double-pole relay.

relay, double-throw *See:* double-throw relay.

relay driving spring The spring that drives the wipers of a stepping relay. (EEC/REE) [87]

relay drop-out *See:* relay release.

relay, dry circuit *See:* dry circuit relay.

relay duty cycle A statement of energized and deenergized time in repetitious operation, as: two seconds on, six seconds off. (EEC/REE) [87]

relay electric bias An electrically produced force tending to move the armature towards a given position. (EEC/REE) [87]

relay, electric reset *See:* electric reset relay.

relay, electromagnetic *See:* electromagnetic relay.

relay, electrostatic *See:* electrostatic relay.

relay, electrostrictive *See:* electrostrictive relay.

relay electrothermal expansion element An actuating element in the form of a wire strip or other shape having a high coefficient of thermal expansion. (EEC/REE) [87]

relay element A subassembly of parts. *Note:* The combination of several relay elements constitutes a relay unit. (SWG/PE/PSR) C37.100-1981s, C37.90-1978s

relay finish lead The outer termination of the coil. (EEC/REE) [87]

relay frame The main supporting portion of a relay. This may include parts of the magnetic structure. (EEC/REE) [87]

relay freezing, magnetic *See:* magnetic relay freezing.

relay fritting Contact erosion in which the electrical discharge makes a hole through the film and produces molten matter that is drawn into the hole by electrostatic forces and solidifies there to form a conducting bridge. (EEC/REE) [87]

relay front contacts Sometimes used for relay contacts, normally open. (EEC/REE) [87]

relay functioning time The time between energization and operation or between de-energization and release. (EEC/REE) [87]

relay functioning value The value of applied voltage, current, or power at which the relay operates or releases. (EEC/REE) [87]

relay header The subassembly that provides support and insulation to the leads passing through the walls of a sealed relay. (EEC/REE) [87]

relay heater A resistor that converts electric energy into heat for operating a thermal relay. (EEC/REE) [87]

relay heel piece The portion of a magnetic circuit of a relay that is attached to the end of the core remote from the armature. (EEC/REE) [87]

relay, high, common, low A type of relay control used in such devices as thermostats and in relays operated by them, in which a momentary contact between the common lead and another lead operates the relay, that then remains operated until a momentary contact between the common lead and a third lead causes the relay to return to its original position. (EEC/REE) [87]

relay hinge The joint that permits movement of the armature relative to the stationary parts of the relay structure. (EEC/REE) [87]

relay hold A specified functioning value at which no relay meeting the specification may release. (EEC/REE) [87]

relay housing An enclosure for one or more relays, with or without accessories, usually providing access to the terminals. (EEC/REE) [87]

relay hum The sound emitted by relays when their coils are energized by alternating current or in some cases by unfiltered rectified current. (EEC/REE) [87]

relaying A function performed at intermediate nodes on an interconnection between communicating end-systems. The relaying function is performed by connecting two independent layer entities. For example, a relaying function at the Data Link Layer connects two Data Link Layer entities to make an interconnection. (LM/C/COM) 8802-9-1996

relay inside lead *See:* relay start lead.

relay, interposing *See:* interposing relay.

relay inverse time A qualifying term applied to a relay indicating that its time of operation decreases as the magnitude of the operating quantity increases. (EEC/REE) [87]

relay just-operate value The measured functioning value at which a particular relay operates. (EEC/REE) [87]

relay just-release value The measured functioning value for the release of a particular relay. (EEC/REE) [87]

relay leakage flux The portion of the magnetic flux that does not cross the armature-to-pole-face gap. (EEC/REE) [87]

relay level A series of contacts served by one wiper in a stepping relay. (EEC/REE) [87]

relay load curves The static force displacement characteristic of the total load of the relay. (EEC/REE) [87]

relay logic A logic network that coordinates the output of measuring units and other inputs to energize output circuits when prescribed conditions and sequences have been met. (PE/PSR) C37.90.1-1989r

relay magnetic bias A steady magnetic field applied to the magnetic circuit of a relay. (EEC/REE) [87]

relay magnetic gap Nonmagnetic portion of a magnetic circuit. (EEC/REE) [87]

relay, manual-reset *See:* manual-reset relay.

relay mechanical bias A mechanical force tending to move the armature towards a given position. (EEC/REE) [87]

relay mounting plane The plane to which the relay mounting surface is fastened. (EEC/REE) [87]

relay-must-operate value A specified functioning value at which all relays meeting the specification must operate. (EEC/REE) [87]

relay must-release value A specified functioning value, at which all relays meeting the specification must release. (EEC/REE) [87]

relay nonfreeze pin Sometimes used for relay armature stop, nonmagnetic. (EEC/REE) [87]

relay nonoperate value A specified functioning value at which no relay meeting the specification may operate. (EEC/REE) [87]

relay normal condition The de-energized condition of the relay. (EEC/REE) [87]

relay operate The condition attained by a relay when all contacts have functioned. *See also:* relay contact actuation time. (EEC/REE) [87]

relay operate time The time interval from coil energization to the functioning time of the last contact to function. Where not otherwise stated the functioning time of the contact in question is taken as its initial functioning time. (EEC/REE) [87]

relay operate time characteristic The relation between the operate time of an electromagnetic relay and the operate power. (EEC/REE) [87]

relay operating frequency The rated alternating-current frequency of the supply voltage at which the relay is designed to operate. (EEC/REE) [87]

relay outside lead *See:* relay finish lead.

relay overtravel Amount of contact wipe. *See also:* relay contact wipe; relay armature overtravel. (EEC/REE) [87]

relay pickup value Sometimes used for relay must-operate value. (EEC/REE) [87]

relay pileup A set of contact arms, assemblies, or springs, fastened one on top of the other with insulation between them. (EEC/REE) [87]

relay pneumatic bellows Gas-filled bellows, sometimes used with plunger-type relays to obtain time delay. (EEC/REE) [87]

relay pole face The part of the magnetic structure at the end of the core nearest the armature. (EEC/REE) [87]

relay pole piece The end of an electromagnet, sometimes separable from the main section, and usually shaped so as to distribute the magnetic field in a pattern best suited to the application. (EEC/REE) [87]

relay pull curves The force-displacement characteristics of the actuating system of the relay. (EEC/REE) [87]

relay pull-in value Sometimes used for relay must-operate value. (EEC/REE) [87]

relay pusher Sometimes used for relay armature stud. *See also:* relay. (EEC/REE) [87]

relay rating A statement of the conditions under which a relay will perform satisfactorily. (EEC/REE) [87]

relay recovery time A cooling time required from heater de-energization of a thermal time-delay relay to subsequent re-energization that will result in a new operate time equal to 85 percent of that exhibited from a cold start. (EEC/REE) [87]

relay recovery time, instantaneous *See:* instantaneous relay recovery time.

relay recovery time, saturated *See:* saturated relay recovery time.

relay release The condition attained by a relay when all contacts have functioned and the armature (where applicable) has reached a fully opened position. (EEC/REE) [87]

relay release time The time interval from coil de-energization to the functioning time of the last contact to function. Where not otherwise stated the functioning time of the contact in question is taken as its initial functioning time. (EEC/REE) [87]

relay reoperate time Release time of a thermal relay. (EEC/REE) [87]

relay reoperate time, instantaneous *See:* instantaneous relay reoperate time.

relay reoperate time, saturated *See:* saturated relay reoperate time.

relay residual gap Sometimes used for relay armature stop, nonmagnetic. (EEC/REE) [87]

relay restoring spring A spring that moves the armature to the normal position and holds it there when the relay is de-energized. (EEC/REE) [87]

relay retractile spring Sometimes used for relay restoring spring. (EEC/REE) [87]

relay return spring Sometimes used for relay restoring spring. (EEC/REE) [87]

relay saturation The condition attained in a magnetic material when an increase in field intensity produces no further increase in flux density. (EEC/REE) [87]

relay sealing Sometimes used for relay seating. (EEC/REE) [87]

relay seating The magnetic positioning of an armature in its final desired location. (EEC/REE) [87]

relay seating time The elapsed time after the coil has been energized to the time required to seat the armature of the relay. (EEC/REE) [87]

relay shading coil Sometimes used for relay shading ring. (EEC/REE) [87]

relay shading ring A shorted turn surrounding a portion of the pole of an alternating-current magnet, producing a delay of the change of the magnetic field in that part, thereby tending to prevent chatter and reduce hum. (EEC/REE) [87]

relay shields, electrostatic spring Grounded conducting members located between two relay springs to minimize electrostatic coupling. (EEC/REE) [87]

relay shim, nonmagnetic *See:* nonmagnetic relay shim.

relay, single-pole *See:* single-pole relay.

relay, single-throw *See:* single-throw relay.

relay sleeve A conducting tube placed around the full length of the core as a short-circuited winding to retard the establishment or decay of flux within the magnetic path. (EEC/REE) [87]

relay slow-release time characteristic The relation between the release time of an electromagnetic relay and the conductance of the winding circuit or of the conductor (sleeve or slug) used to delay release. The conductance in this definition is the quantity $N^2 \cdot R$, where N is the number of turns and R is the resistance of the closed winding circuit. (For a sleeve or slug N, 1). (EEC/REE) [87]

relay slug A conducting tube placed around a portion of the core to retard the establishment or decay of flux within the magnetic path. (EEC/REE) [87]

relay soak The condition of an electromagnetic relay when its core is approximately saturated. (EEC/REE) [87]

relay soak value The voltage, current, or power applied to the relay coil to insure a condition approximating magnetic saturation. (EEC/REE) [87]

relay spool A flanged form upon which a coil is wound. (EEC/REE) [87]

relay spring buffer Sometimes used for relay spring stud. (EEC/REE) [87]

relay spring curve A plot of spring force on the armature versus armature travel. (EEC/REE) [87]

relay spring stop A member that controls the position of a pretensioned spring. (EEC/REE) [87]

relay spring stud An insulating member that transmits the motion of the armature from one movable contact to another in the same pileup. (EEC/REE) [87]

relay stack Sometimes used for relay pileup. (EEC/REE) [87]

relay stagger time The time interval between the actuation of any two contact sets. (EEC/REE) [87]

relay starting switch (rotating machinery) A relay, actuated by current, voltage, or the combined effect of current and voltage, used to perform a circuit-changing function in the primary winding of a single-phase induction motor within a predetermined range of speed as the rotor accelerates: and to perform the reverse circuit-changing operation when the motor is disconnected from the supply line. One of the circuit changes that is usually performed is to open or disconnect the auxiliary-winding circuit. *See also:* starting-switch assembly. (PE) [9]

relay start lead The inner termination of the coil. (EEC/REE) [87]

relay static characteristic The static force-displacement characteristic of the spring system or of the actuating system. (EEC/REE) [87]

relay station (mobile communication) A radio station used for the reception and retransmission of the signals from another radio station. *See also:* mobile communication system. (VT) [37]

relay system (1) (surge withstand capability) An assembly, usually consisting of current and voltage circuits, measuring units, logic, and power supplies, to provide a specific relay scheme such as line, transformer, bus, or generator protection. A relay system may include interfaces with other systems such as data logging, alarm, telecommunications, or other relay systems. (PE/PSR) C37.90.1-1989r, C37.90-1978s
(2) An assembly that usually consists of measuring units, relay logic, communications interfaces, computer interfaces, and necessary power supplies. (SWG/PE) C37.100-1992

relay thermal A relay that is actuated by the heating effect of an electric current. *See also:* relay. (EEC/REE) [87]

relay, three-position *See:* three-position relay.

relay transfer contacts Sometimes used for relay contacts, break-make. (EEC/REE) [87]

relay transfer time The time interval between opening the closed contact and closing the open contact of a break-make contact form. (EEC/REE) [87]

relay unit (general) An assembly of relay elements that in itself can perform a relay function. *Note:* One or more relay units constitutes a relay. (PE/PSR) C37.90-1978s
(2) (A) A subassembly of parts. *Note:* The combination of several relay elements constitutes a relay unit. **(B)** An assembly of relay elements that in itself can perform a relay function. *Note:* One or more relay units constitute a relay. (SWG/PE) C37.100-1992

relay winding Sometimes used for relay coil. (PE/TR) C57.15-1968s

relay wiper The moving contact on a rotary stepping switch or relay. (EEC/REE) [87]

release (1) (A) (telephone switching systems) Disengaging the apparatus used in a connection and restoring it to its idle condition upon recognizing a disconnect signal. **(B)** (of a mechanical switching device) A device, mechanically connected to a mechanical switching device, which releases the holding means and permits the opening or closing of the switching device. *Synonym:* tripping mechanism. **(C) (STEbus)** The ac-

tion of a transmitter in ceasing to hold a signal line in the asserted state. **(D)** The action of applying a logic zero signal to a bus line. **(E)** The state of a bus line when the signal it carries represents a logic zero. (SWG/COM) 312-1977
(2) To stop pressing a mouse button or keyboard key. (C) 1295-1993w
(3) To cease to assert a logic 1 on a bus signal line. (One module's releasing a signal line produces a change in value of the signal line only if no module is asserting the signal.). (TT/C) 1149.5-1995
(4) The formal notification and distribution of an approved version. (C/SE) 828-1998
(5) **(A)** The action of applying a logic 0 signal to a bus line. **(B)** The state of a bus line when the signal it carries represents a logic 0. (C/BA) 896.10-1997

release coil (of a mechanical switching device) A coil used in the electromagnet that initiates the action of a release (trip). *Synonym:* trip coil. (SWG/PE) C37.100-1992

released Having a value equal to logic 0 (said of any signal). Equivalently, in the case of an MTM-Bus signal line, not asserted by any module on the bus. (TT/C) 1149.5-1995

release delay (of a mechanical switching device) Intentional time-delay introduced into contact parting time in addition to opening time. *Note:* In devices employing a shunt release, release delay includes the operating time of protective and auxiliary relays external to the device. In devices employing direct or indirect release, release delay consists of intentional delay introduced into the function of the release. *Synonym:* tripping delay. (SWG/PE) C37.100-1992

release-delay setting (trip delay) A calibrated setting of the time interval between the time when the actuating value reaches the release setting and the time when the release operates. (SWG/PE) C37.100-1992

release-free (as applied to a mechanical device). *See also:* trip-free. (SWG/PE) C37.100-1992

release-free in any position A descriptive term indicating that a switching device is release-free at any part of the closing operation. *Note:* If the release circuit is completed through an auxiliary switch, electrical release will not take place until such auxiliary switch is closed. *Synonym:* trip-free in any position. (SWG/PE) C37.100-1981s

release-free relay *See:* trip-free relay.

release mechanism (mechanical switching device) A device, mechanically connected to the mechanical switching device, that releases the holding means and permits the opening or closing of the switching device. *Synonym:* tripping mechanism. (SWG/PE) C37.100-1981s

release setting A calibrated point at which the release is set to operate. *Synonym:* trip setting. (SWG/PE) C37.100-1992

release signal (telephone switching systems) A signal transmitted from one end of a line or trunk to indicate that the called party has disconnected. (COM) 312-1977w

release time *See:* hang-over time.

release time, relay *See:* relay release time.

relevant failure *See:* failure.

relevant rated step voltage The value of rated step voltage that corresponds to a specific value of rated through current. (PE/TR) C57.131-1995

reliability (1) (relay or relay system) A measure of the degree of certainty that the relay, or relay system, will perform correctly. *Note:* Reliability denotes certainty of correct operation together with assurance against incorrect operation from all extraneous causes. *See also:* security; dependability. (SWG/PE/PSR) C37.100-1992, [6], C37.90-1978s, [56]
(2) (reliability analysis of nuclear power generating station safety systems) The characteristic of an item or system expressed by the probability that it will perform a required mission under stated conditions for a stated mission time. (PE/NP) 352-1987r, 577-1976r
(3) (software) The ability of a system or component to perform its required functions under stated conditions for a specified period of time. (C/BA) 896.9-1994w, 610.12-1990
(4) The probability that a transformer will perform its specified function under specified conditions for a specified period of time. (PE/TR) C57.117-1986r
(5) (power system protective relaying) A combination of dependability and security. (PE/PSC) 487-1992
(6) The probability that an item will perform its intended function for a specified interval under stated conditions. (C/BA) 896.3-1993w
(7) The characteristic of equipment or software that relates to the integrity of the system and ability to maintain trouble-free operations to insure against failure. (C) 610.7-1995
(8) (general) The ability of an item to perform a required function under stated conditions for a stated period of time. *See also:* wearout-failure period; observed reliability; assessed reliability. (R/C/Std100) 1413-1998, [29], 610.7-1995
(9) (general) The probability that a device will function without failure over a specified time period or amount of usage. *Notes:* 1. This definition is most commonly used in engineering applications. In any case where confusion may arise, specify the definition being used. 2. The probability that the system will perform its function over the specified time should be equal to or greater than the reliability. (SMC/C) [63], [20], [85]
(10) The probability that a system will perform its intended functions without failure, within design parameters, under specific operating conditions, and for a specific period of time. (VT/RT) 1475-1999, 1474.1-1999

reliability allocation (nuclear power generating station) The assignment of reliability subgoals to subsystems and elements thereof within a system for the purpose of meeting the overall reliability goal for the system, if each of these subgoals is attained. (PE/NP) 380-1975w, 338-1977s

reliability, assessed *See:* assessed reliability.

reliability assessment (software) The process of determining the achieved level of reliability of an existing system or system component. *See also:* reliability; system; component. (C/SE) 729-1983s

reliability, availability, and maintainability Elements that are considered as unified for reliability enhancement. (PE/NP) 933-1999

reliability-centered maintenance (1) A systematic methodology that establishes initial preventive maintenance requirements or optimizes existing preventive maintenance requirements for equipment based upon the consequences of equipment failure. The failure consequences are determined by the application of the equipment in an operating system. (IA/PSE) 902-1998
(2) A series of orderly steps for identifying system and subsystem functions, functional failures, and dominant failure modes, prioritizing them, and selecting applicable and effective preventive maintenance tasks to address the classified failure modes. (PE/NP) 933-1999

reliability compliance test An experiment used to show whether or not the value of a reliability characteristic of an item complies with its stated reliability requirements. (R) [29]

reliability data (software) Information necessary to assess the reliability of software at selected points in the software life cycle. Examples include error data and time data for reliability models, program attributes such as complexity, and programming characteristics such as development techniques employed and programmer experience. (C/SE) 729-1983s

reliability determination test An experiment used to determine the value of a reliability characteristic of an item. *Note:* Analysis of available data may also be used for reliability determination. (R) [29]

reliability evaluation *See:* reliability assessment.

reliability, extrapolated *See:* extrapolated reliability.

reliability goal (nuclear power generating station) A design objective, stated numerically, applied to reliability or availability. (PE/NP) 380-1975w

reliability growth (software) The improvement in reliability that results from correction of faults. (C) 610.12-1990

reliability, inherent *See:* inherent reliability.

reliability model (1) (software) A model used for predicting, estimating, or assessing reliability. *See also:* reliability assessment; model; reliability. (C/SE) 729-1983s
(2) (modeling and simulation) A model used to estimate, measure, or predict the reliability of a system; for example, a model of a computer system, used to estimate the total down time that will be experienced. (C) 610.3-1989w

reliability modeling (nuclear power generating station) A logical display in a block diagram format and a mathematical representation of component functions as they occur in sequence which is required to produce system success. (PE/NP) 380-1975w, 338-1977s

reliability monitoring Direct monitoring of reliability parameters of a plant, system, or equipment (e.g., failure frequency, downtime due to the maintenance activities, outage rate). (PE/NP) 933-1999

reliability, operational *See:* operational reliability.

reliability, predicted *See:* predicted reliability.

reliability program A description of activities and techniques associated with reliability technology, not necessarily a formalized program or entity unto itself, which may be integrated with design and operations. (PE/NP) 933-1999

reliability targets The reliability goals to be achieved by the plant systems. (PE/NP) 933-1999

reliability, test *See:* test reliability.

reliability unit That portion of a system for which a single reliability model is valid, i.e., for which there is a single mechanism of failure. (PE/NP) 1082-1997

reliable service A communication service in which the received data is guaranteed to be exactly as transmitted. (DIS/C) 1278.1-1995, 1278.2-1995

relief door (rotating machinery) A pressure-operated door to prevent excessive gas pressure within a housing. (PE) [9]

relieving (electroplating) The removal of compounds from portions of colored metal surfaces by mechanical means. *See also:* electroplating. (EEC/PE) [119]

relieving anode (pool-cathode tube) An auxiliary anode that provides an alternative conducting path for reducing the current to another electrode. *See also:* electrode. (EEC/PE) [119]

relocatable Pertaining to code that can be loaded into any part of main memory. The starting address is established by the loader, which then adjusts the addresses in the code to reflect the storage locations into which the code has been loaded. *See also:* relocating loader. (C) 610.12-1990

relocatable address An address that is to be adjusted by the loader when the computer program containing the address is loaded into memory. *Note:* Generally implemented through the use of relative addressing. *Contrast:* absolute address. *See also:* relative address. (C) 610.10-1994w, 610.12-1990

relocatable code Code containing addresses that are to be adjusted by the loader to reflect the storage locations into which the code is loaded. *Contrast:* absolute code. (C) 610.12-1990

relocatable machine code (software) Machine language code that requires relative addresses to be translated into absolute addresses prior to computer execution. *See also:* address; absolute machine code. (C/SE) 729-1983s

relocate (1) (programming) (computers) To move a routine from one portion of storage to another and to adjust the necessary address references so that the routine, in its new location, can be executed. (C) [20], [85]
(2) (software) To move machine code from one portion of main memory to another and to adjust the addresses so that the code can be executed in its new location. (C) 610.12-1990

relocating assembler An assembler that produces relocatable code. *Contrast:* absolute assembler. (C) 610.12-1990

relocating loader A loader that reads relocatable code into main memory and adjusts the addresses in the code to reflect the storage locations into which the code has been loaded. *Synonym:* relative loader. *Contrast:* absolute loader. (C) 610.12-1990

relocation *See:* biasing.

relocation dictionary The part of an object module or load module that identifies the addresses that must be adjusted when a relocation occurs. (C) 610.12-1990

relocation factor *See:* offset.

reluctance (magnetic circuit) The ratio of the magnetomotive force to the magnetic flux through any cross section of the magnetic circuit. (Std100) 270-1966w

reluctance motor A synchronous motor similar in construction to an induction motor, in which the member carrying the secondary circuit has salient poles, without permanent magnets or direct-current excitation. It starts as an induction motor, is normally provided with a squirrel-cage winding, but operates normally at synchronous speed. (PE) [9]

reluctance synchronizing (rotating machinery) Synchronizing by bringing the speed of a salient pole synchronous machine to near-synchronous speed, but without applying excitation to it. (PE) [9]

reluctance torque (synchronous motor) The torque developed by the motor due to the flux produced in the field poles by action of the armature-reaction magnetomotive force. (PE) [9]

reluctivity The reciprocal of permeability. *Note:* In anisotropic media, reluctivity becomes a matrix. (Std100) 270-1966w

REM *See:* ring error monitor.

remanence The magnetic flux density that remains in a magnetic circuit after the removal of an applied magnetomotive force. *Note:* This should not be confused with residual flux density. If the magnetic circuit has an air gap, the remanence will be less than the residual flux density. *See also:* residual flux density. (Std100/PE/PSR) 163-1959w, C37.110-1996

remanent charge The charge remaining when the applied voltage is removed. *Note:* The remanent charge is essentially independent of the previously applied voltage, provided this voltage was sufficient to cause saturation. If the device was not or cannot be saturated, the value of the previously applied voltage should be stated when measurements of remanent charge are reported. *See also:* ferroelectric domain. (UFFC) 180w

remanent induction (magnetic material) The induction when the magnetomotive force around the complete magnetic circuit is zero. *Note:* If there are no air gaps or other inhomogeneities in the magnetic circuit, the remanent induction will equal the residual induction: if there are air gaps or other inhomogeneities, the remanent induction will be less than the residual induction. (Std100) 270-1966w

remanent polarization (primary ferroelectric terms) The value of the polarization P_r that remains after an applied electric field is removed. Remanent polarization can be measured by integrating the compensating surface charge released on heating a poled ferroelectric to a temperature above its Curie point. *Note:* When the magnitude of this electric field is sufficient to saturate the polarization (usually $3E_c$ that is, three times the coercive electric field), the polarization remaining after the field is removed is termed the saturation remanent polarization P_r. In a single-domain ferroelectric material, the saturation remanent polarization is equal to the spontaneous polarization. (UFFC) 180-1986w

remodulator A device located at the headend of a broadband coaxial cable system that receives inbound transmissions and converts them to outbound transmissions via an intermediate step in which the inbound signals are converted to the baseband level. The device may or may not perform operations on the contents of the baseband signal. (LM/C) 802.7-1989r

remote At a distance such that the mutual resistance of the two electrodes is essentially zero. *See also:* ground rod. (PE/PSIM) 81-1983

remote access (test, measurement, and diagnostic equipment) Pertaining to communication with a data processing facility by one or more stations that are distant from that facility. (MIL) [2]

remote-access data processing Data processing in which some or all of the input-output functions are performed at locations away from the primary computer, connected to the primary computer by telecommunication facilities. (C) 610.2-1987

remote alarm indication *See:* yellow alarm.

remote backup A form of backup protection in which the protection is at a station or stations other than that which has the primary protection. (SWG/PE) C37.100-1992

remote batch entry *See:* remote job entry.

Remote Bridge *See:* Remote MAC Bridge.

Remote Bridge Cluster A subset of the Remote Bridges in a single Group, all of which are providing, or preparing to provide, MAC-sublayer interconnection of the attached Locally Bridged Local Area Networks and other Groups. A Remote Bridge Cluster is fully connected, *ie,* it supports communication between any pair of the Remote Bridges that belong to it. Membership of a Remote Bridge Cluster is determined dynamically through protocols operating in support of the Spanning Tree Algorithm. *Note:* A Cluster can—and often will—consist of all the Remote Bridges in the relevant Group. (C/LM) 802.1G-1996

Remote Bridge Group A set of Remote Bridges, capable of communicating with each other over non-LAN communications equipment, which cooperate in providing actual or potential MAC-sublayer interconnection among all the attached Locally Bridged Local Area Networks and other attached Groups. Membership of a Remote Bridge Group is determined statically, as an aspect of the configuration of the Remotely Bridged Local Area Network. *Note:* A Remotely Bridged Local Area Network can contain more than one Remote Bridge Group. (C/LM) 802.1G-1996

Remote Bridge Subgroup A set of Remote Bridges belonging to one Group, such that each Remote Bridge in the Subgroup has a single Virtual Port representing its communication with all the other Remote Bridges in the Subgroup, and with no others. Membership of a Subgroup is determined statically, as an aspect of the configuration of the Group. (C/LM) 802.1G-1996

remote computer system A computer system located at some remote site and connected via a communications network to one or more other systems. *See also:* satellite computer. (C) 610.7-1995, 610.10-1994w

remote concentrator (telephone switching systems) A concentrator located away from a serving system control. (COM) 312-1977w

remote console A console in a remote computer system. *See also:* master console. (C) 610.10-1994w

remote control (1) (general) Control of an operation from a distance: this involves a link, usually electrical, between the control device and the apparatus to be operated. *Note:* Remote control may be over direct wire; other types of interconnecting channels such as carrier-current or microwave; supervisory control; or mechanical means. *See also:* control. (PE/PSE) [54], [84]

(2) (programmable instrumentation) A method whereby a device is programmable via its electrical interface connection in order to enable the device to perform different tasks. (IM/AIN) 488.1-1987r

(3) Control of a device from a distant point. *Note:* Remote control may be over direct wire, or over other types of interconnecting channels such as carrier-current or microwave, or by supervisory control or by (4) mechanical means. (SWG/PE/SUB) C37.100-1992, C37.1-1994

remote-control circuit Any electric circuit that controls any other circuit through a relay or an equivalent device. (NESC/NEC) [86]

remote-cutoff tube *See:* variable-mu tube.

remote data logging An arrangement for the numerical representation of selected telemetered quantities on log sheets or paper or magnetic tape, or the like, by means of an electric typewriter, teletype, or other suitable devices. (SWG/PE) C37.100-1992

remote earth (1) (potential) The location outside the influence of local grounds. Always assumed to be at zero potential. (SPD/PE) C62.23-1995

(2) That distant point on the earth's surface where an increase in the distance from a ground electrode will not measurably increase the impedance between that ground electrode and the new distant point. (PE/PSC) 367-1996

(3) The point beyond which further reduction in ground electrode or grid impedance results in negligible effects. (IA/PSE) 1100-1999

remote error-sensing (power supplies) A means by which the regulator circuit senses the voltage directly at the load. This connection is used to compensate for voltage drops in the connecting wires. (AES) [41]

remote fault The generic ability of a link partner to signal its status even in the event that it may not have an operational receive link. (C/LM) 802.3-1998

remote indication Indication of the position or condition of remotely located devices. *Note:* Remote indication may be over direct wire, or over other types of interconnecting channels such as carrier-current or microwave, or by supervisory indication or by mechanical means. (SWG/PE) C37.100-1992

remote job entry (RJE) (1) (A) Submission of jobs through a remote input device connected to a computer through a data link. *Synonym:* remote batch entry. **(B)** Submission of jobs through an input device that has access to a computer through a communications link. (LM/C/COM) 610.12-1990, 168-1956

(2) A service that allows a user to submit a batch job from a remote site. (C) 610.10-1994w

remotely controlled operation Operation of a device by remote control. (SWG/PE) C37.100-1992

remotely operated (as applied to equipment) Capable of being operated from a position external to the structure in which it is installed or from a protected position within the structure. (NESC) C2-1984s

remote line (electroacoustics) A program transmission line between a remote pickup point and the studio or transmitter site. *See also:* transmission line. (SP) 151-1965w

remote login (rlogin) A login to another computer in a remote location. (C) 610.10-1994w

Remotely Bridged Local Area Network A Bridged Local Area Network of two or more Locally Bridged Local Area Networks interconnected using non-LAN communication technologies, and providing MAC-sublayer interworking between end stations attached to any of the LANs. (C/LM) 802.1G-1996

remotely operable (as applied to equipment) Capable of being operated from a position external to the structure in which it is installed or from a protected position within the structure. (NESC) C2-1997

remote MAC (RMAC) The MAC component at the remote end of the data link as specified by its unique 48-bit address. (C/LM/CC) 8802-2-1998

Remote MAC Bridge A MAC Bridge interconnecting a Locally Bridged Local Area Network and the non-LAN communications equipment of a Remotely Bridged Local Area Network. (C/LM) 802.1G-1996

remote magnetic sensor (navigation aid terms) A magnetic sensor located on a vehicle away from disturbances which provides an electrical signal in synchro format which is proportional to the vehicle heading relative to magnetic north. Often called a flux valve. (AES/GCS) 172-1983w

remote manual operation *See:* indirect manual operation.

remote master *See:* master remote unit.

remote metering *See:* telemetering.

remote on/off control The control over the on/off operation of the unit-under-test (UUT) output power by means initiated externally or away from the UUT. (PEL) 1515-2000

remote operation *See:* remotely controlled operation.

remote release *See:* remote trip.

remote SAP (RSAP) The SAP at the remote end of a data link as specified by its LLC address. (C/LM/CC) 8802-2-1998

remote station (1) (of a supervisory system) A remotely located station wherein units of switchgear or other equipment are controlled by supervisory control or from which supervisory indications or selected telemeter readings are obtained. (SWG/PE) C37.100-1992
(2) (of a supervisory system) The entire complement of devices, functional modules, and assemblies that are electrically interconnected to effect the remote station supervisory functions. The equipment includes the interface with the communication channel, but does not include the interconnecting channel. During communication with a master station, the remote station is hierarchy. *Note:*

- — *Hardwired.* Station supervisory equipment that is comprised entirely of wired-logic elements.
- — *Firmware.* Station supervisory equipment that uses hardware logic programmed routines in a manner similar to a computer. The routines can only be modified by physically exchanging logic memory elements.
- — Station supervisory equipment that uses software routines.
- — *Semiautomatic.* A station that requires both automatic and manual modes to maintain the required character of service.
- — *Submaster.* A station that can perform as a master station on one message transaction and as a remote station on another message transaction.

(SWG/PE/SUB) C37.100-1992, C37.1-1994

remote-station supervisory equipment The part of a (single) supervisory system that includes all supervisory control relays and associated devices located at the remote station for selection, control, indication, and other functions to be performed. (SWG/PE) C37.100-1992

remote switching entity (telephone switching systems) An entity for switching inlets to outlets located away from a serving system control. (COM) 312-1977w

remote terminal The entire complement of devices, functional modules, and assemblies that are electrically interconnected to effect the remote terminal supervisory functions (of a supervisory system). The equipment includes the interface with the communication channel, but does not include the connecting channel. (SUB/PE) 999-1992w

remote terminal unit (RTU) (1) (supervisory control, data acquisition, and automatic control) The remote station equipment of a supervisory system. *See also:* station. (SWG) C37.100-1992
(2) A piece of equipment located at a distance from a master station to monitor and control the state of outlying equipment, and to communicate the information back to the master station or host. (PE/SUB) 1379-1997
(3) *See also:* remote station. (SUB/PE) C37.1-1994

remote trip (remote release) A general term applied to a relay installation to indicate that the switching device is located physically at a point remote from the initiating protective relay, device, or source of release power or all these. *Note:* This installation is commonly called transfer trip when a communication channel is used to transmit the signal for remote tripping. *Synonym:* remote release. (SWG/PE/PSR) C37.100-1992, C37.90-1978s

removable breaker The removable breaker consists of the circuit breaker, disconnecting provisions, network relays, auxiliary panels, current transformers, control devices, other attachments, and all interconnecting wiring, which can be rolled out of the network protector enclosure on rails for maintenance or removal. (PE/TR) C57.12.44-1994

removable conductor link A removable connector between the GIS conductor and the end of the cable termination. (PE/IC) 1300-1996

removable disk A disk that can be removed from the disk drive. *Contrast:* fixed disk. (C) 610.10-1994w

removable element (of a switchgear assembly) The portion that normally carries the circuit-switching and circuit-interrupting devices and the removable part of the primary and secondary disconnecting devices. (SWG/PE) C37.100-1992

removable storage Any storage medium, such as a disk, which can be removed from the storage device and stored or transported somewhere else. *Note:* Some portion of the interface may be included with the medium. For example, some removable disk cartridges include the heads as well as the disk. (C) 610.10-1994w

removal, fault *See:* fault removal.

removal time *See:* recovery/removal timing check.

removal timing check A timing check that establishes only the end of the stable interval for a recovery/ removal timing check. If no recovery timing check is provided for the same arc, transitions, and state, the stable interval is assumed to begin at the reference signal transition, and a negative value for the removal time is not meaningful. *See also:* recovery/ removal timing check. (C/DA) 1481-1999

rendezvous (software) The interaction that occurs between two parallel tasks when one task has called an entry of the other task, and a corresponding accept statement is being executed by the other task on behalf of the calling task. (C/SE) 729-1983s

renegotiation Restart of the Auto-Negotiation algorithm caused by management or user interaction. (C/LM) 802.3-1998

renewable fuse (1) (protection and coordination of industrial and commercial power systems) A fuse in which the element, usually a zinc link, may be replaced after the fuse has opened. Once a very popular item, this fuse is gradually losing popularity due to the possibility of using higher ampere-rated links or multiple links in the field, which can present a hazard. (IA/PSP) 242-1986r
(2) A fuse or fuse unit that, after circuit interruption, may be restored readily for service by the replacement of the renewal element, fuse link, or refill unit. *Synonyms:* field-renewable fuse unit; field-renewable fuse; renewable fuse unit. (SWG/PE) C37.100-1992

renewable fuse unit *See:* renewable fuse.

renewal element (of a low-voltage fuse) The part of a renewable fuse that is replaced after each interruption to restore the fuse to operating condition. (SWG/PE) C37.100-1992

renewal parts Those parts that must be replaced during maintenance as a result of wear. (SWG/PE) C37.100-1992

reoperate time, relay *See:* relay reoperate time.

reorder* *See:* order.

* Deprecated.

reorganization (A) The process of rearranging the contents of a database so that space allocation is minimized and efficiency is maximized. Techniques include pointer optimization and garbage collection. *Synonym:* restructuring. *See also:* concurrent reorganization. **(B)** The process of rearranging the logical schema or physical structure of a database. *Synonym:* reformatting. (C) 610.5-1990

repagination *See:* automatic pagination.

repaint *See:* refresh.

repair (1) (failure data for power transformers and shunt reactors) Any operation that requires the dismantling, modification, or replacement of transformer components that results in restoring the transformer to normal service quality. (PE/TR) C57.117-1986r

(2) (nuclear power quality assurance) The process of restoring a nonconforming characteristic to a condition such that the capability of an item to function reliably and safely is unimpaired, even though that item still does not conform to the original requirement. (PE/NP) [124]

(3) (test, measurement, and diagnostic equipment) The restoration or replacement of parts or components of material as necessitated by wear and tear, damage, failure of parts or the like in order to maintain the specific item of material in efficient operating condition. (MIL) [2]

(4) Work done to restore the component or the circuit breaker to condition for operation. (SWG/PE) C37.10-1995

(5) Includes incoming inspection and test, damage appraisal, cleaning, replacement or fixing of damaged part(s) or both, assembly, postrepair inspection and test, and refinishing. (IA/PC) 1068-1996

repairable item *See:* repaired item.

repaired item An item that is repaired after a failure. *See also:* reliability. (R) [29]

repair facility The entity contracted to make repairs; includes the "on site" repair(s) made by employees of that entity in addition to repair(s) made at a shop operated by or under the supervision of that entity. (IA/PC) 1068-1996

repair rate (nuclear power generating station) The expected number of repair actions of a given type completed on a given item per unit of time. (PE/NP) 352-1987r, 933-1999

repair time The repair time of a failed component or the duration of a failure is the clock time from the occurrence of the failure to the time when the component is restored to service, either by repair of the failed component or by substitution of a spare component for the failed component. It includes time for diagnosing the trouble, locating the failed component, waiting for parts, repairing or replacing, testing, and restoring the component to service. It does not include the time required to restore service to a load by putting alternate circuits into operation. *Synonym:* forced outage duration. (IA/PSE) 399-1997, 493-1997

repair unavailability *See:* unavailability.

repair urgency (electric generating unit reliability, availability, and productivity) When a planned or unplanned outage is initiated, the urgency with which repair activities are carried out is classified according to one of three classes as defined in maximum effort, normal effort, and low-priority effort. (PE/PSE) 762-1987w

repeat The action of receiving a bit stream (for example, frame, token, or fill) and placing it on the medium. Stations repeating the bit stream may copy it into a buffer or modify control bits as appropriate. *Contrast:* transmit. (C/LM) 8802-5-1998

repeatability (1) (electric pipe heating systems) The closeness of agreement among a number of consecutive measurements of the output for the same value of the input under the same operating conditions approaching from the same direction. With respect to electric pipe heating systems, repeatability is usually associated with temperature controllers and is the difference in degrees for repeated operation at a specific temperature setting. (PE/EDPG) 622A-1984r

(2) (electric heat tracing systems) The closeness of agreement among a number of consecutive measurements of the output for the same value of the input under the same operating conditions approaching from the same direction. With respect to electric heat tracing systems, repeatability is usually associated with temperature controllers and is the difference in degrees for repeated operation at a specific temperature setting. (PE/EDPG) 622B-1988r

(3) (measurement) (control equipment) The closeness of agreement among repeated measurements of the same variable under the same conditions. *See also:* measurement system. (PE/PSE) 94-1970w

(4) (supervisory control, data acquisition, and automatic control) The measure of agreement among multiple readings of an output for the same value of input, made under the same operating conditions, approaching from the same direction, using full-range traverses. (SWG/SUB/PE) C37.1-1987s, C37.100-1992

(5) (electrical analog indicating instruments) The ability of an instrument to repeat its readings taken when the pointer is deflected upscale, compared to the readings taken when the pointer is deflected downscale, expressed as a percentage of the full-scale value. *See also:* moving element; measurement system. (EEC/AII) [102]

(6) (analog computer) A quantitative measure of the agreement among repeated operations. (C) 165-1977w

(7) (A) (attenuator, variable in fixed steps) Maximum difference in decibels of residual or incremental characteristic insertion loss for a selected position between the extreme values of a first and second set of ten measurements before and after the specified stepping life. **(B) (attenuator, variable in fixed steps)** (two-port, due to insertion/removal cycle) The maximum difference in decibels between the extreme value of ten measurements before and ten measurements after the number of complete insertion/removal cycles specified in insertion/removal life. **(C) (inertial sensors)** (continuously variable attenuator, due to cycling) Maximum difference in decibels between the extreme values of a first set of ten measurements for a selected calibration point, five of which are approached from the opposite direction, and the extreme values of a similar second set after the specified cycling life. (IM/HFIM/GYAC) 474-1973

(8) (accelerometer) (inertial sensors) (gyros) The closeness of agreement among repeated measurements of the same variable under the same conditions when changes in conditions or nonoperating periods occur between measurements. (AES/GYAC) 528-1994

(9) (nuclear power generating station) The closeness of agreement among a number of consecutive measurements of the output for the same value of the input under the same operating conditions, approaching from the same direction.

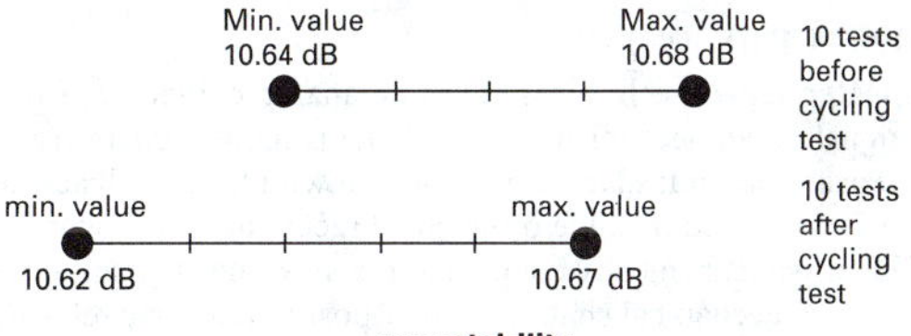

repeatability

(PE/NP) 381-1977w

(10) (software) *See also:* test repeatability. (C) 610.12-1990

repeated selection sort A selection sort in which the set of items to be sorted is divided into subsets; one item that fits specified criteria is selected from each subset, forming a second-level subset; a selection sort is then applied to this second-level subset; the selected item is appended to the sorted set and is replaced in the second-level subset by the next eligible item in the original subset; and the process is repeated until all items are in the sorted set. *See also:* tournament sort. (C) 610.5-1990w

repeater (1) A device used to extend the length, topology, or interconnectivity of the physical medium beyond that imposed by a single segment, up to the maximum allowable end-to-end trunk transmission line length. Repeaters perform the basic actions of restoring signal amplitude, waveform, and timing applied to normal data and collision signals. (C/LM) 802.9a-1995w

(2) (data transmission) A combination of apparatus for receiving either one-way or two-way communication signals and delivering corresponding signals which are either amplified, reshaped, or both. A repeater for one-way communication signals is termed a "one-way repeater" and one for two-way communication signals a "two-way repeater." (PE) 599-1985w

(3) (communication satellite) A receiver-transmitter combination, often aboard a satellite or spacecraft, which receives a signal, performs signal processing (amplification, frequency translation, etc.) and retransmits it. Used in active communication satellite to relay signals between earth stations. *Synonym:* transponder. (COM) [24]
(4) A device that restores signals to their original shape and transmission level at the physical layer only. (C) 610.7-1995
(5) The physical layer coupler of ring segments that provides for physical containment of channels, dividing the ring into segments. A repeater can receive any valid token ring signal and retransmit it with the same characteristics and levels as a transmitting station. (C/LM) 8802-5-1998
(6) A device used to extend the length, topology, or interconnectivity of the physical medium beyond that imposed by a single segment, up to the maximum allowable end-to-end transmission line length. Repeaters perform the basic actions of restoring signal amplitude, waveform, and timing applied to the normal data and collision signals. For wired star topologies, repeaters provide a data distribution function. In 100BASE-T, a device that allows the interconnection of 100BASE-T Physical Layer (PHY) network segments using similar or dissimilar PHY implementations (e.g., 100BASE-X to 100BASE-X, 100BASE-X to 100BASE-T4, etc.). Repeaters are only for use in half duplex mode networks. (C/LM) 802.3-1998
(7) (local area networks) A device used to extend the length, topology, and interconnectivity of the physical medium beyond that imposed by a single segment. Demand-priority repeaters perform the functions of restoring signal amplitude, waveform, and timing. They also arbitrate access to the network from connected end nodes and optionally collect statistics regarding network operations. (C) 8802-12-1998
(8) (fiber optics) *See also:* optical repeater. 812-1984w

repeater medium access control sublayer (local area networks) The sublayer in the repeater that arbitrates packet sequencing and controls packet routing. (C) 8802-12-1998

repeater port *See:* port.

repeater servomechanism (1) In an analog computer, a positional servomechanism in which loop input signals from a transmitting transducer are compared with loop feedback signals from a compatible or identical receiving transducer. The latter is mechanically coupled to the servomechanism to produce a mechanical shaft motion or position linearly related to the motion or position of the transmitting transducer. (C) 610.10-1994w
(2) A positional servomechanism in which loop input signals from a transmitting transducer are compared with loop feedback signals from a compatible or identical receiving transducer mechanically coupled to the servomechanism to produce a mechanical shaft motion or position linearly related to motion or position of the transmitting transducer. *See also:* electronic analog computer. (C) 165-1977w

repeater set (1) A repeater unit plus its associated MAUs and, if present, AU interfaces (AUIs). (LM/C) 8802-3-1990s
(2) A repeater unit plus its associated Physical Layer interfaces [Medium Attachment Units (MAUs) or PHYs] and, if present, Attachment Unit (AU) or Media Independent (MI) interfaces (i.e., AUIs, MIIs). (C/LM) 802.3-1998

repeater station (data transmission) An intermediate point in a transmission system where line signals are received, amplified or reshaped, and retransmitted. (PE) 599-1985w

repeater unit The portion of a repeater that is inboard of its Physical Medium Attachment (PMA)/ Physical Signaling Sublayer (PLS), or PMA/Physical Coding Sublayer (PCS) interfaces. (LM/C) 802.3-1998, 8802-3-1990s

repeating field A field within a record that may have multiple occurrences within a record; for example, the data element "Student Name" may have up to 30 occurrences within the following record structure:

Repeating Field

01	Course Name	20 characters
01	Instructor Name	25 characters
01	Students (30)	
	02 Student Name	25 characters
	02 Student Number	9 characters

(C) 610.5-1990w

repeating group A collection of data elements that may have multiple occurrences within a record; for example, the data elements representing the name and age of each dependent within an employee record. (C) 610.5-1990w

repeating port A transmitting port on a physical layer (PHY) that is repeating a packet from the PHY's receiving port. (C/MM) 1394a-2000

repeat key A key that continues to operate as long as it is held down. (C) 610.10-1994w

repeller (electron tube) (reflector) An electrode whose primary function is to reverse the direction of an electron stream. *See also:* electrode. (ED) 161-1971w

reperforator *See:* perforator.

reperforator switching center A message-relaying center at which incoming messages are received on a reperforator that perforates a storage tape from which the message is retransmitted into the proper outgoing circuit. The reperforator may be of the type that also prints the message on the same tape, and the selection of the outgoing circuit may be manual or under control of selection characters at the head of the message. *See also:* telegraphy. (COM) [49]

repetition equivalent (of a complete telephone connection, including the terminating telephone set) A measure of the grade of transmission experienced by the subscribers using the connection. It includes the combined effects of volume, distortion, noise, and all other subscriber reactions and usages. The repetition equivalent of a complete telephone connection is expressed numerically in terms of the trunk loss of a working reference system when the latter is adjusted to given an equal repetition rate. (EEC/PE) [119]

repetition instruction (1) An instruction that causes one or more instructions to be executed an indicated number of times. (C) [20], [85]
(2) A computer instruction that causes one or more instructions to be executed an indicated number of times, for example:

```
do 10 times:

write a record

add one to a counter

end
```

(C) 610.10-1994w

repetition rate The average number of partial discharge pulses per second measured over a selected period of time. (SWG/PE) 1291-1993r

repetition rate *n* (dry-type transformers) (failure data for power transformers and shunt reactors on electric utility power systems) (partial discharge measurement in liquid-filled power transformers and shunt reactors) The partial discharge pulse repetition rate *n* is the average number of partial discharge pulses per second measured over a selected period of time. (PE/TR) C57.113-1988s, C57.124-1991r

repetitive addressing (1) A method of implied addressing in which the operation field of a computer instruction is understood to address the operands of the last instruction executed. *Contrast:* one-ahead addressing. (C) 610.12-1990
(2) A method of implied addressing, applicable only to zero-address instructions, in which the operation field of an instruction implicitly addresses the operands of the last instruction executed. (C) 610.10-1994w

repetitively pulsed laser (laser maser) A laser with multiple pulses of radiant energy occurring in a sequence. (LEO) 586-1980w

repetitive operation (analog computer) A condition in which the computer operates as a repetitive device; the solution time may be a small fraction of a second or as long as desired, after which the problem is automatically and repetitively cycled through reset, hold, and operate. (C) 165-1977w

repetitive peak forward current (semiconductor) The peak value of the forward current including all repetitive transient currents. (IA) [12]

repetitive peak line voltage (thyristor) The highest instantaneous value of the line voltage including all repetitive transient voltages, but excluding all nonrepetitive transient voltages. (IA/IPC) 428-1981w

repetitive peak OFF-state current (semiconductor) The maximum instantaneous value of the OFF-state current that results from the application of repetitive peak-OFF-state voltage. (IA) [12]

repetitive peak OFF-state voltage The maximum instantaneous value of the OFF-state voltage that occurs across a thyristor, including all repetitive transient voltages, but excluding all nonrepetitive transient voltages. (IA/ED) 223-1966w, [62], [46], [12]

repetitive peak ON-state current (semiconductor) The peak value of the ON-state current including all repetitive transient currents. (IA) [12]

repetitive peak reverse current (semiconductor) The maximum instantaneous value of the reverse current that results from the application of repetitive peak reverse voltage. (IA) [12]

repetitive peak reverse voltage (1) (semiconductor rectifiers) The maximum instantaneous value of the reverse voltage, including all repetitive transient voltages but excluding all nonrepetitive transient voltages, that occurs across a semiconductor rectifier cell, rectifier diode, or rectifier stack. *See also:* semiconductor rectifier stack; principal voltage-current characteristic; rectification. (IA/ED) 59-1962w, [12], [62], [46], 223-1966w
(2) (reverse-blocking thyristor) The maximum instantaneous value of the reverse voltage which occurs across the thyristor, including all repetitive transient voltages, but excluding all non-repetitive transient voltages. (IA) [12]

repetitive peak reverse-voltage rating (rectifier circuit element) The maximum value of repetitive peak reverse voltage permitted by the manufacturer under stated conditions. *See also:* average forward current rating. (IA) 59-1962w, [62], [12]

repetitive surge and follow-current withstand The number of surges of specified voltage and current amplitudes and waveshapes that may be applied to a device without causing degradation beyond specified limits. The repetitive surge and follow-current withstand ratings apply to a device connected to an ac line of specified characteristics and to pulses applied at specified rates and phase angles. The effects of any cumulative heating that may occur are included. (SPD/PE) C62.62-2000

replaceable unit A collection of one or more parts considered as a single part for the purposes of replacement and repair due to physical constraints of the unit under test (UUT). (ATLAS) 1232-1995

replacement part A part for use in place of an existing component of switching equipment. (SWG/PE) C37.30-1971s

replay *See:* reversible execution.

replicate One of multiple aliquants of a sample. (NI) N42.23-1995

replica temperature relay A thermal relay whose internal temperature rise is proportional to that of the protected apparatus or conductor, over a range of values and durations of overloads. (SWG/PE) C37.100-1992

replication (1) (A) Theoretically, repetition of an experiment in exact detail. **(B)** Obtaining similar results from similar experiments. (T&D/PE) 539-1990
(2) The process by which copies of entries are created and maintained. (C/PA) 1328.2-1993w, 1326.2-1993w, 1224.2-1993w, 1327.2-1993w

reply (1) (transponder operation) (navigation aids) A radio-frequency signal or combination of signals transmitted as a result of an interrogation. (AES/GCS) 172-1983w
(2) Messages from the printer to the host. *Synonym:* response. (C/MM) 1284.1-1997
(3) The response sent from a target to an initiator indicating that the target has successfully or unsuccessfully executed the process specified by the command originally sent from the initiator to the target. (C/MM) 1284.4-2000
(4) *See also:* transaction completion. (C/MM) 1212.1-1993

replying agent An agent that participates in a transfer operation with the bus owner. (C/MM) 1296-1987s

reply phase The final phase of a transfer operation that consists of one or more consecutive data and/or status transfers on the parallel system bus. (C/MM) 1296-1987s

report The data objects/elements sent to a master device from slave devices. Used only in connection with slave devices. A slave device may parse requests for objects that it cannot generate or report. (PE/SUB) 1379-1997

report-by-exception The reporting of data (e.g., from RTU to master station) only when the data either changes state (e.g., for a status or digital input point) or exceeds a predefined deadband (e.g., for an analog input point). (SUB/PE) C37.1-1994

Report Generation Language A problem-oriented language designed for file processing and report creation. (C) 610.13-1993w

reporting period A period assumed to be one year unless otherwise stated. (PE/T&D) 1366-1998

reporting period time The duration of the reporting period (equals service time plus outage time). (PE/PSE) 859-1987w

report standard A standard that describes the characteristics of describing results of engineering and management activities. (C) 610.12-1990

report writer (1) A query language that can produce formatted reports using data from a database or other files. (C) 610.5-1990w
(2) A software tool or programming language used specifically for generating reports. (C) 610.13-1993w

repository (A) A collection of all software-related artifacts (e.g., the software engineering environment) belonging to a system. **(B)** The location/format in which such a collection is stored. (C/SE) 1219-1998

repository of last resort In a hierarchical memory (or cache-based) environment, a storage location that "owns" the only, or last remaining, copy of sharable data. *Note:* It may be a unique source, an ultimate destination, or simply a "safe" repository of data that may not be invalidated, unless action is taken to preserve a copy of that data at some higher level in the memory (or cache) hierarchy. In a cache-only Futurebus+ system (e.g., one where even the main DRAM storage is also designed as a hardware cache), the repository of last resort begins life as the binding of an address to a physical location in one of the caches, along with the creation of the data by initialization, a copy from some higher level in the memory hierarchy, or by its arrival from some I/O device. This data may migrate around the system, and be owned by different caches at different times, provided no less than one copy of that data is maintained somewhere. A repository of last resort may end its life by an explicit instruction to "destroy" the data by migration to a higher level in the memory (or cache hierarchy), or by transfer of ownership through some I/O device to another system, storage device, or display. (C/BA) 10857-1994

representation (1) A likeness, picture, drawing, block diagram, description, or symbol that logically portrays a physical,

operational, or conceptual image or situation. (C/SE) 1233-1998
(2) One or more properties used by an algorithm for the realization of a responsibility. (C/SE) 1320.2-1998

representational model *See:* descriptive model.

representation property A property on which an algorithm operates. (C/SE) 1320.2-1998

representation standard A standard that describes the characteristics of portraying aspects of an engineering or management product. (C) 610.12-1990

representative sample (nuclear power generating station) Production/prototype equipment used in a qualification program that is equivalent to that for which qualification is sought in terms of design, function, materials, and manufacturing techniques and processes. (SWG/PE/NP) 649-1980s, C37.100-1992

reproduce *See:* duplicate.

reproducibility (1) The ability of a system or element to maintain its output/input precision over a relatively long period of time. *See also:* precision; accuracy. (IA) [61]
(2) (transmission lines and waveguides) The degree to which a given set of conditions or observations, using different components or instruments each time, can be reproduced. *See also:* measurement system. (IM/HFIM) [40]
(3) (automatic null-balancing electric instrument) The closeness of agreement among repeated measurements by the instrument for the same value of input made under the same operating conditions, over a long period of time, approaching from either direction. *Notes:* 1. It is expressed as a maximum nonreproducibility in percent of span for a specified time. 2. Reproducibility includes drift, repeatability, and dead band. *See also:* measurement system. (EEC/EMI) [112]
(4) (radiation protection) (precision) The degree of agreement of repeated measurements of the same property expressed quantitatively as the standard deviation computed from the results of the series of measurements. (NI) N323-1978r
(5) (supervisory control, data acquisition, and automatic control) The measure of agreement among multiple readings of the output for the same value of input, made under the same operating conditions, approaching from either direction, using full-range traverses. (SUB/PE) C37.1-1987s

reproducing punch *See:* card reproducing punch.

reproducing stylus A mechanical element adapted to following the modulations of a record groove and transmitting the mechanical motion thus derived to the pickup mechanism. *See also:* phonograph pickup. (SP) [32]

reproductibility *See:* repeatability.

reproduction speed (facsimile) The area of copy recorded per unit time. *See also:* recording. (COM) 168-1956w

reprogrammable read-only memory (RPROM) *See:* erasable programmable read-only memory.

reprographics Automated composition, production, and reproduction of printed material. Methods include photocomposition, computer-aided typesetting, and offset printing. *See also:* office automation. (C) 610.2-1987

repulsion-induction motor A motor with repulsion-motor windings and short-circuited brushes, without an additional device for short-circuiting the commutator segments, and with a squirrel-cage winding in the rotor in addition to the repulsion motor winding. (PE) [9]

repulsion motor A single-phase motor that has a stator winding arranged for connection to a source of power and a rotor winding connected to a commutator. Brushes on the commutator are short-circuited and are so placed that the magnetic axis of the rotor winding is inclined to the magnetic axis of the stator winding. This type of motor has a varying-speed characteristic. *See also:* asynchronous machine. (PE) [9]

repulsion-start induction motor A single-phase motor with repulsion-motor windings and brushes, having a commutator-short-circuiting device that operates at a predetermined speed of rotation to convert the motor into the equivalent of a squirrel-cage motor for running operation. For starting operation, this motor performs as a repulsion motor. *See also:* asynchronous machine. (PE) [9]

request (1) Transaction that is generated by a requester, to initiate an action on a responder. For a processor-to-memory read transaction, for example, the request transfers the memory address and command from the processor to memory. In the case of a split transaction, the request would be a separate bus transaction. In the case of a connected transaction, the request would be the connection phase of a bus transaction. (C/BA) 896.3-1993w
(2) (local area networks) (Request_Normal, Request_High) A link control signal indicating that a lower entity has traffic pending for the network. (C) 8802-12-1998
(3) A command, generated by a requester, to initiate an action on a responder. For a processor-to-memory read transaction, for example, the request transfers the memory address and command from the processor to memory. In the case of a split transaction, the request would be a separate bus transaction. In the case of a connected transaction, the request would be the connection phase of a bus transaction. (C/BA) 10857-1994, 896.4-1993w, 1014.1-1994w
(4) A subaction with a transaction code and optional data sent by a node (the requester) to another node (the responder). (C/MM) 1394-1995
(5) A message sent from one object (the sender) to another object (the receiver), directing the receiver to fulfill one of its responsibilities. Specifically, a request may be for the value of an attribute, for the value of a participant property, for the application of an operation, or for the truth of a constraint. *Request* also encompasses sentences of such requests. Logical sentences about the property values and constraints of objects are used for queries, pre-conditions, post-conditions, and responsibility realizations. *See also:* message. (C/SE) 1320.2-1998
(6) A type of primitive in which one layer entity solicits another layer entity to perform a particular function. (EMB/MIB) 1073.4.1-2000
(7) A primary packet (with optional data) sent by one node's link (the requester) to another node's link (the responder). (C/MM) 1394a-2000
(8) *See also:* transaction initiation. (C/MM) 1212.1-1993

request echo The echo packet generated by a responder or agent when it strips the request send packet. (C/MM) 1596-1992

request for proposal (RFP) (1) A request for services, research, or a product prepared by a customer and delivered to prospective developers with the expectation that prospective developers will respond with their proposed cost, schedule, and development approach. (C/SE) 1362-1998
(2) A document used by the acquirer as a means to announce intention to potential bidders to acquire a specified system or software product (which may be part of a system). (C/SE) 1062-1998

requester-capable A term used to describe RamLink slaves that behave as DMA masters in the sense that they generate request packets and receive response packets. The delivery of these request and response packets is done by the controller. (C/MM) 1596.4-1996

requested batch service A service that is either rejected or performed prior to a response from the service to the requester. (C/PA) 1003.2d-1994

requester (1) (VSB) A functional module that resides on the same board as a master and requests use of the DTB whenever its master needs it. When implementing serial arbitration, after requesting use of the DTB, the requester waits for the bus to be granted to it by the arbiter. In the parallel arbitration method, the requester that is associated with the active master initiates an arbitration cycle. This arbitration cycle is used to determine which master will be granted use of the DTB. The

VSB specification calls the requester that is associated with the master the "active requester." Requesters that have a bus request pending and that participate in an arbitration cycle are called "contending requesters." (MM/C) 1096-1988w
(2) **(VMEbus)** A functional module that resides on the same printed-circuit board (pcb) as an interrupt handler or a master and requests use of the data transfer bus (DTB) whenever its interrupt handler or master needs it. (BA/C) 1014-1987
(3) A module that initiates a transaction by sending a request (containing address, command, and sometimes data). (C/BA) 1014.1-1994w, 896.3-1993w, 896.4-1993w, 10857-1994
(4) The node that initiates a transaction, by initiating a request subaction. (C/MM) 1596-1992
(5) A node that initiates a transaction by generating a request subaction (containing address, command, and sometimes data). (C/MM) 1212-1991s

requesting agent An agent that has entered arbitration for bus access. *See also:* arbitration operation. (C/MM) 1296-1987s

request message A message that generates one or more response messages when processed. (C/BA) 896.2-1991w

request packet A packet that is generated by a controller to initiate a directed transaction with a selected slave. (C/MM) 1596.4-1996

request phase The initial phase of a transfer operation in which the bus owner places command and address information on the parallel system bus. (C/MM) 1296-1987s

request send The packet generated by a requester to initiate an action in the responder, containing address, command, and, if appropriate, data. In a processor-to-memory read transaction, for example, the request send transfers the memory address and command from the processor to memory. (C/MM) 1596-1992

request subaction **(1)** A request send and its echo. Often called simply a "request." (C/MM) 1596-1992
(2) A subaction that is generated by a requester to initiate an action on the responder. For a processor-to-memory read transaction, for example, the request subaction transfers the memory address and command from the processor to memory. (C/MM) 1212-1991s

request test (metering) A test made at the request of a customer. (ELM) C12.1-1982s

required feature Either a single facility or behavior, or one of a pair of alternative facilities or behaviors, required by a POSIX standard that is always present on a conformng implementation. (C/PA) 13210-1994, 2003.1-1992

required hyphen In word processing, a hyphen that is to appear in a word or phrase regardless of whether the word or phrase is divided to achieve justification; for example, the hyphen in "computer-aided design." *Note:* A required hyphen is not subject to hyphen drop. *Synonym:* embedded hyphen. *Contrast:* discretionary hyphen. (C) 610.2-1987

required input motion (valve actuators) The input motion in terms of acceleration, velocity, and displacement expressed as a function of frequency that a device being tested shall withstand and still perform its intended function. (PE/NP) 382-1985

required inputs The set of items necessary to perform the minimum V&V tasks mandated within any life cycle activity. (C/SE) 1012-1998

required outputs The set of items produced as a result of performing the minimum V&V tasks mandated within any life cycle activity. (C/SE) 1012-1998

required reserve (power operations) (electric power supply) The system planned reserve capability needed to ensure a specified standard of service. (PE/PSE) 858-1987s, 346-1973w

required response spectrum (RRS) **(1)** **(seismic qualification of Class 1E equipment for nuclear power generating stations)** The response spectrum issued by the user or his agent as part of his specifications for qualification or artificially created to cover future applications. The RRS constitutes a requirement to be met. (SWG/PE/NP/PSR) 344-1987r, C37.98-1977s, C37.81-1989r
(2) The response spectrum issued by the user or the user's agent as part of the specifications for proof testing, or artificially created to cover future applications. The RRS constitutes a requirement to be met. (SWG/PE/NP) C37.100-1992, 382-1985
(3) The response spectrum issued by the user or the user's agent as part of the specifications for qualification. The RRS constitutes a requirement to be met. (PE/SUB) 693-1997

required time (availability) The period of time during which the user requires the item to be in a condition to perform its required function. (R) [29]

requirement **(1)** **(A)** A characteristic that a system or a software item is required to possess in order to be acceptable to the acquirer. **(B)** A binding statement in a standard or another portion of the contract. Requirements are expressed using the word "shall." (C/SE) J-STD-016-1995
(2) A statement that identifies a product or process operational, functional, or design characteristic or constraint, which is unambiguous, testable or measurable, and necessary for product or process acceptability (by consumers or internal quality assurance guidelines). (C/SE) 1220-1998
(3) **(A)** A condition or capability needed by a user to solve a problem or achieve an objective. **(B)** A condition or capability that must be met or possessed by a system or system component to satisfy a contract, standard, specification, or other formally imposed document. **(C)** A documented representation of a condition or capability as in definition (A) or (B). (C/SE) 1233-1998

requirements analysis **(A)** **(software)** The process of studying user needs to arrive at a definition of system, hardware, or software requirements. **(B)** **(software)** The process of studying and refining system, hardware, or software requirements. (C) 610.12-1990

requirements baseline The composite set of operational, functional, and physical requirements that serve to guide development and management decision processes. (C/SE) 1220-1994s

requirements baseline validation The process of evaluating the results of the requirements analysis activities of the systems engineering process to ensure compliance with customer expectations, project and enterprise constraints, and external constraints. (C/SE) 1220-1994s

requirements demonstration metric The result of dividing the total number of separately-identified requirements in the software requirements specification (SRS) that have been successfully demonstrated, by the total number of separately-identified requirements in the SRS. (C/SE) 730-1998

requirements inspection *See:* inspection.

requirements phase **(1)** **(software verification and validation plans)** The period of time in the software life cycle during which the requirements, such as functional and performance capabilities for a software product, are defined and documented. (C/SE) 1012-1986s
(2) **(software)** The period of time in the software life cycle during which the requirements for a software product are defined and documented. (C) 610.12-1990

requirements review A process or meeting during which the requirements for a system, hardware item, or software item are presented to project personnel, managers, users, customers, or other interested parties for comment or approval. Types include system requirements review, software requirements review. *Contrast:* design review; formal qualification review; code review; test readiness review. (C) 610.12-1990

requirements specification (software) A document that specifies the requirements for a system or component. Typically included are functional requirements, performance requirements, interface requirements, design requirements, and development standards. *Contrast:* design description. *See also:* functional specification; performance specification. (C) 610.12-1990

requirements specification language (software) A specification language with special constructs and, sometimes, verification protocols, used to develop, analyze, and document hardware or software requirements. *See also:* design language. (C) 610.12-1990, 610.13-1993w

requirement standard (software) A standard that describes the characteristics of a requirements specification. (C) 610.12-1990

requirements verification *See:* verification.

reradiated field An electromagnetic field resulting from currents induced in a secondary, predominantly conducting, object by electromagnetic waves incident on that object from one or more primary radiating structures or antennas. Reradiated fields are sometimes called "reflected" or more correctly "scattered fields." The scattering object is sometimes called a "re-radiator" or "secondary radiator." *See also:* scattered radiation. (NIR) C95.1-1999

reradiation (1) (A) The scattering of incident radiation. **(B)** The radiation of signals amplified in a radio receiver. *See also:* radio receiver. (EEC/PE) [119]
(2) The process by which an electromagnetic signal induces currents into a structure, which then causes radiation from that structure. (T&D/PE) 1260-1996

rerecording (electroacoustics) The process of making a recording by reproducing a recorded sound source and recording this reproduction. *See also:* dubbing. (SP) [32]

rerecording system (electroacoustics) An association of reproducers, mixers, amplifiers, and recorders capable of being used for combining or modifying various sound recordings to provide a final sound record. *Note:* Recording of speech, music, and sound effects may be so combined. *See also:* dubbing; phonograph pickup. (SP) [32]

rerefining (insulating oil) The use of primary refining processes on used electrical insulating liquids that are suitable for further use as electrical insulating liquids. *Note:* Techniques may include a combination of distillation and acid, clay or hydrogen treating, and other physical and chemical means. (PE/TR) 637-1985r

rering signal (telephone switching systems) A signal initiated by an operator at the calling end of an established connection to recall the operator at the called end or the customer at either end. (COM) 312-1977w

rerunability An attribute of a batch job. If a batch job may be rerun from the beginning after an abnormal termination without affecting the validity of the results, the job is said to be rerunable. (C/PA) 1003.2d-1994

rerun point (computers) The location in the sequence of instructions in a computer program at which all information pertinent to the rerunning of the program is available. (MIL/C) [2], [20], [85]

rerun time That part of operating time that is used for repeating operations or programs whose repetition is due to faults or mistakes in operations. (C) 610.10-1994w

rescue point *See:* restart point.

reseal voltage rating (surge arresters) The maximum arrester recovery voltage permitted for a specified time following one or more unit operation(s) with discharge currents of specified magnitude and duration. (PE) [8]

reservation charge *See:* capacity charge.

reserve (1) (test, measurement, and diagnostic equipment) The setting aside of a specific portion of memory for a storage area. (MIL) [2]
(2) (electric power system) (generating stations electric power system) A qualifying term used to identify equipment and capability that is available and is in excess of that required for the load. *Note:* The reserve may be connected to the system and partially loaded or may be made available by closing switches, contactors, or circuit breakers. Reserve not in operation and requiring switching is sometimes called standby equipment. (PE/EDPG) 505-1977r
(3) (power operations) *See also:* operating reserve; nonspinning reserve; customer generation reserve; spinning reserve; installed reserve; electrical reserve; voltage reduction reserve; interruptible load reserve; required reserve. (PE/PSE) 858-1987s

reserve cell A cell that is activated by shock or other means immediately prior to use. *See also:* electrochemistry. (EEC/PE) [119]

reserved (1) The term used for signals, bits, fields, and code values that are set aside for future standardization. (C/BA) 1496-1993w
(2) Used to describe an instruction field or register field that is reserved for definition by future versions of the architecture. A reserved field should only be written to zero by software. A reserved register field shall read as zero in hardware; software intended to run on future versions of IEEE 1754 should not assume that the field will read as zero. *See also:* ignored; unused. (C/MM) 1754-1994
(3) (FASTBUS acquisition and control) Bus lines, connector pins, codes, bits, etc held for future assignment by the NIM committee. They are not to be used until and except as so assigned. (NID) 960-1993
(4) An object in the delivery, retrieval, or input queue that the client can access without first removing it from that queue, but that no other client can access simultaneously. (C/PA) 1224.1-1993w
(5) Any protocol elements identified as "reserved" are intended for future standardization. Reserved elements shall not be used. Reserved fields or bits shall be set to 0 and shall not be checked. (C/MM) 1284.4-2000

reserved segment interconnect A segment interconnect is said to be reserved if it has gained mastership of the far-side segment and is asserting GK=1 onto that segment. (NID) 960-1993

reserved signal Signals that the application cannot accept and for which the application cannot modify the signal action or masking because the signals are reserved for use by the Ada language implementation. (C) 1003.5-1999

reserved word A word in a programming language whose meaning is fixed by the rules of that language and which, in certain or all contexts, cannot be used by the programmer for any purpose other than its intended one. Examples include IF, THEN, WHILE. (C) 610.12-1990

reserve, electrical *See:* electrical reserve.

reserve equipment The installed equipment in excess of that required to carry peak load. *Note:* Reserve equipment not in operation is sometimes referred to as standby equipment. *See also:* generating station. (T&D/PE) [10]

reserve generation (RG) (1) (electric generating unit reliability, availability, and productivity) The energy that a unit could have produced in a given period but did not, because it was not required by the system. This is the difference between available generation and actual generation. (PE/PSE) 762-1987w
(2) A flexible type of coaxial cable. RG is a military term. (C/CC) 802.7-1989r

reserve, installed *See:* installed reserve.

reserve, nonspinning *See:* nonspinning reserve.

reserve, operating *See:* operating reserve.

reserve, required *See:* required reserve.

reserve shutdown (power system measurement) (electric generating unit reliability, availability, and productivity) The state in which a unit is available but not in service. *Note:* This is sometimes referred to as economy shutdown. (PE/PSE) 762-1987w

reserve shutdown forced derated hours (electric generating unit reliability, availability, and productivity) The reserve shutdown hours during which a Class 1, 2, or 3 unplanned derating was in effect. (PE/PSE) 762-1987w

reserve shutdown hours (electric generating unit reliability, availability, and productivity) The number of hours a unit was in the reserve shutdown state. (PE/PSE) 762-1987w

reserve shutdown maintenance derated hours (electric generating unit reliability, availability, and productivity) The reserve shutdown hours during which a Class 4 unplanned derating was in effect. (PE/PSE) 762-1987w

reserve shutdown planned derated hours (electric generating unit reliability, availability, and productivity) The reserve shutdown hours during which a basic or extended planned derating was in effect. (PE/PSE) 762-1987w

reserve shutdown unit derated hours (electric generating unit reliability, availability, and productivity) The reserve shutdown hours during which a unit derating was in effect. (PE/PSE) 762-1987w

reserve shutdown unplanned derated hours (electric generating unit reliability, availability, and productivity) The reserve shutdown hours during which an unplanned derating was in effect. (PE/PSE) 762-1987w

reservoir operating curve (power operations) A curve, or family of curves (reservoir capability versus time), indicating how a reserve is to be operated under specified conditions to obtain best or predetermined results. (PE/PSE) 858-1987s

reservoir operating rule curve (electric power supply) A curve, or family of curves (reservoir capability versus time), indicating how a reservoir is to be operated under specified conditions to obtain best or predetermined results. (PE/PSE) 346-1973w

reservoir storage (power operations) (electric power system) The volume of water in a reservoir at a given time. (PE/PSE) 858-1987s, 346-1973w

reset (1) (A) (electronic digital computation) To restore a storage device to a prescribed state, not necessarily that denoting zero. **(B) (electronic digital computation)** To place a binary cell in the initial or zero state. *See also:* set. (C/MIL/ICTL) 162-1963, [20], 270-1966, [60], [85], [2], 610.10-1994

(2) (analog computer) The computer control state in which integrators are held constant and the proper initial condition voltages or charges are applied or reapplied. *See also:* initial condition. (C) 165-1977w

(3) (software) To set a variable, register, or other storage location back to a prescribed state. *See also:* initialize; clear. (C) 610.12-1990, 610.10-1994w

(4) An action that occurs when certain error conditions occur, or when error conditions exceed a preset value. Reset causes the Data Link layer to go to the offline state. Reconnection can than be requested by the DCC. (EMB/MIB) 1073.3.1-1994

(5) When describing the operating status of an S-module, the state of the S-module's Status registers produced by execution of the Reset Slave Status command. (TT/C) 1149.5-1995

(6) The state of an inverse-time overcurrent relay when the integral of the function of current *F(I)* that produces a time-current characteristic is zero. (PE/PSR) C37.112-1996

(7) (of a relay) The action of a relay as it makes designated response to decreases in input. As a qualifying term, reset denotes the state of a relay when all response to decrease of input has been completed. Reset is also used to identify the maximum value of an input quantity reached by progressive decreases that will permit the relay to reach the state of complete reset from pickup. *Note:* In defining the designated performance of relays having multiple inputs, reset describes the state when all inputs are zero and also when some input circuits are energized, if the resulting state is not altered from the zero-input condition. (SWG/PE/PSR) C37.100-1992, C37.90-1978s

reset action (process control) A component of control action in which the final control element is moved at a speed proportional to the extent of proportional-position control action. *Note:* This term applies only to a multiple control action including proportional-position control action. *See also:* proportional plus integral control action; positioning control system. (PE/EDPG) [3]

reset, automatic *See:* automatic reset.

reset characteristic The time versus current curve that defines the time required for the integral of the function of current *F(I)* to reach zero for values below current pickup when the integral is initially at the trip value. (PE/PSR) C37.112-1996

reset control action (electric power system) Action in which the controller output is proportional to the input signal and the time integral of the input signal. The number of times per minute that the integral control action repeats the proportional control action is called the reset rate. *Note:* Applies only to a controller with proportional control action plus integral control action. *See also:* speed-governing system. (PE/PSE) 94-1970w

reset current or voltage (faulted circuit indicators) The nominal rms (root-mean-square) value of current or voltage that will cause the indicator of the automatic current or voltage reset FCI (faulted circuit indicator) to change from FAULT to NORMAL indication. (T&D/PE) 495-1986w

reset device A device whereby the brakes may be released after an automatic train-control brake application. (EEC/PE) [119]

reset dwell time The time spent in reset. In cycling the computer from reset, to operate, to hold, and back to reset, this time must be long enough to permit the computer to recover from any overload and to charge or discharge all integrating capacitors to appropriate initial voltages. *See also:* electronic analog computer. (C) 165-1977w

reset interval (1) (automatic circuit recloser) The time required for the counting mechanism to return to the starting position. (SWG/PE) C37.60-1981r

(2) (of an automatic circuit recloser or automatic line sectionalizer) The time required, after a counting operation, for the counting mechanism to return to the starting position of that counting operation. (SWG/PE) C37.100-1992

reset, manual *See:* manual reset.

reset on inertial navigation systems (navigation aid terms) Use of external data (for example, position fix) to refine alignment of and to calibrate the inertial navigation system. (AES/GCS) 172-1983w

reset packet A packet used during initialization to reset the node's CSR state, empty ring buffers, initialize the ring interface and establish that ring closure has been achieved. (C/MM) 1596-1992

reset pulse A drive pulse that tends to reset a magnetic cell. (Std100) 163-1959w

reset rate (process control) (proportional plus reset control action or proportional plus reset plus rate control action) The number of times per minute that the effect of proportional-position control action is repeated. *See also:* integral action rate. (PE/EDPG) [3]

reset switch A machine-operated device that restores normal operation to the control system after a corrective action. *See also:* photoelectric control. (IA/ICTL/IAC) [60]

resettability (1) (electric pipe heating systems) The restoring of a mechanism, electrical circuit, or device to the prescribed state. Resettability is usually associated with temperature controllers and is the difference in degrees when returning to original temperature setting. (PE/EDPG) 622A-1984r, 622B-1988r

(2) (oscillators) The ability of the tuning element to retune the oscillator to the same operating frequency for the same set of input conditions. (ED) 158-1962w

reset test A test or collection of tests that is invoked by a command_reset. Although a reset test is actually a form of initialization test, the term reset test is used to avoid confusing its functionality with the initialization tests that are invoked by writing to the TEST_START register. (C/MM) 1212-1991s

reset time (faulted circuit indicators) The time required for the FCI (faulted circuit indicator) to return automatically to NORMAL indication after its reset current or voltage has been established, or for the elapsed time automatic reset FCI to reset. (T&D/PE) 495-1986w
(2) (A) (of a relay) The time interval from occurrence of specified conditions to reset. *Note:* When the conditions are not specified it is intended to apply to a picked-up relay and to be a sudden change from picked value of input to zero input. **(B)** (of an automatic circuit recloser or automatic line sectionalizer) The time required, after one or more counting operations, for the counting mechanism to return to the starting position. (SWG/PE) C37.100-1992

resident Pertaining to computer programs that remain in a particular storage device or in main storage. (C) 610.10-1994w

resident control program *See:* kernel.

residential-custodial care facility (health care facilities) A building, or part thereof, used for the lodging or boarding of 4 or more persons who are incapable of self-preservation because of age, or physical or mental limitation. This includes facilities such as homes for the aged, nurseries (custodial care for children under 6 years of age), and mentally retarded care institutions. Day care facilities that do not provide lodging or boarding for institutional occupants are not classified as residential custodial care facilities. (NESC/NEC) [86]

residential zone A zone that includes single-family and multifamily residential units, as defined by local ordinances. (PE/SUB) 1127-1998

residual-component telephone-influence factor (three-phase synchronous machine) The ratio of the square root of the sum of the squares of the weighted residual harmonic voltages to three times the root-mean-square no-load phase-to-neutral voltage. (PE) [9]

residual control A microprogramming technique in which the meaning of a field in a microinstruction depends on the value in an auxiliary register. *Contrast:* bit steering. *See also:* two-level encoding. (C) 610.12-1990

residual current (protective relaying) The sum of the three-phase currents on a three-phase circuit. The current that flows in the neutral return circuit of three wye-connected current transformers is residual current. (PE/PSR) C37.95-1973s

residual-current state (thermionics) The state of working of an electronic valve or tube in the absence of an accelerating field from the anode of a diode or equivalent diode, in which the cathode current is due to the nonzero velocity of emission of electrons. *See also:* electron emission; inductive coordination. (ED) [45], [84]

residual element A circuit element connected to a function pin that, for operational reasons, cannot be isolated from the pin in test mode. A residual element can be connected to a power supply pin, or another function pin, or another residual element, provided it can be modelled over a defined working range by a network of ideal resistors, capacitors, and inductors together with independent dc sources. (C/TT) 1149.4-1999

residual error (1) (electronic navigation) The sum of the random errors and the uncorrected systematic errors. *See also:* navigation. (AES/RS) 686-1982s, [42]
(2) (software) The difference between an optimum result derived from experience or experiment and a theoretically exact result. (C) 1084-1986w

residual-error rate *See:* undetected error rate.

residual flux density The magnetic flux density at which the magnetizing force is zero when the material is in a symmetrically cyclically magnetized condition. *See also:* remanence. (Std100/PE/PSR) 163-1959w, C37.110-1996

residual frequency-modulation (frequency modulation) (spectrum analyzer) Short term displayed frequency instability (jitter) of the spectrum analyzer caused by instability of the local oscillators. Given in terms of peak-to-peak frequency deviation (Hz). *Notes:* 1. Any influencing factors such as phase lock on or off, etc. should be given. 2. For the purpose of this standard "short term" shall mean measurements made during a specified period of time. The recommended time duration is 20 s to 20 μs per division. This will accommodate incidental FM from less than one Hz to tens of kHz. The manufacturer shall specify the time to be used. *Synonym:* incidental frequency modulation. (IM) [14], 748-1979w

residual induction (1) (magnetic material) The magnetic induction corresponding to zero magnetizing force in a material that is in a symmetrically cyclically magnetized condition. (Std100) 270-1966w
(2) (residual flux density) (toroidal magnetic amplifier cores) The magnetic induction at which the magnetizing force is zero when the magnetic core is cyclically magnetized with a half-wave sinusoidal magnetizing force of a specified peak magnitude. *Note:* This use of the term residual induction differs from the standard definition that requires symmetrically cyclically magnetized conditions. (MAG) 393-1977s

residual life The remaining period of time during which a system, structure, or component is expected to perform its intended function under specified service conditions. (PE/NP) 1205-1993

residual magnetism (ferromagnetic bodies) A property by which they retain a certain magnetization (induction) after the magnetizing force has been removed. (Std100) 270-1966w, [84]

residual modulation *See:* carrier noise level.

residual probe pickup (slotted line) (constancy of probe coupling) The noncyclical variation of the amplitude of the probe output over its complete range of travel when reflected waves are eliminated on the slotted section by proper matching at the output and the input, discounting attenuation along the slotted section. It is defined by the ratio of one-half of the total variation to the average value of the probe output, assuming linear amplitude response of the probe, at a specified frequency(ies) within the range of usage. *Note:* This quantity consists of two parts of which one is reproducible and the other is not. The repeatable part can be eliminated by subtraction in repeated measurements, while the nonrepeatable part must cause an error. The residual probe pickup depends to some extent on the insertion depth of the probe. *See also:* measurement system; residual standing-wave ratio. (IM/HFIM) [40]

residual reflected coefficient (reflectometer) The erroneous reflection coefficient indicated when the reflectometer is terminated in reflectionless terminations. *See also:* measurement system. (IM/HFIM) [40]

residual relay A relay so applied that its input, derived from external connections of instrument transformers, is proportional to the zero-phase-sequence component of a polyphase quantity. (SWG/PE) C37.100-1992

residual response (1) (non-real time spectrum analyzer) A spurious response in the absence of an input. (IM) [14]
(2) (spectrum analyzer) A spurious response in the absence of an input, not including noise and zero pip. (IM) 748-1979w

residuals The differences between the recorded data and the fitted sine wave for sine-wave curve fitting. (IM/WM&A) 1057-1994w

residual standing-wave ratio (SWR) (slotted line) The standing-wave ratio measured when the slotted line is terminated by a reflectionless termination and fed by a signal source that provides a nonreflecting termination for waves reflected toward the generator. *Note:* Residual standing-wave ratio does not include the residual noncyclical probe pickup or the attenuation encountered as the probe is moved along the line. *See also:* residual probe pickup. (IM/HFIM) [40]

residual voltage (1) (arresters) (discharge voltage) The voltage that appears between the line and ground terminals of an arrester during the passage of discharge current. *See also:* inductive coordination. (PE) [8], [84]

(2) (protective relaying) The sum of the three line-to-neutral voltages on a three-phase circuit. (PE/PSR) C37.95-1973s

residue The value of $\lim_{s \to s_0} (s - s_0) \times F(s)$, where $F(s)$; has the complex pole s_0. *See also:* pole. (C/DA) 1481-1999

residue check (computers) A check in which each operand is accompanied by the remainder obtained by dividing this number by n, the remainder then being used as a check digit or digits. *See also:* modulo *N* check. (C) [20], [85]

resin (rotating machinery) Any of various hard brittle solid-to-soft semisolid amorphous fusible flammable substances of either natural or synthetic origin; generally of high molecular weight, may be either thermoplastic or thermosetting. (PE) [9]

resin-bonded paper-insulated bushing (outdoor electric apparatus) A bushing in which the major insulation is provided by paper bonded with resin. (PE/TR) 21-1976

resin impregnated paper-insulated bushing A bushing in which the internal insulation consists of a condenser wound from untreated paper and subsequently impregnated with a curable resin. *Note:* A resin impregnated paper bushing may be provided with an insulating envelope, in which case the intervening space may be filled with another insulating medium. (PE/TR) C57.19.03-1996

resist (electroplating) Any material applied to part of a cathode or plating rack to render the surface nonconducting. *See also:* electroplating. (EEC/PE) [119]

resistance (1) (A) (network analysis) That physical property of an element, device, branch, network, or system that is the factor by which the mean-square conduction current must be multiplied to give the corresponding power lost by dissipation as heat or as other permanent radiation or loss of electromagnetic energy from the circuit. **(B) (network analysis)** The real part of impedance. *Note:* Definitions (A) and (B) are not equivalent but are supplementary. In any case where confusion may arise, specify definition being used. *See also:* resistor. (IA/IM/IAC/HFIM) 270-1966, [60], [40]
(2) (shunt) The quotient of the voltage developed across the instrument terminals to the current passing between the current terminals. In determining the value, account should be taken of the resistance of the instrument and the measuring cable. The resistance value is generally derived from a direct-current measurement such as by means of a double Kelvin bridge. (PE/PSIM) 4-1978s
(3) (automatic control) A property opposing movement of material, or flow of energy, and involving loss of potential (voltage, temperature, pressure, level). (PE/EDPG) [3]
(4) *See also:* radiation resistance; antenna resistance. (AP/ANT) 145-1993

resistance, apparent *See:* apparent resistance.

resistance, body *See:* body resistance.

resistance box A rheostat consisting of an assembly of resistors of definite values so arranged that the resistance of the circuit in which it is connected may be changed by known amounts. (Std100) 270-1966w

resistance braking A system of dynamic braking in which electric energy generated by the traction motors is dissipated by means of a resistor. *See also:* dynamic braking. (EEC/PE) [119]

resistance bridge smoke detector (fire protection devices) A device that responds to an increase of smoke particles and moisture, present in products of combustion, which fall on an electrical bridge grid. As these conductive substances fall on the grid they reduce the resistance of the grid and cause the detector to respond. (NFPA) [16]

resistance-capacitance characteristic, input (oscilloscopes) The direct-current resistance and parallel capacitance to ground present at the input of an oscilloscope. (IM) 311-1970w

resistance-capacitance coupling Coupling between two or more circuits, usually amplifier stages, by means of a combination of resistance and capacitance elements. *See also:* coupling. (EEC/PE) [119]

resistance-capacitance oscillator Any oscillator in which the frequency is determined principally by resistance-capacitance elements. *See also:* oscillatory circuit. (EEC/PE) [119]

resistance drop (power and distribution transformers) The component of the impedance voltage drop in phase with the current. (PE/TR) C57.12.80-1978r

resistance furnace An electrothermic apparatus, the heat energy for which is generated by the flow of electric current against ohmic resistance internal to the furnace. (EEC/PE) [119]

resistance grading (cr corona shielding) A form of corona shielding embodying high resistance material on the surface of the coil. *Synonym:* corona shielding. (PE) [9]

resistance grounded (1) (power and distribution transformers) Grounded through impedance, the principal element of which is resistance. *Note:* The resistance may be inserted either directly, in the connection to the ground, or indirectly, as for example, in the secondary of a transformer, the primary of which is connected between neutral and ground, or in series with the delta-connected secondary of a wye-delta grounding transformer. (PE/TR) C57.12.80-1978r
(2) (system grounding) Grounded through impedance, the principal element of which is resistance. *Note:* The high-resistance-grounded system is designed to meet the criterion of $R_0 \leq X_{C0}$ in order to limit transient overvoltages due to arcing ground faults. The ground-fault current is usually limited to less than 10 A. X_{C0} is the distributed per-phase capacitive reactance to ground of the system. The low-resistance-grounded system permits a higher ground-fault current (on the order of 25 A to several hundred amperes) to obtain sufficient current for selective relay performance. For the usual system the criterion for limiting transient overvoltages is $R_0/X_0 \geq 2$. (IA/PSE) 142-1982s

resistance lamp An electric lamp used to prevent the current in a circuit from exceeding a desired limit. (EEC/PE) [119]

resistance magnetometer A magnetometer that depends for its operation upon the variation of electrical resistance of a material immersed in the field to be measured. *See also:* magnetometer. (EEC/PE) [119]

resistance method of temperature determination (power and distribution transformers) The determination of the temperature by comparison of the resistance of a winding at the temperature to be determined, with the resistance at a known temperature. (PE/TR) C57.12.80-1978r, C57.15-1999

resistance modulation (bolometric power meters) A change in resistance of the bolometer resulting from a change in power (RF, ac, or dc) dissipated in the element. *Note:* The resistance modulation sensitivity is the (dc) change in resistance per unit (dc) change in power at normal bias and at a constant ambient temperature. Resistance modulation frequency response is the frequency of repetitive (sinusoidal) power change for which the peak-to-peak resistance change is 3 dB lower than the asymptotic, maximum value at zero frequency. (IM) 470-1972w

resistance modulation effect (bolometric power meters) A component of substitution error (for dc power substitution) in bolometer units in which both ac and dc bias is used. *Note:* This component is dependent upon the frequency of the ac bias and the frequency response of the element: it is usually very small, and usually not included in the effective efficiency correction for substitution error. It is caused by resistance modulation of the element, and is more pronounced in barretters than in thermistors. (IM) 470-1972w

resistance-reduction factor A number usually less than or equal to 1.0 used in load and resistance factor design (LRFD). Called strength-reduction factor in 751-1990. (T&D/PE) 751-1990

resistance relay A linear-impedance form of distance relay for which the operating characteristic on an *R-X* diagram is a straight line of constant resistance. *Note:* The operating characteristic may be described by the equation $R = K$ or $Z\cos\theta = K$, where K is a constant, and θ is the angle by which the input voltage leads the input current. (SWG/PE) C37.100-1992

resistance starting A form of reduced-voltage starting employing resistances that are short-circuited in one or more steps to complete the starting cycle. *See also:* resistance starting, motor-armature; resistance starting, generator-field. (IA/ICTL/IAC/APP) [60], [75]

resistance starting, generator-field Field resistance starting provided by one or more resistance steps in series with the shunt field of a generator, the output of which is connected to a motor armature. *See also:* resistance starting; resistance starting, motor-armature. (IA/IAC) [60]

resistance starting, motor-armature Motor resistance starting provided by one or more resistance steps connected in series with the motor armature. *See also:* resistance starting; resistance starting, generator-field. (IA/IAC) [60]

resistance-start motor A form of split-phase motor having a resistance connected in series with the auxiliary winding. The auxiliary circuit is opened when the motor has attained a predetermined speed. *See also:* asynchronous machine. (EEC/PE) [119]

resistance temperature detector (resistance thermometer resistor) (resistance thermometer detector) A resistor made of some material for which the electrical resistivity is a known function of the temperature and that is intended for use with a resistance thermometer. It is usually in such a form that it can be placed in the region where the temperature is to be determined. *Note:* A resistance temperature detector with its support and enclosing envelope, is often called a resistance thermometer bulb. *See also:* electric thermometer; embedded temperature detector. (EEC/PE) [119]

resistance thermometer (resistance temperature meter) An electric thermometer that operates by measuring the electric resistance of a resistor, the resistance of which is a known function of its temperature. The temperature-responsive element is usually called a resistance temperature detector. *Note:* The resistance thermometer is also frequently used to designate the sensor and its enclosing bulb alone, for example, as in platinum thermometer, copper-constantan thermometer, etc. *See also:* instrument; electric thermometer. (PE/PSIM) 119-1974w

resistance times capacitance (RC), RC time constant The product of some resistance and some capacitance (having the dimensions of time) or a time constant computed in some other way. (C/DA) 1481-1999

resistance to ground (surge arresters) The ratio, at a point in a grounding system, of the component of the voltage to ground that is in phase with the ground current, to the ground current that produces it. (PE) [8], [84]

resistance voltage drop (1) The component of voltage drop in phase with the current. (PE/TR) C57.16-1996
(2) The component of the impedance voltage in phase with the current. (PE/TR) C57.15-1999

resistant (1) (rotating machinery) Material or apparatus so constructed, protected or treated, that it will not be injured readily when subjected to the specified material or condition, for example, fire-resistant, moisture-resistant. *See also:* asynchronous machine. (SWG/PE/PSR) C37.30-1971s, C37.90-1978s
(2) (power and distribution transformers) So constructed, protected, or treated that the apparatus will not be damaged when subjected to the specified material or conditions for a specified time. (PE/TR) C57.12.80-1978r
(3) (used as a suffix) So constructed, protected, or treated that damage will not occur readily when the device is subjected to the specified material or condition. (SWG/PE) C37.100-1992, C37.40-1993

resistive attenuator (waveguide) A length of waveguide designed to introduce a transmission loss by the use of some dissipative material. *See also:* waveguide; absorptive attenuator. (AP/ANT) [35], [84]

resistive conductor A conductor used primarily because it possesses the property of high electric resistance. (T&D/PE) [10]

resistive coupling The association of two or more circuits with one another by means of resistance mutual to the circuits. (PE/PSIM) 81-1983

resistive distributor brush Resistive pickup brush in an ignition distributor cap. *See also:* electromagnetic compatibility. (EMC/INT) [53], [70]

resistive feedback preamplifier (germanium gamma-ray detectors) A charge-sensitive preamplifier in which charge that accumulates on the feedback capacitor is continually discharged through a resistor in parallel with the capacitor. (NPS) 325-1986s

resistive ignition cable High-tension ignition cable, the core of which is made of resistive material. *See also:* electromagnetic compatibility. (EMC/INT) [53], [70]

resistive loads Loads for which the current supplied by the low-voltage power supply/battery varies proportionally with the source voltage. *Note:* These loads will demand less current when the source voltage is switched from the low-voltage power supply to the battery. Typically, relays fall into this category. (VT) 1476-2000

resistivity (material) A factor such that the conduction-current density is equal to the electric field in the material divided by resistivity. (PE/PSIM) 81-1983

resistivity, volume *See:* volume resistivity.

resistor (1) An element within a circuit that has specified resistance value designed to restrict the flow of current. *See also:* potentiometer. (C) 610.10-1994w
(2) A device with the primary purpose of introducing resistance into an electric circuit. (A resistor as used in electric circuits for purposes of operation, protection, or control, commonly consists of an aggregation of units. Resistors, as commonly supplied, consist of wire, metal, ribbon, cast metal, or carbon compounds supported by or embedded in an insulating medium. The insulating medium may enclose and support the resistance material as in the case of the porcelain tube type or the insulation may be provided only at the points of support as in the case of heavy duty ribbon or cast iron grids mounted in metal frames.) (IA/MT) 45-1998

resistor, bias *See:* bias resistor.

resistor furnace A resistance furnace in which the heat is developed in a resistor that is not a part of the charge. (EEC/PE) [119]

resistor-start motor A single-phase induction motor with a main winding and an auxiliary winding connected in series with a resistor, with the auxiliary winding circuit opened for running operation. (PE) [9]

resistor-transistor logic (RTL) A family of circuit logic in which the basic circuit element is a network of resistors and transistors. (C) 610.10-1994w

re-solution (electrodeposition) The passing back into solution of metal already deposited on the cathode. (EEC/PE) [119]

resolution (1) (supervisory control, data acquisition, and automatic control) The least value of the measured quantity that can be distinguished. (SWG/PE/SUB) C37.100-1992, C37.1-1994
(2) (A) (data transmission) The result of deriving from a sound, scene, or other form of intelligence, a series of discrete elements wherefrom the original may subsequently be synthesized. **(B) (data transmission)** The degree to which nearly equal values of a quantity can be discriminated. **(C) (data transmission)** The fineness of detail in a reproduced spatial pattern. **(D) (data transmission)** The degree to which a system or a device distinguishes fineness of detail in a spatial pattern. (COM/PE) 599-1985
(3) (storage tubes) A measure of the quantity of information that may be written into and read out of a storage tube. *Notes:* 1. Resolution can be specified in terms of number of bits, spots, lines, or cycles. 2. Since the relative amplitude of the

output may vary with the quantity of information, the true representation of the resolution of a tube is a curve of relative amplitude versus quantity. *See also:* storage tube. (ED) 158-1962w

(4) (television) A measure of ability to delineate picture detail. *Note:* Resolution is usually expressed in terms of a number of lines N (normally alternate black and white lines) the width of each line is $1/N$ times the picture height. In television practice, where the raster has a 4/3 aspect ratio, resolution, measured in either the horizontal or the vertical direction, is the number of test chart lines observable in a distance equal to the vertical dimension of the raster. (BT/AV) 201-1979w

(5) (oscilloscopes) A measure of the total number of trace lines discernible along the coordinate axes, bounded by the extremities of the graticule or other specific limits. *See also:* oscillograph. (IM/HFIM) [40]

(6) (transmission lines and waveguides) The degree to which nearly equal values of a quantity can be discriminated. (IM/HFIM) [40]

(7) The smallest distinguishable increment into which a quantity is divided in a device or system. *See also:* feedback control system. (IA/ICTL/IAC) [60]

(8) (digital delay line) The time spacing between peaks of the doublet. (UFFC) [22]

(9) (acousto-optic deflector) The ratio of the angular swing to the minimum resolvable angular spread of one spot. The minimum spot size depends on the optical beam amplitude and phase distribution, as well as the criteria used to define minimum spot size. When the Rayleigh criteria is used for minimum spot size, resolution, N, is given by $N = 1/\alpha\ \tau\Delta f$, with $\alpha = 1$ rectangular beam, constant amplitude: 1.22 circular beam, constant amplitude: 1.34 circular beam, Gaussian amplitude. For operation in the scanning mode, the resolution will be reduced as the scan time approaches the access time. (UFFC) [17]

(10) (A) (spectrum analyzer) (general). The ability to display adjacent responses discretely (Hz, Hz dB down). The measure of resolution is the frequency separation of two responses that merge with a 3 dB notch. **(B) (spectrum analyzer)** (resolution). As a minimum, instruments will be specified and controls labeled on the basis of two equal amplitude responses under the best operational conditions. **(C) (spectrum analyzer)** (skirt resolution). The frequency difference between two signals of specified unequal amplitude when the notch formed between them is 3 dB down from the smaller signal. **(D) (spectrum analyzer)** (optimum resolution). For every combination of frequency span and sweep time there exists a minimum obtainable value of resolution (R). This is the optimum resolution (R_o), which is defined theoretically as:

$$R_o = K\sqrt{\frac{\text{Frequency Span}}{\text{Sweep Time}}}$$

The factor K shall be unity unless otherwise specified. (IM) [14], 748-1979

(11) (pulse measurement) The smallest change in the pulse characteristic, property, or attribute being measured which can unambiguously be discerned or detected in a pulse measurement process. (IM/WM&A) 181-1977w

(12) (electrothermic power meters) The smallest discrete or discernible change in power that can be measured. In IEEE Std 544-1975w, resolution includes the estimated uncertainty with which the power changes can be determined on the readout scale. (IM) 544-1975w

(13) (plutonium monitoring) The minimum detectable change in instrument response. (NI) N317-1980r

(14) (image processing and pattern recognition) In image processing, the degree to which closely spaced objects in an image can be distinguished from one another. (C) 610.4-1990w

(15) In micrographics, the ability of a photographic system to record fine detail. (C) 610.2-1987

(16) (A) (computer graphics) The smallest distance between two display elements that can be addressed. **(B) (computer graphics)** The fineness of a raster display device expressed in pixels per inch, pixels per screen, number of horizontal lines by number of dots per line, dots per inch, or other ratio. (C) 610.6-1991, 610.10-1994

(17) A measure of the ability to delineate picture detail. (BT/AV) 208-1995

(18) (accelerometer) (gyros) The largest value of the minimum change in input, for inputs greater than the noise level, that produces a change in output equal to some specified percentage (at least 50%) of the change in output expected using the nominal scale factor. (AES/GYAC) 528-1994

(19) A measure of the ability to delineate display detail. (C) 610.10-1994w

(20) The minimum time interval that a clock can measure or whose passage a timer can detect. (C) 1003.5-1999

(21) *See also:* energy resolution. (NPS) 325-1996

resolution bandwidth (spectrum analyzer) The width, in Hz, of the spectrum analyzer's response to a continuous wave (CW) signal. This width is usually defined as the frequency difference at specified points on the response curve, such as the 3 or 6 dB down points. The manufacturer will specify the dB down points to be used. (IM) [14], 748-1979w

resolution cell The one-dimensional or multidimensional region related to the ability of a radar to resolve multiple targets. *Note:* The dimensions that involve resolution can include range, angle, and radial velocity (Doppler frequency). The three dimensional spatial resolution cell is, for example:

$$\theta_a \times \theta_e \times (c\tau/2)$$

where
θ_a = azimuth beamwidth
θ_e = elevation beamwidth
τ = pulsewidth
c = velocity of propagation of electromagnetic waves

Synonym: resolution element. (AES) 686-1997

resolution element *See also:* resolution cell. (AES) 686-1997

resolution, energy *See:* energy resolution.

resolution error (analog computer) The error due to the inability of a transducer to manifest changes of a variable smaller than a given increment. (C) 165-1977w

resolution of output adjustment (inverters) (of any output parameter, voltage, frequency, etc.) The minimum increment of change in setting. *See also:* self-commutated inverters. (IA) [62]

resolution phase The initial phase of an arbitration operation in which all agents requesting access to the bus drive an arbitration ID onto the parallel system bus. Agents mutually resolve requests and allow the agent with the highest priority to gain access to the bus. *See also:* arbitration operation. (C/MM) 1296-1987s

resolution, pulse height *See:* pulse-height resolution.

resolution response In television, the ratio of the peak-to-peak signal amplitude, given by a test pattern consisting of alternate black and white bars of equal widths representing a given TV line number on a test chart, to the peak-to-peak signal amplitude, given by large black areas and large white areas having the same luminance as the black and white bars in the test pattern. (BT/AV) 208-1995

resolution, structural *See:* structural resolution.

resolution test chart In micrographics, a chart containing a number of increasingly smaller horizontal and vertical lines of specific size and spacing, used to measure resolution. *See also:* target. (C) 610.2-1987

resolution time (1) (counter tube or counting system) (radiation counters) The minimum time interval between two distinct events that will permit both to be counted. *See also:* anticoincidence. (ED) [45]

(2) (sequential events recording systems) The minimum time interval between any two distinct events that will permit both to be recorded in sequence of occurrence. *See also:* event. (PE/EDPG) [5], [1]

resolution, time *See:* time resolution.

resolution time correction (radiation counters) Correction to the observed counting rate to allow for the probability of the occurrence of events within the resolution time. *See also:* anticoincidence. (ED) [45]

resolution wedge (television) A narrow-angle wedge-shaped pattern calibrated for the measurement of resolution and composed of alternate contrasting strips that gradually converge and taper individually to preserve equal widths along any given line at right angles to the axis of the wedge. *Note:* Alternate strips may be black and white of maximum contrast or strips of different colors. (BT/AV) 201-1979w

resolver (analog computer) A device or computing element used for vector resolution or composition. The rectangular mode is the mode of operation that produces a transformation from polar to rectangular coordinates or a rotation of rectangular coordinates. The polar mode of operation that produces a transformation from rectangular to polar coordinates. (C) 165-1977w

(2) (A) In an analog computer, a device or computing element used for vector resolution or composition. **(B)** A functional unit whose input analog variables are the polar coordinates of a point and whose output analog variables are the Cartesian coordinates of the same point, or vice-versa. (C) 610.10-1994

resolving power (illuminating engineering) The ability of the eye to perceive the individual elements of a grating or any other periodic pattern with parallel elements measured by the number of cycles per degree that can be resolved. The resolution threshold is the period of the pattern that can be just resolved. Visual acuity, in such a case, is the reciprocal of one half of the period expressed in minutes. The resolution threshold for a pair of points or lines is the distance between them when they can just be distinguished as two, not one, expressed in minutes of arc. (EEC/IE) [126]

resolving time (1) (navigation aids) The minimum time interval by which two events must be separated, to be distinguishable in a navigation system, by the time measurement alone. (AES/GCS/RS) 172-1983w, 686-1990

(2) (liquid-scintillation counting) The minimum time that must exist between successive events if they are to be counted as separate events. (NI) N42.15-1990

(3) The time from the start of a counted pulse to the instant a succeeding pulse can assume the minimum strength to be detected by the counting circuit. *Note:* This quantity pertains to the combination of tube and recording circuit. *See also:* half-amplitude recovery time. (NI/NPS) 309-1999

resonance (1) (seismic design of substations) A dynamic condition which occurs when any input frequency of vibration coincides with one of the natural frequencies of the structure. In a plot of the response of the structure (acceleration, velocity, displacement) versus input frequency for a constant input, as the input frequency approaches one of the natural frequencies of the structure the response increases to a maximum value if damping is less than critical. The response of the structure at resonance may be much greater than the input, if the damping is low. (PE/SUB) 693-1984s

(2) The enhancement of the response of a physical system to a periodic excitation when the excitation frequency is equal to a natural frequency of the system. (CAS) [13]

(3) (automatic control) Of a system or element, a condition evidenced by large oscillatory amplitude, which results when a small amplitude of a periodic input has a frequency approaching one of the natural frequencies of the driven system. *Note:* In a feedback control system, this occurs near the stability limit. (PE/EDPG) [3]

(4) (data transmission) A condition in a circuit containing inductance and capacitance in which the capacitive reactance is equal to the inductive reactance. This condition occurs at only one frequency in a circuit with fixed constants, and the circuit is said to be "tuned" to this frequency. The resonance frequency can be changed by varying the value of the capacitance or inductance of the circuit. (PE) 599-1985w

(5) (mechanical) A dynamic condition that occurs when any forcing frequency of mechanical vibration coincides with one of the natural frequencies of the structure. *Note:* In a plot of the response of the structure (acceleration, velocity, and displacement) vs. forcing frequency for a constant forcing input, as the forcing frequency approaches one of the natural frequencies of the structure, the response increases to a maximum at the natural frequency if damping is less than critical. The response of the structure at resonance may be much greater than the input, depending on the damping. (SWG/SUB/PE) C37.122-1983s, C37.100-1992, C37.122.1-1993

(6) (A) (in an oscillating system) The rapid increase or decrease of the oscillation magnitude as the excitation frequency approaches one of the natural frequencies of the system. **(B)** (of a traveling wave) The change in magnitude as the frequency of the wave approaches or coincides with a natural frequency of the medium (e.g., a plasma frequency). (AP/PROP) 211-1997

resonance bridge A 4-arm alternating-current bridge in which both an inductor and a capacitor are present in one arm, the other three arms being (usually) nonreactive resistors, and the adjustment for balance includes the establishment of resonance for the applied frequency. *Note:* Normally used for the measurement of inductance, capacitance, or frequency. Two general types can be distinguished according as the inductor and capacitor are effectively in series or in parallel. *See also:* bridge.

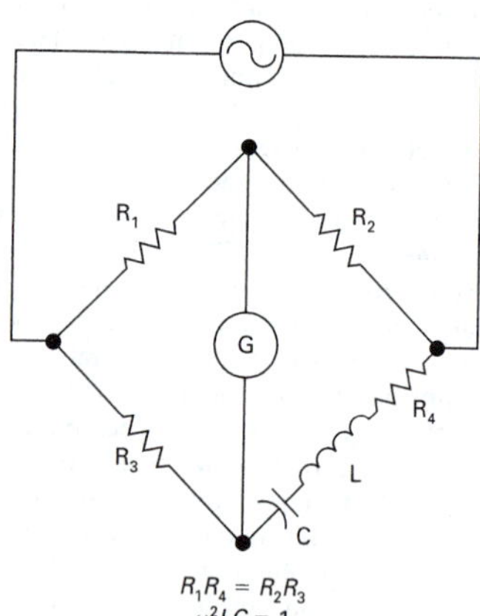

resonance bridge

(EEC/PE) [119]

resonance charging (A) (charging inductors) (direct current) The charging of the capacitance (of a pulse-forming network) to the initial peak value of voltage in an oscillatory series resistance- inductance-capacitance (RLC) circuit, when supplied by a direct voltage. *Note:* in order to provide a pulse train, the network capacitance is repetitively discharged by a synchronous switch at the time when the current through the charging inductor is zero and the peak voltage to which the network capacitance is charged approaches two times the power-supply direct voltage. **(B) (charging inductors)** (alternating current) The charging of the capacitance (of a pulse-forming network) to the peak value of voltage selected, in an oscillatory resistance-inductance-capacitance (RLC) circuit, when supplied by an alternating voltage. *Note:* In order to provide a pulse train, the network capacitance is repetitively discharged at a time in the charging cycle when the current through the charging inductor is zero. At these times the voltage may be essentially:

$\frac{\pi E_p}{2}$, πE_p, $\frac{3\pi E_p}{2}$, etc.

The value chosen depends upon the pulse-repetition rate and the frequency of the alternating voltage. (E_p = peak alternating voltage supply.) (MAG) 306-1969

resonance curve, carrier-current line trap (power-system communication) A graphical plot of the ohmic impedance of a carrier current line trap with respect to frequency at frequencies near resonance. *See also:* power-line carrier. (PE) 599-1985w

resonance frequency (1) (resonant frequency) (networks) Any frequency at which resonance occurs. *Note:* For a given network, resonance frequencies may differ for different quantities, and almost always differ from natural frequencies. For example, in a simple series resistance-inductance-capacitance circuit there is a resonance frequency for current, a different resonance frequency for capacitor voltage, and a natural frequency differing from each of these. *See also:* network analysis. (Std100) 270-1966w
(2) (crystal unit) The frequency for a particular mode of vibration to which, discounting dissipation, the effective impedance of the crystal unit is zero. *See also:* crystal. (EEC/PE) [119]

resonance frequency of charging (charging inductors) The frequency at which resonance occurs in the charging circuit of a pulse-forming network. *Note:* In IEEE Std 306-1969w, it will be assumed to be the frequency determined as follows:

$$f_0 = \frac{1}{2\pi\sqrt{LC_O}}$$

where
f_0 = resonance frequency of charging
C_0 = capacitance of pulse-forming network
L = charging inductance

(MAG) 306-1969w

resonance mode (laser maser) A natural oscillation in a resonator characterized by a distribution of fields which have the same harmonic time dependence throughout the resonator. (LEO) 586-1980w

resonant cavity *See:* optical cavity.

resonant frequency (1) A frequency at which a response peak occurs in a system subjected to forced vibration. This frequency is accompanied by a phase shift of response relative to the excitation. (PE/SUB/NP) 344-1987r, 693-1997
(2) (of an antenna) A frequency at which the input impedance of an antenna is nonreactive. (AP/ANT) 145-1993
(3) The frequency, f, at which a parallel resonant resistive-inductive-capacitive (RLC) load has unity power factor

$$f = 1/(2\pi\sqrt{C \times L})$$

where
f = the resonant frequency
C = effective load capacitance (including shunt capacitors)
L = effective load inductance

Also, the frequency at which the reactive powers P_{qL} and P_{qC} are equal, and hence the parallel RLC load appears equivalent to the load resistance only. (SCC21) 929-2000
(4) (networks) *See also:* resonance frequency.

resonant gap (microwave gas tubes) The small region in a resonant structure interior to the tube, where the electric field is concentrated. (ED) 161-1971w

resonant grounded *See:* ground-fault neutralizer grounded.

resonant grounded neutral system A system in which one or more neutral points are connected to ground through reactors that approximately compensate the capacitive component of a single-phase-to-ground-fault current. *Note:* With resonant grounding of a system, the fault current is limited such that an arc fault in air will be self-extinguishing. (PE/C) 1313.1-1996

resonant grounded system (surge arresters) (arc-suppression coil) A system grounded through a reactor, the reactance being of such value that during a single line-to-ground fault, the power-frequency inductive current passed by this reactor essentially neutralizes the power-frequency capacitive component of the ground-fault current. *Note:* With resonant grounding of a system, the net current in the fault is limited to such an extent that an arc fault in air would be self-extinguishing. *See also:* ground. (PE/SPD) 32-1972r, [8], [84]

resonant iris (waveguide components) An iris designed to have equal capacitive and inductive susceptances at the resonant frequency. (MTT) 147-1979w

resonant line oscillator An oscillator in which the principal frequency-determining elements are one or more resonant transmission lines. *See also:* oscillatory circuit. (AP/BT/ANT) 145-1983s, 182A-1964w

resonant mode (1) (general) A component of the response of a linear device that is characterized by a certain field pattern, and that when not coupled to other modes is representable as a single-tuned circuit. *Note:* When modes are coupled together, the combined behavior is similar to that of the corresponding single-tuned circuits correspondingly coupled. *See also:* waveguide. (EEC/PE) [119]
(2) (cylindrical cavities) When a metal cylinder is closed by two metal surfaces perpendicular to its axis a cylindrical cavity is formed. The resonant modes in this cavity are designated by adding a third subscript to indicate the number of half-waves along the axis of the cavity. When the cavity is a rectangular parallelepiped the axis of the cylinder from which the cavity is assumed to be made should be designated since there are three possible cylinders out of which the parallelepiped may be made. *See also:* guided wave. (MM) 210-1945w

resonating (steady-state quantity or phasor) The maximizing or minimizing of the amplitude or other characteristic provided the maximum or minimum is of interest. *Notes:* 1. Unless otherwise specified, the quantity varied to obtain the maximum or minimum is to be assumed to be frequency. 2. Phase angle is an example of a quantity in which there is usually no interest in a maximum or a minimum. 3. In the case of amplification, transfer ratios, etc., the amplitude of the phasor is maximized or minimized: in the case of currents, voltages, charges, etc., it is customary to think of the amplitude of the steady-state simple sine-wave quantity as being maximized or minimized. *See also:* network analysis. (Std100) 270-1966w

resonating capacitor Provides the capacitance associated with ferroresonant regulating circuits for the purpose of producing ferroresonance. (PEL) 449-1998

resonating capacitor voltamperes The product of the voltage across the resonating capacitor and the current through the resonating capacitor (root-mean-square values) under stated operating conditions. (PEL) 449-1998

resonating winding The winding of the ferroresonant transformer used to connect the resonating capacitance to the circuit. *Note:* It is wound on the secondary section of the core and is separated from the primary winding by a magnetic shunt. It may itself be the output winding or a portion of the output winding. (PEL) 449-1998

resonator A device, the primary purpose of which is to introduce resonance into a system. *See also:* network analysis. (Std100) 270-1966w
(2) (A) A resonating system. **(B)** A device designed to operate in the vicinity of a natural frequency of that device. **(C)** (electrical circuit) An electrical network designed to present a given natural frequency at its terminal. (CAS) [13]

resonator grid (electron tube) An electrode, connected to a resonator, that is traversed by an electron beam and that provides the coupling between the beam and the resonator. *See also:* velocity-modulated tube. (Std100) [84]

resonator mode (oscillators) A condition of operation corresponding to a particular field configuration for which the electron stream introduces a negative conductance into the coupled circuit. *See also:* oscillatory circuit. (ED) 161-1971w

resonator, waveguide *See:* waveguide resonator.

resonant wavelengths (cylindrical cavities) Those given by $\lambda_r = 1[(1/\lambda_c)^2 + (1/2c)^2]^{1/2}$ where λ_c is the cutoff wavelength for the transmission mode along the axis, *l* is the number of half-period variations of the field along the axis, and *c* is the axial length of the cavity. *See also:* guided wave. (MM) 210-1945w

resource **(1)** An attribute of a widget or widget class, represented by a named data value in the defining structure of the widget. (C) 1295-1993w
(2) Any capability that must be scheduled, assigned, or controlled by the underlying implementation to assure nonconflicting usage by processes. (ATLAS) 1232-1995
(3) That part of a LAN/MAN environment for which a managed object provides the management view. The management view of a resource may be limited to a subset of the functionality of the resource; some aspects of the resource may therefore be inaccessible for management purposes. (LM/C) 15802-2-1995

resource allocation The assignment of physical resources to virtual resources such that the virtual resource requirements are satisfied. (SCC20) 1232.1-1997

Resource Description Language (RDL) A standardized computer language used to describe test instrument capabilities and communication sequences. (ATLAS) 1226-1993s

resource lock A type of an attention cycle that indicates to slaves that data items will be referenced in a locked fashion and any nonbus path to referenced data items should be locked out. A null cycle clears this state. (C/MM) 1196-1987w

resource management The identification, estimation, allocation, and monitoring of the means used to develop a product or perform a service. Example is estimating. (C) 610.12-1990

resource manager **(1)** A message-based commander located at logical address 0 that provides configuration management services, such as address map configuration, commander/servant mappings, self-test, and diagnostics management. (C/MM) 1155-1992
(2) A process or activity that initializes and manages the resources in a system. (SCC20) 1226-1998

respecialize A change by an instance from being an instance of its current subclass to being an instance of one of the other subclasses in its current cluster. *Contrast:* specialize; unspecialize. (C/SE) 1320.2-1998

responder **(1)** A module that completes a transaction by sending a response (containing the completion status and sometimes data). (C/BA) 1014.1-1994w, 896.3-1993w
(2) The node that completes a transaction, by returning a response subaction. (C/MM) 1596-1992
(3) A node that completes a transaction by returning a response subaction (containing completion status and sometimes data). (C/MM/BA) 1212-1991s, 10857-1994, 896.4-1993w
(4) The function that completes an I/O transaction-initiation/transaction-completion exchange by sending a completion message to the initiator. (C/MM) 1212.1-1993
(5) The file service user that accepts an FTAM regime establishment requested by the initiator. (C/PA) 1238.1-1994w

responder beacon *See:* transponder.

responding slave *See:* slave; interrupt-acknowledge cycle.

response **(1) (radiation protection)** The instrument reading. (NI) N323-1978r
(2) (airborne radioactivity monitoring) The instrument indication produced as a result of some influence quantity. (NI) N42.17B-1989r
(3) The output, as a function of time or frequency, when a step input voltage or current is applied to the system. (PE/PSIM) 4-1995
(4) A reply generated by a responder, to complete a transaction initiated by a requester. For a processor-to-memory read transaction, for example, the response returns the data and status from the memory to the processor. In the case of a split transaction, the response would be a separate bus transaction. In the case of a connected transaction, the response would be the data and disconnection phases of a bus transaction. (C/BA) 1014.1-1994w, 896.3-1993w, 896.4-1993w, 10857-1994
(5) A pulse, signal, or set of signals indicating a reaction to a preceding transmission. (SUB/PE) 999-1992w
(6) For a dosimeter, the indication (*R*) produced as a result of some influence quantity. (NI) N42.20-1995
(7) In the context of message transmission, the set of packets sent by an S-module during a single message. In the context of the operation of S-modules, an S-module's action that is a direct consequence of the command most recently received by that S-module. (TT/C) 1149.5-1995
(8) A subaction sent by a node (the responder) that sends a response code and optional data back to another node (the requester). (C/MM) 1394-1995
(9) (of a device or system) A quantitative expression of the output as a function of the input under conditions that must be explicitly stated. *Note:* The response characteristic, often presented graphically, gives the response as a function of some independent variable such as frequency or time. (SWG/PE) C37.100-1992
(10) A reply represented in the control field of a response protocol data unit (PDU). It advises the address destination logical link control (LLC) of the action taken by the source LLC to one or more command PDUs. (EMB/MIB) 1073.4.1-2000
(11) A primary packet (with optional data) sent in response to a request subaction. (C/MM) 1394a-2000

response, acceleration-forced *See:* acceleration-forced response.

response data The information sensed from a test subject as the result of an applied stimulus. (SCC20) 1226-1998

response echo The echo packet generated by a requester or agent when it strips the response send packet. (C/MM) 1596-1992

response-expected request The request subaction component of a response-expected transaction. (C/MM) 1596-1992

response-expected transaction A transaction that normally consists of a request subaction and a response subaction. For example, the read, write, and lock transactions are all response-expected transactions. (C/MM) 1596-1992

response, forced *See:* forced response.

response function (linear passive networks) The ratio of response to excitation, both expressed as functions of the complex frequency, $s = \sigma + j\omega$. *Note:* The response function is the Laplace transform of the response due to unit impulse excitation. (CAS) 156-1960w

response, Gaussian *See:* Gaussian response.

response, impulse-forced *See:* impulse-forced response.

response, indicial *See:* indicial response.

response, instrument *See:* instrument response.

responseless request The request subaction component of a responseless transaction. (C/MM) 1596-1992

responseless transaction A transaction that consists of only a request subaction (there is never any response subaction). For example, the move and event transactions are responseless transactions. (C/MM) 1596-1992

response packet A packet that is generated by a slave to return data or status from an address specified by a previous request packet. (C/MM) 1596.4-1996

response protocol data unit (logical link control) All PDUs sent by a logical link control (LLC) in which the command/response (C/R) bit is equal to "1." (LM/PE/C/TR/CC) 799-1987w, 8802-2-1998

response, ramp-forced automatic control *See:* ramp-forced automatic control response.

responses Signals or interrupts generated by a device to notify another device of an asynchronous event. Responses contain the information in the sender's response register. (C/MM) 1155-1992

response send The packet generated by a responder to complete a transaction initiated by a requester. In a processor's memory-read transaction, for example, the response send returns the requested data and related status information from the memory to the processor. (C/MM) 1596-1992

response, sinusoidal *See:* sinusoidal response.

response spectrum (1) (seismic qualification of Class 1E equipment for nuclear power generating stations) (valve actuators) A plot of the maximum response, as a function of oscillator frequency, of an array of single-degree-of-freedom (SDOF) damped oscillators subjected to the same base excitation. (PE/NP) 344-1987r, 382-1985
(2) (seismic testing of relays) (as applied to relays) A plot of the peak acceleration response of damped, single-degree-of-freedom bodies, at a damping value expressed as a percent of critical damping of different natural frequencies, when these bodies are rigidly mounted on the surface of interest. (PE/PSR) C37.98-1977s
(3) (Class 1E metal-enclosed power switchgear) A plot of the maximum response of single-degree-of-freedom bodies of different natural frequencies, at a damping value expressed as a percent of critical damping, when these bodies are rigidly mounted on the surface of interest (that is, on the ground for the ground response spectrum or on the floor of a building for the floor's response spectrum) when that surface is subjected to a given earthquake's motion as modified by any intervening structures. (SWG/PE) C37.81-1989r
(4) (mechanical) A plot of the maximum response of single-degree-of-freedom bodies at a damping value expressed as a percentage of critical damping of different natural frequencies when these bodies are rigidly mounted on the surface of interest (i.e., on the ground for a ground-response spectrum or on the floor for a floor-response spectrum) when that surface is subjected to a given earthquake's motion as modified by intervening structures. (SWG/SUB/PE) C37.122-1983s, C37.100-1992, C37.122.1-1993
(5) A plot of the maximum response of an array of single-degree-of-freedom (SDOF) identically damped oscillators with different frequencies, all subjected to the same base excitation. (PE/SUB) 693-1997

response, steady-state *See:* steady-state response.

response, step-forced *See:* step-forced response.

response subaction (1) A response send and its echo. Often called simply a "response." (C/MM) 1596-1992
(2) A subaction that is returned by a responder to complete a transaction initiated by a requester. In a processor-memory read transaction, for example, the response subaction returns the data and status from the memory to the processor. (C/MM) 1212-1991s

response time (1) (A) (data transmission) (magnetic amplifier). The time (preferably in seconds; may also be in cycles of supply frequency) required for the output quantity to change by some agreed-upon percentage of the differential output quantity in response to a step change in control signal equal to the differential control signal. *Notes:* 1. The initial and final output quantities correspond to the test output quantities. The response time is the maximum obtained including differences arising from increasing or decreasing output quantity or time phase of signal application. **(B) (data transmission)** (turn-ON response time) (control devices). The time required for the output voltage to change from rated OFF voltage to rated ON amplifiers, one serving to amplify the telephone voltage in response to a step change in control signal equal to 120% of the differential trip signal. *Note:* The absolute magnitude of the initial signal condition is the absolute magnitude of the trip OFF control signal plus 10% of the differential trip signal. **(C) (data transmission)** (Turn OFF response time) (control devices). The time required for the output voltage to change from rated ON voltage to rated OFF voltage in response to a step change in control signal equal to 120% of the differential trip signal. *Note:* The absolute magnitude of the initial signal condition is the absolute magnitude of the trip ON control signal minus 10% of the differential trip signal. **(D) (data transmission)** (electrically tuned oscillator). The time following a change in the input to the tuning element required for a characteristic to reach a predetermined range of values within which it remains. **(E) (data transmission)** (instrument). The time required after an abrupt change has occurred in the measured quantity to a new constant value until the pointer, or indicating means, has first come to apparent rest in its new position. *Notes:* 1. Since, in some instruments, the response time depends on the magnitude of the deflection, a value corresponding to an initial deflection from zero to end scale is used in determining response time for rating purposes. 2. The pointer is at apparent rest when it remains within a range on either side of its final position equal to one-half the accuracy rating, when determined as specified above. **(F) (data transmission)** (bolometric power meter). The time required for the bolometric power indication to reach 90% of its final value after a fixed amount of radio-frequency power is applied to the bolometer unit. **(G) (data transmission)** (thermal converter). The time required for the output electromotive force to come to its new value after an abrupt change has occurred in the input quantity (current, voltage, or power) to a new constant value. *Notes:* 1. Since, in some thermal converters, the response time depends upon the magnitude and direction of the change, the value obtained for an abrupt change from zero to rated input quantity is used for rating purposes. 2. The output electromotive force is considered to have come to its new value when all but 1 percent of the change in electromotive force has been indicated. **(H) (data transmission)** (industrial control). The time required, following the initiation of a specified stimulus to a system, for an output going in the direction of necessary corrective action to first reach a specified value. *Note:* The response time is expressed in seconds. **(I) (data transmission)** (electrical conversion). The elapsed time from the initiation of a transient until the output has recovered to 63% of its maximum excursion. **(J) (data transmission)** (arcwelding apparatus). The time required to attain conditions within a specified amount of their final value in an automatically regulated welding circuit after a definitely specified disturbance has been initiated. **(K) (data transmission)** (photoelectric lighting control) (industrial control). The time required for operation following an abrupt change in illumination from 50% above to 50% below the minimum illumination sensitivity. **(L)** (control system or element) (time of response) (control system, feedback) The time required for an output to make the change from an initial value to a large specified percentage of the steady state, either before overshoot or in the absence of overshoot. *Note:* If the term is unqualified, time of response of a first-order system to a unit-step stimulus is generally understood; otherwise the pattern and magnitude of the stimulus should be specified. Usual percentages are 90, 95, or 99. **(M)** (data circuit) The amount of time elapsed between generation of an inquiry at a data communications terminal and receipt of a response at that same terminal. Response time, thus defined includes: transmission time to the computer processing time at the computer, including access time to obtain any file records needed to answer the inquiry; and transmission time back to the terminal. (PE) 599-1985
(2) (station control and data acquisition) The time between initiating some operation and obtaining results. (SWG/PE/SUB) C37.100-1992, C37.1-1987s
(3) (sequential events recording systems) The time interval between receiving a finite input status change and the recognition by the system of the status change. The time interval is usually expressed in milliseconds. (PE/EDPG) [1]
(4) (temperature measurement) The time required for the indication of a thermometer, which has been subjected to an essentially instantaneous change in temperature, to traverse

63% of the temperature interval involved. Following such a temperature change the indication of the thermometer may be expected to traverse 99% of the temperature interval in a period ranging from 5 to 8 time constants so defined, depending on the details of its construction. (PE/PSIM) 119-1974w

(5) (faulted circuit indicators) The time required for the faulted circuit indicator (FCI) to respond to a specified value of fault current.

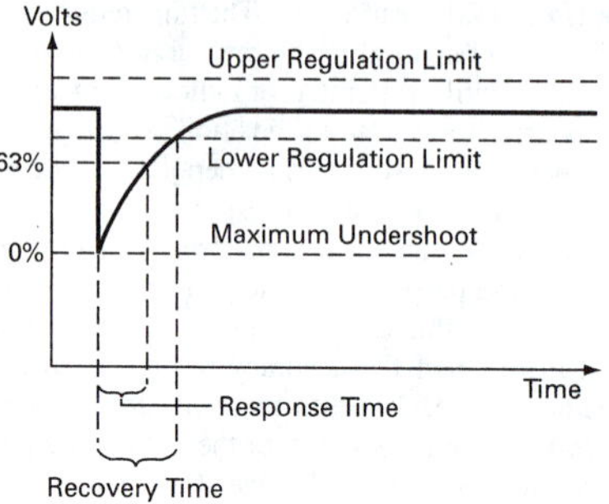

Response and recovery time for a critically damped circuit.

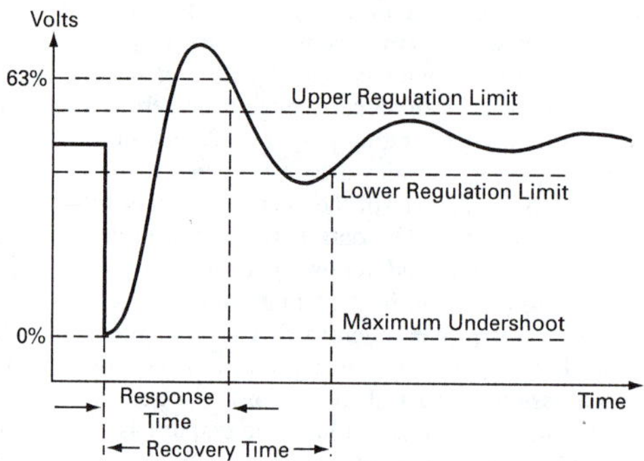

Response and recovery time for an underdamped circuit.

response time

(T&D/PE) 495-1986w

(6) (monitoring radioactivity in effluents) The time interval from a step change in the input concentration at the instrument inlet to a reading of 90% (nominally equivalent to 2.2 time constants) of the ultimate recorded output. (NI) N42.18-1980r

(7) (airborne radioactivity monitoring) The time interval required for the instrument reading to change from 10% to 90% of the final reading following a step change in the radiation field (i.e., signal) at the detector, or, for integrating monitors, 90% of the final value of the first derivative of the indication with respect to time (i.e., rate of change). (NI) N42.17B-1989r

(8) (software) The elapsed time between the end of an inquiry or command to an interactive computer system and the beginning of the system's response. *See also:* turnaround time; port-to-port time; think time. (C) 610.12-1990, 610.10-1994w

(9) A quantity that is indicative of the speed with which a system responds to changing voltages or currents. (PE/PSIM) 4-1995

(10) The time required for a field probe to reach 90% of its steady state value when the field is applied as a step function. The measurement includes test set up response time, thus giving worst case results. (EMC) 1309-1996

(11) The duration from a step change in control signal input until the static var compensator (SVC) output reaches 90% of required output, before any overshoot. (PE/SUB) 1031-2000

(12) (lagged-demand meter) *See also:* time characteristic. (ELM) C12.1-1982s

(13) (lagged-demand meter) *See also:* demand meter—time characteristic. (ELM) C12.1-1981

response timer A timing device within a FASTBUS master or segment interconnect used to terminate an operation that has failed to complete within a given (excessive) period of time. (NID) 960-1993

response time, ramp-forced *See:* ramp-forced response time.

response to signal removal (measuring the performance of tone address signaling systems) The time interval from the end of signal present condition to the time the receiver indication terminates. (COM/TA) 752-1986w

response to signal start (measuring the performance of tone address signaling systems) The time interval from start of a signal present condition to the time at which the appropriate indication occurs in the receiver. (COM/TA) 752-1986w

response to tone removal (measuring the performance of tone address signaling systems) The time interval from the end of tone present condition to the time the receiver indication terminates. (COM/TA) 752-1986w

response to tone start (measuring the performance of tone address signaling systems) The time interval from the start of a tone present condition to the time at which the appropriate indication occurs in the receiver. (COM/TA) 752-1986w

response_timeout An implied split-transaction-error status that is returned when the response subaction is not returned within an expected timeout interval. (C/MM) 1212-1991s

responsibility A generalization of properties (attributes, participant properties, and operations) and constraints. An instance possesses knowledge, exhibits behavior, and obeys rules. These are collectively referred to as the instance's responsibilities. A class abstracts the responsibilities in common to its instances. A responsibility may apply to each instance of the class (instance-level) or to the class as a whole (class-level). (C/SE) 1320.2-1998

responsivity (fiber optics) The ratio of an optical detector's electrical output to its optical input, the precise definition depending on the detector type; generally expressed in amperes per watt or volts per watt of incident radiant power. *Note:* "Sensitivity" is often incorrectly used as a synonym. (Std100) 812-1984w

responsor (1) (navigation aids) The receiving component of an interrogator-responsor. (AES/GCS) 172-1983w

(2) The receiving part of an interrogator-responsor. (AES/RS) 686-1990

rest and de-energized (rotating machinery) The complete absence of all movement and of all electric or mechanical supply. *See also:* asynchronous machine. (PE) [9]

restart (1) (computers) To reestablish the execution of a routine, using the data recorded at a checkpoint. (MIL/C) [2], [20], [85]

(2) (software) To cause a computer program to resume execution after a failure, using status and results recorded at a checkpoint. (C) 610.12-1990

(3) Resume the processing of a job from the point of the last checkpoint. Typically, this is done if the job has been interrupted because of a system failure. (C/PA) 1003.2d-1994

restart instruction An instruction in a computer program at which the program may be restarted. (C) 610.10-1994w

restart point A point in a computer program at which execution can be restarted following a failure. *Synonym:* rescue point. (C) 610.12-1990

resting potential (biological) The voltage existing between the two sides of a living membrane or interface in the absence of stimulation. (EMB) [47]

resting potential The normal potential difference between the inside and the outside of a cell, usually about 80 mV, with the inside negative relative to the outside. (T&D/PE) 539-1990

restorable fire detector (fire protection devices) A device whose sensing element is not ordinarily destroyed by the process of detecting a fire. Restoration may be manual or automatic. (NFPA) [16]

restoration *See:* image restoration.

restore To recover the state of a system, computer program, or database to a specific point. *See also:* roll back; rollforward. (C) 610.5-1990w

restoring force gradient (direct-acting recording instrument) The rate of change, with respect to the displacement, of the resultant of the electric, or of the electric and mechanical, forces tending to restore the marking device to any position of equilibrium when displaced from that position. *Note:* The force gradient may be constant throughout the entire travel of the marking device or it may vary greatly over this travel, depending upon the operating principles and the details of construction. *See also:* accuracy rating. (EEC/PE) [119]

restoring torque gradient (instrument) The rate of change, with respect to the deflection, of the resultant of the electric, or electric and mechanical, torques tending to restore the moving element to any position of equilibrium when displaced from that position. *See also:* accuracy rating. (EEC/PE) [119]

restraint relay A relay so constructed that its operation in response to one input is restrained or controlled by a second input. (SWG/PE) C37.100-1992

restricted character string type A simple type whose values are strings of characters from some defined character set. (C/PA) 1238.1-1994w

restricted radiation frequencies for industrial, scientific, and medical equipment Center of a band of frequencies assigned to industrial, scientific, and medical equipment either nationally or internationally and for which a power limit is specified. *See also:* electromagnetic compatibility. (INT) [53], [70]

restricted-service tone (telephone switching systems) A class-of-service tone that indicates to an operator that certain services are denied the caller. (COM) 312-1977w

restriction *See:* clearance.

restrictive [security] attribute A security attribute that indicates the minimum level of privilege required by an active entity (i.e., subject) in order to gain access to a resource (i.e., object). Commonly, a set of restrictive security attributes are associated with each resource. An active entity may only gain access to a resource if its set of privileges is higher than, or a superset of (i.e., dominates), the attribute set for the resource. (C/LM) 802.10g-1995, 802.10-1998

restrike A resumption of current between the contacts of a switching device during an opening operation after an interval of zero current of $\frac{1}{4}$ cycle at normal frequency or longer. (SWG/PE) C37.100-1992

restrike time (nuclear security systems) The time period during which a momentary loss or reduction of illumination results from the need to cool down to restrike the arc after a momentary loss or reduction of electrical power to a luminaire. (PE/NP) 692-1986s

restriking voltage (1) (gas tube) The anode voltage at which the discharge recommences when the supply voltage is increasing before substantial deionization has occurred. (ED) [45], [84]

(2) The voltage that appears across the terminals of a switching device immediately after the breaking of the circuit. *Note:* This voltage may be considered as composed of two components. One, which subsists in steady-state conditions, is direct current or alternating current at service frequency, according to the system. The other is a transient component that may be oscillatory (single or multifrequency) or nonoscillatory (for example, exponential) or a combination of these depending on the characteristics of the circuit and the switching device. *See also:* switch. (IA/ICTL/IAC) [60], [84]

restructuring *See:* reorganization.

result (1) (binary floating-point arithmetic) The bit string (usually representing a number) that is delivered to the destination. (C/MM) 754-1985r

(2) (radix-independent floating-point arithmetic) The digit string (usually representing a number) that is delivered to the destination. (C/MM) 854-1987r

(3) Information that is returned from an interface operation or a directory operation and that constitutes the outcome of the processing that was performed. (C/PA) 1328.2-1993w, 1327.2-1993w, 1224.2-1993w, 1326.2-1993w

resultant color shift (illuminating engineering) The difference between the perceived color of an object illuminated by a test source and of the same object illuminated by a reference source, taking account of the state of chromatic adaptation in each case; that is, the resultant of colorimetric shift and adaptive color shift. *See also:* state of chromatic adaptation. (EEC/IE) [126]

resultant magnetic field The resultant magnetic field is given by the expression

$$B_R = \sqrt{B_x^2 + B_y^2 + B_z^2}$$

where

B_x, B_y, and B_z are the rms values of the three orthogonal field components.

Notes: 1. The resultant magnetic field is also given by the expression

$$B_R = \sqrt{B_{max}^2 + B_{min}^2}$$

where B_{max} and B_{min} are the rms values of the semimajor and semiminor axes of the magnetic field ellipse, respectively. The resultant B_R is always $\geq B_{max}$. If the magnetic field is linearly polarized, $B_{min} = 0$ and $B_R = B_{max}$. If the magnetic field is circularly polarized, $B_{max} = B_{min}$ and $B_R = 1.41 B_{max}$. 2. A three-axis magnetic field meter simultaneously measures the rms values of the three orthogonal field components and combines them according to the second equation to indicate the resultant magnetic field. Although power line magnetic fields are typically two dimensional in nature, i.e., elliptically polarized, unless two axes of a three-axis probe are in the plane of the ellipse, each of the three probes will sense a component of the rotating magnetic field vector. (T&D/PE) 644-1994, 1308-1994

resuming port A previously suspended port that has observed bias or has been instructed to generate bias. In either case, the resuming port engages in a protocol with its connected peer physical layer (PHY) in order to reestablish normal operations and become active. (C/MM) 1394a-2000

retained image (image burn) A change produced in or on the target that remains for a large number of frames after the removal of a previously stationary light image and that yields a spurious electric signal corresponding to that light image. *See also:* camera tube. (ED) 161-1971w

retainer *See:* separator.

retaining ring (1) (rotating machinery) (steel) A mechanical structure surrounding parts of a rotor to restrain radial movement due to centrifugal action. *See also:* rotor. (PE) [9]

(2) (insulation) The insulation forming a dielectric and mechanical barrier between the rotor end windings and the high-strength steel retaining ring. *See also:* rotor. (PE) [9]

retaining ring liner (rotating machinery) Insulating ring between the end winding and the metallic ring which secures the coil ends against centrifical force. (PE) [9]

retardation (deceleration) The operation of reducing the motor speed from a high level to a lower level or zero. *See also:* electric drive. (IA/ICTL/IAC) [60]

retardation coil *See:* inductor.

retardation test (rotating machinery) A test in which the losses in a machine are deduced from the rate of deceleration of the machine when only these losses are present. *See also:* asynchronous machine; direct-current commutating machine. (PE) [9]

retard coil *See:* inductor.

retarder *See:* bullwheel tensioner.

retarding-field (positive-grid) oscillator An oscillator employing an electron tube in which the electrons oscillate back and forth through a grid maintained positive with respect to the cathode and the plate. The frequency depends on the electron-transit time and may also be a function of the associated circuit parameters. The field in the region of the grid exerts a retarding effect that draws electrons back after passing through it in either direction. Barkhausen-Kurz and Gill-Morell oscillators are examples. *See also:* oscillatory circuit. (AP/ANT) 145-1983s

retarding magnet A magnet used for the purpose of limiting the speed of the rotor of a motor-type meter to a value proportional to the quantity being integrated. *See also:* braking magnet; watthour meter; drag magnet. (EEC/PE) [119]

retard transmitter A transmitter in which a delay period is introduced between the time of actuation and the time of transmission. *See also:* protective signaling. (EEC/PE) [119]

retention (metal-nitride-oxide field-effect transistor) The time period defined by the time elapsed between the instant of writing a metal-nitride-oxide semiconductor (MNOS) transistor into a given high conduction or low conduction (HC or LC) state, and the instant when either state becomes indistinguishable from the other. (ED) 581-1978w

retention characteristic (metal-nitride-oxide field-effect transistor) A plot of both high conduction (HC) and low conduction (LC) threshold voltages v HC or v LC as a function (commonly the logarithm) of the time trd elapsed after the instant of writing. (ED) 581-1978w

retention cycle The length of time specified for data on a data medium to be preserved. *Synonym:* retention period. (C) 610.10-1994w

retention failure The inability to correctly sense the state of a memory cell within the limits of device specifications dependent on the time period of data storage. (ED) 641-1987w

retention longevity The time elapsed between the instant of writing a data pattern into a memory and the time when the read failure rate exceeds some predetermined value. *Note:* This definition allows for soft errors and fits reliability prediction and specification methods. (ED) 641-1987w

retention period *See:* retention cycle.

retention pit A pit designed to retain (hold) oil-contaminated liquids. (SUB/PE) 980-1994

retention time (1) The time interval between the instant of writing a memory pattern into a memory and the first retention failure. (ED) 641-1987w
(2) The time interval between the instant that data is stored and the instant that the data can no longer be read correctly. *Note:* Unless otherwise qualified, the term "read-only memory" implies that the data content is determined by the structure of the memory and in unalterable. (ED) 1005-1998

retention time at maximum read rate The retention tune using the maximum specified read rate on a single address. (ED) 641-1987w

retention time, maximum *See:* maximum retention time.

retentivity (magnetic material) That property that is measured by its maximum residual induction. *Note:* The maximum residual induction is usually associated with a hysteresis loop that reaches saturation, but in special cases this is not so. (Std100) 270-1966w

Re Test Okay (RTOK) The result of a UUT passing at any level of maintenance after failing at a previous level of maintenance. In maintenance, it is a unit determined faulty at one level, but good at the next level of maintenance. (ATLAS) 1232-1995

reticle (navigation aid terms) A system of lines, etc., placed in the focal plane of an optical instrument to serve as a reference. (AES/GCS) 172-1983w

retina (1) (illuminating engineering) A membrane lining the posterior part of the inside of the eye. It comprises photoreceptors (cones and rods) which are sensitive to light, and nerve cells which transmit to the optic nerve the responses of the receptor elements. (EEC/IE) [126]
(2) (laser maser) That sensory membrane which receives the incident image formed by the cornea and lens of the human eye. The retina lines the inside portion of the eye. (LEO) 586-1980w

retirement (A) (software) Permanent removal of a system or component from its operational environment. **(B) (software)** Removal of support from an operational system or component. *See also:* software life cycle; system life cycle. (C) 610.12-1990

retirement phase (software) The period of time in the software life cycle during which support for a software product is terminated. (C) 610.12-1990

retrace (oscillography) Return of the spot on the cathode-ray tube to its starting point after a sweep; also that portion of the sweep waveform that returns the spot to its starting point. *See also:* oscillograph. (IM/HFIM) [40]

retrace blanking *See:* blanking.

retrace interval (television) The interval corresponding to the direction of sweep not used for delineation. *See also:* flyback. (BT/AV) 201-1979w

retrace line The line traced by the electron beam in a cathode-ray tube in going from the end of one line or field to the start of the next line or field. 188-1952w

retraining The process of re-acquiring receiver parameters and synchronizing the scramblers of two connected 100BASE-T2 PHYs. *See also:* receiver training. (C/LM) 802.3-1998

retransmit contacts Auxiliary contacts on an annunciator that provide an output to a remote device to indicate that the annunicator has been actuated. (SUB/PE) C37.123-1996

retrieval *See:* information retrieval.

retrieval code In micrographics, a code used for manual or automatic retrieval of microimages. (C) 610.2-1987

retrieval queue One of two alternative databases that the service uses to convey objects to the client of the MA interface. (C/PA) 1224.1-1993w

retrieve To move data out of a storage device or data medium. *Contrast:* store. *See also:* read. (C) 610.5-1990w

retrodirective antenna An antenna whose monostatic cross section is comparable to the product of its maximum directivity and its area projected in the direction toward the source, and is relatively independent of the source direction. *Note:* Active devices can be added to enhance the return signal. For this case, the term shall be qualified by the word active; that is, active retrodirective antenna system. (AP/ANT) 145-1993

retrofill (handling and disposal of transformer grade insulating liquids containing PCBs) The process of replacing the dielectric liquid in a transformer. (LM/C) 802.2-1985s

retrograde orbit (communication satellite) An inclined orbit with an inclination between 90° and 180°. (COM) [19]

retro-reflector (illuminating engineering) A device designed to reflect light in a direction close to that at which it is incident, whatever the angle of incidence. (EEC/IE) [126]

retrospective trace A trace produced from historical data recorded during the execution of a computer program. *Note:* This differs from an ordinary trace, which is produced cumulatively during program execution. *See also:* execution trace; variable trace; symbolic trace; subroutine trace. (C) 610.12-1990

retry A mechanism whereby a transaction that (for whatever reason) could not complete in the current operation is attempted again at a later time. (C/BA) 1014.1-1994w

retry period The time a master waits after failing to receive a response before trying the operation again. This time should be randomized to avoid system deadlocks. (NID) 960-1993

return (1) (A) (software) To transfer control from a software module to the module that called it. *See also:* return code. **(B) (software)** To assign a value to a parameter that is accessible by a calling module; for example, to assign the value 25 to parameter AGE for use by a calling module. *See also:* return value. **(C) (software)** A computer instruction or process that performs the transfer in definition definition (A). (C) 610.12-1990

(2) (software) *See also:* carriage return. (C) [85]

(3) (local area networks) A secondary link control signal indicating that the pre- empted normal-priority round-robin cycle in a lower repeater is not complete. (C) 8802-12-1998

return air Air returned from the conditioned space. (IA/PSE) 241-1990r

return-beam mode (camera tubes) A mode of operation in which the output current is derived, usually through an electron multiplier, from that portion of the scanning beam not accepted by the target. *See also:* camera tube. (ED) [45]

return beam multiplier gain (diode-type camera tube) The dimensionless ratio between the output signal current at the final anode of the electron multiplier in a return beam camera tube and the modulated portion of the beam current falling on the first dynode of the multiplier. The output signal current is the value of the output current less the dark current. (ED) 503-1978w

return code (1) A code used to influence the execution of a calling module following a return from a called module. (C) 610.12-1990

(2) A value returned by a function indicating whether the function completed successfully. If the function did not complete successfully, it may return a nonzero return code; the exact value may indicate one of several possible severity conditions: informational, warning, error, severe, terminal error, etc. (C/DA) 1481-1999

(3) A value, returned to the caller of an operation, providing information about the completion status of the operation. This standard defines two forms of return codes. (IM/ST) 1451.1-1999

return code register A register used to store a code which is used to influence the carrying out of following programs. (C) 610.10-1994w

return difference (network analysis) One minus the loop transmittance. (CAS) 155-1960w

return interval *See:* retrace interval.

return loss (1) (A) (data transmission) At a discontinuity in a transmission system the difference between the power incident upon the discontinuity. **(B) (data transmission)** The ratio in decibels of the power incident upon the discontinuity to the power reflected from the discontinuity. *Note:* This ratio is also the square of the reciprocal to the magnitude of the reflection coefficient.

Return loss $= 20\log_{10}(1/\Gamma)$.

(C) (data transmission) More broadly, the return loss is a measure of the dissimilarity between two impedances, being equal to the number of decibels that corresponds to the scalar value of the reciprocal of the reflection coefficient, and hence being expressed by the following formula:

$$20\log_{10}\left|\frac{Z_1 + Z_2}{Z_1 - Z_2}\right| \text{ decibel}$$

where Z_1 and Z_2 = the two impedances. (MTT/PE) 146-1980, 599-1985

(2) (waveguide) (or gain) The ratio of incident to reflected power at a reference plane of a network. (MTT) 146-1980w

(3) (transmission characteristics of PCM telecommunications circuits and systems) A measure of power reflected back to the originating end of a channel due to impedance mismatch. This measurement is also used to characterize equipment impedance accuracy vs. a test impedance. (COM/TA) 1007-1991r

(4) (broadband local area networks) The degree of impedance mismatch for an RF component or system. The return loss term expresses the coefficient of reflection in decibels. At the location of an impedance mismatch, part of the incident signal is reflected back toward its source, creating a reflected signal. The return loss is the number of decibels that the reflected signal level is below the incident signal level. (LM/C) 802.7-1989r

(5) In 10BROAD36, the ratio in decibels of the power reflected from a port to the power incident to the port. An indicator of impedance matching in a broadband system. (C/LM) 802.3-1998

(6) The ratio, in dB, of the power incident upon the discontinuity to the power reflected from the discontinuity. *Note:* This ratio is also the square of the reciprocal of the magnitude of the reflection coefficient. (EMC) 1128-1998

return loss, echo *See:* echo return loss.

return loss, singing *See:* singing return loss.

return path Direction towards the headend. *See also:* inbound. (LM/C) 802.7-1989r

return signal (control system feedback) (closed loop) The signal resulting from a particular input signal, and transmitted by the loop and to be subtracted from that input signal. See the figure attached to the definition of **signal, error.** *See also:* feedback control system. (PE/EDPG) 421-1972s

return stroke (lightning) The luminescent, high-current discharge that is initiated after the stepped leader and pilot streamer have established a highly ionized path between charge centers. *See also:* direct-stroke protection. (T&D/PE) [10]

return swing (pulse transformers) (last transition ringing) The maximum amount by which the instantaneous pulse value is below the zero axis in the region following the backswing. It is expressed in amplitude units or as a percentage of AM. (PEL/ET) 390-1987r

return to bias (magnetic tape pulse recorders for electricity meters) A method whereby a recording head current, which results in a magnetic field polarity opposite that of the bias magnet, is applied momentarily in order to record a pulse. (ELM) C12.14-1982r

return-to-reference recording Magnetic recording such that the patterns of magnetization used to represent zeros and ones occupy only part of the storage cell, and the remainder of the cell is magnetized to a reference condition. *Contrast:* nonreturn-to-reference recording. *See also:* return-to-zero recording. (C) 610.10-1994w

return-to-zero recording Return-to-reference recording in which the reference condition is the absence of magnetization. (C) 610.10-1994w

return trace (1) (television) The path of the scanning spot during the retrace interval. (BT/AV) 201-1979w

(2) (oscillography) (television) The path of the scanning spot during the retrace. *See also:* television; oscillograph. (IM) 311-1970w

return-transfer function (control system feedback) (closed loop) The transfer function obtained by taking the ratio of the Laplace transform of the return signal to the Laplace transform of its corresponding input signal. *See also:* feedback control system. (PE/EDPG) [3]

return value A value assigned to a parameter by a called module for access by the calling module. (C) 610.12-1990

reusability The degree to which a software module or other work product can be used in more than one computer program or software system. *See also:* generality. (C) 610.12-1990

(2) (A) The degree to which an asset can be used in more than one software system, or in building other assets. **(B)** In a reuse library, the characteristics of an asset that make it easy to use in different contexts, software systems, or in building different assets. (C/SE) 1517-1999

reusable Pertaining to a software module or other work product that can be used in more than one computer program or software system. *See also:* generality. (C) 610.12-1990

reusable software product A software product developed for one use but having other uses, or one developed specifically to be usable on multiple projects or in multiple roles on one project. Examples include, but are not limited to, commercial off-the-shelf software products, acquirer-furnished software products, software products in reuse libraries, and pre-existing developer software products. Each use may include all or part of the software product and may involve its modification. This term can be applied to any software product (for example, requirements, architectures, etc.), not just to software itself. (C/SE) J-STD-016-1995

reuse The use of an asset in the solution of different problems. (C/SE) 1517-1999

reused source statements Unmodified source statements obtained for the product from an external source. (C/SE) 1045-1992

reuse sponsor A member of the organization's management who authorizes, approves, promotes, and obtains the funding and other resources for the Reuse Program. (C/SE) 1517-1999

revenue service (A) Transit service excluding deadheading or layovers. **(B)** Any service scheduled for passenger trips. (VT/RT) 1474.1-1999

reverberant sound Sound that has arrived at a given location by a multiplicity of indirect paths as opposed to a single direct path. *Notes:* 1. Reverberation results from multiple reflections of sound energy contained within an enclosed space. 2. Reverberation results from scattering from a large number of inhomogeneities in the medium or reflection from bounding surfaces. 3. Reverberant sound can be produced by a device that introduces time delays that approximate a multiplicity of reflections. *See also:* echo. (SP) [32]

reverberation The presence of reverberant sound. (SP) [32]

reverberation chamber An enclosure especially designed to have a long reverberation time and to produce a sound field as diffuse as possible. *See also:* anechoic chamber. (SP) [32]

reverberation room *See:* reverberation chamber.

reverberation time (T_{60}) (1) The time required for the mean-square sound pressure level, or electric equivalent, originally in a steady state, to decrease 60 dB after the source output is stopped. (SP) [32]
(2) The time it takes for sound in a room to decay 60 dB from its initial, steadystate value. (COM/TA) 1329-1999

reverberation-time meter An instrument for measuring the reverberation time of an enclosure. *See also:* instrument. (EEC/PE) [119]

reversal (storage battery) (storage cell) A change in normal polarity of the cell or battery. *See also:* charge. (PE/EEC) [119]

reversal point That point on the input current versus input voltage characteristics where the input current reaches a minimum value and begins to increase. See figure under output voltage versus input voltage characteristics and figure below.

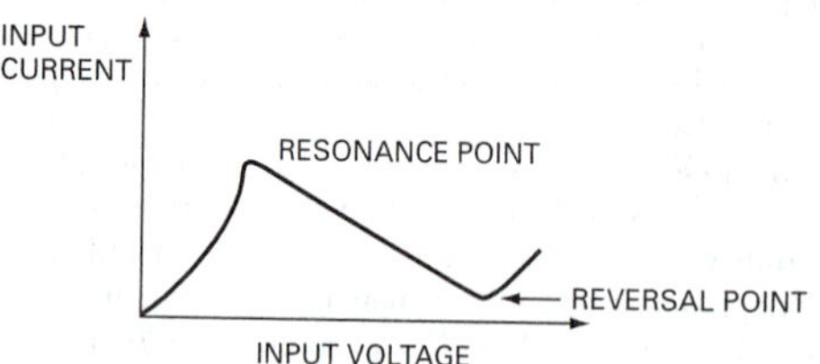

Reversal point without jump resonance

Reversal point

(PEL) 449-1998

reverse The direction of operation that is opposite to forward. (VT) 1475-1999

reverse-battery signaling (telephone switching systems) A method of loop signaling in which the direction of current in the loop is changed to convey on-hook and off-hook signals. (COM) 312-1977w

reverse-battery supervision (telephone switching systems) A form of supervision employing reverse-battery signaling. (COM) 312-1977w

reverse bias (light-emitting diodes) (reverse voltage) The bias voltage that is applied to an LED (light emitting diode) in the reverse direction. (IE/EEC) [126]

reverse-blocking current (reverse-blocking thyristor) The reverse current when the thyristor is in the reverse-blocking state. *See also:* principal current. (IA/ED) 223-1966w, [46], [12], [62]

reverse-blocking diode-thyristor A two-terminal thyristor that switches only for positive anode-to-cathode voltages and exhibits a reverse-blocking state for negative anode-to-cathode voltages. (IA/IPC) 428-1981w

reverse-blocking impedance (reverse-blocking thyristor) The differential impedance between the two terminals through which the principal current flows, when the thyristor is in the reverse-blocking state at a stated operating point. *See also:* principal voltage-current characteristic. (ED) [46]

reverse-blocking state (reverse-blocking thyristor) The condition of a reverse-blocking thyristor corresponding to the portion of the anode-to-cathode voltage-current characteristic for reverse currents of lower magnitude than the reverse-breakdown current. *See also:* principal voltage-current characteristic. (ED) [46]

reverse-blocking triode thyristor (SCR) (1) A three-terminal thyristor that switches only for positive anode-to-cathode voltages and exhibits a reverse-blocking state for negative anode-to-cathode voltages. (IA/IPC) 428-1981w
(2) A monocrystalline reverse-blocking semiconductor device with bistable character in the forward direction normally having three pn junctions and a gate electrode at which a suitable electrical signal will cause switching from the off state to the on state within the first quadrant of the anode to cathode voltage−current characteristics. If cooling means are integrated, they are included. *Note:* In this document the word thyristor means a reverse-blocking triode thyristor. *See also:* *p-n* junction. (IA/IPC) 444-1973w

reverse-breakdown current (reverse-blocking thyristor) The principal current at the reverse-breakdown voltage. *See also:* principal current. (PE/PSR) C37.93-1976s

reverse-breakdown voltage (reverse-blocking thyristor) The value of negative anode-to-cathode voltage at which the differential resistance between the anode and cathode terminals changes from a high value to a substantially lower value. *See also:* principal voltage-current characteristic. (PE/PSR) C37.93-1976s

reverse channel (1) Data path from the peripheral to the host. (C/MM) 1284-1994
(2) *See also:* backward channel. (C) 610.10-1994w

reverse-conducting diode-thyristor A two-terminal thyristor that switches only for positive anode-to-cathode voltages and conducts large currents at negative at negative anode-to-cathode voltages comparable in magnitude to the ON-state voltages. (IA/IPC) 428-1981w

reverse-conducting triode-thyristor A three-terminal thyristor that switches only for positive anode-to-cathode voltages and conducts large currents at negative anode-to-cathode voltages comparable in magnitude to the ON-state voltage. (IA/IPC) 428-1981w

reverse contact A contact that is closed when the operating unit is in the reverse position. (EEC/PE) [119]

reverse coupling The ratio of the spurious signal generated by a signal at some other input to the recorder and the signal recorded at the specified input of the recorder. (IM/WM&A) 1057-1994w

reverse current (1) Current that flows upon application of reverse voltage. (EEC/PE) [119]
(2) (reverse-blocking or reverse-conducting thyristor) The principal current for negative anode-to-cathode voltage. *See also:* principal current. (PE/PSR) C37.93-1976s

(3) **(metallic rectifier)** The current that flows through a metallic rectifier cell in the reverse direction. *See also:* rectification. (PE/EEC) [119]
(4) (semiconductor rectifiers) The total current that flows through a semiconductor rectifier device in the reverse direction. *See also:* rectification. (ED) 216-1960w

reverse-current cleaning *See:* anode cleaning.

reverse-current cutout A magnetically operated direct-current device that operates to close an electric circuit when a predetermined voltage condition exists and operates to open an electric circuit when more than a predetermined current flows through it in the reverse direction. *Fixed-voltage type:* A reverse-current cutout that closes an electric circuit whenever the voltage at the cutout terminal exceeds a predetermined value and is of the correct polarity. It opens the circuit when more than a predetermined current flows through it in the reverse direction. *Differential-voltage type:* A reverse-current cutout that closes an electric circuit when a predetermined differential voltage appears at the cutout terminal, provided this voltage is of the correct polarity and exceeds a predetermined value. It opens the circuit when more than a predetermined current flows through it in the reverse direction. (EEC/PE) [119]

reverse-current relay A relay that operates on a current flow in a dc circuit in a direction opposite to a predetermined reference direction. (SWG/PE) C37.100-1992

reverse-current release (trip) A release that operates upon reversal of the direct current in the main circuit from a predetermined direction. (SWG/PE) C37.100-1992

reverse-current trip *See:* reverse-current release.

reverse-current tripping *See:* reverse-current release.

reverse-power tripping *See:* reverse-current release.

reverse data transfer phase When data transfers from the peripheral to the host. (C/MM) 1284-1994

reverse direction (semiconductor rectifier diode) The direction of higher resistance to steady-state direct-current: that is, from the cathode to the anode. (IA/ED) [12], [127]

reverse-electrode coaxial detector (germanium gamma-ray detectors) Reverse-electrode geometry. A coaxial detector in which the outer contact is a p-type layer. (NPS) 325-1996

reverse emission (vacuum tubes) The inverse electrode current from an anode during that part of a cycle in which the anode is negative with respect to the cathode. *See also:* electron emission. (ED) [45]

reverse engineering The process of extracting software system information (including documentation) from source code. (C/SE) 1219-1998

reverse execution *See:* reversible execution.

reverse gate current (thyristor) The gate current when the junction between the gate region and the adjacent anode or cathode region is reverse biased. *See also:* principal current. (ED) [46]

reverse gate voltage (thyristor) The voltage between the gate terminal and the terminal of an adjacent region resulting from reverse gate current. *See also:* principal voltage-current characteristic. (ED) [46]

reverse host busy data available phase When the peripheral has data to transmit. (C/MM) 1284-1994

reverse host busy data not available phase When the peripheral has no more data to transmit. (C/MM) 1284-1994

reverse leading In photocomposition, the ability of some phototypesetting equipment to allow reverse movement of the photographic medium. *Note:* This technique permits the setting of side-by-side columns of text on the composed page. *See also:* leading. (C) 610.2-1987

reverse offset mho characteristic A modification of a mho characteristic to make it nondirectional so as to encompass the intersection of the *R-X* axes. See figure below.

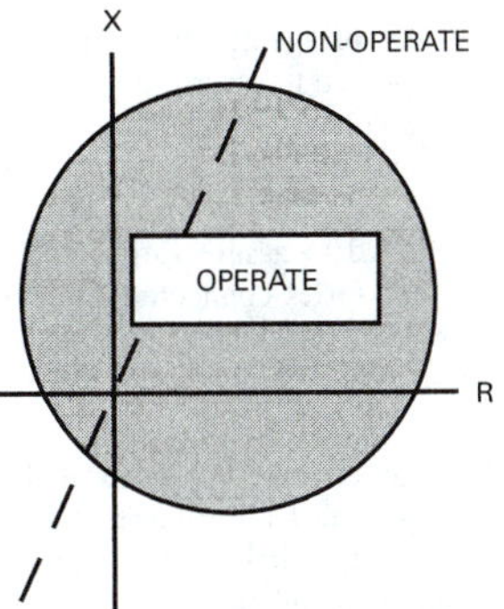

reverse offset mho characteristic

(SWG/PE) C37.100-1992

reverse period (rectifier circuits) (rectifier circuit element) The part of an alternating-voltage cycle during which the current flows in the reverse direction. *See also:* blocking period; rectifier circuit element. (IA) [62]

reverse-phase or phase-balance current relay (power system device function numbers) A relay that functions when the polyphase currents are unbalanced or contain negative phase-sequence components above a given amount. (PE/SUB) C37.2-1979s

reverse Polish notation *See:* postfix notation.

reverse position (device) The opposite of the normal position. (EEC/PE) [119]

reverse power dissipation (semiconductor) The power dissipation resulting from reverse current. (IA) [12]

reverse power loss (semiconductor rectifiers) The power loss resulting from the flow of reverse current. *See also:* rectification; semiconductor rectifier stack. (IA) [12]

reverse printer *See:* bidirectional printer.

reverser A switching device for interchanging electric circuits to reverse the direction of motor rotation. *See also:* multiple-unit control. (EEC/PE) [119]
(2) (A) The portion of the master controller used to change the commanded direction of train movement. **(B)** A circuit device used to change motor connections in order to change the direction of motor rotation and thus train movement. (VT) 1475-1999

reverse recovery current (semiconductor rectifiers) The transient component of reverse current associated with a change from ON state conduction to reserve voltage. *See also:* rectification. (IA) [12]

reverse recovery interval (thyristor) The interval between the instant when the principal ON-state current flowing through a semiconductor passes through zero, and the instant when the reverse current has decayed to 10 percent of the peak reverse value. (IA/IPC) 428-1981w

reverse recovery time (reverse-blocking thyristor or semiconductor diode) The time required for the principal current or voltage to recover to a specified value after instantaneous switching from an ON state to a reverse-voltage or current. *See also:* principal voltage-current characteristic; rectification. (ED) [46]

reverse resistance (metallic rectifier) The resistance measured at a specified reverse voltage or a specified reverse current. *See also:* rectification. (EEC/PE) [119]

reverse scrolling In word processing, the process of moving the text across the display screen in the reverse direction from the normal reading direction. *See also:* scrolling. (C) 610.2-1987

reverse voltage (1) (rectifier) Voltage of that polarity that produces the smaller current. *See also:* rectification; principal voltage-current characteristic. (PE/EEC) [119]
(2) (reverse-blocking or reverse-conducting thyristor) (semiconductor) A negative anode-to-cathode voltage. (IA) [12]

reverse or OFF-state voltage dividers (1) (thyristor) (or OFF-state) Devices employed to assure satisfactory division of reverse or OFF-state voltage among series-connected semiconductor devices under transient or steady state conditions, or both. (IA/IPC) 428-1981w
(2) Devices employed to assure satisfactory division of reverse voltage among series-connected semiconductor rectifier diodes. Transformers, bleeder resistors, capacitors, or combinations of these may be employed. (IA) [62]

reverse wave *See:* reflected wave.

reversibility (Hall generator) The ratio of the change in absolute magnitude of the Hall voltage to the mean absolute magnitude of the Hall voltage, when the control current is kept constant and the magnetic field is changed from a given magnitude of one polarity to the same magnitude of the opposite polarity. (MAG) 296-1969w

reversible capacitance (nonlinear capacitor) The limit, as the amplitude of an applied sinusoidal capacitor voltage approaches zero, of the ratio of the amplitude of the resulting in-phase fundamental-frequency component of transferred charge to the amplitude of the applied voltage, for a given constant bias voltage superimposed on the sinusoidal voltage. *See also:* nonlinear capacitor. (ED) [46]

reversible-capacitance characteristic (nonlinear capacitor) The function relating the reversible capacitance to the bias voltage. *See also:* nonlinear capacitor. (ED) [46]

reversible counter A counter that can be incremented or decremented by a certain amount upon receipt of an appropriate signal. (C) 610.10-1994w

reversible dark current increase (diode-type camera tube) That increase of the target dark current which results from electron bombardment of the charge storage target by the scanning electron beam. This is manifested as a dark current increase which is reversible. (ED) 503-1978w

reversible execution A debugging technique in which a history of program execution is recorded and then replayed under the user's control, in either the forward or backward direction. *Synonyms:* reverse execution; playback; backward execution; replay. (C) 610.12-1990

reversible motor A motor whose direction of rotation can be selected by change in electric connections or by mechanical means but the motor will run in the selected direction only if it is at a standstill or rotating below a particular speed when the change is initiated. *See also:* asynchronous machine; direct-current commutating machine. (EEC/PE) [119]

reversible permeability The limit of the incremental permeability as the incremental change in magnetizing force approaches zero. *Note:* In anisotropic media, reversible permeability becomes a matrix. (Std100) 270-1966w

reversible permittivity The change in displacement per unit field when a very small relatively high-frequency alternating signal is applied to a ferroelectric at any point of a hysteresis loop. *See also:* ferroelectric domain. (UFFC) 180w

reversible potential *See:* equilibrium potential.

reversible power converter (1) An equipment containing thyristor converter assemblies connected in such a way that energy transfer is possible from the alternating-current side to the direct-current side and from the direct-current side to the alternating-current side with or without reversing the current in the direct-current circuit. *See also:* power rectifier. (IA/SPC) [62]
(2) (thyristor converter) A converter in which the transfer of energy is possible both from the ac side to the dc side and vice versa. (IA/IPC) 444-1973w

reversible process An electrochemical reaction that takes place reversibly at the equilibrium electrode potential. *See also:* electrochemistry. (EEC/PE) [119]

reversible target dark current increase (diode-type camera tube) That increase in dark current and dark current nonuniformity and monitoring and is not permanent. It is removable through target operation under special operating procedures or with a nonoperating rest period. (ED) 503-1978w

reversible turbine (power operations) A hydraulic turbine, normally installed in a pumped-storage station, that can be used alternately as a pump or as a prime mover. (PE/PSE) 858-1987s, 346-1973w

reversing The control function of changing motor rotation from one direction to the opposite direction. *See also:* electric drive. (IA/ICTL/IAC) [60]

reversing change-over selector A change-over selector that connects one or the other end of the tap winding to the main winding. (PE/TR) C57.131-1995

reversing device (power system device function numbers) A device that is used for the purpose of reversing a machine field or for performing any other reversing functions. (PE/SUB) C37.2-1979s

reversing motor One the torque and hence direction of rotation of which can be reversed by change in electric connections or by other means. These means may be initiated while the motor is running at full speed, upon which the motor will come to a stop, reverse, and attain full speed in the opposite direction. *See also:* asynchronous machine. (EEC/PE) [119]

reversing starter An electric controller for accelerating a motor from rest to normal speed in either direction of rotation. *See also:* starter; electric controller. (IA/ICTL/IAC) [60]

reversing switch A switch intended to reverse the connections of one part of a circuit. *See also:* switch. (IA/IAC) [60]

reverting call (telephone switching systems) A call between two stations on the same party line. (COM) 312-1977w

reverting-call tone (telephone switching systems) A tone that indicates to a calling customer that the called party is on the same line. (COM) 312-1977w

revertive pulsing (telephone switching systems) A means of pulsing for controlling distant selections whereby the near end receives signals from the far end. (COM) 312-1977w

review (1) (software) A process or meeting during which a work product, or set of work products, is presented to project personnel, managers, users, customers, or other interested parties for comment or approval. Types include code review, design review, formal qualification review, requirements review, test readiness review. (C/SE) 610.12-1990, 1058.1-1987s
(2) A process or meeting during which a software product is presented to project personnel, managers, users, customers, user representatives, or other interested parties for comment or approval. (C/SE) 1028-1997

revision A controlled item with the same functional capabilities as the original plus changes, error resolution, or enhancements. (C/SE) 1074-1995s

revolver track *See:* regenerative track.

rewind (1) (test, measurement, and diagnostic equipment) To return a tape to its beginning or a passed location. (MIL) [2]
(2) To bring a magnetic tape or paper tape back to the beginning of the recording area. (C) 610.10-1994w

rework (nuclear power quality assurance) The process by which an item is made to conform to original requirements by completion or correction. (PE/NP) [124]

rewrite (A) To write again. **(B)** In a destructive-read storage device, to return the data to the state it had prior to reading. *See also:* regenerate. (C) 162-1963, 610.10-1994

REXX A command language used primarily in the IBM VM/CMS environment. *Note:* Supersedes EXEC and EXEC2. (C) 610.13-1993w

Reynolds number A nondimensional number equal to air velocity (V_w) times conductor diameter ($D/_{12}$) divided by kinematic viscosity (μ_f/ρ_f). (T&D/PE) 738-1993

RF *See:* radio frequency.

RF link *See:* radio frequency link.

RH *See:* relative humidity.

RGB display device *See:* red, green, blue display device.

rheobase (medical electronics) The intensity of the steady cathodal current of adequate duration that when suddenly applied just suffices to excite a tissue. (EMB) [47]

rheostat (1) An adjustable resistor so constructed that its resistance may be changed without opening the circuit in which it may be connected. (Std100) 270-1966w
(2) (power system device function numbers) A variable resistance device used in an electric circuit, which is electrically operated or has other electrical accessories, such as auxiliary, position, or limit switches. (SUB/PE) C37.2-1979s

rheostatic brake A form of dynamic brake in which the electrical energy generated by braking is dissipated as heat in onboard resistors during the braking cycle. (VT) 1475-1999

rheostatic braking A form of dynamic braking in which electric energy generated by the traction motors is controlled and dissipated by means of a resistor whose value of resistance may be varied. *See also:* electric drive; dynamic braking. (IA/ICTL/IAC) [60]

rheostatic control (elevators) A system of control that is accomplished by varying resistance and.or reactance in the armature and.or field circuit of the driving-machine motor. *See also:* control. (EEC/PE) [119]

rheostatic-type regulator (rotating machinery) A regulator that accomplishes the regulating function by mechanically varying a resistance. *Note:* Historically, rheostatic-type regulators have been further defined as direct acting and indirect acting. An indirect-acting-type regulator is a rheostatic type that controls the excitation of the exciter by acting on an intermediate device which is not considered part of the regulator or exciter. A direct-acting-type regulator is a rheostatic type that directly controls the excitation of an exciter by varying the input to the exciter field circuit. (PE/EDPG) [9], 421-1972s

rheostat loss (synchronous machines) The I^2R loss in the rheostat controlling the field current. (REM) [115]

rheostriction *See:* pinch effect.

RHI *See:* range-height indicator.

rhombic antenna An antenna composed of long wire radiators arranged in such a manner that they form the sides of a rhombus. *Note:* The antenna usually is terminated in a resistance. The length of the sides of the rhombus, the angle between the sides, the elevation above ground, and the value of the termination resistance are proportioned to give the desired radiation properties. (AP/ANT) 145-1993

rho rho (navigation aids) A generic term referring to navigation systems based on the measurement of two distances for determination of position. (AES/GCS) 172-1983w

rho theta (navigation aids) A generic term referring to polar coordinate navigation systems for determination of position of a vehicle through measurement of distance and direction. (AES/GCS) 172-1983w

rhumbatron (electron tube) (microwave tubes) A resonator, usually in the form of a torus. *See also:* velocity-modulated tube. (ED) [45], [84]

rhumbline (navigation aids) A line on the surface of the earth making the same oblique angle with all meridians. (AES/GCS) 172-1983w

rhythmic light (illuminating engineering) A light which, when observed from a fixed point, has a luminous intensity which changes periodically. (EEC/IE) [126]

ribbon microphone A moving-conductor microphone in which the moving conductor is in the form of a ribbon that is directly driven by the sound waves. *See also:* microphone. (EEC/PE) [119]

ribbon transducer (mechanical recording) A moving-conductor transducer in which the movable conductor is in the form of a thin ribbon. (SP) [32]

Richardson-Dushmann equation (thermionics) An equation representing the saturation current of a metallic thermionic cathode in the saturation-current state:

$$J = A_0(1 - r)T^2 \exp\left(-\frac{b}{T}\right)$$

where
J = density of the saturation current
T = absolute temperature
A_0 = universal constant equal to 120 amperes per centimeter^2 kelvin^2
b = absolute temperature equivalent to the work function
r = reflection coefficient, which allows for the irregularities of the surface.

See also: electron emission; work function. (ED) [45], [84]

Richardson effect *See:* thermionic emission.

ride-through capability The ability of equipment to withstand momentary interruptions or sags. (T&D/PE) 1250-1995

ridged horn (antenna) A horn antenna in which the waveguide section is ridged. (AP/ANT) 145-1993

ridge-pin cover *See:* insulator cover.

rider structure *See:* crossing structure.

ridge waveguide A waveguide with interior projections extending along the length and in contact with the boundary wall. *See also:* waveguide. (AP/ANT) [35]

Rieke diagram (oscillator performance) A chart showing contours of constant power output and constant frequency drawn on a polar diagram whose coordinates represent the components of the complex reflection coefficient at the oscillator load. *See also:* load-impedance diagram; oscillatory circuit. (ED) 161-1971w, [45]

RIF *See:* radio-influence field; routing information field.

rigging An assembly of material used to manipulate or support various tools and equipment in both energized and de-energized line-work. (T&D/PE) 516-1995

right-hand polarization of a field vector *See:* sense of polarization.

right-hand polarization of a plane wave *See:* sense of polarization.

right-hand polarized wave A circularly or an elliptically polarized electromagnetic wave for which the electric field vector, when viewed with the wave approaching the observer, rotates counter-clockwise in space. *Notes:* 1. This definition is consistent with observing a clockwise rotation when the electric field vector is viewed in the direction of propagation. 2. A right-handed helical antenna radiates a right-hand polarized wave. (AP/PROP) 211-1997

right-hand rule Positive rotation is clockwise when viewed toward the positive direction along the axis of rotation. (DIS/C) 1278.1-1995

right justification In text formatting, justification of text such that the right margin is aligned. *Contrast:* ragged right margin. (C) 610.2-1987

right of access (nuclear power quality assurance) The right of a purchaser or designated representative to enter the premises of a supplier for the purpose of inspection, surveillance, or quality assurance audit. (PE/NP) [124]

right-threaded tree A threaded tree in which the right link field in each terminal node is made to point to its successors with respect to a particular order of traversal. *Contrast:* left-threaded tree. (C) 610.5-1990w

rigid-bus structure A bus structure comprised of rigid conductors supported by rigid insulators. (PE/SUB) 605-1998

rigid equipment Equipment, structures, and components whose lowest resonant frequency is greater than the cutoff frequency on the response spectrum. (PE/SUB/NP) 693-1997, 344-1987r

rigid metal conduit A raceway specially constructed for the purpose of the pulling in or the withdrawing of wires or cables after the conduit is in place and made of metal pipe of standard weight and thickness permitting the cutting of standard threads. *See also:* raceway. (EEC/PE) [119]

rigid tower A tower that depends only upon its own structural members to withstand the load that may be placed upon it. *See also:* angle tower; dead-end tower; flexible tower; tower. (T&D/PE) [10]

RII *See:* routing information indicator.

rim (rotating machinery) (spider rim) The outermost part of a spider. A rotating yoke. *See also:* rotor. (PE) [9]

ring (1) (plug) A ring-shaped contacting part, usually placed in back of the tip but insulated therefrom. (EEC/PE) [119]
(2) (data management) *See also:* circularly-linked list. (C) 610.5-1990w
(3) A signal transmitted on a telephone line to indicate an incoming call. (C) 610.7-1995

ring around (A) The undesired triggering of a transponder or repeater by its own transmitter. **(B)** The ring-type (constant-radius echo) presentation on a plan-position indicator (PPI) display that occurs from a very large radar cross section target when the radar has high azimuth sidelobes. (AES) 686-1997

ring array *See:* circular array.

ringback signal (telephone switching systems) A signal initiated by an operator at the called end of an established connection to recall the originating operator. (COM) 312-1977w

ringback tone (telephone switching systems) A tone that indicates to a caller that a ringing signal is being applied to a destination outlet. (COM) 312-1977w

ring circuit (waveguide practice) A hybrid T having the physical configuration of a ring with radial branches. *See also:* waveguide. (PE/EEC) [119]

ring counter A re-entrant multistable circuit consisting of any number of stages arranged in a circle so that a unique condition is present in one stage, and each input pulse causes this condition to transfer one unit around the circle. *See also:* trigger circuit. (EEC/PE) [119]

ringdown In telephone operation, a method of signaling to gain the attention of an operator. (C) 610.10-1994w

ringdown signaling (1) (telephone switching systems) A method of alerting an operator in which ringing is sent over the line to operate a device or circuit to produce a steady indication (normally a visual signal). (COM) 312-1977w
(2) (data transmission) The application of a signal to the line for the purpose of bringing in a line signal or supervisory signal at a switchboard or ringing a user's instrument. (Historically, this was a low frequency signal of about 20 Hz from the user on the line for calling the operator or for disconnect). (PE) 599-1985w

ringer *See:* telephone ringer.

ringer box *See:* bell box.

ring error monitor (REM) A function that collects ring error data from ring stations. The REM may log the received errors, or it may analyze this data and record statistics on the errors. (C/LM) 8802-5-1998

ring feeder *See:* loop-service feeder.

ring head (electroacoustics) A magnetic head in which the magnetic material forms an enclosure with one or more air gaps. The magnetic recording medium bridges one of these gaps and is in contact with or in close proximity to the pole pieces on one side only. (SP) [32]

ring heater, induction *See:* induction ring heater.

ring in A port that receives signals from the main ring path on the trunk cable and transmits signals to the backup path on the trunk cable, and provides connectivity to the immediate upstream ring out port. (C/LM) 8802-5-1998

ringing (1) (pulse transformers) (first transition ringing) (ARI) The maximum amount by which the instantaneous pulse value deviates from the straight-line segment fitted to the top of the pulse in determining AM in the pulse top region following rolloff, or overshoot, or both. It is expressed in amplitude units or as a percentage of AM. (PEL/ET) 390-1987r
(2) (data transmission) The production of an audible or visible signal at a station or switchboard by means of an alternating or pulsating current, or a damped oscillation occurring in the output signal of a system, as a result of a sudden change in input signal. (PE) 599-1985w
(3) (telephone switching systems) An alternating or pulsing current primarily intended to produce a signal at a station or switchboard. (COM) 312-1977w
(4) (facsimile) *See also:* facsimile transient. (IM/WM&A) 194-1977w
(5) (pulse terminology) A distortion in the form of a superimposed damped oscillatory waveform which, when present, usually follows a major transition. *See also:* preshoot. (IM/WM&A) 194-1977w

ringing cycle (telephone switching systems) A recurring sequence made up of ringing signals and the intervals between them. (COM) 312-1977w

ringing key A key whose operation sends ringing current over the circuit to which the key is connected. (EEC/PE) [119]

ring latency In a token ring, the time (measured in bit times) it takes for a signal to propagate once around the ring. The ring latency time includes the signal propagation delay through the ring medium plus the sum of the propagation delays through each station or other element in the data path connected to the token ring. (C/LM) 8802-5-1998

ringless-type meter socket A meter socket that has no provision for a socket sealing ring but has other means of holding a detachable watthour meter in place, such as a cover that is secured in place by a latch. (ELM) C12.7-1993

ringlet (1) The closed path formed by the connection that provides feedback from the output link of a node to its input link. This connection may include other nodes or switch elements. (C/MM) 1596-1992
(2) The concept of RamLink is based on a point-to-point connection of devices circularly connected, starting and ending at a controller, thus forming a ring. Rings are more effective when small, hence the diminutive *ringlet* is emphasized in this standard. (C/MM) 1596.4-1996

RingLink A physical signaling model, consisting of point-to-point connections between nodes in a ring, that supports the RamLink logical protocols. RingLink is optimized for robust longer-distance single-board as well as cross-board communications. (C/MM) 1596.4-1996

ring oscillator An arrangement of two or more pairs of tubes operating as push-pull oscillators around a ring, usually with alternate successive pairs of grids and plates connected to tank circuits. Adjacent tubes around the ring operate in phase opposition. The load is supplied by coupling to the plate circuits. *See also:* oscillatory circuit. (AP/BT/ANT) 145-1983s, 182-1961w

ring out A port that transmits the output signals to the main ring path on the trunk cable and receives from the backup ring path on the trunk cable, and provides connectivity to the immediate downstream ring in port. (C/LM) 8802-5-1998

ring parameter server (RPS) A function that is responsible for initializing a set of operational parameters in ring stations on a particular ring. (C/LM) 8802-5-1998

ring segment A section of transmission path bounded by repeaters or converters. Ring segment boundaries are critical for determining the transmission limits that apply to the devices within the segment. (C/LM) 8802-5-1998

ring shift *See:* circular shift.

ring time The time during which the indicated output of an echo box remains above a specified signal-to-noise ratio. The ring time is used in measuring the performance of radar equipment. (AES/RS) 686-1990

ring topology A topology in which stations are attached to repeaters in a ring fashion. *Note:* Every station has a predecessor and a successor for network transmissions. *Synonym:* loop topology. *See also:* star topology; bus-ring topology; star-ring topology; star-bus topology; tree topology; bus topology. (C) 610.7-1995

ring-type meter socket A meter socket that has a socket rim. (ELM) C12.7-1993

ring wave (100 kHz ring wave). An open-circuit voltage wave characterized by a rapid rise to a defined peak value, followed by a damped oscillation. (SPD/PE) C62.62-2000

R interface The interface provided at the R reference point to allow the connection of non-ISDN terminals using, for example, CCITT V-series, or CCITT X-series interfaces. (LM/C/COM) 8802-9-1996

riometer *See:* relative ionospheric opacity meter.

ripple (1) The alternating-current component from a direct-current power supply arising from sources within the power supply. *Notes:* 1. Unless specified separately, ripple includes unclassified noise. 2. In electrical-conversion technology, ripple is expressed in peak, peak-to-peak, root-mean-square volts, or as percent root-mean-square. 3. Unless otherwise specified, percent ripple is the ratio of the root-mean-square value of the ripple voltage to the absolute value of the total voltage, expressed in percent. *See also:* percent ripple. (Std100) 270-1966w
(2) (high voltage testing) Ripple is the periodic deviation from the arithmetic mean value of the voltage. The amplitude of the ripple is defined as half the difference between the maximum and minimum values. The ripple factor is the ratio of the ripple amplitude to the arithmetic mean value. (PE/PSIM) 4-1978s

ripple amplitude (1) The maximum value of the instantaneous difference between the average and instantaneous value of a pulsating unidirectional wave. *See also:* rectification; power rectifier. (IA/PEL/PCON/CEM/ET) [62], [58], 388-1992r
(2) The fine variations on a frequency plot of an impedance function or of a transfer function are called ripple. The ripple amplitude is the difference between the maximum and the minimum value of the function. (CAS) [13]

ripple content (converter characteristics) (self-commutated converters) The periodic ac (alternating current) function that may be superimposed on a steady zero-frequency (dc) (direct current) voltage or current. (IA/SPC) 936-1987w

ripple current The total harmonic current content superimposed on the dc current. For specific engineering purposes, it is essential to define the harmonic spectrum of the ripple current in terms of amplitude and frequency. For general purposes, the ripple current can be expressed as the root-mean-square (rms) value of the harmonic current at any level of dc current; including the continuous rated dc current. (PE/TR) 1277-2000

ripple factor The ratio of the ripple magnitude to the arithmetic mean value of the voltage. *See also:* radio receiver; interference; power pack. (PE/PSIM) 4-1978s, [55]

ripple filter A low-pass filter designed to reduce the ripple current, while freely passing the direct current, from a rectifier or generator. *See also:* filter. (PE) 599-1985w

ripple voltage (rectifier or generator) The alternating-voltage component of the unidirectional voltage from a direct-current power supply arising from sources within the power supply. *See also:* rectifier; interference. (AP/BT/ANT) 145-1983s, 182A-1964w

ripple voltage or current The alternating component whose instantaneous values are the difference between the average and instantaneous values of a pulsating unidirectional voltage or current. *See also:* rectification. (IA/EEC/PCON) 59-1962w, [110]

RISC *See:* reduced instruction set computer.

rise *See:* travel.

rise-and-fall pendant A pendant, the height of which can be regulated by means of a cord adjuster. (EEC/PE) [119]

riser cable (communication practice) The vertical portion of a house cable extending from one floor to another. In addition, the term is sometimes applied to other vertical sections of cable. *See also:* cable. (PE/EEC) [119]

riser pole type arrester An arrester for pole mounting most often used to protect underground distribution cable and equipment. (SPD/PE) C62.22-1997, C62.11-1999

rise time (1) (pulse transformers) (first transition duration) (t_r) The time interval of the leading edge between the instants at which the instantaneous value first reaches the specified lower and upper limits of 10% and 90% of A_M. Limits other than 10% and 90% may be specified in special cases. (PEL/ET) 390-1987r
(2) The time required for the output of a system (other than first-order) to make the change from a small specified percentage (often 5 or 10) of the steady-state increment to a large specified percentage (often 90 or 95), either before overshoot or in the absence of overshoot. *Note:* If the term is unqualified, response to a step change is understood: otherwise the pattern and magnitude of the stimulus should be specified. *See also:* feedback control system. (IA/ICTL/IAC) [60]
(3) (A) The time required for a voltage or current pulse to increase from 10% to 90% of its maximum value. *Contrast:* fall time. **(B)** In digital logic, the time required to transition from a low state to a high state. (C) 610.10-1994
(4) The interval between the instants at which the instantaneous value first reaches specified lower and upper limits, namely 10 and 90% of the peak pulse value. (NI/NPS) 309-1999

rise time, fall time (amplitude, frequency, and pulse modulation) The time for the light intensity to increase from the 10 to 90% intensity points. The fall time is the time for the light intensity to fall from the 90 to 10 % intensity points. (UFFC) [17]

rise time, pulse *See:* pulse rise time.

rise time tr (of a pulse). The interval on the first transition between the 10% and 90% points, with respect to peak height, unless other levels are specified. *See also:* transition. (NPS) 325-1996

rising edge (1) (test access port and boundary-scan architecture) A transition from a low to a high logic level. In positive logic, a change from logic 0 to logic 1. (TT/C) 1149.1-1990
(2) A transition from a logic zero to a logic one. (TT/C) 1149.5-1995

rising slope *See:* initial slope.

rising-sun magnetron A multicavity magnetron in which resonators of two different resonance frequencies are arranged alternately for the purpose of mode separation. *See also:* magnetron. (ED) 161-1971w

risk (1) (reliability analysis of nuclear power generating station safety systems) A measure of the probability and severity of undesired effects. Often taken as the simple product of probability and consequence. (PE/NP) 352-1987r
(2) (overhead power lines) A measure of the probability of experiencing harm from one or more hazards (e.g., accidents, toxic chemicals). (T&D/PE) 539-1990
(3) The potential for loss (such as the compromising of data confidentiality or data integrity or the denial of service to users) that could result from threats to the system, exploiting vulnerabilities in the system. *See also:* threat. (C/BA) 896.3-1993w
(4) A measure that combines both the likelihood that a system hazard will cause an accident and the severity of that accident. (C/SE) 1228-1994
(5) The combination of the probability of an abnormal event or failure and the consequence(s) of that event or failure to a system's components, operators, users, or environment. (DEI) 1221-1993w
(6) (nuclear power generating station) The expected detriment per unit time to a person or population from a given cause. (PE/NP) 933-1999, 577-1976r

risk analysis A procedure to develop probability estimates of occurrence of each specific hazard. (DEI) 1221-1993w

risk assessment The process and procedures of identifying, characterizing, quantifying, and evaluating risks and their significance. (DEI) 1221-1993w

risk management The activities associated with risk management preparation, risk assessment, risk handling option assessment, and risk control. (C/SE) 1220-1994s

RIV (radio influence voltage) *See:* radio-influence voltage.

RJ-11 A six-pin modular telephone plug. *Notes:* 1. Also called a permissive connection, an RJ-11 plug is generally used on two-wire circuits, but can be used on four-wire circuits. 2. This definition reflects colloquial usage. Standards referencing this term should point to the precise standardized connector specification. (C) 610.7-1995

RJ-45 A eight-pin modular telephone plug. *Notes:* 1. Also called a programmable connection, an RJ-45 plug is generally used on four-wire circuits, but can be used on eight-wire circuits. 2. This definition reflects colloquial usage. Standards referencing this term should point to the precise standardized connector specification. (C) 610.7-1995

RJE *See:* remote job entry.

RLC circuit *See:* simple series circuit.

rlogin *See:* remote login.

RMAC *See:* remote MAC.

RMI *See:* radio magnetic indicator.

r/min Revolutions per minute. (T&D/PE) 957-1987s

rms *See:* root-mean-square value.

rms (effective) burst magnitude *See:* root-mean-square (effective) burst magnitude.

rms detector *See:* root-mean-square detector.

rms deviation *See:* root-mean-square deviation.

rms field The horizontal component of the root-mean-square (rms) field strength in the far field of an array, scaled to an effective value at 1 km. *Synonym:* effective field. (T&D/PE) 1260-1996

rms (effective) pulse amplitude *See:* root-mean-square (effective) pulse amplitude.

rms pulse broadening *See:* root-mean-square pulse broadening.

rms pulse duration *See:* root-mean-square pulse duration.

rms reverse-voltage rating *See:* root-mean-square reverse-voltage rating.

rms ripple *See:* root-mean-square ripple.

rms sensing A term commonly used to indicate the sensing of root-mean-square (rms) value current rather than instantaneous or peak values, as by a circuit-breaker trip unit. (IA/PSP) 1015-1997

rms spectral width The optical wavelength range as measured by ANSI/EIA/TIA 455-127-1991 (FOTP-127). (C/LM) 802.3-1998

rms value *See:* root-mean-square value.

RMW cycle *See:* read-modify-write cycle.

roadband interference (measurement) A disturbance that has a spectral energy distribution sufficiently broad, so that the response of the measuring receiver in use does not vary significantly when tuned over a specified number of receiver bandwidths. *See also:* electromagnetic compatibility. (EMC) [53]

road map A high-level process outline. (C/PA) 1003.23-1998

roadside equipment (RSE) Equipment located at a fixed position along the road transport network, providing communication and data exchange with the onboard equipment (OBE). (SCC32) 1455-1999

roadway The portion of highway, including shoulders, for vehicular use. *Note:* A divided highway has two or more roadways. *See also:* shoulder; traveled way. (NESC/T&D) C2-1997, C2.2-1960

roadway element (track element) That portion of the roadway apparatus associated with automatic train stop, train control, or cab signal systems, such as a ramp, trip arm, magnet, inductor, or electric circuit, to which the locomotive apparatus is directly responsive. *See also:* automatic train control. (EEC/PE) [119]

robbed bit signaling A scheme in which the signaling bits for each channel are assigned to the least significant bit (bit 8) of frames 6 and 12 of superframe format, or frames 6, 12, 18, and 24 of extended superframe format. When a frame is used for signaling, bits 1−7 are used for channel transmission. (COM/TA) 1007-1991r

robot A mechanical device that can be programmed to perform some task of manipulation or locomotion under automatic control. (C) 610.10-1994w

robustness (1) (software) The degree to which a system or component can function correctly in the presence of invalid inputs or stressful environmental conditions. *See also:* fault tolerance; error tolerance. (C) 610.12-1990
(2) A statistical result that is not significantly affected by small changes in parameters, models, or assumptions. (PE/NP) 933-1999

ROC curves *See:* receiver operating characteristic curves.

rock-dust distributor *See:* rock duster.

rock duster (rock-dust distributor) A machine that distributes rock dust over the interior surfaces of a coal mine by means of air from a blower or pipe line or by means of a mechanical contrivance, to prevent coal dust explosions. (PE/EEC) [119]

rock socket A hole drilled in good rock for installing either expanding or grouted guy anchors. (T&D/PE) 751-1990

rodding a duct *See:* duct rodding.

rod, ground *See:* ground rod.

rods (illuminating engineering) Retinal receptors which respond at low levels of luminance even down below the threshold for cones. At these levels there is no basis for perceiving differences in hue and saturation. No rods are found in the center of the fovea. (EEC/IE) [126]

rod storage A type of storage consisting of wires, coated with a nickel-iron alloy, which are cut in such a way as to form stacks of rods. (C) 610.10-1994w

Roebel transposition (rotating machinery) An arrangement of strands occupying two heightwise tiers in a bar (half coil), wherein at regular intervals through the core length, one top strand and one bottom strand cross over to the other tier in such a way that each strand occupies every vertical position in each tier so as to equalize the voltage induced in each of the strands, thereby eliminating current that would otherwise circulate among the strands. Looking from one end of the slot, the strands are seen to progress in a clockwise direction through the core length through what may be interpreted as an angle of 360° so that the strands occupy the same position at both ends of the core. There are several variations of the Roebel transposition in use. In a bar having four tiers of copper, the two pairs of tiers would each have a Roebel transposition. The uninsulated bar, then, would be assembled as two Roebel-transposed bars, side-by-side. In order to transpose against voltages induced by end-winding flux, various modifications of the transposition in the slot, and extension of the Roebel transposition into the end winding have been used. *See also:* rotor; stator. (PE) [9]

roff A text-formatting language. (C) 610.13-1993w

rogue module An unauthorized module introduced into the system to perform malicious activities, or an authorized module corrupted by malicious hardware or software. (C/BA) 896.3-1993w

role The context in which an operation is executed. The utilities in this standard require the ability to perform operations on more than one system, perhaps by more than one person. These operations are separted into distinct roles including developer, packager, manager, source, target, and client. (C/PA) 1387.2-1995

role name (A) A name that more specifically names the nature of a related value class or state class. For a relationship, a role name is a name given to a class in a relationship to clarify the participation of that class in the relationship, i.e., connote the role played by a related instance. For an attribute, a role name is a name used to clarify the sense of the value class in the context of the class for which it is a property. **(B)** A name assigned to a foreign key attribute to represent the use of the foreign key in the entity. (C/SE) 1320.2-1998

roll angle *See:* roll attitude.

roll attitude (navigation aid terms) The angle between the horizontal and the lateral axis of the craft. *Synonym:* roll angle. *See also:* bank. (AES/GCS) 172-1983w

roll back (1) (telecommunications) The procedure by which a central processing unit recovers automatically from a fault that has led to a system malfunction. The complexity of the procedure, and the resulting temporary effect on the service of the system, depend on the nature of the fault. The procedure will usually involve the process of reinitialization. The time required to accomplish roll back is a measure of switching system performance. (COM/TA) 973-1990w
(2) (data management) Backward recovery of a database in which recently applied changes to the current version of a database are reversed. *Note:* A journal or checkpoint file is used to determine which changes must be reversed. *Synonym:* back out. *Contrast:* rollforward. (C) 610.5-1990w

roller *See:* sheave.

roller bearing (rotating machinery) A bearing incorporating a peripheral assembly of rollers. *See also:* bearing. (PE) [9]

roller, hold-down *See:* hold-down block.

roller, uplift *See:* uplift roller.

rollforward Forward recovery of a database in which all or part of a database is restored using data from a backup or snapshot of the database. Changes since the backup are reapplied to the database to restore it to some recently existing state. *Contrast:* roll back. (C) 610.5-1990w

roll in (1) (software) To transfer data or computer program segments from auxiliary storage to main storage. *Contrast:* roll out. *See also:* swap. (C) 610.12-1990
(2) To restore to main storage the sets of data that were previously rolled out. *Contrast:* roll out. (C) 610.10-1994w

rolling contacts A contact arrangement in which one cooperating member rolls on the other. *See also:* contactor. (IA/ICTL/IAC) [60], [84]

rolling interval An interval of time, the beginning of which progresses in steps of sub-intervals and where the interval length is equal to an integral multiple of sub-intervals. (ELM) C12.15-1990

rolling sphere method A simplified technique for applying the electrogeometric theory to the shielding of substations. The technique involves rolling an imaginary sphere of prescribed radius over the surface of a substation. The sphere rolls up and over (and is supported by) lightning masts, shield wires, fences, and other grounded metal objects intended for lightning shielding. A piece of equipment is protected from a direct stroke if it remains below the curved surface of the sphere by virtue of the sphere being elevated by shield wires or other devices. Equipment that touches the sphere or penetrates its surface is not protected. (SUB/PE) 998-1996

rolling transposition A transposition in which the conductors of an open wire circuit are physically rotated in a substantially helical manner. With two wires a complete transposition is usually executed in two consecutive spans. *See also:* open wiring. (EEC/PE) [119]

roll-in-jewel error Error caused by the pivot rolling up the side of the jewel and then falling to a lower position when tapped. This effect is not present when instruments are mounted with the axis of the moving element in a vertical position. (Roll-in-jewel error includes pivot-friction error that is small compared to the roll-in-jewel error.) (EEC/AII) [102]

rolloff (rounding after first transition), A_{RO} (pulse transformers) The amount by which the instantaneous pulse value is less than A_M at the point in time of the intersection of straight-line segments used to determine A_M. It is expressed in amplitude units or as a percentage of A_M. (PEL/ET) 390-1987r

roll out To transfer sets of data, such as files or computer programs of various sizes, from main storage to auxiliary storage for the purpose of freeing main storage for another use. *Contrast:* roll in. *See also:* swap. (C) 610.10-1994w, 610.12-1990

roll-out A movement process by which a snaphook or carabiner unintentionally disengages from another connector or object to which it is coupled. (T&D/PE) 1307-1996

roll over angle (conductor stringing equipment) For tangent stringing, the sum of the vertical angles between the conductor and the horizontal on both sides of the traveler. Resultants of these angles must be considered when stringing through line angles. Under some stringing conditions, such as stringing large diameter conductor, excessive roll over angles can cause premature failure of a conductor splice if it is allowed to pass over the travelers. (T&D/PE) 524-1980s

ROM (1) An abbreviation for read-only memory. The ROM data is maintained through losses of power. In some implementations ROM may actually be writeable, using (normally disabled) vendor-dependent protocols. (C/MM) 1212-1991s
(2) Read-only memory. (C/BA) 14536-1995
(3) *See also:* read-only memory. (C) 610.10-1994w

roof bushing A bushing intended primarily to carry a circuit through the roof, or other grounded barriers of a building, in a substantially vertical position. Both ends must be suitable for operating in air. At least one end must be suitable for outdoor operation. *See also:* bushing. 49-1948w

roof conductor The portion of the conductor above the eaves running along the ridge, parapet, or other portion of the roof. (EEC/PE) [119]

room air velocity The average sustained residual air velocity in the occupied area in the conditioned space. (IA/PSE) 241-1990r

room ambient temperature (electrical insulation tests) 20°C ± 5° (68°F ± 9°). (AES/ENSY) 135-1969w

room bonding point (health care facilities) A grounding terminal or group of terminals which serves as a collection point for grounding exposed metal or conductive building surfaces in a room. (NESC/NEC) [86]

room cavity ratio (illuminating engineering) For a cavity formed by a plane of the luminaires, the work-plane, and the wall surfaces between these two planes, the RCR is computed by using the distance from the work-plane to the plane of the luminaires (hr) as the cavity height in the equations given in the definition for **cavity ratio.** (EEC/IE) [126]

room coefficient, K† (illuminating engineering) A number computed from wall and floor areas. *Note:* The room coefficient is computed from

$$K_r = \frac{\text{height} \times (\text{length} + \text{width})}{2 \times \text{length} \times \text{width}}$$

(This term is retained for reference and literature searches). (EEC/IE) [126]

† Obsolete.

room index† (illuminating engineering) A letter designation for a range of room ratios. (This term is retained for reference and literature searches). (EEC/IE) [126]

† Obsolete.

room ratio† (illuminating engineering) A number indicating room proportions, calculated from the length, width, and ceiling height (or luminaire mounting height) above the work plane. It is used to simplify lighting design tables by expressing the equivalence of room shapes with respect to the utilization of direct or interreflected light. (This term is retained for reference and literature searches). (EEC/IE) [126]

† Obsolete.

room surface dirt depreciation (rsdd) (illuminating engineering) The fractional loss of task illuminance due to dirt on the room surface. (IE/EEC) [126]

room utilization factor (illuminating engineering) The ratio of the luminous flux (lumens) received on the work-plane to that emitted by the luminaire. *Note:* This ratio sometimes is called interflectance. Room utilization factor is based on the flux emitted by a complete luminaire, whereas coefficent of utilization is based on the rated flux generated by the lamps in a luminaire. (EEC/IE) [126]

Root An instance of a subclass of `IEEE1451_Root`. (IM/ST) 1451.1-1999

root *See:* root node.

root arrow segment The arrow segment of a junction from which other arrow segments branch or to which other arrow segments join. *Synonyms:* root; root segment. (C/SE) 1320.1-1998

root cause The underlying or physical cause of problem/failure. (PE/NP) 933-1999

root compiler (software) A compiler whose output is a machine independent, intermediate-level representation of a program. A root compiler, when combined with a code generator, comprises a full compiler. (C) 610.12-1990

root directory (1) A directory, associated with a process, that is used in pathname resolution for pathnames that begin with a slash. (C/PA) 9945-1-1996, 9945-2-1993, 1003.5-1999 **(2)** A region in read-only memory (ROM) specified in ISO/IEC 13213: 1994 whose size is identified in the first location of the root directory and whose contents include ROM entries that may be identified using the ROM key. (C/BA) 896.2-1991w, 896.10-1997

rooted tree *See:* tree.

root locus (1) (control system feedback) (for a closed loop whose characteristic equation is $KG(s)H(s) + 1 = 0$) A plot in the s plane of all those values of s that make $G(s)H(s)$ a negative real number: those points that make the loop transfer function $KG(s)H(s) = -1$ are roots. *Note:* The locus is conveniently sketched from the factored form of $KG(s)H(s)$: each branch starts at a pole of that function with $K = 0$. With increasing K, the locus proceeds along its several branches toward a zero of that function and, often asymptotic to one of several equiangular radial lines, toward infinity. Roots lie at points on the locus for which (1) the sum of the phase angles of component $G(s)H(s)$ vectors totals 180 degrees, and for which (2) $1/K = |G(s)H(s)|$. Critical damping of the closed loop occurs when the locus breaks away from the real axis: instability when it crosses the imaginary axis. *See also:* feedback control system. (PE/EDPG) [3] **(2) (excitation systems)** Consider a linear, stationary, system with closed loop transfer function C(S)/R(S) where R(S) is the Laplace Transform of the excitation (input) driving function of the closed loop system and C(S) is the Laplace Transform of the response (output) function of the closed loop system. When C(S)/R(S) is a function of the gain, K, of one element in either the forward or reverse signal path, the poles of C(S)/R(S) in the S-plane will in general be a function of K. A plot in the S-plane of the loci of poles of the closed loop transfer function as K varies is known as a root locus. (PE/EDPG) 421A-1978s

root-mean-square (rms) The effective value, or the value associated with joule heating, of a periodic electromagnetic wave. The rms value is obtained by taking the square root of the mean of the squared value of a function. (NIR) C95.1-1999

root-mean-square bandwidth The root-mean-square (rms) deviation of the power spectrum of the received signal relative to zero frequency or the spectral center, in units of r/s. This bandwidth, β, is defined as the square root of

$$\beta^2 = \frac{\int_{-\infty}^{\infty} [2\pi(f - f_0)^2] |S(f)|^2 df}{\int_{-\infty}^{\infty} |S(f)|^2 df}$$

where
$S(f)$ = the Fourier transform of the signal.
$s(t - \tau_0)$ with true time delay τ_0 and f_0 is the center frequency of the spectrum.

Note: β^2 is the normalized second moment of the spectrum $|S(f)|^2$ about the mean, and β is sometimes called effective bandwidth. (AES) 686-1997

root-mean-square (effective) burst magnitude (audio and electroacoustics) The square root of the average square of the instantaneous magnitude of the voltage or current taken over the burst duration. See the figure attached to the definition of burst duration. *Synonym:* rms (effective) burst magnitude. *See also:* burst.

root-mean-square detector A detector, the output voltage of which is the rms value of an applied signal or noise. *Note:* The instrument manufacturer must specify a "crest factor" to go along with the rms detector function. Typical crest factors on rms detectors are 20 dB to 26 dB, some are as high as 36 dB, and in rare cases an instrument may have a crest factor as high as 40 dB. *Synonym:* rms detector. (T&D/PE) 539-1990

root-mean-square deviation (fiber optics) A single quantity characterizing a function given, for $f(x)$, by

$$\sigma_{rms} = [1/M_0 \int_{-\infty}^{\infty} (x - M_1)^2 f(x) dx]^{1/2}$$

where

$$M_0 = \int_{-\infty}^{\infty} f(x)dx$$

$$M_1 = 1/M_0 \int_{-\infty}^{\infty} xf(x)dx$$

Note: The term rms deviation is also used in probability and statistics, where the normalization, M_0, is unity. Here, the term is used in a more general sense. *Synonym:* rms deviation. *See also:* spectral width; impulse response; root-mean-square pulse duration; root-mean-square pulse broadening.

root-mean-square (effective) pulse amplitude The square root of the average of the square of the instantaneous amplitude taken over the pulse duration. *Synonym:* rms (effective) pulse amplitude. (IM) 194-1977w

root-mean-square pulse broadening (fiber optics) The temporal rms deviation of the impulse response of a system. *Synonym:* rms pulse broadening. *See also:* root-mean-square pulse duration.

root-mean-square pulse duration (fiber optics) A special case of root-mean-square deviation where the independent variable is time and $f(t)$ is pulse waveform. *Synonym:* rms pulse duration. *See also:* root-mean-square deviation.

root-mean-square reverse-voltage rating (rectifier device) The maximum sinusoidal root-mean-square reverse voltage permitted by the manufacturer under stated conditions. *Synonym:* rms reverse-voltage rating. *See also:* average forward current rating. (IA) 59-1962w, [12]

root-mean-square ripple The effective value of the instantaneous difference between the average and instantaneous values of a pulsating unidirectional wave integrated over a complete cycle. *Note:* The root-mean-square ripple is expressed in percent or per unit referred to the average value of the wave. *Synonym:* rms ripple. *See also:* rectification. (IA/EEC/PCON) [62], [110]

root-mean-square value (1)

$$Y_{rms} = \left[\frac{1}{T}\int_a^{a+T} y^2 dt\right]^{1/2}$$

where y_{rms} is the root-mean-square (rms) value of y, y is an instantaneous value of a period function, a is any instant of time, and T is the period. The rms value of a periodic waveform may also be expressed as the square root of the sum of the squares of the Fourier components of y.

where A_1, A_2, A_n, are the rms values of the fundamental component, second harmonic, and nth harmonic, respectively. (PE/PSIM) 120-1989r

(2) (periodic function) (effective value) The square root of the average of the square of the value of the function taken throughout one period. Thus, if y is a periodic function of t

$$Y_{rms} = \left[\frac{1}{T}\int_a^{a+T} y^2 dt\right]^{1/2}$$

where Y_{rms} is the root-mean-square value of y, a is any value of time, and T is the period. If a periodic function is represented by a Fourier series, then:

$$= \frac{1}{(2)^{1/2}}\left(\frac{1}{2}A_0^2 + A_1^2 + A_2^2 + \ldots + B_n^2 + B_2^2 + \ldots\right)^{1/2}$$

$$= \frac{1}{(2)^{1/2}}\left(\frac{1}{2}A_0^2 + C_1^2 + C_2^2 + \ldots + C_n^2\right)^{1/2}$$

Note: The use of root-mean-square in terms of effective value is deprecated. (Std100) 270-1966w

(3) (high voltage testing) The root mean square value of an alternating voltage is the square root of the mean value of the square of the voltage values during a complete cycle. (PE/PSIM) 4-1978s

root-mean-square (rms) value of alternating voltage The square root of the mean value of the square of the voltage values during a complete cycle. (PE/PSIM) 4-1995

root node In a tree, the single node that is not a member of any subtree. *Note:* All other nodes are descendent nodes of the root node. *Synonym:* root. *See also:* nonterminal node; terminal node.

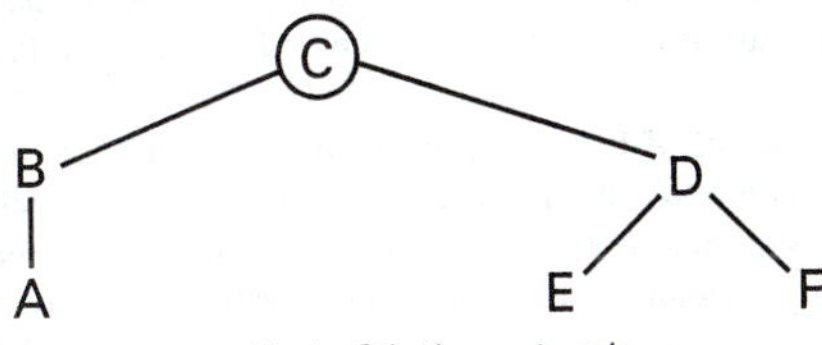

Node C is the root node

root node

(C) 610.5-1990w

(2) The device node that is the root of the device tree. (C/BA) 1275-1994

root repeater (local area networks) The level 1 (topmost) repeater in a cascade. (C) 8802-12-1998

root segment (1) A segment that is the root node in a database. *Contrast:* child segment; dependent segment. *See also:* path; parent; database record. (C) 610.5-1990w
(2) *See also:* root arrow segment. (C/SE) 1320.1-1998

root-sum-square The square root of the sum of the squares. *Note:* Commonly used to express the total harmonic distortion. *See also:* radio receiver. 188-1952w

root_directory A term used to describe the directory at the top level of the hierarchical ROM directory structure. (C/MM) 1212-1991s

rope block A device designed with one or more sheaves, a shell, and an attachment hook or shackle, commonly used in pairs with a rope reeved through the sheaves. The primary purpose of this device is to provide mechanical advantage so as to lift or move equipment *Synonym:* block and tackle. (T&D/PE) 516-1995

rope connector A special high strength steel link used to join two lengths of pulling rope by means of the eye splice at each end. Although designed to pass easily through the grooves of the bullwheels on the puller, it should not be passed under full load. *Synonym:* peanut. (T&D/PE) 524-1992r

roped-hydraulic driving machine (elevators) A machine in which the energy is applied by a piston, connected to the car with wire ropes, that operates in a cylinder under hydraulic pressure. It includes the cylinder, the piston, and the multiplying sheaves if any and their guides. *See also:* roped-hydraulic elevator; driving machine. (EEC/PE) [119]

roped-hydraulic elevator A hydraulic elevator having its piston connected to the car with wire ropes. *See also:* elevator; roped-hydraulic driving machine. (EEC/PE) [119]

rope grab A device which travels on a lifeline and automatically frictionally engages the lifeline and locks so as to arrest the fall of an worker. (T&D/PE) 1307-1996

rope ladder A ladder having vertical synthetic or manila suspension members and wood, fiberglass, or metal rungs. The ladder is suspended from the arm or bridge of a structure to enable workers to work at the conductor level, hang travelers, perform clipping-in operations, etc. *Synonym:* Jacob's ladder. (T&D/PE) 524-1992r

rope-lay conductor or cable A cable composed of a central core surrounded by one or more layers of helically laid groups of wires. *Note:* This kind of cable differs from a concentric-lay conductor in that the main strands are themselves stranded. In the most common type of rope-lay conductor or cable, all wires are of the same size and the central core is a concentric-lay conductor. *See also:* conductor. (T&D/PE) [10]

rosette An enclosure of porcelain or other insulating material, fitted with terminals and intended for connecting the flexible cord carrying apendant to the permanent wiring. *See also:* cabinet. (EEC/PE) [119]

rotary attenuator A variable attenuator in circular waveguide having absorbing vanes fixed diametrically across one section: the attenuation is varied by rotation of this section about the common axis. *See also:* waveguide. (AP/ANT) [35]

rotary converter A machine that combines both motor and generator action in one armature winding connected to both a commutator and slip rings, and is excited by one magnetic field. It is normally used to change alternating-current power to direct-current power. (PE) [9]

rotary dial (telephone switching systems) A type of calling device used in automatic switching that generates pulses by manual rotation and release of a dial, the number of pulses being determined by how far the dial is rotated before being released. (COM) 312-1977w

rotary generator (induction heating) An alternating-current generator adapted to be rotated by a motor or prime mover. (IA) 54-1955w, 169-1955w

rotary hunt An arrangement allowing calls placed to seek an idle circuit in a prearranged multichannel group. (C) 610.7-1995

rotary inverter A machine that combines both motor and generator action in one armature winding. It is excited by one magnetic field and changes direct-current power to alternating-current power. (Usually it has no amortisseur winding.) (PE) [9]

rotary joint (waveguide components) A coupling for efficient transmission of electromagnetic energy between two waveguide or transmission line structures designed to permit unlimited mechanical rotation of one structure. (MTT) 147-1979w

rotary phase changer (waveguide) A phase changer that alters the phase of a transmitted wave in proportion to the rotation of one of its waveguide sections. *Synonym:* rotary phase shifter. (AP/ANT) [35], [84]

rotary phase shifter *See:* rotary phase changer.

rotary relay (A) A relay whose armature moves in rotation to close the gap between two or more pole faces (usually with a balanced armature). **(B)** Sometimes used for stepping relay. *See also:* relay. (EEC/REE) [87]

rotary solenoid relay A relay in which the linear motion of the plunger is converted mechanically into rotary motion. *See also:* relay. (EEC/REE) [87]

rotary switch A bank-and-wiper switch whose wipers or brushes move only on the arc of a circle. *See also:* switch. (EEC/PE) [119]

rotary system An automatic telephone switching system that is generally characterized by the following features: The selecting mechanisms are rotary switches. The switching pulses are received and stored by controlling mechanisms that govern the subsequent operations necessary in establishing a telephone connection. (EEC/PE) [119]

rotatable frame (rotating machinery) A stator frame that can be rotated by a limited amount about the axis of the machine shaft. *See also:* stator. (PE) [9]

rotatable phase-adjusting transformer (phase-shifting transformer) A transformer in which the secondary voltage may be adjusted to have any desired phase relation with the primary voltage by mechanically orienting the secondary winding with respect to the primary. The primary winding of such a transformer usually consists of a distributed symmetrical polyphase winding and is energized from a polyphase circuit. *See also:* auxiliary device to an instrument. (PE/EEC) [119]

rotate *See:* circular shift.

rotating amplifier (excitation systems for synchronous machines) An electric machine in which a small energy change in the field is amplified to a large energy change at the armature terminals. (PE/EDPG) 421.1-1986r

rotating-anode tube (x-ray) An x-ray tube in which the anode rotates. *Note:* The rotation continually brings a fresh area of its surface into the beam of electrons, allowing greater output without melting the target. (ED) [45]

rotating control assembly (rotating machinery) The complete control circuits for a brushless exciter mounted to permit rotation. *See also:* rotor. (PE) [9]

rotating field A variable vector field that appears to rotate with time. (Std100) 270-1966w

rotating-insulator switch A switch in which the opening and closing travel of the blade is accomplished by the rotation of one or more insulators supporting the conducting parts of the switch. (SWG/PE) C37.100-1992

rotating joint (waveguides) A coupling for transmission of electromagnetic energy between two waveguide structures designed to permit mechanical rotation of one structure. *See also:* waveguide.

rotating machinery *See:* electric machine.

rotational position sensing The process of locating a given sector, a desired track, and a specific record by continually comparing the read/write head position with appropriate synchronization signals. (C) 610.10-1994w

rotating storage device A storage device that employs a circular medium that must be rotated in order to access the data. (C) 610.10-1994w

rotation The displacement of one or more display elements about an axis, changing its angular orientation. (C) 610.6-1991w

rotational delay (A) The delay caused by waiting for the read/write head of a rotating storage device such as a disk drive to be positioned over the appropriate storage location on the disk. *Synonym:* latency. *See also:* search time. **(B)** The part of access time that is attributed to the delay as in definition (A). (C) 610.10-1994

rotation plate (rotating machinery) A plaque showing the proper direction of rotor rotation. *See also:* rotor. (PE) [9]

rotation test (rotating machinery) A test to determine that the rotor rotates in the specified direction when the voltage applied agrees with the terminal markings. *See also:* asynchronous machine. (PE) [9]

rotor (1) (watthour meter) That part of the meter that is directly driven by electromagnetic action. (ELM) C12.1-1982s

(2) (rotating machinery) The rotating member of a machine, with shaft. *Note:* In a direct-current machine with stationary field poles, universal, alternating-current series, and repulsion-type motors, it is commonly called the armature. (PE) [9]

rotor angular momentum (gyros) The product of spin angular velocity and rotor moment of inertia about its spin axis. (AES/GYAC) 528-1994

rotor bar (rotating machinery) A solid conductor that constitutes an element of the slot section of a squirrel-cage winding. *See also:* rotor. (PE) [9]

rotor bushing (rotating machinery) A ventilated or nonventilated piece or assembly used for mounting onto a shaft, an assembled rotor core whose inside opening is larger than the shaft. *See also:* rotor. (PE) [9]

rotor coil (rotating machinery) A unit of a rotor winding of a machine. *See also:* rotor. (PE) [9]

rotor core (rotating machinery) That part of the magnetic circuit that is integral with, or mounted on, the rotor shaft. It frequently consists of an assembly of laminations. (PE) [9]

rotor-core assembly (rotating machinery) The rotor core with a squirrel-cage or insulated-conductor winding, put together as an assembly. *See also:* rotor. (PE) [9]

rotor core lamination (rotating machinery) A sheet of magnetic material, containing teeth, slots, or other perforations dictated by design, which forms the rotor core when assembled with other identical or similar laminations. (PE) [9]

rotor displacement angle (rotating machinery) (load angle) The displacement caused by load between the terminal voltage and the armature voltage generated by that component of flux produced by the field current. *See also:* rotor. (PE) [9]

rotor end plate (rotating machinery) An annular disk (ring) fitted at the outer end of the retaining ring. (PE) [9]

rotor end ring (rotating machinery) The conducting structure of a squirrel-cage or amortisseur (damper) winding that short-circuits all of the rotor bars at one end. *See also:* rotor. (PE) [9]

rotor moment-of-inertia (gyros) The moment of inertia of a gyro rotor about its spin axis. (AES/GYAC) 528-1994

rotor-resistance starting (rotating machinery) The process of starting a wound-rotor induction motor by connecting the rotor initially in series with starting resistors that are short-circuited for the running operation. *See also:* asynchronous machine. (PE) [9]

rotor rotation detector (gyros) A device that produces a signal output as a function of the speed of the rotor. (AES/GYAC) 528-1994

rotor slot armor (rotating machinery) (cylindrical-rotor synchronous machine) Main ground insulation surrounding the slot or core portions of a field coil assembled on a slotted rotor. *See also:* rotor. (PE) [9]

rotor-speed sensitivity (dynamically tuned gyro) The change in in-phase spring rate due to a change in gyro rotor speed. (AES/GYAC) 528-1994

rotor spider *See:* spider.

rotor winding (rotating machinery) A winding on the rotor of a machine. *See also:* rotor. (PE) [9]

roughness (navigational system display) (navigation aids) Irregularities resembluing scalloping, but distinguished by their random, noncyclic nature. *Synonym:* course roughness. (AES/GCS) 172-1983w

rough surface An irregular surface separating two media. *See also:* Rayleigh criterion. (AP/PROP) 211-1997

round To delete or omit one or more of the least significant digits in a representation of a number and to adjust the remaining digits according to some specified rule. *Contrast:* truncate. *See also:* round up; round off; round down. (C) 610.5-1990w, 1084-1986w

round conductor Either a solid or stranded conductor of which the cross section is substantially circular. *See also:* conductor. (T&D/PE) [10]

round down To round a number, making no adjustment to the numeral that is retained. For example, the decimal numeral 5.6789, when rounded down to two decimal places, becomes 5.67. *Synonym:* truncate. (C) 610.5-1990w, 1084-1986w

rounding *See:* pulse distortion.

rounding error (1) (test, measurement, and diagnostic equipment) The error resulting from deleting the less significant digits of a quantity and applying some rule of correction to the part retained. A common round-off rule is to take the quantity to the nearest digit. Thus, the value of π, 3.14159265. . ., rounded to four decimals is 3.1416. (MIL) [2]

(2) (mathematics) The error introduced by rounding a number. *Synonym:* round-off error. (C) 1084-1986w

round off (1) To delete the least-significant digit or digits of a numeral and to adjust the part retained in accordance with some rule. (MIL/C) [2], [85], [20]

(2) (A) (data management) To round, adjusting the part of the numeral that is retained by rounding down any digit less than 5, rounding up any digit greater than 5, and rounding 5 up or down to the even digit. For example, 5.5 would be rounded off to 6, and 4.5 rounds off to 4. **(B) (data management)** To round, adjusting the part of the numeral that is retained by rounding down any digit less than 5, rounding up any digit equal or greater than 5. For example, 5.5 rounds off to 6, 4.5 rounds off to 5. (C) 610.5-1990

(3) (mathematics of computing) *See also:* round. (C) 1084-1986w

round-off error *See:* rounding error.

round-power test *See:* circulating-power test.

round robin (1) A bus allocation rule whereby, after a module acquires and uses the bus, it will not be granted use of the bus again until all other modules currently requesting the bus at the same priority level have had a chance to use the bus. (C/BA) 1014.1-1994w

(2) A bus allocation rule where, after a module acquires and uses the bus, it will not be granted use of the bus again until all other modules currently requesting the bus at the same priority level have had control of the bus. (C/BA) 10857-1994, 896.3-1993w, 896.4-1993w

round rotor (rotating machinery) (cyclindrical rotor) A rotor of cylindrical shape in which the coil sides of the winding are contained in axial slots. *See also:* rotor. (PE) [9]

round-trip delay The sum of the absolute delays on an outgoing path and return path. Different methods of measuring round-trip delay may produce somewhat different results. (COM/TA) 743-1995

round-trip envelope delay The sum of the outgoing envelope delay and return path envelope delay, where provision has been made at the far end of the circuit to either loopback or remodulate the envelope delay test signal back to the transmitting measuring set. (COM/TA) 743-1995

round trip time The total time taken for a single packet or datagram to leave one device, reach the other, and return. (C) 610.7-1995, 610.10-1994w

round up To round a number, adjusting the numeral that is retained by adding 1 to its least significant digit and executing any carries required. For example, the decimal numeral 5.6789 when rounded up to two decimal places becomes 5.68. (C) 610.5-1990w, 1084-1986w

route (1) (telephone switching systems) A particular order of a set of switching entities through which call connections may be established. (COM) 312-1977w

(2) Denotes the information employed to generate routing information. It becomes a routing_information parameter when placed in the MAC primitive. The route explicitly describes the path a frame takes through a bridged network. (C/LM/CC) 8802-2-1998

route locking Locking effective when a train passes a signal and adapted to prevent manipulation of levers that would endanger the train while it is within the limits of the route entered. It may be so arranged that a train in clearing each section of the route releases the locking affecting that section. *See also:* interlocking. (EEC/PE) [119]

route query (RQ) An RDE PDU used to explore possible paths between two stations developing a data link. The route query consists of a command PDU (RQC) and a response PDU (RQR). (C/LM/CC) 8802-2-1998

router (1) A functional unit that interconnects two computer networks that use a single Network Layer procedure but may use different Data Link Layer and Physical Layer procedures. (LM/C) 8802-6-1994

(2) In networking, a device that interconnects two networks using the network layer (layer 3) address. *Note:* Routers are protocol dependent because they must be able to identify the address field within a specific network layer protocol. *See also:* hub; gateway; bridge. (C) 610.7-1995

(3) A layer 3 interconnection device that appears as a Media Access Control (MAC) to a CSMA/CD collision domain. (C/LM) 802.3-1998

route selected (RS) An RDE PDU used to announce the selection of a path between two stations developing a data link. (C/LM/CC) 8802-2-1998

route table The list of group addresses recognized by an SI for passing operations to its far-side segment. (NID) 960-1993

route tracing mode A mode of SI operation that generates an error diagnostic response instead of the normal passing of an operation. (NID) 960-1993

routine (software) A subprogram that is called by other programs and subprograms. *Note:* The terms "routine," "subprogram," and "subroutine" are defined and used differently in different programming languages; the preceding definition is advanced as a proposed standard. *See also:* subroutine; coroutine. (C) 610.12-1990

routine entry point (1) (computers) Any place to which control can be passed. (C) [20], [85]

(2) (test, measurement, and diagnostic equipment) One of a set of points in an automatic test equipment program where the test conditions are completely stated and are not dependent on previous tests or setups in any way. Such points are the only ones at which it is permissible to begin part of the complete test program. *See also:* rerun point. (MIL) [2]

(3) (software) A point in a software module at which execution of the module can begin. *Synonyms:* entrance; entry. *Contrast:* exit. *See also:* reentry point. (C) 610.12-1990

routine measurements Radioassays performed on samples by established, validated, verified, and controlled procedures. (NI) N42.23-1995

routine test (1) A test that is carried out by the manufacturer of the heating cable on all cables during or after the production process. (BT/IA/AV/PC) 152-1953s, 515.1-1995

(2) (rotating electric machinery) A test showing that each machine has been run and found to be sound electrically and mechanically, and is essentially identical with those that have been type tested. (PE/EM) 11-1980r

(3) A test made on each completed LTC to establish that the LTC is without manufacturing defects, with the design having been verified by a design test. (PE/TR) C57.131-1995

(4) A test that is carried out by the manufacturer prior to shipment to verify conformance to the manufacturer's specifications. (IA) 515-1997

routine tests (1) (surge arresters) (metal-oxide surge arresters for ac power circuits) Tests made for quality control by the manufacturer on every device or representative samples, or on parts or materials as required to verify during production that the product meets the design specifications. (PE/TR) [57], C57.12.80-1978r, 270-1966w

(2) (rotating machinery) The tests applied to a machine to show that it has been constructed and assembled correctly, is able to withstand the appropriate high-voltage tests, is in

sound working order both electrically and mechanically, and has the proper electrical characteristics. *See also:* asynchronous machine. (PE) [9]
(3) Tests made by the manufacturer on every device or representative samples, or on parts or materials, as required, to verify that the product meets the design specifications. (SPD/PE) C62.11-1999, C62.62-2000
(4) (switchgear) *See also:* production tests. (SWG/PE) C37.100-1992

routing (A) In data communications, a path by which a message reaches its destination. **(B)** A path that network traffic takes from its source to its destination. *See also:* static routing; adaptive routing; stochastic routing; fixed routing. (C) 610.7-1995

routing code (telephone switching systems) A digit or combination of digits used to direct a call towards its destination. (COM) 312-1977w

routing indicator A coded indicator preceding a message showing the transmission routing of the message. (C) 610.7-1995

routing information (1) A field, carried in a frame, used by source routing transparent bridges that provides source routing operation in a bridged LAN. (C/LM) 8802-5-1998
(2) The data that explicitly describes the route a frame takes through a bridged network. The routing_information parameter is included in the MA_UNITDATA request and MA_UNITDATA indication MAC primitives. (C/LM/CC) 8802-2-1998

routing information field (RIF) Denotes the routing information field of the source-routed frame format. (C/LM/CC) 8802-2-1998

routing information indicator (RII) An indication that the frame format contains a routing information field (RIF). (C/LM/CC) 8802-2-1998

routing function Inside a switch, this is function which determines to which numbered node interface a packet is to be sent, based on the information contained in the packet destination. *See also:* switch. (C/BA) 1355-1995

routing pattern (telephone switching systems) The implementation of a routing plan with reference to an individual automatic exchange. (COM) 312-1977w

routing plan (telephone switching systems) A plan for directing calls through a configuration of switching entities. (COM) 312-1977w

roving (rotating machinery) A loose assemblage of fibers drawn or rubbed into a single strand with very little twist. In spun yarn systems, the product of the stage or stages just prior to spinning. (PE) [9]

row (1) (metal nitrite oxide semiconductor arrays) A group of memory cells having a common internal address line. (ED) 641-1987w
(2) (test pattern language) A group of words or bits in a memory, identified by a common X-address. (TT/C) 660-1986w
(3) (data management) A horizontally corresponding set of entries in a table. *Contrast:* column. *See also:* tuple. (C) 610.5-1990w
(4) A horizontal arrangement of characters or other expressions. *See also:* card row; tape row. (C) 610.10-1994w
(5) In a Physical Design Exchange Format (PDEF) datapath cluster, a cluster of cell, spare_cell, and/ or cluster instances placed or contstrained to be placed in the vertical (*Y*-axis) direction. *See also:* column; datapath. (C/DA) 1481-1999

row arrangement Circuit-breaker pole units that are installed in a consecutive mode, thus physically forming a continuous line. The natural expansion of the substation would normally continue in the direction of the row. Arrangements can have two, three, four, or more rows in parallel configuration. (SWG/PE/SUB) C37.100-1992, C37.122-1983s

row binary (1) Pertaining to the binary representation of data on punched cards in which adjacent positions in a row correspond to adjacent bits of data, for example, each row in an 80-column card may be used to represent 80 consecutive bits of two 40-bit words. (C) [20], [85]
(2) (mathematics of computing) Pertaining to the binary representation of data in which adjacent positions in a row correspond to adjacent binary digits. For example, each row in an 80-column card may be used to represent 80 consecutive bits of a binary word. *Contrast:* column binary. *See also:* binary card. (C) 610.10-1994w, 1084-1986w

row enable (semiconductor memory) The input used to strobe in the row address in multiplexed address random access memories (RAM). (TT/C) 662-1980s

row-major order A method for storing the elements of a matrix in computer memory, in which the elements are ordered in a row-by-row manner—that is, all elements of row 1, followed by all elements of row 2, etc. *Contrast:* column-major order. (C) 610.5-1990w

row pitch The distance between corresponding points of adjacent rows measured along a track. (C) 610.10-1994w

row select gate Sometimes called the select gate. The transistor, controlled by the word-line, that isolates the memory transistor from the bit-line so that individual bytes may be altered. (ED) 1005-1998

row select line The line, determined by the row addresses (output of the X decoder), that is used to access the appropriate rows and, when present, the byte select transistors during a read or write. (ED) 1005-1998

row select transistor Sometimes called the select transistor or select gate. The transistor, controlled by the row select line, that isolates the memory transistor from the bit-line so that individual bits are isolated from voltages on the bit-line. (ED) 1005-1998

row vector A matrix with only one row. That is, a matrix of size 1-by-*n*. *Contrast:* column vector. (C) 610.5-1990w

RPE *See:* radial probable error; circular probable error.

RPROM *See:* reprogrammable read-only memory; reprogrammable memory, programmable read-only memory.

RPS *See:* ring parameter server.

RQ *See:* route query.

rrrv *See:* rate-of-rise of restriking voltage.

***r* register** One of the integer registers. (C/MM) 1754-1994

RRS *See:* required response spectrum.

RS (cable systems in power generating stations) Rigid steel conduit. (PE/EDPG) 422-1977

RS-232 *See:* EIA/TIA-232-E.

RS-232-C (1) An EIA/TIA standard, officially known and published as EIA/TIA-232-E. *See also:* EIA/TIA-232-E. (C) 610.7-1995
(2) An EIA standard for asynchronous serial data communications between terminal devices, such as printers; computers; and communications equipment, such as modems. *Note:* This standard defines a 25-pin connector and certain signal characteristics for interfacing computer equipment. Also known as EIA 232-D. (C) 610.10-1994w

RS-422-A (1) An EIA standard, officially known and published as EIA-422-A. *See also:* EIA-422-A. (C) 610.7-1995
(2) An EIA standard that specifies electrical characteristics for balanced transmission in which each of the main circuits has its own ground lead. Also known as EIA 422-A. (C) 610.10-1994w

RS-423-A (1) An EIA standard, officially known and published as EIA-423A. *See also:* EIA-423-A. (C) 610.7-1995
(2) An EIA standard that specifies electrical characteristics for unbalanced circuits using common or shared grounding techniques. Also known as EIA 423-A. (C) 610.10-1994w

RS-449 (1) An EIA standard that has been rescinded. This standard has been replaced by EIA/TIA-530-A. *See also:* EIA/TIA-530-A. (C) 610.7-1995
(2) An EIA standard that specifies cabling and connectors for RS-422-A and RS-423-A interfaces. Where RS-232-C was all inclusive, RS-449 is the equivalent connector and cabling specification. It references, in turn, RS-422-A and RS-423-A

to specify voltage levels. *Note:* Within RS-449, control signals are typically transmitted at RS-423-A levels, and clocks and data at RS-422-A. At lower speeds, RS-423-A may be substituted for RS-422-A for the clocks and data. (C) 610.10-1994w

RSAP *See:* remote SAP.

R scan *See:* R-display.

R-scope A cathode-ray oscilloscope arranged to present an R-display. (AES/RS) 686-1990

R-S flip-flop A flip-flop that has two level-sensitive data inputs; R and S. *Note:* The R input is used to make the output a logical zero (false) and the S input is used to make the output a logical one (true). (C) 610.10-1994w

rs1, rs2, rd The register operands of an instruction. *rs1* and *rs2* are the source registers; *rd* is the destination register. (C/MM) 1754-1994

R_{sig} Signal etch resistance. (C/BA) 896.2-1991w

R_{term} Terminator resistance (shall be placed on the backplane). (C/BA) 896.2-1991w

RTOK *See:* Re Test Okay.

RTL *See:* resistor-transistor logic; register transfer language.

RTT (Round Trip Time) *See:* round trip time.

rubber banding A technique that consists of the real-time display of a line with one endpoint fixed and the location of the other endpoint being controlled by a graphical input device. The displayed line will "stretch" to maintain its connections. (C) 610.6-1991w

rubber tape A tape composed of rubber or rubberlike compounds that provides insulation for joints. (EEC/PE) [119]

rub-out character *See:* delete character.

rudder-angle-indicator system A system consisting of an indicator (usually in the wheel house) so controlled by a transmitter connected to the rudder stock as to show continually the angle of the rudder relative to the center line of the ship. (EEC/PE) [119]

rule **(A)** A series of steps or activities with a single known or anticipated result. **(B)** A guideline for acting or planning action. (PE/NP) 1082-1997

Rule and Constraint Language A declarative specification language that is used to express the realization of responsibilities and to state queries. (C/SE) 1320.2-1998

rule-based language A nonprocedural language that permits the user to state a set of rules and to express queries or problems that use these rules. *See also:* declarative language; interactive language; command language. (C) 610.12-1990, 610.13-1993w

ruling span A calculated span length that will have the same changes in conductor tension due to changes of temperature and conductor loading as will be found in a series of spans of varying lengths between deadends. (T&D/PE) 524-1992r

rumble (electroacoustics) Low-frequency vibration of the recording or reproducing drive mechanism superimposed on the reproduced signal. *See also:* phonograph pickup. (SP) [32]

rumble, turntable *See:* turntable rumble.

run **(1)** **(A)** **(software)** In software engineering, a single, usually continuous, execution of a computer program. *See also:* run time. **(B)** **(software)** To execute a computer program. (C) 610.12-1990

(2) **(image processing and pattern recognition)** In image processing, a sequence of consecutive pixels that all have the same gray level. (C) 610.4-1990w

(3) **(data management)** In sorting, two or more successive items in a set that are in the proper order according to the specified sorting criteria. (C) 610.5-1990w

(4) **(computers)** A single, continuous performance of a computer routine. (MIL/C) [2], [85], [20]

(5) A single and continuous execution of a program by a computer. (C) 610.10-1994w

runaway pipeline temperature (electrical heating systems) The highest equilibrium temperature on the pipeline or the vessel that can occur when the heating system is continuously energized in the maximum ambient temperature. *Synonym:* runaway vessel temperature. (IA/PC) 844-1991

runaway pipe temperature (electrical heat tracing for industrial applications) The highest equilibrium pipe temperature that occurs when the heating cable is continuously energized at the maximum ambient temperature. (IA/BT/AV) 515-1997, 152-1953s

runaway speed The maximum speed obtained when a turbine-generator is operated unloaded with wicket gates fully open at maximum head. (PE/EDPG) 1020-1988r

runaway vessel temperature *See:* runaway pipe temperature.

runback The control of a dc system to reduce power to match loss of generation on the ac network. (PE/SUB) 1378-1997

RUNBIST *See:* run built-in self-test.

run built-in self-test (RUNBIST) A defined instruction for the test logic defined by IEEE Std 1149.1-1990. (TT/C) 1149.1-1990

run-down time (gyros) The time interval required for the gyro rotor to reach a specified speed, or during which the gyro exhibits specified performance, after removal of rotor excitation at a specified speed. (AES/GYAC) 528-1994

run/halt switch (RH) A switch normally operated by the run/halt switch activator bar on crate segments and on the ATC on cable segments which stops bus traffic so that it may be possible to insert or remove modules without affecting other modules on the segment. (NID) 960-1993

run length **(1)** The number of pixels in a run. (C) 610.4-1990w

(2) The maximum number of successive bits of the same value which can occur in the coded bit stream. (C/BA) 1355-1995

(3) The number of consecutive identical bits in a code-group. For example, the pattern 0011111010 has a run length of five. (C/LM) 802.3-1998

run length encoding An image compression technique in which the rows of an image are represented as sequences of runs, each with a given run length and gray level. (C) 610.4-1990w

run-length-limited code Any transmission code that has limited run-length for its transmission. (C/LM) 802.3-1998

runnable process A process that is capable of being a running process, but for which no processor is available. (C/PA) 1003.1b-1993s

runnable thread A thread that is capable of being a running thread, but for which no processor is available. (C/PA) 9945-1-1996

runner The rotating element of a turbine, which converts hydraulic energy into mechanical energy. (PE/EDPG) 1020-1988r

running board (conductor stringing equipment) A pulling device designed to permit stringing more than one conductor simultaneously with a single pulling line. For distribution stringing, it is usually made of lightweight tubing with the forward end curved gently upward to provide smooth transition over pole crossarm rollers. For transmission stringing, the device is either made of sections hinged transversely to the direction of pull or of a hard nose rigid design, both having a flexible pendulum tail suspended from the rear. This configuration stops the conductors from twisting together and permits smooth transition over the sheaves of bundle travelers. *Synonyms:* sled; bird; monkey tail; birdie; alligator. (T&D/PE) 524-1992r

running circuit breaker (power system device function numbers) A device whose principal function is to connect a machine to its source or running or operating voltage. This function may also be used for a device, such as a contactor, that is used in series with a circuit breaker or other fault protecting means, primarily for frequent opening and closing of the circuit. (SUB/PE) C37.2-1979s

running disparity **(1)** The cumulative sum of the disparities of characters transmitted from the start of operation of the link up to the present time. A link has two running disparities, one for each direction. (C/BA) 1355-1995
(2) A binary parameter having a value of + or −, representing the imbalance between the number of ones and zeros in a sequence of 8B/10B code-groups. (C/LM) 802.3-1998

running footer In text formatting, a line of text that is automatically placed at the bottom of each page of a document. *Synonym:* footer. *Contrast:* running header. (C) 610.2-1987

running ground (conductor stringing equipment) A portable device designed to connect a moving conductor or wire rope, or both, to an electrical ground. These devices are normally placed on the conductor or wire rope adjacent to the pulling and tensioning equipment located at either end of a sag section. It is primarily used to provide safety for personnel during construction or reconstruction operations. *Synonym:* ground roller. (T&D/PE) 524a-1993r, 524-1992r, 1048-1990

running header In text formatting, a line of text that is automatically placed at the top of each page of a document. *Synonym:* header. *Contrast:* running footer. (C) 610.2-1987

running-light-indicator panel (telltale) A panel in the wheelhouse providing audible and visible indication of the failure of any running light connected thereto. (EEC/PE) [119]

running lights Lanterns constructed and located as required by navigation laws, to permit the heading and approximate course of a vessel to be determined by an observer on a nearby vessel. *Note:* Usual running lights are port side, starboard side, mast-head, range, and stern lights. (EEC/PE) [119]

running open-phase protection The effect of a device operative on the loss of current in one phase of a polyphase circuit to cause and maintain the interruption of power in the circuit. (IA/ICTL/IAC) [60]

running operation (A) (single-phase motor) (for a motor employing a starting switch or relay) Operation at speeds above that corresponding to the switching operation. **(B) (single-phase motor)** (for a motor not employing a starting switch or relay) Operation in the range of speed that includes breakdown-torque speed and above. *See also:* asynchronous machine. (PE) [9]

running process A process currently executing on a processor. There may be more than one such process in a system at a time in a system with multiple processors. (C/PA) 1003.1b-1993s

running state A node state that is reflected by the value of 0 in the STATE_CLEAR.state field. The running state is the normal operational state in which access to all of the node's CSRs are defined. (C/MM) 1212-1991s

running task The task currently being executed by a processor. (C) 1003.5-1999

running tension control A control function that maintains tension in the material at operating speeds. *See also:* feedback control system. (IA/ICTL/IAC/APP) [60], [75]

running thread A thread currently executing on a processor. There may be more than one such thread in a system at a time in a system with multiple processors. (C/PA) 9945-1-1996

running time *See:* execution time.

run-of-river plant One utilizing stream flow as it occurs and with little or no storage at the project site. (PE/EDPG) 1020-1988r

run-of-river station (power operations) A hydroelectric generating station utilizing limited pondage or the flow of the stream as it occurs. (PE/PSE) 858-1987s, 346-1973w

runout rate The velocity at which the error in register accumulates. (IA/ICTL/CEM) [58]

run stream *See:* job stream.

run time (A) (software) The instant at which a computer program begins to execute. **(B) (software)** The period of time during which a computer program is executing. *See also:* execution time. (C) 610.12-1990, 610.10-1994

run-time test object (RTO) Contains the procedures utilized and data necessary to execute a test on a test subject within a specified context. (SCC20) 1226-1998

run time variable (test, measurement, and diagnostic equipment) An application program condition in which the stimuli is varied under system control based on a measurement result. (MIL) [2]

run-up time (gyros) The time interval required for the gyro rotor to reach a specified speed from standstill. (AES/GYAC) 528-1994

runway alignment indicator (illuminating engineering) A group of aeronautical ground lights arranged and located to provide early direction and roll guidance on the approach to a runway. (EEC/IE) [126]

runway centerline lights (illuminating engineering) Runway lights installed in the surface of the runway along the centerline indicating the location and direction of the runway centerline and are of particular value in conditions of very poor visibility. (EEC/IE) [126]

runway-edge lights (illuminating engineering) Lights installed along the edges of a runway marking its lateral limits and indicating its direction. (EEC/IE) [126]

runway-end identification light (illuminating engineering) A pair of flashing aeronautical ground lights symmetrically disposed on each side of the runway at the threshold to provide additional threshold conspicuity. (EEC/IE) [126]

runway-exit lights (illuminating engineering) Lights placed on the surface of a runway to indicate a path of the taxiway centerline. (EEC/IE) [126]

runway lights (illuminating engineering) Aeronautical ground lights arranged along or on a runway. (EEC/IE) [126]

runway threshold *See:* approach-light beacon.

runway visibility (illuminating engineering) The meteorological visibility along an identified runway. Where a transmissometer is used for measurement, the instrument is calibrated in terms of a human observer; that is, the sighting of dark objects against the horizon sky during daylight and the sighting of moderately intense unfocused lights of the order of 25 candelas at night. (EEC/IE) [126]

runway visual range (1) (navigation aids) The forward distance a human pilot can see along the runway during an approach to landing; this distance is derived from electro-optical instruments operated on the ground and it is improved (increased) by the use of lights (such as high-intensity runway lights). (AES/GCS) 172-1983w
(2) (illuminating engineering in the United States) An instrumentally derived value, based on standard calibrations, that represents the horizontal distance a pilot will see down the runway from the approach end; it is based on the sighting of either high intensity runway lights or on the visual contrast of other targets—whichever yields the greater visual range. (IE/EEC) [126]

rural districts All places not urban. This may include thinly settled areas within city limits. (NESC) C2-1997

rural line A line serving one or more subscribers in a rural area. (EEC/PE) [119]

rust A corrosion product consisting primarily of hydrated iron oxide. *Note:* This term is properly applied only to iron and ferrous alloys. (IA) [59]

rust-resistant So constructed, protected or treated that rust will not exceed a specified limit when subjected to a specified rust resistance test. (PE/TR) C57.12.80-1978r

RVR *See:* runway visual range.

RWM *See:* read/write memory.

RWP *See:* recommended wearing position.

Rytov approximation A mathematical approximation for a scalar wave propagating through an inhomogeneous medium in which the unknown field is expressed as $\exp\{X(r)\}$ and various levels of approximation are developed; the lowest order one is based on an assumed slow spatial variability of $X(r)$ (i.e., $grad[X] = 0$). (AP/PROP) 211-1997

***R-X* diagram** A graphic presentation of the characteristics of a relay unit in terms of the ratio of voltage to current and the phase angle between them. *Note:* For example, if a relay just operates with 10 V and 10 A in phase, one point on the operating curve of the relay would be plotted as 1 Ω on the R axis (i.e., $R = 1, X = 0$, where R is the abscissa and X is the ordinate). (SWG/PE) C37.100-1992

***R–X* plot (protective relaying)** A graphical method of showing the characteristics of a relay element in terms of the ratio of voltage to current and the angle between them. For example, if a relay barely operates with 10 V and 10 A in phase, one point on the operating curve of the relay would be plotted as 1 Ω on the R axis (that is, $R = 1, X = 0$).

RZ(NP) *See:* non-polarized return-to-zero recording.

RZ(P) *See:* polarized return-to-zero recording.

S

S A programming language used widely in the UNIX environment for data analysis and visualization. (C) 610.13-1993w

SA *See:* source address; spin axis.

sabin (audio and electroacoustics) A unit of absorption having the dimensions of area. *Notes:* 1. The metric sabin has dimensions of square meters. 2. When used without a modifier, the sabin is the equivalent of one square foot of a perfectly absorptive surface. (SP) [32]

SAC *See:* slanted array compressor.

SACK *See:* selective acknowledgment.

sacrificial protection Reduction or prevention of corrosion of a metal in an environment acting as an electrolyte by coupling it to another metal that is electrochemically more active in that particular electrolyte. *See also:* stray-current corrosion. (IA) [59]

safe Having acceptable risk of the occurrence of a hazard. (VT/RT) 1483-2000

safe braking model An analytical representation of a train's performance while decelerating to a complete stop, allowing for a combination of worst-case influencing factors and failure scenarios. A communications-based train control equipped train will stop in a distance equal to or less than that guaranteed by the safe braking model. (VT/RT) 1474.1-1999

safeguard (1) (electrolytic cell line working zone) A precautionary measure or stipulation, or a technical contrivance to prevent accidents. (IA/PC) 463-1993w
(2) Security measures for the physical protection of nuclear material and vital equipment at a nuclear power generating station. (PE/NP) 692-1997

safe let-go level The current level passing through a hand grip contact for which 99.5% of the subject population would retain sufficient muscular control to voluntarily release the subject grip and break contact. *Note:* The safe let-go level is a function of the frequency and voltage and varies considerably for various contact areas and pressures. Individual responses vary greatly from the mean level, and different levels are obtained for men, women, and children. (T&D/PE) 539-1990

safe shutdown earthquake (1) (seismic qualification of Class 1E equipment for nuclear power generating stations) An earthquake that is based upon an evaluation of the maximum earthquake potential considering the regional and local geology and seismology and specific characteristics of local subsurface material. It is that earthquake that produces the maximum vibratory ground motion for which certain structures, systems, and components are designed to remain functional. These structures, systems, and components are those necessary to ensure the integrity of the reactor coolant pressure boundary, the capability to shut down the reactor and maintain it in a safe shutdown condition, and the capability to prevent or mitigate the consequences of accidents that could result in potential off-site exposures comparable to the guideline exposures of 10 CFR, Ch 1, Section 100. (PE/NP) 344-1987r
(2) (valve actuators) That earthquake which produces the maximum vibratory ground motion for which certain structural systems are designed to remain functional. These structures, systems, and components are those necessary to ensure:

a) The integrity of the reactor coolant pressure boundary;
b) The capability to shut down the reactor and maintain it in a safe shutdown condition;
c) The capability to prevent or mitigate the consequences of an accident which could result in potential offsite exposures comparable to the exposure guideline of CFR 10, Energy—Nuclear Regulatory Commission, Part 100 (Dec 5, 1973.)

(SWG/PE/NP/PSR) C37.81-1989r, 382-1985, C37.98-1977s
(3) (nuclear power generating station) That earthquake which produces the maximum vibratory ground motion for which certain structures, systems, and components are designed to remain functional. (PE/NP) 649-1980s
(4) That earthquake which produces the maximum vibratory ground motion for which certain structures, systems, and components are designed to remain functional. These structures, systems, and components are those necessary to ensure (1) the integrity of the reactor coolant pressure boundary, and (2) the capability to prevent or mitigate the consequences of accidents that could result in potential offsite exposures comparable to the guideline exposures of Code of Federal Regulations, Title 10, Part 100 (December 5, 1973). (SWG/PE) C37.100-1992

safety assurance A characteristic of the implementation of a system that assures a level of safe operation. (VT/RT) 1483-2000

safety assurance concept A design concept applied to processor-based systems that assures the fail-safe implementation of identified functions, including safe operation in the presence of hardware failures and/or software errors. Examples are: Checked Redundancy; Diversity and Self-Checking; Numerical Assurance; and N-Version Programming. (VT/RT) 1483-2000

safety class features (Class 1E equipment and circuits) Structures design to protect Class 1E equipment against the effects of design basis events. *Note:* For the purposes of this standard, separate safety class structures can be separate rooms in the same building. The rooms may share a common wall. (PE/NP) 384-1981s

safety class structures Structures designed to protect Class 1E equipment against the effects of the design basis events. (PE/NP) 384-1992r, 308-1991

safety, conductor *See:* conductor safety.

safety control feature (deadman's feature) That feature of a control system that acts to reduce or cut off the current to the traction motors or to apply the brakes, or both, if the operator relinquishes personal control of the vehicle. *See also:* multiple-unit control. (EEC/PE) [119]

safety control handle (deadman's handle) A safety attachment to the handle of a controller, or to a brake valve, causing the current to the traction motors to be reduced or cut off, or the brakes to be applied, or both, if the pressure of the operator's hand on the handle is released. *Note:* This function may be applied alternatively to a foot-operated pedal or in combination with attachments to the controller or the brake valve handles, or both. *See also:* multiple-unit control. (EEC/PE) [119]

safety critical A term applied to a system or function, the correct performance of which is critical to safety of personnel and/or equipment; also a term applied to a system or function, the incorrect performance of which may result in an unacceptable hazard. *Note:* A safety-critical designation may require the incorporation of additional special safety design features. *See also:* fail-safe. (VT/RT) 1473-1999, 1483-2000, 1475-1999
(2) (A) A term applied to a system or function, the correct performance of which is critical to safety of personnel and/or equipment. **(B)** A term applied to a system or function, the incorrect performance of which may result in a hazard. *Note:* Vital functions are a subset of safety-critical functions. (VT/RT) 1474.1-1999

safety-critical software Software that falls into one or more of the following categories:

a) Software whose inadvertent response to stimuli, failure to respond when required, response out-of-sequence, or response in combination with other responses can result in an accident.

b) Software that is intended to mitigate the result of an accident.

c) Software that is intended to recover from the result of an accident.

(C/SE) 1228-1994

safety function One of the processes or conditions (for example, emergency negative reactivity insertion, post-accident heat removal, emergency core cooling, post-accident radioactivity removal, and containment isolation) essential to maintain plant parameters within acceptable limits established for a design basis event. *Note:* A safety function is achieved by the completion of all required protective actions by the reactor trip system or the engineered safety features concurrent with the completion of all required protective actions by the auxiliary supporting features, or both.
(PE/NP) 603-1998, 338-1987r, 379-1994

safety ground (1) The connection between a grounding system and metallic parts that are not usually energized but that may become live due to a fault or an accident; often referred to as equipment or frame ground. (PE/EDPG) 665-1995
(2) *See also:* equipment grounding conductor.
(IA/PSE) 1100-1999

safety group A given minimal set of interconnected components, modules, and equipment that can accomplish a safety function. (PE/NP) 603-1998, 308-1991

safety life line A safety device normally constructed from synthetic fiber rope and designed to be connected between a fixed object and the body belt of a worker working in an elevated position when his/her regular safety strap cannot be utilized. *Synonyms:* safety line; life line; scare rope.
(T&D/PE) 524-1992r

safety line *See:* safety life line.

safety outlet *See:* grounding outlet.

safety-related (nuclear power generating station) Any Class IE power or protection system device included in the scope of IEEE 279-1971 or IEEE 308-1974. (PE/NP) 577-1976r

safety ring *See:* write-protect ring.

safety sign A visual alerting device in the form of a sign, label, decal, placard, or other marking that advises the observer of the nature and degree of the potential hazard(s), which can cause injury or death. It can also provide safety precautions or evasive actions to take, or provide other directions to eliminate or reduce the hazard. (NIR/SCC28) C95.2-1999

safety strap *See:* positioning strap.

safety system (1) (nuclear power plants) Those systems (the reactor trip system, an engineered safety feature, or both, including all their auxiliary supporting features and other auxiliary features) that provide a safety function. A safety system is comprised of more than one safety group of which any one safety group can provide the safety function.
(PE/NP) 338-1987r, 600-1983lw, 497-1981w
(2) (nuclear power generating station) (design of control room complex) The collection of systems required to minimize the probability and magnitude of release of radioactive material to the environment by maintaining plant conditions within the allowable limits established for each design basis event. *Note:* The safety system is the aggregate of one or more protective action systems. It includes, but is not necessarily limited to, the engineered safety features, the reactor trip system, and the auxiliary supporting features.
(PE/NP) 567-1980w
(3) A system that is relied upon to remain functional during and following design basis events to ensure: (A) the integrity of the reactor coolant pressure boundary, (B) the capability to shut down the reactor and maintain it in a safe shutdown condition, or (C) the capability to prevent or mitigate the consequences of accidents that could result in potential off-site exposures comparable to the 10 CFR Part 100 guidelines. *Note:* The electrical portion of the safety systems, that perform safety functions, is classified as Class 1E.
(PE/NP) 603-1998

safety validation A structured and managed set of activities that demonstrate that the system, as specified and implemented, performs the intended functions, and that those functions result in overall safe operation. Validation answers the question, "Did we build the right system?" (VT/RT) 1483-2000

safety verification A structured and managed set of activities that identify the vital functions required to be performed by the system, and demonstrate that the system, including its subsystems, inter faces and components, implements the vital functions fail-safely to a level that meets the allocated system safety goals. Verification answers the question, "Did we build the system right?" (VT/RT) 1483-2000

safe working space (electrolytic cell line working zone) The space required to safeguard personnel from hazarous electrical conditions during the conduct of their work in operating and maintaining cells and their attachments. This shall include space allowance for tools and equipment that may be involved. (IA/PC) 463-1977s

safe working voltage to ground (electric recording instrument) The highest safe voltage in terms of maximum peak value that should exist between any circuit of the instrument and its case. *See also:* test. (EEC/ERI) [111]

safe work practices (electrolytic cell line working zone) Those operating and maintenance procedures that are effective in preventing accidents. (IA/PC) 463-1993w

SAFI *See:* semiautomatic flight inspection.

sag (1) The distance measured vertically from a conductor to the straight line joining its two points of support. Unless otherwise stated in the rule, the sag referred to is the sag at the midpoint of the span. *See also:* sag of a conductor at any point in a span; apparent sag at any point in the span; maximum total sag; final unloaded sag; final sag; initial unloaded sag; total sag; apparent sag.
(NESC/T&D/PE) C2-1997, 524-1992r
(2) A decrease in rms voltage or current at the power frequency for durations of 0.5 cycle to 1 min. Typical values are 0.1 to 0.9 pu. *Note:* To give a numerical value to a sag, the recommended usage is "a sag to 20%," which means that the line voltage is reduced down to 20% of the normal value, not reduced by 20%. Using the preposition "of" (as in "a sag of 20%," or "a 20% sag") is deprecated. (SCC22) 1346-1998
(3) An rms reduction in the ac voltage, at the power frequency, for durations from a half cycle to a few seconds. *See also:* notch; undervoltage. (IA/PSE) 1100-1999

sag and appararent sag (apparent sag at any point) The departure of the wire at the particular point in the span from the straight line between the two points of support of the span, at 60°F, with no wind loading. (See the corresponding figure.)

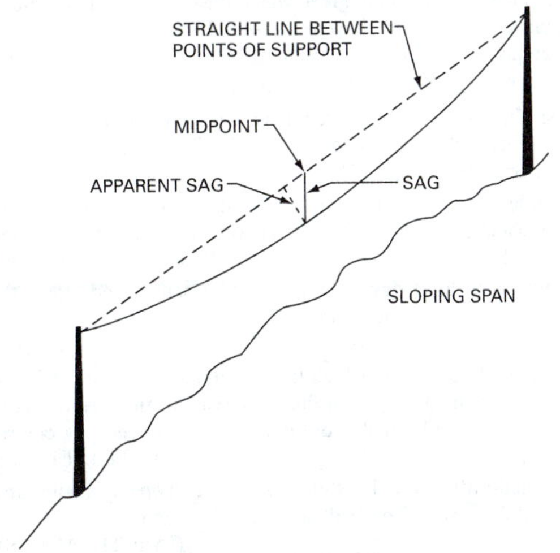

sag and apparent sag

(NESC) C2-1997

sag board *See:* target sag.

sagger *See:* wheel tractor.

Sagnac effect (laser gyro) (interferometric fiber optic gyro) A relativistic rotation-induced optical path length difference between electromagnetic waves that counter-propagate around a closed path. (AES/GYAC) 528-1994

sag of a conductor at any point in a span The distance measured vertically from the particular point in the conductor to a straight line between its two points of support. (T&D/NESC) C2-1997

sag section (conductor stringing equipment) The section of line between snub structures. More than one sag section may be required in order to sag properly the actual length of conductor that has been strung. *Synonyms:* stringing section; pull; setting; pull setting. (T&D/PE) 524-1992r

sag span (conductor stringing equipment) A span selected within a sag section and used as a control to determine the proper sag of the conductor, thus establishing the proper conductor level and tension. A minimum of two, but normally three, sag spans are required within a sag section in order to sag properly. In mountainous terrain or where span lengths vary radically, more than three sag spans could be required within a sag section. *Synonym:* control span. (T&D/PE) 524-1992r

sag target (conductor stringing equipment) A device used as a reference point to sag conductors. It is placed on one structure of the sag span. The sagger, on the other structure of the sag span, can use it as his reference to determine the proper conductor sag. *Synonym:* target. (T&D/PE) 524-1980s

SAID *See:* security association identifier.

sal ammoniac cell A cell in which the electrolyte consists primarily of a solution of ammonium chloride. *See also:* electrochemistry. (EEC/PE) [119]

salient pole (rotating machinery) A field pole that projects from the yoke or hub towards the primary winding core. *See also:* rotor. (PE) [9]

salient-pole machine An alternating-current machine in which the field poles project from the yoke toward the armature and/or the armature winding self-inductance undergoes a significant single cyclic variation for a rotor displacement through one pole pitch. *See also:* asynchronous machine. (PE) [9]

salient pole synchronous induction motor (rotating machinery) A salient pole synchronous motor having a coil winding for starting purposes embedded in the pole shoes. The terminal leads of this coil winding are connected to collector rings. (PE) [9]

salinity indicator system A system, based on measurement of varying electric resistance of the solution, to indicate the amount of salt in boiler feed water, the output of an evaporator plant, or other fresh water. *Note:* Indication is usually in grains per gallon. (EEC/PE) [119]

SAM *See:* sequential access method.

SAM76 A list processing language widely used in artificial intelligence due to its unique suitability for interactive and user-directed applications. (C) 610.13-1993w

sample One or more units of product drawn from a lot, the units of sample being selected at random without regard to their quality. (PE/T&D) C135.61-1997

sample-and-hold device A device that senses and stores the instantaneous value of an analog signal. (C) 610.10-1994w

sampled data Data in which the information content can be, or is, ascertained only at discrete intervals of time. *Note:* Sampled data can be analog or digital. *See also:* feedback control system. (IA/EEC) [61], [74]

sampled-data control system A system that operates with sampled data. *See also:* feedback control system. (IA/ICTL/IAC) [60]

sampled-data system One in which at least one sampled signal is present. *See also:* control system. (CS/IM) [120]

sample description −1, −2, −3, −4 Four 64-character records containing a sample description. These records will provide information on the source of the sample being analyzed. If not used they should be set to spaces. (NPS/NID) 1214-1992r

sampled format (pulse measurement) A waveform which is a series of sample magnitudes taken sequentially or nonsequentially as a function of time. It is assumed that nonsequential samples may be rearranged in time sequence to yield the following samples formats. *See also:* aperiodically sampled equivalent time format; periodically sampled equivalent time format; periodically sampled real time format; aperiodically sampled real time format. (IM/WM&A) 181-1977w

sampled signal The sequence of values of a signal taken at discrete instants. *See also:* control system. (PE/EDPG) [3]

sample edge A time corresponding to a falling edge of the bus clock signal. (C/MM) 1196-1987w

sample equipment (1) (nuclear power generating station) Production equipment tested to obtain data that are valid over a range of ratings and for specific services. (PE/NP) 323-1974s, 650-1979s
(2) (safety systems equipment in nuclear power generating stations) Equipment, representative of a design, used to obtain data that are valid over a range of ratings and for specific service conditions. (PE/NP) 627-1980r

sample instance diagram A form of presenting example instances in which instances are shown as separate graphic objects. The graphic presentation of instances can be useful when only a few instances are presented. *Contrast:* sample instance table. (C/SE) 1320.2-1998

sample instance table A form of presenting example instances in which instances are shown as a tabular presentation. The tabular presentation of instances can be useful when several instances of one class are to be presented. *Contrast:* sample instance diagram. (C/SE) 1320.2-1998

SAMPLE/PRELOAD A defined instruction for the test logic defined by IEEE Std 1149.1-1990. (TT/C) 1149.1-1990

sample size Based on the lot size. Minimum sample sizes are given in the following table.

Lot size	Sample size
1–29	3
30–150	5
151–1200	13
1201–10 000	20
10 001–35 000	32

(PE/T&D) C135.61-1997

sample software Software selected from a current or completed project from which data can be obtained for use in preliminary testing of data collection and metric computation procedures. (C/SE) 1061-1992s

sample time The time of the physical collection of the sample material, as

DD/MM/YR_HH:NN:SS_

where the '_' (underscore character) is an ASCII space; DD is the day; MM is the month; YR is the year; HH is the hours; NN is the minutes; and SS is the seconds. (NPS/NID) 1214-1992r

sample valve actuator A representative unit manufactured in accordance with the manufacturer's quality control system and specifications for production units. (PE/NP) 382-1980s

sample valve operator (nuclear power generating station) A production valve operator type tested to obtain data that are valid over a range of sizes and for the specific services. *Note:* All salient factors must be shown to be common to the sample valve operator and to the intended service valve operator. Commonality of factors such as materials of construction, lubrication, mechanical stresses and clearances, manufacturing processes, and dielectric properties may be established by specification, test, or analyses. (PE/NP) 382-1980s

sampling (1) (pulse terminology) A process in which strobing pulses yield signals that are proportional to the magnitude (typically, as a function of time) of a second pulse or other event. (IM/WM&A) 194-1977w
(2) (image processing) The technique of dividing an image into disjoint regions, selecting a single point in each region to represent the region, and measuring the brightness or color of each of these points. (C) 610.4-1990w
(3) The process of obtaining the values of a function for regularly or irregularly spaced distinct values of an independent variable. (C) 610.10-1994w

sampling circuit (sampler) A circuit whose output is a series of discrete values representative of the values of the input at a series of points in time. (PE/EEC) [119]

sampling control *See:* sampling control system.

sampling control system Control using intermittently observed values of signals such as the feedback signal or the actuating signal. *Note:* The sampling is often done periodically. *See also:* feedback control system. (IM/PE/EDPG) [120], [3]

sampling gate A device that extracts information from the input wave only when activated by a selector pulse or sampling pulse. (AES) 686-1997

sampling, instantaneous *See:* instantaneous sampling.

sampling interval (automatic control) The time between samples in a sampling control system. *See also:* feedback control system. (IM) [120]

sampling period (automatic control) The time interval between samples in a periodic sampling control system. *See also:* feedback control system. (IM) [120]

sampling rate (1) For the purposes of "de-bouncing" and noise rejection, the receiving station (BCC or DCC) of a special function signal shall sample the signal at a specified rate for a specified length of time in order to determine the "filtered" logic sense of the signal. The clock rate used is referred to as the sampling rate. (EMB/MIB) 1073.3.1-1994
(2) The frequency with which the event recorder regularly monitors an input channel to determine its value. (VT) 1482.1-1999

sampling smoke detector (fire protection devices) A device which consists of tubing distributed from the detector unit to the area(s) to be protected. An air pump draws air from the protected area back to the detector through the air sampling ports and piping. At the detector, the air is analyzed for smoke particles. (NFPA) [16]

sampling tests Tests carried out on a few samples taken at random out of one consignment. *See also:* direct-current commutating machine; asynchronous machine. (PE) [9]

sanding Dropping or blowing of sand or similar material on the top of the rail head to increase the coefficient of friction to obtain better adhesion. (VT) 1475-1999

SAP *See:* service access point.

SAR *See:* synthetic-aperture radar.

SAS *See:* Statistical Analysis System.

sash A visual user interface control that separates areas of a window to allow the user to display alternative information. (C) 1295-1993w

satellite (communication satellite) A body which revolves around another body and which has a motion primarily and permanently determined by the force of attraction of this body. *Note:* A body so defined which revolves round the sun is called a planet or planetoid. By extension, a natural satellite of a planet may itself have a satellite. (COM) [19]

satellite computer A processor connected locally or remotely to a larger central processor, and performing certain processing tasks. *See also:* remote computer system. (C) 610.10-1994w

satellite navigation (navigation aid terms) Navigation using artificial earth satellites as an aid. Position is computed by determination of either angles, range and range rate, or range and angle measurements of the vehicle relative to the satellite plus satellite ephemeris data received by the vehicle. Satellite ephemeris data can be determined by tracking stations and transmitted to and stored in the satellite's memory for subsequent transmission to vehicle's receivers. (AES/GCS) 172-1983w

satellite phasing (communication satellite) Maintaining the center of mass of a satellite by propulsion within a prescribed small tolerance in a desired relation with respect to other satellites or a point on the earth or some other point of reference, such as the subsolar point. (COM) [19]

saturable-core magnetometer A magnetometer that depends for its operation on the changes in permeability of a ferromagnetic core as a function of the field to be measured. *See also:* magnetometer. (EEC/PE) [119]

saturable-core reactor *See:* saturable reactor.

saturable reactor (saturable core reactor) (electrical heating applications to melting furnaces and forehearths in the glass industry) A device that provides output voltage modulation by variation of its circuit reactance. This reactance is controlled by changing the saturation of its magnetic core through variation of a superimposed unidirectional flux. (IA) 668-1987w
(2) (A) (power and distribution transformers) A magnetic-core reactor whose reactance is controlled by changing the saturation of the core through variation of a super-imposed unidirection flux. **(B) (power and distribution transformers)** A magnetic-core reactor operating in the region of saturation without independent control means. *Note:* Thus a reactor whose impedance varies cyclically with the alternating current (or voltage). (PE/TR) C57.12.80-1978

saturated relay recovery time Recovery time of a thermal relay measured after temperature saturation has been reached. (EEC/REE) [87]

saturated relay reoperate time Reoperate time of a thermal relay measured when the relay is de-energized after temperature saturation (equilibrium) has been reached. (EEC/REE) [87]

saturated signal *See:* saturating signal.

saturated sleeving A flexible tubular product made from braided cotton, rayon, nylon, glass, or other fibers, and coated or impregnated with varnish, lacquer, a combination of varnish and lacquer, or other electrical insulating materials. The impregnant or coating need not form a continuous film. (EEC/PE) [119]

saturating reactor A magnetic-core reactor operating in the region of saturation without independent control means. *See also:* magnetic amplifier. (EEC/PE) [119]

saturating signal (electronic navigation) A signal of an amplitude greater than can be accommodated by the dynamic range of a circuit. *See also:* navigation. (AES/RS) 686-1982s, [42]

saturation (1) (signal-transmission system) A natural phenomenon or condition in which any further change of input no longer results in appreciable change of output. (IA/IA/APP/IAC) [69], [60]
(2) (automatic control) A condition caused by the presence of a signal or interference large enough to produce the maximum limit of response, resulting in loss of incremental response. *See also:* feedback control system; signal. (IE/PE/EDPG) [43], [3]
(3) (A) (visual) The attribute of a visual sensation which permits a judgment to be made of the proportion of pure chromatic color in the total sensation. *Note:* This attribute is the psychosensorial correlate (or nearly so) of the colorimetric quantity "purity." **(B) (color television)** In a tristimulus reproducer, the degree to which the color lies on the triangle as defined by the three reproducing primaries. *Note:* Full saturation is achieved when one or two of the reproduced primary colors have zero intensity. (BT/AV) 201-1979
(4) (A) (perceived light-source color) The attribute used to describe its departure from a light-source color of the same brightness perceived to have no hue. *See also:* color. **(B) (illuminating engineering)** (of a perceived color) The attribute according to which it appears to exhibit more or less

chromatic color judged in proportion to its brightness. In a given set of viewing conditions, and at luminance levels that result in photopic vision, a stimulus of a given chromaticity exhibits approximately constant saturation for all luminances. (EEC/IE) [126]

(5) **(diode-type camera tube)** The point on the signal transfer characteristic where an increase in the input irradiance signal does not change the resulting output current signal significantly. (ED) 503-1978w

(6) (of a maser, laser, maser material, or laser material) (laser-maser) A condition in which the attenuation or gain of a material or a device remains at a fixed level or decreases as the input signal is increased. (LEO) 586-1980w

(7) **(A)** In a switching device or amplifier, the fully conducting state at which the device is passing the maximum possible current. *Note:* Most commonly used in reference to circuits containing bipolar or field-effect transistors. **(B)** In color graphics and printing, the amount of color in a specified hue. *Note:* The saturation affects the vividness of the image. (C) 610.10-1994

(8) The amount of a particular gas that can be dissolved in a fluid at a given pressure and temperature. The saturation of all gases of interest is linearly proportional to absolute pressure. However, the effect of temperature varies with the specific gas. Some gases exhibit a decrease in saturation with increasing temperature while others tend to increase. The variation with temperature is generally small and can be neglected when evaluating samples in the lab. (PE/IC) 1406-1998

saturation characteristics (nuclear power generating station) A description of the steady state or dynamic conditions or limitations under which a further change in input produces an output response which no longer conforms to the specified steady-state or dynamic input-output relationship. (PE/NP) 381-1977w

saturation current (1) (thermionics) The value of the current in the saturation state. *See also:* electron emission. (ED) [45]

(2) **(semiconductor diode)** That portion of the steady-state reverse current that flows as a result of the transport across the junction of minority carriers thermally generated within the regions adjacent to the junction. *See also:* semiconductor device. (ED) 216-1960w

(3) **(diode-type camera tube)** The value of the output current signal saturation. Units: amperes. (ED) 503-1978w

saturation curve (machine or other apparatus) A characteristic curve that expresses the degree of magnetic saturation as a function of some property of the magnetic excitation. *Note:* For a direct-current or synchronous machine the curve usually expresses the relation between armature voltage and field current for no load or some specified load current, and for specified speed. (PE) [9]

saturation factor (K_s) (1) (rotating machinery) The ratio of the unsaturated value of a quantity to its saturated value. The reciprocal of this definition is also used. (PE) [9]

(2) The ratio of the saturation voltage of a current transformer to the excitation voltage. Saturation factor is an index of how close to saturation a current transformer is in a given application. (PE/PSR) C37.110-1996

saturation flux density (1) (pipelines and vessels) The maximum possible magnetic flux density in a material. (IA/PC) 844-1991

(2) *See also:* saturation induction. (MAG) 393-1977s

saturation induction (magnetic core testing) The maximum intrinsic value of induction possible in a material. *Notes:* 1. This term is often used for the maximum value of induction at a stated high value of field strength where further increase in intrinsic magnetization with increasing field strength is negligible. 2. S.I. unit: Tesla: cgs unit: Gauss (1 Tesla), 10^4 Gauss. 3. Peak induction (Bm) is the magnetic induction corresponding to the peak applied magnetizing force specified in a test. (MAG) 393-1977s

saturation level (storage tubes) The output level beyond which no further increase in output is produced by further writing (then called write saturation) or reading (then called read saturation). *Note:* The word saturation is frequently used alone to denote saturation level. *See also:* storage tube. (ED) 158-1962w, [45]

saturation region of an IGFET *See:* saturation region of an insulated-gate field-effect transistor.

saturation region of an insulated-gate field-effect transistor (metal-nitride-oxide field-effect transistor) A portion of the I_{DS} versus V_{DS} characteristic where I_{DS} is nearly constant regardless of the value of V_{DS}. This is true when $|V_{DS}| \geq |V_{GS} - V_T|$. *Synonym:* saturation region of an IGFET. (ED) 581-1978w

saturation state (thermionics) The state of working of an electron tube or valve in which the current is limited by the emission from the cathode. *See also:* electron emission. (ED) [45]

saturation voltage (V_x) The symmetrical voltage across the secondary winding of the current transformer for which the peak induction just exceeds the saturation flux density. It is found graphically by locating the intersection of the straight portions of the excitation curve on log-log axes. This is not the same as the knee-point voltage which is the point on the curve where the tangent to the curve makes an angle of 45° to the abscissa. (PE/PSR) C37.110-1996

save area An area of main storage in which the contents of registers are saved. (C) 610.10-1994w

saved program state The set of information, necessary to begin or resume the execution of a client program, describing the machine state (including CPU registers) that will be established upon resumption or initiation of client program execution. (C/BA) 1275-1994

saved set-group-ID (1) An attribute of a process that allows some flexibility in the assignment of the effective group ID attribute, when the saved set-user-ID option is implemented. (C/PA/C/PA) 9945-1-1996, 9945-2-1993

(2) When the Saved IDs option is implemented, an attribute of a process that allows some flexibility in the assignment of the effective group ID attribute, as described for `Set_Group_ID`. (C) 1003.5-1999

saved set-user-ID (1) An attribute of a process that allows some flexibility in the assignment of the effective user ID attribute, when the saved set-user-ID option is implemented. (C/PA) 9945-1-1996, 9945-2-1993

(2) When the Saved IDs option is implemented, an attribute of a process that allows some flexibility in the assignment of the effective user ID attribute, as described for `Set_User_ID`. (C) 1003.5-1999

SAW *See:* surface acoustic wave.

SAW filter A filter characterized by the use of surface acoustic waves that are usually generated by an interdigital transducer and that propagate along a substrate surface to a receiving transducer. (UFFC) 1037-1992w

SAW oscillator An oscillator that uses a SAW device (resonator or delay line) as the main frequency-controlling element. (UFFC) 1037-1992w

SAW resonator filter A type of surface acoustic wave filter that offers a high $Q(f_0/\Delta f)$ and incorporates efficient reflective arrays to form a Fabry-Perot resonant cavity. (UFFC) 1037-1992w

sawtooth *See:* sawtooth waveform.

sawtooth sweep A sweep generated by the ramp portion of a sawtooth waveform. *See also:* oscillograph. (IM/HFIM) [40]

sawtooth wave (television) A periodic wave whose instantaneous value varies substantially linearly with time between two values, the interval required for one direction of progress being longer than that for the other. *Note:* In television practice, the waveform during the retrace interval is not necessarily linear, since only the trace interval is used for active scanning. (BT/AV) 201-1979w

sawtooth waveform A waveform containing a ramp and a return to initial value, the two portions usually of unequal duration. *See also:* oscillograph. (IM/HFIM) [40]

S-band A radar-frequency band between 2 GHz and 4 GHz, usually in one of the International Telecommunication Union (ITU) allocated bands 2.3–2.5 GHz or 2.7–3.7 GHz. (AES) 686-1997

S-BASIC A dialect of the BASIC programming language. (C) 610.13-1993w

SBS *See:* system breakdown structure.

SBM *See:* Serial Bus management.

SBus A) The correct spelling of the noun describing the bus defined by this standard. B) The name for the Chip and Module Interconnect Bus described by this standard. (C/BA) 1496-1993w

SBus Card A physical printed circuit assembly that conforms to the single-width or double-width mechanical specifications; meets the connector, power, and signal assignment requirements of this standard; and contains one or more SBus Devices. (C/BA) 1496-1993w

SBus Controller The SBus Device that performs all the centralized services for the SBus, including bias circuitry, arbitration, and address translation for SBus Masters, and selection of and time-outs for SBus Slaves. (C/BA) 1496-1993w

SBus Cycle One complete operation on the SBus, consisting of a set of phases beginning with an optional Arbitration Phase and progressing through the optional Translation Phase, the optional Extended Transfer Information Phase, and the Transfer Phase. (C/BA) 1496-1993w

SBus Device A set of circuitry complying with the electrical and protocol requirements of the SBus and properly implementing all the signals of the SBus. An SBus Device may reside on the computer motherboard or it may be on an SBus Card. See SBus Controller, SBus Master, SBus Slave. (C/BA) 1496-1993w

SBus Master The SBus Device that requests data transfers to be performed by an SBus Slave. (C/BA) 1496-1993w

SBus Master port In an SBus Device that combines both an SBus Master and an SBus Slave, the circuitry that is associated with the SBus Master. (C/BA) 1496-1993w

SBus Slave The SBus Device providing the function of performing the data transfers requested by an SBus Master. The address space for the data transfers may contain data, control registers, or sense registers. (C/BA) 1496-1993w

SBus Slave port In an SBus Device that combines both an SBus Master and an SBus Slave, the circuitry that is associated with the SBus Slave. (C/BA) 1496-1993w

SBus Slot The location on a computer motherboard in which an SBus Card may be installed. The SBus Slot has the connector, the electrical characteristics, and the physical volumes and dimensions that are required by this standard. (C/BA) 1496-1993w

SBus Specification SBus Specification B.0, an earlier specification of SBus, now superseded by the IEEE Std 1496-1993. (C/BA) 1496-1993w

SBus standard IEEE Std 1496-1993, IEEE Standard for a Chip and Module Interconnect Bus: SBus. (C/BA) 1496-1993w

SBus System A computer system containing a motherboard with at least an SBus Controller and some combination of zero or more SBus Slots which may be populated with SBus Cards. The SBus System may additionally have SBus Devices integrated on the motherboard. The SBus System includes the electronic, powering, cooling, and mechanical support functions required by the installed SBus Devices and SBus Slots. (C/BA) 1496-1993w

SC *See:* station-type cubicle switchgear.

SCADA *See:* supervisory control data acquisition system.

scada channel (supervisory control, data acquisition, and automatic control) The communication path between master and remote stations. (SWG/PE/SUB) C37.100-1992, C37.1-1994

scaffolding Computer programs and data files built to support software development and testing, but not intended to be included in the final product. For example, dummy routines or files, test case generators, software monitors, stubs. *See also:* programming support environment. (C) 610.12-1990

scalability The ability to provide functionality up and down a graduated series of application platforms that differ in speed and capacity. (C/PA) 14252-1996

Scalable Coherent Interface (1) The name that refers to IEEE Std 1596-1992. Though functionally behaving as a bus, the SCI's physical implementation is a collection of point-to-point unidirectional links (i.e., a ring). (C/MM) 1212-1991s
(2) The Scalable Coherent Interface standard, IEEE Std 1596-1992. (C/MM) 1596.3-1996

scalable font A font that can be scaled to produce characters in varying sizes. *See also:* derived font; outline font. (C) 610.10-1994w

scalar (1) (computers) A data item used to represent a single number or entity. *Contrast:* vector. (C) 610.5-1990w, 1084-1986w
(2) Quantity that is completely specified by a single number. (Std100) 270-1966w
(3) A value that is atomic, i.e., having no parts. *Contrast:* collection-valued. (C/SE) 1320.2-1998
(4) An integer constant. (C/DA) 1481-1999

scalar approximation The reduction of the vector representation of an electromagnetic field to a scalar description by assuming that the field is identically polarized at every point in space. *Note:* It usually means that cross-polarization effects are ignored. (AP/PROP) 211-1997

scalar field The totality of scalars in a given region represented by a scalar function $S(x, y, z)$ of the space coordinates x, y, z. (Std100) 270-1966w

scalar function A functional relationship that results in a scalar. (Std100) 270-1966w

Scalar Parameter An instance of the class `IEEE1451_ScalarParameter` or of a subclass thereof. (IM/ST) 1451.1-1999

scalar product (dot product) (of two vectors) The scalar obtained by multiplying the product of the magnitudes of the two vectors by the cosine of the angle between them. The scalar product of the two vectors A and B may be indicated by means of a dot A · B. If the two vectors are given in terms of their rectangular components, then

$$\mathrm{A} \cdot \mathrm{B} = A_x B_x + A_y B_y + A_z B_z$$

Example: Work is the scalar product of force and displacement. (Std100) 270-1966w

scalar property *See:* scalar-valued property.

scalar radiative transfer A radiative transfer theory in which the vector nature of the fields is ignored. *Synonym:* scalar radiative transport. (AP/PROP) 211-1997

scalar radiative transport *See:* scalar radiative transfer.

Scalar Series Parameter An instance of the class `IEEE1451_ScalarSeriesParameter` or of a subclass thereof. (IM/ST) 1451.1-1999

scalar solutions Solutions of Maxwell's equations where cross-polarization effects are disregarded, i.e., coupling between transverse electric (TE) and transverse magnetic (TM) fields is ignored. (AP/PROP) 211-1997

scalar unit An arithmetic unit that operates on one data element at a time. (C) 610.10-1994w

scalar-valued class A class in which each instance is a single value. *Contrast:* collection-valued class. (C/SE) 1320.2-1998

scalar-valued property A property that maps to a scalar-valued class. *Contrast:* collection-valued property. (C/SE) 1320.2-1998

scalar wave equation *See:* homogeneous Helmholtz equation.

scale (1) A musical scale is a series of notes (symbols, sensations, or stimuli) arranged from low to high by a specified scheme of intervals, suitable for musical purposes. (SP/ACO) [32]
(2) (computers) To change a quantity by a factor in order to bring its range within prescribed limits. (C) [20], [85]
(3) (instrument scale) *See also:* full scale. (C) 1084-1986w
(4) (mathematics of computing) To multiply the representation of a number by a factor in order to bring its range within prescribed limits. (C) 1084-1986w
(5) (A) (data management) To adjust the representation of a quantity so that its value is brought within a specified range. **(B) (data management)** The difference between the original and resulting adjustment as in definition (A). **(C) (data management)** A system of mathematical notation such as fixed-point or floating point. (C) 610.5-1990
(6) (computer graphics) To change the size of a display element by multiplying its coordinates by a constant value. *Note:* An object can be scaled by the same amount in each of its dimensions (global scaling) or by a different amount in each of its dimensions. (C) 610.6-1991w
(7) A visual user interface control that represents a quantity and its relationship to the range of possible values for that quantity. The user can change the value of the quantity. (C) 1295-1993w

scale class (mechanical demand registers) Denotes, with respect to single-pointer-form, dual-range single-pointer form, or cumulative-form demand registers, the relationship between the full-scale value of the register and the kilovolt ampere (kVA) rating of the meter with which the register is used. (ELM) C12.4-1984

scale factor (1) (high voltage testing) (measuring system) (of a measuring system) The factor by which the output indication is multiplied to determine the measured value of the input quantity or function. (PE/PSIM) 4-1995
(2) In an analog computer, the multiplication factor necessary to transform problem variables into computer variables. *Note:* A problem variable is a variable appearing in the mathematical model of the problem. A computer variable is a dependent variable as represented on the computer. (C) 610.10-1994w, 165-1977w
(3) (mathematics of computing) A number used as a factor in a scaling operation. *See also:* scale. (C) 1084-1986w
(4) (A) (accelerometer) (gyros) The ratio of a change in output to a change in the input intended to be measured. Scale factor is generally evaluated as the slope of the straight line that can be fitted by the method of least squares to input-output data obtained by varying the input cyclically within the input range. **(B) (laser gyro)** The ratio of change in angular displacement about the input axis to a change in output (arc-seconds per pulse). The laser gyro scale factor is directly proportional to the total path length and operating wavelength, and inversely proportional to the effective enclosed ring area. (AES/GYAC) 528-1994

scale factor asymmetry (accelerometer) (gyros) The difference between the scale factor measured with positive input and that measured with negative input, specified as a fraction of the scale factor measured over the input range. Scale factor asymmetry implies that the slope of the input-output function is discontinuous at zero input. It must be distinguished from other nonlinearities. (AES/GYAC) 528-1994

scale-factor potentiometer *See:* parameter potentiometer.

scale length (electric instruments) The length of the path described by the indicating means or the tip of the pointer in moving from one end of the scale to the other. *Notes:* 1. In the case of knife-edge pointers and others extending beyond the scale division marks, the pointer shall be considered as ending at the outer end of the shortest scale division marks. In multiscale instruments the longest scale shall be used to determine the scale length. 2. In the case of antiparallax instruments of the step-scale type with graduations on a raised step in the plane of and adjacent to the pointer tip, the scale length shall be determined by the end of the scale divisions adjacent to the pointer tip. *See also:* instrument. (EEC/AII) [102]

scale model A physical model that resembles a given system, with only a change in scale; for example, a replica of an airplane one tenth the size of the actual airplane. (C) 610.3-1989w

scale-of-two counter A flip-flop circuit in which successive similar pulses, applied at a common point, cause the circuit to alternate between its two conditions of permanent stability. *See also:* trigger circuit. (EEC/PE) [119]

scaler (radiation counters) An instrument incorporating one or more scaling circuits and used for registering the number of counts received. *See also:* anticoincidence. (ED) [45]

scaler, pulse *See:* pulse scaler.

scale span (instrument) The algebraic difference between the values of the actuating electrical quantity corresponding to the two ends of the scale. *See also:* instrument. (EEC/PE) [119]

scaling (A) The formation at high temperatures of thick corrosion product layer(s) on a metal surface. **(B)** The deposition of water-insoluble constituents on a metal surface (as on the interior of water boilers). (IA) [59]

scaling circuit (radiation counters) A device that produces an output pulse whenever a prescribed number of input pulses has been received. *See also:* anticoincidence. (ED) [45]

scalloping (navigation aid terms) The irregularities in the field pattern of the ground facility due to unwanted reflections from obstructions or terrain features, exhibited in flight as cyclical variations in bearing error. *Synonym:* course scalloping. (AES/GCS) 172-1983w

scan (1) (general) To examine sequentially part by part. (C) [20], [85]
(2) (oscillography) The process of deflecting the electron beam. *See also:* uniform luminance area; graticule area; phosphor screen; oscillograph. (IM/HFIM) [40]
(3) (supervisory control, data acquisition, and automatic control) (interrogation) The process by which a data acquisition system interrogates remote stations or points for data. (SWG/PE/SUB) C37.100-1992, C37.1-1994
(4) (data management) To examine a set of items sequentially. (C) 610.5-1990w
(5) The process by which a data acquisition system interrogates remote terminals or points for data. (SUB) 999-1992w
(6) A sampling process of observing attribute values at a specified point in time. (LM/C) 802.1F-1993r
(7) To examine stored information sequentially, part by part. (C) 610.10-1994w

scan angle The angle between the direction of the maximum of the major lobe or a directional null and a reference direction. *Notes:* 1. The term beam angle applies to the case of a pencil beam antenna. 2. The reference boresight is usually chosen as the reference direction. *Synonym:* beam angle. (AP/ANT) 145-1993

scan conversion The process of redefining an image from one that is composed of lines, points, and areas to one that is expressed in an array of pixels. (C) 610.6-1991w

scan converter A device on which a display can be written in refresh line-drawing mode and read out in raster scan mode. (C) 610.10-1994w

scan cycle (supervisory control, data acquisition, and automatic control) The time in seconds required to obtain a collection of data (for example, all data from one remote, all data from all remotes, and all data of a particular type from all remotes). (SWG/PE/PE) C37.100-1992, C37.1-1994

scan design A design technique that introduces shift-register paths into digital electronic circuits and thereby improves their testability. (TT/C) 1149.1-1990

scan function check Accomplished when control function check has been performed with all remotes. A check of master and remote station equipment by exercising a predefined component or capability. (SUB/PE) C37.1-1994

scan head A head within a scanner that sweeps across the item being scanned and transmits the contents of that item to be processed by the scanner. (C) 610.10-1994w

scan input signal A primary signal which may be used to serially precondition the scan register latches of the DUT. (C/TT) 1450-1999

scanner (1) (facsimile) That part of the facsimile transmitter which systematically translates the densities of the subject copy into signal waveform. *See also:* scanning. (COM) 168-1956w

(2) (A) A multiplexing arrangement that sequentially connects one channel to a number of channels. **(B)** An arrangement that progressively examines a surface for information. *See also:* feedback control system. (IA/ICTL/APP/IAC) [69], [60]

(3) (test, measurement, and diagnostic equipment) A device that sequentially samples a number of data points. *See also:* optical scanner; flying spot scanner; visual scanner. (MIL) [2]

(4) (computer graphics) A graphical input device that examines a spatial pattern and generates analog or digital signals, which can be used as input to a computer system. (C) 610.6-1991w

(5) (A) A graphic input device that automatically digitizes images for input to a computer. *See also:* bar code scanner; scan head; magnetic ink scanner; optical scanner. **(B)** Any device that is capable of scanning. (C) 610.10-1994

scanning (1) (navigation aids) A programmed motion given to the major lobe of an antenna for the purpose of searching a larger angular region than can be covered with a single direction of the beam, or for measuring angular location of a target; also, the analogous process using range gates or frequency domain filters. *See also:* supervisory control. (AES/GCS) 172-1983w, 686-1997

(2) (television) The process of analyzing or synthesizing successively, according to a predetermined method, the light values of picture elements constituting a picture area. (BT) 202-1954w

(3) (facsimile) The process of analyzing successively the densities of the subject copy according to the elements of a predetermined pattern. *Note:* The normal scanning is from left to right and top to bottom of the subject copy as when reading a page of print. Reverse direction is from right to left and top to bottom of the subject copy. (COM) 168-1956w

(4) (telephone switching systems) The periodic examination of circuit states under common control. (COM) 312-1977w

(5) (of an antenna beam) A repetitive motion given to the major lobe of an antenna. (AP/ANT) 145-1993

(6) The process of examining information in a systematic manner. (C) 610.10-1994w

scanning, high-velocity *See:* high-velocity scanning.

scanning line (television) A single continuous narrow strip that is determined by the process of scanning. *Note:* In most television systems, the scanning lines that occur during the retrace intervals are blanked. The total number of scanning lines is numerically equal to the ratio of line frequency to frame frequency. (BT/AV) 201-1979w

scanning linearity (television) A measure of the uniformity of scanning speed during the unblanked trace interval. (BT/AV) 201-1979w

scanning line frequency *See:* stroke speed.

scanning line length (facsimile) The total length of scanning line is equal to the spot speed divided by the scanning line frequency. *Note:* This is generally greater than the length of the available line. *See also:* scanning. (COM) 168-1956w

scanning loss (radar system employing a scanning antenna) The reduction in sensitivity, usually expressed in decibels, due to scanning across a target, compared with that obtained when the beam is directed constantly at the target. *See also:* antenna. (AP/ANT) 145-1983s

(2) (A) In a radar using a continuously scanning beam, the reduction in sensitivity due to motion of the beam between transmission and reception of the signal (sometimes called transit-time loss). **(B)** In an electronic scanning radar, the reduction in signal power due to scanning of the beam from broadside (the direction normal to the array face). *See also:* beamshape loss. (AES) 686-1997

scanning, low-velocity *See:* low-velocity scanning.

scanning speed (television) The time rate of linear displacement of the scanning spot. (BT/AV) 201-1979w

scanning spot (1) (television) The area with which the scanned area is being explored at any instant in the scanning process. *See also:* television. (PE/EEC) [119]

(2) (facsimile) The area on the subject copy viewed instantaneously by the pickup system of the scanner. *See also:* scanning. (COM) 168-1956w

scanning spot, *X* dimension (facsimile) The effective scanning-spot dimension measured in the direction of the scanning line on the subject copy. *Note:* The numerical value of this will depend upon the type of system used. *See also:* scanning. (COM) 168-1956w

scanning spot, *Y* dimension (facsimile) The effective scanning-spot dimension measured perpendicularly to the scanning line on the subject copy. *Note:* The numerical value of this will depend upon the type of system used. *See also:* scanning. (COM) 168-1956w

scanning supervisory system (station control and data acquisition) A system in which the master controls all information exchange. The normal state is usually one of repetitive communication with the remote stations. (SWG/PE/SUB) C37.100-1992, C37.1-1994

scanning velocity (spectrum analyzer) Frequency span divided by sweep time. (IM) 748-1979w

scan output signal A primary signal which may be used to serially observe the contents of the scan register latches of the DUT. (C/TT) 1450-1999

scan path The shift-register path through a circuit designed using the scan design technique. (TT/C) 1149.1-1990

scan pitch (facsimile) The number of scanning lines per unit length measured perpendicular to the direction of scanning. (COM) 167-1966w

scan rate (data transmission) The quantity of remote functions or stations that a master station can poll in a given time period. (PE) 599-1985w

scan sector The angular interval over which the major lobe of an antenna is scanned. (AP/ANT) 145-1993

scan test methodology A test methodology that utilizes shift register latches to precondition and observe modeled faults within the DUT. Scan tests typically consist of a serial preconditioning (load via scan inputs), parallel vectors to clock/transition the DUT, and then a serial observation (unload via the scan outputs). (C/TT) 1450-1999

scan time (1) (sequential events recording systems) The time required to examine the state of all inputs. (PE/EDPG) [1]

(2) (acousto-optic deflector) The time for the light beam to be scanned over the angular swing of the deflector. (UFFC) [17]

scan vectors A representation of test information containing lists of states that are to be shifted into or out of the scan pins on the device. *Note:* Scan vectors imply the use of scan test methodology in the design of the device under test. (C/TT) 1450-1999

scare rope *See:* safety life line.

scatterband (navigation aids) (interrogation systems) The total bandwidth occupied by the various received signals from interrogators operating with carriers on the same nominal radio frequency; the scatter results from the individual deviations from the nominal frequency. (AES/GCS) 172-1983w

scattered radiation An electromagnetic field resulting from currents induced in a secondary, conducting or dielectric object by electromagnetic waves incident on that object from one or more primary sources. (NIR) C95.1-1999

scattered wave An electromagnetic wave that results when an incident wave encounters the following:

— One or more discrete scattering objects
— A rough boundary between two media
— Continuous irregularities in the complex constitutive parameters of a medium

(AP/PROP) 211-1997

scatter/gather Data structures in memory that are sequentially ordered in virtual space may be sparcely ordered in physical space. In order to access this data structure with a physical device (such as a DMA controller), the device may need to redirect its address pointer to different physical pages of memory while transferring that data. (C/BA) 1014.1-1994w

scattering (1) (laser maser) The angular dispersal of power from a beam of radiation (or the perturbation of the field distribution of a resonance mode) either with or without a change in frequency, caused for example by inhomogeneities or nonlinearities of the medium or by irregularities in the surfaces encountered by the beam. (LEO) 586-1980w
(2) (fiber optics) The change in direction of light rays or photons after striking a small particle or particles. It may also be regarded as the diffusion of a light beam caused by the inhomogeneity of the transmitting medium. *See also:* leaky mode; unbound mode; Rayleigh scattering; waveguide scattering; refractive index; mode; nonlinear scattering; material scattering. (Std100) 812-1984w
(3) (data transmission) The production of waves of changed direction, frequency, or polarization when radio waves encounter matter. *Note:* The term is frequently used in a narrower sense, implying a disordered change in the incident energy. (PE) 599-1985w
(4) A process in which the energy of a traveling wave is dispersed in direction by means other than reflection and refraction. (AP/PROP) 211-1997

scattering coefficient Element of the scattering matrix. *See also:* scattering matrix. (IM/HFIM) [40]
(2) (A) The scattering cross-section per unit illuminated area of a surface expressed in square meters per square meter:

$$\sigma^0_{pq} = \frac{d\sigma_{pq}}{dA}$$

where p and q are polarization indices. **(B)** The scattering cross-section per unit volume of a medium containing discrete scatterers or random variations of refractive index. It is expressed in meters squared per cubic meter and is often designated σ_V. *Note:* The scattering coefficient may be monostatic (backscatter), when the transmitter and receiver are collocated, or bistatic, when they are not. *See also:* scattering cross section. (AP/PROP) 211-1997

scattering cross section (1) (radio-wave propagation) The projected area required to intercept and isotropically radiate the same power as a scatterer (target) scatters toward the receiver. The scattering cross-section is calculated from the relationship:

$$\sigma_{pq} = \lim_{R\to\infty}\left[4\pi R^2 \frac{\langle|\vec{E}^s_p|^2\rangle}{\langle|\vec{E}^i_q|^2\rangle}\right]$$

where
R = the distance between the scatterer and the receiver
$\vec{E}^s_p$ = the p-polarized component of the scattered electric field at the receiver
$\vec{E}^i_q$ = a q-polarized incident electric field at the scatterer.

The incident field is assumed to be planar over the extent of the target. (AP/PROP) 211-1997
(2) For a scattering object and an incident plane wave of a given frequency, polarization, and direction, an area that, when multiplied by the power flux density of the incident wave, would yield sufficient power that could produce, by isotropic radiation, the same radiation intensity as that in a given direction from the scattering object. *Note:* The scattering cross section is equal to 4π times the ratio of the radiation intensity of the scattered wave in a specified direction to the power flux density of the incident plane wave. *See also:* monostatic cross section; radar cross section; bistatic cross section. (AP/ANT) 145-1993

scattering loss (1) (laser maser) That portion of the loss in received power which is due to scattering. (LEO) 586-1980w
(2) That part of the transmission loss that is due to scattering within the medium or due to roughness of the reflecting surface. (SP/ACO) [32]

scattering matrix (1) (waveguide components) A square array of complex numbers consisting of the transmission and reflection coefficients of a waveguide component. As most commonly used, each of these coefficients relates the complex electric field strength (or voltage) of a reflected or transmitted wave to that of an incident wave. The subscripts of a typical coefficient S_{ij} refer to the output and input ports related by the coefficient. These coefficients, which may vary with frequency, apply at a specified set of input and output reference planes. (MTT/AP/ANT) 148-1959w, [35]
(2) An n × n (square) matrix used to relate incident waves and reflected waves for an n-port network. If the incident wave quantities for the ports are denoted by the vector A and the reflected wave quantities by the vector B then the scattering matrix S is defined such that $B = SA$. where:

$$a_i = \frac{1}{\sqrt{R_e Z_i}}(V_i + Z_i I_i)$$

$$b_i = \frac{1}{\sqrt{R_e Z_i}}(V_i - Z_i I_i)$$

Z_i is the port normalization impedance with $R_e Z_i > 0$. One formula for the scattering matrix is $S = [Z + R]^{-1}[Z - R]$ where Z is the open circuit impedance matrix that describes the network and R is a diagonal matrix representing the source or load resistances at each port. It should be noted that the scattering matrix is defined with respect to a specific set of port terminations. Physical interpretations can be given to the scattering coefficients for example, $|S_{ij}|^2$ is the fraction of available power that is delivered to the port termination at port i due to a source at port j. (CAS) [13]
(3) An $n \times n$ (square) matrix used to relate incident waves and reflected waves for an n-port network. (EMC) 1128-1998
(4) A 2 × 2 complex matrix which characterizes the polarized field scattered by a given object. (AP/PROP) 211-1997

scattering pattern *See:* scattering phase function.

scattering phase function The angular spectrum of a scatterer when illuminated by a plane wave. *Synonym:* scattering pattern. (AP/PROP) 211-1997

scatter read A read operation in which data from an input record is placed into non-adjacent storage areas. *Contrast:* gather write. (C) 610.10-1994w

scatter storage *See:* hashing.

SC device A static configuration device, whose logical address is set manually and cannot be changed by DC protocols. (C/MM) 1155-1992

scenario A set of initial conditions and a sequence of events used to develop, test, or apply a system, model, or simulation. (C) 610.3-1989w
(2) (A) A description of an exercise (initial conditions). A scenario is part of the session database that configures the units and platforms and places them in specific locations with specific missions. **(B)** An initial set of conditions and time line of significant events imposed on trainees or systems to achieve exercise objectives. (DIS/C) 1278.3-1996
(3) (A) A step-by-step description of a series of events that may occur concurrently or sequentially. **(B)** An account or

synopsis of a projected course of events or actions. (C/SE) 1362-1998

scheduled frequency (electric power system) The frequency that a power system or an interconnected system attempts to maintain. (PE/PSE) 94-1970w

scheduled frequency offset (electric power system) Scheduled system frequency minus rated frequency. This offset is usually initiated to correct the system time error. (PE/PSE) 94-1991w

scheduled interruption (1) (electric power system) An interruption caused by a scheduled outage. *See also:* outage. (PE/PSE) [54], 346-1973w

(2) A loss of electric power that results when a component is deliberately taken out of service at a selected time, usually for the purposes of construction, preventative maintenance, or repair. *Notes:* 1. This derives from transmission and distribution applications and does not apply to generation interruptions. 2. The key test to determine if an interruption should be classified as a forced or scheduled interruption is as follows. If it is possible to defer the interruption when such deferment is desirable, the interruption is a scheduled interruption; otherwise, the interruption is a forced interruption. Deferring an interruption may be desirable, for example, to prevent overload of facilities or interruption of service to customers. (PE/T&D) 1366-1998

scheduled maintenance (generation) Capability which has been scheduled to be out of service for maintenance. (PE/PSE) 346-1973w

scheduled net interchange (control area) (electric power system) The net power flow that a control area strives to maintain on its area tie lines in the absence of control biases. (PE/PSE) 94-1991w

scheduled outage (1) (electric power system) A loss of electric power that results when a component is deliberately taken out of service at a selected time, usually for purposes of construction, preventive maintenance, or repair. *Notes:* 1. This derives from transmission and distribution applications and does not necessarily apply to generation outages. 2. The key test to determine if an outage should be classified as forced or scheduled is as follows. If it is possible to defer the outage when such deferment is desirable, the outage is a scheduled outage; otherwise, the outage is a forced outage. Deferring an outage may be desirable, for example, to prevent overload of facilities or an interruption of service to consumers. (PE/PSE) 346-1973w

(2) (electrical transmission facilities) An intentional manual outage that could have been deferred without increasing risk to human life, risk to property, or damage to equipment. *Note:* A manual outage is classified as scheduled if it is possible to defer the outage occurrence when such deferment is desirable. Otherwise, the outage occurrence is a forced outage. Deferring an outage occurrence may be desirable, for example,to prevent overload of facilities or an interruption of service to consumers. (PE/PSE) 859-1987w

(3) An outage that results when a component is deliberately taken out of service at a selected time, usually for purposes of construction, maintenance, or repair. (IA/PSE) 493-1997, 399-1997

scheduled outage duration (1) (electric power system) The period from the initiation of the outage until construction, preventive maintenance, or repair work is completed. (PE/PSE) 346-1973w

(2) The period from the initiation of a scheduled outage until construction, preventive maintenance, or repair work is completed and the affected component is made available to perform its intended function. (IA/PSE) 493-1997, 399-1997

scheduled outage rate (1) (electrical transmission facilities) The number of scheduled outages per unit of service time = number of scheduled outages/service time. In some studies, scheduled outage rate may be defined as the number of outage occurrences per unit of exposure time (including both service time and outage time). (PE/PSE) 859-1987w

(2) The mean number of scheduled outages of a component per unit exposure time. (IA/PSE) 493-1997, 399-1997

scheduled system frequency The frequency that a power system or an interconnected system attempts to maintain. (PE/PSE) 94-1991w

schedule, electric rate *See:* electric rate schedule.

scheduler A computer program, usually part of an operating system, that schedules, initiates, and terminates jobs. (C) 610.12-1990

schedule setter or set-point device (speed governing systems) A device for establishing or setting the desired value of a controlled variable. *See also:* speed-governing system. (PE/PSE) 94-1970w

scheduling The application of a policy to select a runnable thread to become a running thread, or to alter one or more of the thread lists. (C/PA) 9945-1-1996

scheduling allocation domain The set of processors on which an individual thread can be scheduled at any given time. (C/PA) 9945-1-1996

scheduling contention scope A property of a thread that defines the set of threads against which that thread competes for resources. For example, in a scheduling decision, threads sharing scheduling contention scope compete for processor resources. In this standard, a thread has a scheduling contention scope of either PTHREAD_SCOPE_SYSTEM or PTHREAD_SCOPE_PROCESS. (C/PA) 9945-1-1996

scheduling policy A set of rules that is used to determine the order of execution of threads to achieve some goal. In the context of this standard, a scheduling policy affects thread ordering

1) When a thread is a running thread and it becomes a blocked thread
2) When a thread is a running thread and it becomes a preempted thread
3) When a thread is a blocked thread and it becomes a runnable thread
4) When a running thread calls a function that can change the priority or scheduling policy of a thread
5) In other scheduling-policy-defined circumstances

Conforming implementations shall define the manner in which each of the scheduling policies may modify the priorities or otherwise affect the ordering of threads at each of the occurrences listed above. Additionally, conforming implementations shall define at what other circumstances and in what manner each scheduling policy may modify the priorities or affect the ordering of threads. (C/PA) 9945-1-1996

schema (1) A description of the logical structure of a database. *See also:* data model. (C) 610.5-1990w

(2) The set of rules and constraints concerning DIT structure, object class definitions, attribute types, and syntaxes that characterize the DIB. (C/PA) 1328.2-1993w, 1326.2-1993w, 1327.2-1993w, 1224.2-1993w

(3) The structure or framework used to define a data record. This includes each field's name, type, shape, dimension, and mapping. (SCC20) 1226-1998

schema definition language *See:* data definition language.

schema language *See:* data definition language.

schematic diagram (elementary diagram) A diagram that shows, by means of graphic symbols, the electrical connections and functions of a specific circuit arrangement. The schematic diagram facilitates tracing the circuit and its functions without regard to the actual physical size, shape, or location of the component device or parts. (GSD) 315-1975r

SCHEME A dialect of LISP. (C) 610.13-1993w

Scherbius machine (rotating machinery) A polyphase alternating-current commutator machine capable of generator or motor action, intended for connection in the secondary circuit of a wound-rotor induction motor supplied from a fixed-frequency polyphase power system, and used for speed and/or power-factor control. The magnetic circuit components are laminated and may be of the salient-pole type or of the cylindrical-rotor uniformly slotted type, either type having a series-connected armature reaction compensating winding as part of the field system. The control field winding may be separately or shunt-excited with or without an additional series-excited field winding. *See also:* asynchronous machine. (PE) [9]

Schering bridge A four-arm alternating-current bridge in which the unknown capacitor and a standard loss-free capacitor form two adjacent arms, while the arm adjacent to the standard capacitor consists of a resistor and a capacitor in parallel, and the fourth arm is a nonreactive resistor. (See the corresponding figure.) *Note:* Normally used for the measurement of capacitance and dissipation factor. Usually, one terminal of the source is connected to the junction of the unknown capacitor with the standard capacitor. With this connection, if the impedances of the capacitance arms are large compared to those of the resistance arms, most of the applied voltage appears across the former, the maximum test voltage being limited by the rating of the standard capacitor. If the detector and the source of electromotive force are interchanged the resulting circuit is called a conjugate Schering bridge. The balance is independent of frequency. *See also:* bridge.

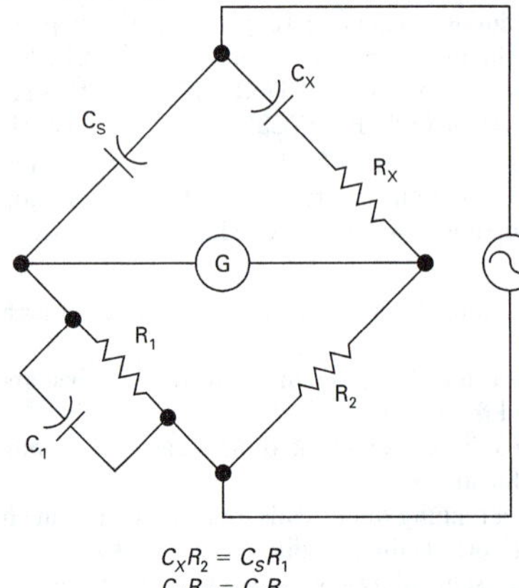

$C_X R_2 = C_S R_1$
$C_X R_X = C_1 R_1$

Schering bridge

(EEC/PE) [119]

Schlieren method The technique by which light refracted by the density variations resulting from acoustic waves is used to produce a visible image of a sound field. (SP/ACO) [32]

Schmitt trigger A solid state element that produces an output when the input exceeds a specified turn-on level, and whose output continues until the input falls below a specified turn-off level. (SWG/PE) C37.100-1981s

Schottky-barrier contact (charged-particle detectors) A metal-semiconductor contact structure in which rectification occurs that is heavily influenced by the difference in the work functions of the materials. The contacts frequently consist of an interfacial metal/semiconductor compound such as a silicide. (NPS) 325-1986s, 300-1988r, 325-1996

Schottky-barrier detector A semiconductor radiation detector in which the blocking contact is of the Schottky barrier type. (NPS) 325-1996, 300-1988r

Schottky effect *See:* Schottky emission.

Schottky emission (electron tube) The increased thermionic emission resulting from an electric field at the surface of the cathode. *See also:* electron emission. (ED) 161-1971w

Schottky noise (electron tube) The variation of the output current resulting from the random emission from the cathode. (ED) [45], [84]

Schuler tuning (inertial navigation system) (navigation aids) The application of parameter values such that accelerations do not deflect the platform system from any vertical to which it has been set; a Schuler-tuned system, if fixed to the mean surface of a nonrotating earth, exhibits a natural period of 84.4 min. (AES/GCS) 172-1983w

SCI *See:* SCI standard.

SCI standard Refers to IEEE Std 1596-1992, which provides computer-bus-like services using a collection of point-to-point unidirectional links. (C/MM) 1596.5-1993, 1596.4-1996, 1596-1992

scientific notation A notation system in which a number is expressed as a coefficient multiplied by a power of ten. (C) 1084-1986w

scintillation (1) (scintillators) The optical photons emitted as a result of the incidence of a particle or photon of ionizing radiation on a scintillator. *Note:* Optical photons unless otherwise specified are photons with energies corresponding to wavelengths between 2000 and 15 000 angstroms. *See also:* ionizing radiation; radiation. (NPS) 398-1972r
(2) (laser maser) The rapid changes in irradiance levels in a cross section of a laser beam. (LEO) 586-1980w
(3) The phenomenon of fluctuation of the amplitude of a wave caused by irregular changes in the transmission path or paths with time. *Note:* The term scintillation is sometimes used to describe fluctuations of phase and angle of arrival. *See also:* fading. (AP/PROP) 211-1997
(4) Random variations in the received signal from a complex target that can occur due to changes in aspect angle or other causes. *Note:* Because this term has been applied variously to target fluctuation and scintillation error, use of one of these more specific terms is recommended to avoid ambiguity. (AES) 686-1997

scintillation counter The combination of scintillation-counter heads and associated circuitry for detection and measurement of ionizing radiation. (NPS) 398-1972r

scintillation-counter cesium resolution The scintillation-counter energy resolution for the gamma ray or conversion electron from cesium-137. *See also:* scintillation counter. (NPS) 398-1972r

scintillation-counter energy resolution A measure of the smallest difference in energy between two particles or photons of ionizing radiation that can be discerned by the scintillation counter. Quantitatively, it is the fractional standard deviation (σ/E_1) of the energy distribution curve. *Note:* The fractional full width at half maximum of the energy distribution curve (FWHM/E_1) is frequently used as a measure of the scintillation-counter energy resolution where E_1 is the mode of the distribution curve. *See also:* scintillation counter. (NPS) 398-1972r

scintillation-counter energy-resolution constant The product of the square of the scintillation-counter energy resolution, expressed as the fractional full width at half maximum (FWHM/E_1), and the specified energy. *See also:* scintillation counter. (NPS) 175-1960w

scintillation counter head The combination of scintillators and phototubes or photocells that produces electric pulses or other electric signals in response to ionizing radiation. *See also:* scintillation counter; phototube. (NPS) 175-1960w

scintillation-counter time discrimination A measure of the smallest interval of time between two individually discernible events. Quantitatively, it is the standard deviation of the time-interval curve. *Note:* The full width at half maximum of the time-interval curve is frequently used as a measure of the time discrimination. *See also:* scintillation counter. 160-1957w

scintillation decay time The time required for the rate of emission of optical photons of a scintillation to decrease from 90% to 10% of its maximum value. *Note:* Optical photons, for the purpose of this Standard, are photons with energies corresponding to wavelengths between 2000 and 15 000 angstroms. *See also:* scintillation counter. (NPS) 398-1972r

scintillation duration The time interval from the emission of the first optical photon of a scintillation until 90% of the optical photons of the scintillation have been emitted. *Note:* Optical photons are photons with energies corresponding to wavelengths between 2000 and 15 000 angstroms. *See also:* scintillation counter. (NPS) 398-1972r

scintillation error Error in radar-derived target position or Doppler frequency caused by interaction of the scintillation spectrum with frequencies used in sequential measurement techniques. *Note:* Not to be confused with glint. (AES) 686-1997

scintillation index The ratio of the second moment to the first moment squared of the intensity. (AP/PROP) 211-1997

scintillation rise-time The time required for the rate of emission of optical photons of a scintillation to increase from 10% to 90% of its maximum value. *Note:* Optical photons are photons with energies corresponding to wavelengths between 2000 and 15 000 angstroms. *See also:* scintillation counter. (NPS) 398-1972r

scintillator The body of scintillator material together with its container. *See also:* scintillation counter. (NPS) 398-1972r

scintillator conversion efficiency The ratio of the optical photon energy emitted by a scintillator to the incident energy of a particle or photon of ionizing radiation. *Note:* The efficiency is generally a function of the type and energy of ionizing radiation. Optical photons are photons with energies corresponding to wavelengths between 2000 and 15 000 angstroms. *See also:* scintillation counter. (NPS) 175-1960w

scintillator material A material that emits optical photons in response to ionizing radiation. *Notes:* 1. There are five major classes of scintillator materials, namely:

a) inorganic crystals such as NaI(Tl) single crystals, ZnS(Ag) screens;
b) organic crystals (such as, anthracene, trans-stilbene);
c) solution scintillators: (1) liquid, (2) plastic, (3) glass;
d) gaseous scintillators;
e) Cerenkov scintillators.

2. Optical photons are photons with energies corresponding to wavelengths between 2000 and 15 000 angstroms. *See also:* scintillation counter. (NPS) 398-1972r

scintillator-material total conversion efficiency The ratio of the optical photon energy produced to the energy of a particle or photon of ionizing radiation that is totally absorbed in the scintillator material. *Note:* The efficiency is generally a function of the type and energy of the ionizing radiation. Optical photons are photons with energies corresponding to wavelengths between 2000 and 15 000 angstroms. *See also:* scintillation counter. (NPS) 398-1972r

scintillator photon distribution (in number) The statistical distribution of the number of optical photons produced in the scintillator by total absorption of monoenergetic particles. *Note:* Optical photons are photons with energies corresponding to wavelengths between 2000 and 15 000 angstroms. *See also:* scintillation counter. (NPS) 398-1972r

scissoring A computer graphics technique in which portions of display elements that lie outside of the physical bounds of a window or view volume are removed. *See also:* wrap-around; clipping. (C) 610.6-1991w

SCN *See:* specification change notice.

scope (1) (navigation aids) The face of a cathode-ray tube or a display of similar appearance. A colloquial abbreviation of oscilloscope. (AES/GCS) 172-1983w
(2) (scheme programming language) The region of a program's source text that is associated with a linguistic construct. Normally used with "variable" to describe the region over which a variable is bound: "the scope of a variable." (C/MM) 1178-1990r
(3) The face of a cathode-ray tube or a display of similar appearance. *Note:* The term *scope* is a colloquial abbreviation of the word *oscilloscope*. (AES) 686-1997
(4) *See also:* transit. (T&D/PE) 524-1992r

scored card A special card that contains one or more scored lines to facilitate precise folding or separation of certain parts of the card. *See also:* processable scored card. (C) 610.10-1994w

scoring system (motion-picture production) (electroacoustics) A recording system used for recording music to be reproduced in timed relationship with a motion picture. (SP) [32]

scotopic spectral luminous efficiency function (light-emitting diodes) (photometric standard observer for scotopic vision) (V′) The ratio of the radiant flux at wavelength m, to that at wavelength λ, when the two fluxes produce the same scotopic luminous sensations under specified photometric conditions, lm, being chosen so that the maximum value of this ratio is unity. Unless otherwise indicated, the values used for the spectral luminous efficiency function relate to scotopic vision by the photometric standard observer having the characteristics laid down by the International Commission on Illumination. (ED) [127]

scotopic vision (illuminating engineering) Vision mediated essentially or exclusively by the rods. It is generally associated with adaptation to a luminance below about 0.034 cd/m^2, (2.2 $\times$ 10^{-5} cd/in^2), (0.01 fL). (EEC/IE) [126]

Scott-connected transformer, interlacing impedance voltage The single-phase voltage applied from the midtap of the main transformer winding to both ends, connected together, that is sufficient to circulate in the supply lines a current equal to the three-phase line current. The current in each half of the winding is 50% of this value. *See also:* efficiency. (IA) [61]

Scott-connected transformer per-unit resistance The measured watts expressed in per-unit on the base of the rated kilovolt-ampere of the teaser winding. (IA) [61]

Scott or T-connected transformer (power and distribution transformers) An assembly used to transfer energy from a three-phase circuit to a two-phase circuit, or vice versa; or from a three-phase circuit to another three-phase circuit. The assembly consists of a main transformer with a tap at its midpoint connected directly between of the phase wires of a three-phase circuit, and of a teaser transformer connected between the mid-tap of the main transformer and a third phase wire of the three-phase circuit. The other windings of the transformers may be connected to provide either a two-phase or a three-phase output. Alternatively, this may be accomplished with an assembly utilizing a three-legged core with main and teaser coil assemblies located on the two outer legs, and with a center leg which has no coil assembly and provides a common magnetic circuit for the two outer legs. *See also:* teaser transformer; interlacing impedance voltage of a Scott-connected transformer; main transformer. (PE/TR) C57.12.80-1978r

SCR *See:* semiconductor controlled rectifier; silicon controlled rectifier; reverse-blocking triode thyristor.

scram (power operations) The rapid shutdown of a nuclear reactor. Usually, a scram is accomplished by rapid insertion of safety or control rods, or both. Emergencies or deviations from normal operation may require scramming the reactor by manual or automatic means. (PE/PSE) 858-1987s

scraper hoist A power-driven hoist operating a scraper to move material (generally ore or coal) to a loading point. (EEC/PE) [119]

scratch (A) To physically erase data from its medium. **(B)** To logically delete the identification of data from its medium. (C) 610.5-1990

scratch file A file used as a work area to hold data temporarily. (C) 610.5-1990w

scratchpad area (SPA) A portion of computer memory shared by a set of computer programs or processes for some special purpose. For example, memory used by two programs for interprocess communication. *Synonym:* scratchpad RAM. (C) 610.5-1990w, 610.10-1994w

scratchpad memory *See:* temporary storage.

scratchpad RAM *See:* scratchpad area.

screen (1) (rotating machinery) A port cover with multiple openings used to limit the entry of foreign objects. (IA/APP) [90]

(2) (cathode-ray tubes) The surface of the tube upon which the visible pattern is produced. *See also:* electrode. (ED) 161-1971w

(3) A rectangular region of columns and lines on a terminal display. A screen may be a portion of a physical display device or may occupy the entire physical area of the display device. (C/PA) 9945-2-1993

(4) The portion of a display that is visible on the display device. A screen may show part of a page, an entire page, or several pages. *See also:* display device. (PE/NP) 1289-1998

(5) *See also:* display screen. (C) 610.10-1994w

screened conductor cable A cable in which the insulated conductor or conductors is/are enclosed in a conducting envelope or envelopes. (PE/IC/TR) C57.15-1968s

screen editor *See:* full-screen editor.

screen factor (electron-tube grid) The ratio of the actual area of the grid structure to the total area of the surface containing the grid. *See also:* electron tube. (ED) [45], [84]

screen font A font designed for use on a display device. *Note:* Usually matches closely the font used when printing. *Synonym:* graphical user interface font. (C) 610.10-1994w

screen grid A grid placed between a control grid and an anode, and usually maintained at a fixed positive potential, for the purpose of reducing the electrostatic influence of the anode in the space between the screen grid and the cathode. *See also:* grid; electrode. (ED) 161-1971w

screen-grid modulation Modulation produced by application of a modulating voltage between the screen grid and the cathode of any multigrid tube in which the carrier is present. (BT) 182A-1964w

screen image *See:* display image.

screening (telephone switching systems) The ability to accept or reject calls by using trunk or line class or trunk or line number information. (C) [85]

screening measurements Measurements made to detect radioactive material under routine conditions, but not used to quantify the amount of a given radionuclide. (NI) N42.23-1995

screening test A test, or combination of tests, intended to remove unsatisfactory items or those likely to exhibit early failures. *See also:* reliability. (R) [29]

screen protected *See:* guarded.

screen size The diameter of a cathode ray tube outside of its housing or, for a non-round tube, the length of the maximum diagonal of the display space after the tube has been mounted inside its housing. (C) 610.6-1991w

screen, viewing *See:* viewing area.

SC resource manager A resource manager that supports static configuration and does not support dynamic configuration of VXIbus devices. (C/MM) 1155-1992

screw machine (elevators) An electric driving machine, the motor of which raises and lowers a vertical screw through a nut with or without suitable gearing, and in which the upper end of the screw is connected directly to the car frame or platform. The machine may be of direct or indirect drive type. (EEC/PE) [119]

SCRIBE A text-formatting language in which formatting commands are embedded in the text, then processed into a formatted document. (C) 610.13-1993w

SCRIPT A text-formatting language in which formatting commands are embedded in the text, then processed into a formatted document. *Note:* SCRIPT is a forerunner to DCF. (C) 610.13-1993w

script An area of nonvolatile memory reserved for user interface commands to be evaluated at particular times during the Open Firmware start-up sequence. (C/BA) 1275-1994

scroll To move the representation of data vertically or horizontally relative to the terminal screen. There are two types of scrolling:

1) The cursor moves with the data
2) The cursor remains stationary while the data moves

(C/PA) 9945-2-1993

scroll bar A visual user interface control, associated with a scrollable area, that indicates to a user that more information is available and can be scrolled into view. (C) 1295-1993w

scrolled window A window that presents information that exceeds the space available for display. The user uses the scroll bar to bring the contents currently outside the display area into view. (C) 1295-1993w

scrolling (1) (word processing) The process of moving text across a display screen to create the effect of a viewing window moving on a large page of a document. An operator may scroll left, right, up, or down in a document. *See also:* reverse scrolling. (C) 610.2-1987

(2) (computer graphics) The process of moving an entire display image in such a manner that new data appears within the viewport as old data disappears, to give a visual impression of vertical movement of the image. *Note:* The term scrolling is sometimes used to mean vertical or horizontal movement. *Contrast:* panning. (C) 610.6-1991w

(3) A method of viewing and moving the data displayed in which the data rolls continuously behind a fixed display frame. (PE/NP) 1289-1998

scrubber The node that marks packets as they go past in a ringlet, and discards any previously marked packet. This prevents damaged or misaddressed packets from circulating indefinitely. The scrubber also performs other housekeeping tasks for the ringlet. There is always exactly one scrubber on a ringlet. Normal nodes may all have scrubber capability built in, but exactly one is enabled as scrubber per ringlet. Often the scrubber will take responsibility for initializing a ringlet, but this could be done by another (unique) node. (C/MM) 1596-1992

SC system A VXIbus system with no DC devices. (C/MM) 1155-1992

sculling error (inertial sensors) (strapdown inertial system) A system error resulting from the combined input of linear vibration along one axis and an angular oscillation, at the same frequency, around a perpendicular axis. In the computer processing, an apparent rectified acceleration is produced along an axis perpendicular to these two axes. (AES/GYAC) 528-1994

scuzzy Colloquial pronunciation for "SCSI." *See also:* small computer systems interface. (C) 610.10-1994w

SDC *See:* self-damping conductor.

SDD *See:* software design description.

SDL *See:* Specification and Description Language; software development library.

SDN *See:* software defined network.

SDP *See:* software development plan.

SDR *See:* system design review.

SDS *See:* sparse data scan; sequential data set.

SDU *See:* service data unit.

SDV (segment delay value (ARCHIVE)) *See:* Segment Delay Value.

SE *See:* segment extender.

seal (window) (in a waveguide) A gastight or watertight membrane or cover designed to present no obstruction to radio-frequency energy. *See also:* waveguide. (AP/ANT) [35]

sealable equipment Equipment enclosed in a case or cabinet that is provided with a means of sealing or locking so that live parts cannot be made accessible without opening the enclosure. The equipment may or may not be operable without opening the enclosure. (NESC/NEC) [86]

seal, double electric conductor (nuclear power generating station) An assembly of two single electric conductor seals in series and arranged in such a way that there is a double pressure barrier seal between the inside and the outside of the containment structure along the axis of the conductors. (IM) [76]

sealed (1) (power and distribution transformers) So constructed that the enclosure will remain hermetically sealed within specified limits of temperature and pressure. (PE/TR) C57.12.80-1978r

(2) (rotating machinery) Provided with special seals to minimize either the leakage of the internal coolant out of the enclosure or the leakage of medium surrounding the enclosure into the machine. *See also:* asynchronous machine. (IA/APP) [90]

sealed-beam headlamp (illuminating engineering) An integral optical assembly designed for headlighting purposes, identified by the name "Sealed Beam" branded on the lens. (EEC/IE) [126]

sealed bushing An oil-filled bushing in which the oil is contained within the bushing and not allowed to mix with the oil of the apparatus on which it is used. (PE/TR) C57.19.03-1996

sealed cell (1) (lead storage batteries) (nuclear power generating station) A cell in which the only passage for the escape of gases from the interior of the cell is provided by a vent of effective spray-trap design adapted to trap and return to the cell particles of liquid entrained in the escaping gases. (PE/EDPG) 484-1987s

(2) A sealed cell (or battery) is one that has no provision for the addition of water or electrolyte or for external measurement of electrolyte specific gravity. (NESC/NEC) [86]

sealed dry-type transformer, self-cooled (power and distribution transformers) (class GA) A dry-type self-cooled transformer with a hermetically sealed tank. *Note:* The insulating gas may be air, nitrogen, or other gases (such as fluorocarbons) with high dielectric strength. (PE/TR) C57.94-1982r, C57.12.80-1978r

sealed end (cable) (shipping seal) The end fitted with a cap for protection against the loss of compound or the entrance of moisture. (EEC/AWM) [91]

sealed refrigeration compressor (hermetic type) A mechanical compressor consisting of a compressor and a motor, both of which are enclosed in the same sealed housing, with no external shaft or shaft seals, the motor operating in the refrigerant atmosphere. *See also:* appliance. (NESC) [86]

sealed relay contacts A contact assembly that is sealed in a compartment separate from the rest of the relay. (EEC/REE) [87]

sealed-tank system (1) (power and distribution transformers) A method of oil preservation in which the interior of the tank is sealed from the atmosphere and in which the gas plus the oil volume remains constant over the temperature range. (PE/TR) C57.12.80-1978r

(2) A method of oil preservation in which the interior of the tank is sealed from the atmosphere and in which the gas volume plus the oil volume remains constant. (PE/TR) C57.15-1999

sealed transformer (power and distribution transformers) A dry-type transformer with a hermetically sealed tank. (PE/TR) C57.12.80-1978r

sealed tube An electron tube that is hermetically sealed. *Note:* This term is used chiefly for pool-cathode tubes. (ED) [45]

sealing current *See:* sealing voltage.

sealing gap The distance between the armature and the center of the core of a magnetic circuit-closing device when the contacts first touch each other. *See also:* electric controller. (IA/ICTL) 74-1958w

sealing voltage (contactors) The voltage (or current) necessary to complete the movement of the armature of a magnetic circuit-closing device from the position at which the contacts first touch each other. *Synonym:* sealing current. *See also:* control switch; contactor. (QUL) 268-1982s

seal-in relay An auxiliary relay that remains picked up through one of its own contacts which bypasses the initiating circuit until deenergized by some other device. (SWG/PE) C37.100-1992

seal, pressure barrier *See:* pressure barrier seal.

seal, single electric conductor *See:* single electric conductor seal.

search (1) (information processing) To examine a set of items for those that have a desired property. *See also:* dichotomizing search; binary search. (C/C) [20], [85]

(2) (test, measurement, and diagnostic equipment) The scanning of information contained on a storage medium by comparing the information of each field with a predetermined standard until an identity is obtained. (IM/WM&A) 194-1977w

(3) (A) (data management) The examination of a set of items to find all those having a desired property or properties. For example, to find all items in a file that meet some search criterion. **(B) (data management)** To examine a set of items as in definition (A). **(C) (data management)** To retrieve the results of an examination as in definition (A). **(D) (data management)** To retrieve the first item witin a set of items as in definition (A). *Note:* The use of the term "search" in place of the term "seek" is deprecated in IEEE Std 610.5-1990. (C) 610.5-1990

search argument In a search, the value compared with the search key of each item in the set being searched. *See also:* condition. (C) 610.5-1990w

search criterion In a search, the relationship that a search key must have to the search argument in order for the search to be successful. For example, "NAME equals 'SMITH;'" "SALARY greater than 10000." (C) 610.5-1990w

search cycle That portion of a search that is repeated for each item in the set being searched. (C) 610.5-1990w

search key In a search, the key within each item in the set being searched that is compared to the search argument. *Synonym:* seek key. (C) 610.5-1990w

search length (A) For a node in a search tree, the number of nodes that must be examined in order to find that node. **(B)** For a search tree, the average search length as in definition (A) for all nodes in the tree. (C) 610.5-1990

searchlight (illuminating engineering) A projector designed to produce an approximately parallel beam of light. *Note:* The optical system of a searchlight has an aperture of greater than 20 cm (8 in). (EEC/IE) [126]

searchlighting The process of projecting a radar beam continuously at a particular object or in a particular direction as contrasted to scanning. (AES) 686-1997

search memory *See:* associative memory.

search radar (1) (navigation aids) A radar used primarily for the detection of targets in a particular volume of interest. (AES/GCS) 172-1983w

(2) A radar used primarily for the initial detection of targets in a particular volume of interest. (AES) 686-1997

search time (A) The time required to locate a particular item of data in a storage medium. **(B)** The time interval required for the read/write head of a rotating storage device to locate a particular record on a track corresponding to a given address or key. *See also:* rotational delay; seek time. (C) 610.10-1994

search tree (A) A tree into which items in a set are placed in order for the set to be searched. The tree is traversed according to some searching algorithm, making key comparisons until the search argument is found or the algorithm is halted. For example, a B-tree. **(B)** A multiways tree of order m in which each nonterminal node may contain $(m-1)$ key values and each terminal node, called a leaf, contains associated data for one of the key values contained in its parent node. Each subtree is used to contain all the items with key values falling in the intervals formed by the key values contained in its root node. *See also:* B-tree; binary search tree; digital search tree. (C) 610.5-1990

sea return (navigation aids) The radar response from the sea surface. (AES/GCS) 686-1997, 172-1983w

season A calendar-specified period used for activation of rate schedules. (AMR/SCC31) 1377-1997

seasonal derated hours (power system measurement) (electric generating unit reliability, availability, and productivity) The available hours during which a seasonal derating was in effect. (PE/PSE) 762-1987w

seasonal derating (electric generating unit reliability, availability, and productivity) The difference between maximum capacity and dependable capacity. (PE/PSE) 762-1987w

seasonal diversity Load diversity between two or more electric systems that occurs when their peak loads are in different seasons of the year. (PE/PSE) 858-1993w, 346-1973w

seasonal unavailable generation (electric generating unit reliability, availability, and productivity) The difference between the energy that would have been generated if operating continuously at maximum capacity and the energy that would have been generated if operating continuously at dependable capacity, calculated only during the time the unit was in the available state.

$$\text{SUG} = \text{equivalent seasonal derated hours} \cdot \text{maximum capacity} = \text{ESDH} \cdot \text{MC}$$

(PE/PSE) 762-1987w

season cracking Cracking resulting from the combined effect of corrosion and internal stress. A term usually applied to stress-corrosion cracking of brass. (IA) [59]

SEC *See:* secondary-electron conduction; secondary-electron conduction camera tube.

second (metric practice) The duration of 9 192 631 770 periods of the radiation corresponding to the transition between the two hyperfine levels of the ground state of the cesium-133 atom. (adopted by 13 General Conference on Weights and Measures 1967). *Note:* This definition supersedes the ephemeris second as the unit of time. (QUL) 268-1982s

secondary (A) Operates after the primary device; for example, secondary arcing contacts. **(B)** Second in preference. **(C)** Referring to auxiliary or control circuits as contrasted with the main circuit; for example, secondary disconnecting devices, secondary and control wiring. **(D)** Referring to the energy output side of transformers or the conditions (voltages) usually encountered at this location; for example, secondary fuse, secondary unit substation. (SWG/PE) C37.100-1992

secondary access method A collection of techniques designed to allow efficient access to all the target data or data records associated with a set of stated secondary key values in a query. (C) 610.5-1990w

secondary address An address for use within a device. It is provided by a secondary address cycle that loads the NTA register of the device following a primary address cycle or a data cycle. (NID) 960-1993

secondary address cycle A data cycle in which a master uses the address/data (AD) lines to load a secondary address into the NTA register of a device. (NID) 960-1993

secondary alarm station (SAS) A continuously manned location that is capable of providing backup security system monitoring and communications functions. (PE/NP) 692-1997

secondary and control wiring Wire used with switchgear assemblies for control circuits and for connections between instrument transformers' secondaries, instruments, meters, relays, or other equipment. *Synonym:* small wiring. (SWG/PE) C37.100-1992

secondary arcing contacts (of a switching device) The contacts on which the arc of the arc-shunting-resistor current is drawn and interrupted. (SWG/PE) C37.100-1992

secondary arrester A surge protective device that is intended to be connected to the low-voltage ac supply mains (1000 V rms and less, frequency between 48 and 62 Hz) at locations between and including the secondary terminals of the distribution transformer and the main service entrance panel. (PE) C62:34-1996

secondary boot program A client program whose purpose is to load and execute another client program. (C/BA) 1275-1994

secondary bus A collection of signals that provides the system with an alternate mechanism for exchanging data between boards as a means to recover from faults in the primary bus. (C/BA) 896.9-1994w, 896.3-1993w

secondary calibration (nuclear power generating station) (monitoring radioactivity in effluents) The determination of the response of a system with an applicable source whose effect on the system was established at the time of a primary calibration. (NI/EEC) N42.18-1980r, [81]

secondary current rating The secondary current existing when the transformer is delivering rated kilovolt-amperes at rated secondary voltage. *See also:* transformer. (PE/TR) [57]

secondary disconnecting devices (of a switchgear assembly) Self-coupling separable contacts provided to connect and disconnect the auxiliary and control circuits between the removable element and the housing. (SWG/PE) C37.100-1992

secondary distribution feeder A feeder operating at secondary voltage supplying a distribution circuit. (EEC/AWM) [91]

secondary distribution mains The conductors connected to the secondaries of distribution transformers from which consumers' services are supplied. *See also:* center of distribution. (EEC/AWM) [91]

secondary distribution network A network consisting of secondary distribution mains. *See also:* center of distribution. (EEC/AWM) [91]

secondary distribution system A low-voltage alternating-current system that connects the secondaries of distribution transformers to the consumers' services. *See also:* alternating-current distribution; center of distribution. (EEC/AWM) [91]

secondary distribution trunk line A line acting as a main source of supply to a secondary distribution system. *See also:* center of distribution. (EEC/AWM) [91]

secondary electric shock An electric shock not sufficiently severe to cause direct physiological harm. Nevertheless, such a shock could result in injury from involuntary muscular response. (T&D/PE) 539-1990

secondary electron (thermionics) An electron detached from a surface during secondary emission by an incident electron. *See also:* electron emission. (ED) [45], [84]

secondary-electron conduction (SEC) The transport of charge under the influence of an externally applied field in low-density structured materials by free secondary electrons traveling in the interparticle spaces (as opposed to solid-state conduction). *See also:* camera tube. (ED) [45]

secondary-electron conduction camera tube (SEC) A camera tube in which an electron image is generated by a photocathode and focused on a target composed of a backplate and a secondary-electron-conduction layer that provides charge amplification and storage. *See also:* camera tube. (ED) [45]

secondary emission Electron emission from solids or liquids due directly to bombardment of their surfaces by electrons or ions. *See also:* electron emission. (ED) 161-1971w

secondary-emission characteristic (thermionics) (surface) The relation, generally shown by a graph, between the secondary-emission rate of a surface and the voltage between the source of the primary emission and the surface. *See also:* electron emission. (ED) [45], [84]

secondary-emission crossover voltage (charge-storage tubes) The voltage of a secondary-emitting surface, with respect to cathode voltage, at which the secondary-emission ratio is unity. The crossovers are numbered in progression with increasing voltage. *Note:* The qualifying phrase secondary-emission is frequently dropped in general usage. *See also:* charge-storage tube. (ED) 158-1962w

secondary-emission ratio (electrons) The average number of electrons emitted from a surface per incident primary electron. *Note:* The result of a sufficiently large number of events should be averaged to ensure that statistical fluctuations are negligible. (ED) 161-1971w

secondary failure *See:* failure.

secondary fault An insulation breakdown occurring as a result of a primary fault. *See also:* center of distribution. (EEC/AWM) [91]

secondary fuse A fuse used on the secondary-side circuits of transformers. *Note:* In high-voltage fuse parlance such a fuse is restricted for use on a low-voltage secondary distribution system that connects the secondaries of distribution transformers to consumers' services. (SWG/PE) C37.100-1992

secondary grid emission Electron emission from a grid resulting directly from bombardment of its surface by electrons or other charged particles. *See also:* electron emission. (ED/ED) 161-1971w, [45]

secondary index (A) A list associated with an inverted file in which entries in the list point to records in the file that contain identical values for the key field on which the file is inverted. **(B)** In a hierarchical database, an index used to establish access to a physical or logical segment by a path different from the one provided by the primary key within the root segment. *Note:* A secondary index allows access on the basis of any field within the segment or any of its dependent segments with secondary indices. *See also:* source segment; secondary processing sequence. (C) 610.5-1990

secondary key (A) In sorting and searching, a key that is given lower priority than the primary key within a group of related keys. That is, after sorting, all items having the same primary key will be in order by the secondary key or keys. *Synonym:* minor key. *Contrast:* primary key. **(B)** Within a record, a key that is used to index that record but which does not necessarily uniquely identify that record. (C) 610.5-1990

secondary neutral grid A network of neutral conductors, usually grounded, formed by connecting together within a given area all the neutral conductors of individual transformer secondaries of the supply system. *See also:* center of distribution. (EEC/AWM) [91]

secondary oil containment A system designed to contain the oil discharged from an oil-filled piece of equipment in situations of primary oil-containment failure. (SUB/PE) 980-1994

secondary outage An outage occurrence that is the result of another outage occurrence. *Notes:* 1. Secondary outages of components or units may be caused by repair of other components or units requiring physical clearance, failure of a circuit breaker to clear a fault, or a protective relay system operating incorrectly and overreaching into the normal tripping zone of another unit. 2. Some secondary outages are solely the result of system configuration; for example, two components connected in series will always go out of service together. These secondary outages may be given special treatment when compiling outage data. 3. At present, primary outages have been referred to in the industry as independent outage occurrences, and secondary outages as dependent or related outage occurrences. *See also:* related multiple outage event. (PE/PSE) 859-1987w

secondary power The excess above firm power to be furnished when, as, and if available. *See also:* generating station. (EEC/AWM) [91]

secondary processing sequence In a hierarchical database, the hierarchical order of segment types in a physical or logical database resulting from a secondary index. (C) 610.5-1990w

secondary radar (1) (A) (navigation aids) A radar technique or mode of operation in which the return signals are obtained from a beacon, transponder, or repeater carried by the target, as contrasted with primary radar in which the return signals are obtained by reflection from the target. **(B) (navigation aids)** A radar, or that portion of a radar, that operates on this principle. *See also:* primary radar. (AES/GCS) 172-1983 **(2)** A cooperative target identification system such as the military identification, friend or foe (IFF) Mark XII or the civil air traffic control radar beacon system (ATCRBS) in which an interrogator transmits a coded signal that asks for a reply. The transponder on the vehicle or platform queried answers with a coded reply. *Notes:* 1. The term secondary radar is more widely used in Europe than in the U.S. 2. The interrogator antenna is often mounted on the radar antenna and the reply from the transponder is often included on the radar display with the echo detection. *See also:* primary radar. (AES) 686-1997

secondary radiator That portion of an antenna having the largest radiating aperture, consisting of a reflecting surface or a lens, as distinguished from its feed. (AP/ANT) 145-1993

secondary representation A second form, an alternative to the primary representation, in which the client may supply an attribute value to the service. (C/PA) 1328-1993w, 1327-1993w, 1224-1993w

secondary ring The alternate paths of the dual ring that are not normally connected to MAC1. It uses links BRx and ATx. Application data may be transmitted on this ring, but not when the primary ring is being used for transmitting application data. (LM/C) 802.5c-1991r

secondary section of the core The section of the ferroresonant transformer on which the output and resonating windings are wound. In steady-state operation, this section of the core is normally driven into magnetic saturation. (PEL) 449-1998

secondary-selective type (low voltage-selective type) A unit substation that has two stepdown transformers each connected to an incoming high-voltage circuit. The outgoing side of each transformer is connected to a separate bus through a suitable switching and protective device. The two sections of bus are connected by a normally open switching and protective device. Each bus has one or more outgoing radial (stub-end) feeders. (PE/TR) C57.12.80-1978r

secondary service area (radio broadcast station) The area within which satisfactory reception can be obtained only under favorable conditions. *See also:* radio transmitter. (PE/EEC) [119]

secondary shock A shock of a magnitude such that it will not produce direct physiological harm, but may cause involuntary muscle reactions. The results of secondary shock are annoyance, alarm, and aversion. (T&D/PE) 524a-1993r, 1048-1990

secondary short-circuit current rating of a high-reactance transformer (power and distribution transformers) One that designates the current in the secondary winding when the primary winding is connected to a circuit of rated primary voltage and frequency and when the secondary terminals are short-circuited. (PE/TR) C57.12.80-1978r

secondary, single-phase induction motor The rotor or stator member that does not have windings that are connected to the supply line. *See also:* induction motor; asynchronous machine. (IA/APP) [90]

secondary socket identifier (SSID) A socket number identifying a particular endpoint on the secondary device. (C/MM) 1284.4-2000

secondary space allocation The amount of space that is reserved for a particular file after the primary space allocation has been exhausted. *Note:* Some systems allow multiple secondary space allocation operations. When a secondary space allocation is granted to a particular file, that file is said to "increase its extents." *Contrast:* primary space allocation. (C) 610.5-1990w

secondary standard (luminous standards) (illuminating engineering) A stable light source calibrated directly or indirectly by comparison with a primary standard. This order of standard also is designated as a reference standard. *Note:* National secondary (reference) standards are maintained at national physical laboratories; laboratory secondary (reference) standards are maintained at other photometric laboratories. (IE/EEC) [126]

secondary station (1) A station that has been temporarily selected to receive a transmission from the primary station. *See also:* primary station. (C) 610.7-1995
(2) As defined by the infrared link access protocol (IrLAP), any station on the data link that does not assume the role of the primary station. It will initiate transmission only as a result of receiving explicit permission to do so from the primary station. (EMB/MIB) 1073.3.2-2000

secondary storage A type of storage which is used to store information for extended periods, while still allowing for online access. *See also:* mass storage; auxiliary storage. (C) 610.10-1994w

secondary unit substation (power and distribution transformers) A substation in which the low-voltage section is rated 1000 V (volts) and below. *See also:* unit substation. (PE/TR) C57.12.80-1978r

secondary voltage (capacitance potential devices) The root-mean-square voltage obtained from the main secondary winding, and when provided, from the auxiliary secondary winding. *See also:* rated secondary voltage; outdoor coupling capacitor. 31-1944w

secondary voltage rating (power and distribution transformers) The load circuit voltage for which the secondary winding is designed. (PE/TR) C57.12.80-1978r

secondary winding (1) (A) (power and distribution transformers) The winding on the energy output side. **(B) (instrument transformers) (power and distribution transformers)** The winding that is intended to be connected to the measuring or control devices. (PE/TR) C57.12.80-1978
(2) (rotating machinery) Any winding that is not a primary winding. *See also:* asynchronous machine; voltage regulator. (IA/APP) [90]
(3) (voltage regulators) The series winding. *See also:* voltage regulator. (PE/TR) C57.15-1968s
(4) The winding intended for connection to the measuring, protection, or control devices. (PE/TR) C57.13-1993

second-channel attenuation *See:* selectance.

second-channel interference Interference in which the extraneous power originates from a signal of assigned (authorized) type in a channel two channels removed from the desired channel. *See also:* interference; radio receiver. 188-1952w

second contingency incremental transfer capability (power operations) The amount of power, incremental above normal base power transfers, that can be transferred over the transmission network in a reliable manner, based on the following conditions:

a) With all transmission facilities in service, all facility loadings are within normal ratings and all voltages are within normal limits.
b) The bulk power system is capable of absorbing the dynamic power swings and remaining stable following a disturbance resulting in the sequential and overlapping outage of two facilities, either being a generating unit, transmission circuit, or transformer with system adjustments made between the two outages as required.
c) After the dynamic power swings following a disturbance resulting in the loss of the second facility, either a generating unit, transmission circuit, or transformer, but before further operator-directed system adjustments are made, all transmission facility loadings are within emergency ratings and all voltages are within emergency limits.

Note: The term second contingency is used to specifically exclude simultaneous outages. Use of the term double contingency has been avoided, since it is often used to include both simultaneous and sequential outages. (PE/PSE) 858-1987s

second generation A period during the evolution of electronic computers in which transistors were used to replace the first generation vacuum tubes. *Note:* Introduced in 1959, thought to have been the state of the art until the introduction of integrated circuits. *See also:* fourth generation; first generation; fifth generation. (C) 610.10-1994w

second generation language *See:* assembly language.

second-level address *See:* n-level address; indirect address.

second normal form One of the forms used to characterize relations; a relation is said to be in second normal form if it is in first normal form and if every nonprime attribute is fully functionally dependent on each candidate key of the relation.

FIRST NORMAL FORM

ORDER1 = {ORDER-NO} + DATE + CUSTOMER-NO
+ CUSTOMER-NAME + CUSTOMER-ADDRESS
+ TOTAL-ORDER-AMOUNT
ITEM1 = {ORDER-NO + SEQUENCE-NO} + ITEM-NO
+ ITEM-DESCRIPTION + QUANTITY-ORDERED
+ UNIT-PRICE + EXTENDED-PRICE

SECOND NORMAL FORM

ORDER2 = {ORDER-NO} + DATE + CUSTOMER-NO
+ CUSTOMER-NAME + CUSTOMER-ADDRESS
+ TOTAL-ORDER-AMOUNT
ORDER-ITEM2 = {ORDER-NO + ITEM-NO}
+ QUANTITY-ORDERED
+ EXTENDED-PRICE
ITEM2 = {ITEM-NO} + ITEM-DESCRIPTION
+ UNIT-PRICE

In first normal form, nonprime attributes ITEM- DESCRIPTION and UNIT-PRICE are not functionally dependent on candidate key SEQUENCE- NO. Keys shown in brackets.

(C) 610.5-1990w

second-order distortion *See:* intermodulation distortion.

second-order lag (automatic control) In a linear system or element, lag which results from changes of energy storage at two separate points in the system, or from effects such as acceleration. *Note:* It is representable by a second-order differential equation, or by a quadratic factor such as $s^2 + 2z\omega_n s + \omega_n^2$ in the denominator of a transfer function. *Synonym:* quadratic lag. *See also:* lag. (PE/EDPG) [3]

second-order nonlinearity coefficient (accelerometer) The proportionality constant that relates a variation of the output to the square of the input, applied parallel to the input reference axis. (AES/GYAC) 528-1994

second source In the event that several vendors offer pin-for-pin compatible components, second-source suppliers are vendors of the component other than the prime source. *See also:* prime source. (TT/C) 1149.1-1990

seconds since the Epoch A value to be interpreted as the number of seconds between a specified time and the Epoch. A Coordinated Universal Time name [specified in terms of seconds (*tm_sec*), minutes (*tm_min*), hours (*tm_hour*), days since January 1 of the year (*tm_yday*), and calendar year minus 1900 (*tm_year*)] is related to a time represented as seconds since the Epoch, according to the expression below. If the year < 1970 or the value is negative, the relationship is undefined. If the year ≥ 1970 and the value is nonnegative, the value is related to a Coordinated Universal Time name according to the expression:

tm_sec + *tm_min**60 + *tm_hour**3600 + tm_yday*86400
+ (*tm_year-70*)*31536000 + ((*tm_year-69*)/4)*86400

(C/PA) 9945-1-1996, 9945-2-1993

second-time-around echo An echo received after a time delay exceeding one pulse-repetition interval but less than two pulse-repetition intervals. *Note:* Third-time-around, etc., echoes are defined in a corresponding manner. The generic term "multiple-time-around" is sometimes used. (AES) 686-1997

second Townsend discharge (gas) A semi-self-maintained discharge in which the additional ionization is due to the secondary electrons emitted by the cathode under the action of the bombardment by the positive ions present in the gas. *See also:* discharge. (ED) [45]

second voltage range *See:* voltage range.

secretary/librarian (software) The software librarian on a chief programmer team. *See also:* software librarian; chief programmer team. (C/SE) 729-1983s

secret key The traditional cryptographic key known only to the communicating parties and used for both encipherment and decipherment. (C/LM) 802.10-1998

section (1) (rectifier unit) A part of a rectifier unit with its auxiliaries that may be operated independently. *See also:* rectification. (IA) [62]
(2) (thyristor converter) Those parts of a thyristor converter unit containing the power thyristors (and when also used, the power diodes) together with their auxiliaries (including individual transformers or cell windings of double converters and circulating current reactors, if any), in which the main direct current when viewed from the converter unit dc terminals always flows in the same direction. A thyristor converter section is supposed to be operated independently. *Note:* A converter equipment may have either only one section or one forward and one reverse section. (IA/IPC) 444-1973w
(3) A length of coaxial cable which forms the transmission medium for a network. *See also:* segment. (C) 610.7-1995
(4) (A) To divide a program into parts such that some portions reside in internal storage and others in auxiliary storage. *See also:* page. **(B)** One of the parts as in (A). **(C)** To divide a program or data into parts of varying lengths, known as sections, such that each section is placed in a main memory area of corresponding size, not necessarily contiguously or in logical order. (C) 610.10-1994

sectional center (1) (telephone switching systems) A toll office to which may be connected a number of primary centers, toll centers, or toll points. Sectional centers are classified as Class 2 offices. *See also:* office class. (C) [85]
(2) Class 2 office in the North American hierarchical routing plan; a control center connecting primary centers of the telephone system together. *See also:* regional center; toll center; end office; primary center. (C) 610.7-1995

sectionalized linear antenna A linear antenna in which reactances are inserted at one or more points along the length of the antenna. *Synonym:* loaded linear antenna. (AP/ANT) 145-1993

sectionalizer *See:* automatic line sectionalizer.

section locking Locking effective while a train occupies a given section of a route and adapted to prevent manipulation of levers that would endanger the train while it is within that section. *See also:* interlocking. (EEC/PE) [119]

section, sag *See:* sag section.

sector *See:* block.

sectoral horn antenna A horn antenna with two opposite sides of the horn parallel and the two remaining sides diverging. (AP/ANT) 145-1983s

sector cable A multiple-conductor cable in which the cross section of each conductor is substantially a sector of a circle, an ellipse, or a figure intermediate between them. *Note:* Sector cables are used in order to obtain decreased overall diameter and thus permit the use of larger conductors in a cable of given diameter. (EEC/AWM) [91]

sector display (1) (continuously rotating radar-antenna system) A range-amplitude display used with a radar set, the antenna system of which is continuously rotating. The screen, which is of the long-persistence type, is excited only while the beam of the antenna is within a narrow sector centered on the object. *See also:* radar. (AES) [42]
(2) A limited display in which only a sector of the total service area of the radar system is shown. *Note:* Usually the sector to be displayed is selectable. (AES) 686-1997

sector impedance relay A form of distance relay that by application and design has its operating characteristic limited to a sector of its operating circle on the *R-X* diagram. (SWG/PE) C37.100-1992

sector scanning (1) A modification of circular scanning in which the direction of the antenna beam generates a portion of a cone or a plane. (AP/ANT) 145-1993
(2) The repeated scanning of a limited volumetric sector by a radar. *See also:* sector display. (AES) 686-1997

sector select line The line, determined by the row addresses (output of the X decoder), that is used to access the appropriate sector select transistor. (ED) 1005-1998

sector select transistor The transistor, controlled by the sector select line, that isolates the sector source from other sectors. (ED) 1005-1998

secure data exchange Layer Manager The SDE portion of the Layer 2 Manager. (C/LM) 802.10-1998

secure modem *See:* port protection system.

secure path *See:* trusted path.

security (1) (software) The protection of computer hardware and software from accidental or malicious access, use, modification, destruction, or disclosure. Security also pertains to personnel, data, communications, and the physical protection of computer installations. *See also:* software; modification; data; protection; hardware. (C/SE) 729-1983s
(2) The protection of computer resources (e.g., hardware, software, and data) from accidental or malicious access, use, modification, destruction, or disclosure. Tools for the maintenance of security are focused on availability, authentication, accountability, confidentiality, and integrity. (C/PA) 14252-1996
(3) (of a relay or relay system) That facet of reliability that relates to the degree of certainty that a relay or relay system will not operate incorrectly. (SWG/PE) C37.100-1992

security association A cooperative relationship between entities formed by the sharing of cryptographic keying information and security management objects. This shared information need not be identical, but it must be compatible. (C/LM) 802.10-1998

security association identifier (SAID) A value placed in the clear header of the SDE PDU that is used to identify the security association. (C/LM) 802.10-1998

security attribute A security-related quality of an object. Security attributes may be represented as hierarchical levels, bits in a bit map, or numbers. Compartments, caveats, and release markings are examples of security attributes. (C/LM) 802.10g-1995, 802.10-1998

security code A group of data bits calculated by a transmitting terminal from the information within its message by use of a prearranged algorithm, appended to the transmitted message, and tested by the receiving terminal to determine the validity of the received message. (SUB/PE) 999-1992w

security dispatch control An automatic generation control subsystem that allocates unit generation levels within a control area based upon system security considerations. (PE/PSE) 94-1991w

security kernel (software) A small, self-contained collection of key security-related statements that works as a privileged part of an operating system, specifying and enforcing criteria that must be met for programs and data to be accessed. (C) 610.12-1990

security label A marking bound to a resource (which may be a data unit) that names or designates the security attributes of that resource. (C/LM) 802.10g-1995

security level (1) The sensitivity of information represented, for example, by a combination of hierarchical classifications and nonhierarchical categories. (C/BA) 896.3-1993w
(2) A hierarchical level whose purpose is to indicate degree of sensitivity to a designated security threat. It indicates a specific level of protection as specified by the security policy being enforced. (C/LM) 802.10g-1995, 802.10-1998

security management In networking, a management function defined for controlling, authenticating, and authorizing access to network resources. (C) 610.7-1995

security management information base (SMIB) A management information base (MIB) that stores security-relevant objects. (C/LM) 802.10a-1999, 802.10-1998

security policy The objectives and mandates for protecting information, services, and other resources in a system, and the philosophy of protection for meeting those objectives. (C/BA) 896.3-1993w

security service (1) A service, provided by a layer of communicating open systems, that ensures adequate security of the systems or of data transfers. Note that these security services need not be directly requested at the (N)- and (N+)-layer boundary as is required for an OSI (N)-service. (LM/C) 802.10-1992
(2) The capability of the system to ensure the security of system resources or data transfers. Access controls, authentication, data confidentiality, data integrity, and nonrepudiation are traditional data communications security services. (C/BA) 896.3-1993w

security system The aggregate assemblage of hardware and associated software that includes all components, equipment, barriers, etc., necessary for the physical protection of nuclear power generating stations against the design basis threat of radiological sabotage. (PE/NP) 692-1997

security tag An information unit containing a representation of certain security-related information (e.g., a restrictive attribute bit map). (C/LM) 802.10g-1995, 802.10-1998

security threat A potential violation of security. (LM/C) 802.10-1992, 802.10g-1995

SED Static Electric Discharge; an alternate name for ESD. (SPD/PE) C62.47-1992r

sedimentation potential (electrobiology) The electrokinetic potential gradient resulting from unity velocity of a colloidal or suspended material forced to move by gravitational or centrifugal forces through a liquid electrolyte. *See also:* electrobiology. (EMB) [47]

sediment separator (rotating machinery) Any device, used to collect foreign material in the lubricating oil. *See also:* oil cup. (IA/APP) [90]

Seebeck coefficient (for homogeneous conductors) (of a couple) The limit of the quotient of: the Seebeck electromotive force by the temperature difference between the junctions as the temperature difference approaches zero: by convention, the Seebeck coefficient of a couple is positive if the first-named conductor has a positive potential with respect to the second conductor at the cold junction. *Note:* The Seebeck coefficient of a couple is the algebraic difference of either the relative or absolute Seebeck coefficients of the two conductors. *See also:* thermoelectric device. (ED) [46]

Seebeck coefficient, absolute *See:* absolute Seebeck coefficient.

Seebeck coefficient, relative *See:* relative Seebeck coefficient.

Seebeck effect The generation of an electromotive force by a temperature difference between the junctions in a circuit composed of two homogeneous electric conductors of dissimilar composition: or, in a nonhomogeneous conductor, the electromotive force produced by a temperature gradient in a nonhomogeneous region. *See also:* thermoelectric device; thermoelectric effect. (ED) [46]

Seebeck electromotive force The electromotive force resulting from the Seebeck effect. *See also:* thermoelectric device. (ED) [46]

Seed In 10BROAD36, the 23 bits residing in the scrambler shift register prior to the transmission of a packet. (C/LM) 802.3-1998

seeding *See:* fault seeding.

SEE-IN *See:* Significant Event Evaluation and Information Network.

seek (1) To position the head or access mechanism of a direct-access device to a specified location. *Synonym:* position. (C) 610.10-1994w
(2) An activity that positions a pointer at a specific location within a data file. (C/MM) 855-1990
(3) (data management) *See also:* search; search cycle. (C) 610.5-1990w

seek key *See:* search key.

seek time The time it takes to position the head or access mechanism of a rotating storage device to a specified location. *Synonym:* positioning time. *See also:* access time; search time. (C) 610.10-1994w

segment (1) (A) (data management) (software) One of the subsystems or combinations of subsystems that make up an overall system; for example, the accounts payable segment of a financial system. **(B) (data management) (software)** In storage allocation, a self-contained portion of a computer program that can be executed without maintaining the entire program in main storage. *See also:* page. **(C) (data management) (software)** A collection of data that is stored or transferred as a unit. **(D) (software) (data management)** In path analysis, a sequence of computer program statements between two consecutive branch points. **(E) (data management) (software)** To divide a system, computer program, or data file into segments as in (A), (B), or (C). **(F) (data management) (software)** A fixed-length unit of data that contains one or more data items. **(G) (software) (data management)** In some databases, the smallest unit of data that can be retrieved or stored. *Synonym:* database segment. *See also:* twin segment; parent segment; dependent segment; logical segment; child segment; physical segment; root segment. (C) 610.5-1990, 610.12-1990
(2) (computer graphics) A logically related collection of display elements with their associated attributes such that the collection can be manipulated as a unit. (See corresponding figure.) *Synonym:* display group; entity; display segment. *See also:* segment attribute.

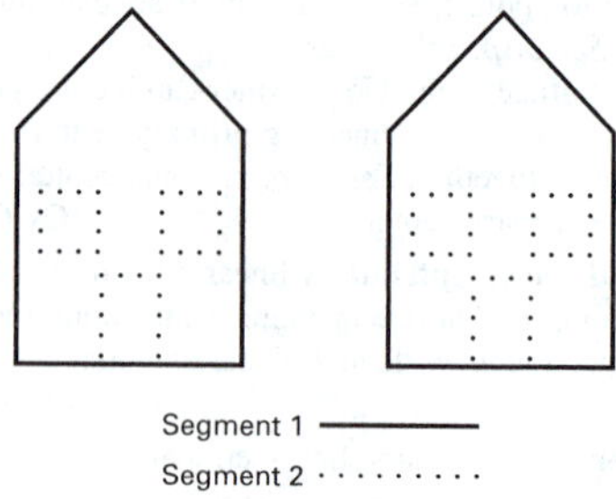

segment

(C) 610.6-1991w
(3) The medium connection, including connectors, between medium dependent interfaces in a LAN. (C/LM) 802.9a-1995w
(4) The portion of a ringlet between the producer and consumer along which a packet is sent. The segment traversed by a send packet is the send segment, and the segment traversed by an echo is the echo segment. (C/MM) 1596-1992
(5) A specific transmission medium that supports the FASTBUS protocol and to which FASTBUS devices may attach. A segment is capable of supporting autonomous operation and communicating with other segments via segment interconnects. (NID) 960-1993
(6) The protocol data unit (PDU) of 52 octets transferred between peer DQDB Layer entities as the information payload of a slot. It contains a segment header of 4 octets and a segment payload of 48 octets. There are two types of segments: Pre-Arbitrated (PA) segments and Queued Arbitrated (QA) segments. (LM/C) 8802-6-1994
(7) One or more sections of coaxial cable that form the transmission medium for a network. (C) 610.7-1995
(8) A portion of a session that is contiguous in simulation time and in wall clock (sidereal) time. (DIS/C) 1278.3-1996
(9) On a magnetic drum or disk, one of a series of addressable segments within a track or a band on which information is stored. *See also:* cluster; storage element. (C) 610.10-1994w

(10) Zero or more contiguous elements of a string. (C/PA) 1328-1993w, 1224-1993w, 1327-1993w
(11) The medium connection, including connectors, between Medium Dependent Interfaces (MDIs) in a CSMA/CD local area network. (C/LM) 802.3-1998
(12) A portion of an interconnect structure treated as a unit for the purposes of extracting or estimating its electrical properties. *See also:* parasitics. (C/DA) 1481-1999

segmental conductor A stranded conductor consisting of three or more stranded conducting elements, each element having approximately the shape of the sector of a circle, assembled to give a substantially circular cross section. The sectors are usually lightly insulated from each other and, in service, are connected in parallel. *Note:* This type of conductor is known as type-M conductor in Canada. *See also:* conductor. (EEC/AWM) [91]

segmental-rim rotor (rotating machinery) A rotor in which the rim is composed of interleaved segmental plates bolted together. *See also:* rotor. (IA/APP) [90]

segmentation (1) The function in the DQDB Layer that fragments a variable length Initial MAC Protocol Data Unit (IMPDU) into fixed-length segmentation units for transfer in Derived MAC Protocol Data Units (DMPDUs) (cf., reassembly). (LM/C) 8802-6-1994
(2) (image processing and pattern recognition) *See also:* image segmentation. (C) 610.4-1990w

segmentation unit The fixed-length data units of 44 octets formed by the fragmentation of an Initial MAC Protocol Data Unit (IMPDU). (LM/C) 8802-6-1994

segment attribute A characteristic that applies to a segment. For example, detectability, highlighting, priority, transformation, visibility. (C) 610.6-1991w

segment, cable *See:* cable segment.

segment, crate *See:* crate segment.

Segment Delay Value (SDV) A number associated with a given segment that represents the delay on that segment including repeaters and end stations, if present, used to assess path delays for 10 Mb/s CSMA/CD networks. (C/LM) 802.3-1998, 802.9a-1995w

segment, extended *See:* extended segment.

segment extender (SE) (FASTBUS acquisition and control) A device for connecting two segments to form an extended segment or part of an extended segment. (NID) 960-1993

segment header The protocol control information in a segment. (LM/C) 8802-6-1994

segmenting A technique used in memory mapping whereby the address space is broken into several various-size blocks; physical addresses are obtained by biasing each of the individual segments. (C) 610.10-1994w

segment interconnect (SI) A device that implements an intersegment connection such that the FBP on the two segments is synchronized. When an operation is passing through an SI, the SI acts as a slave on the near-side and as a master on the far-side. (NID) 960-1993

segment interconnect, active *See:* active segment interconnect.

segment interconnect, reserved *See:* reserved segment interconnect.

segment number A four-character number identifying the subsection of the ADC used for the data. Leading spaces are interpreted as leading zeros. This will be used in a system where the ADC is shared by several data inputs and is multiplexed among the inputs. Each input could be a separate detector, and there is not necessarily any correlation between any two detector inputs. The subsystem identification, ADC number, and segment number should provide sufficient data to be able to trace this data to the apparatus that collected it. This will enable other data, such as calibration spectra or background spectra, which are stored separately, to be identified with this spectrum. In large distributed systems, the ADC and segment numbers are not enough to uniquely identify the data. (NPS/NID) 1214-1992r

segment payload The unit of data carried by a segment. (LM/C) 8802-6-1994

segment shoe (rotating machinery) (bearing shoe) A pad that is part of the bearing surface of a pad-type bearing. *See also:* bearing. (IA/APP) [90]

Segment Variability Value (SVV) A number associated with a given segment that represents the delay variability on that segment (including a repeater) for 10 Mb/s CSMA/CD networks. (C/LM) 802.3-1998, 802.9a-1995w

segregated-phase bus One in which all phase conductors are in a common metal enclosure, but are segregated by metal barriers between phases. (SWG/PE/EDPG) C37.100-1992, 665-1987s

segregated phase comparison Similar to phase comparison, except data on each phase and ground is sent separately to the remote terminal for comparison with the local phase data at that terminal. (PE/PSR) C37.113-1999

seismic category I (nuclear power generating station) The classification assigned to those structures, systems, and components of a nuclear power plant, including foundations and supports, which must be designed to withstand the effects of the Safe Shutdown Earthquake (SSE) and remain functional. (PE/NP) 567-1980w

seismic outline drawing A 280 × 432 mm (11 × 17 in) or 216 × 280 mm (8 1/2 × 11 in) drawing that shows key information concerning the seismic qualification of the equipment. It shows information such as the resonant frequencies of the equipment, important loads, an outline drawing of the equipment, the center of gravity of the equipment, and other key information about the equipment. (PE/SUB) 693-1997

seizure signal (telephone switching systems) A signal transmitted from the sending end of a trunk to the far end to indicate that its sending end has been selected. (C) [85]

select (1) To identify, within a set of items, all items that meet a particular criterion. *See also:* extract; selection. (C) 610.5-1990w
(2) Context determines which of the following applies:

— To establish a particular device node as the active package.
— To establish a particular device as either the console input device or console output device.
— To establish a particular instance as the current instance.

(C/BA) 1275-1994
(3) To identify and fix (for the duration of the current message) a port to which data of a Data Transfer class message are to be directed. (TT/C) 1149.5-1995

selectable element A display element that can be selected by a pick device. (C) 610.6-1991w

selectance (amplitude-modulation broadcast receivers) The ratio of the ordinates of a selectivity graph, described in Section 4.05.03 of IEEE Std 186-1948w, between the resonant frequency and another frequency differing from the resonant frequency by a specified multiple of the width of one channel. (The width of one broadcast channel is 10 kilocycles.) It is expressed in decibels or voltage ratios. The ratio at a frequency n channels above the resonant frequency is denoted by S_{+n} and at a frequency n channels below the resonant frequency is denoted by S_{-n}. The geometric mean of these ratios is denoted by S_n. Expressed in decibels, the value of S_n is the average value of S_{+n} and S_{-n}. The terms adjacent-channel attenuation" (ACA) and "second-channel attenuation" (2ACA) are used to refer to S_1 and S_2, respectively. (CE) 186-1948w

selected emphasis A visual cue that indicates the current selection. (C) 1295-1993w

selected slave (1) A slave that is selected during the connection phase by the master when it recognizes its address on the bus lines. (C/BA/C/BA/C/BA) 10857-1994, 896.4-1993w, 896.3-1993w
(2) *See also:* slave. (C/MM) 1096-1988w

selected test data register A test data register is selected when it is required to operate by an instruction supplied to the test logic. (TT/C) 1149.1-1990

selecting (telephone switching systems) Choosing a particular group of one or more servers in the establishment of a call connection. (C) [85]

selection (A) (data management) A relational operator that extracts specified tuples from a relation and results in a relation containing only those tuples. Also called **select**.

Student No.	Name	Grade	Homeroom
15	Mary	4	26A
21	Harry	4	27

See also: product; projection; coincident-current selection; difference; union; join; amplitude selection; intersection. **(B) (data management)** The process of identifying, within a set of items, all items that meet a particular criterion. (C) 610.5-1990

Selection of Relation *Students* in Fig to entity/attribute matrix where GRADE is ≤ 5

selection box A boxed area on the user interface that presents a list of items to select from and a text field where selected items are displayed.The user can also directly type in an item into the text field. (C) 1295-1993w

selection check (electronic computation) A check (usually an automatic check) to verify that the correct register, or other device, is selected in the interpretation of an instruction. (C) 162-1963w, [85], [20]

selection phase The set of steps performed by software administration utility to process selections and options. (C/PA) 1387.2-1995

selection ratio The least ratio or a magnetomotive force used to select a cell to the maximum magnetomotive force used that is not intended to select a cell. *See also:* coincident-current selection. (Std100) 163-1959w

selection sort A sort in which the items in a set are examined to find an item that fits a specified criterion; for example, the smallest item; this item is appended to the sorted set and removed from further consideration; and the process is repeated until all items are in the sorted set. *Synonym:* straight selection sort. *See also:* repeated selection sort; tree selection sort; heapsort. (C) 610.5-1990w

selective acknowledgment An acknowledgment mechanism used with a sliding window protocol that allows the receiver to acknowledge packets that are received out of order. *Synonym:* extended acknowledgment. (C) 610.7-1995

selective calling The ability of a transmitting station to specify which of several stations on the same line is in condition to receive a message. (C) 610.7-1995

selective choice construct *See:* branch.

selective collective automatic operation (elevators) Automatic operation by means of one button in the car for each landing level served and by UP and DOWN buttons at the landings, wherein all stops registered by the momentary actuation of the car buttons are made as defined under nonselective collective automatic operation, but wherein the stops registered by the momentary actuation of the landing buttons are made in the order in which the landings are reached in each direction of travel after the buttons have been actuated. With this type of operation, all UP landing calls are answered when the car is traveling in the up direction and all DOWN landing calls are answered when the car is traveling in the down direction, except in the case of the uppermost or lowermost calls, which are answered as soon as they are reached irrespective of the direction of travel of the car. *See also:* control. (PE/EEC) [119]

selective coordination *See:* selectivity.

selective dump (1) (computers) A dump of a selected area of storage. (C) [20], [85]
(2) (software) A dump of designated storage location areas only. *See also:* postmortem dump; snapshot dump; dynamic dump; change dump; static dump; memory dump. (C) 610.12-1990

selective erase The removal of one or more display elements or parts of an element without affecting the rest of the display image. *See also:* erase. (C) 610.6-1991w

selective erasing (storage tubes) Erasing of selected storage elements without disturbing the information stored on other storage elements. *See also:* storage tube. (ED) 158-1962w

selective fading *See:* frequency selective fading.

selective listing in combination index (SLIC) An automatic index in which the entries are combinations of terms taken from a set of preselected keywords. (C) 610.2-1987

selective opening (tripping) The application of switching devices in series such that (of the devices carrying fault current) only the device nearest the fault will open and the devices closer to the source will remain closed and carry the remaining load. (SWG/PE) C37.100-1992

selective overcurrent trip *See:* overcurrent release; selective release.

selective overcurrent tripping *See:* overcurrent release; selective opening.

selective pole switching The practice of tripping and reclosing one or more poles of a multipole circuit breaker without changing the state of the remaining pole(s), with tripping being initiated by protective relays that respond selectively to the faulted phases. *Note:* Circuit breakers applied for selective pole switching must inherently be capable of individual pole opening. (SWG/PE) C37.100-1992

selective release (trip) A delayed release with selective settings that will automatically reset if the actuating quantity falls and remains below the release setting for a specified time. (SWG/PE) C37.100-1992

selective retransmission A transmission scheme where the transmitter may send multiple PDUs without waiting for an acknowledgment. If the receiver indicates that an error occurred in a given PDU, the sender will retransmit only the errored PDU. *Note:* In this scheme, the receiver will accept PDUs that are out-of-sequence. *Contrast:* Go-Back-N. (C) 610.7-1995

selective ringing (telephone switching systems) Ringing in which only the ringer at the desired main station on a party line responds. (C) [85]

selective signaling equipment (mobile communication) Arrangements for signaling, selective from a base station, of any one of a plurality of mobile stations associated with the base station for communication purposes. *See also:* mobile communication system. (VT) [37]

selective trace A variable trace that involves only selected variables. *See also:* symbolic trace; subroutine trace; retrospective trace; execution trace; variable trace. (C) 610.12-1990

selectivity (1) The characteristic of a filter that determines the extent to which the filter is capable of altering the frequency spectrum of a signal. A highly selective filter has an abrupt transition between a pass-band region and a stop-band region. (CAS/MTT) 146-1980w
(2) A general term describing the interrelated performance of relays and breakers, and other protective devices; complete selectivity being obtained when a minimum amount of equipment is removed from service for isolation of a fault or other abnormality. (SWG/PE/IA/PSP) C37.100-1992, 1015-1997

selector, amplitude *See:* pulse-height selector.

selector channel An input-output channel that can transfer data to or from only one peripheral device at a time. *See also:* multiplexer. (C) 610.10-1994w

selector field A five-bit field in the Base Link Code Word encoding that is used to encode up to 32 types of messages that define basic abilities. For example, selector field 00001 indicates that the base technology is IEEE 802.3. (C/LM) 802.3-1998

selector pen *See:* light pen.

selector pulse (navigation aids) A pulse which is used to identify, for selection, one event in a series of events. (AES/GCS) 172-1983w

selector, pulse-height *See:* pulse-height selector.

selectors Those who execute the selection process described in this recommended practice. They may also act in other roles (for example, users). (C/SE) 1209-1992w

selector switch (1) A switch arranged to permit connecting a conductor to any one of a number of other conductors. (SWG/PE) C37.30-1971s, C37.100-1992

(2) A manually operated multiposition switch for selecting alternative control circuits. (IA/ICTL/IAC) [60]

self-adapting Pertaining to the ability of a system to change its performance characteristics in response to its environment. (C) [20], [85]

self-adapting computer A computer that can change its performance characteristics in response to its environment. (C) 610.10-1994w

self-aligning bearing (rotating machinery) A sleeve bearing designed so that it can move in the end shield to align itself with the journal of the shaft. *See also:* bearing. (IA/APP) [90]

self-ballasted lamp (illuminating engineering) Any arc discharge lamp of which the current-limiting device is an integral part. (EEC/IE) [126]

self-capacitance (conductor) (total capacitance) (grounded capacitance) In a multiple-conductor system, the capacitance between this conductor and the other $(n - 1)$ conductors connected together. *Note:* The self-capacitance of a conductor equals the sum of its $(n - 1)$ direct capacitances to the other $(n - 1)$ conductors. (Std100) 270-1966w

self-checking (by a relay) Self-testing by microprocessor-based relays that checks operation of the processor software. (PE/PSR) C37.113-1999

self-checking circuit A circuit which is capable of withstanding a specified number of non-fatal failures while continuing to operate. (C) 610.10-1994w

self-checking code (electronic computation) A code that uses expressions such that one (or more) error(s) in a code expression produces a forbidden combination. *See also:* error-detecting code; parity. (C) 162-1963w

self-closing door or gate (elevators) A manually opened hoistway door and/or a car door or gate that closes when released. *See also:* hoistway. (PE/EEC) [119]

self-commutated converter (self-commutated converters) (forced-commutated converter) A converter in which commutation is accomplished by components within the converter. *Note:* In converters using switching devices that can interrupt or turn off current, such as transistors or gate turn-off thyristors, rejection of the current produces a voltage across the device to commutate the current to another device. In converters using circuit-commutated thyristors, the commutating voltages required to transfer current from one device to another are usually supplied by capacitors. (IA/SPC) 936-1987w

self-commutated inverters An inverter in which the commutation elements are included within the power inverter. (IA) [62]

self-complementing code A binary code in which the complement of each decimal digit represented equals the complement of its binary representation. *See also:* excess-three code. (C) 1084-1986w

self-contained Pertaining to a database management system having a programming language that contains all of the necessary facilities for the control and processing of a database. (C) 610.5-1990w

self-contained electromechanical watthour meter A self-contained electromechanical watthour meter is one in which the terminals are arranged for connection to the circuit being measured without using external instrument transformers. (ELM) C12.10-1987

self-contained instrument An instrument that has all the necessary equipment built into the case or made a corporate part thereof. *See also:* instrument. (EEC/AII) [102]

self-contained navigation aid (navigation aids) An aid that consists only of facilities carried by the vehicle. (AES/GCS) 172-1983w

self-contained pressure cable A pressure cable in which the container for the pressure medium is an impervious flexible metal sheath, reinforced if necessary, that is factory assembled with the cable core,. *See also:* oil-filled cable; pressure cable. (EEC/AWM) [91]

self-coupling separable contacts (switchgear assembly disconnecting device) Contacts, mounted on the stationary and removable elements of a switchgear assembly, that align and engage or disengage automatically when the two elements are brought into engagement or disengagement. (SWG/PE) [56]

self-damping (conductor self-damping measurements) Of a conductor subjected to a load T is defined by the power dissipated per unit length of a conductor vibrating in a natural mode, with a loop length l and an antinode displacement amplitude y and a frequency f. The power per unit conductor length P is expressed as a function in the nth mode.

$$P = f_n (T, l, f, y)$$

(T&D/PE) 563-1978r

self-damping conductor (SDC) ACSR that is designed to control aeolian vibration by integral damping. Trapezoidal aluminum wires and annular gaps are utilized. (T&D/PE) 524-1992r

self-descriptiveness The degree to which a system or component contains enough information to explain its objectives and properties. *See also:* maintainability; usability; testability. (C) 610.12-1990

self diagnostics Programs automatically executed, at predetermined intervals, in the master station or RTU, to check the health of the system. (SUB/PE) C37.1-1994

self-discharge The process by which the available capacity of a battery is reduced by internal chemical reactions (local action). (PV) 1013-1990, 1144-1996

self-discharge rate The amount of capacity reduction occurring per unit of time in a battery as the result of self-discharge. (PV) 1013-1990, 1144-1996

self-documented Pertaining to source code that contains comments explaining its objectives, operation, and other information useful in understanding and maintaining the code. (C) 610.12-1990

self-excitation A condition in which an induction generator, operating in an isolated power system, derives its excitation from shunt capacitors or the natural capacitance of the power lines. Applies only to induction machines. (DESG) 1094-1991w

self-excited A qualifying term applied to a machine to denote that the excitation is supplied by the machine itself. *See also:* direct-current commutating machine. (IA/APP) [90]

self-field (Hall generator) The magnetic field caused by the flow of control current through the loop formed by the control current leads and the relevant conductive path through the Hall plate. (MAG) 296-1969w

self-filling bushing An oil-filled bushing in which the oil is allowed to circulate freely between the inside of the bushing and the apparatus on which it is used. (PE/TR) C57.19.03-1996

self-heating The result of exothermic reactions, occurring in some materials under certain conditions, whereby heat is liberated at a rate sufficient to raise the temperature of the material. (DEI) 1221-1993w

Self Identifying Publisher Port An instance of the class `IEEE1451_SelfIdentifyingPublisherPort` or of a subclass thereof. (IM/ST) 1451.1-1999

self-ID packet (1) A special packet sent by a cable PHY during the self-ID phase following a reset. One to four self-ID packets are sent by a given node depending on the maximum number of ports it has. (C/MM) 1394-1995

(2) A physical layer (PHY) packet transmitted by a cable PHY during the self-ID phase or in response to a PHY ping packet. (C/MM) 1394a-2000

self-ignition Ignition resulting from self-heating. (DEI) 1221-1993w

self-impedance (of an array element) The input impedance of a radiating element of an array antenna with all other elements in the array open-circuited. (AP/ANT) 145-1993

self-inductance The property of an electric circuit whereby an electromotive force is induced in that circuit by a change of current in the circuit. *Notes:* 1. The coefficient of self-inductance L of a winding is given by the following expression:

$$L = \frac{\delta\lambda}{\delta i}$$

where λ is the total flux-linkage of the winding and i is the current in the winding. 2. The voltage e induced in the winding is given by the following equation:

$$e = -\left[L\frac{\mathrm{d}i}{\mathrm{d}t} + i\frac{\mathrm{d}L}{\mathrm{d}t}\right]$$

If L is constant

$$e = -L\frac{\mathrm{d}i}{\mathrm{d}t}$$

3. The definition of self-inductance L is restricted to relatively slow changes in i, that is, to low frequencies, but by analogy with the definitions, equivalent inductances may often be evolved in high-frequency applications such as resonators, waveguide equivalent circuits, etc. Such inductances, when used, must be specified. The definition of self-inductance L is also restricted to cases in which the branches are small in physical size compared with a wavelength, whatever the frequency. Thus, in the case of a uniform 2-wire transmission line, it may be necessary even at low frequencies to consider the parameters as distributed rather than to have one inductance for the entire line. (BT) 270-1966w, [33]

self-information *See:* information content.

self-locking gear train A power transmission gear train designed such that the power train will hold its position and not back-drive whenever the primary driving force is removed. A valve actuator with a self-locking gear train will hold the valve in its last position when the VAM is de-energized. (PE/NP) 1290-1996

self-lubricating bearing (rotating machinery) A bearing lined with a material containing its own lubricant such that little or no additional lubricating fluid need be added subsequently to ensure satisfactory lubrication of the bearing. *See also:* bearing. (IA/APP) [90]

self-maintained discharge (gas) A discharge characterized by the fact that it maintains itself after the external ionizing agent is removed. *See also:* discharge. (ED) [45], [84]

self-operated controller (automatic control) A control device in which all the energy to operate the final controlling element is derived from the controlled system through the primary detecting element. (PE/EDPG) [3]

self-organizing Pertaining to the ability of a system to arrange its internal structure. (C) [20], [85]

self-organizing computer A computer that can change its internal structure. (C) 610.10-1994w

self-phasing array antenna system A receiving antenna system that introduces a phase distribution among the array elements so as to maximize the received signal, regardless of the direction of incidence. *Contrast:* retrodirective antenna. (AP/ANT) 145-1993

self-propelled electric car An electric car requiring no external source of electric power for its operation. *Note:* Diesel-electric, gas-electric, and storage-battery-electric cars are examples of self-propelled cars. The prefix self-propelled is also applied to buses. *See also:* electric motor car. (EEC/PE) [119]

self-propelled electric locomotive An electric locomotive requiring no external source of electric power for its operation. *Note:* Storage-battery, diesel-electric, gas-electric and turbine-electric locomotives are examples of self-propelled electric locomotives. *See also:* electric locomotive. (EEC/PE) [119]

self-pulse modulation Modulation effected by means of an internally generated pulse. *See also:* blocking oscillator; oscillatory circuit. (AP/ANT) 145-1983s

self-quenched counter tube A radiation counter tube in which reignition of the discharge is inhibited by internal processes. (EEC/PE) [119]

self-rectifying x-ray tube An x-ray tube operating on alternating anode potential. (ED) [45]

self-refresh A RAM-refresh protocol, where RAM-local hardware schedules the timing and specifies addresses for RAM refresh cycles. (C/MM) 1596.4-1996

self-relative address An address that must be added to the address of the instruction in which it appears to obtain the address of the storage location to be accessed. *See also:* offset; indexed address; base address; relative address. (C) 610.12-1990

self-repairing circuit A circuit capable of automatically correcting for the effects of a failure so that the presence of the failure is not perceptible. (C) 610.10-1994w

self-reset manual release (control) A manual release that is operative only while it is held manually in the release position. *See also:* electric controller. (IA/ICTL/IAC) [60]

self-reset relay (automatically reset relay) (automatic reset relay) A relay that is so constructed that it returns to its reset position following an operation after the input quantity is removed. (SWG/PE/PSR) C37.100-1992, C37.90-1978s

self-restoring fire detector (fire protection devices) A restorable fire detector whose sensing element is designed to be returned to normal automatically. (EEC/LB) [97]

self-restoring insulation (1) Insulation that completely recovers its insulating properties after a disruptive discharge caused by the application of a test voltage; insulation of this kind is generally, but not necessarily, external insulation. (SWG/C/PE/TR) 1313.1-1996, C57.12.80-1978r, C37.100-1992

(2) Insulation that completely recovers its insulating properties after a disruptive discharge. (PE/PSIM) 4-1995

(3) Insulation that completely recovers its insulating properties after a disruptive discharge caused by the application of an overvoltage; insulation of this kind is generally, but not necessarily, external insulation. (SPD/PE) C62.22-1997

self-rest relay (automatically reset relay) A relay that is so constructed that it returns to its reset position following an operation after the input quantity is removed. (SWG/PE) C37.100-1981s

self-retracting lanyard A device which contains a drum-wound line which may be slowly extracted from or retracted onto the drum under slight tension during normal movement of the user. The line has means for attachment to the fall arrest attachment on the body support. After onset of a fall, the device automatically locks the drum and arrests the fall.The device may have integral means for energy absorption. (T&D/PE) 1307-1996

self-revealing component failures Component failures whose effects on system operation are immediately and clearly apparent to a properly trained person. (VT/RT) 1483-2000

self-saturation (magnetic amplifier) The saturation obtained by rectifying the output current of a saturable reactor. (EEC/PE) [119]

self-screening range Range at which a specified target carrying its own specified active jamming [electronic countermeasures (ECM)] can be detected by a specified radar with specified probabilities of detection and false-alarm. *Note:* Also called "burn-through range." (AES) 686-1997

self-supported aerial cable A cable that includes a messenger cable that has an outer jacket that covers the messenger and the shield. The messenger is available for support, gripping, pulling, and tensioning. *See also:* aerial cable. (PE/PSC) 789-1988w

self-supporting aerial cable A cable consisting of one or more insulated conductors factory assembled with a messenger that supports the assemblage, and that may or may not form a part of the electric circuit. *See also:* conductor. (EEC/AWM) [91]

self-surge impedance *See:* surge impedance.

self-test (test, measurement, and diagnostic equipment) A test or series of tests, performed by a device upon itself, which shows whether or not it is operating within designed limits. This includes test programs on computers and automatic test equipment which check out their performance status and readiness. (IM/WM&A) 194-1977w

self-test capability (test, measurement, and diagnostic equipment) The ability of a device to check its own circuitry and operation. The degree of self-test is dependent on the ability to fault detect and isolate. (IM/WM&A) 194-1977w

self-testing circuit A circuit in which for every signal line an error is detected for both stuck-at-zero and stuck-at-one failures. (C) 610.10-1994w

self-triggered gap A bypass gap that is designed to sparkover on the voltage that appears across the gap terminals. The sparkover of the gap is normally initiated by a trigger circuit set at a specified voltage level. A self-triggered bypass gap may be used for the primary protection of the capacitor. The self-triggered gap may sparkover during external as well as internal faults. (T&D/PE) 824-1994

self-ventilated machine A machine that has its ventilating air circulated by means integral with the machine. (IA/MT) 45-1998

SELV Safety extra low voltage. (C/BA) 896.2-1991w

semianechoic absorber-lined chamber (SALC) A room or enclosure (either shielded or unshielded) with all its surfaces, except the floor, lined with radio-frequency (RF) absorber material. The floor is covered with a good conductor. (EMC) 1128-1998

semantic error An error resulting from a misunderstanding of the relationship of symbols or groups of symbols to their meanings in a given language. *Contrast:* syntactic error. (C) 610.12-1990

semantics (1) (software) The relationships of symbols or groups of symbols to their meanings in a given language. (C) [20], 610.12-1990, [85]

(2) The connotative meaning of words within an ATLAS statement. (SCC20) 771-1998

(3) The meaning of the syntactic components of a language. (C/SE) 1320.2-1998

(4) The meaning, including concept(s), associated with a given entity. (SCC32) 1489-1999

semaphore (1) (software) A shared variable used to synchronize concurrent processes by indicating whether an action has been completed or an event has occurred. *See also:* indicator; flag. (C) 610.12-1990

(2) A shareable resource that has a nonnegative integral value. When the value is zero, there is a (possibly empty) set of threads awaiting the availability of the semaphore. (PA/C) 9945-1-1996

(3) A system variable used to synchronize concurrent processes by indicating whether an action has been completed or an event has occurred. (C/MM) 855-1990

(4) A shareable resource that has a nonnegative integer value. When the value is zero, there is a (possibly empty) set of tasks is awaiting the availability of the semaphore. (C) 1003.5-1999

semaphore decrement operation An operation that decrements the value of a semaphore, blocking until this is possible. If, prior to the operation, the value of the semaphore is zero, the semaphore decrement operation shall cause the calling task to be blocked and added to the set of tasks (possibly in different processes) awaiting the semaphore. Otherwise, the semaphore value is decremented. (C) 1003.5-1999

semaphore increment operation An operation that increments the value of a semaphore or unblocks a waiting task. If, prior to the operation, any tasks are awaiting the semaphore, then some task from that set shall be removed from the set and be unblocked. Otherwise, the semaphore value shall be incremented. (C) 1003.5-1999

semaphore lock operation An operation that is applied to a semaphore. If, prior to the operation, the value of the semaphore is zero, the semaphore lock operation shall cause the calling thread to be blocked and added to the set of threads awaiting the semaphore. Otherwise, the value is decremented. (C/PA) 9945-1-1996

semaphore unlock operation An operation that is applied to a semaphore. If, prior to the operation, there are any threads in the set of threads awaiting the semaphore, then some thread from that set shall be removed from the set and become unblocked. Otherwise, the semaphore value is incremented. (C/PA) 9945-1-1996

semiactive guidance A bistatic-radar homing system in which a receiver in the guided vehicle derives guidance information from electromagnetic signals scattered from a target which is illuminated by a transmitter at a third location. *See also:* illuminator. (AES) 686-1997

semiactive homing guidance (navigation aids) Guidance in which a craft is directed toward a destination by means of information received from the destination in response to transmissions from a source other than the craft. (AES/GCS) 172-1983w

semianalytic inertial navigation equipment The same as geometric inertial navigation equipment except that the horizontal measuring axes are not maintained in alignment with a geographic direction. *Note:* The azimuthal orientations are automatically computed. *See also:* navigation. (AES/RS) 686-1982s, [42]

semiautomatic Combining manual and automatic features so that a manual operation is required to supply to the automatic feature the actuating influence that causes the automatic feature to function. (EEC/PE) [119]

semiautomatic controller An electric controller in which the influence directing the performance of some of its basic functions is automatic. *See also:* electric controller. (IA/IAC) [60]

semiautomatic flight inspection (SAFI) (navigation aid terms) A specialized and largely automatic system for evaluating the quality of information in signals from ground-based navigational aids; data from navigational aids along and adjacent to any selected air route are simultaneously received by a specially equipped SAFI aircraft as it proceeds under automatic control along the route, evaluated at once for gross errors, and recorded for subsequent processing and detailed analysis at a computer-equipped central ground facility. *Note:* Flight inspection means the evaluation of performance of navigational aids by means of in-flight measurements. (AES/GCS) 172-1983w

semiautomatic gate (elevators) A gate that is opened manually and that closes automatically as the car leaves the landing. *See also:* hoistway. (PE/EEC) [119]

semiautomatic holdup-alarm system An alarm system in which the signal transmission is initiated by the indirect and secret action of the person attacked or of an observer of the attack. *See also:* protective signaling. (EEC/PE) [119]

semiautomatic load throw-over equipment Equipment that automatically transfers a load to another (emergency) source of power when the original (preferred) source to which it has been connected fails, but that requires manual restoration of the load to the original source. (SWG/PE) C37.100-1992

semiautomatic plating Mechanical plating in which the cathodes are conveyed automatically through only one plating tank. *See also:* electroplating. (EEC/PE) [119]

semiautomatic signal A signal that automatically assumes a stop position in accordance with traffic conditions, and that can be cleared only by cooperation between automatic and manual controls. (EEC/PE) [119]

semiautomatic station (station control and data acquisition) A station that requires both automatic and manual modes to maintain the required character of service. (SWG/PE/SUB) C37.100-1992, C37.1-1994

semiautomatic telephone systems A telephone system in which operators receive orders orally from the calling parties and establish connections by means of automatic apparatus. (EEC/PE) [119]

semiautomatic test equipment (test, measurement, and diagnostic equipment) Any automatic testing device which requires human participation in the decision-making, control, or evaluative functions. (IM/WM&A) 194-1977w

semiconducting jacket A jacket of such resistance that its outer surface can be maintained at substantially ground potential by contact at frequent intervals with a grounded metallic conductor, or when buried directly in the earth. (PE/EEC/AWM) [4], [91]

semiconducting material (1) A conducting medium in which the conduction is by electrons, and holes, and whose temperature coefficient of resistivity is negative over some temperature range below the melting point. *See also:* semiconductor device; semiconductor. (Std100) 270-1966w
(2) A solid material that conducts limited electric current by means of a small number of free electrons and additional electrons that can be freed from their local bonds by the addition of other elements or "doping." For example, silicon is a semiconducting material. *Contrast:* conducting material; insulating material. *See also:* hole. (C) 610.10-1994w

semiconducting paint (rotating machinery) A paint in which the pigment or portion of pigment is a conductor of electricity and the composition is such that when converted into a solid film, the electrical conductivity of the film is in the range between metallic substances and electrical insulators. (IA/APP) [90]

semiconducting tape (power distribution, underground cables) A tape of such resistance that when applied between two elements of a cable the adjacent surfaces of the two elements will maintain substantially the same potential. Such tapes are commonly used for conductor shielding and in conjunction with metallic shielding over the insulation. (PE/TR) C57.15-1968s

semiconductive ignition cable High-tension ignition cable, the core of which is made of semiconductive material. *Note:* Semiconductive is understood here as referring to conductivity and no other physical properties. *See also:* electromagnetic compatibility. (EMC/INT) [53], [70]

semiconductive slot coating The partially conductive paint or tape layer in intimate contact with the groundwall insulation in the slot portion of the stator core. This coating ensures that there is little voltage between the surface of the coil or bar and the grounded stator core. (DEI) 1043-1996

semiconductor (1) An electronic conductor, with resistivity in the range between metals and insulators, in which the electric-charge-carrier concentration increases with increasing temperature over some temperature range. *Note:* Certain semiconductors possess two types of carriers, namely, negative electrons and positive holes. (ED) 216-1960w
(2) A device that is made of semiconducting material. For example: a diode, an integrated circuit, or a transistor. (C) 610.10-1994w
(3) Material in which the conductivity is due to charge carriers of both signs (electrons and holes), is normally in the range between metals and insulators, and in which the charge-carrier density can be changed by external means. (NPS) 325-1996

semiconductor, compensated *See:* compensated semiconductor.

semiconductor controlled rectifier (SCR) An alternative name used for the reverse-blocking triode-thyristor. *Note:* The name of the actual semiconductor material (selenium, silicon, etc.) may be substituted in place of the word semiconductor in the name of the components. *See also:* thyristor. (ED) 216-1960w

semiconductor converters, classification The following designations are intended to describe the functional characteristics of converters, but not necessarily the circuits or components used. *Note:* Forms A through D refer only to the converters. Rotational direction of motors may be changed by field or armature reversal. (form A converter) A single converter unit in which the direct current can flow in one direction only and which is not capable of inverting energy from the load to the ac supply. Operates in quadrant I only (semiconverter). (form B converter) A double converter unit in which the direct current can flow in either direction but which is not capable of inverting energy from the load to the ac supply. Operates in quadrants I and III only. (form C converter) A single converter unit in which the direct current can flow in one direction only and which is capable of inverting energy from the load to the ac supply. Operates in quadrants I and IV. (form D converter) A double converter unit in which the direct current can flow in either direction and which is capable of inverting energy from the load to the ac supply. Operates in quadrants I, II, III, and IV. (IA/IPC) 444-1973w

semiconductor device An electron device in which the characteristic distinguishing electronic conduction takes place within a semiconductor. *See also:* semiconductor. (Std100) 270-1966w

semiconductor device circuit breaker (thyristor) A circuit breaker of special characteristics used to isolate or protect semiconductor devices from overcurrent. (IA/IPC) 428-1981w

semiconductor device fuse (thyristor) A fuse of special characteristics connected in series with one or more semiconductor devices to isolate or protect the semiconductor. (IA/IPC) 428-1981w

semiconductor device lead inductance (nonlinear, active, and nonreciprocal waveguide components) The inductance of a semiconductor device associated with the strap, mesh, or wire connections used to contact the semiconductor chip. In general, a larger cross-sectional contacting area results in decreased lead inductance. (MTT) 457-1982w

semiconductor device, multiple unit A semiconductor device having two or more sets of electrodes associated with independent carrier streams. *Note:* It is implied that the device has two or more output functions that are independently derived from separate inputs, for example, a duo-triode transistor. *See also:* semiconductor. (PE/EDPG) [93]

semiconductor device, single unit A semiconductor device having one set of electrodes associated with a single carrier stream. *Note:* It is implied that the device has a single output function related to a single input. *See also:* semiconductor. (PE/EDPG) [93]

semiconductor diode A two-terminal device formed of a semiconductor junction having a nonlinear characteristic that will conduct electric current more in one direction than in the other. (CAS/MTT) 146-1980w

semiconductor-diode parametric amplifier A parametric amplifier using one or more varactors. *See also:* parametric device. (ED) [46]

semiconductor, extrinsic *See:* extrinsic semiconductor.

semiconductor frequency changer A complete equipment employing semiconductor devices for changing from one alternating-current frequency to another. *See also:* semiconductor rectifier stack. (IA) [62]

semiconductor, intrinsic *See:* intrinsic semiconductor.

semiconductor junction (light-emitting diodes) A region of transition between semiconductor regions of different electrical properties. (ED) [127]

semiconductor laser *See:* injection laser diode.

semiconductor, *n*-type (A) An extrinsic semiconductor in which the conduction electron concentration exceeds the mobile hole concentration. *Note:* It is implied that the net ionized impurity concentration is donor type. *See also:* semiconductor. **(B)** An *n*-type semiconductor in which the excess conduction electron concentration is very large. *See also:* semiconductor. (ED) 216-1960

semiconductor, *n*+-type An *n*-type semiconductor in which the excess conduction electron concentration is very large. *See also:* semiconductor. (ED) 216-1960w

semiconductor, *p*+-type A *p*-type semiconductor in which the excess mobile hole concentration is very large. *See also:* semiconductor. (ED) 216-1960w

semiconductor power converter A complete equipment employing semiconductor devices for the transformation of electric power. *See also:* semiconductor rectifier stack. (IA) [62]

semiconductor, *p*-type An extrinsic semiconductor in which the mobile hole concentration exceeds the conduction electron concentration. *Note:* It is implied that the net ionized impurity concentration is acceptor type. *See also:* semiconductor. (ED) 216-1960w

semiconductor radiation detector (1) (germanium gamma-ray detectors) A semiconductor device that utilizes the production and motion of excess free charge carriers in the semiconductor for the detection and measurement of particles or photons of incident radiation. (NPS/NID) 300-1988r, 759-1984r
(2) A semiconductor device in which the production and motion of excess free carriers is used for the detection and measurement of incident particles or photons. (NPS) 325-1996

semiconductor rectifier A device consisting of a conductor and semiconductor forming a junction. The junction exhibits a difference in resistance to current flow in the two directions through the junction. This results in effective current flow in one direction only. The semiconductor rectifier stack is a single columnar structure of one or more semiconductor rectifier cells. (IA/MT) 45-1998

semiconductor rectifier cell A semiconductor device consisting of one cathode, one anode, and one rectifier junction. *See also:* semiconductor rectifier stack; semiconductor. (IA) 59-1962w, [12]

semiconductor rectifier cell combination The arrangement of semiconductor rectifier cells in one rectifier circuit, rectifier diode, or rectifier stack. The semiconductor rectifier cell combination is described by a sequence of four symbols written in the order 1-2-3-4 with the following significances:

1) Number of rectifier circuit elements.
2) Number of semiconductor rectifier cells in series in each rectifier circuit element.
3) Number of semiconductor rectifier cells in parallel in each rectifier circuit element.
4) Symbol designating circuit. If a semiconductor rectifier stack consists of sections of semiconductor rectifier cells insulated from each other, the total semiconductor rectifier cell combination becomes the sum of the semiconductor rectifier cell combinations of the individual insulated sections. If the insulated sections have the same semiconductor rectifier cell combination, the total semiconductor rectifier cell combination may be indicated by the semiconductor rectifier cell combination of one section preceded by a figure showing the number of insulated sections. Example: 4(4-1-1-B) indicates four single-phase full-wave bridges insulated from each other assembled as one semiconductor rectifier stack.

Notes: 1. The total number of semiconductor rectifier cells in each semiconductor rectifier cell combination is the product of the numbers in the combination. 2. This arrangement can also be applied by analogy to give a semiconductor rectifier diode combination.

Symbol	Circuit	Example
H	half wave	1-1-1-H
C	center tap	2-1-1-C
B	bridge	4-1-1-B 6-1-1-B
Y	wye	3-1-1-Y
S	star	6-1-1-S
D	voltage doubler	2-1-1-D

See also: semiconductor rectifier cell. (IA) 59-1962w, [12]

semiconductor rectifier diode (thyristor) A semiconductor diode having an asymmetrical voltage-current characteristic, used for the purpose of rectification, and including its associated housing, mounting, and cooling attachment if integral with it. (IA/IPC) 428-1981w

semiconductor rectifier stack An integral assembly, with terminal connections, of one or more semiconductor rectifier diodes, and includes its associated mounting and cooling attachments if integral with it. *Note:* It is a subassembly of, but not a complete semiconductor rectifier. (IA) 59-1962w, [12], [62]

semiconductor storage A type of storage whose elements are formed as solid state electronic components on an integrated circuit. *Contrast:* magnetic storage; core memory. (C) 610.10-1994w

semiconverter, bridge *See:* bridge semiconverter.

semi-direct lighting (illuminating engineering) Lighting involving luminaires that distribute 60−90% of the emitted light downward and the balance upward. (EEC/IE) [126]

semienclosed (A) Having the ventilating openings in the case protected with wire screen, expanded metal, or perforated covers. **(B)** Having a solid enclosure except for a slot for an operating handle or small openings for ventilation, or both. (EEC/PE) [119]

semienclosed brake A brake that is provided with an enclosure that covers the brake shoes and the brake wheel but not the brake actuator. *See also:* control. (IA/ICTL/IAC/APP) [60], [75]

semi-flush-mounted device A device in which the body of the device projects in front of the mounting surface a specified distance between the distance specified for flush-mounted and surface-mounted devices. (SWG/PE) C37.100-1992

semiguarded enclosure An enclosure in which all of the openings, usually in the top half, are protected as in the case of a "guarded enclosure," but the others are left open. (IA/MT) 45-1998

semiguarded machine (rotating machinery) One in which part of the ventilating openings, usually in the top half, are guarded as in the case of a guarded machine but the others are left open. (IA/APP) [90]

semi-high-speed low-voltage dc power circuit breaker (1) A low-voltage dc power circuit breaker that, during interruption, limits the magnitude of the fault current so that its crest is passed not later than a specified time after the beginning of the fault current transient, where the system fault current, determined without the circuit breaker in the circuit, falls between specified limits of current at a specified time. *Note:* The specified time in present practice is 0.03 second. (SWG/PE) C37.100-1981s
(2) (low-voltage dc power circuit breakers used in enclosures) A circuit breaker which, when applied in a circuit with the parameter values specified in American National Standard C37.16-1979, Tables 11 and 11A, tests "b" (1.7 A μs initial rate of rise of current), forces a current crest during interruption within 0.030 s after the current reaches the pickup setting of the instantaneous trip device. *Note:* For total performance at other than test circuit parameters values, consult the manufacturer. (SWG/PE) C37.14-1979s

semi-indirect lighting (illuminating engineering) Lighting involving luminaires which distribute 60% to 90% of the emitted light upward and the balance downward. (EEC/IE) [126]

semimagnetic controller An electric controller having only part of its basic functions performed by devices that are operated by electromagnets. (IA/MT) 45-1998

semi-manual hyphenation In text formatting, hyphenation in which most line-ending and word break decisions are made automatically, the user being asked to assist only when a determination cannot be made automatically. *See also:* manual hyphenation; hot zone hyphenation; automatic hyphenation. (C) 610.2-1987

semi-Markov model (modeling and simulation) A Markov chain model in which the length of time spent in each state is randomly distributed. (C) 610.3-1989w

semi-Markov process (modeling and simulation) A Markov process in which the duration of each event is randomly distributed. (C) 610.3-1989w

semioutdoor reactor A reactor suitable for outdoor use provided that certain precautions in installation (specified by the manufacturer) are observed. For example, protection against rain. C57.16-1958w

semiprotected enclosure (electric installations on shipboard) An enclosure in which all of the openings, usually in the top half, are protected as in the case of a "protected enclosure," but the others are left open. (IA/MT) 45-1983s

semi-random-access A mode of data access that, in the search for the desired item, combines a form of direct access with a limited sequential search. (C) 610.10-1994w

semiremote control A system or method of radio-transmitter control whereby the control functions are performed near the transmitter by means of devices connected to but not an integral part of the transmitter. *See also:* radio transmitter. (AP/ANT) 145-1983s

semi-reverberant facility A room with a solid floor and an undetermined amount of sound-absorbing materials on the walls and ceiling. (PE/TR) C57.12.90-1999

semiselective ringing (telephone switching systems) Ringing wherein the ringers at two or more of the main stations on a party line respond simultaneously, differentiation being by the number of rings. (C) [85]

semistop joint (power cable joints) A joint which is designed to restrict movement of the dielectric fluid between cables being joined. (PE/IC) 404-1986s

semistrain insulator (semitension assembly) Two insulator strings at right angles, each making an angle of about 45 degrees with the line conductor. *Note:* These assemblies are used at intermediate points where it may be desirable to partially anchor the conductor to prevent too great movement in case of a broken wire. *See also:* tower. (EEC/AWM) [91]

semit *See:* semitone.

semitone (half-step) (semit) The interval between two sounds having a basic frequency ratio approximately the twelfth root of two. *Note:* In equally tempered semitones, the interval between any two frequencies is 12 times the logarithm to the base 2 (or 39.86 times the logarithm to the base 10) of the frequency ratio. (SP) [32]

semitransparent photocathode (camera tubes or phototubes) A photocathode in which radiant flux incident on one side produces photoelectric emission from the opposite side. *See also:* phototube; electrode. (ED) 161-1971w

senary (A) Pertaining to a selection in which there are six possible outcomes. **(B)** Pertaining to the numeration system with a radix of 6. (C) 1084-1986

send (1) The first of two packets within a request or response subaction (the second packet is an echo). The send packet contains a 16-byte header (containing command and status) and may optionally contain a data component (up to 256 bytes). (C/MM) 1596-1992
(2) The electrical output of a handsfree telephone due to an acoustic input. (COM/TA) 1329-1999

send attenuation during double talk (A_{SDT}) Attenuation in the send path, seen at the send electrical test point, inserted during double talk. The receive talker initiates the double talk. (COM/TA) 1329-1999

send channel *See:* transmit channel.

send electrical test point (SETP) The point in a battery feed circuit, reference codec, or wireless reference base station at which signals coming from the handsfree telephone (HFT) in the send direction are accessed. (COM/TA) 1329-1999

sender (A) The agent that supplies the data for a solicited message. *See also:* solicited messages. **(B) (telephone switching systems)** Equipment that generates and transmits signals in response to information received from another part of the system. (C/MM) 1296-1987, [85]

send front-end syllabic clipping during double talk (T_{SFDT}) The length of time that speech undergoes syllabic clipping, as seen at the send electrical test point, just after the onset of double talk. The send talker initiates the double talk. (COM/TA) 1329-1999

sending-end crossfire The crossfire in a telegraph channel from one or more adjacent telegraph channels transmitting from the end at which the crossfire is measured. *See also:* telegraphy. (EEC/PE) [119]

sending-end impedance (line) The ratio of an applied potential difference to the resultant current at the point where the potential difference is applied. The sending-end impedance of a line is synonymous with the driving-point impedance of the line. *Note:* For an infinite uniform line the sending-end impedance and the characteristic impedance are the same: and for an infinite periodic line the sending-end impedance and the iterative impedance are the same. *See also:* waveguide; self-impedance. (EEC/PE) [119]

send loudness rating directionality (SLRD) Send loudness rating versus angles around the handsfree telephone (HFT), normalized to the loudness rating at the 50 cm test point (50TP). (COM/TA) 1329-1999

send noise level Electrical noise at send electrical test point (SETP), measured in units of decibels relative to 1 mW (dBm), psophometrically weighted (dBmp). (COM/TA) 1329-1999

send-only equipment Data communication channel equipment capable of transmitting signals, but not arranged to receive signals. (COM) [49]

sensation level *See:* level above threshold.

sense (navigation aids) The pointing direction of a vector representing some navigation parameter. (AES/GCS) 172-1983w

sense amplifier voltage window (1) The minimum difference in threshold voltage required by a sense amplifier to differentiate between the low- and high-conductance MNOS thresholds for the two logic states. (ED) 641-1987w
(2) Sense amplifier sensitivity between differential state or reference voltage and sensed state. (ED) 1005-1998

sense and command features The electrical and mechanical components and interconnections involved in generating those signals associated directly or indirectly with the safety functions. The scope of the sense and command features extends from the measured process variables to the execute features' input terminals. (PE/NP) 603-1998

sense finder That portion of a direction-finder that permits determination of direction without 180-degree ambiguity. *See also:* radio receiver. (EEC/PE) [119]

sense of polarization For an elliptical or circularly polarized field vector, the sense of rotation of the extremity of the field vector when its origin is fixed. *Note:* When the plane of polarization is viewed from a specified side, if the extremity of the field vector rotates clockwise [counterclockwise] the sense is right-handed [left-handed]. For a plane wave, the plane of polarization shall be viewed looking in the direction of propagation. (AP/ANT) 145-1993

sense signal The response taken or measured from a test subject. (SCC20) 1226-1998

sense switch A switch found on the front panel or console of a computer. *Note:* The computer can be programmed to check a switch and to take some action depending on whether the switch is on or off. (C) 610.10-1994w

SENSE/INT/SYNC-IN A special function signal from the bedside ommunications controller (BCC) to the device communications controller (DCC) in a BCC-to-DCC interconnection cable used for three purposes. The signal allows the DCC to sense whether it has been physically connected to a BCC port, allows a BCC to send sync pulses to the DCC, and allows the BCC to send interrupt deactivate pulses to the DCC. (EMB/MIB) 1073.3.1-1994, 1073.4.1-2000

SENSE/INT/SYNC-OUT A special function signal from the device communications controller (DCC) to the bedside communications controller (BCC) in a BCC-to-DCC interconnection cable used for three purposes. The signal allows the BCC port to sense whether a DCC has been physically connected, allows a DCC to send sync pulses to the BCC, and allows the DCC to send interrupt activate and interrupt deactivate pulses to the BCC. (EMB/MIB) 1073.3.1-1994, 1073.4.1-2000

sensibility, deflection *See:* deflection sensibility.

sensing (navigation aids) The process of finding the sense, as, for example, in direction finding, the resolution of the 180° ambiguity in bearing indication; and, as in phase or amplitude-comparison systems such as ILS (instrument landing system) and VOR (very high-frequency omnidirectional range), the establishment of a relation between course displacement signal and the proper response in the control of the vehicle. (AES/GCS) 172-1983w

sensing circuit A circuit whose function is to detect the occurrence of some event at its input terminals. (C) 610.10-1994w

sensing coil (interferometric fiber optic gyro) A coil of optical fiber in which counter-propagating light waves differ in phase as a consequence of the Sagnac effect when the coil is rotated about an axis normal to the plane of the coil. (AES/GYAC) 528-1994

sensing element *See:* primary detector; sensor.

sensing station *See:* read station.

sensitive A condition of an object that allows it to accept input events. (C) 1295-1993w

sensitive relay A relay that operates on comparatively low input power, commonly defined as 100 mW or less. *See also:* relay. (EEC/REE) [87]

sensitive volume That portion of the radiation counter gas volume having sufficient potential gradient to operate in the Geiger-Mueller region. (NI/NPS) 309-1999

sensitivity (1) (A) (general comment) Definitions of sensitivity fall into two contrasting categories. In some fields, sensitivity is the ratio of response to cause. Hence increasing sensitivity is denoted by a progressively larger number. In other fields, sensitivity is the ratio of cause to response. Hence increasing sensitivity is denoted by a progressively smaller number. *See also:* sensitivity coefficient. **(B) (electric pipe heating systems)** The ratio of the magnitude of a device response to the magnitude of the quantity measured. In electric pipe heating systems, sensitivity is usually associated with temperature controls and alarms and addresses their response function. **(C) (nuclear power generating station)** The ratio of a change in output magnitude to the change in input which causes it, after the steady-state has been reached. (PE/EDPG/NP) 622A-1984, 381-1977

(2) (electric heat tracing systems) The ratio of the magnitude of a device response to the magnitude of the quantity measured. In electric heat tracing systems, sensitivity is usually associated with temperature controls and alarms and addresses their response function. (PE/EDPG) 622B-1988r

(3) (monitoring radioactivity in effluents) The minimum amount of contaminant that can repeatedly be detected by an instrument. (NI/PE/NP) N42.18-1980r, 381-1977w

(4) (measuring devices) The ratio of the magnitude of its response to the magnitude of the quantity measured. *Notes:* 1. It may be expressed directly in divisions per volt, millimeters per volt, milliradians per microampere, etc., or indirectly by stating a property from which sensitivity can be computed (for example, ohm per volt for a stated deflection. 2. In the case of mirror galvanometers it is customary to express sensitivity on the basis of a scale distance of 1 m. (MIL) [2]

(5) (radio receiver or similar device) Taken as the minimum input signal required to produce a specified output signal having a specified signal-to-noise ratio. *Note:* This signal input may be expressed as power or as voltage, with input network impedance stipulated. (PE) 599-1985w

(6) (transmission lines, waveguides, and nuclear techniques) The least signal input capable of causing an output signal having desired characteristics. *See also:* ionizing radiation. (IM/HFIM) [40]

(7) (camera tubes or phototubes) The quotient of output current by incident luminous flux at constant electrode voltages. *Notes:* 1. The term output current as here used does not include the dark current. 2. Since luminous sensitivity is not an absolute characteristic but depends on the special distribution of the incident flux, the term is commonly used to designate the sensitivity to light from a tungsten-filament lamp operating at a color temperature of 2870 kelvins. *See also:* phototube; cathode luminous sensitivity. (EEC/PE) [119]

(8) (A) (electrothermic unit) (dissipated power) The ratio of the dc output voltage of the electrothermic unit to the microwave power dissipated within the electrothermic unit at a prescribed frequency, power level, and temperature. **(B) (electrothermic unit)** (incident power) The ratio of the dc output voltage of the electrothermic unit to the microwave power incident upon the electrothermic unit at a prescribed frequency, power level, and temperature. **(C) (electrothermic-coupler unit)** The ratio of the dc output voltage of the electrothermic unit on the side arm of the directional coupler to the power incident upon a nonreflecting load connected to the output port of the main arm of the directional coupler at a prescribed frequency, power level, and temperature. If the electrothermic unit is attached to the main arm of the directional coupler, the sensitivity is the ratio of the dc output voltage of the electrothermic unit attached to the main arm of the directional coupler to the microwave power incident upon a nonreflecting load connected to the output port of the side arm of the directional coupler at a prescribed frequency, power level, and temperature. (IM) 544-1975

(9) (non-real time spectrum analyzer) (volts, decibels above or below one milliwatt) Measure of a spectrum analyzer's ability to display minimum level signals. IF (intermediate frequency) bandwidth, display mode, and any other influencing factors must be given. *Notes:* 1. equivalent input noise. The average level of a spectrum analyzer's internally generated noise referenced to the input. 2. input signal level. The input signal level that produces an output equal to twice the value of the average noise alone. This may be power or voltage relationship, but must be so stated. (IM) [14]

(10) (automatic control) Of a control system or element, or combination, the ratio of a change in output magnitude to the change of input which causes it, after the steady state has been reached. *Note:* ASA C85 deprecates use of "sensitivity" to describe smallness of a dead-band. *See also:* gain. (PE/EDPG) [3]

(11) (fiber optics) Imprecise synonym for responsivity. In optical system receivers, the minimum power required to achieve a specified quality of performance in terms of output signal-to-noise ratio or other measure. (Std100) 812-1984w

(12) (radiation protection) The ratio of a change in response to the corresponding change in the field being measured. (NI) N323-1978r

(13) (spectrum analyzer) Measure of a spectrum analyzer's ability to display minimum level signals, (V, dBm). Intermediate frequency (IF) bandwidth, display mode, and any other influencing factors must be given. *See also:* input signal level sensitivity; equivalent input noise sensitivity. (IM) 748-1979w

(14) (accelerometer) (gyros) The ratio of a change in output to a change in an undesirable or secondary input. For example: a scale factor temperature sensitivity of a gyro or accelerometer is the ratio of change in scale factor to a change in temperature. (AES/GYAC) 528-1994

sensitivity analysis (1) (nuclear power generating station) An analysis that determines the variation of a given function caused by changes in one or more parameters about a selected reference value. (PE/NP) 380-1975w

(2) (reliability analysis of nuclear power generating station safety systems) An analysis that assesses the variation in the value of a given function caused by changes in one or more arguments of the function. (PE/NP) 352-1987r

sensitivity, cathode luminous *See:* cathode luminous sensitivity.

sensitivity, cathode radiant *See:* cathode radiant sensitivity.

sensitivity coefficient (1) (automatic control) The partial derivative of a system signal with respect to a system parameter. *See also:* control system. (CS/PE/EDPG) [3]

(2) A coefficient used to relate the change of a system function F due to the variation of one of its parameters x. In some applications (for example control theory) absolute changes are important and the sensitivity coefficient is defined as the $\partial F/\partial x$. In other applications (for example, filter theory) relative changes are important and then sensitivity is defined as $\partial(\text{Ln } F)/\partial(\text{Ln} x) = (\partial F/\partial x)/(F/x)$. (CAS) [13]

sensitivity, deflection *See:* deflection sensitivity.

sensitivity, dynamic *See:* dynamic sensitivity.

sensitivity, illumination *See:* illumination sensitivity.

sensitivity, incremental *See:* incremental sensitivity.

sensitivity label Representation of the security level of an object describing the sensitivity (e.g., classification) of information in the object. (C/BA) 896.3-1993w

sensitivity level (in electroacoustics) (in decibels) (of a transducer) 20 times the logarithm to the base 10 of the ratio of the amplitude sensitivity *SA* to the reference sensitivity S_0, where the amplitude is a quantity proportional to the square root of power. The kind of sensitivity and the reference sensitivity must be indicated. *Note:* For a microphone, the free-field voltage/pressure sensitivity is the kind often used and a common reference sensitivity is S_0, one volt per newton per square meter. The square of the sensitivity is proportional to a power ratio. The free-field voltage sensitivity-squared level, in decibels, is therefore $SA = 10 \log (S_A^2/S_0^2) = 20 \log (S_A/S_0)$. Often, sensitivity-squared level in decibels can be shortened, without ambiguity, to sensitivity level in decibels, or simply sensitivity in decibels. *Synonyms:* sensitivity; response. (SP) [32]

sensitivity, luminous *See:* luminous sensitivity.

sensitivity, quieting *See:* quieting sensitivity.

sensitivity, radiant *See:* radiant sensitivity.

sensitivity response The net number of counts registered by the detector system per unit of time, divided by the activity of the radionuclide. (NI) N42.12-1994

sensitivity, threshold *See:* threshold sensitivity.

sensitivity time control (STC) Programmed variation of the gain (sensitivity) of a radar receiver as a function of time within each pulse-repetition interval or observation time in order to prevent overloading of the receiver by strong echoes from targets or clutter at close ranges. *Note:* Also called swept gain, especially in British usage. (AES) 686-1997

sensitizing (electrostatography) The act of establishing an electrostatic surface charge of uniform density on an insulating medium. *See also:* electrostatography. (ED) 224-1965w, [46]

sensitometry The measurement of the light response characteristics of photographic film under specified conditions of exposure and development. (SP) [32]

sensor (1) (electrical heating applications to melting furnaces and forehearths in the glass industry) A device that responds to a physical stimulus (such as heat and light) and transmits a resulting signal. (IA) 668-1987w

(2) (A) (nuclear power generating station) That portion of a channel which first responds to changes in, and performs the primary measurement of, a plant variable or condition. **(B) (nuclear power generating station)** A device directly responsive to the value of the measured quantity. (PE/NP) 381-1977

(3) (electric heat tracing systems) The first system element that responds quantitatively to the measure and performs the initial measurement operation. Sensors, as used in electric heat tracing systems, respond to the temperature of the system and may be directly connected to controllers, alarms, or both. Sensors can be mechanical (bulb, bimetallic) or electrical (thermocouple, RTD, thermistor). (PE/EDPG) 622A-1984r, 622B-1988r

(4) (temperature measurement) That portion of a temperature-measuring system that responds to the temperature being measured. (PE/PSIM) 119-1974w

(5) (test, measurement, and diagnostic equipment) A transducer which converts a parameter at a test point to a form suitable for measurement by the test equipment. (MIL) [2]

(6) The portion of a channel that responds to changes in a plant variable or condition and converts the measured process variable into an electric, optic, or pneumatic signal. (PE/NP) 603-1998

(7) A transducer that converts a physical, biological, or chemical parameter into an electrical signal. (IM/ST) 1451.2-1997

(8) (as applied to a circuit-breaker with an electronic trip unit) A current sensing element such as a current transformer within a circuit-breaker frame. The sensor will have a current rating less than or equal to the frame size and will provide the sensing function for a specific group of current ratings within the frame size. (IA/PSP) 1015-1997

(9) A component providing a useful output in response to a physical, chemical, or biological phenomenon. This component may already have some signal conditioning associated with it. Examples: platinum resistance temperature detector, humidity sensor with voltage output, light sensor with frequency output, pH probe, and piezoresistive bridge. (IM/ST) 1451.1-1999

sensor, active *See:* active sensor.

sensor-based system An organization of components, including a computer, whose primary source of input is data from sensors and whose output can be used to control the related physical process being sensed. (C) 610.10-1994w

sensor, passive *See:* passive sensor.

sensory saturation (nuclear power generating station) The impairment of effective operator response to an event due to excessive amount of display information that must be evaluated prior to taking action. (PE/NP) 566-1977w

sentinel *See:* flag.

separable insulated connector (1) (separable insulated connectors) A fully insulated and shielded system for terminating and electrically connecting an insulated power cable to electrical apparatus, other power cables, or both, so designed that the electrical connection can be readily established or broken by engaging or separating the connector at the operating interface. (T&D/PE) 386-1995

(2) (power and distribution transformers) A system for terminating and electrically connecting an insulated power cable to electrical apparatus, other power cables, or both, so designed that the electrical connection can be readily established or broken by engaging or separating mating parts of the connector at the operating interface. (PE/TR) C57.12.80-1978r

separate chaining Hashing in which collision resolution is handled by building a linked list, called a collision chain, for each position in the hash table to hold the items whose hash values correspond to that position in the hash table. *Synonyms:* direct chaining; external chaining. *Contrast:* open-address hashing. (C) 610.5-1990w

separate excitation (1) (emergency and standby power) A source of generator field excitation power derived from a source independent of the generator output power. (IA/PSE) 446-1995

(2) (power system device function numbers) A device that connects a circuit, such as the shunt field of a synchronous converter, to a source of separate excitation during the starting sequence; or one that energizes the excitation and ignition circuits of a power rectifier. (SUB/PE) C37.2-1979s

separately excited (rotating machinery) A qualifying term applied to a machine to denote that the excitation is obtained from a source other than the machine itself. (PE) [9]

separately ventilated machine A machine that has its ventilating air supplied by an independent fan or blower external to the machine. (IA/MT) 45-1998

separate parts of a network The parts that are not connected. *See also:* network analysis. (Std100) 270-1966w

separate terminal enclosure (rotating machinery) A form of termination in which the ends of the machine winding are connected to the incoming supply leads inside a chamber that need not be fully enclosed and may be formed by the foundations beneath the machine. (PE) [9]

separating character *See:* information separator.

separation (1) (frequency modulation) The process of deriving individual channel signals (for example, for stereophonic systems) from a composite transmitted signal. *Note:* Separation describes the ability of a receiver to produce left and right stereophonic channel signals at its output terminals and is a measured parameter for stereo receivers only. Left-channel signal separation is defined as the ratio in decibels of the output voltage of the left output of the receiver to that of the right output when an "L"-only signal is received. Right-channel separation is similarly defined. (BT) 185-1975w

(2) (nuclear power generating station) (separation and identification) Physical independence of redundant circuits, components, and equipment. (Physical independence may be achieved by space, barriers, shields, etc.) (PE/EDPG) 690-1984r

(3) The distance between two objects, measured surface to surface, and usually filled with a solid or liquid material. (NESC) C2-1997

separation criteria Curves that relate the frequency displacement to the minimum distance between a receiver and an undesired transmitter to insure that the signal-to-interference ratio does not fall below a specified value. *See also:* electromagnetic compatibility. (EMC) [53]

separation distance Space that has no interposing structures, equipment, or materials that could aid in the propagation of fire or that could otherwise disable Class 1E systems or equipment. (PE/NP) 384-1992r

separation plane A reference plane, used to separate two objects. This plane shall not be encroached upon by the object on either side except in clearly specified interface areas. (C/BA) 1301.4-1996

separation sort *See:* distribution sort.

separator (1) (storage cell) A spacer employed to prevent metallic contact between plates of opposite polarity within the cell. (Perforated sheets are usually called retainers.) *See also:* battery. (PE/EEC) [119]

(2) A visual user interface control consisting of a line boundary that provides a visual distinction between two adjacent areas. The line may be drawn using various graphics styles. (C) 1295-1993w

(3) *See also:* delimiter. (C) 610.5-1990w, 610.12-1990

separator, insulation slot *See:* insulation slot separator.

septenary (A) Pertaining to a selection in which there are seven possible outcomes. **(B)** Pertaining to the numeration system with a radix of 7. (C) 1084-1986

septendecimal (A) Pertaining to a selection in which there are 17 possible outcomes. **(B)** Pertaining to the numeration system with a radix of 17. (C) 1084-1986

septet (1) A group of seven adjacent digits operated upon as a unit. *Synonym:* seven-bit byte. (C) 610.5-1990w, 1084-1986w

(2) A byte composed of seven bits. *Synonym:* seven-bit byte. (C) 610.10-1994w

sequence (1) (A) To place items in a linear arrangement in accordance with the order of the natural numbers. *Note:* Methods or procedures may be specified for other natural linear orders by mapping onto the natural numbers. For example, the sequence may be alphabetic or chronological. *See also:* sort; collating sequence. **(B)** The order in which items are arranged. *See also:* random number sequence; collating sequence; recursively defined sequence. **(C)** A set of items that have been sequenced. *See also:* collating sequence; order. (C) 610.5-1990

(2) (STEbus) An indivisible bus transaction comprising one or more transfers. *See also:* pseudo-random number sequence; calling sequence. (MM/C) 1000-1987r

(3) A set of bits, packets, or messages ordered in time and that are, or that are intended to be, transmitted consecutively without interruption. (TT/C) 1149.5-1995

sequence by merging *See:* sort by merging.

sequence check A check that verifies that a set of items are in a certain sequence. (C) 610.5-1990w

sequence control register *See:* instruction address register.

sequence field *See:* key.

sequence filter *See:* sequence network.

sequence network An electrical circuit that produces an output proportional to one or more of the sequence components of a polyphase system of voltages or currents, e.g., positive sequence network, negative sequence network, or zero-sequence network. (SWG/PE) C37.100-1992

sequence number (1) A number identifying the relative location of blocks or groups of blocks on a tape. (IA) [61]

(2) Each I frame is sequentially numbered with a number listed in the control field, from 0 to 7. The sequence numbers cycle through the entire range. (EMB/MIB) 1073.3.1-1994

sequence-number readout Display of the sequence number punched on the tape. *See also:* block count readout. (IA) [61]

sequence-of-events (SOE) Digital input points that are time tagged to include relative or absolute time of occurrence. (SUB/PE) C37.1-1994

sequence of events function *See:* supervisory control functions.

sequence-of-events point interface Master station or RTU (or both) element(s) that accept(s) a digital input signal to perform the function of time tagging the occurrence of an event. *See also:* sequence-of-events. (SUB/PE) C37.1-1994

sequence-of-events SCADA function The capability of a supervisory system to recognize each predefined event, associate a time of occurrence with each event, and present the event data in order of occurrence of the events. (SUB/PE) C37.1-1994

sequence of operation (packaging machinery) A written detailed description of the order in which electrical devices and other parts of the industrial equipment should function. (IA/PKG) 333-1980w

sequence point A certain point in the execution sequence of a program where all side effects of previous evaluations are complete and no side effects of subsequent evaluations have occurred. (C/DA) 1481-1999

sequencer (1) A mechanical device or computer program that sequences the items in a set. *See also:* sorter. (C) 610.5-1990w

(2) An object that controls the execution flow of programs. (SCC20) 1226-1998

sequence switch A remotely controlled power-operated switching device used as a secondary master controller. *See also:* multiple-unit control. (EEC/PE) [119]

sequence table (electric controller) A table indicating the sequence of operation of contactors, switches, or other control apparatus for each step of the periodic duty. *See also:* multiple-unit control. (VT/LT) 16-1955w

sequence variable A flow-control component involving a public location in System Memory or a CSR, holding the sequence number of the current message. The first message corresponds to sequence number one, etc. This sequence number is operated modulo the variable size (i.e., when the maximum value is reached, it rolls over to zero). (C/MM) 1212.1-1993

sequencing key *See:* sort key.

sequential (1) (formatted system) (telecommunications) If the signal elements are transmitted successively in time over a channel, the transmission is said to be sequential. If the signal elements are transmitted at the same time over a multiwire circuit, the transmission is said to be coincident. *See also:* bit. (COM) [49]
(2) (software) Pertaining to the occurrence of two or more events or activities in such a manner that one must finish before the next begins. *Synonym:* serial. *See also:* consecutive. (C) 610.12-1990
(3) Pertaining to a circuit whose output values, at a given instant, depend upon its input values and internal state at that instant, and whose internal state depends upon the immediately preceding input values and the preceding internal state. *Contrast:* combinational. (C) 610.10-1994w

sequential access (1) (test, measurement, and diagnostic equipment) A system in which the information becomes available in a one after the other sequence only, whether all of it is desired or not. (MIL) [2]
(2) (data management) Pertaining to the process of storing and retrieving data using the sequential access mode. *Synonyms:* serial access; physical sequential access. *Contrast:* random access; direct access. *See also:* indexed access; indexed sequential access. (C) 610.5-1990w
(3) *See also:* nano; access. (C) 610.10-1994w

sequential access method (SAM) A technique for accessing data using sequential access mode. That is, to process a given data record, all data records previous to it must be accessed. *See also:* basic sequential access method. (C) 610.5-1990w

sequential access mode An access mode in which data records are stored and retrieved in such a way that each successive access defines the next record to be retrieved. *Contrast:* indexed sequential access mode; direct access mode. (C) 610.5-1990w

sequential access storage A type of storage that provides only sequential access to data. For example, magnetic tape storage. *Synonym:* serial access storage. (C) 610.10-1994w

sequential circuit A logic circuit whose output values, at a given instant, depend upon its input values and internal state at that instant, and whose internal state depends upon the immediately preceding input values and the preceding internal state. *Contrast:* combinational circuit. *See also:* trigger circuit; toggle. (C) 610.10-1994w

sequential cohesion A type of cohesion in which the output of one task performed by a software module serves as input to another task performed by the module. *Contrast:* temporal cohesion; logical cohesion; coincidental cohesion; procedural cohesion; communicational cohesion; functional cohesion. (C) 610.12-1990

sequential commutation (circuit properties) (self-commutated converters) Commutation occurs from one to the next of three or more principal switching branches arranged as a multipulse group that conduct in cyclic sequential order for usually (but not always) equal time intervals. The commutation may be direct or indirect. (IA/SPC) 936-1987w

sequential computer A computer in which events occur in time sequence, with little or no simultaneity or overlap of events. *Contrast:* simultaneous computer; parallel computer. (C) 610.10-1994w

sequential construct *See:* serial construct.

sequential control (computers) A mode of computer operation in which instructions are executed consecutively unless specified otherwise by a jump. (MIL/C) [2], [20], [85]

sequential data set (SDS) *See:* sequential file.

sequential detection A method of automatic detection in two or more steps. Normally the first step uses a high probability of false alarm and the last uses a low probability of false alarm. *Note:* In radars with controllable scanning, the first detection can be used to order the scan to return to, stop at, or stay longer at the suspected target position. (AES) 686-1997

sequential events recording system A system that monitors bistable equipment operations and process status and records changes of state in the order of detected occurrences. This monitoring may be accomplished using a device dedicated solely to this function, or using a multifunction system such as a data acquisition computer system. (PE/EDPG) [5], [1]

sequential file A file that must be accessed using sequential access; for example, a data file on a magnetic tape. *Synonyms:* serial file; sequential data set. *Contrast:* partitioned data set; direct data set; indexed file. (C) 610.5-1990w

sequential lobing *See:* lobe switching.

sequential logic function A logic function in which there exists at least one combination of input states for which there is more than one possible resulting combination of states at the outputs. *Note:* The outputs are functions of variables in addition to the present states of the inputs, such as time, previous internal states of the element, etc. (GSD) 91-1984r

sequential memory (sequential events recording systems) The memory that stores events in the same order in which they were received by the system. The memory capacity can be expressed as the number of events or levels. *See also:* level; event. (PE/EDPG) [5], [1]

sequential operation Pertaining to the performance of operations one after the other. (C) [20], [85]

sequential precedential database *See:* hierarchical database.

sequential processes (software) Processes that execute in such a manner that one must finish before the next begins. *See also:* concurrent processes; process. (C/SE) 729-1983s

sequential processing *See:* serial processing.

sequential programming (test, measurement, and diagnostic equipment) The programming of a device by which only one arithmetical or logical operation can be executed at one time. (MIL) [2]

sequential relay A relay that controls two or more sets of contacts in a predetermined sequence. *See also:* relay. (EEC/REE) [87]

sequential scanning (television) A rectilinear scanning process in which the distance from center to center of successively scanned lines is equal to the nominal line width. *See also:* television. (PE/EEC) [119]

sequential search A search in which the items in a set are examined in order, starting from the first item in the set, until the search is successful or the end of the set is encountered. *Synonym:* linear search. (C) 610.5-1990w

sequential-stress aging A form of accelerated aging in which two or more individual stresses are applied or intensified in sequence. (DEI/RE) 775-1993w

sequential transfer A transfer operation with multiple data transfers during the reply phase. *See also:* transfer operation; reply phase. (C/MM) 1296-1987s

sequential tripping A situation where one or more relay terminals of a line cannot detect an internal line fault, typically because of infeed, until one or more terminals has already opened and removed the infeed. (PE/PSR) C37.113-1999

serial **(1)** **(A)** Pertaining to the time sequencing of two or more processes. **(B)** Pertaining to the time sequencing of two or more similar or identical processes, using the same facilities for the successive processes. **(C)** Pertaining to the time-sequential processing of the individual parts of a whole, such as the bits of a character, the characters of a word, etc., using the same facilities for successive parts. *See also:* serial-parallel. (C) 162-1963
(2) **(software)** Pertaining to the sequential transfer, occurrence, or processing of the individual parts of a whole, such as the bits of a character, using the same facilities for successive parts. *Contrast:* parallel. *See also:* sequential. (C) 610.12-1990
(3) One bit following another over a single pathway. *Contrast:* parallel. *See also:* bit serial. (C) 610.10-1994w

serial access **(1)** **(computers)** Pertaining to the process of obtaining data from, or placing data into, storage when there is a sequential relation governing the access time to successive storage locations. (C) [20], [85]
(2) **(data management)** *See also:* sequential access. (C) 610.5-1990w

serial access storage *See:* sequential access storage.

serial adder An adder in which addition is performed by adding, digit place after digit place, the corresponding digits of the operands. *Contrast:* parallel adder. (C) 610.10-1994w

serial addition Addition that is performed by adding the corresponding digits of the operands, one digit place at a time. *Contrast:* parallel addition. (C) 1084-1986w

serial by bit *See:* serial transmission.

serial bus **(1)** Intended as a low-cost peripheral connect or an alternate diagnostic and control path. One instantiation of a serial bus is the "Serial Bus" as specified in IEEE P1394. (C/BA) 896.2-1991w
(2) A peripheral interconnect and an alternate diagnostic and control path. (C/BA) 896.10-1997

Serial Bus **(1)** The name that refers to the IEEE project, P1394, which specifies a serial bus intended as a low-cost peripheral connect or an alternate diagnostic and control path. (C/MM) 1212-1991s
(2) A bit-serial interconnect defined by IEEE P1394. (C/MM) 1212.1-1993
(3) Refers to the IEEE P1394 project, which defines an inexpensive serial network that can be used as an alternate control or diagnostic path, as an I/O connection, or in place of a parallel bus in some systems. (C/MM) 1596.5-1993, 1596-1992

Serial Bus management The set of protocols, services, and operating procedures that monitors and controls the various Serial Bus layers: physical, link, and transaction. See figure 34 for the relation of Serial Bus management to the Serial Bus protocol stack. (C/MM) 1394-1995

serial clock driver A functional module that provides a periodic timing signal that synchronizes the operation of IEEE P1132 serial bus. Two backplane signal lines are reserved for use by a serial bus. However, the protocols of the serial bus are completely independent of this standard, and the inclusion of a serial bus is not a required feature of IEEE Std 1014-1987. (C/BA) 1014-1987

serial communication **(1)** **(supervisory control, data acquisition, and automatic control)** A method of transmitting information between devices by sending all bits serially over a single communication channel. (SWG/PE/SUB) C37.100-1992, C37.1-1994
(2) Method of transferring information between devices by transmitting a sequence of individual bits in a prearranged order of significance. (SUB/PE) 999-1992w

serial computer **(A)** A computer that has a single arithmetic and logic unit. **(B)** A computer, some specified characteristic of which is serial; for example, a computer that manipulates all bits of a word serially. *Contrast:* parallel computer. (C) 610.10-1994

serial construct A program construct consisting of a sequence of steps not involving a decision or loop. *Synonym:* sequential construct. (C) 610.12-1990

serial digital computer A digital computer in which the digits are handled serially. Mixed serial and parallel machines are frequently called serial or parallel according to the way arithmetic processes are performed. An example of a serial digital computer is one that handles decimal digits serially although it might handle the bits that comprise a digit either serially or in parallel. *See also:* parallel digital computer. (Std100) 270-1966w

serial file *See:* sequential file.

serial interface An interface that transmits data bit by bit rather than in whole bytes. (C) 610.10-1994w

serialization Serialization is the process of transmitting coded characters one bit at a time. *See also:* deserialization. (C/BA) 1355-1995

serializer A device that converts a set of simultaneous signals into a corresponding time sequence of signals. *Synonyms:* parallel-serial converter; dynamicizer. (C) 610.10-1994w

serially correlated variable *See:* lag variable.

serial medium A medium that contains a POSIX.1 extended tar or extended `cpio` archive. (C/PA) 1387.2-1995

serial mouse A mouse that is connected to a computer system through a serial port. *Contrast:* bus mouse. (C) 610.10-1994w

serial operation **(data transmission)** **(telecommunications)** The flow of information in time sequence, using only one digit, word, line, or channel at a time. (PE) 599-1985w

serial-parallel Pertaining to processing that includes both serial and parallel processing, such as one that handles decimal digits serially but handles the bits that comprise a digit in parallel. (C) 162-1963w

serial port A port that transfers data one bit at a time. *Contrast:* parallel port. (C) 610.10-1994w

serial printer A printer that receives its input data in the form of a serial stream of data. *Contrast:* parallel printer. *See also:* character-at-a-time printer. (C) 610.10-1994w

serial processing Pertaining to the sequential execution of processes in a single device, such as a processing unit or channel. *Synonym:* sequential processing. *Contrast:* parallel processing. (C) 610.10-1994w

serial protocol Any communication protocol in which data is transferred serially to or from a fixed location. (C/MM) 1155-1992

serial transmission **(1)** **(data transmission)** **(telecommunications)** Used to identify a system wherein the bits of a character occur serially in time. Implies only a single transmission channel. *Synonym:* serial by bit. (PE) 599-1985w
(2) In data communications, the conveying of a character of information one bit at a time on a single path. *Contrast:* parallel transmission. (C) 610.7-1995

series In a propulsion system, the motor connection in which all motors are connected in a series circuit for the purpose of applying to them some fraction (usually one half) of the maximum available per-motor voltage. (VT) 1475-1999

series capacitor A device that has the primary purpose of introducing capacitive reactance in series with an electric circuit. (T&D/PE) 824-1994

series capacitor bank **(series capacitor)** An assembly of capacitors and associated auxiliaries, such as structures, support insulators, switches, and protective devices, with control equipment required for a complete operating installation. (T&D/PE) 824-1994

series circuit A circuit supplying energy to a number of devices connected in series, that is, the same current passes through each device in completing its path to the source of supply. *See also:* center of distribution. (T&D/PE) [10]

series circuit lighting transformer **(power and distribution transformers)** Dry-type individual lamp insulating transformer, autotransformer, and group series loop transformers for operation of incandescent or memory lamps on series lighting circuits such as for street and airport lighting. (PE/TR) C57.12.80-1978r

series coil sectionalizer A sectionalizer in which main circuit current impulses above a specified value, flowing through a solenoid or operating coil, provide the energy required to operate the counting mechanism. (SWG/PE) C37.100-1992

series-connected capacitor unit A capacitor unit with the elements connected in series with each other between the line terminals, with more than one such series strings within a capacitor unit (see the below figure).power systems relaying.

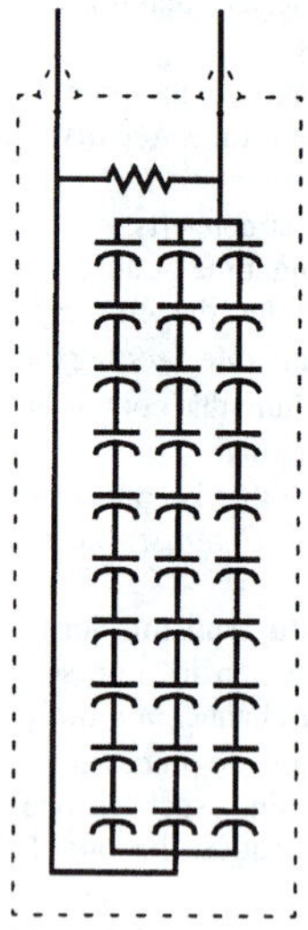

Series-connected capacitor unit with three strings of 10 elements (showing two shorted elements in one string

series-connected capacitor unit

(PE) C37.99-2000

series-connected starting-motor starting (rotating machinery) The process of starting a motor by connecting its primary winding to the supply in series with the primary windings of a starting motor, this latter being short-circuited for the running condition. (PE) [9]

series connection The arrangement of cells in a battery made by connecting the positive terminal of each successive cell to the negative terminal of the next adjacent cell so that their voltages are additive. *See also:* battery. (EEC/PE) [119]

series distribution system A distribution system for supplying energy to units of equipment connected in series. *See also:* direct-current distribution; alternating-current distribution. (T&D/PE) [10]

series elements (A) (networks) Two-terminal elements are connected in series when they form a path between two nodes of a network such that only elements of this path, and no other elements, terminate at intermediate nodes along the path. **(B) (networks)** Two-terminal elements are connected in series when any mesh including one must include the others. *See also:* network analysis. (Std100) 270-1966

series-fed vertical antenna A vertical antenna that is insulated from ground and whose feed line connects between ground and the lower end of the antenna. (AP/ANT) 145-1993

series filter A type of filter that reduces harmonics by putting a high series impedance between the harmonic source and the system to be protected. (IA/SPC) 519-1992

series gap (1) (surge arresters) An intentional gap(s) between spaced electrodes: it is in series with the valve or expulsion element of the arrester, substantially isolating the element from line or ground, or both, under normal line-voltage conditions. (PE/SPD) C62.1-1981s
(2) An intentional gap(s) between spaced electrodes in series with the valve elements across which all or part of the impressed arrester terminal voltage appears. (SPD/PE) C62.22-1997, C62.11-1999
(3) An intentional gap(s) between spaced electrodes. The gap is in series with the valve or expulsion element of the protective device, substantially isolating the element from line or ground, or both, under normal line-voltage conditions. (SPD/PE) C62.62-2000

series heater (electrical heat tracing for industrial applications) Heating elements that are designed to have a specific resistance at a given temperature for a given length. (BT/AV) 152-1953s

series heating cable Heating elements that are electrically connected in series with a single current path and have a specific resistance at a given temperature for a given length. (IA/PC) 515.1-1995, 515-1997

series loading Loading in which reactances are inserted in series with the conductors of a transmission circuit. *See also:* loading. (EEC/PE) [119]

series-mode interference *See:* differential-mode interference.

series modulation Modulation in which the plate circuits of a modulating tube and a modulated amplifier tube are in series with the same plate voltage supply. (EEC/PE) [119]

series noise (of a device) The fraction of electrical noise that can be attributed to a hypothetical white noise generator connected in series with the input of the device. (NPS) 325-1996

series operation (power supplies) The output of two or more power supplies connected together to obtain a total output voltage equal to the sum of their individual voltages. Load current is equal and common through each supply. The extent of series connection is limited by the maximum specified potential rating between any output terminal and ground. For series connection of current regulators, master/slave (compliance extension) or automatic crossover is used. *See also:* isolation voltage. (AES) [41]

series overcurrent tripping *See:* overcurrent release; direct release.

series-parallel connection The arrangement of cells in a battery made by connecting two or more series-connected groups, each having the same number of cells so that the positive terminals of each group are connected together and the negative terminals are connected together in a corresponding manner. *See also:* battery. (EEC/PE) [119]

series-parallel control A method of controlling motors wherein the motors, or groups of them, may be connected successively in series and in parallel. *See also:* multiple-unit control. (EEC/PE) [119]

series-parallel network Any network, containing only two-terminal elements, that can be constructed by successively connecting branches in series and/or in parallel. *Note:* An elementary example is the parallel combination of two branches, one containing resistance and inductance in series, the other containing capacitance. This network is sometimes called a simple parallel circuit. *See also:* network analysis. (Std100) 270-1966w

series-parallel primary current transformer One that has two insulated primary windings that are intended for connection in series or parallel to provide different rated currents. (PE/TR) C57.13-1993, [57]

series-parallel starting (rotating machinery) The process of starting a motor by connecting it to the supply with the primary winding phase circuits initially in series, and changing them over to a parallel connection for running operation. *See also:* asynchronous machine. (PE) [9]

Series Parameter A Scalar Series Parameter or a Vector Series Parameter. (IM/ST) 1451.1-1999

series rating The interrupting rating of a tested combination of a line-side (main) overcurrent protective device and a load-side (branch) circuit-breaker in which the interrupting rating of the combination is greater than the interrupting rating of the branch circuit-breaker. The interrupting rating of the series combination does not exceed the interrupting rating of the main overcurrent protective device. (IA/PSP) 1015-1997

series rectifier circuit A rectifier circuit in which two or more simple rectifier circuits are connected in such a way that their direct voltages add and their commutations coincide. *See also:* rectifier circuit element; rectification. (IA) [12]

series regulator (power supplies) A device placed in series with a source of power that is capable of controlling the voltage or current output by automatically varying its series resistance. (AES) [41]

series relay *See:* relay; current relay.

series resistor (electric instruments) A resistor that forms an essential part of the voltage circuit of an instrument and generally is used to adapt the instrument to operate on some designated voltage or voltages. The series resistor may be internal or external to the instrument. *Note:* Inductors, capacitors, or combinations thereof are also used for this purpose. *See also:* auxiliary device to an instrument. (EEC/AII) [102]

series snubber (ac adjustable-speed drives) Circuit elements, usually including an inductor, connected in series with a switching device to limit the rate of rise or fall of current through the device when switching on or off, respectively. *See also:* snubber. (IA/ID/SPC) 995-1987w, 936-1987w

series street-lighting transformer (power and distribution transformers) A series transformer that receives energy from a current-regulating series circuit and that transforms the energy to another winding at the same or different current from that in the primary. *See also:* specialty transformer. (PE/TR) C57.12.80-1978r, [57]

series system The arrangement in a multielectrode electrolytic cell whereby in each cell an anode connected to the positive bus bar is placed at one end and a cathode connected to the negative bus bar is placed at the other end, with the intervening unconnected electrodes acting as bipolar electrodes. *See also:* electrorefining. (EEC/PE) [119]

series tee junction *See:* E-plane tee junction.

series thyristor converter A thyristor converter in which two or more simple converters are connected in such a way that their direct voltages add and their commutations coincide. (IA/IPC) 444-1973w

series transformer (1) (power and distribution transformers) A transformer with a "series" winding and an "exciting" winding, in which the "series" winding is placed in a series relationship in a circuit to change voltage or phase, or both, in that circuit as a result of input received from the "exciting" winding. *Note:* Applications of series transformers include:

1) Use in a transformer such as a load-tap-changing or regulating transformer to change the voltage or current duty of the load-tap-changing mechanism.
2) Inclusion in a circuit for power factor correction to indirectly insert series capacitance in a circuit by connecting capacitors to the exciting winding.

(PE/TR) C57.12.80-1978r

(2) A transformer in which the primary winding is connected in series with a power-supply circuit, and that transfers energy to another circuit at the same or different current from that in the primary circuit. *See also:* transformer. (PE/TR) [57]

series transformer rating (power and distribution transformers) The lumen rating of the series lamp, or the wattage rating of the multiple lamps, that the transformer is designed to operate. (PE/TR) C57.12.80-1978r

series-trip recloser A recloser in which main-circuit current above a specified value, flowing through a solenoid or operating coil, provides the energy necessary to open the main contacts. (SWG/PE) C37.100-1992

series two-terminal pair networks Two-terminal pair networks are connected in series at the input or at the output terminals when their respective input or output terminals are in series. *See also:* network analysis. (BT) 153-1950w

series undercurrent tripping *See:* direct release; undercurrent release.

series unit (power and distribution transformers) The core and coil unit which has one winding connected in series in the line circuit. (PE/TR) C57.12.80-1978r

series weighting Response weighting by separating a finger into individual elements with capacitive coupling between them; the elements may be separated from the bus bar. (UFFC) 1037-1992w

series winding (1) (A) (autotransformer) (power and distribution transformers) That portion of the autotransformer winding which is not common to both the primary and the secondary circuits, but is connected in series between the input and output circuits. **(B) (power and distribution transformers)** The winding of the series unit which is connected in series in the line circuit. *Note:* If the main unit of a two-core transformer is an autotransformer, both units will have a series winding. In such cases, one is referred to as the series winding of the autotransformer and the other, the series winding of the series unit. (PE/TR) C57.12.80-1978

(2) That portion of the autotransformer winding that is not common to both the primary and secondary circuits, but is connected in series between the input and output circuits. (PE/TR) C57.15-1999

series-wound (rotating machinery) A qualifying term applied to a machine to denote that the excitation is supplied by a winding or windings connected in series with or carrying a current proportional to that in the armature winding. *See also:* asynchronous machine. (PE) [9]

series-wound motor (1) The conductors and equipment for delivering energy from the electricity supply system to the wiring system of the premises served. (NESC/NEC) [86]

(2) A dc motor in which the field circuit and armature circuit are connected in series. Speed is inversely proportional to the square root of load torque. Motor operates at a much higher speed at light load than at full load. (IA/MT) 45-1998

servant A device that is controlled by a commander. There are message-based and register-based servants. (C/MM) 1155-1992

server (1) (telecommunications switching systems) A system component that performs operations required for the processing of a call. *See also:* traffic usage count. (COM/TA) 973-1990w

(2) (MULTIBUS II) An agent that performs a service for clients. *See also:* client. (C/MM) 1296-1987s

(3) In a network, a device or computer system that is dedicated to providing specific facilities to other devices attached to the network. *Contrast:* client. *See also:* mail server; disk server; file server; terminal server; network server; database server; print server. (C) 610.7-1995

(4) The facility in the terminal or work station that provides input (keyboard, mouse) and output (screen graphics) services to the application. *Synonym:* X server. (C) 1295-1993w

(5) The software component on one device that provides services for use by clients on the same or another device. (C/MM) 1284.4-2000

(6) *See also:* batch server.

Server Object Any Object that executes one or more of its operations in response to a request from a Client object. (IM/ST) 1451.1-1999

Server Object Tag An attribute of a Client Port that identifies the Object Tag of the Server Object with which the Port communicates in client-server communications. (IM/ST) 1451.1-1999

Service An instance of a subclass of `IEEE1451_Service`. (IM/ST) 1451.1-1999

service (1) (electric systems) The conductors and equipment for delivering electric energy from the secondary distribution or street main, or other distribution feeder, or from the transformer, to the wiring system of the premises served. *Note:* For overhead circuits, it includes the conductors from the last line pole to the service switch or fuse. The portion of an

overhead service between the pole and building is designated as service drop. (NESC) [86]
(2) (controller) The specific application in which the controller is to be used, for example: general purpose; definite purpose (for example, crane and hoist, elevator, machine tool, etc). *See also:* electric controller. (IA/ICTL/IAC) [60]
(3) A distinct part of the functionality that is provided by an entity on one side of an interface to an entity on the other side of the interface. (C/PA) 14252-1996
(4) The delivery, installation, maintenance, training, and other labor-intensive activities providing life-cycle support associated with the products and processes of the system. (C/SE) 1220-1994s
(5) System behavior as perceived by the system user. (C/BA) 896.9-1994w
(6) Software that implements the interface. (C/PA) 1351-1994w, 1224-1993w, 1224.1-1993w, 1327-1993w, 1328-1993w
(7) Capabilities provided by a tool or user to get a job done. (ATLAS) 1232-1995
(8) A software interface, frequently implemented as a software function, providing a means for communicating information between two applications. (SCC20) 1232.1-1997
(9) Operation or run-time call whose behavior and interface are standardized. *See also:* method. (SCC20) 1226-1998
(10) The operation of the vehicles under normal conditions with or without revenue passengers. (VT) 1475-1999
(11) An action or response initiated by a process (i.e., a server) at the request of some other process (i.e., a client). (SCC20) 1232.2-1998
(12) (local area networks) The capabilities and action provided by one layer for another. (C) 8802-12-1998
(13) The capabilities and features provided by an N-layer to an N-user. (C/LM/CC) 8802-2-1998

service access point (SAP) (1) The point at which services are provided by one layer (or sublayer) to the layer (or sublayer) immediately above it (ISO 7498). (LM/C) 610.7-1995, 8802-6-1994
(2) An address that identifies a user of the services of a protocol entity. (C/EMB/MIB) 610.7-1995, 1073.3.2-2000

service area (1) (navigation) (navigation aids) The area within which a navigational aid provides either generally satisfactory service or a specific quality of service. (AES/GCS) 172-1983w
(2) The territory in which a utility system is required or has the right to supply or make available electric service to the ultimate customer. (PE/PSE) 858-1993w

service band A band of frequencies allocated to a given class of radio service. *See also:* radio transmission. (AP/BT/ANT) 145-1983s, 182-1961w

service bits (telecommunications) Those bits that are neither check nor information bits. *See also:* bit. (COM) [49]

service brake (maximum) A nonemergency brake application that obtains the (maximum) brake rate that is consistent with the design of the brake system, retrievable under the control of master control. (VT) 1475-1999

service braking *See:* service brake.

service cable Service conductors made up in the form of a cable. (NESC/NEC) [86]

service capacity (cell or battery) The electric output (expressed in ampere-hours, watthours, or similar units) on a service test before its working voltage falls to a specified cutoff voltage. *See also:* battery. (EEC/PE) [119]

service circuit (telephone switching systems) A circuit used for signaling purposes connected to and disconnected from a communication path during the progress of a call. (COM) 312-1977w

service circuits Common or shared equipment units or software facilities (e.g., registers) associated with lines or trunks to provide specialized functions. Examples of service circuits include the following:

1) Customer digit receivers
2) Interoffice receivers
3) Interoffice transmitters
4) Ringing circuits
5) Universal announcement circuits
6) Tone circuits
7) Conference circuits
8) Memory registers

(COM/TA) 973-1990w

service class (use in primitives) A parameter used to convey the class of service required or desired. (C/LM/CC) 8802-2-1998

service code (telephone switching systems) Any of the destination codes for use by customers to obtain directory assistance or repair service, or to reach the business office of the telecommunications company. (COM) 312-1977w

service condition (thermal classification of electric equipment and electrical insulation) A combination of factors of influence, which are to be expected in a specific application of electric equipment. (EI) 1-1986r

service conditions (nuclear power generating station) (valve actuators) (safety systems equipment in nuclear power generating stations) Environmental, loading, power and signal conditions expected as a result of normal operating requirements, expected extremes (abnormal) in operating requirements, and postulated conditions appropriate for the design-basis events of the station. (PE/NP/EDPG) 382-1985, 323-1974as, 650-1979s, 317-1983r, 649-1980s, 690-1984r, 627-1980r

service conductors The supply conductors that extend from the street main or from transformers to the service equipment of the premises supplied. (NESC/NEC) [86]

service controls Parameters conveyed as part of a directory operation that constrain various aspects of its performance. (C/PA) 1328.2-1993w, 1327.2-1993w, 1224.2-1993w, 1326.2-1993w

service corrosion (dry cell) The consumption of the negative electrode as a result of useful current delivered by the cell. *See also:* electrolytic cell. (EEC/PE) [119]

service current, continuous *See:* continuous service current.

service data unit (SDU) (1) Information that is delivered as a unit between peer service access points (SAPs). *See also:* service access point. (LM/C/EMB/MIB) 8802-6-1994, 1073.3.2-2000
(2) The 48-byte data payload of an asynchronous transfer mode (ATM) Cell. (C/BA) 1393-1999
(3) The data associated with a service primitive. (SCC32) 1455-1999
(4) Information delivered as a unit between adjacent entities that may also contain a PDU of the upper layer. (C/LM) 8802-5-1998

service date (power system measurement) The date a unit first enters the active state. On this date the reporting of performance data shall begin. *Note:* The service date is not to be confused with the installation date (the date the unit was first electrically connected to the system) or with the commercial operation date (usually related to the satisfactory completion of acceptance tests as specified in the purchase contract). (PE/PSE) 762-1980s

service delay In data communications, the time that elapses from the release of a message by an originator to its receipt by the addressee. (C) 610.7-1995

service discovery The function of providing transport clients with the ability to dynamically query service availability within a peer transport entity. (C/MM) 1284.4-2000

service drop (1) The overhead conductors between the electric supply or communication line and the building or structure being served. (NESC) C2-1997
(2) The overhead service conductors from the last pole or other aerial support to and including the splices, if any, connecting to the service-entrance conductors at the building or other structure. (NEC) [86]

service-entrance cable A single conductor or multiconductor assembly provided with or without an overall covering, primarily used for services and of the following types:

a) Type SE, having a flame-retardant, moisture-resistant covering, but not required to have inherent protection against mechanical abuse.
b) Type USE, recognized for underground use, having a moisture-resistant covering, but not required to have a flame-retardant covering or inherent protection against mechanical abuse. Single-conductor cables having an insulation specifically approved for the purpose do not require an outer covering. Cabled single-conductor Type USE constructions recognized for underground use may have a bare copper conductor cabled with the assembly. Type USE single, parallel, or cabled conductor assemblies recognized for underground use may have a bare copper concentric conductor applied. These constructions do not require an outer overall covering.
c) If Type SE or USE cable consists of two or more conductors, one shall be permitted to be uninsulated.

(NESC/NEC) [86]

service entrance conductors (1) (electric systems) (overhead system) The service conductors between the terminals of the service equipment and a point usually outside the building, clear of building walls, where joined by tap or splice to the service drop. (NESC) [86]
(2) (underground system) The service conductors between the terminals of the service equipment and the point of connection to the service lateral. *Note:* Where service equipment is located outside the building walls, there may be no service-entrance conductors, or they may be entirely outside the building. (NESC/T&D/PE) [10], [86]

service environment (diesel-generator unit) The aggregate of conditions surrounding the diesel-generator unit in its enclosure, while serving the design load during normal, accident, and post-accident operation. (PE/NP) 387-1995, 387-1984s

service equipment The necessary equipment, usually consisting of a circuit breaker or switch and fuses, and their accessories, located near the point of entrance of supply conductors to a building or other structure, or an otherwise defined area, and intended to constitute the main control and means of cutoff of the supply. (NESC/NEC) [86]

service evaluation (telephone switching systems) Determination of the quality of service received by the customer. (COM) 312-1977w

service factor (general-purpose alternating-current motor) A multiplier that, when applied to the rated power, indicates a permissible power loading that may be carried under the conditions specified for the service factor. *See also:* asynchronous machine. (PE/NP) [9], 741-1997

service ground A ground connection to a service equipment or a service conductor or both. *See also:* ground. (T&D/PE) [10]

service hours (power system measurement) (electric generating unit reliability, availability, and productivity) The number of hours a unit was in the in-service state. (PE/EDPG) [3]

service interface The interface as realized, for the benefit of the client, by the service as a whole. (C/PA) 1328-1993w, 1224.1-1993w, 1327-1993w, 1224-1993w

service laboratory A laboratory, either internal to an agency (or company) or commercially contracted, which performs radioassay measurements for the purpose of providing analytical results, exclusive of the purpose of monitoring or testing. The term "service laboratory" is synonymous with "processing laboratory." (NI) N42.23-1995

service lateral The underground service conductors between the street main, including any risers at a pole or other structure or from transformers, and the first point of connection to the service-entrance conductors in a terminal box or meter or other enclosure with adequate space, inside or outside the building wall. Where there is no terminal box, meter, or other enclosure with adequate space, the point of connection shall be considered to be the point of entrance of the service conductors into the building. (NESC/NEC) [86]

service life (1) (primary cell or battery) The period of useful service before its working voltage falls to a specified cutoff voltage. (EEC/PE) [119]
(2) (storage cell or battery) The period of useful service under specified conditions, usually expressed as the period elapsed before the ampere-hour capacity has fallen to a specified percentage of the rated capacity. *See also:* battery; charge. (EEC/PE) [119]
(3) The actual period from initial operation to retirement of a system, structure, or component. (PE/NP) 1205-1993

service life capacity Minimum battery capacity needed to meet design requirements, including temperature correction but excluding margin. (PE/EDPG) 1106-1987s

service life of cable (cable systems) The time during which satisfactory cable performance can be expected for a specific set of service conditions. (PE/SUB/EDPG) 422-1986w, 525-1992r

service order process time A command to a switching system to install, remove, or rearrange trunking, routing, or a customer's service. The time required for input of a service order and its implementation in the switch under specified traffic conditions may be defined separately for manual or computer-generated input. (COM/TA) 973-1990w

service period (illuminating engineering) The number of hours per day for which the daylighting provides a specified illuminance level. It often is stated as a monthly average. (EEC/IE) [126]

service pipe The pipe or conduit that contains underground service conductors and extends from the junction with outside supply wires into the customer's premises. *See also:* service; distributor duct. (T&D/PE) [10]

service-point The point of connection between the facilities of the serving utility and the premises' wiring. *Note:* For clearances of conductors of over 600 V, see the National Electrical Safety Code. (NESC/NEC) [86]

service pole *See:* structure.

service primitive (1) A specific service provided by a particular protocol layer entity. (C/MM) 1394-1995
(2) An abstract, implementation-independent interaction between a service user and the service provider. *Synonym:* primitive. (LM/C) 8802-6-1994
(3) A function in the external interface of a kernel element (KE) that may be invoked to access services provided by the KE. (SCC32) 1455-1999

service provider An implementation of the ACSE and Presentation Layer protocols, to which the APS API provides access. (C/PA) 1351-1994w

service raceway The raceway that encloses the service-entrance conductors. (NESC/NEC) [86]

service rating (rectifier transformer) The maximum constant load that, after a transformer has carried its continuous rated load until there is no further measurable increase in temperature rise, may be applied for a specified time without injury. *See also:* rectifier transformer. (Std100) C57.18-1964w

service recovery Actions or strategies that restore the service capabilities of a switching system from detected troubles, both internal and external to the switching system, in order to protect service with minimal impact on customers, consistent with reliability and service objectives. (COM/TA) 973-1990w

service request A solicitation of services from a client to a server. A service request may entail the exchange of any number of messages between the client and the server. In this amendment, the term *request* denotes a service request. When naming specific types of requests, the term *request* is qualified by the type of request, as in Queue Job Request and Delete Job Request. (C/PA) 1003.2d-1994

service request handler (SRH) A Master responsible for monitoring the service request line, SR, on a segment or a group of segments. When SR=1 the SRH requests bus mastership and after obtaining mastership determines which module(s) is asserting SR, either by polling or by a broadcast operation. The SRH may subsequently service the pending request(s) itself, or may issue interrupt messages to other devices on behalf of the module(s) asserting SR. SR is usually asserted only by modules which lack Mastership capability. (NID) 960-1993

service requirement (thermal classification of electric equipment and electrical insulation) The specified performance to be expected in a specific application under a specified service condition. (EI) 1-1986r

service routine (computers) A routine in general support of the operation of a computer, for example, an input-output, diagnostic, tracing, or monitoring routine. *See also:* utility routine. (MIL/C) [2], [85], [20]

services (1) (logical link control) The capabilities and features provided by an *N*-layer to an *N*-user. (PE/TR) 799-1987w
(2) A set of capabilities provided by one protocol layer entity for use by a higher layer or by management entities. (C/MM) 1394-1995

service specification The formal description of the services provided by an entity of the OSI model to the next higher layer. *See also:* physical layer; sublayer; network layer; presentation layer; data link layer; transport layer; entity layer; session layer; client layer; application layer; logical link control sublayer; medium access control sublayer. (C) 610.7-1995

service, standby *See:* standby service.

service, station *See:* station service.

service test (1) (primary battery) A test designed to measure the capacity of a cell or battery under specified conditions comparable with some particular service for which such cells are used. (EEC/PE) [119]
(2) (meter) A test made during the period that the meter is in service. *Note:* A service test may be made on the consumer's premises without removing the meter from its support, or by removing the meter for test, either on the premises or in a laboratory or meter shop. *See also:* field tests. (ELM) C12.1-1982s
(3) A test, in the "as found" condition, of the battery's capability to satisfy the battery duty cycle. (PE/EDPG) 1106-1995, 450-1995
(4) (battery) A special test of the battery's capability, as found, to satisfy the design requirements (battery duty cycle) of the dc system. (SB) 1188-1996

service time The accumulated time one or more components or units are in the in-service state during the reporting period. (PE/PSE) 859-1987w

service voltage (1) (system voltage ratings) (elecrtic power systems in commercial buildings) The voltage at the point where the electric system of the supplier and the electric system of the user are connected. *See also:* high voltage; maximum system voltage; medium voltage; low voltage; nominal system voltage; system voltage; utilization voltage; service voltage. (IA/APP/PSE) [80], 241-1990r
(2) The root-mean-square phase-to-phase or phase-to-neutral voltage at the point where the electrical system of the supplier and the user are connected. (SPD/PE) C62.62-2000

servicing Planned servicing of the circuit breaker including lubricating and replacing minor parts. (SWG/PE) C37.10-1995

servicing time (electric drive) The portion of down time that is necessary for servicing due to breakdowns or for preventive servicing measures. *See also:* electric drive. (VT/LT) 16-1955w

serving (cable) A wrapping applied over the core of a cable before the cable is leaded, or over the lead if the cable is armored. *Note:* Materials commonly used for serving are jute or cotton. The serving is for mechanical protection and not for insulating purposes. (T&D/PE) [10]

servo *See:* servomechanism.

servo amplifier In an analog computer, an amplifier used as part of a servomechanism that supplies power to the electrical input terminals of a mechanical actuator. (C) 610.10-1994w, 165-1977w

servo function generator A function generator consisting of a position servo driving a function potentiometer. *See also:* electronic analog computer. (C) 165-1977w

servomechanism (A) A feedback control system in which at least one of the systems signals represents mechanical motion. **(B)** Any feedback control system. **(C)** An automatic feedback control system in which the controlled variable is mechanical position or any of its time derivatives. *See also:* feedback control system. (C) [85]
(2) (A) An automatic device that uses feedback to govern the physical position of an element; for example, a tracking servo. *See also:* rate servomechanism; servo potentiometer; repeater servomechanism; positional servomechanism. **(B)** A feedback control system in which at least one of the system signals represents mechanical motion. (C) 610.10-1994

servomechanism, positional *See:* positional servomechanism.

servomechanism, rate *See:* rate servomechanism.

servomechanism, repeater *See:* repeater servomechanism.

servomechanism type number In control systems in which the loop transfer function is

$$\frac{K(1 + a_1 s + a_2 s^2 + \ldots + a_i s^i)}{s^n(1 + b_1 s + b_2 s^2 + \ldots + b_k s^k)}$$

where K, a_1, b_1, b_2, etc., are constant coefficients, the value of the integer n. *Note:* The value of n determines the low-frequency characteristic of the transfer function. The log-gain−log-frequency curve (Bode diagram) has a zero-frequency slope of zero for $n = 0$, slope -1 for $n = 1$, etc. *See also:* feedback control system. (NESC) [86]

servomotor An actuating device used to position turbine wicket gates, runner blades, deflectors, or other turbine control devices. (PE/EDPG) 1020-1988r

servomotor limit (hydraulic turbines) A device that acts on the governor system to prevent the turbine-control servomotor from opening beyond the position for which the device is set. (PE/EDPG) 125-1988r

servomotor position (hydraulic turbines) The instantaneous position of the turbine control servomotor expressed as a percent of the servomotor stroke. This is commonly referred to as gate position, needle position, blade position, or deflector position, although the relationship between servomotor stroke and the position of the controlled device may not always be linear. (PE/EDPG) 125-1977s

servomotor stroke (speed governing systems) Travel of the turbine control servomotor from zero to maximum without overtravel at the maximum position or "squeeze" at the minimum position. *Notes:* 1. For a gate servomotor this shall be established as the change in gate position from no discharge to maximum discharge. 2. For a blade servomotor this shall be established as the change in blade position from "flat" to "steep." 3. For a deflector servomotor this shall be established as the change in deflector position from "no deflection" position to "full flow deflected" position with maximum discharge under maximum specified head including overpressure due to water hammer. (PE/EDPG) [5], 125-1977s

servomotor time (hydraulic turbines) The equivalent elapsed time for one servomotor stroke (either opening or closing) corresponding to maximum servomotor velocity. Servomotor time can be qualified as:

a) gate;
b) blade;
c) deflector;
d) needle.

(PE/EDPG) 125-1977s

servomotor velocity limit (hydraulic turbines) A device that functions to limit the servomotor velocity in either the opening, closing, or both directions exclusive of the operation of the slow closure device. (PE/EDPG) 125-1977s

servomotor velocity limiter A device that functions to limit the servomotor velocity in either the opening, closing, or both directions exclusive of the operation of the slow closure device (above). (PE/EDPG) 125-1988r

servo multiplier An analog multiplier in which one variable is used to position one or more ganged potentiometers across which the other variable voltages are applied. (C) 610.10-1994w, 165-1977w

servo potentiometer A potentiometer driven by a positional servomechanism. *See also:* electronic analog computer. (C) 165-1977w, 166-1977w, 610.10-1994w

session (1) The period of time during which a user of a terminal can communicate with an interactive system, usually equal to elapsed time between logon and logoff. (C) 610.10-1994w **(2)** An execution of a software administration command from initiation to completion on all applicable roles. (C/PA) 1387.2-1995 **(3)** A collection of process groups established for job control purposes. Each process group is a member of a session. A process is considered to be a member of the session of which its process group is a member. A newly created process joins the session of its creator. A process can alter its session membership. Implementations that support the *setpgid*() function can have multiple process groups in the same session. (PA/C) 9945-1-1996, 9945-2-1993 **(4)** A portion of an exercise that is contiguous in wall clock (sidereal) time and is initialized per a session database. (DIS/C) 1278.3-1996 **(5)** A sequence of directory operations requested by a particular user of a particular DUA using the same session OM object. (PA/C) 1328.2-1993w, 1326.2-1993w, 1327.2-1993w, 1224.2-1993w **(6)** A printer state that allows the logical grouping of one or more jobs into a sequential, referenceable collection. (C/MM) 1284.1-1997 **(7)** A portion of an exercise that is contiguous in wall clock (sidereal) time and is initialized by a session database that includes network, entity, and environment initialization and control data. (C/DIS) 1278.4-1997 **(8)** A series of transactions exchanged between the roadside and the vehicle while the vehicle is within a beacon's communications zone. (SCC32) 1455-1999 **(9)** A collection of process groups established for job control purposes. Each process group is a member of a session. A process is considered to be a member of the session of which its process group is a member. A newly created process joins the session of its creator. A process can alter its session membership by the procedure `Create_Session` in the package `POSIX_Process_Identification`. Implementations that support `Set_Process_Group_ID` can have multiple process groups in the same session. (C) 1003.5-1999

session database A database that includes network, entity, and environment initialization and control data. It contains the data necessary to start a session. (DIS/C) 1278.3-1996

session layer (1) (Layer 5) The layer of the ISO Reference Model that provides the mechanisms for organizing and structuring the interaction between two entities. (C/DIS) 1278.2-1995 **(2)** The fifth layer of the seven-layer OSI model, responsible for coordination of the communications in an orderly manner. *See also:* entity layer; client layer; application layer; logical link control sublayer; physical layer; presentation layer; transport layer; network layer; data link layer; sublayer; medium access control sublayer. (C) 610.7-1995

session leader A process that has created a session. (C/PA) 9945-1-1996, 9945-2-1993, 1003.5-1999

session lifetime The period between when a session is created and the end of the lifetime of all the process groups that remain as members of the session. (C/PA) 9945-1-1996, 9945-2-1993, 1003.5-1999

set (1) (A) (test, measurement, and diagnostic equipment) A collection. **(B) (test, measurement, and diagnostic equipment)** To place a storage device into a specified state, usually other than that denoting zero or blank. **(C) (test, measurement, and diagnostic equipment)** To place a binary cell into the one state. *See also:* reset; preset. (MIL/C) [2], 162-1963 **(2) (electric and electronics parts and equipment)** A unit or units and necessary assemblies, subassemblies, and basic parts connected or associated together to perform an operational function. Typical examples: search radar set, radio transmitting set, sound measuring set; these include such parts, assemblies, and units as cables, microphone, and measuring instruments. (GSD) 200-1975w **(3) (A) (data management)** In a CODASYL model or network model, a named collection of records. *Synonym:* CODASYL set. **(B) (data management)** In database design, of entities, or of concepts, that have a given property or properties in common. (C) 610.5-1990 **(4)** To force the contents of one or more storage elements to a logic 1. (TT/C) 1149.5-1995 **(5)** The language-independent syntax for a family of datatypes constructed from a base datatype. A value of a set datatype contains an unordered collection of values of the base datatype, each occurring at most once. (C/PA) 1351-1994w **(6)** To place a binary cell in the true or one state. *See also:* reset. (C) 610.10-1994w **(7)** A datatype constructed from a base datatype; a value of a set datatype contains an unordered collection of values of the base datatype, each occurring only once. The is-member operation returns a Boolean value that depends on whether the specified value is a member of the set. Applying an operation to all members of a set may be supported through either of two programming paradigms. In the first, the sequencing control is provided by the application; in the second, it is provided by the implementation. (C/PA) 1224.1-1993w **(8) (A)** (of *m* phases) A group of *m* interrelated alternating currents, each in a separate phase conductor, that have the same primitive period but normally differ in phase. They may or may not differ in amplitude and waveform. The equations for a set of *m* phase currents, when each is sinusoidal, and has the primitive period, are

$$i_a = (2)^{1/2} I_a \cos(\omega t + \beta_{a1})$$

$$i_b = (2)^{1/2} I_b \cos(\omega t + \beta_{b1})$$

$$i_c = (2)^{1/2} I_c \cos(\omega t + \beta_{c1})$$

$$\cdots$$

$$i_m = (2)^{1/2} I_m \cos(\omega t + \beta_{m1})$$

where the symbols have the same meaning as for the general case given later. The general equations for a set of *m*-phase alternating currents are

$$i_a = (2)^{1/2}[I_a \cos(\omega t + \beta_{a1}) + I_{a2}\cos(2\ \omega t + \beta_{a2}) + \ldots + I_{aq}\cos(q\omega t + \beta_{aq}) + \ldots]$$

$$i_b = (2)^{1/2}[I_b \cos(\omega t + \beta_{b1}) + I_{b2}\cos(2\ \omega t + \beta_{b2}) + \ldots + I_{bq}\cos(q\omega t + \beta_{bq}) + \ldots]$$

$$i_b = (2)^{1/2}[I_b \cos(\omega t + \beta_{b1}) + I_{b2}\cos(2\ \omega t + \beta_{b2}) + \ldots + I_{bq}\cos(q\omega t + \beta_{bq}) + \ldots]$$

$$i_m = (2)^{1/2}[I_m \cos(\omega t + \beta_{m1}) + I_{m2}\cos(2\ \omega t + \beta_{m2}) + \ldots + I_{aq}\cos(q\omega t + \beta_{mq}) + \ldots]$$

here $i_a, i_b, \ldots, i_m$ are the instantaneous values of the currents, and $I_{a1}, I_{a2}, \ldots, I_{aq}$ are the root-mean-square amplitudes of the harmonic components of the individual currents. The first subscript designates the individual current and the second subscript denotes the number of the harmonic component. If there is no second subscript, the quantity is assumed to be sinusoidal. β_{a1} β_{a2}, . . ., β_q are the phase angles of the components of the same subscript determined with relation to a common reference. *Note:* If the circuit has a neutral conductor, the current in the neutral conductor is generally not considered as a separate current of the set, but as the negative of the sum of all the other currents (with respect to the same reference direction). *See also:* network analysis; voltage sets. **(B)** A group of *m* interrelated alternating voltages that have the same primitive period but normally differ in phase. They may or may not differ in amplitude and wave form. The equations for a set of *m*-phase voltages, when each is sinusoidal and has the primitive period, are

$$e_a = (2)^{1/2}E_a\cos(\omega t + \alpha_{a1})$$

$$e_b = (2)^{1/2}E_b\cos(\omega t + \alpha_{b1})$$

$$e_c = (2)^{1/2}E_c\cos(\omega t + \alpha_{c1})$$

$$\ldots$$

$$e_m = (2)^{1/2}E_m\cos(\omega t + \alpha_{m1})$$

where the symbols have the same meaning as for the general case given below. The general equations for a set of *m*-phase alternating voltages are

$$e_a = (2)^{1/2}[E_{a1}\cos(\omega t + \alpha_{a1}) + E_{a2}\cos(2\omega t + \alpha_{a2}) + \ldots + E_{ar}\cos(r\omega t + \alpha_{ar}) + \ldots]$$

$$e_b = (2)^{1/2}[E_{b1}\cos(\omega t + \alpha_{b1}) + E_{b2}\cos(2\omega t + \alpha_{b2}) + \ldots + E_{br}\cos(r\omega t + \alpha_{br}) + \ldots]$$

$$e_m = (2)^{1/2}[E_{m1}\cos(\omega t + \alpha_{m1}) + E_{m2}\cos(2\omega t + \alpha_{m2}) + \ldots + E_{mr}\cos(r\omega t + \alpha_{mr}) + \ldots]$$

where e_a, e_b, . . ., e_m are the instantaneous values of the voltages, and E_{a1}, E_{a2}, . . ., E_{ar} the root-mean-square amplitudes of the harmonic components of the individual voltages. The first subscript designates the individual voltage and the second subscript denotes the number of the harmonic component. If there is no second subscript, the quantity is assumed to be sinusoidal. a_{a1}, a_{a2}, . . ., a_{ar} are the phase angles of the components with the same subscript determined with relation to a common reference. *Note:* This definition may be applied to a two-phase four-wire or five-wire circuit if *m* is considered to be 4 instead of 2. A two-phase three-wire circuit should be treated as a special case. (Std100) 270-1966

(9) (used as a verb) To position the various adjusting devices so as to secure the desired operating characteristic. *Note:* Typical adjustment devices are taps, dials, levers, and scales suitably marked, rheostats that may be adjusted during tests, and switches with numbered positions that refer to recorded operating characteristics. (SWG/PE) C37.100-1992

(10) A kind of collection class with no duplicate members and where order is irrelevant. *Contrast:* bag; list. (C/SE) 1320.2-1998

set difference *See:* difference.

set light (illuminating engineering) The separate illumination of the background or set, other than that provided for principal subjects or areas. (EEC/IE) [126]

set normal response mode (SNRM) A high-level data link control (HDLC) message sent by a bed-side communications controller (BCC) to a device communications controller (DCC) when a successful connection to the network has occurred. (EMB/MIB) 1073.3.2-2000, 1073.3.1-1994

set of commutating groups (rectifier) Two or more commutating groups that have simultaneous commutations. *See also:* rectifier circuit element; rectification. (IA) [62], [12]

set of fours *See:* block.

set of fives *See:* block.

set of sixes *See:* block.

set point (1) (electric pipe heating systems) A fixed or constant (for relatively long time periods) command. With respect to electric pipe heating systems, set points are usually associated with temerature controllers or alarms and are the position of of the dials, taps, levels, scales, etc., so as to secure the desired operating characteristics. (PE/EDPG) 622A-1984r

(2) (electric heat tracing systems) A fixed or constant (for relatively long time periods) command. With respect to electric heat tracing systems, set points are usually associated with temperature controllers or alarms and are the position of the dials, taps, levels, scales, etc., so as to secure the desired operating characteristics. (PE/EDPG) 622B-1988r

(3) (nuclear power generating station) A predetermined point within the range of an instrument where protective or control action is initiated. 336-1980s

set pulse A drive pulse that tends to set a magnetic cell. (Std100) 163-1959w

setting (1) (of circuit breaker) The value of current and/or time at which an adjustable circuit breaker is set to trip. (NESC/NEC) [86]

(2) (used as a noun) The desired characteristic, obtained as a result of having set a device, stated in terms of calibration markings or of actual performance bench marks such as pickup current and operating time at a giving value of input. *Note:* When the setting is made by adjusting the device to operate as desired in terms of a measured input quantity, the procedure may be the same as in calibration. However, since it is for the purpose of finding one particular position of an adjusting device, which in the general case may have several marked positions that are not being calibrated, the term *setting* is to be preferred over the term *calibration.* (SWG/PE) C37.100-1992

setting error The departure of the actual performance from the desired performance resulting from errors in adjustment or from limitations in testing or measuring techniques. (SWG/PE) C37.100-1992

setting limitation The departure of the actual performance from the desired performance resulting from limitations of adjusting devices. (SWG/PE) C37.100-1992

settling time (1) (hybrid computer linkage components) The time required from the instant after the "load" has been completed until the digital-to-analog converter (KDAC) or digital-to-analog multiplier (DAM) output voltage is available within a given accuracy (under the condition of a jam transfer for a double-buffered DAC). (C) 166-1977w

(2) (automatic control) The time required, following the initiation of a specified stimulus to a linear system, for the output to enter and remain within a specified narrow band centered on its steady-state value. *Note:* The stimulus may be a step, impulse, ramp, parabola, or sinusoid. For a step or impulse, the band is often specified as ±2%. For nonlinear behavior, both magnitude and pattern of the stimulus should be specified. (PE/EDPG) [3]

(3) (STEbus) The time taken for a signal line to settle unambiguously to a logical state when making a transition from one state to another. (MM/C) 1000-1987r

(4) Measured from the mesial point (50%) of the output, the time at which the step response enters and subsequently remains within a specified error band around the final value. The final value is defined to occur 1 s after the beginning of the step. (IM/WM&A) 1057-1994w

(5) Time required by channel or terminal equipment to reach an acceptable operating condition. (SWG/PE/SUB) C37.100-1992, C37.1-1994

(6) **(A)** Following the initiation of a specified input signal to a system, the time required for the output signal to enter and remain within a specified narrow range centered on its steady-state value. *Note:* The input may be step, impulse, ramp, parabola, or sinusoidal signal. **(B)** In a hybrid computer, the time required after a load has been completed until the digital-to-analog converter or digital-to-analog multiplier output voltage is available within a given accuracy. *Synonym:* switching time. (C) 610.10-1994

(7) The duration from a step change in control signal input until the static var compensator (SVC) output settles to within ±5% of the required output. (PE/SUB) 1031-2000

setup (television) The ratio between reference black level and reference white level, both measured from blanking level. It is usually expressed in percent. *See also:* television. (BT/AV) [34]

setup/hold timing check A timing check that establishes an interval with respect to a reference signal transition during which some other signal may not change value. This timing check is frequently applied to flip-flops and latches to establish a stable interval for the data input with respect to the active edge of the clock or the active-to-inactive transition of the gate. Two limit values are necessary to define the stable interval. The setup time is the time before the reference signal transition when the stable interval begins and shall be negative if the stable interval begins after the reference signal transition. The hold time is the time after the reference signal transition when the stable interval ends, and shall be negative if the stable interval ends before the reference signal transition. If the data signal changes during the stable interval, the reliability of the resulting state of the flip-flop or latch is unknown. (C/DA) 1481-1999

setup time (1) The period of time during which a system or component is being prepared for a specific operation. *See also:* busy time; down time; idle time. (C/IM/ST) 610.12-1990, 1451.2-1997
(2) *See also:* nochange timing check; setup/hold timing check. (C/DA) 1481-1999

setup timing check A timing check that establishes only the beginning of the stable interval for a setup/hold timing check. If no hold timing check is provided for the same arc, transitions, and state, the stable interval is assumed to end at the reference signal transition and a negative value for the setup time is not meaningful. *See also:* setup/hold timing check. (C/DA) 1481-1999

seven-bit byte *See:* septet.

seven bolt *See:* conductor grip; grip, conductor.

severe lightning current Lightning currents greater than 65 kA, but not greater than 100 kA. (SPD/PE) C62.11-1999

severely errored second A second during which the error ratio is worse than a specified limit, or an OOF event occurs. (COM/TA) 1007-1991r

severity *See:* criticality.

sexadecimal (A) Pertaining to a characteristic or property involving a selection, choice, or condition in which there are sixteen possibilities. **(B)** Pertaining to the numeration system with a radix of sixteen. *Note:* More commonly called hexadecimal. *See also:* hexadecimal; positional notation. (C) [20]

sexagenary (A) Pertaining to a selection in which there are 60 possible outcomes. **(B)** Pertaining to the numeration system with a radix of 60. *Synonym:* sexagesimal. (C) 1084-1986

sexagesimal *See:* sexagenary.

sextant (navigation aids) A double-reflecting instrument for measuring angles—primarily altitudes—of the celestial bodies. (AES/GCS) 172-1983w

sextet (1) A group of six adjacent digits operated upon as a unit. *Synonym:* six-bit byte. (C) 610.5-1990w, 1084-1986w
(2) A byte composed of six bits. *Synonym:* six-bit byte. (C) 610.10-1994w
(3) (local area networks) A contiguous string of six bits. (C) 8802-12-1998

SF_6 *See:* sulfur hexafluoride.

sferics *See:* atmospherics.

SGML *See:* Standard Generalized Markup Language.

shade (illuminating engineering) A screen made of opaque or diffusing material which is designed to prevent a light source from being directly visible at normal angles of view. (EEC/IE) [126]

shaded-pole motor (rotating machinery) A single-phase induction motor with a main winding and one or more short-circuited windings (or shading coils) disposed about the air gap. The effect of the winding combination is to produce a rotating magnetic field which in turn induces the desired motor action. (PE) [9]

shading (1) (A) The rendering of surfaces in the graphical display image of a three-dimensional object by taking into account surface characteristics and the position and orientation of the surfaces with respect to light sources. **(B) (storage tubes)** The type of spurious signal, generated within a tube, that appears as a gradual variation or a small number of gradual variations in the amplitude of the output signal. These variations are spatially fixed with reference to the target area. Note the distinction between this and disturbance. *See also:* storage tube; television. (C/ED) 610.6-1991, 158-1962
(2) (audio and electroacoustics) A method of controlling the directional response pattern of a transducer through control of the distribution of phase and amplitude of the transducer action over the active face. *See also:* television. (SP) [32]
(3) (camera tubes) A brightness gradient in the reproduced picture, not present in the original scene, but caused by the tube. (ED) 161-1971w

shading coil (1) (rotating machinery) The short-circuited winding used in a shaded-pole motor, for the purpose of producing a rotating component of magnetic flux. (PE) [9]
(2) (direct-current motors and generators) A short-circuited winding used on a main (excitation) pole to delay the shift in flux caused by transient armature current. Transient commutation is aided by the use of this coil. *See also:* rotor; stator. (PE) [9]

shading wedge (rotating machinery) A strip of magnetic material placed between adjacent pole tips of a shaded-pole motor to reduce the effective separation between the pole tips. The shading wedge usually has a slot running most of its length to provide some separation effect. *See also:* stator; rotor. (PE) [9]

shadow class A class presented in a view that is specified in some other view. (C/SE) 1320.2-1998

shadow factor (radio-wave propagation) The ratio of the electric field strength which would result from propagation over a convex curved surface to that which would result from propagation over a plane, other factors being the same. (AP) 211-1977s

shadowing (shielding) The interference of any part of an anode, cathode, rack, or tank with uniform current distribution upon a cathode. (EEC/PE) [119]

shadow loss (mobile communication) The attenuation to a signal caused by obstructions in the radio propagation path. *See also:* mobile communication system. (VT) [37]

shadow mask (1) (color picture tubes) A color-selecting-electrode system in the form of an electrically conductive sheet containing a plurality of holes that uses masking to effect color selection. *See also:* television. (ED) 161-1971w
(2) (computer graphics) A metal plate with small holes, positioned behind the display surface of a display device, such that when the electron guns for red, green, and blue colors are focused through each hole, the beam strikes only the phosphor of its associated color. (See the corresponding figure.)

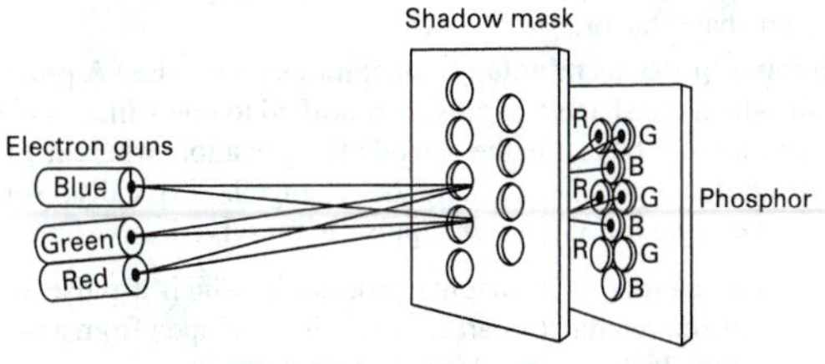

shadow mask

(C) 610.6-1991w

shadow region The region in space that, because of an intervening obstacle, cannot be reached by an incident geometric-optic ray. (AP/PROP) 211-1997

shaft (rotating machinery) That part of a rotor that carries other rotating members and that is supported by bearings in which it can rotate. *See also:* rotor. (PE) [9]

shaft current (rotating machinery) Electric current that flows from one end of the shaft of a machine through bearings, bearing supports, and machine framework to the other end of the shaft, driven by a voltage between the shaft ends that results from flux linking the shaft caused by irregularities in the magnetic circuit. *See also:* rotor. (PE) [9]

shaft extension (rotating machinery) The portion of a shaft that projects beyond the bearing housing and away from the core. *See also:* armature. (PE) [9]

shaft recorder A sensor that is attached to the wheels of an input device such as a mouse; used for delivering electrical pulses as the wheel rotates. (C) 610.10-1994w

shaft revolution indicator A system consisting of a transmitter driven by a propeller shaft and one or more remote indicators to show the speed of the shaft in revolutions per minute, the direction of rotation and (usually) the total number of revolutions made by the shaft. *See also:* electric propulsion system. (EEC/PE) [119]

shaft voltage test (rotating machinery) A test taken on an energized machine to detect the induced voltage that is capable of producing shaft currents. *See also:* rotor. (PE) [9]

shaker-type conveyor A conveyor designed to transport material along a line of troughs by means of a reciprocating or shaking motion. *See also:* conveyor. (EEC/PE) [119]

shallow and deep dose equivalent The dose equivalents (H_s and H_d) at the depths in tissue of 0.007 cm and 1.0 cm, respectively. (NI) N42.20-1995

sham control In an experiment, a group of organisms that is not exposed to the treatment, but is maintained, handled, observed, etc., in an identical manner as the treatment group, and whose overall characteristics are as similar as possible to the treatment organisms. (T&D/PE) 539-1990

shank (cable plowing) A portion of the plow blade to which a removable wear point is fastened. *See also:* wear point. (T&D/PE) 590-1977w

shaped-beam antenna An antenna that is designed to have a prescribed pattern shape differing significantly from that obtained from a uniform-phase aperture of the same size. (AP/ANT) 145-1993

shaped pulse The pulse shape produced by passing the output signal from the preamplifier (approximately a step function) through the pulse-shaping network in the main amplifier (shaping amplifier). (NPS) 325-1996

shaped wire compact conductor (TW) ACSR or AAC that is designed to increase the aluminum area for a given diameter of conductor by the use of trapezoidal shaped aluminum wires. (T&D/PE) 524-1992r

shape factor (1) (spectrum analyzer) A measure of the asymptotic shape of the resolution bandwidth response curve of a spectrum analyzer. Shape factor is defined as the ratio between bandwidths at two widely spaced points on the response curve, such as the 3 dB and 60 dB down points. (IM) 748-1979w
(2) (induction and dielectric heating equipment) *See also:* coil shape factor.

shaping (pulse terminology) (operations on a pulse) A process in which the shape of a pulse is modified to one which is ideal or more suitable for the intended application wherein time magnitude parameters may be changed. Typically, some property(ies) of the original pulse is preserved.

a) (regeneration) A shaping process in which a pulse with desired reference characteristics is developed from a pulse which lacks certain desired characteristics.
b) (stretching) A shaping process in which pulse duration is increased.
c) (clipping) A shaping process in which the magnitude of a pulse is constrained at one or more predetermined magnitudes.
d) (limiting) A clipping process in which the pulse shape is preserved for all magnitudes between predetermined clipping magnitudes.
e) (slicing) A clipping process in which the pulse shape is preserved for all magnitudes less (greater) than a predetermined clipping magnitude.
f) (differentiation) A shaping process in which a pulse is converted to a wave whose shape is or approximates the time derivative of the pulse.
g) (integration) A shaping process in which a pulse is converted to a wave whose shape is or approximates the time integral of the pulse.

(IM) 194-1977w

shaping amplifier *See:* main amplifier.

shaping constant An arbitrary indicator of shaped pulse width. The use of this designation is discouraged. *See also:* shaping index. (NPS) 300-1988r

shaping index (1) The width of a unipolar pulse at 50% of its peak height. (NPS) 300-1988r
(2) In a main amplifier, the width of the shaped unipolar pulse measured at 50% of its peak height, designated as t0.5. (NPS) 325-1996

shaping pulse The intentional processing of a pulse waveform to cause deviation from a reference waveform. *See also:* pulse. (IM/HFIM) [40]

shaping time *See:* amplifier shaping time.

shaping time constant (semiconductor radiation detectors) The time constants of the bandwidth defining CR (capacitance-resistance) differentiators and RC (resistance-capacitance) integrators used in pulse amplifiers. (NID) 301-1976s

Shareable List A DMA queue composed of a linked list of items. Each item contains a pointer to the next item and the message being passed to the consumer. Swap transactions are used to support shared access by multiple producers. (C/MM) 1212.1-1993

shared lock A lock that allows several processes concurrent access to data. *Note:* at most, only one of the processes is allowed to modify the data and the other processes may only read the data. *Contrast:* exclusive lock. (C) 610.5-1990w

shared-logic word processing Word processing performed on a system composed of multiple work stations that share the logic and storage sections of a single central processor. *Contrast:* shared-resource word processing; dedicated word processing; clustered word processing; stand-alone word processing. (C) 610.2-1987

shared memory (1) The address space in the system accessible to all modules. (C/BA) 1014.1-1994w, 896.3-1993w
(2) The address space in the system accessible to all caching modules. (C/BA) 10857-1994, 896.4-1993w

shared memory object (1) An object that represents memory that can be mapped concurrently into the address space of more than one process. (PA/C) 9945-1-1996
(2) An object that represents memory that can be mapped concurrently into the address space of more than one process. These named regions of storage may be independent of the file system and can be mapped into the address space of one or more processes to allow them to share the associated memory. (C) 1003.5-1999

shared port An output or bidirectional port where some other output port of the cell derives its logic function. The output load at a shared port affects not only the delay to that port itself, but also the delay to any ports sharing it. (C/DA) 1481-1999

shared-resource word processing Word processing performed on a system composed on multiple work stations, each with its own processor but sharing certain resources such as printers and disk drives. *Contrast:* shared-logic word processing; dedicated word processing; clustered word processing; stand-alone word processing. (C) 610.2-1987

shared service A CSMA/CD network in which the collision domain consists of more than two DTEs so that the total network bandwidth is shared among them. (C/LM) 802.3-1998

shared systems Structures, systems, and components that can perform functions for more than one unit in multiunit stations. *Note:* This definition includes the following:

1) systems that are simultaneously shared by both units;
2) time sequential sharing or systems that would be shared by two units at different times, according to the sequence of events; and
3) systems that would only be used by one unit, at any given time, but that could be disconnected from that unit and placed in the other unit on demand.

(PE/NP) 379-1994

shared unmodified (SU) An attribute assigned to a cache line if there is an up-to-date copy of the line in the module's cache and the module is to assume that a copy also exists in another module's cache. (C/BA) 896.4-1993w

Shared Virtual Local Area Network (VLAN) Learning (SVL) Configuration and operation of the Learning Process and the Filtering Database such that, for a given set of VLANs, if an individual MAC Address is learned in one VLAN, that learned information is used in forwarding decisions taken for that address relative to all other VLANs in the given set. *Note:* In a Bridge that supports only SVL operation, the "given set of VLANs" is the set of all VLANs. (C/LM) 802.1Q-1998

Shared Virtual Local Area Network (VLAN) Learning (SVL) Bridge A type of Bridge that supports only Shared VLAN Learning. (C/LM) 802.1Q-1998

Shared Virtual Local Area Network (VLAN) Learning (SVL)/Independent Virtual Local Area Network (VLAN) Learning (IVL) Bridge An SVL/IVL Bridge is a type of Bridge that simultaneously supports both Shared VLAN Learning and Independent VLAN Learning. (C/LM) 802.1Q-1998

sharing *See:* time sharing.

sharing transformer and current balancing transformer (current balancing reactor) (electrical heating applications to melting furnaces and forehearths in the glass industry) Two-winding, iron core devices used in paralleled current paths, connected so that any difference in current between the paths causes an induced voltage that opposes the current difference. (IA) 668-1987w

sharp Pertaining to elements in an image that are well defined and readily discernable. *Contrast:* blurred. (C) 610.4-1990w

sharpening Any image enhancement technique in which the effect of blurring in the original image is reduced. *Synonym:* deblurring. *See also:* unsharp masking. (C) 610.4-1990w

shearing machine An electrically driven machine for making vertical cuts in coal. (EEC/PE) [119]

shear pin (1) (rotating machinery) A dowel designed to shear at a predetermined load and thereby prevent damage to other parts. *See also:* rotor; shear section shaft. (PE) [9]
(2) (small hydraulic power plants) Replaceable protective device that fails by shearing when an obstruction prevents a wicket gate from closing. (PE/EDPG) 1020-1988r

shear section shaft (rotating machinery) A section of shaft machined to a controlled diameter, or area, designed to shear at a predetermined load and thereby prevent damage to connected machinery. *See also:* shear pin; rotor. (PE) [9]

shear wave (rotational wave) A wave in an elastic medium that causes an element of the medium to change its shape without a change of volume. *Notes:* 1. Mathematically, a shear wave is one whose velocity field has zero divergence. 2. A shear plane wave in an isotropic medium is a transverse wave. 3. When shear waves combine to produce standing waves, linear displacements may result. (SP/ACO) [32]

sheath (1) (cable systems in power generating stations) (jacket) The overall protective covering for the insulated cable. (PE/EDPG) 422-1977
(2) A continuous covering for a cable. (BT/IA/AV/PC) 152-1953s, 515.1-1995
(3) A uniform and continuous covering, metallic or nonmetallic, enclosing the insulated conductor(s), used to protect the cable against influences from the surroundings (corrosion, moisture, etc.). (IA) 515-1997

sheath/shield sectionalizers A sectionalizer that is used to minimize induced current in the cable metallic shield/shield by electrically interrupting the semiconducting shield and conducting metallic sheath or shield of two cable lengths that are joined together. These sectionalizers are primarily used on cable systems operating at 60 kV and above. (PE/IC) 404-1993

sheath temperature (1) The temperature of the outermost sheath that may be exposed to the surrounding atmosphere. (BT/IA/AV/PC) 152-1953s, 515.1-1995
(2) The temperature of the outermost continuous covering of a heating cable or surface heating device that may be exposed to the surrounding atmosphere. (IA) 515-1997

sheave (A) (grounding of power lines) The grooved wheel of a traveler or rigging block. Travelers are frequently referred to as sheaves. **(B) (grounding of power lines)** A shaft-mounted wheel used to transmit power by means of a belt, chain, band, etc. *See also:* pulley. (T&D/PE) 1048-1990
(2) (A) (conductor stringing equipment) The grooved wheel of a traveler or rigging block. Travelers are frequently referred to as sheaves. *Synonym:* pulley; wheel; roller. **(B)** A shaft-mounted wheel used to transmit power by means of a belt, chain, band, etc. (PE/T&D) 524a-1993, 524-1992

sheet A cut piece of print media, such as a sheet of paper. (C/MM) 1284.1-1997

Sheffer stroke *See:* NAND.

shelf corrosion (dry cell) The consumption of the negative electrode as a result of local action. *See also:* electrolytic cell. (EEC/PE) [119]

shelf depreciation The depreciation in service capacity of a primary cell as measured by a shelf test. *See also:* battery. (EEC/PE) [119]

shelf test A storage test designed to measure retention of service ability under specified conditions of temperature and cutoff voltage. *See also:* battery. (EEC/PE) [119]

Shell, The The Shell Command Language Interpreter, a specific instance of a shell. (C/PA) 9945-2-1993

shell (1) (insulators) A single insulating member, having a skirt or skirts without cement or other connecting devices intended to form a part of an insulator or an insulator assembly. *See also:* insulator. (EEC/IEPL) [89]
(2) (electrolysis) The external container in which the electrolysis of fused electrolyte is conducted. *See also:* fused electrolyte. (EEC/PE) [119]
(3) (electrotyping) A layer of metal (usually copper or nickel) deposited upon, and separated from, a mold. (PE/EEC) [119]
(4) (software) A computer program or routine that provides an interface between the user and a computer system or program. (C) 610.12-1990
(5) (A) A software interface between the user and the operating system in which the shell interprets commands and communicates them to the operating system of the computer. **(B)** Software that allows a kernel program to run under different computing environments. (C) 610.13-1993
(6) A program that interprets sequences of text input as commands. It may operate on an input stream or it may interactively prompt and read commands from a terminal. (C/PA) 9945-2-1993

shell-form transformer (power and distribution transformers) A transformer in which the laminations constituting the iron core surround the windings and usually enclose the greater part of them. (PE/TR) C57.12.80-1978r

shell script A file containing shell commands. If the file is made executable, it can be executed by specifying its name as a simple command. Execution of a shell script causes a shell to execute the commands within the script. Alternatively, a shell can be requested to execute the commands in a shell script by specifying the name of the shell script as the operand to the `sh` utility. (C/PA) 9945-2-1993

Shell's method *See:* diminishing increment sort.

Shell sort *See:* diminishing increment sort.

shell, stator *See:* stator shell.

shell token string A sequence of shell tokens. A shell token string shall be a portable character string. (C/PA) 1387.2-1995

shell-type motor A stator and rotor without shaft, end shields, bearings or conventional frame. *Note:* A shell-type motor is normally supplied by a motor manufacturer to an equipment manufacturer for incorporation as a built-in part of the end product. Separate fans or fans larger than the rotor are not included. *See also:* asynchronous machine. (PE) [9]

sheltered equipment (test, measurement, and diagnostic equipment) Equipment so housed or otherwise protected that the extreme of natural and induced environments are partially or completely excluded or controlled. Examples are laboratory and shop equipment, equipment shielded from sun by a canopy or roof, and so forth. (MIL) [2]

SHF *See:* super high frequency.

shield (1) (nuclear power generating station) (instrumentation cables) Braid copper, metallic sheath, or metallic coated polyester tape (usually copper or aluminum), applied over the insulation of a conductor or conductors for the purpose of reducing elecrostatic coupling between the shielded conductors and others that may be either susceptible to, or generators of, electrostatic fields (noise). When electromagnetic shielding is intended, the term electromagnetic is usually included to indicate the difference in shielding requirement and material. (PE/IA/EDPG/PSE) 690-1984r, 1100-1999
(2) (cable systems in power generating stations) As normally applied to instrumentation cables, refers to metallic sheath (usually copper or aluminum), applied over the insulation of a conductor or conductors for the purpose of providing means for reducing electrostatic coupling between the conductors so shielded and others which may be susceptible to or which may be generating unwanted (noise) electrostatic fields. When electromagnetic shielding is intended, the term "electromagnetic" is usually included to indicate the difference in shielding requirements as well as material. To be effective at power system frequencies, electromagnetic shields would have to be made of high-permeability steel. Such shielding material is expensive and is not normally applied. Other less expensive means for reducing low-frequency electromagnetic coupling, as described herein, are preferred. (PE/EDPG) 422-1977
(3) (power and distribution transformers) A conductive protective member placed in relationship to apparatus or test components to control the shape of magnitude, or both, of electric or magnetic fields, thereby improving performance of apparatus or test equipment by reducing losses, voltage gradients, or interface. (PE/TR) C57.12.80-1978r
(4) (electromagnetic) A housing, screen, or other object, usually conducting, that substantially reduces the effect of electric or magnetic fields on one side thereof, upon devices or circuits on the other side. *See also:* signal; induction heating. (PE/EM) [4], 43-1974s
(5) (rotating machinery) (mechanical protection) An internal part used to protect rotating parts or parts of the electric circuit. In general, the word shield will be preceded by the name of the part that is being protected. (PE) [9]
(6) (induction heating) A material used to suppress the effect of an electric or magnetic field within or beyond definite regions. (IA) 54-1955w
(7) (instrumentation cables) (cable systems) A metallic sheath (usually copper or aluminum), applied over the insulation of a conductor or conductors for the purpose of providing means for reducing electrostatic coupling between the conductors so shielded and others which may be susceptible to or which may be generating unwanted (noise) electrostatic fields. (SUB/PE) 525-1992r
(8) A barrier, usually metallic, within a coaxial cable that is designed to contain the high-powered broadcast signal within the coaxial cable to reduce electromagnetic interference and signal loss. (C) 610.7-1995
(9) (magnetrons) *See also:* end shield. (IA) 54-1955w

shielded conductor cable A cable in which the insulated conductor or conductors is/are enclosed in a conducting envelope or envelopes. 30-1937w

shielded ignition harness A metallic covering for the ignition system of an aircraft engine, that acts as a shield to eliminate radio interference with aircraft electronic equipment. The term includes such items as ignition wiring and distributors when they are manufactured integral with an ignition shielding assembly. (EEC/PE) [119]

shielded insulated splice (power cable joints) An insulated splice in which a conducting material is employed over the full length of the insulation for electric stress control. (PE/IC) 404-1986s

shielded joint A cable joint having its insulation so enveloped by a conducting shield that substantially every point on the surface of the insulation is at ground potential or at some predetermined potential with respect to ground. (T&D/PE) [10]

shielded line A planar transmission line whose cross section is completely enclosed within conducting boundaries. (MTT) 1004-1987w

shielded-loop antenna (probe) An electrically small antenna consisting of a tubular electrostatic shield formed into a loop with a small gap, and containing one or more wire turns for external coupling. (AP/ANT) 145-1993

shielded-loop probe *See:* shielded-loop antenna.

shielded nonmetallic-sheathed cable A factory assembly of two or more insulated conductors in an extruded core of a moisture-resistant, flame-resistant nonmetallic material, covered with an overlapping spiral metal tape and wire shield and jacketed with an extruded moisture-, flame-, oil-, corrosion-, fungus-, and sunlight-resistant nonmetallic material. *Synonym:* SNM cable. (NESC/NEC) [86]

shielded pair (signal-transmission system) A two-wire transmission line surrounded by a sheath of conducting material to protect it from the effects of external fields, or to confine fields produced by the transmission line. *See also:* signal; waveguide. (MTT) 146-1980w

shielded strip transmission line A strip conductor between two ground planes. Some common designations are: Stripline (trade mark); Tri-plate (trade mark); slab line (round conductor); balanced strip line. *See also:* unshielded strip transmission line; strip-type transmission line. (AP/ANT) [35]

shielded transmission line (1) (signal-transmission system) A transmission line surrounded by a sheath of conducting material to protect it from the effects of external fields, or to confine fields produced by the transmission line. *See also:* signal; waveguide. (IE) [43]
(2) (waveguide) A transmission line whose elements essentially confine propagated electrical energy to a finite space inside a conducting sheath. (MTT) 146-1980w

shielded twisted pair (STP) (1) A twisted pair medium surrounded by a metallic shield to minimize electrical interference and noise. *Note:* Specifications are provided in IEEE Std 802.5. *Contrast:* unshielded twisted pair. (C) 610.7-1995
(2) Normally refers to those shielded cables with individual pairs of conductors twisted, or with a group of four conductors in a quad configuration, with any characteristic impedance. Specifically refers to those shielded cables whose pairs have a high-frequency characteristic impedance of 150 Ω and with two pair of conductors shielded from any other individual pairs. (C/LM) 8802-5-1998

shielded twisted-pair (STP) cable An electrically conducting cable, comprising one or more elements, each of which is individually shielded. There may be an overall shield, in which case the cable is referred to as shielded twisted pair cable with an overall shield. Specifically for IEEE 802.3

100BASE-TX, 150 Ω balanced inside cable with performance characteristics specified to 100 MHz (i.e., performance to Class D link standards as per ISO/IEC11801:1995). In addition to the requirements specified in ISO/IEC11801:1995, IEEE 802.3 clauses 23 and 25 provide additional performance requirements for 100BASE-T operation over STP. (LM/C) 802.3u-1995s

shielded-type cable A cable in which each insulated conductor is enclosed in a conducting envelope so constructed that substantially every point on the surface of the insulation is at ground potential or at some predetermined potential with respect to ground under normal operating conditions. (T&D/PE) [10]

shield factor (telephone circuit) The ratio of noise, induced current, or voltage when a source of shielding is present, to the corresponding quantity when the shielding is absent. (PE/EEC) [119]

shield grid (gas tube) A grid that shields the control electrode in a gas tube from the anode or the cathode, or both, with respect to the radiation of heat and the deposition of thermionic activating material and also reduces the electrostatic influence of the anode. It may be used as a control electrode. *See also:* grid; electrode. (ED) 161-1971w

shielding (1) (power cable joints) (screening) A conducting layer, applied to control the dielectric stresses within tolerable limits and minimize voids. It may be applied over the entire joint insulation, on the tapered insulation ends only, or over irregular conductor or connector surfaces. (PE/IC) 404-1986s

(2) (x-radiation limits for ac high-voltage power vacuum interrupters used in power switchgear) Barrier of attenuating material used to reduce radiation hazards. 553-1986

(3) The process of applying a conducting barrier between a potentially disturbing noise source and electronic circuitry. Shielding is used to protect cables (data and power) and electronic circuits. Shielding may be accomplished by the use of metal barriers, enclosures, or wrappings around source circuits and receiving circuits. (IA/PSE) 1100-1999

shielding angle (1) (illuminating engineering) (of a luminaire) The angle between a horizontal line through the light center and the line of sight at which the bare source first becomes visible. (IE/EEC) [126]

(2) The angle between a vertical line through the overhead ground wire and a line connecting the overhead ground wire to the shielded conductor. *See also:* direct-stroke protection (T&D/PE/SPD) [10], C62.23-1995, 1410-1997

(3) (A) (of shield wires with respect to conductors). The angle formed by the intersection of a vertical line drawn through a shield wire and a line drawn from the shield wire to a protected conductor. The angle is chosen to provide a zone of protection for the conductor so that most lightning strokes will terminate on the shield wire rather than on the conductor. **(B)** (of a lightning mast). The angle formed by the intersection of a vertical line drawn through the tip of the mast and another line drawn through the tip to earth at some selected angle with the vertical. Rotation of this angle around the structure forms a cone-shaped zone of protection for objects located within the cone. The angle is chosen so that lightning strokes will terminate on the mast rather than on an object contained within the protective zone so formed. *See also:* negative shielding angle; positive shielding angle. (SUB/PE) 998-1996

shielding effectiveness (SE) (1) For a given external source, the ratio of electric or magnetic field strength at a point before and after the placement of the shield in question. (EMC) [53]

(2) The ratio of the signal received (from a transmitter) without the shield, to the signal received inside the shield; the insertion loss when the shield is placed between the transmitting antenna and the receiving antenna. (EMC/STCOORD) 299-1997

shielding electrode An electrode intended for the reduction of electromagnetic interference signals and that is usually located between input and output transducers. (UFFC) 1037-1992w

shielding enclosure A structure that protects its interior from the effect of an exterior electric or magnetic field, or conversely, protects the surrounding environment from the effect of an interior electric or magnetic field. A high-performance shielding enclosure is generally capable of reducing the effects of both electric and magnetic field strengths by one to seven orders of magnitude depending upon frequency. An enclosure is normally constructed of metal with provisions for continuous electrical contact between adjoining panels, including doors. (EMC/STCOORD) 299-1997

shielding failure (lightning protection) The occurrence of a lightning stroke that bypasses the overhead ground wire and terminates on the phase conductor. *See also:* direct-stroke protection. (T&D/PE) [10]

shielding failure flash-over rate (SFFOR) The annual number of flashovers on a circuit or tower-line length basis caused by shielding failures. (PE/T&D) 1243-1997

shielding failure rate (SFR) The annual number of lightning events on a circuit or tower-line length basis, which bypass the overhead ground/shield wire and terminate directly on the phase conductor. This event may or may not cause flashover. (PE/T&D) 1243-1997

shield wire (1) (electromagnetic fields) A wire employed to reduce the effects on electric supply or communication circuits from extraneous sources. *See also:* inductive coordination. (SPD/PE/EEC) C62.23-1995, [119]

(2) (overheard power line or substation) A wire suspended above the phase conductors positioned with the intention of having lightning strike it instead of the phase conductor(s). *Synonyms:* overhead ground wire; static wire. (SUB/PE) 998-1996

(3) Grounded wire(s) placed near the phase conductors for the purposes of

a) Protecting phase conductors from direct lightning strokes,
b) Reducing induced voltages from external electromagnetic fields,
c) Lowering the self-surge impedance of an OHGW system, or
d) Raising the mutual surge impedance of an OHGW system to the protected phase conductors.

They may be electrically bonded directly to the structure or indirectly through short gaps. (PE/T&D/PE/T&D) 1410-1997, 1243-1997

shift (mathematics of computing) A displacement of an ordered set of characters one or more places to the left or right. If the characters are the digits of a numeral, a shift may be equivalent to multiplying by a power of the base. *See also:* arithmetic shift; logical shift. (C) 1084-1986w

shift character A control character that determines the alphabetic or numeric shift of character codes in a message. (C) 610.5-1990w

shift clock (semiconductor memory) The inputs that when operated in a prescribed manner shift internal data in a serial memory. (TT/C) 662-1980s

shift, direct-current *See:* direct-current shift.

shift-in character (SI) A code extension character, used to terminate a sequence that has been introduced by the shift-out character, that makes effective the graphic characters of the original character set. *Contrast:* shift-out character. (C) 610.5-1990w

shift key A control key that controls the interpretation of other keys. That is, when used in conjunction with another key, the representation of that other key is different from that of the key alone. *Note:* Often used to form uppercase characters. *See also:* alternate key. (C) 610.10-1994w

shift operation An operation for which the VHDL operator is **sll**, **srl**, **sla**, **sra**, **rol**, or **ror**. (C/DA) 1076.3-1997

shift-out character (SO) A code extension character that substitutes, for the graphic characters of the original character set, an alternative set of graphic characters upon which agreement has been reached or that has been designated using code extension procedures. *Contrast:* shift-in character. (C) 610.5-1990w

shift pulse A drive pulse that initiates shifting of characters in a register. (Std100) 163-1959w

shift register **(1)** A register in which the stored data can be moved to the right or left. (C) [20], [85]
(2) A register in which the data bits can be shifted in one direction or both; for example, if the contents are 11010010 and the register is shifted to the right, the result is x1101001; where x is a zero, one, or the bit shifted off the right end, depending on the type of shift register. *See also:* circulating register. (C) 610.10-1994w

shim (A) (rotating machinery) (mechanical) A lamination usually machined to close-tolerance thickness, for assembly between two parts to control spacing. **(B) (rotating machinery)** (magnetic) A lamination added to adjust or change the effective air gap in a magnetic circuit. *See also:* rotor; stator. (PE) [9]

shingle (photoelectric converter) Combination of photoelectric converters in series in a shingle-type structure. *See also:* semiconductor. (AES) [41]

ship control telephone system A system of sound-powered telephones (requiring no external power supply for talking) with call bells, exclusively for communication among officers responsible for control and operation of a ship. *Note:* Call bells are usually energized by hand-cranked magneto generators. (EEC/PE) [119]

shipping brace (rotating machinery) Any structure provided to reduce motion or stress during shipment, that must be removed before operation. (PE) [9]

shipping seal *See:* sealed end.

ship's service electric system On any vessel, all electric apparatus and circuits for power and lighting, except apparatus provided primarily either for ship propulsion or for the emergency system. *Note:* Emergency and interior communication circuits are normally supplied with power from the ship's service system, upon failure of which they are switched to an independent emergency generator or other sources of supply. *See also:* marine electric apparatus. (EEC/PE) [119]

shock excitation (oscillatory systems) The excitation of natural oscillations in an oscillatory system due to a sudden acquisition of energy from an external source or a sudden release of energy stored with the oscillatory system. *See also:* oscillatory circuit. (PE/EEC) [119]

shock motion (mechanical system) Transient motion that is characterized by suddenness, by significant relative displacements, and by the development of substantial internal forces in the system. *See also:* mechanical shock. (SP) [32]

shock, primary *See:* primary shock.

shockproof electric apparatus Electric apparatus designed to withstand, to a specified degree, shock of specified severity. *Note:* The severity is stated in footpounds impact on a special test stand equivalent to shock of gunfire, explosion of mine or torpedo, etc. *See also:* marine electric apparatus. (EEC/PE) [119]

shock, secondary *See:* secondary shock.

shoe (ramp shoe) Part of a vehicle-carried apparatus that makes contact with a ramp. (EEC/PE) [119]

shop instruments Instruments and meters that are used in regular routine shop or field operations. (ELM) C12.1-1982s

shop—meter A place where meters are inspected, repaired, tested, and adjusted. (ELM) C12.1-1988

shop test (laboratory test) A test made upon the receipt of a meter from a manufacturer, or prior to reinstallation. Such tests are made in a shop or a laboratory of a meter department. *See also:* service test. (EEC/PE) [119]

shoran (navigation aid terms) A radio navigation system that provides circular lines of position. The term is derived from the words short-range navigation. (AES/GCS) 172-1983w

shore feeder Permanently installed conductors from a distribution switchboard to a connection box (or boxes) conveniently located for the attachment of portable leads for supply of power to a ship from a source on shore. *See also:* marine electric apparatus. (EEC/PE) [119]

short answer interaction An instruction method employed by some computer-assisted instruction systems, in which the student is asked to provide a word or phrase in response to a question. (C) 610.2-1987

short card A special-purpose punch card that is shorter in length than a standard 80-column punch card; For example, a 51 column card. (C) 610.10-1994w

short circuit (1) (gas-tube surge protective devices) An abnormal connection of relatively low impedance, whether made accidentally or intentionally, between two points of different potential in a circuit. (SPD/PE) C62.31-1987r, C62.32-1981s
(2) An abnormal connection (including an arc) of relatively low impedance, whether made accidentally or intentionally, between two points of different potential. *Note:* The term *fault* or *short-circuit fault* is used to describe a short circuit. (SWG/IA/PE/PSP) 1015-1997, C37.100-1992
(3) The condition in which the output terminals of the power supply are directly connected together, resulting in near-zero output voltage. (PEL) 1515-2000

short circuit, adjustable, waveguide (waveguide components) A longitudinally movable obstacle which reflects essentially all the incident energy. (MTT) 147-1979w

short-circuit current (1) (electric power systems in commercial buildings) An overcurrent resulting from a fault of negligible impedance between live conductors having a difference in potential under normal operating conditions. The fault path may include the path from active conductors via earth to neutral. (IA/PSE) 241-1990r
(2) (protection and coordination of industrial and commercial power systems) An overcurrent usually defined as being in excess of ten times normal continuous rating. *See also:* overload. (IA/PSP) 242-1986r
(3) (overhead power lines) The current between a conductive object and ground through a zero impedance connection or in a closed circuit, as a result of induction or deposition of charge. (T&D/PE) 539-1990
(4) (of a battery charger) The current magnitude at the output terminals, when the terminals are short circuited and with nominal input voltage supplied to the charger. (IA/PSE) 602-1996
(5) (transformer-rectifier system) The steady-state value of the input alternating current that flows when the output direct current terminals are short-circuited and rated line alternating voltage is applied to the line terminals. This current is normally of interest when using current limiting transformers or checking current limiting devices. (PEL/ET) 295-1969r

short-circuit dc current The dc current between a conductive object and ground through a zero impedance, as a result of deposition of charge. (T&D/PE) 539-1990

short-circuit driving-point admittance *See:* admittance, short-circuit driving-point.

short-circuit duration rating (magnetic amplifier) The length of time that a short circuit may be applied to the load terminals nonrecurrently without reducing the intended life of the amplifier or exceeding the specified temperature rise. (MAG) 107-1964w

short-circuiter A device designed to short circuit the commutator bars when the motor has attained a predetermined speed in some forms of single-phase commutator-type motors. *See also:* asynchronous machine. (EEC/PE) [119]

short circuit, external dc *See:* external dc short circuit.

short-circuit feedback admittance *See:* admittance, short-circuit feedback.

short-circuit flux (magnetic sound records) That flux from a magnetic record which flows across a plane normal to the recorded medium, through a magnetic short circuit placed in intimate contact with the record. (SP) 347-1972w

short-circuit flux per unit width (magnetic sound records) The measured short-circuit flux divided by the measured width of the recorded track. *Note:* The term fluxivity has been proposed for the quantity short-circuit flux per unit width. (SP) 347-1972w

short-circuit forward admittance *See:* admittance, short-circuit forward.

short-circuit impedance (1) A qualifying adjective indicating that the impedance under consideration is for the network with a specified pair or group of terminals short-circuited. *See also:* network analysis; self-impedance. (Std100) 270-1966w
(2) (line or four-terminal network) The driving-point impedance when the far-end is short-circuited. *See also:* self-impedance. (PE/EEC) [119]

short-circuit induced current The rms power frequency current between a conductive object and ground through a zero impedance or in a closed circuit, as a result of induction. (T&D/PE) 539-1990

short-circuit inductance The apparent inductance of a winding of a transformer with one or more specified windings short circuited often taken as a means of determining the leakage inductance of a winding. (CAS/CHM) [13], [51]

short-circuiting device (power system device function numbers) A primary circuit switching device that functions to short circuit or to ground a circuit in a response to automatic or manual means. *Synonym:* grounding device. (PE/SUB) C37.2-1979s

short-circuiting relays Telecommunication circuit grounding relays are used to ground an exposed telecommunication or telephone pair, usually on open-wire "joint-use" facilities during periods of severe power system disturbance. *Synonym:* grounding relays. (PE/PSC) 487-1992

short-circuiting switch A single-pole double-throw (make-before-break) transfer switch used to transfer current away from the meter. (ELM) C12.9-1993

short-circuit input admittance *See:* admittance, short-circuit input.

short-circuit input capacitance (*n*-terminal electron device) The effective capacitance determined from the short-circuit input admittance. *See also:* electron-tube admittances. (ED) 161-1971w

short-circuit input voltamperes The product of the input voltage and input current (rms values) with the resonating winding short circuited. (PEL) 449-1998

short circuit, internal *See:* internal short circuit.

short-circuit loss (rotating machinery) The difference in power required to drive a machine at normal speed, when excited to produce a specified balanced short-circuit armature current, and the power required to drive the unexcited machine at the same speed. *See also:* asynchronous machine. (PE) [9]

short-circuit output admittance *See:* admittance, short-circuit output.

short-circuit output capacitance (*n* -terminal electron device) The effective capacitance determined from the short-circuit output admittance. *See also:* electron-tube admittances. (ED) 161-1971w

short-circuit protection (power supplies) (automatic). Any automatic current-limiting system that enables a power supply to continue operating at a limited current, and without damage, into any output overload including short circuits. The output voltage must be restored to normal when the overload is removed, as distinguished from a fuse or circuit-breaker system that opens at overload and must be closed to restore power. (AES) [41]

short-circuit ratio (SCR) (1) For a semiconductor converter, the ratio of the short-circuit capacity of the bus, in MVA, at the point of converter connection to the rating of the converter, in MW. (IA/SPC) 519-1992
(2) The ratio of the ac system three-phase short-circuit MVA (expressing the ac system impedance) to dc power. (PE/T&D) 1204-1997

short-circuit saturation curve (synchronous machines) The relationship between the current in the short-circuited armature winding and the field current. (PE) [9]

short-circuit time constant (rotating machinery) (primary winding) The time required for the direct-current component present in the short-circuit primary-winding current following a sudden change in operating conditions to decrease to 1/e Δ 0.368 of its initial value, the machine running at rated speed. *See also:* asynchronous machine. (PE) [9]

short-circuit transfer admittance *See:* admittance, short-circuit transfer.

short-circuit transfer capacitance (electron tube) The effective capacitance determined from the short-circuit transfer admittance. *See also:* electron-tube admittances. (ED) 161-1971w

short dimension Incremental dimensions whose number of digits is the same as normal dimensions except the first digit is zero, that is 0.XXXX for the example under normal dimension. *See also:* normal dimension; dimension; incremental dimension; long dimension. (IA) [61]

short-distance navigation Navigation utilizing aids usable only at comparatively short distances; this term covers navigation between approach navigation and long-distance navigation, there being no distinct, universally accepted demarcation between them. *See also:* navigation. (AES/RS) 686-1982s, [42]

short field Where two field strengths are required for a series machine, short field is the minimum-strength field connection. *Note:* The use of the term tapped field is deprecated. *See also:* asynchronous machine; direct-current commutating machine. (EEC/PE) [119]

shorting cap A device that provides a closed circuit between line and load when a photocontrol is not used. (RL) C136.10-1996

short-lever relay armature An armature with an armature ratio of 1 : 1 or less. (EEC/REE) [87]

short-line-fault transient recovery voltage The transient recovery voltage obtained when a circuit-switching device interrupts a nearby fault on the line. *Note:* Short-line-fault transient recovery voltage differs from terminal fault conditions in that the length of line adds a high-frequency saw-tooth component to the transient recovery voltage. As the distance to the fault becomes greater, the amplitude of the saw-tooth component increases, the rate of rise of the saw-tooth component decreases, and the fault current decreases. The increased amplitude adversely affects the interrupting capability of the circuit-switching device while the decrease in the rate of rise and the decrease in current makes interruption easier. The effects are not proportional and a distance is reached where interruption is most severe even though the current is less than for a terminal fault. The critical value varies considerably with the type of circuit-switching device (oil, air-blast, gas-blast, etc.), and with the particular design. The critical distance may be in the order of a mile at the higher voltages. The critical distance is less as lower voltages are considered. (SWG/PE) C37.100-1992

short packet A packet with a length of less than 64 bytes. *Synonym:* under-sized packet. *Contrast:* long packet. (C) 610.7-1995, 610.10-1994w

short stack A stack of protocols with less than seven layers. *See also:* thin stack. (C) 610.7-1995

short-time-delay phase or ground trip element A direct-acting trip device element that functions with a purposely delayed action (measured in milliseconds). (SWG/PE) C37.100-1992

short-term exposure Exposure for durations less than the corresponding averaging time. (NIR) C95.1-1999

short-term settling time Measured from the mesial point (50%) of the output, the time at which the step response enters and subsequently remains within a specified error band around the final value. The final value is defined to occur at a specified time less than one second after the beginning of the step. (IM/WM&A) 1057-1994w

short-term timing instability *See:* aperture uncertainty.

short-time current (of an air switch) The current carried by a device, an assembly, or a bus for a specified short-time interval. *See also:* short-time rating; rated short-time withstand current. (SWG/PE/IA/PSP/TR) C37.100-1992, 1015-1997, C57.12.44-1994

short-time current rating The designated rms current that a connector can carry for a specified time under specified conditions. (T&D/PE) 386-1995

short-time current ratings (of a switching device or assembly) The maximum rms total currents, including the transient direct-current component, that the device or assembly is required to carry successfully for specified short-time intervals. (SWG/PE) C37.30-1992s

short-time current tests (high-voltage switchgear) Tests that consist of the application of a current higher than the rated continuous current for specified short periods to determine the adequacy of the device to withstand short-circuit currents for the specified short times. (SWG/PE) C37.40-1981s

short-time delay An intentional time delay in the tripping of a circuit-breaker which is above the overload pickup setting. (IA/PSP) 1015-1997

short-time delay phase A direct-acting trip device element that functions with a purposely delayed action (measured in milliseconds). (IA/PSP) 1015-1997

short-time duty (power and distribution transformers) A duty that demands operation at a substantially constant current for a short and definitely specified time. (NESC/PE/TR) C57.12.80-1978r, [86], C57.16-1996

short-time overload rating (rotating electric machinery) The output that the machine can sustain for a specified time starting hot under the conditions of Section 4 of IEEE Std 11-1980, without exceeding the limits of temperature rise of Section 5. (PE/EM) 11-1980r

short-time pickup The current at which the short-time delay function is initiated. (IA/PSP) 1015-1997

short-time rating (1) (packaging machinery) The rating that defines the load which can be carried for a short and definitely specified time, the machine apparatus, or device being at approximately room temperature at the time the load is applied. (IA/PKG) 333-1980w
(2) (power and distribution transformers) Defines the maximum constant load that can be carried for a specified short time without exceeding established temperature-rise limitations, under prescribed conditions. (PE/TR) C57.12.80-1978r
(3) (of diesel-generator unit) The electric power output capability that the diesel-generator unit can maintain in the service environment for 2 h in any 24 h period, without exceeding the manufacturer's design limits and without reducing the maintenance interval established for the continuous rating. *Note:* Operation at this higher rating does not limit the use of the diesel-generator unit at its continuous rating. (PE/NP) 387-1995
(4) (of a relay) The highest value of current or voltage or their product that the relay can stand, without injury, for specified short-time intervals (for alternating-current circuits, root-mean-square total value including the direct-current component is used). The rating recognizes the limitations imposed by both the thermal and electromagnetic effects. (SWG/PE/SWG-OLD) C37.100-1992
(5) A rating applied to a circuit-breaker that, for reason of system coordination, causes tripping of the circuit-breaker to be delayed beyond the time when tripping would be caused by an instantaneous element. *See also:* rated short-time withstand current; short-time current. (IA/PSP) 1015-1997

short-time test current (thyristor converter) The value of direct current that may be applied to a unit or section for a short period (seconds) following continuous operation at a specified lower dc value under specific conditions. (IA/IPC) 444-1973w

short-time thermal current rating (current transformer) The 1 s thermal current rating of a current transformer is the root-mean-square (rms) symmetrical primary current that can be carried for 1 s with the secondary winding short-circuited without exceeding in any winding the limiting temperature. (PE/TR) [57]

short-time waveform distortion (linear waveform distortion) Distortion of time components from 125 ns to 1 μs; that is, time components of the short-time domain. (PE/BT/NP) 382-1980s, 511-1979w

short-time (symmetrical) withstand current (1) (of an air switch) The current carried by a device, an assembly, or a bus for a specified short-time interval. (SWG/PE) C37.34-1994
(2) An abnormal power-frequency current, the initial portion of which may have a dc offset (expressed in rms symmetrical amperes) that a switch is required to carry. (SWG/PE) 1247-1998

short-time (symmetrical) withstand current duration The maximum duration of short-time (symmetrical) withstand current that a switch is required to carry. (SWG/PE) 1247-1998

short-wave fade-out *See:* sudden ionospheric disturbance.

shot noise (1) (fiber optics) Noise caused by current fluctuations due to the discrete nature of charge carriers and random or unpredictable (or both) of charged particles from an emitter. *Note:* There is often a (minor) inconsistency in referring to shot noise in an optical system: many authors refer to shot noise loosely when speaking of the mean square shot noise current (amp^2) rather than noise power (watts). *See also:* quantum noise. (Std100) 812-1984w
(2) *See also:* noise. (LM/C) 802.7-1989r

shoulder The portion of the roadway contiguous with the traveled way for accommodation of stopped vehicles for emergency use and for lateral support of base and surface course. (NESC) C2-1997

shoulder lobe A radiation lobe that has merged with the major lobe, thus causing the major lobe to have a distortion that is shoulder-like in appearance when displayed graphically. *Synonym:* vestigial lobe. (AP/ANT) 145-1993

show window Any window used or designed to be used for the display of goods or advertising material, whether it is fully or partly enclosed or entirely open at the rear and whether or not it has a platform raised higher than the street floor level. (NESC/NEC) [86]

shrink link (rotating machinery) A bar with an enlarged head on each end for use like a rivet but slipped into place after expansion by heat. It tightens on cooling by shrinkage only. (IA/APP) [90]

Shugart Associates System Interface *See:* small computer systems interface.

shunt (1) (general) A device having appreciable resistance or impedance connected in parallel across other devices or apparatus, and diverting some (but not all) of the current from it. Appreciable voltage exists across the shunted device or apparatus and an appreciable current may exist in it. (Std100) 270-1966w
(2) (air switch) A flexible electrical conductor comprised of braid, cable, or flat laminations designed to conduct current around the mechanical joint between two conductors. (SWG/PE) C37.100-1992

shunt capacitance *See:* auxiliary capacitance.

shunt capacitor bank current Current, including harmonics, supplied to a shunt capacitor bank. *Note:* Current is expressed in rms amperes. (SWG/PE) C37.100-1992

shunt control A method of controlling motors employing the shunt method of transition from series to parallel connections of the motors. *See also:* multiple-unit control. (EEC/PE) [119]

shunt excitation (emergency and standby power) A source of generator field excitation power taken from the generator output, normally through power potential transformers connected directly or indirectly to the generator output terminals. (IA/PSE) 446-1995

shunt-fed vertical antenna (1) A vertical antenna connected to ground at the base and excited (or connected to a receiver) at a point suitably positioned above the grounding point. (T&D/PE) 539-1990

(2) A vertical antenna that is connected directly to ground at its base and whose feed line connects to the antenna between ground and a point suitably positioned above the base. (AP/ANT) 145-1993

shunt filter A type of filter that reduces harmonics by providing a low-impedance path to shunt the harmonics from the source away from the system to be protected. (IA/SPC) 519-1992

shunt gap An intentional gap(s) between spaced electrodes that is electrically in parallel with one or more valve elements. (SPD/PE) C62.11-1999

shunting or discharge switch (power system device function numbers) A switch that serves to open or to close a shunting circuit around any piece of apparatus (except a resistor), such as a machine field, a machine armature, a capacitor, or a reactor. *Note:* This excludes devices that perform such shunting operatings as may be necessary in the process of starting a machine by devices 6 or 42 [a starting circuit breaker or a running circuit breaker], or their equivalent, and also excludes device function 73 [a load-resistor contactor], which serves for the switching of resistors. (SUB/PE) C37.2-1979s

shunting transition *See:* shunt transition.

shunt leads (instrument) Those leads that connect a circuit of an instrument to an external shunt. The resistance of these leads is taken into account in the adjustment of the instrument. *See also:* auxiliary device to an instrument; instrument. (EEC/AII) [102]

shunt loading Loading in which reactances are applied in shunt across the conductors of a transmission circuit. *See also:* loading. (EEC/PE) [119]

shunt noninterfering fire-alarm system A manual fire-alarm system employing stations and circuits such that, in case two or more stations in the same premises are operated simultaneously, the signal from the operated box electrically closest to the control equipment is transmitted and other signals are shunted out. *See also:* protective signaling. (EEC/PE) [119]

shunt reactor (power and distribution transformers) A reactor intended for connection in shunt to an electric system for the purpose of drawing inductive current. *Note:* The normal use for shunt reactors is to compensate for capacitive currents from transmission lines, cable, or shunt capacitors. The need for shunt reactors is most apparent at light load. (PE/TR) C57.117-1986r, C57.21-1981s, C57.12.80-1978r

shunt regulator (power supplies) A device placed across the output that controls the current through a series dropping resistance to maintain a constant voltage or current output. (AES) [41]

shunt release A release energized by a source of voltage. *Note:* The voltage may be derived either from the main circuit or from an independent source. *Synonym:* shunt trip. (SWG/PE/TR) C37.100-1992, C57.12.44-1994

shunt snubber (ac adjustable-speed drives) Circuit elements, usually including a capacitor and a resistor connected in shunt with a switching device to limit the rate of rise of voltage or the peak voltage across the device (or both) when switching from a conducting to a blocking state or when subjected to an external voltage transient. *See also:* snubber. (IA/ID/SPC) 995-1987w, 936-1987w

shunt tee junction *See:* H-plane tee junction.

shunt transformer A transformer in which the primary winding is connected in shunt with a power-supply circuit, and that transfers energy to another circuit at the same or different voltage from that of the primary circuit. *See also:* transformer. (PE/TR) [57]

shunt transition (shunting transition) A method of changing the connection of motors from series to parallel in which one motor, or group of motors, is first shunted or short circuited, then open circuited, and finally connected in parallel with the other motor or motors. *See also:* multiple-unit control. (PE/EEC) [119]

shunt trip *See:* shunt release.

shunt-trip device A trip mechanism energized by a source of voltage that may be derived either from the main circuit or from an independent source. (IA/PSP) 1015-1997

shunt-trip recloser A recloser in which the tripping mechanism, by releasing the holding means, permits the main contacts to open, with both the tripping mechanism and the contact opening mechanism deriving operating energy from other than the main circuit. (SWG/PE) C37.100-1992

shunt winding *See:* common winding.

shunt-wound A qualifying term applied to a direct-current machine to denote that the excitation is supplied by a winding connected in parallel with the armature in the case of a motor, with the load in the case of a generator, or is connected to a separate source of voltage. (EEC/PE) [119]

shunt-wound generator A dc generator in which the entire field excitation is ordinarily derived from one winding consisting of many turns with a relatively high resistance. This winding is connected in parallel with the armature circuit in a self-excited generator. In a separately excited generator, the winding is connected to the load side of another generator or another dc source. (IA/MT) 45-1998

shunt-wound motor A dc motor in which the field circuit is connected either in parallel with the armature circuit or to a separate source of excitation voltage. (IA/MT) 45-1998

Shupe effect A time-variant non-reciprocity due to temperature changes along the length of the fiber. (AES) 952-1997

shutDown The lowest-power operating mode of RamLink slaves, in which all signals, except for a bused *linkOn* signal, may be ignored. (C/MM) 1596.4-1996

shutOff An unpowered state of RamLink slaves, in which a drop in the supply voltage may have caused a loss of volatile memory state. The contents of DRAM storage become undefined; storage on special RAM-devices (such as FLASH memory) may be unaffected by the shutOff state. (C/MM) 1596.4-1996

shutter (1) A protective covering used to close, or to close partially, an opening in a stator frame or end shield. In general, the word shutter will be preceded by the name of the part to which it is attached. As used for an electric machine, a shutter is rigid and hence not adjustable. *See also:* stator. (IA/EM/APP) [90]

(2) (of a switchgear assembly) A device that is automatically operated to completely cover the stationary portion of the primary disconnecting devices when the removable element is either in the disconnected position, test position, or has been removed. (SWG/PE) C37.100-1992

shuttle car A vehicle on rubber tires or caterpillar treads and usually propelled by electric motors, electric energy for which is supplied by a diesel-driven generator, by storage batteries, or by a power distribution system through a portable cable. Its chief function is the transfer of raw materials, such as coal and ore, from loading machines in trackless areas of a mine to the main transportation system. (EEC/PE) [119]

shuttle car, explosion-tested *See:* explosion-tested shuttle car.

SI *See:* International System of Units; segment interconnect; shift-in character; units and letter symbols.

sibling node Relative to a node in a tree, a second node that has the same immediate predecessor or parent node. (See the corresponding figure.) *Synonym:* brother; sister.

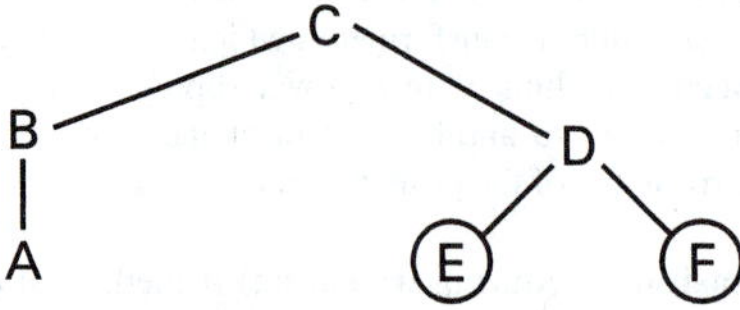

Nodes E and F are sibling nodes
sibling node

(C) 610.5-1990w

side The physical portion of a cartridge that is accessible by a drive after one mount operation. A side contains one or more partitions. (C/SS) 1244.1-2000

side A By convention, the side of the module closest to the connector "A" row, and the same side as the keying pins. (C/BA) 1101.4-1993, 1101.3-1993

side B By convention, the side of the module farthest from the connector "A" row, and the opposite side from the keying pins. (C/BA) 1101.4-1993, 1101.3-1993

side back light (illuminating engineering) Illumination from behind the subject in a direction not parallel to a vertical plane through the optical axis of the camera. (EEC/IE) [126]

sideband attenuation That form of attenuation in which the transmitted relative amplitude of some component(s) of a modulated signal (excluding the carrier) is smaller than that produced by the modulation process. *See also:* wavefront. (AP/ANT) 145-1983s

sideband null (navigation aids) (rectilinear navigation system) The surface of position along which the resultant energy from a particular pair of sideband antennas is zero. (AES/GCS) 172-1983w

sideband-reference glide slope (instrument landing systems) (navigation aids) A modified null reference glide-slope antenna system in which the upper (sideband) antenna is replace with two antennas, both at lower heights, and fed out of phase, so that a null is produced at the desired glide-slope angle. *Note:* This system is used to reduce unwanted reflections of energy into the glide-slope sector at locations where rough terrain exists in front of the approach end of the runway, by producing partial cancellation of energy at low elevation angles. (AES/GCS) 172-1983w

sidebands (1) (A) The frequency bands on both sides of the carrier frequency within which fall the frequencies of the wave produced by the process of modulation. **(B)** The wave components lying within such bands. *Note:* In the process of amplitude modulation with a sine-wave carrier, the upper sideband includes the sum (carrier plus modulating) frequencies: the lower sideband includes the difference (carrier minus modulating) frequencies. *See also:* amplitude modulation; radio receiver. (AP/ANT) 145-1983
(2) (data transmission) A band of frequencies containing components of either the sum (upper side band) or difference (lower sideband) of the carrier and modulation frequencies. (PE) 599-1985w

sideband suppression (power-system communication) A process that removes the energy of one of the sidebands from the modulated carrier spectrum. (AP/ANT) [35]

side-break switch One in which the travel of the blade is in a plane parallel to the base of the switch. (SWG/PE) C37.100-1992

side circuit (data transmission) A circuit arranged for deriving a phantom circuit. *Note:* In the case of two-wire side circuits, the conductors of each side circuit are placed in parallel to form a side of the phantom circuit. In the case of four-wire side circuits, the lines of the two side circuits that are arranged for transmission in the same direction provide a one-way phantom channel for transmission in that same direction, the two conductors of each line being placed in parallel to provide a side for that phantom channel. Similarly, the conductors of the other two lines provide a phantom channel for transmission in the opposite direction. (PE) 599-1985w

side-circuit loading coil A loading coil for introducing a desired amount of inductance in a side circuit and a minimum amount of inductance in the associated phantom circuit. *See also:* loading. (EEC/PE) [119]

side-circuit repeating coil (side-circuit repeat coil) A repeating coil that functions simultaneously as a transformer at a terminal of a side circuit and as a device for superposing one side of a phantom circuit on that side circuit. (PE/EEC) [119]

side effect (1) (software) Processing or activities performed, or results obtained, secondary to the primary function of a program, subprogram, or operation. *See also:* subprogram; program; function. (C/SE) 729-1983s
(2) Loosely speaking, an expression has a side effect if it performs some observable action in addition to returning a value. For example, a variable assignment is a side effect. (C/MM) 1178-1990r

sideflash (lightning) A spark occurring between nearby metallic objects or from such objects to the lightning protection system or to ground. *See also:* direct-stroke protection. (T&D/PE/NFPA) [10], [114]

side flashover (lightning) A flashover of insulation resulting from a direct lightning stroke that bypasses the overhead ground wire and terminates on a phase conductor of a transmission line. *See also:* direct-stroke protection. (T&D/PE) [10]

side frequency One of the frequencies of a sideband. *See also:* amplitude modulation. (AP/ANT) 145-1983s, [123]

side lobe A radiation lobe in any direction other than that of the major lobe. *See also:* minor lobe; maximum relative side lobe level; back lobe; relative side lobe level; relative cross-polar side lobe level; mean side lobe level. (AP/ANT) 145-1993

sidelobe blanker A device that employs an auxiliary wide-angle antenna and receiver to sense whether a received pulse originates in the sidelobe region of the main antenna and if so, to gate it from the output signal. (AES) 686-1997

sidelobe canceler A device that employs one or more auxiliary antennas and receivers to allow linear subtraction of interfering signals from the desired output if they are sensed to originate in the sidelobes of the main antenna. (AES) 686-1997

side lobe level, maximum relative *See:* maximum relative side lobe level.

side lobe level, relative *See:* relative side lobe level.

sidelobes Normally undesired amplitude responses that lie outside a specified frequency or time interval. These amplitude responses are inherently related to the impulse response of the device and not to other effects, such as bulk-wave, harmonic, multiple transit echoes, or direct electromagnetic feed-through responses. (UFFC/FT) 1037-1992w

side lobe suppression Any process, action, or adjustment to reduce the level of the side lobes or to reduce the degradation of the intended antenna system performance resulting from the presence of side lobes. (AP/ANT) 145-1993

side-lock Spurious synchronization in an automatic frequency synchronizing system by a frequency component of the applied signal other than the intended component. *See also:* television. (BT/AV) [34]

sidelooking airborne radar (SLAR) A high-resolution (in both range and angle) airborne imaging radar, without synthetic-aperture radar (SAR) processing, directed sidelooking (perpendicular to the line of flight) using large, narrow-beam-width antennas. (AES) 686-1997

sidelooking radar A ground mapping radar used aboard aircraft involving the use of a fixed antenna beam pointing out the side of an aircraft either abeam or squinted with respect to the aircraft axis. The beam is usually a vertically oriented fan beam having a narrow azimuth width. The narrow azimuth resolution can either be obtained with a long aperture mounted along the axis of the aircraft [sidelooking airborne radar (SLAR)] or by the use of synthetic-aperture radar (SAR) processing. (AES/GCS) 686-1997, 172-1983w

side marker lights (illuminating engineering) Lamps indicating the presence of a vehicle when seen from the front and sometimes serving to indicate its width. When seen from the side they may also indicate its length. (EEC/IE) [126]

side panel (rotating machinery) A structure enclosing or partly enclosing one side of a machine. (PE) [9]

sidereal (navigation aids) Of or pertaining to the stars. (AES/GCS) 172-1983w

sidereal period (communication satellite) The time duration of one orbit measured relative to the stars. (COM) [19]

side relay armature An armature that rotates about an axis parallel to that of the core, with the pole face on a side surface of the core. (EEC/REE) [87]

side stream scrambling A data scrambling technique, used by 100BASE-T2, to randomize the sequence of transmitted symbols and avoid the presence of spectral lines in the signal spectrum. Synchronization of the scrambler and descrambler of connected PHYs is required prior to operation. (C/LM) 802.3-1998

side thrust (disk recording) (skating force) The radial component of force on a pickup arm caused by the stylus drag. *See also:* phonograph pickup. (SP) [32]

sidetone The acoustic output of a telephone set receiver due to an acoustic input to the transmitter of the same telephone set. *Note:* Where the handset is mounted on a test fixture that includes the artificial mouth and artificial ear, the definition includes transmission through the handset proper; there may be also some vibration effect that is expected to be insignificant for handsets of modern design. There are two types of sidetone to be considered: **listener sidetone** and **talker sidetone**. (COM/TA) 269-1992

sidetone objective loudness rating (loudness ratings of telephone connections)

$$\text{SOLR} = -20\log_{10}\frac{S_E}{S_M}$$

where

S_M = sound pressure at the mouth reference point (in pascals)

S_E = sound pressure at the ear reference point (in pascals)

(COM/TA) 661-1979r

sidetone path loss (telephony) The difference in dB of the acoustic output level of the receiver of a given telephone set to the acoustic input level of the transmitter of the same telephone set. (COM/TA) 269-1971w

sidetone telephone set A telephone set that does not include a balancing network for the purpose of reducing sidetone. *See also:* telephone station. (EEC/PE) [119]

sidewalk elevator A freight elevator that operates between a sidewalk or other area exterior to the building and floor levels inside the building below such area, that has no landing opening into the building at its upper limit of travel, and that is not used to carry automobiles. *See also:* elevator. (EEC/PE) [119]

side-wall pressure The crushing force exerted on a cable during installation. (NESC) C2-1997

sideways sum (mathematics of computing) A sum obtained by adding the digits of a numeral without regard to position or significance. (C) 1084-1986w

siemens (metric practice) The electric conductance of a conductor in which a current of one ampere is produced by an electric potential difference of one volt. (QUL) 268-1982s

sievert (metric practice) The dose equivalent when the absorbed dose of ionizing radiation multiplied by the dimensionless factors Q (quality factor) and N (product of any other multiplying factors) stipulated by the International Commission on Radiological Protection is one joule per kilogram. (QUL) 268-1982s

sifting sort *See:* bubble sort.

sigma (σ) The term sigma designates a group of telephone wires, usually the majority or all wires of a line, that is treated as a unit in the computation of noise or in arranging connections to ground for the measurement of noise or current balance ratio. (PE/EEC) [119]

sign (1) (power or energy) Positive, if the actual direction of energy flow agrees with the stated or implied reference direction: negative, if the actual direction is opposite to the reference direction. *See also:* network analysis. (Std100) 270-1966w

(2) (test, measurement, and diagnostic equipment) The symbol that distinguishes positive from negative numbers. (MIL) [2]

(3) *See also:* electric sign. (NESC) [86]

signal (1) (signals and paths) (microcomputer system bus) The physical representation of data. (MM/C) 796-1983r

(2) (signals and paths, 696 interface devices) The physical representation which conveys data from one point to another. For the purpose of IEEE Std 696-1983, this applies to digital electrical signals only. (MM/C) 696-1983w

(3) (A) (data transmission) A visual, audible or other indication used to convey information. **(B) (data transmission)** The intelligence, message or effect to be conveyed over a communication system. **(C) (data transmission)** A signal wave; the physical embodiment of a message. (PE/PSCC) 599-1985

(4) (overhead-power-line corona and radio noise) The intelligence, message, or effect conveyed over a communication system. (T&D/PE) 539-1990

(5) (programmable instrumentation) The physical representation of information. *Note:* For the purposes of IEEE Std 488.1-1987, this term refers to digital electrical signals only. (IM/AIN) 488.1-1987r

(6) (computers) The event or phenomenon that conveys data from one point to another. (C) [20], [85]

(7) Information about a variable that can be transmitted in a system. (IA/ICTL/IAC) [60]

(8) (telephone switching systems) An audible, visual or other indication of information. (C) [85]

(9) A phenomenon (visual, audible, or otherwise) used to convey information. The signal is often coded, such as a modulated waveform, so that it requires decoding to be intelligible. (CAS) [13]

(10) (SBX bus) The physical representation of a logical value. (C/MM) 959-1988r

(11) (STEbus) The physical representation of data. (C/MM) 1000-1987r

(12) Any communication between message-based devices consisting of a write to a signal register. (C/MM) 1155-1992

(13) A measurable quantity (e.g., a voltage) which varies in time in order to transmit information. A signal propagates along a wire or an optic fiber. It is interpreted as a sequence of bits, which is grouped into a sequence of characters by the character layer of the protocol stack. Signals are generated by a link output and are absorbed by a link input. (C/BA) 1355-1995

(14) In networking, an electrical pulse that conveys information through a transmission medium. *See also:* baseband signaling; digital signal; analog signal; broadband signaling; out-of-band signaling. (C) 610.7-1995

(15) (A) A variation of a physical quantity, used to convey data. **(B)** A time-dependent value attached to a physical phenomenon and conveying data. (C/Std100) 610.10-1994

(16) A mechanism by which a process may be notified of, or affected by, an event occurring in the system. Examples of such events include hardware exceptions and specific actions by processes or threads. The term *signal* is also used to refer to the event itself. (C/PA) 9945-1-1996, 9945-2-1993

(17) (A) The behavior controlled or observed by a test resource. **(B)** A visual, audible, or other indication used to convey information. (SCC20) 1226-1998

(18) A point in the design from which a stimulus may be directly applied or a response directly measured. (C/TT) 1450-1999

(19) A mechanism by which a process may be notified of, or affected by, an event occurring in the system. Examples of such events include hardware exceptions and specific actions by processes. The term *signal* is also used to refer to the event itself. (C) 1003.5-1999

signal, actuating *See:* actuating signal.

signal aspect The appearance of a fixed signal conveying an indication as viewed from the direction of an approaching train: the appearance of a cab signal conveying an indication as viewed by an observer in the cab. (EEC/PE) [119]

signal assertion A) The act of driving a signal to the true state. B) The act of driving a bus of signals to the correct pattern of ones and zeros. (C/BA) 1496-1993w

signal back light A light showing through a small opening in the back of an electrically lighted signal, used for checking the operation of the signal lamp. (EEC/PE) [119]

signal charge The charge that flows when the condition of the device is changed from that of zero applied voltage (after having previously been saturated with either a positive or negative voltage) to at least that voltage necessary to saturate in the reverse sense. *Note:* The signal charge Q_s equals the sum of Q_r and Q_i, as illustrated in the corresponding figure. It is dependent on the magnitude of the applied voltage, which should be specified in describing this characteristic of ferroelectric devices. *See also:* ferroelectric domain.

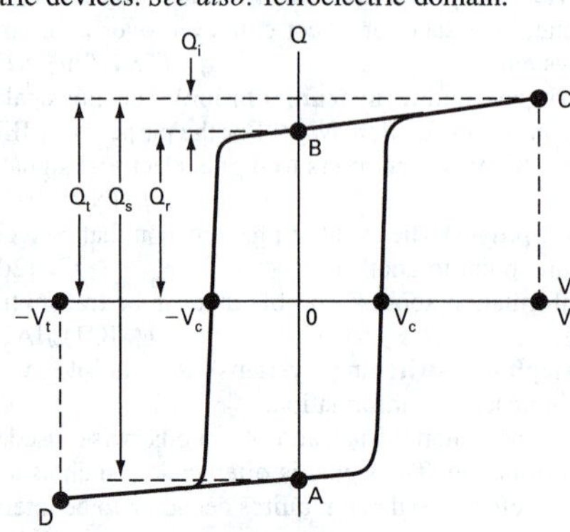

Hysteresis loop for a ferroelectric device.
signal charge

(UFFC) 180w

signal circuit (1) Any electric circuit that supplies energy to an appliance that gives a recognizable signal. Such circuits include circuits for door bells, buzzers, code-calling systems, signal lights, and the like. *See also:* appliance. (NESC) [86]

(2) (protective relay system) Any circuit other than input voltage circuits, input current circuits, power supply circuits, or those circuits that directly or indirectly control power circuit breaker operation. (SWG/PE/PSR) C37.100-1992, C37.90-1978s

(3) (protective relay system) Any circuit other than an input voltage circuit, input current circuit, power supply circuit, or an output circuit. (PE/PSR) C37.90.1-1989r

signal conditioning Sensor signal processing involving operations such as amplification, compensation, filtering, and normalization. (IM/ST) 1451.2-1997

signal contrast (facsimile) The ratio expressed in decibels between white signal and black signal. *See also:* facsimile signal. (COM) 168-1956w

signal converter (test, measurement, and diagnostic equipment) A device for changing a signal from one form or value to another form or value. (MIL) [2]

signal current (diode-type camera tube) The change in target current which occurs when the target is irradiated with photons, or electrons, compared to the case where no radiation is incident on the target. (ED) 503-1978w

signal decay time (measuring the performance of tone address signaling systems) The time interval between the end of the signal present condition and the beginning of the signal off condition at the end of the signal under consideration. (COM/TA) 752-1986w

signal decorrelation time *See:* decorrelation time.

signal delay The transmission time of a signal through a network. The time is always finite, may be undesired, or may be purposely introduced. *See also:* oscillograph; delay line. (IM/HFIM) [40]

signal, difference *See:* differential signal.

signal distance (1) (computers) The number of digit positions in which the corresponding digits of two binary words of the same length are different. *See also:* hamming distance. (COM/C) 312-1977w, [20]

(2) (mathematics of computing) *See also:* hamming distance. (C) 1084-1986w

signal distributing (telephone switching systems) Delivering of signals from a common control to other circuits. (COM) 312-1977w

signal-driven mode A mode of operation in which the signal `POSIX_Signals.Signal_IO` is sent to the owner of a socket whenever an I/O operation becomes possible on that socket. In this mode, `POSIX_Signals.Signal_IO` is sent when additional data could be sent on the socket, when new data arrives to be received on a socket, or a state transition occurs that would allow a send or receive call to return status without blocking. Signal-driven mode is enabled by setting the `POSIX_IO.Signal_When_Socket_Ready` flag on the socket and disabled by resetting the `POSIX_IO.Signal_When_Socket_Ready` flag. The default mode for signal driven mode is disabled. (C) 1003.5-1999

signal duration (measuring the performance of tone address signaling systems) The time interval during which a signal present condition exists continuously. (COM/TA) 752-1986w

signal electrode (camera tubes) An electrode from which the signal output is obtained. *See also:* electrode. (BT/AV) [34]

signal electronics power (thyristor converter) The power used for the analog or digital system power supplies, or both, required for the thyristor converter control and protection systems. (IA/IPC) 444-1973w

signal element (1) (unit interval) (data transmission) The part of a signal that occupies the shortest interval of signaling code. It is considered to be of unit duration in building up signal combinations. (PE) 599-1985w

(2) The logical signal during one half of a bit time which may take on the values of Logic_1 or Logic_0. (C/LM) 8802-5-1998

signal, error *See:* error signal.

signal, feedback *See:* feedback signal.

signal flow graph (network analysis) A network of directed branches in which each dependent node signal is the algebraic sum of the incoming branch signals at that node. *Note:* Thus,

$$x_1 t_{1k} + x_2 t_{2k} + \ldots + x_n t_{nk} = x_k$$

at each dependent node k, where t_{jk} is the branch transmittance of branch jk. (CAS) 155-1960w

signal frequency shift (frequency-shift facsimile system) The numerical difference between the frequencies corresponding to white signal and black signal at any point in the system. *See also:* facsimile signal. (COM) 168-1956w

signal generator A shielded source of voltage or power, the output level and frequency of which are calibrated, and usually variable over a range. *Note:* The output of known waveform is normally subject to one or more forms of calibrated modulation. (IM/HFIM) [40]

signal ground For the purpose of this guide, shall be the grounding system to which signals are referenced. (PE/EDPG) 1050-1996

signal identifier (spectrum analyzer) A means to identify the frequency of the input when spurious responses are possible. A front panel control used to identify the input frequency when spurious responses are present. (IM) 748-1979w

signal indication The information conveyed by the aspect of a signal. (EEC/PE) [119]

signaling (1) (data transmission) The production of an audible or visible signal at a station or switchboard by means of an alternating or pulsating current. In a telephone system, any of several methods used to alert subscribers or operators or to establish and control connections. (PE) 599-1985w
(2) (telephone switching systems) The transmission of address and other switching information between stations and central offices and between switching entities. (COM) 312-1977w
(3) The exchange of information specifically concerned with the establishment and control of connections, and the transfer of user-to-user and management information in a circuit-switched network. (C/LM) 802.9a-1995w
(4) The exchange of information specifically concerned with the establishment and control of connections, and the transfer of user-to-user and management information in a telecommunication network, e.g., in a PPSN. (LM/C/COM) 8802-9-1996

signaling, analog *See:* analog signaling.

signaling and doorbell transformers (power and distribution transformers) Step-down transformers (having a secondary of 30 V or less), generally used for the operation of signals, chimes, and doorbells. (PE/TR) C57.12.80-1978r

signaling, binary *See:* binary signaling.

signaling circuit Any electric circuit that energizes signaling equipment. (NESC/NEC) [86]

signaling light (illuminating engineering) A projector used for directing light signals toward a designated target zone. (EEC/IE) [126]

signal, input *See:* input signal.

signal interface The interface between a test device and the unit under test (UUT) through which signals pass. Stimuli pass from the test device to the UUT. Measured signals generally pass from the UUT to a test device. (SCC20) 993-1997

signal integration The summation of a succession of signals by writing them at the same location on the storage surface. *See also:* storage tube. (ED) 158-1962w

signal interphasing A method of simultaneously overlapping multiple transmission signals to achieve higher transmission rates. (C) 610.7-1995

signal layer The layer of the protocol stack at which signals are specified. (C/BA) 1355-1995

signal level (1) (signals and paths) (696 interface devices) The magnitude of a signal when considered in relation to an arbitrary reference magnitude. (IM/MM/C/AIN) 488.1-1987r, 696-1983w
(2) (SBX bus) The relative magnitude of a signal when compared to a reference. (MM/C) 959-1988r
(3) (STEbus) The relative magnitude of a signal when considered in relation to an arbitrary reference. The unit of representation is the volt. (C/MM/C/MM) 796-1983r, 1000-1987r
(4) (broadband local area networks) The measured voltage or power of a signal usually stated in dBmV. (LM/C) 802.7-1989r

signal line (1) (programmable instrumentation) (696 interface devices) (signals and paths) (microcomputer system bus) One of a set of signal conductors in an interface system used to transfer messages among interconnected devices. (IM/C/MM/AIN) 488.1-1987r, 796-1983r, 696-1983w, 959-1988r
(2) (STEbus) One of a set of signal conductors in an interface system used to transfer data among interconnected boards. (MM/C) 1000-1987r
(3) (NuBus) A conductor on the backplane other than ground, or power. (C/MM) 1196-1987w
(4) An electrical or optical information-carrying facility, such as a differential pair of wires or an optical fiber, with associated transmitter and receiver, carrying binary true/false logic values. (C/MM) 1596-1992
(5) An electrical or optical information-carrying facility, such as a differential pair of wires or an optical fiber, with associated driver and receiver, carrying binary true/false logic values. (C/MM) 1596.3-1996

signal lines The passive transmission lines through which the signal passes from one to another of the elements of the signal transmission system. *See also:* signal. (IE) [43]

signal names (1) Where a group of bus lines are represented by the same letters, the lines within the group are numbered; e.g., AD0*, AD1*, AD2*, etc. In order to represent a group of lines or signals in more convenient form, notation such as AD[63..0]* is used. Also, in these examples, the notation AD[ ]* is used to refer to all of the lines within the group. (C/BA) 896.2-1991w, 10857-1994, 896.4-1993w
(2) The simplified alphanumeric notation used to represent a group of lines or signals. For example, where a group of bus lines are represented by the same letters, the lines within the group are numbered, e.g., **AD0**, **AD1**, **AD2**, etc. In order to represent a group of lines or signals in more convenient form, the notation **AD[31..0]** is used. Also, the notation **AD[ ]** is used to refer to all of the lines within the group. The signal that a module applies to the module-side input of its driver is designated by a lowercase label, e.g., **ai**. The signal that a specific module applies to a bus line is designated by a lowercase label with an asterisk, e.g., **ai***. The signal that appears on a bus line as the result of the combined signals applied to it by all modules is designated by an uppercase label with an asterisk, e.g., **AI***. When appended to a signal name, the suffix "f" (filtered) refers to the bus line signal after it has passed through a receiver and a wire-or glitch filter (integrator). For example, **AIf** refers to the signal on the **AI*** line, after it has passed through an inverting receiver to become **AI** and the wire-or glitch filter to become **AIf**. See the figure below. (C/BA) 896.10-1997

signal negation The act of driving a signal to the false state. (C/BA) 1496-1993w

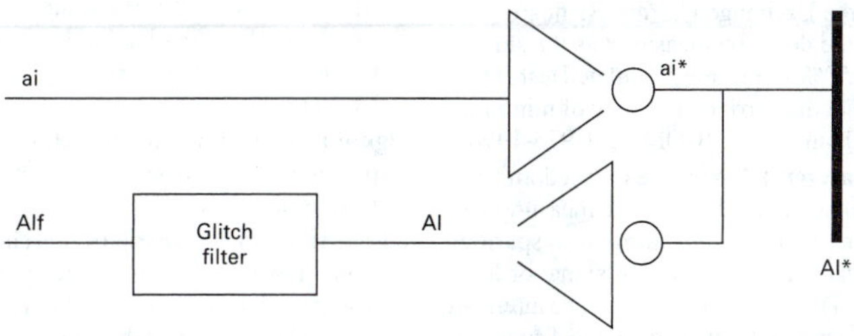

Signal naming convention

Signal names

signal off (measuring the performance of tone address signaling systems) Any condition where all the constituent tones of a tone signaling system are below a specified OFF level for each tone. In a single-tone signaling system, tone off and signal off are synonymous terms. *Note:* During a signal off condition, tones that are not used in the signaling system may be at a higher level. (COM/TA) 752-1986w

signal operation (elevators) Operation by means of single buttons or switches (or both) in the car, and up-or-down direction buttons (or both) at the landings, by which predetermined landing stops may be set up or registered for an elevator or for a group of elevators. The stops set up by the momentary actuation of the car buttons are made automatically in succession as the car reaches those landings, irrespective of its direction of travel or the sequence in which the buttons are actuated. The stops set up by the momentary actuation of the up-and-down buttons at the landing are made automatically by the first available car in the group approaching the landing in the corresponding direction, irrespective of the sequence in which the buttons are actuated. With this type of operation, the car can be started only by means of a starting switch or button in the car. *See also:* control. (EEC/PE) [119]

signal, output *See:* output signal.

signal output current (camera tubes or phototubes) The absolute value of the difference between output current and dark current. *See also:* phototube. (ED/BT/AV) 161-1971w, [34], [45]

signal overload point (electronic navigation) The maximum input signal amplitude at which the ratio of output to input is observed to remain within a prescribed linear operating range. *See also:* navigation. (AES/RS) 686-1982s, [42]

signal parameter (programmable instrumentation) That parameter of an electrical quantity whose values or sequence of values convey information. (IM/MM/C/AIN) 488.1-1987r, 1000-1987r, 796-1983r, 696-1983w

signal phase The initial phase of an exception operation in which all agents are notified of an error condition. *See also:* exception operation. (C/MM) 1296-1987s

signal present (measuring the performance of tone address signaling systems) Any condition where the presence of tone or tones is sufficient to be recognized as a valid digital or supervisory signal. In a single-tone signaling system, tone present and signal present are synonymous terms; in a two-tone signaling system the signal present state exists where two and only two tones each meet the signal present condition, and the two tones represent a valid combination. (COM/TA) 752-1986w

signal processing antenna system An antenna system having circuit elements associated with its radiating element(s) that perform functions such as multiplication, storage, correlation, and time modulation of the input signals. (AP/ANT) 145-1993

signal propagation delay or transmission delay Total time for a signal to pass through the switching system. For digital-to-digital interfaces (DS1) where bit integrity is maintained, the delay is the time elapsed between transmission and reception of the bit that starts a known test pattern. For analog-to-analog, digital-to-analog, or analog-to-digital transmission, T1Q1 (a study group of the Exchange Carriers Standards Association) proposes that the delay be measured as the shift in time of the envelope of a 50% amplitude modulated test signal (or its digital equivalent) at the carrier frequency of minimum delay in the voiceband channel. (COM/TA) 973-1990w

signal purity (network analyzers) A measure of freedom from frequency components other than the desired measurement frequency. It includes harmonics, subharmonics, spurious mixer products, and unwanted components of signal or local oscillator leakage. *Note:* The resulting error in measurement is a function of the detection system and of the frequency response of the network under test, as well as the signal purity. (IM/HFIM) 378-1986w

signal quality error heartbeat A signal from the transceiver to a node peripheral indicating that the transceiver is functioning properly. *Note:* This term is contextually specific to IEEE Std 802.3. (C) 610.7-1995

signal queueing When queueing is enabled for a signal, occurrences of that signal are queued in FIFO order and information is included if the signal is from a source that supplies information. Otherwise, the signal queue may be only one occurrence deep and it is implementation defined whether the data are included. Support for signal queueing is governed by the Realtime Signals option. Not all signals may support queueing. (C) 1003.5-1999

signal, rate *See:* rate signal.

signal, reference input *See:* reference input signal.

signal reference structure A system of conductive paths among interconnected equipment that reduces noise-induced voltages to levels that minimize improper operation. Common configurations include grids and planes. (IA/PSE) 1100-1999

signal relay *See:* alarm relay.

signal release The act of removing electronic drive to a signal thereby placing the driver in a high-impedance condition. Release of an SBus signal will leave the SBus signal in its last state unless the signal is specified to have bias circuits attached that return the signal to a specified state. (C/BA) 1496-1993w

signal repeater lights A group of lights indicating the signal displayed for humping and trimming. (EEC/PE) [119]

signal, return *See:* return signal.

signal rise time (measuring the performance of tone address signaling systems) The time interval between the end of the signal off condition and the beginning of the signal present condition at the beginning of a tone signal. (COM/TA) 752-1986w

signal, sampled *See:* sampled signal.

signal-shaping amplifier (telegraph practice) An amplifier and associated electric networks inserted in the circuit, usually at the receiving end of an ocean cable, for amplifying and improving the waveshape of the signals. *See also:* telegraphy. (EEC/PE) [119]

signal-shaping network (wave-shaping set) An electric network inserted (in a telegraph circuit) for improving the waveshape of the received signals. *See also:* telegraphy. (PE/EEC) [119]

signal shutter (illuminating engineering) A device which modulates a beam of light by mechanical means for the purpose of transmitting intelligence. (EEC/IE) [126]

signal skew The time difference a data bus may have relative to its corresponding strobe signal. Signal skew is a positive quantity when the strobe signal delay exceeds the bus signal delay. (C/BA) 896.2-1991w

signal state *See:* logic state.

signal threshold (dial-pulse address signaling systems) (telephony) The current or voltage value representing a transition between make and break states. The threshold may be expressed either as an absolute value (for example, 15 mA or 21 V) or as a percentage deviation between steady-state conditions (for example, the value which is 70% of the difference between make and break values, added to the off-hook value). Signaling detectors having hysteresis will have different signal threshold values for the make-break and break-make transition. (COM/TA) 753-1983w

signal-to-clutter ratio The ratio of target echo power to the power received from clutter sources lying within the same resolution element. (AES) 686-1997

signal-to-distortion ratio (C-notched noise) The ratio of input signal power to C-message weighted output distortion power. The signal level is measured with a holding tone (typically 1004 Hz) present with a wideband (200 Hz–10 kHz) level detector. The C-notch noise should be measured with a C-message filter after the holding tone has been reduced in amplitude by at least 50 dB with a 1010 Hz notch filter. (COM/TA) 973-1990w

signal-to-interference ratio The ratio of the magnitude of the signal to that of the interference or noise. *Note:* The ratio may be in terms of peak values or root-mean-square values and is often expressed in decibels. The ratio may be a function of the bandwidth of the system. *See also:* signal. (IE) [43]

signal-to-noise ratio (1) (video magnetic-tape recording systems) The ratio of the peak-to-peak amplitude of the video luminance signal from blanking level to reference white level (100 IRE units), 714 megavolts (mV) to the root-mean-square (rms) amplitude of the random noise, expressed in decibels.

$$\mathrm{SNR} = 20\log_{10}\frac{E_V}{E_N}$$

where

E_V = peak-to-peak amplitude of the maximum video luminance component (714 mV)

E_N = rms amplitude of random noise.

Notes: 1. Unless otherwise specified, the definition for signal-to-noise is as defined here. (BT) 618-1984w

(2) (camera tubes) The ratio of peak-to-peak signal output current to root-mean-square noise in the output current. *Note:* Magnitude is usually not measured in tubes where the signal output is taken from target. *See also:* camera tube; television. (PE/EEC) [119]

(3) (television transmission) The signal-to-noise ratio at any point is the ratio in decibels of the maximum peak-to-peak voltage of the video television signal, including synchronizing pulse, to the root-mean-square voltage of the noise. *Note:* The signal-to-noise ratio is defined in this way because of the difficulty of defining the root-mean-square value of the video signal or the peak-to-peak value of random noise. *See also:* television. (EEC/PE) [119]

(4) (mobile communication) The ratio of a specified speech-energy spectrum to the energy of the noise in the same spectrum. *See also:* television. (VT) [37]

(5) (sound recording and reproducing system) The ratio of the signal power output to the noise power in the entire pass band. 191-1953w

(6) (digital delay line) Ratio of the peak amplitude of the output doublet to the maximum peak of any noise response (or signal) outside of the doublet interval. (Includes overshoot.) (UFFC) [22]

(7) (speech quality measurements) In decibels of a speech signal, the difference between its speech level and the noise level. 297-1969w

(8) (overhead-power-line corona and radio noise) The ratio of the value of the signal to that of the noise. *Notes:* 1. This ratio is usually in terms of measured peak values in the case of impulse noise and in terms of the root-mean-square (rms) values in the case of random noise. 2. Where there is a possibility of ambiguity, suitable definitions of the signal and noise should be associated with this term; as, for example, peak signal to peak noise ratio, rms signal to rms noise ratio, peak-to-peak signal to peak-to-peak noise ratio, etc. In measurements of transmission line noise in the AM frequency range, the ratio of average station signal level to quasi-peak line noise level is generally used. 3. This ratio often may be expressed in decibels (dB). 4. This ratio may be a function of the bandwidth of the transmission or measuring system. (PE/T&D/AP/ANT) 539-1990, 599-1985w, 145-1993

(9) The ratio of a signal to the noise. (IM/WM&A) 1057-1994w

(10) The ratio of relative power of the usable signal to the noise present, expressed in decibels. (C) 610.7-1995, 610.10-1994w

(11) The ratio in dB of the power of a single- or multiple-tone test signal to the power of the background noise (after a D Filter) not related to the application of the test signal, plus all spurious signals resulting from the application of the test signal, except for 2nd-order and 3rd-order intermodulation (IMD) distortion. (COM/TA) 743-1995

(12) "Signal" refers to peak signal and "noise" refers to rms noise. (NPS) 325-1996

(13) In radar, the ratio of the power corresponding to a specified target measured at some point in the receiver to the noise power at the same point in the absence of the received signal. (AES) 686-1997

signal-to-total-distortion ratio (S/TD) The ratio in dB of the power of a single or multiple-tone test signal to the power of all spurious signals (after a D Filter) resulting from the application of the test signal plus the power from background noise. (COM/TA) 743-1995

signal transfer characteristic (diode-type camera tube) The relationship between the input image irradiance incident on the camera tube and the resulting output current signal. It is presented as a plot of the logarithm of the output signal as a function of the logarithm of the input signal. (ED) 503-1978w

signal-transfer point (telephone switching systems) A switching entity where common channel signaling facilities are interconnected. (COM) 312-1977w

signal transition distortion The deviation from the ideal time between signal transitions of a signal on a serial communication link. Contributing factors to signal transition distortion include clock oscillator frequency differences, variations in transmitter delay times and rise/fall times, and phase shifts introduced by cables. For a 1 Mb/s Manchester-encoded signal, the ideal times between transitions are 500 ns and 1 μs. A transition is defined as the time that a differential signal passes through 0 V. (EMB/MIB) 1073.4.1-2000

signal transmission system *See:* carrier.

signal, TV waveform *See:* TV waveform signal.

signal, unit-impulse *See:* unit-impulse signal.

signal, unit-ramp *See:* unit-ramp signal.

signal, unit-step *See:* unit-step signal.

signal wave A wave whose shape conveys some intelligence, message, or effect. [53]

signal winding (input winding) (saturable reactor) A control winding to which the independent variable (signal wave) is applied. (EEC/PE) [119]

signal word The word or words that designate a degree of safety alerting. The words shall always be located in a distinctive panel located in the uppermost portion of a safety sign or label. (NIR/SCC28) C95.2-1999

signature (1) Those characteristics of a waveform that help identify an event or conditions. (SWG/PE) C37.100-1992

(2) A statement of what the interface to a responsibility "looks like." A signature consists of the responsibility name, along with a property operator and the number and type of its arguments, if any. A type (class) may be specified for each argument in order to limit the argument values to being instances of that class. (C/SE) 1320.2-1998

signature analysis A technique for compressing a sequence of logic values output from a circuit under test into a small number of bits of data (signature) that, when compared to stored data, will indicate the presence or absence of faults in the circuit. (TT/C) 1149.1-1990

signature diagnosis (test, measurement, and diagnostic equipment) The examination of signature of an equipment for deviation from known or expected characteristics and consequent determination of the nature and location of malfunctions. (IM/WM&A) 194-1977w

sign bit A binary digit used to indicate the algebraic sign of a number. (C) 1084-1986w

sign character A character within a picture specification that represents the sign of a data item. *Note:* S, +, and − are commonly used as sign characters. (C) 610.5-1990w

sign digit (electronic computation) A character used to designate the algebraic sign of a number. (C) 162-1963w, 1084-1986w

signed (1) Pertaining to a representation of a number with which an algebraic sign is associated. (C) 610.5-1990w

(2) The condition of information that has an enciphered summary appended to it that is used to ensure the integrity of the data, the authenticity of the originator, and the unambiguous relationship between the originator and the data. (C/PA) 1328.2-1993w, 1326.2-1993w, 1224.2-1993w, 1327.2-1993w

signed binary arithmetic *See:* sign-magnitude arithmetic.

significance (1) (test, measurement, and diagnostic equipment) The value or weight given to a position, or to a digit in a position, in a positional numeration system. In most positional numeration systems positions are grouped in sequence of significance, usually more significant towards the left. (MIL) [2]
(2) *See also:* weight. (C) 1084-1986w

significand (1) (binary floating-point arithmetic) The component of a binary floating-point number that consists of an explicit or implicit leading bit to the left of its implied binary point and a fraction field to the right. (C/MM) 754-1985r
(2) (mathematics of computing) The component of a floating-point number that consists of an explicit or implicit leading digit to the left of its implied radix point and a fraction field to the right. *Synonyms:* fixed-point part; mantissa. *Contrast:* exponent. (C/MM) 854-1987r, 1084-1986w

significant (nuclear power generating station) Demonstrated to be important by the safety analysis of the station. (PE/NP) 381-1977w, 308-1991l

Significant Event Evaluation and Information Network (SEE-IN) An information database maintained by the Institute of Nuclear Power Operations (INPO). (PE/NP) 933-1999

significant aging mechanism An aging mechanism that, if in the normal and abnormal service environment, causes degradation during the installed life of the equipment that progressively and appreciably renders the equipment vulnerable to failure to perform its safety function(s) during design basis event conditions (DBE). (PE/NP) 1205-1993

significant code A code that identifies a particular item and also yields further information about the properties or classification of the item. *Contrast:* nonsignificant code. (C) 610.5-1990w

significant digit (1) (mathematics of computing) A digit that contributes to the accuracy or precision of a numeral. *See also:* least significant digit; most significant digit. (C) [20], 610.5-1990w, 1084-1986w
(2) (metric practice) Any digit that is necessary to define a value or quantity. (QUL) 268-1982s
(3) Any digit in a number that is necessary to define a numerical value. (SCC14) SI 10-1997

significant-digit arithmetic A method of making calculations using a modified form of floating-point representation in which the number of significant digits in the result is determined by the number of significant digits in the operands, the operations performed, and the degree of precision available. (C) 1084-1986w

significant figure* *See:* significant digit.
* Deprecated.

significant human interface An interface between personnel and equipment, facilities, software, or documentation, where the resulting human performance is a determinant in the achievement of system performance.

sign-magnitude arithmetic Computer arithmetic using numerals expressed in sign-magnitude notation. *Synonym:* signed binary arithmetic. (C) 1084-1986w

sign-magnitude notation A numeration system in which the left-most bit is interpreted as the sign bit and the remaining bits represent the magnitude. *Contrast:* twos-complement notation. (C) 1084-1986w

sign-off *See:* logoff.

sign-on *See:* login.

sign position The position at which the sign of a number is located. (C) [20], [85], 1084-1986w

silent lobing A method for scanning an antenna beam to achieve angle tracking without revealing the scanning pattern on the transmitted signal. (AES) 686-1997

silent zone Part of the skip zone at a distance greater than the range of the ground wave. (AP/PROP) 211-1997

silicon (A) A semiconducting material used in many devices such as integrated circuits and solar cells that in its pure form is a lightweight metal resembling aluminum. **(B)** A colloquial reference to an integrated circuit. (C) 610.10-1994

silicon controlled rectifier (SCR) (thyristor) An alternative name for the reverse blocking triode thyristor. *Note:* Although not an official definition, the term "unidirectional" is sometimes used to describe the single switching class of thyristors consisting of reverse-blocking and reverse-conducting thyristors. This term is useful for comparing or contrasting this class of thyristors with bidirectional thyristors. (IA/IPC) 428-1981w

silicone oil (insulating oil) A generic term for a family of relatively inert liquid organosiloxane polymers used as electrical insulation. (PE/TR) 637-1985r

silvering (electrotyping) The application of a thin conducting film of silver by chemical reduction upon a plastic or wax matrix. (PE/EEC) [119]

silver oxide cell A cell in which depolarization is accomplished by oxide of silver. *See also:* electrochemistry. (EEC/PE) [119]

silver storage battery An alkaline storage battery in which the positive active material is silver oxide and the negative contains zinc. *See also:* battery. (EEC/PE) [119]

silver-surfaced or equivalent Metallic materials having satisfactory long-term performance that operate within the temperature rise limits established for silver-surfaced electrical contact parts and conducting mechanical joints. (SWG/PE) C37.100-1992

SIM++ A programming language used for simulations on distributed computing systems. (C) 610.13-1993w

SIMD *See:* single instruction, multiple data.

similar design bar/coil A similar design bar/coil is a bar/coil of the same design and manufacture using same materials and processes as the actual production bar/coil except that it may be longer in the slot section, and/or larger in the copper and/or groundwall cross section than the actual production bar/coil. The variance in the slot section length and the cross section of the 'similar design bar/coil' and the actual production bar/coil must be identified prior to the start of the thermal cycling test. (PE/EM) 1310-1996

simple arc An arc that does not cross itself. (C) 610.4-1990w

simple buffering A buffering technique in which a buffer is allocated to a computer program for the duration of the program's execution. *Contrast:* dynamic buffering. (C) 610.12-1990

simple circuit (A) A circuit permitting the transmission of signals in either direction, but not in both simultaneously. *Contrast:* two-way circuit. **(B)** A circuit permitting the transmission of signals in one specific direction only. (C) 610.10-1994

simple combination of insulating materials (thermal classification of electric equipment and electrical insulation) A number of insulating materials, which together make possible the evaluation of any interaction between them. (EI) 1-1986r

simple electrical ground and test device A device with one terminal set and a power-operated ground-making switch for connecting the terminal set to the device ground connection system, complete with necessary isolation barriers and suitable interlocking. Voltage test ports may be provided. (SWG/PE) C37.20.6-1997

simple GCL circuit *See:* simple parallel circuit; simple series circuit.

simple interconnect A connection between two or more component pins consisting only of a net. *Contrast:* extended interconnect. (C/TT) 1149.4-1999

simple manual ground and test device A device with one or two terminal sets with provisions for connecting either terminal set through manually installed cables to the device ground connection system, complete with necessary isolation barriers and suitable interlocking. *Note:* Cables are synonymous with manually installed cords of any type. (SWG/PE) C37.20.6-1997

simple message An MTM-Bus message that consists of no more than a HEADER and, optionally, an ACKNOWLEDGE/PACKET COUNT packet pair. (TT/C) 1149.5-1995

simple network management protocol (SNMP) A protocol used for the management of network nodes and devices, used extensively on internet and other networks. The protocol provides for the communication of status and setup information between a management console and a managed device using values of objects defined in the management information base (MIB) for the managed object. (C/MM) 1284.1-1997

simple parallel circuit A linear, constant-parameter circuit consisting of resistance, inductance, and capacitance in parallel. *Synonym:* simple GCL circuit. *See also:* network analysis. (Std100) 270-1966w

simple path A path in which all vertices except the first and last in the sequence are distinct. (C) 610.5-1990w

simple rectifier A rectifier consisting of one commutating group if single-way or two commutating groups if double-way. *See also:* rectification. (IA) [62]

simple rectifier circuit A rectifier circuit consisting of one commutating group if single-way, or two commutating groups if double-way. *See also:* rectifier circuit element; rectification. (IA) [12]

simple scanning (facsimile) Scanning of only one scanning spot at a time during the scanning process. (ELM) C12.1-1982s

simple series circuit A resistance, inductance, and capacitance in series. *See also:* network analysis. (Std100) 270-1966w

simple sine-wave quantity A physical quantity that is varying with time t as either $A\sin(wt + q_A)$ or $A\cos(\omega + \theta_B)$ where A, ω, $_A$, θ_B are constants. (Simple denotes that A, θ_A, θ_B are constants.). *Notes:* 1. It is immaterial whether the sin or cos form is used, so long as no ambiguity or inconsistency is introduced. 2. A is the amplitude or maximum value, $wt + \theta_A$ (or $\omega t + \theta_B$) the phase, θ_A (or θ_B) the phase angle. However, when no ambiguity may arise, phase angle may be abbreviated phase. 3. In certain special applications, for example, modulation, $\omega t + \theta$ is called the angle (of a sine wave), (not phase angle) in order to clarify particular uses of the word "phase." Another permissible term for $(\omega t + \theta)$ is argument (sine wave). (Std100) 270-1966w

simple sound source A source that radiates sound uniformly in all directions under free-field conditions. (SP) [32]

simple target A target which can be represented by only one major scattering center so that its radar cross section is relatively insensitive to viewing aspect over small angular displacements. *Note:* Examples are a sphere or an object shorter than a half wavelength. Also called a point target. (AES) 686-1997

simple thyristor converter A thyristor converter that consists of one commutating group. (IA/IPC) 444-1973w

simple tone (A) (pure tone) A sound wave, the instantaneous sound pressure of which is a simple sinusoidal function of the time. **(B) (pure tone)** A sound sensation characterized by its singleness of pitch. *Note:* Whether or not a listener hears a tone as simple or complex is dependent upon the ability, experience, and listening attitude. *See also:* complex tone. (SP) [32]

simplex (1) A communication channel that permits information transfer in one direction only. Duplex channels consist of two simplex channels simultaneously operating in opposing directions. (SUB/PE) 999-1992w
(2) (local area networks) A link segment configuration capable of transferring signals in one direction only. (C) 8802-12-1998

simplex channel (data transmission) (simplex operation) A method of operation in which communication between two stations takes place in one direction at a time. *Note:* This includes ordinary transmit-receive operation, press-to-talk operation, voice-operated carrier and other forms of manual or automatic switching from transmit to receive. (PE) 599-1985w

simplex circuit (1) (data transmission) A circuit derived from a pair of wires by using the wires in parallel with ground return. (PE) 599-1985w
(2) In networking, a circuit permitting the transmission of signals in one specific direction only. *See also:* four-wire circuit; leased circuit; channel; two-wire circuit; foreign exchange circuit; dial-up circuit. (C) 610.7-1995

Simplex Fiber Optic Link Segment A single fiber path between two Medium Attachment Units (MAUs) or PHYs, including the terminating connectors, consisting of one or more fibers joined serially with appropriate connection devices, for example, patch cables and wall plates. (C/LM) 802.3-1998

simplex lap winding (rotating machinery) A lap winding in which the number of parallel circuits is equal to the number of poles. (PE) [9]

simplex link segment A path between two Medium Dependent Interfaces (MDIs), including the terminating connectors, consisting of one or more segments of twisted pair cable joined serially with appropriate connection devices, for example, patch cords and wall plates. (C/LM) 802.3-1998

simplex operation (1) (radio transmitters) A method of operation in which communication between two stations takes place in one direction at a time. *Note:* This includes ordinary transmit-receive operation, press-to-talk operation, voice-operated carrier and other forms of manual or automatic switching from transmit to receive. *See also:* radio transmission; telegraphy. (AP/BT/ANT) 145-1983s, 182-1961w
(2) *See also:* one-way-only operation. (C) 610.7-1995

simplex signaling (telephone switching systems) A method of signaling over a pair of conductors by producing current flow in the same direction through both of the conductors. (COM) 312-1977w

simplex supervision (telephone switching systems) A form of supervision employing simplex signaling. (COM) 312-1977w

simplex transmission Transmission in which data is sent in one specific direction only. *Contrast:* duplex transmission; half-duplex transmission. (C) 610.7-1995

simplex wave winding (rotating machinery) A wave winding in which the number of parallel circuits is two, whatever the number of poles. (PE) [9]

simplicity (software) The degree to which a system or component has a design and implementation that is straightforward and easy to understand. *Contrast:* complexity. (C) 610.12-1990

simply connected region (two-dimensional space) A region, such that any closed curve in the region encloses points all of which belong to the region. (Std100) 270-1966w

simply mesh-connected circuit A circuit in which two, and only two, current paths extend from the terminal of entry of each phase conductor, one to the terminal of entry that precedes and the other to the terminal of entry that follows the first terminal in the normal sequence, and from which the amplitude of the voltages to the first terminal is normally the smallest (when the number of phases is greater than three). *See also:* network analysis. (Std100) 270-1966w

SIMSCRIPT A high-order language designed for use in performing general-purpose digital simulations. *Note:* Allows for the description of a system in terms of its "attributes," which are properties associated with "entities," which are groups of "sets." *See also:* ECSS II. (C) 610.13-1993w

SIMULA A general-purpose programming language based on ALGOL 60, with special features designed to aid the description and simulation of active processes. (C) 610.13-1993w

simuland The system being simulated by a simulation. (C) 610.3-1989w

simulate (1) (computers) To represent the functioning of one system by another, for example, to represent one computer by another, to represent a physical system by the execution of a computer program, to represent a biological system by a mathematical model. *See also:* electronic analog computer. (C) [20], [85]
(2) (modeling and simulation) To represent a system by a model that behaves or operates like the system. *See also:* emulate. (C) 610.3-1989w

simulated ESD An ESD that originates from an ESD simulator. (EMC) C63.16-1993

simulated fly ash The entrained ash produced by suspension firing in a small-scale pulverized coal combustor designed and operated with the objective of closely approximating certain selected properties of the fly ash produced in the full-scale steam generator of interest. The combustor should have the capability of providing approximately the same time/ temperature profile for combustion as would occur in a full-scale boiler furnace. This process is applicable particularly when coal from a new source has never been burned in a full-scale boiler. (PE/EDPG) 548-1984w

simulated meter A simulated meter is an assembly consisting of a watthour meter cover, base, and jumper bars constructed to represent the thermal characteristics of a specific class of watthour meter to be used in the testing of a meter socket for temperature rise at continuous ampere rating. (ELM) C12.10-1987

simulated source A radioactive source consisting of one or more long-lived radionuclides that are chosen to simulate the radiations from a short-lived or unavailable radionuclide of interest. (NI) N42.12-1994

simulated sources (ionization chambers) ("dose calibrator" ionization chambers) Simulated sources usually contain long-lived radionuclides, alone or in combination, that are chosen to simulate, in terms of photon or particle emission, a short-lived radionuclide of interest. (NI) N42.13-1986

simulated times Time as represented within a simulation. *Synonym:* virtual time. *See also:* real time; fast time; slow time. (C) 610.3-1989w

simulation (1) (analog computer) The representation of an actual or proposed system by the analogous characteristics of some device easier to construct, modify, or understand. (C) 165-1977w
(2) (A) (modeling and simulation) (software) A model that behaves or operates like a given system when provided a set of controlled inputs. *Synonym:* simulation model. *See also:* emulation. **(B) (modeling and simulation) (software)** The process of developing or using a model as in definition (A). (C) 610.3-1989, 610.12-1990
(3) An instruction method employed by some computer-assisted instruction systems, in which a situation is simulated and the student must respond appropriately. *Contrast:* instructional game. (C) 610.2-1987
(4) (mathematical) The use of a model of mathematical equations generally solved by computers to represent an actual or proposed system. (C) 165-1977w

simulation application The executing software on a host computer that models all or part of the representation of one or more simulation entities. The simulation application represents or simulates real-world phenomena for the purpose of training or experimentation. Examples of simulation applications include manned vehicle simulators, computer generated forces, environment simulators and computer interfaces between a DIS network and real equipment. The simulation application receives and processes information concerning entities created by peer simulation applications through the exchange of DIS PDUs. More than one simulation application may simultaneously execute on a host computer. The simulation application is the application layer protocol entity that implements standard DIS protocol. *Note:* The term *simulation application* is used to avoid confusion between protocol entities and simulation entities. The term *simulation* may also be used in place of simulation application. (DIS/C) 1278.1-1995, 1278.3-1996, 1278.2-1995

simulation clock A counter used to accumulate simulated time. (C) 610.3-1989w

simulation entity An element of the synthetic environment that is created and controlled by a simulation application and affected by the exchange of DIS PDUs. Examples of types of simulated entities are: tank, submarine, carrier, fighter aircraft, missiles, bridges, or other elements of the synthetic environment. It is possible that a simulation application may be controlling more than one simulation entity. *Note:* Simulation entities may also be referred to as *entities*. (DIS/C) 1278.1-1995, 1278.3-1996, 1278.2-1995

simulation environment The operational environment surrounding the simulation entities. This environment includes terrain, atmospheric, and oceanographic information. It is assumed that participants in the same DIS exercise will be using environment information that is adequately correlated for the type of exercise to be performed. (DIS/C) 1278.1-1995

simulation exercise An exercise that consists of one or more interacting simulation applications. Simulations participating in the same simulation exercise share a common identifying number called the exercise identifier. These simulations also utilize correlated representations of the synthetic environment in which they operate. *See also:* exercise. (DIS/C) 1278.1-1995, 1278.2-1995, 1278.3-1996

simulation fidelity (A) The similarity, both physical and functional, between the simulation and that which it simulates. **(B)** A measure of the realism of a simulation. **(C)** The degree to which the representation within a simulation is similar to a real world object, feature, or condition in a measurable or perceivable manner. (DIS/C) 1278.3-1996

simulation game A simulation in which the participants seek to achieve some agreed-upon objective within an established set of rules. For example, a management game, a war game. *Note:* The objective may not be to compete, but to evaluate the participants, increase their knowledge concerning the simulated scenario, or achieve other goals. *Synonym:* gaming simulation. (C) 610.3-1989w

simulation host *See:* host computer.

simulation language An application-oriented programming language used to implement simulations. *See also:* continuous simulation language. (C/C) 610.13-1993w, 610.3-1989w

simulation management A process that provides centralized control of the simulation exercise. Functions of simulation management include: start, restart, maintenance, shutdown of the exercise, and collection and distribution of certain types of data. (DIS/C) 1278.1-1995

simulation model *See:* simulation.

Simulation Program with Integrated Circuit Emphasis (SPICE) A simulation language used widely to design electrical circuits. (C) 610.13-1993w

simulation site Location of one or more simulation hosts connected by a LAN. (DIS/C) 1278.2-1995

simulation time The reference time (e.g., UTC) within a simulation exercise. This time is established ahead of time by the simulation management function and is common to all participants in a particular exercise. (DIS/C) 1278.1-1995

simulation time unit (STU) A fixed unit of time that is utilized during simulation for evaluation of data. (SCC20) 1445-1998

simulator (1) (analog computer) A device used to represent the behavior of a physical system by virtue of its analogous characteristics. In this general sense, all computers are, or can be, simulators. However in a more restricted definition, a simulator is a device used to interact with, or to train, a human operator in the performance of a given task or tasks. (C) 165-1977w

(2) (modeling and simulation) (software) A device, computer program, or system that performs simulation. *See also:* emulator. (C) 610.3-1989w, 610.10-1994w, 610.12-1990
(3) (test, measurement, and diagnostic equipment) A device or program used for test purposes that simulates a desired system or condition providing proper inputs and terminations for the equipment under test. (MIL) [2]

simulator approach speed The rate at which an air discharge ESD simulator approaches the EUT or coupling plane. (EMC) C63.16-1993

simultaneous Pertaining to the occurrence of two or more events at the same instant of time. *Contrast:* concurrent. (C) 610.12-1990

simultaneous access *See:* immediate access.

simultaneous computer A parallel computer that contains a separate processing unit to perform each portion of the computation concurrently, allowing the units to be interconnected in a manner determined by the computation. *Contrast:* sequential computer. *See also:* parallel computer. (C) 610.10-1994w

simultaneous line downtime *See:* partial system downtime.

simultaneous lobing (1) (electronic navigation) A direction-determining technique utilizing the received energy of two concurrent and partially overlapped signal lobes: the relative phase, or the relative power, of the two signals received from a target is a measure of the angular displacement of the target from the equiphase or equisignal direction. Compare with lobe switching. (AP/AES/ANT) 149-1979r, [42], [35]
(2) (radar) A direction-determining technique utilizing the signals of overlapping lobes existing at the same time. *Synonym:* monopulse. (AES/AP/RS/ANT) 686-1982s, 145-1993

simultaneous peripheral output on-line *See:* spool.

simultaneous recursion (software) A situation in which two software modules call each other. (C) 610.12-1990

sin^2 *See:* sin-square.

sin^2 pulse *See:* sine-square pulse.

sin^2 step *See:* sine-square step.

SINAD signal plus noise plus distortion to noise plus distortion ratio expressed in decibels (dB), where the "signal plus noise plus distortion" is the audio power recovered from a modulated radio frequency carrier, and the "noise plus distortion" is the residual audio power present after the audio signal is removed. This ratio is a measure of audio output signal quality for a given receiver audio power output level. (EMC) 377-1980r

sinad ratio (mobile communication) A measure expressed in decibels of the ratio of: the signal plus noise plus distortion to noise plus distortion produced at the output of a receiver that is the result of a modulated-signal input. *See also:* mobile communication system. (VT) 184-1969w

sinad sensitivity (receiver performance) The minimum standard modulated carrier-signal input required to produce a specified sinad ratio at the receiver output. (VT) [37]

sine beats A continuous sinusoid of one frequency, the amplitude of which is modulated by a sinusoid of a lower frequency. (PE/SUB/NP) 693-1997, 344-1987r

sine-cosine potentiometer A function potentiometer with movable contacts attached to a rotating shaft so that the voltages appearing at the contacts are proportional to the sine and cosine of the angle of rotation of the shaft, the angle being measured from a fixed referenced position. *See also:* electronic analog computer. (C) 165-1977w

sine-current coercive force (toroidal magnetic amplifier cores) The instantaneous value of sine-current magnetizing force at which the dynamic hysteresis loop passes through zero induction. (Std100) 106-1972

sine-current differential permeability (toroidal magnetic amplifier cores) The slope of the sides of the dynamic hysteresis loop obtained with a sine-current magnetizing force. (Std100) 106-1972

sine-current magnetizing force (toroidal magnetic amplifier cores) The applied magnetomotive force per unit length for a core symmetrically cyclicly magnetized with sinusoidal current. (Std100) 106-1972

sinen shaping In an amplifier, the pulse shape produced by one CR high-pass filter section (differentiator) followed by n RC low-pass filter sections (integrators), all with different time constants, but following a particular pattern related to the differentiating time constant t. If the input signal is a step function and no other high-pass sections are in the signal path, the pulse shape is unipolar and is described by Ke-3t/t sinn(t/t), where K is a constant, t is time, and t is the time constant of the differentiator. (NPS) 325-1996

sine-square pulse (video signal transmission measurement) One cycle of a sine wave, starting and finishing at its negative peaks with an added constant amplitude component of half the peak-to-peak value, thus raising the negative peaks to zero. *Note:* A sin^2 pulse is obtained by squaring a half-cycle of a sine wave. (BT) 511-1979w

sine-square step (video signal transmission measurement) A step function whose transition from zero to the final value is the sum of a ramp and a negative sinusoid of equal durations, with zero slope at both the zero and the final value of the step. *Notes:* 1. A sin^2 step is obtained by integrating a sin^2 pulse. 2. The attractiveness of both the sin^2 pulse and the sin^2 step lies in the fact that their frequency spectra are limited: that is, they are effectively at zero amplitude beyond a given frequency. For the sin^2, pulse this frequency is a function of the half-amplitude duration (HAD) of the pulse: for the sin^2 step the frequency is a function of the 10% to 90% rise time. (Std100) 270-1964w

sine sweep test A sinusoidal input with continuously varying frequency covering the range of interest. (SWG/PE) C37.100-1992, C37.81-1989r

sine wave A wave that can be expressed as the sine of a linear function of time, or space, or both. (AP/ANT) 145-1983s

sine-wave generator An alternating-current generator whose output voltage waveform contains a single main frequency with low harmonic content of prescribed maximum level. *See also:* asynchronous machine. (PE) [9]

sine-wave response *See:* amplitude response.

sine-wave sweep (1) A sweep generated by a sine function. *See also:* oscillograph. (IM/HFIM) [40]
(2) A sweep generated by a sinusoidal function. *See also:* oscillograph. (IM) 311-1970w

singing An undesired self-sustained oscillation existing in a transmission system or transducer. *Note:* Very-low-frequency oscillation is sometimes called motor-boating. (AP/PE/ANT) 145-1983s, 599-1985w

singing margin (gain margin) The excess of loss over gain around a possible singing path at any frequency, or the minimum value of such excess over a range of frequencies. *Note:* Singing margin is usually expressed in decibels. (SP) 151-1965w

singing point (data transmission) For a circuit which is coupled back to itself, the point at which the gain is just sufficient to make the circuit break into oscillation. (PE) 599-1985w

singing point margin (data transmission) The amount of additional gain (dB) which can be inserted into a loop without sustained oscillations developing. (PE) 599-1985w

singing return loss (1) The return loss of a circuit measured with two separately transmitted signals with a flat spectral distribution between 3 dB frequencies of 260 Hz and 500 Hz (SRL low) and 2200 Hz and 3400 Hz (SRL high). The lower of the two return losses (SRL low or SRL high) will be the best measure of the margin of the circuit against singing. (COM/TA) 743-1995
(2) The minimum of SRL-low (SRL-L) and SRL-high (SRL-H). SRL-low is the frequency weighted average of the return losses in a low-frequency band (with 3 dB bandwidth from 260 Hz to 500 Hz). SRL-high is the frequency weighted average of the return loss in a high-frequency band (with 3 dB

bandwidth from 2200 Hz to 3400 Hz). The weightings are given in IEEE Std 743-1984. (COM/TA) 1007-1991r

single-address Pertaining to an instruction that has one address part. In a typical single-address instruction the address may specify either the location of an operand to be taken from storage, the destination of a previously prepared result, the location of the next instruction to be interpreted, or an immediate address operand. *See also:* one-address. (C) [20]

single-address instruction *See:* one-address instruction.

single-anode tank (single-anode tube) An electron tube having a single main anode. *Note:* This term is used chiefly for pool-cathode tubes. (ED) [45]

single aperture seal (electric penetration assemblies) (nuclear power generating station) A single seal between the containment aperture and the electric penetration assembly. (PE/NP) 317-1983r

single automatic operation (elevators) Automatic operation by means of one button in the car for each landing level served and one button at each landing so arranged that if any car or landing button has been actuated, the actuation of any other car or landing operating button will have no effect on the operation of the car until the response to the first button has been completed. *See also:* control. (EEC/PE) [119]

single-blind study A study in which the subject is unaware of his or her role as experimental or control subject in an experiment. (T&D/PE) 539-1990

single-break switch One that opens each conductor of a circuit at one point only. (SWG/PE) C37.100-1992

single-buffered DAC (hybrid computer linkage components) A digital-to-analog converter (DAC) or a digital-to-analog multiplier (DAM) with one dynamic register, which also serves as the holding register for the digital value. *Synonym:* single-buffered DAM. (C) 166-1977w

single-buffered DAM *See:* single-buffered DAC.

single cable system A type of broadband system that uses a single cable for the transmission of the inbound and outbound paths. (LM/C) 802.7-1989r

single capacitance (as applied to interrupter switches) A capacitance is defined to be a single capacitance when the crest of its inrush current does not exceed the switch inrush current capability for single capacitance. (SWG/PE) C37.100-1992

singlecast A mode of operation in which the M-module transmits data to a single S-module. Also, a message transmitted in this mode. (TT/C) 1149.5-1995

single-circuit system (protective signaling) A system of protective wiring that employs only the nongrounded side of the battery circuit, and consequently depends primarily on an open circuit in the wiring to initiate an alarm. *See also:* protective signaling. (EEC/PE) [119]

single-cycle instruction An instruction that is completely executed in one machine cycle. (C) 610.10-1994w

single-degree-freedom gyro A gyro in which the rotor is free to precess (relative to the case) about only the axis orthogonal to the rotor spin axis. *See also:* navigation. (AES/RS) 686-1982s, [42]

single electric conductor seal (1) (electric penetration assemblies) (nuclear power generating station) A mechanical assembly arranged in such a way that there is a single pressure barrier seal between the inside and the outside of the containment structure along the axis of the electric conductor. (PE/NP) 317-1983r

(2) (nuclear power generating station) A mechanical assembly providing a single pressure barrier between the electric conductors and the electric penetration. (IM) [76]

single electron distribution (scintillation counting) The pulse-height distribution associated with single electrons originating at the photocathode. (NPS) 398-1972r

single-electron PHR *See:* single-electron pulse-height resolution.

single-electron pulse-height resolution (scintillation counting) The fractional FWHM (full width at half maximum) of the single-electron distribution of a photomultiplier. (NPS) 398-1972r

single-electron rise time (scintillation counting) The anode-pulse rise time associated with single electrons originating at the photocathode. (NPS) 398-1972r

single-electron transit-time spread (scintillation counting) Transit-time spread measured with single-electron events. (NPS) 398-1972r

single-element fuse A fuse having a current-responsive element comprising one or more parts with a single fusing characteristic. (SWG/PE) C37.100-1992

single-element relay An alternating-current relay having a set of coils energized by a single circuit. (EEC/PE) [119]

single-end control (single-station control) A control system in which provision is made for operating a vehicle from one end or one location only. *See also:* multiple-unit control. (EEC/PE) [119]

single-ended amplifier An amplifier in which each stage normally employs only one active element (tube, transistor, etc.) or, if more than one active element is used, in which they are connected in parallel so that operation is asymmetric with respect to ground. *See also:* amplifier. (AP/ANT) 145-1983s

single-ended push-pull amplifier circuit (electroacoustics) An amplifier circuit having two transmission paths designed to operate in a complementary manner and connected so as to provide a single unbalanced output without the use of an output transformer. *See also:* amplifier. (SP) 151-1965w

single-ended recorder A non-differential waveform recorder, i.e., one that does not subtract the signals at two input terminals. (IM/WM&A) 1057-1994w

single event effect (SEE) Any measurable or observable change in state or performance of a microelectronic device, component, subsystem, or system (digital or analog) resulting from the passing or traversal of a single particle. (C/BA) 1156.4-1997

single-event upset (SEU) The loss of data that is caused by passage of a single ionizing particle through an array. (ED) 1005-1998

single-faced tape Fabric tape finished on one side with rubber or synthetic compound. (T&D/PE) [10]

single feeder A feeder that forms the only connection between two points along the route considered. (T&D/PE) [10]

single-font character recognition Character recognition of one character font. *Contrast:* omni-font character recognition. (C) 610.2-1987

single frequency distortion The production of frequency components other than the applied single frequency; these components may not necessarily be harmonic components. (COM/TA) 1007-1991r

single-frequency pulsing (telephone switching systems) Dial pulsing using the presence or absence of a single frequency to represent break or make intervals, respectively or vice versa. (COM) 312-1977w

single-frequency signaling (telephone switching systems) A method for conveying dial pulse and supervisory signals from one end of a trunk to the other using the presence or absence of a single specified frequency. (COM) 312-1977w

single-frequency signal-to-noise ratio (sound recording and reproducing system) The ratio of the single-frequency signal power output to the noise power in the entire pass band. *See also:* noise. 191-1953w

single-frequency simplex operation (radio communication) The operation of a two-way radio-communication circuit on the same assigned radio-frequency channel, which necessitates that intelligence can be transmitted in only one direction at a time. *See also:* channel spacing. (VT) [37]

single hoistway (elevators) A hoistway for a single elevator or dumbwaiter. *See also:* hoistway. (PE/EEC) [119]

single instruction, multiple data Pertaining to a computer architecture in which all processors receive instructions from a common source but receive data from multiple, disjoint sources. *Contrast:* single instruction, single data; multiple instruction, multiple data. (C) 610.10-1994w

single instruction, single data (SISD) Pertaining to a computer architecture in which the processors receive instructions from a common source and receive data from a common source. *See also:* multiple instruction, single data; single instruction, multiple data. (C) 610.10-1994w

single-layer winding (rotating machinery) A winding in which there is only one actual coil side in the depth of the slot. (Also known as one-coil-side-per-slot winding). *See also:* asynchronous machine. (PE) [9]

single-level encoding (software) A microprogramming technique in which different microoperations are encoded as different values in the same field of a microinstruction. *Contrast:* two-level encoding. (C) 610.12-1990

single-level interrupt A signal causing transfer of processor control to a preassigned memory location at which an interrupt processing routine starts. *Note:* The program must poll all possible sources of interrupt to determine which one requires service. (C) 610.10-1994w

single-level network subject A network subject that causes information to flow through the network at a single security level. *Contrast:* multilevel network subject. (C) 610.7-1995

single-level security The capability of protecting only one security level during processing. (C/BA) 896.3-1993w

single-line diagram *See:* one-line diagram.

single-mode bandwidth The range of frequencies between the cutoff of the dominant mode of propagation and that of the lowest higher order guided mode. (MTT) 1004-1987w

single-mode fiber (interferometric fiber optic gyro) An optical fiber waveguide in which only the lowest-order bound mode (which may consist of a pair of orthogonally polarized fields) can propagate at the wavelength of interest. (AES/GYAC) 528-1994

single-mode optical fiber An optical fiber in which a single concentrated light beam travels straight down the center of the fiber, maximizing its information-carrying capacity. *See also:* multi-mode optical fiber. (C) 610.7-1995

single mode optical waveguide (fiber optics) An optical waveguide in which only the lowest order bound mode (which may consist of a pair of orthogonally polarized fields) can propagate at the wavelength of interest. In step index guides, this occurs when the normalized frequency, V, is less than 2.405. For power-law profiles, single mode operation occurs for normalized frequency, V, less than approximately

$$2.405\sqrt{(g + 2)/g}$$

where g is the profile parameter. *Note:* In practice, the orthogonal polarizations may not be associated with degenerate modes. *Synonym:* monomode optical waveguide. *See also:* normalized frequency; multimode optical waveguide; step index optical waveguide; profile parameter; power-law index profile; mode; bound mode. (Std100) 812-1984w

single-mode surface acoustic wave oscillator A surface acoustic wave oscillator in which there is only one frequency that satisfies the oscillation conditions of having positive excess gain and total phase shift of $N2\pi$ (where N is a positive integer). (UFFC) 1037-1992w

single-office exchange (telephone switching systems) A telecommunications exchange served by one central office. (COM) 312-1977w

single-operand instruction *See:* one-address instruction.

single-operator arc welder An arc-welding power supply designed to deliver current to only one welding arc. (EEC/AWM) [91]

single outage event An outage event involving only one component or one unit. (PE/PSE) 859-1987w

single-phase ac fields Fields whose space components are in phase. These fields will be produced by single-phase power lines. The field at any point can be described in terms of a single direction in space and its time-varying magnitude. *Note:* Such fields are sometimes referred to as being linearly polarized. (T&D/PE) 644-1994

single-phase circuit (1) (power and distribution transformers) An ac circuit consisting of two or three intentionally interrelated conductors that enter (or leave) a delimited region at two or three terminals of entry. If the circuit consists of two conductors, it is intended to be so energized that, in the steady state, the voltage between the two terminals of entry is an alternating voltage. If the circuit consists of three conductors, it is intended to be so energized that, in steady state, the alternating voltages between any two terminals of entry have the same period and are in phase or in phase opposition. (PE/TR) C57.12.80-1978r

(2) A circuit energized by a single alternating electromotive force. *Note:* A single-phase circuit is usually supplied through two conductors. The currents in these two conductors, counted outward from the source, differ in phase by 180° or a half cycle. (IA/MT) 45-1998

single-phase electric locomotive An electric locomotive that collects propulsion power from a single phase of an alternating-current distribution system. *See also:* electric locomotive. (EEC/PE) [119]

single-phase enclosure A metallic enclosure containing the buses and/or devices associated with one phase of a multiple-phase system. *Note:* A single gas-insulated substation need not be composed of all single-phase or all three-phase enclosures. A common compromise is to use buses in three-phase enclosures mated with equipment in single-phase enclosures. (SUB/PE) C37.122.1-1993

single-phase machine A machine that generates or utilizes single-phase alternating-current power. *See also:* asynchronous machine. (PE) [9]

single-phase motor (rotating machinery) A machine that converts single-phase alternating-current electric power into mechanical power, or that provides mechanical force or torque. (PE) [9]

single-phase symmetrical set (A) (polyphase voltages)A symmetrical set of polyphase voltages in which the angular phase difference between successive members of the set is ν radians or odd multiples thereof. The equations of symmetrical set (polyphase voltages) represent a single-phase symmetrical set of polyphase voltages if k/m is 1/2 or an odd multiple thereof. (The symmetrical set of voltages represented by the equations of symmetrical set (polyphase voltages) may be said to have single-phase symmetry if k/m is an odd (positive or negative) multiple of 1/2). *Notes:* 1. A set of polyphase voltages may have single-phase symmetry only if m, the number of members of the set, is an even number. 2. This definition may be applied to a two-phase four-wire or five-wire circuit if m is considered to be 4 instead of 2. It is not applicable to a two- phase three-wire circuit. *See also:* network analysis. **(B) (polyphase currents)** This definition is obtained from the corresponding definitions for voltage by substituting the word current for voltage, and the symbol I for E and b for a wherever they appear in the equations of symmetrical set (polyphase voltages). The subscripts are unaltered. *See also:* network analysis. (Std100) 270-1966

single-phase synchronous generator A generator that produces a single alternating electromotive force at its terminals. It delivers electric power that pulsates at double frequency. (IA/MT) 45-1998

single-phase three-wire circuit A single-phase circuit consisting of three conductors, one of which is identified as the neutral conductor. *See also:* network analysis. (Std100) 270-1966w

single-phase two-wire circuit A single-phase circuit consisting of only two conductors. *See also:* network analysis. (Std100) 270-1966w

single-phasing (rotating machinery) An abnormal operation of a polyphase machine when its supply is effectively single-phase. *See also:* asynchronous machine. (PE) [9]

single-pointer form demand register (metering) An indicating demand register from which the demand is obtained by reading the position of a pointer relative to the markings on a scale. The single pointer is resettable to zero. (ELM) C12.1-1982s

single point failure analysis A reliability analysis that identifies single components or subsystems whose failure results in system failure. (PE/NP) 933-1999

single point of failure With respect to a system, a failure that would result in the inability of that system to perform its intended function. (C/BA) 896.9-1994w, 896.3-1993w

single-polarity pulse A pulse in which the sense of the departure from normal is in one direction only. *See also:* pulse. (EEC/PE) [119]

single-pole relay A relay in which all contacts connect, in one position or another, to a common contact. (EEC/REE) [87]

single-pole switching The practice of tripping and reclosing one pole of a multiple circuit breaker without changing the state of the remaining poles, with tripping being initiated by protective relays that respond selectively to the faulted phase. *Notes:* 1. Circuit breakers used for single-pole switching must inherently be capable of individual pole opening. 2. In most single-pole switching schemes, it is the practice to trip all poles for any fault involving more than one phase. (SWG/PE) C37.100-1992

single precision (data management) (mathematics of computing) Pertaining to the use of a single computer word to represent a number. *Note:* Single precision is implied in number representation and in computer arithmetic unless multiple precision is specified. *Contrast:* triple precision; double precision; multiple precision. (C) 610.5-1990w, 1084-1986w

single-pressure-zone potheads A pressure-type pothead intended to operate with one pressure zone. *See also:* multi-pressure-zone pothead; pressure-type pothead. (PE) 48-1975s

single-pressure zone termination *See:* pressure-type termination.

single processor architecture A computer architecture that uses a single processor. *Contrast:* multiprocessor architecture. (C) 610.10-1994w

single pulse (thyristor) A gate signal applied at the commencement of the conducting interval in the form of a single pulse of predetermined duration, amplitude, and frequency. (IA/IPC) 428-1981w

single quote The character "'", also known as *apostrophe*. (C/PA) 9945-2-1993

single ring station A station that offers one attachment to the network on the primary ring. (LM/C) 802.5c-1991r

single-scan probability of detection *See:* blip-scan ratio.

single scatter approximation An approximation used in the calculation of wave scattering by a surface, volume or a collection of particles. In this approximation, the field that excites the surface element or particle, the exciting field, is assumed to be the same field that would have been present in the absence of all other surface or volume elements or particles (i.e., the exciting field is equal to the incident field). *See also:* Born approximation. (AP/PROP) 211-1997

single service One service only supplying a consumer. *Note:* Either or both lighting and power load may be connected to the service. *See also:* service. (T&D/PE) [10]

single-sheet feed A mechanism enabling a printer to print on individual sheets of paper. *Note:* Usually uses friction feed. *See also:* cut-sheet feed. (C) 610.10-1994w

single-shot blasting unit A unit designed for firing only one explosive charge at a time. *See also:* blasting unit. (EEC/PE) [119]

single-shot blocking oscillator A blocking oscillator modified to operate as a single-shot trigger circuit. *See also:* trigger circuit. (EEC/PE) [119]

single-shot multivibrator (single-trip multivibrator) A multivibrator modified to operate as a single-shot trigger circuit. *See also:* trigger circuit. (EEC/PE) [119]

single-shot trigger circuit (single-trip trigger circuit) A trigger circuit in which a triggering pulse intiates one complete cycle of conditions ending with a stable condition. *See also:* trigger circuit. (PE/EEC) [119]

single-sideband modulation (data transmission) Modulation whereby the spectrum of the modulating function is translated in frequency by a specified amount either with or without inversion. *See also:* modulation. (AP/BT/ANT) 145-1983s, 511-1979w

single-sideband transmission (data transmission) The method of operation in which one sideband is transmitted and the other sideband is suppressed. The carrier wave may be either transmitted or suppressed. (PE) 599-1985w

single-sideband transmitter A transmitter in which one sideband is transmitted and the other is effectively eliminated. (AP/ANT) 145-1983s

single-sided board A board with components on one side only. Often single-sided boards are used with right-angle connectors. (C/BA) 14536-1995

single-sided disk A floppy disk on which information can be stored reliably on only one side. (C) 610.10-1994w

single-speed floating control system (automatic control) A floating control system in which the manipulated variable changes at a fixed rate, increasing or decreasing depending on the sign of the actuating signal. *Note:* A neutral zone of values of the actuating signal, in which no action occurs, may be used. (PE/EDPG) [3]

single-station control *See:* single-end control.

single-stator electromechanical watthour meter A single-stator electromechanical watthour meter (single-phase electromechanical watthour meter or single element electromechanical watthour meter) is an electromechanical watthour meter containing only one stator. (ELM) C12.10-1987

single step (computers) Pertaining to a method of operating a computer in which each step is performed in response to a single manual operation. (C) [20], [85]

single-step execution *See:* single-step operation.

single-step operation A debugging technique in which a single computer instruction, or part of an instruction, is executed in response to an external signal. *Synonyms:* step-by-step operation; single-step execution. (C) 610.12-1990

single-stress aging A form of accelerated aging in which the level of one aging stress is intensified, while the others are minimized or held constant. (DEI/RE) 775-1993w

single-stroke bell An electric bell that produces a single stroke on its gong each time its mechanism is actuated. *See also:* protective signaling. (EEC/PE) [119]

single sweep (spectrum analyzer) Operating mode for a triggered sweep instrument in which the sweep must be reset for each operation, thus preventing unwanted multiple displays. This mode is useful for trace photography. In the interval after the sweep is reset and before it is triggered, it is said to be an armed sweep. (IM) 748-1979w

single-talk (ST) One talker speaking while the opposite transmission direction is silent. (COM/TA) 1329-1999

single throw (switching device) A qualifying term used to indicate that the device has an open and a closed circuit position only. (SWG/PE) C37.100-1992

single-throw relay A relay in which each contact form included is a single contact pair. (EEC/REE) [87]

single thyristor converter unit A thyristor converter unit connected to a dc circuit such that the direct current supplied by the converter is flowing in only one direction. The single converter section is referred to as a forward converter section. *Note:* When used without a reversing switch a single converter can be used in a reversible power sense only in those cases where single-way thyristor connections or symmetrical double-way thyristor connections are used and where the dc circuit can change from accepting energy to giving up energy without the need for current reversal, for example, a heavily inductive load. When used with a reversing switch, a single converter can be used in a reversible power sense in all cases where single-way thyristor connections or uniform double-way thyristor connections are used. (IA/IPC) 444-1973w

single-tone keying (modulation systems) That form of keying in which the modulating function causes the carrier to be modulated by a single tone for one condition, which may be either a mark or a space, the carrier being unmodulated for the other condition. *See also:* telegraphy. (COM/AP/ANT) [49], 145-1983s, 270-1964w

single-track (standard track) (electroacoustics) A variable-density or variable-area sound track in which both positive and negative halves of the signal are linearly recorded. *See also:* phonograph pickup. (SP) [32]

single-transfer read cycle A cycle that is used to transfer 1, 2, 3, or 4 bytes from the responding slave to the active master, and possibly to participating slaves. The cycle begins when the active master broadcasts the addressing information on the address/data lines. Each slave checks the address to see if it is to respond to the cycle. If so, it acknowledges the address and retrieves the data from its internal storage. When the master releases the address/data lines, the responding slave places its data on them and acknowledges the transfer. The master as well as participating slaves capture the data. After all selected slaves signal their agreement, the master terminates the cycle. (C/MM) 1096-1988w

single-transfer write cycle A cycle that is used to transfer 1, 2, 3, or 4 bytes from the active master to the selected slaves(s). The cycle begins when the master broadcasts the addressing information on the address/data lines. Each slave checks the address to see if it is to participate in the cycle. The responding slave acknowledges the address broadcast. The master then switches the address/ data lines to carry data, and places its data on the bus. The selected slaves can then store the data. The responding slave acknowledges the transfer. After all selected slaves signal their agreement, the master terminates the cycle. (C/MM) 1096-1988w

single-transistor cell A memory cell consisting of one transistor. (ED) 641-1987w

single-trip multivibrator *See:* single-shot multivibrator.

single-trip trigger circuit *See:* single-shot trigger circuit.

single-tuned amplifier An amplifier characterized by a resonance at a single frequency as indicated by the s-plane representation of its gain, which is $A(s) = A_0 s/(s^2 + \omega_0 \xi s + \omega_0^2)$. It rejects low and high frequencies while having a peak gain at a center frequency $s = j\omega_0$. *See also:* amplifier. (CAS) [13]

single-tuned circuit A circuit that may be represented by a single inductance and a single capacitance, together with associated resistances. (EEC/PE) [119]

single-valued *See also:* function. (C/SE) 1320.2-1998

single-valued function A function u is single valued when to every value of x (or set of values of $x_1, x_2, \ldots, x_n$) there corresponds one and only one value of u. Thus, $u = ax$ is single valued if a is an arbitrary constant. (Std100) 270-1966w

single-valued property A property with a single-valued mapping. *Contrast:* multivalued property. (C/SE) 1320.2-1998

single-valve unit A single structure comprising only one valve. (SUB/PE) 857-1996

single-way rectifier A rectifier unit that makes use of a single-way rectifier circuit. (IA) [12]

single-way rectifier circuit A rectifier circuit in which the current between each terminal of the alternating voltage circuit and the rectifier circuit element or elements conductively connected to it flows in only one direction. (IA) [12]

single-way thyristor converter A thyristor converter in which the current between each terminal of the alternating-voltage circuit and the thyristor converter circuit element or elements conductively connected to it flows only in one direction. (IA/IPC) 444-1973w

single-winding multispeed motor A type of multispeed motor having a single winding capable of reconnection in two or more pole groupings. *See also:* asynchronous machine. (EEC/PE) [119]

single-wire line (waveguide) A surface-wave transmission line consisting of a single conductor so treated as to confine the propagated energy to the neighborhood of the wire. The treatment may consist of a coating of dielectric. (MTT) 146-1980w

singly linked list *See:* linked list.

singular point Synonymous with equilibrium point. *See also:* control system. (CS/IM) [120]

sink (1) (oscillators) The region of a Rieke diagram where the rate of change of frequency with respect to phase of the reflection coefficient is maximum. Operation in this region may lead to unsatisfactory performance by reason of cessation or instability of oscillations. *See also:* oscillatory circuit. (ED) 161-1971w

(2) A consumer of normal characters at a link interface. *See also:* normal character. (BA/C) 1355-1995

(3) The end of a delay arc; that is, the destination of the logic signal. For arcs across cell instances, the sink is the driver pin; for arcs across interconnect, the sink is the receiver pin. (C/DA) 1481-1999

sink node (network analysis) A node having only incoming branches. (CAS) 155-1960w

sin-square (linear waveform distortion) A step function whose transition from zero to the final value is the sum of a ramp and a negative sinusoid of equal durations, with zero slope at both the zero and the final value of the step. *Notes:* 1. A $\sin^2$ step is obtained by integrating a $\sin^2$ pulse. 2. The attractiveness of both the $\sin^2$ pulse and the $\sin^2$ step lies in the fact that their frequency spectra are limited; that is, they are effectively at zero amplitude beyond a given frequency. For the $\sin^2$ pulse this frequency is a function of the half-amplitude duration (HAD) of the pulse; for the $\sin^2$ step the frequency is a function of the 10% to 90% rise time. (BT) 511-1979w

sin-square pulse *See:* sine-square pulse.

S interface The interface provided at the S reference point for ISDN user-to-network interface. (LM/C/COM) 8802-9-1996

sinusoidal electromagnetic wave (radio-wave propagation) In a homogeneous medium, a wave whose electric field vector is proportional to the sine (or cosine) of an angle that is a linear function of time, or of a distance, or of both. (AP) 211-1977s

sinusoidal field A field in which the field quantities vary as a sinusoidal function of an independent variable, such as space or time. (Std100) 270-1966w

sinusoidal function A function of the form $A\sin(xa)$. A is the amplitude, x is the independent variable, and a the phase angle. Note that $\cos(x)$ may be expressed as sin1x (6.2)0. *See also:* simple sine-wave quantity. (Std100) 270-1966w

sinusoidal response (sine-force) The forced response due to a sinusoidal stimulus. *Note:* A set of steady-state sinusoidal responses for sinusoidal inputs at different frequencies is called the frequency-response characteristic. *See also:* feedback control system. (IM) [120]

siphon recorder A telegraph recorder comprising a sensitive moving-coil galvanometer with a siphon pen that is directed by the moving coil across a traveling strip of paper. *See also:* telegraphy. (EEC/PE) [119]

SISD *See:* single instruction, single data.

sister *See:* sibling node.

site attenuation (1) The ratio of the power input of a matched, balanced, lossless, tuned dipole radiator to that at the output

of a similarly matched, balanced, lossless, tuned dipole receiving antenna for specified polarization, separation, and heights above a flat reflecting surface. It is a measure of the transmission path loss between two antennas. (EMC) C63.4-1991, 1128-1998
(2) The ratio of the power input to a matched balanced lossless tuned dipole radiator to that at the output of a similarly balanced matched lossless tuned dipole receiving antenna for specified polarization, separation, and heights above a flat reflecting surface. (EMC) C63.5-1988
(3) The ratio of the power input to a matched, balanced, lossless, wideband dipole radiator to that at the output of a similarly balanced, matched, lossless, wideband dipole receiving antenna for specified polarization, separation, and heights above a flat reflecting surface. (EMC) 1128-1998

site-attenuation deviation (SAD) The difference, in dB, of the site attenuation and the free-space distance attenuation between the transmit and receive antennas in an open-area test site (OATS), absorber-lined open-area test site (ATS), semi-anechoic chamber, or anechoic chamber. *See also:* absorber-lined open-area test site; open-area test site. (EMC) 1128-1998

site error (navigation) (navigation aids) Error due to the distortion in the electromagnetic field by objects in the vicinity of the navigational equipment. (AES/GCS) 172-1983w

site marker *See:* transit.

site, pull *See:* pull site.

site, tension *See:* tension site.

SI unit of luminance (illuminating engineering) Candela per square meter (cd/m^2); also, lumen per steradian × square meter (lm/(sr/m^2) also, lumen per steradian × square meter (lm/(sr/m^2)). This also is called the nit. (EEC/IE) [126]

SI units Units belonging to the International System of Units. (SCC20) 771-1998

six-bit byte *See:* sextet.

six bolt *See:* conductor grip.

six-phase circuit (power and distribution transformers) A combination of circuits energized by alternating electromotive forces which differ in phase by one-sixth of a cycle, that is, 60°. *Note:* In practice, the phases may vary several degrees from the specified angle. (PE/TR) C57.12.80-1978r

64-bit supportive Uses 64-bit addresses when accessing System Memory. (C/MM) 1212.1-1993

63% response time *See:* apparent time constant.

size distribution (fly ash resistivity) Size distribution of particulate matter is the cumulative frequency of particle diameter, generally expressed on a mass basis. It describes the probability that a particle diameter *x* takes a value equal to or less than probability *P*. Size distribution rather than mean particle size shall be reported. (PE/EDPG) 548-1984w

size metric A value used to estimate properties of interconnect wholly contained in a region. The metric may be freely chosen (for example, square microns or gate sites), but it needs to be consistent between the cells and the wireload models. *See also:* wireload model. (C/DA) 1481-1999

size threshold (illuminating engineering) The minimum perceptible size of an object. It also is defined as the size which can be detected some specific fraction of the times it is presented to an observer, usually 50%. It usually is measured in minutes of arc. (EEC/IE) [126]

sizing (software) The process of estimating the amount of computer storage or the number of source lines required for a software system or component. (C) 610.12-1990

skate machine A mechanism, electrically controlled, for placing on, or removing from, the rails a skate that, if allowed to engage with the wheels of a car, provides continuous braking until the car is stopped and that may be electrically or pneumatically operated. (EEC/PE) [119]

skates Devices used in the climbing of flanged structures. (T&D/PE) 1307-1996

skating force *See:* side thrust.

skeleton frame (rotating machinery) A stator frame consisting of a simple structure that clamps the core but does not enclose it. (PE) [9]

skew (1) (measuring the performance of tone address signaling systems) In a two-tone signal, the time interval from the start of the higher-frequency tone present condition. Skew is negative if the higher-frequency tone starts before the lower-frequency tone. (COM/TA) 752-1986w
(2) (facsimile) The deviation of the received frame from rectangularity due to asynchronism between scanner and recorder. Skew is expressed numerically as the tangent of the angle of the deviation. (COM) 168-1956w
(3) (magnetic storage) The angular displacement of an individual printed character, group of characters, or other data, from the intended or ideal placement. (MIL/C) [2], [20], [85]
(4) (MULTIBUS) The time difference between signals because of timing differences for logic and backplane delays. (C/MM) 1296-1987s
(5) The difference between the propagation delays of two or more signals on any bus line. (BA/C) 10857-1994, 896.4-1993w
(6) The angular or longitudinal deviation of a tape or disk track from a specified reference. (C) 610.10-1994w
(7) The difference in time that is unintentionally introduced between changing signal levels (incident edges) that occur on parallel signal lines. This difference results in an uncertain position with respect to time among parallel signals. (C/MM) 1596.3-1996
(8) The timing ambiguity associated with the occurrence of an automatic test equipment (ATE) Input/Output (I/O) event that is due to the physical limitations of the ATE digital driver and detector electronics. (SCC20) 1445-1998

skew between pairs The difference in arrival times of two initially coincident signals propagated over two different pairs, as measured at the receiving end of the cable. Total skew includes contributions from transmitter circuits as well as the cable. (C/LM) 802.3-1998

skewed slot (rotating machinery) A slot of a rotor or stator of an electric machine, placed at an angle to the shaft so that the angular location of the slot at one end of the core is displaced from that at the other end. Slots are commonly skewed in many types of machines to provide more uniform torque, less noise, and better voltage waveform. *See also:* rotor; stator. (PE) [9]

skewing factor S0.1 or S.02 (of a peak) The ratio (FW.1M)/(1.823 FWHM) or (FW.02M)/(2.376 FWHM). For an ideal Gaussian shape, the factors are 1.000. If the peak is wider at FW.1M or FW.02M than it should be relative to FWHM, the factors are >1.000. (For a perfect Gaussian peak, FW.1M/FWHM = [(ln 0.1)/(ln 0.5)]1/2 = 1.823 and FW.02M/FWHM = [(ln 0.02)/(ln 0.5)]1/2) = 2.376). (NPS) 325-1996

skew ray (fiber optics) A ray that does not intersect the optical axis of a system (in contrast with a meridional ray). *See also:* optical axis; meridional ray; geometric optics; axial ray; hybrid mode; paraxial ray. (Std100) 812-1984w

skew time (FASTBUS acquisition and control) The minimum time that the assertion of a FASTBUS timing signal must be delayed after the assertion of information and/or control signals to allow for differences in propagation time of signals on a FASTBUS segment. (NID) 960-1993

skew timing check A timing check that specifies the maximum time between two signal transitions. This timing check is frequently applied to dual-clock flip-flops to specify the maximum separation of the active edges of the two phases of the clock. (C/DA) 1481-1999

skiatron† **(A)** A dark-trace storage-type cathode-ray tube. **(B)** A display employing an optical system with a dark-trace tube. *See also:* dark-trace tube. (AES/RS) 686-1990

† Obsolete.

skidder *See:* wheel tractor.

skid wire (pipe-type cable) (power distribution, underground cables) Wire or wires, usually D shaped, applied open spiral with curved side outward with a suitable spacing between turns over the outside surface of the cable. Its purpose is to facilitate cable pulling and to provide mechanical protection during installation. (PE) [4]

skill A cognitive and/or physical control process so highly practiced as to require little or no conscious supervision. (PE/NP) 1082-1997

skim tape Filled tape coated on one or both sides with a thin film of uncured rubber or synthetic compound to produce a coating suitable for vulcanization. (T&D/PE) [10]

skin depth (1) (waveguide) Of a conducting material, at a given frequency, the depth at which the surface current density is reduced by one neper. (MTT) 146-1980w
(2) *See also:* penetration depth. (AP/PROP) 211-1997

skin effect (induction heating) Tendency of an alternating current to concentrate in the areas of lowest impedance. (IA) 54-1955w

skin effect heating (electrical heating systems) An electric heating system where a conductor inside a ferromagnetic material generates heat via I^2R losses in the conductor and ferromagnetic material. (IA/PC) 844-1991

skip (1) (computers) To ignore one or more instructions in a sequence of instructions. (C/C) [20], [85]
(2) In Physical Design Exchange Format (PDEF), skip is the spacing between the ordered cell, spare_cell, and/or cluster instances in rows and/or columns of a datapath. (C/DA) 1481-1999

skip distance (1) (data transmission) (navigation aids) The minimum separation for which radio waves of a specified frequency can be transmitted at a specified time (interval) between two points on the earth by reflection from the regular ionized layers of the ionosphere. (PE/AES/GCS) 599-1985w, 172-1983w
(2) For a given frequency, the minimum distance at which the sky wave is returned to the Earth. *Note:* Given frequency is the maximum usable frequency for the skip distance. (AP/PROP) 211-1997

skip distance focusing Ionospheric focusing observed in the vicinity of the skip distance. (AP/PROP) 211-1997

skip zone An area of the surface of the Earth surrounding a transmission point bounded by the skip distance in all directions. (AP/PROP) 211-1997

skiving (A) The process of assembling a fitting to a hose. **(B)** The process of trimming outside of a hose to fit the inside dimensions of a fitting. (T&D/PE) 957-1995

skookum *See:* snatch block.

sky compass (navigation aids) A type of astro compass, designed for use in the arctic during long periods of twilight. (AES/GCS) 172-1983w

sky condition *See:* cloudy sky; overcast sky; partly cloudy sky; clear sky.

sky factor (illuminating engineering) The ratio of the illuminance on a horizontal plane at a given point inside a building due to the light received directly from the sky, to the illuminance due to an unobstructed hemisphere of sky of uniform luminance equal to that of the visible sky. (EEC/IE) [126]

sky light (illuminating engineering) Visible radiation from the sun redirected by the atmosphere. (EEC/IE) [126]

sky noise (communication satellite) Noise contribution of the sky (often the galaxies). *See also:* background noise. (COM) [25]

sky radiometric temperature The observed brightness temperature of the sky, caused by emissions from the Earth's atmosphere as well as cosmic and galactic radiation. (AP/PROP) 211-1997

sky wave A radio wave propagated obliquely toward, and returned from, the ionosphere. *Note:* This term has sometimes been called an ionospheric wave but that term is intended to connote internal waves in ionospheric plasmas. (AP/PE/T&D/PROP) 211-1997, 1260-1996

sky-wave contamination (navigation aids) Degradation of the received ground-wave signal, or of the desired sky-wave signal, by the presence of delayed ionospheric-wave components of the same transmitted signal. (AES/GCS) 172-1983w

sky-wave correction (navigation aids) (navigation) A correction for sky-wave propagation errors applied to measured position data; the amount of the correction is established on the basis of an assumed ionosphere height. (AES/GCS) 172-1983w

sky-wave station-error (sky-wave synchronized loran) (navigation aid terms) The error of station synchronization due to the effect of variations of the ionosphere on the time of transmission of the synchronizing signal from one station to the other. (AES/GCS) 172-1983w

sky wire *See:* overhead ground wire.

sky wire-coupling protector *See:* static wire-coupling protector.

SL *See:* special link.

slabbing or arcwall machine A power-driven mobile-cutting machine that is a single-purpose cutter in that it cuts only a horizontal kerf at variable heights. (EEC/PE) [119]

slab interferometry (fiber optics) The method for measuring the index profile of an optical fiber by preparing a thin sample that has its faces perpendicular to the axis of the fiber, and measuring its index profile by interferometry. *Synonym:* axial slab interferometry. *See also:* interferometer. (Std100) 812-1984w

slab line (waveguide) A uniform transmission line consisting of a round conductor between two extended parallel conducting surfaces, so that the propagating wave is essentially confined between the surfaces. (MTT) 146-1980w

slack-rope switch (elevators) A device that automatically causes the electric power to be removed from the elevator driving-machine motor and brake when the hoisting ropes of a winding-drum machine become slack. *See also:* control. (PE/EEC) [119]

slack stringing The method of stringing conductor slack without the use of a tensioner. The conductor is pulled off the reel by a pulling vehicle and is dragged along the ground, or the reel is carried along the line on a vehicle and the conductor is deposited on the ground. As the conductor is dragged to, or past, each supporting structure, the conductor is placed in the travelers, normally with the aid of finger lines. (T&D/PE) 524-1992r

slant distance (1) (navigation aids) The distance between two points that are not at the same elevation. (AES/GCS) 172-1983w
(2) The distance between two points that are not at the same elevation. Used in contrast to ground distance. (AES) 686-1997

slanted array compressor (SAC) A device containing dispersive transducer arrays that have a length-wise centerline that is inclined at an oblique angle to the incident propagation direction. (UFFC) 1037-1992w

slant range (navigation aids) The slant distance between a radar and a target. (AES/GCS) 686-1997, 172-1983w

slant-voltage-rated (multiple voltage rated) distribution cutout A distribution cutout intended primarily for application on three-phase solidly grounded neutral (multi-grounded) systems where prescribed conditions exist. (SWG/PE) C37.40-1993, C37.100-1992

slant voltage (multiple voltage) ratings of a distribution cutout A pair of maximum voltage ratings assigned to a distribution cutout intended primarily for application on three-phase solidly grounded neutral (multigrounded) systems where construction conditions are such that two cutouts will normally operate in series to clear phase-to-phase faults. In applying these cutouts, the system line-to-line voltage must be equal to or less than the maximum voltage rating to the right of the slant (/), and the system line-to-ground voltage must be equal to or less than the maximum voltage rating to the left of the slant (/). *Note:* Slant voltage rated cutouts may be used in single-phase applications where the normal frequency recovery voltage across the cutout does not exceed the maximum voltage rating to the left of the slant (/). (SWG/PE) C37.100-1992

SLAR *See:* sidelooking airborne radar.

slash The literal character "/". This character is also known as *solidus*. (C/PA) 9945-1-1996, 9945-2-1993, 1003.5-1999

slave (1) (test, measurement, and diagnostic equipment) A device that follows an order given by a master remote control. (MIL) [2]
(2) (VSB) A functional module that detects bus cycles initiated by the active master and, when those cycles select it, transfers data between itself and the master. The VSB specification defines a mechanism through which any number of slaves can participate in a bus cycle. All slaves that are selected by the cycle are called "selected slaves." However, only one of the selected slaves is allowed to acknowledge the cycle and respond to the active master. This slave is called the "responding slave." All other selected slaves are called "participating slaves." Slaves that are not selected by the cycle are called "idle slaves." (MM/C) 1096-1988w
(3) (VMEbus) A functional module that detects data transfer bus (DTB) cycles initiated by a master and, when those cycles specify its participation, transfers data between itself and the master. (BA/C) 1014-1987
(4) (STD bus) A card that responds to a bus transaction. (MM/C) 961-1987r
(5) (NuBus) A bus device that responds to a transaction. (C/MM) 1196-1987w
(6) A module that can be addressed and is capable of participating in bus transactions. (C/BA) 10857-1994, 896.4-1993w, 896.3-1993w
(7) A module that can be addressed and is capable of participating in, but not initiating, bus transactions. (C/BA) 1014.1-1994w
(8) A device that responds to masters according to the FBP. (NID) 960-1993
(9) An input-output device that is driven or controlled by a master unit. (C) 610.10-1994w
(10) The entity that responds to RamLink transactions (the transaction addressing is sufficient to support up to 63 slaves on each RamLink ringlet). (C/MM) 1596.4-1996
(11) A device that gathers data or performs control operations in response to requests from the master, and sends response messages in return. A slave device may also generate unsolicited responses (DNP 3.0 specific). (PE/SUB) 1379-1997

Slave *See:* SBus Slave.

slave, attached *See:* attached slave.

Slave Busy (BSY) bit A bit in the Slave Status register of every S-module that is set by the S-module when the application logic of the S-module is executing a previously transmitted MTM-bus instruction or is executing its power-up sequence. (TT/C) 1149.5-1995

Slave Controller state A state of the fsm required of S-modules that controls S-module Link Layer behavior during message transmission to the module. (TT/C) 1149.5-1995

Slave Data Fault (SDF) bit A bit in the Bus Error register of all S-modules. An S-module sets this bit when it is transmitting on the MSD signal and detects a fault on that signal. (TT/C) 1149.5-1995

slave drive *See:* electric drive; follower drive.

slaved tracking (power supplies) A system of interconnection of two or more regulated supplies in which one (the master) operates to control the others (the slaves). The output voltage of the slave units may be equal or proportional to the output voltage of the master unit. (The slave output voltages track the master output voltage in a constant ratio). *See also:* master/slave operation; complementary tracking. (AES/PE) [41], [78]

slave Physical Layer In a 100BASE-T2 link containing a pair of PHYs, the PHY that recovers its clock from the received signal and uses it to determine the timing of transmitter operations. It also uses the slave transmit scrambler generator polynomial for side stream scrambling. Master and slave PHY status is determined during the Auto-Negotiation process that takes place prior to establishing the transmission link. *See also:* master physical layer. (C/LM) 802.3-1998

slave relay *See:* auxiliary relay; relay.

slave state *See:* problem state.

slave station (navigation) (navigation aids) A station in which some characteristic of its emission is controlled by a master station. (AES/GCS) 172-1983w

Slave Status register A status register that is required to be implemented in the MTM-Bus interface logic of every S-module. Bits in this register are used to record such items of module status as interrupt enable status, whether an error condition has occurred, whether a module application-related error has occurred, whether the module has failed its Built-In Self Test, etc. (TT/C) 1149.5-1995

slave-sweep switching (oscilloscopes) A combination of sweep switching and multiple-trace operation in which a specific channel is displayed with a specific sweep. (IM) 311-1970w

slaving (gyros) The use of a torquer to maintain the orientation of the spin axis relative to an external reference, such as a pendulum or magnetic compass. (AES/GYAC) 528-1994

SLC *See:* software life cycle.

sled *See:* running board.

sleet hood (of a switch) A cover for the contacts to prevent sleet from interfering with successful operation of the switch. (SWG/PE) C37.100-1992

sleetproof (1) (general) So constructed or protected that the accumulation of sleet will not interfere with successful operation. (SWG/PE) C37.30-1971s, C37.100-1981s
(2) (power and distribution transformers) So constructed or protected that the accumulation of sleet (ice) under specified test conditions will not interfere with the successful operation of the apparatus. (PE/TR) C57.12.80-1978r

sleeve (1) (plug) (three-wire telephone-switchboard plug). A cylindrically shaped contacting part, usually placed in back of the tip or ring but insulated therefrom. (EEC/PE) [119]
(2) (rotating machinery) A tubular part designed to fit around another part. *Note:* In a sleeve bearing, the sleeve is that component that includes the cylindrical inner surface within which the shaft journal rotates. (PE) [9]
(3) *See also:* compression joint. (T&D/PE) 524-1992r

sleeve bearing (rotating machinery) A bearing with a cylindrical inner surface in which the journal of a rotor (or armature) shaft rotates. *See also:* rotor. (PE) [9]

sleeve conductor *See:* sleeve wire.

sleeve-dipole antenna A dipole antenna surrounded in its central portion by a coaxial conducting sleeve. *See also:* antenna. (AP/ANT) 145-1993

sleeve-monopole antenna An antenna consisting of half of a sleeve-dipole antenna projecting from a ground plane. *Synonym:* sleeve-stub antenna. (AP/ANT) 145-1993

sleeve-stub antenna *See:* sleeve-monopole antenna.

sleeve supervision The use of the sleeve circuit for transmitting supervisory signals. (EEC/PE) [119]

sleeve-type suppressor A suppressor designed for insertion in a high-tension ignition cable. *See also:* electromagnetic compatibility. (EMC/INT) [53], [70]

sleeve wire (telephone switching systems) That conductor, usually accompanying the tip and ring leads of a switched connection, that provides for miscellaneous functions necessary to the control and supervision of the connection. In cord-type switchboards, the sleeve wire is that conductor associated with the sleeve contacts of the jacks and plugs. (COM) 312-1977w

sleeving trailer *See:* splicing cart.

slew A measure of the shape of the waveform constituting a logic state transition. A slew value can have the dimensions of time, in which case it is a slew time, or the dimensions of voltage-per-time, in which case it is a slew rate. The delay and power calculation system (DPCS) allows either interpretation if used consistently. (C/DA) 1481-1999

slew-dependent delay That part of an input-to-output delay that can be attributed to the signal at the input of the arc taking longer to make a transition than is considered ideal. (C/DA) 1481-1999

slewing (gyros) The rotation of the spin axis about an axis parallel to that of the applied torque causing the rotation. (AES/GYAC) 528-1994

slewing rate (1) (power supplies) A measure of the programming speed or current-regulator-response timing. The slewing rate measures the maximum rate-of-change of voltage across the output terminals of a power supply. Slewing rate is normally expressed in volts per second ($\Delta E/\Delta T$) and can be converted to a sinusoidal frequency-amplitude product by the equation (E_{pp}) = slewing rate/π, where E_{pp} is the peak-to-peak sinusoidal volts. Slewing rate = $\pi f(E_{pp})$. *See also:* high-speed regulator. (AES) [41]
(2) (thyristor) A rate at which the output changes in response to a step change in control signal input. (IA/IPC) 428-1981w

slewing speed (test, measurement, and diagnostic equipment) A continuous speed, usually the maximum at which a tape reader or other rotating device can search for information. (MIL) [2]

slew limit The value of output transition rate of change for which an increased amplitude input step causes no change. (IM/WM&A) 1057-1994w

slew rate (1) Rate of change of (ac voltage) frequency. (IA/PSE) 1100-1999
(2) A measure of how quickly a signal takes to make a transition; that is, a voltage-per-unit time. Slew rate is inversely related to slew time and is sometimes used incorrectly where slew time is intended. (C/DA) 1481-1999

slew time A measure of how long a signal takes to make a transition; that is, the rise time or fall time. Slew time is inversely related to slew rate. The way a slew time value is abstracted from the continuous waveform at a cell pin varies with different cell characterization methods. (C/DA) 1481-1999

SLIC *See:* selective listing in combination index.

slicer (amplitude gate) (clipper-limiter) A transducer that transmits only portions of an input wave lying between two amplitude boundaries. *Note:* The term is used especially when the two amplitude boundaries are close to each other as compared with the amplitude range of the input. (PE/EEC) [119]

slide rail (rotating machinery) A special form of soleplate which is long in the direction of the machine axis to permit sliding the stator frame in the axial direction. (PE) [9]

slide-screw tuner (1) (transmission lines and waveguides) An impedance or matching transformer that consists of a slotted waveguide or coaxial-line section and an adjustable screw or post that penetrates into the guide or line and can be moved axially along the slot. *See also:* waveguide. (IM/HFIM) [40]
(2) (waveguide components) A waveguide or transmission line tuner employing a post of adjustable penetration, adjustable in position along the longitudinal axis of the waveguide. (MTT) 147-1979w

sliding contact An electric contact in which one conducting member is maintained in sliding motion over the other conducting member. *See also:* contactor. (EEC/PE) [119]

sliding load A load sliding inside or along a fixed length of waveguide or transmission line. *See also:* waveguide. (IM/HFIM) [40]

sliding short circuit A short-circuit termination that consists of a section of waveguide or transmission line fitted with a sliding short-circuiting piston (contacting or noncontacting) that ideally reflects all the energy back toward the source. *See also:* waveguide. (IM/HFIM) [40]

sliding window A protocol used to allow a sender to transmit multiple packets before waiting for an acknowledgment. *Note:* The protocol places a small window in the sequence and transmits all packets that lie inside the window. The window size is adjusted based on the successful rate of packet transmission. *See also:* Go-Back-N; selective retransmission. (C) 610.7-1995

sliding window demand The block demand calculated over an integration period that includes sub-intervals of previous demand calculations. (AMR/SCC31) 1377-1997

slime, anode *See:* anode slime.

slinging wire A wire used to suspend and carry current to one or more cathodes in a plating tank. *See also:* electroplating. (EEC/PE) [119]

sling, traveler *See:* traveler sling.

slip (1) (A) (rotating machinery) The quotient of: the difference between the synchronous speed and the actual speed of a rotor, to the synchronous speed, expressed as a ratio, or as a percentage. **(B) (rotating machinery)** The difference between the speed of a rotating magnetic field and that of a rotor, expressed in revolutions per minute. **(C) (rotating machinery)** (electric couplings) The difference between the speeds of the two rotating members. *See also:* asynchronous machine. (PE) [9]
(2) In an induction machine, the difference between its synchronous speed and its operating speed. It may be expressed in the following ways:

a) As a percent of synchronous speed
b) As a decimal fraction of synchronous speed
c) Directly in revolutions per minute

(IA/MT) 45-1998

SLIP *See:* Symmetric List Processing Language.

slip, controlled *See:* controlled slip.

slip-grip *See:* conductor grip.

slip regulator (rotating machinery) A device arranged to produce a reduction in speed below synchronous speed greater than would be obtained inherently. Such a device is usually in the form of a variable impedance connected in the secondary circuit of a slip ring induction motor. (PE) [9]

slip relay A relay arranged to act when one or more pairs of driving wheels increase or decrease in rotational speed with respect to other driving wheels of the same motive power unit. *See also:* multiple-unit control. (EEC/PE) [119]

slip ring *See:* collector ring.

slip-ring induction motor *See:* wound-rotor induction motor.

slip-ring short-circuiting device *See:* brush-operating device.

slip, uncontrolled *See:* uncontrolled slip.

sliver A pulse with a duration less than that specified for that signal (e.g., truncated clock signal). (C/LM) 802.3-1998

slope (1) The gain (or loss) versus frequency characteristic of cable, amplifiers, and other devices. (LM/C) 802.7-1989r
(2) The ratio of the voltage change to the current change over the full (inductive plus capacitive) linearly controlled range of the static var compensator (SVC) at nominal voltage, expressed as a percentage. (PE/SUB) 1031-2000

slope angle *See:* glide-slope angle.

slope compensation The action of a slope-compensated gain control. The gain of the amplifier and the slope of amplifier equalization are changed simultaneously to provide equalization for different lengths of cable; normally specified in terms of cable loss. (LM/C) 802.7-1989r

slope detector (telephony) (dial-pulse address signaling systems) A circuit that provides a means of accurately measuring the open and closed intervals of a contact even though the contact may be shunted by a contact protection network or measured from the far end of a metallic loop, or both. (COM/TA) 753-1983w

slot (1) (rotating machinery) A channel or tunnel opening onto or near the air gap and passing essentially in an axial direction through the rotor or stator core. A slot usually contains the conductors of a winding, but may be used exclusively for ventilation. *See also:* rotor; stator. (PE) [9]
(2) (VMEbus) A position where a printed-circuit board (pcb) can be inserted into the backplane. When the system has a J1 and a J2 backplane (or a combination J1/J2 backplane) each slot provides a pair of 96-pin connectors. When the system has only a J1 backplane, then each slot provides a single 96-pin connector. (C/BA) 1014-1987
(3) (VSB) A position where a board can be inserted into a backplane. Each VSB slot provides at least one 96-pin connector. (C/MM) 1096-1988w
(4) (NuBus) A backplane location that accepts a NuBus module. (C/MM) 1196-1987w
(5) A module-insertion position provided by the backplane and associated card cage. (C/MM) 1596-1992
(6) A position where a module can be inserted into an VXIbus backplane. Each slot provides the 96-pin J connectors to interface with the board P connectors. It may have to provide one, two, or three connectors. (C/MM) 1155-1992
(7) A mechanical location on a backplane. (C/BA) 896.2-1991w, 896.10-1997
(8) A module connector position on a crate segment backplane (see position). (NID) 960-1993
(9) The protocol data unit (PDU) of 53 octets used to transfer segments. It contains a segment of 52 octets and a 1 octet Access Control Field (ACF). There are two type of slots: Pre-Arbitrated (PA) slots and Queued Arbitrated (QA) slots. (LM/C) 8802-6-1994
(10) *See also:* expansion slot. (C) 610.10-1994w

slot antenna (data transmission) A radiating element formed by a slot in a conducting surface. (PE/AP/ANT) 599-1985w, 145-1993

slot array An antenna array formed of slot radiators. *See also:* antenna. (Std100) [84]

slot cell (rotating machinery) A sheet of insulation material used to line a slot before the winding is placed in it. *See also:* rotor; stator. (PE) [9]

slot coupling factor (navigation aid terms) (slot-antenna array) The ratio of the desired slot current to the available slot current, controlled by changing the depth of penetration of the slot probe into the waveguide. (AES/GCS) 172-1983w

slot current ratio (slot-antenna array) (navigation aid terms) The relative slot currents in the slots of the waveguide reading from its center to its end, with the maximum taken as 1; this ratio is dependent upon the slot spacing factor and the slot coupling factor. (AES/GCS) 172-1983w

slot discharge (rotating machine) Sparking between the outer surface of coil insulation and the grounded slot surface, caused by capacitive current between conductors and iron. The resulting current pulses have a fundamental frequency of a few kilohertz. *See also:* asynchronous machine. (PE) [9]

slot-discharge analyzer (rotating machinery) An instrument designed for connection to an energized winding of a rotating machine, to detect pulses caused by slot discharge, and to discriminate between them and pulses otherwise caused. *See also:* asynchronous machine. (PE) [9]

slot insulation (rotating machinery) A sheet or deposit of insulation material used to line a slot before the winding is placed in it. *See also:* asynchronous machine. (PE) [9]

slot line A planar transmission line consisting of two semi-infinite coplanar conductors affixed to the same side of an insulating substrate of arbitrary thickness and separated by a finite gap. (MTT) 1004-1987w

slot liner (rotating machinery) Separate insulation between an embedded coil side and the slot which can provide mechanical and electrical protection. (PE) [9]

slot packing (rotating machinery) Additional insulation used to pack embedded coil sides to ensure a tight fit in the slots. *See also:* stator; rotor. (PE) [9]

slot pitch (rotating machinery) The peripheral distance between fixed points in corresponding positions in two consecutive slots. *Synonym:* tooth pitch. (PE) [9]

slot section Portion of the stator bar/coil which after installation in the rotating electric machine is enclosed in the stator core slot. The outer surface of the slot section of bars/coils is treated with semiconducting (or conducting) materials, commonly referred to as semiconducting (or conducting) tapes or paints. (PE/EM) 1310-1996

slot separator insulation *See:* insulation slot separator.

slot space factor (rotating machinery) The ratio of the cross-sectional area of the conductor metal in a slot to the total cross-sectional area of the slot. *See also:* asynchronous machine. (PE) [9]

slot spacing factor (slot-antenna array) (navigation aids) A value proportional to the size of the angle between the slot location and the null of the internal standing wave; this factor is dependent upon frequency. (AES/GCS) 172-1983w

slotted armature (rotating machine) An armature with the winding placed in slots. *See also:* armature. (PE) [9], [84]

slotted section (slotted waveguide) (slotted line) A section of a waveguide or shielded transmission line the shield of which is slotted to permit the use of a carriage and travelling probe for examination of standing waves. *See also:* auxiliary device to an instrument. (AP/IM/ANT/HFIM) [35], [40], [84]

slot time A multipurpose parameter used in CSMA/CD technique that describes the contention behavior of the MAC sublayer of a LAN. *Note:* This value represents the amount of time during which a collision will occur if two stations transmit simultaneously. It is calculated as a function of the propagation delay of the network. *Synonym:* contention interval. (C) 610.7-1995

slot-type antenna (aircraft) A slot in the normal streamlined metallic surface of an aircraft, excited electromagnetically by a structure within the aircraft. Radiation is thus obtained without projections that would disturb the aerodynamic characteristics of the aircraft. Radiation from a slot is essentially directive. (EEC/PE) [119]

slot wedge (rotating machinery) The element placed above the turns or coil sides in a stator or rotor slot, and held in place by engagement of wedge (slots) grooves along the sides of the coil slot, or by projections from the sides of the slot tending to close the top of the slot. *Note:* A wedge may be a thin strip of material provided solely as insulation or to provide temporary retention of the coils during the manufacturing process. It may be a piece of structural insulating material or high-strength metal to hold the coils in the slot. Slots in laminated cores are normally wedged with insulating material. *See also:* stator; rotor. (PE) [9]

slow-closure device (hydraulic turbines) A cushioning device that retards the closing velocity of the servomotor from a predetermined servomotor position to zero servomotor position. (PE/EDPG) 125-1988r

slow-operate relay A slugged relay that has been specifically designed for long operate time but not for long release time. Caution: The usual slow-operate relay has a copper slug close to the armature, making it also at least partially slow to release. (PE/EM) 43-1974s

slow-operating relay A relay that has an intentional delay between energizing and operation. *See also:* electromagnetic relay. (EEC/PE) [119]

slow-release relay A relay that has an intentional delay between de-energizing and release. *Note:* The reverse motion need not have any intentional delay. *See also:* electromagnetic relay. (PE/EM) 43-1974s

slow release time characteristic, relay *See:* relay slow-release time characteristic.

slow-speed starting A control function that provides for starting an electric drive only at the minimum-speed setting. *See also:* starter. (IA/ICTL/IAC) [60]

slow time (A) Simulated time with the property that a given period of actual time represents less than that period of time in the system being modeled; for example, in a simulation of the internal workings of a computer, running the simulation for one second may result in the model advancing time by only a microsecond; that is, simulated time advances slower than actual time. **(B)** The duration of activities within a simulation in which simulated time advances slower than actual time. *Contrast:* real time; fast time. (C) 610.3-1989

slow wave An electromagnetic wave propagating close to a boundary with a phase velocity less than that which would exist in an unbounded medium having the same electromagnetic properties. *See also:* fast wave. (AP/PROP) 211-1997

slow-wave circuit (microwave tubes) A circuit whose phase velocity is much slower than the velocity of light. For example, for suitably chosen helixes the wave can be considered to travel on the wire at the velocity of light but the phase velocity is less than the velocity of light by the factor that the pitch is less than the circumference. (ED) [45]

slug *See:* connector link.

slug, relay *See:* relay slug.

slug tuning A means for varying the frequency of a resonant circuit by introducing a slug of material into either the electric or magnetic fields or both. *See also:* radio transmission; network analysis. (AP/ANT) 145-1983s

sluice Open water trough, also used to describe operation of a turbine when operated under free discharge conditions to release flood flows. (PE/EDPG) 1020-1988r

slush compound A non-drying oil, grease, or similar organic compound that, when coated over a metal, affords at least temporary protection against corrosion. (IA) [59]

small computer systems interface A data-transfer interface used to connect multiple peripheral devices, such as disk drives, tapes, or printers to computer systems while taking up only one slot in the computer. *Note:* Previously, this was known as Shugart Associates Systems Interface. (C) 610.10-1994w

small ion (dc electric-field strength and ion-related quantities) An ion comprised of molecules or molecular clusters bound together by charge. Mobilities are in the range of 10^{-5} m^2/Vs to 2×10^{-4} m^2/Vs. Typical radius is less than 1×10^{-9} m. *Note:* To avoid confusion with the more general term "ion," the use of the term "small ion" is encouraged. (T&D/PE) 539-1990, 1227-1990r

small-perturbation method An approximate technique for estimating the scattering from a perturbed boundary or from perturbations in the constitutive parameter(s) of a medium applicable when the perturbation is small compared to a reference parameter or scale such as the wavelength in the boundary case. (AP/PROP) 211-1997

small-power producer A non-utility generation source that is a qualifying small-power production facility under PURPA and the Federal Energy Regulatory Commission (FERC). (SUB/PE) 1109-1990w

small scale integration (SSI) (A) Pertaining to an integrated circuit containing less than 100 transistors in its design. *Contrast:* ultra-large scale integration; medium scale integration; very large scale integration; large scale integration. **(B)** Pertaining to an integrated circuit containing fewer than 10 elements. (C) 610.10-1994

small-signal (light-emitting diodes) A signal which when doubled in magnitude does not produce a change in the parameter being measured that is greater than the required accuracy of the measurement. (ED) [127]

small-signal forward transadmittance The value of the forward transadmittance obtained when the input voltage is small compared to the beam voltage. *See also:* electron-tube admittances. (ED) 161-1971w

small signal performance (1) (excitation systems for synchronous machines) The response of an excitation cotrol system, excitation system, or elements of an excitation system to signals that are small enough that nonlinearities can be disregarded in the analysis of the response, and operation can be considered to be linear. (PE/EDPG) 421.1-1986r **(2) (excitation systems)** The response to signals that are small enough so that nonlinearities are insignificant. (PE/EDPG) 421.2-1990

small-signal permittivity The incremental change in electric displacement per unit electric field when the magnitude of the measuring field is very small compared to the coercive electric field. (Measurements are usually made at a frequency of one kilohertz or higher). The small signal relative permittivity κ is equal to the ratio of the absolute permittivity e to the permittivity of free space e_0, that is, $\kappa = \epsilon / \epsilon_0$. *Note:* The value of the small-signal permittivity may depend on the remanent polarization, electric field, mechanical stress, sample history, or frequency of the measuring field. *See also:* remanent polarization; Curie-Weiss temperature. (UFFC) [21]

small-signal resistance (semiconductor rectifiers) The resistive part of the quotient of incremental voltage by incremental current under stated operating conditions. *See also:* rectification. (IA) [12]

SMALLTALK A high-order language based on the metaphor of objects sending messages to one another. *See also:* object-oriented language. (C) 610.13-1993w

small wiring *See:* secondary and control wiring.

smart actuator An actuator version of a smart transducer. (IM/ST) 1451.2-1997

smart sensor A sensor version of a smart transducer. (IM/ST) 1451.2-1997

smart terminal An intelligent terminal that is preprogrammed for a particular application, for example, a word processing workstation with integrated spell checking. (C) 610.10-1994w

Smart Transducer A transducer that provides functions over and above that necessary for generating a correct representation of a sensed or controlled physical quantity. This functionality typically simplifies the integration of the transducer into applications in a networked environment. (IM/ST) 1451.1-1999, 1451.2-1997

Smart Transducer Interface Module (STIM) (1) A module that contains the Transducer Electronic Data Sheet, logic to implement the transducer interface, the transducer(s) and any signal conversion or signal conditioning. This standard expressly requires that no operating mode of the Smart Transducer Interface Module ever permit these components to be physically separated. They may, however, be separated during manufacturing and repair. (IM/ST) 1451.2-1997 **(2)** The supporting electronics on the transducer side of the hardware interface to the NCAP. In IEEE 1451.1, an STIM and an NCAP combined form a Networked Smart Transducer. In the various IEEE 1451.X standards, for example IEEE 1451.2, the term STIM may have a more precise meaning within the scope of the particular standard. (IM/ST) 1451.1-1999

Smart Transducer object model An object model for a Smart Transducer. The model includes an interface to a transducer object and a transducer bus. (IM/ST) 1451.1-1999

smashboard signal A signal so designed that the arm will be broken when passed in the stop position. (EEC/PE) [119]

SME *See:* subject matter expert; systems management entity.

SMFA *See:* specific management functional areas.

SMIB *See:* security management information base.

smog *See:* fog.

smoke The airborne solid and liquid particulates and gases evolved when a material undergoes pyrolysis or combustion. (DEI) 1221-1993w

smoke detector (fire protection devices) A device which detects the visible or invisible particles of combustion. (NFPA) [16]

smooth To apply procedures that decrease or eliminate rapid fluctuations in data. (C) 1084-1986w

smooth current (rotating electric machinery) Current that remains unidirectional and the ripple of which does not exceed 3%. (PE/EM) 11-1980r

smoothing Any image enhancement technique in which the effect of noise in the original image is reduced. *Synonyms:* noise suppression; noise cleaning. (C) 610.4-1990w

smoothing reactor for HVDC transmission A smoothing reactor for HVDC application is a reactor intended for connection in series with an HVDC converter, or an HVDC transmission line or insertion in the intermediate dc circuit of a back-to-back link, for the purpose of

— Reducing harmonics in the dc line.
— Complying, in conjunction with dc filters, with the dc side telephone interference requirements.
— Limiting the surge-current amplitude during faults and disturbances; especially the limitation of cable discharge currents in the case of a long dc cable.
— Providing a high impedance to the flow of harmonics in the case of a cable link (high capacitance of cable).
— Limiting the rate of rise of inverter dc current in the case of inverter ac network disturbances; thus reducing the risk of commutation failures.
— Improving the dynamic stability of the dc transmission system (commutation failures).

Smoothing reactors may be built using either of two designs: dry-type air cooled or oil-immersed. Dry-type smoothing reactors are of air-core design. Oil-immersed smoothing reactors utilize magnetic-core materials as an inherent part of their design. (PE/TR) 1277-2000

smothered-arc furnace A furnace in which the arc or arcs is covered by a portion of the charge. (EEC/PE) [119]

SMT *See:* station management.

S/N *See:* signal-to-noise ratio.

SNA *See:* systems network architecture.

snake *See:* conductor cover; fish tape.

snap-action relay contacts A contact assembly having two or more equilibrium positions, in one of which the contacts remain with substantially constant contact pressure during the initial motion of the actuating member, until a condition is reached at which stored energy snaps the contacts to a new position of equilibrium. (EEC/REE) [87]

snaphook A connector comprised of a hook-shaped member with a normally closed keeper or similar arrangement, which may be opened to permit the hook to receive an object and, when released, automatically closes to retain the object. There are two types of snaphooks: a) The locking type with a self-closing, self-locking keeper which remains closed and locked until unlocked and pressed open for connection or disconnection, or b) The non-locking type with a self-closing keeper which remains closed until pressed open for connection or disconnection. (T&D/PE) 1307-1996

snapover When used in connection with alternating-current testing, a quasi-flashover or quasi-sparkover, characterized by failure of the alternating-current power source to maintain the discharge, thus permitting the dielectric strength of the specimen to recover with the test voltage still applied. (PE/PSIM) [55]

snapshot A copy of all or portions of the data contained in storage or in a database at a particular point in time. *Note:* Considered a "picture" of the data. (C) 610.5-1990w

snapshot dump (A) A dynamic dump of the contents of one or more specified storage areas. *See also:* dynamic dump; change dump; postmortem dump; static dump; selective dump; memory dump. **(B) (computers)** A selective dynamic dump performed at various points in a machine run. (C) 610.12-1990, [20], [85]

snarf The action taken by a a module when it takes a copy of data passing by on the bus, even though it did not request it. (C/BA) 10857-1994, 896.3-1993w, 896.4-1993w

snatch block (power line maintenance) (conductor stringing equipment) A device normally designed with a single sheave, a shell, and an attachment hook or shackle. One side of the shell can be opened to eliminate the need for threading of the line. It is commonly used for lifting loads on a single line, or as a device to control the position or direction, or both, of a fall line or pulling line *Synonyms:* skookum; Washington; Western. (T&D/PE) 516-1995, 524-1992r

Snell's law The relationship between angles of incidence, reflection and transmission, and material constitutive parameters, for a plane wave incident on a planar boundary between media of differing electromagnetic constitutive parameters. Often expressed as:

$\theta_i = \theta_r$

and

$n_i \sin\theta_i = n_t \sin\theta_t$

where
θ_i = the angle of incidence
θ_r = the angle of reflection
θ_t = the angle of transmission
n_i = the real part of the refractive index of the material in which the wave is incident
n_t = the real part of the refractive index of the other material

(AP/PROP) 211-1997

SNM cable *See:* shielded nonmetallic-sheathed cable.

SNOBOL *See:* StriNg-Oriented symBOlic Language.

snoop The action taken by a module on a transaction when it is not the master that originated the transaction or the repository of last resort for the data, but it still monitors the transaction. Cache memories snoop transactions to maintain coherence. (C/BA) 10857-1994, 896.4-1993w, 896.3-1993w

snow (1) (intensity-modulated display) A varying speckled background caused by noise. *See also:* television; radar. (BT/AV) [34]

(2) (overhead power lines) Precipitation composed of white or translucent ice crystals, chiefly in complex branched hexagonal form and often agglomerated into snowflakes. For weather observation purposes, the intensity of snow is characterized as:

a) "Very light," when scattered flakes do not completely cover or wet an exposed surface, regardless of duration;
b) "Light," when the visibility is 1.0 km or more;
c) "Moderate," when the visibility is less than 1.0 km but more than 0.5 km;
d) "Heavy," when the visibility is less than 0.5 km The classification of snowfall according to its intensity is identical to that of rain, where the equivalent amount of water accumulated in millimeters per hour is measured. An easier but less accurate approach uses the depth of the accumulated snow.

(T&D/PE) 539-1990

snow brake A constant application of light friction brake intended to create enough heat to mitigate the buildup of snow and ice, which would interfere with the brake actuators. (VT) 1475-1999

S/N ratio *See:* signal-to-noise ratio.

SNRM *See:* set normal response mode.

snr psoph (data transmission) Signal-to-noise ratio measured with psophometrically weighted receiver; expressed in dB (decibels). (PE) 599-1985w

snub *See:* anchor.

snubber (1) (converter circuit elements) (self-commutated converters) An auxiliary circuit element or combination of elements employed to modify the transient voltage or current of a semiconductor device. *See also:* shunt snubber; series snubber; polarized snubber. (IA/SPC) 936-1987w

(2) (load commutated inverter synchronous motor drives) An auxiliary circuit element or combination of elements employed to modify the transient voltage or current of a semiconductor device during switching. (A) (shunt snubber) Circuit elements, usually including a capacitor and a resistor connected in shunt with a switching device to limit the rate of rise of voltage or the peak voltage across the device (or both) when switching from a conducting to a blocking state or when subjected to an external voltage transient. (B) (series snubber) Circuit elements, usually including an inductor, connected in series with a switching device to limit the rate of rise or fall of current through the device when switching on or off, respectively. (IA/ID) 995-1987w

snub structure A structure located at one end of a sag section and considered as a *zero* point for sagging and clipping offset calculations. The section of line between two such structures is the sag section, but more than one sag section may be required in order to sag properly the actual length of conductor that has been strung. *Synonyms:* zero structure; O structure. (T&D/PE) 524-1992r

SO *See:* shift-out character.

soak, relay *See:* relay soak.

sock *See:* woven wire grip.

socket A file of a particular type that is used as a communications endpoint for process-to-process communication. (C) 1003.5-1999

socket address An address associated with a socket or remote endpoint. The address may include multiple parts, such as a network address associated with a host system and an identifier for a specific endpoint. (C) 1003.5-1999

socket cover The removable portion of the enclosure that provides access to the meter socket wiring. (ELM) C12.7-1993

socket identifier (socket ID) A byte used to uniquely identify the socket. A socket ID can be well known or can be dynamically assigned. (C/MM) 1284.4-2000

socket rim That part of a ring-type meter socket that is required to accommodate the socket sealing ring that holds a detachable watthour meter in place. The socket rim may be a part of the cover that is secured in place by a fastener such as a latch or crossbar. (ELM) C12.7-1993

sockets An interface to transport protocol. (C/PA) 1003.23-1998

socket sealing ring (watthour meter sockets) A ring used to overlap the socket rim and the detachable watthour meter cover ring to hold and provide means for sealing a detachable watthour meter in place. C12.7-1993

sock line *See:* pulling line.

sodium vapor lamp transformers (power and distribution transformers) (multiple-supply type) Transformers, autotransformers, or reactors for operating sodium vapor lamps for all types of lighting applications, including indoor, outdoor area, roadway, and other process and specialized lighting. (PE/TR) C57.12.80-1978r

SOE *See:* sequence-of-events.

sofar (navigation aids) A system of navigation providing hyperbolic lines of position determined by shore listening stations. (AES/GCS) 172-1983w

soft copy A copy of computer output in a form other than a printed page. For example, data displayed on a display device. *Contrast:* hard copy. (C) 610.2-1987, 610.6-1991w

soft error* *See:* transient error.

* Deprecated.

soft failure A failure that permits continued operation of a system with partial operational capability. *Contrast:* hard failure. (C) 610.12-1990

soft font *See:* downloadable font.

soft hyphen *See:* discretionary hyphen.

soft limiting *See:* limiter circuit.

soft region A cluster which does not have a specified physical location in a floorplan. It may have constraints on how closely the cells within the cluster are placed relative to each other. A soft region may be located within a hard region. (C/DA) 1481-1999

soft-sector Pertaining to a magnetic disk that is segmented by recorded data marks on the disk; the location of a sector is determined by the distance from a magnetically or photoelectrically sensed starting mark, known as an index mark. *Note:* This can refer to either a floppy diskette or a hard disk but generally refers to the former, which has one punched hole, known as an index hole, which marks the first sector. *Contrast:* hard-sector. (C) 610.10-1994w

soft start (1) (thyristor) At turn-on, a gradual increase in output at a predetermined rate from zero or a set minimum to a desired maximum. (IA/IPC) 428-1981w

(2) The ability of a controlling device to apply power to a load upon energization in a proportional manner, irrespective of values of the controlling signals. (IA/PC) 844-1991

soft start reset (thyristor) Reset of soft start to initial conditions when ac power is interrupted. (IA/IPC) 428-1981w

software (1) Computer programs, procedures, and possibly associated documentation and data pertaining to the operation of a computer system. (SE/C) 610.12-1990

(2) (programmable digital computer systems in safety systems of nuclear power generating stations) Computer programs and data. 7432-1982w

(3) The programs, procedures, rules, and any associated documentation pertaining to the operation of an information processing system. (C/PA) 14252-1996

(4) A generic term referring to software objects or a structured set of files. This term can refer to the objects forming the hierarchical structure (software objects), or to the actual files and control_files (software files). (C/PA) 1387.2-1995

(5) Computer programs and computer databases. (C/SE) J-STD-016-1995

(6) Computer programs, procedures, and associated documentation and data pertaining to the operation of a computer system. (C/SE) 1062-1998

software accuracy (programmable digital computer systems in safety systems of nuclear power generating stations) The software attribute that provides a quantitative measure of the magnitude of error. 7432-1982w

software acquisition process The period of time that begins with the decision to acquire a software product and ends when the product is no longer available for use. The software acquisition process typically includes nine steps associated with planning the organizational strategy, implementing an organization's process, determining the software requirements, identifying potential suppliers, preparing contract requirements, evaluating proposals and selecting the supplier, managing supplier performance, accepting the software, and using the software. (C/SE) 1062-1998

software characteristic (software) An inherent, possibly accidental, trait, quality, or property of software (for example, functionality, performance, attributes, design constraints, number of states, lines or branches). (C/SE) 610.12-1990, 1008-1987r

software_collection A grouping of software objects that are managed by the software_administration utilities. Software_collections are the sources and targets of these utilities. This standard defines two types of software_collections: installed_software and distributions. (C/PA) 1387.2-1995

software common class The common class describing the common attributes associated with the hierarchical structure of software objects defined by this standard. (C/PA) 1387.2-1995

software component A general term used to refer to a software system or an element, such as module, unit, data, or document. (C/SE) 1061-1998

software configuration item, or subsystem A collection of software elements treated as a unit for the purpose of configuration management. (C/SE) 1016.1-1993w

software configuration management *See:* configuration management.

software consistency (programmable digital computer systems in safety systems of nuclear power generating sta-

tions) The software attribute that provides uniform design and implementation techniques and notation. 7432-1982w

software data base A centralized file of data definitions and present values for data common to, and located internal to, an operational software system. *See also:* data; file. (C/SE) 729-1983s

software data protection (SDP) A means of preventing inadvertent write or access to different operating modes. (ED) 1005-1998

software defined network A network based on a public circuit-switched network that gives the user the appearance of a private network. (C) 610.7-1995

software definition files The files containing the software structure and detailed attributes for distributions, installed_software, bundles, products, subproducts, filesets, files, and control_files. This includes the `INDEX` and `INFO` files and the PSF. To communicate metadata information relating to both distributions and installed software, software definition files serve as input to, or output from, the various software administration utilities. The format used by software administration utilities to store metadata relating to installed software is undefined. (C/PA) 1387.2-1995

software design description (SDD) A representation of software created to facilitate analysis, planning, implementation, and decision making. The software design description is used as a medium for communicating software design information, and may be thought of as a blueprint or model of the system. (C/SE) 1012-1998, 1016-1998

software design document The output of design process in a presentable format, traditionally, a paper-based document. (C/SE) 1016.1-1993w

software design process Organized tasks and activities of design, having appropriate specification. (C/SE) 1016.1-1993w

software design process specification Know-how, technology of design, that specify operationally how to use methodology of design (standardized itself) together with standards for evaluating design, tools to support design automation, and documentation required to represent design information. (C/SE) 1016.1-1993w

software development A set of activities that results in software products. Software development may include new development, modification, reuse, reengineering, maintenance, or any other activities that result in software products. (C/SE) J-STD-016-1995

software development cycle (software) The period of time that begins with the decision to develop a software product and ends when the software is delivered. This cycle typically includes a requirements phase, design phase, implementation phase, test phase, and sometimes, installation and checkout phase. *Notes:* 1. The phases listed above may overlap or be performed iteratively, depending upon the software development approach used. 2. This term is sometimes used to mean a longer period of time, either the period that ends when the software is no longer being enhanced by the developer, or the entire software life cycle. *Contrast:* software life cycle. (C) 610.12-1990

software development file (1) A collection of material pertinent to the development of a given software unit or set of related units. Contents typically include the requirements, design, technical reports, code listings, test plans, test results, problem reports, schedules, and notes for the units. *Synonyms:* software development folder; software development notebook; unit development folder. (C) 610.12-1990
(2) A repository for material pertinent to the development of a particular body of software. Contents typically include (either directly of by reference) considerations, rationale, and constraints related to requirements definition, design, and implementation; developer-internal test information; and schedule and status information. (C/SE) J-STD-016-1995

software development folder *See:* software development file.

software development library (SDL) (1) A software library containing computer readable and human readable information relevant to a software development effort. *Synonyms:* program support library; project library. *Contrast:* system library; master library; software repository; production library. (C) 610.12-1990
(2) A controlled collection of software, documentation, other intermediate and final software products, and associated tools and procedures used to facilitate the orderly development and subsequent maintenance of software. (C/SE) J-STD-016-1995

software development notebook *See:* software development file.

software development plan (SDP) A project plan for a software development project. (C) 610.12-1990

software development process (1) The process by which user needs are translated into a software product. The process involves translating user needs into software requirements, transforming the software requirements into design, implementing the design in code, testing the code, and sometimes, installing and checking out the software for operational use. *Note:* These activities may overlap or be performed iteratively. *See also:* incremental development; spiral model; waterfall model; rapid prototyping. (C) 610.12-1990
(2) An organized set of activities performed to translate user needs into software products. (C/SE) J-STD-016-1995

software diversity A software development technique in which two or more functionally identical variants of a program are developed from the same specification by different programmers or programming teams with the intent of providing error detection, increased reliability, additional documentation, or reduced probability that programming or compiler errors will influence the end results. *See also:* diversity. (C) 610.12-1990

software documentation Technical data or information, including computer listings and printouts, in human-readable form, that describe or specify the design or details, explain the capabilities, or provide operating instructions for using the software to obtain desired results from a software system. *See also:* data; documentation; software; system; system documentation; design; user documentation. (C/SE) 729-1983s

software element A deliverable or in-process document produced or acquired during software development or maintenance. Specific examples include but are not limited to:

a) Project planning documents (for example, software development plans, and software verification and validation plans);
b) Software requirements and design specifications;
c) Test effort documentation;
d) Customer-deliverable documentation;
e) Program source code;
f) Representation of software solutions implemented in firmware;
g) Reports (for example, review, audit, projectect status) and data (for example, defect detection test).

(C/SE) 1028-1988s

software engine An engine characterized by a self-contained software module that performs a set of low-level tasks when called by an application program; for example, a database engine or an inference engine. (C) 610.10-1994w

software engineering (A) The application of a systematic, disciplined, quantifiable approach to the development, operation, and maintenance of software; that is, the application of engineering to software. **(B)** The study of approaches as in definition (A). (C/SE) 1209-1992, 610.12-1990

software engineering environment The hardware, software, and firmware used to perform a software engineering effort. Typical elements include computer equipment, compilers, assemblers, operating systems, debuggers, simulators, emulators, test tools, documentation tools, and database management systems. (C/SE) 610.12-1990, 1348-1995

software error An error in a software element which, when executed, results in unintended system operation. (VT/RT) 1483-2000

software error tolerance (programmable digital computer systems in safety systems of nuclear power generating stations) The software attribute that provides continuity of operation under postulated non-nominal conditions. 7432-1982w

software experience data Data relating to the development or use of software that could be useful in developing models, reliability predictions, or other quantitative descriptions of software. (C/SE) 729-1983s

software feature (1) (software unit testing) A software characteristic specified or implied by requirements documentation (for example, functionality, performance, attributes, or design constraints). (C/SE) 1008-1987r, 610.12-1990
(2) A distinguishing characteristic of a software item (e.g., performance, portability, or functionality). (C/SE) 829-1998

software file A generic term referring to the files and control_files that are contained within software objects and managed by the utilities in this standard. (C/PA) 1387.2-1995

software_file common class The common class that relates the two types of files defined by this standard, namely the actual files that make up the software, plus the control_files that are executed by the utilities when operating on software. (C/PA) 1387.2-1995

software_files A generic term referring to file andcontrol_ file objects (those that share the same software_file common class). (C/PA) 1387.2-1995

software hazard A software condition that is a prerequisite to an accident. (C/SE) 1228-1994

software hierarchy Hierarchical organization of objects that are managed by the software administration utilities. (C/PA) 1387.2-1995

software implementation-defined behavior Behavior, for a correct program construct and correct data, that depends on the software implementation and that each implementation shall document. (C/DA) 1481-1999

software implementation limits Restrictions imposed by an implementation. (C/DA) 1481-1999

software-intensive system A system for which software is a major technical challenge and is perhaps the major factor that affects system schedule, cost, and risk. In the most general case, a software-intensive system is comprised of hardware, software, people, and manual procedures. (C/SE) 1362-1998

software item (1) An aggregation of software, such as a computer program or database, that satisfies an end use function and is designated for purposes of specification, qualification testing, interfacing, configuration management, or other purposes. Software items are selected based on tradeoffs among software function, size, host or target computers, developer, support strategy, plans for reuse, criticality, interface considerations, need to be separately documented and controlled, and other factors. A software item is made up of one or more software units. (C/SE) J-STD-016-1995
(2) Source code, object code, job control code, control data, or a collection of these items. (C/SE) 829-1998

software librarian The person responsible for establishing, controlling, and maintaining a software library. *See also:* software library. (C/SE) 729-1983s

software library (1) A controlled collection of software and related documentation designed to aid in software development, use, or maintenance. Types include master library, production library, software development library, software repository, system library. *Synonym:* program library. (C) 610.12-1990
(2) A collection of object code units that may be linked, either statically or at run-time, with other libraries and/or object code modules to produce a software program. (C/DA) 1481-1999

software life cycle (1) The period of time that begins when a software product is conceived and ends when the software is no longer available for use. The software life cycle typically includes a concept phase, requirements phase, design phase, implementation phase, test phase, installation and checkout phase, operation and maintenance phase, and, sometimes, retirement phase. (See the corresponding figure.) *Note:* These phases may overlap or be performed iteratively.

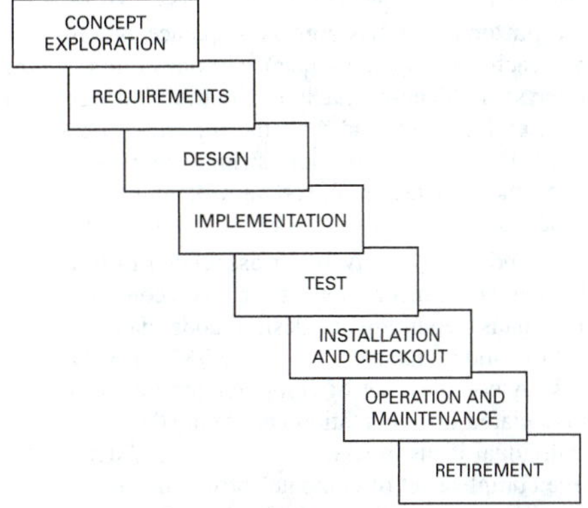

sample software life cycle

(C) 610.12-1990
(2) The various phases for software development from initial conception to final release and maintenance. (PE/EDPG) 1150-1991w
(3) The system or product cycle initiated by a user need or a perceived customer need and terminated by discontinued use of the product. The software life cycle typically includes a concept phase, requirements phase, design phase, implementation phase, test phase, installation and checkout phase, operation and maintenance phase, and, sometimes, retirement phase. These phases may overlap in time or may occur iteratively. (C/SE) 1362-1998
(4) The period of time that begins when a software product is conceived and ends when the software is no longer available for use. (C/SE) 1490-1998

software life cycle process (SLCP) The project-specific description of the process that is based on a project's software life cycle (SLC) and the Organizational Process Assets (OPA). *See also:* Organizational Process Asset. (C/SE) 1074-1997

software location The directory relative to the installed_software root directory where the relocatable files of the software have been located. (C/PA) 1387.2-1995

software maintenance (1) The set of activities that takes place to ensure that software installed for operational use continues to perform as intended and fulfill its intended role in system operation. Software maintenance includes improvements, aid to users, and related activities. (C/SE) J-STD-016-1995
(2) Modification of a software product after delivery to correct faults, to improve performance or other attributes, or to adapt the product to a modified environment. (C/SE) 1219-1998
(3) (software) *See also:* maintenance. (C) 610.12-1990

software model A symbolic model whose properties are expressed in software; for example, a computer program that models the effects of climate on the world economy. *Contrast:* narrative model; mathematical model; graphical model. (C) 610.3-1989w

software modularity (programmable digital computer systems in safety systems of nuclear power generating stations) The software attribute that provides a structure of highly independent computer program units that are discrete and identifiable with respect to compiling, combining with other units, and loading. 7432-1982w

software monitor (software) A software tool that executes concurrently with another program and provides detailed information about the execution of the other program. *See also:* monitor; hardware monitor. (C) 610.12-1990

software object An object that inherits attributes of the software common class, meaning a bundle, product, subproduct, or fileset object. (C/PA) 1387.2-1995

software packaging layout The format for software in a distribution. It contains the metadata for the distribution catalog in a well-defined exported form, as well as the files for the software objects in that distribution. (C/PA) 1387.2-1995

software pattern match string A sequence of one of more strings, each made up of a sequence of one or more characters from the shell "Pattern Matching Notation" strings described in 3.13, of POSIX.2, and with the meaning defined in that clause. If there are two or more strings, the strings are separated by the | character. A software pattern match shall be portable character string. (C/PA) 1387.2-1995

software product (1) Software or associated information created, modified, or incorporated to satisfy a contract. Examples include plans, requirements, design, code, databases, test information, and manuals. (C/SE) J-STD-016-1995
(2) (A) A complete set of computer programs, procedures, and associated documentation and data. **(B)** One or more of the individual items in (A). (C/SE) 1028-1997
(3) The complete set of computer programs, procedures, and associated documentation and data designated for delivery to a user. (C/SE) 1062-1998

software product requirements document The document that describes the full requirements for the software product that is to be developed. *Note:* Equivalent names used in the software industry are: user requirements, segment specification, software product requirements specification, and technical requirements section of the contract. (C/SE) 1298-1992w

software project The set of work activities, both technical and managerial, required to satisfy the terms and conditions of a project agreement. A software project should have specific starting and ending dates, well-defined objectives and constraints, established responsibilities, and a budget and schedule. A software project may be self-contained or may be part of a larger project. In some cases, a software project may span only a portion of the software development cycle. In other cases, a software project may span many years and consist of numerous subprojects, each being a well-defined and self-contained software project. (C/SE) 1058-1998, 1490-1998

software project management The process of planning, organizing, staffing, monitoring, controlling, and leading a software project. (C/SE) 1058.1-1987s

software project management plan The controlling document for managing a software project. A software project management plan defines the technical and managerial project functions, activities, and tasks necessary to satisfy the requirements of a software project, as defined in the project agreement. (C/SE) 1058.1-1987s

software quality (1) (A) The totality of features and characteristics of a software product that bear on its ability to satisfy given needs; for example, conform to specifications. **(B)** The degree to which software possesses a desired combination of attributes. **(C)** The degree to which a consumer or user perceives that software meets his or her composite expectations. **(D)** The composite characteristics of software that determine the degree to which the software in use will meet the expectations of the customer. *See also:* software; specification; software product. (C/SE) 729-1983
(2) The ability of software to satisfy its specified requirements. (C/SE) J-STD-016-1995

software quality assurance *See:* quality assurance.

software quality management That aspect of the overall software management function:Software management that determines and implements the software quality policy. (C/SE) 1074-1995s

software quality metric A function whose inputs are software data and whose output is a single numerical value that can be interpreted as the degree to which software possesses a given attribute that affects its quality. (C/SE) 1061-1998

software quality policy The overall quality intentions and direction of an organization as regards software quality, as expressed by top management. (C/SE) 1074-1995s

software reliability (1) (software reliability) The probability that software will not cause the failure of a system for a specified time under specified conditions. The probability is a function of the inputs to and use of the system as well as a function of the existence of faults in the software. The inputs to the system determine whether existing faults, if any, are encountered. (SE/C) 982.2-1988, 982.1-1988, 729-1983s
(2) The ability of a program to perform a required function under stated conditions for a stated period of time. *See also:* software; program; function; system; failure. (C/SE) 729-1983s

software reliability management The process of optimizing the reliability of software through a program that emphasizes software error prevention, fault detection and removal, and the use of measurements to maximize reliability in light of project constraints such as resources (cost), schedule, and performance. (C/SE) 982.2-1988, 982.1-1988

software repository A software library providing permanent, archival storage for software and related documentation. *Contrast:* system library; software development library; master library; production library. (C) 610.12-1990

software requirements review (SRR) (A) A review of the requirements specified for one or more software configuration items to evaluate their responsiveness to and interpretation of the system requirements and to determine whether they form a satisfactory basis for proceeding into preliminary design of the configuration items. *Note:* This review is called software specification review by the U.S. Department of Defense. *See also:* system requirements review. **(B)** A review as in definition (A) for any software component. (C) 610.12-1990

software requirements specification (SRS) Documentation of the essential requirements (i.e., functions, performance, design constraints, and attributes) of the software and its external interfaces. The software requirements are derived from the system specification. (C/SE) 1012-1998

software reuse The reapplication of previously developed software or information associated with that software. (SCC20) 1226-1998

software reuse resources Reuse libraries, stores of reusable assets, software reuse services, and suppliers. (C/SE) 1430-1996

software safety Freedom from software hazards. (C/SE) 1228-1994

software safety program A systematic approach to reducing software risks. (C/SE) 1228-1994

software sneak analysis A technique applied to software to identify latent (sneak) logic control paths or conditions that could inhibit a desired operation or cause an unwanted operation to occur. *See also:* software. (C/SE) 729-1983s

software_spec A string that is used to identify one or more software objects for input to a software administration utility. (C/PA) 1387.2-1995

software specification review (SSR) *See:* software requirements review.

software system (1) Software that is the subject of a single software project. (C/SE) 1074-1995s
(2) A system consisting solely of software and possibly the computer equipment on which the software operates. (C/SE) J-STD-016-1995
(3) A software-intensive system for which software is the only component to be developed or modified. *See also:* software-intensive system. (C/SE) 1362-1998

software test incident (software unit testing) (software) Any event occurring during the execution of a software test that requires investigation. (C/SE) 610.12-1990, 1008-1987r

software tools **(1)** A computer program used in the development, testing, analysis, or maintenance of a program or its documentation. Examples include comparator, cross reference generator, decompiler, driver, editor, flowcharter, monitor, test case generator, and timing analyzer. (PE/C/NP) 7-4.3.2-1993, 610.12-1990
(2) Computer programs used to aid in the development, testing, analysis, or maintenance of a computer program or its documentation. (C/SE) 730.1-1995

software transition The set of activities that enables responsibility for software development to pass from one organization, usually the organization that performs initial software development, to another, usually the organization that will perform software maintenance. (C/SE) J-STD-016-1995

software unit An element in the design of a software item; for example, a major subdivision of a software item, a component of that subdivision, a class, object, module, function, routine, or database. Software units may occur at different levels of a hierarchy and may consist of other software units. Software units in the design may or may not have a one-to-one relationship with the code and data entities (routines, procedures, databases, data files, etc.) that implement them or with the computer files containing those entities. (C/SE) J-STD-016-1995

software user document Body of material that provides information to users; typically printed or stored on some medium in the format of a printed document. (C/SE) 1063-1987r

software verification and validation plan (SVVP) A plan describing the conduct of software V&V. (C/SE) 1012-1998

software verification and validation report (SVVR) Documentation of V&V results and software quality assessments. (C/SE) 1012-1998

soil structure interaction (SSI) A general concept for effects caused by the influence of the soil dynamic behavior on the response of a structure. (SUB/PE) C37.122.1-1993

solar activity The emission of electromagnetic radiation and particles from the sun, including slowly varying components and transient components caused by phenomena such as solar flares. (AP/PROP) 211-1997

solar activity center A region on the sun containing the sources of variable electromagnetic and corpuscular radiation. (AP/PROP) 211-1997

solar activity index A number characterizing solar activity. Examples are international relative sunspot number, twelve-month running mean sunspot number, and monthly mean solar radio-noise flux. (AP/PROP) 211-1997

solar array (photovoltaic power system) A group of electrically interconnected solar cells assembled in a configuration suitable for oriented exposure to solar flux. (AES) [41]

solar constant **(1) (illuminating engineering)** The irradiance (averaging 1 353 W/m^2 (125.7 W/ft^2), from the sun at its mean distance from the earth 92.9 × 10^6 miles (1.5 × 10^{11}m), before modification by the earth's atmosphere. (EEC/IE) [126]
(2) (electric power systems in commercial buildings) The solar intensity incident on a surface that is oriented normal to the sun's rays and located outside the earth's atmosphere at a distance from the sun that is equal to the mean distance between the earth and the sun. (IA/PSE) 241-1990r

solar cycle The magnitude of slowly varying components of solar activity as a function of time. The solar cycle has a period of approximately 11 years. *Note:* The cycle is not symmetrical. It rises to a maximum in approximately 4 years and declines to a minimum in approximately 7 years. (AP/PROP) 211-1997

solar induced currents (power fault effects) Spurious, quasi-direct currents flowing in grounded power differences due to geomagnetic storms resulting from the particle emission of solar flares erupting from the surface of the sun. *Synonym:* geomagnetically induced currents. *See also:* auroral effects. (PE/PSC) 367-1987s

solar noise (communication satellite) Electrical noise generated by the sun. Exceeds other background noise sources by several orders of magnitude. (COM) [25]

solar panel *See:* solar array.

solar radiation simulator (illuminating engineering) A device designed to produce a beam of collimated radiation having a spectrum, flux density, and geometric characteristics similar to those of the sun outside the earth's atmosphere. (EEC/IE) [126]

solar wind (communication satellite) Energetic particles emitted by the sun and travelling through space. (COM) [19]

solderability That property of a metal surface to be readily wetted by molten solder. (EEC/AWM) [105]

soldered joints The connection of similar or dissimilar metals by applying molten solder, with no fusion of the base metals. (EEC/AWM) [105]

solder projections Icicles, nubs, and spikes are undesirable protrusions from a solder joint. (EEC/AWM) [105]

solder side By convention, the side of the module opposite to the component side. This is the left side when looking at an IEEE 1101.1 system through the front door. (C/MM) 1101.2-1992

solder splatter Unwanted fragments of solder. (EEC/AWM) [105]

solenoid An electric conductor wound as a helix with a small pitch, or as two or more coaxial helixes. *See also:* solenoid magnet. (Std100) 270-1966w

solenoid magnet (solenoid) An electromagnet having an energizing coil approximately cylindrical in form, and an armature whose motion is reciprocating within and along the axis of the coil. (IA/ICTL/IAC) 270-1966w, [60]

solenoid relay *See:* plunger relay.

soleplate (rotating machinery) A support fastened to a foundation on which a stator frame foot or a bracket arm can be mounted. *See also:* slide rail. (PE) [9]

solicited messages A negotiated data transfer in message space. *See also:* message space; data transfer. (C/MM) 1296-1987s

solicited status Information generated by the peripheral in response to a command from the host. (C/MM) 1284-1994

solid angle (ω) (laser maser) The ratio of the area on the surface of a sphere to the square of the radius of that sphere. It is expressed in steradians. (LEO) 586-1980w

solid angle factor (Q) (illuminating engineering) A function of the solid angle (ω) subtended by a source and is given by

$$Q = 20.4\omega = 1.52\omega^{0.2} - 0.075$$

See also: index of sensation. (EEC/IE) [126]

solid-beam efficiency The ratio of the power received over a specified solid angle when an antenna is illuminated isotropically by uncorrelated and unpolarized waves to the total power received by the antenna. *Note:* This term is sometimes used to mean the ratio of the power received corresponding to a particular polarization over the solid angle to the total power received. Equivalently, the term is used to mean the ratio of the power radiated over a specified solid angle by the antenna corresponding to a particular polarization to the total power radiated. (AP/ANT) 145-1993

solid bushing (outdoor electric apparatus) A bushing in which the major insulation is provided by a ceramic or analogous material. (PE/TR) 21-1976

solid conductor A conductor consisting of a single wire. *See also:* conductor. (T&D/PE) [10]

solid contact A contact having relatively little inherent flexibility and whose contact pressure is supplied by another member. (SWG/PE) C37.100-1981s

solid coupling (rotating machinery) A coupling that makes a rigid connection between two shafts. *See also:* rotor. (PE) [9]

solid electrolytic capacitor A capacitor in which the dielectric is primarily an anodized coating on one electrode, with the remaining space between the electrodes filled with a solid semiconductor. (PE/EM) 43-1974s

solid enclosure An enclosure that will neither admit accumulations of flyings or dust nor transmit sparks or flying particles to the accumulations outside. (EEC/PE) [119]

solid insulations (cable-insulation materials) Firm, essentially homogeneous, dielectric materials comprising virtually complete solid-phase structures and having no liquid phase. (PE) 402-1974w

solid-iron cylindrical-rotor generator *See:* cylindrical-rotor generator.

solidly grounded (power and distribution transformers) Grounded through an adequate ground connection in which no impedance has been inserted intentionally. *Note:* Adequate as used herein means suitable for the purpose intended. (PE/TR) C57.12.80-1978r

solid-material fuse unit A fuse unit in which the arc is drawn through a hole in solid material. (SWG/PE) C37.40-1993, C37.100-1992

solid modeling A method of displaying solid constructions on a graphical display device using geometric forms such as cubes, cones, spheres, and cylinders. (C) 610.6-1991w

solid-pole synchronous motor A salient-pole synchronous motor having solid steel pole shoes, and either laminated or solid pole bodies. (PE) [9]

solid rotor (A) (rotating machinery) A rotor, usually constructed of a high-strength forging, in which slots may be machined to accommodate the rotor winding. **(B) (rotating machinery)** A spider-type rotor in which spider hub is not split. *See also:* rotor. (PE) [9]

solid-state component A component whose operation depends on the control of electric or magnetic phenomena in solids, for example, a transistor, crystal diode, ferrite core. (C) [20], [85]

solid state controller An electric controller that utilizes a static power converter as the primary switching device. (IA/MT) 45-1998

solid-state converter static (induction and dielectric heating equipment) A solid state generator or power source that utilizes semiconductor devices to control the switching of currents through inductive and capacitive circuit elements and thus generate a useable alternating current at a desired output frequency. (IA) 54-1955w

solid-state device (control equipment) A device that may contain electronic components that do not depend on electronic conduction in a vacuum or gas. The electrical function is performed by semiconductors or the use of otherwise completely static components such as resistors, capacitors, etc. (PE/PSE) 94-1970w

solid state network relay A solid state relay which performs the combined functions of the master and phasing relays. (PE/TR) C57.12.44-1994

solid-state protector A protective device that employs solid-state circuit elements that provide a combination of high speed voltage and current sensing. These protectors are a combination of voltage clamps (avalanche diodes) and crowbar devices (multilayer diodes similar to SCRs), and are designed to limit the voltage to a specific value and to reduce current flow to low values of milliamperes within nanoseconds. They are usually integrated into the terminal apparatus. (PE/PSC) 487-1992

solid-state relay (or relay unit) A static relay or relay unit constructed exclusively of solid-state components. (SWG/PE/PSR) C37.100-1992, C37.90-1978s

solid-state scanning (facsimile) A method in which all or part of the scanning process is due to electronic commutation of a solid-state array of thin-film photosensitive elements. *See also:* facsimile. (COM) [49]

solid-type paper-insulated cable Oil-impregnated, paper-insulated cable, usually lead covered, in which no provision is made for control of internal pressure variations. (T&D/PE) [10]

solitary wave A propagating wave disturbance where the effects of media dispersion and non-linearity compensate one another to produce a self-preserving wave shape. *Synonym:* soliton. (AP/PROP) 211-1997

soliton *See:* solitary wave.

solution *See:* check solution.

solution domain The environment in which a solution or set of solutions resides. *See also:* problem domain. (C/SE) 1362-1998

solvent cleaning (electroplating) Cleaning by means of organic solvents. *See also:* electroplating. (PE/EEC) [119]

solventless (rotating machinery) A term applied to liquid or semiliquid varnishes, paints, impregnants, resins, and similar compounds that have essentially no change in weight or volume when converted into a solid or semisolid. (PE) [9]

somatic cells *See:* genetic effect.

son *See:* child node.

sonar (navigation aids) A general name for sonic and ultrasonic ranging, sounding and communication systems. (AES/GCS) 172-1983w

son file A file that contains data that have been updated from those in another file, called the father file. *See also:* grandfather file; father file. (C) 610.5-1990w

sonic delay line *See:* acoustic delay line.

sonic depth finder (navigation aids) A direct reading instrument which determines the depth of water by measuring the time interval between emission of sound and the return of its echo from the bottom. (AES/GCS) 172-1983w

sonic pen A pick device that is sensitive to audio signals. *See also:* light pen. (C) 610.6-1991w, 610.10-1994w

sonne (navigation aid terms) A radio navigation aid that provides a number of characteristic signal zones which rotate in a time sequence; a bearing may be determined by observation (by interpolation) of the instant at which transition occurs from one zone to the following zone. *See also:* consol. (AES/GCS) 172-1983w

sonobuoy (navigation aid terms) A buoy with equipment for automatically transmitting a radio signal when triggered by an underwater sound signal. (AES/GCS) 172-1983w

sort To arrange data or items in an ordered sequence by applying specific rules. (MIL/C) [2], [85], [20]

(2) (A) (data management) To arrange items according to a specified order of their sort keys. For example, to arrange the records of a personnel file into alphabetical sequence using the sort key "Employee-name." *See also:* radix sort; internal sort; distribution sort; merge sort; exchange sort; selection sort; insertion sort; external sort. **(B) (data management)** To segregate items into subsets according to specified criteria. **(C) (data management)** A process that achieves the arrangement or segregation described in definition (A) or (B). (C) 610.5-1990

sort by merging (data management) To sort the items of a set by splitting the set into subsets, sorting the subsets, and merging the subsets. *Synonym:* sequence by merging. *See also:* order-by-merging. (C) 610.5-1990w

sorter (1) A person, device, or computer routine that sorts. (C) [20], [85]

(2) (data management) A mechanical device that deposits punched cards in pockets based on the hole patterns in the cards. (C) 610.5-1990w

(3) *See also:* card sorter. (C) 610.10-1994w

sorting item (A) That item of a set that is actively being exchanged or manipulated with other elements during the sorting process. *See also:* sort selection. **(B)** Any element of a set that has a probability of being selected by a sort selection. (C) 610.5-1990

sorting rewind time In a tape merge sort, the length of time needed to rewind a tape to its original position. (C) 610.5-1990w

sorting-sequencing key *See:* sort key.

sorting string A string of characters used as a sort key. (C) 610.5-1990w

sort key A key field whose value is used to determine the position of items within a sorted set. *Synonyms:* sorting-sequencing key; sequencing key. *See also:* sorting string; sort. (C) 610.5-1990w

sort order* *See:* order.

* Deprecated.

sort pass (A) In a sorting algorithm, a single processing of all the items of a set. **(B)** A phase of a merge sort that reads a subset of unsorted data items, orders them, and places the ordered subset on a data medium. This process is repeated until all input data is placed in some subset. The merge phase is then begun to merge the subsets into one ordered set. (C) 610.5-1990

sort selection (A) The choice of a particular sorting algorithm. **(B)** The process of choosing an item to be exchanged with another item as part of a selection sorting process. *See also:* sorting item. (C) 610.5-1990

SOS *See:* radio distress signal.

sound (A) An oscillation in pressure, stress, particle displacement, particle velocity, etc., in a medium with internal forces (for example, elastic, viscous), or the superposition of such propagated oscillations. **(B)** An auditory sensation evoked by the oscillation described above. *Note:* In case of possible confusion, the term sound wave or elastic wave may be used for concept (A) and the term sound sensation for concept (B). Not all sound waves can evoke an auditory sensation, for example, an ultrasonic wave. 2. The medium in which the sound exists is often indicated by an appropriate adjective, for example, air-borne, water-borne, structure-borne. (SP) [32]

sound absorption (A) The change of sound energy into some other form, usually heat, in passing through a medium or on striking a surface. **(B)** The property possessed by material and objects, including air, of absorbing sound energy. (SP) [32]

sound-absorption coefficient (surface) The ratio of sound energy absorbed or otherwise not reflected by the surface, to the sound energy incident upon the surface. Unless otherwise specified, a diffuse sound field is assumed. (SP) [32]

sound analyzer A device for measuring the band pressure level, or pressure spectrum level, of a sound at various frequencies. *Notes:* 1. A sound analyzer usually consists of a microphone, an amplifier and wave analyzer, and is used to measure amplitude and frequency of the components of a complex sound. 2. The band pressure level of a sound for a specified frequency band is the effective root-mean-square sound pressure level of the sound energy contained within the bands. *See also:* instrument. 157-1951w

sound articulation (percent sound articulation) The percent articulation obtained when the speech units considered are fundamental sounds (usually combined into meaningless syllables). *See also:* volume equivalent. (PE/EEC) [119]

sound buoy (navigation aid terms) A buoy equipped with a characteristic sound signal. *See also:* buoy. (AES/GCS) 172-1983w

sound-detection system (protective signaling) A system for the protection of vaults by the use of sound-detecting devices and relay equipment to pick up and convert noise, caused by burglarious attack on the structure, to electric impulses in a protection circuit. *See also:* protective signaling. (EEC/PE) [119]

sound-effects filter (electroacoustics) A filter used to adjust the frequency response of a system for the purpose of achieving special aural effects. *See also:* filter. (SP) 151-1965w

sound energy Of a given part of a medium, the total energy in this part of the medium minus the energy that would exist in the same part of the medium with no sound waves present. (SP) [32]

sound field A region containing sound waves. (SP) [32]

sounding, active *See:* active sounding.

sound intensity (1) (power station noise control) The average rate of sound energy radiated by a source per unit time. (PE/EDPG) 640-1985w

(2) (sound-energy flux density) (sound power density) (in a specified direction at a point) The average rate of sound energy transmitted in the specified direction through a unit area normal to this direction at the point considered. *Notes:* 1. The sound intensity in any specified direction a of a sound field is the sound-energy flux through a unit area normal to that direction. This is given by the expression

$$I_a = \frac{1}{T}\int_0^T pv_a \mathrm{d}t$$

where

T = an integral number of periods or a long time compared to a period

p = the instantaneous sound pressure

v_a = the component of the instantaneous particle velocity in the direction a

t = time

2. In the case of a free plane or spherical wave having an effective sound pressure, p, the velocity of propagation c, in a medium of density ρ, the intensity in the direction of propagation is given by

$$I = \frac{p^2}{\rho c}$$

(SP) [32]

sound level (1) (measurement of sound pressure levels of ac power circuit breakers) Weighted sound pressure level obtained by the use of a metering characteristic and the weightings A, B, C (or other) as specified. The weighting used must be indicated. (SWG/PE) C37.100-1992, C37.082-1982r

(2) (overhead power lines) A weighted sound pressure level, obtained by the use of metering characteristics and the weightings A, B, C, or D specified in ANSI S1.4-1983. The weightings employed must always be stated. The reference pressure is always 20 μPa. *Notes:* 1. The meter reading (in decibels) corresponds to a value of the sound pressure integrated over the audible frequency range with a specified frequency weighting and integration time. 2. A suitable method of stating the weighting is, for example, "The A-weighted sound level was 43 dB," or "The sound level was 490 dB (A)." 3. Weightings are based on psychoacoustically determined time or frequency responses in objective measuring equipment. This is done to obtain data that better predict the subjective listener reaction than would wide-band measurements with a meter having either an instantaneous time response or a slow average or rms response. (T&D/PE) 539-1990

sound level, A-weighted *See:* A-weighted sound level.

sound-level meter An instrument including a microphone, an amplifier, an output meter, and frequency-weighting networks for the measurement of noise and sound levels in a specified manner. *Notes:* 1. The measurements are intended to approximate the loudness level of pure tones that would be obtained by the more-elaborate ear balance method. 2. Loudness level in phons of a sound is numerically equal to the sound pressure level in decibels relative to 0.0002 μbar of a simple tone of frequency 1000 Hz that is judged by the listeners to be equivalent in loudness. 3. Specifications for sound-level meters are given in American National Standard Specification for Sound-Level Meters, S1.4-1971 (or latest revision thereof). *See also:* instrument. (SP) [32]

sound power (power station noise control) The total sound energy radiated by a source per unit time. (PE/EDPG) 640-1985w

sound power level (L_w) (1) (airborne sound measurements on rotating electric machinery) The sound power level, in decibels, is equal to 10 times the logarithm to the base 10 of the ratio of a given power to the reference power, 10^{-12} W.

$$L_W = 10\log_{10}\left(\frac{W}{W_0}\right)$$

where
L_w = sound power level
W = measured sound power in watts
W_0 = reference power

(PE/EM) 85-1973w

(2) (in decibels) Ten times the logarithm to the base ten of the emitted sound power (w) to the reference power of 10^{-12} W, (w_o), or

$$L_W = 10 \times \log_{10}\left(\frac{W}{W_0}\right)$$

(PE/TR) C57.12.90-1999

sound power level, A-weighted *See:* A-weighted sound power level.

sound pressure (1) (power station noise control) The instantaneous pressure measured in a sound wave, that is, the variation in atmospheric pressure. (PE/EDPG) 640-1985w
(2) (transmission performance of telephone sets) The sound pressure at a point, is the total instantaneous pressure at that point, in the presence of a sound wave, minus the static pressure at that point. (COM/TA) 269-1983s

sound pressure, effective *See:* effective sound pressure.

sound pressure, instantaneous *See:* instantaneous sound pressure.

sound pressure level (1) (overhead power lines) Twenty times the logarithm to the base 10 of the ratio of the pressure of a sound to the reference pressure, expressed in decibels. The reference pressure shall be explicitly stated. *Notes:* 1. The following reference pressures are in common use: 20 micropascals (μPa), and (2) 0.1 pascal (Pa). Reference pressure (A) is in general use for measurements concerned with hearing and with sound in air and liquids, while reference pressurehas gained widespread acceptance for calibration of transducers and various kinds of sound measurements in liquids. 2. Unless otherwise explicitly stated, it is to be understood that the sound pressure is the effective (rms) sound pressure. 3. It is to be noted that in many sound fields the sound pressure ratios are not the square roots of the corresponding power ratios. (SWG/T&D/PE) 539-1990, C37.082-1982r
(2) The sound pressure level, in decibels, of a sound is 20 times the logarithm to the base 10 of the ratio of the pressure of this sound to the reference pressure. (COM/TA) 269-1992, 1206-1994
(3) Twenty times the logarithm to the base 10 of the ratio of the pressure of a sound to the reference sound pressure. Unless otherwise specified, the effective rms pressure to be used. The reference sound pressure is 20 μPa. Unit: decibel (dB). (SWG/PE) C37.100-1992
(4) [in decibels (dB)] Twenty times the logarithm to the base 10 of the ratio of the measured sound pressure (p) to a reference pressure (p_o) of 20 (μPa), or

$$L_p = (20 \times \log_{10})\frac{p}{p_o}$$

(PE/TR) C57.12.90-1993s

(5) Twenty times the logarithm to the base 10 of the ratio of the pressure of the sound to the reference pressure. The reference pressure is normally 1 Pascal (Pa), and sound pressure levels are expressed in dB re 1 Pa (dBPa). When a reference pressure of 20 μPa is used, the sound pressure level will be expressed as dBSPL. Unless otherwise indicated, rms values of pressure are used. Most telephony acoustic measurements are referenced to 1 Pa. However, measurements such as receive noise and room noise are generally referenced to 20 uPa. *Note:* 0 dB Pa = 94 dBSPL, 0 dBSPL = 20 μPa, 1 Pa = 1 N/m^2. An A-weighted sound pressure level in dB (dBSPL, A-weighted) is often abbreviated as dBA or dB(A). (COM/TA) 1329-1999

sound probe A device that responds to some characteristic of an acoustic wave (for example, sound pressure, particle velocity) and that can be used to explore and determine this characteristic in a sound field without appreciably altering the field. *Note:* A sound probe may take the form of a small microphone or a small tubular attachment added to a conventional microphone. *See also:* instrument. (SP) [32]

sound recording system A combination of transducing devices and associated equipment suitable for storing sound in a form capable of subsequent reproduction. *See also:* phonograph pickup. (SP) [32]

sound reflection coefficient (surface) The ratio of the sound reflected by the surface to the sound incident upon the surface. Unless otherwise specified, reflection of sound energy in a diffuse sound field is assumed. (SP) [32]

sound reproducing system A combination of transducing devices and associated equipment for reproducing recorded sound. (SP) [32]

sound spectrum analyzer (sound analyzer) A device or system for measuring the band pressure level of a sound as a function of frequency. (SP) [32]

sound tract (electroacoustics) A band that carries the sound record. In some cases, a plurality of such bands may be used. In sound film recording, the band is usually along the margin of the film. *See also:* phonograph pickup. (SP) [32]

sound transmission coefficient (interface or partition) The ratio of the transmitted to incident sound energy. Unless otherwise specified, transmission of sound energy between two diffuse sound fields is assumed. (SP) [32]

source (1) (laser maser) Taken to mean either laser of laser-illuminated reflecting surface. (LEO) 586-1980w
(2) (metal-nitride-oxide field-effect transistor) Region in the device structure of an insulated-gate-field-effect transistor (IGFET) which contains the terminal from which charge carries flow into channel toward the drain. It has the potential which is less attractive than the drain for the carriers in the channel. (ED) 581-1978w
(3) The node that creates a send or echo packet. The source nodeId is contained in the third symbol of the packet. (C/MM) 1596-1992
(4) A generator of normal characters at a link interface. *See also:* normal character. (C/BA) 1355-1995
(5) The specification of a source distribution object for a software administration utility. The source host provides a means to locate the source role and the source path is a path accessible to the source host. (C/PA) 1387.2-1995
(6) A node that initiates a bus transfer. (C/MM) 1394-1995
(7) The start of a delay arc; that is, the origin of the logic signal. For arcs across cell instances, the source is the receiver pin; for arcs across interconnect, the source is the driver pin. (C/DA) 1481-1999

source address (SA) (1) The address of a device or storage location from which data is to be transferred. *Contrast:* destination address; destination. (C) 610.12-1990, 610.10-1994w
(2) (local area networks) A field in the message packet format identifying the sending end node. (C) 8802-12-1998

source, calibrated *See:* calibrated source.

source, check *See:* check source.

source code (1) Computer instructions and data definitions expressed in a form suitable for input to an assembler, compiler, or other translator. *Note:* A source program is made up of source code. *Contrast:* object code. (C) 610.12-1990
(2) When dealing with the Shell Command Language, source code is input to the command language interpreter; the term *shell script* is synonymous with this meaning. When dealing with the C-Language Bindings Option, source code is input to a C compiler conforming to the C Standard. When dealing with another ISO/IEC conforming language, source code is input to a compiler conforming to that ISO/IEC standard. Source code also refers to the input statements prepared for

the following standard utilities: awk, bc, ed, lex, localedef, make, sed, and yacc. Source code can also refer to a collection of sources meeting any or all of these meanings. (C/PA) 9945-2-1993
(3) A piece of software that has not yet been compiled or assembled, and appears in the language used by the programmer, and thus cannot yet run on a machine. (PE/SUB) 1379-1997

source code generator *See:* code generator.

source data card A data card which contains manually or mechanically recorded data that are to be subsequently punched into the same card. (C) 610.10-1994w

source document A document containing information that is to be input to a computer. For example, an original invoice, a library charge-out card, or a machine-readable document. (C) 610.2-1987

source domain The MD that supplies a piece of trace information. (C/PA) 1224.1-1993w

source efficiency (fiber optics) The ratio of emitted optical power of a source to the input electrical power. (Std100) 812-1984w

source ground (signal-transmission system) Potential reference at the physical location of a source, usually the signal source. *See also:* signal. (IE) [43]

source host The host portion of a source specification. (C/PA) 1387.2-1995

source impedance (1) The impedance presented by a source of energy to the input terminals of a device, or network. *See also:* self-impedance; network analysis; input impedance. (SP/IM/HFIM) 151-1965w, [40]
(2) The Thevenin equivalent impedance of an electrical system at the terminal of a transmission line. In network applications, this impedance can vary depending on the location of the fault on the transmission line and the status (i.e., opened or closed) of other terminals associated with the transmission line. (PE/PSR) C37.113-1999
(3) The source impedance is defined to begin at the power source termination and end at the power source return termination. Hence, all cabling is included in the source impedance. The cables and interconnects used in the system should be used during test. Line impedance stabilization networks (LISNs) may be used in series with each input line to provide a uniform standard for source impedance. Different LISNs may be used for different test applications to test the UUT. (PEL) 1515-2000

source language (software) The language in which the input to a machine-aided translation process is represented. For example, the language used to write a computer program. *Contrast:* target language. (C) 610.2-1987, 610.12-1990

source/load impedance (loudness ratings of telephone connections) The source/load impedance used for determining loudness ratings is considered to be 900 Ω resistive. *See also:* source impedance; load impedance. (COM/TA) 661-1979r

source node (1) (network analysis) A node having only outgoing branches. (CAS) 155-1960w
(2) A terminal node which originates data. *See also:* destination node. (C/BA) 1355-1995

source of fault current A terminal that contributes a significant amount of current to a fault on the protected line. Note that it is not necessary for generation to be connected to a terminal for it to be a source of fault current. For instance, large synchronous motor loads can contribute significant amounts of fault current for a few cycles within the duration of fault clearing. Transformers can also be a significat source of zero-sequence currents to unbalanced faults involving ground if they have winding with a grounded neutral connected to the line, and also have a delta or zig-zag winding. (PE/PSR) C37.113-1999

source path The pathname portion of a source specification. (C/PA) 1387.2-1995

source program (software) A computer program that must be compiled, assembled, or otherwise translated in order to be executed by a computer. *Contrast:* object program. (C) 610.12-1990

source quench In networking, a method for controlling congestion in which a device detects the congestion and requests that the source stop transmitting. *See also:* fair queuing. (C) 610.7-1995

source, radioactive *See:* radioactive source.

source, radioactivity standard *See:* radioactivity standard source.

source resistance The resistance presented to the input of a device by the source. *See also:* measurement system. (EEC/EMI) [112]

source resistance rating The value of source resistance that, when injected in an external circuit having essentially zero resistance, will either double the dead band or shift the dead band by one-half its width. *See also:* measurement system. (EEC/EMI) [112]

source role Where the software exists in a form suitable for distribution, forming a context for the establishment of a repository of software from which the manager may choose to distribute to targets. Software exists in the source until it is removed by a task initiated by the manager. The source role provides a repository where software may be stored and provides access for those roles that require the software. (C/PA) 1387.2-1995

source routing (1) A bridging technique where frames contain the list of bridges and networks that must be traversed for the frame to reach the destination. In this scheme, the transmitter must know the route to the destination before sending the frame. *Contrast:* spanning tree. (C) 610.7-1995
(2) A mechanism to route frames through a bridged LAN. Within the source routed frame, the station specifies the route that the frame will traverse. (C/LM) 8802-5-1998
(3) The capability for a source to specify the path that a frame will use to traverse the bridged network. (C/LM/CC) 8802-2-1998

Source Routing Transparent (SRT) The bridging technology defined by ISO/IEC 10038:1993, annex C, as an extension to the transparent bridging rules allowing the source station to specify the path through the bridged network (source routing). (C/LM/CC) 8802-2-1998

source segment In a hierarchical database, a segment that contains the data used to construct a secondary index. (C) 610.5-1990w

source/sink device Source devices originate signals, whereas sink devices terminate signals. Examples of source/sink devices include channel banks and digital crossconnect systems. (COM/TA) 1007-1991r

source statements (SS) The encoded logic of the software product. (C/SE) 1045-1992

source-to-line impedance ratio (SIR) The ratio of the source impedance behind a relay terminal to the line impedance. (PE/PSR) C37.113-1999

SP *See:* space character.

SPA *See:* scratchpad area.

space (1) (data transmission) One of the two possible conditions of an element (bit); an open line in a neutral circuit. In Morse code, a duration of two unit intervals between characters and six unit intervals between words. (PE) 599-1985w
(2) (A) To advance the reading or display position according to a prescribed format. **(B) (data management)** A site intended for the storage of data such as a location in a storage medium. **(C) (data management)** A basic unit of area such as the size of a single character. **(D) (data management)** One or more space characters. (C) [20], 610.5-1990
(3) The absence of a signal; for example, in data communications, the "zero's" state. *Contrast:* mark. (C) 610.10-1994w

(4) The lighter element of a bar code—usually formed by the background between the darker elements of the bar code. (PE/TR) C57.12.35-1996

⟨space⟩ The character defined as ⟨space⟩. The ⟨space⟩ character is a member of the space character class of the current locale, but represents the single character and not all of the possible members of the class. (C/PA) 9945-2-1993

space character (1) A graphic character that is usually represented by a blank site in a series of graphics. The space character, though not a control character, has the function equivalent to that of a format effector that causes the print or display position to move one position forward without producing the printing or display of any graphic. Similarly, the space character may have a function equivalent to that of an information separator. *See also:* null character; space. (C) 610.5-1990w
(2) A byte (hex 20) used in text strings that represents a space. (C/MM) 1284.1-1997

space charge (1) (general) A net excess of charge of one sign distributed throughout a specified volume. (ED) 161-1971w
(2) (thermionics) Electric charge in a region of space due to the presence of electrons and/or ions. *See also:* electron emission. (ED) 161-1971w

space-charge-control tube *See:* density-modulated tube.

space-charge debunching Any process in which the mutual interactions between electrons in the stream disperse the electrons of a bunch. (ED) 161-1971w

space-charge density (thermionics) The space charge per unit volume. *See also:* electron emission. (ED) 161-1971w

space-charge filter (A) A device used to measure net space-charge density in which a filter medium is used to remove the charge from an air stream. **(B)** A device, used to measure net space-charge density, in which a filter medium is used to remove the charge from an airstream. (T&D/PE) 1227-1990, 539-1990

space-charge–free electric field The electric field due to a system of energized electrodes, excluding the effect of space charge present in the interelectrode space. (T&D/PE) 539-1990

space-charge generation (1) (germanium gamma-ray detectors) (charged-particle detectors) (x-ray energy spectrometers) (semiconductor radiation detectors) The thermal generation of free charge carriers in the space-charge region. (NPS/NID) 759-1984r, 301-1976s, 300-1988r
(2) (in a semiconductor radiation detector) The thermal generation of free charge carriers in the depletion region. (NPS) 325-1996

space-charge grid A grid, usually positive, that controls the position, area, and magnitude of a potential minimum or of a virtual cathode in region adjacent to the grid. *See also:* electrode; grid. (ED) 161-1971w

space-charge-limited current (electron vacuum tubes) The current passing through an interelectrode space when a virtual cathode exists therein. *See also:* electrode current. (ED) 161-1971w

space-charge perturbation of the electric field in the interelectrode space A change in the electric field caused by the presence of space charge in the interelectrode space. *Note:* The electric field at ground level under dc transmission lines is generally increased due to the presence of monopolar space charge having the same polarity as the nearest conductor. This increase is generally termed "field enhancement." (T&D/PE) 539-1990

space-charge region (1) (x-ray energy spectrometers) (charged-particle detectors) (of a semiconductor radiation detector) A region in which the net charge density is significantly different from zero. *See also:* depletion region. (NPS/ED/NID) 759-1984r, 216-1960w, 301-1976s, 300-1988r
(2) *See also:* depletion region. (NPS) 325-1996

space correction A method of register control that takes the form of a sudden change in the relative position of the web. (IA/ICTL/IAC) [60]

spacecraft (communication satellite) Any type of space vehicle, including an earth satellite or deep-space probe, whether manned or unmanned, and also rockets and high-altitude balloons which penetrate the earth's outer atmosphere. (COM) [19]

space current (electron tube) Synonym in a diode or equivalent diode of cathode current. *See also:* electrode current; leakage current; load current; quiescent current. (ED) [45], [84]

space diversity *See:* space diversity reception.

space diversity reception (data transmission) That form of diversity reception that utilizes receiving antennas placed in different locations. (PE) 599-1985w

space-division digital switching (telephone switching systems) Digital switching with separate paths for each call. (COM) 312-1977w

space-division switching (1) (telephone switching systems) A method of switching that provides a separate path for each of the simultaneous calls. (COM) 312-1977w
(2) A circuit-switching method in which each connection through the switch takes a physically separate and dedicated path. *See also:* circuit switching; message switching; time multiplexed switching. (C) 610.7-1995

space factor (rotating machinery) The ratio of the sum of the cross-sectional areas of the active or specified material to the cross-sectional area within the confining limits specified. *See also:* asynchronous machine; slot space factor. (PE) [9]

space, head *See:* head space.

space heater (1) A heater that warms occupied spaces. (PE) [9]
(2) (rotating machinery) A device that warms the ventilating air within a machine and prevents condensation of moisture during shut-down periods. (PE) [9]

space pattern (television) A geometrical pattern on a test chart designed for the measurement of geometric distortion. (BT) 202-1954w

space potential The electric potential at any point in space relative to some reference potential, usually ground. It is the electric potential difference between the reference point and the point in question. (T&D/PE) 539-1990

space probe (communication satellite) A spacecraft with a trajectory extending into deep space. (COM) [19]

spacer (insulators) An insulator used to support the inner conductor in the enclosure. (SUB/PE) C37.122-1993, C37.122.1-1993

spacer buggy *See:* conductor car.

spacer cable A type of electric supply line construction consisting of an assembly of one or more covered conductors, separated from each other and supported from a messenger by insulating spacers. (NESC) C2-1997

spacer cart *See:* conductor car.

space-referenced navigation data (navigation aids) Data in terms of a coordinate system referenced to inertial space. (AES/GCS) 172-1983w

spacer insulator As used in a gas-insulated system an insulator used to support the inner conductor in the enclosure. (SWG/SUB/PE) C37.122-1983s, C37.100-1992

spacer shaft (rotating machinery) A separate shaft connecting the shaft ends of two machines. *See also:* armature. (PE) [9]

space, state *See:* state space.

space-tapered array antenna An array antenna whose radiation pattern is shaped by varying the density of driven radiating elements over the array surface. *Synonym:* density-tapered array antenna. (AP/ANT) 145-1993

space-time adaptive processing (STAP) In airborne moving-target indication (MTI), a method of processing that compensates for the adverse effects of platform motion by adaptively placing antenna nulls in the directions of large clutter echoes and/or large noise or jamming sources. *Note:* It simultaneously employs the signals received from the multiple elements of an adaptive phased array antenna (spatial domain) and the signals from multiple pulse repetition periods (time domain) to provide adaptive processing in both the time and spatial domains. (AES) 686-1997

spacing (data transmission) A term which originated with telegraph to indicate an open key condition. Present usage implies the absence of current or carrier on a circuit. It also indicates the binary digit 0 in computer language. (PE) 599-1985w

spacing bicycle *See:* conductor car.

spacing pulse (data transmission) A spacing pulse or space is the signal pulse that, in direct-current neutral operation, corresponds to a circuit open or no current condition. *See also:* pulse. (COM) [49]

spacing wave (telegraph communication) (back wave) The emission that takes place between the active portions of the code characters or while no code characters are being transmitted. *See also:* radio transmitter. (AP/ANT) 145-1983s

spalling Spontaneous separation of a surface layer from a metal. (IA) [59]

span (measuring devices) The algebraic difference between the upper and lower values of a range. *Notes:* 1. For example:

a) Range 0 to 150, span 150;
b) Range −20 to 200, span 220;
c) Range 20 to 150, span 130;
d) Range −100 to −20, span 80.

2. The following compound terms are used with suitable modifications in the units: measured variable span, measured signal span, etc. 3. For multirange devices, this definition applies to the particular range that the device is set to measure. *See also:* instrument. (EEC/EMI) [112]

(2) **(A)** **(overhead conductors)** The horizontal distance between two adjacent supporting points of a conductor. **(B)** **(overhead conductors)** That part of any conductor, cable, suspension strand, or pole line between two consecutive points of support. *See also:* cable. (T&D/PE) [10]

span frequency-response rating The maximum frequency in cycles per minute of sinusoidal variation of measured signal for which the difference in amplitude between output and input represents an error no greater than five times the accuracy rating when the instrument is used under rated operating conditions. The peak-to-peak amplitude of the sinusoidal variation of measured signal shall be equivalent to full span of the instrument. It must be recognized that the span frequency-response rating is a measure of dynamic behavior under the most adverse conditions of measured signal (that is, the maximum sinusoidal excursion of the measured signal). The frequency response for an amplitude of measured signal less than full span is not proportional to the frequency response for full span. The relationship between the frequency response of different instruments at any particular amplitude of measured signal is not indicative of the relationship that will exist at any other amplitude. *See also:* accuracy rating. (EEC/EMI) [112]

span length The horizontal distance between two adjacent supporting points of a conductor. (NESC) C2-1997

spanned record A record that is partially contained in more than one block; that is, it spans a block boundary. *See also:* unblocked record; blocked record. (C) 610.5-1990w

spanning tree A bridging technique where a network of randomly interconnected bridges can automatically build a logical tree structure so as to guarantee a unique path between any pair of stations on the network. In this scheme, the transmitter does not have to know how to route the frame to the destination; that is the job of the bridges. *Contrast:* source routing. (C) 610.7-1995

spanning tree algorithm The abstract distributed algorithm that determines the active topology of an ISO/IEC 10038 Bridged Local Area Network. (C/LM) 802.1G-1996

spanning tree explorer (STE) A type of source-routed frame that will traverse the network following the spanning tree path created by the transparent bridging rules. (C/LM/CC) 8802-2-1998

spanning tree protocol The protocol that MAC Bridges use in exchanging information across Local Area Networks, in order to compute the active topology of a Bridged Local Area Network in accordance with the Spanning Tree Algorithm. (C/LM) 802.1G-1996

spanning tree route (STR) A term used to denote the configuration of transparent bridges such that every segment is connected to the root of the network through exactly one path. A frame sent without routing information (NSR) traverses the network on the spanning tree path according to the rules for transparent bridging. A frame sent with a routing type of STE is forwarded through the network on the spanning tree path, but is forwarded by the rules for SRT bridges (note that a bridge that does not support source routing will not forward STE frames). (C/LM/CC) 8802-2-1998

span, ruling *See:* ruling span.

span, sag *See:* sag span.

span step-response-time rating The time that the step-response time will not exceed for a change in measured signal essentially equivalent to full span when the instrument is used under rated operating conditions. The actual span step-response time shall not be less than 2/3 of the span step-response-time rating. (For example, for an instrument of 3-second span step-response-time rating, the span step-response time, under rated operating conditions, will be between 3 and 2 seconds.) It must be recognized that the step-response time for smaller steps is not proportional to the step-response time for full span. *Note:* The end device shall be considered to be at rest when it remains within a band of plus and minus the accuracy rating from its final position. *See also:* accuracy rating. (EEC/EMI) [112]

span wire An auxiliary suspension wire that serves to support one or more trolley contact conductors or a light fixture and the conductors that connect it to a supply system. (NESC) C2-1997

spare This record is unused at this time. It is reserved for expansion of the standard. (NPS/NID) 1214-1992r

spare_cell A cell instance that is presently not part of the logical function of a design, and therefore is not included in the design's logical netlist. A spare_cell is typically reserved for future logic modifications to be implemented through changes in the interconnect layers of the chip. (C/DA) 1481-1999

spare equipment Equipment complete or in parts, on hand for repair or replacement. *See also:* reserve equipment. (T&D/PE) [10]

spare only point interface Point for which cabinet space only is provided for the future addition of wiring and other necessary plug-in equipment. (SUB/PE) C37.1-1994

spare point (for supervisory control or indication or telemeter selection) A point that is not being utilized but is fully equipped with all of the necessary devices for a point. (SWG/PE) C37.100-1992

spare point interface Point equipment that is not being utilized but is fully wired and equipped. (SUB/PE) C37.1-1994

spark (overhead-power-line corona and radio noise) A sudden and irreversible transition from a stable corona discharge to a stable arc discharge. It is a luminous electrical discharge of short duration between two electrodes in an insulating medium. It is generally brighter and carries more current than corona, and its color is mainly determined by the type of insulating medium. It generates radio noise of wider frequency spectrum (extending into hundreds of megahertz) and wider magnitude range than corona. A spark is not classified as corona. (T&D/PE) 539-1990

spark capacitor A capacitor connected across a pair of contact points, or across the inductance that causes the spark, for the purpose of diminishing sparking at these points. *Note:* The use of the term "spark condenser" for the term spark capacitor is deprecated. (PE) [9]

spark condenser* *See:* spark capacitor.
* Deprecated.

spark gap (1) An air dielectric between two electrodes that may be a combination of several basic shapes that is used to protect telecommunication circuits from damage due to voltage stress in excess of their dielectric capabilities. It may or may not be adjustable. (PE/PSC) 487-1992
(2) Any short-air space between two conductors electrically insulated from or remotely electrically connected to each other. (PE/T&D/NFPA) 1410-1997, [114]

spark-gap converter, mercury-hydrogen *See:* mercury-hydrogen spark-gap converter.

spark-gap modulation A modulation process that produces one or more pulses or energy by means of a controlled spark-gap breakdown for application to the element in which modulation takes place. *See also:* oscillatory circuit. (AP/ANT) 145-1983s

spark gaps (wire-line communication facilities) Spark gaps consist of air dielectric between two electrodes which may be a combination of several basic shapes. Spark gaps are used to protect communication circuits from damage due to voltage stress in excess of their dielectric capabilities. (PE/PSC) 487-1980s

spark killer An electric network, usually consisting of a capacitor and resistor in series, connected across a pair of contact points, or across the inductance that causes the spark, for the purpose of diminishing sparking at these points. *See also:* network analysis. (EEC/PE) [119]

sparkover (1) A disruptive discharge between electrodes in a gas or liquid. (PE/PSIM) 4-1995
(2) A disruptive discharge between electrodes of a measuring gap, voltage-control gap, or gap-type protective device. (SPD/PE) C62.11-1999
(3) A disruptive discharge between electrodes of a measuring gap, voltage-control gap, or protective device. (SPD/PE) C62.62-2000

spark-plug suppressor A suppressor designed for direct connection to a spark plug. *See also:* electromagnetic compatibility. (EMC) [53]

spark transmitter A radio transmitter that utilizes the oscillatory discharge of a capacitor through an inductor and a spark gap as the source of its radio-frequency power. *See also:* radio transmitter. (AP/BT/ANT) 145-1983s, 182-1961w

sparse data scan (SDS) A technique by which arrays of modules with low data occupancy may be scanned efficiently, i.e, without accessing every potential data site. (NID) 960-1993

sparse medium A medium with a low volume fraction of discrete objects, typically less than 1%. In such a medium, multiple scattering is negligibly small. *See also:* homogeneous dense medium; inhomogeneous dense medium. (AP/PROP) 211-1997

spatial average The root mean square of the field over an area equivalent to the vertical cross section of the adult human body, as applied to the measurement of electric or magnetic fields in the assessment of whole-body exposure. The spatial average is measured by scanning (with a suitable measurement probe) a planar area equivalent to the area occupied by a standing adult human (projected area). In most instances, a simple vertical, linear scan of the fields over a 2 m height (approximately 6 ft), through the center of the projected area, will be sufficient for determining compliance with the maximum permissible exposures (MPEs). (NIR) C95.1-1999

spatial coherence (1) (laser maser) (electromagnetic) The correlation between electromagnetic fields at points separated in space. *See also:* coherence area. (LEO) 586-1980w
(2) (fiber optics) *See also:* coherent. 812-1984w

spatial disturbance (diode-type camera tube) In the output signal from a television camera consists of a broad variety of spurious signals, some of which are observable when no optical input is present, while others are input-level dependent. Spatial disturbances are characterized as either independent of time or as having a temporal variation long with respect to a frame interval, provided the operating conditions, including position and temperature, remain fixed. Tolerance for spatial distrubance covers a broad range, depending upon the application. Cosmetic considerations ultimately reduce to a cost decision. Spurious signals have been classified in the following categories:

a) (Fixed pattern) This is a modulation of a uniform background which may be either spatially periodic or random.
b) (Shading) This consists of a broad area continuous variation in the background signal, with or without an optical input. The signal corresponding to uniform irradiance is either curved or tilted, causing a brightness variation in the display.
c) (Moire) This is a periodic amplitude modulation in the output which is not present in the input, usually due to the interaction of two or more periodic tube elements such as the field mesh, scanning raster, and target.
d) (Blemishes) These are bright or dark spots or streaks whose effect is equivalent to viewing the scene through a dirty window. Blemishes affect limited portions of the raster.
e) (Geometric distortion) This includes any skewing, bending, displacement or rotation of the image. It can be localized or include the entire raster.

See also: image storage. (ED) 503-1978w

spatial locality The tendency for a program to reference closely related clusters of memory addresses over short time intervals. (C/BA) 10857-1994

spatially aligned bundle *See:* aligned bundle.

spatially coherent radiation *See:* coherent.

SPCS *See:* stored program control.

SPD *See:* surge-protective device.

special addition* *See:* double-precision addition.
* Deprecated.

special-billing call (telephone switching systems) A call charged to a special number. (COM) 312-1977w

special character (character set) A character that is neither a numeral, a letter, nor a blank, for example, virgule, asterisk, dollar sign, equals sign, comma, period. (C) [20], [85]
(2) (A) (data management) A character that is not in the alphabet, but that is used for punctuation or another special purpose. For example, blank, comma, period, or asterisk. **(B) (data management)** A graphic character in a character set that is not a letter, not a digit, and not a space character. **(C) (data management)** In COBOL, a character that is neither numeric nor alphabetic. (C) 610.5-1990

special color rendering index (illuminating engineering) Measure of the color shift of various standardized special colors including saturated colors, typical foliage, and Caucasian skin. It also can be defined for other color samples when the spectral reflectance distributions are known. (EEC/IE) [126]

special combination protective devices (wire-line communication facilities) (open-wire or hot-line protectors) Combined isolating and drainage transformer type protectors used in conjunction with, but not limited to, horn gaps and grounding relays, are used on open-wire lines to provide protection against lightning, power contacts, or high values or induced voltage. (PE/PSC) 487-1980s

special-date logic Logic found in programs using specific reserved dates to trigger exceptions to normal date data processing. *Note:* The normally valid (abbreviated) date-specifiers 99-12-31 or 99-09-09 might be used as an expiration date on tape archives to mean "never expire" or the normally valid date of 00-01-01 might be used to indicate an unknown or out-of-range date, all being ambiguous reserved dates, since they conflict with actual valid date representations which, when encountered, would trigger the special-date logic erroneously. On the other hand, unconditionally invalid date-specifiers 99-99-99 or 00-00-00 used as reserved dates would not erroneously trigger the special-date logic because such unambiguous reserved dates do not conflict with any real (valid) intrinsic dates. (C/PA) 2000.1-1999

special-dial tone (telephone switching systems) A tone for certain features that indicates that a customer can use his or her calling device. (COM) 312-1977w

special function signal A signal line in a connecting cable from a bedside communications controller (BCC) to a device communications controller (DCC) that conducts other than serial data or power. The two special function signals are $\overline{\text{SENSE/INT/SYNC-IN}}$ and $\overline{\text{SENSE/INT/SYNC-OUT}}$. Special function signals convey indications of physical connection and disconnection, use of interrupts, frame transmission, interrupt requests, and synchronization pulses. (EMB/MIB) 1073.4.1-2000

specialId A reserved nodeId value associated with special-send packets. (C/MM) 1596-1992

specialize A change by an instance from being an instance of its current class to being additionally an instance of one (or more) of the subclasses of the current subclass. A specialized instance acquires a different (lower) lowclass. *Contrast:* respecialize; unspecialize. (C/SE) 1320.2-1998

specialized common carrier (1) A company that provides private line communications services, for example, voice, teleprinter, data, facsimile transmission. *See also:* value-added service; common carrier. (LM/COM) 168-1956w
(2) A common carrier providing a limited set of services. For example, only private line services. (C) 610.7-1995

special link (SL) A transmission system that replaces the normal medium. (C/LM) 802.3-1998

special logic Many programs use specific dates to trigger exceptions to normal date processing. A common example is expiration date on tape archives. Rather than adding another flag to the tape header, the date 1999-9-9 is used by many systems to mean never expire. Other dates have been used to indicate an unknown or out-of-range date. Since there is no standard for this practice, the specific dates used may be difficult to trace. The potential for dramatic failures increases in 1999 due to date code flags and various other Year 2000 problems. (C/PA) 2000.2-1999

special parameter In the shell command language, a parameter named by a single character from the following list:

* @ # ? ! − $ 0

(C/PA) 9945-2-1993

special permission The written consent of the authority having jurisdiction. (NESC/NEC) [86]

special process (replacement parts for Class 1E equipment in nuclear power generating stations) (nuclear power quality assurance) A process, the results of which are highly dependent on the control of the process or the skill of the operators, or both, and in which the specified quality cannot be readily determined by inspection or test of the product. (PE/NP) 934-1987w, [124]

special-purpose computer A computer designed to solve a restricted class of problems. *Contrast:* general-purpose computer. *See also:* incremental computer; dedicated computer. (C) [20], [85], 610.10-1994w

special-purpose electronic test equipment *See:* special-purpose test equipment.

special-purpose motor A motor with special operating characteristics or special mechanical construction, or both designed for a particular application and not falling within the definition of a general-purpose or definite-purpose motor. *See also:* direct-current commutating machine; asynchronous machine. (PE/IA/APP) [9], [82]

special-purpose test equipment (test, measurement, and diagnostic equipment) Equipment used for test, repair and maintenance of a specified system, subsystem or module, having application to only one or a very limited number of systems. (MIL) [2]

special send A packet having one of a particular set of special addresses and a special format used for initialization, such as "reset" or "clear." (C/MM) 1596-1992

specialty transformer (power and distribution transformers) A transformer generally intended to supply electric power for control, machine tool, Class 2, signaling, ignition, luminous-tube, cold-cathode lighting series street-lighting, low-voltage general purpose, and similar applications. See the following types of transformers: individual-lamp; series street-lighting; energy-limiting; high-reactance; non-energy-limiting; high power factor; low power factor; insultating; individual-lamp insulating; group-series loop insulating; luminous tube; ignition; series circuit lighting; signaling and doorbell; control; machine tool control; general-purpose; mercury vapor lamp (multiple-supply type); sodium vapor lamp (multiple-supply type); saturable reactor (saturable-core reactor); electronic. *See also:* secondary voltage rating; Class 2 transformer; secondary short-circuit current rating of a high-reactance transformer; IR-drop compensation transformer; kVA or volt-ampere short-circuit input rating of a high-reactance transformer; series transformer rating. (PE/TR) C57.12.80-1978r

special unit capacity purchases (electric power supply) That capacity that is purchased or sold in transactions with other utilities and that is from a designated unit on the system of the seller. It is understood that the seller does not provide reserve capacity for this type of capacity transaction. *See also:* generating station. (PE/PSE) [54]

specific With respect to a class, the attribute types that may appear in an instance of the class but not in an instance of its superclasses. (C/PA) 1328-1993w, 1224-1993w, 1327-1993w

specific absorption (SA) The quotient of the incremental energy (*dW*) absorbed by (dissipated in) an incremental mass (*dm*) contained in a volume (*dV*) of a given density (ρ).

$$SA = \frac{dW}{dm} = \frac{dW}{\rho dV}$$

The specific absorption is expressed in units of joules per kilogram (J/kg). (NIR) C95.1-1999

specific absorption rate (SAR) The time derivative of the incremental energy (*dW*) absorbed by (dissipated in) an incremental mass (*dm*) contained in a volume element (*dV*) of given density (ρ).

$$SAR = \frac{d}{dt}\left(\frac{dW}{dm}\right) = \frac{d}{dt}\left(\frac{dW}{\rho dV}\right)$$

SAR is expressed in units of watts per kilogram (W/kg). (NIR) C95.1-1999

specific acoustic impedance (unit area acoustic impedance) (at a point in the medium) The complex ratio of sound pressure to particle velocity. (SP) [32]

specific acoustic reactance The imaginary component of the specific acoustic impedance. (SP) [32]

specific acoustic resistance The real component of the specific acoustic impedance. (SP) [32]

specific address *See:* absolute address.

specifically routed frame (SRF) A frame sent with a routing information field that describes the exact path that the frame will take through the bridged network. (C/LM/CC) 8802-2-1998

specification (1) (software) A document that specifies, in a complete, precise, verifiable manner, the requirements, design, behavior, or other characteristics of a system or component, and, often, the procedures for determining whether these provisions have been satisfied. *See also:* formal specification; requirements specification; product specification. (C) 610.12-1990
(2) A document that prescribes, in a complete, precise, verifiable manner, the requirements, design, behavior, or characteristics of a system or system component. (C/PA) 14252-1996

(3) A document that fully describes a design element or its interfaces in terms of requirements (functional, performance, constraints, and design characteristics) and the qualification conditions and procedures for each requirement. (C/SE) 1220-1998

Specification and Description Language (SDL) A specification language for telecommunications and distributed systems that provides both textual and graphic description techniques. (C) 610.13-1993w

specification change notice A document used in configuration management to propose, transmit, and record changes to a specification. *See also:* engineering change; notice of revision; configuration control. (C) 610.12-1990

specification element A product, subsystem, assembly, component, subcomponent, subassembly, or part of the specification tree described by a specification. (C/SE) 1220-1994s

specification language (1) An application-oriented computer language, often a machine-processible combination of natural and formal language, used to express the requirements, design, behavior, or other characteristics of a system or component. For example, a design language or requirements specification language. *Contrast:* compiler specification language; programming language; query language. *See also:* design language. (C) 610.13-1993w, 610.12-1990
(2) *See also:* Rule and Constraint Language. (C/SE) 1320.2-1998

specification, system *See:* system specification.

specification tree (1) A diagram that depicts all of the specifications for a given system and shows their relationships to one another. (C) 610.12-1990
(2) A hierarchy of specification elements and their interface specifications that identifies the elements and the specifications related to design elements of the system configuration which are to be controlled. (C/SE) 1220-1998

specification verification *See:* verification.

specific code *See:* absolute code.

specific coordinated methods Those additional methods applicable to specific situations where general coordinated methods are inadequate. *See also:* inductive coordination. (EEC/PE) [119]

specific creepage distance For bushings for pure dc application, the specific creepage distance is the creep distance divided by the rated voltage for the DC system where the bushing is intended to be used. For bushings for combined voltage application, the specific creepage distance is the creep distance divided by Z * Vd where:

Z = number of six pulse bridges in series

V_d = dc rated voltage per valve bridge.

(PE/TR) C57.19.03-1996

specific detectivity *See:* D*.

specific emission The rate of emission per unit area. (Std100) [84]

specific heat (electric power systems in commercial buildings) The ratio of the quantity of heat required to raise the temperature of a given mass of a substance 1° to the heat required to raise the temperature of an equal amount of water by 1°. (IA/PSE) 241-1990r

specific inductive capacitance *See:* relative capacitivity.

specific intensity (*I*) A positive real quantity *I*, in general a function of position *r*, direction *s*, frequency *f*, and time *t*, representing the quantity of power *dP* flowing outward through an elemental area *dA* at a particular location *r*, within an elemental solid angle $d\Omega$ containing a particular direction *s*, within a frequency interval $(f, f+df)$:

$$dP = I(r, s, f, t)s\ dA\ d\Omega\ df$$

(AP/PROP) 211-1997

specific management functional areas A category of systems management user requirements. (C) 610.7-1995

specific repetition frequency (navigation aids) (loran) One of a set of closely-spaced pulse repetition frequencies derived from the basic repetition frequency and associated with a specific set of synchronized stations. (AES/GCS) 172-1983w

specific repetition rate *See:* specific repetition frequency.

specific unit capacity (power operations) Capacity which is purchased, or sold, in transactions with other systems and which is from a designated unit on the system of the seller. (PE/PSE) 858-1987s

specified achromatic lights (A) Light of the same chromaticity as that having an equi-energy spectrum. **(B)** The standard illuminants of colorimetry A, B, and C, the spectral energy distributions of which were specified by the International Commission on Illumination (CIE) in 1931, with various scientific applications in view. Standard A: incandescent electric lamp of color temperature 2854 K. Standard B: Standard A combined with a specified liquid filter to give a light of color temperature approximately 4800 K. Standard C: Standard A combined with a specified liquid filter to give a light of color temperature approximately 6500 K. **(C)** Any other specified white light. *See also:* color. (BT/AV) [34], [84]

specified breakaway torque (rotating machinery) The torque which a motor is required to develop to break away its load from rest to rotation. (PE) [9]

specified mechanical load (SML) The bending moment load at which irreversible visible damage may be evident. SML is supplied from the manufacturer. (PE/SUB) 693-1997

speckle (1) The random distribution of intensity in space. (AP/PROP) 211-1997
(2) A mottled effect in coherent radar images, such as those from synthetic-aperture radar (SAR) and laser radar caused by random additive and subtractive interference of signals from individual scatterers within each resolution cell. *Note:* This is the same as target fluctuation for isolated targets. (AES) 686-1997

speckle noise *See:* modal noise.

speckle pattern (fiber optics) A power intensity pattern produced by the mutual interference of partially coherent beams that are subject to minute temporal and spatial fluctuations. *Note:* In a multimode fiber, a speckle pattern results from a superposition of mode field patterns. If the relative modal group velocities change with time, the speckle pattern will also change with time. If, in addition, differential mode attenuation is experienced, modal noise results. *See also:* modal noise. (Std100) 812-1984w

spectral bandwidth (light-emitting diodes) The difference between the wavelengths at which the spectral radiant intensity is 50% (unless otherwise stated) of the maximum value. The term spectral linewidth is sometimes used. (ED) [127]

spectral brightness (of an object) The total power radiated by an object per unit solid angle per unit projected area per unit bandwidth. *Note:* In radiative transfer theory it is called the spectral specific intensity. In infrared radiometry it is called the spectral radiance. (AP/PROP) 211-1997

spectral characteristic (A) (television) The set of spectral responses of the color separation channels with respect to wavelength. *Notes:* 1. The channel terminals at which the characteristics apply must be specified, and an appropriate modifier, such as pickup spectral characteristic or studio spectral characteristic may be added to the term. 2. Because of nonlinearity, some spectral characteristics depend on the magnitude of radiance used in the measurement. 3. Nonlinearizing and matrixing operations may be performed within the channels. 4. The spectral taking characteristics are uniquely related to the chromaticities of the display primaries. **(B)** (camera tube) A relation, usually shown by a graph, between wavelength and sensitivity per unit wavelength interval. **(C)** (luminescent screen) The relation, usually shown by a graph, between wavelength and emitted radiant power per unit wavelength interval. *Note:* The radiant power is commonly expressed in arbitrary units. **(D)** (phototube) A relation, usually shown by a graph, between the radiant sensitivity and the wavelength of the incident radiant flux. (BT/AV) 201-1979

spectral-conversion luminous gain (optoelectronic device) The luminous gain for specified wavelength-intervals of both incident and emitted luminous flux. *See also:* optoelectronic device. (ED) [46]

spectral-conversion radiant gain (optoelectronic device) The radiant gain for specified wavelength intervals of both incident and emitted radiant flux. *See also:* optoelectronic device. (ED) [46]

spectral data The channel data are stored with the channel number at the beginning of each record. The channel number is six characters. The first channel is channel 0. The channel data is 10 characters per number, separated by a space. There are 5 channels per line giving a total line length of 61 characters, including the end-of-record character. Leading spaces are interpreted as zeros. (NPS/NID) 1214-1992r

spectral-directional emissivity $\varepsilon(\theta, \tau, \lambda, T)$ (of an element of surface of a temperature radiator at any wavelength and in a given direction) The ratio of its spectral radiance at that wavelength and in the given direction to that of a black body at the same temperature and wavelength.

$$\varepsilon(\lambda, \theta, \phi, T) = L_{\lambda}(\lambda, \theta, \phi, T)/L_{\text{blackbody}}(\lambda, T)$$

(EEC/IE) [126]

spectral emissivity (element of surface of a temperature radiator at any wavelength) The ratio of its radiant flux density per unit wavelength interval (spectral radiant exitance) at that wavelength to that of a blackbody at the same temperature. *See also:* radiant energy. (IE/EEC) [126]

spectral-hemispherical emissivity, $\varepsilon(\lambda, T)$ (of an element of surface of a temperature radiator) The ratio of its spectral radiant exitance to that of a blackbody at the same temperature. *Note:* Hemispherical emissivity is frequently called "total" emissivity. "Total" by itself is ambiguous, and should be avoided since it may also refer to "spectral-total" (all wavelengths) as well as "directional-total" (all directions). (EEC/IE) [126]

spectral irradiance (fiber optics) Irradiance per unit wavelength interval at a given wavelength, expressed in watts per unit area per unit wavelength interval. *See also:* irradiance; radiometry. (Std100) 812-1984w

spectral line (1) (fiber optics) A narrow range of emitted or absorbed wavelengths. *See also:* line spectrum; monochromatic; spectral width; line source. (Std100) 812-1984w
(2) A sharply peaked portion of the spectrum that represents a specific feature of the incident radiation, usually the full energy of a monoenergetic radiation. (NPS) 300-1988r

spectral luminous efficacy (illuminating engineering) (of radiant flux) $(K(\lambda) = \Phi^{-}_{\omega\lambda}/\Phi_{\varepsilon\lambda})$ The quotient of the luminous flux at a given wavelength by the radiant flux at that wavelength. It is expressed in lumens per watt. *Note:* This term formerly was called "luminosity factor." The reciprocal of the maximum luminous efficacy of radiant flux is sometimes called "mechanical equivalent of light;" that is, the ratio between radiant and luminous flux at the wavelength of maximum luminous efficacy. The most probable value is 0.00146 W/lm, corresponding to 683 lm/W as the maximum possible luminous efficacy. For scotopic vision values (13.7) the maximum luminous efficacy is 1754 "scotopic" lm/W. (EEC/IE) [126]

spectral luminous efficiency (illuminating engineering) (of radiant flux) The ratio of the luminous efficacy for a given wavelength to the value at the wavelength of maximum luminous efficacy. It is dimensionless. *Note:* The term "spectral luminous efficiency" replaces the previously used terms "relative luminosity" and "relative luminosity factor." (EEC/IE) [126]

spectral luminous flux (light-emitting diodes) The luminous flux per unit wavelength interval at wavelength l that is, lumens per nanometer. (ED) [127]

spectral luminous gain (optoelectronic device) Luminous gain for a specified wavelength interval of either the incident or the emitted flux. *See also:* optoelectronic device. (ED) [46]

spectral luminous intensity (light-emitting diodes) The luminous intensity per unit wavelength (at wavelength λ), that is, candela per nanometer. (ED) [127]

specific MAC service The service provided by the MAC protocol and procedures of a specific Local Area Network technology (which can contain features not present in other specific MAC services or in the ISO/IEC 10039 MAC service). (C/LM) 802.1G-1996

spectral-noise density (sound recording and reproducing system) The limit of the ratio of the noise output within a specified frequency interval to the frequency interval, as that interval approaches zero. *Note:* This is approximately the total noise within a narrow frequency band divided by that bandwidth in hertz. *See also:* noise. 191-1953w

spectral power density Power per unit bandwidth, in watts per Hertz. (AP/PROP) 211-1997

spectral power flux density The power density per unit bandwidth in watts per square meter per Hertz. (AP/PROP) 211-1997

spectral quantum efficiency (diode-type camera tube) ($\eta\lambda$) The average number of electrons produced in the output signal per photon incident on the camera tube faceplate at a particular photon energy or wavelength. It is a dimensionless quantity that can be conveniently calculated from the spectral response Rl through the relation

$$\eta\lambda = \frac{1241\ R_{\lambda}}{\lambda}$$

where $R\lambda$ is in amperes per watt and λ in nanometers. (ED) 503-1978w

spectral quantum yield (photocathodes) The average number of electrons photoelectrically emitted from the photocathode per incident photon of a given wavelength. *Note:* The spectral quantum yield may be a function of the angle of incidence and of the direction of polarization of the incident radiation. *See also:* phototube. (NPS) 175-1960w

spectral radiance (1) (fiber optics) Radiance per unit wavelength interval at a given wavelength, expressed in watts per steradian per unit area per wavelength interval. *See also:* radiance; radiometry. (Std100) 812-1984w
(2) (laser maser) The power transmitted in a radiation field per unit frequency (or wavelength) interval unit solid angle unit area normal to a given direction ($W \cdot nm^{-1} \cdot sr^{-1} \cdot m^{-2}$). (LEO) 586-1980w

spectral radiant energy (light-emitting diodes) ($Q_{\lambda} = dQ_e/d_{\lambda}$) Radiant energy per unit wavelength interval at wavelength l; that is, joules per nanometer. (ED) [127]

spectral radiant flux (light-emitting diodes) ($\phi\lambda = d\phi_e/d\lambda$) Radiant flux per unit wavelength interval at wavelength l: that is watts per nanometer. (IE/EEC/ED) [126], [127]

spectral radiant gain (optoelectronic device) Radiant gain for a specified wavelength interval of either the incident or the emitted radiant flux. *See also:* optoelectronic device. (ED) [45]

spectral radiant intensity (light-emitting diodes) ($I\lambda = dI_e/d\lambda$) The radiant intensity per unit wavelength interval: for example watts per (steradian-nanometer). (IE/EEC/ED) [126], [127]

spectral range (acoustically tunable optical filter) The wavelength region over which the dynamic transmission is greater than some specified minimum value. (UFFC) [17]

spectral reflectance (illuminating engineering) ($\rho(\lambda) = \Phi_{r\lambda}/\Phi_{i\lambda}$) The ratio of the reflected flux to the incident flux at a particular wavelength, l, or within a small band of wavelengths, Dl, about l. *Note:* The various geometrical aspects of reflectance may each be considered restricted to a specific region of the spectrum and may be so designated by the addition of the adjective "spectral." (EEC/IE) [126]

spectral response (diode-type camera tube) The spectral response (R_μ) of a camera is the current produced in the output signal per incident radiant power in the input signal as a function of the photon energy frequency, or wavelength. Units: amperes watt^{-1} (AW^{-1}). (ED) 503-1978w

spectral response characteristic *See:* spectral sensitivity characteristic.

spectral responsivity (fiber optics) Responsivity per unit wavelength interval at a given wavelength. *See also:* responsivity. (Std100) 812-1984w

spectral selectivity (photoelectric device) The change of photoelectric current with the wavelength of the irradiation. *See also:* photoelectric effect. (ED) [45], [84]

spectral sensitivity characteristic (camera tubes or phototubes) The relation between the radiant sensitivity and the wavelength of the incident radiation, under specified conditions of irradiation. *Note:* Spectral sensitivity characteristic is usually measured with a collimated beam at normal incidence. *See also:* phototube. (BT/ED/AV) [34], [45]

spectral temperature (laser maser) (of a radiation field) The temperature of a black body which produces the same spectral radiance as the radiation field at a given frequency and in a given direction. (LEO) 586-1980w

spectral-total directional emissivity (illuminating engineering) (ε, ϕ, T) (of an element of surface of a temperature radiator in a given direction) The ratio of its radiance to that of a blackbody at the same temperature.

$$\varepsilon(\theta, \phi, T) == L(\theta, \phi, T)/L_{\text{blackbody}}) (T)$$

where θ and ϕ are directional angles and T is temperature. (EEC/IE) [126]

spectral-total hemispherical emissivity, ε (of an element of surface of a temperature radiator) The ratio of its radiant exitance to that of a blackbody at the same temperature.

$$\varepsilon = \frac{1}{\pi}\int \varepsilon(\theta, \phi) \cdot \cos\theta \cdot d\omega = \frac{1}{\pi}\int\int \varepsilon$$

$$(\lambda, \theta, \phi) \cdot \cos\theta \cdot d\omega \cdot d\lambda == M(T)/M_{\text{blackbody}} (T)$$

(EEC/IE) [126]

spectral transmittance (illuminating engineering) ($\tau(\lambda) = \Phi_{t\lambda}/\Phi_{i\lambda}$) The ratio of the transmitted flux to the incident flux at a particular wavelength, l, or within a small band of wavelengths, Dl, about l. *Note:* The various geometrical aspects of transmittance may each be considered restricted to a specific region of the spectrum and may be so designated by the addition of the adjective "spectral." (EEC/IE) [126]

spectral tristimulus values Values per unit wavelength interval and unit spectral radiant flux. *Note:* Spectral tristimulus values have been adopted by the International Commission on Illumination (CIE). They are tabulated as functions of wavelength throughout the spectrum and are the basis for the evaluation of radiant energy as light. (EEC/IE) [126]

spectral width (fiber optics) A measure of the wavelength extent of a spectrum. *Notes:* 1. One method of specifying the spectral linewidth is the full width at half maximum (FWHM), specifically the difference between the wavelengths at which the magnitude drops to one half of its maximum value. This method may be difficult to apply when the line has a complex shape. 2. Another method of specifying spectral width is a special case of root-mean-square (rms) deviation where the independent variable is wavelength (λ), and $f(\lambda)$ is a suitable radiometric quantity. 3. The relative spectral width ($\Delta\lambda/\lambda$) is frequently used, where $\Delta\lambda$; is obtained according to Note 1 or Note 2. *See also:* root-mean-square deviation; coherence length; line spectrum; material dispersion. (Std100) 812-1984w

spectral width, full-width half maximum (FWHM) The absolute difference between the wavelengths at which the spectral radiant intensity is 50% of the maximum. (C/LM) 802.3-1998

spectral window (fiber optics) A wavelength region of relatively high transmittance, surrounded by regions of low transmittance. *Synonym:* transmission window. (Std100) 812-1984w

spectrophotometer (illuminating engineering) An instrument for measuring the transmittance and reflectance of surfaces and media as a function of wavelength. (EEC/IE) [126]

spectroradiometer (illuminating engineering) An instrument for measuring radiant flux as a function of wavelength. (EEC/IE) [126]

spectrum (1) (germanium gamma-ray detectors) (radiation) (x-ray energy spectrometers) (charged-particle detectors) A distribution of the intensity of radiation as a function of energy or its equivalent electric analog (such as charge or voltage) at the output of a radiation detector. (NPS/NID) 325-1986s, 759-1984r, 301-1976s

(2) (fiber optics) *See also:* optical spectrum. 812-1984w

(3) (data transmission) The distribution of the amplitude (and sometimes phase) of the components of a wave as a function of frequency. Spectrum is also used to signify a continuous range of frequencies, usually wide in extent, within which waves have some specified common characteristic. (PE) 599-1985w

(4) (radiation) A distribution of the intensity of radiation as a function of energy, or a distribution of the amplitudes of the electrical pulses caused by the radiation. (NPS) 300-1988r

spectrum amplitude (impulse strength and impulse bandwidth) The voltage spectrum of a pulse can be expressed as

$$V(\omega) = R(\omega) + jX(\omega) = \int_{-\infty}^{+\infty} v(t)\, e^{-j\omega t} dt$$

where

$$R(\omega) = \int_{-\infty}^{+\infty} v(t) \cos \omega t \, dt$$

$$X(\omega) = \int_{-\infty}^{+\infty} v(t) \sin \omega t \, dt$$

and

$$\omega = 2\pi f$$

Notes: 1. See IEEE Std 263-1965, Measurement of Radio Noise Generated by Motor Vehicles and Affecting Mobile Communications Receivers in the Frequency Range 25 to 1000 megahertz. The spectrum then has the amplitude characteristic

$$A(\omega) = \sqrt{R^2(\omega) + X^2(\omega)} \qquad (V/\text{rad})/\text{s}$$

and the phase characteristic

$$\varphi(\omega) = \tan^{-1}\frac{X(\omega)}{R(\omega)}$$

The inverse transform can be written

$$v(t) = \frac{1}{\pi}\int_0^{\infty} A(\omega) \cos\,[\omega t + \varphi(\omega)]\, d\omega$$

for real $v(t)$. The spectrum amplitude is also expressible in volts per hertz (volt-seconds) as follows:

$$S(f) = 2A(\omega)$$

It is this form that is used as the basis for calibration of commercially available impulse generators. A practical impulse is a function of time duration short compared with the reciprocals of all frequencies of interest. 2. For a rectangular pulse, the spectrum is flat within about 1 dB up to a frequency for which the pulse duration is equal to 1/4 cycle. Its spectrum amplitude $S(f)$ is substantially uniform in this frequency range and is equal to twice the area under the impulse time function or 2s. At frequencies higher than this it is still of interest to define the spectrum amplitude that will usually be less than 2σ. In most broadband impulse generators a dc voltage is used to charge a calibrated coaxial transmission line. The pulses are produced when the line is discharged into its terminating impedance through mechanically activated contacts. These mechanical contacts may be parts of either a vibrating diaphragm or mercury wetted relay switches. By proper choice of transmission line length and resistive termination, it is possible to produce impuls es having a predictable uniform spectrum amplitude range. The advent of solid-state switches has made it possible to switch on a sine wave for a precisely measurable time interval (τ), producing in the frequency band in the vicinity of the sine wave a spectrum simulating that produced by an impulse. The spectrum amplitude at that particular frequency can be measured in terms of a measurement of the amplitude of the sine wave when not switched, and a measurement of the on time (τ_0) for the switch. (EMC) 376-1975r

spectrum analyzer (1) An instrument generally used to display the power distribution of an incoming signal as a function of frequency. *Notes:* 1. Spectrum analyzers are useful in analyzing the characteristics of electrical waveforms in general since, by repetitively sweeping through the frequency range of interest, they display all components of the signal. 2. The display format may be a cathode ray tube or chart recorder. (IM) [14]

(2) An instrument that measures the power of a complex signal in many bands. The frequency bands can be either constant absolute bandwidth or constant percentage bandwidth. (COM/TA) 269-1992

(3) An instrument that measures the power of a signal in multiple frequency bands. The frequency bands may be constant bandwidth (e.g., fast Fourier transform (FFT) analyzer), or constant percentage bandwidth (e.g., real-time filter analyzer). (COM/TA) 1329-1999

(4) A test instrument which measures and displays the measurements of amplitude versus frequency for a given signal. Its frequency range should be at least 100 Hz to 1500 MHz; resolution bandwidth should be at least 10 Hz to 3 MHz; the video bandwidth should be at least 1 Hz to 3 MHz, and amplitude range should be at least −135 to +35 dBm. (PEL) 1515-2000

spectrum, angular *See:* angular spectrum.

spectrum, angular power *See:* angular power spectrum.

spectrum, energy *See:* energy spectrum.

spectrum intensity (impulse strength and impulse bandwidth) (For spectra which have a continuous distribution of components components are not discrete over the frequency range of interest). The spectrum intensity is the ratio of the power contained in a given frequency range to the frequency range as the frequency range approaches zero. It has the dimensions watt-seconds or joules and is usually stated quantitatively in terms of watts per hertz. (EMC) 376-1975r

spectrum level (spectrum density level) (specified signal at a particular frequency) The level of that part of the signal contained within a band 1 Hz wide, centered at the particular frequency. Ordinarily, this has significance only for a signal having a continuous distribution of components within the frequency range under consideration. The words spectrum level cannot be used alone but must appear in combination with a prefatory modifier: for example, pressure, velocity, voltage. *Note:* For illustration, if L_{ps} is a desired pressure spectrum level, p the effective pressure measured through the filter system, p_0 reference sound pressure, Δf the effective bandwidth of the filter system, and $\Delta_0 f$ the reference bandwidth (1 Hz), then

$$L_{ps} = L_p - 10\log_{10}\frac{\Delta df}{\Delta_0 f}$$

For computational purposes, if L_{ps} is the band pressure level observed through a filter of bandwidth Δf, the above relation reduces to

$$L_{ps} = \log_{10}\frac{p^2/\Delta f}{p_0^2/\Delta_0 f}$$

(SP/ACO) [32]

spectrum locus (1) (color) The locus of points representing the chromaticities of spectrally pure stimuli in a chromaticity diagram. (See the corresponding figure.)

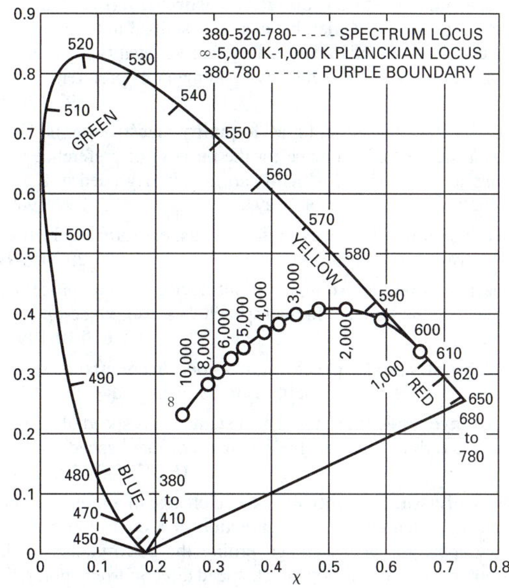

Chromaticity diagram

spectrum locus

(BT/AV) 201-1979w

(2) (illuminating engineering) The locus of points representing the colors of the visible spectrum in a chromaticity diagram. (EEC/IE) [126]

spectrum mask A graphic representation of the required power distribution as a function of frequency for a modulated transmission. (C/LM) 802.3-1998

specular angle (illuminating engineering) That angle between the perpendicular to the surface and the reflected ray that is numerically equal to the angle of incidence and that lies in the same plane as the incident ray and the perpendicular but on the opposite side. (EEC/IE) [126]

specular reflection (1) (laser maser) A mirrorlike reflection. (LEO) 586-1980w

(2) The reflection of a wave when incident on an infinite planar surface. The angle of incidence is equal to the angle of reflection. (EMC) 1128-1998

(3) The process by which all or part of a wave, incident on a smooth surface, is returned to the original medium, in accordance with Snell's law of reflection. (AP/PROP) 211-1997

(4) (fiber optics) *See also:* reflection. 812-1984w

specular surface (illuminating engineering) A surface from which the reflection is predominantly regular. *See also:* regular reflection. (EEC/IE) [126]

(2) (A) A surface, smooth enough that all energy is reflected from it or transmitted across it in those directions specified by Snell's law. **(B)** A planar interface separating two media. (AP/PROP) 211-1997

speech interpolation The method of obtaining more than one voice channel per voice circuit by giving each subscriber a speech path in the proper direction only at times when his speech requires it. (EEC/PE) [119]

speech level (speech quality measurements) The speech level defined and measured subjectively by comparison of the speech signal with a signal obtained by passing pink noise through a filter with A-weighting characteristics that has been judged to be equal to it in loudness. *Note:* The value of the speech level is defined to be the A-weighted sound pressure level of this noise [dB(A)]. 297-1969w

speech network (transmission performance of telephone sets) An electrical circuit that connects the transmitter and the receiver to a telephone line or telephone test loop and to each other. (COM/TA) 269-1983s

speech quality (speech quality measurements) A characteristic of a speech signal that can be described in terms of subjective and objective parameters. Speech quality is evaluated only in terms of the subjective parameter of preference. 297-1969w

speech reference signal (speech quality measurements) Used as a standard of reference for the purpose of preference testing, a speech signal which is artificially degraded in a measurable and reproducible way. 297-1969w

speech signals Utterances in their acoustical form or electrical equivalent. 297-1969w

speech synthesizer An input-output device that can process or generate the sound of human speech. *See also:* voice-operated device. (C) 610.10-1994w

speech test signal (speech quality measurements) A speech signal whose speech quality is to be evaluated. 297-1969w

speed (hydraulic turbines) The instantaneous speed of rotation of the turbine expressed as a percent of rated speed. (PE/EDPG) 125-1977s

speed adjustment (control) A speed change of a motor accomplished intentionally through action of a control element in the apparatus or system governing the performance of the motor. *Note:* For an adjustable-speed direct-current motor, the speed adjustment is expressed in percent (or per unit) of base speed. Speed adjustment of all other motors is expressed in percent (or per unit) of rated full-load speed. *See also:* base speed; electric drive; adjustable-speed motor. (IA/IAC) [60]

speed changer (1) (hydraulic turbines) A device for changing the governor speed reference. (PE/EDPG) 125-1988r
(2) (gas turbines) A device by means of which the speed-governing system is adjusted to change the speed or power output of the turbine during operation. *See also:* asynchronous machine. (PE/EDPG) 282-1968w, [5]

speed-changer high-speed stop (gas turbines) A device that prevents the speed changer from moving in the direction to increase speed or power output beyond the position for which the device is set. *See also:* asynchronous machine. (PE/EDPG) 282-1968w, [5]

speed code (1) The code used to indicate various bit rates for Serial Bus: S25 indicates 24.576 Mbit/s for TTL backplanes; S50 indicates 49.152 Mbit/s for BTL and ECL backplanes; S100 indicates the 98.304 Mbit/s base rate for cable; S200 and S400 indicate 196.608 Mbit/s and 393.216 Mbit/s for the cable. (C/MM) 1394-1995
(2) The code used to indicate bit rates for Serial Bus. (C/MM) 1394a-2000

speed controller (control systems for steam turbine-generator units) Includes only those components and control elements that are responsive to speed and speed reference, and that supplies an input signal to the control mechanism for the purpose of controlling speed. (PE/EDPG) 122-1985s

speed-control mechanism (electric power system) All equipment such as relays, servomotors, pressure or power-amplifying devices, levers, and linkages between the speed governor and the governor-controlled valves. (PE/PSE) 94-1991w

speed deadband (hydraulic turbines) The total magnitude of the change in steady-state speed, expressed in percent of rated speed, required to reverse the direction of travel of the turbine control servomotor. (See the corresponding figure.) One half of the governor speed deadband is termed the governor speed insensitivity.

$$DB_S = \Delta n$$

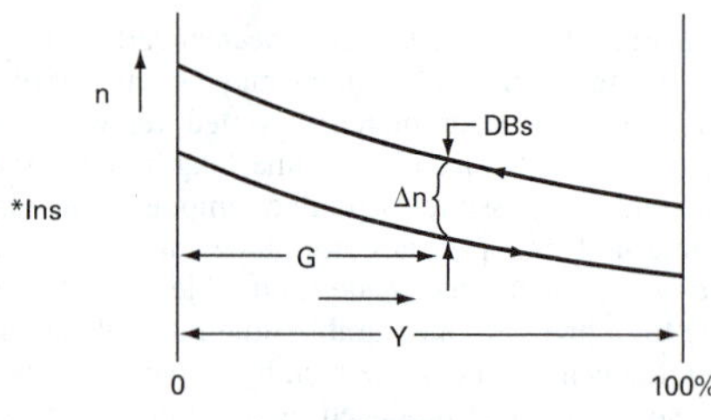

speed deadband

(PE/EDPG) 125-1977s, [5]

speed deviation (hydraulic turbines) The instantaneous difference between the actual speed and a reference speed. (PE/EDPG) 125-1977s

speed droop (hydraulic turbines) The speed droop and speed regulation graphs may indicate a nonlinear relationship between the two measured variables depending on the adjustment of the governor speed changer and the quantity (servomotor position or generator power output) used to develop the feedback signal used in the governor system. Speed droop and speed regulation are considered positive when speed increases with a decrease in gate position or power output. (See the corresponding figure.) The slope of the speed droop graph at a specified point of operation *G*. The change in a steady-state speed expressed in percent of rated speed corresponding to the 100% turbine servomotor stroke with no change in setting of any governor adjustments and with the turbine supplying power to a load independently of any other power source.

$$D_S = \left(\frac{-\Delta n}{\Delta G}\right) \cdot (100)$$

Speed droop is classified as either permanent or temporary:

a) *(permanent speed droop).* The speed droop that remains in steady state after the decay action of the damping device has been completed.

b) *(temporary speed droop).* The speed droop in steady state that would occur if the decay action of the damping device were blocked and the permanent speed droop were made inactive.

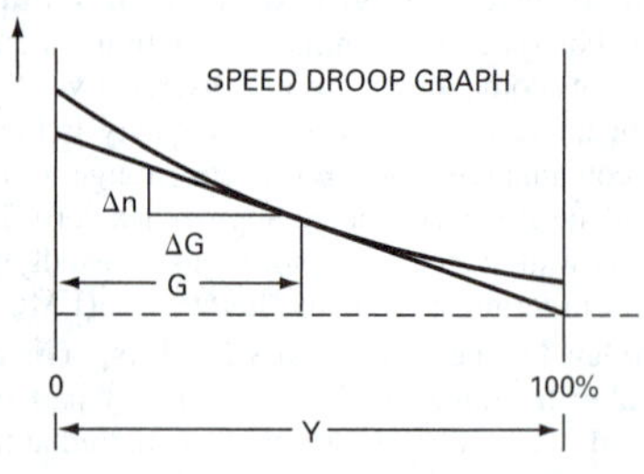

speed droop graph

(PE/EDPG) 125-1977s

speed-droop changer (hydraulic turbines) A device for changing the speed droop. (PE/EDPG) 125-1988r

speed-governing system (1) (automatic generation control on electric power systems) The control system used to maintain the rotational speed of a turbine-generator unit. The system consists of the speed governor, the speed-control mechanism, and the governor-controlled valves on a steam turbine or gates on a hydro turbine. (PE/PSE) 94-1991w
(2) Control elements and devices for the control of the speed or power output of a gas turbine. This includes a speed governor, speed changer, fuel-control mechanism, and other devices and control elements. (PE/EDPG) 282-1968w, [4]

speed governor (1) (electric power system) Includes only those elements that are directly responsive to speed and that position or influence the action of other elements of the speed-governing system. *See also:* asynchronous machine; speed-governing system. (PE/EDPG) 282-1968w, [5]

(2) (electric power system) A device that varies the governor-controlled valves to adjust energy input levels to maintain a uniform speed of a rotating machine. (PE/PSE) 94-1991w

speed limit A control function that prevents a speed from exceeding prescribed limits. Speed-limit values are expressed as percent of maximum rated speed. If the speed-limit circuit permits the limit value to change somewhat instead of being a single value, it is desirable to provide either a curve of the limit value of speed as a function of some variable, such as load, or to give limit values at two or more conditions of operation. *See also:* feedback control system. (IA/ICTL/APP/IAC) [69], [60]

speed-limit indicator A series of lights controlled by a relay to indicate the speeds permitted corresponding to the track conditions. (EEC/PE) [119]

speed/load control system (control systems for steam turbine-generator units) A system that controls the speed and load of a steam turbine-generator. The system typically includes the speed and load sensing and referencing elements, the controller(s), the control mechanism(s), and the control valve(s). (PE/EDPG) 122-1985s

speed/load reference changer (control systems for steam turbine-generator units) A device or devices by means of which the control system reference may be adjusted to change the speed or load of the turbine while the turbine is in operation. (PE/EDPG) 122-1985s

speed of transmission (data transmission) The instantaneous rate of which information is processed by a transmission facility. This quantity is usually expressed in characters per unit time or bits per unit time. (Rate of transmission is in more common use.) (PE) 599-1985w

speed of transmission, effective *See:* effective speed of transmission.

speed of vision (illuminating engineering) The reciprocal of the duration of the exposure time required for something to be seen. (EEC/IE) [126]

speed or frequency matching device (power system device function numbers) A device that functions to match and hold the speed or the frequency of a machine or of a system equal to, or approximately equal to, that of another machine, source or system. (PE/SUB) C37.2-1979s

speed range All the speeds that can be obtained in a stable manner by action of part (or parts) of the control equipment governing the performance of the motor. The speed range is generally expressed as the ratio of the maximum to the minimum operating speed. *See also:* electric drive. (IA/ICTL/IAC) [60]

speed ratio (1) (high-voltage switchgear) The ratio between 0.1s and 300 s or 600 s minimum melting currents, whichever is specified, which designates the relative speed of the fuse link. (SWG/PE) C37.40-1981s

(2) *See also:* clearing time; fuse tube; melting-speed ratio. (SWG/PE) C37.100-1992

speed ratio control A control function that provides for operation of two drives at a preset ratio of speed. *See also:* feedback control system. (IA/ICTL/IAC) [60]

speed-regulating rheostat A rheostat for the regulation of the speed of a motor. *See also:* control. (IA/ICTL/IAC) [60], [84]

speed regulation (speed governing of hydraulic turbines) The slope of the speed regulation graph at a specified point P of operation. The change in steady-state speed expressed in percent of rated speed when the power output of the unit is reduced from rated power output to zero power output under rated head with no change in setting of any governor adjustments and with the unit supplying power to a load independently of any other power source. (See the corresponding figure.)

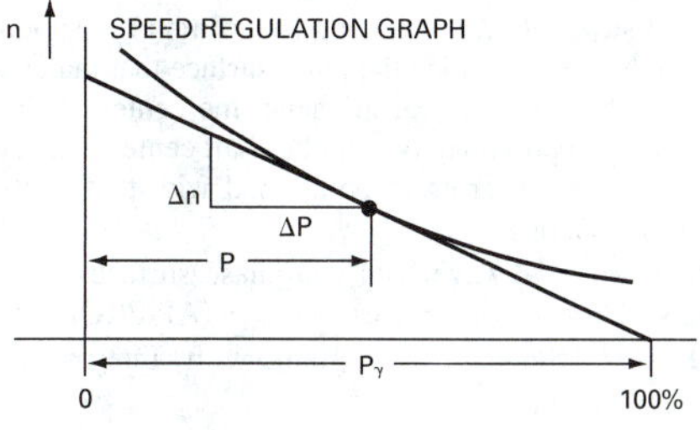

$$R_S = \left(\frac{-\Delta n/100}{\Delta P/P_r}\right) \cdot (100)$$

speed regulation graph

speed regulation changer (hydraulic turbines) A device for changing the speed regulation. (PE/EDPG) 125-1988r

speed regulation characteristic (rotating machinery) The relationship between speed and the load of a motor under specified conditions. *See also:* asynchronous machine. (PE) [9]

speed regulation of a constant-speed direct-current motor The change in speed when the load is reduced gradually from the rated value to zero with constant applied voltage and field-rheostat setting, expressed as a percent of speed at rated load. (EEC/PE) [119]

speed-sensing elements (hydraulic turbines) The speed-responsive elements that determine speed and influence the action of other elements of the governing system. Included are the means used to transmit a signal proportional to the speed of the turbine to the governor. (PE/EDPG) 125-1988r

speed variation Any change in speed of a motor resulting from causes independent of the control-system adjustment, such as line-voltage changes, temperature changes, or load changes. *See also:* electric drive. (IA/ICTL/IAC) [60]

SPeedy ImplemenTation of SNOBOL A version of SNOBOL that requires less memory and provides faster execution than SNOBOL. (C) 610.13-1993w

sphere illumination (illuminating engineering) The illumination on a task from a source providing equal illuminance in all directions about that task, such as a uniformly illuminated sphere with the task located at the center. (EEC/IE) [126]

spherical array A two-dimensional array of elements whose corresponding points lie on a spherical surface. (AP/ANT) 145-1993

spherical diffraction Transhorizon propagation due to diffraction by the spherical surface of the Earth, or more generally by any rounded obstacle which is extremely large in relation to the wavelength. (AP/PROP) 211-1997

spherical hyperbola (navigation aid terms) The locus of the points on the surface of a sphere having a specified constant difference in great circle distances from two fixed points on the sphere. (AES/GCS) 172-1983w

spherical propagation function The function given by:

$$f(r) = \frac{e^{-jkr}}{r}$$

where

r = the range from the source

jk = the propagation constant of the medium

(AP/PROP) 211-1997

spherical reduction factor (illuminating engineering) The ratio of the mean spherical luminous intensity to the mean horizontal intensity. (EEC/IE) [126]

spherical reflector A reflector that is a portion of a spherical surface. *See also:* antenna. (AP/ANT) 145-1993

spherical-seated bearing (rotating machinery) (self-aligning bearing) A journal bearing in which the bearing liner is supported in such a manner as to permit the axis of the journal to be moved through an appreciable angle in any direction. *See also:* bearing. (PE) [9]

spherical support seat (rotating machinery) A support for a journal bearing in which the inner surface that mates with the bearing shell is spherical in shape, the center of the sphere coinciding approximately with the shaft centerline, permitting the axis of the bearing to be aligned with that of the shaft. *See also:* bearing. (PE) [9]

spherical wave A wave with equiphase surfaces that form a family of concentric spheres. (AP/PROP) 211-1997

SPICE *See:* Simulation Program with Integrated Circuit Emphasis.

spider (rotating machinery) (rotor spider) A structure supporting the core or poles of a rotor from the shaft, and typically consisting of a hub, spokes, and rim, or some modified arrangement of these. (PE) [9]

spider rim *See:* rim.

spider web (rotating machinery) The component of a rotor that provides radial separation between the hub or shaft and the rim or core. *See also:* rotor. (PE) [9]

spike (1) (pulse terminology) A distortion in the form of a pulse waveform of relatively short duration superimposed on an otherwise regular or desired pulse waveform. *See also:* preshoot. (IM/WM&A) 194-1977w
(2) (as applied to relaying) An output signal of short duration and limited crest derived from an alternating input of specified polarity. *Note:* The duration of a spike usually does not exceed 1 ms. (SWG/PE) C37.100-1992

spike leakage energy (microwave gas tubes) The radio-frequency energy per pulse transmitted through the tube before and during the establishment of the steady-state radio-frequency discharge. *See also:* gas tube. (ED) 161-1971w

spikes *See:* transient overvoltages.

spike train (electrotherapy) (courant iteratif) A regular succession of pulses of unspecified shape, frequency, duration, and polarity. *See also:* electrotherapy. (EMB) [47]

spill (1) (charge-storage tubes) The loss of information from a storage element by redistribution. (ED) 161-1971w
(2) (liquid-filled power transformers) Spills, leaks, and other uncontrolled discharges of polychlorinated biphenyls (PCBs) constitute the disposal of PCBs. *See also:* disposal. (LM/C) 802.2-1985s

spillover In the transmit mode of a reflector antenna, the power from the feed that is not intercepted by the reflecting elements. (AP/ANT) 145-1993

spillover loss (radar) In a transmitting antenna having a focusing device such as a reflector or lens illuminated by a feed, spillover loss is the reduction in gain due to the portion of the power radiated by the feed in directions that do not intersect the focusing device. By reciprocity, the same loss occurs when the same antenna is used for reception. (AES/RS) 686-1982s

spillway Section of dam, or structure near dam, for flow of excess water. (PE/EDPG) 1020-1988r

spinaxis (navigation aids) The axis of rotation of a gyroscope. (AES/GCS) 172-1983w

spin axis (SA) (gyros) The axis of rotation of the rotor. (AES/GYAC) 528-1994

spin-axis-acceleration detuning error (dynamically tuned gyro) The error in a dynamically tuned gyro whereby deflection of the flexure, resulting from acceleration along the spin axis, can cause a shift in the tuning frequency. This will result in a change in the gyro output when there also exists a pickoff or capture loop offset. (AES/GYAC) 528-1994

spindle A device within a disk drive that maintains the axis of rotation and the force to rotate the disk. (C) 610.10-1994w

spindle speed (numerically controlled machines) The rate of rotation of the machine spindle usually expressed in terms of revolutions per minute. (IA) [61]

spindle wave (electrobiology) A sharp, rather large wave considered of diagnostic importance in the electroencephalogram. *See also:* electrocardiogram. (EEC/PE) [119]

spin-input-rectification drift rate (gyros) The drift rate in a single-degree-of-freedom gyro resulting from coherent oscillatory rates about the spin reference axis (SRA) and input reference axis (IRA). It occurs only when gyro and loop dynamics allow the gimbal to move away from null in response to the rate about the input reference axis, resulting in a cross coupling of the spin reference axis rate. This drift rate is a function of the input rate amplitudes and the phase angle between them. (AES/GYAC) 528-1994

spinner† Rotating part of a radar antenna, together with directly associated equipment, used to impart any subsidiary motion, such as conical scanning, in addition to the primary slewing of the beam. (AES/RS) 686-1990

† Obsolete.

spinning reserve Unloaded generation that is synchronized and ready to serve additional demand. (PE/PSE) 858-1993w

spin-offset coefficient (accelerometer) The constant of proportionality between bias change and the square of angular rate for an accelerometer that is spun about an axis parallel to its input reference axis and that passes through its effective center of mass for angular velocity. (AES/GYAC) 528-1994

spin-output-rectification drift rate (gyros) The drift rate in a single-degree-of-freedom gyro resulting from coherent oscillatory rates about the spin reference axis (SRA) and output reference axis (ORA). It occurs only when gyro and loop dynamics allow the float motion to lag case motion when subjected to a rate about the output reference axis, resulting in a cross coupling of the spin reference axis rate. This drift rate is a function of the input rate amplitudes and the phase angle between them. (AES/GYAC) 528-1994

spin reference axis (SRA) (gyros) An axis normal to the input reference axis and nominally parallel to the spin axis when the gyro output has a specified value, usually null. (AES/GYAC) 528-1994

spin-wait A condition during multi-unit synchronization where a processor remains in a tight loop, retesting a variable while waiting for the value to be changed by another unit. This condition may waste processor time and cost considerable bus bandwidth if the test variable must be read from system memory in another unit. (C/MM) 1212.1-1993

spiral antenna An antenna consisting of one or more conducting wires or tapes arranged as a spiral. *Note:* Spiral antennas are usually classified according to the shape of the surface to which they conform (for example, conical or planar spirals), and according to the mathematical form (for example, equiangular or archimedean). (AP/ANT) 145-1993

spiral distortion (camera tubes or image tubes using magnetic focusing) A distortion in which image rotation varies with distance from the axis of symmetry of the electron optical system. (ED) 161-1971w

spiral four (1) A quad in which the four conductors are twisted about a common axis, the two sets of opposite conductors being used as pairs. *Synonym:* star quad. *See also:* cable. (EEC/PE) [119]
(2) A cable element that comprises four insulated connectors twisted together. Two diametrically facing conductors form a transmission pair. *Note:* Cables containing star quads can be used to interchangeably with cables consisting of pairs, provided the electrical characteristics meet the same specifications. *Synonym:* star quad. (LM/C) 802.3u-1995s

spiral model A model of the software development process in which the constituent activities, typically requirements analysis, preliminary and detailed design, coding, integration, and testing, are performed iteratively until the software is complete. *Contrast:* waterfall model. *See also:* prototyping; incremental development. (C) 610.12-1990

spiral scanning (electronic navigation) Scanning in which the direction of maximum response describes a portion of a spiral. *See also:* antenna. (AP/ANT) [35]

SPL (sound-pressure level) *See:* sound pressure level.

s-plane In the Laplace transform, the notation σ = σ + φω is introduced. The s-plane is a coordinate system with s as the abscissa and w as the ordinate. The letter "p" is sometimes used instead of "s." (CAS) [13]

splashproof (packaging machinery) So constructed and protected that external splashing will not interfere with successful operation. *See also:* traction motor. (SWG/PE/VT/IA/ICTL/LT/PKG) C37.30-1971s, C37.100-1981s, 16-1955w, 333-1980w

splashproof enclosure An enclosure in which the openings are so constructed that drops of liquid or solid particles falling on the enclosure or coming towards it in a straight line at any angle not greater than 100° from the vertical cannot enter the enclosure either directly or by striking and running along a surface. (IA/MT) 45-1998

splashproof machine An open machine in which the ventilating openings are so constructed that drops of liquid or solid particles falling on the machine or coming towards it in a straight line at any angle not greater than 100° downward from the vertical cannot enter the machine either directly or by striking and running along a surface. *See also:* asynchronous machine. (PE) [9]

splice (1) (power cable joints) The physical connection of two or more conductors to provide electrical continuity. (PE/IC) 404-1986s

(2) (optical waveguide) *See also:* optical waveguide splice 812-1984w

(3) *See also:* compression joint. (T&D/PE) 524-1992r

splice box (mine type) An enclosed connector permitting short sections of cable to be connected together to obtain a portable cable of the required length. *See also:* mine feeder circuit.

splice loss *See:* insertion loss.

splice release block *See:* hold-down block.

splice site The location along the line where the conductors are temporarily anchored to join the conductors together to form a splice. (T&D/PE) 524a-1993r

splice, wire rope *See:* wire rope splice.

splicing cart (conductor stringing equipment) A unit which is equipped with a hydraulic pump and compressor (press) and all other necessary equipment for performing splicing operations on conductors. *Synonyms:* splicing truck; splicing trailer; sleeving trailer. (T&D/PE) 524a-1993r, 524-1992r

splicing chamber *See:* manhole.

splicing trailer *See:* splicing cart.

splicing truck *See:* splicing cart.

split acceptor A master module that splits the current transaction upon detecting SR* asserted. (C/BA) 896.4-1993w

split-anode magnetron A magnetron with an anode divided into two segments, usually by slots parallel to its axis. *See also:* magnetron. (ED) 161-1971w

split-beam cathode-ray tube (double-beam cathode-ray tube) A cathode-ray tube containing one electron gun producing a beam that is split to produce two traces on the screen. (ED) [45]

split brush Either an industrial or fractional-horsepower brush consisting of two pieces that are used in place of one brush. The adjacent sides of the split brush are parallel to the commutator bars. *Note:* A split brush is normally mounted so that the plane formed by the adjacent contacting brush sides is parallel to or passes through the rotating axis of the rotor. *See also:* brush; asynchronous machine. (EEC/EM/LB) [101]

split collector ring (rotating machinery) A collector ring that can be separated into parts for mounting or removal without access to a shaft end. *See also:* rotor. (PE) [9]

split-conductor cable A cable in which each conductor is composed of two or more insulated conductors normally connected in parallel. *See also:* segmental conductor. (T&D/PE) [10]

split-core-type current transformer *See:* current transformer.

split fitting A conduit split longitudinally so that it can be placed in position after the wires have been drawn into the conduit, the two parts being held together by screws or other means. *See also:* raceway. (EEC/PE) [119]

split-gate tracker A form of range tracker using a pair of time gates called an *early gate* and a *late gate*, contiguous or partly overlapping in time. *Note:* When tracking is established, the pair of gates straddles the received pulse that is being tracked. The position of the pair of gates then gives a measure of the time of arrival of the pulse (i.e., the range of the target from which the echo is received). Deviation of the pair of gates from the proper tracking position increases the signal energy in one gate and decreases it in the other, thus producing an error signal that moves the pair of gates so as to reestablish equilibrium. (AES) 686-1997

split hub (rotating machinery) A hub that can be separated into parts for ease of mounting on removal from a shaft. *See also:* rotor. (PE) [9]

split hydrophone *See:* split transducer.

split initiator A slave module that asserts sr* to require that the current transaction be split. (C/BA) 896.4-1993w

split key A foreign key containing two or more attributes, where at least one of the attributes is a part of the entity's primary key and at least one of the attributes is not a part of the primary key. (C/SE) 1320.2-1998

split node (network analysis) A node that has been separated into a source node and a sink node. *Notes:* 1. Splitting a node interrupts all signal transmission through that node. 2. In splitting a node, all incoming branches are associated with the resulting sink node, and all outgoing branches with the resulting source node. (CAS) 155-1960w

split operation A request to exchange data or an event that cannot be satisfied while all its participants remain interlocked. Rather, the participants record the requested action and, at a later time, when the exchange can be completed, reestablish a communication path to complete it. (C/BA) 1014.1-1994w

split-phase electric locomotive A single-phase electric locomotive equipped with electric devices to change the single-phase power to polyphase power without complete conversion of the power supply. *See also:* electric locomotive. (EEC/PE) [119]

split-phase motor A single-phase induction motor having a main winding and an auxiliary winding, designed to operate with no external impedance in either winding. The auxiliary winding is energized only during the starting operation of the auxiliary-winding circuits and is open-circuited during running operation. *See also:* asynchronous machine. (PE) [9]

split projector *See:* split transducer.

split rotor (rotating machinery) A rotor that can be separated into parts for mounting or removal without access to a shaft end. *See also:* rotor. (PE) [9]

split-sleeve bearing (rotating machinery) A journal bearing having a bearing sleeve that is split for assembly. *See also:* bearing. (PE) [9]

splitter Splitters divide or combine power. The power division causes an insertion loss and a small amount of internal loss that contributes to the attenuation of the signals passing through the device. The splitter has a common port and split port(s). The signals between the common and split port(s) has an insertion loss of 10 log *n*, where *n* equals the number of power splits. The splitter also has an isolation that attenuates signals passing between port(s). (LM/C) 802.7-1989r

splitter, optical *See:* optical directional coupler.

split-throw winding (rotating machinery) A winding wherein the conductors that constitute one complete coil side in one slot do not all appear together in another slot. *See also:* asynchronous machine. (PE) [9]

split transaction (1) An operation in which the request is transmitted in one bus transaction and the response is transmitted in a separate subsequent bus transaction. (C/BA) 896.3-1993w, 896.4-1993w

(2) A transaction that consists of separate request and response subactions. On a backplane bus, for example, a split transaction is one in which bus mastership is relinquished between the request and response subactions. Few buses permit split transactions. *See also:* unified transaction. (C/MM) 1596-1992

(3) A transaction that consists of separate request and response subactions. On a backplane bus, for example, bus ownership is relinquished between the request and response subactions. A transaction that is not split is called a unified transaction. (C/MM) 1212-1991s

(4) A system transaction in which the request is transmitted in one bus transaction and the response is transmitted in a separate subsequent bus transaction. (C/BA) 10857-1994

(5) A transaction where the responder releases control of the bus after sending the acknowledge and then some time later starts arbitrating for the bus so it can start the response subaction. Other subactions may take place on the bus between the request and response subactions for the transaction. (C/MM) 1394-1995

(6) A transaction where unrelated subactions may take place on the bus between its request and response subactions. (C/MM) 1394a-2000

split transducer (audio and electroacoustics) A directional transducer in which electroacoustic transducing elements are so divided and arranged that each division is electrically separate. (SP) [32]

split-winding protection A form of differential protection in which the current in all or part of the winding is compared to the normally proportional current in another part of the winding. (SWG/PE) C37.100-1992

SPMP *See:* software project management plan.

spoiler resistors (power supplies) Resistors used to spoil the load regulation of regulated power supplies to permit parallel operation when not otherwise provided for. (AES) [41]

s pole *See:* junction pole.

sponge (electrodeposition) A loose cathode deposit that is fluffy and of the nature of a sponge, contrasted with a reguline metal. (EEC/PE) [119]

spontaneous emission (1) (fiber optics) Radiation emitted when the internal energy of a quantum mechanical system drops from an excited level to a lower level without regard to the simultaneous presence of similar radiation. *Note:* Examples of spontaneous emission include:

1) radiation from a light emitting diode (LED), and
2) radiation from an injection laser below the lasing threshold.

See also: superradiance; light-emitting diode; injection laser diode; stimulated emission. (Std100) 812-1984w

(2) (laser maser) The emission of radiation from a single electron, atom, molecule, or ion in an excited state at a rate independent of the presence of applied external fields. (LEO) 586-1980w

spontaneous polarization (primary ferroelectric terms) Magnitude of the polarization within a single ferroelectric domain in the absence of an external electric field. A spontaneous polarization P_s is a fundamental property of all pyroelectric crystals, although it is reversible or reorientable only in ferroelectrics. Most ferroelectric phases originate from a nonpolar prototypic phase and all of the polarization is reorientable. However, if the prototypic phase is polar, only a portion of the total spontaneous polarization may be reoriented. This reorientable or reversible portion is commonly called the spontaneous polarization. The corresponding figures illustrate an example of a case where the prototypic phase is tetragonal (4*mm*) and the ferroelectric phase is monoclinic (*m*), one of the two special polar groups. The spontaneous polarization, P_s in the second part of the figure, is composed of a switchable part, P_1, and a nonswitchable component, P_4; thus P_s is reorientable when P_1 switches between any of its four allowed states. In this example the pyroelectric vector is not collinear with the polar axis P_s, and both P_1 and P_s may independently change direction with temperature. The unit cell is the smallest group of atoms within a crystal whose repetition in space generates the whole crystal. It is electrically neutral in all states of the crystal, but on this microscopic scale, P_s is associated with a polar displacement of the ionic and electronic charges within the crystalline unit cell, and this, in turn, gives rise to a microscopic electric dipole moment. In ferroelectric crystals, dipoles in adjacent unit cells are aligned in the same direction, resulting in a net P_s within a macroscopic volume (much larger than a unit cell) called a domain.

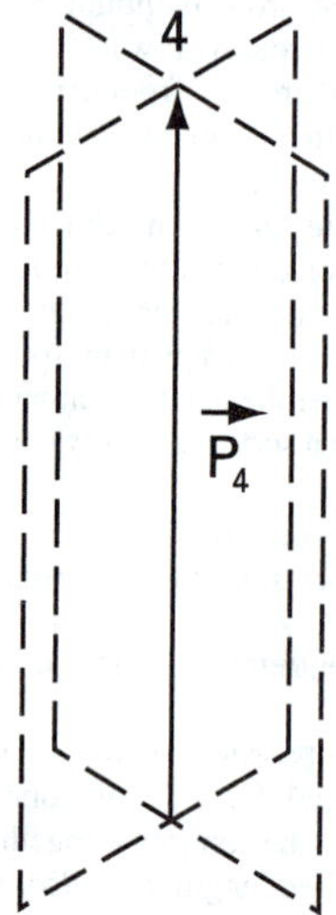

Note: The polar axis lies along the four-fold axis of symmetry.

Polar prototype phase, tetragonal (4 mm)

spontaneous polarization

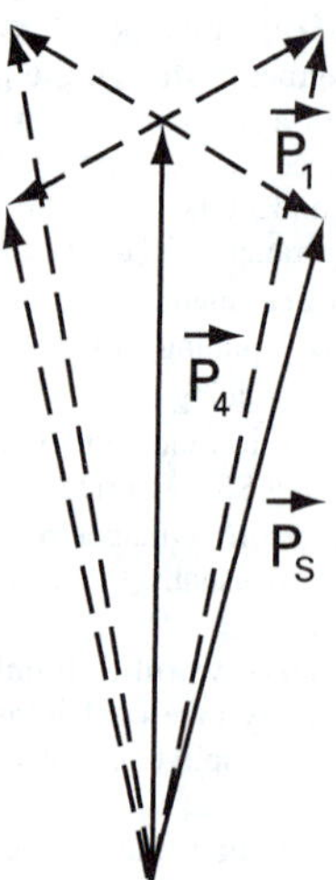

Note: The spontaneous polarization is not collinear with the prototype four-fold axis.

Ferroelectric phase in the monoclinic symmetry (m), derived from the tetragonal prototype phase

(UFFC) 180-1986w

spontaneous strain (primary ferroelectric terms) The summation of all the strains necessary to convert a ferroelastic crystal from the nonferroelastic prototype state to one of the ferroelastic orientation states. The prototype state, by definition, has zero spontaneous strain. A ferroelastic crystal can be switched from one ferroelastic orientation state to another by mechanical stress. Any two of the states are identical or enantiomorphous in crystal structure but different in mechanical strain tensor at zero mechanical stress (and at zero electrical field). (UFFC) 180-1986w

spool To read input data, or write output data, to auxiliary or main storage for later processing or output, in order to permit input/output devices to operate concurrently with job execution. Derived from the acronym SPOOL for Simultaneous Peripheral Output On Line. (C) 610.12-1990, 610.10-1994w
(2) **(A)** Secondary storage used as an interim holding area for output waiting to be printed as in definition (1) above. **(B)** A cylinder without flanges on which tape may be wound. *Synonym:* hub. *Contrast:* reel. *See also:* bore. (C) 610.10-1994

spooler A program that initiates and controls spooling. (C) 610.12-1990

spool insulator An insulating element of generally cylindrical form having an axial mounting hole and a circumferential groove or grooves for the attachment of a conductor. *See also:* insulator; tower. (T&D/PE/EEC/IEPL) [10], [89]

sporadic E layer An ionospheric layer of the E region which is thin, transient and of limited geographical extent. *Note:* An equatorial sporadic E layer occurs regularly during the day in association with the equatorial electrojet. *Synonym:* Es layer. (AP/PROP) 211-1997

sporadic ionization Ionization of the upper atmosphere, irregularly distributed in space and time, and abnormally high relative to the average ionization level of the region in which it is produced. (AP/PROP) 211-1997

sporadic tasks Asynchronous tasks that may have hard (rigid) deadlines that must be met, but have a minimum interarrival duration between instances (typically by enforcement). (C/BA) 896.3-1993w

spot (oscilloscopes) (cathode-ray tubes) The illuminated area that appears where the primary electron beam strikes the phosphor screen of a cathode-ray tube. *Note:* The effect of the impact on this small area of the screen is practically instantaneous. *See also:* oscillograph; cathode-ray tube. (IM/ED/HFIM) [40], [45], [84]

spotlight (illuminating engineering) A form of floodlight usually equipped with lens and reflectors to give a fixed or adjustable narrow beam. (EEC/IE) [126]

spotlight synthetic-aperture radar A form of SAR in which very high resolution is obtained by steering the real antenna beam to dwell longer on a scene or target than allowed by a fixed antenna. (AES) 686-1997

spot measurement (point-in-time measurement) A measurement that is performed at some instant and point in space that does not provide information regarding temporal or spatial variations of the field. (T&D/PE) 1308-1994

spot network A small network, usually at one location, consisting of two or more primary feeders, with network units and one or more load service connections. (PE/TR) C57.12.44-1994

spot-network type A unit substation which has two stepdown transformers, each connected to an incoming high-voltage circuit. The outgoing side of each transformer is connected to a common bus through circuit breakers equipped with relays which are arranged to trip the circuit breaker on reverse power flow to the transformer and to reclose the circuit breaker upon the restoration of the correct voltage, phase angle and phase sequence at the transformer secondary. The bus has one or more outgoing radial (stub end) feeders. (PE/TR) C57.12.80-1978r

spot noise figure (spot noise factor) (transducer at a selected frequency) The ratio of the output noise power per unit bandwidth to the portion thereof attributable to the thermal noise in the input termination per unit-bandwidth, the noise temperature of the input termination being standard (290 kelvins). The spot noise figure is a point function of input frequency. *See also:* signal-to-noise ratio; noise figure. (PE/EEC) [119]

spot projection (facsimile) The optical method of scanning or recording in which the scanning or recording spot is defined in the path of the reflected or transmitted light. *See also:* recording; scanning. (COM) 168-1956w

spot punch A punch device used for punching one hole at a time into a punch card. (C) 610.10-1994w

spot size The diameter of a pixel on a display surface. *Synonym:* beam spot size. (C) 610.6-1991w

spot speed (facsimile) The speed of the scanning or recording spot within the available line. *Note:* This is generally measured on the subject copy or on the record sheet. *See also:* recording; scanning. (COM) 168-1956w

spotting (electroplating) The appearance of spots on plated or finished metals. (PE/EEC) [119]

spot-type fire detector (fire protection devices) A device whose detecting element is concentrated at a particular location. (NFPA) [16]

spot wobble (television) A process wherein a scanning spot is given a small periodic motion transverse to the scanning lines at a frequency above the picture signal spectrum. (BT/AV) 201-1979w

spray plume *See:* positive prebreakdown streamers.

spread, Doppler *See:* Doppler spread.

spread F A phenomenon observed on ionograms displaying a wide range of delays of echo pulses, near the F region critical frequencies. *Notes:* 1. The echoes usually are spread in the frequency and virtual height domains on an ionogram. 2. Spread F commonly occurs at night at low latitudes (e.g., near the magnetic dip equator) and at high latitudes. (AP/PROP) 211-1997

spreading factor For propagation in isotropic unbounded media, that amplitude factor that accounts for geometric spreading of the field intensity. *Note:* In the far field region of plane, cylindrical, and spherical waves this factor is 1, $r^{-1/2}$, and r^{-1} respectively, where r is the distance from the source to the observation point. (AP/PROP) 211-1997

spreading loss (1) (wave propagation) The reduction in radiant-power surface density due to spreading. (Std100) 270-1966w
(2) The decrease in power or power density due to divergence of the outward energy flow of cylindrical and spherical waves. (AP/PROP) 211-1997

spread sheet *See:* electronic spread sheet.

spread spectrum A modulation technique for multiple access, or for increasing immunity to noise and interference. (C/COM) 610.7-1995, [19]

spread, time delay *See:* time delay spread.

spring-actuated stepping relay A stepping relay that is cocked electrically and operated by spring action. *See also:* relay. (EEC/REE) [87]

spring attachment (burglar-alarm system) (trap) A device designed for attachment to a movable section of the protected premises, such as a door, window, or transom, so as to carry the electric protective circuit in or out of such section, and to indicate an open- or short-circuit alarm signal upon opening of the movable section. *Synonym:* spring contact. *See also:* protective signaling. (PE/EEC) [119]

spring barrel The part that retains and locates the short-circuiter. *See also:* rotor. (EEC/PE) [119]

spring buffer A buffer that stores in a spring the kinetic energy of the descending car or counterweight. *See also:* elevator. (EEC/PE) [119]

spring-buffer load rating (elevators) The load required to compress the spring an amount equal to its stroke. *See also:* elevator. (EEC/PE) [119]

spring-buffer stroke (elevators) The distance the contact end of the spring can move under a compressive load until all coils are essentially in contact. *See also:* elevator. (PE/EEC) [119]

spring contact An electric contact that is actuated by a spring. (EEC/PE) [119]

spring-loaded bearing (rotating machinery) A ball bearing provided with a spring to ensure complete angular contact between the balls and inner and outer races, thereby removing the effect of diametral clearance in both bearings of a machine provided with ball bearing at each end. *See also:* bearing. (PE/EEC) [119]

spring operation Stored-energy operation by means of spring-stored energy. (SWG/PE) C37.100-1992

sprinkler supervisory system A supervisory system attached to an automatic sprinkler system that initiates signal transmission automatically upon the occurrence of abnormal conditions in valve positions, air or water pressure, water temperature or level, the operability of power sources necessary to the proper functioning of the automatic sprinkler, etc. *See also:* protective signaling. (EEC/PE) [119]

sprite engine A graphics controller that supports sprites; small, high-resolution objects that can be moved about the display surface. (C) 610.10-1994w

sprocket feed *See:* tractor feed.

sprocket hole (1) (test, measurement, and diagnostic equipment) The hole in a tape that is used for electrical timing or mechanically driving the tape. (MIL) [2]
(2) *See also:* feed hole. (C) 610.10-1994w

sprocket track *See:* feed track.

SPS *See:* Symbolic Programming Systems.

SPSS *See:* Statistical Package for Social Sciences.

spurious bulk-wave signals Unwanted signals caused by bulk wave excitation existing at the surface acoustic wave filter output. (UFFC) 1037-1992w

spurious components Persistent sine waves at frequencies other than the harmonic frequencies. *See also:* harmonic distortion. (IM/WM&A) 1057-1994w

spurious count (1) (nuclear techniques) A count from a scintillation counter other than one purposely generated or one due directly to ionizing radiation. *See also:* scintillation counter. (NPS) 398-1972r
(2) A count caused by any event other than the passage into or through the counter tube of the ionizing radiation to which it is sensitive. (NI/NPS) 309-1999

spurious emission power (land-mobile communications transmitters) Any part of the radio frequency output that is not a component of the theoretical output, as determined by the type of modulation and specified bandwidth limitations. (EMC) 377-1980r

spurious emission power radiation field (land-mobile communications transmitters) That portion of the spurious emission power which may be radiated from a transmitter enclosure and which can be measured in the near or far field regions. (EMC) 377-1980r

spurious emissions (transmitter performance) Any part of the radio-frequency output that is not a component of the theoretical output, as determined by the type of modulation and specified bandwidth limitations. *See also:* audio-frequency distortion. (VT) [37]

spurious initiation Attempts to establish a communication connection or association, using a false identity or through the replay of a previous, legitimate initiation sequence. Spurious initiation includes spoofing or masquerading attempts in the communication system, and coupled with other attacks, could result in unauthorized disclosure or modification of information, unauthorized receipt of services, or denial of service to legitimate users or critical functions. (C/BA) 896.3-1993w

spurious output (signal generators) (nonharmonic) Those signals in the output of a source that have a defined amplitude and frequency and are not harmonically related to the fundamental frequency. This definition excludes sidebands due to residual and intentional modulation. *See also:* signal generator. (IM/HFIM) [40]

spurious pulse (1) (nuclear techniques) A pulse in a scintillation counter other than one purposely generated or one due directly to ionizing radiation. *See also:* scintillation counter. (NPS) 398-1972r
(2) (dial-pulse address signaling systems) (telephony) The intermittent and undesired change of state in a circuit from its on-hook condition (spurious make) or off-hook condition (spurious break) lasting more than 1ms. (COM/TA) 753-1983w

spurious pulse mode An unwanted pulse mode, formed by the chance combination of two or more pulse modes, that is indistinguishable from a pulse interrogation or pulse reply. (IM/WM&A) 194-1977w

spurious radiation (radio-noise emissions) Any emission from an electronic communications equipment at frequencies outside its occupied bandwidth. (EMC) C63.4-1988s

spurious reflections Unwanted signals caused by reflection of surface acoustic wave or bulk waves from substrate edges or electrodes. (UFFC) 1037-1992w

spurious response (1) (general) Any response, other than the desired response, of an electric transducer or device. (PE) 599-1985w
(2) (mobile communication or electromagnetic compatibility) Output, from a receiver, due to a signal or signals having frequencies other than that to which the receiver is tuned. *See also:* electromagnetic compatibility. (VT) [37]
(3) (spectrum analyzer) A characteristic of a spectrum analyzer wherein the displayed frequency does not conform to the input frequency. (IM) 748-1979w
(4) (frequency-modulated mobile communications receivers) Any receiver response that occurs because of frequency conversions other than the desired frequency translations in the receiver. (VT) 184-1969w

spurious-response ratio (radio receivers) The ratio of: A) the field strength at the frequency that produces the spurious response to B) the field strength at the desired frequency, each field being applied in turn, under specified conditions, to produce equal outputs. *Note:* Image ratio and intermediate-frequency-response ratio are special forms of spurious response ratio. *See also:* radio receiver. 188-1952w

spurious transmitter output (1) (A) (general) Any part of the radio-frequency output that is not implied by the type of modulation (amplitude modulation, frequency modulation, etc.) and specified bandwidth. *See also:* radio transmitter. **(B) (conducted)** Any spurious output of a radio transmitter conducted over a tangible transmission path. *Note:* Power lines, control leads, radio-frequency transmission lines and waveguides are all considered as tangible paths in the foregoing definition. Radiation is not considered as tangible path in the foregoing definition. Radiation is not considered a tangible path in this definition. *See also:* radio transmitter. **(C) (inband)** Spurious output of a transmitter within its specified band of transmission. *See also:* radio transmitter. **(D) (radiated)** Any spurious output radiated from a radio transmitter. *Note:* The radio transmitter does not include the associated antenna and transmission lines. *See also:* radio transmitter. (BT) 182-1961
(2) (extraband) *See also:* extraband spurious transmitter output. (EMC) 377-1980r

spurious tube counts (radiation counter tubes) Counts in radiation-counter tubes, other than background counts and those caused by the source measured. *Note:* Spurious counts are caused by failure of the quenching process, electric leakage, and the like. Spurious counts may seriously affect measurement of background counts. (ED) 161-1971w

sputtering (cathode sputtering) (electroacoustics) A process sometimes used in the production of the metal master wherein the original is coated with an electric conducting layer by means of an electric discharge in a vacuum. *Note:* This is done prior to electroplating a heavier deposit. *See also:* phonograph pickup. (SP) [32]

SQL *See:* Structured Query Language.

SQL engine A relational engine that accepts SQL commands and accesses the database in order to obtain the requested data. (C) 610.10-1994w

square-law detection The form of detection of an amplitude-modulated signal in which the output voltage is a linear function of the square of the envelope of the input wave. (IT) [123]

squareness ratio (magnetic storage) (material in a symmetrically cyclically magnetized condition) The ratio of the flux density at zero magnetizing force to the maximum flux density; the ratio of the flux density when the magnetizing force has changed halfway from zero toward its negative limiting value, to the maximum flux density. *Note:* Both these ratios are functions of the maximum magnetizing force. (C) [20]

square wave (1) (data transmission) A periodic wave that alternately for equal lengths of time assumes one of two fixed values, the time of transition being negligible in comparison. (PE) 599-1985w

(2) (pulse terminology) A periodic rectangular pulse train with a duty factor of 0.5 or an on-off ratio of 1.0. (IM/WM&A) 194-1977w

square-wave response (1) (camera tubes) The ratio of the peak-to-peak signal amplitude given by a test pattern consisting of alternate black and white bars of equal widths to the difference in signal between large-area blacks and large-area whites having the same illuminations as the black and white bars in the test pattern. *Note:* Horizontal square-wave response is measured if the bars run perpendicular to the direction of horizontal scan. Vertical square-wave response is measured if the bars run parallel to the direction of horizontal scan. *See also:* amplitude response. (ED) 161-1971w

(2) (diode-type camera tube) Square-wave spatial inputs may be used, in which case the response curve is called the contrast transfer function or square-wave amplitude response. Units: lines per picture height. (ED) 503-1978w

square-wave response characteristic (camera tubes) The relation between square-wave response and the ratio of a raster dimension to the bar width in the square-wave response test pattern. *Note:* Unless otherwise specified, the raster dimension is the vertical height. *See also:* amplitude response characteristic; television. (ED) 161-1971w

squaring amplifier (as applied to relaying). A circuit that produces a block. (SWG/PE) C37.100-1992

squeezable waveguide A variable-width waveguide for shifting the phase of the radio-frequency wave traveling through it by mechanically squeezing the dimensions of a rectangular waveguide. (AES) 686-1997

squeeze section (transmission lines and waveguides) A length of rectangular waveguide so constructed as to permit alteration of the broad dimension with a corresponding alteration in electrical length. *See also:* waveguide. (IM/HFIM) [40]

squeeze trace (electroacoustics) A variable-density sound track wherein, by means of adjustable masking of the recording light beam and simultaneous increase of the electric signal applied to the light modulator, a track having variable width with greater signal-to-noise ratio is obtained. *See also:* phonograph pickup. (SP) [32]

squelch (1) (radio receivers) A circuit function that acts to suppress the audio output of a receiver when noise power that exceeds a predetermined level is present. (VT) 184-1969w

(2) Facility incorporated in radio receivers to disable their signal output while the received carrier signal level is less than a preset value. (SUB/PE) 999-1992w

squelch circuit (1) (data transmission) A circuit for preventing a radio receiver from producing audio-frequency output in the absence of a signal having predetermined characteristics. A squelch circuit may be operated by signal energy in the receiver pass band, by noise quieting, or by a combination of the two (ratio squelch). It may also be operated by a signal having special modulation characteristics (selective squelch). (PE) 599-1985w

(2) A circuit for preventing production of an unwanted output in the absence of a signal having predetermined characteristics. (SWG/PE) C37.100-1992

squelch clamping (frequency-modulated mobile communications receivers) The characteristic of the receiver, when receiving a normal signal, in which the squelch circuit under certain conditions of modulation will cause suppression of the audio output. (VT) 184-1969w

squelch selectivity (frequency-modulated mobile communications receivers) The characteristic that permits the receiver to remain squelched when a radio-frequency signal not on the receiver's tuned frequency is present at the input. (VT) 184-1969w

squelch sensitivity (frequency-modulated mobile communications receivers) The minimum radio-frequency signal input level, with standard test modulation required to increase the audio power output from the reference threshold squelch adjustment condition to within 6 dB of the reference audio power output. (VT) 184-1969w

squiggle A short "s"-shaped line attached at one end to an arrow label and at the other end to an arrow segment. A squiggle binds an object type set (arrow label) to an object set (arrow segment). (C/SE) 1320.1-1998

squint A condition in which a specified axis of an antenna, such as the direction of maximum directivity or of a directional null, departs slightly from a specified reference axis. *Notes:* 1. Squint is often the undesired result of a defect in the antenna; but in certain cases, squint is intentionally designed in in order to satisfy an operational requirement. 2. The reference axis is often taken to be the mechanically defined axis of the antenna; for example, the axis of a paraboloidal reflector. (AP/ANT) 145-1993

(2) **(A)** The angle between the major lobe axis of each lobe and the central axis in a lobe-switching or simultaneous-lobing (monopulse) antenna. **(B)** The angular difference between the axis of antenna radiation and a selected geometric axis, such as the axis of the reflector, the center of the cone formed by movement of the radiation axis, or the broadside direction of a moving vehicle. (AES) 686-1997

squint angle The angle between a specified axis of an antenna, such as the direction of maximum directivity or a directional null, and the corresponding reference axis. (AP/ANT) 145-1993

squint-mode synthetic-aperture radar A SAR in which the beam is pointed other than at right angles to the flight path of the airborne radar platform. (AES) 686-1997

squirrel-cage induction motor A motor in which the secondary circuit consists of a squirrel-cage winding suitably disposed in slots in the secondary core. (IA/MT) 45-1998

squirrel-cage rotor (rotating machinery) A rotor core assembly having a squirrel-cage winding. *See also:* rotor. (PE) [9]

squirrel-cage winding (1) (rotating machinery) A winding, usually on the rotor of a machine, consisting of a number of conducting bars having their extremities connected by metal rings or plates at each end. (PE) [9]

(2) A permanently short-circuited winding, usually uninsulated (primarily used in induction machines) having its conductors uniformly distributed around the periphery of the machine and joined by continuous end rings. (IA/MT) 45-1998

squitter Random output pulses from a transponder caused by ambient noise or by an intentional random triggering system, but not by the interrogation pulses. (AES) 686-1997

SR *See:* symbol rate.

SRA *See:* spin reference axis.

SRAM *See:* static random-access memory.

SRF *See:* specifically routed frame.

SRH *See:* service request handler.

SRM *See:* standard reference material.

SRR *See:* software requirements review; system requirements review.

SRS *See:* standard response spectrum; software requirements specification.

SRT *See:* Source Routing Transparent.

SS *See:* source statements.

SSAC *See:* steel supported aluminum conductor.

SSAP *See:* address fields.

SSB *See:* single-sideband modulation.

SSD *See:* start of stream delimiter.

SSI *See:* small scale integration; soil structure interaction.

SSR *See:* software specification review.

ST *See:* symbol time.

stability **(1)** An aspect of system behavior associated with systems having the general property that bounded input perturbations result in bounded output perturbations. *Notes:* 1. A stable system will ultimately attain a steady state. 2. Deviations from this steady state due to component aging or environmental changes do not indicate instability, but a change in the system. *See also:* transient stability; steady-state stability. (CAS) [13]

(2) (perturbations) For convenience in defining various stability concepts, only those parameters or signals that are perturbed are explicitly exhibited, or mentioned, that is, for perturbations in initial states, a perturbed solution is denoted

$$\varphi(\mathbf{x}(t_0) + \Delta\mathbf{x}(t_0);t)$$

where $\Delta\mathbf{x}(t_0)$ represents the perturbation in initial state. Finally, the perturbed-state solution is denoted $\Delta\varphi = \varphi(\mathbf{x}(t_0) + \Delta\mathbf{x}(t_0);t) - \varphi(\mathbf{x}(t_0);t)$. (IM) [120]

(3) (power system stability) In a system of two or more synchronous machines connected through an electric network, the condition in which the difference of the angular positions of the rotors of the machines either remains constant while not subjected to a disturbance, or becomes constant following an aperiodic disturbance. *Note:* If automatic devices are used to aid stability, their use will modify the steady-state and transient stability terms to: steady-state stability with automatic devices; transient stability with automatic devices. Automatic devices as defined for this purpose are those devices that are operating to increase stability during the period preceding and following a disturbance as well as during the disturbance. Thus relays and circuit breakers are excluded from this classification and all forms of voltage regulators included. Devices for inserting and removing shunt or series impedance may or may not come within this classification depending upon whether or not they are operating during the periods preceding and following the disturbance. *See also:* transient stability; steady-state stability. (T&D/PE) [10]

(4) (oscilloscopes) The property of retaining defined electrical characteristics for a prescribed time and environment. *Notes:* 1. Deviations from a stable state may be called drift if it is slow, or jitter or noise if it is fast. In triggered-sweep systems, triggering stability may refer to the ability of the trigger and sweep systems to maintain jitter-free displays of high- frequency waveforms for long (seconds to hours) periods of time. 2. Also, the name of the control used on some oscilloscopes to adjust the sweep for triggered, free-running, or synchronized operation. *See also:* sweep mode control. (IM) 311-1970w

(5) (hydraulic turbines) Characteristics of the governing system pertaining to limitation of oscillations of speed or power under sustained conditions, to damping of oscillations of speed following rejection of load, and to damping of speed oscillations under isolated load conditions following sudden load changes. (PE/EDPG) 125-1977s

(6) (electrothermic power meters) For a constant input rf power, constant ambient temperature and constant power line voltage, the variation in rf power indication over stated time intervals. *Note:* Long term stability or drift is the maximum acceptable change in 1 h. Short-term stability or fluctuation is the maximum (peak) change in 1 min. (IM) 544-1975w

(7) (nuclear power generating station) The ability of a module to attain and maintain a steady state. (PE/NP) 381-1977w

(8) (A) (software) The ability to continue unchanged despite disturbing or disruptive events. **(B) (software)** The ability to return to an original state after disturbing or disruptive events. (C/SE) 729-1983

(9) (accelerometer) (gyros) A measure of the ability of a specific mechanism or performance coefficient to remain invariant when continuously exposed to a fixed operating condition. (This definition does not refer to dynamic or servo stability.) (AES/GYAC) 528-1994

stability, absolute *See:* absolute stability.

stability, asymptotic *See:* asymptomatic stability.

stability, bounded-input-bounded-output *See:* bounded-input-bounded-output stability.

stability, conditional *See:* conditional stability.

stability drift (electric conversion) Gradual shift or change in the output over a period of time due to change or aging of circuit components. (All other variables held constant). (T&D/PE) [26]

stability, driven *See:* driven stability.

stability, equiasymptotic *See:* equiasymptotic stability.

stability, excitation-system *See:* excitation-system stability.

stability factor The ratio of a stability limit (power limit) to the nominal power flow at the point of the system to which the stability limit is referred. *Note:* In determining stability factors it is essential that the nominal power flow be specified in accordance with the one of several bases of computation, such as rating or capacity of, or average or maximum load carried by, the equipment or the circuits. *See also:* alternating-current distribution. (T&D/PE) [10]

stability, finite-time *See:* finite-time stability.

stability, global *See:* global stability.

stability in-the-whole *See:* control system; global stability.

stability, Lagrange *See:* Lagrange stability.

stability limit (electric systems) (power limit) The maximum power flow possible through some particular point in the system when the entire system or the part of the system to which the stability limit refers is operating with stability. *See also:* alternating-current distribution. (T&D/PE) [10]

stability, long-term *See:* long-term stability.

stability, Lyapunov *See:* Lyapunov stability.

stability of a limit cycle Synonymous with orbital stability. *See also:* control system. (CS/IM) [120]

stability of the speed-governing system (gas turbines) A characteristic of the system that indicates that the speed-governing system is capable of actuating the turbine fuel-control valve so that sustained oscillations in turbine speed, or rate of energy input to the turbine, are limited to acceptable values by the speed-governing system. (PE/EDPG) 282-1968w, [5]

stability of the temperature-control system (gas turbines) A characteristic of the system that indicates that the temperature-control system is capable of actuating the turbine fuel-control valve so that sustained oscillations in rate of energy input to the turbine are limited to acceptable values by the temperature-control system during operation under constant system frequency. (PE/EDPG) 282-1968w, [5]

stability, orbital *See:* orbital stability.

stability, practical *See:* finite-time stability.

stability, quasi-asymptotic *See:* quasi-asymptotic stability.

stability, relative *See:* relative stability.

stability, short-time *See:* finite-time stability.

stability, synchronous-machine regulating-system *See:* synchronous-machine regulating-system stability.

stability, total *See:* total stability.

stability, trajectory *See:* trajectory stability.

stability, uniform-asymptotic *See:* uniform-asymptotic stability.

stabilization (1) (control system feedback) Act of attaining stability or of improving relative stability. *See also:* feedback control system. (PE/EDPG) [3]

(2) (navigation aids) (navigation) Maintenance of a desired orientation of a vehicle or device with respect to one or more reference directions. (AES/GCS) 172-1983w

(3) (direct-current amplifier) *See also:* drift stabilization.

stabilization, drift *See:* drift stabilization.

stabilization network (analog computer) As applied to operational amplifiers, a network used to shape the transfer characteristics to eliminate or minimize oscillations when feedback is provided. (C) 165-1977w

stabilized feedback Feedback employed in such a manner as to stabilize the gain of a transmission system or section thereof with respect to time or frequency or to reduce noise or distortion arising therein. *Note:* The section of the transmission system may include amplifiers only, or it may include modulators. *See also:* feedback. (AP/ANT) 145-1983s

stabilized flight (navigation aids) That type of flight which obtains control information from devices which sense orientation with respect to external references. (AES/GCS) 172-1983w

stabilized shunt-wound generator Same as the shunt-wound type, except that a series field winding is added. The series field winding is proportioned such that it does not require equalizers for satisfactory parallel operation. The voltage regulation of this type of generator should be the same as shunt-wound generators. (IA/MT) 45-1998

stabilized shunt-wound motor A shunt-wound motor that has a light series winding added to prevent a rise in speed, or to obtain a slight reduction in speed, with increase of load. (IA/MT) 45-1998

stabilized-variable model A model in which some of the variables are held constant and the others are allowed to vary; for example, a model of a controlled climate in which humidity is held constant and temperature is allowed to vary. (C) 610.3-1989w

stabilizer, excitation control system *See:* excitation control system stabilizer.

stabilizer, power system *See:* power system stabilizer.

stabilizing winding (power and distribution transformers) A delta connected auxiliary winding used particularly in Y-connected three-phase transformers for such purposes as the following:

a) To stabilize the neutral point of the fundamental frequency voltages;
b) to minimize third-harmonic voltage and the resultant effects on the system;
c) to mitigate telephone influence due to third-harmonic currents and voltages;
d) to minimize the residual direct-current magnetomotive force on the core;
e) to decrease the zero-sequence impedance of transformers with Y-connected windings.

Note: A winding is regarded as a stabilizing winding if its terminals are not brought out for connection to an external circuit. However, one or two points of the winding which are intended to form the same corner point of the delta may be brought out for grounding, or grounded internally to the tank. For a three-phase transformer, if other points of the winding are brought out, the winding should be regarded as a normal winding as otherwise defined. (PE/TR) C57.12.80-1978r

stable (1) (excitation systems) Possessing stability, where, for a feedback control system or element, stability is the property such that its output is asymptotic, that is, will ultimately attain a steady-state, within the linear range and without continuing external stimuli. For certain nonlinear systems or elements, the property such that the output remains bounded, that is, in a limit cycle of continued oscillation, when the input is bounded. *See also:* conditional stability; asymptomatic stability; bounded-input-bounded-output stability. (PE/EDPG) 421A-1978s

(2) Pertaining to a state of a circuit in which the circuit will remain until an input signal causes a change to another state. *Contrast:* unstable. *See also:* monostable; bistable. (C) 610.10-1994w

stable circuit A circuit that alternates between its two unstable states. (C) 610.10-1994w

stable element (navigation aids) (navigation) An instrument or device which maintains a desired orientation independently of the motion of the vehicle. (AES/GCS) 172-1983w

stable limit cycle One that is approached asymptotically by a state trajectory for all initial states sufficiently close. (CS/PE/EDPG) [3]

stable oscillation A response that does not increase indefinitely with increasing time: an unstable oscillation is the converse. *Note:* The response must be specified or understood: a steady current in a pure resistance network would be stable, although the total charge passing any cross section of a network conductor would be increasing continuously. (Std100) 270-1966w

stable platform (navigation aids) A gimbal-mounted platform, usually containing gyros and accelerometers, whose purpose is to maintain a desired orientation in inertial space, independent of the motion of the vehicle. (AES/GCS) 172-1983w

stack (1) (A) (data management) A list in which items are appended to and retrieved from the same end of the list, known as the top. That is, the next item to be retrieved is the item that has been in the list for the shortest time. *Synonyms:* storage stack; push-down stack. *Contrast:* pushdown list; queue; pushdown storage. **(B) (data management)** A line formed by items waiting for service in a system in which the next item to exit the line is the item that has been in the line for the shortest time. **(C) (data management)** To arrange in, or to form a stack as in definition (A). (C) 610.5-1990

(2) (software) A list that is accessed in a last-in, first-out manner. *See also:* list; queue. (C/SE) 729-1983s

(3) A last-in, first-out (LIFO) data structure. This document sometimes uses the phrase "the stack" to mean "the Forth data stack". (C/BA) 1275-1994

(4) An area in memory for the temporary storage of data. *Note:* Can be implemented using either last-in-first-out or first-in-first out. *Synonym:* pushdown storage. *See also:* evaluation stack; pushup storage. (C) 610.10-1994w

stack architecture A computer architecture whose design relies on a push-down stack to store data and process operands. *Contrast:* register architecture. (C) 610.10-1994w

stack diagram A notational convention used to show the effect of a Forth word on the data stack and, where applicable, the input buffer and return stack. See ANSI X3.215-1994 for syntactic details. (C/BA) 1275-1994

stacked-beam radar A radar that forms two or more simultaneous receive beams at the same azimuth but at different elevation angles. *Note:* The beams are usually contiguous or partly overlapping. Each stacked beam feeds an independent receiver channel. (AES) 686-1997

stacker *See:* card stacker.

stack height A dimension used to define the spacing between adjacent surfaces of mezzanine cards or a mezzanine card and its host. (C/BA) 1301.4-1996

stack indicator *See:* stack pointer.

stack, insulator *See:* insulator stack.

stack pointer A data item that specifies the address of the data item most recently stored in a stack. *Synonym:* stack indicator. (C) 610.5-1990w

stack storage *See:* pushdown storage.

staff-hour An hour of effort expended by a member of the staff. (C/SE) 1045-1992

stage (1) (communication practice) One step, especially if part of a multistep process, or the apparatus employed in such a step. The term is usually applied to an amplifier. *See also:* amplifier. (EEC/PE) [119]

(2) (thermoelectric device) One thermoelectric couple or two or more similar thermoelectric couples arranged thermally in parallel and electrically connected. *See also:* thermoelectric device. (ED) [46], 221-1962w

stage efficiency The ratio of useful power delivered to the load (alternating current) and the plate power input (direct current). *See also:* network analysis. (AP/ANT) 145-1983s

stagger (facsimile) Periodic error in the position of the recorded spot along the recorded line. *See also:* recording. (COM) 168-1956w

staggered-repetition-interval moving-target indicator A moving-target indicator with multiple interpulse intervals. The interval may vary either from pulse to pulse or from scan to scan. (AES/RS) 686-1990

staggered-repetition-interval waveform A waveform in which the pulse repetition interval (PRI) changes from pulse to pulse, to fill blind speeds or to distinguish echoes having ambiguous range or Doppler shifts. *Note:* Not to be confused with changing PRI from scan to scan or from one group of pulses to another, which can be described as multiple-PRI or PRI-diversity waveforms. (AES) 686-1997

staggering The offsetting of two channels of different carrier systems from exact sideband frequency coincidence in order to avoid mutual interference. (EEC/PE) [119]

staggering advantage The effective reduction, in decibels, of interference between carrier channels, due to staggering. (EEC/PE) [119]

stagger time, relay *See:* relay stagger time.

stagger-tuned amplifier An amplifier consisting of two or more single-tuned stages that are tuned to different frequencies. *See also:* amplifier. (EEC/PE) [119]

stain spots (electroplating) Spots produced by exudation, from pores in the metal, of compounds absorbed from cleaning, pickling plating solutions. The appearance of stain spots is called spotting out. (PE/EEC) [119]

staircase (pulse terminology) Unless otherwise specified, a periodic and finite sequence of steps of equal magnitude and of the same polarity. (IM/WM&A) 194-1977w

staircase signal (television) A waveform consisting of a series of discrete steps resembling a staircase. (BT/AV) 201-1979w

STAIRS A nonprocedural computer language used in manipulation of textual data. (C) 610.13-1993w

stairstepping The jagged effect that results from representing a diagonal line or curve by pixels arranged in horizontal or vertical rows and columns on a display surface. *Synonym:* jaggies. *See also:* aliasing. (C) 610.6-1991w

stairstep sweep (oscilloscopes) An incremental sweep in which each step is equal. The electric deflection waveform producing a stairstep sweep is usually called a staircase or stairstep waveform. *See also:* incremental sweep; oscillograph. (IM/HFIM) [40], 311-1970w

stalled tension control A control function that maintains tension in the material at zero speed. *See also:* electric drive. (IA/ICTL/IAC) [60]

stalled torque control A control function that provides for the control of the drive torque at zero speed. *See also:* electric drive. (IA/ICTL/IAC) [60]

stalo (STALO) Acronym for stable local oscillator. A highly stable radio-frequency local oscillator used for heterodyning signals to produce an intermediate frequency (IF). (AES) 686-1997

stamper (electroacoustics) A negative (generally made of metal by electroforming) from which finished pressings are molded. *See also:* phonograph pickup. (SP) [32]

stand-alone (1) Pertaining to hardware or software that is capable of performing its function without being connected to other components; for example, a stand-alone word processing system. (C) 610.12-1990
(2) Pertaining to a system that is self-contained and not connected to other systems or system components. (C) 610.10-1994w

stand-alone data dictionary *See:* passive data dictionary.

stand-alone testing A test of a component performed before it is assembled onto a board or other substrate; for example, using ATE. (TT/C) 1149.1-1990

stand-alone word processing Word processing performed on a system that does not depend on the resources of other equipment to perform word processing activities. *Contrast:* dedicated word processing; shared-resource word processing; shared-logic word processing. (C) 610.2-1987

standard (1) (A) (radiation protection) (instrument or source) (national standard) An instrument, source, or other system or device maintained and promulgated by the U.S. National Bureau of Standards as such. **(B) (radiation protection) (instrument or source)** (derived or secondary standard) A calibrated instrument, source, or other system or device directly relatable (that is, with no intervening steps) to one or more U.S. National Standards. **(C) (radiation protection) (instrument or source)** (laboratory standard) A calibrated instrument, source, or other system or device without direct one-step relatability to the U.S. National Bureau of Standards, maintained and used primarily for calibration and standardization. (NI) N323-1978
(2) (transmission lines and waveguides) A device having stable, precisely defined characteristics that may be used as a reference. (IM/HFIM) [40]
(3) (test, measurement, and diagnostic equipment) A laboratory type device which is used to maintain continuity of value in the units of measurement by periodic comparison with higher echelon or national standards. They may be used to calibrate a standard of lesser accuracy or to calibrate test and measurement equipment directly. (MIL) [2]
(4) (A) (instrument or source) (National Standard) An instrument, source, or other system or device maintained and promulgated by the U.S. National Institute of Standards and Technology (NIST). **(B) (instrument or source)** (Transfer Standard) A physical measurement standard that has been compared directly or indirectly with the national standard. This standard is typically a measurement instrument or a radiation source used as a laboratory standard. **(C) (instrument or source)** (Laboratory Standard) An instrument, source, or other system or device calibrated by comparisons with a standard other than a U.S. National Standard. (NI) N42.17B-1989
(5) A document, established by consensus and approved by an accredited standards development organization, that provides, for common and repeated use, rules, guidelines, or characteristics for activities or their results, aimed at the achievement of the optimum degree of order and consistency in a given context. (C/PA) 14252-1996
(6) (A) A set of detailed technical guidelines, used as a means of establishing uniformity in an area of computing development. **(B)** Pertaining to the set of guidelines, as in (A). For example, a standard interface or a standard definition. **(C)** An agreement among any number of organizations that defines certain characteristics, specifications, or parameters related to a particular aspect of computer technology. For example, ANSI, ISO, and IEEE are standards-making bodies. *Note:* Such organizations may include industrial, academic, or governmental entities. **(D)** In software engineering, mandatory requirements employed and enforced to prescribe a disciplined uniform approach to software development, that is, mandatory conventions and practices are in fact standards. *See also:* standard language; convention; de facto standard; language standard. (C) 610.7-1995, 610.10-1994
(7) An agreement among any number of organizations that defines certain characteristics, specification, or parameters related to a particular aspect of computer technology. For example, ANSI, ISO, and IEEE are standards-making bodies. *Note:* Such organizations may include industrial, academic, or governmental entities. (C) 610.10-1994w

standard adapter A two-port device having standard connectors for joining together two waveguides or transmission lines with nonmating standard connectors. (IM/HFIM) 474-1973w

standard antenna (amplitude-modulation broadcast receivers) An open single-wire antenna (including the lead-in wire) having an effective height of 4 m. (CE) 186-1948w

standard antenna calibration site A flat, open area site, devoid of nearby scatterers such as trees, power lines, and fences, that has a large metallic groundplane. (EMC) C63.5-1988, C63.4-1988s

standard antenna input voltages (amplitude-modulation broadcast receivers) Four standard antenna input voltages are specified for the purpose of certain tests, as follows:

1) A "distant signal voltage" is taken as 86 dB below 1 V, or 50 μV.
2) A "mean-signal voltage" is taken as 46 dB below 1 V, or 5 000 μV.
3) A "local signal voltage" is taken as 20 dB below 1 V, or 100 000 μV.
4) A "strong-signal voltage" is taken as 1 V.

(CE) 186-1948w

standard atmosphere for refraction An atmosphere for which the refractivity is determined by the equation:

$$N(h) = 315 \exp(-0.136h)$$

where h is the altitude in kilometers above mean sea level. *Note:* The standard atmosphere for refraction is almost identical to the standard radio atmosphere up to a height of one kilometer. *Synonym:* reference atmosphere for refraction. *See also:* refractive index gradient. (AP/PROP) 211-1997

standard binary *See:* binary.

standard bus An abbreviated notation used throughout this document, rather than the more exact "physical bus standard that claims conformance to this specification." (C/MM) 1212-1991s

standard cable The standard cable formerly used for specifying transmission losses had, in American practice, a linear series resistance and linear shunt capacitance of 88 ohms and 0.054 microfarad, respectively, per loop mile, with no inductance or shunt conductance. (EEC/PE) [119]

standard cell A cell that serves as a standard of electromotive force. *See also:* unsaturated standard cell; electrochemistry; auxiliary device to an instrument; Weston normal cell. (EEC/PE) [119]

standard chopped lightning impulse A standard lightning impulse chopped by an external gap after 2–5 μs. (PE/PSIM) 4-1995

standard chopped wave impulse voltage shape A standard lightning impulse that is intentionally interrupted on the tail by spark over of a gap or other equivalent chopping device. Usually the time to chop is 2–3 μs. (PE/C) 1313.1-1996

standard code The operating, block signal, and interlocking rules of the Association of American Railroads. (EEC/PE) [119]

standard compass A magnetic compass so located that the effect of the magnetic mass of the vessel and other factors that may influence compass indication is the least practicable. (EEC/PE) [119]

standard connector (fixed and variable attenuators) A connector, the critical mating dimensions of which have been standardized to assure nondestructive mating. *Notes:* 1. It butts against its mating standard connector only in the mechanical reference plane. 2. It joins to its waveguide or transmission line with a minimum discontinuity. 3. All its discontinuities are to the maximum extent possible, self-compensated, not within the mating connector. (IM/HFIM) 474-1973w

standard connector pair (fixed and variable attenuators) Two standard connectors designed to mate with each other. (IM/HFIM) 474-1973w

standard de-emphasis characteristic (frequency modulation) A falling response with modulation frequency, complementary to the standard pre-emphasis characteristic and equivalent to an RC circuit with a time constant of 75 μs. *Note:* The de-emphasis characteristic is usually incorporated in the audio circuits of the receiver. (BT) 185-1975w

standard development organization An accredited organization that formally develops and coordinates standards for use by a community. (C/PA) 14252-1996

standard deviation The square root of the variance of a random variable. For this application, the variance is a measure of the variation of the observations within a measurement set. The standard deviation is often estimated using a set of measurements of the random variable. The standard deviation has the same units as the measured quantity, and therefore is particularly convenient when describing the variability of the measured quantity. This parameter may also be expressed as a relative standard deviation (i.e., as a percentage of the measured quantity). (NI) N42.23-1995

standard directivity The maximum directivity from a planar aperture of area A, or from a line source of length L, when excited with a uniform amplitude, equiphase distribution. *Notes:* 1. For planar apertures in which $A >> \lambda^2$. The value of the standard directivity is $4\pi A/\lambda^2$, with λ the wavelength and with radiation confined to a half space. 2. For line sources with $L >> \lambda$, the value of the standard directivity is $2L/\lambda$. *Synonym:* reference directivity. (AP/ANT) 145-1993

standard electrode potential An equilibrium potential for an electrode in contact with an electrolyte, in which all of the components of a specified electrochemical reaction are in their standard states. The standard state for a gas is the pressure of one atmosphere, for an ionic constituent it is unit ion activity, and it is a constant for a solid. *See also:* electrochemistry. (EEC/PE) [119]

standard error An output stream usually intended to be used for diagnostic messages. (C/PA) 9945-2-1993

standard form *See:* normalized form.

standard frequency (electric power system) A precise frequency intended to be used for a frequency reference. *Note:* In the U.S. a frequency of 50 Hz is recognized as a standard for all ac lighting and power systems. (IA/PE/MT/PSE) 45-1983s, 94-1991w

standard full impulse voltage wave (1) (insulation strength) An impulse that rises to crest value of volage in 1.2 μs (virtual time) and drops to 0.5 crest value of voltage in 50 μs (virtual time), both time being measured from the same origin and in accordance with established standards of impulse testing techniques. *Note:* The virtual value for the duration of the wavefront is 1.67 times the time taken by the voltage to increase from 30% to 90% of its crest value. The origin from which time is measured is the intersection with the zero axis of a straight line drawn through points on the front of the voltage wave at 30%and 90% crest value. (EEC/LB) [100] **(2) (mercury lamp transformers)** An impulse that rises to crest value of voltage in 1.5 μs (nominal time) and drops to 0.5 crest value of voltage in 40 μs (nominal time), both times being measured from the same time origin and in accordance with established standards of impulse testing techniques. *See also:* basic impulse insulation level. (EEC/LB) [98]

Standard Generalized Markup Language (SGML) A text-formatting language. (C) 610.13-1993w

standard gravity By international agreement, the standard value of gravity acceleration magnitude, or $g_o = 9.80665$ m/s^2, which corresponds closely to the plumb-bob-gravity-acceleration magnitude measured at 45° latitude and sea level. In the calibration of accelerometer scale factor, the local value g of plumb-bob-gravity-acceleration magnitude is used as a reference. If an absolute measurement of scale factor is required, a conversion to the standard gravity unit g_o or to m/s2 has to be made. (AES/GYAC) 1293-1998

standard illuminant (illuminating engineering) A hypothetical light source of specified relative spectral power distribution. *Note:* The International Commission on Illumination has specified spectral power distributions for standard illuminants A, B, and C, and several D-illuminants. (EEC/IE) [126]

standard illuminant A (illuminating engineering) A blackbody at a temperature of 2856 K. It is defined by its relative spectral power distribution over the range from 300 nm to 830 nm. (EEC/IE) [126]

standard illuminant B (illuminating engineering) A representation of noon sunlight with a correlated color temperature of approximately 4900 K. It is defined by its relative spectral power distribution over the range from 320 nm to 770 nm. *Note:* It is anticipated that at some future date, that is yet to be decided, illuminant B will be dropped from the list of recommended standard illuminants. (EEC/IE) [126]

standard illuminant C (illuminating engineering) A representation of daylight having a correlated color temperature of approximately 6800 K. It is defined by its relative spectral power distribution over the range from 320 nm to 770 nm. *Note:* It is anticipated that at some future date, that is yet to be decided, illuminant C will be dropped from the list of recommended standard illuminants. (EEC/IE) [126]

standard illuminant D65 (illuminating engineering) A representation of daylight at a correlated color temperature of approximately 6500 K. It is defined by its relative spectral power distribution over the range from 300 nm to 830 nm. *Note:* At present, no artificial source for matching this illuminant has been recommended. (EEC/IE) [126]

standard input An input stream usually intended to be used for primary data input. (C/PA) 9945-2-1993

standard insulation class (instrument transformers) Denotes the maximum voltage in kilovolts that the insulation of the primary winding is designed to withstand continuously. *See also:* instrument transformer. (ELM) C12.1-1982s

standardization *See:* laboratory reference standards; echelon; laboratory working standards.

standardization coefficient A factor used for the direct conversion of a net area counting rate of a gamma-ray peak of a given energy, *E*, and from a specific radionuclide, *i*, to the activity of that radionuclide. (NI) N42.14-1991

standardize *See:* check; normalize.

standardized profile A balloted, formal, harmonized document that specifies a profile. (C/PA) 14252-1996

standard language Any language that conforms to an existing language standard. For example, ALGOL 60 and ALGOL 68 are considered standard languages.
(C) 610.13-1993w, 610.10-1994w

standard lightning impulse (1) (power and distribution transformers) An impulse that rises to crest value of voltage in 1.2 μs (virtual time) and drops to 0.5 crest value of voltage in 50 μs (virtual time), both times being measured from the same origin and in accordance with established standards of impulse testing techniques. It is described as a 1.2/50 μs impulse. *Note:* The virtual value for the duration of the wavefront is 1.67 times the time taken by the voltage to increase from 30% to 90% of its crest value. The origin from which time is measured is the intersection with the zero axis of a straight line drawn through points on the front of the voltage wave at 30% and 90% crest value.
(PE/TR) C57.12.80-1978r

(2) A full lightning impulse having a virtual front time of 1.2 μs and a virtual time to half-value of 50 μs.
(PE/PSIM) 4-1995

(3) The wave shape of the standard impulse used is 1.2/50 μs (when not in conflict with products standards).
(SPD/PE) C62.22-1997

(4) A unidirectional surge having a 30–90% equivalent rise time of 1.2 μs and a time to half value of 50 μs.
(PE/T&D) 1243-1997

standard lightning impulse voltage shape An impulse that rises to crest value of voltage in 1.2 μs (virtual time) and drops to 0.5 crest value of voltage in 50 μs (virtual time), both times being measured from the same origin and in accordance with established standards of impulse testing techniques. It is described as a 1.2/50 impulse.
(PE/C) 1313.1-1996

standard logic type The type STD_ULOGIC defined by IEEE Std 1164-1993, or any type derived from it, including, in particular, one-dimensional arrays of STD_ULOGIC or of one of its subtypes. (C/DA) 1076.3-1997

standard loop input signals (A) (amplitude-modulation broadcast receivers) A "distant-signal" loop input is taken as 86 dB below 1 V/m, or 5 000 μV/m. **(B) (amplitude-modulation broadcast receivers)** A "mean-signal" loop input is taken as 46 dB below 1 V/m, or 5 000 μV/m. **(C) (amplitude-modulation broadcast receivers)** A "local-signal" loop input is taken a 26 dB below 1 V/m, or 50 000 μV/m. **(D) (amplitude-modulation broadcast receivers)** A "strong-signal" loop input is taken as 14 dB below 1 V/m, or 200 000 μV/m. *Note:* The above loop field intensities are not equivalent to the standard antenna input voltages for the corresponding class of service. For example, the "mean-signal" voltage for antenna operation is 5 000 μV. This corresponds to a field intensity if 1 250 μV/m assuming a standard 4-meter antenna, whereas the mean-signal voltage for loop receivers is arbitrarily taken as 5 000 μV/m. (CE) 186-1948

standard maximum usable frequency *See:* maximum usable frequency.

standard M gradient *See:* refractive index gradient.

standard microphone A microphone the response of which is accurately known for the condition under which it is to be used. *See also:* instrument. (EEC/PE) [119]

standard N gradient *See:* standard refractive index gradient.

standard noise temperature (interference terminology) The temperature used in evaluating signal transmission systems for noise factor 290 K (27°C). *See also:* interference.
(IE) [43]

standard observer (television) (color) (CIE 1931) Receptor of radiation whose colorimetric characteristics correspond to the distribution coefficients $\bar{x}_\lambda$, $\bar{y}_\lambda$, $\bar{z}_\lambda$ adopted by the International Commission on Illumination (CIE) in 1931.
(BT/AV) 201-1979w

standard operating duty *See:* operating duty.

standard output An output stream usually intended to be used for primary data output. (C/PA) 9945-2-1993

standard pitch *See:* standard tuning frequency.

standard potential (standard electrode potential) The reversible potential for an electrode process when all products and reactants are at unit activity on a scale in which the potential for the standard hydrogen half-cell is zero.
(GSD) [71], 315-1975r

standard power-frequency short-duration voltage shape A sinusoidal voltage with frequency between 48 Hz and 62 Hz, and duration of 60 s. *Note:* Some apparatus standards (e.g., transformers) use a modified wave shape when practical test considerations or particular dielectric strength characteristics make such modification necessary. (PE/C) 1313.1-1996

standard propagation The propagation of radio waves over a smooth spherical Earth of uniform dielectric constant and conductivity, under conditions of standard refraction in the atmosphere. *See also:* refractive index gradient.
(AP/PROP) 211-1997

standard radio atmosphere An atmosphere whose vertical refractivity gradient is equal to the standard refractive index gradient. *See also:* refractive index gradient.
(AP/PROP) 211-1997

standard radio horizon The radio horizon corresponding to propagation through the standard radio atmosphere. *See also:* refractive index gradient. (AP/PROP) 211-1997

standard reference material Material characterized by the U.S. National Institute of Standards and Technology (NIST) for the activity of radionuclides and issued with a certificate that gives the results of the characterization. (NI) N42.23-1995

standard reference position (of a contact) The nonoperated or de-energized position of the associated main device to which the contact position is referred. *Note:* Standard reference positions of typical devices are shown in the following table:

Device	Standard Reference Position
Circuit breaker	Main contacts open
Disconnecting switch	Main contacts open
Relay	De-energized position
Contactor	De-energized position
Valve	Closed position

(SWG/PE) C37.100-1992

standard refraction *See:* refractive index gradient.

standard refractive index gradient A standard value of vertical gradient of refractivity, namely 39.2?N-Units/km, used in studies of the refraction of radio waves in the troposphere. *Note:* This value corresponds, approximately, to the median value of the gradient in the first kilometer of altitude in temperate regions. *Synonym:* standard N gradient. *See also:* refractive index gradient. (AP/PROP) 211-1997

standard refractive index modulus gradient *See:* refractive index gradient.

standard register (motor meter) (dial register) A four- or five-dial register, each dial of which is divided into ten equal parts, the division marks being numbered from zero to nine, and the gearing between the dial pointers being such that the reltive movements of the adjacent dial pointers are in opposite directions and in a 10-to-1 ratio. *See also:* watthour meter. (PE/EEC) [119]

standard resistor (resistance standard) A resistor that is adjusted with high accuracy to a specified value, is but slightly affected by variations in temperature, and is substantially constant over long periods of time. *See also:* auxiliary device to an instrument. (PE/EEC) [119]

standard response spectrum (SRS) A required response system (RRS) that is artificially created to cover the standard testing of relays and whose shape is defined. The SRS may be terminated at any convenient frequency above 35 Hz. (SWG/PE) C37.100-1992

standard rod gap A gap between the ends of the two one-half-inch square rods cut off squarely and mounted on suppots so that a length of rod equal to or greater than one-half the gap spacing overhangs the inner edge of each support. It is intended to be used for the approximate measurement of crest voltages. *See also:* instrument. (EEC/PE) [119]

standards Mandatory requirements employed to prescribe a disciplined, uniform approach to software development, maintenance, and operation. (C/SE) 730.1-1995

standards—basic reference Those standards with which the values of the electrical units are maintained in the laboratory, and that serve as the starting point of the chain of sequential measurements carried out in the laboratory. (ELM) C12.1-1988

standards—dc-ac transfer Instruments used to establish the equality of an rms current or voltage (or the average values of alternating power) with the corresponding steady-state dc quantity. (ELM) C12.1-1988

standards—laboratory reference (metering) Standards that are used to assign and check the values of laboratory secondary standards. (ELM) C12.1-1988, C12.1-1982s

standards—laboratory secondary (metering) Standards that are used in the routine calibration tasks of the laboratory. (ELM) C12.1-1988, C12.1-1982s

standards—national Those standards of electrical measurements that are maintained by the National Institute of Standards and Technology. (ELM) C12.1-1988

standards—transport Standards of the same nominal value as the basic reference standards of a laboratory (and preferably of equal quality) that are regularly intercompared with the basic group but are reserved for periodic interlaboratory comparison tests to check the stability of the basic reference group. (ELM) C12.1-1988

standard source (illuminating engineering) An artificial source having the same spectral distribution as a specified standard illuminant. (EEC/IE) [126]

standard source A (illuminating engineering) A tungsten filament lamp operated at a color temperature of 2856 K (International Practical Temperature Scale, 1968) and approximating the relative spectral power distribution of standard illuminant A. (EEC/IE) [126]

standard source B (illuminating engineering) An approximation of standard illuminant B obtained by a combination of Source A and a special filter. (EEC/IE) [126]

standard source C (illuminating engineering) An approximation of standard illuminant C obtained by a combination of Source A and a special filter. (EEC/IE) [126]

standard source diameter (x-ray energy spectrometers) The diameter of the x-ray emission source which is used to measure the response characteristics of the spectrometer. Unless otherwise specified, this is assumed to be a point source. (NPS/NID) 759-1984r

standard sources ("dose calibrator" ionization chambers) A general term used to refer to the standard sources listed below:

- National radioactivity standard source. A calibrated radioactive source prepared and distributed as a standard reference material by the US Bureau of Standards.
- Certified radioactivity standard source. A calibrated radioactive source, with stated accuracy, whose calibration is certified by the source supplier as traceable to the National Radioactive Measurements System.

(NI) N42.13-1986

standard sphere gap (high voltage testing) A peak-voltage device constructed and arranged in accordance with this document. It consists of two metal spheres of the same diameter, D, with their shanks, operating gear, insulating supports, supporting frame, and leads for connections to the point at which the voltage is to be measured. Standard values of D are 625 mm, 125 mm, 250 mm, 500 mm, 750 mm, 1000 mm, 1500 mm, and 2000 mm. The spacing between the spheres is designated as S. The points on the two spheres that are closest to each other are called the sparking points. In practice, the disruptive discharge may occur between other neighboring points. The corresponding figures A and B show two arrangements; one of which is typical of sphere gaps with a vertical axis, and the other, of sphere gaps with a horizontal axis.

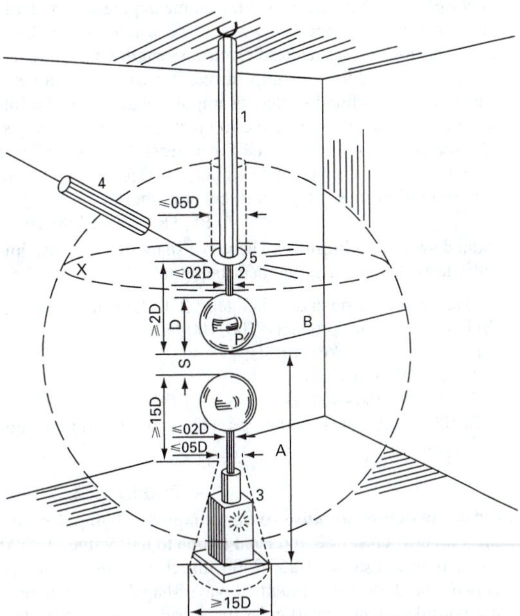

Key: 1. Insulating support; 2. Sphere shank; 3. Operating gear, showing maximum dimensions; 4. High-voltage connection with series resistor; 5. Stress distributor, showing maximum dimensions; P Sparking point of high-voltage sphere; A Height of P above ground plane; B Radius of space free from external structures; X item 4 not to pass through this plane within a distance from B from P. Note: The figure is drawn to scale for a 100 cm sphere gap at radius spacing.

standard sphere gap (figure A)

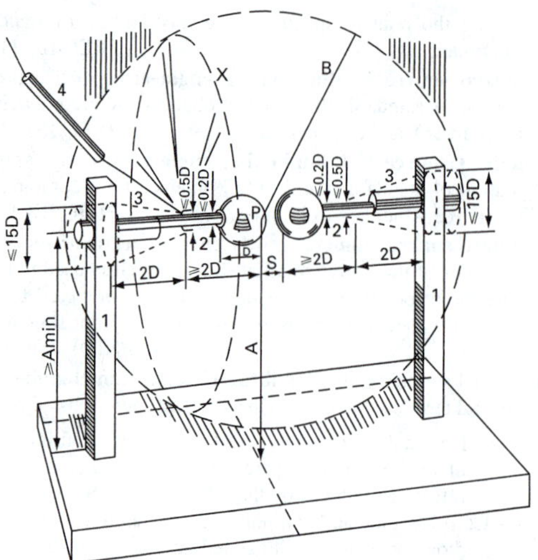

Key: 1. Insulating support; 2. Sphere shank; 3. Operating gear, showing maximum dimensions; 4. High-voltage connection with series resistor; P Sparking point of high-voltage sphere; A Height of P above ground plane; B Radius of space free from external structures; X item 4 not to pass through this plane within a distance from B from P. Note: This figure is drawn to scale for a 25 cm sphere gap at radius spacing.

standard sphere gap (figure B)

(PE/PSIM) 4-1978s

standard structure A particular C structure, defined in `std_stru.h`, that contains fields used to pass data over the procedural interface (PI) (thus avoiding large numbers of arguments). Most functions of the PI have a pointer to a *standard structure* as their first argument. (C/DA) 1481-1999

standard switching impulse (power and distribution transformers) A full impulse having a front time of 250 μs and a time to half value of 2500 μs. It is described as a 250/2500 impulse. *Note:* It is recognized that some apparatus standards may have to use a modified wave shape where practical test considerations or particular dielectric strength characteristics make some modification imperative. Transformers, for example, use a modified switching impulse wave with the following characteristics: 1) Time to crest greater than 100 μs; 2) Exceeds 90% of crest value for at least 200 μs; 3) Time to first voltage zero on tail not less than 1000 μs, except where core saturation causes the tail to become shorter.

(PE/TR) C57.12.80-1978r

standard switching impulses The wave shapes of standard impulse tests depend on equipment being tested:

a) For air insulation and switchgear: 250/2500 μs
b) For transformer products: 100/1000 μs
c) For arrester sparkover tests:
  1) 30–60/90–180 μs
  2) 50–300/400–900 μs
  3) 1000–2000/3000–6000 μs (The tail duration is not critical)

(SPD/PE) C62.22-1997

standard switching impulse voltage shape A full impulse having a time-to-crest of 250 μs and a time to half value of 2500 μs. It is described as a 250/2500 impulse. *Note:* Some apparatus standards use a modified wave shape where practical test considerations or particular dielectric strength characteristics make some modification imperative.

(PE/C) 1313.1-1996

standard systems (electric installations on shipboard) The following systems of distribution are recognized as standard:

a) Two-wire with single-phase alternating current, or direct current.
b) Three-wire with single-phase alternating current, or direct current.
c) Three-phase three-wire, alternating current.
d) Three-phase, four-wire, alternating current.

(IA/MT) 45-1983s

standard television signal A signal that conforms to certain accepted specifications. *See also:* television.

(EEC/PE) [119]

standard test fiber A silica graded index multi-mode optical fiber with a core diameter of 100 μm, an outside diameter of 140 μm, a numerical aperture of 0.29, a bandwidth of 400 MHz-km at 1300 nm and terminated in a standard subminiature assembly (SMA) connector. (C/BA) 1393-1999

standard test frequencies in the broadcast band (amplitude-modulation broadcast receivers) The standard group of seven carrier frequencies for testing is 540, 600, 800, 1 000, 1 200, 1 400, and 1 600 kilocycles. The standard group of three carrier frequencies for testing is 600, 1 000, and 1 400 kilocycles. (CE) 186-1948w

standard test interface language (STIL) A syntax for the description of device stimulus and expected response used for stimulus development, as well as input to automated test equipment (ATE). (C/TT) 1450-1999

standard test modulation (frequency-modulated mobile communications receivers) Sixty percent of the rated system deviation at a frequency of 1 kilohertz. (VT) 184-1969w

standard test position (STP) The default test position for a headset that does not have a fixed spatial relationship between the location of their transmitter and receiver sound ports. See the figure below.

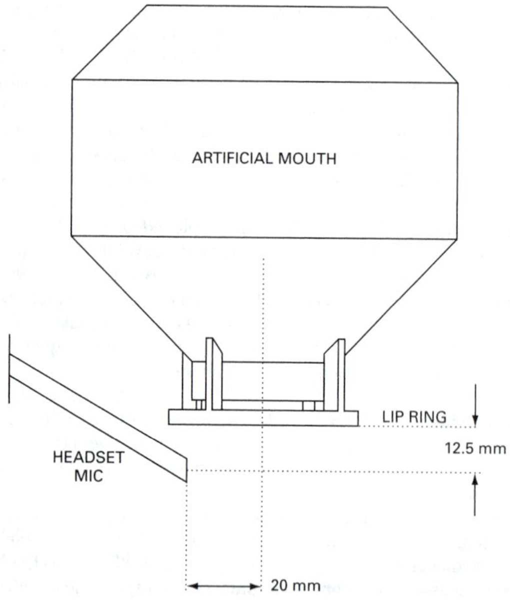

headset transmitter port in standard test position (STP)

(COM/TA) 1206-1994

standard test problem (test, measurement, and diagnostic equipment) An evaluation of the performance of a system, or any part of it, conducted by setting parameters into the system; the parameters are operated on and the results obtained from system read outs. (MIL) [2]

standard test tone (data transmission) A 1mW (0 dBm) 1000 Hz signal applied to the 600 Ω audio portion of a circuit at a zero transmission level reference point. If referred to a point with a relative level other than 0, the absolute power of the tone shall be adjusted to suit the relative level at the point of application. (PE) 599-1985w

standard track *See:* single-track.

standard transmitter test modulation (land-mobile communications transmitters) The standard test modulation shall be 60% of the maximum rated deviation at 1 kKz. (EMC) 377-1980r

standard tuning frequency (standard musical pitch) The frequency for the note A4, namely, 440 Hz. (SP) [32]

standard voltages (electric installations on shipboard) The following voltages are recognized as standard:

	Alternating Current (volts)	Direct Current (volts)
Lighting	115	115
Power	115–200–220–440	115 and 230
Generators	120–208–230–450	120 and 240

Note: Satisfactory to use 120 V lamps

standard volume indicator (volume measurements of electrical speech and program waves) A device for the indication of volume, and having the characteristic described in IEEE Std 152-1953w. *Note:* A standard volume indicator consists of at least two parts: 1) A meter; and 2) an attenuator (adjustable loss) or pad (fixed loss).

standard watthour meter *See:* portable standard watthour meter; reference standard watthour meter.

standard-wave error (navigation aids) (direction finder measurements) The bearing error produced by a wave whose vertically and horizontally polarized electric fields are equal and phased so as to give maximum error in the DF, and whose incidence direction is arranged to be 45°. (AES/GCS) 172-1983w

standard working axis (of a semiconductor x-ray energy spectrometer) A straight line drawn between the center of the entrance window on the detector and the specified location of the source of x rays. (NPS/NID) 759-1984r

standBy A lower-power operating mode of RamLink slaves, in which a change in the flag line is sufficient to quickly reactivate attached chips. (C/MM) 1596.4-1996

standby *See:* reserve; alternative.

standby current The current flowing in any specific conductor (including a conductive case) when the device is connected as intended to the energized power system at rated voltage with no connected load. (SPD/PE) C62.62-2000

standby current, dc *See:* direct-current standby current.

standby equipment Equipment not normally in operation that is available on demand to perform a specific function. (PE/NP) 933-1999

standby failure rate (reliability data for pumps and drivers, valve actuators, and valves) The probability (per hour) of failure for those components which are normally dormant or in a standby state until tested or required to operate to perform their function. (PE/NP) 500-1984w

standby losses The losses produced with the HVDC converter station energized, but with the valves blocked. (SUB/PE) 1158-1991r

standby monitor A station on the ring that is not in active monitor mode. The function of the standby monitor in normal ring operation is to assure that an active monitor is operating. (C/LM) 8802-5-1998

standBy packet An event packet that initiates the transition into the standBy state. (C/MM) 1596.4-1996

standby power The power consumption while the chip is not performing any read or write operation. (ED) 641-1987w

standby power, ac *See:* alternating-current standby power.

standby power supply (nuclear power generating station) (diesel-generator unit) The power supply that is selected to furnish electric energy when the preferred power supply is not available. (PE/NP) 387-1995, 308-1991

standby power system (emergency and standby power) An independent reserve source of electric energy that, upon failure or outage of the normal source, provides electric power of acceptable quality so that the user's facilities may continue in satisfactory operation. (IA/PSE) 446-1995

standby redundancy (1) (software) In fault tolerance, the use of redundant elements that are left inoperative until a failure occurs in a primary element. *Contrast:* active redundancy. (C) 610.12-1990

(2) That redundancy wherein the alternative means for performing a given function are inoperative until needed. (R) [29]

standby service (electric power utilization) Service through a permanent connection not normally used but available in lieu of, or as a supplement to, the usual source of supply. (PE/PSE) 346-1973w, 858-1993w

standby time *See:* idle time.

ST and CBT *See:* sharing transformer and current balancing transformer.

standing-on-nines carry (mathematics of computing) A carry process in which a carry digit transferred to a given digit place is further transferred to the next higher digit place if the current sum in the given digit place is nine. *See also:* carry. (C) [20], 1084-1986w, [85]

standing wave (1) (overhead power-line corona and radio noise) A wave in which, for any component of the field, the ratio of its instantaneous value at one point to that at any other point does not vary with time. *Notes:* 1. A standing wave is most frequently produced by reflection. The sum of the incident and reflected waves, if they are periodic, will produce a standing wave. 2. Commonly, a standing wave is a periodic wave in which the amplitude of the displacement in the medium is a periodic function of the distance in the direction of any line of propagation of the waves. (T&D/PE) 539-1990

(2) A wave formed by the interference of two oppositely traveling plane waves having the same frequency and polarization. (AP/PROP) 211-1997

standing-wave antenna An antenna whose excitation is essentially equiphase, as the result of two feeding waves that traverse its length from opposite directions, their combined effect being that of a standing wave. (AP/ANT) 145-1993

standing-wave detector *See:* standing-wave meter.

standing wave dissipation factor (waveguide) The ratio of the transmission loss in an unmatched waveguide to that in the same waveguide when matched. (MTT) 146-1980w

standing-wave indicator *See:* standing-wave meter.

standing-wave loss factor The ratio of the transmission loss in an unmatched waveguide to that in the same waveguide when matched. *See also:* waveguide. (MTT) 146-1980w

standing-wave machine *See:* standing-wave meter.

standing-wave meter (standing-wave indicator) (standing-wave machine) (standing-wave detector) An instrument for measuring the standing-wave ratio in a transmission line. In addition, a standing-wave meter may include means for finding the location of maximum and minimum amplitudes. See table on previous page. *See also:* instrument. (PE/EEC) [119]

standing wave ratio (1) (data transmission) The ratio of the amplitude of a standing wave at an antinode to the amplitude of a node. *Note:* The standing wave ratio in a uniform transmission line is

$$\frac{1 + p}{1 - p}$$

where $p =$ the reflection coefficient. (T&D/PE) 539-1990, 599-1985w

(2) (waveguide) At a given frequency in a uniform transmission line or waveguide, the ratio of the maximum to the minimum amplitudes of corresponding components of the field (or the voltage or current) along the waveguide in the direction of propagation. *Note:* The standing wave ratio is occasionally expressed as the reciprocal of the ratio defined above. (MTT) 148-1959w, 146-1980w

standing-wave-ratio indicator (standing-wave-ratio meter) A device or part thereof used to indicate the standing-wave ratio. *Note:* In common terminology, it is the combination of amplifier and meter as a supplement to the slotted line or bridge, etc., when performing impedance or reflection measurements. (IM/WM&A) 181-1977w

stand, reel *See:* reel stand.

standstill locking (rotating machinery) The occurrence of zero or unusably small torque in an energized polyphase induction motor, at standstill, for certain rotor positions. *See also:* asynchronous machine. (PE) [9]

Stanford Artificial Intelligence Language A dialect of LISP that was developed at Stanford's Artificial Intelligence Laboratory. (C) 610.13-1993w

STAP *See:* space-time adaptive processing.

star-bus topology A topology where the stations are physically star-wired to a hub but which logically act like a bus. *Note:* This is a common wiring scheme when using traditional point-to-point media such as twisted pair and optical fiber in a bus network. *See also:* bus topology; star-ring topology; star topology; bus-ring topology; loop topology; ring topology; tree topology. (C) 610.7-1995

star chain (navigation aids) A radio navigation transmitting system comprising a master station about which three (or more) slave stations are symmetrically located. (AES/GCS) 172-1983w

star-connected circuit A polyphase circuit in which all the current paths of the circuit extend from a terminal of entry to a common terminal or conductor (which may be the neutral conductor). *Note:* In a three-phase system this is sometimes called a Y (or wye) connection. (IA/PSE) 1100-1999

star connection *See:* Y connection.

star coupler (fiber optics) A passive device in which power from one or several input waveguides is distributed amongst a larger number of output optical waveguides. *See also:* tee coupler; optical combiner. (Std100) 812-1984w

star-delta starter A switch for starting a three-phase motor by connecting its windings first in star and then in delta. *See also:* starter. (IA/ICTL/IAC) [60], [84]

star-delta starting The process of starting a three-phase motor by connecting it to the supply with the primary winding initially connected in star, then reconnected in delta for running operation. (PE) [9]

star ground *See:* radial ground.

star network A set of three or more branches with one-terminal of each connected at a common node. *See also:* network analysis. (Std100) 270-1966w

star quad A cable element that comprises four insulated connectors twisted together. Two diametrically facing conductors form a transmission pair. *Note:* Cables containing star quads can be used interchangeably with cables consisting of pairs, provided the electrical characteristics meet the same specifications. (C/LM) 802.3-1998

star rectifier circuit A circuit that employs six or more rectifying elements with a conducting period of 60 electrical degrees plus the commutating angle. *See also:* rectification. (EEC/PE) [119]

star-ring topology A topology having a logical arrangement of a ring with a physical implementation of a star. This results in a system with relatively short cables, as in a ring network, and allows maintenance to be performed from a single point, as in a star network. *Note:* This is accomplished by connecting each node over a cable to a wiring closet and connecting all cables in a ring topology within the wiring closet. This is the common way in which IEEE 802.5 token rings are built. *See also:* bus-ring topology; star topology; bus topology; star-bus topology; tree topology; loop topology; ring topology. (C) 610.7-1995

start (1) An electric controller for accelerating a motor from rest to normal speed and to stop the motor. (IA/ICTL/MT/PKG) 45-1983s, 333-1980w
(2) (gas tube) A control electrode, the principal function of which is to establish sufficient ionization to reduce the anode breakdown voltage. *Note:* This has sometimes been referred to as a "trigger electrode." (ED) 161-1971w

start and stop characters Distinct bar/space patterns used at the beginning and end of each bar code symbol that provide initial timing references and direction-of-read information to the coding logic. (PE/TR) C57.12.35-1996

start bit (1) In asynchronous transmission, a signal that lasts a single bit time, indicating the beginning of a character. *Contrast:* stop bit. (C) 610.7-1995
(2) For the low-speed version of the Physical layer, a bit that is encoded identically to a logic "0" data bit that is used to delineate the beginning of each individual octet transmission. (EMB/MIB) 1073.4.1-2000

Start Current The current taken from the line when the motor is producing rated start torque at rated voltage and frequency. (PE/NP) 1290-1996

start cycle A cycle that initiates a transaction. The address and transfer type are valid during this cycle. (C/MM) 1196-1987w

start-dialing signal (semiautomatic or automatic working) (telecommunications) A signal transmitted from the incoming end of a circuit, following the receipt of a seizing signal, to indicate that the necessary circuit conditions have been established for receiving the numerical routing information. (COM) [49]

start-diesel signal That input signal to the diesel-generator unit start logic that initiates a diesel-generator unit start sequence. (PE/NP) 387-1995

start element (1) (data transmission) In a character transmitted in a start-stop system, the first element in each character, which serves to prepare the receiving equipment for the reception and registration of the character. The start element is a spacing signal. (PE) 599-1985w
(2) *See also:* start signal. (C) 610.7-1995

starter (1) (illuminating engineering) A device used in conjunction with a ballast for the purpose of starting an electric-discharge lamp. (EEC/IE) [126]
(2) An electric controller that is used to accelerate a motor from rest to normal speed and to stop the motor. (A device designed for starting a motor in either direction of rotation includes the additional function of reversing and should be designated a controller.) (IA/MT) 45-1998

starter gap (gas tube) The conduction path between a starter and the other electrode to which starting voltage is applied. *Note:* Commonly used in the glow-discharge cold-cathode tube. (ED) 161-1971w

starters (fluorescent lamps) Devices that first connect a fluorescent or similar discharge lamp in a circuit to provide for cathode preheating and then open the circuit so that the starting voltage is applied across the lamp to establish an arc. Starters also include a capacitor for the purpose of assisting the starting operation and for the suppression of radio interference during lamp starting and lamp operation. They may also include a circuit-opening device arraned to disconnect the preheat circuit if the lamp fails to light normally. (NPS) 325-1972s

starter voltage drop (glow-discharge cold-cathode tube) The starter-gap voltage drop after conduction is established in the starter gap. (ED) 161-1971w

starting (rotating machinery) The process of bringing a motor up to speed from rest. *Note:* This includes breaking away, accelerating and if necessary, synchronizing with the supply. (PE) [9]

starting address The address of the first instruction of a computer program in main storage. *Note:* This address may or may not be the same as the program's origin, depending upon whether there are data preceding the first instruction. *Contrast:* origin. *See also:* loaded origin; assembled origin. (C) 610.12-1990

starting amortisseur An amortisseur, the primary function of which is the starting of the synchronous machine and its connected load. (EEC/PE) [119]

starting anode An electrode that is used in establishing the initial arc. *See also:* rectification. (EEC/PE) [119]

starting attempt (electric generating unit reliability, availability, and productivity) The action to bring a unit from shutdown to the in-service state. Repeated initiations of the starting sequence without accomplishing corrective repairs are counted as a single attempt. (PE/PSE) 762-1987w

starting capacitance (capacitor motor) The total effective capacitance in series with the auxiliary winding for starting operation. *See also:* asynchronous machine.

starting circuit breaker (power system device function numbers) A device whose principal function is to connect a machine to its source of starting voltage. (SUB/PE) C37.2-1979s

starting current (1) (rotating machinery) The current drawn by the motor during the starting period. (A function of speed or slip). *See also:* asynchronous machine. (PE) [9]
(2) (oscillators) The value of electron-stream current through an oscillator at which selfsustaining oscillations will start under specified conditions of loading. *See also:* magnetron. (ED) 161-1971w

starting failure (electric generating unit reliability, availability, and productivity) The inability to bring a unit from some unavailable state or reserve shutdown state to the in-service state within a specified period. The specified period may be different for individual units. Repeated failures within the specified starting period are counted as a single starting failure. (PE/PSE) 762-1987w

starting motor An auxiliary motor used to facilitate the starting and accelerating of a main machine to which it is mechanically connected. *See also:* asynchronous machine. (PE) [9]

starting open-phase protection The effect of a device operative to prevent connecting the load to the supply unless all conductors of a polyphase system are energized. (IA/IAC) [60]

starting operation (A) (single-phase motor) The range of operation between locked rotor and switching for a motor employing a starting-switch or relay. **(B) (single-phase motor)** The range of operation between locked rotor and a point just below but not including breakdown-torque speed for a motor not employing a starting switch or relay. *See also:* asynchronous machine. (PE) [9]

starting point (A) (for common channel signaling [CCS] incoming trunk). Receipt of the initial address message (IAM). **(B)** For non-centralized automatic message accounting (CAMA) p*er-trunk-signaling incoming trunk)*. End of valid called-number digit reception. **(C)** (for CAMA per-trunk-signaling incoming trunk). End of reception of first automatic number identification (ANI) digit [exclusive of start of dialing or key pulse (KP) signal and information digit] or first operator number identification (ONI) digit. *See also:* ending point; cross-office delay. (COM/TA) 973-1990

starting reactor (power and distribution transformers) A current-limiting reactor for decreasing the starting current of a machine or device. *See also:* reactor. (PE/TR) C57.12.80-1978r, [57]

starting resistor (rotating machinery) A resistor connected in a secondary or field circuit to modify starting performance of an electric machine. *See also:* rotor; stator. (PE) [9]

starting rheostat A rheostat that controls the current taken by a motor during the period of starting and acceleration, but does not control the speed when the motor is running normally. (IA/ICTL/IAC) [60]

starting sheet (electrorefining) A thin sheet of refined metal introduced into an electrolytic cell to serve as a cathode surface for the deposition of the same refined metal. *See also:* electrorefining. (EEC/PE) [119]

starting-sheet blank (electrorefining) A rigid sheet of conducting material designed for introduction into an electrolytic cell as a cathode for the deposition of a thin temporarily adherent deposit to be stripped off as a starting sheet. *See also:* electrorefining. (EEC/PE) [119]

starting success (electric generating unit reliability, availability, and productivity) The occurrence of bringing a unit from some unavailable state or the reserve shutdown state to the in-service state within a specified period. The specified period may be different for individual units. (PE/PSE) 762-1987w

starting-switch assembly The make-and-break contacts, mechanical linkage, and mounting parts necessary for starting or running, or both starting and running, split-phase and capacitor motors. *Note:* The starting-switch assembly may consist of a stationary-contact assembly and a contact that moves with the rotor. (EEC/PE) [119]

starting switch, centrifugal *See:* centrifugal starting switch.

starting switch, relay *See:* relay starting switch.

starting temperature (grounding device) The winding temperature at the start of the flow of thermal current. (PE/SPD) 32-1972r

starting test (rotating machinery) A test taken on a machine while it is accelerating from standstill under specified conditions. *See also:* asynchronous machine. (PE) [9]

starting torque (1) (electric coupling) The minimum torque of an electric coupling developed with the output member stationary and the input member rotating, with excitation applied. *Note:* Starting torque is usually specified with rated speed of rotation and rated excitation applied. (EM/PE) 290-1980w
(2) (synchronous motor) The torque exerted by the motor during the starting period. (A function of speed or slip). (PE) [9]

starting-to-running transition contactor (power system device function numbers) A device that operates to initiate or cause the automatic transfer of a machine from the starting to the running power connection. (PE/SUB) C37.2-1979s

starting voltage (radiation counters) The voltage applied to a Geiger-Mueller tube at which pulses of 1 V amplitude appear across the tube when irradiated. *See also:* anticoincidence. (ED) [45]

starting winding (rotating machinery) A winding, the sole or main purpose of which is to set up or aid in setting up a magnetic field for producing the torque to start and accelerate a rotating electric machine. (PE) [9]

startle shock An electric shock from a steady-state or a discharge current that, if it occurred unexpectedly, would produce an unintentional muscular reflex. (T&D/PE) 539-1990

Start_of_Packet Delimiter (SPD) In 1000BASE-X, a single code-group 8B/10B ordered_set used to delineate the starting boundary of a data transmission sequence for a single packet. (C/LM) 802.3-1998

start of packet byte A single byte (hex A5, decimal 165) that is used by both the printer device and the host to quickly determine whether or not they are synchronized. (C/MM) 1284.1-1997

start of stream delimiter (ssd) (1) A pattern of defined code words used to delineate the boundary of a data transmission sequence on the Physical Layer stream. The SSD is unique in that it may be recognized independent of previously defined code-group boundaries and it defines subsequent code-group boundaries for the stream it delimits. For 100BASE-T4, SSD is a pattern of three predefined sosb code-groups (one per wire pair) indicating the positions of the first data code-group on each wire pair. For 100BASE-X, SSD consists of the code-group sequence /J/K/. For 100BASE-T2, the SSD is indicated by two consecutive pairs of predefined PAM5×5 symbols (±2, ±2) (±2, 0) which are generated using unique SSD/ESD coding rules. (C/LM) 802.3-1998
(2) (local area networks) Reserved code patterns that identify the beginning of the MII channel transmission frame. The ssd indicates the transmission priority of the packet. (C) 8802-12-1998

star topology A topology in which stations are connected to a single central switching facility. *See also:* bus topology; star-ring topology; ring topology; bus-ring topology; star-bus topology; tree topology; loop topology. (C) 610.7-1995

start-pulsing signal (telephone switching systems) A signal transmitted from the receiving end to the sending end of a trunk to indicate that the receiving end is in a condition to receive pulsing. (COM) 312-1977w

star tracker *See:* astrotracker.

start-record signal A signal used for starting the process of converting the electric signal to an image on the record sheet. *See also:* facsimile signal. (COM) 168-1956w

start signal (1) (start-stop system) Signal serving to prepare the receiving mechanism for the reception and registration of a character, or for the control of a function. (COM) [49]
(2) (facsimile) A signal that initiates the transfer of a facsimile equipment condition from standby to active. *See also:* facsimile signal. (COM) 168-1956w
(3) (telephone switching systems) In multifrequency and key pulsing, a signal used to indicate that all digits have been transmitted. (COM) 312-1977w
(4) (data management) A signal at the beginning of a start-stop character that prepares the receiving device for the reception of the code elements. *Note:* A start signal is limited to one signal element generally having the duration of unit interval. (C) 610.5-1990w
(5) In asynchronous transmission, a signal preceding a character that prepares the receiving device for the reception of code elements. *Synonym:* start element. *Contrast:* stop signal. (C) 610.7-1995

start-stop character A character including one start signal at the beginning and one or two stop signals at the end. (C) 610.5-1990w

start-stop printing telegraphy That form of printing telegraphy in which the signal-receiving mechanisms are started in operation at the beginning and stopped at the end of each character transmitted over the channel. *See also:* telegraphy. (EEC/PE) [119]

start-stop signal A signal composed of a sequence or group of signal elements, each group representing a character or block, having a duration equal to the duration of an integral number of unit intervals and which are separated by time intervals for which the duration is not fixed. (C) 610.7-1995

start-stop system (data transmission) A system in which each group of code elements corresponding to a character is preceded by a start element which serves to prepare the receiving equipment for the reception and registration of a character, and is followed by a stop element during which the receiving equipment comes to rest in preparation for the reception of the next character. (PE) 599-1985w

start-stop tape drive A tape drive capable of coming to a complete stop and restarting in the gap between two recorded data blocks. *Contrast:* streaming tape drive. (C) 610.10-1994w

start-stop transmission (1) (data transmission) A synchronous transmission in which a group of code elements corresponding to a character signal is preceded by a start signal which serves to prepare the receiving mechanism for the reception and registration of a character and is followed by a stop signal which serves to bring the receiving mechanism to rest in preparation for the reception of the next character. (PE) 599-1985w
(2) *See also:* asynchronous transmission. (C) 610.7-1995

start time *See:* acceleration time.

start transition (data transmission) In a character transmitted in a start-stop system, the mark-to-space transition at the beginning of the start element. (PE) 599-1985w

startup (of a relay) The action of a relay as it just departs from complete reset. Startup is also used as a qualifying term to identify the minimum value of the input quantity that will permit this condition. (SWG/PE) C37.100-1992

startup current (1) The transient current of a heating cable immediately following energization. (IA/PC) 515.1-1995
(2) The current response of a heating cable or surface heating device following energization. (IA) 515-1997

startup testing All testing of the generating unit from initial-powered rotation to verify suitability for operation. (PE/EDPG) 1248-1998

starvation A condition that occurs when one or more modules perform no useful work for an indefinite period of time due to lack of access to the bus or other system resources. (C/BA) 1014.1-1994w, 10857-1994, 896.3-1993w

starved electrolyte cell *See:* absorbed electrolyte cell.

statcoulomb The unit of charge in the centimeter-gram-second electrostatic system. It is that amount of charge that repels an equal charge with a force of one dyne when they are in a vacuum, stationary, and one centimeter apart. One statcoulomb is approximately 3.335×10^{-10} C. (Std100) 270-1966w

State An unordered, finite datatype. Each state value is identified by an associated name. (C/PA) 1224.1-1993w

state (1) (high-level microprocessor language) The condition of the target microprocessor, given in terms of the contents of its registers, internal flags, local memory, etc. (C/MM) 755-1985w
(2) (A) (modeling and simulation) (software) A condition or mode of existence that a system, component, or simulation may be in; for example, the pre-flight state of an aircraft navigation program or the input state of given channel. **(B) (modeling and simulation) (software)** The values assumed at a given instant by the variables that define the characteristics of a system, component, or simulation. *Synonym:* system state. *See also:* steady state. (C) 610.3-1989, 610.12-1990
(3) (power outages) Component or unit state is a particular condition or status of a component or a unit which is important for outage reporting purposes. (PE/PSE) 859-1987w
(4) The language-independent syntax for a family on unordered, finite datatypes. Each state value is identified by an associated name. (C/PA) 1351-1994w
(5) The input to and information stored in a circuit or device. *Note:* A full description of the state of a device allows its future behavior to be predicted for any combination of inputs. (C) 610.10-1994w
(6) A condition that characterizes the behavior of a function/subfunction or element at a point in time. (C/SE) 1220-1998

state class A kind of class that represents a set of real or abstract objects (people, places, events, ideas, things, combinations of things, etc.) that have common knowledge or behavior. A state class represents instances with changeable state. The constituent instances of a state class can come and go and can change state over time, i.e., their property values can change. (C/SE) 1320.2-1998

state data (software unit testing) Data that defines an internal state of the test unit and is used to establish that state or compare with existing states. (SE/C) 1008-1987r, 610.12-1990

state diagram (software) A diagram that depicts the states that a system or component can assume, and shows the events or circumstances that cause or result from a change from one state to another. (C) 610.12-1990

state element (high-level microprocessor language) A microprocessor component containing a distinguishable part of the state information, such as a single register. (MM/C) 755-1985w

state machine A model of a system in which all values are discrete, as in a digital computer. (C) 610.3-1989w

statement (1) (computer programming) A meaningful expression or generalized instruction in a source language. (C) [20], [85]
(2) (software) In a programming language, a meaningful expression that defines data, specifies program actions, or directs the assembler or compiler. *See also:* control statement; assignment statement; declaration. (C) 610.12-1990

statement of work A document used by the acquirer as a means to identify, describe, and specify the tasks to be performed under the contract. (C/SE) 1062-1998

statement testing (software) Testing designed to execute each statement of a computer program. *Contrast:* path testing; branch testing. (C) 610.12-1990

state-of-charge factor Actual capacity of a battery expressed as a percentage of a fully-charged capacity. *Note:* This is based on experience, application (cycling/float service), and charging parameters. (VT) 1476-2000

state of chromatic adaptation (illuminating engineering) The condition of the chromatic properties of the visual system at a specified moment as a result of exposure to the totality of colors of the visual field currently and in the past. *See also:* chromatic adaptation. (EEC/IE) [126]

state of polarization (of a plane wave [field vector]) At a given point in space, the condition of the polarization of a plane wave [field vector] as described by the axial ratio, tilt angle, and sense of polarization. *Synonym:* polarization state. (AP/ANT) 145-1993

state of statistical control (pulse measurement process) That state wherein a degree of consistency among repeated measurements of a characteristic, property, or attribute is attained. (IM/WM&A) 181-1977w

State Sequence Error (SSE) bit A bit in the Bus Error register of all S-modules. An addressed S-module sets this bit to indicate that the S-module's Slave Link Layer Controller has entered the ERROR Slave Controller State. (TT/C) 1149.5-1995

state space (1) (automatic control) A space which contains the state vectors of a system. *Note:* The number of state variables in the system determines the dimension of the state space. *See also:* control system. (PE/EDPG) [3]
(2) Memory that is used to store the parameters, variables, workspace, etc., related to an I/O transaction that is currently being processed. (C/MM) 1212.1-1993

state, system *See:* system state.

state trajectory (automatic control) The vector function describing the dependence of the state on time and initial state. *Note:* If ϕ is the state trajectory, then

$$\phi(t_0,x_0) = x_0$$

$$\phi(t_2,x_0) = \phi[t_2,\phi(t_1,x_0)]$$

(PE/EDPG) [3]

state transition A change from one state to another in a system, component, or simulation. (C) 610.3-1989w

state transition diagram A graphical means of expressing the allowed states of an object and the allowed transitions from one state to another. (IM/ST) 1451.1-1999

state variable A variable that defines one of the characteristics of a system, component, or simulation. The values of all such variables define the state of the system, component, or simulation. (C) 610.3-1989w

state variable formulation (excitation systems) (eigenvalue, eigenvector, characteristic equation) A system may be mathematically modeled by assigning variables $x_1, x_2, \ldots, x_n$ to system parameters: when these *x*s comprise the minimum number of parameters which completely specify the system, they are termed "states" or "state variables." System states arranged in a *n*-vector form a state vector. The mathematical model of the system may be manipulated into the form

$$(dx)/(dt) = \dot{H} = AX + bu$$

$$Y = CX + bu$$

where X is the system state vector, u is the input vector, Y is the output vector, and A, b, C, d are matrices of appropriate dimension which specify the system. Such a model is known as a state variable or modern control formulation.

$$\det(A - \lambda I) = 0$$

is called the characteristic equation and has n roots which are called eigenvalues (det (·) denotes determinant). When eigenvalues are real, they are the negative inverses of closed loop system time constants. Eigenvalues are also the pole locations of the closed loop transfer function. Any vector e_i such that

$$(A - \lambda_i I)e_i = 0$$

$$\|e_i\| \neq 0$$

is called an eigenvector of the eigenvalue λ_i($\|\cdot\|$) denotes the square root of the sum of the squares of all entries of a vector. All *n* eigenvectors of a system form a modal matrix of matrix A when arranged side-by-side in a square matrix. The modal matrix is used in certain analytic procedures in modern control theory whereby large, complex systems are decoupled into many first order systems. (PE/EDPG) 421A-1978s

state variables (automatic control) Those whose values determine the state. (PE/EDPG) [3]

state vector (automatic control) One whose components are the state variables. (PE/EDPG) [3]

static (1) (atmospherics) Interference caused by natural electric disturbances in the atmosphere, or the electromagnetic phenomena capable of causing such interference. *See also:* radio transmitter. (PE/EEC) [119]
(2) (adjective) (automatic control) Referring to a state in which a quantity exhibits no appreciable change within an arbitrarily long time interval. (PE/EDPG) [3]
(3) (software) Pertaining to an event or process that occurs without computer program execution; for example, static analysis, static binding. *Contrast:* dynamic. (C) 610.12-1990

static accuracy (analog computer) Accuracy determined with a constant output. (C) 165-1977w

static analysis (software) The process of evaluating a system or component based on its form, structure, content, or documentation. *Contrast:* dynamic analysis. *See also:* walkthrough; inspection. (C) 610.12-1990

static analyzer (software) A software tool that aids in the evaluation of a computer program without executing the program. Examples include syntax checkers, compilers, cross-reference generators, standards enforcers, and flowcharters. *See also:* dynamic analyzer; computer program; syntax; compiler; program. (C/SE) 729-1983s

static binding (software) Binding performed prior to the execution of a computer program and not subject to change during program execution. *Contrast:* dynamic binding. (C) 610.12-1990

static breakpoint A breakpoint that can be set at compile time, such as entry into a given routine. *See also:* data breakpoint; epilog breakpoint; code breakpoint; prolog breakpoint; programmable breakpoint. (C) 610.12-1990

static breeze *See:* convective discharge.

static characteristic (electron tube) A relation, usually represented by a graph, between a pair of variables such as electrode voltage and electrode current, with all other voltages maintained constant. (ED) [45]

static characteristic, relay *See:* relay static characteristic.

static charge Any electric charge at rest, e.g., charge on capacitor. Static charge is often loosely used to describe discharge conditions resulting from electric field coupling. (T&D/PE) 524a-1993r, 1048-1990

static check *See:* problem check.

static compensator (STATCOM) A static synchronous generator operated as a shunt-connected static var compensator (SVC), whose capacitive or inductive output current can be controlled independently of the ac system voltage. (PE/SUB) 1031-2000

static converter A unit that employs solid state devices such as semiconductor rectifiers or controlled rectifiers (thyristors), gated power transistors, electron tubes, or magnetic amplifiers to change ac power to dc power, dc power to ac power, or fixed frequency ac power to variable frequency ac power. (IA/MT) 45-1998

static decay (charge-storage tubes) Decay that is a function only of the target properties, such as lateral and transverse leakage. *See also:* charge-storage tube. (ED) 158-1962w

static dissipative Having a level of resistivity that typically leads to charge dissipation. (SPD/PE) C62.47-1992r

static dose rate test Test of the permanent changes induced by radiation that are obtained by a comparison of characteristics before and after exposure at a given dose rate. (ED) 1005-1998

static dump (1) (software) A dump that is produced before or after the execution of a computer program. *Contrast:* dynamic dump. *See also:* selective dump; memory dump; snapshot dump; postmortem dump; change dump. (C) 610.12-1990 **(2) (computers)** A dump that is performed at a particular point in time with respect to a machine run, frequently at the end of a run. (C) [20], [85]

static electrode potential The electrode potential that exists when no current is flowing between the electrode and the electrolyte. *See also:* electrolytic cell. (EEC/PE) [119]

static error (software) An error that is independent of the time-varying nature of an input. *Contrast:* dynamic error. (C) 610.12-1990, 165-1977w

static exciter Nonrotating source of direct current for the synchronous generator field, utilizing controlled rectifiers. (PE/EDPG) 1020-1988r

static friction *See:* stiction.

static induced current The charging and discharging current of a pair of Leyden jars or other capacitors, which current is passed through a patient. *See also:* electrotherapy. (EMB) [47]

staticize (A) (electronic digital computation) To convert serial or time-dependent parallel data into static form. **(B) (electronic digital computation)** Occasionally, to retrieve an instruction and its operands from storage prior to its execution. (C) 162-1963

staticizer (electronic computation) A storage device for converting time-sequential information into static parallel information. (Std100) 270-1966w

static Kraemer system (rotating machinery) A system of speed control below synchronous speed for wound-rotor induction motors. Slip power is recovered through the medium of a static converter equipment electrically connected between the secondary winding of the induction motor and a power system. *See also:* asynchronous machine. (PE) [9]

static load line The locus of all simultaneous average values of output electrode current and voltage, for a fixed value of direct-current load resistance. (ED) [45]

static magnetic cell *See:* magnetic cell.

static method A method that can be executed without an instance of its package. (C/BA) 1275-1994

static model (1) A model of a system in which there is no change; for example, a scale model of a bridge, studied for its appearance rather than for its performance under varying loads. *Contrast:* dynamic model. (C) 610.3-1989w **(2)** A kind of model that describes an interrelated set of classes (and/or subject domains) along with their relationships and responsibilities. *Contrast:* dynamic model. (C/SE) 1320.2-1998

static noise (atmospherics) (telephone practice) Interference caused by natural electric disturbances in the atmosphere, or the electromagnetic phenomena capable of causing such interference. *See also:* static. (PE/PSR) C37.93-1976s

static optical transmission (acousto-optic device) The ratio of the transmitted zero order intensity, I_0, to the incident light intensity, I_{in}, when the acoustic drive power is off: thus $T = I_0/I_n$. (UFFC) [23]

static overvoltage (surge arresters) An overvoltage due to an electric charge on an isolated conductor or installation. (PE) [8]

static patterns A set of controlled, time-invariant patterns. (SCC20) 1445-1998

static phase offset The constant difference between the phase of the recovered clock and the optimal sampling position of the received data. (LM/C) 802.5-1989s

static plow (cable plowing) A plowing unit that depends upon drawbar pull only for its movement through the soil. (T&D/PE) 590-1977w

static power converter Any static power converter with control, protection, and filtering functions used to interface an electric energy source with an electric utility system. Sometimes referred to as power conditioning subsystems, power conversion systems, solid-state converters, or power conditioning units. (DESG) 1035-1989w

static pressure (audio and electroacoustics) (at a point in a medium) The pressure that would exist at that point in the absence of sound waves. (SP/ACO) [32]

static radiation test (metal nitride oxide semiconducter arrays) A test of the permanent changes induced by radiation obtained by a comparison of characteristics before and after exposure. (ED) 641-1987w, 581-1978w

static random-access memory (SRAM) A static form of random-access memory that does not require periodic refresh to retain data. *Contrast:* dynamic random-access memory. (C) 610.10-1994w

static regulation Expresses the change from one steady-state condition to another as a percentage of the final steady-state condition.

$$\text{Static Regulation} = \frac{E_{\text{initial}} - E_{\text{final}}}{E_{\text{final}}}$$

(AES) [41]

static regulator A transmission regulator in which the adjusting mechanism is in self-equilibrium at any setting and requires control power to change the setting. *See also:* transmission regulator. (EEC/PE) [119]

static relay (or relay unit) A relay or relay unit in which the designed response is developed by electronic, solid-state, magnetic or other components without mechanical motion. *Note:* A relay that is composed of both static and electromechanical units in which the designed response is accomplished by static units may be referred to as a static relay. (SWG/PE) C37.100-1992

static resistance (semiconductor rectifier device) (forward or reverse) The quotient of the voltage by the current at a stated point on the static characteristic curve. *See also:* rectification. (IA) [12]

static routing A routing strategy that determines the path to be followed by network traffic using the information and algorithms fixed at the time of network generation. (C) 610.7-1995

static short-circuit ratio (arc-welding apparatus) The ratio of the steady-state output short-circuit current of a welding power supply at any setting to the output current at rated load voltage for the same setting. (EEC/AWM) [91]

static, solid-state converter *See:* solid-state converter static.

static storage A type of storage that does not require periodic refreshment for retention of data. *Contrast:* dynamic storage. *See also:* static random-access memory. (C) 610.10-1994w

static test (1) (A) (test, measurement, and diagnostic equipment) A test of a non-signal property, such as voltage and current, of an equipment or of any of its constituent units, performed while the equipment is energized. **(B) (test, measurement, and diagnostic equipment)** A test of a device in a stationary of helddown position as a means of testing and measuring its dynamic reactions. (MIL) [2] **(2)** The computer-control state that applies a predetermined set of voltages and conditions to the analog computer, allowing a static check to be performed. (C) 610.10-1994w, 165-1977w

static timing error The constant part of the difference in time between the ideal sampling point for the received data and the actual sampling point. (C/LM) 8802-5-1998

static torque (electric coupling) The minimum torque an electric coupling will transmit or develop with no relative motion between the input and output members, with excitation applied. *Note:* Static torque is usually specified for rated excitation. (EM/PE) 290-1980w

static value (light-emitting diodes) A non-varying value or quantity of measurement at a specified fixed point, or the slope of the line from the origin to the operating point on the appropriate characteristic curve. (ED) [127]

static var compensator (SVC) A shunt-connected static var generator or absorber whose output is adjusted to exchange capacitive or inductive current to maintain or control specific parameters of the electrical power system (typically bus voltage). (PE/SUB) 1031-2000

static var system (SVS) A combination of different static and mechanically switched var compensators whose outputs are coordinated. (PE/SUB) 1031-2000

static volt-ampere characteristic (arc-welding apparatus) The curve or family of curves that gives the terminal voltage of a welding power supply as ordinate, plotted against output load current as abscissa, is the static volt-ampere characteristic of the power supply. (EEC/AWM) [91]

static wave current (electrotherapy) The current resulting from the sudden periodic discharging of a patient who has been raised to a high potential by means of an electrostatic generator. *See also:* electrotherapy. (EMB) [47]

static wire *See:* shield wire; overhead ground wire.

static wire-coupling protector A device for protecting carrier terminals that are used in conjunction with overhead, insulated ground wires (static wires) of a power transmission line. *Synonym:* sky wire-coupling protector. (PE/PSC) 487-1992

station (1) One of the input or output devices on a communications network. *Synonym:* data station. *See also:* secondary station; device; server; primary station. (C) 610.7-1995
(2) A facility where several components of a system are located. (PE/PSE) 858-1993w
(3) Any device that contains an IEEE 802.11 conformant medium access control (MAC) and physical layer (PHY) interface to the wireless medium (WM). (C/LM) 8802-11-1999
(4) A physical device that may be attached to a shared medium local area network (LAN) to transmit and receive information on that shared medium. A data station is identified by a destination address. *Synonym:* data station. (EMB/MIB) 1073.4.1-2000
(5) (generating station grounding) *See also:* generating station. (PE/EDPG) 665-1987s
(6) *See also:* semiautomatic station; automatic station; remote station; master station. (SWG/PE/SUB) C37.100-1992, C37.1-1994

Station When capitalized, Station refers to DTR station or a C-Port in Station Emulation mode. (C/LM) 802.5t-2000

stationarity (seismic qualification of Class 1E equipment for nuclear power generating stations) A condition that exists when a waveform is stationary and when its amplitude distribution, frequency content, and other descriptive parameters are statistically constant with time. (PE/NP) 344-1987r

stationary appliance (electric systems) An appliance that is not easily moved from one place to another in normal use. *See also:* appliance. (NESC) [86]

stationary battery A storage battery designed for service in a permanent location. *See also:* battery. (EEC/PE) [119]

stationary-contact assembly The fixed part of the starting-switch assembly. *See also:* starting-switch assembly. (EEC/PE) [119]

stationary contact member A conducting part having a contact surface that remains substantially stationary. (SWG/PE) C37.100-1992

stationary-mounted device One that cannot be removed except by the unbolting of connections and mounting supports. *See also:* drawout-mounted device. (SWG/PE) C37.100-1992

stationary phase approximation A technique for evaluating or estimating integrals whose integrands have rapid variations in phase everywhere except near stationary phase points. (AP/PROP) 211-1997

stationary phase point Point in space near which the phase of a function is slowly varying. (AP/PROP) 211-1997

stationary relay contact The member of a contact pair that is not moved directly by the actuating system. (EEC/REE) [87]

stationary satellite (communication satellite) A synchronous satellite with an equatorial, circular and direct orbit. A stationary satellite remains fixed in relation to the surface of the primary body. *Note:* A geo-stationary satellite is a stationary earth satellite. (COM) [19]

stationary system (excitation systems) (time invarient) Let a system have zero input response Z(t), then the system is stationary (time invarient) if the response to input R(t) is C(t) + Z(t) and the response to input R(t + T) is C(t + T) + Z(t). Otherwise the system is nonstationary. *Note:* A stationary system is modelled mathematically by a stationary differential equation the coefficients of which are not functions of time. (PE/EDPG) 421A-1978s

stationary wave *See:* standing wave.

station auxiliary (generating station) (generating stations electric power system) An auxiliary at a generating station not assigned to a specific unit. (PE/EDPG) 505-1977r

station auxiliary losses The electric power required to feed the HVDC station auxiliary loads. (SUB/PE) 1158-1991r

station basic rate A data transfer rate belonging to the extended service set (ESS) basic rate set that is used by a station for specific transmissions. The station basic rate may change dynamically as frequently each medium access control (MAC) protocol data unit (MPDU) transmission attempt, based on local considerations at that station. (C/LM) 8802-11-1999

station blackout The complete loss of ac electric power to the essential and nonessential switchgear buses in a nuclear power plant (i.e., loss of offsite electric power system concurrent with turbine trip and unavailability of the onsite emergency ac power system). Station blackout does not include the loss of available ac power to buses fed by station batteries through inverters or by alternate ac sources. (PE/NP) 765-1995

station changing (communication satellite) The changeover of service from one earth station to another, especially in a system using satellites that are not stationary. (COM) [19]

station check (supervisory control, data acquisition, and automatic control) (supervisory check, status update) The automatic selection, in a definite order, of all the supervisory alarm and indication points associated with one remote station or all remote stations of a system, and the transmission of all the indications to the master station. (SWG/SUB/PE) C37.1-1987s, C37.100-1992

station-control error (electric power system) The actual station generation minus assigned station generation. (PE/PSE) 94-1991w

station equipment (data transmission) A broad term used to denote equipment located at the customer's premises. The equipment may be owned by the telephone company or the customer. If the equipment is owned by the customer it is referred to as the customer's equipment. (PE) 599-1985w

station ground A ground grid or any equivalent system of grounding electrodes buried beneath or adjacent to the gas-insulated substation that determines the rise of ground voltage level relative to remote earth and controls the distribution of voltage gradients within the gas-insulated substation area during a fault. (SWG/SUB/PE) C37.122-1983s, C37.122.1-1993, C37.100-1992

station identification (supervisory control, data acquisition, and automatic control) A sequence of signal elements used to identify a station. (SWG/PE/SUB) C37.1-1987s, C37.100-1992

station line (telephone switching systems) Conductors carrying direct current between a central office and a main station, private branch exchange, or other end equipment. (COM) 312-1977w

station lobe The wiring that connects a LAN station or other device to a hub, excluding equipment and station attachment cables. (C) 610.7-1995

station-loop resistance (telephone switching systems) The series resistance of the loop conductors, including the resistance of an off-hook station. (COM) 312-1977w

station management (SMT) The conceptual control element of a station that interfaces with all of the layers of the station and is responsible for the setting and resetting of control parameters, obtaining reports of error conditions, and determining if the station should be connected to or disconnected from the medium. (C/LM) 8802-5-1998

station number (subroutines for CAMAC) The number n represents an integer which is the station number component of a CAMAC address. (NPS) 758-1979r

station, peaking *See:* peaking station.

station, pumped storage *See:* pumped storage station.

station ringer *See:* telephone ringer.

station, run-of-river *See:* run-of-river station.

station service (SS) (1) (power operations) Facilities that provide power for station use in a generating, switching, converting, or transforming station. (PE/PSE) 858-1987s, 346-1973w
(2) The set of services that support transport of medium access control (MAC) service data units (MSDUs) between stations within a basic service set (BSS). (C/LM) 8802-11-1999

station service power The power used to operate a station. (PE/PSE) 94-1991w

station service transformer (generating stations electric power system) A transformer that supplies power from a station high-voltage bus to the station auxiliaries and also to the unit auxiliaries during unit startup and shutdown or when the unit auxiliaries transformer is not available, or both. (PE/TR/EDPG) C57.116-1989r, 505-1977r

station, steam-electric *See:* steam-electric station.

station, storage *See:* storage station.

station-to-station call (telephone switching systems) A call intended for a designated main station. (COM) 312-1977w

station-type cubicle switchgear (SC) Metal-enclosed power switchgear characterized by the following required features:

— The main switch and interrupting device is of the stationary mounted type, composed of a primary circuit compartment and a secondary or mechanism compartment; arranged with gang-operated isolating switches that are mechanically interlocked with the main switching and interrupting device.

— Each phase for the major parts of the primary circuit switching or interrupting devices, buses, and line-to-ground potential transformers is completely enclosed (or segregated) by grounded metal barriers that have no intentional openings between compartments. Specifically included are mechanically interlocked doors in front of or a part of the primary circuit compartment of the circuit switching and interrupting device so that when the group operated isolating switches are closed, no primary parts can be exposed by the attempted opening of the interlocked doors.

— All live parts are enclosed within grounded metal compartments.

— Primary bus conductor and connections are bare.

— Mechanical interlocks are provided for proper operating sequence under normal operating conditions.

— Secondary control devices and their wiring are isolated by grounded metal barriers from all primary circuit elements with the exception of short lengths of wire, such as at instrument transformer terminals.

— The doors to the secondary or mechanism compartment of the primary switching or interrupting device are to provide access to the secondary or control equipment within the housing without danger of exposure to the primary circuit parts.

Note: Auxiliary vertical sections may be required for mounting devices or for use as bus transition. (SWG/PE) C37.100-1992, C37.20.2-1993

station-type regulator A regulator designed for ground-type installations in stations or substations. (PE/TR) C57.15-1999

station-type transformer (power and distribution transformers) A transformer designed for installation in a station or substation. (PE/TR) C57.12.80-1978r

Statistical Analysis System (SAS) A programming language used for statistical analysis, data manipulation, and application development. (C) 610.13-1993w

statistical BIL The crest values of a standard lightning impulse for which the insulation exhibits a 90% probability of withstand (or a 10% probability of failure) under specified conditions, applicable specifically to self-restoring insulations. (SPD/PE/C) C62.22-1997, 1313.1-1996

statistical BSL The crest value of a standard switching impulse for which the insulation exhibits a 90% probability of withstand (or a 10% probability of failure), under specified conditions, applicable to self-restoring insulations. (PE/SPD/C) C62.22-1997, 1313.1-1996

statistical delay (gas tube) The time lag from the application of the specified voltage to initiate the discharge to the beginning of breakdown. *See also:* gas tube. (ED) [45]

statistical descriptors Many sounds have sound-pressure levels that are not constant in time and cannot, without qualification, be adequately characterized by a single value of sound level. One method for dealing with fluctuating or intermittent sounds is to examine the sound level statistically as a function of time. Statistical descriptors are often applied to A-weighted sound levels. They are called exceedance levels or L-levels. For example, the L_{10} is the A-weighted sound level exceeded for 10% of the time over a specified time period. The other 90% of the time, the sound level is less than the L_{10}. Similarly, the L_{50} is the sound level exceeded 50% of the time; the L_{90} is the sound level exceeded 90% of the time; etc. (T&D/PE) 656-1992

statistical indicators Parameters based on past plant-specific or generic experience used to predict the failure of identical or similar equipment based on time or stress histories. (PE/NP) 933-1999

statistically homogeneous Having statistical characteristics that are independent of the specific locations at which those characteristics are measured. (AP/PROP) 211-1997

statistically isotropic Having statistical characteristics that are independent of the directions along which those characteristics are measured. (AP/PROP) 211-1997

statistical multiplexer A multiplexer that uses time division multiplexing technique by dynamically allocating telecommunication line time to each of the various attached terminals, according to whether a terminal is active or inactive at a particular moment. (C) 610.7-1995

Statistical Package for Social Sciences (SPSS) A nonprocedural language used for statistical analysis of research results, particularly data collected in polls and surveys. (C) 610.13-1993w

statistical pattern recognition An approach to pattern recognition that uses probability and statistical methods to assign patterns to pattern classes. (C) 610.4-1990w

statistical sparkover voltage A transient overvoltage level that produces a 97.72% probability of sparkover (i.e., two standard deviations above the 50% sparkover voltage value). *Note:* IEC uses 90%. (T&D/PE) 516-1995

statistical terms related to corona effects Terms applied to the procedures of data collection, classification, and presentation relating to corona effects. (T&D/PE) 539-1990

statistical test model (software) A model that relates program faults to the input data set (or sets) which cause them to be encountered. The model also gives the probability that these faults will cause the program to fail. *See also:* model; data. (C/SE) 729-1983s

statistical withstand voltage (1) A transient overvoltage level that produces a 0.14% probability of sparkover (i.e., three standard deviations below the 50% sparkover voltage value). *Note:* IEC uses 2%. (T&D/PE) 516-1995
(2) The voltage that an insulation is capable of withstanding with a given probability of failure, corresponding to a specified probability of failure (e.g., 10%, 0.1%). (PE/SPD/C) C62.22-1997, 1313.1-1996

stat mux *See:* statistical multiplexer.

stator (1) (watthour meter) An assmbly of an induction watthour meter, which consists of a voltage circuit, one or more current circuits, and a combined magnetic circuit so arranged that their joint effect, when energized, is to exert a driving torque on the rotor by the reaction with currents induced in an individual or common conducting disk. (ELM) C12.1-1982s
(2) (rotating machinery) The portion that includes and supports the stationary active parts. The stator includes the stationary portions of the magnetic circuit and the associated winding and leads. It may, depending on the design, include a frame or shell, winding supports, ventilation circuits, coolers, and temperature detectors. A base, if provided, is not ordinarily considered to be part of the stator. (PE) [9]

stator bar A unit of winding on the stator of a machine. *See also:* stator coil; bar. (PE/EM) 1129-1992r

stator coil (rotating machinery) A unit of a winding on the stator of a machine. *See also:* stator. (PE) [9]

stator coil pin (rotating machinery) A rod through an opening in the stator core, extending beyond the faces of the core, for the purpose of holding coils of the stator winding to a desired position. *See also:* stator. (PE) [9]

stator core (rotating machinery) The stationary magnetic-circuit of an electric machine. It is commonly an assembly of laminations of magnetic steel, ready for winding. *See also:* stator. (PE) [9]

stator-core lamination (rotating machinery) A sheet of material usually of magnetic steel, containing teeth and winding slots, or containing pole structures, that forms the stator core when assembled with other identical or similar laminations. *See also:* stator. (PE) [9]

stator frame (rotating machinery) The supporting structure holding the stator core or core assembly. *Note:* In certain types of machines, the stator frame may be made integral with one end shield. *See also:* stator. (PE) [9]

stator iron (rotating machinery) A term commonly used for the magnetic steel material or core of the stator of a machine. *See also:* stator. (PE) [9]

stator mounting lug (rotating machinery) A part attached to the outer surface of stator core or a stator shell to provide a means for the bolting or equivalent attachment to the appliance, machine, or other foundation. *See also:* stator. (PE) [9]

stator resistance starting (rotating machinery) The process of starting a motor at reduced voltage by connecting the primary winding initially in series with starting resistors that are short-circuited for the running condition. (PE) [9]

stator shell (rotating machinery) A cylinder in tight assembly around the wound stator core, all or a portion of which is machined or otherwise made to a specific outer dimension so that the stator may be mounted into an appliance, machine, or other end product. *See also:* stator. (PE) [9]

stator winding (rotating machinery) A winding on the stator of a machine. *See also:* stator. (PE) [9]

stator winding copper (rotating machinery) A term commonly used for the material or conductors of a stator winding. *See also:* stator. (PE) [9]

stator winding terminal (rotating machinery) The end of a lead cable or a stud or blade of a terminal board to which connections are normally made during installation. *See also:* stator. (PE) [9]

status (1) (supervisory control, data acquisition, and automatic control) Information describing a logical state of a point or equipment. (SWG/PE/SUB) C37.100-1992, C37.1-1994
(2) A term used generally to describe data generated by the peripheral that reflects the current operating state of the peripheral. (C/MM) 1284-1994
(3) (A) The condition at a particular time of a system or system component. **(B)** Pertaining to the condition as in (A), for example a status bit containing a bit that represents the status of a system. (C) 610.10-1994

status bit A bit used to indicate a non-error condition important to S-module operation. Status bits are located in an S-module's Slave Status register and Bus Error register (the BMR bit) and may be located in the optional Module Status register or an Additional Status register Status bits in the Module Status register or in an Additional Status register are permitted to affect the value of the EVO bit of the Slave Status register. (TT/C) 1149.5-1995

status code A code used to indicate the results of a computer program operation. For example, a code indicating a carry, an overflow, or a parity error. *Synonym:* condition code. (C) 610.12-1990

status codes Information used to indicate the state or condition of system components. (SUB/PE) 999-1992w

status datatype An abstract datatype whose values may be bound to "control" values as well as "data" values. (C/PA) 1328-1993w, 1327-1993w, 1224.1-1993w, 1224-1993w, 1351-1994w

status flag (radix-independent floating-point arithmetic) (binary floating-point arithmetic) A variable that may take two states, set and clear. A user may clear a flag, copy it, or restore it to a previous state. When set, a status flag may contain additional system-dependent information, possibly inaccessible to some users. The operations of IEEE Std 754-1985 and IEEE Std 854-1987 may as a side effect set some of the following flags: inexact result, underflow, overflow, divide by zero, and invalid operation. (C/MM) 854-1987r, 754-1985r

status lines Unidirectional signals from the peripheral to the host, defined in Compatibility Mode to handshake data and to report error conditions. In other IEEE 1284 interface modes defined in this standard, these lines are used for control, data, and/or status. (C/MM) 1284-1994

status memory (sequential events recording systems) The memory that contains the most recently scanned status of all inputs. (PE/EDPG) [5], [1]

status point, supervisory control *See:* status.

status register A register in an S-module by means of which current operating conditions of the S-module (e.g., interrupt enabled, module pass/fail status, multicast address of the S-module, etc.) and event occurrence (e.g., detection of an error condition during transmission of a message) can be recorded either for later interrogation by the M-module or to record the necessity of particular S-module activity at a later time. (TT/C) 1149.5-1995

status transfer The passing of information over the system control signal group, between the bus owner and the replying agent, during the reply phase of a transfer operation. *See also:* agent status. (C/MM) 1296-1987s

status word Together with the contents of the processor's registers, this defines the state or condition of the processor at any given moment. *Note:* If the processor is interrupted, it must save the status word so it can return to its former task. (C) 610.10-1994w

STC *See:* sensitivity time control.

S/TD *See:* signal-to-total-distortion ratio.

STD *See:* subscriber trunk dialing.

STE *See:* spanning tree explorer.

steady current A current that does not change with time. (Std100) 270-1966w

steady state (1) The condition of a specified variable at a time when no transients are present. *Note:* For the purpose of this definition, drift is not considered to be a transient. *See also:* feedback control system. (IA/ICTL/IAC) [60]
(2) (cable insulation materials) Conditions of current in the material attained when the difference between the maximum and minimum current observed during four consecutive hourly readings is less than 5% of the minimum current. (PE) 402-1974w
(3) (excitation systems) That in which some specified characteristic of a condition, such as value, rate, periodicity, or amplitude, exhibits only negligible change over an arbitrarily long interval of time. *Note:* It may describe a condition in which some characteristics are static, others dynamic. (PE/EDPG) 421A-1978s, 421-1972s
(4) (data management) A situation in which a model, process, or device exhibits stable behavior independent of time. *Synonym:* equilibrium. (C) 610.3-1989w
(5) The operating condition of a system wherein the observed variable has reached an equilibrium condition in response to an input or other stimulus in accordance with the definition of the system transfer function. This may involve a system output being at some constant voltage or current values in the case of power supplies. Referring to a subsystem operating parameter such as a thermal base-plate, it may be refer to a temperature that has reached stability as a function of the system operating inputs, load, and ambient environment. (PEL) 1515-2000

steady-state condition *See:* equilibrium mode distribution.

steady-state current perception threshold The current at which stimulation is perceptible for 50% of the subject population. *Note:* The threshold is a function of the frequency and voltage and varies considerably for various contact areas and pressures. Individual responses vary greatly from the mean threshold, and different levels are obtained for men, women, and children. (T&D/PE) 539-1990

steady-state deviation (control) The system deviation after transients have expired. *Note:* For the purpose of this definition, drift is not considered to be a transient. *See also:* deviation. (IM/IA/IAC) [120], [60]

steady-state governing load band (hydraulic turbines) The magnitude of the envelope of cyclic load variations caused by the governing system, expressed as a percent of rated power output, when the generating unit is operating in parallel with other generators and under steady-state load demand. (PE/EDPG) 125-1977s

steady-state governing speed band (hydraulic turbines) The magnitude of the envelope of the cyclic speed variations caused by the governing system, expressed as a percent of rated speed when the generating unit is operating independently and under steady-state load demand. (PE/EDPG) 125-1977s

steady-state incremental speed regulation (gas turbines) (excluding the effects of deadband) At a given steady-state speed and power, the rate of change of the steady-state speed with respect to the power output. It is the slope of the tangent to the steady-state speed versus power curve at the point of power output under consideration. It is the difference in steady-state speed, expressed in percent of rated speed, for any two points on the tangent, divided by the corresponding difference in power output, expressed as a fraction of the rated power output. For the basis of comparison, the several points of power output at which the values of steady-state incremental speed regulation are derived are based upon rated speed being obtained at each point of power output. (PE/PSE/EDPG) 94-1970w, 282-1968w, [5]

steady-state induced current The rms power-frequency current in any circuit, as a result of induction. (PE/T&D) 539-1990

steady-state oscillation A condition in which some aspect of the oscillation is a continuing periodic function. (SP) [32]

steady-state response (system or element) The part of the time response remaining after transients have expired. *Note:* The term steady-state may also be applied to any of the forced-response terms: for example steady-state sinusoidal response. *See also:* feedback control system; sinusoidal response. (IM) [120]

steady-state short-circuit current (synchronous machines) The steady-state current in the armature winding when short-circuited. (PE) [9]

steady-state speed regulation (A) (straight condensing and noncondensing steam turbines, nonautomatic extraction turbines, hydro turbines, and gas turbines) The percent change in rated speed as the power output is gradually reduced from rated power to zero while all speed-governing system settings remain unchanged. **(B) (straight condensing and noncondensing steam turbines, nonautomatic extraction turbines, hydro-turbines, and gas turbines)** The change in steady-state speed, expressed in percent of rated speed, when the power output of the turbine operating isolated is gradually reduced from rated power output to zero power output with unchanged settings of all adjustments of the speed-governing system. *Note:* Speed regulation is considered positive when the speed increases with a decrease in power output. *See also:* asynchronous machine; speed-governing system. (PE/PE/PSE/PSE/EDPG) 94-1991, 94-1970, 282-1968, [5]

steady-state stability A condition that exists in a power system if it operates with stability when not subjected to an aperiodic disturbance. *Note:* In practical systems, a variety of relatively small aperiodic disturbances may be present without any appreciable effect upon the stability, as long as the resultant rate of change in load is relatively slow in comparison with the natural frequency of oscillation of the major parts of the system or with the rate of change in field flux of the rotating machines. (PE/T&D) [10]

steady-state stability factor (system or part of a system) The ratio of the steady-state stability limit to the nominal power flow at the point of the system to which the stability limit is referred. *See also:* stability factor. (T&D/PE) [10]

steady-state stability limit (steady-state power limit) The maximum power flow possible through some particular point in the system when the entire system or the part of the system to which the stability limit refers is operating with steady-state stability. (T&D/PE) [10]

steady-state temperature rise (grounding device) The maximum temperature rise above ambient which will be attained by the winding of a device as the result of the flow of rated continuous current under standard operating conditions. It may be expressed as an average or a hot-spot winding rise. (PE/SPD) 32-1972r

steady state thermal rating The constant electrical current that would yield the maximum allowable conductor temperature for specified weather conditions and conductor characteristics under the assumption that the conductor is in thermal equilibrium (steady state). (T&D/PE) 738-1993

steady-state value The value of a current or voltage after all transients have decayed to a negligible value. For an alternating quantity, the root-mean-square value in the steady state does not vary with time. *See also:* asynchronous machine. (PE) [9]

steady voltage *See:* steady current.

steam capability (power operations) The maximum net capability of steam generating units which can be obtained under normal operating practices for a given period of time as calculated based on design or test data or as demonstrated by total plant tests. The limitation on steam capability may be electrical or mechanical in nature. (PE/PSE) 858-1987s

steam-electric station (power operations) An electric generating station utilizing steam for the motive force of its prime movers. (PE/PSE) 858-1987s, 346-1973w

steam turbine-electric drive A self-contained system of power generation and application in which the power generated by a steam turbine is transmitted electrically by means of a generator and a motor (or multiples of these) for propulsion purposes. *Note:* The prefix steam turbine-electric is applied to ships, locomotives, cars, buses, etc., that are equipped with this drive. *See also:* electric propulsion system; electric locomotive. (EEC/PE) [119]

steel container (storage cell) The container for the element and electrolyte of a nickel-alkaline storage cell. This steel container is sometimes called a can. *See also:* battery. (PE/EEC) [119]

steel supported aluminum conductor (SSAC) ACSR with the aluminum wires annealed. (T&D/PE) 524-1992r

steerable-beam antenna system An antenna with a non-moving aperture for which the direction of the major lobe can be changed by electronically altering the aperture excitation or by mechanically moving a feed of the antenna. (AP/ANT) 145-1993

steering compass A compass located within view of a steering stand, by reference to which the helmsman holds a ship on the set course.

Stefan-Boltzmann law (illuminating engineering) The statement that the radiant exitance of a blackbody is proportional to the fourth power of its absolute temperature; that is,

$$M = \sigma T^4$$

Note: The currently recommended value of the Stefan-Boltzmann constant s is 5.67032×10^{-8} W $\times$ m^{-2} $\times$ K^{-4}. (EEC/IE) [126]

stellar guidance (navigation aids) Guidance by means of celestial bodies, particularly the stars. (AES/GCS) 172-1983w

stellar-inertial navigation equipment *See:* celestial-inertial navigation equipment.

step (1) (pulse techniques) A waveform that, from the observer's frame of reference, approximates a Heaviside (unit step) function. *See also:* unit-step signal. (IM/HFIM) [40] **(2) (A) (computers)** One operation in a computer routine. **(B) (computers)** To cause a computer to execute one operation. *See also:* single step. (C/C) [20], [85] **(3) (pulse terminology) (single transition)** A transition waveform that has a transition duration that is negligible relative to the duration of the waveform epoch or to the duration of its adjacent first and second nominal states. (IM/WM&A) 194-1977w

step-back relay A relay that operates to limit the current peaks of a motor when the armature or line current increases. A step-back relay may, in addition, operate to remove such limitations when the cause of the high current has been removed. (IA/MT) 45-1998

step-by-step operation *See:* single-step operation.

step-by-step switch (1) A bank-and-wiper switch in which the wipers are moved by electromagnet ratchet mechanisms individual to each switch. *Note:* This type of switch may have either one or two types of motion. (EEC/PE) [119] **(2)** A switch that moves in synchronism with a pulse device such as a rotary telephone dial. *Synonym:* line switch. *See also:* crossbar switch. (C) 610.7-1995

step-by-step system (1) An automatic telephone switching system that is generally characterized by the following features:

a) The selecting mechanisms are step-by-step switches;
b) The switching pulses may either actuate the successive selecting mechanisms directly or may be received and stored by controlling mechanisms that, in turn, actuate the selecting mechanisms.

(EEC/PE) [119]

(2) A type of line-switching system which uses step-by-step switches. *Synonym:* line switching system. *See also:* crossbar system; electronic switching system. (C) 610.7-1995

step change (control) (step function) An essentially instantaneous change of an input variable from one value to another. *See also:* feedback control system. (IA/ICTL/IAC) [60]

step compensation (correction) The effect of a control function or a device that will cause a step change in an other function when a predetermined operating condition is reached. (IA/ICTL/IAC) [60]

step control system (automatic control) A system in which the manipulated variable assumes discrete predetermined values. *Note:* The condition for change from one predetermined value to another is often a function of the value of the actuating signal. When the number of values of the manipulated variable is two, it is called a two-step control system; when more than two, a multi-step control system. (PE/EDPG) [3]

step distance A non-pilot distance relay scheme using multiple zones with time delay to differentiate between the zones. (PE/PSR) C37.113-1999

step-down transformer (power and distribution transformers) A transformer in which the power transfer is from a higher voltage source circuit to a lower voltage circuit. (PE/TR) C57.12.80-1978r

step-forced response (automatic control) The total (transient plus steady-state) time response resulting from a sudden change from one constant level of input to another. (PE/EDPG) [3]

step index optical waveguide (fiber optics) An optical waveguide having a step index profile. *See also:* step index profile. (Std100) 812-1984w

step index profile (fiber optics) A refractive index profile characterized by a uniform refractive index within the core and a sharp decrease in refractive index at the core-cladding interface. *Note:* This corresponds to a power-law profile with profile parameter, g, approaching infinity. *See also:* total internal reflection; multimode optical waveguide; normalized frequency; graded index profile; critical angle; dispersion; refractive index; mode volume; optical waveguide. (Std100) 812-1984w

stepless (electrical heating applications to melting furnaces and forehearths in the glass industry) Power modulation by means of a device, such as a saturable reactor or thyristor that provides essentially infinite resolution in output voltage, current, or power. (IA) 668-1987w

step line-voltage change (power supplies) An instantaneous change in line voltage (for example, 105−125 volts alternating current): for measuring line regulation and recovery time. (AES) [41]

step load change (power supplies) An instantaneous change in load current (for example, zero to full load) for measuring the load regulation and recovery time. (AES) [41]

stepped (electrical heating applications to melting furnaces and forehearths in the glass industry) Power modulation by means of discrete voltage steps, such as with a tapped transformer. (IA) 668-1987w

stepped antenna *See:* zoned antenna.

stepped-gate structure (metal-nitride-oxide field-effect transistor) Also source-drain protected structure: a variant of the metal-nitride-oxide semiconductor (MNOS) transistor whose gate dielectric along the channel is divided into two or three parts. One portion has the standard MNOS layer sequence of the memory device. On one or either side of the memory portion, particularly covering the lines where source and drain junction emerge at the silicon surface, is a gate dielectric that is used for the threshold insulated-gate field-effect transistor (IGFET) in a given technology. (ED) 581-1978w

stepped leader (1) (lightning) A series of discharges emanating from a region of charge concentration at short time intervals. Each discharge proceeds with a luminescent tip over a greater distance than the previous one. *See also:* direct-stroke protection. (T&D/PE) [10] **(2)** Static discharge that propagates from a cloud into the air. Current magnitudes that are associated with stepped leaders are small (on the order of 100 A) in comparison with the final stroke current. The stepped leaders progress in a random direction in discrete steps from 10 to 80 m in length. Their most frequent velocity of propagation is about 0.05% of the speed of light, or approximately 500 000 ft/s (150 000 m/s). It is not until the stepped leader is within the striking distance of the point to be struck that the stepped leader is positively directed toward this point. (SUB/PE) 998-1996

stepped wave (converter characteristics) (self-commutated converters) The waveform obtained from the summation of any number of square waves of the same frequency, each displaced in time from the others. The square waves are often uniformly displaced in time, but are not necessarily of equal amplitudes. An example is shown below.

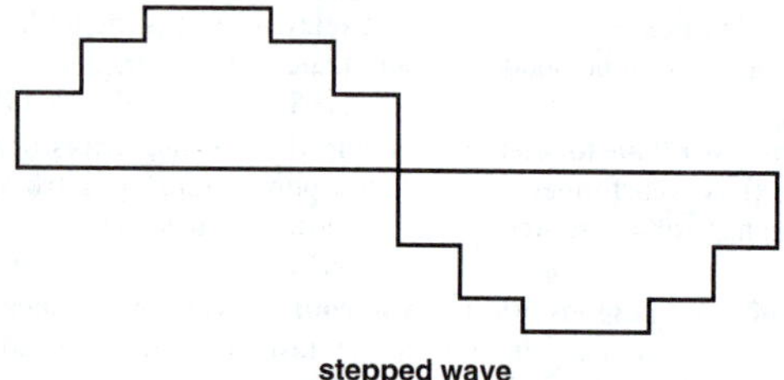

stepped wave

(IA/SPC) 936-1987w

stepping life (for attenuator variable in fixed steps) Number of times to switch from any selected position to any other selected positions, after which the residual and incremental characteristic insertion loss remain within the specified repeatability. (IM/HFIM) 474-1973w

stepping relay (1) A multiposition relay in which moving wiper contacts mate with successive sets of fixed contacts in a series of steps, moving from one step to the next in successive operations of the relay. *See also:* relay. (EEC/REE) [87]
(2) (rotary type) A relay having many rotary positions, ratchet actuated, moving from one step to the next in successive operations, and usually operating its contacts by means of cams. There are two forms:

a) directly driven, where the forward motion occurs on energization; and
b) indirectly (spring) driven, where a spring produces the forward motion on pulse cessation.

Note: The term is also incorrectly used for stepping switch. (PE/EM) 43-1974s

stepping relay, spring-actuated *See:* spring-actuated stepping relay.

step potential (1) The potential difference between two points on the earth's surface separated by a distance of one pace (assumed to be one meter) in the direction of maximum potential gradient. This potential difference could be dangerous when current flows through the earth or material upon which a worker is standing, particularly under fault conditions. *Synonym:* step voltage. (T&D/PE) 1048-1990
(2) *See also:* step voltage. (T&D/PE) 524-1992r

step response (1) The recorded output response for an ideal input step with designated baseline and topline. (IM/WM&A) 1057-1994w
(2) (high voltage testing) $g(t)$ The normalized output as a function of time t when the input is a voltage or current step. (PE/PSIM) 4-1995

step-response time The time required for the end device to come to rest in its new position after an abrupt change to a new constant value has occurred in the measured signal. *See also:* accuracy rating. (EEC/EMI) [112]

step restoration The restoration of service to blocks of customers in an area until the entire area or feeder is restored. (PE/T&D) 1366-1998

step speed adjustment The speed drive can be adjusted in rather large and definite steps between minimum and maximum speed. *See also:* electric drive. (IA/ICTL/IAC) [60]

step-stress test A test consisting of several stress levels applied sequentially, for periods of equal duration, to a (one) sample. During each period a stated stress-level is applied and the stress level is increased from one step to the next. *See also:* reliability. (R) [29]

step twist, waveguide *See:* waveguide step twist.

step-up transformer (power and distribution transformers) A transformer in which the power transfer is from a lower voltage source circuit to a higher voltage circuit. (PE/TR) C57.12.80-1978r

step voltage (1) (conductor stringing equipment) The potential difference between two points on the earth's surface separated by a distance of one pace (assumed to be 1 m) in the direction of maximum potential gradient. This potential difference could be dangerous when current flows through the earth or material upon which a worker is standing, particularly under fault conditions. *Synonym:* step potential. (T&D/PE/PSIM) 524a-1993r, 81-1983, 524-1992r
(2) The difference in surface potential experienced by a person not in contact with any grounded object and whose feet are spaced 1 m apart. (PE/SUB) 1268-1997
(3) The difference in surface potential experienced by a person bridging a distance of 1m with the feet without contacting any grounded object. (PE/SUB) 80-2000

step-voltage regulator (1) (power and distribution transformers) A regulating transformer in which the voltage of the regulated circuit is controlled in steps by means of taps and without interrupting the load. *Note:* Such units are generally 833 kVA (output) and below, single-phase; or 2500 kVA (output) and below, three-phase. (PE/TR) C57.12.80-1978r
(2) (transformer type) An induction device having one or more windings in shunt with, and excited from, the primary circuit, and having one or more windings in series between the primary circuit and the regulated circuit, all suitably adapted and arranged for the control of the voltage, or of the phase angle, or of both, of the regulated circuit in steps by means of taps without interrupting the load. (PE/TR) C57.15-1999

step-voltage test (rotating machinery) A controlled overvoltage test in which designated voltage increments are applied at designated times. Time increments may be constant or graded. *See also:* asynchronous machine; graded-time step-voltage test. (PE) [9]

step wedge* *See:* gray scale.
* Deprecated.

stepwise refinement A software development technique in which data and processing steps are defined broadly at first and then further defined with increasing detail. *See also:* transaction analysis; data structure-centered design; transform analysis; structured design; object-oriented design; modular decomposition; input-process-output. (C) 610.12-1990

steradian (1) (metric practice) The solid angle which, having its vertex in the center of a sphere, cuts off an area of the surface of the sphere equal to that of a square with sides of length equal to the radius of the sphere. (QUL) 268-1982s
(2) (laser maser) The unit of measure for a solid angle. There are 4π sr in a sphere. (LEO) 586-1980w

stereophonic (frequency modulation) (adjective) Pertains to audio information carried by a plurality of channels arranged to afford the listener a sense of the spatial distribution of the sound sources. *Note:* A stereophonic receiver responds to both the L + R main channel and the L − R subcarrier channel of a composite stereophonic signal, so that the one output contains substantially only L information, and the other only R. In addition to the main channel, stereophonic program modulation requires transmission of a 19 kHz pilot signal and the sidebands of a suppressed 38 kHz subcarrier carrying L − R information. This combination is called the composite signal, and it may be used alone or with other subcarrier (SCA) signals to frequency modulate the RF carrier. After pre-emphasis, the left and right channels are added for main channel information. The right-channel program material is subtracted from the left to derive a difference signal that then amplitude modulates a 38 kHz subcarrier. The subcarrier is suppressed, divided by two, and transmitted as a 19 kHz pilot signal to facilitate demodulation of the suppressed carrier information at the receiver. (BT) 185-1975w

stereopsis The three-dimensional effect achieved by simultaneously viewing two two-dimensional images of the same object projected from slightly different viewpoints. *See also:* stereoscopic projection. (C) 610.6-1991w

stereoscopic projection The projection of two two-dimensional images onto a two-dimensional display by use of stereopsis. (C) 610.6-1991w

ST-506 interface A data-transfer interface used in many early personal computers with hard disk capacities less than 40MB; characterized by a 34-pin control cable, a 20-pin data cable and an modest data-transfer rate. (C) 610.10-1994w

stick A type of insulating tool used in various operations of live-line work *Synonyms:* work stick; hot stick; pole; work pole; live-line tool. (T&D/PE) 516-1995

stick circuit A circuit used to maintain a relay or similar unit energized through its own contact. (EEC/PE) [119]

stickiness The condition caused by physical interference with the rotation of the moving element. *See also:* moving element. (EEC/AII) [102]

sticking voltage (luminescent screen) The voltage applied to the electron beam below which the rate of secondary emission from the screen is less than unity. The screen then has a negative charge that repels the primary electrons. *See also:* cathode-ray tube. (ED) [45], [84]

stick operation Manual operation of a switching device by means of a switch stick. *Synonym:* hook operation. (SWG/PE) C37.100-1992

stick printer An element printer in which a stick moves from left to right, printing one character at a time. (C) 610.10-1994w

stiction (1) The force in excess of the coulomb friction required to start relative motion between two surfaces in contact. (IM) [120]

(2) (static friction) The total friction that opposes the start of relative motion between elements in contact. *See also:* feedback control system. (IA/ICTL/IAC) [60]

stiffness The ability of a system or element to resist deviations resulting from loading at the output. *See also:* feedback control system. (PE/IA/ICTL/EDPG/IAC) [3], [60]

stiffness coefficient The factor K (also called spring constant) in the differential equation for oscillatory motion $M\ddot{x} + B\dot{x} + Kx = 0$. (PE/EDPG) [3]

stilb (illuminating engineering) A centimeter-gram-second (CGS) unit of luminance. One stilb equals one candela per square centimeter. The use of this term is deprecated. (EEC/IE) [126]

Stiles-Crawford effect (illuminating engineering) The reduced luminous efficiency of rays entering the peripheral portion of the pupil of the eye. This effect applies only to cones and not to rod visual cells. Hence, there is no Stiles-Crawford effect in scotopic vision. (EEC/IE) [126]

stimulate To provide input to a system in order to observe or evaluate the system's response. (C) 610.3-1989w

stimulated emission (1) (fiber optics) Radiation emitted when the internal energy of a quantum mechanical system drops from an excited level to a lower level when induced by the presence of radiant energy at the same frequency. An example is the radiation from an injection laser diode above lasing threshold. *See also:* spontaneous emission. (Std100) 812-1984w

(2) (laser maser) The emission of radiation at a given frequency caused by an applied external radiation field of the same frequency. (LEO) 586-1980w

stimulus (1) Any change in signal that affects the controlled variable: for example, a disturbance or a change in reference input. (MIL/IA/ICTL/APP/IAC) [2], [69], [60]

(2) The logic states within a pattern that drives a circuit model in simulation, or a unit under test (UUT) on an automatic test equipment (ATE). (SCC20) 1445-1998

stimulus data The information associated with stimuli. (SCC20) 1226-1998

stipple To change the appearance of an object by covering it with a pattern or regularly spaced small dots, spaced sufficiently so that the underlying image or text is still recognizable to the user. Stippling is frequently used to denote an inaccessible, or busy, object. (C) 1295-1993w

stirring effect (induction heater usage) The circulation in a molten charge due to the combined forces of motor and pinch effects. *See also:* pinch effect; induction heating. (IA) 54-1955w, 169-1955w

stochastic (1) (computer modeling and simulation) Pertaining to a process, model, or variable whose outcome, result, or value depends on chance. *Contrast:* deterministic. (C) 610.3-1989w

(2) (mathematics of computing) Pertaining to variables that are probabilistic in nature. (C) 1084-1986w

stochastic model A model in which the results are determined by using one or more random variables to represent uncertainty about a process or in which a given input will produce an output according to some statistical distribution; for example, a model that estimates the total dollars spent at each of the checkout stations in a supermarket, based on probable number of customers and probable purchase amount of each customer. *Synonym:* probabilistic model. *Contrast:* deterministic model. *See also:* Markov chain model. (C) 610.3-1989w

stochastic routing A routing strategy in which the results of individual decisions vary according to the conditions in the network at decision time. (C) 610.7-1995

Stokes matrix *See:* Mueller matrix.

Stokes parameters Elements of the Stokes vector. *See also:* Stokes vector. (AP/PROP) 211-1997

Stokes vector A 4 × 1 vector of real numbers called the Stokes parameters, representing the polarization state of a propagating wave:

$$\bar{I} = \begin{bmatrix} I_0 \\ Q \\ U \\ V \end{bmatrix} = \frac{1}{2\eta} \begin{bmatrix} |E_v|^2 + |E_h|^2 \\ |E_v|^2 - |E_h|^2 \\ 2\mathrm{Re}\{E_v E^*_h\} \\ 2\mathrm{Im}\{E_v E^*_h\} \end{bmatrix}$$

where

E_v is the vertical electric field component of the wave

E_h is the horizontal electric field component of the wave

η is the intrinsic impedance of the medium

* indicates the complex conjugate

(AP/PROP) 211-1997

stop (1) (limit stop) A mechanical or electric device used to limit the excursion of electromechanical equipment. *See also:* limiter circuits. (C) 165-1977w

(2) (software) To terminate the execution of a computer program. *Synonym:* halt. *Contrast:* pause. (C) 610.12-1990

stop band A band of frequencies that pass through a filter with a substantial amount of loss (relative to other frequency bands such as a pass band). (CAS) [13]

stop-band ripple The difference between maxima and minima of loss in a filter stop band. (CAS) [13]

stop bit (1) In asynchronous transmission, a bit that signals the end of a character. *Contrast:* start bit. (C) 610.7-1995

(2) For the low-speed version of the Physical layer, a bit that is encoded identically to a logic "1" that is used to delineate the end of each individual octet transmission. (EMB/MIB) 1073.4.1-2000

stop character A word processing control character that interrupts the sequence of output processing to provide the ability to make changes in the text formatting parameters, the text itself, the character font on the output device, or other items. *Synonym:* stop code. (C) 610.2-1987

stop code *See:* stop character.

stop dowel (rotating machinery) A pin fitted into a hole to limit motion of a second part. (PE) [9]

stop element (1) (data transmission) In a character transmitted in a start-stop system, the last element in each character, to which is assigned a minimum duration, during which the receiving equipment is returned to its rest condition in preparation for the reception of the next character. The stop element is a marking signal. (PE) 599-1985w
(2) *See also:* stop signal. (C) 610.7-1995

stop-go pulsing (telephone switching systems) A method of pulsing control wherein the pulsing operation may take place in stages, and the sending end is arranged to pulse the digits continuously unless or until the stop-pulsing signal is received. *Note:* When this occurs, the pulsing of the remaining digits is suspended until the sending end receives a start-pulsing signal. (COM) 312-1977w

stop instruction A computer instruction that specifies the termination of the execution of a computer program. *See also:* pause instruction. (C) 610.10-1994w

stop joint (power cable joints) A joint that is designed to prevent any transfer of dielectric fluid between the cables being joined. (PE/IC) 404-1986s

stop lamp (illuminating engineering) A lighting device giving a steady warning light to the rear of a vehicle or train of vehicles, to indicate the intention of the operator to diminish speed or to stop. (EEC/IE) [126]

stop list In automatic indexing, a list of terms, words, or roots of words that are considered insignificant for purposes of information retrieval, and are excluded from being keywords in an index. *Synonym:* stopword list. *Contrast:* go list. (C) 610.2-1987

stop-motion switch *See:* machine final-terminal stopping device.

stopping off The application of a resist to any part of a cathode or plating rack. *See also:* electroplating. (EEC/PE) [119]

stop-pulsing signal (telephone switching systems) A signal transmitted from the receiving end to the sending end of a trunk to indicate that the receiving end is not in a condition to receive pulsing. (COM) 312-1977w

stop-record signal (facsimile) A signal used for stopping the process of converting the electrical signal to an image on the record sheet. *See also:* facsimile signal. (COM) 168-1956w

stop signal (1) (facsimile) A signal that initiates the transfer of a facsimile equipment condition from active to standby. *See also:* facsimile signal. (COM) 168-1956w
(2) (data management) A signal at the end of a start-stop character that prepares the receiving device for the reception of a subsequent character. *Note:* A stop signal is usually limited to one signal element having any duration equal to or greater than a specified minimum value. (C) 610.5-1990w
(3) In asynchronous transmission, a signal following a character that prepares the receiving device for the reception of a subsequent character or block. *Synonym:* stop element. *Contrast:* start signal. (C) 610.7-1995

stop time *See:* deceleration time.

stop valve (1) (control systems for steam turbine-generator units) [throttle valve(s)] Those valve(s) that normally provide fast interruption of the main energy input to the turbine. Throttle valves are sometimes used for turbine control during start-up. *Note:* The term stop valve is defined as an open or closed valve. A throttle valve has some portion of its opening through which it can modulate flow. (PE/EDPG) 122-1985s
(2) (power system device function numbers) A control device used primarily to shut down an equipment and hold it out of operation. This device may be manually or electrically actuated, but excludes the function of electrical lockout on abnormal conditions. *See also:* lockout relay. (SUB/PE) C37.2-1979s

stopword list *See:* stop list.

storable swimming or wading pool A pool with a maximum dimension of 15 ft and a maximum wall height of 3 ft and is so constructed that it may be readily disassembled for storage and reassembled to its original integrity. (NESC/NEC) [86]

storage (1) (A) (electronic computation) The act of storing information. **(B) (electronic computation)** Any device in which information can be stored, sometimes called a memory device. **(C) (electronic computation)** In a computer, a section used primarily for storing information. Such a section is sometimes called a memory or store (British). *Notes:* 1. The physical means of storing information may be electrostatic, ferroelectric, magnetic, acoustic, optical, chemical, electronic, electric, mechanical, etc., in nature. 2. Pertaining to a device in which data can be entered, in which it can be held, and from which it can be retrieved at a later time. *See also:* store. (MIL/C) [2], [85], [20]
(2) (data management) In a computer, one or more bytes that are used to store data. (C) 610.5-1990w
(3) (A) The retention of data in a storage device. **(B)** The action of placing data into a storage device. **(C)** A storage device. **(D)** Any medium in which data can be retained. (C) 610.10-1994

storage access *See:* access.

storage allocation (1) (computers) The assignment of sequences of data or instructions to specified blocks of storage. (C) [20], [85]
(2) (software) An element of computer resource allocation, consisting of assigning storage areas to specific jobs and performing related procedures, such as transfer of data between main and auxiliary storage, to support the assignments made. *See also:* paging; buffer; contiguous allocation; cyclic search; virtual storage; overlay; memory compaction. (C) 610.12-1990

storage assembly (storage tubes) An assembly of electrodes (including meshes) that contains the target together with electrodes used for control of the storage process, those that receive an output signal, and other members used for structural support. *See also:* storage tube. (ED) 158-1962w

storage battery A battery comprised of one or more rechargeable cells of the lead-acid, nickel-cadmium, or other rechargeable electrochemical types. (NESC/NEC) [86]

storage breakpoint *See:* data breakpoint.

storage capacitor A low leakage capacitor on which a data value can be stored. (C) 610.10-1994w

storage capacity (1) The amount of data that can be contained in a storage device. *Notes:* 1. The units of capacity are bits, characters, words, etc. For example, capacity might be "32 bits," "10 000 decimal digits," "16 384 words with 10 alphanumeric characters each." 2. When comparisons are made among devices using different character sets and word lengths, it may be convenient to express the capacity in equivalent bits, which is the number obtained by taking the logarithm to the base 2 of the number of usable distinguishable states in which the storage can exist. 3. The storage (or memory) capacity of a computer usually refers only to the internal storage section. (C) 162-1963w
(2) (software) The maximum number of items that can be held in a given storage device; usually measured in words or bytes. (C) 610.12-1990
(3) The amount of data that can be contained in a storage device measured in binary characters, bytes, words, or other units of data. (C) 610.10-1994w
(4) The amount of data that can be contained in a storage device. (ED) 1005-1998

storage cell (1) (electric energy) (secondary cell or accumulator) A galvanic cell for the generation of electric energy in which the cell, after being discharged, may be restored to a fully charged condition by an electric current flowing in a direction opposite to the flow of current when the cell discharges. (EEC/PE) [119]

(2) (A) One or more storage elements considered as a unit. (B) The smallest subdivision of storage into which a unit of data can be placed, retained, and with which the unit can be retrieved. *Synonym:* data cell. *See also:* binary cell; magnetic cell. (C) 610.10-1994
(3) An elementary unit of storage (e.g., a binary cell or a decimal cell). (ED/C) 1005-1998, [85], [20]

storage channel A channel that can be used to access a storage device. (C) 610.10-1994w

storage device (1) A device in which data can be stored and from which it can be copied at a later time. The means of storing data may be chemical, electrical, mechanical, etc. *See also:* storage. (C) 162-1963w
(2) A device into which data can be placed, in which they can be retained, and from which they can be retrieved. *See also:* store. (C) 610.10-1994w

storage display *See:* storage tube display device.

storage efficiency The degree to which a system or component performs its designated functions with minimum consumption of available storage. *See also:* execution efficiency. (C) 610.12-1990

storage element (1) (storage tubes) An area of a storage surface that retains information distinguishable from that of adjacent areas. *Note:* The storage element may be a portion of a continuous storage surface or a discrete area such as a dielectric island. *See also:* storage tube. (ED) 158-1962w, 161-1971w
(2) The basic unit of a storage device, such as a sector, or a track. (C) 610.10-1994w

storage-element equilibrium voltage (storage tubes) A limiting voltage toward which a storage element charges under the action of primary electron bombardment and secondary emission. At equilibrium voltage the escape ratio is unity. *Note:* Cathode equilibrium voltage, second-crossover equilibrium voltage, and gradient-established equilibrium voltage are typical examples. *See also:* charge-storage tube. (ED) 158-1962w

storage-element equilibrium voltage, cathode (storage tubes) The storage element equilibrium voltage near cathode voltage and below first-crossover voltage. *See also:* charge-storage tube. (ED) 158-1962w

storage-element equilibrium voltage, collector *See:* charge-storage tube.

storage-element equilibrium voltage, gradient established (storage tubes) The storage-element equilibrium voltage, between first- and second-crossover voltages, at which the escape ratio is unity. *See also:* charge-storage tube. (ED) 158-1962w

storage-element equilibrium voltage, second-crossover (storage tubes) The storage-element equilibrium voltage at the second-crossover voltage. *See also:* charge-storage tube. (ED) 158-1962w

storage error An error in which the data retrieved from storage is different from that which was originally stored in that location. *See also:* soft error; hard error; transient error. (C) 610.10-1994w

storageid (microprocessor operating systems parameter types) An identifier for a block of data. The identifier is not guaranteed to be valid outside the allocating process and should not be passed between processes. (C/MM) 855-1985s

storage integrator In an analog computer, a device used to store a voltage in the hold condition for future use. *See also:* electronic analog computer. (C) 610.10-1994w, 165-1977w

storage life (accelerometer) (gyros) (inertial sensors) The nonoperating time interval under specified conditions, after which a device will still exhibit a specified operating life and performance. *See also:* operating life. (AES/GYAC) 528-1994

storage light A light found on a storage device indicating that a parity check error has occurred on a character as it was read into storage. (C) 610.10-1994w

storage light-amplifier *See:* image-storage panel.

storage location (1) An area in a storage device that can be explicitly and uniquely specified by means of an address. (C) 610.5-1990w
(2) A location in a storage device that is uniquely specified by means of an address. (C) 610.10-1994w

storage medium Any device or recording medium into which data can be stored and held until some later time, and from which the entire original data can be obtained. (IA) [61]

storage protection (computers) An arrangement for preventing access to storage for either reading or writing, or both. (C) [20]

storage rate The frequency with which sampled signals are recorded in crashworthy nonvolatile memory. The event recorder may store any signal less often than it samples. (VT) 1482.1-1999

storage, reservoir *See:* reservoir storage.

storage schema In a CODASYL database, statements expressed in data storage definition language that describe storage areas, stored records, and any associated indices and access paths supporting the records and sets defined by a given schema. *See also:* CODASYL database. (C) 610.5-1990w

storage stack *See:* stack.

storage station (power operations) A hydroelectric generating station associated with a water storage reservoir. (PE/PSE) 858-1987s, 346-1973w

storage structure (A) The manner in which data structures are represented in storage. **(B)** The configuration of a database resident on computer storage devices after mapping the data elements of the logical structure of the database onto their respective physical counterparts. *Note:* The relationships and associations that provide the physical means for accessing the information stored in the database are preserved. (C) 610.5-1990

storage surface (storage tubes) The surface upon which information is stored. *See also:* storage tube. (ED) 158-1962w

storage temperature (1) (power supply) The range of environmental temperatures in which a power supply can be safely stored (for example, −40°C to +85°C). (AES/IA) [41], [12]
(2) (light-emitting diodes) The temperature at which the device, without any power applied, is stored. (ED) [127]

storage temperature range The range of temperatures over which the Hall generators may be stored without any voltage applied, or without exceeding a specified change in performance. (MAG) 296-1969w

storage time *See:* decay time; maximum retention time.

storage tube An electron tube into which information can be introduced and read at a later time. *Note:* The output may be an electric signal and.or a visible image corresponding to the stored information. (ED) 161-1971w, 158-1962w

storage tube display device A type of cathode ray tube display device that retains a display image on its surface in the form of a pattern of electric charges. *Synonyms:* storage display; display storage tube; direct-view storage tube. *Contrast:* refresh display device. (C) 610.10-1994w

storage unit The length of an addressable element of storage in the machine, measured in bits. (Every storage element has the same size.). *Note:* The storage unit is very likely to be one byte, but this is not a requirement. For example, it might be 32 or 64 bits. (C) 1003.5-1999

store (A) A device into which data can be placed, in which they can be retained, and from which they can be retrieved. *Note:* This term is the equivalent of the term storage in British (U.K.) usage. **(B)** To place data into a device as in definition (A). **(C)** To retain data in a device as in definition (A). (C) 162-1963, 610.10-1994
(2) (A) To place or retain data in a storage device. **(B) (software) (data management)** To copy computer instructions or data from a register to internal storage or from internal storage to external storage. *Contrast:* retrieve; load. *See also:* move; fetch. (C) 610.12-1990, 610.5-1990

store-and-forward Pertaining to communications where a message is received completely before beginning transmission onto the next node. (C) 610.7-1995

store-and-forward buffer (local area networks) A first-in-first-out (FIFO) buffer in the network repeater that can provide temporary storage for an entire message packet prior to retransmission. The buffer acts as a shift register and must hold an entire, full-length packet. *See also:* elasticity buffer. (C) 8802-12-1998

store-and-forward switched network A switched network in which the store-and-forward principle is used to handle transmissions between the sender and the recipient. (C) 610.7-1995

store-and-forward switching A method of switching whereby messages are transferred directly or with interim storage, each in accordance with its own address. *See also:* packet switching. (LM/COM) 168-1956w

store-and-forward switching system (telephone switching systems) A switching system for the transfer of messages, each with its own address or addresses, in which the message can be stored for subsequent transmission. (COM) 312-1977w

stored-energy indicator An indicator that visibly shows that the stored-energy mechanism is in the charged or discharged position. (SWG/PE) C37.100-1992

stored-energy operation Operation by means of energy stored in the mechanism itself prior to the completion of the operation and sufficient to complete it under predetermined conditions. *Note:* This kind of operation may be subdivided according to: (1) how the energy is stored (spring, weight, etc.), (2) how the energy originates (manual, electric, etc.), and (3) how the energy is released (manual, electric, etc.). (SWG/PE/IA/PSP) C37.100-1992, 1015-1997

stored logic (telephone switching systems) Instructions in memory arranged to direct the performance of predetermined functions in response to readout. (COM) 312-1977w

stored paragraph *See:* boilerplate text.

stored program (telephone switching systems) A program in memory that a processor can execute. (COM) 312-1977w

stored-program computer **(1)** A digital computer that, under control of internally stored instructions, can synthesize, alter, and store instructions as though they were data and can subsequently execute these new instructions. (C) [20], [85]
(2) A computer that is controlled by internally stored instructions that are treated as though they were data, and that can subsequently be executed. (C) 610.10-1994w

stored program control (telephone switching systems) A system control using stored logic. (COM) 312-1977w

stored record *See:* internal record.

stored-program switching system (telephone switching systems) An automatic switching system having stored program control. (COM) 312-1977w

storm guys Anchor guys, usually placed at right angles to direction of line, to provide strength to withstand transverse loading due to wind. *See also:* tower. (T&D/PE) [10]

storm loading The mechanical loading imposed upon the components of a pole line by the elements, that is, wind and/or ice, combined with the weight of the components of the line. *Note:* The United States has been divided into three loading districts, light, medium, and heavy, for which the amounts of wind and.or ice have been arbitrarily defined. *See also:* cable. (EEC/PE) [119]

STP (local area networks) A 150 Ω shielded balanced cable meeting the specifications in ISO/IEC 11801:1995. (C) 8802-12-1998

STR *See:* spanning tree route.

straggling, energy *See:* energy straggling.

straight air brake An arrangement of brakes whereby air is admitted from the main reservoir through a brake valve to the straight air pipe to the brake cylinders in the operating unit. *Note:* In most rail transit vehicle applications, an electropneumatic overlay is utilized to assist in the straight air brake command transmission. (VT) 1475-1999

straight air pipe A method of transmitting a pneumatic command from the active cab to the straight air brake equipment on each vehicle in the operating unit. (VT) 1475-1999

straightaway A one-way measurement requiring no path or connection in the opposite direction. (COM/TA) 743-1995

straight binary *See:* binary.

straight condensing turbine (control systems for steam turbine-generator units) All the steam enters the turbine at one pressure and all the steam leaves the turbine exhaust at a pressure below atmospheric pressure. (PE/EDPG) 122-1985s

straight-cut control system (numerically controlled machines) A system in which the controlled cutting action occurs only along a path parallel to linear, circular, or other machine ways. (IA) [61]

straightforward trunking (manual telephone switchboard system) That method of operation in which the A operator gives the order to the B operator over the trunk on which talking later takes place. (PE/EEC) [119]

straight insertion sort *See:* insertion sort.

straight joint (power cable joints) A cable joint used for connecting two lengths of cable, each of which consists of one or more conductors. (PE/IC) 404-1986s

straight-line code A sequence of computer instructions in which there are no loops. (C) 610.12-1990

straight-line coding **(1) (computers)** Coding in which loops are avoided by the repetition of parts of the coding when required. (C) [20], [85]
(2) (software) A programming technique in which loops are avoided by stating explicitly and in full all of the instructions that would be involved in the execution of each loop. *See also:* unwind. (C) 610.12-1990

straight line sort *See:* linear sort.

straight noncondensing turbine (control systems for steam turbine-generator units) All the steam enters the turbine at one pressure and all the steam leaves the turbine exhaust at a pressure equal to or greater than atmospheric pressure. (PE/EDPG) 122-1985s

straight radix sort A radix sort in which items are sorted repeatedly on successive digits within the numeric representation of the sort key, starting with the least significant digit. (C) 610.5-1990w

straight-seated bearing (rotating machinery) (cylindrical bearing) A journal bearing in which the bearing liner is constrained about a fixed axis determined by the supporting structure. *See also:* bearing. (PE) [9]

straight selection sort *See:* selection sort.

straight storage system (electric power supply) A system in which the electrical requirements of a car are supplied solely from a storage battery carried on the car. *See also:* axle-generator system. (PE/EEC) [119]

straight two-way merge sort A variation of the natural two-way merge sort in which the set to be sorted is repeatedly divided into two ordered subsets of length 2 to the power of *k*, where *k* is the number of passes made so far. *Contrast:* natural two-way merge sort. (C) 610.5-1990w

strain attachment *See:* dead-end.

strain-bus structure A bus structure comprised of flexible conductors supported by strain insulators. (PE/SUB) 605-1998

strain element (of a fuse) That part of the current-responsive element that is connected in parallel with the fusible element in order to relieve it of tensile strain. *Note:* The fusible element melts and severs first, and then the strain element melts during circuit interruption. *Synonym:* strain wire. (SWG/PE/SWG-OLD) C37.40-1993, C37.100-1992

strain insulator An insulator generally of elongated shape, with two transverse holes or slots. (EEC/IEPL) [89]

strain stick An insulating support tool used primarily to relieve mechanical loading at suspension and dead-end configurations so as to replace damaged insulators or hardware. (T&D/PE) 516-1995

strain wire *See:* strain element.

STRAND A concurrent logic programming language. (C) 610.13-1993w

strand (A) One of the wires, or groups of wires, of any stranded conductor. **(B)** One of a number of paralleled uninsulated conducting elements of a conductor which is stranded to provide flexibility in assembly or in operation. **(C)** One of a number of paralleled insulated conducting elements that constitute one turn of a coil in rotating machinery. The strands are usually separated electrically through all the turns of a multi-turn coil. Various types of transposition are commonly employed to reduce the circulation of current among the strands. A strand has a solid cross section, or it may be hollow to permit the flow of cooling fluid in intimate contact with the conductor (one form of "conductor cooling"). *See also:* rotor; conductor; stator. (PE) [9]

stranded conductor A conductor composed of a group of wires or of any combination of groups of wires. *Note:* The wires in a stranded conductor are usually twisted or braided together. *See also:* conductor. (T&D/PE) [10]

stranded wire *See:* stranded conductor.

strand insulation (rotating machinery) The insulation on a strand or lamination or between adjacent strands or laminations that comprise a conductor. (PE) [9]

strand restraining clamp An adjustable circular clamp commonly used to keep the individual strands of a conductor in place and to prevent them from spreading when the conductor is cut. *Synonyms:* plier clamp; vise grip; hose clamp. (T&D/PE) 524-1992r

strand-to-strand test (rotating machinery) A test that is designed to apply a voltage of specified amplitude and waveform between the strands of a coil for the purpose of determining the integrity of the strand insulation. (PE) [9]

strap, anode *See:* anode strap.

strapdown (accelerometer) (gyros) Direct-mounting of inertial sensors (without gimbals) to a vehicle to sense the linear and angular motion of the vehicle. (AES/GYAC) 528-1994

strapdown inertial navigation equipment (navigation aid terms) Inertial navigation equipment wherein the inertial sensors, (for example, gyros and accelerometers) are directly mounted to the vehicle, (eliminating the stable platform and gimbal system) to sense the linear and angular motion of the vehicle. *Notes:* 1. In this equipment, a computer utilizes gyro information to resolve the accelerations that are sensed along the carrier axes, and to refer these accelerations to an inertial frame of reference. Navigation is then accomplished in the same manner as in systems using a stable platform. 2. Also called strapped down. (AES/GCS) 172-1983w

strap key A pushbutton circuit controller that is biased by a spring metal strip and is used for opening or closing a circuit momentarily. (EEC/PE) [119]

strapping *See:* jumper; anode strap.

stratified language A language that cannot be used as its own metalanguage. Examples include FORTRAN, COBOL. *Contrast:* unstratified language. (C) 610.12-1990, 610.13-1993w

stratopause The upper boundary of the stratosphere. (AP/PROP) 211-1997

stratosphere That part of the Earth's atmosphere located above the troposphere in which the temperature remains constant or increases slightly with increasing height. The stratosphere extends to a height of around 50 km. (AP/PROP) 211-1997

straw line *See:* pilot line.

stray An element or occurrence usually not desired in a theoretical design, but unavoidable in a practical realization. For example, the relative proximity of wires can cause stray capacitance. *See also:* parasitic element. (CAS) [13]

stray capacitance (electric circuits) Capacitance arising from proximity of component parts, wires, and ground. *Note:* It is undesirable in most circuits, although in some high-frequency applications it is used as the tuning capacitance. In bridges and other measuring equipment, its effect must be eliminated by preliminary balancing out, or known and included in the results of any measurement performed. *See also:* measurement system. (IM/HFIM) [40]

stray current Currents or components that do not constitute information desired for measurement. Examples are currents due to the stray capacitance of an object to the ground plane, walls, etc. (T&D/PE) 516-1995

stray-current corrosion Corrosion caused by current through paths other than the intended circuit or by an extraneous current in the earth. *See also:* long-line current; noble potential; sacrificial protection; cathodic corrosion. (IA) [59]

stray light (illuminating engineering) (in the eye) Light from a source which is scattered onto parts of the retina lying outside the retinal image of the source. (EEC/IE) [126]

stray load loss (synchronous machines) The losses due to eddy currents in copper and additional core losses in the iron, produced by distortion of the magnetic flux by the load current, not including that portion of the core loss associated with the resistance drop. (PE) [9], [84]

stray losses (electronic power transformer) Those occurring in the core and case structure that result from the leakage flux of a transformer when supplying rated load current. (PEL/ET) 295-1969r

strays† Electromagnetic disturbances in radio reception other than those produced by radio transmitting systems. *See also:* radio transmitter. (EEC/PE) [119]

† Obsolete.

stream (1) An ordered sequence of characters, as described by the C Standard. (C/PA) 9945-2-1993
(2) The Physical Layer (PHY) encapsulation of a Media Access Control (MAC) frame. Depending on the particular PHY, the MAC frame may be modified or have information appended or prepended to it to facilitate transfer through the Physical Medium Attachment (PMA). Any conversion from a MAC frame to a PHY stream and back to a MAC frame is transparent to the MAC. (C/LM) 802.3-1998

streamer *See:* streamer mode.

streamer mode A repetitive corona discharge characterized by luminous filaments extending into the low electric field strength region near either a positive or a negative electrode, but not completely bridging the gap. (T&D/PE) 539-1990

stream flow The quantity rate of water passing a given point. *See also:* generating station. (T&D/PE) [10]

streaming (audio and electroacoustics) Unidirectional flow currents in a fluid that are due to the presence of acoustic waves. (SP) [32]

streaming cassette A magnetic tape cassette for a streaming tape drive. (C) 610.10-1994w

streaming potential (electrobiology) The electrokinetic potential gradient resulting from unit velocity of liquid forced to flow through a porous structure or past an interface. *See also:* electrobiology. (EMB) [47]

streaming tape drive A tape drive that does not come to a stop at each interrecord gap; rather the tape moves continuously past the read/write heads. *Note:* This type of tape drive is particularly appropriate for performing nonstop dumps or for restoring magnetic disks. *Synonym:* streamer. *Contrast:* start-stop tape drive. (C) 610.10-1994w

streetlighting luminaire (illuminating engineering) A complete lighting device consisting of a light source together with its direct appurtenances such as globe, reflector, refractor, housing, and such support as is integral with the housing. The pole, post, or bracket is not considered part of the luminaire. *Note:* Modern streetlighting luminaires contain the ballasts for high intensity discharge lamps where they are used. (EEC/IE) [126]

streetlighting unit (illuminating engineering) The assembly of a pole or lamp post with a bracket and a luminaire. (EEC/IE) [126]

strength-duration curve (medical electronics) (time-intensity) A graph of the intensity curve of applied electrical stimuli as a function of the duration just needed to elicit response in an excitable tissue. (EMB) [47]

strength of a sound source (strength of a simple source) The maximum instantaneous rate of volume displacement produced by the source when emitting a wave with sinusoidal time variation. *Note:* The term is properly applicable only to sources of dimension small with respect to the wavelength. (SP) [32]

strength-reduction factor *See:* resistance-reduction factor.

strength test A test to ensure that components of the fall protection system will not fail when subjected to a 22.2 kN (5000 pound) force per worker or to the design load of an engineered system. (T&D/PE) 1307-1996

STRESS *See:* STRuctural Engineering Systems Solver.

stress The nonspecific response of an organism to any demand upon it, whether pleasant or unpleasant, that results in certain biochemical changes. In popular usage, the harmful connotation is often assumed; i.e., excessive stress or distress. (T&D/PE) 539-1990

stress-accelerated corrosion Corrosion that is accelerated by stress. (IA) [59]

stress analysis (Class 1E battery chargers and inverters) An electrical and thermal design analysis of component applications in specific circuits under the specified range of service conditions. (PE/NP) 650-1979s

stress control coating The paint or tape on the outside of the groundwall insulation which extends several centimeters beyond the semiconductive slot coating in high voltage stator bars and coils. The stress control coating often contains silicon carbide particles which tend to linearize the electric field distribution along the coil or bar endturn. The stress control coating overlaps the semiconductive slot coating to provide electrical contact between them. (DEI) 1043-1996

stress corrosion cracking Spontaneous cracking produced by the combined action of corrosion and static stress (residual or applied). (IA) [59]

stressor An agent or stimulus that stems from fabrication or preservice and service conditions and can produce immediate degradation or aging degradation of a system, structure, or component. (PE/NP) 1205-1993

stress relief A predetermined amount of slack to relieve tension in component or lead wires. (EEC/AWM) [105]

stress test (Class 1E battery chargers and inverters) A type test performed on a sample equipment which "stresses" the equipment to the specified range of service conditions. (PE/NP) 650-1979s

stress testing Testing conducted to evaluate a system or component at or beyond the limits of its specified requirements. *See also:* boundary value. (C) 610.12-1990

strike deposit (A) (electroplating) A thin film of deposited metal to be followed by other coatings. *See also:* electroplating. **(B) (electroplating)** (bath) An electrolyte used to deposit a thin initial film of metal. *See also:* electroplating. (EEC/PE) [119]

striking (1) (arc) (spark) (gas) The process of establishing an arc or a spark. *See also:* discharge. (ED) [45], [84]
(2) (electroplating) The electrode position of a thin initial film of metal, usually at a high current density. *See also:* electroplating. (EEC/PE) [119]

striking current (gas tube) The starter-gap current required to initiate conduction across the main gap for a specified anode voltage. (ED) [45], [84]

striking distance (1) The shortest distance, measured through air, between parts of different polarities. (SWG/PE) C37.100-1992, [56]
(2) The shortest unobstructed distance measured through a dielectric medium such as liquid, gas, or vacuum between parts of different electric potential. (PE/TR) C57.12.80-1978r
(3) (outdoor apparatus bushings) The shortest tight string distance measured externally over the weather casing between the metal parts which have the operating line to ground voltage between them. (PE/TR) 21-1976
(4) The length of the final jump of the stepped leader as its potential exceeds the breakdown resistance of this last gap; found to be related to the amplitude of the first return stroke. (SUB/PE) 998-1996

String The IEEE 1451.1 representation of a sequence of human readable characters. (IM/ST) 1451.1-1999

string (1) (microprocessor operating systems parameter types) A sequence of characters. (C/MM) 855-1985s
(2) (software) A linear sequence of entities such as characters or physical elements. (C/SE) 729-1983s
(3) (A) (data management) A sequence of bits, characters, or other entities; for example, the bit string 0101010 or the character string XYZ. **(B) (data management)** Pertaining to data that contains a sequence as in definition (A). *Contrast:* arithmetic. *See also:* character string; bit string. (C) 610.5-1990
(4) An ordered sequence of zero or more bits, octets, or characters, accompanied by the length of the string. (C/PA) 1328-1993w, 1224-1993w, 1327-1993w
(5) (of capacitors) Capacitors connected in series between the line terminals.power systems relaying. (PE) C37.99-2000

string device (1) A logical input device used to provide a character string to a graphics system. (C) 610.6-1991w
(2) An input device that is used to specify or detect a character string. For example, an alphanumeric keyboard. (C) 610.10-1994w

stringing (conductor stringing equipment) The pulling of pilot lines, pulling lines, and conductors over travelers supported on structures of overhead transmission lines. Quite often, the entire job of stringing conductors is referred to as stringing operations, beginning with the planning phase and terminating after the conductors have been installed in the suspension clamps. (T&D/PE) 524a-1993r, 1048-1990, 524-1992r

stringing block *See:* traveler.

stringing section *See:* sag section.

stringing sheave *See:* traveler.

stringing, slack *See:* slack stringing.

stringing traveler *See:* traveler.

stringing, tension *See:* tension stringing.

StriNg-Oriented symBOlic Language (SNOBOL) A programming language designed for use in string manipulation tasks such as language translation, program compilation and combinatorial problems. (C) 610.13-1993w

string-shadow instrument An instrument in which the indicating means is the shadow (projected or viewed through an optical system) of a filamentary conductor, the position of which in a magnetic or an electric field depends upon the measured quantity. *See also:* instrument. (EEC/PE) [119]

strip (1) (electroplating) A solution used for the removal of a metal coating from the base metal. *See also:* electroplating. (EEC/PE) [119]

(2) To replace a received nonidle symbol by an idle symbol and hence to remove it from transmission on a ringlet. For example, a send packet is stripped by the receiving port of an agent or the target and replaced by idles (most of which may be consumed in the process of emptying the bypass FIFO) and an echo. Similarly an echo is stripped by the receiving port of an agent or the source and replaced by idles. (C/MM) 1596-1992

strip-chart recorder (analog computer) (hybrid computer linkage components) A recorder in which one or more records are made simultaneously as a function of time. *See also:* electronic analog computer. (C) 165-1977w, 166-1977w

stripe-to-gap ratio The ratio of the metallized surface to the free surface within the interdigital transducer. (UFFC) 1037-1992w

stripline A class of planar transmission line characterized by one or more thin conducting strips of finite width parallel to and approximately midway between two extended conducting ground planes. The space between the strips and the ground planes is filled by a homogeneous insulating medium. (MTT) 1004-1987w

strip line (waveguide) A transmission line consisting of a strip conductor above or between extended parallel conducting surfaces. Some common examples of such transmission lines are:

a) Partially-shielded strip transmission line: a strip conductor above a single ground plane;
b) Shielded strip transmission line: a strip conductor between two ground planes.

See also: slab line. (MTT) 146-1980w

stripper tank (electrorefining) An electrolytic cell in which the cathode deposit, for the production of starting sheets, is plated on starting-sheet blanks. *See also:* electrorefining. (EEC/PE) [119]

stripping (1) (electroplating) (mechanical) The removal of a metal coating by mechanical means. (PE/EEC) [119]
(2) (chemical) The removal of a metal coating by dissolving it. (EEC/PE) [119]
(3) (electrolytic) The removal of a metal coating by dissolving it or an underlying coating anodically with the aid of a current. *See also:* electroplating. (EEC/PE) [119]
(4) The action of a station removing the frames it has transmitted from the ring. (C/LM) 8802-5-1998

stripping compound (electrometallurgy) Any suitable material for coating a cathode surface so that the metal electro deposited on the surface can be conveniently stripped off in sheets. *See also:* electrowinning. (EEC/PE) [119]

strip terminals (rotating machinery) A form of terminal in which the ends of the machine winding are brought out to terminal strips mounted integral with the machine frame or assembly. (PE) [9]

strip-type transmission line (waveguides) A transmission line consisting of a conductor above or between extended conducting surfaces. *See also:* unshielded strip transmission line; shielded strip transmission line. (AP/ANT) [35]

strobe (1) (A) A pulse used to cause a register to assume and retain the state indicated by its data inputs. **(B)** A pulse used as an input to a trigger circuit. (C) 610.10-1994
(2) To record or measure the state of a particular node at an instant in time. Strobing will have a skew associated with it. (SCC20) 1445-1998

strobing (pulse terminology) A process in which a first pulse of relatively short duration interacts with a second pulse or other event of relatively longer duration to yield a signal which is indicative (typically, proportional to) the magnitude of the second pulse during the first pulse. (IM/WM&A) 194-1977w

stroboscopic lamp (illuminating engineering) A flashtube designed for repetitive flashing. (EEC/IE) [126]

stroboscopic tube A gas tube designed for the periodic production of short light flashes. *See also:* gas tube. (ED/C) [45], [85]

stroke (1) (cable plowing) Peak to peak displacement of the plow blade tip. (T&D/PE) 590-1977w
(2) In character recognition, a straight line or arc used as a segment of a graphic character. (C) 610.2-1987
(3) A straight line or arc that is a segment of a graphic character. *See also:* keystroke. (C) 610.10-1994w, 610.6-1991w

stroke centerline In character recognition, a line midway between two stroke edges. (C) 610.2-1987

stroke character generator A character generator that creates characters composed of line segments. *Contrast:* matrix character generator. (C) 610.6-1991w

stroke device (1) A logical input device used to provide a sequence of data points as the input data to a graphics system. (C) 610.6-1991w
(2) An input device that provides a set of coordinates that record the path of the device. (C) 610.10-1994w

stroke edge In character recognition, the line of discontinuity between a side of a stroke and the background, obtained by averaging, over the length of the stroke, the irregularities resulting from the printing and detecting process. (C) 610.2-1987

stroke font *See:* vector font.

stroker display *See:* random-scan display device.

stroke speed (facsimile) (scanning or recording line frequency) The number of times per minute, unless otherwise stated, that fixed line perpendicular to the direction of scanning is crossed in one direction by a scanning or recording spot. *Note:* In most conventional mechanical systems this is equivalent to drum speed. In systems in which the picture signal is used while scanning in both directions, the stroke speed is twice the above figure. *See also:* scanning; recording. (COM) 168-1956w

stroke width (1) The thickness of the lines that make up a character, usually expressed as a proportion of the character's height. (PE/NP) 1289-1998
(2) In character recognition, the distance between two stroke edges, measured perpendicular to the stroke centerline. (VT) 1477-1998

strong sequential consistency A state exhibited by a system when each participating cache in the system observes all modifications to lines within itself in the same order as all other participating caches in the system. *Note:* Futurebus+ allows modules to issue transactions that dynamically choose between following a weakly consistent behavior model (which implies greater concurrency and hence higher performance) and strongly consistent behavior models (which may be necessary to ensure correct operation of an algorithm written by a programmer not cognizant of the concurrency possible in different parts of a system). *See also:* weak sequential consistency. (C/BA) 10857-1994

strong typing (software) A feature of some programming languages that requires the type of each data item to be declared, precludes the application of operators to inappropriate data types, and prevents the interaction of data items of incompatible types. (C) 610.12-1990

STRUctural Design language An extension of STRESS, used for analysis and design of structures. (C) 610.13-1993w

STRuctural Engineering Systems Solver (STRESS) A problem-oriented programming language used in structural engineering. *See also:* ICES. (C) 610.13-1993w

structurally dual networks A pair of networks such that their branches can be marked in one-to-one correspondence so that any mesh of one corresponds to a cut-set of the other. Each network of such a pair is said to be the dual of the other. *See also:* network analysis.

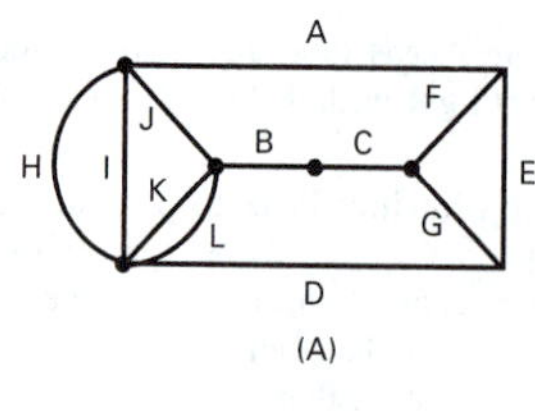

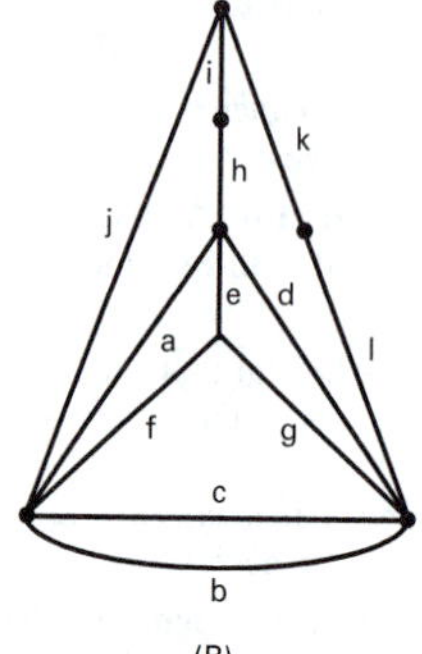

structurally dual networks

(Std100) 270-1966w

structurally symmetrical network A network that can be arranged so that a cut through the network produces two parts that are mirror images of each other. *See also:* network analysis. (Std100) 270-1966w

structural model A model of the physical or logical structure of a system; for example, a model that represents a computer network as a set of boxes connected by communication lines. *Contrast:* process model. (C) 610.3-1989w

structural pattern recognition An approach to pattern recognition in which patterns are represented in terms of primitives and relationships among those primitives in order to describe and classify pattern structure. *See also:* syntactic pattern recognition. (C) 610.4-1990w

structural resolution (color picture tubes) The resolution as limited by the size and shape of the screen elements. *See also:* television. (BT/AV) [34]

structural return loss A term used to describe the structural integrity of the coaxial cable. Structural return loss defines impedance variation due to deformed coaxial cable concentricity. (LM/C) 802.7-1989r

structural testing Testing that takes into account the internal mechanism of a system or component. Types include branch testing, path testing, statement testing. *Synonyms:* glass-box testing; white-box testing. *Contrast:* functional testing. (C) 610.12-1990

structural vector A pattern generated to exercise a device's structural elements (e.g., scan-based ATPG test generation). *Contrast:* functional vector. (C/TT) 1450-1999

structure Material assembled to support conductors or associated apparatus, or both, used for transmission and distribution of electricity (e.g., service pole, tower). (T&D/PE) 516-1995

structure base ground (conductor stringing equipment) A portable device designed to connect (bond) a metal structure to an electrical ground. Primarily used to provide safety for personnel during construction, reconstruction or maintenance operations. *Synonyms:* ground chain; butt ground. (PE/T&D) 524a-1993r, 524-1992r, 1048-1990

structure chart A diagram that identifies modules, activities, or other entities in a system or computer program and shows how larger or more general entities break down into smaller, more specific entities. *Note:* The result is not necessarily the same as that shown in a call graph. *Synonym:* hierarchy chart. *Contrast:* call graph.

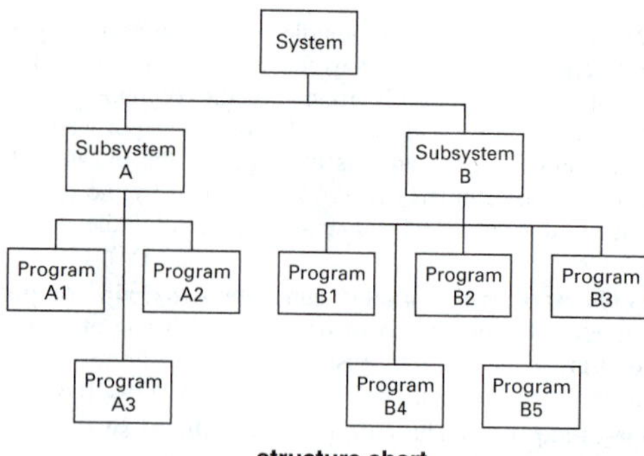

structure chart

(C) 610.12-1990

structure clash In software design, a situation in which a module must deal with two or more data sets that have incompatible data structures. *See also:* order clash; data structure-centered design. (C) 610.12-1990

structure conflict A line so situated with respect to a second line that the overturning of the first line will result in contact between its supporting structures or conductors and the conductors of the second line, assuming that no conductors are broken in either line. (NESC/T&D) C2-1997, C2.2-1960

structure constant ($C^2{}_n$) A measure of the turbulent fluctuations of the refractive index of the atmosphere. (AP/PROP) 211-1997

structure, crossing *See:* crossing structure.

structured A postal O/R address that specifies the postal address of a user by means of several attributes. Its structure is prescribed in some detail. (C/PA) 1224.1-1993w

structured design (A) (software) Any disciplined approach to software design that adheres to specified rules based on principles such as modularity, top-down design, and stepwise refinement of data, system structures, and processing steps. *See also:* object-oriented design; input-process-output; transform analysis; modular decomposition; stepwise refinement; rapid prototyping; transaction analysis; data structure-centered design. **(B) (software)** The result of applying the approach in definition (A). (C) 610.12-1990

structure designer A party who designs the structure based on criteria given by a line designer. The structure designer could be an owner, an agent acting for the owner, or a manufacturer or fabricator. (T&D/PE) 1025-1993r, 951-1996

structured program (software) A computer program constructed of a basic set of control structures, each having one entry and one exit. The set of control structures typically includes: sequence of two or more instructions, conditional selection of one of two or more sequences of instructions, and repetition of a sequence of instructions. *See also:* structured design. (C) 610.12-1990

structured programming (software) Any software development technique that includes structured design and results in the development of structured programs. (C) 610.12-1990

structured programming language A programming language that provides structured program constructs such as single-entry-single-exit sequences, branches, and loops, and facilitates the development of structured programs. *See also:* block-structured language. (C) 610.13-1993w, 610.12-1990

Structured Query Language (SQL) A query language designed for accessing data and performing queries on relational databases, standardized by ASC ×3. (C) 610.13-1993w

structure, snub *See:* snub structure.

structures, safety class *See:* safety class structures.

stub (1) (A) (software) A skeletal or special-purpose implementation of a software module, used to develop or test a module that calls or is otherwise dependent on it. **(B) (software)** A computer program statement substituting for the body of a software module that is or will be defined elsewhere. (C) 610.12-1990

(2) The signal path on the module connecting the BTL transceiver to the Futurebus+ connector. (C/BA) 896.2-1991w

stub antenna A short, thick monopole. (AP/ANT) 145-1993

stub card A card that has a separable stub attached to a regular punch card. *Note:* May also be scored. (C) 610.10-1994w

stub feeder A feeder that connects a load to its only source of power. *Synonym:* radial feeder. (SWG/PE) C37.100-1992, [56]

stub length Stub length is measured from the point at which the connector assembly connects to the PC board to the pad where the transceiver lead is soldered to the PC board. (C/BA) 14536-1995

stub-multiple feeder A feeder that operates as either a stub or a multiple feeder. (SWG/PE) C37.100-1992

stub shaft (rotating machinery) A separate shaft not carried in its own bearings and connected to the shaft of a machine. *See also:* rotor. (PE) [9]

stub-supported coaxial A coaxial whose inner conductor is supported by means of short-circuited coaxial stubs. *See also:* waveguide. (EEC/PE) [119]

stub tuner A stub that is terminated by movable short-circuiting means and used for matching impedance in the line to which it is joined as a branch. *See also:* waveguide. (AP/ANT) [35]

stub, waveguide *See:* waveguide stub.

stuck-at fault A failure in a logic circuit that causes a signal connection to be fixed at 0 or 1 regardless of the operation of the circuitry that drives it. (TT/C) 1149.1-1990

stud (of a switching device) A rigid conductor between a terminal and a contact. (SWG/PE) C37.100-1992

stuffing bits Bits inserted into a frame to compensate for timing differences in constituent lower rate signals. (COM/TA) 1007-1991r

stuffing box (watertight gland) A device for use where a cable passes into a junction box or other piece of apparatus and is so designed as to render the joint watertight. (T&D/PE) [10]

style Set of editorial conventions covering grammar, terminology, punctuation, capitalization, etc, of a software user document. (C/SE) 1063-1987r

stylus (1) (electroacoustics) A mechanical element that provides the coupling between the recording or the reproducing transducer and the groove of a recording medium. *See also:* phonograph pickup. (SP) [32]
(2) (computer graphics) A pointing device, resembling a pencil, that is used with a data tablet to determine locations using an electrical sensing mechanism. (C) 610.6-1991w
(3) A pointing device used with a data tablet as a locator. Examples include light pens, sonic pens, and voltage pencils. *See also:* twinkle box. (C) 610.10-1994w

stylus drag (electroacoustics) An expression used to denote the force resulting from friction between the surface of the recording medium and the reproducing stylus. *See also:* phonograph pickup.

stylus force (electroacoustics) The vertical force exerted on a stationary recording medium by the stylus when in its operating position. *See also:* phonograph pickup. (SP) [32]

SU *See:* shared unmodified; system unit.

SUB *See:* substitute character.

subaction (1) A component of a transaction; a request or a response. (C/MM) 1596-1992
(2) One of the two components in a transaction; a transaction consists of request and response subactions. (C/MM) 1212-1991s
(3) A complete link layer operation: arbitration, packet transmission and acknowledgment. The arbitration may be missing when a node already controls the bus, and the acknowledge is not present for subactions with broadcast addresses or for isochronous subactions. (C/MM) 1394-1995
(4) A complete link layer operation: optional arbitration, packet transmission, and optional acknowledgment. (C/MM) 1394a-2000

subaction gap (1) The period of idle bus between subactions. There is no gap between the request and response subaction of a concatenated split transaction. (C/MM) 1394-1995
(2) For an asynchronous subaction, the period of idle bus that precedes arbitration. (C/MM) 1394a-2000

subaddress (subroutines for CAMAC) The symbol a represents an integer which is the subaddress component of a CAMAC address. (NPS) 758-1979r

subassembly (1) (A) One or more compartments that comprise the gas-insulated substation assembly. **(B) (electric and electronics parts and equipment)** Two or more basic parts which form a portion of an assembly or a unit, replaceable as a whole, but having a part or parts which are individually replaceable. *Notes:* 1. The application, size, and construction of an item may be factors in determining whether an item is regarded as a unit, an assembly, a subassembly, or a basic part. A small electric motor might be considered as a part if it is not normally subject to disassembly. 2. The distinction between an assembly and a subassembly is not always exact: an assembly in one instance may be a subassembly in another where it forms a portion of an assembly. Typical examples: filter network, terminal board with mounted parts. (SWG/SUB/PE/GSD) C37.122-1983, C37.100-1992, 200-1975
(2) An element of the physical or system architecture, specification tree, and system breakdown structure that is subordinate to a complex component, and is comprised of two or more subcomponents. (C/SE) 1220-1994s
(3) Items that have an identifiable function. Subassemblies are not completed equipment or individual components. (SPD/PE) C62.38-1994r

subcarrier (facsimile) A carrier that is applied as a modulating wave to modulate another carrier. (COM) 168-1956w

subclass (1) A class that is defined by specializing another class. The subclass typically adds features to the other class, called the superclass, or uses the facilities of the superclass to implement a more specific set of interfaces. (C) 1295-1993w
(2) One of the classes, designated as such, whose attribute types are a superset of those of another class. (C/PA) 1328-1993w, 1327-1993w, 1224-1993w
(3) A specialization of one or more superclasses. Each instance of a subclass is an instance of each superclass. A subclass typically specifies additional, different responsibilities to those of its superclasses or overrides superclass responsibilities to provide a different realization. (C/SE) 1320.2-1998

subclass cluster (A) A set of one or more generalization structures in which the subclasses share the same superclass and in which an instance of the superclass is an instance of no more than one subclass. A cluster exists when an instance of the superclass can be an instance of only one of the subclasses in the set, and each instance of a subclass is an instance of the superclass. **(B)** A set of one or more mutually exclusive specializations of the same generic entity. (C/SE) 1320.2-1998

subclass responsibility A designation that a property of a class must be overridden in its subclasses, i.e., the designation given to a property whose implementation is not specified in this class. A property that is a subclass responsibility is a specification in the superclass of an *interface* that each of its subclasses must provide. A property that is designated as a subclass responsibility has its *realization* deferred to the subclass(es) of the class. (C/SE) 1320.2-1998

subclutter visibility The ratio by which the target echo power may be weaker than the coincident clutter echo power and still be detected with specified detection and false-alarm probabilities. *Note:* Target and clutter powers are measured on a single pulse return and all target radial velocities are assumed equally likely. (AES) 686-1997

subcomponent (1) A part or portion of a component which is relevant for quantifying exposure to outage occurrences, or failures, or both, or for identifying the cause of an outage occurrence or failure. (PE/PSE) 859-1987w
(2) An element of the physical or system architecture, specification tree, and system breakdown structure that is subordinate to a noncomplex component, or a subassembly, and is comprised of two or more parts. (C/SE) 1220-1994s

subcontractor A party having a direct contract with the constructor for performing work covered by the Contract Documents, when the constructor is not the owner. (T&D/PE) 951-1996

subdatabase A subset of the data contained in a database as used for a specific type of application or system. (C) 610.5-1990w

subdirectory entry A read-only memory (ROM) entry that specifies the address of another ROM subdirectory. (C/BA) 896.10-1997, 896.2-1991w

subdivided capacitor (condenser box) A capacitor in which several capacitors known as sections are so mounted that they may be used individually or in combination. (Std100) 270-1966w

subfactor *See:* quality subfactor.

subfeeder A feeder originating at a distribution center other than the main distribution center and supplying one or more branch-circuit distribution centers. *See also:* feeder. (EEC/PE) [119]

subframe overhead (OH) 12-byte SFODB Overhead structure used to transfer configuration and status information between the control fiber-optic bus interface unit (CFBIU) and fiber-optic bus interface units (FBIUs) on the network. (C/BA) 1393-1999

subgraph (data management) A graph consisting of a subset of nodes from a larger graph. See the figure below.

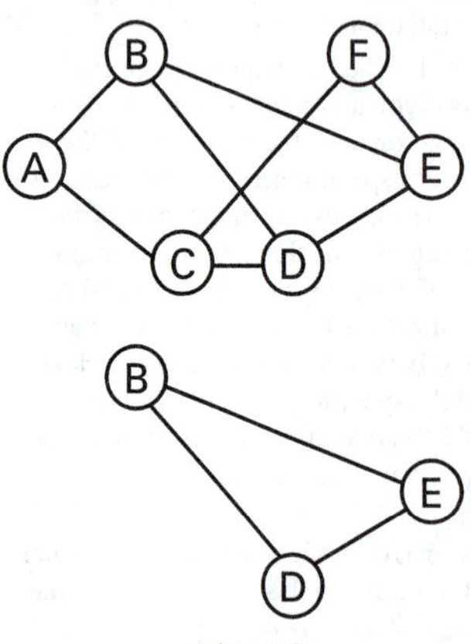

subgraph

(C) 610.5-1990w

Subgroup *See:* Remote Bridge Subgroup.

Subgroup Port A Virtual Port by which a Remote Bridge attaches to a Group consisting of two or more Subgroups. *Note:* Every Virtual Port is either a Virtual LAN Port or a Subgroup Port. In a Group that is not a Virtual LAN, every Bridge attaches to the Group by at least two Subgroup Ports. (C/LM) 802.1G-1996

subharmonic (data transmission) A sinusoidal quantity having a frequency which is an integral submultiple of the fundamental frequency of a periodic quantity to which it is related. For example, a wave the frequency of which is half the fundamental frequency of another wave is called the second subharmonic of that wave. (PE) 599-1985w

subharmonic detector A device that detects subharmonic current of specified frequency and duration and initiates an alarm signal or corrective action. (T&D/PE) 824-1994

subharmonic protector (series capacitor) A device to detect subharmonic current of a specified frequency and duration to initiate closing of the capacitor bypass switches. (T&D/PE) [26]

subject An active entity in a system that causes information to flow between objects or otherwise affects the system state. Typically, subjects include users, processes, and processing elements. (C/BA) 896.3-1993w

subject copy (facsimile) The material in graphic form that is to be transmitted for facsimile reproduction. *See also:* facsimile. (COM) 168-1956w

subject domain (1) The MD that contains the MTA embodied by the client and service of the MT interface. (C/PA) 1224.1-1993w
(2) An area of interest or expertise. The responsibilities of a subject domain are an aggregation of the responsibilities of a set of current or potential named classes. A subject domain may also contain other subject domains. A subject domain encapsulates the detail of a view. (C/SE) 1320.2-1998

subject domain responsibility A generalized concept that the analyst discovers by asking "in general, what do instances in this subject domain need to be able to do or to know?" The classes and subject domains in a subject domain together supply the knowledge, behavior, and rules that make up the subject. These notions are collectively referred to as the subject domain's responsibilities. Subject domain responsibilities are not distinguished as sub-domains or classes during the early stages of analysis. (C/SE) 1320.2-1998

subjective brightness (illuminating engineering) The subjective attribute of any light sensation giving rise to the percept of luminous magnitude, including the whole scale of qualities of being bright, light, brilliant, dim, or dark. *Note:* The term brightness often is used when referring to the measurable luminance. While the context usually makes it clear as to which meaning is intended, the preferable term for the photometric quality is luminance, thus reserving brightness for the subjective sensation. *See also:* luminance; saturation. (EEC/IE) [126]

subject matter expert (SME) An individual knowledgeable in the subject area being trained or tested. (DIS/C) 1278.3-1996

subject message When used in reference to a communique, the communique, if it is a message; or any of the messages denoted by the communique, if it is a probe. (C/PA) 1224.1-1993w

sublayer A subdivision of a layer in the OSI reference model. *See also:* physical layer; presentation layer; data link layer; session layer; entity layer; logical link control sublayer; application layer; client layer; network layer; transport layer; medium access control sublayer. (LM/C) 610.7-1995, 8802-6-1994

submarine cable A cable designed for service under water. *Note:* Submarine cable is usually a lead-covered cable with a steel armor applied between layers of jute. (T&D/PE) [10]

submaster *See:* master remote unit.

submaster station A station that can perform as a master station on one message transaction and as a remote station on another message transaction. *Note:* Examples of station equipments include

- *Hardwired.* Station supervisory equipment that is comprised entirely of wired-logic elements.
- *Firmware.* Station supervisory equipment that uses hardware logic programmed routines in a manner similar to a computer. the routines can only be modified by physically exchanging logic memory elements.
- *Programmable.* Station supervisory equipment that uses software routines.

(SWG/PE/SUB) C37.100-1992, C37.1-1994

submerged-resistor induction furnace A device for melting metal comprising a melting hearth, a depending melting channel closed through the hearth, a primary induction winding, and a magnetic core which links the melting channel and the primary winding. *See also:* induction heating. (IA) 54-1955w

submersible (rotating machinery) So constructed as to be successfully operable when submerged in water under specified conditions of pressure and time.
(SWG/IA/PE/ICTL/MT/TR) 45-1983s, [9], C37.100-1992, C57.12.80-1978r

submersible enclosure An enclosure constructed so that the equipment within it will operate successfully when submerged in water under specified conditions of submergence depth and time. (IA/MT) 45-1998

submersible entrance terminals (cable-heads) (of distribution oil cutouts) A hermetically sealable entrance terminal for the connection of cable having a submersible sheathing or jacket.
(SWG/PE) C37.100-1992, C37.40-1993

submersible fuse (1) (subway oil cutout) (high-voltage switchgear) A fuse that is so constructed that it will operate successfully when submerged in water under specified conditions of pressure and time. (SWG/PE) C37.40-1993
(2) *See also:* submersible; fuse. (SWG/PE) C37.100-1992

submersible transformer (power and distribution transformers) A transformer so constructed as to be successfully operable when submerged in water under predetermined conditions of pressure and time. (PE/TR) C57.12.80-1978r

submillimeter 300 GHz to 3 THz. *See also:* radio spectrum.
(AP/PROP) 211-1997

submission The process by which a client requests that a batch server create a job via a Queue Job Request to perform a specified computational task. (C/PA) 1003.2d-1994

submission queue The database that the client of the MA interface uses to convey objects to the service.
(C/PA) 1224.1-1993w

submodel *See:* subschema.

subnetwork (1) A functional unit comprised of a single dual bus pair and those access units (AUs) attached to it. Subnetworks are physically formed by connecting adjacent nodes with transmission links.
(LM/EMB/C/MIB) 1073.3.1-1994, 8802-6-1994
(2) A set of one or more intermediate open systems that provide relaying and through which end systems may communicate. It is a representation within the OSI model of a real network such as a carrier network, a provider network, or a LAN. It can also be defined as a collection of equipment and physical media that forms an autonomous whole and that can be used to interconnect real systems for the purpose of communication. (C/LM) 802.9a-1995w

subnetwork configuration The topological arrangement of nodes to form a subnetwork. In normal operation a DQDB subnetwork can have one of two configurations, open Dual Bus or looped Dual Bus. (LM/C) 8802-6-1994

subnetwork service The service supported by the subnetwork access protocol.
(C/LM/COM) 802.9a-1995w, 8802-9-1996

subnormality (electrical depression) (electrobiology) The state of reduced electrical sensitivity after a response or succession of responses. *See also:* excitability. (EMB) [47]

subnormal number (mathematics of computing) A non-zero floating-point number whose exponent is the precision's minimum and whose leading significant digit is zero.
(C) 1084-1986w

subobject An immediate subobject of an object or of one of its subobjects applied recursively.
(C/PA) 1328-1993w, 1327-1993w, 1224-1993w

subordinate The converse of superior.
(C/PA) 1328.2-1993w, 1224.2-1993w, 1326.2-1993w, 1327.2-1993w

subpanel (1) (photoelectric converter) Combination of photoelectric converters in parallel mounted on a flat supporting structure. *See also:* semiconductor. (AES) [41]
(2) (solar cells) A combination of solar cells in series.parallel matrix to provide current at array (bus) voltage.
(AES/SS) 307-1969w

subpost car frame (elevators) A car frame all of whose members are located below the car platform. *See also:* hoistway.
(PE/EEC) [119]

subproduct A software object that is a grouping of software filesets and other subproducts within a product.
(C/PA) 1387.2-1995

subprogram (1) (software) A separately compilable, executable component of a computer program. *Note:* The terms "routine," "sub-program," and "subroutine" are defined and used differently in different programming languages; the preceding definition is advanced as a proposed standard. *See also:* coroutine; main program; routine; subroutine.
(C) 610.12-1990
(2) (test, measurement, and diagnostic equipment) A part of a larger program which can be converted into machine language independently. (MIL) [2]

subrack (VSB) A rigid framework that provides mechanical support for boards inserted into the backplane, ensuring that the connectors mate properly and that adjacent boards do not contact each other. It also guides the cooling airflow through the system, and ensures that inserted boards are not disengaged from the backplane due to vibration.
(C/MM/BA) 1096-1988w, 1014-1987

subreflector A reflector other than the main reflector of a multiple-reflector antenna. (AP/ANT) 145-1993

subrefraction Refraction for which the vertical gradient of refractivity is greater (less negative) than the standard gradient of refractivity. (AP/PROP) 211-1990s

subrefractive atmosphere *See:* refractive index gradient.

subremote unit (SRU) A physical device (for example, peripheral boards, RTUs, meters, or other intelligent electronic devices) that collects data, processes it in some way, and communicates it to an MRU. SRUs are able to respond to commands from MRUs. (PE/SUB) 1379-1997

subroutine (1) (A) (electronic computation) In a routine, a portion that causes a computer to carry out a well-defined mathematical or logic operation. **(B) (electronic computation)** A routine that is arranged so that control may be transferred to it from a master routine and so that at the conclusion of the subroutine, control reverts to the master routine. *Note:* Such a subroutine is usually called a closed subroutine. A single routine may simultaneously be both a subroutine with respect to another routine and a master routine with respect to a third. Usually control is transferred to a single subroutine from more than one place in the master routine and the reason for using the subroutine is to avoid having to repeat the same sequence of instructions in different places in the master routine. (MIL) [2], 270-1966
(2) (software) A routine that returns control to the program or subprogram that called it. *Note:* The terms "routine," "sub-program," and "subroutine" are defined and used differently in different programming languages; the preceding definition is advanced as a proposed standard. *Contrast:* coroutine. *See also:* open subroutine; closed subroutine. (C) 610.12-1990
(3) *See also:* procedure. (C/MM) 1178-1990r

subroutine, open *See:* open subroutine.

subroutine trace A record of all or selected subroutines or function calls performed during the execution of a computer program and, optionally, the values of parameters passed to and returned by each subroutine or function. *See also:* subroutine trace; retrospective trace; symbolic trace; variable trace; execution trace. (C) 610.12-1990

subschema (A) A subset of a schema that defines a view of the database that is needed by one or more application programs. *Synonym:* submodel. **(B)** A description of the logical structure of a record in a database. (C) 610.5-1990

subscriber carrier (telephone loop performance) A system that multiplexes customer signals to achieve pair gain in the loop plant. Usually, it consists of (1) An end office terminal (EOT); it interfaces with the end office (EO) through analog line appearances, one per each integrated loop. If the carrier is a digital system integrated in a digital end office, this terminal and its interfaces are replaced by much simpler all-digital equipment that may be integrated into the switching system. (2) A remote terminal (RT); it interfaces with the cable pairs to the customers' premises through analog interfaces, one per each implemented loop. (3) A transmission medium between the EOT (or EO in a digital integrated system) and the RT; it provides a control channel for internal EOT/EO-to-RT communication and communciation channels for customer traffic. A nonconcentrated system has as many customer channels as implemented loops, with fixed loop-channel assignments. A concentrated system has fewer channels than implemented loops, and changes loop-channel assignments to accomodate changing traffic patterns. (COM/TA) 820-1984r

subscriber equipment (protective signaling) That portion of a system installed in the protected premises or otherwise supervised. *See also:* protective signaling. (EEC/PE) [119]

subscriber line (data transmission) A telephone line between a central office and a telephone station, private branch exchange, or other end equipment. (PE) 599-1985w

subscriber loop *See:* subscriber's line.

subscriber multiple A bank of jacks in a manual switchboard providing outgoing access to subscriber lines, and usually having more than one appearance across the face of the switchboard. (EEC/PE) [119]

Subscriber object Any object that receives publications from the network via an Object of class `IEEE1451_SubscriberPort`. (IM/ST) 1451.1-1999

Subscriber Port An instance of the class `IEEE1451_SubscriberPort` or of a subclass thereof. (IM/ST) 1451.1-1999

subscriber's drop A wire that runs from a cable terminal or distribution point to the subscriber's premises. (C) 610.7-1995

subscriber set (customer set) An assembly of apparatus for use in originating or receiving calls on the premises of a subscriber to a communication or signaling service. *See also:* voice-frequency telephony. (EEC/PE) [119]

subscriber's line (telephony) A link between a local exchange and a telephone set, a private telephone system, or another terminal using signals compatible with the telephone network. According to evolving practice in North America, the subscriber loop may be referred to as a local exchange access line. *Note:* Most subscriber loops are physical pairs. Subscriber loops may also be provided by means of a radio link, associated "go" and "return" channels in a multiplex system, or line sections allocated by a line concentrator. For purposes of this standard, only that part of the line connecting the telephone set and the feeding bridge is considered to be part of the local sending or local receiving system. (COM/TA) 823-1989w

subscriber's loop *See:* local loop.

subscriber's telephone line *See:* subscriber's line.

subscriber trunk dialing *See:* direct distance dialing.

subscript A symbol that is associated with the name of a set to identify a particular subset or element of the set. (C) 610.5-1990w

Subscription Domain A Domain identified by a subscriber for a specific publication. *See also:* Domain. (IM/ST) 1451.1-1999

Subscription Qualifier A configurable identifier used to help define the selection of received publications on the basis of publication's Publication Topic in a publish-subscribe communication. Specifically, a value having datatype `SubscriptionQualifier`. For a Subscriber Port, the operation `GetSubscriptionQualifier` returns a value, `subscription_qualifier`, that has the same value as Port's Subscription Qualifier. (IM/ST) 1451.1-1999

subset A dialect of a particular language that varies from its referenced standard language such that its capabilities include some, but not all, the capabilities of the referenced language. For example, TINT is a subset of JOVIAL. *Contrast:* extension. (C) 610.13-1993w

subset entity Entities that are related to and dependent on other primary entity sets called parent entity sets. (PE/EDPG) 1150-1991w

subshell A shell execution environment, distinguished from the main or current shell execution environment by the attributes described in 3.12. (C/PA) 9945-2-1993

subsidence *See:* attenuation; damping.

subsidence ratio (automatic control) A measure of the damping of a second-order linear oscillation, resulting from step or ramp forcing, expressed as the greater divided by the lesser of two successive excursions in the same direction from an ultimate steady-state value. *Note:* The term is also used loosely to describe the ratio of the first two consecutive peaks of any damped oscillation. (PE/EDPG) [3]

subsidiary communications authorization subcarrier modulation The FCC permits broadcasters to transmit privileged information and control signals on subcarriers as specified under the SCA but only when transmitted in conjunction with broadcast programming. *Notes:* 1. With monophonic broadcasting, the SCA service may use from 20 to 75 kHz with no restriction on the number of subcarriers, but the total SCA modulation of the RF (radio frequency) carrier must not exceed 30 percent and the crosstalk into the main channel must be at least 60 dB down. 2. With stereophonic broadcasting, the SCA service is limited to 53−75 kHz, 10 percent modulation of the carrier, and must still comply with the 60 dB crosstalk ratio. A 67 kHz subcarrier with ±6 kHz modulation is often used. (BT) 185-1975w

subsidiary conduit (lateral) A terminating branch of an underground conduit run, extending from a manhole or handhole to a nearby building, handhole, or pole. *See also:* cable. (EEC/PE) [119]

subsplit A frequency division scheme that allows two-way traffic on a single cable. Inbound path signals come to the headend from 5 to 30 MHz. Outbound path signals go from the headend from 54 to the upper frequency limit. The guardband is located from 30 to 54 MHz. (LM/C) 802.7-1989r

substantial So constructed and arranged as to be of adequate strength and durability for the service to be performed under the prevailing conditions. (T&D) C2.2-1960

substation (1) (generating stations electric power system) An area or group of equipment containing switches, circuit breakers, buses, and transformers for switching power circuits and to transform power from one voltage to another or from one system to another. (PE/EDPG) 505-1977r

(2) An assemblage of equipment for purposes other than generation or utilization, through which electric energy in bulk is passed for the purpose of switching or modifying its characteristics. Service equipment, distribution transformer installations, or other minor distribution or transmission equipment are not classified as substations. *Note:* A substation is of such size or complexity that it incorporates one or more buses, a multiplicity of circuit breakers, and usually is either the sole receiving point of commonly more than one supply circuit, or it sectionalizes the transmission circuits passing through it by means of circuit breakers. *See also:* alternating-current distribution; direct-current distribution. (T&D/PE) [10]

(3) An enclosed assemblage of equipment (e.g., switches, circuit breakers, buses, and transformers) under the control of qualified persons, through which electric energy is passed for the purposes of switching or modifying its characteristics. (PE/EDPG) 665-1995

(4) An enclosed assemblage of equipment, e.g., switches, circuit breakers, buses, and transformers, under the control of qualified persons, through which electric energy is passed for the purpose of switching or modifying its characteristics. (NESC) C2-1997

substitutability A principle stating that, since each instance of a subclass is an instance of the superclass, an instance of the subclass should be acceptable in any context where an instance of the superclass is acceptable. Any request sent to an instance receives an acceptable response, regardless of whether the receiver is an instance of the subclass or the superclass. (C/SE) 1320.2-1998

substitute character (SUB) A control character used in the place of a character that is recognized to be invalid or in error, or that cannot be represented on a given device. (C) 610.5-1990w

substitution The introduction unauthorized, potentially malicious components to intercept communications, generate incorrect or misleading information, masquerade as a legitimate component, or perform other undesirable functions. Substitution may be perpetrated throughout a system (e.g., hardware and software components), and could result in unauthorized disclosure or modification of information, unauthorized receipt of services, or denial of service to legitimate users or critical functions. (C/BA) 896.3-1993w

substitution error, direct-current-radio-frequency (bolometers) The error arising in the bolometric measurement technique when a quantity of direct-current or audio-frequency power is replaced by a quantity of radio-frequency power with the result that the different current distributions generate different temperature fields that give the bolometer element different values of resistance for the same amonts of power. This error is expressed as where e is the effective efficiency of the bolometer units and h is the efficiency of the bolometer unit. *See also:* bolometric power meter. (IM/HFIM) [40]

substitution error, dual-element A substitution error peculiar to dual-element bolometer units that results from a different division of direct-current (or audio-frequency) and radio-frequency powers between the two elements. (IM/HFIM) [40]

substitution power (bolometers) The difference in bias power required to maintain the resistance of a bolometer at the same value before and after radio-frequency power is applied. Commonly, a bolometer is placed in one arm of a Wheatstone bridge that is balanced when the bias current (direct current and.or audio frequency) holds the bolometer at its nominal operating resistance. Following the application of the radio-frequency signal, the reduction in bias power is taken as a measure of the radio-frequency power. This reduction in the bias power is the substitution power and is given by

$$P = I_1^2R - I_2^2R$$

where I_1 and I_2 are the bias currents before and after radio-frequency power is applied and R is the nominal operating resistance of the bolometer. *See also:* bolometric power meter. (IM) 470-1972w

substrate (1) (integrated circuit) The supporting material upon or within which an integrated circuit is fabricated or to which an integrated circuit is attached. (ED) 274-1966w
(2) (photovoltaic power system) Supporting material or structure for solar cells in a panel assembly. Solar cells are attached to the substrate. (AES) [41]
(3) (planar transmission lines) The supporting material upon or within which a planar transmission line is fabricated or to which it is attached. A substrate can be composed of one or more nonconducting layers. (MTT) 1004-1987w
(4) The base material upon which or in which a transistor or integrated circuit is fabricated; for example, materials such as glass-ceramic or silicon oxide. (C) 610.10-1994w

subsurface corrosion Formation of isolated particles of corrosion product(s) beneath the metal surface. This results from the preferential reaction of certain alloy constituents by inward diffusion of oxygen, nitrogen, sulfur, etc. (internal oxidation). (IA) [59]

subsurface switch A submersible switching assembly suitable for application in a below-grade enclosure that does not allow space for personnel access. (SWG/PE) C37.71-1984r, C37.100-1992

subsurface transformer (power and distribution transformers) A transformer utilized as part of an underground distribution system, connected below ground to high-voltage and low-voltage cables, and located below the surface of the ground. (PE/TR) C57.12.80-1978r

subsynchronous reluctance motor A form of reluctance motor that has the number of salient poles greater than the number of electrical poles of the primary winding, thus causing the motor to operate at a constant average speed that is a submultiple of its apparent synchronous speed. *See also:* asynchronous machine. (PE) [9]

subsynchronous satellite (communication satellite) A satellite, for which the sidereal period of rotation of the primary body about its own axis is an integral multiple of the mean sidereal period of revolution of the satellite about the primary body. (COM) [19]

subsystem (1) (unique identification in power plants) A portion of a system containing two or more integrated components which, while not completely performing the specific function of a system, may be isolated for design, test, or maintenance. (PE/EDPG) 804-1983r, 803-1983r
(2) (nuclear power generating station protective systems) That part of the system which effects a particular protective function. These subsystems may include, but are not limited to those actuating: reactor shutdown: safety injection: containment isolation: emergency core cooling: containment pressure and temperature reduction: containment air cleaning. (PE/NP) 380-1975w
(3) (software) A secondary or subordinate system with a larger system. (C) 610.12-1990
(4) An element of the physical or system architecture, specification tree, or system breakdown structure that is a subordinate element to a product and is comprised of one or more assemblies and their associated life-cycle processes. (C/SE) 1220-1994s
(5) An interconnected, interrelated group of equipment intended to serve a single basic purpose within a larger installation or facility. (SUB/PE) 1303-1994
(6) An element in a hierarchical division of an open system that interacts directly only with elements in the next higher division or the next lower division of that open system (ISO 7498). (LM/C) 8802-6-1994

subsystem identification An eight-character label further describing the system. Leading spaces are not interpreted as leading zeros. This is used with the system identification to uniquely describe the data. (NPS/NID) 1214-1992r

subsystem verification Testing to verify that all of the systems required for a main generating unit startup have been tested and are operational. (PE/EDPG) 1248-1998

subtracter A device whose output data is the arithmetic difference of the two or more quantities presented as input data. *Contrast:* adder. *See also:* half subtracter; full subtracter; adder-subtracter. (C) 610.10-1994w

subtract time The elapsed time required to perform one subtraction operation, not including the time required to obtain the operands or to return the result to storage. *Contrast:* multiply time; add time. (C) 610.10-1994w

subtrahend A number to be subtracted from another number (the minuend) to produce a result (the difference). (C) 1084-1986w

subtransient current (rotating machinery) The initial alternating component of armature current following a sudden short circuit. *See also:* armature. (PE) [9]

subtransient internal voltage (synchronous machines) (specified operating condition) The fundamental-frequency component of the voltage of each armature phase that would appear at the terminals immediately following the sudden removal of the load. *Note:* The subtransient internal voltage, as shown in the phasor diagram, is related to the terminal-voltage and phase-current phasors by the equation:

$$E''_1 = E_a + RI_a + jX''_dI_{ad} + jX''_qI_{aq}$$

For a machine subject to saturation, the reactances should be determined for the degree of saturation applicable to the specified operating conditions. (PE) [9]

subtransient reactance (1) (electrical power systems in commercial buildings) The apparent reactance of the stator winding at the instant the short circuit occurs. (IA/PSE) 241-1990r
(2) Reactance of a generator at the initiation of a fault. This reactance is used in calculations of the initial symmetrical fault current. The current continuously decreases, but it is assumed to be steady at this value as a first step, lasting approximately 0.05 s after an applied fault. (PE/SUB/PSC) 80-2000, 367-1996

subtrate (metal-nitride-oxide field-effect transistor) This insulated-gate field-effect transistor (IGFET) region separates source from drain and is of opposite conductivity type. The potential on the substrate terminal can only be equally, or less attractive to the carriers in the channel than the source terminal. (ED) 581-1978w

subtree A tree whose root node is part of a larger tree. *Note:* A subtree is made up of a node and all of its hierarchical descendants. *Synonym:* branch. (C) 610.5-1990w

subtype (1) A subset of a data type, obtained by constraining the set of possible values of the data type. *Note:* The operations applicable to the subtype are the same as those of the original data type. *See also:* derived type. (C) 610.12-1990
(2) *See also:* subclass. (C/SE) 1320.2-1998

subunit A logical subcomponent of a unit that is accessed by a largely independent subcomponent of I/O driver software. For example, a terminal multiplexer unit could have multiple subunits (two for each full-duplex connection). (C/MM) 1212-1991s

subvoice-band channel A channel with a bandwidth narrower than that of a voice-band channel. *Note:* It is generally used in telegraphy. (C) 610.7-1995

subway transformer (power and distribution transformers) A submersible-type distribution transformer suitable for installation in an underground vault. (PE/TR) C57.12.80-1978r

successfully transferred For a write operation to a regular file, when the system ensures that all data written is readable on any subsequent open of the file (even one that follows a system or power failure) in the absence of a failure of the physical storage medium. For a read operation, when an image of the data on the physical storage medium is available to the requesting process. (C/PA) 9945-1-1996, 1003.5-1999

successful test A completed test that is invoked by a write to the TEST_START register, in which no errors are detected. (C/MM) 1212-1991s

sudden failure *See:* failure.

sudden ionospheric disturbance (SID) An ionospheric disturbance with a duration of from a few minutes to a few hours, characterized by the sudden increase in the ionization of the D region in the daylight hemisphere as a result of electromagnetic radiation from a solar flare. *Note:* This effect is sometimes called the Mägel-Delinger effect. (AP/PROP) 211-1997

sudden-pressure relay A relay that operates by the rate of rise in pressure of a liquid or gas. (SWG/PE) C37.100-1992

sudden short-circuit test (synchronous machines) A test in which a short-circuit is suddenly applied to the armature winding of the machine under specified operating conditions. (PE) [9]

Suez Canal searchlight A searchlight constructed to the specifications of the Canal Administration that by regulation of the Administration, must be carried by every ship traversing the canal, so located as to illuminate the banks. (EEC/PE) [119]

suffix notation *See:* postfix notation.

suicide control (adjustable-speed drive) A control function that reduces and automatically maintains the generator voltage at approximately zero by negative feedback. *See also:* feedback control system. (IA/IAC) [60]

suitable test (faulted circuit indicators) Where a condition or a set of conditions are so variable from one utility to another or even within the utility itself that no test can be properly specified for all conditions, it is left to the user to determine their individual test needs. A suitable test and anticipated service life are mutually agreed to between manufacturer and user. (T&D/PE) 495-1986w

suitcase *See:* conductor grip.

sulfur hexafluoride (SF_6) A gaseous dielectric for high-voltage power applications having characteristics as specified in ASTM D2472-92. (SWG/SUB/PE) C37.122-1983s, C37.122.1-1993, C37.100-1992

sum (mathematics of computing) The result of an addition operation. (C) 1084-1986w

sum check *See:* summation check.

sum frequency (parametric device) The sum of a harmonic (nf_p) of the pump frequency (f_p) and the signal frequency (f_s), where n is a positive integer. *Note:* Usually n is equal to one. *See also:* parametric device. (ED) [46]

sum-frequency parametric amplifier *See:* parametric device; noninverting parametric device.

summary punch (1) (test, measurement, and diagnostic equipment) A tape or card punch operating in conjunction with another machine to punch data which have been summarized or calculated by the other machine. (MIL) [2]
(2) A card punch used to record data that were calculated or summarized by another device. (C) 610.10-1994w

summation check (1) (computers) A check based on the formation of the sum of the digits of a numeral. The sum of the individual digits is usually compared with a previously computed value. (C) [20], [85]
(2) (mathematics of computing) A check in which a group of digits is summed, usually without regard to overflow, and that sum is checked against a previously computed value to verify that no digits have been changed. *Synonym:* sum check. (C) 1084-1986w

summer *See:* summing amplifier.

summer outage rate The number of outage occurrences per unit of service time during the summer. Summer outage rate = number of outages during the summer/service time during summer. *See also:* outage rate. (PE/PSE) 859-1987w

summing amplifier An operational amplifier whose output analog variable is the integral of a weighted sum of the input analog variables with respect to time or with respect to another input analog variable. (C) 610.10-1994w, 165-1977w

summing junction In an analog computer, the junction common to the input and feedback impedances used with an operational amplifier. (C) 610.10-1994w, 165-1977w

summing point (1) Any point at which signals are added algebraically. *Note:* For example the null junction of a power supply is a summing point because, as the input to a high-gain direct-current amplifier, operational summing can be performed at this point. As a virtual ground, the summing point decouples all inputs so that they add linearly in the output, without other interaction. *See also:* operational programming; null junction. (PE/EDPG) [3]
(2) The point in a feedback control system at which the algebraic sum of two or more signals is obtained. *See also:* feedback control system. (IA/ICTL/IAC) [60]

sum pattern A radiation pattern characterized by a single main lobe whose cross section is essentially elliptical, and a family of side lobes, the latter usually at a relatively low level. *Note:* Antennas that produce sum patterns are often designed to produce a difference pattern and have application in acquisition and tracking radar systems. *Contrast:* difference pattern. (AP/ANT) 145-1993

sun bearing (illuminating engineering) The angle measured in the plane of the horizon between a vertical plane at a right angle to the window wall and the position of this plane after it has been rotated to contain the sun. (EEC/IE) [126]

sun light (illuminating engineering) Direct visible radiation from the sun. (EEC/IE) [126]

superclass (1) A class that is the ancestor of another class in the class hierarchy. (C) 1295-1993w
(2) One of the classes, designated as such, whose attribute types are a subset of those of another class. (C/PA) 1328-1993w, 1224-1993w, 1327-1993w
(3) A class whose instances are specialized into one or more subclasses. *See also:* partial cluster; total cluster. (C/SE) 1320.2-1998

supercomputer Any of the group of computers that have the fastest processing speeds available at a given time. (C) 610.10-1994w

superconducting The state of a superconductor in which it exhibits superconductivity. Example: Lead is superconducting below a critical temperature and at sufficiently low operating frequencies. *See also:* normal; superconductivity. (ED) [46]

superconductive Pertaining to a material or device that is capable of exhibiting superconductivity. Example: Lead is a superconductive metal regardless of temperature. The cryotron is a superconductive computer component. *See also:* superconductivity. (ED) [46]

superconductivity A property of a material that is characterized by zero electric resistivity and, ideally, zero permeability. (ED) [46]

superconductor Any material that is capable of exhibiting superconductivity. Example: Lead is a superconductor. *See also:* superconductivity. (ED) [46]

superdirectivity The condition that occurs when the antenna illumination efficiency significantly exceeds 100%. *Note:* Superdirectivity is only obtained at a cost of a large increase in the ratio of average stored energy to energy radiated per cycle. (AP/ANT) 145-1993

superframe A structure that consists of 12 DS1 frames (2316 bits). The DS1 frame comprises 193 bit positions, the first of which is the frame overhead bit position. Frame overhead bit positions are used for terminal frame (F_t) and signaling frame (F_s) alignment only. (COM/TA) 1007-1991r

supergroup *See:* channel supergroup.

superheterodyne reception A method of receiving radio waves in which the process of heterodyne reception is used to convert the voltage of the received wave into a voltage of an intermediate, but usually superaudible, frequency, that is then detected. *See also:* radio receiver. (EEC/PE) [119]

super high frequency (SHF) 3–30 GHz. *See also:* radio spectrum. (AP/PROP) 211-1997

superimposed ringing (telephone switching systems) Selective ringing that utilizes direct current polarity to obtain selectivity. (COM) 312-1977w

superior (applying to entry or directory object) Immediately superior, or superior to one that is immediately superior (recursively). (C/PA) 1326.2-1993w, 1224.2-1993w, 1327.2-1993w, 1328.2-1993w

super-large scale integration *See:* ultra-large scale integration; very large scale integration.

superluminescent LED *See:* superluminescent light-emitting diode.

superluminescent light-emitting diode (SLED, SLD) (1) (fiber optics) An emitter based on stimulated emission with amplification but insufficient feedback for oscillation to build up. *Synonym:* superluminescent LED. *See also:* spontaneous emission; stimulated emission. (Std100) 812-1984w
(2) A p-n junction semiconductor emitter based on stimulated emission with amplification, but insufficient for feedback oscillation to build up. (AES/GYAC) 528-1994

superobject The immediate superobject of an object, or one of its superobjects applied recursively. (C/PA) 1328-1993w, 1224-1993w, 1327-1993w

superposed circuit An additional channel obtained from one or more circuits, normally provided for other channels, in such a manner that all the channels can be used simultaneously without mutual interference. *See also:* transmission line. (EEC/PE) [119]

superposition theorem States that the current that flows in a linear network, or the potential difference that exists between any two points in such a network, resulting from the simultaneous application of a number of voltages distributed in any manner whatsoever throughout the network is the sum of the component currents at the first point, or the component potential differences between the two points, that would be caused by the individual voltages acting separately. (EEC/PE) [119]

superradiance (fiber optics) Amplification of spontaneously emitted radiation in a gain medium, characterized by moderate line narrowing and moderate directionality. *Note:* This process is generally distinguished from lasing action by the absence of positive feedback and hence the absence of well-defined modes of oscillation. *See also:* stimulated emission; spontaneous emission; laser. (Std100) 812-1984w

super-refraction (1) Refraction for which the vertical gradient of refractivity is less (more negative) than the standard gradient of refractivity. (AP/PROP) 211-1990s
(2) *See also:* ducting. (AES) 686-1997

super-refractive atmosphere *See:* refractive index gradient.

superregeneration A form of regenerative amplification, frequently used in radio receiver detecting circuits, in which oscillations are alternately allowed to build up and are quenched at a superaudible rate. *See also:* radio receiver. (EEC/PE) [119]

supersonic frequency *See:* ultrasonic frequency.

supersynchronous satellite (communication satellite) A satellite with mean sidereal period of revolution about the primary body which is an integral multiple of the sidereal period of rotation of the primary body about its axis. (COM) [19]

supertype *See:* superclass.

supervised circuit (protective signaling) A closed circuit having a current-responsive device to indicate a break in the circuit, and, in some cases, to indicate an accidental ground. *See also:* protective signaling. (EEC/PE) [119]

supervision (telephone switching systems) The function of indicating and controlling the status of a call. (COM) 312-1977w

supervisor *See:* supervisory program.

supervisor mode A processor state that is active when the S bit of the PSR is set (PSR.S = 1). *See also:* privileged. (C/MM) 1754-1994

supervisor software Software that executes when the processor is in **supervisor mode.** (C/MM) 1754-1994

supervisor state In the operation of a computer system, a state in which the supervisory program is executing. This state usually has higher priority than, and precludes the execution of, application programs. *Synonyms:* executive state; master state; privileged state. *Contrast:* problem state. (C) 610.12-1990

supervisory control (1) (supervisory control, data acquisition, and automatic control) An arrangement for operator control and supervision of remotely located apparatus using multiplexing techniques over a relatively small number of interconnecting channels. (PE/SUB) 999-1992w, C37.1-1994
(2) A form of remote control comprising an arrangement for the selective control of remotely located units by electrical means over one or more common interconnecting channels. (SWG/PE) C37.100-1992

supervisory control data acquisition system (supervisory control, data acquisition, and automatic control) A system

operating with coded signals over communication channels so as to provide control of remote equipment (using typically one communication channel per remote station). The supervisory system may be combined with a data acquisition system, by adding the use of coded signals over communication channels to acquire information about the status of the remote equipment for display or for recording functions.
(SWG/PE/SUB) C37.100-1992, C37.1-1994

supervisory control functions Equipment governed by this standard that comprises one or more of the following functions:

1) **alarm function.** The capability of a supervisory system to accomplish a predefined action in response to an alarm condition.
2) **analog function.** The capability of a supervisory system to accept, record, or display, or do all of these, an analog quantity as presented by a transducer or external device. The transducer may or may not be a part of the supervisory control system.
3) **control function.** The capability of a supervisory system to selectively perform manual operation, automatic operation, or both (singularly or in selected groups), of external devices. Control may be either analog (magnitude or duration) or digital.
4) **indication (status) function.** The capability of a supervisory system to accept, record, or display, or do all of these, the status of a device. The status of a device may be derived from one or more inputs giving two or more states of indication.
   a) **two-state indication.** Only one of the two possible positions of the supervised device is displayed at a time. Such display may be derived from a single set of contacts.
   b) **three-station indication.** Indication in which the transitional state or security indication as well as the terminal positions of the supervised device is displayed. Such a display is derived from at least two sets of initiating contacts.
   c) **multistate indication.** Only one of the predefined states (transitional or discrete, or both) is indicated at a time. Such a display is derived from multiple inputs.
   d) **indication with memory.** An indication function with the additional capability of storing single or multiple changes of status that occur between scans.
   e) **accumulator function.** The capability of a supervisory system to accept and totalize digital pulses and make them available for display or recording or both.
5) **sequence of events function.** The capability of a supervisory system to recognize each predefined event, associate a time of occurrence with each event, and present the event data in order of occurrence of the events.

See also: alarm condition.
(SWG/PE/SUB) C37.100-1992, C37.1-1987s

supervisory format The format used to perform data link supervisory control such as acknowledging I frames, requesting retransmission of I frames, and requesting a temporary suspension of transmission of I frames.
(EMB/MIB) 1073.3.1-1994

supervisory indication A form of remote indication comprising an arrangement for the automatic indication of the position or condition of remotely located units by electrical means over one or more common interconnecting channels.
(SWG/PE) C37.100-1992

supervisory program (software) A computer program, usually part of an operating system, that controls the execution of other computer programs and regulates the flow of work in a computer system. *Synonyms:* executive; control program; executive program; supervisor. *See also:* supervisor state.
(C) 610.12-1990

supervisory relay A relay that, during a call, is generally controlled by the transmitter current supplied to a subscriber line in order to receive, from the associated station, directing signals that control the actions of operators or switching mechanisms with regard to the connection. (EEC/PE) [119]

supervisory routine *See:* executive routine.

supervisory scanning cycle (station control and data acquisition) The time interval to start and complete a supervisory scan. (SUB/PE) C37.1-1979s

supervisory sequence In data communications, a sequence of communication control characters, and possibly other characters, that define control function. (C) 610.7-1995

supervisory signal (telephone switching systems) Any signal used to indicate or control the states of the circuits involved in a particular connection. (COM) 312-1977w

supervisory station check The automatic selection in a definite order, by means of a single initiation of the master station, of all of the supervisory points associated with one remote station of a system; and the transmission to the master station of indications of positions or conditions of the individual equipment or device associated with each point.
(SWG/PE) C37.100-1992

supervisory system (supervisory control, data acquisition, and automatic control) All control indicating and associated with telemetering equipment at the master station and all of the complementary devices at the remote station, or stations. *See also:* scanning supervisory system; continuous update supervisory system; quiescent supervisory system; polling supervisory system.
(SWG/PE/SUB) C37.100-1992, C37.1-1994

supervisory system check The automatic selection in a definite order, by means of a single initiation at the master station, of all supervisory points associated with all of the remote stations in a system; and the transmission to the master station of indications of positions or conditions of the individual equipment or device associated with each point.
(SWG/PE) C37.100-1992

supervisory telemeter selection A form of remote telemeter selection comprising an arrangement for the selective connection of telemeter transmitting equipment to an appropriate telemeter receiving equipment over one or more common interconnecting channels. (SWG/PE) C37.100-1992

supervisory tone (telephone switching systems) A tone that indicates to equipment, an operator or a customer that a particular state in the call has been reached, and which may signify the need for action to be taken. The terms used for the various supervisory tones are usually self-explanatory.
(COM) 312-1977w

supplementary control Any control action that is superimposed upon normal governor action. (PE/PSE) 94-1991w

supplementary equipment ground (generating station grounding) A grounding conductor used to connect the equipment frame to local grounding system to minimize potential difference. (PE/EDPG) 665-1987s

supplementary group ID **(1)** An attribute of a process used in determining file access permissions. A process has up to {NGROUPS_MAX} supplementary group IDs in addition to the effective group ID. The supplementary group IDs of a process are set to the supplementary group IDs of the parent process when the process is created. Whether the effective group ID of the process is included in or omitted from its list of supplementary group IDs is unspecified.
(C/PA) 9945-1-1996, 1003.2-1992s

(2) An attribute of a process, used in determining file access permissions. A process has group IDs in addition to the effective group ID. The size of this list of supplementary group IDs is specified at compile time by `Groups_Maxima` in package `POSIX_Limits`, or at run time by the value of the function `Groups_Maximum` in package `POSIX_Configurable_System_Limits`. The supplementary group IDs of a process are set to the supplementary group IDs of the parent process when the process is created. Whether the effective group ID of a process is included in or omitted from its list of supplementary group IDs is unspecified. (C) 1003.5-1999

supplementary insulation Independent insulation applied in addition to basic insulation to protect against electric shock if the basic insulation fails. (EMB/MIB) 1073.4.1-2000

supplementary lighting (illuminating engineering) Lighting used to provide an additional quantity and quality of illuminance which cannot readily be obtained by a general lighting system and which supplements the general lighting level, usually for specific work requirements. (EEC/IE) [126]

supplementary standard illuminant D_{55} (illuminating engineering) A representation of a phase of daylight at a correlated color temperature of approximately 5500 K. (EEC/IE) [126]

supplementary standard illuminant D_{75} (illuminating engineering) A representation of a phase of daylight at a correlated color temperature of approximately 7500 K. (EEC/IE) [126]

supplier (1) (nuclear power quality assurance) Any individual or organization who furnishes items or services to a procurement document. An all inclusive term used in place of any of the following: vendor, seller, contractor, subcontractor, fabricator, consultant, and subtier levels. (PE/NP) [124]
(2) An organization that develops some or all of the project deliverables for an acquirer. Suppliers may include organizations that have primary responsibility for project deliverables and subcontractors that deliver some part of the project deliverables to a primary supplier. In the latter case, the primary supplier is also an acquirer. (C/SE) 1058-1998
(3) The person, or persons, who produce a product for a customer. In the context of this recommended practice, the customer and the supplier may be members of the same organization. (C/SE) 830-1998
(4) The entity that contractually acts as the source of a product. *Note:* The supplier may or may not be the actual builder. (VT) 1475-1999, 1476-2000
(5) A person or organization that enters into a contract with the acquirer for the supply of a software product (which may be part of a system) under the terms of the contract. (C/SE) 1062-1998
(6) *See also:* developer. (C/SE) 1362-1998

suppliers Those who build and/or sell the CASE tools, or intermediate distributors of the CASE tools. (C/SE) 1209-1992w

supply circuit (household electric ranges) The circuit that is the immediate source of the electric energy used by the range. *See also:* appliance outlet. (IA/APP) [90]

supply equipment *See:* electric-supply equipment.

supply impedance (1) (converters having dc input) (self-commutated converters) The impedance appearing in the input lines to the converter. (IA/SPC) 936-1987w
(2) (inverters) The impedance appearing across the input lines to the power inverter with the power inverter disconnected. *See also:* self-commutated inverters. (IA) [62]

supply line, motor *See:* motor supply line.

supply lines *See:* electric supply lines.

supply short-circuit current (self-commutated converters) (converters having dc input) The steady-state current that the dc (direct current) supply system can deliver into a short-circuit across the terminals to which the converter is to be connected. (IA/SPC) 936-1987w

supply station *See:* electric supply station.

supply transient energy (converters having dc input) (self-commutated converters) The energy that the dc (direct current) supply system, due to a transient, is capable of delivering at the terminals to which the terminal is to be connected. (IA/SPC) 936-1987w

supply transient overvoltage (self-commutated converters) (converters having dc input) The peak instantaneous voltage that may appear between the input lines to the converter with the converter disconnected. (IA/SPC) 936-1987w

supply transient voltage (inverters) The peak instantaneous voltage appearing across the input lines to the power inverter with the inverter disconnected. *See also:* self-commutated inverters. (IA) [62]

supply voltage (electron tube) (electrode) The voltage, usually direct, applied by an external source to the circuit of an electrode. *See also:* electrode voltage. (ED) [45], [84]

support (1) (raceway systems for Class 1E circuits for nuclear power generating stations) (raceway) An assembly of structural members whose function is to restrain and provide structural stability for raceways. (PE/NP) 628-1987r
(2) (software) The set of activities necessary to ensure that an operational system or component fulfills its original requirements and any subsequent modifications to those requirements. For example, software or hardware maintenance, user training. *See also:* system life cycle; software life cycle. (C) 610.12-1990

support components The components that give additional strength and rigidity or both to the bus enclosure and are basic subassemblies of the enclosure. (SWG/PE) C37.100-1992, C37.23-1987r

supported A condition regarding optional functionality. (C/PA) 1326.2-1993w, 1003.1-1988s, 1003.5-1999

supported transaction A transaction whose returned data value and side effects are defined by the hardware architecture that is addressed. For example, a write 4 transactions to the 4-byte STATE_CLEAR register is supported. (C/MM) 1212-1991s

support equipment (test, measurement, and diagnostic equipment) Equipment required to make an item, system or facility operational in its environment. This includes all equipment required to maintain and operate the item, system or facility and the computer programs related thereto. (MIL) [2]

supporting data item Data used to describe an anomaly and the environment in which it was encountered. (C/SE) 1044-1993, 1044.1-1995

supporting operations area(s) (nuclear power generating station) Functional area(s) allocated for controls and displays that support plant operation. (PE/NP) 566-1977w

supporting structure The main supporting unit (usually a pole or tower). (NESC/T&D) C2-1997, C2.2-1960

supporting process A collection of work activities that span the entire duration of a software project. Examples of supporting processes include software documentation, quality assurance, configuration management, software reviews, audit processes, and problem resolution activities. (C/SE) 1058-1998

support manual A document that provides the information necessary to service and maintain an operational system or component throughout its life cycle. Typically described are the hardware and software that make up the system or component and procedures for servicing, repairing, or reprogramming it. *Synonym:* maintenance manual. *See also:* installation manual; operator manual; programmer manual; user manual; diagnostic manual. (C) 610.12-1990

support package A package, residing in the `/packages` node, that provides a service to assist in the implementation of a particular device type. (C/BA) 1275-1994

support ring (rotating machinery) A structure for the support of a winding overhang: either constructed of insulating material, carrying support-ring insulation, or separately insulated before assembly. *See also:* stator. (PE) [9]

support-ring insulation (rotating machinery) Insulation between the winding overhang or end winding and the winding support rings. *See also:* rotor; stator. (PE) [9]

support software Software that aids in the development or maintenance of other software, for example, compilers, loaders, and other utilities. *Contrast:* application software. *See also:* system software. (C) 610.12-1990

support staff-hour An hour of effort expended by a member of the staff who does not directly define or create the software product, but acts to assist those who do. (C/SE) 1045-1992

support test system (test, measurement, and diagnostic equipment) A measurement system used to assess the quality of operational equipments and may include: test equipment; ancillary equipment; supporting documentation; operating personnel. (MIL) [2]

suppressed-carrier modulation Modulation in which the carrier is suppressed. *Note:* By carrier is meant that part of the modulated wave that corresponds in a specified manner to the unmodulated wave. (Std100) 270-1964w

suppressed-carrier operation (data transmission) That form of amplitude-modulation carrier transmission in which the carrier wave is suppressed. (PE) 599-1985w

suppressed carrier transmission A method of transmission in which the carrier frequency is suppressed partially or fully. (C) 610.7-1995

suppressed time delay (navigation aids) A deliberate displacement of the zero of the time scale with respect to the time of emission of a pulse. (AES/GCS) 172-1983w

suppressed-zero instrument An indicating or recording instrument in which the zero position is below the end of the scale markings. *See also:* instrument. (EEC/PE) [119]

suppressed-zero range A range where the zero value of the measured variable, measured signal, etc., is less than the lower range value. Zero does not appear on the scale. *Note:* For example: 20 to 100. (EEC/EMI) [112]

suppression *See:* zero suppression.

suppression characteristic (thyristor) Predicated on a device's ability to block voltage at higher than rated junction temperatures (Ts) when either the voltage or the rate of application of the principal blocking voltage, or both, are below the rated voltage of the silicon controlled rectifier (SCR). (IA/IPC) 428-1981w

suppression corrugation Grooves or surface roughness intentionally placed in the nonactive side of the substrate for reflected bulk-wave signals. (UFFC) 1037-1992w

suppression distributor rotor Rotor of an ignition distributor with a built-in suppressor. *See also:* electromagnetic compatibility. (INT) [53], [70]

suppression rating (thyristor) Repetitive surge ON-state current. A specified ON-state current of short time duration resulting in a specified junction temperature, above rated, immediately prior to supporting a specified principal voltage without turning on (gate signal removed, gate impedance specified). *Note:* Proper coordination with this rating permits a thyristor power controller to limit fault currents without fuse blowing or circuit breaker action. For a given silicon controlled rectifier (SCR) its suppression characteristic may be defined in one of two ways: (1) T_L and I_f together with shape of fault I waveform may be specified together with time $t_2 - t_1$. This then determines maximum V_{line} and shape of reapplied V at time t_2, that is, dv/dt. Alternately, ac frequency and peak V may be given for sinusoidal waveforms. (2) T_s may be specified together with shape and magnitude of reapplied voltage at time t_2. Criteria (2) serves as well as (1) since the magnitude and shape of the fault current determine T_2 together with the time $(t_2 - t_1)$. (IA/IPC) 428-1981w

suppression ratio (suppressed-zero range) The ratio of the lower range-value to the span. *Note:* For example: Range 20 to 100

$$\text{suppression ratio} = \frac{20}{80} = 0.25$$

(EEC/EMI) [112]

suppressive wiring techniques (coupling in control systems) Those wiring techniques which result in the reduction of electric or magnetic fields in the vicinity of the wires which carry current without altering the value of the current. Wires which are candidates for suppressive techniques are generally connected to a noise source, may couple noise into a susceptible circuit by induction. Example: twisting or transposing of alternating-current power lines to reduce the intensity of magnetic field produced by current in these lines. *See also:* compensatory wiring techniques; barrier wiring techniques. (IA/ICTL) 518-1982r

suppressor grid A grid that is interposed between two positive electrodes (usually the screen grid and the plate), primarily to reduce the flow of secondary electrons from one electrode to the other. *See also:* grid; electrode. (ED) 161-1971w

suppressor spark plug A spark plug with a built-in interference suppressor. *See also:* electromagnetic compatibility. [53]

supra-aural receiver A receiver that rests upon the pinna of the ear. (For example, conventional telephone handsets use receivers of the supra-aural type.) (COM/TA) 1206-1994

supra-concha receiver A receiver that rests upon the ridges of the concha cavity. (COM/TA) 1206-1994

surface acoustic wave (SAW) An acoustic or Rayleigh wave, propagating along a surface of an elastic substrate whose amplitude decays exponentially with substrate depth. See the corresponding figure. (UFFC) 1037-1992w

surface acoustic wave diffraction A phenomenon (analogous to optical diffraction due to the finite aperture of the source) that causes surface acoustic wave beam spreading and wavefront distortion. (UFFC) 1037-1992w

surface active agent *See:* wetting agent.

surface-barrier contact (1) (x-ray energy spectrometers) (semiconductor radiation detectors) A rectifying contact that is characterized by a potential barrier associated with an inversion or accumulation layer; said inversion or accumulation layer being caused by surface charge resulting from the presence of surface states and work function differences, or both. (NPS/NID) 759-1984r, 301-1976s **(2) (charged-particle detectors)** A metal-insulator-semiconductor contact structure in which the rectification properties are dominated or heavily influenced by charge trapped at the

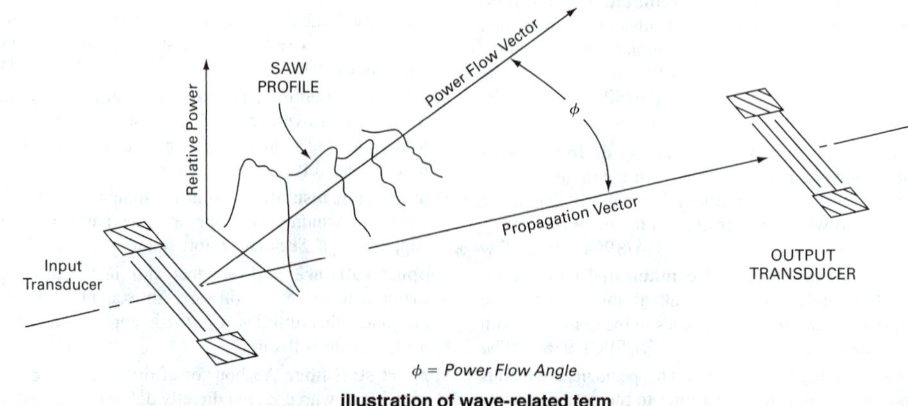

illustration of wave-related term

interfaces and in the insulator. (NPS) 300-1988r, 325-1996

surface barrier detector A radiation detector in which the principal rectifying junction is a surface barrier contact. (NPS) 325-1996

surface barrier radiation detector (1) (charged-particle detectors) A radiation detector for which the blocking contact is a surface barrier contact. (NPS) 300-1988r
(2) (germanium gamma-ray detectors) (x-ray energy spectrometers) (semiconductor radiation detectors) A radiation detector for which the principal rectifying junction is a surface barrier contact. (NPS/NID) 325-1986s, 759-1984r, 301-1976s

surface channel (semiconductor radiation detectors) A thin region at a semiconductor surface of *p* or *n*-type conductivity created by the action of an electric field; for example, that due to trapped surface charge. (NPS) 300-1988r, 325-1996

surface connecting cable (electric submersible pump cable) Power cable connecting the ESP (electric submersible pump) cable to surface equipment. (IA/PC) 1017-1985s

surface contamination (plutonium monitoring) Radioactive material deposited on the surface of facilities (floor surfaces, workbench tops, machines, etc.), equipment, or personnel. (NI) N317-1980r

surface density *See:* recording density.

surface duct An atmospheric radio duct for which the lower boundary is the Earth's surface. (AP/PROP) 211-1997

surface flame spread The propagation of a flame away from the source of ignition across the surface of a liquid or a solid. (DEI) 1221-1993w

surface heating device A heater comprising series or parallel connected elements having sufficient flexibility to conform to the shape of the surface to be heated. (IA) 515-1997

surface impedance For a monochromatic electromagnetic wave incident on a locally planar boundary, the complex ratio of the total orthogonal electric to magnetic field components tangent to the surface. The surface impedance is taken as having a positive real part. (AP/PROP) 211-1997

surface leakage The passage of current over the surface of a material rather than through its volume. (Std100) 270-1966w

surface leakage current (I_L) A current that is constant with time, and which usually exists over the surface of the end-turns of the stator winding or between exposed conductors and the rotor body in insulated rotor windings. The magnitude of the surface leakage current is dependent upon temperature and the amount of conductive material, i.e., moisture or contamination on the surface of the insulation. (PE/EM) 43-2000

surface material A material installed over the soil consisting of, but not limited to, rock or crushed stone, asphalt, or man-made materials. The surfacing material, depending on the resistivity of the material, may significantly impact the body current for touch and step voltages involving the person's feet. (PE/SUB) 80-2000

surface metal raceway (metal molding) A raceway consisting of an assembly of backing and capping. *See also:* raceway. (EEC/PE) [119]

surface-mounted (illuminating engineering) A luminaire that is mounted directly on the ceiling. (EEC/IE) [126]

surface-mounted device A device, the entire body of which projects in front of the mounting surface. (SWG/PE) C37.100-1992

surface navigation (navigation aids) Navigation of a vehicle on the surface of the earth. (AES/GCS) 172-1983w

surface noise (mechanical recording) The noise component in the electric output of a pickup due to irregularities in the contact surface of the groove. *See also:* phonograph pickup. (SP) [32]

surface of position (navigation aids) Any surface defined by a constant value of some navigation quantity. (AES/GCS) 172-1983w

surface operable (1) A term indicating that the switch and its accessories are operable from above grade. (SWG/PE) C37.71-1984r
(2) A term indicating that an underground switch and its accessories are operable from above grade. (SWG/PE) C37.100-1992

surface-potential gradient The slope of a potential profile, the path of which intersects equipotential lines at right angles. (PE/PSIM) 81-1983

surface, prescribed *See:* prescribed surface.

surface search radar *See:* navigational radar.

surface-skimming (shallow-bulk) acoustic wave A predominantly horizontally polarized bulk shear wave that propagates in a direction almost parallel to and at a depth just below the substrate surface. (UFFC) 1037-1992w

surface, specular *See:* specular surface.

surface state coefficient (m) (overhead-power-line corona and radio noise) A coefficient ($0 < m < 1$) by which the nominal corona inception gradient must be multiplied to obtain the actual corona inception gradient on overhead power lines. *Note:* Examples of conditions that affect the surface state are given in the definition of corona, overhead power lines. *See also:* corona, overhead power lines. (T&D/PE) 539-1990

surface wave (1) (planar transmission lines) A mode of propagation where the energy is concentrated near the interface of two media having different electric or magnetic properties, or both, and whose field amplitude decays in a direction normal to the interface. (MTT) 1004-1987w
(2) (fiber optics) A wave that is guided by the interface between two different media or by a refractive index gradient in the medium. The field components of the wave may exist (in principle) throughout space (even to infinity) but become negligibly small within a finite distance from the interface. *Note:* All guided modes, but not radiation modes, in an optical waveguide belong to a class known in electromagnetic theory as surface waves. (Std100) 812-1984w
(3) A wave guided by a boundary with a surface impedance whose reactive part exceeds the resistive part. A surface wave is generally characterized as a slow wave having a magnitude that exponentially decreases with distance from the interface but may be modified by curvature. (AP/PROP) 211-1997

surface wave antenna An antenna that radiates power from discontinuities in the structure that interrupt a bound wave on the antenna surface. (AP/ANT) 145-1993

surface-wave transmission line (waveguide) A transmission line in which propagation in other than a TEM mode is constrained to follow the external face of a guiding structure. (MTT) 146-1980w

surge (1) A transient voltage or current, which usually rises rapidly to a peak value and then falls more slowly to zero, occurring in electrical equipment or networks in service. (PE/PSIM) 4-1995
(2) A transient wave of voltage or current. (The duration of a surge is not tightly specified, but it is usually less than a few milliseconds.) (T&D/PE/SPD) 1250-1995, C62.34-1996, C62.48-1995
(3) A transient wave of current, potential, or power in an electric circuit. (SPD/PE) C62.22-1997, C62.11-1999, C62.62-2000
(4) *See also:* transient. (IA/PSE) 1100-1999

surge arrester (1) (electrical heating applications to melting furnaces and forehearths in the glass industry) A protective device for limiting surge voltages on equipment by discharging or bypassing surge current. It prevents continued flow of current to ground and is capable of repeating these functions, as specified. As surge protective devices, arresters are connected from sensitive circuit points to ground, thus limiting dangerous surge voltage below damaging levels. (IA) 668-1987w

(2) (ac power circuits) A protective device for limiting surge voltages on equipment by discharging or bypassing surge current; it prevents continued flow of follow current to ground, and is capable of repeating these functions as specified. *Notes:* 1. The term "arrester" as used in IEEE Stds 28-1974 and C62.1-1981 shall be understood to mean "surge arrester." 2. Use of the term "lightning arrester" is deprecated. (SPD/PE) 28-1974, C62.1-1981s

(3) (broadband local area networks) A device that protects electronic equipment against surge voltage and transient signals on trunk and distribution lines. (LM/C) 802.7-1989r

(4) A device that guards against dielectric failure of protection apparatus due to lightning or surge voltages in excess of their dielectric capabilities and serves to interrupt power follow current. (PE/PSC) 487-1992

(5) A protective device for limiting surge voltages on equipment by discharging or bypassing surge current; it limits the flow of power follow current to ground, and is capable of repeating these functions as specified. (SPD/PE) C62.22-1997

(6) A protective device for limiting surge voltages on equipment by diverting surge current and returning the device to its original status. It is capable of repeating these functions as specified. (SPD/PE) C62.11-1999

surge breakdown voltage *See:* impulse sparkover voltage.

surge capacitor (electrical heating applications to melting furnaces and forehearths in the glass industry) Capacitors used to decrease the slope of the surge voltage wave fronts. They help to reduce the voltage stresses on protected apparatus by spreading the impressed voltage over a greater time span. (IA) 668-1987w

surge-crest ammeter A special form of magnetometer intended to be used with magnetizable links to measure the crest value of transient electric currents. *See also:* instrument. (EEC/PE) [119]

surge diverter *See:* surge arrester.

surge electrode current *See:* fault electrode current.

surge energy The energy (in joules) contained in a surge. It can be calculated if the current and voltage wave shape are known:

where
E is the energy
t is the time
I is the instantaneous current
V is the instantaneous voltage
T is the time duration of the pulse
$\int_{dt}$ is the time integral

(RL) C136.10-1996

surge generator (impulse generator) An electric apparatus suitable for the production of surges. *Notes:* 1. Surge generator types common in the art are: transformer-capacitor; transformer-rectifier; transformer-rectifier-capacitor; parallel charging; series discharging. 2. Use of the term lightning generator is deprecated. (T&D/PE) [10], [8]

surge ground The point of external connection to the relay system reference or common bus for surge protection. (PE/PSR) C37.90.1-1989r

surge impedance (1) (self-surge impedance) The ratio between voltage and current of a wave that travels on a line of infinite length and of the same characteristics as the relevant line. *See also:* characteristic impedance. (PE) [8], [84]

(2) The impedance of an electrical circuit under surge conditions (which may differ significantly from the impedance of a circuit under steady state conditions). (PE/IC) 1143-1994r

(3) The ratio between voltage and current of a wave that travels on a conductor. (SUB/PE) 998-1996

surge let-through That part of the surge that passes by a surge-protective device with little or no alteration. *See also:* surge remnant. (SPD/PE) C62.45-1992r, C62.62-2000

surge life The number of surges of specified voltage and current amplitudes and waveshapes that may be applied to a device without causing degradation beyond specified limits. The pulse life applies to a device connected to an ac line of specified characteristics and to surges sufficiently spaced in time to preclude the effects of cumulative heating. (SPD/PE) C62.62-2000

surge protection *See:* rate-of-change protection.

surge-protective device (1) The generic term used to describe a device by its protective function, regardless of technology used, ratings, packaging, point of application, etc. (SPD/PE) C62.45-1992r

(2) A device intended to limit transient overvoltages, divert surge currents, or both. It contains at least one nonlinear component. (T&D/PE/IA/PSE) 1250-1995, C62.34-1996, 1100-1999

(3) An assembly of one or more components intended to limit or divert surges. The device contains at least one nonlinear component. (SPD/PE) C62.48-1995

(4) The generic term used to describe a device by its protective function, regardless of technology used, ratings, packaging, point of application, etc. It contains at least one nonlinear component. (SPD/PE) C62.62-2000

surge protective level (surge arresters) The highest value of surge voltage that may appear across the terminals under the prescribed conditions. (PE) [8]

surge protector (1) (gas-tube surge protective devices) A protective device, consisting of one or more surge arresters, a mounting assembly, optional fuses and short-circuiting devices, etc, which is used for limiting surge voltages on low-voltage ($\leq$1000 V rms or 1200 V dc) electrical and electronic equipment or circuits. (SPD/PE) C62.31-1987r

(2) An assembly of protective devices consisting of one or more series, parallel, or any combination of elements used to limit surge voltages, currents, or both to a specified level. (SPD/PE) C62.36-1994

(3) The term used to refer to a specific complete device [generally the equipment under test (EUT) in the context of the present guide], as opposed to a component of the surge protector or a generic surge-protective device. (SPD/PE) C62.45-1992r

surge rating (thyristor) Rated values for surge forward current is given for two time regions:

a) For times smaller than one-half cycle (at 50 hertz (Hz) or 60 Hz) down to approximately one millisecond (ms), the value is given in terms of maximum rated $\int dt_i^2$. They may be given by means of a curve or by single values. No immediate subsequent application of reverse blocking voltage is assumed.
b) Maximum values of surge forward current versus time up to at least 10 cycles. The frequency, the conducting period length, the current waveshape and the reverse blocking voltage capability including the rate-of-rise of voltage for the intervals after and between the surges are specified. In either case, a previous application of rate maximum junction temperature is assumed if not otherwise specifically mentioned.

(IA/IPC) 428-1981w

surge reference equalizer A surge protective device used for connecting equipment to external systems whereby all conductors connected to the protected load are routed, physically and electrically, through a single enclosure with a shared reference point between the input and output ports of each system. (IA/PSE) 1100-1999

surge remnant (surge testing for equipment connected to low-voltage ac power circuits) That part of an applied surge that remains downstream of one or several protective devices. *See also:* surge let-through. (SPD/PE) C62.45-1992r, C62.62-2000

surge-response current The current flowing in a surge-protective device during its diverting function upon occurrence of an impinging surge. (SPD/PE) C62.62-2000

surge-response voltage (1) The voltage profile appearing at the output terminals of a surge-protective device and applied to downstream loads, during and after a specified impinging surge, until normal, stable conditions are reached. (SPD/PE) C62.48-1995, C62.45-1992r

(2) The voltage that appears at the output terminals of a surge-protective device during and after a specified impinging surge, until normal stable conditions are reached. (SPD/PE) C62.62-2000

surge suppressor A device operative in conformance with the rate of change of current, voltage, power, etc., to prevent the rise of such quantity above a predetermined value. (IA/ICTL/IAC) [60]

surge, switching *See:* switching surge.

surge voltage recorder *See:* Lichtenberg figure camera.

surveillance (1) (diesel-generator unit) The determination of the state or condition of a system or subsystem. (PE/NP) 387-1995, 338-1987r

(2) (nuclear power quality assurance) The act of monitoring or observing to verify whether an item or activity conforms to specified requirements. (PE/NP) [124]

(3) The act of monitoring or observing whether an item or activity conforms to specific requirements. A surveillance is less extensive than an audit and concentrates on a single activity or item. It is usually conducted more frequently than an audit. Reports are issued to cognizant personnel or groups with a request for corrective action if required. (NI) N42.23-1995

surveillance radar (1) (navigation aid terms) A search radar used to maintain cognizance of selected traffic within a selected area, such as an airport terminal area or air route. (AES/GCS) 172-1983w

(2) A radar used to detect, locate, and track targets over a large volume of space. (AES) 686-1997

surveillance test The test that can determine the state or condition of a system or subsystem. (PE/NP) 933-1999

surveillance testing Periodic testing to verify that safety systems continue to function or are in a state of readiness to perform their safety function. (PE/NP) 338-1987r

survey (plutonium monitoring) The examination of an area for the purpose of detecting the presence of radioactive materials and determining the quantity of that radioactivity. (NI) N317-1980r

survey contamination control (plutonium monitoring) A survey conducted to determine the presence of unwanted contaminants, normally conducted with alpha or gamma, or both, sensitive instruments. (NI) N317-1980r

survey dose rate (plutonium monitoring) A survey conducted to determine the dose rate at some specified location or area and usually conducted with gamma exposure rate survey instruments. Neutron surveys may also be required frequently. (NI) N317-1980r

survey meter A lightweight battery operated meter that can be held conveniently by hand in order to conduct survey type measurements. (T&D/PE) 1308-1994

susceptance The imaginary part of admittance. (IM/HFIM) 270-1966w, [40]

susceptance relay A mho type of distance relay for which the center of the operating characteristic on the *R-X* diagram is on the *X* axis. *Note:* The equation that describes such a characteristic is $Z = K\sin\theta$, where K is a constant and θ is the phase angle by which the input voltage leads the input current. (SWG/PE) C37.100-1992

susceptibility (1) (grounding in generating stations) The property of an equipment that describes its capability to function acceptably when subjected to unwanted electromagnetic energy. (PE/EDPG) 1050-1996

(2) The inability of a device, equipment, or system to resist an electromagnetic disturbance. *Note:* Susceptibility is the lack of immunity. (SPD/PE) C62.45-1992r

(3) The property of equipment that describes its capability to function acceptably when subjected to unwanted interfering energy. (PE/PSC) 367-1996

(4) (electromagnetic) The characteristic of any equipment that results in an undesired response to an electromagnetic field. (SWG/PE) C37.100-1992

susceptiveness The characteristics of a communication circuit, including its connected apparatus, that determine the extent to which it is adversely affected by inductive fields. (NESC/T&D) C2-1997, C2.2-1960

susceptor Energy absorbing device generally used to transfer heat to another load. (IA) 54-1955w

suspended (illuminating engineering) (pendant) A luminaire which is hung from a ceiling by supports. (EEC/IE) [126]

suspended domain One or more suspended nodes linked by suspended connection(s). Two nodes are part of the same suspended domain if there is a physical connection between them and all ports on the path are suspended. A boundary node is adjacent to one or more suspended domain(s) but not part of the suspended domain(s). (C/MM) 1394a-2000

suspend initiator An active port that transmits the TX_SUSPEND signal and engages in a protocol with its connected peer physical layer (PHY) to suspend the connection. (C/MM) 1394a-2000

suspended job A job that has received a SIGSTOP, SIGTSTP, SIGTTIN, or SIGTTOU signal that caused the process group to stop. A suspended job is a background job, but a background job is not necessarily a suspended job. (C/PA) 9945-2-1993

suspended node An isolated node with at least one port that is suspended. (C/MM) 1394a-2000

suspended port A connected port not operational for normal Serial Bus arbitration, but otherwise capable of detecting both a physical cable disconnection and received bias. (C/MM) 1394a-2000

suspended stripline A type of stripline in which the major dielectric is empty space. The strip conductor is located on a thin dielectric substrate supported between two ground planes. The conductor can be either a single strip or two strips in double registration acting electrically as a single conductor. (MTT) 1004-1987w

suspended substrate microstrip A compound planar transmission line consisting of one or more thin conducting strips of finite width affixed to an insulating substrate of finite thickness and suspended above a single extended conducting ground plane with the strips facing the ground plane and separated from it by free space. The semi-infinite space above the substrate is also free space. (MTT) 1004-1987w

suspended-type handset telephone *See:* hang-up hand telephone set.

suspend target An active port that observes the RX_SUSPEND signal. A suspend target requests all of the physical layer's (PHY's) other active ports to become suspend initiators while the suspend target engages in a protocol with its connected peer PHY to suspend the connection. (C/MM) 1394a-2000

suspension (accelerometer) (inertial sensors) (gyros) A means of supporting and positioning a float (floated gyro), rotor (dynamically tuned gyro, electrically suspended gyro), or proof mass (accelerometer) with respect to the case. (AES/GYAC) 528-1994

suspension insulator (1) One or a string of suspension-type insulators assembled with the necessary attaching members and designed to support in a generally vertical direction the weight of the conductor and to afford adequate insulation from tower or other structure. *See also:* tower; insulator. (T&D/PE) [10]

(2) **(composite insulators)** As used in IEEE Std 987-1985, any insulator intended primarily to carry tension loads. It includes tangent, deadend, and vee-string installations. (T&D/PE) 987-1985w

suspension-insulator unit An assembly of a shell and hardware, having means for nonrigid coupling to other units or terminal hardware. (EEC/IEPL) [89]

suspension-insulator weights Devices, usually cast iron, hung below the conductor on a special spindle supported by the conductor clamp. *Note:* Suspension insulator weights will limit the swing of the insulator string, thus maintaining adequate clearances. In practice, weights of several hundreds of pounds are sometimes used. *See also:* tower. (T&D/PE) [10]

suspension of reclosing To make inoperative automatic reclosing equipment *Synonyms:* live-line permit; hold order; hold out; hold off. (T&D/PE) 516-1995

suspension strand (messenger) A stranded group of wires supported above the ground at intervals by poles or other structures and employed to furnish within these intervals frequent points of support for conductors or cables. *See also:* cable. (EEC/PE) [119]

sustained When used to quantify the duration of a voltage interruption, refers to the time frame associated with a long duration variation (i.e., greater than 1 min). (SCC22) 1346-1998

sustained bypass current detection (series capacitor) A means to detect prolonged current flow through the protective device and to initiate closing of the bypass device. (T&D/PE) 824-1985s

sustained gap-arc protection (series capacitor) A means to detect prolonged arcing of the protective power gap or arcing of the backup gap if included to initiate closing of the capacitor bypass switch. (T&D/PE) [26]

sustained interruption (1) (electric power system) Any interruption not classified as a momentary interruption. *See also:* outage. (PE/PSE) [54], 346-1973w
(2) Any interruption not classified as a momentary event. Any interruption longer than 5 min. (PE/T&D) 1366-1998
(3) A type of long duration variation. The complete loss of voltage (<0.1 pu) on one or more phase conductors for a time greater than 1 min. (SCC22) 1346-1998
(4) (power quality monitoring) The complete loss of voltage for a time period greater than 1 min. (IA/PSE) 1100-1999

sustained-operation influence The change in the recorded value, including zero shift, caused solely by energizing the instrument over extended periods of time, as compared to the indication obtained at the end of the first 15 min of the application of energy. It is to be expressed as a percentage of the full-scale value. *Note:* The coil used in the standard method shall be approximately 80 in in diameter, not over 5 in long, and shall carry sufficient current to produce the required field. The current to produce a field to an accuracy of ±1% in air shall be calculated without the instrument in terms of specific dimensions and turns of the coil. In this coil, 800 ampere-turns will produce a field of approximately 5 oersteds. The instrument under test shall be placed in the center of the coil. (EEC/ERI) [111]

sustained oscillation (1) (system) (sustained vibration) The oscillation when forces controlled by the system maintain a periodic oscillation of the system. Example: Pendulum actuated by a clock mechanism. (Std100) 270-1966w
(2) (gas turbines) Those oscillations in which the amplitude does not decrease to zero, or to a negligibly small, final value. (PE/EDPG) 282-1968w, [5]

sustained overvoltage detection device (series capacitor) A device that detects capacitor voltage above rating but below the operation level of the protective device and initiates an alarm signal or corrective action. (T&D/PE) 824-1985s

sustained overvoltage protection device (series capacitor) A device to detect capacitor voltage that is above rating or predetermined value but is below the sparkover of the protective power gaps, and to initiate the closing of the capacitor bypass switch according to a predetermined voltage-time characteristic. (T&D/PE) [26]

sustained short-circuit test (synchronous machines) A test in which the machine is run as a generator with its terminals short-circuited. (PE) [9]

SVC *See:* switched virtual circuit.

SVV *See:* Segment Variability Value.

swamp buggy *See:* off-road vehicle.

swap (A) An exchange of the contents of two storage areas, usually an area of main storage with an area of auxiliary storage. *See also:* roll in; roll out. **(B)** To perform an exchange as in definition (A). (C) 610.12-1990

sweep A traversing of a range of values of a quantity for the purpose of delineating, sampling, or controlling another quantity. *Notes:* 1. Examples of swept quantities are: the displacement of a scanning spot on the screen of a cathode-ray tube; and the frequency of a wave. 2. Unless otherwise specified, a linear time function is implies; but the sweep may also vary in some other controlled and desirable manner. (BT/IM/AV/HFIM) [34], [40]

sweep accuracy (oscilloscopes) Accuracy of the horizontal (vertical) displacement of the trace compared with the reference independent variable, usually expressed in terms of average rate error as a percent of full scale. *See also:* oscillograph. (IM) 311-1970w

sweep-delay accuracy (oscilloscopes) Accuracy of indicated sweep delay, usually specified in error terms. (IM) 311-1970w

sweep, delayed *See:* delayed sweep.

sweep duration (sawtooth sweep) The time required for the sweep ramp. *See also:* oscillograph. (IM/HFIM) [40]

sweep duty factor For repetitive sweeps, the ratio of the sweep duration to the interval between the start of one sweep and the start of the next. *See also:* oscillograph. (IM/HFIM) [40]

sweep, expanded *See:* magnified sweep.

sweep, external (oscilloscopes) A sweep generated external to the instrument. (IM) 311-1970w

sweep, free-running *See:* free-running sweep.

sweep frequency (oscilloscopes) The sweep repetition rate. *See also:* oscillograph. (IM/HFIM) [40]

sweep gate (oscilloscopes) Rectangular waveform used to control the duration of the sweep; usually also used to unblank the cathode-ray tube for the duration of the sweep. *See also:* oscillograph. (IM/HFIM) [40]

sweep, gated *See:* gated sweep.

sweep generator (oscilloscopes) A circuit that generates a signal used as an independent variable; the signal is usually a ramp, changing value at a constant rate. (IM) 311-1970w

sweep holdoff interval (oscilloscopes) The interval between sweeps during which the sweep and/or trigger circuits are inhibited. (IM) 311-1970w

sweep linearity (oscilloscopes) Maximum displacement error of the independent variable between specified points on the display area. (IM) 311-1970w

sweep mode control (oscilloscopes) The control used on some oscilloscopes to set the sweep for triggered, free-running, or synchronized operation. (IM) 311-1970w

sweep oscillator An oscillator in which the output frequency varies continuously and periodically between two frequency limits. *See also:* telephone station. (COM) [50]

sweep-out time, charge *See:* charge collection time.

sweep range (oscilloscopes) The set of sweep-time/division settings provided. *See also:* oscillograph. (IM/HFIM) [40]

sweep recovery time (oscilloscopes) The minimum possible time between the completion of one sweep and the initiation of the next, usually the sweep holdoff interval. *See also:* oscillograph. (IM/HFIM) [40]

sweep, recurrent *See:* recurrent sweep.

sweep reset (oscilloscopes) In oscilloscopes with single-sweep operation, the arming of the sweep generator to allow it to cycle once. *See also:* oscillograph. (IM/HFIM) [40]

sweep, sine-wave *See:* sine-wave sweep.

sweep, stairstep *See:* stairstep sweep.

sweep switching (automatic) Alternate display of two or more time bases or other sweeps using a single-beam cathode-ray tube: comparable to dual- or multiple-trace operation of the deflection amplifier. (IM) 311-1970w

sweep time (acoustically tunable optical filter) The time to continuously tune the filter over its spectral range. (UFFC) [17]

sweep time division (spectrum analyzer) The nominal time required for the spot in the reference coordinate to move from one graticule division to the next. Also the name of the control used to select this time. (IM) 748-1979w

swell (1) A momentary increase in the power frequency voltage delivered by the mains, outside of the normal tolerances, with a duration of more than one cycle and less than a few seconds. *See also:* surge. (SPD/PE) C62.48-1995, C62.41-1991r
(2) An rms increase in the ac voltage, at the power frequency, for durations from a half-cycle to a few seconds. *See also:* overvoltage; surge. (PE/T&D) 1250-1995
(3) An increase in rms voltage or current at the power frequency for durations from 0.5 cycles to 1 min. Typical values are 1.1 to 1.8 pu. See the figure below. (SCC22/IA/PSE) 1346-1998, 1100-1999

swellable powder A powder that swells upon contact with moisture. A jelly like material is formed to block the longitudinal transmission of moisture. (PE/IC) 1142-1995

swim The visual misrepresentation that occurs when images on a display surface appear to move about their normal positions. (C) 610.6-1991w

swing A transient power flow due to change in relative angles of generation on the system caused by a change in transmission or generation configuration. (PE/PSR) C37.113-1999

swinging compass (navigation aid terms) An accurate, portable magnetic compass used to indicate magnetic headings during aircraft magnetic compass calibration. (AES/GCS) 172-1983w

swingout panel (packaging machinery) A panel that is hinge-mounted in such a manner that the back of the panel may be made accessible from the front of the enclosure. (IA/PKG) 333-1980w

swing rack cabinet An assembly enclosed at the top, side, and rear with front hinged door for front access having a swing open frame for equipment mounting (e. g., nominal 19-inch wide chassis and subpanel assemblies). (SWG/PE) C37.100-1992, C37.21-1985r

switch (1) (telephone loop performance) (switching system) A system that establishes communication channels among two or more of its interfaces at customers' demand. (COM/TA) 820-1984r
(2) (high-voltage switchgear) A device designed to close or open, or both, one or more electric circuits. *See also:* switching device. (SWG/PE) C37.40-1993
(3) (computers) A device or programming technique for making a selection, for example, a toggle, a conditional jump. (C) [20], [85]
(4) (electric and electronics parts and equipment) A device for making, breaking, or changing the connections in an electric circuit. *Note:* a switch may be operated by manual, mechanical, hydraulic, thermal, barometric, or gravitational means, or by electromechanical means not falling within the definition of "relay." (GSD) 200-1975w
(5) A device that connects ringlets and has queues. It can behave as a consumer (when accepting remote subactions) and as a producer (when forwarding the subaction to another ringlet). It may be visible as a node, with a nodeId, or be transparent, with no nodeId. A switch differs from a bridge in that a switch may connect more than two ringlets, but a bridge connects only two. A switch is generally assumed to connect multiple instances of the same bus standard, while a bridge may connect different bus standards. (C/MM) 1596-1992
(6) A routing device (for example, a box or board) providing a set of numbered node interfaces, constructed from one or more switch chips (or by other methods). *See also:* switch chip; fabric; node interface. (C/BA) 1355-1995
(7) (A) An electrical or mechanical device used for opening, closing, or changing the connection of a circuit. *Synonym:*

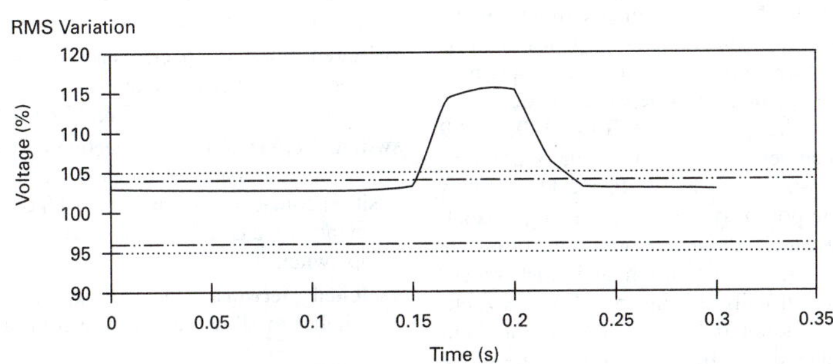

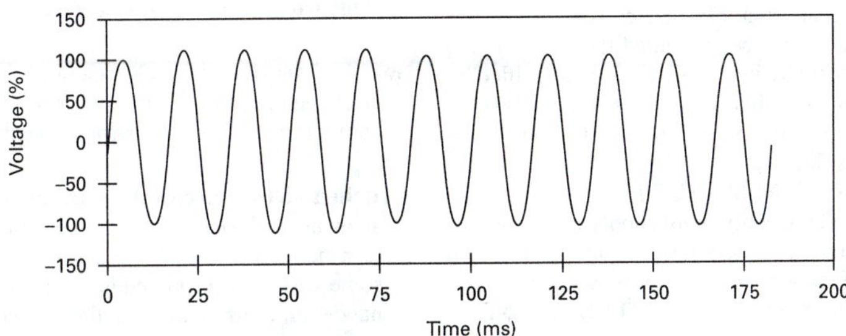

Swells occurring upon recovery from a remote system fault

swell

switchpoint. *See also:* DIP switch; display switch; sense switch; function switch; relay. **(B)** To open, close, or change the connection of a circuit as in definition (A). **(C)** A device used for making a selection, as in a toggle. (C) 610.10-1994

(8) A device for opening and closing or for changing the connection of a circuit. In these rules, a switch is understood to be manually operable, unless otherwise stated. (NESC/T&D) C2-1997, C2.2-1960

(9) In a propulsion system, the historic name for the lowest level of positive tractive effort and power; so called because it is typically utilized for slow-speed switching movements, such as yard moves, train makeup, etc. (VT) 1475-1999

(10) A layer 2 interconnection device that conforms to the ISO/IEC 10038 [ANSI/IEEE 802.1D-1990] International Standard. *Synonym:* bridge. (C/LM) 802.3-1998

(11) An electronic device connected between two data lines. A switch can exist in one of two states, referred to as "open" and "closed." The state at any time depends on a digital control variable. When the switch is open, the pathway between the two data lines has a very high impedance (ideally infinite) so that signals appearing on the data lines should be completely independent. When the switch is closed, the pathway between the two data lines has a very low impedance (ideally zero) so that signals on the two data lines should be identical. *Notes:* 1. Practical electronic switches implemented in silicon depart from the ideal in at least three ways.

a) In the "on" state, the pathway between the two data lines may have significant impedance, or the relationship between voltage and current may be nonlinear (e.g., a voltage-dependent "impedance").
b) In the "off" state, there may be significant interaction between the signals on the two data lines due to, for example, stray capacitance.
c) In either state there may be significant leakage pathways through which current can pass from the data lines to the surrounding circuitry or vice versa.

The effects of all these characteristics will need to be considered as part of the detailed implementation, especially in a system containing multiple-switch networks. 2. A switching action effectively in series with the function signal pathway can sometimes be obtained without a physically separate device by incorporating a high-Z or enable facility into the functional circuitry. 3. Data transmission through a switch is normally assumed to be bidirectional (as with electromechanical devices such as relays or semiconductor switches such as transmission gates). Some forms of switch can implement only unidirectional voltage or current dependence. *See also:* conceptual switch; high-Z. (C/TT) 1149.4-1999

switch base The main members to which the insulator units are attached. (SWG/PE) C37.30-1992s

switchboard (1) (electric power system) A large single panel, frame, or assembly of panels, on which are mounted, on the face or back or both, switches, overcurrent and other protective devices, buses, and usually instruments. *Note:* Switchboards are generally accessible from the rear as well as from the front and are not intended to be installed in cabinets. *See also:* panelboard; center of distribution; distribution center. (NESC) [86]

(2) A type of switchgear assembly that consists of one or more panels with electric devices mounted thereon, and associated framework. *Note:* Switchboards may be classified by function, that is, power switchboards or control switchboards. Both power and control switchboards may be further classified by construction as defined. (SWG/PE/NESC) C37.100-1992, C2-1997

(3) When referred to in connection with supply of electricity, a large single panel, frame, or assembly of panels, on which are mounted (on the face, or back, or both) switches, fuses, buses, and usually instruments. (T&D) C2.2-1960

switchboard cord A cord that is used in conjunction with switchboard apparatus to complete or build up a telephone connection. (EEC/PE) [119]

switchboard lamp (switchboard) A small electric lamp associated with the wiring in such a way as to give a visual indication of the status of a call or to give information concerning the condition of trunks, subscriber lines, and apparatus. (EEC/PE) [119]

switchboard position (telephone switching systems) That portion of a manual switchboard normally provided for the use of one operator. (COM) 312-1977w

switchboards and panels (electric installations on shipboard) A generator and distribution switchboard receives energy from the generating plant and distributes directly or indirectly to all equipment supplied by the generating plant. A subdistribution switchboard is essentially a section of the generator and distribution switchboard (connected thereto by a bus-feeder and remotely located for reasons of convenience or economy) that distributes energy for lighting, heating, and power circuits in a certain section of the vessel. A distribution panel receives energy from a distribution or subdistribution switchboard and distributes energy to energy-consuming devices or other distribution panels or panelboards. A panelboard is a distribution panel enclosed in a metal cabinet. (IA/MT) 45-1983s

switchboard section (telephone switching systems) A structural unit providing for one or more operator positions. A complete switchboard may consist of one or more sections. (COM) 312-1977w

switchboard supervisory lamp (cord circuit or trunk circuit) A lamp that is controlled by one or other of the users to attract the attention of the operator. (EEC/PE) [119]

switchboard supervisory relay A relay that controls a switchboard supervisory lamp. (EEC/PE) [119]

switch chip A VLSI integrated circuit with two or more link interfaces, between which it provides packet routing. *See also:* link; switch. (C/BA) 1355-1995

switch compartment (metal-enclosed interrupter switchgear) That portion of the switchgear assembly that contains one switching device, such as an interrupter switch, power fuse interrupter switch combination, etc., and the associated primary conductors. (SWG/PE) C37.20.3-1996

switch core A magnetic core in which the core material generally has a high residual flux density and a high ratio of residual to saturated flux density; Switching does not occur when the magnetic force imposed on the core is below a threshold value. (C) 610.10-1994w

switched bank A capacitor bank designed for controlled operation.power systems relaying. (T&D/PE) 1036-1992, C37.99-2000

switched current The prospective current to be broken during a switching operation by each set of main switching or transition contacts (resistance-type LTC) or transfer contacts (reactance-type LTC) incorporated in the arcing switch or arcing tap switch. (PE/TR) C57.131-1995

switched network (1) A computer interconnect that uses switches to allow intermodule communications. (C/BA) 14536-1995

(2) A network, using a switching technique, to direct messages from the sender to the ultimate recipient. *See also:* circuit-switched network; store-and-forward switched network. (C) 610.7-1995

switched-service network (telephone switching systems) An arrangement of dedicated switching facilities to provide telecommunications services for a specific customer. (COM) 312-1977w

switched virtual circuit A virtual circuit that is established on an as-needed basis to interconnect any two end users attached to a network. *Note:* SVC service requires the definition of some call control procedures for the establishment, maintenance, and termination of the virtual circuit. An SVC may not be available when the user wants if too many SVCs are open at once. *See also:* permanent virtual circuit. (C) 610.7-1995

switched way (1) A way connected to the bus through a three-pole, group operated switch. (SWG/PE) C37.71-1984r
(2) A way connected to the bus through a switch. (SWG/PE) C37.100-1992
(3) A way connected to the bus through a three-phase group-operated switch or single-phase switch. (SWG/PE) C37.73-1998

switchgear (1) A general term covering switching and interrupting devices and their combination with associated control, instrumentation, metering, protective and regulating devices, also assemblies of these devices with associated interconnections, accessories and supporting structures used primarily in connection with the generation, transmission, distribution, and conversion of electric power. (SWG/PE/IA/PSP) C37.20.3-1996, 1015-1997, C37.20.2-1993, C37.20.1-1993r, C37.100-1992
(2) (hydroelectric power plants) An assembly of equipment used to switch and control electrical power. (PE/EDPG) 1020-1988r

switchgear assembly An assembled equipment (indoor or outdoor) including, but not limited to, one or more of the following categories: switching, interrupting, control, instrumentation, metering, protective, and regulating devices; together with their supporting structures, enclosures, conductors, electrical interconnections, and accessories. (SWG/PE) C37.20.1-1993r, C37.20.2-1993, C37.20.3-1996, C37.100-1992

switchgear pothead A pothead intended for use in a switchgear where the inside ambient air temperature may exceed 40°C. It may be an indoor or outdoor pothead that has been suitably modified by silver surfacing (or the equivalent) the current-carrying parts and incorporates sealing materials suitable for the higher operating temperatures. *See also:* pothead. (PE) 48-1975s

switchgear, protective *See:* protective switchgear.

switch hook (hookswitch) A switch on a telephone set, associated with the structure supporting the receiver or handset. It is operated by the removal or replacement of the receiver or handset on the support. *See also:* telephone station; switch stick. (EEC/PE) [119]

switch indicator (1) A device used at a noninterlocked switch to indicate the presence of a train in a block. (EEC/PE) [119]
(2) An indicator that shows the setting of a switch. *See also:* flag. (C) 610.10-1994w

switching (1) The process by which the remanent polarization is reversed (or reoriented) to a new value of P_r (generally equal and opposite). Switching can be produced by electric fields or mechanical stresses. (UFFC) 180-1986w
(2) (single-phase motor) The point in the starting operation at which the stator-winding circuits are switched from one connection arrangement to another. *See also:* asynchronous machine. (PE) [9]
(3) (test, measurement, and diagnostic equipment) The act of manually, mechanically or electrically actuating a device for opening or closing an electrical circuit. (MIL) [2]
(4) In networking, pertaining to a connection that is established by closing switches. *See also:* packet switching; message switching; digital switching; circuit switching. (C) 610.7-1995
(5) The process of using a switch. (C) 610.10-1994w

switching amplifier An amplifier that is designed to be applied so that its output is sustained at one of two specified states dependent upon the presence of specified inputs. *See also:* feedback control system. (IA/ICTL/IAC) [60]

switching array (telephone switching systems) An assemblage of multipled crosspoints. (COM) 312-1977w

switching branch (synchronous motor drives) A part of the circuit, including at least one switching element, bounded by two principal terminals. *Note:* A switching branch may include one or more simultaneously conducting converter switching elements connected together, commutating reactor windings, and other devices intended to protect the semiconductor devices or to ensure their proper function, such as voltage and current dividers, and snubbers. In the simplest case, a switching branch may consist of only the switching element, which may be a single semiconductor device. The adjective "switching" may be omitted when the context of converter circuits is clear. (IA/ID/SPC) 995-1987w, 936-1987w

switching card (test, measurement, and diagnostic equipment) A plug-in device that provides the necessary interconnection to the unit under test. (MIL) [2]

switching circuit (data transmission) Term applied to the method of handling traffic through a switching center, either from a local user or from other switching centers, whereby additional electrical connection is established between the calling and the called station. (PE) 599-1985w

switching coefficient The derivative of applied magnetizing force with respect to the reciprocal of the resultant switching time. It is usually determined as the reciprocal of the slope of a curve of reciprocals of switching times versus values of applied magnetizing forces. The magnetizing forces are applied as step functions. (C) [20]

switching computer A communications computer designed to handle switching messages or packets in a network. *See also:* gate. (C) 610.7-1995, 610.10-1994w

switching control center (telephone switching systems) A place where maintenance analysis and control activities are centralized for switching entities situated in different locations. (COM) 312-1977w

switching current The value of current expressed in rms symmetrical amperes that the power circuit breaker element of the circuit protector interrupts at the rated maximum voltage and rated frequency under the prescribed test conditions. (SWG/PE) C37.100-1992, C37.29-1981r

switching current rating The designated rms current that a load-break connector can connect and disconnect for a specified number of times under specified conditions. (T&D/PE) 386-1995

switching device (switch) A device designed to close or open, or both, one or more electric circuits. *Note:* The term *switch* in international (IEC) practice refers to a mechanical switching device capable of opening and closing rated continuous load current. *See also:* nonmechanical switching device; mechanical switching device. (SWG/PE) C37.100-1992

switching entity (telephone switching systems) A switching network and its control. (COM) 312-1977w

switching function A function that has only a finite number of possible values and whose independent variables each have only a finite number of possible values. (C) 1084-1986w

switching impulse (1) Ideally, an aperiodic transient impulse voltage that rises rapidly to a maximum value and falls, usually less rapidly, to zero. Switching impulses generally have front times of the order of tens to thousands of microseconds, in contrast to lightning impulses, which have front times from fractions of a microsecond to tens of microseconds. 332-1972w
(2) An impulse with a front duration of some tens to thousands of microseconds. (PE/PSIM) 4-1995

switching impulse insulation level (power and distribution transformers) An insulation level expressed in kilovolts of the crest value of a switching impulse withstand voltage. (PE/TR) C57.12.80-1978r

switching impulse protective level (of a surge protective device) The maximum switching impulse expected at the terminals of a surge protective device under specified conditions of operation. *Note:* The switching impulse protective levels given by the higher of either: a) the switching impulse discharge voltage for a specified current magnitude and wave shape, or b) the switching impulse sparkover voltage for a specified voltage wave shape. (C/PE/TR) 1313.1-1996, C57.12.80-1978r

switching-impulse sparkover voltage (arresters) The impulse sparkover voltage with an impulse having a virtual duration of wavefront greater than 30 μs. (PE) [8]

switching impulse test (power and distribution transformers) Application of the "standard switching impulse," a full wave having a front time of 250 μ and a time to half value of 2500 μ, described as a 250/2500 impulse. *Note:* It is recognized that some apparatus standards may have to use a modified wave shape where practical test considerations or particular dielectric strength characteristics make some modification imperative. Transformers, for example, use a modified switching impulse wave with the following characteristics:

— Time to crest greater than 100 μ.
— Exceeds 90% of crest value for at least 200 μ.
— Time to first voltage zero on tail not less than 1000 μ, except where core saturation causes the tail to become shorter.

(PE/TR) C57.12.80-1978r

switching impulse withstand voltage The crest value of a voltage impulse with a front duration from tens to thousands of microseconds that, under specified conditions, can be applied without causing flashover or puncture. (SWG/PE) C37.34-1994

switching network (telephone switching systems) Switching stages and their interconnections. Within a switching system there may be more than one switching network. (COM) 312-1977w

switching-network plan (telephone switching systems) The switching stages and their interconnections within a specific switching system. (COM) 312-1977w

switching node The intelligent interface point where the customer's equipment is connected to a public packet switching network. (C) 610.7-1995

switching overvoltage (1) A transient overvoltage in which a slow front, short duration, unidirectional or oscillatory, highly damped voltage is generated (usually by switching or faults). (PE/C) 1313.1-1996

(2) Any combination of switching surge(s) and temporary overvoltage(s) associated with a single switching episode. (SPD/PE) C62.22-1997

switching plan (telephone switching systems) A plan for the interconnection of switching entities. (COM) 312-1977w

switching, slave-sweep *See:* slave-sweep switching.

switching stage (telephone switching systems) An assemblage of switching arrays within each inlet that can be connected through a single crosspoint to its associated outlet. (COM) 312-1977w

switching station (power operations) A station where transmission lines are connected without power transformers. (PE/PSE) 858-1987s

switching structure An open framework supporting the main switching and associated equipment, such as instrument transformers, buses, fuses, and connections. It may be designed for indoor or outdoor use and may be assembled with or without switchboard panels carrying the control equipment. (SWG/PE) C37.100-1992

switching surge (1) (conductor stringing equipment) A transient wave of overvoltage in an electrical circuit caused by a switching operation. When this occurs, a momentary voltage surge could be induced in a circuit adjacent and parallel to the switched circuit in excess of the voltage induced normally during steady-state conditions. If the adjacent circuit is under construction, switching operations should be minimized to reduce the possibility of hazards to the workers. (T&D/PE) 524a-1993r, 524-1992r, 1048-1990

(2) A heavily damped transient electrical disturbance associated with switching. System insulation flashover may precede or follow the switching in some cases but not all. (SPD/PE) C62.22-1997

switching-surge protective level (arresters) The highest value of switching-surge voltage that may appear across the terminals under the prescribed conditions. *Note:* The switching-surge protective levels are given numerically by the maximums of the following quantities: (1) discharge voltage at a given discharge current, and (2) switching-impulse sparkover voltage. (PE) [8]

switching system (telephone switching systems) A system in which connections are established between inlets and outlets either directly or with intermediate storage. (COM) 312-1977w

switching-system processor (telephone switching systems) Circuitry to perform a series of switching system operations under control of a program. (COM) 312-1977w

switching time (T_S) (1) (A) (magnetic storage cells) T_s, the time interval between the reference time and the last instant at which the instantaneous voltage response of a magnetic cell reaches a stated fraction of its peak value. **(B) (magnetic storage cells)** T_x, the time interval between the reference time and the first instant at which the instantaneous integrated voltage response reaches a stated fraction of its peak value. (C) [20]

(2) (hybrid computer linkage components) (settling time) That time required from the time at which a channel is addressed until the selected analog signal is available at the output within a given accuracy. (C) 166-1977w

(3) The time required for a device to change from one state to another. (C) 610.10-1994w

(4) (reliable industrial and commercial power systems planning and design) The period from the time a switching operation is required due to a component failure until that switching operation is completed. Switching operations include such operations as throwover to an alternate circuit, opening or closing a sectionalizing switch or circuit breaker, reclosing a circuit breaker following a trip-out due to a temporary fault, etc. (PE/IA/PSE) [54], 493-1997, 399-1997

(5) Time taken to switch from one transmission direction to the other in alternating single talk conversation on a handsfree telephone (HFT). (COM/TA) 1329-1999

switching torque (1) (motor having an automatic connection change during the starting period). The minimum external torque developed by the motor as it accelerates through switch operating speed. *Note:* It should be noted that if the torque on the starting connection is never less than the switching torque, the pull-up torque is identical with the switching torque; however, if the torque on the starting connection falls below the switching torque at some speed below switch operating speed, the pull-up and switching torques are not identical. *See also:* asynchronous machine. (EEC/PE) [119]

(2) (single-phase motor) The minimum torque which a motor will provide at switching at normal operating temperature, with rated voltage applied at rated frequency. *See also:* asynchronous machine. (PE) [9]

switching transients (radiation survey instruments) Sudden excursions of the meter which occur when the range switch is changed from one position to the next. (NI) N13.4-1971w

switching variable A variable that may take only a finite number of possible values or states. *Synonym:* logical variable. (C) 1084-1986w

switch inrush current capability for single capacitance (as applied to interrupter switches) This capability is a function of the rated switching current, for single capacitance, the rated differential capacitance voltage (minimum) and the maximum design voltage of the switch. *Note:* This can be calculated from the equation:

$$\text{Capability, in Peak Amperes} = \sqrt{2}I_C\sqrt{1 + \frac{0.816E_m}{\Delta V_{min}}}$$

where

I_C = Rated switching current for single capacitance
ΔV_{min} = Rated differential capacitance voltage, minimum
E_m = Switch rated maximum voltage, in volts, rms

(SWG/PE) C37.100-1992

switch, load matching *See:* load-matching switch.

switch, load transfer *See:* load transfer switch.

switch machine A quick-acting mechanism, electrically controlled, for positioning track switch points, and so arranged that the accidental trailing of the switch points does not cause damage. A switch machine may be electrically or pneumatically operated. *See also:* car retarder. (EEC/PE) [119]

switch machine lever lights A group of lights indicating the position of the switch machine. (EEC/PE) [119]

switch-machine point detector *See:* point detector.

switch mode (thyristor) The starting instant of the controller ON-state interval is nonperiodic. This instant may be random (analogous to contactor operation), or it may be selected, for example, at voltage zero. (IA/IPC) 428-1981w

switch onto fault protection This provides tripping in the event that the breaker is closed into a zero voltage bolted fault, such as occurs if the grounding chains were left on the line following maintenance. (PE/PSR) C37.113-1999

switch, optical *See:* optical switch.

switch or contactor, load *See:* load switch or contactor.

switch part class designation A code which identifies the curve that relates the loadability factor (LF) of a switch part material and function to the ambient temperature θ_A. (SWG/PE) C37.30-1992s, C37.37-1996

switch point (watthour meters) The transition from one time-of-use period to another. (ELM) C12.13-1985s

switchpoint *See:* switch.

switch register A register made up of a number of manual switches, typically equal to the number of bits in the computer, and generally located on the computer control panel. *Note:* Used to manually enter addresses and data into main storage and to manually intervene in program execution. (C) 610.10-1994w

switchroom (telephone switching systems) That part of a building that houses an assemblage of switching equipment. (COM) 312-1977w

switch signal A low two-indication horizontal color light signal with electric lamps for indicating position of switch or derail. (EEC/PE) [119]

switch sleeve A component of the linkage between the centrifugal mechanism and the starting-switch assembly. *See also:* starting-switch assembly. (EEC/PE) [119]

switch starting *See:* preheat-starting.

switch stick A device with an insulated handle and a hook or other means for performing stick operation of a switching device. *Synonym:* switch hook. (SWG/PE) C37.100-1992

switch train A series of switches in tandem. (EEC/PE) [119]

switch-type function generator A function generator using a multitap switch rotated in accordance with the input and having its taps connected to suitable voltage sources. *See also:* electronic analog computer. (C) 165-1977w

swivel link (conductor stringing equipment) A swivel device designed to connect pulling lines and conductors together in series or connect one pulling line to the drawbar of a pulling vehicle. The device will spin and help relieve the torsional forces which build up in the line or conductor under tension. (T&D/PE) 524-1992r

SWR *See:* residual standing-wave ratio; standing-wave-ratio indicator.

syllabic companding (modulation systems) Companding in which the gain variations occur at a rate comparable to the syllabic rate of speech; but do not respond to individual cycles of the audio-frequency signal wave. (IT) [7]

syllable articulation (percent syllable articulation) The percent articulation obtained when the speech units considered are syllables (usually meaningless and usually of the consonant-vowel-consonant type). *See also:* volume equivalent; articulation (percent articulation) and intelligibility (percent intelligibility). (EEC/PE) [119]

syllable hyphen *See:* discretionary hyphen.

symbol (1) A representation of something by reason of relationship, association, or convention. *See also:* logic symbol. (C) [20], [85]
(2) (packaging machinery) A sign, mark, or drawing agreed upon to represent an electrical device of component part thereof. (IA/PKG) 333-1980w
(3) (computer graphics) A conventional representation of an object, composed of one or more display elements that is expressed as a unit. (C) 610.6-1991w
(4) A 16 bit unit of data accompanied by flag information. The flag information may be explicitly present as a 17th bit, or implied by the context. Symbols are transmitted one after another to form SCI packets or idles. The particular physical layer used to transmit these symbols is not visible to the logical layer. (C/MM) 1596-1992
(5) Refers to data within an SCI packet. A 16-bit unit of data accompanied by flag information. The flag information may be explicitly present as a 17th bit, or implied by the context. Symbols are transmitted one after another to form SCI packets or idles. The particular physical layer used to transmit these symbols is not visible to the logical layer. (C/MM) 1596.3-1996
(6) A 10-bit, 8B/10B encoded byte. (C/BA) 1393-1999
(7) Two signal elements. Four symbols are defined: data_zero, data_one, non-data_J, and non-data_K. (C/LM) 8802-5-1998
(8) The smallest unit of data transmission on the medium. Symbols are unique to the coding system employed. 100BASE-T4 uses ternary symbols; 10BASE-T and 100BASE-X use binary symbols or code bits; 100BASE-T2 uses quinary symbols. (C/LM) 802.3-1998
(9) The radio-frequency (RF) energy and/or the RF current advisory symbols. (NIR/SCC28) C95.2-1999
(10) (data management) *See also:* code. (C) 610.5-1990w

symbol for a quantity (abbreviation) (quantity symbol) A letter (which may have letters or numbers, or both, as subscripts or superscripts, or both), used to represent a physical quantity or a relationship between quantities. *Compare with:* abbreviation; functional designation; mathematical symbol; reference designation; symbol for a unit. *See also:* abbreviation. (GSD) 267-1966

symbol for a unit (unit symbol) (abbreviation) A letter, a character, or combinations thereof, that may be used in place of the name of the unit. With few exceptions, the letter is taken from the name of the unit. *Compare with:* abbreviation; mathematical symbol; symbol for a quantity. *See also:* abbreviation. (GSD) 267-1966

symbolic address (1) (computers) An address expressed in symbols convenient to the programmer. (C) [20], [85]
(2) (software) An address expressed as a name or label that must be translated to the absolute address of the device or storage location to be accessed. *Contrast:* absolute address. (C) 610.12-1990
(3) An address, expressed in symbols convenient to the computer programmer, that will be translated to an absolute or virtual address before it can be interpreted by the computer. (C) 610.10-1994w

symbolic addressing An addressing mode in which the address field of an instruction contains a symbolic address. (C) 610.10-1994w

symbolic coding (computers) Coding that uses machine instructions with symbolic addresses. (C) [20], [85]

symbolic execution (software) A software analysis technique in which program execution is simulated using symbols, such as variable names, rather than actual values for input data, and program outputs are expressed as logical or mathematical expressions involving these symbols. (C) 610.12-1990

symbolic image A digital image in which the value associated with each pixel is a symbol, rather than a gray level. (C) 610.4-1990w

symbolic language A programming language that expresses operations and addresses in symbols convenient to humans rather than in machine language. Examples are assembly language, high-order language. *Contrast:* machine language. *See also:* list processing language. (C) 610.12-1990, 610.13-1993w

symbolic link A type of file that contains a pathname. Rather than containing data itself, this type of file will resolve to another, as defined by the contained pathname. The way in which this type of file is handled by implementations of this standard is undefined. (C/PA) 1387.2-1995

symbolic logic The discipline that treats formal logic by means of a formalized artificial language or symbolic calculus whose purpose is to avoid the ambiguities and logical inadequacies of natural languages. (C) [20], [85]

symbolic model A model whose properties are expressed in symbols. Examples include graphical models, mathematical models, narrative models, software models, and tabular models. *Contrast:* physical model. (C) 610.3-1989w

symbolic processor (A) A computer which manipulates data at the algorithm level, typically not reducing computed equation values to a numerical resultant value. **(B)** A processor optimized to manipulate character strings and other symbolic data. *Note:* This is often done in the LISP or Prolog programming languages. (C) 610.10-1994

Symbolic Programming Systems (SPS) A programming language in which terms may represent quantities and locations. (C) 610.13-1993w

symbolic quantity *See:* mathematico-physical quantity.

symbolic trace A record of the source statements and branch outcomes that are encountered when a computer program is executed using symbolic, rather than actual, values for input data. *See also:* subroutine trace; variable trace; retrospective trace; execution trace. (C) 610.12-1990

symbol manipulation language *See:* list processing language.

symbol rank *See:* digit place.

symbol rate (1) The total number of symbols per second transferred to or from the Medium Dependent Interface (MDI) on a single wire pair. For 100BASE-T4, the symbol rate is 25 MBd; for 100BASE-X, the symbol rate is 125 MBd; for 100BASE-T2, the symbol rate is 25 MBd. (C/LM) 802.3-1998
(2) (local area networks) The number of symbols transmitted per second and expressed in baud (e.g., 1 Mbd = 1 000 000 symbols per second). (C) 8802-12-1998

symbol table A table that presents program symbols and their corresponding addresses, values, and other attributes. (C) 610.12-1990

symbol time (ST) The duration of one symbol as transferred to and from the Medium Dependent Interface (MDI) via a single wire pair. The symbol time is the reciprocal of the symbol rate. (C/LM) 802.3-1998

symmetrical The shape of the ac current waves about the zero axis (when both sides have equal value and configuration). (IA/PSE) 241-1990r

symmetrical alternating current A periodic alternating current in which points one-half a period apart are equal and have opposite signs. *See also:* network analysis; alternating function. (Std100) 270-1966w

symmetrical channel One of a pair of channels in which the transmit and receive directions of transmission have the same data signaling rate. (C) 610.10-1994w

symmetrical component (ac component) (of a total current) That portion of the total current that constitutes the symmetry. (SWG/PE) C37.100-1992

symmetrical components (A) (set of polyphase alternating voltages) The symmetrical components of an unsymmetrical set of sinusoidal polyphase alternating voltages of m phases are the m symmetrical sets of polyphase voltages into which the unsymmetrical set can be uniquely resolved, each component set having an angular phase lag between successive members of the set that is a different integral multiple of the characteristic angular phase difference for the number of phases. The successive component sets will have phase differences that increase from zero for the first set to $(m - 1)$ times the characteristic angular phase difference for the last set. The phase sequence of each component set is identified by the integer that denotes the number of times the angle of lag between successive members of the component set contains the characteristic angular phase difference. If the members of an unsymmetrical set of alternating polyphase voltages are not sinusoidal, each voltage is first resolved into its harmonic components, then the harmonic components of the same period are grouped to form unsymmetrical sets of sinusoidal voltages, and finally each harmonic set of sinusoidal voltages is uniquely resolved into its symmetrical components. Because the resolution of a set of polyphase voltages into its harmonic components is also unique, it follows that the resolution of an unsymmetrical set of polyphase voltages into its symmetrical components is unique. There may be a symmetrical-component set of voltages for each of the possible phase sequences from zero to $(m - 1)$ and for each of the harmonics present from 1 to r, where r may approach infinity in particular cases. Each member of a set of symmetrical component voltages of kth phase sequence and rth harmonic may be denoted by

$$e_{ski} = (2)^{1/2} E_{akr} \cos\left(r\omega t + \alpha_{akr} - (s - 1) K \frac{2\pi}{m} \right)$$

where e_{skr} is the instantaneous voltage component of phase sequence k and harmonic r in phase s. E_{akr} is the root-mean-square amplitude of the voltage component of phase sequence k and harmonic r, using phase a as reference, a_{akr} is the phase angle of the first member of the set, selected as phase a, with respect to a common reference. The letter s as the first subscript denotes the phase identification of the individual member, a, b, c, etc., for successive members, and a denotes that the first phase, a, has been used as a reference from which other members are specified. The second subscript, k, denotes the phase sequence of the component, and may run from 0 to $m - 1$. The third subscript denotes the order of the harmonic, and may run from 1 to ∞. The letter s as an algebraic quantity denotes the member of the set and runs from 1 for phase a to m for the last phase. Of the m symmetrical component sets for each harmonic, one will be of zero phase sequence, one of positive phase sequence, and one of negative phase sequence. If the number of phases m $(m > 2)$ is even, one of the symmetrical component sets for $k = m/2$ will be a single-phase symmetrical set (polyphase voltages). The zero-phase-sequence component set will constitute a zero-phase symmetrical set (polyphase voltages), and the remaining sequence components will constitute polyphase symmetrical sets (polyphase voltages). *See also:* network analysis. **(B) (set of polyphase alternating currents)** Obtained from the corresponding definition for symmetrical components (set of polyphase alternating voltages) by substituting the word current for voltage wherever it appears. *See also:* network analysis. (Std100/OLD TERMS) 270-1966

symmetrical fractional-slot winding (rotating machinery) A distributed winding in which the average number of slots per pole per phase is not integral, but in which the winding pattern repeats after every pair of poles, for example, $3\frac{1}{2}$ slots per pole per phase. *See also:* rotor; stator. (PE) [9]

symmetrical grid current That portion of the symmetrical ground fault current that flows between the grounding grid and surrounding earth. It may be expressed as

$$I_g = S_f \times I_f$$

where

I_g = the rms symmetrical grid current in A
I_f = the rms symmetrical ground fault current in A
S_f = the fault current division factor

(PE/SUB) 80-2000

symmetrical ground fault current The maximum rms value of symmetrical fault current after the instant of a ground fault initiation. As such, it represents the rms value of the symmetrical component in the first half-cycle of a current wave that develops after the instant of fault at time zero. For phase-to-ground faults

$I_{f(0+)} = 3I''_0$

where

$I_{f(0+)}$ = the initial rms symmetrical ground fault current

I''_0 = the rms value of zero-sequence symmetrical current that develops immediately after the instant of fault initiation, reflecting the subtransient reactances of rotating machines contributing to the fault

This rms symmetrical fault current is shown in an abbreviated notation as I_f, or is referred to only as $3I_0$. The underlying reason for the latter notation is that, for purposes of this guide, the initial symmetrical fault current is assumed to remain constant for the entire duration of the fault. (PE/SUB) 80-2000

symmetrically cyclically magnetized condition A condition of a magnetic material when it is in a cyclically magnetized condition and the limits of the applied magnetizing forces are equal and of opposite sign, so that the limits of flux density are equal and of opposite sign. (Std100) 270-1966w

symmetrical network *See:* structurally symmetrical network.

symmetrical periodic function A function having the period 2π is symmetrical if it satisfies one or more of the following identities.

(1) $f(x) = -f(-x)$

(2) $f(x) = -f(\pi + x)$

(3) $f(x) = -f(\pi - x)$

(4) $f(x) = f(-x)$

(5) $f(x) = f(\pi + x)$

(6) $f(x) = f(\pi - x)$

See also: network analysis. (Std100) 270-1966w

symmetrical set (A) (polyphase voltages) A symmetrical set of polyphase voltages of m phases is a set of polyphase voltages in which each voltage is sinusoidal and has the same amplitude, and the set is arranged in such a sequence that the angular phase difference between each member of the set and the one following it, and between the last member and the first, can be expressed as the same multiple of the characteristic angular phase difference $2\pi/m$ radians. A symmetrical set of polyphase voltages may be expressed by the equations

$$e_a = (2)^{1/2}E_{ar}\cos(r\omega t + \alpha_{ar})$$

$$e_b = (2)^{1/2}E_{ar}\cos\left(r\omega t + \alpha_{ar} - k\frac{2\pi}{m}\right)$$

$$e_c = (2)^{1/2}E_{ar}\cos\left(r\omega t + \alpha_{ar} - 2k\frac{2\pi}{m}\right)$$

$$e_m = (2)^{1/2}E_{ar}\cos\left(r\omega t + \alpha_{ar} - (m-1)\,k\frac{2\pi}{m}\right)$$

where E_{ar} is the root-mean-square amplitude of each member of the set, r is the order of the harmonic of each member, with respect to a specified period. a_{ar} is the phase angle of the first member of the set with respect to a selected reference. k is an integer that denotes the phase sequence. *Notes:* 1. Although sets of polyphase voltages that have the same amplitude and waveform but that are not sinusoidal possess some of the characteristics of a symmetrical set, only in special cases do the several harmonics have the same phase sequence. Since phase sequence is an important feature in the use of symmetrical sets, the definition is limited to sinusoidal quantities. This represents a change from the corresponding definition in the 1941 edition of the American Standard Definitions of Electrical Terms. 2. This definition may be applied to a two-phase four-wire or five-wire circuit if m is considered to be 4 instead of 2. The concept of symmetrical sets is not directly applicable to a two-phase three-wire circuit. **(B) (polyphase currents)** This definition is obtained from the corresponding definitions for voltage by substituting the word current for voltage, and the symbol I for E and b for a wherever they appear. The subscripts are unaltered. *See also:* network analysis. (Std100) 270-1966

symmetrical terminal voltage Terminal voltage measured in a delta network across the mains lead. *See also:* electromagnetic compatibility. (EMC) [53]

symmetrical transducer (specified terminations in general) A transducer in which all possible pairs of specified terminations may be interchanged without affecting transmission. *See also:* transducer. (Std100) 270-1966w

symmetric channel *See:* binary symmetric channel; symmetrical channel.

Symmetric List Processing Language (SLIP) A high-order list processing language using a structure reader, a parser that can traverse a data structure. (C) 610.13-1993w

symmetric multiprocessor A multiprocessor system in which each processor is equal to all others. *Contrast:* asymmetric multiprocessor. (C) 610.10-1994w

symmetric system Tunnel lighting system or luminaires having a symmetric light distribution with respect to the direction of travel. (RL) C136.27-1996

symmetric traversal *See:* inorder traversal.

synapse The junction between two neural elements, which has the property of one-way propagation. (EMB) [47]

sync (1) (television) Abbreviation for synchronizing signal extensively used in speech and writing. *Note:* This abbreviation is so commonly used that it has achieved the status of a word. (BT/AV) 201-1979w

(2) A function in which a bedside communications controller (BCC) may precisely synchronize a common fiduciary (now) maintained by multiple device communications controllers (DCCs) to the value of its own real-time clock, to an accuracy of approximately 1 ms. The BCC implements this function by transmitting a 3 μs pulse to one or more DCCs. A sync pulse may also be transmitted by a BCC or a DCC for reasons other than time synchronization. (EMB/MIB) 1073.4.1-2000

synchro control transformer (synchro or selsyn devices) A transformer with relatively rotatable primary and secondary windings. The primary inputs is a set of two or more voltages from a synchro transmitter that define an angular position relative to that of the transmitter. The secondary output voltage varies with the relative angular alignment of primary and secondary windings, of the control transformer and the position of the transmitter. The output voltage is substantially zero in value at a position known as correspondence. *See also:* synchro system. (PE) [9]

synchro differential receiver (synchro or selsyn devices) (motor) A transformer identical in construction to a synchro differential transmitter but used to develop a torque increasing with the difference in the relative angular displacement (up to about 90 electrical degrees) between the two sets of voltage input signals to its primary and secondary windings, the torque being in a direction to reduce this difference to zero. *See also:* synchro system. (PE) [9]

synchro differential transmitter (rotating machinery) (generator) A transformer with relatively rotatable primary and secondary windings. The primary input is a set of two or more voltages that define an angular position. The secondary output is a set of two or more voltages that represent the sum or difference, depending upon connections, of the position defined by the primary input and the relative angular displacement between primary and secondary windings. *See also:* synchro system. (PE) [9]

synchronism The state where connected alternating-current systems, machines, or a combination operate at the same frequency and where the phase angle displacement between voltages in them are constant, or vary about a steady and stable average value. *See also:* asynchronous machine. (PE/PSR) 1344-1995, [9]

synchronism-check relay A verification relay whose function is to operate when two input voltage phasors are within predetermined limits. (SWG/PE) C37.100-1992

synchronization (data transmission) A means of ensuring that both transmitting and receiving stations are operating together (equal scanning line frequencies) in a fixed phase relationship. (PE) 599-1985w

synchronization bit One or more bits that are added to a string of data to allow a receiving circuit to align its clocks with the data. *See also:* clock track. (C) 610.10-1994w

synchronization error (navigation) (navigation aids) The error due to imperfect timing of two operations; this may or may not include signal transmission time. (AES/GCS) 172-1983w

synchronization internal The time period between clock synchronization cycles. (C/BA) 896.2-1991w

synchronization time (gyros) The time interval required for the gyro rotor to reach synchronous speed from standstill. (AES/GYAC) 528-1994

synchronized I/O completion The state of an I/O operation that has either been successfully transferred or diagnosed as unsuccessful. (C/PA) 1003.5-1999, 9945-1-1996

synchronized I/O data integrity completion A degree of completion for an I/O operation that occurs when:

1) For read, the operation has been completed or diagnosed as unsuccessful. The read is complete only when an image of the data has been successfully transferred to the requesting task. If there were any pending write requests affecting the data to be read at the time that the synchronized read operation was requested, these write requests shall be successfully transferred prior to reading the data.
2) For write, the operation has been completed or diagnosed as unsuccessful. The write is complete only when the data specified in the write request is successfully transferred, and all file system information required to retrieve the data is successfully transferred.

File attributes that are not necessary for data retrieval (***Last Access Time***, ***Last Modification Time***, ***Last Status Change Time***) need not be successfully transferred prior to returning to the calling task. (C/PA) 1003.5-1999, 9945-1-1996

synchronized I/O file integrity completion (1) Identical to a synchronized I/O data integrity completion with the addition that all file attributes relative to the I/O operation (including ***Last Access Time***, ***Last Modification Time***, ***Last Status Change Time***) shall be successfully transferred prior to returning to the calling task. (C) 1003.5-1999
(2) Identical to a synchronized I/O data integrity completion with the addition that all file attributes relative to the I/O operation (including access time, modification time, status change time) shall be successfully transferred prior to returning to the calling process. (C/PA) 9945-1-1996

synchronized I/O operation An I/O operation performed on a file that provides the application assurance of the integrity of its data and files. *See also:* synchronized I/O file integrity completion; synchronized I/O data integrity completion. (C/PA) 1003.5-1999, 9945-1-1996

synchronous I/O operation (1) An I/O operation that causes the task requesting the I/O to be blocked from further use of the processor until that I/O operation completes. *Note:* A synchronous I/O operation does not imply synchronized I/O data integrity completion or synchronized I/O file integrity completion. (C) 1003.5-1999
(2) An I/O operation that causes the process requesting the I/O to be blocked from further use of the processor until that I/O operation completes. *Note:* A synchronous I/O operation does not perforce imply synchronized I/O data integrity completion or synchronized I/O file integrity completion. (C/PA) 9945-1-1996

synchronized operation (1) An operating mode where system facilities are connected and controlled to operate at the same frequency. (PE/PSE) 858-1993w
(2) (power operations) An operation wherein power facilities are electrically connected and controlled to operate at the same frequency. (PE/PSE) 346-1973w

synchronized phasor A phasor calculated from data samples using a standard time signal as the reference for the sampling process. In this case, the phasors from remote sites have a defined common phase relationship. (PE/PSR) 1344-1995

synchronized sweep (spectrum analyzer) (non-real time spectrum analyzer) (oscilloscopes) A sweep that would free run in the absence of an applied signal but in the presence of the signal is synchronized by it. *See also:* oscillograph. (IM) 748-1979w

synchronizing (1) (rotating machinery) The process whereby a synchronous machine, with its voltage and phase suitably adjusted, is paralleled with another synchronous machine or system. *See also:* asynchronous machine. (PE) [9]
(2) (facsimile) The maintenance of predetermined speed relations between the scanning spot and the recording spot within each scanning line. *See also:* facsimile. (COM) 168-1956w
(3) (television) Maintaining two or more scanning processes in phase. (BT/AV) [34]
(4) (pulse terminology) The process of rendering a first pulse train or other sequence of events synchronous with a second pulse train. (IM/WM&A) 194-1977w
(5) (hydroelectric power plants) Process of paralleling and connecting a synchronous generator to another source. (PE/EDPG) 1020-1988r

synchronizing coefficient (rotating machinery) The quotient of the shaft power and the angular displacement of the rotor. *Note:* It is expressed in kilowatts per electrical radian. Unless otherwise stated, the value will be for rated voltage, load, power-factor, and frequency. *See also:* asynchronous machine. (PE) [9]

synchronizing or synchronism-check device (power system device function numbers) A device that operates when two ac circuits are within the desired limits of frequency, phase angle, and voltage, to permit or to cause the paralleling of these two circuits. (PE/SUB) C37.2-1979s

synchronizing reactor (power and distribution transformers) A current-limiting reactor for connecting momentarily across the open contacts of a circuit-interrupting device for synchronizing purposes. *See also:* reactor. (PE/TR) C57.12.80-1978r, [57]

synchronizing relay A programming relay whose function is to initiate the closing of a circuit breaker between two ac sources when the voltages of these two sources have a predetermined relationship of magnitude, phase angle, and frequency. (SWG/PE) C37.100-1992

synchronizing signal (1) (television) The signal employed for the synchronizing of scanning. *Note:* In television, this signal is composed of pulses at rates related to the line and field frequencies. The signal usually originates in a central synchronizing generator and is added to the combination of picture signal and blanking signal, comprising the output signal from the pickup equipment, to form the composite picture signal. In a television receiver, this signal is normally separated from the picture signal and is used to synchronize the deflection generators. (BT/AV) [34]
(2) (facsimile) A signal used for maintenance of predetermined speed relations between the scanning spot and recording spot within each scanning line. *See also:* facsimile signal. (COM) 168-1956w
(3) (oscillograph) A signal used to synchronize repetitive functions. *See also:* oscillograph. (IM/HFIM) [40]
(4) (telecommunications) A special signal which may be sent to establish or maintain a fixed relationship in synchronous systems. (COM) [49]

synchronizing signal compression (television) The reduction in gain applied to the synchronizing signal over any part of its amplitude range with respect to the gain at a specified reference level. *Notes:* 1. The gain referred to in the definition is for a signal amplitude small in comparison with the total peak-to-peak composite picture signal involved. A quantitative evaluation of this effect can be obtained by a measurement of differential gain. 2. Frequently the gain at the level of the peaks of synchronizing pulses is reduced with respect to the gain at the levels near the bases of the synchronizing pulses. Under some conditions, the gain over the entire synchronizing signal region of the composite picture signal may be reduced with respect to the gain in the region of the picture signal. *See also:* television. (BT/AV) [34]

synchronizing signal level (television) The level of the peaks of the synchronizing signal. *See also:* television. (BT/AV) [34]

synchronizing torque (synchronous machines) The torque produced, primarily through interaction between the armature currents and the flux produced by the field winding, tending to pull the machine into synchronism with a connected power system or with another synchronous machine. (PE) [9]

synchronous (1) A mode of transmission in which the sending and receiving terminal equipment are operating continuously at the same rate and are maintained in a desired phase relationship by an appropriate means. (COM/TA) 1007-1991r

(2) Protocol operation in which only one exchange between a given pair of entities can be handled at any moment in time. The current exchange must complete before the next can be initiated. (LM/C) 15802-2-1995

(3) Describes an activity specified by a function that is expected to be complete when the function returns. (C/MM) 855-1990

synchronous assignment An assignment that takes place when a signal or variable value is updated as a direct result of a clock edge expression evaluating as true. (C/DA) 1076.6-1999

synchronous booster converter A synchronous converter having a mechanically connected alternating-current reversible booster connected in series with the alternating-current supply circuit for the purpose of adjusting the output voltage. *See also:* converter. (EEC/PE) [119]

synchronous booster inverter An inverter having a mechanically connected reversible synchronous booster connected in series for the purpose of adjusting the output voltage. *See also:* converter. (EEC/PE) [119]

synchronous capacitor (rotating machinery) A synchronous machine running without mechanical load and supplying or absorbing reactive power to or from a power system. *See also:* converter; synchronous condenser. (PE) [9]

synchronous circuit A circuit in which clock pulses synchronize the operations of the elements. *Contrast:* asynchronous circuit. (C) 610.10-1994w

synchronous communication In the IEEE 1451.1 client-server model, refers to a stateless communication in which the client blocks until the return is received from the server. That is, the client can perform no activities until unblocked by the return of the result. (IM/ST) 1451.1-1999

synchronous computer (1) A computer in which each event, or the performance of each operation, starts as a result of a signal generated by a clock. *Contrast:* asynchronous computer. (C) [20], [85], 610.10-1994w

(2) A computer in which each event or operation is performed upon receipt of a signal generated by the completion of a previous event or operation, or upon availability of the system resources required by the event or operation. (C) 610.10-1994w

synchronous condenser* (1) (electric installations on shipboard) A synchronous phase modifier running without mechanical load, the field excitation of which may be varied so as to modify the power factor of the system: or through such modification, to influence the load voltage. (IA/MT) 45-1983s

(2) A synchronous machine running without mechanical load and supplying or absorbing reactive power. *See also:* synchronous capacitor. (PE) [9]

* Deprecated.

synchronous converter A converter that combines both motor and generator action in one armature winding and is excited by one magnetic field. It is normally used to change ac power to dc power, or to create an isolated ac power source. (IA/MT) 45-1998

synchronous coupling (1) (electric coupling) An electric coupling in which torque is transmitted by attraction between magnetic poles on both rotating members which revolve at the same speed. The magnetic poles may be produced by direct current excitation, permanent magnet excitation, or alternating current excitation, and those on one rotating member may be salient reluctance poles. (EM/PE) 290-1980w

(2) (rotating machinery) A type of electric coupling in which torque is transmitted at zero slip, either between two electromagnetic members or like number of poles, or between one electromagnetic member and a reluctance member containing a number of saliencies equal to the number of poles. *Note:* Synchronous couplings may have induction members or other means for providing torque during nonsynchronous operation such as starting. *See also:* electric coupling. (PE) [9]

synchronous detector A device whose output is proportional to the amplitude of a vector component of an input radio frequency (RF) or intermediate frequency (IF) signal measured with respect to an externally supplied reference signal. (AES) 686-1997

synchronous device (data transmission) A device whose speed of operation is related to the rest of the system to which the device is connected. (PE) 599-1985w

synchronous errored second A one-second interval during which one or more errors are received, which is measured by triggering the one-second time period on a detected error. (COM/TA) 1007-1991r

synchronous gate A time gate wherein the output intervals are synchronized with an incoming signal. (AP/ANT) 145-1983s

synchronous generator (1) (hydroelectric power plants) A generator that produces power with rotor speed exactly proportional to the frequency of the system. The generator has field poles excited by direct current. (PE/EDPG) 1020-1988r

(2) A synchronous ac machine that transforms mechanical power into electric power. (A synchronous machine is one in which the average speed of normal operation is exactly proportional to the frequency of the system to which it is connected.) (IA/MT) 45-1998

synchronous impedance (1) (per unit direct-axis) The ratio of the field current at rated armature current on sustained symmetrical short-circuit to the field current at normal open-circuit voltage on the air-gap line. *Note:* This definition of synchronous impedance is used to a great extent in electrical literature and corresponds to the definition of direct-axis synchronous reactance as determined from open-circuit and sustained short-circuit tests. *See also:* positive-phase-sequence reactance. (EEC/PE) [119]

(2) (rotating machinery) The ratio of the value of the phasor difference between the synchronous internal voltage and the terminal voltage of a synchronous machine to the armature current under a balanced steady-state condition. *Note:* This definition is of rigorous application to turbine type machines only, but it gives a good degree of approximation for salient pole machines. *See also:* synchronous reactance. (PE) [9]

synchronous induction motor (rotating machinery) A cylindrical rotor synchronous motor having a secondary coil winding similar to that of a wound rotor induction motor. *Note:* This winding is used for both starting and excitation purposes. (PE) [9]

synchronous (interdigital) transducer An interdigital transducer that has uniform electrode center-to-center spacing. (UFFC) 1037-1992w

synchronous internal voltage (synchronous machine for any specified operating conditions) The fundamental-frequency component of the voltage of each armature phase that would be produced by the steady (or very slowly varying) component of the current in the main field winding (or field windings) acting alone provided the permeance of all parts of the magnetic circuit remained the same as for the specified operating condition. *Note:* The synchronous internal voltage, as shown in the phasor diagram, is related to the terminal-voltage and phase-current phasors by the equation. For a machine subject to saturation, the reactances should be determined for the degree of saturation applicable to the specified operating condition. (PE) [9]

synchronous inverter An inverter that combines both motor and generator action in one armature winding. It is excited by one magnetic field and changes direct-current power to alternating-current power. *Note:* Usually it has no amortisseur winding. *See also:* converter. (EEC/PE) [119]

synchronously generated signal A signal that is attributable to a specific thread. For example, a thread executing an illegal instruction or touching invalid memory causes a synchronously generated signal. Being synchronous is a property of how the signal was generated and not a property of the signal number. (C/PA) 9945-1-1996

synchronous machine A machine in which the average speed of normal operation is exactly proportional to the frequency of the system to which it is connected. (IA/MT) 45-1998

synchronous-machine regulating-system stability The property of a synchronous-machine-regulating system in which a change in the controlled variable, resulting from a stimulus, decays with time if the stimulus is removed. (PE) [9]

synchronous machine, ideal *See:* ideal synchronous machine.

synchronous-machine regulating system An electric-machine regulating system consisting of one or more principal synchronous electric machines and the associated excitation system. (PE) [9]

synchronous machine regulator (1) (excitation systems for synchronous machines) A regulator that couples the output variables of the synchronous machine to the output of the exciter through feedback and forward controlling elements for the purpose of regulating the synchronous machine output variables. (PE/EDPG) 421.4-1990, 421.1-1986r
(2) (excitation systems) One that couples the output variables of the synchronous machine to the input of the exciter through feedback and forward controlling elements for the purpose of regulating the synchronous machine output variables. *Note:* In general, the regulator is assumed to consist of an error detector, preamplifier, power amplifier, stabilizers, auxiliary inputs, and limiters. (PE/EDPG) 421-1972s
(3) (rotating machinery) An electric-machine regulator that controls the excitation of a synchronous machine. (PE) [9]

synchronous motor A polyphase ac motor with separately supplied dc field and an auxiliary (amortisseur) winding for starting purposes. The operating speed is fixed by the frequency (f) of the system and the number of poles (p) of the motor. (Synchronous speed (r/min) = $120\,f/p$). Thus the speed of the motor can be varied by varying the frequency of the power source. The synchronous motor generally operates at unity power factor and can be used to improve the system power factor. It is generally the motor of choice for ac propulsion systems. (IA/MT) 45-1998

synchronous operation (1) An operation that causes the process requesting the operation to be blocked until it is complete. (C/PA) 1327.2-1993w, 1224.2-1993w
(2) (opening or closing) Operation of a switching device in such a manner that the contacts are closed or opened at a predetermined point on a reference voltage or current wave. *Note:* Synchronous operation applied on multiphase circuits may require that closing or opening of the contacts of each pole be responsive to a different reference. (SWG/PE) C37.100-1992

synchronous reactance (1) (rotating machinery) (effective) An assumed value of synchronous reactance used to represent a machine in a system study calculation for a particular operating condition. *Note:* The synchronous internal voltage, as shown in the phasor diagram, is related to the terminal-voltage and phase-current phasors by the equation

$$E'_i = E_a + RI_a + jX_{\mathrm{eff}}I_a$$

See also: synchronous internal voltage; synchronous impedance. (PE) [9]
(2) (power fault effects) The steady-state reactance of a generator during fault conditions used to calculate the steady-state fault current. The current so calculated excludes the effect of the automatic voltage regulator or governor. (PE/PSC) 367-1996
(3) (electric power systems in commercial buildings) The reactance that determines the current flow when a steady-state condition is reached. (IA/PSE) 241-1990r

synchronous satellite (navigation aids) An equatorial satellite orbiting the earth in a west-to-east direction at an altitude of approximately 35 900 km. At this altitude the satellite makes one revolution in 24 h, synchronous with the earth's rotation. (AES/GCS) 172-1983w

synchronous speed (rotating machinery) The speed of rotation of the magnetic flux, produced by or linking the primary winding. (PE) [9]

synchronous-speed device (power system device function numbers) A device such as a centrifugal-speed switch, a slip-frequency relay, a voltage relay, an undercurrent relay, or any type of device that operates at approximately the synchronous speed of a machine. (SUB/PE) C37.2-1979s

synchronous system (data transmission) A system in which the sending and receiving instruments are operating continuously at substantially the same rate and are maintained by means of correction if necessary, in a fixed relationship. (PE) 599-1985w

synchronous time division multiplexing A method of time division multiplexing in which time slots on a shared communication channel are assigned to devices on a fixed, predetermined basis. (C) 610.7-1995

synchronous transmission (1) (data transmission) A mode of data transmission in which the sending and receiving data processing terminal equipments are operating continuously at substantially the same frequency and are maintained in a desired phase relationship by an appropriate means. (PE) 599-1985w
(2) A transmission in which information and control characters are sent at regular clocked intervals so that sending and receiving stations are operating continuously in step with each other. *Synonym:* synchronous communication. *Contrast:* asynchronous transmission. (C) 610.7-1995

synchronous-vibration sensitivity (dynamically tuned gyro) The functions that relate drift rates to linear or angular vibrations that are phase coherent with spin frequency or its harmonics. *See also:* two-N (2N) translational sensitivity; one-N (1N) translational sensitivity; two-N (2N) angular sensitivity. (AES/GYAC) 528-1994

synchronous voltage (traveling-wave tubes) The voltage required to accelerate electrons from rest to a velocity equal to the phase velocity of a wave in the absence of electron flow. *See also:* magnetron. (ED) [45]

synchro receiver (rotating machinery) (or motor) A transformer electrically similar to a synchro transmitter and that, when the secondary windings of the two devices are interconnected, develops a torque increasing with the difference in angular alignment of the transmitter and receiver rotors and in a direction to reduce the difference toward zero. *See also:* synchro system. (PE) [9]

synchroscope An instrument for indicating whether two periodic quantities are synchronous. It usually embodies a continuously rotatable element the position of which at any time is a measure of the instantaneous phase difference between the quantities: while its speed of rotation indicates the frequency difference between the quantities: and its direction of rotation indicates which of the quantities is of higher frequency. *Note:* This term is also used to designate a cathode-ray oscilloscope providing either: 1) a rotating pattern giving indications similar to that of the conventional synchroscope; or 2) a triggered sweep, giving an indication of synchronism. *See also:* instrument. (EEC/PE) [119]

synchro system (alternating current) An electric system for transmitting angular position or motion. It consists of one or more synchro transmitters, one or more synchro receivers or synchro control transformers and may include differential synchro machines. (PE) [9]

synchro transmitter (rotating machinery) (or generator) A transformer with relatively rotatable primary and secondary windings, the output of the secondary winding being two or more voltages that vary with and completely define the relative angular position of the primary and secondary windings. *See also:* synchro system. (PE) [9]

synchrotron A device for accelerating charged particles (for example, electrons) to high energies in a vacuum. The particles are guided by a changing magnetic field while they are accelerated many times in a closed path by a radio-frequency electric field. (ED) [45]

SyncLink A physical interconnect model, consisting of shared input and output links, that supports the RamLink logical protocols. SyncLink is optimized for short-distance single-board communications using a bused connection. The signal-levels for this interconnect model have been left for a future extension to this standard. (C/MM) 1596.4-1996

sync packet (1) A special packet that is used heavily during initialization and occasionally during normal operation for the purpose of checking and adjusting receiver circuit timing. (C/MM) 1596.3-1996, 1596-1992

(2) A packet consisting of four high and four low flags, as well as four data bytes whose bits are the complement of the flag values, typically generated when leaving shutdown (to synchronize the device and controller receiver circuits). (C/MM) 1596.4-1996

sync signal *See:* synchronizing signal.

sync slip An error condition in serial communication channels in which the receiving terminal incorrectly recognizes the start of a new message. (SUB/PE) 999-1992w

syndrome A particular group of symptoms that occur together and that define a particular disease or abnormality. (T&D/PE) 539-1990

synergism The presence of a performance or aging effect, produced by a combination of aging factors, that is different from the effect predicted by the simple summations of the effects of these factors acting separately. (DEI/RE) 775-1993w

synonym In hashing, an item whose hash value is identical to that of another item. *See also:* collision resolution. (C) 610.5-1990w

syntactic error A violation of the structural or grammatical rules defined for a language; for example, using the statement $B + C = A$ in Fortran, rather than the correct $A = B + C$. *Synonym:* syntax error. *Contrast:* semantic error. (C) 610.12-1990

syntactic pattern recognition A type of structural pattern recognition that identifies primitives and relationships in natural or artificial language patterns. (C) 610.4-1990w

syntax (1) (A) (computers) The structure of expressions in a language. **(B) (computers)** The rules governing the structure of a language. (C/C) [20], [85]

(2) (software) The structural or grammatical rules that define how the symbols in a language are to be combined to form words, phrases, expressions, and other allowable constructs. (C) 610.12-1990

(3) A category into which an attribute value is placed on the basis of its form. (C/PA) 1328-1993w, 1327-1993w, 1224-1993w

(4) The structural components or features of a language and rules that define the ways in which the language constructs may be assembled together to form sentences. (C/SE) 1320.2-1998

(5) The grammatical rules pertaining to the structure of an ATLAS statement. (SCC20) 771-1998

(6) The structure of expressions in a language and the rules governing the structure of a language. (SCC32) 1489-1999

(7) *See also:* attribute syntax. (C/PA) 1328.2-1993w

syntax error *See:* syntactic error.

syntax template A lexical construct containing an asterisk from which several attribute syntaxes can be derived by substituting text for the asterisk. (C/PA) 1328-1993w, 1224-1993w, 1327-1993w

synthetic address *See:* generated address.

synthetic-aperture radar (SAR) (1) A radar system that generates the effect of a long antenna by signal processing means rather than by the actual use of a long physical antenna. (AES/GCS) 172-1983w

(2) A coherent radar system that generates a narrow cross range impulse response by signal processing (integrating) the amplitude and phase of the received signal over an angular rotation of the radar line of sight with respect to the object (target) illuminated. *Note:* Due to the change in line-of-sight direction, a synthetic aperture is produced by the signal processing that has the effect of an antenna with a much larger aperture (and hence a much greater angular resolution). (AES) 686-1997

synthetic-aperture radar-moving-target indication (SAR-MTI) A synthetic-aperture imaging radar that also detects moving targets (especially slow-moving ground vehicles) and displays them on the SAR image. (AES) 686-1997

synthetic benchmark program A benchmark program that consists of a small program constructed especially for benchmarking purposes, but does not necessarily perform any useful function. (C) 610.10-1994w

synthetic environment The integrated set of data elements that define the environment within which a given simulation application operates. The data elements include information about the initial and subsequent states of the terrain including cultural features, and atmospheric and oceanographic environments throughout a DIS exercise. The data elements include databases of externally observable information about instantiable DIS entities, and are adequately correlated for the type of exercise to be performed. (DIS/C) 1278.1-1995, 1278.2-1995

synthetic test (1) A test in which the major part of, or the total current, is obtained from one source (current circuit), and the major part of, or all of the transient recovery voltage from a separate source or sources (voltage circuit). (SWG/PE) C37.100-1992, C37.081-1981r

(2) A test in which the major part of, or the total current, is obtained from a source or sources (current circuit), and the major part of, or all of the transient recovery voltage from a separate source or sources (voltage circuit). (SWG/PE) C37.083-1999

synthesis tool Any system, process, or tool that interprets RTL VHDL source code as a description of an electronic circuit and derives a netlist description of that circuit. (C/DA) 1076.6-1999, 1076.3-1997

SyRS *See:* System Requirements Specification.

sysgen *See:* system generation.

system (1) (microcomputer system bus) (general system) A set of interconnected elements that achieve a given objective through the performance of a specified function. (C/MM/IM/AIN) 796-1983r, 488.1-1987r, 696-1983w

(2) (monitoring radioactivity in effluents) The entire assembled equipment excluding only the sample collecting pipe. (NI) N42.18-1980r

(3) (reliability data for pumps and drivers, valve actuators, and valves) A collection of components arranged to provide a desired function (for example, containment spray system, residual heat removal system, high pressure coolant injection system). (PE/NP) 500-1984w
(4) (seismic design of substations) A group of components operating together to perform a function (for example, disconnect switch, support structure and foundation, relay protection system, and telemetering system). (PE/SUB) 693-1984s
(5) (unique identification in power plants and related facilities) A combination of two or more interrelated components that perform a specific function related to plant operation and safety. A system may peform a function such as control, monitoring, electrical, mechanical, or structural. (PE/EDPG) 804-1983r
(6) An integrated whole even though composed of diverse, interacting, specialized structures and subjunctions. *Notes:* 1. Any system has a number of objectives and the weights placed on them may differ widely from system to system. 2. A system performs a function not possible with any of the individual parts. Complexity of the combination is implied. (SMC) [63]
(7) (computers) A collection of components organized to accomplish a specific function or set of functions. (C) 610.3-1989w, 610.5-1990w, 610.12-1990
(8) (electric and electronics parts and equipment) A combination of two or more sets, generally physically separated when in operation, and such other units, assemblies, and basic parts necessary to perform an operational function or functions. Typical examples: telephone carrier system, ground-controlled approach (GCA) electronic system, telemetering system, facsimile transmission system. (GSD) 200-1975w
(9) (test access port and boundary-scan architecture) Pertaining to the nontest function of the circuit. (TT/C) 1149.1-1990
(10) (measurement of radio-noise emissions) An arrangement of interconnected devices and their cables designed to perform a particular function or functions. (EMC) C63.4-1991
(11) A group of devices and a controller interconnected with a system interface. (IM/AIN) 488.2-1992r
(12) Hardware and software collectively organized to achieve an operational objective. (SUB/PE) 999-1992w
(13) A set of interconnected boards that achieve a specified objective by performing designated functions. (C/BA/MM) 896.9-1994w, 1000-1987r
(14) One or more mainframes that are connected, having a common resource manager. Each device in a system has a unique logical address. (C/MM) 1155-1992
(15) (A) A collection of entities to be processed by applying a top-down, hierarchical approach. **(B)** A collection of interacting, interrelated, or interdependent elements forming a collective, functioning entity. **(C)** A set of objects or phenomena grouped together for classification or analysis. **(D)** A collection of hardware or software components necessary for performing a high-level function. (ATLAS) 1232-1995
(16) (power operations) A group of components connected or associated in a fixed configuration to perform a specified function of distributing power. (PE/IA/PSE) 858-1987s, 493-1997, 399-1997
(17) An integration of parts or constituents, their relationships, their mutual behavior or possible states, and the laws or rules that determine their behavior. (PE/NP) 1082-1997
(18) A set or arrangement of elements [people, products (hardware and software) and processes (facilities, equipment, material, and procedures)] that are related and whose behavior satisfies customer/operational needs, and provides for the life cycle sustainment of the products. (C/SE) 1220-1998
(19) (A) A collection of interacting components organized to accomplish a specific function or set of functions within a specific environment. **(B)** A group of people, objects, and procedures constituted to achieve defined objectives of some operational role by performing specified functions. A complete system includes all of the associated equipment, facilities, material, computer programs, firmware, technical documentation, services, and personnel required for operations and support to the degree necessary for self-sufficient use in its intended environment. (C/SE) 1362-1998
(20) A set of interlinked units organized to accomplish one or several specific functions. (C/SE) 1219-1998
(21) An interdependent group of people, objects, and procedures constituted to achieve defined objectives or some operational role by performing specified functions. A complete system includes all of the associated equipment, facilities, material, computer programs, firmware, technical documentation, services, and personnel required for operations and support to the degree necessary for self-sufficient use in its intended environment. (C/SE) 1233-1998
(22) An integrated composite that consists of one or more of the processes, hardware, software, facilities, and people, that provides a capability to satisfy a stated need or objective. (C/SE) 1517-1999
(23) (controlling) *See also:* controlling system. (SPD/PE) 32-1972r
(24) *See also:* batch system.
(25) *See also:* insulation system. (PE/TR) 1276-1997

system acceptance The formal approval of system operation parameters. (LM/C) 802.7-1989r

system access Attacks on computer system resources (e.g., files, directories, devices) from an intruder having access to the application software or operating system. This threat area assumes that anintruder has access to some form of man-machine interface in the system (e.g., terminal, operator console, workstation). (C/BA) 896.3-1993w

system architecture (1) (software) The structure and relationship among the components of a system. The system architecture may also include the system's interface with its operational environment. (C/SE) 729-1983s
(2) The composite of the design architectures for products and their life cycle processes. (C/SE) 1220-1998

system assured capability (power operations) The dependable capability of all power sources available to a system under short range conditions, including firm power contracts, less that reserve assigned to provide for planned outages, equipment and operating limitations, and unplanned outages of power sources. (PE/PSE) 858-1987s

systematic drift rate (gyros) That component of drift rate that is correlated with specific operating conditions. It is composed of acceleration-sensitive drift rate and acceleration-insensitive drift rate. It is expressed as angular displacement per unit time. (AES/GYAC) 528-1994

systematic error (1) (electrothermic power meters) The inherent bias (off-set) of a measurement process or of one of its components. (IM/HFIM) 544-1975w, 314-1971w
(2) (electronic navigation) Error capable of identification due to its orderly character. *See also:* navigation. (AES/RS) 686-1982s, [42]
(3) Errors where the magnitudes and directions are constant throughout the calibration process. (PE/PSIM) 4-1995

systematic reuse The practice of reuse according to a well-defined, repeatable process. (C/SE) 1517-1999

system breakdown structure (SBS) A hierarchy of elements, related life cycle processes, and personnel used to assign development teams, conduct technical reviews, and to partition out the assigned work and associated resource allocations to each of the tasks necessary to accomplish the objectives of the project. It also provides the basis for cost tracking and control. (C/SE) 1220-1998

system bus The IEEE 488.1 bus and protocols that interconnect the **devices** and **controllers** in a **system.** The content of this standard applies to device-dependent traffic over this bus. (IM/AIN) 488.2-1992r

system bus bridge The interface between a Profile B Futurebus+ and the system CPU/main memory which uses Futurebus+ to communicate to I/O subsystems. Typically, the bridge couples the Profile B I/O bus to an internal system bus, which links CPU and main memory. A distinction should be made between the two types of bus bridges defined in this clause: the *bus bridge* must comply with Profile B requirements in all respects, while the *system bus bridge* may be exempt from certain Profile requirements, such as accessibility of its CSRs from the Futurebus+, and mechanical requirements. Where different or reduced constraints apply to system bus bridges, they are called out in the appropriate section of the profile. (C/BA) 896.2-1991w

system clock driver A functional module that provides a 16 MHz timing signal on the utility bus. (C/BA) 1014-1987

system configuration process The software that intializes the system. The monarch processor executes the system configuration process. (C/BA) 896.2-1991w, 896.10-1997

system control (telephone switching systems) The means for collecting and processing pulsing and supervisory signals in a switching system. (COM) 312-1977w

system controller board A board that resides in slot I of the backplane and has a system clock driver, an arbiter, an lack daisy-chain driver, and a bus timer. Some also have a serial clock driver, a power monitor, or both. (C/BA) 1014-1987

system control signal group A set of ten (10) signals, including two parity bits, which supply command and status information between the bus owner and the replying agent. (C/MM) 1296-1987s

system crash (1) An interval initiated by an unspecified circumstance that causes all processes (possibly other than special system processes) to be terminated in an undefined manner, after which any changes to the state and contents of files created or written to by a Conforming POSIX.1 Application prior to the interval are undefined, except as required elsewhere in this standard. (C/PA) 9945-1-1996
(2) An event initiated by an unspecified circumstance that causes all processes (possibly other than special system processes) to be terminated in an undefined manner, after which any changes to the state and contents of files created or written to by a conforming POSIX.5 application prior to the interval are undefined. (C) 1003.5-1999

system delay time (mobile communication) The time required for the transmitter associated with the system to provide rated radio-frequency output after activation of the local control (push to talk) plus the time required for the system receiver to provide useful output. *See also:* mobile communication system. (VT) [37]

system demand factor *See:* demand factor.

system design (A) (software) The process of defining the hardware and software architectures, components, modules, interfaces, and data for a system to satisfy specified system requirements. **(B) (software)** The result of the system design process. *See also:* system; data; component; software; hardware; requirement; module. (C/SE) 729-1983

system design review A review conducted to evaluate the manner in which the requirements for a system have been allocated to configuration items, the system engineering process that produced the allocation, the engineering planning for the next phase of the effort, manufacturing considerations, and the planning for production engineering. *See also:* critical design review; preliminary design review. (C) 610.12-1990

system development cycle (1) The period of time that begins with the decision to develop a system and ends when the system is delivered to its end user. *Note:* This term is sometimes used to mean a longer period of time, either the period that ends when the system is no longer being enhanced, or the entire system life cycle. *Contrast:* system life cycle. *See also:* software development cycle. (C) 610.12-1990
(2) The life cycle through which a system is developed, which consists of the following:

a) Concept development
b) Design
c) Test and construction
d) Operation
e) Maintenance

(PE/NP) 845-1999

system, directly controlled *See:* directly controlled system.

system, discrete *See:* discrete system.

system, discrete-state *See:* discrete-state system.

system diversity factor *See:* diversity factor.

system documentation (1) (software) Documentation conveying the requirements, design philosophy, design details, capabilities, limitations, and other characteristics of a system. *See also:* documentation; user documentation; requirement; system; design. (C/SE) 729-1983s
(2) All documentation provided with an implementation, except the conformance document. Electronically distributed documents for an implementation are considered part of the system documentation.
(C/PA) 1326.2-1993w, 2003.2-1996, 1003.5-1999

system effectiveness A measurement of the ability of a system to satisfy its intended operational uses as a function of how the system performs under anticipated environmental conditions, and the ability to produce, test, distribute, operate, support, train, and dispose of the system throughout its life cycle. (C/SE) 1220-1998

system element (1) One or more basic elements with other components and necessary parts to form all or a significant part of one of the general functional groups into which a measurement system can be classified. While a system element must be functionally distinct from other such elements it is not necessarily a separate measurement device. Typical examples of system elements are: a thermocouple, a measurement amplifier, a millivoltmeter. *See also:* measurement system. (EEC/PE) [119]
(2) A product, subsystem, assembly, component, subcomponent, subassembly, or part of the system breakdown structure that includes the specifications, configuration baseline, budget, schedule, and work tasks. (C/SE) 1220-1994s
(3) One or more software, hardware, or firmware components that perform a specified task. *Note:* A system element may be a composed of a combination of system elements. A system element may be a system. (C/PA) 2000.1-1999

system failure Malfunctions in the hardware and software that could compromise the security of the system (for example, non-security-related failures and design flaws are not considered). The malfunctions include both intentional and inadvertent design or implementation flaws (including malicious hardware and software) and component failures. For intentional attacks, this threat area assumes that an intruder has access to the design or implementation processes of the system or to the operational system in such a way as to be able to cause a failure in a component. For inadvertent attacks, there may not be a specific intruder.
(C/BA) 896.9-1994w, 896.3-1993w

system, finite-state *See:* discrete-state system.

system flowchart *See:* flowchart.

system frequency (electric power system) The prevailing frequency in hertz (Hz) of the alternating current and voltage throughout a power system. (PE/PSE) 94-1991w

system frequency stability (mobile communication) (radio system) The measure of the ability of all stations, including all transmitters and receivers, to remain on an assigned frequency-channel as determined on both a short-term and long-term basis. *See also:* mobile communication system.
(VT) [37]

system generation The process of using an operating system to assemble and link together all the parts that constitute another operating system. (C) 610.10-1994w

system ground (surge arresters) The connection between a grounding system and a point of an electric circuit (for example, a neutral point). (PE/EDPG) 665-1995, [84], [8]

system grounding conductor An auxiliary solidly grounded conductor that connects together the individual grounding conductors in a given area. *Note:* This conductor is not normally a part of any current-carrying circuit including the system neutral. *See also:* ground. (T&D/PE) [10]

system handshake A handshake in a broadcast operation where the handshake signal is from the last segment of the addressed system rather than from individual devices. (NID) 960-1993

system hazard A system condition that is a prerequisite to an accident. (C/SE) 1228-1994

system high The highest security level supported by a system. (C/BA) 896.3-1993w

system, idealized *See:* idealized system.

system identification An eight-character label describing the system. Leading spaces are not interpreted as leading zeros. This is intended to be used to identify the laboratory or experimental apparatus where the data were collected. This, along with the subsystem identification, can be used to uniquely describe the source of the data. (NPS/NID) 1214-1992r

system incremental cost The additional cost of delivering another megawatt of power to the load center. This cost is commonly called lambda (λ). (PE/PSE) 94-1991w

system interconnection The connecting together of two or more power systems. *See also:* direct-current distribution; alternating-current distribution. (T&D/PE) [10]

system interface An interface that connects a **device** or **controller** to the **system bus.** A "non-IEEE 488.2 system interface" is any interface other than **the system interface** that may happen to be connected to a **device** or **controller.** (IM/AIN) 488.2-1992r

system integrator A person who combines software and hardware into a working ATE. (SCC20) 993-1997

system library (software) A software library containing system-resident software that can be accessed for use or incorporated into other programs by reference; for example, a macro library. *Contrast:* master library; production library; software repository; software development library. (C) 610.12-1990

system life cycle The period of time that begins when a system is conceived and ends when the system is no longer available for use. *See also:* software life cycle; system development cycle. (C) 610.12-1990

system load (1) (power operations) Equal to internal load plus pumping load plus firm sales for resale. (PE/PSE) 858-1987s

(2) (electric power system) Total loads within the system including transmission and distribution losses. (PE/PSE) 346-1973w

system logic (1) (test access port and boundary-scan architecture) Any item of logic that is dedicated to realizing the nontest function of the component or is at the time of interest configured to achieve some aspect of the nontest function. (TT/C) 1149.1-1990

(2) That equipment that monitors the output of two or more channels and supplies output signals in accordance with a prescribed combination rule (for example, two of three, two of four, etc.). (PE/NP) 379-1994, 308-1980s

system loss (L_s) (of a radio system) The ratio of the input power to the terminals of the transmitting antenna to the available output power at the terminals of the receiving antenna. Usually expressed in decibels as a positive number. (AP/PROP) 211-1997

system low The lowest security level supported by a system. (C/BA) 896.3-1993w

system management stack The protocols residing above SDE that request services via an SDE SAP that is supported by the use of a bootstrap SAID with either of the two values reserved for system management. (C/LM) 802.10-1998

system margin capability (power operations) The difference between system capability and system load. (PE/PSE) 858-1987s

system matrix A matrix of transfer functions that relate the Laplace transforms of the system outputs and of the system inputs. *See also:* control system. (CS/IM/PE/EDPG) [120], [3]

system maximum hourly load (power operations) (electric power system) The maximum hourly integrated system load. This is an energy quantity usually expressed in kilowatt-hours per hour (kWh/h). (PE/PSE) 858-1987s, 346-1973w

system memory Read/write memory accessible by both the Processor and I/O Unit in the system address space. It is usually thought of as being in one or more third-party nodes, but it could alternatively be wholly or partially located in the communicating Processor or I/O nodes. (C/MM) 1212.1-1993

system model In computer performance evaluation, a representation of a system depicting the relationships between workloads and performance measures in the system. *See also:* workload model. (C) 610.12-1990

system monitor and control subsystem (terrestrial photovoltaic power systems) Logic and control circuitry that supervises the overall operation of the system by controlling the interaction between all subsystems. *See also:* array control. (PV) 928-1986r

system, multidimensional *See:* multidimensional system.

system, multivariable *See:* multivariable system.

system noise (sound recording and reproducing system) The noise output that arises within or is generated by the system or any of its components, including the medium. *See also:* noise.

system nominal response The rate of increase of the excitation system output voltage determined from the excitation system voltage response curve, divided by the rated field voltage. This rate, if maintained constant, would develop the same voltage-time area as obtained from the actual curve over the first half-second interval (unless a different time interval is specified). (PE/EDPG) 421.2-1990

system of units A set of interrelated units for expressing the magnitudes of a number of different quantities. (Std100) 270-1966w

system operator (1) A person designated to operate the system or parts thereof. (NESC) C2-1977s

(2) Electric utility personnel in charge of system-wide coordination of generation, interchange, and transmission security. (PE/PSE) 858-1993w

system overshoot (control) The largest value of system deviation following the first dynamic crossing of the ideal value in the direction of correction, after the application of a specified stimulus. *See also:* feedback control system. (IA/ICTL/IAC) [60]

system performance testing (nuclear power generating station) Tests performed on completed systems, including all their electric, instrumentation, controls, fluid, and mechanical subsystems under normal or simulated normal process conditions of temperature, flow, level, pressure, etc. 336-1980s

system pin A component pin that feeds, or is fed from, the on-chip system logic. (TT/C) 1149.1-1990

system process (1) An object, other than a process executing an application, that is defined by the system and has a process ID. (C/PA) 9945-1-1996

(2) An object, other than a process executing an application, that is defined by the system and has a process ID. An implementation shall reserve at least one process ID for system processes. (C) 1003.5-1999

system production time The part of operating time that is actually used by a user. *Contrast:* system test time. (C) 610.10-1994w

system profile A set of measurements used in computer performance evaluation, describing the proportion of time each of the major resources in a computer system is busy, divided by the time that resource is available. (C) 610.12-1990

system, quantized *See:* quantized system.

system reboot An implementation defined sequence of events that may result in the loss of transitory data, i.e., data that is not saved in permanent storage. This includes message queues, shared memory, semaphores, and processes. (C/PA) 9945-1-1996, 1003.5-1999

system recovery time (mobile communication) The elapsed time from deactivation of the local transmitter control until the local receiver is capable of producing useful output. *See also:* mobile communication system. (VT) [37]

system-related outage A forced outage that results from system effects or conditions and is not caused by an event directly associated with the component or unit being reported. *Note:* line outage occurrences due to cascading, out-of-step conditions, and so forth, are all examples of system-related outages. (PE/PSE) 859-1987w

system reliability (software) The probability that a system, including all hardware and software subsystems, will perform a required task or mission for a specified time in a specified environment. *See also:* system; operational reliability; software reliability; hardware. (C/SE) 729-1983s

system reliability service (SRS) A United Kingdom reliability information and database service. (PE/NP) 933-1999

system requirements review A review conducted to evaluate the completeness and adequacy of the requirements defined for a system; to evaluate the system engineering process that produced those requirements; to assess the results of system engineering studies; and to evaluate system engineering plans. *See also:* software requirements review. (C) 610.12-1990

System Requirements Specification (SyRS) A structured collection of information that embodies the requirements of the system. (C/SE) 1233-1998

system reserve The capacity, in equipment and conductors, installed on the system in excess of that required to carry the peak load. *See also:* generating station. (T&D/PE) [10]

system reset An initialization event intiated when any module asserts the re* signal for 100 to 200 ms. (C/BA) 896.2-1991w, 896.10-1997

system resources chart *See:* block diagram.

system routing code (telephone switching systems) In World Zone 1, a three-digit code consisting of a country code and two additional numerals that uniquely identifies an international switching center. *See also:* world-zone number. (COM) 312-1977w

system safety **(1)** Freedom from system hazards. (C/SE) 1228-1994

(2) The application of engineering and management principles, criteria, and techniques to optimize all aspects of safety within the constraints of operational effectiveness, time, and cost throughout all phases of the system life cycle. (VT/RT) 1474.1-1999, 1483-2000

system safety goals—quantitative A quantitative limit of the probability and/or frequency with which any vital function fails to be implemented safely. (VT/RT) 1483-2000

system safety program The combined tasks and activities of system safety management and system safety engineering that enhance operational effectiveness by satisfying the system safety requirements in a timely, cost-effective manner throughout the system life cycle. (VT/RT) 1483-2000, 1474.1-1999

system safety program plan A formal document that fully describes the planned safety tasks required to meet the system requirements, including organizational responsibilities, methods of accomplishment, milestones, depth of effort, and integration with other program engineering and management functions. (VT/RT) 1483-2000

system, sampled-data *See:* sampled-data system.

system scheduling priority A number used as advice to the system to alter process scheduling priorities. Raising the value of the system scheduling priority should give a process additional preference when scheduling a process to run. Lowering the value should reduce the preference and make a process less likely to run. Typically, a process with higher system scheduling priority will run to completion more quickly than an equivalent process with lower system scheduling priority. A scheduling priority of zero specifies the default policy of the system. (C/PA) 9945-2-1993

system science The branch of organized knowledge dealing with systems and their properties, the systematized knowledge of systems. *See also:* system; cybernetics; learning system; systems engineering; adaptive system; tradeoff. (SMC) [63]

systems engineering **(1)** The application of the mathematical and physical sciences to develop systems that utilize economically the materials and forces of nature for the benefit of mankind. *See also:* system science. (SMC) [63]

(2) An interdisciplinary collaborative approach to derive, evolve, and verify a life-cycle balanced system solution that satisfies customer expectations and meets public acceptability. (C/SE) 1220-1994s

systems management In networking, functions in the application layer related to the management of various OSI resources and their status across all layers of the OSI architecture. (LM/C) 610.7-1995, 8802-6-1994, 802.10-1992

systems management entity An entity that carries out communications to perform systems management functions such as monitoring, controlling and coordination in the OSI environment. (C) 610.7-1995

systems network architecture A network architecture used widely by IBM and its compatible products for transmitting information units through and controlling the configuration and operation of a network. (C) 610.10-1994w

system software (software) Software designed to facilitate the operation and maintenance of a computer system and its associated programs; for example, operating systems, assemblers, utilities. *Contrast:* application software. *See also:* support software. (C) 610.12-1990

system-source short-circuit current The short-circuit current when the source of the short-circuit current is from the power system through at least one transformation. (SWG/PE) C37.013-1997

system specification Agreed description of the requirements of the system. (C/BA) 896.9-1994w

system state (modeling and simulation) (software) The values assumed at a given instant by the variables that define the characteristics of a system, component, or simulation. *Synonym:* system state. *See also:* steady state; final state. (C/BA) 896.9-1994w

system structure (unique identification in power plants) A combination of two or more integrated components, generally physically remote or occupying a large area, interacting to perform a specific function important to plant operation or safety, or both. A system may be civil/structural, that is, a building or structure, mechanical/fluid, or electrical/control. A system, for the purpose of IEEE Std 803-1983, will not be considered a subsystem of another system. (PE/EDPG) 803-1983r

systemtag (microprocessor operating systems parameter types) A "tag" returned by one function for use by another. Its contents may not be examined or changed. Its form is system dependent. A systemtag is valid system-wide (that is, global) and may be passed between processes. (C/MM) 855-1985s

system test A test, or collection of tests, that may use an external memory buffer and may require cooperation of other nodes. For example, identical unit architectures may collaborate to test cache coherence or to generate background "noise" traffic for other nodes being tested. A system test is invoked by writing to the TEST_START register. (C/MM) 1212-1991s

system testing (1) (software) Testing conducted on a complete, integrated system to evaluate the system's compliance with its specified requirements. *See also:* interface testing; integra-

tion testing; component testing; unit testing. (C/PE/NP) 610.12-1990, 7-4.3.2-1993
(2) The activities of testing an integrated hardware and software system to verify and validate whether the system meets its original objectives. (C/SE) 1012-1998
(3) Preoperational testing to verify components of the system, and operational testing to verify the coordination of the system components for proper design operation of the system functions. (PE/EDPG) 1248-1998

system test time The part of operating time during which the computer is tested for proper system operation. *Contrast:* system production time. (C) 610.10-1994w

system time (1) (supervisory control, data acquisition, and automatic control) A coordinated value of time maintained at stations throughout the power system. (SWG/PE) C37.100-1992
(2) A coordinated value of time maintained throughout the control and data acquisition equipment. (SUB/PE) C37.1-1994
(3) The state of any system element that is used to synchronize system events with real-world events based on a date, time, or combination of the two. (C/PA) 2000.2-1999, 2000.1-1999

system transaction A complete operation, such as a memory read or write, as viewed from the initiating unit. A system transaction can be translated into one or more bus transactions by the Futurebus+ interface to complete the operation. (C/BA) 10857-1994

system-transfer function (automatic control) The transfer function obtained by taking the ratio of the Laplace transform of the signal corresponding to the ultimately controlled variable to the Laplace transform of the signal corresponding to the command. *See also:* feedback control system. (IM/PE/EDPG) [120], [3]

system under test (SUT) The computer system hardware and software on which the implementation under test operates. (C/PA) 2003-1997

system unit (SU) (1) The equipment reference increment mounting pitch equal to 25 mm. (C/BA) 1101.3-1993
(2) The equipment reference increment mounting pitch, 1 SU = 25.0 mm, for height, width, and depth. (C/MM) 1301.3-1992r, 1301.1-1991
(3) The equipment reference increment or mounting pitch (mp_1), used for length and width (1 SU = 25 mm = mp_1). (C/BA) 1301.4-1996

system utilization factor *See:* utilization factor.

system validation *See:* validation.

system verification *See:* verification.

system voltage (1) (electric power systems in commercial buildings) The root-mean square phase-to-phase voltage of a portion of an ac electric system. Each system voltage pertains to a portion of the system that is bounded by transformers or utilization equipment. (IA/PSE) 241-1990r
(2) Phase-to-phase voltage of the circuit(s). When phase-to-ground voltage is the intention, it should be so noted. (PE/T&D) 957-1995
(3) (power and distribution transformers) The root-mean-square (rms) phase-to-phase power frequency voltage on a three-phase alternating-current electric system. (PE/SPD/TR) C57.12.80-1978r, C62.22-1997
(4) The rms power-frequency voltage from line-to-line, as distinguished from the voltage from line-to-neutral. (SPD/PE) C62.11-1999, C62.62-2000

T

T Letter symbol for the duration of a half-period of the nominal upper cut-off frequency of a transmission system. Therefore

$$T = \frac{1}{2f_c}$$

Note: for the TV system M

$$T = \frac{1}{2 \times 4\ \text{(MHz)}} = 125'' \text{ (ns)}$$

The duration T is commonly referred to as the Nyquist interval. The concept of T is employed not only when the frequency cut-off is a physical property of a given system but also when the system is flat and there is no interest in the performance of the system beyond a given frequency. (BT) 511-1979w

TA *See:* watthour meter—test current; terminal adapter; technical advisory; test amperes.

⟨`tab`⟩ The horizontal tab character. (C/PA) 9945-2-1993

table **(1) (A) (software)** An array of data, each item of which may be unambiguously identified by means of one or more arguments. **(B) (software)** A collection of data in which each item is uniquely identified by a label, by its position relative to the other items, or by some other means. *See also:* label; data. (SE/C) 729-1983
(2) (data management) A two-dimensional array. (See below for an example.)

Table

State	Abbreviation	Zone
Alabama	AL	2
Alaska	AK	9
. . .		
West Virginia	WV	3
Wisconsin	WI	4
Wyoming	WY	3

See also: code-decode table. (C) 610.5-1990w
(3) Functionally related utility application data elements, grouped together into a single data structure for transport. (AMR/SCC31) 1377-1997

table ESD test An indirect test in which ESD is applied to the HCP. (EMC) C63.16-1993

table lamp (illuminating engineering) A portable luminaire with a short stand suitable for standing on furniture. (EEC/IE) [126]

table lookup (TLU) **(1) (mathematics of computing)** A procedure for obtaining the value of a function corresponding to a given argument from a table of function values. *See also:* look-up table. (C) 1084-1986w
(2) (data management) The process of obtaining the value y corresponding to an argument x from a two-dimensional table of (x,y) pairs. *See also:* direct lookup; associative lookup. (C) 610.5-1990w

tablet *See:* graphic tablet; acoustic tablet; data tablet.

table-top device A device designed for and normally placed on the raised surface of a table; e.g., most personal computers are considered to be table-top devices. (EMC) C63.4-1991

tab sequential format (numerically controlled machines) A means of identifying a word by the number of tab characters preceding the word in the block. The first character in each word is a tab character. Words must be presented in a specific order but all characters in a word, except the tab character, may be omitted when the command represented by that word is not desired. (IA) [61]

tabular model A symbolic model whose properties are expressed in tabular form; for example, a truth table that represents the logic of an OR gate. *Contrast:* narrative model; software model; mathematical model; graphical model. (C) 610.3-1989w

tabulate **(A)** To form data into a table. **(B)** To print totals. (C) [20], [85]

tabulation character A format effector character that causes the print or display position to move to the next corresponding horizontal or vertical position in a series of predetermined positions. *See also:* horizontal tabulation character; vertical tabulation character. (C) 610.5-1990w

tabulator A device that reads data from some medium such as punch cards or punched tape, and which produces lists, totals, or calculations. (C) 610.10-1994w

tacan (tactical air navigation) (navigation aids) A complete ultra-high frequency (uhf), polar coordinate (rho theta) navigation system using pulse techniques. The distance (rho) function operates as DME (distance measuring equipment) and the bearing function is derived by rotating the ground transponder antenna so as to obtain a rotating multilobe pattern for coarse and fine bearing information. (AES/GCS) 172-1983w

TACCAR *See:* time-averaged-clutter coherent airborne radar.

tachometer A device to measure speed or rotation. *See also:* rotor. (PE) [9]

tachometer electric indicator A device that provides an indication of the speed of an aircraft engine, of a helicopter rotor, of a jet engine, and of similar rotating apparatus used in aircraft. Such tachometer indicators may be calibrated directly in revolutions per minute or in percent of some particular speed in revolutions per minute. (EEC/PE) [119]

tachometer generator (rotating machinery) A generator, mechanically coupled to an engine, whose main function is to generate a voltage, the magnitude or frequency of which is used either to determine the speed of rotation of the common shaft or to supply a signal to a control circuit to provide speed regulation. (PE) [9]

tachometric relay A relay in which actuation of the contacts is effected at a predetermined speed of a moving part. *See also:* relay. (IA/ICTL/IAC) [60]

taffrail log A device that indicates distance traveled based on the rotation of a screw-type rotor towed behind a ship which drives, through the towing line, a counter mounted on the taffrail. *Note:* An electric contact made (usually) each tenth of a mile causes an audible signal to permit ready calculation of speed. (EEC/PE) [119]

tag **(1) (supervisory control, data acquisition, and automatic control)** A visual indication, usually at the master station, to indicate that a device has been cleared for field maintenance/construction purposes and is not available for control or data acquisition. (SWG/PE/SUB) C37.100-1992, C37.1-1994
(2) (A) Same as flag. **(B)** Same as label. (MIL/C) [2], [20]
(3) (data management) One or more characters associated with a set of data, containing information about the set. (C) 610.5-1990w
(4) (computer graphics) A unique textual identifier associated with a display element. (C) 610.6-1991w
(5) Accident prevention tag (DANGER, PEOPLE AT WORK, etc.) of a distinctive appearance used for the purpose of personnel protection to indicate that the operation of the device to which it is attached is restricted. (NESC) C2-1997

tag address *See:* symbolic address.

tagged architecture A computer architecture in which each word is "tagged" as either an instruction or a unit of data. *Contrast:* Von Neumann architecture. (C) 610.10-1994w

tagged frame A frame that contains a tag header immediately following the Source MAC Address field of the frame or, if the frame contained a Routing Information field, immediately following the Routing Information field. There are two types of tagged frames: *VLAN-tagged frames* and *priority-tagged frames*. (C/LM) 802.1Q-1998

tag header Allows user priority information, and optionally, VLAN identification information, to be associated with a frame. (C/LM) 802.1Q-1998

tag line A control line, normally manila or synthetic fiber rope, attached to a suspended load to enable a worker to control its movement. *Synonym:* tag rope. (T&D/PE) 524-1992r

tag rope *See:* tag line.

tags Men-at-work tags of distinctive appearance, indicating that the equipment or lines so marked are being worked on. (T&D) C2.2-1960

Tag Set Name A numeric identifier associated with a set of security tags. (C/LM) 802.10g-1995

tag sort A sort that uses the address table sorting technique. (C) 610.5-1990w

tail (x-ray energy spectrometers) (on a monoenergetic peak) Any peak shape distortion that does not comply with the limits defining the full energy peak intensity and that does not come from a source of radiation other than the monoenergetic source in question. (NPS/NID) 759-1984r

tailing (1) (hangover) (facsimile) The excessive prolongation of the decay of the signal. *See also:* facsimile signal. (COM) 168-1956w

(2) (hydrometallurgy and ore concentration) (electrometallurgy) The discarded residue after treatment of an ore to remove desirable minerals. *See also:* electrowinning. (PE/EEC) [119]

tailing time (charged-particle detectors) (of an amplifier output pulse) The time interval between the top centerline of a unipolar pulse and the one percent level on the last transition. (NPS) 300-1988r

tail lamp (illuminating engineering) A lighting device used to designate the rear of a vehicle by a warning light. (EEC/IE) [126]

tail-of-wave (chopped wave) impulse test voltage (insulation strength) The crest voltage of a standard impulse wave that is chopped by flashover at or after crest. (PE/TR) 21-1976

tailrace The exit channel of water from the powerhouse. (PE/EDPG) 1020-1988r

tail recursive A property of the implementation of a programming language. In a tail-recursive implementation, iterative processes can be expressed by means of procedure calls. (The process described by a program is iterative if and only if the order of its space growth is constant, aside from that used for the values of the program's variables.) (C/MM) 1178-1990r

tailwater The water in the tailrace. (PE/EDPG) 1020-1988r

take-over time (T_T) Time taken to switch from one transmission direction to the other in double talk conversation on a handsfree telephone (HFT). The signal in the first direction is continuously applied while the interrupting signal is applied in the opposite direction. T_T is measured from the application of the interrupting signal until the output level due to the interrupting signal reaches 3 dB below its final value. (COM/TA) 1329-1999

takeup reel *See:* reel winder.

talker A node that sends an isochronous subaction for an isochronous channel. (C/MM) 1394-1995

talker echo Echo that reaches the ear of the talker. (EEC/PE) [119]

talker sidetone The sidetone acoustic output of the telephone caused by an input from the artificial mouth. (COM/TA) 269-1992

talking key A key whose operation permits conversation over the circuit to which the key is connected. (EEC/PE) [119]

talk-ringing key (listening and ringing key) A combined talking key and ringing key operated by one handle. (EEC/PE) [119]

tamper To interfere with the performance of a security sensor or the electrical connections within an alarm communications system. (PE/NP) 692-1997

tampering Attacks to physical and logical system components (i.e., hardware, software, media), including substituting rogue components in place of legitimate components, physical theft of components, damage to components, and electrical probing within the system. This threat area assumes that an intruder has physical or electrical access to the system components and their internal structure (e.g., processor and memory boards, software modules). (C/BA) 896.3-1993w

tamper protection The aspect of physical security concerned with the protection of system resources from unauthorized modification, especially where modifying or altering a component degrades system security. *See also:* quadrant. (C/BA) 896.3-1993w

tamper-resistant enclosure A metal-enclosure for a power switchgear assembly that is designed to resist damage to or improper operation of the switchgear from willful acts of destruction and which is designed to provide reasonably safe protection against tampering by unauthorized persons who may attempt to gain entry by forcible means, to insert foreign substances into, or otherwise tamper with the assembly. (SWG/PE) C37.20-1968w

tandem (cascade) (networks) Networks are in tandem when the output terminals of one network are directly connected to the input terminals of the other network. *See also:* network analysis. (CAS) [13]

tandem blocking The first and final failure matching loss for trunk-to-trunk connections. *See also:* matching loss. (COM/TA) 973-1990w

tandem central office (tandem office) A central office used primarily as a switching point for traffic between other central offices. (SWG/PE) C37.20-1968w

tandem-completing trunk A trunk, extending from a tandem office to a central office, used as part of a telephone connection between stations. (EEC/PE) [119]

tandem control (electric power system) A means of control whereby the area control error of an area or areas A, connected to the interconnected system B only through the facilities of another area C, is included in control of area C generation. (PE/PSE) 94-1970w

tandem drive Two or more drives that are mechanically coupled together. *See also:* electric drive. (IA/ICTL/APP/IAC) [69], [60]

tandem office (telephone switching systems) An intermediate office used primarily for interconnecting end offices with each other and with toll connecting trunks. (COM) 312-1977w

tandem trunk (data transmission) A trunk extending from a central office or a tandem office to a tandem office and used as part of a telephone connection between stations. (PE) 599-1985w

tangential plane The plane that is tangential to the surface of the VDT screen at the center-center point. (EMC) 1140-1994r

tangential wave path (data transmission) In radio wave propagation over the earth, a path of propagation of a direct wave, which is tangential to the surface of the earth. The tangential wave path is curved by atmospheric refraction. (PE) 599-1985w

tangent-plane approximation *See:* Kirchhoff approximation.

tank (storage cell) A lead container, supported by wood, for the element and electrolyte of a storage cell. *Note:* This is restricted to some relatively large types of lead-acid cells. *See also:* battery. (EEC/PE) [119]

tank circuit (signal-transmission system) A circuit consisting of inductance and capacitance, capable of storing electric energy over a band of frequencies continuously distributed about a single frequency at which the circuit is said to be resonant, or tuned. *Note:* The selectivity of the circuit is proportional to the ratio of the energy stored in the circuit to the energy dissipated. The ratio is often called the Q of the circuit. *See also:* oscillatory circuit; signal. (AP/ANT) 145-1983s

tank, single-anode *See:* single-anode tank.

tank vessel A vessel that carries liquid or gaseous cargo in bulk. (IA/MT) 45-1998

tank voltage The total potential drop between the anode and cathode bus bars during electrodeposition. *See also:* electroplating. (EEC/PE) [119]

TAP *See:* test access port.

tap (1) (fiber optics) A device for extracting a portion of the optical signal from a fiber. (Std100) 812-1984w
(2) (power and distribution transformers) (in a transformer) A connection brought out of a winding at some point between its extremities, to permit changing the voltage, or current, ratio. (PE/TR) C57.12.80-1978r
(3) (power and distribution transformers) An available connection that permits changing the active portion of the device in the circuit. *See also:* grounding device. (SPD/PE) 32-1972r
(4) (reactor) A connection brought out of a winding at some point between its extremities, to permit changing the impedance. *See also:* reactor. C57.16-1958w
(5) (rotating machinery) A connection made at some intermediate point in a winding. *See also:* stator; voltage regulator; rotor. (PE) [9]
(6) (broadband local area networks) A passive device in the feeder system that provides a connection between the drop cable and the feeder. The tap is the principal means of access to the cable system by the user. It removes a portion of the signal power from the distribution line and delivers it to the drop line. The amount of power tapped off the main line depends on the input power to the tap and the attenuation value of the tap. Only the information signal (and not 60 Hz power) goes to the outlet ports. *See also:* multi-tap. (LM/C) 802.7-1989r
(7) (A) In a baseband system, a component or connector that attaches a transceiver to a cable. **(B)** In a broadband system, a passive device used to remove a portion of the signal power from the distribution line and deliver it onto the drop line. *See also:* fan-out box. **(C)** In the security environment, the term is used for a breach of security on a telecommunication line or channel. (C) 610.7-1995
(8) A connection brought out of a winding at some point between its extremities to permit the changing of the voltage ratio. (PE/TR) C57.15-1999

tap change operation A complete sequence of events from the initiation to the completion of the transition of the through current from one tap position to an adjacent one. (PE/TR) C57.131-1995

tap-changer, for deenergized operation (power and distribution transformers) A selector switch device used to change transformer taps with the transformer de-energized. (PE/TR) C57.12.80-1978r

tape (1) (rotating machinery) A relatively narrow, long, thin, flexible fabric, mat, or film, or a combination of them with or without binder, not over 20 cm in width. *See also:* rotor; stator. (PE) [9]
(2) (electronic computation) *See also:* magnetic tape. (PE) [9]
(3) *See also:* punch tape; magnetic tape; chadless tape; carriage control tape; perforated tape. (C) 610.10-1994w

tape block (test, measurement, and diagnostic equipment) A group of frames or tape lines. (MIL) [2]

taped A joint that is constructed in the field with the use of one or more tapes that are applied over the cable in layers. Heat may or may not be applied as part of the installation procedure. (PE/IC) 404-1993

tape deck *See:* tape drive.

taped insulation Insulation of helically wound tapes applied over a conductor or over an assembled group of insulated conductors. A) When successive convolutions of a tape overlie each other for a fraction of the tape width, the taped insulation is lap wound. This is also called positive lap wound. B) When a tape is applied so that there is an open space between successive convolutions, this construction is known as open butt or negative lap wound. C) When a tape is applied so that the space between successive convolutions is too small to measure with the unaided eye, it is a closed butt taping. (T&D/PE) [10]

taped joint (power cable joints) A joint with hand-applied tape insulation. (PE/IC) 404-1986s

tape drive A device that moves tape past a head. (C) [20], [85]
(2) (A) An input device that reads magnetic tape. *Contrast:* disk drive. **(B)** A mechanism for moving magnetic tape and controlling its movement. *Note:* This mechanism is used to move magnetic tape past a read head or write head, or used to allow automatic rewinding. *Synonym:* transport; magnetic tape drive. *See also:* incremental tape drive; hypertape drive. (C) 610.10-1994

tape frame *See:* tape row; tape drive.

tape line *See:* frame.

tape merge sort An external merge sort in which the auxiliary storage used is a magnetic tape. *See also:* direct-access merge sort. (C) 610.5-1990w

tape preparation The act of translating command information into punched or magnetic tape. (IA) [61]

tape punch *See:* perforator.

taper (communication practice) A continuous or gradual change in electrical properties with length, as obtained, for example, by a continuous change of cross-section of a waveguide. *See also:* transmission line. (Std100) 270-1966w

tape recorder *See:* magnetic recorder.

tapered fiber waveguide (fiber optics) An optical waveguide whose transverse dimensions vary monotonically with length. *Synonym:* tapered transmission line. (Std100) 812-1984w

tapered hose *See:* leader cone.

tapered key (rotating machinery) A wedge-shaped key to be driven into place, in a matching hole or recess. (PE) [9]

tapered potentiometer A function potentiometer that achieves a prescribed functional relationship by means of a nonuniform winding. *See also:* electronic analog computer. (C) 165-1977w

tapered transmission line *See:* tapered waveguide; tapered fiber waveguide.

tapered waveguide (waveguide terms) A waveguide in which a physical or electrical characteristic increases or decreases continuously with distance along the axis of the guide. (MTT) 147-1979w, 146-1980w

tape reproducer A device that prepares one tape from another tape by copying all or part of the data from the tape that is read. (C) 610.10-1994w

tape row A group of binary characters recorded or sensed in parallel on a line perpendicular to the reference edge of a tape. *Synonym:* tape frame. *See also:* skew; row pitch. (C) 610.10-1994w

taper, waveguide *See:* waveguide taper.

tape station *See:* tape unit; tape drive.

tape thickness The lesser of the cross-sectional dimensions of a length of ferromagnetic tape. *See also:* tape-wound core. (Std100) 163-1959w

tape to card Pertaining to equipment or methods that transmit data from either magnetic tape or punched tape to punched cards. (C) [20], [85]

tape transmitter (telegraphy) A machine for keying telegraph code signals previously recorded on tape. *See also:* telegraphy. (COM) [49]

tape transport (1) (test, measurement, and diagnostic equipment) A device which moves magnetic or punched tape past the tape reader. Reels for storage of the tape are usually provided. *See also:* tape drive. (MIL) [2]
(2) *See also:* tape drive. (C) 610.10-1994w

tape unit **(1)** A device containing a tape drive, together with reading and writing heads and associated controls. (C) [20], [85]
(2) *See also:* tape drive. (C) 610.10-1994w

tape width The greater of the cross-sectional dimensions of a length of ferromagnetic tape. *See also:* tape-wound core. (Std100) 163-1959w

tape-wound core A length of ferromagnetic tape coiled about an axis in such a way that one convolution falls directly upon the preceding convolution. *See also:* wrap thickness. (Std100) 163-1959w

tap outlet An F-type connector port on a tap used to attach a drop cable to an outlet. (LM/C) 802.7-1989r

tapped field *See:* short field.

tapped field control A system of regulating the tractive force of an electrically driven vehicle by changing the number of effective turns of the traction motor series-field windings by means of an intermediate tap or taps in those windings. *See also:* multiple-unit control. (EEC/PE) [119]

tapped potentiometer A potentiometer, usually a servo potentiometer, that has a number of fixed contacts (or taps) to the resistance element in addition to the end and movable contacts. *See also:* electronic analog computer. (C) 165-1977w

tapped-secondary current or voltage transformer One with two ratios obtained by use of a tap on the secondary winding. (PE/TR) C57.13-1993, [57]

tapped way A way solidly connected to the bus. (SWG/PE) C37.71-1984r, C37.100-1992

tapper bell A single-stroke bell having a gong designed to produce a sound of low intensity and relatively high pitch. *See also:* protective signaling. (EEC/PE) [119]

tap selector A device designed to carry, but not to make or break current, used in conjunction with an arcing switch to select tap connections. (PE/TR) C57.131-1995

target **(1)** **(camera tubes)** A structure employing a storage surface that is scanned by an electron beam to generate a signal output current corresponding to a charge-density pattern stored thereon. *Note:* The structure may include the storage surface that is scanned by an electron beam, the backplate, and the intervening dielectric. *See also:* radar; television. (ED/BT/AV) 161-1971w, [34], [45]
(2) **(storage tubes)** The storage surface and its immediate supporting electrodes. *See also:* radar; storage tube. (ED) 158-1962w, [45]
(3) In micrographics, any document or chart containing identification information or a resolution test chart. *See also:* flash card. (C) 610.2-1987
(4) The node addressed by the first symbol of a packet; i.e., the final destination of the packet. (C/MM) 1596-1992
(5) The specification of a target distribution object, or installed software object, for a software administration utility. The target host provides a means to locate the target role and the target path is a path accessible to the target host. (C/PA) 1387.2-1995
(6) **(operation indicator)** (of a relay) A supplementary device operated either mechanically or electrically, to indicate visibly that the relay has operated or completed its function. *Notes:* 1. A mechanically operated target indicates the physical operation of the relay. 2. An electrically operated target, when not further described, is actuated by the current in the control circuit associated with the relay and hence indicates not only that the relay has operated but also that it has completed its function by causing current to flow in the associated control circuit. 3. A shunt-energized target only indicates operation of the relay contact and does not necessarily show that current has actually flowed in the associated control circuit. (SWG/PE) C37.100-1992
(7) Broadly, any discrete object that scatters energy back to the radar. (AES/RS) 686-1990
(8) *See also:* target sag. (T&D/PE) 524-1992r
(9) **(A)** Specifically, an object of radar search or tracking. **(B)** Broadly, any discrete object that scatters energy back to the radar. (AES/GCS) 686-1997, 172-1983

target capacitance (camera tubes) The capacitance between the scanned area of the target and the backplate. *See also:* television. (ED/BT/AV) 161-1971w, [45], [34]

target classification *See:* target recognition.

target cutoff voltage (camera tubes) The lowest target voltage at which any detectable electric signal corresponding to a light image on the sensitive surface of the tube can be obtained. *See also:* television. (ED) 161-1971w

target fluctuation Variation in the amplitude of the echo from a complex target caused by changes in target aspect angle, motion of target-scattering sources, or changes in radar wavelength (i.e., the amplitude component of target noise). *Note:* Rapid fluctuation is usually modeled as independent from pulse to pulse within a scan and independent from scan to scan. The terms *scintillation* and *amplitude noise* have been used in the past as synonyms for target fluctuation and also to denote location errors caused by target fluctuation, and should be avoided because of this ambiguity. (AES) 686-1997

target glint *See:* glint.

target host The host portion of a target specification. (C/PA) 1387.2-1995

target identification Target identification means identifying a particular target such as the name painted on the side of a ship, an aircraft's side-number, or the flight number of a commercial aircraft. *Note:* Primary radar cannot usually provide the identity of a target, but secondary radar systems including transponders can be used for such cooperative target identification. *See also:* target recognition. (AES) 686-1997

target language (software) The language in which the output from a machine-aided translation process is represented. For example, the language output by an assembler or compiler. *Synonym:* object language. *Contrast:* source language. (C) 610.2-1987, [85], [20], 610.12-1990

target machine **(A)** **(software)** The computer on which a program is intended to execute. **(B)** **(software)** A computer being emulated by another computer. *Contrast:* host machine. (C) 610.12-1990

target noise Random variations in observed amplitude, location, and/or Doppler of a target, caused by changes in target aspect angle, rotation or vibration of target-scattering sources, or by changes in wavelength. *See also:* glint; scintillation error; target fluctuation. (AES) 686-1997

target path The pathname portion of a target specification. (C/PA) 1387.2-1995

target program A program written in a target language. *See also:* object program. (C) [20], [85]

target recognition The use of a radar to recognize one class of target from another. Also known as noncooperative target recognition (NCTR) or target classification. *See also:* target identification. (AES) 686-1997

target role Where software is installed, removed, listed, and otherwise operated on by the utilities. For example, when installing software, the target is where software is installed after having been delivered from a source. As another example, the target for a copy operation command refers to the distribution to which products are added. For management operations like removing software, the target refers to either the installed_software objects or the distributions from which software is being removed. (C/PA) 1387.2-1995

target sag A device used as a reference point to sag conductors. It is placed on one structure of the sag span. The sagger, on the other structure of the sag span, can use it as a reference to determine the proper conductor sag. *Synonyms:* target; sag board. (T&D/PE) 524-1992r

target system The combination of the computer system on which the PCTS is executed and the parts of the development system that are used to generate the executable code of a PCTS. (C/PA) 13210-1994, 2003.1-1992

target transmitter (electronic navigation) (navigation aids) A source of radio-frequency energy suitable for providing test signals at a test site. *See also:* navigation.
(AES/GCS/RS) 173-1959w, [42], 686-1982s, 172-1983w

target user The name of a user on the destination batch server. The *target user* is the *user_name* under whose account the job is to execute on the destination batch server.
(C/PA) 1003.2d-1994

target voltage (camera tube with low-velocity scanning) The potential difference between the thermionic cathode and the backplate. *See also:* television.
(ED/BT/AV) 161-1971w, [34], [45]

tariff (1) (data transmission) The published rate for a particular approved commercial service of a common carrier.
(PE) 599-1985w
(2) (electric power utilization) (power operations) A published volume of rate schedules and general terms and conditions. (PE/PSE) 858-1987s, 346-1973w
(3) A published list of rate schedules and terms and conditions. (AMR/SCC31) 1377-1997

tarnish Surface discoloration of a metal caused by formation of a thin film of corrosion product. (IA) [59]

TASI *See:* time assigned speech interpolation.

task (1) (A) (software) A sequence of instructions treated as a basic unit of work by the supervisory program of an operating system. **(B) (software)** In software design, a software component that can operate in parallel with other software components. (C) 610.12-1990
(2) The smallest unit of work subject to management accountability. A task is a well-defined work assignment for one or more project members. Related tasks are usually grouped to form activities. (C/SE) 1074-1995s
(3) An Ada object with a thread of control. The execution of an Ada program consists of the execution of one or more tasks. Each task represents a separate thread of control that proceeds independently and concurrently between the points where it interacts with other tasks. (C) 1003.5-1999

task-ambient lighting A concept involving a component of light directed toward tasks from appropriate locations by luminaires located close to the task for energy efficiency.
(IA/PSE) 241-1990r

task illumination (health care facilities) Provision for the minimum lighting required to carry out necessary tasks in the described areas, including safe access to supplies and equipment, and access to exits. (NESC/NEC) [86]

tasking activity The person(s) or organization that directs a performing activity to accomplish the work specified in this standard. (C/SE) 1220-1994s

taut-band suspension (electric instruments) A mechanical arrangement whereby the moving element of an instrument is suspended by means of ligaments, usually in the form of a thin flat conducting ribbon, at each of its ends. The ligaments normally are in tension sufficient to restrict the lateral motion of the moving element to within limits that permit freedom of useful motion when the instrument is mounted in any position. A restoring torque is produced within the ligaments with rotation of the moving element. *See also:* moving element. (EEC/AII) [102]

taxi-channel lights (illuminating engineering) Aeronautical ground lights arranged along a taxi-channel of a water aerodrome to indicate the route to be followed by taxiing aircraft.
(EEC/IE) [126]

taxi light (illuminating engineering) An aircraft aeronautical light designed to provide necessary illumination for taxiing.
(EEC/IE) [126]

taxiway-centerline lights (illuminating engineering) Taxiway lights placed along the centerline of a taxiway except that on curves or corners having fillets, these lights are placed a distance equal to half the normal width of the taxiway from the outside edge of the curve or corner. (EEC/IE) [126]

taxiway-edge lights (illuminating engineering) Taxiway lights placed along or near the edges of a taxiway.
(EEC/IE) [126]

taxiway holding-post light (illuminating engineering) A light or group of lights installed at the edge of a taxiway near an entrance to a runway, or to another taxiway, to indicate the position at which the aircraft should stop and obtain clearance to proceed. (EEC/IE) [126]

taxiway lights (illuminating engineering) Aeronautical ground lights provided to indicate the route to be followed by taxiing aircraft. (EEC/IE) [126]

taxonomy (1) A scheme that partitions a body of knowledge and defines the relationships among the pieces. It is used for classifying and understanding the body of knowledge.
(C) 610.12-1990, 610.10-1994w
(2) A classification scheme, or its results.
(PE/NP) 1082-1997

Taylor distribution, circular A continuous distribution of a circular planar aperture that is equiphase, with the amplitude distribution dependent only on distance from the center of the aperture and such as to produce a pattern with a main beam plus side lobes. The side lobe structure is rotationally symmetric, with a specified number of inner side lobes at a quasi-uniform height, the remainder of the side lobes decaying in height with their angular separation from the main beam. *Note:* Taylor distributions are often sampled to obtain the excitation for a planar array. (AP/ANT) 145-1993

Taylor distribution, linear A continuous distribution of a line source that is symmetric in amplitude, has a uniform progressive phase, and yields a pattern with a main beam plus side lobes. The side lobe structure is symmetrical, with a specified number of inner side lobes at a quasi-uniform height, the remainder of the side lobes decaying in height with their angular separation from the main beam. *Note:* Taylor distributions are often sampled to obtain the excitation for a planar array. (AP/ANT) 145-1993

TB cell *See:* transmitter-blocker cell.

Tbyte Terabyte. Indicates 2^{40} bytes. (MM/C) 1212-1991s

T-carrier system A hierarchy of high-speed digital transmission facility designed to carry speech and other signals in digital form according to their transmission capacity. *See also:* T1; T2; T3; T1C; T4. (C) 610.7-1995

TCB *See:* trusted computing base.

TCBH *See also:* traffic engineering limits; time-consistent busy hour. (COM/TA) 973-1990w

TCF *See:* transformer correction factor.

TCM *See:* time compression multiplexing.

T-connected transformer (power and distribution transformers) (or tee-connected) A three-phase to three-phase transformer, similar to a Scott-connected transformer. *See also:* Scott or T-connected transformer.
(PE/TR) C57.12.80-1978r

TC traffic measures *See:* time-consistent traffic measures.

TCR *See:* thyristor-controlled reactor.

TCU *See:* trunk coupling unit.

TDD *See:* total demand distortion.

TDM *See:* time-division multiplexing.

TDMA *See:* time division multiple access.

TDR *See:* time domain reflectometer.

TE *See:* transverse electric mode.

tearing (television) An erratic lateral displacement of some scanning lines of a raster caused by disturbance of synchronization. *See also:* television. (BT/AV) [34]

teaser transformer (power and distribution transformers) As applied to two single-phase Scott-connected units for the three-phase to two-phase or two-phase to three-phase operation, designates the transformer that is connected between the midpoint of the main transformer and the third-phase wire of the three-phase system. (PE/TR) C57.12.80-1978r

technical advisory (TA) A telephone company publication intended to disclose information and request comments regard-

ing network services.
(AMR/SCC31) 1390-1995, 1390.2-1999, 1390.3-1999

technical effort The total engineering, testing, manufacturing, and specialty engineering effort associated with the development of a product that encompasses all of the system, equipment, facilities, etc., necessary for the enterprise to develop, produce, distribute, operate, test, support, train, and dispose of the product. (C/SE) 1220-1994s

technical management The application of technical and administrative resources to plan, organize, and control engineering functions. (C) 610.12-1990

technical requirements (TR) A telephone company publication intended to disclose information and operation regarding network services.
(AMR/SCC31) 1390-1995, 1390.2-1999, 1390.3-1999

technical review A systematic evaluation of a software product by a team of qualified personnel that examines the suitability of the software product for its intended use and identifies discrepancies from specifications and standards. Technical reviews may also provide recommendations of alternatives and examination of various alternatives. (C/SE) 1028-1997

technical standard A standard that describes the characteristics of applying accumulated technical or management skills and methods in the creation of a product or performing a service.
(C) 610.12-1990

techniques (1) (software) Technical and managerial procedures that aid in the evaluation and improvement of the software development process. (C/SE) 610.12-1990, 983-1986w
(2) Technical and managerial procedures used to achieve a given objective. (C/SE) 730.1-1995

technology Scientific knowledge used to achieve a practical purpose. (C/PA) 1003.23-1998

Technology Ability Field An eight-bit field in the Auto-Negotiation base page that is used to indicate the abilities of a local station, such as support for 10BASE-T, 100BASE-T4, 100BASE-TX, and 100BASE-T2, as well as full duplex capabilities. (C/LM) 802.3-1998

technology component model The assembled IT services required to deliver one or more IS services to support the BSRs.
(C/PA) 1003.23-1998

technology data Data used to calculate the timing properties of a cell instance based on its context in the design. This term includes information that is not cell type specific or data specific for each cell type in the library. The kind of data used varies with the timing calculation methodology. General data and cell data may be contained in the same file or in separate files. Cell data also may be merged with the timing models of each cell, for example, when a tool performs its own timing calculation. (C/DA) 1481-1999

technology library A technology library is a program written in delay calculation language (DCL) consisting of one or more subrules, each of which may contain references to other subrules (yet to be loaded). There is no hierarchical limit to the nesting of subrules within the scope of a technology library. Subrules can also be segmented into technology families, which alters the way they are made available to the application. (C/DA) 1481-1999

TEDL *See:* Test Equipment Description Language.

tee connection Connection of heater in series or parallel to accommodate a branch on a pipe or equipment.
(IA) 515-1997

tee coupler (fiber optics) A passive coupler that connects three ports. *See also:* star coupler. (Std100) 812-1984w

teed feeder A feeder that supplies two or more feeding points. *See also:* center of distribution. (PE/TR) [57]

tee junction (waveguide components) A junction of waveguides or transmission lines in which the longitudinal guide axes form a tee. (MTT) 147-1979w

teleautograph A telegraphic writing instrument, in which movement of a pen at the transmitting end causes corresponding movement of a pen at the remote receiving instrument. *Synonym:* telewriter. *See also:* telegraphy.
(C/PE/EEC) 610.2-1987, [119]

telecommunication (1) The transmission of signals over long distance, such as by telegraph, radio, or television. *See also:* computer conferencing; office automation.
(SWG/C/PE) 610.2-1987, C37.20.1-1987s
(2) (data transmission) The transmission of information from one point to another. (PE) 599-1985w
(3) The transmission of signals by electrical, electromagnetic, optical acoustic, or mechanical means. (C) 610.7-1995

telecommunication access program A software program located in a front-end communications processor that handles tasks associated with the routing, scheduling, and movement of messages between remote terminals and the host computer.
(C) 610.7-1995

telecommunication circuit A circuit that is designed to handle remote transmission of information. *See also:* wideband circuit. (C) 610.7-1995, 610.10-1994w

telecommunication line A medium, such as wire or circuit, that connects equipment which enables data to be sent and received. (C) 610.7-1995

telecommunication loop (telephone switching systems) A channel between a telecommunications station and a switching entity. (COM) 312-1977w

telecommunication monitor *See:* teleprocessing monitor.

telecommunications Any transmission, emission, and reception of signs, signals, writings, images, and sounds, i.e., information of any nature, by cable, radio, optical, or other electromagnetic systems. (IA/PSE) 1100-1999

telecommunications customer (telephone switching systems) One for whom telecommunications service is provided (formerly referred to as a "subscriber"). (COM) 312-1977w

telecommunications equipment room (TER) A centralized space for telecommunications equipment that serves the occupants of the building. (IA/PSE) 1100-1999

telecommunications exchange (telephone switching systems) A means of providing telecommunications services to a group of users within a specified geographical area.
(COM) 312-1977w

Telecommunications Industries Association A sister organization of EIA that establishes and maintains standards for the telecommunications industries in the United States.
(C) 610.7-1995

telecommunications interface equipment A portion of a relay system that transmits or receives information from a telecommunications system; eg, audio tone equipment or carrier transmitter-receiver included as an integral part of the relay system. (PE/PSR) C37.90.1-1989r

telecommunications switchboard (telephone switching systems) A manual means of interconnecting telecommunications lines, trunks, and associated circuits, and including signaling facilities. (COM) 312-1977w

telecommunications switching (telephone switching systems) The function of selectively establishing and releasing connections among telecommunication transmission paths.
(COM) 312-1977w

telecommunications system (1) (telephone switching systems) An assemblage of telecommunications stations, lines, and channels, and switching arrangements for their interconnection, together with all the accessories for providing telecommunications services. (COM) 312-1977w
(2) (surge withstand capability tests for protective relays and relay systems) Any of the telecommunication media; eg, microwave, power-line carrier, wire line.
(PE/PSR) C37.90.1-1989r

telecommuting An employment alternative involving working at home using a computer and telecommunication system instead of commuting between home and workplace.
(SWG/C/PE) 610.2-1987, C37.20.1-1987s

teleconferencing A form of communication that uses telephones, computer networks, and television to allow participants at different geographical locations to confer. (SWG/C/PE) 610.2-1987, C37.20.1-1987s

telecopier A device used for facsimile transmission. (C) 610.2-1987, 610.10-1994w

telefax *See:* facsimile transmission.

telegraph A mechanized or electric device for the transmission of stereotyped orders or information from one fixed point to another. *Note:* The usual form of telegraph is a transmitter and a receiver, each having a circular dial in sectors upon which are printed standard orders. When the index of the transmitter is placed at any order, the pointer of the receiver designates that order. dual mechanism is generally provided to permit repeat back or acknowledgment of orders. (EEC/PE/MT) [119]

telegraph channel (data transmission) A channel suitable for the transmission of telegraph signals. *Note:* Three basically different kinds of telegraph channels used in multichannel telegraph transmission are

- One of a number of paths for simultaneous transmission in the same frequency range as in bridge duplex, differential duplex, and quadruplex telegraphy.
- One of a number of paths for simultaneous transmission in the same frequency range as in bridge duplex, differential duplex, and quadruplex telegraphy.
- One of a number of paths for successive transmission as in multiplex printing telegraphy.

Combinations of these three types may be used on the same circuit. (PE) 599-1985w

telegraph concentrator A switching arrangement by means of which a number of branch or subscriber lines or station sets may be connected to a lesser number of trunk lines or operating positions or instruments through the medium of manual or automatic switching devices in order to obtain more efficient use of facilities. *See also:* telegraphy. (COM) [49]

telegraph distortion (data transmission) The condition in which the significant intervals have not all exactly their theoretical durations. The reference point used when measuring telegraph distortion is the initial space-to-mark transition of each character which occurs at the beginning of each "start" element. The slicing level for all measurements is at the 50% point on the rising or falling current waveforms. Percent distortion is expressed by

$$\text{percent distortion} = \frac{\Delta t}{t_e} \times 100$$

where

t = time difference between the actual slicing point and the ideal crossover point

t_e = time interval of one signal element.

(PE) 599-1985w

telegraph distributor A device that effectively associates one direct-current or carrier telegraph channel in rapid succession with the elements of one or more signal sending or receiving devices. *See also:* telegraphy. (EEC/PE) [119]

telegraph key A hand-operated telegraph transmitter used primarily in Morse telegraphy. *See also:* telegraphy. (EEC/PE) [119]

telegraph repeater An arrangement of apparatus and circuits for receiving telegraph signals from one line and retransmitting corresponding signals into another line. *See also:* telegraphy. (EEC/PE) [119]

telegraph selector A device that performs a switching operation in response to a definite signal or group of successive signals received over a controlling circuit. *See also:* telegraphy. (EEC/PE) [119]

telegraph sender A transmitting device for forming telegraph signals. Examples are a manually operated telegraph key and a printer keyboard. *See also:* telegraphy. (EEC/PE) [119]

telegraph signal (telecommunications) The set of conventional elements established by the code to enable the transmission of a written character (letter, figure, punctuation sign, arithmetic sign, etc.) or the control of a particular function (spacing, shift, line-feed, carriage return, phase correction, etc.): this set of elements being characterized by the variety, the duration and the relative position of the component elements or by some of these features. (COM) [49]

telegraph signal distortion Time displacement of transitions between conditions, such as marking and spacing, with respect to their proper relative positions in perfectly timed signals. *Note:* The total distortion is the algebraic sum of the bias and the characteristic and fortuitous distortions. *See also:* telegraphy. (EEC/PE) [119]

telegraph sounder A telegraph receiving instrument by means of which Morse signals are interpreted aurally (or read) by noting the intervals of time between two diverse sounds. *See also:* telegraphy. (EEC/PE) [119]

telegraph transmission speed The rate at which signals are transmitted, and may be measured by the equivalent number of dot cycles per second or by the average number of letters or words transmitted, and received per minute. *Note:* A given speed in dot cycles per second (often abbreviated to dots per second) may be converted to bauds by multiplying by two. The baud is the unit of signaling transmission speed recommended by the International Consultative Committee on Telegraph Communication. Where words per minute are used as a measure of transmission speed, five letters and a space per word are assumed. *See also:* telegraphy. (EEC/PE) [119]

telegraph transmitter A device for controlling a source of electric power so as to form telegraph signals. *See also:* telegraphy. (EEC/PE) [119]

telegraph word (conventional) A word comprising five letters together with one letter-space, used in computing telegraph speed in words per minute or traffic capacity. *See also:* telegraphy. (COM) [49]

telegraphy (1) (data transmission) A system of telecommunication for the transmission of graphic symbols, usually letters or numerals, by the use of a signal code. It is used primarily for record communication. The term may be extended to include any system of telecommunication for the transmission of graphic symbols or images for reception in record form, usually without gradation of shade values. (PE) 599-1985w

(2) The communication of textual messages through a telecommunication medium at speeds of 150 baud or less. (C) 610.7-1995

teleinformatics Data transfer via telecommunication systems. (C) 610.2-1987

telemetering (1) (A) (supervisory control, data acquisition, and automatic control) Transmission of measurable quantities using telecommunication techniques. *See also:* pulse-type telemeter; current-type telemeter; frequency-type telemeter; ratio-type telemeter. **(B) (supervisory control, data acquisition, and automatic control)** (Analog) Telemetering in which some characteristic of the transmitter signal is proportional to the quantity being measured. **(C) (supervisory control, data acquisition, and automatic control)** (Digital) Telemetering in which a numerical representation is generated and transmitted, the number being representative of the quantity being measured. (SWG/PE/SUB) C37.1-1987, C37.100-1992

(2) (data transmission) Measurement with the aid of intermediate means that permit the measurement to be interpreted at a distance from the primary detector. *Note:* The distinctive feature of telemetering is the nature of the translating means, which includes provision for converting the measure into a representative quantity of another kind that can be transmitted conveniently for measurement at a distance. The actual distance is irrelevant. (SWG/PE) 599-1985w

telemetering selection point interface Master station or RTU (or both) element(s) for the selective connection of telemetering transmitting equipment to appropriate telemetering receiving equipment over an interconnecting communication channel. This type of point is more commonly used in electromechanical or stand-alone type of supervisory control. (SUB/PE) C37.1-1994

telemeter service Metered telegraph transmission between paired telegraph instruments over an intervening circuit adapted to serve a number of such pairs on a shared-time basis. *See also:* telegraphy. (EEC/PE) [119]

telemetry interface unit (TIU) A customer premise equipment (CPE) device that provides a network gateway function and an interface to one or more meters (water, gas, and electric) or other telemetry/control devices or to a local area network. The TIU may be placed in series with or bridged onto the local loop assigned to the end user. Because the TIUs are not network elements but CPE, they are connected to the end user's line (tip/ring) of the local loop at the network interface. In existing systems, these units are also known as meter interface units (MIUs). (AMR/SCC31) 1390-1995, 1390.2-1999, 1390.3-1999

teleordering Use of a telecommunication system to accept orders from customers at remote locations. *Synonyms:* online ordering; teleshopping. (C) 610.2-1987

telephone air-to-air input-output characteristic The acoustical output level of a telephone set as a function of the acoustical input level of another telephone set to which it is connected. The output is measured in an artificial ear, and the input is measured free-field at a specified location relative to the reference point of an artificial mouth. *See also:* telephone station. (COM) [50]

telephone booth A booth, closet, or stall for housing a telephone station. *See also:* telephone station. (EEC/PE) [119]

telephone central office (data transmission) A telephone switching unit, installed in a telephone system providing service to the general public, having the necessary equipment and operating arrangements for terminating and interconnecting lines and trunks. *Note:* There may be more than one central office in the same building. (PE) 599-1985w

telephone channel (data transmission) A channel suitable for the transmission of telephone signals. (PE) 599-1985w

telephone connection A two-way telephone channel completed between two points by means of suitable switching apparatus and arranged for the transmission of telephone currents, together with the associated arrangements for its functioning with the other parts of a telephone system in switching and signaling operations. *Note:* The term is also sometimes used to mean a two-way telephone channel permanently established between two telephone stations. (EEC/PE) [119]

telephone electrical impedance The complex ratio of the voltage to the current at the line terminals at any given single frequency. *See also:* telephone station. (COM) [50]

telephone equalization A property of a telephone circuit that ideally causes both transmit and receive responses to be inverse functions of current, thus tending to equalize variations in loop loss. (COM) [50]

telephone exchange (1) (data transmission) A unit of a telephone communication system for the provision of communication service in a specified area which usually embraces a city, town, or village, and its environs. Incoming lines are connected to outgoing lines as required by the individual caller dial code. (PE) 599-1985w

(2) *See also:* central office. (C) 610.7-1995

telephone feed circuit An arrangement for supplying direct-current power to a telephone set and an alternating-current path between the telephone set and a terminating circuit. (COM) [50]

telephone frequency characteristics Electrical and acoustical properties as functions of frequency. (COM) [50]

telephone handset A telephone transmitter and receiver combined in a unit with a handle. *See also:* telephone station. (COM) [50]

telephone influence factor (TIF) (1) (high-voltage direct-current systems) A dimensionless quantity which includes C-message weighting and is used to express the effect of the deviation of a voltage or current wave shape from a pure sinusoidal wave on a voice-frequency communication network caused by electromagnetic or electrostatic induction, or both. The frequencies and amplitudes of harmonics present on the power circuit, among other factors, determine a power circuit's inductive influence on a voice communications circuit. TIF expressed in terms of $I \cdot T$ product current and voltage TIF (that is, kV $\cdot$ T product per kilovolt) is a measure of this influence. TIF of a voltage or current wave is the ratio of the square root of the sum of the squares (rss) of the weighted root-mean-square (rms) values of all the sine-wave components (including in ac waves both fundamental and harmonics to the root-mean-square value (unweighted) of the entire wave. C-message weighting is derived from listening tests to indicate the relative annoyance of speech impairment by an interfering signal of frequency f as heard through a modern (since 1960) telephone set. The result, called C-message weighting, is shown in graphical and tubular form in the figure on the next page in terms of relative interfering effect P_f at frequency f.

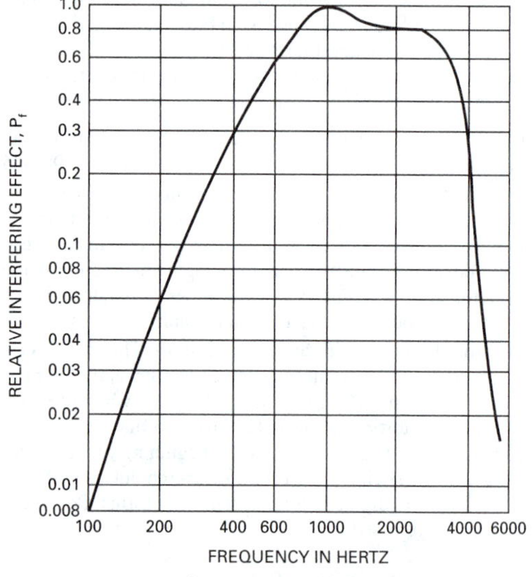

C-message weighting

telephone influence factor

(COM/TA) 368-1977w

(2) (thyristor) Of a voltage or current wave in an electric supply circuit, the ratio of the square root of the sum of the square of the weighted root-mean-square (rms) values of all sine-wave components (including in alternating waves both the fundamental and harmonics) to the rms (unweighted) values of the entire wave. (The weightings are applied to the individual components of different frequencies according to a prescribed curve). (IA/IPC) 428-1981w

(3) (voice-frequency electrical-noise test) The ratio of the square root of the sum of the squares of the weighted root-mean-square values of all the sine-wave components (including, in alternating-current waves, both fundamental and harmonics) to the root-mean-square value (unweighted) of the entire wave. The TIF represents the relative interfering effect of voltages and currents at the various harmonic frequencies that appear in power supply circuits. It is a dimensionless quantity indicative of waveform and not of amplitude. TIF takes into consideration the characteristics of the telephone receiver and the ear (all represented by *c*-message weighting) and the assumption that the coupling between the electric supply circuit and the telephone circuit is directly proportional to the interfering frequency. TIF is also shown as T for convenience and is expressed as

1960 SINGLE FREQUENCY TIF VALUES							
FREQ	TIF	FREQ	TIF	FREQ	TIF	FREQ	TIF
60	0.5	1020	5100	1860	7820	3000	9670
180	30	1080	5400	1980	8330	3180	8740
300	225	1140	5630	2100	8830	3300	8090
360	400	1260	6050	2160	9080	3540	6730
420	650	1380	6370	2220	9330	3660	6130
540	1320	1440	6650	2340	9840	3900	4400

1960 SINGLE FREQUENCY TIF VALUES							
FREQ	TIF	FREQ	TIF	FREQ	TIF	FREQ	TIF
660	2260	1500	6680	2460	10340	4020	3700
720	2760	1620	6970	2580	10600	4260	2750
780	3360	1740	7320	2820	10210	4380	2190
900	4350	1800	7570	2940	9820	5000	840
1000	5000						

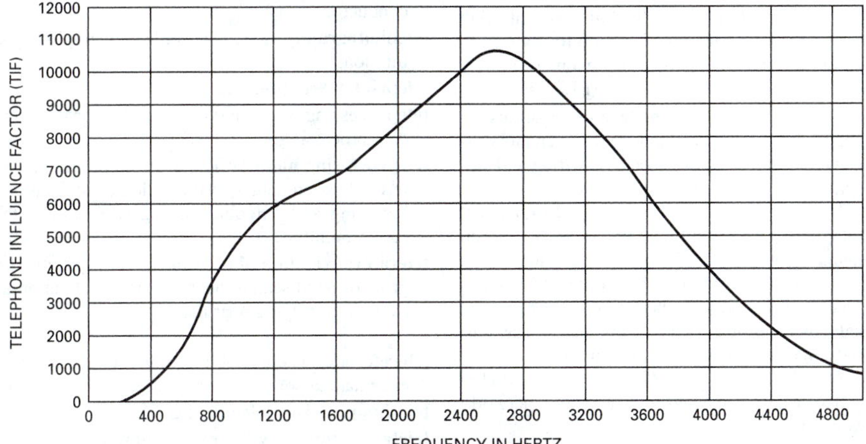

TIF weighting characteristic

telephone influence factor

$$T = \frac{\sqrt{\Sigma(X_f \cdot W_f)^2}}{X_t}$$

or

$$T = \sqrt{\Sigma\left(\frac{X_t \cdot W_f}{X_t}\right)^2}$$

where

X_t = The total effective or rms current (*I*) or voltage (kV)

X_f = The single-frequency effective current (*I*) or voltage (kV) at frequency *f*, including the fundamental

W_f = The single-frequency TIF weighting at frequency *f*. The TIF contribution of power-circuit voltage or current at frequency *f* may be expressed as follows:

$$T = \frac{X_f \cdot W_f}{X_t}$$

The 1960 TIF weighting characteristic represents the relative interfering effect of a voltage or current in a supply circuit at frequency *f*. The weighting takes into account the relative subjective effect of frequency *f* as heard through a telephone set (that is, the *c*-message weighting) and the coupling between the power and telephone circuit, assumed to be directly proportional to frequency. It is defined as

$$W_f = 5P_f\,f$$

where

5 = A constant

P_f = The c-message weighting at frequency *f*

f = The frequency under consideration.

The 1960 TIF weighting characteristic is shown in the corresponding figure. (COM/TA) 469-1988w

(4) (power and distribution transformers) Of a voltage or current wave in an electric supply circuit, the ratio of the square root of the sum of the squares of the weighted root-mean-square values of all the sine-wave components (including in alternating-current waves both fundamental and harmonics) to the root-mean-square value (unweighted) of the entire wave. *Note:* This factor was formerly known as telephone interference factor, which term is still used occasionally when referring to values based on the original (1919) weighting curve. (PE/TR) C57.12.80-1978r

(5) For a voltage or current wave in an electric supply circuit, the ratio of the square root of the sum of the squares of the weighted root-mean-square values of all the sine-wave components (including alternating current waves both fundamental and harmonic) to the root-mean-square value (unweighted) of the entire wave. (IA/SPC) 519-1992

telephone line (data transmission) A general term used in communication practice in several different senses, the more important of which are:

— The conductor or conductors and supporting or containing structures extending between telephone stations and central offices or between central offices whether they be in the same or in different communities.

— The conductors and circuit apparatus associated with a particular communication channel.

(PE) 599-1985w

telephone modal distance The distance between the center of the grid of a telephone handset transmitter and the center of the lips of a human talker (or the reference point of an artificial mouth) when the handset is in the modal position. *See also:* telephone station. (COM/TA) 269-1971w, [50]

telephone modal position The position a telephone handset assumes when the receiver of the handset is held in close contact with the ear of a person with head dimensions that are modal for a population. *See also:* telephone station. (COM) [50]

telephone network A telecommunication network primarily intended for telephony. (COM/TA) 823-1989w

telephone operator A person who handles switching and signaling operations needed to establish telephone connections between stations or who performs various auxiliary functions associated therewith. (COM) [48]

telephone receive input-output characteristic The acoustical output level of a telephone set as a function of the electric input level. The output is measured in an artificial ear, and the input is measured across a specified termination connected to the telephone feed circuit. *See also:* telephone station. (COM) [50]

telephone receiver An earphone for use in a telephone system. (SP) [32]

telephone repeater A repeater for use in a telephone circuit. *See also:* repeater. (EEC/PE) [119]

telephone ringer (ringer) (station ringer) An electric bell designed to operate on low-frequency alternating or pulsating current and associated with a telephone station for indicating a telephone call to the station. *See also:* telephone station. (EEC/PE) [119]

telephone set (1) (telephone) An assemblage of apparatus including a telephone transmitter, a telephone receiver, and usually a switch, and the immediately associated wiring and signaling arrangements. *See also:* telephone station. (PE/EEC) [119]
(2) (speech telephony) An assembly of apparatus for speech telephony, including at least a telephone transmitter and a telephone receiver in a handset, and the wiring and components immediately associated with these transducers. A telephone set usually includes other components such as a switchhook; it may also include a telephone bell and a dial. (COM/TA) 823-1989w
(3) A device that, when connected to a telephone network, allows two-way voice communication. (COM/TA) 269-1992

telephone sidetone The ratio of the acoustical output of the receiver of a given telephone set to the acoustical input of the transmitter of the same telephone set. *See also:* telephone station; telephone air-to-air input-output characteristic. (COM) [50]

telephone speech network An electric circuit that connects the transmitter and the receiver to a telephone line or telephone test loop and to each other. *See also:* telephone station. (COM) [50]

telephone station An installed telephone set and associated wiring and apparatus, in service for telephone communication. *Note:* As generally applied, this term does not include the telephone sets employed by central-office operators and by certain other personnel in the operation and maintenance of a telephone system. (EEC/PE) [119]

telephone subscriber A customer of a telephone system who is served by the system under a specific agreement or contract. (COM) [48]

telephone switchboard A switchboard for interconnecting telephone lines and associated circuits. (COM) [48]

telephone system An assemblage of telephone stations, lines, channels, and switching arrangements for their interconnection, together with all the accessories for providing telephone communication. (COM) [48]

telephone test circuit An assembly consisting of a telephone set(s) and interface(s) as may be required to realize simulated partial and overall telephone connections. (COM/TA) 269-1992

telephone test connection Two telephone sets connected together by means of telephone test loops and a telephone feed circuit. (COM) [50]

telephone test loop A circuit that is interposed between a telephone set and a telephone feed circuit to simulate a real telephone line. (COM) [50]

telephone transmit input-output characteristic The electric output level of a telephone set as a function of the acoustical input level. The output is measured across a specified impedance connected to the telephone feed circuit, and the input is measured free-field at a specified location relative to the reference point of an artificial mouth. *See also:* telephone station. (COM) [50]

telephone transmitter A microphone for use in a telephone system. *See also:* telephone station. (SP) [32]

telephone-type relay A type of electromechanical relay in which the significant structural feature is a hinged armature mechanically separate from the contact assembly. This assembly usually consists of a multiplicity of stacked leaf-spring contacts. (SWG/PE) C37.100-1992

telephony (1) (speech telephony) A form of telecommunication primarily intended for the exchange of information in the form of speech. (COM/TA) 823-1989w
(2) (speech telephony) *See also:* sleeve wire; sleeve conductor.

telephotography *See:* facsimile telegraphy; picture transmission.

teleprinter *See:* printer.

teleprocessing *See:* remote-access data processing; distributed data processing.

teleprocessing monitor The software program, usually located in the host computer, that handles various tasks required for incoming and outgoing messages. *Synonym:* telecommunication monitor. (C) 610.7-1995

teleran (navigation aids) A navigation system which employs ground-based search radar equipment along an airway to locate aircraft flying near that airway. (AES/GCS) 172-1983w

telereference The use of a telecommunication system to reference data at some remote location. (C) 610.2-1987

teleshopping *See:* teleordering.

teletext A form of videotex that allows users to receive textual or pictorial material via broadcast signals interpreted by a special decoder attached to a television set. *Contrast:* viewdata. (C) 610.2-1987

teletype exchange A service that permits the transmission of data using commercial telecommunication facilities. *Synonym:* telex. (C) 610.2-1987

teletypesetting (TTS) Use of a telecommunication system to allow typesetting to be done at remote locations. (C) 610.2-1987

teletypewriter *See:* printer.

television (TV) The electric transmission and reception of transient visual images. (EEC/PE) [119]

television broadcast station A radio station for transmitting visual signals, and usually simultaneous aural signals, for general reception. *See also:* television. (EEC/PE) [119]

television camera A pickup unit used in a television system to convert into electric signals the optical image formed by a lens. *See also:* television. (EEC/PE) [119]

television channel A channel suitable for the transmission of television signals. The channel for associated sound signals may or may not be considered a part of the television channel. *See also:* channel. (EEC/PE) [119]

television interference (overhead-power-line corona and radio noise) A radio interference occurring in the frequency range of television signals. (T&D/PE) 539-1990

television line number The ratio of the raster height to the half period of a periodic test pattern. Example: In a test pattern composed of alternate equal-width black and white bars, the television line number is the ratio of the raster height to the width of each bar. *Note:* Both quantities are measured at the camera-tube sensitive surface. *See also:* television. (ED) 158-1962w

television lines (TVL) The number of television lines for the measurement of camera resolution is defined as the total number of alternate, equal width, black and white, horizontally oriented lines that can be drawn between the top edge and bottom edge of the picture and that will fill the complete height of the resolution chart. Specification and measurement of the number of television lines always refers to *lines per picture height. Note:* Total Lines per Picture Width = (Picture Width ÷ Picture Height) × (TVL). For a given value of television lines (TVL), the total number of alternate, equal width, vertically oriented black or white lines that can be drawn between the left edge and right edge of the picture, and that will fill the complete width of the resolution chart, can be determined by multiplying the number of horizontally oriented lines *N*, by the aspect ratio of the resolution chart, which is the ratio of the chart width to the chart height. (BT/AV) 208-1995

television lines per raster height (diode-type camera tube) The number of half-cycles of a uniform periodic array referred to a unit length equal to the raster height. The array may be sinusoidal or comprised of equal width alternating light and dark bars (lines). For a given array, the TVL/RH value is numerically twice the spatial frequency in line pairs per raster height (LP/RH) units. *Note:* While the unit TVL/RH has had wide usage throughout the television industry, it is recommended that the more accurately descriptive unit LP/RH be adopted. (ED) 503-1978w

television picture tube *See:* picture tube.

television receiver A radio receiver for converting incoming electric signals into television pictures and customarily associated sound. *See also:* television. (EEC/PE) [119]

television repeater A repeater for use in a television circuit. *See also:* television; repeater. (EEC/PE) [119]

television transmitter The aggregate of such radio-frequency and modulating equipment as is necessary to supply to an antenna system modulated radio-frequency power by means of which all the component parts of a complete television signal (including audio, video, and synchronizing signals) are concurrently transmitted. *See also:* television. (AP/ANT) 145-1983s

telewriter *See:* teleautograph.

telex *See:* teletype exchange.

Tell-a-graf A computer language used to develop presentation and business graphics. (C) 610.13-1993w

telltale *See:* running-light-indicator panel.

telluric (power fault effects) Currents circulating in the earth or in conductors connecting two grounded points due to voltages in the earth. (PE/PSC) 367-1979, 367-1996

telluric currents (power fault effects) Currents circulating in the earth or in conductors connecting two grounded points due to voltages in the earth. (PE/PSC) 367-1979, 367-1996

telluric effects Currents circulating in the earth or in conductors connecting two grounded points due to voltages in the earth. (PE/PSC) 367-1987s

TEM *See:* transverse-electromagnetic wave.

TEM mode (1) (waveguide terms) A waveguide mode in which the longitudinal components of the electric and magnetic fields are everywhere zero. (MTT) 146-1980w
(2) (transverse electromagnetic) A mode of propagation characterized by frequency-independent electric and magnetic-field patterns that are purely transverse with respect to the axis of the transmission line; that is, patterns that possess no field component in the direction of propagation. A transmission line cannot possess TEM modes unless there are at least two disjoint conductors in its cross section, and unless the medium filling the cross section is homogeneous and normally isotropic. (MTT) 1004-1987w
(3) (fiber optics) *See also:* transverse-electromagnetic mode 812-1984w

TE_{mn} mode (A) (waveguide) (H_{mn} mode) In a rectangular waveguide, the subscripts $_m$ and $_n$ denote the number of half-period variations in the electric field parallel to the broad and narrow sides, respectively, of the guide. *Note:* In the United Kingdom, the reverse order is preferred. **(B) (waveguide)** (H_{mn} mode) In a circular waveguide, a mode that has $_m$ diametral planes in which the longitudinal component of the magnetic field is zero, and $_n$ cylindrical surfaces of nonzero radius (including the wall of the guide) at which the tangential component of the electric field is zero. **(C) (waveguide)** (H_{mn} mode) In a resonant cavity consisting of a length of rectangular or circular waveguide, a third subscript is used to indicate the number of half-period variations of the field along the waveguide axis. (MTT) 146-1980

TE mode (1) (waveguide) (H mode) A waveguide mode in which the longitudinal component of the electrical field is everywhere zero and the longitudinal component of the magnetic field is not. (MTT) 146-1980w
(2) (fiber optics) *See also:* transverse electric mode. 812-1984w

temperature, ambient air *See:* ambient air temperature.

temperature class (1) (evaluation of thermal capability) (thermal classification of electric equipment and electrical insulation) A standardization designation of the temperature capability of the insulation in electric equipment, as defined by the appropriate technical committee. It may be determined by experience or test and expressed by letters or numbers. (EI) 1-1986r
(2) One of the values of temperature allocated to electrical heating devices derived from a system of classification according to the maximum surface temperature of the heater. Also referred to as T-class, identification number, T-rating, and temperature code. (IA) 515-1997

temperature classification (solid electrical insulating materials) The term is reserved for insulation systems as used in specific equipment and is no longer recognized as a description of the temperature capability of individual insulating materials. Historically, the term has been used in reference to insulation systems and to electrical equipment. In the future the term may be reserved for use in rating electrical equipment, while thermal identification may be used in the specification of insulation systems for particular applications. (EI) 98-1984r

temperature class ratings insulation These temperatures are and have been, in most cases over a long period of time, benchmarks descriptive of the various classes of insulating materials, and various accepted test procedures have been or are being developed for use in their identification. They should not be confused with the actual temperatures at which these same classes of insulating materials may be used in the various specific types of equipment, nor with the temperatures on which specified temperature rise in equipment standards are based. 1) In the following definitions the words "accepted tests" are intended to refer to recognized test procedures established for the thermal evaluation of materials by themselves or in simple combinations. Experience or test data, used in classifying insulating materials, are distinct from the experience or test data derived for the use of materials in complete insulation systems. The thermal endurance of complete systems may be determined by test procedures specified by the responsible technical committees. A material that is classified as suitable for a given temperature may be found suitable for a different temperature, either higher or lower, by an insulation system test procedure. For example, it has been found that some materials suitable for operation at one temperature in air may be suitable for a higher temperature when used in a system operated in an inert gas atmosphere. Likewise some insulating materials when operated in dielectric liquids will have lower or higher thermal endurance than in air. 2) It is important to recognize that other characteristics, in addition to thermal endurance, such as mechanical strength, moisture resistance, and corona endurance, are required in varying degrees in different applications for the successful use of insulating materials. *class 105 insulation system.* Materials or combinations of materials such as cotton, silk, and paper when suitably impregnated or coated or when immersed in a dielectric liquid. *Notes:* 1. Other materials or combinations may be included in this class if by experience or accepted tests the insulation system can be shown to have comparable thermal life at 105°C. *class 120 insulation system.* Materials or combinations of materials such as cotton, silk, and paper when suitably impregnated or coated or when immersed in a dielectric liquid: and which possess a degree of

thermal stability which allows them to be operated at a temperature 15°C higher than Class 105 insulation materials. 2. Other materials or combinations may be included in this class if by experience or accepted tests the insulation system can be shown to have comparable thermal life at 120°C. *class 150 insulation system.* Materials or combinations of materials such as mica, glass fiber, asbestos, etc., with suitable bonding substances. 3. Other materials or combinations of materials may be included in this class if by experience or accepted tests the insulation system can be shown to have comparable life at 150 degrees Celsius. *class 185 insulation system.* Materials or combinations of materials such as silicone elastomer, mica, glass fiber, asbestos, etc., with suitable bonding substances such as appropriate silicone resins. 4. Other materials or combinations of materials may be included in this class if by experience or accepted tests the insulation system can be shown to have comparable thermal life at 185°C. *class 220 insulation system.* Materials or combinations of materials such as silicone elastomer, mica, glass fiber, asbestos, etc., with suitable bonding substances such as appropriate silicone resins. 5. Other materials or combinations of materials may be included in this class if by experience or accepted tests, the system can be shown to have comparable thermal life at 220°C. *class over-220 insulation system.* Materials consisting entirely of mica, porcelain, glass quartz, and similar inorganic materials. 6. Other materials or combinations of materials may be included in this class if by experience or accepted tests the insulation system can be shown to have the required thermal life at temperatures over 220°C. *class O insulation.* (nonpreferred term). *class A insulation.* (nonpreferred term). *See also:* Class 105 insulation system.
(PE/TR) C57.12.80-1978r

temperature coefficient (1) (power supplies) (ferroresonant voltage regulators) The percent change in the output voltage or current as a result of a 1°C change in the ambient operating temperature (percent per degree Celsius). *See also:* environmental coefficient; overall regulation.
(AES/PEL/ET) [41], 449-1990s
(2) (rotating machinery) The variation of the quantity considered, divided by the difference in temperature producing it. Temperature coefficient may be defined as an average over a temperature range or an incremental value applying to a specified temperature. *See also:* asynchronous machine.
(PE) [9]
(3) (variable or fixed attenuator) Maximum temporary and reversible change of insertion loss in decibels per degree Celsius over operating temperature range.
(IM/HFIM) 474-1973w

temperature coefficient of capacity (storage cell or battery) The change in delivered capacity (ampere-hour or watt hour capacity) per degree Celsius relative to the capacity of the cell or battery at a specified temperature. (EEC/PE) [119]

temperature coefficient of electromotive force (storage cell or battery) The change in open-circuit voltage per degree Celsius relative to the electromotive force of the cell or battery at a specified temperature. (EEC/PE) [119]

temperature coefficient of resistance (rotating machinery) The temperature coefficient relating a change in electric resistance to the difference in temperature producing it. *See also:* asynchronous machine. (PE) [9]

temperature coefficient of sensitivity (electrothermic power meters) The change in rf sensitivity (microvolts/milliwatts) resulting from a specified temperature change of the electrothermic unit at a specified power level. Expressed in percent per degree celsius. (IM) 544-1975w

temperature coefficient of voltage drop (glow-discharge tubes) The quotient of the change of tube voltage drop (excluding any voltage jumps) by the change of ambient (or envelope) temperature. *Note:* It must be indicated whether the quotient is taken with respect to ambient or envelope temperature. *See also:* gas tube. (ED) 161-1971w

temperature compensated overload relay A device that functions at any current in excess of a predetermined value essentially independent of the ambient temperature.
(IA/MT) 45-1998

temperature control (packaging machinery) A control device responsive to temperature. (IA/PKG) 333-1980w

temperature control device (power system device function numbers) A device that functions to raise or lower the temperature of a machine or other apparatus, or of any medium, when its temperature falls below, or rises above, a predetermined value. *Note:* An example is a thermostat that switches on a space heater in a switchgear assembly when the temperature falls to a desired value as distinguished from a device that is used to provide automatic temperature regulation between close limits and would be designated as device function 90T [regulating device T]. (SUB/PE) C37.2-1979s

temperature control system (gas turbines) The devices and elements, including the necessary temperature detectors, relays, or other signal-amplifying devices and control elements, required to actuate directly or indirectly the fuel-control valve, speed of the air compressor, or stator blades of the compressor so as to limit or control the rate of fuel input or air flow inlet to the gas turbine. By this means, the temperature in the combustion system or the temperatures in the turbine stages or turbine exhaust may be limited or controlled.
(PE/EDPG) [5]

temperature conversion (tolerance requirements) (International System of Units (SI)) Standard practice for converting tolerances from degrees Fahrenheit to kelvins or degrees Celsius is:

Conversion of Temperature Tolerance Requirements

Tolerance °F	Tolerance °K or °C
±1	±0.5
±2	±1
±5	±3
±10	±5.5
±15	±8.5
±20	±11
±25	±14

Normally, temperatures expressed in a whole number of degrees Fahrenheit should be converted to the nearest 0.5 kelvin (or degrees Celcius). As with other quantities, the number of significant digits to retain will depend upon implied accruacy of the original dimension; for example:

100 ± 5°F implied accuracy estimated total 2°F

37.7777 ± 2.7777°C rounds to 38 ± 3°C

1000 ± 50°F implied accuracy estimated total 20°F

537.7777 ± 27.7777°C rounds to 540 ± 30°C

See also: units and letter symbols. (QUL) 268-1982s

temperature derating (semiconductor devices) The reduction in reverse-voltage or forward-current rating, or both, assigned by the manufacturer under stated conditions of higher ambient temperatures. *See also:* average forward current rating; semiconductor rectifier stack. (IA) [62]

temperature detectors (gas turbines and rotating electric machinery) The primary temperature-sensing elements that are directly responsive to temperature. *See also:* asynchronous machine; electric thermometer. (PE) [9]

temperature, dew point *See:* dew point temperature.

temperature, dry bulb *See:* dry bulb temperature.

temperature, effective *See:* effective temperature.

temperature, equilibrium *See:* equilibrium temperature.

temperature index (1) (power and distribution transformers) An index that allows relative comparisons of the temperature capability of insulating materials or insulation systems based on specified controlled test conditions. Preferred values of temperature index numbers are shown in the table below. (PE/TR) C57.12.80-1978r

(2) (thermal classification of electric equipment and electrical insulation) (evaluation of thermal capability) The number that corresponds to the temperature in °C, derived mathematically or graphically from the thermal endurance relationship at a specified time (often 20 000 h). The temperature index (TI) may be reported for materials and insulation systems. However, for insulation systems it may be preferable to make comparisons at a particular temperature, for example, 130°C, 155°C, or over a range of temperatures. (The TI is not used for equipment). *See also:* thermal endurance graph. (EI) 1-1986r
(3) (solid electrical insulating materials) This is the number corresponding to the temperature in degrees Celsius derived from the thermal endurance graph at a given time.

Number Range	Preferred Temperature Index
90–104	90
105–129	105
130–154	130
155–179	155
180–199	180
200–219	200

For 220 and above, no preferred indices established.

See also: thermal endurance graph. (EI) 98-1984r

temperature inversion (in the troposphere) An increase of temperature with height in the troposphere. (AP/PROP) 211-1997

temperature meter *See:* electric thermometer.

temperature, operating *See:* operating temperature.

temperature radiator (illuminating engineering) An ideal radiator whose radiant flux density (radiant exitance) is determined by its temperature and the material and character of its surface, and is independent of its previous history. (EEC/IE) [126]

temperature-regulating equipment (rectifier) Any equipment used for heating and cooling a rectifier, together with the devices for controlling and indicating its temperature. *See also:* rectification. (IA) [62]

temperature relay A relay whose operation is caused by specified external temperature. *See also:* thermal relay. (SWG/PE) C37.100-1992

temperature relays (gas turbines) Devices by means of which the output signals of the temperature detectors are enabled to control directly or indirectly the rate of fuel energy input, the air flow input, or both, to the combustion system. *Note:* Operation of a temperature relay is caused by a specified external temperature: whereas operation of a thermal relay is caused by the heating of a part of the relay. *See also:* thermal relay. (SWG/PE/PSR) C37.90-1978s, C37.100-1981s

temperature rise (1) The difference between the temperature of the part under consideration [commonly the *average winding rise* or the *maximum (hottest-spot) winding temperature rise*] and the ambient temperature. (PE/TR) C57.134-2000
(2) The difference between the temperature of the part under consideration and the ambient temperature. (PE/EM/TR) 67-1990r, C57.12.80-1978r

temperature-rise tests (1) Tests to determine the temperature rise, above ambient, of various parts of the tested device when subjected to specified test quantities. *Note:* The test quantities may be current, load, etc. *See also:* allowable continuous current. (SWG/PE) C37.40-1981s, C37.100-1992
(2) A test in which rated current at rated frequency is applied to equipment to determine its temperature rise. (SWG/PE/SWG-OLD) [9], C37.34-1971s, [56]

temperature sensor (sensing element) A device that responds to temperature and provides an electrical signal or mechanical operation. (IA/BT/AV/PC) 515-1997, 152-1953s, 515.1-1995

temperature stability (electrical conversion) Static regulation caused by a shift or change in output that was caused by temperature variation. This effect may be produced by a change in the ambient or by self-heating. (AES) [41]

temperature, wet bulb *See:* wet bulb temperature.

tempest The investigation, study, and control of spurious electromagnetic signals emitted by electronic equipment. *See also:* emanations security. (C/BA) 896.3-1993w

template An asset with parameters or slots that can be used to construct an instantiated asset. *See also:* construction. (C/SE) 1517-1999

template matching (A) An image processing technique in which patterns or shapes are detected by comparison with prespecified patterns or shapes called templates. *See also:* image matching. **(B)** A pattern recognition technique using the principle described in definition (A). (C) 610.4-1990

temporal coherence (1) (laser maser) (electromagnetic) The correlation in time of electromagnetic fields at a point in space. (LEO) 586-1980w
(2) (fiber optics) *See also:* coherent. 812-1984w

temporal cohesion A type of cohesion in which the tasks performed by a software module are all required at a particular phase of program execution; for example, a module containing all of a program's initialization tasks. *Contrast:* coincidental cohesion; logical cohesion; sequential cohesion; procedural cohesion; functional cohesion; communicational cohesion. (C) 610.12-1990

temporal locality The tendency for a program to reference the same memory locations over short time intervals. (C/BA) 10857-1994

temporally coherent radiation *See:* coherent.

temporally weighted terminal coupling loss (TCL_T) The terminal coupling loss, weighted in both time and frequency domains to account for subjective perception. (COM/TA) 1329-1999

temporal noise (diode-type camera tube) The varying amplitude portion of what should be a fixed amplitude video signal. It is statistical in nature, being random in both time and amplitude. (ED) 503-1978w

temporary Intermittent or transient. (C/BA) 896.3-1993w

temporary emergency circuits Circuits arranged for instantaneous automatic transfer to a storage-battery supply upon failure of a ship's service supply. *See also:* emergency electric system. (EEC/PE/MT) [119]

temporary emergency lighting The lighting of exits and passages to permit passengers and crew, upon failure of a ship's service lighting, readily to find their way to the lifeboat embarkation deck. *See also:* emergency electric system. (EEC/PE/MT) [119]

temporary fault One that may be self-clearing, or may be cleared if the faulted circuit is rapidly de-energized by opening of a protective device, such as a circuit breaker or recloser. (T&D/PE) 1250-1995

temporary forced outage A forced outage where the unit or component is undamaged and is restored to service by manual switching operations without repair but possibly with on-site inspection. (PE/PSE) 859-1987w

temporary ground A connection between a grounding system and parts of an installation that are normally alive, applied temporarily so that work may be safely carried out in them. (PE) [8]

temporary interruption (1) A short-duration outage that interferes with call processing but does not affect established connections. SPCSs may have frequent outages of short duration due to system reinitialization. Although established calls may remain connected during these outages, new calls may be delayed and calls in the dialing state may be lost. Most customers do not perceive these short outages because they are not likely to be using their telephones when they occur. However, an excessive number of short outages can lead to degradation of service and can cause delay in dial tone or ineffective attempts. (COM/TA) 973-1990w
(2) A type of short duration variation. The complete loss of voltage (<0.1 pu) on one or more phase conductors for a time period between 3 s and 1 min. (SCC22) 1346-1998

(3) (A) **(power quality monitoring)** A type of short-duration variation. **(B) (power quality monitoring)** The complete loss of voltage (< 0.1 pu) on one or more phase conductors for a time period between 3 s and 1 min. (IA/PSE) 1100-1999

temporary master *See:* master.

temporary overvoltage (1) An oscillatory phase-to-ground or phase-to-phase overvoltage that is at a given location of relatively long duration (seconds, even minutes) and that is undamped or only weakly damped. Temporary overvoltages usually originate from switching operations or faults (for example, load rejection, single-phase fault, fault on a high-resistance grounded or ungrounded system) or from nonlinearities (ferroresonance effects, harmonics), or both. They are characterized by the amplitude, the oscillation frequencies, the total duration, or the decrement. (C/PE/TR) 1313.1-1996, C57.12.80-1978r
(2) An oscillatory overvoltage, associated with switching or faults (for example, load rejection, single-phase faults) and/or nonlinearities (ferroresonance effects, harmonics), of relatively long duration, which is undamped or slightly damped. (SPD/PE) C62.22-1997

temporary storage (programming) Storage locations reserved for intermediate results. *See also:* working storage. (MIL/C) [2], 610.10-1994w, [85], [20]

temporary structure *See:* crossing structure.

TEM wave *See:* transverse-electromagnetic wave.

10BASE-T (1) ISO/IEC 8802-3 Physical Layer specification for Ethernet over two pairs of unshielded twisted pair (UTP) media at 10 Mbit/s. (C/LM) 802.9a-1995w
(2) IEEE 802.3 Physical Layer specification for a 10 Mb/s CSMA/CD local area network over two pairs of twisted-pair telephone wire. (C/LM) 802.3-1998

10BASE2 IEEE 802.3 Physical Layer specification for a 10 Mb/s CSMA/CD local area network over RG 58 coaxial cable. (C/LM) 802.3-1998

10BASE-F IEEE 802.3 Physical Layer specification for a 10 Mb/s CSMA/CD local area network over fiber optic cable. (C/LM) 802.3-1998

10BASE-FB port A port on a repeater that contains an internal 10BASE-FB Medium Attachment Unit (MAU) that can connect to a similar port on another repeater. (C/LM) 802.3-1998

10BASE-FB segment A fiber optic link segment providing a point-to-point connection between two 10BASE-FB ports on repeaters. (C/LM) 802.3-1998

10BASE5 IEEE 802.3 Physical Layer specification for a 10 Mb/s CSMA/CD local area network over coaxial cable (i.e., thicknet). (C/LM) 802.3-1998

10BASE-FL segment A fiber optic link segment providing point-to-point connection between two 10BASE-FL Medium Attachment Units (MAUs). (C/LM) 802.3-1998

10BASE-FP segment A fiber optic mixing segment, including one 10BASE-FP Star and all of the attached fiber pairs. (C/LM) 802.3-1998

10BASE-FP Star A passive device that is used to couple fiber pairs together to form a 10BASE-FP segment. Optical signals received at any input port of the 10BASE-FP Star are distributed to all of its output ports (including the output port of the optical interface from which it was received). A 10BASE-FP Star is typically comprised of a passive-star coupler, fiber optic connectors, and a suitable mechanical housing. (C/LM) 802.3-1998

10BROAD36 IEEE 802.3 Physical Layer specification for a 10 Mb/s CSMA/CD local area network over single broadband cable. (C/LM) 802.3-1998

ten high day *See:* ten high day busy-hour load; time-consistent traffic measures.

ten high day busy-hour load To calculate the THDBH load, traffic data for the time-consistent busy hour is processed all year to identify the 10 highest traffic days of the year. The 10-day average traffic level for this time-consistent busy hour is the THDBH load. *Synonym:* THDBH load. *See also:* time-consistent traffic measures. (COM/TA) 973-1990w

ten-minute reserve An additional amount of operating reserve sufficient to reduce area control error to zero within ten minutes following the loss of generating capacity that would result from the most severe single contingency. (PE/PSE) 858-1993w

tens complement (mathematics of computing) The radix complement of a decimal numeral, which may be formed by subtracting each digit from 9, then adding 1 to the least significant digit and executing any required carries. For example, the tens complement of 4830 is 5170. *Synonym:* complement on ten. (C) 1084-1986w

tension *See:* final unloaded conductor tension; conductor.

tensioner *See:* bullwheel tensioner.

tensioner, bullwheel *See:* bullwheel tensioner.

tension site (conductor stringing equipment) The location on the line where the tensioner, reel stands, and anchors (snubs) are located. This site may also serve as the pull or tension site for the next sag section. *Synonyms:* payout site; conductor payout station; payout site; reel setup; conductor payout station; reel setup. (T&D/PE) 524a-1993r, 524-1992r

tension stringing The use of pullers and tensioners to keep the conductor under tension and positive control during the stringing phase, thus keeping it clear of the earth and other obstacles that could cause damage. (T&D/PE) 524-1992r

tension, unloaded *See:* unloaded tension.

tenth-power width (in a plane containing the direction of the maximum of a lobe) The full angle between the two directions in that plane about the maximum in which the radiation intensity is one-tenth the maximum value of the lobe. *See also:* antenna. (AP/ANT) 145-1983s

tenuous medium A medium in which the spatial variations of constitutive parameters, either continuous or discrete, are small relative to their mean values. (AP/PROP) 211-1997

tenure (1) (STEbus) The time during which a master has control of the bus. (C/MM) 1000-1987r
(2) (NuBus) Time period of unbroken ownership of the bus by a particular module. May consist of one or more transactions or attention cycles. (C/MM) 1196-1987w

teratology The study of developmental abnormalities in the fetus. (T&D/PE) 539-1990

terdenary (A) Pertaining to a selection in which there are 13 possible outcomes. **(B)** Pertaining to the numeration system with a radix of 13. (C) 1084-1986

terminal (1) (A) (supervisory control, data acquisition, and automatic control) A point in a system or communication network at which data can either enter or leave. *See also:* virtual terminal. **(B) (supervisory control, data acquisition, and automatic control)** An input/output device capable of transmitting entries to and obtaining output from the system of which it is a part, for example cathode-ray tube (crt) terminal. **(C) (power and distribution transformers)** A conducting element of an equipment or a circuit intended for connection to an external conductor. **(D) (power and distribution transformers)** A device attached to a conductor to facilitate connection with another conductor. **(E)** An input-output peripheral device capable of transmitting entries to and obtaining output from a system. *See also:* link-attached terminal; intelligent terminal; logical terminal; master terminal; smart terminal; output terminal; input terminal; video display terminal; graphic user terminal; facsimile terminal; dumb terminal; channel-attached terminal; job-oriented terminal; remote terminal; local terminal. (SWG/SUB/PE/C/TR) C37.1-1987, C37.100-1992, 610.7-1995, 610.10-1994, C57.12.80-1978
(2) (packaging machinery) A point of connection in an electric circuit. (IA/PKG) 333-1980w
(3) (terminal connector) A connector for attaching a conductor to electrical apparatus. (SWG/PE) C37.40-1993, C37.100-1992

(4) (networks) A point at which any element may be directly connected to one or more other elements. *See also:* network analysis. (BT) 153-1950w

(5) (light-emitting diodes) (semiconductor devices) An externally available point of connection to one or more electrodes or elements within the device. *See also:* semiconductor rectifier cell; semiconductor; anode. (ED/ICTL) 216-1960w

(6) (telegraph circuits) A general term referring to the equipment at the end of a telegraph circuit, modems, input-output and associated equipment. *See also:* telegraph. (COM) [49]

(7) (rotating machinery) A conducting element of a winding intended for connection to an external electrical conductor. *See also:* stator. (PE) [9]

(8) (power outages) A functional facility (substation, generating station, or load center) that includes components such as bus sections, circuit breakers, and protection systems where transmission units terminate. (PE/PSE) 859-1987w

(9) A master or remote terminal connected to a communication channel. (SUB/PE) 999-1992w

(10) A character special file that obeys the specifications of the POSIX.1 General Terminal Interface. *Synonym:* terminal device. (C/PA/C/PA) 9945-2-1993, 9945-1-1996

(11) (terminal device) A character special file that obeys the specifications of IEEE Std 1003.5b-1995. (PA/C) 1003.5b-1995

terminal adapter (TA) An adapter required to map one specified interface to another. An example is the adapting of the R interface in the ITU-T ISDN to the S/T interface. The TA may be an integral functional entity as part of the terminal or may be a separate physical unit connected between an R interface and the S/T interface. (C/LM/COM) 802.9a-1995w, 8802-9-1996

terminal block An insulating base equipped with terminals for connecting secondary and control wiring. *Synonym:* terminal board. (SWG/PE) C37.100-1992

terminal board (power and distribution transformers) A plate of insulating material that is used to support terminations of winding leads. *Notes:* 1. The terminations, which may be mounted studs or blade connectors, are used for making connections to the supply line, the load, other external circuits, or among the windings of the machine. 2. Small terminal boards may also be termed terminal blocks, or terminal strips. (PE/TR) C57.12.80-1978r

terminal board cover (rotating machinery) A closure for the opening which permits access to the terminal board and prevents accidental contact with the terminals. (PE) [8]

terminal box (rotating machinery) A form of termination in which the ends of the machine winding are connected to the incoming supply leads inside a box that virtually encloses the connections, and is of minimum size consistent with adequate access and with clearance and creepage-distance requirements. The box is provided with a removable cover plate for access. *See also:* stator. (PE) [9]

terminal chamber A metal-enclosed container that includes all necessary mechanical and electrical items to complete the connections to other equipment. (SWG/PE) C37.100-1992, C37.23-1969s

terminal charge *See:* apparent charge.

terminal conformity *See:* conformity.

terminal connection (battery) Connections made between cells or rows of cells or at the positive and negative terminals of the battery, which may include terminal plates, cables with lugs, and connectors. (SB) 1188-1996

terminal connection detail (1) (nickle-cadmium storage batteries for generating stations and substation) Connections made between cells, rows of cells, and at positive and negative terminals of the battery, which may include nickel- or cadmium-plated terminal plates, cables with nickel- or cadmium-plated lugs, and nickel- or cadmium-plated solid copper or steel connectors. (PE/EDPG) 1106-1987s

(2) Connections made between rows of cells or at the positive and negative terminals of the battery, which may include terminal plates, cables with lugs, and connectors. (PE/EDPG) 450-1995

terminal connector (1) (power and distribution transformers) A connector for attaching a conductor to a lead, terminal block, or stud of electric apparatus. (PE/TR) C57.12.80-1978r

(2) *See also:* terminal. (SWG/PE) C37.100-1981s

terminal corona charge (corona measurement) A charge equal to the product of the capacitance of the insulation system and the terminal corona-pulse voltage. (MAG/ET) 436-1977s

terminal corona-pulse voltage (corona measurement) The pulse voltage resulting from a corona discharge that is represented as a voltage source suddenly applied in series with the capacitance of the insulation system under test, and which would appear at the terminals of the system under open-circuit conditions. (MAG/ET) 436-1977s

terminal coupling loss (TCL) The loss in the echo path from the receive electrical test point to the send electrical test point. TCL can be weighted in time, frequency, or both. *Synonym:* echo return loss. (COM/TA) 1329-1999

terminal device *See:* terminal.

terminal emulator Software that makes a frame-buffer appear to have the characteristics of a cursor-addressable text terminal. (C/BA) 1275-1994

terminal guidance (navigation aids) Guidance from an arbitrary point, at which midcourse guidance ends, to the destination. (AES/GCS) 172-1983w

terminal interface processor (TIP) A computer that connects terminals directly to a network, eliminating the need for a host computer. (C) 610.7-1995

terminal interference voltage *See:* terminal voltage.

terminal linearity *See:* conformity.

terminal, master *See:* master terminal.

terminal node (1) In a tree, a node that has no subtrees. *Synonym:* external node. *See also:* root node; leaf. (C) 610.5-1990w

(2) A node with one or more link interfaces which are used to originate or consume data across an interconnect complying with this standard. *See also:* source; sink. (C/BA) 1355-1995

terminal of entry (for a conductor entering a delimited region) That cross section of the conductor that coincides with the boundary surface of the region and that is perpendicular to the direction of the electric field intensity at its every point within the conductor. In a conventional circuit, in which the conductors have a cross section that is uniform and small by comparison of the largest dimension with the length, the terminal of entry is a cross section perpendicular to the axis of the conductor. If the cross section of the conductor is infinitesimal, the terminal of entry becomes the point at which the conductor cuts the surface. *Notes:* 1. It follows from this definition and delimited region that the algebraic sum of the currents directed into a delimited region through all the terminals of entry is zero at every instant. 2. The term terminal of entry has been introduced because of the need in precise definitions of indicating definitely the terminations of the paths along which voltages are determined. The terms phase conductor and neutral conductor refer to a portion of a conductor rather than to a particular cross section although they may be considered by a practical engineer as representing a portion along which the integral of the electric intensity is negligibly small. Hence he may treat these terms as synonymous with terminal of entry in particular cases. *See also:* network analysis. (Std100) 270-1966w

terminal pad A usually flat conducting part of a device to which a terminal connector is fastened. (SWG/PE) C37.40-1993, C37.100-1992

terminal pair (networks) An associated pair of accessible terminals, such as input pair, output pair, and the like. *See also:* network analysis. (IA/PSE) 141-1976s, 270-1966w

terminal-per-line system (telephone switching systems) A switching entity having an outlet corresponding to each line. (COM) 312-1977w

terminal-per-station system (telephone switching systems) A switching entity having an outlet corresponding to each main-station code. (COM) 312-1977w

terminal, remote *See:* remote terminal.

terminal repeater (data transmission) A repeater for use at the end of a trunk or line. (PE) 599-1985w

terminal room (telephone switching systems) That part of a building that contains distributing frames, relays and similar apparatus associated with switching equipment. (COM) 312-1977w

terminals (1) (storage battery) (storage cell) The parts to which the external circuit is connected. *See also:* battery. (PE/EEC) [119]

(2) The conducting parts provided for connecting the arrester across the insulation to be protected. (SPD/PE) C62.11-1999

(3) The conducting parts provided for connecting the surge-protective device across the circuit to be protected. Terminal designations could be phase(s), neutral or ground with line and/or load designations. (SPD/PE) C62.62-2000

terminal server On a network, a server that provides access to a central computer from one or more terminals. *See also:* disk server; file server; mail server; database server; print server; network server. (C) 610.7-1995

terminal set A set of three primary terminals with primary disconnecting devices typically called "upper" or "lower," "front" or "back," or "bus" or "line," depending upon design configuration. (SWG/PE) C37.20.6-1997

terminal screw *See:* binding screw.

terminal, stator winding *See:* stator winding terminal.

terminal strip *See:* terminal board.

terminal trunk (data transmission) A trunk circuit connecting switching centers used in conjunction with local switching only in these centers. (PE) 599-1985w

terminal unit (programmable instrumentation) An apparatus that terminates the considered interface system and by means which a connection (and translation, if required) is made between the considered interface system and another external interface system. (IM/AIN) 488.1-1987r

terminal voltage (terminal interference voltage) Interference voltage measured between two terminals of an artificial mains network. *See also:* electromagnetic compatibility. (EMC/INT) [53], [70]

terminate (a connection) To dissolve an association established between two or more endpoints for the transfer of data. (C) 1003.5-1999

terminating (line or transducer) The closing of the circuit at either end by the connection of some device thereto. Terminating does not imply any special condition, such as the elimination of reflection. (PE/EEC) [119]

terminating power meter or measuring system A device which terminates a waveguide or transmission line in a prescribed manner and contains provisions for measuring the incident of absorbed power. (IM) 470-1972w, 544-1975w

terminating test circuit (measuring longitudinal balance of telephone equipment operating in the voice band) A network connected to a transmission port of a circuit to terminate it in a suitable balanced termination for longitudinal balance testing. This circuit is used when a driving test circuit is connected to one such port and the test specimen has additional transmission ports. (COM/TA) 455-1985w

terminating toll center code (telephone switching systems) In operator distance dialing, the three digits used for identifying the toll center within the area to which a call is routed. (COM) 312-1977w

terminating traffic (telephone switching systems) Traffic delivered directly to lines. (COM) 312-1977w

termination (1) (A) (power outages) A facility where a transmission line ends within a terminal, typically at a circuit breaker. **(B) (metal-enclosed bus and calculating losses in isolated-phase bus) (terminal chamber)** A metal enclosure that contains all necessary and mechanical and electrical items to complete the connections to other equipment. *See also:* dead-end; cable terminal. (SWG/PE/PSE) 859-1987

(2) A one-port load that terminates a section of a transmission system in a specified manner. *See also:* transmission line. (NESC/IM/HFIM) [40]

(3) (rotating machinery) The arrangement for making the connections between the machine terminals and the external conductors. *See also:* stator. (PE) [9]

(4) (waveguide components) A one port load in a waveguide or transmission line. (MTT) 147-1979w

(5) A constant impedance and digital logic state that a signal is held at during some or all of a test. (C/TT) 1450-1999

termination capacity (lines and trunks) The number of lines and trunks that can be terminated and maintained on a switching system. (COM/TA) 973-1990w

termination charge (electric power utilization) The amount paid by a customer when service is terminated at the customer's request. (PE/PSE) 346-1973w

termination, conjugate *See:* conjugate termination.

termination construct A program construct that results in a halt or exit. (C) 610.12-1990

termination failure A failure in the portion of the cable, which does not have a metallic shield covering. (PE/IC) 1407-1998

termination insulator (cable terminations) An insulator used to protect each cable conductor passing through the device and provide complete external leakage insulation between the cable conductor(s) and ground. (PE/IC) 48-1996

termination, matched *See:* matched termination.

termination proof (software) In proof of correctness, the demonstration that a program will terminate under all specified input conditions. *See also:* proof of correctness; program. (C/SE) 729-1983s

termination, reflectionless *See:* reflectionless termination.

terminations charge (power operations) The amount paid by a customer when service is terminated at the customer's request. (PE/PSE) 858-1987s

termination sequence The process by which the AS/AK lock is broken. (NID) 960-1993

terminator A single-port, 75-Ω device that is used to absorb energy from a transmission line or RF device. Terminators prevent energy from reflecting back into a cable plant by absorbing the RF signals. A terminator is usually shielded, which also prevents ingress and egress from an unused port. (LM/C) 802.7-1989r

terminology bank *See:* automated glossary.

ternary (A) (mathematics of computing) Pertaining to the numeration system with a radix of 3. *See also:* base; positional notation; radix. **(B) (mathematics of computing)** Pertaining to a selection in which there are three possible outcomes. (C) 1084-1986

ternary code A code whose alphabet consists of three symbols. *See also:* ternary. (IT) [7]

ternary incremental representation Representation of changes in variables in which the value of an increment is plus one, zero, or minus one. *Synonym:* incremental ternary representation. (C) 1084-1986w

ternary relation A relation with three attributes. (C) 610.5-1990w

ternary symbol In 100BASE-T4, a ternary data element. A ternary symbol can have one of three values: −1, 0, or +1. (C/LM) 802.3-1998

ternary torquing (1) (digital accelerometer) System with three stable torquing states (for example, positive, negative, and off). (AES/GYAC) 530-1978r
(2) (accelerometer) (gyros) A torquing mechanization that utilizes three levels of torquer current, usually positive and negative of the same magnitude, and a zero current or off condition. The positive and negative torque conditions can be either discrete pulses or pulse-duration-modulated current periods. In both implementations, the case of zero input (acceleration or angular rate) will result in zero torquer current. Ternary torquer power is proportional to the input (acceleration or angular rate), resulting in minimum power as compared to binary torquing. (AES/GYAC) 528-1994

terrain-avoidance radar (1) A radar that provides assistance to a pilot for flight around obstacles by sensing obstacles at or above his or her altitude. (AES/GCS) 172-1983w
(2) A radar that provides information about the ground environment so that an aircraft can fly around high ground or obstacles. (AES) 686-1997

terrain-clearance indicator (navigation aids) An absolute altimeter using the measurement of height above terrain to alert the pilot of danger. (AES/GCS) 172-1983w

terrain echoes *See:* ground clutter.

terrain error (navigation aids) The error resulting from the use of a wave which has become distorted by the terrain over which it has propagated. (AES/GCS) 172-1983w

terrain-following radar (1) (navigation aids) A radar that works with the aircraft flight control system to provide low-level flight following the contour of the earth's surface at some given altitude. (AES/GCS) 172-1983w
(2) An airborne radar that works with the aircraft flight control system to achieve flight that follows the contour of the earth's surface at some given altitude. (AES) 686-1997

terrestrial-reference flight (navigation aids) That type of stabilized flight which obtains control information from terrestrial phenomena, such as earth's magnetic field, atmospheric pressure, etc. (AES/GCS) 172-1983w

tertiary winding (power and distribution transformers) An additional winding in a transformer which can be connected to a synchronous condenser, a reactor, an auxiliary circuit, etc. For transformers with Y-connected primary and secondary windings, it may also help to: stabilize voltages to the neutral, when delta connected; reduce the magnitude of third harmonics when delta connected; control the value of the zero-sequence impedance; serve load. (PE/TR) C57.12.80-1978r

tesla The unit of magnetic induction in the International System of Units (SI). The tesla is a unit of magnetic induction equal to 1 weber per square meter. (Std100) 270-1966w

Tesla current (coagulating current) (electrotherapy) A spark discharge having a drop of 5–10 kV in air, from monopolar or bipolar electrodes, generated by a special arrangement of transformers, spark gaps, and capacitors, delivered to a tissue surface, and dense enough to precipitate and oxidize (char) tissue proteins. *Note:* The term Tesla current is appropriate if the emphasis is on the method of generation: a coagulating current, if the emphasis is on the physiological effects. *See also:* electrotherapy. (EMB) [47]

test (1) An action or group of actions performed on a particular unit under test (UUT) to evaluate a parameter or characteristic. (ATLAS) 771-1989s
(2) (A) (electronic digital computation) To ascertain the state or condition of an element, device, program, etc. **(B) (electronic digital computation)** Sometimes used as a general term to include both check and diagnostic procedures. **(C) (electronic digital computation)** Loosely, same as check. *See also:* check problem; check. **(D) (supervisory control, data acquisition, and automatic control)** *See also:* data test; point test; certified design test. (SWG/C/Std100) 162-1963
(3) (instrument or meter) To ascertain its performance characteristics while functioning under controlled conditions. (MIL) [2]
(4) (A) (software) An activity in which a system or component is executed under specified conditions, the results are observed or recorded, and an evaluation is made of some aspect of the system or component. **(B) (software)** To conduct an activity as in definition (A). (C/Std100) 610.12-1990
(5) (airborne radioactivity monitoring) A procedure whereby the instrument, component, or circuit is evaluated for performance or satisfactory operation. (NI) N42.17B-1989r, N323-1978r
(6) (physical interpretation) Zero or more actions or groups of actions performed on a particular system to evaluate a parameter or characteristic. (Abstract interpretation) A source of information about the behavior of a system. In the abstract sense, an observation (symptom) can be interpreted as a test. (ATLAS) 1232-1995
(7) A set of stimuli, either applied or known, combined with a set of observed responses and criteria for comparing these responses to a known standard. (SCC20) 1232.1-1997
(8) An observed activity that may be caused to occur (e.g., stimulus-response) in order to obtain information about the behavior of a test subject. (SCC20) 1226-1998
(9) (A) A set of one or more test cases. **(B)** A set of one or more test procedures. **(C)** A set of one or more test cases and procedures. (C/SE) 829-1998

testability (1) (software) The degree to which a system or component facilitates the establishment of test criteria and the performance of tests to determine whether those criteria have been met. (C/Std100) 610.12-1990
(2) The degree to which a requirement is stated in terms that permit establishment of test criteria and performance of tests to determine whether those criteria have been met. (C/SE) 1233-1998

test—acceptance A test to demonstrate the degree of compliance of a device with the purchaser's requirement. (ELM) C12.1-1988

test access port (TAP) A general-purpose port that can provide access to many support functions built into a component, including the test logic defined by IEEE Std 1149.1-1990. It is composed as a minimum of the three input connections and one output connection required by the test logic in IEEE Std 1149.1-1990. An optional fourth input connection provides for asynchronous initialization of the test logic defined by IEEE Std 1149.1-1990. (TT/C) 1149.1-1990

test adapter *See:* adapter.

test amperes *See:* watthour meter—test current.

test analysis (test, measurement, and diagnostic equipment) The examination of the test results to determine whether the device is in a go or no-go state or to determine the reasons for or location of a malfunction. (MIL) [2]

test application framework (TAF) A structure for organizing the test objects related to a specific testing requirement. (SCC20) 1226-1998

test—approval A test of one or more meters or other items under various controlled conditions to ascertain the performance characteristics of the type of which they are a sample. (ELM) C12.1-1988

test asset An assemblage of instruments, interconnect devices, supporting software, and manual procedures that enable one or more test objectives to be achieved. *See also:* automatic test system. (SCC20) 1226-1998

test bed (software) An environment containing the hardware, instrumentation, simulators, software tools, and other support elements needed to conduct a test. (C) 610.12-1990

test bench (test, measurement, and diagnostic equipment) An equipment specifically designed to provide a suitable work surface for testing a unit in a particular test setup under controlled conditions. (MIL) [2]

test block *See:* test switch.

test block cabinet (watthour meter) An enclosure to house a test block and wiring for a bottom-connected watthour meter. (ELM) C12.8-1981r

test board A switchboard equipped with testing apparatus so arranged that connections can be made from it to telephone lines or central-office equipment for testing purposes. (COM) [48]

test bus interface circuit (TBIC) A circuit module that allows an internal analog test bus in an integrated circuit to be isolated from or connected to the pins in the analog test access port (ATAP). *See also:* analog test access port. (C/TT) 1149.4-1999

test bypass A mode of testing whereby the safety group under test is designed to permit any one channel or load group to be maintained, tested or calibrated during power operation, without initiating a protective action of the safety group. (PE/NP) 338-1987r

test cabinet (for a switchgear assembly) An assembly of a cabinet containing permanent electric connections, with cable connections to a contact box arranged to make connection to the secondary contacts on an electrically operated removable element, permitting operation and testing of the removable element when removed from the housing. It includes the necessary control switch and closing relay, if required. (SWG/PE) C37.100-1992

test call (telephone switching systems) A call made to determine if circuits or equipment are performing properly. (COM) 312-1977w

test cap A protective structure that is placed over the exposed end of the cable to seal the sheath or other covering completely against the entrance of dirt, moisture, air, or other foreign substances. *Note:* Test caps are often provided with facilities for vacuum treatment, oil filling, or other special field operations. *See also:* live cable test cap. (T&D/PE) [10]

test case (1) (software) A set of test inputs, execution conditions, and expected results developed for a particular objective, such as to exercise a particular program path or to verify compliance with a specific requirement. (C/Std100) 610.12-1990
(2) Documentation that specifies inputs, predicted results, and a set of execution conditions for a test item. (C/SE) 1012-1998, 610.12-1990

test case generator (software) A software tool that accepts as input source code, test criteria, specifications, or data structure definitions; uses these inputs to generate test input data; and, sometimes, determines expected results. *Synonyms:* test generator; test data generator. *See also:* automated test generator. (C) 610.12-1990

test case specification (1) (software) A document that specifies the test inputs, execution conditions, and predicted results for an item to be tested. *Synonyms:* test specification; test description. *See also:* test procedure; test report; test plan; test item transmittal report; test incident report; test log. (C) 610.12-1990
(2) A document specifying inputs, predicted results, and a set of execution conditions for a test item. (C/SE) 829-1998

test circuit breaker (ac high-voltage circuit breakers) The circuit breaker under test. (SWG/PE) C37.081-1981r, C37.083-19992, C37.100-1992

test connection (telephony) Two telephone sets connected together by means of test loops and a feed circuit. (COM/TA) 269-1971w

test control The functionality that directs and facilitates the execution of tests and the collection of data. (SCC20) 1226-1998

test coordinator A person typically responsible for organizing and scheduling tests; deciding how, when, and where the system and its components will be tested; and determining the test equipment that is needed. (SUB/PE) 1303-1994

test coverage The degree to which a given test or set of tests addresses all specified requirements for a given system or component. (C) 610.12-1990

test criteria The criteria that a system or component must meet in order to pass a given test. *See also:* acceptance criteria; pass/fail criteria. (C) 610.12-1990

test current *See:* watthour meter—test current.

test current, continuous *See:* continuous test current.

test current in alternating-current circuits (insulation tests) The normal current flowing in the test circuit as the result of insulation leakage and, in alternating-current circuits, is the vector sum of the inphase leakage currents and quadrature capacitive currents. (AES/ENSY) 135-1969w

test current, long-time *See:* long-time test current.

test current, short-time *See:* short-time test current.

test data (1) Data from observations during tests. *Note:* All conditions should be stated in detail, for example, time, stress conditions and failure or success criteria. (R) [29]
(2) (software) Data developed to test a system or system component. *See also:* test case; component; data; system. (C/SE) 729-1983s
(3) (A) (station control and data acquisition) The recorded results of test. **(B) (station control and data acquisition)** A set of data developed specifically to test the adequacy of a computer run or system. They may be actual data taken from previous operations or artificial data created for this purpose. (SWG/PE/SUB) C37.100-1992, C37.1-1994, C37.1-1979
(4) Data that are entered into an electronic system of any kind (component, printed circuit assembly, subsystem, system) to verify the integrity of part or all of the system. Test data may be entered through function pins or test pins or both. (C/TT) 1149.4-1999

test data generator *See:* automated test generator; test case generator.

test description *See:* test case specification.

test design Documentation that specifies the details of the test approach for a software feature or combination of software features and identifying the associated tests. (C/SE) 1012-1998

test design specification A document specifying the details of the test approach for a software feature or combination of software features and identifying the associated tests. (C/SE) 829-1998

test desk (telephone switching systems) A position equipped with testing apparatus so arranged that connections can be made from it to telephone lines or central office equipment for testing purposes. (COM) 312-1977w

test documentation Documentation describing plans for, or results of, the testing of a system or component. Types include test case specification, test incident report, test log, test plan, test procedure, test report. (C) 610.12-1990

test driver (software) A software module used to invoke a module under test and, often, provide test inputs, control and monitor execution, and report test results. *Synonym:* test harness. (C) 610.12-1990

test duration (nuclear power generating station) The elapsed time between the test initiation and the test termination. (PE/NP) 338-1987r

test enclosure (for low-voltage ac power circuit breakers) A single-unit enclosure used for test purposes for a specific frame-size circuit breaker, which conforms to the manufacturer's recommendation for minimum volume, minimum electrical clearances, effective areas and locations of ventilation openings, and configuration of connections to terminals. (SWG/PE) C37.100-1992

Test Equipment Description Language (TEDL) A standardized computer language used to describe the configuration of ATE systems. (ATLAS) 1226-1993s

tester cycle *See:* vector.

test event An action or group of actions performed on a particular unit under test (UUT) to evaluate a parameter or characteristic. It is the process of initialization, stimulus, and measurement of the UUT. It includes one or more tests and occurs during a continuous period of time. Test events are defined by the test procedure or test software requirements. A Parametric Data Log (PDL) file corresponds to a single event and would be a test of a single device or unit. (SCC20) 1545-1999

test event data The necessary and sufficient set of information that allows the parametric data to be understood in the context within which it was obtained. It includes the identification of the unit under test (UUT), test set hardware and procedure, the start time and stop time when the data was taken, and any other information needed to interpret the parametric data. (SCC20) 1545-1999

test executive The part of the test system within the AI-ESTATE architectural concept that controls the test resources. (ATLAS) 1232-1995

test foundation framework (TFF) A comprehensive set of object classes that supports the use of product data, the development of test programs, and the utilization of diagnostic data elements. The classes defined in the TFF are based wholly on the fundamental functionality required for test. (SCC20) 1226-1998

test frequency (reliability analysis of nuclear power generating station safety systems) The number of tests of the same type per unit time interval; the reciprocal of the test interval. (PE/NP) 352-1987r

test gas phase (fly ash resistivity) The gaseous environment to which the ash layer being tested is exposed in a test cell used for the laboratory measurement of electrical resistivity of fly ash.

test generator *See:* test case generator.

test handset (telephone) A handset used for test purposes in a central office or in the outside plant. It may contain in the handle other components in addition to the transducer, as for example a dial, keys, capacitors, and resistors. *See also:* telephone station. (EEC/PE) [119]

test harness *See:* test driver.

test incident report (1) (software) A document that describes an event that occurred during testing that requires further investigation. *See also:* test log; test plan; test case specification; test item transmittal report; test report; test procedure. (C) 610.12-1990

(2) A document reporting on any event that occurs during the testing process which requires investigation. (C/SE) 829-1998

testing (1) (nuclear power quality assurance) An element of verification for the determination of the capability of an item to meet specified requirements by subjecting the item to a set of physical, chemical, environmental, or operating conditions. (PE/NP) [124]

(2) Dynamic verification performed with valued inputs. (C/BA) 896.9-1994w

(3) The process of analyzing a software item to detect the differences between existing and required conditions (that is, bugs) and to evaluate the features of the software item. (C/SE) 829-1998

testing agency The organization that actually performs the tests and records the data. (EMC/STCOORD) 299-1997

testing constant A constant that is not specified in the standard being tested but is required by an assertion test to test an assertion. (C/PA) 2003-1997

testing state A node state that is reflected by the value of 2 in the STATE_CLEAR.state field. The testing state is an optional transient state that is entered immediately after a write to the TEST_START register. The node remains in the testing state until the active test completes. (C/MM) 1212-1991s

testing unavailability *See:* unavailability.

test initiation (nuclear power generating station) The application of a test input or removal of equipment train from service to perform a test. (PE/NP) 338-1987r

test input (nuclear power generating station) A real or simulated, but deliberate action that is imposed upon a sensor, channel, train, load group, or other system or device for the purpose of testing. (PE/NP) 338-1987r

test—in-service A test made during the period that the meter is in service. It may be made on the customer's premises without removing the meter from its mounting, or by removing the meter for test either on the premises or in a laboratory or meter shop. (ELM) C12.1-1988

test instruction A computer instruction that checks the condition of data and sets status or overflow flag bits for a subsequent branch instruction. For example:

test x (sets flag to zero, negative or

overflow, depending on value of x

)

branch p (if flag is TRUE, then branch to p)

n: . . . .

p:

(C) 610.10-1994w

test interval (nuclear power generating station) The elapsed time between the initiation (or successful completion) of tests on the same sensor, channel, load group, safety group, safety system, or other specified system or device. (PE/NP) 338-1987r, 352-1987r

test item A software item which is an object of testing. (C/SE) 829-1998

test item transmittal report A document identifying test items. It contains current status and location information. (C/SE) 829-1998

test jack A spring-jaw receptacle in the current element of a test switch that provides a bipolar test connection in the metering current circuit without interruption of the current circuit. (ELM) C12.9-1993

test jack switch A single-pole single-throw disconnect switch used in conjunction with a test jack to provide a parallel current path during normal operating conditions. (ELM) C12.9-1993

test language A computer language used in testing components of hardware or of software. Examples include ATLAS, ATOLL, DETOL, and DMAD. (C) 610.13-1993w

test log A chronological record of relevant details about the execution of tests. (C/SE) 829-1998

test logic (1) (test access port and boundary-scan architecture) Any item of logic that is a dedicated part of the test logic architecture defined by this standard or is at the time of interest configured as a part of the test logic architecture defined by 1149.1-1990. (TT/C) 1149.1-1990

(2) (test, measurement, and diagnostic equipment) The logical, systematic examination of circuits and their diagrams to identify and analyze the probability and consequence of potential malfunctions for determining related maintenance or maintainability design requirements. (MIL) [2]

test loop (transmission performance of telephone sets) A circuit that is interposed between a telephone set and a telephone feed circuit to simulate a telephone line. (COM/TA) 269-1983s

test, measurement, and diagnostic equipment Any system or device used to evaluate the operational condition of a system or equipment to identify and isolate or both any actual or potential malfunction. (MIL) [2]

test method (1) The software, procedures, or other means specified by a standard to measure conformance. (C/PA) 1003.10-1995, 2003.1-1992

(2) The software, procedures, or other means specified by a POSIX standard to measure conformance. Test methods may include a PCTS, PCTP, or an audit of a PCD. (C/PA) 13210-1994

(3) A testing approach, philosophy, or strategy. (SCC20) 771-1998

(4) A specification that defines the algorithm, procedures, and required controllable inputs and potential behavior (nominal or anomalous) of a test object. (SCC20) 1226-1998

(5) The software, procedures, or other means specified to measure conformance to a specification.
(C/PA) 1328.2-1993w, 1224-1993w, 1224.1-1993w, 1327-1993w, 1326.1-1993w, 1328-1993w, 1326.2-1993w

test method implementation The software, procedures, or other means used to measure conformance. For PASC, test method implementations may include a CTS, a CTP, or an audit of a CD. (C/PA) 2003-1997

test method specification A document that expresses the required functionality and behavior of a base standard as assertions and provides the complete set of conforming test result codes. (C/PA) 2003-1997

test method standard A test method specification that has been adopted as a standard. (C/PA) 2003-1997

test model (thermal classification of electric equipment and electrical insulation) (evaluation of thermal capability) A representation of equipment, a component or part of equipment, or the equipment itself, that is suitable for use in a functional test. (EI) 1-1986r

test mode select input pin (TMS) The test mode select input pin contained in the test access port (TAP) defined by IEEE Std 1149.1-1990. *See also:* test access port.
(TT/C) 1149.1-1990

test node Any physical location(s) of relevance to a test.
(ATLAS) 1232-1995

test object (1) An encapsulated stand-alone, executable test procedure. (ATLAS) 1226.2-1993w
(2) Any object defined for use within the domain of test representing an encapsulated view of a test method with interfaces to a test system. (SCC20) 1226-1998

test objective (software unit testing) (software) An identified set of software features to be measured under specified conditions by comparing actual behavior with the required behavior described in the software documentation.
(C/SE) 610.12-1990, 1008-1987r

test operating cycle (valve actuators) The movement of an actuator through its required operations travel under specified loading conditions, terminating with a return to the starting position. (PE/NP) 382-1985

test-oriented language (test, measurement, and diagnostic equipment) A computer language utilizing English mnemonics that are commonly used in testing. Examples are measure, apply, connect, disconnect, and so forth. (MIL) [2]

test outcome A mapping from an observation to one of a set of discrete possibilities. (SCC20) 1226-1998

test phase (software verification and validation plans) (software) The period of time in the software life cycle during which the components of a software product are evaluated and integrated, and the software product is evaluated to determine whether or not requirements have been satisfied.
(C/SE) 1012-1986s, 610.12-1990

test pin A pin on a component or a printed circuit assembly that is provided solely or primarily for use during test or maintenance operations. (C/TT) 1149.4-1999

test plan (1) (safety systems equipment in nuclear power generating stations) A document that identifies the equipment to be qualified, defines the acceptance criteria and the total scope of the testing activities required for qualification to a specified set of conditions. (PE/NP) 600-1983w
(2) Documentation that specifies the scope, approach, resources, and schedule of intended testing activities.
(C/SE) 1012-1998
(3) A document describing the scope, approach, resources, and schedule of intended testing activities. It identifies test items, the features to be tested, the testing tasks, who will do each task, and any risks requiring contingency planning.
(C/SE) 829-1998

test plug A bipolar mating plug to a **test jack** for inserting instrumentation into the metering current circuit.
(ELM) C12.9-1993

test point (1) (separable insulated connectors) A capacitively coupled terminal for use with voltage sensing devices.
(T&D/PE) 386-1995
(2) (station control and data acquisition) A predefined location within equipments or routines at which known result should be present if the equipment or routine is operating properly. (SUB/PE) C37.1-1979s
(3) (test, measurement, and diagnostic equipment) A convenient, safe access to a circuit or system so that a significant quantity can be measured or introduced to facilitate maintenance, repair, calibration, alignment, and checkout.
(MIL) [2]
(4) A geographic location which has been selected for the measurement of field strength. (T&D/PE) 1260-1996

test point selector (test, measurement, and diagnostic equipment) A device capable of selecting test points on an item being tested in accordance with instructions from the programmer. (MIL) [2]

test position (of a switchgear assembly removable element) That position in which the primary disconnecting devices of the removable element are separated by a safe distance from these in the housing, and some or all of the secondary disconnecting devices are in operating contact. *Notes:* 1. A set of test jumpers or mechanical movement of secondary disconnecting devices may be used to complete all secondary connections for test in the test position. This may correspond with the disconnected position. 2. Safe distance, as used here, is a distance at which the equipment will meet its withstand ratings, both power frequency and impulse, between line and load stationary terminals and phase-to-phase and phase-to-ground on both line and load stationary terminals with the switching device in the closed position.
(SWG/PE) C37.100-1992

test procedure (1) (safety systems equipment in nuclear power generating stations) A document that defines the implementation of the test plan and describes the methodology for performing the specific test. (PE/NP) 600-1983w
(2) (test, measurement, and diagnostic equipment) A document that describes step by step the operation required to test a specific unit with a specific test system. (MIL) [2]
(3) A description of the tests, test methods, and test sequences to be performed on a unit under test (UUT) to verify conformance with its test specification with or without fault diagnosis and without reference to specific test equipment.
(SCC20) 771-1998
(4) Documentation that specifies a sequence of actions for the execution of a test. (C/SE) 1012-1998
(5) The implementation of a test method.
(SCC20) 1226-1998

test procedure specification (1) A document specifying a sequence of actions for the execution of a test.
(C/SE) 829-1998
(2) (software) *See also:* test procedure. (C) 610.12-1990

test program (TP) (1) (test, measurement, and diagnostic equipment) A program specifically intended for the testing of a unit under test (UUT). (MIL) [2]
(2) An implementation of the tests, test methods, and test sequences to be performed on a unit under test (UUT) to verify conformance with its test specification with or without fault diagnosis and designed for execution on a specific test system.
(SCC20) 771-1998
(3) A test program implements the tests, test methods, and test sequences to be performed on a unit under test (UUT) to verify conformance with its test specification with or without fault diagnosis and designed for execution on a specific test system. (SCC20) 993-1997
(4) A program specifically intended for the testing of a test subject. (SCC20) 1226-1998

test program documentation *See:* test programming procedures.

test programming procedures (test, measurement, and diagnostic equipment) Documents which explain in detail the composition of test programs including definitions and logic used to compose the program. Provides instructions to implement changes in the program. (MIL) [2]

test program set (TPS) **(1)** The complete set of hardware, software, and documentation needed to evaluate a unit under test (UUT) on a given test system. (ATLAS) 1232-1995

(2) The complete set of hardware, software, and documentation needed to evaluate a unit under test (UUT) on a given test system. The test program set includes the test program, adapter devices, ancillary hardware and software, operator initiated procedures, and supporting documentation (source data, adapter device schematics, parts lists, etc.). (SCC20) 993-1997

(3) An assembly of items necessary to test a test subject on a piece of automatic test equipment (ATE). This includes the electrical, mechanical, instructional, and logical decision elements. The individual elements of the TPS are the TP, the adapter, and the TPS documentation (TPSD). (SCC20) 1226-1998

test provisions (test, measurement, and diagnostic equipment) The capability included in the design for conveniently evaluating the performance of a prime equipment, module, assembly, or part. (MIL) [2]

test radius A circle, with the center of the antenna array as its origin, on which the test points for field strength measurements ideally should be located. (T&D/PE) 1260-1996

test readiness review (TRR) **(A)** A review conducted to evaluate preliminary test results for one or more configuration items; to verify that the test procedures for each configuration item are complete, comply with test plans and descriptions, and satisfy test requirements; and to verify that a project is prepared to proceed to formal testing of the configuration items. **(B)** A review as in definition (A) for any hardware or software component. *Contrast:* formal qualification review; requirements review; design review; code review. (C) 610.12-1990

test—referee A test made by or in the presence of one or more representatives of a regulatory body or other impartial agency. (ELM) C12.1-1988

test reliability The assessed reliability of an item based on a particular test with stated stress and stated failure criteria. *See also:* reliability. (R) [29]

test repeatability (software) An attribute of a test, indicating that the same results are produced each time the test is conducted. (C) 610.12-1990

test report (software) A document that describes the conduct and results of the testing carried out for a system or component. *Synonym:* test summary report. *See also:* test log; test incident report; test case specification; test item transmittal report; test plan; test procedure. (C) 610.12-1990

test—request A test made at the request of a customer. (ELM) C12.1-1988

test requirement **(1)** A definition of the tests and test conditions required to be performed on a unit under test (UUT) to verify conformance with its performance specification. (SCC20) 771-1998

(2) A specification of a particular test action giving the necessary power input conditioning, the stimulus and load applications, the measurements to be taken, and any special operator actions. The complete collection of all test requirements defines the actions necessary to validate proper operation of a unit under test (UUT) in accordance with a predetermined design or product specification. (SCC20) 993-1997

(3) A specification of the test methods and test conditions needed to evaluate and diagnose a test subject. (SCC20) 1226-1998

test requirement analysis (test, measurement, and diagnostic equipment) The examination of documents such as schematics, assembly drawings and specifications for the purpose of deriving test requirements for a unit. (MIL) [2]

test requirement document (test, measurement, and diagnostic equipment) The document that specifies the tests and test conditions required to test and fault isolate a unit under test. (MIL) [2]

Test Requirement Specification Language (TRSL) A standardized computer language used to specify test requirements. (ATLAS) 1226-1993s

test response spectrum (TRS) **(1)** **(nuclear power generating station) (seismic qualification of Class 1E equipment)** The response spectrum that is constructed using analysis or derived using spectrum analysis equipment based on the actual motion of the shake table. (SWG/PE) C37.100-1992

(2) **(seismic qualification of Class 1E metal-enclosed power switchgear assemblies)** The response spectrum that is developed from the actual time history of the actual motion of the shake table. *Note:* When qualifying equipment bu utilizing response spectra, the TRS is to be compared with the RRS (required response spectrum) using the methods described in 7.6.2 and 7.6.3 of IEEE Std 344-1987. (SWG/PE/NP) C37.81-1989r, 344-1987r

(3) **(valve actuators)** The response spectrum that is constructed using analysis or derived using spectrum analysis equipment based on the actual input test table motion to the device. (PE/NP) 382-1985

(4) **(seismic testing of relays)** (as applied to relays) The acceleration response spectrum that is constructed using analysis or derived using spectrum analysis equipment based on the actual motion of the shake table. (PE/PSR) C37.98-1977s

(5) The calculated response spectrum that is developed from the actual time history of the motion of the shake table (not any point on the equipment or equipment structure) for a particular damping value. (PE/SUB) 693-1997

test result code A value that describes the result of an assertion test. (C/PA) 13210-1994, 2003-1997, 2003.1-1992

test routine **(A)** Usually a synonym for check routine. **(B)** Sometimes used as a general term to include both check routine and diagnostic routine. (Std100) 270-1966

tests after delivery Those tests made by the purchaser after delivery of the circuit breaker, which supplement inspection, to determine whether the circuit breaker has arrived in good condition. These tests may consist of timing tests on closing, opening, close-open no-load operations, and power frequency voltage withstand tests at 75% of the rated power frequency withstand voltage. (SWG/PE) C37.013-1997

test schedule (reliability analysis of nuclear power generating station safety systems) The pattern of testing applied to systems or the parts of a system. In general, there are two patterns of interest:

a) (simultaneous) Redundant items or systems are tested at the beginning of each test interval, one immediately following the other.

b) (perfectly staggered) Redundant items or systems are tested such that the test interval is divided into equal subintervals.

(PE/NP) 352-1987r

test script *See:* test procedure.

test sequence **(1)** **(A)** **(test, measurement, and diagnostic equipment)** A unique setup of measurements, and. **(B)** **(test, measurement, and diagnostic equipment)** A specific order of related tests. (MIL) [2]

(2) A specified order of related tests. (SCC20) 771-1998

test sequence number (test, measurement, and diagnostic equipment) Identification of a test sequence. (MIL) [2]

test set architecture (software) The nested relationships between sets of test cases that directly reflect the hierarchic decomposition of the test objectives. (C/SE) 610.12-1990, 1008-1987r

test site A site meeting specified requirements suitable for measuring radio interference fields radiated by an appliance under test. *See also:* electromagnetic compatibility. (EMC/INT) [53], [70]

test software (test, measurement, and diagnostic equipment) Maintenance instructions which control the testing operations and procedures of the automatic test equipment. This software is used to control the unique stimuli and measurement parameters used in testing the unit under test. (MIL) [2]

test specification (1) (safety systems equipment in nuclear power generating stations) A document that defines the test requirements including test levels and performance requirements. (PE/NP) 600-1983w
(2) (software) *See also:* test case specification. (C) 610.12-1990
(3) Describes the test criteria and the methods to be used in a specific test to assure that the performance and design specifications have been satisfied. The test specification identifies the capabilities or program functions to be tested, and identifies the test environment. (C/SE) 1298-1992w
(4) A definition of the tests to be performed on a unit under test (UUT) to verify conformance with its performance specification and without reference to any specific test equipment or test method. (SCC20) 771-1998
(5) A document that defines the tests to be performed on a test subject to verify conformance with its performance specification, without reference to any specific test equipment or test method. (SCC20) 1226-1998

test specimen (insulators) An insulator that is representative of the product being tested: it is a specimen that is undamaged in any way that would influence the result of the test. (EEC/IEPL) [89]

test spectrum (test, measurement, and diagnostic equipment) A range of test stimuli and measurements based on analysis of prime equipment test requirements. (MIL) [2]

test stand (test, measurement, and diagnostic equipment) An equipment specifically designed to provide suitable mountings, connections, and controls for testing electrical, mechanical, or hydraulic equipment as an entire system. (MIL) [2]

test stimulus (electrical) A single shock or succession of shocks, used to characterize or determine the state of excitability or the threshold of a tissue. (EMB) [47]

test strategy (A) The arrangement of specific tester types to achieve optimum throughput and diagnostic capability at the least possible cost given the fault spectrum, process yield, production rate, and product mix for a particular environment. (Adapted from MIL-STD-1309D). **(B)** A selection of test methods to achieve some diagnostic test within execution time and test resource constraints. (SCC20) 1226-1998
(2) (A) An approach taken to combine factors including constraints, goals, and other considerations to be applied to the testing of a unit under test (UUT). **(B)** The approach taken to the evaluation of a UUT by which a result is obtained. **(C)** The requirements and constraints to be reflected in test and diagnostic strategies. (ATLAS) 1232-1995

test subject (1) The entity to be tested. It may range from a simple to a complex system, e.g., a unit under test or a human patient. (SCC20) 1232.1-1997
(2) The specific product design that is the focus of attention or target for the development of tests and diagnostics. (SCC20) 1226-1998

test summary report A document summarizing testing activities and results. It also contains an evaluation of the corresponding test items. (C/SE) 829-1998

test support Those facilities not specified by the standard(s) being tested, or specified but not required, that need to be provided by the SUT in order to perform an assertion test. (C/PA) 2003-1997

test support software (test, measurement, and diagnostic equipment) Computer programs used to prepare, analyze, and maintain test software. Test software includes automatic test equipment (ATE) compilers, translation/analysis programs and punch/print programs. (MIL) [2]

test switch A combination of connection studs, jacks, plugs, or switch parts arranged conveniently to connect the necessary devices for testing instruments, meters, relays, etc. *Synonym:* test block. (SWG/PE) C37.100-1992

test system One system within the AI-ESTATE architecture. This system handles the execution of tests. (ATLAS) 1232-1995

test temperature The temperature of the heater plates mounted on the stator coil or bar, as measured by a temperature sensor embedded within the heater plate. (DEI) 1043-1996

test termination The removal of a test input with results of the test being known, or the committal of the equipment for repair based on the results of the test. (PE/NP) 380-1975w

test testboard (telephone switching systems) A position equipped with testing apparatus so arranged that connections can be made from it to trunks for testing purposes. (COM) 312-1977w

test unit (software unit testing) A set of one or more computer program modules together with associated control data, (for example, tables), usage procedures, and operating procedures that satisfy the following conditions:

a) All modules are from a single computer program
b) At least one of the new or changed modules in the set has not completed the unit test.
c) The set of modules together with its associated data and procedures are the sole object of a testing process.

Notes: 1. A test unit may occur at any level of the design hierarchy from a single module to a complete program. Therefore, a test unit may be a module, a few modules, or a complete computer program along with associated data and procedures. 2. A test unit may contain one or more modules that have already been unit tested. (C/SE) 1008-1987r, 610.12-1990

test validity (software) The degree to which a test accomplishes its specified goal. (C/SE) 729-1983s

test voltage (electrical insulation tests) The voltage applied across the specimen during a test. (AES/ENSY) 135-1969w

test voltage, partial discharge-free A specified voltage applied in a specified test procedure, at which the test object is free from partial discharges exceeding a specified level. This voltage is expressed as a peak value divided by the square root of two. *Note:* The term corona-free test voltage has frequently been used with this connotation. It is recommended that such usage be discontinued in favor of the term partial discharge-free test voltage. (PE/PSIM) 454-1973w

test voltage related to partial discharges The phase-to-ground alternating voltage whose value is expressed by its peak value divided by $\sqrt{2}$. (SWG/PE) 1291-1993r

test withstand voltage The maximum value of a test voltage at which a new valve, with unimpaired integrity, does not show any disruptive discharge, nor suffer component failures above permissible levels, when subjected to a specified number of applications of the test voltage, under specified conditions. (SUB/PE) 857-1996

tetanizing current (electrotherapy) The current that, when applied to a muscle or to a motor nerve connected with a muscle stimulates the muscle with sufficient intensity and frequency to produce a smoothly sustained contraction as distinguished from a succession of twitches. *See also:* electrotherapy. (EMB) [47]

tetrad A group of four closely related items or digits. (C) 1084-1986w

tetrode A four-electrode electron tube containing an anode, a cathode, a control electrode, and one additional electrode that is ordinarily a grid. (ED) [45]

TEX A public-domain word processing language. *See also:* teletype exchange. (C) 610.2-1987

TeX A page description language used widely for formatting text containing mathematical symbols. (C) 610.13-1993w

text (1) In word processing, information that is intended for presentation for human comprehension in a two-dimensional form. Text may consist of symbols, phrases, sentences in natural or artificial language, pictures, diagrams, and tables. (C) 610.2-1987
(2) (computer graphics) A display element that consists of a character string. (C) 610.6-1991w

text attribute A characteristic of text. For example, color index, font, path, precision. (C) 610.6-1991w

text-based user interface *See:* character-based user interface.

text column A roughly rectangular block of characters capable of being laid out side-by-side next to other text columns on an output page or terminal screen. The widths of text columns are measured in column positions. (C/PA) 9945-2-1993

text cursor A screen object that indicates the insertion point for text input. (C) 1295-1993w

text editing The process of entering, altering, and viewing text. (C) 610.2-1987

text editor A computer program, often part of a word processing system, that allows a user to enter, alter, and view text. *Synonym:* editor. *See also:* program editor; document editor; full-screen editor; line editor. (C) 610.2-1987, 610.12-1990

text end adjustment The ability of a text formatter to automatically reformat text to comply with specified line lengths and page sizes. *See also:* adjust line mode. (C) 610.2-1987

text field A visual user interface control into which a user types or places alphanumeric text on one line. Its boundaries are usually indicated. (C) 1295-1993w

text file A file that contains characters organized into one or more lines. The lines shall not contain NUL characters and none shall exceed {LINE_MAX} bytes in length, including the ⟨newline⟩. Although POSIX.1 does not distinguish between text files and binary files (see the C Standard), many utilities only produce predictable or meaningful output when operating on text files. The standard utilities that have such restrictions always specify *text files* in their Standard Input or Input Files subclauses. (C/PA) 9945-2-1993

text formatter A computer program, often part of a word processing system, that interprets formatting commands embedded in text and performs the indentation, pagination, tabulation, underscoring, and other formatting procedures indicated by the commands. *Synonym:* print formatter. (C) 610.2-1987

text formatting In word processing, the process of interpreting formatting commands embedded in text and performing the indentation, pagination, tabulation, underscoring, and other formatting procedures indicated by the formatting commands. *Synonym:* print formatter. (C) 610.2-1987

text-formatting language A computer language used to format text documents. Examples include Bookmaster, Cyphertext, DCF, SCRIPT, PAGE, and SCRIBE. *See also:* page description language. (C) 610.13-1993w

text page A model page that contains textual material related to a specific diagram. (C/SE) 1320.1-1998

text processing *See:* word processing.

texture (image processing and pattern recognition) An attribute representing the spatial arrangement of the gray levels of the pixels in a region. (C) 610.4-1990w

text window That portion of a display screen that is being used to display text (human-readable characters and words). (C/BA) 1275-1994

T4 A carrier facility that transmits digital signal level four. The data rate is 274.176 Mb/s, the equivalent of 4032 voice-band channels. (C) 610.7-1995

THD *See:* total harmonic distortion.

THDBH load *See:* ten high day busy-hour load.

theft Unauthorized removal of system components (i.e., hardware, software, media). Theft could result in unauthorized disclosure of information or sensitive technology and denial of service conditions. (C/BA) 896.3-1993w

theoretical cutoff frequency (theoretical cutoff) (electric structure) A frequency at which, disregarding the effects of dissipation, the attenuation constant changes from zero to a positive value or vice versa. *See also:* cutoff frequency. (PE/EEC) [119]

theory *See:* information theory.

therapeutic high-frequency diathermy equipment (health care facilities) Therapeutic high-frequency diathermy equipment is therapeutic induction and dielectric heating equipment. (NESC/NEC) [86]

therm A quantity of heat that is equal to 100 000 Btu. (IA/PSE) 241-1990r

thermal aging (1) (rotating machinery) Normal load/temperature deteriorating influence on insulation. (PE/EM) 432-1976s
(2) (thermal classification of electric equipment and electrical insulation) The aging that takes place at an elevated temperature. (EI) 1-1986r

thermal burden rating of a voltage transformer The voltampere output that the voltage transformer will provide continuously at rated secondary voltage without exceeding the specified temperature limits. (PE/TR) C57.13-1993, C57.12.80-1978r

thermal capability (solid electrical insulating materials) Includes the ability to withstand without failure the maximum short time operating temperatures and the long time integrated degradative effect of temperature and time. It constitutes a design limitation on the use of insulating materials in electrical and electronic equipment to the extent that both thermal softening (or other short term effects) and long term aging affect functional properties. (EI) 98-1984r

thermal cell A reserve cell that is activated by the application of heat. *See also:* electrochemistry. (EEC/PE) [119]

thermal conduction The transport of thermal energy by processes having rates proportional to the temperature gradient and excluding those processes involving a net mass flow. *See also:* thermoelectric device. (ED) [46]

thermal conductivity (1) (electric power systems in commercial buildings) The time rate of heat flow through a unit area of a homogeneous substance under steady conditions when a unit temperature gradient is maintained in the direction that is normal to the area. (IA/PSE) 241-1990r
(2) The quotient of the conducted heat through unit area per unit time by the component of the temperature gradient normal to that area. *See also:* thermoelectric device. (ED) [46]

thermal conductivity, electronic *See:* electronic thermal conductivity.

thermal converter (thermoelement) (electric instruments) (thermocouple converter) A device that consists of one or more thermojunctions in thermal contact with an electric heater or integral therewith, so that the electromotive force developed at its output terminals by thermoelectric action gives a measure of the input current in its heater. *Note:* The combination of two or more thermal converters when connected with appropriate auxiliary equipment so that its combined direct-current output gives a measure of the active power in the circuit is called a thermal watt converter. (EEC/PE) [119]

thermal current converter (electric instruments) A type of thermal converter in which the electromotive force developed at the output terminals gives a measure of the current through the input terminals. *See also:* thermal converter. (EEC/AII) [102]

thermal current rating (A) (neutral grounding devices) (electric power) The root-mean-square neutral current in amperes that it will carry under standard conditions for its rated time without exceeding standard temperature limitations, unless otherwise specified. *See also:* grounding device.
(B) (resistors) The initial root-mean-square symmetrical value of the current that will flow when rated voltage is applied. (SPD/PE) 32-1972

thermal cutout An overcurrent protective device that contains a heater element in addition to and affecting a renewable fusible member which opens the circuit. It is not designed to interrupt short-circuit currents. (NESC/NEC) [86]

thermal diffusivity Thermal conductivity divided by the product of density and specific heat. (IA/PSE) 241-1990r

thermal duty cycle (nuclear power generating station) The percentage of time that heat producing electrical current flows in equipment over a specific period of time. (PE/NP) 649-1980s

thermal electromotive force Alternative term for Seebeck electromotive force. *See also:* thermoelectric device. (ED) [46], 221-1962w

thermal endurance (1) (rotating machinery) The relationship, between temperature and time spent at that temperature, required to produce such degradation of an electrical insulation that it fails under specified conditions of stress, electric or mechanical, in service or under test. For most of the chemical reactions encountered, this relationship is a straight line when plotted with ordinates of logarithm of time against abscissae of reciprocal of absolute temperature (Arrhenius plot). *See also:* asynchronous machine. (PE/EI) [9]
(2) (solid electrical insulating materials) Related to the rate at which important properties deteriorate as a function of temperature and time. It is determined by accelerated testing. (EI) 98-1984r

thermal endurance graph (thermal classification of electric equipment and electrical insulation) (evaluation of thermal capability) The graphical expression of the thermal endurance relationship in which time to failure is plotted against the reciprocal of the absolute test temperature. (EI) 1-1986r

thermal endurance relationship (thermal classification of electric equipment and electrical insulation) (evaluation of thermal capability) The expression of aging time to failure as a function of test temperature in an aging test. (EI) 1-1986r

thermal equilibrium (rotating machinery) The state reached when the observed temperature rise of the several parts of the machine does not vary by more than 2°C over a period of one hour. *See also:* asynchronous machine. (PE) [9]

thermal fire hazard A hazard resulting from the generation of heat in a fire. (DEI) 1221-1993w

thermal flow switch *See:* flow relay.

thermal insulation (1) (electrical heating systems) Material having air- or gas-filled pockets, void spaces, or heat-reflective surfaces that, when properly applied, will reduce the transfer of heat with reasonable effectiveness under ordinary conditions. (BT/IA/AV/PC) 152-1953s, 515.1-1995, 844-1991, 515-1997
(2) (electric power systems in commercial buildings) A material having a high resistance to heat flow and used to retard the flow of heat to the outside. (IA/PE/PSE/EDPG) 241-1990r, 622-1979s

thermal limit curves for large squirrel-cage motors Plots of maximum permissible time versus percent of rated current flowing in the motor winding under specified emergency conditions. These curves can be used in conjunction with the motor time-current curve for a normal start to set protective relays and breakers for motor thermal protection during starting and running conditions. (EM/PE) 620-1987w

thermally delayed overcurrent trip *See:* thermally delayed release; overcurrent release.

thermally delayed release A release delayed by a thermal device. *Synonym:* thermally delayed trip. (SWG/PE) C37.100-1992

thermally delayed trip *See:* thermally delayed release.

thermally protected (as applied to motors) The words "Thermally Protected" appearing on the nameplate of a motor or motor-compressor indicate that the motor is provided with a thermal protector. (NESC/NEC) [86]

thermal-mechanical cycling (rotating machinery) The experience undergone by rotating-machine windings, and particularly their insulation, as a result of differential movement between copper and iron on heating and cooling. Also denotes a test in which such actions are simulated for study of the resulting behavior of an insulation system, particularly for machines having a long core length. *See also:* asynchronous machine. (PE) [9]

thermal noise (1) (telephone practice) Noise occurring in electric conductors and resistors and resulting from the random movement of free electrons contained in the conducting material. The name derives from the fact that such random motion depends on the temperature of the material. Thermal noise has a flat power spectrum out to extremely high frequencies. (PE/PSR) C37.93-1976s
(2) (electron tube) The noise caused by thermal agitation in a dissipative body. *Note:* The available thermal noise power N, from a resistor at temperature T, is $N = kT\Delta f$, where k is Boltzmann's constant and Δf is the frequency increment. (ED) 161-1971w
(3) (resistance noise) (data transmission) Random noise in a circuit associated with the thermodynamic interchange of energy necessary to maintain thermal equilibrium between the circuit and its surroundings. *Note:* The average square of the open-circuit voltage across the terminals of a passive two-terminal network of uniform temperature, due to thermal agitation, is given by:

$$V_T^2 = 4kT \int R(f)\, df$$

where T is the absolute temperature in degrees Celsius, R is the resistance component Ω in ohms of the network impedance at the frequency f measured in hertz, and k is the Boltzmann constant, 1.38×10^{-23}. (PE) 599-1985w
(4) *See also:* noise. (LM/C) 802.7-1989r

thermal-overload detection (series capacitor) A means to detect excessive heating of series capacitor bank components and to initiate an alarm signal, or the closing of the associated bypass device, or both. (T&D/PE) 824-1985s

thermal-overload protection (series capacitor) A means to detect excessive heating of capacitor units as a result of a combination of current, ambient temperature, and solar radiations, and to initiate an alarm signal or the closing of the associated capacitor bypass switch, or both. (T&D/PE) [26]

thermal power converter (thermal watt converter) (electric instruments) A complex type of thermal converter having both potential and current input terminals. It usually contains both current and potential transformers or other isolating elements, resistors, and a multiplicity of thermoelements. The electromotive force developed at the output terminals gives a measure of the power at the input terminals. *See also:* thermal converter. (EEC/AII) [102]

thermal printer A nonimpact printer in which the characters are produced by applying heated elements to heat-sensitive paper directly or by melting ink from a ribbon onto normal paper. *Synonym:* thermal transfer printer. (C) 610.10-1994w

thermal protection (motors) The words thermal protection appearing on the nameplate of a motor indicate that the motor is provided with a thermal protector. *See also:* contactor. (NESC) [86]

thermal protector (1) (as applied to motors) A protective device for assembly as an integral part of a motor or motor-compressor and which, when properly applied, protects the motor against dangerous overheating due to overload and failure to start. The thermal protector may consist of one or more sensing elements integral with the motor or motor-compressor and an external control device. (NESC/NEC) [86]
(2) (rotating machinery) A protective device, for assembly as an integral part of a machine, that protects the machine against dangerous overheating due to overload or any other

reason. *Notes:* 1. It may consist of one or more temperature-sensing elements integral with the machine and a control device external to the machine. 2. When a thermal protector is designed to perform its function by opening the circuit to the machine and then automatically closing the circuit after the machine cools to a satisfactory operating temperature, it is an automatic-reset thermal protector. 3. When a thermal protector is designed to perform its function by opening the circuit to the machine but must be reset manually to close the circuit, it is a manual-reset thermal protector. *See also:* contactor. (PE) [9]

thermal relay (1) A relay in which the displacement of the moving contact member is produced by the heating of a part of the relay under the action of electric currents. *See also:* relay. (SWG/PE/IA/ICTL/IAC) C37.100-1992, [60], [84] **(2)** A relay whose operation is caused by heat developed within the relay as a result of specified external conditions. *See also:* temperature relay. (SWG/PE) C37.100-1992

thermal residual voltage (Hall effect devices) That component of the zero field residual voltage caused by a temperature gradient in the Hall plate. (MAG) 296-1969w

thermal resistance (1) (cable) The resistance offered by the insulation and other coverings to the flow of heat from the conductor or conductors to the outer surface. *Note:* The thermal resistance of the cable is equal to the difference of temperature between the conductor or conductors and the outside surface of the cable divided by rate of flow of heat produced thereby. It is preferably expressed by the number of degrees Celsius per watt per foot of cable. (T&D/PE) [10] **(2) (Hall generator)** The difference between the mean Hall plate temperature and the temperature of an external reference point, divided by the power dissipation in the Hall plate. (MAG) 296-1969w

thermal resistance case-to-ambien (light-emitting diodes) The thermal resistance (steady-state) from the device case to the ambient. (ED) [127]

thermal resistance, effective *See:* effective thermal resistance.

thermal resistance junction-to-ambient ($R\theta$CA) (light-emitting diodes) (formerly θ_{J-C}) The thermal resistance (steady-state) from the semiconductor junction(s) to the ambient. (ED) [127]

thermal resistance junction-to-case ($R\theta$CA) (light-emitting diodes) (formerly θ_{J-C}) The thermal resistance (steady-state) from the semiconductor junction(s) to a stated location on the case. (ED) [127]

thermal runaway A condition that is caused by a battery charging current that produces more internal heat than the battery can dissipate. This condition ultimately causes cell venting and premature failure. (PE/EDPG) 1184-1994

thermal short-circuit rating The maximum steady-state short-circuit rms current that can be carried for a specified time, the reactor being approximately at rated temperature rise and maximum ambient at the time the load is applied, without exceeding the specified temperature limits, and within the limitations of established standards. (PE/TR) C57.16-1996

thermal short-time current rating (current transformer) The root-mean-square symmetrical primary current that may be carried for a stated period (five seconds or less) with the secondary winding short-circuited, without exceeding a specified maximum temperature in any winding. *See also:* instrument transformer. (PE/TR) [57]

thermal subsystem (terrestrial photovoltaic power systems) The subsystem that receives thermal energy from the array subsystem. The thermal energy may be utilized for a thermal load application or dissipated. *See also:* array control. (PV) 928-1986r

thermal telephone receiver (thermophone) A telephone receiver in which the temperature of a conductor is caused to vary in response to the current input, thereby producing sound waves as a result of the expansion and contraction of the adjacent air. (EEC/PE) [119]

thermal time constant The time required for the conductor temperature to accomplish 63.2% of a change in initial temperature to the final temperature when the electrical current going through a conductor undergoes a step change. (T&D/PE) 738-1993

thermal transfer printer *See:* thermal printer.

thermal transmittance (U factor) The time rate of heat flow per unit temperature difference. (IA/PSE) 241-1990r

thermal tuning The process of changing the operating frequency of a system by using a controlled thermal expansion to alter the geometry of the system. *See also:* oscillatory circuit. (ED) 161-1971w

thermal tuning rate The initial time rate of change in frequency that occurs when the input power to the tuner is instantaneously changed by a specified amount. *Note:* This rate is a function of the power input to the tuner as well as the sign and magnitude of the power change. *See also:* oscillatory circuit. (ED) 161-1971w

thermal tuning sensitivity The rate of change of resonator equilibrium frequency with respect to applied thermal tuner power. (ED) 161-1971w

thermal tuning time constant The time required for the frequency to change by a fraction $(1 - 1/e)$ of the change in equilibrium frequency after an incremental change of the applied thermal tuner power. *Notes:* 1. If the behavior is not exponential, the initial conditions must be stated. 2. Here e is the base of natural logarithms. *See also:* oscillatory circuit. (ED) 161-1971w

thermal voltage converter (electric instruments) A thermoelement of low-current input rating with an associated series impedance or transformer, such that the electromotive force developed at the output terminals gives a measure of the voltage applied to the input terminals. *See also:* thermal converter. (EEC/AII) [102]

thermal watt converter *See:* thermal converter; thermal power converter.

thermionic arc (gas) An electric arc characterized by the fact that the thermionic cathode is heated by the arc current itself. *See also:* discharge. (ED) [45], [84]

thermionic emission (Edison effect) (Richardson effect) The liberation of electrons or ions from a solid or liquid as a result of its thermal energy. *See also:* electron emission. (ED) 161-1971w, 160-1957w

thermionic generator A thermoelectric generator in which a part of the circuit, across which a temperature difference is maintained, is a vacuum or a gas. *See also:* thermoelectric device. (ED) [46]

thermionic grid emission Current produced by electrons thermionically emitted from a grid. *See also:* electron emission. (ED) 161-1971w, 160-1957w

thermionic tube An electron tube in which the heating of one or more of the electrodes is for the purpose of causing electron or ion emission. *See also:* hot-cathode tube. (ED) 161-1971w

thermistor (1) (general) An electron device that makes use of the change of resistivity of semiconductor with change in temperature. *See also:* bolometric detector; semiconductor. (PE/ED/PSIM) 119-1974w, 216-1960w

(2) (power semiconductor) A semiconductor device whose electric resistance is dependent upon temperature. (IA) [12]

(3) (waveguide components) A form of bolometer element having a negative temperature coefficient of resistivity which typically employs a semiconductor bead. (MTT) 147-1979w

thermistor mount (waveguide) (bolometer mount) A waveguide termination in which a thermistor (bolometer) can be incorporated for the purpose of measuring electromagnetic power. *See also:* waveguide. (AP/ANT) [35], [84]

thermochromeric Pertaining to heat-sensitive materials that change color when heated to different temperatures. (C) 610.10-1994w

thermochromeric display device A display device that uses thermochromeric materials to form images on the display surface. (C) 610.10-1994w

thermocouple A pair of dissimilar conductors so joined at two points that an electromotive force is developed by the thermoelectric effects when the junctions are at different temperatures. *See also:* electric thermometer; thermoelectric effect. (IM/PE/PSIM) 544-1975w, 119-1974w

thermocouple converter *See:* thermal converter.

thermocouple extension wire A pair of wires having such electromotive-force-temperature characteristics relative to the thermocouple with which the wires are intended to be used that, when properly connected to the thermocouple, the reference junction is in effect transferred to the other end of the wires. (PE/PSIM) 119-1974w

thermocouple instrument An electrothermic instrument in which one or more thermojunctions are heated directly or indirectly by an electric current or currents and supply a direct current that flows through the coil of a suitable direct-current mechanism, such as one of the permanent-magnet moving-coil type. *See also:* instrument. (EEC/PE) [119]

thermocouple leads A pair of electrical conductors that connect the thermocouple to the electromotive force measuring device. One or both leads may be simply extensions of the thermoelements themselves or both may be of copper, dependent on the thermoelements in use and upon the physical location of the reference junction or junctions relative to the measuring device. (PE/PSIM) 119-1974w

thermocouple thermometer A temperature-measuring instrument comprising a device for measuring electromotive force, a sensing element called a thermocouple that produces an electromotive force of magnitude directly related to the temperature difference between its junctions, and electrical conductors for operatively connecting the two. (PE/PSIM) 119-1974w

thermocouple vacuum gauge A vacuum gauge that depends for its operation on the thermal conduction of the gas present, pressure being measured as a function of the electromotive force of a thermocouple the measuring junction of which is in thermal contact with a heater that carries a constant current. It is ordinarily used over a pressure range of 10^{-1} to 10^{-3} conventional millimeter of mercury. *See also:* instrument. (EEC/PE) [119]

thermodynamic equilibrium A situation in which the net thermal radiation exchanged by members of a system is zero. (AP/PROP) 211-1997

thermoelectric arm The part of a thermoelectric device in which the electric-current density and temperature gradient are approximately parallel or antiparallel and that is electrically connected only at its extremities to a part having the opposite relation between the direction of the temperature gradient and the electric-current density. *Note:* The term thermoelement is ambiguously used to refer to either a thermoelectric arm or to a thermoelectric couple, and its use is therefore not recommended. *See also:* thermoelectric device. (ED) [46], 221-1962w

thermoelectric cooling device A thermoelectric heat pump that is used to remove thermal energy from a body. *See also:* thermoelectric device. (ED) [46], 221-1962w

thermoelectric device A generic term for thermoelectric heat pumps and thermoelectric generators. (ED) [46]

thermoelectric effect *See:* Seebeck effect.

thermoelectric effect error (bolometric power meters) An error arising in bolometric power meters that employ thermistor elements in which the majority of the bias power is alternating current and the remainder direct current. The error is caused by thermocouples at the contacts of the thermistor leads to the metal oxides of the thermistors. *See also:* bolometric power meter. (IM/HFIM) [40]

thermoelectric generator A device that converts thermal energy into electric energy by direct interaction of a heat flow and the charge carriers in an electric circuit, and that requires for this process the existence of a temperature difference in the electric circuit. *See also:* thermoelectric device. (ED) [46]

thermoelectric, graded arm A thermoelectric arm whose composition changes continuously along the direction of the current density. *See also:* thermoelectric device. (ED) [46], 221-1962w

thermoelectric heating device A thermoelectric heat pump that is used to add thermal energy to a body. *See also:* thermoelectric device. (ED) [46]

thermoelectric heat pump A device that transfers thermal energy from one body to another by the direct interaction of an electric current and the heat flow. *See also:* thermoelectric device. (ED) [46]

thermoelectric power *See:* thermoelectric device; Seebeck coefficient.

thermoelectric thermometer (thermocouple thermometer) An electric thermometer that employs one or more thermocouples of which the set of measuring junctions is in thermal contact with the body, the temperature of which is to be measured, while the temperature of the reference junctions is either known or otherwise taken into account. *See also:* electric thermometer. (EEC/PE) [119]

thermoelement (electric instruments) The simplest type of thermal converter. It consists of a thermocouple, the measuring junction of which is in thermal contact with an electric heater or integral therewith. *See also:* thermal converter. (EEC/AII) [102]

thermogalvanic corrosion Corrosion resulting from a galvanic cell caused primarily by a thermal gradient. *See also:* electrolytic cell. (IA) [59]

thermographic printer A nonimpact printer that creates images on paper through heat impressions. (C) 610.10-1994w

thermojunction One of the surfaces of contact between the two conductors of a thermocouple. The thermojunction that is in thermal contact with the body under measurement is called the measuring junction, and the other thermojunction is called the reference junction. *See also:* electric thermometer. (EEC/PE) [119]

thermometer An instrument for determining the temperature of a body or space. (SWG/PE) C37.30-1971s

thermometer method of temperature determination **(1) (power and distribution transformers)** The determination of the temperature by mercury, alcohol, resistance, or thermocouple thermometer, any of these instruments being applied to the hottest accessible part of the device. (PE/TR) C57.12.80-1978r
(2) This method consists of the determination of the temperature by thermocouple or suitable thermometer, with either being applied to the hottest accessible part of the equipment. (PE/TR) C57.15-1999

thermophone An electroacoustic transducer in which sound waves of calculable magnitude result from the expansion and contraction of the air adjacent to a conductor whose temperature varies in response to a current input. *Note:* When used for the calibration of pressure microphones, a thermophone is generally used in a cavity the dimensions of which are small compared to a wavelength. *See also:* microphone. (SP) [32]

thermopile A group of thermocouples connected in series aiding. This term is usually applied to a device used either to measure radiant power or energy or as a source of electric energy. *See also:* electric thermometer. (IM/PE/PSIM) 544-1975w, 119-1974w

thermoplastic **(A)** A plastic that is thermoplastic in behavior. **(B)** Having the quality of softening when heated above a certain temperature range and of returning to its original state when cooling below that range. (PE) [9]

thermoplastic insulating tape A tape composed of a thermoplastic compound that provides insulation for joints. (EEC/PE) [119]

thermoplastic insulations and jackets (power distribution, underground cables) Insulations and jackets made of materials that are softened by heat for application to the cable and then become firm, tough and resilient upon cooling. Subsequent heating and cooling will reproduce similar changes in the physical properties of the material. (PE) [4]

thermosphere That part of the Earth's atmosphere located above the mesosphere in which temperature increases and then remains constant with increasing height and from which there is virtually no further escape of particles to free space. The thermosphere extends to an altitude of 500–600 km. (AP/PROP) 211-1997

thermostat A device that responds to temperature and, directly or indirectly, controls temperature in a building. (IA/PSE) 241-1990r

thermostatic switch (thermostat) A form of temperature-operated switch that receives its operating energy by thermal conduction or convection from the device being controlled or operated. *See also:* switch. (IA/IAC) [60]

theta polarization (θ polarization) The state of the wave in which the E vector is tangential to the meridian lines of a given spherical frame of reference. *Note:* The usual frame of reference has the polar axis vertical and the origin at or near the antenna. Under these conditions, a vertical dipole will radiate only theta (θ) polarization and the horizontal loop will radiate only phi (ϕ) polarization. *See also:* antenna. (AP) 149-1979r, [84]

Thevenin's theorem States that the current that will flow through an impedance Z', when connected to any two terminals of a linear network between which there previously existed a voltage E and an impedance Z, is equal to the voltage E divided by the sum of Z and Z'. (EEC/PE) [119]

thickener (hydrometallurgy) (electrometallurgy) A tank in which suspension of solid material can settle so that the solid material emerges from a suitable opening with only a portion of the liquid while the remainder of the liquid overflows in clear condition at another part of the thickener. *See also:* electrowinning. (EEC/PE) [119]

thick film technology A technology in which a thick film (about 1 mil) is screen-printed onto an insulating substrate and then fused to the substrate by firing. *Note:* Resistors, capacitors, and conductors are commonly made by this technology. (CAS) [13]

thimble A print element shaped like a sewing thimble, used for letter quality printing, with type slugs arranged around its perimeter. (C) 610.10-1994w

thin film (1) Loosely, magnetic thin film. (C) [20], [85]
(2) *See also:* magnetic thin film. (C) 610.10-1994w

thin film storage *See:* magnetic thin film storage.

thin film technology A technology in which a thin film (a few hundred to a few thousand angstroms in thickness) is applied by vacuum deposition to an insulating substrate. Resistors, capacitors, and conductors are commonly made by this technology. (CAS) [13]

thin film waveguide (fiber optics) A transparent dielectric film, bounded by lower index materials, capable of guiding light. *See also:* optical waveguide. (Std100) 812-1984w

think time The elapsed time between the end of a prompt or message generated by an interactive system and the beginning of a human user's response. *See also:* response time; turnaround time; port-to-port time. (C) 610.12-1990

thinned array antenna An array antenna that contains substantially fewer driven radiating elements than a conventional uniformly spaced array with the same beamwidth having identical elements. Interelement spacings in the thinned array are chosen such that no large grating lobes are formed and side lobes are minimized. (AP/ANT) 145-1993

thinning An image processing technique in which regions are reduced to sets of thin curves. (C) 610.4-1990w

thin phase screen approximation An approximation in which the cumulative effects of phase distortion take place in an equivalent thin layer and amplitude effects are neglected. (AP/PROP) 211-1997

thin stack A less than fully featured protocol stack. *See also:* short stack. (C) 610.7-1995

thin-wall counter (radiation counters) A counter tube in which part of the envelope is made thin enough to permit the entry of radiation of low penetrating power. *See also:* anticoincidence. (ED) [45]

third generation A period during the evolution of electronic computers in which integrated circuits, core memory technology and miniaturized components replaced transistors and discrete passive components. *Note:* Introduced in 1964, thought to have been the state of the art until the introduction of large scale integration, as is found in many microcomputers. *See also:* second generation; first generation; fourth generation; fifth generation. (C) 610.10-1994w

third generation language *See:* high-order language.

third-level address *See:* n-level address.

third normal form One of the forms used to characterize relations; a relation is said to in third normal form if it is in second normal form and if no nonprime attribute is transitively dependent on the primary key. See the table below.

Third Normal Form

SECOND NORMAL FORM
ORDER2 = ORDER-NO + DATE + CUSTOMER-NO
+CUSTOMER-NAME
+CUSTOMER-ADDRESS
+TOTAL-ORDER-AMOUNT
ORDER-ITEM2 = ORDER-NO + ITEM-NO
+QUANTITY-ORDERED
+EXTENDED-PRICE
ITEM2 = ITEM-NO + ITEM-DESCRIPTION
+UNIT-PRICE
THIRD NORMAL FORM
ORDER3 = ORDER-NO + DATE + CUSTOMER-NO
+TOTAL-ORDER-AMOUNT
CUSTOMER3 = CUSTOMER-NO
+CUSTOMER-NAME
+CUSTOMER-ADDRESS
ORDER-ITEM3 = ORDER-NO + ITEM-NO
+QUANTITY-ORDERED + EXTENDED-PRICE
ITEM3 = ITEM-NO + ITEM-DESCRIPTION
+UNIT-PRICE

In second normal form, nonprime attributes CUSTOMER-NAME and CUSTOMER-ADDRESS are transitively dependent on CUSTOMER-NO. Keys shown in brackets.
See also: Boyce/Codd Normal form. (C) 610.5-1990w

third-order distortion *See:* intermodulation distortion.

third-order nonlinearity coefficient (accelerometer) The proportionality constant that relates a variation of the output to the cube of the input, applied parallel to the input reference axis. (AES/GYAC) 528-1994

third-rail clearance line (railroads) The contour that embraces all cross sections of third rail and its insulators, supports, and guards located at an elevation higher than the top of the running rail. *See also:* electric locomotive. (PE/EEC) [119]

third-rail electric car An electric car that collects propulsion power through a third-rail system. *See also:* electric motor car. (EEC/PE) [119]

third-rail electric locomotive An electric locomotive that collects propulsion power from a third-rail system. *See also:* electric locomotive. (EEC/PE) [119]

third voltage range *See:* voltage range.

32-bit supportive Uses 32-bit addresses when accessing System Memory. (C/MM) 1212.1-1993

Thomson bridge *See:* Kelvin bridge.

Thomson coefficient The quotient of the rate of Thomson heat absorption per unit volume of conductor by the scalar product of the electric current density and the temperature gradient. The Thomson coefficient is positive if Thomson heat is absorbed by the conductor when the component of the electric current density in the direction of the temperature gradient is positive. *See also:* thermoelectric device. (ED) [46]

Thomson effect The absorption or evolution of thermal energy produced by the interaction of an electric current and a temperature gradient in a homogeneous electric conductor. *Notes:* 1. An electromotive force exists between two points in a single conductor that are at different temperatures. The magnitude and direction of the electromotive force depend on the material of the conductor. A consequence of this effect is that if a current exists in a conductor between two points at different temperatures, heat will be absorbed or liberated depending on the material and on the sense of the current. 2. In a nonhomogeneous conductor, the Peltier effect and the Thomson effect cannot be separated. *See also:* thermoelectric device. (ED) [46]

Thomson heat The thermal energy absorbed or evolved as a result of the Thomson effect. *See also:* thermoelectric device. (ED) [46]

thrashing A state in which a computer sys-tem is expending most or all of its resources on overhead operations, such as swapping data between main and auxiliary storage, rather than on intended computing functions. (C) 610.12-1990

thread (1) (control) A control function that provides for maintained operation of a drive at a preset reduced speed such as for setup purposes. *See also:* electric drive. (IA/ICTL/IAC) [60]

(2) (data management) In a tree, a set of link fields, one in each node, each of which points to the successor or predecessor of that node with respect to a particular traversal order. (C) 610.5-1990w

(3) A single sequential flow of control within a process. (C/PA) 1328.2-1993w, 1326.2-1993w, 1224.2-1993w, 1327.2-1993w, 14252-1996

(4) A single flow of control within a process. Each thread has its own thread ID, scheduling priority and policy, *errno* value, thread-specific key/value bindings, and the required system resources to support a flow of control. Anything whose address may be determined by a thread, including but not limited to static variables, storage obtained via *malloc*(), directly addressable storage obtained through implementation-supplied functions, and automatic variables shall be accessible to all threads in the same process. (C/PA) 9945-1-1996

threaded coupling (rigid steel conduit) An internally threaded steel cylinder for connecting two sections of rigid steel conduit. (EEC/CON) [28]

threaded tree A tree whose nodes contain link fields for one or more threads, allowing nonrecursive traversal of the tree. *See also:* left-threaded tree; triply-threaded tree; doubly-threaded tree; right-threaded tree. (C) 610.5-1990w

thread ID A unique value of type *pthread_t* that identifies each thread during its lifetime in a process. (C/PA) 9945-1-1996

threading line (conductor stringing equipment) A lightweight flexible line, normally manila or synthetic fiber rope, used to lead a conductor through the bullwheels of a tensioner or pulling line through a bull wheel puller. *Synonyms:* threading rope; bull line. (T&D/PE) 524-1992r

threading rope *See:* threading line.

thread list An ordered set of runnable threads that all have the same ordinal value for their priority. The ordering of threads on the list is determined by a scheduling policy or policies. The set of thread lists includes all runnable threads in the system. (C/PA) 9945-1-1996

thread of control A sequence of instructions executed by a conceptual sequential subprogram, independent of any programming language. More than one thread of control may execute concurrently, interleaved on a single processor, or on separate processors. The conceptual threads of control in an Ada application are Ada tasks. They may, but need not, correspond to the POSIX threads defined in POSIX.1. (C) 1003.5-1999

thread-safe A function that may be safely invoked concurrently by multiple threads. Each function defined by this standard is thread-safe unless explicitly stated otherwise. An example is any "pure" function (a function that holds a mutex locked while it is accessing static storage or objects shared among threads). (C/PA) 9945-1-1996

thread-specific data key A process global handle of type *pthread_key_t* that is used for naming thread-specific data. Although the same key value may be used by different threads, the values bound to the key by *pthread_setspecific*() and accessed by *pthread_getspecific*() are maintained on a per-thread basis and persist for the life of the calling thread. (C/PA) 9945-1-1996

threat (1) A potential violation of security. (LM/C) 802.10g-1995, 802.10-1992

(2) Means by which a system may be adversely affected. Threats include both inadvertent and malicious actions. (C/BA) 896.3-1993w

three-address Pertaining to an instruction code in which each instruction has three address parts. Also called triple-address. In a typical three-address instruction the addresses specify the location of two operands and the destination of the result, and the instructions are taken from storage in a preassigned order. *See also:* two-plus-one address. (C) 162-1963w

three-address instruction (1) A computer instruction that contains three address fields. For example, an instruction to add the contents of locations A and B, and place the results in location C. *Contrast:* four-address instruction; one-address instruction; zero-address instruction; two-address instruction. (C) 610.12-1990

(2) An instruction containing three addresses. *Synonym:* triple-address instruction. *See also:* address format. (C) 610.10-1994w

three-bit byte *See:* triplet.

three-conductor bundle *See:* bundle.

three-dimensional graphics The presentation of data on a two-dimensional display surface so that it appears to represent a three-dimensional model, and can be viewed from any position. *Note:* Each coordinate of the model contains a triplet of information; for example, *x*, *y*, and *z* in the Cartesian coordinate system. (C) 610.6-1991w

three-dimensional hardware A graphical display processor that accepts three-dimensional information as input and generates an image directly rather than using a projection transformation. (C) 610.6-1991w

three-dimensional priority The property possessed by a line or surface that is in front of another line or surface from the viewer's perspective. (C) 610.6-1991w

three-dimensional radar (navigation aid terms) A radar capable of producing three-dimensional position data on a multiplicity of targets. (AES/GCS) 686-1997, 172-1983w

3GL *See:* high-order language.

three-input adder *See:* full adder.

three-level address *See:* n-level address.

3-of-9 bar code A variable length, bidirectional, discrete, self-checking, alpha-numeric bar code. Its basic data character set contains 43 characters: 0 to 9, A to Z, −, ., /, +, $, %, and space. Each character is composed of 9 elements: 5 bars and 4 spaces. Three of the nine elements are wide (binary value 1) and six are narrow (binary value 0). A common character (*) is used exclusively for both a start and stop character. (PE/TR) C57.12.35-1996

three-phase ac fields (electric and magnetic fields from ac power lines) Three-phase transmission lines generate a three-phase field whose space components are not in phase. The field at any point can be described by the field ellipse, that is, by the magnitude and direction of the semi-major axis and the magnitude and direction of its semi-major axis. In a three-phase field, the electric field at large distances $\geq$ 15 m away from the outer phases (conductors) can frequently be considered a single-phase field because the minor axis of the electric field ellipse is only a fraction (less than 10%) of the major axis when measured at a height of 1 m. Similar remarks apply for the magnetic field. *See also:* electric field strength. (T&D/PE) 644-1979s

three-phase circuit A combination of circuits energized by alternating electromotive forces that differ in phase by one-third of a cycle (120°). *Note:* In practice, the phases may vary several degrees from the specified angle. (IA/MT) 45-1998

three-phase dry-type air-core reactor Dry-type air core rectors are single phase devices. In a three-phase reactor the single phase reactors are stacked and magnetically coupled. Depending on the application the self inductance may be modified to compensate for mutual coupling effects. (PE/TR) C57.16-1996

three-phase electric locomotive An electric locomotive that collects propulsion power from three phases of an alternating-current distribution system. *See also:* electric locomotive. (EEC/PE) [119]

three-phase enclosure A metallic enclosure containing the buses and/or devices of all phases of a three-phase system. (SUB/PE) C37.122.1-1993

three-phase four-wire system A system of alternating-current supply comprising four conductors, three of which are connected as in a three-phase three-wire system, the fourth being connected to the neutral point of the supply, which may be grounded. *See also:* alternating-current distribution. (T&D/PE) [10]

three-phase seven-wire system A system of alternating-current supply from groups of three single-phase transformers connected in Y so as to obtain a three-phase four-wire grounded-neutral system for lighting and a three-phase three-wire grounded-neutral system of a higher voltage for power, the neutral wire being common to both systems. *See also:* alternating-current distribution. (T&D/PE) [10]

three-phase three-wire system A system of alternating-current supply comprising three conductors between successive pairs of which are maintained alternating differences of potential successively displaced in phase by one-third of a period. *See also:* alternating-current distribution. (T&D/PE) [10]

three-plus-one address Pertaining to a four-address code in which one address part always specifies the location of the next instruction to be interpreted. (C) 162-1963w

three-plus-one address instruction A computer instruction that contains four address fields, the fourth containing the address of the instruction to be executed next. For example, an instruction to add the contents of locations A and B, place the results in location C, then execute the instruction at location D. *Contrast:* four-plus-one address instruction; two-plus-one address instruction; one-plus-one address instruction. (C) 610.12-1990

three-position relay A relay that may be operated to three distinct positions. (EEC/PE) [119]

three-state A type of bus driver. Either drives high, low, or not at all. (C/MM) 1196-1987w

three-state circuit A digital circuit which has three output states: logical one (false), logical zero (true) and a high impedance output to isolate itself from the circuit. (C) 610.10-1994w

three-state indication *See:* supervisory control functions.

3-state pin A component output pin where the drive may be either active or inactive (for example, at high impedance). (TT/C) 1149.1-1990

three-terminal capacitor Two conductors (the active electrodes) insulated from each other and from a surrounding third conductor that constitutes the shield. When the capacitor is provided with properly designed terminals and used with shielded leads, the direct capacitance between the active electrodes is independent of the presence of other conductors. (Specialized usage.) (Std100) 270-1966w

three-tone slope A measure of attenuation distortion at 404 Hz and 2804 Hz relative to loss at 1020Hz (or 1004 Hz). Three-tone slope is specified for many telecommunication services. (COM/TA) 743-1995

three-wire control A control function that utilizes a momentary-contact pilot device and a holding-circuit contact to provide undervoltage protection. *See also:* undervoltage protection; relay. (IA/ICTL/IAC) [60]

three-wire system (direct current or single-phase alternating current) A system of electric supply comprising three conductors, one of which (known as the neutral wire) is maintained at a potential midway between the potential of the other two (referred to as the outer conductors). *Note:* Part of the load may be connected directly between the outer conductors, the remainder being divided as evenly as possible into two parts each of which is connected between the neutral and one outer conductor. There are thus two distinct voltages of supply, the one being twice the other. *See also:* alternating-current distribution; direct-current distribution. (T&D/PE) [10]

three-wire type current transformer (1) (power and distribution transformers) One which has two primary windings each completely insulated for the rated insulation level of the transformer. This type of current transformer is for use on a three-wire single-phase service. *Note:* The primary windings and secondary windings are permanently assembled on the core as an integral structure. The secondary current is proportional to the phasor sum of the primary currents. (PE/TR) C57.12.80-1978r

(2) One that has two insulated primary windings and one secondary winding and is for use on a three-wire, single-phase service. *Note:* The primary windings and the secondary winding are permanently assembled on the core as an integral structure. The secondary current is proportional to the phasor sum of the primary currents. (PE/TR) C57.13-1993

threshold (1) (A) (mathematics of computing) A logic operator having the property that if P is a statement, Q is a statement, R is a statement, . . ., then the threshold of P, Q, R, . . . is true if at least *N* statements are true, false if less than *N* statements are true, where *N* is a specified non-negative integer called the threshold condition. **(B) (mathematics of computing)** The threshold condition as in definition (A). (C) 1084-1986

(2) (image processing) A specified gray level used for producing a binary image. *See also:* thresholding. (C) 610.4-1990w

(3) (illuminating engineering) The value of a variable of a physical stimulus (such as size, luminance, contrast or time) which permits the stimulus to be seen a specific percentage of the time or at a specific accuracy level. In many psychophysical experiments, thresholds are presented in terms of 50% accuracy or accurately 50% of the time. However, the threshold also is expressed as the value of the physical variable which permits the object to be just barely seen. The threshold may be determined by merely detecting the presence of an object or it may be determined by discriminating certain details of the object. (EEC/IE) [126]

(4) (of a maser or laser) The condition of a maser or laser wherein the gain of its medium is just sufficient to permit the start of oscillation. (LEO) 586-1980w

(5) (accelerometer) (gyros) The largest absolute value of the minimum input that produces an output equal to at least 50% of the output expected using the nominal scale factor. (AES/GYAC) 528-1994

(6) A value of voltage or other measure that a signal must exceed in order to be detected or retained for further processing. (AES) 686-1997

threshold audiogram *See:* audiogram.

threshold center voltage The algebraic average of the (HC) and (LC) threshold voltages; that is, $(V_{LC} + V_{HC})/2$.

threshold current (1) (protection and coordination of industrial and commercial power systems) The magnitude of

current at which a fuse becomes current limiting, specifically the symmetrical root-mean-square (rms) available current at the threshold of the current-limiting range, where the fuse total clearing time is less than half-cycle at rated voltage and rated frequency, for symmetrical closing, and a power factor of less than 20%. Refer to various peak let-through current curves for each type of fuse. The threshold ratio is the relationship of the threshold current to the fuse's continuous-current rating. (IA/PSP) 242-1986r
(2) (fiber optics) The driving current corresponding to lasing threshold. *See also:* lasing threshold. (Std100) 812-1984w
(3) (of a current-limiting fuse) A current magnitude of specified wave shape at which the melting of the current-responsive element occurs at the first instantaneous peak current for that wave shape. *Note:* The current magnitude is usually expressed in rms amperes. (SWG/PE) C37.100-1992

threshold element (1) (A) A combinational logic element such that the output channel is in its one state if and only if at least n input channels are in their one states, where n is a specified fixed nonnegative integer, called the threshold of the element. **(B)** By extension, a similar element whose output channel is in its one state if and only if at least n input channels are in states specified for them, not necessarily the one state but a fixed state for each input channel. (C) 162-1963
(2) A device that performs the logic threshold operation but in which the truth of each input statement contributes to the output determination a weight associated with that statement. (C) [20], [85]
(3) *See also:* threshold gate. (C) 610.10-1994w

threshold field The least magnetizing force in a direction that tends to decrease the remanence, that, when applied either as a steady field of long duration or as a pulsed field appearing many times, will cause a stated fractional change of remanence. (C) [20]

threshold frequency (photoelectric tubes) (photoelectric device) The frequency of incident radiant energy below which there is no photoemissive effect. *See also:* photoelectric effect. (ED) [45], [84]

threshold function A two-value switching function of one or more not necessarily Boolean arguments that take the value 1 if a specified mathematical function of the arguments exceeds a given threshold value, and zero otherwise. *See also:* threshold operation. (C) 610.10-1994w

threshold gate A combinational circuit that performs a threshold operation. *Synonym:* threshold element. (C) 610.10-1994w

thresholding The process of producing a binary image from a gray scale image by assigning each output pixel the value 1 if its corresponding input pixel is at or above a specified gray level (the threshold) and the value 0 if the input pixel is below that threshold. (C) 610.4-1990w

threshold level (L_{TH}) The minimum signal level necessary for removing insertion loss on a handsfree telephone (HFT). (COM/TA) 1329-1999

threshold lights (illuminating engineering) Runway lights so placed as to indicate the longitudinal limits of that portion of a runway, channel, or landing path usable for landing. (EEC/IE) [126]

threshold limit value—short term exposure limit (TLV-STEL) as defined by the American Conference of Governmental Industrial Hygienists: The maximum concentration to which workers can be exposed for a period of up to 15 min continuously without suffering adverse effects, or materially reduced work efficiency, provided that no more than four 15 min excursions per day are permitted with at least 60 min between exposure periods, and provided that the TLV-TWA is not exceeded. (In most jurisdictions in North America, the TLVs are legislated limits to exposure.) (SUB/PE) C37.122.1-1993

threshold limit value—time weighted average (TLV-TWA) as defined by the American Conference of Governmental Industrial Hygienists: The time-weighted average concentration for a normal 8h work day and 40 h work week to which nearly all workers may be exposed repeatedly, day after day, without adverse effect. (SUB/PE) C37.122.1-1993

threshold of audibility (specified signal) (threshold of detectability) The minimum effective sound pressure level of the signal that is capable of evoking an auditory sensation in a specified fraction of the trials. The characteristics of the signal, the manner in which it is presented to the listener, and the point at which the sound pressure is measured must be specified. *Notes:* 1. Unless otherwise specified, the ambient noise reaching the ears is assumed to be negligible. 2. The threshold is usually given as a sound pressure level in decibels relative to 20 micronewtons per square meter. 3. Instead of the method of constant stimuli, which is implied by the phrase in a specified fraction of the trials, another psychophysical method (which should be specified) may be employed. (SP) [32]

threshold of discomfort (audio and electroacoustics) (for a specified signal) The minimum effective sound pressure level at the entrance to the external auditory canal that, in a specified fraction of the trials, will stimulate the ear to a point at which the sensation of feeling becomes uncomfortable. (SP) [32]

threshold of feeling (audio and electroacoustics) (tickle) (for a specified signal) The minimum effective sound pressure level at the entrance to the external auditory canal that, in a specified fraction of the trials, will stimulate the ear to a point at which there is a sensation of feeling that is different from the sensation of hearing. (SP) [32]

threshold of pain (audio and electroacoustics) (for a specified signal) The minimum effective sound pressure level at the entrance to the external auditory canal that, in a specified fraction of the trials, will stimulate the ear to a point at which the discomfort gives way to definite pain that is distinct from the mere nonnoxious feeling of discomfort. (SP) [32]

threshold of perception The level of stimulation at which 50% of the population is just able to consciously detect the presence of the stimulus. (T&D/PE) 539-1990

threshold operation An operation that evaluates the threshold function of the operands. *See also:* majority operation. (C) 610.10-1994w

threshold ratio (of a current-limiting fuse) The ratio of the threshold current to the fuse current rating. (SWG/PE) C37.100-1992

threshold sensitivity (test, measurement, and diagnostic equipment) The smallest quantity that can be detected by a measuring instrument or automatic control system. (MIL) [2]

threshold signal (navigation aid terms) (navigation) The smallest signal capable of effecting a recognizable change in navigational information. (AES/GCS) 172-1983w

threshold signal-to-interference ratio (TSI) The minimum signal to interference power, described in a prescribed way, required to provide a specified performance level. *See also:* electromagnetic compatibility. (EMC) [53]

threshold voltage (1) (metal-nitride-oxide field-effect transistor) Minimum gate voltage necessary for onset of current flow between source and drain of an insulated-gate field-effect transistor (IGFET). This is a serviceable general definition. There are three more specific definitions possible.

(1) $V_T = V_{GS}$ at $I_{DS} = 10\ \mu A$, when $V_{GS} = V_{DS}$

that is, V_T is the gate voltage necessary to result in a defined low current level;

(2) $V_T = V_{GS} - \left[\frac{I_{DS}\ell}{k'W}\right]^{1/2}$

that is, V_T is that value that gives the best fit to the I–V relationship of the IGFET;

(3) $V_T = \phi_{ms} + \phi_s - \frac{Q_B}{C_G} - \frac{Q_I}{C_G}$

that is, V_T is derived from inherent structural parameters only. *Note:* Equation (1) is commonly used in practice. While normally I_{DS} is set at 10 μA, a value of I_{DS} independent of lateral geometry is I_{DS} 0.5 (W/l) μA. Equation (3) suffers from the fact that in packaged devices, none of the terms can be varified by independent measurement. *See also:* insulated-gate field-effect transistor symbols. (ED) 581-1978w

(2) (semiconductor rectifiers) The zero-current−voltage intercept of a straight-line approximation of the forward current-voltage characteristic over the normal operating range. (IA) [62]

(3) The minimum voltage considered to be a high state or the maximum voltage considered to be a low state. (SCC20) 1445-1998

threshold voltage saturation (metal-nitride-oxide field-effect transistor) For a given gate-to-source voltage, the transistor metal-nitride-oxide semiconductor (MNOS) threshold voltage achieved in either of the two written states for which an order of magnitude increase in pulse width causes less than a 100-mV change in threshold voltage. Thus pulse width can also be achieved by sequence of shorter pulses of the same polarity. (ED) 581-1978w

threshold voltage window (metal-nitride-oxide field-effect transistor) The algebraic difference between the two threshold voltages (the threshold voltage after a write-high operation minus the threshold voltage after a write-low operation). (ED) 581-1978w

threshold wavelength (photoelectric tubes) (photoelectric device) The wavelength of the incident radiant energy above which there is no photoemission effect. *See also:* photoelectric effect. (ED) [45], [84]

throat microphone A microphone normally actuated by mechanical contact with the throat. *See also:* microphone. (EEC/PE) [119]

through-board This refers to the implementation of dual-in-line packages, axial-leaded devices, etc., into plated through-holes in a PWB. *Synonym:* through-hole. (C/BA) 1101.3-1993

through bolt (rotating machinery) A bolt passing axially through a laminated core, that is used to apply pressure to the end plates. (PE) [9]

through-hole *See:* through-board.

through loss *See:* insertion loss.

throughput (1) (data transmission) The total capability of equipment to process or transmit data during a specified time period. (PE) 599-1985w

(2) (software) The amount of work that can be performed by a computer system or component in a given period of time; for example, number of jobs per day. *See also:* turnaround time; workload model. (C) 610.12-1990

(3) (automatic control) *See also:* capacity.

through supervision (1) (communication switching) The automatic transfer of supervisory signals through one or more trunks in a manual telephone switchboard. (COM) [48]

(2) (telephone switching systems) The capability of apparatus within a switched connection to pass or repeat signaling. (COM) 312-1977w

throwing power (electroplating) (of a solution) A measure of its adaptability to deposit metal uniformly upon a cathode of irregular shape. In a given solution under specified conditions it is equal to the improvement (in percent) of the metal distribution ratio above the primary-current distribution ratio. *See also:* electroplating. (EEC/PE) [119]

throw-over equipment *See:* automatic transfer equipment.

thrust bearing (rotating machinery) A bearing designed to carry an axial load so as to prevent or to limit axial movement of a shaft, or to carry the weight of a vertical rotor system. *See also:* bearing. (PE) [9]

thrust block (rotating machinery) A support for a thrust-bearing runner. *See also:* bearing. (PE) [9]

thrust collar (rotating machinery) The part of a shaft or rotor that contacts the thrust bearing and transmits the axial load. *See also:* rotor. (PE) [9]

thumbwheel An input device consisting of a dial or wheel, inset into a surface so that only a portion of its rim protrudes, that can be moved with one degree of freedom to provide coordinate input data. *Note:* It is usually used in pairs to control the display of crosshairs on a display surface. (C) 610.6-1991w, 610.10-1994w

thump A low-frequency transient disturbance in a system or transducer characterized audibly by the onomatopoeic connotation of the word. *Note:* In telephony, thump is the noise in a receiver connected to a telephone circuit on which a direct-current telegraph channel is superposed caused by the telegraph currents. *See also:* signal-to-noise ratio. (SP) 151-1965w

thunder The sound that follows a flash of lightning and is caused by the sudden expansion of the air in the path of electrical discharge. (SUB/PE) 998-1996

thunderstorm day (1) A day during which thunder is heard at least once at a specified observation point. (PE/PSC) 487-1992

(2) A day on which thunder can be heard, and hence when lightning occurs. (SUB/PE) 998-1996

thunderstorm hour An hour during which thunder can be heard, and hence when lightning occurs. (SUB/PE) 998-1996

Thury transmission system A system of direct-current transmission with constant current and a variable high voltage. *Note:* High voltage used on this system is obtained by connecting series direct-current generators in series at the generating station: and is utilized by connecting series direct-current motors in series at the substations. *See also:* direct-current distribution. (T&D/PE) [10]

thyratron A hot-cathode gas tube in which one or more control electrodes initiate but do not limit the anode current except under certain operating conditions. (ED) [45]

thyristor (1) (thyristor ac power controllers) A bistable semiconductor device comprising three or more junctions that can be switched from the OFF state to the ON state or vice versa, such switching occurring within at least one quadrant of the principal voltage-current characteristic. (IA/IPC) 428-1981w

(2) (electrical heating applications to melting furnaces and forehearths in the glass industry) A bistable semiconductor device comprising three or more junctions that can be switched from an off (nonconducting) to an on (conducting) condition, or vice versa, by the application of a small electric signal. Such switching occurs within at least one quadrant of the principal voltage-current characteristic. (IA) 668-1987w

thyristor ac power controller A power electronic equipment for the control or switching of ac power where switching, multicycle control and phase control are included. The only power controlling element is the thyristor, although other power elements may be included. (IA/IPC) 428-1981w

thyristor assembly An electrical and mechanical functional assembly of thyristors in combination with diodes, if any, or thyristor stacks, complete with all its connections and auxiliary components, including trigger equipment, together with means for cooling, if any, in its own mechanical structure, but without the controller transformers and other switching devices. *Note:* A thyristor assembly may be combined of several subassemblies, which are made and traded as mechanically combined units, for example, thyristor stacks or other combinations or one or more thyristors with control devices, protective devices, etc. (IA/IPC) 428-1981w

thyristor-controlled reactor (TCR) (1) A series connection thyristor controller, typically connected between two halves of a reactor, that forms one leg of the connected circuit. The thyristor controller consists of antiparallel phase angle con-

trolled thyristors for vernier control of the reactor susceptance (current). (SUB/PE) 1303-1994
(2) A shunt-connected thyristor-controlled inductor whose effective reactance is varied in a continuous manner by partial conduction of the thyristor valve. (SUB/PE) 1031-2000
(3) The effective value of the reactor is changed by using thyristors to control the flow of current by phase-controlling the turn-on signal to the thyristors.power systems relaying. (PE) C37.99-2000

thyristor converter (thyristor converter unit, thyristor converter equipment) An operative unit comprising one or more thyristor sections together with converter transformers, essential switching devices, and other auxiliaries, if any of these items exist. System control equipments are optionally included. (IA/IPC) 444-1973w

thyristor converter, bridge *See:* bridge thyristor converter.

thyristor converter, cascade *See:* cascade thyristor converter.

thyristor converter circuit element A group of one or more thyristors, connected in series or parallel or any combination of both, bounded by no more than two circuit terminals and conducting forward current in the same direction between these terminals. *Note:* A circuit element is also referred to as a leg or arm, and in the case of paralleled thyristors each path is referred to as a branch. (IA/IPC) 444-1973w

thyristor converter, multiple *See:* multiple thyristor converter.

thyristor converter, parallel *See:* parallel thyristor converter.

thyristor converter, series *See:* series thyristor converter.

thyristor converter, simple *See:* simple thyristor converter.

thyristor converter, single-way *See:* single-way thyristor converter.

thyristor converter transformer A transformer that operates at the fundamental frequency of the ac system and is designed to have one or more output windings conductively connected to the thyristor converter elements. (IA/IPC) 444-1973w

thyristor converter unit, double *See:* double thyristor converter unit.

thyristor converter unit, single *See:* single thyristor converter unit.

thyristor fuses Fuses of special characteristics connected in series with one or more thyristors to protect the thyristor or other circuit components, or both. (IA/IPC) 444-1973w

thyristor-level A single thyristor, or thyristors if the valve has parallel connected thyristors, and associated components for control, voltage grading, protection, and monitoring that constitute a single voltage level within the valve. (SUB/PE) 857-1996

thyristor, reverse-blocking triode *See:* reverse-blocking triode thyristor.

thyristor stack A single structure of one or more thyristors with its (their) associated mounting(s), cooling attachments, if any, connections whether electrical or mechanical, and auxiliary components, if any. A thyristor stack may consist of thyristors and semiconductor rectifier diodes in combination and in this case it may be referred to as a non-uniform thyristor stack. Trigger equipments are not included in this definition. (IA/IPC) 428-1981w

thyristor-switched capacitor (TSC) **(1)** A series connection thyristor switch, typically connected between a capacitor bank and a current limiting reactor, that forms one leg of the connected circuit. The thyristor switch consists of antiparallel thyristors that are blocked or fired for full conduction (on/off control). (SUB/PE) 1303-1994
(2) A shunt-connected thyristor-switched capacitor whose effective reactance is varied in a stepwise manner by full- or zero-conduction operation of the thyristor valve. (PE/SUB) 1031-2000
(3) **(power systems relaying)** A capacitor switched on and off by thyristor control action. (PE) C37.99-2000

thyristor-switched reactor (TSR) **(1)** A series connection thyristor switch, typically connected between two halves of a reactor, that forms one leg of the connected circuit. The thyristor switch consists of antiparallel thyristors that are blocked or fired for full conduction (on/off control). (SUB/PE) 1303-1994
(2) A shunt-connected thyristor-switched inductor whose effective reactance is varied in a stepwise manner by full- or zero-conduction operation of the thyristor valve. (PE/SUB) 1031-2000

thyristor trigger circuit A circuit for the conversion of a control signal to suitable trigger signals for the thyristors in a thyristor ac power controller including phase shifting circuits, pulse generating circuits, and power supply circuits. (IA/IPC) 428-1981w

TIA *See:* Telecommunications Industries Association.

tick A time interval that is equal to the transmission of one RamLink data byte. (C/MM) 1596.4-1996

ticker A form of receiving-only printer used in the dissemination of information such as stock quotations and news. *See also:* telegraphy. (COM) [49]

tick-pair A time interval that is equal to the transmission time of two RamLink data bytes. (C/MM) 1596.4-1996

tie (rotating machinery) A binding of the end turns used to hold a winding in place or to hold leads to windings for purpose of anchoring. *See also:* stator; rotor. (PE) [9]

tie feeder A feeder that connects together two or more independent sources of power and has no tapped load between the terminals. *Note:* If a feeder has any tapped load between the two sources, it is designated as a multiple feeder. (SWG/PE) C37.100-1992

tie line (electric power system) A transmission line connecting two power systems. (PE/PSE) 94-1991w

tie-line bias control A mode of load-frequency control in which the area control error is a function of the net interchange error and frequency-related biases. (PE/PSE) 858-1993w, 94-1991nw

tie point (electric power system) The location of the switching facilities of a tie line. (PE/PSE) 94-1991w

tier (rotating machinery) A concentric winding is said to have one, two or more tiers according to whether the periphal extremities of the end windings of groups of coils at each end of the machine form one, two or more solids of revolution around the axis of the machine. (PE) [9]

tier chart *See:* call graph.

tie trunk (data transmission) A telephone line or channel directly connecting two private branch exchanges. (COM/PE) 312-1977w, 599-1985w

tie wire A short piece of wire used to bind an overhead conductor to an insulator or other support. *See also:* conductor; tower. (T&D/PE) [10]

TIF *See:* telephone influence factor.

tight **(1)** **(packaging machinery)** Used as a suffix, indicating that apparatus is so constructed that the enclosing case will exclude the specified material. (IA/PKG) 333-1980w
(2) **(power and distribution transformers)** (used as a suffix) Apparatus is designed as watertight, dusttight, etc, when so constructed that the enclosing case will exclude the specified material under specified conditions. (PE/TR) C57.12.80-1978r
(3) (used as a suffix) So constructed that the specific material is excluded under specified conditions. (SWG/PE) C37.100-1992, C37.40-1993

tight coupling *See:* close coupling.

tightly coupled A condition that exists when simulation entities are involved in very close interaction such that every action of an entity must be immediately accounted for by the other entities. Several tanks in close formation involving rapid, complicated maneuvers over the terrain is an example of a tightly coupled situation. (DIS/C) 1278.2-1995

tile The pixmap used repetitively to pattern an area on the screen. (C) 1295-1993w

tilde **(1)** The character "~". (C/PA) 9945-2-1993
(2) Negation. (C/BA) 14536-1995

tilt (1) (navigation aids) (directional antenna) The angle that the antenna axis forms with the horizontal. (AES/GCS) 172-1983w

(2) (pulse transformers) (A_D) (droop) The difference between A_M and A_T. It is expressed in amplitude units or in percentage of A_M. *See also:* input pulse shape. (PEL/ET) 390-1987r

(3) (pulse terminology) A distortion of a pulse top or pulse base wherein the overall slope over the extent of the pulse top or the pulse base is essentially constant and other than zero. Tilt may be of either polarity. *See also:* preshoot. (IM/WM&A) 194-1977w

(4) (broadband local area networks) The relative level of multiplexed carriers with respect to a designated reference carrier. The gross difference in level between signals at the upper and lower frequency of the bandwidth of interest. *See also:* cable tilt. (LM/C) 802.7-1989r

tilt angle (1) (navigation aids) The vertical angle between the axis of measurement and a reference axis; the reference is normally horizontal. (AES/GCS) 172-1983w

(2) (of a polarization ellipse) When the plane of polarization is viewed from a specified side, the angle measured clockwise from a reference line to the major axis of the ellipse. *Notes:* 1. For a plane wave, the plane of polarization shall be viewed looking in the direction of propagation. 2. The tilt angle is only defined up to a multiple of π radians and is usually taken in the range $(-\pi/2, +\pi/2)$ or $(0, \pi)$. (AP/ANT) 145-1993

(3) (of polarization) Angle of major axis of the polarization ellipse relative to horizontal. (AP/PROP) 211-1997

tilted bushing A bushing intended to be mounted at an angle of 20° to 70° from the vertical. (PE/TR) C57.19.03-1996

tilt error *See:* ionospheric tilt error.

tilting-insulator switch One in which the opening and closing travel of the blade is accomplished by a tilting movement of one or more of the insulators supporting the conducting parts of the switch. (SWG/PE) C37.100-1992

tilting-pad bearing (rotating machinery) (kingsbury bearing) A pad-type bearing in which the pads are capable of moving in such a manner as to improve the flow of lubricating fluid between the bearing and the shaft journal or collar (runner). *See also:* bearing. (PE) [9]

timbering machine An electrically driven machine to raise and hold timbers in place while supporting posts are being set after being cut to length by the machine's power-driven saw. (EEC/PE) [119]

timbre The attribute of auditory sensation in terms of which a listener can judge that two sounds similarly presented and having the same loudness and pitch are dissimilar. *Note:* Timbre depends primarily upon the spectrum of the stimulus, but it also depends upon the waveform, the sound pressure, the frequency location of the spectrum, and the temporal characteristics of the stimulus. (SP) [32]

time (1) Any duration of observations of the considered items either in actual operation or in storage, readiness, etc., but excluding down time due to a failure. *Note:* In definitions where time is used, this parameter may be replaced by distance, cycles, or other measures of life as may be appropriate. This refers to terms such as acceleration factor, wear-out failure, failure rate, mean life, mean time between failures, mean time to failure, reliability, and useful life. *See also:* reliability. (R) [29]

(2) (International System of Units (SI)) The SI unit of time is the second. This unit is preferred and should be used if practical, particularly when technical calculations are involved. In cases where time relates to life customs or calendar cycles, the minute, hour, day, and other calendar units may be necessary. For example, vehicle velocity will normally be expressed in kilometers per hour. *See also:* units and letter symbols. (QUL) 268-1982s

(3) (A) The measured or measurable period during which an action, process, or condition exists or continues. **(B)** The instant at which an event occurs. (C) 610.10-1994

(4) (electronic computation) *See also:* reference time; access time; switching time; word time; real time. (QUL) 268-1982s

time above 90 percent (switching impulse testing) The time interval T_d during which the switching impulse exceeds 90% of its crest value. 332-1972w

time-and-charge-request call (telephone switching systems) A call for which a request is made to be informed of its duration and cost upon its completion. (COM) 312-1977w

time assigned speech interpolation (TASI) The sending of two or more voice calls on the same telephone circuit simultaneously by interleaving the active signals of one conversation with the periods of silence of other conversations. (C) 610.7-1995

time-averaged-clutter coherent airborne radar (TACCAR) An airborne moving-target indication (MTI) radar that uses a technique to compensate for the changing Doppler frequency from fixed clutter due to the motion of the aircraft (or other vehicle) carrying the radar or as the moving radar antenna scans in angle. The clutter is sampled over some range interval, the average Doppler frequency of the clutter in the range interval is used to set the frequency of a voltage-controlled oscillator (VCO) in a phase-lock loop to cause the mean Doppler frequency of the clutter echo to coincide with a null of the MTI Doppler frequency response over the range of observation. *Note:* Also known as clutter-locked MTI. (AES) 686-1997

time-averaged Poynting vector ($\vec{S}$) Of a periodic electromagnetic wave, the time average of the instantaneous Poynting vector over the wave period. For time harmonic waves, it is equal to:

$$(1/2)\mathrm{Re}(\vec{E} \times \vec{H}^*)$$

where
Re indicates the real part
$\vec{E}$ = the electric field vector in phasor notation
$\vec{H}$ = the magnetic field vector in phasor notation
* indicates the complex conjugate

(AP/PROP) 211-1997

time-bandwidth product The product of the device time delay and the chirp bandwidth. (UFFC) 1037-1992w

time base (1) (oscilloscopes) The sweep generator in an oscilloscope. *See also:* oscillograph. (IM/HFIM) [40]

(2) A stable, periodic signal, usually a square wave, used to synchronize and to provide power to circuits. (C) 610.10-1994w

time base primary The principal means of establishing timing relationships. (ELM) C12.15-1990

time base—primary (watthour meters) A timing system established from the power line-source. (ELM) C12.13-1985s

time base—secondary (watthour meters) A timing system established from an alternate source when the line source is not available or not used. (ELM) C12.13-1985s

time bias (electric power system) An offset in the scheduled net interchange power of a control area that varies in proportion to the time deviation. This offset is in a direction to assist in restoring the time deviation to zero. *See also:* power system. (PE/PSE) 94-1970w, [54]

time bias setting (electric power system) A coefficient that, when multiplied by time error, yields time error bias. (PE/PSE) [54], 94-1991w

time, build-up *See:* build-up time.

time characteristic *See:* demand meter—time characteristic.

time coherence *See:* coherent.

time compression multiplexing (TCM) A multiplexing technique that provides full-duplex digital data transmission over a single twisted pair. *Synonym:* ping-pong transmission technique. (C) 610.7-1995

time-consistent busy hour (TCBH) The hour having the highest average traffic for the three highest traffic months. A "busy hour" determination study uses only about two weeks worth of hour-by-hour data collected just in advance of the expected high-traffic months. *Synonym:* busy hour. *See also:* traffic engineering limits; time-consistent traffic measures. (COM/TA) 973-1990w

time-consistent traffic measures Three time frames in common use for PTS today are termed as follows: average busy season, 10 high day, and high day, each with busy hour appended (but sometimes omitted as understood). These will be defined in terms of load volumes and, implicitly, there is always a corresponding service criterion. In the definitions of average busy season busy-hour (ABSBH) load, ten high day busy-hour (THDBH) load, and high day busy-hour (HDBH) load, "hour" refers to 60 contiguous minutes starting at a clock hour or half-hour, and "month" refers to a service observing month, which is not generally a calendar month. Traffic intensity or event data, typically expressed for local switching systems as CCS or calls or call attempts per hour, is collected and processed to determine the candidate busy hours for each of the switching system components to be engineered. The hour having the highest average traffic for the three highest traffic months is defined as the "busy hour" or "time consistent busy hour." However, a "busy hour" determination study uses only about two weeks worth of hour-by-hour data collected just in advance of the expected high-traffic months. *Synonym:* TC traffic measures. *See also:* average busy season busy-hour load; high day busy-hour load; ten high day busy-hour load. (COM/TA) 973-1990w

time constant (1) (electrothermic unit) The time required for the dc electrothermic output voltage to reach $1 - (1/e)$, or 63% of its final value after a fixed amount of power is applied to the electrothermic unit. (IM) 544-1975w
(2) (excitation systems) The value T in an exponential response term $A^{-t/T}$ or in one of the transform factors $1 + sT$, $1 + j\omega T$, $1/(1 + sT)$, $1/(1 + j\omega T)$. *Note:* For the output of a first-order (lag or lead) system forced by a step or an impulse, T is the time required to complete 63.2% of the total rise or decay; at any instant during the process, T is the quotient of the instantaneous rate of change divided into the change still to be completed. In higher order systems, there is a time constant for each of the first-order components of the process. In a Bode diagram, breakpoints occur at $\omega = 1/T$. (PE/EDPG) 421A-1978s
(3) The time required for a field probe output to reach a stable, repeatable reading. The measurement includes test set up, metering unit, cables, etc., thus is a worst case. The measurements assume an exponential response of the field probe. The time constant is used specifically for burst peak field strength measurements. (EMC) 1309-1996

time constant, derivative action (automatic control) A parameter whose value is equal to $1/2\pi f_a$ where f_a is the frequency (cycles.unit time) on a Bode diagram of the lowest frequency gain corner resulting from derivative control action. (PE/EDPG) [3]

time constant of an exponential function $1/b$, if t represents time and b is real. (Std100) 270-1966w

time constant of fall (data transmission) (pulse) The time required for the pulse to fall from 70.7% to 26.0% of its maximum amplitude excluding spike. (PE) 599-1985w

time constant of integrator (for each input) The ratio of the input to the corresponding time rate of change of the output. *See also:* electronic analog computer. (C) 165-1977w

time constant of rise (data transmission) (pulse) The time required for the pulse to rise from 26.0% to 70.7% and of its maximum amplitude excluding spike. (PE) 599-1985w

time constant of the damping device (hydraulic turbines) A time constant which describes the decay of the output signal from the damping device. (PE/EDPG) 125-1977s

time correction (A) (manual) A change made to the system frequency in order to correct for system time error. **(B) (automatic)** A component of the area control error formula that adds a bias to correct interconnection time error. (PE/PSE) 858-1993

time critical Applications where the communication delay is bound to a fixed upper limit, independent of the load conditions. (VT) 1473-1999

time-current characteristic *See:* fuse time-current characteristic.

time-current tests *See:* fuse time-current tests.

timed acceleration A control function that accelerates the drive by automatically controlling the speed change as a function of time. *See also:* electric drive. (IA/ICTL/APP/IAC) [69], [60]

timed deceleration A control function that decelerates the drive by automatically controlling the speed change as a function time. *See also:* feedback control system. (IA/ICTL/APP/IAC) [69], [60]

time delay (1) (analog computer) The time interval between the manifestation of a signal at one point and the manifestation or detection of the same signal at another point. *Notes:* 1. Generally, the term time delay is used to describe a process whereby an output signal has the same form as an input signal causing it, but is delayed in time; that is, the amplification of all frequency components of the output are related by a single constant to those of corresponding input frequency components but each output component lags behind the corresponding input component by a phase angle proportional to the frequency of the component. 2. Transport delay is synonymous with time delay but usually is reserved for applications that involve the flow of material. (C) 165-1977w
(2) (protection and coordination of industrial and commercial power systems) Meaningless unless defined. This term is now used by National Electrical Manufacturers Association (NEMA), American National Standards Institute (ANSI), and Underwriters Laboratories (UL) to mean, in Classes H, K, J, and R cartridge fuses, a minimum opening time of 10 s on an overload current five times the ampere rating of the fuse. Such a delay is particularly useful in allowing the fuse to pass the momentary starting current of a motor, yet not hindering the opening of the fuse should the overload persist. In Class G, CC, and plug fuses, the phrase "time delay" is required by UL to be a minimum opening time of 12 s on an overload of twice the fuse's ampere rating. The time-delay characteristic does not affect the fuse's short-circuit current clearing ability. (IA/PSP) 242-1986r
(3) The time interval between the manifestation of a signal at one point and the manifestation or detection of the same signal at another point. *Synonym:* transport delay. *See also:* propagation delay. (C) 610.7-1995, 610.10-1994w
(4) A time interval purposely introduced in the performance of a function. *See also:* feedback control system. (C/IA/IAC) 610.10-1994w, [60]

time delay register *See:* delay-line storage.

time-delay relay *See:* delay relay; relay.

time delay spread ($\sigma\tau$) A measure of the differential propagation times due to multipath propagation. Specifically, time delay spread is the rms width of the signal received when a very narrow pulse has been transmitted. *Note:* The time delay spread is inversely proportional to the frequency selective bandwidth (f_τ):

$$\sigma_\tau = (2\pi f_\tau)^{-1}$$

(AP/PROP) 211-1997

time-delay starting or closing relay (power system device function numbers) A device that functions to give a desired amount of time delay before or after any point of operation in a switching sequence or protective relay system, except as specifically provided by incomplete sequence relay, time-delay stopping or opening relay, and alternating current (ac) reclosing relay, device functions 48, 62, and 79. (PE/SUB) C37.2-1979s

time-delay stopping or opening relay (power system device function numbers) A time-delay relay that serves in conjunction with the device that initiates the shutdown, stopping, or opening operating in an automatic sequence or protective relay system. (SUB/PE) C37.2-1979s

time-dependent event An event that occurs at a predetermined point in time or after a predetermined period of time has elapsed. *See also:* conditional event. (C) 610.3-1989w

time derived channel A channel that is obtained from multiplexing a channel by time division. (C) 610.7-1995

time deviation ($x(t)$) (1) (power system) The integrated or accumulated difference in cycles between system frequency and rated frequency. This is usually expressed in seconds by dividing the deviation in cycles by the rated frequency. (PE/PSE) 94-1970w
(2) Instantaneous time departure from a nominal time. (SCC27) 1139-1999

time dial (1) The control that determines the value of the integral at which the trip output is actuated and hence controls the time scale of the time-current characteristic produced by the relay. In the induction type relay, the time dial sets the distance the disk must travel which is the integral of the velocity with respect to time. (PE/PSR) C37.112-1996
(2) (time lever) (of a relay) An adjustable, graduated element of a relay by which, under fixed input conditions, the prescribed relay operating time can be varied. (SWG/PE) C37.100-1992

time difference (navigation aids) (loran) The difference in the time of reception of the two signals of a loran rate. (AES/GCS) 172-1983w

time-difference-of-arrival *See:* time-of-arrival location.

time discriminator (electronic navigation) A circuit in which the sense and magnitude of the output is a function of the time difference of the occurrence, and relative time sequence, of two pulses. *See also:* navigation. (AES/RS) 686-1982s, [42]

time distortion (broadband local area networks) Time distortion (group delay) is the difference in transmission time between frequencies of a service. The broadband service usually resides in a single channel, but the delay distortion may be specified over a bandwidth that is different than the bandwidth of the channel. Video specifies the time delay distortion to be less than a channel bandwidth. Video channels (6 MHz) normally specify the group delay between the video and color carriers (3.58 MHz). The delay distortion in video services may influence color rendition. In data services, group delay may influence the bit error rate. The specification for group delay must always be applied across a referenced bandwidth to be valid. This distortion is most prominent at the frequency band-edges of a diplex filter, but may also be observed in band-pass, band-stop, and equalizing filters. (LM/C) 802.7-1989r

time distribution analyzer (nuclear techniques) An instrument capable of indicating the number or rate of occurrence of time intervals falling within one or more specified time interval ranges. The time interval is delineated by the separation between pulses of a pulse pair. *See also:* ionizing radiation. (NPS) 175-1960w

time-division analog switching (telephone switching systems) Analog switching with common time-divided paths for simultaneous calls. (COM) 312-1977w

time-division digital switching (telephone switching systems) Digital switching with common time-divided paths for simultaneous calls. (COM) 312-1977w

time division multiple access (TDMA) (1) (communication satellite) A technique whereby earth stations communicate with each other on the basis of non-overlapping time sequenced bursts of transmissions through a common satellite repeater. (COM) [19]
(2) A multiplexing technique in which a channel is divided among different users allocating to each of them a time slot in a repeating cycle. (C) 610.7-1995

time-division multiplex (data transmission) The process or device in which each modulating wave modulates a separate pulse subcarrier, the pulse subcarriers being spaced in time so that no two pulses occupy the same time interval. *Note:* Time division permits the transmission of two or more signals over a common path by using different time intervals for the transmission of the intelligence of each message signal. (AP/PE/ANT) 145-1983s, 599-1985w

time-division multiplexing (TDM) (1) Sharing a communication channel among several users by allowing each to use the channel for a given period of time in a defined, repeated sequence. (LM/C) 802.7-1989r
(2) A method by which two or more channels of information are transmitted over the same link by allocating a different time interval for the transmission of each channel. *See also:* synchronous time division multiplexing; wave-division multiplexing. (C) 610.7-1995

time division multiplexing bus switching A method of time division switching in which time slots are used to transfer data over a shared bus between transmitter and receiver. (C) 610.7-1995

time division switching (1) The switching of inputs to outputs using time-division multiplexing techniques. *See also:* time division multiplexing bus switching. (C) 610.7-1995
(2) A method of switching that provides a common path with separate time intervals assigned to each of the simultaneous calls. (COM) 312-1977w

time domain A function in which the signals are represented as a function of time. (EMC) 1128-1998

time domain calibration A result which is the impulse response function of the sensor or probe in the time domain. (EMC) 1309-1996

time domain reflectometer (TDR) (1) Test equipment that verifies proper functioning of the physical components of the network with a sequence of time-delayed electrical pulses. (C) 610.7-1995
(2) An instrument designed to indicate and to measure reflection characteristics of a transmission system connected to the instrument by monitoring the step-formed signals entering the test object and the superimposed reflected transient signals on an oscilloscope that is equipped with a suitable time-based sweep. The measuring system basically consists of a fast-rise function generator, a tee coupler, and an oscilloscope connected to the probing branch of the coupler. *See also:* instrument. (EMC/IM/HFIM) 1128-1998, [40]

timed release (telephone switching systems) Release accomplished after a specified delay. (COM) 312-1977w

time, electrification *See:* electrification time.

time error Power system time minus a reference time. This quantity is derived by integrating frequency error over time and dividing it by rated frequency. (PE/PSE) 858-1993w, 94-1991w

time error bias An offset of the scheduled net interchange of a control area that varies in proportion to time error and that assists in restoring time error to zero. (PE/PSE) 858-1993w, 94-1991w

time gain control *See:* differential gain-control circuit.

time gate A transducer that gives output only during chosen time intervals. (AP/ANT) 145-1983s

time history (1) (gas-insulated substations) The trace of acceleration, velocity, or displacement as a function of time that the ground, the floor of a building, or a point of support experiences due to an earthquake. (SWG/SUB/PE) C37.122-1983s, C37.100-1992
(2) A record of motion, usually in terms of acceleration, as a function of time. (PE/SUB) 693-1997

time-insensitivity A type of year-insensitivity in which year, date, day indicator, and time-of-day are not maintained or represented. (C/PA) 2000.1-1999

time instability ($S_x(f)$) One-sided spectral density of the time deviation. (SCC27) 1139-1999

time-interval error (TIE) The variation of the time difference between a real clock and an ideal uniform time scale following a time period t after perfect synchronization. (SCC27) 1139-1999

time-interval selector (nuclear techniques) A circuit that produces a specified output pulse when and only when the time interval between two pulses lies between specified limits. *See also:* scintillation counter. (NPS) 175-1960w

time-interval simulation *See:* time-slice simulation.

time-invariant filtering (germanium gamma-ray detectors) Pulse shaping in which the filter response does not change with respect to time. [CR-(RC)n shaping is an example of time-invariant filtering.] (NPS) 325-1986s

time lag *See:* lag.

time lag of impulse flashover (surge arresters) The time between the instant when the voltage of the impulse wave first exceeds the power-frequency flashover crest voltage and the instant when the impulse flashover causes the abrupt drop in the testing wave. (T&D/PE) [10], [8]

time-load withstand strength (of an insulator) The mechanical load that, under specified conditions, can be continuously applied without mechanical or electrical failure. *See also:* insulator. (EEC/IEPL) [89]

time locking A method of locking, either mechanical or electric, that, after a signal has been caused to display an aspect to proceed, prevents, until after the expiration of a predetermined time interval after such signal has been caused to display its most restrictive aspect, the operation of any interlocked or electrically locked switch, movable-point frog, or derail in the route governed by that signal, and that prevents an aspect to proceed from being displayed for any conflicting route. *See also:* interlocking. (EEC/PE) [119]

time meridian (navigation aids) Any meridian used as a reference for reckoning time, particularly a zone. (AES/GCS) 172-1983w

time-multiplexed bus A bus which uses time-division multiplexing techniques to share its data paths between a number of devices. (C) 610.10-1994w

time multiplexed switching (TMS) A form of space-division switching in which each input line is a time division multiplexing strea. At the receiving end, the different signals are divided out and merged back into single streams. *See also:* message switching; circuit switching; space-division switching. (C) 610.7-1995

time-of-arrival location (TOA) A process whereby the position of a radiating transmitter can be located by means of the relative time delay between its signals as received in multiple receivers of known relative position. *Synonym:* time-difference-of-arrival. (AES) 686-1997

time-of-day clock A clock that indicates the actual time of the day. *Synonym:* real-time clock. *See also:* wall clock. (C) 610.10-1994w

time of death The term used to describe a field within a send packet that is used to determine when a send packet is stale and should be discarded. (C/MM) 1596-1992

time of decay of video pulses The duration of the decaying portion of a pulse measured between specified levels. *See also:* pulse timing of video pulses. (BT) 207-1950w

time-of-flight The time delay between a signal leaving a driving pin or primary input port and reaching a receiving pin or primary output. Time-of-flight is generally dominated by the time taken to charge the distributed capacitance of the interconnect and the capacitance of the driven pins through the distributed impedance of the interconnect. The internal impedance of the driving port affects the load-dependent delay but *not* (directly) the time-of-flight. (C/DA) 1481-1999

time of response *See:* response time.

time of rise of video pulses (television) (decay) The duration of the rising (decaying) portion of a pulse measured between specified levels. *See also:* pulse timing of video pulses. (BT) 207-1950w

time-of-use metering Metering equipment that separately records metered or measured quantities according to a time schedule. (AMR/SCC31) 1377-1997

time-of-use period (watthour meters) A selected period of time during which a specified rate will apply to the energy usage or demand. (ELM) C12.13-1985s

time-of-use register (watthour meters) That portion of a watthour meter that, for selected periods of time, accumulates and may display amounts of electric energy, demand, or other quantities measured or calculated. (ELM) C12.13-1985s

time origin line (pulse terminology) A line of constant and specified time which, unless otherwise specified, has a time equal to zero and passes through the first datum time, t_0, of a waveform epoch. *See also:* waveform epoch. (IM/WM&A) 194-1977w

timeout (1) A mechanism for terminating requested activity that, at least from the requester's perspective, does not complete within the time specified by the timeout's "value." (IM/ST) 1451.1-1999
(2) A method of error checking whereby an expected event is tested to occur within a specified period of time. (C) 1003.5-1999

time-out (1) (A) A condition that occurs when a predetermined amount of time elapses without the occurrence of an expected event. For example, the condition that causes termination of an on-line process if no user input is received within a specified period of time. **(B)** To experience the condition in definition (A). (C/Std100) 610.12-1990, 610.10-1994
(2) A time-out occurs when a protective timer completes its assigned time without the expected event occurring. Time-outs prevent the system from waiting indefinitely in case of error or failure. (NID) 960-1993

time-out of tone If the calling party reaches a call progress tone or announcement and does not abandon the call within a specified length of time, called the timeout of ringing interval, the switch may release the call. (COM/TA) 973-1990w

time-overcurrent relay An overcurrent relay in which the input current and operating time are inversely related throughout a substantial portion of the performance range. (SWG/PE) C37.100-1992

Time Parameter An instance of the class `IEEE1451_TimeParameter` or of a subclass thereof. (IM/ST) 1451.1-1999

time parameters and references *See:* time reference lines; pulse start time; transition duration; pulse duration.

time pattern (television) A picture-tube presentation of horizontal and vertical lines or dot rows generated by two stable frequency sources operating at multiples of the line and field frequencies. (BT) 202-1954w

time per point (multiple-point recorders) The time interval between successive points on printed records. *Note:* For some instruments this interval is variable and depends on the magnitude of change in measured signal. For such instruments, time per point is specified as the minimum and maximum time intervals. (EEC/EMI) [112]

time proportioning (electrical heating applications to melting furnaces and forehearths in the glass industry) An operation in which variable length bursts of full cycles of output voltage are alternated with variable length off periods to produce modulation of output. (IA) 668-1987w

timer (1) A register or storage location whose value is changed at regular intervals in such a manner as to measure time. *Synonyms:* clock register; time register. *See also:* watchdog timer; interval timer. (C) 610.10-1994w
(2) An object that can notify a process when the time as measured by a particular clock has reached or passed a specified value, or when a specified amount of time, as measured by a particular clock, has passed. (C/PA) 9945-1-1996
(3) An object that can notify a process when the time as measured by a particular clock has reached or passed a specified value or when a specified amount of time as measured by a particular clock has passed. Timers are per process; that is, they cannot be shared between processes. (C) 1003.5-1999

time rate (storage cell) The current in amperes at which a storage battery will be discharged in a specified time, under specified conditions of temperature and final voltage. *See also:* battery. (PE/EEC) [119]

time rating (of a VAM) The maximum amount of time the motor can be operated at rated running load without exceeding the allowable temperature rise for the insulation class being used. (PE/NP) 1290-1996

time, real *See:* real time.

timer EEPROM Typically a byte-alterable electrically erasable programmable read-only memory (EEPROM) with on-chip latches for all address, data, and control lines to internally control the duration of a write cycle. (ED) 1005-1998

time referenced point (pulse terminology) A point at the intersection of a time reference line and a waveform. (IM/WM&A) 194-1977w

time reference line (pulse terminology) A line parallel to the time origin line at a specified instant. (IM/WM&A) 194-1977w

time reference lines (A) (pulse terminology) (pulse start [stop]) line. The time reference line at pulse start (stop) time. *See also:* waveform epoch. **(B) (pulse terminology)** (top center line) The time reference line at the average of pulse start time and pulse stop time. *See also:* waveform epoch. (IM/WM&A) 194-1977

time register *See:* timer.

time-related adjectives (A) (pulse terminology) (periodic [aperiodic]) Of or pertaining to a series of specified waveforms or features which repeat or recur regularly (irregularly in time. **(B) (pulse terminology)** (coherent [incoherent]) Of or pertaining to two or more repetitive waveforms whose constituent features have (lack) time correlation. **(C) (pulse terminology)** (synchronous [asynchronous]) Of or pertaining to two or more repetitive waveforms whose sequential constituent features have (lack) time correlation. (IM/WM&A) 194-1977

time-related definitions *See:* cycle; frequency; duration; interval; period.

time release A device used to prevent the operation of an operative unit until after the expiration of a predetermined time interval after the device has been actuated. (EEC/PE) [119]

time resolution The minimum time interval that a clock can measure or whose passage a timer can detect. (PA/C/C/PA) 1003.5b-1995, 9945-1-1996

time response (1) (control system feedback) An output, expressed as a function of time, resulting from the application of a specified input under specified operating conditions. *Note:* It consists of a transient component that depends on the initial conditions of the system, and a steady-state component that depends on the time pattern of the input. *Synonym:* dynamic response. (IA/IAC) [60]
(2) (excitation systems) An output expressed as a function of time, resulting from the application of a specified input under specified operating conditions. See the corresponding figure for a typical time response of a system to step increase of input and for identification of the principle characteristics of interest.

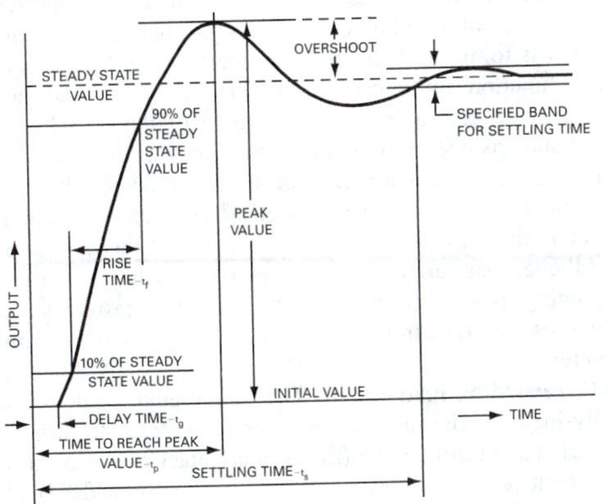

[Typical time response of a feedback control system to a step change in input.]

time response

(PE/EDPG) 421A-1978s

(3) (synchronous-machine regulator) The output of the synchronous-machine regulator (that is, voltage, current, impedance, or position) expressed as a function of time following the application of prescribed inputs under specfied conditions. (PE/EDPG) 421A-1978s

time, response *See:* response time.

time rise tone (measuring the performance of tone address signaling systems) The time interval between the end of the tone off condition and the beginning of the tone present condition at the beginning of the tone under consideration. (COM/TA) 752-1986w

timer overrun (1) A condition that occurs each time a timer, for which there is already an expiration signal queued to the process, expires. (C/PA) 9945-1-1996
(2) A condition that occurs each time a timer for which there is already an expiration signal queued to the process expires. (C) 1003.5-1999

time scale *See:* time.

time schedule controller (process control) A controller in which the command (or reference input signal) automatically adheres to a pre-determined time schedule. *Note:* The time schedule mechanism may be programmed to switch motors or other devices. (PE/EDPG) [3]

time sharing (software) A mode of operation that permits two or more users to execute computer programs concurrently on the same computer system by interleaving the execution of their program. *Note:* Time sharing may be implemented by time slicing, priority-based interrupts, or other scheduling methods. (C) 610.12-1990, 610.10-1994w

time signal (navigation aids) An accurate signal marking a specified time or time interval. (AES/GCS) 172-1983w

time skew (analog-to-digital converter) In an analog to digital conversion process, the time difference between the conversion of one analog channel and any other analog channel, such that the converted (digital) representations of the analog signals do not correspond to values of the analog variables that existed at the same instant of time. Time skew is eliminated, where necessary, by the use of a multiplexor with a sample/hold feature, allowing all input channels to be simultaneously sampled and stored for later conversion. *See also:* switching time; analog-to-digital converter. (C) 165-1977w, 166-1977w
(2) (A) In a conversion from analog to digital, the time difference between the conversion of one analog channel and any other analog channel, such that the converted (digital) representations of the analog signals do not correspond to values of the analog variables that existed at the same instant of time. **(B)** The time interval between two events which are intended to be simultaneous. (C) 610.10-1994

time-slice simulation (A) A discrete simulation that is terminated after a specific amount of time has elapsed; for example, a model depicting the year-by-year forces affecting a volcanic eruption over a period of 100 000 years. *Synonym:* time-interval simulation. *See also:* critical event simulation. **(B)** A discrete simulation of continuous events in which time advances by intervals chosen independent of the simulated events; for example, a model of a time multiplexed communication system with multiple channels transmitting signals over a single transmission line in very rapid succession. (C) 610.3-1989

time slicing A mode of operation in which two or more processes are each assigned a small, fixed amount of continuous processing time on the same processor, and the processes execute in a round-robin manner, each for its allotted time, until all are completed. (C) 610.12-1990

time slot (1) In time division multiplexing, when time is divided into slots to route data from input to output. (C) 610.7-1995
(2) Any cyclic time interval that can be recognized and defined uniquely. (COM/TA) 1007-1991r

time sorter *See:* time distribution analyzer.

time-to-amplitude converter (scintillation counting) An instrument producing an output pulse whose amplitude is proportional to the time difference between start and stop pulses. (NPS) 398-1972r

time to chopping (switching impulse testing) The time interval T_c between actual zero and the instant when the chopping occurs. 332-1972w

time to crest The time interval T_{cr} between actual zero and the instant when the voltage has reached its crest value. 332-1972w

time-to-crest value (T_r) The time that an impulse rises to crest value. (PE/C) 1313.1-1996

time to failure The amount of time remaining until the event horizon. (C/PA) 2000.1-1999

time to first voltage zero on the tail of the wave The time interval from the start of the transient to the time when the first voltage zero occurs on the tail of the wave. (PE/TR) C57.12.90-1999

time to half value (1) The time interval T_h between actual zero and the instant on the tail when the impulse has decreased to half its crest value. (Std100) 332-1972w **(2)** The time that an impulse drops to 0.5 crest value. (PE/C) 1313.1-1996

time to half-value on the wavetail *See:* virtual time to half-value.

time to impulse flashover The time between the initial point of the voltage impulse causing flashover and the point at which the abrupt drop in the voltage impulse takes place. (T&D/PE) [10]

time to impulse sparkover (1) The time between virtual zero of the voltage impulse that causes sparkover and the point on the voltage wave at which sparkover occurs. The voltage across the terminals of the surge-protective-device during the flow of discharge current and contributes to the limitation of follow current at normal power-frequency voltage. (SPD/PE) C62.62-2000 **(2)** The time between virtual zero of the voltage impulse causing sparkover and the point on the voltage wave at which sparkover occurs. (SPD/PE) C62.11-1999

time to repair (TTR) Time required to accomplish corrective maintenance or repair successfully. It includes all of the time required for diagnosis, set-up, replacement, reassembly, and test, but does not include logistics scheduling and approval. (PE/NP) 933-1999

time-to-saturation The time during which the secondary current is a faithful replica of the primary current. *Note:* The core does not saturate suddenly. Beyond the saturation flux level, the exciting current increases more rapidly than the secondary current, causing distortion in the secondary waveform. (PE/PSR) C37.110-1996

time, turnaround *See:* turnaround time.

time-undervoltage protection A form of undervoltage protection that disconnects the protected equipment upon a deficiency of voltage after a predetermined time interval. (SWG/PE) C37.100-1992

time unit (TU) A measurement of time equal to 1024 μs. (C/LM) 8802-11-1999

time update (sequential events recording systems) The correction or resetting of a real time clock to match a time standard. *See also:* real time. (PE/EDPG) [5], [1]

time variable A variable whose value represents simulated time or the state of the simulation clock. (C) 610.3-1989w

time-variant filtering Pulse shaping in which the filter response varies with time. (NPS) 325-1996

time zone diversity Load diversity between two or more electric systems that occurs when their peak loads are in different time zones. (PE/PSE) 858-1993w, 346-1973w

timing accuracy The maximum timing error allowable in message accounting records. Different tolerances may be required for recording time of day and for call duration. Time of day would reflect the time zone of the originating station and affects billing rates. Tolerances on call duration are usually chosen to avoid charging a customer for time that a call is not connected. (COM/TA) 973-1990w

timing ambiguity The period of time in a nodal transition during which the state of the node cannot be guaranteed. (SCC20) 1445-1998

timing analyzer (software) A software tool that estimates or measures the execution time of a computer program or portion of a computer program, either by summing the execution times of the instructions along specified paths or by inserting probes at specified points in the program and measuring the execution time between probes. (C) 610.12-1990

timing annotation The annotation of a design in one tool with timing data computed by another tool. If timing calculation is performed as an off-line process (separately from the application using the timing data), the process of reading the timing data into the tool is known as timing annotation. A timing annotation file stores the data written by the timing calculator and is later read by an application. *Synonym:* back-annotation. (C/DA) 1481-1999

timing arc A pair of ports, pins, or nodes possess some timing relationship, such as the propagation delay of a signal from one to the other or a timing check between them. Delay arcs may be between two distinct ports or nodes of a cell or over the interconnect from driver pins to receiver pins. (C/DA) 1481-1999

timing calculation The process of calculating values for the delays and timing checks associated with the physical primitives (cells) of an integrated circuit design, or part of an integrated circuit design, and their interconnections. (C/DA) 1481-1999

timing check A timing property of a circuit (frequently a cell) that describes a relationship in time between two input signal events. This relationship needs to be satisfied for the circuit to function correctly. (C/DA) 1481-1999

timing deviation demand meter (metering) The difference between the elapsed time indicated by the timing element and the true elapsed time, expressed as a percentage of the true elapsed time. (ELM) C12.1-1982s

timing discriminator A class of discriminators in which the initiation of the output signal is keyed to the instant when the input signal crosses the discriminator threshold. *See also:* constant-fraction discriminator; discriminator. (NPS) 325-1996

timing generator The function in the automatic test equipment (ATE) that stores and produces timing sets, or its analogous construct in the simulation process. (SCC20) 1445-1998

timing jitter Short-term deviations of the significant instants of a digital signal from their ideal positions in time. 1007-1991r

timing mechanism (1) (demand meter) That mechanism through which the time factor is introduced into the result. The principal function of the timing mechanism of a demand meter is to measure the demand interval, but it has a subsidiary function, in the case of certain types of demand meters, to provide also a record of the time of day at which any demand has occurred. A timing mechanism consists either of a clock or its equivalent, or of a lagging device that delays the indications of the electric mechanism. In thermally lagged meters the time factor is introduced by the thermal time lag of the temperature responsive elements. In the case of curve-drawing meters, the timing element merely provides a continuous record of time on a chart or graph. *See also:* demand meter. (EEC/PE) [119] **(2) (recording instrument)** The time-regulating device usually includes the motive power unit necessary to propel the chart at a controlled rate (linear or angular). *See also:* moving element. (EEC/ERI) [111]

timing model The timing behavior of a cell for applications, such as simulation and timing analysis. For black-box timing behavior, it represents the definition of pin-to-pin delays between any pair of pins as well as internal nodes. In addition, for sequential cells it provides the definition of timing checks and constraints on any pair of pins and/or internal nodes. (C/DA) 1481-1999

timing offset The difference between two physical units' fundamental clock sources; those sources being the timing basis from which signals and sampling are derived and analyzed (usually expressed proportionally in parts per million). Timing offset will cause a uniform percentage change in signal frequencies. (COM/TA) 743-1995

timing phase noise *See:* aperture uncertainty.

timing pulse *See:* clock signal.

timing relay An auxiliary relay or relay unit whose function is to introduce one or more time delays in the completion of an associated function. *Synonym:* relay unit. (SWG/PE) C37.100-1992

timing sequence Sequence of enable, coding, and data pulses to permit writing or reading of information. (ED) 1005-1998

timing set (TSET) An automatic test equipment (ATE) timing-cycle during which stimuli are applied and unit under test (UUT) responses are measured. A timing set includes the specification of the pattern period, UUT input pin groupings that will transition at a specific time within a pattern, and UUT output pin groupings that share the same window. (SCC20) 1445-1998

timing table That portion of central-station equipment at which means are provided for operators' supervision of signal reception. *See also:* protective signaling. (EEC/PE) [119]

timing track *See:* clock track.

tinning (electrotyping) The melting of lead-tin foil or tin plating upon the back of shells. (PE/EEC) [119]

tinsel cord A flexible cord in which the conducting elements are thin metal ribbons wound helically around a thread core. *See also:* transmission line.

TINT A subset of JOVIAL designed for simplified time-sharing programming. (C) 610.13-1993w

TIP *See:* terminal interface processor.

tip (1) (plug) The contacting part at the end of the plug. (EEC/PE) [119]

(2) (electron tube) (pip) A small protuberance on the envelope resulting from the sealing of the envelope after evacuation. (ED) [45], [84]

tip and ring wires (1) (telephone switching systems) A pair of conductors associated with the transmission portions of circuits and apparatus. Tip or ring designation of the individual conductors is arbitrary except when applied to cord-type switchboard wiring in which case the conductors are designated according to their association with tip or ring contacts of the jacks and plugs. (COM) 312-1977w

(2) (communication and control cables) The pair of conductors associated with the transmission portions of telephone cables, circuits, and apparatus. (PE/PSC) 789-1988w

tip switch A button on the end of a light pen or stylus that is depressed as the pen is touched to a data tablet, determining the position of a display element. (C) 610.6-1991w

TIU *See:* telemetry interface unit.

T junction (waveguide) A junction of waveguides in which the longitudinal guide axes form a T. *Note:* The guide that continues through the junction is the main guide: the guide that terminates at a junction is the branch guide. *See also:* waveguide. (AP/ANT) [35]

TLP *See:* transmission level point.

TLU *See:* table lookup.

TLV-STEL *See:* threshold limit value—short term exposure limit.

TLV-TWA *See:* threshold limit value—time weighted average.

T matrix Relates the scattered field to the exciting field. (AP/PROP) 211-1997

TM_{mn}mode (A) (E_{mn}mode) In a rectangular waveguide, the subscripts $_m$ and $_n$ denote the number of half-period variation in the magnetic field parallel to the broad and narrow sides, respectively, of the guide. *Note:* In the United Kingdom, the reverse order is preferred. **(B) (E_{mn}mode)** In a circular waveguide, a mode that has $_m$ diametral planes and $_n$ cylindrical surfaces of nonzero radius (including the wall of the guide) at which the longitudinal component of the electric field is zero. **(C) (E_{mn}mode)** In a resonant cavity consisting of a length of rectangular or circular waveguide, a third subscript is used to indicate the number of half-period variations of the field along the waveguide axis. (MTT) 146-1980

TM mode (1) (E mode) A waveguide mode in which the longitudinal component of the magnetic field is everywhere zero and the longitudinal component of the electric field is not. (MTT) 146-1980w

(2) (fiber optics) *See also:* transverse magnetic mode. 812-1984w

TMS *See:* time multiplexed switching; test mode select input pin.

TNA *See:* transient network analyzer.

T network A network composed of three branches with one end of each branch connected to a common junction point, and with the three remaining ends connected to an input terminal, an output terminal, and a common input and output terminal, respectively. *See also:* network analysis.

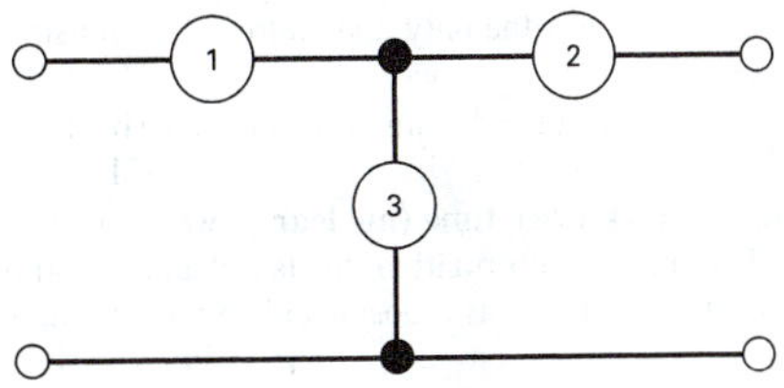

One end of each of the branches 1, 2, and 3 is connected to a common point. The other ends of branches 1 and 2 form, respectively, an input and an output terminal, and the other end of branch 3 forms a common input and output terminal.

T network

(BT) 153-1950w, 270-1966w

TOA location *See:* time-of-arrival location.

toe and shoulder (photographic techniques) [of a Hurter and Driffield (H and D) curve] The terms applied to the nonlinear portions of the H and D curve that lie, respectively, below and above the straight portion of this curve. (SP) [32]

to-from indicator (navigation aids) (omnirange receiver) A supplementary device used with an omnibearing selector to resolve the ambiguity of measured omnibearings. (AES/GCS) 172-1983w

toggle (1) Pertaining to any device having two stable states. *See also:* flip-flop. (C) [20], [85]

(2) A switching action performed on an object with two states. (C) 1295-1993w

(3) The action of changing state in a sequential circuit. *See also:* flip-flop. (C) 610.10-1994w

toggle bit An end-of-write indicator. (ED) 1005-1998

token (1) In a local area network, a control mechanism that is passed among stations to indicate which station is currently in control. *See also:* token passing; token ring; token bus; token access. (C) 610.7-1995

(2) In the shell command language, a sequence of characters that the shell considers as a single unit when reading input. A token is either an operator or a word. (C/PA) 9945-2-1993

(3) The 3-bit field of authority that is passed between data hosts using a token access method to indicate which data host is currently in control of the medium. (C/BA) 1393-1999

(4) A signal sequence passed from station to station that is used to control access to the medium. (C/LM) 8802-5-1998

token access (1) A means of transmitting data over a local area network that employs a token, a special bit pattern, to which a station attaches its data. (C) 610.7-1995

(2) The asynchronous data transmission process utilizing tokens to indicate which fiber-optic bus interface unit (FBIU) is currently in control of a token group. (C/BA) 1393-1999

token bus A network with a physical bus and logical ring topology where token passing is used to determine which node is allowed to transmit next. *See also:* bus-ring topology. (C) 610.7-1995

token group There are four token groups. Each fiber-optic bus interface unit (FBIU) can be a member of up to four token groups. Members of each token group share access to the same Tx Data Slots defined by the Token Arbitrated Transmit Slot Masks of FBIU configuration and status registers 16 through 19 [FCSR-(16–19)]. That is, FCSR-(16–19) for each member of a token group must be identical. This shared access is controlled using a simple token passing protocol. (C/BA) 1393-1999

tokenizer A development tool that converts FCode source code into a (binary) FCode program. (C/BA) 1275-1994

token passing A local area network access method in which a terminal can transmit only after it has acquired the network's token. (C) 610.7-1995

token ring A network in a logical ring configuration around which a token is periodically passed. The node which has the token at any time is the only node allowed to transmit on the network. (C) 610.7-1995

tokens The content of a document as characterized by words, ideograms, and graphics. (C/SE) 1045-1992

tolerable out-of-service time (nuclear power generating station) The time an information display channel is allowed to be unavailable for use as a post accident monitoring display. (PE/NP) 497-1981w

tolerable voltage difference (generating station grounding) The maximum potential difference that would cause a body current to flow of such value as not to cause ventricular fibrillation. (PE/EDPG) 665-1987s

tolerance (1) (nuclear power generating station) The allowable deviation from a specified or true value. 41-1982
(2) (software) The ability of a system to provide continuity of operation under various abnormal conditions. *See also:* system. (C/SE) 729-1983s
(3) (test, measurement, and diagnostic equipment) The total permissible variation of a quantity from a designated value. (MIL) [2]
(4) (metric practice) The amount by which the value of a quantity is allowed to vary; thus, the tolerance is the algebraic difference between the maximum and minimum limits. (SCC14/QUL) SI 10-1997, 268-1982s

tolerance band (1) (self-commutated converters) (converter characteristics) The range of steady-state values of a stabilized output quantity lying between the limits of operating error. *Notes:* 1. Tolerance band describes the permissible deviation of a stabilized output quantity from a rated or preset value. 2. A statement of tolerance band is useful when a subdivision into output effects and intrinsic errors is not of interest. (IA/SPC) 936-1987w
(2) (thyristor) The range of values specified in terms of permissible deviations of the steady state value of a parameter from a specified nominal value. (IA/IPC) 428-1981w

tolerance chart *See:* instrument tolerance chart.

tolerance, fault *See:* fault tolerance.

tolerance field (A) (fiber optics) In general, the region between two curves (frequently two circles) used to specify the tolerance on component size. **(B) (fiber optics)** When used to specify fiber cladding size, the annular region between the two concentric circles of diameter $D + \Delta D$ and $D - \Delta D$. The first circumscribes the outer surface of the homogeneous cladding; the second (smaller) circle is the largest circle that fits within the outer surface of the homogeneous cladding. **(C) (fiber optics)** When used to specify the core size, the annular region between the two concentric circles of diameter $d + \Delta d$ and $d - \Delta d$. The first circumscribes the core area; the second (smaller) circle is the largest circle that fits within the core area. *Note:* The circles of definition B need not be concentric with the circles of definition C. *See also:* core; cladding; homogeneous cladding; concentricity error. (Std100) 812-1984

toll board A switchboard used primarily for establishing connections over toll lines. (EEC/PE) [119]

toll call (telephone switching systems) A call for a destination outside the local-service area of the calling station. (COM) 312-1977w

toll center (1) (telephone switching systems) A toll office where trunks from end offices are connected to intertoll trunks and where operator's assistance is provided in completing incoming calls and where other traffic operating functions are performed. Toll centers are classified as Class 4C offices. *See also:* office class. (COM) 312-1977w
(2) Class 4 office in the North American hierarchical routing plan; a control center connecting end offices of the telephone system together. *See also:* primary center; end office; regional center; sectional center. (C) 610.7-1995

toll circuit *See:* trunk circuit.

toll connecting trunk (telephone switching systems) A trunk between a local office and a toll office or switchboard. (COM) 312-1977w

toll line A telephone line or channel between two central offices in different telephone exchanges. (EEC/PE) [119]

toll office (telephone switching systems) An intermediate office serving toll calls. (COM) 312-1977w

toll point (telephone switching systems) A toll office where trunks from end offices are connected to the distance dialing network and where operators handle only outward calls or where there are no operators present. Toll points are classified as Class 4P offices. (COM) 312-1977w

toll restriction (telephone switching systems) A method that prevents private automatic branch exchange stations from completing certain or any toll calls or reaching a toll operator, except through the attendant. (COM) 312-1977w

toll station A public telephone station connected directly to a toll telephone switchboard. *See also:* telephone station. (EEC/PE) [119]

toll switching trunk (telephone switching systems) A trunk for completing calls from a toll office or switchboard to a local office. (COM) 312-1977w

toll switch train A switch train that carries a connection from a toll board to a subscriber line. *Synonym:* toll train. *See also:* switching system. (EEC/PE) [119]

toll terminal loss (toll connection) That part of the over-all transmission loss that is attributable to the facilities from the toll center through the tributary office to and including the subscriber's equipment. *Note:* The toll terminal loss at each end of the circuit is ordinarily taken as the average of the transmitting loss and the receiving loss between the subscriber and the toll center. *See also:* transmission loss. (PE/EEC) [119]

toll train *See:* toll switch train.

toll transmission selector A selector in a toll switch train that furnishes toll-grade transmission to the subscriber and controls the ringing. (EEC/PE) [119]

T1 A carrier facility that transmits digital signal level one. The data rate is 1.544 Mb/s, the equivalent of 24 voice-band channels. *Note:* T1 is used to provide long-distance telephone service and also to provide voice and data communications to individual subscribers. (C) 610.7-1995

tone (1) (A) (general) A sound wave capable of exciting an auditory sensation having pitch. **(B) (general)** A sound sensation having pitch. (SP) [32]
(2) (telephone switching systems) An audible signal transmitted over the telecommunications network. (COM) 312-1977w

T1C A carrier facility that transmits digital signal level 1C. The data rate is 3.152 Mb/s, the equivalent of 48 voice-band channels. (C) 610.7-1995

tone, call *See:* call tone.

tone control A means for altering the frequency response at the audio-frequency output of a circuit, particularly of a radio receiver or hearing aid, for the purpose of obtaining a quality more pleasing to the listener. *See also:* radio receiver; amplifier. (EEC/PE) [119]

tone decay time (measuring the performance of tone address signaling systems) The time interval between the end of the tone present condition and the beginning of the tone off condition at the end of the tone under consideration. (COM/TA) 752-1986w

tone duration (measuring the performance of tone address signaling systems) The time interval during which a tone present condition exists continuously. (COM/TA) 752-1986w

tone leak (measuring the performance of tone address signaling systems) The occurrence of any address signaling tone during the signal present or signal off intervals when such tone is not intended. (COM/TA) 752-1986w

tone localizer *See:* equisignal localizer.

tone-modulated waves Waves obtained from continuous waves by amplitude modulating them at audio frequency in a substantially periodic manner. *See also:* telegraphy. (EEC/PE) [119]

tone off (measuring the performance of tone address signaling systems) Any condition where the tone under consideration is below a specified OFF level. (COM/TA) 752-1986w

tone-operated net-loss adjuster *See:* tonlar.

tone present (measuring the performance of tone address signaling systems) Any condition where the tone under consideration is equal to or greater than a specified threshold value. (COM/TA) 752-1986w

toner (electrostatography) The image-forming material in a developer that, deposited by the field of an electrostatic-charge pattern, becomes the visible record. *See also:* electrostatography. (ED) [46]

tone-to-C-Notched noise ratio The ratio in dB of the incoming holding tone power to the C-Notched noise power at the point of measurement. The incoming holding tone power is reduced in power by at least 50 dB by the 1010 Hz Notch. (COM/TA) 743-1995

tone-to-D-Notched noise ratio The ratio in dB of the incoming holding tone power to the D-Notched noise power at the point of measurement. The incoming holding tone power is reduced in power by at least 50 dB by the 1010 Hz Notch. (COM/TA) 743-1995

tongs, fuse *See:* fuse tongs.

tonlar A system for stabilizing the net loss of a telephone circuit by means of a tone transmitted between conversations. The name is derived from the initial letters of the expression tone-operated net-loss adjuster. (EEC/PE) [119]

ton of refrigeration Is equal to 12 000 Btu/hour. (IA/PSE) 241-1990r

tool (software) A hardware device used to analyze software or its performance. *See also:* performance; hardware. (C/SE) 729-1983s

tool function (numerically controlled machines) A command identifying a tool and calling for its selection either automatically or manually. The actual changing of the tool may be initiated by a separate tool-change command. (IA) [61]

toolkit An implementation of IEEE Std 1003.5b-1995. (C/PA) 1295-1993w, 1387.2-1995, 1003.5b-1995, 9945-1-1996, 1003.5-1992r, 9945-2-1993

tool offset (numerically controlled machines) A correction for tool position parallel to a controlled axis. (IA) [61]

tool or equipment current The total current delivered to the tool or equipment. (T&D/PE) 516-1995

tools, access *See:* access tools.

tooth (1) (rotating machinery) A projection from a core, separating two adjacent slots, the tip of which forms part of one surface of the air gap. *See also:* stator; rotor. (PE) [9]

(2) (cable plowing) *See also:* wear point. (T&D/PE) 590-1977w

tooth pitch *See:* slot pitch.

tooth tip (rotating machinery) That portion of a tooth that forms part of the inner or outer periphery of the air gap. It is frequently considered to be the section of a tooth between the radial location of the wedge and the air gap. *See also:* stator; rotor. (PE) [9]

top (1) (pulse terminology) The portion of a pulse waveform which represents the second nominal state of a pulse. (IM/WM&A) 194-1977w

(2) (data management) In a queue or a stack, the position of the next item to be retrieved. *Contrast:* bottom. (C) 610.5-1990w

(3) By convention, the edge of the module seen clockwise from the faceplate when viewing the component side. In IEEE 1101.1 systems, the P1 connector is closer to the top than the bottom edge. (C/MM) 1101.2-1992

top box The box in the A-0 context diagram that models the top-level function of an IDEF0 model. (C/SE) 1320.1-1998

top cap (electron tube) (side contact) A small metal shell on the envelope of an electron tube or valve used to connect one electrode to an external circuit. *See also:* electron tube. (ED) [45]

top car clearance (elevators) The shortest vertical distance between the top of the car crosshead, or between the top of the car where no car crosshead is provided, and the nearest part of the overhead structure or any other obstruction when the car floor is level with the top terminal landing. *See also:* hoistway. (PE/EEC) [119]

top center line (of a pulse) In a peaked pulse, the ordinate that passes through the peak; in a flat-topped pulse, the ordinate that bisects the nominally flat-topped region. (NPS) 325-1996

top centerline (of a pulse) In a peaked pulse, the ordinate that passes through the peak. In a flat-topped pulse, the ordinate that bisects the nominally flat-topped region. (NPS) 300-1988r

top coil side (rotating machinery) (radially inner coil side) The coil side of a stator slot nearest the bore of the stator or nearest the slot wedge. *See also:* stator. (PE) [9]

top counterweight clearance (elevators) (elevator counterweight) The shortest vertical distance between any part of the counterweight structure and the nearest part of the overhead structure or any other obstruction when the car floor is level with the bottom terminal landing. *See also:* hoistway. (EEC/PE) [119]

top-down Pertaining to an activity that starts with the highest level component of a hierarchy and proceeds through progressively lower levels; for example, top-down design; top-down testing. *Contrast:* bottom-up. *See also:* critical piece first. (C) 610.12-1990

top-down design (software) The process of designing a system by identifying its major components, decomposing them into their lower level components, and iterating until the desired level of detail is achieved. *See also:* system; bottom-up design; level; component. (C/SE) 729-1983s

top-down testing (software) The process of checking out hierarchically organized programs, progressively, from top to bottom, using simulation of lower level components. *See also:* simulation; component; program. (C/SE) 729-1983s

top half bearing (rotating machinery) The upper half of a split sleeve bearing. *See also:* bearing. (PE) [9]

top-level environment The environment used to resolve free variable references in a Scheme program. (C/MM) 1178-1990r

top-level function The function modeled by the single box in the A-0 context diagram of an IDEF0 model. (C/SE) 1320.1-1998

top-loaded vertical antenna A vertical monopole with an additional metallic structure at the top intended to increase the effective height of the antenna and to change its input impedance. (AP/ANT) 145-1993

top magnitude (pulse terminology) The magnitude of the top as obtained by a specified procedure or algorithm. *See also:* waveform epoch. (IM/WM&A) 194-1977w

topology (1) (A) The interconnection pattern of nodes on a network. **(B)** The logical and/or physical arrangement of stations on a network. *See also:* bus-ring topology; star topology; star-bus topology; tree topology; loop topology; ring topology; bus topology; star-ring topology. (C) 610.7-1995
(2) The geometric pattern or configuration of intelligent devices and how they are linked together for communications. (VT) 1473-1999

topside ionospheric sounding Vertical incidence ionospheric sounding made from an artificial Earth satellite above the height of the maximum electron density of the F region. (AP/PROP) 211-1997

top side sounding (communication satellite) Ionospheric sounding from medium altitude satellites for measuring ionospheric densities at high altitudes. (COM) [25]

top terminal landing (elevators) The highest landing served by the elevator that is equipped with a hoistway door and hoisting-door locking device that permits egress from the hoistway side. *See also:* elevator landing. (EEC/PE) [119]

torchere (illuminating engineering) An indirect floor lamp which sends all or nearly all of its light upward. (EEC/IE) [126]

toroid (doughnut) A toroidal-shaped vacuum envelope in which electrons are accelerated. *See also:* electron device. (ED) [45]

toroidal coil A coil wound in the form of a toroidal helix. (IM) [120]

toroidal reflector A reflector formed by rotating a segment of plane curve about a nonintersecting co-planar line. *Note:* The plane curve segment is called the torus cross section and the co-planar line is called the toroidal axis. (AP/ANT) 145-1993

torque (instrument) The turning moment on the moving element produced by the quantity to be measured or some quantity dependent thereon acting through the mechanism. This is also termed the deflecting torque and in many instruments is opposed by the controlling torque, which is the turning moment produced by the mechanism of the instrument tending to return it to a fixed position. *Note:* Full-scale torque is the particular value of the torque for the condition of full-scale deflection and as an index of performance should be accompanied by a statement of the angle corresponding to this deflection. *See also:* accuracy rating; energy and torque. (PE/EEC) [119]

torque (force) balance accelerometer A device that measures acceleration by applying a torquer (force) rebalance. (AES/GYAC) 528-1994

torque buildup time constant (electric coupling) The time constant applicable when excitation voltage is changed from zero to full value. (EM/PE) 290-1980w

torque-coil magnetometer A magnetometer that depends for its operation on the torque developed by a known current in a coil that can turn in the field to be measured. *See also:* magnetometer. (EEC/PE) [119]

torque-command storage (gyros) The transient deviation of the output of a rate-integrating gyro from that of an ideal integrator when the gyro is subjected to a torquer command signal. It is a function of the gyro's characteristic time and the torquer time constant. *See also:* float storage; attitude storage. (AES/GYAC) 528-1994

torque control (1) (protective relaying of utility-consumer interconnections) A means of supervising the operation of one relay element with another. For example, an overcurrent relay cannot operate unless the lag coil circuit is closed. It may be closed by the contact of an undervoltage element. (PE/PSR) C37.95-1973s
(2) (of a relay) A method of constraining the pickup of a relay by preventing the torque-producing element from developing operating torque until another associated relay unit operates. (SWG/PE) C37.100-1992

torque decay time constant (electric coupling) The time constant applicable when the excitation voltage is changed from full value to zero. (EM/PE) 290-1980w

torque-generator reaction torque *See:* torquer reaction torque.

torque margin The increase in torque above rated torque to which a motor may be subjected without the motor pulling out of step. This is of particular concern with electric propulsion systems. (IA/MT) 45-1998

torque motor A motor designed primarily to exert torque through a limited travel or in a stalled position. *Note:* Such a motor may be capable of being stalled continuously or only for a limited time. *See also:* asynchronous machine. (PE) [9]

torquer (accelerometer) (or forcer) (gyros) A device that exerts a torque (or force) on a gimbal, a gyro rotor, or a proof mass, in response to a command signal. (AES/GYAC) 528-1994

torquer axis (accelerometer) (inertial sensors) (gyros) The axis about which a force couple is produced by a torquer. (AES/GYAC) 528-1994

torquer-current rectification (accelerometer) (gyros) (inertial sensors) An apparent drift rate (or bias) in an inertial sensor resulting from effects such as torquer nonlinearity or capture loop asymmetry. (AES/GYAC) 528-1994

torquer reaction torque (accelerometer) (gyros) The usually undesired reaction torque that is a function of the frequency and amplitude of the command torque signal. (AES/GYAC) 528-1994

torque seating A control scheme that uses the torque switch as the primary control for operation of the VAM. The torque switch controls the VAM by interrupting power to the motor contactor when the valve actuator output torque exceeds a predetermined value. (PE/NP) 1290-1996

torque time constants (electric coupling) Torque time constants define the time required for the coupling torque to reach 63.2% of its total excursion, whenever the magnitude of excitation voltage is instantly changed between specified values. This does not imply that the torque time constant of a given coupling under certain conditions of slip, speed, temperature, and environmental conditions will be the same under other conditions. (EM/PE) 290-1980w

torquing (accelerometer) (gyros) The application of torque to a gimbal or a gyro rotor about an axis-of-freedom for the purpose of precessing, capturing, slaving, or slewing. (AES/GYAC) 528-1994

torquing rate (navigation aids) (inertial navigation) The angular rate at which the orientation of a gyro, with respect to inertial space, is changed in response to a command. (AES/GCS) 172-1983w

torsional critical speed (rotating machinery) The speed at which the amplitudes of the angular vibrations of a machine rotor due to shaft torsional vibration reach a maximum. *See also:* rotor. (PE) [9]

torsional mechanism An operating mechanism that transfers rotary motion by torsion through a pipe or shaft from the operating means to open or close the switching device. (SWG/PE) C37.100-1992, C37.30-1971s

torsionmeter A device to indicate the torque transmitted by a propeller shaft based on measurement of the twist of a calibrated length of the shaft. *See also:* electric propulsion system. (EEC/PE) [119]

total A complete mapping. The mapping `M` from a set `D` to a set `R` is *total* if for every `X` in `D`, there is at least one `Y` in `R` and pair [`X`, `Y`] in `M`. A property of a class is total, meaning that it will have a value for every instance of the class, unless it is explicitly declared partial. *Contrast:* partial. *See also:* mandatory; mapping completeness. (C/SE) 1320.2-1998

total cluster A subclass cluster in which each instance of a superclass must be an instance of at least one of the subclasses of the cluster. *Contrast:* partial cluster. *See also:* superclass. (C/SE) 1320.2-1998

total current The combination of the symmetrical component and the dc component of the current. (SWG/PE) C37.100-1992

total average power dissipation (semiconductor) The sum of the full cycle average forward and full cycle average reverse power dissipation. (IA) [12]

total break time *See:* interrupting time.

total capability for load (electric power supply) The capability available to a system from all sources including purchases. *See also:* generating station. (PE/PSE) [54]

total capacitance *See:* self-capacitance.

total charge One-half of the charge that flows as the condition of the device is changed from that of full applied positive voltage to that of full negative voltage (or vice versa). *Note:* Total charge is dependent on the amplitude of the applied voltage which should be stated when measurements of total charge are reported. *See also:* ferroelectric domain. (UFFC) 180w

total clearing time (1) (protection and coordination of industrial and commercial power systems) The total time between the beginning of the specified overcurrent and the final interruption of the circuit, at rated voltage. It is the sum of the minimum melting time plus tolerance and the arcing time. For clearing times in excess of half-cycle, the clearing time is substantially the maximum melting time for low-voltage fuses. *See also:* clearing time. (IA/PSP) 242-1986r
(2) *See also:* clearing time; fuse tube; melting-speed ratio. (SWG/PE) C37.100-1992

total correctness In proof of correctness, a designation indicating that a program's output assertions follow logically from its input assertions and processing steps, and that, in addition, the program terminates under all specified input conditions. *Contrast:* partial correctness. *See also:* input assertion; output assertion; proof of correctness; assertion. (C) 610.12-1990

total-current regulation (axle generator) That type of automatic regulation in which the generator regulator controls the total current output of the generator. *See also:* axle-generator system.

total cyanide (electroplating) (in a solution for metal deposition) The total content of the cyanide radical (CN), whether present as the simple or complex cyanide of an alkali or other metal. *See also:* electroplating. (PE/EEC) [119]

total detection efficiency (germanium detectors) The ratio of the total (peak plus Compton) counting rate to the gamma-ray emission rate. *Note:* The terms standard source and radioactivity standard are general terms used to refer to the sources and standards of National Radioactivity Standard Source and Certified Radioactivity Standard Source. (PE/EDPG) 485-1983s

total demand distortion (TDD) (1) The total root-sum-square harmonic current distortion, in percent of the maximum demand load current (15 or 30 min demand). (IA/SPC) 519-1992
(2) The total rms current distortion in percent of maximum demand current. (T&D/PE) 1250-1995

total distortion (1) A measure of the difference between a pure sine waveform of a specified frequency and a test voltage waveform. Usually measured by an audio distortion analyzer. (DESG) 1035-1989w
(2) The ratio in dB of the energy of a fundamental test signal, to all the other energy appearing in the band of interest. (COM/TA) 1007-1991r
(3) The summation of noise and distortion resulting from the application of a test signal (quantizing noise, phase jitter, intermodulation distortion, etc.), and the noise not related to the application of a test signal (background noise). (COM/TA) 743-1995

total dose (1) (metal-nitride-oxide field-effect transistor) The total amount of ionizing radiation deposited in the active area of a device over a given period of time. The unit of measure for total dose most commonly used in the present context is rads (SI). (ED) 581-1978w, 1005-1998
(2) The total amount of ionizing radiation that is deposited in the active area of a device over a given period of time. The unit of measure for total dose that is most commonly used in the present context is rads(Si).

total efficiency The ratio of the number of pulses in the entire energy spectrum due to an X or gamma ray of a given energy to the number of X- or gamma-ray photons emitted by the source for a specified source-to-detector distance. (NI) N42.14-1991

total electric current density At any point, the vector sum of the conduction-current density vector, the convection-current density vector, and the displacement-current density vector at that point. (Std100) 270-1966w

total electrode capacitance (electron tube) The capacitance of one electrode to all other electrodes connected together. (ED) [45]

total electron content (TEC) The total number of free electrons in a tube (generally with a vertical axis) of unit transverse cross-section passing through the ionosphere. *Note:* The units for TEC are 10^{16} electrons/m^2 (or 10^{12} electrons/cm^2). (AP/PROP) 211-1997

total emissivity (element of surface of a temperature radiator) The ratio of its radiant-flux density (radiant exitance) to that of a blackbody at the same temperature. (EEC/IE) [126]

total fall distance The maximum vertical distance between the person's fall arrest attachment point at the onset of a fall and after the fall is arrested, including free fall distance and maximum deceleration distance. Total fall distance excludes dynamic elongation. (T&D/PE) 1307-1996

total for load capability (power operations) The dependable capability available to a system from all sources including purchases. (PE/PSE) 858-1987s

total harmonic distortion (1) The root sum square of all harmonic distortion components including their aliases. (IM/WM&A) 1057-1994w
(2) The ratio of the rms value of the sum of the squared individual harmonic amplitudes to the rms value of the fundamental frequency of a complex waveform. *Synonym:* distortion factor. (T&D/PE/DESG) 1250-1995, 1035-1989w
(3) The ratio in dB of the energy of a fundamental test frequency to the energy of its harmonics appearing in the band of interest. (COM/TA) 1007-1991r
(4) The ratio, expressed as a percent, of the rms value of the ac signal after the fundamental component is removed and inter-harmonic components are ignored, to the rms value of the fundamental. The formula defining total harmonic distortion (THD) is provided below. The variables 'X1' and xn may represent either voltage or current, and may be expressed either as rms or peak values, so long as all are expressed in the same fashion.

$$D_x = \frac{\sqrt{\sum_{n=2} x_n^2}}{X_1} \cdot 100\%$$

where
X_1 = fundamental value of current or voltage;
x_n = n^{th} harmonic value of current or voltage.
(PEL) 1515-2000
(5) *See also:* distortion factor. (IA/SPC) 519-1992
(6) *See also:* distortion factor. (IA/PSE) 1100-1999

total hazard current (health care facilities) The hazard current of a given isolated system with all devices, including the line isolation monitor, connected. *See also:* hazard current. (EMB) [47]

total internal reflection (fiber optics) The total reflection that occurs when light strikes an interface at angles of incidence (with respect to the normal) greater than the critical angle. *See also:* step index optical waveguide; critical angle. (Std100) 812-1984w

total ionizing dose (TID) The radiation dose accumulated in a component, device, or module. (C/BA) 1156.4-1997

totalizing pulse relay (metering) A device used to receive and totalize pulses from two or more sources for proportional transmission to another totalizing relay or to a receiver. (ELM) C12.1-1982s

totalizing relay A device used to receive and totalize pulses from two or more sources for proportional transmission to another totalizing relay or to a receiver. *See also:* auxiliary device to an instrument. (ELM) C12.1-1982s

total loss (rotating machinery) The difference between the active electrical power (mechanical power) input and the active electrical power (mechanical power) output. (PE) [9]

total losses (1) (transformer or regulator) The sum of the no-load and load losses, excluding losses due to accessories. (PE/TR) C57.12.80-1978r
(2) (shunt reactors over 500 kVA) (of a shunt reactor) The sum of the conductor loss, magnetic circuit loss, cooling loss, shielding loss, and any other stray losses in the shunt reactor. (PE/TR) C57.21-1981s
(3) The sum of the no-load losses and the load losses. (PE/TR) C57.12.90-1999
(4) Those losses that are the sum of the no-load losses and the load losses. Power required for cooling fans, oil pumps, space heaters, and other ancillary equipment is not included in the total loss. When specified, loss data on such ancillary equipment shall be furnished. (PE/TR) C57.15-1999

totally-depleted detector (charged-particle detectors) (germanium gamma-ray detectors) (semiconductor radiation detectors) (x-ray energy spectrometers) A detector in which the thickness of the depletion region is essentially equal to the thickness of the semiconductor material. (NPS/AES/NID/GCS) 759-1984r, 325-1996, 172-1983w, 301-1976s, 300-1988r

totally enclosed (rotating machinery) A term applied to apparatus with an integral enclosure that is constructed so that while it is not necessarily airtight, the enclosed air has no deliberate connection with the external air except for the provision for draining and breathing. (PE) [9]

totally enclosed fan-cooled A term applied to a totally enclosed apparatus equipped for exterior cooling by means of a fan or fans, integral with the apparatus but external to the enclosing parts. *Synonym:* totally enclosed fan-ventilated. *See also:* asynchronous machine. (IA/PE/MT) 45-1983s, [9]

totally enclosed fan-cooled machine (TEFC) A totally enclosed machine equipped for exterior cooling by means of a fan or fans integral with the machine but external to the enclosing parts. (IA/MT) 45-1998

totally enclosed fan-ventilated *See:* totally enclosed fan-cooled.

totally-enclosed fan-ventilated air-cooled (rotating machinery) Applied to a totally-enclosed machine having an air-to-air heat exchanger in the internal air circuit, the external air being blown through the heat exchanger by a fan mechanically driven by the machine shaft. (PE) [9]

totally enclosed machine (electric installations on shipboard) A machine so enclosed as to prevent the exchange of air between the inside and outside of the case, but not sufficiently enclosed to be termed airtight. (IA/MT) 45-1983s

totally enclosed nonventilated (rotating machinery) A term applied to a totally enclosed apparatus that is not equipped for cooling by means external to the enclosing parts. *See also:* asynchronous machine. (PE) [9]

totally enclosed nonventilated machine (TENV) A machine enclosed to prevent the free exchange of air between the inside and outside of the case, but not sufficiently enclosed to be airtight. (IA/MT) 45-1998

totally enclosed pipe-ventilated machine A totally enclosed machine except for openings so arranged that inlet and outlet ducts or pipes may be connected to them for the admission and discharge of the ventilating air. This air may be circulated by means integral with the machine or by means external to and not a part of the machine. In the latter case, these machines shall be known as separately ventilated or forced ventilated machines. *See also:* closed air circuit; asynchronous machine. (EEC/PE) [119]

totally enclosed ventilated apparatus Apparatus totally enclosed in which the cooling air is carried through the case and apparatus by means of ventilating tubes and the air does not come in direct contact with the windings of the apparatus. (EEC/PE) [119]

totally enclosed water/air cooled machine (TEWAC) A totally enclosed machine with integral water-to-air heat exchanger and internal fan to provide closed-loop air cooling of the windings. (IA/MT) 45-1998

totally unbalanced currents (balanced line) Push-push currents. *See also:* waveguide. (MTT) 146-1980w

total motor temperature The motor temperature rise plus the ambient temperature. (PE/NP) 1290-1996

total number of customers served The total number of customers served on the last day of the reporting period. If a different customer total is used, it must be clearly defined within the report. (PE/T&D) 1366-1998

total operating losses The total station losses produced with the converter station energized and the valves operating. (SUB/PE) 1158-1991r

total power The total (or apparent) power (S) is the product of rms voltage and current (VA). (PEL) 1515-2000

total power factor The ratio of the total power input, in watts, to the total volt-ampere input. *Note:* This definition includes the effect of harmonic components of current and voltage and the effect of phase displacement between current and voltage. (IA/PSE/SPC) 1100-1999, 519-1992

total power loss (semiconductor rectifiers) The sum of the forward and reverse power losses. *See also:* rectification. (IA) [12]

total propagated uncertainty An estimate or approximation of the accuracy of a measured value by propagation of individual uncertainties in accordance with NIST recommendations. (NI) N42.22-1995

total range (instrument) The region between the limits within which the quantity measured is to be indicated or recorded and is expressed by stating the two end-scale values. *Notes:* 1. If the span passes through zero, the range is stated by inserting zero or 0 between the end-scale values. 2. In specifying the range of multiple-range instruments, it is preferable to list the ranges in descending order, for example, 750/300/150. *See also:* instrument. (EEC/ERI) [111]

total sag The distance measured vertically from the conductor to the straight line joining its two points of support, under conditions of ice loading equivalent to the total resultant loading for the district in which it is located. (NESC/T&D) C2-1997, C2.2-1960

total scattering cross-section The average over 4π steradians of the bistatic scattering cross-section for a specific illumination, given by

$$\sigma_\tau = \frac{1}{4\pi}\int_\Omega \sigma d\Omega$$

(AP/PROP) 211-1997

total stability (solution, $\phi = \phi(x(t_0);t)$ of the system $x = f(x,t)$) Implies that for every given $\varepsilon > 0$ there exist a $\delta_1 > 0$ and a $\delta_2 > 0$ (both of which, in general, may depend on ε and t_0) such that $\|\Delta x(t_0)\| \leq \delta_1$ and $\|g(x,t)\| \leq \delta_2$ imply $\|\phi - \Psi\| \leq \varepsilon$ for $t \geq t_0$, where $\Psi = \Psi(x(t_0) + \Delta x(t_0);t)$ is a solution of the system $x = f(x,t) + g(x,t)$. *See also:* control system. (CS/IM) [120]

total start-stop telegraph distortion Refers to the time displacement of selecting-pulse transitions from the beginning of the start pulse expressed in percent of unit pulse. (AP/ANT) 145-1983s

total switching time The time required to reverse the signal charge. *Note:* Total switching time is measured from the time of application of the voltage pulse, which must have a rise time much less than and a duration greater than the total switching time. The magnitude of the applied voltage pulse should be specified as part of the description of this characteristic. *See also:* ferroelectric domain. (UFFC) 180w

total system downtime (switching system) The time interval over which the entire switching system is down and cannot process any calls. It is the long-term mean time out of service in minutes per year for all outages greater than 30 s. (COM/TA) 973-1990w

total telegraph distortion Telegraph transmisson impairment, expressed in terms of time displacement of mark-space and space-mark transitions from their proper positions relative to one another, in percent of the shortest perfect pulses called the unit pulse. (Time lag affecting all transitions alike does not cause distortion). Telegraph distortion is specified in terms of its effect on code and terminal equipment. Total Morse telegraph distortion for a particular mark or space pulse is expressed as the algebraic sum of time displacements of space-mark and mark-space transitions determining the beginning and end of the pulses, measured in percent of unit pulse. Lengthening of mark is positive, and shortening, negative. *See also:* distortion. (AP/ANT) 145-1983s

total varactor capacitance The capacitance between the varactor terminals under specified conditions. (ED) 318-1971w

total voltage regulation (rectifier) The change in output voltage, expressed in volts, that occurs when the load current is reduced from its rated value to zero or light transition load with rated sinusoidal alternating voltage applied to the alternating-current line terminals, but including the effect of the specified alternating-current system impedance as if it were inserted between the line terminals and the transformer, with the rectifier transformer on the rated tap. *Note:* The measurement shall be made with zero phase control and shall exclude the corrective action of any automatic voltage-regulating means, but not impedance. *See also:* rectification; power rectifier. (IA) [62]

touchdown zone lights (illuminating engineering) Barettes of runway lights installed in the surface of the runway between the runway edge lights and the runway centerline lights to provide additional guidance during the touchdown phase of a landing in conditions of very poor visibility. (EEC/IE) [126]

touch panel A touch-sensitive input device that allows users to interact with a computer system by touching an area on the panel. *See also:* touch screen. (C) 610.6-1991w, 610.10-1994w

touch potential (1) The potential difference between a grounded metallic structure and a point on the earth's surface separated by a distance equal to the normal maximum horizontal reach, approximately one meter. This potential difference could be dangerous and could result from induction or fault conditions, or both. *Synonym:* touch voltage. (T&D/PE) 1048-1990
(2) *See also:* touch voltage. (T&D/PE) 524-1992r

touch screen A display screen equipped with a touch panel in front of it such that users may interact with a computer system by touching an area on the panel. *See also:* touch panel. (C) 610.10-1994w

touch-sensitive Pertaining to an input device that can detect when a user touches its surface with a finger, pencil or other object. *See also:* light-sensitive. (C) 610.10-1994w

touch symbol Refers to the overall design and shape shown in the figure below. This symbol is normally shown under a red circle with bar to show the action (touching) is prohibited.

Touch symbol

touch symbol

(NIR/SCC28) C95.2-1999

touch voltage (1) The potential difference between a grounded metallic structure and a point on the earth's surface separated by a distance equal to the normal maximum horizontal reach, approximately 1 m (3 ft). *Note:* This potential difference could be dangerous and could result from induction or fault conditions, or both. *Synonym:* touch potential. (PE/T&D/PSIM) 81-1983, 524a-1993r, 524-1992r
(2) The potential difference between the ground potential rise (GPR) and the surface potential at the point where a person is standing while at the same time having a hand in contact with a grounded structure. (PE/SUB) 80-2000, 1268-1997

tournament sort A repeated selection sort in which each of the subsets that make up the set to be sorted consists of no more than two items. (C) 610.5-1990w

tower (1) A broad-base latticed steel support for line conductors. (T&D/PE) [10]
(2) (mainenance of energized power lines) *See also:* structure. (T&D/PE) 516-1987s

tower footing resistance (lightning protection) The resistance between the tower grounding system and true ground. *See also:* direct-stroke protection. (T&D/PE) [10]

tower ladder A ladder complete with hooks and safety chains attached to one end of the side rails. These units are normally fabricated from fiberglass, wood, or metal. The ladder is suspended from the arm or bridge of a structure to enable workers to work at the conductor level, to hang travelers, perform clipping-in operations, etc. In some cases, these ladders are also used as lineperson's platforms. *Synonym:* hook ladder. (T&D/PE) 524-1992r

tower loading The load placed on a tower by its own weight, the weight of the wires with or without ice covering, the insulators, the wind pressure normal to the line acting both on the tower and the wires and the pull from the wires in the direction of the line. *See also:* tower. (T&D/PE) [10]

towing light A lantern or lanterns fixed to the mast or hung in the rigging to indicate that a ship is towing another vessel or other objects. (EEC/PE) [119]

Townsend coefficient (gas) The number of ionizing collisions per centimeter of path in the direction of the applied electric field. *See also:* discharge. (ED) [45]

TPC *See:* trigger pulse converter.

T-pin broadcast A method of using the pattern of ones on the A/D lines to select devices on a segment that should participate in a broadcast. (NID) 960-1993

TPS *See:* test program set.

T pulse (linear waveform distortion) A $\sin^2$ pulse with a half-amplitude duration (HAD) of 125 ns. The amplitude of the envelope of the frequency spectrum at 4 MHz is 0.5 of the amplitude at zero frequency and effectively zero at and beyond 8 MHz. (See the corresponding figures.)

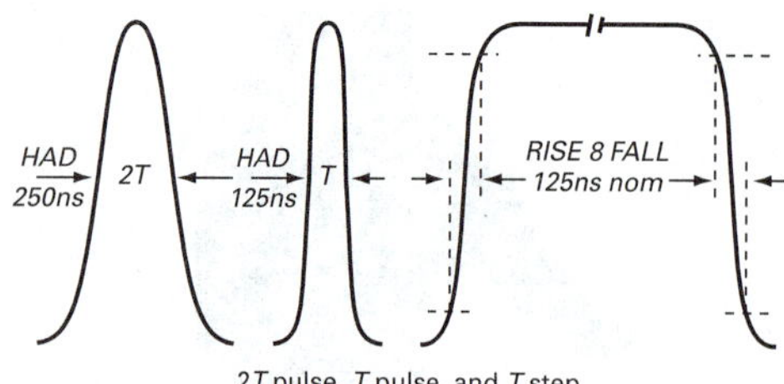

2*T* pulse, *T* pulse, and *T* step

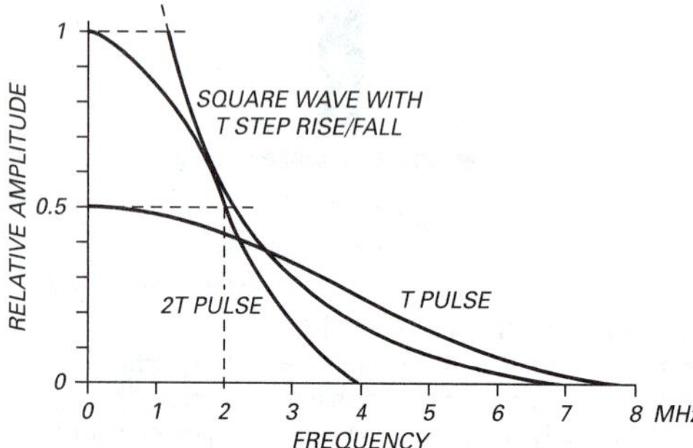

Envelope of frequency spectrum of 2*T* pulse, *T* pulse and square wave with *T* step rise and fall

T pulse

(BT) 511-1979w

TRSL *See:* Test Requirement Specification Language.

TRV *See:* transient recovery voltage; actual transient recovery voltage.

trace (1) The cathode-ray-tube display produced by a moving spot. (IM/HFIM) [40]
(2) (A) (software) A record of the execution of a computer program, showing the sequence of instructions executed, the names and values of variables, or both. Types include execution trace, retrospective trace, subroutine trace, symbolic trace, variable trace. **(B) (software)** To produce a record as in definition (A). **(C) (software)** To establish a relationship between two or more products of the development process; for example, to establish the relationship between a given requirement and the design element that implements that requirement. (C) 610.10-1994, 610.12-1990
(3) To execute the component steps of a computer program, displaying the state of selected system resources after each step. (C/BA) 1275-1994
(4) A diagnostic fault isolation program that uses a probe on a tester. (SCC20) 1445-1998

traceable reference material or standard A NIST prepared standard reference material or a sample of known activity concentration prepared from a NIST traceable reference material (derived standard material). Analogous to certified reference material. (NI) N42.23-1995

traceability (1) (nuclear power quality assurance) The ability to trace the history, application, or location of an item and like items or activities by means of recorded identification. (PE/NP) [124]
(2) (test, measurement, and diagnostic equipment) Process by which the assigned value of a measurement is compared, directly or indirectly, through a series of calibrations to the value established by the U.S. national standard. (MIL) [2]
(3) (software) The degree to which each element in a software development product establishes its reason for existing; for example, the degree to which each element in a bubble chart references the requirement that it satisfies. (C) 610.12-1990
(4) Demonstrated lineage of measurement process quality to the national physical standards. (NI) N42.23-1995
(5) The identification and documentation of derivation paths (upward) and allocation or flowdown paths (downward) of work products in the work product hierarchy. Important kinds of traceability include: to or from external sources to or from system requirements; to or from system requirements to or from lowest level requirements; to or from requirements to or from design; to or from design to or from implementation; to or from implementation to test; and to or from requirements to test. (C/SE) 1362-1998
(6) The degree to which a relationship can be established between two or more products of the development process, especially products having a predecessor-successor or master-subordinate relationship to one another; e.g., the degree to which the requirements and design of a given system element match. (C/SE) 1233-1998

traceability matrix A matrix that records the relationship between two or more products of the development process; for example, a matrix that records the relationship between the requirements and the design of a given software component. (C) 610.12-1990

traced tube bundle Pretraced and thermally insulated instrument tubing that is used for fluid transport, containment, or conditioning system. The bundle is factory fabricated and consists of tubing, heating cable, thermal insulation, and weatherproof jacket. (IA) 515-1997

trace finder *See:* beam finder.

trace interval (television) The interval corresponding to the direction of sweep used for delineation. (BT/AV) 201-1979w

tracer (software) A software tool used to trace. *See also:* trace. (C/SE) 729-1983s

trace, return *See:* return trace.

trace width (oscilloscopes) The distance between two points on opposite sides of a trace perpendicular to the direction of motion of the spot, at which luminance is 50% of maximum. With one setting of the beam controls, the width of both horizontally and vertically going traces within the quality area should be stated. *See also:* oscillograph. (IM/HFIM) [40]

tracing distortion The nonlinear distortion introduced in the reproduction of mechanical recording because the curve traced by the motion of the reproducing stylus is not an exact replica of the modulated groove. For example, in the case of a sine-wave modulation in vertical recording the curve traced by the center of the tip of a stylus is a poid. *See also:* phonograph pickup. (SP) [32]

tracing domain The MD that produces an external trace entry. (C/PA) 1224.1-1993w

tracing MTA The MTA that produces an internal trace entry. (C/PA) 1224.1-1993w

tracing routine (computers) A routine that provides a historical record of specified events in the execution of a program. (C) [20], [85]

track (1) (A) (navigation) (navigation aids) The resultant direction of actual travel projected in the horizontal plane and expressed as a bearing. **(B) (navigation) (navigation aids)** The component of motion that is in the horizontal plane and represents the history of accomplished travel. (AES/GCS) 172-1983
(2) (in electronic computers) The portion of a moving-type storage medium that is accessible to a given reading station: for example, as on film, drum, tapes, or discs. *See also:* band. (C) [85], 338
(3) (test, measurement, and diagnostic equipment) *See also:* channel. (MIL) [2]
(4) (A) A path that is to be followed. For example, the track followed by the read or write head in a storage device during access to a storage medium. *See also:* band; address track; feed track; recording density; card track; storage element; clock track; alternate track; regenerative track. **(B)** One consecutive stream of recorded data on a storage medium. (C) 610.10-1994

track and hold unit A device whose input analog variable is equal to either the input analog variable or a sample of this variable selected by the action of an external Boolean signal. *Synonyms:* track store; track and store unit. (C) 610.10-1994w

track and store unit *See:* track and hold unit.

track angle (navigation aid terms) Track measured from 0° at the reference direction. (AES/GCS) 172-1983w

track ball *See:* control ball.

track brake A magnetic friction brake that compresses against the running rail and is activated by an electrical signal. (VT) 1475-1999

track circuit An electric circuit that includes the rails of a track relay as essential parts. (EEC/PE) [119]

track density The number of tracks per unit length of a data medium, measured in a direction perpendicular to the tracks. (C) 610.10-1994w

tracked munition A munition for which tracking data is required. By necessity, a tracked munition becomes a simulation entity during its flight; its flight path is represented, therefore, by Entity State PDUs. (DIS/C) 1278.1-1995

track element *See:* roadway element.

track homing (navigation aids) The process of following a line of position known to pass through an objective. (AES/GCS) 172-1983w

track indicator chart A maplike reproduction of railway tracks controlled by track circuits so arranged as to indicate automatically for defined sections of track whether such sections are or are not occupied. (EEC/PE) [119]

tracking (1) A motion given to the major lobe of an antenna with the intent that a selected moving target be contained within the major lobe. *Synonym:* angle tracking. (AP/ANT) 145-1993

(2) (A) (data transmission) (radar) The process of following a moving object or a variable input quantity using a servomechanism. *Note:* In radar, tracking is carried out by keeping a narror beam or angle cursor centered on the target angle, a range mark or gate on the delayed echo, or a narrowband filter of the signal frequency. **(B) (data transmission)** (electric) The maintenance of proper frequency relations in circuits designed to be simultaneously varied by gang operation. **(C) (data transmission)** (phonographic technique). The accuracy with which the stylus of a phonograph pickup follows a prescribed path. **(D) (data transmission)** (instrument) The ability of an instrument to indicate, at the division line being checked, when energized by corresponding proportional value of actual end-scale excitation, expressed as a percentage of actual end-scale value. **(E) (data transmission)** (communication satellite) (1) The determination of the orbit and the ephemeris to a satellite or spacecraft. (2) Maintaining the point of a high gain antenna at a moving spacecraft. **(F) (data transmission)** (antenna) A motion given to the major lobe of an antenna so that a selected moving target is contained within the major lobe. (PE/AES/GCS) 599-1985, 172-1983

(3) (image processing and pattern recognition) An image segmentation technique in which arcs are detected by searching sequentially from one arc pixel to the next. (C) 610.4-1990w

(4) (computer graphics) The process of moving a tracking symbol across a display surface in response to coordinate data from an input device. (C) 610.6-1991w

(5) Irreversible degradation of surface material from the formation of conductive carbonized paths. (SPD/PE) C62.11-1999

(6) The process of following a moving object or a variable input quantity. In radar, target tracking in angle, range, or Doppler frequency is accomplished by keeping a beam or angle cursor on the target angle, a range mark or gate on the delayed echo, or a narrowband filter on the signal frequency, respectively. *Note:* This process may be carried out manually or automatically for one or more of the above input quantities. The beam, range gate, or filter can be either centered on the input quantity or can be coarsely placed, with interpolation measurements providing accurate data to a computer that does the fine tracking. *See also:* automatic tracking; tracking radar; track-while-scan. (AES) 686-1997

(7) (computer graphics) *See also:* tracking level. (COM/TA) 1007-1991r

tracking error (1) The deviation of a dependent variable with respect to a reference function. *Note:* As applied to power inverters, tracking error may be the deviation of the output volts per hertz from a prescribed profile or the deviation of the output frequency from a given input synchronizing signal or others. *See also:* self-commutated inverters. (IA) [62]

(2) (lateral mechanical recording) (phonographic techniques) The angle between the vibration axis of the mechanical system of the pickup and a plane containing the tangent to the unmodulated record groove that is perpendicular to the surface of the recording medium at the point of needle contact. *See also:* phonograph pickup. (SP) [32]

(3) The portion of alignment error due to failure to track receiver jitter. (C/LM) 8802-5-1998

tracking level Deviation of the gain or loss as the input level to a codec is varied. Usually the reference level is 0 dBm0 at 1004 Hz. (COM/TA) 1007-1991r

tracking radar (navigation aids) A radar whose primary function is the automatic tracking of targets. *See also:* automatic tracking; tracking. (AES/GCS) 686-1997, 172-1983w

tracking servo (A) A servomechanism that allows a device to follow the path of a target; for example, a telescope or a radar device. **(B)** A mechanism in a rotating storage device that keeps the head centered on a track by following recorded signals on the medium. (C) 610.10-1994

tracking symbol A cross or other predefined symbol, appearing on a display surface, that represents the position of an object being tracked by a computer system. (C) 610.6-1991w

track instrument A device in which the vertical movement of the rail or the blow of a passing wheel operates a contact to open or close an electric circuit. (EEC/PE) [119]

trackless trolley coach *See:* trolley coach.

track pitch The distance between adjacent tracks, measured in a direction perpendicular to the tracks. *See also:* track density; row pitch. (C) 610.10-1994w

track relay A relay receiving all or part of its operating energy through conductors of which the track rails are an essential part and that responds to the presence of a train on the track. (EEC/PE) [119]

track store (1) (analog computer) A component, controlled by digital logic signals, whose output equals the input, when in the "track" mode, and whose output becomes constant and is held (stored) at the value it possessed at the instant its mode was switched to the "store" mode. (C) 165-1977w

(2) *See also:* track and hold unit. (C) 610.10-1994w

track-while-scan (TWS) An automatic target tracking process in which the radar antenna and receiver provide periodic video data from a search scan, together with interpolation measurements, as inputs to computer channels that follow individual targets. *See also:* tracking. (AES) 686-1997

traction machine (elevators) A direct-drive machine in which the motion of a car is obtained through friction between the suspension ropes and a traction sheave. *See also:* driving machine. (EEC/PE) [119]

traction motor An electric propulsion motor used for exerting tractive force through the wheels of a vehicle. (EEC/PE) [119]

traction system *See:* propulsion system.

tractive effort The force generated at the wheel-rail interface as a result of the action of the propulsion system. It may be either positive, indicating motoring/powering, or negative, indicating brake. (VT) 1475-1999

tractive force (electrically propelled vehicle) The total propelling force measured at the rims of the driving wheels, or at the pitch line of the gear rack in the case of a rack vehicle. *Note:* Tractive force of an electrically propelled vehicle is commonly qualified by such terms as: maximum starting tractive force; short-time-rating tractive force; continuous-rating tractive force. *See also:* electric locomotive. (PE/EEC) [119]

tractor *See:* crawler tractor; wheel tractor.

tractor, crawler *See:* crawler tractor.

tractor feed A method for feeding paper or preprinted forms into a printer using an attachment that guides the paper using advancing sprockets that fit into specially prepared guide holes in the paper. *Synonyms:* sprocket feed; form feed; pin feed. *Contrast:* friction feed. (C) 610.10-1994w

tractor, wheel *See:* wheel tractor.

trademark A symbol, word, or phrase used to denote a particular source of goods or services. (C/SE) 1420.1b-1999

tradeoff Parametric analysis of concepts or components for the purpose of optimizing the system or some trait of the system. *See also:* system science. (SMC) [63]

tradeoff analysis An analytical evaluation of design options/alternatives against performance, design-to-cost objectives, and life cycle quality factors. (C/SE) 1220-1998

trade secret Any formula, process, design, or intellectual property interest that is protected by secrecy. (C/SE) 1420.1b-1999

traffic (A) Messages that are transmitted and received over a communication channel. **(B)** A quantitative measure of network load. *Note:* Generally refers to the packet transmission rate, frames/second or frames/hour. (C) 610.7-1995, 610.10-1994

traffic analysis The collection, analysis, and interpretation of communication patterns to inferoperational and logistics-related information. Traffic analysis may be perpetrated using the same techniques as wiretapping, and could result in the unauthorized disclosure of sensitive operations or logisticsinformation. (C/BA) 896.3-1993w

traffic beam *See:* lower beams.

traffic-control system A block signal system under which train movements are authorized by block signals whose indications supersede the superiority of trains for both opposing and following movements on the same track. *See also:* centralized traffic-control system. (EEC/PE) [119]

traffic engineering limits Two engineering methodologies are in common use for PTS. These are time-consistent busy hour (TCBH) and extreme value engineering (EVE). In each methodology, two or three capacities are stated. The lowest capacity is meant to correspond to peak or busy-hour daily loads and is designed to protect the network as several switching systems interact under moderate-to-heavy load. The highest capacity corresponds to peak yearly loads and defines the boundary between normal load and overload. The switching system should maintain throughput at peak capacity, but some of its interactions with other systems may be slightly degraded. Greater loads than peak load may require network management actions. (COM/TA) 973-1990w

traffic flow security (A) The concealment of valid messages on a communication circuit, usually by causing the circuit to appear busy at all times or by encrypting the source and destination addresses of valid messages. **(B)** The state of protection that results from (A). (C) 610.7-1995

traffic intensity A measure of load volume, typically expressed for local switching systems as CCS or calls or call attempts per hour. *Synonym:* event data. *See also:* time-consistent traffic measures. (COM/TA) 973-1990w

traffic locking Electric locking adapted to prevent the manipulation of levers or other devices for changing the direction of traffic on a section of track while that section is occupied or while a signal is displayed for a train to proceed into that section. *See also:* interlocking. (EEC/PE) [119]

traffic peg count A measure of the number of occurrences of particular events; for example, recognition of a call by the system. Accuracy of peg counts is usually not an issue, since under normal load it is relatively easier to get good peg count data than usage data. Accuracy of peg count data should be expressed in percentages at peak system load. *Note:* It is very important to collect accurate data exactly when it may become most difficult—when the switching system is in overload. (COM/TA) 973-1990w

traffic service (telephone switching systems) The services rendered to customers by telephone company operators. (COM) 312-1977w

traffic usage capacity The capacity of the switching network of a switching system depends on the customer usage that can be supported while meeting the network service requirements. The usage of a customer line is composed of originating and terminating usage. (COM/TA) 973-1990w

traffic usage count A measure of server occupancy, usually expressed in Erlangs or hundreds of call seconds per hour (CCS). Measurements may be required for individual servers or for a set of servers performing the same function. Usage may be measured directly, (e.g., by measuring the server holding time of each call) and then summing for all calls. The direct method can be made to be as accurate as the precision of the data register (e.g., tenths of a second). It is usually simpler to measure usage by the scan method. In the scan method, servers are scanned periodically and the number of busy servers is added to a usage register. The scan period is typically every second, 10 s, or 100 s, depending on the desired accuracy. Accuracy will also depend on the service holding time, the number of servers, and the total load being measured. Accuracy of the scan method was the subject of much early statistical analysis in telephony. Accuracy of usage data collected by the scan method should be expressed as the probability that error exceeds a certain percentage of the total usage. *See also:* server. (COM/TA) 973-1990w

traffic usage recorder (telephone switching systems) A device or system for sampling and recording the occupancy of equipment. (COM) 312-1977w

trailer (1) Identification or control information placed at the end of a file or message. *Contrast:* header. (C) 610.12-1990 **(2)** The contiguous control bits following a transmission that contain information used for such purposes as bit error detection and end-of-transmission indication. *Contrast:* header. (C) 610.7-1995 **(3)** The portion of tape that follows the end-of-tape marker. *Contrast:* leader. (C) 610.10-1994w

trailer card A punch card that contains information identifying data on the preceding cards. *Note:* Usually the last card in a deck of cards *Contrast:* header card. (C) 610.10-1994w

trailer label *See:* end-of-file label.

trailer plow (cable plowing) (static or vibratory plows) A unit that is self-contained except for drawbar pull that is furnished by a prime mover. (T&D/PE) 590-1977w

trailing decision A loop control that is executed after the loop body. *Contrast:* leading decision. *See also:* UNTIL. (C) 610.12-1990

trailing edge (pulse transformers) That portion of the pulse occurring between the time of intersection of straight-line segments used to determine A_T and the time at which the instantaneous value reduces to zero. *Synonym:* last transition. (PEL/ET) 390-1987r

trailing edge amplitude (pulse transformers) That quantity determined by the intersection of a line passing through the points on the trailing edge where the instantaneous value reaches 90% and 10% of A_T, and the straight-line segment fitted to the top of the pulse in determining A_M. (PEL/ET) 390-1987r

trailing edge, pulse *See:* pulse trailing edge.

trailing-edge pulse time The time at which the instantaneous amplitude last reaches a stated fraction of the peak pulse amplitude. (IM/WM&A) 194-1977w

trailing-type antenna (aircraft) A flexible conductor usually wound on a reel within the aircraft passing through a fairlead to the outside of the aircraft, terminated in a streamlined weight or wind sock and fed out to the proper length for the desired radio frequency of operation. It has taken other forms such as a capsule that when exploded releases the antenna. (EEC/PE) [119]

trailing zero A zero that comes after the last digit in a numeric representation that is non-zero, and that is to the right of the decimal point; for example, the two zeros in "324.600." *Contrast:* leading zero. (C) 610.5-1990w

train (1) (illuminating engineering) The angle between the vertical plane through the axis of the searchlight drum and the plane in which this plane lies when the searchlight is in a position designated as having zero train. (EEC/IE) [126]
(2) A consist of one or more basic operating units. (VT/RT) 1477-1998, 1473-1999, 1475-1999, 1474.1-1999

train control system The system for controlling train movement, enforcing train safety, and directing train operations. (VT) 1477-1998

train-control territory That portion of a division or district equipped with an automatic train-control system. *See also:* automatic train control. (EEC/PE) [119]

train describer An instrument used to give information regarding the origin, destination, class, or character of trains, engines, or cars moving or to be moved between given points. (EEC/PE) [119]

trained listening group (speech-quality measurements) Six to ten listeners who understand thoroughly the purpose of the speech quality test and respond properly throughout the test. All persons of the group shall meet the requirements on auditory acuity as described by USAS S3.2-1960 (Monosyllabic Word Intelligibility). The training of the listeners will depend on the special type of tests to be conducted. 297-1969w

training (1) The process of synchronizing the receiver circuit of a linc to the incoming data stream during initialization. (C/MM) 1596-1992
(2) (local area networks) (Training_Up, Training_Down) A link control signal indicating that the sending entity is either requesting or giving permission to train (initialize) the link. (C) 8802-12-1998

trainline interoperability The ability of the basic operating units that constitute a train to communicate successfully with each other through coupler interface(s), without limitation as to the sequence or orientation of the basic operating units within the train, and without requirement for manual configuration other than optional manual confirmation of basic operating unit sequence within the train. (VT) 1473-1999

trainlines Wires routed through and/or between vehicles or units by means of couplers, jumpers, or other means so that power or signals may be transmitted to all vehicles of the train. (VT) 1475-1999

train printer An impact printer in which the type slugs are moved around on a circular track, known as a print train. (C) 610.10-1994w

trajectory stability Orbital stability where the solution curve is not closed. *See also:* control system. (CS/IM) [120]

trajectory, state *See:* state trajectory.

trans-μ-factor (multibeam electron tubes) The ratio of the magnitude of an infinitesimal change in the voltage at the control grid of any one beam to the magnitude of an infinitesimal change in the voltage at the control grid of a second beam. The current in the second beam and the voltage of all other electrodes are maintained constant. (ED) 161-1971w

transaction (1) An event that requires data contained in a master file to be processed. *See also:* change transaction; null transaction; update transaction; delete transaction; add transaction. (C) 610.2-1987
(2) A data element, control element, signal, event, or change of state that causes, triggers, or initiates an action or sequence of actions. (SE/C) 610.12-1990
(3) (supervisory control, data acquisition, and automatic control) That sequence of messages between master and remote stations required to perform a specific function (for example, acquire specific data or control a selected device). (SUB/PE) C37.1-1994
(4) (STEbus) The combination of data transfer sequences controlled by a master during a single bus tenure. (MM/C) 1000-1987r
(5) (NuBus) A sequence of cycles beginning with a start cycle and ending with an ack cycle that is used to convey data between a master and a slave. (C/MM) 1196-1987w
(6) A sequence of messages between cooperating terminals to perform a specific function. Usually a minimum of one message in each direction that is comprised of a command followed by a response. (SUB/PE) 999-1992w
(7) A unit of work consisting of an arbitrary number of individual operations, all of which will either complete successfully or abort with no effect on the intended resources. A transaction has well-defined boundaries. A transaction starts with a request from the application program and either completes successfully (commits) or has no effect (abort). Both the commit and abort signify completion of a transaction. (C/PA) 14252-1996
(8) An information exchange between two nodes. A transaction consists of a request subaction and a response subaction. The request subaction transfers commands (and possibly data) between a requester and a responder. The response subaction returns status (and possibly data) from the responder to the requester. (C/MM) 1596.5-1993, 1596-1992
(9) A transfer between requester and responder consisting of a request and response subaction. The request subaction transfers a command (and sometimes data) between a requester and responder. The response subaction returns status (and sometimes data) from the responder to the requester. A transaction may be either unified or split. (C/MM) 1212-1991s
(10) A single use of a service. (ATLAS) 1232-1995
(11) A transaction is a sequence of packets sent between two or more terminal nodes to perform some function. *See also:* transaction layer. (C/BA) 1355-1995
(12) A request and the corresponding response. The response may be null for transactions with broadcast destination addresses. This is the PDU for the transaction layer. (C/MM) 1394-1995
(13) An event initiated with a connection phase and terminated with a disconnection phase. Data may or may not be transferred during a transaction. Often used instead of the more precise phrase "bus transaction" for the sake of brevity. *See also:* bus transaction; system transaction. (C/BA) 10857-1994
(14) A functionally continuous and complete exchange of information between the roadside equipment (RSE) and the vehicle transponder. (SCC32) 1455-1999

transaction analysis A software development technique in which the structure of a system is derived from analyzing the transactions that the system is required to process. *Synonym:* transaction-centered design. *See also:* transform analysis; structured design; object-oriented design; modular decomposition; input-process-output; rapid prototyping; data structure-centered design; stepwise refinement. (C) 610.12-1990

transaction bystander A module that is not participating in the current transaction. A transaction bystander monitors, asserts, and releases the synchronization signals, even though it is not the initiator of a transaction or the target of it. (C/BA) 896.4-1993w

transaction-centered design *See:* transaction analysis.

transaction code An identifier associated with a transaction and representing the operation to be carried out by that transaction. For example, "A" for an add transaction, "D" for a delete transaction. (C) 610.2-1987

transaction completion (reply) A reply generated by a function in response to one or more I/O transaction initiations. The completion returns status and sometimes data. In a disk read I/O transaction, for example, the completion returns the data and status from the disk and the function in the I/O Unit to the Processor. (C/MM) 1212.1-1993

transaction file An organized collection of transaction records. *Synonym:* detail file. *Contrast:* master file. (C) 610.2-1987

transaction_ID A value selected by an initiator to designate a given I/O transaction. It is included either explicitly or implicitly in a transaction-initiation message and returned in a transaction-completion message. (C/MM) 1212.1-1993

transaction initiation (request) A request generated by the initiator to start an action by the responder. An initiation message usually transfers a command and sometimes data. For a disk read I/O transaction, for example, the initiation transfers the address and command. (C/MM) 1212.1-1993

transaction, I/O *See:* I/O transaction.

transaction layer (1) The layer above the packet layer for use by applications. It is unspecified in this standard. *See also:* transaction. (C/BA) 1355-1995
(2) The layer, in a stack of three protocol layers defined for the Serial Bus, that defines a request-response protocol to perform bus operations of type read, write, and lock. (C/MM) 1394-1995

transaction matrix A matrix that identifies possible requests for database access and relates each request to information categories or elements in the database. (C) 610.12-1990

transaction record A record, representing one transaction, used to process data stored in a master file. *See also:* update transaction; null transaction; change transaction; delete transaction; add transaction. (C) 610.2-1987

transactor A magnetic device with an air-gapped core having an input winding which is energized with an alternating current and having an output winding which produces a voltage that is a function of the input current. *Note:* The term "transactor" is a contraction of the words "transformer" and "reactor." (SWG/PE/PSR) C37.110-1996, C37.100-1992

transadmittance For harmonically varying quantities at a given frequency, the ratio of the complex amplitude of the current at one pair of terminals of a network to the complex amplitude of the voltage across a different pair of terminals. *See also:* interelectrode transadmittance. (IM/HFIM) [40]

transadmittance compression ratio (electron tube) The ratio of the magnitude of the small-signal forward transadmittance of the tube to the magnitude of the forward transadmittance at a given input signal level. (ED) 161-1971w

transadmittance, forward *See:* forward transadmittance.

transceiver (1) (data transmission) The combination of radio transmitting and receiving equipment in a common housing, usually for portable or mobile use, and employing common circuit components for both transmitting and receiving. (PE) 599-1985w
(2) (navigation aids) A combination transmitter and receiver in a single housing, with some components being used by both parts. *See also:* transponder. (AES/GCS) 172-1983w
(3) (A) A device that both transmits and receives data. **(B)** A device that connects a host interface to a network. **(C)** A device that applies electronic signals to the cable and may sense collisions. *Note:* Definition (C) is contextually specific to IEEE Std 802.3. (C) 610.7-1995

transceiver cable A four-pair, shielded cable which interconnects a workstation to a transceiver or fan-out box. *Note:* This term is contextually specific to IEEE Std 802.3. *See also:* coaxial cable; trunk cable; drop cable; attachment unit interface cable. (C) 610.7-1995

transceiver chatter *See:* chatter.

transconductance The real part of the transadmittance. *Note:* Transconductance is, as most commonly used, the interelectrode transconductance between the control grid and the plate. At low frequencies, transconductance is the slope of the control-grid-to-plate transfer characteristic. *See also:* interelectrode transconductance; electron-tube admittances. (ED) 161-1971w

transconductance meter (mutual-conductance meter) An instrument for indicating the transconductance of a grid-controlled electron tube. *See also:* instrument. (EEC/PE) [119]

transcribe (electronic computation) To convert data recorded in a given medium to the medium used by a digital computing machine or vice versa. (C) 162-1963w

transcriber (electronic computation) Equipment associated with a computing machine for the purpose of transferring input (or output) data from a record of information in a given language to the medium and the language used by a digital computing machine (or from a computing machine to a record of information). (Std100) 270-1966w

transducer (1) (electrical heating applications to melting furnaces and forehearths in the glass industry) A device that is actuated by power from one system and supplies power in any other form to a second system. (IA) 668-1987w
(2) (communication and power transmission) A device by means of which energy can flow from one or more transmission systems or media to one or more other transmission systems or media. *Note:* The energy transmitted by these systems or media may be of any form (for example, it may be electric, mechanical, or acoustical), and it may be of the same form or different forms in the various input and output systems or media. (MIL/C/AP/ANT) [2], [85], 145-1983s
(3) (metering) A device to receive energy from one system and supply energy (of either the same or of a difference kind) to another system, in such a manner that the desired characteristics of the energy input appear at the output. (ELM) C12.1-1988
(4) (thyristor) A device which under the influence of a change in energy level of one form or in one system, produces a specified change in energy level of another form or in another system. (IA/IPC) 428-1981w
(5) A device for converting energy from one form to another. (C) 610.10-1994w
(6) A device converting energy from one domain into another. The device may either be a sensor or an actuator. (IM/ST) 1451.2-1997
(7) A device converting energy from one domain into another, calibrated to minimize the errors in the conversion process. A sensor or an actuator. (IM/ST) 1451.1-1999

transducer, active *See:* active transducer.

Transducer Block An instance of a subclass of `IEEE1451_TransducerBlock`. (IM/ST) 1451.1-1999

transducer conversion loss The ratio of the SAW power generated in the substrate at the transducer output to the power available in the circuit at the transducer input in decibels. (UFFC) 1037-1992w

Transducer Electronic Data Sheet (TEDS) (1) A data sheet describing a transducer stored in some form of electronically readable memory. (IM/ST) 1451.2-1997
(2) Several of the IEEE 1451.X standards use TEDS to provide a machine-readable specification of the characteristics of the transducer interface. (IM/ST) 1451.1-1999

transducer gain (1) The ratio of the power that the transducer delivers to the specified load under specified operating conditions to the available power of the specified source. *Notes:* 1. If the input and/or output power consist of more than one component, such as multifrequency signals or noise, then the particular components used and their weighting must be specified. 2. This gain is usually expressed in decibels. *See also:* transducer. (Std100) 270-1966w
(2) (two-port linear transducer) At a specified frequency, the ratio of the actual signal power transferred from the output port of the transducer to its load, to the available signal power from the source driving the transducer. (ED) 161-1971w

transducer, ideal *See:* ideal transducer.

Transducer Independent Interface The digital interface used to connect a Smart Transducer Interface Module to a Network Capable Application Processor. (IM/ST) 1451.2-1997

transducer interface The physical connection by which a transducer communicates with the control or data systems that it is a member of, including the physical connector, the signal wires used and the rules by which information is passed across the connection. (IM/ST) 1451.2-1997

transducer, line *See:* line transducer.

transducer loss The ratio of the available power of the specified source to the power that the transducer delivers to the specified load under specified operating conditions. *Notes:* 1. If the input and/or output power consist of more than one component, such as multifrequency signals or noise, then the particular components used and their weighting must be specified. 2. This loss is usually expressed in decibels. *See also:* transducer. (Std100) 270-1966w

transducer, passive *See:* passive transducer.

transfer (1) (telephone switching systems) A feature that allows a customer to instruct the switching equipment or operator to transfer his call to another station. (COM) 312-1977w

(2) (A) (electronic computation) To transmit, or copy, information from one device to another. **(B) (electronic computation)** To jump. **(C) (electronic computation)** The act of transferring. *See also:* transmit; jump. (C) 162-1963

(3) (electrostatography) The act of moving a developed image, or a portion thereof, from one surface to another, as by electrostatic or adhesive forces, without altering the geometric configuration of the image. *See also:* electrostatography. (ED) [46]

(4) (data management) (software) To send data from one place and receive it at another. *See also:* transmit. (C) 610.5-1990w, 610.12-1990

(5) (software) To relinquish control by one process and assume it at another, either with expectation of return (call) or without such expectation (jump). *See also:* call; jump. (C) 610.12-1990

(6) (STEbus) The movement of a single byte of data from the current master to the addressed slave(s) or from the addressed slave to the master. (C/MM) 1000-1987r

(7) The successful movement of a bit or bits between an MTM-Bus Master module and one or more modules co-connected by the MTM-Bus. (TT/C) 1149.5-1995

(8) To transmit, or copy, information from one device to another. (IM/ST) 1451.2-1997

transfer admittance (1) (linear passive networks) A transmittance for which the excitation is a voltage and the response is a current. (CAS) 156-1960w

(2) (from the *i*th terminal to the *j*th terminal of an *n*-terminal network) The (complex) current flowing to the *i*th terminal divided by the (complex) voltage applied between the *j*th terminal with respect to the reference point when all other terminals have arbitrary terminations. For example, for a 3-terminal network terminated in short circuits,

$$y_{12} = \frac{I}{V}\bigg|_{V_1} = 0$$

transfer alignment (navigation aids) A method of transfer of reference coordinates to an inertial navigation system for initial alignment. Accomplished by way of: structure to structure mating, simultaneous measurement of acceleration patterns, or by optical measurement techniques. (AES/GCS) 172-1983w

transfer capability The capacity and ability of a transmission network to allow for the reliable movement of electric power from an area of supply to an area of need. (PE/PSE) 858-1993w

transfer characteristic (1) (electron tube) A relation, usually shown by a graph, between the voltage of one electrode and the current to another electrode, all other electrode voltages being maintained constant. *See also:* electrode. (ED) 161-1971w

(2) (camera tubes) A relation between the illumination on the tube and the corresponding signal output current, under specified conditions of illumination. *Note:* The relation is usually shown by a graph of the logarithm of the signal output current as a function of the logarithm of the illumination. *See also:* illumination; television; sensitivity. (ED) 161-1971w

transfer check (electronic computation) A check (usually an automatic check) on the accuracy of a data transfer. *Note:* In particular, a check on the accuracy of the transfer of a word. (C) 162-1963w

transfer constant *See:* image transfer constant.

transfer contacts For reactance-type LTCs, a set of contacts that makes and breaks current. *Note:* In cases where no bypass contacts are provided, the transfer contact is a continuous current carrying contact. (PE/TR) C57.131-1995

transfer control *See:* jump.

transfer current (gas tube) The current to one electrode required to initiate breakdown to another electrode. *Note:* The transfer current is a function of the voltage of the second electrode. (ED) 161-1971w

transfer-current ratio (linear passive networks) A transmittance for which the variables are currents. *Note:* The word transfer is frequently dropped in present usage. (CAS) 156-1960w

transfer function (1) (seismic qualification of Class 1E equipment for nuclear power generating stations) A complex frequency response function that defines the dynamic characteristics of a constant parameter linear system. For an ideal system, the transfer function is the ratio of the Fourier transform of the output to that of a given input. (PE/NP) 344-1987r

(2) (control system feedback) A mathematical, graphic, or tabular statement of the influence that a system or element has on a signal or action compared at input and at output terminals. *Note:* For a linear system, general usage limits the transfer function to mean the ratio of the Laplace transform of the output to the Laplace transform of the input in the absence of all other signals, and with all initial conditions zero. *See also:* transfer function; feedback control system. (IM/PE/EDPG) [120], [3]

(3) (low-power wide-band transformers) The complex ratio of the output of the device to its input. It is also the combined phase and frequency responses. (MAG/PEL/ET) 264-1977w, 111-1984w

(4) (nuclear power generating station) A mathematical, graphical, or tabular statement of the influence which a module has on a signal or action compared at input and at output terminals. This should be specified as to whether it is transient or steady state. (PE/NP) 381-1977w

(5) (excitation systems) A mathematical, graphical, or tabular statement of the influence which a system or element has on a signal or action compared at input and output terminals. *Note:* For a linear system, general usage limits the transfer function to mean the ratio of the Laplace transform of the output to the Laplace transform of the input in the absence of all other signals, and with all initial conditions zero. (PE/EDPG) 421A-1978s

(6) The relationship between the input and output signals of a circuit, especially when expressed as a continuous mathematical function. (C) 610.10-1994w

(7) The ratio of the device output signal (voltage, current, frequency, meter reading, etc.) to the incident field or field vector of interest in the Frequency Domain. The transfer function is the Laplace (or Fourier) transform of the impulse response function. (EMC) 1309-1996

(8) [$H(f)$] The quantity $Y(f)$ divided by $X(f)$, where $Y(f)$ and $X(f)$ are the frequency domain representations of the output and input signals respectively. (PE/PSIM) 4-1995

(9) (fiber optics) (of a device) The complex function, $H(f)$, equal to the ratio of the output to input of the device as a function of frequency. The amplitude and phase responses are, respectively, the magnitude of $H(f)$ and the phase of $H(f)$. *Notes:* 1. For an optical fiber, $H(f)$ is taken to be the ratio of output optical power to input optical power as a function of modulation frequency. 2. For a linear system, the transfer function and the impulse response $h(t)$ are related through the Fourier transform pair, a common form of which is given by

$$H(f) = \int_{-\infty}^{\infty} h(t)^{(i2\pi ft)}dt$$

$$h(t) = \int_{-\infty}^{\infty} H(f)^{(-2\pi ft)}df$$

where f is frequency. Often $H(f)$ is normalized to $H(0)$ and $h(t)$ to

$$\int_{-\infty}^{\infty} h(t)\mathrm{d}t,$$

which by definition is $H(0)$. *Synonyms:* baseband response function; frequency response. *See also:* impulse response. (Std100) 812-1984w

transfer immittance *See:* transmittance.

transfer impedance (linear passive networks) A transmittance for which the excitation is a current and the response is a voltage. *Note:* It is therefore the impedance obtained when the response is determined at a point other than that at which the driving force is applied, all terminals being terminated in any specified manner. In the case of an electric circuit, the response would be determined in any branch except that in which the driving force is. *See also:* network analysis; self-impedance. (CAS) 156-1960w
(2) **(A)** (linear passive networks) (general). A transmittance for which the excitation is a current and the response is a voltage. **(B)** (from the ith terminal to the jth terminal of an n- terminal network). The (complex) voltage measured between the ith terminal and the reference point divided by the (complex) current applied to the jth terminal when all other terminals have arbitrary terminations. For example, for a 3-terminal network terminated in open circuits

$$Z_{12} = \frac{V_1}{I_2}\bigg|_{I_1} = 0$$

(CAS/PE/EM) 95-1977

transfer instruction *See:* branch instruction.

transfer instrument (radiation protection) Instrument or dosimeter exhibiting high precision which has been standardized against a national or derived standardized source. (NI) N323-1978r

transfer interpreter A device that prints on a punch card the characters corresponding to hole patterns punched in another card. *See also:* interpreter. (C) 610.10-1994w

transfer lag *See:* first-order lag; multiorder lag.

transfer line size The size of the block of data transferred to or from main memory in a caching environment. (C/BA) 896.4-1993w

transfer locus (linear system or element) A plot of the transfer function as a function of frequency in any convenient coordinate system. *Note:* A plot of the reciprocal of the transfer function is called the inverse transfer locus. *See also:* phase locus; feedback control system. (IM) [120]

transfer of control *See:* jump.

transfer operation The bus operation in which a bus owner transfers data on the parallel system bus. *See also:* bus operation; bus owner. (C/MM) 1296-1987s

transfer rate The average number of bits, characters, or blocks per unit time passing between corresponding devices in a data transmission system. It is expressed in terms of bits, characters, or blocks per second, minute, or hour. *Synonym:* data rate. (C) 610.7-1995

transfer ratio A dimensionless transfer function. (Std100) 270-1966w

transfer ratio correction (correction to setting) The deviation of the output phasor from nominal, in proportional parts of the input phasor.

$$\frac{\text{Output}}{\text{Input}} = A + \alpha + j\beta$$

INPUT A OUTPUT

A = setting
α = in-phase transfer ratio correction
β = quadrature transfer ratio correction

transferred charge The net electric charge transferred from one terminal of a capacitor to another via an external circuit. *See also:* nonlinear capacitor. (ED) [46]

transferred-charge characteristic (nonlinear capacitor) The function relating transferred charge to capacitor voltage. *See also:* nonlinear capacitor. (ED) [46]

transferred information *See:* transinformation.

transferred jitter The amount of jitter in the recovered clock of the upstream PHY which is subsequently transferred to the downstream PHY which in turn is transferred to the next downstream PHY. Transferred jitter is important because each PHY must both limit the amount of jitter it generates and track the jitter delivered by the upstream PHY. (C/LM) 8802-5-1998

transferred voltage **(1)** That voltage between points of contact, hand to foot or feet, where the grounded surface touched is intentionally grounded at a remote point (or unintentionally touching at a remote point a conductor connected to the station ground system). Here the voltage rise encountered due to ground fault conditions may equal or exceed the ground potential rise of the ground grid discharging the fault current (and not a fraction of this total as is encountered in the usual touch contact). (PE/EDPG) 665-1995
(2) A special case of the touch voltage where a voltage is transferred into or out of the substation from or to a remote point external to the substation site. (PE/SUB) 1268-1997, 80-2000

transferring (as applied to fall protection) The act of moving from one distinct object to another (e.g., between an aerial device and a structure). (NESC/T&D/PE) C2-1997, 1307-1996

transfer standard A term that refers to an electrically small field probe or field sensor. This can be a short dipole for sensing E-fields or a small loop for H-fields, which has a known response over a given range of frequency and amplitude. This known response can be either accurately calculable quasi-static response parameters or a calibration performed to some specified accuracy and precision by an accredited calibration facility. (EMC) 1309-1996

transfer standards, alternating-current–direct-current Devices used to establish the equality of a root-mean-square current or voltage (or the average value of alternating power) with the corresponding steady-state direct-current quantity that can be referred to the basic standards through potentiometric techniques. *See also:* auxiliary device to an instrument. (ELM) C12.1-1982s

transfer state A state of a Link Layer Controller the name of which begins with the letters "XFER". Such states in the MTM-Bus Master Link Layer Controller are called M-transfer states and in the MTM-Bus Slave Link Layer Controller are called S-transfer states. (TT/C) 1149.5-1995

transfer switch **(1)** **(emergency and standby power)** A device for transferring one or more load conductor connections from one power source to another. (IA) [18]
(2) (a high-voltage switch) A switch arranged to permit transferring a conductor connection from one circuit to another without interrupting the current.

1) A tandem transfer switch is a switch with two blades, each of which can be moved into or out of only one contact.
2) A double-blade double-throw transfer switch is a switch with two blades, each of which can be moved into or out of either of two contacts.

Note: In contrast to high-voltage switches, many low-voltage, control and instrument transfer switches interrupt current during transfer. *See also:* automatic transfer equipment; selector switch. (SWG/PE) C37.100-1992

transfer switch, load *See:* load transfer switch.

transfer time **(1)** **(A)** **(gas tube surge arresters)** The time required for the voltage across a conducting gap to drop into the arc region after the gap initially begins to conduct. **(B)** The time duration of the transverse voltage. (PE/SPD) [8], C62.31-1987
(2) The part of access time attributed to the time between the beginning of a transfer of data to or from storage and its completion. (C) 610.10-1994w

(3) **(uninterruptible power supply)** The time that it takes an uninterruptible power supply to transfer the critical load from the output of the inverter to the alternate source, or back again. (IA/PSE) 1100-1999

transfer time, relay *See:* relay transfer time.

transfer trip (1) A form of remote trip in which a communication channel is used to transmit a trip signal from the relay location to a remote location. (SWG/PE) C37.100-1992
(2) The sending of a TRIP signal via a communication channel to a remote line terminal. (PE/PSR) C37.113-1999

transform analysis A software development technique in which the structure of a system is derived from analyzing the flow of data through the system and the transformations that must be performed on the data. *Synonyms:* transformation analysis; transform-centered design. *See also:* rapid prototyping; transaction analysis; data structure-centered design; structured design; stepwise refinement; modular decomposition; input-process-output; object-oriented design. (C) 610.12-1990

transformation A segment attribute that determines the translation, scaling, and rotation applied to a segment when it is displayed on a display surface. (C) 610.6-1991w

transformation analysis *See:* transform analysis.

transformation function A mapping function that performs graphical coordinate transformations such as scaling, rotation, and translation. (C) 610.6-1991w

transform-centered design *See:* transform analysis.

transformer (1) A device, which when used, will raise or lower the voltage of alternating current of the original source. (NESC/NEC) [86]
(2) (power and distribution transformers) A static electric device consisting of a winding, or two or more coupled windings, with or without a magnetic core, for introducing mutual coupling between electric circuits. Transformers are extensively used in electric power systems to transfer power by electromagnetic induction between circuits at the same frequency, usually with changed values of voltage and current. (PE/TR) C57.12.80-1978r
(3) (failure data for power transformers and shunt reactors) A static electric device consisting of a winding, or two or more coupled windings, with or without a magnetic core, for introducing mutual coupling between electric circuits. *Note:* The transformer includes all transformer-related components, such as bushings, LTCs, fans, temperature gauges, etc, and excludes all system-related components, such as surge arresters, grounding resistors, high voltage switches, low-voltage switches, and house service equipment. (PE/TR) C57.117-1986r
(4) An inductive electrical device which uses electromagnetic energy to transform voltage and current levels within a circuit. (C) 610.10-1994w
(5) *See also:* transformer coupled; dry-type encapsulated water-cooled transformer; liquid-filled, or liquid-cooled transformer; dry-type transformer. (IA) 668-1987w

transformer, alternating-current arc welder *See:* alternating-current arc welder transformer.

transformerboard Pressboard specifically manufactured for use as transformer dielectric insulation. (PE/TR) 1276-1997

transformer category definitions (distribution, power and regulating transformers) n/a. *Note:* All kVA ratings are minimum nameplate kVA for the principal windings. Category I includes distribution transformers manufactured in accordance with ANSI C57.12.20-1974, Requirements for Overhead-Type Distribution Transformers 67 000 Volts and Below; 500 kVA and Smaller, up through 500 kVA, single phase or three phase. In addition, autotransformers of 500 equivalent two-winding kVA or less that are manufactured as distribution transformers in accordance with ANSI C57.12.20-1974 are included in Category I, even through their nameplate kVAs may exceed 500. (PE/TR) C57.12.00-1987s

transformer class designations *See:* oil-immersed transformer.

transformer, constant-voltage *See:* constant-voltage transformer.

transformer correction factor (TCF) The ratio of the true watts or watthours to the measured secondary watts or watthours, divided by the marked ratio. *Note:* The transformer correction factor for a current or voltage transformer is the ratio correction factor multiplied by the phase angle correction factor for a specified primary circuit power factor. The true primary watts or watthours are equal to the watts or watthours measured, multiplied by the transformer correction factor and the marked ratio. The true primary watts or watthours, when measured using both current and voltage transformers, are equal to the current transformer ratio correction factor multiplied by the voltage transformer ratio correction factor multiplied by the marked ratios of the current and voltage transformers multiplied by the observed watts or watthours. It is usually sufficiently accurate to calculate true watts or watthours as equal to the product of the two transformer correction factors multiplied by the marked ratios multiplied by the observed watts or watthours. (PE/TR) C57.13-1993, C57.12.80-1978r, [57]

transformer coupled (electrical heating applications to melting furnaces and forehearths in the glass industry) The power modulation device is connected in the primary circuit of a transformer whose secondary circuit is connected to the glass. (IA) 668-1987w

transformer, dry-type *See:* dry-type transformer.

transformer, energy-limiting *See:* energy-limiting transformer.

transformer equipment rating A volt-ampere output together with any other characteristics, such as voltage, current, frequency, and power factor, assigned to it by the manufacturer. *Note:* It is regarded as a test rating that defines an output that can be taken from the item of transformer equipment without exceeding established temperature-rise limitations, under prescribed conditions of test and within the limitations of established standards. *See also:* duty. (PE/TR) [57]

transformer, grounding *See:* grounding transformer.

transformer grounding switch and gap (capacitance potential devices) Consists of a protective gap connected across the capacitance potential device and transformer unit to limit the voltage impressed on the transformer and the auxiliary or shunt capacitor, when used; and a switch that when closed removes voltage from the potential device to permit adjustment of the potential device without interrupting high-voltage line operation and carrier-current operation when used. *See also:* outdoor coupling capacitor. (PE/EM) 43-1974s

transformer, group-series loop insulating *See:* group-series loop insulating transformer.

transformer, high-power-factor *See:* high-power−factor transformer.

transformer, high-reactance *See:* high-reactance transformer.

transformer, ideal *See:* ideal transformer.

transformer, individual-lamp insulating *See:* individual-lamp insulating transformer.

transformer, insulating *See:* insulating transformer.

transformer insulation life For a given temperature of the transformer insulation, the total time between the initial state for which the insulation is considered new and the final state for which dielectric stress, short circuit stress, or mechanical movement, which could occur in normal service, and would cause an electrical failure. (PE/TR) C57.91-1995

transformer integrally mounted cable terminating box A weatherproof air-filled compartment suitable for enclosing the sidewall bushings of a transformer and equipped with any one of the following entrance devices:

a) Single or multiple-conductor potheads with couplings or wiping sleeves;
b) Wiping sleeves;

c) Couplings with or without stuffing boxes for conduit-enclosed cable, metallic-sheathed cable, or rubber-covered cable.

(PE/TR) [108]

transformer, interphase *See:* interphase transformer.

transformer, isolating *See:* isolating transformer.

transformer, line *See:* line transformer.

transformer loss The ratio of the signal power that an ideal transformer would deliver to a load, to the power delivered to the same load by the actual transformer, both transformers having the same impedance ratio. *Note:* Transformer loss is usually expressed in decibels. *See also:* transmission loss. (COM/SP) 151-1965w

transformer-loss compensator (metering) A passive electric network that adds to or subtracts from the meter registration to compensate for predetermined iron and copper losses of transformers and transmission lines. (ELM) C12.1-1988

transformer, low-power factor *See:* low-power factor transformer.

transformer, matching *See:* matching transformer.

transformer, network *See:* network transformer.

transformer, nonenergy-limiting *See:* nonenergy-limiting transformer.

transformer, oil-immersed *See:* oil-immersed transformer.

transformer, outdoor *See:* outdoor transformer.

transformer overcurrent tripping *See:* overcurrent release; indirect release.

transformer, phase-shifting *See:* phase-shifting transformer.

transformer, pole-type *See:* pole-type transformer.

transformer, protected outdoor *See:* protected outdoor transformer.

transformer-rated electromechanical watthour meter A transformer-rated electromechanical watthour meter is one in which the terminals are arranged for connection to the secondary windings of external instrument transformers.

(ELM) C12.10-1987

transformer-rectifier, alternating-current–direct-current arc welder A combination of static rectifier and the associated isolating transformer, reactors, regulators, control, and indicating devices required to produce either direct or alternating current suitable for arc-welding purposes.

(EEC/AWM) [91]

transformer-rectifier, direct-current arc welder A combination of static rectifiers and the associated isolating transformer, reactors, regulators, control, and indicating devices required to produce direct current suitable for arc welding.

(EEC/AWM) [91]

transformer relay A relay in which the coils act as a transformer. (EEC/PE) [119]

transformer removable cable-terminating box A weatherproof air-filled compartment suitable for enclosing the side-wall bushings of a transformer and equipped with mounting flange(s) (one or two) to accommodate either single-conductor or multiconductor potheads or entrance fittings, depending upon the type of cable termination to be used and the number of three-phase cable circuits (one or two) to be terminated.

(PE/TR) [107]

transformer secondary current rating *See:* secondary current rating.

transformer, series *See:* series transformer.

transformer, series street-lighting *See:* series street-lighting transformer.

transformer, series street-lighting, rating The lumen rating of the series lamp, or the wattage rating of the multiple lamps, that the transformer is designed to operate. *See also:* specialty transformer. (PE/TR) [57]

transformer short-circuit impedance **(A)** For Category I and Category II transformers, the transformer impedance, expressed in percent on the transformer's rated voltage and rated base kilovoltamperes. **(B)** For Category III and Category IV transformers, the sum of transformer impedance and system short-circuit impedance at the transformer location, expressed in percent on the transformer's rated voltage and rated base kilovoltamperes. (PE/TR) C57.109-1993

transformer, shunt *See:* shunt transformer.

transformer, specialty *See:* specialty transformer.

transformer, station-type *See:* station-type transformer.

transformer undercurrent tripping *See:* undercurrent release; indirect release.

transformer vault An isolated enclosure either above or below ground with fire-resistant walls, ceiling, and floor, in which transformers and related equipment are installed, and which is not continuously attended during operation. *See also:* vault.

(NESC) C2-1997

transformer, vault-type *See:* vault-type transformer.

transformer voltage (of a network protector) The voltage between phases or between phase and neutral on the transformer side of a network protector. (SWG/PE) C37.100-1992

transforming station (power operations) A station where power is transformed from one voltage level to another.

(PE/PSE) 858-1987s

transhorizon tropospheric propagation Tropospheric propagation between two points, the reception point being beyond the radio horizon of the transmission point. Transhorizon propagation includes a variety of possible propagation mechanisms such as diffraction, scattering, ducting, refraction and reflection. *See also:* tropospheric scatter propagation.

(AP/PROP) 211-1997

transient **(1)** **(cable systems in substations)** A change in the steady-state condition of voltage or current, or both. As used in this guide, transients occurring in control circuits are a result of rapid changes in the power circuits to which they are coupled. The frequency, damping factor, and magnitude of the transients are determined by resistance, inductance, and capacitance of the power and control circuits and the degree of coupling. Voltages as high as 10 kV in the frequency range of 0.3–3.0 MHz have been observed where little or no protection was provided. Transients may be caused by a lightning stroke, a fault, or by switching operation, such as the opening of a disconnect, and may readily be transferred from one conductor to another by means of electrostatic or electromagnetic coupling. 382-1987

(2) **(industrial power and control)** That part of the change in a variable that disappears during transition from one steady-state operating condition to another. *Note:* Using the term to mean the total variation during the transition between two steady states is deprecated. (IA) [18]

(3) **(excitation systems)** In a variable observed during transition from one steady-state operating condition to another, that part of the variation which ultimately disappears. *Note:* ANSI C85 deprecates using the term to mean the total variable during the transition between two steady-states.

(PE/EDPG) 421A-1978s

(4) Any disturbance with a duration of less than a few cycles. *See also:* swell; notch; sag; surge. (T&D/PE) 1250-1995

(5) A fault or error resulting from temporary environmental conditions. (C/BA) 896.9-1994w

(6) That part of the change in a variable, such as voltage, current, or speed, which may be initiated by a change in steady-state conditions or an outside influence, that decays and/or disappears following its appearance.

(IA/PSE) 446-1995

(7) Pertaining to or designating a phenomenon or a quantity which varies between two consecutive steady states during a time interval that is short compared to the time scale of interest. A transient can be a unidirectional impulse of either polarity or a damped oscillatory wave with the first peak occurring in either polarity. (SCC22) 1346-1998

(8) A subcycle disturbance in the ac waveform that is evidenced by a sharp, brief discontinuity of the waveform. May be of either polarity and may be additive to, or subtractive from, the nominal waveform. *See also:* notch; overvoltage; swell. (IA/PSE) 1100-1999

(9) A momentary departure of a characteristic from steady-state conditions and back to steady state conditions as a result of a system disturbance. Normal transients occur as a result of normal disturbances such as load or line changes. Abnormal transients result from abnormal disturbances such as a power interruption or wire fault. (PEL) 1515-2000

transient adaptation factor (illuminating engineering) A factor which reduces the equivalent contrast due to readaptation from one luminous background to another. (EEC/IE) [126]

transient analyzer An electronic device for repeatedly producing in a test circuit a succession of equal electric surges of small amplitude and of adjustable waveform, and for presenting this waveform on the screen of an oscilloscope. *See also:* oscillograph. (EEC/PE) [119]

transient blanking *See:* chopping transient blanking.

transient blocking A circuit function that blocks tripping during the interval in which an external fault is being cleared. (SWG/PE) C37.100-1992

transient-cause forced outage (electric power system) A component outage whose cause is immediately self-clearing so that the affected component can be restored to service either automatically or as soon as a switch or circuit breaker can be reclosed or a fuse replaced. *Note:* An example of a transient-cause forced outage is lightning flashover that does not permanently disable the flashed component. *See also:* outage. (PE/PSE) [54]

transient-cause forced outage duration (electric power system) The period from the initiation of the outage until the affected component is restored to service by switching or fuse replacement. *See also:* outage. (PE/PSE) [54]

transient critical component temperatures The temperature that a semiconductor critical component may reach for a short (transient) period of time. (C/BA) 14536-1995

transient current (A) (rotating machinery) The current under nonsteady conditions. **(B) (rotating machinery)** The alternating component of armature current immediately following a sudden short-circuit, neglecting the rapidly decaying component present during the first few cycles. (PE) [9]

transient-decay current (photoelectric device) The decreasing current flowing in the device after the irradiation has been abruptly cut off. *See also:* phototube. (ED) [45]

transient deviation (control) The instantaneous value of the ultimately controlled variable minus its steady-state value. *Synonym:* transient overshoot. *See also:* deviation. (PE/IA/EDPG/IAC) 421-1972s, [60]

transient discharge An electric discharge of momentary nature, resulting from a sudden change in the electric-circuit voltage or current. The discharge may be energized via electric, magnetic, or electromagnetic field induction. *Note:* In many cases, the sudden change in the electric circuit is the result of an insulation breakdown of a small gap, such as between an energized object and a person attempting to grasp it. When the open-circuit voltage is high, the transient discharge may be initiated by a spark. For low open-circuit voltage, physical contact may produce the transient discharge without any associated spark. (T&D/PE) 539-1990

transient discharge perception threshold The level of transient discharge that is perceptible for 50% of the subject population. *Note:* The threshold varies considerably for various contact areas and transient discharge characteristics. Individual responses vary greatly from the mean thresholds, and different levels are obtained for men, women, and children. (T&D/PE) 539-1990

transient electrical noise An electrical disturbance that occurs in a time interval separated from other interferences. Transient electrical noise may be superimposed on other transients or on continuous waves. Transient electrical noise may be of several types, such as pulse, step, or oscillatory. It may occur as a response to a network to one of these types. Sources of transients affecting low voltage circuits are impulse noise, power switching, or lightning. (PE/IC) 1143-1994r

transient enclosure voltage (TEV) Very fast transient phenomena, which are found on the grounded enclosure of GIS systems. Typically, ground leads are too long (inductive) at the frequencies of interest to effectively prevent the occurrence of TEV. The phenomenon is also known as transient ground rise (TGR) or transient ground potential rise (TGPR). (PE/SUB) 80-2000

transient error (1) An error that occurs once, or at unpredictable intervals. *See also:* intermittent fault; random failure. (C) 610.12-1990

(2) A storage error in which data is retrieved incorrectly by the first read operation, but a second read operation is successful. *Contrast:* soft error; hard error. (C) 610.10-1994w

transient fault (1) (surge arresters) A fault that disappears of its own accord. (PE) [9], [84]

(2) A nonrecurring temporary error caused by temporary environmental conditions. (C/BA) 896.3-1993w

transient forced outage (1) A forced outage where the unit or component is undamaged and is restored to service automatically. (PE/PSE) 859-1987w

(2) (electric power system) An outage whose cause is immediately self-clearing so that the affected component can be restored to service either automatically or as soon as a switch or circuit breaker can be reclosed or a fuse replaced. *Notes:* 1. An example of a transient forced outage is a lightning flashover which does not permanently disable the flashed component. 2. This definition derives from transmission and distribution applications and does not necessarily apply to generation outages. (PE/PSE) 346-1973w

transient forced outage duration (electric power system) The period from the initiation of the outage until the component is restored to service by switching or fuse replacement. *Note:* Thus transient forced outage duration is really switching time. (PE/PSE) 346-1973w

transient inrush current Current that results when a switching device is closed to energize a capacitance or an inductive circuit. *Note:* Current is expressed by the highest peak value in amperes and frequency in hertz. (SWG/PE) C37.100-1992

transient insulation level (power and distribution transformers) An insulation level expressed in kilovolts of the crest value of the withstand voltage for a specified transient wave shape; that is, lightning or switching impulse. (PE/TR) C57.12.80-1978r

transient internal voltage (synchronous machines) (for any specified operating condition) The fundamental-frequency component of the voltage of each armature phase that would be determined by suddenly removing the load, without changing the excitation voltage applied to the field, and extrapolating the envelope of the voltage back to the instant of load removal, neglecting the voltage components of rapid decrement that may be present during the first few cycles after removal of the load. *Note:* The transient internal voltage, as shown in the phasor diagram, is related to the terminal-voltage and phase-current phasors by the equation

$$E_i' = E_a + RI_a + jX_d'I_{ad} + jX_q'I_{aq}$$

For a machine subject to saturation, the reactances should be determined for the degree of saturation applicable to the specified operating condition. *See also:* phasor diagram; direct-axis synchronous reactance. (PE) [9]

transient motion (audio and electroacoustics) Any motion that has not reached or that has ceased to be a steady state. (SP) [32]

transient network analyzer (TNA) An analog test circuit representing a scaled down version of the pertinent power circuit components, used mainly for control response and performance testing. (SUB/PE) 1303-1994

transient overshoot An excursion beyond the final steady-state value of output as the result of a step-input change. *Note:* It is usually referred to as the first such excursion; expressed as a percent of the steady-state output step. *See also:* accuracy rating; feedback control system. (EEC/EMI) [112]

transient overvoltage (1) A short-duration highly damped, oscillatory or nonoscillatory overvoltage, having a duration of few milliseconds or less. Transient overvoltage is classified as one of the following types: lightning, switching and very fast front, short duration. (PE/C) 1313.1-1996
(2) The peak voltage during the transient conditions resulting from the operation of a switching device. *Note:* The location and units of measurement are specified in apparatus standards. *See also:* transient overvoltage ratio. (SWG/PE) C37.100-1992

transient overvoltage ratio (factor) The ratio of the transient overvoltage to the closed-switching device operating line-to-neutral peak voltage with the load connected. *Note:* The location of measurement is specified in the apparatus standards. (SWG/PE) C37.100-1992

transient overvoltages Momentary excursions of voltage outside of the normal 60 Hz voltage wave. *Synonym:* spikes. (IA/PSE) 241-1990r

transient performance (synchronous-machine regulating system) The performance under a specified stimulus, before the transient expires. (PE) [9]

transient phenomena (rotating machinery) Phenomena appearing during the transition from one operating condition to another. (PE) [9], [84]

transient reactance (1) (electric power systems in commercial buildings) Determines the current flowing during the period when the subtransient reactance is the controlling value. (IA/PSE) 241-1990r
(2) (power fault effects) The reactance of a generator between the subtransient and synchronous states. This reactance is used for the calculation of the symmetrical fault current during the period between the subtransient and steady states. The current decreases continuously during this period but is assumed to be steady at this value for approximately 0.25 s. (PE/PSC) 367-1996

transient read *See:* dirty read.

transient recovery voltage (TRV) (1) The voltage transient that occurs across the terminals of a pole of a switching device upon interruption of the current flowing through the pole. *Note:* It is the difference between and in some cases the sum of the transient voltages to ground occurring on the terminals. The term "transient recovery voltage" is usually designated as TRV, and may refer to inherent TRV, modified inherent TRV, or actual TRV, as defined elsewhere. In a multipole switching device, the term is usually applied to the voltage across the first pole to interrupt in a three-phase ungrounded test, but not necessarily the first phase to interrupt when tested with a three-phase or multigrounded fault. For switching devices having several interrupting units in series, the term may be applied to the voltage across units or groups of units. (SWG) C37.04E-1985w, C37.4D-1985w, C37.100B-1981w
(2) The voltage transient that occurs across the terminals of a pole of a circuit switching device upon interruption of the current. *Note:* TRV is the difference between the transient voltages to ground occurring on the terminals. The term may refer to a circuit TRV, a modified circuit TRV, or an actual TRV. (SWG/PE) C37.40-1993
(3) The voltage transient that occurs across the terminals of a pole of a switching device upon interruption of the current. *Note:* TRV is the difference between transient voltages to ground occurring on the terminals. The term *transient recovery voltage* is usually designated as TRV, and may refer to inherent TRV, modified inherent TRV, or actual TRV as defined elsewhere. In a multiple switching device, the term is usually applied to the voltage across the first pole to interrupt. For switching devices having several interrupting units in series, the term may be applied to the voltage across units or groups of units. (SWG/PE/IA/PSP) C37.100-1992, 1015-1997

transient recovery voltage rate The rate at which the voltage rises across the terminals of a pole of a circuit-switching device upon interruption of the current. *Note:* The transient recovery voltage rate is usually determined by dividing the voltage at one of the crests of the TRV by the time from current zero to that crest. In case no definite crest exists, the rate may be taken to some stated value usually arbitrarily selected as a certain percentage of the crest value of the normal-frequency recovery voltage. In case the transient is an exponential function, the rate may also be taken at the point of zero voltage. It is the rate of rise of the algebraic difference between the transient voltages occurring on the terminals of the switching device upon interruption of the current. The transient recovery voltage rate may be a circuit transient recovery voltage rate or a modified circuit transient recovery voltage rate, or an actual transient recovery voltage rate according to the type of transient from which it is obtained. When giving actual transient recovery voltage rates, the points between which the rate is measured should be definitely stated. (SWG/PE) C37.100-1992

transient response (1) (excitation systems) A typical transient response of a feedback control system is shown below. The principal characteristics of interest are the rise time, overshoot, and settling time as indicated. *Note:* In some applications, the time to attain 10% of steady-state value is of interest. This time may be appreciable even though the delay time may be very small or even zero. (PE/EDPG) 421A-1978s
(2) (oscilloscopes) Time-domain reactions to abruptly varying inputs. (IM) 311-1970w
(3) (of a relay) The manner in which a relay, relay unit, or relay system responds to a sudden change in the input. (SWG/PE) C37.100-1992

transient speed deviation (A) (gas turbines) *load decrease.* The maximum instantaneous speed above the steady-state speed occurring after the sudden decrease from one specified steady-state electric load to another specified steady-state electric load having values within limits of the rated output of the gas-turbine-generator unit. It is expressed in percent of rated speed. **(B) (gas turbines)** *load increase.* The minimum instantaneous speed below the steady-state speed occurring after the sudden increase from one specified steady-state electric load having values within the limits of rated output of the gas-turbine-generator unit. It is expressed in percent of rated speed. (PE/EDPG) 282-1968, [5]

transient stability A condition that exists in a power system if, after an aperiodic disturbance, the system regains steady-state stability. *See also:* alternating-current distribution. (T&D/PE) [10]

transient stability factor (system or part of a system) The ratio of the transient stability limit to the nominal power flow at the point of the system to which the stability limit is referred. *See also:* stability factor; alternating-current distribution. (T&D/PE) [10]

transient stability limit (transient power limit) The maximum power flow possible through some particular point in the system when the entire system or the part of the system to which the stability limit refers is operating with transient stability. *See also:* alternating-current distribution. (T&D/PE) [10]

transient suppression networks Capacitors, resistors, or inductors so placed as to control the discharge of stored energy banks. They are commonly used to suppress transients caused by switching. (PEL/ET) 295-1969r

transient thermal impedance (semiconductor devices) The change in the difference between the virtual junction temperature and the temperature of a specified reference point or region at the end of a time interval divided by the step function change in power dissipation at the beginning of the same time interval which causes the change of temperature-difference. *Note:* It is the thermal impedance of the junction under conditions of change and is generally given in the form of a curve as a function of the duration of an applied pulse. *See also:* semiconductor rectifier stack; principal voltage-current characteristic. (ED) [46]

transient thermal rating The transient thermal rating is that final current (I_f) that yields the maximum allowable conductor temperature (T_{max}) in a specified time after a step change in electrical current from some initial current, I_i. (T&D/PE) 738-1993

transient voltage capability (thyristor) Rated nonrepetitive peak reverse voltage. The maximum instantaneous value of any nonrepetitive transient reverse voltage that may occur across a thyristor without damage. (IA/IPC) 428-1981w

transient voltage surge suppressor (TVSS) A device that functions as a surge protective device (SPD) or surge suppressor. (IA/PSE) 1100-1999

transimpedance (of a magnetic amplifier) The ratio of differential output voltage to differential control current. (MAG) 107-1964w

transinformation (of an output symbol about an input symbol) The difference between the information content of the input symbol and the conditional information content of the input symbol given the output symbol. *Notes:* 1. If x_i is an input symbol and y_j is an output symbol, the transinformation is equal to

$$[-\log p(x_i)] - [-\log p(x_i|y_j)]$$

$$= \log \frac{p(x_i|y_j)}{p(x_i)} = \log \frac{p(x_i,y_j)}{p(x_i)p(y_j)}$$

where $p(x_i|y_j)$ is the conditional probability that x_i was transmitted when y_j is received, and $p(x_i,y_j)$ is the joint probability of x_i and y_j. 2. This quantity has been called transferred information, transmitted information, and mutual information. *See also:* information theory. (IT) [123]

transistor (1) An active semiconductor device with three or more terminals. It is an analog device. (ED) 216-1960w
(2) A semiconducting device for controlling the flow of current between two terminals, the emitter and the collector, by means of variations in the current flow between a third terminal, the base, and one of the other two. *See also:* logic gate. (C) 610.10-1994w

transistor, conductivity-modulation *See:* conductivity-modulation transistor.

transistor equivalent (A) A model approximating the behavior of an electronic component using only transistors, resistors, capacitors and inductors. **(B)** An approximation of the size of an integrated circuit, counting all circuit elements as transistors or portions thereof. (C) 610.10-1994

transistor, filamentary *See:* filamentary transistor.

transistor, junction *See:* junction transistor.

transistor, point-contact *See:* point-contact transistor.

transistor, point-junction *See:* point-junction transistor.

transistor reset preamplifier An integrating preamplifier in which the charge that accumulates on the feedback capacitor is drained off through a transistor when the charge exceeds a predetermined value. (NPS) 325-1996

transistor-transistor logic (TTL) A family of bipolar integrated circuit logic in which the multiple inputs on gates are provided by multiple transistors. (C) 610.10-1994w

transistor, unipolar *See:* unipolar transistor.

transit (1) (navigation aids) A radio navigation system using low orbit satellites to provide world-wide coverage, with transmissions from the satellites at vhf (very high frequency) and uhf (ultra high frequency), in which fixes are determined from measurements of the Doppler shift of the continuous wave signal received from the moving satellite. (AES/GCS) 172-1983w
(2) An instrument primarily used during construction of a line to survey the route, to set hubs and point on tangent (POT) locations, to plumb structures, to determine downstrain angles for locations of anchors at the pull and tension sites, and to sag conductors. *Synonyms:* site marker; level; scope. (T&D/PE) 524-1992r

transit angle The product of angular frequency and the time taken for an electron to traverse a given path. *See also:* electron emission. (ED) 161-1971w, [45]

transition (1) (A) (data transmission) (signal transmission) The change from one circuit condition to the other, that is, to change from mark to space or from space to mark. **(B) (data transmission)** (waveform) (pulse techniques) A change of the instantaneous amplitude from one amplitude to another amplitude level. **(C) (data transmission)** (transition frequency) (disk recording system) (crossover frequency) (turnover frequency) The frequency corresponding to the point of intersection of the asymptotes to the constant-amplitude and the constant-velocity portions of its frequency response curve. This curve is plotted with output voltage ratio in decibels as the ordinate and the logarithm of the frequency as the abscissa. (PE) 599-1985
(2) (A) (pulse) A portion of a wave or pulse between a first nominal state and a second nominal state. Throughout the remainder of this document the term transition is included in the term pulse and wave. **(B) (pulse)** The region of a pulse in which a major change in amplitude occurs, such as the leading edge (first transition) or final trailing edge (last transition). (NPS) 300-1988
(3) A joint that connects two cable types. A joint on extruded cable rated 46–138 kV connecting an ethylene propylene rubber (EPR) insulated cable to a crosslinked polyethylene (XLPE) or high-molecular-weight polyethylene (HMWPE) insulated cable should be considered a transition joint. (PE/IC) 404-1993
(4) (of a pulse) The region of a pulse in which a major change in amplitude occurs, such as at the leading edge (first transition) or final trailing edge (last transition). (NPS) 325-1996
(5) The change of a logic signal from one state to another (as in ". . .a transition at the input shall cause. . .") or the pair of logic states between which a transition may occur (as in ". . .the delay for a low-to- high transition. . ."). (C/DA) 1481-1999
(6) *See also:* software transition. (C/SE) J-STD-016-1995

transitional mode The change from the nonoperating to the operating mode, caused by switching the input to the relay from the nonoperating to the operating input, or vice versa. (SWG/PE/PSR) C37.100-1992, C37.98-1977s

transition compartment The compartment specifically designed for joining gas-insulated substation equipment of different design or manufacture. This compartment provides the necessary transition for the current-carrying conductor and the gas enclosure. (SWG/PE/SUB) C37.100-1992, C37.122.1-1993, C37.122-1993

transition contacts For resistance-type LTCs, a set of contacts that is connected in series with a transition impedance and makes and breaks current. (PE/TR) C57.131-1995

transition current The current required at a given temperature and duration to cause a current-protective device to change state. (SPD/PE) C62.36-1994

transition density The number of times the stream of bits within an 8B/10B code-group changes its value. (C/LM) 802.3-1998

transition duration (1) (pulse terminology) The duration between the proximal point and the distal point on a transition waveform. (IM/WM&A) 194-1977w
(2) (of a step response) The duration between the proximal point (10%) and the distal point (90%) on the recorded output response transition, for an ideal input step with designated baseline and topline. (IM/WM&A) 1057-1994w

transition frequency (disk recording) The frequency corresponding to the point of intersection of the asymptotes to the constant-amplitude and the constant-velocity portions of its frequency response curve. This curve is plotted with output voltage ratio in decibels as the ordinate and the logarithm of the frequency as the abscissa. *Synonyms:* turnover frequency;

crossover frequency. *See also:* phonograph pickup. (SP) [32]

transition impedance A resistor or reactor consisting of one or more units that bridge adjacent taps for the purpose of transferring load from one tap to the other without interruption or appreciable change in the load current, at the same time limiting the circulating current for the period that both taps are used. Normally, reactance-type LTCs use the bridging position as a service position and, therefore, the reactor is designed for continuous loading. (PE/TR) C57.131-1995

transitioning (as applied to fall protection) The act of moving from one location to another on equipment or a structure. (NESC/T&D/PE) C2-1997, 1307-1996

transition joint (power cable joints) A cable joint which connects two different types of cable. (PE/IC) 404-1986s

transition load (1) (rectifier circuits) The load at which a rectifier unit changes from one mode of operation to another. *Note:* The load current corresponding to a transition load is determined by the intersection of extensions of successive portions of the direct-current voltage-regulation curve where the curve changes shape or slope. *See also:* rectifier circuit element; rectification. (IA) [12]
(2) The load at which a thyristor converter changes from one mode of operation to another. *Note:* The load current corresponding to a transition load is determined by the intersection of extensions of successive portions of the direct-voltage regulation curve where the curve changes shape or slope. (IA/IPC) 444-1973w

transition loss (1) (A) (wave propagation) At a transition or discontinuity between two transmission media, the difference between the power incident upon the discontinuity and the power transmitted beyond the discontinuity that would be observed if the medium beyond the discontinuity were match-terminated. **(B) (wave propagation)** The ratio in decibels of the power incident upon the discontinuity to the power transmitted beyond the discontinuity that would be observed if the medium beyond the discontinuity were match terminated. *See also:* waveguide. (MTT) 146-1980
(2) (junction between a source and a load) The ratio of the available power to the power delivered to the load. Transition loss is usually expressed in decibels. *See also:* waveguide; transmission loss. (MTT) 146-1980w

transition matrix A matrix which maps the state of a linear system at one instant of time into another state at a later instant of time, provided that the system inputs are zero over the closed time interval between the two instants of time. *Note:* This is also the matrix of solutions of the homogeneous equations. *Synonym:* fundamental matrix. (CS/PE/EDPG) [3]

transition point (1) A point in a transmission system at which there is change in the surge impedance. (CAS/PE) [8], [84]
(2) The input value that causes 50% of the output codes to be greater than or equal to the upper code of the transition, and 50% to be less than the upper code of the transition. (IM/WM&A) 1057-1994w

transition pulse (pulse waveform) That segment comprising a change from one amplitude level to another amplitude level. *See also:* pulse. (IM/HFIM) [40]

transition region (semiconductor) The region, between two homogeneous semiconductor regions, in which the impurity concentration changes. *See also:* semiconductor; transistor. (AES/SS) 307-1969w

transition shape (A) (pulse terminology) For descriptive purposes a transition waveform may be imprecisely described by any of the adjectives, or combinations thereof, in descriptive adjectives, major (minor); polarity related adjectives; geometrical adjectives, round; and functional adjectives. When so used, these adjectives describe general shape only, and no precise distinctions are defined. **(B) (pulse terminology)** For tutorial purposes, a hypothetical transition waveform may be precisely defined by the further addition of the adjective ideal. **(C) (pulse terminology)** For measurement or comparison purposes a transition waveform may be precisely defined by the further addition of the adjective reference. (IM/WM&A) 194-1977

transition time (gas-tube surge protective devices) The time required for the voltage across a conducting gap to drop into the arc region after the gap initially begins to conduct. (SPD/PE) C62.31-1987r

transitive dependency A type of dependency among attributes in a relation, in which a nonprime attribute A is said to be transitively dependent on another attribute B if and only if there is another attribute C that is functionally dependent on B and functionally determining A but not B. *Contrast:* nontransitive dependency. (C) 610.5-1990w

transitron oscillator A negative-transconductance oscillator employing a screen-grid tube with negative transconductance produced by a retarding field between the negative screen grid and the control grid that serves as the anode. *See also:* oscillatory circuit. (AP/ANT) 145-1983s

transit time (1) (electron tube) The time taken for a charge carrier to traverse a given path. *See also:* electron emission. (ED) [45]
(2) (multiplier-phototube) The time interval between the arrival of a delta-function light pulse at the entrance window of the tube and the time at which the output pulse at the anode terminal reaches peak amplitude. *See also:* electron emission; phototube. (ED) 158-1962w

transit-time mode (electron tube) A condition of operation of an oscillator corresponding to a limited range of drift-space transit angle for which the electron stream introduces a negative conductance into the coupled circuit. (ED) 161-1971w

transit-time spread (1) (electron tube) The time interval between the half-amplitude points of the output pulse at the anode terminal, arising from a delta function of light incident on the entrance window of the tube. *See also:* phototube. (ED) 158-1962w
(2) (scintillation counting) The FWHM (full-width-at-half-maximum) of the time distribution of a set of pulses each of which corresponds to the photomultiplier transit time for that individual event. (NPS) 398-1972r

translate (1) (A) To convert expressions in one language to synonymous expressions in another language. **(B)** To encode or decode. *See also:* matrix; translator. (C) 162-1963
(2) (data management) To transform data from one language to another. (C) 610.5-1990w

translation (1) (telecommunications) The process of converting information from one system of representation into equivalent information in another system of representation. (COM) [49]
(2) (computer graphics) The displacement of one or more display elements without rotation, maintaining its orientation. (C) 610.6-1991w
(3) In a single-cable 10BROAD36 system, the process by which incoming transmissions at one frequency are converted into another frequency for outgoing transmission. The translation takes place at the headend. (C/LM) 802.3-1998

translation buffer A set of registers in a memory management unit in which virtual addresses are converted to physical addresses. *Note:* Typically the complete map of translations will not fit into the memory management unit at one time so only a portion are buffered there while the entire map is in main storage. (C) 610.10-1994w

translation loss (playback loss) (reproduction of a mechanical recording) The loss whereby the amplitude of motion of the reproducing stylus differs from the recorded amplitude in the medium. *See also:* phonograph pickup. (SP) [32]

translation manager A facility that maps X event sequences (such as keyboard actions) into widget-supplied functionality (action procedures). (C) 1295-1993w

translator (1) (software) A computer program that transforms a sequence of statements expressed in one language into an

equivalent sequence of statements expressed in another language. *See also:* assembler; compiler. (C) 610.12-1990
(2) (telephone switching systems) Equipment capable of interpreting and converting information from one form to another form. (COM) 312-1977w
(3) (test, measurement, and diagnostic equipment) An automatic means, usually a program, to translate machine language mnemonic symbols for computer operations into true machine language. Memory locations and input-output lines must be written in numerical code, not symbolically. (MIL) [2]
(4) (broadband local area networks) A frequency conversion device located at the headend. Its sole purpose is to provide gain and convert inbound signal frequencies to the outbound frequency range. (LM/C) 802.7-1989r

transliterate (1) (data management) To convert data character-by-character from one character set to another. (C) 610.5-1990w
(2) (data management) To convert the characters of one alphabet to the corresponding characters of another alphabet. (C) [20], [85]

transmissibility Ratio of the response at any one point in the equipment to the input of the equipment at a single frequency. (SWG/PE) C37.100-1992, C37.81-1989r

transmission (1) (data transmission) The electrical transfer of a signal, message, or other form of intelligence from one location to another. (PE) 599-1985w
(2) (laser maser) Passage of radiation through a medium. (LEO) 586-1980w
(3) (illuminating engineering) A general term for the process by which incident flux leaves a surface or medium on a side other than the incident side, without change in frequency. *Note:* Transmission through a medium is often a combination of regular and diffuse transmission. *See also:* regular transmission; diffuse transmission. (EEC/IE) [126]
(4) The propagation of a signal, message, or other form of intelligence by any means, such as optical fiber, wire, or visual means. (C) 610.7-1995, 610.10-1994w

transmission band (uniconductor waveguide) The frequency range above the cutoff frequency. *See also:* waveguide. (MTT) 146-1980w

transmission block character *See:* end of transmission block character.

transmission coefficient (1) (waveguide) (of a network) At a given frequency and for a given mode, the ratio of some quantity associated with the transmitted wave at a specified reference plane to the corresponding quantity in the incident wave at a specified reference plane. *Notes:* 1. The transmission coefficient may be different for different associated quantities, and the chosen quantity must be specified. The voltage transmission coefficient is commonly used and is defined as the complex ratio of the resultant electric field strength (or voltage) to that of the incident wave. Examples of other quantities are power or current. 2. An interface is a special case of a network where the reference planes associated with the incident and transmitted waves become coincident; in this case the voltage transmission coefficient is equal to one plus the voltage reflection coefficient. (MTT) 146-1980w
(2) (multiport) Ratio of the complex amplitude of the wave emerging from a port of a multiport terminated by reflectionless terminations to the complex amplitude of the wave incident upon another port. *See also:* reflection coefficient; scattering coefficient. (IM/HFIM) [40]
(3) *See also:* Fresnel coefficients. (AP/PROP) 211-1997

transmission control character (1) (A) Any control character used to control or facilitate transmission of data. **(B)** Any character transmitted that is not part of the message being transferred, but that is used to control or to facilitate the transfer. *Synonym:* communication control character. (C) 610.5-1990
(2) A control character used to control or facilitate transmission of data between DTEs. (C) 610.7-1995

transmission delay or propagation delay *See:* absolute delay.

transmission detector (1) (charged-particle detectors) (semiconductor radiation detectors) (x-ray energy spectrometers) A totally depleted detector whose thickness including its entrance and exit windows is sufficiently small to permit radiation to pass completely through the detector. (PE/NID/NP) 301-1976s, [124]
(2) (charged-particle detectors) A totally depleted detector in which the thickness, including entrance and exit windows, is sufficiently thin to permit radiation to pass completely through it. (NPS) 300-1988r

transmission error control The process that ensures no errors are introduced while transmitting data between sender and receiver. (C) 610.7-1995

transmission facility (data transmission) The transmission medium and all the associated equipment required to transmit a message. (PE) 599-1985w

transmission factor *See:* Fresnel coefficients.

transmission feeder A feeder forming part of a transmission circuit. *See also:* center of distribution. (T&D/PE) [10]

transmission frequency meter (waveguide) A cavity frequency meter that, when tuned, couples energy from a waveguide into a detector. *See also:* waveguide. (AP/ANT) [35], [84]

transmission format The specified arrangement of delimiter symbols or start and stop symbols and of data bit symbols that constitute a complete transmitted signal frame (PhPDU). Data bit symbols are always arranged in octets (8 b groupings). The definition of the transmission format also includes the selected encoding scheme, which is binary for low-speed operation and Manchester biphase-L for high-speed operation. (EMB/MIB) 1073.4.1-2000

transmission gain (data transmission) General term used to denote an increase in signal power in transmission from one point to another. Gain is usually expressed in decibels and is widely used to denote transducer gain. (PE) 599-1985w

transmission level (data transmission) The ratio of the signal power at any point in a transmission system to the power at some point in the system chosen as a reference point. This ratio is usually expressed in decibels. The transmission level at the transmitting switchboard is frequently taken as the zero level reference point. (PE) 599-1985w

transmission level point (TLP) (1) For a particular point in a transmission system, the design signal level in dB relative to the level at the zero TLP reference point. (COM/TA) 1007-1991r
(2) a point in a transmission system at which the ratio is specified in dB of the power of a test signal at that point to the power of a signal at a reference point. The reference level point, called the zero transmission level point (0 TLP), is an arbitrary established point relative to which transmission levels at all other points are specified. A signal level of X dBm at the 0 TLP is designated X dBm0. (COM/TA) 743-1995

transmission line (1) (A) (data transmission) (signal-transmission system) The conductive connections between system elements which carry signal power; A waveguide consisting of two or more conductors. **(B) (data transmission)** (electric power) A line used for electric power transmission. **(C) (data transmission)** (electromagnetic wave guidance) A system of material boundaries or structures for guiding electromagnetic waves, in the TEM (transverse electromagnetic) mode. Commonly a two-wire or coaxial system of conductors. (PE) 599-1985
(2) (planar transmission lines) A structure designed to guide the propagation of electromagnetic energy in a well-defined direction. For purposes of definition and description relating to wave propagation, planar transmission lines are usually assumed to be of invariant cross section along the direction of propagation. *See also:* load leads. (MTT) 1004-1987w

(3) (waveguide) A system of material boundaries or structures for guiding electromagnetic waves. Frequently, such a system is used for guiding electromagnetic waves, in the TEM mode. Commonly, a two-wire or coaxial system of conductors. *See also:* waveguide. (MTT) 146-1980w

(4) Any overhead line used for electric power transmission with a phase-to-phase voltage exceeding 69 kV and an average conductor height of more than 10 m. (PE/T&D) 1243-1997

transmission-line capacity (electric power supply) The maximum continuous rating of a transmission line. The rating may be limited by thermal considerations, capacity of associated equipment, voltage regulation, system stability, or other factors. *See also:* generating station. (PE/PSE) [54]

transmission line, coaxial *See:* coaxial transmission line.

transmission link The physical unit of a DQDB subnetwork that provides the transmission connection between adjacent nodes. Each transmission link accommodates both buses of the dual bus pair between the adjacent nodes. (LM/C) 8802-6-1994

transmission loss (*L*) (1) (data transmission) In communication, a general term used to denote a decrease in power in transmission from one point to another. Transmission loss is usually expressed in decibels. *Synonym:* loss. (PE) 599-1985w

(2) (A) (electric power system) The power lost in transmission between one point and another. It is measured as the difference between the net power passing the first point and the net power passing the second. **(B) (electric power system)** The ratio in decibels of the net power passing the first point to the net power passing the second. (MTT) 146-1980

(3) (fiber optics) Total loss encountered in transmission through a system. *See also:* attenuation; reflection; transmittance. (Std100) 812-1984w

(4) (of a radio system) The ratio of the power radiated from the transmitting antenna to the resultant power that would be available from a loss-free (but otherwise identical) receiving antenna. (AP/PROP) 211-1997

transmission-loss coefficients (electric power system) Mathematically derived constants to be combined with source powers to provide incremental transmission losses from each source to the composite system load. These coefficients may also be used to calculate total system transmission losses. (PE/PSE) 94-1991w

transmission measuring set (data transmission) A measuring instrument comprising a signal source and a signal receiver having known impedances, that is designed to measure the insertion loss or gain of a network or transmission path connected between those impedances. (PE) 599-1985w

transmission media The physical facility utilized for the interconnection and transmission of messanges between a user station and network device; for example, twisted pair wire, coaxial cable, optical fiber, microwave, and infrared light beams. (C) 610.10-1994w

transmission medium (1) A means of transporting electrical or optical signals. *See also:* signal. (C/BA) 1355-1995

(2) The material on which information signals may be carried; e.g., optical fiber, coaxial cable, and twisted-wire pairs. (LM/C) 8802-6-1994

(3) The physical facility utilized for the interconnection and transmission of messages between a user station and network device. For example: coaxial cable; optical fiber. (C) 610.7-1995

transmission mode A form of propagation along a transmission line characterized by the presence of any one of the elemental types of TE (transverse electric), TM (transverse magnetic), or TEM (transverse electromagnetic) waves. *Note:* Waveguide transmission modes are designated by integers (modal numbers) associated with the orthogonal functions used to describe the waveform. These integers are known as waveguide mode subscripts. They may be assigned from observations of the transverse field components of the wave and without reference to mathematics. A waveguide transmission mode is commonly described as a $TE_{m,n}$ or $TM_{m,n}$ mode, $_{m,n}$ being numerics according to the following system:

a) (waves in rectangular waveguides). If a single wave is transmitted in a rectangular waveguide, the field that is everywhere transverse may be resolved into two components, parallel to the wide and narrow walls respectively. In any transverse section, these components vary periodically with distance along a path parallel to one of the walls. m = the total number of half-period variations of either component of field along a path parallel to the wide walls. n = the total number of half-period variations of either component of field along a path parallel to the narrow walls.

b) (waves in circular waveguides). If a single wave is transmitted in a circular waveguide, the transverse field may be resolved into two components, radial and angular, respectively. These components vary periodically along a circular path concentric with the wall and vary in a manner related to the Bessel function of order m along a radius, where m the total number of full-period variations of either component of field along a circular path concentric with the wall. n = one more than the total number of reversals of sign of either component of field along a radial path. This system can be used only if the observed waveform is known to correspond to a single mode.

See also: waveguide. (EEC/PE) [119]

transmission-mode photocathode A photocathode in which radiant flux incident on one side produces photoelectric emission from the opposite side. (NPS) 398-1972r

transmission modulation (storage tubes) Amplitude modulation of the reading-beam current as it passes through apertures in the storage surface, the degree of modulation being controlled by the charge pattern stored on that surface. *See also:* storage tube. (ED) 158-1962w

transmission network A group of interconnected transmission lines or feeders. *See also:* transmission line. (T&D/PE) [10]

transmission performance (in telephony) The effectiveness of a complete telephone connection for transmitting and reproducing speech under actual conditions. *Note:* The specification of transmission generally requires the consideration of more than one attribute or test method. (COM/TA) 823-1989w

transmission primaries (color television) The set of three colorimetric primaries that, if used in a display and controlled linearly and individually by a corresponding set of three channel signals generated in the color television camera, would result in exact colorimetric rendition (over the gamut defined by the primaries) of the scene viewed by the camera. *Note:* Ideally the primaries used at the receiver display would be identical with the transmission primaries, but this is not usually possible since developments in display phosphors occurring since the setting of transmission standards, for example, may result in the use of receiver display primaries that differ from the transmission primaries. Within a linear part of the overall system, it is always possible to compensate for differences existing between transmission and display primaries by means of matrixing. Because of the capability afforded by matrixing, the transmission primaries need not be real. There exists a unique relationship between the chromaticity coordinates of the transmission primaries and the spectral taking characteristics used at the camera to generate the three respective channel signals. (BT/AV) 201-1979w

transmission quality (mobile communication) The measure of the minimum usable speech-to-noise ratio, with reference to the number of correctly received words in a specified speech sequence. *See also:* mobile communication system. (VT) [37]

transmission regulator (electric communication) A device that functions to maintain substantially constant transmission over a transmission system. (PE/EEC) [119]

transmission route The route followed by a transmission circuit. *See also:* transmission line. (T&D/PE) [10]

transmission service charge The amount paid to a system for the use of its transmission facilities. (PE/PSE) 858-1993w

transmission system (1) (power operations) An interconnected group of electric transmission lines and associated equipment for the movement or transfer of electric energy in bulk between points of supply and points for delivery. (PE/PSE) 858-1987s

(2) (data transmission) In communication practice, an assembly of elements capable of functioning together to transmit signal waves. (PE) 599-1985w

(3) The interface and *transmission medium* through which peer *Physical Layer* entities transfer bits. (LM/C) 8802-6-1994

transmission test *See:* end-to-end test.

transmission throughput *See:* effective speed of transmission.

transmission time (data transmission) The absolute time interval from transmission to reception of a signal. (PE) 599-1985w

transmission window *See:* spectral window.

transmissivity (fiber optics) The transmittance of a unit length of material, at a given wavelength, excluding the reflectance of the surfaces of the material; the intrinsic transmittance of the material, irrespective of other parameters such as the reflectances of the surfaces. No longer in common use. *See also:* transmittance. (Std100) 812-1984w

(2) (A) (of a boundary) The ratio of the normal component of the power density transmitted across the boundary between two media to the normal component of the incident power density. **(B)** (of a layer) The ratio of the normal component of the power density transmitted through the layer to the normal component of the incident power density. (AP/PROP) 211-1997

transmissivity matrix A 4 × 4 matrix of dimensionless real numbers which, when multiplied by the Stokes vector incident upon a boundary or through a medium, yields the Stokes vector that is propagated across that boundary or medium. (AP/PROP) 211-1997

transmissometer (illuminating engineering) A photometer for measuring transmittance. *Note:* Transmissometers may be visual or physical instruments. (EEC/IE) [126]

transmit (1) (computers) To move data from one location to another location. *See also:* transfer. (C) [20], [85]

(2) (data management) To send data from one place for reception elsewhere. *See also:* transfer. (C) 610.5-1990w

(3) The electrical output of a telephone set or connecting test circuit due to an acoustic input to the telephone set. (COM/TA) 269-1992

(4) The electrical output of a handset, headset, or connecting test circuit, due to an acoustic input to the device. (COM/TA) 1206-1994

(5) The action of a station generating a frame, token, abort sequence, or fill and placing it on the medium. *Contrast:* repeat. (LM/C/LM) 8802-5-1998

transmit channel A channel used within a data circuit to transmit data. *Synonym:* send channel. *Contrast:* receive channel. (C) 610.10-1994w

transmit characteristic (telephony) The electrical output level of a telephone set as a function of the acoustic input level. The output is measured across a specified impedance connected to the telephone feed circuit, and the input is measured in free field at a specified location relative to the reference point of an artificial mouth. (COM/TA) 269-1971w

transmit-receive box *See:* transmit-receive switch.

transmit-receive cavity The resonant portion of a transmit-receive switch. (AES/RS) 686-1990

transmit-receive cell (waveguide) (tube) A gas-filled waveguide cavity that acts as a short circuit when ionized but is transparent to low-power energy when un-ionized. It is used in a transmit-receive switch for protecting the receiver from the high power of the transmitter but is transparent to low-power signals received from the antenna. *See also:* waveguide. (AP/ANT) [35]

transmit-receive module An active TR electronic module, usually with integrated circuits, consisting of an antenna (or direct connection thereto), transmitter, receiver, duplexer, phase shifters, and power conditioner employed at the radiating elements of a phased array radar. (AES) 686-1997

transmit-receive switch (TR switch) (1) An automatic device employed in a radar for substantially preventing the transmitted energy from reaching the receiver but allowing the received energy to reach the receiver without appreciable loss. *See also:* radar. (EEC/PE) [119]

(2) An RF switch, frequently of the gas discharge type, which automatically decouples the receiver from the antenna during the transmitting period. *Note:* Employed when a common antenna is used for transmission and reception. *Synonym:* transmit-receive box. (AES) 686-1997

transmit-receive switch, duplexer A switch, frequently of the gas discharge type, employed when a common transmitting and receiving antenna is used, that automatically decouples the receiver from the antenna during the transmitting period. *See also:* navigation. (AES/RS) 686-1982s, [42]

transmit-receive tube A gas-filled radio-frequency switching tube used to protect the receiver in pulsed radio-frequency systems. *See also:* gas tube. (ED) 161-1971w

transmittal of a CCS message *See:* transmittal of a CCS signal.

transmittal of a CCS signal Occurs when the signal or complete message becomes available for transmission (that is, stored in the output buffer). *Synonym:* transmittal of a CCS message. (COM/TA) 973-1990w

transmittal of a per-trunk-signaling supervisory signal Occurs when the state transition that begins the signal occurs at the M lead (or M-lead equivalent). (COM/TA) 973-1990w

transmittance (1) (fiber optics) The ratio of transmitted power to incident power. *Note:* In optics, frequently expressed as optical density or percent; in communications applications, generally expressed in decibels (dB). Formerly called "transmission." *See also:* transmission loss; antireflection coating. (Std100) 812-1984w

(2) (photovoltaic power system) The fraction of radiation incident on an object that is transmitted through the object. *See also:* photovoltaic power system. (AES) [41]

(3) (linear passive networks) (transfer function) A response function for which the variables are measured at different ports (terminal pairs). (CAS) 156-1960w

(4) (laser maser) (τ) The ratio of total transmitted radiant power to total incident radiant power. (LEO) 586-1980w

(5) (illuminating engineering) ($\tau = \Phi_t/\Phi_i$) (of a medium) The ratio of the transmitted flux to the incident flux. *Notes:* 1. It should be noted that transmittance refers to the ratio of flux emerging to flux incident; therefore; reflections at the surface as well as absorption within the material operate to reduce the transmittance. Transmittance is a function of:

a) geometry (i) of the incident flux (ii) of collection for the transmitted flux;
b) spectral distribution (i) characteristic of the incident flux (ii) weighting function for the collected flux; and
c) polarization (i) of the incident flux (ii) component defined for the collected flux.

2. Unless the state of polarization for the incident flux and the polarized component of the transmitted flux are stated, it shall be considered that the incident flux is unpolarized and that the total transmitted flux is evaluated. 3. Unless qualified by the term "spectral" (see "spectral reflectance") or other modifying adjectives, luminous transmittance (see "luminous transmittance") is meant. 4. If no qualifying geometric adjec-

tive is used, transmittance for hemispherical collection is meant. For other modifying adjectives see listing in reflectance factor entry. 5. In each case of conical incidence or collection, the solid angle is not restricted to a right circular cone, but may be of any cross section including rectangular, a ring, or a combination of two or more solid angles. 6. These concepts must be applied with care, if the area of the transmitting element is not large compared to its thickness, due to internal transmission across the boundary of the area. 7. The following breakdown of transmittance quantities is applicable only to the transmittance of thin films with negligible internal scattering so that the transmitted radiation emerges from a point that is not significantly separated from the point of incidence of the incident ray that produces the transmitted ray(s). The governing considerations are similar to those for application of the bidirectional reflectance-distribution function (BRDF), rather than the bidirectional scattering-surface reflectance-distribution function (BSSRDF). *See also:* luminous transmittance; spectral reflectance. (EEC/IE) [126]

transmittance, thermal *See:* thermal transmittance.

transmitted-carrier operation That form of amplitude-modulation carrier transmission in which the carrier wave is transmitted. *See also:* amplitude modulation. (EEC/PE) [119]

transmitted harmonics (electrical conversion) (induced harmonics) Harmonics that are transformed or pass through the conversion device from the input to the output. (AES) [41]

transmitted information *See:* transinformation.

transmitted light scanning The scanning of changes in the magnitude of light transmitted through a web. *See also:* photoelectric control. (IA/ICTL/IAC) [60]

transmitted wave (1) (waveguide) At a transverse plane in a transmission line or waveguide, a wave transmitted past a discontinuity in the same direction as the incident wave. *See also:* reflected wave. (MTT) 146-1980w
(2) A wave (or waves) produced by an incident wave that continue(s) beyond the transition point. (CAS) [84]
(3) **(A)** The wave launched by a transmitting antenna. **(B)** *See also:* refracted wave. (AP/PROP) 211-1997

transmitter (1) (protective signaling) A device for transmitting a coded signal when operated by any one of a group of actuating devices. *See also:* protective signaling. (EEC/PE) [119]
(2) (radio) A device or circuit that generates high-frequency electric energy, controlled or modulated, which can be radiated by an antenna. (SWG/PE/PSR) C37.90.2-1995, C37.100-1992

transmitter-blocker cell (antitransmit-receive tube) (with reference to a waveguide). A gas-filled waveguide cavity that acts as a short circuit when ionized but as an open circuit when un-ionized. It is used in a transmit-receive switch for directing the energy received from the aerial to the receiver, no matter what the transmitter impedance may be. *See also:* waveguide. (AP/ANT) [35]

transmitter, facsimile *See:* facsimile transmitter.

transmitter on/transmitter off An asynchronous protocol that synchronizes the receiving terminal with the sending terminal. (C) 610.7-1995

transmitter performance *See:* audio input signal; audio input power.

transmitter, telephone *See:* telephone transmitter.

transmitting (transmission performance of telephone sets) The electric output level of a telephone set or connecting test circuit due to an acoustic input to the telephone set. The acoustic input may be varied either in frequency or level. The output is measured across a specified impedance and the input is measured at the calibration point of an artificial mouth. (COM/TA) 269-1983s

transmitting converter (facsimile) (amplitude-modulation to frequency-shift-modulation converter) A device which changes the type of modulation from amplitude to frequency shift. *See also:* facsimile transmission. (COM) 168-1956w

transmitting current response (electroacoustic transducer used for sound emission) The ratio of the sound pressure apparent at a distance of one meter in a specified direction from the effective acoustic center of the transducer to the current flowing at the electric input terminals. *Note:* The sound pressure apparent at a distance of one meter can be found by multiplying the sound pressure observed at a remote point (where the sound field is spherically divergent) by the number of meters from the effective acoustic center of the transducer to that point. (SP) [32]

transmitting efficiency (electroacoustic transducer) (projector efficiency). The ratio of the total acoustic power output to the electric power input. *Note:* In computing the electric power input, it is customary to omit any electric power supplied for polarization or bias. (SP) [32]

transmitting loop loss That part of the repetition equivalent assignable to the station set, subscriber line, and battery supply circuit that are on the transmitting end. *See also:* transmission loss. (EEC/PE) [119]

transmitting objective loudness rating (loudness ratings of telephone connections)

$$\mathrm{TOLR} = -20 \log_{10} \frac{V_T}{S_M}$$

where
S_M = sound pressure at the mouth reference point (in pascals)
V_T = output voltage of the transmitting component (in millivolts).

Note: Normally occurring TOLRs will be in the -30 to -55 (dB) range. These numbers are a result of the units chosen and have no physical significance. (COM/TA) 661-1979r

transmitting power response (electroacoustic transducer used for sound emission) (projector power response) The ratio of the mean-square sound pressure apparent at a distance of one meter in a specified direction from the effective acoustic center of the transducer to the electric power input. *Note:* The sound pressure apparent at a distance of one meter can be found by multiplying the sound pressure observed at a remote point (where the sound field is spherically divergent) by the number of meters from the effective acoustic center of the transducer to that point. (SP) [32]

transmitting voltage response (electroacoustic transducer used for sound emission) The ratio of the sound pressure apparent at a distance of one meter in a specified direction from the effective acoustic center of the transducer to the signal voltage applied at the electric input terminals. *Note:* The sound pressure apparent at a distance of one meter can be found by multiplying the sound pressure observed at a remote point (where the sound field is spherically divergent) by the number of meters from the effective acoustic center of the transducer to that point. (SP) [32]

transobuoy (navigation aids) A free floating or moored automatic weather station providing weather reports from the open ocean. (AES/GCS) 172-1983w

transparency A capability of a communications medium to pass within specified limits a range of signals having one or more defined properties, for example, a channel may be code transparent, or an equipment may be bit pattern transparent. (LM/COM) 168-1956w

transparent (1) (A) In data transmission, pertaining to information that does not contain transmission control characters. **(B)** To perform in a manner that is invisible to, and of no concern to a user. For example, a computer program may perform file allocation, database operations, and housekeeping operations transparent to its user. (C) 610.5-1990
(2) The state of a protocol when all of the following conditions are met:

a) Previously existing protocol implementations are able to recover when receiving packets formed by this new protocol.

b) The implementations of this protocol are able to process packets formed by previously existing protocols without problems.
c) The protocol does not affect the operations of the (N+1) and (N-1)-layer implementations. (C/LM) 802.10-1998

transparent bridging A bridging mechanism in a bridged LAN that is transparent to the end stations. (C/LM) 8802-5-1998

transparent latch A latch that has a level sensitive trigger input such that when the trigger signal is in the 'enable' state the outputs follow the inputs, and when the trigger signal goes to the 'latch' state the outputs retain the data then at the inputs. (C) 610.10-1994w

transponder (1) (navigation aid terms) A transmitter-receiver facility, the function of which is to transmit signals automatically when the proper interrogation is received. (AES/GCS) 172-1983w
(2) (communication satellite) A receiver-transmitter combination, often aboard a satellite, or spacecraft, which receives a signal and retransmits it at a different carrier frequency. Transponders are used in communication satellites for reradiating signals to earth stations or in spacecraft for returning ranging signals. *See also:* repeater. (COM) [24]
(3) (broadband local area networks) A device that responds to a physical or electrical stimulus and emits an electrical signal in response to the stimulus. (LM/C) 802.7-1989r
(4) The onboard component of a dedicated short-range communications (DSRC) system. (SCC32) 1455-1999
(5) Receiver-transmitter equipment, the function of which is to transmit signals automatically when the proper interrogation is received from a radar or an interrogator. (AES) 686-1997

transponder beacon *See:* transponder.

transponder, crossband *See:* crossband transponder.

transponder reply efficiency (navigation aids) The ratio of the number of replies emitted by a transponder to the number of interrogations which the transponder recognizes as valid. The interrogations recognized as valid include those accidentally combined to form recognizable codes, a statistical computation of them normally being made. (AES/GCS) 172-1983w

transport *See:* tape transport; tape drive.

transportability *See:* portability.

transportable (x-ray) X-ray equipment to be installed in a vehicle or that may be readily disassembled for transport in a vehicle. (NEC/NESC) [86]

transportable computer A personal computer that weighs more than 21 lb, yet is designed and configured to permit easy transportation. *See also:* portable computer. (C) 610.2-1987, 610.10-1994w

transportable transmitter A transmitter designed to be readily carried or transported from place to place, but which is not normally operated while in motion. *Note:* This has been commonly called a portable transmitter, but the term transportable transmitter is preferred. *See also:* radio transmitter; radio transmission. (AP/BT/ANT) 145-1983s, 182A-1964w

transportation and storage conditions The conditions to which a device may be subjected between the time of construction and the time of installation. Also included are the conditions that may exist during shutdown. *Note:* No permanent physical damage or impairment of operating characteristics shall take place under these conditions, but minor adjustments may be needed to restore performance to normal. (EEC/EMI) [112]

transportation lag* *See:* lag.

* Deprecated.

transport delay *See:* time delay.

transport lag* *See:* lag.

* Deprecated.

transport layer (1) (Layer 4) The layer of the ISO Reference Model that accomplishes the transparent transfer of data over the established link, providing an end-to-end service with high data integrity. (DIS/C) 1278.2-1995
(2) The middle layer in the ISO seven-layer open system communications reference model, and the boundary between the communication subnet layers (physical, data link, and network) and the host process layers (session, presentation, and application). (C/MM) 1284.1-1997
(3) The fourth layer of the seven-layer OSI model, responsible for error-free end-to-end communication. *See also:* application layer; physical layer; data link layer; presentation layer; session layer; entity layer; client layer; logical link control sublayer; network layer; sublayer; medium access control sublayer. (C) 610.7-1995

transport standards Standards of the same nominal value as the basic reference standards of a laboratory (and preferably of equal quality), which are regularly intercompared with the basic group but are reserved for periodic interlaboratory comparison tests to check the stability of the basic reference group. (ELM) C12.1-1982s

transport time (feedback system) The time required to move an object, element or information from one predetermined position to another. *See also:* feedback control system. (IA/ICTL/APP/IAC) [69], [60]

transport vehicle (handling and disposal of transformer grade insulating liquids containing PCBs) A motor vehicle or rail car used for the transportation of cargo by any mode, each cargo carrier (for trailer, freight car) is a separate vehicle. (LM/C) 802.2-1985s

transposed file A file in which corresponding fields in corresponding records are stored contiguously, in contrast to the usual practice of storing entire records contiguously. (C) 610.5-1990w

transposition (1) (A) (data transmission) An interchange of positions of the several conductors of a circuit between successive lengths. *Notes:* 1. It is normally used to reduce inductive interference on communication or signal circuits by cancellation. 2. The term is most frequently applied to open wire circuits. **(B) (data transmission)** The ordered permutation of the pattern of the multiple of a switching stage to improve traffic carrying characteristics and reduce crosstalk. (PE) 599-1985
(2) (transmission lines) An interchange of positions of the several conductors of a circuit between successive lengths. *Notes:* 1. It is normally used to reduce inductive interference on communication or signal circuits by cancellation. 2. The term is most frequently applied to open-wire circuits. *See also:* signal; tower. (T&D/PE) [10]
(3) (rotating machinery) An arrangement of the strands or laminations of a conductor or of the conductors comprising a turn or coil whereby they take different relative positions in a slot for the purpose of reducing eddy current losses. (PE) [9]

transposition section A length of open wire line to which a fundamental transposition design or pattern is applied as a unit. (EEC/PE) [119]

transreactance The imaginary part of the transimpedance. (IM/HFIM) [40]

transrectification factor The quotient of the change in average current of an electrode by the change in the amplitude of the alternating sinusoidal voltage applied to another electrode, the direct voltages of this and other electrodes being maintained constant. *Note:* Unless otherwise stated, the term refers to cases in which the alternating sinusoidal voltage is of infinitesimal magnitude. *See also:* rectification factor. (ED) 161-1971w

transrectifier A device, ordinarily a vacuum tube in which rectification occurs in one electrode circuit when an alternating voltage is applied to another electrode. *See also:* rectifier. (EEC/PE) [119]

transresistance The real part of the transimpedance. (IM/HFIM) [40]

transsusceptance The imaginary part of the transadmittance. (IM/HFIM) [40]

transverse-beam traveling-wave tube A traveling-wave tube in which the direction of motion of the electron beam is transverse to the average direction in which the signal wave moves. (ED) 161-1971w

transverse crosstalk coupling (between a disturbing and a disturbed circuit in any given section) The vector summation of the direct couplings between adjacent short lengths of the two circuits, without dependence on intermediate flow in other nearby circuits. *See also:* coupling. (EEC/PE) [119]

transverse-electric hybrid wave (radio-wave propagation) An electromagnetic wave in which the electric field vector is linearly polarized normal to the plane of propagation and the magnetic field vector is elliptically polarized in this plane. (AP) 211-1977s

transverse electric mode (TE) (1) (fiber optics) A mode whose electric field vector is normal to the direction of propagation. *Note:* In an optical fiber, TE and transverse magnetic (TM) modes correspond to meridional rays. *See also:* mode; meridional ray. (Std100) 812-1984w
(2) A mode in which the longitudinal components of the electric and magnetic fields are everywhere zero. *Synonym:* TEM mode. *See also:* waveguide. (AP/ANT) [35], [84]

transverse electric resonant mode ($TE_{m,n,p}$) (cylindrical cavity) In a hollow metal cylinder closed by two plane metal surfaces perpendicular to its axis, the resonant mode whose transverse field pattern is similar to the $TE_{m,n}$ wave in the corresponding cylindrical waveguide and for which p is the number of half-period field variations along the axis. *Note:* When the cavity is a rectangular parallelepiped, the axis of the cylinder from which the cavity is assumed to be made should be designated since there are three such axes possible. *See also:* waveguide. (MTT) 146-1980w

transverse electric wave (1) (A) (circular waveguide) (hollow circular metal cylinder). The transverse electric wave for which *m* is the number of axial planes along which the normal component of the electric vector vanishes, and *n* is the number of coaxial cylinders (including the boundary of the waveguide) along which the tangential component of the electric vector vanishes. *Notes:* 1. $TE_{0,n}$ waves are circular electric waves of order *n*. The $TE_{0,1}$ wave is the circular electric wave with the lowest cutoff frequency. 2. The $TE_{1,1}$ wave is the dominant wave. Its lines of electric force are approximately parallel to a diameter. *Synonym:* TE. *See also:* waveguide. **(B) (rectangular waveguide)** (hollow rectangular metal cylinder) The transverse electric wave for which *m* is the number of half-period variations of the field along the *x* coordinate, which is assumed to coincide with the larger transverse dimension, and *n* is the number of half-period variations of the field along the *y* coordinate, which is assumed to coincide with the smaller transverse dimension. *Note:* The dominant wave in a rectangular waveguide is $TE_{1,0}$: its electric lines are parallel to the shorter side. *See also:* waveguide; guided wave. **(C)** In a homogeneous isotropic medium, an electromagnetic wave in which the electric field vector is everywhere perpendicular to the direction of propagation. *See also:* waveguide. (MTT) 146-1980
(2) For waves propagating in homogeneous space, an electromagnetic wave whose electric field is perpendicular to the direction of propagation. For waves incident on a scatterer, the wave whose electric field is perpendicular to the plane of incidence. (AP/PROP) 211-1997

transverse-electromagnetic cell A rectangular transmission line segment that produces a transverse electromagnetic (TEM). Cables, cable/connector assemblies, and/or electronic devices are placed inside the cell. Alternatively, the cell can be used as a detector to measure radiation emitted by a cable or device inside the cell. *Synonym:* Crawford cell. *See also:* absorbing clamp. (PE/IC) 1143-1994r

transverse-electromagnetic mode (1) (fiber optics) A mode whose electric and magnetic field vectors are both normal to the direction of propagation. *Synonym:* TEM mode. *See also:* mode. (Std100) 812-1984w
(2) (waveguide) A mode in which the longitudinal components of the electric and magnetic fields are everywhere zero. *Synonym:* TEM mode. *See also:* waveguide. (AP/ANT) [35], [84]
(3) (planar transmission lines) *Synonym:* TEM mode. (MTT) 1004-1987w

transverse-electromagnetic wave (1) In a homogeneous isotropic medium, an electromagnetic wave in which both the electric and magnetic field vectors are everywhere perpendicular to the direction of propagation. *Synonym:* TEM wave. *See also:* radio-wave propagation; waveguide. (MTT) 146-1980w
(2) A wave that propagates with the electric field and magnetic field vectors transverse (at right angles to) the direction of propagation. (At high frequencies waves may also propagate in transverse electric [TE] or transverse magnetic [TM] waves that bounce back and forth between guiding structures such as waveguides.) (PE/IC) 1143-1994r
(3) (radio-wave propagation) An electromagnetic wave in which both the electric and magnetic field vectors are everywhere perpendicular to the direction of propagation. *Synonym:* TEM wave. (AP/EMC/PROP) 211-1997, 1128-1998

transverse-field traveling-wave tube A traveling-wave tube in which the traveling electric fields that interact with electrons are essentially transverse to the average motion of the electrons. (ED) 161-1971w

transverse interference *See:* signal; differential-mode interference; normal-mode interference.

transverse interferometry (fiber optics) The method used to measure the index profile of an optical fiber by placing it in an interferometer and illuminating the fiber transversely to its axis. Generally, a computer is required to interpret the interference pattern. *See also:* interferometer. (Std100) 812-1984w

transverse-magnetic hybrid wave (radio-wave propagation) An electromagnetic wave in which the magnetic field vector is linearly polarized normal to the plane of propagation and the electric field vector is elliptically polarized in this plane. (AP) 211-1977s

transverse magnetic mode A mode whose magnetic field vector is normal to the direction of propagation. *Note:* In a planar dielectric waveguide (as within an injection laser diode), the field direction is parallel to the core-cladding interface. In an optical waveguide, transverse electric (TE) and TM modes correspond to meridional rays. *Synonym:* TM mode. *See also:* mode; meridional ray. (Std100) 812-1984w

transverse-magnetic polarization *See:* parallel polarization.

transverse magnetic resonant mode (cylindrical cavity) In a hollow metal cylinder closed by two plane metal surfaces perpendicular to its axis, the resonant mode whose transverse field pattern is similar to the $TM_{m,n}$ wave in the corresponding cylindrical waveguide and for which p is the number of half-period field variations along the axis. *Note:* When the cavity is a rectangular parallelepiped, the axis of the cylinder from which the cavity is assumed to be made should be designated since there are three such axes possible. *See also:* waveguide. (MTT) 146-1980w

transverse-magnetic wave (1) (A) In a homogeneous isotropic medium, an electromagnetic wave in which the magnetic field vector is everywhere perpendicular to the direction of propagation. *See also:* waveguide. **(B) (hollow circular metal cylinder) (circular waveguide)** The transverse magnetic wave for which *m* is the number of axial planes along which the normal component of the magnetic vector vanishes, and *n* is the number of coaxial cylinders to which the electric vector is normal. *Note:* $TM_{0,n}$ waves are circular magnetic waves of order *n*. The $TM_{0,1}$ wave is the circular magnetic wave with the lowest cutoff frequency. *See also:* circular magnetic wave; waveguide; guided wave. **(C) (hollow rectangular metal cylinder) (rectangular waveguide)** The transverse magnetic wave for which *m* is the number of half-period variations of the magnetic field along the longer transverse dimension, and *n* is the number of half-period variations of the magnetic field along the shorter transverse dimension. *See also:* guided wave; circular magnetic wave; waveguide. (MTT) 146-1980

(2) For waves propagating in homogeneous space, an electromagnetic wave whose magnetic field is perpendicular to the direction of propagation. For waves incident on a scatterer, the wave whose magnetic field is perpendicular to the plane of incidence. (AP/PROP) 211-1997

transverse magnetization Magnetization of the recording medium in a direction perpendicular to the line of travel and parallel to the greatest cross-sectional dimension. *See also:* phonograph pickup. (SP/MR) [32]

transverse mode (laser maser) A mode which is detected by measuring one or more maxima in transverse field intensity in the cross-section of a beam. (LEO) 586-1980w

transverse (differential) mode voltage (low-voltage air-gap surge-protective devices) (gas-tube surge protective devices) The voltage at a given location between two conductors of a group. *Synonym:* differential-mode voltage. (SWG/PE/SPD) C37.100-1992, C62.31-1987r, C62.32-1981s

transverse-mode interference *See:* differential-mode interference.

transverse-mode noise (with reference to load device input ac power) Noise signals measurable between or among active circuit conductors feeding the subject load, but not between the equipment grounding conductor or associated signal reference structure and the active circuit conductors. (IA/PSE) 1100-1999

transverse-mode voltage (1) (gas-tube surge protective devices) (surge withstand capability tests) (low-voltage air-gap surge-protective devices) The voltage at a given location between two conductors of a group. *Synonym:* differential-mode voltage. (SWG/SPD/PE/PSR) C62.31-1987r, C62.32-1981s, C37.100-1992, C37.90-1978s

(2) (protective relays and relay systems) The voltage between two conductors of a circuit at a given location. (PE/PSR) C37.90.1-1989r

transverse offset loss *See:* lateral offset loss.

transverse propagation constant (fiber optics) The propagation constant evaluated along a direction perpendicular to the waveguide axis. *Note:* The transverse propagation constant for a given mode can vary with the transverse coordinates. *See also:* propagation constant. (Std100) 812-1984w

transverse scattering (fiber optics) The method for measuring the index profile of an optical fiber or preform by illuminating the fiber or preform coherently and transversely to its axis, and examining the far-field irradiance pattern. A computer is required to interpret the pattern of the scattered light. *See also:* scattering. (Std100) 812-1984w

transverse wave A wave in which the direction of displacement at each point of the medium is perpendicular to the direction of propagation. *Note:* In those cases where the displacement makes an acute angle with the direction of propagation, the wave is considered to have longitudinal and transverse components. (Std100) 270-1966w

trap (1) (burglar-alarm system) An automatic device applied to a door or window frame for the purpose of producing an alarm condition in the protective circuit whenever a door or window is opened. *See also:* protective signaling. (EEC/PE) [119]

(2) (computers) An unprogrammed conditional jump to a known location, automatically activated by hardware, with the location from which the jump occurred recorded. (C) [20], [85]

(3) (A) (software) A conditional jump to an exception or interrupt handling routine, often automatically activated by hardware, with the location from which the jump occurred recorded. **(B) (software)** To perform the operation in definition (A). (C) 610.12-1990

(4) A vectored transfer of control to supervisor software through a table, the address of which is specified by a privileged **integer unit** register referred to as the trap base register (TBR). (C/MM) 1754-1994

trap circuit A circuit used at locations where it is desirable to protect a section of track on which it is impracticable to maintain a track circuit. It usually consists of an arrangement of one or more stick circuits so connected that when a train enters the trap circuit the stick relay drops and cannot be picked up again until the train has passed through the other end of the trap circuit. (EEC/PE) [119]

trapezium distortion (cathode-ray tubes) A fault characterized by a variation of the sensitivity of the deflection parallel to one axis (vertical or horizontal) as a function of the deflection parallel to the other axis and having the effect of transforming an image that is a rectangle into one which is a trapezium. (EEC/ACO) [109], [84]

trapped flux (superconducting material) Magnetic flux that links with a closed superconducting loop. *See also:* superconductivity. (ED) [46]

trapped inverted microstrip A type of inverted microstrip where the ground plane below the substrate completely encloses the strip conductor. (MTT) 1004-1987w

trapped mode *See:* bound mode.

trapped ray *See:* guided ray.

trapping *See:* ducting.

travel (1) (elevators) (rise) Of an elevator, dumbwaiter, escalator, or of a private-residence inclined lift, the vertical distance between the bottom terminal landing and the top terminal landing. *See also:* elevator. (PE/PSR) C37.90-1978s

(2) (of a relay) The amount of movement in either direction (towards pickup or reset) of a responsive element. *Note:* Travel may be specified in linear, angular, or other measure. (SWG/PE) C37.100-1992

traveled way The portion of the roadway for the movement of vehicles, exclusive of shoulders and full-time parking lanes. (NESC) C2-1997

traveler (conductor stringing equipment) A sheave complete with suspension arm or frame used separately or in groups and suspended from structures to permit the stringing of conductors. These devices are sometimes bundled with a center drum, or sheave and another traveler, and used to string more than one conductor simultaneously. For protection of conductors that should not be nicked or scratched, the sheaves are often lined with nonconductive or semiconductive neoprene or with nonconductive urethane. Any one of these materials acts as a padding or cushion for the conductor as it passes over the sheave. Traveler grounds must be used with lined travelers in order to establish an electrical ground. *Synonyms:* sheave; dolly; stringing block; stringing traveler; stringing sheave; block. (T&D/PE) 524a-1993r, 524-1992r, 1048-1990

traveler ground (conductor stringing equipment) A portable device designed to connect a moving conductor or wire rope, or both, to an electrical ground. Primarily used to provide safety for personnel during construction or reconstruction operations. This device is placed on the traveler (sheave, block, etc.) at a strategic location where an electrical ground is required. (T&D/PE) 524a-1993r, 524-1992r, 1048-1990

traveler, hold-down *See:* hold-down block.

traveler rack A device designed to protect, store, and transport travelers. It is normally designed to permit efficient use of transporting vehicles, spotting by helicopters on the line, and stacking during storage to utilize space. The exact design of each rack is dependent upon the specific travelers to be stored. *Synonym:* dolly car. (T&D/PE) 524-1992r

traveler sling A sling of wire rope, sometimes utilized in place of insulators, to support the traveler during stringing operations. Normally, it is used when insulators are not readily available or when adverse stringing conditions might impose severe downstrains and cause damage or complete failure of the insulators. *Synonym:* choker. (T&D/PE) 524-1992r

traveling cable (elevators) A cable made up of electric conductors that provides electric connection between an elevator or dumbwaiter car and fixed outlet in the hoistway. *See also:* control. (PE/EEC) [119]

traveling ionospheric disturbance (TID) A localized disturbance in the electron density distribution propagating in the ionosphere. *Note:* A TID is the signature in the ionosphere of an atmospheric gravity wave (AGW) in the neutral thermosphere. (AP/PROP) 211-1997

traveling overvoltage (surge arresters) A surge propagated along a conductor. (PE) [8], [84]

traveling plane wave (1) (radio-wave propagation) A plane wave each of whose frequency components has an exponential variation of amplitude and a linear variation of phase with distance. (AP) 211-1977s
(2) (waveguide) A plane wave each of whose components have an exponential variation of amplitude and a linear variation of phase in the direction of propagation. (MTT) 146-1980w

traveling wave The resulting wave when an electrical variation in a circuit such as a transmission line takes the form of translation of energy along a conductor, such energy being always equally divided between current and potential forms. (SPD/PE) C62.22-1997

traveling-wave antenna An antenna whose excitation has a quasi-uniform progressive phase, as the result of a single feeding wave traversing its length in one direction only. (AP/ANT) 145-1993

traveling-wave magnetron A traveling-wave tube in which the electrons move in crossed static electric and magnetic fields which are substantially normal to the direction of wave propagation. (ED) 161-1971w

traveling-wave magnetron oscillations Oscillations sustained by the interaction between the space-charge cloud of a magnetron and a traveling electromagnetic field whose phase velocity is approximately the same as the mean velocity of the cloud. (ED) [45]

traveling-wave parametric amplifier A parametric amplifier that has a continuous or iterated structure incorporating nonlinear reactors and in which the signal, pump, and difference-frequency wave are propagated along the structure. *See also:* parametric device. (ED) [46]

traveling-wave tube An electron tube in which a stream of electrons interacts continuously or repeatedly with a guided electromagnetic wave moving substantially in synchronism with it, and in such a way that there is a net transfer of energy from the stream to the wave. *See also:* transverse-field traveling-wave tube; transverse-beam traveling-wave tube. (ED) 161-1971w

traveling-wave-tube interaction circuit An extended electrode arrangement in a traveling-wave tube designed to propagate an electromagnetic wave in such a manner that the traveling electromagnetic fields are retarded and extended into the space occupied by the electron stream. *Note:* traveling-wave tubes are often designated by the type of interaction circuit used, as in helix traveling-wave tube. (ED) 161-1971w

travel trailer A vehicular unit mounted on wheels, designed to provide temporary living quarters for recreational, camping, or travel use, of such size or weight as not to require special highway movement permits when drawn by a motorized vehicle, and with a living area of less than 220 sq ft, excluding built-in equipment such as wardrobes, closets, cabinets, kitchen units or fixtures) and bath and toilet rooms. *See also:* recreational vehicle. (NESC/NEC) [86]

traversable widget A widget that can receive input focus. (C) 1295-1993w

traversal The process of enumerating or visiting each of the nodes of an ordered tree exactly once. *See also:* postorder traversal; traversal order; inorder traversal; converse postorder traversal; converse inorder traversal; converse preorder traversal; preorder traversal; traverse. (C) 610.5-1990w

traversal order The order in which the nodes of a tree are visited in a traversal. (C) 610.5-1990w

traverse To enumerate or to visit each of the nodes of an ordered tree exactly once. *See also:* traversal. (C) 610.5-1990w

tray (storage battery) (storage cell) A support or container for one or more storage cells. *See also:* battery. (PE/EEC) [119]

TR box *See:* transmit-receive box.

treated fabric (rotating machinery) A fabric or mat in which the elements have been essentially coated but not filled with an impregnant such as a compound or varnish. *Synonym:* treated mat. *See also:* stator; rotor. (PE) [9]

treated mat *See:* treated fabric.

treatment The systematic application of some agent (e.g., chemical, electric field) to a sample of organisms in an experimental setting for the purpose of determining the biological effect(s) of the agent. (T&D/PE) 539-1990

treble boost An adjustment of the amplitude-frequency response of a system or transducer to accentuate the higher audio frequencies. (SP) 151-1965w

tree (1) A set of connected branches including no meshes. *See also:* network analysis. (NIR) C95.1-1982s
(2) (software) An abstract hierarchical structure consisting of nodes connected by branches, in which: each branch connects one node to a directly subsidiary node; there is a unique node called the root which is not subsidiary to any other node; and every node besides the root is directly subsidiary to exactly one other node. (C/SE) 729-1983s
(3) (data management) A nonlinear data structure consisting of a finite set of nodes in which one node is called the root node and the remaining nodes are partitioned into disjoint sets, called subtrees, each of which is itself a tree. (See the corresponding figure.) *Note:* The nodes are connected by pointers. *Synonyms:* tree structure; rooted tree. *See also:* subtree; ordered tree; threaded tree; height-balanced tree; *n–m* tree; search tree; unordered tree; null tree; n-ary tree.

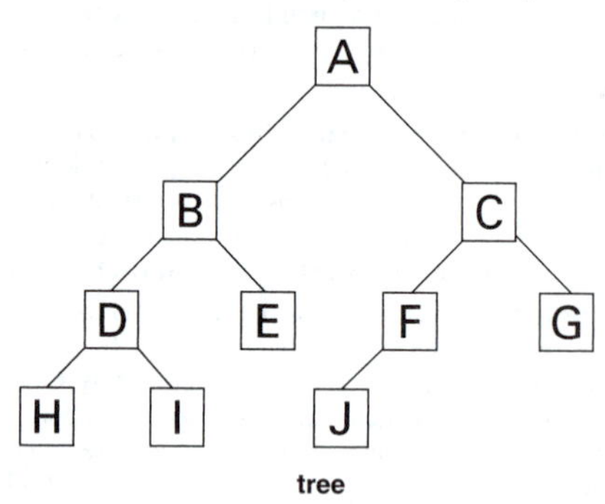

tree

(C) 610.5-1990w

treeing (composite insulators) Irreversible internal degradation by the formation of conductive carbonized paths. (T&D/PE) 987-1985w

tree insertion sort An insertion sort in which the items in the set to be sorted are treated as nodes on a tree. *Contrast:* tree selection sort. (C) 610.5-1990w

tree machine A multiprocessor whose processing elements are connected in an n-ary tree arrangement. (C) 610.10-1994w

trees and nodules Projections formed on a cathode during electrodeposition. Trees are branched whereas nodules are rounded. (EEC/PE) [119]

tree selection sort A selection sort in which the items in the set to be sorted are treated as nodes on a tree. *Contrast:* tree insertion sort. (C) 610.5-1990w

tree structure (1) (data management) *See also:* tree. (C) 610.5-1990w

(2) A set of connected segments with no loops (cross connections). (NID) 960-1993

tree topology A topology in which stations are attached in a tree layout fashion on a shared transmission medium. The tree layout begins at the head/end and each of these may have branches. The branches in turn may have additional branches to allow quite complex layouts. *See also:* bus topology; star-ring topology; loop topology; star-bus topology; bus-ring topology; star topology; ring topology. (C) 610.7-1995

tree wire A conductor with an abrasion-resistant outer covering, usually nonmetallic, and intended for use on overhead lines passing through trees. *See also:* armored cable; conductor. (T&D/PE) [10]

triad A group of three closely related items or digits. (C) 1084-1986w

trial operation (1) A period during which the equipment or system is placed under service conditions and is also monitored for stable, smooth, and reliable performance. (SUB/PE) 1303-1994

(2) A period of normal operation of the dc system near the end of commissioning. (PE/SUB) 1378-1997

triangular array *See:* triangular grid array.

triangular grid array A regular arrangement of array elements, in a plane, such that lines connecting corresponding points of adjacent elements form triangles, usually equilateral. (AP/ANT) 145-1993

triaxial Testing or analysis in the two horizontal orthogonal directions and the vertical direction simultaneously. (PE/SUB) 693-1997

triboelectric charging The generation of electrostatic charges when two pieces of material are brought into intimate contact and are then separated. *Synonym:* triboelectrification. (SPD/PE) C62.47-1992r

triboelectric series A list of substances in an order of relative positive to negative charging as a result of the triboelectric charging effect. (SPD/PE) C62.47-1992r

triboelectrification (electrification by friction) The mechanical separation of electric charges of opposite sign by processes such as: the separation (as by sliding) of dissimilar solid objects; interaction at a solid-liquid interface; breaking of a liquid-gas interface. (Std100) 270-1966w

tri-bundle *See:* bundle; bundle, two-conductor, three-conductor, four-conductor, multiconductor.

tributary office A telephone central office that passes toll traffic to, and receives toll traffic from, a toll center. (EEC/PE) [119]

tributary trunk (data transmission) A trunk circuit connecting a local exchange with a toll center or other toll office through which access to the long-distance network is achieved. (PE) 599-1985w

Trichel streamers (overhead power-line corona and radio noise) Streamers occurring at a negative electrode with electric field strengths at and above the corona inception voltage gradient. A Trichel streamer appears as a small, constantly moving purple fan. The current pulse is of small amplitude, short duration (in the range of a hundred nanoseconds), and high repetition rate (in the range of tens of kilohertz or more). (T&D/PE) 539-1990

trickle charge (storage battery) (storage cell) A continuous charge at a low rate approximately equal to the internal losses and suitable to maintain the battery in a fully charged condition. *Note:* This term is also applied to very low rates of charge suitable not only for compensating for internal losses but to restore intermittent discharges of small amount delivered from time to time to the load circuit. *See also:* charge; floating. (EEC/PE) [119]

triclinic system (piezoelectricity) A triclinic crystal has neither symmetry axes nor symmetry planes. The lengths of the three axes are in general unequal; and the angles α, β, and g between axes *b* and *c*, *c* and *a*, and *a* and *b*, respectively, are also unequal. The *a* axis has the direction of the intersection of the faces *b* and *c* (extend the faces to intersection if necessary), the *b* axis has the direction of the intersection of faces *c* and *a*, the *c* axis has the direction of the intersection of faces *a* and *b*. The *X*, *Y*, *Z* axes are associated as closely as possible with the *a*, *b*, *c* axes, respectively. The *Z* axis is parallel to *c*, *Y* is normal to the *ac* plane, and *X* is thus in the *ac* plane. The +*Z* and +*X* axes are chosen so that d_{33} and and d_{11} are positive. The +*Y* axis is chosen so that it forms a right-handed system with +*Z* and +*X*. *See also:* crystal systems. (UFFC) 176-1978s

trie An *n*-ary tree each of whose nonterminal nodes is the parent of a sequence of subtrees, where the *k*-th subtree represents the *k*-th digit or character in an *n*-character alphabet. *Notes:* 1. A sequence of nodes (length *p*) from the root of a trie to the root of a subtree represents the first *p* digits or characters of the keys of the elements represented by that subtree. 2. The term is pronounced "try" and is derived from the word "re-trie-val." *See also:* binary radix trie search; radix trie search; multiway radix trie search. (C) 610.5-1990w

Triex *See:* explosives.

trigatron (1) A triggered spark-gap switch on which control is obtained by a voltage applied to a trigger electrode. *Note:* This voltage distorts the field between the two main electrodes converting the sphere-to-sphere gap to a point to sphere gap. *See also:* electron device. (ED) [45], [84]

(2) (radar) An electronic switch in which conduction is initiated by the breakdown of an auxiliary gap. (AES/RS) 686-1990, [42]

trigger (1) To start action in another circuit which then functions for a period of time under its own control. (EEC/PE) [119]

(2) A pulse used to initiate some function, for example, a triggered sweep or delay ramp. *Note:* Trigger may loosely refer to a waveform of any shape used as a signal from which a trigger pulse is derived as in trigger source, trigger input, etc. *See also:* triggering signal; oscillograph. (IM/HFIM) [40]

(3) (thyristor) The act of causing a thyristor to switch from the off-state to the on-state. *See also:* gate trigger current. (IA) [12]

(4) A signal to start an action. A trigger is a signal from the Network Capable Application Processor serving as a command to the Smart Transducer Interface Module for an action to occur. (IM/ST) 1451.2-1997

(5) (A) To cause a circuit or device to change state or to perform some other operaton. *See also:* clock. **(B)** A signal that causes a circuit or device to change state as in (A). (C) 610.10-1994

trigger circuit (1) A circuit that has two conditions of stability, with means for passing from one to the other when certain conditions are satisfied, either spontaneously or through application of an external stimulus. (EEC/PE) [119]

(2) A sequential circuit that has a number of states, at least one of which is stable, and has one or more inputs that allow external signals to force a change of state. *See also:* multivibrator; strobe; flip-flop. (C) 610.10-1994w

trigger countdown A process that reduces the repetition rate of a triggering signal. *See also:* oscillograph. (IM/HFIM) [40]

trigger cycle A complete cycle comprising the assertion of the trigger signal by the Network Capable Application Processor followed by the acknowledgment by the Smart Transducer Interface Module. (IM/ST) 1451.2-1997

trigger delay The elapsed time from the occurrence of a trigger pulse at the trigger input connector to the time at which the first or a specified data sample is recorded. (IM/WM&A) 1057-1994w

triggered sweep A sweep that can be initiated only by a trigger signal, not free running. *See also:* oscillograph. (IM/HFIM) [40]

trigger, external *See:* external trigger.

trigger gap (series capacitor) Enclosed electrodes that initiate the sparkover of the bypass gap. (T&D/PE) 824-1985s

triggering (pulse terminology) A process in which a pulse initiates a predetermined event or response. (IM/WM&A) 194-1977w

triggering level The instantaneous level of a triggering signal at which a trigger is to be generated. Also, the name of the control that selects the level. *See also:* oscillograph. (IM/HFIM) [40]

triggering, line *See:* line triggering.

trigger minimum rate of change The slowest rate of change of the leading edge of a pulse of a specified level that will trigger the recorder. (IM/WM&A) 1057-1994w

triggering signal The signal from which a trigger is derived. *See also:* oscillograph. (IM/HFIM) [40]

trigger signal coupling The ratio of the spurious signal level (that is recorded by an input to the recorder) to the trigger signal level. (IM/WM&A) 1057-1994w

triggering slope The positive-going (+slope) or negative-going (−slope) portion of a triggering signal from which a trigger is to be derived. Also, the control that selects the slope to be employed. *Note:* + and − slopes apply to the slope of the waveform only and not to the absolute polarity. *See also:* oscillograph. (IM/HFIM) [40]

triggering stability *See:* stability.

trigger jitter The standard deviation in the trigger delay time over multiple records. (IM/WM&A) 1057-1994w

trigger level (1) (navigation aids) (transponder) The minimum input to the receiver that is capable of causing the transmitter to emit a reply. (AES/GCS) 172-1983w
(2) In a transponder, the minimum input to the receiver that is capable of causing the transmitter to emit a reply. (AES/RS) 686-1990

trigger pickoff A process or a circuit for extracting a triggering signal. *See also:* oscillograph. (IM/HFIM) [40]

trigger pulse converter (TPC) A device in the control system that converts the control signal to a signal that can be transmitted to the thyristor valve. (SUB/PE) 1303-1994

trigger-starting systems (fluorescent lamps) Applied to systems in which hot-cathode electric discharge lamps are started with cathodes heated through low-voltage heater windings built into the ballast. Sufficient voltage is applied across the lamp and between the lamp and fixture to initiate the discharge when the cathodes reach a temperature high enough for adequate emission. The ballast is so designed that the cathode-heating current is greatly reduced as soon as the arc is struck. (EEC/LB) [94]

trigger tube A cold-cathode gas-filled tube in which one or more electrodes initiate, but do not control, the anode current. (ED) [45]

trigonal and hexagonal systems (piezoelectricity) These systems are distinguished by an axis of sixfold (or threefold) symmetry. This axis is always called the *c* axis. According to the Bravais-Miller axial system, which is most commonly used, there are three equivalent secondary axes, a_1, a_2, and a_3, lying 120 degrees apart in a plane normal to *c*. These axes are chosen as being either parallel to a twofold axis or perpendicular to a plane of symmetry, or if there are neither twofold axes perpendicular to *c* nor planes of symmetry parallel to *c*, the a axes are chosen so as to give the smallest unit cell. The *Z* axis is parallel to *c*. The *X* axis coincides in direction and sense with any one of the *a* axes. The *Y* axis is perpendicular to *Z* and *X*, so oriented as to form a right-handed system. Positive-sense rules for $+Z$, $+X$, $+Y$ are listed in the table below for the piezoelctric trigonal and hexagonal crystals. *Note:* "Positive" and "negative" may be checked using a carbon-zinc flashlight battery. The carbon anode connection will have the same effect on meter deflection as the + end of the crystal axis upon release of compression. *See also:* crystal systems. (UFFC) 176-1978s

trilateration *See:* multilateration.

trimmer capacitor (trimming capacitor) A small adjustable capacitor associated with another capacitor and used for fine adjustment of the total capacitance of an element or part of a circuit. (IM) [120]

trimmer signal A signal that gives indication to the engineman concerning movements to be made from the classification tracks into the switch and retarder area. (EEC/PE) [119]

triode A three-electrode electron tube containing an anode, a cathode, and a control electrode. (ED) 161-1971w

triode region of an insulated-gate field-effect transistor (metal-nitride-oxide field-effect transistor) The same as non-saturation region. (ED) 581-1978w

trip (A) A release that initiates either an opening or a closing operation or other specified action. **(B)** A release that initiates an opening operation only. **(C)** A complete opening operation. **(D)** The action associated with the opening of a circuit breaker or other interrupting device. **(E)** To release in order to initiate either an opening or a closing operation or other specified action. **(F)** To release in order to initiate an opening operation only. **(G)** To initiate and complete an opening operation. **(H)** Pertaining to a release that initiates either an opening or a closing operation or other specified action. *Synonym:* tripping. **(I)** Pertaining to a release that initiates an opening operation only. *Synonym:* tripping. **(J)** Pertaining to a complete opening operation. *Synonym:* tripping. (SWG/PE) C37.100-1992

trip arm *See:* mechanical trip.

trip/close A type of digital output that stops or starts an action, usually affecting actual electric power circuits. (PE/SUB) 1379-1997

trip coil *See:* release coil.

trip current (faulted circuit indicators) The actual value of current in amperes rms (root-mean-square) that will cause the FCI (faulted circuit indicator) to indicate FAULT. (T&D/PE) 495-1986w

trip current rating (faulted circuit indicators) The published rms (root-mean-square) sinusoidal fault current in amperes that causes the FCI (faulted circuit indicator) to indicate FAULT. (T&D/PE) 495-1986w

trip delay setting *See:* release-delay setting.

trip device, impulse *See:* impulse trip device.

trip device (opening release), impulse A trip device that is designed to operate only by the discharge of a capacitor into its release (trip) coil and is utilized on high speed circuit breakers to produce tripping times that are independent of *di/dt*. (SWG/PE) C37.100-1992, C37.14-1992s

trip-free The capability of a switching device to have the moving contacts return to and remain in the opening position when the opening operation is initiated after the initiation of the closing operation, even if the closing force and command are maintained. *Notes:* 1. To ensure proper breaking of the current that may be established, it may be necessary for the contacts to momentarily reach the closed position. 2. If the release circuit is completed through an auxiliary switch, electrical release will not take place until such auxiliary switch is closed. *Synonym:* release-free. (SWG/PE) C37.100-1992

trip-free in any position *See:* release-free.

trip-free relay An auxiliary relay whose function is to open the closing circuit of an electrically operated switching device so that the opening operation can prevail over the closing operation. *Synonym:* release-free relay. (SWG/PE) C37.100-1992

trip lamp A removable self-contained mine lamp, designed for marking the rear end of a train (trip) of mine cars. (EEC/PE) [119]

triple-address Same as three-address.

triple-address instruction *See:* three-address instruction.

triple detection *See:* double-superheterodyne reception.

triple-length register Three registers that function as a single register. *Note:* Typically used in display controllers to store x, y, z information. *Synonym:* triple register. *See also:* n-tuple length register; double-length register. (C) 610.10-1994w

triplen (rotating machinery) An order of harmonic that is a multiple of three. *See also:* asynchronous machine. (PE) [9]

triple precision Pertaining to the use of three computer words to represent a number in order to preserve or gain precision. *Contrast:* double precision; single precision. *See also:* multiple precision. (C) 610.5-1990w, 1084-1986w

triple-precision arithmetic Computer arithmetic performed with operands that are expressed in triple-precision representation. (C) 1084-1986w

triple register *See:* triple-length register.

triplet (1) (mathematics of computing) (data management) A group of three adjacent digits operated upon as a unit. *Synonym:* three-bit byte. (C) 610.5-1990w, 1084-1986w
(2) (navigation systems) (navigation aids) Three radio stations, operated as a group, for the determination of positions. (AES/GCS) 172-1983w
(3) An ordered set of three adjacent bytes. (C/BA) 1014.1-1994w
(4) A byte composed of three bits. *Synonym:* three-bit byte. (C) 610.10-1994w

triple transit echo (TTE) Normally unwanted signals in a surface acoustic wave device that arise from reflection of the wave at the output and input transducers and subsequent transduction at the output; the propagation path between input and output interdigital transducer is traversed three times. (UFFC) 1037-1992w

triplex cable A cable composed of three insulated single-conductor cables twisted together. *Note:* The assembled conductors may or may not have a common covering of binding or protecting material. (T&D/PE) [10]

triply-threaded tree A binary tree in which each node contains three link fields: one for its parent node and one for each of its left child and right child nodes. *Contrast:* doubly-threaded tree. (C) 610.5-1990w

trip OFF control signal (magnetic amplifier) The final value of signal measured when the amplifier has changed from the ON to the OFF state as the signal is varied so slowly that an incremental increase in the speed with which it is varied does not affect the measurement of the trip OFF control signal. That is, the change in trip OFF control signal is below the sensitivity of the measuring instrument. (MAG) 107-1964w

trip ON control signal (magnetic amplifier) The final value of signal measured when the amplifier has changed from the OFF to the ON state as the signal is varied so slowly than an incremental increase in the speed with which it is varied does not affect the measurement of the trip ON control signal. That is, the change in trip ON control signal is below the sensitivity of the measuring instrument. (MAG) 107-1964w

tripping *See:* automatic opening.

tripping delay *See:* release delay.

tripping mechanism *See:* release.

tripping or trip-free relay (power system device function numbers) A relay that functions to trip a circuit breaker, contactor, or equipment, or to permit immediate tripping by other devices; or to prevent immediate reclosure of a circuit interrupter if it should open automatically even though its closing circuit is maintained closed. (SUB/PE) C37.2-1979s

trip-point repeatability (magnetic amplifier) The change in trip point (either trip OFF or trip ON, as specified) control signal due to uncontrollable causes over a specified period of time when all controllable quantities are held constant. (MAG) 107-1964w

trip-point repeatability coefficient (magnetic amplifier) The ratio of: the maximum change in trip-point control signal due to uncontrollable causes, to the specified time period during which all controllable quantities have been held constant. *Note:* The units of this coefficient are the control signal units per the time period over which the coefficient was determined. (MAG) 107-1964w

trip setting *See:* release setting.

trip switch A device mounted on the truck of a vehicle, responding to a raised arm on the way-side, used to cause an emergency brake application if a train attempts to pass a mandatory stop signal. (VT) 1475-1999

tristimulus values (television) (of a light) The amounts of the three reference or matching stimuli required to give a match with the light considered, in a given trichromatic system. *Notes:* 1. In the standard colorimetric system, CIE (1931), the symbols, X, Y, Z are recommended for the tristimulus values. 2. These values may be obtained by multiplying the spectral concentration of the radiation at each wavelength by the distribution coefficients $\bar{x}_\lambda$, $\bar{y}_\lambda$, $\bar{z}_\lambda$ and integrating these products over the whole spectrum. (BT/AV) 201-1979w

tristimulus values of a light, *X*, *Y*, *Z* (illuminating engineering) The amounts of each of three specific primaries required to match the color of the light. *See also:* color matching functions. (EEC/IE) [126]

troff A text-formatting language used widely in the UNIX environment. (C) 610.13-1993w

troffer (illuminating engineering) A long recessed lighting unit usually installed with the opening flush with the ceiling. The term is derived from "trough" and "coffer." (EEC/IE) [126]

troland (illuminating engineering) A unit of retinal illuminance which is based upon the fact that retinal illuminance is proportional to the product of the luminance of the distal stimulus and the area of entrance pupil. One troland is the retinal illuminance produced when the luminance of the distal stimulus is one candela per square meter and the area of the pupil is one square millimeter. *Note:* The troland makes no allowance for interocular attenuation or for the Stiles-Crawford effect. *See also:* Stiles-Crawford effect. (EEC/IE) [126]

TROLL A nonprocedural computer language. (C) 610.13-1993w

trolley A current collector, the function of which is to make contact with a contact wire. *See also:* contact conductor. (VT/LT) 16-1955w

trolley bus *See:* trolley coach.

trolley car An electric motor car that collects propulsion power from a trolley system. *See also:* electric motor car. (EEC/PE) [119]

trolley coach An electric bus that collects propulsion power from a trolley system. *Synonyms:* trackless trolley coach; trolley bus. *See also:* electric bus. (EEC/PE) [119]

trolley locomotive An electric locomotive that collects propulsion power from a trolley system. *See also:* electric locomotive. (EEC/PE) [119]

trombone line (transmission lines and waveguides) A U-shaped length of waveguide or transmission line of adjustable length. *See also:* waveguide. (IM/HFIM) [40]

tropopause The upper boundary of the troposphere. (AP/PROP) 211-1997

troposcatter *See:* tropospheric scatter propagation.

troposphere (1) (data transmission) That part of the earth's atmosphere in which temperature generally decreases with altitude, clouds form, and convection is active. *Note:* Experiments indicate that the troposphere occupies the space above the earth's surface up to a height ranging from about 6 km (kilometers) at the poles to about 18 km at the equator. (PE) 599-1985w
(2) The lower part of the Earth's atmosphere, situated immediately above the surface of the Earth and in which the temperature decreases with increasing altitude except in certain local temperature inversion layers. The troposphere extends to an altitude of around 9 km at the poles and 17 km at the equator. (AP/PROP) 211-1997

tropospheric layer An elevated portion of the troposphere having radio propagation properties that are clearly distinguished from those of the surrounding atmosphere. Horizontal dimensions are generally in excess of 100 km, and vertical dimensions are on the order of 1 km. (AP/PROP) 211-1997

tropospheric propagation Propagation within the troposphere. (AP/PROP) 211-1997

tropospheric radio duct *See:* atmospheric radio duct.

tropospheric scatter propagation Propagation of radio waves through the atmosphere caused by scattering from inhomogeneities in the refractive index of the troposphere. *Note:* Troposcatter enables propagation beyond the radio horizon. *Synonym:* troposcatter. (AP/PROP) 211-1997

tropospheric wave (1) (data transmission) A radio wave that is propagated by reflection from a place of abrupt change in the dielectric constant or its gradient in the troposphere. *Note:* In some cases the ground wave may be so altered that new components appear to arise from reflections in regions of rapidly changing dielectric constant. When these components are distinguishable from the other components, they are called tropospheric waves. (PE) 599-1985w
(2) A radio wave that propagates in the troposphere. (AP/PROP) 211-1997

trouble Equipment malfunction or loss of power. (PE/NP) 692-1997

trouble recorder (telephone switching systems) A device or system associated with one or more switching systems for automatically recording data on calls encountering trouble. (COM) 312-1977w

troubleshoot (supervisory control, data acquisition, and automatic control) Action taken by operating or maintenance personnel, or both, to isolate a malfunctioned component of a system. Actions may be supported by printed procedures, diagnostic circuits, test points, and diagnostic routines. *See also:* debug; fault isolation. (PE/SUB) C37.1-1994

troughing An open channel of earthenware, wood, or other material in which a cable or cables may be laid and protected by a cover. (T&D/PE) [10]

TRR *See:* test readiness review.

TRS *See:* test response spectrum.

TR switch *See:* transmit-receive switch.

truck A rail vehicle component that consists of a frame—normally two axles, brakes, suspension, and other parts—and supports the vehicle body and can swivel under it on curves. If powered, it may also contain traction motors and associated drive mechanisms. (VT) 1475-1999

truck camper A portable unit constructed to provide temporary living quarters for recreational, travel, or camping use, consisting of a roof, floor, and sides, designed to be loaded onto and unloaded from the bed of a pick-up truck. *See also:* recreational vehicle. (NESC/NEC) [86]

truck generator suspension A design of support for an axle generator in which the generator is supported by the vehicle truck. (EEC/PE) [119]

true air speed (navigation aids) The actual speed of an aircraft relative to the surrounding air. (AES/GCS) 172-1983w

true air-speed indicator (navigation aids) An instrument for measuring indicated true air speed. (AES/GCS) 172-1983w

true bearing (navigation aids) Bearing relative to true north. (AES/GCS) 172-1983w

true complement (1) A number representation that can be derived from another by subtracting each digit from one less than the base and then adding one to the least significant digit and executing all carries required. Tens complements and twos complements are true complements. (C) 162-1963w
(2) *See also:* radix complement. (C) 1084-1986w

true course (navigation aids) Course relative to true north. (AES/GCS) 172-1983w

true density (fly ash resistivity) The weight of the particles divided by the solid volume of the particles. (PE/EDPG) 548-1984w

true heading (navigation aids) Heading relative to true north. (AES/GCS) 172-1983w

true-motion display A display in a vehicle- or ship-mounted radar that shows the motions of the radar and of targets tracked by that radar, relative to a fixed background. True-motion display is accomplished by inserting compensation for the motion of the vehicle carrying the radar. (AES) 686-1997

true neutral point (at terminals of entry) Any point in the boundary surface that has the same voltage as the point of junction of a group of equal nonreactive resistors placed in the boundary surface of the region and connected at their free ends to the appropriate terminals of entry of the phase conductors of the circuit, provided that the resistance of the resistors is so great that the voltages are not appreciably altered by the introduction of the resistors. *Notes:* 1. The number of resistances required is two for direct-current or single-phase alternating-current circuits, four for two-phase four-wire or five-wire circuits, and is equal to the number of phases when the number of phases is three or more. Under normal symmetrical conditions the number of resistors may be reduced to three for six- or twelve-phase systems when the terminals are properly selected, but the true neutral point may not be obtained by this process under all abnormal conditions. The concept of a true neutral point is not considered applicable to a two-phase, three-wire circuit. 2. Under abnormal conditions the voltage of the true neutral point may not be the same as that of the neutral conductor. *See also:* network analysis. (Std100) 270-1966w

true north (navigation aids) The direction of the north geographical pole. (AES/GCS) 172-1983w

true power factor For user equipment, the true power factor is the ratio of the active, or real, power (P) consumed in watts to the apparent power (S) drawn in volt-amperes, with

$$PF = P/S$$

and

$$S = \sqrt{P^2 + Q^2}$$

where
PF = power factor;
P = active power in watts;
Q = reactive power in vars;
S = total power in volt-amperes.

This definition of power factor includes the effect of both displacement and distortion in the input current (and/or voltage) waveform. Alternatively, if there are no interharmonics, the previous equation can be simplified to

$$F = PF_{dp} \cdot PF_d$$

(PEL) 1515-2000

true ratio (1) (power and distribution transformers) The ratio of the root-mean-square (rms) primary value of the rms secondary value under specified conditions. (PE/TR) C57.12.80-1978r
(2) The ratio of the root-mean-square (rms) primary voltage or current to the rms secondary voltage or current under specified conditions. (PE/TR) C57.13-1993

true rms detector A detector that contains a circuit component that performs the mathematical operation

$$\sqrt{\frac{1}{T}\int_0^T ([v(t)]^2 dt)}$$

to a periodic signal, $v(t)$

where
T is the period of the signal.

Note: If there are harmonics in the field and $v(t)$ is proportional to the time-derivative of the field, the detector circuit must also contain a stage of integration prior to the rms operation in order to avoid error. This type of detector gives the true rms value of a field containing harmonics provided that the frequency response of the detector is flat over the frequency range of interest. If significant levels of harmonics are present in $v(t)$, particular attention should be given to the possibility of amplifier saturation effects if the integration follows one or more stages of amplification. *See also:* average sensing rms detector. (T&D/PE) 1308-1994

true time *See:* real time.

true value (measured quantity) The actual value of a precisely defined quantity under the conditions existing during its measurement. (IM/HFIM) 314-1971w

truncate (computers) To terminate a computational process in accordance with some rule; for example, to end the evaluation of a power series at a specified term. *See also:* round down. (C) [20], [85], 1084-1986w
(2) **(A)** To remove the beginning or ending entities in a string; for example, the string 'PINEAPPLE,' when truncated on the right to six characters, is 'PINEAP.'. **(B)** To delete or omit one or more of the digits in a representation of a number; for example, the numbers 57.5634 and 25.437, when truncated to two decimal digits, become 57.56 and 25.43. *Contrast:* round. (C) 610.5-1990

truncation The process of truncating. (C) 610.5-1990w

truncation error An error caused by truncation. (C) 1084-1986w

truncation loss In a modulated data waveform, the power difference before and after implementation filtering necessary to constrain its spectrum to a specified frequency band. (C/LM) 802.3-1998

trunk (1) (analog computer) A connecting line between one analog computer and another, or between an analog compute and an external point, allowing the input (or output) of an analog component to communicate directly with the output (or input) or another component which is located outside of the analog computer. (C) 165-1977w
(2) (data transmission) A telephone line or channel between two central offices or switching devices, which is used in providing telephone connections between subscribers. (PE) 599-1985w
(3) (telephone switching systems) A channel provided as a common traffic artery between switching entities. (COM) 312-1977w
(4) (broadband local area networks) (system) That portion of a broadband coaxial cable system that serves as the RF signal path between the headend and the feeders. (LM/C) 802.7-1989r
(5) **(A)** A transmission path between exchanges or central offices. **(B)** A telephone exchange line that ends in a PBX. (C) 610.7-1995

trunk amplifier station A low-distortion amplifier that amplifies RF signals for long-distance transport. An active device designed to compensate for cable losses in the trunk system. (LM/C) 802.7-1989r

trunk cable (1) (medium attachment units and repeater units) The trunk coaxial cable system.
(2) (broadband local area networks) Coaxial cable used for distribution of RF signals over long distances throughout a cable system. Usually the largest rigid cable used in the system. (C) 802.7-1989r
(3) **(A)** A cable circuit between two switching centers or two individual distribution points. **(B)** The main (large-diameter) cable of a broadband coaxial cable system. *See also:* drop cable. (C) 610.7-1995
(4) The transmission medium for interconnection of concentrators providing a main signal path and a back-up signal path, exclusive of the lobe cabling. (C/LM) 8802-5-1998
(5) The main (often large diameter) cable of a coaxial cable system. *See also:* drop cable. (C/LM) 802.3-1998

trunk circuit (1) (telephone switching systems) An interface circuit between a trunk and a switching system. (COM) 312-1977w
(2) A pair of complementary circuits with associated equipment terminating in two switching centers. *Synonym:* toll circuit. (C) 610.7-1995

trunk circuit, combined line and recording (CLR) **(A)** Name given to a class of trunk circuits that provide access to operator positions generally referred to by abbreviation only. **(B)** Recording-completing trunk circuit for operator recording and completing of toll calls originated by subscribers of central offices. (COM) [48]

trunk concentrator (telephone switching systems) A concentrator in which all inlets and outlets are trunks. (COM) 312-1977w

trunk conditioning *See:* carrier group alarm.

trunk coupling unit (TCU) (1) A physical device that enables a data terminal equipment (DTE) to connect to a trunk cable. *Note:* The trunk coupling unit may be a passive connector, or may contain active elements. A drop cable may be used between the trunk coupling unit and the DTE to facilitate communication. (C/Std100) 610.7-1995
(2) A device that couples a station to the main ring path. A TCU provides the mechanism for insertion of a station into the ring and removal of it from the ring. (C/LM) 8802-5-1998, 610.7-1995

trunk failure rate The expected frequency of outages a trunk can experience due to switching system and subsystem malfunctions. The trunk failure rate may be given for hardware faults alone. (COM/TA) 973-1990w

trunk feeder A feeder connecting two generating stations or a generating station and an important substation. *See also:* center of distribution. (T&D/PE) [10]

trunk group (telephone switching systems) A number of trunks that can be used interchangeably between two switching entities. (COM) 312-1977w

trunk hunting The operation of a selector or other similar device, to establish connection with an idle circuit of a chosen group. This is usually accomplished by successively testing terminals associated with this group until a terminal is found that has an electrical condition indicating it to be idle. (EEC/PE) [119]

trunk line The major cable from the headend to downstream branches. *Synonym:* main trunk. (LM/C) 802.7-1989r

trunk-line conduit A duct-bank provided for main or trunk-line cables. (T&D/PE) [10]

trunk loss That part of the repetition equivalent assignable to the trunk used in the telephone connection. *See also:* transmission loss. (EEC/PE) [119]

trunk multifrequency pulsing (telephone switching systems) A means of pulsing embodying a simultaneous combination of two out of six frequencies to represent each digit or character. (COM) 312-1977w

trunks Interoffice facilities. (COM/TA) 973-1990w

trunk transmission line A transmission line acting as a source of main supply to a number of other transmission circuits. *See also:* transmission line. (T&D/PE) [10]

trussed blade (of a switching device) A blade that is reinforced by truss construction to provide stiffness. (SWG/PE) C37.100-1992

trusted communications path A path by which a network user, program, process, or device can communicate directly with the trusted network base. (C) 610.7-1995

trusted computing base (TCB) The totality of security controls and mechanisms within an information system. (C/BA) 896.3-1993w

trusted functionality That which is perceived to be correct with respect to some criteria, e.g., as established by a security policy. (LM/C) 802.10-1992

trusted identification forwarding In networks, an identification method in which a sending host transmits user authentication information to the receiving host and the receiving host can verify that the user is authorized for access to its systems. (C) 610.7-1995

trusted network base The totality of security mechanisms within a network that are responsible for enforcing a security policy on the network. (C) 610.7-1995

trusted network component base The totality of the security mechanisms within a network component that are responsible for enforcing the component security policy. (C) 610.7-1995

trusted path A path by which a user at a terminal can communicate directly with the trusted network base in a computer system. *Synonym:* secure path. *See also:* trusted communications path. (C) 610.7-1995

truth function A function that may take one of two possible values: true or false. (C) 1084-1986w

truth table (1) (computers) A table that describes a logic function by listing all possible combinations of input values and indicating, for each combination, the true output values. (MIL/C) [2], [85], [20]
(2) (mathematics of computing) An operation table that describes a truth function by listing all possible combinations of input values and giving the corresponding output values. *Synonym:* Boolean operation table. *See also:* NOT; AND; OR. (C) 1084-1986w
(3) An operation table for a logic operation. (C) 610.10-1994w

TSC *See:* thyristor-switched capacitor.

TSI *See:* threshold signal-to-interference ratio.

TSR *See:* thyristor-switched reactor.

T step (linear waveform distortion) A $\sin^2$ step with a 10% to 90% rise (fall) time of nominally 125 ns (nanoseconds). The amplitude of the envelope of the frequency spectrum is effectively zero at and beyond 8 MHz. *Note:* In practice, the T step is part of a square wave (or line bar), so that there is a T step rise and fall. *See also:* line bar. (BT) 511-1979w

TTE *See:* triple transit echo.

T3 A carrier facility that transmits digital signal level three. The data rate is 44.736 Mb/s, the equivalent of 672 voice-band channels. (C) 610.7-1995

TTL *See:* transistor-transistor logic.

TTS *See:* teletypesetting.

T2 A carrier facility that transmits digital signal level two. The data rate is 6.312 Mb/s, the equivalent of 96 voice-band channels. (C) 610.7-1995

T-2 A two-conductor twisted construction designed to control wind-induced motion. (T&D/PE) 524-1992r

tube (1) (protection and coordination of industrial and commercial power systems) The cylindrical enclosure of a fuse. Such a tube may be made of laminated paper, special fiber, melamine impregnated glass cloth, bakelite, ceramic, glass, plastic, or other materials. (IA/PSP) 242-1986r
(2) (interior wiring) A hollow cylindrical piece of insulating material having a head or shoulder at one end, through which an electric conductor is threaded where passing through a wall, floor, ceiling, joist, stud, etc. *See also:* raceway. (EEC/PE) [119]
(3) (primary cell) A cylindrical covering of insulating material, without closure at the bottom. *See also:* electrolytic cell. (EEC/PE) [119]
(4) A generic term for any kind of vacuum or electron tube. *See also:* cathode ray tube display device; storage tube display device; Nixie tube display device. (C) 610.10-1994w
(5) *See also:* fuse tube; melting-speed ratio; clearing time. (SWG/PE) C37.100-1992

tube count A terminated discharge produced by an ionizing event in a radiation-counter tube. (NI/NPS) 309-1999

tube current averaging time The time interval over which the current is averaged in defining the operating capability of the tube. *See also:* rectification. (EEC/PCON) [110]

tube, display *See:* display tube.

tube, electron *See:* electron tube.

tube fault current The current that flows through a tube under fault conditions, such as arc-back or short circuit. *See also:* rectification. (EEC/PCON) [110]

tube, fuse *See:* fuse tube.

tube heating time (mercury-vapor tube) The time required for the coolest portion of the tube to attain operating temperature. *See also:* electronic controller; preheating time; gas tube. (ED) 161-1971w

tubelet *See:* eyelet.

tuberculation The formation of localized corrosion products scattered over the surface in the form of knoblike mounds. (IA) [59]

tube scintillation pulses (photomultipliers) Dark pulses caused by scintillations within the photomultiplier structure. Example: cosmic-ray-induced events. (NPS) 398-1972r

tube-type plate (storage cell) A plate of an alkaline storage battery consisting of an assembly of metal tubes filled with active material. *See also:* battery. (EEC/PE) [119]

tube voltage drop (electron tube) The anode voltage during the conducting period. *See also:* electronic controller; electrode voltage. (ED) 161-1971w

tubing (rotating machinery) A tubular flexible insulation, extruded or made of layers of film plastic, into which a conductor is inserted to provide additional insulation. Tubing is frequently used to insulate connections and crossovers. *See also:* asynchronous machine. (PE) [9]

Tudor plate (storage cell) A lead storage battery plate obtained by molding and having a large area. *See also:* battery. (EEC/PE) [119]

tugger *See:* drum puller; two-drum, three-drum puller.

tugger setup *See:* pull site.

tumbling (gyros) The loss of reference in a two-degree-of-freedom gyro due to gimbal lock or contact between a gimbal and a mechanical stop. This is not to be confused with "tumble testing," which is a method of evaluating gyro performance. (AES/GYAC) 528-1994

tuned filter (harmonic control and reactive compensation of static power converters) A filter consisting generally of combinations of capacitors, inductors, and resistors which have been selected in such a way as to present a relative minimum (maximum) impedance to one or more specific frequencies. For a shunt (series) filter the impedance is a minimum (maximum). Tuned filters generally have a relatively high $Q(X/R)$. (IA/SPC) 519-1981s

tuned-grid oscillator An oscillator whose frequency is determined by a parallel-resonance circuit in the grid circuit coupled to the plate to provide the required feedback. *See also:* oscillatory circuit. (AP/ANT) 145-1983s

tuned grid-tuned plate oscillator An electron tube circuit in which both grid and plate circuits are tuned to resonance where the feedback voltage normally is developed through the inter-electrode capacity of the tube. *See also:* radio-frequency generator. (IA) 54-1955w

tuned-plate oscillator An oscillator whose frequency is determined by a parallel-resonance circuit in the plate circuit coupled to the grid to provide the required feedback. *See also:* oscillatory circuit. (AP/ANT) 145-1983s

tuned rotor gyro *See:* dynamically tuned gyro.

tuned speed (dynamically tuned gyro) The rotor spin velocity at which the dynamically induced spring rate is equal in magnitude, and of opposite sign, to the physical spring rate of the rotor suspension. (AES/GYAC) 528-1994

tuned transformer A transformer, the associated circuit elements of which are adjusted as a whole to be resonant at the frequency of the alternating current supplied to the primary, thereby causing the secondary voltage to build up to higher values than would otherwise be obtained. *See also:* power pack. (IM) [120]

tuner (1) (radio receivers) In the broad sense, a device for tuning. Specifically, in radio receiver practice, it is (A) a packaged unit capable of producing only the first portion of the functions of a receiver and delivering either radio-frequency, intermediate-frequency, or demodulated information to some other equipment, or (B) that portion of a receiver that contains the circuits that are tuned to resonance at the received-signal frequency. *See also:* radio receiver. (EEC/PE) [119]

(2) (transmission lines) (waveguide) An ideally lossless, fixed or adjustable, network capable of transforming a given impedance into a different impedance. *See also:* transmission loss; waveguide. (AP/IM/ANT/HFIM) [35], [40]

tuner, waveguide *See:* waveguide tuner.

tungsten-halogen lamp (illuminating engineering) A gas filled tungsten filament incandescent lamp containing a certain proportion of halogens. *Note:* The tungsten-iodine lamp (UK) and quartz-iodine lamp (USA) belong to this category. (EEC/IE) [126]

tuning (data transmission) The adjustment in relation to frequency of a circuit or system to secure optimum performance; commonly the adjustment of a circuit or circuits to resonance. (PE) 599-1985w

tuning creep (oscillators) The change of an essential characteristic as a consequence of repeated cycling of the tuning element. (ED) 158-1962w

tuning, electronic *See:* electronic tuning.

tuning hysteresis (oscillators) (microwave) The difference in a characteristic when a tuner position, or input to the tuning element, is approached from opposite directions. (ED) 158-1962w

tuning indicator An electron-beam tube in which the signal supplied to the control electrode varies the area of luminescence of the screen. (ED) [45]

tuning probe (waveguides) An essentially lossless probe of adjustable penetration extending through the wall of the waveguide or cavity resonator. *See also:* waveguide.

tuning range (1) (switching tubes) The frequency range over which the resonance frequency of the tube may be adjusted by the mechanical means provided on the tube or associated cavity. *See also:* gas tube. (ED) 161-1971w
(2) (oscillators) The frequency range of continuous tuning within which the essential characteristics fall within prescribed limits. (ED) 158-1962w

tuning range, electronic *See:* electronic tuning range.

tuning rate, thermal *See:* thermal tuning rate.

tuning screw (waveguide technique) An impedance-adjusting element in the form of a rod whose depth of penetration through the wall into a waveguide or cavity is adjustable by rotating the screw. *See also:* waveguide. (AP/ANT) [35]

tuning sensitivity (oscillators) The rate of change of frequency with the control parameter (for example, the position of a mechanical tuner, electric tuning voltage, etc.) at a given operating point. (ED) 158-1962w

tuning sensitivity, electronic *See:* electronic tuning sensitivity.

tuning sensitivity, thermal *See:* thermal tuning sensitivity.

tuning susceptance (anti-transmit-receive tube) The normalized susceptance of the tube in its mount due to the deviation of its resonance frequency from the desired resonance frequency. *Note:* Normalization is with respect to the characteristic admittance of the transmission line at its junction with the tube mount. (ED) 161-1971w

tuning, thermal *See:* thermal tuning.

tuning time constant, thermal *See:* thermal tuning time constant.

tuning time thermal (1) (cooling) The time required to tune through a specified frequency range when the tuner power is instantaneously changed from the specified maximum to zero. *Note:* The initial condition must be one of equilibrium. *See also:* electron emission. (ED) 161-1971w
(2) (heating) The time required to tune through a specified frequency range when the tuner power is instantaneously changed from zero to the specified maximum. *Note:* The initial condition must be one of equilibrium. *See also:* electron emission. (ED) 161-1971w

tunneled arrow An arrow left undrawn between its attachment to an ancestral box and its appearance as a boundary arrow on some hierarchically consecutive descendent diagram. (C/SE) 1320.1-1998

tunneling The act of applying tunnel notation to an arrow segment. (C/SE) 1320.1-1998

tunneling mode *See:* leaky mode.

tunneling ray *See:* leaky ray.

tunnel notation A pair of short shallow arcs, resembling a pair of left and right parentheses characters, that bracket the arrowhead or the arrowtail of an arrow segment. (C/SE) 1320.1-1998

tuple (A) A suffix meaning "an ordered set of items," as in *n*-tuple. **(B)** In a relational data model, a set of values of related attributes. *Note:* Often thought of as a row in a table. *Synonym:* row. *See also:* relation; attribute. (C) 610.5-1990

turbine-control servomotor (hydraulic turbines) The actuating element that moves the turbine-control mechanism in response to the action of the governor control actuator. Turbine-control servomotors are designated as: gate servomotor; blade servomotor; deflector servomotor; needle servomotor. (PE/EDPG) 125-1988r

turbine-driven generator An electric generator driven by a turbine. (EEC/PE) [119]

turbine end (rotating machinery) The driven or power-input end of a turbine-driven generator. (PE) [9]

turbine-generator *See:* cylindrical-rotor generator.

turbine/generator testing Consists of four testing phases: subsystem verification, prestartup, startup, and performance testing. (PE/EDPG) 1248-1998

turbine-generator unit An electric generator with its driving turbine. (EEC/PE) [119]

turbine-nozzle control system (gas turbines) A means by which the turbine diaphragm nozzles are adjusted to vary the nozzle angle or area, thus varying the rate of energy input to the turbine(s). (PE/EDPG) [5]

turbine, reversible *See:* reversible turbine.

turbine-type (rotating machinery) Applied to alternating-current machines designed for high-speed operation and having an excitation winding embedded in slots in a cylindrical steel rotor made from forgings or thick disks. *See also:* asynchronous machine. (PE) [9]

turbo-machine (rotating machinery) (turbo-generator) A machine of special design intended for high-speed operation. Turbo-generators usually are directly connected to gas or steam turbines. (PE) [9]

turbulence Random movements within a liquid or gaseous medium inducing heterogeneous values of certain characteristics of the medium. (AP/PROP) 211-1997

turbulence scale A length representative of the average size of the irregularities of a specified property of a medium subject to turbulence. (AP/PROP) 211-1997

Turing machine A mathematical model of a device that changes its internal state and reads from, writes on, and moves a potentially infinite tape, all in accordance with its present state, thereby constituting a model for computerlike behavior. *See also:* universal Turing machine. (C) [20], [85]

turn (rotating machinery) The basic coil element which forms a single conducting loop comprising one insulated conductor. *Note:* The conductor may consist of a number of strands or laminations. Each strand or lamination is in the form of a wire, rod, strip or bar, depending on its cross-section, and may be either uninsulated or insulated for the sole purpose of reducing eddy currents. (PE) [9]

turnaround time (1) The elapsed time between the submission of a job to a batch processing system and the return of completed output. *See also:* port-to-port time; response time; think time. (C) 610.12-1990
(2) In data communications, the amount of time required to reverse the direction of transmission from send to receive or vice-versa in a half duplex transmission. (C) 610.7-1995

(3) The elapsed time between the submission of a job and the return of the completed output. (C) 610.10-1994w

(4) (test, measurement, and diagnostic equipment) The time needed to service or check out an item for recommitment. (MIL) [2]

turnbuckle A threaded device inserted in a tension member to provide minor adjustment of tension or sag. *See also:* tower. (T&D/PE) [10]

turn error (gyros) An error in gyro output due to cross coupling and acceleration encountered during vehicle turns. (AES/GYAC) 528-1994

turning center A numerical control machine capable of performing lathe-oriented operations, such as boring, facing, turning, and threading. (C) 610.2-1987

turning gear (rotating machinery) A separate drive to rotate a machine at very low speed for the purpose of thermal equilization at a time when it would otherwise be at rest. *See also:* rotor. (PE) [9]

turn insulation (rotating machinery) Insulation applied to provide electrical separation between turns of a coil. *Note:* In the usual case, the insulation encircles each turn. However, in the case of edgewise-wound field coils for salient pole synchronous machines, the outer edges may be left bare to facilitate cooling. (PE) [9]

turnkey Pertaining to a hardware or software system delivered in a complete, operational state. (C) 610.12-1990

turnkey system A complete computer system that is fully operational and supplied to the user in a ready-to-run condition. (C) 610.10-1994w

turn-off and/or turn-on response time The time between a rapid change in light level and the switching of the load. Turn-off delay may be different than turn-on delay. The response time is measured at 25°C and rated voltage. (RL) C136.10-1996

turn-off branch (self-commutated converters) (converter circuit elements) An auxiliary branch intended to take over the current transiently from a previously conducting principal branch. (IA/SPC) 936-1987w

turn-off thyristor (gate controlled switch) A thyristor that can be switched from the ON state to the OFF state and vice versa by applying control signals of appropriate polarities to the gate terminal, with the ratio of triggering power appreciably less than one. (IA/IPC) 428-1981w

turn-off time (1) (self-commutated converters) (circuit properties) (applies to converters that use circuit-commutated thyristor devices) The time interval between that instant when the principal current of a thyristor device has been reduced to zero and that instant when the same thyristor device is again subjected to voltage that could cause conduction. *Note:* for proper operation, the turn-off time, t_o, made available by the action of the circuit must exceed the turn-off time, t_q, required by the thyristor device for recovery of its voltage-blocking ability. Both the available and required turn-off times depend on the operating conditions of the converter. (See the corresponding figures.)

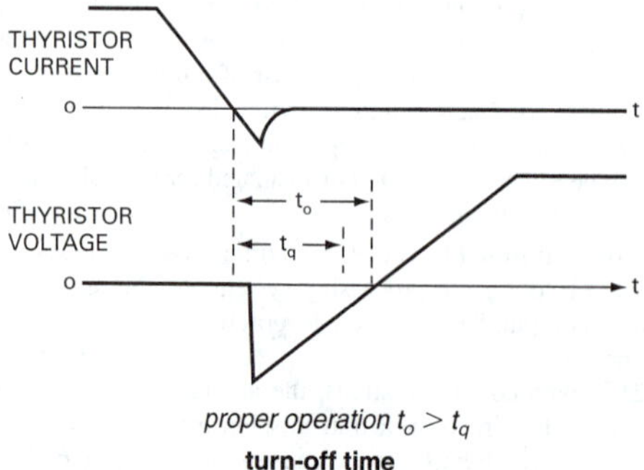

proper operation $t_o > t_q$

turn-off time

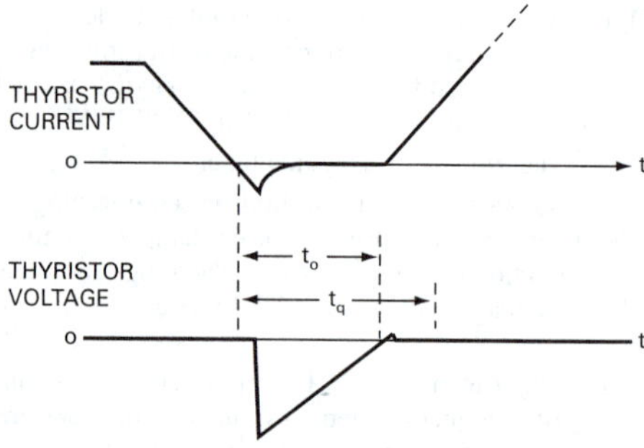

(IA/SPC) 936-1987w

(2) (thyristor) *See also:* gate-controlled turn-off time; circuit-commutated turn-off time.

turn-on time (accelerometer) (inertial sensors) (gyros) The time from the initial application of power until a sensor produces a specified useful output, though not necessarily at the accuracy of full specification performance. (AES/GYAC) 528-1994

turn-off/turn-on ratio The turn-off light level divided by the turn-on light level. (RL) C136.10-1996

turnover frequency *See:* transition frequency.

turn ratio (1) (constant-current transformer) The ratio of the number of turns in the primary winding to that in the secondary winding. *Note:* In case of a constant-current transformer having taps for changing its voltage ratio, the turn ratio is based on the number of turns corresponding to the normal rated voltage of the respective windings, to which operation and performance characteristics are referred. (PE/TR) [117]

(2) (current transformer) The ratio of the secondary winding turns to the primary winding turns. (PE/TR/PSR) C57.12.80-1978r, C57.13-1978s, C57.13-1993, C37.110-1996

(3) (potential transformer) (voltage transformer) The ratio of the primary winding turns to the secondary winding turns. (PE/TR) [57], C57.13-1978s, C57.13-1993, C57.12.80-1978r

(4) (rectifier transformer) The ratio of the number of turns in the alternating-current winding to that in the direct-current winding. *Note:* The turn ratio is based on the number of turns corresponding to the normal rated voltage of the respective windings to which operating and performance characteristics are referred. *See also:* rectifier transformer. (Std100) C57.18-1964w

(5) (power and distribution transformers) The ratio of the number of turns in a higher voltage winding to that in a lower voltage winding. *Note:* In the case of a constant-voltage transformer having taps for changing its voltage ratio, the nominal turn ratio is based on the number of turns corresponding to the normal rated voltage of the respective windings, to which operating and performance characteristics are referred. (PE) C57.12.80-1978r

turn separator (rotating machinery) An insulation strip between turns: a form of turn insulation. *See also:* rotor; stator. (PE) [9]

turns factor (magnetic core testing) Under stated conditions the number of turns that a coil of specified shape and dimensions placed on the core in a given position should have to obtain a given unit of self inductance. When measured with a measuring coil of the specified shape and dimensions and placed in the same position, it is defined as:

$$\alpha = \frac{N}{\sqrt{L}}$$

where

α = turns factor

N = number of turns of the measuring coil

L = self-inductance in henrys of the measuring coil placed on the core.

turn-signal operating unit (illuminating engineering) That part of a signal system by which the operator of a vehicle indicates the direction a turn will be made, usually by a flashing light. (EEC/IE) [126]

turns per phase, effective *See:* effective turns per phase.

turns ratio (1) (electronic power transformer) The number of turns of a given secondary divided by the number of primary turns. Thus a ratio less than one is a step-down transformation, a ratio greater than one is a step-up transformation, and a ratio equal to one is unity ratio. (PEL/ET) 295-1969r
(2) (potential transformer) (voltage transformer) (of a voltage transformer) The ratio of the primary winding turns to the secondary winding turns. (PE/TR) C57.13-1993, [57], C57.12.80-1978r, C57.13-1978s
(3) (current transformer) (of a current transformer) The ratio of the secondary winding turns to the primary winding turns. (PE/TR/PSR) C57.13-1993, C37.110-1996, C57.12.80-1978r, C57.13-1978s

turnstile antenna An antenna composed of two dipole antennas, perpendicular to each other, with their axes intersecting at their midpoints. Usually, the currents on the two dipole antennas are equal and in phase quadrature. (AP/ANT) 145-1993

turntable rumble (audio and electroacoustics) Low-frequency vibration mechanically transmitted to the recording or reproducing turntable and superimposed on the reproduction. *See also:* rumble; phonograph pickup. 188-1952w

turn-to-turn test (rotating machinery) A test for applying or more often introducing between adjacent turns of an insulated component, a voltage of predetermined amplitude, for the purpose of checking the integrity of the interturn insulation. *Synonym:* interturn test. (PE) [9]

turn-to-turn voltage (rotating machinery) The voltage existing between adjacent turns of a coil. *See also:* rotor; stator. (PE) [9]

turtle graphics *See:* LOGO.

tutorial simulation *See:* instructional simulation.

TV *See:* television.

TV broadcast band (overhead-power-line corona and radio noise) Any one of the frequency bands assigned for the transmission of audio and video signals for television (TV) broadcasting to the general public. *Note:* In the United States and Canada, the frequency bands are 54–72 MHz; 76–88 MHz; 174–216 MHz, and 400–890 MHz. (T&D/PE) 539-1990

TVL *See:* television lines.

TV waveform signal (linear waveform distortion) An electrical signal whose amplitude varies with time in a generally nonsinusoidal manner and whose shape (that is, duration and amplitude) carries the TV signal information. (BT) 511-1979w

TW *See:* shaped wire compact conductor.

TWA *See:* two-way alternating.

twelve punch A zone punch in punch row twelve (top row) of a twelve-row punch card. *Synonym:* Y punch. *See also:* eleven punch; zone punch. (C) 610.10-1994w

twelve-row punch card A punch card with twelve rows. (C) 610.10-1994w

21-type repeater *See:* twenty-one-type repeater.

twenty-one-type repeater (data transmission) A two-wire telephone repeater in which there is one amplifier serving to amplify the telephone current in both directions, the circuit being arranged so that the input and output terminals of the amplifier are in one pair of conjugate branches, while the lines in the two directions are in another pair of conjugate branches. (PE) 599-1985w

22-type repeater *See:* twenty-two-type repeater.

twenty-two-type repeater (data transmission) A two-wire telephone repeater in which there are two amplifiers, one serving to amplify the telephone current being transmitted in one direction and the other serving to amplify the telephone currents in the other direction. (PE) 599-1985w

twinaxial cable A cable consisting of two conductors, insulated from each other, within and insulated from another conductor of larger diameter. *Contrast:* coaxial cable. (C) 610.7-1995

twin-bundle *See:* bundle; bundle, two-conductor, three-conductor, four-conductor, multiconductor.

twin cable A cable composed of two insulated conductors laid parallel and either attached to each other by the insulation or bound together with a common covering. (T&D/PE) [10]

twinkle box An input device employing light sensors, rotating disks, and a stylus, used to measure three-dimensional positions by angular light sensing. (C) 610.10-1994w

twin segment In a hierarchical database, a child segment N that shares a common parent segment with another child segment M. Segments N and M are said to be twin segments. *See also:* logical twin segment; physical twin segment. (C) 610.5-1990w

twin-T network *See:* parallel-T network.

twin wire A cable composed of two small insulated conductors laid parallel, having a common covering. *See also:* conductor. 30-1937w

twist (measuring the performance of tone address signaling systems) In a two-tone signal, during the signal present condition, the level of the higher-frequency tone relative to the level of the lower-frequency tone, expressed in decibels. Twist is negative if the higher-frequency tone level is below the lower-frequency tone level. (COM/TA) 752-1986w

twisted-lead transposition (rotating machinery) A form of transposition used on a distributed armature winding wherein the strands comprising each turn are kept insulated from each other throughout all the coils in a phase belt, and the last half turn of each coil is given a 180-degree twist prior to connecting it to the first half turn of the next coil in the series. *See also:* rotor; stator. (PE) [9]

twisted pair (1) A cable composed of two small insulated conductors, twisted together without a common covering. *Note:* The two conductors of a twisted pair are usually substantially insulated, so that the combination is a special case of a cord. *See also:* conductor. (T&D/PE) [10]
(2) A medium consisting of two insulated wires arranged in a regular spiral pattern. (C) 610.7-1995
(3) A cable element that consists of two insulated conductors twisted together in a regular fashion to form a balanced transmission line. (C/LM) 802.3-1998
(4) (local area networks) Two continuous, insulated copper conductors twisted around each other in a helical manner. Twisted-pair cable may be unshielded (UTP) or shielded (STP). (C) 8802-12-1998

twisted-pair cable (1) A group of twisted pairs within a single protective sheath. (C) 610.7-1995
(2) A bundle of multiple twisted pairs within a single protective sheath. (C/LM) 802.3-1998
(3) (local area networks) A group of two or four twisted pairs within a single sheath. (C) 8802-12-1998

twisted-pair cable binder group (1) A group of twisted pairs within a cable that are bound together. Large telephone cables have multiple binder groups with high interbinder group near-end crosstalk loss. (C/LM) 802.3-1998
(2) (local area networks) A group of twisted pairs within a cable that are bound together. (C) 8802-12-1998

twisted-pair link (1) A twisted-pair cable plus connecting hardware. (LM/C) 802.3u-1995s

(2) **(local area networks)** A link segment consisting of a twisted-pair cable and two attached MDI connectors. (C) 8802-12-1998

twisted-pair link segment In 100BASE-T, a twisted-pair link for connecting two Physical Layers (PHYs). (C/LM) 802.3-1998

twist, waveguide *See:* waveguide twist.

two-address Pertaining to an instruction code in which each instruction has two address parts. Some two-address instructions use the addresses to specify the location of one operand and the destination of the result, but more often they are one-plus-one-address instructions. (C) 162-1963w

two-address instruction (1) A computer instruction that contains two address fields. For example, an instruction to add the contents of A to the contents of B. *Synonym:* double-operand instruction. *Contrast:* four-address instruction; three-address instruction; one-address instruction; zero-address instruction. (C) 610.12-1990
(2) An instruction containing two addresses. *Synonym:* double-address instruction. *See also:* address format. (C) 610.10-1994w

two-bit byte *See:* doublet.

two-conductor bundle *See:* bundle.

two-degree-freedom gyro A gyro in which the rotor axis is free to move in any direction. *See also:* navigation. (AES/RS) 686-1982s, [42]

two-dimensional radar A radar that provides information in range and one angle coordinate, as in a 2-D air-surveillance radar that uses a fan-beam antenna to obtain range and azimuth angle. (AES) 686-1997

two-dimensional scanning Scanning the beam of a directive antenna using two degrees of freedom to provide solid angle coverage. (AP/ANT) 145-1993

two-directional signal line A signal line that may be defined in either direction across an interface, and that cannot be defined in both directions simultaneously. The direction of operation for a two-directional signal line in a system is a configuration option. (C/MM) 959-1988r

two-drum, three-drum puller The definition and application for this unit is essentially the same as that for the drum puller. It differs in that this unit is equipped with two or three drums and thus can pull one, two, or three conductors individually or simultaneously. *Synonyms:* tugger; hoist, double drum; hoist, triple drum; winch, double-drum; winch, triple-drum; winch, three-drum; winch, two-drum. *See also:* drum puller. (T&D/PE) 524-1992r

two-element relay An alternating-current relay that is controlled by current from two circuits through two cooperating sets of coils. (EEC/PE) [119]

two-family dwelling A building consisting solely of two dwelling units. (NESC/NEC) [86]

two-fluid cell A cell having different electrolytes at the two electrodes. *See also:* electrochemistry. (EEC/PE) [119]

two-frequency mutual coherence function The correlation between two fields at two frequencies measured at the same point in space and time. (AP/PROP) 211-1997

two-frequency simplex operation (radio communication) The operation of a two-way radio-communication circuit utilizing two radio-frequency channels, one for each direction of transmission, in such manner that intelligence can be transmitted in only one direction at a time. *See also:* channel spacing. (VT) [37]

2GL *See:* assembly language.

two-input adder *See:* half adder.

two-layer winding A winding in which there are two coil sides in the depth of a slot. *See also:* rotor; stator. (PE) [9]

two-level address (1) An indirect address that specifies the storage location containing the address of the desired operand. *See also:* n–level address. (C) 610.12-1990
(2) *See also:* n-level address; indirect address. (C) 610.10-1994w

two-level encoding A microprogramming technique in which different microoperations may be encoded identically into the same field of a microinstruction, and the one that is executed depends upon the value in another field internal or external to the microinstruction. *Contrast:* single-level encoding. *See also:* residual control; bit steering. (C) 610.12-1990

two-N (2N) angular sensitivity (dynamically tuned gyro) The coefficient that relates drift rate to angular vibration at twice spin frequency applied about an axis perpendicular to the spin axis. It has the dimensions of angular displacement per unit time, per unit of angle of the input vibration. *See also:* angular-case-motion sensitivity. (AES/GYAC) 528-1994

two-N (2N) translational sensitivity (dynamically tuned gyro) The coefficient that relates drift rate to linear vibrations at twice spin frequency applied perpendicularly to the spin axis. It has the dimensions of angular displacement per unit time, per unit of acceleration of the input vibration. (AES/GYAC) 528-1994

two-out-of-five code (A) A BCD code in which each decimal digit is represented by a five-digit numeral of which two bits are in one state (usually ones) and three are in the other state. *See also:* m–out–of–n code. **(B)** A code in which each decimal digit is represented by five binary digits of which two are one kind (for example, ones) and three are the other kind (for example, zeros). (C) 1084-1986, [20], [85]

two-phase circuit (power and distribution transformers) A polyphase circuit of three, four, or five distinct conductors intended to be so energized that in the steady state the alternating voltages between two selected pairs of terminals of entry, other than the neutral terminal when one exists, have the same periods, are equal in amplitude, and have a phase difference of 90 degrees. When the circuit consists of five conductors, but not otherwise, one of them is a neutral conductor. *Note:* A two-phase circuit as defined here does not conform to the general pattern of polyphase circuits. Actually a two-phase, four-wire, or five-wire circuit could more properly be called a four-phase circuit, but the term two-phase is in common usage. A two-phase three-wire circuit is essentially a special case, as it does not conform to the general pattern of other polyphase circuits. (PE/TR) C57.12.80-1978r

two-phase five-wire system A system of alternating-current supply comprising five conductors, four of which are connected as in a four-wire two-phase system, the fifth being connected to the neutral points of each phase. *Note:* The neutral is usually grounded. Although this type of system is usually known as the two-phase five-wire system, it is strictly a four-phase five-wire system. *See also:* network analysis; alternating-current distribution. (T&D/PE) [10]

two-phase four-wire system A system of alternating-current supply comprising two pairs of conductors between one pair of which is maintained an alternating difference of potential displaced in phase by one-quarter of a period from an alternating difference of potential of the same frequency maintained between the other pair. *See also:* network analysis; alternating-current distribution. (T&D/PE) [10]

two-phase three-wire system A system of alternating-current supply comprising three conductors between one of which (known as the common return) and each of the other two are maintained alternating differences of potential displaced in phase by one quarter of a period with relation to each other. *See also:* alternating-current distribution; network analysis. (T&D/PE) [10]

two-plus-one address (electronic computation) Pertaining to an instruction that contains two operand addresses and a control address. *See also:* operand; instruction. (C) 162-1963w

two-plus-one address format *See:* address format.

two-plus-one address instruction A computer instruction that contains three address fields, the third containing the address of the instruction to be executed next. For example, an instruction to add the contents of A to the contents of B, then execute the instruction at location C. *Contrast:* four-plus-one address instruction; three-plus-one address instruction; one-plus-one address instruction. (C) 610.12-1990

two-port surge protective device A surge protective device with two sets of terminals, input and output. A specific series impedance is inserted between these terminals. (PE) C62.34-1996

two-quadrant DAM (hybrid computer linkage components) A digital-to-analog multiplier (DAM) that multiplies with a single sign only for the digital value. (C) 166-1977w

two-quadrant multiplier (1) (analog computer) A multiplier in which operation is restricted to a single sign of one input variable only. *See also:* electronic analog computer. (C) 165-1977w, 166-1977w

(2) A multiplier in which the multiplication operation is restricted to a single sign of one input variable. *Contrast:* four-quadrant multiplier; one-quadrant multiplier. (C) 610.10-1994w

two-rail logic *See:* double-rail logic.

two-range Decca *See:* lambda.

two-rate watthour meter A meter having two sets of register dials, with a changeover arrangement such that integration of the quantity will be registered on one set of dials during a specified time each day, and on the other set of dials for the remaining time. (ELM) C12.1-1982s

two-sample deviation ($\sigma_y(\tau)$) The square root of the two-sample variance. *Synonym:* Allan deviation. (SCC27) 1139-1999

two-sample variance ($\sigma_y{}^2(\tau)$) Time average over the sum of the squares of the differences between successive readings of the normalized frequency departure sampled over the sampling time τ, under the assumption that there is no dead time between the normalized frequency departure samples. *Synonym:* Allan deviation. (SCC27) 1139-1999

two-scale *See:* binary notation.

twos complement (mathematics of computing) The radix complement of a binary numeral, which may be formed by subtracting each digit from 1, then adding 1 to the least significant digit and executing any required carries. For example, the twos complement of 1101 is 0011. *Synonym:* complement on two. (C) 1084-1986w

twos-complement arithmetic Computer arithmetic performed with operands that are expressed in twos-complement notation. (C) 1084-1986w

twos-complement notation A binary numeration system in which negative numbers are represented by their twos complement and positive numbers are expressed in their usual binary form. *Contrast:* sign-magnitude notation. (C) 1084-1986w

two-sided z transform (data processing) The two-sided z transform of $f(t)$ is

$$F(z) = \sum_{n=\infty}^{-1} f(nT)z^{-n} + \sum_{n=0}^{\infty} f(nT)z^{-n}$$

where the first summation is for $f(t)$ over all negative time and the second summation is for $f(t)$ over all positive time. (IM) [52]

two-signal selectivity (frequency-modulated mobile communications receivers) The characteristic that determines the extent to which the receiver is capable of differentiating between the desired signal and disturbances of signals at other frequencies. It is expressed as the amplitude ratio of the modulated desired signal and the unmodulated disturbing signal when the reference sensitivity sinad of the desired signal is degraded six decibels. (VT) 184-1969w

two-source frequency keying That form of keying in which the modulating wave abruptly shifts the output frequency between predetermined values, where the values of output frequency are derived from independent sources. *Note:* Therefore, the output wave is not coherent and, in general, will have a phase discontinuity. *See also:* telegraphy. (AP/ANT) 145-1983s

two-speed alternating-current control A control for two-speed driving-machine induction motor that is arranged to run near two different synchronous speeds by connecting the motor windings so as to obtain different numbers of poles. *See also:* control. (EEC/PE) [119]

two-state indication *See:* supervisory control functions.

two-state variable *See:* binary variable.

two-step control system A control system in which the manipulated variable alternates between two predetermined values. *Note:* A control system in which the manipulated variable changes to other predetermined value whenever the actuating signal passes through zero is called a two-step single-point control system. A two-step neutral-zone control system is one in which the manipulated variable changes to the other predetermined value when the actuating signal passes through a range of values known as the neutral zone. The neutral zone may be produced by a mechanical differential gap. The neutral zone is also called overlap, and two-step neutral-zone control overlap control. *See also:* feedback control system. (IM/PE/EDPG) [120], [3]

two-step neutral zone control system *See:* two-step control system.

two-step single-point control system *See:* two-step control system.

two-terminal capacitor Two conductors separated by a dielectric. The construction is usually such that one conductor essentially surrounds the other and therefore the effect of the presence of other conductors, except in the immediate vicinity of the terminals, is eliminated. (Specialized usage). (Std100) 270-1966w

two-terminal pair network (quadripole) (four-pole) A network with four accessible terminals grouped in pairs, for example, input pair, output pair. (CAS) [13]

two-tone keying That form of keying in which the modulating wave causes the carrier to be modulated with a single tone for the marking condition and modulated with a different single tone for the spacing condition. *See also:* telegraphy. (AP/ANT) 145-1983s

2T pulse (television) (waveform test signals) A $\sin^2$ pulse with a half-amplitude duration (HAD) of 250 ns. The amplitude of the envelope of the frequency spectrum at 2 MHz is 0.5 of the amplitude at zero frequency and effectively zero at and beyond 4 MHz. *Note:* The 2T pulse is mentioned here for the sake of completeness. The short-time domain may be tested by the 2T pulse in conjunction with the T pulse and a reference signal. This method is not used in this standard since the T step alone tests the short-time domain in a simpler and more direct manner. (BT) 511-1979w

two-transistor cell A memory cell consisting of two MNOS transistors. (ED) 641-1987w

two-value capacitor motor A capacitor motor using different values of effective capacitance for the starting and running conditions. *See also:* asynchronous machine. (PE) [9]

two-valued variable *See:* binary variable.

two-way alternate operation A mode of operation of a data link in which data may be transmitted in both directions, one direction at a time. *Synonym:* either-way operation. *See also:* two-way simultaneous operation; one-way-only operation. (C) 610.7-1995

two-way alternating (TWA) A subset of HDLC defining half-duplex, rather than simultaneous, two-way communications. (EMB/MIB) 1073.3.1-1994

two-way automatic maintaining leveling device A device that corrects the car level on both underrun and overrun, and maintains the level during loading and unloading. *See also:* elevator car-leveling device. (EEC/PE) [119]

two-way automatic nonmaintaining leveling device A device that corrects the car level on both underrun and overrun, but will not maintain the level during loading and unloading. *See also:* elevator car-leveling device. (EEC/PE) [119]

two-way chain *See:* doubly-linked list.

two-way circuit A circuit in which the transmission of signals is permitted in both directions. *Contrast:* simple circuit. (C) 610.10-1994w

two-way correction A method of register control that effects a correction in register in either direction. (IA/ICTL/CEM) [58]

two-way insertion sort An insertion sort in which each item in the set to be sorted is inserted in its proper position in the sorted set such that the first item is placed in the middle of the output set and space is made for subsequent items by moving the previously-inserted items to the right or left. *Contrast:* binary insertion sort. (C) 610.5-1990w

two-way merge sort A merge sort in which the set to be sorted is divided into two subsets, the items in each subset are sorted, and the subsets are merged by comparing the smallest items of each subset, outputting the smallest of those, then repeating the process. *See also:* natural two-way merge sort; multiway merge sort; straight two-way merge sort. (C) 610.5-1990w

two-way simultaneous operation A mode of operation of a data link in which data may be transmitted over a link simultaneously in both directions. *See also:* one-way-only operation; two-way alternate operation. (C) 610.7-1995

two-way trunk (telephone switching systems) A trunk between two switching entities used for calls that originate from either end. (COM) 312-1977w

two-wire channel (1) (data transmission) (two-wire circuit) A metallic circuit formed by two adjacent conductors insulated from each other. *Note:* Also used in contrast with four-wire circuit to indicate a circuit using one line or channel for transmission of electric waves in both directions. (PE) 599-1985w
(2) (telephone loop performance) A transmission medium that simultaneously carries, without multiplexing, two signals traveling in opposite directions. (COM/TA) 820-1984r

two-wire circuit (1) (data transmission) A metallic circuit formed by two adjacent conductors insulated from each other. *Note:* Also used in contrast with four-wire circuit to indicate a circuit using one line or channel for transmission of electric waves in both directions. (PE) 599-1985w
(2) (transmission performance of telephone sets) A metallic circuit formed by two conductors insulated from each other. The electric waves are transmitted in both directions over the path provided by the two-wire circuit. (COM/TA) 269-1983s
(3) A leased circuit in which two conductors are used, each for a one-way transmission path. *See also:* foreign exchange circuit; dial-up circuit; simplex circuit; four-wire circuit. (C) 610.7-1995
(4) A circuit formed by a pair of conductors that are insulated from one another and that each feed a load in one direction at a time. (C) 610.10-1994w

two-wire control A control function that utilizes a maintained-contact type of pilot device to provide undervoltage release. (See the corresponding figure.) *See also:* undervoltage release; control.

two-wire control

(IA/ICTL/IAC) [60]

two-wire repeater (data transmission) A telephone repeater that provides for transmission in both directions over a two-wire telephone circuit. (PE) 599-1985w

two-wire switching (telephone switching systems) Switching using the same path, frequency, or time interval for both directions of transmission. (COM) 312-1977w

two-wire system *See:* two-wire circuit.

two-wire transmission A transmission scheme where the send and receive signals are carried in one pair of wires. (COM/TA) 269-1992, 1329-1999

TWS *See:* track-while-scan.

type (1) A category into which attribute values are placed on the basis of their purpose. (C/PA) 1224-1993w
(2) *See also:* class. (C/SE) 1320.2-1998

Type A step-voltage regulator A step-voltage regulator in which the primary circuit is connected directly to the shunt winding of the regulator. The series winding is connected to the shunt winding and, in turn, via taps, to the regulated circuit, per the figure below. In a Type A step-voltage regulator, the core excitation varies because the shunt winding is connected across the primary circuit.

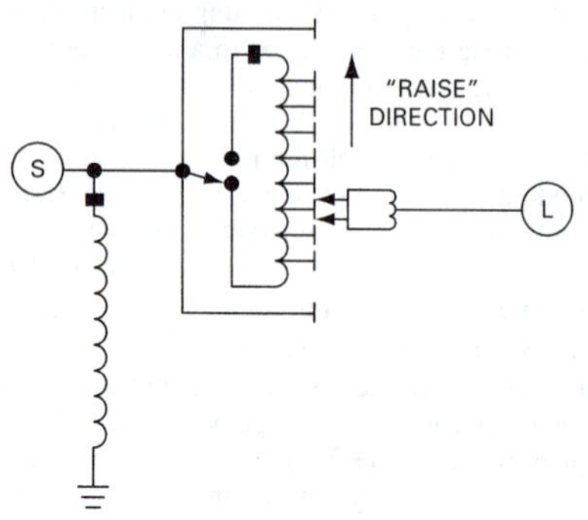

Schematic diagram of single-phase, Type A step-voltage regulator

Type A step-voltage regulator

(PE/TR) C57.15-1999

type attributes The classification of each source statement as either executable, data declaration, compiler directive, or comment. (C/SE) 1045-1992

type bar In a bar printer, a print element in the form of a bar that holds type slugs. *Synonym:* print bar. (C) 610.10-1994w

Type B step-voltage regulator A step-voltage regulator in which the primary circuit is connected, via taps, to the series winding of the regulator. The series winding is connected to the shunt winding, which is connected directly to the regulated circuit, per the figure below. In a Type B step-voltage regulator, the core excitation is constant because the shunt winding is connected across the regulated circuit.

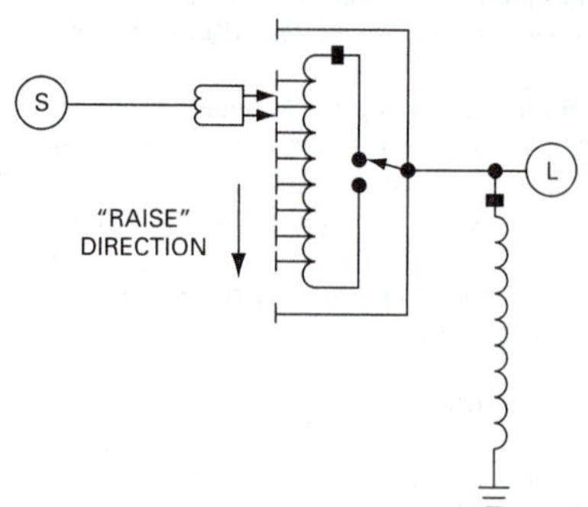

Schematic diagram of single-phase, Type B step-voltage regulator

Type B step-voltage regulator

(PE/TR) C57.15-1999

Type DB (1) (cable systems in power generating stations) (formerly Type II) Conduit designed for underground installation without encasement in concrete. (C) 166-1977w
(2) Duct designed for direct burial without encasement in concrete (also referred to as Type II duct), fabricated from PVC or ABS. (SUB/PE) 525-1992r

Type EB (1) (cable systems in power generating stations) (formerly Type I) Conduit designed to be encased in concrete when installed. (C) 166-1977w
(2) Duct designed to be encased in concrete when installed (also referred to as Type I duct), fabricated from PVC or ABS. (SUB/PE) 525-1992r

type element *See:* print element.

type error An error that occurs when a node is encountered with improper protocol information. (C) 610.7-1995

type_error A status code that is returned when the transaction is directed to an existing address, but the transaction command (for example, a read64 directed to a quadlet register) is not implemented. (C/MM) 1212-1991s

type font (1) A type face of a given size and design; for example, 10-point Bodoni Book Medium, or 9-point Gothic. (C) [20], [85]
(2) *See also:* character font. (C) 610.2-1987
(3) *See also:* font. (C) 610.10-1994w

Type IV Duct designed for heavy-duty applications above grade. (SUB/PE) 525-1992r

type of emission (mobile communication) A system of designating emission, modulation, and transmission characteristics of radio-frequency transmissions, as defined by the Federal Communications Commission. *See also:* mobile communication system. (VT) [37]

type of piezoelectric crystal cut The orientation of a piezoelectric crystal plate with respect to the axes of the crystal. It is usually designated by symbols. For example, *GT*, *AT*, *BT*, *CT*, and *DT* identify certain quartz crystal cuts having very low temperature coefficients. *See also:* crystal. (EEC/PE) [119]

type of service The specific type of application in which the controller is to be used; for example: general purpose; special purpose—namely, crane and hoist, elevator, steel mill, machine tool, printing press, etc. *See also:* electric controller. (IA/ICTL/IAC) [60]

Type I Duct designed to be encased in concrete. (SUB/PE) 525-1992r

type I error *See:* misidentification.

type slug A type element, usually with two characters arranged one above the other, for mounting on a type bar. (C) 610.10-1994w

types of metal-enclosed bus assemblies (metal-enclosed bus and calculating losses in isolated-phase bus) In general, three basic types of construction are used: nonsegregated-phase, segregated-phase, and isolated-phase.

1) (nonsegregated-pase bus) One in which all phase conductors are in a common metal enclosure without barriers between the phases. When associated with metal-clad switchgear, the primary bus and connections are covered with insulating material equivalent to the switchgear insulating system.
2) (segregated-phase bus) One in which all phase conductors are in a common metal enclosure but are segregated by metal barriers between phases.
3) (isolated-phase bus) One in which each phase conductor is enclosed by an individual metal housing separated from adjacent conductor housing by an air space. The bus may be self-cooled or may be force-cooled by means of circulating a gas or liquid.

(SWG/PE) C37.23-1987r

type test (1) (electrical heat tracing for industrial applications) A test or series of tests carried out on equipment, representative of a type, to determine compliance of the design, construction, and manufacturing methods within the requirements of IEEE Std 515-1983. (BT/AV) 152-1953s
(2) (valve actuators) Tests made on one or more sample actuators to verify adequacy of design and the manufacturing processes. (PE/NP) 382-1985
(3) (rotating electric machinery) A test made by the manufacturer on a machine that is identical in all essential respects with those supplied on an order, to demonstrate that it complies with this standard. (PE/EM) 11-1980r
(4) A test or series of tests carried out on heating cables and accessories, representative of a type, to determine compliance of the design, construction, and manufacturing methods within the requirements of the specified standard (in this case, IEEE Std 515.1-1995). (IA/PC) 515.1-1995
(5) A test or series of tests carried out on heating cables or surface heating devices and accessories, representative of a type, to determine compliance of the design, construction, and manufacturing methods within the requirements of this standard. (IA) 515-1997

type testing Evaluation or measurement of all identified performance characteristics of a representative sample of production model instruments. (NI) N42.17B-1989r

type tests (1) (Class 1E battery chargers and inverters) (nuclear power generating station) Tests made on one or more sample equipment to verify adequacy of design and the manufacturing processes. (PE/NP) 323-1974s, 650-1979s
(2) (rotating machinery) The performance tests taken on the first machine of each type of design. *See also:* asynchronous machine. (PE) [9]
(3) Tests made on representative samples that are intended to be used as a part of routine production. The applicable portions of these type tests may also be used to evaluate modifications of a previous design and to ensure that performance has not been adversely affected. (SUB/PE) C37.122-1993
(4) Tests made on representative samples that are intended to be used as part of routine production. The applicable portions of these type tests may also be used to evaluate modifications of a previous design and to ensure that performance has not been adversely affected. (SUB/PE) C37.122.1-1993

Type III Duct designed for normal-duty applications above grade. (SUB/PE) 525-1992r

Type II Duct designed for underground installation without encasement in concrete. (SUB/PE) 525-1992r

type wheel *See:* print wheel.

typewriter key *See:* typing key.

typing key Any key on a keyboard that represents a printable character, an alphanumeric or special character. *Synonym:* typewriter key. *Contrast:* control key. (C) 610.10-1994w

T0 Pronounced "tee-zero." A reference to a MASTER clock that synchronizes all events across all signals to a common starting point. Initiates the start of each test vector. (C/TT) 1450-1999

U

UART *See:* universal asynchronous receiver/transmitter.

UAT *See:* unit auxiliaries transformer.

UC *See:* utility controller.

UDF *See:* unit development folder; software development file.

UDT *See:* unidirectional transducer.

U_{50} A transient overvoltage level that produces a 50% probability of sparkover. (T&D/PE) 516-1995

ufer ground *See:* concrete-encased ground electrode.

UHF *See:* ultra-high frequency.

UHF radar *See:* ultra-high-frequency radar.

uhv *See:* ultra-high voltage.

UI *See:* unscheduled interrupt; unit interval; user interface.

UIB *See:* unit_initialization_block.

ULF *See:* ultra-low frequency.

ULSI *See:* ultra-large scale integration.

ultimate deformation or displacement (raceway systems for Class 1E circuits for nuclear power generating stations) The maximum deformation or displacement an element can undergo without failure. (PE/NP) 628-1987r

ultimate load (raceway systems for Class 1E circuits for nuclear power generating stations) The maximum load an element can carry without failure as obtained from failure load tests or manufacturer's recommendations, whichever is less. (PE/NP) 628-1987r

ultimately controlled variable (control) The variable the control of which is the end purpose of the automatic control system. *See also:* feedback control system. (IM/IA/ICTL/IAC) [120], [60]

ultimate mechanical strength (insulators) The load at which any part of the insulator fails to perform its function of providing a mechanical support without regard to electrical failure. *See also:* insulator. (EEC/IEPL) [89]

ultimate mechanical strength-static (UMS-static) The load at which any part of the surge arrester fails to perform its mechanical function. (SPD/PE) C62.11-1999

ultimate period *See:* undamped frequency.

ultimate strength (1) (power distribution) The tensile load at which any part of the insulator fails to perform its function of providing mechanical support based on a short term test. (T&D/PE) 1024-1988w
(2) (power distribution) The rated breaking strength of a material determined by the results of tests to destruction. (T&D/PE) 751-1990

ultimate strength rating The minimum tensile strength allowed on a test of five insulators. (T&D/PE) 1024-1988w

ultra-audible frequency *See:* ultrasonic frequency.

ultra-audion oscillator *See:* Colpitts oscillator.

ultrafiche In micrographics, microfiche with images reduced more than ninety times. (C) 610.2-1987

ultra-high frequency (UHF) 300 MHz to 3 GHz. *See also:* radio spectrum. (AP/PROP) 211-1997

ultra-high-frequency radar (UHF radar) A radar operating at frequencies between 300 MHz and 1000 MHz, usually in one of the International Telecommunication Union (ITU) bands allocated for radiolocation: 420–450 MHz or 890–942 MHz. *Note:* Radars between 1 GHz and 3 GHz, although within the UHF band as defined by the ITU, are described as L-band or S-band radars, as appropriate. (AES) 686-1997

ultra-high voltage (uhv) A term applied to voltage levels that are higher than 800 000 V. (T&D/PE) 516-1995

ultra-high-voltage system An electric system having a maximum rms ac (root-mean-square alternating current) voltage above 800 000 V to 2 000 000 V. (PE/TR) C57.12.80-1978r

ultra-large scale integration Pertaining to an integrated circuit containing more than 106 elements. *Contrast:* medium scale integration; large scale integration; very large scale integration; small scale integration. (C) 610.10-1994w

ultra-low frequency (ULF) Lower than 3 Hz. *See also:* radio spectrum. (AP/PROP) 211-1997

ultrasonic cross grating (grating) A space grating resulting from the crossing of beams of ultrasonic waves having different directions of propagation. *Note:* The grating may be two- or three-dimensional. (SP) [32]

ultrasonic delay line A transmission device, in which use is made of the propagation time of sound to obtain a time delay of a signal. (SP) [32]

ultrasonic depth finder (navigation aids) A direct reading instrument that determines the depth of water by measuring the time interval between the emission of an ultrasonic signal and the return echo from the bottom. (AES/GCS) 172-1983w

ultrasonic frequency (supersonic frequency) (ultra-audible frequency) A frequency lying above the audio-frequency range. The term is commonly applied to elastic waves propagated in gases, liquids, or solids. *Note:* The word ultrasonic may be used as a modifier to indicate a device or system employing or pertaining to ultrasonic frequencies. The term supersonic, while formerly applied to frequency, is now generally considered to pertain to velocities above those of sound waves. Its use as a synonym of ultrasonic is now deprecated. *See also:* signal wave. (SP) [32]

ultrasonic generator A device for the production of sound waves of ultrasonic frequency. (EEC/PE) [119]

ultrasonic grating constant The distance between diffracting centers of the sound wave that is producing particular light diffraction spectra. (SP) [32]

ultrasonic light diffraction Optical diffraction spectra or the process that forms them when a beam of light is passed through the field of a longitudinal wave. (SP) [32]

ultrasonic space grating (grating) A periodic spatial variation of the index of refraction caused by the presence of acoustic waves within the medium. (SP) [32]

ultrasonic stroboscope A light interrupter whose action is based on the modulation of a light beam by an ultrasonic field. (SP) [32]

ultraviolet (fiber optics) The region of the electromagnetic spectrum between the short wavelength extreme of the visible spectrum (about 0.4 μm) and 0.04 μm. *See also:* light; infrared. (Std100) 812-1984w

ultraviolet-erasable programmable read-only memory (UV-EPROM) *See:* erasable programmable read-only memory.

ultraviolet flame detector (fire protection devices) A device whose sensing element is responsive to radiant energy outside the range of human vision (below approximately 4000 Angstroms). (NFPA) [16]

ultraviolet radiation (1) (illuminating engineering) For practical purposes any radiant energy within the wavelength 10 to 380 nm (nanometers) is considered ultraviolet radiation. *Note:* On the basis of practical applications and the effect obtained, the ultraviolet region often is divided into the following bands:

a) ozone-producing: 180–220 nm
b) bactericidal (germicidal): 220–300 nm
c) erythemal: 280–320 nm
d) "black light": 320–400 nm

There are no sharp demarcations between these bands, the indicated effects usually being produced to a lesser extent by longer and shorter wavelengths. For engineering purposes, the "black light" region extends slightly into the visible portion of the spectrum. *See also:* regions of electromagnetic spectrum. (EEC/IE) [126]

(2) (laser maser) Electromagnetic radiation with wavelengths smaller than those for visible radiation; for the purposes of IEEE Std 586-1980, 0.2 to 0.4 μm. (LEO) 586-1980w

umbrella antenna A type of top-loaded short vertical antenna in which the top-loading structure consists of elements sloping down toward the ground but not connected to it. (AP/ANT) 145-1993

umbrella reflector antenna An antenna constructed in a form similar to an umbrella that can be folded for storage or transport and unfolded to form a large reflector antenna for use. (AP/ANT) 145-1993

UNA *See:* upstream neighbor's address.

unaligned A term that refers to the constraints placed on the address of the data; the address is unconstrained and may be any integer value. (C/MM) 1596.5-1993

unaligned address An unaligned address is a noninteger multiple of the data block size. The maximum data block size that can be transferred by an IUT Master is the product of data width and data length. (C/BA) 896.4-1993w

unannounced unavailability *See:* unavailability.

unary operation *See:* monadic operation.

unary operator *See:* monadic operator.

unary relation A relation with one attribute. (C) 610.5-1990w

unasserted The state of a signal line. Since all signal lines are active low, this state is the high state for all bus lines. (C/MM) 1196-1987w

unassigned Describes a value (for example, an **address space identifier [ASI]** number), the semantics of which are not architecturally mandated and may be determined independently by each implementation (preferably within any guidelines given). (C/MM) 1754-1994

unattainable limit A limit that is undefined for a target system or that has a magnitude exceeding the POSIX.1 {3} specified minimum and that would require an unreasonable amount of time or system resources to test. (C/PA) 2003.1-1992

unattended automatic exchange A normally unattended telephone exchange, wherein the subscribers, by means of calling devices, set up in the central office the connections to other subscribers or to a distant central office. (EEC/PE) [119]

unauthorized data modification Alteration of data not consistent with the defined security policy. (C/LM) 802.10-1998

unauthorized disclosure The process of making information available to unauthorized individuals, entities, or processes. (C/LM) 8802-11-1999

unauthorized resource use Use of a resource not consistent with the defined security policy. (C/LM) 8802-11-1999

unavailability (1) (nuclear power generating station) The probability that an item or system will not be operational at a future instant in time. Unavailability may be a result of the item being repaired (repair unavailability) or it may occur as a result of malfunctions. Unavailability is the complement of availability. (PE/NP) 352-1987r

(2) (the numerical complement of availability) Unavailability may occur as a result of the item being repaired (repair unavailability), tested (testing unavailability), or it may occur as a result of undetected malfunctions (unannounced unavailability). (PE/NP) 577-1976r

(3) (nuclear power plants) The numerical complement of availability. Unavailability may occur as a result of the item being repaired or a detected malfunction (repair unavailability), tested (testing unavailability), or it may occur as a result of undetected malfunctions (unannounced unavailability). (PE/NP) 338-1987r

(4) (power outages) Unavailability = outage time/reporting period time. *Note:* Some examples are:

1) Forced unavailability = forced outage time/reporting period time.
2) Scheduled unavailability = scheduled outage time/reporting period time.

(PE/PSE) 859-1987w

(5) (telecommunications) A state of nonservice that occurs when one or more of the following is true:

a) A bit-error ratio worse than one in ten to the *n* power for a specific number of consecutive observation periods of fixed duration has occurred;
b) A block-error ratio worse than one in ten to the *n* power, for a specific number of consecutive observation periods of fixed duration, has occurred;
c) More than a specific number of consecutive severely errored units of time has occurred;
d) An LOS event is detected.

(COM/TA) 1007-1991r

(6) The long-term average fraction of time that a component or system is out of service due to failures or scheduled outages. An alternative definition is the steady-state probability that a component or system is out of service due to failures or scheduled outages. Mathematically, unavailability = (1− availability). (IA/PSE) 493-1997, 399-1997

(7) The numerical complement of availability. Unavailability may occur as a result of the item being repaired (that is, repair unavailable) or as a result of undetected malfunctions (that is, unannounced unavailability). (PE/NP) 933-1999

unavailability margin (nuclear power generating station) The favorable difference between the desired goal and the calculated or observed unavailability. (PE/NP) 577-1976r

unavailable (electric generating unit reliability, availability, and productivity) The state in which a unit is not capable of operation because of operational or equipment failures, external restrictions, testing, work being performed, or some adverse condition. The unavailable state persists until the unit is made available for operation, either by being synchronized to the system (in-service state) or by being placed in the reserve shut-down state. (PE/PSE) 762-1987w

unavailable generation (electric generating unit reliability, availability, and productivity) The difference between the energy that would have been generated if operating continuously at dependable capacity and the energy that would have been generated if operating continuously at available capacity. This is the energy that could not be generated by a unit due to planned and unplanned outages and unit deratings.

$$\begin{aligned} UG &= (\text{planned outage hours} + \text{unplanned outage hours} \\ &\quad + \text{equivalent unit deratedhours}) \cdot \text{maximum capacity} \\ &= (POH + UOH + EUNDH) \cdot MC \end{aligned}$$

(PE/PSE) 762-1987w

unavailable hours (electric generating unit reliability, availability, and productivity) The number of hours a unit was in the unavailable state. *Note:* Unavailable hours are the sum of planned outage hours and unplanned outage hours, or the sum of planned outage hours, forced outage hours, and maintenance outage hours. (PE/PSE) 762-1987w

unavailable seconds The time interval in seconds, starting with the first of ten or more consecutive SES and ending at the beginning of ten consecutive non-SES. *See also:* unavailability. (COM/TA) 1007-1991r

unavailable time The time during which a device cannot be accessed or used. *Contrast:* available time. (C) 610.10-1994w

unbalance (data transmission) A differential mutual impedance or mutual admittance between two circuits that ideally would have no coupling. (PE) 599-1985w

unbalanced (1) (to ground) The state of impedance on a two-wire circuit when the impedance-to-ground of one wire is different from the impedance-to-ground of the other wire. *Contrast:* balanced. (C) 610.7-1995

(2) Pertaining to a relationship between two or more objects that are not alike or unsymmetrical in some respect. *Contrast:* balanced. (C) 610.10-1994w

unbalanced circuit A circuit, the two sides of which are inherently electrically unlike with respect to a common reference point, usually ground. *Note:* Frequently, unbalanced signifies a circuit, one side of which is grounded. (IE) [43]

unbalanced error (A) A set of error values in which the maximum and minimum are not necessarily opposite in sign and equal in magnitude. **(B)** A set of error values whose average is not zero. *Contrast:* balanced error. (C) 1084-1986

unbalanced merge A merge in which the subsets to be merged are unequally distributed among half of the available auxiliary storage devices, then the subsets are merged onto the other half of the auxiliary storage devices. *Contrast:* balanced merge. (C) 610.5-1990w

unbalanced merge sort A merge sort in which the sorted subsets created by internal sorts are unequally distributed among some of the available storage, the subsets are merged onto the remaining available storage, and this process is repeated until all the items are in one sorted set. *Contrast:* balanced merge sort. *See also:* polyphase merge sort. (C) 610.5-1990w

unbalanced modulator *See:* signal.

unbalanced load regulation A specification that defines the maximum voltage difference between the three output phases that will occur when the loads on the three are of different levels. (IA/PSE) 1100-1999

unbalanced phase components (thyristor) In multiphase systems unbalance of the phases can be expressed in terms of negative, positive, and zero sequence components. *Note:* Defined for the load only under conditions of balanced lines and balanced loads. (IA/IPC) 428-1981w

unbalanced strip line *See:* strip-type transmission line.

unbalanced three-phase system (self-commutated converters) (converters having ac output) A three-phase system in which the rms (root-mean-square) value of at least one phase voltage (or current) or line-to-line voltage is significantly different from the others, or in which the phase angle displacement between any pair of phases significantly differs from 120 degrees. *Note:* In an unbalanced three-phase system, negative or zero-sequence components exist. (IA/SPC) 936-1987w

unbalanced wire circuit (data transmission) One whose two sides are inherently electrically unlike. (PE) 599-1985w

unbalance factor (self-commutated converters) (converters having ac output) The ratio of the negative sequence component to the positive sequence component. (IA/SPC) 936-1987w

unbalance ratio (converters having ac output) (self-commutated converters) The difference between the highest and the lowest fundamental rms (root-mean-square) values in a three-phase system, referred to the average of the three fundamental rms values of current or voltages, respectively. (IA/SPC) 936-1987w

unbiased A measurement of a random variable is called unbiased if the expected value of the measurement is equal to the stated value of the property being measured. (NI) N42.23-1995

unbiased rounding A rounding process in which the rules for adjusting the retained numeral ensure that the average rounding error is zero. (C) 1084-1986w

unbiased telephone ringer A telephone ringer whose clapper-driving element is not normally held toward one side or the other, so that the ringer will operate on alternating current. Such a ringer does not operate reliably on pulsating current. *Note:* A ringer that is weakly biased so as to avoid tingling when dial pulses pass over the lines may be referred to as an unbiased ringer. *See also:* telephone station. (EEC/PE) [119]

unbind To remove the association between a network address and an endpoint. (C) 1003.5-1999

unblanking Turning on of the cathode-ray-tube beam. *See also:* oscillograph. (IM/HFIM) [40]

unblock* *See:* deblock.

* Deprecated.

unblocked mode A function that behaves like a blocked function, except that when it returns, the function may be incomplete and the application process may have to invoke the interface again. (C) 1003.5-1999

unblocked record A record that is contained in exactly one entire block. *See also:* blocked record; spanned record. (C) 610.5-1990w

unblocking Logic that will allow a permissive pilot scheme to trip for an internal fault within a time window, even though the pilot TRIP signal is not present when the signal is lost due to the fault. (PE/PSR) C37.113-1999

unbound mode (fiber optics) Any mode that is not a bound mode; a leaky or radiation mode of the waveguide. *See also:* bound mode; cladding mode; leaky mode. (Std100) 812-1984w

unbundle The separation of arrow meanings, expressed by branching arrow segments, i.e., the separation of object types from an object type set. (C/SE) 1320.1-1998

uncached data-access operation A data-access operation, when used to access data that is not cached. (C/MM) 1596.5-1993

uncertainty (1) (radiation protection) The estimated bounds of the deviation from the mean value, generally expressed as a percent of the mean value. Ordinarily taken as the sum of the random errors at the 95% confidence level and the estimated upper limit of the systematic error. (NI) N323-1978r

(2) (general) The estimated amount by which the observed or calculated value of a quantity may depart from the true value. *Note:* The uncertainty is often expressed as the average deviation, the probable error, or the standard deviation. *See also:* measurement system; measurement uncertainty. (MIL/IM/HFIM) [2], [40]

(3) (electrothermic power meters) The assigned allowance for the systematic error, together with the random error attributed to the imprecision of the measurement process. (IM) 470-1972w, 544-1975w

(4) (germanium spectrometers) The likely inaccuracy of a reported value, expressed in terms of estimated standard deviations. *See also:* combined uncertainty. (NI) N42.14-1991

(5) (mathematics of computing) The upper bound on an absolute error or relative error. (C) 1084-1986w

(6) An estimated limit based on an evaluation of the various sources of error. (PE/PSIM) 4-1995

unconditional branch* *See:* unconditional jump.

* Deprecated.

unconditional jump (1) An instruction that interrupts the normal process of obtaining instructions in an ordered sequence and specifies the address from which the next instruction must be taken. *See also:* jump. (C) 162-1963w

(2) A jump that takes place regardless of execution conditions. *Contrast:* conditional jump. (C) 610.12-1990

unconditional jump instruction A computer instruction that specifies an unconditional jump. *Contrast:* conditional jump instruction. (C) 610.10-1994w

unconditionally invalid date-component value A date-component value that is improperly produced or improperly accepted by a system element independent of other date-component values. *Note:* In the Julian and Gregorian calendars, the following are the unconditionally invalid date-component values:

— Values of the day-of-the-month less than 1 or greater than 31.
— Values of numeric-month less than 1 or greater than 12.
— Values of a non-numeric month (text string) outside its normal list of culturally-accepted values.
— Values of a day-of-the-year less than 1 or greater than 366.

Normalization of invalid date-component values to valid date-component values does not constitute improper acceptance of a date-component. (C/PA) 2000.1-1999

unconditional transfer of control *See:* unconditional jump.

uncontrolled ESD environment One in which no attempt is made to maintain charge levels on humans and objects below a certain level. (EMC) C63.16-1993

uncontrolled environment Locations where there is the exposure of individuals who have no knowledge or control of their exposure. (NIR) C95.1-1999

uncontrolled slip The loss or gain of one or more digit positions or a set of consecutive digit positions in a digital signal that is not a controlled slip of the timing processes associated with transmission or switching of the digital signal, and in which either the magnitude or the instant of that loss or gain is not controlled. (COM/TA) 1007-1991r

uncontrolled variable A factor affecting the outcome of an experiment that is designed to assess other factors and which is unknown to, or unaccounted for, by the experimenter. *Synonym:* confounding variable. (T&D/PE) 539-1990

uncorrelated jitter The portion of the total jitter that is independent of the data pattern. This jitter is generally caused by noise that is uncorrelated among stations and therefore grows in a nonsystematic way along the ring. *Synonyms:* noise jitter; nonsystematic jitter. (C/LM) 8802-5-1998

undamped frequency (A) (frequency, natural) Of a second-order linear system without damping, the frequency of free oscillation in radians per unit time or in hertz. **(B) (frequency, natural)** Of any system whose transfer function contains the quadratic factor $s^2 + 2\zeta\omega_n s + \omega_n^2$ in the denominator, the value ω_n ($0 < \zeta < 1$). **(C) (frequency, natural)** Of a closed loop control system or controlled system, a frequency at which continuous oscillation (hunting) can occur without periodic stimuli. *Note:* In linear systems, the undamped frequency is the phase crossover frequency. With proportional control action only, the undamped frequency of a linear system may be obtained by raising (in most cases) the proportional gain until hunting occurs. This value of gain has been called the "ultimate gain" and the undamped period the "ultimate period." *Synonym:* natural frequency. (IM/PE/EDPG) [120], [3]

undefined (1) A value or behavior is undefined if the standard imposes no portability requirements on applications for erroneous program construct, erroneous data, or use of an indeterminate value. Implementations (or other standards) may specify the result of using that value or causing that behavior. An application using such behaviors is using extension. (C/PA) 1003.1-1988s

(2) Describes an aspect of the architecture that has deliberately been left unspecified. Software should have no expectation of, nor make any assumptions about, such an architectural feature or behavior. Use of such a feature may deliver random results, may or may not cause a trap, may vary among implementations, and may vary with time on a given implementation. Notwithstanding any of the above, undefined aspects of the architecture shall not cause security holes, such as allowing user software to access supervisor state, put the processor into supervisor state, or put the processor into an unrecoverable state. (C/MM) 1754-1994

underbilling error Occurs when a call is billed less that it should be due to shortened time interval or wrong class of service or time zone given by the switch. (COM/TA) 973-1990w

underbuilt shield wires Shield wires arranged among or below the average height of the protected phase conductors for the purposes of lowering the OHGW system impedance and improving coupling. Underbuilt shield wires may be bonded to the structure directly or indirectly through short gaps. Insulated earth return conductors on HVDC transmission lines and/or faulted phases both function as underbuilt shield wires. (PE/T&D) 1243-1997

underbunching A condition representing less than optimum bunching. (ED) 161-1971w

undercounter dumbwaiter A dumbwaiter that has its top terminal landing located underneath a counter and that serves only this landing and the bottom terminal landing. (EEC/PE) [119]

undercurrent or underpower relay (power system device function numbers) A relay that functions when the current or power flow decreases below a predetermined value. (SUB/PE) C37.2-1979s

undercurrent relay (1) A relay that operates when the current through the relay is equal to or less than its setting. *See also:* relay. (PE/PSR) [6]

(2) A relay that operates when the current is less than a predetermined value. (SWG/PE) C37.100-1992

undercurrent release A release that operates when the current in the main circuit is equal to or less than the release setting. *Synonym:* undercurrent trip. (SWG/PE) C37.100-1992

undercurrent trip *See:* undercurrent release.

undercurrent tripping *See:* undercurrent release.

underdamped Damped insufficiently to prevent oscillation of the output following an abrupt input stimulus. *Note:* In an underdamped linear second-order system, the roots of the characteristic equation have complex values. *See also:* damped harmonic system. (IA/IAC) [60]

underdamped period (instrument) (periodic time) The time between two consecutive transits of the pointer or indicating means in the same direction through the rest position, following an abrupt change in the measurand. (PE/EEC) [119]

underdamping (periodic damping) The special case of damping in which the free oscillation changes sign at least once. A damped harmonic system is underdamped if F^2 less MS. *See also:* damped harmonic system. (Std100) 270-1966w

underdome bell A bell whose mechanism is mostly concealed within its gong. *See also:* protective signaling. (EEC/PE) [119]

underfilm corrosion Corrosion that occurs under films in the form of randomly distributed hairlines (filiform corrosion). (IA) [59]

undefined behavior Behavior for which the standard imposes no requirements (e.g., use of an erroneous program construct). Permissible undefined behavior ranges from:

— Ignoring a situation completely with unpredictable results.
— Behaving during translation or program execution in a documented manner characteristic of the environment (with or without the issuance of a diagnostic message).
— Terminating a translation or execution (with the issuance of a diagnostic message).

Note: Many erroneous program constructs do not engender undefined behavior; they are required to be diagnosed. (C/DA) 1481-1999

underfloor raceway A raceway suitable for use in the floor. *See also:* raceway. (EEC/PE) [119]

underflow (mathematics of computing) The condition that arises when the result of a floating-point arithmetic operation is smaller than the smallest non-zero number that can be represented in a digital computer. *Synonym:* arithmetic underflow. (C) 1084-1986w

underflow error (mathematics of computing) The error caused by an underflow condition in computer arithmetic. (C) 1084-1986w

underflow exception (software) An exception that occurs when the result of an arithmetic operation is too small a fraction to be represented by the storage location designated to receive it. *See also:* operation exception; protection exception; data exception; addressing exception; overflow exception. (C) 610.12-1990

underground cable (1) A cable for installation below the surface of the earth in ducts or conduits so it can readily be removed without disturbing the surrounding earth, and that is designed to withstand submersion in ground waters. (PE/PSC) 789-1988w

(2) A cable installed below the surface of the ground. *Note:* This term is usually applied to cables installed in ducts or conduits or under other conditions such that they can readily be removed without disturbing the surrounding ground. *See also:* cable; tower. (T&D/PE) [10]

underground collector or plow A current collector, the function of which is to make contact with an underground contact rail. *See also:* contact conductor. (VT/LT) 16-1955w

underground duct system (raceway systems for Class 1E circuits for nuclear power generating stations) Metallic or nonmetallic conduit enclosed in reinforced concrete or directly buried, including access points. (PE/NP) 628-1987r

underground system service-entrance conductors The service conductors between the terminals of the service equipment and the point of connection to the service lateral. Where service equipment is located outside the building walls, there may be no service-entrance conductors, or they may be entirely outside the building. (NESC/NEC) [86]

under-jacket type A moisture barrier applied under the jacket and over the metallic shield or concentric neutral of a cable. Also, a combination moisture barrier and shield. (PE/IC) 1142-1995

underlap, *X* (facsimile) The amount by which the center-to-center spacing of the recorded spots exceeds the recorded spot *X* dimension. *Note:* This effect arises in that type of equipment which responds to a constant density in the subject copy by a succession of discrete recorded spots. *See also:* recording. (COM) 168-1956w

underlap, *Y* (facsimile) The amount by which the nominal line width exceeds the recorded spot *Y* dimension. *See also:* recording. (COM) 168-1956w

underreaching protection A form of protection in which the relays at a given terminal do not operate for faults at remote locations on the protected equipment, the given terminal being cleared either by other relays with different performance characteristics or by a transferred trip signal from a remote terminal similarly equipped with underreaching relays. (SWG/PE) C37.100-1992

undershoot (1) (television) (rounding) That part of the distorted wave front characterized by a decaying approach to the final value. *Note:* Generally, undershoots are produced in transfer devices having insufficient transient response. (BT/AV) 201-1979w

(2) (oscilloscopes) In the display of a step function (usually of time), that portion of the waveform that, following any overshoot or rounding that may be present, falls below its nominal or final value. (IM) 311-1970w

(3) The peak value of an impulse voltage or current that passes through zero in the opposite polarity of the initial peak. (PE/PSIM) 4-1995

under-sized packet *See:* short packet.

underslung car frame A car frame to which the hoisting-rope fastenings or hoisting rope sheaves are attached at or below the car platform. *See also:* hoistway. (EEC/PE) [119]

underspeed (hydraulic turbines) Any speed below rated speed expressed as a percent of rated speed. (PE/EDPG) 125-1977s

underspeed device (power system device function numbers) A device that functions when the speed of a machine falls below a predetermined value. (SUB/PE) C37.2-1979s

undervoltage (1) When used to describe a specific type of long duration variation, refers to a measured voltage having a value less than the nominal voltage for a period of time greater than 1 min. Typical values are 0.8–0.9 pu. (SCC22) 1346-1998

(2) When used to describe a specific type of long duration variation, refers to an RMS decrease in the ac voltage, at the power frequency, for a period of time greater than 1 min. Typical values are 0.8-0.9 pu. (IA/PSE) 1100-1999

undervoltage protection (1) The effect of a device, operative on the reduction or failure of voltage, to cause and maintain the interruption of power in the main circuit. (IA/MT/PKG) 45-1998, 333-1980w

(2) A form of protection that operates when voltage is less than a predetermined value. *Synonym:* low-voltage protection. (SWG/PE) C37.100-1992

undervoltage relay A relay that operates when its voltage is less than a predetermined value. (SWG/PE/SUB) C37.100-1992, C37.2-1979s

undervoltage release (1) (trip) A release that operates when the voltage of the main circuit is equal to less than the release setting. (SWG/PE) C37.100-1992

(2) The effect of a device, operative on the reduction or failure of voltage, to cause the interruption of power to the main circuit, but not to prevent the re-establishment of the main circuit on return of voltage. (IA/MT) 45-1998

undervoltage tripping *See:* undervoltage release.

underwater log A device that indicates a ship's speed based on the pressure differential, resulting from the motion of the ship relative to the water, as developed in a Pitot tube system carried by a retractable support extending through the ship's hull. Continuous integration provides indication of total distance travelled. The ship's draft is indicated, based on static pressure. (EEC/PE) [119]

underwater sound projector A transducer used to produce sound in water. *Notes:* 1. There are many types of underwater sound projectors whose definitions are analogous to those of corresponding loudspeakers, for example, crystal projector, magnetic projector, etc. 2. Where no confusion will result, the term underwater sound projector may be shortened to projector. *See also:* microphone. (SP) [32]

undesirable response rate The percentage of undesirable ESD responses exhibited by the EUT when subjected to a specific number of ESD events. (EMC) C63.16-1993

undesired conducted power (frequency-modulated mobile communications receivers) Radio-frequency power that is present at the antenna, power terminals, or any other interfacing terminals. (VT) 184-1969w

undesired radiated power (frequency-modulated mobile communications receivers) Radio-frequency power radiated from the receiver that can be measured outside a specified area. (VT) 184-1969w

undetected error rate (data transmission) The ratio of the number of bits, unit elements, characters, blocks incorrectly received but undetected or uncorrected by the error-control equipment, to the total number of bits, unit elements, characters, blocks sent. (COM) [49]

undeveloped stage The time prior to the installation of permanent structures, site preparation, preliminary surveying, surface stripping, fence erection, road building, equipment and material staging, furnishing construction power, etc. (PE/SUB) 1402-2000

undirected graph A graph in which no direction is implied in the internode connections. *Contrast:* directed graph. (C) 610.5-1990w, 610.12-1990

undisturbed-ONE output (magnetic cell) A ONE output to which no partial-read pulses have been applied since that cell was last selected for writing. *See also:* coincident-current selection. (Std100) 163-1959w

undisturbed-ZERO output (magnetic cell) A ZERO output to which no partial-write pulses have been applied since that cell was last selected for reading. *See also:* coincident-current selection. (Std100) 163-1959w

undressed timber Rough unsurfaced lumber. (T&D/PE) 751-1990

undulating current (rotating electric machinery) Current that remains unidirectional, but the ripple of which exceeds that defined for smooth current. (PE/EM) 11-1980r

unexposed side (cable penetration fire stop qualification test) The side of a fire-rated wall, floor-ceiling assembly, or floor that is opposite to the fire side. *Synonym:* cold side. (PE) 634-1978w

unfired tube (microwave gas tubes) The condition of the tube during which there is no radio-frequency glow discharge at either the resonant gap or resonant window. *See also:* gas tube. (ED) 161-1971w

Unformatted Page (UP) A Next Page encoding that contains an unformatted 12-bit message field. Use of this field is defined through Message Codes and information contained in the UP. (C/LM) 802.3-1998

unfused capacitor A capacitor without any internal fuses.power systems relaying. (PE) C37.99-2000

unfused capacitor bank Any capacitor bank without fuses, internal or external.power systems relaying. (PE) C37.99-2000

unfused switch A switch which has no fuses directly attached or in close proximity to the switch. (SWG/PE) C37.20.4-1996

ungrounded (electric power) A system, circuit, or apparatus without an intentional connection to ground except through potential-indicating or measuring devices or other very-high-impedance devices. *Note:* Though called ungrounded, this type of system is in reality coupled to ground through the distributed capacitance of its phase windings and conductors. In the absence of a ground fault, the neutral of an ungrounded system under reasonably balanced load conditions will usually be close to ground potential, being held there by the balanced electrostatic capacitance between each phase conductor and ground. (IA/PE/PSE/TR) 142-1982s, C57.12.80-1978r

ungrounded potentiometer (analog computer) A potentiometer with neither end terminal attached directly to ground. *See also:* electronic analog computer. (C) 165-1977w, 166-1977w

ungrounded system (systems grounding) A system, circuit, or apparatus without an intentional connection to ground, except through potential-indicating or measuring devices or other very-high-impedance devices. (PE/C/IA/PSE) 1313.1-1996, 142-1982s

unguarded release (telephone switching systems) A condition during the restoration of a circuit to its idle state when it can be prematurely seized. (COM) 312-1977w

uniconductor waveguide A waveguide consisting of a cylindrical metallic surface surrounding a uniform dielectric medium. *Note:* Common cross-sectional shapes are rectangular and circular. *See also:* waveguide. (MTT) 148-1959w, 146-1980w

unicast A transmission mode in which a single message is sent to a single network destination, (i.e., one-to-one). (DIS/C) 1278.1-1995, 1278.2-1995

unicast address (local area networks) An individual address identifying an individual end node. (C) 8802-12-1998

unicast frame A frame that is addressed to a single recipient, not a broadcast or multicast frame. *Synonym:* directed address. (C/LM) 8802-11-1999

unidirectional A connection between telegraph sets, one of which is a transmitter and the other a receiver. (COM) [49]

unidirectional antenna An antenna that has a single well-defined direction of maximum gain. *See also:* antenna. (AP/ANT) 145-1983s

unidirectional bus (1) (programmable instrumentation) A bus used by any individual device for one-way transmission of messages only; that is, either input only or output only. (IM/AIN) 488.1-1987r
(2) (696 interface devices) (signals and paths) A bus used by a device for one-way transmission of messages, that is, either input only or output only. (MM/C) 696-1983w

unidirectional current A current that has either all positive or all negative values. (Std100) 270-1966w

unidirectional microphone A microphone that is responsive predominantly to sound incident from a single solid angle of one hemisphere or less. *See also:* microphone. (SP) [32]

unidirectional operation When the peripheral and host communicate data in one direction only. Compatibility Mode is unidirectional in the forward direction; Byte and Nibble Modes are unidirectional in the reverse direction. (C/MM) 1284-1994

unidirectional pulse train (signal-transmission system) Pulses in which pertinent departures from the normally constant value occur in one direction only. *See also:* pulse. (IM/WM&A) 194-1977w

unidirectional transducer (UDT) (1) (unilateral transducer) A transducer that cannot be actuated at its output by waves in such a manner as to supply related waves at its input. *See also:* transducer. (Std100) 270-1966w
(2) A transducer capable of radiating and receiving surface acoustic waves in or from a single direction. (UFFC) 1037-1992w

unified atomic mass unit The unit equal to the fraction 1/12 of the mass of an atom of the nuclide ^{12}C: 1 u = 1.660 53 × 10^{-27} kg approximately. (QUL) 268-1982s

unified s-band system (communication satellite) A communication system using an s-band carrier (2000−2300 megahertz) combining all links into one spectrum. The functions of spacecraft command, data transmission, tracking, ranging, etc., are transmitted on separate carrier frequencies for earth-space and space-earth links. (COM) [19]

unified transaction (1) A transaction in which the request and response subactions are completed in an indivisible sequence; i.e., no other subactions may be performed on the bus until this response subaction is complete. Most buses use unified transactions, but SCI uses only split transactions. The concept of a unified transaction is only relevant to SCI in the context of bridges to other buses. (C/MM) 1596-1992
(2) A transaction in which the request and response subactions are completed as an indivisible sequence. Between the initiation of the request and the completion of the response, other subactions are blocked. The Futurebus+ standard also calls this a connected transaction. A transaction that is not unified is called a split transaction. (C/MM) 1212-1991s
(3) A transaction that is completed in a single subaction. (C/MM) 1394-1995

uniform-asymptotic stability Asymptotic stability where the rate of convergence to zero of the perturbed-state solution is independent of the initial time t_0. *Note:* An example of a solution that is asymptotically stable but not uniformly asymptotically stable is the solution

$\varphi(x(t_0);t) = x(t_0)t_0/t$

of the system $\dot{x} = -x/t, t_0$. Note that the initial rate of decay,

$\dot{x}t_0)/x(t_0) = -1/t_0$

is clearly a function of t_0. Compare with the time-invariant system $\dot{x} = ax$ where $\dot{x}(t_0)/x(t_0) = a$ is independent of t_0. The concept of uniformity with respect to the initial time t_0 applies only to time-varying systems. All stable time-invariant systems are uniformly stable. *See also:* control system. (CS/IM) [120]

uniform current density A current density that does not change (either in magnitude or direction) with position within a specified region. (A uniform current density may be a function of time.) (Std100) 270-1966w

uniform field A field whose magnitude and direction are uniform at each instant in time at all points within a defined region. (T&D/PE) 644-1994, 539-1990

uniform line A line that has substantially identical electrical properties throughout its length. *See also:* transmission line. (EEC/PE) [119]

uniform linear array A linear array of identically oriented and equally spaced radiating elements having equal current amplitudes and equal phase increments between excitation currents. *See also:* antenna. (AP/ANT) [35], 145-1993

uniform luminance area The area in which a display on a cathode-ray tube retains 70% or more of its luminance at the center of the viewing area. *Note:* The corners of the rectangle formed by the vertical and horizontal boundaries of this area may be below the 70% luminance level. *See also:* oscillograph. (IM/HFIM) [40]

uniform plane wave *See:* homogeneous plane wave.

uniform probing Open-address hashing in which collision resolution is handled by selecting positions at uniform distances from the original position in the hash table until an available position is found. *Contrast:* random probing; linear probing. (C) 610.5-1990w

uniform random number Any member of a random number sequence that has a uniform statistical distribution. (C) 1084-1986w

uniform waveguide A waveguide in which the physical and electrical characteristics do not change with distance along the axis of the guide. (MTT) 147-1979w, 146-1980w

unilateral area track A sound track in which one edge only of the opaque area is modulated in accordance with the recorded signal. There may, however, be a second edge modulated by a noise-reduction device. *See also:* phonograph pickup. (SP) [32]

unilateral connection (control system feedback) A connection through which information is transmitted in one direction only. *See also:* feedback control system. (IM) [120]

unilateral network A network in which any driving force applied at one pair of terminals produces a nonzero response at a second pair but yields zero response at the first pair when the same driving force is applied at the second pair. *See also:* network analysis. (Std100) 270-1966w

unilateral transducer *See:* unidirectional transducer.

unimpaired observation Conditions that enable an unobstructed view to ensure direct visual or closed-circuit television (CCTV) surveillance of individuals or vehicles. (PE/NP) 692-1997

uninhibited oil (power and distribution transformers) Mineral transformer oil to which no synthetic oxidation inhibitor has been added. *See also:* oil-immersed transformer. (PE/TR) C57.12.80-1978r, [57]

unintentional disconnect Any disconnection from the network, at either end, that is not preceded by an intentional disconnect primitive, within the specified time period. (EMB/MIB) 1073.3.1-1994

unintentional radiator A device that generates radio-frequency energy for use within the device, or sends radio-frequency signals by conduction to associated equipment via connecting wiring, but which is not intended to emit radio-frequency energy by radiation or induction. (EMC) C63.4-1991

uninterruptible power supply (UPS) (1) (electric power systems in commercial buildings) A device or system that provides quality and continuity of an ac power source. (IA/PSE) 241-1990r

(2) A system designed to provide power automatically, without delay or transients, during any period when the normal power supply is incapable of performing acceptably. (IA/PSE) 446-1995

uninterruptible power supply module (electric power systems in commercial buildings) The power conversion portion of the uninterruptible power system. *Synonym:* UPS module. (IA/PSE) 241-1990r

union (data management) A relational operator that combines two relations of the same degree and results in a relation containing all of the tuples that are in either of the original relations. (See corresponding figure.) *See also:* intersection; projection; selection; product; join; difference.

A B U X Y Z = A B X Y Z SUT

S T

union

(C) 610.5-1990w

(2) (mathematics of computing) *See also:* OR. (C) 1084-1986w

unipolar (power supplies) Having but one pole, polarity, or direction. Applied to amplifiers or power supplies, it means that the output can vary in only one polarity from zero and, therefore, must always contain a direct-current component. (AES) [41]

unipolar electrode system (monopolar electrode system) (electribiology) Either a pickup or a stimulating system, consisting of one active and one dispersive electrode. *See also:* electrobiology. (EMB) [47]

unipolar pulse A signal pulse having a single lobe above (or below) the baseline. (NPS) 325-1996

unipolar transistor A transistor that utilizes charge carriers of only one polarity. *See also:* transistor; semiconductor. (ED) 216-1960w

unipole *See:* antenna.

uniprocessor (A) A computer that can execute only one program at a time. *Contrast:* multiprocessor. **(B)** A computer system with one central processing unit. (C) 610.10-1994

unique identification code (unique identification in power plants) A code applied at the component function level to uniquely distinguish a specific function within a specific system from all other similar or different functions occurring within the system or facility. The basic code format described in IEEE 803-1983 may also be applied, with appropriate field identifiers, for project software and project control elements (schedule and budget items). (PE/EDPG) 803-1983r

uniquely addressed Said of an MTM-Bus S-module participating in a singlecast. (TT/C) 1149.5-1995

uniqueness constraint A kind of constraint stating that no two distinct instances of a class may agree on the values of all the properties that are named in the uniqueness constraint. (C/SE) 1320.2-1998

UNIRAM A modeling methodology and software for the performance of reliability, availability, and maintainability (RAM) analysis of power production systems. (PE/NP) 933-1999

unit (1) (nuclear power generating station) One independent portion of a motor control center vertical section. It is normally a plug-in module which connects to the motor control center vertical bus. (PE/NP) 649-1980s

(2) That portion of the switch−gear assembly which contains one switching device such as a circuit breaker, interrupter switch, power fuse interrupter switch combination, etc. and the associated primary conductors. *See also:* relay unit. (SWG/PE) C37.20-1968w

(3) (electric and electronics parts and equipment) A major building block for a set or system, consisting of a combination of basic parts, subassemblies, and assemblies packaged together as a physically independent entity. The application, size, and construction of an item may be factors in determining whether an item is regarded as a unit, an assembly, a subassembly, or a basic part. A small electric motor might be considered as a part if it is not normally subject to disassembly. Typical examples are: radio receiver, radio transmitter, electronic power supply, antenna. (GSD) 200-1975w

(4) The generator or generators, associated prime mover or movers, auxiliaries, and energy supply or supplies that are normally operated together as a single source of electric power. (PE/TR/EDPG) C57.116-1989r, 505-1977r

(5) (power outages) A group of components that are functionally related and are regarded as an entity for purposes of recording and analyzing data on outage occurrences. *Notes:* 1. A unit can be defined in a number of ways. For example, it may be:

a) A group of components which constitute an operating entity bounded by automatic fault interrupting devices which isolate it from other such entities for faults on any component within the group.

b) A group of components protected by and within the sensing zone of a particular system of protective relays. Examples include a transformer or an overhead line and associated terminal facilities switched with it.

c) A group of components including a transmission line, one or more transformers supplied by the line, and a subtransmission or distribution network radially supplied from the transformer. These components are so configured that the subtransmission network is in the outage state during outage occurrences of the transmission line.

2. A unit may be single-terminal, two-terminal or multi-terminal. A multi-terminal unit is connected to three or more terminals. 3. It is recognized that certain components (for example, circuit breakers) may be part of more than one unit. 4. Different types of units include transmission unit (overhead or cable), transformer unit, bus unit, and special units that consist of any equipment protected by separate breakers, such as shunt capacitors. (PE/PSE) 859-1987w

(6) **(A)** **(software)** A separately testable element specified in the design of a computer software component. **(B)** **(software)** A logically separable part of a computer program. **(C)** **(software)** A software component that is not subdivided into other components. *Note:* The terms "module," "component," and "unit" are often used interchangeably or defined to be subelements of one another in different ways depending upon the context. The relationship of these terms is not yet standardized. *See also:* test unit. (C/Std100) 610.12-1990

(7) A logically separable part of a program. (C/SE) 1074-1995s

(8) A portion of a computer that constitutes the means of accomplishing some inclusive operation or function as; for example, an arithmetic unit. *See also:* execution unit; processing unit; logic unit; arithmetic unit; control unit; functional unit. (C) 610.10-1994w

(9) A unit is a logical component of a node that is accessed by I/O driver software. After the node is initialized and configured, the units normally operate independently. Note that one node could have multiple units (for example, processor, memory, and SCSI controller). (C/BA) 896.4-1993w

(10) Multiple cells in a single jar. (SB) 1188-1996

(11) A nuclear steam supply system, its associated turbine-generator, auxiliaries, and engineered safety features. (PE/NP) 308-1991

(12) **(A)** An aggregation of entities. **(B)** A basis of measurement. (DIS/C) 1278.3-1996

(13) *See also:* relay unit. (SWG/PE) C37.100-1992

(14) *See also:* basic operating unit. (VT/RT) 1473-1999, 1475-1999, 1474.1-1999

(15) A subcomponent of a node that provides a processing, memory, or I/O functionality. After the node has been initialized (typically by generic software), the unit provides the register interface that is accessed by I/O driver software. The units normally operate independently of each other, and do not affect the operation of the node upon which they reside. *Note:* One node could have multiple units (for example, processor, memory, and SCSI controller). (C/MM/BA) 1212-1991s, 14536-1995, 896.2-1991w, 896.10-1997

unit address The component of a node name that indicates the device node's position within the address space defined by its parent node. (C/BA) 1275-1994

unit architecture **(1)** The specification document describing the format and function of the unit's software-visible registers. (C/MM) 1212-1991s

(2) The specification document describing the format and function of the software-visible resources of the unit. (C/MM) 1394-1995

unit-area capacitance (electrolytic capacitor) The capacitance of a unit area of the anode surface at a specified frequency after formation at a specified voltage. (PE/EEC) [119]

unitary code A code having only one digit; the number of times it is repeated determines the quantity it represents. (C) 1084-1986w

unit auxiliaries transformer (UAT) (generating stations electric power system) A transformer intended primarily to supply all or a portion of the unit auxiliaries. (PE/TR/EDPG) C57.116-1989r, 505-1977r

unit auxiliary (generating stations electric power system) An auxilary intended for a specific generating unit. (PE/EDPG) 505-1977r

unit cable construction That method of cable manufacture in which the pairs of the cable are stranded into groups (units) containing a certain number of pairs and these groups are then stranded together to form the core of the cable. *See also:* cable. (EEC/PE) [119]

unit-control error (electric power system) The unit generation minus assigned unit generation. (PE/PSE) 94-1991w

unit data *See:* datagram.

unit-dependent A term used to describe parameters that may vary between different unit architectures. Although the CSR Architecture may specify the size and location of these fields, their format and most of their definition is provided by the appropriate unit architecture specification. (C/MM) 1212-1991s

unit derated generation (power system measurement) The unavailable generation resulting from unit derating. (PE/PSE) 762-1980s

unit derated hours (electric generating unit reliability, availability, and productivity) The available hours during which a unit derating was in effect. (PE/PSE) 762-1987w

unit derating (electric generating unit reliability, availability, and productivity) The difference between dependable capacity and available capacity. (PE/PSE) 762-1987w

unit development folder (UDF) *See:* software development file.

unit-distance code (mathematics of computing) A code in which the Hamming distance between consecutive numerals is 1. *Synonyms:* continuous-progression code; cyclic permuted code. (C) 1084-1986w

uniterm indexing A variation of derivative indexing in which each keyword must be a single word. (C) 610.2-1987

unit function *See:* function.

unit-impulse function *See:* unit-impulse signal.

unit-impulse signal (automatic control) A signal that is an impulse having unity area. *See also:* feedback control system. (PE/EDPG) [3]

unit_initialization_block (UIB) A contiguous buffer in System Memory used to pass Unit-dependent parameters to the I/O Unit during initialization or on command during normal DMA operation. The Processor writes the address of this block in a Unit-global CSR or passes it in a message, respectively. (C/MM) 1212.1-1993

unit interval (UI) **(1)** **(local area networks)** One half of a bit time. 125 ns for 4 Mbit/s transmission and 31.25 ns for 16 Mbit/s transmission. UI is used in the specification of jitter. (LM/C) 802.5-1989s

(2) **(telecommunications)** The nominal difference in time between consecutive significant instants of an isochronous signal. (COM/TA) 1007-1991r

(3) *See also:* signal element. (PE) 599-1985w

unitized equipment (packaging machinery) Electrical controls so constructed that separate panels are provided for each working station, or section as specified, of a multiple-station transfer-type machine. (IA/PKG) 333-1980w

unit of acceleration (digital accelerometer) The symbol *g* denotes a unit of acceleration equal in magnitude to the local value of gravity at the test site unless otherwise specified. (AES/GYAC) 530-1978r

unit operation **(1)** An interrupting operation followed by a closing operation. The final operation is also considered one unit operation. (SWG/PE) C37.60-1981r

(2) Discharge of a surge through an arrester while the arrester is energized. (SPD/PE) C62.22-1997, C62.11-1999

(3) Discharging a surge through the surge-protective-device while the device is energized. (SPD/PE) C62.62-2000

unit-ramp function *See:* unit-ramp signal.

unit-ramp signal (automatic control) A signal that is zero for all values of time prior to a certain instant and equal to the time measured from that instant. *Note:* The unit-ramp signal is the integral of the unit-step signal. *See also:* feedback control system. (PE/EDPG) [3]

unit rate-limiting controller (electric power system) A controller that limits rate of change of generation of a generating unit to an assigned value or values. (PE/PSE) 94-1991w

unit requirements documentation (software unit testing) Documentation that sets forth the functional, interface, performance, and design constraint requirements for a test unit. (C/SE) 610.12-1990, 1008-1987r

units The units of a measured value of a physical variable define the standard quantity of the measure of that variable used to express the value. The representation of the units of an Object is the datatype `Units`. (IM/ST) 1451.1-1999

units and letter symbols (International System of Units (SI)) The three classes of SI units are: 1) Base units, regarded by convention as dimensionally independent:

Quantity	Unit	Symbol
length	meter	m
mass	kilogram	kg
time	second	s
electric current	ampere	A
thermodynamic temperature	kelvin	K
amount of substance	mole	mol
luminous intensity	candela	cd

(2) Supplementary units, regarded as either base units or as derived units:

Quantity	Unit	Symbol
plane angle	radian	rad
solid angle	steradian	sr

(3) Derived units, formed by combining base elements, supplementary units, and other derived units according to the algebraic relations linking the corresponding quantitiies. The sysmbols for derived units are obtained by means of the mathematical signs for multiplication, division, and use of exponents.

unit sequence starting relay (power system device function numbers) A relay that functions to start the next available unit in a multiple-unit equipment upon the failure or nonavailability of the normally preceding unit. (PE/SUB) C37.2-1979s

unit sequence switch (power system device function numbers) A switch that is used to change the sequence in which units may be placed in and out of service in multiple-unit equipments. (SUB/PE) C37.2-1979s

unit service power The power used to operate a unit. (PE/PSE) 94-1991w

units of luminance (light emitting diodes) The luminance (photometric brightness) of a surface in a specified direction may be expressed in luminous intensity per unit of projected area of surface. *Note:* Typical units in this system are the candela per square meter. *Synonym:* units of photometric brightness. (ED) [127]

units of luminous exitance (illuminating engineering) Lumens per square meter (lam/m^2) and lumens per square foot (lm/ft^2) are preferred practice for the SI and English (USA) systems respectively. (EEC/IE) [126]

units of photometric brightness *See:* units of luminance.

units of wavelength The distance between two successive points of a periodic wave in the direction of propagation, in which the oscillation has the same phase. The three commonly used units are listed in the following table:

Name	Symbol	Value
micrometer	μm	$1 \mu m = 10^{-3}$ millimeters
nanometer	nm	$1 \text{ nm} = 10^{-6}$ millimeters
angstrom	Å = 10^{-7} millimeters	

See also: radiant energy. (EEC/IE) [126]

units position In a positional notation system, the position corresponding to the zero power of the radix. This is the rightmost position in a numeral representing an integer. (C) 1084-1986w

unit state (electric generating unit reliability, availability, and productivity). A particular unit condition that is important for purposes of collecting data on performance. *Note:* The state definitions are related as shown in the figure below. The transitions between states are described in Appendix B of IEEE Std 762-1987. The correlation between these definitions and those in use by the industry is shown in Appendix A of IEEE Std 762-1987. (PE/PSE) 762-1987w

unit-step function *See:* unit-step signal.

unit-step signal (automatic control) A signal that is zero for all values of time prior to a certain instant and unity for all values of time following. *Note:* The unit-step signal is the integral of the unit-impulse signal. *See also:* feedback control system. (PE/EDPG) [3]

unit string A string consisting of only one entity. (C) 610.5-1990w

unit substation A substation consisting primarily of one or more transformers mechanically and electrically connected and coordinated in design with one or more switchgear or motor control assemblies or combination thereof. *Note:* A unit substation may be described as *primary* or *secondary* depending on the voltage rating of the low-voltage section: *primary,* more than 1000 V; *secondary,* 1000 V and below. (SWG/PE/TR) C37.100-1992, C57.12.80-1978r

unit-substation transformer (power and distribution transformers) A transformer that is mechanically and electrically connected to, and coordinated in design with, one or more switchgear or motor-control assemblies, or combinations thereof. *See also:* integral unit substation; articulated unit substation; secondary unit substation; primary unit substation; unit substation. (PE/TR) C57.12.80-1978r

unit symbol *See:* symbol for a unit.

unit test A test performed on a single unit or group of units. *Note:* one widespread use of such tests is extrapolation of test results for the purpose of representing overall performance of a device composed of several units. (SWG/PE) C37.20-1968w, C37.100-1981s

unit testing Testing of individual hardware or software units or groups of related units. *See also:* system testing; component testing; interface testing; integration testing. (C) 610.12-1990

unit transformer (UT) (generating stations electric power system) A power system supply transformer that transforms all or a portion of the unit power from the unit to the power system voltage. (PE/TR/EDPG) C57.116-1989r, 505-1977r

unit under test (UUT) (1) The entity to be tested. It may range from a simple diagnostic unit to a complete system. (ATLAS) 1232-1995
(2) The entity to be tested. It may range from a simple component to a complete system. (SCC20/SCC20) 771-1998, 993-1997

unit-under-test-oriented language (test, measurement, and diagnostic equipment) A computer language used to program automatic test equipment to test units under test (UUTs), whose characteristics are directed to the test needs of the UUTs and therefore do not imply the use of a specific ATE (automatic test equipment) system or family of ATE systems. *Synonym:* UTT-oriented language. (MIL) [2]

unit vector A vector whose magnitude is unity. (Std100) 270-1966w

unit warmup time (power supply) The interval between the time of application of input power to the unit and the time at which the regulated power supply is supplying regulated power at rated output voltage. *See also:* regulated power supply. 209-1950w

unit years (power system measurement) For any unit or for a group of units, unit years is the total period hours accumulated, divided by 8760: PH UY = 8760. (PE/PSE) 762-1980s

unity gain (broadband local area networks) A design principle wherein amplifiers supply enough signal gain at appropriate frequencies to compensate for the system's cable loss and flat loss: cable loss + flat loss = amplifier gain. (LM/C) 802.7-1989r

unity-gain bandwidth (power supplies) A measure of the gain-frequency product of an amplifier. Unity-gain bandwidth is the frequency at which the open-loop gain becomes unity, based on a 6-decibel-per-octave crossing. (See corresponding figure.)

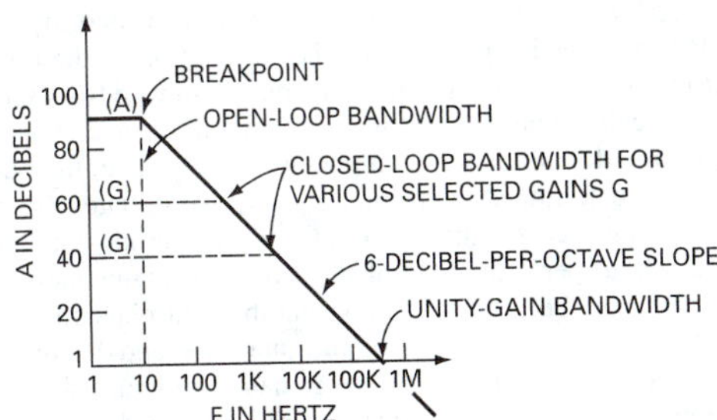

Typical gain-frequency (Bode) plot, showing unity-gain bandwidth.

unity-gain bandwidth

(AES) [41]

unity power-factor test (synchronous machines) A test in which the machine is operated as a motor under specified operating conditions with its excitation adjusted to give unity power factor. (PE) [9]

univalent function If to every value of u there corresponds one and only one value of x (or one and only one set of values of $x_1, x_2, \ldots, x_n$) then u is a univalent function. Thus $u^2 = ax + b$ is univalent, within the interval of definition. (Std100) 270-1966w

universal asynchronous receiver/transmitter A universal receiver/transmitter device used in asynchronous transmission applications. *Synonym:* asynchronous receiver/transmitter. (C) 610.7-1995

universal demand register (mechanical demand registers) A demand register of specific ratio used in conjunction with all ratings of any type of integrating electricity meter designed to accommodate it. The register constant of a universal demand register is proportional to the watthour constant K_1h of the meter on which it is mounted. (ELM) C12.4-1984

universal fuse links Fuse links that, for each rating, provide mechanical and electrical interchangeability within prescribed limits over the specified time-current range. (SWG/PE) C37.40-1993, C37.100-1992

universally unique identifier (UUID) Various versions of UUIDs exist, and references for generation of UUIDs abound. (C/SS) 1244.1-2000

universal motor A series-wound or a compensated series-wound motor designed to operate at approximately the same speed and output on either a direct- or single-phase alternating current of a frequency not greater than 60 Hz and of approximately the same rms voltage. (IA/MT) 45-1998

universal-motor parts (rotating machinery) A term applied to a set of parts of a universal motor. Rotor shaft, conventional stator frame (or shell), end shields, or bearings may not be included, depending on the requirements of the end product into which the universal-motor parts are to be assembled. *See also:* asynchronous machine. (PE) [9]

universal-numbering plan (telephone switching systems) A numbering plan employing nonconflicting codes so arranged that all main stations can be reached from any point within a telecommunications system. (COM) 312-1977w

universal or arcshear machine A power-driven cutter that will not only cut horizontal kerfs, but will also cut vertical kerfs or at any angle, and is designed for operation either on track, caterpillar treads, or rubber tires. (EEC/PE) [119]

universal product code (UPC) A bar code appearing on many retail products to uniquely identify the product. The code is designed to be read by an optical scanner attached to an electronic cash register. (See corresponding figure.)

universal product code

(SWG/C/PE) C37.20.1-1987s

universal receiver/transmitter A circuit used in data communication applications to provide the necessary logic to recover data in a serial-in/parallel-out fashion and to transmit data in a parallel-in/serial-out fashion. (C) 610.7-1995

universal stick A stick, or type of insulating tool, with an end to which universal tools can be attached. (T&D/PE) 516-1995

universal synchronous receiver/transmitter A universal receiver/transmitter that is used in synchronous communication applications. (C) 610.7-1995

universal tool An accessory designed to attach to a universal stick allowing one insulated stick to be used to perform many different operations. (T&D/PE) 516-1995

universal Turing machine A Turing machine that can simulate any other Turing machine. (C) [20], [85]

unloaded applicator impedance (dielectric heating) The complex impedance measured at the point of application, without the load material in position, at a specified frequency. (IA) 54-1955w

unloaded delay The conceptual delay value for a delay arc of a cell when the output pin is unloaded (unconnected) and the signal at the input pin conforms to some ideal waveform. (C/DA) 1481-1999

unloaded labor rate Variable costs of labor per hour, excluding all forms of benefits such as vacations, medical insurance, retirement, etc. (SCC22) 1346-1998

unloaded *Q* (switching tubes) (intrinsic) The Q of a tube unloaded by either the generator or the termination. *Note:* As here used, Q is equal to 2ν times the energy stored at the resonance frequency divided by the energy dissipated per cycle in the tube or, for cell-type tubes, in the tube and its external resonant circuit. *See also:* gas tube. (ED) 161-1971w

unloaded sag (conductor or any point in a span) The distance measured vertically from the particular point in the conductor to a straight line between its two points of support, without any external load. (IA/APP) [90]

unloaded tension (A) (initial) The longitudinal tension in a conductor prior to the application of any external load. **(B)** (final) The longitudinal tension in a conductor after it has been subjected for an appreciable period to the loading prescribed for the loading district in which it is situated, or equivalent loading, and the loading removed. Final unloaded tension includes the effect of inelastic deformation (creep). (NESC/T&D) C2-1984, C2.2-1960

unloading amplifier An amplifier that is capable of reproducing or amplifying a given voltage signal while drawing negligible current from the voltage source. *Note:* The term buffer amplifier is sometimes used as a synonym for unloading amplifier, in an incorrect sense, since a buffer amplifier draws significant current, but at a constant load impedance (seen at the input). (C) 610.10-1994w, 165-1977w

unloading circuit (1) (analog computer) In an analog computer, a computing element or combination of computing elements capable of reproducing or amplifying a given voltage signal while drawing negligible current from the voltage source, thus eliminating any possible loading errors. *See also:* unloading amplifier. (C) 165-1977w

(2) In an analog computer, a circuit that is capable of reproducing or amplifying a given voltage signal while drawing negligible current from the voltage source, thus eliminating possible load errors. (C) 610.10-1994w

unloading point (electric transmission system used on self-propelled electric locomotives or cars) The speed above or below which the design characteristics of the generators and traction motors or the external control system, or both, limit the loading of the prime mover to less than its full capacity. *Note:* The unloading point is not always a sharply defined point, in which case the unloading point may be taken as the useful point at which essentially full load is provided. *See also:* traction motor. (EEC/PE) [119]

unmodulated groove (mechanical recording) A groove made in the medium with no signal applied to the cutter. *Synonym:* blank groove. *See also:* phonograph pickup. (SP) [32]

unnormalized form The form assumed by data that have not been normalized. *Contrast:* normalized form. (C) 610.5-1990w

unnormalized relation A relation that is not in normal form. *Contrast:* normalized relation. (C) 610.5-1990w

unnumbered (U) format The format used to provide additional data link control functions and unnumbered information transfer. This format shall contain no sequence numbers, but shall include a P/F bit that may be set to "1" or "0." (EMB/MIB) 1073.3.1-1994

unodecimal (A) Pertaining to a selection in which there are 11 possible outcomes. **(B)** Pertaining to the numeration system with a radix of 11. (C) 1084-1986

unordered access (communication satellite) A system in which access to a radio frequency channel is gained without determining channel availability. This method is useful in common spectrum or random access discrete address systems. (COM) [19]

unordered list A list in which data items are not arranged in any specific order. *Synonym:* random-ordered list. *Contrast:* ordered list. (C) 610.5-1990w

unordered tree A tree in which the left-to-right order of the subtrees of a given node is not significant. *Contrast:* ordered tree. (C) 610.5-1990w

unpack (1) To separate various sections of packed data. (C) [20], [85]

(2) (data management) (software) To recover the original form of one or more data items from packed form. *Contrast:* pack. (C) 610.5-1990w, 610.12-1990

unpacked decimal data *See:* zoned decimal data.

unplanned derated hours (electric generating unit reliability, availability, and productivity) The available hours during which an unplanned derating was in effect. (PE/PSE) 762-1987w

unplanned derating (electric generating unit reliability, availability, and productivity) That portion of the unit derating that is not a planned derating. Unplanned derating events are classified according to the urgency with which the derating needs to be initiated. Class 1 (immediate). A derating that requires an immediate action for the reduction of capacity. Class 2 (delayed). A derating that does not require an immediate reduction of capacity, but requires a reduction of capacity within 6 h (hours). Class 3 (postponed). A derating that can be postponed beyond 6 h, but requires a reduction of capacity before the end of the next weekend. Class 4 (deferred). A derating that can be deferred beyond the end of the next weekend, but requires a reduction of capacity before the next planned outage. (PE/PSE) 762-1987w

unplanned outage (electric generating unit reliability, availability, and productivity) The state in which a unit is unavailable but is not in the planned outage state. *Notes:* 1. When an unplanned outage is initiated, the outage is classified according to one of five classes, as defined in Class 0 unplanned outage, Class 1 unplanned outage, Class 2 unplanned outage, Class 3 unplanned outage, and Class 4 unplanned outage. Unplanned outage Class 0 applies to a start-up failure and Class 1 applies to a condition requiring immediate outage. Also, unplanned outage starts when planned outage ends but is extended due to unplanned work. Classes 2, 3, and 4 apply to outages where some delay is possible in time of removal of the unit from service. The class (2, 3, or 4) of outage is to be determined by the amount of delay that can be exercised in the time of removal of the unit. The class of outage is not made more urgent if the time of removal is advanced due to favorable conditions of system reserves or availability of replacement capacity for the predicted duration of the outage. However, outage starts when the unit is removed from service or is declared unavailable when it is not in service. 2. During the time the unit is in the unplanned outage state, the outage class is determined by the outage class that initiates the state. 3. In some cases, the opportunity exists during unplanned outages to perform some of the repairs or maintenance that would have been performed during the next planned outage. If the additional work extends the outage beyond that required for the unplanned outage, the remaining outage should be reported as a planned outage. 4. Unlike planned outages, unplanned outages do not have a fixed duration that can be estimated each year. *See also:* Class 1 unplanned outage; Class 0 unplanned outage; Class 3 unplanned outage; Class 2 unplanned outage; Class 4 unplanned outage. (PE/PSE) 762-1987w

unplanned outage hours (electric generating unit reliability, availability, and productivity) The number of hours a unit was in a Class 0, 1, 2, 3, or 4 unplanned outage state. (PE/PSE) 762-1987w

unpolarized *See:* randomly polarized.

unpowered retention time The retention time with an terminals of the memory at ground voltage, except for an occasional read cycle to test the device state. (ED) 641-1987w

unprecedented system A system for which design examples do not exist so that the design architecture alternatives are unconstrained by previous system descriptions. (C/SE) 1220-1998

unpredictable Describes an aspect of the architecture that in nondeterministic. This term is used only to describe the targets of branches and the occurrence of certain traps in unusual situations. *See also:* undefined. (C/MM) 1754-1994

unpropagated potential (electrobiology) An evoked transient localized potential not necessarily associated with changed excitability. *See also:* excitability. (EEC/PE) [119]

unprotected field On a display device, a field in which a user can enter, modify or erase data. *Contrast:* protected field. (C) 610.10-1994w

unqualified climber A worker that does not meet the requirements of a qualified climber. (T&D/PE) 1307-1996

unquenched sample (1) (liquid-scintillation counters) A counting sample (material of interest plus liquid-scintillation solution) that contains a minimum of colored species and chemical impurities that would reduce the photon output from the vial. (NI) N42.16-1986

(2) (liquid-scintillation counting) A counting sample (material of interest plus liquid-scintillation solution) that contains a minimum of colored species and chemical impurities that would reduce the light output to the photomultiplier tubes. (NI) N42.15-1990

unrecoverable light loss factors (illuminating engineering) Factors which give the fractional light loss that cannot be recovered by cleaning or lamp replacement. (EEC/IE) [126]

unregulated voltage (electronically regulated power supply) The voltage at the output of the rectifier filter. *See also:* regulated power supply. 209-1950w

unrepresented channel A channel that is not represented by a Public Transducer. (IM/ST) 1451.1-1999

unsafe Having unacceptable risk of the occurrence of a hazard. (VT/RT) 1483-2000

unsaturated standard cell A cell in which the electrolyte is a solution of cadmium sulphate at less than saturation at ordinary temperatures. (This is the commercial type of cadmium standard cell commonly used in the United States). *See also:* electrochemistry. (EEC/PE) [119]

unscheduled interrupt (UI) An interrupt caused by the occurrence of an event within the computer that is not associated with normal functional operation. (C) 610.10-1994w

unselected slave A slave that does not recognize its address on the bus lines during the connection phase of a bus transaction. (C/BA) 10857-1994, 896.4-1993w

unsharp masking In image processing, a sharpening technique in which an intentionally blurred version of the image is subtracted from the image itself. (C) 610.4-1990w

unshielded strip transmission line A strip conductor above a single ground plane. Some common designations are, microstrip (flat-strip conductor), unbalanced strip line. *See also:* strip-type transmission line; shielded strip transmission line; waveguide. (AP/ANT) [35]

unshielded twisted pair (UTP) **(1)** A twisted pair medium consisting of only a pair of conductors exposed to outside electrical interferences and noise. *Contrast:* shielded twisted pair. (C) 610.7-1995

(2) Normally refers to those cables with individual pairs of conductors twisted, or with a group of four conductors in a star quad configuration, with any characteristic impedance. When used in this document, the term specifically refers to those cables whose pairs have a high-frequency characteristic impedance of 100 Ω. Shielded cables with the same high-frequency characteristic impedance are included within this definition. (C/LM) 8802-5-1998

unshielded twisted-pair cable (UTP) An electrically conducting cable, comprising one or more pairs, none of which is shielded. There may be an overall shield, in which case the cable is referred to as unshielded twisted-pair with overall shield. (C/LM) 802.3-1998

unsigned byte A byte that represents positive integers in the decimal range 0–255. (C/MM) 1284.1-1997

unsigned dword A dword that represents positive integers in the decimal range 0–4 294 967 295. (C/MM) 1284.1-1997

unsigned word A word that represents positive integers in the decimal range 0–65 535. (C/MM) 1284.1-1997

unsigned packed decimal data Integer data in which each decimal digit is represented in binary, occupying four bits. *Note:* Since no sign is stored, only non-negative integers can be represented.

decimal 75_{10}

unsigned packed decimal $0111\ 0101_2$

See also: packed decimal data. (C) 610.5-1990w

unsolicited status Information generated by the peripheral that has not been asked for by the host, yet is important enough that the peripheral desires to send it to the host. (C/MM) 1284-1994

unspecialize A change by an instance from being an instance of its current subclass within a cluster to being an instance of none of the subclasses in the cluster. *Contrast:* respecialize; specialize. (C/SE) 1320.2-1998

unspecified **(1)** A value or behavior is unspecified if the standard imposes no portability requirements on applications for a correct program construct or correct data. Implementations (or other standards) may specify the result of using that value or causing that behavior. An application requiring a specific behavior, rather than tolerating any behavior when using that functionality, is using extensions. (C/PA) 1003.1-1988s

(2) An indication that this standard imposes no portability requirements on applications for correct program constructs or correct data regarding a value or behavior. Implementations (or other standards) may specify the result of using that value or causing that behavior. An application requiring a specific behavior, rather than tolerating any behavior when using that functionality, is using extensions. (PA/C) 1238.1-1994w, 2003.2-1996, 1326.2-1993w

unspecified behavior Behavior (for a correct program construct and correct data) that depends on the implementation. The implementation is not required to document which behavior occurs. Usually the range of possible behaviors is delineated by the standard. (C/DA) 1481-1999

unstable **(1) (control system feedback)** Not possessing stability. *See also:* feedback control system. (IM/PE/EDPG) [120], [3]

(2) Pertaining to circuit or device in which the circuit will remain for a limited time, after which the circuit will change to another state without any external stimulus. *Note:* Often used to describe an undesirable or unexpected circuit behavior. *Contrast:* stable. (C) 610.10-1994w

unstable limit cycle One from which state trajectories recede for all initial states sufficiently close. (CS/PE/EDPG) [3]

unstratified language A language that can be used as its own metalanguage; for example English, German. *Contrast:* stratified language. *See also:* natural language. (C) 610.13-1993w, 610.12-1990

unstructured A postal O/R address that specifies the postal address of a user in a single attribute. Its structure is left largely unspecified. (C/PA) 1224.1-1993w

unsuccessful test A completed test that is invoked by a write to the TEST_START register and detects one or more errors. (C/MM) 1212-1991s

unsupported command A word-serial protocol error that occurs when a servant receives a command that it does not support. (C/MM) 1155-1992

unsupported transaction A transaction whose returned data value or side effects are not defined by the hardware architecture that is addressed. For example, a write64 transactions to the 4-byte STATE_CLEAR register is unsupported. (C/MM) 1212-1991s

untagged frame An *untagged frame* is a frame that does not contain a tag header immediately following the Source MAC Address field of the frame or, if the frame contained a Routing Information field, immediately following the Routing Information field. *See also:* tagged frame. (C/LM) 802.1Q-1998

UNTIL A single-entry, single-exit loop, in which the loop control is executed after the loop body. (See the corresponding figure.) *Contrast:* WHILE; closed loop. *See also:* trailing decision.

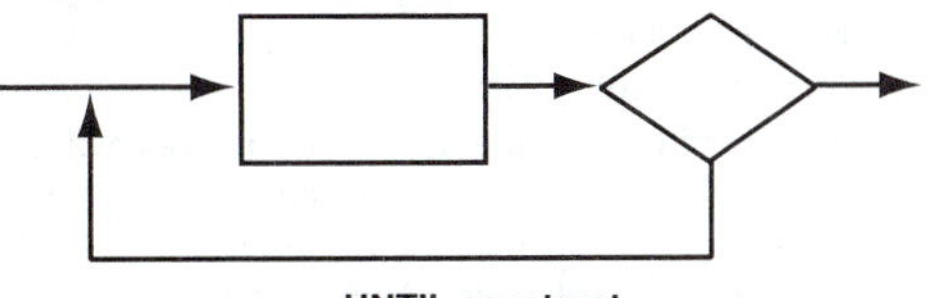

UNTIL construct

(C) 610.12-1990

untransposed Refers to the physical positions of the phase conductors of a transmission line, which are not interchanged periodically to balance the mutual impedances between phases. (PE/PSR) C37.113-1999

unused Used to describe an instruction field or register field that is not currently defined by the architecture. When read by software, the value of an unused register field is undefined. However, since an unused field could be defined by a future version of the architecture, an unused field should only be written to zero by software. *See also:* reserved; ignored. (C/MM) 1754-1994

unwind In programming, to state explicitly and in full all of the instructions involved in multiple executions of a loop. *See also:* straight-line coding. (C) 610.12-1990

unsupported length Unbraced length of a column. (PE/T&D) 751-1990

unusual service conditions Environmental conditions that may affect the constructional or operational requirements of a machine. This includes the presence of moisture and abrasive, corrosive, or explosive atmosphere. It also includes external structures that limit ventilation, unusual conditions relating to the electrical supply, the mechanical loading, and the position of the machine. (PE) [9]

unwanted radiation (radiation protection) Any ionizing radiation other than that which the instrument is designed to measure. (NI) N323-1978r

unwanted signal A signal that may impair the measurement or reception of a wanted signal. (T&D/PE) 539-1990

UP *See:* Unformatted Page.

up (1) Pertaining to a system or component that is operational and in service. Such a system is either busy or idle. *Contrast:* down. *See also:* idle; busy. (C) 610.12-1990
(2) A colloquial expression used in reference to a system or system component that is functioning and ready to use. *Contrast:* down. (C) 610.10-1994w

UPC *See:* universal product code.

updateable argument The designation given to an operation argument that identifies an instance to which a request may be sent that will change the state of the instance. An argument not designated as "updatable" means that there will be no requests sent that will change the state of the instance identified by the argument. (C/SE) 1320.2-1998

update (1) (supervisory control, data acquisition, and automatic control) The process of modifying or reestablishing data with more recent information.
(SWG/PE/SUB) C37.1-1987s, C37.100-1992
(2) (A) (data management) To change information in accordance with information that is more recent than that which was available previously. For example, a master file containing account balances might be updated nightly to reflect transactions precessed the previous day. **(B) (data management)** To replace data in a storage device or on a data medium. *See also:* read; delete; write. (C) 610.5-1990
(3) Installing a newer version of software than one that is currently installed, into the same location. This is also referred to as upgrading. (C/PA) 1387.2-1995

updateable microfilm Microfilm that permits the addition or deletion of images. (C) 610.2-1987

update access A type of access to data in which the data can be updated. *See also:* read/write access; write access; read-only access; delete access. (C) 610.5-1990w

update transaction A transaction that modifies a master file by adding, deleting, or changing data to make it more current. *See also:* null transaction; change transaction; add transaction; delete transaction. (C) 610.2-1987

uplift plates *See:* bearing plates.

uplift roller (conductor stringing equipment) A small single-grooved wheel designed to fit in or immediately above the throat of the traveler and keep the pulling line in the traveler groove when uplift occurs due to stringing tensions.
(T&D/PE) 524-1992r

up link (communication satellite) A ground to satellite link, very often the command link. (COM) [24]

uplink (local area networks) The transmission medium between an end node or repeater and a connected higher-level repeater, as viewed from the local entity. *Contrast:* downlink.
(C) 8802-12-1998

upload (A) To transfer some collection of data from some storage location to a computer memory. **(B)** To transfer some collection of data from the memory of a small computer to the memory of a relatively larger computer; for example, to transfer data from a microcomputer to a mainframe computer.
(C) 610.5-1990

upper (driving) beams (illuminating engineering) One or more beams intended for distant illumination and for use on the open highway when not meeting other vehicles. Formerly "country beam." (EEC/IE) [126]

upper bracket (rotating machinery) A bearing bracket mounted above the core of a vertical machine. (PE) [9]

upper burst reference (audio and electroacoustics) A selected multiple of the long-time average magnitude of the quantity mentioned in the definition of burst. *See also:* burst duration; burst. (SP) 257-1964w, [32]

upper coil support (rotating machinery) A coil support to restrain field-coil motion in the direction toward the air gap. *See also:* stator; rotor. (PE) [9]

upper frequency limit (coaxial transmission line) The limit determined by the cutoff frequency of higher-order waveguide modes of propagation, and the effect that they have on the impedance and transmission characteristics of the normal TEM coaxial-transmission-line mode. The lowest cutoff frequency occurs with the TE1,1 mode, and this cutoff frequency in air dielectric line is the upper frequency limit of a practical transmission line. How closely the TE1,1 mode cutoff frequency can be approached depends on the application. *See also:* waveguide. (EEC/REWS) [92]

upper guide bearing (rotating machinery) A guide bearing mounted above the core of a vertical machine. *See also:* bearing. (PE) [9]

upper half bearing bracket (rotating machinery) The top half of a bracket that can be separated into halves for mounting or removal without access to a shaft end. *See also:* bearing.
(PE) [9]

upper limit (test, measurement, and diagnostic equipment) The maximum acceptable value of the characteristic being measured. (MIL) [2]

upper range-value The highest quantity that a device is adjusted to measure. *Note:* The following compound terms are used with suitable modifications in the units: measured variable upper range-value, measured signal upper range-value, etc. *See also:* instrument. (EEC/EMI) [112]

upper sideband (data transmission) The higher of two frequencies or groups of frequencies produced by a modulation process. (PE) 599-1985w

upper-sideband parametric down-converter A noninverting parametric device used as a parametric downconverter. *See also:* parametric device. (ED) 254-1963w, [46]

upper-sideband parametric up-converter A noninverting parametric device used as a parametric up-converter. *See also:* parametric device. (ED) [46]

UPS *See:* uninterruptible power supply.

upset Malfunction of a system because of electrical disturbances. (SPD/PE) C62.62-2000

upset duplex system A direct-current telegraph system in which a station between any two duplex equipments may transmit signals by opening and closing the line circuit, thereby causing the signals to be received by upsetting the duplex balance. *See also:* telegraphy. (EEC/PE) [119]

UPS module *See:* uninterruptible power supply module.

upstream The direction along a bus that is towards the head of bus function. This is opposite to the direction of data flow along a bus. (LM/C) 8802-6-1994

upstream neighbor's address (UNA) The address of the station functioning upstream from a specific station.
(C/LM) 8802-5-1998

up time (1) (supervisory control, data acquisition, and automatic control) The time during which a device or system is capable of meeting performance requirements.
(SWG/PE/SUB) C37.100-1992, C37.1-1994
(2) (availability) The period of time during which an item is in a condition to perform its required function. (R) [29]
(3) The period of time during which a system or component is operational and in service; that is, the sum of busy time and idle time. *Contrast:* down time. *See also:* idle time; busy time; mean time between failures; setup time. (C) 610.12-1990

upward compatible (1) Pertaining to hardware or software that is compatible with a later or more complex version of itself; for example, a program that handles files created by a later version of itself. *Contrast:* downward compatible. (C) 610.12-1990

(2) Pertaining to hardware or software that is compatible with a later or more complex version of itself; for example, a new version of a program that handles files created by an earlier version of that program is said to be "upwardly compatible." *Contrast:* downward compatible. (C) 610.10-1994w

upward component (illuminating engineering) That portion of the luminous flux from a luminaire which is emitted at angles above the horizontal. (EEC/IE) [126]

upward compression In software design, a form of demodularization in which a subordinate module is copied in-line into the body of a superordinate module. *Contrast:* downward compression; lateral compression. (C) 610.12-1990

urban districts Thickly settled areas (whether in cities or suburbs) or where congested traffic often occurs. A highway, even though in thinly settled areas, on which the traffic is often very heavy, is considered as urban. (NESC) C2-1997

usability (1) The ease with which a user can learn to operate, prepare inputs for, and interpret outputs of a system or component. (C) 610.12-1990

(2) A measure of an executable software unit's or system's functionality, ease of use, and efficiency. *See also:* reusability. (C/SE) 1517-1999

usable levels (storage tubes) The output levels, each related to a different input, that can be distinguished from one another regardless of location on the storage surface. *Note:* The number of usable levels is normally limited by shading and disturbance. *See also:* storage tube. (ED) 158-1962w

usable sensitivity The minimum standard modulated carrier-signal power required to produce usable receiver output. (VT) [37]

usage attributes The classification of software as delivered to a user of the final product, or as nondelivered when created only to support the development process. (C/SE) 1045-1992

usage count *See:* traffic usage count.

usage mode Primary manner in which the document issuer expects that document to be used. IEEE Std 1063-1987 recognizes two usage modes, instructional and reference. (C/SE) 1063-1987r

USASCII* *See:* USA Standard Code for Information Interchange.

* Deprecated.

USA Standard Code for Information Interchange* (USASCII) (nuclear power quality assurance) A disposition permitted for a nonconforming item when it can be established that the item is satisfactory for its intended use. (PE/NP) [124]

* Deprecated.

useful active dimension (charged-particle detectors) (of a position-sensitive detector) A dimension (length, width) of that region of a position-sensitive detector over which the specifications of resolution and linearity are met. (NPS) 300-1988r

useful energy range The set or range of continuous energies for a specific type of radiation in which the instrument meets specified criteria. (NI) N42.17B-1989r

useful life (1) The period from a stated time, during which, under stated conditions, an item has an acceptable failure rate, or until an unrepairable failure occurs." (R) [29]

(2) (nuclear power generating station) The time to failure for a specific service condition. (PE/NP) 380-1975w

useful line *See:* available line.

useful output power That part of the output power that flows into the load proper. (ED) [45]

useful service life (thermal classification of electric equipment and electrical insulation) The length of time (usually in hours) for which an insulating material, insulation system, or electric equipment performs in an adequate or specified fashion. (EI) 1-1986r

user (1) (radix-independent floating-point arithmetic) (binary floating-point arithmetic) Any person, hardware, or program not itself specified by IEEE Std 754-1985 or IEEE Std 854-1987 or both, having access to and controlling those operations of the programming environment specified in these standards. (MM/C) 854-1987r, 754-1985r

(2) One who uses the services of a computer system. (C) 610.2-1987, 610.10-1994w

(3) (broadband local area networks) An individual whose principal concern is the transfer of information through the system, and to whom the system is transparent. The user is assumed to be in possession of a device that is capable of one- or two-way communication through the system. (LM/C) 802.7-1989r

(4) (software user documentation) Person who uses software to perform a task. (C/SE) 1063-1987r

(5) (repair and rewinding of motors) The owner of the motor or an authorized agent of the owner. (IA/PC) 1068-1996

(6) The independent party that may be a purchaser of utility electric power or a producer of electric energy for sale to an electric utility, or both. (SUB/PE) 1109-1990w

(7) (power systems) The owner of the transformer. (PE/TR) C57.117-1986r

(8) The ultimate human interface or top-most application program. For example, text typed into a terminal interface can be referred to as "user data." This document does not use the ISO Open Systems "layer-user" concept, in which each module in a vertical stack is the "user" of the adjacent, lower module. (MM/C) 1212.1-1993

(9) The entity or person that accesses the Directory, used in this standard to refer to the application program that is using the interface. (C/PA) 1328.2-1993w, 1224.2-1993w, 1327.2-1993w, 1326.2-1993w

(10) Those who use the CASE tools. They are not necessarily those who will execute this evaluation and selection process. (C/SE) 1209-1992w

(11) A person, system, process, or tool that generates the VHDL source code that a synthesis tool processes. (C/DA) 1076.3-1997

(12) The person, or persons, who operate or interact directly with the product. The user(s) and the customer(s) are often not the same person(s). (C/SE) 830-1998

(13) (A) An individual or organization who uses a software-intensive system in their daily work activities or recreational pursuits. **(B)** The person (or persons) who operates or interacts directly with a software-intensive system. (C/SE) 1362-1998

(14) The person or persons operating or interacting directly with the system. (C/SE) 1219-1998

(15) The source of the business drivers that the User Organization OSE Profile must address and support. (C/PA) 1003.23-1998

(16) A person, system, process, or tool that generates the VHDL source code that a synthesis tool. (C/DA) 1076.6-1999

user application program *See:* application program.

user certificate The public keys of a user, together with some other information, rendered unforgeable by encipherment with the secret key of the certification authority that issued it. *Synonym:* certificate. (C/PA) 1328.2-1993w, 1327.2-1993w, 1224.2-1993w, 1326.2-1993w

USERCODE *See:* user identity code.

user command A command that may be issued by the user. For example, "sort" or "print." (C) 610.2-1987

user-definable key (1) A key on a computer keyboard that initiates operations or functions that have been defined by the user. *Synonyms:* user-programmable key. (C) 610.2-1987

(2) A function key on a keyboard that initiates operations or functions that have been defined by the user by programming the terminal or keyboard. *Synonym:* programmable function key; user-programmable key. (C) 610.10-1994w

user defined These 12 records can be used to contain any text data that needs to go with the spectrum. (NPS/NID) 1214-1992r

user-defined data type A non-standard data type determined to meet the needs of a particular user or to solve a particular problem. (C) 610.5-1990w

user-defined date systems Some users may implement locally developed date systems to track user-specific events. For example, manufacturing, production, or just-in-time (JIT) dates. Methodologies developed for verification of Year 2000 issues associated with the Gregorian calendar may need to be adapted to apply to user-defined date systems. (C/PA) 2000.2-1999

user documentation (software) Documentation describing the way in which a system or component is to be used to obtain desired results. *See also:* data input sheet; user manual. (C) 610.12-1990

user-driven computing *See:* end user computing.

user friendly Pertaining to a computer system, device, program, or document designed with ease of use as a primary objective. *Synonym:* user-oriented. (C/C) 610.2-1987, 610.12-1990

user group An organization of users of a particular class of computer systems, designed to allow the users to share knowledge about and programs for those systems and to formulate feedback for the systems' manufacturers. *Synonym:* user's group. (C) 610.2-1987

user guide *See:* user manual.

user hotline Telephone access to a specialist who provides users with answers to questions concerning some product, system, or application. (C) 610.2-1987

user ID (1) A nonnegative integer, which can be contained in an object of type *uid_t,* that is used to identify a system user. When the identity of a user is associated with a process, a user ID value is referred to as a real user ID, an effective user ID, or an (optional) saved set-user-ID. (C/PA) 9945-1-1996, 9945-2-1993
(2) A value identifying a system user. A User ID is a value of the type `User_ID` defined in the package `POSIX_Process_-Identification`. When the identity of a user is associated with a process, a user ID value is referred to as a real user ID, an effective user ID, or an (optional) saved set-user-ID. (C) 1003.5-1999

user identity code A defined instruction for the test logic defined by 1149.1-1990. (TT/C) 1149.1-1990

user interaction Communication between a computer system and a user in which each user entry causes a response from the system. (C) 610.2-1987

user interface (1) The portion of a firmware system that process commands entered by a human. (The user interface defined by this standard consists of a Forth command interpreter plus a set of Forth words for interactively performing various Open Firmware functions. In its fully elaborated form, the Open Firmware user interface gives interactive access to all Firmware capabilities.) (C/BA) 1275-1994
(2) A physical interface between the operator and the system equipment. (SUB/PE) C37.1-1994
(3) An interface that enables information to be passed a human user and the hardware or software components of a computer system. *Synonym:* human interface. *See also:* character-based user interface. (C) 610.10-1994w, 610.12-1990
(4) The part of the application that permits the user and application to communicate with each other to perform certain tasks. (SCC20) 1226-1998

user manual A document that presents the information necessary to employ a system or component to obtain desired results. Typically described are system or component capabilities, limitations, options, permitted inputs, expected outputs, possible error messages, and special instructions. *Note:* A user manual is distinguished from an operator manual when a distinction is made between those who operate a computer system (mounting tapes, etc.) and those who use the system for its intended purpose. *Synonym:* user guide. *See also:* user manual; operator manual; data input sheet; programmer manual; installation manual; support manual; diagnostic manual. (C) 610.12-1990

user mode A processor state that is active when the S bit of the PSR is not set (when PSR.S = 0). (C/MM) 1754-1994

user name (1) A string that is used to identify a user. (PA/C) 9945-1-1996, 9945-2-1993
(2) A value of `POSIX_String` that is used to identify a user. (C) 1003.5-1999

user need (1) A user requirement for a system that a user believes would solve a problem experienced by the user. (C/SE) 1362-1998
(2) The user's set of qualitative and quantitative requirements in a particular problem domain. (C/SE) 1209-1992w

user-oriented *See:* user friendly.

user outlet port (A) (broadband local area networks) A connection port, located at a user location, that provides user access to the cable system. **(B) (broadband local area networks)** A broadband attachment location that provides connection access to the broadband coaxial cable system. (LM/C) 802.7-1989

user program (mathematics of computing) (data management) A computer program written specifically for or by a particular user. (C) 610.2-1987, 610.5-1990w

user-programmable computer A computer that can be programmed by the user. *Contrast:* fixed-instruction computer. *See also:* microprogrammable computer. (C) 610.10-1994w

user-programmable key *See:* user-definable key.

user's group *See:* user group.

user stack The protocols residing above SDE that request services from any SDE SAP except those supported by the use of a bootstrap SAID. (C/LM) 802.10-1998

user state *See:* problem state.

user terminal An input-output device by which a user communicates with a computer. (C) 610.10-1994w

user-user protocol A protocol that is adopted between two or more users to ensure communication between them. (C) 610.7-1995

user working area (data management) A work area used by a database management system to load and unload data in response to a call by some application program for data. *Synonym:* workspace. (C) 610.5-1990w

USRT *See:* universal synchronous receiver/transmitter.

usual service conditions Environmental conditions in which standard machines are designed to operate. The temperature of the cooling medium does not exceed 40 degrees Celsius and the altitude does not exceed 3300 feet. (PE) [9]

UT *See:* unit transformer.

UTP (1) (local area networks) A 100 Ω unshielded balanced cable meeting the specifications in ISO/IEC 11801:1995. (C) 8802-12-1998
(2) *See also:* unshielded twisted pair. (C) 610.7-1995

utilance *See:* room utilization factor.

utility (1) (software) A software tool designed to perform some frequently used support function. For example, a program to copy magnetic tapes. (C) 610.12-1990
(2) A program that can be called by name from a shell to perform a specific task or a related set of tasks. This program shall either be an executable file, such as might be produced by a compiler/linker system from computer source code, or a file of shell source code, directly interpreted by the shell. The program may have been produced by the user, provided by the implementor of this standard, or acquired from an independent distributor. The term *utility* does not apply to the special built-in utilities provided as part of the shell command language. The system may implement certain utilities as shell

functions or built-ins, but only an application that is aware of the command search order described or of performance characteristics can discern differences between the behavior of such a function or built-in and that of a true executable file. (C/PA) 9945-2-1993

(3) An organization responsible for the installation, operation, or maintenance of electric supply or communications systems. (NESC) C2-1997

(4) A provider of electricity, gas, water, telecommunications, or related services to a community. (AMR/SCC31) 1377-1997

utility bus This bus includes signals that provide periodic timing and coordinate the power-up and power-down of the systems. It is one of the four buses provided by the backplane. (C/BA) 1014-1987

utility controller (UC) A controller resident on a utility/enhanced service provider (ESP) premises, that connects, via the telephone network, to the telemetry interface unit (TIU) (using the direct dial network access method), to the central office service unit (COSU) (using the COSU network access method or the no-test trunk network access method). (AMR/SCC31) 1390-1995, 1390.2-1999, 1390.3-1999

utility interactive system An electric power production system that is operating in parallel with and capable of delivering energy to a utility electric supply system. (NESC) C2-1997

utility interconnection point Point of interconnection between the utility-owned equipment (conductors) and that of the SPC owner. This is usually the metering location. (DESG) 1035-1989w

utility-interface disconnect switch A switch that may be required at the interface between the photovoltaic (PV) system and the utility system. This terminology is used to distinguish this switch from others that may be installed for other reasons, such as to satisfy requirements of the National Electrical Code®. (SCC21) 929-2000

utility power *See:* commercial power.

utility routine *See:* service routine.

utility signals A set of discrete lines in the backplane that provide communications among modules for which a bus is not adequate because of latency or other reasons. (C/BA) 14536-1995

utility software Computer programs or routines designed to perform some general support function required by other application software, by the operating system, or by system users. *See also:* computer program; operating system; routine; application software; function; system. (C/SE) 729-1983s

utility telemetry trunk (UTT) A two-way telephone company facility connecting the central office service unit (COSU) to the switch. This facility allows a utility or enhanced service provider (ESP), via the telephone network (COSU), to automatically invoke/ignore certain telephone network capabilities as well as provide suppressed or abbreviated ringing access to a telemetry interface unit(s) [TIU(s)] on an end user's line. The TIU may also originate calls, through the telephone network (COSU and switch), which will automatically invoke/ignore certain telephone network capabilities and provide a connection to the utility or ESP. (AMR/SCC31) 1390-1995, 1390.2-1999, 1390.3-1999

utilization (software) In computer performance evaluation, a ratio representing the amount of time a system or component is busy divided by the time it is available. *See also:* busy time; idle time. (C) 610.12-1990

utilization equipment (1) Equipment that utilizes electric energy for mechanical, chemical, heating, lighting, or similar purposes. (NESC/NEC) [86]

(2) (electric power systems in commercial buildings) Electrical equipment that converts electric power into some other form of energy, such as light, heat, or mechanical motion. (IA/PSE) 241-1990r

(3) Equipment, devices, and connected wiring that utilize electric energy for mechanical, chemical, heating, lighting, testing, or similar purposes and are not a part of supply equipment, supply lines, or communication lines. (NESC) C2-1997

utilization factor (system utilization factor) The ratio of the maximum demand of a system to the rated capacity of the system. *Note:* The utilization factor of a part of the system may be similarly defined as the ratio of the maximum demand of the part of the system to the rated capacity of the part of the system under consideration. *See also:* direct-current distribution; alternating-current distribution. (T&D/PE) [10]

utilization time (A) (medical electronics) (hauptnutzzeit) The minimum duration that a stimulus of rheobasic strength must have to be just effective. **(B) (medical electronics)** (hauptnutzzeit) The shortest latent period between stimulus and response obtainable by very strong stimuli. **(C) (medical electronics)** (hauptnutzzeit) The latent period following application of a shock of theobasic intensity. (EMB) [47]

utilization voltage (1) (system voltage ratings) The root-mean-square phase-to-phase or phase-to-neutral voltage at the line terminals of utilization equipment. *See also:* high voltage; medium voltage; low voltage; maximum system voltage; service voltage; nominal system voltage; system voltage. (IA/SPD/PE/APP) [80], C62.62-2000

(2) (elecrtic power systems in commercial buildings) The voltage at the line terminals of utilization equipment. (IA/PSE) 241-1990r

UTT *See:* utility telemetry trunk.

UTT-oriented language *See:* unit-under-test-oriented language.

UUID *See:* universally unique identifier.

UUT *See:* unit under test.

uvh *See:* ultra-high voltage.

V

vacant code (telephone switching systems) A digit or a combination of digits that is unassigned. (COM) 312-1977w

vacant-code tone (telephone switching systems) A tone that indicates that an unassigned code has been dialed. (COM) 312-1977w

vacant number (telephone switching systems) An unassigned or unequipped directory number. (COM) 312-1977w

vacuum column In a tape drive, a cavity in which a low air pressure is maintained so as to attract a tape loop between the spool and the driving mechanism. (C) 610.10-1994w

vacuum envelope (electron tube) The airtight envelope that contains the electrodes. *See also:* electrode. (EEC/PE) [119]

vacuum-tube amplifier An amplifier employing electron tubes to effect the control of power from the local source. *See also:* amplifier. (AP/ANT) 145-1983s

vacuum-tube radio frequency generator *See:* radio-frequency generator, electron tube type.

vacuum-tube transmitter A radio transmitter in which electron tubes are utilized to convert the applied electric power into radio-frequency power. *See also:* radio transmitter. (AP/ANT) 145-1983s

vacuum valve A device for sealing and unsealing the passage between two parts of an evacuated system. *See also:* rectification. (EEC/PE) [119]

valance (illuminating engineering) A longitudinal shielding member mounted across the top of a window or along a wall, to conceal light sources, giving both upward and downward distributions. (EEC/IE) [126]

valance lighting (illuminating engineering) Lighting comprising light sources shielded by a panel parallel to the wall at the top of a window. (EEC/IE) [126]

valence band The range of energy states in the spectrum of a solid crystal in which lie the energies of the valence electrons that bind the crystal together. *See also:* semiconductor. (ED) 216-1960w

validated export license A license for which the exporter must submit an application requesting specific Bureau of Export Administration (BXA) approval. (C/SE) 1420.1b-1999

validated metric A metric whose values have been statistically associated with corresponding quality factor values. (C/SE) 1061-1998

validation (1) (programmable digital computer systems in safety systems of nuclear power generating stations) The test and evaluation of the integrated computer system to ensure compliance with the functional, performance, and interface requirements. 7432-1982w
(2) (software verification and validation) The process of evaluating software at the end of the software development process to ensure compliance with software requirements. (SE/C) 1012-1986s
(3) (test, measurement, and diagnostic equipment) That process in the production of a test program by which the correctness of the program is verified by running it on the automatic test equipment together with the unit under test. The process includes the identification of run-time errors, procedure errors, and other non-compiler errors, not uncovered by pure software methods. The process is generally performed with the customer or designated representative as a witness. (MIL) [2]
(4) The process of testing an application or system to ensure that it conforms to its specification. (C/PA) 14252-1996
(5) The process of evaluating a system or component during or at the end of the development process to determine whether a system or component satisfies specified requirements. (C/SE) 1233-1998
(6) The process of determining the degree to which a distributed simulation is an accurate representation of the real world from the perspective of the intended use(s) as defined by the requirements. Validation also refers to the process of determining the confidence that should be placed on this assessment. (C/DIS) 1278.4-1997

valid bid for service An originating or incoming call attempt for which the switching system receives the expected number of digits. *See also:* ineffective attempts. (COM/TA) 973-1990w

valid call Occurs if enough digits are received to complete the call through the office or to give the subscriber an appropriate tone or signal. (COM/TA) 973-1990w

valid compare A condition on output response when the precise state of the response is not important to the test, but the fact that the output is a valid state value is pertinent. (C/TT) 1450-1999

valid input A condition on input stimulus when the state of that stimulus will not affect the current test. In the simulator perspective, this condition is often identified as an unknown, or X, state. (C/TT) 1450-1999

validity check A consistency check that is based upon known limits relating to particular data. For example, a month may not be numbered greater than 12, and a week cannot have more than 168 hours. (C) 610.5-1990w

valley (pulse terminology) A portion of a pulse waveform between two specified peak magnitudes of the same polarity. *See also:* preshoot. (IM/WM&A) 194-1977w

valley point (tunnel-diode characteristic) The point on the forward current-voltage characteristic corresponding to the second-lowest positive (forward) voltage at which $di/dV = 0$. *See also:* peak point. (ED) 253-1963w, [46]

valley-point current (tunnel-diode characteristic) The current at the valley point. *See also:* peak point. (ED) 253-1963w, [46]

valley-point voltage (tunnel-diode characteristic) The voltage at which the valley point occurs. *See also:* peak point. (ED) 253-1963w, [46]

valuator A logical input device used to input a scalar value in a graphics system. A typical physical device is a control dial. (C) 610.6-1991w
(2) (A) An input device that provides a scalar value; for example, a thumb wheel or a potentiometer. **(B)** A logical input device used to input a scalar value in a graphics system. *Note:* A corresponding physical device is a control dial. (C) 610.10-1994

value (1) (automatic control) The quantitative measure of a signal or variable. *See also:* feedback control system. (IM) [120]
(2) (direct-current through test object) The arithmetic mean value. (PE/PSIM/PSR) 4-1978s, [6]
(3) (test direct voltages) The arithmetic mean value: that is, the integral of the voltage over a full period of the ripple divided by the period. *Note:* The maximum value of the test voltage may be taken approximately as the sum of the arithmetic mean value plus the ripple magnitude. (PE/PSIM) 4-1978s, [55]
(4) (alternating test voltage) The peak value divided by $(2)^{1/2}$. (PE/PSIM) [55]
(5) An arbitrarily complex information item that can be viewed as a characteristic or property of an object. (C/PA) 1328-1993w, 1224-1993w, 1327-1993w
(6) *See also:* OM attribute value; attribute value. (C/PA) 1328.2-1993w
(7) *See also:* data value. (C) 610.5-1990w

value-added network A communications network that provides enhanced services, such as character set conversion, protocol conversion, and message storing and forwarding. (C) 610.7-1995

value-added service A communications service utilizing communications common carrier networks for transmission and providing added data services with separate additional equipment. Such added service features may be store-and-forward switching, terminal interfacing and host interfacing. (LM/COM) 168-1956w

value class A kind of class that represents instances that are pure values. The constituent instances of a value class do not come and go and cannot change state. (C/SE) 1320.2-1998

value, desired *See:* ideal value.

value domain An expression of a specific and explicit representational value of some information about something of interest within the Intelligent Transportation Systems (ITS) domain. *Note:* An example of a value domain is numbers as applied to freeway lanes, which have Integer Type as their data type and a Valid Value Rule expressing the range 1–99. (SCC32) 1489-1999

value, ideal *See:* ideal value.

value list constraint A kind of constraint that specifies the set of all acceptable instance values for a value class. (C/SE) 1320.2-1998

value, Munsell *See:* Munsell value.

value of the test current (high voltage testing) The value of the test current is normally defined by the crest value. With some test circuits, overshoot or oscillations may be present on the current. The appropriate apparatus standard should specify whether the value of the test current should be defined by the actual crest or by a smooth curve drawn through the oscillations. (PE/PSIM) 4-1978s

value of test voltage (1) (high voltage testing) The voltage value that is to be applied in a test. (PE/PSIM) 4-1978s
(2) (lightning impulse tests, general applicability) The value of the test voltage is, for a smooth lightning impulse, the crest value. With some test circuits, oscillations or an overshoot may occur at the crest of the impulse [see figure below, (a)–(d)]. If the frequency of such oscillations is not less than 0.5MHz or the duration of overshoot not over one μ, a mean curve should be drawn as in Fig (a) and (b) and, for the purpose of measurement, the maximum amplitude of this curve defines the value of the test voltage. Permissible amplitude limits for the oscillations of overshoot, on standard lightning impulses, are given in IEEE Std 4-1978. For other impulse shapes [see, for example, (e) and (h)], the appropriate apparatus standard should define the value of the test voltage, taking account of the type of test and of test object. (PE/PSIM) 4-1978s
(3) The value of the test voltage is defined by its arithmetic mean value. (PE/PSIM) 4-1978s
(4) The value of the test voltage is defined by its peak value divided by

$\sqrt{2}$.

Note: The appropriate apparatus standards may require a measurement of the rms value of the test voltage instead of the peak value for cases where the rms value may be of importance. Such cases are, for instance, when thermal effects are under investigation. (PE/PSIM) 4-1978s

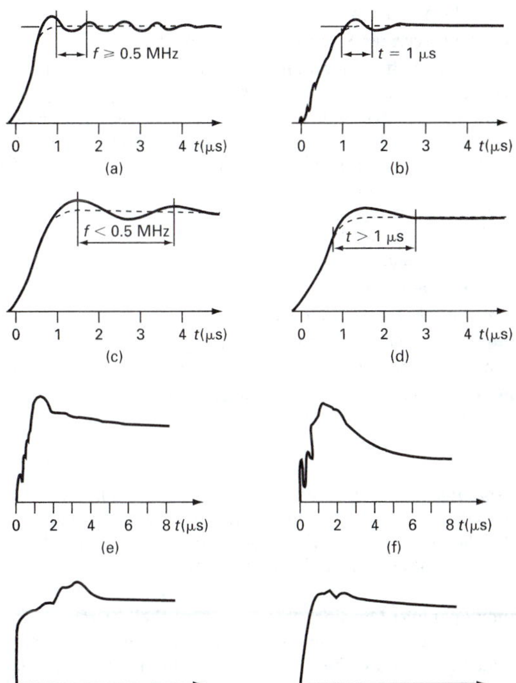

[Examples of lighting impulses with oscillations or overshoot. (a), (b)—The value of the test voltage is determined by a mean curve (broken line). (c), (d)—The value of the test voltage is determined by the crest value. (e), (f), (g), (h)—No general guidance can be given for the determination of the value of the test voltage.]

value of test voltage

value of the test voltage for alternating voltage The peak value divided by the square root of 2, or the rms value as defined by the appropriate apparatus standard. (PE/PSIM) 4-1995

value of the test voltage for lightning impulse voltage The peak value when the impulse is without overshoot or oscillations. (PE/PSIM) 4-1995

value range constraint A kind of constraint that specifies the set of all acceptable instance values for a value class where the instance values are constrained by a lower and/or upper boundary. An example of the value range constraint is `Azimuth`, which is required to be between −180° to +180°. A range constraint only makes sense if there is a linear ordering specified. (C/SE) 1320.2-1998

values of spectral luminous efficiency for photopic vision (illuminating engineering) Values at 10-nm intervals were adopted by the International Commission on Illumination in 1924 and were adopted in 1933 by the International Committee for Weights and Measures as a basis for the establishment of photometric standards of types of sources differing from the primary standard in spectral distribution of radiant flux. These values are given in the second column of the accompanying table; the intermediate values given in the other columns have been interpolated. *Note:* These standard values of spectral luminous efficiency were determined by observations with a two-degree photometric field having a moderately high luminance, photometric evaluations based upon them consequently do not apply exactly to other conditions of observation. Watts weighted in accord with these standard values are often referred to as light-watts. (EEC/IE) [126]

values of spectral luminous efficiency for scotopic vision (illuminating engineering) Values at 10-nm intervals were provisionally adopted by the International Commission on Illumination in 1951. *Note:* These values of spectral luminous efficiency were determined by observation by young dark-adapted observers using extra-foveal vision at near-threshold luminance. (EEC/IE) [126]

value trace *See:* variable trace.

value, true *See:* true value.

`value` word A Forth word created by the defining word `value`. (A `value` word, when executed by itself, places a numeric value on the data stack, much like a `constant`. The numeric value of a `value` word is changed by preceding it with the Forth word `to`.). (C/BA) 1275-1994

valve A converter arm in a three-phase, six pulse bridge converter connection. (SUB/PE) 857-1996

valve action (electrochemical) The process involved in the operation of an electrochemical valve. *See also:* electrochemical valve. (EEC/PE) [119]

valve actuator (valve actuators) An electric, pneumatic, hydraulic, or electrohydraulic power-driven mechanism for positioning two-position or modulating valves, and dampers. Included are those components required to control valve action and to provide valve position output signals, as defined in the actuator specification. (PE/NP) 382-1985

valve actuator specification (valve actuators) A document to be provided to the valve actuator manufacturer which contains technical requirements for a specific application. (PE/NP) 382-1980s

valve arrester (1) (surge arresters) An arrester that includes a valve element. (SPD/PE) C62.1-1981s
(2) An arrester that includes one or more valve elements. (SPD/PE) C62.22-1997, C62.11-1999

valve base That assembly that mechanically supports, and electrically insulates the valves from ground. *Note:* A part of a valve that is clearly identifiable in a discrete form to be a valve base may not exist in all designs of valves. A valve base could be a separate platform, insulated from ground by post-type insulators, that carries a live-tank valve unit, or a steel framework insulated from ground by post-type insulators on which the various modules of a multiple valve unit are mounted, or a raised platform of insulating material that is integral to the valve structure and forms the base. (SUB/PE) 857-1996

valve base electronics (VBE) Electronic circuitry that directs gate pulses into the thyristor valve. (SUB/PE) 1303-1994

valve electronics (VE) Electronic circuitry associated with the thyristors and mounted at thyristor level potential. (SUB/PE) 1303-1994

valve element (1) A resistor that, because of its nonlinear current-voltage characteristic, limits the voltage across the arrester terminals during the flow of discharge current and contributes to the limitation of follow current at normal power-frequency voltage. (SPD/PE) C62.22-1997, C62.11-1999
(2) A resistor that, because of its nonlinear current-voltage characteristic, limits the voltage across the terminals of the surge-protective-device during the flow of discharge current and contributes to the limitation of follow current at normal power-frequency voltage. (SPD/PE) C62.62-2000

valve module The smallest assembly, comprising a number of thyristors and their immediate auxiliaries for firing and protection, voltage dividing components, and distributed or lumped valve reactors, from which the valve is built up and which exhibits the same electrical properties as the complete valve but can withstand only a portion of the full voltage blocking capability of the valve. (SUB/PE) 857-1996

valve point A camshaft (turbine input valve control mechanism) position at which point one of the turbine input valves is fully or near fully open and the next valve remains closed. (PE/PSE) 94-1991w

valve point loading An economic control strategy that dispatches generation based on minimizing production costs by considering the effects of turbine input valve points on incremental heat rate. (PE/PSE) 94-1991w

valve-point loading control (electric power system) A control means for making a unit operate in the more efficient portions of the range of the governor-controlled valves. (PE/PSE) 94-1991w

valve position limiter (control systems for steam turbine-generator units) (load limit) A device that acts on the speed/load-control system to prevent the control valve(s) from opening beyond a preset limit. (PE/EDPG) 122-1985s

valve protector A protective device that includes a valve element. (SPD/PE) C62.62-2000

valve ratio (electrochemical valve) The ratio of the impedance to current flowing from the valve metal to the compound or solution, to the impedance in the opposite direction. *See also:* electrochemical valve. (EEC/PE) [119]

valve-regulated battery (1) A battery in which the venting of the products of electrolysis is controlled by a reclosing pressure-sensitive valve. (PV) 1013-1990
(2) A battery that is sealed with the exception of a valve that opens to the atmosphere when the internal gas pressure exceeds the atmospheric pressure by a preselected amount. Valve-regulated batteries provide a means for recombination of internally generated oxygen. (PV) 1144-1996

valve-regulated lead-acid (VRLA) cell (1) A lead-acid cell that is sealed with the exception of a valve that opens to the atmosphere when the internal gas pressure in the cell exceeds atmospheric pressure by a pre-selected amount. (IA/PSE) 446-1995
(2) A lead-acid cell that is sealed with the exception of a valve that opens to the atmosphere when the internal gas pressure in the cell exceeds atmospheric pressure by a pre-selected amount. VRLA cells provide a means for recombination of internally generated oxygen and the suppression of hydrogen gas evolution to limit water consumption. (SCC29/SCC21) 485-1997, 937-2000
(3) A cell that is sealed with the exception of a valve that opens to the atmosphere when the internal gas pressure in the cell exceeds atmospheric pressure by a preselected amount. VRLA cells provide a means for recombination of internally generated oxygen and the suppression of hydrogen gas evolution to limit water consumption. (SB/PE/EDPG) 1188-1996, 1184-1994, 1189-1996

valve section An electrical assembly comprising a number of thyristor levels and other components that exhibits prorated electrical properties of a complete valve. (SUB/PE) 857-1996

Valve Stroke Complete travel of a valve from either fully open to fully closed position or vice-versa. (PE/NP) 1290-1996

valve tube *See:* kenotron.

valve-type arrester *See:* valve-type arrester.

Van Allen belts (communication satellite) The belts of charged particles (electrons and protons) trapped by the earth's (external) magnetic field and which surround the earth at altitudes from 1000 to 6000 kilometers. The paths of the particles are determined by the directions of the (external) lines of force of the earth's magnetic field. The particles migrate from the region above earth's equator toward the North Pole, then toward the South Pole, then return to the region above the equator. (COM) [19]

VAN *See:* value-added network.

V&V *See:* verification and validation.

vane (navigation aids) A device to indicate the direction from which the wind blows. (AES/GCS) 172-1983w

vane-type relay A type of alternating-current relay in which a light metal disc or vane moves in response to a change of the current in the controlling circuit or circuits. (EEC/PE) [119]

V antenna A V-shaped arrangement of two conductors, balanced-fed at the apex, with included angle, length, and apex height above the earth chosen so as to give the desired directive properties to the radiation pattern. (AP/ANT) 145-1993

vapor openings Openings through a tank shell or roof above the surface of the stored liquid. Such openings may be provided for tank breathing, tank gauging, fire fighting, or other operating purposes. (NFPA) [114]

vapor-safe electric equipment A unit so constructed that it may be operated without hazard to its surroundings in an atmosphere containing fuel, oil, alcohol, or other vapors that may occur in aircraft: that is, the unit is capable of so confining any sparks, flashes, or explosions of the combustible vapors within itself that ignition of the surrounding atmosphere is prevented. *Note:* This definition closely parallels that given for explosionproof: however, it is believed that the new term is needed in order to avoid the connotation of compliance with Underwriter's standards that are now associated with explosionproof in the minds of most engineers who are familiar with the use of that term applied to industrial motors and control equipment. (EEC/PE) [119]

vaportight So enclosed that vapor will not enter the enclosure. (SWG/PE) C37.100-1992

vapor-tight luminaire (illuminating engineering) A luminaire designed and approved for installation in damp or wet locations. It is also described as "enclosed and gasketed." (EEC/IE) [126]

VAR *See:* visual-aural range.

var (electric power circuits) The unit of reactive power in the International System of Units (SI). The var is the reactive power at the two points of entry of a single-phase, two-wire circuit when the product of the root-mean-square value in amperes of the sinusoidal current by the root-mean-square value in volts of the sinusoidal voltage and by the sine of the angular phase difference by which the voltage leads the current is equal to one. (Std100) 270-1966w

varactor A two-terminal semiconductor device in which the electrical characteristic of primary interest is a voltage-dependent capacitance. (ED) [46]

var-aligned Alignment for the storage of a Forth variable. (C/BA) 1275-1994

varhour The unit of a quadrature-energy (quadergy) in the International System of Units (SI). The varhour is the quadrature energy that is considered to have flowed past the points of entry of a reactive circuit when a reactive power of one var has been maintained at the terminals of entry for one hour. (Std100) 270-1966w

varhour constant (metering) The registration, expressed in varhours, corresponding to one revolution of the rotor. (ELM) C12.1-1988

varhour meter (metering) An electricity meter that measures and registers the integral, with respect to time, of the reactive power of the circuit in which it is connected. The unit in which this intergral is measured is usually the kilovarhour. (ELM) C12.1-1988

variable (1) (electrical heating applications to melting furnaces and forehearths in the glass industry) A quantity or condition that is subject to change. (IA) 668-1987w
(2) (modeling and simulation) (software) A quantity or data item whose value can change; for example, the variable Current time. (C) 610.3-1989w, 610.12-1990
(3) In the shell command language, a named parameter. (C/PA) 9945-2-1993
(4) A quantity, the value of which is assigned at program runtime. (SCC20) 771-1998
(5) An instance whose identity is unknown at the time of writing. A variable is represented by an identifier that begins with an upper-case letter. (C/SE) 1320.2-1998

variable address *See:* indexed address.

variable-area track (electroacoustics) A sound track divided laterally into opaque and transparent areas, a sharp line of demarcation between these areas forming an oscillographic trace of the wave shape of the recorded signal. *See also:* phonograph pickup. (SP) [32]

variable assignment In the shell command language, a word consisting of the following parts

varname = *value*

When used in a context where assignment is defined to occur and at no other time, the value (representing a word or field) shall be assigned as the value of the variable denoted by *varname*. The *varname* and *value* parts meet the requirements for a name and a word, respectively, except that they are delimited by the embedded unquoted equals sign in addition to the delimiting described. In all cases, the variable shall be created if it did not already exist. If *value* is not specified, the variable shall be given a null value. An alternative form of variable assignment

symbol = *value*

(where *symbol* is a valid word delimited by an equals sign, but not a valid name) produces unspecified results. *Synonym:* assignment. (C/PA) 9945-2-1993

variable-block format A format that allows the number of words in successive blocks to vary. (IA/EEC) [61], [74]

variable carrier *See:* controlled carrier.

variable, complex *See:* complex variable.

variable-density track (electroacoustics) A sound track of constant width, usually but not necessarily of uniform light transmission on any instantaneous transverse axis, on which the average light transmission varies along the longitudinal axis in proportion to some characteristic of the applied signal. *See also:* phonograph pickup. (SP) [32]

variable, directly controlled *See:* directly controlled variable.

variable field One that varies with time. (Std100) 270-1966w

variable format A file organization in which logical records are of variable length. *Synonym:* V format. *Contrast:* fixed format. (C) 610.5-1990w

variable-frequency telemetering (electric power system) A type of telemetering in which the frequency of the alternating-voltage signal is varied as a function of the magnitude of the measured quantity. *See also:* telemetering. (PE/PSE) 94-1970w

variable, indirectly controlled *See:* indirectly controlled variable.

variable inductor *See:* continuously adjustable inductor.

variable, input *See:* input variable.

variable length Pertaining to a record or field that does not have a constant length, but whose length depends on the length of the specific data contained in it. *Contrast:* fixed length. *See also:* variable format. (C) 610.5-1990w

variable-length field A field whose length may vary according to data stored. *Contrast:* fixed-length field. *See also:* variable format. (C) 610.5-1990w

variable, manipulated *See:* manipulated variable.

variable modulation (navigation aids) (very high-frequency omnidirectional range) That modulation of the ground station radiation which produces a signal in the airborne receiver whose phase with respect to a radiated reference modulation corresponds to the bearing of the receiver. (AES/GCS) 172-1983w

variable-mu tube (remote-cutoff tube) (variable-μ tube) An electron tube in which the amplification factor varies in a predetermined way with control-grid voltage. (ED) 161-1971w

variable name data element A data element whose name can vary depending upon the particular data item represented; for example, a data element named "Population of *X* in *Y*," where *X* takes on the name of a city and *Y* represents a given year. (C) 610.5-1990w

variable operating cost Cost that varies or fluctuates with operating or use of plant. (PE/PSE) 858-1993w

variable, output *See:* output variable.

variable point (data management) (mathematics of computing) Pertaining to a numeration system in which the position of the radix point is indicated by a special character at that position. *Contrast:* fixed point; floating point. (C) 610.5-1990w, 1084-1986w

variable-reluctance microphone (magnetic microphone) A microphone that depends for its operation on variations in the reluctance of a magnetic circuit. *See also:* microphone. (EEC/PE) [119]

variable-reluctance pickup (magnetic pickup) A phonograph pickup that depends for its operation on the variation in the reluctance of a magnetic circuit. *See also:* phonograph pickup. (EEC/PE) [119]

variable-reluctance transducer An electroacoustic transducer that depends for its operation on the variation in the reluctance of a magnetic circuit. (SP) [32]

variable scope *See:* scope.

variable-speed axle generator An axle generator in which the speed of the generator varies directly with the speed of the car. *See also:* axle-generator system. (EEC/PE) [119]

variable speed constant frequency generator (VSCF) An ac generator designed to have a constant frequency output with a variable speed input. This may be accomplished with an induction generator having an ac/ac converter feedback circuit that excites the wound rotor at a frequency to produce a constant frequency output. This may also be accomplished by a synchronous generator whose variable output frequency is fed into a frequency changer that produces a constant output frequency. Basic frequency changers may be of the cycloconverter or dc link type. (IA/MT) 45-1998

variable-speed drive An electric drive so designed that the speed varies through a considerable range as a function of load. *See also:* electric drive. (IA/ICTL/IAC) [60]

variable speed motor (rotating machinery) A motor with a positively damped speed-torque characteristic which lends itself to controlled speed applications. (PE) [9]

variables, state *See:* state variables.

variable threshold transistor (metal-nitride-oxide field-effect transistor) An insulated-gate field-effect transistor (IGFET) whose threshold voltage can be varied electrically to predetermined levels. The memory metal-nitride-oxide semiconductor (MNOS) memory transistor is a specific example of this type. (ED) 581-1978w

variable-torque motor (A) A multispeed motor whose rated load torque at each speed is proportional to the speed. Thus the rated power of the motor is proportional to the square of the speed. **(B)** An adjustable-speed motor in which the specified torque increases with speed. It is common to provide a variable-torque adjustable-speed motor in which the torque varies as the square of the speed and hence the power output varies as the cube of the speed. *See also:* asynchronous machine. (PE) [9]

variable trace A record of the name and values of variables accessed or changed during the execution of a computer program. *Synonyms:* data trace; value trace. *See also:* execution trace; subroutine trace; retrospective trace; symbolic trace; data flow trace. (C) 610.12-1990

variable, ultimately controlled *See:* ultimately controlled variable.

variable-voltage transformer (power and distribution transformers) An autotransformer in which the output voltage can be changed (essentially from turn to turn) by means of a movable contact device sliding on the shunt winding turns. (PE/TR) C57.12.80-1978r

variant (1) In fault tolerance, a version of a program resulting from the application of software diversity. (C) 610.12-1990

(2) *See also:* dialect. (C) 610.13-1993w

variation (1) (navigation aids) The angle between the magnetic and geographical meridians at any place. (AES/GCS) 172-1983w

(2) *See also:* dialect. (C) 610.13-1993w

varindor An inductor whose inductance varies markedly with the current in the winding. (EEC/PE) [119]

variocoupler (radio practice) A transformer, the self-impedance of whose windings remains essentially constant while the mutual impedance between the windings is adjustable. (IM) [120]

variolosser A device whose loss can be controlled by a voltage or current. (EEC/PE) [119]

variometer A variable inductor in which the change of inductance is effected by changing the relative position of two or more coils. (PE/EM) 43-1974s

varioplex A telegraph switching system that establishes connections on a circuit-sharing basis between a multiplicity of telegraph transmitters in one locality and respective corresponding telegraph receivers in another locality over one or more intervening telegraph channels. Maximum usage of channel capacity is secured by momentarily storing the signals and allocating circuit time in rotation among those transmitters having intelligence in storage. *See also:* telegraphy. (EEC/PE) [119]

varistor (A) A two-terminal resistive element, composed of an electronic semiconductor and suitable contacts, that has a markedly nonlinear volt-ampere characteristic. **(B)** A two-terminal semiconductor device having a voltage-dependent nonlinear resistance. *Note:* Varistors may be divided into two groups, symmetrical and nonsymmetrical, based on the symmetry or lack of symmetry of the volt-ampere curve. *See also:* semiconductor. (ED) 216-1960

varistor capacitance (low voltage varistor surge arresters) Capacitance between the two terminals of the varistor measured at a specified frequency and bias. (PE) [8]

varistor resistance (low voltage varistor surge arresters) Static resistance of the varistor at a given operating point, described as the ratio of varistor voltage to varistor current. (PE) [8]

varistor voltage (low voltage varistor surge arresters) Voltage across the varistor measured at a given current. (PE) [8]

varmeter (reactive volt-ampere meter) An instrument for measuring reactive power. It is provided with a scale usually graduated in either vars, kilovars, or megavars. If the scale is graduated in kilovars or megavars, the instrument is usually designated as a kilovarmeter or megavarmeter. *See also:* instrument. (PE/EEC) [119]

varnish (rotating machinery) A liquid composition that is converted to a transparent or translucent solid film after application as a thin layer. (PE) [9]

varnished fabric (rotating machinery) A fabric or mat in which the elements and interstices have been essentially coated and filled with an impregnant such as a compound or varnish and that is relatively homogeneous in structure. *Synonym:* varnished mat. *See also:* rotor; stator. (PE) [9]

varnished mat *See:* varnished fabric.

varnished tubing A flexible tubular product made from braided cotton, rayon, nylon, glass, or other fibers, and coated, or impregnated and coated, with a continuous film or varnish, lacquer, a combination of varnish and lacquer, or other electrical insulating materials. (EEC/PE) [119]

varying duty (1) Operation at loads, and for intervals of time, both of which may be subject to wide variation. (NESC/NEC) [86]

(2) (rating of electric equipment) A requirement of service that demands operation at loads, and for periods of time, both of which may be subject to wide variation. *See also:* voltage regulator; asynchronous machine. (PE/IA/EI/TR/PKG) 96-1969w, C57.12.80-1978r, 333-1980w, C57.15-1968s

(3) A requirement of service that demands operation at intermittent current loading, and for periods of time, both of which may be subject to wide variation. (PE/TR) C57.16-1996

varying parameter *See:* linear varying parameter.

varying-speed motor A motor whose speed varies with the load, ordinarily decreasing when the load increases, such as a series-wound or repulsion motor. (IA/MT) 45-1998

varying-voltage control A form of armature-voltage control obtained by impressing on the armature of the motor a voltage that varies considerably with change in load, with a consequent change in speed, such as may be obtained from a differentially compound-wound generator or by means of resistance in the armature circuit. *See also:* control. (IA/IAC) [60]

vary off To make a device, control unit, or line unavailable for its normal intended use. *Contrast:* vary on. (C) 610.10-1994w

vary off-line To place a device in a state where it is not available for use by the system. *Contrast:* vary on-line. (C) 610.10-1994w

vary on To make a device, control unit, or line available for its normal intended use. *Contrast:* vary off. (C) 610.10-1994w

vary on-line To restore a device to a state where it is available for use by the system. *Contrast:* vary off-line. (C) 610.10-1994w

VASIS *See:* visual approach slope indicator system.

vault (1) A structurally solid enclosure above or below ground with access limited to personnel qualified to install, maintain, operate, or inspect the equipment or cable enclosed. The enclosure may have openings for ventilation, personnel access, cable entrance, and other openings required for operation of equipment in the vault. (NESC) C2-1997
(2) A non-automated (manual) library, i.e., a shelf. (C/SS) 1244.1-2000

vault-type transformer (power and distribution transformers) A transformer that is so constructed as to be suitable for occasional submerged operation in water under specified conditions of time and external pressure. (PE/TR) C57.12.80-1978r

VBA *See:* vibrating beam accelerometer.

V-band A radar-frequency band between 40 GHz and 75 GHz, usually in the International Telecommunication Union (ITU) allocated band 59–64 GHz. Included within the definition of millimeter-wave radar. *See also:* millimeter-wave radar. (AES) 686-1997

VBE *See:* valve base electronics.

V-beam radar A ground-based, three-dimensional radar system for the determination of range, bearing, and, uniquely, the height or elevation angle of the target. It uses two fan-shaped beams, one vertical and the other inclined, that rotate together in azimuth so as to give two responses from the target. The time difference between these responses, together with range, are used in determining the height of the target. (AES) 686-1997

V-channel metal-oxide semiconductor (VMOS) A type of n-channel metal-oxide semiconductor in which a V-shaped notch is used to increase the density. (C) 610.10-1994w

VCI *See:* virtual channel identifier.

VCO *See:* voltage-controlled oscillator.

VCP Vertical coupling plane. *See also:* coupling plane. (EMC) C63.16-1993

V curve (synchronous machines) The load characteristic giving the relationship between the armature current and the field current for constant values of load, power, and armature voltage. *See also:* asynchronous machine. (PE) [9]

VDD *See:* version description document.

VDM *See:* Vienna Development Method.

VDT *See:* video display terminal; visual display terminal.

VDU *See:* video display terminal; video display unit.

VE *See:* valve electronics.

vector (1) A mathematico-physical quantity that represents a vector quantity. (Std100) 270-1966w
(2) (data management) A quantity represented by an ordered set of numbers; for example, a one-dimensional array. (C) 610.5-1990w
(3) (computer graphics) A directed line segment. (C) 610.6-1991w
(4) A packet of data that is passed, by an interrupt handler, to a processor in response to an interrupt acknowledge. This data is typically used to identify to an operating system which interrupt subroutine handler should be dispatched in response to the asserted interrupt. (C/BA) 1014.1-1994w
(5) A one-dimensional array. (C/DA) 1076.3-1997, 1076.6-1999
(6) Every signal's stimuli/response to be applied/observed in the smallest integral "step" of a device test. Contains a collection of waveforms to be applied to the primary signals. *See also:* T0. (C/TT) 1450-1999

vector display device *See:* random-scan display device.

vector electrocardiogram (vectorcardiogram) (electrobiology) The 2-dimensional or 3-dimensional presentation of cardiac electric activity that results from displaying lead pairs against each other rather than against time. More strictly, it is a loop pattern taken from leads placed orthogonally. *See also:* electrocardiogram. (EMB) [47]

vector field The totality of vectors in a given region represented by a vector function $V(x,y,z)$ of the space coordinates x,y,z. (Std100) 270-1966w

vector font A scalable font that is stored as a series of geometric objects such as line or curve segments. *Synonym:* raster font. *See also:* outline font; bit map font. (C) 610.10-1994w

vector function A functional relationship that results in a vector. (Std100) 270-1966w

vector generator A component of the display processor hardware that generates directed line segments from end-point coordinates. (C) 610.6-1991w

vector graphics (1) The representation of an image by a collection of vectors. *Contrast:* raster graphics. *See also:* random-scan; random-scan display device. (C) 610.6-1991w
(2) *See also:* random-scan display device. (C) 610.10-1994w

vector norm The measure of the size of a vector, with the usual norm properties. *Notes:* 1. Vector norm of x is denoted by $\|x\|$. 2. Norm properties are:

$$\|x\| > 0 \qquad \text{for } x \neq 0$$

$$\|0\| = 0$$

$$\|\alpha x\| = |\alpha| \times \|x\|$$

$$\|x_1 + x_2\| \leq \|x_1\| + \|x_2\|$$

(CS/PE/EDPG) [3]

vector operator del (∇) A differential operator defined as follows in terms of Cartesian coordinates:

$$\nabla = \mathbf{i}\frac{\partial}{\partial x} + \mathbf{j}\frac{\partial}{\partial y} + \mathbf{k}\frac{\partial}{\partial z}$$

(Std100) 270-1966w

Vector Parameter An instance of the class `IEEE1451_VectorParameter` or of a subclass thereof. (IM/ST) 1451.1-1999

vector processor *See:* array processor.

vector product (cross product) The vector product of vector A and a vector B is a vector C that has a magnitude obtained by multiplying the product of the magnitudes of A and B by the sine of the angle between them: the direction of C is that traveled by a right-hand screw turning about an axis perpendicular to the plane of A and B, in the sense in which A would move into B by a rotation of less than 180 degrees: it is assumed that A and B are drawn from the same point. The vector product of two vectors A and B may be indicated by using a small cross: A × B. The direction of the vector product depends on the order in which the vectors are multiplied, so that A × B = −B × A. If the two vectors are given in terms of their rectangular components, then

$$\mathbf{A} \times \mathbf{B} = \begin{vmatrix} \mathbf{ijk} \\ A_xA_yA_z \\ B_xB_yB_z \end{vmatrix}$$

$$= \mathbf{i}(A_yB_z - A_zB_y) + \mathbf{j}(A_zB_x - A_xB_z) + \mathbf{k}(A_xB_y - A_yB_x)$$

Example: The linear velocity V of a particle in a rotating body is the vector product of the angular velocity ω and the radius vector r from any point on the axis to the point in question, or

$$V = \omega \times r = -r \times \omega$$

(Std100) 270-1966w

vector quantity Any physical quantity whose specification involves both magnitude and direction and that obeys the parallelogram law of addition. (Std100) 270-1966w

vector radiative transport An attempt to incorporate the vector nature of electromagnetic waves into the energy conserving transport theory. *See also:* radiative transfer theory. (AP/PROP) 211-1997

Vector Series Parameter An instance of the class `IEEE1451_VectorSeriesParameter` or of a subclass thereof. (IM/ST) 1451.1-1999

vector, state *See:* state vector.

vector unit An arithmetic unit that operates on multiple data elements at the same time. *Contrast:* scalar unit. (C) 610.10-1994w

vehicle (1) (navigation aids) That in or on which a person or thing is being or may be carried. (AES/GCS) 172-1983w **(2)** A land conveyance assembly for carrying or transporting people or objects, capable of traversing a guideway, having structural integrity, and general mechanical completeness, but not necessarily designed for independent operation. (VT/RT) 1477-1998, 1473-1999, 1475-1999, 1474.1-1999, 1476-2000

vehicle-derived navigation data (navigation aids) Data obtained from measurements made at a vehicle. (AES/GCS) 172-1983w

vehicle maneuver effects Gyro output errors due to vehicle maneuvers. (AES/GYAC) 528-1984s

vehicle service table (VST) The answer of the onboard equipment (OBE) application layer in response to a received beacon service table (BST). The VST contains the identifier of applications supported by the OBE and profiles used for further point-to-point communication. (SCC32) 1455-1999

veiling brightness (illuminating engineering) A brightness superimposed on the retinal image which reduces its contrast. It is this veiling effect produced by bright sources or areas in the visual field which results in decreased visual performance and visibility. (EEC/IE) [126]

veiling reflections (1) (illuminating engineering) Regular reflections that are superimposed upon diffuse reflections from an object which partially or totally obscure the details to be seen by reducing the contrast. This sometimes is called reflected glare. *See also:* reflected glare. (EEC/IE) [126] **(2) (electric power systems in commercial buildings)** Reflected light from a task that reduces visibility because the light is reflected specularly from shiny details of the task, which brightens those details and reduces the contrast with the background. (IA/PSE) 241-1990r

Veitch chart *See:* Veitch diagram.

Veitch diagram A variation of the Karnaugh map in which the rows and columns are headed with combinations of the variables in a straight binary sequence. *Synonyms:* Veitch-Karnaugh diagram; Veitch chart. (C) 1084-1986w

Veitch-Karnaugh diagram *See:* Veitch diagram.

velocity correction A method of register control that takes the form of a gradual change in the relative velocity of the web. (IA/ICTL/IAC) [60]

velocity level in decibels of a sound Twenty times the logarithm to the base 10 of the ratio of the particle velocity of the sound to the reference particle velocity. The reference particle velocity shall be stated explicitly. *Note:* In many sound fields the particle velocity ratios are not proportional to the square root of corresponding power ratios and hence cannot be expressed in decibels in the strict sense: however, it is common practice to extend the use of the decibel to these cases. (SP/ACO) [32]

velocity microphone A microphone in which the electric output substantially corresponds to the instantaneous particle velocity in the impressed sound wave. *Note:* A velocity microphone is a gradient microphone of order one, and it is inherently bidirectional. *See also:* gradient microphone; microphone. (SP) [32]

velocity-modulated amplifier (velocity-variation amplifier) An amplifier that employs velocity modulation to amplify radio frequencies. *See also:* amplifier. (EEC/PE) [119]

velocity-modulated oscillator An electron-tube structure in which the velocity of an electron stream is varied (velocity-modulated) in passing through a resonant cavity called a buncher. Energy is extracted from the bunched electron stream at a higher energy level in passing through a second cavity resonator called the catcher. Oscillations are sustained by coupling energy from the catcher cavity back to the buncher cavity. *See also:* oscillatory circuit. (AP/ANT) 145-1983s

velocity-modulated tube An electron-beam tube in which the velocity of the electron stream is alternately increased and decreased with a period comparable with the total transit time. (ED) [45], [84]

velocity modulation (velocity variation) (of an electron beam) The modification of the velocity of an electron stream by the alternate acceleration and deceleration of the electrons with a period comparable with the transit time in the space concerned. *See also:* velocity-modulated tube; velocity-modulated oscillator. (ED) 161-1971w

velocity response The clutter filter frequency response defined by the ratio of power gain at a specific target Doppler frequency to the average power gain over all target Doppler frequencies of interest. *Note:* Applied to moving-target indication (MTI) radars. (AES) 686-1997

velocity, room air *See:* room air velocity.

velocity shock A mechanical shock resulting from a nonoscillatory change in velocity of an entire system. (SP) [32]

velocity sorting (electronics) Any process of selecting electrons according to their velocities. (ED) 161-1971w

velocity storage (accelerometer) The velocity information that is stored in the accelerometer as a result of its dynamics. (AES/GYAC) 528-1994, 530-1978r

velocity storage, normal *See:* normal velocity storage.

velocity storage, overrange *See:* overrange velocity storage.

velocity variation *See:* velocity modulation.

velocity-variation amplifier *See:* velocity-modulated amplifier.

vendor A supplier of packaged software. (C/PA) 1387.2-1995

vendor-defined (1) Parameters that identify and particularize a specific vendor implementation. *Synonym:* vendor-dependent. (C/BA) 896.3-1993w, 896.4-1993w **(2)** An item, such as a nonstandard attribute, that is defined by the vendor that created (packaged) the software. (C/PA) 1387.2-1995

vendor-dependent (1) A term used to describe parameters that may vary between vendors supplying the same node or unit architectures. Although the CSR Architecture may constrain the definition of these fields, their format and definition is provided by the module vendor. Note that vendor-dependent fields may be standardized or left implementation-specific, depending on the vendor's needs. (C/MM) 1212-1991s **(2)** *See also:* vendor-defined. (C/BA) 896.4-1993w

vendor-supplied An item, such as a control file, that is supplied by the creator (packager) of the software. (C/PA) 1387.2-1995

Venn diagram A diagram in which sets are represented by closed regions. (See the corresponding figure.)

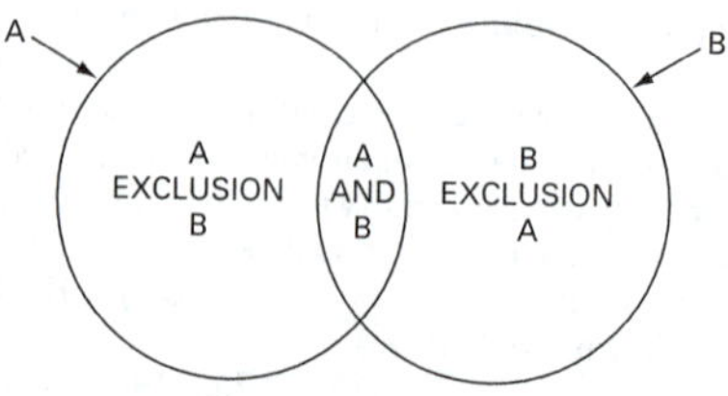

Venn diagram

(C) [20], [85], 1084-1986w

vent (1) (rotating machinery) An opening that will permit the flow of air. (PE) [9]

(2) (of a fuse) The means provided for the escape of the gases developed during circuit interruption. *Note:* In distribution oil cutouts, the vent may be an opening in the housing, or an accessory attachable to a vent opening in the housing with suitable means to prevent loss of oil.
(SWG/PE) C37.100-1992, C37.40-1993
(3) An intentional opening for the escape of gases to the outside. (SPD/PE) C62.11-1999

vented battery A battery in which the products of electrolysis and evaporation are allowed to escape freely to the atmosphere. These batteries are commonly referred to as "flooded." (SCC29) 485-1997

vented cell (1) A cell in which the products of electrolysis and evaporation are allowed to escape to the atmosphere as they are generated. These cells are commonly referred to as "flooded."
(PE/SB/EDPG) 1106-1995, 450-1995, 1184-1994, 1189-1996
(2) A cell in which the products of electrolysis and evaporation are allowed to escape to the atmosphere as they are generated. These batteries are commonly referred to as "flooded." (SCC21) 937-2000
(3) A cell design that is characterized by an excess of free electrolyte, and in which the products of electrolysis and evaporation can freely exit the cell through a vent. *Synonyms:* flooded cell; wet cell. (IA/PSE) 446-1995
(4) *See also:* vented battery. (PE/EDPG) 484-1996

vented fuse (or fuse unit) A fuse with provision for the escape of arc gases, liquids, or solid particles to the surrounding atmosphere during circuit interruption.
(SWG/PE) C37.100-1992, C37.40-1993

vented power fuse (installations and equipment operating at over 600 volts, nominal) D A fuse with provision for the escape of arc gases, liquids, or solid particles to the surrounding atmosphere during circuit interruption.
(NEC/NESC) [86]

vent finger *See:* duct spacer.

ventilated (power and distribution transformers) Provided with a means to permit circulation of the air sufficiently to remove an excess of heat, fumes, or vapors.
(PE/EM/TR) C57.12.80-1978r, 86-1987w

ventilated dry-type transformer (dry-type general purpose distribution and power transformers) A dry-type transformer which is so constructed that the ambient air may circulate through its enclosure to cool the transformer core and windings. (PE/TR) C57.94-1982r, C57.12.80-1978r

ventilated enclosure (1) (metal-enclosed bus and calculating losses in isolated-phase bus) An enclosure so constructed as to provide for the circulation of external air through the enclosure to remove heat, fumes, or vapors.
(SWG/PE) C37.23-1987r
(2) An enclosure provided with means to permit circulation of sufficient air to remove an excess of heat, fumes, or vapors. *Note:* For outdoor applications ventilating openings or louvres are usually filtered, screened, or restricted to limit the entrance of dust, dirt, or other foreign objects.
(SWG/PE) C37.100-1992

ventilating and cooling loss (synchronous machines) Any power required to circulate the cooling medium through the machine and cooler (if used) by fans or pumps that are driven by external means (such as a separate motor) so that their power requirements are not included in the friction and windage loss. It does not include power required to force ventilating gas through any circuit external to the machine and cooler. (PE) [9], [84]

ventilating duct (rotating machinery) A passage provided in the interior of a magnetic core in order to facilitate circulation of air or other cooling agent. *Synonym:* cooling duct.
(PE) [9], [84]

ventilating passage (rotating machinery) A passage provided for the flow of cooling medium. (PE) [9]

ventilating slot (rotating machinery) A slot provided for the passage of cooling medium. (PE) [9]

verbal-descriptive model *See:* narrative model.

verb phrase (A) A part of the label of a relationship that names the relationship in a way that a sentence can be formed by combining the first class name, the verb phrase, the cardinality expression, and the second class name or role name. A verb phrase is ideally stated in active voice. For example, the statement "*each project funds one or more tasks*" could be derived from a relationship showing "`project`" as the first class, "`task`" as the second class with a "one or more" cardinality, and "funds" as the verb phrase. **(B)** A phrase used to name a relationship, which consists of a verb and words that constitute the object of the phrase.
(C/SE) 1320.2-1998

verge-punched card *See:* edge-punched card.

verification (1) (programmable digital computer systems in safety systems of nuclear power generating stations) The process of determining whether or not the product of each phase of the digital computer system development process fulfills all the requirements imposed by the previous phase.
7432-1982w
(2) (software verification and validation plans) The process of determining whether or not the products of a given phase of the software development cycle fulfill the requirements established during the previous phase. (C/SE) 1012-1986s
(3) (nuclear power quality assurance) The act of reviewing, inspecting, testing, checking, auditing, or otherwise determining and documenting whether items, processes, services, or documents conform to specified requirements.
(PE/NP) [124]
(4) (A) (software) The process of evaluating a system or component to determine whether the products of a given development phase satisfy the conditions imposed at the start of that phase. **(B) (software)** Formal proof of program correctness. *See also:* proof of correctness. (C) 610.12-1990
(5) The process of evaluating a system or component to determine whether the system of a given development phase satisfies the conditions imposed at the start of that phase.
(C/SE) 1233-1998
(6) Confirmation by examination (testing) with evidence that specified requirements have been met. (NI) N42.22-1995
(7) The process of determining that an implementation of a distributed simulation accurately represents the developer's conceptual description and specifications.
(C/DIS) 1278.4-1997

verification and validation (V&V) The process of determining whether the requirements for a system or component are complete and correct, the products of each development phase fulfill the requirements or conditions imposed by the previous phase, and the final system or component complies with specified requirements. *See also:* independent verification and validation. (C/PE/NP) 610.12-1990, 7-4.3.2-1993

verification relay A monitoring relay restricted to functions pertaining to power-system conditions and not involving opening circuit breakers during fault condition. *Note:* Such a relay is sometimes referred to as a check or checking relay.
(SWG/PE) C37.100-1992

verification system *See:* automated verification system.

verified frame A valid frame, addressed to the station, for which the information field has met the validity requirements.
(C/LM) 8802-5-1998

verify To check, usually automatically, one typing or recording of data against another in order to minimize human and machine errors in the punching of tape or cards.
(C) 162-1963w

vernier control A method for improving resolution. The amount of vernier control is expressed as either the percent of the total operating range or of the actual operating value, whichever is appropriate to the circuit in use. *See also:* feedback control system. (PE/ICTL/PSE) [54]

version **(1)** **(A)** An initial release or re-release of a computer software configuration item, associated with a complete compilation or recompilation of the computer software configuration item. **(B)** An initial release or complete re-release of a document, as opposed to a revision resulting from issuing change pages to a previous release. *See also:* version description document; configuration control. (C) 610.12-1990
(2) A unique identification of software based on the attributes of the software. Version differentiates software objects with the same value of the *tag* attribute. Versions of bundles or products have the same value of the *tag* attribute and will differ by the value of at least one of *revision*, *architecture*, *vendor_tag*, *location*, or *qualifier* attributes. The *location* and *qualifier* attributes only apply to software in installed_software software_collections. A fileset is considered a version of another fileset if they have the same fileset.tag and their respective products have the same *product.tag*. (C/PA) 1387.2-1995
(3) *See also:* dialect. (C) 610.13-1993w

version description document (VDD) A document that accompanies and identifies a given version of a system or component. Typical contents include an inventory of system or component parts, identification of changes incorporated into this version, and installation and operating information unique to the version described. (C) 610.12-1990

vertex *See:* node.

vertex plate (of a reflector antenna) A small auxiliary reflector placed in front of the main reflector near its vertex for the purpose of reducing the standing waves in the feed due to reflected waves from the main reflector. (AP/ANT) 145-1993

vertical amplifier (oscilloscopes) An amplifier for signals intended to produce vertical deflection. *See also:* oscillograph. (IM/HFIM) [40]

vertical antenna An antenna consisting of a vertically arranged conductor. *Synonym:* whip antenna. (T&D/PE) 539-1990

vertical-break switch A switch in which the travel of the blade is in a plane perpendicular to the plane of the mounting base. The blade in the closed position is parallel to the mounting base. (SWG/PE) C37.100-1992

vertical bushing A bushing intended to be mounted vertically or at an angle not exceeding 20° from the vertical. (PE/TR) C57.19.03-1996

vertical component of the electric field strength (measurement of power frequency electric and magnetic fields from ac power lines) The root-mean-square (rms) value of the component of the electric field along the vertical line passing through the point of measurement. This quantity is often used to characterize electric field induction effects in objects close to ground level. (T&D/PE) 539-1990, 644-1994

vertical deflection axis (oscilloscopes) The vertical trace obtained when there is a vertical deflection signal and no horizontal deflection signal. (IM) 311-1970w

vertical feed Pertaining to the motion of a punch card along a card feed path with the short edge first. *Contrast:* horizontal feed. (C) 610.10-1994w

vertical footcandles (VFC) Illuminance measured in a vertical plane. (RL) C136.10-1996

vertical, gravity *See:* mass-attraction vertical.

vertical gyro A two-degree-of-freedom gyro with provision for maintaining the spin axis vertical. In this gyro, output signals are produced by gimbal angular displacements that correspond to angular displacements of the case surrounding two nominally orthogonal, horizontal axes. (AES/GYAC) 528-1994

vertical-hold control (television) A synchronizing control that adjusts the free-running period of the vertical-deflection oscillator. (BT/AV) 201-1979w

vertical justification In text formatting, justification of text by adding small increments of vertical space between paragraphs and lines to create a well-spaced output page or a series of pages with equal top and bottom margins. (C) 610.2-1987

vertical linearity (oscilloscopes) The change in deflection factor of an oscilloscope as the display is positioned vertically within the graticule area. *See also:* compression; expansion. (IM) 311-1970w

vertically integrated microprocessor A microprocessor in which vertical microinstructions can be performed. *Contrast:* horizontally integrated microprocessor. (C) 610.10-1994w

vertically polarized field vector A linearly polarized field vector whose direction is vertical. (AP/ANT) 145-1993

vertically polarized plane wave A plane wave whose electric field vector is vertically polarized. (AP/ANT) 145-1993

vertically polarized wave (radio-wave propagation) A linearly polarized wave whose electric field vector is vertical. *Notes:* 1. See parallel polarization. 2. The term "vertical polarization" is commonly employed to characterize groundwave propagation in the medium-frequency broadcast band; these waves, however, have a small component of electric field in the direction of propagation due to finite ground conductivity. (AP) 211-1977s

vertical machine (rotating machinery) A machine whose axis is rotation is approximately vertical. (PE) [9]

vertical magnetic recording *See:* perpendicular magnetic recording.

vertical microinstruction A microinstruction that specifies one of a sequence of operations needed to carry out a machine language instruction. *Note:* Vertical microinstructions are relatively short, 12 to 24 bits, and are called "vertical" because a sequence of such instruction, normally listed vertically on a page, are required to carry out a single machine language instruction. *Contrast:* diagonal microinstruction; horizontal microinstruction. (C) 610.12-1990, 610.10-1994w

vertical plane of a searchlight (illuminating engineering) The plane through the axis of the searchlight drum which contains the elevation angle. (EEC/IE) [126]

vertical polarization *See:* parallel polarization.

vertical reach switch A switch in which the stationary contact is supported by a structure separate from the hinge-mounting base. The blade in the closed position is perpendicular to the hinge-mounting base. (SWG/PE) C37.100-1992, C37.30-1971s

vertical recording A mechanical recording in which the groove modulation is in a direction perpendicular to the surface of the recording medium. (EEC/PE) [119]

vertical redundancy check A parity check performed on each character of a transmitted block of data as the block is received. *Note:* This method can use even or odd parity, and it may be used on non-ASCII characters. *See also:* longitudinal redundancy check. (C) 610.7-1995

vertical riser cable Cable designed for use in long vertical runs, as in tall buildings. (PE) [4]

vertical rod or shaft A component of a switch-operating mechanism designed to transmit motion from an operating handle or power operator to a switch offset bearing or bell crank. (SWG/PE) C37.100-1992, C37.30-1971s

vertical section **(1)** **(metal-clad and station-type cubicle switchgear) (metal-enclosed interrupter switchgear) (metal-enclosed low-voltage power circuit-breaker switchgear)** That portion of the switchgear assembly between two successive vertical delineations and may contain one or more circuit breakers, auxiliary compartments, and associated primary conductors. (SWG/PE) C37.20.1-1993r, C37.20.2-1993
(2) **(nuclear power generating station)** A portion of the motor control center normally containing one vertical bus assembly. (PE/NP) 649-1980s
(3) That portion of the switchgear assembly between two successive vertical delineations and may contain one or more switch compartments and associated primary conductors. (SWG/PE) C37.20.3-1996

(4) That portion of the switchgear assembly between two successive vertical delineations. It may contain one or more units. (SWG/PE) C37.100-1992

vertical switchboard A control switchboard composed only of vertical panels. *Note:* This type of switchboard may be enclosed or have an open rear. An enclosed vertical switchboard has an overall sheet-metal enclosure (not grille) covering back and ends of the entire assembly, access to which is usually provided by doors or removable covers. (SWG/PE) C37.100-1992, C37.21-1985r

⟨`vertical-tab`⟩ The vertical tab character. (C/PA) 9945-2-1993

vertical tabulation (A) On an impact printer or typewriter, movement of the imprint position to another writing line. **(B)** On a display device, movement of the cursor to another display line. *Contrast:* horizontal tabulation. (C) 610.10-1994

vertical tabulation character (VT) A format effector character that causes the print or display position to move to the corresponding position on the next of a series of predetermined lines. (C) 610.5-1990w

very fast front, short duration overvoltage A transient overvoltage in which a short duration, usually unidirectional, voltage is generated (often by GIS disconnect switch operation or when switching motors). High frequency oscillations are often superimposed on the unidirectional wave. (PE/C) 1313.1-1996

very fast front voltage shape This category has not been standardized at this time. (PE/C) 1313.1-1996

very fast transients (VFT) (1) Switching- or breakdown-induced transients with rise times of 3–10 ns that propagate as traveling waves throughout the GIS and cause overvoltage waveforms that vary as a function of position throughout a substation, and that couple to the external enclosure of SF_6-to-air terminations and can thereby cause external sparking between the enclosure and the support structure. (SUB/PE) C37.122.1-1993
(2) A class of transients generated internally within a gas-insluated substation (GIS) characterized by short duration and very high frequency. VFT is generated by the rapid collapse of voltage during breakdown of the insulating gas, either across the contacts of a switching device or line-to-ground during a fault. These transients can have rise times in the order of nanoseconds implying a frequency content extending to about 100 MHz. However, dominant oscillation frequencies, which are related to physical lengths of GIS bus are usually in the 20–40 MHz range. (PE/SUB) 80-2000

very fast transients overvoltage (VFTO) System overvoltages which result from generation of VFT. While VFT is one of the main constituents of VFTO, some lower frequency (≅ 1 MHz) component may be present as a result of the discharge of lumped capacitance (voltage transformers). Typically, VFTO will not exceed 2.0 per unit although higher magnitudes are possible in specific instances. (PE/SUB) 80-2000

very high frequency (VHF) 30–300 MHz. *See also:* radio spectrum. (AP/PROP) 211-1997

very high-frequency omnidirectional range (VOR) (navigation aids) A navigation aid operating at VHF (very high frequency) and providing radial lines of position in any direction as determined by bearing selection within the receiving equipment; it emits a (variable) modulation whose phase, relative to a reference modulation, is different for each bearing of the receiving point from the station. (AES/GCS) 172-1983w

very-high-frequency radar A radar operating at frequencies between 30 MHz and 300 MHz, usually in one of the International Telecommunication Union (ITU) allocated bands 138–144 MHz or 216–225 MHz. (AES) 686-1997

very-high-speed integrated circuit (VHSIC) An integrated circuit designed to operate at extremely high speeds. (C) 610.10-1994w

very large scale integration (VLSI) (A) Pertaining to an integrated circuit containing between 2×104 and 106 transistors in its design. *See also:* ultra-large scale integration; medium scale integration; large scale integration; small scale integration. **(B)** Pertaining to an integrated circuit containing between 5000 and 106 elements. (C) 610.10-1994

very long instruction word (VLIW) An instruction word of uniform length, in excess of 128 bits. (C) 610.10-1994w

very low frequency (VLF) 3–30 kHz. *See also:* radio spectrum. (AP/PROP) 211-1997

very-low-frequency high-potential test An alternating-voltage high-potential test performed at a frequency equal to or less than 1 hertz. *See also:* asynchronous machine. (PE) [9]

very-low-frequency test A test made at a frequency considerably lower than the normal operating frequency. *Note:* In order to facilitate communication and comparison among investigators, this document recommends that the very low frequency used be 0.1 Hz + 25 percent. (PE/EM) 433-1974r

vessel A container such as a barrel, a drum, or a tank for holding fluids or other material. (NESC/IA/PC) 844-1991, [86]

vestigial lobe *See:* shoulder lobe.

vestigial sideband (data transmission) The transmitted portion of the sideband that has been largely suppressed by a transducer having a gradual cutoff in the neighborhood of the carrier frequency, the other sideband being transmitted without much suppression. (PE) 599-1985w

vestigial-sideband modulation A modulation process involving a prescribed partial suppression of one of the two sidebands. (Std100) 270-1964w

vestigial-sideband transmission (facsimile) That method of signal transmission in which one normal sideband and the corresponding vestigial sideband are utilized. *See also:* amplitude modulation; facsimile transmission. (COM) 168-1956w

vestigial-sideband transmitter A transmitter in which one sideband and a portion of the other are intentionally transmitted. *See also:* radio transmitter. (AP/ANT) 145-1983s

VHDL *See:* VHSIC Hardware Description Language.

VHF *See:* very high frequency.

VHSIC Hardware Description Language (VHDL) (1) A language format for a simulatable product description for digital systems. (ATLAS) 1226-1993s
(2) A standard language of the United States Department of Defense, used in the description, design and simulation of very high-speed integrated circuits (VHSIC) and computer logic systems; standardized by IEEE. (C) 610.13-1993w

VF *See:* voice frequency; forward bias.

VFC *See:* vertical footcandles.

V format *See:* variable format.

VFT *See:* very fast transients.

VHF radar *See:* very-high-frequency radar.

VHSIC *See:* very-high-speed integrated circuit.

via In Physical Design Exchange Format (PDEF), a physical connection between two different levels of interconnect, or between a level of interconnect and a physical or logical pin. (C/DA) 1481-1999

vial (liquid-scintillation counters) (liquid-scintillation counting) A glass or plastic sample container that meets the dimensional specifications of International Electrotechnical Commission (IEC) Pub 582-1977. (NI) N42.15-1990, N42.16-1986

via net loss (vnl) (data transmission) The net losses of trunks in the long distance switched telephone network of North America. The trunk is said to be in a via condition when it is an intermediate trunk in a longer switched connection. (PE) 599-1985w

vibrating beam accelerometer (VBA) (inertial sensors) A linear accelerometer whose proof mass is mechanically constrained by a force-sensitive beam resonator. The resultant oscillation frequency is a function of input acceleration. (AES/GYAC) 528-1994

vibrating bell A bell having a mechanism designed to strike repeatedly when and as long as actuated. *See also:* protective signaling. (EEC/PE) [119]

vibrating circuit (telegraph circuits) An auxiliary local timing circuit associated with the main line receiving relay for the purpose of assisting the operation of the relay when the definition of the incoming signals is indistinct. *See also:* telegraphy. (EEC/PE) [119]

vibrating-contact machine regulator A regulator that varies the excitation of an electric machine by changing the average time of engagement of vibrating contacts in the field circuit. (SWG/PE) C37.100-1992

vibrating probe (measurement of dc electric field strength) A device in which a plate is modulated below an aperture of a face plate in the electric field to be measured. *Note:* The meter responds to the oscillating displacement current from the induced charge on the vibrating plate by generating a negative feedback voltage on the face plate to null the signal from the vibrating plate. The electric-field strength is proportional to the feedback voltage. (T&D/PE) 539-1990, 1227-1990r

vibrating-reed relay A relay in which the application of an alternating or a self-interrupted voltage to the driving coil produces an alternating or pulsating magnetic field that causes a reed to vibrate and operate contacts. *See also:* relay. (EEC/REE) [87]

vibrating string accelerometer (VSA) A device that employs one or more vibrating strings whose natural frequencies are affected as a result of acceleration acting on one or more proof masses. (AES/GYAC) 528-1994

vibrating-type conveyor A conveyor consisting of a movable bed mounted at an angle to the horizontal, that vibrates in such a way that the material advances. *See also:* conveyor. (EEC/PE) [119]

vibration An oscillation wherein the quantity is a parameter that defines the motion of a mechanical system. *See also:* oscillation. (SP) [32]

vibration detection system (protective signaling) A system for the protection of vaults by the use of one or more detector buttons firmly fastened to the inner surface in order to pick up and convert vibration, caused by burglarious attack on the structure, to electric impulses in a protection circuit. *See also:* protective signaling. (EEC/PE) [119]

vibration meter An apparatus including a vibration pickup, calibrated amplifier, and output meter for the measurement of displacement, velocity, and acceleration of vibrations. *See also:* instrument. (EEC/PE) [119]

vibration relay A relay that responds to the magnitude and frequency of a mechanical vibration. (SWG/PE) C37.100-1992

vibration test (rotating machinery) A test taken on a machine to measure the vibration of any part of the machine under specified conditions. (PE) [9]

vibrato A family of tonal effects in music that depend upon periodic variations in one or more characteristics of the sound wave. *Note:* When the particular characteristics are known, the term vibrato should be modified accordingly, for example, frequency vibrato; amplitude vibrato; phase vibrato and so forth. (SP) [32]

vibrator (cable plowing) That device which induces the vibration in a vibratory plow. *See also:* vibratory plow. (T&D/PE) 590-1977w

vibratory isolation (cable plowing) Percentage reduction in force transmitted from vibration source to receiver by use of flexible mounting(s) (amount of isolation for a given unit varies with plow blade frequency). (T&D/PE) 590-1977w

vibratory plow (cable plowing) A plow utilizing induced periodic motion(s) of the blade in conjunction with drawbar pull for its movement through the soil. *Note:* Orbital and oscillating plows are types of vibratory plows that are commercially available. (T&D/PE) 590-1977w

vibropendulous error (accelerometer) A cross-coupling rectification error caused by angular motion of the pendulum in a pendulous accelerometer in response to a linear vibratory input. The error varies with frequency and is maximum when the vibratory acceleration is applied in a plane normal to the output axis and at 45° to the input axis. (AES/GYAC) 528-1994

vicenary (A) Pertaining to a selection in which there are 20 possible outcomes. **(B)** Pertaining to the numeration system with a radix of 20. (C) 1084-1986

video (1) (television) A term pertaining to the bandwidth and spectrum position of the signal resulting from television scanning. *Note:* In present usage, video means a bandwidth of the order of several megahertz, and a spectrum position that goes with a direct-current carrier. *See also:* signal wave. (PE/EEC) [119]
(2) Refers to the signal after envelope or phase detection, which in early radar was the displayed signal. Contains the relevant radar information after removal of the carrier frequency. (AES) 686-1997

video board *See:* graphics adapter.

video conferencing A form of teleconferencing that uses television to allow participants to see one another. *See also:* computer conferencing. (C) 610.2-1987

video disk An optical disk used to store visual images that are to appear on a display device. (C) 610.10-1994w

video display *See:* video display terminal.

video display device *See:* display device; video display terminal.

video display terminal (VDT) (1) A device for the presentation of information by controlled excitation of a CRT screen for visual observation designed for interactive use by an operator. (EMC) 1140-1994r
(2) The visual equipment used as a user interface. *See also:* user interface. (SUB/PE) C37.1-1994
(3) A terminal in which a CRT, liquid-crystal, or plasma display device is used for the visual presentation of data. *Synonyms:* video display unit; video terminal. (C) 610.10-1994w

video display unit *See:* video display terminal.

video filter (non-real time spectrum analyzer) (spectrum analyzer) A post detection low-pass filter. (IM) 748-1979w

video-frequency amplifier A device capable of amplifying such signals as comprise periodic visual presentation. *See also:* television. (COM) 167-1966w

video integration A method of utilizing the redundancy of repetitive video signals to improve the output signal-to-noise ratio, by summing successive signals. Also called post-detection integration or noncoherent integration. (AES) 686-1997

video look-up table *See:* color look-up table.

video mapping The electronic superposition of geographic or other data on a radar display. (AES) 686-1997

video mixing The formation of a graphical display image by the merging of two images, one from a display buffer and one from a video signal. (C) 610.6-1991w

video monitor *See:* video display terminal.

video RAM (VRAM) (A) A special type of RAM used to hold and transfer an image onto a display device. *See also:* image memory. **(B)** A dual-port semiconductor memory that is specially designed for raster display devices. *Note:* One port is connected directly to the processor; the other to the display device. (C) 610.10-1994

video stretching The increasing of the duration of a video pulse. (AES) 686-1997

video-telephone call (telephone switching systems) A call between stations equipped to provide video-telephone service. (COM) 312-1977w

video terminal *See:* video display terminal.

videotex A telecommunication system that allows users to interact with a computer by using a specially equipped television set and a keyboard to access remote data banks and to obtain consumer services such as electronic mail, teleordering, and bank services. (C) 610.2-1987, 610.6-1991w

video unit *See:* video display terminal.

vidicon A camera tube in which a charge-density pattern is formed by photoconduction and stored on that surface of the photoconductor that is scanned by an electron beam, usually of low-velocity electrons. *See also:* television. (BT/ED/AV) [34], [45]

Vienna Definition Language A metalanguage used to formally define the syntax and semantics of PL/1. (C) 610.13-1993w

Vienna Development Method (VDM) A specification language developed by IBM; widely used in Europe. (C) 610.13-1993w

view (A) (data management) A subset of a relational database, formed by applying relational operations to the base relations represented. *See also:* logical database. **(B) (data management)** A subset of a data model. *See also:* external schema. (C) 610.5-1990
(2) (A) A collection of subject domains, classes, relationships, responsibilities, properties, constraints, and notes assembled or created for a certain purpose and covering a certain scope. A view may cover the entire area being modeled or only a part of that area. **(B)** A collection of entities and assigned attributes (domains) assembled for some purpose. (C/SE) 1320.2-1998

view area (computer graphics) A rectangular region in a world coordinate system, containing a subset of the model. (C) 610.6-1991w

viewdata A form of videotex that allows users to access remote data banks via telephone and cable lines. *Contrast:* teletext. (C) 610.2-1987

view diagram A graphic representation of the underlying semantics of a view. (C/SE) 1320.2-1998

viewing area (oscilloscopes) The area of the phosphor screen of a cathode-ray tube that can be excited to emit light by the electron beam. *See also:* oscillograph. (IM) [39]

viewing operation *See:* viewing transformation.

viewing time (storage tubes) The time during which the storage tube is presenting a visible output corresponding to the stored information. *See also:* storage tube. (ED) 158-1962w

viewing time, maximum usable *See:* maximum usable viewing time.

viewing transformation (computer graphics) The process of mapping positions in world coordinates to positions in normalized device coordinates. (See the corresponding figure.) *Synonym:* normalization transformation; viewing operation.

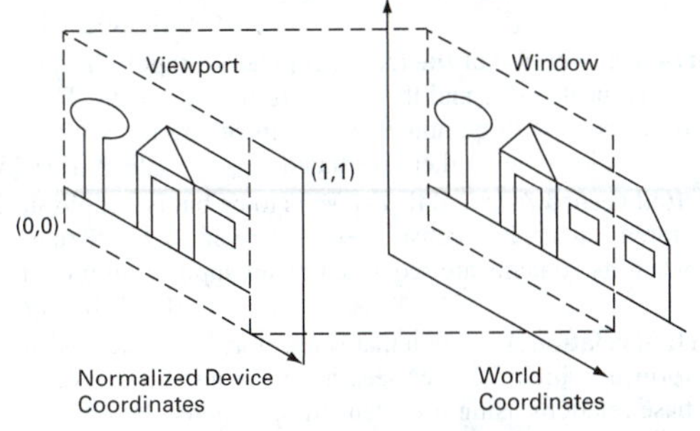

viewing transformation

(C) 610.6-1991w

view integration (data management) The integration of two or more logical views into a single logical view. *Note:* This is generally done in the normalization stage of database design. (C) 610.5-1990w

view plane (computer graphics) A two-dimensional display surface onto which a three-dimensional image is projected. (C) 610.6-1991w

viewpoint statement A brief statement of the perspective of an IDEF0 model that is presented in the A-0 context diagram of the model. (C/SE) 1320.1-1998

viewport (computer graphics) A rectangular portion of a display surface onto which the contents of a display image are mapped. (C) 610.6-1991w

view surface The medium on which graphical display images appear. For example, plotter paper or the screen of a cathode ray tube display device. (C) 610.6-1991w

view volume (computer graphics) A region of a three-dimensional world coordinate system that is to be visible as the display image. (C) 610.6-1991w

vignette A self-contained portion or a scenario. (DIS/C) 1278.3-1996

virgin bar/coil A virgin bar/coil is a new, completely manufactured bar/coil with all armor tapes and coatings which, except for any semiconducting (or conducting) rubber coating, would have been installed in the stator core, but is now used in the thermal cycle tests. (PE/EM) 1310-1996

virgin medium (1) (data management) A data medium in or on which data have never been recorded. (C) 610.5-1990w
(2) A data medium on which neither marks of reference, nor user data, are or have ever been recorded; for example, paper that is unmarked, or magnetic tape that has never recorded information. *See also:* blank medium; empty medium. (C) 610.10-1994w

virtual address (1) (software) In a virtual storage system, the address assigned to an auxiliary storage location to allow that location to be accessed as though it were part of main storage. *Contrast:* real address. (C) 610.12-1990
(2) The address that a program uses to access a memory location or memory-mapped device register. (Depending on the presence or absence of memory mapping hardware in the system, and whether or not that mapping hardware is enabled, a virtual address may or may not be the same as the physical address that appears on an external bus.) (C/BA) 1275-1994
(3) In a virtual storage system, the address of a storage location. *See also:* physical address; direct reference address; address translator. (C) 610.10-1994w

virtual address space The set of all possible virtual addresses that a process can use to identify an instruction. (C) 610.10-1994w

virtual attribute (data management) An attribute that is derived from stored data by means of user-defined operations rather than being stored. (C) 610.5-1990w

Virtual Bridged Local Area Network (LAN) A Bridged LAN in which the existence of one or more VLAN-aware Bridges allows the definition, creation, and maintenance of VLANs. (C/LM) 802.1Q-1998

virtual cathode (electron tube) (potential-minimum surface) A region in the space charge where there is a potential minimum that, by reason of the space charge density, behaves as a source of electrons. (ED) [45], [84]

virtual channel A channel that behaves as a transducer from the point of view of the Network Capable Application Processor even though nothing outside of the Smart Transducer Interface Module is sensed or changed. Virtual channels are useful for setting or reading operating parameters of other channels. (IM/ST) 1451.2-1997

virtual channel identifier A label that is used to distinguish between the different virtual channels. A virtual channel is a logical association between entities that enables unidirectional transfer of *segments* between the entities. In the context of this part of ISO/IEC 8802, the VCI label can be used to allow a transmitter to distinguish between different outgoing *protocol data units (PDUs),* and is used to allow a receiver to determine whether to receive an incoming segment as well as to distinguish between incoming PDUs. (LM/C) 8802-6-1994

virtual circuit (1) The generic concept of a logical connection. A virtual circuit may be implemented by means of a frame-switched service or a packet-switched service. (C/LM/COM) 802.9a-1995w, 8802-9-1996
(2) In networking, a circuit connecting a source and a sink that may be physically accomplished by using different circuit configurations during transmission of a message. *Note:* A virtual circuit looks like a permanent connection to the user. A switched virtual circuit requires call control and can be established and terminated by the user at will. *See also:* permanent virtual circuit; data circuit; switched virtual circuit; virtual data connection. (C) 610.7-1995

virtual circuit service *See:* connection-oriented service.

virtual data connection A data connection in which one or more of the data circuits are interconnected by a virtual circuit. (C) 610.7-1995

virtual disk *See:* RAM disk.

virtual duration (1) (surge arresters) (of a peak of a rectangular-wave current or voltage impulse) The time during which the amplitude of the wave is greater than 90 percent of its peak value. (PE/PSIM) [8], [84], 4-1978s
(2) (impulse) (of wave front) The virtual value of the duration of the wave front is as follows:

a) For voltage waves with a wave front duration of less than 30 microseconds, either full or chopped on the front, crest, or tail, 1.67 times the time it takes for the voltage to increase from 30% to 90% of its crest value.
b) For voltage waves with a wave front duration of the 30 microseconds or greater, the time it takes for the voltage to increase from actual zero to maximum crest value.
c) For current waves, 1.25 times the time it takes for the current to increase from 10% to 90% of crest value.

(SPD/PE) C62.62-2000

virtual duration of wavefront (of an impulse) The virtual value for the duration of the wavefront is as follows:

— For voltage waves with wavefront duration less than 30 μs, either full or chopped on the front, crest, or tail, 1.67 times the time for the voltage to increase from 30−90% of its crest value;
— For voltage waves with wavefront duration of 30 μs or more, the time taken by the voltage to increase from actual zero to maximum crest value;
— For current waves, 1.25 times the time for the current to increase from 10%−90% of crest value.

(SPD/PE) C62.11-1999

virtual field (data management) A field that appears to be but is not physically stored; rather, it is constructed or derived from existing data when its contents are requested by an application program. (C) 610.5-1990w

virtual filestore An abstract model for describing files and filestores, and the possible actions on them. (C/PA) 1238.1-1994w

virtual front time (T_1) (of a lightning impulse) The time interval between the instants when a smooth impulse is 30% and 90% of the peak value multiplied by 1.67. (PE/PSIM) 4-1995

virtual height (1) (data transmission) The apparent height of an ionized layer determined from the time interval between the transmitted signal and the ionospheric echo at vertical incidence, assuming that the velocity of propagation is the velocity of light over the entire path. (PE) 599-1985w
(2) The apparent height of reflection of a radio wave from an ionized layer. It is determined from the time interval between the transmitted pulse and the ionospheric echo at vertical incidence, assuming that the velocity of propagation is the velocity of light (in vacuum) over the entire path. (AP/PROP) 211-1997

virtual instant of chopping (voltage testing) The instant preceding point C on the figures (under virtual front time), by 0.3 times the (estimated) virtual time of voltage collapse during chopping. (PE/PSIM) [55]

virtual instrument software architecture (VISA) The general name given to the VPP (VXI Plug & Play) 4 Specification and its associated architecture. The architecture consists of two main VISA components: the VISA resource manager and the VISA instrument control resources. (SCC20) 1226-1998

virtual junction temperature The temperature of the active semiconductor element of a semiconductor device based on a simplified representation of the thermal and electrical behavior of the device. It is particularly applicable to multi-junction semiconductor devices. (IA) [12]

Virtual Local Area Network (VLAN) (1) A subset of the active topology of a Bridged Local Area Network. Associated with each VLAN is a VLAN Identifier (VID). (C/LM) 802.1Q-1998
(2) A Group to which each member Remote Bridge attaches by a single Virtual Port. *Note:* Such a Group comprises a single Subgroup. (C/LM) 802.1G-1996

Virtual LAN Port A Virtual Port by which a Remote Bridge attaches to a Virtual LAN. (C/LM) 802.1G-1996

virtual machine (software) A functional simulation of a computer and its associated devices. *See also:* simulation; computer. (C/SE) 729-1983s

virtual medical device (VMD) An abstract representation of a medical device system (or a patient care system) described in Medical Device Data Language-specific terminology. (EMB/MIB) 1073-1996

virtual memory *See:* virtual storage.

Virtual Mesh A Group to which each member Remote Bridge attaches by a set of Individual Virtual Ports, one for each other Remote Bridge in the Group. *Note:* Such a Group comprises $n \times (n-1)/2$ two-member Subgroups, where n is the number of Bridges in the Group. (C/LM) 802.1G-1996

virtual origin The intersection with the time axis of a straight line drawn as a tangent to the steepest portion of the impulse or response curve. (PE/PSIM) 4-1995

virtual peak value *See:* peak value.

virtual point picture character *See:* radix point character.

Virtual Port An abstraction of a Remote Bridge's point(s) of attachment to the non-LAN communications equipment of a single Group. A Virtual Port represents the capability for bi-directional communication with one, some, or all of the other Remote Bridges in that Group. *Note:* A Bridge can attach by two or more Virtual Ports to a single Group. (C/LM) 802.1G-1996

virtual rate of rise of the front (impulse voltage) The quotient of the peak value and the virtual front time. *Note:* The term peak value is to be understood as including the term virtual peak value unless otherwise stated. (PE/PSIM) [55]

virtual record A record that appears to be but is not physically stored; rather, it is constructed or derived from existing data when its contents are requested by an application program. (C) 610.5-1990w

virtual relation A relation that is not stored in a database in the form in which the user sees it, but is instead derived from base relations using user-defined operations. (C) 610.5-1990w

virtual resource (1) An abstract representation of a test capability, independent of its physical realization. (ATLAS) 1226-1993s

(2) An abstract representation of a UUT-directed, signal-oriented test capability. It is represented in ALTPI by instantiation of an Ada generic package with links to the test capability and path description objects. (ATLAS) 1226.2-1993w
(3) A notional test resource, the performance characteristics of which conform to a summary of related test requirements. (SCC20) 771-1998

virtual sequential access method (VSAM) An access method for direct or sequential access to data records on storage devices in which auxiliary storage can be addressed as though it were part of main storage. Pages of data are transferred as needed between auxiliary and main storage. *See also:* indexed sequential access method; basic sequential access method. (C) 610.5-1990w

virtual steepness of voltage during chopping (surge arresters) The quotient of the estimated voltage at the instant of chopping and the virtual time of voltage collapse. (PE/PSIM) [55], [84], [8]

virtual steepness of wavefront of an impulse (surge arresters) The slope of the line that determines the virtual-zero time. It is expressed in kilovolts per microsecond or kiloamperes per microsecond. (PE) [8], [84]

virtual storage (1) (software) A storage allocation technique in which auxiliary storage can be addressed as though it were part of main storage. Portions of a user's program and data are placed in auxiliary storage, and the operating system automatically swaps them in and out of main storage as needed. *Synonyms:* virtual memory; multilevel storage. *Contrast:* real storage. *See also:* virtual address; paging. (C) 610.12-1990
(2) The storage space that may be regarded as addressable main storage by the user of a computer system in which virtual addresses are mapped into real addresses. *Note:* The size of virtual storage is limited by the addressing scheme of the computer system and by the amount of auxiliary storage allocated to such use, but not by the actual number of main storage locations. (C) 610.10-1994w

virtual terminal A terminal that is defined as a standard on a network that can handle diverse terminals. (C) 610.7-1995

virtual time *See:* simulated times.

virtual time of voltage collapse during chopping 1.67 times the time interval between points C and D on the figures attached to the definition of virtual instant of chopping. (PE/PSIM) [55]

virtual time to chopping (impulse voltage) The time interval between the virtual origin and the virtual instant of chopping. (PE/PSIM) [55]

virtual time to half-value (T_2) The time interval between the virtual origin and the instant on the tail when the voltage has decreased to half of the peak value. (PE/PSIM) 4-1995

virtual total duration of a rectangular impulse current (high voltage testing) The virtual total duration of a rectangular impulse current is the time during which the amplitude of the impulse is greater than 10 percent of its peak value. If oscillations are present on the front, a mean curve should be drawn in order to determine the time at which the 10 percent value is reached. (PE/PSIM) 4-1978s

virtual zero point (1) (surge arresters) (of an impulse) The intersection with the zero axis of a straight line drawn through points on the front of the current wave at 10% and 90% crest value, or through points on the front of the voltage wave at 30% and 90% crest value. (PE/SPD) C62.1-1981s, C62.62-2000
(2) (of an impulse) The intersection with the time axis of a straight line drawn through points on the front of the current wave at 10% and 90% crest value or through points on the front of the voltage wave at 30% and 90% crest value. (SPD/PE) C62.22-1997, C62.11-1999

virtual zero time (surge arresters) (impulse voltage or current in a conductor) The point on a graph of voltage-time or current-time determined by the intersection with the zero voltage or current axis, of a straight line drawn through two points on the front of the wave: for full voltage waves and voltage waves chopped on the front, peak, or ail, the reference points shall be 30% of the peak value, and 82 for current waves the reference points shall be 10% and 90% of the peak value. *Synonym:* virtual origin. (PE) [8], [84]

viscous friction The component of friction that is due to the viscosity of a fluid medium, usually idealized as a force proportional to velocity, and that opposes motion. *See also:* feedback control system. (IA/IM/ICTL/APP/IAC) [69], [60], [120]

vise grip *See:* strand restraining clamp.

visibility (1) (meteorological) (illuminating engineering) A term that denotes the greatest distance that selected objects (visibility markers) or lights of moderate intensity of the order of 25 candles can be seen and identified under specified conditions of observation. The distance may be expressed in kilometers or miles in the USA until the metric system becomes more widely used. (IE/EEC) [126]
(2) (light-emitting diodes) The quality or state of being perceivable by the eye. In many outdoor applications, visibility is defined in terms of the distance at which an object can be just perceived by the eye. In indoor applications it usually is defined in terms of the contrast or size of a standard test object, observed under standardized viewing conditions, having the same threshold as the given object. *See also:* visual field. (IE/EEC) [126]
(3) (computer graphics) A segment attribute that determines if a display element is or is not to be displayed. (C) 610.6-1991w
(4) The specification, for a property, of "who can see it?"—i.e., whose methods can reference the property. Visibility is either private, protected, or public. (C/SE) 1320.2-1998

visibility factor (A) (pulsed radar) The ratio of single-pulse signal energy to noise power per unit bandwidth that provides stated probabilities of detection and false alarm on a display, measured in the intermediate-frequency portion of the receiver under conditions of optimum bandwidth and viewing environment. **(B) (continuous-wave radar)** The ratio of single-look signal energy to noise power per unit bandwidth using a filter matched to the time on target. The equivalent term for radar using automatic detection is detectability factor; for operation in a clutter environment a clutter visiblity factor is defined. (AES) [42]
(2) (A) (pulsed radar) The ratio of single-pulse signal energy to noise power per unit bandwidth that provides stated probability of detection for a given false alarm probability on a display, measured in the intermediate-frequency portion of the receiver under conditions of optimum bandwidth and viewing environment. **(B)** (continuous-wave radar) The ratio of single-look signal energy to noise power per unit bandwidth using a filter matched to the time on target. *Note:* The equivalent term for radar using automatic detection is detectability factor; for operation in clutter environment, a clutter visibility factor is defined. (AES) 686-1997

visibility level (illuminating engineering) A contrast multiplier to be applied to the visibility reference function to provide the luminance contrast required at different levels of task background luminance to achieve visibility for specified conditions relating to the task and observer. (EEC/IE) [126]

visibility performance criteria function (illuminating engineering) A function representing the luminance contrast required to achieve 99% visual certainty for the same task used for the visibility reference function, including the effects of dynamic presentation and uncertainty in task location. (EEC/IE) [126]

visibility reference function (illuminating engineering) A function representing a luminance contrast required at different levels of task background luminance to achieve visibility threshold for the visibility reference task consisting of a 4 min disk exposed for 1/5s. (EEC/IE) [126]

visible corona (1) (as applies to an air switch) A luminous discharge due to ionization of the air surrounding an air switch, caused by a voltage gradient exceeding a certain critical value. (SWG/PE) C37.34-1994
(2) A luminous discharge due to ionization of the air surrounding a device, caused by voltage gradient exceeding a certain critical value.
(SWG/PE) C37.100-1992, C37.30-1971s

visible radiation (light) (laser maser) Electromagnetic radiation which can be detected by the human eye. It is commonly used to describe wavelengths which lie in the range between 0.4 μm and 0.7 μm. (LEO) 586-1980w

visible radiation emitting diode (light-emitting diodes) A semiconductor device containing a semiconductor junction in which visible light is nonthermally produced when a current flows as a result of an applied voltage. (ED) [127]

visible range For the case in which the field pattern of a continuous line source, L_λ wavelengths long, is expressed as a function of ψ ($\psi = L_\lambda \cos \theta$, the angle θ is measured from an axis coincident with the line source), that part of the infinite range of ψ that corresponds to a variation in the directional angle θ from π to 0 radians; that is, $-L_\lambda < \psi < L_\lambda$. *Notes:* 1. All values of ψ outside the visible range are said to be in the invisible range. 2. The formulation of the field pattern as a function of ψ is useful because the side lobes in the invisible range are a measure of the Q of the antenna. 3. This concept of a visible range can be extended to other antenna types.
(AP/ANT) 145-1993

visible spectrum *See:* light.

visit To access the node of a tree during a traversal.
(C) 610.5-1990w

visual A visual user interface control that provides a type indicating the specification for the color handling of a screen. A visual type may support multiple variations of depths and colormaps. Examples are monochrome, gray-scale, and several types of color. A visual is specified by a *depth* (for example, 8 b) and a *visual class* (for example, `Gray-Scale` or `PseudoColor`). A `Shell` widget or a `DrawArea` widget can have a nondefault visual. Other widgets use the visuals of their nearest shell or `DrawArea` ancestor. An application in which eligible widgets have nondefault visuals is termed a *multivisual* application. (C) 1295-1993w

visual acuity (illuminating engineering) A measure of the ability to distinguish fine details. Quantitatively, it is the reciprocal of the minimum angular separation in minutes of two lines of width subtending one minute of arc when the lines are just resolvable as separate. (EEC/IE) [126]

visual angle (illuminating engineering) The angle which an object or detail subtends at the point of observation. It usually is measured in minutes of arc. (EEC/IE) [126]

visual approach slope indicator system (VASIS) (illuminating engineering) The system of angle-of-approach lights accepted as a standard by the International Civil Aviation Organization, comprising two bars of lights located at each side of the runway near the threshold and showing red or white or a combination of both (pink) to the approaching pilot depending upon his position with respect to the glide path.
(EEC/IE) [126]

visual-aural radio range *See:* visual-aural range.

visual-aural range (navigation aids) A special type of VHF (very high frequency) radio range which provides: 1) two reciprocal radio lines of position presented to the pilot visually on a course deviation indicator; and 2) two reciprocal radial lines of position presented to the pilot as interlocked and alternate A and N aural code signals. The aural lines of position are displaced 90° from the visual and either may be used to resolve the ambiguity of the other.
(AES/GCS) 172-1983w

visual comfort probability (A) (illuminating engineering) The rating of a lighting system expressed as a percent of people who, when viewing from a specified location and in a specified direction, will be expected to find it acceptable in terms of discomfort glare. Visual Comfort Probability is related to Discomfort Glare Rating. *See also:* discomfort glare rating. **(B)** A rating of a lighting system expressed as a percentage of people who, if seated at the center of the rear of a room, will find the lighting visually acceptable in relation to the perceived glare. (EEC/IE/IA/PSE) [126], 241-1990

visual display terminal *See:* video display terminal.

visual display unit *See:* video display terminal.

visual field (illuminating engineering) The locus of objects or points in space which can be perceived when the head and eyes are kept fixed. Separate monocular fields for the two eyes may be specified or the combination of the two.
(EEC/IE) [126]

visual inspection Qualitative observation of physical characteristics utilizing the unaided eye or with stipulated levels of magnification. (EEC/AWM) [105]

visual perception (illuminating engineering) The interpretation of impressions transmitted from the retina to the brain in terms of information about a physical world displayed before the eye. *Note:* Visual perception involves any one or more of the following: recognition of the presence of something (object, aperture or medium); identifying it; locating it in space; noting its relation to other things; identifying its movement, color, brightness or form. (EEC/IE) [126]

visual performance (illuminating engineering) The quantitative assessment of the performance of a task taking into consideration speed and accuracy. (EEC/IE) [126]

visual photometer (illuminating engineering) One in which the equality of brightness of two surfaces is established visually. *Note:* The two surfaces usually are viewed simultaneously side by side. This is satisfactory when the color difference between the test source and comparison source is small. However, when there is a color difference, a flicker photometer provides more precise measurements. In this type of photometer the two surfaces are viewed alternately at such a rate that the color sensations either nearly or completely blend and the flicker due to brightness difference is minimized by adjusting the comparison source.
(EEC/IE) [126]

visual radio range (navigation aids) Any radio range (such as VOR [very high-frequency omnidirectional range]) whose primary function is to provide lines of position to be flown by visual reference to a course deviation indicator.
(AES/GCS) 172-1983w

visual range (illuminating engineering) (of a light or object) The maximum distance at which that particular light (or object) can be seen and identified. (EEC/IE) [126]

visual scanner *See:* optical scanner.

visual signal device (protective signaling) A general term for pilot lights, annunciators, and other devices providing a visual indication of the condition supervised. *See also:* protective signaling. (EEC/PE) [119]

visual surround (illuminating engineering) Includes all portions of the visual field except the visual task.
(EEC/IE) [126]

visual task (A) (illuminating engineering) Conventionally designates those details and objects which must be seen for the performance of a given activity, and includes the immediate background of the details or objects. *Note:* The term visual task as used is a misnomer because it refers to the visual display itself and not the task of extracting information from it. The task of extracting information also has to be differentiated from the overall task performed by the observer. **(B)** Work that requires illumination in order for it to be accomplished. (EEC/IE/IA/PSE) [126], 241-1990

visual task evaluator (illuminating engineering) A contrast reducing instrument which permits obtaining a value of luminance contrast, called the equivalence contrast C of a standard visibility reference task giving the same visibility as that of a task whose contrast has been reduced to threshold when the background luminances are the same for the task and the reference task. *See also:* equivalent contrast. (EEC/IE) [126]

visual terminal *See:* video display terminal.

visual transmitter All parts of a television transmitter that handle picture signals, whether exclusively or not. *See also:* television. (AP/ANT) 145-1983s

visual transmitter power The peak power output during transmission of a standard television signal. *See also:* television. (EEC/PE) [119]

vital *See:* safety critical.

vital area An area that contains vital equipment. (PE/NP) 692-1997

vital circuit Any circuit the function of which affects the safety of train operation. (EEC/PE) [119]

vital equipment Any equipment, system, device, or material, the failure of which could directly or indirectly endanger the public health and safety by exposure to radiation. Equipment or systems that would be required to function to protect public health and safety following such failure, destruction, or release are also considered to be vital. (PE/NP) 692-1997

vital function A function in a safety-critical system that is required to be implemented in a fail-safe manner. *Note:* Vital functions are a subset of safety-critical functions. (VT/RT) 1483-2000, 1474.1-1999

vital services Services normally considered to be essential for the safety of the ship and its passengers and crew. These usually include propulsion, steering, navigation, fire fighting, emergency power, emergency lighting, electronics, and communications functions. The identification of all vital services in a particular vessel is generally specified by the government regulatory agencies. (IA/MT) 45-1998

vitreous silica (fiber optics) Glass consisting of almost pure silicon dioxide (SiO_2). *Synonym:* fused silica. *See also:* fused quartz. (Std100) 812-1984w

VLAN-aware A property of Bridges or end stations that recognizes and supports VLAN-tagged frames. (C/LM) 802.1Q-1998

VLAN-tagged frame A tagged frame whose tag header carries both VLAN identification and priority information. (C/LM) 802.1Q-1998

VLAN-unaware A property of Bridges or end stations that do not recognize VLAN-tagged frames. (C/LM) 802.1Q-1998

VLF *See:* very low frequency.

VLIW *See:* very long instruction word.

VLSI *See:* very large scale integration.

V_{max} input The maximum allowable input voltage rating at which a unit under test can operate to specifications. (PEL) 1515-2000

VMD *See:* virtual medical device.

VMEbus Refers to IEEE Std 1014-1987, which defines a 32-bit backplane bus. (C/MM) 1596.5-1993

V_{min} input The minimum allowable input voltage rating at which a unit under test can operate to specifications. (PEL) 1515-2000

VMOS *See:* V-channel metal-oxide semiconductor.

V-network An artificial mains network of specified disymmetric impedance used for two-wire mains operation and comprising resistors in V formation connected between each conductor and earth. *See also:* electromagnetic compatibility. (EMC) [53]

vnl *See:* via net loss.

V_{nom} input The stated or objective value of the input voltage, which may not be the actual value measured. The value should be between the minimum and maximum input value. (PEL) 1515-2000

V number *See:* normalized frequency.

vodas A system for preventing the over-all voice-frequency singing of a two-way telephone circuit by disabling one direction of transmission at all times. The name is derived from the initial letters of the expression voice-operated device antising. *See also:* voice-frequency telephony. (EEC/PE) [119]

vogad A voice-operated device is used to give a substantially constant volume output for a wide range of inputs. The name is derived from the initial letters of the expression voice-operated gain-adjusted device. *See also:* voice-frequency telephony. (EEC/PE) [119]

voice band (measuring longitudinal balance of telephone equipment operating in the voice band) That part of the audio-frequency range that is employed for the transmission of speech. For the purpose of IEEE Std 455-1985, the voice band extends from 50 hertz (Hz) to 4000 Hz. (COM/C/TA) 455-1985w, 610.7-1995

voice-band channel A channel that is suitable for transmission of speech, or analog data and has a maximum usable frequency range of 300–3400 cycles per second. *See also:* sub-voice-band channel; wideband channel. (C) 610.7-1995

voice channel (mobile communication) A transmission facility defined by the constraints of the human voice. For mobile-communication systems, a voice channel may be considered to have a range of approximately 250 to 3000 hertz; since the Rules and Regulations of the Federal Communications Commission do not authorize the use of modulating frequencies higher than 3000 hertz for radiotelephony or tone signaling on radio frequencies below 500 megahertz. *See also:* channel spacing. (VT) [37]

voice-coil actuator An access arm that moves the head in relation to a magnetic field produced by a coil of wire in the manner of a speaker voice coil. (C) 610.10-1994w

voice frequency (1) (data transmission) A frequency lying within that part of the audio range which is employed for the transmission of speech. *Note:* Voice frequencies used for commercial transmission of speech usually lie within the range 200 to 3500 Hz (hertz). (PE) 599-1985w
(2) The analog signal bandwidth of approximately 300–3400 Hz used in telephone circuits. (SUB/PE) 999-1992w

voice-frequency carrier telegraph (data transmission) A telegraph transmission system which provides several narrow-band individual channels in the voice-frequency range. (PE) 599-1985w

voice-frequency telephony (data transmission) That form of telephony in which the frequencies of the components of the transmitted electric waves are substantially the same as the frequencies of corresponding components of the actuating acoustical waves. (PE) 599-1985w

voice-grade A channel suitable for the transmission of speech, digital or analog data, or facsimile. (SUB/PE) 999-1992w

voice-grade channel (1) (data transmission) A channel suitable for the transmission of speech, digital or analog data, or facsimile, generally with a frequency range of about 300 to 3000 Hz (hertz). (PE) 599-1985w
(2) *See also:* voice-band channel. (C) 610.7-1995

voice-operated device A device that can be controlled by human speech commands. *See also:* speech synthesizer. (C) 610.10-1994w

voice processing Information processing in which the human voice is the data input. *See also:* office automation. (C) 610.2-1987

void volume ratio The volume of the void spaces between stones divided by the total volume occupied by the stones in a stone-filled collecting pit. (SUB/PE) 980-1994

volatile (electronic data processing) Pertaining to a storage device in which data cannot be retained without continuous power dissipation, for example, an acoustic delay line. *Note:* Storage devices or systems employing nonvolatile media may or may not retain data in the event of planned or or accidental power removal. (C/MIL) 162-1963w, [2]

volatile flammable liquid A flammable liquid having a flash point below 38°C (100°F) or whose temperature is above its flash point. (NESC/NEC) [86]

volatile storage A type of storage in which information cannot be retained without continuous power application. *Contrast:* nonvolatile storage. (C) 610.10-1994w

volcas A voice-operated device that switches loss out of the transmitting branch and inserts loss in the receiving branch under control of the subscriber's speech. The name is derived from the initial letters of the expression voice-operated loss control and suppressor. *See also:* voice-frequency telephony. (EEC/PE) [119]

volt (metric practice) (unit of electric potential difference and electromotive force) The difference of electric potential between two points of a conductor carrying a constant current of one ampere, when the power dissipated between these points is equal to one watt. (QUL) 268-1982s

volta effect *See:* contact potential.

voltage (1) (electromotive force) (general) (along a specified path in an electric field). The dot product line integral of the electric field strength along this path. *Notes:* 1. Voltage is a scalar and therefore has no spatial direction. 2. As here defined, voltage is synonymous with potential difference only in an electrostatic field. 3. In cases in which the choice of the specified path may make a significant difference, the path is taken in an equiphase surface unless otherwise noted. 4. It is often convenient to use an adjective with voltage, for example, phase voltage, electrode voltage, line voltage, etc. The basic definition of voltage applies and the meaning of adjectives should be understood or defined in each particular case. *See also:* reference voltage. (Std100) 270-1966w
(2) **(A)** (voltage of circuit not effectively grounded) The highest nominal voltage available between any two conductors of the circuit. *Note:* If one circuit is directly connected to and supplied from another circuit of higher voltage (as in the case of an autotransformer), both are considered to be of the higher voltage, unless the circuit of the lower voltage is effectively grounded, in which case its voltage is not determined by the circuit of the higher voltage. Direct connection implies electric connection as distinguished from connection merely through electromagnetic or electrostatic induction. **(B)** (voltage of a constant current circuit) The highest normal full-load voltage of the current. **(C)** (voltage of an effectively grounded circuit) The highest nominal voltage available between any conductor of the circuit and ground unless otherwise indicated. **(D)** The effective (rms) potential difference between any two conductors or between a conductor and ground. Voltages are expressed in nominal values unless otherwise indicated. The nominal voltage of a system or circuit is the value assigned to a system or circuit of a given voltage class for the purpose of convenient designation. The operating voltage of the system may vary above or below this value. (NESC) C2-1997
(3) (surge arresters) (electromotive force) The voltage between a part of an electric installation connected to a grounding system and points on the ground at an adequate distance (theoretically at an infinite distance) from any earth electrodes. (PE) [8], [84]
(4) (of a circuit) The greatest root-mean-square (effective) difference of potential between any two conductors of the circuit concerned. Some systems, such as 3-phase 4-wire, single-phase 3-wire, and 3-wire direct-current may have various circuits of various voltages. (NESC/NEC) [86]

voltage amplification (1) An increase in signal voltage magnitude in transmission from one point to another or the process thereof. *See also:* amplifier. (Std100) 270-1966w
(2) (transducer) The scalar ratio of the signal output voltage to the signal input voltage. Warning: By incorrect extension of the term decibel, this ratio is sometimes expressed in decibels by multiplying its common logarithm by 20. It may be correctly expressed in decilogs. *Note:* If the input and.or output power consist of more than one component, such as multifrequency signal or noise, then the particular components used and their weighting must be specified. *See also:* transducer. (Std100) 270-1966w
(3) (magnetic amplifier) The ratio of differential output voltage to differential control voltage. (MAG) 107-1964w

voltage and power directional relay (power system device function numbers) A relay that permits or causes the connection of two circuits when the voltage difference between them exceeds a given value in a predetermined direction and causes these two circuits to be disconnected from each other when the power flowing between them exceeds a given value in the opposite direction. (SUB/PE) C37.2-1979s

voltage attenuation (1) (data transmission) An adjustable device for reducing the amplitude of a wave without introducing distortion. An adjustable passive network that reduces the power level of a signal without introducing appreciable distortion. (PE) 599-1985w
(2) (analog computer) A device for reducing the amplitude of a signal without introducing appreciable distortion. (C) 165-1977w

voltage at the instant of chopping The voltage at the instant of the initial discontinuity. (PE/PSIM) 4-1995

voltage balance relay A balance relay that operates by comparing the magnitudes of two voltage inputs. (SWG/PE) C37.100-1992

voltage buildup (rotating machinery) The inherent establishment of the excitation current and induced voltage of a generator. (PE) [9]

voltage circuit (1) (A) (ac high-voltage circuit breakers) That part of the synthetic test circuit from which the major part of the test voltage is obtained. **(B)** An input circuit to which is applied a voltage or a current that is a measure of primary voltage. (SWG/PE) C37.081-1981
(2) (instrument) That combination of conductors and windings of the instrument to which is applied the voltage of the circuit in which a given electrical quantity is to be measured, or a definite fraction of that voltage, or a voltage or current dependent upon it. *See also:* moving element; instrument; watthour meter. (EEC/AII) [102]
(3) That part of the synthetic test circuit from which the major part of the test voltage is obtained. (SWG/PE) C37.083-1999

voltage clamp (converter circuit elements) (self-commutated converters) A clamp that limits the peak voltage across a semiconductor device. (IA/SPC) 936-1987w

voltage clamping ratio (low voltage varistor surge arresters) A figure of merit measure of the varistor voltage clamping effectiveness as determined by the ratio of clamping voltage to rated root-mean-square (rms) voltage, or by the ratio of clamping voltage to rated direct-current (dc) voltage. (PE) [8]

voltage class *See:* medium-voltage power cable; control cable; low-level analog signal cable; low-level digital signal circuit cable; low-voltage power cable (s).

voltage classes Voltage classes are as shown in the corresponding figure.

VOLTAGE CLASSES
NOMINAL SYSTEM VOLTAGE

Voltage class	Two Wire	Three Wire	Four Wire	Maximum Voltage[3]
	Single-Phase Systems			
LOW VOLTAGE SYSTEMS (IEEE Std for Industrial & Commercial Power Systems[1])	*(120)*			*127*
		120/240		*127/254*
	Three-Phase Systems			
			208Y/120	220Y/127
		(240)	240/120	245/127
		480	480Y/277	508Y/293
		(600)		635
MEDIUM VOLTAGE (IEEE Std for Industrial & Commercial Power Systems[1]); ANSI C84.1–1977 no voltage class stated		(2400)		2540
		4160	4160Y/2400	4400Y/2540
		(4800)		5080
		(6900)		7260
			(8320Y/4800)	8800Y/5080
			(12000Y/6930)	12700Y/7330
			12470Y/7200	13200Y/7620
			13200Y/7620	13970Y/8070
		13800	(13800Y/7970)	14520Y/8380
			(20780Y/12000)	22000Y/12700
			(22860Y/13200)	24200Y/13970
		(23000)		24340
			24940Y/14400	26400Y/15240
		(34500)	34500/19920	36510Y/21080
HIGH VOLTAGE (IEEE Std for Industrial & Commercial Power Systems[1]); ANSI C84.1–1977 HIGHER VOLTAGE SYSTEMS		(46 kV)		48.3 kV
		69 kV		72.5 kV
		115 kV		121 kV
		138 kV		145 kV
		(161 kV)		169 kV
		230 kV		242 kV
EHV (ANSI C92.2–1978[4])		345 kV[2]		362 kV
EHV (ANSI C92.2–1978[4])		500 kV[2]		550 kV
EHV (ANSI C92.2–1978[4])		765 kV[2]		800 kV
ANSI C92.2–1978		1100kV[2]		1200 kV

[Preferred nominal voltages as shown without parentheses ( ).
1. Voltage class designations applicable to industrial and commercial power systems, adapted by IEEE Standards Board (LB 100A–April 23, 1975).
2. Typical nominal system voltage.
3. A comprehensive list of minimum and maximum voltage ranges is given in ANSI C84.1-1977.]

voltage classes

voltage class, rated nominal *See:* rated nominal voltage class.

voltage coefficient of capacitance (nonlinear capacitor) The derivative with respect to voltage of a capacitance characteristic, such as a differential capacitance characteristic or a reversible capacitance characteristic, at a point, divided by the capacitance at that point. *See also:* nonlinear capacitor. (ED) [46]

voltage, common-mode *See:* common-mode voltage.

voltage-controlled oscillator (VCO) An oscillator whose frequency is a function of the voltage of a control signal. (AES) 686-1997

voltage corrector (power supplies) An active source of regulated power placed in series with an unregulated supply to sense changes in the output voltage (or current): also to correct for the changes by automatically varying its own output in the opposite direction, thereby maintaining the total output voltage (or current) constant. (See the corresponding figure.)

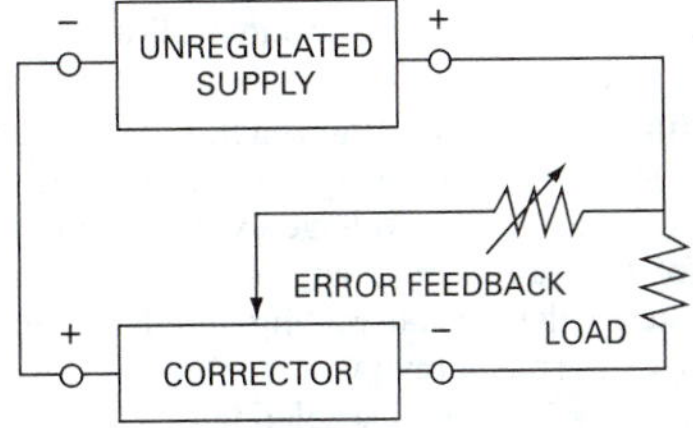

Circuit used to sense output voltage changes.

voltage corrector

(AES/PE) [41], [78]

voltage/current (V/I) characteristic The relationship between the steady-state current of the static var compensator (SVC) and the voltage at its point of connection. (PE/SUB) 1031-2000

voltage deviation (A) (self-commutated converters) (converters having ac output) (transient) The instantaneous difference between the actual instantaneous voltage and the corresponding value of the previously undisturbed wave form. *Note:* Voltage deviation amplitude is expressed in percent or per unit referred to the peak value of the previously undisturbed voltage. **(B) (electromagnetic site survey)** The ratio of the root-mean-squared envelope voltage to the average envelope of a signal expressed in decibels. (IA/EMC/SPC) 936-1987, 473-1985

voltage dip *See:* sag.

voltage directional relay (power system device function numbers) A relay that operates when the voltage across an open circuit breaker or contactor exceeds a given value in a given direction. (SUB/PE) C37.2-1979s

voltage distortion Any deviation from the nominal sine waveform of the ac line voltage. (IA/T&D/PE/PSE) 1100-1999, 1250-1995

voltage divider A network consisting of impedance elements connected in series, to which a voltage is applied, and from which one or more voltages can be obtained across any portion of the network. *Notes:* 1. Dividers may have parasitic impedances affecting the response. These impedances are, in general, the series inductance and the capacitance to ground and to neighboring structures at ground or at other potentials. 2. An adjustable voltage divider of the resistance type is frequently referred to as a potentiometer. (PE/PSIM/EM) 4-1978s, 43-1974s, [55]

voltage doubler A voltage multiplier that separately rectifies each half cycle of the applied alternating voltage and adds the two rectified voltages to produce a direct voltage whose amplitude is approximately twice the peak amplitude of the applied alternating voltage. *See also:* rectifier. (EEC/PE) [119]

voltage drop (1) The difference of voltages at the two terminals of a passive impedance. (PE) [9]
(2) (supply system) The difference between the voltages at the transmitting and receiving ends of a feeder, main, or service. *Note:* With alternating current, the voltages are not necessarily in phase and hence the voltage drop is not necessarily equal to the algebraic sum of the voltage drops along the several conductors. *See also:* alternating-current distribution. (T&D/PE) [10]

voltage efficiency (specified electrochemical process) The ratio of the equilibrium reaction potential to the bath voltage. (EEC/PE) [119]

voltage endurance (1) (rotating machinery) A characteristic of an insulation system, obtained by plotting voltage against time to failure, for a number of samples tested to destruction at each of several sustained voltages. Constant conditions of frequency, waveform, temperature, mechanical restraint, and ambient atmosphere are required. Ordinate scales of arithmetical or logarithmic voltage, and abscissa scales of multicycle logarithmic time, normally give approximately linear characteristics. *See also:* asynchronous machine. (PE) [9]

(2) The time-to-failure of the groundwall insulation under a high electrical stress. (DEI) 1043-1996

voltage endurance test (rotating machinery) A test designed to determine the effect of voltage on the useful life of electric equipment. When this test voltage exceeds the normal design voltage for the equipment, the test is voltage accelerated. When the test voltage is alternating and the frequency of alternation exceeds the normal voltage frequency for the equipment, the test is frequency accelerated. *See also:* asynchronous machine. (PE) [9]

voltage, equivalent test alternating *See:* equivalent test alternating voltage.

voltage, exciter-ceiling *See:* exciter-ceiling voltage.

voltage factor (electron tube) The magnitude of the ratio of the change in one electrode voltage to the change in another electrode voltage, under the conditions that a specified current remains unchanged and that all other electrode voltages are maintained constant. *See also:* ON period. (ED) [45], [84]

voltage/frequency function (self-commutated converters) (converters having ac output) The ratio of output voltage to the fundamental frequency of the output as a function of that frequency. (IA/SPC) 936-1987w

voltage generator (network analysis and signal-transmission system) A two-terminal circuit element with a terminal voltage substantially independent of the current through the element. *Note:* An ideal voltage generator has zero internal impedance. *See also:* network analysis; signal. (ED) 161-1971w

voltage gradient (overhead-power-line corona and radio noise) Corona work particularly emphasizes the property that the voltage gradient is equal to and is in the direction of the maximum space rate of change of the voltage at the specified point. The voltage gradient is obtained as a vector field by applying the operator ∇ to the scalar potential function, u. Thus, if $u = f(x, y, z)$,

$$\vec{E} = -\nabla u = -\text{grad } u = -\left(\vec{a}_x \frac{\partial u}{\partial x} + \vec{a}_y \frac{\partial u}{\partial y} + \vec{a}_z \frac{\partial u}{\partial z}\right)$$

Note: For alternating voltage, the voltage gradient is expressed as the peak value divided by the square root of two. For sinusoidal voltage, this is the rms value. *Synonyms:* potential gradient; field strength; gradient; electric field strength. (T&D/PE) 539-1990

voltage gradient stylus *See:* voltage pencil.

voltage imbalance (unbalance), polyphase systems The ratio of the negative or zero sequence component to the positive sequence component, usually expressed as a percentage. (SCC22) 1346-1998

voltage impulse A voltage pulse of sufficiently short duration to exhibit a frequency spectrum of substantially uniform amplitude in the frequency range of interest. As used in electromagnetic compatibility standard measurements, the voltage impulse has a uniform frequency spectrum over the frequency range 25 to 1000 megahertz. (EMC) 263-1965w

voltage influence (electric instruments) In instruments, other than indicating voltmeters, wattmeters, and varmeters, having voltage circuits, the percentage change (of full-scale value) in the indication of an instrument that is caused solely by a voltage departure from a specified reference voltage. *See also:* accuracy rating. (EEC/ERI/AII) [111], [102]

voltage-injection method (ac high-voltage circuit breakers) A synthetic test method in which the voltage circuit is applied to the test circuit breaker after power frequency current zero. (SWG/PE) C37.081-1981r, C37.100-1992

voltage jump (glow-discharge tubes) An abrupt change or discontinuity in tube voltage drop during operation. *Note:* This may occur either during life under constant operating conditions or as the current or temperature is varied over the operating range. *See also:* gas tube. (ED) 161-1971w

voltage level (data transmission) At any point in a transmission system, the ratio of the voltage existing at that point to an arbitrary value of voltage used as a reference. Specifically, in systems such as television systems, where wave shapes are not sinusoidal or symmetrical about a zero axis and where the arithmetical sum of the maximum positive and negative excursions of the wave is important in system performance, the voltage level is the ratio of the peak-to-peak voltage existing at any point in the transmission system to an arbitrary peak-to-peak voltage used as a reference. This ratio is usually expressed in dBV, signifying decibels referred to one V (volt) peak-to-peak. (PE) 599-1985w

voltage limit A control function that prevents a voltage from exceeding prescribed limits. Voltage limit values are usually expressed as percent of rated voltage. If the voltage-limit circuit permits the limit value to increase somewhat instead of being a single value, it is desirable to provide either a curve of the limit value of voltage as a function of some variable such as current or to give limit values at two or more conditions of operation. *See also:* feedback control system. (IA/ICTL/IAC) [60]

voltage-limiting-type SPD An SPD that has a high impedance when no surge is present, but will reduce it continuously with increased surge current and voltage. Common examples of components used as nonlinear devices are varistors and suppressor diodes. These SPDs are sometimes called "clamping-type" SPDs. (SPD/PE) C62.48-1995

voltage loss (current circuits) (electric instruments) In a current-measuring instrument, the value of the voltage between the terminals when the applied current corresponds to nominal end-scale deflection. In other instruments the voltage loss is the value of the voltage between the terminals at rated current. *Note:* By convention, when an external shunt is used, the voltage loss is taken at the potential terminals of the shunt. The overall voltage drop resulting may be somewhat higher owing to additional drop in shunt lugs and connections. *See also:* accuracy rating. (EEC/AII) [102]

voltage multiplier A rectifying circuit that produces a direct voltage whose amplitude is approximately equal to an integral multiple of the peak amplitude of the applied alternating voltage. *See also:* rectifier. (EEC/PE) [119]

voltage, nominal *See:* nominal system voltage.

voltage of a constant current circuit *See:* voltage.

voltage of an effectively grounded circuit *See:* voltage.

voltage of circuit not effectively grounded *See:* voltage.

voltage or current balance relay (power system device function numbers) A relay that operates on a given difference in voltage, or current input or output, of two circuits. (SUB/PE) C37.2-1979s

voltage overshoot (1) (arc-welding apparatus) The ratio of transient peak voltage substantially instantaneously following the removal of the short circuit to the normal steady-state voltage value. *See also:* voltage recovery time. (EEC/AWM) [91]

(2) (low voltage varistor surge arresters) The excess voltage above the clamping voltage of the device for a given current that occurs when current waves of less than 8μs virtual front duration are applied. This value may be expressed as a percent of the clamping voltage for an $8 \times 20\mu s$ current wave. (PE) [8]

voltage overshoot, effective *See:* effective voltage overshoot.

voltage pattern *See:* radiation pattern.

voltage, peak working *See:* peak working voltage.

voltage pencil A stylus whose position is detected by voltage ratios measured on a resistive grid. *Synonym:* voltage gradient stylus. (C) 610.10-1994w

voltage phase-angle method (electric power system) (economic dispatch) Considers the actual measured phase-angle difference between the station bus and a reference bus in the determination of incremental transmission losses. (PE/PSE) 94-1970w

voltage-phase-balance protection A form of protection that disconnects or prevents the connection of the protected equipment when the voltage unbalance of the phases of a normally balanced polyphase system exceed a predetermined amount. (SWG/PE) C37.100-1992

voltage probe A connecting device, usually consisting of a two-conductor shielded cable and frequency-compensating network, with a hand-held tip, for use with an oscilloscope to measure the amplitude and waveshape of a dc, ac, or composite signal. It should include a ground reference. The measurement bandwidth should be at least 10 times greater than the frequency of interest. The impedance should be at least 50 times greater than the node impedance under measurement. A low impedance probe should be used for measurement purposes. (PEL) 1515-2000

voltage protection level A parameter that characterizes the performance of the surge protective device in limiting the voltage across its terminals. This value shall be equal to or greater than the highest value measured in measured limiting voltage tests. (PE) C62.34-1996

voltage range (electrically propelled vehicle) Divided into five voltage ranges, as follows. first voltage range: 30 volts or less; second voltage range: over 30 volts to and including 175 volts; third voltage range: over 175 volts to and including 250 volts; fourth voltage range: over 250 volts to and including 660 volts; fifth voltage range: over 660 volts. (PE/EEC) [119]

voltage range multiplier (instrument multiplier) A particular type of series resistor or impedor that is used to extend the voltage range beyond some particular value for which the measurement device is already complete. It is a separate component installed external to the measurement device. (EEC/ERI) [111]

voltage, rated maximum *See:* rated maximum voltage.

voltage, rated maximum interrupting of main contacts *See:* rated maximum interrupting of main contacts voltage.

voltage, rated short-time of main contacts *See:* rated short-time of main contacts voltage.

voltage rating (1) (protection and coordination of industrial and commercial power systems) The root-mean-square (rms) alternating current (or the direct current) voltage at which the fuse is designed to operate. All low-voltage fuses will function on any lower voltage, but use on higher voltages than rated is hazardous. For high short-circuit currents, the magnitude of applied voltage will affect the arcing and clearing times and increase the clearing I^2t values. (IA/PSP) 242-1986r

(2) (surge arresters) The designated maximum permissible operating voltage between its terminals at which an arrester is designed to perform its duty cycle. It is the voltage rating specified on the nameplate. (PE/SPD) C62.1-1981s

(3) (household electric ranges) The voltage limits within which the range is intended to be used. *See also:* appliance outlet. (Std100) [84]

(4) (grounding transformer) (of a grounding transformer) The maximum "line-to-line" voltage at which it is designed to operate continuously from line to ground without damage to the grounding transformer. (PE/TR) C57.12.80-1978r

(5) (relay) The voltage at a specified frequency that may be sustained by the relay for an unlimited period without causing any of the prescribed limitations to be exceeded. (SWG/PE/PSR) C37.90-1978s, C37.100-1992

(6) The voltage specified on the nameplate. (SPD/PE) C62.62-2000

voltage rating, maximum *See:* maximum voltage rating.

voltage ratio (1) (capacitance potential device, in combination with its coupling capacitor or bushing) The overall ratio between the root-mean-square primary line-to-ground voltage and the root-mean-square secondary voltage. *Note:* It is not the turn ratio of the transformer used in the network. *See also:* outdoor coupling capacitor. (PE/EM) 43-1974s

(2) (power and distribution transformers) (of a transformer) The ratio of the rms terminal voltage of a higher voltage winding to the rms terminal voltage of a lower voltage winding, under specified conditions of the load. (PE/TR) C57.12.80-1978r

(3) (of a voltage divider) The factor by which the output voltage is multiplied to determine the measured value of the input voltage. (PE/PSIM) 4-1995

voltage recovery time (arc-welding apparatus) With a welding power supply delivering current through a short-circuiting resistor whose resistance is equivalent to the normal load at that setting on the power supply, and measurement being made when the short circuit is suddenly removed, the time measured in seconds between the instant the short circuit is removed and the instant when voltage has reached 95% of its steady-state value. *See also:* effective voltage overshoot; voltage overshoot. (EEC/AWM) [91]

voltage reduction A means of achieving a reduction of system demand and energy by reducing the customer supply voltage. (PE/PSE) 858-1993w

voltage reduction reserve (power operations) The operating reserve available through voltage reduction of a specified percentage. (PE/PSE) 858-1987s

voltage reference (power supplies) A separate, highly regulated voltage source used as a standard to which the output of the power supply is continuously referred. (AES) [41]

voltage-reference tube A gas tube in which the tube voltage drop is approximately constant over the operating range of current and relatively stable with time at fixed values of current and temperature. (ED) 161-1971w

voltage reflection coefficient The ratio of the complex number (phasor) representing the phase and magnitude of the electric field of the backward-traveling wave to that representing the forward-traveling wave at a cross section of a waveguide. The term is also used to denote the magnitude of this complex ratio. *See also:* waveguide. (AP/ANT) [35]

voltage regulating adjuster (excitation systems for synchronous machines) A device associated with a synchronous machine voltage regulator by which adjustment of the synchronous machine terminal voltage can be made. (PE/EDPG) 421.1-1986r

voltage regulating device A voltage sensitive device that is used on an automatically operated voltage regulator to control the voltage of the regulated circuit. (PE/TR) C57.15-1999

voltage-regulating relay (power and distribution transformers) A voltage-sensitive device that is used on an automatically operated voltage regulator to control the voltage of the regulated circuit. (PE/TR) C57.12.80-1978r

voltage-regulating transformer (step-voltage regulator) A voltage regulator in which the voltage and phase angle of the regulated circuit are controlled in steps by means of taps and without interrupting the load. *See also:* voltage regulator. (PE/TR) [57]

voltage-regulating transformer, two-core A voltage-regulating transformer consisting of two separate core and coil units in a single tank. *See also:* voltage regulator. (PE/TR) [57]

voltage-regulating transformer, two-core, excitation-regulating winding In some designs, the main unit will have one winding operating as an autotransformer that performs both functions listed under regulating winding and excitation winding. Such a winding is called the excitation-regulating winding. *See also:* voltage regulator. (PE/TR) [57]

voltage-regulating transformer, two-core, excitation winding The winding of the main unit that draws power from the system to operate the two-core transformer. *See also:* voltage regulator. (PE/TR) [57]

voltage-regulating transformer, two-core, excited winding The winding of the series unit that is excited from the regulating winding of the main unit. *See also:* voltage regulator. (PE/TR) [57]

voltage-regulating transformer, two-core, regulating winding The winding of the main unit in which taps are changed to control the voltage or phase angle of the regulated circuit through the series unit. *See also:* voltage regulator. (PE/TR) [57]

voltage-regulating transformer, two-core, series unit The core and coil unit that has one winding connected in series in the line circuit. *See also:* voltage regulator. (PE/TR) [57]

voltage-regulating transformer, two-core, series winding The winding of the series unit that is connected in series in the line circuit. *Note:* If the main unit of a two-core transformer is an autotransformer, both units will have a series winding. In such cases, one is referred to as the series winding of the autotransformer and the other, the series winding of the series unit. *See also:* voltage regulator. (PE/TR) [57]

voltage regulation (1) (constant-voltage transformer) The change in output (secondary) voltage which occurs when the load (at a specified power factor) is reduced from rated value to zero, with the primary impressed terminal voltage maintained constant. *Note:* In case of multi-winding transformers, the loads on all windings, at specified power factors, are to be reduced from rated kVA to zero simultaneously. The regulation may be expressed in per unit, or percent, on the base of the rated output (secondary) voltage at full load.
(PE/TR) C57.12.80-1978r
(2) (outdoor coupling capacitor) The variation in voltage ratio and phase angle of the secondary voltage of the capacitance potential device as a function of primary line-to-ground voltage variation over a specified range, when energizing a constant, linear impedance burden. *See also:* outdoor coupling capacitor. (PE/EM) 43-1974s
(3) (A) (direct-current generator) The final change in voltage with constant field-rheostat setting when the specified load is reduced gradually to zero, expressed as a percent of rated-load voltage, the speed being kept constant. *Note:* In practice it is often desirable to specify the over-all regulation of the generator and its driving machine thus taking into account the speed regulation of the driving machine. **(B)** (induction frequency converter). The rise in secondary voltage when the rated load at rated power factor is reduced to zero, expressed in percent of rated secondary voltage, the primary voltage, primary frequency, and the speed being held constant. *See also:* asynchronous machine. (EEC/PE) [119]
(4) (synchronous generator) The rise in voltage with constant field current, when, with the synchronous generator operated at rated voltage and rated speed, the specified load at the specified power factor is reduced to zero, expressed as a percent of rated voltage. (PE) [9]
(5) (thyristor converter) The change in output voltage that occurs when the load current is reduced from its rated value to zero, or light transition load, with rated sinusoidal alternating voltage applied to the thyristor power converter with the transformer on its rated tap, but excluding the corrective action of any voltage regulating means. *Note:* The regulation may be expressed in volts or in percent of rated volts.
(IA/IPC) 444-1973w
(6) The degree of control or stability of the rms voltage at the load. Often specified in relation to other parameters, such as input voltage changes, load changes, or temperature changes.
(IA/PSE) 1100-1999
(7) (line regulator circuits) *See also:* pulse-width modulation; Zener diode.

voltage regulation curve (synchronous generator) (voltage regulation characteristic) The relationship between the armature winding voltage and the load on the generator under specified conditions and constant field current. (PE) [9]

voltage regulation of a constant-voltage transformer (power and distribution transformers) The change in output (secondary) voltage which occurs when the load (at a specified power factor) is reduced from rated value to zero, with the primary impressed terminal voltage maintained constant. *Note:* In case of multiwinding transformers, the loads of all windings, at specified power factors, are to be reduced from rated kVA to zero simultaneously. The regulation may be expressed in per unit, or percent, on the base of the rated output (secondary) voltage at full load.
(PE/TR) C57.12.80-1978r

voltage regulator (1) (excitation systems for synchronous machines) A synchronous machine regulator that functions to maintain the terminal voltage of a synchronous machine at a predetermined value, or to vary it according to a predetermined plan. *Note:* Historical term, included for reference only. The preferred term is synchronous machine regulator.
(PE/EDPG) 421.1-1986r
(2) (transformer type) An induction device having one or more windings in shunt with and excited from the primary circuits, and having one or more windings in series between the primary circuits and the regulated circuit, all suitably adapted and arranged for the control of the voltage, or of the phase angle, or of both, of the regulated circuit.
(PE/TR) [57], C57.15-1986s

voltage regulator, continuously acting type (rotating machinery) A regulator that initiates a corrective action for a sustained infinitesimal change in the controlled variable.
(PE) [9]

voltage regulator, direct-acting type (rotating machinery) A rheostatic-type regulator that directly controls the excitation of an exciter by varying the input to the exciter field circuits.
(PE) [9]

voltage regulator, dynamic type (rotating machinery) A continuously acting regulator that does not require mechanical acceleration of parts to perform the regulating function. *Note:* Dynamic-type voltage regulators utilize magnetic amplifiers, rotating amplifiers, electron tubes, semiconductor elements, and/or other static components. (PE) [9]

voltage regulator, indirect-acting type (rotating machinery) A rheostatic-type regulator that controls the excitation of the exciter by acting on an intermediate device not considered part of the voltage regulator or exciter. (PE) [9]

voltage regulator, noncontinuously acting type (rotating machinery) A regulator that requires a sustained finite change in the controlled variable to initiate corrective action.
(PE) [9]

voltage regulator, synchronous-machine (rotating machinery) A synchronous-machine regulator that functions to maintain the voltage of a synchronous machine at a predetermined value, or to vary it according to a predetermined plan.
(PE) [9]

voltage-regulator tube A glow-discharge cold-cathode tube in which the voltage drop is approximately constant over the operating range of current, and that is designed to provide a regulated direct-voltage output. (ED) 161-1971w, [45]

voltage related to partial discharges (1) (A) (liquid-filled power transformers) The phase to ground alternating voltage whose value is expressed by its peak divided by the square root of two. **(B)** (partial discharge measurement in liquid-filled power transformers and shunt reactors) The phase to ground alternating voltage whose value is expressed by its peak divided by the square root of two.
(PE/TR) C57.113-1988
(2) (dry-type transformers) Voltage within the terms of C57.124-1991 is the phase-to-ground alternating voltage for applied tests or terminal to terminal alternating voltage for induced voltage tests. Its value is expressed by its peak value divided by the square root of two.
(PE/TR) C57.124-1991r

voltage relay (1) A relay that functions at a predetermined value of voltage. *Note:* It may be an overvoltage relay, an undervoltage relay, or a combination of both. *See also:* relay.
(IA/ICTL/IAC) [60]
(2) A relay that responds to voltage.
(SWG/PE) C37.100-1992

voltage response The ratio of the open-circuit output voltage to the applied sound pressure, measured by a laboratory standard microphone placed at a stated distance from the plane of the opening of the artificial voice. *Note:* The voltage response is usually measured as a function of frequency. *See also:* close-talking pressure-type microphones. (SP) 258-1965w

voltage response, exciter *See:* exciter voltage response.

voltage response ratio, excitation-system (rotating machinery) The numerical value that is obtained when the excitation-system voltage response in volts per second, measured over the first 1/2 second interval unless otherwise specified, is divided by the rated-load field voltage of the synchronous machine. *Note:* This response, if maintained constant, would develop, in 1/2 second, the same excitation voltage-time area as attained by the actual response. (PE) [9]

voltage response, synchronous-machine excitation-system The rate of increase or decrease of the excitation-system output voltage, determined from the synchronous machine excitation-system voltage-time response curve, that if maintained constant would develop the same excitation-system voltage-time areas as are obtained from the curve for a specified period. The starting point for determining the rate of voltage change shall be the initial value of the synchronous-machine excitation-system voltage-time response curve. (PE) [9]

voltage restraint A method of restraining the operation of a relay by means of a voltage input that opposes the typical response of the relay to other inputs. (SWG/PE) C37.100-1992

voltage sensing relay (A) A term correctly used to designate a special-purpose voltage-rated relay that is adjusted by means of a voltmeter across its terminals in order to secure pickup at a specified critical voltage without regard to coil or heater resistance and resulting energizing current at that voltage. **(B)** A term erroneously used to describe a general-purpose relay for which operational requirements are expressed in voltage. (PE/EM) 43-1974

voltage-sensitive preamplifier An amplifier, preceding the main amplifier, in which the amplitude of the output signal is proportional to the signal voltage appearing across the capacitance that exists at the input of the preamplifier. *See also:* charge-sensitive preamplifier. (NPS) 325-1996

voltage sensitivity *See:* voltage coefficient of capacitance; nonlinear capacitor.

voltage sets (polyphase circuit) The voltages at the terminals of entry to a polyphase circuit into a delimited region are usually considered to consist of two sets of voltages: the line-to-line voltages, and the line to-neutral voltages. If the phase conductors are identified in a properly chosen sequence, the voltages between the terminals of entry of successive pairs of phase conductors form the set of line-to-line voltages, equal in number to the number of phase conductors. The voltage from the successive terminals of entry of the phase conductors to the terminal of entry of the neutral conductor, if one exists, or to the true neutral point, form the set of line-to-neutral voltages, also equal in number to the number of phase conductors. In case of doubt, the set intended must be identified. In the absence of other information, stated or implied, the line-to-neutral-conductor set is understood. *Notes:* 1. Under abnormal conditions the voltage of the neutral conductor and of the true neutral point may not be the same. Therefore it may become necessary to designate which is intended when the line-to-neutral voltages are being specified. 2. The set of line-to-line voltages may be determined by taking the differences in pairs of the successive line-to-neutral voltages. The line-to-neutral voltages can be determined from the line-to-line voltages by an inverse process only when the voltage between the neutral conductor and the true neutral point is completely specified, or equivalent additional information is available. If instantaneous voltages are used, algebraic differences are taken, but if root-mean-square voltages are used, information regarding relative phase angles must be available, so that the voltages may be expressed in phasor form and the phasor differences taken. 3. This definition may be applied to a two-phase, four-wire or five-wire circuit. A two-phase, three-wire circuit should be treated as a special case. *See also:* network analysis. (Std100) 270-1966w

voltage shape A waveform of a voltage impulse that has been standardized to define insulation strength. The standardized voltage shapes are as follows: power-frequency short-duration, standard switching impulse, standard lightning impulse, very fast front, standard chopped wave impulse, and front-of-wave lightning impulse). (PE/C) 1313.1-1996

voltage spread The difference between maximum and minimum voltages. (IA/APP) [80]

voltage-stabilizing tube *See:* voltage-regulator tube.

voltage standing-wave ratio (VSWR) (mode in a waveguide) The ratio of the magnitude of the transverse electric field in a plane of maximum strength to the magnitude at the equivalent point in an adjacent plane of minimum field strength. *See also:* waveguide. (AP/ANT) [35]

voltage surge, internal *See:* internal voltage surge.

voltage surge suppressor (semiconductor rectifiers) A device used in the semiconductor rectifier to attenuate surge voltages of internal or external origin. Capacitors, resistors, nonlinear resistors, or combinations of these may be employed. Nonlinear resistors include electronic and semiconductor devices. *See also:* semiconductor rectifier stack. (IA) [62]

voltage switch (test switches for transformer-rated meters) A single-pole single-throw switch used to open or close a voltage circuit. (ELM) C12.9-1993

voltage-switching-type SPD An SPD that has a high impedance when no surge is present, but can have a sudden change in impedance to a low value in response to a voltage surge. Common examples of components used as nonlinear devices are spark gaps, gas tubes, and silicon-controlled rectifiers. These SPDs are sometimes called crowbar-type SPDs. (SPD/PE) C62.48-1995

voltage test *See:* controlled overvoltage test.

voltage/time curve for impulses of constant prospective shape (high voltage testing) The curve relating the disruptive discharge voltage of a test object to the time to chopping, which may occur on the front, at the crest or on the tail. The curve is obtained by applying impulse voltages of constant shape, but with different peak values. (PE/PSIM) 4-1978s

voltage/time curve for linearly rising impulses (high voltage testing) The voltage/time curve for impulses with fronts rising linearly is the curve relating the voltage at the instant of chopping to the rise time Tr. The curve is obtained by applying impulses with approximately linear fronts of different steepnesses. (PE/PSIM) 4-1978s

voltage-time product (pulse transformers) The time integral of a voltage pulse applied to a transformer winding. (PEL/ET) 390-1987r

voltage-time product rating (pulse transformers) (of a transformer winding) Considered as being a constant and is the maximum voltage-time product of a voltage pulse that can be applied to the winding before a specified level of core saturation-region effects is reached. The level of core saturation-region effects is determined by observing either the shape of the output voltage pulse for a specified degradation (for example, a maximum tilt [droop]), or the shape of the exciting current pulse for a specified departure from linearity (for example, deviation from a linear ramp by a given percentage). (PEL/ET) 390-1987r

voltage-time response, synchronous-machine excitation-system The output voltage of the excitation system, expressed as a function of time, following the application of prescribed inputs under specified conditions. (PE) [9]

voltage-time response, synchronous-machine voltage-regulator The voltage output of the synchronous-machine voltage regulator expressed as a function of time following the application of prescribed inputs under specified conditions. (PE) [9]

voltage to ground (1) For grounded circuits, the voltage between the given conductor and that point or conductor of the circuit that is grounded; for ungrounded circuits, the greatest voltage between the given conductor and any other conductor of the circuit. (NESC/NEC) [86]

(2) **(A)** (of a grounded circuit) The highest nominal voltage available between any conductor of the circuit and that point or conductor of the circuit that is grounded. **(B)** (of an ungrounded circuit) The highest nominal voltage available between any two conductors of the circuit concerned. (NESC) C2-1997

(3) **(power and distribution transformers)** The voltage between any live conductor of a circuit and the earth. *Note:* Where safety considerations are involved, the voltage to ground which may occur in an ungrounded circuit is usually the highest voltage normally existing between the conductors of the circuit, but in special circumstances, higher voltages may occur. (PE/SPD/TR) 32-1972r, C57.16-1996, C57.12.80-1978r

voltage to ground of a conductor **(A)** (of a grounded circuit) The nominal voltage between such conductor and that point or conductor of the circuit that is grounded. **(B)** (of an ungrounded circuit) The highest nominal voltage between such conductor and any other conductor of the circuit concerned. (NESC) C2-1997

voltage to luminaire factor (illuminating engineering) The fractional loss of task illuminance due to improper voltage at the luminaire. (EEC/IE) [126]

voltage, touch *See:* touch voltage.

voltage transformer (VT) **(1)** **(instrument transformers) (power and distribution transformers)** An instrument transformer intended to have its primary winding connected in shunt with a power supply circuit, the voltage of which is to be measured or controlled. (PE/TR) C57.12.80-1978r

(2) **(metering)** An instrument transformer designed for use in the measurement or control of voltage. *Note:* Its primary winding is connected across the supply circuit. *See also:* instrument transformer. (ELM) C12.1-1982s

(3) An instrument transformer intended to have its primary connected in shunt with the voltage to be measured or controlled. (PE/TR) [57]

(4) An instrument transformer intended to have its primary winding connected in shunt with the voltage to be measured or controlled. (PE/TR) C57.13-1993

voltage transient suppression (thyristor) Reduction of the effects of voltage transients on controller components by reducing the voltage or energy of the transients to tolerable levels. (IA/IPC) 428-1981w

voltage, transverse mode *See:* transverse-mode voltage.

voltage-tunable magnetron (microwave tubes) A magnetron in which the resonant circuit is heavily loaded (QL = 1 to 10) and in which the supply of electrons to the interaction space is restricted whereby the frequency of oscillation becomes proportional to the plate voltage. *See also:* magnetron. (ED) [45]

voltage-type telemeter A telemeter that employs the magnitude of a single voltage as the translating means. (SWG/PE/SUB) C37.100-1992, C37.1-1994

voltage winding for regulating equipment (power and distribution transformers) (or transformer) The winding (or transformer) which supplies voltage within close limits of accuracy to instruments, such as contact-making voltmeters. (PE/TR) C57.12.80-1978r

voltage-withstand test **(1)** **(insulation materials)** The application of a voltage higher than the rated voltage for a specified time for the purpose of determining the adequacy against breakdown of insulation materials and spacing under normal conditions. *See also:* dielectric tests. (EEC/PE) [119]

(2) **(rotating machinery)** *See also:* overvoltage test; asynchronous machine.

voltage-withstand tests Tests made to determine the ability of insulating materials and spacings to withstand specified overvoltages for a specified time without flashover or puncture. (ELM) C12.1-1981

voltameter *See:* coulometer; instrument.

volt-ammeter An instrument having circuits so designed that the magnitude either of voltage or of current can be measured on a scale calibrated in terms of each of these quantities. *See also:* instrument. (EEC/PE) [119]

voltampere The unit of apparent power in the International System of Units (SI). The voltampere is the apparent power at the points of entry of a single-phase, two-wire system when the product of the root-mean-square value in amperes of the current by the root-mean-square value in volts of the voltage is equal to one. (Std100) 270-1966w

voltampere loss *See:* apparent-power loss.

voltampere meter An instrument for measuring the apparent power in an alternating-current circuit. It is provided with a scale graduated in volt-amperes or in kilovolt-amperes. *See also:* instrument. (EEC/PE) [119]

volt efficiency (storage battery) (storage cell) The ratio of the average voltage during the discharge to the average voltage during the recharge. *See also:* charge; electrochemistry. (PE/EEC) [119]

voltmeter An instrument for measuring the magnitude of electric potential difference. It is provided with a scale, usually graduated in either volts, millivolts, or kilovolts. If the scale is graduated in millivolts or kilovolts the instrument is usually designated as a millivoltmeter or a kilovoltmeter. *See also:* instrument. (EEC/PE) [119]

voltmeter-ammeter The combination in a single case, but with separate circuits, of a voltmeter and an ammeter. *See also:* instrument. (EEC/PE) [119]

volts per hertz relay A relay whose pickup is a function of the ratio of voltage to frequency. (SWG/PE) C37.100-1992

volt-time area The area under a curve plotted with voltage versus time with areas of positive and negative polarities added algebraically. The volt-time area that is generally of concern consists of the net accumulated volt-time area that occurs during a certain number of power frequency cycles of the ground potential rise (GPR). The area is a function of the magnitude and decay rate of the dc offset. (PE/PSC) 367-1996

volt-time curve **(1)** **(surge arresters)** (impulses with fronts rising linearly) The curve relating the disruptive-discharge voltage of a test object to the virtual time to chopping. The curve is obtained by applying voltages that increase at different rates in approximately linear manner. (PE) [8], [84]

(2) **(standard impulses)** A curve relating the peak value of the impulse causing disruptive discharge of a test object to the virtual time to chopping. The curve is obtained by applying standard impulse voltages of different peak values. (PE) [8], [84]

volume **(1)** **(information transfer)** One collection of data commencing with a Volume ID and containing a number of bytes specified in the volume size of the volume director. (C/MM) 949-1985w

(2) **(volume measurements of electrical speech and program waves)** The magnitude of a complex audio-frequency wave in an electrical circuit as measured on a standard volume indicator. (BT/AV) 152-1953s

(3) **(data transmission)** In general, volume is the intensity or loudness of sound. In a telephone or other audio-frequency circuit, a measure of the power corresponding to an audio-frequency wave at that point [expressed in decibels (dB)]. (PE) 599-1985w

(4) **(electric circuits)** The magnitude of a complex audio-frequency wave as measured on a standard volume indicator. *Notes:* 1. Volume is expressed in volume units (vu). 2. The term volume is used loosely to signify either the intensity of a sound or the magnitude of an audiofrequency wave. (SP) 151-1965w

(5) **(International System of Units (SI))** The SI unit of volume is the cubic meter. This unit, or one of the regularly formed multiples such as the cubic centimeter, is preferred for all applications. The special name liter has been approved for the cubic decimeter, but use of this unit is restricted to the measurement of liquids and gases. No prefix other than milli- should be used with liter. *See also:* units and letter symbols. (QUL) 268-1982s

(6) (A) (data management) A portion of data that, together with it's data carrier, can be handled as a unit. *Synonym:* physical volume. (B) (data management) A data carrier that is mounted and demounted as a unit; for example, a disk pack or a reel of magnetic tape. (C) 610.5-1990
(7) (A) A data carrier that is mounted and demounted as a unit; for example, a spool of magnetic tape, or a disk pack. (B) A storage medium, together with its data carrier, that can be handled conveniently as a unit; for example, a reel of magnetic tape or a disk pack. (C) The portion of a single unit of storage that is accessible to a single read/write head. (C) 610.10-1994

volume density of magnetic pole strength At any point of the medium in a magnetic field, the negative of the divergence of the magnetic polarization vector there. (Std100) 270-1966w

volume equivalent (complete telephone connection, including the terminating telephone sets) A measure of the loudness of speech reproduced over it. The volume equivalent of a complete telephone connection is expressed numerically in terms of the trunk loss of a working reference system when the latter is adjusted to give equal loudness. *Note:* For engineering purposes, the volume equivalent is divided into volume losses assignable to: 1) the station set, subscriber line, and battery-supply circuit that are on the transmitting end 2) the station set, subscriber line, and battery supply that are on the receiving end 3) the trunk and 4) interaction effects arising at the trunk terminals. (EEC/PE) [119]

volume fraction The ratio of the volume of inclusions to the total volume (inclusions plus host material). (AP/PROP) 211-1997

volume header *See:* beginning-of-volume label.

volume indicator (standard volume indicator) A standardized instrument having specified electric and dynamic characteristics and read in a prescribed manner, for indicating the volume of a complex electric wave such as that corresponding to speech or music. *Notes:* 1. The reading in volume units is equal to the number of decibels above a reference volume. The sensitivity is adjusted so that the reference volume or zero volume unit is indicated when the instrument is connected across a 600-ohm resistor in which there is dissipated a power of 1 milliwatt at 1000 hertz. 2. Specifications for a volume indicator are given in American National Standard Volume Measurements of Electrical Speech and Program Waves, C16.5. *See also:* volume unit; instrument. (COM/SP/TA) 269-1971w, [50], [32]

volume label *See:* beginning-of-volume label; end-of-volume label.

volume lifetime (semiconductor) The average time interval between the generation and recombination of minority carriers in a homogeneous semiconductor. *See also:* semiconductor device; semiconductor. (ED) 216-1960w

volume limiter A device that automatically limits the output volume of speech or music to a predetermined maximum value. *See also:* peak limiter. (EEC/PE) [119]

volume-limiting amplifier An amplifier containing an automatic device that functions when the input volume exceeds a predetermined level and so reduces the gain that the output volume is thereafter maintained substantially constant notwithstanding further increase in the input volume. *Note:* The normal gain of the amplifier is restored when the input volume returns below the predetermined limiting level. *See also:* amplifier. (AP/ANT) 145-1983s

volume mixing ratio The ratio defined by $N(z)/N(\text{air})$ where $N(z)$ is the number density (number of molecules per unit volume) of a particular species and $N(\text{air})$ is the number density of air. (AP/PROP) 211-1997

volume range (1) (transmission system) The difference, expressed in decibels between the maximum and minimum volumes that can be satisfactorily handled by the system. (EEC/PE) [119]
(2) (complex audio-frequency signal) The difference, expressed in decibels, beween the maximum and minimum volumes occurring over a specified period of time. (EEC/PE) [119]

volume resistivity The reciprocal of volume conductivity, measured in siemens per centimeter, which is a steady-state parameter. (PE) 402-1974w

volume scattering Scattering from inhomogeneities distributed throughout a volume. The inhomogeneities can be discrete particles or structures or continuous spatial variations of refractive index. (AP/PROP) 211-1997

volumetric radar A surveillance radar with coverage extending over a significant sector in two angular coordinates. (AES) 686-1997

volume unit The unit in which the standard volume indicator is calibrated. *Note:* One volume unit equals one decibel for a sine wave but volume units should not be used to express results of measurements of complex waves made with devices having characteristics differing from those of the standard volume indicator. *See also:* volume indicator. (SP) 151-1965w

Von Neumann architecture A computer architecture characterized by a processor, memory and input-output devices interconnected with a single bus, thus allowing a single path to main storage for instructions and data. *Note:* This is the classic architecture and the basis for most modern computers. *Synonym:* control flow architecture. *Contrast:* Harvard class architecture; tagged architecture. (C) 610.10-1994w

VOR *See:* very high-frequency omnidirectional range.

vortac (navigation aids) A designation applied to certain navigation stations (primarily in the United States) in which both VOR (very high-frequency omnidirectional range) and tacan (tactical air navigation) are used; the distance function in tacan is used with VOR to provide VOR/DME (rho theta) navigation. (AES/GCS) 172-1983w

voter *See:* voting circuit.

voting circuit A logic circuit whose result is true only if the number of its inputs in the true state exceeds a predetermined amount. *Synonym:* voter. (C) 610.10-1994w

voting computer A fault tolerant computer with three or more processing elements, all computing the same operation, the final output of which is produced by the majority of the elements. (C) 610.10-1994w

vowel articulation The percent articulation obtained when the speech units considered are vowels (usually combined with consonants into meaningless syllables). *See also:* articulation (percent articulation) and intelligibility (percent intelligibility); volume equivalent. (SP) [32]

voxel The smallest three-dimensional element of a display space whose characteristics are independently assigned. *Note:* This term is derived from the term "volume element." *See also:* pixel. (C) 610.6-1991w

VRAM *See:* video RAM.

Vref *See:* reference voltage.

VRLA cell *See:* valve-regulated lead-acid (VRLA) cell.

VSA *See:* vibrating string accelerometer.

VSAM *See:* virtual sequential access method.

VSB arbitration bus The second of the two sub-buses defined in the VSB specification. It allows arbiter modules and/or requester modules to coordinate the use of the DTB by VSB masters. The VSB defines two arbitration methods—a serial arbitration method and a parallel arbitration method. (C/MM) 1096-1988w

VSB backplane An assembly that includes a printed circuit (pc) board and 96-pin connectors. The backplane buses the 64 pins on the two outer rows of the VSB connectors, providing the signal paths needed for VSB operation. (C/MM) 1096-1988w

VSB bus cycle A sequence of level transitions on the signal lines of the DTB that results in the transfer of an address and (in most cases) data between the active master and selected slaves. The protocols of the VSB are fully asynchronous. The active master asserts a strobe signal indicating that a cycle is in progress. The responding slave acknowledges the master's signal. However, the responding slave can delay its acknowledgment for as long as it needs. The DTB cycle is generally divided into three phases: an address broadcast, zero or more data transfers, and then cycle termination. (C/MM) 1096-1988w

V-series (ITU-TSS) A CCITT family of recommendations describing the connection of digital equipment to the analog public telephone network. (C) 610.7-1995, 610.10-1994w

VSWR *See:* voltage standing-wave ratio.

VT *See:* vertical tabulation character; voltage transformer.

V_{term} Terminator source voltage to be maintained under all load conditions. (C/BA) 896.2-1991w

V-terminal voltage Terminal voltage measured with a V network between each mains conductor and earth. *See also:* electromagnetic compatibility. (EMC/INT) [53], [70]

V2F A transaction that is originated at the VME64 (the Bridge acts as a VME64 slave), and has its destination at the Futurebus+ (the Bridge acts as a Futurebus+ master). (C/BA) 1014.1-1994w

vu (volume measurements of electrical speech and program waves) (pronounced "vee-you" and customarily written with lower case letters.) A quantitative expression for volume in an electric circuit. *Notes:* 1. The volume in vu is numerically equal to the number of decibels (dB) which expresses the ratio of the magnitude of the waves to the magnitude of reference volume. 2. The term vu should not be used to express results of measurements of complex waves made with devices having characteristics differing from those of the standard volume indicator. *See also:* standard volume indicator. (BT/AV) 152-1953s

vulnerability (surge testing for equipment connected to low-voltage ac power circuits) The characteristic of a device for being damaged by an external influence, such as a surge. (SPD/PE) C62.45-1992r

VXIbus VMEbus extensions for instrumentation. (C/MM) 1155-1992

VXIbus instrument A message-based device that supports the VXIbus instrument protocols. (C/MM) 1155-1992

VXIbus subsystem A central timing module referred to as Slot 0, with up to 12 additional adjacent VXIbus modules. The VXIbus subsystem bus defines the lines on the P2 and P3 connectors. (C/MM) 1155-1992

W

WADEX *See:* word and author index.

wait (test, measurement, and diagnostic equipment) A programmed instruction that causes an automatic test system to remain in a given state for a predetermined period. (MIL) [2]

waiting, call *See:* call waiting.

waiting-passenger indicator (elevators) An indicator that shows at which landings and for which direction elevator-hall stop or signal calls have been registered and are unanswered. *See also:* control. (EEC/PE) [119]

wait packet A packet with four high flags and four undefined data bytes, typically output by the controller after a *standBy* packet has been output. (C/MM) 1596.4-1996

wait state A condition in which a device or component is idle; for example, a central processor that is waiting for some event and not executing instructions. (C) 610.10-1994w

wait timeout period The time a master will wait after recognizing WT (wait) before terminating the connection. (NID) 960-1993

waiver (1) (nuclear power quality assurance) Documented authorization to depart from specified requirements. (PE/NP) [124]
(2) (software) A written authorization to accept a configuration item or other designated item that, during production or after having been submitted for inspection, is found to depart from specified requirements, but is nevertheless considered suitable for use as is or after rework by an approved method. *Contrast:* deviation; engineering change. *See also:* configuration control. (C) 610.12-1990

wake packet A packet with two low and two high flags as well as four undefined data bytes, typically generated by the processing of idle packets when leaving the standBy mode. (C/MM) 1596.4-1996

walk (1) (shaped pulse) The change in a timing point that occurs when the pulse height changes. (NPS) 300-1988r
(2) (in a pulse) The change in the timing (with respect to the initiating event) of a reference point on a pulse caused by a change in pulse amplitude. (NPS) 325-1996

walkie-talkie A two-way radio communication set designed to be carried by one person, usually strapped over the back, and capable of operation while in motion. *See also:* radio transmission. (EEC/PE) [119]

walk-off mode (laser maser) A mode characterized by successive shifts per reflection in the location of a maximum in the transverse field intensity. *See also:* transverse mode. (LEO) 586-1980w

walkout of reverse drain breakdown (metal-nitride-oxide field-effect transistor) An effect where the reverse current/voltage characteristic of the drain junction changes with time toward larger voltages as a function of applied bias. This effect is generally reversible. (ED) 581-1978w

walk-through (1) (software) A static analysis technique in which a designer or programmer leads members of the development team and other interested parties through a segment of documentation or code, and the participants ask questions and make comments about possible errors, violation of development standards, and other problems. (C) 610.12-1990
(2) A static analysis technique in which a designer or programmer leads members of the development team and other interested parties through a software product, and the participants ask questions and make comments about possible errors, violation of development standards, and other problems. (C/SE) 1028-1997

wall bushing (1) A bushing intended primarily to carry a circuit through a wall or other grounded barrier in a substantially horizontal position. Both ends must be suitable for operating in air. *See also:* bushing. (Std100) 49-1948w
(2) A bushing intended to be mounted on the wall (roof) of a building such as a Converter Valve Hall. *Note:* A wall or roof bushing may be outdoor, outdoor-indoor, or indoor bushing. (PE/TR) C57.19.03-1996

wall clock A clock that is on the wall. *Note:* A wall clock is typically referred to in order to demonstrate the difference between system time and real time. *Synonym:* wall time. *See also:* time-of-day clock. (C) 610.10-1994w

wall-mounted oven An oven for cooking purposes designed for mounting in or on a wall or other surface and consisting of one or more heating elements, internal wiring, and built-in or separately mountable controls. *See also:* counter-mounted cooking unit. (NESC/NEC) [86]

wall telephone set A telephone set arranged for wall mounting. *See also:* telephone station. (EEC/PE) [119]

wall time *See:* wall clock.

WAN *See:* wide area network.

wander (telecommunications) The long-term variations of the significant instants of a digital signal from their ideal position in time. The phrase "long-term" implies that these variations are of low frequency, less than 10 Hz. *See also:* scintillation. (COM/TA) 1007-1991r

wanted signal A signal that constitutes the object of the particular measurement or reception. (T&D/PE) 539-1990

warble-tone generator (alarm monitoring and reporting systems for fossil-fueled power generating stations). An audio-frequency oscillator, the frequency of which is varied cyclically at a subaudio rate over a fixed range. (PE/EDPG) 676-1986w

war game (computer graphics) A simulation game in which participants seek to achieve a specified military objective given preestablished resources and constraints; for example, a simulation in which participants make battlefield decisions and a computer determines the results of those decisions. *See also:* management game. (C) 610.3-1989w

warm-start A sequence of events performed to reset a running system. (C/MM) 1296-1987s

warm-up time (1) (power supplies) The time (after power turn on) required for the output voltage or current to reach an equilibrium value within the stability specification. (AES) [41]
(2) (nuclear power generating station) The time, following power application to a module, required for the output to stabilize within specifications. (PE/NP) 381-1977w
(3) (accelerometer) (gyros) The time interval required for a gyro or accelerometer to reach specified performance from the instant that it is energized, under specified operating conditions. (AES/GYAC) 528-1994

warning Advisory in a software user document that performing some action will lead to serious or dangerous consequences. *Contrast:* caution. (C/SE) 1063-1987r

warning whistle *See:* audible cab indicator.

WAROM *See:* word-alterable read-only memory.

warp (navigation aids) (loran) Variations of the propagation times for loran signals due to the variations of conductivity over land. Causes errors in the determination of absolute position. (AES/GCS) 172-1983w

washer *See:* collar.

Washington *See:* snatch block.

watchdog timer (1) (supervisory control, data acquisition, and automatic control) (station control and data acquisition) A form of interval timer that is used to detect a possible malfunction. (SWG/PE/SUB) C37.1-1987s, C37.100-1992
(2) A timer that prevents a computer program from looping endlessly or becoming idle because of program errors or equipment faults. *Note:* This is typically implemented by resetting the computer if the timer is not refreshed often enough. (C) 610.10-1994w

watchman's reporting system A supervisory system arranged for the transmission of a patrolling watchman's regularly recurrent report signals to a central supervisory agency from stations along his patrol route. *See also:* protective signaling. (EEC/PE) [119]

water-air-cooled machine A machine that is cooled by circulating air that in turn is cooled by circulating water. *Note:* The machine is so enclosed as to prevent the free exchange of air between the inside and outside of the enclosure, but not sufficiently to be termed airtight. It is provided with a water-cooled heat exchanger for cooling the ventilating air and a fan or fans, integral with the rotor shaft or separate, for circulating the ventilating air. *See also:* asynchronous machine. (PE) [9]

water-blocking tape A nonwoven synthetic tape impregnated with a swellable powder. Also called a water-swellable tape. Such tapes may also provide cushioning to absorb expansion of the cable core and may also be semiconducting. (PE/IC) 1142-1995

water conditions *See:* median water conditions; adverse water conditions; average water conditions.

water-cooled (A) (rotating machinery) A term applied to apparatus cooled by circulating water, the water or water ducts coming in direct contact with major parts of the apparatus. **(B) (rotating machinery)** In certain types of machine, it is customary to apply this term to the cooling of the major parts by enclosed air or gas ventilation, where water removes the heat through an air-to-water or gas-to-water heat exchanger. *See also:* asynchronous machine. (PE) [9]

water cooler (rotating machinery) A cooler using water as one of the fluids. (PE) [9]

water failure A failure in the active, shielded cable length that is below the waterline and which did not occur as a result of mechanical damage. (PE/IC) 1407-1998

waterfall model (software) A model of the software development process in which the constituent activities, typically a concept phase, requirements phase, design phase, implementation phase, test phase, and installation and checkout phase, are performed in that order, possibly with overlap but with little or no iteration. *Contrast:* spiral model; rapid prototyping; incremental development. (C) 610.12-1990

waterflow-alarm system (protective signaling) An alarm system in which signal transmission is initiated automatically by devices attached to an automatic sprinkler system and actuated by the flow through the sprinkler system pipes of water in excess of a predetermined maximum. *See also:* protective signaling. (EEC/PE) [119]

water hazard The abnormal presence of a quantity of water, either in the form of condensation, accumulation, flow, or spray; not postulated to be associated with any design basis event; and considered likely to cause the loss of desired function of electric equipment in nuclear power plants if it enters such equipment. (PE/NP) 833-1988r

water inertia time (hydraulic turbines) A characteristic time, usually taken at rated conditions, due to inertia of the water in the water passages from intake to exit defined as:

$$T_w = \frac{Q_r}{gH_r}\int\frac{dL}{A} \approx \frac{Q_r}{gH_r}\sum\frac{L}{A}$$

where

A = area of each section
L = corresponding length
Q_r = rated discharge
H_r = rated head
g = acceration due to gravity

(PE/EDPG) 125-1977s

waterline failure A failure at the interface between air and the tank water to include the distance of the total water line variation. (PE/IC) 1407-1998

water load (high-frequency circuits) A matched termination in which the electromagnetic energy is absorbed in a stream of water for the purpose of measuring power by continuous-flow calorimetric methods. *See also:* waveguide. (AP/ANT) [35], [84]

Waterloo FORTRAN A programming language based on FORTRAN, characterized by its fast compilation, excellent diagnostic messages and debugging aids. *See also:* WATFIV. (C) 610.13-1993w

Waterloo FORTRAN V (WATFIV) An extension of WATFOR characterized by additional simplifications in the control (test and looping) constructs and the input/output structure. (C) 610.13-1993w

water-motor bell A vibrating bell operated by a flow of water through its water-motor striking mechanism. *See also:* protective signaling. (EEC/PE) [119]

waterproof electric blasting cap A cap specially insulated to secure reliability of firing when used in wet work. *See also:* blasting unit. (EEC/PE) [119]

waterproof enclosure An enclosure constructed so that any moisture or water leakage that may occur into the enclosure will not interfere with its successful operation. In the case of motor or generator enclosures, leakage that may occur around the shaft may be considered permissible provided it is prevented from entering the oil reservoir and provision is made for automatically draining the motor or generator enclosure. (IA/MT) 45-1998

waterproof machine (rotating machinery) A machine so constructed that water directed on it under prescribed conditions cannot cause interference with satisfactory operation. *See also:* asynchronous machine. (PE) [9]

water resistivity Resistance of water is expressed in Ω cm or Ω in. (T&D/PE) 957-1995

watertight (1) So constructed that moisture will not enter the enclosure. (NESC/NEC) [86]
(2) (power and distribution transformers) So constructed that water will not enter the enclosing case under specified conditions. *Note:* A common form of specification for watertight is: "So constructed that there shall be no leakage of water into the enclosure when subjected to a stream from a hose with a one-in nozzle and delivering at least 65 gal/min, with the water directed at the enclosure from a distance of not less than 10 ft for a period of five min, during which period the water may be directed in one or more directions as desired." (PE/TR) C57.12.80-1978r

watertight door-control system A system of control for power-operated watertight doors providing individual local control of each door and, at a remote station in or adjoining the wheelhouse, individual control of any door, collective control of all doors, and individual indication of open or closed condition. *See also:* marine electric apparatus. (EEC/PE) [119]

watertight enclosure An enclosure constructed so that a stream of water from a hose not less than 25 mm (1 in) in diameter under a head of 10 m (35 ft) from a distance of 3 m (10 ft) can be played on the enclosure from any direction for a period of 15 min without leakage. The hose nozzle shall have a uniform inside diameter of 25 mm (1 in). (IA/MT) 45-1998

water treatment equipment (thyristor) Any apparatus such as deionizers, electrolytic targets, filters, or other devices employed to control electrolysis, corrosion, scaling, or clogging in water systems. (IA/IPC) 428-1981w

water vapor The amount of water in parts per million by volume that is in the gaseous state and mixed with the insulating gas. Water vapor content may vary with temperature. *Synonym:* moisture content. (SWG/PE) C37.100-1992

water vapor content The amount of water in parts per million by volume (ppmv) that is in the gaseous state and mixed with the insulating gas. *Synonym:* moisture content. (SWG/SUB/PE) C37.122-1993, C37.122.1-1993, C37.100-1992

WATFIV *See:* Waterloo FORTRAN V.

WATS *See:* wide area telecommunications service.

watt (W) (1) (general) The unit of power in the International System of Units (SI). The watt is the power required to do work at the rate of one joule per second. (Std100) 270-1966w

(2) (laser maser) The unit of power, or radiant flux. (LEO) 586-1980w

watt density (electrical heat tracing for industrial applications) Thermal output of heating cable in watts per unit area. (BT/AV) 152-1953s

watthour 3600 joules. (Std100) 270-1966w

watthour capacity (storage battery) (storage cell) The number of watthours that can be delivered under specified conditions as to temperature, rate of discharge, and final voltage. *See also:* battery. (EEC/PE) [119]

watthour constant (watthour meter) The registration, expressed in watthours, corresponding to one revolution of the rotor. *Note:* It is commonly denoted by the symbol Kh. When a meter is used with instrument transformers, the watthour constant is expressed in terms of primary watthours. For a secondary test of such a meter, the constant is the primary watthour constant divided by the product of the nominal ratios of transformation. (ELM) C12.1-1982s

watthour-demand meter A watthour meter and a demand meter combined as a single unit. *See also:* electricity meter. (EEC/PE) [119]

watthour efficiency (storage battery) (storage cell) The energy efficiency expressed as the ratio of the watthours output to the watthours of the recharge. *See also:* charge. (PE/EEC) [119]

watthour meter An electricity meter that measures and registers the integral, with respect to time, of the active power of the circuit in which it is connected. This power integral is the energy delivered to the circuit during the interval over which the integration extends, and the unit in which it is measured is usually the kilowatthour. *See also:* class designation; heavy load; rated voltage; load range; percentage error; rated current; basic current range; stator; register ratio; light load; reference performance; test current; watthour constant; load current; motor-type watthour meter; adjustment; two-rate watthour meter; registration; register constant; basic voltage range; portable standard watthour meter; rotor; form designation; induction; percentage registration; creep; gear ratio; register. (ELM) C12.1-1988

watthour meter—adjustment Adjustment of internal controls to bring the percentage registration of the meter to within specified limits. (ELM) C12.1-1988

watthour meter—basic current range The current range of a multirange standard watthour meter designated by the manufacturer for the adjustment of the meter (normally the 5 A range). (ELM) C12.1-1988

watthour meter—basic voltage range The voltage range of a multirange standard watthour meter designated by the manufacturer for the adjustment of the meter (normally the 120 V range). (ELM) C12.1-1988

watthour meter—class designation The maximum of the load range in amperes. *See also:* watthour meter—load range. (ELM) C12.1-1988

watthour meter—creep A continuous motion of the rotor of a meter with normal operating voltage applied and the load terminals open-circuited. (ELM) C12.1-1988

watthour meter—form designation An alphanumeric designation denoting the circuit arrangement for which the meter is applicable and its specific terminal arrangement. The same designation is applicable to equivalent meters of all manufacturers. (ELM) C12.10-1987

watthour meter—gear ratio The number of revolutions of the rotor for one revolution of the first dial pointer, commonly denoted by the symbol R_g. (ELM) C12.1-1988

watthour meter—heavy load *See:* watthour meter—test current.

watthour meter—induction A motor-type meter in which currents induced in the rotor interact with a magnetic field to produce the driving torque. (ELM) C12.1-1988

watthour meter—light load The current at which the meter is adjusted to bring its response near the lower end of the load range to the desired value. It is usually 10% of the test current for a revenue meter and 25% for a standard meter. (ELM) C12.1-1988

watthour meter—load current *See:* watthour meter—test current.

watthour meter—load range The maximum range in amperes over which the meter is designed to operate continuously with a specified accuracy under specified conditions. (ELM) C12.1-1988

watthour meter—motor type A motor in which the speed of the rotor is proportional to the power, with a readout device that counts the revolutions of the rotor. (ELM) C12.1-1988

watthour meter—percentage error The difference between its percentage registration and 100%. A meter whose percentage registration is 95% is said to be 5% slow, or its error is −5%. A meter whose percentage registration is 105% is 5% fast, or its error is +5%. (ELM) C12.1-1988

watthhour meter—percentage registration The percentage registration of a meter is the ratio of the actual registration of the meter to the true value of the quantity measured in a given time, expressed as a percentage. (ELM) C12.1-1981

watthour meter—portable standard A portable meter, principally used as a standard for testing other meters. It is usually provided with several current and voltage ranges and with a readout indicating revolutions and fractions of a revolution of the rotor. *Note:* Electronic portable standards not using a rotor may have a readout indicating equivalent revolutions and fractions of revolutions or other units, such as percentage registration. (ELM) C12.1-1988

watthour meter—rated current The nameplate current for each range of a standard watthour meter. *Note:* The main adjustment of the meter is ordinarily made with rated current on the basic current range. (ELM) C12.1-1988

watthour meter—rated voltage The nameplate voltage for a meter or for each range of a standard watthour meter. *Note:* The main adjustment of the standard meter is ordinarily made with rated voltage on the basic voltage range. (ELM) C12.1-1988

watthour meter—reference performance Performance at specified reference conditions for each test, used as a basis for comparison with performances under other conditions of the test. (ELM) C12.1-1988

watthour meter—reference standard A meter used to maintain the unit of electric energy. It is usually designed and operated to obtain the highest accuracy and stability in a controlled laboratory environment. (ELM) C12.1-1988

watthour meter—register That part of the meter that registers the revolutions of the rotor, or the number of pulses received from or transmitted to a meter, in terms of units of electric energy or other quantity measured. (ELM) C12.1-1988

watthour meter—register constant The multiplier used to convert the register reading to kilowatthours (or other suitable units). *Note:* This constant, commonly denoted by the symbol, K_r, takes into consideration the watthour constant, gear ratio, and instrument transformer ratios. (ELM) C12.1-1988

watthour meter—register ratio The number of revolutions of the first gear of the register, for one revolution of the first dial pointer. *Note:* This is commonly denoted by the symbol, R_r. (ELM) C12.1-1988

watthour meter—registration The registration of a meter is the apparent amount of electric energy (or other quantity being measured) that has passed through the meter, as shown by the register reading. It is equal to the product of the register reading and the register constant. The registration during a given period is equal to the product of the register constant and the difference between the register readings at the beginning and the end of the period. (ELM) C12.1-1988

watthour meter—rotor That part of the meter that is directly driven by electromagnetic action. (ELM) C12.1-1988

watthour meter—standard *See:* watthour meter—portable standard; watthour meter—reference standard.

watthour meter—stator An assembly of an induction watthour meter, which consists of a voltage circuit, one or more current circuits, and a combined magnetic circuit so arranged that their joint effect, when energized, is to exert a driving torque on the rotor by the reaction with currents induced in an individual or common conducting disk. (ELM) C12.1-1988

watthour meter—test current (TA) The current specified by the manufacturer for the main adjustment of the meter (heavy- or full-load adjustment). *Notes:* 1. It has been identified as "TA" on revenue meters manufactured since 1960. 2. The main adjustment of a meter used with a current transformer may be made either at the test current or at the rated secondary current of the transformer. (ELM) C12.1-1988

watthour meter—two-rate A meter having two sets of register dials with a changeover arrangement such that integration of the quantity will be registered on one set of dials during a specified time each day and on the other set of dials for the remaining time. (ELM) C12.1-1988

watthour meter—watthour constant The registration, expressed in watthours, corresponding to one revolution of the rotor. *Note:* It is commonly denoted by the symbol K_h. When a meter is used with instrument transformers, the watthour constant is expressed in terms of primary watthours. For a secondary test of such a meter, the constant is the primary watthour constant divided by the product of the nominal ratios of transformation. (ELM) C12.1-1988

watt loss *See:* power loss.

wattmeter An instrument for measuring the magnitude of the active power in an electric circuit. It is provided with a scale usually graduated in either watts, kilowatts, or megawatts. If the scale is graduated in kilowatts or megawatts, the instrument is usually designated as a kilowattmeter or megawattmeter. *See also:* instrument. (EEC/PE) [119]

wattsecond constant (meter) The registration in wattseconds corresponding to one revolution of the rotor. *Note:* The wattsecond constant is 3600 times the watthour constant and is commonly denoted by the symbol Ks. *See also:* electricity meter. (EEC/PE) [119]

wave (1) (A) (data transmission) A disturbance that is a function of time or space or both. **(B) (data transmission)** A disturbance propagated in a medium or through space. *Notes:* 1. Any physical quantity that has the same relationship to some independent variable (usually time) that a propagated disturbance has, at a particular instant, with respect to space, may be called a wave. 2. Disturbance, in this definition, is used as a generic term indicating not only mechanical displacement but also voltage, current, electric field strength, temperature, etc. **(C) (data transmission)** (electric circuit). The variation of current, potential, or power at any point in the electric circuit. *See also:* wave tilt; ground-reflected wave; tropospheric wave; continuous wave; ground wave; wave filter; square wave; wave interference; rectangular wave; tangential wave path. (PE/AP/ANT) 599-1985, 145-1983
(2) (overhead-power-line corona and radio noise) A disturbance propagated in a medium. *Note:* "Disturbance" in this definition is used as a generic term indicating not only mechanical displacement but also voltage, current, electric field strength, temperature, etc. Any physical quantity that has the same relationship to some independent variable (usually time) as a propagated disturbance, (at a particular instance) with respect to space, may be called a wave. (T&D/PE) 539-1990
(3) (pulse terminology) A modification of the physical state of a medium that propagates in the medium as a function of time as a result of one or more disturbances. (IM/WM&A) 194-1977w
(4) The variation with time of current, potential, or power at any point in an electric circuit. (SPD/PE) C62.11-1999, C62.62-2000

wave analyzer An electric instrument for measuring the amplitude and frequency of the various components of a complex current or voltage wave. *See also:* instrument. (EEC/PE) [119]

wave antenna *See:* Beverage antenna.

wave clutter Clutter caused by echoes from waves of the sea. *See also:* radar. (EEC/PE) [119]

wave-division multiplexing (WDM) A multiplexing technique used in optical fiber transmission systems that defines multiple paths on the fiber by using different wavelengths (colors) of light for each channel. *See also:* time-division multiplexing. (C) 610.7-1995

wave filter (data transmission) A transducer for separating waves on the basis of their frequency. *Note:* A filter introduces relatively small insertion loss to waves in one or more frequency bands and relatively large insertion loss to waves of other frequencies. (PE) 599-1985w

waveform (1) (pulse waveform) (transition waveform) (pulse terminology) A manifestation or representation (that is, graph, plot, oscilloscope presentation, equation(s), table of coordinate or statistical data, etc.) or a visualization of a wave, pulse, or transition. *Notes:* 1. The term pulse waveform is included in the term waveform. 2. The term transition waveform is included in the terms pulse waveform and waveform. (IM/WM&A) 194-1977w
(2) A stream of defined events containing both state and timing information. (C/TT) 1450-1999

waveform-amplitude distortion Nonlinear distortion in the special case where the desired relationship is direct proportionality between input and output. *Note:* Also sometimes called "amplitude distortion." *See also:* nonlinear distortion. (Std100) 154-1953w

Waveform and Vector Exchange Specification (WAVES) A standardized computer language used to describe digital test vectors. (ATLAS) 1226-1993s

waveform distortion (1) (oscilloscopes) A displayed deviation from the representation of the input reference signal. *See also:* oscillograph. (IM/HFIM) [40]
(2) (oscilloscopes) A displayed deviation from the correct representation of the input reference signal. *See also:* oscillograph. (IM) 311-1970w

waveform distortion—percent The ratio of the root-means-square value of the harmonic content (excluding the fundamental) to the root-mean-square value of the nonsinusoidal quantity, expressed as a percentage. (ELM) C12.1-1988

waveform epoch (pulse terminology) The span of time for which waveform data are known or knowable. A waveform epoch manifested by equations may extend in time from minus infinity to plus infinity or, like all waveform data, may extend from a first datum time, t_0, to a second datum time, t_1. (See the corresponding figure.)

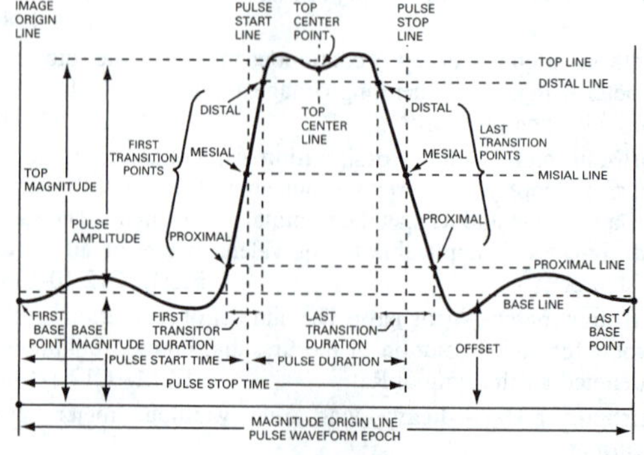

The single pulse.

waveform epoch

(IM/WM&A) 194-1977w

waveform epoch contraction (pulse measurement) A technique for the determination of the characteristics of individual pulse waveforms (or pulse waveform features) wherein the waveform epoch (or pulse waveform epoch) is contracted in time to a pulse waveform epoch (or transition waveform epoch) for the determination of time or magnitude characteristics. In any waveform epoch contraction procedure two or more sets of time or magnitude reference lines may exist, and the set of reference lines being used in any pulse measurement process shall be specified. (IM/WM&A) 181-1977w

waveform epoch expansion (pulse measurement) A technique for the determination of the characteristics of a transition waveform (or pulse waveform) wherein the transition waveform epoch (or pulse waveform epoch) is expanded in time to a pulse waveform epoch (or waveform epoch) for the determination of magnitude or time reference lines. The reference lines determined by analysis of the pulse waveform (or waveform) are transferred to the transition waveform (or pulse waveform) for the determination of characteristics. In any waveform epoch expansion procedure two or more sets of reference lines may exist, and the set of reference lines being used in any pulse measurement process shall be specified. (IM/WM&A) 181-1977w

waveform formats Waveforms may exist, be recorded, or be stored in a variety of formats. It is assumed that: A) waveform formats are in terms of Cartesian coordinates, or some transform thereof; B) conversion from one waveform format to any other is possible; and C) such waveform format conversions can be made with precision, accuracy, and resolution that is consistent with the accuracy desired in the pulse measurement process. (IM/WM&A) 181-1977w

waveform influence of root-mean-square responding instruments The change in indication produced in an RMS responding instrument by the presence of harmonics in the alternating electrical quantity under measurement. In magnitude, it is the deviation between an indicated RMS value of an alternating electrical quantity and the indication produced by the measurement of a pure sine-wave form of equal RMS value. *See also:* instrument. (EEC/AII) [102]

waveforms produced by continuous time superposition of simpler waveforms *See:* square wave; pulse train time-related definitions; pulse train.

waveforms produced by magnitude superposition *See:* offset; composite waveform; offset waveform.

waveforms produced by noncontinuous time superposition of simpler waveforms *See:* pulse burst; pulse burst time-related definitions.

waveforms produced by operations on waveforms (pulse terminology) All envelope definitions in this section are based on the cubic natural spline (or its related approximation, the draftsman's spline) with knots at specified points. All burst envelopes extend in time from the first to the last knots specified, the remainder of the waveform being: A) that portion of the waveform that precedes the first knot; and B) that portion of the waveform that follows the last knot. Burst envelopes and their adjacent waveform bases, taken together, comprise a continuous waveform that has a continuous first derivative except at the first and last knots of the envelope. *See also:* pulse burst top envelope; pulse-train top (base) envelope; pulse burst base envelope.
(IM/WM&A) 194-1977w

waveform pulse A waveform or a portion of a waveform containing one or more pulses or some portion of a pulse. *See also:* pulse. (IM/HFIM) [40]

waveform reference A specified waveform, not necessarily ideal, relative to which waveform measurements, derivations, and definitions may be referred. (IM/HFIM) [40]

waveform test (rotating machinery) A test in which the waveform of any quantity associated with a machine is recorded. *See also:* asynchronous machine. (PE) [9]

wavefront (1) (fiber optics) The locus of points having the same phase at the same time. (Std100) 812-1984w
(2) (impulse in a conductor) That part (in time or distance) between the virtual-zero point and the point at which the impulse reaches its crest value. (T&D/PE) [10]
(3) (of a surge or impulse) The part that occurs prior to the crest value. (SPD/PE) C62.11-1999, C62.62-2000
(4) (of an impulse) That part of an impulse that occurs prior to the crest value. (SPD/PE) C62.22-1997

waveguide (1) (waveguide terms) A system of material boundaries or structures for guiding electromagnetic waves. Usually such a system is used for guiding waves in other than TEM modes. Often, and originally, a hollow metal pipe for guiding electromagnetic waves. *See also:* transmission line.
(MTT) 146-1980w
(2) (A) (data transmission) Broadly, a system of material boundaries capable of guiding electromagnetic waves. **(B) (data transmission)** More specifically, a transmission line comprising a hollow conducting tube within which electromagnetic waves may be propagated or a solid dielectric or dielectric-filled conductor for the same purpose. **(C) (data transmission)** A system of material boundaries or structures for guiding transverse-electromagnetic mode, often and originally a hollow metal pipe for guiding electromagnetic waves.
(PE) 599-1985
(3) A metal tube used to transmit microwaves.
(C) 610.7-1995
(4) Metallic or dielectric structures, usually uniform in the longitudinal direction, that are capable of guiding waves.
(AP/PROP) 211-1997

waveguide adapter (waveguide components) A structure used to interconnect two waveguides that differ in size or type. If the modes of propagation also differ, the adapter functions as a mode transducer. (MTT) 147-1979w

waveguide attenuator (waveguide components) A waveguide component that reduces the output power relative to the input by any means, including absorption and reflection. *See also:* waveguide. (MTT) 147-1979w

waveguide bend (waveguide components) A section of waveguide or transmission line in which the direction of the longitudinal axis is changed. In common usage the waveguide corner formed by an abrupt change in direction is considered to be a bend. (MTT) 147-1979w

waveguide calorimeter (waveguide components) A waveguide or transmission line structure that uses the temperature rise in a medium as a measure of absorbed power. The medium, typically water or a thermoelectric element, is either the power-absorbing agent or has heat transferred to it from a power-absorbing element. (MTT) 147-1979w

waveguide circulator (nonlinear, active, and nonreciprocal waveguide components) A passive waveguide device of three or more ports in which the ports can be numbered in such an order that, when power is fed into any port, the power is transferred to the next sequentially numbered port. The first port is counted as following the last in order.
(MTT) 457-1982w

waveguide component A device designed to be connected at specified ports in a waveguide system. (MTT) 148-1959w

waveguide connector (fixed and variable attenuators) A mechanical device, excluding an adapter, for electrically joining separable parts of a waveguide or transmission-line system.
(IM/HFIM) 474-1973w

waveguide corner *See:* waveguide bend.

waveguide cutoff frequency *See:* cutoff frequency.

waveguide differential phase circulator (nonlinear, active, and nonreciprocal waveguide components) A waveguide circulator based on the use of at least one nonreciprocal differential insertion phase element or gyrator, usually in connection with other waveguide components such as microwave hybrid junctions. (MTT) 457-1982w

waveguide dispersion (fiber optics) For each mode in an optical waveguide, the term used to describe the process by which an electromagnetic signal is distorted by virtue of the dependence of the phase and group velocities on wavelength as a consequence of the geometric properties of the waveguide. In particular, for circular waveguides, the dependence is on the ratio (a/λ), where a is core radius and λ is wavelength. *See also:* profile dispersion; dispersion; multimode distortion; material dispersion; distortion. (Std100) 812-1984w

waveguide Faraday rotation circulator (nonlinear, active, and nonreciprocal waveguide components) A waveguide circulator based on the use of a Faraday rotation element in conjunction with other waveguide components such as dual-mode transducers. (MTT) 457-1982w

waveguide ferrite isolator (nonlinear, active, and nonreciprocal waveguide components) A waveguide two-port device, using gyromagnetic material, in which the attenuation in one direction of propagation is much greater than in the opposite direction. (MTT) 457-1982w

waveguide gasket (waveguide components) A resilient insert usually between flanges intended to serve one or more of the following primary purposes: A) to reduce gas leakage affecting internal waveguide pressure; B) to prevent intrusion of foreign material into the waveguide; or C) to reduce power leakage and arcing. (MTT) 147-1979w

waveguide iris (waveguide components) A partial obstruction at a transverse cross-section formed by one or more metal plates of small thickness compared with the wavelength. (MTT) 147-1979w

waveguide joint A connection between two sections of waveguide. *See also:* waveguide. (IM/HFIM) [40]

waveguide junction circulator (nonlinear, active, and nonreciprocal waveguide components) A waveguide circulator based on the use of gyromagnetic material at the common junction of several waveguides. (MTT) 457-1982w

waveguide matched termination *See:* matched termination.

waveguide mode (waveguide) In a uniform waveguide, a wave that is characterized by exponential variation of the fields along the direction of the guide. *Note:* In other types of waveguides, such as radial, spherical, toroidal, etc., some particular variation will have to be specified according to the geometry. (MTT) 146-1980w

waveguide modes Those spurious modes of a planar transmission line that are guided by the enclosure or shielding of this line rather than the transmission line itself. (MTT) 1004-1987w

waveguide phase shifter (waveguide components) An essentially loss-less device for adjusting the phase of a forward-traveling electromagnetic wave at the output of the device relative to the phase at the input. (MTT) 147-1979w

waveguide post (waveguide components) A cylindrical rod placed in a transverse plane of the waveguide and behaving substantially as a shunt susceptance. (MTT) 147-1979w

waveguide resonator (waveguide components) A waveguide or transmission-line structure which can store oscillating electromagnetic energy for time periods that are long compared with the period of the resonant frequency, at or near the resonant frequency. (MTT) 147-1979w

waveguide scattering (fiber optics) Scattering (other than material scattering) that is attributable to variations of geometry and index profile of the waveguide. *See also:* scattering; material scattering; Rayleigh scattering; nonlinear scattering. (Std100) 812-1984w

waveguide short circuit, adjustable (waveguide components) A longitudinally movable obstacle that reflects essentially all the incident energy. (MTT) 147-1979w

waveguide step twist (waveguide components) A waveguide twist formed by abruptly rotating about the waveguide longitudinal axis one or more waveguide sections, each nominally a quarter wavelength long. (MTT) 147-1979w

waveguide stub (waveguide components) A section of waveguide or transmission line joined to the main guide or transmission line and containing an essentially nondissipative termination. (MTT) 147-1979w

waveguide switch (waveguide system) A device for stopping or diverting the flow of high-frequency energy as desired. *See also:* waveguide. (AP/ANT) [35], [84]

waveguide taper (waveguide components) A section of tapered waveguide. (MTT) 147-1979w

waveguide termination *See:* cavity; unloaded applicator impedance.

waveguide-to-coaxial transition A mode changer for converting coaxial line transmission to rectangular waveguide transmission. *See also:* waveguide. (AP/ANT) [35]

waveguide transformer (waveguide components) A structure added to a waveguide or transmission line for the purpose of impedance transformation. (MTT) 147-1979w

waveguide tuner (waveguide components) An adjustable waveguide transformer. (MTT) 147-1979w

waveguide twist (waveguide components) A waveguide section in which there is progressive rotation of the cross-section about the longitudinal axis. *See also:* waveguide step twist. (MTT) 147-1979w

waveguide wavelength (1) (waveguide terms) For a traveling wave in a uniform waveguide at a given frequency and for a given mode, the distance along the guide between corresponding points at which a field component (or the voltage or current) differs in phase by 2π radians. (MTT) 146-1980w

(2) (data transmission) The distance along a uniform guide between points at which a field component (or the voltage or current) differs in phase by 2π radians. *Note:* It is equal to the quotient of phase velocity divided by frequency. For a waveguide with air dielectric, the waveguide wavelength is given by the formula:

$$\lambda_g = \frac{\lambda}{(1 - (\lambda^2/\lambda_c))^{1/2}}$$

where λ is the free space wavelength and λ_c is the cutoff wavelength of the guide. (PE) 599-1985w

waveguide window (waveguide components) A gas- or liquid-tight barrier or cover designed to be essentially transparent to the transmission of electromagnetic waves. (MTT) 147-1979w

wave heater (dielectric heating) A heater in which heating is produced by energy absorption from a traveling electromagnetic wave. (IA) 54-1955w

wave heating The heating of a material by energy absorption from a traveling electromagnetic wave. *See also:* induction heating. (IA) 54-1955w

wave impedance (overhead power lines) The complex factor relating the transverse component of the magnetic field to the transverse component of the electric field at every point in any specified plane, for a given mode. (T&D/PE/MTT) 539-1990, 146-1980w

wave impedance, characteristic *See:* characteristic wave impedance.

wave interference (1) (data transmission) The variation of wave amplitude with distance or time, caused by the superposition of two or more waves. *Notes:* 1. As most commonly used, the term refers to the interference of waves of the same or nearly the same frequency. 2. Wave interference is characterized by a spatial or temporal distribution of amplitude of some specified characteristic differing from that of the individual superposed waves. (PE) 599-1985w

(2) The variation of wave amplitude with distance or time, caused by the superposition of two or more waves of the same (or very nearly the same) frequency. *Note:* If the waves have very nearly the same frequency, they are said to "beat with each other." (AP/PROP) 211-1997

wavelength (λ) (1) (laser maser) The distance between two points in a periodic wave that have the same phase. (LEO) 586-1980w

(2) (overhead power lines) The distance between points of corresponding phase of two consecutive cycles of a sinusoidal wave. The wavelength, λ, is related to the phase velocity, v, and the frequency, f, by $\lambda = v/f$. (T&D/PE) 539-1990

(3) (of a monochromatic wave) The distance between two points of corresponding phase of two consecutive cycles in the direction of the wave normal. The wavelength, λ, is related to the magnitude of the phase velocity, v_p, and the frequency, f, by the equation:

$$\lambda = v_p/f$$

(AP/PROP) 211-1997

(4) (of a monochromatic wave) The distance between two points of corresponding phase of two consecutive cycles in the direction of propagation. The wavelength (λ) of an electromagnetic wave is related to the frequency (f) and velocity (v) by the expression $v = f\lambda$. In free space the velocity of an electromagnetic wave is equal to the speed of light, i.e., approximately 3×10^8 m/s. (NIR) C95.1-1999

wavelength constant *See:* phase constant.

wavelength division multiplexing (fiber optics) The provision of two or more channels over a common optical waveguide, the channels being differentiated by optical wavelength. (Std100) 812-1984w

wavelength shifter (scintillator) A photofluorescent compound used with a scintillator material to absorb photons and emit related photons of a longer wavelength. *Note:* The purpose is to cause more efficient use of the photons by the phototube or photocell. *See also:* phototube. (NPS) 175-1960w

wavemeter *See:* cavity resonator frequency meter.

wave normal (1) (waveguide) A unit vector normal to an equiphase surface with its positive direction taken on the same side of the surface as the direction of propagation. In isotropic media, the wave normal is in the direction of propagation. (MTT) 146-1980w

(2) (of a traveling wave) The direction normal to an equiphase surface taken in the direction of increasing phase. *See also:* direction of propagation. (AP/PROP) 211-1997

wave number (k) 2π divided by the wavelength in the medium. (AP/PROP) 211-1997

WAVES *See:* Waveform and Vector Exchange Specification.

wave shape (1) (of an impulse test wave) The graph of an impulse test wave as a function of time. (SPD/PE) C62.22-1997

(2) (of an impulse test wave) The graph of the wave as a function of time. (SPD/PE) C62.11-1999

(3) A stream of defined states or transitions with no associated timing. (C/TT) 1450-1999

wave shape designation (1) (of an impulse)

a) The wave shape of an impulse (other than rectangular) of a current or voltage is designated by a combination of two numbers. The first, an index of the wave front, is the virtual duration of the wave front in microseconds. The second, an index of the wave tail, is the time in microseconds from virtual zero to the instant at which one-half of the crest value is reached on the wave tail. Examples are 1.2/50 and 8/20 waves.

b) The wave shape of a rectangular impulse of current or voltage is designated by two numbers. The first designates the minimum value of current or voltage that is sustained for the time in microseconds designated by the second number. An example is the 75 A × 2000 μs wave.

(SPD/PE) C62.22-1997

(2) (A) (of an impulse) The wave shape of an impulse (other than rectangular) of a current or voltage is designated by a combination of two numbers. The first, an index of the wavefront, is the virtual duration of the wavefront in microseconds. The second, an index of the wave tail, is the time in microseconds from virtual zero to the instant at which one-half of the crest value is reached on the wave tail. Examples are 1.2/50 and 8/20 waves. **(B)** (of an impulse) The wave shape of a rectangular impulse of current or voltage is designated by two numbers. The first designates the minimum value of current or voltage that is sustained for the time in microseconds designated by the second number. An example is the 75 A 1000 μs wave. (SPD/PE) C62.11-1999

(3) The shape of a nonrectangular impulse of current or voltage is designated by a combination of two numbers: The first, an index of the wave front, is the virtual duration of the wave front in microseconds (virtual duration of wavefront). The second, an index of the wave tail, is the time in microseconds from virtual zero to the instant at which one-half of the crest value is reached on the wave tail. Examples are 1.2/50 and 8/20 waves. (SPD/PE) C62.62-2000

wave, square *See:* square wave.

wave tail (of an impulse) That part between the crest value and the end of the impulse. (SPD/PE) C62.22-1997, C62.11-1999, C62.62-2000

wave tilt (1) (data transmission) The forward inclination of a radio wave due to its proximity to ground. (PE) 599-1985w

(2) (of a monochromatic electromagnetic wave propagating near the interface between two media) The complex ratio of the electric (or magnetic) field component that is tangent to the interface to that which is normal to the interface, both field components lying in the plane of propagation. *Note:* Wave tilt is generally associated with ground wave propagation over the Earth's surface. (AP/PROP) 211-1997

wave train A limited series of wave cycles caused by a periodic disturbance of short duration. *See also:* pulse train. (PE/EDPG) [3]

wave vector *See:* propagation vector.

wave vector in physical media The complex vector 15$\bar{k}$ in plane wave solutions of the form e^{-jkr}, for an e^{jvt} time variation and 15r> the position vector. *See also:* propagation vector in physical media. (AP/ANT) 145-1983s

wave winding A winding that progresses around the armature by passing successively under each main pole of the machine before again approaching the starting point. In commutator machines, the ends of individual coils are not connected to adjacent commutator bars. *See also:* direct-current commutating machine; asynchronous machine. (EEC/PE) [119]

way (1) A three-phase circuit entrance to a switching assembly. (SWG/PE) C37.71-1984r

(2) A three-phase circuit entrance to a switch or bus; or for single-phase switches, single-phase entrance to a switch or bus. (SWG/PE) C37.100-1992

(3) A three-phase or single-phase circuit section containing a switch, fuse, combination switch and fuse, or bus. (SWG/PE) C37.73-1998

way point (navigation) (navigation aids) A selected point on or near a course line and having significance with respect to navigation or traffic control. (AES/GCS) 172-1983w

way station (data transmission) A telegraph term for one of the stations on a multipoint circuit. (PE) 599-1985w

W-band A radar-frequency band between 75 GHz and 110 GHz, usually in one of the International Telecommunication Union (ITU) allocated bands 76–81 GHz or 92–100 GHz. Included within the definition of millimeter-wave radar. (AES) 686-1997

WBC (wide band channel (ARCHIVE)) *See:* wideband channel.

WDM *See:* wave-division multiplexing.

weak ac system *See:* high-impedance ac system.

weak field In a propulsion system, a motor connection or operating mode in which the exciting field current is less than the full field value. (VT) 1475-1999

weakly guiding fiber (fiber optics) A fiber for which the difference between the maximum and the minimum refractive index is small (usually less than 1 percent). (Std100) 812-1984w

weakly perturbed field (1) (overhead power lines) At a given point, a field whose magnitude does not change by more than 5% or whose direction does not vary by more than 5°, or both, when an object is introduced into the region. (T&D/PE) 539-1990
(2) At a given point, a field whose magnitude does not change by more than 5% or whose direction does not vary by more than 5 degrees when an object is introduced into the region. (T&D/PE) 644-1994

weak sequential consistency A state exhibited by a system when references to global synchronizing variables exhibit strong sequential consistency, and if no reference to a synchronizing variable is issued by any processor until all previous modifications to global data have been observed by all caches, and if no reference to global data is issued by any processor until all previous modifications to synchronizing variables have been observed by all caches. *See also:* strong sequential consistency. (C/BA) 10857-1994

wearout The state of a component in which the failure rate increases with time as a result of a process characteristic of the population. (PE/NP) 933-1999

wear-out failure *See:* failure.

wearout-failure period (1) That possible period during which the failure rate increases rapidly in comparison with the preceding period. (R) [29]
(2) (software) The period in the life cycle of a system or component during which hardware failures occur at an increasing rate due to deterioration. *Contrast:* constant-failure period; early-failure period. *See also:* bathtub curve. (C) 610.12-1990

wear-out failures (station control and data acquisition) The pattern of failures experienced when equipments reach their period of deterioration. Wear-out failure profiles may be approximated by a Gaussian (bell curve) distribution centered on the nominal life of the equipment. (PE/SUB) C37.1-1994

wearout of reverse drain breakdown (metal-nitride-oxide field-effect transistor) An effect where the reverse current voltage characteristic of the drain junction changes progressively and irreversibly toward larger (leakage) currents at the same reverse voltages. (ED) 581-1978w

wearout period (reliability analysis of nuclear power generating station safety systems) The time interval, following the period of constant failure rate, during which failures occur at an increasing rate. (PE/NP) 352-1975s

wear point (cable plowing) A removable tip on the end of some shanks or plow blades. (T&D/PE) 590-1977w

weather (outage occurrences and outage states of electrical transmission facilities) Exposure, whether measured in time or operations, may be subdivided according to the type of weather to which a component or components within a unit is exposed. (PE/PSE) 859-1987w

weather, adverse *See:* adverse weather.

weather barrier (1) (electrical heating systems) A material that protects thermal insulation from environmental conditions such as rain, sleet, snow, wind, contamination, and physical damage. (IA/PC) 844-1991
(2) (electrical heat tracing for industrial applications) Material that, when installed on the outer surface of thermal insulation, protects the insulation from weather, such as rain, snow, sleet, wind, solar radiation, or atmospheric contamination, and physical damage. (BT/AV) 152-1953s
(3) Material that, when installed on the outer surface of thermal insulation, protects the insulation from water or other liquids; physical damage caused by sleet, wind, or mechanical abuse; and deterioration caused by solar radiation or atmospheric contamination. (IA/PC) 515.1-1995, 515-1997

weather, normal *See:* normal weather.

weatherproof (1) So constructed or protected that exposure to the weather will not interfere with successful operation. Rainproof, raintight, or watertight equipment can fulfill the requirements for weatherproof where varying weather conditions other than wetness, such as snow, ice, dust, or temperature extremes, are not a factor. (NESC/NEC) [86]
(2) (outside exposure) So constructed or protected that exposure to the weather will not interfere with successful operation. *See also:* outdoor. (SWG/PE) C37.30-1971s, C37.100-1981s

weatherproof enclosure An enclosure for outdoor application designed to protect against weather hazards such as rain, snow, or sleet. *Note:* Condensation is minimized by use of space heaters. (SWG/PE) C37.100-1981s

weather-protected machine A guarded machine whose ventilating passages are so designed as to minimize the entrance of rain, snow, and airborne particles to the electric parts. *See also:* asynchronous machine; direct-current commutating machine. (PE) [9]

weathershed The external part of the termination insulator that protects the core and provides the wet electrical strength and leakage distance. (PE/T&D/IC) 48-1996, 987-1985w

weathertight *See:* raintight.

weather vane *See:* vane.

web (1) That portion of the cold wall that extends between modules to permit heat transfer. (C/MM) 1101.2-1992
(2) That portion of the chassis cold wall that extends between modules forming the module slot and providing a heat transfer path (for conduction-cooled modules only). (C/BA) 1101.3-1993

weber The unit of magnetic flux in the International System of Units (SI). The weber is the magnetic flux whose decrease to zero when linked with a single turn induces in the turn a voltage whose time integral is one volt-second. (Std100) 270-1966w

Web page A digital multimedia object as delivered to a client system. A Web page may be generated dynamically from the server side, and may incorporate applets or other elements active on either the client or server side. (C) 2001-1999

Web site A collection of logically connected Web pages managed as a single entity. A Web site may contain one or more subordinate Web sites. (C) 2001-1999

wedge (rotating machinery) A tapered shim or key. *See also:* rotor; stator; slot wedge. (PE) [9]

wedge groove (rotating machinery) (wedge slot) A groove, usually in the side of a coil slot, to permit the insertion of and to retain a slot wedge. (PE) [9]

wedge, slot *See:* wedge groove.

wedge washer (rotating machinery) (salient pole) Insulation triangular in cross section placed underneath the inner ends of field coils and spanning between field coils. (PE) [9]

weekend processing The operations required to complete a weekly cycle. (C) 610.2-1987

weekly cycle One complete execution of a data processing function that must be performed once a week. For example, a weekly payroll system. *See also:* monthly cycle; annual cycle; daily cycle. (C) 610.2-1987

weight (1) (mathematics of computing) In positional representation of numbers, the value of a given digit position. *Synonym:* significance. (C) 1084-1986w
(2) (data management) For a given node in a tree, the number of terminal nodes in the subtree for that node. (C) 610.5-1990w

weight-2 code (local area networks) An unbalanced code sextet containing exactly two 1's. (C) 8802-12-1998

weight 2/4 code alternation A rule requiring successive unbalanced-code symbols in a data stream to be alternately chosen between weight-2 and weight-4 code groups. (LM/C) 802.12-1995s

weight-4 code (local area networks) An unbalanced code sextet containing exactly four 1's. (C) 8802-12-1998

weight-balanced tree (data management) A binary tree in which the ratio of the weight of the left subtree to the weight of the right subtree is between the square root of two plus and minus one. *Contrast:* height-balanced tree. (C) 610.5-1990w

weight coefficient (thermoelectric generator couple) (thermoelectric generator) The quotient of the electric power output by the device weight. *See also:* thermoelectric device. (ED) [46]

weighted average cost of capital The average interest rate used in financial analysis by business for capital projects. (SCC22) 1346-1998

weighted average quantum efficiency (diode-type camera tube) (η) The spectral quantum efficiency η_λ, integrated over a spectral band λ_1 to λ_2; and weighted by a particular input spectral distribution $N(\lambda)$.

$$\eta \equiv \frac{\int_{\lambda_1}^{\lambda_2} \eta_\lambda N(\lambda) d\lambda}{\int_{\lambda_1}^{\lambda_2} N(\lambda) d\lambda}$$

Since the input spectral distribution appears in both numerator and denominator, it can have dimensions of radiant power, or irradiance, or it can be a relative number, normalized, for example, to the peak value of the input spectral distribution. (ED) 503-1978w

weighted peak flutter Flutter and wow indicated by the weighted peak flutter measuring equipment specified in IEEE Std 193-1971 (withdrawn), Method for Measurement of Weighted Peak Flutter of Sound Recording and Reproducing Equipment. *Note:* The meter indicates one-half the peak-to-peak demodulated signal. (SP) 193-1971w

weighted response transducer A transducer intended to produce a surface wave with spatial distribution corresponding to a weighted-impulse response by designing the structure so that the finger lengths, finger locations, or electrical connections may vary. (UFFC) 1037-1992w

weighted sound level A-weighted sound-pressure level, obtained by the use of metering characteristics and the weightings A, B, C, or D specified in ANSI S1.4-1983. The weightings employed must always be stated. The reference pressure is always 20 μPa. *Notes:* 1. The meter reading (in decibels) corresponds to a value of the sound pressure integrated over the audible frequency range with a specified frequency weighting and integration time. 2. A suitable method of stating the weighting is, for example, "The A-weighted sound level was 43 dB," or "The sound level was 490 dB (A)." 3. Weightings are based on psychoacoustically determined time or frequency responses in objective measuring equipment. This is done to obtain data that better predict the subjective listener reaction than would wide-band measurements with a meter having either an instantaneous time response or a slow average or rms response. Standard weighting characteristics indicating relative response as a function of frequency are designated A, B, C, and D. (T&D/PE) 656-1992

weighting (data transmission) The artificial adjustment of measurements in order to account for factors that in the normal use of the device, would otherwise be different from the conditions during measurement. For example, background noise measurements may be weighted by applying factors or by introducing networks to reduce measured values in inverse ratio to their interfering effects. (AP/PE/ANT) 145-1983s, 599-1985w

weighting function (control system feedback) A function representing the time response of a linear system, or element to a unit-impulse forcing function: the derivative of the time response to a unit-step forcing function. *Notes:* 1. The Laplace transform of the weighting function is the transfer function of the system or element. 2. The time response of a linear system or element to an arbitrary input is described in terms of the weighting function by means of the convolution integral. *See also:* feedback control system. (IM/PE/EDPG) [120], [3]

weight of 6T code group The algebraic sum of the logical ternary symbol values listed in the 100BASE-T4 8B6T code table. (C/LM) 802.3-1998

weight transfer compensation A system of control wherein the tractive forces of individual traction motors may be adjusted to compensate for the transfer of weight from one axle to another when exerting tractive force. *See also:* multiple-unit control. (EEC/PE) [119]

weight 2/4 code alternation (local area networks) A rule requiring successive unbalanced-code symbols in a data stream to be alternately chosen between weight-2 and weight-4 code groups. (C) 8802-12-1998

weld decay Localized corrosion at or adjacent to a weld. (IA) [59]

welding arc voltage The voltage across the welding arc. (EEC/AWM) [91]

well-defined Containing no metalogical or high-impedance element values. (C/DA) 1076.3-1997, 1076.6-1999

well-formed requirement A statement of system functionality (a capability) that can be validated, and that must be met or possessed by a system to solve a customer problem or to achieve a customer objective, and is qualified by measurable conditions and bounded by constraints. (C/SE) 1233-1998

well-type coaxial detector A coaxial detector that is mounted and encapsulated in such a way that a radioactive sample may be placed within the inner cylindrical electrode such that the sample is essentially surrounded by active detector material. (NPS) 325-1996

Western *See:* snatch block.

Weston normal cell A standard cell of the cadmium type containing a saturated solution of cadmium sulphate as the electrolyte. *Note:* Strictly speaking this cell contains a neutral solution, but acid cells are now in more common use. *See also:* electrochemistry. (EEC/PE) [119]

wet bulb temperature The temperature at which liquid or solid water, by evaporating into the air, can bring the air into saturation adiabatically at the same temperature. (IA/PSE) 241-1990r

wet cell (1) A cell whose electrolyte is in liquid form. *See also:* electrochemistry. (EEC/PE) [119]
(2) *See also:* vented cell. (IA/PSE) 446-1995

wet contact (telephone switching systems) A contact through which direct current flows. *Note:* The term has significance because of the healing action of direct current flowing through contacts. (COM) 312-1977w

wet-dry signaling (telephone switching systems) Two-state signaling achieved by the application and removal of battery at one end of a trunk. (COM) 312-1977w

wet electrolytic capacitor A capacitor in which the dielectric is primarily an anodized coating on one electrode, with the remaining space between the electrodes filled with a liquid electrolytic solution. (PE/EM) 43-1974s

wetlands Any land that has been so designated by governmental agencies. Characteristically, such land contains vegetation associated with saturated types of soil. (PE/SUB) 1127-1998

wet location Installations underground or in concrete slabs or masonry in direct contact with the earth, and locations subject to saturation with water or other liquids, such as vehicle washing areas, and locations exposed to weather and unprotected. (NESC/NEC) [86]

wet location, health care facility A patient care area, that is normally subject to wet conditions, including standing water on the floor, or routine dousing or drenching of the work area. Routine housekeeping procedures and incidental spillage of liquids do not define a wet location. (NESC/NEC) [86]

Wet-Niche lighting fixture A lighting fixture intended for installation in a metal forming shell mounted in a swimming pool structure where the fixture will be completely surrounded by pool water. (NESC/NEC) [86]

wet snow Deposited snow that contains a great deal of liquid water. If free water entirely fills the air space in the snow, it is classified as "very wet" snow. *Note:* This condition causes water drops similar to rain to form on the conductors. (T&D/PE) 539-1990

wetting The free flow of solder alloy, with proper application of heat and flux, on a metallic surface to produce an adherent bond. (EEC/AWM) [105]

wetting agent (electroplating) (surface active agent) A substance added to a cleaning, pickling or plating solution to decrease its surface tension. *See also:* electroplating. (EEC/PE) [119]

wet-wound (rotating machinery) A coil in which the conductors are coated with wet resin in passage to the winding form, or on to which a bonding or insulating resin is applied on each successive winding layer to produce an impregnated coil. *See also:* rotor; stator. (PE) [9]

what-if-analysis An exercise that determines what capabilities an overall system would have if a changed capability were added (e.g., larger fuel tanks). (DIS/C) 1278.3-1996

Wheatstone bridge A 4-arm bridge, all arms of which are predominantly resistive. (See the corresponding figure.) *See also:* bridge.

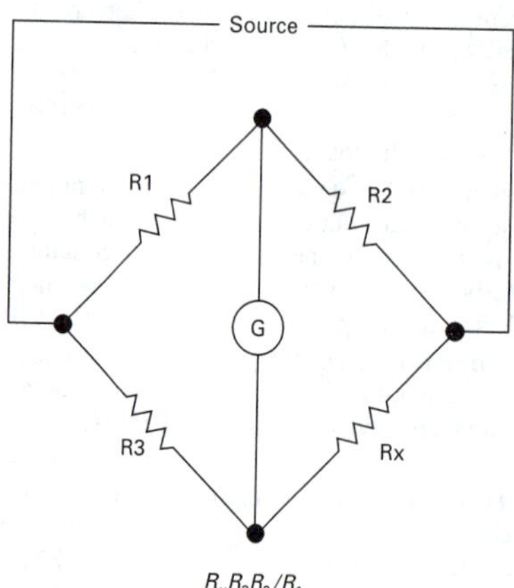

Wheatstone bridge

(EEC/PE) [119]

wheel *See:* sheave.

wheel diameter compensation A function that corrects for either the wear of the wheel(s) or the difference(s) in rolling diameter between different wheels on the vehicle or both. (VT) 1475-1999

wheeling The use of the transmission facilities of one or more parties to transmit electricity for another party. (PE/PSE) 858-1993w

wheeling charge (power operations) The amount paid to an intervening system for the use of its transmission facilities. (PE/PSE) 858-1987s, 346-1973w

wheel printer An element printer in which a set of type slugs, carried on the rim of a print wheel, is made available for each printing position. *See also:* daisy wheel printer. (C) 610.10-1994w

wheel slide During braking, the condition existing when the rotational speed of the wheel is slower than that for pure rolling contact between tread and rail/running surface. (VT) 1475-1999

wheel slip The condition existing when the rotational speed of the wheel does not correspond with pure rolling contact between tread and rail/running surface. (VT) 1475-1999

wheel-speed sensitivity *See:* rotor-speed sensitivity.

wheel spin During acceleration, the condition existing when the rotational speed of the wheel is faster than that for pure rolling contact between tread and rail/running surface. (VT) 1475-1999

wheel tractor A wheeled unit employed to pull pulling lines, sag conductor, and miscellaneous other work. Sagging winches on this unit are usually arranged in a horizontal configuration. It has some advantages over crawler tractors in that it has a softer footprint, travels faster, and is more maneuverable. *Synonyms:* sagger; logger; tractor; skidder. (T&D/PE) 524-1992r

WHILE (software) A single-entry, single-exit loop in which the loop control is executed before the loop body. (See the corresponding figure.) *Synonym:* UNTIL; pre-tested iteration. *Contrast:* closed loop. *See also:* leading decision.

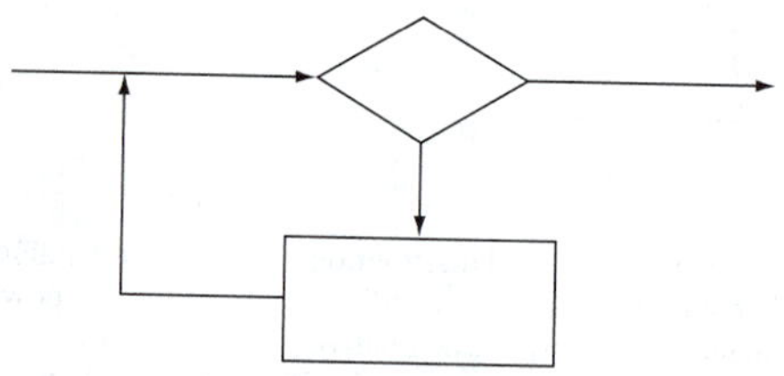

WHILE construct

(C) 610.12-1990

whip antenna A thin, flexible monopole antenna. (AP/ANT) 145-1993

whistle operator A device to provide automatically the timed signals required by navigation laws when underway in fog, and also manual control of electrical operation of a whistle or siren, or both, for at-will signals. (EEC/PE) [119]

whistler A form of radio energy in the extremely low frequency/very low frequency portion of the spectrum, usually originating from lightning strokes and characterized by a whistling tone of decreasing pitch that may last for several seconds. *Note:* Propagation of this energy is in the whistler mode, which is strongly guided along the Earth's magnetic field. *See also:* whistler mode. (AP/PROP) 211-1997

whistler mode The propagation mode of any right-hand polarized electromagnetic wave propagating along a magnetic field line in a plasma at a frequency less than the electron gyrofrequency but greater than the ion gyrofrequency. (AP/PROP) 211-1997

white (color television) Used most commonly in the nontechnical sense. More specific usage is covered by the term achromatic locus, and this usage is explained in the note under the term achromatic locus. *See also:* reference white. (BT/AV) 201-1979w

white box *See:* glass box.

white-box model *See:* glass box model.

white-box testing *See:* structural testing.

white compression (television) (white saturation) The reduction in gain applied to a picture signal at those levels corresponding to light areas in a picture with respect to the gain at that level corresponding to the midrange light value in the picture. *Notes:* 1. The gain referred to in the definition is for a signal amplitude small in comparison with the total peak-to-peak picture signal involved. A quantitative evaluation of this effect can be obtained by a measurement of differential gain. 2. The overall effect of white compression is to reduce contrast in the highlights of the picture as seen on a monitor. *See also:* television. (BT/AV) [34]

white light photocathode response (diode-type camera tube) The ratio of the output signal current to the total input radiant power from a tungsten filament source at a 2854K color temperature. Units: amperes watt $^{-1}$ (AW^{-1}). (CIE illuminant A). (ED) 503-1978w

white noise (1) (data transmission) (overhead-power-line corona and radio noise) Noise, either random or impulsive, that has a flat frequency spectrum in the frequency range of interest. (T&D/PE) 539-1990, 599-1985w
(2) (telephone practice) Noise, either random or impulsive type, that has a flat frequency spectrum at the frequency range of interest. This type of noise is used in the evaluation of systems on a theoretical basis and is produced for testing purposes by a white-noise generator. The use of the term should be limited and is not good usage in describing message circuit noise. (PE/PSR) C37.93-1976s
(3) (broadband local area networks) *See also:* noise. (LM/C) 802.7-1989r
(4) A type of noise that has a uniform power spectral density across a specified frequency spectrum. (C) 610.7-1995

white object (television) (color) An object that reflects all wavelengths of light with substantially equal high efficiencies and with considerable diffusion. (BT/AV) 201-1979w

white peak (television) A peak excursion of the picture signal in the white direction. *See also:* television. (BT/AV) [34]

white recording (frequency-modulation facsimile system) That form of recording in which the lowest received frequency corresponds to the minimum density of the record medium. *See also:* recording. (COM) 168-1956w

white saturation *See:* white compression.

white signal (at any point in a facsimile system) The signal produced by the scanning of a minimum-density area of the subject copy. *See also:* facsimile signal. (COM) 168-1956w

white space (1) A sequence of one or more characters that belong to the space character class as defined via the LC_CTYPE category in the current locale. In the POSIX Locale, white space consists of one or more ⟨`blank`⟩s (⟨`space`⟩s and ⟨`tab`⟩s), ⟨`newline`⟩s, ⟨`carriage-return`⟩s, ⟨`form-feed`⟩s, and ⟨`vertical-tab`⟩s. (C/PA) 9945-2-1993
(2) The nondisplaying formatting characters such as spaces, tabs, etc., that are embedded within a block of free text. (C/SE) 1320.2-1998

white space expansion *See:* kerning.

white space reduction *See:* kerning.

white space string A sequence of one or more white space characters including ⟨`space`⟩, ⟨`tab`⟩, and ⟨`newline`⟩. Within software definition files of exported catalogs, all such strings shall be encoded using IRV. (C/PA) 1387.2-1995

white transmission (1) (amplitude-modulation facsimile system) That form of transmission in which the maximum transmitted power corresponds to the minimum density of the subject copy. (COM) 168-1956w
(2) (frequency-modulation facsimile system) That form of transmission in which the lowest transmitted frequency corresponds to the minimum density of the subject copy. *See also:* facsimile transmission. (COM) 168-1956w

whole body irradiation (electrobiology) Pertains to the case in which the entire body is exposed to the incident electromagnetic energy or in which the cross section of the body is smaller than the cross section of the incident radiation beam. *See also:* electrobiology. (NIR) C95.1-1982s

wicket gates Series of overlapping adjustable guide vanes that regulate the amount of water flowing through a reaction turbine.

wicking The flow of solder along the strands and under the insulation of stranded lead wires. (EEC/AWM) [105]

wick-lubricated bearing (A) (rotating machinery) A sleeve bearing in which a supply of lubricant is provided by the capillary action of a wick that extends into a reservoir of free oil or of oil-saturated packing material. **(B) (rotating machinery)** A sleeve bearing in which the reservoir and other cavities in the bearing region are packed with a material that holds the lubricant supply and also serves as a wicking. *See also:* bearing. (PE) [9]

wide-angle diffusion (illuminating engineering) That in which flux is scattered at angles far from the direction that the flux would take by regular reflection or transmission. *See also:* regular reflection. (EEC/IE) [126]

wide-angle luminaire (illuminating engineering) A luminaire that distributes the light through a comparatively large solid angle. (EEC/IE) [126]

wide area network (WAN) (1) A communications network designed for large geographic areas. Sometimes called *long-haul network.* (DIS/C) 1278.2-1995
(2) A network that connects hosts across large geographic regions such as cities, states, and countries. *See also:* local area network; metropolitan area network; long haul network. (C) 610.7-1995

wide area telecommunications service (WATS) Telephone service that permits customers to make or receive long distance voice or telephone calls and have them billed on a bulk rather than individual call basis. (C) 610.7-1995

wideband channel (WBC) (1) A channel that is wider in bandwidth than a voice-band channel. (C/PE) 610.7-1995, 599-1985w
(2) A 6.144 Mbit/s isochronous channel. (C/LM) 802.9a-1995w

wideband circuit A telecommunication circuit capable of transferring data at speeds from 19 200–2 000 000 b/s. (C) 610.7-1995

wideband improvement The ratio of the signal-to-noise ratio of the system in question to the signal-to-noise ratio of a reference system. *Note:* In comparing frequency-modulation and amplitude-modulation systems, the reference system usually is a double-sideband amplitude-modulation system with a carrier power, in the absence of modulation, that is equal to the carrier power of the frequency-modulation system. (AP/ANT) 145-1983s

wideband ratio The ratio of the occupied frequency bandwidth to the intelligence bandwidth. (AP/ANT) 145-1983s

widget A specific instance of a widget class, providing a control in the user interface, such as a menu, pushbutton, or text fields. (C) 1295-1993w

widget class A collection of code and data structures that provides a generic implementation of a part of the user interface. (C) 1295-1993w

widow prevention The ability of a text formatter to avoid placing a title or the first one or two lines of a paragraph at the end of a page. *See also:* orphan prevention. (C) 610.2-1987

width By convention, the width axis is perpendicular to the PWB. (C/MM) 1101.2-1992

width line (illuminating engineering) The radial line (the one that makes the larger angle with the reference line) that passes through the point of one-half maximum candlepower on the lateral candlepower distribution curve plotted on the surface of the cone of maximum candlepower. (EEC/IE) [126]

Wiedemann-Franz ratio The quotient of the thermal conductivity by the electric conductivity. *See also:* thermoelectric device. (ED/ED) [46], 221-1962w

Wien bridge oscillator An oscillator whose frequency of oscillation is controlled by a Wien bridge. *See also:* oscillatory circuit. (EEC/PE) [119]

Wien capacitance bridge A four-arm alternating-current bridge characterized by having in two adjacent arms capacitors respectively in series and in parallel with resistors, while the other two arms are normally nonreactive resistors. (See the corresponding figure.) *Note:* Normally used for the measurement of capacitance in terms of resistance and frequency. The balance depends upon frequency, but from the balance conditions the capacitance of either or both capacitors can be computed from the resistances of all four arms and the frequency. *See also:* bridge.

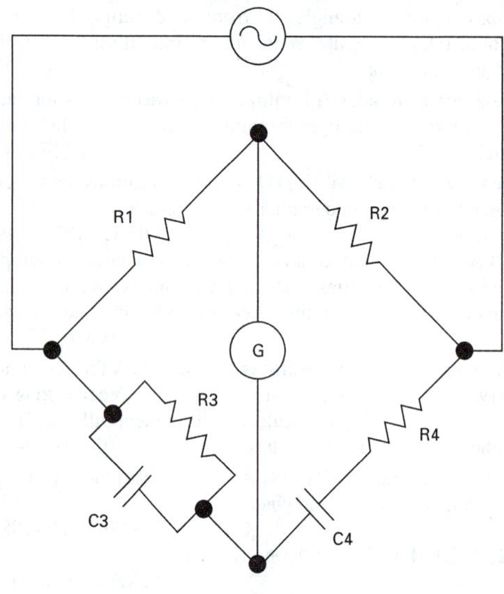

Wien capacitance bridge

(EEC/PE) [119]

Wien displacement law (illuminating engineering) An expression representing, in a functional form, the spectral radiance Lλ of a blackbody as a function of the wavelength λ and the temperature *T*.

$$L_\lambda = I_\lambda/A' = c_1\lambda^{-5}f(\lambda T)$$

where the symbols are those used in the definition of **Planck radiation law.** The two principal corollaries of this law are:

$$\lambda_m T = b$$

$$L_m/T^5 = b'$$

which show how the maximum spectral radiance L_m and the wavelength λ_m at which it occurs are related to the absolute temperature *T*. *Note:* The currently recommended value of b is 2.8978×10^{-3} m · K or 2.8978×10^{-1} cm · K. From the definition of the Planck radiation law, and with the use of the value of b, as given above b', is found to be 4.10×10^{-12} W · cm^{-3} · K^{-5} · sr^{-1}. (EEC/IE) [126]

Wien inductance bridge A 4-arm alternating-current bridge characterized by having in two adjacent arms inductors respectively in series and in parallel with resistors, while the other two arms are normally nonreactive resistors. (See the corresponding figure.) *Note:* Normally used for the measurement of inductance in terms of resistance and frequency. The balance depends upon frequency, but from the balance conditions the inductances of either or both inductors can be computed from the resistances of the four arms and the frequency. *See also:* bridge.

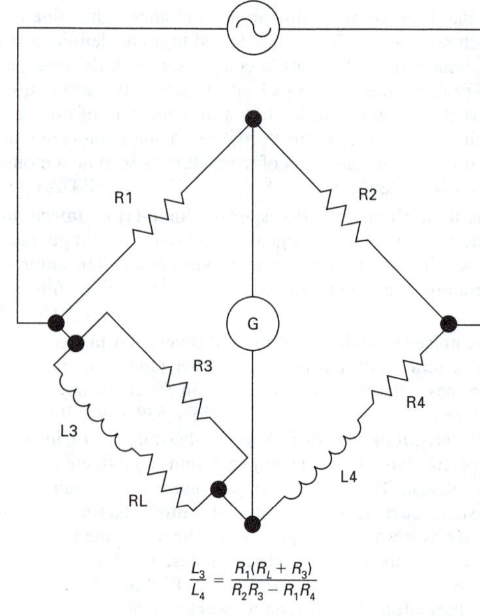

Wien inductance bridge

(EEC/PE) [119]

Wien radiation law (illuminating engineering) An expression representing approximately the spectral radiance of a blackbody as a function of its wavelength and temperature. It commonly is expressed by the formula

$$L_\lambda = I_\lambda/A' = c_{1L}\lambda^{-5}e^{-}(c_2/\lambda T)$$

where the symbols are those used in the definition of Planck radiation law. This formula is accurate to one percent or better for values of lT less than 3000 micrometer kelvins. *See also:* radiant energy. (EEC/IE) [126]

wigwag signal A railroad-highway crossing signal, the indication of which is given by a horizontally swinging disc with or without a red light attached. (EEC/PE) [119]

wildcard character One of *?[(asterisk, question mark, open bracket). Such characters are used in software pattern match strings. (C/PA) 1387.2-1995

Williams-tube storage (1) (electronic computation) A type of electrostatic storage. (Std100) 270-1966w
(2) A type of electrostatic storage that employs a cathode-ray tube. (C) 610.10-1994w

Wilson center (medical electronics) (electrocardiography) (V potential) (limb center) An electric reference contact: the junction of three equal resistors to the limb leads. (EMB) [47]

Wilson plate (measurement of dc electric field strength and ion-related quantities) A conducting plate that is grounded through an ammeter; it is used to collect the ion current, which is measured as it flows through the ammeter. The plate is sensitive to both ion-current density and changes in electric field (displacement current). *Note:* Long integration times are used to minimize the effects of the changes in the electric field (displacement current). If the power-line voltage and geometry are constant with time, the average displacement current is zero. (T&D/PE) 539-1990, 1227-1990r

winch, double-drum *See:* two-drum, three-drum puller.

Winchester disk A hard disk in which the magnetic heads and platter are contained within a sealed unit so that contaminants such as dust particles cannot interfere with the close tolerance between the disk and the head. *Note:* The entire assembly may be removable or fixed. (C) 610.10-1994w

winch, single-drum *See:* drum puller.

winch, three-drum *See:* two-drum, three-drum puller.

winch, triple-drum *See:* two-drum, three-drum puller.

winch, two-drum *See:* two-drum, three-drum puller.

wind direction The direction of the movement of air relative to the conductor axis. The wind direction and the conductor axis are assumed to be in a plane parallel to the earth. When the wind is blowing parallel to the conductor axis it is termed "parallel wind." When the wind is blowing perpendicularly to the conductor axis it is termed "perpendicular wind." (T&D/PE) 738-1993

wind-driven generator for aircraft A generator used on aircraft that derives its power from the air stream applied on its own air screw or impeller during flight. (EEC/PE) [119]

winder, pilot line A device designed to payout and rewind pilot lines during stringing operations. It is normally equipped with its own engine, which drives a drum or a supporting shaft for a reel mechanically, hydraulically, or through a combination of both. These units are usually equipped with multiple drums or reels, depending upon the number of pilot lines required. The pilot line is payed out from the drum or reel, pulled through the travelers in the sag section, and attached to the pulling line on the reel stand or drum puller. It is then rewound to pull the pulling line through the travelers. A pilot line winder can be a unit similar to a bullwheel puller and often has the reelwinder as an integral part of the machine. (T&D/PE) 524-1992r

winder, reel *See:* reel winder.

winding (data processing) A conductive path, usually of wire, inductively coupled to a magnetic core or cell. *Note:* When several windings are employed, they may be designated by the functions performed. Examples are: sense, bias, and drive windings. Drive windings include read, write, inhibit, set, reset, input, shift, and advance windings. (Std100) 163-1959w

winding, ac *See:* ac winding.

winding, autotransformer series *See:* series winding.

winding, control-power *See:* control-power winding.

winding, dc *See:* direct-current winding.

winding-drum machine (elevators) A geared-drive machine in which the hoisting ropes are fastened to and wind on a drum. *See also:* driving machine. (EEC/PE) [119]

winding end wire (rotating machinery) The portion of a random-wound winding that is not inside the core. *See also:* rotor; stator. (PE) [9]

winding factor (rotating machinery) The product of the distribution factor and the pitch factor. *See also:* stator; rotor. (PE) [9]

winding hottest spot temperature (power and distribution transformers) The highest temperature inside the transformer winding. It is greater than the measured average temperature (using the resistance change method) of the coil conductors. (PE/TR) C57.12.80-1978r

winding impregnation (rotating machinery) The process of applying an insulating varnish to a winding and, when required, baking to cure the varnish. (PE) [9]

winding inductance *See:* air-core inductance.

winding loss (electronic power transformer) The power losses of all windings involved, expressed in watts, in an inductor or transformer with the values measured at or corrected to the rated load current, frequency, and waveshape and stabilized at the maximum ambient temperature. *Synonym:* copper losses. (PEL/ET) 295-1969r

winding overhang (rotating machinery) That portion of a winding extending beyond the ends of the core. (PE) [9]

winding pitch *See:* coil pitch.

winding, primary *See:* primary winding.

winding, secondary *See:* secondary winding.

winding shield (rotating machinery) A shield secured to the frame to protect the windings but not to support the bearing. (PE) [9]

windings, high-voltage and low-voltage The terms high-voltage and low-voltage are used to distinguish the winding having the greater from that having the lesser voltage rating. (PE/TR) C57.12.80-1978r

winding, stabilizing *See:* stabilizing winding.

winding, tertiary *See:* tertiary winding.

winding voltage rating The voltage for which the winding is designed. *See also:* duty. (PE/TR) [116]

window (1) (counter tube) (radiation counter tubes) That portion of the wall that is made thin enough for radiation of low penetrating power to enter. (ED) [45]

(2) (charged-particle detectors) *See also:* dead layer thickness. (NPS) 300-1988r

(3) (computer graphics) A region of a two-dimensional world coordinate system that is to be visible as the display image. (C) 610.6-1991w

(4) A work area on the screen used by an application. (C) 1295-1993w

(5) A contiguous unit of addressing space that one bus utilizes to provide access to data on another bus. (C/BA) 1014.1-1994w

(6) In applications and graphical user interfaces, a defined portion of the display screen that is separated by a frame from the rest of the screen and which may be opened, closed, resized, and moved. (C) 610.10-1994w

(7) The measured difference in voltage or current between the erased and programmed states. (ED) 1005-1998

(8) The period of time during a pattern cycle when a primary output is actively monitored by an automatic test equipment (ATE) channel. (SCC20) 1445-1998

window amplifier *See:* biased amplifier.

window annunciator (alarm monitoring and reporting systems for fossil-fueled power generating stations) A visual signal device consisting of a number of backlighted windows, each one indicating a condition that exists or has existed in a monitored circuit, and being identified accordingly. (PE/EDPG) 676-1986w

window-type current transformer One that has a secondary winding insulated from and permanently assembled on the core, but has no primary winding as an integral part of the structure. Primary insulation is provided in the window, through which one turn of the line conductor can be passed to provide the primary winding. (PE/TR/PSR) C57.13-1993, C37.110-1996, C57.12.80-1978r

window, waveguide *See:* waveguide window.

window width The difference between the upper-level and lower-level discriminator settings. (NI) N42.12-1994

windshield wiper for aircraft A motor-driven device for removing rain, sleet, or snow from a section of an aircraft windshield, window, navigation dome, or turret. (EEC/PE) [119]

wind speed (navigation aids) The rate of motion of air. (AES/GCS) 172-1983w

windup Lost motion in a mechanical system that is proportional to the force or torque applied. (IA) [61]

wind velocity (navigation aids) The speed and direction of wind. (AES/GCS) 172-1983w

wing clearance lights (illuminating engineering) A pair of aircraft lights provided at the wing tips to indicate the extent of the wing span when the navigation lights are located an appreciable distance inboard of the wing tips. (EEC/IE) [126]

wink (1) A momentary off-hook condition in telephone trunk signaling. A wink may have different meanings depending on where it is used in the signaling stream (i.e., start or connect). (AMR/SCC31) 1390-1995, 1390.2-1999

(2) A momentary off-hook condition in telephone trunk signaling. A wink may have different meanings depending on where it is used in the signaling stream (i.e., start or connect). (SCC31) 1390.3-1999

wink-start pulsing (telephone switching systems) A method of pulsing control and trunk integrity check wherein the sender delays the sending of the address pulses until it receives a momentary off-hook signal from the far end. (COM) 312-1977w

wiper (brush) That portion of the moving member of a selector or other similar device, that makes contact with the terminals of a bank. (EEC/PE) [119]

wiper relay *See:* relay wiper.

wiping gland A projecting sleeve on a junction box, pothead, or other piece of apparatus serving to make a connection to the lead sheath of a cable by means of a plumber's wiped joint. *Synonyms:* wiping sleeve; transformer removable cable-terminating box. *See also:* tower. (T&D/PE) [10]

wiping sleeve *See:* wiping gland.

wire (1) A slender rod or filament of drawn metal. *Note:* The definition restricts the term to what would be ordinarily understood by the term solid wire. In the definition, the word slender is used in the sense that the length is great in comparison with the diameter. If a wire is covered with insulation, it is properly called an insulated wire: while primarily the term wire refers to the metal, nevertheless when the context shows that the wire is insulated, the term wire will be understood to include the insulation. *See also:* car-wiring apparatus. (VT/LT) 16-1955w

(2) *See also:* conductor. (T&D/PE) 516-1987s

(3) *See also:* conductor. (T&D/PE) 524-1992r

wire antenna An antenna composed of one or more conductors, each of which is long compared to the transverse dimensions, and with transverse dimensions of each conductor so small compared to a wavelength that for the purpose of computation the current can be assumed to flow entirely longitudinally and to have negligible circumferential variation. (AP/ANT) 145-1993

wire-band serving (power distribution, underground cables) A short closed helical serving of wire applied tightly over the armor of wire-armored cables spaced at regular intervals, such as on vertical riser cables, to bind the wire armor tightly over the core to prevent slippage. (PE) [4]

wire broadcasting The distribution of programs over wire circuits to a large number of receivers, using either voice frequencies or modulated carrier frequencies. (EEC/PE) [119]

wire center (1) (telephone loop performance) A central point from which loop feeder networks extend in a tree-like manner into the serving areas associated with the center. One or more end offices may be located at a wire center. (COM/TA) 820-1984r

(2) *See also:* end office. (C) 610.7-1995

wired equivalent privacy (WEP) The optional cryptographic confidentiality algorithm specified by IEEE 802.11 used to provide data confidentiality that is subjectively equivalent to the confidentiality of a wired local area network (LAN) medium that does not employ cryptographic techniques to enhance privacy. (C/LM) 8802-11-1999

wired logic (telephone switching systems) A fixed pattern of interconnections among a group of devices to perform predetermined functions in response to input signals. (COM) 312-1977w

wired OR A technique employed in circuit design in which separate circuits are connected to a common point so that the combination of their outputs results in an OR function, that is, the point at which the circuits are wired together will be true if any circuit feeding it is true. (C) 610.10-1994w

wired point interface Point for which all common equipment, wiring, and space are provided. To activate the point requires only the addition of plug-in hardware for the specific point. (SUB/PE) C37.1-1994

wired program (telephone switching systems) A program embodied in a pattern of fixed physical interconnections among a group of devices. (COM) 312-1977w

wired program control (telephone switching systems) A system control using wired logic. (COM) 312-1977w

wire frame representation (computer graphics) A technique for displaying a three-dimensional object as a series of lines outlining its shape without removing hidden surfaces. (See the corresponding figure.)

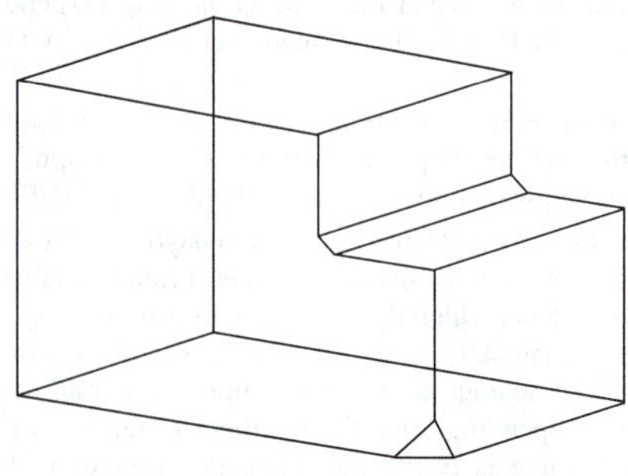

wire frame representation

(C) 610.6-1991w

wire gages Throughout these rules the American Wire Gage (AWG), formerly known as Brown & Sharpe (B&S), is the standard gage for copper, aluminum, and other conductors, excepting only steel conductors, for which the Steel Wire Gage (Stl WG) is used. *Note:* The Birmingham Wire Gage is obsolete. (NESC) C2-1997

wire-grid lens antenna A lens antenna constructed of wire grids, in which the effective index of refraction (and thus the path delay) is locally controlled by the dimensions and the spacings of the wire grid. *Contrast:* geodesic lens antenna; Luneburg lens antenna. (AP/ANT) 145-1993

wire holder (insulators) An insulator of generally cylindrical or pear shape, having a hole for securing the conductor and a screw or bolt for mounting. *See also:* insulator. (EEC/IEPL) [89]

wire insulation (rotating machinery) The insulation that is applied to a wire before it is made into a coil or inserted in a machine. *See also:* stator; rotor. (PE) [9]

wireless connection diagram The general physical arrangement of devices in a control equipment and connections between these devices, terminals, and terminal boards for outgoing connections to external apparatus. Connections are shown in tabular form and not by lines. An elementary (or schematic) diagram may be included in the connection diagram. (IA/ICTL/IAC) 270-1966w, [60]

wireless medium (WM) The medium used to implement the transfer of protocol data units (PDUs) between peer physical layer (PHY) entities of a wireless local area network (LAN). (C/LM) 8802-11-1999

wireload model A statistical model for the estimation of interconnect properties as a function of the geometric measures available before the completion of layout and routing. Typical model properties include fanout, capacitance, length, and resistance. *See also:* size metric. (C/DA) 1481-1999

wire mesh grip *See:* woven wire grip.

wire, overhead ground *See:* overhead ground wire.

wire-pilot protection Pilot protection in which an auxiliary metallic circuit is used for the communicating means between relays at the circuit terminals. (SWG/PE/PSR) C37.90-1978s, C37.100-1981s

wire printer *See:* dot matrix printer.

wire rope splice The point at which two wire ropes are joined together. The various methods of joining (splicing) wire ropes together include *hand tucked* woven splices, compression splices that utilize compression fittings but do not incorporate loops (eyes) in the ends of the ropes, and mechanical splices that are made through the use of loops (eyes) in the ends of the ropes held in place by either compression fittings or wire rope clips. The latter are joined together with connector links or steel bobs and, in some cases, are rigged *eye to eye*. Woven splices are often classified as short or long. A short splice varies in length from 7 to 17 ft (2 to 5 m) for 0.25 to 1.5 in (6 to 38 mm) diameter ropes, respectively, while a long splice varies from 15 to 45 ft (4 to 14 m) for the same size ropes. (T&D/PE) 524-1992r

wire spring relay A relay design in which the contacts are attached to round wire springs instead of the conventional flat or leaf spring. (PE/EM) 43-1974s

wire storage *See:* plated wire storage.

wiretapping Passive surveillance of communication channels to gain access to information transmitted over those channels. Wiretapping may be perpetrated through physical, electrical, and radio-frequency taps into the communication channel, and could result in the unauthorized disclosure of information transmitted over communication channels. Also known as eavesdropping. (C/BA) 896.3-1993w

wireway (1) (packaging machinery) A rigid rectangular raceway provided with a cover. (IA/PKG) 333-1980w
(2) (raceway systems for Class 1E circuits for nuclear power generating stations) Sheet-metal troughs with hinged or removable covers to house or protect wires and cables external to panelboards and cabinets. (PE/NP) 628-1987r

wireways Sheet-metal troughs with hinged or removable covers for housing and protecting electric wires and cable and in which conductors are laid in place after the wireway has been installed as a complete system. (NESC/NEC) [86]

wire-wrapped board A circuit board in which electrical connections between components are accomplished by wrapping wire around contact posts on the board. *Contrast:* printed circuit board. (C) 610.10-1994w

wiring closet A central point at which all the circuits in a wiring system begin or end, allowing cross-connection. *Synonym:* main distribution frame. (C) 610.7-1995

wiring or busing terminal, screw and/or lead That terminal, screw or lead to which a power supply will be connected in the field. (PE/TR) C57.12.80-1978r

wiring panel *See:* patch bay.

with Ada language construct to make the contents of an external Ada library unit visible within another Ada compilation unit. (ATLAS) 1226.2-1993w

withdrawal weighting Response weighting by omission of selected fingers, or weighting by changing the connections of selected fingers from one bus bar to the other. (UFFC) 1037-1992w

withholder (microprocessor architectures) A potential master that requires control of the bus module and is fairness inhibited. (MM/C) 896.1-1987s

withstand current (1) (surge) The crest value attained by a surge of a given wave shape and polarity that does not cause disruptive discharge on the test specimen. (T&D/PE) [10]
(2) *See also:* rated short-time withstand current. (IA/PSP) 1015-1997

withstand probability The probability that one application of a prospective voltage of a given shape and type will not cause a disruptive discharge. (PE/PSIM) 4-1995

withstand rating *See:* rated short-time withstand current.

withstand test voltage The voltage that the device must withstand without flashover, disruptive discharge, puncture, or other electrical failure when voltage is applied under specified conditions. *Note:* For power frequency voltages, the values specified are RMS values and for a specified time. For lightning or switching impulse voltages, the values specified are crest values of a specified wave. For direct voltages, the values specified are average values and for a specified time. (PE/IC) 48-1996

withstand voltage (1) (impulse) (electric power) The crest value attained by an impulse of any given wave shape, polarity, and amplitude, that does not cause disruptive discharge on the test specimen. (SPD/PE) 32-1972r
(2) (surge arresters) A specified voltage that is to be applied to a test object in a withstand test under specified conditions. During the test, in general no disruptive discharge should occur. (PE) [8], [84]
(3) The prospective value of the test voltage that equipment is capable of withstanding when tested under specified conditions. (PE/PSIM) 4-1995
(4) The specified voltage that, under specified conditions, can be applied to insulation without causing flashover or puncture. (SWG/PE/T&D) C37.100-1992, 386-1995
(5) The voltage that an insulation is capable of withstanding. In terms of insulation, this is expressed as either conventional withstand voltage or statistical withstand voltage. (PE/C/SPD) 1313.1-1996, C62.11-1999
(6) The voltage that an insulation is capable of withstanding with a given probability of failure. In terms of insulation, this is expressed as either conventional withstand voltage or statistical withstand voltage. (SPD/PE) C62.22-1997

withstand voltage test A high-voltage test that the armature winding must withstand without flashover or other electric failure at a specified voltage for a specified time and under specified conditions. (PE/EM) 433-1974r

word (1) (mathematics of computing) A sequence of bits or characters that is stored, addressed, transmitted, and operated on as a unit within a given computer. (C) 610.7-1995, 1084-1986w
(2) (microprocessor operating systems) An ordered set of bytes or bits that is the normal unit in which information may be stored, transmitted, or operated on within a given computer. (C/MM) 162-1963w, 855-1990
(3) (signals and paths) (microcomputer system bus) Two bytes or sixteen bits operated on as a unit. (C/MM) 796-1983r
(4) (696 interface devices) A set of bit-parallel signals corresponding to binary digits and operated on as a unit. For IEEE Std 696-1983 word connotes a group of 16 bits where the most significant bit carries the subscript 15 and the least significant bit carries the subscript 0. (MM/C) 696-1983w
(5) (mathematics of computing) (software) (data management) A sequence of bits or characters that is stored, addressed, transmitted, and operated on as a unit within a given computer. (C) 610.5-1990w, 610.12-1990, 1084-1986w
(6) (software) (data management) An element of computer storage that can hold a sequence of bits or characters as in the following definition for "word:" A sequence of bits or characters that is stored, addressed, transmitted, and operated on as a unit within a given computer. (C) 610.5-1990w, 610.12-1990
(7) (software) A sequence of bits or characters that has meaning and is considered an entity in some language; for example, a reserved word in a computer language. (C) 610.12-1990
(8) (SBX bus) Two bytes operated on as a unit. (MM/C) 959-1988r
(9) (NuBus) For the purpose of IEEE Std 1196-1987, 32-bit data item taken as a unit. (C/MM) 1196-1987w
(10) A group of adjacent binary digits operated on as a unit. Usually an integral number of octets. (SUB/PE) 999-1992w
(11) Four bytes or 32 bits operated on as a unit. The most significant byte carries the index value 0 and the least significant byte carries the index value 3. (C/BA) 1496-1993w
(12) An aligned **quadlet.** *Note:* The definition of this term is architecture-dependent, and so may differ from that used in other processor architectures. (C/MM) 1754-1994
(13) An ordered set of 16 bits operated on as a unit. The most significant bit is labeled bit 15 and the least significant bit is labeled bit 0. *Note:* When a word of data is embedded in a MTM-Bus packet, bit <0> of the data word is placed in bit ⟨1⟩ of the 17-bit packet. (TT/C) 1149.5-1995
(14) An element of computer storage that can hold a sequence of bits or characters as in (1). (C) 610.7-1995
(15) In the shell command language, a token other than an operator. In some cases a word is also a portion of a word token: in the various forms of parameter expansion (3.6.2),

such as $ {*name-word*}, and variable assignment, such as *name* = *word,* the word is the portion of the token depicted by *word*. The concept of a word is no longer applicable following word expansions—only fields remain. (C/PA) 9945-2-1993
(16) A character string or bit string that is considered as an entity. (C/ED) 610.10-1994w, 1005-1998
(17) A field composed of two eight-bit bytes. In a byte serial message, the most significant byte is transmitted/received first (big endian). It is capable of describing integers in the decimal range −32 768 to 32 767. (C/MM) 1284.1-1997
(18) *See also:* Forth word. (C/BA) 1275-1994

word address format Addressing each word of a block by one or more characters that identify the meaning of the word. (IA) [61]

word-alterable read-only memory (WAROM) A memory that permits erasing and writing to memory cells constituting a computer word without disturbing any other word. (ED) 641-1987w

word and author index (WADEX) A variation of a keyword out of context (KWOC) index in which author and keyword entries are combined and presented in a KWOC format. *Contrast with:* author and keyword in context index. (C) 610.2-1987

word clear Operation that sets all bits of a word to a common logic "1" state. (ED) 1005-1998

word erase The operation of removing the electrons from all bits in a word. (ED) 1005-1998

word index An automatic index containing an alphabetical list of the words found in a given text and indicating the number of times each word occurs in the text and each word's position in the text. (C) 610.2-1987

word length (1) (hybrid computer linkage components) (analog-to-digital converter) The number of data bits, including sign, that form the digital representation of the analog input in a prescribed voltage range. (C) 166-1977w
(2) (digital-to-analog converter) The number of data bits, including sign, in the digital register of a digital-to-analog converter, or a digital-to-analog multiplier. (C) 166-1977w
(3) The number of characters or bits in a word. (C) 610.10-1994w

word-line The line, determined by the row addresses (output of the X decoder), that is used to access the appropriate memory transistors, pass gate, or byte and row select transistors during a read or write. (ED) 1005-1998

word mark A mark that indicates the beginning or end of a word. *Note:* Used when word length is not fixed by the architecture but can vary under software control. (C) 610.10-1994w

word name A text string denoting a particular Forth word. (C/BA) 1275-1994

word-organized storage A type of storage in which data can be stored or from which data can be retrieved in units of computer words. (C) 610.10-1994w

word processing (WP) The use of computers to enter, view, edit, store, retrieve, manipulate, organize, transmit, and print textual material. A word processor system typically includes text editing and text formatting. *Synonym:* text processing. *See also:* dedicated word processing; shared-logic word processing; shared-resource word processing; word processor; office automation; stand-alone word processing; clustered word processing. (C) 610.2-1987

word processing output microfilm (WPOM) Microimages produced by a word processor. (C) 610.2-1987

word processor (WP) (A) A computer capable of performing word processing functions. **(B)** A computer program capable of performing word processing functions. *See also:* text editor; text formatter. (C) 610.2-1987

word serial The simplest required communication protocol supported by message-based devices in the VXIbus system. It utilizes the A16 communication registers to transfer data using a simple polling handshake method. (C/MM) 1155-1992

word time (electronic computation) In a storage device that provides serial access to storage locations, the time interval between the appearance of corresponding parts of successive words. *See also:* minor cycle. (C) [20], [85], 610.10-1994w

word wrap The ability of a word processing system to divide text into lines that fit into the horizontal space available on a display device without leaving broken words or requiring explicit carriage returns. (C) 610.2-1987

work The work done by a force is the dot-product line integral of the force. *See also:* line integral. (Std100) 270-1966w

work activity A collection of work tasks spanning a fixed duration within the schedule of a software project. Work activities may contain other work activities, as in a work breakdown structure. The lowest-level work activities in a hierarchy of activities are work tasks. Typical work activities include project planning, requirements specification, software design, implementation, and testing. (C/SE) 1058-1998

work area The region of a window where controls such as buttons, settings, and text fields are displayed. (C) 1295-1993w

work coil *See:* load, work, or heater coil.

worker certification The act of documenting the training and demonstrated proficiency of the worker for the task to be performed. (T&D/PE) 1307-1996

work file (A) (data management) A file used to provide storage space for data that is needed only during the duration of a particular event, such as the execution of a computer program. **(B) (data management)** In sorting, an intermediate file used for temporary storage of data between phases of the sort. (C) 610.5-1990

work function The minimum energy required to remove an electron from the Fermi level of a material into field-free space. *Note:* Work function is commonly expressed in electron volts. (ED) 161-1971w

working (electrolysis) The process of stirring additional solid electrolyte or constituents of the electrolyte into the fused electrolyte in order to produce a uniform solution thereof. *See also:* fused electrolyte. (PE/EEC) [119]

working area *See:* working space.

working directory A directory, associated with a process, that is used in pathname resolution for pathnames that do not begin with a slash. *Synonym:* current working directory. (C/PA) 9945-2-1993, 9945-1-1996, 1003.5-1999

working distance (x-ray energy spectrometers) The distance, measured along the working axis, between the source of x-rays and the outermost window on the detector. (NPS/NID) 759-1984r

working ground *See:* personal ground.

working level Any combination of short-lived radon daughters in air that will result in the ultimate emission of 1.3×10^5 MeV of alpha-particle energy. (NI) N42.17B-1989r

working level monitor Monitors used to measure the alpha-energy deposition from the decay of radon daughters. Calibrated in units of working levels. (NI) N42.17B-1989r

working optical aperture (acousto-optic device) That aperture which is equal to the size of the acoustic column that the light will encounter. (UFFC) [23]

working point *See:* operating point.

working pressure The pressure, measured at the cylinder of a hydraulic elevator, when lifting the car and its rated load at rated speed. *See also:* elevator. (EEC/PE) [119]

working reference system A secondary reference telephone system consisting of a specified combination of telephone sets, subscriber lines, and battery supply circuits connected through a variable distortionless trunk and used under specified conditions for determining, by comparison, the transmission performance of other telephone systems and components. (EEC/PE) [119]

working set (software) In the paging method of storage allocation, the set of pages that are most likely to be resident in main storage at any given point of a program's execution. (C) 610.12-1990

working space (software) That portion of main storage that is assigned to a computer program for temporary storage of data. *Synonyms:* working storage; working store; working area. (C) 610.12-1990, 610.10-1994w

working standard (luminous standards) (illuminating engineering) A standardized light source for regular use in photometry. (EEC/IE) [126]

working storage *See:* temporary storage; working space.

working store *See:* working space.

working stress *See:* acting stress.

working value The electrical value that when applied to an electromagnetic instrument causes the movable member to move to its fully energized position. This value is frequently greater than pick-up. *See also:* pickup. (EEC/PE) [119]

working voltage to ground (electric instruments) The highest voltage, in terms of maximum peak value, that should exist between any terminal of the instrument proper on the panel, or other mounting surface, and ground. *See also:* instrument. (EEC/AII) [102]

work in progress Production units in a semifinished state, either being processed or waiting in buffer inventories between processing steps. (SCC22) 1346-1998

workload (software) The mix of tasks typically run on a given computer system. Major characteristics include input/output requirements, amount and kinds of computation, and computer resources required. *See also:* workload model. (C) 610.12-1990

workload model (software) A model used in computer performance evaluation, depicting resource utilization and performance measures for anticipated or actual workloads in a computer system. *See also:* system model. (C) 610.12-1990

work package A specification of the work that must be accomplished to complete a work task. A work package should have a unique name and identifier, preconditions for initiating the work, staffing requirements, other needed resources, work products to be generated, estimated duration, risks factors, predecessor and successor work tasks, any special considerations for the work, and the completion criteria for the work package—including quality criteria for the work products to be generated. (C/SE) 1058-1998

work permit The authorization to perform work on a circuit. *Synonyms:* guarantee; clearance. (T&D/PE) 516-1995

work plane (1) (illuminating engineering) The plane on which work is usually done, and on which the illuminance is specified and measured. Unless otherwise indicated, this is assumed to be a horizontal plane 0.76 m (30 inches) above the floor. (EEC/IE) [126]
(2) (electric power systems in commercial buildings) The plane in which visual tasks are located. (IA/PSE) 241-1990r

work pole *See:* stick.

work positioning system A system of equipment or hardware that, when used with a line-worker's body belt or full body harness, allows a worker to be supported on an elevated vertical surface, such as a pole or tower, and work with both hands free. The primary difference between a positioning device system and a fall arrest system is that the positioning device supports a worker to prevent a fall, while a fall arrest system is used to stop the descent of a worker who has actually fallen from an elevated surface. (T&D/PE) 1307-1996

work product Any tangible item produced during the process of developing or modifying software. Examples of work products include the project plan, supporting process requirements, design documentation, source code, test plans, meeting minutes, schedules, budgets, and problem reports. Some subset of the work products will be baselined and some will form the set of project deliverables. (C/SE) 1058-1998

worksite (as applied to fall protection) The location on the structure or equipment where, after the worker has completed the climbing (horizontally and vertically), the worker is in position to perform the assigned work or task. (NESC/T&D/PE) C2-1997, 1307-1996

workspace (1) A repository for instances of OM classes in the closures of one or more packages associated with the workspace. (C/PA) 1327.2-1993w, 1224.2-1993w
(2) A repository for instances of classes in the closures of one or more packages associated with the workspace. (C/PA) 1238.1-1994w, 1224-1993w
(3) *See also:* user working area. 610.5-1990w

workspace interface The interface as realized, for the benefit of the dispatcher, by each workspace individually. (C/PA) 1328-1993w, 1327-1993w

workstation (A) An input-output device employed to perform applications such as data processing, software development, or computer-aided design. *See also:* data input station. **(B)** A single-user computer system that is dedicated to a particular task. *Note:* This term is commonly used in reference to an extremely powerful personal computer. *See also:* diskless workstation. **(C)** A device used to perform tasks such as data processing and word processing. (C) 610.10-1994, 610.2-1987

work stick *See:* stick.

work task The smallest unit of work subject to management accountability. A work task must be small enough to allow adequate planning and control of a software project, but large enough to avoid micro-management. The specification of work to be accomplished in completing a work task should be documented in a work package. Related work tasks should be grouped to form supporting processes and work activities. (C/SE) 1058-1998

world coordinate system (1) (computer graphics) A device-independent Cartesian coordinate system used to define a model in a two-dimensional or three-dimensional world. *See also:* normalized device coordinate system; viewing transformation. (C) 610.6-1991w
(2) The right-handed geocentric Cartesian system. The shape of the world is described in DMA TR 8350.2, 1987. The origin of the world coordinate system is the centroid of the earth. The axes of this system are labeled *X, Y,* and *Z,* with the positive *X*-axis passing through the prime meridian at the equator, with the positive *Y*-axis passing through 90° east longitude at the equator and the positive *Z*-axis passing through the north pole. (DIS/C) 1278.1-1995

world-numbering plan (telephone switching systems) The arrangement whereby, for the purpose of international distance dialing, every telephone main station in the world is identified by a unique number having a maximum of twelve digits representing a country code plus a national number. (COM) 312-1977w

world-zone number (telephone switching systems) The first digit of a country code. In the world-numbering plan, this number identifies one of the larger geographical areas into which the world is arranged, namely:

Zone 1—North America (includes areas operating with unified regional numbering).
Zone 2—Africa
Zone 3 & 4—Europe
Zone 5—South America, Cuba, Central America including part of Mexico
Zone 6—South Pacific (Australia-Asia)
Zone 7—Union of Soviet Socialist Republics
Zone 8—North Pacific (Eastern Asia)
Zone 9—Far East and Middle East
Zone 0—Spare

(COM) 312-1977w

WORM *See:* write-once/read-many; write-once/read-multiple; write-once/read-mostly.

WORM drive A disk drive that uses write-once/read-many technology to store and retrieve data. (C) 610.10-1994w

worm-geared machine (elevators) A direct-drive machine in which the energy from the motor is transmitted to the driving sheave or drum through worm gearing. *See also:* driving machine. (EEC/PE) [119]

worst-case modal bandwidth (WCMB) The lowest value of the modal bandwidth found when measured using either an overfilled launch (OFL) or a radial overfilled launch (ROFL). (C/LM) 802.3-1998

worst-case retention failure The change of state of any memory cell from the last-written state, even if the state later returns to that last-written state. (ED) 641-1987w

worst-case retention time The time interval between the instant of writing a memory pattern into a memory and the first worst-case retention failure. (ED) 641-1987w

wound rotor (rotating machinery) A rotor core assembly having a winding made up of individually insulated wires. *See also:* asynchronous machine. (PE) [9]

wound-rotor induction motor (1) (rotating machinery) An induction motor in which a primary winding on one member (usually the stator) is connected to the alternating-current power source and a secondary polyphase coil winding on the other member (usually the rotor) carries alternating current produced by electromagnetic induction. *Note:* The terminations of the rotor winding are usually connected to collector rings. The brush terminals may be either short-circuited or closed through suitable adjustable circuits. *See also:* asynchronous machine. (PE) [9]
(2) An induction motor in which the secondary circuit consists of polyphase winding or coils whose terminals are either short-circuited or closed through suitable circuits. (When provided with collector or slip rings, it is also known as a slip-ring induction motor.) (IA/MT) 45-1998

wound stator core (rotating machinery) A stator core into which the stator winding, with all insulating elements and lacing has been placed, including any components imbedded in or attached to the winding, and including the lead cable when this is used. *See also:* stator. (PE) [9]

wound-type current transformer A current transformer that has a primary winding consisting of one or more turns mechanically encircling the core or cores. The primary and secondary windings are insulated from each other and from the core(s) and are assembled as an integral structure. (PE/PSR/TR) C37.110-1996, C57.12.80-1978r, C57.13-1993

woven wire grip A device designed to permit the temporary joining or pulling of conductors without the need of special eyes, links, or grips *Synonyms:* Chinese finger; sock; basket; wire mesh grip; Kellem. (T&D/PE) 524-1992r

wow (sound recording and reproducing equipment) Frequency modulation of the signal in the range of approximately 0.5 Hz to 6 Hz, resulting in distortion that may be perceived as a fluctuation of pitch of a tone or program. *Note:* Measurement of unweighted wow only is not covered by IEEE Std 193-1971w. (SP) 193-1971w

WP *See:* word processing; word processor.

WPOM *See:* word processing output microfilm.

wrap One convolution of a length of ferromagnetic tape about the axis. *See also:* tape-wound core. (Std100) 163-1959w

wrap-around (computer graphics) A situation in which a display element goes off one side of the display surface and reappears on the opposite side. *Note:* Clipping is often used to prevent this. *See also:* word wrap; clipping; scissoring. (C) 610.6-1991w

wrapper (A) (rotating machinery) A relatively thin flexible sheet material capable of being formed around the slot section of a coil to provide complete enclosure. **(B) (rotating machinery)** The outer cylindrical frame component used to contain the ventilating gas. *See also:* stator; rotor. (PE) [9]

wrapping Reconfiguration function that involves dual ring stations using contra-rotating links to avoid a failed link or node. (LM/C) 802.5c-1991r

wrap thickness The distance between corresponding points on two consecutive wraps, measured parallel to the ferromagnetic tape thickness. *See also:* tape-wound core. (Std100) 163-1959w

wrap width *See:* tape-wound core.

write (1) To introduce data, usually into some form of storage. (ED/MIL/C) 158-1962w, [2], 162-1963w, [85]
(2) (charge-storage tubes) To establish a charge pattern corresponding to the input. (ED) 161-1971w
(3) (software) (data management) To record data in a storage device or on a data medium. (C) 610.5-1990w, 610.12-1990, 610.10-1994w
(4) To output characters to a file, such as standard output or standard error. Unless otherwise stated, standard output is the default output destination for all uses of the term *write.* (C/PA) 9945-2-1993
(5) The process of an *access unit (AU)* sending data downstream on a bus by logically ORing its outgoing data with the data pattern (normally all zeros) arriving from *upstream* on that bus. (LM/C) 8802-6-1994
(6) The operation by which a data state is entered into one or more memory cells. This may be a program or an erase operation. (ED) 1005-1998

writeable control store (WCS) A control store implemented in read/write memory to allow the processor instruction set to be redefined or extended at a later date. (C) 610.10-1994w

write access (data management) A type of access to data in which data may be written. *Synonym:* write-only access. *See also:* delete access; read/write access; read-only access; update access. (C) 610.5-1990w

write-after-read To write recently-read data back into storage after completion of the read cycle in order to prevent data loss. *Note:* Some media lose data by the mere act of being read and must be rewritten with the data. (C) 610.10-1994w

write barrier *See:* barrier transaction.

write circuitry Section of the memory that is used in the alteration of the stored data during the write operation. (ED) 1005-1998

write cycle (1) (write) A cycle in which the direction of data flow is from a master to slave(s). (NID) 960-1993
(2) A cycle in which data are transferred to some storage location from the device that requested the write. *Contrast:* read cycle. (C) 610.10-1994w
(3) A data transfer bus (DTB) cycle that is used to transfer 1, 2, 3, or 4 bytes from a master to a slave. The cycle begins when the master broadcasts an address and an address modifier and places data on the data transfer bus (DTB). Each slave captures the address and the address modifier and verifies if it will respond to the cycle. If it is intended to respond, it stores the data and then acknowledges the transfer. The master then terminates the cycle. (C/BA) 1014-1987

write cycle time The minimum time interval between the starts of successive write cycles of a storage device that has separate reading and writing cycles. *Contrast:* read cycle time. (C) 610.10-1994w

write data transfer One or more data transfers from the bus owner to a replying agent(s), with uninterrupted bus ownership. (C/MM) 1296-1987s

write disturb The corruption of data in one location caused by the writing of data at another location. (ED) 1005-1998

write enable (semiconductor memory) The inputs that when true enable writing data into the memory. The data sheet must define the effect of both states of this input on the reading of data and the condition of the output. (TT/C) 662-1980s

write-enable ring *See:* write ring.

write frame The transfer of data from a Network Capable Application Processor to a Smart Transducer Interface Module. (IM/ST) 1451.2-1997

write head A head capable of writing information on the medium. *Contrast:* read/write head; read head. (C) 610.10-1994w

write high (metal-nitride-oxide field-effect transistor) Process of generating a threshold voltage condition that increases source to drain current (high-conductance state) for a given gate to source voltage. (ED) 581-1978w

write low (metal-nitride-oxide field-effect transistor) Process of generating a threshold voltage condition that decreases source to drain current (low-conductance state) for a given gate to source voltage. (ED) 581-1978w

write-once/read-many (WORM) Pertaining to a storage medium which, once written to, cannot be changed or updated. *Synonyms:* write-once/read-mostly; write-once/read-multiple. (C) 610.10-1994w

write-once/read-mostly (WORM) *See:* write-once/read-many.

write-once/read-multiple (WORM) *See:* write-once/read-many.

write one to clear A method used to clear specific bits in a register. For example, if a write one to clear register contained 0xFFFFFFFF and the value 0x00800000 was written into it, the contents of the register would become 0xFF7FFFFF. (C/BA) 896.2-1991w, 896.10-1997

write-only access *See:* write access.

write-protect label A removable label, the presence of which on a diskette prevents writing on the diskette. *Note:* Generally used only on floppy disks that are flexible. *Contrast:* write-protect tab. *See also:* write-protect notch. (C) 610.10-1994w

write-protect mechanism Any mechanism employed to prevent accidentally destroying data on a data medium. For example, a write ring on a magnetic tape, or a write-protect notch on a floppy disk. (C) 610.10-1994w

write-protect notch A write-protect mechanism on flexible magnetic disks consisting of a notch on the side of the disk. *Note:* When the notch is not covered by a write-protect label, the disk is unprotected and may be written upon; when it is covered, the disk is write-protected. (C) 610.10-1994w

write-protect ring *See:* write ring.

write-protect tab A write-protect mechanism used on rigid floppy disks consisting of a small plastic tab that slides back and forth over a hole. *Note:* When the hole is covered, the disk is unprotected and may be written upon. *Contrast:* write-protect label. (C) 610.10-1994w

write pulse *See:* ONE state.

write ready violation A word-serial protocol error that occurs when data is written to a servant while its *write ready* bit is zero (0). (C/MM) 1155-1992

write ring A removable plastic or metal ring that can be inserted within a tape reel to permit writing on the tape. *Note:* If the tape is mounted without a write ring, data may not be written on the tape; the tape is said to be "write protected." *Synonyms:* write-enable ring; safety ring; file-protection ring; write-protect ring. *See also:* write-protect mechanism. (C) 610.10-1994w

write transaction A transaction that passes an address, size parameter, and data values from the requester to the responder. The size parameter specifies the number of bytes that are transferred. (C/MM) 1212-1991s

writing characteristic (metal-nitride-oxide field-effect transistor) The collection of high-conduction and low-conduction threshold voltage data as a function of the writing pulse width of both writing voltage polarities. (ED) 581-1978w

writing line An imaginary line on which the bottom of a displayed, printed, or typed character, excluding descenders, rests. *See also:* printing line; display line. (C) 610.10-1994w

writing rate (1) (storage tubes) The time rate of writing on a storage element, line, or area to change it from one specified level to another. Note the distinction between this and writing speed. *See also:* storage tube. (ED) 158-1962w
(2) (oscilloscopes) *See also:* writing time/division.

writing speed (storage tubes) Lineal scanning rate of the beam across the storage surface in writing. Note the distinction between this and writing rate. *See also:* storage tube. (ED) 158-1962w, 161-1971w

writing speed, maximum usable *See:* maximum usable writing speed.

writing tablet *See:* data tablet.

writing time/division (oscilloscopes) The minimum time per unit distance required to record a trace. The method of recording must be specified. (IM) 311-1970w

writing time, minimum usable *See:* minimum usable writing time.

Wullenweber antenna An antenna consisting of a circular array of radiating elements, each having its maximum directivity along the outward radial, and a feed system that provides a steerable beam that is narrow in the azimuth plane. (AP/ANT) 145-1993

wye connection (power and distribution transformers) So connected that one end of each of the windings of a polyphase transformer (or of each of the windings for the same rated voltage of single-phase transformers associated in a polyphase bank) is connected to a common point (the neutral point) and the other end to its appropriate line terminal. *Synonym:* Y connection. (PE/TR) C57.12.80-1978r

wye junction (waveguide components) A junction of waveguides or transmission lines in which the longitudinal guide axes form a Y. (MTT) 147-1979w

wye rectifier circuit A circuit employing three or more rectifying elements with a conducting period of 120 electrical degrees plus the commutating angle. *See also:* rectification. (EEC/PE) [119]

X

X.25 (ITU-TSS) A CCITT family of recommendations describing packet-switching protocols. (C) 610.7-1995

X.75 (ITU-TSS) A CCITT family of recommendations specifying interconnections between public data networks, including signaling, satellite usage, and multiple physical circuits of different nations. (C) 610.7-1995

X.200 (ITU-TSS) A CCITT family of recommendations describing OSI protocols and service definitions. (C) 610.7-1995

X.400 (ITU-TSS) **(1)** A CCITT family of recommendations describing message handling systems. (C) 610.7-1995 **(2)** The set of CCITT Recommendations on message handling systems. This term covers both the X.400 (1984) and X.400 (1988) recommendations. (PA/C) 1224.1-1993w

X.400 (1984) The set of CCITT Recommendations on message handling systems approved in 1984. (C/PA) 1224.1-1993w

X.400 (1988) The set of CCITT Recommendations on message handling systems approved in 1988. (C/PA) 1224.1-1993w

X.400 Application API The interface that makes the functionality of the MTS accessible to an MS or a UA, or the functionality of a simple MS accessible to a UA. (C/PA) 1224.1-1993w

X.400 Gateway API The interface that divides an MTA into two software components, a mail system gateway and an X.400 gateway service. (C/PA) 1224.1-1993w

X.400 gateway service Software that implements the MT interface (the service). (C/PA) 1224.1-1993w

X-address (test pattern language) The coordinates by which a row of a memory is specified. (TT/C) 660-1986w

X-axis amplifier *See:* horizontal amplifier.

X-band A radar-frequency band between 8 GHz and 12 GHz, usually in the International Telecommunication Union (ITU) allocated band 8.5–10.68 GHz. (AES) 686-1997

X-band radar (radar) A radar operating at frequencies between 8 and 12 GHz, usually in the International Telecommunications Union (ITU) assigned band 8.5 to 10.68 GHz. (AES/RS) 686-1982s

X-datum line An imaginary line along the top edge of a punch card, used as a reference edge for mark sensing or scanning. (C) 610.2-1987

xerographic printer A page printer used to print optical images using electrostatic technology. *See also:* laser printer. (C) 610.10-1994w

xerography The branch of electrostatic electrophotography that employs a photoconductive insulating medium to form, with the aid of infrared, visible, or ultraviolet radiation, latent electrostatic-charge patterns for producing a viewable record. *See also:* electrostatography. (ED) [46]

xeroprinting The branch of electrostatic electrography that employs a pattern of insulating material on a conductive medium to form electrostatic-charge patterns for duplicating purposes. *See also:* electrostatography. (ED) [46]

xeroradiography The branch of electrostatic electrophotography that employs a photoconductive insulating medium to form, with the aid of x rays or gamma rays, latent electrostatic-charge patterns for producing a viewable record. *See also:* electrostatography. (ED) [46]

XID *See:* eXchange IDentification.

Xmodem A protocol used for file transfer employing an eight-bit error checking protocol with a block size of 128 B. *Note:* Xmodem was developed by Ward Christensen. (C) 610.7-1995

XNOR *See:* exclusive NOR.

X-on/X-off *See:* transmitter on/transmitter off.

XOR *See:* exclusive OR.

***x* percent disruptive discharge voltage (high voltage testing)** The *x* percent disruptive discharge voltage is the prospective voltage value which has *x* percent probability of producing a disruptive discharge. (PE/PSIM) 4-1978s

x-position register A register within a display controller which controls the position of the electron beam in the x, or horizontal, direction on the display device. (C) 610.10-1994w

X punch *See:* eleven punch.

x-ray tube A vacuum tube designed for producing x-rays by accelerating electrons to a high velocity by means of an electrostatic field and then suddenly stopping them by collision with a target. (ED) [45]

X/R ratio **(1)** The ratio of the system inductive reactance to resistance. It is proportional to the time constant L/R and is, therefore, indicative of the rate of decay of any dc offset. A large X/R ratio corresponds to a large time constant and a slow rate of decay, whereas a small X/R ratio indicates a small time constant and a fast rate of decay of the dc offset. (PE/PSC) 367-1996 **(2)** Ratio of the system reactance to resistance. It is indicative of the rate of decay of any dc offset. A large X/R ratio corresponds to a large time constant and a slow rate of decay. (PE/SUB) 80-2000

X-series (ITU-TSS) A CCITT family of recommendations describing public digital data networks. (C) 610.7-1995, 610.10-1994w

X server *See:* server.

X wave *See:* extraordinary wave.

***X-Y* display** A rectilinear coordinate plot of two variables. *See also:* oscillograph. (IM/HFIM) [40]

X-Y plotter A plotter used to plot coordinate points in the form of a graph. (C) 610.10-1994w

***x-y* recorder (plotting board) (analog computer)** A recorder that makes a record of any one voltage with respect to another. (C) 165-1977w

XY switch A remotely controlled bank-and-wiper switch arranged in a flat manner, in which the wipers are moved in a horizontal plane, first in one direction and then in another. (EEC/PE) [119]

Y

Y-address (test pattern language) The coordinates by which a column of a memory is specified. (TT/C) 660-1986w

Yagi antenna* *See:* Yagi-Uda antenna.

* Deprecated.

Yagi-Uda antenna A linear end-fire array consisting of a driven element, a reflector element, and one or more director elements. (AP/ANT) 145-1993

Y amplifier *See:* vertical amplifier.

yaw angle (A) (navigation aids) The horizontal angular displacement of the longitudinal axis of a vehicle from its neutral position. **(B) (navigation aids)** The angle between a line in the direction of the relative wind and a plane through the longitudinal and vertical axes of the vehicle. (AES/GCS) 172-1983

Y-axis amplifier *See:* vertical amplifier.

Y-connected circuit A three-phase circuit that is star connected. *See also:* network analysis. (Std100) 270-1966w

Y connection (power and distribution transformers) So connected that one end of each of the windings of a polyphase transformer (or of each of the windings for the same rated voltage of single-phase transformers associated in a polyphase bank) is connected to a common point (the neutral point) and the other end to its appropriate line terminal. *Synonym:* wye connection. (PE/TR) C57.12.80-1978r

Y-datum line An imaginary line along the right edge of a punch card, used as a reference edge for mark sensing or scanning. (C) 610.2-1987

year-end processing The operations required to complete an annual cycle. (C) 610.2-1987

year-insensitivity The property of a system element in which no year-digits are maintained or represented; therefore, no century-digits ambiguity can exist and no leap year determination can be made. *See also:* date-insensitivity; day-insensitivity; time-insensitivity. (C/PA) 2000.1-1999

yearly cycle *See:* annual cycle.

yellow alarm An indication provided to a source device indicating a signal failure condition in a sink device. *Synonym:* remote alarm indication. (COM/TA) 1007-1991r

Yet Another Compiler Compiler A compiler specification language used to express the characteristics of a compiler by defining a set of grammar rules that YACC uses to generate a table-driven lookahead LR(1) (LALR) parser that can be used in a compiler. *See also:* LEX. (C) 610.13-1993w

Y junction *See:* wye junction.

Ymodem A protocol for file transfer employing a CRC Xmodem with a packet size of 1024 B. *Note:* Ymodem was developed by Chuck Forsburg. (C) 610.7-1995

Y network A star network of three branches. *See also:* network analysis. (Std100) 270-1966w

yoke (1) (rotating machinery) (magnetic) The element of ferromagnetic material, not surrounded by windings, used to connect the cores of an electromagnet, or of a transformer, or the poles of a machine, or used to support the teeth of stator or rotor. *Note:* A yoke may be of solid material or it may be an assembly of laminations. *See also:* rotor; stator. (PE) [9]

(2) A system of electromagnetic coils (for focus and deflection) employed with an electromagnetic cathode ray tube to provide the necessary control of focusing and deflection of the electron beam. *Note:* The focus coil is wound on an iron core which may be moved along the neck of the tube to focus the electron beam. The deflection coils are mounted at right angles to each other around the neck of the tube and may be rotated around the axis of the tube. (C) 610.10-1994w

yoked variable (modeling and simulation) One of two or more variables that are dependent on each other in such a manner that a change in one automatically causes a change in the others. (C) 610.3-1989w

y-position register A register within a display controller which controls the position of the electron beam in the y, or vertical, direction on the display device. (C) 610.10-1994w

Y punch *See:* twelve punch.

Y-T display An oscilloscope display in which a time-dependent variable is displayed against time. *See also:* oscillograph. (IM/HFIM) [40]

Y2K This is a popular shorthand designation for "Year 2000" (where "Y" = year and "2K" = two thousand) and is often used in conjunction with "problem" or "bug," as in "Y2K problem" or "Y2K bug," to denote the Year 2000 problem. (C/PA) 2000.1-1999

Year 2000 compliant Describes a system element that:

a) Performs correct date data processing (including leap year determination) for the date interval it supports.

b) Supports a continuous date interval of at least 10 years which must include 1998-12-31T23:59:59Z to 2001-01-01T00:00:01Z, inclusive.

c) Has associated conformance documentation that describes a supported continuous date interval.

d) Has associated conformance documentation that describes the formats of date data it accepts and produces for interchange.

The associated conformance documentation shall be available and readily identifiable to users of the system element and those dealing with remediation and integration efforts. *Note:* The combination of two or more Year 2000 compliant system elements is not necessarily Year 2000 compliant. Multiple system elements may be integrated into more complex system elements. Even if all of the system elements were Year 2000 compliant, the resulting integrated system element is not necessarily Year 2000 compliant. This can result from inappropriate interchange of date data, or from incompatible remediation techniques. Evaluation and potential remediation are needed with every level of integration. (C/PA) 2000.1-1999

Year 2000 rollover (1) The instant when a system's year changes from 1999 to 2000. In a system that uses a six-character date format, this is a transition from 99-12-31 (YY-MM-DD) to 00-01-01 (YY-MM-DD). (C/PA) 2000.2-1999

(2) The instant when a system element's year changes from 1999 to 2000. (C/PA) 2000.1-1999

Year 2000 user ready Describes a system element that has been determined by the user to be suitable for continued use into the year 2000 even though it might not be Year 2000 compliant. *Note:* A claim of Year 2000 user ready cannot be used as a basis for establishing that a system element is Year 2000 compliant. (C/PA) 2000.1-1999

Z

Z *See:* operational impedance.

Z-address (test pattern language) The coordinates by which a matrix in a memory is sp ecified. (TT/C) 660-1986w

Z-axis amplifier (oscilloscopes) An amplifier for signals controlling a display perpendicular to the X-Y plane (commonly intensity of the spot). *See also:* oscillograph; intensity amplifier. (IM) 311-1970w

ZBASIC A dialect of the BASIC programming language. (C) 610.13-1993w

Zeeman effect If an electric discharge tube, or other light source emitting a bright-line spectrum, is placed between the poles of a magnet, each spectrum line is split by the action of the magnetic field into three or more close-spaced but separate lines. The amount of splitting or the separation of the lines, is directly proportional to the strength of the magnetic field. (Std100) 270-1966w

zeitgebers Biological triggers that respond to external stimuli and that influence the circadian rhythm. *See also:* circadian rhythm. (T&D/PE) 539-1990

Zener breakdown (semiconductor devices) A breakdown that is caused by the field emission of charge carriers in the depletion layer. *See also:* semiconductor device; semiconductor. (ED) 216-1960w

Zener diode (semiconductor) A class of silicon diodes that exhibit in the avalanche-breakdown region a large change in reverse current over a very narrow range of reverse voltage. *Note:* This characteristic permits a highly stable reference voltage to be maintained across the diode despite a relatively wide range of current through the diode. *See also:* Zener breakdown; avalanche breakdown. (AES) [41]

Zener diode regulator A voltage regulator that makes use of the constant-voltage characteristic of the Zener diode to produce a reference voltage that is compared with the voltage to be regulated to initiate correction when the voltage to be regulated varies through changes in either load or input voltage. (See the corresponding figure.) *See also:* Zener diode.

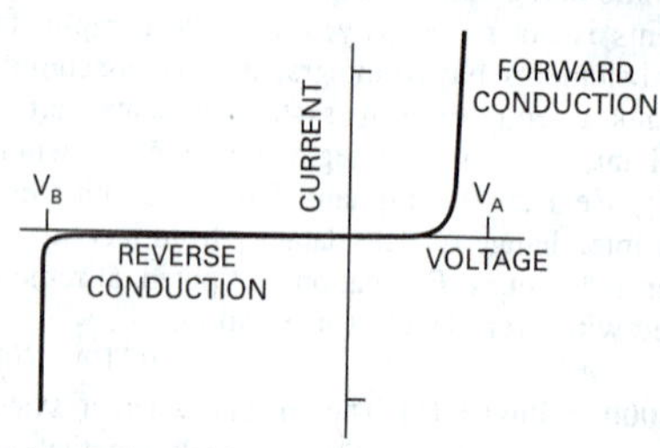

Current and voltage characteristics for a typical Zener diode regulator $|V_A| \gg |V_B|$

Zener diode regulator

(AES) [41]

Zener impedance *See:* breakdown impedance; semiconductor.

Zener voltage *See:* semiconductor; breakdown voltage.

zero (1) (function) (root of an equation) A zero of a function $f(x)$ is any value of the argument X for which $f(x) = 0$. *Note:* Thus the zeros of sin x are $x_1 = 0$, $x_2 = \pi$, $x_3 = 2\pi$, $x_4 = 3\pi, \ldots, x_n = (n-1)\pi, \ldots$. The roots of the equation $f(x) = 0$ are the zeros of $f(x)$. (Std100) 270-1966w
(2) (A) (transfer function in the complex variables) A value of s that makes the function zero. **(B)** The corresponding point in the s plane. *See also:* pole; feedback control system. (IM) [120]
(3) (network function) Any value of p, real or complex, for which the network function is zero. *See also:* network analysis. (Std100) 270-1966w
(4) A false logic state or a false condition of a variable. (C/BA) 1496-1993w

zero-address instruction (1) (software) A computer instruction that contains no address fields. *Contrast:* one-address instruction; three-address instruction; four-address instruction; two-address instruction. (C) 610.12-1990
(2) An instruction that has no address field because the address is implied or no address is required. *Synonyms:* implicit address instruction; no-address instruction; addressless instruction. *See also:* repetitive addressing. (C) 610.10-1994w

zero adjuster A device for bringing the indicator of an electric instrument to a zero or fiducial mark when the electrical quantity is zero. *See also:* moving element. (EEC/PE) [119]

zero-based linearity *See:* linearity.

zero-beat reception *See:* homodyne reception.

zero bias retention (metal-nitride-oxide field-effect transistor) This is the retention inherent in the metal-nitride-oxide semiconductor (MNOS) transistor when all terminals are grounded during information storage. The time period is defined by an (extrapolated) zero window between the two high-conduction (HC) and low-conduction (LC) threshold voltage curves plotted versus the logarithm of trd, the time elapsed between writing and threshold voltage measurement. *Synonym:* relaxation time. (ED) 581-1978w

zero-byte timeslot interchange A method of coding in which a variable address code is exchanged for any zero octet. The address information describes where, in the serial bitstream, zero octets originally occurred. It is a five-step process where data enters a buffer, zero octets are identified and removed, the nonzero bytes move to fill in the gaps, the first gap is identified, and a transparent flag bit is set in front of the message to indicate that one or more bytes originally contained zeros. (COM/TA) 1007-1991r

zero carryover (1) (bolometric power meters) A characteristic of multirange direct reading bolometer bridges that is a measure of the ability of the meter to maintain a zero setting from range to range without readjustment after initially being set to zero on the most sensitive range. (IM) 544-1975w
(2) (electrothermic power meters) A characteristic of multirange direct reading electrothermic power indicators which is a measure of the ability of the meter to maintain a zero setting from range to range without readjustment after initially being set to zero on the most sensitive range. Expressed in terms of percentage of full scale. (IM) 544-1975w

zero complement *See:* radix complement.

zero compression *See:* zero suppression.

zero control current residual voltage (Hall effect devices) The voltage across the Hall terminals that is caused by a time-varying magnetic field when there is no control current. (MAG) 296-1969w

zero direct current voltage test *See:* direct-current side short-circuit test.

zero dispersion wavelength That wavelength where the chromatic dispersion of a fiber is at its minimum. (C/LM) 802.3-1998

zero drift (analog computer) Drift with zero input. (C) 165-1977w

zero elimination *See:* zero suppression.

zero-error (device operating under the specified conditions of use) The indicated output when the value of the input presented to it is zero. *See also:* feedback control system. (IM) [120]

zero-error reference *See:* linearity.

zero field residual voltage (Hall effect devices) The voltage across the Hall terminals that exists when control current flows but there is zero applied magnetic field. (MAG) 296-1969w

zero field residual voltage temperature drift (Hall generator) The maximum change in output voltage per degree Celsius over a given temperature range when operated with zero external field and a given magnitude of control current. (MAG) 296-1969w

zero field resistive residual voltage (Hall effect devices) That component of the zero field residual voltage that remains proportional to the voltage across the control current terminals of the Hall generator for a specified temperature. (MAG) 296-1969w

zero fill (mathematics of computing) (data management) To fill the digit positions of a storage medium with the representation of the character zero. *Synonym:* zeroize. (C) 610.5-1990w, 1084-1986w

zero guy A line guy installed in a horizontal position between poles to provide clearance and transfer strain to an adjacent pole. *See also:* tower. (T&D/PE) [10]

zero inertia system An isolated ac system having no local generation. (PE/T&D) 1204-1997

zeroize *See:* zero fill.

zero-latency storage A type of storage that has an extremely small rotational delay, or latency. *See also:* disk cache. (C) 610.10-1994w

zero lead *See:* bioelectric null.

zero-level address *See:* n-level address; immediate address.

zero-minus call (telephone switching systems) A call for which the digit zero is dialed alone to indicate that operator assistance is desired. (COM) 312-1977w

zero-modulation medium noise (sound recording and reproducing system) The noise that is developed in the scanning or reproducing device during the reproducing process when a medium is scanned in the zero-modulation state. *Note:* For example, zero-modulation medium noise is produced in magnetic recording by undesired variations of the magnetomotive force in the medium, that are applied across the scanning gap of a demagnetized head, when the medium moves with the desired motion relative to the scanning device. Medium noise can be ascribed to nonuniformities of the magnetic properties and to other physical and dimensional properties of the medium. *See also:* noise. 191-1953w

zero-modulation state (sound recording medium) The state of complete preparation for playback in a particular system except for omission of the recording signal. *Notes:* 1. Magnetic recording media are considered to be in the zero-modulation state when they have been subjected to the normal erase, bias, and duplication printing fields characteristic of the particular system with no recording signal applied. 2. Mechanical recording media are considered to be in the zero-modulation state when they have been recorded upon and processed in the customary specified manner to form the groove with no recording signal applied. 3. Optical recording media are considered to be in the zero-modulation state when all normal processes of recording and processing, including duplication, have been performed in the customary specified manner, but with no modulation input to the light modulator. *See also:* noise. 191-1953w

zero offset (1) A control function for shifting the reference point in a control system. *See also:* feedback control system. (IA/ICTL/IAC) [60]

(2) (numerically controlled machines) A characteristic of a numerical machine control permitting the zero point on an axis to be shifted readily over a specified range. The control retains information on the location of the permanent zero. *See also:* floating zero. (IA) [61]

(3) (rate gyros) (restricted to rate gyros) The gyro output when the input rate is zero, generally expressed as an equivalent input rate. It excludes outputs due to hysteresis and acceleration. *See also:* input-output characteristic. 528-1994

zero-period acceleration (1) (seismic design of substations) The peak time history acceleration that can be determined from response spectra by the merging of response spectra, for all damping values, in the high-frequency range (usually above 30 Hz) in which no change of acceleration occurs with frequency. (PE/SUB) C37.122.1-1993

(2) (seismic qualification of Class 1E equipment for nuclear power generating stations) The acceleration level of the high frequency, nonamplified portion of the response spectrum. This acceleration corresponds to the maximum peak acceleration of the time history used to derive the spectrum. (PE/NP) 344-1987r

(3) (valve actuators) The acceleration that appears as a constant portion of a response spectrum in the highest frequency range. It is the maximum acceleration in the time history from which that response spectrum was developed. (PE/NP) 382-1985

(4) (seismic testing of relays) The peak acceleration of the motion/time history that corresponds to the high-frequency asymptote on the response spectrum. (SWG/PE/PSR) C37.98-1977s, C37.81-1989r

(5) (gas-insulated substations) The peak acceleration experienced by a rigid, single-degree-of-freedom oscillator when it is subjected to the design earthquake either directly to its base or through an intervening structure. *Note:* Generally, a body can be considered rigid, for seismic excitation purposes, if its natural frequency is greater than 30 Hz. The peak acceleration for a body with a natural frequency greater than 30 Hz is the same as that of a body with an infinitely high natural frequency, or conversely, a body with an infinitely small period (zero period). On a response spectrum, the zero-period acceleration is also equal to the asymptotic value of acceleration. (SUB/PE/SWG-OLD) C37.122-1983s, C37.100-1992

(6) The acceleration level of the high-frequency, nonamplified portion of the response spectrum (e.g., above the cut-off frequency). This acceleration corresponds to the maximum (peak) acceleration of the time history used to derive the spectrum. (PE/SUB) 693-1997

zero-phase-sequence relay A relay that responds to the zero-phase-sequence component of a polyphase input quantity. (SWG/PE) C37.100-1992

zero-phase-sequence symmetrical components (unsymmetrical set of polyphase voltages or currents of m phases) That set of symmetrical components that have zero phase sequence. That is, the angular phase lag from each member to every other member is 0 radians. The members of this set will all reach their positive maxima simultaneously. The zero-phase-sequence symmetrical components for a three-phase set of unbalanced sinusoidal voltages ($m = 3$) having the primitive period are represented by the equations

$$e_{a0} = e_{b0} = e_{c0} = (2)^{1/2} E_{a0} \cos(\omega t + \alpha_{a0})$$

derived from the equation of symmetrical components (set of polyphase alternating voltages). Since in this case $r = 1$ for every component (of first harmonic), the third subscript is omitted. Then k is 0 for the zero sequence, and s takes on the values 1, 2, and 3 corresponding to phases a, b, and c. These voltages have no phase sequence since they all reach their positive maxima simultaneously. (Std100) 270-1966w

zero-phase symmetrical set (1) (polyphase voltage) A symmetrical set of polyphase voltages in which the angular phase difference between successive members of the set is zero or a multiple of 2π radians. The equations of symmetrical set (polyphase voltages) represent a zero-phase symmetrical set of polyphase voltages if k/m is zero or an integer. (The symmetrical set of voltages represented by the equations of symmetrical set of polyphase voltages may be said to have zero-phase symmetry if k/m is zero or an integer (positive or negative).) *Note:* This definition may be applied to a two-phase four-wire or five-wire system if m is considered to be 4 instead of 2. (Std100) 270-1966w

(2) (polyphase currents) This definition is obtained from the corresponding definitions for voltage by substituting the word current for voltage, and the symbol I for E and β for α wherever they appear. The subscripts are unaltered. (Std100) 270-1966w

zero pip (spectrum analyzer) An output indication that corresponds to zero input frequency. (IM) 748-1979w

zero-plus (telephone switching systems) A call in which the digit zero is dialed as a prefix where operator intervention is necessary. (COM) 312-1977w

zero-power-factor saturation curve (synchronous machines) (zero-power-factor characteristic) The saturation curve of a machine supplying constant current with a power-factor of approximately zero, overexcited. (PE) [9]

zero-power-factor test (synchronous machines) A no-load test in which the machine is overexcited and operates at a power-factor very close to zero. (PE) [9]

zero proof (mathematics of computing) A method of checking computations by adding positive and negative values so that if all computations are accurate the total will be zero. (C) 1084-1986w

zero punch A zone punch in punch row 10 (third from the top) in a twelve-row punch card. *See also:* eleven punch; twelve punch. (C) 610.10-1994w

zero-sequence impedance (1) (power and distribution transformers) An impedance voltage measured between a set of primary terminals and one or more sets of secondary terminals when a single-phase voltage source is applied between the three primary terminals connected together and the primary neutral, with the secondary line terminals shorted together and connected to their neutral. *Note:* (if one exists).

1) For two-winding transformers, the other winding is short-circuited. For multiwinding transformers, several tests are required, and the zero-sequence impedance characteristics are represented by an impedance network.
2) In some transformers, the test must be made at a voltage lower than that required to circulate rated current in order to avoid magnetic core saturation or to avoid excessive current in other windings.
3) Zero-sequence impedances are usually expressed in per unit or percent on a suitable voltage and kVA base.

(PE/TR) C57.12.80-1978r

(2) (rotating machinery) The quotient of the zero-sequence component of the voltage, assumed to be sinusoidal, supplied to a synchronous machine, and the zero-sequence component of the current at the same frequency. *See also:* direct-axis synchronous reactance. (PE) [9]

zero-sequence reactance (rotating machinery) The ratio of the fundamental component of reactive armature voltage, due to the fundamental zero-sequence component of armature current, to this component at rated frequency, the machine running at rated speed. *Note:* Unless otherwise specified, the value of zero-sequence reactance will be that corresponding to a zero-sequence current equal to rated armature current. *See also:* direct-axis synchronous reactance. (PE) [9]

zero-sequence resistance The ratio of the fundamental in-phase component of armature voltage, resulting from fundamental zero-sequence current, to this component of current at rated frequency. (EEC/PE) [119]

ZERO shift error Error measured by the difference in deflection as between an initial position of the pointer, such as at zero, and the deflection after the instrument has remained deflected upscale for an extended length of time, expressed as a percentage of the end-scale deflection. *See also:* moving element. (EEC/AII) [102]

zero span (spectrum analyzer) A mode of operation in which the frequency span is reduced to zero. (IM) 748-1979w

0-state The logic state represented by the binary number 0 and usually standing for an inactive or false logic condition. (GSD) 91-1984r

zero structure *See:* snub structure.

zero-subcarrier chromaticity (color television) The chromaticity that is intended to be displayed when the subcarrier amplitude is zero. *Note:* This chromaticity is also known as reference white for the display. (BT/AV) 201-1979w

zero suppression (1) The elimination of nonsignificant zeros in a numeral. (C) [20], [85]

(2) (mathematics of computing) The elimination of zeros that have no significance or use, such as zeros to the left of the integral part of a numeral or zeros to the right of the fractional part. *Synonyms:* zero elimination; zero compression. (C) 1084-1986w

zero-suppression character (data management) A character within a picture specification that represents a decimal digit in which a blank character is used in place of a zero. *Note:* Z, Y, and * are commonly used. (C) 610.5-1990w

zero synchronization (numerically controlled machines) A technique that permits automatic recovery of a precise position after the machine axis has been approximately positioned by manual control. (IA) [61]

zero vector A vector whose magnitude is zero. (Std100) 270-1966w

zero voltage fired (electrical heating applications to melting furnaces and forehearths in the glass industry) A circuit in which antiparallel connected thyristors are fired at points of voltage zero in the alternating current voltage wave. (IA) 668-1987w

0x A numerical prefix indicating that the number following is a hexadecimal number. (PE/SUB) 1379-1997

0.0.x talker An application at a node that transmits a stream packet. (C/MM) 1394a-2000

0.0.x terabyte A quantity of data equal to 2^{40}, or 1099511627776, bytes. (C/MM) 1394a-2000

0.0.x transaction A request and the optional, corresponding response. (C/MM) 1394a-2000

0.0.x transaction layer The Serial Bus protocol layer that defines a request-response protocol for read, write, and lock operations. (C/MM) 1394a-2000

0.0.x transmitting port Any port transmitting clocked data or an arbitration state. A transmitting port is further characterized as either originating or repeating. (C/MM) 1394a-2000

0.0.x unit A component of a Serial Bus node that provides processing, memory, input/output (I/O), or some other functionality. Once the node is initialized, the unit provides a Command and Status Register (CSR) interface. A node may have multiple units, which normally operate independently of each other. (C/MM) 1394a-2000

0.0.x unit architecture The specification document that describes the interface to, and the behaviors of, a unit implemented within a node. (C/MM) 1394a-2000

zeta potential *See:* electrokinetic potential.

Z-fold paper *See:* continuous form.

zigzag connection (power and distribution transformers) A polyphase transformer with *Y*-connected windings, each one of which is made up of parts in which phase-displaced voltages are induced. (PE/TR) C57.12.80-1978r

zig-zag connection of polyphase circuits (zig-zag or interconnected star) The connection in star of polyphase windings, each branch of which is made up of windings that generate phase-displaced voltage. *See also:* connections of polyphase circuits; polyphase circuit. (PE) [9], [84]

zig-zag fold paper *See:* continuous form.

zig-zag leakage flux The high-order harmonic air-gap flux attributable to the location of the coil sides in discrete slots. *See also:* stator; rotor. (PE) [9]

Z marker *See:* zone marker.

z-marker beacon *See:* zone marker beacon.

Zobel filters A filter designed according to image parameter techniques. (CAS) [13]

zonal-cavity interreflectance method (illuminating engineering) A procedure for calculating coefficients of utilization, wall luminance coefficients, and ceiling cavity luminance coefficients taking into consideration the luminaire intensity distribution, room size and shape (cavity ratio concepts), and room reflectances. It is based on flux transfer theory. (EEC/IE) [126]

zonal constant (illuminating engineering) A factor by which the mean intensity emitted by a source of light in a given angular zone is multiplied to obtain the lumens in the zone. (EEC/IE) [126]

zonal factor interflection method† (illuminating engineering) A procedure for calculating coefficients of utilization based on integral equations which takes into consideration the ultimate disposition of luminous flux from every 10 degree zone from luminaires. (This term is retained for reference and literature searches). (EEC/IE) [126]

† Obsolete.

zonal factor method (illuminating engineering) A procedure for predetermining, from typical luminaire photometric data in discrete angular zones, the proportion of luminaire output which would be incident initially (without interreflections) on the work-plane, ceiling, walls, and floor of a room. (EEC/IE) [126]

zone *See:* reach.

zone comparison protection A form of pilot protection in which the response of fault-detector relays, adjusted to have a zone of response commensurate with the protected line section, is compared at each line terminal to determine whether a fault exists within the protected line section. (SWG/PE) C37.100-1981s

zoned antenna A lens or reflector antenna having various portions (called zones or steps) that form a discontinuous surface such that a desired phase distribution of the aperture illumination is achieved. *Synonym:* stepped antenna. (AP/ANT) 145-1993

zoned decimal data (data management) Integer data in which each decimal digit occupies one byte, the first four bits of which is called the zone portion and the second four bits, the data portion. The zone portion of the lowest-order byte contains the sign of the integer (hexadecimal A, C, E, or F for positive; B or D for negative); otherwise the zone portion contains the binary value 1111.

decimal	75_{10}
zoned decimal	$0000\ 0111\ 1111\ 0101_2$
decimal	-91_{10}
zoned decimal	$0000\ 1001\ 1011\ 0001_2$

Synonym: unpacked decimal data. (C) 610.5-1990w

zone leveling (semiconductor processing) The passage of one or more molten zones along a semiconductor body for the purpose of uniformly distributing impurities throughout the material. *See also:* semiconductor device. (Std100) 102-1957w

zone marker (electronic navigation) A marker used to define a position above a radio range station. *Synonym:* Z marker. (AES/RS/GCS) 686-1982s, 172-1983w

zone marker beacon (navigation aids) A vertical beam—horizontal cross section in the shape of a circle. *Synonym:* z-marker beacon. (AES/GCS) 172-1983w

zone of influence An area around a ground electrode bounded by points of specified equal potential resulting from the voltage drop through the earth between the ground electrode and remote earth. (PE/PSC) 367-1996

zone of protection (1) The adjacent space provided by a grounded air terminal, mast, or overhead ground wire that is protected against most direct lightning strikes. (PE/EDPG) 665-1995
(2) (for relays) That segment of a power system in which the occurrence of assigned abnormal conditions should cause the protective relay system to operate. (SWG/PE) C37.100-1992

zone of silence (navigation aids) A local region in which the signals of a given radio transmitter cannot be received satisfactorily. (AES/GCS) 172-1983w

Zone 1 Terminology used for classification of an industrial area in which an explosive atmosphere is likely to exist under normal operation. (IA) 515-1997

zone-plate lens antenna *See:* Fresnel lens antenna.

zone punch (1) A punch in the 0, 11, or 12 row on a Hollerith punched card. (C) [20]
(2) A punch (2) in one of the upper three rows (0, 11, 12) of a twelve-row punch card. *Contrast:* digit punch. *See also:* eleven punch; zero punch; twelve punch. (C) 610.10-1994w

zone purification (semiconductor processing) The passage of one or more molten zones along a semiconductor for the purpose of reducing the impurity concentration of part of the ingot. *See also:* semiconductor device. (Std100) 102-1957w

zone selective interlocking A function provided for rapid clearing while retaining coordination. The function is a communication interconnection between the electronic trip units of two or more circuit-breakers connected in series on multiple levels. By means of intercommunication between the short-time delay and/or ground fault elements, the one nearest the fault trips with minimum time delay while signaling the supply-side circuit-breaker(s) to delay for a predetermined period. (IA/PSP) 1015-1997

Zone 2 Terminology used for classification of an industrial area in which an explosive atmosphere is not likely to occur in normal operation, and if it does, it will exist only for a short period. (IA) 515-1997

Zone 0 Terminology used for classification of an industrial area in which an explosive atmosphere is present continuously, or present for long periods of time during normal operation. (IA) 515-1997

zoning (1) (electrical heating systems) A division of circuits to minimize different conditions in any one circuit, so the temperature at the location of a sensor is typical for the complete circuit. (IA/PC) 844-1991
(2) (lens or reflector) The displacement of various portions (called zones or steps) of the lens or surface of the reflector so that the resulting phase front in the near field remains unchanged. *See also:* antenna. (AP/ANT) [35]

zoom (computer graphics) To display all or part of a graphical display image at a scale different from the scale of the previous image. *Note:* In zoom in, the scale is larger; in zoom out, the scale is smaller. (C) 610.6-1991w

z-position register A register in a display controller which simulates the beam position (on the display) in the z-direction. *Note:* The z-axis represents depth into and out of the display screen. This illusion is achieved by varying the intensity, or color, of the z-vector in proportion to the value of the z-coordinate. (C) 610.10-1994w

Z_{stub} Characteristic impedance of signal line stub on the module. (C/BA) 896.2-1991w

z transform, advanced *See:* advanced z transform.

z transform, delayed *See:* delayed z transform.

z transform, modified *See:* modified z transform.

z transform, one-sided *See:* one-sided z transform.

z transform, two-sided *See:* two-sided z transform.

Z_0 Characteristic imedance of signal lines, without including connector vias, on the unpopulated backplane. (C/BA) 896.2-1991w

Z_0' Characteristic impedance of signal lines on backplane populated by male connectors, without termination resistors and with no modules inserted. (C/BA) 896.2-1991w

Abstracts and Sources

Abstracts

IEEE Std 1-1986 (R1992). *IEEE Standard General Principles for Temperature Limits in the Rating of Electric Equipment and for the Evaluation of Electrical Insulation.* These principles are intended to serve as a guide in the preparation of IEEE and other standards that deal with the selection of temperature limits and the measurement of temperature for specific types of electric equipment. Fundamental considerations are outlined, and the elements to be considered in applying the principles to specific cases are reviewed. Guiding principles are included for the development of test procedures for thermal evaluation of electrical insulating materials, thermal evaluation of insulation systems, and thermal classification of insulation systems for use in rating electric equipment.

IEEE Std 4-1995. *IEEE Standard Techniques for High-Voltage Testing.* This standard establishes standard methods to measure high-voltage and basic testing techniques, so far as they are generally applicable, to all types of apparatus for alternating voltages, direct voltages, lightning impulse voltages, switching impulse voltages, and impulse currents. This revision implements many new procedures to improve accuracy, provide greater flexibility, and address practical problems associated with high-voltage measurements.

IEEE Std 7-4.3.2-1993. *IEEE Standard Criteria for Digital Computers in Safety Systems of Nuclear Power Generating Stations.* Additional computer specific requirements to supplement the criteria and requirements of IEEE Std 603-1991 are specified. Within the context of this standard, the term computer is a system that includes computer hardware, software, firmware, and interfaces. The criteria contained herein, in conjunction with criteria in IEEE Std 603-1991, establish minimum functional and design requirements for computers used as components of a safety system.

IEEE Std 11-1980 (R2000). *IEEE Standard for Rotating Electric Machinery for Rail and Road Vehicles.* Rotating electric machinery that forms part of the propulsion and major auxiliary equipment on internally and externally powered electrically propelled rail and road vehicles and similar large transport and haulage vehicles, and their trailers where specified in a contract, is covered. Major auxiliary equipment includes equipment such as blower and compressor motors, motor-generator and motor-alternator sets, auxiliary generators, and exciters, usually larger than 3 kW. Ratings, tests, and calculation procedures are defined to permit comparison among machines for similar use, and to enable suitability of machines for a given use to be evaluated. The following are covered: ratings, temperature rises and temperature-rise tests, temperature measurements, high-potential tests, commutation tests, overspeed requirements and tests, characteristic curves and tests, external power systems, terminal markings, mechanical measurements.

IEEE Std 18-1992. *IEEE Standard for Shunt Power Capacitors.* Capacitors rated 216 V or higher, 2.5 kvar or more, and designed for shunt connection to alternating-current transmission and distribution systems operating at a nominal frequency of 50 or 60 Hz are considered. Service conditions, ratings, manufacturing, and testing are covered. A guide to the application and operation of power capacitors is included.

IEEE Std 32-1972 (R1997). *IEEE Standard Requirements, Terminology, and Test Procedures for Neutral Grounding Devices.* Devices used for the purpose of controlling the ground current or the potentials to ground of an alternating-current system are covered. These devices can be grounding transformers, ground-fault neutralizers, resistors, reactors, capacitors, or combinations of these. Rating, insulation classes and dielectric-withstand levels, temperature limitations, testing, and construction are considered.

IEEE Std 43-2000. *IEEE Recommended Practice for Testi Insulation Resistance of Rotating Machinery.* This docume describes the recommended procedure for measuring insul tion resistance of armature and field windings in rotating m chines rated 1 hp, 750 W or greater. It applies to synchrono machines, induction machines, dc machines, and synchrono condensers. Contained within this document is the gene theory of insulation resistance (IR) and polarization ind (P.I.), as well as factors affecting the results, test procedure methods of interpretation, test limitations, and recommend minimum values.

IEEE Std 48-1996. *IEEE Standard Test Procedures and R quirements for Alternating-Current Cable Terminatio 2.5 kV Through 765 kV.* All indoor and outdoor cable te minations used on alternating-current cables having lam nated or extruded insulation rated 2.5 kV through 765 kV a covered, except for separable insulated connectors, which a covered by IEEE Std 386-1995.

IEEE Std 62-1995. *IEEE Guide for Diagnostic Field Testi of Electric Power Apparatus—Part 1: Oil Filled Pow Transformers, Regulators, and Reactors.* Diagnostic tests ar measurements that are performed in the field on oil-immers power transformers and regulators are described. Whenev possible, shunt reactors are treated in a similar manner transformers. Tests are presented systematically in categori depending on the subsystem of the unit being examined. diagnostic chart is included as an aid to identify the vario subsystems. Additional information is provided regardir specialized test and measuring techniques.

IEEE Std 67-1990 (R1995). *IEEE Guide for Operation an Maintenance of Turbine Generators.* General recommend tions for the operation, loading, and maintenance of turbin driven synchronous generators having cylindrical rotors a provided. The manufacturer's and user's responsibility is di cussed, and the classification of generators and the basis c which they are rated are covered. Mechanical consideratio are also addressed.

IEEE Std 80-2000. *IEEE Guide for Safety in AC Substatic Grounding.* Outdoor ac substations, either conventional gas-insulated, are covered in this guide. Distribution, tran mission, and generating plant substations are also include With proper caution, the methods described herein are als applicable to indoor portions of such substations, or to sul stations that are wholly indoors. No attempt is made to cov the grounding problems peculiar to dc substations. A quar titative analysis of the effects of lightning surges is also b yond the scope of this guide.

IEEE Std 81-1983 (R1991). *IEEE Guide for Measuring Ear Resistivity, Ground Impedance, and Earth Surface Potentia of a Ground System.* The present state of the technique c measuring ground resistance and impedance, earth resistivit and potential gradients from currents in the earth, and th prediction of the magnitude of ground resistance and potenti gradients from scale-model tests are described and discusse Factors influencing the choice of instruments and the tech niques for various types of measurements are covered. Thes include the purpose of the measurement, the accuracy re quired, the type of instruments available, possible sources c error, and the nature of the ground or grounding system unde test. The intent is to assist the engineer or technician in ob taining and interpreting accurate, reliable data. The test pro cedures described promote the safety of personnel an property and prevent interference with the operation of neigh boring facilities.

IEEE Std 81.2-1991. *IEEE Guide for Measurement of Im pedance and Safety Characteristics of Large, Extended o*

Interconnected Grounding Systems. Practical instrumentation methods are presented for measuring the ac characteristics of large, extended, or interconnected grounding systems. Measurements of impedance to remote earth, step and touch potentials, and current distributions are covered for grounding systems ranging in complexity from small grids (less than 900 m^2) with only a few connected overhead or direct-burial bare concentric neutrals, to large grids (greater than 20 000 m^2) with many connected neutrals, overhead ground wires (sky wires), counterpoises, grid tie conductors, cable shields, and metallic pipes. This standard addresses measurement safety; earth-return mutual errors; low-current measurements; power-system staged faults; communication and control cable transfer impedance; current distribution (current splits) in the grounding system; step, touch, mesh, and profile measurements; the foot-equivalent electrode earth resistance; and instrumentation characteristics and limitations.

IEEE Std 82-1994. *IEEE Standard Test Procedure for Impulse Voltage Tests on Insulated Conductors.* A test procedure for impulse testing of insulated conductors (cables) and cables with accessories installed (cable systems) is provided. This procedure can be used as a design or qualification test for cables or for cable systems. This test procedure is not intended to replace any existing or future standards covering cable or cable accessories, impulse generators, impulse testing or voltage measurements. It is intended to supplement such standards by indicating specific procedures for a specific type of cable system or cable system component.

IEEE Std 91-1984 (R1994). *IEEE Standard Graphic Symbols for Logic Functions (bound with IEEE Std 91a-1991).* An international language by which it is possible to determine the functional behavior of a logic or circuit diagram with minimal reference to supporting documentation is defined; as such, it is designed to allow a single concept to be expressed in one of several different ways, according to the demands of a particular situation. Consequently, this standard does not attempt, nor intend, to establish single correct symbols for particular devices. The symbols for representing logic functions or devices enable users to understand the logic characteristics of these functions or devices without specific knowledge of their internal characteristics. Definitions and an explanation of symbol construction are provided. Information is presented on: qualifying symbols associated with inputs, outputs, and other connections; dependency notation; combinational and sequential elements; and symbols for highly complex functions. The symbols and representation techniques are compatible with IEC Pub. 617, Part 12.

IEEE Std 91a-1991 (R1994). *Supplement to IEEE Standard Graphic Symbols for Logic Functions (bound with IEEE Std 91-1984).* Graphic symbols for representing logic functions or physical devices capable of carrying out logic functions are presented. Descriptions of logic functions, the graphic representation of these functions, and examples of their applications are given. The symbols are presented in the center of electrical applications, but most may also be applied to nonelectrical systems (for example, pneumatic, hydraulic, or mechanical). This supplement provides additional internationally approved graphic symbols and makes corrections as needed to IEEE Std 91-1984.

IEEE Std 95-1977 (R1991). *IEEE Recommended Practice for Insulation Testing of Large AC Rotating Machinery with High Direct Voltage.* Recommendations are made to aid in the selection of metric units, so as to promote uniformity in the use of metric units and to limit the number of different metric units that will be used in electrical and electronics science and technology. The recommendations can cover units for space and time, periodic and related phenomena, mechanics, heat, electricity and magnetism, light and related electromagnetic radiations, and acoustics. This document does not cover how metric units are to be used nor does it offer guidance concerning correct metric practice.

IEEE Std 98-1984 (R1993). *IEEE Standard for the Preparation of Test Procedures for the Thermal Evaluation of Solid Electrical Insulating Materials.* Principles are given for the development of test procedures to evaluate the thermal endurance of solid electrical insulating materials and simple combinations of such materials. The results of accelerated thermal endurance tests, which are conducted according to prescribed procedures, may be used to establish temperature indexes for insulating materials. The test procedures apply to materials before they are fabricated into insulating structures identified with specific parts of electric equipment. Tests for specific insulating materials are not covered. The procedures may or may not apply to the aging characteristics of dielectric fluids or of porous materials impregnated with dielectric fluids.

IEEE Std 99-1980 (R1992). *IEEE Recommended Procedures for the Preparation of Test Procedures for the Thermal Evaluation of Insulation Systems for Electric Equipment.* A general form is provided for the preparation of test procedures. Points to be considered by technical committees in the preparation of specific instructions for the thermal evaluation of insulation systems for electric equipment are suggested. The test procedures involve accelerated thermal aging of insulation systems and specify tests that the committees deem pertinent, based on conditions of use. The objective of the procedures is to provide for the functional evaluation, by test, of insulation systems electric equipment

IEEE Std 101-1987 (R1995). *IEEE Guide for the Statistical Analysis of Thermal Life Test Data.* Statistical analyses of data from thermally accelerated aging tests are described. The basis and use of statistical calculations are explained. Data analysis, estimation of the relationship between life and temperature, and the comparison between two sets of data are covered.

IEEE Std 112-1996. *IEEE Standard Test Procedure for Polyphase Induction Motors and Generators.* Instructions are given for conducting and reporting the more generally applicable and acceptable tests to determine the performance characteristics of polyphase induction motors and generators. Electrical measurements, performance testing, temperature tests, and miscellaneous tests are covered.

IEEE Std 115-1995. *IEEE Guide: Test Procedures for Synchronous Machines.* Instructions are given for conducting the more generally applicable and accepted tests to determine the performance characteristics of synchronous machines. Although the tests described are applicable in general to synchronous generators, synchronous motors (larger than fractional horsepower), synchronous condensers, and synchronous frequency changers, the descriptions make reference primarily to synchronous generators and synchronous motors. Alternative methods of making many of the tests covered in this guide are described and are suitable for different sizes and types of machines under different conditions. This guide covers miscellaneous tests; saturation curves, segregated losses, and efficiency; load excitation and voltage regulation; temperature tests; torque tests; synchronous machine quantities; and sudden short-circuit tests.

IEEE Std 117-1974 (R1991). *IEEE Standard Test Procedure for Evaluation of Systems of Insulating Materials for Random-Wound AC Electric Machinery.* Useful methods for the evaluation of systems of insulation for random-wound stators of rotating electric machines are given. The chief purpose is to classify insulation systems in accordance with their temperature limits by test, rather than by chemical composition. The procedure is intended to evaluate insulation systems for use in usual service conditions with air cooling. It has also been a useful tool for evaluating systems for special requirements where machines are enclosed in gas atmospheres, subjected to strong chemicals, metal dusts, or submersion in liquids, although these special requirements are beyond the scope of this test procedure.

IEEE Std 118-1978 (R1992). *IEEE Standard Test Code for Resistance Measurements.* Methods of measuring electrical resistance that are commonly used to determine the characteristics of electric machinery and equipment are presented. The methods are limited to those using direct-current or commercial power frequencies of 60 Hz or below, and to those measurements required to determine performance characteristics. The choice of method in any given case depends on the degree of accuracy required and the nature of the circuit to be measured. A guide for selecting the appropriate method is given.

IEEE Std 120-1989 (R1997). *IEEE Master Test Guide for Electrical Measurements in Power Circuits.* Instructions are given for measuring electrical quantities that are commonly needed to determine the performance characteristics of electric machinery and equipment. Methods are given for measuring voltage, current, power, energy, power factor, frequency, impedance, and magnetic quantities, with either analog or digital indicating or integrating instruments, in dc or ac rotating machines; transformers; induction apparatus; arc and resistance heating equipment; and mercury arc, thermionic, or solid-state rectifiers and inverters. Ancillary instruments and equipment are discussed. Computer-based techniques and the use of optical fibers in instrumentation are considered.

IEEE Std 122-1991 (R1997). *IEEE Recommended Practice for Functional Performance Characteristics of Control Systems for Steam Turbine Generator Units.* Minimum functional and performance characteristics related to speed/load-control systems for steam turbine generator units that may be interconnected on a power system are recommended. The recommendations apply to the following types of steam turbines, rated at 500 kW and larger, intended to drive electric generators at constant speed: (1) condensing or noncondensing turbines without initial or exhaust steam-pressure control, or both, including turbines used with reheat or regenerative feed-water heaters, or both; (2) condensing or noncondensing turbines with initial or exhaust steam-pressure control, or both, including turbines used with reheat or regenerative feed-water heaters, or both; (3) automatic extraction, or induction, or both, and mixed-pressure turbines. Emergency governors, or other overspeed control devices, and, in general, devices that are not responsive to speed are excluded. This recommended practice can be included in prime-mover purchase specifications.

IEEE Std 125-1988 (R1996). *IEEE Recommended Practice for Preparation of Equipment Specifications for Speed-Governing of Hydraulic Turbines Intended to Drive Electric Generators.* Terms, functions, and characteristics as commonly used in North America for preparing equipment specifications for speed-governing of hydraulic turbine-driven generators are defined. Specific components that may be included in a governor system are described. The performance characteristics of a good governor system and adjustments and tests to obtain and confirm the desired performance are delineated. Information to be provided by the manufacturer, so that the purchaser can be assured that the governor equipment will interface properly with other equipment, is specified. The intent is also to provide adequate information for maintenance purposes. The criteria for acceptance tests are given, and the data that will be furnished by the purchaser are listed.

IEEE Std 139-1988 (R1999). *IEEE Recommended Practice for the Measurement of Radio Frequency Emission from Industrial, Scientific, and Medical (ISM) Equipment Installed on User's Premises.* Equipment inspection and RF electromagnetic field measurement procedures are described for equipment that generates RF energy for purposes other than radio communications, to cause physical, chemical, or biological changes. The procedures are designed to help ensure that the equipment does not interfere with radio communications, navigation, and other essential radio services. The reporting of RF field measurements is covered.

IEEE Std 140-1990 (R1995). *IEEE Recommended Practice for Minimization of Interference from Radio-Frequency Heating Equipment.* Procedures that may be applied in the design and construction of radio-frequency heating equipment used for heating in industrial settings and for other purposes are described. These procedures are intended to reduce the levels of radio-frequency energy leaks that can interfere with other equipment and broadcast services. They may also be used as remedial measures when harmful interference occurs. Applications in the field of telecommunication and information technology are excluded.

IEEE Std 141-1993 (R1999). *IEEE Recommended Practice for Electric Power Distribution for Industrial Plants (IEEE Red Book).* A thorough analysis of basic electric-systems considerations is present. Guidance is provided in design, construction, and continuity of an overall system to achieve safety of life and preservation of property; reliability; simplicity of operation; voltage regulation in the utilization of equipment within the tolerance limits under all load conditions; care and maintenance; and flexibility to permit development and expansion. Recommendations are made regarding system planning; voltage considerations; urge voltage protection; system protective devices; fault calculations; grounding; power switching, transformation, and motor-control apparatus; instruments and meters; cable systems; busways; electrical energy conservation; and cost estimation.

IEEE Std 142-1991. *IEEE Recommended Practice for Grounding of Industrial and Commercial Power Systems (IEEE Green Book).* The problems of system grounding, that is, connection to ground of neutral, of the corner of the delta, or of the midtap of one phase are covered. The advantages and disadvantages of grounded versus ungrounded systems are discussed. Information is given on how to ground the system, where the system should be grounded, and how to select equipment for the grounding of the neutral circuits. Connecting the frames and enclosures of electric apparatus, such as motors, switchgear, transformers, buses, cables, conduits, building frames, and portable equipment, to a ground system is addressed. The fundamentals of making the interconnection or ground-conductor system between electric equipment and the ground rods, water pipes, etc., are outlined. The problems of static electricity—how it is generated, what processes may produce it, how it is measured, and what should be done to prevent its generation or to drain the static charges to earth to prevent sparking—are treated. Methods of protecting structures against the effects of lightning are also covered. Obtaining a low-resistance connection to the earth, using of ground rods, connections to water pipes, etc., is discussed.

IEEE Std 145-1993. *IEEE Standard Definitions of Terms for Antennas.* Definitions of terms in the field of antennas are provided.

IEEE Std 149-1979 (R1990). *IEEE Standard Test Procedures for Antennas.* Procedures for the measurement of antenna properties are presented. It is assumed that the antenna to be measured can be treated as a passive, linear, and reciprocal device, and that its radiation properties can therefore be measured in either the transmitting or the receiving mode. However, many of the procedures can be adapted for use in the measurement of antenna systems containing circuit elements that may be active, nonlinear, or nonreciprocal. The measurement of radiation patterns on an antenna range is addressed. The instrumentation required for the antenna range, directions for the evaluation of an existing range, and the operation of ranges are discussed. A variety of special measurement techniques are included.

IEEE Std 167A.1-1995. *IEEE Standard Facsimile Test Chart: Bi-level (Black & White).* A facsimile test chart for assessing performance of document facsimile systems, including any compatible combination of facsimile equipment, computers, transmission facilities, and image storage, is provided. The chart is composed solely of high-resolution, high contrast black-and-white patterns. Although the chart is designed for Group 3 and Group 4 facsimile, it is also expected to be useful in testing other imaging systems. The received image may be recorded or displayed. This standard offers a means of assessing various technical quality parameters, detecting defects produced in received images, and evaluating the readability of text when the original is black and white.

IEEE Std 167A.2-1996. *IEEE Standard Facsimile Test Chart: High Contrast (Gray Scale).* A means of assessing performance of document facsimile systems, including any compatible combination of facsimile equipment, computers, transmission facilities, and image storage, is provided.

IEEE Std 167A.3-1997. *IEEE Standard Facsimile Color Test Chart.* A facsimile test chart for assessing performance of document facsimile systems, including any compatible combination of facsimile equipment, computers, transmission facilities, and image storage, is provided.

IEEE Std 187-1990 (R1995). *IEEE Standard on Radio Receivers: Open Field Method of Measurement of Spurious Radiation from FM and Television Broadcast Receivers.* The potential sources of spurious radiation from frequency modulation (FM) and television broadcast receivers, and methods of measurement, are described. This standard is not intended to apply to equipment other than FM and television broadcast receivers.

IEEE Std 208-1995. *IEEE Standard on Video Techniques: Measurement of Resolution of Camera Systems, 1993 Techniques.* The methods for measuring the resolution of camera systems are described. The primary application is for users and manufacturers to quantify the limit where fine detail contained in the original image is no longer reproduced by the camera system. The techniques described may also be used for laboratory measurements and for proof-of-performance specifications for a camera.

IEEE Std 211-1997. *IEEE Standard Definitions of Terms for Radio Wave Propagation.* Terms and definitions used in the context of electromagnetic wave propagation relating to the fields of telecommunications, remote sensing, radio astronomy, optical waves, plasma waves, the ionosphere, the magnetosphere, and magnetohydrodynamic, acoustic, and electrostatic waves are supplied.

IEEE Std 213-1987 (R1998). *IEEE Standard Procedure for Measuring Conducted Emissions in the Range of 300 kHz to 25 MHz from Television and FM Broadcast Receivers to Power Lines.* Procedures for testing television and FM broadcast receivers are included. The user is cautioned that this method might not be appropriate for conducted emissions testing of systems or products other than televisions or FM receivers. Other more general methods exist and it is suggested that they be used for review. These include, but are not limited to, ANSI C63.4-1981, American National Standard Methods of Measurement of Radio Noise Emissions from Low-Voltage Electrical and Electronic Equipment in the Range of 10 kHz to 1 GHz. A method for measuring the emissions conducted by the power line from these receivers in the frequency range of 300 kHz to 25 MHz is defined. Standard input signals, the equipment setup, and measurement techniques are described.

IEEE Std 241-1990 (R1997). *IEEE Recommended Practice for Electric Power Systems in Commercial Buildings (IEEE Gray Book).* This recommended practice is intended to promote the use of sound engineering principles in the design of electrical systems for commercial buildings. It covers load characteristics; voltage considerations; power sources and distribution systems; power distribution apparatus; controllers; services, vaults, and electrical equipment rooms; wiring systems; systems protection and coordination; lighting; electric space conditioning; transportation; communication systems planning; facility automation; expansion, modernization, and rehabilitation; special requirements by occupancy; and energy conservation. Although directed to the power-oriented engineer with limited commercial-building experience, it can be an aid to all engineers responsible for the electrical design of commercial buildings. This standard is not intended to be a complete handbook; however, it can direct the engineer to texts, periodicals, and references for commercial buildings and act as a guide through the myriad of codes, standards, and practices published by the IEEE and other professional associations and governmental bodies.

IEEE Std 242-1986 (R1991). *IEEE Recommended Practice for Protection and Coordination of Industrial and Commercial Power Systems (IEEE Buff Book).* The selection, application, and coordination of the components that constitute system protection for industrial plants and commercial buildings is presented. Complete information on protection and coordination principles designed to protect industrial and commercial power systems against any abnormalities that could reasonably be expected to occur in the course of system operation is presented. Design features are provided for: quick isolation of the affected portion of the system while maintaining normal operation elsewhere; reduction of the short-circuit current to minimize damages to the system, its components, and the utilization equipment it supplies; and provision of alternate circuits, automatic throwovers, and automatic reclosing devices. The following are covered: basic principles; calculation of short-circuit currents; instrument transformers; selection and application of protective relays; fuses; low-voltage circuit breakers; ground-fault protection; conductor protection; motor protection; transformer protection; generator protection; bus and switchgear protection; service supply line protection; overcurrent coordination; and maintenance, testing, and calibration.

IEEE Std 252-1995. *IEEE Standard Test Procedure for Polyphase Induction Motors Having Liquid in the Magnetic Gap.* Instructions for conducting and reporting the more generally applicable and acceptable tests to determine the performance characteristics of polyphase induction motors having liquid in the magnetic gap are given. Constants in several equations and forms apply to three-phase motors only and require modification for application to motors having another number of phases. It is not intended that the procedure cover all possible tests or tests of a research nature. The procedure shall not be interpreted as requiring the making of any or all of the tests described herein in any given transaction.

IEEE Std 259-1999. *IEEE Standard Test Procedure for Evaluation of Systems of Insulation for Dry-Type Specialty and General-Purpose Transformers.* A uniform method by which the thermal endurance of electrical insulation systems for dry-type specialty and general-purpose transformers can be compared is established. Covered are insulation systems intended for use in the types of transformers described in NEMA ST 1-1988 and NEMA ST 20-1992.

ANSI 260.1-1993. *American National Standard Letter Symbols for Units of Measurement (IS Units, Customary Inch-Pound Units, and Certain Other Units).* General principles of letter symbol standardization are discussed. Symbols are given for general use and for use with limited character sets. The symbols given are intended for all applications, including use in text and equations; in graphs and diagrams; and on panels, labels, and nameplates.

ANSI 260.3-1993 (R2000). *American National Standard Mathematical Signs and Symbols for Use in Physical Sciences and Technology.* Signs and symbols used in writing mathematical text are defined. Special symbols peculiar to certain branches of mathematics, such as non-Euclidean Geometry's, Abstract Algebra's, Topology, and Mathematics of Finance, which are not ordinarily applied to the physical sciences and engineering, are omitted.

ANSI 260.4-1996 (R2000). *American National Standard Letter Symbols and Abbreviations for Quantities Used in Acoustics.* Letter symbols for physical quantities used in the science and technology of acoustics are covered. Abbreviations for a number of acoustical levels and related measures that are in common use are also given. The symbols given in this standard are intended for all applications.

IEEE Std 267-1966. *IEEE Recommended Practice for the Preparation and Use of Symbols.* Guidelines to be used in developing and applying those symbols that are employed in the electrical and electronics fields are provided. These include abbreviations, functional designations, graphic symbols, letter combinations, mathematical symbols, reference designations, symbols for quantities, and symbols for units. The guidelines should be useful to any committee engaged in developing standards publications in the areas mentioned.

IEEE Std 269-1992. *IEEE Standard Methods for Measuring Transmission Performance of Analog and Digital Telephone Sets.* Practical methods for measuring the transmission characteristics of both digital and conventional to-wire analog telephone sets by means of objective measurements on a test connection are described. The test results thus obtained may be used as a means of evaluating or specifying the transmission performance of a telephone set on a standardized basis. The measurements are applicable to telephone sets incorporating carbon or linear transmitters. Measurements are over the frequency range most useful for speech: 100–5000 Hz. The test methods are not intended to be applicable to special devices, such as noise-exclusion transmitters, distant-talking transmitters, insert-type receivers, or noise-exclusion receivers equipped with large ear pads.

IEEE Std 275-1992 (R1998). *IEEE Recommended Practice for Thermal Evaluation of Insulation Systems for Alternating-Current Electric Machinery Employing Form-Wound Preinstalled Stator Coils for Machines Rated 6900 V and Below.* A test procedure for comparing two or more insulation systems in accordance with their expected life at rated temperature is described. The procedure is limited to insulation systems for ac electric machines using form-wound preinstalled stator coils and rated 6900 V and below. This procedure is intended to evaluate insulation systems for use in usual service conditions with air cooling. It does not cover such special requirements as machines that are enclosed in gas atmospheres, or that are subjected to strong chemicals, to metal dust, or to submersion in liquids, etc. The procedure includes instructions for testing candidate systems in comparison with known systems having a proven record of service experience and interpretation of the results.

IEEE Std 277-1994. *IEEE Recommended Practice For Cement Plant Power Distribution.* Electrical distribution systems in cement plants that would result in satisfactory equipment utilization, reliability, performance, safety, and low maintenance—all at a reasonable cost are recommended.

IEEE Std 280-1985 (R1997). *IEEE Standard Letter Symbols for Quantities Used in Electrical Science and Electrical Engineering.* Letter symbols used to represent physical quantities in the field of electrical science and electrical engineering are defined. The symbols are independent of the units employed or special values assigned. Also included are selected symbols for mathematics and for physical constants.

IEEE Std 290-1980 (R1986). *IEEE Standard for Electric Couplings: Part I—General, Rating, Performance Characteristics; Part II—Test Procedures.* The more generally applicable characteristics, and how to conduct and report on the tests for determining them, are covered. Service conditions, rating, temperature and temperature use, torque characteristics, speed, losses, and markings are discussed. Methods for electrical measurements, preliminary tests, performance determination, temperature tests, high-potential tests, and miscellaneous tests are given.

IEEE Std 291-1991. *IEEE Standards Methods for Measuring Electromagnetic Field Strength of Sinusoidal Continuous Wave, 30 Hz to 30 Hz.* Two standard methods for field-strength measurement are described. The standard-antenna method consists of measuring the voltage developed in a standard antenna by the field to be measured and computing the field strength from the measured voltage and the dimensions and form of the standard antenna. The standard-field method consists of comparing voltages produced in an antenna by the field to be measured and by a standard field, the magnitude of which is computed from the dimensions of the transmitting antenna, its current distribution, the distance of separation, and effect of the ground. The measurement procedures are outlined, including calibration of commercial field strength and extension of the methods to microwave frequencies. Methods for measuring power radiated from an antenna under several different conditions are briefly presented, and the important considerations for securing useful and accurate measurements are described.

IEEE Std 292-1969 (R2000). *IEEE Specification Format for Single-Degree-of-Freedom Spring-Restrained Rate Gyros.* A guide for the preparation of a single-degree-of-freedom spring-restrained rate gyro specification is given. The format used provides a common meeting ground of terminology and practice for manufacturers and users.

IEEE Std 293-1969 (R2000). *IEEE Test Procedure for Single-Degree-of-Freedom Spring-Restrained Rate Gyros.* Recommended rate gyro test procedures derived from those currently in use, including test conditions to be considered, are compiled. In some cases alternate methods for measuring a performance characteristic have been included. This standard is intended to be a guide in the preparation of Section 4 of a specification that follows the format of IEEE Std 292-1969, Specification Format for Single-Degree-of-Freedom Spring-Restrained Rate Gyros.

IEEE Std 295-1969 (R2000). *IEEE Standard for Electronics Power Transformers.* Application guidance and test procedures are given for power transformers and inductors that are used in electronic equipment and supplied by power lines or generators of essentially sine wave or polyphone voltage. Provision is made for relating the characteristics of transformers to the associated rectifiers and circuits. This Standard includes, but is not limited to, the following transformers and inductors: rectifier supply transformers for either high- or low-voltage supplies, filament and cathode heater transformers, transformers for alternating current resonant charging circuits, inductors used in rectifier filters, and autotransformers with fixed taps.

IEEE Std 299-1997. *IEEE Standard Method for Measuring the Effectiveness of Electromagnetic Shielding Enclosures.* Uniform measurement procedures and techniques are provided for determining the effectiveness of electromagnetic shielding enclosures at frequencies from 9 kHz to 18 GHz (extendable to 50 Hz and 100 GHz, respectively) for enclosures having no dimension less than 2.0 m. The types of enclosures covered include, but are not limited to, single-shield or double-shield structures of various construction, such as bolted demountable, welded, or integral with a building; and made of materials such as steel plate, copper or aluminum sheet, screening, hardware cloth, metal foil, or shielding fabrics.

IEEE Std 300-1988 (R1999). *IEEE Standard Test Procedures for Semiconductor Charged-Particle Detectors.* Test procedures for semiconductor charged-particle detectors for ionizing radiation are provided. They apply to detectors that are used for the detection and high-resolution spectroscopy of charged particles. The measurement techniques were selected to be readily available to all manufacturers and users of charged-particle detectors. Some superior techniques are not included because the methods are too complex or require equipment (such as particle accelerators) that may not be readily available. The standard covers measurement of resolution, noise, sensitivity to ambient conditions, current-voltage characteristics, dead-layer energy loss, sensitive arc, detector thickness (for transmission detectors), and capacitance-voltage characteristics.

IEEE Std 301-1988 (R1999). *IEEE Standard Test Procedures for Amplifiers and Preamplifiers used with Detectors of Ionizing Radiation.*

IEEE Std 303-1991 (R1996). *IEEE Recommended Practice for Auxiliary Devices for Motors in Class 1, Groups A, B, C, and D, Division 2 Locations.* Installation procedures and wiring methods and materials are recommended. Termination boxes, motor surge protection, and power-factor-correction capacitors are discussed. The aim is to promote consistent application of the devices covered.

IEEE Std 304-1977 (R1991). *IEEE Test Procedure for Evaluation and Classification of Insulation Systems for Direct-Current Machines.* Insulation systems for direct-current machines are classified in accordance with their limiting temperatures as determined by test rather than by chemical composition. The intention is to classify according to the recognized A, B, F, and H categories by determining thermal capability in accordance with machine temperature-rise standards. This test procedure has been prepared to indicate accepted tests. It is applicable to insulation systems for use in usual service conditions. This standard does not cover special requirements such as for machines in gas atmospheres being subjected to strong chemicals, metal dusts, or submersion in liquids.

IEEE Std 308-1991. *IEEE Standard Criteria for Class 1E Power Systems for Nuclear Power Generating Stations.* Class 1E portions of ac and dc power systems and instrumentation and control power systems in single-unit and multiunit nuclear power generating stations are covered. Not included are the preferred power supply; unit generator(s) and their buses; generators breaker; step-up, auxiliary, and start-up transformers; connections to the station switchyard; switchyard; transmission lines; and the transmission network. The intent is to provide criteria for the determination of Class 1E power system design features, criteria for sharing Class 1E power systems in multiunit stations, and the requirements for their testing and surveillance.

IEEE Std 309-1999/ANSI N42.3-1999. *IEEE Standard Test Procedures and Bases for Geiger-Mueller Counters.* Test procedures for Geiger-Mueller counters that are used for the detection of ionizing radiation are presented so that they have the same meaning to both manufacturers and users. Also included is information on bases (i.e., connections) for the counters.

IEEE Std 315-1975 (R1993). *IEEE Standard Graphic Symbols for Electrical and Electronics Diagrams.* A list of graphic symbols and class designation letters for use on electrical and electronics diagrams is provided. All of the symbols are designed so that their connection points fall on a modular grid to help those who use a grid basis for the preparation of diagrams. A substantial effort has been made to make this standard compatible with approved International Electrotechnical Commission (IEC) Recommendations (IEC Publication 117, in various parts).

IEEE Std 315A-1986 (R1993). *IEEE Standard Supplement to Graphic Symbols for Electrical and Electronics Diagrams.* Symbols approved by the International Electrotechnical Commission since 1975, or for which there is now a greater need in the US arising from international commerce, are provided. Besides the addition of new symbols, some updating of the information in IEEE Std 315-1975 has been undertaken.

IEEE Std 317-1983 (R1996). *IEEE Standard for Electric Penetration Assemblies in Containment Structures for Nuclear Power Generating Stations.* Requirements for the design, construction, qualification, test, and installation of electric penetration assemblies in nuclear containment structures for stationary nuclear power generating stations are presented. Quality control and quality assurance requirements and requirements for purchaser's specification are included. The requirements for external circuits that connect to penetration assemblies and for operation, maintenance, or periodic testing after installation are not covered.

IEEE Std 323-1983 (R1996). *IEEE Standard for Qualifying Class 1E Equipment for Nuclear Power Generating Stations.* The basic requirements for qualifying Class 1E equipment with interfaces that are to be used in nuclear power generating stations are described. The principles, procedures, and methods of qualification are covered. These qualification requirements, when met, will confirm the adequacy of the equipment design under normal, abnormal, design basis event, post design basis event, and in-service test conditions for the performance of safety functions. The methods are to be used for qualifying equipment, extending qualification, and updating qualification if the equipment is modified.

IEEE Std 325-1996. *IEEE Standard Test Procedures for Germanium Gamma-Ray Detectors.* Terminology and standard test procedures are established for germanium radiation detectors that are used for the detection and high-resolution spectrometry of gamma rays, X rays, and charged particles that produce hole-electron pairs in the crystal lattice so that these items have the same meaning to both manufacturers and users. Not all tests described in this standard are mandatory, but tests that are used to specify performance shall be performed in accordance with this standard. Detector endcap and reentrant (Marinelli) beaker standards are discussed; measurements that depend upon phonon production are not covered in this standard.

IEEE Std 334-1994 (R1999). *IEEE Standard for Qualifying Continuous Duty Class 1E Motors for Nuclear Power Generating Stations.* Methods and requirements for qualifying continuous duty Class 1E motors for use in nuclear power generating stations are provided. The methods are used for qualifying motors, extending the qualification, and updating the qualification if the motor's design or specified service conditions are modified. The requirements include the principles, procedures, and methods of qualification as they relate to continuous duty Class 1E polyphase squirrel cage ac motors.

IEEE Std 336-1985 (R1991). *IEEE Standard Installation, Inspection, and Testing Requirements for Power, Instrumentation, and Control Equipment at Nuclear Facilities.* Requirements for installation, inspection, and testing of power, instrumentation, and control equipment and systems during the construction phase of a nuclear facility are set forth. These requirements also cover modifications and those operating phase activities that are comparable in nature and extent to related initial construction activities of the facility. The intent is to establish requirements for safety systems equipment. However, this standard may also be applied to nonsafety systems equipment.

IEEE Std 337-1972 (R1992). *IEEE Standard Specification Format Guide and Test Procedure for Linear, Single-Axis, Pendulous, Analog Torque Balance Accelerometer.* A format guide for the preparation of an accelerometer specification that provides a common meeting ground of terminology and practice for manufacturers and users is given. It covers performance; mechanical, electrical, and environmental requirements; quality assurance, preparation for delivery, and use of notes. A compilation of recommended procedures for testing an accelerometer is presented. These procedures, including test conditions to be considered, are derived from those currently in use. Not all tests outlined in this document need be included, nor are additional tests precluded. In some cases, alternative methods for measuring performance characteristics have been included or indicated. The torque balance electronics are not considered to be part of the instrument.

IEEE Std 338-1987 (R2000). *IEEE Standard Criteria for the Periodic Surveillance Testing of Nuclear Power Generating Station Safety Systems.* Design and operational criteria are provided for the performance of periodic testing as part of the surveillance program of nuclear power generating station safety systems. Such testing consists of functional tests and checks, calibration verification and time response measurements, as required, to verify that the safety system performs to meet its defined safety function. The system status, associated system documentation, test intervals, and test procedures during operation are also addressed.

IEEE Std 344-1987 (1993). *IEEE Recommended Practice for Seismic Qualification of Class 1E Equipment for Nuclear Power Generating Stations.* Recommended practices for establishing procedures that will yield data that verify that the Class 1E equipment can meet its performance requirements during and following one safe shutdown earthquake preceded by a number of operating basis earthquakes are provided. This recommended practice may be used to establish tests or analyses that will yield data to substantiate performance claims or to evaluate and verify performance of representative devices and assemblies as part of an overall qualification effort. Two approaches to seismic analysis are described, one based on dynamic analysis and the other on static coefficient analysis. Common methods currently in use for seismic qualification by test are presented.

IEEE Std 352-1987 (R1999). *IEEE Guide for General Principles of Reliability Analysis of Nuclear Power Generating Station Safety Systems.* The basic principles that are needed to conduct a reliability analysis of safety systems are provided for designers and operators of nuclear power plant safety systems and the concerned regulatory groups. By applying the principles given, systems may be analyzed, results may be compared with reliability objectives, and the basis for decisions may be suitably documented. The quantitative principles are applicable to the analysis of the effects of component failures on safety system reliability. Although they have their greatest value during the design phase, the principles are applicable during any phase of the system's lifetime. They may also be applied during the preoperational phase or at any time during the normal lifetime of a system.

IEEE Std 367-1996. *IEEE Recommended Practice for Determining the Electric Power Station Ground Potential Rise and Induced Voltage From a Power Fault.* Guidance for the calculation of power station ground potential rise (GPR) and longitudinal induction (LI) voltages is provided, as well as guidance for their appropriate reduction from worst-case values, for use in metallic telecommunication protection design.

IEEE Std 376-1975 (R1998). *IEEE Standard for the Measurement of Impulse Strength and Impulse Bandwidth.* The use of the impulse generator for calibration purposes in electromagnetic compatibility measurements is addressed. In particular, basic information relating to the use of this device is provided, and interpretation of measurements made using instruments based on it is considered. Two methods of measurement of spectrum amplitude and impulse bandwidth are described in detail. The first method uses a video pulse technique. The second uses a substitution method in which the reference is a pulse-modulated sine wave generator whose parameters are measured. Both techniques are capable of about equal accuracy.

IEEE Std 377-1980 (R1997). *IEEE Recommended Practice for Measurement of Spurious Emission from Land-Mobile Communication Transmitters.* Controlled test conditions, test apparatus, test methods, and data presentation, all of which form the basis for establishing the energy levels of spurious emissions of mobile communication transmitters designed to generate FM signals in the frequency range of 25 to 1000 MHz, are covered. The purpose is to enable design and system engineers engaged in a variety of development projects to achieve uniform results in recognizing the sources and nature of RF spurious emissions emanating from vehicular communications transmitters. Procedures for measuring both broadband and narrowband spectra are provided for both conducted and radiated emissions. Specified limits are not included. However, reference values that are not limited by the state of the art are provided. Transmitter test conditions, apparatus, and method are based on standard instrumentation and measuring techniques and do not require any special apparatus other than necessary terminal simulators. The procedures do not cover the associated antenna and transmission lines.

IEEE Std 379-1994 (R1997). *IEEE Standard Application of the Single-Failure Criterion to Nuclear Power Generating Station Safety Systems.* The application of the single-failure criterion to the electrical power, instrumentation, and control portions of nuclear power generating station safety systems is covered. Conformance with the requirements of IEEE Std 603-1991 and the single-failure criterion as stated in that document is established. Interpretation and guidance in the application of the single-failure criterion, a discussion of the failures, and an acceptable method of single-failure analysis are presented.

IEEE Std 382-1996. *IEEE Standard for Qualification of Actuators for Power-Operated Valve Assemblies With Safety-Related Functions for Nuclear Power Plants.* The qualification of all types of power-driven valve actuators, including damper actuators, for safety-related functions in nuclear power generating stations, is described. This standard may also be used to separately qualify actuator components. The minimum requirements for, and guidance regarding, the methods and procedures for qualification of power-driven valve actuators with safety-related functions are provided.

IEEE Std 383-1974 (R1992). *IEEE Standard for Type Test of Class 1E Electric Cables, Field Splices, and Connections for Nuclear Power Generating Stations.* Directions for establishing type tests that may be used in qualifying Class 1E electric cables, field splices, and other connections for service in nuclear power generating stations are provided. Though intended primarily for cable for field installation, this guide may also be used for the qualification of internal wiring of manufactured devices. It does not cover cables for service within the reactor vessel.

IEEE Std 384-1992 (R1998). *IEEE Standard Criteria for Independence of Class 1E Equipment and Circuits.* The independence requirements of the circuits and equipment comprising or associated with Class 1E systems are described. Criteria for the independence that can be achieved by physical separation and electrical isolation of circuits and equipment that are redundant are set forth. The determination of what is to be considered redundant is not addressed.

IEEE Std 386-1995. *IEEE Standard for Separable Insulated Connectors System for Power Distribution Systems Above 600 V.* Definitions, service conditions, ratings, interchangeable construction features, and tests are established for load-break and dead-break separable insulated connector systems rated 601 V and above, 600 A or less, for use on power distribution systems.

IEEE Std 387-1995. *IEEE Standard Criteria for Diesel-Generator Units Applied as Standby Power Supplies for Nuclear Power Generating Stations.* The criteria for the application and testing of diesel-generator units as Class 1E standby power supplies in nuclear power generating stations is described.

IEEE Std 388-1992 (R1998). *IEEE Standard for Transformers and Inductors in Electronic Power Conversion Equipment.* Transformers and inductors of both the saturating and nonsaturating type are covered. The power-transfer capability of the transformers and inductors covered ranges from less than 1 W to the multikilowatt level. The purpose is to provide a common basis between the engineers designing power-conversion circuits and the engineers designing the transformers and inductors used in those circuits. Apparatus used in equipment for high-voltage power conversion for distribution by electric utilities is not covered.

IEEE Std 389-1996. *IEEE Recommended Practice for Testing Electronics Transformers and Inductors.* A number of tests are presented for use in determining the significant parameters and performance characteristics of electronics transformers and inductors. These tests are designed primarily for transformers and inductors used in all types of electronics applications, but they may apply to the other types of transformers of large apparent-power rating used in the electric power utility industry.

IEEE Std 390-1987 (R1998). *IEEE Standard for Pulse Transformers.* Pulse transformers for use in electronic equipment are considered. This standard applies to the following transformer types: power output (drivers), impedance matching, interstage coupling, current sensing, and blocking-oscillator transformers. For these transformers, the peak power transmitted ranges from a few milliwatts to kilowatts, and the peak voltage transmitted ranges from a few volts to many kilovolts. Symbols, performance tests, equivalent circuits, preferred test methods, marking, and service conditions are covered.

IEEE Std 393-1991 (R1998). *IEEE Standard Test Procedures for Magnetic Cores.* Test methods useful in the design, analysis, and operation of magnetic cores in many types of applications are presented. Tests for specifying and/or measuring permeability, core loss, apparent core loss, induction, hysteresis, thermal characteristics, and other properties are given. Most of the test methods described include specific parameter ranges, instrument accuracies, core sizes, etc., and may be used in the specification of magnetic cores for industrial and military applications. More generalized test procedures are included for the benefit of the R & D engineer and university student. Although the primary concern is with cores of the type used in electronic transformers, magnetic amplifiers, inductors, and related devices, many of the tests are adaptable to cores used in many other applications.

IEEE Std 399-1997. *IEEE Recommended Practice for Industrial and Commercial Power Systems Analysis (IEEE Brown Book).* This Recommended Practice is a reference source for engineers involved in industrial and commercial power systems analysis. It contains a thorough analysis of the power system data required, and the techniques most commonly used in computer-aided analysis, in order to perform specific power system studies of the following: short-circuit, load flow, motor-starting, cable ampacity, stability, harmonic analysis, switching transient, reliability, ground mat, protective coordination, dc auxiliary power system, and power system modeling.

IEEE Std 400-1991. *IEEE Guide for Making High-Direct-Voltage Tests on Power Cable Systems in the Field.* Procedures and test-voltage values for acceptance and maintenance high-direct-voltage testing of power cable systems are presented. The procedures apply to all types of insulated cable systems rated between 2000 V and 69 kV and intended primarily for the transmission or distribution of power. They are not intended to apply to communication cables, control cables, high-frequency or other special-purpose cables, although information of some value may be obtained thereby. The aim of this standard is to provide uniform procedures and to provide guidelines for evaluation of the test results.

IEEE Std 404-1993. *IEEE Standard for Cable Joints for Use with Extruded Dielectric Cable Rated 5000 to 138 000 V and Cable Joints for Use with Laminated Dielectric Cable Rated 2500–500 000 V.* This standard establishes electrical ratings and test requirements for cable joints for use with extruded dielectric shielded cable rated in preferred voltage steps from 5000–138 000 V and cable joints for use with laminated dielectric cable rated in preferred voltage steps from 2500–500 000 V. It also defines a variety of common joint constructions. This standard is designed to provide uniform testing procedures that can be used by manufacturers and users to evaluate the ability of underground power cable splices to perform reliably in service.

IEEE Std 420-1982 (R1999). *IEEE Standard for the Design and Qualification of Class 1E Control Boards, Panels, and Racks Used in Nuclear Power Generating Stations.* Design requirements that are unique to Class 1E control boards, panels, and racks are specified. Standards for qualification tests to verify that these design requirements have been satisfied are provided. This standard is not intended to define the selection, design, or qualification of piping, modules, or other equipment mounted on the Class 1E control boards, panels, or racks. It is concerned, however, with the effect such mounted equipment has on the design and qualifications. Qualification and testing of individual Class 1E control board components and modules and external field-run cables are not covered.

IEEE Std 421.1-1986 (R1996). *IEEE Standard Definitions for Excitation Systems for Synchronous Machines.* Elements and commonly used components in excitation control systems and for excitation systems as applied to synchronous machines are defined. The primary purpose of the standard is to provide a vocabulary for writing excitation systems specifications, evaluating excitation system performance, specifying methods for excitation system tests, and preparing excitation system standards. It is also intended to serve as an educational aid for those becoming acquainted with excitation systems.

IEEE Std 421.2-1990 (R1996). *IEEE Guide for Identification, Testing, and Evaluation of the Dynamic Performance of Excitation Control Systems.* Criteria, definitions, and test procedures for evaluating the dynamic performance of excitation control systems as applied by electric utilities are provided. Since an excitation control system, including the synchronous machine and its excitation system, is a feedback control system, many definitions and performance criteria that are common to all feedback control systems have been adopted. Others specifically related to excitation control systems have been derived. The primary purposes of this guide are to provide a basis for evaluating closed-loop performance of excitation control systems (including both the synchronous machine and its excitation system) for both large and small signal disturbances; to confirm the adequacy of mathematical models of excitation systems for use in analytical studies of power systems; to specify methods for performing tests of excitation control systems and their components; and to prepare excitation system specifications and additional standards. Portions of this standard can also serve as a tutorial for people becoming acquainted with excitation control systems.

IEEE Std 421.3-1997. *IEEE Standard for High-Potential Test Requirements for Excitation Systems for Synchronous Machines.* High-potential test voltages for excitation systems used with synchronous machines are established. Test voltages are established based on whether equipment is connected to the exciter power circuit or is electrically isolated from the exciter power circuit.

IEEE Std 421.4-1990 (R1999). *IEEE Guide for the Preparation of Excitation System Specifications.* This guide is intended to provide to the specification writer the necessary material to prepare a specification for the procurement of an excitation system for a synchronous machine. The information is given in narrative form, with descriptions and functions of particular items that should be examined in preparing the specifications. Excitation systems for synchronous machines rated 5000 kVA or larger are covered.

IEEE Std 421.5-1992 (R1996). *IEEE Recommended Practice for Excitation System Models for Power System Stability Studies.* Excitation system models suitable for use in large-scale system stability studies are presented. With these models, most of the excitation systems currently in widespread use on large, system-connected synchronous machines in North America can be represented. They include updates of models published in the Transactions on Power Apparatus and Systems in 1981, as well as models for additional control features, such as discontinuous excitation controls.

IEEE Std 429-1994. *IEEE Recommended Practice for Thermal Evaluation of Sealed Insulation Systems for AC Electric*

Machinery Employing Form-Wound Preinsulated Stator Coils for Machines Rated 6900 V and Below. A test procedure for comparing two or more sealed insulation systems in accordance with their expected life at rated temperature is outlined. The procedure is limited to insulation systems for alternating-current (ac) electrical machines using form-wound preinsulated stator coils rated 6900 V and below.

IEEE Std 432-1992 (R1998). *IEEE Guide for Insulation Maintenance for Rotating Electric Machinery (5 hp to less than 10 000 hp).* Information necessary to permit an effective evaluation of the insulation systems of medium and small rotating electrical machines is presented. The guide is intended to apply in general to industrial air-cooled machines rated from 5 hp to less than 10 000 hp. However, the procedures may be found useful for other types of machines.

IEEE Std 433-1974 (R1991). *IEEE Recommended Practice for Insulation Testing of Large AC Rotating Machinery with High Voltage at Very Low Frequency.* Terms that have a specific meaning in VLF testing are defined, and VLF test equipment and wave shape are described. A uniform procedure for testing the armature insulation of large ac machines with VLF voltage is provided. Constants for relating VLF tests to power-frequency and direct-voltage tests to obtain equally effective test levels are recommended.

IEEE Std 434-1973 (R1991). *IEEE Guide for Functional Evaluation of Insulation Systems for Large High-Voltage Machines.* Classification test methods that may be used to compare insulation systems in use, or proposed for use, in large high-voltage rotating machines are described. Thermal aging, voltage endurance, thermomechanical forces, and electromechanical forces are addressed.

IEEE Std 442-1981 (R1991). *IEEE Guide for Soil Thermal Resistivity Measurement.* A method for measurement of soil thermal resistivity that is based on the theory that the rate of temperature rise of a line heat source is dependent upon the thermal constants of the medium in which it is placed is given. This information will enable the user to properly install and load underground cables. The aim is to provide sufficient information to enable the user to select useful commercial test equipment, or to manufacture equipment that is not readily available on the market, and to make meaningful resistivity measurements with this equipment in the field or on soil samples in the laboratory. Designs for both laboratory and field thermal needles are described.

IEEE Std 446-1995 (R2000). *IEEE Recommended Practice for Emergency and Standby Power Systems for Industrial and Commercial Applications (IEEE Orange Book).* This Recommended Practice addresses the uses, power sources, design, and maintenance of emergency and standby power systems. Chapter 3 is a general discussion of needs for and the configuration of emergency and standby systems. Chapter 9 lists the power needs for specific industries. Chapters 4 and 5 deal with the selection of power sources. Chapter 6 provides recommendations for protecting both power sources and switching equipment during fault conditions. Chapter 7 provides recommendations for design of system grounding, and Chapter 10 provides recommendations for designing to reliability objectives. Chapter 8 provides recommended maintenance practices.

IEEE Std 449-1998. *IEEE Standard for Ferroresonant Voltage Regulators.* Ferroresonant transformers used as regulators in electronic power supplies and in other equipment are covered. Guides to application and test procedures are included.

IEEE Std 450-1995 (R2000). *IEEE Recommended Practice for Maintenance, Testing, and Replacement of Vented Lead-Acid Batteries for Stationary Applications.* Maintenance, test schedules, and testing procedures that can be used to optimize the life and performance of permanently installed, vented lead-acid storage batteries used for standby power applications are provided. This recommended practice also provides guidance to determine when batteries should be replaced. This recommended practice is applicable to all stationary applications. However, specific applications, such as emergency lighting units and semiportable equipment, may have other appropriate practices and are beyond the scope of this recommended practice.

IEEE Std 473-1985 (R1997). *IEEE Recommended Practice for an Electromagnetic Site Survey (10 kHz to 10 GHz).* Guidelines for the systematic, documented investigation of the amplitudes of RF electromagnetic fields found at one or more locations with respect to frequency, time, and position are provided. Periodic and random radiated electric and magnetic fields and conducted interference within the frequency range of 10 kHz to 10 GHz are considered. Although several aspects of radio-emission investigation are not addressed directly, including signal identification and discrimination; field emissions from regularly occurring, low-frequency, pulsed sources; and test enclosure fields, much information pertinent to these areas is provided.

IEEE Std 475-2000. *IEEE Standard Measurement Procedure for Field Disturbance Sensors 300 MHz to 40 GHz.* Test procedures for microwave field disturbance sensors to measure radio frequency (RF) radiated field strength of the fundamental frequency, harmonic frequencies, near field power flux density, and nonharmonic spurious emissions of sensors operating within the frequency range of 300 MHz to 40 GHz are defined.

IEEE Std 484-1996. *IEEE Recommended Practice for Installation Design and Installation of Vented Lead-Acid Batteries for Stationary Applications.* Recommended design practices and procedures for storage, location, mounting, ventilation, instrumentation, preassembly, assembly, and charging of vented lead-acid batteries are provided. Required safety practices are also included. These recommended practices are applicable to all stationary applications. However, specific applications, such as emergency lighting units and semiportable equipment, and alternate energy applications, may have other appropriate practices and are beyond the scope of this recommended practice.

IEEE Std 485-1997. *IEEE Recommended Practice for Sizing Lead-Acid Batteries for Stationary Applications.* Methods for defining the dc load and for sizing a lead-acid battery to supply that load for stationary battery applications in full float operations are described. Some factors relating to cell selection are provided for consideration. Installation, maintenance, qualification, testing procedures, and consideration of battery types other than lead-acid are beyond the scope of this recommended practice. Design of the dc system and sizing of the battery charger(s) are also beyond the scope of this recommended practice.

IEEE Std 487-1992 (R1994). *IEEE Recommended Practice for the Protection of Wire-Line Communication Facilities Serving Electric Power Stations.* Workable methods for protecting wire-line communication circuits entering power stations are presented. This standard covers: the electric power station environment; protection apparatus; service types, reliability, service performance objective classifications, and transmission considerations; protection theory and philosophy; protection configurations; installation and inspection; and safety.

IEEE Std 488.1-1987 (R1994). *IEEE Standard Digital Interface for Programmable Instrumentation.* Interface systems used to interconnect both programmable and nonprogrammable electronic measuring apparatus with other apparatus and accessories necessary to assemble instrumentation systems are considered. The standard applies to the interface of instrumentation systems, or portions of them, in which the data exchanged among the interconnected apparatus is digital, the number of devices that may be interconnected by one contiguous bus does not exceed 15, total transmission path lengths over the interconnecting cables do not exceed 20 m, and the data rate across the interface on any signal line does not exceed 1 Mb/s. The basic functional specifications of this standard may also be used in digital interface applications that require longer distances, more devices, increased noise immunity, or combinations of these.

IEEE Std 488.2-1992 (R1998). *IEEE Standard Codes, Formats, Protocols, and Common Commands for Use With IEEE Std 488.1-1987, IEEE Standard Digital Interface for Programmable Instruction.* A set of codes and formats to be used by devices connected via the IEEE 488.1 bus is specified. This standard also defines communication protocols necessary to effect application-independent and device-dependent message exchanges, and further defines common commands and characteristics useful in instrument system applications. It is intended to apply to small- to medium-scale instrument systems comprised mainly of measurement, stimulus, and interconnect devices with an instrumentation controller. The standard may also apply to certain devices outside the scope of the instrument system environment. IEEE 488.1 subsets, standard message-handling protocols including error handling, unambiguous program and response-message syntactic structures, common commands useful in a wide range of instrument system applications, standard status reporting structures, and system configuration and synchronization protocols are covered.

IEEE Std 492-1999. *IEEE Guide for Operation and Maintenance of Hydro-Generators.* General recommendations for the operation, loading, and maintenance of synchronous hydro-generators and generator/motors are covered. This guide does not apply to synchronous machines having cylindrical rotors. In this guide, the term hydro-generator is used to describe a synchronous machine coupled to a hydraulic turbine or pump-turbine. This guide is not intended to apply in any way to the prime mover.

IEEE Std 493-1997. *IEEE Recommended Practice for the Design of Reliable Industrial and Commercial Power Systems (IEEE Gold Book).* The fundamentals of reliability analysis as it applies to the planning and design of industrial and commercial electric power distribution systems are presented. Included are basic concepts of reliability analysis by probability methods, fundamentals of power system reliability evaluation, economic evaluation of reliability, cost of power outage data, equipment reliability data, examples of reliability analysis. Emergency and standby power, electrical preventive maintenance, and evaluating and improving reliability of the existing plant are also addressed. The presentation is self-contained and should enable trade-off studies during the design of industrial and commercial power systems design, installation, and maintenance practices for electrical power and grounding (including both power-related and signal-related noise control) of sensitive electronic processing equipment used in commercial and industrial applications are presented.

IEEE Std 499-1997. *IEEE Recommended Practice for Cement Plant Electric Drives and Related Electrical Equipment.* All electric drives, including motors and control wiring associated with machinery or equipment commonly used in the manufacturing areas of cement plants are covered. Recommendations are not intended to apply to power distribution circuits. These recommendations apply to electrical equipment having a supply voltage of 13 800 V or less.

IEEE Std 502-1985 (R1998). *IEEE Guide for Protection, Interlocking, and Control of Fossil-Fuel Unit-Connected Steam Stations.* Information regarding the essential subsystems that make up a fossil-fueled unit-connected boiler-turbine-generator (BTG) station is presented. Typical interlocking, control, and protection for operating the subsystems in a coordinated order to ensure proper start-up and safe shutdown are described. The primary purpose is to provide a basis for qualitative evaluation of overall design of a unit-connected fossil-fuel plant, and for writing general operating guides of an educational nature to aid in acquainting personnel with boiler-turbine-generator systems.

IEEE Std 505-1977 (R1996). *IEEE Standard Nomenclature for Generating Station Electric Power Systems.* Electric power systems in stationary generating stations that provide electric power to the power system are covered. Nomenclature is included for the following interrelated systems: generating unit power system, generating unit auxiliaries power system, station auxiliaries power system, generating unit dc auxiliaries power system, and station dc auxiliaries power system. Nomenclature for instrumentation, controls, or auxiliaries is not included.

IEEE Std 510-1983 (R1992). *IEEE Recommended Practices for Safety in High-Voltage and High-Power Testing.* Safety practices for those who are involved with making measurements on high-voltage sources or with high-power sources of various types, including power-system lines, 60-Hz test transformers, direct-voltage supplies, lightning-impulse generators, and switching-impulse generators are recommended. Electrical hazards involved in temporary measurements, as opposed to metering, relaying, or routine line work, are considered. Safety is considered in connection with testing in laboratories, in the field, in substations, and on lines, and with the test equipment utilized. Cable-fault location, large-capacitance-load testing, high-current testing, and direct connection to power lines are treated separately.

IEEE Std 515-1997. *IEEE Standard for the Testing, Design, Installation, and Maintenance of Electrical Resistance Heat Tracing for Industrial Applications.* The specific test requirements for qualifying electrical resistance heating cables for industrial service are provided, as well as the basis for electrical and thermal design. Heating device characteristics are addressed, and installation and maintenance requirements are detailed. Heating cable and surface heating device application recommendations and requirements are made for ordinary (unclassified) and hazardous (classified) potentially flammable atmospheres and locations.

IEEE Std 515.1-1995. *IEEE Recommended Practice for the Testing, Design, Installation, and Maintenance of Electrical Resistance Heat Tracing for Industrial Applications.* This standard provides the specific test requirements for qualifying electrical resistance heating cables for industrial service, and provides the basis for electrical and thermal design. Heater characteristics are addressed, and installation and maintenance requirements are detailed. Recommendations and requirements are made for unclassed, Class I, Division 2, and Class I, Division 1 heating cable applications.

IEEE Std 516-1995. *IEEE Guide for Maintenance Methods on Energized Power Lines.* General recommendations for performing maintenance work on energized power lines are provided. Technical explanations as required to cover certain laboratory testing of tools and equipment, field maintenance and care of tools and equipment, and work methods for the maintenance of energized lines and for persons working in the vicinity of energized lines are included.

IEEE Std 517-1974 (R2000). *IEEE Standard Specification Format Guide and Test Procedure for Single-Degree-of-Freedom Rate-Integrating Gyros.* A specification format guide for the preparation of a rate-integrating gyro specification that provides a common meeting ground of terminology and practice for manufacturers and users is presented. A compilation of recommended procedures for testing a rate-integrating gyro is given.

IEEE Std 518-1982 (R1996). *IEEE Guide for the Installation of Electrical Equipment to Minimize Electrical Noise Inputs to Controllers from External Sources.* Techniques for the installation and operation of industrial controllers, so as to minimize the disturbing effects of electrical noise on these controllers, are addressed. The identification of noise in control circuits and the classification of noise are discussed. A systems approach to noise reduction is presented. Installation recommendations and wiring practices are covered.

IEEE Std 519-1992 (R1996). *IEEE Recommended Practices And Requirements For Harmonic Control In Electric Power*

Systems. This guide applies to all types of static power converters used in industrial and commercial power systems. The problems involved in the harmonic control and reactive compensation of such converters are addressed, and an application guide is provided. Limits of disturbances to the ac power distribution system that affect other equipment and communications are recommended. This guide is not intended to cover the effect of radio frequency interference.

IEEE Std 522-1992 (R1998). *IEEE Guide for Testing Turn-to-Turn Insulation on Form-Wound Stator Coils for Alternating-Current Rotating Electric Machines*. Suggestions are made for testing the dielectric strength of the insulation separating the various turns from each other within multiturn form-wound coils to determine their acceptability. Typical ratings of machines employing such coils normally lie within the range of 200 kW to 100 MW. The test-voltage levels described do not evaluate the ability of the turn insulation to withstand abnormal voltage surges, as contrasted to surges associated with normal operation. The suggestions apply to: (1) individual stator coils after manufacture; (2) coils in completely wound stators of original manufacture; (3) coils and windings for rewinds of used machinery; and (4) windings of machines in service to determine their suitability for further service (preventive-maintenance testing). Coil service conditions, test devices, and test sequence are discussed. High-frequency test levels for new coils, as well as procedures for maintenance tests or tests after installation of machines, are proposed. Specific test procedures for wound machines, for coils during winding, and for applying surge tests to complete windings are given in the appendixes.

IEEE Std 524-1992 (R1997). *IEEE Guide to the Installation of Overhead Transmission Line Conductors*. General recommendations for the selection of methods, equipment, and tools that have been found practical for the stringing of overhead transmission line conductors and overhead ground wires are provided. The aim is to present in one document sufficient details of present-day methods, materials, and equipment to outline the basic considerations necessary to maintain safe and adequate control of conductors during stringing operations.

IEEE Std 524a-1993 (R1998). *IEEE Guide to Grounding During the Installation of Overhead Transmission Line Conductors—Supplement to IEEE Guide to the Installation of Overhead Transmission Line Conductors*. General recommendations for the selection of methods and equipment found to be effective and practical for grounding during the stringing of overhead transmission line conductors and overhead ground wires are provided. The guide is directed to transmission voltages only. The aim is to present in one document sufficient details of present day grounding practices and equipment used in effective grounding and to provide electrical theory and considerations necessary to safeguard personnel during the stringing operations of transmission lines.

IEEE Std 525-1992 (R1999). *IEEE Guide for the Design and Installation of Cable Systems in Substations*. Guidance for the design, installation, and protection of wire and cable systems in substations with the objective of minimizing cable failures and their consequences is provided. The design of wire and cable systems in generating stations is not covered.

IEEE Std 528-1994. *IEEE Standard for Inertial Sensor Terminology*. Terms and definitions relating to inertial sensors are presented. Usage as understood by the inertial sensor community is given preference over general technical usage of the terms herein. The criterion for inclusion of a term and its definition in this document is usefulness as related to inertial sensor technology.

IEEE Std 529-1980 (R2000). *IEEE Supplement for Strapdown Applications to IEEE Standard Specification Format Guide and Test Procedure for Single-Degree-of-Freedom Rate-Integrating Gyros*. A specification format guide for the preparation of a rate-integrating gyroscope specification is presented. Recommended procedures for testing a rate-integrating gyroscope are compiled. This standard, when combined with IEEE Std 517-1974 (R1994), defines the requirements and test procedures in terms of characteristics unique to the gyroscope or those applications in which the dynamic angular inputs are significantly greater than the limitations identified in IEEE Std 517-1974.

IEEE Std 530-1978 (R1992). *IEEE Standard Specification Format Guide and Test Procedure for Linear, Single-Axis, Digital, Torque Balance Accelerometer*. A guide for the preparation of a digital accelerometer specification and test procedure is provided. It is intended to provide common terminology and practice for manufacturers and users. The accelerometer considered utilizes a linear, single-axis, nongyroscopic accelerometer sensor with a permanent magnet torquer. The torquing electronics are considered part of the accelerometer. General design, performance, environmental, and reliability requirements are covered. Information on classification of tests, acceptance tests, qualification tests, reliability tests, standard test conditions, test equipment, test methods, and data submittal is given.

IEEE Std 532-1993. *IEEE Guide for Selecting and Testing Jackets for Underground Cables*. This guide covers corrosion protection, properties of commonly used jackets, electrical characteristics of jackets, physical requirements for jackets referenced in industry standards, and selection and testing of jackets. It is written for those responsible for optimizing underground cable installations. The purpose is to present a reasonably complete picture of the role of jackets so that the subject can be approached in an orderly and organized manner. An effort has been made to shun the highly technical language and theory commonly used by electrical engineers, corrosion engineers, and chemists to discuss the more detailed application of jackets.

IEEE Std 535-1986 (R1994). *IEEE Standard Qualification of Class 1E Lead Storage Batteries for Nuclear Power Generating Stations*. Qualification methods for Class 1E lead storage batteries and racks to be used in nuclear power generating stations outside of primary containment are described. Principles and methods of qualification, qualification information, qualification by type testing, type tests and analysis procedures, and documentation are covered. Battery sizing, maintenance, capacity testing, installation, charging equipment, and other types of batteries that are beyond the scope of this standard are not considered.

IEEE Std 539-1990 (R1994). *IEEE Standard Definitions of Terms Relating to Corona and Field Effects of Overhead Power Lines*. The most widely used terms specific to or associated with overhead power-line corona and electromagnetic fields are defined. This includes terms related to electric and magnetic fields, ions, radio frequency propagation, electromagnetic signals and noise, audible noise, coupled voltages and current, shock and perception, weather and related statistical terms, and measurements and measuring devices.

IEEE Std 563-1978 (R1996). *IEEE Guide on Conductor Self-Damping Measurements*. Methods for measuring the inherent vibration damping characteristics of overhead conductors are presented. The intent is to obtain information in a compatible and consistent form that will provide a reliable basis for studying the vibration and damping of conductors in the future, and for comparing data of various investigators. The methods and procedures recommended are not intended for quality-control test purposes.

IEEE Std 572-1985 (R1992). *IEEE Standard for Qualification of Class 1E Connection Assemblies for Nuclear Power Generating Stations*. General requirements, direction, and methods for qualifying Class 1E connection assemblies for service in nuclear power generating stations are provided. Connectors, terminations, and environmental seals in combination with related cables or wires as assemblies are covered. Emphasis is placed on multipin, quick, disconnect-type connection assemblies primarily utilized for instrumentation, control, and power. This standard does not apply to containment electric penetrations, fire stops, in-line splices, or components for service within the reactor vessel.

IEEE Std 576-1989 (R1992). *IEEE Recommended Practice for Installation, Termination, and Testing of Insulated Power Cable as Used in the Petroleum and Chemical Industry.* A guide to installation, splicing, termination, and field-proof testing of cable systems is provided. The aim is to avoid premature cable failure due to improper installation and mechanical damage during installation, and to provide a reference that can be specified for cable installations. This standard is not intended to be a design document; many of the problems of installation can be avoided by designing cable layouts with the installation limits of this recommended practice.

IEEE Std 577-1976 (R1992). *IEEE Standard Requirements for Reliability Analysis in the Design and Operation of Safety Systems for Nuclear Power Generating Stations.* Uniform minimum acceptable requirements for the performance of reliability analyses for safety-related systems found in nuclear power generating stations are provided. The requirements can be applied during design, fabrication, testing, maintenance, and repair of systems and components in nuclear power plants. The timing of the analysis depends upon the purpose for which it is performed.

IEEE Std 583-1982 (R1999). *IEEE Standard Modular Instrumentation and Digital Interface System (CAMAC).* This standard is intended to serve as a basis for a range of modular instrumentation capable of interfacing transducers and other devices to digital controllers for data and control. It consists of mechanical standards and signal standards that are sufficient to ensure physical and operational compatibility between units regardless of source. The standard fully specifies a data bus (Dataway) by means of which instruments and other functional modules can communicate with each other, with peripherals, with computers, and with other external controllers.

IEEE Std 592-1990 (R1996). *IEEE Standard for Exposed Semiconducting Shields on High-Voltage Cable Joints and Separable Insulated Connectors.* Design tests for shield resistance and a simulated fault-current initiation are provided for exposed semiconducting shields used on cable accessories, specifically joints and separable insulated connectors rated 15 kV through 35 kV. The shield is intended to protect the insulation, provide voltage stress relief, maintain the accessory surface at or near ground potential under normal operating conditions, and initiate fault-current arcing if the accessory insulation should fail. A maximum shield-resistance performance is specified to ensure that the accessory shield provides stress relief, and that the shield surface is maintained at or near ground potential. The shield fault-current initiation test demonstrates the ability of the accessory shield to initiate fault-current arcs to ground that will cause overcurrent protective devices to operate should the accessory insulation fail. In this test, special connections and procedures are specified to ensure that full-circuit voltage will be applied to the shield during the test. The test specifications do not, however, attempt to simulate all service conditions or field assembly.

IEEE Std 595-1982 (R1999). *IEEE Standard Serial Highway Interface System (CAMAC).* A serial highway (SH) system using byte-organized messages and configured as a unidirectional loop, to which are connected a system controller and up to sixty-two CAMAC crate assemblies, is defined. In the primary application, the controlled devices are CAMAC crate assemblies with serial crate controllers that conform to a defined message structure. In other applications, some or all of the controlled devices connected to the SH can be equipment that conforms to a subset of the full specification and is not necessarily constructed in CAMAC format or controlled by CAMAC commands.

IEEE Std 596-1982 (R1999). *IEEE Standard Parallel Highway Interface System (CAMAC).* The CAMAC parallel highway interface system for interconnecting up to seven CAMAC crates (or other devices) and a system controller is defined. In particular, the signals, timing, and logical organization of the connections from crate controllers and parallel highway drivers to the parallel highway through a standard connector are defined. The internal structures of crate controllers and parallel highway drivers, and the physical construction of the parallel highway system, are defined only as they affect compatibility between parts of the system.

IEEE Std 602-1996. *IEEE Recommended Practice for Electric Systems in Health Care Facilities (IEEE White Book).* A recommended practice for the design and operation of electric systems in health care facilities is provided. The term "health care facility," as used here, encompasses buildings or parts of buildings that contain hospitals, nursing homes, residential custodial care facilities, clinics, ambulatory health care centers, and medical and dental offices. Buildings or parts of buildings within an industrial or commercial complex, used as medical facilities, logically fall within the scope of this book.

IEEE Std 603-1998. *IEEE Standard Criteria for Safety Systems for Nuclear Power Generating Stations.* Minimum functional and design criteria for the power, instrumentation, and control portions of nuclear power generating station safety systems are established. The criteria are to be applied to those systems required to protect the public health and safety by functioning to mitigate the consequences of design basis events. The intent is to promote safe practices for design and evaluation of safety system performance and reliability. Although the standard is limited to safety systems, many of the principles may have applicability to equipment provided for safe shutdown, post-accident monitoring display instrumentation, preventative interlock features, or any other systems, structures, or equipment related to safety.

IEEE Std 605-1998. *IEEE Guide for Design of Substation Rigid-Bus Structures.* Rigid-bus structures for outdoor and indoor, air-insulated, and alternating-current substations are covered. Portions of this guide are also applicable to strain-bus structures or direct-current substations, or both. Ampacity, radio influence, vibration, and forces due to gravity, wind, fault current, and thermal expansion are considered. Design criteria for conductor and insulator strength calculations are included.

IEEE Std 610-1990 (R1992). *IEEE Standard Computer Dictionary—A Compilation of IEEE Standard Computer Glossaries.* This dictionary is a compilation of IEEE standard glossaries covering the fields of mathematics of computing, computer applications, modeling and simulation, image processing and pattern recognition, data management, and software engineering. Every effort has been made to include all terms within the designated subject areas. Terms were excluded if they were considered to be parochial to one group or organization; company-proprietary or trademarked; multiword terms whose meaning could be inferred from the definitions of the component words; or terms whose meaning in the computer field could be directly inferred from their standard English meaning.

IEEE Std 610.7-1995. *IEEE Standard Glossary of Computer Networking Terminology.* Terms that pertain to data communications and networking, from the following areas, are defined: Data transmission, general communications, general networks, local area networks, network communications security, network errors, networking hardware, network management, network nodes, network signaling, open system architecture, packet, protocols, standards, and standards organizations, telephony. The glossary is primarily a compilation of terms defined in individual IEEE standards, but also includes a number of common terms.

IEEE Std 610.12-1990. *IEEE Standard Glossary of Software Engineering Terminology.* Terms currently in use in the computer field are identified, and standard definitions are established for them. Topics covered include: addressing; assembling, compiling, linking, and loading; computer performance evaluation; configuration management; data types; errors, faults, and failures; evaluation techniques; instruction types; language types; libraries; microprogramming; operating systems; quality attributes; software documentation; software and system testing; software architecture; software development processes; software development techniques; and software tools. This glossary is intended to serve as a useful reference both for those in the computer field and for those who come into contact with computers either through their work or in their everyday lives.

IEEE Std 620-1996. *IEEE Guide for the Presentation of Thermal Limit Curves for Squirrel Cage Induction Machines.* Thermal limit curves for induction machines are defined. A procedure is established for the presentation of these curves, and guidance for the interpretation and use of these curves for machine thermal protection is provided.

IEEE Std 622-1987 (R1994). *IEEE Recommended Practice for the Design and Installation of Electric Heat Tracing Systems for Nuclear Power Generating Stations.* Recommended practices for designing, installing, and maintaining electric heat tracing systems are provided. These electric heat tracing systems are applied, both for critical process temperature control and for process temperature control, on mechanical piping systems that carry borated water, caustic soda, and other solutions. Electric heat tracing systems are also applied on water piping systems to prevent them from freezing in cold weather. The recommendations include identification of requirements, heater design considerations, power systems design considerations, temperature control considerations, alarm considerations, finished drawings and documents, installation of materials, start-up testing, temperature tests, and maintenance of electric pipe heating systems.

IEEE Std 622A-1984 (R1999). *IEEE Recommended Practice for the Design and Installation of Electric Pipe Heating Control and Alarm Systems for Power Generating Stations.* Recommended practices for designing and installing electric pipe heating control and alarm systems, as applied to mechanical piping systems that require heat, are provided. The recommendations include selection of control and alarm systems, accuracy considerations, local control usage, centralized control usage, qualification criteria of controls and alarms, and calibration and testing of controls and alarms. The intent is to ensure design consistency and reliable operation of electric pipe heating control and alarm systems, which in turn will ensure that piping system fluids will be available for use not only during station operation but also during normal shutdown.

IEEE Std 622B-1988 (R2000). *IEEE Recommended Practice for Testing and Start-up Procedures for Electric Heat Tracing Systems for Power Generating Stations.* Recommendations that may be used to ensure that an electric heat tracing system is installed correctly, is properly tested and commissioned, and is functioning correctly are provided. The recommendations cover the sequence for testing materials and components of the electric heat tracing system, installation, preoperational testing of the system, verification of system performance, and the necessary records to be filed. Although this standard is written for power generating stations, the techniques presented can be used on electric heat tracing systems in any application.

IEEE Std 625-1990. *IEEE Recommended Practice to Improve Electrical Maintenance and Safety in the Cement Industry.* Assists in the effective application of relays and other devices for the protection of shunt capacitors used in substations. It covers the protective considerations, along with recommended and alternate methods of protection for the most commonly used capacitor bank configurations. Capacitor bank design trade-offs are also discussed. This guide covers protection of filter tanks and very large EHV capacitor banks, but does not include a discussion of pole-mounted capacitor banks on distribution circuits or application of capacitors connected to rotating apparatus.

IEEE Std 627-1980 (R1996). *IEEE Standard for Design Qualification of Safety Systems Equipment Used in Nuclear Power Generating Stations.* Basic principles for design qualification of safety systems equipment used in nuclear power generating stations are provided. Specification criteria, the development of a qualification program, and documentation are addressed. All types of safety systems equipment—mechanical and instrumentation as well as electrical—are covered. Principles and procedures for preparing specific safety systems equipment standards are established.

IEEE Std 628-1987 (R1992). *IEEE Standard Criteria for the Design, Installation, and Qualification of Raceway Systems for Class 1E Circuits for Nuclear Power Generating Stations.* Criteria for the minimum requirements in the selection, design, installation, and qualification of raceway systems for Class 1E circuits for nuclear power generating stations are provided. Methods for the structural qualification of such raceway systems are prescribed. Since aging and radiation have no known detrimental effect upon metallic raceway systems, and since nonmetallic raceway systems are limited to underground or embedded applications, these two environmental conditions are not considered, nor are the embedments or structural members to which a support is attached.

IEEE Std 635-1989 (R1994). *IEEE Guide for Selection and Design of Aluminum Sheaths for Power Cables.* Requirements are outlined and design guidelines are established for the selection of aluminum sheaths for extra-high, high-, medium-, and low-voltage cables. Basic installation parameters for aluminum-sheathed cables are also established. In addition, references to industry standards and codes incorporating design and installation requirements of aluminum-sheathed cables and a comprehensive bibliography on the subject are provided.

IEEE Std 637-1985 (R1992). *IEEE Guide for the Reclamation of Insulating Oil and Criteria for Its Use.* Detailed procedures are provided for reclaiming used mineral insulating oils (transformer oils) by chemical and mechanical means to make them suitable for reuse as insulating fluids. Reclamation procedures are described, as are the test methods used to evaluate the progress and end point of the reclamation process, and the essential properties required for reuse in each class of equipment. Suitable criteria for the use of reclaimed oils are identified. The use of oil in new apparatus under warranty is not covered.

IEEE Std 638-1992 (R1999). *IEEE Standard for Qualification of Class 1E Transformers for Nuclear Power Generating Stations.* Procedures for demonstrating the adequacy of new Class 1E transformers, located in a mild environment of a nuclear power generating station, to perform their required safety functions under postulated service conditions are presented. Single- and three-phase transformers rated 601 V to 15 000 V for the highest voltage winding, and up to 2500 kVA (self-cooled rating), are covered. Because of the conservative approach used in the development of this standard for new transformers, the end-point criteria cannot be used for in-service transformers.

IEEE Std 643-1980 (R1992). *IEEE Guide for Power-Line Carrier Applications.* Application information is provided to users of carrier equipment as applied on power-transmission lines. Material on power line carrier channel characteristics is presented, along with discussions on intrabundle conductor systems and insulated shield wire systems. Procedures for the calculation of channel performance are given. Data for the calculations are drawn from various sections of the guide. Coupling components are discussed, covering line traps, coupling capacitors, line tuners, coaxial cables, hybrids, and filters. Frequency selection practices are discussed. Future trends are examined with respect to electronic equipment, system improvements, and applications.

IEEE Std 644-1994. *IEEE Standard Procedures for Measurement Frequency Electric and Magnetic Fields from AC Power Lines.* Uniform procedures for the measurement of power frequency electric and magnetic fields from alternating current (ac) overhead power lines and for the calibration of the meters used in these measurements are established. The procedures apply to the measurement of electric and magnetic fields close to ground level. The procedures can also be tentatively applied (with limitations, as specified in the standard) to electric fields near an energized conductor or structure.

IEEE Std 647-1995. *IEEE Standard Specification Format Guide and Test Procedure for Single-Axis Laser Gyros.* The specification and test requirements for a single-axis laser gyro for use as a sensor in attitude control systems, angular displacement measuring systems, and angular rate measuring systems is defined. A standard specification format guide for the preparation of a single-axis laser gyro is provided. A complication of recommended procedures for testing a laser gyro, derived from those presently used in the industry, is also provided.

IEEE Std 649-1991 (R1999). *IEEE Standard for Qualifying Class 1E Motor Control Centers for Nuclear Power Generating Stations.* The basic principles, requirements, and methods for qualifying Class 1E motor control centers for outside containment applications in nuclear power generating stations are described. In addition to defining specific qualification requirements that are in accordance with the more general qualification requirements of IEEE Std 323-1974, this standard is intended to provide guidance in establishing a quantification program for demonstrating the design adequacy of Class 1E motor control centers.

IEEE Std 650-1990 (R1998). *IEEE Standard for Qualification of Class 1E Static Battery Chargers and Inverters for Nuclear Power Generating Stations.* Methods for qualifying static battery chargers and inverters for Class 1E installations for mild-environment outside containment in nuclear power generating stations are described. The qualification methods set forth employ a combination of type testing and analysis, the latter including a justification of methods, theories, and assumptions used. These procedures meet the requirements of IEEE Std 323-1983 (R1990), IEEE Standard for Qualifying Class 1E Equipment for Nuclear Power Generating Stations.

IEEE Std 656-1992 (R2000). *IEEE Standard for the Measurement of Audible Noise From Overhead Transmission Lines.* Uniform procedures are established for manual and automatic measurement of audible noise from overhead transmission lines. Their purpose is to allow valid evaluation and comparisons of the audible noise performance of various overhead lines. Definitions are provided, and instruments are specified. Measurement procedures are set forth, and precautions are given. Supporting data that should accompany the measurement data are specified, and methods for presenting the latter are described.

IEEE Std 662-1992 (R1998). *IEEE Standard Terminology for Semiconductor Memory.* Guidelines under which data sheets for new semiconductor memories are to be generated are provided. Adherence to these guidelines is intended to produce data sheets that are concise and that consistently define the operation and characteristics of semiconductor memory devices. Terminology relevant to product description, product specification, and user information is covered.

IEEE Std 664-1993 (R2000). *IEEE Guide for Laboratory Measurement of the Power Dissipation Characteristics of Aeolian Vibration Dampers for Single Conductors.* The current methodologies, including apparatus, procedures, and measurement accuracies, for determining the dynamic characteristics of vibration dampers and damping systems are described. Some basic guidance is provided regarding a given method's strengths and weaknesses. The methodologies and procedures described are applicable to indoor testing only.

IEEE Std 665-1995. *IEEE Guide for Generating Station Grounding.* Grounding practices that have generally been accepted by the electric utility industry as contributing to effective grounding systems for personnel safety and equipment protection in generating stations are identified. A guide for the design of generating station grounding systems and for grounding practices applied to generating station indoor and outdoor structures and equipment, including the interconnection of the station and substation grounding systems, is provided.

IEEE Std 666-1991 (R1996). *IEEE Design Guide for Electric Power Service Systems for Generating Stations.* A listing of typical power plant auxiliaries and criteria for their power service are given, as well as examples of one-line diagrams for a typical plant. Tables of typical power service parameters are included to illustrate the range of typical values for each parameter, and the approximate effect of the minimum and maximum value of each parameter on the load is identified. This guide applies to all types of power generating stations, but it is particularly applicable where the electric power service system is required to perform continuously.

IEEE Std 671-1985 (R1997). *IEEE Standard Specification Format Guide and Test Procedure for Nongyroscopic Inertial Angular Sensors: Jerk, Acceleration, Velocity, and Displacement.* A guide is presented for the preparation of a specification and test procedure for an inertial angular sensor that provides a common meeting ground of terminology and practice for manufacturers and users of an array of sensors that have been developed to meet needs not easily met by traditional spinning rotor gyroscopes. A test procedure for verifying that the specifications have been met is given. The standard is not intended to compete with existing standards for specific devices with highly specific models and error sources, such as spring restrained rate gyros, but to provide a uniform guide for those inertial angular sensors that have not been covered elsewhere.

IEEE Std 675-1982 (R1999). *IEEE Standard Multiple Controllers in a CAMAC Crate.* A method for incorporating more than one source of control into a CAMAC crate is defined. The aim is to provide for the use of auxiliary controllers in order to extend the capabilities and fields of application of the CAMAC modular instrumentation and interface system of IEEE Std 583-1982 (R1994).

IEEE Std 683-1976 (R1999). *IEEE Recommended Practice for Block Transfers in CAMAC Systems.* The recommended block-transfer algorithms are discussed, and those given in the basic CAMAC specification are described. These algorithms are well established and are supported by existing hardware. Some new algorithms are then discussed. Compatibility, hardware design, and software considerations are addressed.

IEEE Std 686-1997. *IEEE Standard Radar Definitions.* Definitions for the purpose of promoting clarity and consistency in the use of radar terminology are provided. The definitions represent the consensus of a panel of radar experts.

IEEE Std 692-1997. *IEEE Standard Criteria for Security Systems for Nuclear Power Generating Stations.* Criteria are provided for the design of an integrated security system for nuclear power generating stations. Requirements are included for the overall system, interfaces, subsystems, and individual electrical and electronic equipment. This standard addresses equipment for security-related detection, surveillance, access control, communication, and data acquisition.

IEEE Std 693-1997. *IEEE Recommended Practice for Seismic Design of Substations.* Recommendations for seismic design of substations, including qualification of each equipment type, are discussed. Design recommendations consist of seismic criteria, qualification methods and levels, structural capacities, performance requirements for equipment operation, installation methods, and documentation.

IEEE Std 694-1985 (R1994). *IEEE Standard for Microprocessor Assembly Language.* A common set of instructions used by most general-purpose microprocessors is presented. Rules for the naming of new instructions and the derivation of new mnemonics are provided. Assembly language conventions are established. This standard does not prescribe programming style, specify or restrict the number of instructions or directives, prescribe or restrict the type of instructions or directives, specify or restrict machine architectures, or specify source or object file formats.

IEEE Std 716-1995. *IEEE Standard C/ATLAS Test Language for All Systems—Common/Abbreviated Test Language for All Systems (C/ATLAS).* A high order language for testing is defined. This language is designed to describe tests in terms that are independent of any specific test system, and has been constrained to ensure that it can be implemented on automatic test equipment.

IEEE Std 726-1982 (R1999). *IEEE Standard, Real-Time BASIC for CAMAC.* This standard defines ANSI Standard Real-Time BASIC, in which the declarations and real-time statements are defined for use with CAMAC hardware. It covers real-time capabilities, declarations, parallel activities, CAMAC input and output, the CAMAC Q and X signals, CAMAC LAM handling, message passing, shared data, and bit manipulation. The aim is to achieve maximum compatibility between different implementations of ANSI BASIC for use with CAMAC.

IEEE Std 730-1998. *IEEE Standard for Software Quality Assurance Plans.* Uniform, minimum acceptable requirements for preparation and content of Software Quality Assurance Plans (SQAPs) are provided. This standard applies to the development and maintenance of critical software. For noncritical software, or for software already developed, a subset of the requirements of this standard may be applied.

IEEE Std 730.1-1995 (Redesignation of IEEE Std 938). *IEEE Guide for Software Quality Assurance Planning.* Approaches to good Software Quality Assurance practices in support of IEEE Std 730-1989, IEEE Standard for Software Quality Assurance Plans, are identified. These practices are directed toward the development and maintenance of critical software, that is, where failure could impair safety or cause large financial losses.

IEEE Std 738-1993. *IEEE Standard for Calculating the Current-Temperature Relationship of Bare Overhead Conductors.* A simplified method of calculating the current–temperature relationship of bare overhead lines, given the weather conditions, is presented. Along with a mathematical method, sources of the values to be used in the calculation are indicated. This standard does not undertake to list actual temperature–ampacity relationships for a large number of conductors in a large number of conditions.

IEEE Std 739-1995 (R2000). *IEEE Recommended Practice for Energy Management in Industrial and Commercial Facilities (IEEE Bronze Book).* This recommended practice serves as an engineering guide for use in electrical design for energy conservation. It provides a standard design practice to assist engineers in evaluating electrical options from an energy standpoint. It establishes engineering techniques and procedures to allow efficiency optimization in the design and operation of an electrical system considering all aspects (safety, costs, environment, those occupying the facility, management needs, etc.).

IEEE Std 741-1997. *IEEE Standard Criteria for the Protection of Class 1E Power Systems and Equipment in Nuclear Power Generating Stations.* Criteria that establish protection requirements for Class 1E power systems and equipment are prescribed. The purpose of and the means for obtaining protection from electrical and mechanical damage or failures that can occur within a time period that is shorter than that required for operator action are described. Testing and surveillance requirements are included. Plant physical design requirements to protect against certain events are not included.

IEEE Std 743-1995. *IEEE Standard Equipment Requirements and Measurement Techniques for Analog Transmission Parameters for Telecommunications.* Performance requirements for test equipment that measures the analog transmission parameters of subscriber loops, message trunks, PBX trunks, and ties lines are specified. Requirements for these measurements with DS1 bit stream access are also provided. The measurement of loss, noise, and impulse noise on non-loaded cable pairs used for digital subscriber lines is addressed.

IEEE Std 751-1990 (R1992). *IEEE Trial-Use Design Guide for Wood Transmission Structures.* This standard discusses the structural design and application of wood transmission structures. The guide includes definitions, application of loads, structure application, characteristics of natural wood and laminated wood members, design stresses, fabrication of laminated wood members, connections, nonwood members, erection and framing, and quality assurance.

IEEE Std 754-1985 (R1990). *IEEE Standard for Binary Floating-Point Arithmetic.* A family of commercially feasible ways for new systems to perform binary floating-point arithmetic is defined. This standard specifies basic and extended floating-point number formats; add, subtract, multiply, divide, square root, remainder, and compare operations; conversions between integer and floating-point formats; conversions between different floating-point formats; conversions between basic-format floating-point numbers and decimal strings; and floating-point exceptions and their handling, including nonnumbers.

IEEE Std 758-1979 (R1999). *IEEE Standard, Subroutines for CAMAC.* A set of standard subroutines that provide access to CAMAC facilities in a variety of computer programming languages is described. The subroutines are specifically intended to be suitable for use with FORTRAN, although they are not restricted to that language. The subroutines have been grouped into three subsets in order to provide different standard levels of implementation. The lowest level requires only two subroutines, but, nevertheless, gives access to most of the facilities that can be found in CAMAC. In higher levels of implementation, subroutines are added that permit procedures to be written in more mnemonic terminology, provide better handling of LAMs, permit procedures to be independent of the type of CAMAC highway used, and provide efficient block-transfer capability.

IEEE Std 759-1984 (R1999). *IEEE Test Procedures for Semiconductor X-Ray Energy Spectrometers.* Test procedures for X-ray spectrometers consisting of a semiconductor radiation detector assembly and signal processing electronics interfaced to a pulse-height analyzer/computer are presented. Energy resolution, spectral distortion, pulse-height linearity, counting rate effects, overload effects, pulse-height stability, and efficiency are covered. Test procedures for pulse-height analyzers and computers are not covered.

IEEE Std 765-1995. *IEEE Standard for Preferred Power Supply (PPS) for Nuclear Power Generating Stations.* The design criteria of the preferred power supply (PPS) and its interfaces with the Class 1E power system, switchyard, transmission system, and alternate ac (AAC) source are described. This standard provides PPS requirements for nuclear power plants and guidance in the areas of AAC power source interfaces with PPS, physical independence of the PPS power and control circuits, and expanded PPS criteria for multiunit stations.

IEEE Std 771-1998. *IEEE Guide to the Use of the ATLAS Specification.* Guidance in the use of ATLAS test languages is provided. ATLAS may be used to describe test requirements independent of any specific test equipment, and examples of best practice.

IEEE Std 776-1992 (R1998). *IEEE Recommended Practice for Inductive Coordination of Electric Supply and Communication Lines.* The inductive environment that exists in the vicinity of electric power and wire-line telecommunications systems and the interfering effects that may be produced are addressed. An interface that permits either party, without need to involve the other, to verify the induction at the interface by use of a probe wire is presented. This recommended practice does not apply to railway signal circuits.

IEEE Std 790-1989 (R1996). *IEEE Guide for Medical Ultrasound Field Parameter Measurements.* Information is provided to assist in selecting measurement procedures and implementing 'cookbook' descriptions for building and using devices that measure medical ultrasound field parameters such as pressure, power, and intensity. It is intended for use by persons involved in measurement of acoustic fields produced by medical ultrasound instruments and is divided into three parts. Hydrophones are discussed, with regard to types, calibration and evaluation techniques, and measurement techniques using hydrophones. Fifteen radiation force techniques that are commercially available for purchase or are currently in routine use in established research laboratories are described. Three thermal techniques that utilize acoustic absorption and the measurement of temperature and an acousto-optical approach are presented.

IEEE Std 792-1995. *IEEE Recommended Practice for the Evaluation of the Impulse Voltage Capability of Insulation Systems for AC Electric Machinery Employing Form-Wound Stator Coils.* A test procedure for the evaluation of the impulse voltage capability of insulation systems of form-wound ac rotating electrical machinery is outlined. The procedure is primarily directed toward providing a qualification test for the turn insulation in regard to its ability to withstand impulses that might be impressed on the terminals of a machine and that result from switching surges, lightning, or other disturbances. The standard also presents information on the ability of the ground insulation to withstand impulses. The procedure provides a basis for the accumulation, analysis, and reporting of information concerning impulse-voltage withstand strength of ground and turn insulation, both new and aged. The use of multifactor aging tests, combining thermal and electrical aging in order to address the withstand capability of micaceous insulation, is recommended.

IEEE Std 802-1990 (R1992). *IEEE Standards for Local and Metropolitan Area Networks—Overview and Architecture.* This document serves as the foundation for the family of IEEE 802® standards for local area networks (LANs) and metropolitan area networks (MANs) that deals with the physical and data link layers as defined by the International Organization for Standardization (ISO) Open Systems Interconnection Basic Reference Model. Descriptions of the networks considered as well as a reference model for protocol standards are provided. Compliance with the family of IEEE 802® standards is defined, and a standard for the identification of public, private, and standard protocols is included. Universal addresses and protocol identifiers are considered.

IEEE Std 802.1F-1993 (R1998). *IEEE Standard* for *Local and Metropolitan Area Networks—Common Definitions and Procedures for IEEE 802® Management Information.* Management information and procedures applicable across the entire family of IEEE 802® LAN/MAN standards within the architectural framework for LAN/MAN Management specified in IEEE Std 802-1990 are identified. Common management information, such as attributes to represent MAC address and managed objects to represent configurable gauges, are specified. The need of developers of LAN/MAN management specifications for common procedures to develop, describe, and register management information is addressed.

IEEE Std 802.1Q-1998. *IEEE Standard for Virtual Bridged Local Area Networks.* This standard defines an architecture for Virtual Bridged LANs, the services provided in Virtual Bridged LANs, and the protocols and algorithms involved in the provision of those services.

IEEE Std 802.3, 1998 Edition. (Incorporating ANSI/IEEE Std 802.3, 1996 Edition, IEEE Std 802.3r-1996, IEEE Std 802.3u-1995, IEEE Std 802.3x&y-1997, IEEE Std 802.3z-1998, and IEEE Std 802.3aa-1998). *IEEE Standard for Information technology—Telecommunications and information exchange between systems—Local and metropolitan areas networks—Specific requirements—Part 3: Carrier sense multiple access with collision detection (CSMA/CD) access method and physical layer specifications.* The media access control characteristics for the Carrier Sense Multiple Access with Collision Detection (CSMA/CD) access method for shared medium local area networks are described. The control characteristics for full duplex dedicated channel use are also described. Specifications are provided for MAU types 1BASE5 at 1 Mb/s; Attachment Unit Interface (AUI) and MAU types 10BASE5, 10BASE2, FOIRL (fiber optic inter-repeater link), 10BROAD36, 10BASE-T, 10BASE-FL, 10BASE-FB, and 10BASE-FP at 10 Mb/s; Media Independent Interface (MII) and PHY types 100BASE-T4, 100BASE-TX, 100BASE-FX, and 100BASE-T2 at 100 Mb/s; and the Gigabit MII (GMII) and 1000BASE-X PHY types, 1000BASE-SX, 1000BASE-LX, and 1000BASE-CX, which operate at 1000 Mb/s (Gigabit Ethernet). Repeater specifications are provided at each speed. Full duplex specifications are provided at the Physical Layer for 10BASE-T, 10BASE-FL, 100BASE-TX, 100BASE-FX, 100BASE-T2, and Gigabit Ethernet. System considerations for multisegment networks at each speed and management information base (MIB) specifications are also provided.

IEEE Std 802.3ab-1999. (Supplement to IEEE Std 802.3, 1998 Edition). *IEEE Standard for Information technology—Telecommunications and information exchange between systems—Local and metropolitan area networks—Specific requirements—Supplement to Carrier Sense Mulitple Access with Collision Detection (CSMA/CD) Access Method and Physical Layer Specifications—Physical Layer Parameters and Specifications for 1000 Mb/s Operation Over 4-Pair of Category 5 Balanced Copper Cabling, Type 1000BASE-T.* Type 1000BASE-T PCS, type 1000BASE-T PMA sublayer, and type 1000BASE-T Medium Dependent Interface (MDI) are defined. This supplement provides fully functional, electrical and mechanical specifications for the type 1000BASE-T PCS, PMA, and MDI. This supplement also specifies the baseband medium used with 1000BASE-T.

IEEE Std 802.3ac-1998. (Supplement to IEEE Std 802.3, 1998 Edition). *IEEE Standard for Information technology—Telecommunications and information exchange between systems—Local and metropolitan area networks—Specific requirements—Supplement to Carrier Sense Multiple Access with Collision Detection (CSMA/CD) Access Method and Physical Layer Specifications—Frame Extensions for Virtual Bridged Local Area Network (VLAN) Tagging on 802.3 Networks.* Changes and additions to IEEE Std 802.3, 1998 Edition, to support Virtual Bridged Local Area Networks (VLANs) as specified in IEEE P802.1Q, Draft Standard for Local and Metropolitan Area Networks: Virtual Bridged Local Area Networks, are provided.

IEEE Std 802.3ad-2000. (Amendment to IEEE Std 802.3, 1998 Edition). *IEEE Standard for Information technology—Telecommunications and information exchange between systems—Local and metropolitan area networks—Specific requirements—Amendment to Carrier Sense Multiple Access with Collision Detection (CSMA/CD) Access Method and Physical Layer Specifications—Aggregation of Multiple Link Segments.* An optional Link Aggregation sublayer for use with CSMA/CD MACs is defined. Link Aggregation allows one or more links to be aggregated together to form a Link Aggregation Group, such that a MAC Client can treat the Link Aggregation Group as if it were a single link. To this end, it specifies the establishment of DTE to DTE logical links, consisting of N parallel instances of full duplex point-to-point links operating at the same data rate.

IEEE Std 802.4h-1997. (Supplement to ISO/IEC 8802-4: 1990 [ANSI/IEEE Std 802.4-1990]). *IEEE Standards for*

Local and Metropolitan Area Networks: Supplement to Token-Passing Bus Access Method and Physical Layer Specifications Alternative Use of BNC Connectors and Manchester-Encoded Signaling Methods for Single-Channel Bus Physical Layer Entities. This supplement to ISO/IEC 8802-4:1990 [ANSI/IEEE Std 802.4-1990] provides the functional, electrical, and mechanical characteristics of single-channel differential and Manchester-data-encoded bus Physical Layer Entities (PLEs).

IEEE Std 802.5c-1991 (R1997). *IEEE Standards for Local and Metropolitan Area Networks: Supplement to Token Ring Access Method and Physical Layer Specifications: Recommended Practice for Dual Ring Operation with Wrapback Configuration.* Extensions to the IEEE 802.5 Token-Passing Ring standard are defined. These extensions implement a Dual Ring local area network (LAN) topology that provides full interoperability between stations conforming to IEEE Std 802.5, including coexistence on the same ring, and recovery from all single media failures with full capability restored. The Dual Ring Topology and operation described are intended for applications that require very high availability and recovery from media and station failures.

IEEE Std 802.5t-2000. (Amendment to ANSI/IEEE Std 802.5, 1998 Edition; ANSI/IEEE Std 802.5r, 1998 Edition; and ANSI/IEEE Std 802.5j, 1998 Edition). *IEEE Standard for Information technology—Telecommunications and information exchange between systems—Local and metropolitan area networks—Specific requirements: Amendment to Part 5: Token Ring Access Method and Physical Layer Specifications.* This supplement specifies the changes required to ANSI/IEEE Std 802.5, 1998 Edition, (Base standard) and ANSI/IEEE 802.5r, 1998 Edition, and ANSI/IEEE Std 802.5j, 1998 Edition, (Amendment 1 standard) to support 100 Mbit/s Dedicated Token Ring (DTR) operation. The Base standard, together with the Amendment 1 standard, specifies shared and dedicated (point-to-point) Token Ring operation at both 4 Mbit/s and 16 Mbit/s using either the TKP Access Protocol or the TXI Access Protocol. This supplement extends Token Ring operation to 100 Mbit/s for the DTR C-Port and Station using the TXI Access Protocol. Extensions to the medium access control (MAC) have been made to accommodate the requirements for high media rates (100 Mbit/s and above).

IEEE Std 802.6-1994 (R1997). *IEEE Standard for Information Technology Telecommunications and information exchange between systems—Local And Metropolitan Area Networks Specific Requirements—Part 6: Distributed Queue Dual Bus (DQDB) Access Method And Physical Layer Specifications.* This standard is part of a family of standards for local area networks (LANs) and metropolitan area networks (MANs) that deals with the Physical and Data Link Layers as defined by the ISO Open Systems Interconnection Reference Model. It defines a high-speed shared medium access protocol for use over a dual, counterflowing, unidirectional bus subnetwork. The Physical Layer and Distributed Queue Dual Bus (DQDB) Layer are required to support a Logical Link Control (LLC) Sublayer by means of a connectionless Medium Access Control (MAC) Sublayer service in a manner consistent with other IEEE 802® networks. Additional DQDB Layer functions are specified as a framework for other services. These additional functions will support Isochronous Service Users and Connection-Oriented Data Service Users, but their implementation is not required for conformance.

IEEE Std 802.6j-1995 (R1997). *IEEE Standard for Local and Metropolitan Area Networks: Supplement to 802.6: Connection-Oriented Service on a Distributed Queue Dual Bus (DQDB) Subnetwork of a Metropolitan Area Network (MAN).* Enhanced Queued Arbitrated (QA) Functions, which can support applications requiring bandwidth guarantees and delay limits on a DQDB subnetwork, are specified. Connection-Oriented Convergence Functions (COCFs) using the enhanced QA Functions, which are necessary to support connection-oriented service, are also specified

IEEE Std 802.7-1989 (R1997). *IEEE Recommended Practices for Broadband Local Area Networks.* The physical, electrical, and mechanical characteristics of a properly designed IEEE 802.7 broadband cable medium are specified. The medium supports the communication of IEEE 802.3b, IEEE 802.4, video, and narrow-band radio frequency (RF) modem devices. The broadband bus topology consists of amplifiers, coaxial cable, and directional couplers that create a full duplex directional medium. The characteristics described are intended as the minimum acceptable parameters for the design, installation, and test of an IEEE 802.7 cable plant. Single and dual cable systems are specified for the support of existing ISO 8802-3 and IEEE 802.4 broadband devices.

IEEE Std 802.10-1998. *IEEE Standards for Local and Metropolitan Area Networks: Standard for Interoperable LAN/MAN Security (SILS).* IEEE Std 802.10 provides specifications for an interoperable data link layer security protocol and associated security services. The Secure Data Exchange (SDE) protocol is supported by an application layer Key Management Protocol (KMP) that establishes security associations for SDE and other security protocols. A security label option is specified that enables rule-based access control to be implemented using the SDE protocol. A method to allow interoperability with type-encoded Medium Access Control (MAC) clients is also provided, as well as a set of managed object classes to be used in the management of the SDE sublayer and its protocol exchanges.

IEEE Std 802.10a-1999. (Supplement to IEEE Std 802.10-1998). *IEEE Standards for Local and Metropolitan Area Networks: Supplement to Standard for Interoperable LAN/MAN Security (SILS)—Security Architecture Framework.* An architectural description of the functions and location of SILS components is provided. The SILS components and their relationships to applications, communications protocols, system management, and security management are described.

IEEE Std 802.10c-1998. *IEEE Standards for Local and Metropolitan Area Networks: Supplement to Standard for Interoperable LAN/MAN Security (SILS)—Key Management (Clause 3).* A cryptographic key management model and a key management OSI Basic Reference Model Application Layer protocol are specified.

IEEE Std 802.10h-1997. *IEEE Standards for Local and Metropolitan Area Networks: Supplement to Interoperable LAN/MAN Security (SILS)—Secure Data Exchange (SDE): Protocol Implementation Conformance Statement (PICS) Proforma (Annex 2L).* The secure data exchange (SDE) protocol implementation conformance statement (PICS) proforma is provided. The SDE PICS proforma defines the information to be supplied by protocol implementors claiming conformance with IEEE Std 802.10, Clause 2, Secure Data Exchange (SDE).

IEEE Std 802.11a-1999. (Supplement to IEEE Std 802.11-1999). *IEEE Supplement to IEEE Standard for Information technology—Telecommunications and information exchange between systems—Local and metropolitan area networks—Specific Requirements—Part 11: Wireless LAN Medium Access Control (MAC) and Physical Layer (PHY) Specifications: High-speed Physical Layer in the 5 GHZ Band.* Changes and additions to IEEE Std. 802.11-1999 are provided to support the new high-rate physical layer (PHY) for operation in the 5 GHz band.

IEEE Std 802.11b-1999. (Supplement to ANSI/IEEE Std 802.11, 1999 Edition). *Supplement to IEEE Standard for Information technology—Telecommunications and information exchange between systems—Local and metropolitan area networks—Specific requirements—Part 11: Wireless LAN Medium Access Control (MAC) and Physical Layer (PHY) Specifications: Higher-Speed Physical Layer Extension in the 2.4 GHz Band.* Changes and additions to IEEE Std

802.11, 1999 Edition are provided to support the higher rate physical layer (PHY) for operation in the 2.4 GHz band.

IEEE Std 802.12-1995. *IEEE Standards for Local and Metropolitan Area Networks: Demand Priority Access Method, Physical Layer and Repeater Specification for 100 Mb/s Operation.* The media access control characteristics for the Demand Priority access method are specified. The layer management, physical layers, and media that support this access method are also specified. Layer and sublayer interface specifications are aligned to the ISO Open Systems Interconnection Basic Reference Model and ISO/IEC 8802 models. Specifications for 100 Mb/s operation over 100 Ω balanced cable (twisted-pair) category 3 through 5, 150 Ω shielded balanced cable, and fibre optic media are included.

IEEE Std 802.12c-1998. *Supplement to Information technology—Local and metropolitan area networks—Specific requirements—Part 12: Demand-priority access method, physical layer and repeater specifications: Full-Duplex Operation.* Optional MAC capabilities are defined to allow direct link connection between two end nodes with provision for both half-duplex and full-duplex operation; burst-mode packet transmission from an end node to a repeater where the end node may send one or more packets each time it is granted permission to transmit; and implementation of the MAC Control sublayer to allow the exchange of control requests between peer MAC entities across the network when in 8802-3 compatibility mode. Full interoperability is maintained with existing ISO/IEC 8802-12 products.

IEEE Std 803-1983 (R1999). *IEEE Recommended Practice for Unique Identification in Power Plants and Related Facilities—Principles and Definitions.* This recommended practice provides unique identification principles and definitions that, when used with related recommended practices concerning component function identifiers, implementation instructions, and system descriptions, provides a basis for uniquely identifying systems, structures, and components of nuclear and fossil fueled power plant projects (electric power generating stations) and related facilities. Hydro and other types of power plant projects are not included. The standard is part of a series of recommended practices, entitled the Energy Industry Identification System (EIIS), the purpose of which is to present a common language that will permit a user to correlate a system, structure, or component with that of another organization for the purposes of reporting, comparison, or general communication. A significant feature of the concept is that the unique identification code identifies the function at the component level and not the hardware itself.

IEEE Std 803.1-1992 (R2000). *IEEE Recommended Practice for Unique Identification in Power Plants and Related Facilities—Component Function Identifiers.* This recommended practice provides component function identifiers that, when used with related recommended practices concerning unique identification principles and definitions, implementation instructions, and system descriptions, provide a basis for uniquely identifying systems, structures, and components of nuclear and fossil-fueled power plant projects (electric power generating stations) and related facilities. Hydro and other types of power plant projects are not included. The standard is part of a series of recommended practices, entitled the Energy Industry Identification System (EIIS), the purpose of which is to present a common language of communication that will permit a user to correlate a system, structure, or component with that of another organization for the purpose of reporting, comparison, or general communication. A significant feature of this concept is that the unique identification code identifies the function at the component level and not the hardware itself.

IEEE Std 805-1984 (R2000). *IEEE Recommended Practice for System Identification in Nuclear Power Plants and Related Facilities.* This recommended practice provides a single source of nuclear power plant system descriptions which, along with related recommended practices concerning unique identification principles and definitions, component function identifiers, and implementation instructions, provides a basis for uniquely identifying systems, structures, and components of light water nuclear power plant projects (electric power generating stations) and related facilities. The system descriptions concentrate on system function and include such internal details as is necessary to clearly support the system function description. They are not intended to serve as design input. Fossil, hydro, and other types of power plants are not included. This standard is part of a series of recommended practices, entitled the Energy Industry Identification Systems (EIIS), the purpose of which is to present a common language of communication that will permit a user to correlate a system, structure, or component with that of another organization for the purposes of reporting, comparison, or general communication. A significant feature of this concept is that the unique identification code identifies the function at the component level and not the hardware itself.

IEEE Std 810-1987. (R1994) *IEEE Standard for Hydraulic Turbine and Generator Integrally Forged Shaft Couplings and Shaft Runlet Tolerances.* The dimensions of integrally forged shaft couplings and the shaft runlet tolerances are specified. The shafts and couplings covered are used for both horizontal and vertical connections between generators and turbines in hydroelectric installations. Data on fabricated shafts, shaft stresses, and bolt tensioning are not given.

IEEE Std 813-1988 (R2000). *IEEE Specification Format Guide and Test Procedure for Two-Degree-of-Freedom Dynamically Tuned Gyros.* A format guide for the preparation of a two-degree-of-freedom dynamically tuned gyro (DTG) specification is given that provides a common ground of terminology and practice for manufacturers and users. A compilation of recommended procedures for testing a DTG is also given. The requirements and test procedures are defined in terms unique to the DTG. They cover applications of the gyro as an angular motion sensor in navigation and control systems. They apply to two modes of use: (1) as a strapdown sensor in operating environments typical of aircraft and missile applications, and (2) as a sensor in gimballed platform applications in which the dynamic angular inputs to which the gyro is subjected are benign relative to the accuracy required. In the case of the strapdown DTG, the characteristics of the external capture loops are considered to the extent necessary to define the gyro performance.

IEEE Std 824-1994. *IEEE Standard for Series Capacitors in Power Systems.* Capacitors and assemblies of capacitors, insulation means, switching, protective equipment, and control accessories that form a complete bank for inserting in series with a transmission line are applied. Included are requirements for safety, rating, and protective device levels. Functional requirements for alarm devices, maintenance, design and production tests, and a guide for operation are included. Functional requirements for protective devices are addressed, including varistors and bypass gaps.

IEEE Std 828-1998. *IEEE Standard for Software Configuration Management Plans.* The minimum required contents of a Software Configuration Management Plan (SCMP) are established, and the specific activities to be addressed and their requirements for any portion of a software product's life cycle are defined.

IEEE Std 829-1998. *IEEE Standard for Software Test Documentation.* A set of basic software test documents is described. This standard specifies the form and content of individual test documents. It does not specify the required set of test documents.

IEEE Std 830-1998. *IEEE Recommended Practice for Software Requirements Specifications.* The content and qualities of a good software requirements specification (SRS) are described and several sample SRS outlines are presented. This recommended practice is aimed at specifying requirements of software to be developed but also can be applied to assist in the selection of in-house and commercial software products. Guidelines for compliance with IEEE/EIA 12207.1-1997 are also provided.

IEEE Std 833-1988 (1994). *IEEE Recommended Practice for the Protection of Electric Equipment in Nuclear Power Generating Stations from Water Hazards.* This document recommends methods and design features that, if implemented, would provide water-hazard protection to class-1E and non-class-1E systems and equipment from direct sources of water (for example, water spray from decontamination activities) and indirect sources of water (for example, water running along cables and raceways). It does not classify water-hazard protection features as nuclear-safety-related or non-nuclear-safety-related. Protection of equipment by choice of location, equipment design, and sealing are shielding are considered. The following are covered: design and construction features for electric equipment rooms; protection of electric equipment located in open areas subject to water hazards; electric equipment enclosures; electric equipment installation practices; sealing methods; and maintenance, surveillance, and testing activities.

IEEE 835-1994 (R2000). *IEEE Standard Power Cable Ampacity Tables.* Over 3000 ampacity tables for extruded dielectric power cables rated through 138 kV and laminar dielectric power cables rated through 500 kV are provided.

IEEE Std 836-1991 (R1997). *IEEE Recommended Practice for Precision Centrifuge Testing of Linear Accelerometers.* A guide to the conduct and analysis of precision centrifuge tests of linear accelerometers is provided, covering each phase of the tests beginning with the planning. Possible error sources and typical methods of data analysis are addressed. The intent is to provide those involved in centrifuge testing with a detailed understanding of the various factors affecting accuracy of measurement, both those factors associated with the centrifuge and those in the data collection process. Model equations are discussed, both for the centrifuge and for a typical linear accelerometer, with each equation having the complexity needed to accommodate the various identified characteristics and error sources in both. A new iterative matrix equation solution for deriving from the centrifuge test data the various model equation coefficients for the accelerometer under test is presented.

IEEE Std 837-1989 (R1996). *IEEE Standard for Qualifying Permanent Connections Used in Substation Grounding.* Directions and methods for qualifying permanent connections used for substation grounding are provided. Particular attention is given to the connectors used within the grid system, connectors used to join ground leads to the grid system, and connectors used to join the ground leads to equipment and structures. The purpose is to give assurance to the user that connectors meeting the requirements of this standard will perform in a satisfactory manner over the lifetime of the installation provided, that the proper connectors are selected for the application, and that they are installed correctly. Parameters for testing grounding connections on aluminum, copper, steel, copper-clad steel, galvanized steel, stainless steel, and stainless-clad steel are addressed. Performance criteria are established, test procedures are provided, and mechanical, current-temperature cycling, freeze-thaw, corrosion, and fault-current tests are specified.

IEEE Std 841-1994 (R1996). *IEEE Standard for Petroleum and Chemical Industry—Severe Duty Totally Enclosed Fan-Cooled (TEFC) Squirrel Cage Induction Motors—Up to and Including 500 hp.* This standard applies to high-efficiency TEFC, horizontal and vertical, single-speed, squirrel cage polyphase induction motors, up to and including 500 hp, in NEMA frame sizes 143T and larger, for petroleum, chemical, and other severe duty applications (commonly referred to as severe duty motors). Excluded from the scope of this standard are motors with sleeve bearings and additional specific features required for explosion-proof motors.

IEEE Std 844-1991 (R1996). *IEEE Recommended Practice for Electrical Impedance, Induction, and Skin Effect Heating of Pipelines and Vessels.* Recommended practices are provided for the design, installation, testing, operation and maintenance impedance, induction, and skin-effect heating systems. Thermal insulation and control and monitoring are addressed. General considerations for heating systems are discussed, covering selection criteria, design guidelines and considerations, power systems, receiving and storage, installation, testing, operations, and maintenance. These aspects are then discussed for each of the above types of systems, along with special considerations particular to each. These recommended practices are intended to apply to the use of these heating systems in general industry.

IEEE Std 845-1999. *IEEE Guide for the Evaluation of Human-System Performance in Nuclear Power Generating Stations.* Guidance for evaluating human-system performance related to systems, equipment, and facilities in nuclear power generating stations is provided. Specific evaluation techniques and rationale for their application within the integrated systems approach to plant design, operations, and maintenance described in IEEE Std 1023-1988 are summarized.

IEEE Std 848-1996. *IEEE Standard Procedure for the Determination of the Ampacity Derating of Fire-Protected Cables.* A detailed test procedure is provided for determining the ampacity or derating factor in the following cable installation configurations: block-out or sleeve type cable penetration fire stops; conduits covered with a protective material; tray covered with a protective material; cable directly covered or coated with a fire-retardant material; and free-air drops enclosed with a protective material.

IEEE Std 854-1987 (R1994). *IEEE Standard for Radix-Independent Floating-Point Arithmetic.* A family of commercially feasible ways for new systems to perform floating-point arithmetic is defined. This standard specifies constrains on parameters defining values of basic and extended floating-point numbers; add, subtract, multiply, divide, square root, remainder and compare operations; conversions between integers and floating-point numbers; conversions between different floating-point precisions; conversion between basic precision floating-point numbers and decimal strings; and floating-point exceptions and their handling, including non-numbers. It is intended that an implementation of a floating-point system conforming to this standard can be realized entirely in software, entirely in hardware, or in any combination of software and hardware. Retrofitting issues are not considered.

IEEE Std 857-1996. *IEEE Recommended Practice for Test Procedures for High-Voltage Direct-Current Thyristor Valves.* Information and recommendations for the type testing of thyristor valves for high-voltage direct-current (HVDC) power transmission systems are provided. These tests cover only the principal tests on the valves and do not include tests of auxiliary equipment associated with the valves.

IEEE Std 896.10-1997. *IEEE Standard for Futurebus+® Spaceborne Systems—Profile S.* In the Futurebus+ series of standards, tools with which high-performance bus-based systems may be developed are provided. This architecture provides a wide range of performance scalability over both cost and time for multiple generations of single- and multiple-bus multiprocessor systems. This document, a companion standard to the ISO/IEC 10857:1994 (896.1, 1994 Edition) Futurebus+ Logical Layer Specification, builds on the logical layer by adding requirements for a spaceborne profile. It is to this profile that products will claim conformance. Other specifications may be required in conjunction with this standard.

IEEE Std 902-1998. *IEEE Guide for Maintenance, Operation, and Safety of Industrial and Commercial Power Systems (IEEE Yellow Book).* Guidelines for the numerous personnel who are responsible for safely operating and maintaining industrial and commercial electric power facilities are provided.

This guide provides plant engineers with a reference source for the fundamentals of safe and reliable maintenance and operation of industrial and commercial electric power distribution systems.

IEEE Std 928-1986 (R1991). *IEEE Recommended Criteria for Terrestrial Photovoltaic Power Systems.* General performance criteria for terrestrial photovoltaic (PV) systems are established, and an overall framework for all detailed terrestrial photovoltaic power system performance standards is provided. Criteria for subsystem performance and standard test methods to be used for performance measurements are recommended. System installation, operation, and maintenance are covered. Since thermal conditioning elements may be part of the system design, some consideration is given to the thermal subsystem. The criteria apply to all terrestrial photovoltaic power systems.

IEEE Std 929-2000. *IEEE Recommended Practice for Utility Interface of Photovoltaic (PV) Systems.* This recommended practice contains guidance regarding equipment and functions necessary to ensure compatible operation of photovoltaic (PV) systems that are connected in parallel with the electric utility. This includes factors relating to personnel safety, equipment protection, power quality, and utility system operation. This recommended practice also contains information regarding islanding of PV systems when the utility is not connected to control voltage and frequency, as well as techniques to avoid islanding of distributed resources.

IEEE Std 930-1987 (R1995). *IEEE Guide for the Statistical Analysis of Voltage Endurance Data for Electrical Insulation.* A description is given, with examples of statistical methods for analyzing the data, for time-to-failure from constant-stress voltage endurance tests or breakdown voltage from progressive-stress tests on specimens or systems of electrical insulation. Methods to compare test data are also given. The methods are principally applied to data from tests on solid insulation, but they may also apply to the analysis of data from tests on gas, liquid, and composite systems. The statistical methods discussed do not take into consideration the physical mechanism of voltage aging. They assume that the only aging stress is alternating voltage of constant frequency. The methods may not apply if there is more than one aging stress. Methods to ascertain the short-time withstand voltage or operating voltage of an insulation system are not included, and the mathematical techniques may not directly apply to the estimation of equipment life.

IEEE Std 933-1999. *IEEE Guide for the Definition of Reliability Program Plans for Nuclear Power Generating.* Guidelines for the definition of a reliability program at nuclear power generating stations are developed. Reliability programs during the operating phase of such stations are emphasized; however, the general approach applies to all phases of the nuclear power generating station life cycle (e.g., design, construction, start-up, operating, and decommissioning).

IEEE Std 935-1989 (R1995). *IEEE Guide on Terminology for Tools and Equipment to Be Used in Live Line Working.* Terminology for tools and equipment used in live line working is given to permit identification of the tools and equipment and to standardize their names. Detailed definitions are not given for all the terms used in live line working; only the necessary details, without indications of their components and their methods of use, are provided. The following are covered: insulating sticks; universal tool fittings; insulating covers and similar assemblies; bypassing equipment; small individual hand tools; personal equipment; equipment for positioning a worker; handling and anchoring equipment; measuring and testing equipment, and hydraulic and miscellaneous equipment.

IEEE Std 937-2000. *IEEE Recommended Practice for Installation and Maintenance of Lead-Acid Batteries for Photovoltaic (PV) Systems.* Design considerations and procedures for storage, location, mounting, ventilation, assembly, and maintenance of lead-acid storage batteries for photovoltaic power systems are provided. Safety precautions and instrumentation considerations are also included. Even though general recommended practices are covered, battery manufacturers may provide specific instructions for battery installation and maintenance.

IEEE Std 943-1986 (R1992). *IEEE Guide for Aging Mechanisms and Diagnostic Procedures in Evaluating Electrical Insulation Systems.* Background information necessary for proper construction of aging mechanisms and selection of diagnostic procedures when designing tests for functional evaluation of insulation systems for electrical equipment is presented. Aging mechanisms of insulation systems and methods for ascertaining correlation of aging during testing and aging during actual service are described. Diagnostic techniques for use in functional tests are also listed. The intent is primarily to aid committees in standardizing tests within the scope of their responsibilities.

IEEE Std 944-1986 (R1996). *IEEE Application and Testing of Uninterruptible Power Supplies for Power Generating Stations.* The application and performance requirements for a low-voltage uninterruptible power supply (UPS) system used for service in power generating stations are defined. Service conditions and requirements for design application, procurement documents, and testing are covered. The recommendations apply only to semiconductor ac-to-ac converter systems (static) with dc electric energy storage capability. Equipment or component design requirements, safety-related design criteria, or requirements for equipment qualification and preoperational/surveillance testing are not addressed.

IEEE Std 945-1984 (R1997). *IEEE Standard Preferred Metric Units for Use in Electrical and Electronics Science and Technology.* Recommendations are made to aid in the selection of metric units, so as to promote uniformity in the use of metric units and to limit the number of different metric units that will be used in electrical and electronics science and technology. The recommendations can cover units for space and time, periodic and related phenomena, mechanics, heat, electricity and magnetism, light and related electromagnetic radiations, and acoustics. This document does not cover how metric units are to be used, nor does it offer guidance concerning correct metric practice.

IEEE Std 946-1992 (R1997). *IEEE Recommended Practice for the Design of DC Auxiliary Power Systems for Generating Stations.* Guidance for the design of the dc auxiliary power systems for nuclear and large fossil-fueled power generating stations is provided. The components of the dc auxiliary power system, including lead storage batteries, static battery chargers, and distribution equipment, are addressed. Guidance for selecting the quantity and types of equipment, the equipment ratings, interconnections, instrumentation, control and protection is also provided.

IEEE Std 951-1996. *IEEE Guide to the Assembly and Erection of Metal Transmission Structures.* Various good practices that will enable users to improve their ability to assemble and erect self-supporting and guyed steel or aluminum lattice and tubular steel structures are presented. Construction considerations after foundation installation, and up to the conductor stringing operation, are also covered. The guide focuses on the design and construction considerations for material delivery, assembly and erection of metal transmission structures, and the installation of insulators and hardware. This guide is intended to be used as a reference source for parties involved in the ownership, design, and construction of transmission structures.

IEEE Std 952-1997. *IEEE Standard Specification Format Guide and Test Procedure for Single-Axis Interferometric Fiber Optic Gyros.* Specification and test requirements for a single-axis interferometric fiber optic gyro (IFOG) for use as a sensor in attitude control systems, angular displacement measuring systems, and angular rate measuring systems are defined. A standard specification format guide for the preparation of a single-axis IFOG is provided. A compilation of recommended procedures for testing a fiber optic gyro, derived from those presently used in the industry, is also provided.

IEEE Std 957-1995. *IEEE Guide for Cleaning Insulators.* Procedures for cleaning contaminated electrical insulators (excluding nuclear, toxic, and hazardous chemical contaminants) of all types, using various equipment and techniques, are provided.

IEEE Std 959-1988 (R1995). *IEEE Specifications for an I/O Expansion Bus: SBX Bus.* An I/O expansion bus for microcomputers that is independent of processor or board type is specified. Each expansion interface supports up to 16 8-bit I/O ports directly. Enhanced addressing capability is available using slave processors or FIFO devices. In addition, each expansion interface may optionally support a DMA channel capable of data rates up to 2 16-bit Mwords/sec. These features are supported for both 8- and 16-bit data paths. The specification has been prepared for those users who intend to design or evaluate products that will be compatible with the bus. For this purpose, functional, electrical, and mechanical specification is covered in detail. The intent of the specification is to guarantee compatibility between baseboards and expansion modules while not restricting the actual designs any more than necessary.

IEEE Std 960-1993 (R1999). *IEEE Standard FASTBUS Modular High-Speed Data Acquisition and Control System and IEEE FASTBUS Standard Routines.* Mechanical, signal, electrical, and protocol specifications are given for a modular data bus system, which, while allowing equipment designers a wide choice of solutions, ensure compatibility of all designs that obey the mandatory parts of the specification. This standard applies to systems consisting of modular electronic instrument units that process or transfer data or signals, normally in association with computers or other automatic data processors. Standard software routines for use with the system in IEEE Std 960-1993 are defined.

IEEE Std 961-1987 (R1994). *IEEE Standard for an 8-Bit Microcomputer Bus System: STD Bus.* An 8-bit microcomputer bus system derived from the industry bus known as the STD bus is described. The STD bus is a modular packaging and interconnect scheme for 8-bit microprocessor card systems. The bus size and bus organization were selected to serve the interface between any 8-bit microprocessor and a variety of memory and I/O functions. Logical, timing, electrical, and mechanical specifications are provided. The body of the standard provides a core specification for the device-independent parameters. Appendixes provide device-dependent parameters for various processors. This document also contains IEEE Std 1101-1987, IEEE Standard for Mechanical Core Specifications for Microcomputers.

IEEE Std 977-1991 (R1997). *IEEE Guide to Installation of Foundations for Transmission Line Structures.* Various approaches to good construction practices, which could improve the installation of transmission-line structure foundations, are presented. Spread foundations, drilled shaft foundations, pile foundations, and anchors are treated. This guide is intended to be used as a reference for those involved in the ownership, design, and construction of transmission structures.

IEEE Std 979-1994. *IEEE Guide for Substation Fire Protection.* Guidance is provided to substation engineers in determining the design, equipment, and practices deemed necessary for the fire protection of substations. A list of publications that can be used to acquire more detailed information for specific substations or substation components is presented.

IEEE Std 980-1994. *IEEE Guide for Containment and Control of Oil Spills in Substations.* The significance of oil-spillage regulations and their applicability to electric supply substations are discussed; the sources of oil spills are identified; typical designs and methods for dealing with oil containment and control of oil spills are discussed; and guidelines for preparation of a typical Spill Prevention Control and Countermeasures (SPCC) plan are provided. This guide excludes polychlorinated biphenyl (PCB) handling and disposal considerations.

IEEE Std 982.1-1988. *IEEE Standard Dictionary of Measures to Produce Reliable Software.* A set of measures indicative of software reliability that can be applied to the software product as well as to the development and support processes is provided. The measures can be applied early in the development process to indicate the reliability of the delivered product. The aim is to provide a common set of definitions that allows a meaningful exchange of data and evaluations to occur, and that serves as the foundation on which researchers and practitioners can build consistent methods. The standard is designed to assist management in directing product development and support toward specific reliability goals.

IEEE Std 982.2-1988. *IEEE Guide for the Use of IEEE Standard Dictionary of Measures to Produce Reliable Software.* This guide provides the underlying concepts and motivation for establishing a measurement process for reliable software, utilizing IEEE Std 982.1-1988, IEEE Standard Dictionary of Measures to Produce Reliable Software. The guide contains information necessary for application of measures to a project. It includes guidance for the following: applying product and process measures throughout the software life cycle, providing the means for continual self-assessment and reliability improvement; optimizing the development of reliable software, beginning at the early development stages with respect to constraints such as cost and schedule; maximizing the reliability of software in its actual use environment during the operation and maintenance phases; and developing the means to manage reliability in the same manner that cost and schedule are managed. The guide is intended for design, development, evaluation (e.g., auditing or procuring agency), and maintenance personnel; software quality and software reliability personnel; and operations and acquisition support managers. It is organized to provide input to the planning process for reliability management.

IEEE Std 991-1986 (R1994). *IEEE Standard for Logic Circuit Diagrams.* Guidelines for preparation of diagrams depicting logic functions are provided. Definitions, requirements for assignment of logic levels, application of logic symbols, presentation techniques, and labeling requirements are included, with typical examples where appropriate. The techniques are presented in the context of electrical and electronic systems, but they also may be applied to nonelectrical systems (e.g., pneumatic, hydraulic, or mechanical).

IEEE Std 993-1997. *IEEE Standard for Test Equipment Description Language (TEDL).* A language useful for describing Automatic Test Equipment (ATE) instrumentation and configurations, as well as Interface Test Adapters (ITA), is defined. Principally intended for testing environments using the ATLAS test language, TEDL can also be used to describe instrumentation in non-ATLAS environments.

IEEE Std 998-1996. *IEEE Guide for Direct Lightning Stroke Shielding of Substations.* Design information for the methods historically and typically applied by substation designers to minimize direct lightning strokes to equipment and buswork within substations is provided. Two approaches, the classical empirical method and the electrogeometric model, are presented in detail. A third approach involving the use of active lightning terminals is also briefly reviewed.

IEEE Std 1000-1987 (R1994). *IEEE Standard Specification for a Standard 8-Bit Backplane Interface: STEbus.* A bus that can be used to implement general-purpose, high-performance, 8-bit microcomputer systems is defined. Such a system may be used in a stand-alone configuration or in larger multiple-bus architectures as a private (or secondary) bus or a high-speed I/O channel. The standard is applicable to systems and system elements with the common commercial designation STE Bus and is intended for users who plan to evaluate, implement, or design various system elements that are compatible with the IEEE 1000 Std Bus system structure. It provides a functional description and covers signal lines, arbitration, the data transfer protocol, interboard signaling, and electrical specifications. The physical attributes and method of interconnect utilized by boards and modules conforming to this standard are derived from several IEC standards, which, when implemented jointly in a systems environment, result in a mechanical configuration commonly referred to as Eurocard. This document also contains IEEE Std 1101-1987, IEEE Standard for Mechanical Core Specifications for Microcomputers.

IEEE Std 1003.0-1995. *IEEE Guide to the POSIX® Open System Environment (OSE).* This guide presents an overview of open system concepts and their applications. Information is provided to persons evaluating systems based on the existence of, and interrelationships among, application software standards, with the objective of enabling application portability and system interoperability. A framework is presented that identifies key information system interfaces involved in application portability and system interoperability and describes the services offered across these interfaces. Standards or standards activities associated with the services are identified where they exist or are in progress. Gaps are identified where POSIX Open System Environment services are not currently being addressed by formal standards. Finally, the concept of a profile is discussed with examples from several application domains.

IEEE Std 1003.1/2003.1/INT (March 1994 Edition). *IEEE Standards Interpretations for IEEE Std 1003.1-1990 and IEEE Std 2003.1-1992 (March 1994 Edition).* The Portable Applications Standards Committee of the IEEE Computer Society carried out a series of analyses of various problems encountered by users of IEEE Std 1003.1-1990, IEEE Standard for Information Technology—Portable Operating System Interface (POSIX®)—Part 1: System Application Program Interface (API) and IEEE Std 2003.1-1992, IEEE Standard for Information Technology—Test Methods for Measuring Conformance to POSIX—Part 1: System Interfaces. The results of its deliberations are presented in this document. The intent is to give the POSIX community reasonable ways of interpreting unclear portions of these standards.

IEEE Std 1003.1d-1999. *IEEE Standard for Information Technology—Portable Operating System Interface (POSIX®)—Part 1: System Application Program Interface (API)—Amendment d: Additional Realtime Extensions [C Language].* This standard is part of the POSIX series of standards for applications and user interfaces to open systems. It defines the applications interface to system services for spawning a process, timeouts for blocking services, sporadic server scheduling, execution time clocks and timers, and advisory information for file management. This standard is stated in terms of its C binding.

IEEE Std 1003.5, 1999 Edition. *IEEE Standard for Information Technology—POSIX® Ada Language Interfaces—Part 1: Binding for System Application Program Interface (API) Includes Amendment 1: Realtime Extensions and Amendment 2: Protocol-Independent.* This standard is part of the POSIX® series of standards for applications and user interfaces to open systems. It defines the Ada language bindings as package specifications and accompanying textual descriptions of the application program interface (API). This standard supports application portability at the source code level through the binding between ISO 8652:1995 (Ada) and ISO/IEC 9945-1:1996 (IEEE Std 1003.1-1996) (POSIX) as amended by IEEE P1003.1g/D6.6.Terminology and general requirements, process primitives, the process environment, files and directories, input and output primaries, device- and class-specific functions, language-specific services for Ada, system databases, synchronization, memory management, execution scheduling, clocks and timers, message passing, task management, the XTI and socket detailed network inter-faces, event management, network support functions, and protocol-specific mappings are covered. It also specifies behavior to support the binding that must be proviced by the Ada.

IEEE Std 1003.9-1992 (R1997). *IEEE Standard for Information Technology—POSIX® FORTRAN 77 Language Interfaces—Part 1: Binding for System Application Program Interface (API).* This standard provides a standardized interface for accessing the system services of ISO/IEC 9945:1990 (IEEE Std 1003.1-1990, also known as POSIX.1), and support routines to access constructs not directly accessible with FORTRAN 77. This standard supports application portability at the source level through the binding between ANSI X3.9-1978 and POSIX.1, and a standardized definition of language-specific services. The goal is to provide standardized interfaces to the POSIX.1 system services via a FORTRAN 77 language interface. Terminology and general requirements, process primitives, the process environment, files and directories, input and output primitives, device- and class-specific functions, the FORTRAN 77 language library, and system databases are covered.

IEEE Std 1003.10-1995 (R1997). *IEEE Standard for Information Technology POSIX®—Based Supercomputing Application Environment Profile.* This standard is related to the POSIX series of standards for applications and user interfaces to open systems. It specifies the set of standards and the requirements needed for portability of supercomputing applications, users, and system administrators.

IEEE Std 1005-1998. *IEEE Standard Definitions and Characterization of Floating Gate Semiconductor Arrays.* This standard describes the underlying physics and the operation of floating gate memory arrays, specifically, UV erasable EPROM, byte rewritable E2PROMs, and block rewritable flash EEPROMs. In addition, reliability hazards are covered with focus on retention, endurance, and disturb. There are also clauses on the issues of testing floating gate arrays and their hardness to ionizing radiation.

IEEE Std 1007-1991 (R1997). *IEEE Standard Methods and Equipment for Measuring the Transmission Characteristics of Pulse-Code Modulation (PCM) Telecommunications Circuits and Systems.* Test equipment requirements and methods for testing the transmission characteristics of PCM telecommunications equipment, circuits, and systems are set forth. The requirements are intended for certification, installation, preservice, out-of-service operational, and in-service operational tests of the PCM transmission facilities. The PCM equipment that may be tested includes primary multiplex equipment containing analog–digital conversion devices, digital multiplex equipment, digital links, and digital sections. This standard is limited to testing at the analog interfaces of the primary multiplex equipment and the digital interfaces at DS1, DS1C, DS2, and DS3 levels of the North American digital hierarchy. Synchronous multiplex equipment and equipment offering other than 64 kbps coded voiceband channels are not covered in this standard, nor is signaling parameter measurement.

IEEE Std 1008-1987 (R1993). *IEEE Standard for Software Unit Testing.* An integrated approach to systematic and documented unit testing is defined. The approach uses unit design and unit implementation information, in addition to unit requirements, to determine the completeness of the testing. The testing process described is composed of a hierarchy of phases, activities, and tasks and defines a minimum set of tasks for each activity. The standard can be applied to the unit testing of any digital computer software or firmware and to the testing of both newly developed and modified units. The software engineering concepts and testing assumptions on which this standard approach is based, and guidance and resource information to assist with the implementation and usage of the standard unit testing approach, are provided in appendixes.

IEEE Std 1010-1987 (R1992). *IEEE Guide for Control of Hydroelectric Power Plants.* The control and monitoring requirements for equipment and systems associated with conventional and pumped-storage hydroelectric plants are described. Typical methods of local and remote control, details of the control interfaces for plant equipment, requirements for centralized and off-site control, and trends in control systems are included. The various categories that affect the levels of control for a plant, namely, location, mode, and supervision, are described. Block diagrams and descriptions of the control and monitoring requirements for major plant systems and equipment are given. Control sequencing of generating and pumped storage units, centralized control, and off-site control are covered. The information is directed toward practicing engineers in the field of power plant design who have a basic knowledge of hydroelectric facilities.

IEEE Std 1012-1998. *IEEE Standard for Software Verification and Validation.* Software verification and validation (V&V) processes, which determine whether development products of a given activity conform to the requirements of that activity, and whether the software satisfies its intended use and user needs, are described. This determination may include analysis, evaluation, review, inspection, assessment, and testing of software products and processes. V&V processes assess the software in the context of the system, including the operational environment, hardware, interfacing software, operators, and users.

IEEE Std 1012a-1998. *IEEE Standard for Software Verification and Validation: Content Map to IEEE/EIA 12207.1-1997.* The relationship between the two sets of requirements on plans for verification and validation of software, found in IEEE Std 1012-1998 and IEEE/EIA Std 12207.1-1997, is explained so that users may produce documents that comply with both standards.

IEEE Std 1013-1990. *IEEE Recommended Practice for Sizing Lead-Acid Batteries for Photovoltaic (PV) Systems.* Methods for sizing both vented and valve-regulated lead-acid batteries used with terrestrial photovoltaic (PV) systems, regardless of size, are described. The purpose is to assist system designers in sizing batteries for residential, commercial, and industrial PV systems. Sizing examples are given for various representative system applications. Iterative techniques to optimize battery costs, which include consideration of the interrelationship between battery size, PV array size, and weather, are not covered. A worksheet with examples of its use is included to facilitate the battery sizing process.

IEEE Std 1014-1987 (R1992). *IEEE Standard for a Versatile Backplane Bus: VMEbus.* A high-performance backplane bus for use in microcomputer systems that employ single or multiple microprocessors is specified. This interfacing system, which is used to interconnect data processing, data storage, and peripheral control devices in a tightly coupled hardware configuration, is based on the VMEbus specification. The bus includes four subbuses: data transfer bus, priority interrupt bus, arbitration bus, and utility bus. Specifications are given for each of these, and overall electrical and mechanical specifications are given as well. Signal line description, use of the SERCIK and SERDAT lines, metastability and synchronization, and permissible capability subsets are covered in the appendixes.

IEEE Std 1015-1997. *IEEE Recommended Practice for Applying Low-Voltage Circuit Breakers Used in Industrial and Commercial Power Systems. (IEEE Blue Book).* Information is provided for selecting the proper circuit breaker for a particular application. This recommended practice helps the application engineer specify the type of circuit breaker, ratings, trip functions, accessories, acceptance tests, and maintenance requirements. It also discusses circuit breakers for special applications, e.g., instantaneous only and switches. In addition, it provides information for applying circuit breakers at different locations in the power system, and for protecting specific components. Guidelines are also given for coordinating combinations of line-side and load-side devices.

IEEE Std 1016-1998. *IEEE Recommended Practice for Software Design Descriptions.* The necessary information content and recommendations for an organization for Software Design Descriptions (SDDs) are described. An SDD is a representation of a software system that is used as a medium for communicating software design information. This recommended practice is applicable to paper documents, automated databases, design description languages, or other means of description.

IEEE Std 1020-1988 (R1994). *IEEE Guide for Control of Small Hydroelectric Power Plants.*

IEEE Std 1023-1988 (R1995). *IEEE Guide for the Application of Human Factors Engineering to Systems, Equipment, and Facilities of Nuclear Power Generating Stations.* Guidance is provided to management and engineers who wish to develop an integrated program for the application of human factors engineering (HFE) in the design, operation, and maintenance of nuclear power generating stations. The standard covers the program organization and applicability, the plant design aspects to consider, the HFE methodologies that may be used, and a typical program plan for the application of HFE. It is applicable to new facilities or modifications to existing facilities.

IEEE Std 1025-1993 (R1999). *IEEE Guide to the Assembly and Erection of Concrete Pole Structures.* Good practice that will improve the ability to assemble and erect self-supporting and guyed concrete pole structures for overhead transmission lines is presented. Construction aspects after foundation installation and up to the conductor stringing operation are covered. Some aspects of construction related to other materials use in concrete pole structures are covered, but the treatment is not complete. The guide is intended to be used as a reference source for parties involved in the ownership, design, and construction of transmission structures.

IEEE Std 1026-1995. *IEEE Recommended Practice for Test Methods for Determination of Compatibility of Materials with Conductive Polymeric Insulation Shields and Jackets.* A test method is provided to qualify various essentially nonvolatile, highly viscous fluids or solid materials at 90 °C, for use with high-voltage cable shields and jackets. A suggested alternative test method for more fluid and more volatile materials is also provided.

IEEE Std 1027-1996. *IEEE Standard for Measurement of the Magnetic Field in the Vicinity of a Telephone Receiver.* The methodology for measuring the magnetic field strength in the vicinity of a telephone receiver is discussed.

IEEE Std 1028-1997. *IEEE Standard for Software Reviews.* This standard defines five types of software reviews, together with procedures required for the execution of each review type. This standard is concerned only with the reviews; it does not define procedures for determining the necessity of a review, nor does it specify the disposition of the results of the review. Review types include management reviews, technical reviews, inspections, walk-throughs, and audits.

IEEE Std 1029.1-1998. *IEEE Standard for VHDL Waveform and Vector Exchange to Support Design and Test Verification (WAVES) Language Reference Manual.* This standard is a formal notation intended for use in all phases of the development of electronic systems. Because it is both machine-readable and human-readable, it supports the verification and testing of hardware designs; the communication of hardware design and test verification data; and the maintenance, modification, and procurement of hardware systems. This standard provides the syntactic and semantic framework for the unambiguous expression and aggregation of digital test data and timing information necessary to completely describe a test or set of tests for a digital system. WAVES digital test data (stimulus and expected responses) is described at the logic level. Voltage and current values are not described by WAVES and are beyond the scope of this standard.

IEEE Std 1031-2000. *IEEE Guide for the Functional Specification of Transmission Static Var Compensators.* This guide documents an approach to preparing a specification for a transmission static var compensator. The document is intended to serve as a base specification with an informative annex provided to allow users to modify or develop specific clauses to meet a particular application.

IEEE Std 1036-1992. *IEEE Guide for Application of Shunt Power Capacitors.* Guidelines for the application protection, and ratings of equipment for the safe and reliable utilization of shunt power capacitors are provided. This guide applies to the use of 50 and 60 Hz shunt power capacitors rated 2400 V ac and above, and assemblies of capacitors. Applications that range from simple unit utilization to complex bank situations are covered.

IEEE Std 1043-1996. *IEEE Recommended Practice for Voltage-Endurance Testing of Form-Wound Bars and Coils.* The voltage endurance testing of form-wound bars and coils for use in large rotating machines is covered. Such testing is defined for machines with a nominal voltage rating up to 30 000 V.

IEEE Std 1044-1993. *IEEE Standard Classification for Software Anomalies.* A uniform approach to the classification of anomalies found in software and its documentation is provided. The processing of anomalies discovered during any software life cycle phase are described, and comprehensive lists of software anomaly classifications and related data items that are helpful to identify and track anomalies are provided. This standard is not intended to define procedural or format requirements for using the classification scheme. It does identify some classification measures and does not attempt to define all the data supporting the analysis of an anomaly.

IEEE Std 1044.1-1995. *IEEE Guide to Classification of Software Anomalies.* This guide provides supporting information to assist users who are applying IEEE Std 1044-1993, IEEE Standard Classification for Software Anomalies, to decide whether to conform completely to or just extract ideas from IEEE Std 1044-1993. This guide will enable users of IEEE Std 1044-1993 to implement and customize that standard for their organization in an effective and efficient manner.

IEEE Std 1045-1992. *IEEE Standard for Software Productivity Metrics.* A consistent way to measure the elements that go into computing software productivity is defined. Software productivity metrics terminology are given to ensure an understanding of measurement data for both source code and document production. Although this standard prescribes measurements to characterize the software process, it does not establish software productivity norms, nor does it recommend productivity measurements as a method to evaluate software projects or software developers. This standard does not measure the quality of software. This standard does not claim to improve productivity, only to measure it. The goal of this standard is for a better understanding of the software process, which may lend insight to improving it.

IEEE Std 1046-1991 (R1996). *IEEE Application Guide for Distributed Digital Control and Monitoring for Power Plants.* Alternate approaches to applying a digital control system, for both new construction and existing plant modernization projects, are described, and their advantages and disadvantages are compared. Criteria to be used to judge the suitability of commercially available systems for use in the power generation industry are provided. Terminology is defined, and the objectives of distributed control and monitoring systems are described. The following system application issues are addressed: integrated versus segregated systems functional and geographic distribution, hierarchical architecture and automation, control and protection functions, input/output systems, environmental considerations, and documentation. The data communications structure and the functions that support it are considered. Data acquisition and monitoring (the man/machine interfaces) are discussed. Reliability, availability, and fault tolerance of distributed control and monitoring systems are addressed.

IEEE Std 1048-1990 (R1996). *IEEE Guide for Protective Grounding of Power Lines.* Guidelines are provided for safe protective grounding methods for persons engaged in de-energized overhead transmission and distribution line maintenance. They comprise state-of-the-art information on protective grounding as currently practiced by power utilities in North America. The principles of protective grounding are discussed. Grounding practices and equipment, power-line construction, and ground electrodes are covered.

IEEE Std 1050-1996 (R1998). *IEEE Guide for Instrumentation and Control Equipment grounding in Generating Stations.* Information about grounding methods for generating station instrumentation and control (I & C) equipment is provided. The identification of I & C equipment grounding methods to achieve both a suitable level of protection for personnel and equipment is included, as well as suitable noise immunity for signal ground references in generating stations. Both ideal theoretical methods and accepted practices in the electric utility industry are presented.

IEEE Std 1058-1998. *IEEE Standard for Software Project Management Plans.* The format and contents of software project management plans, applicable to any type or size of software project, are described. The elements that should appear in all software project management plans are identified.

IEEE Std 1061-1998. *IEEE Standard for a Software Quality Metrics.* A methodology for establishing quality requirements and identifying, implementing, analyzing, and validating the process and product software quality metrics is defined. The methodology spans the entire software life cycle.

IEEE Std 1062, 1998 Edition. *IEEE Recommended Practice for Software Acquisition.* A set of useful quality practices that can be selected and applied during one or more steps in a software acquisition process is described. This recommended practice can be applied to software that runs on any computer system regardless of the size, complexity, or criticality of the software, but is more suited for use on modified-off-the-shelf software and fully developed software.

IEEE Std 1063-1987 (R1993). *IEEE Standard for Software User Documentation.* Minimum requirements for the structure and information content of user documentation are provided. The requirements apply primarily to technical substance rather than to style. Editorial and stylistic considerations are addressed only when they impact structure and content. Only traditional documentation, either printed on paper or stored in some other medium in the format of a printed document and used in a manner analogous to the way a printed document is used, is addressed.

IEEE Std 1067-1996. *IEEE Guide for In-Service Use, Care, Maintenance, and Testing of Conductive Clothing for Use on Voltages up to 765 kV ac and ± 750 kV dc.* Recommendations are provided for the in-service use, care, maintenance, and electrical testing of conductive clothing, including suits, gloves, socks, and boots, for use on voltages up to 765 kV ac and ± 750 kV dc.

IEEE Std 1068-1996. *IEEE Recommended Practice for the Repair and Rewinding of Motors for the Petroleum and Chemical Industry.* General recommendations are provided for owners (users?) of motors that need repair as well as owners and operators of establishments that offer motor repair services. The use of this recommended practice is expected to result in higher quality, more cost-effective, and timely repairs. Guidelines are also provided for evaluating repairs and facilities.

IEEE Std 1069-1991 (R1996). *IEEE Recommended Practice for Precipitator and Baghouse Hopper Heating Systems.* Recommendations on hopper heating system performance and equipment requirements necessary to provide an economical and effective hopper heating system are presented. System characteristics are described, and heat transfer analysis is covered. Heating module design considerations are presented. Control, monitoring, and alarm systems are discussed. Insulation, installation, operation, and maintenance are addressed.

IEEE Std 1070-1995. *IEEE Guide for the Design and Testing of Transmission Modular Restoration Structure Components.* A generic specification, including design and testing, for transmission modular restoration structure components used by electric utilities is provided.

IEEE Std 1073-1996. *IEEE Standard for Medical Device Communications—Overview and Framework.* An overall definition of the IEEE 1073 family of standards is provided, describing the interconnection and interoperation of medical devices with computerized healthcare information systems in a manner suitable for the clinical environment.

IEEE Std 1073.3.1-1994. *IEEE Standard for Medical Device Communications—Transport Profile—Connection Mode.* A local area network (LAN) for the interconnection of computers and medical devices is defined by the specifications and guidelines set forth in this standard. The functions, features, and protocols of the intra-room communications subnet of a bedside communications network known as the Medical Information Bus (MIB) are defined. This communications subnet is the functional equivalent for the MIB of the Transport, Network, Data Link, and Physical layers of the Organization for International Standards (ISO) Reference Model for Open Systems Interconnection (OSI). This standard defines the services and protocols for the MIB Transport, Network, and Data Link layers.

IEEE Std 1073.3.2-2000. *IEEE Standard for Medical Device Communications—Transport Profile—IrDA Based—Cable Connected.* A connection-oriented transport profile and physical layer suitable for medical device communications in legacy devices is established. Communications services and protocols consistent with specifications of the Infrared Data Association are defined. These communication services and protocols are optimized for use in patient-connected bedside medical devices.

IEEE Std 1073.4.1, 2000 Edition. *IEEE Standard for Medical Device Communications—Physical Layer Interface—Cable Connected.* A physical interface for the interconnection of computers and medical devices in the IEEE 1073 family of standards is defined. This interface is intended to be highly robust in an environment where devices are frequently connected to and disconnected from the network. The physical and electrical characteristics of the connector and signals necessary to exchange digital information between cable-connected medical devices and host computer systems are specified.

IEEE Std 1073.4.1a-2000. *IEEE Standard for Medical Device Communications—Physical Layer Interface—Cable Connected.* A physical interface for the interconnection of computers and medical devices in the IEEE 1073 family of standards is defined. This interface is intended to be highly robust in an environment where devices are frequently connected to and disconnected from the network. The physical and electrical characteristics of the connector and signals necessary to exchange digital information between cable-connected medical devices and host computer systems are specified.

IEEE Std 1074-1997. *IEEE Standard for Developing Software Life Cycle Processes.* A process for creating a software life cycle process is provided. Although this standard is directed primarily at the process architect, it is useful to any organization that is responsible for managing and performing software projects.

IEEE Std 1076-1993. *VHDL Interactive Tutorial—A CD-ROM Learning Tool for IEEE Std 1076 (VHDL).* Aiding in the comprehension and use of IEEE VHDL, this unique product offers a comprehensive & reliable tutorial on VHDL not available anywhere else. An enhancement to IEEE Std 1076-1993, the interactive tutorial is organized into four modules designed to incrementally add to the user's understanding of VHDL and it's applications. This hands-on tutorial shows clear links between the many levels and layers of VHDL and provides actual examples of VHDL implementation, making it an indispensable tool for VHDL product development and users.

IEEE Std 1076a-2000. *IEEE Standard VHDL Language Reference Manual.* VHSIC Hardware Description Language (VHDL) is defined. VHDL is a formal notation intended for use in all phases of the creation of electronic systems. Because it is both machine readable and human readable, it supports the development, verification, synthesis, and testing of hardware designs; the communication of hardware design data; and the maintenance, modification, and procurement of hardware. Its primary audiences are the implementors of tools supporting the language and the advanced users of the language.

IEEE Std 1076.1-1999. *IEEE Standard VHDL Analog and Mixed-Signal Extensions.* This standard defines the IEEE 1076.1 language, a hardware description language for the description and the simulation of analog, digital, and mixed-signal systems. The language, also informally known as VHDL-AMS, is built on IEEE Std 1076-1993 (VHDL) and extends it with additions and changes to provide capabilities of writing and simulating analog and mixed-signal models.

IEEE Std 1076.2-1996. *IEEE Standard VHDL Language Math Package.* The MATH_REAL package declaration, the MATH_COMPLEX package declaration, and the semantics of the standard mathematical definition and conventional meaning of the functions that are part of this standard are provided. Ways for users to implement this standard are given in an informative annex. Samples of the MATH_REAL and MATH_COMPLEX package bodies are provided in an informative annex as guidelines for implementors to verify their implementation of this standard. Implementors may choose to implement the package bodies in the most efficient manner available to them.

IEEE Std 1076.3-1997. *IEEE Standard VHDL Synthesis Packages.* The current interpretation of common logic values and the association of numeric values to specific VHDL array types is described. This standard provides semantic for the VHDL synthesis domain, and enables formal verification and simulation acceleration in the VHDL based design. The standard interpretations are provided for values of standard logic types defined by IEEE Std 1164-1993, and of the BIT and BOOLEAN types defined in IEEE Std 1076-1993. The numeric types SIGNED and UNSIGNED and their associated operators define integer and natural number arithmetic for arrays of common logic values. Two's compliment and binary encoding techniques are used. The numeric semantic is conveyed by two VHDL packages. This standard also contains any allowable modifications.

IEEE Std 1076.4-1995. *IEEE Standard VITAL Application-Specific Integrated Circuit (ASIC) Modeling Specification.* The VITAL (VHDL Initiative Towards ASIC Libraries) ASIC Modeling Specification is defined. It creates a methodology that promotes the development of highly accurate, efficient simulation models for ASIC (Application-Specific Integrated Circuit) components in VHDL.

IEEE Std 1076.6-1999. *IEEE Standard for VHDL Register Transfer Level (RTL) Synthesis.* A standard syntax and semantics for VHDL register transfer level (RTL) synthesis is de-fined. The subset of IEEE 1076 (VHDL) that is suitable for RTL synthesis is defined, along with the semantics of that subset for the synthesis domain.

IEEE Std 1082-1997. *IEEE Guide for Incorporating Human Action Reliability Analysis for Nuclear Power Generating*

Stations. A structured framework for the incorporation of human/system interactions into probabilistic risk assessments is provided.

IEEE Std 1095-1989 (R1994). *IEEE Guide for Installation of Vertical Generators and Generator/Motors for Hydroelectric Applications.* Installation procedures are given for all types of synchronous generators and generator/motors rated 5000 kVA and above to be coupled to hydraulic turbines or hydraulic pump/turbines having vertical shafts. The standard covers tools and facilities; personnel; generator construction; preparation of generator and turbine shafts in the factory; installation precautions; receiving, storing, and unpacking; erection procedures; mechanical run; balancing; insulation testing and drying out; initial operation; and field tests.

IEEE Std 1100-1999. *IEEE Recommended Practice for Powering and Grounding Electronic Equipment (IEEE Emerald Book).* Recommended design, installation, and maintenance practices for electrical power and grounding (including both power-related and signal-related noise control) of sensitive electronic processing equipment used in commercial and industrial applications are presented. The main objective is to provide a consensus of recommended practices in an area where conflicting information and confusion, stemming primarily from different viewpoints of the same problem, have dominated. Practices herein address electronic equipment performance issues while maintaining a safe installation. A brief description is given of the nature of power quality problems, possible solutions, and the resources available for assistance in dealing with problems. Fundamental concepts are reviewed. Instrumentation and procedures for conducting a survey of the power distribution system are described. Site surveys and site power analyses are considered. Case histories are given to illustrate typical problems.

IEEE Std 1101.1-1998. *IEEE Standard for Mechanical Core Specifications for Microcomputers Using IEC 60603-2 Connectors.* The basic dimensions of a range of modular subracks conforming to 60297-3 (1984-01) and 60297-4 (1995-03) for mounting in equipment according to 60297-1 (1986-09) and 310-D-1992, together with the basic dimensions of a compatible range of plug-in units, printed boards, and backplanes, are covered. The dimensions and tolerances necessary to ensure mechanical function compatibility are provided. This standard offers total system integration guidelines with attendant advantages, such as reduction in design and development time, manufacturing cost savings, and distinct marketing advantages.

IEEE Std 1101.2-1992 (R1994). *IEEE Standard for Mechanical Core Specifications for Conduction-Cooled Eurocards.* Mechanical characteristics of conduction-cooled versions of Eurocard-based circuit card assemblers are described. This specification is applicable to, but not limited to, the VMEbus standard, an internal interconnect (backplane) bus intended for connecting processing elements to their immediate fundamental resources. The aim is to ensure mechanical interchangeability of conduction-cooled circuit card assemblers in a format suitable for military and rugged applications, and to ensure their compatibility with commercial, double-height 160-mm Eurocard chassis.

IEEE Std 1101.3-1993. *IEEE Mechanical Standard for Conduction-Cooled and Air-Cooled 10 SU Modules.* The mechanical design requirements for conduction-cooled and air-cooled modules of the 10SU by 6.375 in (161.9 mm) format are established. The specification of dimensions and tolerances is intended to ensure the mechanical intermateability of modules within associated subracks. The basic dimension, frames, PWBs, materials, assembly, and chassis interface of single-sided and double-sided modules are covered.

IEEE Std 1101.4-1993. *IEEE Standard for Futurebus+®, Profile M (Military).* Futurebus+ standards provide systems developers with a set of tools with which high performance bus-based systems may be developed. This architecture provides a wide range of performance scalability over both cost and time for multiple generations of single- and multiple-bus multiprocessor systems. This document, a companion standard to IEEE Std 896.1-1991, builds on the logical layer by adding requirements for three military profiles. It is to these profiles that products will claim conformance. Other specifications that may be required in conjunction with this standard are IEEE Std 896.1-1991, IEEE Std 896.2-1991, IEEE Std 896.3-1993, IEEE Std 896.4-1993, IEEE Std 1101.3-1993, IEEE Std 1101.4-1993, IEEE Std 1212-1991, IEEE Std 1194.1-1991, IEEE P1394, IEEE Std 1301-1991, and IEEE Std 1301.1-1991.

IEEE Std 1101.7-1995. *IEEE Standard for Space Applications Module, Extended Height Format E Form Factor.* The design requirements for a module designated for use in spacecraft, boosters, and other highly rugged, conductively cooled environments are established in this standard. The requirements herein serve to specify the mechanical design of the module. Dimensions and tolerances for racks, modules, printed wiring boards, backplanes, and other connector-related dimensions that are specific to the use of 300-pin and 396-pin connectors are given. These dimensions and tolerances are designed to ensure mechanical function and interoperability.

IEEE Std 1101.10-1996. *IEEE Standard for Additional Mechanical Specifications for Microcomputers Using the IEEE Std 1101.1-1991 Equipment Practice.* A generic standard that may be applied in all fields of electronics where equipment and installations are required to conform to the 482.6 mm (19 in) equipment practice based on IEEE Std 1101.1-1991, IEC 297-3 (1984), and IEC 297-4 (1995). Dimensions are provided that will ensure mechanical interchangeability of subracks and plug-in units.

IEEE Std 1101.11-1998. *IEEE Standard for Mechanical Rear Plug-in Units Specifications for Microcomputers Using IEEE Std 1101.1 and IEEE Std 1101.10 Equipment Practice.* Additional dimensions that will ensure mechanical interchangeability of subracks and plug-in units based on IEEE P1101.1 (D1.0, 1997), 1101.10-1996, and the environmental requirements of IEC 61587-1 (May 1998-Draft) and IEC 61587-3 (May 1998-Draft) are specified.

IEEE Std 1106-1995. *IEEE Recommended Practice for Maintenance, testing and Replacement of Nickel-Cadmium Storage Batteries for Generating Stations and Substations.* Installation design, installation, maintenance and testing procedures, and test schedules that can be used to optimize the life and performance of vented nickel-cadmium batteries used for continuous-float operations are provided. Guidance for determining when these batteries should be replaced is also provided. This recommended practice is applicable to all stationary applications. However, specific applications, such as alternative energy, emergency lighting units, and semiportable equipment, may have other appropriate practices and are beyond the scope of this recommended practice. Sizing, qualification, other battery types, and battery application are beyond the scope of this recommended practice.

IEEE Std 1107-1996. *IEEE Recommended Practice For Thermal Evaluation Of Sealed Insulation Systems for AC Electric Machinery Employing Random-Wound Stator Cells.* A test procedure for comparing expected life, at rated temperature, of two or more sealed insulation systems is outlined. The procedure is limited to insulation systems for ac electric machines using random-wound stator coils. It is the intent of this procedure to evaluate insulation systems for use with air cooling under severe environmental conditions where the insulation is exposed to conducting contaminants. It does not cover special requirements such as those for machines enclosed in gas atmospheres, subjected to strong chemicals, or to submersion in liquids.

IEEE Std 1110-1991 (R1994). *IEEE Guide for Synchronous Generator Modeling Practices In Stability Analyses.* Three direct-axis and four quadrature-axis models are categorized, along with the basic transient reactance model. Some of the assumptions made in using various models and presents the fundamental equations and concepts involved in generator/system interfacing are discussed. The various attributes of power system stability are generally covered, recognizing two basic approaches. The first is categorized under large-disturbance nonlinear analysis; the second approach considers small disturbances, where the corresponding dynamic equations are linearized. Applications of a range of generator models are discussed and treated. The manner in which generator saturation is treated in stability studies, both in the initialization process, as well as during large or small disturbance stability analysis procedures, is addressed. Saturation functions that are derived, whether from test data or by the methods of finite elements are developed. Different saturation algorithms for calculating values of excitation and internal power angle, depending upon generator terminal conditions, are compared. The question of parameter determination is covered. Two approaches in accounting for generator field and excitation system base quantities are identified. Conversion factors are given for transferring field parameters from one base to another for correct generator/excitation system interface modeling. Suggestions for modeling of negative field currents and other field circuit discontinuities are included.

IEEE Std 1115-1992 (R1994). *IEEE Recommended Practice for Sizing Nickel-Cadmium Batteries for Stationary Applications.* Methods for defining the dc load and for sizing a nickel-cadmium battery to supply that load are described. Installation, maintenance, qualification, testing procedures, and consideration of battery types other than nickel-cadmium batteries are not included. Design of the dc system and sizing of the battery charger(s) are also not included.

IEEE Std 1120-1990 (R1995). *IEEE Guide to the Factors to Be Considered in the Planning, Design, and Installation of Submarine Power and Communications Cables.* A checklist of factors relating to power and communications cables installed in a submarine environment, including the shore ends of such cables, is presented. These factors should be considered in the planning, design, and installation of submarine cable systems in a safe and environmentally acceptable manner. Special requirements of communications cables such as repeaters, etc., are not covered.

IEEE Std 1122-1998. *IEEE Standard for Digital Recorders for Measurements in High-Voltage Impulse Tests.* This standard defines the terms specifically related to the digital recorders used for monitoring high-voltage and high-current impulse tests, specifies the necessary performance characteristics for such digital recorders to ensure their compliance with the requirements for high-voltage and high-current impulse tests, and describes the tests and procedures that are necessary to show that these performance characteristics are within the specified limits.

IEEE Std 1125-1993 (R2000). *IEEE Guide for Measurement and Control in SF6 Gas-Insulated Equipment.* Guidelines for moisture level measurement, moisture data interpretation, and moisture control in gas-insulated transmission class equipment (GIE) are provided.

IEEE Std 1127-1998. *IEEE Guide for the Design, Construction, and Operation of Electric Power Substations for Community Acceptance and Environmental Compatibility.* Significant community acceptance and environmental compatibility items to be considered during the planning and design phases, the construction period, and the operation of electric supply substations are identified, and ways to address these concerns to obtain community acceptance and environmental compatibility are documented. On-site generation and telecommunication facilities are not considered.

IEEE Std 1128-1998. *IEEE Recommended Practice for Radio-Frequency (RF) Absorber Evaluation in the Range of 30 MHz to 5 GHz.* Realistic and repeatable criteria, as well as recommended test methods, for characterizing the absorption properties of typical anechoic chamber linings applied to a metallic surface are described. Parameters and test procedures are described for the evaluation of RF absorbers to be used for radiated emissions and radiated susceptibility testing of electronic products, in the absorber manufacturer and/or absorber user environment, over the frequency range of 30 MHz to 5 GHz.

IEEE Std 1129-1992 (R1998). *IEEE Recommended Practice for Monitoring and Instrumentation of Turbine Generators.* A basic philosophy and guidelines are established for the design and implementation of monitoring systems for cylindrical-rotor, synchronous turbine generators. Monitoring systems are used to display the status of the generator and auxiliary systems while these systems are operating online. The basic information needed to choose monitoring schemes best suited for each application is provided. This standard does not specify actual equipment or instrumentation, but it does indicate some critical areas where it is important to provide monitoring capability.

IEEE Std 1137-1991 (R1998). *IEEE Guide for the Implementation of Inductive Coordination Mitigation Techniques and Application.* Guidance is provided for controlling or modifying the inductive environment and the susceptibility of affected wire-line telecommunications facilities in order to operate within the acceptable levels of steady-state or surge-induced voltages of the environmental interface (probe wire) defined by IEEE Std 776-1987. Procedures for determining the source of the problem are given. Mitigation theory and philosophy are discussed, and mitigation devices are described. The application of typical mitigation apparatus and techniques and installation, maintenance, and inspection of mitigation apparatus are addressed. Advice for determining the best engineering solution is offered, and general safety considerations are discussed.

IEEE Std 1138-1994. *IEEE Standard Construction of Composite Fiber Optic Groundwire (OPGW) for Use on Electric Utility Power Lines.* The construction, mechanical and electrical performance, installation guidelines, acceptance criteria, and test requirements for a composite overhead ground wire with optical fibers, commonly known as OPGW, are discussed.

IEEE Std 1139-1999. *IEEE Standard Definitions of Physical Quantities for Fundamental Frequency and Time Metrology-Random Instabilities.* Methods of describing random instabilities of importance to frequency and time metrology is covered in this standard. Quantities covered include frequency, amplitude, and phase instabilities; spectral densities of frequency, amplitude, and phase fluctuations; and time-domain variances of frequency fluctuations. In addition, recommendations are made for the reporting of measurements of frequency, amplitude and phase instabilities, especially as regards the recording of experimental parameters, experimental conditions, and calculation techniques.

IEEE Std 1140-1994 (R1999). *IEEE Standard Test procedures for the Measurement of Electric and Magnetic Fields from Video Display Terminals (VDTs) from 5 Hz to 400 kHz.* Procedures for the measurement of electric and magnetic fields in close proximity to video display terminals (VDTs) in the frequency range of 5 Hz to 400 kHz are provided. Existing international measurement technologies and practices are adapted to achieve a consistent and harmonious VDT measurement standard for testing in a laboratory controlled environment.

IEEE Std 1142-1995. *IEEE Guide for the Design, Testing, and Application of Moisture-Impervious, Solid Dielectric 5-35 kV Power Cable Using Metal-Plastic Laminates.* The user of underground cables is provided with information on the design, testing, and application of moisture-impervious, medium-voltage, solid dielectric power cable using metal-plastic laminates as moisture barriers. Information is also provided on selection of jacketing materials and installation practices. Other types of moisture barriers, such as extruded metal sheaths and bare metallic tapes with sealed seams, are beyond the scope of this guide.

IEEE Std 1143-1994 (R1999). *IEEE Guide on Shielding Practice for Low Voltage Cables.* A concise overview of shielding options for various types of interference and recommendations on shielding practices, including suggestions on terminating and grounding methods, are provided.

IEEE Std 1144-1996. *IEEE Recommended Practice for Sizing Nickel-Cadmium Batteries for Photovoltaic (PV) Systems.* Methods for sizing nickel-cadmium batteries used in residential, commercial, and industrial photovoltaic (PV) systems are described.

IEEE Std 1145-1990 (R1999). *IEEE Recommended Practice for Installation and Maintenance of Nickel-Cadmium Batteries for Photovoltaic (PV) Systems.* Safety precautions, installation design considerations, and procedures for receiving, storing, commissioning, and maintaining pocket and fiber-plate nickel-cadmium storage batteries for photovoltaic power systems are provided. Disposal and recycling recommendations are also discussed. This recommended practice applies to all terrestrial photovoltaic power systems, regardless of size or application, that contain nickel-cadmium battery storage subsystems.

IEEE Std 1147-1991 (R1996). *IEEE Guide for the Rehabilitation of Hydroelectric Power Plants.* This guide is directed to the practicing engineer in the field of hydroelectric power plant design for the purpose of providing guidance in the decision-making processes and design for rehabilitation of hydroelectric power plants. It covers general assessment considerations, rehabilitation of waterways, and rehabilitation of equipment. An extensive bibliography is included.

IEEE Std 1149.1-1990 (R1996). *IEEE Standard Test Access Port and Boundary-Scan Architecture.* A test access port and boundary-scan architecture for digital integrated circuits and for the digital portions of mixed analog/digital integrated circuits are discussed. These facilities seek to provide a solution to the problem of testing assembled printed circuit boards and other products based on highly complex digital integrated circuits and high-density, surface-mounting assembly techniques. The facilities also provide a means of accessing and controlling design-for-test features built into the digital integrated circuits themselves. The circuitry includes a standard interface through which instructions and test data are communicated. A set of test features is defined, including a boundary-scan register, so that the component is able to respond to a minimum set of instructions designed to assist with testing of assembled printed circuit boards.

IEEE Std 1149.1a-1993 (R1996). *IEEE Supplement to Standard Test Access Port and Boundary-Scan Architecture.* A test access port and boundary-scan architecture for digital integrated circuits and for the digital portions of mixed analog/digital integrated circuits are discussed. These facilities seek to provide a solution to the problem of testing assembled printed circuit boards and other products based on highly complex digital integrated circuits and high-density, surface-mounting assembly techniques. The facilities also provide a means of accessing and controlling design-for-test features built into the digital integrated circuits themselves. The circuitry includes a standard interface through which instructions and test data are communicated. A set of test features is defined, including a boundary-scan register, so that the component is able to respond to a minimum set of instructions designed to assist with testing of assembled printed circuit boards. (This publication includes IEEE 1149.1a-1993.)

IEEE Std. 1149.1b-1994 (R1996). *IEEE Supplement to Standard Test Access Port and Boundary-Scan Architecture.* A language to describe components that conform to IEEE Std 1149.1-1990 is described in this supplement. The language is based on the VHSIC Hardware Description Language (VHDL). General characteristics, the overall structure of a boundary-scan description language (BSDL) description, special cases, and example packages are included.

IEEE Std 1149.4-1999. *IEEE Standard for a Mixed-Signal Test Bus.* The testability structure for digital circuits described in IEEE Std 1149.1-1990 has been extended to provide similar facilities for mixed-signal circuits. The architecture is described, together with the means of control of and access to both analog and digital test data. Sample implementation and application details (which are not part of the standard) are included for illustration.

IEEE Std 1149.5-1995. *IEEE Standard for Module Test and Maintenance Bus (MTM-Bus) Protocol.* This Standard specifies a serial, backplane, test and maintenance bus (MTM-Bus) that can be used to integrate modules from different design teams or vendors into testable and maintainable subsystems. Physical, link, and command layers are specified. Standard interface protocol and commands can be used to provide the basic test and maintenance features needed for a module as well as access to on-module assets (memory, peripherals, etc.) and IEEE Std 1149.1 boundary-scan. Standard commands and functions support fault isolation to individual modules and test of backplane interconnect between modules.

IEEE Std 1150-1991 (R1998). *IEEE Trial-Use Recommended Practice for Integrating Power Plant Computer-Aided Engineering (CAE) Applications.* A data model, called the plant information network, that standardizes categories of generating plant data and data relationships is presented. Guidelines are provided for using the model to integrate computer-aided engineering (CAE) applications across the spectrum of plant work activities during the complete cycle of the plant from site selection through decommissioning. Instructions are given to aid the utility's engineering, construction, and operating groups in specifying integrated CAE applications. The information engineering concepts that are the basis for integrated CAE development are covered.

IEEE Std 1155-1992 (R1998). *IEEE Standard for VMEbus Extensions for Instrumentation: VXIbus.* A technically sound modular instrument standard based on IEEE Std 1014-1987, IEEE Standard for a Versatile Backplane Bus: VMEbus, which is open to all manufacturers and is compatible with present industry standards, is defined. The VXIbus specification details the technical requirements of VXIbus compatible components, such as mainframes, backplanes, power supplies, and modules.

IEEE Std 1156.1-1993 (R1998). *IEEE Standard Microcomputer Environmental Specifications for Computer Modules.* Fundamental information on minimum environmental withstand conditions is provided. The information is intended to be used in those cases in which a generic or detail specification for a certain module has been prepared. The intent is to achieve uniformity and reproducibility in the test conditions for all modules that may make up larger systems and are purported to have a rated environmental performance level. The specifications pertain to both the natural and artificial environments to which modules may be exposed. These conditions include, but are not limited to, thermal, mechanical, electrical, and atmospheric stresses

IEEE Std 1156.2-1996. *IEEE Standard for Environmental Specifications for Computer Systems.* This standard is designed for use in conjunction with other documents such as the IEEE 1101 group of standards, the IEEE 896 group of standards, the IEEE 1596 group of standards, the IEEE 1014 group of standards, and ISO/IEC 10861:1994. This standard is one of the IEEE P1156.x series for environmental specifications. It is intended to be used as a core specification. It contains minimum environmental withstand conditions applicable to computer systems and all of their associated components. It has been created to provide general environmental withstand conditions for one or more of the above listed computer busses or interconnect standards, and electronic equipment in general.

IEEE Std 1156.4-1997. *IEEE Standard for Environmental Specifications for Spaceborne Computer Modules.* Fundamental information on minimum environmental withstand conditions for space electronics is provided. The intent is to achieve uniformity and reproducibility in the test conditions for all spaceborne computer modules that may make up larger systems and are purported to have a rated environmental performance level. The specifications pertain to both the natural and artificial environments to which spaceborne computer modules may be exposed. These conditions include, but are not limited to, thermal, mechanical, electrical, and radiation stresses.

IEEE Std 1158-1991 (R1996). *IEEE Recommended Practice for Determination of Power Losses in High-Voltage Direct-Current (HVDC) Converter Stations.* A set of standard procedures for determining and verifying the total losses of a high-voltage direct-current (HVDC) converter station is recommended. The procedures are applicable to all parts of the converter station and cover standby, partial load, and full load losses and methods of calculation and measurement. All line commutated converter stations used for power exchange in utility systems are covered. Loss determination procedures for synchronous compensators or static VAR compensators are not included.

IEEE Std 1159-1995. *IEEE Recommended Practice on Monitoring Electrical Power Quality.* The monitoring of electric power quality of ac power systems, definitions of power quality terminology, impact of poor power quality on utility and customer equipment, and the measurement of electromagnetic phenomena are covered.

IEEE Std 1160-1993 (R1999). *IEEE Standard Test Procedures for High-Purity Germanium Crystals for Radiation Detectors.* This standard applies to the measurement of bulk properties of high-purity germanium as they relate to the fabrication and performance of germanium detectors for gamma rays and x rays. Such germanium is monocrystalline and has a net concentration of fewer than 1011 electrically active impurity center per cm3, usually on the order of 1010 cm^{-3}.

IEEE Std 1164-1993 (R1999). *IEEE Standard Multivalue Logic System for VHDL Model Interoperability (Std_logic_1164).* This standard is embodied in the Std_logic_1164 package declaration and the semantics of the Std_logic_1164 body. An annex is provided to suggest ways in which one might use this package.

IEEE Std 1175-1992 (R1999). *IEEE Trial-Use Standard Reference Model for Computing System Tool Interconnections.* Reference models for tool-to-organization interconnections, tool-to-platform interconnections, and information transfer among tools are provided. The purpose is to establish agreements for information transfer among tools in the contexts of human organization, a computer system platform, and a software development application. To make the transfer of semantic information among tools easier, a semantic transfer language (STL) is also provided. Interconnections that must be considered when buying, building, testing, or using computing system tools for specifying behavioral descriptions or requirements of system and software products are described.

IEEE Std 1177-1993 (R1997). *IEEE Standard FASTBUS Modular High-Speed Data Acquisition and Control System and IEEE FASTBUS Standard Routines.* Mechanical, signal, electrical, and protocol specifications are given for a modular data bus system, which, while allowing equipment designers a wide choice of solutions, ensure compatibility of all designs that obey the mandatory parts of the specification. This standard applies to systems consisting of modular electronic instrument units that process or transfer data or signals, normally in association with computers or other automatic data processors. Standard software routines for use with the system in IEEE Std 960-1993 are defined.

IEEE Std 1178-1990 (R1995). *IEEE Standard for the Scheme Programming Language.* The form and meaning of programs written in the Scheme programming language, in particular, their syntax, the semantic rules for interpreting them, and the representation of data to be input or output by them, are specified. The fundamental ideas of the language and the notational conventions used for describing and writing programs in the language are presented. The syntax and semantics of expressions, programs, and definitions are specified. Scheme's built-in procedures, which include all of the language's data manipulation and input/output primitives, are described, and a formal syntax for Scheme written in extended Backus-Naur form is provided. Formal denotational semantics for Scheme and some issues in the implementation of Scheme's arithmetic are covered in the appendixes.

IEEE Std 1184-1994 (R1997). *IEEE Guide for the Selection and Sizing of Batteries for Uninterruptible Power Systems.* The characteristics of the various battery energy systems available are described so that users can select the system best suited to their requirements. This guide also describes how the rectifier and the inverter components of the uninterruptible power system (UPS) can relate to the selection of the battery system.

IEEE Std 1185-1994 (R2000). *IEEE Guide for Installation Methods for Generating Station Cables.* Installation methods to improve cable installation practices in generating stations are provided. These include cable lubrication methods, conduit-cable pulling charts, pull rope selection criteria, pulling attachment methods, and alternative methods to traditional cable pulling tension monitoring. This guide supplements IEEE Std 422-1986 and IEEE Std 690-1984, which provide specific cable installation limits. This guide may also be of benefit to cable pulling crews in commercial and industrial facilities when similar cable types and raceways are used.

IEEE Std 1187-1996. *IEEE Recommended Practice for Installation Design and Installation of Valve-Regulated Lead-Acid Storage Batteries for Stationary Applications.* Recommended design practices and procedures for storage, location, mounting, ventilation, instrumentation, preassembly, assembly, and charging of valve-regulated lead-acid (VRLA) storage batteries are provided. Recommended safety practices are also included. This recommended practice applies to all VRLA battery stationary installations.

IEEE Std 1188-1996. *IEEE Recommended Practice for Maintenance, Testing, and Replacement of Valve-Regulated Lead-Acid (VRLA) Batteries for Stationary Applications.* Maintenance, test schedules and testing procedures that can be used to optimize the life and performance of valve-regulated lead-acid (VRLA) batteries for stationary applications are covered. Guidance to determine when batteries should be replaced is also provided.

IEEE Std 1189-1996. *IEEE Guide for Selection of Valve-Regulated Lead-Acid (VRLA) Batteries for Stationary Applications.* Methods for selecting the appropriate type of valve-regulated, immobilized-electrolyte, recombinant lead-acid battery for any of a variety of potential stationary float applications are described.

IEEE Std 1193-1994. *IEEE Guide for Measurement of Environmental Sensitivities of Standard Frequency Generators.* Standard frequency generators that include all atomic frequency standards and precision quartz crystal oscillators are addressed.

IEEE Std 1194.1-1991 (R2000). *IEEE Standard for Electrical Characteristics of Backplane Transceiver Logic (BTL) Interface Circuits.* The electrical characteristics of digital interface circuits (drivers, receivers, or transceivers), used to drive a backplane bus that appears as a transmission line to the interface circuit, are specified in order to ensure proper electrical functioning with respect to timing and noise constraints. The performance requirements of buses using these interface circuits make it necessary to impose constraints beyond those normally encountered in a specification for backplane buses. These constraints are imposed to: (1) optimize the system to accommodate the physical behavior of high-speed signals traveling between boards in a transmission-line environment; (2) minimize backplane propagation delay and backplane propagation skew between these signals to attain the required performance; and (3) minimize cross talk to attain the required reliability of information transfer.

IEEE Std 1202-1991 (R1996). *IEEE Standard for Flame Testing of Cables for Use in Cable Tray in Industrial and Commercial Occupancies.* A test protocol and the performance criteria used to determine the flame propagation tendency of cables in a vertical cable tray are established. The standard applies to single-insulated and multiconductor cables. The test consists of exposing cable samples to a theoretical 20 kW (70 000 Btu/h) flaming ignition source for a 20 min duration. The test facility, test sample requirements, test procedure, and evaluation of results are covered.

IEEE Std 1204-1997. *IEEE Guide for Planning DC Links Terminating at AC Locations Having Low Short-Circuit Capacities.* Guidance on the planning and design of dc links terminating at ac system locations having low short-circuit capacities relative to the dc power infeed is provided in this guide. This guide is limited to the aspects of interactions between ac and dc systems that result from the fact that the ac system is weak compared to the power of the dc link (i.e., ac system appears as a high impedance at the ac/dc interface bus). This guide contains two parts: Part I, AC/DC Interaction Phenomena, classifies the strength of the ac/dc system, provides information about interactions between ac and dc systems, and gives guidance on design and performance; and Part II, Planning Guidelines, considers the impact of ac/dc system interactions and their mitigation on economics and overall system performance and discusses the studies that need to be performed.

IEEE Std 1205-1993. *IEEE Guide for Assessing, Monitoring, and Mitigating Aging Effects on Class 1E Equipment Used in Nuclear Power Generating Stations.* The guidelines are provided for assessing, monitoring, and mitigating degradation of Class 1E equipment used in nuclear power generating stations due to aging. The methods described can be used to identify the performance capability of Class 1E equipment beyond its qualified life. A discussion of stressors and aging mechanisms is included. If aging considerations have been satisfactorily addressed through other means (e.g., equipment qualification), then use of this guide may not be warranted. For some equipment, only partial application of this guide may be warranted.

IEEE Std 1206-1994. *IEEE Standard Methods for Measuring Transmission Performance of Telephone Handsets and Headsets.* Practical methods for measuring the transmission characteristics of a telephone handset or headset by means of using a test connection to obtain objective measurements are provided. The obtained test results may be used as a means of evaluating or specifying the transmission performance of a handset or headset on a standardized basis.

IEEE Std 1210-1996. *IEEE Standard Tests for Determining Compatibility of Cable-Pulling Lubricants with Wire and Cable.* Criteria and test methods for determining the compatibility of cable-pulling lubricants (compounds) with cable jacket or other exterior cable covering are described in this standard. Cable-pulling lubricants are used to lower the tension on cable as it is pulled into conduit, duct, or directionally bored holes. Compatibility is important because lubricants should not negatively interact with the cables they lubricate. Compatibility of lubricants with a variety of common cable coverings is considered.

IEEE Std 1212.1-1993 (R1996). *IEEE Standard for Communicating Among Processors and Peripherals Using Shared Memory (Direct Memory Access—DMA).* Primitive yet high-performance means are defined for passing messages across the bus between the Processor and some form of intelligence in the I/O Unit's Function. This message-passing scheme makes minimal demands on the instruction set and hardware required. In addition, several simple conventions are defined for the structure of the messages passed. The intent is to provide a standard architectural framework that supports the detailed definition of application-dependent I/O Unit and Function interface standards. The algorithms and definitions themselves are useful in the design of integrated circuits for I/O.

IEEE Std 1214-1992 (R1999). *IEEE Standard Multichannel Analyzer (MCA) Histogram Data Interchange Format for Nuclear Spectroscopy.* A standard format for data interchange used to transfer pulse height data on magnetic media between laboratories is provided. The terms used in file records are defined. The contents consist only of ASCII characters and can be transmitted over networks and other direct links. Example programs to read data in FORTRAN, BASIC and C are provided.

IEEE Std 1219-1998. *IEEE Standard for Software Maintenance.* The process for managing and executing software maintenance activities is described.

IEEE Std 1220-1998. *IEEE Standard for Application and Management of the Systems Engineering Process.* The interdisciplinary tasks, which are required throughout a system's life cycle to transform customer needs, requirements, and constraints into a system solution, are defined. In addition, the requirements for the systems engineering process and its application throughout the product life cycle are specified. The focus of this standard is on engineering activities necessary to guide product development while ensuring that the product is properly designed to make it affordable to produce, own, operate, maintain, and eventually to dispose of, without undue risk to health or the environment.

IEEE Std 1226-1998. *IEEE Standard for a Broad-Based Environment for Test (ABBET®: Overview and Architecture).* The overall concept of A Broad-Based Environment for Test (ABBET) is defined, and mandatory requirements for implementation of ABBET are specified. The elements of ABBET and their interrelationships are described. Guidelines and requirements governing the elements of the ABBET set of standards and guides are established, and common terms to be used throughout the set are defined.

IEEE Std 1226.3-1998. *IEEE Standard for Software Interface for Resource Management for A Broad-Based Environment for Test (ABBET®).* The services needed to access and manage descriptive information about resources in an automatic test system (ATS) are covered. This information includes data about the automatic test equipment (ATE) instruments, switching, and the test subject adapter. This standard is a component of the ABBET set of standards.

IEEE Std 1226.6-1996. *IEEE Guide for the Understanding of A Broad-Based Environment for Test (ABBET®) Standard.* As a part of the family of IEEE ABBET standards, this guide facilitates an understanding of the relationships of IEEE ABBET 1226-1993 and its component standards, as well as the relationship of an ABBET implementation with the design, production, support, and operational environments with which it may be used.

IEEE Std 1227-1990 (R1995). *IEEE Guide for the Measurement of DC Electric-Field Strength and Ion Related Quantities.* Guidance is provided for measuring the electric-field strength, ion-current density, conductivity, monopolar space-charge density, and net space-charge density in the vicinity of high-voltage dc (HVDC) power lines, in converter substations, and in apparatus designed to simulate the HVDC power-line environment. The interrelationship between electrical parameters and the operating principles of measuring instruments are described. Methods of calibration are suggested where applicable, and measurement procedures are given. Significant sources of measurement error are identified.

IEEE Std 1228-1994. *IEEE Standard For Software Safety Plans.* The minimum acceptable requirements for the content of a software safety plan are established. This standard applies to the software safety plan used for the development, procurement, maintenance, and retirement of safety-critical software. This standard requires that the plan be prepared within the context of the system safety program. Only the safety aspects of the software are included. This standard does not contain special provisions required for software used in distributed systems or in parallel processors.

IEEE Std 1232-1995. *IEEE Trial-Use Standard for Artificial Intelligence and Expert System Tie to Automatic Test Equipment (AI-ESTATE): Overview and Architecture.* This document is the base standard for the AI-ESTATE set of standards. The overall concept of AI-ESTATE, which is a set of specifications for data interchange and for standard services for the test and diagnostic environment, is defined; mandatory requirements for implementing AI-ESTATE are specified; the elements of AI-ESTATE and their interrelationships are described; guidelines and requirements to govern the documents in the AI-ESTATE set of standards are established; and the terminology used throughout the set is defined. The purpose of the AI-ESTATE set of standards is to standardize interfaces between functional elements of an intelligent test environment and representations of knowledge and data for the functional elements of the intelligent test environment.

IEEE Std 1232.1-1997. *IEEE Trial-Use Standard for Artificial Intelligence Exchange and Service Tie to All Test Environments (AI-ESTATE): Data and Knowledge Specification.* Formal models for information used in system diagnosis are defined. As part of the AI-ESTATE set of standards, this standard includes several models that form the basis for a format to facilitate exchange of persistent diagnostic information between two reasoners, and also provides a formal typing system for the services defined in the AI-ESTATE service specification.

IEEE Std 1232.2-1998. *IEEE Trial-Use Standard for Artificial Intelligence Exchange and Service Tie to All Test Environments (AI-ESTATE): Service Specification.* Formal software interfaces to system diagnosis tools and applications are defined. As part of the AI-ESTATE set of standards, this standard defines services to manipulate information models as defined in IEEE Std 1232.1-1997 and to control a diagnostic reasoner. This standard includes a new information model to manipulate dynamic information obtained during the process of system diagnosis. Service bindings to ANSI C and ANSI Ada are also provided.

IEEE Std 1233-1998 Edition. (Includes IEEE Std 1233-1996 and IEEE Std 1233a-1998). *IEEE Guide for Developing System Requirements Specifications.* Guidance for the development of the set of requirements, System Requirements Specification (SyRS), that will satisfy an expressed need is provided. Developing an SyRS includes the identification, organization, presentation, and modification of the requirements. Also addressed are the conditions for incorporating operational concepts, design constraints, and design configuration requirements into the specification. This guide also covers the necessary characteristics and qualities of individual requirements and the set of all requirements.

IEEE Std 1242-1999. *IEEE Guide for Specifying and Selecting Power, Control, and Special-Purpose Cable for Petroleum and Chemical Plants.* Information on the specification and selection of power, control, and special-purpose cable, as typically used in petroleum, chemical, and similar plants, is provided in this guide. Materials, design, testing, and applications are addressed. More recent developments, such as strand filling, low-smoke, zero-halogen materials, and chemical-moisture barriers have been included.

IEEE Std 1243-1997. *IEEE Guide for Improving the Lightning Performance of Transmission Lines.* The effects of routing, structure type, insulation, shielding, and grounding on transmission lines are discussed. The way these transmission-line choices will improve or degrade lightning performance is also provided. An additional section discusses several special methods that may be used to improve lightning performance. Finally, a listing and description of the FLASH program is presented.

IEEE 1246-1997. *IEEE Guide for Temporary Protective Grounding Systems Used in Substations.* The design, performance, use, testing, and installation of temporary protective grounding systems, including the connection points, as used in permanent and mobile substations are covered.

IEEE Std 1247-1998. *IEEE Standard for Interrupter Switches for Alternating Current, Rated Above 1000 Volts.* The basic requirements of interrupter switches used indoors, outdoors, and in enclosures are covered. This standard does not apply to load-break separable insulated connectors.

IEEE Std 1248-1998. *IEEE Guide for the Commissioning of Electrical Systems in Hydroelectric Power Plants.* Inspection procedures and tests for use following the completion of the installation of components and systems through to commercial operation are provided. This guide is directed to the plant owners, designers, and contractors involved in the commissioning of electrical systems of hydroelectric plants.

IEEE Std 1249-1996. *IEEE Guide for Computer-Based Control for Hydroelectric Power Plant Automation.* The application, design concepts, and implementation of computer-based control systems for hydroelectric power plant automation is addressed. Functional capabilities, performance requirements, interface requirements, hardware considerations, and operator training are discussed. Recommendations for system testing and acceptance are provided, and case studies of actual computer-based control applications are presented.

IEEE Std 1250-1995. *IEEE Guide for Service to Equipment Sensitive to Momentary Voltage Disturbances.* Computers, computer-like products, and equipment using solid-state power conversion have created entirely new areas of power quality considerations. There is an increasing awareness that much of this new user equipment is not designed to withstand the surges, faults, and reclosing duty present on typical distribution systems. Momentary voltage disturbances occurring in ac power distribution and utilization systems, their potential effects on this new, sensitive, user equipment, and guidance toward mitigation of these effects are described. Harmonic distortion limits are also discussed.

IEEE Std 1260-1996. *IEEE Guide on the Prediction, Measurement and Analysis of AM Broadcast Reradiation by Power Lines.* A set of procedures to be followed to cope with reradiation of AM broadcast signals from power lines and other large metallic structures is provided. Reradiation may be described as electromagnetic waves radiated from a structure that has parasitically picked up a signal from the environment. A simplified prediction technique called a survey is described to determine which structures could possibly cause a problem. Guidelines for measurements and data analysis are included.

IEEE Std 1262-1995. *IEEE Recommended Practice for Qualification of Photovoltaic (PV) Modules.* Recommended procedures and specifications for qualification tests that are structured to evaluate terrestrial flat-plate photovoltaic non-concentrating modules intended for power generation applications are established.

IEEE Std 1264-1993. *IEEE Guide for Animal deterrents for Electrical Power Supply Substations.* Methods and designs to mitigate interruptions and equipment damage resulting from animal intrusions into electric power supply substations thereby improving reliability and minimizing the associated revenue loss are addressed.

IEEE Std 1267-1999. *IEEE Trial-Use Guide for Development of Specification for Turnkey Substation Projects.* The technical requirements to engineer, design, specify, fabricate, manufacture, furnish, install, test, commission, and provide as-built documents for air-insulated substations are covered. This guide investigates the methods, practices, and requirements of both users and suppliers in order to promogate a systematic and coordinated approach for development of specifications for turnkey substation projects.

IEEE Std 1268-1997. *IEEE Guide for the Safe Installation of Mobile Substation Equipment.* Information pertaining to the installation of mobile substation equipment up to 230 kV is provided.

IEEE Std 1275.4-1995. *IEEE Standard for Boot (Initialization Configuration) Firmware: Bus Supplement for IEEE 896 (FutureBus+®).* Firmware is the read-only-memory (ROM) based software that controls a computer between the time it is turned on and the time the primary operating system takes control of the machine. Firmware's responsibilities include testing and initializing the hardware, determining the hardware configuration, loading (or booting) the operating system, and providing interactive debugging facilities in case of faulty hardware or software. The core requirements and practices specified by IEEE Std 1275-1994 must be supplemented by system-specific requirements to form a complete specification for the firmware for a particular system. This standard establishes such additional requirements pertaining to the bus architecture defined by the IEEE Futurebus+ standards: ISO/IEC 10857:1994 [ANSI/IEEE Std 896.1, 1994 Edition], Information technology—Microprocessor systems—Futurebus+ —Logical protocol specification; and IEEE Std 896.2-1991, IEEE Standard for Futurebus+ —Physical Layer and Profile Specification.

IEEE Std 1276-1997. *IEEE Trial-Use Guide for the Application of High-Temperature Insulation Materials in Liquid-Immersed Power Transformers.* Technical information is provided related to liquid-immersed power transformers insulated with high-temperature materials. Guidelines for applying existing qualified high-temperature materials to certain insulation systems, recommendations for loading high-temperature liquid-immersed power transformers, and technical information on insulation-system temperature ratings and test procedures for qualifying new high-temperature materials are included.

IEEE Std 1277-2000. *IEEE Trial-Use General Requirements and Test Code for Dry-Type and Oil-Immersed Smoothing Reactors for DC Power Transmission.* The electrical, mechanical, and physical requirements of oil-immersed and dry-type air core smoothing reactors for high-voltage direct current (HVDC) applications are specified. Test code is defined and appropriate technical background information is presented or identified.

IEEE 1278.1-1995. *IEEE Standard for Distributed Interactive Simulation—Applications Protocols.* Data messages, known as protocol data units (PDUs), that are exchanged on a network between simulation applications are defined. These PDUs are for interactions that take place within specified domains called protocol families, which include Entity Information/Interaction, Warfare, Logistics, Simulation Management, Distributed Emission Regeneration, and Radio Communications.

IEEE Std 1278.1a-1998. *IEEE Standard for Distributed Interactive Simulation—Application Protocols.* Data messages, known as protocol data units (PDUs), that are exchanged on a network between simulation applications are defined. These PDUs are for interactions that take place within specified domains called protocol families, which include Entity Information/Interaction, Warfare, Logistics, Simulation Management, Distributed Emission Regeneration, Radio Communications, Entity Management, Minefield, Synthetic Environment, Simulation Management with Reliability, Live Entity Information/Interaction, and Non-Real Time.

IEEE Std 1278.2-1995. *IEEE Standard for Distributed Interactive Simulation—Communication Services and Profiles.* Communication services to support information exchange between simulation applications participating in the Distributed Interactive Simulation (DIS) environment are defined. These communication services describe a connectionless information transfer that supports real-time, as well as non-real-time, exchange. Several communication profiles specifying communication services are provided.

IEEE Std 1278.3-1996. *IEEE Recommended Practice for Distributed Interactive Simulation-Exercise Management and Feedback.* Guidelines are established for exercise management and feedback in Distributed Interactive Simulation (DIS) exercises. Guidance is provided to sponsors, providers, and supporters of DIS compliant systems and exercises as well as to developers of DIS exercise management and feedback stations. The activities of the organizations involved in a DIS exercise and the top-level processes used to accomplish those activities are addressed. The functional requirements of the exercise management and feedback process are also addressed. This standard is one of a series of standards developed for DIS to assure interoperability between dissimilar simulations for currently installed and future simulations developed by different organizations.

IEEE Std 1278.4-1997. *IEEE Trial-Use Recommended Practice for Distributed Interactive Simulation—Verification, Validation, and Accreditation.* Guidelines are established for the verification, validation, and accreditation (VV&A) of distributed interactive simulation (DIS) exercises. "How-to" procedures for planning and conducting DIS exercise VV&A are provided. Intended for use in conjunction with IEEE Std 1278.3-1996, this recommended practice presents data flow and connectivity for all proposed verification and validation activities and provides rationale and justification for each step. VV&A guidance is provided to exercise users/sponsors and developers.

IEEE Std 1284-1994. *IEEE Standard Signaling Method for a Bidirectional Parallel Peripheral Interface for Personal Computers.* A signaling method for asynchronous, fully interlocked, bidirectional parallel communications between hosts and printers or other peripherals is defined. A format for a peripheral identification string and a method of returning this string to the host outside of the bidirectional data stream is also specified.

IEEE Std 1284.1-1997. *IEEE Standard for Information Technology—Transport Independent Printer/System Interface (TIP/SI).* A protocol and methodology for software developers, computer vendors, and printer manufacturers to facilitate the orderly exchange of information between printers and host computers are defined in this standard. A minimum set of functions that permit meaningful data exchange is provided. Thus a foundation is established upon which compatible applications, computers, and printers can be developed, without compromising an individual organizations desire for design innovation.

IEEE Std 1289-1998. *IEEE Guide for the Application of Human Factors Engineering in the Design of Computer-Based Monitoring and Control Displays for Nuclear Power Generating Stations.* System design considerations, information display and control techniques for use with computer-based displays, and human factors engineering guidance for the use of these techniques in nuclear power generating stations are provided.

IEEE Std 1290-1996 (R2000). *IEEE Guide for Motor Operated Valve (MOV) Motor Application, Protection, Control, and Testing in Nuclear Power Generating Stations.* Motors used to drive valve operators in nuclear power generating stations are discussed. Guidelines to evaluate the adequacy of motors used to drive valve operators; to provide recommendations for motor application; and to provide methods for protection, control, and testing of motors used for valve operation are presented.

IEEE Std 1291-1993 (R1998). *IEEE Guide for Partial Discharge Measurement in Power Switchgear.* The IEEE Guide for Partial Discharge Measurement in Power Switchgear defines methods of measuring partial discharges that may occur in energized power switchgear apparatus in flaws, voids and interfaces of non-self restoring insulation which may then result in dielectric failure of the switchgear. Guidance on instrumentation and calibration technique are also given. This guide defines methods of measuring partial discharges that may occur in energized power switchgear apparatus in flaws, voids, and interfaces of non-self-restoring insulation that may then result in dielectric failure of the switchgear. Guidance on instrumentation and calibration technique is also given.

IEEE Std 1299/C62.22.1-1996. *IEEE Guide for the Connection of Surge Arresters to Protect Insulated, Shielded Electric Power Cable Systems.* This guide suggests surge arrester installation methods at distribution cable terminal poles in order to minimize the total impressed transient voltage on medium-voltage distribution cables. Grounding electrode techniques, pole ground values, and system ground grid values are not addressed or considered in this document.

IEEE Std 1300-1996. *IEEE Guide for Cable Connections for Gas-Insulated Substations.* The coordination of design, material supply, installation, and test procedures required for the connection of a gas-insulated substation (GIS) is described. Preferred dimensions for mechanical and electrical interchangeability for voltage classes of 69 kV and above are established.

IEEE Std 1301-1991 (R1994). *IEEE Standard for a Metric Equipment Practice for Microcomputers—Coordination Document.* The metric mechanical coordination of microcomputer components, including the cabinet, rack, subracks, printed boards, and common connector-dependent dimensions for connector pitches of 2.5, 2.0, 1.5, 1.0, and 0.5 mm is addressed. This generic standard may be applied in all fields of electronics where equipment and installations are required to conform to a metric modular order. The choice of coordination dimensions for the mechanical structure for heights, widths, and depths lies within a homogeneous, metric modular three-dimensional grid as specified in IEC 917. The intent is to provide a single metric equipment practice for worldwide use.

IEEE Std 1301.1-1991 (R1994). *IEEE Standard for a Metric Equipment Practice for Microcomputers—Convection-Cooled With 2 mm Connectors.* The metric mechanical coordination of microcomputer components, including the cabinet, rack, subracks, printed boards, and common connector-dependent dimensions for connector pitches of 2.5, 2.0, 1.5, 1.0, and 0.5 mm is addressed. This generic standard may be applied in all fields of electronics where equipment and installations are required to conform to a metric modular order. The choice of coordination dimensions for the mechanical structure for heights, widths, and depths lies within a homogeneous, metric, modular, three-dimensional grid as specified in IEC 917. The intent is to provide a single metric equipment practice for worldwide use.

IEEE Std 1301.2-1993 (R1997). *IEEE Recommended Practices for the Implementation of a Metric Equipment Practice.* Recommendations provide guidance in the implementation of the generic standard, IEEE Std 1301-1991, and the connector-related standards, such as IEEE Std 1301.1-1991 and IEEE Std 1301.3-1992. This recommended practice may be applied in all fields of electronics where equipment and installations are required to conform to a metric modular order. The IEEE 1301 metric equipment practices are in accordance with IEC 917 (1988) IEC 917-0 (1989), including cabinet, rack, subracks, printed boards, and common connector-dependent dimensions for connector pitches of 2.5, 2.0, 115, 1.0, and 0.5 mm.

IEEE Std 1301.3-1992 (R1997). *IEEE Standard for Metric Practice for Microcomputers—Convection-Cooled with 2.5 mm Connectors.* Dimension requirements are presented for subracks, plug-in units, printed boards, and backplanes to be used in conjunction with IEEE Std 1301-1991 and with a 2.5 mm connector as defined in IEC 48B (Central Office) 245. The general arrangement, dimensions, and environmental requirements are covered. This standard may be used with other IEEE Std 1301.x connector implementations in the subracks.

IEEE Std 1301.4-1996. *IEEE Standard for a Metric Equipment Practice for Microcomputers—Coordination Document for Mezzanine Cards.* This standard establishes the metric modular order and coordination dimensions for mezzanine cards for use on host modules.

IEEE Std 1302-1998. *IEEE Guide for the Electromagnetic Characterization of Conductive Gaskets in the Frequency Range of DC to 18 GHz.* Information to assist users of gaskets in evaluating gasket measurement techniques to determine which reveal the properties critical to the intended application, to highlight limitations and sources of error of the competing measurement techniques, and to provide a basis for comparing the techniques is provided. emphasis is placed on those measurement techniques that have been adopted through incorporation into standards, both commercial and military, or that have been used extensively.

IEEE Std 1303-1994 (R2000). *IEEE Guide for Static Var Compensator Field Tests.* General guidelines and criteria for the field testing of static var compensators (SVCs), before they are placed in-service, for the purpose of verifying their specified performance are described. The major elements of a commissioning program are identified so that the user can formulate a specific plan that is most suited for his or her own SVC.

IEEE Std 1307-1996. *IEEE Trial-Use Guide for Fall Protection for the Utility Industry.* General recommendations for fall protection and worker protection are provided. Sufficient details of the methods, equipment, and training requirements necessary to provide safe and adequate procedures for personnel working at elevated worksites are presented.

IEEE Std 1308-1994. *IEEE Recommended Practice for Instrumentation: Specifications for Magnetic Flux Density and Electric Field Strength Meters—10 Hz to 3 kHz.* Specifications that should be provided to characterize instrumentation used to measure the steady state rms value of magnetic and electric fields with sinusoidal frequency content in the range 10 Hz to 3 kHz in residential and occupational settings as well as in transportation systems are identified. The instrumentation, recommended calibration methods, and sources of measurement uncertainty are also described.

IEEE Std 1309-1996. *IEEE Standard for Calibration of Electromagnetic Field Sensors and Probes, Excluding Antennas from 9 kHz to 40 GHz.* Consensus calibration methods for electromagnetic field sensors and field probes are provided. Data recording and reporting requirements are given, and a method for determining uncertainty is specified.

IEEE Std 1310-1996. *IEEE Trial-Use Recommended Practice for Thermal Cycle Testing of Form-Wound Stator Bars and Coils for Large Generators.* A test method to determine the relative ability of high-voltage, form-wound stator bars and coils of large rotating machines to resist deterioration due to rapid heating and cooling resulting from machine load cycling is described.

IEEE Std 1312-1993. (Reaffirmation and redesignation of ANSI C92.2-1987). *IEEE Standard Preferred Voltage Ratings for Alternating-Current Electrical Systems and Equipment Operating at Voltages Above 230 kV.* Preferred voltage ratings above 230 kV nominal for alternating-current (ac) systems and equipment are provided, along with definitions of various types of system voltages.

IEEE Std 1313.1-1996. *IEEE Standard for Insulation Coordination—Definitions, Principles, and Rules.* The procedure for selection of the withstand voltages for equipment phase-to-ground and phase-to-phase insulation systems is specified. A list of standard insulation levels, based on the voltage stress to which the equipment is being exposed, is also identified. This standard applies to three-phase ac systems above 1 kV.

IEEE Std 1313.2-1999. *IEEE Guide for the Application of Insulation Coordination.* The calculation method for selection of phase-to-ground and phase-to-phase insulation withstand votlages for equipment is presented. This guide gives methods for insulation coordination of different air-insulated systems like transmission lines and substations. The methods of analysis are illustrated by practical examples.

IEEE Std 1320.1-1998. *IEEE Standard for Functional Modeling Language—Syntax and Semantics for IDEF0.* IDEF0 function modeling is designed to represent the decisions, actions, and activities of an existing or prospective organization or system. IDEF0 graphics and accompanying texts are presented in an organized and systematic way to gain understanding, support analysis, provide logic for potential changes, specify requirements, and support system-level design and integration activities. IDEF0 may be used to model a wide variety of systems, composed of people, machines, materials, computers, and information of all varieties and structured by the relationships among them, both automated and nonautomated. For new systems, IDEF0 may be used first to define requirements and to specify functions to be carried out by the future system. As the basis of this architecture, IDEF0 may then be used to design an implementation that meets these requirements and performs these functions. For existing systems, IDEF0 can be used to analyze the functions that the system performs and to record the means by which these are done.

IEEE Std 1320.2-1998. *IEEE Standard for Conceptual Modeling Language Syntax and Semantics for IDEF1X97 (IDEFobject).* IDEF1X97 consists of two conceptual modeling languages. The key-style language supports data/information modeling and is downward compatible with the US government's 1993 standard, FIPS PUB 184. The identity-style language is based on the object model with declarative rules and constraints. IDEF1X97 identity style includes constructs for the distinct but related components of object abstraction: interface, requests, and realization; utilizes graphics to state the interface; and defines a declarative, directly executable Rule and Constraint Language for requests and realizations. IDEF1X97 conceptual modeling supports implementation by relational databases, extended relational databases, object databases, and object programming languages. IDEF1X97 is formally defined in terms of first order logic. A procedure is given whereby any valid IDEF1X97 model can be transformed into an equivalent theory in first order logic. That procedure is then applied to a meta model of IDEF1X97 to define the valid set of IDEF1X97 models.

IEEE Std 1325-1996. *IEEE Recommended Practice for Reporting Field Failure Data for Power Circuit Breakers.* A format is presented that provides a concise and meaningful method for recording pertinent information on power circuit breaker field failures. It is recommended that this format be utilized in record keeping and directing corrective action to improve field reliability of power circuit breakers.

IEEE Std 1329-1999. *IEEE Standard Method for Measuring Transmission Performance of Handsfree Telephone Sets.* Techniques for objective measurement of electroacoustic and voice switching characteristics of analog and digital handsfree telephones (HFTs) are provided. Due to the various characteristics of HFTs and the environments in which they operate, not all of the test procedures in this standard are applicable to all HFTs. Application of the test procedures to atypical HFTs should be determined on an individual basis.

IEEE Std 1332-1998. *IEEE Standard Reliability Program for the Development and Production of Electronic Systems and Equipment.* Guidance for providing products that satisfy the customer is given. This standard guides suppliers in planning a program that suits their design philosophy, the product concept, and the resources at their disposal, so that every activity adds value. This standard encourages suppliers and customers to cooperatively integrate their reliability processes. Requirements are written to properly establish the contractual or obligatory relationship between the supplier and customer in a product program.

IEEE Std 1333-1994 (R2000). *IEEE Guide for Installation of Cable Using the Guided Boring Method.* The method and equipment involved in proper and economical installation of insulated conductors and/or conduits using the guided boring method are covered. The method addresses installations of: insulated cable, cable preinstalled in conduit (CIC), and conduit alone.

IEEE Std 1344-1995 (R2000). *IEEE Standard for Synchrophasors for Power Systems.* The synchronizing input and the data output for phasor measurements made by substation computer systems is discussed. Processes involved in computing phasors from sampled data, data-to-phasor conversions, and formats for timing imputs and phasor data output from a Phasor Measurement Unit (PMU) are also addressed.

IEEE Std 1346-1998. *IEEE Recommended Practice for Evaluating Electric Power System Compatibility With Electronic Process Equipment.* A standard methodology for the technical and financial analysis of voltage sag compatibility between process equipment and electric power systems is recommended. The methodology presented is intended to be used as a planning tool to quantify the voltage sag environment and process sensitivity. It shows how technical and financial alternatives can be evaluated. Performance limits for utility systems, power distribution systems, or electronic process equipment are not included.

IEEE Std 1348-1995 (R2000). *IEEE Recommended Practice For The Adoption Of Computer-Aided Software Engineering (CASE) Tools.* Difficulties that may be encountered, and how they can be avoided, by organizations intending to adopt CASE tools are addressed. An overview of the adoption process, including analysis of the organization's needs and readiness for automation, use of a pilot project, and definition of activities necessary to integrate the new technology into the organization's standard software engineering practice, is provided.

IEEE Std 1355-1995 (R2000). *IEEE Standard for Heterogeneous InterConnect (HIC), (Low-Cost, Low-Latency Scalable Serial Interconnect for Parallel System Construction).* Enabling the construction of high-performance, scalable, modular, parallel systems with low system integration cost is discussed. Complementary use of physical connectors and cables, electrical properties, and logical protocols for point-to-point serial scalable interconnect, operating at speeds of 10–200 Mb/s and at 1 Gb/s in copper and optic technologies, is described.

IEEE Std 1362-1998. *IEEE Guide for Information Technology—System Definition—Concept of Operations (ConOps) Document.* The format and contents of a concept of operations (ConOps) document are described. A ConOps is a user-oriented document that describes system characteristics for a proposed system from the users viewpoint. The ConOps document is used to communicate overall quantitative and qualitative system characteristics to the user, buyer, developer, and other organizational elements (for example, training, facilities, staffing, and maintenance). It is used to describe the user organization(s), mission(s), and organizational objectives from an integrated systems point of view.

IEEE Std 1364-1995 (R2000). *IEEE Standard Hardware Description Language Based on the Verilog® Hardware Description Language.* The Verilog Hardware Description Language (HDL) is defined. Verilog HDL is a formal notation intended for use in all phases of the creation of electronic systems. Because it is both machine readable and human readable, it supports the development, verification, synthesis, and testing of hardware designs; the communication of hardware design data; and the maintenance, modification, and procurement of hardware. The primary audiences for this standard are the implementors of tools supporting the language and advanced users of the language.

IEEE Std 1366-1998. *IEEE Trial-Use Guide for Electric Power Distribution Reliability Indices.* Useful distribution reliability indices, and factors that affect their calculation, are identified. This guide includes indices that are useful today as well as ones that may be useful in the future. The indices are intended to apply to distribution systems, substations, circuits, and defined regions.

IEEE Std 1374-1998. *IEEE Guide for Terrestrial Photovoltaic Power System Safety.* The design, equipment applicability, and hardware installation of electrically safe, stand-alone, and grid-connected PV power systems operating at less than 50 kW output are addressed. Storage batteries and other generating equipment are discussed briefly.

IEEE Std 1375-1998. *IEEE Guide for the Protection of Stationary Battery Systems.* Guidance in the protection of stationary battery systems is provided. For the purposes of this guide, stationary battery systems include the battery and dc components to and including the first protective device downstream of the battery terminals. This guide does not set requirements; rather, it presents a number of options to the dc system designer of the different types of stationary battery system protection available.

IEEE Std 1377-1997. *IEEE Standard for Utility Industry End Device Data Tables.* Functionally related utility application data elements, grouped into a single data structure for transport are described. Data may be utilized peer-to-peer or upstream to readers or billing systems by being carried by one lower layered protocol to another stack of lower layered protocol. The data structure does not change from end device to the user of the data.

IEEE Std 1378-1997. *IEEE Guide for Commissioning High-Voltage Direct-Current (HVDC) Converter Stations and Associated Transmission Systems.* General guidelines for commissioning high-voltage direct-current (HVDC) converter stations and associated transmission systems are provided. These guidelines apply to HVDC systems utilizing 6-pulse or 12-pulse thyristor-valveconverter units operated as a two-terminal HVDC transmission system or an HVDC back-to-back system.

IEEE Std 1379-1997. *IEEE Trial-Use Recommended Practice for Data Communications Between Intelligent Electronic Devices and Remote Terminal Units in a Substation.* A uniform set of guidelines for communications and interoperations of Intelligent Electronic Devices (IEDs) and Remote Terminal Units (RTUs) in an electric utility substation is provided. A mechanism for adding data elements and message structures to this recommended practice is described.

IEEE Std 1387.2-1995 (R2000). *IEEE Standard for Information Technology—Portable Operating Interface System Interface (POSIX®) System Administration—Part 2: Software Administration.* This standard is part of the POSIX series of standards for applications and user interfaces to open systems. It defines a software packaging layout, a set of information maintained about software, and a set of utility programs to manipulate that software and information.

IEEE Std 1387.3-1996 (R2000). *IEEE Standard for Information Technology—Portable Operating System Interface (POSIX®) System Administration — Part 3: User and Group Account Administration.* IEEE Std 1387.3-1996 System Administration Interface/User and Group Administration for Computer Operating System Environments, is part of the POSIX® series of standards for applications and user interfaces to open systems. The purpose of this standard is to provide a common set of utility programs, for the administration of the User and Group Account entities described in the ISO/IEC 9945-1:1996 (IEEE Std 1003.1-1996) and ISO/IEC 9945-2:1993 (IEEE Std 1003.2-1992) standards.

IEEE Std 1390-1995 (R2000). *IEEE Standard for Utility Telemetry Service Architecture for Switched Telephone Network.* This standard describes a utility telemetry service architecture operated over the telephone network. The architecture described is a basic transport architecture capable of supporting many different applications. The text is described in terms of a utility meter reading application, but any enhanced service provider (ESP) communication can be transported. Telemetry calls may be initiated by either the utility/service provider (outbound) or the telemetry interface unit (TIU)/CPE (inbound) on the end user's premise.

IEEE Std 1390.2-1999. *IEEE Standard for Automatic Meter Reading Via Telephone—Network to Telemetry Interface Unit.* The telephone network interface to a telemetry interface unit operating under the utility telemetry service architecture is described. The interface is described in terms of a utility meter reading application, but any enhanced service provider communication can be transported. Telemetry calls may be initiated by either the utility/enhanced service provider (outbound) or the telemetry Interface unit/customer premise equipment (inbound) on the end user's premise.

IEEE Std 1390.3-1999. *IEEE Standard for Automatic Meter Reading Via Telephone—Network to Utility Controller.* The telephone network interface to a utility controller operating under the utility telemetry service architecture is described. The interface is described in terms of a utility meter reading application but any enhanced service provider communication can be transported. Telemetry calls may be initiated by either the utility/service provider (outbound) or the telemetry interface unit (TIU)/CPE (inbound) on the end user's premises.

IEEE Std 1393-1999. *IEEE Standard for Spaceborne Fiber-Optic Data Bus.* The design requirements for a fiber-optic serial interconnect protocol, topology, and media is established. The application target for this standard is the interconnection of multiple aerospace sensors, processing resources, bulk storage resources, and communications resources onboard aerospace platforms. This standard is for subsystem interconnection, as opposed to intra-backplane connection.

IEEE Std 1394-1995 (R1994). *IEEE Standard for a High Performance Serial Bus.* A high-speed serial bus that integrates well with most IEEE standard 32-bit and 64-bit parallel buses, as well as such nonbus interconnects as the IEEE Std 1596-1992, Scalable Coherent Interface, is specified. It is intended to provide a low-cost interconnect between cards on the same backplane, cards on other backplanes, and external peripherals. This standard follows the IEEE Std 1212-1991 Command and Status Register (CSR) architecture.

IEEE Std 1394a-2000. *IEEE Standard for a High Performance Serial Bus—Amendment 1.* Amended information for a high-speed Serial Bus that integrates well with most IEEE standard 32-bit and 64-bit parallel buses is specified. This amendment is intended to extend the usefulness of a low-cost interconnect between external peripherals, as described in IEEE Std 1394-1995. This amendment to IEEE Std 1394-1995 follows the ISO/IEC 13213:1994 Command and Status Register (CSR) Architecture.

IEEE Std 1402-2000. *IEEE Guide for Electric Power Substation Physical and Electronic Security.* Security issues related to human intrusion upon electric power supply substations are identified and discussed. Various methods and techniques presently being used to mitigate human intrusions are also presented in this guide.

IEEE Std 1404-1998. *IEEE Guide for Microwave Communications System Development: Design, Procurement, Construction, Maintenance, and Operation.* The needs and requirements specific to the design, procurement, construction, maintenance, and operation of a microwave system are addressed. Steps for a variety of applications have been included in this guide; however, users should select only those steps that apply to their particular system(s) and their procurement policies.

IEEE Std 1406-1998. *IEEE Trial-Use Guide to the Use of Gas-In-Fluid Analysis for Electric Power Cable Systems.* The application of the analysis of gases dissolved in the fluids of fluid-filled cable systems is discussed with respect to the procedures for sampling, obtaining the dissolved gas data, and analyzing the results.

IEEE Std 1407-1998. *IEEE Trial-Use Guide for Accelerated Aging Tests for Medium-Voltage Extruded Electric Power Cables Using Water-Filled Tanks.* Accelerated aging tests on extruded medium-voltage cables using water-filled tanks are addressed. Information on the equipment, cable samples, test conditions, and measurements to perform the aging tests is provided. Techniques on how to analyze the test data are also included. The implementation of this guide will allow a better description of the test data obtained by different laboratories.

IEEE Std 1410-1997. *IEEE Guide for Improving the Lightning Performance of Electric Power Overhead Distribution Lines.*

IEEE Std 1413-1998. *IEEE Standard Methodology for Reliability Prediction and Assessment for Electronic Systems and Equipment.* The framework for the reliability prediction process for electronic systems and equipment, including hardware and software predictions at all levels, is covered.

IEEE Std 1416-1998. *IEEE Recommended Practice for the Interface of New Gas-Insulated Equipment in Existing Gas-Insulated Substations.* Recommendations for the connection of a gas-insulated substation to another gas-insulated substation of a different make are given.

IEEE Std 1420.1-1995. *IEEE Standard for Information Technology—Software Reuse—Data Model for Reuse Library Interoperability: Basic Interoperability Data Model (BIDM).* The minimal set of information about assets that reuse libraries should be able to exchange to support interoperability is provided.

IEEE Std 1420.1a-1996. *(Supplement to IEEE Std 1420.1-1995), Supplement to IEEE Standard for Information Technology—Software Reuse—Data Model for Reuse Library Interoperability: Asset Certification Framework.* A consistent structure for describing a reuse library's asset certification policy in terms of an Asset Certification Framework is defined, along with a standard interoperability data model for interchange of asset certification information.

IEEE Std 1420.1b-1999. *IEEE Standard for Information Technology—Software Reuse—Data model for Reuse Library Interoperability: Intellectual Property Rights Framework.* This extension to the Basic Interoperability Data Model (IEEE Std 1420.1-1995) incorporates intellectual property rights issues into software asset descriptions for reuse library interoperability.

IEEE Std 1430-1996. *IEEE Guide for Information Technology—Software Reuse—Concept of Operations for Interoperating Reuse Libraries.* This document describes the concepts necessary and appropriate for Networks of Interoperating Reuse Libraries (NIRLs). The purpose is to provide a context for standardization efforts toward the goal of supporting and enhancing interoperability.

IEEE Std 1445-1998. *IEEE Standard for Digital Test Interchange Format (DTIF).* The information content and the data formats for the interchange of digital test program data between digital automated test program generators (DATPGs) and automatic test equipment (ATE) for board-level printed circuit assemblies are defined. This information can be broadly grouped into data that defines the following: UUT Model, Stimulus and Response, Fault Dictionary, and Probe.

IEEE Std 1450-1999. *IEEE Standard Test Interface Language (STIL) for Digital Test Vector Data.* Standard Test Interface Language (STIL) provides an interface between digital test generation tools and test equipment. A test description language is defined that: (a) facilitates the transfer of digital test vector data from CAE to ATE environments; (b) specifies pattern, format, and timing information sufficient to define the application of digital test vectors to a DUT; and (c) supports the volume of test vector data generated from structured tests.

IEEE Std 1451.2-1997. *IEEE Standard for a Smart Transducer Interface for Sensors and Actuators Transducer to Microprocessor Communication Protocols and Transducer Electronic Data Sheet (TEDS) Formats.* A digital interface for connecting transducers to microprocessors is defined. A TEDS and its data formats are described. An electrical interface, read and write logic functions to access the TEDS and a wide variety of transducers are defined. This standard does not specify signal conditioning, signal conversion, or how the TEDS data is used in applications.

IEEE Std 1455-1999. *IEEE Standard for Message Sets for Vehicle/Roadside Communications.* Those characteristics of a dedicated short-range communications (DSRC) system that are independent of the Physical and Data Link Layers (ISO model Layers 1 and 2) are specified. The required and optional features of the roadside equipment (RSE) and the onboard equipment (OBE) are specified. In addition, the Applications Layer (ISO model Layer 7) services and protocols, the RSE resource manager, the corresponding OBE command interpreter, and the application-specific messages are all specified. Standard supports and guidelines are provided for implementing secure DSRC systems.

IEEE Std 1459-2000. *IEEE Trial-Use Standard Definitions for the Measurement of Electric Power Quantities Under Sinusoidal, Nonsinusoidal, Balanced, or Unbalanced Conditions.* This is a trial-use standard for definitions used for measurement of electric power quantities under sinusoidal, nonsinusoidal, balanced, or unbalanced conditions. It lists the mathematical expressions that were used in the past, as well as new expressions, and explains the features of the new definitions.

IEEE Std 1460-1996. *IEEE Guide for the Measurement of Quasi-Static Magnetic and Electric Fields.* A listing of possible measurement goals related to characterizing quasi-static magnetic and electric fields and possible methods for their accomplishment is provided.

IEEE Std 1462-1998. (Adoption of International Standard ISO/IEC 14102:1995). *IEEE Standard for Information Technology—Guideline for the Evaluation and Selection of CASE.* IEEE Std 1462-1998 is an adoption of International Standard ISO/IEC 14102:1995. The International Standard deals with the evaluation and selection of CASE tools, covering a partial or full portion of the software engineering life cycle. The adoption of the International Standard by IEEE includes an implementation note, which explains terminology differences, identifies related IEEE standards, and provides interpretation of the International Standard.

IEEE Std 1465-1998. [Adoption of International Standard ISO/IEC 12119:1994(E)]. *IEEE Standard for Information Technology—Software packages—Quality Requirements and Testing.* Quality requirements for software packages and instructions on how to test a software package against these requirements are established. The requirements apply to software packages as they are offered and delivered, not to the production process (including activities and intermediate products, such as specifications).

IEEE Std 1473-1999. *IEEE Standard for Communications Protocol Aboard Trains.* Communications protocols to be used for intercar and intracar serial data communications between subsystems aboard passenger trains are defined by this standard. Minimum acceptable parameters for a network that can simultaneously handle monitoring and control traffic from multiple systems are set forth. While the network is not vital, it is intended to be capable of carrying vital messages.

IEEE Std 1474.1-1999. *IEEE Standard for Communications-Based Train Control (CBTC) Performance and Functional Requirements.* Performance and functional requirements for a communications-based train control (CBTC) system are established in this standard. A CBTC system is a continuous, automatic train control system utilizing high-resolution train location determination, independent of track circuits; continuous, high-capacity, bidirectional train-to-wayside data communications; and trainborne and wayside processors capable of implementing automatic train protection (ATP) functions, as well as optional automatic train operation (ATO) and automatic train supervision (ATS) functions. In addition to CBTC functional requirements, this standard also defines headway criteria, system safety criteria, and system availability criteria for a CBTC system.

IEEE Std 1475-1999. *IEEE Standard for the Functioning of and Interfaces Among Propulsion, Friction Brake, and Trainborne Master Control on Rail Rapid Transit Vehicles.* The interfaces between and among functional systems on rail rapid transit vehicles is prescribed. The systems themselves are treated as black boxes; requirements for the input signals and the output response are given. For each category of interface, three types are listed in increasing technical sophistication.

IEEE Std 1476-2000. *IEEE Standard for Passenger Train Auxiliary Power Systems Interfaces.* The electrical interfaces among the components comprising the auxiliary power systems and their electrical interface with other train-borne systems are described. As such, this standard treats the auxiliary power system components (e.g., static inverters and converters, low-voltage dc power supplies, back-up battery systems, and battery chargers) as black boxes and addresses only their interface requirements.

IEEE Std 1477-1998. *IEEE Standard for Passenger Information System for Rail Transit Vehicles.* Rail transit vehicle passenger information system interfaces with the vehicles carbody, train crew, control system, power system, and passengers are described in this standard. The physical, logical, and electrical interfaces of the passenger information system for rail transit vehicle systems and subsystems are specified.

IEEE Std 1481-1999. *IEEE Standard for Integrated Circuit (IC) Delay and Power Calculation System.* Ways for integrated circuit designers to analyze chip timing and power consistently across a broad set of electric design automation (EDA) applications are covered in this standard. Methods by which integrated circuit vendors can express timing and power information once per given technology are also covered. In addition, this standard covers means by which EDA vendors can meet their application performance and capacity needs.

IEEE Std 1482.1-1999. *IEEE Standard for Rail Transit Vehicle Event Recorders.* On-board device systems, with crashworthy memory, that record data to support accident incident analysis for rail transit vehicles, are covered. The requirements of this standard are limited to event recorder functions and interfaces. Data transmission methods are excluded. The information in this standard is independent of the hardware and or software employed for other vehicle systems.

IEEE Std 1483-2000. *IEEE Standard for Verification of Vital Functions in Processor-Based Systems Used in Rail Transit.* A set of standard verification tasks for processor-based equipment used in safety-critical applications on rail and transit systems is covered. This standard also covers processes that verify the level of safety achieved in the implementation of safety-critical functions that are required to be fail-safe. Quality assurance or validation processes that affect the overall level of system safety are not covered.

IEEE Std 1488-2000. *IEEE Trial-Use Standard for Message Set Template for Intelligent Transportation Systems.* The expanding use of digital communications among subsystems of the transportation infrastructure has spawned the development of message sets for the communications between these subsystems. A format for Intelligent Transportation System (ITS) message sets, including common terms (e.g., object identifier), as well as attributes necessary to document ITS data messages, is addressed in this standard.

IEEE Std 1489-1999. *IEEE Standard for Data Dictionaries.* for Intelligent Transportation Systems. The expanding use of digital communications among subsystems of the transportation infrastructure has spawned the development of data dictionaries for the communications between these subsystems. A format for Intelligent Transportation System (ITS) data dictionaries, including common terms (e.g., time, date, location), as well as meta-attributes necessary to document ITS data concepts is addressed in this standard.

IEEE Std 1490-1998. (Adoption of PMI Guide to PMBOK). *IEEE Guide—Adoption of PMI Standard—A Guide to the Project Management Body of Knowledge.* The subset of the Project Management Body of Knowledge that is generally accepted is identified and described in this guide. "Generally accepted" means that the knowledge and practices described are applicable to most projects most of the time, and that there is widespread consensus about their value and usefulness. It does not mean that the knowledge and practices should be applied uniformly to all projects without considering whether they are appropriate.

IEEE Std 1498-1995. *IEEE Standard for Information Technology—Software Life Cycle Processes—Software Development—Acquirer-Supplier Agreement (Issued for Trial Use).* This standard defines a set of software development activities and resulting software products. It provides a framework for software development planning and engineering. It is also intended to merge commercial and Government software development requirements within the framework of the software life cycle process requirements of the Electronic Industries Association (EIA), Institute of Electrical and Electronics Engineers (IEEE) and International Organization for Standardization (ISO). The term "software development" is used as an inclusive term encompassing new development, modification, reuse, reengineering, maintenance, and all other processes or activities resulting in software products.

IEEE Std 1499-1998. *IEEE Standard Interface for Hardware Description Models of Electronic Components.* The standard interface for hardware description models of electronic components is defined. The primary audiences of this standard are model developers and implementors of software supporting this interface.

IEEE Std 1512-2000. *IEEE Standard for Common Incident Management Message Sets for Use by Emergency Management Centers.* This standard addresses the exchange of vital data about transportation-related incidents among emergency management centers through common incident management message sets. Message sets specified are consistent with the National Intelligent Transportation Systems Architecture and are described using Abstract Syntax Notation One syntax. This standard comprises the base standard of a family of incident management standards; specific incident management message sets for traffic, public safety, and HAZMAT centers may be found in forthcoming companion volumes which build upon and augment this base standard.

IEEE Std 1517-1999. *IEEE Standard for Information Technology—Software Life Cycle Processes—Reuse Processes.* A common framework for extending the software life cycle processes of IEEE/EIA Std 12207.0-1996 to include the systematic practice of software reuse is provided. This standard specifies the processes, activities, and tasks to be applied during each phase of the software life cycle to enable a software product to be constructed from reusable assets. It also specifies the processes, activities, and tasks to enable the identification, construction, maintenance, and management of assets supplied.

IEEE Std 1545-1999. *IEEE Standard for Parametric Data Log Format.* A language and file format for describing parametric test data is defined. Data types, data formats, and file formats are included.

IEEE Std 1596-1992. *IEEE Standard for Scalable Coherent Interface (SCI).* The Scalable Coherent Interface (SCI), which provides computer-bus-like services but uses a collection of fast point-to-point links instead of a physical bus in order to reach far higher speeds than any bus could, is described. The packets and protocols that implement transactions are defined, and the formal specification of the SCI packet protocols is given. In addition to the usual read and write transactions, SCI supports efficient multiprocessor lock transactions, cache coherence in a shared-distributed memory model, noncoherent caching, and message passing. A mechanical package and several physical links that may be used to implement the logical protocols and the cache coherence protocols are defined. Background information for understanding the protocols used by two or more SCI nodes to maintain coherence between cached copies of shared data is provided.

IEEE 1596.3-1996 (R2000). *IEEE Standard for Low-Voltage Differential Signals (LVDS) for Scalable Coherent Interface (SCI).* Scalable Coherent Interface (SCI), specified in IEEE Std 1596-1992, provides computer-bus-like services but uses a collection of fast point-to-point links instead of a physical bus in order to reach far higher speeds. The base specification defines differential ECL signals, which provide a high transfer rate (16 bits are transferred every 2 ns), but are inconvenient for some applications. IEEE Std 1596.3-1996, an extension to IEEE Std 1596-1992, defines a lower-voltage differential signal (as low as 250 mV swing) that is compatible with low-voltage CMOS, BiCMOS, and GaAs circuitry. The power dissipation of the transceivers is low, since only 2.5 mA is needed to generate this differential voltage across a 100 W termination resistance. Signal encoding is defined that allows transfer of SCI packets over data paths that are 4-, 8-, 32-, 64-, and 128-bits wide. Narrow data paths (4 to 8 bits) transferring data every 2 ns can provide sufficient bandwidth for many applications while reducing the physical size and cost of the interface. The wider paths may be needed for very-high-performance systems.

IEEE Std 1596.4-1996 (R2000). *IEEE Standard for High-Bandwidth Memory Interface Based on Scalable Coherent Interface (SCI) Signaling Technology (RamLink).* A high-bandwidth interface optimized for interchanging data between a memory controller and one or more dynamic RAMs is specified. RamLink is an applicable interface for other RAM-like devices as well.

IEEE Std 1596.5-1993 (R2000). *IEEE Standard for Shared-Data Formats Optimized for Scalable Coherent Interfaces (SCI) Processors.* Formats for interchanging integer, bit-field, and floating-point data between heterogeneous multiprocessors in a Scalable Coherent Interface (SCI) system are specified. The defined data formats can also be used to share data among multiprocessors on other bus standards that support the read, write, and lock transactions set defined by IEEE Std 1212-1991 CSR Architecture. The intent is to support efficient data transfers among heterogeneous workstations within a distributed computing environment.

IEEE Std 1754-1994 (R2000). *IEEE Standard for a 32-bit Microprocessor Architecture.* A 32-bit microprocessor architecture, available to a wide variety of manufacturers and users, is defined. The standard includes the definition of the instruction set, register model, data types, instruction op-codes, and coprocessor interface.

IEEE Std 1802.3-1991 (R2000). *Conformance Test Methodology for IEEE Standards for Local and Metropolitan Area Networks: Carrier Sense Multiple Access with Collision Detection (CSMA/CD) Access Method and Physical Layer Specifications [Currently contains Attachment Unit Interface (AUI) Cable (Section 4)].* This standard is part of a standards series on conformance test methodology for the family of local area network (LAN) and metropolitan area network (MAN) standards dealing with the physical and data link layers as defined by the International Organization for Standardization (ISO) Open Systems Interconnection Basic Reference Model. Methods for the conformance testing of AUI cable implementations to satisfy conformance requirements arising from the ISO/IEC 8802-3 AUI cable specification are defined. The conformance test suite is intended to detect incorrect implementations of the ISO/IEC 8802-3 standard. It is comprised of two categories of test groups. The first category relates to basic interconnection testing, and the second to capability and behavior testing. The test setups, adapters, and instruments used are described.

IEEE Std 1802.3d-1993 (R2000). *Supplement to IEEE Std 1802.3-1991, Type 10BASE-T Medium Attachment unit (MAU) Conformance Test Methodology (Section 6).* Methods for conformance testing to satisfy requirements arising from the ISO/IEC 8802-3 [ANSI/IEEE Std 802.3] standard are defined. The conformance test suite is intended to detect incorrect implementations of the ISO/IEC 8802-3 standard, clause 14. It comprises two categories of test groups. The first category relates to basic interconnection testing and the second to capability and behavior testing. The test setups, adapters, and instruments used are described.

IEEE Std 2000.1-1999. *IEEE Standard for Year 2000 Terminology.* This standard revises IEEE Std 2000.1-1998. It provides a detailed set of definitions. In addition, it addresses calendar information that is helpful in understanding the time-line issues surrounding the year 2000 rollover. The definitions section remains the core of the standard. With this expanded set of definitions, the standard now addresses areas that are relevant to both engineering and business environments. An increased degree of specificity has been added to the definition of "Year 2000 compliance," making it more precise in its meaning and application.

IEEE Std 2000.2-1999. *IEEE Recommended Practice for Information Technology—Year 2000 Test Methods.* This document provides users of computer hardware, firmware, software, or data systems with recommended practices for assessing and demonstrating the system elements within their organization that may be at risk of failure due to the Year 2000 problem and related date-specific issues. This recommended practice provides the framework for detailed planning and execution of all steps and tasks involved in testing for Year 2000 compliance. The resulting plan will outline the testing approach and identify system elements that are at risk of failure when crossing into the Year 2000 or using data that includes dates after 2000-01-01.

IEEE Std 2001-1999. *IEEE Recommended Practice for Internet Practices—Web Page Engineering—Intranet/Extranet Applications.* This standard defines recommended practices for Web page engineering. It addresses the needs of Webmasters and managers to effectively develop and manage World Wide Web projects (internally via an intranet or in relation to specific communities via an extranet). This standard discusses life cycle planning: identifying the audience, the client environment, objectives, and metrics, and continues with recommendations on server considerations, and specific Web page content. IEEE Std 2001-1999 defines conformance for both Web pages and tools that generate Web pages. This document is intended to reduce site-management costs, reduce legal risks, facilitate user satisfaction, and increase the productivity of Web applications for both maintainers and users.

IEEE Std 2003-1999. *IEEE Standard for Information Technology—Requirements and Guidelines for Test Methods Specifications and Test Method Implementations for Measuring Conformance to POSIX® Standards.* This International Standard defines the requirements and guidelines for test method specifications and test method implementations for measuring conformance to POSIX standards. Test specification standard developers for other Application Programming

Interface (API) standards are encouraged to use this standard. This document is aimed primarily at developers and users of test method specifications and implementations.

IEEE Std 2003.1-1992 (R2000). *IEEE Standard for Information Technology—Test Methods for Measuring Conformance to POSIX®—Part 1: System Interfaces.* This standard provides a definition of the requirements placed upon providers of POSIX test methods for POSIX.1 (IEEE Std 1003.1-1990; ISO/IEC 9945-1:1990). These requirements consist of a POSIX.1-ordered list of assertions defining those aspects of POSIX.1 that are to be tested and the associated test methods that are to be used in performing those tests. This standard is aimed primarily at POSIX.1 test suite providers and POSIX.1 implementors. This standard specifies those aspects of POSIX.1 that shall be verified by conformance test methods.

IEEE Std 2003.2-1996. *IEEE Standard for Information Technology—Test Methods for Measuring Conformance to POSIX® Part 1: Shell and Utilities Interfaces.* This standard defines the test methods to be used to measure conformance to IEEE 1003.2 (Shell and Utility Application Interface for Computer Operating System Environments). A definition of the requirements placed upon providers of a POSIX Conformance Test Suite for the POSIX.2 standard (ISO/IEC 9945-2: 1993, IEEE/ANSI Std 1003.2-1992) is provided. These requirements consist of a list of assertions defining those aspects of POSIX.2 that are to be tested and the associated test methods that are to be used in performing those tests. This standard is primarily aimed at test suite providers, but it also defines to POSIX.2 implementors those aspects of the standard that will be verified by a conformance test suite.

IEEE/ASTM SI 10-1997. *IEEE/ASTM Standard for Use of the International System of Units (SI): The Modern Metric System.* Guidance for the use of the modern metric system is given. Known as the International System of Units (abbreviated SI), the system is intended as a basis for worldwide standardization of measurement units. Information is included on SI, a list of units recognized for use with SI, and a list of conversion factors, together with general guidance on proper style and usage.

ISO/IEC 8802-2:1998 (R2000). (ANSI/IEEE Std 802.2, 1998 Edition). *Information technology—Telecommunications and information exchange between systems—Local and metropolitan area networks—Specific requirements—Part 2: Logical Link Control.* This standard is part of a family of standards for local area networks (LANs) and metropolitan area networks (MANs) that deals with the physical and data link layers as defined by the ISO Open Systems Interconnection Basic Reference Model. The functions, features, protocol, and services of the Logical Link Control (LLC) sublayer, which constitutes the top sublayer in the data link layer of the ISO/IEC 8802 LAN protocol, are described. The services required of, or by, the LLC sublayer at the logical interfaces with the network layer, the medium access control (MAC) sublayer, and the LLC sublayer management function are specified. The protocol data unit (PDU) structure for data communication systems is defined using bit-oriented procedures, as are three types of operation for data communication between service access points. In the first type of operation, PDUs are exchanged between LLCs without the need for the establishment of a data link connection. In the second type of operation, a data link connection is established between two LLCs prior to any exchange of information-bearing PDUs. In the third type of operation, PDUs are exchanged between LLCs without the need for the establishment of a data link connection, but stations are permitted to both send data and request the return of data simultaneously.

ISO/IEC 8802-3:1996 (R2000). (ANSI/IEEE Std 802.3, 1996 Edition). *Information technology—Telecommunications and information exchange between systems—Local and metropolitan area networks—Part 3: Carrier sense multiple access with collision detection (CSMA/CD) access method and physical layer specifications.* [Supplements ANSI/IEEE 802.3b-1985, ANSI/IEEE 802.3c-1985, ANSI/IEEE 802.3d-1987, and ANSI/IEEE 802.3e-1987 have been incorporated into this edition.] This standard is part of a family of local area network (LAN) and metropolitan area network (MAN) standards dealing with the physical and data link layers as defined by the International Organization for Standardization (ISO) Open Systems Interconnection Reference Model. Media access control characteristics for the Carrier Sense Multiple Access with Collision Detection (CSMA/CD) access method are specified. The media, Medium Attachment Unit (MAU), and physical layer repeater unit for 10 Mb/s baseband and broadband systems are also specified and a 1 Mb/s baseband implementation is provided. Specifications for MAU types 10BASE5, 10BASE2, FOIRL (fiber-optic inter-repeater link), 10BROAD36, and 1BASE5 are included. Layer and sublayer interface specifications are aligned to the ISO Open Systems Interconnection Basic Reference Model and 8802 models. The 8802-3 internal model is defined and used.

ISO/IEC 8802-4:1990 [ANSI/IEEE Std 802.4-1990 (R1995)]. *Information processing systems—Telecommunications and information exchange between systems—Local area networks—Part 4: Token-passing bus access method and physical layer specifications.* This standard is part of a family of local area network (LAN) and metropolitan area network (MAN) standards dealing with the physical and data link layers as defined by the International Organization for Standardization (ISO) Open Systems Interconnection Reference Model. The following are specified in this standard: the electrical and/or optical and physical characteristics of the transmission medium; the electrical or optical signaling method used; the frame formats transmitted; the actions of a station upon receipt of a frame; the services provided at the conceptual interface between the medium access control (MAC) sublayer and the Logical Link Control (LLC) sublayer above it; and the actions, entities, and values used to manage the MAC sublayer and physical layer entity.

ISO/IEC 8802-5:1998 (ANSI/IEEE 802.5, 1998 Edition) *Information processing systems—Telecommunications and information exchange between systems—Local area networks—Part 5: Token ring access method and physical layer specifications.* This Local and Metropolitan Area Network standard, ISO/IEC 8802-5:1998, is part of a family of local area network (LAN) standards dealing with the physical and data link layers as defined by the ISO/IEC Open Systems Interconnection Basic Reference Model. Its purpose is to provide compatible interconnection of data processing equipment by means of a LAN using the token-passing ring access method. The frame format, including delimiters, addressing, and priority stacks, is defined. The medium access control (MAC) protocol is defined. The finite state machine and state tables are supplemented with a prose description of the algorithms. The physical layer (PHY) functions of symbol encoding and decoding, symbol time, and latency buffering are defined. The services provided by the MAC to the station management (SMT) and the services provided by the PHY to SMT and the MAC are described. These services are defined in terms of service primitives and associated parameters. The 4 and 16 Mbit/s, shielded twisted pair attachment of the station to the medium, including the medium interface connector (MIC), is also defined. The applications environment for the LAN is intended to be commercial and light industrial. The use of token ring LANs in home and heavy industrial environments, while not precluded, has not been considered in the development of the standard. A Protocol Implementation Conformance Statement (PICS) proforma is provided as an annex to the standard.

ISO/IEC 8802-5:1998/Amd 1. (ANSI/IEEE 802.5 and 802.5j, 1998 Edition). *Information technology—Telecommunications and information exchange between systems—Local and metropolitan area networks—Specific requirements—Part 5: Token ring access method and physical layer specifications—*

Amendment 1: Dedicated token ring operation and fibre optic media. This amendment to Local and Metropolitan Area Network standard, ISO/IEC 8802-5:1998, is part of a family of local area network (LAN) standards dealing with the physical and data link layers as defined by the ISO/IEC Open Systems Interconnection Basic Reference Model. The requirements for dedicated token ring (DTR) operation are specified, including the changes and additions to the Medium Access Control (MAC) layer to provide for an additional full-duplex mode of operation (switching), and for interconnection of shared LAN segments to switch ports. Also specified are the characteristics of a fibre optic interface for connecting a 4 Mbit/s or 16 Mbit/s token ring station to the trunk coupling unit (TCU) of a token ring, including station, port, and channel requirements. Fibre optic trunk signaling recommendations are also made.

ISO/IEC 8802-6:1994. (ANSI/IEEE 802.6, 1994 Edition). *Information technology—Telecommunications and information exchange between systems—Local and metropolitan area networks—Specific requirements—Part 6: Distributed queue dual bus (DQDB) access method and physical layer specifications.* This standard is part of a family of standards for local area networks (LANs) and metropolitan area networks (MANs) that deals with the Physical and Data Link Layers as defined by the ISO Open Systems Interconnection Reference Model. It defines a high-speed shared medium access protocol for use over a dual, counterflowing, unidirectional bus subnetwork. The Physical Layer and Distributed Queue Dual Bus (DQDB) Layer are required to support a Logical Link Control (LLC) Sublayer by means of a connectionless Medium Access Control (MAC) Sublayer service in a manner consistent with other IEEE 802® networks. Additional DQDB Layer functions are specified as a framework for other services. These additional functions will support Isochronous Service Users and Connection-Oriented Data Service Users, but their implementation is not required for conformance.

ISO/IEC 8802-9:1996. (ANSI/IEEE Std 802.9, 1996 Edition). *Information technology—Telecommunications and information exchange between systems—Local and metropolitan area networks—Specific requirements—Part 9: Integrated services (IS) LAN interface at the medium access control (MAC) and physical (PHY) layers.* A unified access method that offers integrated services (IS) to the desktop for a variety of publicly and privately administered backbone networks (e.g., ANSI FDDI, IEEE 802.x, and ISDN) is defined. In addition, the interface at the MAC sublayer and the PHY Layer is specified.

ISO/IEC 8802-11:1999. (IEEE Std 802.11-1999). *Information technology—Telecommunications and information exchange between systems—Part 11: Wireless LAN medium access control (MAC) and physical layer (PHY) specifications.* The medium access control (MAC) and physical characteristics for wireless local area networks (LANs) are specified in this standard, part of a series of standards for local and metropolitan area networks. The medium access control unit in this standard is designed to support physical layer units as they may be adopted dependent on the availability of spectrum. This standard contains three physical layer units: two radio units, both operating in the 2400–2500 MHz band, and one baseband infrared unit. One radio unit employs the frequency-hopping spread spectrum technique, and the other employs the direct sequence spread spectrum technique.

ISO/IEC 8802-12:1998. (ANSI/IEEE Std 802.12, 1998 Edition). *Information technology—Telecommunications and information exchange between systems—Local and metropolitan area networks—Part 12: Demand-priority access method, physical layer and repeater specifications.* The media access control characteristics for the demand-priority access method are specified. The layer management, physical layers, and media that support this access method are also specified. Layer and sublayer interface specifications are aligned to the ISO Open Systems Interconnection Basic Reference Model and ISO/IEC 8802 models. Specifications for 100 Mb/s operation over 100 balanced cable (twisted-pair) Categories 3 through 5, 150 shielded balanced cable, and fibre-optic media are included. Optional implementation of redundant links to facilitate automatic recovery of network connectivity in case of link or repeater failure any where in the network path is specified. Rules for connecting redundant links within a network are defined.

ISO/IEC 9945-1:1996 (IEEE Std 1003.1-1996). *Information technology—Portable Operating System Interface (POSIX®)—Part 1: System Application Interface (API) [C Language].* A standard operating system interface and environment based on the UNIX® operating system documentation to support application portability at the source level is defined. Intended for use by both application developers and system implementors, the standard focuses on a C language interface, although future revisions are expected to contain bindings for other programming languages as well. Information is provided on: terminology, concepts, and definitions and specifications that govern structures, headers, environment variables, and related requirements; definitions for system service and subroutines; language-specific system services for the C programming language; and interface issues, including portability, error handling, and error recovery.

ISO/IEC 9945-2:1993 (R1995). (IEEE Std 1003.2-1992). *Information technology—Portable Operating System Interface (POSIX®)—Part 2: Shell and Utilities.* This standard is part of the POSIX series of standards for applications and user interfaces to open systems. It defines the applications interface to a shell command language and a set of utility programs for complex data manipulation. When the User Portability Utilities Option is included, the standard also defines a common environment for general-purpose time-sharing users on character-oriented display terminals. Included in this standard is ANSI/IEEE Std 1003.2a-1992.

ISO/IEC 10857:1994 (ANSI/IEEE 896.1, 1994 Edition). *Information technology—Microprocessor Systems-Futurebus+®—Logical Protocol Specification.* This International Standard provides a set of tools with which to implement a Futurebus+ architecture with performance and cost scalability over time, for multiple generations of single- and multiple-bus multiprocessor systems. Although this specification is principally intended 64-bit address and data operation, a fully compatible 32-bit subset is provided, along with compatible extensions to support 128- and 256-bit data highways. Allocation of bus bandwidth to competing modules is provided by either a fast centralized arbiter, or a fully distributed, one or two pass, parallel contention arbiter. Bus allocation rules are provided to suit the needs of both real-time (priority based) and fairness (equal opportunity access based) configurations. Transmission of data over the multiplexed address/data highway is governed by one of two intercompatible transmission methods: a) a technology-independent, compelled-protocol, supporting broadcast, broadcall, and transfer intervention (the minimum requirement for all Futurebus+ systems), and b) a configurable transfer-rate, source-synchronized protocol supporting only block transfers and source-synchronized broadcast for systems requiring the highest possible performance. Futurebus+ takes its name from its goal of being capable of the highest possible transfer rate consistent with the technology available at the time modules are designed, while ensuring compatibility with all modules designed to this standard both before and after. The plus sign (+) refers to the extensible nature of the specification, and the hooks provided to allow further evolution to meet unanticipated needs of specific application architectures. It is intended that this International Standard be used as a key component of an approved IEEE Futurebus+ profile.

ISO/IEC 10861:1994 (ANSI/IEEE Std 1296-1994). *Information technology—High Performance Synchronous 32-Bit Bus: MULTIBUS II.* The operation, functions, and attributes of a parallel system bus (PSB), called MULTIBUS II, are defined. A high-performance backplane bus intended for use in multiple processor systems, the PSB incorporates synchronous, 32-bit multiplexed address/data, with error detection, and uses a 10 MHz bus clock. This design is intended to provide reliable state-of-the-art operation and to allow the implementation of cost-effective, high-performance VLSI for the bus interface. Memory, I/O, message, and geographic address spaces are defined. Error detection and retry are provided for messages. The message-passing design allows a VLSI implementation, so that virtually all modules on the bus will utilize the bus at its highest performance—32 to 40 Mbyte/s. An overview of PSB, signal descriptions, the PSB protocol, electrical characteristics, and mechanical specifications are covered.

ISO/IEC 11802-5:1997(E). ISO/IEC Technical Report 11802-5:1997 [ANSI/IEEE Std 802.1H, 1997 Edition]. *Information technology—Telecommunications and information exchange between systems—Local and metropolitan area networks—Technical reports and guidelines—Part 5: Media access control (MAC) bridging of ethernet V2.0 in local area networks.* Extensions to the behavior of ISO/IEC 10038 (IEEE 802.1D) media access control (MAC) Bridges, in order to facilitate interoperability in bridged local area networks (LANs) comprising CSMA/CD networks interconnected with other types of LAN using MAC Bridges, where the CSMA/CD networks contain a mixture of ISO/IEC 8802-3 and Ethernet V2.0 end stations, are specified. Additionally, guidelines are provided for the of nonstandard 802® protocols, with particular emphasis on conversion of existing Ethernet protocols and the behavior to be expected from a Bridge, for the purpose of avoiding future incompatibilities.

IEEE/EIA 12207.0-1996. *IEEE/EIA Standard for Industry Implementation of International Standard ISO/IEC 12207: 1995 (ISO/IEC 12207) for Information Technology—Software Life Cycle Processes.* ISO/IEC 12207 provides a common framework for developing and managing software. IEEE/EIA 12207.0 consists of the clarifications, additions, and changes accepted by the Institute of Electrical and Electronics Engineers (IEEE) and the Electronic Industries Association (EIA) as formulated by a joint project of the two organizations. IEEE/EIA 12207.0 contains concepts and guidelines to foster better understanding and application of the standard. Thus this standard provides industry a basis for software practices that would be usable for both national and international business.

IEEE/EIA 12207.1-1997. *IEEE/EIA Guide for Industry Implementation of International Standard ISO/IEC 12207:1995 (ISO/IEC 12207) Standard for Information Technology—Software Life Cycle Processes—Life Cycle.* ISO/IEC 12207 provides a common framework for developing and managing software. IEEE/EIA 12207.0 consists of the clarifications, additions, and changes accepted by the Institute of Electrical and Electronics Engineers (IEEE) and the Electronic Industries Association (EIA) as formulated by a joint project of the two organizations. IEEE/EIA 12207.1 provides guidance for recording life cycle data resulting from the life cycle processes of IEEE/EIA 12207.0.

IEEE/EIA 12207.2-1997. *IEEE/EIA Guide for Industry Implementation of International Standard ISO/IEC 12207:1995 (ISO/IEC 12207) Standard for Information Technology—Software life cycle processes—Implementation considerations.* ISO/IEC 12207 provides a common framework for developing and managing software. IEEE/EIA 12207.0 consists of the clarifications, additions, and changes accepted by the Institute of Electrical and Electronics Engineers (IEEE) and the Electronic Industries Association (EIA) as formulated by a joint project of the two organizations. IEEE/EIA 12207.2 provides implementation consideration guidance for the normative clauses of IEEE/EIA 12207.0. The guidance is based on software industry experience with the life cycle processes presented in IEEE/EIA 12207.0.

ISO/IEC 13210:1994 (ANSI/IEEE 1003.3-1991). *Information technology—Test Methods for Measuring Conformance to POSIX®.* The general requirements and test methods for measuring conformance to POSIX standards are defined. This document is aimed primarily at working groups developing test methods for POSIX standards, developers of POSIX test methods, and users of POSIX test methods.

ISO/IEC 13213:1994. [ANSI/IEEE Std 1212, 1994 Edition] (Incorporates ANSI/IEEE Std 1212-1991). *Information technology—Microprocessor systems—Control and Status Registers (CSR) Architecture for microcomputer.* The document structure and notation are described, and the objectives and scope of the CSR Architecture are outlined. Transition set requirements, node addressing, node architectures, unit architectures, and CSR definitions are set forth. The ROM specification and bus standard requirements are covered.

IEEE Std 14143.1-2000. (Adoption of ISO/IEC 14143-1: 1998). *Implementation Note for IEEE Adoption of ISO/IEC 14143-1:1998 Information Technology—Software Measurement—Functional Size Measurement—Part 1: Definition of Concepts.* Implementation notes that relate to the IEEE interpretation of ISO/IEC 14143-1:1998 are described.

ISO/IEC 14536:1995 (ANSI/IEEE Std 896.5-1993). *IEEE Standard for Futurebus+®, Profile M (Military).* Futurebus+ standards provide systems developers with a set of tools with which high performance bus-based systems may be developed. This architecture provides a wide range of performance scalability over both cost and time for multiple generations of single- and multiple-bus multiprocessor systems. This document, a companion standard to IEEE Std 896.1-1991, builds on the logical layer by adding requirements for three military profiles. It is to these profiles that products will claim conformance. Other specifications that may be required in conjunction with this standard are IEEE Std 896.1-1991, IEEE Std 896.2-1991, IEEE Std 896.3-1993, IEEE Std 896.4-1993, IEEE Std 1101.3-1993, IEEE Std 1101.4-1993, IEEE Std 1212-1991, IEEE Std 1194.1-1991, IEEE Std 1394-1995, IEEE Std 1301-1991, and IEEE Std 1301.1-1991.

ISO/IEC 15802-3:1998. (ANSI/IEEE Std 802.1D, 1998 Edition). *Information technology—Telecommunications and information exchange between systems—Local and metropolitan networks area networks—Common specifications—Part 3: Media access control (MAC) bridges.* The concept of Media Access Control (MAC) Bridging. Introduced in the 1993 edition of this standard, has been expanded to define additional capabilities in Bridged LANs aimed at providing for expedited traffic capabilities, to support the transmission of time-critical information in a LAN environment; and providing filtering services that support the dynamic use of Group MAC Addresses in a LAN environment.

ISO/IEC 15802-5:1998. (ANSI/IEEE Std 802.1G, 1998 Edition). *Information technology—Telecommunications and information exchange between systems—Local and metropolitan networks area networks—Common specifications—Part 5: Remote media access control (MAC) bridging.* Extensions to the behavior of ISO/IEC 10038 (IEEE 802.1D) media access control (MAC) bridges, including the aspects of operation of remote MAC bridges that are observable on the interconnected LANs, are specified. A protocol for (optional) use between remote MAC bridges, across the non-LAN communications equipment that interconnects them, to configure the remote bridges within the bridged LAN in accordance with the spanning tree algorithm of ISO/IEC 10038: 1993, is also provided.

ANSI C2-1997. *American National Standard Electrical Safety Code®—1997 Edition.* This standard covers basic provisions for safeguarding of persons from hazards arising from the installation, operation, or maintenance of 1) conductors and equipment in electric supply stations, and 2) overhead and underground electric supply and communication lines. It also includes work rules for the construction, maintenance, and operation of electric supply and communication lines and equipment. The standard is applicable to the systems and equipment operated by utilities, or similar systems and equipment, of an industrial establishment or complex under the control of qualified persons. This standard consists of the introduction, definitions, grounding rules, list of referenced and bibliographic documents, and Parts 1, 2, 3, and 4 of the 1997 Edition of the National Electrical Safety Code.

ANSI C12.1-1988. *American National Standard Code for Electricity Metering.* Acceptable performance criteria for new types of ac watt-hour meters, demand meters, demand registers, pulse devices, instrument transformers, and auxiliary devices are established. Acceptable in-service performance levels for meters and devices used in revenue metering are stated. Information on related subjects, such as recommended measurement standards, installation requirements, test methods, and test schedules, is included. Some of the provisions are applicable to dc watthour meters as well, and acceptable in-service performance levels of such meters are given in an appendix.

ANSI C12.4-1984 (R1995). *American National Standard for Mechanical Demand Registers.* The voltage and frequency ratings, full-scale values, scale classes, demand intervals, multiplying constants, timing mechanisms, and other general features of mechanical demand registers required for use on watthour meters are covered. Single-pointer-form, cumulative-form, and multiple-pointer-form registers are included. Although mechanical demand registers are designed for use as accessories in watthour meters, items relating to watthour meters are not covered.

ANSI C12.6-1987 (R1992). *American National Standard for Marking and Arrangement of Terminals for Phase-Shifting Devices Used in Metering.* Phase-shifting devices designed to provide the proper lagged voltages required for kVAR and kVA measurement are covered. Terminal marking for devices for specific types of services as well as universal devices is considered. The number of terminals and the provision of diagrams of internal connections are specified.

ANSI C12.7-1993. *American National Standard Requirements for Watthour Meter Sockets.* The general requirements and pertinent dimensions applicable to watthour meter sockets rated up to and including 600 V and up to and including 320 A continuous duty per socket opening are covered.

ANSI C12.8-1981 (R1991). *American National Standard for Test Blocks and Cabinets for Installation of Self-Contained "A" Base Watthour Meters.* The dimensions and functions of test blocks and cabinets used with self-contained A-base watthour meters are covered. Standard ratings are defined, and general requirements are addressed.

ANSI C12.9-1993. *American National Standard for Test Switches for Transformer-Rated Meters.* This standard is intended to encompass the dimensions and functions of meter test switches used with transformer-rated watthour meters in conjunction with instrument transformers.

ANSI C12.10-1987 (R1991). *American National Standard for Electromechanical Watthour Meters.* Class designations, voltage and frequency ratings, test-current values, internal wiring arrangements, pertinent dimensions, rotor markings, register requirements, and other general specifications are covered for both detachable and bottom-connected electromechanical watthour meters. Combination devices, the essential elements of watthour meters, are also covered insofar as their application is practicable. The terminal arrangements and mounting dimensions covered by this standard are essentially those adopted by the watthour meter industry during 1928 to 1936.

ANSI C12.11-1987 (R1991). *American National Standard for Instrument Transformers for Revenue Metering 10 kV BIL Through 350 kV BIL (0.6 kV NSV Through 69 kV NSV).* The general requirements, metering accuracy, thermal ratings, and dimensions are established for current and inductively coupled voltage transformers for revenue metering. Both indoor and outdoor types are covered.

ANSI C12.13-1991. *American National Standard for Time-of-Use Registers for Electricity Meters.* Physical requirements and test procedures for time-of-use registers are set forth. The following features of the register are covered: number and format of displays; voltage frequency and temperature ratings; demand intervals; multiplying constants; timing systems; communication requirements; nameplate information; finish; rain-tightness; and other general requirements. Test requirements and conditions and performance requirements for the registers are specified.

ANSI C12.14-1982 (R1987). *American National Standard for Magnetic Tape Pulse Recorders for Electricity Meters.* Minimum requirements for magnetic tape pulse recorders for electricity meters are recommended. The voltage, frequency ratings, recording format, enclosure requirements, and other general specifications are covered. The intent is to assure recorder reliability to the extent that such a quality can be demonstrated by laboratory testing.

ANSI C12.15-1990. *American National Standard for Solid-State Demand Registers for Electromechanical Watthour Meters.* Solid-state demand registers designed for use as accessories with electromechanical watthour meters are covered. Requirements are set forth regarding number and format of displays; voltage, frequency, and temperature ratings; demand intervals; multiplying constants; timing systems; and other general features. Test conditions for evaluating register performance are stated. Items relating to the watthour meters themselves are not covered.

ANSI C12.16-1991. *American National Standard for Solid-State Electricity Meters.* Acceptable performance criteria for solid-state electricity meters are established. Detachable socket, type S, and bottom-connected, type A, as well as any other arrangement agreed upon between the manufacturer and the user are included. Class designations, voltage and frequency ratings, test current values, service connection arrangements, pertinent dimensions, form designations, and environmental tests are covered.

IEEE Std C37.37-1996. *IEEE Loading Guide for AC High-Voltage Air Switches (in Excess of 1000 V).* An aid to users to determine (1) the allowable continuous current class (ACCC), (2) the continuous load current capabilities of air switches under various conditions of ambient temperature, and (3) the emergency load current capabilities of air switches under various conditions of ambient temperature, is provided. This guide does not apply to switches used in enclosures covered by IEEE Std C37.20.2-1993, IEEE Std C37.20.3-1996, IEEE Std C37.23-1987, IEEE Std C37.71-1984, and ANSI C37.72-1987.

IEEE Std C37.04-1999. *IEEE Standard Rating Structure for AC High-Voltage Circuit Breakers.* This standard covers the rating structure for all high-voltage circuit breakers, which include all voltage ratings above 1000 V ac and comprise both indoor and outdoor types having the preferred ratings as listed in ANSI C37.06-1997. Typical circuit breakers covered by these standards have maximum voltage ratings ranging from 4.76 kV through 800 kV, and continuous current ratings of 600 A, 1200 A, 2000 A, and 3000 A associated with the various maximum voltage ratings. The rating structure establishes the basis for all assigned ratings, including continuous current, dielectric withstand voltages, short-circuit current, transient recovery voltage, and capacitor switching, plus associated capabilities such as mechanical endurance, load current, and out-of-phase switching. This standard does not cover generator circuit breakers, which are covered in IEEE Std C37.013-1997.

ANSI C37.06-1997. *American National Standard for Switchgear—AC High-Voltage Circuit Breakers Rated on a Sym-*

metrical Current Basis—Preferred Ratings and Related Required Capabilities.

ANSI C37.06.1-1997. *American National Standard Trial-Use Guide for High-Voltage Circuit Breakers Rated on a Symmetrical Current Basis Designated Definite Purpose for Fast Transient Recovery Voltage Rise Times.*

IEEE Std C37.09-1999 *IEEE Standard Test Procedure for AC High-Voltage Circuit Breakers Rated on a Symmetrical Current Basis.* The testing procedures for all high-voltage circuit breakers that include all voltage ratings above 1000 V ac and comprise both indoor and outdoor types having the preferred ratings as listed in ANSI C37.06-1997 are covered. Typical circuit breakers covered by these standards have maximum voltage ratings from 4.76 kV through 800 kV, and continuous current ratings of 600 A, 1200 A, 2000 A, and 3000 A associated with the various maximum voltage ratings. The test procedures verify all assigned ratings, including continuous current, dielectric withstand voltages, short-circuit current, transient recovery voltage, and capacitor switching, plus associated capabilities such as mechanical endurance, load current, and out-of-phase switching. Production test procedures are also included. This standard does not cover generator circuit breakers as these are covered in IEEE Std C37.013-1993.

IEEE Std C37.010-1999. *IEEE Application Guide for AC High-Voltage Circuit Breakers Rated on a Symmetrical Current.* This guide covers the application of indoor and outdoor high-voltage circuit breakers rated above 1000 V for use in commercial, industrial, and utility installations. It deals with usage under varied service conditions, temperature conditions affecting continuous current compensation, reduced dielectrics, reclosing derating as applicable, calculation of system short-circuit current, compensation at different X/R ratios, detailed calculations with application curves, out-of-phase switching, and general application.

IEEE Std C37.011-1994. *IEEE Application Guide for Transient Recovery Voltage for AC High-Voltage Circuit Breakers Rated on a Symmetrical Current Basis.* Procedures and calculations necessary to apply the standard transient recovery voltage (TRV) ratings for ac high-voltage circuit breakers rated above 1000 V and on a symmetrical current basis are covered. The capability limits of these circuit interrupting devices are determined largely by the TRV. TRV ratings are compared with typical system TRV duties.

IEEE Std C37.013-1997. *IEEE Standard for AC High-Voltage Generator Circuit Breakers Rated on a Symmetrical Current Basis.* Ratings, performance requirements, and compliance test methods are provided for ac high-voltage generator circuit breakers rated on a symmetrical current basis that are installed between the generator and the transformer terminals. Guidance for applying generator circuit breakers is given. Pumped storage installations are considered a special application, and their requirements are not completely covered by this standard.

IEEE Std C37.015-1993. *IEEE Application Guide for Shunt Reactor Switching* . Guidance for the application of ac high voltage circuit breakers for shunt reactor switching is provided. Overvoltage generation for the three cases of directly grounded, ungrounded, and neutral reactor grounded shunt reactors is addressed in terms of derivation and limitation methods. Circuit breaker specification for the purpose and the use of laboratory test results to predict field performance is also covered by this guide.

IEEE Std C37.081-1981 (R1988). *IEEE Guide for Synthetic Fault Testing of AC High-Voltage Circuit Breakers Rated on a Symmetrical Current Basis.* Guidelines are established for synthetic testing of circuit breakers, as well as test criteria for demonstrating the short-circuit current rating of circuit breakers on a single-phase basis. Criteria for evaluating results are also provided. The standard covers short-circuit current interruption process; basic principles of synthetic test; synthetic test circuits; requirements for synthetic test methods; parameters, test procedures, and tolerances; short line fault; multiple loops; circuit breakers equipped with parallel impedance; duty cycle; and test records.

IEEE Std C37.081a-1997 *Supplement to IEEE Guide for Synthetic Fault Testing of AC High-Voltage Circuit Breakers Rated on a Systemmetrical Current Basis—8.3.2: Recovery Voltage for Terminal Faults; Asymmetrical Short-Circuit Current.* The transient recovery voltage needs to be modified when interrupting asymmetrical currents. The voltage rate R, the peak voltage E2 and the rate of change of current di/dt all change with the asymmetrical current zero. Guidance is provided on how to make these corrections when compared to the symmetrical case.

IEEE Std C37.082-1982 (R1988). *IEEE Standard Methods for the Measurement of Sound Pressure Levels of AC Power Circuit Breakers.* Guidelines for uniform measurement and reporting of sound produced by ac power circuit breakers are established. The methods are intended for measuring the sound produced by outdoor circuit breakers in a free-field environment. The methods may be used indoors or in a restricted field, provided that precautions are observed in measurement and interpretation of results. Three types of tests are described: design tests, conformance tests, and field tests. The methods are intended to provide data that can be used in evaluating the effects of circuit breaker sound on human observers, but the evaluation itself is not covered.

IEEE Std C37.083-1999. *IEEE Guide for Synthetic Capacitive Switching Tests of AC High-Voltage Circuit Breakers.* As an aid in testing circuit breakers under conditions of switching capacitive currents synthetic test circuits may be used. The design of the circuit should simulate the stress of actual service conditions as closely as possible. A number of circuits are given as examples. The limitation of the use of synthetic test methods is that the breaker under test must not display evidence of reignition or restriking. The known circuits do not properly represent the interaction between the source and the capacitive load under this condition. Such breakers must be tested using direct circuits.

IEEE Std C37.1-1994. *IEEE Standard Definition, Specification, and Analysis of Systems Used for Supervisory Control, Data Acquisition, and Automatic Control.* Distributed multicomputer master stations and distributed remote terminal units (RTUs) are introduced. Submaster RTUs used in an automated distribution system with downstream feeder RTUs is defined. Local area networks with master stations are discussed. Intelligent electronic devices (IEDs) with respect to their interface to RTUs and master stations are defined. New surge withstand capability (SWC) standards and their applicability to SCADA is shown. An example channel loading calculation is provided.

IEEE Std C37.10-1995. *IEEE Guide for Diagnostics and Failure Investigation of Power Circuit Breakers.* Procedures to be used to perform failure investigations of power circuit breakers are recommended. Although the procedure may be used for any circuit breaker, it is mainly focused on high-voltage ac power circuit breakers used on utility systems. Recommendations are also made for monitoring circuit breaker functions as a means of diagnosing their suitability for service condition.

IEEE Std C37.11-1997. *IEEE Standard Requirements for Electrical Control for AC High-Voltage Circuit Breakers Rated on a Symmetrical Current Basis.* Standard requirements for all types of electrical control circuits for ac high-voltage breakers rated above 1000 V are given. This standard is applicable for any type of power-operated mechanism and for both ac and dc control power. Only basic control elements of the circuit breaker, including reclosing where required, are included in this standard. Devices or circuits for protective relaying, special interlocking, etc., are not included.

IEEE Std C37.13-1990 (R1995). *IEEE Standard for Low-Voltage AC Power Circuit Breakers Used in Enclosures.* This standard covers enclosed low-voltage ac power circuit breakers of the stationary or draw-out type of two- or three-pole construction; with one or more rated maximum voltages of 635 V (600 V for units incorporating fuses), 508 V, and 254 V, for application on systems having nominal voltages of 600 V, 480 V, and 240 V; with unfused or fused circuit breakers; manually or power operated; and with or without electromechanical or solid-state trip devices. The standard deals with service conditions, ratings, functional components, temperature limitations and classifications of insulating materials, insulation-withstand (dielectric) voltage requirements, test procedures, and application.

IEEE Std C37.14-1999. *IEEE Standard for Low-Voltage DC Power Circuit Breakers Used in Enclosures.* This standard covers enclosed low-voltage dc power circuit breakers of the stationary or draw-out type of single- or two-pole construction with one or more rated maximum voltages of 300 V, 325 V, 800 V, 1200 V, 1600 V, or 3200 V for applications on dc systems having nominal voltages of 250 V, 275 V, 750 V, 1000 V, 1500 V, or 3000 V, with general-purpose, high-speed, semi-high-speed and rectifier circuit breakers; manually or power-operated; and with or without electro-mechanical or electronic trip devices. It deals with service conditions, ratings, functional components, temperature limitations and classification of insulating materials, dielectric withstand voltage requirements, test procedures, and application.

IEEE Std C37.18-1979 (R1996). *IEEE Standard for Enclosed Field Discharge Circuit Breakers for Rotating Electric Machinery.* Low-voltage power-circuit breakers that are intended for use in field circuits of apparatus such as generators, motors, synchronous condensers, or exciters, and embodying contacts for establishing field discharge circuits, are covered. Service conditions, ratings, and functional components are discussed. Temperature limitations and classification of insulating materials, insulation (dielectric) withstand voltage requirements, and test requirements are addressed. An application guide is included.

IEEE Std C37.2-1996. *IEEE Standard Electrical Power System Device Function Numbers and Contact Designations.* The definition and application of function numbers for devices used in electrical substations and generating plants and in installations of power utilization and conversion apparatus are covered. The purpose of the numbers is discussed, and 94 numbers are assigned. The use of prefixes and suffixes to provide a more specific definition of the function is considered. Device contact designation is also covered.

IEEE Std C37.20.1-1993 (R1998). *IEEE Standard for Metal-Enclosed Low-Voltage Power Circuit-Breaker Switchgear.* Low-voltage metal-enclosed switchgear, which can contain either stationary or drawout, manually or electrically operated low-voltage ac or dc power circuit breakers in individual grounded metal compartments, in three-pole, two-pole, or single-pole construction is covered. Rated maximum voltage levels can be 254 V, 508 V, or 635 V ac and 300/325 V, 800 V, 1000 V, 1600 V, or 3200 V dc. The continuous current ratings of the main bus in ac designs can be 1600 A, 2000 A, 2500 A, 4000 A, 6000 A, 8000 A, 10 000 A, or 12 000 A. The switchgear can also contain associated control, instruments, metering, protective, and regulating devices as necessary. The standard deals with service conditions, ratings, temperature limitations, and classification of insulating materials, insulation (dielectric) withstand voltage requirements, test procedures, and application.

IEEE Std C37.20.2-1999. *IEEE Standard for Metal-Clad Switchgear.* Metal-clad (MC) medium-voltage switchgear that contains drawout electrically operated circuit breakers is covered. MC switchgear is compartmentalized to isolate all components such as instrumentation, main bus, and both incoming and outgoing connections with grounded metal barriers. Rated maximum voltage levels for metal-clad switchgear range from 4.76 kV to 38 kV with main bus continuous current ratings of 1200 A, 2000 A, and 3000 A. MC switchgear also contains associated control, instruments, metering, relaying, protective, and regulating devices, as necessary. Service conditions, ratings, temperature limitations and classification of insulating materials, insulation (dielectric) withstand voltage requirements, test procedures, and applications are discussed.

IEEE Std C37.20.3-1996. *IEEE Standard for Metal-Enclosed Interrupter Switchgear.* Metal-enclosed interrupter switchgear assemblies containing but not limited to such devices as interrupter/switches, selector switches, power fuses; control, instrumentation and metering; and protective equipment is covered. It includes, but is not specifically limited to, equipment for the control and protection of apparatus used for distribution of electrical power.

IEEE Std C37.20.4-1996. *IEEE Trial-Use Standard for Indoor AC Switches (1 kV–38 kV) for Use in Metal-Enclosed Switchgear.* Indoor ac medium-voltage switches for use in enclosures for application in power circuits at voltages above 1 kV through 38 kV are covered. These include stationary or drawout, manual or power operation, fused or unfused.

IEEE Std C37.20.6-1997. *IEEE Standard for 4.76 kV to 38 kV Rated Grounding and Testing Devices Used in Enclosures.* Drawout type grounding and testing (GT) devices for use in medium-voltage metal-clad switchgear rated above 4.76 kV through 38 kV are covered. The description, design, and testing of these accessory devices that are inserted in place of drawout circuit breakers for the purpose of grounding and testing are also covered.

IEEE Std C37.21-1985 (R1998). *IEEE Standard for Control Switchboards.* Ratings, construction, and testing of dead-front control switchboards containing, but not limited to, devices such as switches, control devices, instrumentation, metering, monitoring, protective and auxiliary relays, and regulating devices and accessories are covered. Switchboards for the control and protection of apparatus used for, or associated with, power generation, conversion, transmission, and distribution are included, but the Standard is not limited to these. Industrial controls, communication equipment, switchboards for use onboard ships, Class 1E switchboards for use in nuclear generating stations, and human factors are not considered.

IEEE Std C37.23-1987 (R1991). *IEEE Standard for Metal-Enclosed Bus and Calculating Losses in Isolated-Phase Bus.* Assemblies of metal-enclosed conductors and their associated interconnections, enclosures, supporting structures, switches, and disconnecting links are addressed. Ratings, tests, construction, miscellaneous accessories, and loss calculation for isolated-phase buses are covered. Specifically excluded are busways or bus assemblies for distribution of electric power less than 600 V consisting of enclosed sectionalized prefabricated bus bars or associated structures and fittings, such as feeder busways (indoor or outdoor) and plug-in busways (indoor only) and bus assemblies utilized at voltages in excess of 38.0 kV.

IEEE Std C37.24-1986 (R1998). *IEEE Guide for Evaluating the Effect of Solar Radiation on Outdoor Metal-Enclosed Switchgear.* This standard applies to all forms of outdoor metal-enclosed switchgear. It covers operating limitations; the effect of ambient temperature, solar radiation, and wind on internal operating temperatures; ventilation and condensation control; enclosure color and finish considerations; current-carrying capabilities of switchgear; and suggested modifications of standard designs.

IEEE Std C37.26-1972 (R1996). *IEEE Guide for Methods of Power-Factor Measurement for Low-Voltage Inductive Test Circuits.* Methods used to measure the power factor in low-voltage test circuits are covered. Since the power factor measurement for high-capacity test circuits is particularly difficult and different methods may yield different results, the methods that are least likely to yield error are recommended for particular circuit conditions. The ratio method is recommended for fast clearing devices that may have total interrupting times of 0.5 cycle or less. The dc decrement method is recommended for circuits of 30% power factor or less when the device to be tested interrupts at a point in time more than one-half cycle from the initiation of the current. The phase relationship method, using current and voltage waves, is recommended for circuits having power factors over 30%.

IEEE Std C37.27-1987 (R1998). *IEEE Application Guide for Low-Voltage AC Non-Integrally Fused Power Circuit Breakers (Using Separately Mounted Current-Limiting Fuses).* Low-voltage power circuit breakers of the 600 V insulation class with separately mounted current-limiting fuses, for use on ac circuits with available short-circuit current of 200 000 A (rms symmetrical) or less, are covered. Guidance is provided respecting coordination of circuit breaker and fuse, location of fuses, open fuse trip devices, addition of fuses to existing installations, protection of connected equipment, and tested combinations of circuit breakers and fuses.

IEEE Std C37.29-1981 (R1985). *IEEE Standard for Low-Voltage AC Power Circuit Protectors Used in Enclosures.* This standard covers enclosed low-voltage ac power circuit protectors of the stationary type with 2-pole or 3-pole construction, having one or more rated maximum voltages of 508 V and 254 V rms for application on systems having nominal voltages of 480 V and 240 V rms, that are manually operated or power operated. The circuit protectors considered are furnished with current limiting fuses such that the entire device is suitable for application on circuits capable of delivering not more than 200 000 A rms symmetrical short-circuit current. Service conditions and ratings are discussed, and the functional components of the circuit protectors are described. Temperature limitations and classification of insulating materials are covered. Insulation (dielectric) withstand voltage requirements are specified, and an application guide is given. Test procedures are also specified.

IEEE Std C37.30-1997. *IEEE Standard Requirements for High-Voltage Switches.* Required ratings and constructional requirements for switches above 1000 V are described.

IEEE Std C37-1996. *IEEE Standard Electrical Power System Device Function Numbers and Contact Designations.* The definition and application of function numbers for devices used in electrical substations and generating plants and in installations of power utilization and conversion apparatus are covered. The purpose of the numbers is discussed, and 94 numbers are assigned. The use of prefixes and suffixes to provide a more specific definition of the function is considered. Device contact designation is also covered.

IEEE Std C37.34-1994. *IEEE Standard Test Code for High-Voltage Air Switches.* Design test requirements for all high-voltage enclosed indoor and outdoor and non-enclosed indoor and outdoor air switches rated above 1000 V are specified. This includes requirements for such switches as disconnecting, selector, horn-gap, grounding, interrupter, etc., for manual and power operation, except for distribution-enclosed single-pole air switches and distribution cutouts fitted with disconnecting blades.

IEEE Std C37.35-1995. *IEEE Guide for the Application, Installation, Operation, and Maintenance of High-Voltage Air Disconnecting and Interrupter Switches.* Guidance for users in the application, installation, operation, and maintenance of high-voltage air switches and interrupter switches is provided.

IEEE Std C37.36b-1990 (1996). *IEEE Guide to Current Interruption with Horn-Gap Air Switches.* Ameans for determining the magnitude of excitation as well as resistive and capacitive currents that may be successfully interrupted with horn-gap, vertical-break air switches in outdoor locations is provided for users of air switches. The practices suggested apply only to switches mounted in the normal horizontal-upright position and not equipped with interrupting aids. It is assumed that the switches are applied to an effectively grounded wye system.

IEEE Std C37.38-1989. *IEEE Standard for Gas-Insulated, Metal-Enclosed Disconnecting, Interrupter, and Grounding Switches.* Requirements for switches rated 72.5 kV and above intended for use in metal-enclosed, gas insulated substations are presented. These switches are characterized by grounded, leak-tight metal enclosures that are filled with a gas (most commonly SF_6) at some pressure above atmospheric, with live parts contained within the housing and insulated therefrom by the gas and by suitable solid insulation that supports the live parts in their proper position. Gas-insulated switches are normally electrically connected to and structurally joined to other gas-insulated components such as buses, gas-to-air bushings, circuit breakers, instrument transformers, cable terminations, etc. Switches may be manually or power operated. Service conditions, ratings, supporting structures, and nameplates are covered. Testing of disconnecting and grounding switches is covered.

IEEE Std C37.40-1993. *IEEE Standard Service Conditions and Definitions for High-Voltage Fuses, Distribution Enclosed Single-Pole Air Switches, Fuse Disconnecting Switches, and Accessories.* Service conditions and definitions for high-voltage fuses (above 1000 V), distribution enclosed single-pole air switches, fuse disconnecting switches, and accessories for ac distribution systems are covered. These include enclosed, open, and open-link types of distribution cutouts and fuses; distribution current-limiting fuses; distribution oil cutouts; distribution enclosed single-pole air switches; power fuses, including current-limiting types; outdoor and indoor fuse disconnecting switches; fuse supports, mountings, hooks, and links, all of the type used exclusively with the above; and removable switch blades for certain products among the above.

IEEE Std C37.40b-1996. *IEEE Standard Service Conditions and Definitions for External Fuses for Shunt Capacitors Supplement to IEEE Std C37.40-1993.* Definitions for high-voltage external capacitor fuses (above 1000 V) used for the protection of shunt capacitor banks are covered in this supplement.

IEEE Std C37.41-1994. *IEEE Standard Design Tests for High-Voltage Fuses, Distribution Enclosed Single-Pole Air Switches, Fuse Disconnecting Switches, and Accessories.* Required procedures for performing design tests for high-voltage distribution-class and power-class fuses, as well as for fuse disconnecting switches and enclosed single-pole air switches are specified. These design tests, as appropriate to a particular device, include the following test types—dielectric, interrupting, load-break, making-current, radio-influence, short-time current, temperature-rise, time-current, mechanical, and liquid-tightness.

IEEE Std C37.48-1997. *IEEE Guide for the Application, Operation, and Maintenance of High-Voltage Fuses, Distribution Enclosed Single-Pole Air Switches, Fuse Disconnecting Switches, and Accessories.* Information on the application, operation, and maintenance of high-voltage fuses (above 1000 V), distribution enclosed single-pole air switches, fuse disconnecting switches, and accessories for use on ac distribution systems is provided. This guide is one of a series of complementary standards covering various types of high-voltage fuses and switches, so arranged that two of the standards apply to all devices while each of the other standards provides additional specifications for a particular device. For each device, IEEE Std C37.40-1993, IEEE Std C37.41-1994, plus the standard covering that device, constitute a complete set of standards for each device. In addition, IEEE Std C37.48-1997 is an application, operation, and maintenance guide for all the devices.

IEEE Std C37.59-1996. *IEEE Standard Requirements for Conversion of Power Switchgear Equipment.* Power switchgear equipment that is converted from the original manufacturer's designs, whether the conversion is performed in manufacturing plants or at installation sites, is covered.

IEEE Std C37.60-1981 (R1992). *IEEE Standard Requirements for Overhead, Pad Mounted, Dry Vault, and Submersible Automatic Circuit Reclosers and Fault Interrupters for AC Systems.* The requirements set forth apply to all overhead, pad mounted, dry vault and submersible single- or multipole ac automatic circuit reclosers and fault interrupters for rated maximum voltages above 1000 V. Service conditions and ratings are discussed. Conditions and procedures are specified for design tests, including dielectric, interruption, current, partial-discharge, radio-influence-voltage, surge-current, temperature-rise, time-current, mechanical-operations, and surge-withstand tests. Production tests and construction requirements are covered.

IEEE Std C37.61-1973 (R1992). *IEEE Guide for the Application, Operation, and Maintenance of Automatic Circuit Reclosers.* Information on the selection, application, operation, and maintenance of single- or multipole ac automatic circuit reclosers is provided. The principal characteristics of reclosers are identified, and the necessary system information is indicated. Step-by-step procedures for selecting reclosers for specific applications are given.

IEEE Std C37.63-1997. *IEEE Standard Requirements for Overhead, Pad-Mounted, Dry-Vault, and Submersible Automatic Line Sectionalizers for AC Systems.* Required definitions (for cutout type sectionalizers), ratings, procedures for performing design tests and production tests, constructional requirements, and application considerations for overhead and pad-mounted, dry-vault, and submersible automatic line sectionalizers for ac systems are specified.

IEEE Std C37.71-1984 (R1990). *IEEE Standard for Three-Phase, Manually Operated Subsurface Load-Interrupting Switches for AC Systems.* This standard applies to three-phase, group-operated, 60 Hz, subsurface, load-interrupting switches with maximum ratings of 600 A and 38 kV and utilizing separable insulated connectors. It covers service conditions; ratings and test requirements; design, production, and conformance tests; construction requirements; and shipping requirements.

IEEE Std C37.73-1998. *IEEE Standard Requirements for Pad-Mounted Fused Switchgear.* Requirements for assemblies of single-phase and three-phase, dead-front and live-front, pad-mounted, load-interrupter switches with expulsion, current-limiting, and other types of fuses in enclosures up to 38 kV rated maximum voltage are given. Definitions are given, and service conditions and ratings are discussed. Design tests, production tests, and construction requirements are included.

IEEE Std C37.81-1989 (R1999). *IEEE Guide for Seismic Qualification of Class 1E Metal-Enclosed Power Switchgear Assemblies.* Requirements and guidance are provided for the seismic qualification of metal-enclosed power switchgear assemblies including switching, interrupting, control, instrumentation, metering, and protective and regulating devices mounted therein. Seismic criteria are discussed, performance requirements are established, and qualification by testing alone and by combined testing and analysis is considered. Documentation is addressed. Although the primary purpose of this guide is for the application of metal-enclosed power switchgear assemblies in nuclear power generating stations, it may be used in other applications in which the seismic response of metal-enclosed power switchgear assemblies is a consideration.

IEEE Std C37.82-1987 (R1998). *IEEE Standard for the Qualification of Switchgear Assemblies for Class 1E Applications in Nuclear Power Generating Stations.* Methods and requirements for qualifying switchgear assemblies for indoor areas outside of the containment in nuclear power generating stations are described. These assemblies include metal-enclosed low-voltage power circuit breaker switchgear assemblies, metal-clad switchgear assemblies, and metal-enclosed interrupter switchgear assemblies. This standard amplifies the general requirements of IEEE Std 323-1983, IEEE Standard for Qualifying Class 1E Equipment for Nuclear Power Generating Stations, as they apply to Class 1E switchgear assemblies.

IEEE Std C37.90-1989 (R1994). *IEEE Standard for Relays and Relay Systems Associated with Electric Power Apparatus.* Standard service conditions, standard ratings, performance requirements, and requirements for testing of relays and relay systems associated with power apparatus are established. Test requirements cover temperature rise limits for foils, dielectric tests, and surge withstand capability tests. Relays designed primarily for industrial control, for switching communication or other low-level signals, or any other equipment not intended for the control of power apparatus are not covered.

IEEE Std C37.90.1-1989 (R1994). *IEEE Standard Surge Withstand Capability (SWC) Tests for Protective Relays and Relay Systems.* Design tests intended for protective relays and relay systems, including those incorporating digital processors, are specified. The tests are intended to be applied to a complete relay system under simulated operating conditions. Oscillatory and fast transient test wave shapes and characteristics are defined. The equipment to be tested and the test conditions are described, and the points of application of the test wave are shown. Acceptance is defined, and the requisite test data are specified.

IEEE Std C37.90.2-1995. *IEEE Standard Withstand Capability of Relay Systems to Radiated Electromagnetic Interference from Transceivers.* A design test to evaluate the susceptibility of protective relays to single-frequency electromagnetic fields in the radio frequency domain, such as those generated by portable or mobile radio transceivers is established.

IEEE Std C37.91-1985 (R1990). *IEEE Guide for Protective Relay Applications to Power Transformers.* A guide to the effective application of relays and other devices for the protection of power transformers is provided. Emphasis is placed on practical applications. The general philosophy and economic considerations involved in transformer protection are reviewed, the types of faults experienced are described, and technical problems with such protection, including current transformer behavior during fault conditions, are discussed. Various types of electrical, mechanical, and thermal protective devices are described, and associated problems such as fault clearing and re-energizing considerations are discussed.

IEEE Std C37.93-1987 (R1999). *IEEE Guide for Power System Protective Relay Applications of Audio Tones over Telephone Channels.* Information and recommendations are provided for applying, installing, and testing audio tones over telephone channels for power system relaying. A basic introduction to and description of leased telephone channels is provided. Also included are typical interface requirements and the transmission line characteristics of three channel offerings along with examples. The intent is to provide a reference for equipment manufacturers engaged in the design and application of relaying equipment, and for telephone personnel engaged in providing telecommunications channels for audio-tone protective relay schemes. The guide has been prepared not only for those considering audio-tone relaying for the first time, but also as a reference for the experienced user.

IEEE Std C37.95-1989 (R1994). *IEEE Guide for Protective Relaying of Utility-Consumer Interconnections.* Information on a number of different protective relaying practices for the utility-consumer interconnection is provided. The following are covered: establishing consumer service requirements and supply method, typical utility-consumer interconnection configurations, protection theory, system studies, and interconnection examples. The information is provided only for applications involving service to a consumer that normally requires a transformation between the utility's supply voltage and the consumer's utilization voltage. Interconnections supplied at the ultimate utilization voltage are not covered.

IEEE Std C37.97-1979 (R1990). *IEEE Guide for Protective Relay Applications to Power System Buses.* The effective application of relays for protection of power system electrical buses is addressed. Common bus arrangements and some special arrangements used in the United States are covered; not all bus protection systems or all possible bus arrangements are included. Factors which determine the need and type of bus protection and basic principles of bus protection operation are discussed. Relay input sources are covered. Bus protection systems and common bus arrangements with relay input sources are described. Also discussed are current transformer locations, wiring and grounding, location of the bus on the system, bus construction, problems associated with switching and by-passing, auxiliary tripping relays, reclosing of breakers after a bus differential operation, testing of bus differential relaying, and bus backup protection.

IEEE Std C37.98-1987 (R1999). *IEEE Standard for Seismic Testing of Relays.* The procedures to be used in the seismic testing of relays used in power system facilities are specified. The concern is with determining the seismic fragility level of relays. Recommendations for proof testing are given. Documentation and generalization of test results are discussed.

IEEE Std C37.99-2000. *IEEE Guide for the Protection of Shunt Capacitor Banks.* The protection of shunt power capacitor and filter banks are covered. Guidelines for reliable applications of protection methods intended for use in many shunt capacitor applications and designs are included. The protection of pole-mounted capacitor banks on distribution circuits and the application of capacitors connected directly to routing apparatus are not included.

IEEE Std C37.100-1992. *IEEE Standard Definitions for Power Switchgear.* Terms that encompass the products within the scope of the C37 project are defined. These include power switchgear for switching, interrupting, metering, protection, and regulating purposes, as used primarily in connection with generation, transmission, distribution, and conversion of electric power. The definitions do not purport to embrace other meanings that the terms may properly have when used in connection with other subjects.

IEEE Std C37.101-1993 (R2000). *IEEE Guide for Generator Ground Protection.* Guidance in the application of relays and relaying schemes for protection against stator ground faults on high-impedance grounded generators is provided.

IEEE Std C37.102-1995. *IEEE Guide for AC Generator Protection.* A review of the generally accepted forms of relay protection for the synchronous generator and its excitation system is presented. This is guide primarily concerned with protection against faults and abnormal operating conditions for large hydraulic, steam, and combustion-turbine generators.

IEEE Std C37.105-1987 (R1999). *IEEE Standard for Qualifying Class 1E Protective Relays and Auxiliaries for Nuclear Power Generating Stations.* The basic principles, requirements, and methods for qualifying Class 1E protective relays and auxiliaries such as test and control switches, terminal blocks, and indicating lamps for applications in nuclear power generating stations are described. The qualification procedure is generic in nature can be used to demonstrate the design adequacy of such equipment under normal, abnormal, design-basis-event, and post-design-basis-event conditions. Protective relays and auxiliaries located inside primary containment in a nuclear power generating station are not covered.

IEEE Std C37.106-1987 (R1992). *IEEE Guide for Abnormal Frequency Protection for Power Generating Plants.* This guide has been prepared to assist the protection engineer in applying relays for the protection of generating plant equipment from damage caused by operation at abnormal frequencies including overexcitation. Emphasis is placed on the protection of the major generating station components at steam generating stations, nuclear stations, and on combustion-turbine installations. Consideration is also given to the effect of abnormal frequency operation on those associated station auxiliaries whose response can affect plant output. The guide also presents background information regarding the hazards caused by operating generation equipment at abnormal frequencies. It documents typical equipment capabilities and describes acceptable protective schemes. Recommended methods for coordinating the underfrequency protective scheme with system load shielding schemes are also included. Sufficient information is provided to apply suitable coordinated protection for given specific situations.

IEEE Std C37.108-1989 (R1994). *IEEE Guide for the Protection of Network Transformers.* This guide is intended to aid those engineers who have reevaluated the risks associated with faults within network vaults, particularly for those network vaults located within or near high-rise buildings. Currently available devices that are being used in network transformer protection schemes are identified. The fault-detection capabilities of these devices are described.

IEEE Std C37.109-1988 (R1999). *IEEE Guide for the Protection of Shunt Reactors.* Protection of shunt reactors used typically to compensate for capacitive shunt reactance of transmission lines is covered. Two basic shunt-reactor configurations are considered: dry-type, connected ungrounded wye, which is connected to the impedance-grounded tertiary of a power transformer; and oil-immersed, wye-connected, with a solidly grounded or impedance-grounded neutral, connected to the transmission system. Reactor construction and characteristics are discussed. Other arrangements or special applications of reactors such as harmonic filter banks, static VAR compensation (SVC), high-voltage direct current (HVDC), or current-limiting reactors are not specifically addressed; however, the protective methods described in this guide are usually applicable to this equipment.

IEEE Std C37.110-1996. *IEEE Guide For The Application of Current Transformers Used for Protective Relaying Purposes.* The characteristics and classification of current transformers (cts) used for protective relaying are described. This guide also describes the conditions that cause the ct output to be distorted and the effects on relaying systems of this distortion. The selection and application of cts for the more common protection schemes are also addressed.

IEEE Std C37.111-1999. *IEEE Standard Common Format for Transient Data Exchange (COMTRADE) for Power Systems.* A common format for data files and exchange medium used for the interchange of various types of fault, test, or simulation data for electrical power systems is defined. Sources of transient data are described, and the case of diskettes as an exchange medium is recommended. Issues of sampling rates, filters, and sample rate conversions for transient data being exchanged are discussed. Files for data exchange are specified, as is the organization of the data. A sample file is given.

IEEE Std C37.112-1996 (R1999). *IEEE Standard Inverse-Time Characteristic Equations for Overcurrent Relays.* The inverse-time characteristics of overcurrent relays are defined in this standard. Operating equations and allowances are provided in the standard. The standard defines an integral equation for microprocessor relays that ensures coordination not only in the case of constant current input but for any current condition of varying magnitude. Electromechanical inverse-time overcurrent relay reset characteristics are defined in the event that designers of microprocessor based relays and computer relays want to match the reset characteristics of the electromechanical relays.

IEEE Std C37.113-1999. *IEEE Guide for Protective Relay Applications to Transmission Lines.* This newly developed guide compiles information on the application considerations of protective relays to ac transmission lines. The guide describes accepted transmission line protection schemes and the different electrical system parameters and situations that affect their application. Its purpose is to provide a reference for the selection of relay schemes and to assist less experienced protective relaying engineers in their application.

IEEE Std C37.122-1993 (R1999). *IEEE Standard for Gas-Insulated Substations.* The technical requirements for the design, fabrication, testing, and installation of a gas-insulated substation (GIS) are covered. The parameters to be supplied by the purchaser are set, and the technical requirements for the design, fabrication, testing, and installation to be furnished by the manufacturer are established.

IEEE Std C37.122.1-1993. *IEEE Guide for Gas-Insulated Substations.* The technical requirements for the design, fabrication, testing, and installation of a gas-insulated substation (GIS) are covered. Parameters to be supplied by the purchaser are suggested, and technical requirements for the design, fabrication, testing, and installation to be furnished by the manufacturer are established.

IEEE Std C37.123-1996 (R1996). *IEEE Guide to Specifications for Gas-Insulated Electric Power Substation Equipment.* IEEE Std C37.123-1996 covers the technical requirements for the design, fabrication, testing and installation of a gas-insulated substation (GIS); its intent is advisory. This guide discusses parameters to be supplied by the purchaser and technical requirements for the design, fabrication, testing, and installation to be furnished by the manufacturer. Environmental conditions, general and specific equipment requirements, and a proposal data sheet form are included to aid the user.

ANSI C50.13-1977 (R1999). *American National Standard Requirements for Cylindrical-Rotor Synchronous Generators.* Requirements for 60Hz cylindrical-rotor synchronous generators, except those covered in standard C50.14-1977, are set forth. The standard covers classification, usual service conditions, rating, temperature rise, abnormal conditions, efficiency, overspeed, telephone influence factor, tests, direction of rotation, nameplate marking, and performance specification forms.

ANSI C50.14-1977 (R1999). *American National Standard Requirements for Combustion Gas Turbine Driven Cylindrical Rotor Synchronous Generators.* Requirements are provided for 60 Hz open-ventilated air-cooled cylindrical rotor synchronous generators rated 10 000 kVA and above. Classification, service conditions, output rating and capabilities, temperature, abnormal and short-circuit requirements, efficiency, overspeed, telephone influence factor, tests, direction of rotation, and nameplate marking are covered. A performance specification form is shown.

IEEE Std C57.12.00-2000. *IEEE Standard General Requirements for Liquid-Immersed Distribution, Power, and Regulating Transformers.* Electrical, mechanical, and safety requirements are set forth for liquid-immersed distribution and power transformers, and autotransformers and regulating transformers; single and polyphase, with voltages of 601 V or higher in the highest voltage winding. This standard is a basis for the establishment of performance, limited electrical and mechanical interchangeability, and safety requirements of equipment described; and for assistance in the proper selection of such equipment. The requirements in this standard apply to all liquid-immersed distribution, power, and regulating transformers except the following: instrument transformers, step-voltage and induction voltage regulators, arc furnace transformers, rectifier transformers, specialty transformers, grounding transformers, mobile transformers, and mine transformers.

IEEE Std C57.12.01-1998. *IEEE Standard General Requirements for Dry-Type Distribution and Power Transformers Including Those with Solid-Cast and/or Resin-Encapsulated Windings.* Electrical, mechanical, and safety requirements of ventilated, nonventilated, and sealed dry-type distribution and power transformers or autotransformers, single and polyphase, with a voltage of 601 V or higher in the highest voltage winding, are described. Information that can be used as a basis for the establishment of performance, interchangeability, and safety requirements of equipment described, and for assistance in the proper selection of such equipment, is given.

ANSI C57.12.20-1997. *American National Standard for Transformers Standard For Overhead Type Distribution Transformers, 500 kVA and Smaller: High Voltage, 34500 Volts and Below; Low Voltage, 7970/13800Y Volts and Below.*

IEEE Std C57.12.23-1992 (R1999). *IEEE Standard for Transformers—Underground-Type, Self-Cooled, Single-Phase Distribution Transformers with Separable, Insulated, High-Voltage Connectors; High Voltage (24 940 GrdY/14 400 V and Below) and Low-Voltage (240/120 V, 167 kVA and Smaller).* Electrical, dimensional, and mechanical characteristics and certain safety features of single-phase, 60 Hz, mineral-oil-immersed, self-cooled, distribution transformers with separable insulated high-voltage connectors are covered. Ratings, testing, and construction are discussed. These transformers are generally used for step-down purposes from an underground primary cable supply and are suitable for occasional submerged operation. The intent is to provide a basis for determining their performance, interchangeability, and safety, and for their selection. This standard does not cover the electrical and mechanical requirements of accessory devices that may be supplied with the transformer.

IEEE Std C57.12.35-1996. *IEEE Standard for Bar Coding for Distribution Transformers.* This standard sets forth bar code label requirements for overhead, padmounted, and underground-type distribution transformers. Included herein are requirements for data content, symbology, label layout, print quality, and label life expectancy.

IEEE Std C57.12.44-1994. *IEEE Standard Requirements for Secondary Network Protectors.* The performance, electrical and mechanical interchangeability as well as the safety of the equipment are covered. The proper selection of such equipment is established as a basis for use in this standard. Certain electrical, dimensional, and mechanical characteristics are described, and certain safety features of three-phase, 60 Hz, low-voltage 600 V and below network protectors are taken into consideration. They are used for automatically connecting and disconnecting a network transformer from a secondary spot or grid network.

IEEE Std C57.12.56-1986 (R1998). *IEEE Standard Test Procedure for Thermal Evaluation of Insulation Systems for Ventilated Dry-Type Power and Distribution Transformers.* A test procedure for determining the temperature classification of ventilated dry-type power and distribution transformer insulation systems by test rather than by chemical composition is established. The intent is to provide a uniform method for providing data for selection of the temperature classification of the insulation system, for providing data which may be used as a basis for a loading guide, and for comparative evaluation of different insulation systems. Voltage withstand endpoint criteria are related to the impulse voltage distribution within the coil or to the initial-voltage withstand of the coil. A relationship between impulse withstand of the insulation and short-term 60 Hz withstand is identified so that 50/60 Hz testing of model coils is possible.

IEEE Std C57.12.58-1991 (R1996). *IEEE Guide for Conducting a Transient Voltage Analysis of a Dry-Type Transformer Coil.* General recommendations for measuring voltage transients in dry-type distribution and power transformers are provided. Recurrent surge-voltage generator circuitry, instrumentation, the test sample, test point location, mounting the test coil, conducting the test, and reporting results are covered.

IEEE Std C57.12.60-1998. *IEEE Guide for Test Procedures for Thermal Evaluation of Insulation Systems for Solid-Cast and Resin-Encapsulated Power and Distribution Transformers.* A uniform method is established for determining the temperature classification of solid-cast and resin-encapsulated power and distribution transformer insulation systems by testing rather than by chemical composition. These insulation systems are intended for use in transformers covered by C57.12.01-1989 and C57.12.91-1995 as they apply to solid-cast and resin-encapsulated transformers whose highest voltages exceed nominal 600 V.

IEEE Std C57.12.80-1978 (R1992). *IEEE Standard Terminology for Power and Distribution Transformers.* This standard is a compilation of terminology and definitions primarily related to electrical transformers and associated apparatus included within the scope of ANSI Committee C57, Transformers, Regulators, and Reactors. It also includes similar data relating to power systems and insulation that is commonly involved in transformer technology.

IEEE Std C57.12.90-1999. *IEEE Standard Test Code for Liquid-Immersed Distribution, Power, and Regulating Transformers.* Methods for performing tests specified in IEEE Std C57.12.00-1993 and other standards applicable to liquid-immersed distribution, power, and regulating transformers are described. Instrument transformers, step-voltage and induction voltage regulators, arc furnace transformers, rectifier transformers, specialty transformers, grounding transformers, and mine transformers are excluded. This standard covers resistance measurements, polarity and phase-relation tests, ratio tests, no-load-loss and excitation current measurements, impedance and load loss measurements, dielectric tests, temperature tests, short-circuit tests, audible sound level measurements, calculated data, and certified test data.

IEEE Std C57.12.91-1995. *IEEE Test Code for Dry-Type Distribution and Power Transformers.* Methods for performing tests specified in IEEE Std C57.12.01-1989 and other referenced standards applicable to dry-type distribution and power transformers are described. This standard is intended for use as a basis for performance, safety, and the proper testing of dry-type distribution and power transformers. This standard applies to all dry-type transformers except instrument transformers, step-voltage and induction voltage regulators, arc furnace transformers, rectifier transformers, specialty transformers, and mine transformers.

IEEE Std C57.13-1993. *IEEE Standard Requirements for Instrument Transformers.* Electrical, dimensional, and mechanical characteristics are covered, taking into consideration certain safety features, for current and inductively coupled voltage transformers of types generally used in the measurement of electricity and the control of equipment associated with the generation, transmission, and distribution of alternating current. The aim is to provide a basis for performance, interchangeability, and safety of equipment covered and to assist in the proper selection of such equipment. Accuracy classes for metering service are provided. The test code covers measurement and calculation of ratio and phase angle, demagnetization, impedance and excitation measurements, polarity determination, resistance measurements, short-time characteristics, temperature rise tests, dielectric tests, and measurement of open-circuit voltage of current transformers.

IEEE Std C57.13.1-1981 (R1999). *IEEE Guide for Field Testing of Relaying Current Transformers.* A description is given of field test methods that will assure that the current transformers used as a source of relay input current are connected properly, are of marked ratio and polarity, and are in condition to perform as designed both initially and after a period of service. The standard covers safety considerations; current transformer types and construction, and the effect of these on test methods; insulation resistance tests; ratio tests; polarity check; winding and lead resistance (internal resistance) excitation test, burden measurements, and specialized situations.

IEEE Std C57.13.3-1983 (R1990). *IEEE Guide for the Grounding of Instrument Transformer Secondary Circuits and Cases.* General and specific recommendations for grounding current and voltage transformer secondary circuits and cases of connected equipment are provided. The practices recommended apply to all transformers of this type, including capacitive voltage transformers and linear couplers, irrespective of primary voltage or whether the primary windings are connected to, or are in, power circuits or are connected in the secondary circuits of other transformers as auxiliary current or voltage transformers. The primary emphasis is personnel safety and proper performance of relays at power-line frequencies. The grounding and shielding of cables and other grounding considerations are not addressed.

IEEE Std C57.15-1999. *IEEE Standard Requirements, Terminology, and Test Code for Step-Voltage Regulators.* Electrical, mechanical, and safety requirements of oil-filled, single- and three-phase voltage regulators not exceeding regulation of 2500 kVA (for three-phase units) or 833 kVA (for single-phase units) are covered.

IEEE Std C57.16-1996. *IEEE Standard Requirements, Terminology, and Test Code for Dry-Type Air-Core Series-Connected Reactors.* Series-Connected dry-type air-core single-phase and three-phase outdoor or indoor reactors of distribution and transmission voltage class that are connected in the power system to control power flow under steady-state conditions and/or limit fault current under short-circuit conditions are covered. Dry-Type air-core reactors covered by this standard are self-cooled by natural air convection. With some restrictions, other reactors, including filter reactors, shunt capacitor reactors (used with shunt capacitor banks), and discharge current limiting reactors (used with series capacitor banks), are also covered.

IEEE Std C57.18.10-1998. *IEEE Standard Practices and Requirements for Semiconductor Power Rectifier Transformers.* Practices and requirements for semiconductor power rectifier transformers for dedicated loads rated single-phase 300 kW and above and three-phase 500 kW and above are included. Static precipitators, high-voltage converters for dc power transmission, and other nonlinear loads are excluded. Service conditions, both usual and unusual, are specified, or other standards are referenced as appropriate. Routine tests are specified. An informative annex provides several examples of load loss calculations for transformers when subjected to non-sinusoidal currents, based on calculations provided in the standard.

IEEE Std C57.19.00-1991 (R1997). *IEEE Standard General Requirements and Test Procedure for Outdoor Power Apparatus Bushings.* Service conditions, rating, general requirements, and test procedures for outdoor apparatus bushings are set forth. They apply to outdoor power apparatus bushings that have basic impulse insulation levels of 110 kV and above for use as components of oil-filled transformers, oil-filled reactors, and oil circuit breakers. The following are not covered: high-voltage cable terminations (potheads), bushings for instrument transformers, bushings for test transformers, bushings in which the internal insulation is provided by a gas, bushings applied with gaseous insulation (other than air at atmospheric pressure) external to the bushing, bushings for distribution-class circuit breakers and transformers, bushings for automatic circuit reclosures and line sectionalizers, and bushings for oil-less and oil-poor apparatus.

IEEE Std C57.19.01-2000. *IEEE Standard Performance Characteristics and Dimensions for Outdoor Apparatus Bushings.* Electrical, dimensional, and related requirements for outdoor power apparatus bushings that have basic impulse insulation levels (BILs) of 200 kV and above are covered. Specific values for dimensional and related requirements that are to be interpreted, measured, or tested, in accordance with IEEE Std C57.19.00-1991, are provided.

IEEE Std C57.19.03-1996. *IEEE Standard Requirements, Terminology, and Test Code for Bushings for DC Applications.* This standard applies to outdoor and indoor power apparatus dc bushings of condenser type that have basic impulse insulation levels of 110 kV and above for use as components of oil-filled converter transformers and smoothing reactors, as well as air-to-air dc bushings. This standard defines the special terms used, service conditions, rating, general requirements, electrical insulation characteristics, and test procedures for the bushings for dc application.

IEEE Std C57.19.100-1995. *IEEE Guide for Application of Power Apparatus Bushings.* Guidance on the use of outdoor power apparatus bushings is provided. The bushings are limited to those built in accordance with IEEE Std C57.19.00-1991. General information and recommendations for the application of power apparatus bushings, when incorporated as part of power transformers, power circuit breakers, and isolated-phase bus, are provided.

IEEE Std C57.21-1990 (R1995). *IEEE Standard Requirements, Terminology, and Test Code for Shunt Reactors Rated Over 500 kVA.* An oil-immersed or dry-type, single-phase or three-phase, outdoor or indoor shunt reactors rated over 500 kVA are covered. Terminology and general requirements are stated, and the basis for rating shunt reactors is set forth. Routine, design, and other tests are described, and methods for performing them are given. Losses and impedance, temperature rise, dielectric tests, and insulation levels are covered. Construction requirements for oil-immersed reactors and construction and installation requirements for dry-type reactors are presented.

IEEE Std C57.91-1995. *IEEE Guide for Loading Mineral-Oil-Immersed Transformers.* General recommendations for loading 65 °C rise mineral-oil-immersed distribution and power transformers are covered.

IEEE Std C57.93-1995. *IEEE Guide for Installation of Liquid-Immersed Power Transformers.* Guidance is given for the shipping, installation, and maintenance of liquid-immersed power transformers rated 501 kVA and above with secondary voltage of 1000 V and above. The entire range of power transformers is covered, including EHV transformers, with distinctions as required for various sizes, voltage ratings, and liquid insulation types.

IEEE Std C57.94-1982 (R2000). *IEEE Recommended Practice for Installation, Application, Operation, and Maintenance of Dry-Type General Purpose Distribution and Power Transformers.* The application, installation, operation and maintenance of single- and polyphase dry-type general purpose, distribution, power, and autotransformers are covered. The following types are included: ventilated, indoor and outdoor, self- or forced-air cooled; nonventilated, indoor and outdoor, self- or forced-air cooled; and sealed, indoor and outdoor, and self-cooled. Instrument transformers, step voltage and induction voltage regulators, arc furnace transformers, rectifier transformers, and specialty transformers are not covered.

IEEE Std C57.96-1999. *IEEE Guide for Loading Dry-Type Distribution and Power Transformers.* General recommendations for the loading of dry-type distribution and power transformers that have 80 °C, 115 °C, and 150 °C average winding rises and insulation systems limited to 150 °C, 180 °C, and 220 °C maximum hottest-spot operating temperatures, respectively, are covered in this guide. Recommendations for ventilated, nonventilated, and sealed dry-type transformers having impregnated insulation systems are included.

IEEE Std C57.98-1993 (R1999). *IEEE Guide for Transformer Impulse Tests.* Transformer connections, test methods, circuit configurations, and failure analysis of lightning impulse and switching impulse testing of power transformers are addressed. This guide is also generally applicable to distribution and instrument transformers.

IEEE Std C57.100-1999. *IEEE Standard Test Procedure for Thermal Evaluation of Liquid-Immersed Distribution and Power Transformers.* A test procedure is established to provide a uniform method for investigating the effect of operating temperature on the life expectancy of liquid-immersed transformers. The test procedures are intended to provide data for the selection of a limiting hottest-spot temperature for rating purposes, provide data which may serve as the basis for a guide for loading, and permit the comparative evaluation of a proposed insulation system with reference to a system that has proven to be acceptable in service.

IEEE Std C57.104-1991. *IEEE Guide for the Interpretation of Gases Generated in Oil-Immersed Transformers.* Detailed procedures for analyzing gas from gas spaces or gas-collecting devices, as well as gas dissolved in oil, are described. The procedures cover: (1) the calibration and use of field instruments for detecting and estimating the amount of combustible gases present in gas blankets above oil, or in gas-detector relays; (2) the use of fixed instruments for detecting and determining the quantity of combustible gases present in gas-blanketed equipment; (3) methods for obtaining samples of gas and oil from the transformer for laboratory analysis; (4) laboratory methods for analyzing the gas blanket and the gases extracted from the oil; and (5) methods for interpreting the results in terms of transformer serviceability. The intent is to provide the operator with positive and useful information concerning the serviceability of the equipment. An extensive bibliography on gas evolution, detection, and interpretation is included.

IEEE Std C57.105-1978 (R1999). *IEEE Guide for Application of Transformer Connections in Three-Phase Distribution Systems.* The characteristics of the various transformer connections and possible operating problems under normal or abnormal conditions are treated for three-phase distribution systems. These systems are characterized by primary voltages up to and including 34.5 kV, usually have a preponderance of connected transformers with low-voltage windings below 1000 V, and furnish electric service to consumers. All combinations of D and Y, grounded and ungrounded, T-connected, zigzag, and certain special connections are considered. Only two-winding transformers are included. Phasing procedures and loading practices are not covered.

IEEE Std C57.106-1991 (R1998). *IEEE Guide for Acceptance and Maintenance of Insulating Oil in Equipment.* Recommendations are made regarding oil tests and evaluation procedures, methods of reconditioning and reclaiming conventional petroleum (mineral) dielectric oils, the levels at which these become necessary, and the routines for restoring oxidation resistance, where required, by the addition of inhibitors. The intent is to assist the power equipment operator in evaluating the serviceability of oil received in equipment, oil as received from the refiner for filling new equipment at the installation site, and oil as processed into such equipment, and to assist the operator in maintaining the oil in serviceable condition. The mineral oil covered is used in transformers, switchgear, reactors, and current breakers.

IEEE Std C57.109-1993 (R2000). *IEEE Guide for Liquid-Immersed Transformer Through-Fault-Current Duration.* Recommendations believed essential for the application of overcurrent protective devices applied to limit the exposure time of transformers to short circuit current are set forth. Transformer coordination curves are presented for four categories of transformers. There is no intent to imply overload capability.

IEEE Std C57.110-1998. *IEEE Recommended Practice for Establishing Transformer Capability When Supplying Nonsinusoidal Load Currents.* Methods are developed to conservatively evaluate the feasibility of supplying additional nonsinusoidal load currents from an existing installed dry-type or liquid-filled transformer, as a portion of the total load. Clarification of the necessary application information is provided to assist in properly specifying a new transformer expected to carry a load, a portion of which is composed of nonsinusoidal load currents. A number of examples illustrating these methods and calculations are presented. Reference annexes make a comparison of the document calculations to calculations found in other industry standards and suggested temperature rise methods are detailed for reference purposes.

IEEE Std C57.111-1989 (R1995). *IEEE Guide for Acceptance of Silicone Insulating Fluid and Its Maintenance in Transformers.* Tests and evaluation procedures for silicone transformer fluid are recommended. Criteria for maintenance and methods of reconditioning of silicone fluid are described. The aim is to assist the transformer operator in evaluating the silicone insulating fluids in transformers, fluid received from the manufacturer for filling transformers at the installation site, and fluid processed into such transformers as well as in maintaining the properties of silicone fluid in operating transformers.

IEEE Std C57.113-1991 (R1995). *IEEE Guide for Partial-Discharge Measurement in Liquid-Filled Power Transformers and Shunt Reactors.* The detection and measurement by the wideband apparent charge method of partial discharges occurring in liquid-filled power transformers and shunt reactors during dielectric tests are covered. The measuring instrument, calibrator characteristics, test circuits, calibration procedure, and partial discharge measurement during induced-voltage tests are covered.

IEEE Std C57.116-1989 (R2000). *IEEE Guide for Transformers Directly Connected to Generators.* The selection, application, and specification considerations for the unit and unit auxiliaries transformers are described, taking into account their connections, voltage and kilovoltampere ratings, and excitation and through-fault capabilities during possible operating conditions, both normal and abnormal. Consideration is given to direct connections and connections through generator breakers and load-break switches. Both hydroelectric and thermal electric generating stations are covered. Phasing procedures, basic impulse insulation level selection, and loading practices are not covered.

IEEE Std C57.117-1986 (R1998). *IEEE Guide for Reporting Failure Data for Power Transformers and Shunt Reactors on Electric Utility Power Systems.* The reporting and statistical analysis of reliability of power transformers and shunt reactors used on electric utility power systems are addressed. The following types and applications of transformers are covered: power transformers, autotransformers, regulating transformers, phase-shifting transformers, shunt reactors, HVDC converter transformer, substation transformers, transmission tie transformers, unit transformers, unit auxiliary transformers, and grounding transformers. The format for the collection and reporting of data is presented, and the kinds of reports that may be useful to both users and manufacturers of transformers are illustrated.

IEEE Std C57.120-1991 (R2000). *IEEE Standard Loss Evaluation Guide for Power Transformers and Reactors.* A method for establishing the dollar value of the electric power needed to supply the losses of a transformer or reactor is provided. Users can use this loss evaluation to determine the relative economic benefit of a high-first-cost, low-loss unit versus one with a lower first cost and higher losses, and to compare the offerings of two or more manufacturers to aid in making the best purchase choice. Manufacturers can use the evaluation to optimize the design and provide the most economical unit to bid and manufacture. The various types of losses are reviewed.

IEEE Std C57.121-1998. *IEEE Guide for Acceptance and Maintenance of Less Flammable Hydrocarbon Fluid in Transformers.* The evaluation and handling procedures for less flammable hydrocarbon transformer insulating fluids are covered. The guide's purpose is to assist the transformer operator in receiving new fluids, filling transformers, and maintaining the fluids in serviceable condition.

IEEE Std C57.124-1991 (R1996). *IEEE Recommended Practice for the Detection of Partial Discharge and the Measurement of Apparent Charge in Dry-Type Transformers.* The detection of partial discharges occurring in the insulation of dry type transformers or their components, and the measurement of the associated apparent charge at the terminals when an alternating test voltage is applied, are covered. The wideband method is used. The detection system and calibrator characteristics are described, and the test procedure is established.

IEEE Std C57.125-1991 (R1998). *IEEE Guide for Failure Investigation, Documentation, and Analysis for Power Transformers and Shunt Reactors.* A procedure to be used to perform a failure analysis is recommended. The procedure is primarily focused on power transformers used on electrical utility systems, although it may be used for an investigation into any ac transformer failure. This document provides a methodology by which the most probable cause of any particular transformer failure may be determined. This document is also intended to encourage the establishment of routine and uniform data collection procedures, consistency of nomenclature and compatibility with similar efforts by other organizations, and cooperative effects by users and manufacturers during the failure analysis.

IEEE Std C57.129-1999. *IEEE Trial Use General Requirements and Test Code for Oil Immersed HVDC Converter Transformers.* The electrical, mechanical, and physical requirements of oil-immersed single-phase and three-phase converter transformers are specified. Tests are described and test code defined. Devices such as arc furnace transformers and rectifier transformers for industrial or locomotive applications are not covered.

IEEE Std C57.131-1995 (R1998). *IEEE Standard Requirements for Load Tap Changers.* Electrical and mechanical performance and test requirements for load tap changers installed in power transformers and voltage regulating transformers of all voltage and kVA ratings are covered.

IEEE Std C57.134-2000. *IEEE Guide for Determination of Hottest-Spot Temperature in Dry-Type Transformers.* Methodologies for determination of the steady-state winding hottest-spot temperature in dry-type distribution and power transformers with ventilated, sealed, solid cast, and encapsulated windings built in accordance with IEEE Std C57.12.01-1998 and IEC 60726 (1982-01) are described in this guide. Converter transformers are not included in this guide.

IEEE Std C57.138-1998. *IEEE Recommended Practice for Routine Impulse Test for Distribution Transformers.* General test procedures for performing routine quality control test that is suitable for high-volume, production line testing. Transformer connections, test methods, circuit configurations, and failure detection methods are addressed. This recommended practice covers liquid-immersed, single- and three-phase distribution transformers.

IEEE Std C62.11-1999. *IEEE Standard for Metal-Oxide Surge Arresters for AC Power Circuits (1 kV).* Metal-oxide surge arresters designed to repeatedly limit the voltage surges on 4862 Hz power circuits (1 kV) are covered in this standard. These devices operate by discharging surge current. Devices for separate mounting and those supplied integrally with other equipment are also discussed.

IEEE Std C62.22-1997. *IEEE Guide for the Application of Metal-Oxide Surge Arresters for Alternating-Current Systems.* The application of metal-oxide surge arresters to safeguard electric power equipment against the hazards of abnormally high voltage surges of various origins is covered. Step-by-step directions toward proper solutions of various applications are provided. In many cases, the prescribed steps are adequate. More complex and special solutions requiring study by experienced engineers are described, but specific solutions are not always given. The procedures are based on theoretical studies, test results, and experience.

IEEE Std C62.23-1995. *IEEE Application Guide for Surge Protection of Electric Generating Plants.* This standard consolidates most electric utility power industry practices, accepted theories, existing standards/guides, definitions, and technical references as they specifically pertain to surge protection of electric power generating plants. Where technical information is not readily available, guidance is provided to aid toward proper surge protection and to reduce interference to communication, control, and protection circuits due to surges and other overvoltages. It has to be recognized that this application guide approaches the subject of surge protection from a common or generalized application viewpoint. Complex applications of surge protection practices may require specialized study by experienced engineers.

IEEE Std C62.31-1987 (R1998). *IEEE Standard Test Specifications for Gas-Tube Surge-Protective Devices.* Gas-tube surge-protective devices for application on systems with voltages = 1000 V rms or 1200 V dc are covered. These protective devices are designed to limit voltage surges on balanced or unbalanced communication circuits and on power circuits operating from dc to 420 Hz. Test criteria for determining the electrical characteristics of these devices are provided.

IEEE Std C62.32-1987 (R1997). *IEEE Standard Test Specifications for Low-Voltage Air Gap Surge-Protective Devices.* Air gaps for over-voltage protection applications on systems with operating voltages equal to or less than 600 V rms are covered. These protective devices are designed for limiting the voltages on balanced or unbalanced communication, power, and signaling circuits. A series of standard design tests for determining the electrical characteristics of these air gap devices is specified. The tests provide a means of comparison among various air gap surge-protective devices.

IEEE Std C62.33-1982 (R1994). *IEEE Standard Test Specifications for Varistor Surge-Protective Devices.* Varistors for surge-protective applications on systems with dc to 420 Hz frequency and voltages equal to or less than 1000 V rms, or 1200 V dc, are covered. Definitions, service conditions, and a series of test criteria for determining the electrical characteristics of the varistors are provided. The tests are intended as design tests and provide a means of comparing various surge-protective devices.

IEEE Std C62.34-1996. *IEEE Standard for Performance of Low-Voltage Surge-Protective Devices (Secondary Arresters).* Surge-protective devices designed for application on the low-voltage ac supply mains (1000 V rms and less, frequency between 48 Hz and 62 Hz) are covered.

IEEE Std C62.35-1987 (R2000). *IEEE Standard Test Specifications for Avalanche Junction Semiconductor Surge-Protective Devices.* A two-terminal avalanche junction surge suppressor for surge protective application on systems with dc to 420 Hz frequency and voltages equal to or less than 1000 V rms or 1200 V dc is considered. The device is a single package that may be assembled from any combination of series and/or parallel diode chips. Definitions, service conditions, and a series of test criteria for determining its electrical characteristics are provided. These devices are used as a surge diverter for limiting transient overvoltages in power and communications circuits.

IEEE Std C62.36-1994. *IEEE Standard Test Methods for Surge Protectors Used in Low-Voltage Data, Communications, and Signaling Circuits.* Methods are established for testing and measuring the characteristics of surge protectors used in low-voltage data, communications, and signaling circuits with voltages less than or equal to 1000 V rms or 1200 V dc. The surge protectors are designed to limit voltage surges, current surges, or both. The surge protectors covered are multiple-component series or parallel combinations of linear or nonlinear elements. Tests are included for characterizing standby performance, surge-limiting capabilities, and surge lifetime. Packaged single gas-tube, air-gap, varistor, or avalanche junction surge-protective devices are not covered, nor are test methods for low-voltage power circuit applications.

IEEE Std C62.37-1996. *IEEE Standard Test Specification for Thyristor Diode Surge Protective Devices.* This standard applies to two or three terminal, four or five layer, thyristor surge protection devices (SPDs) for application on systems with voltages equal to or less than 1000 V rms or 1200 V dc.

IEEE Std C62.38-1994 (R1999). *IEEE Guide on Electrostatic Discharge (ESD): ESD Withstand Capability Evaluation Methods for Electronic Equipment Subassemblies.* This guide establishes test methods for the evaluation of ESD withstand capability for electronic equipment subassemblies. It includes information about test conditions, test equipment, and test procedures for ESD tests of printed circuit boards and other subassemblies.

IEEE Std C62.41-1991 (R1995). *IEEE Recommended Practice for Surge Voltages in Low-Voltage AC Power Circuits.* A practical basis is provided for the selection of voltage and current tests to be applied in evaluating the surge withstand capability of equipment connected to utility power circuits, primarily in residential, commercial, and light industrial applications. The standard covers the origin of surge voltages, rate of occurrence and voltage levels in unprotected circuits, waveshapes of representative surge voltages, energy and source, and impedance. AC power circuits with rated voltages up to 277 V line to ground are addressed, although some of the conclusions offered could apply to higher voltages and also to some dc power systems. The data have been recorded primarily on 120, 220/380, or 277/480 V systems. The general conclusions may be valid for 600 V systems, but more data are needed for the higher voltages.

IEEE Std C62.42-1992 (R1999). *IEEE Guide for the Application of Gas Tube and Air Gap Arrester Low-Voltage (Equal to or Less than 1000 V rms or 1200 V dc) Surge-Protective Devices.* Assistance in selecting the most appropriate type of low-voltage surge-protection device (either gas tube or air gap) for a particular application is provided. Evaluation of the characteristics of each device to meet specific service requirements is also given.

IEEE Std C62.43-1999. *IEEE Guide for the Application of Surge Protectors Used in Low-Voltage (Equal to or Less than 1000 V rms or 1200 V dc) Data, Communications, and Signaling Circuits.* Assistance is provided for the selection of the most appropriate type of low-voltage data, communications, and/or signalling circuit surge protector for a particular application or set of conditions. Surge protector functions and characteristics are also explained and evaluated. AC power circuit applications are not addressed in this document.

IEEE Std C62.45-1992 (R1997). *IEEE Guide on Surge Testing for Equipment Connected to Low-Voltage AC Power Circuits.* Guidance is provided for applying surge testing to ac power interfaces of equipment connected to low-voltage ac power circuits that are subject to transient overvoltages. Signal and data lines are not addressed in this document, nor are any specifications stated on the withstand levels that might be assigned to specific assignments. An important objective of the document is to call attention to the safety aspects of surge testing.

IEEE Std C62.47-1992 (R1997). *IEEE Guide on Electrostatic Discharge (ESD)—Characterization of the ESD Environment.* This guide describes the electromagnetic threat posed to electronic equipment and subassemblies by actual Electrostatic Discharge (ESD) events from humans and mobile furnishings. This guide organizes existing data on the subject of ESD in order to characterize the ESD surge environment. This guide is not an ESD test standard. It is intended to be a resource for equipment designers, and for preparers and users of ESD test standards. The manufacturing, handling, packaging, and transportation of individual electronic components, including integrated circuits, are not discussed, and this guide does not deal with mobile items such as automobiles, aircraft, or other masses of comparable size.

IEEE Std C62.48-1995 (R2000). *IEEE Guide on Interactions Between Power System Disturbances and Surge-Protective Devices.* Information is provided to users and manufacturers of surge-protective devices (SPDs) about the interactions that may occur between SPDs and power system disturbances. This guide applies to SPDs manufactured to be connected to 50 or 60 Hz ac power circuits rated at 100–1000 V rms. The effects and side effects of the presence and operation of SPDs on the quality of power available to the connected loads are described. The interaction between multiple SPDs on the same circuit is also described.

IEEE Std C62.62-2000. *IEEE Standard Test Specifications for Surge-Protective Devices for Low-Voltage AC Power.* This standard establishes methods for testing and measuring the performance characteristics for surge-protective devices used in low-voltage ac power circuits. Definitions are stated that apply specifically to surge-protective devices. The testing requirements are categorized into two groups, in which a minimum set of basic tests (BTs) are prescribed for all surge-protective devices within the scope of its documents, supplemented by additional tests (ATs) that might be needed to establish particular application requirements.

IEEE Std C62.64-1997. *IEEE Standard Specifications for Surge Protectors Used in Low-Voltage Data, Communications, and Signaling Circuits.* This standard applies to surge protectors for application on multiconductor and coaxial, balanced or unbalanced, data, communications, and signaling circuits with voltages less than or equal to 1000 V rms, or 1200V dc. These surge protectors are intended to limit voltage surges, current surges, or both.

IEEE Std C62.92.1-1987 (R1993). *IEEE Guide for the Application of Neutral Grounding in Electrical Utility Systems Part I—Introduction.* Some basic considerations for the selection of neutral grounding parameters that will provide for the control of ground-fault current and overvoltage on all portions of three-phase electrical utility systems are presented. These considerations apply specifically to electric utility systems and do not recognize the neutral grounding requirements for dispersed storage and generation. They are intended to serve as an introduction to a series of standards on neutral grounding in electrical utility systems.

IEEE Std C62.92.2-1989 (R2000). *IEEE Guide for the Application of Neutral Grounding in Electrical Utility Systems Part II—Grounding of Synchronous Generator Systems.* General considerations for grounding synchronous generator systems are summarized, focusing on the objectives of generator grounding. The factors to be considered in the selection of a grounding class and the application of grounding methods are discussed. Four generator grounding types are considered: unit-connected generation systems, common-bus generators without feeders, generators with feeders directly connected at generated voltage, and three-phase, four-wire connected generators.

IEEE Std C62.92.3-1993 (R2000). *IEEE Guide for the Application of Neutral Grounding in Electrical Utility Systems, Part III—Generator Auxiliary Systems.* Basic factors and general considerations in selecting the class and means of neutral grounding for electrical generating plant auxiliary power systems are given in this guide. Apparatus to be used to achieve the desired grounding are suggested, and methods to specify the grounding devices are given. Sensitivity and selectivity of equipment ground-fault protection as affected by selection of the neutral grounding device are discussed, with examples.

IEEE Std C62.92.4-1991 (R1996). *IEEE Guide for the Application of Neutral Grounding in Electrical Utility Systems, Part IV—Distribution.* The neutral grounding of single- and three-phase ac electric utility primary distribution systems with nominal voltages in the range of 2.4 to 34.5 kV is addressed. Classes of distribution systems grounding are defined. Basic considerations in distribution system grounding concerning economics, control of temporary overvoltages, control of ground-fault currents, and ground relaying are addressed. Also considered are use of grounding transformers, grounding of high-voltage neutral of Wye-Delta distribution transformers, and interconnection of primary and secondary neutrals of distribution transformers.

IEEE Std C62.92.5-1992 (R1997). *IEEE Guide for the Application of Neutral Grounding in Electrical Utility Systems, Part V—Transmission Systems and Subtransmission Systems.* Basic factors and general considerations in selecting the class and means of neutral grounding for a particular ac transmission or subtransmission system are covered. Apparatus to be used to achieve the desired grounding are suggested, and methods for specifying the grounding devices are given. Transformer tertiary systems, equipment neutral grounding, and the effects of series compensation on grounding are discussed. The document includes references and an extensive bibliography on the subject of Transmission and Subtransmission Grounding.

ANSI C63.022-1996. *American National Standard for Limits and Methods of Measurement of Radio Disturbance Characteristics of Information Technology Equipment.* Emission limits are provided that are an are acceptable alternative for limits of the current issue of FCC Part 15, Subpart B, for Information Technology Equipment (ITE). This document republishes CISPR 22 (1993) and Amendment 1 (1995) as an American National Standard, ANSI C63.022-1996, which is recognized within the U.S.

ANSI C63.4-1991. *American National Standard for Methods of Measurement of Radio-Noise Emissions from Low-Voltage Electrical and Electronic Equipment in the Range of 9 kHz to 40 GHz.* Uniform methods of measurement of radio-frequency (RF) signals and noise from both unintentional and intentional emitters of RF energy in the frequency range of 9 kHz to 40 GHz are set forth. Methods for the measurement of radiated and ac power-line conducted radio noise are covered and may be applied to any such equipment unless otherwise specified by individual equipment requirements. Measurement of licensed transmitters is not covered, nor is certification/approval of avionic equipment or industrial, scientific, and medical (ISM) equipment.

ANSI C63.5-1998. *American National Standard for Calibration of Antennas Used for Radiated Emission Measurements in Electromagnetic Interference (EMI) Control.* Methods for determining antenna factors of antennas used for radiated emission measurements of electromagnetic interference (EMI) from 30 MHz to 1000 MHz are provided. Antennas included are linearly polarized antennas such as tuned dipoles, biconical dipoles, log-periodic arrays, etc. The methods include standard site, reference antennas, standard antenna and standard field methods. The latter two methods are incorporated by reference.

ANSI C63.6-1996. *American National Standard Guide for the Computation of Errors in Open-Area Test Site Measurements.* The basis for the acceptability criterion of ± 4 dB for the site attenuation measurements required by ANSI C63.4-1988, American National Standard Methods of Measurements of Emissions from Low-Voltage Electrical and Electronics Equipment in the Range of 10 kHz to 1 GHz is shown.

ANSI C63.7-1992 (R1997). *American National Standard Guide for Construction of Open-Area Test Sites for Performing Radiated Emission Measurements.* Information that is useful in constructing an open-area test site (OATS) is used to perform radiated emission measurements in the frequency range of 30-1000 MHz is provided. Final validity of the test site can only be made by performing site attenuation measurements as described in ANSI C63.4-1992.

IEEE Std C37.63-1997 (R1997). *IEEE Standard Requirements for Overhead, Pad-Mounted, Dry-Vault, and Submersible Automatic Line Sectionalizers for AC Systems.* Required definitions (for cutout type sectionalizers), ratings, procedures for performing design tests and production tests, constructional requirements, and application considerations for overhead and pad-mounted, dry-vault, and submersible automatic line sectionalizers for ac systems are specified.

ANSI C63.12-1999. *American National Standard Recommended Practice for Electromagnetic Compatibility Limits.* This recommended practice presents a rationale for developing limits and recommends sets of limits that are representative of current practice. These limits may be adjusted in particular applications as circumstances dictate.

ANSI C63.13-1991 (R1997). *American National Standard Guide on the Application and Evaluation of EMI Power-Line Filters for Commercial Use.* A basic understanding of the application, evaluation, and safety considerations of electromagnetic interference (EMI) power-line filters used in both ac and dc applications is provided. The construction of an EMI power-line filter and its functions in providing suppression of conducted noise are described. The functions and performance of the filter components, particularly the capacitors and inductors, are discussed. It is explained why seemingly identical filters may not give the same performance in a particular application. No-load insertion-loss test methods are presented. Proper installation of the filters in equipment is discussed. Safety regulations are briefly addressed.

ANSI C63.14-1998. *American National Standard Dictionary for Technologies of Electromagnetic Compatibility (EMC), Electromagnetic Pulse (EMP), and Electrostatic Discharge (ESD).* Terms associated with electromagnetic compatibility (EMC), electromagnetic pulse (EMP), and electrostatic discharge (ESD) are defined. Quantities, units, multiplying factors, symbols, and abbreviations are covered.

ANSI C63.16-1993. *American National Standard Guide for Electrostatic Discharge Test Methodologies and Criteria for Electronic Equipment.* Based upon ESD events on electronic equipment in actual use environments, a process to establish ESD test criteria is provided. Test procedures for highly repeatable ESD immunity evaluation of tabletop and floor-standing equipment are described. Simulator characteristics for hand/metal and furniture ESD testing are specified both for air and contact discharge methods. Statistical criteria is given to determine the number of test trials required, based on the confidence factor desired and various pass/fail categories. This ANSI ESD guide has been harmonized with other international ESD standards except where other standards have technical approaches that would reduce equipment quality or result in degraded product operation.

ANSI C63.17-1998. *American National Standard for Methods of Measurement of the Electromagnetic and Operational Compatibility of Unlicensed Personal Communications Services (UPCS) Devices.* Specific test procedures are established for verifying the compliance of unlicensed personal communications services (UPCS) devices with applicable regulatory requirements regarding radio-frequency (RF) emission levels and spectrum access procedures.

ANSI C63.18-1997. *American National Standard Recommended Practice for an On-Site, Ad Hoc Test Method for Estimating Radiated Electromagnetic Immunity of Medical Devices to Specific Radio-Frequency Transmitters.* Guidance is provided for health-care organizations in evaluating the radiated RF electromagnetic immunity of their existing inventories of medical devices to their existing inventories of RF transmitters, as well as to RF transmitters that are commonly available. This recommended practice can also be used for newly purchased medical devices and RF transmitters, as well as for pre-purchase evaluation. It applies to medical devices used in health-care facilities and to portable transmitters with a rated power output of 8 W or less. It does not apply to implantable medical devices, transport environments such as ambulances and helicopters, or to RF transmitters rated at more than 8 W.

IEEE Std C95.1-1999 Edition. *IEEE Standard for Safety Levels with Respect to Human Exposure to Radio Frequency Electromagnetic Fields, 3 kHz to 300 GHz.* IEEE Std C95.1-1991 gives recommendations to prevent harmful effects in human beings exposed to electromagnetic fields in the frequency range from 3 kHz to 300 GHz. The recommendations are intended to apply to exposures in controlled, as well as uncontrolled, environments. They are not intended to apply to the purposeful exposure of patients under the direction of practitioners of the healing arts. The induced and contact current limits of C95.1-1991 are modified in this edition. In addition, field strengths below which induced and contact currents do not have to be measured are specified, spatial averaging and measurement distance requirements are clarified, and more precise definitions for averaging volume and radiated power are provided.

IEEE Std C95.2-1999. *IEEE Standard for Radio-Frequency Energy and Current-Flow Symbols.* Symbols to inform people about the presence of potentially hazardous levels of radio-frequency energy or the presence of contact current hazards in the frequency range of 3 kHz to 300 GHz are specified. Guidance is given about how these symbols should be used on warning signs and labels.

IEEE Std C95.3-1991 (R1997). *IEEE Recommended Practice for the Measurement of Potentially Hazardous Electromagnetic Fields—RF and Microwave.* Techniques and instrumentation for the measurement of potentially hazardous electromagnetic fields are specified. The recommendations apply to hazards to personnel. However, the measurement techniques and instruments described are also applicable to the measurement of fields in the neighborhood of flammable materials and explosive devices, even though exposure standards for these situations have not been established.

IEEE Std C135.1-1999. (Revision of ANSI C135.1-1979). *IEEE Standard for Zinc-Coated Steel Bolts and Nuts for Overhead Line Construction.* The requirements for inch-based carriage bolts, machine bolts, double-arming bolts, and double-end bolts and nuts, commonly used in overhead line construction and where the applied load is primarily a tensile load, are covered.

IEEE Std C135.2-1999. (Revision of ANSI C135.2-1987). *IEEE Standard for Threaded Zinc-Coated Ferrous Strand-Eye Anchor Rods and Nuts for Overhead Line Construction.* Requirements for threaded zinc-coated ferrous strand-eye anchor rods and nuts commonly used in overhead line construction are covered in this standard.

IEEE Std C135.20-1998. *IEEE Standard for Zinc-Coated Ferrous Insulator Clevises for Overhead Line Construction.* Zinc-coated ferrous clevises for spool-type insulators commonly used for supporting or dead-ending conductors in line construction are covered. The specifications for spool-type insulators used with these clevises are covered in C29.3-1986.

IEEE Std C135.61-1997. *IEEE Standard for the Testing of Overhead Transmission and Distribution Line.* Hardware Requirements for mechanically testing load-rated line hardware for use on transmission and distribution facilities are described. Items specifically addressed in this standard include clevis and eye fittings, Y-clevis fittings, socket fittings, ball fittings, chain links, shackles, triangular and rectangular yoke plates, suspension clamps, and strain clamps. This standard is intended to cover routine acceptance testing. It is not intended for initial design tests.

IEEE Std C135.63-1998. *IEEE Standard for Shoulder Live Line Extension Links for Overhead Line Construction.* Dimensions and strength requirements for shoulder live line extension links used in overhead transmission and distribution hardware are covered.

ANSI C136.2-1996. *American National Standard for Roadway Lighting Equipment—Luminaires Voltage Classification.* Three voltage classifications for luminaires used in roadway lighting are covered. General testing methods for determining the dielectric withstand and the transient voltage withstand are given. This standard applies to luminaire electrical insulation between ungrounded current-carrying members and noncurrent-carrying members that may be grounded by design or accident.

ANSI C136.3-1995. *American National Standard for Roadway Lighting Equipment—Luminaire Attachments.* Attachment features of luminaires used in roadway lighting equipment are covered. The features covered apply to luminaires that are side- or post-top-mounted.

ANSI C136.4-1995. *American National Standard for Roadway Lighting Equipment—Series Sockets and Series Sockets Receptacles.* Equipment for luminaires for lighting roadways is covered in the following categories: series sockets having medium impact strength and intended for service at high temperatures, series sockets having high impact strength and intended for service at limited temperatures, and series-socket receptacles in the 5000 V classification.

ANSI C136.5-1996. *American National Standard for Roadway Lighting Equipment—Film Cutouts (Reaffirmation of C136.5-1969).* Operating and dimensional features of single-shot film cutouts used with series roadway lighting equipment and circuits are covered. The film cutouts function by dielectric breakdown and subsequent partial fusing of components.

ANSI C136.6-1997. *American National Standard for Roadway Lighting Equipment—Metal Heads and Reflector Assemblies—Mechanical and Optical Interchangeability.* Dimensional features of luminaires with metal heads that permit mechanical and optical interchangeability of both head and reflector assemblies are covered. The features covered in this standard apply to metal heads that are slipfitter mounted. The reflector assembly is of the latched collar type and may be part of an open or enclosed optical assembly.

ANSI C136.10-1996. *American National Standard for Roadway Lighting Equipment—Locking-Type Photocontrol Devices and Mating Receptacle Physical and Electrical Interchangeability and Testing.* Equipment that may be physically and electrically interchanged to operate within established values is covered in this standard, such as locking-type photocontrol devices, locking-type mating receptacles, and shorting and nonshorting caps.

ANSI Std C136.11-1995 (R1997). *American National Standard for Roadway Lighting Equipment—Multiple Sockets.* Medium and mogul multiple sockets as used in luminaires designed and intended for use in lighting roadways and other areas open to general use by the pubic are covered. This standard provides interchangeability of lamps, minimum safety standards for operating personnel, and minimum performance criteria.

ANSI C136.12-1996. *American National Standard for Roadway Lighting Equipment—Mercury Lamps—Guide for Selection.* Medium and mogul multiple sockets as used in luminaires designed and intended for use in lighting roadways and other areas open to general use by the pubic are covered. This standard provides interchangeability of lamps, minimum safety standards for operating personnel, and minimum performance criteria.

ANSI C136.15-1997. *American National Standard for Roadway Lighting Equipment—High-Intensity-Discharge and Low-Pressure Sodium Lamps in Luminaires—Field Identification.* A simple, uniform method for identifying the type and wattage rating of a high-intensity-discharge or a low-pressure sodium lamp installed in a luminaire is provided.

ANSI C136.16-1997. *American National Standard for Roadway Lighting Equipment Enclosed Post Top-Mounted Luminaires.* Dimensional, maintenance, and light distribution features that will permit interchange of post top-mounted luminaires whose center of mass is approximately over the mounting tenon are covered.

ANSI C136.17-1997. *American National Standard for Roadway Lighting Equipment—Enclosed Side-Mounted Luminaires for Horizontal-Burning High-Intensity-Discharge Lamps—Mechanical Interchangeability of Refractors.* The dimensional features and the material of refractors as described in ANSI C136.14-1988, American National Standard for Roadway Lighting Equipment—Enclosed Side-Mounted Luminaires for Horizontal-Burning High-Intensity Discharge Lamps, are covered

ANSI C136.18-1999. *American National Standard for High-Mast Side-Mounted Luminaires for Horizontal- or Vertical-Burning High-Intensity Discharge Lamps.* Used in Roadway Lighting Equipment. Physical, operational, maintenance, and light-distribution features that permit use of high-mast luminaires in roadway applications when so specified are covered. It is not intended that compliance with this standard will permit interchangeability with existing roadway equipment without thorough engineering review and evaluation.

ANSI C136.19-1997. *American National Standard for Roadway Lighting Equipment High-Pressure Sodium Lamps—Guide for Selection.* The selection of high-pressure sodium lamps recommended for use in roadway lighting equipment is covered.

ANSI C136.23-1997. *American National Standard for Roadway Lighting Equipment Enclosed Architectural Luminaires.* Physical, operating, maintenance, and light distribution features that permit use of architectural luminaires in roadway applications when so specified are covered. Specific features for horizontal, pendant, and vertical architectural luminaires, together with various types of lamps to meet the individual needs of special architectural roadway lighting applications, are included.

ANSI C136.27-1996. *American National Standard for Roadway Lighting Equipment—Tunnel Lighting Luminaires.* Luminaires used for illuminating roadway tunnels are covered. The requirements in this standard are limited to general attributes of tunnel luminaires due to the wide variety of designs possible.

ANSI C136.32-1999. *American National Standard for Roadway Lighting Equipment—Enclosed Setback Luminaires and Directional Floodlights for High-Intensity-Discharge Lamps Accredited Standards.* Dimensional, maintenance, and electrical features that permit the interchange of similar style enclosed luminaires having the same light distribution classification or type for high-intensity-discharge lamps used in roadway lighting equipment are covered. Luminaires covered by this standard are generally yoke, trunnion, or tenon mounted.

ANSI N42.4-1971 (R1991). *American National Standard for High Voltage Connectors for Nuclear Instruments.* Coaxial high-voltage connectors on nuclear instruments for dc applications up to 5000 V and ac applications up to 3500 V rms at 60 Hz are covered. The connectors may also be used at higher frequencies provided the operating voltage is appropriately reduced to provide for interchangeability of safe high-voltage connectors in nuclear instrument applications. The connectors are safe in that the pin and socket contacts are well and securely recessed in the connector housing so that hand or body contact of the unmated connector with rated voltage applied will not result in electrical shock.

ANSI N42.5-1965 (R1991)/N42.6-1980 (R1991). *Bases for GM Counter Tubes and American National Standard Interrelationship of Quartz-Fiber Electrometer Type Exposure Meters and Companion Exposure Meter Chargers.* This document contains two standards. ANSI N42.5 specifies bases for Geiger-Mueller counter tubes. ANSI N42.6 specifies interrelating mechanical and electrical properties so that quartz-fiber exposure meters may be used with any charger. Characteristics peculiar to these devices but not affecting the interrelationship between chargers and exposure meters are omitted.

ANSI N42.12-1994. *American National Standard Calibration and Usage of Thallium-Activated Sodium Iodide Detector Systems for Assay of Radionuclides.* This standard establishes methods for performance testing, calibration, and usage of NaI(Tl) detector systems for the measurement of gamma ray emission rates of radionuclides; the assay for radioactivity; and the determination of gamma ray energies and intensities. It covers both energy calibration and efficiency calibration.

ANSI N42.13-1986 (R1993). *American National Standard for Calibration and Usage of "Dose Calibrator" Ionization Chambers for the Assay of Radionuclides.* A technique for the quantification of the activity of identified radionuclides using any of a variety of ionization chambers currently available for this purpose is presented. Application of the standard is limited to instruments that incorporate well-type ionization chambers as detectors. The method provides measurements that are accurate to within ± 10% and reproducible to within ± 5%. The standard is also intended to assure continuing performance of the apparatus within these specifications.

ANSI N42.14-1999. *American National Standard for Calibration and Use of Germanium Spectrometers for the Measurement of Gamma-Ray Emission Rates of Radionuclides.* Methods for the calibration and use of germanium spectrometers for the measurement of gamma-ray energies and emission rates over the energy range from 59 keV to approximately 3000 keV, and for the calculation of source activities from these measurements, are established. Minimum requirements for automated peak finding are stated. Methods for measuring the full-energy peak efficiency with calibrated sources are given. Performance tests that ascertain the proper functioning of the Ge spectrometer and evaluate the limitations of the algorithms used for locating and fitting single and multiple peaks are described. Methods for the measurement of, and the correction for pulse pileup are suggested. Techniques are recommended for the inspection of spectral-analysis results for large errors resulting from summing of cascade gamma rays in the detector. Suggestions are provided for the establishment of data libraries for radionuclide identification, decay corrections, and the conversion of gamma-ray rates to decay rates.

ANSI N42.15-1997. *American National Standard Check Sources for and Verification of Liquid Scintillation Counting Systems.* Tests and procedures to ensure that a liquid-scintillation counting system is producing reliable data are provided for designers and users. This standard does not cover the calculation of sample activity for quenched unknown samples, sample preparation, efficiency correlation (quench correction) procedures, or identification of unknown radionuclides.

ANSI N42.17A-1989. *Performance Specifications for Health Physics Instrumentation—Portable Instrumentation for Use in Normal Environmental Conditions.* Minimum acceptable performance criteria for health physics instrumentation for use in ionizing radiation fields are established. Included are testing methods to establish the acceptability of each type of instrumentation. This standard does not specify which instruments or systems are required, nor does it consider the number of specific applications of such instruments.

ANSI N42.17B-1989. *American National Standard Performance Specifications for Health Physics Instrumentation—Occupational Airborne Radioactivity Monitoring Instrumentation.* Performance criteria and testing procedures for instruments and instrument systems designed to continuously sample and quantify concentrations of radioactivity in ambient air in the workplace are specified. General test procedures, general criteria, electronic criteria, radiation response, interfering responses, environmental criteria, air circuit criteria, and documentation are covered. This standard does not specify which instruments or systems are required, nor does it address the specific locations or applications of such instruments.

ANSI N42.17C-1989. *American National Standard for Performance Specifications for Health Physics Instrumentation—Portable Instrumentation for Use in Extreme Environmental Conditions.* Minimum acceptable performance criteria for health physics instrumentation for use in ionizing radiation fields under extreme environmental conditions are established. Included are testing methods to establish the acceptability of each type of instrumentation. Performance testing criteria for use in generic (type) tests of new instrument models are given. This standard covers general test procedures, general characteristics, electronic and mechanical requirements and tests, radiation response, interfering responses, environmental factors, and documentation. It does not specify which instruments or systems are required, nor does it consider the number of specific applications of such instruments.

ANSI N42.18-1980 (R1991) (Redesignation of ANSI N13.10-1974). *American National Standard Specification and Performance of On-Site Instrumentation for Continuously Monitoring Radioactivity in Effluents.* Installed instrumentation for measuring the quantity or rate, or both, of the release of radionuclides in the effluent streams, and to provide documentation useful for scientific and legal purposes is covered. Recommendations for the selection of instrumentation are provided. This standard applies to continuous monitors that measure normal releases, detect inadvertent releases, show general trends, and annunciate radiation levels that have exceeded predetermined values.

ANSI N42.20-1995. *American National Standard Performance Criteria for Active Personnel Radiation Monitors.* This standard provides performance and design criteria for monitors that are worn on the trunk of the body to measure the personal dose equivalent or the dose equivalent rate from external sources of ionizing radiation

ANSI N42.22-1995. *American National Standard Traceability of Radioactive Sources to NIST and Associated Instrument Quality Control.* A mechanism for manufacturers to establish traceability of radionuclide sources that are certified for radionuclide activity; concentration; or alpha, beta, x-, or gamma-ray emission rate to the National Institute of Standards and Technology (NIST) is described.

ANSI N42.23-1996. *American National Standard Measurement and Associated Instrumentation Quality Assurance for Radioassay Laboratories.* A framework that can be used to create a national or an organizational NIST-traceable measurement quality assurance (MQA) program that will optimize the quality of radioassays performed by service laboratories is presented. This standard serves as a guide for MQA programs developed for specialized sectors of the radioassay laboratory community, i.e., bioassay, routine environmental monitoring, environmental restoration and waste management, radiopharmaceuticals, nuclear power radiochemistry, and other areas involved in radioassays.

ANSI N42.25-1997. *American National Standard Calibration and Usage of Alpha/Beta Proportional Counters.* This standard establishes methods for the calibration and use of gas proportional counters with and without active guard detectors. This standard also establishes methods for measuring the alpha and beta counting plateau, crosstalk factors, background, alpha and beta efficiency from prepared standards, correction factors for samples whose self-attenuation or mass differs from that of the standard, and calculation of the sample activities together with their random and total uncertainties. Correction for pulse pileup due to high count rate is also discussed. Although many principles articulated in this standard apply to the counting of radionuclides emitting a maximum beta energy below 100 keV as well, the counting of these low-energy beta emitters requires a higher degree of attention to detail in sample preparation, instrument calibration, and measurement correction factors than addressed in this standard. Therefore, this standard is intended for measuring radionuclides with maximum beta energies above 100 keV.

ANSI N42.27-1999. *American National Standard for Determination of Uniformity of Solid Gamma-Emitting Flood Sources.* Minimum informational requirements for a Test and Measurement Report for flood sources used with scintillation cameras are provided. It is not intended to specify the means by which such information is obtained although it does place requirements and limitations on the methodology. In addition, it is not intended to cover the use of the source in the determination of the operating characteristics or correction factors for a scintillation camera.

ANSI N317-1980 (R1991). *American National Standard Performance Criteria for Instrumentation Used for Inplant Plutonium Monitoring.* Performance criteria are defined, and plutonium radiation is characterized. The specifications apply to plutonium handling and storage facilities, excluding reactors and irradiated fuel reprocessing facilities. This standard does not apply to the construction of specific instruments nor does it specify instrumentation to be employed for each survey to be conducted, other than in generic terms. It does not define specifications for personnel dosimeters, effluent monitoring systems, or instruments needed in bioassay programs, nor does it define those requirements that may be needed to monitor emergency conditions.

ANSI N320-1979 (R1993). *American National Standard Performance Specifications for Reactor Emergency Radiological Monitoring Instrumentation.* The essential performance parameters and general placement for monitoring the release of radionuclides associated with a postulated serious accident at a reactor facility are defined for various types of instrumentation. The predominant consideration in the assessment of radiation emergencies is the measurement of fission products made promptly enough to permit timely emergency decisions. This standard does not specify which of the instruments or systems are required, nor does it consider the number or specific locations of such instruments. This standard also does not address single failure criteria associated with nuclear safety instrumentation.

ANSI N322-1997. *American National Standard Inspection, Test, Construction, and Performance Requirements for Direct Reading Electrostatic/Electroscope Type Dosimeters.* Inspection, test, construction and performance requirements for direct reading electrostatic/electroscope type dosimeters designed to measure the personal dose equivalent or ambient exposure delivered by external sources of ionizing radiation (X-rays or gamma-rays) are given.

ANSI N323-1978 (R1983). *American National Standard for Radiation Protection Instrumentation Test and Calibration.* Calibration methods for portable (hand-carried) radiation protection instruments used for detection and measurement of levels of ionizing radiation fields or levels of radioactive surface contamination are established. Included are conditions, equipment, and techniques for calibration as well as the degree of precision and accuracy required. Alpha, beta, photon, and neutron radiation are considered. Passive integrating dosimetric devices such as film, thermoluminescent, and chemical dosimeters are not covered, although the basic principles and intent may apply to them as well as to nonportable radiation detection instrumentation in general.

ANSI N323A-1997. *American National Standard Radiation Protection Instrumentation Test and Calibration, Portable Survey Instruments.* Specific requirements are established for portable radiation protection instruments used for detection and measurement of levels of ionizing radiation fields or levels of radioactive surface contamination.

ANSI N449.1-1978 (R1983). *American National Standard Procedures for Periodic Inspection of Cobalt-60 and Cesium-137 Teletherapy Equipment.* Procedures for the inspection of cobalt-60 and cesium-137 teletherapy equipment are suggested. Their purpose is to enable users to identify and quantify malfunctions or maladjustments of the safety and radiation defining components. Methods and equipment are listed for each procedure.

ANSI Y32.9-1972 (R1989). *American National Standard for Graphic Symbols for Electrical Wiring and Layout Diagrams Used in Architecture and Building Construction.* A basis is provided for showing the general physical location and arrangement of the sections of the required wiring system and identifying the physical requirements for various types of materials needed to provide the electrical installation in buildings. In some instances, the symbols may indicate the function or electrical characteristics of the system; however, that is not their primary purpose. The required installation is shown on the drawing by the use of the various applicable outlet and equipment symbols, together with interconnecting circuit or feeder run lines, supplemented with necessary notations. In general, basic symbols have been included in the symbol schedule.

J-STD-016-1995. *Standard for Information Technology—Software Life Cycle Processes—Software Development—Acquirer-Supplier Agreement (Issued for Trial-Use).* This standard defines a set of software development activities and resulting software products. It provides a framework for software development planning and engineering. It is also intended to merge commercial and Government software development requirements within the framework of the software life cycle process requirements of the Electronic Industries Association (EIA), Institute of Electrical and Electronics Engineers (IEEE) and International Organization for Standardization (ISO). The term “software development” is used as an inclusive term encompassing new development, modification, reuse, reengineering, maintenance, and all other processes or activities resulting in software products.

Non-IEEE Standard Sources

[1] Sequential events recording systems terms prepared by the Power Generation Committee of the Power Engineering Society in 1974. (Terms approved for use in IEEE Std 100 only).

[2] Mil. Std. 1309B; Automated Instrumentation 9.8 Terms for Test Measurement, and Diagnostic Equipment, Definitions of.

[3] ANSI Std C85.1-1963, (a) 1966 (b) 1972 Terminology for Automatic Control.

[4] IEEE Power Engineering Society Committee on Insulated Conductors.

[5] IEEE Power Engineering Society Committee on Power Generation.

[6] IEEE Power Engineering Society Committee on Power System Relaying.

[7] IEEE Information Theory Group.

[8] IEEE Power Engineering Society Committee on Surge Protective Devices. See IEEE Std 28-1974 and IEEE Std 32-1973 (R1984).

[9] IEEE Power Engineering Society Committee on Rotating Machinery.

[10] IEEE Power Engineering Society Committee on Transmission and Distribution.

[11] IEEE Industry Applications Society Committee on Petroleum and Chemical Industry. Definitions taken from the NFPA (National Fire Protection Association)

[12] IEEE Industry Application Society Committee on Static Power Converters.

[13] IEEE Circuits and Systems Society. Network Applications of Circuits and Systems.

[14] IEEE Instrumentation and Measurement Society, Nonreal Time Spectrum Analyzer.

[15] IEEE Instrumentation and Measurement Society, Test, Measurement, and Diagnostic Equipment. See source [2].

[16] ANSI Std SE3.13-1974; NFPA 72E-1974, Standard on Automatic Fire Detectors.

[17] IEEE Ultrasonics, Ferroelectrics, and Frequency Control Society. Definitions for specific (acoustic-optical) devices, delay lines, and ferroelectric material terms. See sources [21], [22], and [23].

[18] IEEE Industry Applications Society, Subcommittee 2-447-02 on Emergency and Standby Power Systems. See IEEE Std 446-1987.

[19] IEEE Communications Society, Committee on Space Communications. Definitions of Communication Satellite Terms.

[20] IEEE Computer Society, Computing Systems.

[21] IEEE Ultrasonics, Ferroelectrics, and Frequency Control Society. Definitions replaced by those in IEEE Std 180-1986.

[22] IEEE Ultrasonics, Ferroelectrics, and Frequency Control Society. Definitions for Delay Lines, Dispersive and Nondispersive.

[23] IEEE Ultrasonics, Ferroelectrics, and Frequency Control Society. Definitions for Acousto-optic Devices.

[24] IEEE Communications Society, Space Communications Committee. Component parts of communications systems; Communications satellite terms.

[25] IEEE Communications Society, Space Communications Committee. Transmission and Propagation Terms.

[26] ANSI Std C55.2-1974. See IEEE Std 18-1980 and IEEE Std 824-1985.

[27] ANSI Std C104.2-1968; EIA RS 330-1966 Closed Circuit Television Camera 525/60 Inter face 2:1, Electrical Performance of.

[28] ANSI Std C80.1-1971, Rigid Steel Conduit, Zinc Coated, Specification for.

[29] IEEE Reliability Society. Availability, Reliability, and Maintainability Terms.

[30] IEEE Electromagnetic Compatibility Society.

[31] IEV entry. Document 7.

[32] IEEE Acoustics, Speech, and Signal Processing Society.

[33] IEEE Broadcast Technology Society—Television.

[34] IEEE Broadcast Technology Society—Video Techniques.

[35] IEEE Antennas and Propagation Society—Antennas and Waveguides.

[36] IEEE Antennas and Propagation Society—Wave Propagation.

[37] IEEE Vehicular Technology Society—Mobile, Communications Systems.

[38] IEEE Instrumentation and Measurement Society—Electromagnetic Measurement State-of-the-Art.

[39] IEEE Instrumentation and Measurement Society—Fundamental Electrical Standards.

[40] IEEE Instrumentation and Measurement Society—High-Frequency Instrumentation and Measurements.

[41] IEEE Aerospace and Electronic Systems Society—Energy Conversion.

[42] IEEE Aerospace and Electronic Systems Society—Navigational Aids. See IEEE Std 686-1990 and IEEE Std 172-1983.

[43] IEEE Industrial Electronics Society.

[44] IEEE Electron Devices Society—Solid-State Devices.

[45] IEEE Electron Devices Society—Standards on Electron Tubes.

[46] IEEE Electron Devices Society—Standards on Solid State Devices.

[47] IEEE Engineering in Medicine and Biology Society.

[48] IEEE Communications Society—Communications Switching.

[49] IEEE Communications Society—Data Communication Systems.

[50] IEEE Communications Society—Wire Communication.

[51] IEEE Components, Hybrids, and Manufacturing Technology Society.

[52] IEEE Instrumentation and Measurement Society—Control Systems.

[53] IEEE Electromagnetic Compatibility Society.

[54] IEEE Power Engineering Society—Power System Engineering.

[55] IEEE Power Engineering Society—Power System Instrumentation and Measurement.

[56] IEEE Power Engineering Society—Switchgear.

[57] IEEE Power Engineering Society—Transformers.

[58] IEEE Industry Applications Society—Cement Industry.

[59] IEEE Industry Applications Society—Corrosion and Cathodic Protection.

[60] IEEE Industry Applications Society—Industrial Control.

[61] IEEE Industry Applications Society—Machine Tools Industry.

[62] IEEE Industry Applications Society—Static Power Converters.

[63] IEEE Systems, Man and Cybernetics Society.

[64] AD8—American Society for Testing and Materials Publication D8.

[65] AD16—American Society for Testing and Materials Publication D16.

[66] AD123—American Society for Testing and Materials Publication D123.

[67] AD883—American Society for Testing and Materials Publication 883.

[68] ADl566—American Society for Testing and Materials Publication D1566.

[69] National Electrical Manufacturers Association Publication AS 1.

[70] CISPR—International Special Committee on Radio Interference.

[71] CM—Corrosion Magazine.

[72] CTD—Chambers Technical Dictionary.

[73] CV 1—National Electrical Manufacturers Association Publication CV 1.

[74] Electronic Industries Association Publication 3B.

[75] IC 1—National Electrical Manufacturers Association Publication IC 1.

[76] 15A-Instrument Society of America.

[77] International Telecommunications Union.

[78] KPSH—Kepco Power Supply Handbook.

[79] LA 1—National Electrical Manufacturers Association Publication LA 1.

[80] MA 1—National Electrical Manufacturers Association Publication MA 1.

[81] MDE—Modern Dictionary of Electronics.

[82] MG 1—National Electrical Manufacturers Association Publication MG 1.

[83] SCC—IEEE Standards Coordinating Committee. See source [123].

[84] IEC—International Electrotechnical Commission.

[85] ANSI Std X3.12-1970; Std 2382/V, VI (150) Vocabulary for Information Processing.

[86] NFPA No.70-1978 (previously Std C1-1978). National Electrical Code.

[87] ANSI Std C83.16-1971, Relays and Electronic Equipment, Definitions and Terminology for.

[88] ANSI Std C84.1-1970 (revised in 1977); IEC 38 and 71 Voltage Ratings for Electric Power Systems and Equipment (60 Hz), including Supplement C84.1A-1973.

[89] ANSI Std C29.1-1961 (R1974), Electrical Power Insulators, Test Methods for, including Addendum C29.2A (reaffirmed 1974).

[90] ANSI Std C71.1-1972, Household Electric Ranges (AHAM ER-1), including Supplements C71.1A-1975 and C71.1B-1975.

[91] ANSI Std C87.1-1971; NEMA Publication EW 1-1970. Electric Arc Welding Apparatus.

[92] ANSI Std C83.14-1963 (R1969); EIA RS 225-1959; IEC 339-1. Requirements for Rigid Coaxial Transmission Lines—50 ohms.

[93] ANSI Std C85.1-1963, Automatic Control, Terminology for, including Supplements C85.1A-1966 and C85.1B-1972.

[94] ANSI Std C82.1-1972, Fluorescent Lamp Ballasts, including Supplement C82.1A-1973, Specifications for.

[95] ANSI Std C82.4-1974 (ANSI); IEC 262. Mercury Lamp Ballasts (Multiple Supply Type), Specifications for.

[96] ANSI Std C82.3-1972 (ANSI); IEC 82. Fluorescent Lamp Reference Ballasts, Specifications for.

[97] ANSI Std C82.9-1971, High-Intensity Discharge Lamp Ballasts and Transformers, Definitions for.

[98] ANSI Std C82.7-1971 (ANSI); IEC 262. Mercury Lamp Transformers, Constant Current (Series) Supply Type, Specifications for.

[99] ANSI Std C82.8-1963 (R1971), Incandescent Filament Lamp Transformers, Constant Current (Series) Supply Type, Specifications for.

[100] ANSI Std C92.1-1971, Voltage Values for Preferred Transient Insulation Levels.

[101] ANSI Std C64.1-1970 (ANSI); IEC 136-1; IEC 136-2; IEC 276. Brushes for Electrical Machines.

[102] ANSI Std C39.1-1972, Electrical Analog Indicating Instruments, Requirements for.

[103] ANSI Std C37.1 (redesignated C37.90). See IEEE Std C37.90-1989.

[104] ANSI Std C78.385-1961, Electric Lamps.

[105] ANSI Std C99.1, Highly Reliable Soldered Connections in Electronic and Electrical Application.

[106] ANSI Std C79.1-1971, Glass Bulbs Intended for Use with Electron Tubes and Electric Lamps, Nomenclature for.

[107] ANSI Std C57.12.75, Removable Air-Filled Junction Boxes for Cable Termination for Power Transformers.

[108] ANSI Std 51.1-1960 (R1971). Integral Air-Filled Junction Boxes for Cable Termination for Power Transformers.

[109] ANSI Std 51.1-1960 (R1971); ISO 131; ISO 16; IEC 50-08. Acoustical Terminology (including Mechanical Shock and Vibration).

[110] ANSI Std C31.4-1958 (R1975). Pool-Cathode Mercury-Arc Power Converters, Practices and Requirements for.

[111] ANSI Std C39.2-1964 (R1969). Direct Acting Electrical Recording Instruments, Requirements for.

[112] ANSI Std C39.4-1966 (R1972), Automatic Null-Balancing Electrical Measuring Instruments, Specifications for.

[113] ANSI Std C80.4-1963 (R1974), Fittings for Rigid Metal Conduit and Electrical Metallic Tubings, Specifications for.

[114] ANSI Std C5.1-1969; NFPA No. 70-1968. Lightning Protection Code.

[115] ANSI Std C50.10-1977; IEC 34-1. Synchronous Machines, General Requirements for.

[116] ANSI Std C89.1-1974, Specialty Transformers except General-Purpose Type.

[117] ANSI Std C57.14, Constant-Current Transformers of the Moving Coil Type.

[118] ANSI Std C67.1, Preferred Nominal Voltages, 100 Volts and Under.

[119] This definition was derived from a standard previously listed in the ANSI category C42.

[120] IEEE Committee on Automatic Control, now the IEEE Instrumentation and Measurement Society.

[121] Office definition prepared by the staff of IEEE Std 100.

[122] IEEE Committee on Sonics and Ultrasonics.

[123] SCC—IEEE Standards Coordinating Committee. See source [83].

[124] ANSI/ASME Std NQA-1-1979, Quality Assurance Program Requirements for Nuclear Power Plants. These definitions are reprinted here with the permission of the American Society of Mechanical Engineers (ASME).

[125] Std 545 (unapproved IEEE Project Standard; as of May 1988 to be issued for trial use).

[126] ANSI/IES Std RP-16-1980, Nomenclature and Definitions for Illuminating Engineering. A revision of ANSI Z7.1-1967 (R1973).

[127] P347 (IEEE Committee draft) (withdrawn). Task group for solid-state displays of the Standardization Committee of the IEEE Group on Electron Devices.